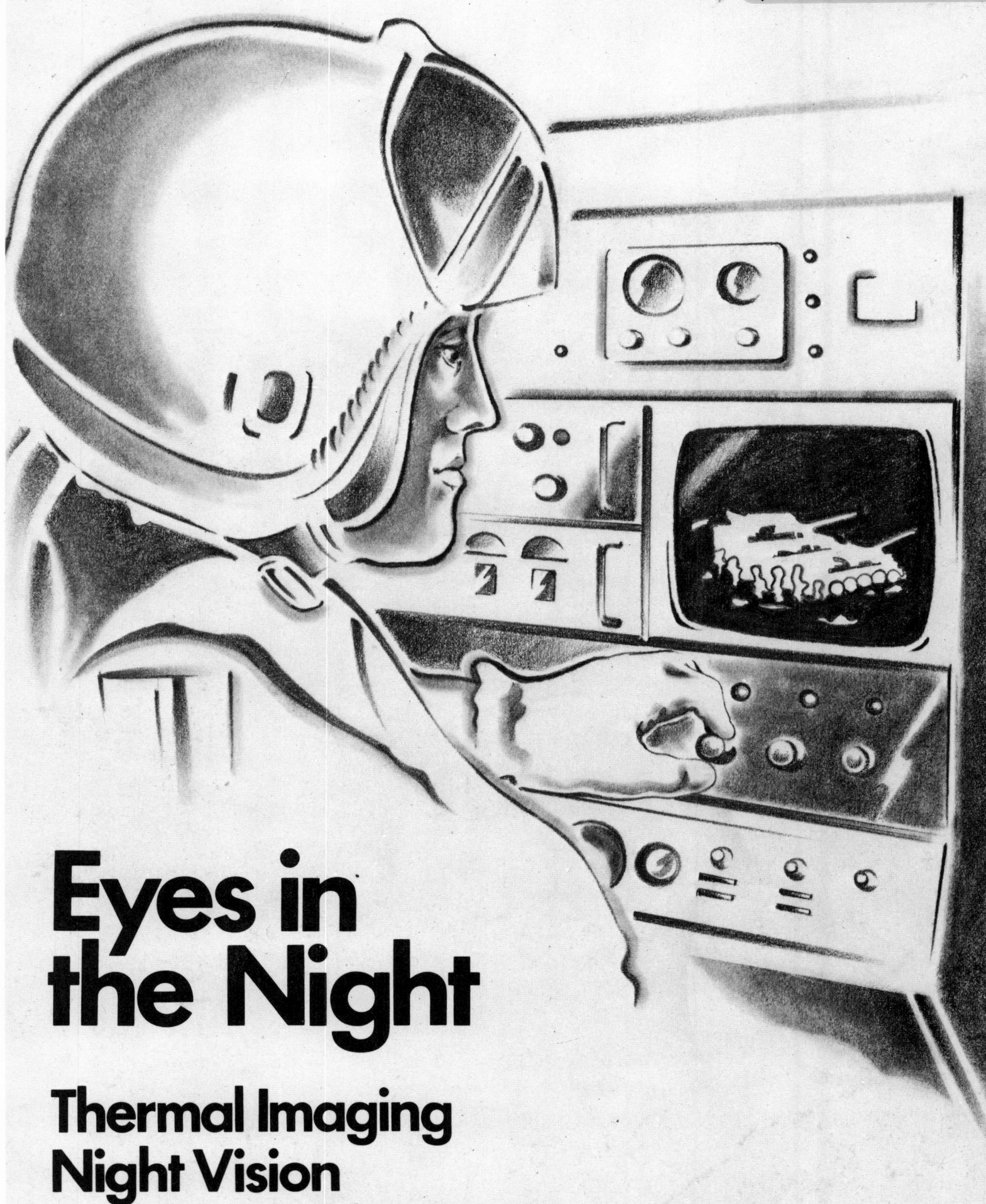

Eyes in the Night

Thermal Imaging
Night Vision
Laser Rangefinding
and other Optronic Systems

Glasgow and London

ZK-DBV
MOUNT COOK AIRLINES

PROVINCETOWN-BOSTON AIRLINE

PEACE AIR LTD

7P-LAC
LESOTHO AIRWAYS

PT-KRP

F-OCSB
F-OCSB

JANE'S ALL THE WORLD'S AIRCRAFT

Edited by **John W. R. Taylor**
FRHistS, MRAeS, FSLAET

Order of Contents

World Sales Distribution

Jane's Yearbooks
Paulton House, 8 Shepherdess Walk
London, N1 7LW, England

All the World
except

United States of America and Canada:
Franklin Watts Inc.
730 Fifth Avenue
New York, NY 10019

Editorial communication to:

The Editor, Jane's All The World's Aircraft
Jane's Yearbooks, Paulton House, 8 Shepherdess Walk
London N1 7LW, England
Telephone 01-251 1666

Advertisement communication to:

The Advertisement Manager
Jane's Yearbooks, Paulton House, 8 Shepherdess Walk
London N1 7LW, England
Telephone 01-251 1666

***Classified List of Advertisers**

The various products available from the advertisers in this edition are listed alphabetically in about 350 different headings.

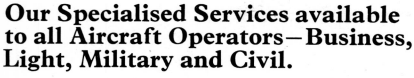

JANE'S ALL THE WORLD'S AIRCRAFT
ALPHABETICAL LIST OF ADVERTISERS
1976/77 EDITION

JANE'S ALL THE WORLD'S AIRCRAFT
CLASSIFIED LIST OF ADVERTISERS
1976/77 EDITION

The companies advertising in this publication have informed us that they are invloved in the fields of manufacture indicated below:

AC MOTORS
Aviaexport
Garrett Corporation
Marvin Tomkins

ACCELEROMETERS
Aeritalia
Aviaexport
Marvin Tomkins
S.F.E.N.A.

ACCESSORIES
Aviaexport
Garrett Corporation
Woodward Governor

ACCUMULATORS—CADMIUM-NICKEL
Aviaexport
S.A.F.T.

ACTUATORS—ELECTRIC
Aviaexport
Garrett Corporation
Marvin Tomkins
S.F.E.N.A.

AERIAL SURVEY INSTRUMENTS
Marvin Tomkins

AERIALS—AIRCRAFT
Aeritalia
British Aircraft Corporation
Dornier
Marvin Tomkins
Messerschmitt Bölkow-Blohm
Siai-Marchetti
S.N.E.C.M.A.

AERO AUXILIARY EQUIPMENT
Garrett Corporation
Marvin Tomkins
Siai Marchetti

AERO-ENGINE TEST PLANT
Avco Lycoming
John Curran
Marvin Tomkins
Siai Marchetti
S.N.E.C.M.A.

AERO ENGINES
Avco Lycoming
Aviaexport
Garrett Corporation
Klöckner-Humboldt-Deutz
Marvin Tomkins
Meteor
Motoren-und Turbinen-Union
Rinaldo Piaggio
S.N.E.C.M.A.

AERONAUTICAL ENGINEERS AND CONSULTANTS
Aviaexport
Boeing
British Aircraft Corporation
Siai Marchetti

AEROSYSTEMS
Boeing
British Aircraft Corporation

AGRICULTURAL AIRCRAFT SPRAY AND DUST SYSTEMS AND COMPONENTS
Aviaexport
Dornier
Pilatus
Siai Marchetti

AIR COMPRESSORS
Garrett Corporation
Marvin Tomkins

AIR COMPRESSORS—CABIN
Garrett Corporation

AIR COMPRESSORS FOR ENGINE STARTING
Garrett Corporation

AIR-CONDITIONING EQUIPMENT
Aviaexport
Garrett Corporation
Marvin Tomkins
M.L. Aviation

AIR-CONDITIONING SYSTEMS
Aviaexport
Garrett Corporation

AIR CONTROL EQUIPMENT FOR CABINS
Aviaexport
Garrett Corporation
Marvin Tomkins

AIR DATA COMPUTER SYSTEMS
Aviaexport
Garrett Corporation
Meteor

AIR TRAFFIC CONTROL EQUIPMENT
Aeritalia
S. G. Brown Communications
Eltro
Hollandse Signaalapparaten
Selenia

AIRCRAFT—AGRICULTURAL (DUSTERS AND SPRAYERS)
Aviaexport
Dornier
Pilatus
Siai Marchetti

AIRCRAFT—AMBULANCE
Aviaexport
Dornier
Fokker-VFW
Hawker Siddeley Aviation
Israel Aircraft Industries
Pilatus
Siai Marchetti

AIRCRAFT—COMMERCIAL
Aeritalia
Aérospatiale
Aviaexport
Boeing
British Aircraft Corporation
Dornier
Fokker-VFW
Hawker Siddeley Aviation
Israel Aircraft Industries
Marvin Tomkins
Messerschmitt-Bölkow-Blohm
Rinaldo Piaggio
Siai Marchetti

AIRCRAFT—EXECUTIVE
Aérospatiale
Aviaexport
Boeing
British Aircraft Corporation
Dornier
Fokker-VFW
Garrett Corporation
Hawker Siddeley Aviation
Israel Aircraft Industries
Marvin Tomkins
Messerschmitt-Bölkow-Blohm
Rinaldo Piaggio
Siai Marchetti

AIRCRAFT INTEGRATED DATA SYSTEMS
Eltro
Garrett Corporation
Hawker Siddeley Aviation

AIRCRAFT—MILITARY
Aeritalia
Aeronautica Macchi
Aérospatiale
Boeing
British Aircraft Corporation
Dornier
Fairey Britten-Norman
Fokker-VFW
G.I.A.T.
Hawker Siddeley Aviation
Marvin Tomkins

AIRCRAFT—MILITARY (contd.)
Messerschmitt Bölkow-Blohm
Panavia
Rinaldo Piaggio
Siai Marchetti
Vought Corporation

AIRCRAFT—NAVAL
Aeritalia
Boeing
Fairey Britten-Norman
Hawker Siddeley Aviation
Marvin Tomkins
Rinaldo Piaggio
Siai Marchetti
Vought Corporation

AIRCRAFT—PRIVATE
Dornier
Fairey Britten-Norman
Hawker Siddeley Aviation
Marvin Tomkins
Omnipol
Siai Marchetti

AIRCRAFT—RADIO CONTROLLED
Boeing
British Aircraft Corporation
Meteor

AIRCRAFT—SUPERSONIC
Aeritalia
British Aircraft Corporation

AIRCRAFT—TRAINING
Aeritalia
Aeronautica Macchi
Aérospatiale
Boeing
British Aircraft Corporation
Dornier
Fokker-VFW
Hawker Siddeley Aviation
Marvin Tomkins
Omnipol
Pilatus
Siai Marchetti
Vought Corporation

AIRCRAFT—TRANSPORT
Aeritalia
Aérospatiale
Aviaexport
Boeing
British Aircraft Corporation
Dornier
Fairey Britten-Norman
Fokker-VFW
Hawker Siddeley Aviation
Israel Aircraft Industries
Marvin Tomkins
Messerschmitt-Bölkow-Blohm
Omnipol

AIRCRAFT—V/STOL
Aviaexport
Boeing
British Aircraft Corporation
Dornier
Fairey Britten-Norman

AIRCRAFT—V/STOL (contd.)
Fokker-VFW
Hawker Siddeley Aviation
Pilatus
Siai Marchetti
Vought Corporation

AIRCRAFT ARRESTING GEAR
S.N.E.C.M.A.

AIRCRAFT CANOPIES
Aeritalia
Aeronautica Macchi
Goodyear Tyre & Rubber Co.
Marvin Tomkins
Siai Marchetti

AIRCRAFT DEVELOPMENT
Aeritalia
Boeing
Hawker Siddeley Aviation
Messerschmitt-Bölkow-Blohm
Siai Marchetti
Vought Corporation

AIRCRAFT ESCAPE SYSTEMS
Garrett Corporation

AIRCRAFT FIELD OPERATIONS AND SUPPORT
Boeing
Vought Corporation

AIRCRAFT FLOATS
Dornier
Garrett Corporation
Meteor

AIRCRAFT FREIGHT HANDLING EQUIPMENT
Aviaexport
Dornier
Fokker-VFW

AIRCRAFT MECHANICAL HANDLERS
Aviaexport
Dornier

AIRCRAFT MODIFICATIONS
Boeing
Garrett Corporation
Israel Aircraft Industries
Siai Marchetti

AIRCRAFT PISTON ENGINE CYLINDERS
Marvin Tomkins

AIRCRAFT PROPELLER GOVERNORS
Aviaexport
Marvin Tomkins
Woodward Governor

AIRCRAFT PROPELLERS
Aviaexport
Marvin Tomkins
Snia Viscosa

AIRCRAFT SEATS
Aérospatiale
Aviaexport
Marvin Tomkins
Messerschmitt-Bölkow-Blohm

AIRCRAFT WIRE AND CABLE
Marvin Tomkins
Standard Wire & Cable

AIR CYCLE REFRIGERATION PACKAGES
Aviaexport
Garrett Corporation

AIRFIELD LIGHTING
Aviaexport
Omnipol

AIRLINE TECHNICAL ASSISTANCE
Boeing

AIRPORT MAINTENANCE EQUIPMENT
Israel Aircraft Industries
S.N.E.C.M.A.

AIRSPEED INDICATORS
Aeritalia
Aviaexport
Dornier
Marvin Tomkins

ALTERNATORS
Aviaexport
Garrett Corporation
Marvin Tomkins

ALTITUDE CONTROL SYSTEMS
Aviaexport
Eltro
Marvin Tomkins
Meteor
S.F.E.N.A.

ALTIMETERS—ENCODING
Aeritalia
Marvin Tomkins

ANTENNAE
S.G. Brown Communications

ANTI-SKID SYSTEMS
Aviaexport
Goodyear Tyre & Rubber Co.
S.N.E.C.M.A.

ARMAMENTS FOR AIRCRAFT
G.I.A.T.
Marvin Tomkins
M.L. Aviation
Snia Viscosa
Vought Corporation

ASTRO-INERTIAL NAVIGATION SYSTEMS
British Aircraft Corporation

The TempAir Force
A force to be reckoned with. Remember we are the wet leasing specialists—the only airline in the world that operates exclusively as a wet lease carrier to the airlines of the world.
Let us fill your capacity shortages or get your new routes into operation ahead of your competitors. Our Boeings can be operating in your colours within days of contract signature.
Tempair International Airlines Limited
THE AIRLINES' AIRLINE
Thames Side House, Thames Side, Windsor, Berkshire SL4 1QN, England. Telephone: Windsor 53361. Telex: 848967. Cables: Tempair Windsor.
Tempair in Hong Kong: 1105A Prosperity House, 8A/10 Granville Road, Kowloon, Hong Kong. Telephone: Kowloon 685062. Telex: 74558.
Tempair in Manila: P.O. Box 7434, Air Mail Exchange Office, Manila International Airport 3120, Philippines. Telephone: Manila 807011. Telex: 45352.

AUDIO ANCILLARY TEST SETS
S.G. Brown Communications

AUTOMATIC CHECKOUT SYSTEMS
Aviaexport
British Aircraft Corporation
Messerschmitt-Bölkow-Blohm
Selenia
Siai Marchetti

AUTOMATIC PARACHUTE OPENERS
Aviaexport

AUTOMATIC PILOTS
Aviaexport
Marvin Tomkins
Meteor
S.F.E.N.A.

AUTOMATIC VOLTAGE AND CURRENT REGULATORS
Aviaexport
Marvin Tomkins

AUXILIARY POWER PLANT
Aviaexport
Garrett Corporation

BARS—STAINLESS STEEL AND HEAT-RESISTING STEEL
Aviaexport

BATTERIES
Aviaexport
Marvin Tomkins
S.A.F.T.

BATTERIES—AVIATION
Aviaexport
Marvin Tomkins
S.A.F.T.

BATTERY CHARGERS
Aviaexport
Marvin Tomkins
S.A.F.T.

BATTERY TESTING EQUIPMENT
Marvin Tomkins
M.L. Aviation
S.A.F.T.

BELTS—SAFETY
Aviaexport

BINOCULARS
Aeritalia
Barr & Stroud

BLADES—GAS TURBINE
Avoc Lycoming
Aviaexport
Marvin Tomkins
S.N.E.C.M.A.

BOMB CARRIERS
Marvin Tomkins
M.L. Aviation

BOMBSIGHTS
Marvin Tomkins
Saab Scania

BRAKE LININGS
Goodyear Tyre & Rubber Co.
Marvin Tomkins

BRAKES FOR AIRCRAFT
Aviaexport
Goodyear Tyre & Rubber Co.
Israel Aircraft Industries
Marvin Tomkins
S.N.E.C.M.A.

CABIN COOLING (TROPICAL AIRFIELD EQUIPMENT)
Aviaexport
Garrett Corporation

CABIN PRESSURE CONTROL SYSTEMS
Aviaexport
Garrett Corporation
Marvin Tomkins
Saab Scania

CABIN PRESSURE CONTROLS
Aviaexport
Garrett Corporation
Marvin Tomkins

CABIN PRESSURISING TEST EQUIPMENT
Aviaexport
Garrett Corporation

CABLES—ELECTRIC
Aviaexport
Marvin Tomkins
M.L. Aviation
Standard Wire & Cable

CABLES—R.F.
Standard Wire & Cable

CENTRAL AIR DATA COMPUTERS
Garrett Corporation
Saab Scania

COATINGS—EROSION RESISTANT
Avco Lycoming
Goodyear Tyre & Rubber Co.
Israel Aircraft Industries

COMMUNICATIONS CONTROL SYSTEMS—AIRBORNE
S.G. Brown Communications

COMPONENTS
Aviaexport
Dornier
Flight Refuelling
Fokker-VFW
Garrett Corporation
Israel Aircraft Industries
Marvin Tomkins

COMPUTERS
Dornier
Garrett Corporation
Hollandse Signaalapparaten
Saab Scania
Selenia

COMPUTERS—AERODYNAMIC ANALOGUE AND DIGITAL
British Aircraft Corporation
Dornier
Garrett Corporation
Israel Aircraft Industries
S.F.E.N.A.

CONNECTORS
Aviaexport
Garrett Corporation
Marvin Tomkins

CONSTANT SPEED ALTERNATOR DRIVE UNITS
Garrett Corporation
S.F.E.N.A.

CONTROL EQUIPMENT FOR AIRCRAFT
Aviaexport
Dornier
Garrett Corporation
Marvin Tomkins
Saab Scania
S.F.E.N.A.
S.N.E.C.M.A.

CONTROLS—COCKPIT
Aviaexport
Marvin Tomkins
Saab Scania

CONTROLS—MAIN ENGINE FUEL
Aviaexport
Marvin Tomkins
Woodward Governor

COOLING COMPRESSORS
Aviaexport
Garrett Corporation

COOLING TURBINES
Garrett Corporation

CRYOGENIC TURBINES
Garrett Corporation

DATA PROCESSING EQUIPMENT
S.G. Brown Communications
Dornier
Garrett Corporation
Saab Scania
Selenai
S.N.E.C.M.A.

DATA PROCESSING EQUIPMENT FOR ATC
S.G. Brown Communications
Hollandse Signaalapparaten
Selenia

Nobody else can offer you this range.

Ours is a hard working family. A range of three aircraft that gets on with the job and makes little noise doing it. So effectively in fact that it covers every short-haul need in the book and at the same time keeps well inside FAA noise regulations. Whatever the traffic pattern, route network or operating conditions, we have an aircraft to take the task in its stride. And do it profitably. Consider each member of this remarkable family. You'll see why we lead the short-haul field.

The F 27 Friendship. The world's most successful turboprop airliner.

Economy. Versatility. The two foremost reasons why the F 27 has been chosen by 144 operators throughout the world. It's economical because despite high cruising speed it goes easy on the fuel. While low structure weight means it can carry 40-56 passengers or 12,000 lb. of cargo over ranges up to 1,100 n. m. Versatile? Well, apart from passenger and cargo services, it's being used by governments and international organisations, as a business aircraft and for rescue and survey missions.

The VFW 614. The most economical way to join the jet set.

The VFW 614. A feeder jet that will rapidly replace propeller aircraft on short-haul, low density routes. It can land on any airstrip, including semi-prepared fields. It needs no special ground facilities. Uses little fuel and has a low break-even point. For up to 44 passengers it has all the advantages of big jet comfort. And for you, all the advantages of small jet economy.

F 28. The jetliner that makes a profit where others no longer can. With room for 65-85 passengers in spacious comfort the F 28 is a profitable addition to any fleet.

But unlike other jets it can operate economically with load factors as low as 30%.

It offers high speed transportation on secondary routes. Self-sufficiency gives quick turnround times on airfields with limited ground facilities. Low fuel consumption and low maintenance cost ensure high profit potential.

VFW-FOKKER
The short-haul specialists.

Fokker-VFW International bv, P. O. Box 7600, Schiphol-Oost, the Netherlands, Telex: FINT 11526.

DATA TRANSMISSION EQUIPMENT
S.G. Brown Communications
Dornier
Eltro
Saab Scania
Selenia
S.N.E.C.M.A.

DC MOTORS
Aviaexport
Garrett Corporation
Marvin Tomkins

DE-ICING EQUIPMENT
Flight Refuelling
Garrett Corporation
Goodyear Tyre & Rubber Co.
Marvin Tomkins

DIRECTION FINDING EQUIPMENT (TRIANGULATION)
Aviaexport
Selenia

DROGUE GUNS
M.L. Aviation

DRONES
Dornier
Meteor
Selenia

EJECTION SEATS
Saab Scania
S.N.E.C.M.A.

EJECTOR RELEASE UNITS
M.L. Aviation

ELECTRIC AUXILIARIES
Aviaexport
Garrett Corporation
Marvin Tomkins
Woodward Governor

ELECTRICAL EQUIPMENT
Aviaexport
Garrett Corporation
Marvin Tomkins
M.L. Aviation

ELECTRICAL PLUGS AND SOCKETS (WATERPROOF)
Marvin Tomkins

ELECTRICAL WIRING ASSEMBLIES
Aviaexport
Fokker-VFW
M.L. Aviation
Standard Wire & Cable

ELECTRICAL WIRE CABLE CORD OF ALL TYPES
Standard Wire & Cable

ELECTRO-OPTICAL SYSTEMS
Barr & Stroud
Eltro
Saab Scania
Selenia

ELECTRONIC EQUIPMENT
Aeritalia
Aviaexport
Boeing
British Aircraft Corporation
S.G. Brown Communications
Garrett Corporation
Israel Aircraft Industries
Marvin Tomkins
Messerschmitt-Bölkow-Blohm
Meteor
M.L. Aviation
Saab Scania
Selenia
S.F.E.N.A.
S.N.E.C.M.A.

ELECTRONIC FLOWMETERS
Marvin Tomkins

ELECTRONICS AND GUIDANCE
Israel Aircraft Industries
Meteor
Selenia
S.F.E.N.A.

ENGINE HANDLING EQUIPMENT
John Curran

ENGINE COMPRESSOR CLEANING RIGS
John Curran

ENGINE PARTS FABRICATION
Avco Lycoming
Marvin Tomkins

ENGINE STARTING EQUIPMENT
Marvin Tomkins

ENGINE TESTING EQUIPMENT
Avco Lycoming
Aviaexport
John Curran
Garrett Corporation
S.N.E.C.M.A.

ENGINES—AIRCRAFT
Avco Lycoming
Aviaexport
Garrett Corporation
Klöckner-Humboldt-Deutz
Marvin Tomkins
Motoren-und Turbinen-Union
S.N.E.C.M.A.

ENGINES—AUXILIARY
Aviaexport
Garrett Corporation
Klöckner-Humboldt-Deutz
Motoren-und Turbinen-Union

ENGINES—V/STOL
Avco Lycoming
Aviaexport
Garrett Corporation
S.N.E.C.M.A.

ENVIRONMENTAL CONTROL SYSTEMS
British Aircraft Corporation
Garrett Corporation
Selenia

EXPERIMENTAL ASSEMBLIES
John Curran
M.L. Aviation

ELECTRIC TRACTORS
M.L. Aviation

FIBRE OPTICS
Barr & Stroud

FILTERS—AIR
Aviaexport

FILTERS—ELECTRONIC
Aviaexport
Barr & Stroud
Selenia

FILTERS—FUEL AND OIL
Aviaexport
Flight Refuelling

FLIGHT INSTRUMENT TEST SETS
Garrett Corporation
Israel Aircraft Industries
S.F.E.N.A.
S.N.E.C.M.A.

FLOTATION GEAR
Garrett Corporation

FLOW GAUGES
Aviaexport
Marvin Tomkins

FLOWMETERS
Marvin Tomkins

FLYING CLOTHING
M.L. Aviation

FORGINGS—STEEL
S.N.E.C.M.A.

FUEL FLOW PROPORTIONERS
Flight Refuelling
Garrett Corporation
Marvin Tomkins

FUEL PUMPS
Aviaexport
Garrett Corporation
Marvin Tomkins
Siai Marchetti

FUEL SYSTEMS PROTECTION
PRB

FUEL SYSTEMS AND REFUELLING EQUIPMENT
Aviaexport
Flight Refuelling
Israel Aircraft Industries
Saab Scania

FUEL TANK PRESSURISATION EQUIPMENT
Flight Refuelling
Garrett Corporation
Israel Aircraft Industries
Sabb Scania

FURNISHINGS FOR AIRCRAFT CABINS
Aviaexport
Garrett Corporation

FEEL SIMULATOR CONTROLS AND JACKS
Garrett Corporation
Saab Scania

GAS TURBINE STARTER SYSTEMS
Aviaexport
Garrett Corporation
Klöckner-Humboldt-Deutz

GAS TURBINES
Avco Lycoming
Aviaexport
Garrett Corporation
Klöckner-Humboldt-Deutz
S.N.E.C.M.A.

GAS TURBINES AND ACCESSORIES
Avco Lycoming
Aviaexport
Garrett Corporation
Klöckner-Humboldt-Deutz
S.N.E.C.M.A.

GAUGES
Aviaexport
Marvin Tomkins

GENERATORS
Aviaexport
Garrett Corporation
Marvin Tomkins

GROUND REFUELLING EQUIPMENT
Flight Refuelling
Goodyear Tyre & Rubber Co.

GROUND WORKSHOP AND HANGAR EQUIPMENT
Israel Aircraft Industries
Marvin Tomkins

GUIDED MISSILE GROUND HANDLING EQUIPMENT
Garrett Corporation
Messerschmitt-Bölkow-Blohm

GUIDED MISSILE GROUND HANDLING EQUIPMENT (contd.)
M.L. Aviation
Saab Scania
Selenia

HIGH PRESSURE COUPLINGS
Flight Refuelling

HYDRAULIC EQUIPMENT
Aviaexport
Flight Refuelling
Garrett Corporation
Israel Aircraft Industries
Marvin Tomkins
Saab Scania
S.A.M.M.
S.N.E.C.M.A.

HYDRAULIC PRESSURE SUITS
Aviaexport
Garrett Corporation
Marvin Tomkins
S.A.M.M.

HYDRAULIC TEST UNITS—STATIC AND MOBILE
Aeronautica Macchi

INDICATORS—FAULT ISOLATION
Minelco

INERTIAL NAVIGATION SYSTEMS
S.F.E.N.A.

INFLATABLE STRUCTURES
Garrett Corporation

INFRA-RED LINESCAN
Eltro

INFRA-RED MATERIALS
Barr & Stroud
Selenia

INFRA-RED SYSTEMS
Barr & Stroud
Eltro
Garrett Corporation
Saab Scania
Selenia

INSTRUMENT COMPONENTS (MECHANICAL)
Marvin Tomkins

INSTRUMENTS—AIRCRAFT
Aeritalia
Israel Aircraft Industries
Marvin Tomkins

INSTRUMENTS—ELECTRONIC
Aeritalia
Aviaexport
Barr & Stroud
British Aircraft Corporation

INSTRUMENTS—ELECTRONIC (contd.)
S.G. Brown Communications
Israel Aircraft Industries
Marvin Tomkins
S.N.E.C.M.A.

INSTRUMENTS—NAVIGATION
Aeritalia
Aérospatiale
Aviaexport
British Aircraft Corporation
Marvin Tomkins
S.F.E.N.A.

GUIDED MISSILES
Aérospatiale
British Aircraft Corporation
Dornier
Messerschmitt-Bölkow-Blohm
Saab Scania
Selenia
Vought Corporation

GUNNERY TRAINING APPARATUS
Saab Scania

HANGAR TEST STANDS
John Curran
Garrett Corporation
Israel Aircraft Industries

HEADPHONES
S.G. Brown Communications
Marvin Tomkins

HEAT EXCHANGERS
Garrett Corporation
S.N.E.C.M.A.

HEAT TRANSFER SYSTEMS
Garrett Corporation

HEATED WINDOWS
Barr & Stroud

HEATED WINDSCREEN CONTROLLERS
Garrett Corporation

HELICOPTER WINCHES ETC.
Garret Corporation

HELICOPTERS—COMMERCIAL
Aérospatiale
Agusta
Aviaexport
Bell Helicopter
Boeing
Dornier
Messerschmitt-Bölkow-Blohm
Siai Marchetti

HELICOPTERS—MILITARY
Aérospatiale
Agusta
Boeing
Messerschmitt-Bölkow-Blohm
Siai Marchetti

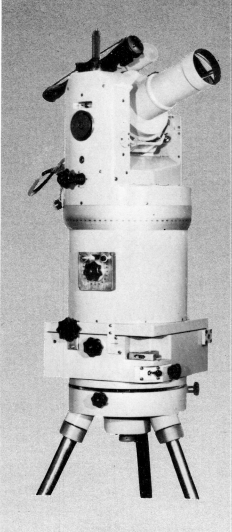

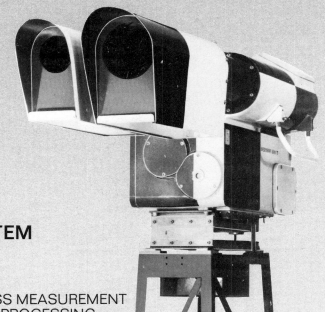

HELICOPTER PARTS AND COMPONENTS
Aviaexport
Dornier
Israel Aircraft Industries
Marvin Tomkins
Messerschmitt-Bölkow-Blohm
Siai Marchetti

HELICOPTER SEARCHLIGHTS
Garrett Corporation

HIGH ALTITUDE TESTING PLANT
Garrett Corporation

HIGH ALTITUDE PRESSURE SUITS
Garrett Corporation

INSTRUMENTS—PRECISION
British Aircraft Corporation
Israel Aircraft Industries
Marvin Tomkins
Rolex

INSTRUMENTS—TEST EQUIPMENT
Aérospatiale
Aviaexport
British Aircraft Corporation
S.G. Brown Communications
Garrett Corporation
Israel Aircraft Industries
Marvin Tomkins
S.N.E.C.M.A.
Woodward Governor

INTEGRATED TOTAL PNEUMATIC SYSTEMS
Garrett Corporation

INTERCOMMUNICATION EQUIPMENT
S.G.Brown Communications

JACKS
Marvin Tomkins
Saab Scania

JET ENGINE PARTS
Avco Lycoming
Klöckner-Humboldt-Deutz
Marvin Tomkins
S.N.E.C.M.A.

JET ENGINE TEST PLANT
Avco Lycoming
John Curran
Garrett Corporation
S.N.E.C.M.A.

JET FUEL STARTERS
Garrett Corporation
Marvin Tomkins

JET PROPULSION ENGINES
Avco Lycoming
Garrett Corporation
Marvin Tomkins
S.N.E.C.M.A.

JOINTING COMPOUND
Goodyear Tyre & Rubber Co.

LAMPS—AIRSTRIP/OBSTRUCTION
Aviaexport

LAMPS—COCKPIT
Aviaexport
Marvin Tomkins

LAMPS FOR GROUND STATIONS
Marvin Tomkins

LANDING LAMPS
Aviaexport
Marvin Tomkins

LASERS
Aérospatiale
Barr & Stroud
Eltro
Garrett Corporation
Messerschmitt-Bölkow-Blohm
Selenia

LASER RANGEFINDERS
Barr & Stroud
Eltro
Saab Scania
Selenia

LIFE-SAVING EQUIPMENT
Garrett Corporation
Marvin Tomkins
Saab Scania

LIGHTS—AIRCRAFT
Aviaexport
Marvin Tomkins

LIGHTS—IDENTIFICATION
Marvin Tomkins

LIGHTS—LANDING
Aviaexport
Marvin Tomkins

LIGHTS—NAVIGATION
Aviaexport
Marvin Tomkins

LINEAR ACTUATORS
Garrett Corporation
Marvin Tomkins

LININGS—BRAKES
Goodyear Tyre & Rubber Co.
Marvin Tomkins

MACH NUMBER TRANSDUCERS
Garrett Corporation

MACHINE TOOLS
Marvin Tomkins

MARINE ENGINES
Garrett Corporation
Klöckner-Humboldt-Deutz

MATERIALS TECHNOLOGY
Boeing
Vought Corporation

METAL FITTINGS
Aviaexport

MICROPHONES
Aviaexport
S.G. Brown Communications

MISSILES—GUIDED
Aérospatiale
British Aircraft Corporation
Dornier
Messerschmitt-Bölkow-Blohm
Saab Scania
Selenia
Vought Corporation

MOTORS—ELECTRIC
Aviaexport
Garrett Corporation
Marvin Tomkins

MOTOR GENERATORS
Aviaexport
Garrett Corporation
Marvin Tomkins

MOTORS—HYDRAULIC
Aviaexport
Garrett Corporation
Marvin Tomkins

NIGHT VISION EQUIPMENT
Barr & Stroud
Eltro
Selenia

NON-DESTRUCTIVE INSPECT EQUIPMENT
Fokker-VFW

OIL VALVES
Flight Refuelling
Garrett Corporation
Marvin Tomkins

OPTICAL EQUIPMENT
Barr & Stroud
Marvin Tomkins

OPTICAL GUN SIGHTS
Eltro
G.I.A.T.
Marvin Tomkins
Saab Scania
Selenia

OXYGEN APPARARUS
Aviaexport
Marvin Tomkins

VOUGHT
UNITED STATES OF AMERICA
U.S.ARMY
U.S.ARMY
U.S.ARMY
AF 66431
WHO'S KEEPING THINGS MOVING IN ADVANCED TECHNOLOGY?
We've had a lot of practice getting from one point to another. With the Space Shuttle's leading edge. The Lance artillery missile. The Corsair tactical fighter. Airtrans people mover. And more.
All hard-working solutions to tough problems.
But one of the things we've learned over the years is never to be content with our past success.
Because we keep some fast moving company.
VOUGHT CORPORATION/An LTV Company

OXYGEN BREATHING APPARATUS
Garrett Corporation
Marvin Tomkins
Saab Scania

OXGEN BREATHING SYSTEMS
Garrett Corporation
Marvin Tomkins
Saab Scania

OVERHAUL AND MODIFICATION KITS
British Aircraft Corporation
Dornier
Garrett Corporation
Siai Marchetti
Woodward Governor

PARACHUTES
Aviaexport

PARACHUTES—SPECIAL PURPOSE
Aviaexport

PARTS FOR U.S. BUILT AIRCRAFT
Garrett Corporation
Marvin Tomkins
Vought Corporation

PASSENGER BRIDGES (AVIOBRIDGE)
Fokker-VFW

PERISCOPES
Barr & Stroud
Marvin Tomkins

PHOTOGRAPHIC EQUIPMENT
Fokker-VFW
Marvin Tomkins

PLASTICS FABRICATIONS
Aeritalia
Israel Aircraft Industries

PLASTIC FABRICATIONS (RE-INFORCED WITH FIBREGLASS)
Aeritalia
Boeing
Fokker-VFW
Israel Aircraft Industries
Messerschmitt-Bölkow-Blohm

PLASTIC MOULDINGS
Fokker-VFW

PNEUMATIC CONTROLS
Garrett Corporation

POWER CONTROL FOR AIRCRAFT
Garrett Corporation
S.A.M.M.

PRESSURE CONTROL EQUIPMENT
Garrett Corporation

PRESSURE RADIO TRANSDUCERS
Garrett Corporation

PRESSURE REGULATING VALVES—FLUIDS AND GASES
Flight Refuelling
Garrett Corporation
Saab Scania

PRESSURE SWITCHES
Marvin Tomkins
S.A.M.M.

PRESSURE TRANSDUCERS
Garrett Corporation
Marvin Tomkins

PROPELLER GOVERNORS
Marvin Tomkins
Woodward Governor

PROPELLER HUBS
Marvin Tomkins
Meteor

PROPELLER TEST STANDS
John Curran

PROPELLERS
Marvin Tomkins

PROPOSAL FOR AIRCRAFT GROUND SUPPORT OPERATIONS
Boeing
M.L. Aviation
Siai Marchetti
Vought Corporation

PROTECTIVE CLOTHING
Marvin Tomkins

PUMPS—AIR COMPRESSOR
Garrett Corporation
Marvin Tomkins
S.N.E.C.M.A.

PUMPS—FUEL AND OIL
Garrett Corporation
Israel Aircraft Industries
Marvin Tomkins
Siai Marchetti

PUMPS—HYDRAULIC
Garrett Corporation
Israel Aircraft Industries
Marvin Tomkins
Saab Scania
S.A.M.M.
S.N.E.C.M.A.

PLATFORM TRUCKS
Fokker-VFW
M.L. Aviation

PROVISIONING PARTS BREAKDOWN LIST
John Curran
Marvin Tomkins

PUMPS—AGRICULTURAL SPRAY
Dornier

RADAR FOR NAVIGATION, WARNING INTERCEPTION, FIRE CONTROL AND AIRFIELD SUPERVISION
Aviaexport
Hollandse Signaalapparaten
Israel Aircraft Industries
Omnipol
Selenia
S.N.E.C.M.A.

RADAR REFLECTORS
John Curran
Fokker-VFW
Selenia

RADAR TOWERS
John Curran

RADAR TURNING GEARS AND EQUIPMENT
Aviaexport
John Curran

RADIO ALTIMETERS
Aviaexport

RADIO EQUIPMENT
Aviaexport
Israel Aircraft Industries
Marvin Tomkins
S.N.E.C.M.A.

RADIO EQUIPMENT—GROUND HF AND AIRBORNE HF/VHF
S.G. Brown Communications

RADIO NAVIGATION EQUIPMENT
Aérospatiale
Aviaexport
Israel Aircraft Industries
Marvin Tomkins

RAMJET PROPULSION ENGINES
Garrett Corporation
Messerschmitt-Bölkow-Blohm

RANGEFINDERS
Barr & Stroud
Eltro
Selenia

REFRIGERATION COMPRESSORS
Garrett Corporation
M.L. Aviation

REPAIR AND MAINTENANCE OF AIRCRAFT
Aeritalia
Avco Lycoming

Partner in International Programs

Civil Aircraft
MBB together with European partners is building the Airbus A300, the world's most economic and ecologically benign wide-bodied commercial plane, and is handling the overall spare parts supply. MBB is a partner in the development and construction of the commercial aircraft VFW 614 and Fokker F. 28.

Military Aircraft
MBB is engaged in the development, construction and support of military aircraft. The main program comprises the combat aircraft Tornado developed in European cooperation together with BAC and Aeritalia within the scope of PANAVIA. Further activities: F-4 Phantom, F-104 G Starfighter, C.160 Transall.

Helicopters
MBB developed and built the first German production helicopter, the BO 105. This multi-purpose helicopter has proved itself in 14 countries over four continents in executive, rescue, offshore and police operations.

Messerschmitt-Bölkow-Blohm GmbH
Postfach 80 11 09
D-8000 München 80
Germany

ZA-9/76 E

REPAIR AND MAINTENANCE OF AIRCRAFT (contd.)
Aviaexport
Boeing
Dornier
Fokker-VFW
Garrett Corporation
Israel Aircraft Industries
Messerschmitt-Bölkow-Blohm
Meteor
Pilatus
Saab Scania
S.F.E.N.A.
Siai Marchetti
Vought Corporation

REPAIR AND OVERHAUL OF AERO ENGINES
Garrett Corporation
Marvin Tomkins
Meteor
S.N.E.C.M.A.

REPAIR OF AIRCRAFT INSTRUMENTS
Aviaexport
Fokker-VFW
Israel Aircraft Industries
Marvin Tomkins
Pilatus
S.F.E.N.A.
Siai Marchetti

ROCKET ENGINE TEST PLANT
John Curran
Messerschmitt-Bölkow-Blohm

ROCKET PROPULSION
Messerschmitt-Bölkow-Blohm
Snia Viscosa

ROCKET SOUNDING
British Aircraft Corporation
Dornier
Saab Scania

ROTARY ACTUATORS
Garrett Corporation
Marvin Tomkins

RUNWAY FRICTION MEASURING EQUIPMENT
M.L. Aviation

SEAT BELTS
Aviaexport

SERVO ACTUATORS
Garrett Corporation
Marvin Tomkins
S.A.M.M.
S.F.E.N.A.

SHEET METAL WORK
Fokker-VFW
Israel Aircraft Industries

SHEET METAL WORKING MACHINES
Fokker-VFW

SIMULATORS
Dornier
Saab Scania
Selenia
Siai Marchetti
Vought Corporation

SOLENOID VALVES
Flight Refuelling
Marvin Tomkins
Saab Scania

SPACE HARDWARE RECOVERY
Dornier

SPACE SYSTEMS
Aeritalia
Aérospatiale
Boeing
British Aircraft Corporation
Dornier
Fokker-VFW
Garrett Corporation
Messerschmitt-Bölkow-Blohm
Selenia
Vought Corporation

SPACECRAFT
Aeritalia
British Aircraft Corporation
Dornier
Selenia
Vought Corporation

SPACE SATELLITES
Aeritalia
Aérospatiale
British Aircraft Corporation
Dornier
Fokker-VFW
Messerschmitt-Bölkow-Blohm
Saab Scania
Selenia
Snia Viscosa

SPARE PARTS FOR U.S. BUILT AIRCRAFT
Garrett Corporation
Marvin Tomkins
Siai Marchetti
Vought Corporation

STABILITY AUGMENTATION SYSTEMS
Aérospatiale
Dornier
S.F.E.N.A.

STALL WARNING SYSTEMS
Fokker-VFW
Saab Scania

STARTER PODS—AIRBORNE
Garrett Corporation

STARTING SYSTEMS—AIRBORNE
Garrett Corporation
Marvin Tomkins

STATIC INVERTERS
Flite-Tronics International

STATION BOXES
S.G. Brown Communications

STEEL AND STEEL ALLOYS
Aviaexport

STORAGE TANKS
Garrett Corporation

SURVEILLANCE SYSTEMS
S.G. Brown Communications
Selenia

SURVIVAL EQUIPMENT
Garrett Corporation
Saab Scania

SWITCHGEAR
Aviaexport
Marvin Tomkins

SWITCHES—MINIATURE ELECTRICAL
Marvin Tomkins
Minelco
S.A.M.M.

TACHOMETERS
Aviaexport

TARGET TOWING WINCHES
Flight Refuelling
Garrett Corporation
Saab Scania

TECHNICAL PUBLICATIONS
John Curran
Dornier
Flight Refuelling
Fokker-VFW
Israel Aircraft Industries
Siai Marchetti

TECHNICAL PUBLICATIONS—SPECIAL STUDIES
John Curran
Dornier

TEMPERATURE CONTROL EQUIPMENT
Garrett Corporation
Marvin Tomkins

TEST EQUIPMENT
Aeronautica Macchi
Avco Lycoming
Aviaexport
S.G. Brown Communications
John Curran
Garrett Corporation
Israel Aircraft Industries
Marvin Tomknis
M.L. Aviation
Saab Scania
S.F.E.N.A.
S.N.E.C.M.A.
Woodward Governor

TEST EQUIPMENT—AIRBORNE RADIO
S.G. Brown Communications

TEST EQUIPMENT—AIRFIELD RADIO
S.G. Brown Communications
S.N.E.C.M.A.

TEST EQUIPMENT METAL BONDING
Aviaexport
Siai Marchetti

THERMO COUPLE CABLES
Israel Aircraft Industries
Standard Wire & Cable

TRAINING DEVICES
Vought Corporation

TRANSFORMER RECTIFIER UNITS
Israel Aircraft Industries

TUBES—STAINLESS STEEL
Aviaexport

TURBINES—RAM AIR
Garrett Corporation

TURBOFAN ENGINES
Avco Lycoming
Garrett Corporation
S.N.E.C.M.A.

TYRES FOR AIRCRAFT
Goodyear Tyre & Rubber Co.

UNDERCARRIAGE EQUIPMENT
Marvin Tomkins
S.N.E.C.M.A.

UNDERCARRIAGE GEAR—RETRACTABLE
S.N.E.C.M.A.

VALVES
Flight Refuelling
Garrett Corporation
Israel Aircraft Industries
Marvin Tomkins
Saab Scania
Siai Marchetti

VALVES AND MINIATURE RELAYS
Garrett Corporation
Marvin Tomkins

VALVES—CONTROL HYDRAULIC
Garrett Corporation
Israel Aircraft Industries
Marvin Tomkins
Saab Scania

VALVES—ELECTRONIC
Garrett Corporation
Marvin Tomkins

VALVES—FUSES HYDRAULIC
Marvin Tomkins

VALVES—NON-RETURN FUEL
Flight Refuelling
Garrett Corporation
Marvin Tomkins
Saab Scania

VALVES—NON-RETURN HYDRAULIC
Flight Refuelling
Garret Corporation
Marvin Tomkins
Saab Scania
S.A.M.M.

VALVES—SEQUENCE HYDRAULIC
Marvin Tomkins
Saab Scania

VALVES—RELIEF HYDRAULIC
Garrett Corporation
Marvin Tomkins
Saab Scania

VAPOUR CYCLE REFRIGERATION PACKAGES
Garrett Corporation

VERTICAL TAKE-OFF AIRCRAFT
Dornier
Vought Corporation

VISIBILITY MEASURING EQUIPMENT
Eltro
S.N.E.C.M.A.

VOLTAGE AND CURRENT REGULATORS
Garrett Corporation
Marvin Tomkins

WATER SEPARATORS
Flight Refuelling
Garrett Corporation
Marvin Tomkins

WHEELS FOR AIRCRAFT
Aviaexport
Goodyear Tyre & Rubber Co.
Marvin Tomkins
S.N.E.C.M.A.

WIND TUNNEL TESTING PLANT
Aeronautica Macchi
Boeing
John Curran
Dornier
Vought Corporation

F-BVHX
AIR FRANCE
N612AC
From the Rallye light aircraft to the supersonic Concorde
AEROSPATIALE-FRANCE / BRITISH AIRCRAFT CORPORATION...
helicopters in operation throughout the world...
A major share in multi-national space programmes...
395000 tactical missiles sold...
QUALITY :
the hallmark of Aerospatiale products.
aerospatiale
aerospatiale
37, boulevard de Montmorency
75781 PARIS Cedex 16 - FRANCE

SNIA
DEFENCE AND AEROSPACE DIVISION
COMPLETE ROUNDS FOR ARTILLERY AND MORTARS • CARTRIDGES FOR SMALL ARMS • PROPELLING POWDERS AND BURSTING EXPLOSIVES
AIR-TO-AIR AND AIR-TO-GROUND ROCKETS • FIELDS ROCKETS • WARHEADS FOR ROCKETS AND MISSILES • SOLID PROPELLANT ROCKET ENGINES FOR MILITARY AND SPACE USE • DOUBLE BASE AND COMPOSITE SOLID PROPELLANTS
RESEARCH AND DEVELOPMENT IN THE MISSILE AND SPACE FIELD
00187 ROMA VIA LOMBARDIA 31
TELEF. 4680 • TELEX 61114

on
8 out of 10 airliners
there is
a SFENA
standby
horizon
The wide range of models, clear presentation
and exceptional reliability of SFENA gyroscopic
horizons have already secured their unrivalled
reputation as standby instruments.
Today over 140 airlines and 30 airforces
rely on SFENA horizons.
(manufactured by SFENA France and under license,
in particular in the U.S.A.)
SFENA
Aérodrome de Villacoublay B.P. n° 59
78140 VELIZY VILLACOUBLAY France
630.23.85
260046 SFENA VELIZ.
SFENA OF AMERICA
1730 Rhode Island Ave. N.W.
WASHINGTON, D.C. 20036
Tel. (202) 296.7734
SFENA
A research centre of international calibre for
advanced technology systems
A production division with over 35 000 ft²
of clean rooms
An after-sales division with branches all
over the world
SFENA LEADS THE WAY
VOLK-00.76.3

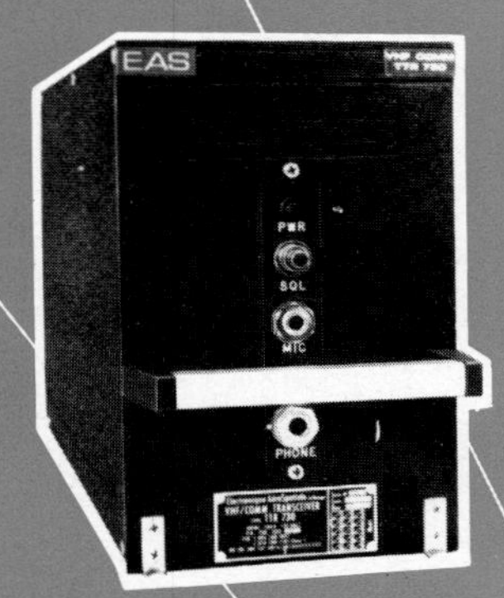

VHF TTR 730
ARINC 566 A
- 118.00 to 135.975 MHZ
- 116.00 to 149.975 MHZ

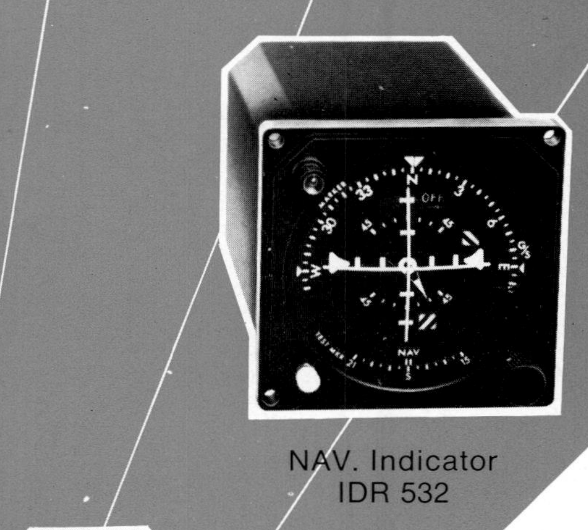

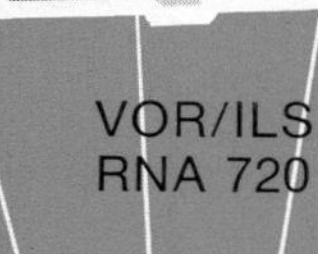

NAV. Indicator
IDR 532

VOR/ILS
RNA 720

MICRO-TACAN
Indicator
IMT 565
- ARINC 582 I/P

COM/NAV Control unit
BCN 3.671

une certaine idée de l'avionique.

Electronique AeroSpatiale

B.P. 4 . 93350 Aeroport du Bourget - France
Tel. 834.4999 - Telex. 220809 F

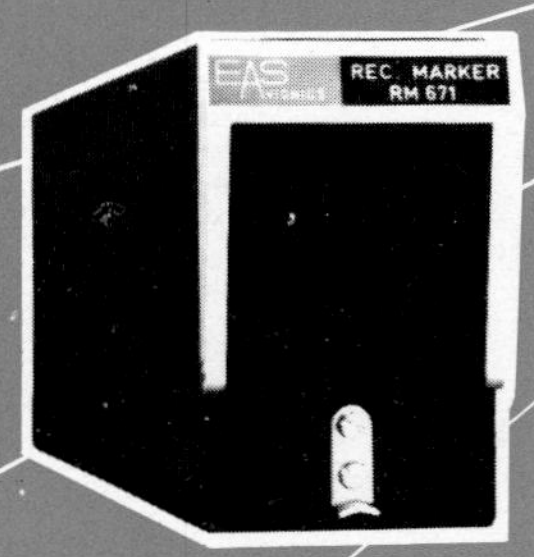

MARKER
beacon Receiver
RM 671

VHF ERC 741
- 116.00 to 149.975 MHZ
- 100.00 to 149.975 MHZ (optional)
- 20 W power O/P
- 1/4 ATR

UHF TRU 750
- 225.00 to 399.975 MHZ
- 20 W Power O/P
- 1/4 ATR

V/UHF TVU 740
- 116.00 to 149.975 MHZ
- 100.00 to 149.975 MHZ (option)
- 225.00 to 399.975 MHZ
- VHF Power O/P : 20 W
- VHF Power O/P : 20 W

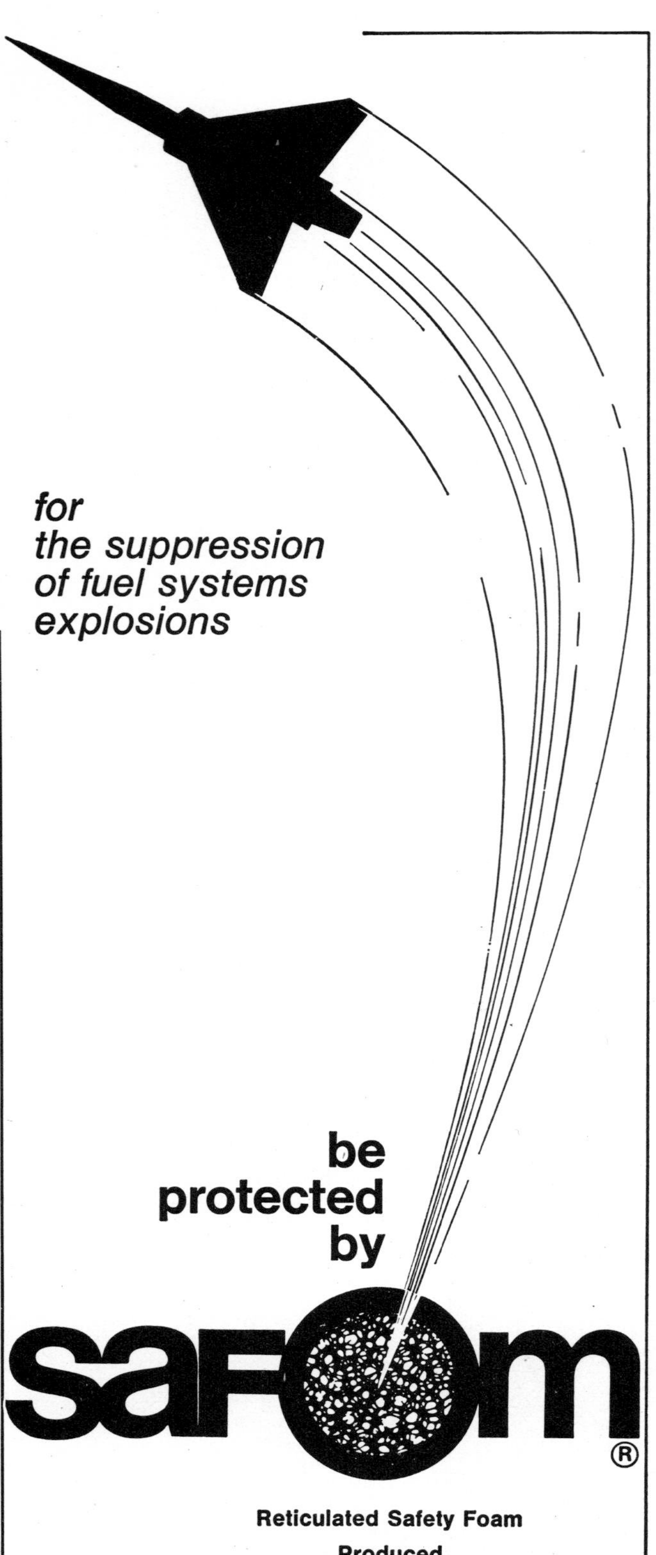

Reticulated Safety Foam

**Produced
to military specifications**

When fitted into a fuel tank it will
completely suppress the explosion
which follows the ignition
of a fuel/air mixture
in a confined space.

Weight: 30 kg per m3.

Reduction of tank capacity: 3%.

Can be fitted during construction or retrofitted on existing
aircraft fuel-system by our specialised personnel.

PRB s.a.

Division POLYURETHANE
Damstraat, 2
9200 Wetteren/BELGIUM
Tel. 091 69 2801
Telex: 11 363

BRAZIL: a growing aerospace market.

Brazil will produce roughly 700 aircraft of a dozen different types in 1976. The Armed Forces are undergoing substancial modernization and reequipment programs. The Government is giving impetus to the establishment of a comprehensive network of 3rd-level airline services to complement the country's vast domestic trunklines. More and more airports are needed to cope with the increasing traffic demand. Military and scientific rockets are being series manufactured. Space technology programs have put the country in the foremost position among all Latin American nations.

This alone shows the potential market for suppliers from all over the world planning to make business (and money) in Brazil. RELATÓRIO AERONÁUTICO has the widest circulation among top-level civil and military officials in the country's aerospace market, to whom it has for long been the source of reliable, first-hand information.

If you are planning to advertise in Brazil, why not use the pages of RELATÓRIO AERONÁUTICO? After all, you have the right to get the most out of your advertising budget.

CAIXA POSTAL 4928 – 20000 RIO DE JANEIRO, RJ – BRAZIL – TEL.(021)722-6741

our trail in the sky
The G222 is today the only medium range military transport aircraft available on the market today. Wide operational capabilities and high reliability even in critical flight conditions are guaranteed by the advanced G.E. T64 P4D turboengines. The most sophisticated avionics systems and airborne instrumentation enable the aircraft to operate in all weather conditions independently of ground assistance. The G222 is certified for operation with a crew of two and can take-off and land on grass strips. Its large volumetric capacity allows to carry a wide range of loads (up to 20,000 lbs.) or 44 fully equipped soldiers and to parachute 32 paratroopers or to air drop heavy loads up to 11,000 lbs. The G222 can be rapidly converted to fulfill rescue missions, aeromedical transport, aerophotogrammetry, fire fithting and radio calibration. And all the above at a very low operational cost.
AERITALIA
HEAD OFFICE
PIAZZALE TECCHIO, 51
80125 NAPOLI, ITALY
PHONE 619.522 TELEX 71370

Turboshaft engine for helicopters

Engine assembly line

Development test rig

Jet engine with afterburner

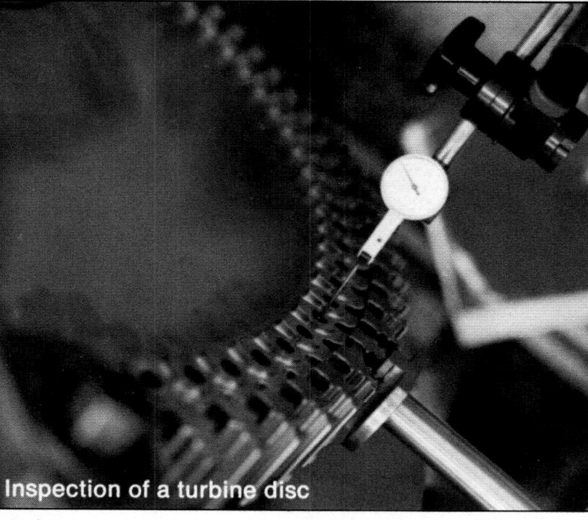
Inspection of a turbine disc

SAMM air equipments

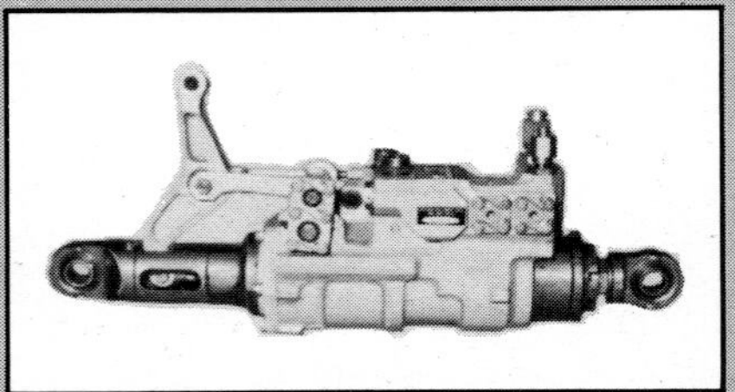

Powered flying control
Pilot's handgrips
Electro pump sets
Pressure switches ....
are fitted to :

MIRAGE F1, G, III, IV et V
JAGUAR A, E, B et S
FOUGA CM170 - ALPHA JET
MACCHI MB326 K - FALCON
AIRBUS A300 B - CORVETTE
MERCURE - CONCORDE
SUPER FRELON - SA321
PUMA - SA330
GAZELLE - SA341
DAUPHIN - SA360
SA350 - SA365
ALOUETTE II et III

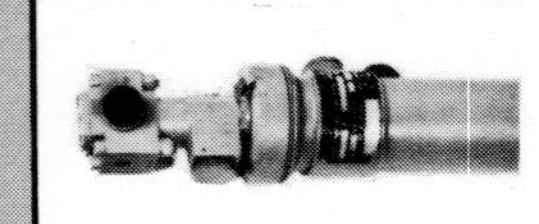

224, Quai de Stalingrad
92130 ISSY-les-MOULINEAUX
Tél. PARIS (1) 644.55.44 Télex 200091

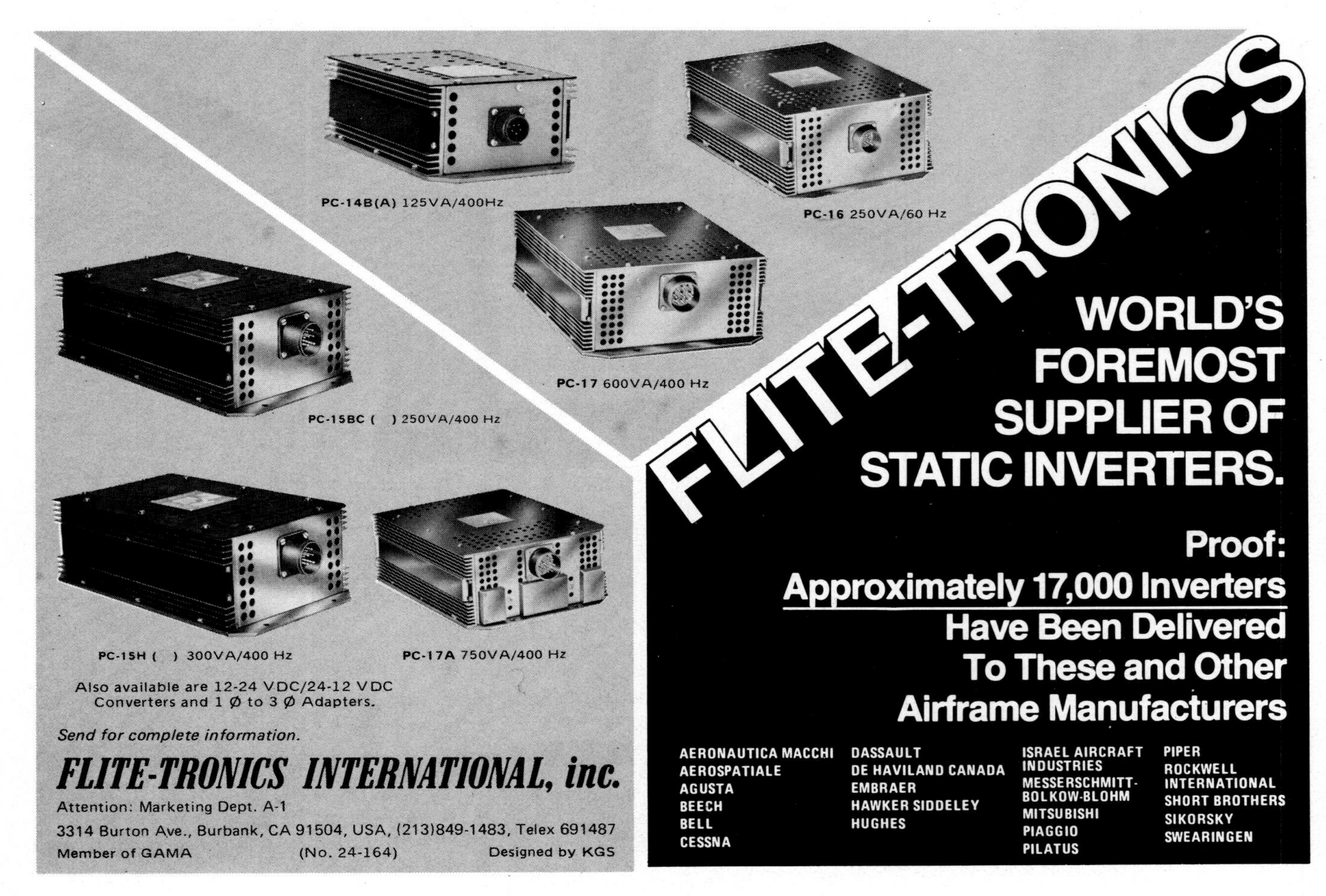

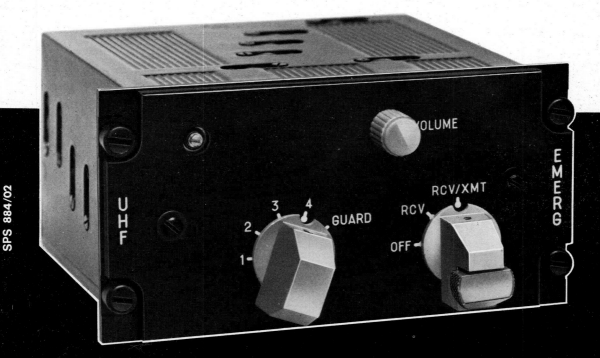

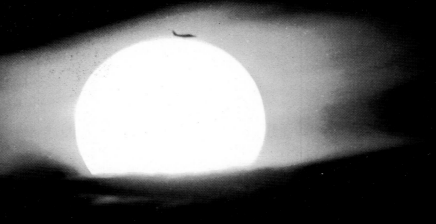

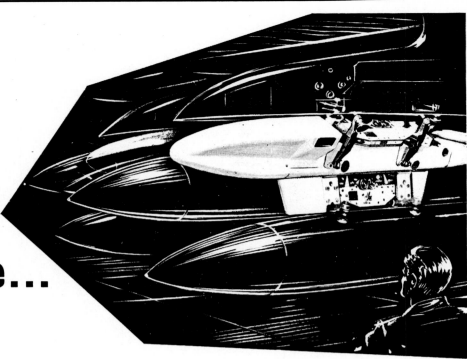

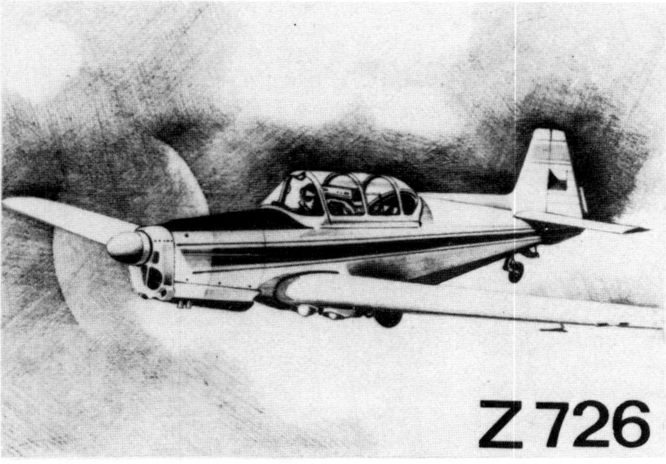

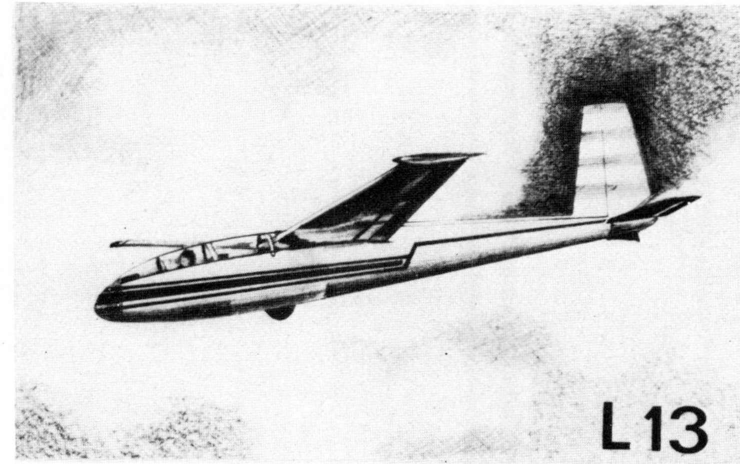

True to its tradition the Czechoslovak aviation industry manufactures highly economical, high performance and highly reliable aircraft for various purposes:

Z-42: all-metal low-wing monoplane powered by a M137 engine rated at 180 h.p. for basic and advanced training, aerobatic training and general flying.

Z-43: four-seater powered by a M337 engine rated at 210 h.p. similar in design to the former model for business trips, general flying, navigation training, IFR training etc.

Z-726: universal trainer for all stages of training—an all-metal low-wing monoplane seating a crew of two. It is powered by a M137 engine or alternatively, in the Z-726K model, by a supercharged M337 engine.

L-13: the all-metal two-seater sailplane manufactured for many years and demanded throughout the world.

Exporter:
OMNIPOL
Washingtonova 11, Praha 1
Czechoslovakia

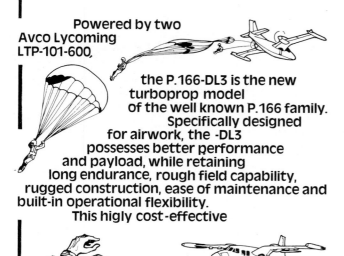

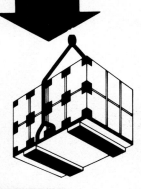

PORTRAIT OF A DEFENC

Rockwell International B-1 strategic bomber skimming over the Mojave Desert at about Mach 0·85 during a test mission, using natural terrain to hide from ground radar

JANE'S ALL THE WORLD'S AIRCRAFT

FOUNDED IN 1909 BY FRED T. JANE

COMPILED AND EDITED BY
JOHN W. R. TAYLOR FRHistS, MRAeS, FSLAET
ASSISTANT EDITOR
KENNETH MUNSON, Associate RAeS

1976-77

I.S.B.N. 0-531 03260 4

L. of C. Cat. No. 75-15174

JANE'S YEARBOOKS

FRANKLIN WATTS INC.
NEW YORK

First Published in the United Kingdom 1976 by Jane's Yearbooks
First American Publication 1976 by Franklin Watts Inc.

JANE'S ALL THE WORLD'S AIRCRAFT 1976-77

The Editor has been assisted in the compilation of this edition as follows:

Kenneth Munson	AIRCRAFT SECTION, THE ARGENTINE TO FINLAND, GREECE TO ISRAEL, JAPAN TO TURKEY, THE UNITED KINGDOM (190-205); RPVS; SAILPLANES.
David Mondey	AIRCRAFT SECTION, GERMANY, THE UNITED KINGDOM (160-190), UNITED STATES OF AMERICA; INDEX
W. T. Gunston	AERO-ENGINES
Michael Taylor	HOMEBUILTS; METRIC CONVERSIONS
Maurice Allward	AIRCRAFT SECTION, ITALY; SPACEFLIGHT AND RESEARCH ROCKETS
The Lord Ventry	AIRSHIPS

CONTENTS

We build more than just products

We build leadership. We are one of the handful of international aerospace companies whose work shapes technological progress.

We build confidence. Our products, chosen by over half of all the countries in the world, are backed by after-sales services second to none.

We build international partnerships. We have greater experience than any other company in the world of collaboration with other nations on high-technology programmes.

We build technological capability. We are constantly probing the frontier of knowledge on programmes which cover the whole span of aerospace technology.

We build experience. For more than 60 years we have been continuously involved in the design, development, production and world-wide marketing of military and civil aircraft and, more recently, missiles, space satellites, and associated equipment and systems.

We build Concorde *(with Aerospatiale of France)*
Tornado *(with MBB of Germany and Aeritalia of Italy)*
Jaguar *(with Dassault/Breguet of France)*
BAC One-Eleven
Strikemaster
Rapier
Swingfire
Seawolf
Intelsat IV and IVA satellites and sub-systems
(with Hughes Aircraft Company of USA)
GEOS *(as leader of the 14-company STAR consortium)* and a constantly developing range of high-technology products for use in the air, on land, at sea, and in space.

FOREWORD

The final quarter of our century has begun with a truly momentous year in aerospace history. A new era of exploration was heralded by the touchdown of the first Viking Lander on Mars, on 20 July 1976. Supersonic air transportation became available to all who could afford it six months earlier, on 21 January. The world's absolute air speed record was raised for the first time in more than a decade on 27 July. The first operational manned air and space craft emerged from its assembly hall in California on 17 September. On the 6th day of the same month, a defecting pilot presented his country's potential enemies with what some experts regard as the most valuable intelligence gift of the postwar period.

Of the five headline-making events listed, two were provided by aircraft that were designed in the early sixties but remain the fastest ever to enter production and service. The speed record was set by a USAF Lockheed SR-71A unarmed reconnaissance aircraft, a type first flown in December 1964, which averaged 1,901 knots (subject to confirmation) over a 15-25 km straight course. The aircraft flown to Hakodate airport, Japan, by the defecting Soviet pilot, Lt Viktor Belenko, was a MiG-25 fighter. Known to NATO as *Foxbat-A,* this type of aeroplane set a speed record as long ago as the Spring of 1965 yet was still rated by America's 1973 Air Force Secretary as "probably the best interceptor in production in the world today".

If the USAF's technical staff and the US aircraft industry had been offered their choice of any aircraft in existence, to dismantle and evaluate, the MiG-25 would have been on the final short list of two, with Tupolev's *Backfire-B* variable-geometry bomber. It had long been regarded as the standard against which the effectiveness of Western air defences had to be assessed. The urgency of acquiring a means of dealing with it was underlined when Israeli Phantoms failed to get anywhere near *Foxbat-B* reconnaissance aircraft during the October 1973 Yom Kippur War, and when the Soviet aircraft began overflying Northern Iran and parts of Western Europe with impunity.

Some newspapers and technical journals on both sides of the Atlantic have implied that the Japanese and US technicians who made a very thorough inspection of Lt Belenko's MiG were surprised by what they discovered. One told its readers "Japanese call Soviet MiG-25 a 'Flying Tank'". Another headlined its report: "Probers discover MiG-25 isn't a 'Miracle Plane'". An important and widely-read magazine referred to spots of brownish rust on the "old-fashioned steel" body and wings of what had been considered the world's hottest warplane, and the use of vacuum tubes instead of transistors in the 1960-vintage electronics.

In its summary of newly-discovered information, one much-respected European aviation journal commented that the thrust of about 24,500 lb per engine which some sources were continuing to report was lower than the 31,000 lb that had been estimated. In fact, the correct thrust of 24,250 lb has appeared annually in *Jane's* since 1968, when the Soviet authorities released the rating in documents relating to a new record set by the MiG.

Another important journal stated that world time-to-climb records held for two years by *Foxbat* had been reclaimed by the USAF McDonnell Douglas F-15, without apparently realising that the F-15's two best records were recaptured subsequently by the MiG, which went on to climb in 4 min 11 sec to 35,000 m (114,830 ft), a height the F-15 pilots had not attempted to reach. This particular Soviet record-breaker was described as an E-266M, whereas earlier MiG-25 record-breakers had been referred to in Soviet documents by the designation E-266. Assuming that the suffix stands for *modfikatsirovanny,* as in other, known Soviet designations, it is interesting to conjecture what modifications were made to the basic MiG-25 to boost its rate of climb to such an extent.

The journalists who prepared these reports, and the people who published them, did so in good faith; but the dangers inherent in comparing the MiG-25 that landed in Japan with US fighters of the mid-seventies are obvious. Bearing in mind that it must have been designed in the early sixties, is it surprising that its radar embodies vacuum tubes? Are there none in the early F-4 Phantoms that were contemporary with the MiG?

It is equally worrying to read scathing remarks about the fact that much of the MiG-25's airframe is made of nickel steel, with about 3 per cent titanium in areas such as the wing leading-edges and engine nozzles. There is little wrong with steel, provided the designer is prepared to compensate for the weight penalty. The outcry which followed the revelation in the Autumn of 1976 that key parts of the UK/German/Italian Tornado combat aircraft are made from titanium 'sponge' purchased from the Soviet Union hardly suggests any shortage of the metal in that country. And anyone who doubts Soviet competance to fashion titanium should visit the metallurgical displays which highlight Soviet static participation in the Paris air shows.

Nor should it have caused surprise that the MiG's Machmeter was redlined at 2·8. No other combat aircraft flies at such a speed with four large missiles on pylons under its wings; and there is ample evidence that the reconnaissance *Foxbat-B*—with the same basic airframe and power plant, but no missiles—can and does exceed Mach 3 as routine.

Any dismay felt by members of the Soviet defence ministry at having the MiG-25's secrets revealed—particularly those concerning its radar and ECM—must have been mitigated by the conclusions and reactions published in the USA. These people are familiar with the generations of combat aircraft built since the MiG-25 was designed, whereas the only Soviet types designed since the last Soviet Aviation Day display, in 1967, of which the public can have the scantiest knowledge are the Tupolev *Backfire* bomber, Sukhoi Su-19 *(Fencer)* fighter-bomber and Yakovlev Yak-36 *(Forger)* VTOL carrier-based combat aircraft which put in a first appearance during the Summer 1976 cruise of the carrier/cruiser *Kiev* through the Mediterranean and North Atlantic, en route to Murmansk.

If the Soviet Navy was prepared to show off the Yak-36 so blatantly, one must assume that it is merely a first step towards something better. It would be wrong to imagine that Soviet designers can produce nothing more versatile than the Yak, which appears to lack both the STOL capability of the British Hawker Siddeley Harrier and this aircraft's ability to use thrust vectoring in forward flight (VIFF) to enhance its combat manoeuvrability.

The value of the Harrier's VIFF was demonstrated dramatically when the US Navy decided to investigate the potential of its F-14A Tomcat interceptor, which has a unique AN/AWG-9 radar fire control system able to guide six large Phoenix air-to-air missiles simultaneously against six targets.

There was no doubt of the Tomcat's superiority when it was matched in a series of simulated air combats against Mirage F1s of the French Air Force, over the Mediterranean, and against F-5E Tiger II 'aggressor' aircraft flown by US pilots using known Soviet tactics. But against Harriers of the US Marine Corps the results were startlingly different. Using to the full the V/STOL aircraft's low-speed manoeuvrability, and rapid acceleration and deceleration, the Marine pilots outfought F-14s in six of the 16 engagements, losing only three, with the other seven declared indecisive. There could be no better incentive for speeding development of the McDonnell Douglas AV-8B advanced version of the Harrier.

The most alarming possibility for the Western powers—and, indeed, for the world as a whole—is that their politicians might grasp eagerly at implied shortcomings in Soviet aircraft like the MiG-25 as an excuse for penny-pinching defence economies. The best argument in support of the USAF's demand for B-1 strategic bombers is that the Soviet Union believes it needs the Tupolev *Backfire* in a missile age and already has more than 100 in air force and naval service.

Soviet delegates to the latest SALT (Strategic Arms Limitation Talks) negotiations have insisted that *Backfire* should be regarded as a tactical bomber. How, then, does one define a strategic bomber? Is it solely an aircraft able to attack the USA from the Soviet Union, and vice versa? Does it cease to be strategic if potential targets for its weapons are in less-distant allied nations, such as the UK or East Germany? Or is the test whether or not it will cover the distance between the US and USSR without being flight refuelled on either the outward or return journey?

Such play with words and definitions is ludicrous. On 20 July 1976, Air Force Secretary Reed stated that there is "absolutely no question" as to whether or not *Backfire* is an intercontinental strategic weapon. "With no refuelling" he said, "*Backfire* could be launched from Soviet soil against targets in the US and then fly on to Cuba for recovery. With only one refuelling, the Soviet bomber could be launched from Russia against all areas of the US, except for some parts of Florida, and return to the Soviet Union." It has long been known, of course, that *Backfire* is fitted with flight refuelling equipment.

On such evidence, any SALT agreement that was signed at the cost of accepting *Backfire* as a tactical aircraft would lessen the hope of lasting peace. This can only be a product of precisely balanced strength, on any scale from extravagant overkill to commonsense basic self defence. The three immediate requirements for the US, as cornerstone of the West, are to recognise that *Backfire* is a strategic bomber, to build the B-1 as its wholly essential and uniquely flexible counterpart, and to order as a matter of urgency replacements for Aerospace Defense Command's time-expired F-106 Delta Dart interceptors.

Opposition to such a programme is bound to be intense, but the Defense Department is unlikely ever to encounter a more ridiculous argument than that put forward last May by the Council on Economic Priorities. According to the *Chicago Tribune,* this worthy body pointed out that the State of Illinois paid $95·5 million more in taxes toward B-1 construction than it got back in B-1 contracts. One can only feel thankful, as an Englishman, that British counties without fighter factories continued to pay their Income Tax in 1940, before the Battle of Britain.

There was a certain logic in the US decision to deactivate in 1976 the modest Safeguard anti-ballistic missile defence system that had been permitted under the SALT I agreement. The number of ABMs that could be emplaced around Washington and a single ICBM site was clearly no more than a token force, maintained at enormous cost; but defence

against air attack is different. If there is any merit in plans to restrict a future conflict to low-key non-nuclear or localised tactical nuclear level for some days, as NATO intends, it must be shortsighted to deploy no more than 12 squadrons of 1956-model F-106s and one F-4 squadron, plus three Army Nike Hercules surface-to-air missile batteries in Alaska, as the total dedicated defence force for 48 home States *and* to support overseas missions. The Soviet Union considers that it needs 2,600 manned interceptors—many of them new—and 10,000 surface-to-air missile launchers.

Twenty years ago, a disastrous British Defence White Paper stated that there would be no need for new interceptors and strategic bombers for the Royal Air Force beyond those in service or under development at the time. Britain soon learned the utter stupidity of such thinking, based on the assumption that any war would generate an immediate nuclear holocaust. The United States had the same wrong thought somewhat later, and has yet to take the positive step of rebuilding its air defence. It is fortunate in having three excellent interceptors available for immediate increased production, in the shape of the F-14 Tomcat, F-15 Eagle and F-16, with the Navy's F-18 not far behind.

At the moment the F-14 is favoured for this important additional application, because of its unrivalled fire control system and missile armament. Far beyond it, the US has already caught a glimpse of the future by destroying two target drones with a high-energy laser mounted on an armoured vehicle at Redstone Arsenal, Alabama. This modest initial success by something akin to the 'death rays' of science fiction must send a cold shiver down the spine of anyone who still believes that all men, made in the image of God, have an inalienable right to life, liberty and the pursuit of happiness—a battered promise of which we have been reminded often during America's bicentennial year.

Study of the Soviet sections of this edition will show that no category of combat aircraft is being neglected in Eastern defence budgets. The *Flogger-D* version of what began as the MiG-23, and the formidable Sukhoi Su-19 *(Fencer),* make up a large proportion of the aircraft to which Air Chief Marshal Sir Andrew Humphrey, former RAF Chief of Air Staff, was referring when he said in December 1975: "Russia is now making more than 1,700 military aircraft each year, and of these more than 700 are of the most advanced types of high-performance combat aircraft. She is replacing her older aircraft at least one for one with new and vastly more advanced types. To take a civilian analogy, it is as though an airline was replacing its aged fleet of Comets and early 707s, one for one, with Concordes and 747s—a gigantic increase in capability.

Gone is the one-time Soviet emphasis on smallest possible size, light weight, simplicity and maximum manoeuvrability. *Flogger* may be in the compact class of what the USAF referred to as its LWF (lightweight fighter) prototypes, with everything hung outside an airframe of minimum cross-section; but it is a far more effective weapon than the original MiG-21, which was woefully short on fuel, radar and weapons. Nor is it lacking in sophistication, with its swing-wings, Gatling-type gun, laser rangefinder and many still-unidentified antennae and dielectric panels.

Fencer is even more thought-provoking. In tracing its pedigree, it is helpful to begin by recognising that the Su-15 *(Flagon)* interceptor is a stretched Su-11 *(Fishpot-C),* with the same basic airframe, adapted to take two turbojets side by side in the rear fuselage and a large radar in an ogival nosecone which necessitated the use of side intakes. The more one studies known details of *Fencer,* the more it looks like a *Flagon* fuselage and tail unit, widened to take a crew of two side by side and fitted with variable-geometry wings. The nose may be similar to that of *Flogger-D* rather than ogival as drawn on page 434. It is clearly a tactical strike aircraft of a quality hitherto beyond the technological reach of the Soviet industry.

Bearing in mind that that industry provides all the combat aircraft deployed by the Warsaw Pact nations, and many of those operated by their allies and friends as far afield as Libya and Peru, it is easy to understand suggestions that European members of NATO should be equally willing to acquire all their first-line equipment from the USA. As subcontractors to US industry, it is suggested that their national industries could maintain high levels of employment and know-how without any of the cost and economic risk of designing and developing their own, often-competing aircraft. This argument overlooks one key factor. . . . .

Good though the aerospace industries of the smaller Warsaw Pact nations are, they do not have the experience and capability of, for example, the UK industry. Alone, or in partnership with its continental neighbours, Britain has produced the Harrier—still the only operational V/STOL fixed-wing combat aircraft in the world; the Concorde—still the only supersonic airliner in scheduled passenger service; and the Tornado multi-role combat aircraft, described by the UK aerospace minister as "the supreme European aircraft".

At the time this Foreword was being written, Britain's projected airborne early warning version of the Nimrod was still competing strongly with the Boeing E-3A AWACS as a Shackleton replacement to fill a vital gap in Western Europe's defences in the eighties. Of more immediate significance, the Finnish Air Force has selected the Hawker Siddeley Hawk as its next jet training aircraft, against strong competition from Czechoslovakia, France, Germany, Italy and Sweden. Up to 50 Hawks are expected to be covered by a firm contract to be signed in 1977; the only sad feature is that so many different European aircraft should be chasing the same contracts in this category, with yet another to come soon from Spain.

More hopeful, as a pointer to future collaboration rather than wasteful competition, is current preliminary discussion of the next generation of battlefield support aircraft. RAF thought on the subject led to the formulation of Air Staff Target (AST) 403, defining the basic parameters of the kind of aircraft that might replace both the Jaguar and the Harrier before the end of the eighties. France needs something similar to partner its future Delta Mirage 2000 high-performance interceptors. Belgium, the Netherlands and Germany have parallel requirements. So the five nations have formed a sub-group of the European Programme Group to progress the project, under the chairmanship of the UK.

AST 403 was aimed at a Mach 1·6 close support/air combat type able to destroy any battlefield target in a single pass and to match the agility of anything encountered in the air. This led, inevitably, to the suggestion that what the RAF wanted was the F-16, of which more than 1,000 had already been ordered by the air forces of Belgium, Denmark, Iran, the Netherlands, Norway and the USA. Experience in operating the Harrier in Germany, and in working with European partners on the Tornado, prompted other thoughts.

A major problem confronting every modern air force is how to survive a pre-emptive strike by an enemy which destroyed all its runways in the opening minutes of a confrontation. Harrier squadrons have no such problems, as they need no runways or even dirt strips from which to fly. However, the cost of taking off vertically is so high in terms of reduced payload/range that they normally operate in a short take-off (STO) mode in order to lift a greater weight of fuel and weapons.

On the other hand, all operational experience by the RAF and USMC points to the importance of being able to land vertically. It would be feasible to touch down at around 100 knots into some form of mobile arrester gear; but the risk would be high if circumstances compelled the use of a narrow cambered road, surrounded by natural or structural obstacles, subject to crosswinds, cluttered with ground equipment, vehicles and other aircraft. Hence what has become known as STOVL—short take-off/vertical landing—making the best of both worlds.

STOVL combined with Mach 1·6, all-weather instrumentation, and the ability to carry a wide range of air-to-ground and air-to-air weapons would seem to meet most anticipated needs for the remainder of the present century. Add thrust vectoring in forward flight, and the end product begins to sound too costly for nations with economic problems; but an entirely practical solution to this problem could be worked out if NATO ceased thinking of itself as a sort of mother hen with a family of small helpless chicks.

It is too early to estimate whether or not the Future Tactical Combat Aircraft (FTCA) being discussed by the European five-nation group could be reconciled with AST 403, and whether the result might have STOVL capability. Even less can one suggest that the USAF technical staff responsible for defining the future needs of their service might be prepared to consider the European project as an answer to one of their urgent needs for the eighties. In this respect, all the nations concerned could benefit from further study of the methods adopted by Warsaw Pact countries, as the Soviet Union has long been content to acquire its jet trainers from Czechoslovakia and its light helicopters from Poland.

NATO has just as much to gain from taking advantage of the considerable skills and experience offered by its individual members. Harmony is not promoted by US newspaper headlines such as "New Europe Warplane Perils US Aircraft Firms"—suggesting that Britain, Germany and Italy are presumptuous in developing the Tornado to meet their own needs, instead of buying American F-15s or F-16s and so helping to keep the US manufacturers profitable. It may not have occurred to the journalist responsible for that story that the F-15 and F-16 do not do the same job as the Tornado.

A more insidious weapon being used against the UK industry by its competitors is the suggestion that impending nationalisation of the larger companies will make them inefficient and unreliable. It could happen, although it is really of little consequence whether capital is provided from private or public funds. Success or failure depends on the competence and imagination of the management, the skill and will to work of the labour force, enthusiastic and consistent government support through a well-conceived programme of work, and a responsible attitude from trade unions.

Nobody would pretend that the already-state-owned Rolls-Royce has achieved perfection, but it has recorded impressive successes during the past year. Its RB.211 turbofans enabled a new version of the Boeing 747 to take off at a heavier weight than that at which any previous aeroplane has flown; and a £100 million ($162 million) licence agreement with China is expected to lead to a new generation of Chinese military aircraft built around the well-proven Spey. Availability of such an engine may well provide a basis for progress as sound as that which availability of Rolls-Royce Nenes for the MiG-15 and Il-28 bomber gave the Soviet Union in the mid-forties. One can only hope that it will prove more profitable for the UK than did the earlier deal.

It must be remembered, of course, that China's aerospace industry already has a well-earned reputation for efficiency. The MiGs it has exported to countries such as Pakistan and Tanzania reflect first-class workmanship; it has manufactured aircraft as large and complex as the Tu-16 *(Badger)* twin-jet strategic bomber; and it has put a succession of satellites into orbit. When the Soviet Union refused to supply spares to keep Egypt's combat aircraft serviceable, following President Sadat's

decision to dispense with the help of Russian advisers, China stepped in with the items needed for certain types that are operated also by the Chinese Air Force.

Egypt is not the only country to learn that aircraft supplied by other nations arrive sometimes with political strings attached. To ensure a higher degree of independence, Romania and Yugoslavia are developing jointly a Jaguar-like attack aircraft, known to the Romanians as the IAR-93 and to the Yugoslavs as the Orao (Eagle). Iran, Turkey and Greece are all setting up new aircraft industries, with foreign assistance. There is no better way to begin, but it will be surprising if they do not progress to designs of their own within a few years.

The Aero Industry Development Center in Taiwan has progressed in seven years from licence-built Bell helicopters and Pazmany piston-engined trainers to a turboprop trainer that owes much to the North American T-28 and, now, a twin-turboprop transport of original design. Israel Aircraft Industries has satisfied its Air Force's desperate need for more fighters by marrying an airframe based on the French Mirage to an engine of the type used in the Air Force's Phantoms, and then adding canard foreplanes which confer a performance that no ordinary Mirage 5 could match. Already the resulting fighter, known as the Kfir-C2, is challenging another highly independent and competent manufacturer, Saab-Scania, in the Swedish company's traditional markets.

With the tandem-delta Viggen scheduled for production throughout the eighties, in new roles, Saab-Scania is assured of one major production programme. It is also nibbling at the commercial market with studies for a Skyvan-like utility transport; but the primary new project is a light attack/trainer, known as Attack Aircraft System 85, intended as an eventual replacement for the Swedish Air Force's Saab 105s (SK 60s).

Most astonishing of all has been the rapid growth of the Brazilian industry. A decade ago it produced only single-engined light aircraft and trainers, which were described and illustrated in four pages of the contemporary *Jane's*. Today, EMBRAER claims to rank eighth among aircraft manufacturers in the non-Communist world, in terms of numbers of aircraft produced annually. Its products include a family of impressive twin-turboprop general-purpose aircraft, and the entire Brazilian industry fills a total of more than seven pages in the main Aircraft section.

Industries in places like Brazil, Sweden and Israel prosper because they do not have to rely on a steady flow of export orders for their survival. The situation has been very different for the major manufacturers of airliners for the world market. During the period of recession, most airlines decided that they could not afford or did not need new equipment. Those that placed orders in 1976 usually supplemented their fleets with additional aircraft of existing types, offering little encouragement to the industry, which regards as essential the continual evolution and production of new designs.

There has been no shortage of proposals, and this edition of *Jane's* would have been twice as bulky had all of them been included. Airline operators must be thoroughly confused by a seemingly never-ending succession of designations such as 7X7, 7N7, DC-X-200 and X-Eleven, applied often to designs available in a range of 'take-your-pick' sizes, configurations and power plants. In November 1976, the only one that seems to have any immediate prospect of advancing from project to prototype construction is the Dassault-Breguet/McDonnell Douglas ASMR (Advanced Short/Medium-Range) airliner, which began life as the 'stretched' Mercure 200 and is intended to carry about 160 passengers. It is, therefore, the only one of which details can be found in this edition, as *Jane's* has always done its best to avoid becoming *All the World's Paper Aircraft*.

Biggest disappointment of the year in commercial aviation has been the non-acceptance of Concorde by airlines other than British Airways and Air France. This is no fault of the aircraft, which continues to perform brilliantly, doing all that had ever been expected, and staying usually within the predicted noise levels when monitored by US officials on the approach and take-off paths at Dulles, Washington. Refusal of the New York authorities to permit landings at John F. Kennedy Airport has restricted Air France utilisation to a rate of 1,000 hours per aircraft per year, against the 2,500 hours with 70% load factor required to show a profit. As a consequence, the French airline expects to lose at least Fr160 million ($32 million) on Concorde operations in 1976.

Meanwhile, in the Soviet Union, the Tu-144 continues to accumulate hours on freight and mail runs before being cleared for that nation's first supersonic passenger services. In other respects, Aeroflot is able to report truly astronomical figures concerning its operations. In 1971-75 it carried more than 420 million passengers and 11 million tonnes of cargo, over routes totalling nearly 432,000 nm, linking over 3,500 cities and other localities. In the same period, its agricultural aircraft treated more than 1,075 million acres of farmland and forests.

Looking ahead fifty years, to 2026 AD, United Air Lines Vice President Andy DeVoursney has become even more astronomic, foreseeing 1,700-passenger airliners, each costing a billion dollars; UAL captains earning $480,000 a year, carrying a million passengers every day; and profits for this one airline in the $11 billion range. Maybe, but one wonders what those huge airliners will burn as fuel. The Space Shuttle Orbiter *Enterprise,* rolled out at Palmdale on 17 September 1976, may represent far more than just the start of a new era of space research. It should not be overlooked that it will use solid boosters and onboard liquid-propellant rocket engines to thrust it to orbital speed—making it, perhaps, the prototype for all future transport aircraft, built for operation after the last gallons of hydrocarbon fuel have been burned.

The growing frequency with which entries in *Jane's* refer to supercritical wings, 'winglets' and the switch to turbofans in business jets underlines the seriousness with which designers are beginning to view both diminishing reserves of fuel and its growing cost. Lockheed has even proposed a 200-passenger Mach 0·8 medium-range airliner for the mid-eighties powered by four turboprops, driving eight-blade variable-pitch propfans. It claims that this aircraft would use up to 18% less fuel and reduce direct operating costs by 5-8% compared with a jet of similar performance.

It is difficult to envisage a return to turboprops for such aircraft, but strange things happen in aviation. A glance through this book will show that the vintage Taylorcraft lightplane is now back in production for pilots who would like to return to the simplicity of thirty years ago. An enthusiast with more time can even buy a Piper Cub kit and build the aircraft himself. What is more, the long and fruitless search for a genuine 'DC-3 replacement' may be over at last, as a company named Tamco is proposing to market conversions of existing DC-3 airframes fitted with Rolls-Royce Dart turboprops, a tricycle landing gear and other changes. So the 1977-78 *Jane's* may well contain a slightly modified version of the most familiar aeroplane shape of them all.

* * * * * *

Like its predecessors for 67 years, the 1976-77 *Jane's* is intended to provide its professional readers with a completely reliable and up-to-date reference work on the products of the world's aircraft industry. In doing so, it embodies far more changes than have been made in any previous edition.

It has long been apparent that a change from letterpress to modern litho printing would enable the publisher to produce economically, and to even higher standards, the large number of copies now sold annually. When the decision was taken to make the change this year—including the re-setting of 1½ million words of text and production of some 1,500 new plates for the illustrations—it seemed logical to give precedence to metrics rather than the old Imperial units that are being superseded in most countries, including the USA and Great Britain.

This was done only after seeking the advice of leading manufacturers in the aerospace industry, and of organisations such as the British CAA and NASA. All recommended adoption of the International System of Units (designated SI in all languages), as accepted by 41 of the principal industrial nations of the world. This has been done, with the sole exception that bars are retained as units of pressure (mechanical stress) instead of the Pascal specified as the SI unit. The change to kiloPascals for items such as tyre pressures will be made when this is considered appropriate by tyre manufacturers and aircraft operators.

Another change that has taken place this year has involved separation of homebuilt aircraft from those manufactured professionally. This was suggested some years ago by designers of the homebuilts, and by organisations such as the Experimental Aircraft Association, to facilitate reference and selection of a particular design by would-be constructor/pilots. The fact that this new section fills 93 pages indicates the scale of this branch of the private flying movement. It may be significant to quote the words of one well-known kit manufacturer who wrote to the Editor in July 1976: "We get lots of letters from all over the world from people who see our aeroplanes in *Jane's*. So you are being widely read, in places other than just the USA." If anyone doubts the significance of these aircraft, he need only glance at the statistics which reveal that many designs are flying or being built in hundreds.

An aircraft is normally put in the Homebuilts section if it is intended for the personal use of its designer/constructor or for construction by other amateurs. If, like the Lockspeiser LDA-1, the prototype is built privately but the type is intended for factory production, it is included in the main Aircraft section. Racing and man-powered aircraft, being usually amateur-built 'one-offs', are put with the homebuilts. Types like the Cranwell A1, built by professional aerospace training or research establishments, are regarded normally as products of the aerospace industry.

Last year, *All the World's Aircraft* included several pages of hang gliders. This sub-section proved popular, but seemed inappropriate—even potentially hazardous to inexperienced would-be pilots—at a time when the regulations governing hang gliding are being revised and given more authority in several countries, following accidents. When the new regulations are published, and it can be ensured that gliders described and illustrated in *Jane's* conform with them, the sub-section will be re-introduced—hopefully in 1977-78.

Offsetting this loss, the 1976-77 *Jane's* includes for the first time a sub-section devoted to hot-air balloons, the oldest practical aircraft of all. Requests that balloons should be added to airships have been received repeatedly since *All the World's Air-Ships* (sic) first appeared in 1909. Considerable help in tracking down the types listed this year, and in deciding what data should be included, came from Mr P. G. Dunnington, Administration Officer, British Balloon and Airship Club, and from Directors of the two major British manufacturers of hot-air balloons. Nor should it be overlooked that the number of airship entries is growing, with a potentially large export order reported by Aerospace Developments.

Removal of homebuilt aircraft from the main Aircraft section has made some national sub-sections look rather thin. For example, the French

can you afford not to have this kind of joker when the game gets tough?
JOKER
The SELENIA ASPIDE missile is a truly multirole weapon, optimized from the design stage for use in air-to-air ground-to-air and ship-to-air weapon systems. In its air-to-air role, ASPIDE consitutes the all aspect all-weather weapon for high performance interceptors. The missile's launch envelope, both as regards range and altitude, its snap-down/snap-up capabilities its high manoeuvrability and effectice guidance system endow the aircraft with superior intercept and dog-fight potential under the most adverse operational conditions. In its ground-to-air role, ASPIDE constitutes the medium range, low-altitude air defence weapon for the SPADA ground based fixed and mobile missile batteries. In its ship-to-air role, ASPIDE has been adopted for the ALBATROS naval air defence weapon system and endows this system with superior capabilities especially as regards SSKP, defended volume, ECCM and performance against low flying targets.
MULTIROLE MEANS REDUCED PRODUCTION, TRAINING, MAINTENANCE AND LOGISTICS COSTS
SELENIA
INDUSTRIE ELETTRONICHE ASSOCIATE S.p.A.
RADAR AND MISSILE SYSTEMS DIVISION
Via Tiburtina Km 12.400, 00131 ROME - ITALY
P.O.Box 7083, 00100 ROME, Cables SELENIA ROMA,
Telex 61106 SELENIAT

section has diminished from last year's 47 pages to 32, and the UK from 45 to 35. The US section has shrunk from 243 pages to 198; but the Soviet section has increased slightly, from 44 to 47. The inclusion later in the book of five Soviet homebuilt aircraft not recorded in previous editions of *Jane's* may come as a surprise. It is also interesting to note that the Antonov An-2 biplane (listed under its current Polish manufacturer) is in its 30th year of continuous production, which must be a record for longevity.

* * * * * *

As well as being completely re-set, the 1976-77 *Jane's* contains 24 more pages of main text than did its predecessor. Its compilation, checking and production demanded an immense effort from everyone concerned, and the sincere thanks of the Publisher and Editor go to each one of them—particularly Assistant Editor Kenneth Munson, who met cheerfully every demand made upon him as both a compiler and the team-member responsible for a major part of the proof checking.

The Editorial team is unchanged from previous years, but there has been some redistribution of tasks. Michael Taylor assumed responsibility for the new Homebuilts section, in addition to the onerous task of converting the entire text to SI metric units. Maurice Allward added the Italian aircraft to his usual commitment on the Spaceflight section. David Mondey agreed to add hot-air balloons to the other sections which he updates annually. Bill Gunston again took charge of the Aero-engines, and has begun planning the Glossary that will be included in future editions at the request of numerous readers, at home and overseas.

In London, those responsible for preparing the pages for camera worked meticulously, for long periods, to ensure the highest standard of accurate alignment—so important to the appearance of the book. In Huddersfield, Netherwood Dalton & Co Ltd made the change to computer setting and litho printing seem deceptively easy after 66 years of letterpress work on *Jane's*. It is a privilege and a pleasure for an Editor to work with such loyal and thoroughly professional colleagues.

* * * * * *

The team is completed by many hundreds of men and women in the aerospace industries of the world who supply facts, figures and illustrations each year. Special thanks go, as always, to Norman Polmar and our friends of *Air Force Magazine,* in Washington, without whose ever-willing assistance the US sections could not be so comprehensive and up-to-date; to Delden Badcock (Australia), Ronaldo Olive (Brazil), Vico Rosaspina (Italy), Eiichiro Sekigawa (Japan), Stuart H. Morison (Romanian sailplanes), Javier Taibo (Spain), Dr Ulrich Haller (Switzerland), Wolfgang Wagner of *Deutscher Aerokurier* and Peter Pletschacher of *Flug Revue* (Germany), friends on the staff of the *Biuletyn Informacyjny Instytutu Lotnictwa* (Poland), William Green and Gordon Swanborough of *Air International* (UK), R. Moulton of Model Aeronautical Press (UK), Alan Hall of *Aviation News* (UK), the editorial staffs of *Flight International* (UK), *Aviation Magazine International* (France), *Repules* (Hungary), *Flieger-Revue* (German Democratic Republic), *FLYGvapenNYTT* (Sweden) and *de Vliegende Hollander* (the Netherlands). The new photographs received from industry and official sources have again been supplemented by often-exclusive prints from Howard Levy, Neil Macdougall, Jean Seele, Peter M. Bowers, Gordon S. Williams, Norman E. Taylor and other photographers and friends whose names are included in the captions to illustrations. Our three-view drawings were provided by Dennis Punnett of Pilot Press, Roy Grainge and Michael Badrocke, whose skill at converting scraps of information and poor photographs into superb art-work, when necessary, is unrivalled.

PHOTOGRAPHS AND THREE-VIEW DRAWINGS

The Editor and Publishers receive many requests for prints of photographs that appear in *Jane's*. It is not possible for them to offer any form of photographic service; but photographs of a high proportion of the aircraft described in this edition, as well as of many earlier types, are available at normal trade rates from:

Air Portraits, 131 Welwyndale Road, Sutton Coldfield, West Midlands, B72 1AL

Flight International, Dorset House, Stamford Street, London SE1 9LU

Stephen Peltz, 9 Cambridge Square, London W.2

Three-view drawings are available to the press from:

Pilot Press Ltd, PO Box 16, Bromley, Kent BR2 7RB

Roy J. Grainge, 12 Bonaly Gardens, Colinton, Edinburgh EH13 0EX

November 1976 JWRT

SOME FIRST FLIGHTS MADE DURING THE PERIOD 1 JUNE 1975 — 30 SEPTEMBER 1976

June 1975
2 Fairchild A-10A, second DT&E aircraft (USA)
3 Mitsubishi FS-T2-KAI, second prototype (first to fly; 59-5107) (Japan)
7 Mitsubishi FS-T2-KAI, first prototype (second to fly; 59-5106) (Japan)
15 Akaflieg Stuttgart FS-29 sailplane (D-2929) (Germany)
18 Chasle LMC-1 Sprintair (F-WXKD) (France)
18 Atlas C4M Kudu, first military prototype for SAAF evaluation (961) (South Africa)
25 Sperry (Convair) PQM-102 (USA)
26 DHC-7 Dash 7, second pre-production (C-GNCA-X) (Canada)
29 Jeffair Barracuda (N19GS) (USA)

July 1975
1 Valmet Leko-70, first prototype (Finland)
2 Bowers Namu II (N75PA) (USA)
3 Sawyer Skyjacker II (USA)
4 Boeing 747SP, first production (N530PA) (USA)
8 Shorts SD3-30, second prototype (G-BDBS) (UK)
9 Hawker Siddeley Super Trident 3B, first production (UK)
17 Piper PA-32R-300 Cherokee Lance, first production (USA)
18 Zlin Z 50 L (Czechoslovakia)
21 IAI 1124 Westwind, development aircraft (4X-CJA) (Israel)
21 Sikorsky XH-59A, second prototype (21942) (USA)
25 Boeing E-3A, second 'full-scale' development aircraft (USA)
25 McDonnell Douglas DC-10 Srs 40 with JT9D-59A engines (USA)

August 1975
1 Emair MA-1B (USA)
5 Panavia Tornado, third prototype (first trainer) (XX947) (International)
5 Boeing Vertol Model 179 (N179BV) (USA)
6 Boeing Vertol CH-46E/SLEP (Service Life Extension Program) (USA)
14 LMSC Aquila mini-RPV (½ scale test model) (USA)
26 Cessna Model 441 Conquest, first prototype (N441CC) (USA)
26 McDonnell Douglas YC-15, first prototype (01875) (USA)
28 Leisure Sport S.5 'replica' (G-BDFF) (UK)
28 Robinson R22 (USA)

September 1975
2 Panavia Tornado, fourth prototype (D-9592) (International)
27 Williams W-18 Falcon (N28LL) (USA)
29 Grumman Gulfstream II, first for Space Shuttle training (USA)
30 Hughes YAH-64, first prototype (22248) (USA)

October 1975
1 Bell YAH-63, first prototype (22246) (USA)
9 Hughes Model 500D, first production (USA)
21 Fairchild A-10A, first production (75-00258) (USA)
25 Concorde, fifth production (first for Air France) (F-BVFA) (International)
31 Beechcraft Baron Model 58TC (USA)
31 Boeing E-3A, second development aircraft, first with production electronics (USA)
31 Harmon 1-2 Mister America (USA)

November 1975
5 Concorde, sixth production (first for British Airways) (G-BOAA) (International)
13 Fuji FA-300/Rockwell Commander 700, first prototype (Japan/USA)
19 AmEagle American Eaglet powered sailplane (N101EA) (USA)
22 Hughes YAH-64, second prototype (22249) (USA)
23 Schleicher ASW 19 sailplane (D-1909) (Germany)
28 Crutchley Special (ZS-UHH) (South Africa)

December 1975
5 Panavia Tornado, fifth (first Italian) prototype (X-586) (International)
5 Grumman F-14A Tomcat, first for Iran (3-863) (USA)
5 McDonnell Douglas YC-15, second prototype (USA)
8 Sikorsky CH-53E, first pre-production (USA)
12 Saunders ST-28, first production (Canada)
15 Grumman (General Dynamics) EF-111A, first development aircraft (USA)
18 RRAFAGA J-1 Martin Fierro (Argentina)
20 Panavia Tornado, sixth prototype, first with gun (XX948) (International)
20 EEUFMG CB-2 Minuano sailplane (PP-ZPZ) (Brazil)
21 Bell YAH-63, second prototype (USA)
23 Aeritalia G222, first production (Italy)
25 Nippi NP-100A (NP-100) (Japan)
31 EMBRAER EMB-820 Navajo (PT-EBN) (Brazil)

January 1976
15 C.A.A.R.P. CAP 20L (F-WVKY) (France)
20 Vought A-7, with composite wing (USA)
22 Northrop TEDS RPV (USA)

February 1976
4 Rolladen-Schneider LS3 sailplane (Germany)
10 Westland/Aérospatiale Lynx HAS. Mk 2, first production (UK/France)
25 Fuji FA-300/Rockwell Commander 700, second (first US-assembled) prototype (N9901S) (Japan/USA)

March 1976
4 Fournier RF-6B Club, first pre-production (F-BVKS) (France)
12 Nihon University NM-75 Stork man-powered aircraft (Japan)
24 General Dynamics CCV YF-16 (72-01567) (USA)
25 Fokker-VFW F27 Maritime (PH-FCX) (Netherlands)

April 1976
1 Rockwell International B-1, third prototype (second to fly) (USA)
9 Kortenbach & Rauh Kora I powered sailplane, second prototype (Germany)
19 PIK-20D sailplane (Finland)
21 Zinno Olympian ZB-1 man-powered aircraft (USA)

May 1976
17 Shin Meiwa PS-1, water bomber version (Japan)
20 UTVA-75, first prototype (Yugoslavia)
26 Dassault Mirage F1-B (France)
27 Boeing Model 727-200, JT8D-17R engines (USA)

June 1976
14 Rockwell International B-1, second prototype (third to fly) (USA)
16 Practavia Sprite, first example (G-BDDB) (UK)
28 Hawker Siddeley HS 125 Series 700, prototype (G-BFAN) (UK)

July 1976
2 Lockheed US-3A Viking, development aircraft (USA)
3 Piaggio P.166-DL3 (I-PJAG) (Italy)
15 Panavia Tornado, eighth prototype (International)
19 Westland/Aérospatiale Lynx, PT6B-34 engines (G-BEAD) (UK/France)

August 1976
6 DSI/NASA oblique wing RPRA (USA)
9 Boeing YC-14, first prototype (01873) (USA)
12 Aermacchi M.B.339, first prototype (I-NOVE) (Italy)
13 Bell Model 222 (N9988K) (USA)
18 Lockheed JetStar II, first production (N5527L) (USA)
20 Scottish Aviation Bullfinch prototype (G-BDOG) (UK)
23 Flight Invert Cranfield A1, prototype (UK)

September 1976
3 Boeing 747-200B, RB.211-524 engines (USA)

Boeing 747-200B with RB.211-524 engines, which flew for the first time on 3 September 1976. Subsequently, it set a world record for the greatest mass lifted to a height of 2,000 m (see under Boeing heading in Addenda)

OFFICIAL RECORDS

Corrected to September 1976

ABSOLUTE WORLD RECORDS

Seven records are classed as Absolute World Records for aeroplanes by the Fédération Aéronautique Internationale, as follows:

Distance in a straight line (USA)
Major Clyde P. Evely, USAF, in a Boeing B-52H Stratofortress, on 10-11 January 1962, from Okinawa to Madrid, Spain. 10,890·27 nm (20,168·78 km; 12,532·3 miles).

Distance in a closed circuit (USA)
Captain William M. Stevenson, USAF, in a Boeing B-52H Stratofortress, on 6-7 June 1962. Seymour Johnson AFB-Bermuda-Sondrestrom (Greenland)-Anchorage (Alaska)-March AFB-Key West-Seymour Johnson AFB. 9,851·54 nm (18,245·05 km; 11,337 miles).

Height (USSR)
Alexander Fedotov in an E-266 (MiG-25) on 25 July 1973. 36,240 m (118,898 ft).

Height in sustained horizontal flight (USA)
Col Robert L. Stephens and Lt Col Daniel Andre (USAF) in a Lockheed YF-12A, on 1 May 1965, over a 15/25 km course at Edwards AFB, California. 24,462·596 m (80,257·86 m).
Awaiting confirmation is a new record of 26,213 m (86,000 ft) set by Captain Robert C. Helt and Major Larry A. Elliott in a Lockheed SR-71A on 27 July 1976 (USA).

Height, after launch from a 'mother-plane' (USA)
Major R. White, USAF, in the North American X-15A-3 on 17 July 1962, at Edwards AFB, California. 95,935·99 m (314,750 ft).

Speed in a straight line (USA)
Col Robert L. Stephens and Lt Col Daniel Andre (USAF) in a Lockheed YF-12A, on 1 May 1965, over a 15/25 km course at Edwards AFB, California. 1,798·87 knots (3,331·507 km/h; 2,070·102 mph).
Awaiting confirmation is a new record of 1,901 knots (3,523 km/h; 2,189 mph) set by Captain Eldon W. Joersz and Major George T. Morgan Jr in a Lockheed SR-71A on 27 July 1976 (USA).

Speed in a closed circuit (USSR)
M. Komarov in a Mikoyan E-266 (MiG-25), on 5 October 1967, at Podmoskovnœ, over a 500 km (310·7 mile) closed circuit. 1,609·88 knots (2,981·5 km/h; 1,852·62 mph).
Awaiting confirmation is a new record of 1,812 knots (3,357 km/h; 2,086 mph), over a 1,000 km circuit, set by Major Adolphus H. Bledsoe Jr and Major John T. Fuller in a Lockheed SR-71A on 27 July 1976 (USA).

Seven records are classed as Absolute World Records for manned spacecraft, by the Fédération Aéronautique Internationale, as follows:

Endurance in Earth orbit (USA)
Gerald P. Carr, Edward G. Gibson and William R. Pogue in Skylab 3, from 16 November 1973 to 8 February 1974. 84 days 1 hr 15 min 30.8 sec.

Altitude (USA)
F. Borman, J. A. Lovell and W. Anders in Apollo 8, on 21-27 December 1968. 203,925 nm (377,668·9 km; 234,673 miles).

Greatest mass lifted to altitude (USA)
F. Borman, J. A. Lovell and W. Anders in Apollo 8, on 21-27 December 1968. 127,980 kg (282,147 lb).

Distance in Earth orbit (USA)
Gerald P. Carr, Edward G. Gibson and William R. Pogue in Skylab 3, from 16 November 1973 to 8 February 1974. 29,953,582 nm (55,474,039 km; 34,469,960 miles).

Extravehicular Duration (USA)
Eugene A. Cernan, from the Apollo 17 lunar module *Challenger*, on 12, 13 and 14 December 1972, during mission of 7-19 December 1972. 21 hr 31 min 44 sec.

Number of Astronauts remaining simultaneously outside Spacecraft (USSR)
A. Eliseiev and E. Khrounov, from Soyuz 4 and 5, on 14-18 January 1969. 37 min.

Accumulated Time in Spaceflight (USA)
Gerald P. Carr in Skylab 3, launched by Saturn IB vehicle, 16 November 1973 to 8 February 1974. 84 days 1 hr 15 min 30.8 sec.

WORLD CLASS RECORDS

Following are details of some of the more important world class records confirmed by the FAI:

CLASS C, GROUP I (Aeroplanes with piston engines)

Distance in a straight line (USA)
Cdr Thomas D. Davies, USN, and crew of three in a Lockheed P2V-1 Neptune, on 29 September-1 October 1946, from Perth, Western Australia, to Columbus, Ohio, USA. 9,763·49 nm (18,081·99 km; 11,235·6 miles).

Distance in a closed circuit (USA)
James R. Bede in the Bede BD-2, on 7-10 November 1969, between Columbus, Ohio, and Toledo, Ohio, USA. 7,797·66 nm (14,441·26 km; 8,973·38 miles).

Height (Italy)
Mario Pezzi, in a Caproni 161*bis*, on 22 October 1938. 17,083 m (56,046 ft).

Speed in a straight line (USA)
Darryl Greenamyer in a modified Grumman F8F-2 Bearcat, on 16 August 1969, at Edwards AFB, California. 419·249 knots (776·449 km/h; 482·463 mph).

CLASS C, GROUP II (Aeroplanes with turboprop engines)

Distance in a straight line (USA)
Lt Col E. L. Allison and crew in a Lockheed HC-130H Hercules, on 20 February 1972. 7,587·99 nm (14,052·95 km; 8,732·098 miles).

Distance in a closed circuit (USA)
Cdr Philip R. Hite and crew in a Lockheed RP-3D Orion, on 4 November 1972. 5,455·46 nm (10,103·51 km; 6,278·03 miles).

Height (USA)
Donald R. Wilson in an LTV Electrosystems L450F, on 27 March 1972, at Majors Field, Greenville, Texas. 15,549 m (51,014 ft).

Speed in a straight line (USA)
Cdr Donald H. Lilienthal and crew in a Lockheed P-3C Orion, on 27 January 1971. 435·26 knots (806·10 km/h; 500·89 mph).

Speed in a closed circuit (USSR)
Ivan Sukhomlin and crew in a Tupolev Tu-114, on 9 April 1960, carrying a 25,000 kg payload over a 5,000 km circuit. 473·66 knots (877·212 km/h; 545·07 mph).

CLASS C, GROUP III (Aeroplanes with jet engines)

Distance in a straight line, distance in a closed circuit, height, speed in straight line and speed in 500 km closed circuit
See Absolute World Records.

Speed in a 100 km (62.14 mile) closed circuit (USSR)
Alexander Fedotov in a Mikoyan E-266 (MiG-25), on 8 April 1973. 1,406·641 knots (2,605·1 km/h; 1,618·734 mph).

Speed in a 1,000 km (621.4 mile) closed circuit (USSR)
P. Ostapenko in a Mikoyan E-266 (MiG-25), on 27 October 1967, at Podmoskovnœ. 1,577·036 knots (2,920·67 km/h; 1,814·82 mph).
Awaiting confirmation is a new record of 1,812 knots (3,357 km/h; 2,086 mph) set by Major Adolphus H. Bledsoe Jr and Major John T. Fuller in a Lockheed SR-71A on 27 July 1976 (USA).

Speed around the World (USA)
Walter H. Mullikin and crew of four, in a Boeing 747SP of Pan American, on 1-3 May 1976, from New York City, via Delhi and Tokyo, back to New York, in 1 day 22 hr 50 sec. 436·95 knots (809·24 km/h; 502·84 mph).

CLASS C.2, ALL GROUPS (Seaplanes)

Distance in a straight line (UK)
Capt D. C. T. Bennett and First Officer I. Harvey, in the Short-Mayo Mercury, on 6-8 October 1938, from Dundee, Scotland, to the Orange River, South Africa. 5,211·66 nm (9,652 km; 5,997·5 miles).

Height (USSR)
Georgi Buryanov and crew of two in a Beriev M-10, on 9 September 1961, over the Sea of Azov. 14,962 m (49,088 ft).

Speed in a straight line (USSR)
Nikolai Andrievsky and crew of two in a Beriev M-10, on 7 August 1961, at Joukovski-Petrovskœ, over a 15/25 km course. 492·44 knots (912 km/h; 566·69 mph).

CLASS D, GROUP I (Single-seat sailplanes)

Distance in a straight line (Federal Germany)
Hans W. Grosse in a Schleicher AS-W 12, on 25 April 1972. 788·77 nm (1,406·8 km; 907·70 miles).

Height (USA)
Paul F. Bickle, in a Schweizer SGS 1-23E, on 25 February 1961, at Mojave-Lancaster, California. 14,102 m (46,266 ft).

CLASS D, GROUP II (Two-seat sailplanes)

Distance in straight line (Australia)
Ingo Renner and Hilmer Geissler in a Caproni Vizzola Calif A-21, on 27 January 1975, from Bendigo Aerodrome to Langley Station, Australia. 523·97 nm (970·4 km; 602·98 miles).

Height (USA)
L. E. Edgar and H. E. Klieforth in a Pratt-Read sailplane, on 19 March 1952, at Bishop, California. 13,489 m (44,256 ft).

CLASS E.1 (Helicopters)

Distance in a straight line (USA)
R. G. Ferry in a Hughes OH-6A, on 6-7 April 1966, 1,923·08 nm (3,561·55 km; 2,213 miles).

Height (France)
Jean Boulet in an Aérospatiale SA 315B Lama on 21 June 1972. 12,442 m (40,820 ft).

Speed in a straight line (USA)
Kurt Cannon in a Sikorsky S-67 Blackhawk, on 19 December 1970, over a 15/25 km course. 191·947 knots (355·485 km/h; 220·888 mph).

Speed in a 100 km closed circuit (USSR)
Boris Galitsky and crew of five in a Mil Mi-6, on 26 August 1964, at Podmoskovnœ. 183·67 knots (340·15 km/h; 211·36 mph).

CLASS E.2 (Convertiplanes)

Height (USSR)
D. Efremov and crew of two, in the Kamov Ka-22 Vintokryl, on 24 November 1961 at Bykovo. 2,588 m (8,491 ft).

Speed in a straight line (USSR)
D. Efremov and crew of five, in the Kamov Ka-22 Vintokryl, on 7 October 1961, at Joukovski-Petrovskœ, over a 15/25 km course. 192·39 knots (356·3 km/h; 221·4 mph).

Speed in a 100 km closed circuit (New Zealand)
Sqd Ldr W. R. Gellatly and J. G. P. Morton, in the Fairey Rotodyne, on 5 January 1959, White Waltham-Wickham-Radley Bottom-Kintbury-White Waltham. 165·89 knots (307·22 km/h; 190·90 mph).

CLASS E.3 (Autogyros)

Height (UK)
Wing Cdr K. H. Wallis, in a Wallis WA-116/Mc, on 11 May 1968. 4,639 m (15,220 ft).

Distance in a closed circuit (USA)
Igor Bensen, in a Bensen B-8M Gyro-Copter, on 15 May 1967. 64·57 nm (119·58 km; 74·30 miles).
Awaiting confirmation is a new record of 362 nm (670 km; 416 miles) set up in July 1974 by Wing Cdr K. H. Wallis in a Wallis WA-116/F.

Speed in a straight line (UK)
Wing Cdr K. H. Wallis, in a Wallis WA-116/Mc, over a 3 km course, on 12 May 1969. 96·65 knots (179 km/h; 111·225 mph).

Lockheed SR-71A, fastest production aeroplane ever built, with three absolute world records awaiting confirmation

THE ARGENTINE REPUBLIC

AERO BOERO

AERO BOERO SRL

Head Office:
Hipolito Irigoyen 505, Morteros, Córdoba
Telephone: Morteros 409
Directors:
Cesar E. Boero
Hector A. Boero

This company is producing and developing the Aero Boero 115 BS and 180 series of light monoplanes, and is developing the AG.260 agricultural aircraft.

AERO BOERO 115 BS

Earlier versions of this aircraft, no longer in production, were the Aero Boero 95 (1969-70 *Jane's*) and Aero Boero 95/115 (1972-73 *Jane's*). The current production version, first flown in February 1973, is known as the Aero Boero 115 BS. This has a sweptback fin and rudder, increased wing span and greater fuel capacity than the AB 95/115, to which it is otherwise generally similar.

The description which follows applies to the AB 115 BS, of which 20 had been completed by the Spring of 1976. A further five were then under construction.

Type: Three-seat light aircraft.

Wings: Braced high-wing monoplane. Wing section NACA 23012. Dihedral 1° 45'. Incidence 3° at root, 1° at tip. Light alloy structure, including skins. Streamline-section Vee bracing strut each side. Aluminium alloy ailerons and flaps.

Fuselage: SAE 4130 steel tube structure, Ceconite-covered.

Tail Unit: Wire-braced welded steel tube structure, Ceconite-covered. Sweptback vertical surfaces.

Landing Gear: Non-retractable tailwheel type. Shock-absorption by helicoidal springs inside fuselage. Mainwheel tyres size 6·00-6, pressure 1·66 bars (24 lb/sq in). Hydraulic disc brakes. Fully-castoring steerable tailwheel.

Power Plant: One 85·5 kW (115 hp) Lycoming O-235-C2A flat-four engine, driving either a McCauley 1C90-7345 or a Sensenich 72CK-050 fixed-pitch propeller. Two wing fuel tanks, total capacity 130 litres (28·5 Imp gallons).

Accommodation: Normal accommodation for pilot and two passengers in enclosed cabin. Baggage compartment on port side, aft of cabin. Ambulance version can accommodate one stretcher in place of the two passengers.

Electronics and Equipment: One 40A alternator and one 12V battery. VHF radio standard. Provision for dual controls, and night or blind-flying equipment, at customer's option.

Dimensions, external:
Wing span 10·72 m (35 ft 2 in)
Wing chord (constant) 1·61 m (5 ft 3½ in)
Wing aspect ratio 7·05
Length overall 7·273 m (23 ft 10¼ in)
Height overall 2·10 m (6 ft 10½ in)
Wheel track 2·05 m (6 ft 8¾ in)
Wheelbase 5·10 m (16 ft 8¾ in)

Dimensions, internal:
Cabin: Length 1·90 m (6 ft 3 in)
Max width 0·84 m (2 ft 9 in)
Max height 1·20 m (3 ft 11¼ in)

Areas:
As for Aero Boero 180 RV and RVR

Weights and Loadings:
Weight empty, equipped 530 kg (1,168 lb)
Max T-O weight 770 kg (1,697 lb)
Max wing loading 47·1 kg/m² (9·65 lb/sq ft)
Max power loading 9·0 kg/kW (14·77 lb/hp)

Performance (at max T-O weight, except where indicated):
Max level speed at S/L 113 knots (210 km/h; 130 mph)
Max cruising speed at S/L 102 knots (188 km/h; 117 mph)
Stalling speed, flaps down 39 knots (72 km/h; 45 mph)
Max rate of climb at S/L 300 m (1,000 ft)/min
T-O run, full load 115 m (380 ft)
T-O to 15 m (50 ft), two persons 185 m (607 ft)
Landing from 15 m (50 ft) 150 m (500 ft)
Landing run, heavy braking 45 m (150 ft)
Range with max fuel 429 nm (800 km; 495 miles)

AERO BOERO 180 RV and RVR

The first production three-seat Aero Boero 180 (1972-73 *Jane's*) was delivered in December 1969.

This version was followed by the Aero Boero 180 RV (standard version) and 180 RVR (glider-towing version), the first of which flew for the first time in October 1972. These current versions have extended-span all-metal wings, similar to those of the AB 115 BS, increased fuel capacity, a recontoured fuselage and sweptback vertical tail surfaces.

The description which follows applies to the AB 180 RV and 180 RVR. Twenty-nine current-model AB 180s had been built and seven more had been ordered by the Spring of 1976.

Type: Three-seat light aircraft.

Wings: Strut-braced high-wing monoplane. Streamline-section Vee bracing strut each side. Wing section NACA 23012. Dihedral 1° 45'. Incidence 3° at root, 1° at tip. Light alloy structure, including skins. Ailerons and flaps of aluminium alloy construction.

Fuselage: Welded steel tube structure (SAE 4130), covered with Ceconite.

Tail Unit: Wire-braced welded steel tube structure, covered with Ceconite. Sweptback vertical surfaces. Ground-adjustable tab on rudder.

Landing Gear: Non-retractable tailwheel type, with shock-absorption by helicoidal springs inside fuselage. Main wheels and tyres size 6·00-6, pressure 1·66 bars (24 lb/sq in). Hydraulic disc brakes. Tailwheel steerable and fully castoring.

Power Plant: One 134 kW (180 hp) Lycoming O-360-A1A flat-four engine, driving (according to customer's choice) either a Hartzell constant-speed or McCauley 1A200 or Sensenich 76EM8 fixed-pitch propeller. Three wing fuel tanks, total capacity 201 litres (53 US gallons; 44 Imp gallons).

Accommodation: Normal accommodation for pilot and two passengers in enclosed cabin. Baggage compartment on port side, aft of cabin. Transparent roof panel in 180 RVR.

Electronics and Equipment: One 40A alternator and one 12V battery. VHF radio standard. Provision for night or blind-flying instrumentation at customer's option. Towing hook in 180 RVR.

Dimensions, external:
Wing span 10·72 m (35 ft 2 in)
Wing chord (constant) 1·61 m (5 ft 3½ in)
Wing aspect ratio 7·05
Length overall 7·273 m (23 ft 10¼ in)
Height overall 2·10 m (6 ft 10½ in)
Wheel track 2·05 m (6 ft 8¾ in)
Wheelbase 5·10 m (16 ft 8¾ in)

Areas:
Wings, gross 16·47 m² (177·3 sq ft)
Ailerons (total) 1·84 m² (19·81 sq ft)
Flaps (total) 1·94 m² (20·9 sq ft)
Fin 0·93 m² (10·01 sq ft)
Rudder, incl tab 0·41 m² (4·41 sq ft)
Tailplane 1·40 m² (15·07 sq ft)
Elevators 0·97 m² (10·44 sq ft)

Weights and Loadings:
Weight empty, equipped 550 kg (1,212 lb)
Max T-O weight 844 kg (1,860 lb)
Max wing loading 52·0 kg/m² (10·7 lb/sq ft)
Max power loading 6·30 kg/kW (10·36 lb/hp)

Performance (at max T-O weight, except where indicated):
Never-exceed speed 134 knots (249 km/h; 155 mph)
Max level speed at S/L:
RV 132 knots (245 km/h; 152 mph)
RVR 122 knots (225 km/h; 140 mph)
Max cruising speed at S/L 114 knots (211 km/h; 131 mph)
Stalling speed, flaps down 41·5 knots (77 km/h; 48 mph)
Max rate of climb at S/L 360 m (1,180 ft)/min
Time to 600 m (1,970 ft), 75% power, with Blanik two-seat sailplane 3 min 10 sec
Service ceiling 6,700 m (22,000 ft)
T-O run 100 m (330 ft)
T-O to 15 m (50 ft), two persons 188 m (615 ft)
Landing from 15 m (50 ft) 160 m (525 ft)
Landing run 60 m (195 ft)
Range with max fuel 636 nm (1,180 km; 733 miles)

AERO BOERO 180 Ag

This version of the Aero Boero 180 is certificated in the Restricted category for use as an agricultural aircraft.

The description of the Aero Boero 180 RV and RVR applies also to the 180 Ag, except in the following respects:

Wings: Incidence 3° 30' at root, 2° at tip.

Equipment: Flush-fitting underfuselage pod containing agricultural chemical. Spraybars fitted along rear bar of Vee strut and horizontally below wings. Electrically-operated rotary atomisers (two each side) fitted to rear bar of Vee strut.

Performance (at max T-O weight):
Never-exceed speed 117 knots (217 km/h; 135 mph)
Max level speed at S/L 109 knots (201 km/h; 125 mph)
Max cruising speed at S/L 100 knots (185 km/h; 115 mph)
Econ cruising speed at S/L 96 knots (177 km/h; 110 mph)
Stalling speed, flaps down 48 knots (89 km/h; 55 mph)
Max rate of climb at S/L 107 m (350 ft)/min
T-O run 213 m (700 ft)
T-O to 15 m (50 ft) 335 m (1,100 ft)
Landing from 15 m (50 ft) 229 m (750 ft)
Landing run 152 m (500 ft)
Range with max fuel 434 nm (804 km; 500 miles)

Aero Boero 115 BS three-seat light aircraft (Lycoming O-235 engine)

Aero Boero 180 Ag, with underwing spraybars and ventral chemical pod

AERO BOERO AG.260

Aero Boero began the design of this single-seat agricultural monoplane in mid-1971, at which time it was known as the AG.235/260. Construction of a prototype began in October 1971, and this aircraft flew for the first time on 23 December 1972. A static test airframe has also been completed.

The static test and flight certification programmes were due to be completed in 1976.

TYPE: Single-seat agricultural aircraft.

WINGS: Low-wing monoplane. Wing section NACA 23012. Dihedral 5°. Construction, including trailing-edge flaps and ailerons, is of aluminium alloy, with inverted Vee bracing struts on each side.

FUSELAGE: Welded SAE 4130 steel tube structure with plastics covering.

TAIL UNIT: Wire-braced welded steel tube structure with plastics covering.

LANDING GEAR: Non-retractable tailwheel type, with coil spring shock-absorbers. Hydraulic disc brakes on main wheels.

POWER PLANT: One 194 kW (260 hp) Lycoming O-540 flat-six engine, driving a McCauley P235/AFA 8456 two-blade propeller. Four wing fuel tanks, total capacity 268 litres (70·8 US gallons; 59 Imp gallons).

ACCOMMODATION: Pilot only, in enclosed cabin. Door on starboard side, which can be jettisoned in an emergency. Cabin heated, and ventilated by adjustable cool-air vents. Utility compartment on port side, aft of cabin.

ELECTRONICS AND EQUIPMENT: VHF radio standard. Non-corrosive glassfibre tank installed forward of cockpit, with capacity of 500 litres (110 Imp gallons) of liquid or 500 kg (1,102 lb) of dry chemical. Quick-dump valve, to jettison contents of tank in an emergency. Engine-driven pump.

DIMENSIONS, EXTERNAL:
- Wing span: 10·90 m (35 ft 9 in)
- Wing chord (constant over most of span): 1·61 m (5 ft 3 ½ in)
- Wing aspect ratio: 6·8

Aero Boero AG.260 agricultural aircraft (Lycoming O-540 engine) *(Alex Reinhard)*

- Length overall (tail up): 7·45 m (24 ft 5¼ in)
- Height overall (tail up): 1·90 m (6 ft 2¾ in)
- Tailplane span: 3·04 m (9 ft 11¾ in)
- Propeller diameter: 2·13 m (7 ft 0 in)

DIMENSION, INTERNAL:
- Hopper volume: 0·5 m³ (17·66 cu ft)

AREAS:
- Wings, gross: 16·47 m² (177·28 sq ft)
- Ailerons (total): 1·84 m² (19·81 sq ft)
- Flaps (total): 1·94 m² (20·88 sq ft)
- Fin: 0·93 m² (10·01 sq ft)
- Rudder: 0·41 m² (4·41 sq ft)
- Tailplane: 1·40 m² (15·07 sq ft)
- Elevators: 0·97 m² (10·44 sq ft)

WEIGHTS AND LOADINGS:
- Weight empty: 720 kg (1,587 lb)
- Max T-O weight: 1,350 kg (2,976 lb)
- Max wing loading: 77·28 kg/m² (15·83 lb/sq ft)
- Max power loading: 6·96 kg/kW (11·44 lb/hp)

PERFORMANCE (at max T-O weight):
- Never-exceed speed: 117 knots (217 km/h; 135 mph)
- Max cruising speed at S/L: 109 knots (201 km/h; 125 mph)
- Econ cruising speed: 95·5 knots (177 km/h; 110 mph)
- Stalling speed, flaps down: 52·5 knots (97 km/h; 60 mph)
- Max rate of climb at S/L: 410 m (1,345 ft)/min
- Service ceiling: 6,400 m (21,000 ft)
- T-O to 15 m (50 ft): 200 m (656 ft)
- Landing from 15 m (50 ft): 120 m (394 ft)
- Range with max fuel: 593 nm (1,100 km; 683 miles)

CHINCUL

CHINCUL S.A.C.A.I.F.I.

HEAD OFFICE:
Mendoza S/N, Calle 6 y 7, Departamento Pocito, Casilla Correo·80, San Juan (San Juan)

WORKS:
25 de Mayo 489, 60 Piso, Buenos Aires

PRESIDENT:
José Maria Beraza

VICE-PRESIDENT:
Juan José Beraza

This company, a wholly-owned subsidiary of La Macarena SRL, Piper's Argentine distributor, concluded an agreement with Piper Aircraft Corporation on 22 November 1971, for manufacture of a broad range of Piper products in Argentina. The proposed plan called for a progression through five manufacturing phases of increasing complexity, designed to permit the gradual assimilation of aircraft manufacturing technology by Chincul.

The programme, officially inaugurated on 20 December 1972, scheduled the completion of 1,000 single-engined and 340 twin-engined Piper aircraft by Chincul. Phase 1, following the delivery of four Seneca and 15 Cherokee kits, was carried out in 1973. Under a programme sponsored by the Comando de Regiones Aéreas, 40 Cherokees were to be built during the first two to three years for use as trainers by Argentine flying clubs, but no information regarding the progress of this programme has been made available, either by Chincul or by Piper Aircraft Corporation.

A new assembly facility in San Juan was inaugurated on 13 December 1972. This facility, occupying a covered area of 2,845 m² (30,623 sq ft), is part of a 12,000 m² (129,165 sq ft) plant which was eventually to be devoted to the assembly of all models of Piper aircraft. Finished aircraft were to be test-flown and certificated by Argentine personnel.

CICARÉ

CICARÉ AERONÁUTICA SC

ADDRESS:
Ave Ibañez Frocham s/n, CC24, Saladillo, Provincia de Buenos Aires

ENQUIRIES TO:
Comodoro Antonio R. Mantel, Santa Fé 1256, Buenos Aires

Telephone: Buenos Aires 41-5260

PARTNERS:
Augusto Ulderico Cicaré
Comodoro Ildefonso Domingo Durana
Comodoro Antonio Raúl Mantel

This company was formed in 1972 to undertake the development and construction of small aero-engines and light helicopters. Sr Cicaré, originally an engine designer, has designed and constructed two experimental helicopters, the Cicaré I and Cicaré II, brief details of which appeared in the 1970-71, 1973-74 and 1974-75 *Jane's*. A description follows of the more recently-designed CH-III Colibri:

CICARÉ CH-III COLIBRÍ

Design of the CH-III began in August 1973, and prototype construction started in 1974. This work is being done, under contract from the Argentine Air Force, to evolve a light helicopter suitable for training and agricultural duties. First flight was scheduled to take place in early 1976.

TYPE: Two/three-seat light helicopter.

ROTOR SYSTEM: Four-blade rigid main rotor and two-blade tail rotor. Blade section NACA 0015. All blades are of glassfibre construction. No rotor brake or blade folding.

ROTOR DRIVE: Ten Vee-belts, via a reduction gearbox, with freewheel system for autorotation. Main rotor/engine rpm ratio 1 : 6; tail rotor/engine rpm ratio 1 : 1.

FUSELAGE: Steel tube structure, with glassfibre cabin and aluminium tailboom.

TAIL UNIT: Glassfibre horizontal and vertical fixed stabilisers.

LANDING GEAR: Tubular steel skid type.

POWER PLANT: One 142 kW (190 hp) Lycoming HIO-360-D1A flat-four engine, mounted horizontally.

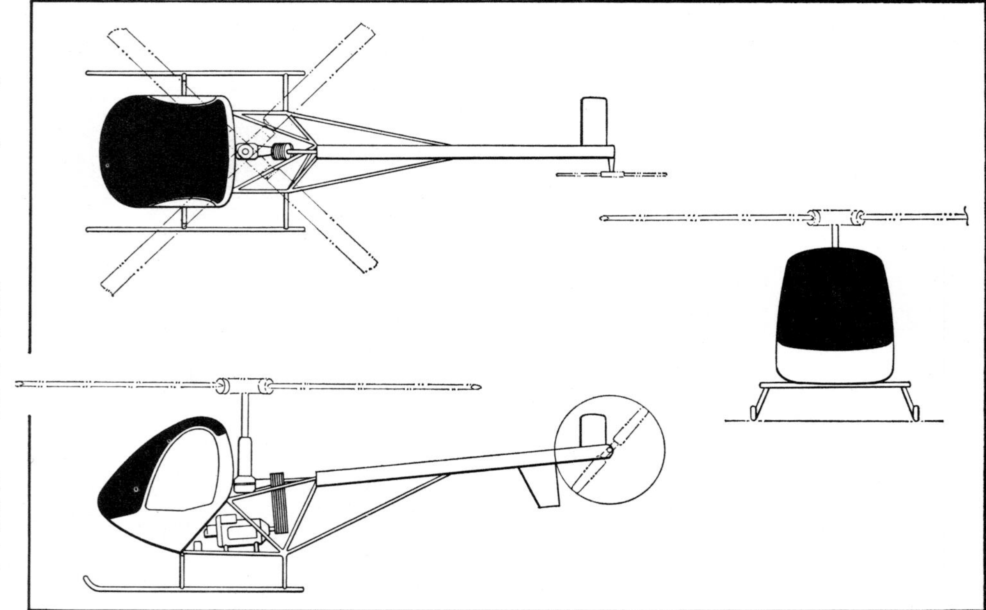

Cicaré CH-III Colibri two/three-seat light helicopter *(Roy J. Grainge)*

Single glassfibre fuel tank, capacity 135 litres (29·5 Imp gallons). Optional auxiliary tank, capacity 75 litres (16·5 Imp gallons).

ACCOMMODATION: Two or three seats side by side in enclosed cabin (instructor and pupil only in training version). Door on each side of cabin. Space for up to 45 kg (100 lb) of baggage. Cabin heated and ventilated.

ELECTRONICS AND EQUIPMENT: VHF radio standard. Mission equipment includes spraying or dusting gear and cargo sling.

DIMENSIONS, EXTERNAL:
- Diameter of main rotor: 7·47 m (24 ft 6 in)
- Main rotor blade chord: 152 mm (6 in)
- Diameter of tail rotor: 1·10 m (3 ft 7·2 in)
- Distance between rotor centres: 4·27 m (14 ft 0 in)
- Length overall: 8·53 m (28 ft 0 in)
- Height overall: 2·47 m (8 ft 1·2 in)
- Skid track: 2·01 m (6 ft 7·2 in)

AREAS:
- Main rotor disc: 43·8 m² (471·43 sq ft)
- Tail rotor disc: 0·95 m² (10·18 sq ft)

WEIGHTS AND LOADINGS:
- Weight empty, equipped: 469 kg (1,034 lb)
- Max payload: 226 kg (500 lb)
- Max T-O weight: 800 kg (1,764 lb)
- Max disc loading: 18·3 kg/m^2 (3·74 lb/sq ft)
- Max power loading: 5·63 kg/kW (9·28 lb/hp)

PERFORMANCE (estimated, at max T-O weight):
- Max level and cruising speed at S/L: 88 knots (163 km/h; 101 mph)
- Econ cruising speed: 65 knots (120 km/h; 74·5 mph)
- Max rate of climb at S/L: 360 m (1,180 ft)/min
- Service ceiling: 3,900 m (12,800 ft)
- Hovering ceiling out of ground effect: 1,700 m (5,575 ft)
- Range with internal fuel 259 nm (480 km; 298 miles)

FMA (AREA DE MATERIAL CÓRDOBA)

AGRUPACIÓN AVIONES-DEPARTAMENTO INGENIERÍA, GUARNICION AÉREA CÓRDOBA

ADDRESS:

Avenida Fuerza Aérea Argentina Km 5½, Córdoba

Telephone: 45011, 37048 and 44732

DIRECTOR:

Brigadier César F. Ferrante

CHIEF DESIGNER AND ENGINEER:

Vicecomodoro Héctor Eduardo Ruiz

The original Fábrica Militar de Aviones (Military Aircraft Factory) was founded in 1927 as a central organisation for aeronautical research and production in the Argentine. Its name was changed to Instituto Aerotécnico in 1943 and then to Industrias Aeronáuticas y Mecánicas del Estado (IAME) in 1952. In 1957 it became a State enterprise under the title of Dirección Nacional de Fabricaciones e Investigaciónes Aeronáuticas (DINFIA), but reverted to its original title in 1968. It is now a component of the Area de Material Córdoba division of the Argentine Air Force.

FMA comprises two large divisions. The Instituto de Investigaciónes Aeronáuticas y Espacial (IIAE) is responsible for the design, manufacture and testing of rockets, sounding equipment and other equipment. The Fábrica Militar de Aviones itself controls the aircraft manufacturing facilities situated in Córdoba. The laboratories, factories and other aeronautical division buildings occupy a total covered area of 148,557 m^2 (1,599,059 sq ft); the Area de Material Córdoba employs 3,500 persons, of whom about 1,500 are in the FMA.

FMA's head offices are situated in Buenos Aires. It also controls the Centro de Ensayos en Vuelo (Flight Test Centre), to which all aircraft produced in the Argentine are sent for certification tests.

The major aircraft of national design in current production is the IA 58 Pucará counter-insurgency aircraft. The IA 60 twin-turbofan trainer, based on the Pucará, is in the design stage. Production of the IA 50 GII, described in the 1974-75 *Jane's*, has ended.

FMA is also producing Cessna single-engined aircraft under licence, under a renewed and extended form of the agreement announced in October 1965. First phase called for assembly of 80 aircraft from major assemblies supplied by Cessna. Phase 2 involved assembly of 100 aircraft from detail parts provided by Cessna. Phase 3 involves an estimated 320 aircraft, for which FMA is manufacturing or acquiring in the Argentine as many parts as possible. All aircraft are repurchased by Cessna for sale through its distributors and dealers in Latin America or sold directly by FMA to Argentine government agencies. Forty Cessna Model 150s have been ordered by the Comando de Regiones Aéreas for use as trainers by Argentine flying clubs.

The first A182J (Argentine 182) was completed in August 1966 from the initial batch of twelve sets of components supplied by Cessna, and was delivered to its owner on 2 September 1966. The renewed and extended agreement, announced in April 1971, provided for continued production of the Cessna 182 and, in addition, for the range to be extended to include the Model 150 trainer and the AGwagon agricultural aircraft.

By February 1975, FMA had completed 136 Cessna Model A182s, 27 Model A150 trainers, 8 Model A-A150 Aerobats and 23 Model A188 AGwagons.

IA 58 PUCARÁ

This twin-turboprop counter-insurgency aircraft was developed to meet an Argentine Air Force requirement. Originally known as the Delfin, it was later renamed Pucará. An unpowered aerodynamic prototype was described in the 1968-69 *Jane's*. The first powered prototype, designated AX-01, flew for the first time on 20 August 1969 with AiResearch TPE 331 engines. It was described in the 1971-72 *Jane's*.

A second prototype, designated AX-02, flew for the first time on 6 September 1970 with 1,022 ehp Turboméca Astazou XVIG turboprop engines, as now fitted to production Pucarás.

An order for 30 Pucarás has been placed by the Argentine Air Force, and the first of these (A-501) flew for the first time on 8 November 1974. It is anticipated that this order will be increased to 70 aircraft. Interest in the Pucará has also been expressed by the air forces of Bolivia, Libya and Peru.

The following description applies to the production version, of which the first five had been delivered to operational units of the Argentine Air Force by May 1976:

TYPE: Twin-turboprop counter-insurgency aircraft.

WINGS: Cantilever low-wing monoplane. Wing section NACA 64_2A215 at root, NACA 64_1A212 at tip. Dihedral 7° on outer wing panels. Incidence 2°. No sweepback. Conventional semi-monocoque fail-safe structure of duralumin. Frise-type fabric-covered duralumin ailerons and all-dural slotted trailing-edge flaps. No slats. Balance tab in starboard aileron, electrically-operated trim tab in port aileron. Kléber-Colombes pneumatic de-icing boots on leading-edges.

FUSELAGE: Conventional semi-monocoque fail-safe structure of duralumin. Door-type airbrakes at rear which form tailcone when closed.

TAIL UNIT: Cantilever semi-monocoque structure of duralumin. Fixed-incidence tailplane and elevators mounted near top of fin. Trim tab in rudder and each elevator. Kléber-Colombes pneumatic de-icing boots on leading-edges.

LANDING GEAR: Retractable tricycle type, all units retracting forward hydraulically. Shock-absorbers of Kronprinz Ring-Feder type, designed by Vicecomodoro Ruiz. Single wheel on nose unit, twin wheels on main units, all with Dunlop tubeless Type III tyres size 7·50-10. Tyre pressures: 2·82 bars (41 lb/sq in) on main units, 2·41 bars (35 lb/sq in) on nose unit. Dunlop hydraulic disc brakes. No anti-skid units.

POWER PLANT: Two 761 kW (1,022 ehp) Turboméca Astazou XVIG turboprop engines, each driving a Hamilton Standard 33LF/1015-0 three-blade metal propeller. Fuel in two fuselage tanks and one self-sealing tank in each wing, with total capacity of 1,422 litres (313 Imp gallons). Attachment point beneath each wing at junction of centre and outer panels for external weapons or jettisonable auxiliary fuel tank of 300 litres (66 Imp gallons). Oil capacity 11·75 litres (2·6 Imp gallons).

ACCOMMODATION: Crew of two in tandem on Martin-Baker Mk AP06A ejection seats beneath transparent moulded canopy. Rear seat slightly elevated. Bullet-proof windscreen.

SYSTEMS: Hydraulic system, pressure 207 bars (3,000 lb/sq in), supplied by two engine-driven pumps, actuates landing gear, flaps, wheel brakes and airbrakes. Wing and tail unit de-icing by engine bleed air. Electrical system includes two 300A 28V starter/generators for DC power and two 500/750VA rotary inverters for 115V AC power. One 36Ah SAFT Voltabloc 4006 battery. No APU.

ELECTRONICS AND EQUIPMENT: Blind-flying instrumentation standard. Radio equipment includes Bendix DFA-73A-1 ADF, Bendix RTA-42A VHF communications system, Bendix RNA-2bc VHF navigation system, Northern N-420 HF 55B communications system, amplifier and audio-selector system with AS-A-31 panel. Optional equipment includes weather radar, IFF and VHF/FM tactical communications system.

ARMAMENT AND OPERATIONAL EQUIPMENT: Two 20 mm Hispano cannon and four 7·62 mm FN machine-guns in fuselage. One attachment point beneath centre of fuselage and one beneath each wing outboard of engine nacelle for a variety of external stores, including auxiliary fuel tanks. Librascope 335336 gunsight and one AN/AWE programmer.

DIMENSIONS, EXTERNAL:
- Wing span: 14·50 m (47 ft 6¾ in)
- Wing chord at root: 2·24 m (7 ft 4¼ in)
- Wing chord at tip: 1·60 m (5 ft 3 in)

First production IA 58 Pucará for the Argentine Air Force, equipped with underwing and underfuselage weapons

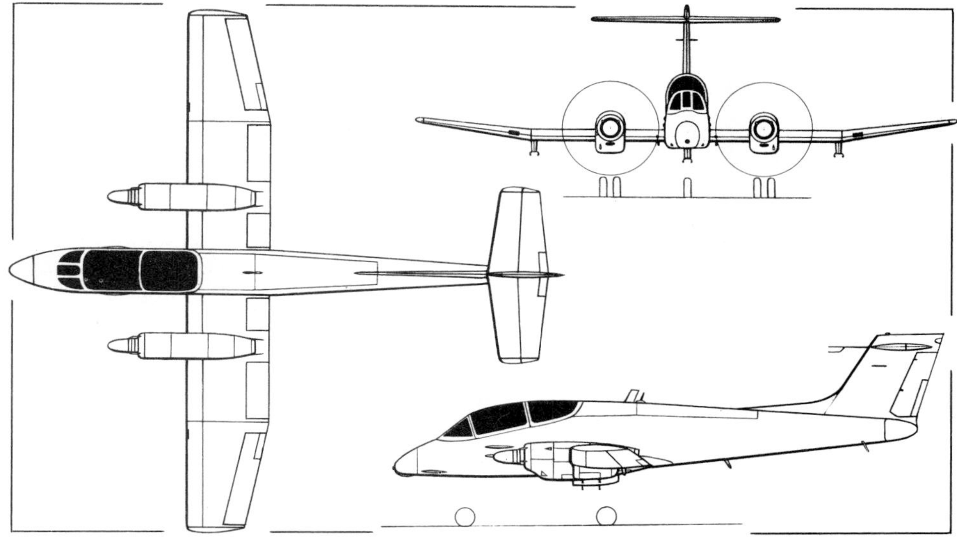

FMA IA 58 Pucará twin-turboprop counter-insurgency aircraft *(Pilot Press)*

Wing aspect ratio	6·95
Length overall	14·10 m (46 ft 3 in)
Length of fuselage	13·32 m (43 ft 8½ in)
Fuselage: Max width	1·24 m (4 ft 0¾ in)
Height overall	5·36 m (17 ft 7 in)
Tailplane span	4·70 m (15 ft 5 in)
Wheel track (c/l of shock-absorbers)	4·20 m (13 ft 9¼ in)
Wheelbase	3·48 m (11 ft 5 in)
Propeller diameter	2·59 m (8 ft 6 in)

DIMENSIONS, INTERNAL:

Cabin: Floor area	2·90 m² (31·2 sq ft)
Volume	2·74 m³ (96·8 cu ft)

AREAS:

Wings, gross	30·30 m² (326·1 sq ft)
Ailerons (total)	3·29 m² (35·41 sq ft)
Trailing-edge flaps (total)	3·58 m² (38·53 sq ft)
Fin	3·465 m² (37·30 sq ft)
Rudder, incl tab	1·565 m² (16·84 sq ft)
Tailplane	4·60 m² (49·51 sq ft)
Elevators, incl tabs	2·612 m² (28·11 sq ft)

WEIGHTS AND LOADINGS:

Weight empty	4,037 kg (8,900 lb)
Max T-O weight	6,486 kg (14,300 lb)
Max landing weight	5,806 kg (12,800 lb)
Max wing loading	214·8 kg/m² (44 lb/sq ft)
Max power loading	4·26 kg/kW (7 lb/hp)

PERFORMANCE (at max T-O weight, except where indicated):

Never-exceed speed	404 knots (750 km/h; 466 mph)
Max level speed at 3,000 m (9,840 ft)	281 knots (520 km/h; 323 mph)
Max cruising speed	261 knots (485 km/h; 301 mph)
Econ cruising speed	232 knots (430 km/h; 267 mph)
Stalling speed, flaps down, at 4,790 kg (10,560 lb) AUW	77·5 knots (142·5 km/h; 89 mph)
Max rate of climb at S/L	1,080 m (3,543 ft)/min
Service ceiling at 6,200 kg (13,668 lb) AUW, 0° flap	8,280 m (27,165 ft)
Service ceiling, one engine out, at 4,960 kg (10,934 lb) AUW, 0° flap	5,344 m (17,533 ft)
T-O run	300 m (985 ft)
T-O to 15 m (50 ft)	705 m (2,313 ft)
Landing from 15 m (50 ft) at 5,100 kg (11,243 lb) AUW	603 m (1,978 ft)
Landing run at 5,100 kg (11,243 lb) AUW	200 m (656 ft)
Range with max fuel at 5,000 m (16,400 ft)	1,641 nm (3,042 km; 1,890 miles)

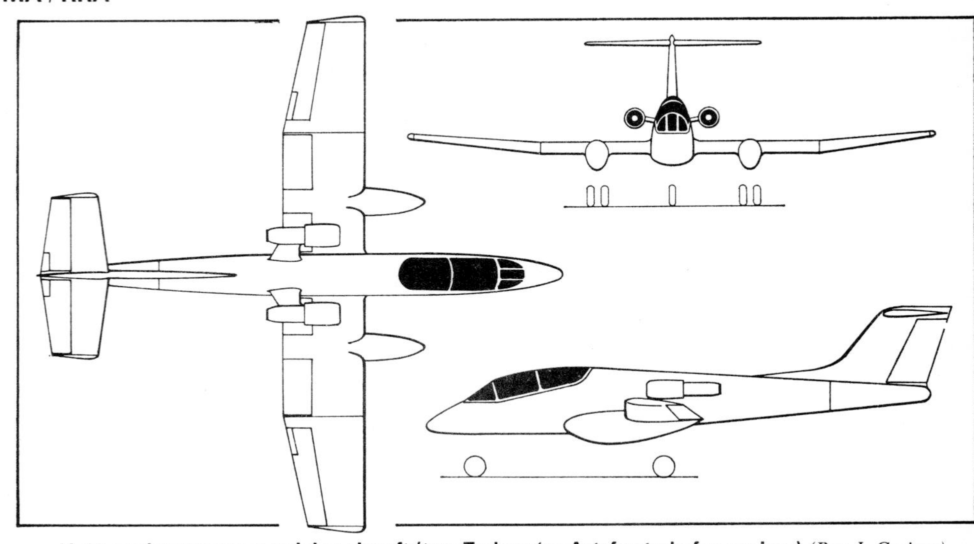

IA 60 tandem two-seat training aircraft (two Turboméca Astafan turbofan engines) *(Roy J. Grainge)*

IA 60

Based on the airframe of the Pucará, Vicecomodoro Ruiz has designed the IA 60 tandem two-seat trainer to be powered by two Turboméca Astafan turbofan engines. As shown in the accompanying three-view drawing, these engines will be mounted in pods on the fuselage sides above the wing, with streamlined landing gear fairings replacing the turboprop engine nacelles of the standard Pucará. The following details are provisional:

DIMENSIONS, EXTERNAL: As IA 58 Pucará

WEIGHTS:

Weight empty, equipped	3,800 kg (8,377 lb)
Max T-O weight	6,500 kg (14,330 lb)

PERFORMANCE:

Max critical Mach number	0·73
Service ceiling	10,000 m (32,800 ft)

RACA
REPRESENTACIONES AERO COMERCIALES ARGENTINAS SA

HEAD OFFICE:
Lavalle 715, 5° Piso, Buenos Aires
Telephone: 392-1334 and 392-9488
Telex: 012-2844

WORKS:
Aeródromo San Fernando, Provincia Buenos Aires

PRESIDENT:
J. R. Fernández Racca

SALES MANAGER:
Francisco Fournier

This company is the representative or dealer in Argentina for a number of world aerospace companies, and is the exclusive national distributor for the Concorde (jointly with the French distributor), BAC One-Eleven, Shorts Skyvan Srs 3 and Canadair CL-215 aircraft, and the MBB BO 105 and Hughes helicopters. Under a licence agreement concluded in December 1972 RACA is undertaking, with Argentine government approval (granted in mid-1973), the progressive local manufacture of a minimum of 120 Hughes Model 500 helicopters (see US section) from knock-down components. These are known locally as RACA-Hughes 500s, and are identical to the Hughes-built version described in the US section.

The programme is covering, in three phases, a period of eight years, to supply military and civil customers in Argentina and neighbouring countries. In anticipation of this programme, RACA expanded its workshop facilities at San Fernando aerodrome to a covered area of 4,600 m² (49,514 sq ft).

By January 1976 the programme had entered its second phase, in which an average of 3,258 man-hours is spent by RACA on the assembly of each aircraft, and kit number 48 had been ordered from Hughes.

First RACA-assembled Hughes 500C helicopter to be delivered

RRA
RONCHETTI, RAZZETTI AVIACIÓN SA

ADDRESS:
Aeropuerto Internacional Rosario, Casilla Correo 7, Funes, Santa Fé Province
Telephone: Rosario 58251 or Funes 93276

PRESIDENT:
Julio E. Razzetti

HEAD OF DESIGN AND DEVELOPMENT GROUP (RRAFAGA): Ing Norberto S. Cobelo

HEAD OF PRODUCTION: Julio Di Giuseppe

RRAFAGA J-1 MARTIN FIERRO

The J-1 Martin Fierro is a single-seat agricultural aircraft, designed in Argentina by a team led by Ing Norberto S. Cobelo. Design began in September 1972, and construction of the first of two prototypes started three months later. This aircraft made its first flight on 18 December 1975. Five more J-1s are under construction, including one for structural testing.

TYPE: Single-seat agricultural aircraft.

WINGS: Cantilever low-wing monoplane. Thickness/chord ratio 15%. Dihedral 7°. Incidence 2° at root. All-metal single box-spar structure, with detachable leading-edges. All-metal Frise-type ailerons and semi-Fowler trailing-edge flaps. No tabs.

Prototype RRA J-1 Martin Fierro agricultural aircraft (Lycoming IO-540-K1JG engine)

FUSELAGE: Welded tube structure with detachable metal skin.

TAIL UNIT: Cantilever all-metal two-spar structure. Fixed-incidence tailplane. Trim tab in each elevator.

LANDING GEAR: Non-retractable tailwheel type, with spring steel shock-absorption. Cleveland wheels and main-wheel brakes. Goodyear tyres, size 8·50-10 (8 ply) on main units, 2·80-8 (4 ply) on tail unit.

POWER PLANT: One 224 kW (300 hp) Lycoming IO-540-K1JG flat-six engine, driving a Hartzell variable-pitch constant-speed propeller with spinner. Fuel in two wing-root tanks, total capacity 162 litres (35·6 Imp gallons). Refuelling point on top of each tank. Oil capacity 12 litres (2·64 Imp gallons).

ACCOMMODATION: Single seat in heated and ventilated framed cockpit. Downward-opening window/door on each side.

SYSTEMS AND EQUIPMENT: 60A 12V Prestolite generator and 12V 30Ah Prestolite battery. Single hopper, forward of cockpit, of 850 litres (187 Imp gallons) capacity. Can be fitted with dusting or spraying systems, for dry or liquid chemicals.

DIMENSIONS, EXTERNAL:
- Wing span: 13·00 m (42 ft 7¾ in)
- Wing chord (constant): 1·60 m (5 ft 3 in)
- Wing aspect ratio: 8·12
- Length overall: 7·00 m (22 ft 11¾ in)
- Height overall: 4·00 m (13 ft 1½ in)
- Wheelbase: 2·40 m (7 ft 10½ in)
- Propeller diameter: 2·13 m (7 ft 0 in)
- Propeller ground clearance: 0·50 m (1 ft 7¾ in)

AREAS:
- Wings, gross: 20·80 m² (223·9 sq ft)
- Ailerons (total): 2·76 m² (29·71 sq ft)
- Trailing-edge flaps (total): 2·76 m² (29·71 sq ft)
- Fin: 1·44 m² (15·50 sq ft)
- Rudder: 0·96 m² (10·33 sq ft)
- Tailplane: 2·64 m² (28·42 sq ft)
- Elevators (total, incl tabs): 1·76 m² (18·94 sq ft)

WEIGHTS AND LOADINGS:
- Basic operating weight with spray equipment: 920 kg (2,028 lb)
- Max payload: 850 kg (1,874 lb)
- Max T-O and landing weight (Restricted category): 2,000 kg (4,409 lb)
- Max wing loading: 96·14 kg/m² (19·69 lb/sq ft)
- Max power loading: 8·93 kg/kW (14·70 lb/hp)

PERFORMANCE (at max FAR Pt 23 Aerobatic T-O weight):
- Never-exceed speed: 188 knots (349 km/h; 217 mph)
- Max level speed at S/L, 'clean': 130 knots (241 km/h; 150 mph)
- Econ cruising speed at S/L, 'clean': 114 knots (211 km/h; 131 mph)
- Stalling speed, flaps up: 55 knots (102 km/h; 63 mph)
- Stalling speed, flaps down: 43 knots (79 km/h; 49 mph)
- Max rate of climb at S/L: 290 m (950 ft)/min
- Service ceiling: 3,660 m (12,000 ft)
- T-O run: 168 m (550 ft)
- T-O to 15 m (50 ft): 251 m (824 ft)
- Landing from 15 m (50 ft): 244 m (800 ft)
- Landing run: 85 m (280 ft)
- Max range, 45 min reserves: 265 nm (490 km; 305 miles)

AUSTRALIA

CAC
COMMONWEALTH AIRCRAFT CORPORATION LTD

HEAD OFFICE AND WORKS:
304 Lorimer Street, Port Melbourne, Victoria 3207
Telephone: 64 0771
Telex: 30721

DIRECTORS:
N. F. Stevens (Chairman)
M. L. Baillieu
Sir Ian McLennan
L. C. Bridgland
R. R. Law-Smith, CBE, AFC
A. W. Stewart
R. T. M. Rose
K. G. Wilkinson

GENERAL MANAGER: R. L. Abbott
SECRETARY: E. W. Stodden

Commonwealth Aircraft Corporation Pty Ltd was formed in 1936 to establish an aircraft industry that would make Australia independent of outside supplies. On 18 August 1975 it became a public company, changing its name to Commonwealth Aircraft Corporation Ltd.

The Corporation has an authorised capital of $6,000,000. Shareholders include BHP Nominees Pty Ltd; North Broken Hill Ltd; B. H. South Ltd; Electrolytic Zinc Co of Australasia Ltd; Nobel (Australasia) Pty Ltd; Rolls-Royce (1971) Ltd; and P & O Australian Holdings Pty Ltd.

Under a co-production agreement between the Australian government and Bell Helicopter of the USA, announced in February 1971, CAC is the prime contractor in a programme to provide 56 Bell 206B-1 JetRanger II helicopters for the Australian Army. The Army received its first 206B-1 in April 1973, and a total of 29 plus 2 civil JetRangers had been delivered by the end of January 1975. Two Bell 206B-1s were delivered in the Spring of 1973 to the Royal Australian Navy.

A constant overhaul programme for Atar engines used by the RAAF is maintained, together with an overhaul programme for Sabre aircraft and Avon engines operated by the air forces of Malaysia and Indonesia.

Other contracts include a variety of offset work for Boeing, Sikorsky, Pratt & Whitney and Hawker Siddeley Aviation; manufacture of components for the Government Aircraft Factories Nomad and New Zealand Aerospace Industries CT4 Airtrainer (which see); and aircraft furnishings, including galleys for airlines.

Rex Aviation Ltd, Bankstown Aerodrome, New South Wales, a wholly-owned subsidiary of CAC, is the distributor of Cessna aircraft and Hughes 300 and 500 helicopters in Australia and New Guinea.

Details of the company's aero-engine activities can be found in the Aero-engines section.

GOVERNMENT OF AUSTRALIA
DEPARTMENT OF INDUSTRY AND COMMERCE

ADDRESS:
Anzac Park West Building, Constitution Avenue, Parkes, Canberra ACT 2600
Telephone: 48 2111
Telex: 62063

SECRETARY:
N. S. Currie, OBE, BA

DIRECTOR OF PUBLIC RELATIONS:
A. L. Witsenhuysen, BA

Government Aircraft Factories

HEADQUARTERS:
Fishermen's Bend, Private Bag No. 4, Post Office, Port Melbourne, Victoria 3207
Telephone: 64 0661

AIRFIELD AND FINAL ASSEMBLY WORKSHOPS:
Avalon Airfield, Beach Road, Lara, Victoria 3212
Telephone: Lara 82 1202

MANAGER: J. H. Dolphin

The Government Aircraft Factories are units of the Defence Production facilities owned by the Australian government and operated by the Department of Industry and Commerce. Their functions include the design, development, manufacture, assembly, maintenance and modification of aircraft and guided weapons. At Avalon airfield, subassembly of components, final assembly, modification, repair and test-flying of jet and other aircraft are undertaken.

Current activity includes development and production of the Nomad twin-turboprop STOL aircraft, the Ikara anti-submarine missile, and the Jindivik and Turana target drones. The Jindivik and Turana are described in the RPVs and Targets section of this edition.

The GAF are producing elevators for the Boeing 727 under contract to The Boeing Company, and wing flaps for the F28 Fellowship under contract to Fokker-VFW; and are subcontractors to Commonwealth Aircraft Corporation in manufacturing bonded structures for Australian-produced examples of the Bell 206B-1 light observation helicopter.

GOVERNMENT AIRCRAFT FACTORIES NOMAD

This small, twin-turboprop utility aircraft is in production at the GAF. The first of two Model N2 prototypes (VH-SUP) was flown for the first time on 23 July 1971; it was followed by the second aircraft (VH-SUR) on 5 December 1971.

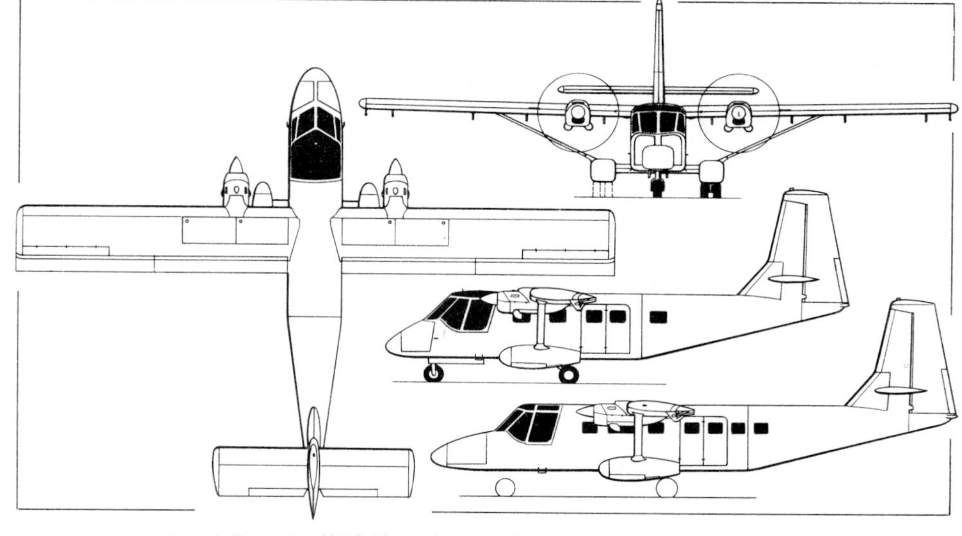

Government Aircraft Factories N22B Nomad, with additional side view *(bottom)* **of N24** *(Pilot Press)*

Design certification for the Nomad is to US FAR 23 requirements administered by the Australian Dept of Transport; the domestic Type Certificate for the N2 was issued on 11 August 1972. The basic design incorporates features of common interest to military and civil operators, including quick role-change capabilities and the ability to operate from short fields and unprepared surfaces.

Two versions have so far been announced, as follows:

N22B. Short-fuselage version, currently in production for the Australian Army Aviation Corps and commercial operators. The two prototypes were built to this standard. The military version is known as the **Mission Master**. Commercial version entered service with Aero Pelican (operating a leased aircraft) on 18 December 1975. The type certificate for the N22B production aircraft was issued on 29 April 1975, and deliveries began in that month.

N24. Higher-capacity version, with lengthened fuselage. Design includes the insertion of a 1·14 m (3 ft 9 in) section in the cabin, and increased forward baggage capacity.

Production of 95 Nomads has been authorised by the Australian government. This number includes 11 N22Bs for the Australian Army, 6 for the Indonesian Navy, 12 for the Philippine Air Force, 2 for Sabah Air and 1 for the Australian Dept of National Mapping; and 6 of the N24 version for the Northern Territory Aeromedical Service.

The following description applies generally to both versions, except where a specific model is indicated:

TYPE: Twin-turboprop STOL utility aircraft.

WINGS: Braced high-wing monoplane. Basic NACA 23018 wing section, modified to incorporate increased nose radius and camber. Dihedral 1° from roots. Incidence 2°. No sweepback. Two-spar fail-safe torsion-box structure of riveted light alloy. Full-span double-slotted trailing-edge flaps. All-metal ailerons, which droop with the flaps and transfer their motion progressively to slot-lip ailerons as the flaps extend, resulting in full-span flap. Controls actuated manually by cables and push-rods. Pneumatic de-icing of leading-edges optional. Small stub wings at cabin floor level support the main landing gear fairings from which a single strut on each side braces the main wing.

Government Aircraft Factories N22B Mission Master (Nomad) STOL utility aircraft in the insignia of the Australian Army

Fuselage: Conventional semi-monocoque riveted light alloy structure of stringers and frames.

Tail Unit: Cantilever all-metal structure. One-piece all-moving tailplane, with inset trim and anti-balance tab. Tailplane and rudder actuated manually by cables. Trim tab in rudder. Pneumatic de-icing of leading-edges optional.

Landing Gear: Retractable tricycle type, with electrical retraction by means of single actuator in the fuselage. GAF oleo-pneumatic shock-absorbers. Single rearward-retracting steerable nosewheel, tyre size 8·00-6, pressure 3·17 bars (46 lb/sq in). Twin wheels, tyre size 8·00-6, pressure 2·34 bars (34 lb/sq in), on each main unit. Main wheels retract forward into streamlined fairings at outer ends of stub wings. Dual hydraulically-operated single-disc brakes on main units. No anti-skid units.

Power Plant: Two 298 kW (400 shp) Allison 250-B17B turboprop engines, each driving a Hartzell three-blade constant-speed fully-feathering reversible-pitch metal propeller. Standard fuel capacity 813 kg (1,794 lb) plus 11·3 kg (25 lb) unusable in flexible bag tanks; or 767 kg (1,692 lb) plus 25 lb unusable in self-sealing bag tanks. Provision for internal auxiliary tanks for ferry purposes. An additional fuel capacity of 270 kg (595 lb) is provided by two optional integral tanks, one in each wingtip. Gravity refuelling via overwing point above each pair of tanks. Oil capacity 8·5 litres (1·9 Imp gallons) per engine.

Accommodation (N22B): Designed for single-pilot operation, but can accommodate crew of two on side-by-side seats. Access to flight deck by forward-opening door on each side. Main cabin has individual seats for up to 12 passengers, at 787 mm (31 in) pitch, with continuous seat tracks and readily-removable seats which allow rapid rearrangement of the cabin to suit alternative loads. Access to main cabin via double doors on port side, with single emergency exit on starboard side. Baggage compartments in nose (with door on each side) and optionally in rear of fuselage (with internal access). Whole interior, including flight deck, is heated and ventilated.

Accommodation (N24): Flight deck accommodation and access as for N22B. Lengthened main cabin, with similar internal provision to N22B for up to 15 passengers, and access via double port-side doors as in N22B. Enlarged nose baggage compartment. Rear baggage compartment of same capacity as N22B. Ventilation and heating system with individual adjustable outlets.

Systems: No air-conditioning, hydraulic or pneumatic systems normally, but air-conditioning is proposed for future models and pneumatic airframe de-icing is available optionally. Electrical system comprises a 28V 150A DC starter/generator on each engine, and a 22Ah battery with AC inverters. Other optional systems include oxygen demand system for crew and continuous-flow system for passengers; electrical de-icing for propellers, cabin floor hatch and underwing pylon racks.

Electronics and Equipment: Provision is made for a wide range of nav/com equipment to meet specific customer requirements. Other optional items include full IFR instrumentation and a lightweight weather radar.

Armament and Operational Equipment (N22B): The military variant has been designed to have four underwing hardpoints capable of accepting up to 227 kg (500 lb) loads, including gun and rocket pods. The nose bay can be utilised to accommodate surveillance and night vision aid equipment. Removable seat armour and self-sealing fuel tanks can be fitted for added protection.

Dimensions, external:

Wing span	16·46 m (54 ft 0 in)
Wing chord (constant)	1·81 m (5 ft 11¼ in)
Wing aspect ratio	9·11
Length overall:	
N22B	12·56 m (41 ft 2·4 in)
N24	14·36 m (47 ft 1¼ in)
Height overall	5·52 m (18 ft 1½ in)
Tailplane span	5·39 m (17 ft 8·4 in)
Wheel track	3·23 m (10 ft 7 in)
Wheelbase: N22B	3·65 m (11 ft 11·6 in)
Propeller diameter	2·29 m (7 ft 6 in)
Propeller ground clearance	1·22 m (4 ft 0 in)
Distance between propeller centres	4·36 m (14 ft 3·6 in)
Crew doors (each):	
Height	0·86 m (2 ft 10 in)
Width	0·69 m (2 ft 3 in)
Passenger double doors (port):	
Height	1·32 m (4 ft 4 in)
Width	1·22 m (4 ft 0 in)
Height to sill	0·89 m (2 ft 11 in)
Emergency exit (stbd):	
Height	0·58 m (1 ft 11 in)
Width	0·63 m (2 ft 1 in)

Dimensions, internal:

Cabin, excl flight deck and rear baggage compartment:	
Length:	
N22B	5·18 m (17 ft 0 in)
N24	6·27 m (20 ft 7 in)
Max width	1·30 m (4 ft 3 in)
Max height	1·58 m (5 ft 2·4 in)
Floor area:	
N22B	6·53 m² (70·25 sq ft)
N24	8·08 m² (87·0 sq ft)
Volume:	
N22B	10·19 m³ (360·0 cu ft)
N24	12·46 m³ (440·0 cu ft)
Baggage compartment volume (nose):	
N22B	0·76 m³ (27·0 cu ft)
N24	1·13 m³ (40·0 cu ft)
Baggage compartment volume (optional, rear):	
N22B, N24	0·79 m³ (28·0 cu ft)

Areas:

Wings, gross	30·10 m² (324·0 sq ft)
Ailerons (total net)	2·55 m² (27·4 sq ft)
Trailing-edge flaps (total net)	9·81 m² (105·6 sq ft)
Fin	3·63 m² (39·1 sq ft)
Rudder, incl tab	2·89 m² (31·1 sq ft)
Tailplane, incl tabs	7·25 m² (78·0 sq ft)

Weights and Loadings:

Manufacturer's basic weight empty:	
N22B	2,019 kg (4,451 lb)
N24	2,063 kg (4,549 lb)
Typical operating weight empty:	
N22B	2,116 kg (4,666 lb)

N24 extended-fuselage version of the GAF Nomad

Max disposable load:
N22B 1,739 kg (3,834 lb)
Max T-O and landing weight:
N22B, N24 3,855 kg (8,500 lb)
Max wing loading:
N22B, N24 127·9 kg/m² (26·2 lb/sq ft)
Max power loading:
N22B, N24 6·47 kg/kW (10·625 lb/shp)

PERFORMANCE (at max T-O weight, ISA at S/L, except where indicated otherwise):
Normal cruising speed:
N22B, N24 168 knots (311 km/h; 193 mph)
Stalling speed, power off, flaps up, at AUW of 3,402 kg (7,500 lb):
N22B, N24 65 knots (121 km/h; 75 mph)
Stalling speed, power off, flaps down, at AUW of 3,402 kg (7,500 lb):
N22B, N24 47 knots (88 km/h; 54·5 mph)
Max rate of climb at S/L, both engines, T-O rating for 5 min:
N22B, N24 445 m (1,460 ft)/min
N22B, N24 (ISA+20°C) 396 m (1,300 ft)/min
Rate of climb at S/L, one engine out, max continuous rating:
N22B, N24 70 m (230 ft)/min
N22B, N24 (ISA+20°C) 52 m (170 ft)/min
Service ceiling, both engines, climbing at 30·5 m (100 ft)/min, max cruise rating:
N22B, N24 6,860 m (22,500 ft)
Min ground turning radius:
N22B, N24 11·66 m (38 ft 3 in)
Runway LCN at max T-O weight:
N22B, N24 2·3
T-O run:
N22B, N24 (FAR 23) 244 m (800 ft)
N22B, N24 (STOL) 183 m (600 ft)
N22B, N24 (FAR 23), ISA+20°C 296 m (970 ft)
N22B, N24 (STOL), ISA +20°C 213 m (700 ft)
T-O to 15 m (50 ft):
N22B, N24 (FAR 23) 411 m (1,350 ft)
N22B, N24 (STOL) 384 m (1,260 ft)
N22B, N24 (FAR 23), ISA+20°C 497 m (1,630 ft)
Landing from 15 m (50 ft), AUW of 3,630 kg (8,000 lb):
N22B, N24 (FAR 23) 366 m (1,200 ft)
N22B, N24 (STOL) 194 m (635 ft)
N22B, N24 (FAR 23), ISA+20°C 384 m (1,260 ft)
Landing run, AUW of 3,630 kg (8,000 lb):
N22B, N24 (FAR 23) 212 m (695 ft)
N22B, N24 (STOL) 76 m (250 ft)
N22B, N24 (FAR 23), ISA+20°C 226 m (740 ft)
Max range at 90% power, reserves for 45 min hold:
N22B, N24 at S/L 580 nm (1,074 km; 668 miles)
N22B, N24 at 3,050 m (10,000 ft) 730 nm (1,352 km; 840 miles)

HAWKER DE HAVILLAND

HAWKER DE HAVILLAND AUSTRALIA PTY, LTD (Member Company of HAWKER SIDDELEY GROUP)

HEAD OFFICE:
PO Box 78, Lidcombe, NSW 2141
Telephone: 649-0111
Telex: 20214

DIRECTORS:
R. Kingsford-Smith (Chairman and Managing Director)
L. R. Jones (Deputy Managing Director)
C. D. MacQuaide
B. S. Price (Commercial Director)
S. S. Schaetzel (Director and Chief Designer)
I. S. Gregg

PUBLICITY OFFICER:
Miss J. Wilson

This company has its head office in Sydney, and branch offices throughout Australia and in Djakarta, Singapore, Kuala Lumpur and Hong Kong.

Aero-Industrial Division has two plants in Sydney concerned primarily with the overhaul, modification and repair of aircraft, engines, propellers, engine accessories and helicopter transmission systems for the defence forces; and at Perth International Airport also services and overhauls aircraft engines and accessories for the defence forces and commercial operators. This Division is also a substantial subcontractor to overseas aerospace manufacturers, including Boeing, Lockheed, McDonnell Douglas and Westland.

Aviation & Systems Sales Division sells aircraft, aviation equipment, missile and electronic systems on behalf of overseas principals including Hawker Siddeley Aviation, Hawker Siddeley Dynamics, de Havilland Canada, Airbus Industrie, Westland Helicopters, British Hovercraft Corporation, Raytheon and Cossor.

Research Division is engaged in a number of design and development contracts for Australian government agencies.

Distributor Products Group, comprising General Aviation Division, Spares Division and Servicing Department, provides sales and support services for civil and government aircraft operators and agencies throughout Australia, the south-west Pacific and south-east Asia. These activities include distributorship for Beech and Britten-Norman aircraft, and Allison 250, Lycoming and Continental aero-engines.

One of HDH Australia's latest products is the Enmoth target drone, described in the RPVs & Targets section.

TRANSAVIA

TRANSAVIA CORPORATION PTY LTD

ADDRESS:
73 Station Road, Seven Hills, NSW 2147
Telephone: 624-4244
Telex: 21396

SERVICE DIVISION:
Hangar 120, Bankstown Aerodrome, NSW
Telephone: 70-6968

CHAIRMAN:
F. Belgiorno-Nettis

DIRECTOR:
C. Salteri

GENERAL MANAGER:
G. Forrester

Transavia Corporation was formed in 1964 as a subsidiary of Transfield Pty Ltd, one of Australia's largest construction companies.

Its current product is the multi-purpose Airtruk.

TRANSAVIA PL-12 AIRTRUK

The Airtruk, designed by Mr Luigi Pellarini, was originally type-certificated on 10 February 1966, for spreading fertiliser and for seeding. Swath width is up to 32 m (35 yd) and of unusual uniformity. A liquid-spraying conversion, developed in 1968, is capable of covering a 30·2 m (33 yd) swath. This version has an engine-driven spray pump and a liquid chemical capacity of 818 litres (180 Imp gallons). The PL-12's unconventional layout keeps the tails clear of chemicals, and also permits rapid loading by a vehicle which approaches the aircraft between the tails.

The three-seat prototype Airtruk flew for the first time on 22 April 1965. Delivery of production Airtruks began in December 1966, and a total of 72 had been built by December 1974, for customers in Australia, Denmark, India, New Zealand, Thailand and East and South Africa.

Production of the PL-12 was continuing in 1976, together with that of the **PL-12-U**, a multi-purpose cargo/passenger/ambulance/aerial survey version of which a prototype flew for the first time in December 1970. Certification of this version was granted in February 1971, by which time two production aircraft had been completed, and deliveries began later in the year.

Airtruks are being assembled by Flight Engineers Ltd in New Zealand (which see).

The following description applies to both the PL-12 and PL-12-U, except where a particular version is indicated:

TYPE: Single-engined agricultural (PL-12) or multi-purpose (PL-12-U) aircraft.

WINGS: Strut-braced sesquiplane. Wing section NACA 23012. Dihedral 1° 30′ on upper wings. Incidence (upper wings) 3° 0′. Conventional all-metal structure, covered with Alclad sheet. All-metal trailing-edge flaps and ailerons, covered with ribbed Alclad sheet, and operated manually. Small stub wings below fuselage, braced to cabin by a single strut and to upper wings by a Vee strut on each side.

FUSELAGE: Pod-shaped structure, of 4130 welded steel tube construction with 2024 Alclad covering and glassfibre tailcone.

Transavia PL-12 Airtruk agricultural aircraft (Continental IO-520-D engine)

TAIL UNIT: Twin units, each comprising a fin, rudder and separate T tailplane and elevator, and each carried on a cantilever tubular Alclad boom extending from the upper wings. Small bumper fairing underneath each fin. Manually-operated control surfaces. Adjustable tab in each elevator. No tabs on rudder.

LANDING GEAR: Non-retractable tricycle type, each of the three wheels being carried on a pivoted trailing leg. Shock-absorbers of Transavia patented type, of bonded rubber block moulded within four hinged plates forming a diamond shape, loaded at the long axis and deformed by loads to exchange long and short axes. All wheels and tyres same size, 8·00-6. Nosewheel tyre pressure 1·38 bars (20 lb/sq in); main-wheel tyre pressure 2·21 bars (32 lb/sq in). Cleveland hydraulic disc brakes with parking lock.

POWER PLANT: One 224 kW (300 hp) Rolls-Royce Continental IO-520-D flat-six engine, driving a McCauley D2A34C58/90AT-2 two-blade constant-speed metal propeller with spinner. Two upper-wing fuel tanks, total capacity 189 litres (50 US gallons; 41·5 Imp gallons). Optional long-range installation of second tank in each upper mainplane, increasing total capacity to 379 litres (100 US gallons; 83·4 Imp gallons). Refuelling point above each upper wing. Oil capacity 11·4 litres (2·5 Imp gallons).

ACCOMMODATION (PL-12): Single-seat cockpit, with door on starboard side. Two-seat cabin aft of chemical hopper/tank for carriage of ground crew, with door at rear of lower deck. Accommodation heated and ventilated.

Accommodation (PL-12-U): Single-seat cockpit as in PL-12. By removing the central hopper or tank, passenger cabin is enlarged to seat one passenger on upper deck (back to back with pilot's seat) and four more passengers on lower deck. Doors on upper deck (starboard side) and lower deck (port side). Lower-deck cabin is heated.
Systems: 12V electrical system standard.
Electronics and Equipment: Optional equipment for PL-12-U includes VHF (also available optionally for PL-12), HF, ADF, artificial horizon and directional gyro.
Dimensions, external:
- Wing span 11·98 m (39 ft 3½ in)
- Wing chord (constant) 1·75 m (5 ft 9 in)
- Length overall 6·40 m (21 ft 0 in)
- Length of fuselage 3·96 m (13 ft 0 in)
- Height overall 2·74 m (9 ft 0 in)
- Tailplane span (each) 2·13 m (7 ft 0 in)
- Distance between tailplanes 3·48 m (11 ft 5 in)
- Wheel track 3·05 m (10 ft 0 in)
- Wheelbase 1·91 m (6 ft 3 in)
- Propeller diameter 2·23 m (7 ft 4 in)
- Min propeller ground clearance 0·30 m (1 ft 0 in)
- Passenger door (PL-12, rear):
  - Height 0·97 m (3 ft 2 in)
- Passenger doors (PL-12-U, stbd upper and port lower, each):
  - Height 0·91 m (3 ft 0 in)

Dimensions, internal (PL-12):
- Rear passenger cabin:
  - Length 1·83 m (6 ft 0 in)
  - Max width 0·97 m (3 ft 2 in)
  - Max height 2·03 m (6 ft 8 in)
  - Floor area 0·37 m² (4 sq ft)
  - Volume 0·85 m³ (30 cu ft)

Dimensions, internal (PL-12-U):
- Passenger cabin:
  - Length 2·74 m (9 ft 0 in)
  - Max width 0·97 m (3 ft 2 in)
  - Max height 2·11 m (6 ft 11 in)
  - Floor area 1·67 m² (18 sq ft)
  - Volume 2·10 m³ (74 cu ft)

Areas:
- Wings, gross 23·8 m² (256 sq ft)
- Ailerons, total 1·67 m² (18·0 sq ft)
- Trailing-edge flaps, total 1·67 m² (18·0 sq ft)
- Fins, total 1·30 m² (14·0 sq ft)
- Rudders, total 0·56 m² (6·0 sq ft)
- Tailplanes, total 2·60 m² (28·0 sq ft)
- Elevators, total, incl tabs 1·30 m² (14·0 sq ft)

Weights and Loadings:
- Weight empty:
  - PL-12 775 kg (1,710 lb)
  - PL-12-U 830 kg (1,830 lb)
- Max T-O weight:
  - PL-12 (normal category) 1,723 kg (3,800 lb)
  - PL-12 (agricultural category) 1,855 kg (4,090 lb)
  - PL-12-U 1,723 kg (3,800 lb)
- Max landing weight (both) 1,723 kg (3,800 lb)
- Max wing loading:
  - PL-12 79 kg/m² (16·2 lb/sq ft)
  - PL-12-U 73 kg/m² (15·0 lb/sq ft)
- Max power loading:
  - PL-12 8·28 kg/kW (13·7 lb/hp)
  - PL-12-U 7·69 kg/kW (12·7 lb/hp)

Performance (at max T-O weight except where indicated):
- Never-exceed speed:
  - PL-12 180 knots (333 km/h; 207 mph)
  - PL-12-U 150 knots (276·5 km/h; 172 mph)
- Max level speed at S/L, ISA:
  - PL-12 103 knots (192 km/h; 119 mph)
  - PL-12-U 112 knots (208 km/h; 129 mph)
- Max cruising speed (75% power) at S/L, ISA:
  - PL-12 95 knots (175 km/h; 109 mph)
  - PL-12-U 102 knots (188 km/h; 117 mph)
- Stalling speed, flaps up:
  - PL-12 55 knots (103 km/h; 64 mph)
  - PL-12-U 52 knots (97 km/h; 60 mph)
- Stalling speed, flaps down:
  - PL-12 52 knots (97 km/h; 60 mph)
  - PL-12-U 50 knots (94 km/h; 58 mph)
- Max rate of climb at S/L:
  - PL-12 183 m (600 ft)/min
  - PL-12-U 244 m (800 ft)/min
- Service ceiling (both versions) 3,200 m (10,500 ft)
- *T-O run:
  - PL-12 334 m (1,095 ft)
  - PL-12-U 274 m (900 ft)
- *T-O to 15 m (50 ft):
  - PL-12 564 m (1,850 ft)
  - PL-12-U 457 m (1,500 ft)
- Landing run (both versions, at max landing weight) 183 m (600 ft)
- Normal range with standard fuel 286 nm (531 km; 330 miles)
- Ferry range, standard fuel 330 nm (611 km; 380 miles)

**DCA Australia technique*

TRANSAVIA T-320 AIRTRUK

This new version of the Airtruk, certificated in January 1976, is a development of the PL-12, from which it differs principally in having a Continental Tiara 6-320-2B engine and a larger-diameter propeller; oleo-pneumatic shock-absorbers; and a 28V electrical system. A T-320-U utility version is under development.

Type, Wings, Fuselage and Tail Unit: Similar to those of PL-12. Upper-wing fence added on each side of each tailboom.
Landing Gear: Similar to PL-12, but with oleo-pneumatic shock-absorbers and size 7·00-6 (8 ply) tyres and tubes on all units.
Power Plant: One 242 kW (325 hp) Continental Tiara 6-320-2B flat-six engine, driving (on prototype) a two-blade Hartzell HC-C2YF-1BF constant-speed propeller. Fuel and oil capacities as for PL-12.
Accommodation: As PL-12. Two-man rear cabin can carry up to 155 kg (342 lb) of equipment with seats removed.
Systems: 28V electrical system and 28V battery standard.
Equipment: Standard 907 kg (2,000 lb) capacity hopper aft of cockpit, with twin nozzles for dry chemical dispersal and seeding. Optional Powermist spray system (up to 454 litres; 120 US gallons; 100 Imp gallons/min) for liquid chemical (max capacity 816 litres; 216 US gallons; 180 Imp gallons). Hopper can be filled in 15-20 sec and contents jettisoned in 3·2-4·5 sec. Other optional equipment includes spray nozzles; additional wing fuel tanks; 8·00-6 (6 ply) tyres; radio; cockpit heating; navigation, landing, taxi and cockpit lights; lighting system, including iodine quartz lamps, for night spraying; windscreen demister.
Dimensions, external:
- Wing span 11·98 m (39 ft 3½ in)
- Length overall 6·35 m (20 ft 10 in)
- Height overall 2·79 m (9 ft 2 in)
- Distance between tailplanes 3·48 m (11 ft 5 in)
- Propeller diameter (prototype) 2·41 m (7 ft 11 in)

Dimension, internal:
- Hopper volume 1·02 m³ (36·0 cu ft)

Area:
- Wings, gross 23·48 m² (252·7 sq ft)

Weights and Loadings:
- Typical weight empty 816 kg (1,800 lb)
- Max T-O weight:
  - Normal category 1,723 kg (3,800 lb)
  - agricultural category 1,855 kg (4,090 lb)
- Max wing loading 79·1 kg/m² (16·2 lb/sq ft)
- Max power loading 7·66 kg/kW (12·6 lb/hp)

Performance (A: at 1,855 kg; 4,090 lb AUW; B: at unloaded weight, S/L, ISA):
- Max level speed:
  - A 120 knots (222 km/h; 138 mph)
  - B 125 knots (232 km/h; 144 mph)
- Cruising speed (75% power):
  - A 111 knots (206 km/h; 128 mph)
  - B 116 knots (216 km/h; 134 mph)
- Stalling speed, flaps up, power off:
  - A 50·5 knots (93·5 km/h; 58 mph)
  - B 40 knots (74 km/h; 46 mph)
- Stalling speed, flaps down, power off:
  - A 47 knots (87 km/h; 54 mph)
  - B 35 knots (64·5 km/h; 40 mph)
- Max rate of climb at S/L:
  - A 254 m (833 ft)/min
  - B 547 m (1,795 ft)/min
- Service ceiling:
  - A 5,260 m (17,250 ft)
  - B 6,005 m (19,700 ft)
- T-O run, grass, zero wind:
  - A 311 m (1,020 ft)
  - B 75 m (246 ft)
- T-O to 15 m (50 ft):
  - A 443 m (1,452 ft)
  - B 166 m (543 ft)
- Landing run, grass, zero wind:
  - B 77 m (255 ft)
  - at 1,723 kg (3,800 lb) AUW 169 m (555 ft)
- Swath width:
  - liquid spray 27·5 m (90 ft)
  - dry fertiliser 29 m (96 ft)
  - insecticide dusting 30·5 m (100 ft)

T-320 version of the Transavia Airtruk, with Continental Tiara engine

VTOL

VTOL AIRCRAFT CO PTY LTD

Address:
PO Box 5195C, Newcastle West, New South Wales 2302
Telephone: 435348
Chairman of Directors:
D. A. Phillips

PHILLIPS PHILLICOPTER Mk 1

Mr Phillips, assisted by Mr P. Gerakiteys, designed and built a prototype two-seat helicopter known as the Phillicopter Mk 1. Design work began in 1962, construction started in 1967, and the prototype flew for the first time in 1971. Flight trials were continuing in early 1975. Orders for the Phillicopter have been held in abeyance pending the completion of these trials, the results of which had not been learned at the time of closing for press.

Type: Two-seat light helicopter.
Rotor System: Two-blade main and tail rotors. Main rotor blades, of NACA 0012 section and fully-extruded hollow-section construction, are attached to hub by solid grips. Fixed tab on each trailing-edge. Tail rotor blade construction similar to main rotor blades.
Rotor Drive: Via three gearboxes: one transfer box, one reduction box and one tail rotor box. Main rotor/engine rpm ratio 1 : 5·66; tail rotor/engine rpm ratio 1 : 1.
Fuselage: Tubular steel airframe, with aluminium and glassfibre covering.
Tail Unit: Tubular steel open space-frame. Tailplane incidence adjustable manually on ground.
Landing Gear: Tubular skid type. Shock-absorption through bending and torsion of cross-members. Ground handling wheels. Float gear available optionally.
Power Plant: One 108 kW (145 hp) Rolls-Royce Continental O-200-C flat-four engine. Single fuel tank, capacity 82 litres (18 Imp gallons). Oil capacity 5·7 litres (1·25 Imp gallons).
Accommodation: Side-by-side seating for pilot and one passenger. Door on each side of cabin. Accommodation ventilated.
Systems and Equipment: Battery for electrical power. Radio fitted.
Dimensions, external:
- Main rotor diameter 7·77 m (25 ft 6 in)
- Tail rotor diameter 1·22 m (4 ft 0 in)
- Distance between rotor centres 4·19 m (13 ft 9 in)
- Main rotor blade chord (each) 203 mm (8 in)
- Length overall, rotors fore and aft 8·79 m (28 ft 10 in)
- Length of fuselage 6·65 m (21 ft 10 in)
- Height to top of rotor hub 2·44 m (8 ft 0 in)
- Height overall 2·54 m (8 ft 4 in)
- Skid track 1·93 m (6 ft 4 in)

DIMENSIONS, INTERNAL:
- Cabin: Length — 1·63 m (5 ft 4 in)
- Max width — 1·17 m (3 ft 10 in)
- Max height — 1·27 m (4 ft 2 in)

AREAS:
- Main rotor blades (each) — 0·89 m² (9·56 sq ft)
- Tail rotor blades (each) — 0·046 m² (0·50 sq ft)
- Main rotor disc — 47·38 m² (510·00 sq ft)
- Tail rotor disc — 1·17 m² (12·56 sq ft)

WEIGHTS AND LOADINGS:
- Weight empty — 476 kg (1,050 lb)
- Max T-O and landing weight — 748 kg (1,650 lb)
- Max disc loading — 15·82 kg/m² (3·24 lb/sq ft)
- Max power loading — 6·93 kg/kW (11·35 lb/hp)

PERFORMANCE (at max T-O weight):
- Max level speed — 78 knots (145 km/h; 90 mph)
- Max cruising speed — 74 knots (137 km/h; 85 mph)
- Econ cruising speed — 60·5 knots (112·5 km/h; 70 mph)
- Max rate of climb at S/L — 365 m (1,200 ft)/min
- Vertical rate of climb at S/L — 91 m (300 ft)/min
- Service ceiling — 4,880 m (16,000 ft)
- Hovering ceiling in ground effect — 2,440 m (8,000 ft)

Phillips Phillicopter Mk 1 prototype (Rolls-Royce Continental O-200-C engine) *(S. J. Cherz)*

- Hovering ceiling out of ground effect (optimum) — 1,830 m (6,000 ft)
- Range with max fuel — 200 nm (370 km; 230 miles)

BELGIUM

FAIREY
FAIREY SA

HEAD OFFICE, WORKS AND AIRPORT:
Route Nationale 5, B-6200 Gosselies
Telephone: Charleroi (071) 35 01 90
Telex: 51241

DIRECTORS:
A. Talbott (Chairman and Managing Director)
R. W. Holder
I. G. Tylee (General Manager)
A. W. Hicks

DEPARTMENT HEADS:
A. W. Hicks (Finance)
J. Lapiere (Personnel)
J. Hoebeke (Procurement)
L. Gosset (Industrial Engineering)
E. Dumont (Production)
M. Leleux (Technical)
R. Bouniton (Commercial)
J. Maitre (Quality Control)

PUBLICITY AND PRESS:
C. Loriaux

This company, known formerly as Avions Fairey SA, was formed in 1931 as a subsidiary of the English Fairey Aviation Co Ltd, later expanding its facilities considerably to manufacture Gloster Meteor, Hawker Hunter and Lockheed F-104G fighters post-war for NATO and the Belgian Air Force.

Following completion of this work, the company contributed to the Breguet Atlantic programme. In addition, a contract was signed in 1969 whereby Fairey undertook a major share in building Mirage 5s for the Belgian Air Force. Details of this programme were given in the 1973-74 *Jane's*. The agreement also covers the manufacture of rear fuselage sections for all Mirage F1 aircraft ordered from Dassault-Breguet, no matter by whom they are ordered. Other aspects of the agreement provide for participation by Fairey in production programmes for other Dassault aircraft.

The company is collaborating on an international basis on many other projects, including the VFW 614 transport aircraft.

Overhaul and repair of military and civil aircraft constitute an important part of the company's activities. Major subcontract work currently in hand includes the manufacture of glassfibre components for Aérospatiale helicopters and the glass-reinforced plastics airframe for the MBLE Epervier (see RPVs and Targets section).

Since October 1972 Britten-Norman in the UK (which see) has been a member of the Fairey group of companies, as a result of which Fairey SA is now the principal constructor of the Britten-Norman Islander and Trislander light transport aircraft. At the end of 1975, output was at the rate of eight Islanders and three Trislanders per month.

Fairey's works cover an area of some 46,450 m² (499,985 sq ft) and employ 1,700 people.

SABCA
SOCIÉTÉ ANONYME BELGE DE CONSTRUCTIONS AÉRONAUTIQUES

HEAD OFFICE:
Chaussée de la Hulpe 185, B-1170 Brussels
Telephone: Brussels (02) 660 00 64
Telex: SABDG 23 244

WORKS:
Haren-Brussels, Chaussée de Haecht 1470, B-1130 Brussels
Telephone: Brussels (02) 216 80 10
Telex: SABUSH 21 237
Aéroport de Gosselies/Charleroi, B-6200 Gosselies
Telephone: Charleroi (071) 35 01 70
Telex: SABGO 51 251

CHAIRMAN:
A. Dubuisson

DIRECTOR, GENERAL MANAGER:
P. G. Willekens

SABCA has been since 1920 the largest aircraft manufacturer in Belgium. In addition to building aircraft and aero-engines under licence pre-war for the Belgian government and the Sabena company, it also built aircraft of its own design.

Dassault-Breguet (France) and Fokker-VFW (Netherlands/Germany) have parity holdings in SABCA, through which the Belgian company now participates in various European projects.

At Haren, the company is working on components for the Dassault-Breguet/Dornier Alpha Jet; Dassault Mirage F1, Mirage III and Mirage 5; VFW 614; Fokker-VFW F27 and F28; and Aérospatiale/Westland SA 330 Puma. SABCA is also manufacturing hydraulic components and tanks for F-104G and TF-104G aircraft in European service and airstairs for the Lockheed JetStar II, and is preparing to contribute to European production of the General Dynamics F-16, which will succeed the F-104G in the Belgian Air Force.

At Gosselies, SABCA is continuing to maintain and overhaul Mirage 5 and F-104G aircraft for the Belgian Air Force. This works is also engaged on overhaul and repair of other military aeroplanes and helicopters for the Belgian and foreign armed forces.

SABCA's Electronic Division, Cobelda, is manufacturing IFF equipment, SATT electronic equipment (all-weather landing monitoring system), Doppler equipment, and a variety of aircraft electronic ground equipment. It undertakes also work on electronic equipment under F-104G overhaul contracts. The division is undertaking series production of a tank fire control system for the Belgian Army and other armed forces. It is also engaged in the production of some of the equipment for this system, and of a simulator which allows the tank crews to be trained in firing.

Its work on the Belgian Mirage 5 and F-104G aircraft included functional electronic test procedures in the laboratory before installation of equipment in the aircraft and checkout after such installation.

SABCA is involved in space projects, including the Ariane space launcher and Spacelab. It is a member of various European industrial consortia, and of national and European aircraft associations.

In 1976 the Haren and Gosselies works occupied a total area of approx 66,000 m² (710,420 sq ft) and between them employed about 1,750 people.

BRAZIL

AEROTEC
SOCIEDADE AEROTEC LTDA

HEAD OFFICE AND WORKS:
Caixa Postal 286, 12200 São José dos Campos, São Paulo State
Telephone: (0123) 21-8011 and 21-8877

GENERAL AND INDUSTRIAL DIRECTOR:
Eng Carlos Gonçalves

COMMERCIAL DIRECTOR:
Wlademir Monteiro Carneiro

ADMINISTRATIVE DIRECTOR:
Almir Medeiros

This company was formed in 1968. It designed and built the Uirapuru light aircraft, which, under the military designation T-23, has been ordered by the Brazilian, Bolivian and Paraguayan air forces and for civil flying clubs.

Aerotec is engaged in the manufacturing programme for the EMBRAER Ipanema agricultural aircraft (which see), being responsible for building the wings under contract from EMBRAER; and is manufacturing, under sub-contract, Microturbo starter pods for the Xavante jet trainer and light attack aircraft. The company is also participating in EMBRAER's general aviation aircraft construction programme, by building components for the EMB-720 Minuano, EMB-721 Sertanejo and EMB-810 Seneca II. As a result of this commitment, it has abandoned the A-144 version of the Uirapuru, described briefly in the Addenda to the 1975-76 *Jane's*.

In 1976 Aerotec employed 250 persons and its premises occupied approx 6,800 m² (73,195 sq ft) of covered space.

AEROTEC A-122 UIRAPURU
Brazilian Air Force designation: T-23

The Uirapuru was designed as a private venture by Engs José Carlos de Souza Reis and Carlos Gonçalves. The prototype (PP-ZTF), with an 80·5 kW (108 hp) Lycoming O-235-C1 engine, flew for the first time on 2 June 1965 and was described and illustrated in the 1966-67 *Jane's*. It was followed by a second Uirapuru (PP-ZTT), with a 112 kW (150 hp) Lycoming O-320-A engine.

Two military pre-production models (0940 and 0941) were completed in early 1968, making their first flights on 23 January and 11 April respectively. These both had 150 hp engines.

In early 1968, the Brazilian Air Force placed an order for 30 **A-122A** production aircraft (Brazilian Air Force designation T-23), powered by 119 kW (160 hp) Lycoming O-320-B2B engines, increasing the total to 70 in April 1969. Delivery of these has been completed. A further 20 were ordered in February 1973, and Aerotec is supplying 10 additional aircraft per year to the Brazilian Air Force to offset losses and compensate for aircraft undergoing IRAN. In Brazilian Air Force service the T-23 Uirapurus are employed at the Academia da Força Aérea (Air Force Academy) in Pirassununga, São Paulo State.

Eighteen A-122A military Uirapurus were ordered by the Bolivan Air Force; delivery of these took place between April and September 1974. The Paraguayan Air Force ordered eight, deliveries of which took place during 1975.

Aerotec has developed a version of the Uirapuru, the **A-122B**, for the civil market. This is basically similar to the military version, except for having a cockpit canopy like that of the second prototype (see 1967-68 *Jane's*). Eighteen were delivered in 1975 to civil flying clubs supported by the Ministry of Aeronautics.

A total of 148 Uirapurus had been sold by 20 April 1976, at which time 120 had been delivered and production was continuing at the rate of four per month.

The following description applies to the standard A-122A military version, except where indicated:

TYPE: Two-seat primary trainer.

WINGS: Cantilever low-wing monoplane. Wing section NACA 43013. Dihedral 5°. Incidence 2°. Light alloy structure of centre-section and two outer panels. Glassfibre wingtips. All-metal ailerons and trailing-edge split flaps. Flap settings 0°, 20° and 40°; ailerons deflect 20° up and 13° down.

FUSELAGE: All-metal semi-monocoque structure in 2024-T-3 aluminium, with 4130 steel for critical areas.

TAIL UNIT: Cantilever all-metal construction. Fin, rudder, tailplane and elevator tips of glassfibre. Trim tab in starboard half of elevator. Statically and aerodynamically balanced elevator, aerodynamically balanced rudder. Fixed ventral fin. Vertical and horizontal surfaces of NACA 0009 section. Sweepback 30° on fin leading-edge. Elevator deflects 30° up, 23° down; trim tab 22° up, 40° down; rudder 25° to left and right.

LANDING GEAR: Non-retractable tricycle type, with nosewheel steerable 22° to each side. Rubber-cushioned shock-absorbers on main units; oleo shock-absorber on nose unit. All wheels have Goodyear 6·00-6 tyres, pressure 1·79 bars (26 lb/sq in) on main units and 1·65 bars (24 lb/sq in) on nose unit. Independent hydraulic disc brakes on main units. Parking brake. Legs of 4130 steel, with small fairings on main-wheel legs.

POWER PLANT: One 119 kW (160 hp) Lycoming O-320-B2B flat-four engine, driving a Sensenich M-74-DM-6-060 two-blade fixed-pitch metal propeller. Variable-pitch propeller optional. Two integral fuel tanks in wing leading-edges, with total capacity of 140 litres (31 Imp gallons). Refuelling points above tanks. Optional 40 litre (8·8 Imp gallon) wingtip tanks.

ACCOMMODATION: Two fully-adjustable seats side by side under rearward-sliding transparent canopy. Two-piece windscreen. For emergency jettison, canopy separates into two pieces. Seats permit the use of either back-type or seat-type parachutes. Dual controls. Baggage compartment, capacity 30 kg (66 lb), aft of seats with access from cockpit only.

SYSTEMS: Hydraulic system for brakes. Electrical system includes 24V 50A generator, 24V 24Ah battery (12V 40Ah in A-122B) and electric starter.

ELECTRONICS AND EQUIPMENT: Conventional VFR equipment. Optional items include VHF transceiver, ADF, artificial horizon and directional gyro. Adjustable-angle 100W landing light in each wing leading-edge.

Aerotec T-23 Uirapuru two-seat primary training aircraft of the Brazilian Air Force *(Ronaldo S. Olive)*

Aerotec A-122B, civil version of the Uirapuru, of the Brasilia Aero Club *(Ronaldo S. Olive)*

DIMENSIONS, EXTERNAL:

Wing span	8·50 m (27 ft 10¾ in)
Wing chord (constant)	1·53 m (5 ft 0½ in)
Wing aspect ratio	5·5
Length overall	6·60 m (21 ft 8 in)
Length of fuselage	6·40 m (21 ft 0 in)
Width of fuselage	1·08 m (3 ft 6½ in)
Height overall	2·70 m (8 ft 10 in)
Tailplane span	2·80 m (9 ft 2¼ in)
Wheel track	2·36 m (7 ft 9 in)
Wheelbase	1·47 m (4 ft 9¾ in)
Propeller diameter	1·87 m (6 ft 1½ in)
Propeller ground clearance	0·27 m (10¾ in)

AREAS:

Wings, gross	13·50 m² (145·3 sq ft)
Ailerons (total)	1·19 m² (12·81 sq ft)
Flaps (total)	0·95 m² (10·23 sq ft)
Fin	0·60 m² (6·46 sq ft)
Rudder	0·50 m² (5·38 sq ft)
Tailplane	1·50 m² (16·15 sq ft)
Elevator, incl tab	1·10 m² (11·84 sq ft)

WEIGHTS AND LOADINGS:

Weight empty	540 kg (1,191 lb)
Max T-O weight	840 kg (1,825 lb)
Max wing loading	63·0 kg/m² (13·90 lb/sq ft)
Max power loading	7·06 kg/kW (11·41 lb/hp)

PERFORMANCE (at max T-O weight, S/L, ISA):

Never-exceed speed	165 knots (307 km/h; 190 mph)
Max level speed:	
A-122A	122 knots (227 km/h; 141 mph)
A-122B	128 knots (238 km/h; 148 mph)
Max cruising speed (75% power):	
A-122A	100 knots (185 km/h; 115 mph)
A-122B	105 knots (195 km/h; 121 mph)
Econ cruising speed (65% power):	
A-122A	89 knots (164 km/h; 102 mph)
A-122B	94 knots (174 km/h; 108 mph)
Stalling speed, flaps up (both versions)	56·5 knots (104 km/h; 65 mph)
Stalling speed, flaps down (both versions)	39 knots (72 km/h; 45 mph)
Max rate of climb at S/L:	
A-122A	255 m (836 ft)/min
A-122B	273 m (895 ft)/min
Service ceiling	4,500 m (14,760 ft)
T-O run (zero wind)	200 m (656 ft)
Landing run (zero wind)	180 m (590 ft)
Max range	429 nm (800 km; 495 miles)
Endurance	4 hr 30 min

CTA
CENTRO TÉCNICO AEROESPACIAL

HEADQUARTERS:
São José dos Campos, São Paulo State

DIRECTOR OF CTA:
Major Brigadeiro Hugo de Miranda e Silva

The Centro Técnico Aeroespacial (Aerospace Technical Centre) is a Ministry of Aeronautics establishment for training aeronautical and aerospace personnel and for conducting aeronautical and aerospace research and development. It is composed of five institutes: the Instituto Tecnológico de Aeronáutica (ITA); the Instituto de Pesquisas e Desenvolvimento (IPD); the Instituto de Atividades Espaciais (IAE); the Instituto de Ensaios e Padrões (IEP); and the Instituto de Fomento e Coordenação Industrial (IFI).

The **ITA** is the college of engineering for aeronautical and aerospace personnel. It provides training for BS, MS and PhD degrees in aeronautical, electronic and mechanical engineering.

The **IPD** conducts aeronautical research and development. Its major activities are in the fields of aeronautics, electronics, energy, materials, propulsion and fuels.

The **IAE** conducts space research and development, mainly in the fields of aerospace vehicles, astrophysics, atmospheric sciences, space technology, control, guidance and navigation.

The **IEP** is devoted to the development of aeronautical and aerospace standards in Brazil.

The **IFI** is responsible for the fostering and co-ordination of the Brazilian aerospace industry. Its main activities are devoted to increasing the development rate of aeronautical and aerospace activities, by transferring the most appropriate technology to the aerospace industry and by providing incentives for the adoption of such technology by the industrial community.

The IFI is also responsible for the certification of civil aircraft and approval of other aeronautical products.

EMBRAER
EMPRESA BRASILEIRA DE AERONÁUTICA SA

HEAD OFFICE AND WORKS:
Av Brig Faria Lima, Caixa Postal 343, 12200 São José dos Campos, São Paulo State
Telephone: (0123) 21-5400
Telex: 391-01122445 São José dos Campos

RIO OFFICE:
Av Nilo Pecanha 50, Sala 2405, 20000 Rio de Janeiro GB
Telephone: (021) 231-3652

PRESIDENT:
Aldo B. Franco

SUPERINTENDENT DIRECTOR:
Ozires Silva

PRODUCTION DIRECTOR:
Ozilio Carlos da Silva

TECHNICAL DIRECTOR:
Guido Fontegalante Pessotti

FINANCIAL DIRECTOR:
Alberto Franco Faria Marcondes

INDUSTRIAL RELATIONS DIRECTOR:
Antonio Garcia da Silveira

COMMERCIAL DIRECTOR:
Renato José da Silva

EMBRAER was created in August 1969, and came into

operation on 2 January 1970 to promote the development of the Brazilian aircraft industry. It now has an authorised capital of Cr $250 million (about US $25 million), of which Cr $198·3 million (US $19·83 million) had been subscribed by early 1976. The Brazilian government owns 51% of the shares, the remaining 49% being held by private shareholders. EMBRAER has a work-force of some 3,600 persons and a factory area of 112,360 m² (1,209,430 sq ft).

In August 1974, EMBRAER signed a comprehensive co-operative agreement with Piper Aircraft Corporation involving the assembly and eventual manufacture in Brazil of the Seneca II and Navajo Chieftain twin-engined aircraft and worldwide distribution of the products of both companies. At the beginning of 1975, this agreement was extended to include the single-engined Cherokee series. In consequence, with the co-operation of other Brazilian companies (Neiva, Aerotec, Aeromot and Motortec), EMBRAER now produces and markets these aircraft in Brazil. The two companies are also studying a similar arrangement for reciprocal marketing of each other's turboprop-engined aircraft.

In early 1975, EMBRAER negotiated with Northrop Corporation a contract for manufacturing in Brazil components for the US company's F-5E Tiger II combat aircraft. This work was scheduled to begin in 1976.

EMBRAER has in current production the EMB-110 Bandeirante twin-turboprop transport aircraft, the EMB-326GB Xavante licence-built version of the Italian Aermacchi M.B.326GC jet trainer and ground attack aircraft, the EMB-201 Ipanema agricultural aircraft, and various EMBRAER-built versions of the Piper Cherokee Pathfinder, Arrow II, Cherokee SIX, Lance, Seneca and Navajo Chieftain. The principal current design and development programme, known as Project 12X, is for three related designs, the EMB-120 Araguaia, EMB-121 Xingu and EMB-123 Tapajós. Other programmes concern the EMB-111 coastal patrol version of the Bandeirante, and an EMB-201A developed version of the Ipanema. Details of all of these follow.

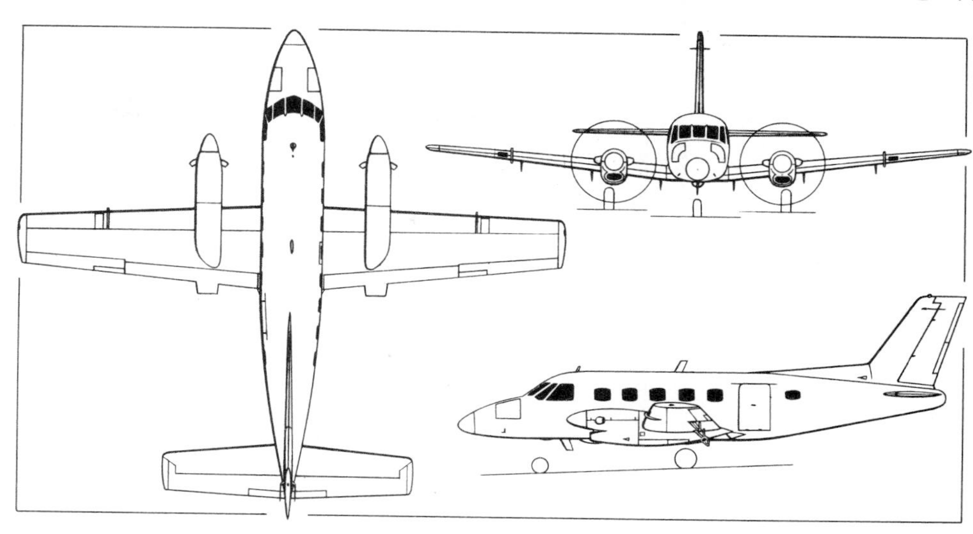

EMBRAER EMB-110 Bandeirante twin-turboprop light transport *(Pilot Press)*

EMBRAER EMB-110 BANDEIRANTE (PIONEER)

Brazilian Air Force designations: C-95, EC-95 and RC-95

The Bandeirante twin-turboprop light transport was developed to a Ministry of Aeronautics specification calling for a general-purpose aircraft capable of carrying out missions such as transport, navigation training and aeromedical evacuation. Its Brazilian design team was, initially, under the leadership of M Max Holste, the well-known French aircraft designer.

The first YC-95 prototype flew for the first time on 26 October 1968, followed by the second on 19 October 1969, and the basically similar third YC-95 on 26 June 1970. These prototypes were described in the 1970-71 *Jane's*.

The first production EMB-110 Bandeirante flew for the first time on 9 August 1972, and was test-flown until December 1972 as part of the certification programme. Following the completion of testing to FAR 23, the aircraft was granted a type certificate by the Aerospace Technical Centre of the Brazilian Air Ministry, and the first three Bandeirantes were delivered to the Brazilian Air Force on 9 February 1973.

By March 1976, a total of 135 Bandeirantes of various models had been sold to the Brazilian Air Force (88); Transbrasil (6); VASP (10); Taxi Aéreo Sagres (3); Transportes Aéreos da Bacia Amazônica (5); the Uruguayan Air Force (5); DNOCS (1), SUDECO (1) and FUNAI (1) governmental agencies of the Ministry of the Interior; Furnas Centrais Elétricas (2); the Chilean Navy (3); and other operators (10). By then a total of 74 Bandeirantes had been delivered and production was scheduled to continue during 1976-77 at a rate of four per month.

The Bandeirante is available in the following versions:

EMB-110. Basic 12-seat aircraft; 60 ordered by Brazilian Air Force and three by Chilean Navy.

EMB-110A. Navaid checking and calibration version. Two operated by Brazilian Air Force as EC-95.

EMB-110B. Aerial photogrammetric version, with cabin floor apertures permitting the use of aerial cameras (Zeiss RMK A8·5/23, RMK A15/23, RMK A30/23 and Wild RC-10), a Zeiss IRU regulator, and Zeiss NT-1 navigation visors. Other equipment includes Decca 72 Doppler navigation system. Crew includes three equipment operators. Six ordered by Brazilian Air Force as RC-95s, and one by VASP Aerofotogrametria S/A.

EMB-110C. Standard 15-passenger commercial transport version, to which the detailed description applies. Entered commercial service with Transbrasil on 16 April 1973.

EMB-110E. Executive transport version with accommodation for seven passengers, four in individual seats and three on a sideways-facing sofa. Other features include a galley, wardrobe, and stereo AM/FM and tape deck. Nine sold in 1975, four to J. P. Martins and one each to Cacique, Bradesco, Banco Noroeste, Zwigal, and Frigus.

EMB-110F. All-cargo version.

EMB-110K. Developed from EMB-110C with enlarged fuselage door 1·35 m (4 ft 5¼ in) high × 1·80 m (5 ft 10¾ in) wide; 20 to be delivered to Brazilian Air Force.

EMB-110K1. Proposed version of K with 0·84 m (2 ft 9 in) longer fuselage, 1·78 m × 1·30 m (5 ft 10 in × 4 ft 3¼ in) cargo door and additional crew door aft of flight deck.

EMB-110P. Commercial third-level commuter version for 18 passengers, developed from EMB-110C. Seats in six rows of three, at 775 mm (30·5 in) pitch. Rear baggage hold volume increased to 2·0 m³ (70·63 cu ft). Two over-wing emergency exits. Five delivered to TABA (Transportes Aéreos da Bacia Amazônica) by early 1976.

EMB-110S. Geophysical survey version. Equipment includes Geometrics proton magnetometers, gamma ray spectrometers and data recording systems.

EMB-111. Maritime patrol version, described separately.

The following description applies to the standard production EMB-110C:

TYPE: Twin-turboprop general-purpose transport.

WINGS: Cantilever low-wing monoplane. Wing section NACA 23016 (modified) at root, NACA 23012 (modified) at tip. Sweepback 19′ 48″ at quarter-chord. Dihedral 7° at 28% chord. Incidence 3°. All-metal two-spar structure, of 2024-T3 and -T4 aluminium alloy, with detachable glassfibre wingtips. Glassfibre wing/fuselage fairing. All-metal statically-balanced ailerons and double-slotted flaps. Trim tab in port aileron.

FUSELAGE: All-metal semi-monocoque structure of 2024-T3 aluminium alloy. Two upward-hinged doors, one on each side of nose, provide access to avionics equipment.

TAIL UNIT: Cantilever all-metal structure, with sweptback vertical surfaces. Glassfibre dorsal fin. Trim tabs in rud-

EMB-110P commuter transport version of the Bandeirante, with high-density seating for up to 18 passengers

der and port elevator. Tab in starboard elevator linked to flaps, to offset pitching moment during flap extension.

LANDING GEAR: Hydraulically-retractable tricycle type, of ERAM manufacture, with single wheel and oleo-pneumatic shock-absorber on each unit. Main wheels have Kléber-Colombes tyres, size 670 × 210-12, pressure 2·76-3·93 bars (40-57 lb/sq in). Steerable, forward-retracting nosewheel unit has Goodyear tyre, size 6·50-8.

POWER PLANT: Two 507 kW (680 shp) Pratt & Whitney Aircraft of Canada PT6A-27 turboprop engines, each driving a Hartzell HC-B3TN-3C/T10178H-8R constant-speed three-blade metal propeller with autofeathering and full reverse-pitch capability. Four integral fuel tanks in wings, with total capacity of 1,720 litres (378 Imp gallons). Gravity refuelling point on top of each wing.

ACCOMMODATION: Two seats side by side on flight deck, which is separated from main cabin by door. Cabin seats up to 15 passengers. Conversion into an ambulance for four stretcher patients takes ten minutes. Downward-hinged door on port side, aft of wing, with built-in airstairs. Cabin floor stressed for loads of up to 400 kg/m² (82 lb/sq ft). Emergency exit over wing on starboard side. Baggage compartment at rear of cabin, with total capacity of 1·30 m³ (46 cu ft). Toilet/lavatory standard.

SYSTEMS: Air-cycle-type air-conditioning system with cooling capacity of 20,000 BTU/hr and engine bleed heating. Hydraulic system, pressure 207 bars (3,000 lb/sq in), for landing gear actuation, independent braking system, nosewheel steering and parking brake. No pneumatic system. Electrical system utilises two 200A starter/generators and an MS-24498-1 24V 34Ah alkaline battery with two 250VA inverters to supply 115/26V 400Hz AC power. External power receptacle on port side of forward fuselage. Oxygen system for crew and passengers, using oxygen bottle in rear of fuselage with capacity of 3·3 m³ (115 cu ft) at 124 bars (1,800 lb/sq in) pressure.

ELECTRONICS AND EQUIPMENT: Standard equipment includes one Brazilian-built Whinner CY04A03C 140-channel VHF transceiver, one Collins 618 M-2B 360-channel VHF transceiver, one HF-AM/SSB Sunair Type ASB-100 transceiver, one Collins 51 R-7A VOR/ILS receiver, two Bendix DFA-73 A1 ADF receivers, one Collins 51Z-6 marker beacon receiver and one Collins 51V-5 glideslope receiver. 450W landing light in each wing leading-edge, and 250W GE landing and taxying light on nosewheel unit. RCA AVQ-47 weather radar and Bendix M-4C autopilot optional.

DIMENSIONS, EXTERNAL:
- Wing span: 15·32 m (50 ft 3 in)
- Wing chord at root: 2·32 m (7 ft 7½ in)
- Wing chord at tip: 1·35 m (4 ft 5 in)
- Wing aspect ratio: 8·1
- Length overall: 14·23 m (46 ft 8¼ in)
- Length of fuselage: 13·74 m (45 ft 1 in)
- Height overall: 4·13 m (13 ft 6½ in)
- Fuselage: Max width: 1·70 m (5 ft 7 in)
- Tailplane span: 7·54 m (24 ft 9 in)
- Propeller diameter: 2·36 m (7 ft 9 in)
- Distance between propeller centres: 4·80 m (15 ft 9 in)
- Propeller ground clearance: 0·345 m (1 ft 1½ in)
- Wheel track: 4·94 m (16 ft 2½ in)
- Wheelbase: 4·56 m (14 ft 11½ in)
- Passenger door (rear, port):
  - Height: 1·30 m (4 ft 3¼ in)
  - Width: 0·85 m (2 ft 9½ in)
- Emergency exit (stbd, over wing):
  - Height: 0·80 m (2 ft 7½ in)
  - Width: 0·63 m (2 ft 1 in)

DIMENSIONS, INTERNAL:
- Cabin: Max length: 8·65 m (28 ft 4½ in)
  - Width: 1·60 m (5 ft 3 in)
  - Height: 1·60 m (5 ft 3 in)
  - Floor area: 11·60 m² (124·9 sq ft)

AREAS:
- Wings, gross: 29·00 m² (312 sq ft)
- Ailerons (total): 2·18 m² (23·5 sq ft)
- Flaps (total): 5·04 m² (54·3 sq ft)
- Fin, incl dorsal fin: 2·07 m² (22·3 sq ft)
- Rudder, incl tab: 1·68 m² (18·1 sq ft)
- Tailplane: 5·43 m² (58·4 sq ft)
- Elevators, incl tabs: 4·40 m² (47·3 sq ft)

WEIGHTS AND LOADINGS:
- Weight empty, equipped: 3,380 kg (7,451 lb)
- Max T-O weight: 5,600 kg (12,345 lb)
- Max landing weight: 5,300 kg (11,684 lb)
- Max zero-fuel weight: 5,180 kg (11,420 lb)
- Max wing loading: 193·1 kg/m² (39·55 lb/sq ft)
- Max power loading: 5·52 kg/kW (9·08 lb/shp)

PERFORMANCE (at max T-O weight, ISA, except where indicated):
- Never-exceed speed: 300 knots (558 km/h; 346 mph)
- Max level speed at 4,575 m (15,000 ft): 234 knots (434 km/h; 270 mph)
- Max Mach No.: 0·592
- Max cruising speed at 4,575 m (15,000 ft): 228 knots (422 km/h; 262 mph)
- Econ cruising speed at 4,575 m (15,000 ft): 192 knots (356 km/h; 221 mph)
- Stalling speed:
  - at max landing weight: 71·5 knots (132 km/h; 82 mph) IAS
  - at 3,570 kg (7,870 lb): 60 knots (111 km/h; 69 mph) IAS
- Max rate of climb at S/L: 518 m (1,700 ft)/min
- Rate of climb at S/L, one engine out: 91 m (300 ft)/min
- Time to 2,440 m (8,000 ft): 5 min
- Time to 4,575 m (15,000 ft): 13 min
- Service ceiling: 8,000 m (26,250 ft)
- Service ceiling, one engine out: 2,440 m (8,000 ft)
- T-O run: 380 m (1,245 ft)
- T-O to 15 m (50 ft): 540 m (1,770 ft)
- Landing from 15 m (50 ft): 680 m (2,230 ft)
- Landing run: 350 m (1,150 ft)
- Range, 30 min reserves:
  - with max fuel: 1,197 nm (2,220 km; 1,379 miles)
  - with max payload: 132 nm (246 km; 153 miles)

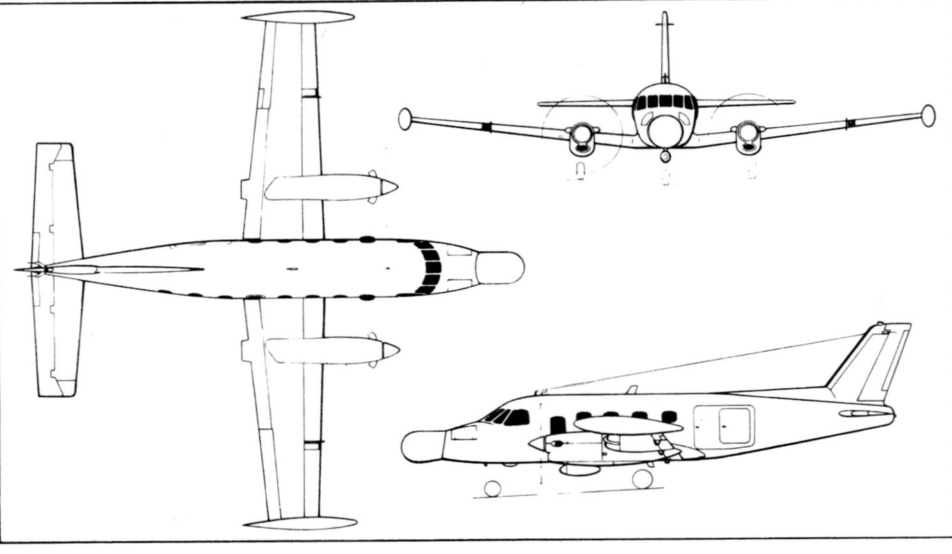

EMB-111 patrol version of the Bandeirante, developed by EMBRAER *(Roy J. Grainge)*

EMBRAER EMB-111

This designation has been given to a shore-based patrol aircraft, based on the EMB-110 Bandeirante, which EMBRAER has designed to meet specifications issued by the Comando Costeiro, the Brazilian Air Force's Coastal Command. It is powered by two 559 kW (750 shp) Pratt & Whitney Aircraft of Canada PT6A-34 turboprop engines, and carries 610 litres (134 Imp gallons) of additional fuel in wingtip tanks. The main external difference in this version is the large nose radome, which houses an AIL AN/APS-128 search radar. Other electronics and equipment include a Litton LN-33 inertial navigation system, Collins VIR-30A VOR/ILS marker beacon receiver, Collins DF-301E VHF/DF, Collins DME-40 DME, two Bendix DFA-74 ADF, Bendix ALA-51 radio altimeter, and Sperry C-14 gyro-magnetic compass. There is provision for eight 5 in air-to-surface rockets to be carried. A ventrally-mounted searchlight of 10 million candlepower is fitted for night operations. For target marking, six Brazilian-built Mk 6 smoke grenades are carried, as well as a Motorola SST-121 transponder. Flares of 1 million candlepower are also available for illumination of targets at night.

Twelve EMB-111s were ordered by the Brazilian Air Force in December 1975. A six-month evaluation programme will follow the first flight, which is scheduled to take place in about July 1977.

DIMENSIONS, EXTERNAL: As EMB-110, except:
- Wing span over tip-tanks: 15·69 m (51 ft 5¾ in)
- Length overall: 14·71 m (48 ft 3¼ in)
- Length of fuselage: 14·22 m (46 ft 7¾ in)
- Height overall: 4·73 m (15 ft 6¼ in)

WEIGHTS:
- Basic operating weight with 6 crew: 4,200 kg (9,259 lb)
- Max T-O weight: 6,150 kg (13,558 lb)
- Max landing weight: 5,300 kg (11,684 lb)

PERFORMANCE (estimated, at max T-O weight except where indicated):
- Cruising speed at 3,000 m (9,845 ft): 250 knots (464 km/h; 288 mph)
- Max rate of climb at S/L: 534 m (1,752 ft)/min
- Service ceiling: 8,550 m (28,050 ft)
- T-O run at S/L: 651 m (2,136 ft)
- T-O to 15 m (50 ft): 835 m (2,740 ft)
- Landing from 15 m (50 ft) at max landing weight: 636 m (2,087 ft)
- Landing run at S/L at max landing weight: 412 m (1,352 ft)
- Endurance: approx 8 hr

EMBRAER PROJECT 12X

Under the general series designation Project 12X, EMBRAER is developing from the Bandeirante a family of three medium-sized pressurised twin-engined aircraft, all named after Brazilian rivers. Three different fuselage lengths are employed, but all have the same cross-section and modular construction, of fail-safe design and built of chemically milled panels; the same flight deck layout; and substantially the same systems, electronics and equipment. Each will have a normal cabin pressure differential of 0·414 bars (6·0 lb/sq in) and a maximum differential of 0·425 bars (6·17 lb/sq in). A T-tail configuration is common to all three models, and each will have baggage compartments in the rear of the cabin and in the nose, the latter accessible via two upward-hinged doors.

The three versions are designated as follows:

EMB-121 Xingu. First version to be developed. Pro-

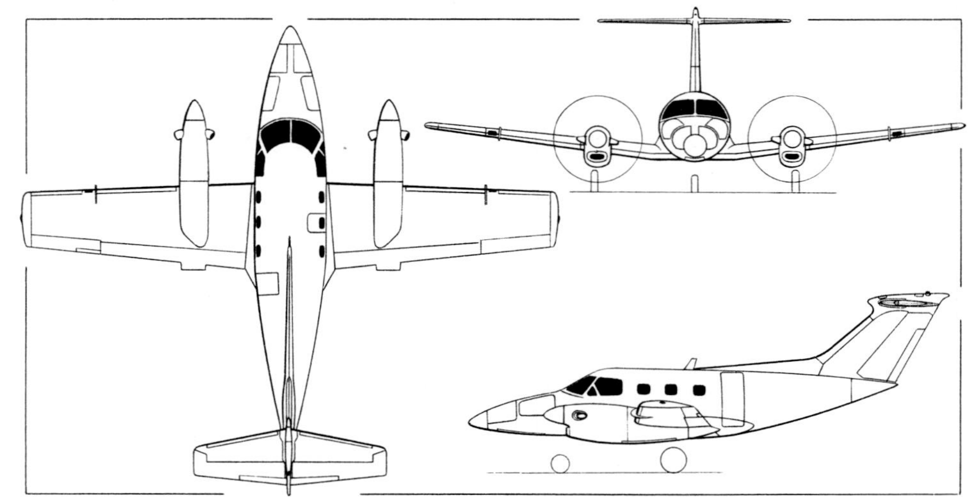

EMB-121 Xingu six/nine-passenger twin-turboprop transport *(Pilot Press)*

totype, built with production jigs from unpressurised Bandeirante, scheduled to fly in May 1976. Utilises same engine nacelles, landing gear and (with reduced span) wings as the Bandeirante. Accommodation for 6-9 passengers. Powered by two 507 kW (680 shp) Pratt & Whitney Aircraft of Canada PT6A-28 turboprop engines, each driving a Hartzell HC-B3TN-3D/T10178HB-8R three-blade constant-speed metal propeller with autofeathering and full reverse-pitch capability.

EMB-123 Tapajós. Second version to be developed. Compared with EMB-121, will have new landing gear and new supercritical wings of GAW.1 section with full-span Fowler flaps. Accommodation for 10 passengers. Powered by two 835 kW (1,120 shp) Pratt & Whitney Aircraft of Canada PT6A-45 turboprop engines, each driving a Hartzell five-blade reversible-pitch propeller. Wingtip fuel tanks optional.

EMB-120 Araguaia. Third version to be developed. Same wings and power plant as EMB-123, but lengthened fuselage seating up to 24 passengers. Constant-speed reversible-pitch propellers, similar to those of EMB-123, with electrical de-icing. Wingtip fuel tanks optional.

Electronics and Equipment (all models): One Bendix RDR-1200 weather radar, two Collins VHF-20A, one Sunair ASB-100A HF, two Collins VIR-30A automatic VOR/ILS, two Collins DF-206 ADF, two RCA AVA-310 audio control panels, two Sperry SPZ-200 flight directors/autopilots, one Collins DME-40 DME, one Collins ALT-50 radio altimeter, one Collins TDR-90 transponder, and one Garrett Rescue 88 emergency transmitter (ELT).

Dimensions, external:
Wing span:
121 — 14·14 m (46 ft 4¾ in)
120, 123 (over tip-tanks) — 14·40 m (47 ft 3 in)
Wing chord at root:
121 — 2·46 m (8 ft 0¾ in)
120, 123 — 2·62 m (8 ft 7¼ in)
Wing chord at tip:
121 — 1·50 m (4 ft 11 in)
120, 123 — 0·96 m (3 ft 1¾ in)
Wing mean aerodynamic chord:
120 — 1·919 m (6 ft 3½ in)
123 — 1·805 m (5 ft 11 in)
Wing aspect ratio:
121 — 7·15
120, 123 — 9
Length overall:
120 — 19·45 m (63 ft 9¾ in)
121 — 12·32 m (40 ft 5 in)
123 — 15·62 m (51 ft 3 in)
Length of fuselage:
120 — 18·13 m (59 ft 5¾ in)
121 — 11·01 m (36 ft 1½ in)
123 — 14·31 m (46 ft 11½ in)
Fuselage max width:
120, 121, 123 — 1·86 m (6 ft 1¼ in)
Height overall:
120 — 5·25 m (17 ft 2¾ in)
121 — 4·94 m (16 ft 2½ in)
123 — 5·20 m (17 ft 0¾ in)
Tailplane span:
120, 121, 123 — 5·58 m (18 ft 3¾ in)
Wheel track:
121 — 5·24 m (17 ft 2¼ in)
120, 123 — 5·10 m (16 ft 8¾ in)
Wheelbase:
120 — 6·40 m (21 ft 0 in)
121 — 2·86 m (9 ft 4½ in)
123 — 5·10 m (16 ft 8¾ in)
Propeller diameter:
121 — 2·36 m (7 ft 9 in)
120, 123 — 2·64 m (8 ft 8 in)
Distance between propeller centres:
120, 121, 123 — 5·10 m (16 ft 8¾ in)
Passenger door (rear, port):
Height (120, 121, 123) — 1·31 m (4 ft 3½ in)
Width (120, 121, 123) — 0·63 m (2 ft 0¾ in)
Emergency exits (2 overwing, each):
Height (120, 121, 123) — 0·85 m (2 ft 9½ in)
Width (120, 121, 123) — 0·51 m (1 ft 8 in)

Dimensions, internal:
Pressurised cabin: Max length:
120 — 12·27 m (40 ft 3 in)
121 — 5·18 m (17 ft 0 in)
123 — 8·45 m (27 ft 8¾ in)
Max width (120, 121, 123) — 1·74 m (5 ft 8½ in)
Max height (120, 121, 123) — 1·52 m (4 ft 11¾ in)

Areas:
Wings, gross:
121 — 27·50 m^2 (296·0 sq ft)
120, 123 — 25·00 m^2 (269·1 sq ft)
Ailerons (total):
121 — 1·42 m^2 (15·28 sq ft)
Trailing-edge flaps (total):
121 — 5·04 m^2 (54·25 sq ft)
120, 123 — 6·61 m^2 (71·15 sq ft)
Spoilers (total):
120, 123 — 0·25 m^2 (2·69 sq ft)

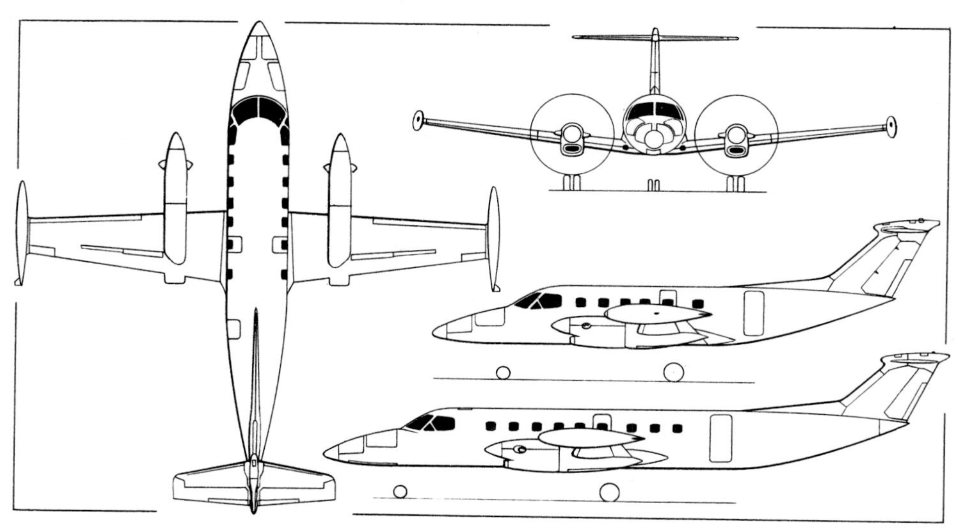

EMB-123 Tapajós, with additional side view (bottom) of EMB-120 Araguaia *(Pilot Press)*

Fin, excl dorsal fin:
120, 121, 123 — 2·18 m^2 (23·46 sq ft)
Rudder, incl tab:
120, 121, 123 — 1·78 m^2 (19·16 sq ft)
Tailplane:
120, 121, 123 — 3·33 m^2 (35·84 sq ft)
Elevator, incl tabs:
120, 121, 123 — 2·51 m^2 (27·02 sq ft)

Weights and Loadings:
Weight empty, equipped:
120 — 4,300 kg (9,480 lb)
121 — 3,175 kg (7,000 lb)
123 — 3,875 kg (8,543 lb)
Max T-O weight:
120 — 8,000 kg (17,637 lb)
121 — 5,600 kg (12,346 lb)
123 — 7,000 kg (15,432 lb)
Max landing weight:
120 — 7,600 kg (16,755 lb)
121 — 5,300 kg (11,684 lb)
123 — 6,650 kg (14,660 lb)
Max zero-fuel weight:
120 — 7,300 kg (16,094 lb)
123 — 6,650 kg (14,660 lb)
Max wing loading:
120 — 320 kg/m^2 (65·54 lb/sq ft)
121 — 204 kg/m^2 (41·78 lb/sq ft)
123 — 280 kg/m^2 (57·35 lb/sq ft)
Max power loading:
120 — 4·79 kg/kW (7·88 lb/shp)
121 — 5·52 kg/kW (9·10 lb/shp)
123 — 4·19 kg/kW (6·89 lb/shp)

Performance (estimated, 121 at AUW of 5,200 kg; 11,464 lb, 120 and 123 at max T-O weight, except where indicated):
Never-exceed speed:
120 — 365 knots (676 km/h; 420 mph)
121 — 316 knots (586 km/h; 364 mph)
123 — 364 knots (675 km/h; 419 mph)
Max level speed:
120 — 305 knots (565 km/h; 351 mph)
121 at 4,575 m (15,000 ft) — 252 knots (467 km/h; 290 mph)
123 — 310 knots (574 km/h; 356 mph)
Max Mach No:
121 — 0·635
120, 123 — 0·67
Max cruising speed at 4,575 m (15,000 ft):
120 — 292 knots (541 km/h; 336 mph)
121 — 252 knots (467 km/h; 290 mph)
123 at 6,000 kg (13,227 lb) AUW — 300 knots (556 km/h; 345 mph)
Econ cruising speed at 4,575 m (15,000 ft):
120 — 251 knots (465 km/h; 289 mph)
121 — 210 knots (389 km/h; 242 mph)
123 — 240 knots (445 km/h; 276 mph)
Stalling speed at max landing weight, full flap:
120 — 80 knots (148 km/h; 92 mph) IAS
121 — 70·5 knots (130 km/h; 81 mph) IAS
123 — 76·5 knots (141 km/h; 88 mph) IAS
Stalling speed at max T-O weight, flaps up:
120 — 106 knots (196 km/h; 122 mph)
121 — 92 knots (170 km/h; 106 mph)
123 — 100 knots (185 km/h; 115 mph)
Max rate of climb at S/L:
120 — 487 m (1,597 ft)/min
121 — 579 m (1,900 ft)/min
123 — 624 m (2,047 ft)/min
Rate of climb at S/L, one engine out:
120 — 183 m (600 ft)/min
121 — 165 m (541 ft)/min
123 — 274 m (899 ft)/min
Service ceiling:
120 — 8,535 m (28,000 ft)
121 — 8,230 m (27,000 ft)
123 — 9,390 m (30,800 ft)
Service ceiling, one engine out:
120 — 4,270 m (14,000 ft)
121 — 3,960 m (13,000 ft)
123 — 5,945 m (19,500 ft)
T-O run:
120 — 455 m (1,493 ft)
121 at max T-O weight — 520 m (1,706 ft)
123 — 341 m (1,119 ft)
T-O to 10·7 m (35 ft):
120 — 850 m (2,789 ft)
123 — 396 m (1,299 ft)
T-O to 15 m (50 ft):
121 — 715 m (2,346 ft)
Landing from 15 m (50 ft):
120 — 710 m (2,329 ft)
121 — 520 m (1,706 ft)
123 — 631 m (2,070 ft)
Landing run at max landing weight:
120 — 410 m (1,345 ft)
121 — 315 m (1,033 ft)
123 — 363 m (1,191 ft)
Range at 6,100 m (20,000 ft), 45 min reserves:
120 with max payload of 2,720 kg (6,000 lb) — 300 nm (556 km; 345 miles)
121 with max payload — 1,300 nm (2,410 km; 1,497 miles)
123 with payload of 2,040 kg (4,500 lb) — 630 nm (1,167 km; 725 miles)
120 with max fuel and 1,500 kg (3,300 lb) payload — 1,590 nm (2,946 km; 1,830 miles)
121 with max fuel 1,400 nm (2,595 km; 1,612 miles)
123 with max fuel and 893 kg (1,970 lb) payload — 1,850 nm (3,428 km; 2,130 miles)

EMBRAER (AERMACCHI) EMB-326GB XAVANTE

Brazilian Air Force designation: AT-26

In accordance with an agreement signed in May 1970, EMBRAER is assembling under licence, from Italian-built components, Aermacchi M.B. 326GB jet trainer/ground attack aircraft for the Brazilian Air Force, by whom the type is known as the AT-26 Xavante, the name of a Brazilian Indian tribe.

The initial order called for the manufacture of 112 aircraft, at a rate of two per month; a further 40 were ordered in December 1975. The first Brazilian-completed Xavante made its first flight on 3 September 1971, and the first two aircraft were handed over to the Brazilian Air Force at an official ceremony three days later.

By 1 January 1976 a total of 94 Xavantes had been delivered to the Brazilian Air Force. These are in service with the 3° Esquadrão Misto de Reconhecimento e Ataque (3rd Mixed Reconnaissance and Attack Squadron) at Santa Cruz AFB, Rio de Janeiro; the 4° Esquadrão Misto de Reconhecimento e Ataque at Santa Maria AFB, Rio Grande do Sul State; the 4° Grupo de Aviação (4th Aviation Group) at Fortaleza, Ceará State; the CATRE (Centro de Aplicações Táticas e Recompletamento de Equipagens: Tactical and Aircrew Training Centre) at Natal, Rio Grande do Norte State; the 10° Grupo de Aviação at Cumbica AFB, São Paulo State; and other units.

A full description of the standard M.B. 326GB appears in the Italian section of this edition; the version for the Brazilian Air Force is basically similar, except in the following respects:

Electronics and Equipment: Two Collins Type 618M-2B 360-channel VHF transceivers, Collins CIA-102A interphone system, Bendix DFA 73A-1 ADF, and a complete VOR/ILS system using a Collins 51V-1 VOR/LOC/glideslope receiver, Collins 51Z-4 marker beacon receiver and AN/APX-72 IFF transponder.

ARMAMENT: Six underwing points for bombs, gun pods or other stores. Typical loads include six 250 lb bombs; two 500 lb bombs; two 500 lb bombs and two twin 7·62 mm gun pods; four 250 lb bombs and two twin 7·62 mm gun pods; two twin 7·62 mm gun pods and two underwing drop-tanks; two twin 7·62 mm gun pods and four LM-70/7 rocket pods (each with seven SBAT 70 mm folding-fin air-to-ground projectiles); two twin 7·62 mm gun pods and two LM-37/36 rocket pods (each with thirty-six SBAT 37 mm air-to-ground rockets); six LM-70/7 rocket pods (each with seven SBAT 70 mm air-to-ground rockets); or two LM-70/19 rocket pods (each with nineteen SBAT 70 mm air-to-ground rockets); or photographic reconnaissance pods. All armament loads are designed and manufactured in Brazil.

EMBRAER (Aermacchi) AT-26 Xavante jet trainer/ground attack aircraft of the Brazilian Air Force

EMBRAER EMB-201 IPANEMA

The original EMB-200 version of this agricultural aircraft was designed and developed to specifications laid down by the Brazilian Ministry of Agriculture. Design was started in May 1969, and the prototype (PP-ZIP) made its first flight on 30 July 1970. A CTA type certificate was granted on 14 December 1971.

Production of the EMB-200 and EMB-200A initial versions, described in the 1973-74 *Jane's,* ended in mid-1974 after the completion of 73 aircraft. The version in current production, at a rate of 12 aircraft per month, is the further-developed EMB-201, to which the following description applies. The 100th Ipanema (PT-GFP) made its first flight on 24 January 1975.

TYPE: Single-seat agricultural aircraft.

WINGS: Cantilever low-wing monoplane. Wing section NACA 23015. Dihedral 7° from roots. Incidence 3°. All-metal single-spar structure of 2024 aluminium alloy with all-metal Frise-type ailerons outboard and all-metal slotted flaps on trailing-edge, and all-detachable leading-edges. Flap settings 0°, 8° and 30° (max); ailerons deflect 22° up, 14° down. No tabs. Variable-camber wingtips standard.

FUSELAGE: Rectangular-section all-metal safe-life structure, of welded 4130 steel tube with removable skin panels of 2024 aluminium alloy. Structure is specially treated against chemical corrosion.

TAIL UNIT: Cantilever two-spar all-metal structure of 2024 aluminium alloy. Slight sweepback on fin and rudder. Fixed-incidence tailplane. Elevators deflect 35° up, 20° down; rudder deflects 25° each side. Trim tab in starboard elevator.

LANDING GEAR: Non-retractable tailwheel type, with oleo shock-absorbers on main units. Tailwheel has tapered spring shock-absorber. Goodyear main wheels and tyres, size 8·50-10. Scott tailwheel, diameter 250 mm (10 in). Goodyear hydraulic disc brakes on main units.

POWER PLANT: One 224 kW (300 hp) Lycoming IO-540-K1D5 flat-six engine, driving a two-blade constant-speed metal propeller with spinner. Integral fuel tank in each wing leading-edge, with total capacity of 230 litres (50·6 Imp gallons). Refuelling point on top of each tank. Oil capacity 12 litres (2·6 Imp gallons).

ACCOMMODATION: Single horizontally/vertically-adjustable seat for pilot, in fully-enclosed cabin with bottom-hinged window/door on each side. Ventilation system in cabin. Provision for inertial type shoulder harness.

SYSTEMS: 28V DC electrical system supplied by a 24Ah AN-3151 battery and a Bosch K.1-28V-35A24 alternator. Power receptacle for external battery (AN-2552-3A type) on port side of forward fuselage.

ELECTRONICS AND EQUIPMENT: Standard VFR equipment, including VHF radio transceiver (Brazilian-made 14-channel Whinner Model 601 or 360-channel Bendix RT 241A) and ADF receiver (Brazilian-made Pontes & Moraes Model ADF-101/CP-101 or Bendix Model T-12C). Hopper for agricultural chemicals has capacity of 680 litres (149·5 Imp gallons). Transland dusting system below centre of fuselage. Transland or Micronair spraybooms aft of and above wing trailing-edges.

DIMENSIONS, EXTERNAL:

Wing span	11·20 m (36 ft 9 in)
Wing chord (constant)	1·60 m (5 ft 3 in)
Wing aspect ratio	7
Length overall	7·43 m (24 ft 4½ in)
Height overall (tail down)	2·20 m (7 ft 2½ in)
Fuselage: Max width	0·93 m (3 ft 0½ in)
Tailplane span	3·46 m (11 ft 4¼ in)
Wheel track	2·20 m (7 ft 2½ in)
Wheelbase	5·20 m (17 ft 7¼ in)
Propeller diameter	2·13 m (7 ft 0 in)

DIMENSIONS, INTERNAL:

Cockpit: Max length	1·20 m (3 ft 11¼ in)
Max width	0·85 m (2 ft 9½ in)
Max height	1·34 m (4 ft 4¾ in)

AREAS:

Wings, gross	18·00 m² (193·75 sq ft)
Ailerons (total)	1·60 m² (17·21 sq ft)
Flaps (total)	2·30 m² (24·76 sq ft)
Fin	0·58 m² (6·24 sq ft)
Rudder	0·63 m² (6·78 sq ft)
Tailplane	3·00 m² (32·29 sq ft)
Elevators, incl tab	1·50 m² (16·15 sq ft)

One of ten EMB-201 Ipanema agricultural aircraft supplied to Uruguayan Ministry of Agriculture and Fishing's Air Service *(Ronaldo S. Olive)*

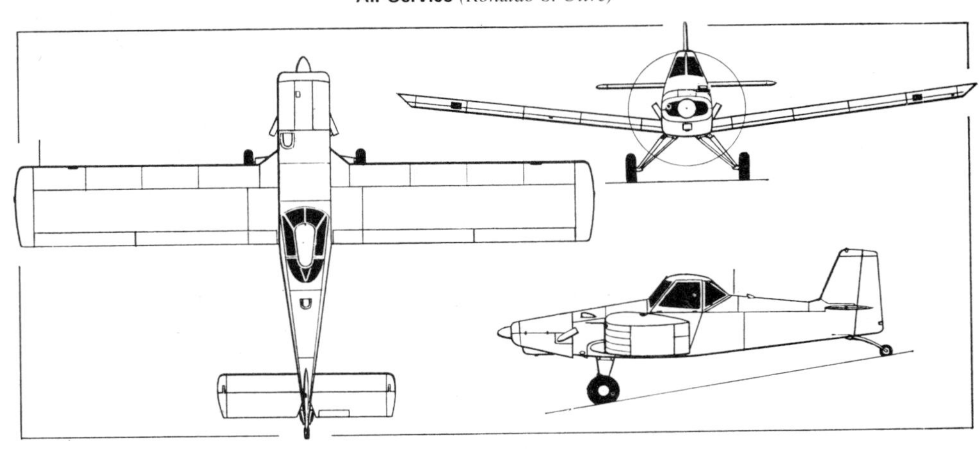

EMBRAER EMB-201 Ipanema single-seat agricultural aircraft *(Pilot Press)*

WEIGHTS AND LOADINGS (N: Normal; R: Restricted category):

Max payload	
N	550 kg (1,212 lb)
R	800 kg (1,763 lb)
Max T-O and landing weight:	
N	1,550 kg (3,417 lb)
R	1,800 kg (3,968 lb)
Max wing loading:	
N	86 kg/m² (17·6 lb/sq ft)
R	100 kg/m² (20·5 lb/sq ft)
Max power loading:	
N	6·92 kg/kW (11·39 lb/hp)
R	8·04 kg/kW (13·22 lb/hp)

PERFORMANCE (at max T-O weight, Normal category 'clean' configuration, ISA):

Never-exceed speed
165 knots (305 km/h; 190 mph) IAS

Max level speed at S/L
128 knots (238 km/h; 148 mph)

Max cruising speed (75% power) at 1,830 m (6,000 ft)
121 knots (224 km/h; 139 mph)

Econ cruising speed (65% power) at 1,830 m (6,000 ft)
106 knots (196 km/h; 122 mph)

Stalling speed, power off:

flaps up	68 knots (125 km/h; 78 mph)
T-O	64·5 knots (119 km/h; 74 mph)
landing	57·5 knots (106 km/h; 66 mph)
Max rate of climb at S/L	320 m (1,050 ft)/min
Service ceiling	5,180 m (17,000 ft)

T-O run at S/L, 8° flap, asphalt runway
466 m (1,530 ft)

T-O to 15 m (50 ft), conditions as above
876 m (2,875 ft)

Landing from 15 m (50 ft) at S/L, 30° flap, asphalt runway
400 m (1,310 ft)

Landing run, conditions as above 185 m (605 ft)

Range at 1,830 m (6,000 ft), no reserves:

at 1,800 kg (3,968 lb) 339 nm (629 km; 391 miles)
at 1,550 kg (3,417 lb) 386 nm (717 km; 445 miles)

Endurance at 1,830 m (6,000 ft), no reserves:

at 1,800 kg (3,968 lb)	3 hr 10 min
at 1,550 kg (3,417 lb)	3 hr 27 min

EMBRAER EMB-201A IPANEMA

Now under development, this new version of the Ipanema differs from the EMB-201 in the following respects:

Wings: New leading-edges and wingtips.

Accommodation: New control column.

Area:
- Wings, gross — 19·00 m² (204·5 sq ft)

Loadings:
- Max wing loading:
  - Normal — 81·54 kg/m² (16·7 lb/sq ft)
  - Restricted — 94·23 kg/m² (19·3 lb/sq ft)

Performance (at max T-O weight, Normal category, 'clean' configuration):
- Stalling speed, power off:
  - flaps up — 57·5 knots (106 km/h; 66 mph)
  - T-O — 54·5 knots (100 km/h; 62·5 mph)
  - landing — 50 knots (92 km/h; 57·5 mph)
- Max rate of climb at S/L — 426 m (1,400 ft)/min
- T-O run at S/L, 8° flap, asphalt runway — 354 m (1,160 ft)
- T-O to 15m (50 ft), conditions as above — 666 m (2,185 ft)

EMBRAER-PIPER LIGHT AIRCRAFT PROGRAMME

Detailed descriptions of the Piper aircraft built by EMBRAER (see introductory copy) can be found in the US section. EMBRAER names and designations are as follows:

EMB-710 Carioca. Piper PA-28-235 Cherokee Pathfinder, named after inhabitants of Rio de Janeiro. Fixed landing gear, four seats. Total of 57 produced by 1 January 1976.

EMB-711 Corisco (Lightning). Piper PA-28R-200 Cherokee Arrow II. Retractable landing gear. Total of 32 produced by 1 January 1976.

EMB-720 Minuano. Piper PA-32-300 Cherokee SIX, named after a wind of southern Brazil. Fixed landing gear, six seats. Total of 21 produced by 1 January 1976.

EMB-721 Sertanejo. Piper PA-32R-300 Cherokee Lance, named after a farming people of the Brazilian interior. Retractable landing gear, six seats. Two produced by 1 January 1976.

EMB-810 Seneca II. Piper PA-34-200T Seneca II. Total of 21produced by 1 January 1976.

EMB-820 Navajo. Piper PA-31-350 Navajo Chieftain. Six produced by 1 January 1976.

EMB-710 Carioca (Cherokee Pathfinder)

EMB-711 Corisco (Cherokee Arrow II)

EMB-720 Minuano (Cherokee SIX)

EMB-721 Sertanejo (Cherokee Lance)

EMB-810 Seneca II

EMB-820 Navajo (Navajo Chieftain)

NEIVA
SOCIEDADE CONSTRUTORA AERONÁUTICA NEIVA LTDA

Head Office and Works:
Estrada Velha Rio-São Paulo 2076, São José dos Campos, SP, Caixa Postal 247, Código 12200
Telephone: (0123) 216333
Telegrams: Aeroneiva
Other Works:
Av Brigadeiro Faria Lima s/n, São José dos Campos, SP, Caixa Postal 363, Código 12200
Rua Nossa Senhora de Fátima 360, Botucatu, SP, Caixa Postal 10, Código 18600
Directors:
José Carlos de Barros Neiva (Director General)
Breno A. B. Junqueira

Neiva completed in early 1975 the production and delivery of 150 N621 (T-25) Universal basic trainers for the Brazilian Air Force.

The company participates in EMBRAER's general aviation aircraft production programme, partially manufacturing and assembling the EMB-710 Carioca and EMB-711 Corisco, and building components for the EMB-720 Minuano, EMB-721 Sertanejo and EMB-810 Seneca II. Neiva also participates in EMBRAER's agricultural aircraft production programme, building at Botucatu the fuselage structure of the EMB-201 Ipanema.

NEIVA LANCEIRO

The Lanceiro was a civil development of the Neiva Regente described in the 1974-75 and earlier editions of *Jane's*. It was based on the L-42 Regente, and an aerodynamic prototype (PP-ZAH) began flying in 1970; the first production-standard aircraft (PP-ZCL) flew on 5 September 1973.

In view of its current involvement in the EMBRAER general aviation aircraft programme (see introductory copy), Neiva has discontinued manufacture of the Lanceiro. A full description can be found in the 1975-76 *Jane's*.

NEIVA N621 UNIVERSAL

Brazilian Air Force designation: T-25

The Universal was designed by Mr Joseph Kovacs to meet a Brazilian Air Force requirement for a trainer to replace the Fokker S-11/S-12 Instructor and North American T-6 Texan.

Initial design work was started in January 1963, and construction of the prototype began in May 1965.

This aircraft (PP-ZTW), flown for the first time on 29 April 1966, had side-by-side seating to conform with Ministry of Aeronautics preference.

The Brazilian Air Force ordered 150 Universals under the designation T-25. Production of these took place at the Botucatu and São José dos Campos factories, the former manufacturing the fuselage and the latter the wings, with final assembly at São José dos Campos. Wingtips, tailplanes and elevators were manufactured under subcontract by Motortec (formerly Avitec) in Rio de Janeiro. The first production T-25 was flown on 7 April 1971, and deliveries to the Brazilian Air Force began in the Autumn of 1971. The 150th T-25 was completed in early 1975. An additional Brazilian Air Force order was imminent in March 1976.

The T-25 is currently in service, for various duties, with 17 units of the Brazilian Air Force including the Centro de Aplicacões Táticas e Recompletamento de Equipagens (Tactical and Aircrew Training Centre) at Natal, Rio Grande do Norte State; the Academia da Força Aérea (Air Force Academy) at Pirassununga, São Paulo State; the 2a Esquadrilha de Ligação e Observação (2nd Liaison and Observation Flight) at São Pedro da Aldeia, Rio de Janeiro State; the EMRA 1 (1st Reconnaissance and Attack Squadron) at Belem, Para State; and the EMRA 2 (2nd Reconnaissance and Attack Squadron) at Recife, Pernambuco State.

Ten Universals were built and delivered to the Chilean Army during 1975. Negotiations with other countries were under way in early 1976.

Type: Two/three-seat basic trainer.

Wings: Cantilever low-wing monoplane. Wing section NACA 63_2A315 at root, NACA $63_1$212 at tip. Dihedral 6°. Incidence 2°. Single-spar structure of riveted aluminium alloy. All-metal dynamically-balanced slotted ailerons. All-metal split flaps.

Fuselage: Welded steel tube centre fuselage with aluminium skin panels. Semi-monocoque tailcone of riveted aluminium alloy.

Tail Unit: Cantilever all-metal structure, with electrically-actuated tab in port elevator.

Landing Gear: Retractable tricycle type. Hydraulic retraction, main units inward, nosewheel rearward. ERAM oleo shock-absorbers. Main wheels fitted with Goodyear tyres size 6·50-8 and Goodyear or OLDI disc brakes. Nosewheel steerable and fitted with Goodyear tyre size 6·00-6. Tyre pressure 2·28 bars (33 lb/sq in) on main units, 1·79 bars (26 lb/sq in) on nose unit.

Power Plant: One 224 kW (300 hp) Lycoming IO-540-K1D5 flat-six engine, driving a Hartzell HC-C2YK-4/C8475-A2 non-feathering two-blade constant-speed metal propeller. Six aluminium fuel tanks in wings, total capacity 332 litres (73 Imp gallons). Refuelling points above wings. Oil capacity 11·5 litres (2·5 Imp gallons).

Accommodation: Two seats side by side, with full dual controls, and optional third seat at rear. Large rearward-sliding transparent canopy. Baggage compartment aft of rear seat.

Systems: Electrically-actuated hydraulic system, pressure 103 bars (1,500 lb/sq in), for flaps and landing gear. Manual emergency pump. 28V electrical system.

Electronics and Equipment: 140-channel Brazilian-made VHF radio, ADF and VOR/LOC. Complete IFR instrumentation.

Armament: Two underwing hardpoints for the attachment of 7·62 mm machine-gun pods.

Dimensions, external:

Wing span	11·00 m (36 ft 1 in)
Wing chord at root	2·00 m (6 ft 6½ in)
Wing chord at tip	1·08 m (3 ft 6½ in)
Wing aspect ratio	7·1
Length overall	8·60 m (28 ft 2½ in)
Height overall	3·00 m (9 ft 9¾ in)
Tailplane span	3·95 m (12 ft 11½ in)
Wheel track	2·65 m (8 ft 8¼ in)
Wheelbase	2·33 m (7 ft 7¾ in)
Propeller diameter	2·13 m (7 ft 0 in)
Propeller ground clearance	0·37 m (1 ft 2½ in)

Neiva T-25 Universal two/three-seat basic training aircraft of the Chilean Army

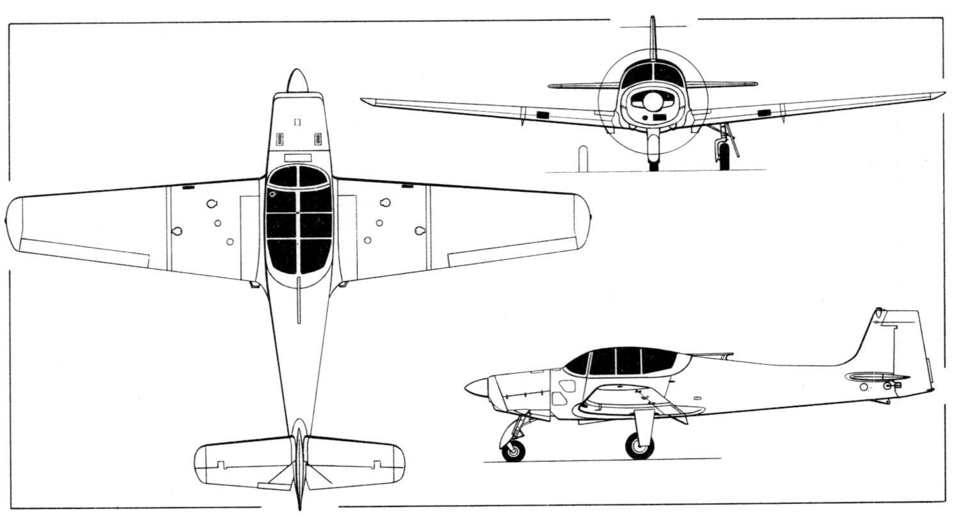

Neiva T-25 Universal basic training aircraft of the Brazilian and Chilean Services *(Pilot Press)*

Dimensions, internal:

Cabin:	
Length	2·20 m (7 ft 2½ in)
Max width	1·25 m (4 ft 1 in)
Max height	1·25 m (4 ft 1 in)
Floor area	3·0 m^2 (32 sq ft)
Volume	4·00 m^3 (141 cu ft)
Baggage compartment volume	0·35 m^3 (12·5 cu ft)

Areas:

Wings, gross	17·20 m^2 (185·14 sq ft)
Ailerons (total)	1·47 m^2 (15·82 sq ft)
Trailing-edge flaps (total)	1·34 m^2 (14·42 sq ft)
Fin	0·82 m^2 (8·83 sq ft)
Rudder	0·90 m^2 (9·69 sq ft)
Tailplane	1·72 m^2 (18·51 sq ft)
Elevators, incl tab	1·35 m^2 (14·53 sq ft)

Weights and Loadings (A: Aerobatic; U: Utility):

Weight empty, equipped:	
A, U	1,150 kg (2,535 lb)
Max T-O weight:	
A	1,500 kg (3,306 lb)
U	1,700 kg (3,747 lb)
Max wing loading:	
A	88·2 kg/m^2 (18·1 lb/sq ft)
U	100·0 kg/m^2 (20·5 lb/sq ft)
Max power loading:	
A	6·70 kg/kW (11·02 lb/hp)
U	7·59 kg/kW (12·49 lb/hp)

Performance (at max T-O weight. A: Aerobatic; U: Utility):

Never-exceed speed:	
A, U	269 knots (500 km/h; 310 mph)
Max level speed at S/L:	
A	162 knots (300 km/h; 186 mph)
U	160 knots (296 km/h; 184 mph)
Max cruising speed (75% power) at S/L:	
A	153 knots (285 km/h; 177 mph)
U	151 knots (280 km/h; 174 mph)
Stalling speed, flaps up:	
A	63·5 knots (117 km/h; 73 mph)
U	66 knots (122 km/h; 76 mph)
Stalling speed, flaps down:	
A	56·5 knots (104 km/h; 65 mph)
U	59·5 knots (110 km/h; 68·5 mph)
Max rate of climb at S/L:	
A	400 m (1,312 ft)/min
U	320 m (1,050 ft)/min
Service ceiling:	
A	6,100 m (20,000 ft)
U	5,000 m (16,400 ft)
T-O run at S/L:	
A	350 m (1,148 ft)
U	455 m (1,493 ft)
T-O to 15 m (50 ft) at S/L:	
A	510 m (1,673 ft)
U	650 m (2,133 ft)
Landing from 15 m (50 ft) at S/L:	
A	600 m (1,970 ft)
U	760 m (2,493 ft)
Range (75% power) at 2,000 m (6,550 ft), 10% reserves:	
A	539 nm (1,000 km; 621 miles)
U	809 nm (1,500 km; 932 miles)

NEIVA 621A UNIVERSAL II

As a production successor to the T-25 Universal, Neiva has proposed the N621A Universal II as a follow-on aircraft, differing from the earlier model principally in having a 298 kW (400 hp) Lycoming IO-720 series flat-eight engine in place of the 224 kW (300 hp) six-cylinder engine employed in the T-25. The nose has been redesigned to accommodate the larger power plant, thus increasing the aircraft's overall length to 8·78 m (28 ft 9¾ in); other dimensions would remain unchanged. Four underwing pylons would be able to carry light bombs and rocket pods of Brazilian manufacture.

The Esquadrilha da Fumaça (Smoke Flight) is studying the possibility of adopting the Universal II to replace the ageing North American T-6 Texans of the Brazilian Air Force's aerobatic team.

Performance (estimated, at AUW of 1,800 kg; 3,968 lb, 'clean'):

Never-exceed speed	269 knots (500 km/h; 310 mph)
Max level speed	173 knots (320 km/h; 199 mph)
Cruising speed at 3,000 m (9,845 ft)	163 knots (302 km/h; 188 mph)
Stalling speed, flaps down	57 knots (105 km/h; 66 mph)
Max rate of climb at S/L	540 m (1,770 ft)/min
Service ceiling	5,000 m (16,400 ft)
T-O to 15 m (50 ft)	600 m (1,970 ft)
Landing from 15 m (50 ft)	500 m (1,640 ft)
Range (75% power at 1,500 m; 5,000 ft, standard fuel)	277 nm (515 km; 320 miles)

CANADA

CANADAIR
CANADAIR LIMITED

HEAD OFFICE AND WORKS:
Cartierville Airport, St Laurent, Montreal, Quebec
POSTAL ADDRESS:
PO Box 6087, Station 'A', Montreal, Quebec H3C 3G9
Telephone: (514) 744-1511
CHAIRMAN OF THE BOARD:
Léo Lavoie
PRESIDENT AND CHIEF EXECUTIVE OFFICER:
Frederick R. Kearns
EXECUTIVE VICE-PRESIDENT:
Harry Halton
VICE-PRESIDENTS:
Peter J. Aird (Finance)
Frank M. Francis (Marketing)
Andreas Throner (Operations)
Robert A. Wohl (Administration)
DIRECTORS:
Harold V. Buhr (Contracts)
Bernard Langlois (Industrial Relations)
Harry Louis (Support)
Robert R. Ross (Surveillance Systems)
Harry H. Whiteman (Engineering)
PUBLIC RELATIONS MANAGER:
Benoit Kerub

Canadair Limited, formerly the Canadian subsidiary of General Dynamics Corporation and now owned by the Canadian government, has been engaged in the development and manufacture of military and commercial aircraft since 1944. It has also been employed in the research, design, development and production of missile components, drone surveillance systems and a variety of non-aerospace products.

Canadair Limited is located at Cartierville Airport, Montreal, and has 232,257 m² (2·5 million sq ft) of covered floor space.

In current production are the CL-215 tanker/utility amphibious aircraft, the Canadair (Northrop) CF-5D fighter and the AN/USD-501 drone surveillance system. Components for the General Dynamics F-111, McDonnell Douglas F-15, Boeing 747SP and Dassault-Breguet Mercure aircraft are manufactured under subcontracts. Production of aircraft spares, and the modification, repair and overhaul of aircraft, are other activities included in the current work programme.

Canadair Flextrac, a Canadair subsidiary located in Calgary, Alberta, is engaged in the design, development and production of military and commercial off-road vehicles.

The Canadian government announced in January 1975 that negotiations had been completed with General Dynamics Corporation for the option to purchase Canadair Limited, as a further step in the implementation of government policy to rationalise the Canadian aerospace industry and to increase Canadian ownership. The sale was completed on 5 January 1976.

LEARSTAR 600

Canadair has acquired world manufacturing and marketing rights in the LearStar 600 twin-turbofan executive transport aircraft (see US section), and in the Spring of 1976 was about to initiate an extensive assessment, in consultation with Mr William P. Lear, of the aircraft and its potential market.

CANADAIR CL-215

The Canadair CL-215 is a twin-engined amphibian, intended primarily for firefighting but adaptable to a wide variety of other duties. It is designed for simplicity of operation and maintenance, and can operate from small airstrips, lakes, ocean bays etc.

The CL-215 made its first flight on 23 October 1967, and its first water take-off on 2 May 1968.

The Protection Civile of France has operated CL-215s since June 1969. Following an initial purchase of 10 aircraft, it took delivery of a further four and in early 1976 had three more on order. The French aircraft have seen considerable action fighting forest fires in southern France, Corsica, the Federal Republic of Germany and Italy.

The Province of Quebec operates 15 aircraft, mainly in a firefighting role. Several of these were converted to wide-swath liquid sprayers for a massive campaign which began in Spring 1973 to protect huge tracts of valuable timberland from budworm infestation.

The Spanish government, which has operated two aircraft since February 1971, took delivery of eight more in 1974. The later aircraft, although equipped for search and rescue, are capable of firefighting and other roles.

The Greek government took delivery of two CL-215s in 1974; three more have been purchased, which were due for delivery before the start of the 1976 fire season.

The CL-215 offers fire protection agencies three methods of attacking fires in grass, brush or forest: (1) with pre-mixed long-term chemical retardants ground-loaded at a land base; (2) with short-term retardants mixed automatically during the water scooping operation; and (3) with plain water scooped from any 1,200 m (¾ mile) stretch of lake or ocean near the fire. It carries a maximum water or retardant load of 5,455 litres (1,200 Imp gallons). The tanks can be ground filled in 90 sec or scooped filled in 16-20 sec with the original scooping system, while the aircraft skims the water at about 60 knots (111 km/h; 69 mph). Pickup distance in still air, from 15 m (50 ft) above the water during landing to 15 m (50 ft) on take-off, is 1,660 m (5,450 ft) with the original installation. All new production aircraft now incorporate an improved scooping system which reduces the time to fill the tanks to 10 sec and the 15 m/15 m (50 ft/50 ft) pickup distance to 1,070 m (3,500 ft).

On a number of occasions single CL-215s have made over 100 drops totalling more than 545,520 litres (120,000 Imp gallons) in one day. Full loads have been scooped from the Mediterranean in wave heights of up to 2 m (6 ft).

An on-board mixing system for long-term retardants is being developed. A CL-215 equipped with this system will be able to mix and drop 20,910 litres (4,600 Imp gallons) of long-term retardant in a single flight. The prototype system was to be tested early in the 1976 fire season.

Tests conducted at Canadair have shown that the CL-215 can be used to extinguish oil fires when loaded with a foaming material.

TYPE: Twin-engined multi-purpose amphibian.

WINGS: Cantilever high-wing monoplane. No dihedral. All-metal one-piece fail-safe structure, with front and rear spars at 16% and 49% chord. Spars of conventional construction, with extruded caps and web stiffened by vertical members. Aluminium alloy skin, with riveted spanwise extruded stringers, is supported at 762 mm (30 in) pitch by interspar ribs. Leading-edge consists of aluminium alloy skin attached to pressed nose-ribs and spanwise stringers. Hydraulically-operated all-metal single-slotted flaps, supported by four external hinges on interspar ribs on each wing. Trim tab and geared tab in port aileron, rudder/aileron interconnect tab in starboard aileron. Detachable glassfibre wingtips.

FUSELAGE: All-metal single-step flying-boat hull of conventional fail-safe construction.

TAIL UNIT: Cantilever all-metal fail-safe structure with horizontal surfaces mounted midway up fin. Structure of aluminium alloy sheet, honeycomb panels, extrusions and fittings. Elevators and rudder fitted with dynamic balance, trim tab (port elevator only) and spring tabs and geared tabs. Provision for de-icing of leading-edges.

LANDING GEAR: Hydraulically-retractable tricycle type. Fully-castoring, self-centering twin-wheel nose unit retracts rearward into hull and is fully enclosed by doors. Main gear support structures retract into wells in sides of hull. A plate mounted on each main gear assembly encloses bottom of wheel well. Main-wheel tyre pressure 5·31 bars (77 lb/sq in); nosewheel tyre pressure 6·55 bars (95 lb/sq in). Hydraulic disc brakes. Non-retractable stabilising floats are each carried on a pylon cantilevered from wing box structure, with breakaway provision.

POWER PLANT: Two 1,566 kW (2,100 hp) Pratt & Whitney R-2800-83AM2AH, -83AM12AD or -CA3 eighteen-cylinder radial engines, each driving a Hamilton Standard Hydromatic constant-speed fully-feathering three-blade propeller, with 43E60 hub and type 6903 blades. First 30 aircraft have two fuel tanks, each of six flexible cells, in wing spar box, with total usable capacity of 4,336 litres (954 Imp gallons). Next 20 aircraft have two tanks each of eight flexible cells, with total usable capacity of 5,910 litres (1,300 Imp gallons). Gravity refuelling through two points above each tank. Oil in two tanks, with total capacity of 272·75 litres (60 Imp gallons), aft of engine firewalls.

ACCOMMODATION (water bomber version): Crew of two side by side on flight deck. Dual controls standard. Two 2,673 litre (588 Imp gallon) water tanks in main fuselage compartment, with retractable pickup probe in each side of hull bottom. Water-drop door in each side of hull bottom. Doors on port side of fuselage forward and aft of wings, of sliding type on first 30 aircraft, flush type on next 35. Emergency exit on starboard side aft of wing trailing-edge. Emergency hatch above starboard cockpit. Mooring hatch on top of hull nose below flight deck windows.

ACCOMMODATION (utility versions): Basic aircraft is equipped with canvas folding seats for eight passengers. With the tank headers in situ, 15 passengers can be carried. Removing the tank headers provides space for a total of 19 passengers. In the search and rescue configuration the utility version has, in addition, a navigator's station on the starboard side, aft of the pilot's bulkhead; a flight engineer's station between the pilot and co-pilot; two observer's stations in the aft cabin, forward of the rear door; and provision for four seats or two banks of three stretchers each, one on the port side forward of the wheel well and one on the starboard side aft of the wheel well. Provision for up to 18 seats or nine stretchers in casualty evacuation/supply role.

SYSTEMS: Hydraulic system, pressure 207 bars (3,000 lb/sq in), utilises two engine-driven pumps to actuate landing gear, flaps, water-drop doors and pickup probes, and wheel brakes. Electric pump in system provides power for emergency actuation of landing gear and brakes and closure of water doors. Electrical system includes two 250VA 115V 400Hz single-phase inverters, two 28V 200A DC generators, one 34Ah nickel-cadmium battery and one aircooled petrol engine-driven 28V 200A generator GPU. In the SAR version, two 800VA inverters are installed.

ELECTRONICS AND EQUIPMENT: Standard installation includes HF, VHF and FM communications equipment, VOR/ILS, glideslope receiver, ADF and marker beacon. For the SAR version, radar, radio altimeter and DME are added to the navigation equipment; other equipment for this role includes IFF/SIF, UHF intercom, and crash location communications equipment. Optional equipment includes UHF, DME and radar.

DIMENSIONS, EXTERNAL:

Wing span	28·60 m (93 ft 10 in)
Wing chord (constant)	3·54 m (11 ft 7½ in)
Wing aspect ratio	8·15
Length overall	19·82 m (65 ft 0½ in)
Beam	2·59 m (8 ft 6 in)
Length/beam ratio	7·5
Height overall (on land)	8·98 m (29 ft 5½ in)
Tailplane span	10·97 m (36 ft 0 in)
Wheel track	5·25 m (17 ft 2¾ in)
Wheelbase	7·23 m (23 ft 8½ in)
Propeller diameter	4·34 m (14 ft 3 in)
Forward door:	
Height	1·37 m (4 ft 6 in)
Width	1·03 m (3 ft 4 in)
Rear door:	
Height	1·12 m (3 ft 8 in)
Width	1·03 m (3 ft 4 in)
Water-drop door:	
Length	1·60 m (5 ft 3 in)
Width	0·81 m (2 ft 8 in)
Emergency exit:	
Height	0·91 m (3 ft 0 in)
Width	0·51 m (1 ft 8 in)

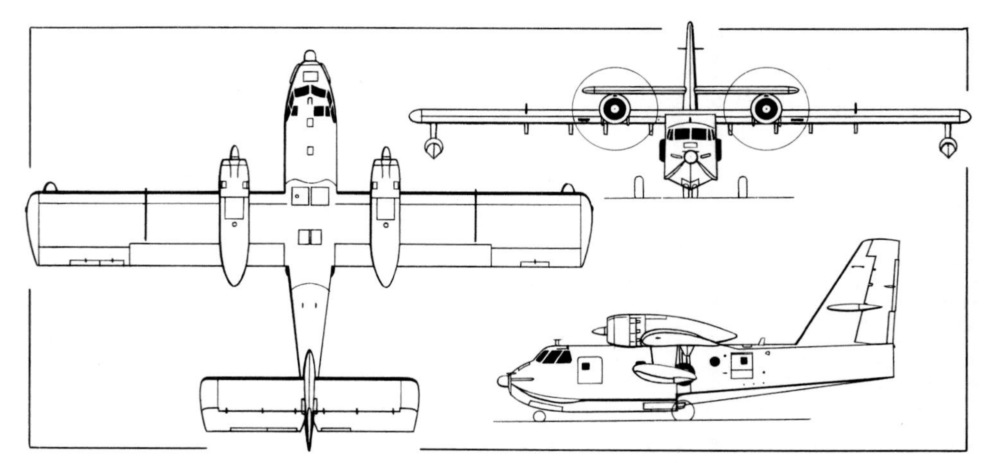

Canadair CL-215 twin-engined multi-purpose amphibian *(Pilot Press)*

Canadair CL-215 amphibian (two Pratt & Whitney R-2800 radial engines) in service with the Spanish Air Force *(Javier Taibo)*

DIMENSIONS, INTERNAL:
Cabin, excl flight deck:
- Length: 9·38 m (30 ft 9½ in)
- Max width: 2·39 m (7 ft 10 in)
- Max height: 1·90 m (6 ft 3 in)
- Floor area: 19·69 m² (212 sq ft)
- Volume: 35·03 m³ (1,237 cu ft)

AREAS:
- Wings, gross: 100·33 m² (1,080 sq ft)
- Ailerons (total): 8·05 m² (86·6 sq ft)
- Flaps (total): 22·39 m² (241 sq ft)
- Vertical tail surfaces (total): 17·23 m² (185·5 sq ft)
- Rudder, incl tabs: 6·02 m² (64·75 sq ft)
- Horizontal tail surfaces (total): 28·43 m² (306 sq ft)
- Elevators, incl tabs: 7·88 m² (84·8 sq ft)

WEIGHTS AND LOADINGS (A: aircraft Nos. 1-30; B: Nos. 31-50; C: Nos. 51-65):
Manufacturer's weight empty:
- A: 11,793 kg (26,000 lb)
- B, C: 12,065 kg (26,600 lb)

Typical operating weight empty:
- A: 12,247 kg (27,000 lb)
- B, C: 12,587 kg (27,750 lb)

Max payload:
- Water bomber: 5,443 kg (12,000 lb)
- Utility version (A): 3,062 kg (6,750 lb)
- Utility version (B, C): 2,839 kg (6,260 lb)

Max T/O weight (land): 19,731 kg (43,500 lb)
Max T-O weight (water):
- all versions: 17,100 kg (37,700 lb)

Max zero-fuel weight:
- A: 17,235 kg (38,000 lb)
- B: 18,145 kg (40,000 lb)
- C: 19,275 kg (42,500 lb)

Max landing weight (land and water):
- all versions: 16,780 kg (37,000 lb)

Cabin floor loading: 732 kg/m² (150 lb/sq ft)
Max wing loading: 196·66 kg/m² (40·3 lb/sq ft)
Max power loading: 6·23 kg/kW (10·36 lb/hp)

PERFORMANCE:
Cruising speed (max recommended power) at AUW of 18,595 kg (41,000 lb) at 3,050 m (10,000 ft)
157 knots (291 km/h; 181 mph)
Stalling speed, 15° flap, AUW of 19,731 kg (43,500 lb)
75 knots (139 km/h; 86 mph)
Stalling speed, 25° flap, AUW of 15,603 kg (34,400 lb), power off
66 knots (123 km/h; 76 mph)
Max rate of climb at S/L at AUW of 19,731 kg (43,500 lb) at max continuous power 305 m (1,000 ft)/min
Rate of climb at S/L, one engine out, at AUW of 17,100 kg (37,700 lb) at T-O power 75 m (245 ft)/min
T/O to 15 m (50 ft):
- on land at AUW of 19,731 kg (43,500 lb): 811 m (2,660 ft)
- on water at AUW of 17,100 kg (37,700 lb): 800 m (2,620 ft)

Landing from 15 m (50 ft):
- on land at AUW of 15,603 kg (34,400 lb): 732 m (2,400 ft)
- on water at AUW of 16,780 kg (37,000 lb): 835 m (2,740 ft)

Range with 1,587 kg (3,500 lb) payload:
- at max cruise power: 1,000 nm (1,853 km; 1,151 miles)
- at long-range cruise power: 1,220 nm (2,260 km; 1,405 miles)

DE HAVILLAND CANADA

THE DE HAVILLAND AIRCRAFT OF CANADA LTD

HEAD OFFICE AND WORKS:
Downsview M3K 1Y5, Ontario
Telephone: (416) 633-7310
Telex: 06-22128

CHAIRMAN: J. H. Smith

MANAGING DIRECTOR AND CHIEF EXECUTIVE OFFICER:
R. Bannock

VICE-PRESIDENTS:
D. B. Annan (Operations)
W. T. Heaslip (Engineering)
F. A. Johnson (Customer Support)
J. A. Timmins (Marketing and Sales)

DIRECTOR OF MARKET DEVELOPMENT: R. B. McIntyre

CHIEF DESIGNER: F. H. Buller

The de Havilland Aircraft of Canada Ltd was established in early 1928 as a subsidiary of The de Havilland Aircraft Co Ltd, and became subsequently a member of the Hawker Siddeley Group. On 26 June 1974 ownership was transferred to the Canadian government, which plans to operate the company only until responsible Canadian investors are found to purchase and operate de Havilland.

Facilities in 1976 covered a total area of 97,215 m² (1,046,430 sq ft), comprising a 77,022 m² (829,070 sq ft) main plant on the southern border of Downsview airport and 20,193 m² (217,360 sq ft) of leased space on the northern boundary of the airport. The company also has a product support facility, known as de Havilland Canada Inc, at Chicago in the USA.

Until the beginning of the second World War, de Havilland Canada acted principally as a sales and servicing organisation for products of the parent company. It became a manufacturing unit during the war and has since produced several original designs.

Of these the DHC-1 Chipmunk two-seat ab initio trainer, DHC-2 Beaver STOL utility aircraft, DHC-3 Otter and DHC-4/4A Caribou have been described in previous editions of *Jane's*. In production and service are the twin-engined DHC-5 Buffalo, evolved from the Caribou, and the DHC-6 Twin Otter STOL utility transport. Latest design is the DHC-7 Dash 7 'quiet STOL' transport, of which flight testing began in the Spring of 1975.

DHC-5D BUFFALO

CAF designation: CC-115
USAF designation: C-8A

In early May 1962, the US Army invited 25 companies to submit proposals for a new STOL tactical transport aircraft. De Havilland Canada won the competition with a developed version of the Caribou known as the Buffalo (originally Caribou II) with an enlarged fuselage and two General Electric T64 turboprop engines. Development costs were shared equally by the US Army, the Canadian government and de Havilland Canada.

Four evaluation DHC-5s were built initially, of which the first flew for the first time on 9 April 1964. Delivery of these aircraft to the US Army, as YAC-2s, began in April 1965. Fifteen DHC-5As were ordered by the Canadian Ministry of Defence in December 1964, deliveries of which began in 1967 and were completed at the end of 1968. These aircraft, currently designated **C-8A (DHC-5)** and **CC-115 (DHC-5A)**, have been described in earlier editions of *Jane's*.

Twenty-four Buffalos were ordered in 1967 by the Brazilian government. Twelve were delivered in 1969 and

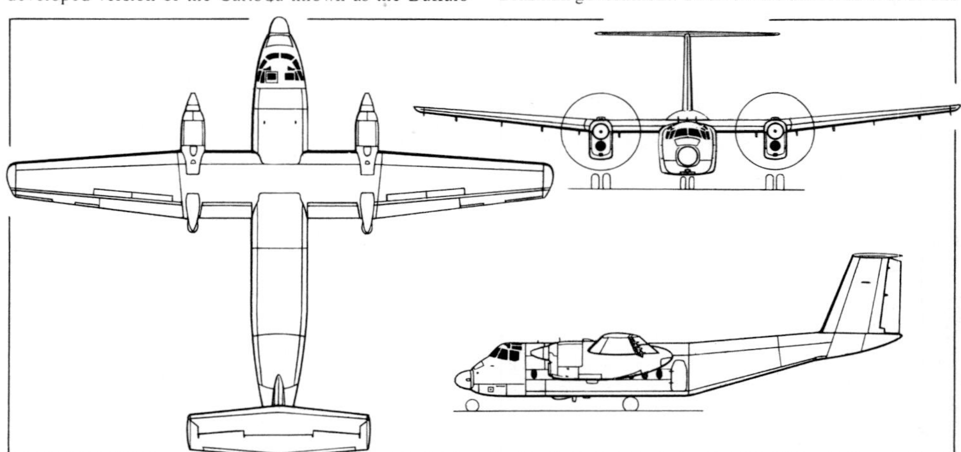

de Havilland Canada DHC-5D Buffalo twin-turboprop STOL utility transport *(Pilot Press)*

the remainder in 1970. Sixteen were ordered for use by Grupo Aéreo No. 8 of the Peruvian Air Force, based at Lima. The first was handed over on 16 June 1971; delivery was completed in mid-1972. Buffalos supplied to Brazil and Peru are designated DHC-5A/C-115, and are generally similar to the CC-115 Canadian version.

Two NASA-owned C-8As, one modified to evaluate the augmentor wing concept devised by de Havilland Canada and the other for quiet short-haul research (QSRA) and evaluation of propulsive-lift technology, are described under the NASA entry in the US section.

To meet an Indian Air Force requirement, DHC-5B and C versions were proposed in 1974 with CT64-P4C and Rolls-Royce Dart RDa.12 engines respectively.

One CC-115, modified by de Havilland Canada, has been loaned by the Canadian Department of Defence to Bell Aerospace for tests with an air cushion landing system. This aircraft is described under the Bell Aerospace heading in the US section.

In 1974 the air forces of Ecuador (two), Zaîre (six) and Zambia (seven) ordered Buffalos, and 19 aircraft are being built to cover these and other anticipated orders. They are of a later model known as the **DHC-5D**, to which the following description applies. In February 1976 a DHC-5D set up, subject to confirmation, new time-to-height records of 2 min 12·75 sec to 3,000 m; 4 min 27·5 sec to 6,000 m; and 8 min 3·5 sec to 9,000 m. These times qualify both in the class for turboprop-powered aircraft of unlimited weight category, and for those in the new 12,000-16,000 kg weight category.

Type: Twin-turboprop STOL utility transport.

Wings: Cantilever high-wing monoplane. Wing section NACA 64_3A417·5 (mod) at root, NACA 63_2A615 (mod) at tip. Dihedral 0° inboard of nacelles, 5° outboard. Incidence 2° 30′. Sweepback at quarter-chord 1° 40′. Conventional fail-safe multi-spar structure of high-strength aluminium alloys. Full-span double-slotted aluminium alloy flaps, outboard sections functioning as ailerons. Aluminium alloy slot-lip spoilers, forward of inboard flaps, are actuated by Jarry Hydraulics unit. Spoilers coupled to manually-operated ailerons for lateral control, uncoupled for symmetrical ground operation. Electrically-actuated trim tab in starboard aileron. Geared tab in each aileron. Rudder/aileron interconnect tab on port aileron. Outer wing leading-edges fitted with electrically-controlled flush pneumatic rubber de-icing boots.

Fuselage: Fail-safe structure of high-strength aluminium alloy. Cargo floor supported by longitudinal keel members.

Tail Unit: Cantilever structure of high-strength aluminium alloy, with fixed-incidence T-tailplane. Elevator aerodynamically and mass balanced. Fore and trailing serially-hinged rudders are powered by tandem jacks operated by two independent hydraulic systems manufactured by Jarry Hydraulics. Trim tab in port half of elevator, spring tab in starboard half. Electrically-controlled flush pneumatic rubber de-icing boot on tailplane leading-edge.

Landing Gear: Retractable tricycle type, with twin wheels on each unit. Hydraulic retraction, nose unit aft, main units forward. Jarry Hydraulics oleo-pneumatic shock-absorbers. Goodrich main wheels and tyres, size 37 × 15-12, pressure 3·10 bars (45 lb/sq in). Goodrich nosewheels and tyres, size 8·9 × 12·5, pressure 2·62 bars (38 lb/sq in). Goodrich multi-disc brakes.

Power Plant: Two General Electric CT64-820-4 turboprop engines, each rated at 2,336 kW (3,133 shp) and driving a Hamilton Standard 63E60-21 three-blade reversible-pitch fully-feathering propeller. Fuel in one integral tank in each inner wing, capacity 2,423 litres (533 Imp gallons), and rubber bag tanks in each outer wing, capacity 1,527 litres (336 Imp gallons). Total fuel capacity 7,900 litres (1,738 Imp gallons). Refuelling points above wings and in side of fuselage for pressure refuelling. Total oil capacity 45·5 litres (10 Imp gallons).

Accommodation: Crew of three, comprising pilot, co-pilot and crew chief. Main cabin can accommodate roll-up troop seats or folding forward-facing seats for 41 troops or 35 paratroops, or 24 stretchers and six seats. Provision for toilet in forward part of cabin. Door on each side at rear of cabin. Loading height with rear cargo loading door up and ramp down 2·90 m (9 ft 6 in).

Systems: AiResearch bleed air cabin heating and cooling system. Two independent hydraulic systems, each of 207 bars (3,000 lb/sq in), actuate landing gear, flaps, spoilers, rudders, brakes, nosewheel steering, winch and APU starting. 3·45 bar (50 lb/sq in) pneumatic system for engine starting, de-icing and environmental control. Two Lucas Aerospace engine-driven variable-frequency 3-phase 20kVA AC generators with 28V DC and 400Hz conversion subsystems. Solar T-62T-40 gas turbine APU in port engine nacelle provides electric (10kVA generator), hydraulic and pneumatic power. Brooks & Perkins rail-type cargo handling system, with hydraulic winch and floor rollers.

Electronics and Equipment: Radio and radar to customer's specification. Blind-flying instrumentation standard.

Dimensions, external:
- Wing span 29·26 m (96 ft 0 in)
- Wing chord at root 3·59 m (11 ft 9¼ in)
- Wing chord at tip 1·19 m (5 ft 11 in)
- Wing aspect ratio 9·75
- Length overall 24·08 m (79 ft 0 in)
- *Height overall 8·73 m (28 ft 8 in)
- Tailplane span 9·75 m (32 ft 0 in)
- Wheel track 9·29 m (30 ft 6 in)
- Wheelbase 8·48 m (27 ft 10 in)
- Propeller diameter 4·42 m (14 ft 6 in)
- Cabin doors (each side):
  - Height 1·68 m (5 ft 6 in)
  - Width 0·84 m (2 ft 9 in)
  - *Height to sill 1·17 m (3 ft 10 in)
- Emergency exits (each side, below wing leading-edge):
  - Height 1·02 m (3 ft 4 in)
  - Width 0·66 m (2 ft 2 in)
  - *Height to sill approx 1·52 m (5 ft 0 in)
- Rear cargo loading door and ramp:
  - Height 6·33 m (20 ft 9 in)
  - Width 2·34 m (7 ft 8 in)
  - *Height to ramp hinge 1·17 m (3 ft 10 in)

**will vary with aircraft configuration and loading conditions*

Dimensions, internal:
- Cabin, excl flight deck:
  - Length, cargo floor 9·58 m (31 ft 5 in)
  - Max width 2·67 m (8 ft 9 in)
  - Max height 2·08 m (6 ft 10 in)
  - Floor area 22·63 m² (243·5 sq ft)
  - Volume 48·56 m³ (1,715 cu ft)

Areas:
- Wings, gross 87·8 m² (945 sq ft)
- Ailerons (total) 3·62 m² (39 sq ft)
- Trailing-edge flaps (total, incl ailerons) 26·01 m² (280 sq ft)
- Spoilers (total) 2·34 m² (25·2 sq ft)
- Fin 8·55 m² (92 sq ft)
- Rudder 5·57 m² (60 sq ft)
- Tailplane 14·07 m² (151·5 sq ft)
- Elevator, incl tabs 7·57 m² (81·5 sq ft)

Weights and Loadings (A: STOL assault mission from unprepared airfield; B: STOL transport mission, firm smooth airfield surface):
- Operational weight empty (incl 3 crew and 680 kg; 1,500 lb allowance for options and electronics):
  - A, B 11,362 kg (25,050 lb)
- Max payload:
  - A 5,443 kg (12,000 lb)
  - B 8,164 kg (18,000 lb)
- Max normal fuel:
  - A, B 6,212 kg (13,696 lb)
- Max unit load for air drop:
  - A, B 2,721 kg (6,000 lb)
- Manoeuvring limit load factor:
  - A 3·0
  - B 2·5
- Max T-O weight:
  - A 18,597 kg (41,000 lb)
  - B 22,316 kg (49,200 lb)
- Max landing weight:
  - A 17,735 kg (39,100 lb)
  - B 21,273 kg (46,900 lb)
- Max zero-fuel weight:
  - A 16,782 kg (37,000 lb)
  - B 19,731 kg (43,500 lb)
- Max wing loading:
  - A 322·2 kg/m² (66 lb/sq ft)
  - B 268·5 kg/m² (55 lb/sq ft)

Performance (at max T-O weight except where indicated. A: STOL assault mission from unprepared airfield; B: STOL transport mission from firm smooth airfield surface):
- Max cruising speed at 3,050 m (10,000 ft):
  - A 250 knots (463 km/h; 288 mph) TAS
  - *B 227 knots (420 km/h; 261 mph) TAS

de Havilland Canada CC-115 (DHC-5A) Buffalo twin-turboprop STOL utility transport aircraft in Canadian Armed Forces insignia

Stalling speed, 40° flap:
A at 17,690 kg (39,000 lb) AUW
66 knots (122·5 km/h; 76 mph)
B at 21,273 kg (46,900 lb) AUW
71 knots (132 km/h; 82 mph)
Max rate of climb at S/L, normal rated power:
A 710 m (2,330 ft)/min
B 555 m (1,820 ft)/min
Rate of climb at S/L, one engine out:
A, max power 201 m (660 ft)/min
B, max power 113 m (370 ft)/min
†Service ceiling, normal rated power:
A, B 7,620 m (25,000 ft)
Service ceiling, one engine out:
A, max power 5,575 m (18,300 ft)
*B, max power 4,235 m (13,900 ft)
**STOL T-O run:
A 289 m (950 ft)
B 701 m (2,300 ft)
**STOL T-O to 15 m (50 ft), mid-CG:
A 381 m (1,250 ft)
B 876 m (2,875 ft)
**STOL landing from 15 m (50 ft):
A 346 m (1,135 ft)
B 613 m (2,010 ft)
**STOL landing run:
A 183 m (600 ft)
B 259 m (850 ft)
Range at 3,050 m (10,000 ft):
A, max payload 350 nm (648 km; 403 miles)
B, max payload 600 nm (1,112 km; 691 miles)
A, B, zero payload
1,770 nm (3,280 km; 2,038 miles)

† *Recommended max operating altitude; climb capability has been demonstrated up to 9,450 m (31,000 ft) at AUW of 18,597 kg (41,000 lb), ISA*
* *at 21,320 kg (47,000 lb) AUW*
***with 5,533 kg (12,200 lb) payload*

DHC-5 BUFFALO AUGMENTOR WING JET STOL RESEARCH AIRCRAFT

In co-operation with the Canadian Department of Industry, Trade and Commerce, a NASA-owned C-8A Buffalo has been modified as a flying testbed for the 'augmentor wing' concept devised by de Havilland Canada. A full description of the modified aircraft appears under the NASA heading in the US section of this edition.

In 1975 the flight programme continued with evaluation of a new longitudinal stability augmentation system. Handling qualities were assessed with the aid of various modes of control integration. The aerial navigation and automatic terminal guidance system (STOLAND) was checked and accepted by NASA following successful autoland trials in June 1975.

The flight programme is scheduled to continue into 1977 with 4-D navigation, autoland evaluations and flight studies for STOL certification criteria. At the beginning of 1976 a total of 250 flying hours and 340 engine hours had been accumulated.

DHC-5 BUFFALO QUIET SHORT-HAUL RESEARCH AIRCRAFT (QSRA)

A second C-8A has been acquired by NASA for evaluation of an alternative powered lift system. It is to have a new wing, four overwing Lycoming YF102 turbofan engines, and a redesigned tail assembly furnished by NASA. The QSRA prototype is being converted by Boeing Commercial Airplane Co under a $20 million NASA contract. Further details can be found under the NASA heading in the US section.

The DHC-5D Buffalo which set three new time-to-height records in early 1976

DHC-5 BUFFALO ACLS RESEARCH AIRCRAFT

Under a programme sponsored jointly by the Canadian Department of Industry, Trade and Commerce and the US Air Force Flight Dynamics Laboratory, a CAF CC-115 Buffalo was delivered to Bell Aerospace after modification by DHC to accept an air cushion landing system (ACLS). In this role the aircraft is designated XC-8A. Flight testing with the ACLS installed began in 1973.

A full description of this programme appears under the Bell Aerospace heading in the US section of this edition.

DHC-6 TWIN OTTER

CAF designation: CC-138
US Army designation: UV-18A

First announced in 1964, the Twin Otter is a STOL transport powered by two Pratt & Whitney Aircraft of Canada PT6A series turboprop engines. Design work was started in January 1964, and construction of an initial batch of five aircraft began in November of the same year. The first of these (CF-DHC-X), powered by two 432 kW (579 ehp) PT6A-6 engines, flew for the first time on 20 May 1965.

The fourth and subsequent aircraft of the initial Series 100 version were fitted with PT6A-20 engines, and the first delivery of a production aircraft, to the Ontario Department of Lands and Forests, was made in July 1966, shortly after the Twin Otter received FAA Type Approval. All Series are certificated to FAR 23 for Pt 135 operation.

By 1 January 1976, 490 Twin Otters had been sold, and operating hours totalled about 3½ million. The 400th Twin Otter was delivered on 18 December 1973, and production was continuing in 1976 at a rate of four aircraft per month.

Military operators of Twin Otters include the Argentine Air Force (six) and Army (two); Chilean Air Force (five); Ecuadorean Air Force (three); Jamaica Defence Force (one); Panamanian Air Force (one); Paraguayan Air Force (one); Peruvian Air Force (twelve); Royal Norwegian Air Force (four); the Canadian Armed Forces (eight CC-138 for SAR and utility duties); and the US Army (two).

Four versions of the Twin Otter have so far been announced, of which the Series 100 (115 built) and Series 200 (115 built) were described in the 1967-68 and 1970-71 *Jane's* respectively. The others are:

Series 300. Current production version, to which the following description applies. Deliveries began in the Spring of 1969 with the 231st Twin Otter off the line. Available, with short nose, as floatplane. Ten of the 12 aircraft supplied to Peru are fitted with floats, and are operated by Grupo Aéreo No. 42 of the Peruvian Air Force, based at Iquitos.

Series 300S. First announced in 1973. Improved Series 300 with added operational safety features associated with FAR Pt 25 (Transport Category) regulations and technical refinements to enhance the aircraft's STOL capability. Improvements include high-capacity brakes; anti-skid braking system; wing spoilers; electrical and hydraulic systems improvements; emergency brakes; propeller autofeather time delay; and improved power plant fire protection. The number of passenger seats has been reduced to 11, to provide an improved level of passenger comfort, and full airline-standard transport category electronics are installed.

Six Series 300S built, for operation by Airtransit Canada, an Air Canada subsidiary, on a government-funded experimental air service between Ottawa and Montreal. Using STOLports in the two cities, the service began on 24 July 1974 and was run until April 1976.

In September 1969, the Twin Otter was certificated by the MoT in the Normal category with a new-type external fire-bombing tank. This completely new forest firefighting concept, known as the Membrane Tank System, was designed and built by Field Aviation Company Ltd. A

DHC-6 Twin Otter Series 300 aircraft operated by Merpati Nusantara Airlines of Indonesia

rectangular tank of two 3·66 × 0·58 m (12 ft 0 in × 1 ft 11 in) sections, capable of holding 1,818 litres (480 US gallons; 400 Imp gallons), is mounted on the underside of the aircraft. An expendable fabric membrane supports the fluid, and is jettisoned with the load. It is designed for use on the landplane Twin Otter, using chemical fire retardants.

In 1971, a 3·05 m (10 ft) long, 1·4 m³ (50 cu ft) capacity ventral pod, carrying up to 272 kg (600 lb) of baggage or freight, was designed for the Twin Otter Series 300 by Field Aviation of Toronto and tested in service by Rocky Mountain Airlines of Denver.

TYPE: Twin-turboprop STOL transport.

WINGS: Strut-braced high-wing monoplane. Wing section NACA 6A series mean line; NACA 0016 (modified) thickness distribution. Dihedral 3°. No sweepback. All-metal safe-life structure, each wing being attached to the fuselage by two bolts at the front and rear spar fitting and braced by a single streamline section strut on each side. Light alloy riveted construction is used throughout except for the upper skin panels, which have spanwise corrugated stiffeners bonded to them. All-metal double-slotted full-span trailing-edge flaps. Spoilers fitted to Series 300S aircraft only. All-metal ailerons which also droop for use as flaps. Electrically-actuated tab in port aileron; geared trim tabs in both port and starboard ailerons. Optional pneumatic-boot de-icing equipment.

FUSELAGE: Conventional semi-monocoque safe-life structure, built in three sections. Primary structure of frames, stringers and skin of aluminium alloy. Windscreen and cabin windows of acrylic plastics. Cabin floor is of low-density aluminium-faced sandwich construction and is designed to accommodate distributed loads of up to 976·5 kg/m² (200 lb/sq ft).

TAIL UNIT: Cantilever all-metal structure of high-strength aluminium alloys. Fin and fixed-incidence tailplane are bolted to rear fuselage. Manually-operated trim tabs in rudder and elevators. A geared tab is fitted to the rudder to lighten control forces, and a tab fitted to the starboard elevator is linked to the flaps to control longitudinal trim during flap retraction and extension. Optional pneumatic-boot de-icing of tailplane leading-edge.

LANDING GEAR: Non-retractable tricycle type, with single wheel on each unit. Fully-steerable nosewheel. Urethane compression-block shock-absorption on main units. Oleo-pneumatic nosewheel shock-absorber. Goodyear main-wheel tyres size 11·00-12, pressure 2·62 bars (38 lb/sq in). Goodyear nosewheel tyre size 8·90-12·50, pressure 2·28 bars (33 lb/sq in). Goodrich independent, hydraulically-operated disc brakes on main wheels. Anti-skid braking system in Series 300S. Alternatively, high-flotation wheels and tyres, for operation in soft-field conditions, are available at customer's option, size 15·0-12·0 for nosewheel and main wheels. Provision for alternative wheel/ski landing gear. Twin-float gear available for short-nose Srs 300, with added wing fences and small auxiliary fins.

POWER PLANT: Two 486 kW (652 ehp) Pratt & Whitney Aircraft of Canada PT6A-27 turboprop engines, each driving a Hartzell HC-B3TN-3D three-blade reversible-pitch fully-feathering metal propeller. Two underfloor fuel tanks (eight cells), total capacity of 1,446 litres (318 Imp gallons). Refuelling point for each tank on port side of fuselage. Oil capacity 9·1 litres (2 Imp gallons) per engine. Optional electrical de-icing system for propellers and air intakes.

ACCOMMODATION: Side-by-side seats for one or two pilots on flight deck, access to which is by a forward-opening car-type door on each side or via the passenger cabin. Dual controls standard. Windscreen demisting and defrosting standard. Cabin divided by bulkhead into main passenger or freight compartment and baggage compartment. Seats for up to 20 passengers in main cabin. Standard interior is 20-seat commuter layout, with Douglas track, carpets, double windows, individual air vents and reading lights, and airstair door. Optional layouts include 18- or 19-seat commuter versions, 13/20-passenger utility version with foldaway seats and double cargo doors with ladder, and 11-passenger layout in Series 300S. Access to cabin by door on each side of rear fuselage; airstair door on the port side. Optional double door for cargo on port side instead of airstair door. Compartments in nose and aft of main cabin, each with upward-hinged door on port side, for 136 kg (300 lb) and 227 kg (500 lb) of baggage respectively; rear baggage hold accessible from cabin in emergency. Emergency exits near front of cabin on each side. Heating of flight deck and passenger cabin by engine bleed air; ventilation via a ram-air intake on the port side of the fuselage nose. Oxygen system for crew and passengers optional. Executive, survey or ambulance interiors can be fitted at customer's option. Tie-down cargo rings are installed as standard for the freighter role.

SYSTEMS: Hydraulic system, pressure 103 bars (1,500 lb/sq in), for flaps, brakes, nosewheel steering and (where fitted) ski retraction mechanism. A hand pump in the crew compartment provides emergency pressure for standby or ground operation if the electric pump is inoperative. Accumulators smooth the system pressure pulses and provide pressure for parking and emergency braking. Optional low-pressure pneumatic system (1·24 bars; 18 lb/sq in) for operation of autopilot or wing and tail de-icing boots, if fitted. Primary electrical system is 28V DC, with one 200A starter/generator on each engine. One 40Ah nickel-cadmium battery (optionally a 36Ah lead-acid battery) for emergency power and engine starting. Separate 3·6Ah battery supplies independent power for engine starting relays and ignition. 250VA main and standby static inverters provide 400Hz AC power for instruments and electronics. External DC receptacle aft of port side cabin door permits operation of complete system on the ground.

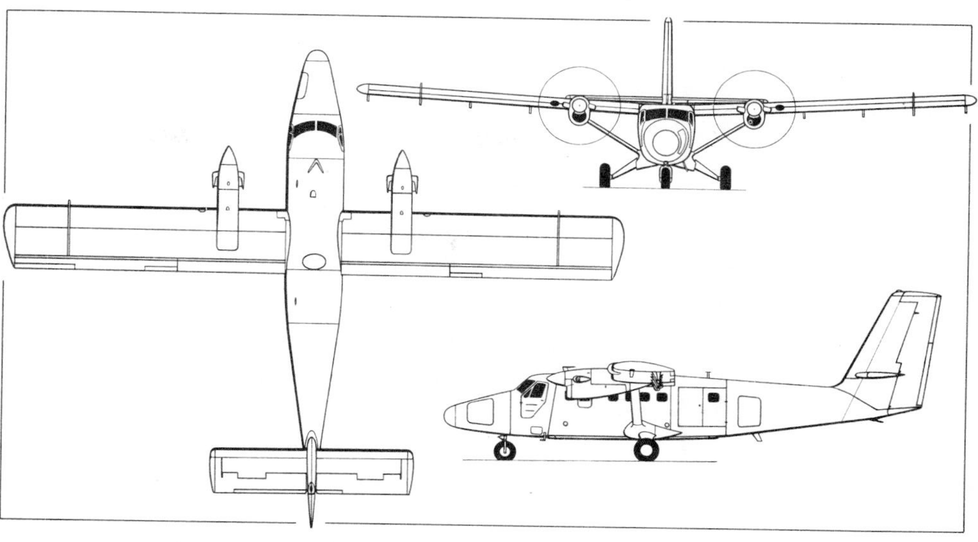
de Havilland Canada DHC-6 Twin Otter Series 300 STOL utility transport *(Pilot Press)*

DHC-6 Twin Otter Series 300 in the insignia of Deutsche Luft Transport of Germany

ELECTRONICS AND EQUIPMENT: Navigation and communications equipment, including weather radar, to customer's specification. Blind-flying instrumentation standard.

DIMENSIONS, EXTERNAL:
Wing span 19·81 m (65 ft 0 in)
Wing chord (constant) 1·98 m (6 ft 6 in)
Wing aspect ratio 10
Length overall 15·77 m (51 ft 9 in)
*Height overall 5·94 m (19 ft 6 in)
Tailplane span 6·30 m (20 ft 8 in)
Wheel track 3·71 m (12 ft 2 in)
Wheelbase 4·53 m (14 ft 10½ in)
Propeller diameter 2·59 m (8 ft 6 in)
Passenger door (port side):
Height 1·27 m (4 ft 2 in)
Width 0·76 m (2 ft 6 in)
*Height to sill 1·32 m (4 ft 4 in)
Passenger door (starboard side):
Height 1·15 m (3 ft 9½ in)
Width 0·77 m (2 ft 6¼ in)
*Height to sill 1·32 m (4 ft 4 in)
Baggage compartment door (nose):
Mean height 0·69 m (2 ft 3¼ in)
Width 0·76 m (2 ft 5¾ in)
*Height to sill 1·32 m (4 ft 4 in)
Baggage compartment door (port, rear):
Max height 0·97 m (3 ft 2 in)
Width 0·65 m (2 ft 1½ in)
Cargo double door (port, rear):
Height 1·27 m (4 ft 2 in)
Width 1·42 m (4 ft 8 in)
*Height to sill 1·32 m (4 ft 4 in)

**will vary with aircraft configuration and loading conditions*

DIMENSIONS, INTERNAL:
Cabin, excl flight deck, galley and baggage compartment:
Length 5·64 m (18 ft 6 in)
Max width 1·61 m (5 ft 3¼ in)
Max height 1·50 m (4 ft 11 in)
Floor area 7·45 m² (80·2 sq ft)
Volume 10·87 m³ (384 cu ft)
Baggage compartment (nose):
Volume 1·08 m³ (38 cu ft)
Baggage compartment (rear):
Length 1·88 m (6 ft 2 in)
Volume 2·49 m³ (88 cu ft)

AREAS:
Wings, gross 39·02 m² (420 sq ft)
Ailerons (total) 3·08 m² (33·2 sq ft)
Trailing-edge flaps (total) 10·42 m² (112·2 sq ft)
Fin 4·46 m² (48·0 sq ft)
Rudder, incl tabs 3·16 m² (34·0 sq ft)
Tailplane 9·29 m² (100·0 sq ft)
Elevator, incl tabs 3·25 m² (35·0 sq ft)

WEIGHTS:
Typical operating weight (20-seat commuter, incl 2 crew and 59 kg; 130 lb of electronics) 3,320 kg (7,320 lb)
Max payload for 100 nm (185 km; 115 miles) 2,004 kg (4,420 lb)
Max T-O weight 5,670 kg (12,500 lb)

Max landing weight:
- wheels and skis: 5,579 kg (12,300 lb)
- floats: 5,670 kg (12,500 lb)

PERFORMANCE (at max T-O weight, ISA):
- Max cruising speed at 3,050 m (10,000 ft): 182 knots (338 km/h; 210 mph)
- Stalling speed, flaps up: 74 knots (137·5 km/h; 85·5 mph)
- Stalling speed, flaps down: 58 knots (108 km/h; 67 mph)
- Max rate of climb at S/L: 488 m (1,600 ft)/min
- Rate of climb at S/L, one engine out: 104 m (340 ft)/min
- Service ceiling: 8,140 m (26,700 ft)
- Service ceiling, one engine out: 3,530 m (11,600 ft)
- T-O run:
  - STOL: 213 m (700 ft)
  - CAR Pt 3: 262 m (860 ft)
- T-O to 15 m (50 ft):
  - STOL: 366 m (1,200 ft)
  - CAR Pt 3: 457 m (1,500 ft)
- Landing from 15 m (50 ft):
  - STOL: 320 m (1,050 ft)
  - CAR Pt 3: 591 m (1,940 ft)
- Landing run:
  - STOL: 157 m (515 ft)
  - CAR Pt 3: 290 m (950 ft)
- Range at max cruising speed with 1,156 kg (2,550 lb) payload: 690 nm (1,277 km; 794 miles)
- Range at max cruising speed with 966 kg (2,131 lb) payload and wing tanks: 958 nm (1,775 km; 1,103 miles)

DHC-7 DASH 7

The Dash 7 'quiet STOL' airliner project was begun by de Havilland Canada in late 1972, following a worldwide market survey of short-haul transport requirements. The DHC-7 is designed to inaugurate Metroflight STOL service between downtown STOLports having 610 m (2,000 ft) runways. Pratt & Whitney Aircraft of Canada has participated in the development of a quiet engine/propeller combination which will limit external noise to 95 EPNdB at 152 m (500 ft) from the aircraft during take-off and landing.

De Havilland Canada built two pre-production aircraft for the flight test programme. The first of these (C-GNBX-X) was rolled out on 5 February 1975, and made its first flight at Downsview on 27 March 1975. The second (C-GNCA-X) made its first flight on 26 June 1975. A third airframe is undergoing static testing, and a fourth will be used for fatigue testing. Funds have been allocated for an initial production quantity of 50 Dash 7s.

Certification of the Dash 7 is scheduled for early 1977. Production deliveries will start in the Spring of 1977, and output will reach a rate of four aircraft per month by 1978. Orders for the Dash 7 up to 1 January 1976 were as follows:

Air Alpes (France)	4
AirWest Airlines (Canada)	2
Eastern Provincial Airways (Canada)	3
Ethiopian Airlines	5
Greenlandair (Denmark)	2
Nordair (subject to Canadian government approval)	8
Quebecair	2
Rocky Mountain Airways (USA) (plus 1 on option)	2
Widerøe's Flyveselskap (Norway)	4

The Dash 7 will be certificated by the Canadian Ministry of Transport to FAR 25; STOL performance will be approved under conventional FAR 25 and FAR 121 regulations apart from a 7° 30′ glideslope and 10·7 m (35 ft) landing reference height adopted by the FAA for STOL aircraft.

TYPE: Four-engined short/medium-range quiet STOL transport.

WINGS: Cantilever high-wing monoplane, with 4° 30′ dihedral from centre-section. Wing section NACA 63A418 (modified) at root, NACA 63A415 (modified) at tip. Incidence 3° at root. Conventional all-metal two-spar bonded skin/stringer structure. Double-slotted flaps, extending over approx 80% of trailing-edge, are actuated mechanically for take-off, by irreversible screwjacks, and hydraulically for landing. Two inboard ground spoilers/lift dumpers and two outboard air spoilers in each upper surface, forward of flaps. Outboard sections can be operated symmetrically, or differentially in combination with the ailerons. Trim tab in each aileron. Pneumatic-boot de-icing of leading-edges outboard of the inner nacelles.

FUSELAGE: Conventional all-metal stressed-skin pressurised structure, of bonded skin/stringer construction. Basically circular cross-section, with flattened profile under floor level.

TAIL UNIT: Cantilever all-metal T-tail, with large dorsal fin. Fixed-incidence tailplane, and one-piece horn-balanced elevator with trim tabs. Two-piece vertically-split rudder, actuated hydraulically. Pneumatic-boot de-icing of tailplane leading-edge.

LANDING GEAR: Menasco retractable tricycle type, with twin wheels on all units. Oleo-pneumatic shock-absorbers. Hydraulic retraction, main units forward into inboard engine nacelles, steerable nose unit rearward into fuselage. Main-wheel tyres size 30 × 9·00-15, pressure 6·90 bars (100 lb/sq in); nosewheel tyres size 6·50-10, pressure 5·52 bars (80 lb/sq in). Larger, low-pressure tyres optional, with pressures of 4·83 bars (70 lb/sq in) on main units, 4·14 bars (60 lb/sq in) on nose unit. Anti-skid hydraulic braking system for all units. Small retractable tailskid under rear fuselage.

POWER PLANT: Four 835 kW (1,120 shp) Pratt & Whitney Aircraft of Canada PT6A-50 turboprop engines, each driving a Hamilton Standard 24PF-303 fully-feathering reversible-pitch four-blade glassfibre propeller of slow-turning type (1,210 rpm) to reduce noise level. Fuel in two integral tanks in each wing, total capacity 5,602 litres (1,480 US gallons; 1,232 Imp gallons). Single pressure refuelling/defuelling point on underside of rear fuselage, aft of pressure dome. Pneumatic de-icing of engine air intakes; electrical de-icing for propellers. Oil capacity 23 litres (6 US gallons; 5 Imp gallons).

ACCOMMODATION: Flight crew of two, plus one or two cabin attendants. Seats for 50 passengers at 813 mm (32 in) pitch, in pairs on each side of centre aisle, with generous provision for underseat carry-on baggage. Outward-opening airstair door at rear on port side. Emergency exits on each side at front of cabin and on starboard side at rear. Baggage compartment in rear fuselage (capacity 998 kg; 2,200 lb), with external access on starboard side and internal access from cabin. Galley, coat rack and toilet at rear of cabin. Optional arrangements include movable bulkhead for mixed freight/passenger loads with forward freight door on port side. Up to five standard pallets can be accommodated in an all-cargo role. Brownline quick-change cargo handling system available optionally. Entire accommodation pressurised and air-conditioned.

SYSTEMS: Cabin pressure differential 0·294 bars (4·26 lb/sq in). Two air-cycle systems, driven by engine bleed air, for cabin air-conditioning. Two independent hydraulic systems, each of 207 bars (3,000 lb/sq in). No. 1 system actuates flaps, rudder, wing spoilers, main-wheel brakes and elevator boost: No. 2 system actuates landing gear, nosewheel and backup main-wheel brakes, parking brakes, nosewheel steering, rudder, outboard wing spoilers and elevator boost. Primary DC power provided by four Lucas 28V 250A 7·5kW starter/generators. 115/200V three-phase AC power at 400Hz from four 10kVA Lucas brushless generators for propeller and windscreen de-icing and standby fuel pumps. Lucas static inverters supply constant-frequency 400Hz loads, including engine instrumentation and navigational systems. Nickel-cadmium batteries for engine starting.

ELECTRONICS AND EQUIPMENT: Standard equipment includes crew interphone system; cabin PA system; flight data recorder; flight compartment voice recorder; emergency locator transmitter; two independent VHF communications systems; two independent VHF (VOR/ILS) radio navigation systems; one LF (ADF) radio navigation system; one ATC transponder; one DME; one RCA AVQ-21 weather radar; one marker beacon; Sperry Flight Systems SPZ-200A autopilot/flight director system, incorporating Z-500 flight computer and ADC-200 central air data computer; Sperry STARS ADI and HSI; Sperry AA-215 radio altimeter; and two Sperry C-14 slaved gyro compasses and VG-14 vertical gyros. Provision for variety of optional equipment to customer's requirements. Standard options include Sperry STARS CT-107 communications transceiver, NR-106A navigation receiver, GS-100B glideslope/marker receiver, and TP-114B transponder.

DIMENSIONS, EXTERNAL:
- Wing span: 28·35 m (93 ft 0 in)
- Wing chord at root: 3·81 m (12 ft 6 in)
- Wing chord at tip: 1·68 m (5 ft 6 in)
- Wing mean aerodynamic chord: 2·99 m (9 ft 9¾ in)
- Wing aspect ratio: 10
- Length overall: 24·58 m (80 ft 7·7 in)
- *Height overall: 7·98 m (26 ft 2 in)
- Tailplane span: 9·45 m (31 ft 0 in)
- Fuselage: Max diameter: 2·79 m (9 ft 2 in)
- Wheel track: 7·16 m (23 ft 6 in)
- Wheelbase: 8·38 m (27 ft 6 in)
- Propeller diameter: 3·43 m (11 ft 3 in)
- *Propeller ground clearance (inboard engines): 1·60 m (5 ft 3 in)
- Min propeller/fuselage clearance: 0·75 m (2 ft 5·4 in)
- Passenger door (rear, port):
  - Height: 1·78 m (5 ft 10 in)
  - Width: 0·76 m (2 ft 6 in)
  - Height to sill: 1·09 m (3 ft 7 in)
- Emergency exit door (rear, stbd):
  - Height: 1·35 m (4 ft 5 in)
  - Width: 0·61 m (2 ft 0 in)
  - Height to sill: 1·09 m (3 ft 7 in)

Second pre-production example of the DHC-7 Dash 7 quiet STOL transport (four PT6A-50 turboprop engines)

Emergency exit doors (fwd, each):
Height 0·91 m (3 ft 0 in)
Width 0·51 m (1 ft 8 in)
Height to sill 1·55 m (5 ft 1 in)
Baggage hold door (rear, stbd):
Height 0·97 m (3 ft 2 in)
Width 0·79 m (2 ft 7 in)
Height to sill 1·47 m (4 ft 10 in)
Cargo door (fwd, port, optional):
Height 1·78 m (5 ft 10 in)
Width 2·31 m (7 ft 7 in)
will vary with aircraft configuration and loading conditions

DIMENSIONS, INTERNAL:
Cabin, excl flight deck:
Length 12·19 m (40 ft 0 in)
Max width 2·62 m (8 ft 7 in)
Floor width 2·13 m (7 ft 0 in)
Max height 1·98 m (6 ft 6 in)
Height under wing 1·85 m (6 ft 1 in)
Volume 54·1 m³ (1,910 cu ft)
Baggage compartment (rear fuselage):
Max length 2·64 m (8 ft 8 in)
Volume 6·8 m³ (240 cu ft)

AREAS:
Wings, gross 79·90 m² (860·0 sq ft)
Ailerons (total) 2·16 m² (23·22 sq ft)
Trailing-edge flaps (total) 27·33 m² (294·20 sq ft)
Spoilers (total) 3·63 m² (39·04 sq ft)
Vertical tail surfaces (total, excl dorsal fin) 15·79 m² (170·0 sq ft)
Horizontal tail surfaces (total) 20·16 m² (217·0 sq ft)

WEIGHTS AND LOADINGS:
Basic weight empty (standard 50-passenger layout) 11,453 kg (25,250 lb)
Operating weight empty 11,730 kg (25,860 lb)
Max payload (50 passengers or cargo) 5,280 kg (11,640 lb)
Max T-O weight 19,504 kg (43,000 lb)
Max zero-fuel weight 17,009 kg (37,500 lb)
Max landing weight 18,597 kg (41,000 lb)
Max cabin floor loading 366 kg/m² (75 lb/sq ft)
Max wing loading 244 kg/m² (50 lb/sq ft)
Max power loading 5·84 kg/kW (9·60 lb/shp)

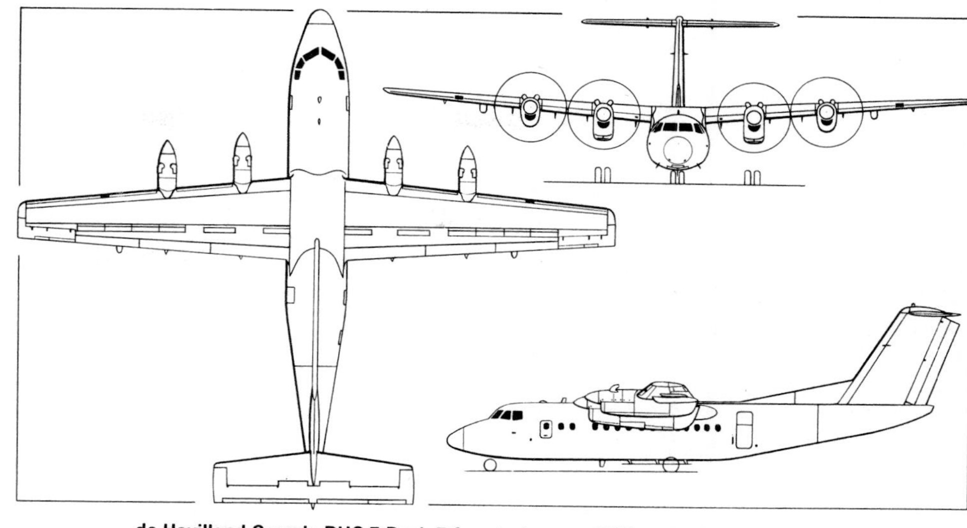

de Havilland Canada DHC-7 Dash 7 four-turboprop STOL transport *(Pilot Press)*

PERFORMANCE (estimated, at max T-O weight except where indicated, FAR 25 at S/L, ISA):
Max cruising speed at 4,570 m (15,000 ft) 244 knots (452 km/h; 281 mph)
En-route rate of climb, flaps and landing gear up:
four engines, max climb power 399 m (1,310 ft)/min
three engines, max continuous power 231 m (760 ft)/min
Service ceiling 6,770 m (22,200 ft)
Service ceiling, one engine out 4,330 m (14,200 ft)
Accelerate/stop distance 710 m (2,330 ft)
T-O to 10·7 m (35 ft) 710 m (2,330 ft)
Landing from 10·7 m (35 ft) at max landing weight 408 m (1,338 ft)
Landing field length (factored) at max landing weight 680 m (2,230 ft)
Min ground turning radius 8·84 m (29 ft 0 in)
Runway LCN with 32 × 11·50-15 low-pressure tyres, 30 movements 8
Range at 4,570 m (15,000 ft) with max passenger payload at 80% max cruise rating, IFR reserves 812 nm (1,504 km; 935 miles)
Max range at 4,570 m (15,000 ft) with standard fuel and 3,211 kg (7,080 lb) payload, 80% max cruise rating 1,237 nm (2,293 km; 1,425 miles)

OPERATIONAL NOISE CHARACTERISTICS (FAR 36, estimated):
T-O noise level 74 EPNdB
Approach noise level on 3° glideslope 91 EPNdB
Approach noise level on 7° 30′ glideslope 82 EPNdB
Sideline noise level 81 EPNdB

HAWKER SIDDELEY

HAWKER SIDDELEY CANADA LTD (Member Company of HAWKER SIDDELEY GROUP)

HEAD OFFICE:
7 King Street East, Toronto M5C 1A3, Ontario
Telephone: (416) 362-2941
Telex: 02-2605

CHAIRMAN:
Sir Arnold Hall, FRS
VICE-CHAIRMEN:
R. G. Smith
R. R. Kenderdine
PRESIDENT AND CHIEF EXECUTIVE OFFICER:
R. S. Faulkner
DIRECTOR OF PUBLIC RELATIONS: J. F. A. Painter

Known as A. V. Roe Canada Ltd until 1962, this company controls operating units and subsidiaries in Canada employing about 8,500 people.

The company's chief aviation unit is Orenda Division (see Aero-engines section), which manufactures aircraft jet engines under licence and carries out repairs and overhauls.

NWI

NORTHWEST INDUSTRIES LTD (Division of CAE Industries Ltd)

ADDRESS:
Industrial Airport, PO Box 517, Edmonton, Alberta
Telephone: (304) 455-3161

This company, other recent activities of which were described in the 1972-73 and 1974-75 *Jane's,* received two important aircraft modification contracts during 1975.

The first of these, valued at $1 million (Canadian), was for the modification of four CAF C-130E Hercules transport aircraft for use at the Air Navigation and Instrument Rating School at Winnipeg. The modification involves the fitting of two palletised training modules, each accommodating an instructor and two trainees, into the main hold of the aircraft. The modules are designed to be quickly and easily removable, to permit normal transportation missions to be flown when the Hercules are not required for training flights.

The second contract, valued at some $2 million (Canadian), is for the modernisation of 113 Canadair CT-114 Tutor training aircraft of the Canadian Armed Forces. Improvements include some new equipment and associated antennae, provision of external auxiliary fuel tanks, relocation of the engine ice detection probe, and a more efficient electrical canopy jettison system. NWI was due to have delivered the first 12 modified Tutors to the CAF by the end of May 1976, and to continue delivery thereafter at a rate of eight aircraft per month.

SAUNDERS

SAUNDERS AIRCRAFT CORPORATION LTD

HEAD OFFICE AND WORKS:
PO Box 1230, Gimli, Manitoba
Telephone: (204) 642-5101
Telex: 07-587-850
PRESIDENT:
S. R. Kersey

Saunders Aircraft Corporation was formed in 1968 to design and manufacture the ST-27, a conversion of the de Havilland Heron transport aircraft incorporating major design changes, including turboprop engines. A full description of the ST-27 appeared in the 1974-75 *Jane's.*

The company employed more than 500 personnel in 1975, and had begun to produce the ST-28, a successor to the ST-27, before financial support was halted in early 1976.

The company's debts were to be written off by the Manitoba government; its assets, which included two further ST-27s completed for Aero Trades (Winnipeg) and Tropic Air Services (Barbados), and the prototype for the ST-28, were to be sold; and its status reduced to that of a product support organisation. At the time of the announcement, the work force had already been reduced to 200, and was eventually to be reduced to 24 pending a decision on the company's long-term future.

SAUNDERS ST-28

The ST-28 is a new-build aircraft, based on the ST-27, and until 1 February 1975 was known as the ST-27B. It features extensive structural and systems redesign to meet the requirements of FAR Pts 25 and 36. The prototype (CF-YBM-X), a converted Heron designated ST-27A, flew for the first time on 18 July 1974 and by the end of 1975 was nearing the completion of its certification programme. By that time the first production ST-28 had flown (on 12 December), and orders and options totalled 34.

On 3 January 1976, however, further financial support was withdrawn by the Manitoba government, as noted in the introductory copy.

The following description applies to the production ST-28:

TYPE: Twin-turboprop light transport.

WINGS: Cantilever low-wing monoplane. Thickness/chord ratio 18·3% at root, 14·5% at tip. Dihedral 6° ±10′ at chord line. Incidence 2° ±10′. Conventional three-spar stressed-skin fail-safe metal structure. Single-spar mass-balanced ailerons and three-section flaps, with metal skins. Flying controls operated manually. Two-position flaps (10° for T-O, 41·5° for landing), operated by Airite hydraulic actuators. Interconnected servo trim tab in each aileron. Pneumatic rubber-boot de-icing of leading-edges.

FUSELAGE: Conventional semi-monocoque light alloy fail-safe structure. Wing triple-spar carry-through structure integral with fuselage. Main cabin floor of sandwich construction.

TAIL UNIT: Cantilever light alloy structure. Horizontal surfaces, which have marked dihedral, comprise one-piece three-spar fixed-incidence tailplane and elevators, each with an inset trim tab. Anti-servo trim tab in rudder. Rubber-boot de-icing of fin and tailplane leading-edges.

LANDING GEAR: Retractable tricycle type. Hydraulic retraction, nosewheel rearward into fuselage, main wheels outward into wings. Oleo-pneumatic shock-absorbers on all units. Nosewheel is castoring and self-centering. Single wheel and tyre on each unit, size 700 × 230 mm on main units, 500 × 200 mm on nose unit. Tyre pressure 5·52 bars (80 lb/sq in) on main units, 2·90 bars (42 lb/sq in) on nose unit. Dunlop hydraulic disc brakes on main units.

POWER PLANT: Two 584 kW (750 shp; 783 ehp) Pratt & Whitney Aircraft of Canada PT6A-34 turboprop engines, each driving a Hartzell HC-B4TN-3A four-blade fully-feathering reversible-pitch propeller with spinner. One main and one auxiliary fuel tank in each wing, each tank consisting of two cells composed of a flexible liner housed in a sealed sheet metal liner. Full cross-feed capability. Total fuel capacity 1,900 litres (418 Imp gallons), of which 1,864 litres (410 Imp gallons) are usable. One overwing gravity refuelling point for each tank. Oil capacity 8·6 litres (1·9 Imp gallons).

ACCOMMODATION: Crew of two on flight deck. Non-pressurised main cabin can seat up to 23 persons, or 22 and a cabin attendant. Individual seats, with underseat baggage restraint, on each side of centre aisle at 762 mm (30 in) pitch. Washroom, wardrobe and galley optional. Alternative layouts for 21 passengers, one attendant, and washroom; or for 20 passengers, attendant, wardrobe and washroom. Special role adaptions (eg cargo, passenger/cargo, executive, aerial photography, ambulance) are possible. Main passenger/crew door on starboard side at front, opening outward and downward, with integral stairs. Passenger/service door on port side at rear, opening outward and forward. Outward-opening emergency exit over wing on each side. Baggage/freight holds at rear (aft of cabin bulkhead) and in nose (beneath crew compartment). Cabin heated, by engine bleed air, and ventilated. Air-conditioning system optional.

SYSTEMS: Two independent hydraulic systems, pressure 138 bars (2,000 lb/sq in) each, for landing gear, brake and flap actuation. No pneumatic system. Electrical system includes two 300A starter/generators for 27V DC supply, with two 24V 40Ah nickel-cadmium batteries for engine starting and emergency supply. AC power supplied by four single-phase static inverters. Airframe de-icing utilises pneumatic boots supplied from engine bleed air/vacuum system. Engine intakes utilise direct bleed air de-icing; propellers and windscreens are de-iced electrically.

ELECTRONICS AND EQUIPMENT: Standard equipment includes dual Collins VHF-20 VHF com radio; dual Collins VIR-30A VHF nav radio, with single marker; one Collins DF-206 ADF; one Collins TDR-90 ATC transponder; one Collins DME-40 DME; one Collins PN-101 compass; one Collins 346D-1B PA system; and dual Smith M1035 K2 audio. Blind-flying instrumentation standard for both pilots. Optional equipment includes RCA AVQ-21 radar; radio altimeter; flight director: microwave landing system; and area navigation equipment.

DIMENSIONS, EXTERNAL:

Wing span	21·79 m (71 ft 6 in)
Wing chord at root	3·22 m (10 ft 6¾ in)
Wing chord at tip	0·86 m (2 ft 10 in)
Wing aspect ratio	10·3
Length overall	17·93 m (58 ft 10 in)
Height overall	4·75 m (15 ft 7 in)
Tailplane span	6·81 m (22 ft 4 in)
Wheel track	5·08 m (16 ft 8 in)
Wheelbase	7·09 m (23 ft 3 in)
Propeller diameter	2·23 m (7 ft 4 in)
Propeller ground clearance	0·36 m (1 ft 2 in)
Passenger door (fwd, stbd):	
Height	1·30 m (4 ft 3 in)
Width	0·71 m (2 ft 3¾ in)
Passenger door (aft, port):	
Height	1·30 m (4 ft 3 in)
Width	0·84 m (2 ft 9 in)
Baggage door (rear, stbd):	
Height	0·65 m (2 ft 1·6 in)
Width	1·02 m (3 ft 4 in)
Baggage door (fwd, port):	
Height	0·44 m (1 ft 5½ in)
Width	0·69 m (2 ft 3 in)
Emergency exits (each):	
Height	0·69 m (2 ft 3 in)
Width	0·69 m (2 ft 3 in)

DIMENSIONS, INTERNAL:

Cabin (excl flight deck):	
Max length	9·14 m (30 ft 0 in)
Max width	1·40 m (4 ft 7 in)
Max height	1·75 m (5 ft 9 in)
Volume	20·47 m³ (723 cu ft)
Baggage holds, volume:	
forward	0·40 m³ (14 cu ft)
rear	3·85 m³ (136 cu ft)

AREAS:

Wings, gross	46·36 m² (499·0 sq ft)
Ailerons (total)	2·98 m² (32·04 sq ft)
Trailing-edge flaps (total)	6·60 m² (71·0 sq ft)
Fin	2·83 m² (30·5 sq ft)
Rudder, incl tab	2·09 m² (22·5 sq ft)
Tailplane	5·35 m² (57·6 sq ft)
Elevators, incl tabs	3·53 m² (38·0 sq ft)

WEIGHTS AND LOADINGS:

Weight empty, equipped	4,002 kg (8,824 lb)
Max payload (incl crew)	2,121 kg (4,676 lb)
Max T-O weight	6,577 kg (14,500 lb)
Max zero-fuel weight	6,123 kg (13,500 lb)
Max landing weight	6,418 kg (14,150 lb)
Max wing loading	141·8 kg/m² (29·06 lb/sq ft)
Max power loading	5·64 kg/kW (9·26 lb/ehp)

PERFORMANCE (estimated, at max T-O weight except where indicated):

Never-exceed speed	253 knots (470 km/h; 292 mph)
Max cruising speed at 2,135 m (7,000 ft)	200 knots (370 km/h; 230 mph)
Econ cruising speed at 2,135 m (7,000 ft)	182 knots (338 km/h; 210 mph)
T-O to 10·7 m (35 ft), FAR 25	1,143 m (3,750 ft)
Landing from 15 m (50 ft), FAR 25 at max landing weight	1,067 m (3,500 ft)
Landing run at max landing weight	305 m (1,000 ft)
Min ground turning radius	5·08 m (16 ft 8 in)
Range with max fuel, IFR reserves, 100 nm (185 km; 115 miles) diversion plus 45 min	860 nm (1,594 km; 990 miles)
Range with max payload, reserves as above	110 nm (204 km; 126 miles)

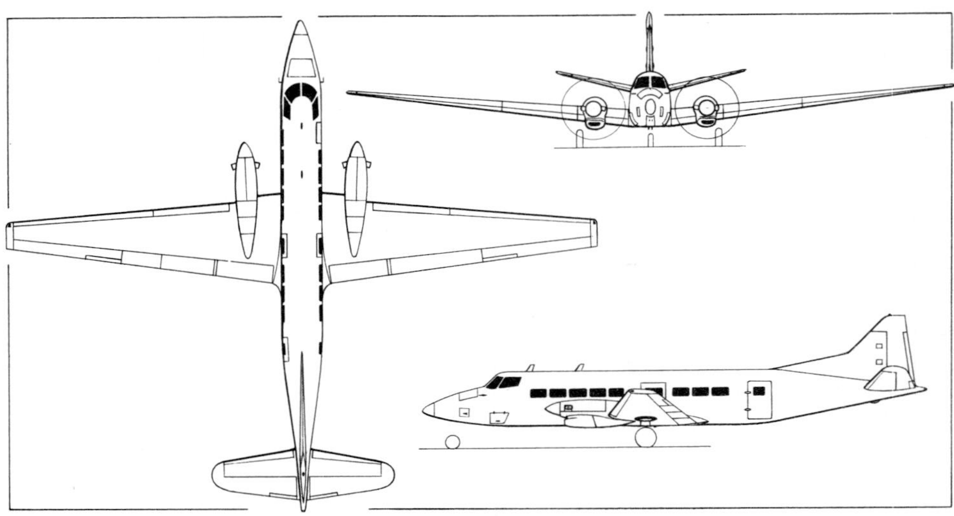

Saunders ST-28 twin-turboprop light transport *(Pilot Press)*

TRIDENT

TRIDENT AIRCRAFT LIMITED

ADDRESS:
261 Viscount Way, Richmond, British Columbia
Telephone: (604) 278-6204
PRESIDENT: D. A. Hazlewood, PEng
CHIEF ENGINEER: J. C. Galizia
PROJECT MANAGER: P. S. Masterton

TRIDENT TR-1 TRIGULL-320

The TR-1 incorporates a number of significant differences from earlier aircraft of its type. It is claimed that, compared with other similar aircraft, it provides better low-speed handling (including stall) and improved general performance.

Design of the Trigull-320 was started in Canada in July 1971. The first prototype (CF-TRI-X) flew for the first time on 5 August 1973. By February 1976, more than 300 flying hours had been logged. A second prototype was scheduled to fly in early 1976; a third airframe has completed static testing. Certification was anticipated by mid-1976.

Production of the Trigull-320 has been delayed, pending completion of the necessary funding arrangements. Having failed in 1975 to raise by its own efforts the amount of capital required as a condition of financial assistance from the Canadian government, Trident was obliged to slow down the Trigull programme. By the beginning of 1976, orders for 83 Trigull-320s had been received. The airframes will be built by Canadian Aircraft Products Ltd and assembled by Trident.

TYPE: Six-seat light amphibian.

WINGS: Cantilever high-wing monoplane. Wing section NACA 23015R-4 (modified). Dihedral 2° from roots. Incidence 2° 15′. No sweepback. Two-spar aluminium stressed-skin fail-safe structure, of constant chord, with drooped leading-edges. Hydraulically-operated single-slotted aluminium Fowler flap and Frise-type aileron on each trailing-edge. Fixed tabs.

FUSELAGE: Flying-boat type, with single-step hull and rear boom to support tail unit. Conventional aluminium stressed-skin construction.

TAIL UNIT: Cantilever type, of aluminium stressed-skin construction, with single sweptback fin and rudder. Variable-incidence tailplane, actuated by screwjack. No tabs.

Trident TR-1 Trigull-320 prototype (Teledyne Continental Tiara 6-320 engine)

LANDING GEAR: Fuselage hull and independently retractable wingtip floats for landing on water. Manually-retractable water rudder, extending from air rudder. Retractable tricycle-type wheeled gear for operation on land, with single wheel on each unit. Hydraulic retraction, both of floats and of wheeled gear. Main wheels retract outward into wings, nosewheel (which is steerable) upward to lie semi-recessed in nose to act as bumper. Oleo-pneumatic shock-absorbers and Cleveland brakes. Main wheels and tube-type tyres size 7·00-6, nosewheel 6·00-6.

POWER PLANT: One 239 kW (320 hp) Teledyne Continental Tiara 6-320 flat-six fuel-injection engine, driving a Hartzell three-blade constant-speed reversible-pitch metal pusher propeller. Fuel in single bag-type tank in lower hull, capacity 378 litres (100 US gallons). Refuelling point in hull. Oil capacity 11·4 litres (3 US gallons).

ACCOMMODATION: Seating for pilot and up to five passengers, in three pairs, in enclosed, heated and ventilated glassfibre cabin. Access via large forward-hinged door on each side and forward-hinged bow door on starboard side. Space for 68 kg (150 lb) of baggage in rear of cabin and below floor. Alternative layouts for use as ambulance (one stretcher and one medical attendant) or freighter.

SYSTEMS AND EQUIPMENT: Hydraulic system for landing gear, wingtip floats and flap actuation. Electrical system includes 24V 50A alternator. Basic blind-flying instrumentation standard. Radio and other equipment to customer's specification.

DIMENSIONS: EXTERNAL:

Wing span:	
floats up	12·73 m (41 ft 9 in)
floats down	11·73 m (38 ft 6 in)
Wing chord (constant)	1·83 m (6 ft 0 in)
Length overall	8·69 m (28 ft 6 in)
Height overall	3·81 m (12 ft 6 in)
Wheel track	3·66 m (12 ft 0 in)
Wheelbase	3·35 m (11 ft 0 in)
Propeller diameter	2·13 m (7 ft 0 in)
Passenger doors (each):	
Height	1·14 m (3 ft 9 in)

Width 1·09 m (3 ft 7 in)
Bow door (stbd):
Height 0·51 m (1 ft 8 in)
Width 0·51 m (1 ft 8 in)
DIMENSIONS, INTERNAL:
Cabin: Max length 3·00 m (9 ft 10 in)
Max width 1·22 m (4 ft 0 in)
Max height 1·22 m (4 ft 0 in)
Baggage compartments: Volume:
rear of cabin 0·79 m³ (28 cu ft)
underfloor (total) 1·22 m³ (43 cu ft)
AREAS:
Wings, gross:
floats up 22·85 m² (246·0 sq ft)
floats down 21·18 m² (228·0 sq ft)
Ailerons (total) 0·89 m² (9·6 sq ft)
Trailing-edge flaps (total) 4·11 m² (44·2 sq ft)
Fin 2·91 m² (31·3 sq ft)
Rudder 0·91 m² (9·8 sq ft)
Tailplane 2·75 m² (29·6 sq ft)
Elevators (total) 2·44 m² (26·3 sq ft)
WEIGHTS AND LOADINGS:
Basic operating weight, empty 1,134 kg (2,500 lb)
Max T-O weight 1,723 kg (3,800 lb)
Max wing loading:
floats up 75·40 kg/m² (15·45 lb/sq ft)
floats down 81·35 kg/m² (16·67 lb/sq ft)
Max power loading 7·21 kg/kW (11·88 lb/hp)
PERFORMANCE (at max T-O weight):
Never-exceed speed 183 knots (339 km/h; 211 mph)
Max level speed at S/L
146 knots (270 km/h; 168 mph)
Max cruising speed (75% power) at 1,980 m (6,500 ft)
143 knots (265 km/h; 165 mph)
Cruising speed (65% power)
135 knots (249 km/h; 155 mph)
Stalling speed at S/L, flaps and landing gear down
49 knots (91 km/h; 56·5 mph)
Max rate of climb at S/L 384 m (1,260 ft)/min
Service ceiling 5,485 m (18,000 ft)
T-O run at S/L, 20° flap, 15°C:
from land 158 m (520 ft)
from water 241 m (790 ft)
T-O to 15 m (50 ft), conditions as above:
from land 320 m (1,050 ft)
from water 427 m (1,400 ft)
Landing from 15 m (50 ft), conditions as above:
on land 396 m (1,300 ft)
on water 366 m (1,200 ft)
Landing run, conditions as above:
on land 174 m (570 ft)
on water 149 m (490 ft)
Range:
75% power 690 nm (1,279 km; 795 miles)
65% power 736 nm (1,364 km; 848 miles)
55% power 802 nm (1,486 km; 924 miles)
45% power 848 nm (1,572 km; 976 miles)

CHINA
(PEOPLE'S REPUBLIC)

STATE AIRCRAFT FACTORY

ADDRESS:
Shenyang (formerly Mukden)
DIRECTOR: Professor Hsue Shen Tsien

This factory had its origin in the Mukden plant of the Manshu Aeroplane Manufacturing Company, one of several aircraft and aero-engine manufacturing facilities established in Manchukuo (Manchuria) by the Japanese invaders in 1938. After the Communist regime became responsible for the whole of mainland China in 1949 the Manchurian factories were re-established and re-equipped with Soviet assistance. Today the factories at Shenyang, Chungking and Harbin are the main centres of Chinese aircraft and aero-engine production, with design and development centres at Shenyang, Peking and Harbin.

In the middle and late 1950s the Shenyang factory produced in large numbers under licence several aircraft types, including the Soviet An-2, Yak-12 and Yak-18 and the Mi-4 helicopter, and the Czech Super Aero 45 light transport.

First combat aircraft manufactured at Shenyang, under licence, was the Soviet MiG-17 fighter, with deliveries to the Chinese Air Force beginning in 1956. Well over a thousand MiG-17s were built, under the Chinese designation F-4, plus several hundred MiG-15UTI (F-2) fighter/trainers, before production was completed in the mid-1960s. F-4s were exported to Albania (30), Cambodia and North Vietnam. These types have been followed by Chinese versions of the MiG-19 (F-6) and MiG-21 (F-8), and by the Il-28 bomber, production of which may still continue at a modest rate. The Chairman of the US Joint Chiefs of Staff told the Senate Armed Services Committee in early 1975 that the Chinese Air Force of the People's Liberation Army had an operational home defence fighter force of more than 3,500 MiG-17s, -19s and -21s, and had more than 400 Il-28s still in operational service. He added that China may plan to equip some of the Il-28s for a nuclear attack role, and that the fighter force had a severe shortage of all-weather interceptors. A copy of the Tupolev Tu-16 twin-jet medium bomber was also produced in China, and a twin-jet fighter designated F-9, embodying experience gained with the F-6, has been developed in China.

The capability of China's aircraft industry has been revealed most openly by study of the F-6 single-seat day fighters supplied to Pakistan. Generally similar to the Soviet MiG-19SF, the F-6s equipped three first-line squadrons of the Pakistan Air Force (Nos. 11, 23 and 25) at the time of the 1971 war with India. They were credited with the destruction of twelve Indian aircraft, made up of one MiG-21, eight Su-7s and three Hunters, for the loss of three F-6s.

An assessment of the F-6 by a western observer described the general standard of workmanship of the airframe as very good. At low altitudes this fighter was said to outmanoeuvre any type of combat aircraft in service in Asia except the F-86, and to outclimb the MiG-21 and F-104 Starfighter. The potential of the Pakistani F-6s has been much enhanced by supplementing their standard cannon armament with two Sidewinder missiles.

Il-28 bomber of the Chinese Air Force taxying out for a night exercise

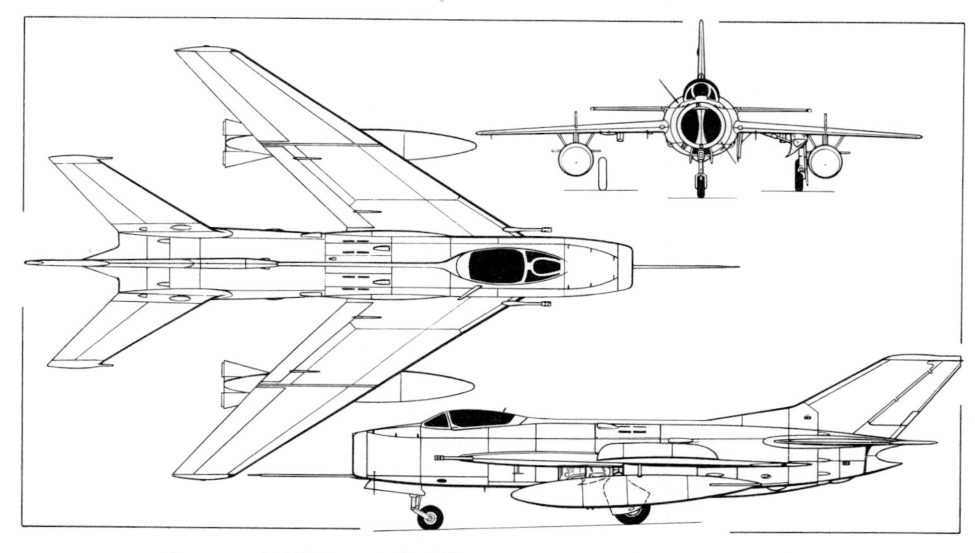

Shenyang F-6 (Chinese-built MiG-19SF) single-seat day fighter *(Pilot Press)*

SHENYANG F-6

NATO reporting names: *Farmer-C* **(MiG-19SF) and** *Farmer-D* **(MiG-19PF)**

The F-6 is basically a MiG-19 fighter built in the Chinese State Aircraft Factory. Its original design was initiated by the Mikoyan bureau in the early 'fifties, with the aim of producing the first Soviet fighter able to exceed Mach 1 in level flight. Construction of a prototype, designated I-350 at the time, was authorised on 30 July 1951. Powered by two 19·62 kN (4,410 lb st) Mikulin AM-5 turbojets, this aircraft was flown for the first time by Major Grigori Sedov in September 1953. It achieved its maximum speed of Mach 1·1 in level flight on several occasions before being handed over for state trials in early 1954.

The initial production MiG-19 day fighter began to enter service with the Soviet air defence force in early 1955. Before long an all-moving tailplane replaced the former, ineffective, elevators. This new version, which also had three 30 mm guns (instead of the original armament of one 37 mm and two 23 mm cannon) and introduced an attachment under each wing for a bomb or air-to-surface rocket, was known as the MiG-19S (for *Stabilisator*).

Meanwhile, Vladimir Klimov's bureau had been developing a new turbojet designated R-9. Of similar overall dimensions to the small-diameter AM-5, it had a considerably higher rating, and was adopted as the standard power plant of the MiG-19 in 1957. Again the aircraft's designation was changed, to MiG-19SF (*Forsirovanny*; increased power). At the same time, another version with limited all-weather capability was put into production as the MiG-19PF (*Perekhvatchik*; interceptor), with a small *Izumrud* (Emerald) radar scanner inside its engine air intake and a ranging unit in the intake top lip. The later MiG-19PM (*Modifikatsirovanny*; modified) differed from the PF in having four first-generation radar-homing missiles (NATO *Alkali*) instead of guns.

In the Soviet Union the MiG-19 was phased out of production by the end of the 'fifties, although many SFs and PMs remain in service in the Warsaw Pact countries, Cuba, Iraq and Egypt. Some had been delivered to China before the deterioration of Moscow-Peking relations and, with great skill, these were copied down to the last detail so that assembly lines of MiG-19s and their R-9B turbojets could be set up at Shenyang. The designation F-6 was given to the resulting fighter, which first flew in December 1961 and from the following year became standard equipment in the Chinese Air Force of the People's Liberation Army. Many hundreds (perhaps as many as 1,500) were built subsequently and, as the accompanying Chinese official photograph shows, included counterparts of both the MiG-19PF and MiG-19SF versions. Production of the F-6 is believed to be continuing.

Immediately after the Indo-Pakistan war of September 1965, China offered F-6s to Pakistan, which had an urgent need of replacements. Forty were supplied initially and

Two Chinese counterparts of the Soviet MiG-19PF all-weather fighter taxi past standard F-6s (MiG-19SFs), the nearest of which has large underwing weapon carriers

despite problems such as poor component interchangeability resulting from hand manufacture, and the fact that the spares and servicing handbooks were in Chinese, the first PAF squadron was operational within a year. Subsequent deliveries brought to about 90 the total of F-6s acquired by Pakistan, and by the Spring of 1974 the Air Wing of the Tanzanian People's Defence Force had received sufficient F-6s for a single squadron.

The following description is based on known details of the basic MiG-19SF, modified where possible to apply specifically to the Chinese F-6:

TYPE: Single-seat day interceptor and air-superiority fighter.

WINGS: Cantilever mid-wing monoplane of all-metal construction. Wing section TsAGI S-12S at root, SR-7S at tip. Anhedral 4° 30'. Sweepback at quarter-chord 55°. Entire trailing-edge of each wing formed by aileron (outboard) and large Fowler-type flap, both hydraulically powered. Compressed-air emergency extension system for flaps. Trim tab in port aileron. Large full-chord boundary layer fence above each wing at mid-span to enhance aileron effectiveness.

FUSELAGE: Conventional all-metal semi-monocoque structure of circular section, with divided air intake in nose and side-by-side twin orifices at rear. Top and bottom 'pen-nib' fairings aft of nozzles. Entire rear fuselage detaches at wing trailing-edge for engine servicing. Forward-hinged door-type airbrake, operated hydraulically, on each side of fuselage aft of wing trailing-edge. Forward-hinged perforated door-type airbrake under centre-fuselage. Shallow ventral fin strake under rear fuselage. Upward-hinged pitot boom mounted on lower lip of nose intake.

TAIL UNIT: Conventional all-metal structure. Hydraulically-actuated one-piece horizontal surfaces, with electrical emergency actuation in the event of hydraulic failure. Anti-flutter weight projecting forward from each tailplane tip. Stick-to-tailplane gearing, via electro-mechanical linkage, reduces required stick forces during high-*g* manoeuvres. Sweepback on vertical surfaces 57° 30'. Electrically-actuated trim tab in rudder. Large dorsal fin between fin and dorsal spine enclosing actuating rods for tail control surfaces.

LANDING GEAR: Wide-track tricycle type, with single wheel on each unit. Hydraulic actuation, nosewheel forward, main units inward into wing roots. Pneumatic emergency extension system. All units of levered-suspension type, with oleo-pneumatic shock-absorbers. Main-wheel tyres size 660 × 200 mm; max pressure 9·8 bars (142 lb/sq in). Nosewheel tyre size 500 × 180; pressure 6·9 bars (100 lb/sq in). Pneumatically-operated brakes on main wheels, with pneumatic emergency backup. Pneumatically-deployed brake parachute housed in bottom of rear fuselage above ventral fin strake. Small tail bumper.

POWER PLANT: Two Chinese-built versions of Klimov R-9B axial-flow turbojet, each rated at 25·5 kN (5,730 lb st) dry and 31·9 kN (7,165 lb st) with afterburning. Hydraulically-actuated nozzles. Two main fuel tanks in tandem between cockpit and engines, and two smaller tanks under forward end of engine tailpipes, with total capacity of 2,170 litres (477 Imp gallons). Provision for two 800 litre (176 Imp gallon) underwing drop-tanks, raising max total fuel capacity to 3,770 litres (829 Imp gallons).

ACCOMMODATION: Pilot only, on ejection seat (Martin-Baker Mk 10 zero-zero rocket-assisted type in Pakistani aircraft), under rearward-sliding blister canopy. In emergency canopy is jettisoned by an explosive charge at the lock, after which it is carried away by the slipstream. Fluid anti-icing system for windscreen. Cockpit pressurised, heated and air-conditioned.

SYSTEMS: Cockpit pressurised by air-conditioning system mounted in top of fuselage aft of cockpit, using compressor bleed air. Constant temperature maintained by adjustable electric thermostat. Two independent hydraulic systems. Main system, powered by pump on starboard engine, actuates landing gear retraction and extension, flaps, airbrakes and afterburner nozzle mechanism. System for tailplane and aileron boosters is powered by a pump on the port engine, and can also be supplied by the main system should the booster system fail. Electrical system powered by two DC starter/generators, supplemented by a battery, providing 27V DC, and 115V 400Hz and 36V 400Hz AC.

ELECTRONICS AND EQUIPMENT: Standard equipment includes VHF radio, blind-flying equipment, radio compass, radio altimeter, tail-warning system, navigation lights, taxying light on nosewheel unit and landing light in bottom of front fuselage.

ARMAMENT: Installed armament of two or three 30 mm NR-30 guns, one in each wing root and (not on MiG-19PF) one under starboard side of nose. Aircraft supplied to Pakistan have an attachment under each wing for a Sidewinder air-to-air missile, outboard of drop-tank. Alternatively, an attachment inboard of each tank for a bomb weighing up to 250 kg (or 500 lb), a rocket of up to 212 mm calibre, or a pack of eight air-to-air rockets. Optical gunsight. Gun camera in top lip of air intake of MiG-19SF; *Izumrud* airborne interception radar bullet in centre of nose intake of MiG-19PF, with ranging unit in top lip of intake.

DIMENSIONS, EXTERNAL:

Wing span	9·00 m (29 ft 6½ in)
Wing chord, mean	3·02 m (9 ft 10¾ in)
Wing aspect ratio	3·24
Thickness/chord ratio, mean	8·24%
Length overall (MiG-19SF):	
incl nose probe	14·90 m (48 ft 10½ in)
excl nose probe	12·60 m (41 ft 4 in)
Length of fuselage	11·82 m (38 ft 9½ in)
Height overall	4·02 m (13 ft 2¼ in)
Tailplane span	5·00 m (16 ft 4¾ in)
Wheel track	4·15 m (13 ft 7½ in)

AREAS:

Wings, gross	25·00 m² (269 sq ft)
Airbrakes (three, total)	1·50 m² (16·15 sq ft)
Ventral fin	0·614 m² (6·61 sq ft)

WEIGHTS AND LOADINGS:

Weight empty, nominal	5,760 kg (12,700 lb)
Normal T-O weight	7,600 kg (16,755 lb)
Max T-O weight	8,700 kg (19,180 lb)
Max wing loading	348 kg/m² (71·28 lb/sq ft)
Max power loading	136·4 kg/kN (1·34 lb/lb st)

PERFORMANCE:

Max level speed at 10,000 m (32,800 ft)	783 knots (1,452 km/h; 902 mph)
Cruising speed	512 knots (950 km/h; 590 mph)
Stalling speed, flaps up	189 knots (350 km/h; 218 mph)
Landing speed	127 knots (235 km/h; 146 mph)
Max rate of climb at S/L	6,900 m (22,635 ft)/min
Time to service ceiling	8 min 12 sec
Service ceiling	17,900 m (58,725 ft)
Absolute ceiling	19,870 m (65,190 ft)
T-O run, with afterburning	515 m (1,690 ft)
T-O run, with underwing tanks, no afterburning	900 m (2,953 ft)
T-O to 25 m (82 ft), with afterburning	1,525 m (5,000 ft)
T-O to 25 m (82 ft), with underwing tanks, no afterburning	1,880 m (6,170 ft)
Landing from 25 m (82 ft), with brake-chute	1,700 m (5,580 ft)
Landing from 25 m (82 ft), without brake-chute	1,980 m (6,495 ft)
Landing run, with brake-chute	600 m (1,970 ft)
Landing run, without brake-chute	890 m (2,920 ft)
Combat radius with external tanks	370 nm (685 km; 426 miles)
Normal range at 14,000 m (46,000 ft)	750 nm (1,390 km; 863 miles)
Max range with external tanks	1,187 nm (2,200 km; 1,366 miles)
Max endurance at 14,000 m (46,000 ft)	2 hr 38 min

SHENYANG F-8

Design of this Chinese version of the Mikoyan MiG-21 fighter was based initially on that of a number of Soviet-built MiG-21Fs *(Fishbed-Cs)* that had been delivered to China prior to the political break in 1960. The difficult task of copying the airframe, R-11 afterburning turbojet and equipment was completed so quickly and efficiently that the F-8 made its first flight in December 1964 and began to enter service with the Chinese Air Force in 1965. The design has been updated by reference to later-model Soviet-built aircraft (including the MiG-21PF *Fishbed-D*) despatched to North Vietnam via China.

Exports of the F-8 to Albania, and of 16 to the Air Wing of the Tanzanian People's Defence Force, were reported in 1974. In early 1975 General George S. Brown, USAF, told the Senate Armed Services Committee that China "has produced a number of MiG-21s, but for reasons which are not yet fully clear . . .production was suspended and only a small number of PRC (People's Republic of China) produced MiG-21s are operational with the PRC Air Force. The balance of the operational MiG-21s were Soviet-provided some years ago".

Aircrew of the Chinese Air Force with their Il-28U (NATO reporting name *Mascot*) tandem-cockpit training version of the Il-28 bomber

SHENYANG F-9

NATO reporting name: *Fantan-A*

In the FY 1977 report of the US Defense Department, Secretary Donald H. Rumsfeld remarked that "Tactical aviation in the PRCNAF (People's Republic of China Naval Air Force) also plays an air defense role relative to naval forces, with the *Beagle* bomber (Il-28) and *Fantan-A* fighter-bomber being the principal tactical aircraft."

Fantan-A is known to be the NATO reporting name for the F-9, a twin-engined fighter embodying technology derived from the F-6/MiG-19. Reports suggest that it first flew in the early 'seventies, and has lateral air intakes to permit use of a pointed nose radome. The F-9 is said to be somewhat larger overall than the F-6, with a wing span of about 10·20 m (33 ft 5 in), overall length of about 15·25 m (50 ft 0 in) and T-O weight of about 10,000 kg (22,050 lb). Combat radius is thought to be up to 430 nm (800 km; 500 miles), and max level speed almost Mach 2.

It is likely that a future version of the F-9, or a development of this aircraft, will be powered by the Rolls-Royce Spey turbofans that China is to manufacture under licence. Most F-9s delivered by 1976 serve with strike squadrons of the Chinese Air Force; others are operated by the Naval Air Force.

SHENYANG (TUPOLEV) Tu-16

The first Chinese-built Tu-16 is believed to have been flown in 1968, and a production rate of six per month was reported to have been achieved by the Spring of 1972. According to a statement in early 1975 by General George S. Brown, USAF, Chairman of the US Joint Chiefs of Staff, Tu-16 production in China "apparently ceased after some 60 . . .had been built, and it is too early to determine whether production will be resumed".

CHINESE HELICOPTER

It has been confirmed that at least one type of helicopter is currently being produced by the Chinese aircraft industry. It is not yet clear whether this is entirely of Chinese design, or whether it is a development of the Soviet Mi-4 which was built under licence in China some years ago.

CHINESE TRANSPORT AIRCRAFT (?)

A $2 million order for aero-engines, placed with Pratt & Whitney Aircraft of Canada by Machimpex, the Chinese foreign trading organisation, includes about six or seven examples of the 11·13 kN (2,500 lb st) JT15D turbofan engine. This has given rise to speculation that China may plan to build a twin-engined VIP or business transport aircraft with such a power plant, possibly in a rear-engine installation.

COLOMBIA

URDANETA Y GALVEZ
URDANETA Y GALVEZ LTDA

ADDRESS:
Carrera 9, No. 16-51, Of. 406, Apartado Aéreo No. 6876, Bogotá
Telephone: Bogotá 428301 or 430230
PRESIDENT:
Antonio Urdaneta
MANAGER:
Hector Galvez

This company, which was established in the 1950s and has since 1961 been a South American distributor for Cessna aircraft, is currently also assembling and partly building selected Cessna types under licence. Sixty-five aircraft were so produced in 1973, and a further 93 in 1974, some 40% of these being Cessna Model A188B AGwagons (see US section). Other types assembled and sold include the twin-engined Cessna 310. About 80% of production has been for customers in Colombia, the remainder being exported, chiefly to Ecuador and Peru.

In early 1975 the company had a work force of about 300 persons, and an output of 130 aircraft for the year was planned, to be increased to 200 in 1976. Production in 1974 was limited to wings, tail units and seats for the Cessna range, but welding and other techniques are being learnt, and the company hoped to be qualified to manufacture complete airframes by the end of 1975. Meanwhile, it is already qualified to repair and overhaul airframes, engines and propellers for all Cessna piston-engined aircraft and also has a limited capability to repair and overhaul electronics equipment for these aircraft.

CZECHOSLOVAKIA

Central direction of the Czechoslovak aircraft industry is by a body known as the Generální Reditelstvi Aero—Ceskoslovenské Letecke Podniky; Trust Aero—Czechoslovak Aeronautical Works, Prague-Letnany, whose General Manager is Josef Skaromlid. Principal factories concerned with aircraft manufacture are the Aero Vodochody National Corporation, Let National Corporation and Zlin Aircraft-Moravan National Corporation, whose current products appear under the appropriate headings in this section. Other Czechoslovak factories engaged in the production of aero-engines and sailplanes are listed in the relevant sections of this edition.

Sales of all aircraft products outside Czechoslovakia are handled by the Omnipol Foreign Trade Corporation, whose address is given below:

OMNIPOL
FOREIGN TRADE CORPORATION

ADDRESS:
Washingtonova 11, Prague 1
Telephone: 2126
Telex: 121489, 121808 and 121077

GENERAL MANAGER:
Tomás Marecek, GE
SALES MANAGER:
Maroslav Vesely
PUBLICITY MANAGER:
Jirí Matula

This concern handles the sales of products of the Czechoslovak aircraft industry outside Czechoslovakia and furnishes all information requested by customers with regard to export goods.

About 29,000 people are employed by the Czechoslovak aircraft industry.

AERO
AERO VODOCHODY NÁRODNÍ PODNIK (Aero Vodochody National Corporation)

ADDRESS:
Vodochody, p. Odelená Voda, near Prague
MANAGING DIRECTOR:
Jirí Chmelícek
VICE-DIRECTORS:
Ing Josef Sedlácek (Technical)
Jan Spára (Production)
Ing Otakar Stella (Sales)
Ing Václav Klouda (Works Economy)
CHIEF DESIGNER:
Dipl Ing Jan Vlcek
CHIEF PILOT:
Vlastimil David

This factory perpetuates the name of one of the three founder companies of the Czechoslovak aircraft industry, which began activities shortly after the first World War with the manufacture of Austrian Phönix fighters. Subsequent well-known products included the A 11 military general-purpose biplane and its derivatives, and licence manufacture of the French Bloch 200 twin-engined bomber. The present works was established on 1 July 1953, since when it has seven times received the Red Banner award of the Ministry of Engineering and UVOS, as well as many other awards including those of Exemplary Exporting Corporation and the Order of Labour.

Aero's major product from 1963-74 was the L-29 Delfin jet basic and advanced trainer, of which more than 3,000 were built. A full description of this can be found in the 1974-75 edition of *Jane's*. It has now been superseded in production by the L-39, a description of which follows:

AERO L-39

The L-39 basic and advanced jet trainer was developed in the Aero works at Vodochody by a team led by the chief designer, Dipl Ing Jan Vlcek. Two prototype airframes had been completed by 4 November 1968 when the 02 aircraft flew for the first time. The 01 airframe was utilised for structural testing. By the end of 1970, five flying prototypes and two for ground testing had been completed. Slightly larger and longer air intake trunks were fitted after preliminary flight tests.

A pre-production batch of 10 aircraft began to join the flight test programme in 1971, and series production started in late 1972, following official selection of the L-39 to succeed the L-29 (1974-75 *Jane's*) as the standard jet trainer of all Warsaw Pact countries except Poland. Service acceptance trials, in Czechoslovakia and the USSR, took place in 1973, and by the Spring of 1974 the L-39 had begun to enter service with the Czech Air Force.

The L-39 forms part of a comprehensive training system which includes a specially designed pilot training flight simulator (TL-39), a pilot ejection ground training simulator (NK-TL-29/39), and vehicle-mounted mobile automatic test equipment (AKZ-KL-39). The aircraft is capable of operation from unpaved or unprepared runways.

TYPE: Two-seat basic and advanced jet trainer.

WINGS: Cantilever low-wing monoplane, with 2° 30′ dihedral from roots. Wing section NACA 64A012 mod. 5. Incidence 2°. Sweepback at quarter-chord 1° 45′. One-piece all-metal stressed-skin structure, with all-metal hydraulically-operated double-slotted trailing-edge flaps. Small fence above and below each trailing-edge between flap and aileron. Trim tab in each aileron. Control surfaces actuated by pushrods. Flaps deflect 25° for take-off, 44° for landing; ailerons deflect 17° up or down; airbrakes deflect 55° downward. Non-jettisonable wingtip fuel tanks, incorporating landing lights.

FUSELAGE: All-metal semi-monocoque structure, built in two portions. Front portion consists of three sections, the first of which houses electrical and radio equipment and nose landing gear. Next comes the pressurised compartment for the crew. The third section contains fuel tanks and the engine bay. The rear fuselage, carrying the tail unit, can be removed quickly to provide access for engine servicing. Two airbrakes side by side under fuselage, just forward of wing leading-edge.

TAIL UNIT: Conventional all-metal cantilever structure, with sweepback on vertical surfaces. Variable-incidence tailplane. Control surfaces actuated by pushrods. Trim tab in each elevator.

LANDING GEAR: Retractable tricycle type, with single wheel on each unit. Hydraulic retraction, main wheels inward into wings, nosewheel rearward into fuselage. Oleo-pneumatic shock-absorbers and low-pressure tyres on all units. Hydraulic disc brakes on main wheels. Pneumatic ram-air system for emergency extension.

POWER PLANT: One 16·87 kN (3,792 lb st) Walter Titan (Motorlet-built Ivchenko AI-25-TL) turbofan engine mounted in rear fuselage, with semi-circular lateral air intake, fitted with splitter plate, on each side of fuselage above wing centre-section. Fuel in rubber bag-type main tanks aft of cockpit, capacity 824 kg (1,816 lb), and two non-jettisonable wingtip tanks with total capacity of 156 kg (344 lb).

ACCOMMODATION: Crew of two in tandem on zero-height ejection seats beneath individual transparent canopies which hinge sideways to starboard. Seats ensure safe ejection at speeds above 81 knots (150 km/h; 94 mph). Dual controls standard.

SYSTEMS: Cabin pressurised (differential 0·23 bars; 3·4 lb/sq in) and air-conditioned. High-pressure hydraulic system for landing gear retraction and control of flaps, airbrakes and wheel brakes. Pneumatic, electrical, fire extinguishing and de-icing systems.

ELECTRONICS AND EQUIPMENT: Standard equipment includes RTL-11 VHF com, crew intercom, RKL-41 ADF, RV-5 radio altimeter, MRP-56-P/S marker beacon receiver, and IFF.

ARMAMENT: Electrically-controlled ASP-3-NMU-39 gunsight and FKP-2-2 gun camera standard. Provision for underwing bombs, rocket pods or air-to-air missiles.

DIMENSIONS, EXTERNAL:

Wing span	9·46 m (31 ft 0½ in)
Wing chord (mean)	2·15 m (7 ft 0½ in)
Wing aspect ratio (geometric)	4·4
Length overall	12·32 m (40 ft 5 in)
Height overall	4·72 m (15 ft 5½ in)
Tailplane span	4·40 m (14 ft 5 in)
Wheel track	2·44 m (8 ft 0 in)
Wheelbase	4·39 m (14 ft 4¾ in)

AREAS:

Wings, gross	18·8 m² (202·4 sq ft)
Ailerons (total)	1·23 m² (13·26 sq ft)
Flaps (total)	2·68 m² (28·89 sq ft)
Airbrakes (total)	0·50 m² (5·38 sq ft)
Fin	2·77 m² (29·78 sq ft)
Rudder	0·71 m² (7·68 sq ft)
Tailplane	3·93 m² (42·30 sq ft)
Elevators, incl tabs	1·14 m² (12·27 sq ft)

WEIGHTS AND LOADINGS:

Weight empty	3,330 kg (7,341 lb)
Normal T-O weight	4,100 kg (9,039 lb)
Max T-O weight	4,600 kg (10,141 lb)
Max wing loading	219 kg/m² (44·85 lb/sq ft)
Max power loading	272·7 kg/kN (2·67 lb/lb st)

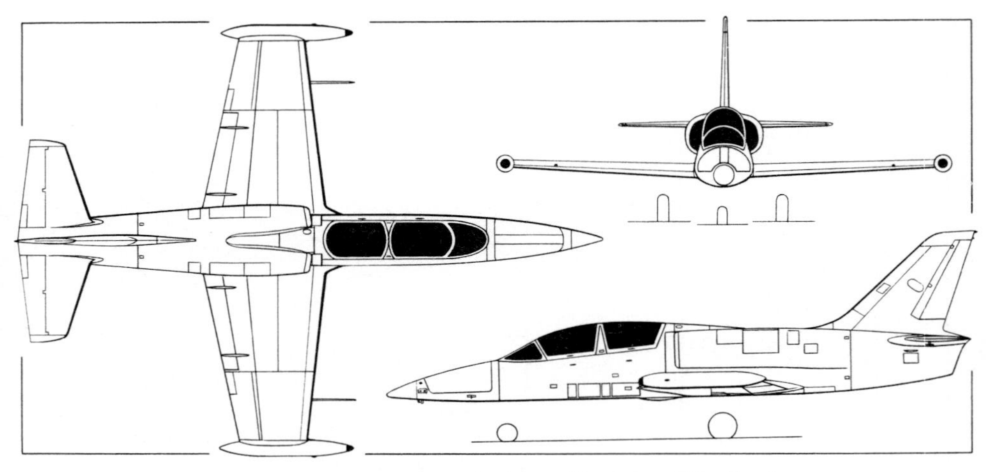

Aero L-39 two-seat basic and advanced jet trainer *(Pilot Press)*

Above and below: two views of Aero L-39 jet trainers in service with the Czechoslovakian Air Force

PERFORMANCE (at normal T-O weight):

Max limiting Mach number	0·80
Max level speed at 5,000 m (16,400 ft)	405 knots (750 km/h; 466 mph)
Max level speed at S/L	378 knots (700 km/h; 435 mph)
Cruising speed at 5,000 m (16,400 ft)	367 knots (680 km/h; 423 mph)
Stalling speed, flaps up	97 knots (180 km/h; 112 mph)
Stalling speed, 25° flap	90 knots (165 km/h; 103 mph)
Stalling speed, 44° flap	84 knots (155 km/h; 97 mph)
Max rate of climb at S/L	1,320 m (4,330 ft)/min
Optimum climbing speed	210 knots (390 km/h; 242 mph)
Service ceiling	11,300 m (37,075 ft)
T-O run (25° flap): on concrete	450 m (1,475 ft)
on grass	560 m (1,835 ft)
T-O to 15 m (50 ft) on concrete	665 m (2,180 ft)
Landing from 15 m (50 ft) on concrete	880 m (2,885 ft)
Landing run (44° flap): on concrete	620 m (2,035 ft)
Max range at 5,000 m (16,400 ft) with 824 kg (1,816 lb) fuel	491 nm (910 km; 565 miles)
Max endurance	1 hr 55 min
g limits	+8; −4

LET

LET NÁRODNÍ PODNIK (Let National Corporation)

Address:
Uherské Hradiste-Kunovice
Telephone: Uherské Hradiste 5121
Telex: 060180 and 060181

Managing Director:
Ing Josef Kurz

Chief Designer:
Ing Ladislav Smrcek

Chief Pilot:
Vladimír Vlk

The Let plant at Kunovice was established in 1950, its early activities including licence production of the Soviet Yak-11 piston-engined trainer under the Czechoslovak designation C-11. It also contributed to the production of the Aero 45 and L 200 Morava twin-engined air taxi aircraft, and is currently responsible for development and manufacture of the L-410 twin-turboprop light transport, the Z-37 Cmelák agricultural aircraft, and the L 13 Blanik sailplane.

The factory also produces apparatus and equipment for radar and computer technology.

Three of the fleet of L-410A Turbolets in service with the Czech domestic operator Slov-Air

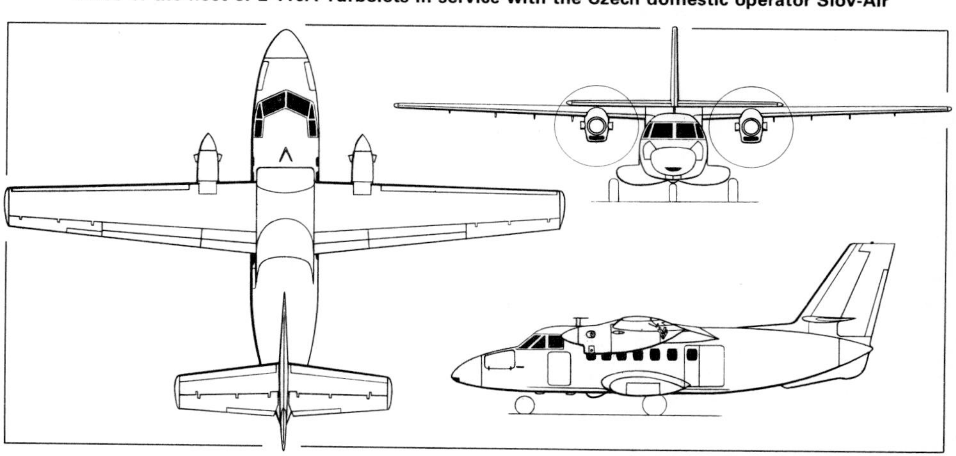

Let L-410A Turbolet twin-turboprop 15/19-passenger light transport *(Pilot Press)*

LET L-410 TURBOLET

The L-410 is a twin-turboprop light transport, intended primarily for use on local passenger and freight services. It is suitable also for executive, aerial survey, radio/navigation training, ambulance and other duties, and can operate from airfields with a natural grass surface.

Design of the L-410 was started in 1966, by a team led by Ing Ladislav Smrcek. The first prototype (OK-YKE), powered by Pratt & Whitney Aircraft of Canada PT6A-27 turboprop engines, flew for the first time on 16 April 1969. Three additional PT6A-engined prototypes were completed subsequently.

Forty Turbolets, including the four prototypes, had been built by the Summer of 1975.

The following versions have been announced:

L-410A. Initial passenger/cargo production version, powered by Pratt & Whitney Aircraft of Canada PT6A-27 engines. First deliveries were to the domestic operator Slov-Air, with which it entered service at Bratislava in late 1971 on scheduled, non-scheduled and charter services. This airline has 12 L-410As. Five underwent hot and cold weather trials, and route evaluation, in the USSR between Spring and Autumn 1973. Intended primarily for western markets.

L-410AF. Aerial photography/survey version, announced in mid-1974 and displayed at the 16th International Engineering Fair at Brno during that year. Generally similar to L-410A, but has a larger, wider and extensively glazed nose section in which are a vertically-mounted camera and an inward-facing seat for the navigator/camera operator. Modifications prevent nose-wheel from being retracted in flight. One example of this version exported to Hungary in 1974.

L-410M. Version with Motorlet M 601 A turboprop engines. Scheduled to enter service in 1975; primarily for non-western markets. Two aircraft used for certification programme in 1974-75.

The following description applies to the current standard L-410A version:

Type: Twin-turboprop light passenger and freight transport.

Wings: Cantilever high-wing monoplane. Wing section NACA 63A418 at root, NACA 63A412 at tip. Dihedral 1° 45′. Incidence 2° at root, −0° 30′ at tip. No sweepback at front spar. Conventional all-metal two-spar structure, attached to fuselage by four-point mountings. Chemically-machined skin with longitudinal reinforcement. All-metal ailerons with electrically-controlled trim tab in port aileron. No spoilers. Double-slotted hydraulically actuated metal flaps, with both slots variable. Kléber-Colombes pneumatic de-icing of leading-edges.

Fuselage: Conventional all-metal semi-monocoque structure.

Tail Unit: Cantilever all-metal structure of conventional semi-monocoque type. Sweptback vertical surfaces, with small dorsal fin and curved ventral fin. One-piece tailplane, mounted part-way up fin. Manually-controlled trim tab in each elevator. Rudder has electrically-actuated trim tab. Kléber-Colombes pneumatic de-icing of leading-edges.

Landing Gear: Retractable tricycle type with single wheel on each unit. Hydraulic retraction, nosewheel forward, main wheels inward to lie flat in fairing on each side of fuselage. Technometra Radotin oleo-pneumatic shock-absorbers. Non-braking nosewheel, with servo-assisted steering, fitted with 548 × 221 mm (9·00-6) tubeless tyre, pressure 2·74 bars (39·8 lb/sq in). Main wheels fitted with 718 × 306 mm (12·50-10) tubeless tyres, pressure 3·14 bars (45·5 lb/sq in). All wheels manufactured by Moravan Otrokovice, tyres by Rudy Rijen, Gottwaldov. Moravan Otrokovice hydraulic disc brakes on main wheels. No anti-skid units. Metal ski landing gear, with plastics undersurface, optional.

Power Plant: Two 533 kW (715 ehp) Pratt & Whitney Aircraft of Canada PT6A-27 turboprop engines in L-410A, each driving a Hamilton Standard LF-23 Type 343 or Hartzell HC-B3TN-3D reversible-pitch fully-feathering three-blade propeller. De-icing for propeller blades (electrical) and lower intakes. Six (optionally eight) bag-type fuel tanks in wings, with total capacity (eight tanks) of 1,300 litres (286 Imp gallons). Four standard refuelling points above wings, with provision for two extra points when all eight tanks are fitted. Usable oil capacity 5·6 litres (1·25 Imp gallons) for each engine. Alternative installation, in L-410 M, of two 549 kW (736 ehp) M 601 A turboprop engines, each driving an Avia V 508 A three-blade hydraulically-adjustable and reversible-pitch constant-speed propeller. Fuel system as for L-410A. Oil capacity 10 litres (2·2 Imp gallons).

Accommodation: Crew of one or two on flight deck. Dual controls standard. Standard accommodation in main cabin for 15 to 19 passengers, with pairs of adjustable seats on starboard side of aisle and single seats opposite, all at 762 mm (30 in) pitch. Alternative layouts include de luxe seating for 12 passengers in individual chairs; an executive layout with eight individual seats, four work desks, and a wardrobe; and an ambulance version accommodating six stretchers, five sitting patients and a medical attendant. Baggage compartment in nose with two separate doors; toilet and additional baggage compartment at rear. Double upward-opening doors aft on port side, right hand door serving as passenger entrance and exit; both doors open for cargo loading. Both doors can be removed for paratroop training missions. Downward-opening crew door, forward on starboard side, serves also as emergency exit. All-cargo version has protective floor covering, crash nets on each side of cabin, and tiedown provisions; floor is at truckbed height. Standard passenger version can be quickly and easily converted to all-cargo configuration, and vice versa. Cabin heated and ventilated by engine bleed air.

Systems: No air-conditioning or pressurisation systems. Isopressure duplicated hydraulic system, pressure 147 bars (2,133 lb/sq in), for flap and landing gear actuation. Electrical system includes two 28V 6kW DC starter/generators, three 36V 400Hz rotary inverters and two Varley 25Ah lead-acid storage batteries. No APU or oxygen systems.

Electronics and Equipment: Standard equipment includes cockpit, instrument and passenger cabin lights, navigation lights, cabin-mounted fire extinguisher, three landing lights in nose (one in L-410AF, plus second light in port main-wheel well), and windscreen wipers. Optional equipment includes two Mesit (LUN 3524) VHF; two King KDF 800 or Collins DF 203 ADF; one Collins 51Z6 marker; two Collins 51RV2B or RCA AVN 210 VOR/ILS; anti-collision lights; wing and tail de-icing system (Kléber-Colombes pneumatic); and windscreen de-icing.

Dimensions, external:

Wing span	17·48 m (57 ft 4¼ in)
Wing chord at root	2·534 m (8 ft 3¾ in)
Wing chord at tip	1·267 m (4 ft 1¾ in)
Wing aspect ratio	9·3
Length overall:	
L-410A and M	13·61 m (44 ft 7¾ in)
L-410AF	13·54 m (44 ft 5¼ in)
Length of fuselage	12·89 m (42 ft 2¾ in)
Height overall	5·65 m (18 ft 6½ in)
Tailplane span	6·77 m (22 ft 2½ in)
Wheel track	3·65 m (11 ft 11½ in)
Wheelbase	3·67 m (12 ft 0½ in)
Propeller diameter:	
PT6A-27	2·59 m (8 ft 6 in) *or* 2·49 m (8 ft 2 in)
M 601	2·49 m (8 ft 2 in)
Propeller ground clearance	1·12 m (3 ft 8 in)
Distance between propeller centres	4·82 m (15 ft 9¾ in)
Passenger/cargo door (port, aft):	
Height	1·30 m (4 ft 3¼ in)
Width overall	1·25 m (4 ft 1¼ in)
Width (passenger door only)	0·75 m (2 ft 5½ in)
Height to sill	0·80 m (2 ft 7½ in)
Crew door/emergency exit door (stbd, fwd):	
Height	1·05 m (3 ft 5¼ in)
Width	0·66 m (2 ft 2 in)
Height to sill	0·80 m (2 ft 7½ in)
Nose baggage compartment doors (each):	
Height	0·43 m (1 ft 5 in)
Width	0·74 m (2 ft 5 in)
Height to sill	1·30 m (4 ft 3¼ in)

DIMENSIONS, INTERNAL:
Cabin, excl flight deck:
Length 6·25 m (20 ft 6 in)
Max width 1·92 m (6 ft 3½ in)
Max height 1·658 m (5 ft 5¼ in)
Floor area 9·69 m² (104·3 sq ft)
Volume 18·00 m³ (635 cu ft)
Baggage compartment volume (nose)
1·1 m³ (38·8 cu ft)
Baggage compartment volume (rear):
15 passengers 2·5 m³ (88·3 cu ft)
19 passengers 1·6 m³ (56·5 cu ft)

AREAS:
Wings, gross 32·86 m² (353·70 sq ft)
Ailerons (total) 2·248 m² (24·2 sq ft)
Trailing-edge flaps (total) 5·92 m² (63·7 sq ft)
Fin 3·74 m² (40·26 sq ft)
Rudder, incl tab 2·78 m² (29·92 sq ft)
Tailplane 6·60 m² (71·04 sq ft)
Elevators, incl tabs 2·96 m² (31·86 sq ft)

WEIGHTS AND LOADINGS:
Basic weight empty 3,400 kg (7,495 lb)
Operating weight empty (cargo version)
3,570 kg (7,870 lb)
Max fuel load 1,100 kg (2,425 lb)
Max payload (cargo version) 1,850 kg (4,078 lb)
Max T-O weight 5,700 kg (12,566 lb)
Max landing weight 5,500 kg (12,125 lb)
Max zero-fuel weight 5,290 kg (11,662 lb)
Max wing loading 173·5 kg/m² (35·53 lb/sq ft)
Max power loading:
PT6A-27 5·35 kg/kW (8·79 lb/ehp)
M 601 5·19 kg/kW (8·54 lb/ehp)

PERFORMANCE (at max T-O weight, ISA, except where indicated):
Never-exceed speed
216 knots (400 km/h; 248 mph) EAS
Max cruising speed at 3,000 m (9,845 ft)
200 knots (370 km/h; 230 mph) TAS
Econ cruising speed (80% power) at 3,000 m (9,845 ft) 194 knots (360 km/h; 224 mph)TAS
Stalling speed, flaps up
83 knots (152 km/h; 95 mph) EAS
Stalling speed, flaps down, at max landing weight
64 knots (118 km/h; 73 mph) EAS
Max rate of climb at S/L 492 m (1,615 ft)/min
Rate of climb at S/L, one engine out
96 m (315 ft)/min
Service ceiling 7,100 m (23,300 ft)
Service ceiling, one engine out 2,800 m (9,175 ft)
Min ground turning radius 5·40 m (17 ft 8¾ in)
T-O run 496 m (1,627 ft)
T-O to 15 m (50 ft) 564 m (1,850 ft)
Landing from 15 m (50 ft) at max landing weight
592 m (1,942 ft)
Landing run at max landing weight 273 m (896 ft)
Range with max fuel (eight tanks), 30 min reserves
701 nm (1,300 km; 807 miles)
Range with max payload (two pilots, IMC conditions) 161 nm (300 km; 186 miles)

LET Z-37A CMELÁK (BUMBLE-BEE)

Design of the Cmelák began in August 1961, and the first XZ-37 prototype of this agricultural aircraft flew for the first time on 29 March 1963. Ten prototypes were built altogether. Certification in the Normal category, Aerial Work Class D, BCAR, was awarded on 20 June 1966. Additional applications for the production Cmelák include mail and cargo transport during the Winter season. The aircraft has been certificated for operation with ski landing gear and for glider towing.

The improved Z-37A version, with fixed instead of adjustables louvres, was introduced in early 1971 and is now the standard production version. It features some structural reinforcement, increased use of corrosion-resistant materials and other modifications designed to extend operational life (some agricultural operators claim to have achieved as much as 4,000 flying hours before overhaul). All main structural members are wrapped in textile material for added protection against chemical corrosion. Duralumin fittings on the hopper and wing flaps, and covering between the hopper and wing centre-section, are replaced by fittings and coverings of stainless steel.

A total of 600 Cmeláks of all versions had been built by the beginning of 1975, for customers in Bulgaria, Czechoslovakia, Finland, Germany (Democratic Republic), Hungary, India, Iraq, Poland, the UK and Yugoslavia. This total included 27 examples of the two-seat training version, which is designated **Z-37A-2.**

The following description applies to the standard Z-37A:

TYPE: Agricultural monoplane.

WINGS: Cantilever low-wing monoplane. Wing section NACA 33015 at root, NACA 43012A at tip. Dihedral 7° on outer panels only. Incidence 3° at root, 0° at tip. All-metal single-spar fail-safe structure, with auxiliary rear spar, comprising centre-section, built integrally with fuselage, and two outer panels. Centre-section is strengthened in the vicinity of the main landing gear by comparison with original Z-37. Fabric-covered hermetically-sealed aluminium slotted ailerons. Pneumatically-operated double-slotted aluminium flaps. Leading-edge fixed slats of aluminium alloy on outer wings, in line with ailerons.

L-410AF aerial survey version of the Turbolet, with modified nose

FUSELAGE: Welded steel tube fail-safe structure. Engine, cockpit and underfuselage covered in dural sheet, remainder fabric-covered. Rear end of fuselage is detachable to facilitate cleaning.

TAIL UNIT: Cantilever aluminium alloy structure. Fin and tailplane metal-covered; rudder and elevator fabric-covered and hermetically sealed. Trim tabs in elevator and rudder.

LANDING GEAR: Non-retractable tailwheel type. Technometra N. C. Semily oleo-pneumatic shock-absorbers. Moravan Otrokovice wheels and Rudy Rijen low-pressure tyres and tubes. Main-wheel tyres size 556 × 163, pressure 2·21 bars (32 lb/sq in); anti-shimmy tail-wheel, tyre size 290 × 110, pressure 2·79 bars (40·5 lb/sq in). Moravan Otrokovice hydraulic shoe-type brakes on main wheels. Chemical deflector fitted to starboard main wheel, eliminating the need for rubber protection for the shock-absorbers and undercarriage bracing struts. Provision for fitting wooden skis, with pneumatically-actuated hydraulic brakes.

POWER PLANT: One 235 kW (315 hp) M 462 RF nine-cylinder radial aircooled engine, driving an Avia V 520 two-blade constant-speed metal propeller. Aluminium alloy fuel tank in port wing centre-section is standard; provision for optional tank in starboard side of centre-section to give total capacity of 250 litres (55 Imp gallons). For carrying replenishment fuel to operating site, two externally-suspended tanks, each of 125 litres (27·5 Imp gallons) capacity, can be carried. Similar tanks can also be used to transport chemicals in concentrated form. Refuelling points above port wing centre-section. Oil capacity 13·5 litres (2·97 Imp gallons). Anti-dust filter in front of carburettor air intake. Provision for pre-heating intake air.

ACCOMMODATION: Pilot in enclosed cockpit forward of hopper. One auxiliary seat behind hopper for mechanic or loader. Cabin ventilated and heated by ram air and heat exchanger. Forward-opening door on starboard side.

SYSTEMS: Pneumatic system for engine starter, flaps, parking brake and hopper actuation. Electrical power provided by 28V 1·5kW generator, 24V 10Ah battery and 36V 400Hz converter.

ELECTRONICS AND EQUIPMENT: ADF and Mesit LUN 3522 optional. Hopper for 650 litres (143 Imp gallons) of spray or 600 kg (1,323 lb) of dust. Spray system and distributor for dry chemicals interchangeable. Total volume available for chemical hopper or cargo 1·8 m³ (63·5 cu ft). Effective swath width with the aircraft flying 5 m (16 ft) above the ground is 35 m (115 ft) for oily spray, 20 m (66 ft) for aqueous spray; at flying height of 15-20 m (50-65 ft), effective swath width is 20-25 m (66-82 ft) for granules and 40 m (130 ft) for dust.

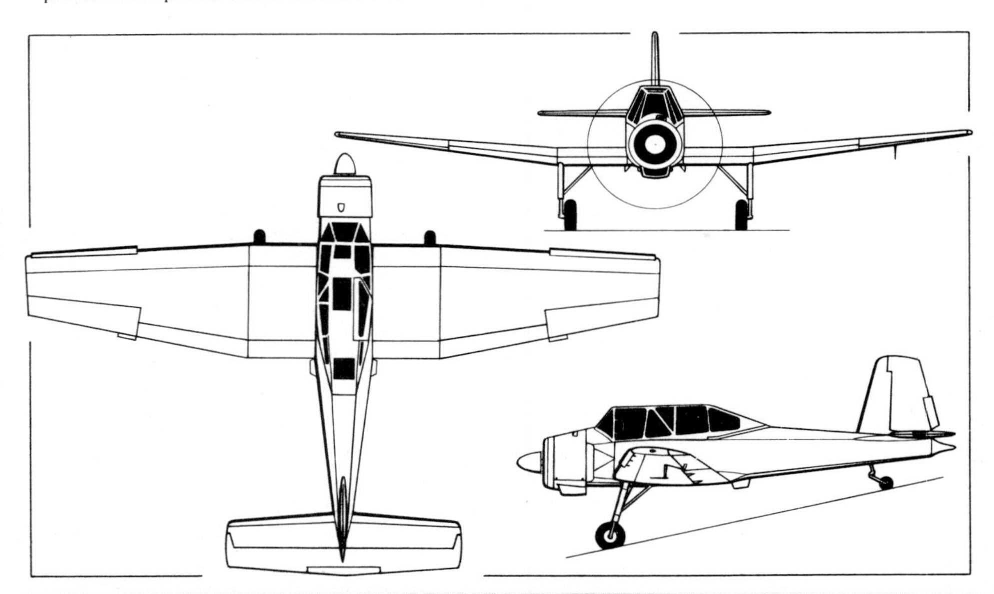

Photograph and three-view drawing *(Pilot Press)* **of the Let Z-37A-2 two-seat training version of the Z-37A Cmelák agricultural aircraft**

DIMENSIONS, EXTERNAL:

Wing span	12·22 m (40 ft 1¼ in)
Wing chord at root	2·32 m (7 ft 7¼ in)
Wing chord at tip	1·16 m (3 ft 9¾ in)
Wing aspect ratio	6·3
Length overall	8·55 m (28 ft 0½ in)
Height overall	2·90 m (9 ft 6 in)
Tailplane span	4·53 m (14 ft 10½ in)
Wheel track	3·30 m (10 ft 9¾ in)
Wheelbase	5·50 m (18 ft 0½ in)
Propeller diameter	2·70 m (8 ft 10½ in)
Propeller ground clearance	0·35 m (1 ft 1¾ in)

DIMENSIONS, INTERNAL:
Pilot's cockpit:

Width at chest level	0·97 m (3 ft 2¼ in)
Height (seat to roof)	1·01 m (3 ft 3¾ in)
Height (floor to roof)	1·22 m (4 ft 0 in)

Mechanic's compartment:

Width	0·915 m (3 ft 0 in)
Height (seat to roof)	1·11 m (3 ft 7¾ in)

AREAS:

Wings, gross	23·8 m² (256·2 sq ft)
Ailerons (total)	2·05 m² (22·07 sq ft)
Trailing-edge flaps (total)	4·37 m² (47·04 sq ft)
Vertical tail surfaces (total)	2·05 m² (22·07 sq ft)
Horizontal tail surfaces (total)	5·03 m² (54·14 sq ft)

WEIGHTS AND LOADINGS:

Weight empty, standard equipment, without agricultural equipment	1,043 kg (2,295 lb)
Max chemicals	600 kg (1,323 lb)
Max T-O weight: freight version	1,750 kg (3,855 lb)
agricultural version	1,850 kg (4,080 lb)
Max wing loading (agricultural)	77·7 kg/m² (15·93 lb/sq ft)
Max power loading (agricultural)	7·87 kg/kW (12·95 lb/hp)

Let Z-37A Cmelák agricultural aircraft, photographed prior to delivery to a customer in India

PERFORMANCE (at max T-O weight. A: freight version; B: agricultural version):

Never-exceed speed	145 knots (270 km/h; 167 mph)
Max level speed (without application equipment)	113 knots (210 km/h; 130 mph)
Cruising speed at 1,500 m (4,920 ft): A	99 knots (183 km/h; 114 mph)
B	92 knots (170 km/h; 106 mph)
Operating speed, agricultural operations: B	65 knots (120 km/h; 75 mph)
Stalling speed, flaps up	49 knots (90 km/h; 56 mph)
Stalling speed, flaps down	45 knots (81 km/h; 51 mph)
Max rate of climb at S/L: A	282 m (925 ft)/min
B	222 m (728 ft)/min
Service ceiling: A	4,000 m (13,125 ft)
Min ground turning radius	5·68 m (18 ft 7¾ in)
T-O run: A	125 m (410 ft)
B	150 m (492 ft)
Landing run: A	100 m (328 ft)
B	122 m (400 ft)
Range with reserves for 1 hour's flying plus 10%: A	345 nm (640 km; 398 miles)

VZLU
VYZKUMNY A ZKUSEBNÍ LETECKY USTAV (Aeronautical Research and Test Institute)

ADDRESS:
Beranovych 130, 19905, Prague 9-Letnany

Telephone: Prague 827041 and 826541
Telex: Prague 1493

MANAGING DIRECTOR:
Ing J. Havlicek

This Institute, whose title is self-explanatory, was founded in 1922 and undertakes a range of activities corresponding broadly to those carried out by the RAE in Britain. Details of its principal facilities appeared in the 1970-71 and 1972-73 *Jane's*. It is a member of the Czechoslovak aircraft manufacturing group, under the general management of Aero (which see).

ZLIN
MORAVAN NÁRODNÍ PODNIK (Zlin Aircraft Moravan National Corporation)

ADDRESS:
Otrokovice
Telephone: Gottwaldov 92 2041-44
Telex: Gottwaldov 067 334

MANAGING DIRECTOR:
Frantisek Klapil

VICE-DIRECTORS:
Ing Stanislav Machálka (Technical)
Jan Munclinger (Production)
Frantisek Muzny (Sales)
Ing Adolf Dolezal (Works Economy)

CHIEF DESIGNER:
Ing Jirí Navrátil

CHIEF PILOT:
Vlastimil Berg

The Moravan works, responsible for production of the famous range of Zlin aerobatic and light touring aircraft, was formed originally on 8 July 1935 as Zlinská Letecká Akciová Spolecnost (Zlin Aviation Joint Stock Co) in Zlin, although manufacture of Zlin aircraft was actually started two years earlier by the Masarykova Letecká Liga (Masaryk League of Aviation). The factory was renamed Moravan after the second World War. In 1967 it was awarded the FAI Diploma of Honour in recognition of its work in the design and manufacture of training and aerobatic aircraft. At present, in addition to production of the Zlin 42 M, Zlin 43 and Zlin 726, Moravan is building fuselages for the Let Z-37A Cmelák and items of aircraft equipment.

ZLIN 42 M

The prototype Zlin 42 (OK-41) was the first of a new series of small sporting and touring aircraft developed by the Moravan works at Otrokovice. Intended for basic and advanced training, aerobatic training (solo or dual), navigation training, sport, touring and glider towing, it was first flown on 17 October 1967, and in 1969 began undergoing flight trials prior to certification. Standard power plant of the initial version, which entered production in 1971 and was described in the 1973-74 *Jane's*, was the 134 kW (180 hp) M 137 A engine, with which the Z 42 conforms to FAR Pt 23 airworthiness specifications in the Aerobatic category and is suitable for service in climates with temperatures between +40° and −20°C. The principal customer for this version was the German Democratic Republic, which acquired several dozen.

The following description applies to the subsequent **Zlin 42 M** production version, of which a prototype was first flown in November 1972, certification under FAR Pt 23 in the Aerobatic (+6*g* to −3·5*g*) and Normal categories being obtained in 1973. This differs from the initial version in having a constant-speed propeller and a tail fin identical to that of the Z 43. Production of the Z 42 M began in 1974, and was continuing in 1976. By early 1976, total Zlin 42 production (all versions) was in excess of 120.

Zlin 42 M two-seat light training and touring aircraft (Avia M 137 AZ engine)

TYPE: Two-seat light training and touring aircraft.

WINGS: Cantilever low-wing monoplane. Wing section NACA $63_2416{\cdot}5$. Dihedral 6° from roots. Sweepforward 4° 20′ at quarter-chord. All-metal structure with single main spar. All-metal slotted ailerons and flaps all have same dimensions. Mass-balanced flaps and ailerons, operated mechanically by control rods. Ground-adjustable tab in each aileron.

FUSELAGE: Engine cowlings of sheet metal. Centre fuselage of welded steel tube truss construction, covered with laminated glassfibre panels. Rear fuselage is all-metal semi-monocoque structure.

TAIL UNIT: Cantilever all-metal structure. Control surfaces have partial mass and aerodynamic balance. Trim tabs on elevator and rudder. Rudder actuated by control cables, elevator by control rods.

LANDING GEAR: Non-retractable tricycle type, with nosewheel offset to port. Oleo-pneumatic nosewheel shock-absorber. Main wheels carried on flat spring steel legs. Nosewheel steering by means of rudder pedals. Single wheel on each unit. Main wheels and Barum tyres size 420 × 150, pressure 1·86 bars (27 lb/sq in); nosewheel and Barum tyre size 350 × 135, pressure 2·45 bars (35·6 lb/sq in). Hydraulic brakes on main wheels can be operated from either seat. Parking brake standard. Wheel fairings and skis optional.

POWER PLANT: One 134 kW (180 hp) Avia M 137 AZ inverted six-cylinder aircooled in-line engine, with low-pressure injection pump, driving a two-blade Avia V 503 A fully-automatic constant-speed propeller. Fuel tanks in each wing leading-edge, with total capacity of 130 litres (28·5 Imp gallons). Fuel and oil systems permit inverted flying for up to 3 minutes.

ACCOMMODATION: Individual side-by-side seats for two

persons, the pilot's seat being to port. Both are adjustable for height and permit the use of back-type parachutes. Baggage space aft of seats. Cabin and windscreen heating and ventilation. Forward-opening door on each side of cabin. Dual controls standard.

SYSTEMS: Electrical system includes a 600W 27V engine-driven generator and 25Ah 27V Varley battery. External power source can be used for starting the engine.

ELECTRONICS AND EQUIPMENT: VHF radio and IFR instrumentation optional.

DIMENSIONS, EXTERNAL:
- Wing span 9·11 m (29 ft 10¾ in)
- Wing chord (constant) 1·42 m (4 ft 8 in)
- Length overall 7·07 m (23 ft 2¼ in)
- Height overall 2·69 m (8 ft 10 in)
- Tailplane span 2·90 m (9 ft 6 in)
- Wheel track 2·33 m (7 ft 7¾ in)
- Wheelbase 1·66 m (5 ft 5¼ in)
- Propeller diameter 2·00 m (6 ft 6¾ in)

DIMENSIONS, INTERNAL:
- Cabin:
  - Length 1·80 m (5 ft 10¾ in)
  - Width 1·12 m (3 ft 8 in)
  - Height 1·20 m (3 ft 11¼ in)
- Baggage space 0·2 m³ (7·1 cu ft)

AREAS:
- Wings, gross 13·15 m² (141·5 sq ft)
- Ailerons (total) 1·408 m² (15·16 sq ft)
- Trailing-edge flaps (total) 1·408 m² (15·16 sq ft)
- Fin 0·54 m² (5·81 sq ft)
- Rudder, incl tab 0·81 m² (8·72 sq ft)
- Tailplane 1·23 m² (13·24 sq ft)
- Elevator, incl tab 1·36 m² (14·64 sq ft)

WEIGHTS AND LOADINGS:
- Basic weight, empty 645 kg (1,422 lb)
- Max T-O weight:
  - Aerobatic 920 kg (2,028 lb)
  - Normal 970 kg (2,138 lb)
- Max wing loading:
  - Aerobatic 70 kg/m² (14·3 lb/sq ft)
  - Normal 74 kg/m² (15·2 lb/sq ft)
- Max power loading:
  - Aerobatic 6·87 kg/kW (11·27 lb/hp)
  - Normal 7·24 kg/kW (11·88 lb/hp)

PERFORMANCE (at max Aerobatic T-O weight):
- Max level speed at 600 m (1,975 ft), ISA 122 knots (226 km/h; 140 mph)
- Cruising speed at 600 m (1,975 ft), ISA 116 knots (215 km/h; 134 mph)
- Stalling speed, flaps down, power off 49 knots (89 km/h; 56 mph)
- Max rate of climb at S/L 312 m (1,025 ft)/min
- Service ceiling 4,250 m (13,950 ft)
- T-O run 250 m (820 ft)
- T-O to 15 m (50 ft) 380 m (1,245 ft)
- Landing from 15 m (50 ft) 410 m (1,345 ft)
- Landing run 135 m (443 ft)
- Range with max standard fuel 286 nm (530 km; 329 miles)
- g limits:
  - Aerobatic +6·0; −3·5
  - Normal +3·8; −1·5

ZLIN 43

The Zlin 43 was first flown in prototype form on 10 December 1968.

It is designed primarily for advanced navigation, night and all-weather flying training, but is also suitable for sports and competitive flying, touring and aerial taxi flying, ambulance work (Z 43 S), basic aerobatics (solo or dual) and (with a wooden propeller) glider towing.

The Z 42 M and Z 43 have some 80% of their structural components in common, the Z 43 differing principally in power plant and in having an enlarged centre section in the fuselage to accommodate a bigger, four-seat cabin with more comprehensive instrumentation.

Certification under FAR 23, in the Utility and Normal categories, has been obtained. Production began in 1972, and 60 Zlin 43s had been completed by the end of 1973. Recent deliveries have included aircraft for military service in the Czechoslovak and East German Air Forces. Production continues in 1976.

TYPE: Two/four-seat light training and touring aircraft.

WINGS: Cantilever low-wing monoplane. Wings are of greater span and area than those of Z 42 M, but are otherwise similar except that they have a flat centre-section and no sweep. Tab on each aileron.

FUSELAGE: Similar to that of Z 42 M, but with additional steel tube section inserted in centre to permit incorporation of larger cabin.

TAIL UNIT: As Z 42 M.

LANDING GEAR: As Z 42 M, but with some reinforcement of the nosewheel unit and strengthened spring steel legs on main units. Wheel and tyre sizes as Z 42 M; tyre pressure 2·45 bars (35·6 lb/sq in) on all units. Hydraulic disc brakes on main wheels. Optional streamline wheel fairings for all units.

POWER PLANT: One 157 kW (210 hp) Avia M 337 A inverted six-cylinder aircooled in-line engine, with supercharger for take-off and climb, normally driving an Avia V 500 A two-blade constant-speed metal propeller. Wooden propeller on glider towing version. Fuel tanks in each wing leading-edge, with total capacity of 130 litres (28·5 Imp gallons). Standard additional tanks in each wingtip, each of 55 litres (12 Imp gallons) capacity. Fuel and oil systems permit inverted flight (restricted to maximum of 5 consecutive seconds at negative load factors).

ACCOMMODATION: Individual side-by-side seats for two persons in front of cabin, the pilot's seat being to port. Both are adjustable longitudinally and for height, and have tilting backs. Bench seat in rear of cabin for two additional passengers, with baggage space to rear of this seat. Forward-opening door on each side of cabin. Cabin and windscreen heating (by heat exchange system) and ventilation standard. Additional baggage compartment in rear of fuselage, with external access. Dual controls optional.

SYSTEMS: As Z 42 M.

ELECTRONICS AND EQUIPMENT: Standard Z 43 is equipped with instrumentation for day and night flying under VMC conditions. Optional items include full radio/navigation equipment, and instrumentation for various training roles, and for flight under IFR conditions.

DIMENSIONS, EXTERNAL:
- Wing span 9·76 m (32 ft 0¼ in)
- Wing chord (constant) 1·42 m (4 ft 8 in)
- Length overall 7·75 m (25 ft 5 in)
- Height overall 2·91 m (9 ft 6½ in)
- Tailplane span 3·00 m (9 ft 10 in)
- Wheel track 2·44 m (8 ft 0 in)
- Wheelbase 1·75 m (5 ft 9 in)
- Propeller diameter 1·95 m (6 ft 4¾ in)

DIMENSIONS, INTERNAL:
- Cabin:
  - Length 2·50 m (8 ft 2½ in)
  - Width 1·12 m (3 ft 8 in)
  - Height 1·20 m (3 ft 11¼ in)
- Baggage space (inside cabin) 0·2 m³ (7·1 cu ft)
- Baggage compartment (rear) 0·25 m³ (8·8 cu ft)

AREAS:
- As for Z 42 M, except:
  - Wings, gross 14·50 m² (156·1 sq ft)
  - Tailplane 2·59 m² (27·88 sq ft)
  - Elevator, incl tab 1·36 m² (14·64 sq ft)

WEIGHTS AND LOADINGS:
- Basic weight empty, equipped:
  - Normal and Utility 730 kg (1,609 lb)
- Max T-O weight:
  - Normal 1,350 kg (2,976 lb)
  - Utility 1,000 kg (2,204 lb)
- Max wing loading:
  - Normal 93·2 kg/m² (19·1 lb/sq ft)
  - Utility 69·0 kg/m² (14·1 lb/sq ft)
- Max power loading:
  - Normal 8·60 kg/kW (14·15 lb/hp)
  - Utility 6·37 kg/kW (10·49 lb/hp)

PERFORMANCE (at max Normal category T-O weight except where indicated):
- Never-exceed speed:
  - Normal 147 knots (273 km/h; 169·5 mph)
  - Utility 157 knots (292 km/h; 181 mph)
- Max level speed at S/L 127 knots (235 km/h; 146 mph)
- Max permissible manoeuvring speed:
  - Utility 120 knots (223 km/h; 138·5 mph)
- Cruising speed 113 knots (210 km/h; 130 mph)
- Stalling speed, flaps up 63·5 knots (117 km/h; 72·5 mph)
- Stalling speed, flaps down 56 knots (103 km/h; 64 mph)
- Max rate of climb at S/L 210 m (689 ft)/min
- Service ceiling 3,800 m (12,465 ft)
- T-O to 15 m (50 ft) 700 m (2,297 ft)
- Landing from 15 m (50 ft) 590 m (1,936 ft)
- Max range (standard fuel) 325 nm (610 km; 375 miles)

Z 43 S ambulance version of the Zlin 43 two/four-seat light aircraft

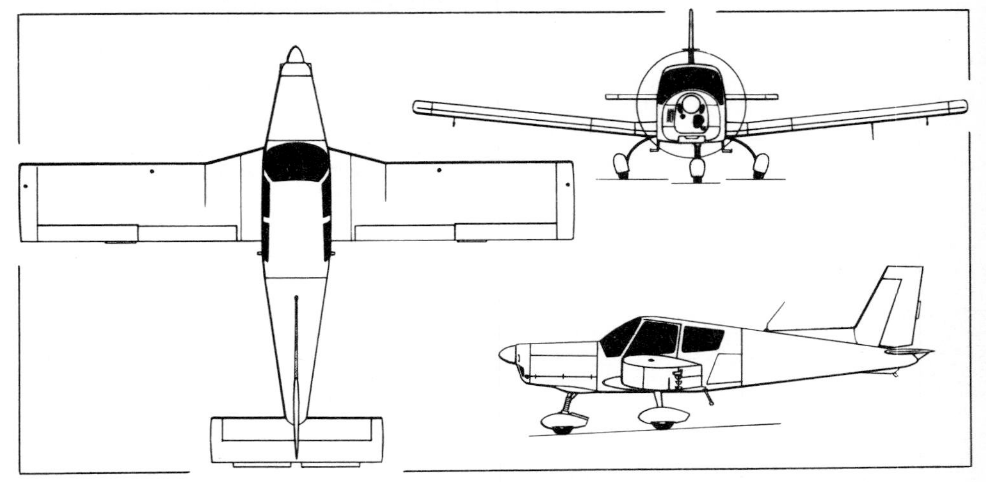

Zlin 43 two/four-seat light training and touring aircraft *(Pilot Press)*

Max range (with wingtip tanks)
620 nm (1,150 km; 714 miles)

g limits:
- Normal: +3·8; −1·52
- Utility: +4·4; −1·76

ZLIN Z 50 L

The Z 50 L is a fully-aerobatic single-seat competition light aircraft, the design of which began in 1973. Construction of a prototype was started in 1975, and this aircraft flew for the first time on 18 July 1975. Two additional prototypes and seven production aircraft had been completed by 1 March 1976.

TYPE: Single-seat aerobatic aircraft.

WINGS: Cantilever low-wing monoplane. Wing section NACA 0018 at root, NACA 0012 at tip. Dihedral 1° 7′ 24″. All-metal structure, with single continuous main spar, rear auxiliary spar, and aluminium-clad duralumin skin. All-metal mass-balanced ailerons, actuated by pushrods, occupy most of each trailing-edge, and have 20° travel (± 1°) up and down. Ground-adjustable tabs. No flaps. Provision for fitting wingtip fuel tanks for cross-country flights.

FUSELAGE: All-metal semi-monocoque stressed-skin structure, with duralumin skin.

TAIL UNIT: Conventional metal structure. Braced tailplane and fin duralumin-covered, balanced elevators and rudder fabric-covered. Mechanically-adjustable tab in port elevator, ground-adjustable tab on rudder. Elevators actuated by pushrods, rudder by cables. Elevator travel 30° (± 1°) up and down, rudder 30° (± 2°) to left and right.

LANDING GEAR: Non-retractable tailwheel type. Main wheels carried on flat-spring titanium cantilever legs. Mechanical main-wheel brakes actuated by rudder pedals. Fully-castoring tailwheel, with flat-spring shock-absorption, has automatic locking device to maintain aircraft on a straight track during taxying, take-off and landing. Main wheels size 350 × 135 mm, pressure 2·5 bars (36 lb/sq in); tailwheel size 122 × 60 mm. Streamlined main-wheel fairings optional.

POWER PLANT: One 194 kW (260 hp) Lycoming AEIO-540-D4B5 flat-six engine, without reduction gear, driving a Hoffmann HO-V123/K/200AH three-blade constant-speed variable-pitch wooden propeller with spinner. Single main fuel tank in fuselage, aft of firewall, capacity 60 litres (13·2 Imp gallons). Auxiliary 50 litre (11 Imp gallon) tank can be attached to each wingtip for cross-country flights only. Fuel and oil systems designed for full aerobatic manoeuvres, including inverted flight.

ACCOMMODATION: Single seat under fully-transparent sideways-opening bubble canopy, which can be jettisoned in an emergency. Seat and backrest are adjustable, and permit the use of a back-type parachute. Cockpit ventilated by sliding panel in canopy.

SYSTEM: Electrical single-conductor system only, utilising an alternator as main power source and a storage battery for standby power. External power socket in fuselage side for engine starting.

ELECTRONICS AND EQUIPMENT: VHF radio optional.

DIMENSIONS, EXTERNAL:

Wing span	8·58 m (28 ft 1¾ in)
Wing span over tip-tanks	9·03 m (29 ft 7½ in)
Wing chord at root	1·73 m (5 ft 8¼ in)
Wing chord at tip	1·21 m (3 ft 11¾ in)
Wing mean aerodynamic chord	1·4853 m (4 ft 10½ in)
Wing aspect ratio	5·88
Length overall (tail up)	6·512 m (21 ft 4½ in)
Length overall (tail down)	6·62 m (21 ft 8¾ in)
Height over tail (static)	1·86 m (6 ft 1¼ in)
Elevator span	3·44 m (11 ft 3½ in)
Wheel track	1·90 m (6 ft 2¾ in)
Propeller diameter	2·00 m (6 ft 6¾ in)

AREAS:

Wings, gross	12·50 m² (134·55 sq ft)
Ailerons (total)	2·80 m² (30·14 sq ft)
Fin	0·59 m² (6·35 sq ft)
Rudder, incl tab	0·81 m² (8·72 sq ft)
Tailplane	1·66 m² (17·87 sq ft)
Elevators (total, incl tabs)	1·20 m² (12·92 sq ft)

WEIGHTS AND LOADINGS (Aerobatic category):

Weight empty, equipped	570 kg (1,256 lb)
Max T-O weight	720 kg (1,587 lb)
Max wing loading	57·6 kg/m² (11·80 lb/sq ft)
Max power loading	3·71 kg/kW (6·11 lb/hp)

PERFORMANCE (at max Aerobatic T-O weight):

Never-exceed speed	181·5 knots (337 km/h; 209 mph)
Max level speed at 500 m (1,640 ft)	158 knots (293 km/h; 182 mph)
Max cruising speed at 500 m (1,640 ft)	143 knots (265 km/h; 165 mph)
Econ cruising speed at 500 m (1,640 ft)	129 knots (240 km/h; 149 mph)
Stalling speed	53 knots (98 km/h; 61 mph)
Max rate of climb at S/L	720 m (2,360 ft)/min
Service ceiling	6,000 m (19,675 ft)
T-O run	100 m (330 ft)
T-O to 15 m (50 ft)	220 m (722 ft)
Landing from 15 m (50 ft)	500 m (1,640 ft)
Landing run	200 m (656 ft)
Range with max fuel (incl wingtip tanks)	345 nm (640 km; 397 miles)

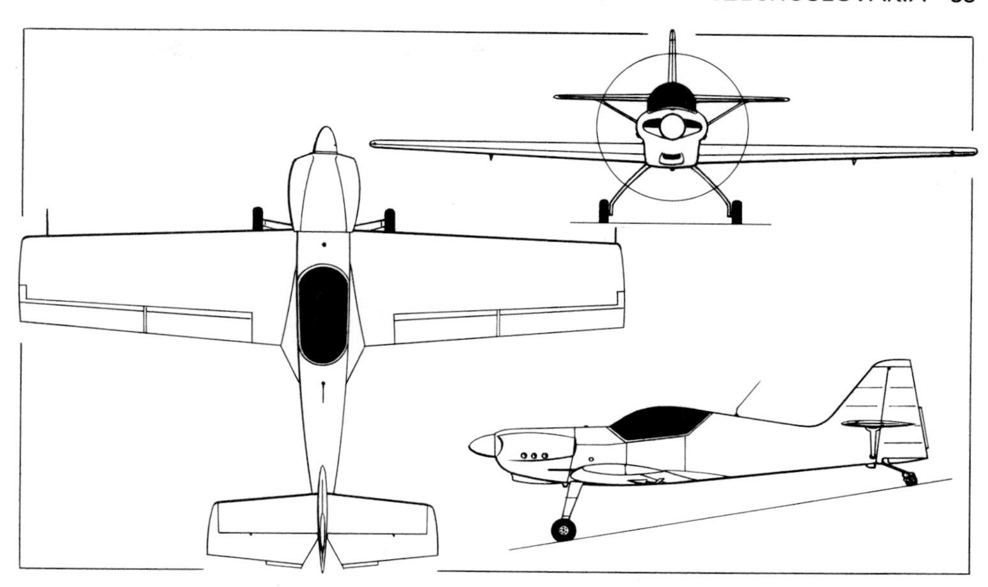

Three-view drawing *(Pilot Press)* **and photograph of second prototype of the Zlin Z 50 L**

g limits: +9; −6

ZLIN Z 526 F TRENER

The Z 526 F is one of the aircraft of the Trener/Akrobat series. Major modification, compared with the basic Z 526 (see 1968-69 *Jane's*), is the installation of an Avia M 137 A engine in place of the Walter Minor 6-III engine in the Z 526.

The prototype was flown in the Autumn of 1968 and the Z 526 F was certificated in 1969. More than 150 had been built by the beginning of 1974; production ended in 1975. A full description and illustration can be found in the 1975-76 *Jane's*.

ZLIN Z 726 UNIVERSAL

The Z 26/126/226/326/526/726 series has been built at Otrokovice since 1947 and has operated in 40 countries. Total production of the Zlin Trener and Akrobat series is more than 1,600, including 163 Z 26s, 166 Z 126s, 360 Z 226s, 421 Z 326s, and more than 300 of the Z 526 series. Sporting successes have included first place in the First, Second, Third and Fifth World Aerobatic Championships in 1960, 1962, 1964 and 1968, first place in the Lockheed Trophy aerobatic competition in Britain in 1957, 1958, 1961, 1963, 1964 and 1965, and first place in the Léon Biancotto Trophée aerobatic competition in France in 1965, 1967 and 1969.

The Z 726, of which the first of two prototypes flew for the first time in March 1973, is generally similar to the Z 526 F (1975-76 *Jane's*) except for having a different power plant, shorter-span wings with new root fillets, and metal-covered rudder and elevators. It was certificated to FAR Pt 23 in the Aerobatic and Normal categories in February 1974, and is now in production.

Two versions are available, as follows:

Z 726. Standard version, with 134 kW (180 hp) Avia M 137 AZ engine for basic and advanced training, aerobatic training and glider towing.

Z 726 K. More powerful version with 157 kW (210 hp) Avia M 337 AK supercharged engine and V 500 A variable-pitch propeller.

The following description applies to the standard Z 726, except where indicated:

TYPE: Two-seat basic trainer.

WINGS: Cantilever low-wing monoplane. Wings of combined NACA 2418 and NACA 4412 section. Dihedral 4° 30′ from roots. Sweepback 9° at quarter-chord. All-metal two-spar structure with flush-riveted light alloy stressed skin. All-metal ailerons, statically and aerodynamically balanced, are operated differentially. All-metal trailing-edge flaps. Flaps and ailerons actuated mechanically by control rods. Ground-adjustable tabs on ailerons.

FUSELAGE: Welded steel tube structure. Forward portion covered with easily-removable duralumin panels and remainder with fabric.

TAIL UNIT: Cantilever type. Removable tailplane and fin of all-metal stressed-skin construction. Elevators and rudder also of all-metal construction. Trim tabs on rudder and each elevator. Rudder actuated by control cables, elevator by control rods.

LANDING GEAR: Tailwheel type, with retractable main units. Electrical retraction. Oleo-pneumatic shock-absorbers. Main wheels retract backward into wings. Tyres protrude in retracted position to reduce damage in event of wheels-up landing. Fully-castoring self-centering tailwheel, steerable 30° to either side of centreline. Barum tyres, size 420 × 150, pressure 2·16 bars (31·3 lb/sq in) on main wheels; size 260 × 85, pressure 2·45 bars (35·6 lb/sq in) on tailwheel. Hydraulic brakes on main wheels, actuated from either cockpit.

POWER PLANT: One 134 kW (180 hp) Avia M 137 AZ inverted six-cylinder aircooled in-line engine, with low-pressure injection pump, driving an Avia V 503 A fully-automatic two-blade constant-speed propeller. One fuel tank of 45 litres (9·9 Imp gallons) capacity in each wing root. Fuel and oil installation, designed for aerobatics, permits inverted flying for 3 minutes. Provision for fitting wingtip fuel tanks, with total capacity of 70 litres (15·4 Imp gallons).

ACCOMMODATION: Tandem seats under continuous sliding canopy which is jettisonable in an emergency. Adjustable seats and rudder pedals in both cockpits. Seat cushions may be replaced by seat-type parachutes. Windscreen frame reinforced as crash pylon. Can be flown solo from either cockpit. Instructor occupies rear seat in training version. Complete dual controls and instrumentation. Cabin ventilation.

SYSTEMS: Electrical system includes a 28V 600W engine-driven generator and 24V 25Ah Varley battery. External power source can be used for starting the engine.

ELECTRONICS AND EQUIPMENT: Optional equipment includes VKDC-1 (or alternative) radio, and glider towing gear.

DIMENSIONS, EXTERNAL:

Wing span	9·875 m (32 ft 4¾ in)

Wing span over tip-tanks 10·335 m (33 ft 11 in)
Wing chord at root 1·545 m (5 ft 0¾ in)
Length overall 7·975 m (26 ft 2 in)
Height overall 2·06 m (6 ft 9 in)
Tailplane span 3·00 m (9 ft 10 in)
Wheel track 1·76 m (5 ft 9¼ in)
Wheelbase 4·33 m (14 ft 2½ in)
Propeller diameter 2·00 m (6 ft 6¾ in)

DIMENSIONS, INTERNAL:
Cabin:
Max length 2·30 m (7 ft 6½ in)
Max width 0·65 m (2 ft 1½ in)
Max height 1·50 m (4 ft 11 in)

AREAS:
Wings, gross 14·89 m² (160·3 sq ft)
Ailerons (total) 1·25 m² (13·45 sq ft)
Trailing-edge flaps (total) 1·376 m² (14·81 sq ft)
Fin 0·49 m² (5·27 sq ft)
Rudder, incl tab 0·94 m² (10·12 sq ft)
Tailplane 1·42 m² (15·28 sq ft)
Elevators, incl tabs 1·07 m² (11·52 sq ft)

WEIGHTS AND LOADINGS:
Weight empty:
726 (Aerobatic), 726 K (Utility) 690 kg (1,521 lb)
726 (Normal), 726 K (Normal) 700 kg (1,543 lb)
Max T-O weight:
726 (Aerobatic), 726 K (Utility) 940 kg (2,072 lb)
726 (Normal), 726 K (Normal) 1,000 kg (2,204 lb)
Max wing loading:
726 (Aerobatic), 726 K (Utility) 63 kg/m² (12·9 lb/sq ft)
726 (Normal), 726 K (Normal) 67·5 kg/m² (13·82 lb/sq ft)
Max power loading:
726 (Aerobatic) 7·01 kg/kW (11·51 lb/hp)
726 K (Utility) 5·99 kg/kW (9·87 lb/hp)
726 (Normal) 7·46 kg/kW (12·24 lb/hp)
726 K (Normal) 6·37 kg/kW (10·50 lb/hp)

PERFORMANCE (A: Z 726 at max T-O weight of 1,000 kg; 2,204 lb. B: Z 726 K at T-O weight of 940 kg; 2,072 lb. All figures for ISA at S/L):
Max level speed:
A 127 knots (236 km/h; 147 mph)
B 147 knots (272 km/h; 169 mph)
Max cruising speed:
A 116 knots (216 km/h; 134 mph)
B 133 knots (247 km/h; 153 mph)
Stalling speed, flaps up:
A, B 58 knots (107 km/h; 67 mph)
Stalling speed, flaps down:
A, B 53 knots (98 km/h; 61 mph)
Max rate of climb at S/L:
A 300 m (985 ft)/min
B 420 m (1,380 ft)/min
Service ceiling:
A 4,500 m (14,775 ft)
B 5,500 m (18,050 ft)
T-O to 15 m (50 ft):
A 395 m (1,295 ft)
B 350 m (1,150 ft)
Landing from 15 m (50 ft):
A, B 440 m (1,445 ft)
Max range with standard fuel:
A 237 nm (440 km; 273 miles)
B 269 nm (500 km; 310 miles)
Max range with wingtip tanks:
A 425 nm (790 km; 490 miles)
B 485 nm (900 km; 559 miles)
g limits:
726 (Aerobatic) +6·0; −3·0
726 K (Utility) +4·4; −2·2
726 (Normal), 726 K (Normal) +3·8; −1·5

Zlin Z 726 Universal two-seat training aircraft (Avia M 137 AZ engine)

EGYPT

Negotiations between the Egyptian and British governments, concluded in mid-1975, resulted in approval for the Westland/Aérospatiale Lynx helicopter and its Rolls-Royce Gem turboshaft engine to be manufactured under licence in Egypt. Westland is to deliver 20 Lynx from British production before the start of licence production; the Anglo-Egyptian agreement covers a total of 250 Lynx and their engines, and the provision of British technical assistance in setting up the Egyptian production line.

Payment for Lynx/Gem production will be made by the Arab Military Industrialisation Organisation, from funds contributed by Kuwait, Qatar, Saudi Arabia and the United Arab Emirates. New premises to accommodate the Lynx/Gem production lines are understood to be under construction at Helwan, near Cairo. Use may also be made of existing facilities at Cairo and Heliopolis.

In the longer term, as part of a large-scale armaments modernisation programme, the AMIO is expected to finance the production of other types of military aircraft and aero-engines in Egypt, subject to suitable agreements being reached. Among those reportedly under consideration are the Hawker Siddeley Hawk, SEPECAT Jaguar, Dassault Mirage F1 and Dassault-Breguet/Dornier Alpha Jet combat aircraft and the Rolls-Royce/Turboméca Adour turbofan engine.

FINLAND

PIK—*See 'Sailplanes' section*

VALMET
VALMET OY KUOREVEDEN TEHDAS

OFFICE AND WORKS:
35600 Halli
Telephone: 942-82210 Exchange
Telex: 28269 Valku SF
MANAGER:
Heikki Mäntylä

Valmet Oy Kuoreveden Tehdas (Kuorevesi Works) is affiliated to Valmet Oy, a State-owned company consisting of several metal-working factories.

Kuoreveden Tehdas continues the traditions of Ilmailuvoimien Lentokonetehdas, which was established in 1921. It was formerly a part of the factory group Valmet Oy Tampere, from which it was separated in 1974. It is now an independent factory directly responsible to Valmet's Head Office in Helsinki, and is currently the largest aircraft industry establishment in Finland. Since 1922, Valmet Oy Kuoreveden Tehdas and its predecessors have built 29 different types of aircraft, of which 18 have been of Finnish design.

In addition to aircraft manufacture, the present activities of Valmet Oy Kuoreveden Tehdas include the overhaul and repair of military and civil aircraft, and the periodical maintenance and repair of piston engines and instruments. The factory had a covered area in 1976 of approximately 14,000 m² (150,695 sq ft).

Valmet Oy Linnavuoren Tehdas, at Siuro, also directly subordinate to Valmet's Head Office in Helsinki, is concerned primarily with the overhaul and repair of aircraft jet engines.

In April 1970 it was announced that Valmet Oy would be responsible for assembly of the 12 Saab 35XS Drakens ordered by Finland. The first of these (DK-201) was completed and handed over to the Finnish Air Force on 25 April 1974; the last was delivered, on schedule, on 28 July 1975.

The latest aircraft of Finnish design to be built by Valmet is the Leko-70 piston-engined trainer, flown for the first time in 1975. It was designed by Ilmailuteollisuuden Kehitysosasto (IKO), which was formed on 15 September 1970 following an agreement between Valmet and the Finnish Ministry of Commerce and Industry to establish an aeronautical research and development group at Valmet Oy Tampere.

VALMET LEKO-70

In late 1970 an aeronautical research and development group was established in Finland, its first major task being to study a Finnish Air Force requirement for a basic training aircraft to replace the Saab 91 Safir. After considering various alternatives it was decided to produce an entirely Finnish design to fulfil this need, and a development contract was placed with Valmet by the Finnish Air Force on 23 March 1973. The aircraft is known as the Leko-70, the name being an abbreviation of 'Lentokone', the Finnish word for 'aeroplane'.

The Leko-70 made its first flight, lasting 1 hr 5 min, at Kuorevesi on 1 July 1975, and initial test flights were described as "very encouraging". A second prototype is being used for static and fatigue testing; a full-size cockpit mockup and components for a third aircraft have also been completed.

The flight test programme is divided into a preliminary phase, after which a decision will be made on the aircraft's suitability for Finnish Air Force service, and on its development potential; and a second, certification, phase. If the Leko-70 is accepted, it is estimated that at least 30 will be required by the Finnish Air Force.

The Leko-70 is designed for aerobatic flying as a two-seater. In civil use, in Normal or Utility category, it is capable of seating two or three persons, depending upon the amount of baggage carried.

TYPE: Two-seat training or two/three-seat touring light aircraft.

WINGS: Cantilever low-wing monoplane. Wing section NACA 63_2A615 (modified). Dihedral 6° from roots. Single-spar structure, of constant chord except for forward-swept wing-root leading-edges, attached to fuselage by steel fittings. Riveted aluminium alloy skin. Electrically-operated slotted flaps, and slotted ailerons, on trailing-edges. Ailerons actuated by stainless steel control cables. Flaps and ailerons have fluted skins.

Ailerons on prototype have adjustable geared tabs.

FUSELAGE: Conventional aluminium alloy semi-monocoque structure of frames and longerons, with riveted skin. Welded steel tube engine mount and stainless steel firewall. Cockpit floor panels of bonded sandwich.

TAIL UNIT: Cantilever aluminium alloy structure, with fluted and riveted skin. Slight sweepback on vertical surfaces; shallow dorsal fin from rear of canopy to base of fin. Elevators and rudder are aerodynamically and mass balanced, and are actuated by stainless steel control cables. Geared trim and balance tabs in elevators, and geared trim tab in rudder.

LANDING GEAR: Non-retractable tricycle type. Cantilever sprung main legs. Telescopic nosewheel strut. Disc brakes. Provision for ski gear.

POWER PLANT: One 149 kW (200 hp) Lycoming IO-360-A1B6 flat-four engine, driving a two-blade constant-speed propeller with spinner; or the 200 hp AEIO-360-A1B6 version of this engine, in which case a Christen-801 inverted oil system is also fitted. Two bonded sandwich fuel tanks, one in each wing root ahead of main spar; total normal capacity 150 litres (33 Imp gallons), max capacity 190 litres (41·8 Imp gallons).

ACCOMMODATION: Side-by-side seats for instructor and pupil in trainer version, with integral longitudinal central console which serves also to reinforce fuselage floor. Dual controls standard, but instructor's or pupil's control column can be removed if desired. Windscreen and one-piece rearward-sliding fully-transparent canopy, with steel tube turnover frame. Provision for third seat at rear, which can be removed to make room for additional baggage. Cockpit heated and ventilated.

SYSTEM: 28V DC electrical system.

ELECTRONICS AND EQUIPMENT: Two VHF, one ADF and VOR/ILS standard.

DIMENSIONS, EXTERNAL:

Wing span	9·30 m (30 ft 6¼ in)
Wing chord (constant over most of span)	1·53 m (5 ft 0¼ in)
Wing aspect ratio	6
Length overall	7·30 m (23 ft 11½ in)
Tailplane span	3·60 m (11 ft 9¾ in)
Wheel track	2·30 m (7 ft 6½ in)
Wheelbase	1·60 m (5 ft 3 in)

AREAS:

Wings, gross	14·00 m² (150·70 sq ft)
Ailerons (total)	1·40 m² (15·07 sq ft)
Trailing-edge flaps (total)	2·20 m² (23·68 sq ft)
Fin	0·90 m² (9·69 sq ft)
Rudder, incl tab	0·60 m² (6·46 sq ft)
Tailplane	1·90 m² (20·45 sq ft)
Elevators, incl tabs	1·00 m² (10·76 sq ft)

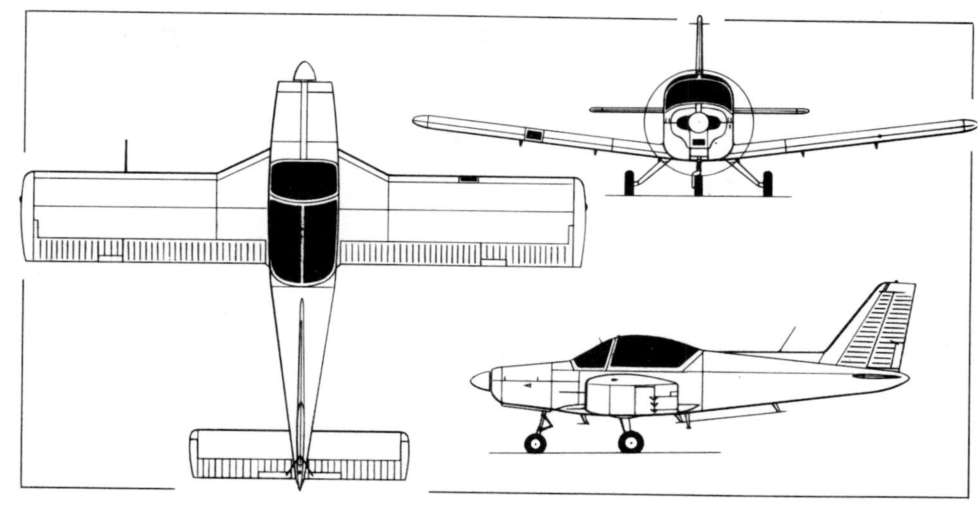

Valmet Leko-70 two/three-seat training and touring aircraft. The name is a military abbreviation of the word 'Lentokone', meaning 'aeroplane' *(Pilot Press)*

Prototype Valmet Leko-70 (Lycoming IO-360-A1B6 engine)

WEIGHTS:

Weight empty, equipped, without fuel	720 kg (1,587 lb)
Max T-O weight	1,200 kg (2,645 lb)

PERFORMANCE (estimated, at max T-O weight*):

Max level speed at S/L	129·5 knots (240 km/h; 149 mph)
Stalling speed, flaps up	57 knots (105 km/h; 66 mph)
Max rate of climb at S/L	360 m (1,180 ft)/min

**Confirmed performance figures to be released on completion of flight testing*

FRANCE

AÉROSPATIALE
SOCIÉTÉ NATIONALE INDUSTRIELLE AÉROSPATIALE

HEAD OFFICE:
37 boulevard de Montmorency, 75781-Paris Cédex 16

Telephone: 524.43.21
Telex: AISPA X 620059 F

BOARD OF DIRECTORS

CHAIRMAN:
Jacques Mitterrand

REPRESENTATIVES OF THE SHAREHOLDERS:
René Ravaud, Ingénieur Général de l'Armement (President and Director General of SNECMA)
André Giraud (General Non-executive Director, Government Delegate to the Atomic Energy Commission)
Pierre Jouven (Honorary President of Péchiney Ugine Kuhlmann)
Le Crédit Lyonnais, represented by Alain Bizot (Director of Crédit Lyonnais)

REPRESENTATIVES OF THE STATE:
Jean Martre, Ingénieur Général de l'Armement (representing the Ministry of Defence)
Jacques de Larosière (Treasury Director, representing the Ministry of Economy and Finance)
Claude Abraham (Director General of Civil Aviation, representing the Secretary of State for Transport)

REPRESENTATIVES OF THE EMPLOYEES:
Paul Bienfait (representing executives)
Georges Girard and Guy Carraz (representing workmen, office staff, technicians and supervisors)

GENERAL MANAGEMENT

PRESIDENT AND DIRECTOR GENERAL:
Jacques Mitterrand

DIRECTOR GENERAL DELEGATE:
Roger Chevalier

DEPUTY DIRECTOR GENERAL:
Yves Barbé

GENERAL SECRETARY:
Marc Robert

DIRECTOR OF THE PRESIDENT'S OFFICE:
René Dor

INSPECTOR GENERAL:
Jean Soissons

ADVISER TO THE PRESIDENT:
Serge Bisone

DIRECTOR OF INTERNATIONAL AFFAIRS:
Michel Thomas

DIRECTOR OF PUBLIC RELATIONS:
Jean Calmel

PRODUCTION DIRECTOR:
Jean Coupain

DIRECTOR OF LONG-RANGE PLANNING:
Jacques de Montravel

CENTRAL FINANCIAL DIRECTOR:
Michel Euvrard

DIRECTOR (INDUSTRIAL RELATIONS):
André Escoulin

DIRECTOR (ORGANISATION AND TRAINING):
Michael Mézaize

DIRECTOR (HEADQUARTERS ESTABLISHMENT):
Jean René Signori

AIRCRAFT DIVISION

DIVISION DIRECTOR:
André Etesse

DIRECTOR, ASSISTANT TO DIVISION MANAGER:
Edouard Debout

TECHNICAL DIRECTOR:
Pierre Lecomte

COMMERCIAL DIRECTOR:
Jean-Charles Poggi

PRODUCTION DIRECTOR:
René Puydebois

CONCORDE PROGRAMME DIRECTOR:
Pierre Gautier

AIRBUS PROGRAMME DIRECTOR:
Maurice de Charnacé

DIRECTOR (MISCELLANEOUS AIRCRAFT PROGRAMMES):
Jacques Hablot

DIRECTOR (ADMINISTRATION AND FINANCE):
Georges Roche

FLIGHT TEST DIRECTOR:
Henri Perrier

DIRECTOR OF AIRCRAFT DESIGN, TOULOUSE:
Gilbert Cormery

WORKS AND FACILITIES:
Toulouse. DIRECTOR: Bernard Dufour
Nantes-Bouguenais. DIRECTOR: Gilbert Colas
Saint-Nazaire. DIRECTOR: Jean Renon
Méaulte. DIRECTOR: Jean-Paul Chandez
Suresnes. DIRECTOR: Roger Berthier

HELICOPTER DIVISION

DIVISION DIRECTOR:
François Legrand

TECHNICAL DIRECTOR:
Georges Petit

DIRECTOR OF DESIGN:
René Mouille

COMMERCIAL DIRECTOR:
Jean-Claude Rebuffel

DIRECTOR (PRODUCT SUPPORT):
Marc Fourcade

FLIGHT TEST DIRECTOR:
Jean-Marie Besse

DIRECTOR (QUALITY CONTROL):
André Breton

WORKS AND FACILITIES:
Marignane. DIRECTOR: Fernand Carayon
La Courneuve. DIRECTOR; Lucien Fournier

Aérospatiale Frégate (two Turboméca Bastan VII turboprop engines) in service with the French Air Force

TACTICAL MISSILES DIVISION
DIVISION DIRECTOR:
Michel Allier
DEPUTY DIVISION DIRECTOR:
Philippe Girard
WORKS AND FACILITIES:
Châtillon. DIRECTOR: Jean-Claude Renaut
Bourges. DIRECTOR: Georges Barroy

BALLISTIC AND SPACE SYSTEMS DIVISION
DIVISION DIRECTOR:
Pierre Usunier
DEPUTY DIVISION DIRECTOR:
Louis Marnay
WORKS AND FACILITIES:
Aquitaine. DIRECTOR: Robert Laurentjoye
Cannes. DIRECTOR: Louis Marnay
Les Mureaux. DIRECTOR: Séverin Golbert

SUBSIDIARIES

Société Girondine d'Entretien et de Réparation de Matériel Aéronautique (SOGERMA)
WORKS: Bordeaux-Mérignac Airport
PRESIDENT AND DIRECTOR GENERAL:
Raymond Brohon

Société de Construction d'Avions de Tourisme et d'Affaires (Socata)
WORKS: Tarbes-Ossun Airport
PRESIDENT AND DIRECTOR GENERAL:
I. G. Jean Soissons
DIRECTOR GENERAL:
Jean Pierson

Société d'Exploitation et de Constructions Aéronautiques (SECA)
WORKS: Le Bourget Airport, Paris
PRESIDENT AND DIRECTOR GENERAL:
Raymond Brohon

Saint-Gramond-Granat
WORKS: Courbevoie and St-Ouen-l'Aumône
PRESIDENT AND DIRECTOR GENERAL:
I. G. Jean Soissons

Electronique Aérospatiale (EAS)
WORKS: Le Bourget Airport, Paris
PRESIDENT AND DIRECTOR GENERAL:
Bernard de Royer

Société Charentaise d'Equipements Aéronautiques (SOCEA)
WORKS: Rochefort and Suresnes
PRESIDENT AND DIRECTOR GENERAL:
Pierre Marion

Aérospatiale Helicopter Corporation (USA)
European Aerospace Corporation (USA)

The Société Nationale Industrielle Aérospatiale was formed on 1 January 1970, by decision of the French government, as a result of the merger of the former Sud-Aviation, Nord-Aviation and SEREB companies. It is the biggest aerospace company in the Common Market countries on the Continent of Europe, with a registered capital of 497,250,000 francs, facilities extending over a total area of 8,005,000 m² (86,165,800 sq ft), of which 1,490,500 m² (16,043,750 sq ft) are covered, and a staff (including subsidiary companies) of 40,267 persons at the beginning of 1976.

In the aircraft field, major products include the Concorde supersonic transport, developed in co-operation with BAC; the short-to-medium-range large-capacity subsonic A300 European Airbus, in co-operation with Deutsche Airbus GmbH, Hawker Siddeley Aviation, VFW-Fokker and CASA; the N 262/Frégate twin-turboprop transport and the SN 601 Corvette twin-turbofan business and third-level transport aircraft.

Aérospatiale produces a range of light piston-engined aircraft through its subsidiary, Socata (which see).

Helicopter activities, concentrated at Marignane, involve the development and production of a wide range of turbine-powered types, described under this entry, including the new 5/6-seat AS 350 Ecureuil, the ten-seat single-engined SA 360 Dauphin and the twin-engined SA 365 Dauphin 2. Agreements concluded with Westland in the UK covered joint development and production of the Puma and Gazelle, and the Westland-designed Lynx, after all three types had been chosen to equip the French and British armed forces. Commercial versions of these helicopters are being produced or developed.

Tactical missiles and pilotless aircraft produced by Aérospatiale include the first-generation surface-to-surface and air-to-surface Entac, SS.11, SS.12M, AS.12, AS.20 and AS.30; target missiles; RPVs; the second-generation Harpon, Milan, Hot, Roland and Exocet anti-tank, surface-to-air and ship-to-ship missiles; and the Pluton nuclear-warhead surface-to-surface missile.

Aérospatiale also produces the SSBS (surface-to-surface) and MSBS (submarine-launched) strategic ballistic missiles; and a range of research rockets. It made major contributions to development and production of the Diamant, Europa II and Ariane launch vehicles, and the Peole, D2-A, Eole, Symphonie, Cos-B, Meteosat, Aerosat and ECSS satellites; and is participating in the post-Apollo Spacelab programme.

AÉROSPATIALE/BAC CONCORDE

Full details of the Concorde programme can be found in the International section of this edition.

AÉROSPATIALE N 262 and FRÉGATE

Design of the N 262 began in the Spring of 1961, and the prototype (F-WKVR) flew for the first time on 24 December 1962. It was followed by three pre-production aircraft, which were built at Châtillon-sous-Bagneux and assembled at Melun-Villaroche flight test centre. Final assembly of production models is undertaken at Bourges. The following versions have been produced:

N 262 Series A. Preceded by Series B. Standard early production version, with 805 kW (1,080 ehp Bastan) VIC turboprop engines. Received FAA Type Approval on 15 March 1965. First production Series A was airframe number 9 (F-WLHX), delivered to Lake Central Airlines (now Allegheny) on 17 August 1965. Production continues.

N 262 Series B. Designation of first four production aircraft only, built for Air Inter. Same power plant as Series A. First Series B (F-BLHS, airframe number 4) flown for first time on 8 June 1964. Recieved SGAC certification on 16 July 1964. Entered service 24 July 1964.

Frégate (formerly N 262 Series C and D). Version for both civil and military use, with more powerful Bastan VII turboprop engines, having improved single-engine ceiling, cruising speed and T-O performance at 'hot and high' airfields. Power plant dispenses with water-methanol system of Series A and B and has higher initial TBO. An N 262 (airframe number 36) began flying experimentally with Bastan VIIA engines in July 1968, and was also test-flown with different wingtips (see general arrangement drawing) which bestow improved low-speed handling. The Frégate entered production in 1970, alongside the Series A, from the 74th aircraft. Certification granted 24 December 1970.

Orders received for the N 262 and Frégate by the early Spring of 1976 were as follows:

Series A Total 71
Original purchasers listed in 1974-75 *Jane's*

Series B Total 4
For Air Inter (Rousseau Aviation)

Frégate

East African Community	1
Gabon government	3
SFA (France)	1
Upper Volta government	2
French Air Force	24
Congo Air Force	1

TYPE: Twin-engined light transport.

WINGS: Cantilever high-wing monoplane. Wing section NACA 23016 (modified) at root, NACA 23012 (modified) at tip. Dihedral 3° from root. Incidence 3°. No sweepback. All-metal two-spar fail-safe structure in conventional light alloys. Sealed all-metal ailerons. Balance tab in starboard aileron. Electrically-controlled hydraulically-actuated all-metal three-position flaps in inner and outer sections on each trailing-edge. Kléber-Colombes (Goodrich licence) pneumatic de-icing boots on outer leading-edges.

FUSELAGE: Semi-monocoque light alloy fail-safe structure, built up from 39 circular main and secondary frames, covered with skin panels arranged circumferentially in sets of four.

TAIL UNIT: Cantilever metal structure, built as separate unit and bolted to rear fuselage frame. Fixed-incidence tailplane. Control surfaces fabric-covered. One controllable tab and one balance tab in rudder and each elevator. Kléber-Colombes (Goodrich licence) pneumatic de-icing system on leading-edges.

LANDING GEAR: Retractable tricycle type, designed and manufactured by ERAM, with single wheel on each unit. Electro-hydraulic retraction, nosewheel forward, main wheels rearward into fairings on sides of fuselage. ERAM oleo-pneumatic nitrogen-filled shock-absorbers. Main wheels have Dunlop or Kléber-Colombes tyres size 12·50-16, pressure 4·07 bars (59 lb/sq in). Nosewheel has Dunlop or Kléber-Colombes Type 06 tyre, size 9·00-6, pressure 3·24 bars (47 lb/sq in). Goodyear hydraulic disc brakes, with anti-skid units. Self-centering nosewheel is fitted with shimmy damper and is steerable hydraulically.

POWER PLANT (Frégate): Two 854 kW (1,145 ehp) Turboméca Bastan VII turboprop engines, each driving a Ratier Forest FH 206-1 four-blade constant-speed fully-feathering metal propeller. Six bag-type flexible fuel tanks between wing spars, forming two groups of three tanks with provision for cross-feed and having a total usable capacity of 2,000 litres (440 Imp gallons). Provision for two additional optional bag tanks in wing centre-section, each of 285 litres (62·5 Imp gallons) usable capacity, giving a max usable capacity of 2,570 litres (565 Imp gallons). Refuelling point above outer wing tank on each side. Pressure refuelling point at front of starboard side main landing gear fairing. No fuel dump system. Oil capacity 23 litres (5 Imp gallons). Electrical anti-icing of engine intakes, spinners and propellers, with additional anti-icing of intakes by engine bleed air.

ACCOMMODATION: Crew of two on flight deck, with central jump-seat at rear for a third crew member if carried. Standard airline version has seating for 26 passengers at 810 mm (32 in) pitch; maximum seating for 29 at 710 mm (28 in) pitch, in three-abreast rows, with two seats on starboard side of aisle and single seat on port side. Movable forward bulkhead, to cater for variable mixed cargo (in front)/passenger (at rear) layouts. Bulkhead can be located in two intermediate positions, to provide 20 or 14 seats at 810 mm (32 in) pitch in rear of cabin, with 9·7 m³ (342 cu ft) or 13·2 m³ (467 cu ft) of cargo space respectively in front part of cabin. Galley, toilet and (on 26-seat version) separate coat space at rear of cabin. For quick-change passenger/cargo operation, foldaway seats can be installed which, when folded, give an available width for cargo of 1·68 m (5 ft 6 in) throughout entire cabin length. Alternative layouts include a six-person executive suite forward with 10 passengers aft; ambulance version with accommodation for 12 stretchers and two medical attendants; or aerial survey version with wide range of cameras and survey equipment and fully-equipped darkroom. Military versions can be fitted out to carry 18 paratroops or 29 troops, or as 22-seat utility transports. Naval versions (Series A) can be equipped for target towing, artillery and missile observation, radar calibration or crew training duties. Standard transport versions have two-section passenger door at rear on port side, the lower half of which has built-in airstairs, and a large cargo door at front on the port side. Emergency exits at front of cabin on each side, at rear on starboard side, and on port side of flight deck. Standard baggage compartments between flight deck and cabin on each side. All accommodation is pressurised, soundproofed and air-conditioned. Windscreen has electrical anti-icing.

Systems: SEMCA air-conditioning system using bleed air from engine. Max pressure differential 0·29 bars (4·20 lb/sq in). Auxiliary ventilation via ram-air inlet at front of port main landing gear fairing. Hydraulic system, operated by two engine-driven pumps at pressure of 207 bars (3,000 lb/sq in), actuates landing gear, nosewheel steering, flaps, brakes and gust locks. Electrically-driven (27V DC) backup pump and 100 bars (1,450 lb/sq in) surge accumulator. Hand pump for emergency operation of flaps, landing gear and gust locks. Pneumatic system for de-icing only. Two 24/27V 40Ah nickel-cadmium batteries, in rear fuselage, and two 9kVA engine-driven starter/generators provide 28V DC electrical supply for engine starting, feathering pumps and rotary inverters. External 28V DC power receptacle. AC system includes two engine-driven 12 kVA three-phase alternators providing 115/200V 400Hz power for engine anti-icing, windscreen heating and anti-icing, and heating for galley. Two single-phase 750VA rotary inverters provide continuous 115V 400Hz AC supply for flight deck instruments. System also includes four 115/26V 400Hz auto-transformers. Optional APU, in port landing gear fairing, provides power for electrical services, engine starting and cabin air-conditioning.

Electronics and Equipment: Standard equipment includes two Collins 618 M 1 VHF, two Collins 51 RV 1 VOR/ILS, Collins 51 Z 4 marker beacon receiver, Collins DF 203 ADF, Collins 331 A6A course indicator, SFIM A 213 flight recorder, Sperry C 14 gyro compass, two Allen RMI, one Bendix OMI, interphone and public address systems. Emergency equipment includes oxygen masks and cylinders, fire extinguishers, life rafts and radio set. Optional equipment includes HF radio, autopilot, second gyro compass, second ADF, weather radar, ATC transponder, radio altimeter and DME; and choice of flight director/recorder, VHF, VOR/ILS and marker beacon receiver.

Dimensions, external (Frégate):
Wing span 22·60 m (74 ft 1¾ in)
Wing chord at root 3·10 m (10 ft 2 in)
Wing chord at tip 1·80 m (5 ft 11 in)
Wing aspect ratio 9·10
Length overall 19·28 m (63 ft 3 in)
Height overall 6·21 m (20 ft 4½ in)
Fuselage: Max diameter 2·45 m (8 ft 0½ in)
Tailplane span 8·80 m (28 ft 10½ in)
Wheel track 3·13 m (10 ft 3 in)
Wheelbase 7·23 m (23 ft 9 in)
Propeller diameter 3·20 m (10 ft 6 in)
Distance between propeller centres 5·91 m (19 ft 4¾ in)
Passenger door (rear, port):
Height 1·66 m (5 ft 5¼ in)
Width 0·68 m (2 ft 3 in)
Height to sill 1·08 m (3 ft 6½ in)
Cargo door (fwd, port):
Height 1·53 m (5 ft 0¼ in)
Width 1·28 m (4 ft 2½ in)
Height to sill 1·08 m (3 ft 6½ in)
Emergency exit doors (fwd, port and stbd):
Height 1·38 m (4 ft 6¼ in)
Width 0·51 m (1 ft 8 in)
Emergency exit door (aft, stbd):
Height 0·92 m (3 ft 0¼ in)
Width 0·51 m (1 ft 8 in)

Dimensions, internal:
Cabin, incl baggage space and toilet:
Length 10·61 m (34 ft 10 in)
Max width 2·15 m (7 ft 1 in)
Width at floor 1·66 m (5 ft 5¼ in)
Max height 1·80 m (5 ft 11 in)
Floor area 17·0 m² (183 sq ft)
Volume 32·5 m³ (1,146 cu ft)
Baggage hold (port) 1·9 m³ (67 cu ft)
Baggage hold (stbd) 2·6 m³ (92 cu ft)

Areas (Frégate):
Wings, gross 55·79 m² (601 sq ft)
Ailerons (total) 4·07 m² (43·8 sq ft)

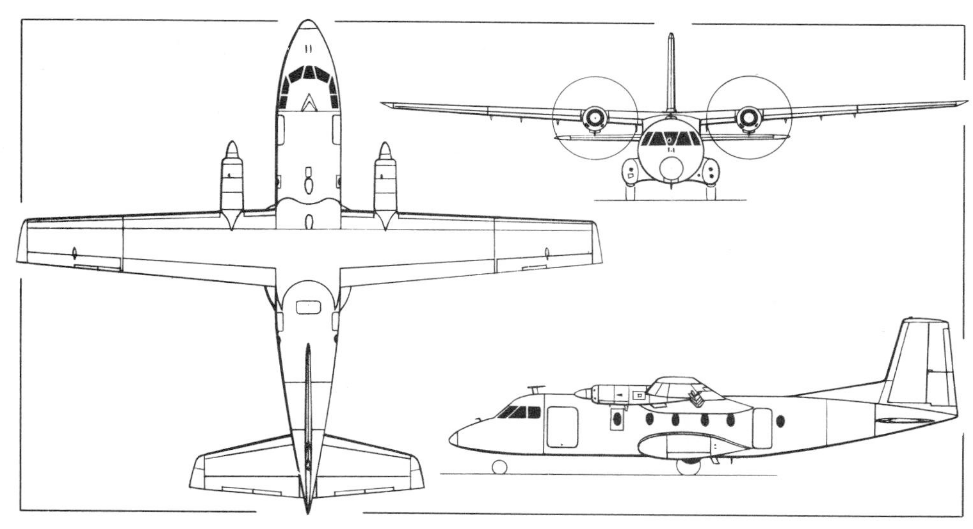

Aérospatiale Frégate twin-turboprop pressurised light transport *(Pilot Press)*

Trailing-edge flaps (total) 8·98 m² (96·6 sq ft)
Fin 10·1 m² (108·7 sq ft)
Rudder, incl tabs 3·75 m² (40·4 sq ft)
Tailplane 18·0 m² (193·7 sq ft)
Elevators, incl tabs 4·54 m² (48·8 sq ft)

Weights and Loadings (Frégate):
Basic weight empty 6,200 kg (13,668 lb)
Manufacturer's weight empty, equipped 6,959 kg (15,342 lb)
Basic operating weight 7,225 kg (15,928 lb)
Max payload 3,075 kg (6,779 lb)
Max T-O weight 10,800 kg (23,810 lb)
Max ramp weight 10,850 kg (23,920 lb)
Max landing weight 10,450 kg (23,040 lb)
Max zero-fuel weight 10,300 kg (22,710 lb)
Max wing loading 193 kg/m² (39·5 lb/sq ft)
Max power loading 6·32 kg/kW (10·2 lb/ehp)

Performance (Frégate, at max T-O weight except where indicated):
Max level speed 225 knots (418 km/h; 260 mph)
Max and econ cruising speed 220 knots (408 km/h; 254 mph)
Normal operating limit speed 214 knots (397 km/h; 247 mph)
Max speed with landing gear extended 154 knots (285 km/h; 177 mph)
Max speed with 15° flap 143 knots (265 km/h; 165 mph)
Max speed with 35° flap 126 knots (235 km/h; 146 mph)
Final approach speed 90 knots (167 km/h; 104 mph)
Stalling speed, flaps up, at max landing weight 86 knots (159 km/h; 99 mph)
Stalling speed, wheels and flaps down, at max landing weight 74 knots (136 km/h; 85 mph)
Max rate of climb at S/L 420 m (1,380 ft)/min
Service ceiling 8,690 m (28,500 ft)
Service ceiling, one engine out, at AUW of 9,525 kg (21,000 lb) 4,920 m (15,000 ft)
Min ground turning radius 8·0 m (26 ft 3 in)
Runway LCN at max weight 8
T-O run 570 m (1,870 ft)
T-O to 10·7 m (35 ft) 1,070 m (3,510 ft)
Landing from 15 m (50 ft) 530 m (1,740 ft)
Range with max fuel, no reserves 1,295 nm (2,400 km; 1,490 miles)
Range with max fuel, FAA reserves 985 nm (1,825 km; 1,135 miles)
Range with 26 passengers and baggage, no reserves 780 nm (1,450 km; 900 miles)
Range with 26 passengers and baggage, FAA reserves 550 nm (1,020 km; 633 miles)

MOHAWK 298

This version of the Frégate is powered by two Pratt & Whitney Aircraft of Canada PT6A-45 turboprop engines. A prototype, converted from an N 262, was produced in the USA by Frakes Aviation Inc, under contract from Mohawk Air Services (see US section), and flew for the first time on 7 January 1975. A similar engine retrofit scheme is offered to existing operators of the N 262/Frégate, the converted aircraft being known as Mohawk 298s.

AÉROSPATIALE SN 601 CORVETTE

The Corvette was designed to fulfil a variety of roles, including executive transport, air taxi, ambulance, freighter or training aircraft. It can be equipped for radio aids calibration or aerial photography.

Two versions have flown, as follows:

SN 600. Prototype only (F-WRSN), with two rear-mounted Pratt & Whitney Aircraft of Canada JT15D-1 turbofan engines, each rated at 9·81 kN (2,200 lb st). First flew on 16 July 1970, and completed more than 270 flying hours before being lost in a crash on 23 March 1971.

SN 601. Initial production Corvette, with two JT15D-4 turbofan engines and a longer fuselage than the prototype. The first SN 601 (F-WUAS) was completed in 1972, together with two airframes for static and fatigue testing. It flew for the first time on 20 December 1972, and was followed by the second SN 601 (F-WRNZ) on 7 March 1973, the third (F-WUQN) on 9 November 1973 and the fourth (F-WUQP) on 12 January 1974. Type certification of this version was received from the SGAC on 28 May

Mohawk 298, an Aérospatiale N 262 re-engined by Frakes Aviation with PT6A-45 turboprops

1974 and from the FAA on 24 September 1974. Deliveries began in September 1974.

By 1 January 1976, orders totalled 23, plus options on 6 more aircraft. Eleven of the aircraft delivered were being operated by third-level airlines. Production will be suspended after delivery of 40 Corvettes.

The following description applies to the SN 601:

TYPE: Multi-purpose twin-turbofan aircraft.

WINGS: Cantilever low-wing monoplane of all-metal construction. Thickness/chord ratio 13·65% at root, 11·5% at tip. Dihedral 3° on outer panels. Sweepback 22° 32′ on leading-edge. Conventional two-spar fail-safe structure, of aluminium alloy. Manually-operated aluminium alloy ailerons and electrically-operated double-slotted long-travel trailing-edge flaps of aluminium alloy and honeycomb construction. Three-section spoiler forward of each outer flap. Hydraulically-actuated airbrakes inboard of spoilers, above and below each wing. Electrically-actuated trim tab in port aileron. TKS-type de-icing of leading-edges.

FUSELAGE: Aluminium alloy semi-monocoque fail-safe structure of circular cross-section.

TAIL UNIT: Cantilever aluminium alloy structure, with tailplane mounted on fin. Sweepback on all surfaces. Electrically-actuated variable-incidence tailplane. Manually-operated elevators and rudder. Electrically-actuated trim tab in rudder.

LANDING GEAR: Hydraulically-retractable tricycle type, with hydraulic shock-absorbers and single wheel on each unit. Main wheels retract inward, nosewheel forward, into fuselage. Low-pressure tyres of 254 mm (10 in) diameter on main wheels and 152 mm (6 in) diameter on nosewheel. Main-wheel tyre pressure 5·45 bars (79 lb/sq in); nosewheel tyre pressure 4·48 bars (65 lb/sq in). Hydraulic brakes and anti-skid units. Nosewheel steerable.

POWER PLANT: Two 11·12 kN (2,500 lb st) Pratt & Whitney Aircraft of Canada JT15D-4 turbofan engines, mounted in pod on each side of rear fuselage. Two integral wing fuel tanks, with total capacity of 1,660 litres (365 Imp gallons; 439 US gallons). Provision for tip-tanks, of approx 700 litres (154 Imp gallons; 185 US gallons) total capacity.

ACCOMMODATION: Crew of one or two on flight deck. Normal seating for 6 to 14 passengers in single seats on each side of centre aisle. Galley, toilet and baggage compartments available to customer's requirements. Two-part door, with built-in airstairs, at front on port side; upper part of door is hinged at top, lower part is hinged at bottom of doorway.

SYSTEMS: Cabin air-conditioning and pressurisation by engine bleed air; max differential 0·59 bars (8·6 lb/sq in). Hydraulic system for actuating landing gear, nosewheel steering, wheel brakes and airbrakes. Main electrical system includes two 10·5kW 28·5V DC starter/generators, one 36Ah battery and two inverters for 400Hz AC supply.

ELECTRONICS AND EQUIPMENT: Blind-flying instrumentation standard. Radio, radar or other special equipment to customer's specification.

DIMENSIONS, EXTERNAL:
- Wing span: 12·87 m (42 ft 2½ in)
- Wing aspect ratio: 7·45
- Length overall: 13·83 m (45 ft 4½ in)
- Length of fuselage: 12·90 m (42 ft 4 in)
- Fuselage diameter: 1·70 m (5 ft 7 in)
- Height overall: 4·23 m (13 ft 10½ in)
- Tailplane span: 5·00 m (16 ft 4¾ in)
- Wheel track: 2·57 m (8 ft 5¼ in)
- Wheelbase: 5·22 m (17 ft 1½ in)
- Passenger door:
  - Height: 1·31 m (4 ft 3½ in)
  - Width: 0·71 m (2 ft 4 in)
  - Mean height to sill: 0·85 m (2 ft 9½ in)

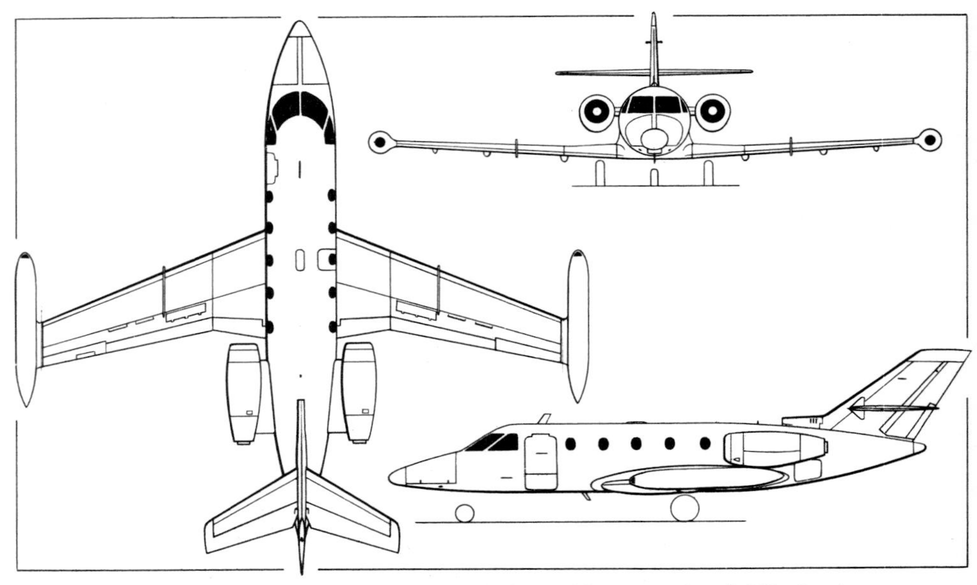

Aérospatiale SN 601 Corvette twin-turbofan multi-purpose aircraft *(Pilot Press)*

DIMENSIONS, INTERNAL:
- Cabin, excl flight deck:
  - Max length: 5·73 m (18 ft 9½ in)
  - Max width: 1·56 m (5 ft 1½ in)
  - Max height: 1·52 m (5 ft 0 in)
  - Floor area: 6·60 m² (71 sq ft)
  - Volume: 9·93 m³ (351 cu ft)
- Baggage compartment volume (10-passenger layout): 1·08 m³ (38·1 cu ft)

AREAS:
- Wings, gross: 22·00 m² (236·8 sq ft)
- Vertical tail surfaces (total): 4·22 m² (45·4 sq ft)
- Horizontal tail surfaces (total): 5·47 m² (58·9 sq ft)

WEIGHTS:
- Manufacturer's weight empty: 3,510 kg (7,738 lb)
- Max T-O weight: 6,600 kg (14,550 lb)
- Max ramp weight: 6,650 kg (14,660 lb)
- Max landing weight: 5,700 kg (12,550 lb)
- Max zero-fuel weight: 5,600 kg (12,345 lb)

PERFORMANCE (at max T-O weight, except where indicated):
- Max cruising speed at 9,000 m (30,000 ft): 410 knots (760 km/h; 472 mph)
- Econ cruising speed at 11,900 m (39,000 ft): 306 knots (566 km/h; 352 mph)
- Stalling speed, flaps and landing gear down, at max landing weight: 91 knots (168 km/h; 105 mph)
- Max rate of climb at S/L: 823 m (2,700 ft)/min
- Service ceiling: 12,500 m (41,000 ft)
- T-O balanced field length (FAR 25): 1,390 m (4,560 ft)
- Landing distance at max landing weight: 755 m (2,480 ft)
- Max range with tip-tanks, 45 min reserves:
  - Max cruise power: 1,290 nm (2,390 km; 1,485 miles)
  - Econ cruise power: 1,380 nm (2,555 km; 1,588 miles)
- Range with 12 passengers, 45 min reserves:
  - Max cruise power: 800 nm (1,480 km; 920 miles)
  - Econ cruise power: 840 nm (1,555 km; 967 miles)

OPERATIONAL NOISE CHARACTERISTICS:
- T-O noise level at 3·5 nm (6·5 km; 4 miles) from start of T-O roll:
  - Without power reduction: 81·2 EPNdB
  - With power reduction: 74 EPNdB
- Approach noise level at 1·08 nm (2 km; 1·24 miles) from landing threshold on 3° glideslope: 90 EPNdB

AÉROSPATIALE SA 318C ALOUETTE II ASTAZOU

Production of the Alouette II ended in 1975. Altogether, 1,305 were built, with Turboméca Artouste and Astazou turboshaft engines. The Artouste-powered SE 313B was described in the 1967-68 and 1968-69 editions of *Jane's*; the SA 318C Alouette II Astazou was described in the 1975-76 edition.

AÉROSPATIALE SA 315B LAMA

Indian Army name: Cheetah

Design of the SA 315B Lama began in late 1968, initially to meet a requirement of the Indian armed forces, and a prototype was flown for the first time on 17 March 1969. French certification was granted on 30 September 1970 and FAA Type Approval on 25 February 1972.

The Lama combines features of the Alouette II and III, having the airframe (with some reinforcement) of the former and the dynamic components, including the Artouste power plant and rotor system, of the SA 316 Alouette III.

During demonstration flights in the Himalayas in 1969 a Lama, carrying a crew of two and 140 kg (308 lb) of fuel, made the highest landings and take-offs ever recorded, at a height of 7,500 m (24,600 ft).

On 21 June 1972, a Lama set a helicopter absolute height record of 12,442 m (40,820 ft). The pilot was Jean Boulet, holder of the previous record in an SE 3150 Alouette.

The production Lama is capable of transporting an external load of 1,135 kg (2,500 lb) at an altitude of more than 2,500 m (8,200 ft). In an agricultural role, it is fitted with spraybars and an underbelly tank of 1,135 litres (300 US gallons) capacity, developed jointly by Aérospatiale Helicopter Corporation and Simplex Manufacturing Company. The tank is equipped with an electrical emergency dump system.

A total of 185 Lamas had been ordered by 56 operators in 22 countries by the Spring of 1976. In addition to manufacture by Aérospatiale, the SA 315 is produced under licence by HAL for the Indian Army under the name of Cheetah.

Aérospatiale SN 601 Corvette (two Pratt & Whitney Aircraft of Canada JT15D-4 turbofan engines) in service with Air Alpes

TYPE: Turbine-driven general-purpose helicopter.

ROTOR SYSTEM: Three-blade main and anti-torque rotors. All-metal main rotor blades, of constant chord, are on articulated hinges, with hydraulic drag-hinge dampers. Rotor brake standard.

ROTOR DRIVE: Main rotor driven through planetary gearbox, with freewheel for autorotation. Take-off drive for tail rotor at lower end of main gearbox, from where a torque shaft runs to a small gearbox which supports the tail rotor and houses the pitch-change mechanism. Cyclic and collective pitch controls are powered.

FUSELAGE: Glazed cabin has light metal frame. Centre and rear of fuselage have a triangulated steel tube framework.

LANDING GEAR: Skid type, with removable wheels for ground manoeuvring. Pneumatic floats for normal operation from water, and emergency flotation gear, inflatable in the air, are available.

POWER PLANT: One 649 kW (870 shp) Turboméca Artouste IIIB turboshaft engine, derated to 410 kW (550 shp). Fuel tank in fuselage centre-section, with capacity of 575 litres (126·5 Imp gallons), of which 573 litres (126 Imp gallons) are usable.

ACCOMMODATION: Glazed cabin seats pilot and passenger side by side in front and three passengers behind. Provision for external sling for loads of up to 1,135 kg (2,500 lb). Can be equipped for rescue (hoist capacity 160 kg; 352 lb), liaison, observation, training, agricultural, photographic and other duties. As an ambulance, can accommodate two stretchers and a medical attendant internally.

DIMENSIONS, EXTERNAL:
Main rotor diameter 11·02 m (36 ft 1¼ in)
Tail rotor diameter 1·91 m (6 ft 3¼ in)
Main rotor blade chord (constant) 0·35 m (13·8 in)
Length overall, both rotors turning 12·92 m (42 ft 4¾ in)
Length of fuselage 10·26 m (33 ft 8 in)
Height overall 3·09 m (10 ft 1¾ in)
Skid track 2·38 m (7 ft 9¾ in)

WEIGHTS:
Weight empty 1,014 kg (2,235 lb)
Normal max T-O weight 1,950 kg (4,300 lb)
Max T-O weight with externally-slung cargo 2,300 kg (5,070 lb)

PERFORMANCE (at AUW of 2,200 kg; 4,850 lb, with slung load):
Max cruising speed 65 knots (120 km/h; 75 mph)
Max rate of climb at S/L 250 m (820 ft)/min
Service ceiling 4,000 m (13,125 ft)
Hovering ceiling in ground effect 3,750 m (12,300 ft)
Hovering ceiling out of ground effect 2,800 m (9,185 ft)

AÉROSPATIALE SA 316B ALOUETTE III

The Alouette III helicopter was evolved from the Alouette II, with larger cabin, greater power, improved equipment and higher performance. The prototype flew for the first time on 28 February 1959, and a total of 1,357 Alouette IIIs had been sold to 186 operators in 71 countries by the Spring of 1976.

Those delivered up to the end of 1969 were designated **SE 3160**. The subsequent Artouste-engined **SA 316B** has strengthened main and rear rotor transmissions, higher AUW and increased payload; first deliveries were made in 1970, and this version received FAA Type Approval on 25 March 1971. The **SA 319B**, with Astazou engine, is described separately, but is included in the total sales figures above.

The sale of Alouette IIIs to India, Romania and Switzerland included a licence agreement for manufacture of the aircraft in those countries. Quantities involved were 80 in India, 50 in Romania and 60 in Switzerland.

TYPE: Turbine-driven general-purpose helicopter.

ROTOR SYSTEM: Three-blade main and anti-torque rotors. All-metal main rotor blades, on articulated hinges, with hydraulic drag-hinge dampers. Rotor brake standard.

ROTOR DRIVE: Main rotor driven through planetary gearbox, with freewheel for autorotation. Take-off drive for tail rotor at lower end of main gearbox, from where a torque shaft runs to a small gearbox which supports the tail rotor and houses the pitch-change mechanism. Cyclic and collective pitch controls are powered.

FUSELAGE: Welded steel tube centre-section, carrying the cabin at the front and a semi-monocoque tailboom.

TAIL UNIT: Cantilever all-metal fixed tailplane, with twin endplate fins, mounted on tailboom.

LANDING GEAR: Non-retractable tricycle type, manufactured by Messier-Hispano. Nosewheel is fully-castoring. Provision for pontoon landing gear.

POWER PLANT: One 649 kW (870 shp) Turboméca Artouste IIIB turboshaft engine, derated to 425 kW (570 shp). Fuel in single tank in fuselage centre-section, with capacity of 575 litres (126·5 Imp gallons), of which 573 litres (126 Imp gallons) are usable.

ACCOMMODATION: Normal accommodation for pilot and six persons, with three seats in front and a four-person folding seat at the rear of the cabin. Two baggage holds in centre-section, on each side of the welded structure and enclosed by the centre-section fairings. Provision for carrying two stretchers athwartships at rear of cabin, and two other persons, in addition to pilot. All passenger seats removable to enable aircraft to be used for freight-carrying. Provision for external sling for loads of up to 750 kg (1,650 lb). One forward-opening door on each side, immediately in front of two rearward-sliding doors. Dual controls and cabin heating optional.

Aérospatiale SA 315B Lama (Turboméca Artouste IIIB turboshaft)

OPERATIONAL EQUIPMENT (military version): In the assault role, the Alouette III can be equipped with a wide range of weapons. A 7·62 mm AA52 machine-gun (with 1,000 rds) can be mounted athwartships on a tripod behind the pilot's seat, firing to starboard, either through a small window in the sliding door or through the open doorway with the door locked open. The rear seat is removed to allow the gun mounting to be installed. In this configuration, max accommodation is for pilot, co-pilot, gunner and one passenger, although normally only the pilot and gunner would be carried. Alternatively, a 20 mm MG 151/20 cannon (with 480 rds) can be carried on an open turret-type mounting on the port side of the cabin. For this installation all seats except that of the pilot are removed, as is the port side cabin door, and the crew consists of pilot and gunner. Instead of these guns, the Alouette III can be equipped with four AS.11 or two AS.12 wire-guided missiles on external jettisonable launching rails, with an APX-Bézu 260 gyro-stabilised sight, or 68 mm rocket pods. Firing trials with Hot missiles have also been completed successfully.

DIMENSIONS, EXTERNAL:
Diameter of main rotor 11·02 m (36 ft 1¾ in)
Main rotor blade chord (each) 0·35 m (13·8 in)
Diameter of tail rotor 1·91 m (6 ft 3¼ in)
Length overall, rotors turning 12·84 m (42 ft 1½ in)
Length overall, blades folded 10·03 m (32 ft 10¾ in)
Width overall, blades folded 2·60 m (8 ft 6¼ in)
Height to top of rotor head 3·00 m (9 ft 10 in)
Wheel track 2·60 m (8 ft 6¼ in)

WEIGHTS:
Weight empty 1,143 kg (2,520 lb)
Max T-O weight 2,200 kg (4,850 lb)

PERFORMANCE (standard version, at max T-O weight):
Max level speed at S/L 113 knots (210 km/h; 130 mph)
Max cruising speed at S/L 100 knots (185 km/h; 115 mph)
Max rate of climb at S/L 260 m (850 ft)/min
Service ceiling 3,200 m (10,500 ft)
Hovering ceiling in ground effect 2,880 m (9,450 ft)
Hovering ceiling out of ground effect 1,520 m (5,000 ft)
Range with max fuel at S/L 258 nm (480 km; 298 miles)
Range at optimum altitude 290 nm (540 km; 335 miles)

AÉROSPATIALE SA 319B ALOUETTE III ASTAZOU

Indian military name: Chetak

The SA 319B Alouette III Astazou is a direct development of the SA 316B, from which it differs principally in having an Astazou XIV turboshaft engine (649 kW; 870 shp, derated to 447 kW; 600 shp) with increased thermal efficiency and a 25% reduction in fuel consumption.

A prototype SA 319 was completed in 1967. The production total to the Spring of 1976 is included in the figures given under the SA 316B entry.

OPERATIONAL EQUIPMENT (naval version): The Alouette III can fulfil a variety of shipborne roles; features common to all naval configurations include a quick-mooring harpoon to ensure instant and automatic mooring on landing and before take-off, a nosewheel locking device, and folding main rotor blades. For detecting and destroying small surface craft such as torpedo-boats, it can be equipped with a SFENA three-axis stabilisation system, OMERA ORB 31 radar, APX-Bézu 260 gyro-stabilised sight and two AS.12 wire-guided missiles. For the ASW role, it can carry two Mk 44 homing torpedoes beneath the fuselage, or one torpedo and MAD (magnetic anomaly detection) gear in a streamlined container which is towed behind the helicopter on a 50 m (150 ft) cable. The aircraft can be used for air/sea rescue when the cabin floor is protected by an anti-corrosion covering to prevent sea water from reaching vital components. Rescue hoist (capacity 225 kg; 500 lb) mounted on port side of fuselage.

WEIGHTS:
Weight empty 1,140 kg (2,513 lb)
Max T-O weight 2,250 kg (4,960 lb)

PERFORMANCE (at max T-O weight):
Max level speed at S/L 118 knots (220 km/h;136 mph)
Max cruising speed at S/L 106 knots (197 km/h; 122 mph)
Max rate of climb at S/L 270 m (885 ft)/min
Hovering ceiling in ground effect 3,100 m (10,170 ft)
Hovering ceiling out of ground effect 1,700 m (5,575 ft)
Range with 6 passengers (80 kg; 176 lb each), T-O at S/L 325 nm (605 km; 375 miles)

Aérospatiale SA 319B Alouette III Astazou equipped for search and rescue operations in Spain *(Javier Taibo)*

AÉROSPATIALE SA 321 SUPER FRELON

The Super Frelon is a three-engined multi-purpose helicopter derived from the smaller SA 3200 Frelon (see 1961-62 *Jane's*).

Under a technical co-operation contract, Sikorsky Aircraft, USA, provided assistance in the development of the Super Frelon, in particular with the detail specifications, design, construction and testing of the main and tail rotor systems. Under a further agreement, the main gearcase and transmission box are produced in Italy by Fiat.

The first prototype of the Super Frelon (originally designated SA 3210-01) flew on 7 December 1962, powered by three 985 kW (1,320 shp) Turmo IIIC₂ engines, and represented the troop transport version. In July 1963 this aircraft set up several international helicopter records, including a speed of 184 knots (341 km/h; 212 mph) over a 3 km course, and a speed of 189·115 knots (350·47 km/h; 217·77 mph) over a 15/25 km course.

The second prototype, flown on 28 May 1963, was representative of the naval version, with stabilising floats on the main landing gear supports. Four pre-production aircraft followed, and the French government ordered an initial production series of 17, designated SA 321G, in October 1965. By the beginning of 1976, a total of 91 Super Frelons had been sold to 11 operators in 9 countries, with production continuing to fill an order for 13 for China. Rate of manufacture was then one aircraft per month.

Passenger and utility versions of the Super Frelon are available, and the main differences between the current versions are summarised as follows:

SA 321F. Commercial airliner, designed to carry 34-37 passengers in a standard of comfort comparable to that of fixed-wing airliners, over 94 nm (175 km; 108 mile) stage lengths at a cruising speed of 124 knots (230 km/h; 143 mph), with 20 min reserve fuel. The prototype was designed in accordance with US FAR 29 regulations and flew for the first time on 7 April 1967. Type certification was granted by the SGAC on 27 June 1968 and by the FAA on 29 August 1968.

SA 321G. Anti-submarine helicopter. First version of the SA 321 to enter production. The first SA 321G flew on 30 November 1965 and deliveries began in early 1966. Twenty-four built. In service with Flottille 32F of Aéronavale, which was commissioned at Lanvéoc-Poulmic on 5 May 1970. Duties of this squadron include patrols in support of *Redoutable* class nuclear submarines entering and leaving their base on the Île Longue. The SA 321G can also be operated from the French helicopter carrier *Jeanne d'Arc*.

SA 321H. Version for air force and army service, without stabilising floats or external fairings on each side of lower fuselage. Turmo IIIE₆ engines instead of Turmo IIIC₆ in other versions. No de-icing equipment fitted.

SA 321Ja. Utility and public transport version, intended to fulfil the main roles of personnel and cargo transport. Designed to carry a maximum of 27 passengers. External loads of up to 5,000 kg (11,023 lb) can be suspended from the cargo sling and carried 27 nm (50 km; 31 miles), the aircraft returning to base without load. An internal payload of 4,000 kg (8,818 lb) can be carried over 100 nm (185 km; 115 miles) at 124 knots (230 km/h; 143 mph) with 20 min fuel reserves. The SA 321J prototype flew for the first time on 6 July 1967. A French certificate of airworthiness was granted in December 1971.

The following description applies generally to all current models of the Super Frelon, except where specific variants are indicated:

TYPE: Three-engined heavy-duty helicopter.

ROTOR SYSTEM: Six-blade main rotor and five-blade anti-torque tail rotor. Main rotor head consists basically of two six-armed star-plates carrying the drag and flapping hinges for each blade. The root of each blade carries a fitting for pitch control and each blade has an individual hydraulic damper to govern movement in the drag plane. Each main blade is 8·60 m (28 ft 2½ in) long, with constant chord and NACA 0012 section. All-metal construction, with D-section main spar forming leading-edge. Tail rotor of similar construction to main rotor, with blades 1·60 m (5 ft 3 in) long. Rearward folding of all six main rotor blades of SA 321G is accomplished automatically by hydraulic jacks, simultaneously with automatic folding of the tail rotor pylon.

ROTOR DRIVE: The driveshaft from the rear engine is geared directly to the shaft from the port forward engine. The two forward engines have a common reduction gear from which an output shaft drives the main rotor shaft through helical gearing. There are two reduction gear stages on the main rotor shaft. The tail rotor shaft is driven by gearing from the shaft linking the rear and port forward engines and incorporates two-stage reduction. The rotor can be stopped within 40 sec by a boosted disc-type rotor brake fitted to this shaft. Main rotor rpm 207 and 212. Tail rotor rpm 990.

FUSELAGE: Boat-hull fuselage of conventional metal semi-monocoque construction, with watertight compartments inside planing bottom. On the SA 321G, there is a small stabilising float attached to the rear landing gear support structure on each side. The tail section of the SA 321G folds for stowage. Small fixed stabiliser on starboard side of tail rotor pylon on all versions. The SA 321F does not have stabilising floats, but large external fairings on each side of the centre fuselage serve a similar purpose and also act as baggage containers.

LANDING GEAR: Non-retractable tricycle type, by Messier-Hispano. Twin wheels on each unit. Oleo-pneumatic shock-absorbers can be shortened on the SA 321G to reduce height of aircraft for stowage. Magnesium alloy wheels, all of same size. Tyre pressure 6·9 bars (100 lb/sq in). Optionally, low-pressure (3·45 bars; 50 lb/sq in) tyres may be fitted. Hydraulic disc brakes on main wheels. Nosewheel unit is steerable and self-centering.

POWER PLANT: Three 1,156 kW (1,550 shp) Turboméca Turmo IIIC₆ turboshaft engines (IIIE₆ in SA 321H); two mounted side by side forward of main rotor shaft and one aft of rotor shaft. Fuel in flexible tanks under floor of centre fuselage, with total standard capacity of 3,975 litres (874 Imp gallons) in SA 321G/H and 3,900 litres (858 Imp gallons) in SA 321Ja. Optional auxiliary fuel tankage comprises two 500 litre (110 Imp gallon) external tanks on all models, two 500 litre (110 Imp gallon) internal tanks in the SA 321G, and three 666 litre (146·5 Imp gallon) internal tanks in the SA 321H/Ja.

ACCOMMODATION (military versions): Crew of two on flight deck, with dual controls and advanced all-weather equipment. Equipment in the SA 321G, which carries a flight crew of five, includes a tactical table and a variety of devices for anti-submarine detection and attack, towing, minesweeping and other duties. This version also has provision for carrying 27 passengers. SA 321H transport accommodates 27-30 troops, 5,000 kg (11,023 lb) of internal or external cargo, or 15 stretchers and two medical attendants. Rescue hoist of 275 kg (606 lb) capacity. Main cabin is ventilated and soundproofed. Sliding door on starboard side of front fuselage. Rear loading ramp is actuated hydraulically and can be opened in flight.

ACCOMMODATION (SA 321F): Airliner seats for up to 37 passengers (34 if toilets are installed) in three-abreast rows with centre aisle. Alternative layouts for 8, 14 or 23 passengers, with toilets, or 11, 17 or 26 passengers without toilets, the remainder of the cabin space being blanked off by movable partitions and used for the carriage of freight; with these configurations, unused seats are folded against the cabin wall. All seats and interior furnishings are designed for quick removal when the helicopter is to be used for all-freight services. To cater for operations over marshland or water, the hull and lateral cargo compartments are sealed sufficiently to permit an occasional landing on water.

ACCOMMODATION (SA 321Ja): Seating for up to 27 passengers in the personnel transport role. As a cargo transport, external loads of up to 5,000 kg (11,023 lb) can be suspended from the cargo sling. Loading of internal cargo (up to 5,000 kg; 11,023 lb) is effected via the rear ramp-doors, with the assistance of a Tirefor hand winch.

OPERATIONAL EQUIPMENT (SA 321G): This version operates normally in tactical formations of three or four aircraft, each helicopter carrying the full range of detection, tracking and attack equipment, including a central navigational system, Doppler radar, radio altimeter, Sylphe panoramic radar with IFF capability and dipping sonar. Four homing torpedoes can be carried in pairs on each side of the main cabin.

DIMENSIONS, EXTERNAL:

Diameter of main rotor	18·90 m (62 ft 0 in)
Main rotor blade chord (each)	0·54 m (1 ft 9¼ in)
Diameter of tail rotor	4·00 m (13 ft 1½ in)
Tail rotor blade chord (each)	0·30 m (11¾ in)
Length overall, rotors turning	23·03 m (75 ft 6⅝ in)
Length of fuselage, incl tail rotor	20·08 m (65 ft 10¾ in)
Length of fuselage	19·40 m (63 ft 7¾ in)
Length overall:	
SA 321G, blades and tail folded	17·07 m (56 ft 0 in)
Width overall:	
SA 321G, blades and tail folded	5·20 m (17 ft 0¾ in)
SA 321F, incl baggage containers	5·04 m (16 ft 6⅜ in)
Width of fuselage	2·24 m (7 ft 4¼ in)
Height at tail rotor (normal)	6·66 m (21 ft 10¼ in)
Height overall:	
SA 321G, blades and tail folded	4·94 m (16 ft 2½ in)
Wheel track	4·30 m (14 ft 1 in)
Wheelbase	6·56 m (21 ft 6¼ in)
Cabin door:	
Height	1·55 m (5 ft 1 in)
Width	1·20 m (3 ft 11¼ in)
Rear loading ramp:	
Length	1·90 m (6 ft 2¾ in)
Width	1·90 m (6 ft 2¾ in)

Aérospatiale SA 321Ja Super Frelon heavy-duty helicopter, specially equipped for servicing offshore oil rigs

DIMENSIONS, INTERNAL:
Cabin:
Length:
SA 321F 9·67 m (31 ft 9 in)
SA 321G and Ja 7·00 m (22 ft 11½ in)
Width:
SA 321F 1·96 m (6 ft 5 in)
SA 321G and Ja, at floor 1·90 m (6 ft 2¾ in)
Height:
SA 321F 1·80 m (5 ft 11 in)
SA 321G and Ja 1·83 m (6 ft 0 in)
Usable volume:
SA 321G and Ja 25·3 m³ (893 cu ft)

WEIGHTS:
Weight empty, standard aircraft:
SA 321G 6,863 kg (15,130 lb)
SA 321H 6,702 kg (14,775 lb)
SA 321Ja 6,868 kg (15,141 lb)
Max T-O weight 13,000 kg (28,660 lb)

PERFORMANCE (at max T-O weight):
Never-exceed speed at S/L 148 knots (275 km/h; 171 mph)
Cruising speed at S/L 135 knots (249 km/h; 155 mph)
Cruising speed at S/L, one engine out 113 knots (210 km/h; 130 mph)
Max rate of climb at S/L 400 m (1,312 ft)/min
Rate of climb at S/L, one engine out 146 m (479 ft)/min
Service ceiling 3,150 m (10,325 ft)
Service ceiling, one engine out 1,200 m (3,940 ft)
Hovering ceiling in ground effect 2,170 m (7,120 ft)
Normal range at S/L 442 nm (820 km; 509 miles)
Normal range at S/L, one engine out 496 nm (920 km; 572 miles)
Range at S/L with 3,500 kg (7,716 lb) payload 549 nm (1,020 km; 633 miles)
Endurance in ASW role 4 hr

AÉROSPATIALE/WESTLAND SA 330 PUMA

The twin-engined SA 330 Puma was developed initially to meet a French Army requirement for a medium-sized *hélicoptère de manoeuvre,* able to operate by day or night in all weathers and all climates. In 1967, the SA 330 was selected for the RAF Tactical Transport Programme, and was included in the joint production agreement between Aérospatiale and Westland in the UK.

The first of two SA 330 prototypes flew on 15 April 1965, and the last of six pre-production models on 30 July 1968, followed in September 1968 by the first production aircraft.

The following versions of the Puma have been announced:

SA 330B. For French Army (ALAT) and French Air Force. First flown January 1969; deliveries began Spring 1969; became operational with the Groupe de l'Aviation Légère of the 7th French Division at Habsheim Base, Mulhouse, France, in June 1970. Turmo IIIC4 engines, each of 990 kW (1,328 shp) for T-O and 884 kW (1,185 shp) max continuous rating.

SA 330C/H. Military export versions. First flown September 1968. For engine details see Power Plant paragraph.

SA 330E. For Royal Air Force, by whom it is designated Puma HC. Mk 1. Forty built, with Turmo IIIC4 engines. First production example (XW198) flown on 25 November 1970. First RAF squadron (No. 33) formed in 1971, followed by No. 230 in 1972. Deliveries completed.

SA 330F/G. Civil passenger or cargo versions. First flown 26 September 1969. Awarded French certification on 12 October 1970, FAA Type Approval for IFR operation (FAR Pt 29, category A and B) on 23 June 1971 and UK certification on 4 June 1975. For engine details see Power Plant paragraph.

SA 330J/L. Civil (J) and military (L) versions introduced in 1976 with main rotor blades of composite materials. Increased max T-O weight. For engine details see Power Plant paragraph.

SA 330Z. This is the current designation of the fifth pre-production Puma (F-ZWWR), which has been flying at Marignane since September 1975 with an eleven-blade 'fenestron' anti-torque tail rotor and T-type horizontal stabiliser, as part of the Super Puma development programme.

By the Spring of 1976, a total of 460 Pumas had been sold to 38 operators in 33 countries. Deliveries were increased to eight a month in early 1976.

TYPE: Medium-sized transport helicopter.

ROTOR SYSTEM (except SA 330J/L): Four-blade main rotor, with a fully-articulated hub and integral rotor brake. The blade cuffs, equipped with horns, are connected by link-rods to the swashplate, which is actuated by three hydraulic twin-cylinder servo-control units. The blades, which are of constant chord, NACA 00 series section and twisted, consist of an aluminium alloy extruded spar, milled on the outside to form the leading-edge, and a series of sheet metal pockets hotbonded to the rear of the spar to form the trailing-edge. Attachment of the blades to their sleeve by means of two pins enables them to be folded back quickly by manual methods. The five-blade tail rotor has flapping hinges only, and is located on the starboard side of the tailboom.

Aérospatiale SA 330G Puma (two Turboméca Turmo IVC turboshaft engines)

ROTOR SYSTEM (SA 330J/L): Generally similar to other versions, but main rotor blades of composite construction and increased chord. Each blade has a roving spar and trailing-edge, with glassfibre skin. The interior is filled with honeycomb inside an inner skin of carbon fibre. The leading-edge is sheathed in glassfibre with an outer stainless steel protective section.

ROTOR DRIVE: Mechanical shaft and gear drive. Main gearbox, mounted on top of cabin behind engines, has two separate inputs from the engines and five reduction stages. The first stage drives, from each engine, an intermediate shaft directly driving the alternator and the ventilation fan, and indirectly driving the two hydraulic pumps. At the second stage the action of the two units becomes synchronised on a single main driveshaft by means of freewheeling spur gears. If one or both engines are stopped, this enables the drive gears to be rotated by the remaining turbine or the autorotating rotor, thus maintaining drive to the ancillary systems when the engines are stopped. Drive to the tail rotor is via shafting and an intermediate angle gearbox, terminating at a right-angle tail rotor gearbox. Turbine output 23,000 rpm, main rotor shaft 265 rpm. Tail rotor shaft 1,278 rpm. The hydraulically-controlled rotor brake, installed on the main gearbox, permits stopping of the rotor 15 seconds after engine shutdown.

FUSELAGE: Conventional all-metal semi-monocoque structure. Local use of titanium alloy under engine installation, which is outside the main fuselage shell. Monocoque tailboom supports the tail rotor on the starboard side and a horizontal stabiliser on the port side.

LANDING GEAR: Messier-Hispano semi-retractable tricycle type, with twin wheels on each unit. Main units retract upward hydraulically into fairings on sides of fuselage; self-centering nose unit retracts rearward. When landing gear is down, the nosewheel jack is extended and the main-wheel jacks are telescoped. Dual-chamber oleo-pneumatic shock-absorbers. All tyres same size (7·00-6), of Dunlop or Kléber-Colombes tubeless type, pressure 4·83 bars (70 lb/sq in) on all units. Hydraulic differential disc brakes, controlled by foot pedals. Lever-operated parking brake. Emergency pop-out flotation units can be mounted on rear landing gear fairings and forward fuselage.

POWER PLANT: Initial civilian version had two Turboméca Turmo IVA turboshaft engines, each with max rating of 1,070 kW (1,435 shp) and equipped for air intake anti-icing. Initial military export versions had two Turmo IVB engines, each with max rating of 1,044 kW (1,400 shp) and not equipped for air intake anti-icing. From the end of 1973 SA 330G and H versions have been delivered with 1,175 kW (1,575 shp) Turmo IVC engines, with intake anti-icing; these engines are fitted also in the SA 330J and L. Engines are mounted side by side above cabin forward of the main rotor assembly and separated by a firewall. They are coupled to the main rotor transmission box, with shaft drive to tail rotor, and form a completely independent system from the fuel tanks up to the main gearbox inputs. Fuel in four flexible tanks and one auxiliary tank beneath cargo compartment floor, with total capacity of 1,544 litres (339·5 Imp gallons). Provision for additional 1,900 litres (418 Imp gallons) in four auxiliary ferry tanks installed in cabin. External auxiliary tanks (two, each 350 litres; 77 Imp gallons capacity) are available. Each engine is supplied by a pair of interconnected tanks, the lower halves of which have self-sealing walls for protection against small-calibre projectiles. RAF version has fuel flow meters and fuel jettison system. Refuelling point on starboard side of main cabin. Oil capacity 22 litres (4·8 Imp gallons) for engines, 25·5 litres (5·6 Imp gallons) for transmission.

ACCOMMODATION: Crew of one or two side by side on anti-crash seats on flight deck, with jump-seat for third crew member if required. Pilot's door on starboard side and jettisonable door for co-pilot on port side. Internal doorway connects flight deck to cabin, with folding seat in doorway for an extra crew member or cargo supervisor. Dual controls standard. Accommodation in main cabin for 16 individually-equipped troops, six stretchers and six seated patients, or equivalent freight. The number of troops can be increased to 20 in the high-density version. Strengthened floor for cargo-carrying, with lashing points. Jettisonable sliding door on each side of main cabin; or port-side door with built-in steps and starboard-side double door in VIP or airline configurations. Removable panel on underside of fuselage, at rear of main cabin, permits longer loads to be accommodated and also serves as emergency exit on SA 330C and H versions. Removable door with integral steps for access to baggage racks on SA 330F and G versions. A hatch in the floor below the centreline of the main rotor is provided for carrying loads of up to 3,000 kg (6,600 lb) on an internally-mounted cargo sling. A fixed or retractable rescue hoist (capacity 275 kg; 606 lb) can be mounted externally on the starboard side of the fuselage and is standard on the RAF version, together with an abseiling beam, cargo hook and full-width main cabin steps. The cabin can be equipped in 8/9/12-seat VIP, 17-seat commuter or 20-seat high-density layouts, with baggage compartment and/or toilet facilities in rear of cabin. Cabin and flight deck are heated, ventilated and soundproofed. Demisting, de-icing, washers and wipers for pilots' windscreens.

SYSTEMS: Two independent hydraulic systems, each 172 bars (2,500 lb/sq in), supplied by self-regulating pumps driven by the main gearbox. Each system supplies one set of servo unit chambers, the left-hand system supplying in addition the autopilot, landing gear, rotor brake and wheel brakes. Freewheels in main gearbox ensure that both systems remain in operation, for supplying the servo-controls, if the engines are stopped in flight. Other hydraulically-actuated systems can be operated on the ground from the main gearbox, or by external power through the ground power receptacle. There is also an independent auxiliary system, fed through a hand pump, which can be used in an emergency to lower the landing gear and pressurise the accumulator for the parking brake on the ground. Three-phase 200V AC electrical power supplied by two 20kVA 400Hz alternators, driven by the port side intermediate shaft from the main gearbox and available on the ground under the same conditions as the hydraulic ancillary systems. 28·5V 10kW DC power provided from the AC system by two transformer-rectifiers. Main aircraft battery used for self-starting and emergency power in flight. For the latter purpose, an emergency 400VA inverter can supply the essential navigation equipment from the battery, permitting at least 20 min continued flight in the event of a main power failure. SEMCA air-conditioning system in SA 330G and H. De-icing of engines and engine air intakes by warm air bled from compressor. Anti-snow shield for Winter operations.

ELECTRONICS AND EQUIPMENT: Optional communications equipment includes VHF, UHF, tactical HF and HF/SSB radio installations and intercom system. Navigational equipment includes radio compass, radio altimeter, Decca navigator and flight log, Doppler, and

VOR/ILS with glidepath. Autopilot, with provision for coupling to self-contained navigation and microwave landing systems. Full IFR instrumentation available optionally. Standard equipment in the RAF version includes VHF/UHF radio, standby UHF, UHF homing, intercom, IFF/SSR, ICS, radio altimeter, and Decca navigation system with flight log. The search and rescue version has nose-mounted Bendix RDR 1200 search radar, Doppler, and Decca self-contained navigation system, including navigation computer, polar indicator, roller-map display, hover indicator, route mileage indicator and ground speed and drift indicator.

ARMAMENT (optional): A wide range of armament can be carried, including side-firing 20 mm cannon, axial-firing 7·62 mm machine-guns and missiles.

DIMENSIONS, EXTERNAL:
Diameter of main rotor 15·00 m (49 ft 2½ in)
Diameter of tail rotor 3·04 m (9 ft 11½ in)
Distance between rotor centres 9·20 m (30 ft 2¼ in)
Blade chord, main rotor:
Except SA 330J/L 0·54 m (1 ft 9 in)
SA 330J/L 0·60 m (1 ft 11½ in)
Ground clearance of tail rotor 2·00 m (6 ft 6¾ in)
Length overall 18·15 m (59 ft 6½ in)
Length of fuselage 14·06 m (46 ft 1½ in)
Length, blades folded 14·80 m (48 ft 6¾ in)
Width, blades folded 3·50 m (11 ft 5¾ in)
Height overall 5·14 m (16 ft 10½ in)
Height to top of rotor hub 4·38 m (14 ft 4½ in)
Width over wheel fairings 3·00 m (9 ft 10 in)
Wheel track 2·38 m (7 ft 10¾ in)
Wheelbase 4·045 m (13 ft 3 in)
Passenger cabin doors, each:
Height 1·35 m (4 ft 5 in)
Width 1·35 m (4 ft 5 in)
Height to sill 1·00 m (3 ft 3½ in)
Floor hatch, rear of cabin:
Length 0·98 m (3 ft 2¼ in)
Width 0·70 m (2 ft 3½ in)

DIMENSIONS, INTERNAL:
Cabin: Length 6·05 m (19 ft 10½ in)
Max width 1·80 m (5 ft 11 in)
Max height 1·55 m (5 ft 1 in)
Floor area 7·80 m² (84 sq ft)
Usable volume 11·40 m³ (403 cu ft)

AREAS:
Main rotor blades (each) 4·00 m² (43 sq ft)
Tail rotor blades (each) 0·28 m² (3·01 sq ft)
Main rotor disc 177·0 m² (1,905 sq ft)
Tail rotor disc 7·30 m² (78·6 sq ft)
Horizontal stabiliser 1·34 m² (14·4 sq ft)

WEIGHTS:
Weight empty, basic aircraft:
SA 330H 3,536 kg (7,795 lb)
Max T-O and landing weight:
SA 330H 7,000 kg (15,430 lb)
SA 330J/L 7,400 kg (16,315 lb)

PERFORMANCE (SA 330H, Turmo IVC engines, at AUW of 6,700 kg; 14,770 lb):
Never-exceed speed at S/L 147 knots (273 km/h; 169 mph)
Max cruising speed at S/L 138 knots (257 km/h; 159 mph)
Econ cruising speed at S/L 134 knots (248 km/h; 154 mph)
Max rate of climb at S/L 426 m (1,400 ft)/min
Hovering ceiling in ground effect 2,230 m (7,315 ft)
Hovering ceiling out of ground effect 1,350 m (4,430 ft)
Max range at S/L, standard fuel, no reserves 313 nm (580 km; 360 miles)

PERFORMANCE (SA 330J/L: A at 6,000 kg; 13,230 lb AUW, B at 7,400 kg; 16,315 lb AUW):
Never-exceed speed
A 158 knots (294 km/h; 182 mph)
B 142 knots (263 km/h; 163 mph)
Normal cruising speed:
A 146 knots (271 km/h; 168 mph)
B 139 knots (258 km/h; 160 mph)
Max rate of climb at S/L:
A 552 m (1,810 ft)/min
B 366 m (1,200 ft)/min
Service ceiling (30 m; 100 ft/min rate of climb):
A 6,000 m (19,680 ft)
B 4,800 m (15,750 ft)
Hovering ceiling in ground effect:
A, ISA 4,450 m (14,600 ft)
A, ISA +20°C 3,700 m (12,135 ft)
B, ISA 2,300 m (7,545 ft)
B, ISA +20°C 1,500 m (4,920 ft)
Hovering ceiling out of ground effect:
A, ISA 4,250 m (13,940 ft)
A, ISA +20°C 3,600 m (11,810 ft)
B, ISA 1,700 m (5,575 ft)
B, ISA +20°C 950 m (3,115 ft)
Max range at normal cruising speed, no reserves:
A 309 nm (572 km; 355 miles)
B 297 nm (550 km; 341 miles)

AÉROSPATIALE SA 331 SUPER PUMA

This developed version of the Puma is under development to French government contract. Few details are available except that the Super Puma will have a 1,342 kW (1,800 shp) Turboméca Makila turboshaft engine, driving a glassfibre main rotor and 'fenestron' tail rotor of the kind being flight tested on the SA 330Z Puma, and a lengthened main cabin to accommodate an increased payload.

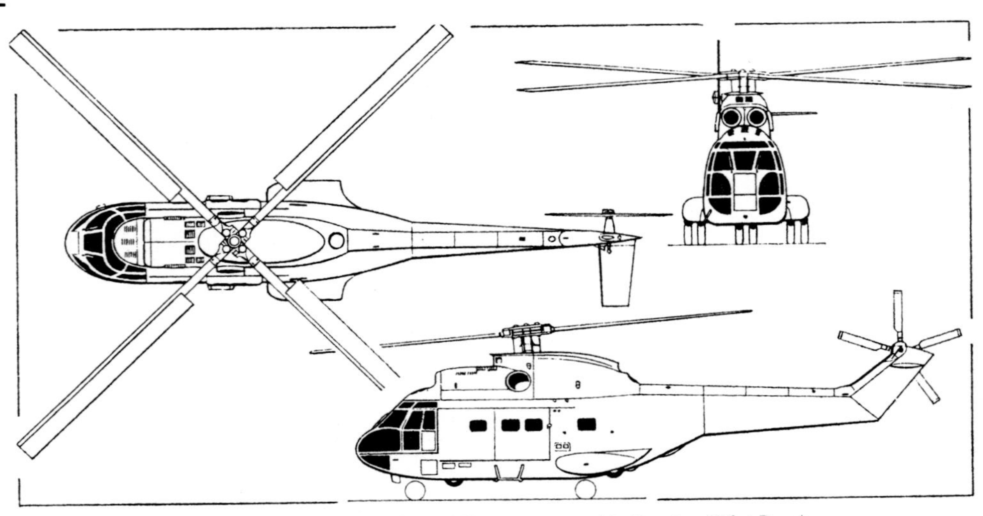

Aérospatiale/Westland SA 330 Puma transport helicopter *(Pilot Press)*

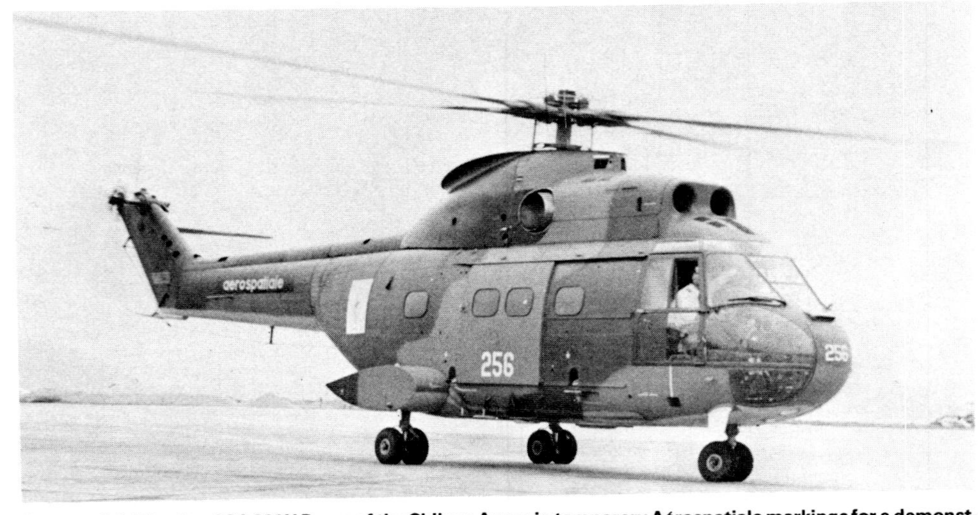

Aérospatiale/Westland SA 330H Puma of the Chilean Army, in temporary Aérospatiale markings for a demonstration tour *(Ronaldo S. Olive)*

AÉROSPATIALE/WESTLAND SA 341/342 GAZELLE

The Gazelle all-purpose lightweight helicopter is a five-seat aircraft, with a Turboméca Astazou turboshaft engine. Under an Anglo-French agreement signed in 1967, it is produced jointly with Westland Helicopters Ltd, and is also built under licence in Yugoslavia.

The first prototype (designated SA 340) made its first flight on 7 April 1967, powered by an Astazou III engine, and the second on 12 April 1968. It was followed by four pre-production SA 341 Gazelles, of which the third was equipped to British Army requirements and given the British military serial number XW276.

The first production SA 341 Gazelle flew for the first time on 6 August 1971, with a longer cabin than its predecessors, enlarged tail unit, additional door on the starboard side at rear (optional on production aircraft) and uprated Astazou IIIA engine.

The second pre-production Gazelle (F-ZWRL) was later re-engined with a 485 kW (650 shp) Turboméca Arriel free-turbine turboshaft engine, as a flying testbed for this power plant. It flew for the first time in this form on 7 December 1974.

Ten versions of the Gazelle have been announced, as follows:

SA 341B. British Army version, with Astazou IIIN engine. Designated Gazelle AH. Mk 1.

SA 341C. British Navy version. Designated Gazelle HT. Mk 2.

SA 341D. Royal Air Force training version. Designated Gazelle HT. Mk 3.

SA 341E. Projected Royal Air Force communications version. Designated Gazelle HCC. Mk 4.

SA 341F. French Army version, with Astazou IIIC engine; 166 procured.

SA 341G. Civil version, with Astazou IIIA engine. Certificated by SGAC on 7 June 1972 and by the FAA on 18 September 1972. In January 1975, it was announced that the SA 341G had become the first helicopter in the world authorised to be flown by a single pilot under IFR Cat I conditions, with a ceiling of 61 m (200 ft) and 550 m (1,800 ft) forward visibility. Equipment fitted to the aircraft which qualified for this FAA certification comprised a Sperry flight director coupled to SFENA servo-dampers.

SA 341H. Military export version, with Astazou IIIB engine.

SA 342J. Similar to SA 342K, for commercial operators. Higher max T-O weight. Certificated by DGAC on 27 April 1976. Deliveries to begin in 1977.

SA 342K. Military version, supplied initially to Kuwait. 650 kW (870 shp) Astazou XIVH engine, with momentum-separation shrouds over intakes.

SA 342L. Similar to SA 342J/K but with improved 'fenestron' tail rotor.

A two-stretcher ambulance configuration has received FAA Standard Type Certification. No major modification is necessary to convert the aircraft to carry two patients longitudinally on the port side of the cabin, one above the other, leaving room for the pilot and a medical attendant in tandem on the starboard side. The dual spine board arrangement weighs 27 kg (60 lb) and stows into the baggage compartment when not in use.

A total of 719 Gazelles had been sold to 89 operators in 26 countries by the Spring of 1976.

Three Class E1c records were set up by the SA 341-01 at Istres on 13 and 14 May 1971. These were: 167·28 knots (310·00 km/h; 192·62 mph) in a straight line over a 3 km course; 168·36 knots (312·00 km/h; 193·87 mph) in a straight line over a 15/25 km course; and 159·72 knots (296·00 km/h; 183·93 mph) over a 100 km closed circuit.

The following details apply to all SA 341 variants:

TYPE: Five-seat light utility helicopter.

ROTOR SYSTEM: Three-blade semi-articulated main rotor and 13-blade shrouded-fan anti-torque tail rotor (known as a 'fenestron' or 'fan-in-fin'). Rotor head and rotor mast form a single unit. The main rotor blades are of NACA 0012 section, attached to NAT hub by flapping hinges. There are no drag hinges. Each blade has a single leading-edge spar of plastics material reinforced with glassfibre, a laminated glass-fabric skin and honeycomb filler. Tail rotor blades are of die-forged light alloy, with articulation for pitch change only. Main rotor blades can be folded manually for stowage. Rotor brake optional.

ROTOR DRIVE: Main reduction gearbox forward of engine, which is mounted above the rear part of the cabin. Intermediate gearbox beneath engine, rear gearbox supporting the tail rotor. Main rotor/engine rpm ratio

378·3 : 6,179. Tail rotor/engine rpm ratio 5,774 : 6,179.

FUSELAGE: Cockpit structure is based on a welded light alloy frame which carries the windows and doors. This is mounted on a conventional semi-monocoque lower structure consisting of two longitudinal box sections connected by frames and bulkheads. Central section, which encloses the baggage hold and fuel tank and supports the main reduction gearbox, is constructed of light alloy honeycomb sandwich panels. Rear section, which supports the engine and tailboom, is of similar construction. Honeycomb sandwich panels are also used for the cabin floors and transmission platform. Tailboom is of conventional sheet metal construction, as are the horizontal tail surfaces and the tail fin.

TAIL UNIT: Small horizontal stabiliser on tailboom, ahead of tail rotor fin.

LANDING GEAR: Steel tube skid type. Wheel can be fitted at rear of each skid for ground handling. Provision for alternative float or ski landing gear.

POWER PLANT (SA 341): One Turboméca Astazou IIIA turboshaft, installed above fuselage aft of cabin and delivering 440 kW (590 shp) for take-off (max continuous rating also 440 kW; 590 shp). Main fuel tank in fuselage, usable capacity 445 litres (98 Imp gallons). Provision for 90 litre (19·8 Imp gallon) auxiliary tank beneath baggage compartment and/or 200 litre (44 Imp gallon) ferry tank inside rear cabin. Total possible usable fuel capacity 735 litres (161 Imp gallons). Refuelling point on starboard side of cabin. Oil capacity 13 litres (2·8 Imp gallons) for engine, 3·5 litres (0·77 Imp gallons) for gearbox.

ACCOMMODATION: Crew of one or two on side-by-side seats in front of cabin, with bench seat to the rear for a further three persons. The bench seat can be folded into floor wells to leave a completely flat cargo floor. Access to baggage compartment via rear cabin bulkhead, or via optional door on starboard side. Cargo tie-down points in cabin floor. Forward-opening car-type door on each side of cabin, immediately behind which are rearward-opening auxiliary cargo loading doors. Baggage compartment at rear of cabin. Ventilation standard. Dual controls optional.

SYSTEMS: Hydraulic system, pressure 39·25 bars (569 lb/sq in), serves three pitch change jacks for main rotor head and one for tail rotor. 28V DC electrical system supplied by 4kW engine-driven generator and 40Ah battery. Optional 26V AC system, supplied by 0·5kVA alternator at 115/200V 400Hz.

ELECTRONICS AND EQUIPMENT: Optional communications equipment includes UHF, VHF, HF, intercom systems and homing aids. Optional navigation equipment includes radio compass, radio altimeter and VOR. Blind-flying instrumentation standard on SA 341B and F, optional on other versions. A variety of operational equipment can be fitted, according to role, including a 700 kg (1,540 lb) cargo sling, 135 kg (300 lb) rescue hoist, one or two stretchers (internally), or photographic and survey equipment.

ARMAMENT: Military loads can include two pods of 36 mm rockets, four AS.11 or Hot wire-guided missiles or two AS.12s with APX-Bézu 334 gyro-stabilised sight, four TOW missiles with XM 26 sight, two forward-firing 7·62 mm machine-guns, reconnaissance flares or smoke markers, cabin-mounted side-firing GE Minigun or 7·62 mm machine-gun or Emerson Minitat or chin turret mounting with pantograph sight system.

DIMENSIONS, EXTERNAL:

Diameter of main rotor	10·50 m (34 ft 5½ in)
Diameter of tail rotor	0·695 m (2 ft 3⅜ in)
Distance between rotor centres	5·85 m (19 ft 2¼ in)
Main rotor blade chord (constant)	0·30 m (11·8 in)
Length overall	11·97 m (39 ft 3⁵⁄₁₆ in)
Length of fuselage	9·53 m (31 ft 3³⁄₁₆ in)
Width, rotors folded	2·015 m (6 ft 7⁵⁄₁₆ in)
Height to top of rotor hub	2·72 m (8 ft 11⅛ in)
Height overall	3·15 m (10 ft 2⅝ in)
Skid track	2·015 m (6 ft 7⁵⁄₁₆ in)
Main cabin doors, each:	
Height	1·05 m (3 ft 4⁹⁄₁₆ in)
Width	1·00 m (3 ft 3¼ in)
Height to sill	0·63 m (2 ft 0¾ in)
Auxiliary cabin doors, each:	
Height	1·05 m (3 ft 4⁹⁄₁₆ in)
Width	0·48 m (1 ft 6¾ in)
Height to sill	0·63 m (2 ft 0¾ in)

DIMENSIONS, INTERNAL:

Cabin: Length	2·20 m (7 ft 2⁹⁄₁₆ in)
Max width	1·32 m (4 ft 4 in)
Max height	1·21 m (3 ft 11⅝ in)
Floor area	1·50 m² (16·1 sq ft)
Volume	1·80 m³ (63·7 cu ft)
Baggage hold volume	0·45 m³ (15·9 cu ft)

AREAS:

Main rotor blades, each	1·57 m² (16·9 sq ft)
Tail rotor blades, each	0·007 m² (0·075 sq ft)
Main rotor disc	86·5 m² (931 sq ft)
Tail rotor disc	0·37 m² (3·98 sq ft)
Fin	0·45 m² (4·84 sq ft)
Tailplane	1·80 m² (19·4 sq ft)

Aérospatiale/Westland SA 341 Gazelle operated by Public Service Electricity & Gas (New Jersey) *(Howard Levy)*

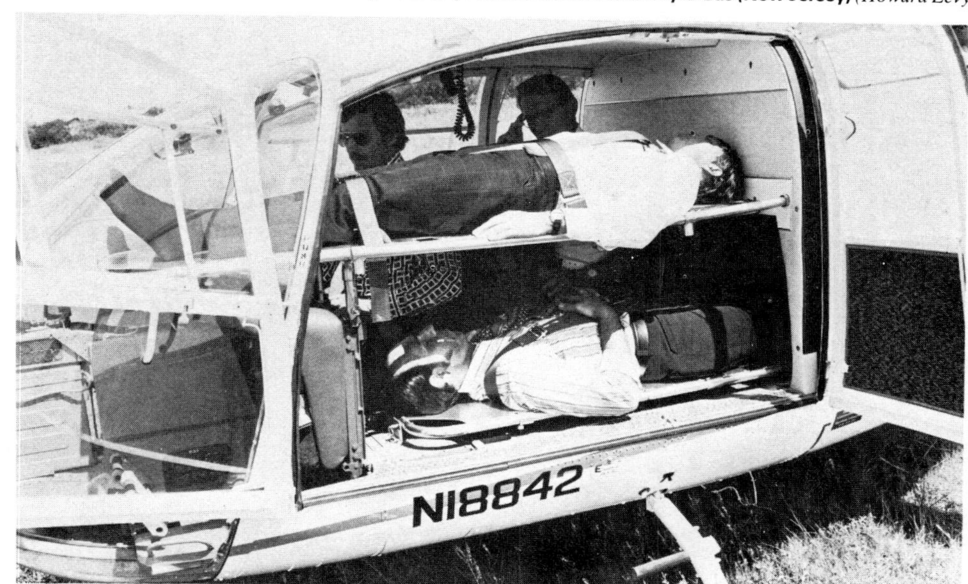

Stretcher installation in ambulance version of Aérospatiale/Westland SA 341 Gazelle

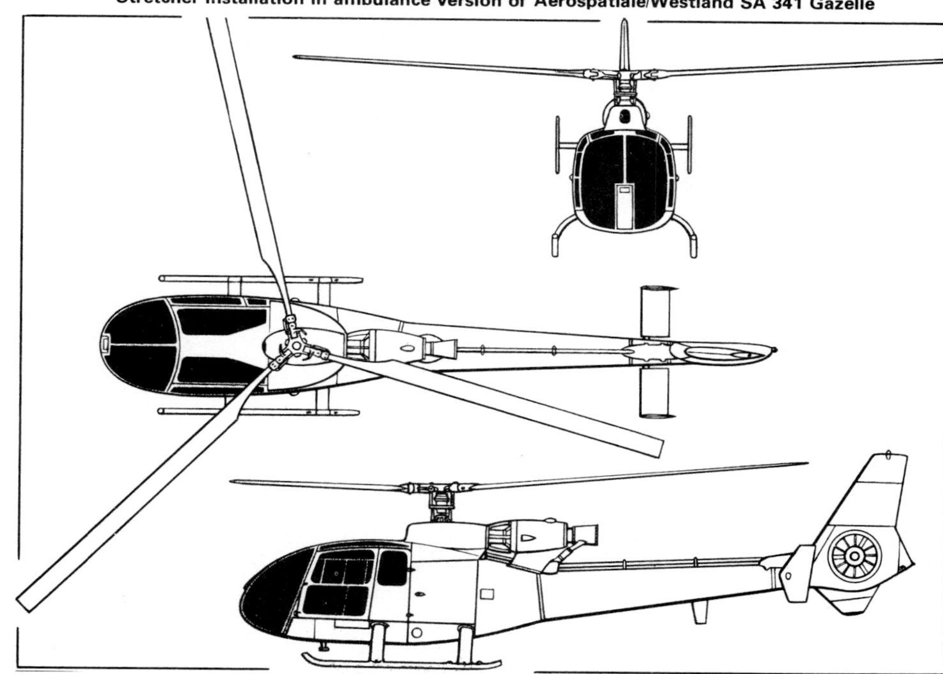

Aérospatiale/Westland SA 341 Gazelle five-seat light utility helicopter *(Pilot Press)*

WEIGHTS AND LOADING:

Weight empty:	
341G	917 kg (2,022 lb)
341H	908 kg (2,002 lb)
Max T-O and landing weight:	
341G/H	1,800 kg (3,970 lb)
342J	1,900 kg (4,190 lb)
Max disc loading:	
341G/H	19·5 kg/m² (4 lb/sq ft)

PERFORMANCE (SA 341, at max T-O weight):

Never-exceed speed at S/L	167 knots (310 km/h; 192·5 mph)
Max cruising speed at S/L	142 knots (264 km/h; 164 mph)
Econ cruising speed at S/L	126 knots (233 km/h; 144 mph)
Max rate of climb at S/L	540 m (1,770 ft)/min
Service ceiling	5,000 m (16,400 ft)
Hovering ceiling in ground effect	2,850 m (9,350 ft)
Hovering ceiling out of ground effect	2,000 m (6,560 ft)
Range at S/L with max fuel	361 nm (670 km; 416 miles)
Range with pilot and 500 kg (1,102 lb) payload	193·5 nm (360 km; 223 miles)

AÉROSPATIALE AS 350 ECUREUIL (SQUIRREL)

Intended as a successor to the Alouette, the AS 350 Ecureuil was designed with an emphasis on low operating and maintenance costs, and low noise and vibration levels. It embodies Aérospatiale's new Starflex type of main rotor hub, made of glassfibre, with elastomeric spherical stops and oleo-elastic frequency matchers.

The decision to build prototypes of the Ecureuil was taken in April 1973. The first of these (F-WVKH) flew on 27 June 1974, powered by an Avco Lycoming LTS 101 turboshaft engine. It was followed on 14 February 1975 by a second prototype (F-WVKI) with a Turboméca Arriel turboshaft. Both engines will be offered in production helicopters, of which the first is expected to fly in mid-1977, permitting deliveries to begin in early 1978.

TYPE: Five/six-seat light general-purpose helicopter.

ROTOR SYSTEM: Three-blade main rotor, with Starflex glassfibre hub in which the three conventional hinges for each blade are replaced by a single ball-joint of rubber/steel sandwich construction, requiring no maintenance. Glassfibre blades, with stainless steel leading-edge sheath, produced by an entirely mechanised process. Two-blade tail rotor; each blade comprises a sheet metal skin around a glassfibre spar, the flexibility of which obviates the need for hinges.

ROTOR DRIVE: Simplified transmission, with single epicyclic main gear train. By comparison with Alouette II, number of gear wheels is reduced from 22 to 9 and number of bearings from 23 to 9. Tail rotor drive-shaft coupling on engine.

FUSELAGE: Basic structure of light alloy pressings, with skin mainly of thermoformed plastic, including doors to cabin and to baggage compartment.

TAIL UNIT: Horizontal stabiliser, of inverted aerofoil section, mid-mounted on tailboom. Sweptback fin, in two sections above and below tailboom.

LANDING GEAR: Steel tube skid type.

POWER PLANT: One 478 kW (641 shp) Turboméca Arriel or 441 kW (592 shp) Avco Lycoming LTS 101 turboshaft engine, mounted above fuselage to rear of cabin. Plastics fuel tanks with total capacity of 530 litres (116·5 Imp gallons).

ACCOMMODATION: Two individual bucket seats at front of cabin; dual controls. Three individual armchair seats, or four-place bench seat at rear of cabin. Large forward-hinged door on each side. Baggage compartment aft of cabin, with full-width upward-hinged door on starboard side. Top of baggage compartment reinforced to provide platform on each side for inspecting and servicing rotor head. Provision for underfuselage cargo sling, capacity 800 kg (1,763 lb).

DIMENSIONS, EXTERNAL:
- Diameter of main rotor: 10·69 m (35 ft 0¾ in)
- Diameter of tail rotor: 1·86 m (6 ft 1¼ in)
- Length overall: 13·00 m (42 ft 8 in)
- Length of fuselage: 10·91 m (35 ft 9½ in)
- Width of fuselage: 1·80 m (5 ft 10¾ in)
- Height overall: 2·94 m (9 ft 7¾ in)
- Wheel track: 2·10 m (6 ft 10¾ in)

DIMENSION, INTERNAL:
- Baggage compartment volume: 1·00 m³ (35·31 cu ft)

WEIGHTS:
- Weight empty: 950 kg (2,094 lb)
- Max T-O weight: 1,900 kg (4,190 lb)

PERFORMANCE (at max T-O weight, with LTS 101 engine):
- Never-exceed speed below 500 m (1,640 ft): 144 knots (268 km/h; 166 mph)
- Max cruising speed at S/L: 124 knots (230 km/h; 143 mph)
- Max cruising speed at 1,500 m (5,000 ft): 118 knots (220 km/h; 136 mph)
- Vertical rate of climb at S/L: 480 m (1,575 ft)/min
- Vertical rate of climb at 1,500 m (5,000 ft): 288 m (945 ft)/min
- Max rate of climb at S/L: 540 m (1,770 ft)/min
- Max rate of climb at 1,500 m (5,000 ft): 408 m (1,340 ft)/min
- Service ceiling: 5,800 m (19,025 ft)
- Hovering ceiling in ground effect: 2,750 m (9,025 ft)
- Hovering ceiling out of ground effect: 2,000 m (6,560 ft)
- Max range at S/L: 432 nm (800 km; 497 miles)
- Max range at 1,500 m (5,000 ft): 496 nm (920 km; 571 miles)

AÉROSPATIALE SA 360 DAUPHIN

The SA 360 Dauphin was developed, with the twin-engined SA 365 variant (described separately), as a replacement for the Alouette III. The first of two SA 360 prototypes (F-WSQL) flew for the first time on 2 June 1972, powered by a 730 kW (980 shp) Turboméca Astazou XVI turboshaft engine. After 180 flights, it was re-engined with an Astazou XVIIIA turboshaft and modified in certain respects, including the addition of small weights to the rotor blades, to eliminate ground resonance and reduce vibration to an unprecedented level, even at high speed. The aircraft flew for the first time in its modified form on 4 May 1973, having been joined by the second prototype (F-WSQX) on 29 January 1973.

DGAC certification of the SA 360 was awarded on 18 December 1975, followed by FAA certification on 31 March 1976. At that time, Aérospatiale had received a total of 50 orders for the SA 360 and SA 365, from 11 operators in four countries, including the USA. Deliveries were scheduled to begin during 1976.

Aérospatiale AS 350 Ecureuil (Avco Lycoming LTS 101 turboshaft engine)

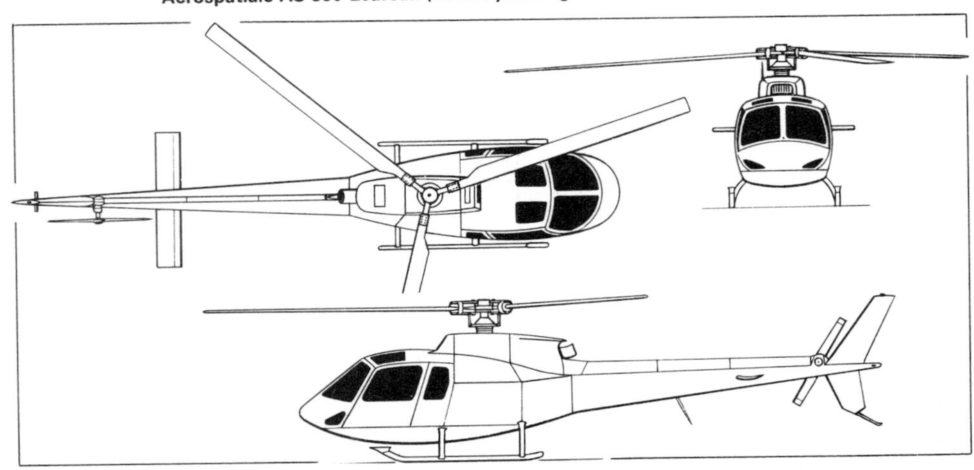
Aérospatiale AS 350 Ecureuil five/six-seat light helicopter *(Pilot Press)*

Three helicopter speed records in Class Eld (1,750 to 3,000 kg weight) were set up at Istres by the first prototype of the SA 360 on 15, 16 and 17 May 1973, piloted by Roland Coffignot. Carrying a payload equivalent to eight persons and fuel for one hour's flying, the SA 360 achieved, successively, 161·4 knots (299 km/h; 185·8 mph) over a 100 km closed circuit; 168·4 knots (312 km/h; 193·9 mph) over a 3 km course; and 163·5 knots (303 km/h; 188·3 mph) over a 15 km course.

As part of the continuing development programme, an SA 360 was to be flight tested with an Astazou XX turboshaft engine during 1976.

TYPE: Turbine-powered general-purpose helicopter.

ROTOR SYSTEM: Four-blade semi-articulated main rotor and 13-blade shrouded-fan anti-torque tail rotor (known as a 'fenestron' or 'fan-in-fin'). Main rotor blades are of symmetrical NACA 0012 section, with a theoretical twist of 8° and constant chord, and are attached to the NAT hub via flapping hinges. There are no drag hinges. Each blade has a single leading-edge spar of polyester plastics, extending back to about 30% chord at top and bottom. The outer skin is of glassfibre, with an inner skin of carbon fibre, and the entire blade is filled with Nomex honeycomb. The leading-edge is formed by a layer of Vulkollan plastics with an outer protective shield of thin-gauge stainless steel. Tail rotor blades are of die-forged light alloy, with articulation for pitch change only. Main rotor blades can be folded manually for stowage. Rotor brake and main rotor blade de-icing optional.

ROTOR DRIVE: Main reduction gearbox forward of engine, which is mounted above the fuselage to the rear of the cabin. Output shaft enters main transmission box above the driveshaft to the tail rotor. Self-lubricating bearings. Main rotor rpm: 348 normal; 393 in autorotation. Tail rotor rpm: 4,700.

FUSELAGE: Conventional all-metal assembly of cabin and semi-monocoque tailboom. Cabin built on a strong box structure embodying two transverse frames and the cabin floor.

TAIL UNIT: Horizontal stabiliser mid-set on tailboom, forward of shrouded tail rotor, with endplate fins. Tailboom terminates in large fin of unsymmetrical section, housing the tail rotor. The section of this fin is such that in cruising flight it counters the torque of the main rotor; the tail rotor is thus required to provide only yaw control, with minimal variation of pitch, requiring only

Second prototype of the Aérospatiale SA 360 Dauphin ten-seat general-purpose helicopter

small power intake.

Landing Gear: Prototypes have Eram non-retractable tailwheel-type landing gear, with single wheel on each unit. Main legs embody hydraulic shock-absorbers. Tailwheel carried on anti-shimmy leg which can be locked manually in central position. Dunlop main-wheel tyres size 355 × 150-4, pressure 5 bars (73 lb/sq in). Dunlop tailwheel tyre size 260 × 80-4, pressure 5 bars (73 lb/sq in). Disc brakes on main wheels. Wheel fairings standard. Two main wheels will retract forward into cabin underfloor structure on production aircraft. Provision for emergency floats or skis.

Power Plant: One Turboméca Astazou XVIIIA turboshaft engine, delivering 783 kW (1,050 shp) for take-off. Two Kléber-Colombes bag-type fuel tanks under cabin floor, total normal capacity 475 litres (104 Imp gallons). Provision for larger tanks, capacity 660 litres (145 Imp gallons) and for two ferry tanks, one of 275 litres (60·5 Imp gallons) capacity on the cabin floor and another of 200 litres (44 Imp gallons) capacity at the back of the cabin. Tanks can be of self-sealing type in military versions.

Accommodation: Standard ten-seat version has seats for pilot (to starboard) and co-pilot or passenger in front, and two rows of four seats to the rear. Interior of the cabin is clear except for a vertical duct, housing the flying control rods, positioned centrally aft of the centre row of seats. Two large forward-hinged doors on each side. Compartment for hand baggage or coats aft of rear row of seats. Separate main baggage compartment aft of cabin, with door on starboard side. Alternative 13-seat layout has an extra row of three seats between the four-seat rows, and no space for hand baggage or coats. Ambulance version carries four stretcher patients, a medical attendant and two crew. Mixed-traffic version carries six persons at front of cabin, with 2·50 m³ (88·3 cu ft) of cargo space to the rear. The floor in this area will support a loading of 610 daN/m² (125 lb/sq ft). Executive versions are available with VIP interiors for four or five passengers. Cabin is heated and ventilated. Provision for 1,250 kg (2,755 lb) capacity cargo sling, 272 kg (600 lb) capacity rescue hoist, and a wide range of other civil and military equipment, including six or eight Hot missile launchers and underfuselage Minitat gun turret.

Dimensions, external:
- Diameter of main rotor: 11·50 m (37 ft 8¾ in)
- Main rotor blade chord (constant): 0·35 m (1 ft 1¾ in)
- Diameter of tail rotor: 0·90 m (2 ft 11⁷/₁₆ in)
- Length overall: 13·20 m (43 ft 3½ in)
- Length of fuselage: 10·98 m (36 ft 0 in)
- Height overall: 3·50 m (11 ft 6 in)
- Stabiliser span: 3·15 m (10 ft 4 in)
- Wheel track: 1·95 m (6 ft 4¾ in)
- Wheelbase: 7·23 m (23 ft 8¾ in)
- Cabin doors (fwd, each):
  - Height: 1·16 m (3 ft 9½ in)
  - Width: 1·14 m (3 ft 9 in)
- Cabin doors (aft, each):
  - Height: 1·16 m (3 ft 9½ in)
  - Width: 0·87 m (2 ft 10¼ in)
- Freight compartment door:
  - Height: 0·56 m (1 ft 10 in)
  - Width: 0·75 m (2 ft 5½ in)

Dimensions, internal:
- Cabin: Usable length: 2·30 m (7 ft 6½ in)
  - Height at front: 1·40 m (4 ft 7 in)
  - Height at rear: 1·06 m (3 ft 5¾ in)
  - Width at front: 1·92 m (6 ft 3½ in)
  - Width at rear: 1·60 m (5 ft 3 in)
  - Floor area: 4·2 m² (45·20 sq ft)
  - Volume: 5·0 m³ (176 cu ft)
- Baggage compartment volume: 1·00 m³ (35·31 cu ft)

Weights:
- Basic operating weight: 1,555 kg (3,428 lb)
- Max payload:
  - internal: 1,150 kg (2,500 lb)
  - slung: 1,500 kg (3,300 lb)
- Max T-O weight: 3,000 kg (6,613 lb)

Performance (at max T-O weight):
- Never-exceed speed at S/L: 170 knots (315 km/h; 196 mph)
- Max cruising speed at S/L: 150 knots (278 km/h; 172 mph)
- Econ cruising speed at S/L: 132 knots (245 km/h; 152 mph)
- Max rate of climb at S/L: 540 m (1,770 ft)/min
- Hovering ceiling in ground effect: 2,250 m (7,380 ft)
- Range at S/L with max fuel: 351 nm (650 km; 405 miles)
- Endurance at S/L: 4 hr

AÉROSPATIALE SA 365 DAUPHIN 2

Announced in early 1973, the SA 365 is a twin-engined version of the SA 360, powered by Turboméca Arriel turboshaft engines, each rated at 485 kW (650 shp). The prototype (F-WVKE) flew for the first time on 24 January 1975. Delivery of production SA 365s is scheduled to begin in early 1978.

In mid-1976, an SA 365 was flying with a Starflex glassfibre rotor hub, as described for the AS 350 Ecureuil. This will be standard on production aircraft.

The SA 365 is designed for single-pilot IFR flight. It differs from the SA 360 in the following details:

Rotor Drive: The installation of free-turbine engines has eliminated the need for a clutch in the output drive from each engine into the main gearbox.

Fuselage: The profile of the firewall between the two turboshaft engines is extended rearward in the form of a curved fairing which blends into the dorsal spine fairing over the tail rotor driveshaft.

Tail Unit: Horizontal stabiliser has inverted-camber aerofoil section, and the fixed vertical tail fins are offset to produce a lateral component which enhances the anti-torque function of the unsymmetrical surfaces. These features increase the efficiency of the 'fenestron' tail rotor, notably during hover.

Power Plant: Two Turboméca Arriel free-turbine turboshaft engines, each rated at 485 kW (650 shp), mounted side by side above the fuselage, aft of the main rotor driveshaft. Four separate bag-type fuel tanks, filling full width of fuselage under cabin floor. Two refuelling points aft of rear cabin door on port side.

Weights:
- Weight empty: 1,823 kg (4,018 lb)
- Max T-O weight: 3,400 kg (7,495 lb)

Performance ('clean' aircraft. A at AUW of 2,600 kg; 5,730 lb; B at AUW of 3,200 kg; 7,055 lb, zero wind, ISA at S/L, unless otherwise stated):
- Never-exceed speed:
  - A, B: 170 knots (315 km/h; 196 mph)
- Max cruising speed:
  - A: 151 knots (280 km/h; 173 mph)
  - B: 144 knots (268 km/h; 165 mph)
- Econ cruising speed:
  - A: 146 knots (270 km/h; 168 mph)
  - B: 140 knots (260 km/h; 161 mph)
- Max rate of climb at S/L:
  - A: 570 m (1,870 ft)/min
  - B: 390 m (1,280 ft)/min
- Service ceiling (30 m; 100 ft/min climb):
  - A, B: 4,575 m (15,000 ft)
- Hovering ceiling in ground effect:
  - A (ISA to ISA + 20°C): 4,900 m (16,070 ft)
  - B (ISA): 2,900 m (9,510 ft)
  - B (ISA + 20°C): 2,050 m (6,725 ft)
- Hovering ceiling out of ground effect:
  - A (ISA): 4,300 m (14,105 ft)
  - A (ISA + 20°C): 3,600 m (11,810 ft)
  - B (ISA): 2,900 m (9,510 ft)
  - B (ISA + 20°C): 1,910 m (6,275 ft)
- Max range at econ cruising speed, no reserves:
  - A: 313 nm (580 km; 360 miles)
  - B: 296 nm (548 km; 340 miles)
- Max endurance at 70 knots (130 km/h; 81 mph), no reserves:
  - A: 3 hr 18 min
  - B: 3 hr 0 min

Operating Limits:
- Max pressure altitude: 4,575 m (15,000 ft)
- Max temperature: 45°C
- Min temperature: −40°C

AÉROSPATIALE SA 366 DAUPHIN

This variant of the Dauphin 2, with two 441 kW (592 shp) Avco Lycoming LTS 101 turboshaft engines, flew for the first time on 28 January 1975. Similar to the SA 365 except for its power plant, it will not be developed further.

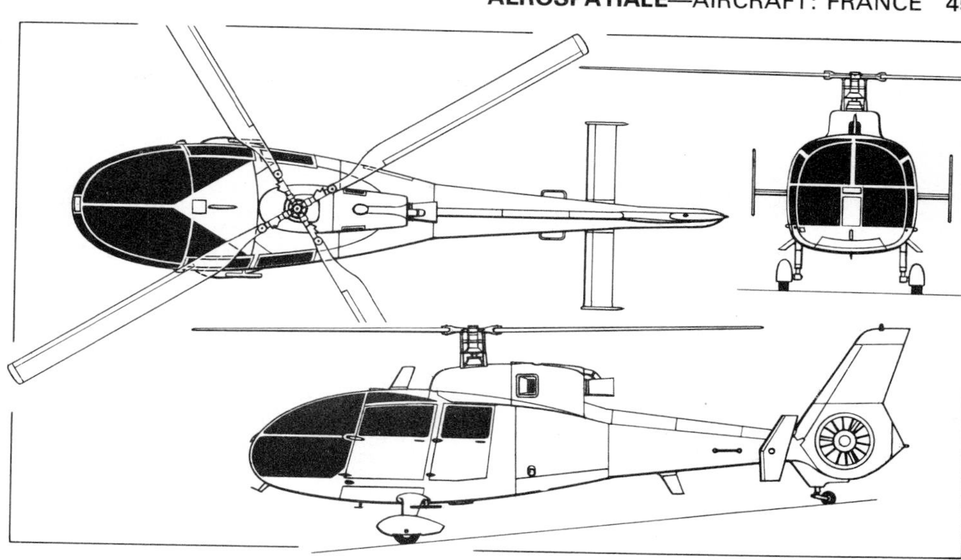

Aérospatiale SA 360 Dauphin (Turboméca Astazou XVIIIA turboshaft engine) *(Pilot Press)*

Prototype Aérospatiale SA 365, a version of the Dauphin with twin Turboméca Arriel turboshaft engines

C.A.A.R.P.
COOPÉRATIVE DES ATELIERS AÉRONAUTIQUES DE LA RÉGION PARISIENNE

HEAD OFFICE AND WORKS:
Aérodrome, 78650-Beynes
Telephone: 489-10-69
DIRECTOR: Auguste Mudry

This company specialised at first in aircraft modification and repair. It then began the manufacture, under subcontract, of components for sailplanes, and in 1965 took over from Scintex-Aviation production of the Super Emeraude light aircraft. It also built a prototype of the C.P. 100 side-by-side two- seat aerobatic version of the Emeraude.

C.A.A.R.P. is now associated with Avions Mudry et Cie (which see) in production of the CAP 10 and CAP 20 aerobatic aircraft, and acts as design and development centre for aircraft produced by the two companies. It is responsible for complete manufacture of the CAP 20, described below, and for fuselages for the CAP 10. Final assembly of the CAP 10 is undertaken by Mudry at Bernay.

C.A.A.R.P./MUDRY CAP 20

The CAP 20, developed in parallel with the CAP 10 (see entry under 'Mudry' in this section), is essentially a single-seat derivative of the latter aircraft, although of almost completely new design. Construction of a prototype was financed by the SGAC, and this aircraft (F-WPXU) flew for the first time on 29 July 1969. It was followed by eight more CAP 20s, of which six were delivered to the Equipe de Voltige of the Armée de l'Air (EVAA).

Under the terms of an agreement between the Armée de l'Air and the Mudry group, several significant modifications were made and tested on two of the production aircraft, to improve the crispness of their manoeuvres, their pull-outs, rate of roll and overall performance when inverted.

The progressive design changes are covered by the following designations:

CAP 20A. This is the second of six production aircraft delivered to the EVAA, as modified and flown for the first time in its new form in March 1973. Its landing gear was lightened by removal of the leg and wheel fairings, and its wings were remounted without dihedral. Subsequently, the wingtips were shortened in span (CAP 20A1); the tips were then fitted with wooden endplates (CAP 20A2); finally, the endplates were removed, and larger slotted ailerons were fitted in place of the former type with automatic tabs, which had required excessive balance weights (CAP 20A3). Extensive flight testing of the A3 model, including inverted flight and measurement of accelerations and vibration, showed considerable improvement by comparison with the basic CAP 20, except that the absence of dihedral made it difficult to hold precise positions in manoeuvres such as four-point or eight-point hesitation rolls.

CAP 20B. First flown in January 1974, this represented a further modification of the CAP 20A with 1° 30′ dihedral to improve rolling aerobatics; a new wing spar of spruce and glassfibre which is both stronger (+20*g*) and more rigid because of its symmetrical form; and ailerons similar to those of the CAP 20A3.

CAP 20C. This is a modification of the second prototype, which also belongs to the Armée de l'Air. It differs from the standard CAP 20 only in having a 194 kW (260 hp) Lycoming engine, adapted in Switzerland, with modified oil and fuel feed for inverted flight.

CAP 20D. This version has a 194 kW (260 hp) Lycoming AEIO-540-D4B5 engine, wings without dihedral, and a structure stressed for +12*g* load factors. Static testing began on 27 February 1975, and the prototype went to the CEV for certification in February 1976.

CAP 20L. Redesigned lightweight version. Described separately.

Following development testing of the modified CAP 20s and of the lightweight CAP 20L, all aircraft were undergoing modification in the Spring of 1976. Wider-chord ailerons, as fitted to the CAP 20L, are now standard. The wingtips have been clipped and fitted with new tip fairings, reducing the span to 7·64 m (25 ft 0¾ in).

The description below applies to the prototype CAP 20:

TYPE: Single-seat aerobatic light aircraft.

WINGS: Cantilever low-wing monoplane. All-wood single-spar wings, of NACA 23012 section, similar in construction and planform to those of CAP 10 but with only ailerons on trailing-edge, and hydraulically-actuated airbrakes. Dihedral 5° from roots.

FUSELAGE: Conventional all-wood structure, of basically triangular section with rounded top-decking. Wooden covering, except for laminated plastics engine cowling.

TAIL UNIT: Cantilever all-wood structure. Trim tab in rudder and each elevator.

LANDING GEAR: Non-retractable tailwheel type. Streamline fairings on main wheels and legs.

POWER PLANT: One 149 kW (200 hp) Lycoming AIO-360-B1B flat-four engine, driving a Hartzell two-blade constant-speed metal propeller. Fuel tank aft of cockpit, with system modified to permit periods of inverted flight. Provision for 75 litre (16·5 Imp gallon) under-fuselage auxiliary tank for ferrying.

ACCOMMODATION: Single seat under transparent moulded canopy which opens sideways to starboard.

DIMENSIONS, EXTERNAL:
Wing span:
Prototype 8·04 m (26 ft 4¾ in)
Current aircraft 7·64 m (25 ft 0¾ in)
Wing aspect ratio:
Prototype 5·96
Length overall 7·21 m (23 ft 7¾ in)
Height overall 1·55 m (5 ft 1 in)
Propeller diameter 1·83 m (6 ft 0 in)

AREA:
Wings, gross:
Prototype 10·85 m² (116·79 sq ft)

WEIGHTS AND LOADING:
Weight empty 640 kg (1,410 lb)
Max T-O weight (Aerobatic) 760 kg (1,675 lb)
Max wing loading (prototype) 70·0 kg/m² (14·3 lb/sq ft)

PERFORMANCE (at max Aerobatic T-O weight):
Never-exceed speed 202 knots (376 km/h; 233 mph)
Max cruising speed 183 knots (340 km/h; 211 mph)
Max speed for aerobatics 146 knots (270 km/h; 168 mph)
Stalling speed 52·5 knots (96 km/h; 60 mph)
g limits +8; −6

C.A.A.R.P. CAP 20L

First flown on 15 January 1976, the CAP 20L ('léger') is a lightweight development of the basic CAP 20. It is intended as both a comparatively inexpensive high-performance type for private individuals to fly in competitive aerobatics, and as an aircraft superior to any previous version of the CAP 20 for the Armée de l'Air and for international competition against the best aerobatic types of foreign design. To this end, it is offered in two versions:

CAP 20L-180. With 134 kW (180 hp) Lycoming AEIO-360 engine and constant-speed propeller. Designed for load factors of +8*g* and −6*g*. Shown in accompanying photograph.

CAP 20L-200. With 149 kW (200 hp) Lycoming engine and constant-speed propeller.

A pre-production batch of five CP 20L-180s was under construction in mid-1976. The following data apply to this version:

DIMENSIONS, EXTERNAL:
Wing span 7·43 m (24 ft 4½ in)
Length overall 7·05 m (23 ft 1½ in)

AREA:
Wings, gross 10·40 m² (111·9 sq ft)

WEIGHTS:
Weight empty 460 kg (1,014 lb)
Max T-O weight (Aerobatic) 600 kg (1,322 lb)

PERFORMANCE (estimated):
Max cruising speed 146 knots (270 km/h; 168 mph)
Stalling speed, flaps down 43·5 knots (80 km/h; 50 mph)
Max rate of climb at S/L 780 m (2,560 ft)/min

C.A.A.R.P./Mudry CAP 20 single-seat aerobatic aircraft (Lycoming AIO-360-B1B engine)

C.A.A.R.P. CAP 20L-180 single-seat aerobatic aircraft, with new wings and other modifications

CERVA (G.I.E.)
CONSORTIUM EUROPÉEN DE RÉALISATION ET DE VENTES D'AVIONS (GROUPEMENT D'INTÉRÊTS ÉCONOMIQUES)

ADDRESS:
13 rue Saint-Honoré, 78000-Versailles
Telephone: 950-63-95

This company was formed in 1971 by Siren SA and Wassmer-Aviation SA (which see), each of which has a 50% holding, to build and market an all-metal version of the Wassmer Super 4/21, known as the CE.43 Guépard. It is also developing two more powerful versions of the same design, named CE.44 Couguar and CE.45 Léopard.

CERVA CE.43 GUÉPARD (CHEETAH)

The CE.43 Guépard is basically an all-metal derivative of the WA Super 4/21 (see 1972-73 *Jane's*), retaining the general features of that aircraft. The prototype (F-WSNJ) flew for the first time on 18 May 1971 and was exhibited at the Paris Air Show later that month. It was followed by a second flying prototype, which was delivered to the Service de la Formation Aéronautique (SFA), and a further airframe for static testing by the CEAT at Toulouse.

Following certification by the SGAC, on 1 June 1972, the Guépard was put into production. Five were delivered to the SFA in 1975; subsequent deliveries have been to private owners.

Like the prototypes, and the WA Super 4/21, these initial production aircraft have a 186 kW (250 hp) Lycoming engine and are basically four/five-seaters. Development of the design to have six seats and a more powerful engine is under way (see separate entries on the CERVA Couguar and Léopard), and a light cargo-carrying version is projected with the rear seats removed.

The basic airframe of the Guépard is manufactured by Siren at Argenton-sur-Creuse. Equipment installation, final assembly and flight testing are performed by Wassmer at Issoire.

TYPE: Four/five-seat all-metal light aircraft.

WINGS: Cantilever low-wing monoplane of constant

chord, except at roots, where leading-edge is swept forward. Wing section NACA 63-618. Dihedral 6° from roots. No sweep. All-metal structure, with I-section main spar at 33% chord, light plate front spar at 3·2% chord, and light rear spar at 65% chord to carry ailerons and flaps. Each wing contains 13 ribs, four top-surface stringers and three bottom-surface stringers, and is covered with AU4G alloy sheet. All-metal mass-balanced unslotted ailerons, with top hinges. Electrically-actuated slotted flaps of all-metal construction. No trim tabs. Landing and navigation lights in wingtips. De-icing system optional.

Fuselage: All-metal 'boat-type' cabin structure of heavy frames, stringers and skin. Conventional metal semi-monocoque rear fuselage. Engine cowling, cabin top and cabin door of polyester plastics.

Tail Unit: Cantilever metal structure, except for Klégécel ribs in rudder, with vertical surfaces swept back at 37°. All-moving mass-balanced horizontal surfaces, with anti-tab at root on each side. Controllable tab on rudder.

Landing Gear: Retractable tricycle type, with steerable nosewheel. Main wheels retract inward into wing roots, nosewheel rearward. Electrical retraction. Oleo-pneumatic shock-absorbers. Main-wheel tyres size 420-150. Nosewheel tyre size 360-125·7. Differentially-operated hydraulic brakes. Parking brake. Small tail bumper.

Power Plant: One 186 kW (250 hp) Lycoming IO-540-C4B5 flat-six engine, driving a Hartzell HC 22YK two-blade variable-pitch propeller. Laminated plastics main fuel tank in centre of each wing, aft of main spar; laminated plastics auxiliary tank in each wing, outboard of main tank. Total capacity of main fuel tanks 220 litres (48·4 Imp gallons). Total capacity with auxiliary tanks 440 litres (96·8 Imp gallons). Refuelling point above each tank.

Accommodation: Two adjustable seats side by side at front, with dual controls and adjustable rudder bars. EFA 602 M2 seat harness. Rear bench seat for two persons, or three or four individual adjustable rear seats. Upward-hinged door on starboard side, raised by pneumatic jacks. Baggage compartment aft of cabin, with upward-hinged door on starboard side. Foul-weather window. Front and rear of cabin heated and ventilated.

Electronics and Equipment: Standard equipment by King Radio Corp includes dual N1 KX 175B VHF/VOR, KN 77 VOR/LOC converter, KN 73 glide-slope receiver, KR 85 ADF, KT 76 transponder, OBS KNI 520 indicator, and KMA 20, including marker beacon receiver. Alternative standard installation by Narco comprises 2 Comm 11A VHF, Nav 12 VOR, Nav 11 VOR, ADF 31B radio compass, UGR 2 glideslope receiver, MBT marker beacon receiver, KT 75R ATC, and associated control equipment and antennae. DME and HF are available. Standard equipment includes IFR instrumentation, anti-collision beacon, dual landing lights, navigation lights, interior lights, heated pitot and fire extinguisher. Optional equipment includes IFR instrumentation for co-pilot.

CERVA CE.43 Guépard four/five-seat light aircraft (Lycoming IO-540-C4B5 engine)

Systems: Electrical system includes Prestolite 12V 80A alternator, regulator and starter, and Fulmen 40Ah battery.

Dimensions, external:

Wing span	10·00 m (32 ft 9½ in)
Wing chord (constant)	1·60 m (5 ft 3 in)
Length overall	7·85 m (25 ft 9 in)
Height overall	2·90 m (9 ft 6 in)
Tailplane span	3·46 m (11 ft 4 in)
Wheel track	3·30 m (10 ft 10 in)
Wheelbase	2·10 m (6 ft 10½ in)

Dimension, internal:

Cabin: Max width	1·09 m (3 ft 7 in)

Area:

Wings, gross	16·0 m² (172 sq ft)

Weights and Loading:

Weight empty	890 kg (1,962 lb)
Max T-O weight: Utility	1,460 kg (3,220 lb)
Wing loading: Utility	91·25 kg/m² (18·69 lb/sq ft)

Performance (at max T-O weight, Utility category):

Never-exceed speed	189 knots (350 km/h; 217 mph)
Max level speed	172 knots (320 km/h; 198 mph)
Normal operating speed	150 knots (280 km/h; 174 mph)
Min flying speed	50 knots (93 km/h; 58 mph)
Max rate of climb at S/L	330 m (1,080 ft)/min
Service ceiling	5,300 m (17,400 ft)
Range with max fuel	1,565 nm (2,900 km; 1,800 miles)
g limits	+4·4; −2·2

CERVA CE.44 COUGUAR

The Couguar is generally similar to the Guépard, but has a 212 kW (285 hp) Continental Tiara engine, driving a three-blade Hoffmann propeller. Built under French government contract, the prototype flew for the first time on 24 October 1974. The second Couguar was delivered to the SFA in 1975.

Weight:

Max T-O weight	1,655 kg (3,650 lb)

CERVA CE.45 LÉOPARD

Flight testing of this further development of the Guépard, with a 231 kW (310 hp) Lycoming TIO-540 flat-six engine, began at the end of 1975. No details are available.

DASSAULT-BREGUET

AVIONS MARCEL DASSAULT/BREGUET AVIATION

Head Office: 27 rue du Professeur Victor Pauchet, 92420-Vaucresson

Postal Address:
BP 32, 92420-Vaucresson

Telephone: 970-38-50 and 970-75-21

Telex: 60755 Brevau

Press Information Office: 46 avenue Kléber, 75116-Paris

Works: 92210-Saint-Cloud, 77000-Melun-Villaroche, 95100-Argenteuil, 92100-Boulogne/Seine, 78140-Vélizy-Villacoublay, 33610-Martignas, 33700-Bordeaux-Mérignac, 33400-Talence, 33630-Cazaux, 31770-Toulouse-Colomiers, 64600-Biarritz-Anglet, 64200-Biarritz-Parme, 13800-Istres, 74370-Argonay, 59113-Lille-Seclin, 86000-Poitiers

Founder: Marcel Dassault

Chairman: B. C. Vallières

Deputy General Manager: X. D'Iribarne

General Secretary: C. Edelstenne

General Technical Manager: J. Cabrière

Technical Adviser: H. Deplante

Export Technical Manager: Y. Thiriet

Military Aircraft Manager: F. Serralta

Military Aircraft Manager: P. E. Jaillard

Civil Aircraft Manager: B. Latreille

Production Manager: C. Barrière

Flight Test Manager: J. F. Cazaubiel

Press Information Manager: A. Segura

Avions Marcel Dassault/Breguet Aviation resulted from the merger on 14 December 1971 of the Avions Marcel Dassault and Breguet Aviation companies. It is engaged in the development and production of military and civil aircraft, guided missiles and servo control equipment.

The company's principal current products are the Mirage III multi-purpose fighter; Mirage 5 ground-support aircraft; Mirage F1 fighter; Jaguar tactical support aircraft and advanced trainer (under Anglo-French collaborative programme; see SEPECAT in the International section of this edition); Falcon 20 executive transport and its scaled-down development, the Falcon 10.

Under development are the Delta Mirage 2000 and Delta Super-Mirage multi-role combat aircraft; Alpha Jet basic and advanced training and strike aircraft (under Franco-German programme); Super Étendard carrier-based combat aircraft; Atlantic Mk II second-generation version of the Atlantic maritime patrol aircraft for operational service at the end of the present decade; and Falcon 50 long-range executive transport.

The company is also engaged in the development and manufacture of equipment, notably hydraulic and electro-hydraulic powered aircraft controls with 'feel' simulation. Its subsidiary, Electronique Marcel Dassault, is engaged on a variety of projects, including research and production of weapon system equipment for air-to-air, surface-to-air and surface-to-surface missiles; tracking radar for use with missiles; and equipment for supersonic jet aircraft.

Series production of Avions Marcel Dassault/Breguet Aviation aircraft is undertaken under a wide-spread subcontracting programme, with final assembly and flight testing being handled by the company. Its 17 separate works and facilities cover more than 700,000 m² (7,535,000 sq ft), with a total of 15,000 employees, including 3,000 engineers.

The principal works is at Saint-Cloud, where the design office is situated and where most of the company's prototypes are built. Mérignac is the flight test centre for production aircraft. Istres is the flight test centre for all prototypes, and Cazaux the flight test centre for armament. Biarritz is devoted to production, modification and overhaul.

Avions Marcel Dassault/Breguet Aviation has established close links with the industries of other countries. The Atlantic programme associated manufacturers in Belgium, France, Germany, Italy and the Netherlands under the overall responsibility of their respective governments. In the same way the British and French governments are associated in the SEPECAT concern, formed to control the Dassault-Breguet/BAC Jaguar programme; and the German and French governments are associated in the Dassault-Breguet/Dornier Alpha Jet programme.

Purchase of Mirage fighters by Belgium and Spain led to Belgian and Spanish participation in Mirage III/5 and Mirage F1 production.

Production by Dassault-Breguet in 1975 totalled 116 military aircraft and 63 civil aircraft.

DASSAULT MIRAGE III

The Mirage III was designed initially as a Mach 2 high-altitude all-weather interceptor, capable of performing ground support missions and requiring only small airstrips. Developed versions include a two-seat trainer, long-range fighter-bomber and reconnaissance aircraft, and a total of more than 1,300 Mirage IIIs of all types had been ordered by 1 January 1976, including licence production abroad. Production in France is at the rate of five Mirage IIIs and Mirage 5s (described separately) each month.

The experimental prototype flew for the first time on 17 November 1956, powered by a SNECMA Atar 101G turbojet with afterburner (44·1 kN; 9,900 lb st).

Production versions of the Mirage III are as follows:

Mirage III-A. Pre-series of ten aircraft, with SNECMA Atar 9B turbojet (58·8 kN; 13,225 lb st). First Mirage III-A flew on 12 May 1958. Last six equipped to production standard, with CSF Cyrano *Ibis* air-to-air radar.

Mirage III-B. Two-seat version of III-A, with tandem seating under one-piece canopy; radar deleted, but fitted with radio beacon equipment. Fuselage 0·6 m (23·6 in) longer than that of III-A. Intended primarily as a trainer, but suitable for strike sorties, carrying same air-to-surface armament as Mirage III-C. Prototype flew for first time on 20 October 1959, and first production model on 19 July 1962. Total of 174 two-seaters ordered, including III-B and III-D variants for 19 countries.

Mirage III-BE. Two-seat version of the III-E for French Air Force. Similar model also supplied to foreign air forces.

Mirage III-C. All-weather interceptor and day ground attack fighter. Production version of III-A with SNECMA Atar 9B turbojet engine, optional SEPR 841 rocket engine and CSF Cyrano *Ibis* air-to-air radar. Initial series of 95 for French Air Force, of which the first flew on 9 October 1960. One supplied to Swiss Air Force. Total of 244 built, including III-CJ for Israel and III-CZ for South Africa. Full description in 1968-69 *Jane's*.

Mirage III-D. Two-seat version of the Mirage III-O, built in Australia for the RAAF. First of ten ordered for Mirage OCU was assembled in Australia and delivered in November 1966. Similar, French-built models ordered by 12 countries, including six more for Australia.

Mirage III-D2Z. For South Africa. Generally similar to

III-D but with SNECMA Atar 9K-50 turbojet. Delivered 1974-75.

Mirage III-E. Long-range fighter-bomber/intruder version, of which 453 have been built for the French Air Force and for the air forces of Argentina (III-EA), Brazil (III-EBR), Lebanon (III-EL), Pakistan (III-EP), South Africa (III-EZ), Spain (III-EE) and Venezuela (III-EV). First of three prototypes flew on 5 April 1961, and the first delivery of a production III-E was made in January 1964. Length increased by 0·3 m (11·8 in) compared with III-C.

Mirage III-O. Version of the Mirage III-E manufactured under licence in Australia. Main differences compared with the standard III-E are fitment of a Sperry twin gyro platform and PHI 5CI navigation unit. First two III-Os assembled in France; first of these handed over on 9 April 1963. Further 98 built in Australia, details of which were given under entry for Commonwealth of Australia, Government Aircraft Factories, in 1969-70 *Jane's*.

Mirage III-R. Reconnaissance version of III-E for French Air Force. Set of five OMERA type 31 cameras, in place of radar in nose, can be focused in four different arrangements for very low altitude, medium altitude, high altitude and night reconnaissance missions. Self-contained navigation system and same air-to-surface armament as Mirage III-C. Two prototypes, converted from III-As, of which the first flew in November 1961. Total of 153 production models ordered, including variants for Pakistan (III-RP), South Africa (III-RZ and R2Z) and Switzerland (III-RS).

Mirage III-R2Z. For South Africa. Generally similar to III-R but with SNECMA Atar 9K-50 turbojet. Delivered 1974-75.

Mirage III-RD. Similar to III-R but with improved Doppler navigation system in fairing under front fuselage, gyro gunsight and automatic cameras. Provision for carrying SAT Cyclope infra-red tracking equipment in ventral fairing, and two 1,700 litre (374 Imp gallon) underwing auxiliary fuel tanks. Twenty ordered for French Air Force.

Mirage III-S. Developed from the Mirage III-E, with a Hughes TARAN electronics fire-control system and armament of HM-55 Falcon missiles. Thirty-six supplied to Swiss Air Force, of which the first two were built in France and the remainder by the Federal Aircraft Factory in Switzerland.

The following description refers to the Mirage III-E, but is generally applicable to all versions:

Type: Single-seat fighter-bomber/intruder aircraft.

Wings: Cantilever low-wing monoplane of delta planform, with conical camber. Thickness/chord ratio 4·5% to 3·5%. Anhedral 1°. No incidence. Sweepback on leading-edge 60° 34'. All-metal torsion-box structure with stressed skin of machined panels with integral stiffeners. Elevons are hydraulically powered by Dassault twin-cylinder actuators with artificial feel. Airbrakes, comprising small panels hinged to upper and lower wing surfaces, near leading-edge.

Fuselage: All-metal structure, 'waisted' in accordance with the area rule.

Tail Unit: Cantilever fin and hydraulically-actuated powered rudder only. Dassault twin-cylinder actuators with artificial feel.

Landing Gear: Retractable tricycle type, with single wheel on each unit. Hydraulic retraction, nosewheel rearward, main units inward. Messier-Hispano shock-absorbers and disc brakes. Main-wheel tyre pressure 5·9-9·8 bars (85·5-142 lb/sq in). Braking parachute.

Power Plant: One SNECMA Atar 9C turbojet engine (60·8 kN; 13,670 lb st with afterburning), fitted with an overspeed system which is engaged automatically from Mach 1·4 and permits a thrust increase of approx 8 per cent in the high supersonic speed range. Optional and jettisonable SEPR 844 single-chamber rocket motor (14·7 kN; 3,300 lb st) or interchangeable fuel tank. Movable half-cone centrebody in each air intake. Total internal fuel capacity 3,330 litres (733 Imp gallons) when rocket motor is not fitted. Provision for this to be augmented by two 600, 1,300 or 1,700 litre (132, 285 or 374 Imp gallon) underwing drop-tanks.

Accommodation: Single seat under rearward-hinged canopy. Hispano-built Martin-Baker Type RM.4 zero-altitude ejection seat.

Systems: Two separate air-conditioning systems for cockpit and electronics. Two independent hydraulic systems, pressure 207 bars (3,000 lb/sq in), for flying controls, landing gear and brakes. Power for DC electrical system from 24V 40Ah batteries and a 26·5V 9kW generator. AC electrical system power provided by one 200V 400Hz transformer and one 200V 400Hz 9kVA alternator.

Electronics and Equipment: Duplicated UHF, Tacan, Doppler, CSF Cyrano II fire-control radar in nose, navigation computer, bombing computer, automatic gunsight.

The Mirage III-E has a normal magnetic detector mounted in the fin, and a central gyro and other electronics to provide accurate and stabilised heading information. The pilot's equipment determines at any instant the geographical co-ordinates of the aircraft and compares them with the co-ordinates of the target, the differences between the two being presented to the pilot as a 'course to steer' and 'distance to run'. Associated with this facility is a rotative magazine in the cockpit in which it is possible to insert up to twelve plastics punch-cards. Each card represents the co-ordinates of a geographical position. Therefore it is possible before take-off at point A to select point B on the rotating magazine. During take-off, ie after reaching 150 knots (278 km/h; 173 mph), the computer will switch on and the heading and distance to point B will be presented to the pilot. When overhead point B (assuming a pure navigational sortie) he can either select point A or the next turning point, or if required this sequence can continue until a maximum of twelve pre-set turning points have been used. Another facility available in the computer is known as the 'additional base'. Assuming that between points A and B the pilot receives instructions by radio to go to point C (and that there is no punch-card in the magazine for point C) the pilot can, by means of setting knobs, wind on the bearing and distance of point C from point B; then, when he selects the switch 'additional base', the heading to steer and distance to run to point C will be indicated.

Marconi Doppler equipment provides the ground speed and drift information for the above, while Tacan is presented as a 'bearing and distance' on the navigation indicator located on the starboard side of the instrument panel.

The Cyrano II installation in the aircraft's nose provides orthodox air-to-air interception radar, and has the additional mode available of control from the ground. In the latter case the pilot simply obeys his gunsight instructions, and radio silence is maintained. Cyrano II also functions in an air-to-ground role for high-level navigation, presenting a radar picture of the ground; for low-level navigation, presenting the obstacles above a preselected altitude; for blind descent, presenting obstacles that intercept the descent path; for anti-collision, presenting the obstacles that can be avoided by applying a 0·1g pull-up; and for distance measuring, by presenting in the sight the oblique aircraft-to-ground distance.

Allied to the Cyrano II installation is the CSF 97 sighting system, of illuminated points, dots, bars and figures, giving air-to-air facility for cannon and missiles, air-to-ground facility for dive-bombing or LABS, and navigation facility for horizon and heading.

Armament: Ground attack armament consists normally of two 30 mm DEFA cannon in fuselage, each with 125 rounds of ammunition, and two 1,000 lb bombs, or an AS.30 air-to-surface missile under the fuselage and 1,000 lb bombs under the wings. Alternative underwing stores include JL-100 pods, each with 18 rockets, and 250 litre (55 Imp gallon) fuel tanks. For interception duties, one Matra R.530 air-to-air missile can be carried under fuselage, with optional guns and two Sidewinder missiles.

Dassault Mirage III-E fighter-bomber/intruder of the French Air Force

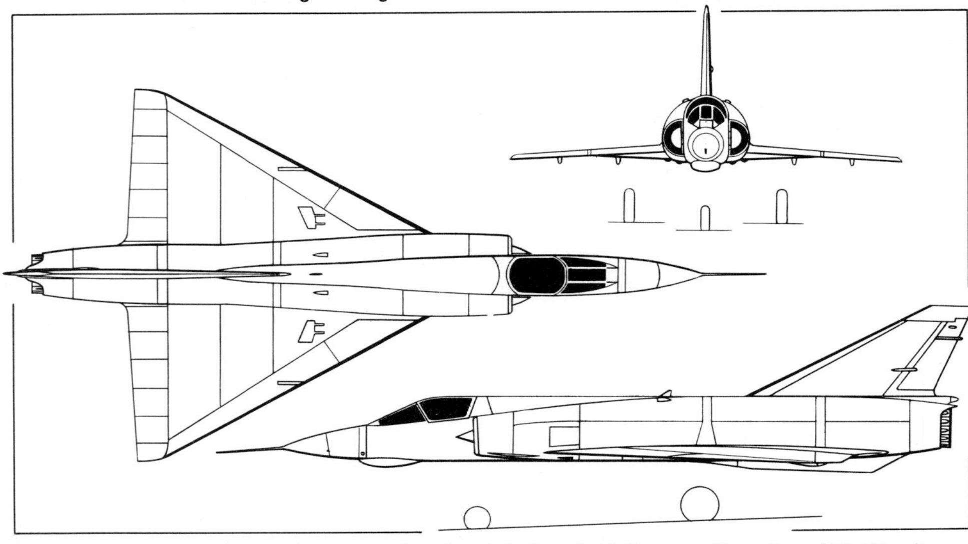

Dassault Mirage III-E single-seat combat aircraft in French Air Force configuration *(Pilot Press)*

Dimensions, external:
- Wing span: 8·22 m (26 ft 11½ in)
- Wing aspect ratio: 1·94
- Length overall:
  - III-B: 15·40 m (50 ft 6¼ in)
  - III-E: 15·03 m (49 ft 3½ in)
  - III-R: 15·50 m (50 ft 10¼ in)
- Height overall: 4·25 m (13 ft 11¼ in)
- Wheel track: 3·15 m (10 ft 4 in)
- Wheelbase:
  - III-E: 4·87 m (15 ft 11¾ in)

Areas:
- Wings, gross: 34·85 m² (375 sq ft)
- Vertical tail surfaces (total): 4·5 m² (48·4 sq ft)

Weights and Loading:
- Weight empty:
  - III-B: 6,270 kg (13,820 lb)
  - III-E: 7,050 kg (15,540 lb)
  - III-R: 6,600 kg (14,550 lb)
- Max T-O weight:
  - III-B: 12,000 kg (26,455 lb)
  - III-E, R: 13,500 kg (29,760 lb)
- Max wing loading:
  - III-E, R: 370 kg/m² (75·85 lb/sq ft)

Performance (Mirage III-E, in 'clean' condition with guns installed, except where indicated):
- Max level speed at 12,000 m (39,375 ft): Mach 2·2 (1,268 knots; 2,350 km/h; 1,460 mph)
- Max level speed at S/L: 750 knots (1,390 km/h; 863 mph)
- Cruising speed at 11,000 m (36,000 ft): Mach 0·9
- Approach speed: 183 knots (340 km/h; 211 mph)
- Landing speed: 162 knots (300 km/h; 187 mph)
- Time to 11,000 m (36,000 ft), Mach 0·9: 3 min
- Time to 15,000 m (49,200 ft), Mach 1·8: 6 min 50 sec
- Service ceiling at Mach 1·8: 17,000 m (55,775 ft)
- Ceiling, using rocket motor: 23,000 m (75,450 ft)
- T-O run, according to mission (up to max T-O weight): 700-1,600 m (2,295-5,250 ft)
- Landing run, using brake parachute: 700 m (2,295 ft)
- Combat radius, ground attack: 647 nm (1,200 km; 745 miles)

DASSAULT MIRAGE 5

The Mirage 5 is a ground attack aircraft using the same airframe and engine as the Mirage III-E. The basic VFR version has simplified electronics, 500 litres (110 Imp gallons) greater fuel capacity than III-E and considerably extended stores carrying capability. It combines the full Mach 2+ capability of the Mirage III, and its ability to operate from semi-prepared airfields, with simpler maintenance. In ground attack configuration, up to 4,000 kg (8,820 lb) of weapons and 1,000 litres (220 Imp gallons) of fuel can be carried externally on seven wing and fuselage attachment points. The Mirage 5 can also be flown as an interceptor, with two Sidewinder air-to-air missiles and 4,700 litres (1,034 Imp gallons) of external fuel. At customer's option, any degree of IFR/all-weather operation can be provided for, with reduced fuel or weapons load. The Mirage 5 was flown for the first time on 19 May 1967.

Up to 1 January 1976, a total of 430 Mirage 5s had been ordered. The 106 aircraft for the Belgian Air Force were assembled in Belgium by SABCA.

Versions announced so far are as follows:

Mirage 5-AD and 5-RAD. For Abu Dhabi.

Mirage 5-DAD. Two-seat version of Mirage 5-AD for Abu Dhabi.

Mirage 5-BA. Single-seat ground attack model, with more advanced navigation system than basic Mirage 5, for Belgian Air Force. First (Dassault-built) 5-BA flew on 6 March 1970.

Mirage 5-BD. Two-seat version of Mirage 5-BA for Belgian Air Force.

Mirage 5-BR. Single-seat reconnaissance version of 5-BA for Belgian Air Force, with five Vinten type 360 cameras installed in nose. Provision to install infra-red photographic equipment is under consideration.

Mirage 5-COA, 5-COD and 5-COR. For Colombian Air Force.

Mirage 5-D, 5-DD, 5-DE and 5-DR. For Libyan Air Force. Deliveries began January 1971.

Mirage M5-F. Former Mirage 5-J aircraft repurchased from the Israeli Air Force by the French government. In service with French Air Force tactical squadrons since mid-1973.

Mirage 5-G. For Gabon.

Mirage 5-DG. Two-seat version of Mirage 5-G, for Gabon.

Mirage 5-M. For Zaïre.

Mirage 5-DM. Two-seat version of Mirage 5-M, for Zaïre.

Mirage 5-P. Ordered by Peru in April 1968.

Mirage 5-DP. Two-seat version of Mirage 5-P for Peruvian Air Force.

Mirage 5-PA. For Pakistan.

Mirage 5-SDE. For Saudi Arabia.

Mirage 5-SDD. Two-seat version of Mirage 5-SDE, for Saudi Arabia.

Mirage 5-V. For Venezuela.

Mirage 5-DV. Two-seat version of Mirage 5-V, for Venezuela.

The structural description of the Mirage III-E is generally applicable to the Mirage 5, with the following exceptions:

ARMAMENT: Seven attachment points for external loads, with multiple launchers permitting a max load of more than 4 tons. Ground attack armament consists normally of two 30 mm DEFA cannon in fuselage, each with 125 rounds of ammunition, and two 1,000 lb bombs or an AS.30 air-to-surface missile under the fuselage and 1,000 lb bombs under the wings. Alternative underwing stores include tank/bomb carriers, each with 500 litres (110 Imp gallons) of fuel and four 500 lb or two 1,000 lb bombs, and JL-100 pods, each with eighteen 68 mm rockets and 250 litres (55 Imp gallons) of fuel. For interception duties, two Sidewinder missiles can be carried under the wings.

EQUIPMENT: Can have Aïda II radar rangefinder in nose.

DIMENSIONS, EXTERNAL:
As III-E, except:
Length overall 15·55 m (51 ft 0¼ in)

WEIGHTS AND LOADING:
As III-E, except:
Weight empty 6,600 kg (14,550 lb)

PERFORMANCE (in 'clean' condition, with guns installed, except where indicated):
As III-E, plus:
Combat radius with 907 kg (2,000 lb) bomb load:
hi-lo-hi 699 nm (1,300 km; 805 miles)
lo-lo-lo 347 nm (650 km; 400 miles)
Ferry range with three external tanks 2,158 nm (4,000 km; 2,485 miles)

DASSAULT MIRAGE 50

Among the Dassault fighters displayed at the 1975 Paris Air Show was a prototype identified as the Mirage 50. This retains the basic airframe of the Mirage III/5 series, but is powered by the higher-rated SNECMA Atar 9K-50 turbojet, as fitted in the Mirage F1-Cs of the French Air Force and Mirage III-R2Zs of the South African Air Force. This gives 70·6 kN (15,873 lb st) with afterburning, representing a 16% thrust increase compared with standard Mirage III/5s.

The Mirage 50 is a multi-mission fighter, suitable for air superiority duties with guns and dogfight missiles, air patrol and supersonic interception, and ground attack combined with self-defence capability. It can carry the full range of operational stores, armament and equipment developed for the Mirage III/5 series, and is available in reconnaissance configuration. A two-seat training version is also projected. Improvements compared with other delta-wing Mirages include better take-off performance, higher rate of climb, faster acceleration and better manoeuvrability.

Dassault Mirage 5-DM two-seat trainer/combat aircraft supplied to Zaïre

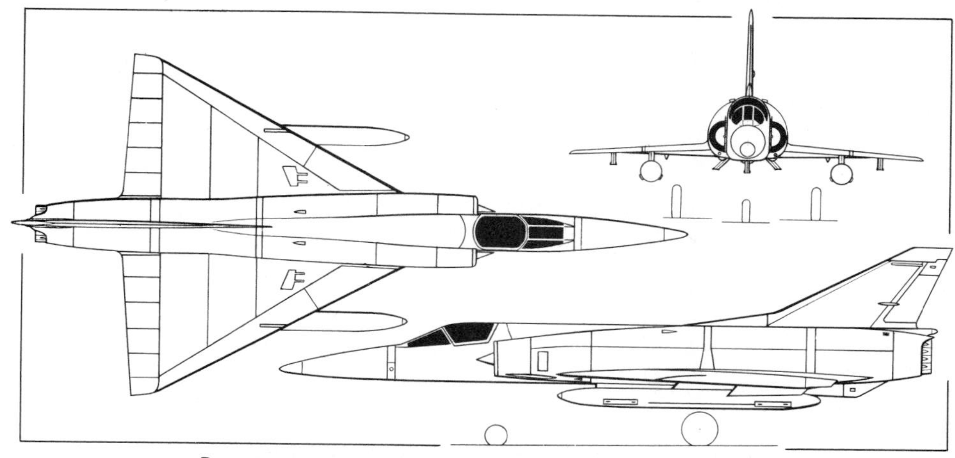

Dassault Mirage 5 single-seat ground attack aircraft *(Pilot Press)*

DIMENSIONS, EXTERNAL:
Wing span 8·22 m (27 ft 0 in)
Length overall 15·56 m (51 ft 0½ in)
Height overall 4·50 m (14 ft 9 in)

AREA:
Wings, gross 35·00 m² (376·7 sq ft)

WEIGHTS:
Ramp weight, 'clean' 9,500 kg (20,945 lb)
Max T-O weight 13,500 kg (29,760 lb)

PERFORMANCE:
Max speed at altitude above Mach 2 (750 knots; 1,390 km/h; 863 mph IAS)

DASSAULT MIRAGE F1

In early 1964 Dassault was awarded a French government contract to develop a replacement for the Mirage III, followed shortly afterwards by an order for a prototype aircraft which was designated Mirage F2. This was a two-seat fighter, powered by a SNECMA (Pratt & Whitney) TF 306 turbofan engine. It first flew on 12 June 1966 and was described and illustrated in the 1967-68 *Jane's*.

Concurrently with work on the Mirage F2 Dassault also developed, as a private venture, a much smaller single-seat aircraft, the Mirage F1, with a SNECMA Atar 9K turbojet engine. The prototype Mirage F1-01 flew for the first time on 23 December 1966, and exceeded Mach 2 during its fourth flight on 7 January 1967.

In September 1967, three pre-series F1 aircraft and a structural test airframe were ordered by the French government. The first pre-series aircraft, the Mirage F1-02, reached Mach 1·15 during its first flight on 20 March 1969, and Mach 2·03 during its third flight on 24 March. It completed the first phase of its flight test programme on 27 June 1969. This comprised 62 flights, during which the aircraft was flown at speeds of up to Mach 2·12 (1,200 knots; 2,260 km/h; 1,405 mph) at 11,000 m (36,000 ft) and up to 702 knots (1,300 km/h; 808 mph) at low level; at altitudes of more than 15,250 m (50,000 ft); and with various external military loads, including air-to-air missiles and drop-tanks.

The F1-02, during its initial flight tests, was powered by an Atar 9K-31 turbojet engine developing 65·7 kN (14,770 lb st) with afterburning. It was re-engined in 1969 with the more powerful Atar 9K-50 turbojet; this engine was also fitted in the two later pre-series aircraft, and is the standard power plant of the current production versions.

The Mirage F1-03, which flew for the first time on 18 September 1969, had its wing leading-edges extended for a greater proportion of the overall span than the preceding aircraft. It was followed by the final pre-series aircraft, the F1-04, on 17 June 1970. This had a complete electronics system and, after modification of its wing leading-edges to be similar to those of the F1-03, became representative of the initial production version.

The Mirage F1 is dimensionally similar to the Mirage III series, and its swept wing is virtually a scaled-down version of that fitted to the F2 prototype, with improved high-lift devices which help to make possible take-offs and landings within 500-800 m (1,600-2,600 ft) at average combat mission weight. Operation from semi-prepared, or even sod, runways is possible, and aircraft systems have been improved by comparison with the Mirage III, for increased efficiency and easy servicing. Compared with the Mirage III, internal fuel capacity is some 45 per cent greater, trebling the endurance of the F1 for patrol or high-altitude supersonic interception missions and doubling the possible combat radius in the attack role. Performance during flight testing met or exceeded all expectations, and included reductions of 22 per cent in approach speed and 28 per cent in take-off distance compared with the Mirage III. Manoeuvrability is claimed to have been increased by as much as 80 per cent.

Dassault Mirage 50 fighter (SNECMA Atar 9K-50 turbojet engine)

The primary role of the Mirage F1 is that of all-weather interception at any altitude, and the **F1-C** production version, to which the detailed description applies, utilises initially weapon systems similar to those of the Mirage III-E, with more advanced systems to follow. It is equally suitable for attack missions, carrying a variety of external loads beneath the wings and fuselage. A 'utility' version, the **F1-A,** is also in production for operation only under VFR conditions, with much of the more costly electronic equipment deleted and the space so vacated occupied by an additional fuel tank. Further versions include the **F1-B** two-seat trainer, the first of which made its first flight on 26 May 1976.

By January 1976, a total of 300 Mirage F1s had been ordered for the French Air Force and for service with foreign air forces. Production is at the rate of six aircraft each month in France.

The Mirage F1 is produced by Dassault-Breguet in co-operation with the Belgian companies SABCA, in which Dassault-Breguet has a parity interest, and Fairey SA, which is building rear fuselage sections for all Mirage F1s ordered. Dassault-Breguet also has a technical and industrial co-operation agreement with the Armaments Development and Production Corporation of South Africa Ltd, whereby the latter company has rights to build the Mirage F1 under licence.

The first production Mirage F1 flew on 15 February 1973 and was delivered officially to the French Air Force on 14 March 1973. The first unit to receive the F1 was the 30e Escadre at Reims, which became operational in early 1974. It was followed by the 5e Escadre at Orange and the 12e Escadre at Cambrai.

By the beginning of 1976, the 90 pre-series and production Mirage F1s then completed had logged a total of more than 20,000 flying hours.

Dassault has built a prototype of the Mirage **F1-E,** powered by a SNECMA M53 afterburning turbofan engine instead of the Atar 9K-50 of the basic Mirage F1. This is described separately. As part of the continuing development programme, a pre-production Mirage F1 has been flying since 7 May 1976 with a tailplane made of boron fibre.

The following description applies to the F1-C production version for the French Air Force:

Type: Single-seat multi-mission fighter and attack aircraft.

Wings: Cantilever shoulder-wing monoplane. Anhedral from roots. Sweepback 47° 30′ on leading-edges, with extended chord (saw-tooth) on approx the outer two-thirds of each wing. All-metal two-spar torsion-box structure, making extensive use of mechanically or chemically milled components. Trailing-edge control surfaces of honeycomb sandwich construction. Entire leading-edge can be drooped hydraulically (manually for T-O and landing, automatic in combat). Two differentially-operating double-slotted flaps and one aileron on each trailing-edge, actuated hydraulically by servo controls. Ailerons are compensated by trim devices incorporated in linkage. Two spoilers on each wing, ahead of flaps.

Fuselage: Conventional all-metal semi-monocoque structure. Primary frames are milled mechanically, secondary frames and fuel tank panels chemically. Electrical spot-welding for secondary stringers and sealed panels, remainder titanium flush-riveted or bolted and sealed. Titanium alloy also used for landing gear trunnions, engine firewall and certain other major structures. High-tensile steel wing attachment points. Nosecone over radar, and antennae fairings on fin, are of plastics. Large hydraulically-actuated door-type airbrake in forward underside of each intake trunk.

Tail Unit: Cantilever all-metal two-spar structure, with sweepback on all surfaces. All-moving tailplane mid-set on fuselage, and actuated hydraulically by electrical or manual control. Tailplane trailing-edge panels are of honeycomb sandwich construction. Auxiliary fin beneath each side of rear fuselage.

Landing Gear: Retractable tricycle type, by Messier-Hispano. Hydraulic retraction, nose unit rearward, main units upward into rear of intake trunk fairings. Twin wheels on each unit. Nose unit steerable and self-centering. Oleo-pneumatic shock-absorbers. Mainwheel tyre pressure 8·83 bars (128 lb/sq in), permitting operation from semi-prepared airfields. Messier-Hispano brakes and anti-skid units. Brake parachute in bullet fairing at base of rudder.

Power Plant: One SNECMA Atar 9K-50 turbojet engine, rated at 70·6 kN (15,873 lb st) with afterburning. Movable semi-conical centrebody in each intake. Fuel in integral tanks in wings and fuselage, on each side of intake trunks. Provision for three jettisonable auxiliary fuel tanks (each 1,200 litres; 264 Imp gallons) to be carried under fuselage and on inboard wing pylons.

Accommodation: Single SEMMB (Martin-Baker Mk 4) ejection seat for pilot, under rearward-hinged canopy. Cockpit is air-conditioned, and is heated by warm air bled from engine which also heats the radar compartment and certain equipment compartments. Intertechnique liquid oxygen system for pilot.

Systems: Two independent hydraulic systems, for landing gear retraction, flaps and flying controls, supplied by pumps similar to those fitted in Mirage III. Electrical system includes two Auxilec 15kVA variable-speed alternators, either of which can supply all functional and operational requirements. Emergency and standby power provided by SAFT Voltabloc 40Ah nickel-cadmium battery and EMD static converter. DC power provided by transformer-rectifiers operating in conjunction with battery.

Electronics and Equipment: Thomson-CSF Cyrano IV fire-control radar in nose. This permits all-sector interception at any altitude and incorporates a system to eliminate 'fixed' echoes when following low-flying aircraft. Two UHF transceivers (one UHF/VHF), Socrat 6200 VOR/ILS with Socrat 5600 marker, LMT Tacan, LMT NR-AI-4-A IFF, remote-setting interception system, three-axis generator, central air data computer, Bézu Sphere with ILS indicator, Crouzet Type 63 navigation indicator and SFENA 505 autopilot. CSF head-up display, with magnifying lens, provides all necessary data for flying and fire control. Equipment for attack role can include Doppler radar and bombing computer, navigation computer, position indicator, laser rangefinder and terrain-avoidance radar.

Armament and Operational Equipment: Standard fixed armament of two 30 mm DEFA 553 cannon, with 125 rds/gun, mounted in lower front fuselage. Two Alkan universal stores attachment pylons under each wing and one under centre fuselage, plus provision for carrying one air-to-air missile at each wingtip. Max external combat load 4,000 kg (8,820 lb). Externally-mounted weapons for interception role include Matra R.530 or Super 530 radar homing or infra-red homing air-to-air missiles on underfuselage and inboard wing pylons, and/or a Sidewinder or Matra 550 Magic infra-red hom-

Dassault Mirage F1 of the South African Air Force, carrying two Matra 550 Magic air-to-air missiles on wingtip launchers, four Matra 155 rocket packs and an underbelly fuel tank

Prototype Dassault Mirage F1-B two seat combat trainer

Dassault Mirage F1-C of the Spanish Air Force, by which it is designated C-14

ing air-to-air missile at each wingtip station. For ground attack duties, typical loads may include one AS.37 Martel anti-radar missile or AS.30 air-to-surface missile, eight 450 kg bombs, four launchers each containing 18 air-to-ground rockets, or six 600 litre (132 Imp gallon) napalm tanks. Other possible external loads include three 1,200 litre (264 Imp gallon) auxiliary fuel tanks, or two photoflash containers and a reconnaissance pod incorporating an SAT Cyclope infra-red system and EMI side-looking radar.

DIMENSIONS, EXTERNAL:
- Wing span: 8·40 m (27 ft 6¾ in)
- Length overall: 15·00 m (49 ft 2½ in)
- Height overall: 4·50 m (14 ft 9 in)
- Wheel track: 2·50 m (8 ft 2½ in)
- Wheelbase: 5·00 m (16 ft 4¾ in)

AREA:
- Wings, gross: 25·00 m² (269·1 sq ft)

WEIGHTS AND LOADING:
- Weight empty: 7,400 kg (16,314 lb)
- T-O weight, 'clean': 10,900 kg (24,030 lb)
- Max T-O weight: 14,900 kg (32,850 lb)
- Max wing loading: 596 kg/m² (122·2 lb/sq ft)

PERFORMANCE:
- Max level speed (high altitude): Mach 2·2
- Max level speed (low altitude): Mach 1·2
- Approach speed: 141 knots (260 km/h; 162 mph)
- Landing speed: 124 knots (230 km/h; 143 mph)
- Max rate of climb at S/L (with afterburning): 12,780 m (41,930 ft)/min
- Max rate of climb at high altitude (with afterburning): 14,580 m (47,835 ft)/min
- Service ceiling: 20,000 m (65,600 ft)
- Stabilised supersonic ceiling: 18,500 m (60,700 ft)
- T-O run (AUW of 11,500 kg; 25,355 lb): 450 m (1,475 ft)
- T-O run (typical interception mission): 640 m (2,100 ft)
- Landing run (AUW of 8,500 kg; 18,740 lb): 500 m (1,640 ft)
- Landing run (typical interception mission): 610 m (2,000 ft)
- Endurance: 3 hr 45 min

DASSAULT MIRAGE F1-E

Developed with French government support, the prototype of this multi-mission version of the Mirage F1 flew for the first time on 22 December 1974.

Major features of the F1-E are as follows:

POWER PLANT: One SNECMA M53 turbofan engine, rated at 54·4 kN (12,235 lb st) dry and 83·4 kN (18,740 lb st) with afterburning. Fuel tanks in centre and rear fuselage, around and between intake trunks and around engine; also in inboard half of each wing, between spars. Internal fuel capacity 4,300 litres (946 Imp gallons). Max fuel capacity 8,320 litres (1,830 Imp gallons), including two 1,200 litre (264 Imp gallon) underwing tanks and one tank under fuselage.

EQUIPMENT: Thomson-CSF Cyrano IV Srs 100 modular fire control radar in nose, with downward-look capability. Inertial navigation system, with digital computer. SFENA 505 autopilot. UHF transceiver, VOR/ILS digital receiver, digital OBS and VHF/UHF equipment.

ARMAMENT: Installed armament of two 30 mm DEFA 553 cannon, with 270 rds/gun, in lower fuselage. Seven external attachments for stores: one under fuselage with capacity of 2,040 kg (4,500 lb); two under each wing, with capacity of 1,270 kg (2,800 lb) and 500 kg (1,100 lb) respectively; and one at each wingtip, capacity 127 kg (280 lb). Armament for the interception role comprises two Matra 550 Magic or Sidewinder close-range missiles on the wingtips, plus two Matra 530 or Super 530 missiles under the wings. Attack armament includes an air-to-surface missile in the class of the AM39 Exocet or Martel, packs of 18 or 36 SNEB 68 mm rockets, and more than three tonnes of bombs of various kinds.

DIMENSIONS, EXTERNAL:
- Wing span: 8·45 m (27 ft 8½ in)
- Wing aspect ratio: 2·8
- Wing thickness/chord ratio: 4·5 to 3·5%
- Length overall: 15·53 m (50 ft 11½ in)
- Height overall: 4·56 m (14 ft 11½ in)

AREA:
- Wings, gross: 25·00 m² (269·1 sq ft)

WEIGHTS:
- Weight empty, equipped, incl pilot: 8,100 kg (17,857 lb)
- T-O weight, clean: 11,550 kg (25,460 lb)
- Max T-O weight: 15,200 kg (33,510 lb)

PERFORMANCE:
- Design limit speed at S/L: Mach 1·2 (800 knots; 1,480 km/h; 920 mph IAS)
- Max continuous speed at high altitude: Mach 2·2
- Landing speed: 127 knots (235 km/h; 146 mph)
- Max rate of climb from S/L to Mach 2 and 10,000 m (33,000 ft): approx 18,000 m (59,000 ft)/min
- T-O run, 'clean': 500 m (1,640 ft)
- Landing run, 'clean': 600 m (1,970 ft)

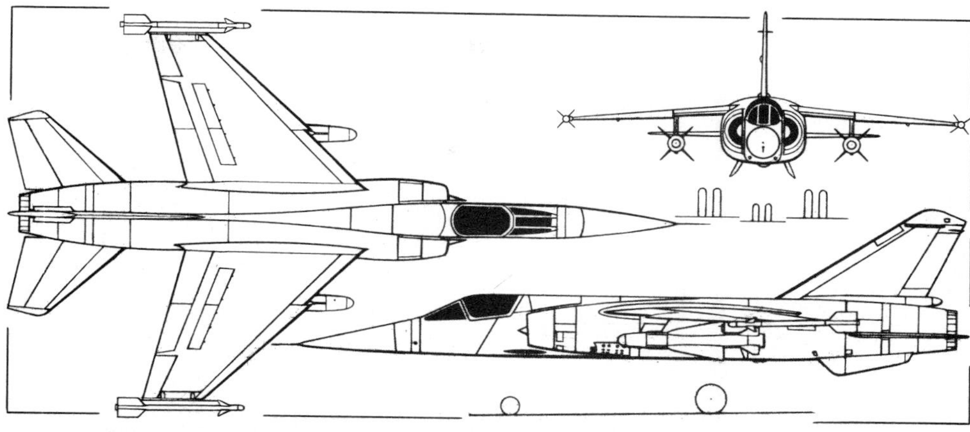

Dassault Mirage F1 single-seat multi-mission fighter and attack aircraft *(Pilot Press)*

Range with six bombs:
- at low altitude: 400 nm (740 km; 460 miles)
- hi-lo-hi: 650 nm (1,200 km; 750 miles)

DASSAULT DELTA MIRAGE 2000

The Delta Mirage 2000, which is being developed under French Air Force contract, is basically an interceptor and air superiority fighter powered by a single SNECMA M53 turbofan engine. It will be equally suitable for reconnaissance, close support and low-altitude attack missions in areas behind a battlefield. Dassault claims that its performance will be markedly superior to that of current combat aircraft in the class of the Mirage F1-E.

Series production of the Delta Mirage 2000 is planned to be under way within four years, to take the place of the now-abandoned ACF (Avion de Combat Futur) swept-wing Super Mirage, described briefly in the 1975-76 *Jane's*. No details are yet available.

DASSAULT DELTA SUPER MIRAGE

The Delta Super Mirage, of which a prototype will be produced as a private venture by Dassault, is intended as an export fighter powered by two SNECMA M53 turbofan engines. It will perform the same combat missions as the Delta Mirage 2000, and will also be able to make low-altitude penetration attacks on targets situated at considerable distances from its base. A prototype is scheduled to fly in the Summer of 1978. No details are yet available.

DASSAULT SUPER ETENDARD

The Super Etendard is developed from the Dassault Etendard IV-M carrier-based fighter, which has served with operational squadrons of the French Navy since 1962 and was last described in the 1965-66 *Jane's*. It is a transonic single-seat strike fighter, for low and medium altitude operations from ships of the *Clémenceau* and *Foch* class.

Very comprehensive high-lift devices are fitted to suit it for shipboard use. There is a Thomson-CSF/EMD Agave radar in the enlarged nose. Armament includes two 30 mm guns and a wide variety of external stores on five attachment points; equipment includes a highly sophisticated and accurate nav/attack integrated electronic system. Inherent long range is increased by flight refuelling capability and, like the Etendard, the new fighter is able to operate as a tanker for other aircraft.

The Super Etendard is powered by a SNECMA Atar 8K-50 turbojet engine, which is a non-afterburning version of the Atar 9K-50 giving 50·1 kN (11,265 lb st). It has a lower specific fuel consumption than the Atar 8 of the Etendard IV-M. The thrust increase of about 10% allows a significant increase of AUW for catapulting and, hence, makes possible an increase in fuel load and range.

A prototype of the Super Etendard, produced by conversion of a standard Etendard IV-M, flew for the first time on 28 October 1974. A second prototype followed on 28 March 1975. Deliveries of production Super Etendards

Dassault Mirage F1-E prototype, with SNECMA M53 afterburning turbofan engine

Dassault Super Etendard, photographed during deck landing trials

to the French Navy are expected to begin in the Summer of 1977.

DIMENSIONS, EXTERNAL:

Wing span	9·60 m (31 ft 6 in)
Length overall	14·31 m (46 ft 11½ in)
Height overall	3·85 m (12 ft 8 in)

AREA:

Wings, gross	28·4 m² (305·7 sq ft)

WEIGHTS:

Weight empty	6,300 kg (13,890 lb)
Mission T-O weight	9,200-11,500 kg (20,280-25,350 lb)

PERFORMANCE (estimated):

Max level speed at 11,000 m (36,000 ft) above Mach 1
Max level speed at low altitude
637 knots (1,180 km/h; 733 mph)
Approach speed for shipboard landing
135 knots (250 km/h; 155 mph)
Radius, anti-ship mission with air-to-surface missile
350 nm (650 km; 403 miles)

DASSAULT-BREGUET/DORNIER ALPHA JET

Details of the Alpha Jet programme can be found in the International section of this edition.

DASSAULT-BREGUET/BAC JAGUAR

Details of the Jaguar programme can be found under 'SEPECAT' in the International section of this edition.

BREGUET 1150 ATLANTIC

On 19 July 1974, Dassault-Breguet delivered the 18th Breguet 1150 Atlantic maritime patrol aircraft ordered by the Italian government. This completed the production programme for 87 operational Atlantics of the basic type, made up of 40 aircraft for the French Navy (3 of these have been passed on to Pakistan), 20 for the German Navy, 9 for the Royal Netherlands Navy and 18 for the Italian Navy. Manufacture was shared by companies in all four countries, with additional airframe components supplied by the Belgian ABAP group and some equipment from the USA and UK. Details can be found in the 1974-75 *Jane's*.

The French government has authorised development of an **Atlantic Mk II** to meet the anticipated Aéronavale requirement for 50 uprated aircraft of this type for service by the late 'seventies. Design studies for the Mk II have been underway for some time, and the 04 prototype of the Atlantic was returned to the manufacturer in late 1974 for conversion to Mk II standard. Following a general refurbishment, the aircraft's entire weapon system will be removed and replaced by new equipment. This is expected to place primary emphasis on the location and destruction of surface targets, while retaining anti-submarine capability.

Atlantic Mk II equipment will include an inertial navigator, Doppler, an advanced radar, digital processing of sonobuoy information, an improved magnetic anomaly detector, passive electronics surveillance devices, and a central data processor. Current anti-submarine weapons will be supplemented by the AS.37 Martel anti-radiation missile and the AM39 Exocet.

The prototype Atlantic Mk II was scheduled to fly for the first time in 1976. Aéronavale is expected to require about 50, of which deliveries could begin in 1977.

Dassault-Breguet is also studying an Atlantic Mk II-B, which would have the standard Tyne turboprop engines supplemented by two Rolls-Royce/SNECMA M45H turbofans, mounted in underwing pods outboard of the turboprop nacelles. These would permit an increase of max T-O weight to 52,500 kg (115,745 lb), including 6,000 kg (13,225 lb) of additional fuel to extend the aircraft's patrol endurance and range.

The following details apply to the basic version currently in service:

TYPE: Twin-engined maritime patrol aircraft.

WINGS: Cantilever mid-wing monoplane. Wing section NACA 64 series. Dihedral on outer wings only. All-metal three-spar fail-safe structure, with bonded light alloy honeycomb skin panels on torsion box and on main landing gear doors. Conventional all-metal ailerons actuated by SAMM twin-cylinder jacks. All-metal slotted flaps, with bonded light alloy honeycomb filling, over 75% of span. Three hinged spoilers on upper surface of each outer wing, forward of flaps. Metal airbrake above and below each wing. No trim tabs. Kléber-Colombes pneumatic de-icing boots on leading-edges.

FUSELAGE: All-metal 'double-bubble' fail-safe structure, with bonded honeycomb sandwich skin on pressurised central section of upper fuselage, weapons bay doors and nosewheel door.

TAIL UNIT: Cantilever all-metal structure with bonded honeycomb sandwich skin panels on torsion boxes. Fixed-incidence tailplane. Control surfaces operated through SAMM twin-cylinder jacks. No trim tabs. Kléber-Colombes pneumatic de-icing boots on leading-edges.

LANDING GEAR: Retractable tricycle type, supplied by Messier-Hispano, with twin wheels on each unit. Hydraulic retraction, nosewheels rearward, main units forward into engine nacelles. Kléber-Colombes dimpled tyres, size 39 × 13-16 on main wheels, 26 × 7·75-13 on nosewheels. Tyre pressures: main 9·52 bars (138 lb/sq in), nose 6·07 bars (88 lb/sq in). Messier-Hispano disc brakes with Maxaret anti-skid units.

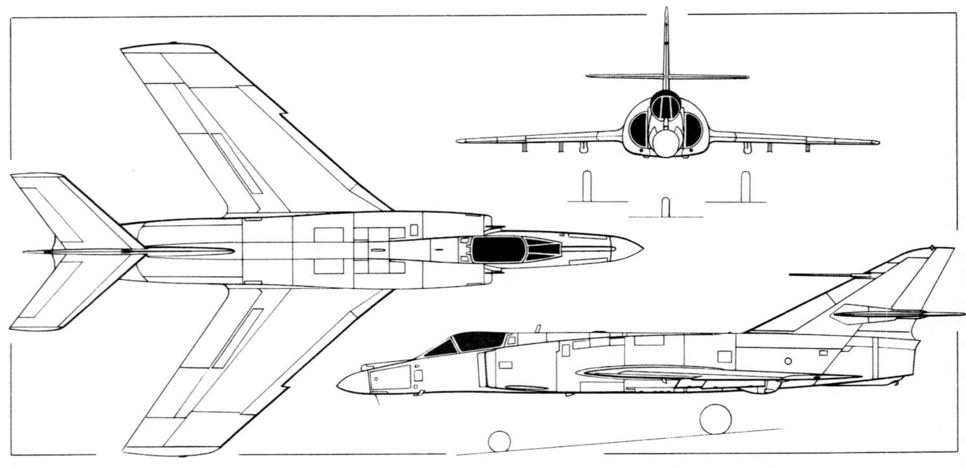

Dassault Super Etendard naval fighter (SNECMA Atar 8K-50 turbojet engine) *(Pilot Press)*

POWER PLANT: Two 4,553 kW (6,106 ehp) SNECMA-built Rolls-Royce Tyne RTy.20 Mk 21 turboprop engines, each driving a Ratier-built HSD four-blade constant-speed propeller. Six integral fuel tanks with total capacity of 21,000 litres (4,619 Imp gallons). Provision for wingtip tanks to be fitted.

ACCOMMODATION: Normal flight crew of 12, comprising observer in nose; pilot and co-pilot on flight deck; a tactical co-ordinator, navigator, two sonobuoy operators, and radio, radar and ECM/MAD/Autolycus operators in tactical compartment; and two observers in beam positions. On long-range patrol missions a further 12 men can be carried as relief crew. The upper, pressurised section of the fuselage, from front to rear, comprises the nose observer's compartment, flight deck, tactical operations compartment, rest compartment for crew, and beam observers' compartment.

SYSTEMS: SEMCA air-conditioning and pressurisation system. Hydraulic system pressure 207 bars (3,000 lb/sq in). Electrical system provides 28·5V DC, 115/200V variable-frequency AC and 115/200V stabilised-frequency AC. AiResearch GTCP 85-100 APU in starboard side of front fuselage, adjacent to radar compartment, for engine starting and ground air-conditioning, can also power one 20 kVA AC alternator and one 4kW DC generator for emergency electrical power supply.

ARMAMENT AND OPERATIONAL EQUIPMENT: Main weapons carried in bay in unpressurised lower fuselage. Weapons include all NATO standard bombs, 175 kg (385 lb) US or French depth charges, HVAR rockets, homing torpedoes, including types such as the Mk 44 Brush or LX.4 with acoustic heads, or four underwing air-to-surface missiles with nuclear or high-explosive warheads. Electronic equipment includes a retractable CSF radar installation, an MAD tailboom and an electronic countermeasures pod at the top of the tail-fin. Sonobuoys are carried in a compartment aft of the main weapons bay, while the whole of the upper and lower rear fuselage acts as a storage compartment for sonobuoys and marker flares. Compartment for retractable CSF radar 'dustbin' forward of main weapons bay. Forward of this, the lower nose section acts as additional storage for military equipment and the APU. Weapons system includes Plotac optical tactical display, 80 × 80 cm (31·5 × 31·5 in) in size, consisting of separate tables for search display and localisation and attack display. At 1:30,000 scale, this gives coverage of an area 21,950 × 21,950 m (24,000 × 24,000 yd) to an accuracy of 1 mm (*ie* less than 30·5 m; 100 ft at that scale). Heading references provided by duplicated gyroscopic platforms of the 3-gyro (1° of freedom) 4-gimbals type, with magnetic compasses as backup system. Janus-type Doppler has stabilised antenna and works in the Ke band to provide direct indication of ground speed and drift. In case of failure an automatic switch is made to the air data system. The analogue-type navigation computer is accurate to 0·25%. The MAD is of the atomic resonance type and uses light-stimulation techniques. Plotac system has provision to accept additional detectors. Radar has 'sea-return' circuits and stabilised antenna enabling it to detect a submarine snorkel at up to 40 nm (75 km; 46 miles) even in rough seas.

DIMENSIONS, EXTERNAL:

Wing span	36·30 m (119 ft 1 in)
Wing aspect ratio	10·94
Length overall	31·75 m (104 ft 2 in)
Height overall	11·33 m (37 ft 2 in)

Breguet 1150 Atlantic maritime patrol aircraft of the Royal Netherlands Navy, with radar extended and weapons bay open *(Dr Alan Beaumont)*

Fuselage:
Max width 2·90 m (9 ft 6 in)
Max depth 4·00 m (13 ft 1½ in)
Tailplane span 12·31 m (40 ft 4½ in)
Wheel track 9·00 m (29 ft 6¼ in)
Wheelbase 9·44 m (31 ft 0 in)
Propeller diameter 4·88 m (16 ft 0 in)

DIMENSIONS, INTERNAL:
Tactical compartment:
Length 8·60 m (28 ft 2½ in)
Height 1·93 m (6 ft 4 in)
Max width 2·70 m (8 ft 10½ in)
Rest compartment:
Length 5·10 m (16 ft 8¾ in)
Height 1·93 m (6 ft 4 in)
Max width 2·70 m (8 ft 10½ in)
Beam observers' compartment:
Length 1·00 m (3 ft 3¼ in)
Main weapons bay:
Length 9·00 m (29 ft 6¼ in)
Height 1·55 m (5 ft 1 in)
Height under wing 1·00 m (3 ft 3¼ in)
Max width 2·20 m (7 ft 2½ in)

AREAS:
Wings, gross 120·34 m² (1,295 sq ft)
Ailerons (total) 5·40 m² (58·0 sq ft)
Trailing-edge flaps (total) 26·80 m² (288·4 sq ft)
Spoilers (total) 1·66 m² (17·8 sq ft)
Fin 16·64 m² (179·1 sq ft)
Rudder 5·96 m² (64·1 sq ft)
Tailplane 32·5 m² (349·7 sq ft)
Elevators 8·28 m² (89·1 sq ft)

WEIGHTS:
Useful load 18,551 kg (40,900 lb)
Max zero-fuel weight 34,473 kg (76,000 lb)
Max T-O weight 43,500 kg (95,900 lb)

PERFORMANCE (at max T-O weight):
Max level speed at high altitudes 355 knots (658 km/h; 409 mph)
Cruising speed 300 knots (556 km/h; 345 mph)
Service ceiling 10,000 m (32,800 ft)
T-O to 10·7 m (35 ft), ISA 1,500 m (4,925 ft)
T-O to 10·7 m (35 ft), ISA+17°C, 15° flap 1,700 m (5,575 ft)
Max range 4,854 nm (9,000 km; 5,590 miles)
Max endurance at patrol speed of 169 knots (320 km/h; 195 mph) 18 hr

DASSAULT MYSTÈRE 20/FALCON 20

The Dassault Mystère 20/Falcon 20 is a light twin-turbofan executive transport, with standard accommodation for 8-10 passengers and a crew of two. An alternative layout offers seats for 14 passengers, and the aircraft can be used for a variety of alternative duties.

Its development was undertaken jointly with Aérospatiale (then Sud-Aviation) and construction of the prototype began in January 1962. The fuselage of the prototype was built by Dassault and the wings and tail unit by Sud-Aviation. Dassault was responsible for final assembly. For production aircraft, Dassault builds the wings and Aérospatiale the fuselages and tail units.

The prototype flew for the first time on 4 May 1963, with Pratt & Whitney JT12A-8 turbojets (each 14·6 kN; 3,300 lb st). It was re-engined subsequently with General Electric CF700 turbofans, which are standard on subsequent aircraft, and flew for the first time with these engines on 10 July 1964.

In August 1963 the Business Jets Division of Pan American World Airways ordered 54 production aircraft, with an option on 106 more for distribution in the western hemisphere. From the start, these aircraft were marketed by Pan American under the name Fan Jet Falcon, but the original name of Mystère 20 continues to be used in France. The first production aircraft flew on 1 January 1965.

The Mystère 20/Falcon 20 received French and US Transport Category Type Approval on 9 June 1965. On the following day, Mme Jacqueline Auriol established a Class C1g speed record of 463·80 knots (859·51 km/h; 534·075 mph) over a 1,000 km closed circuit in a Mystère 20/Falcon 20. On 15 June she set up a second Class C1g record of 442·00 knots (819·13 km/h; 508·98 mph) over a 2,000 km circuit.

By 1 January 1976, total sales of Mystère 20/Falcon 20s had reached 357, of which 141 were for customers in the areas marketed by Dassault with options on 120 more; of these, 338 had been delivered, including 210 to the Business Jets Division of Pan American and its successor, Falcon Jet Corporation, formed jointly by Dassault-Breguet and Pan American in 1973.

Current production version is the **Mystère 20/Falcon Series F,** which introduced high-lift devices to improve T-O and landing performance, more powerful engines than earlier Falcons and increased wing fuel tank capacity. The prototype was displayed at the Paris Air Show in June 1969 and deliveries began in July 1970. During 1970 the Series F became the first aircraft to receive type approval under FAA FAR Pt 36 anti-noise regulations. This approval was subsequently extended to other versions.

The new **Mystère 20/Falcon 20 Series G,** announced in the late Spring of 1976, is described in the Addenda to this edition.

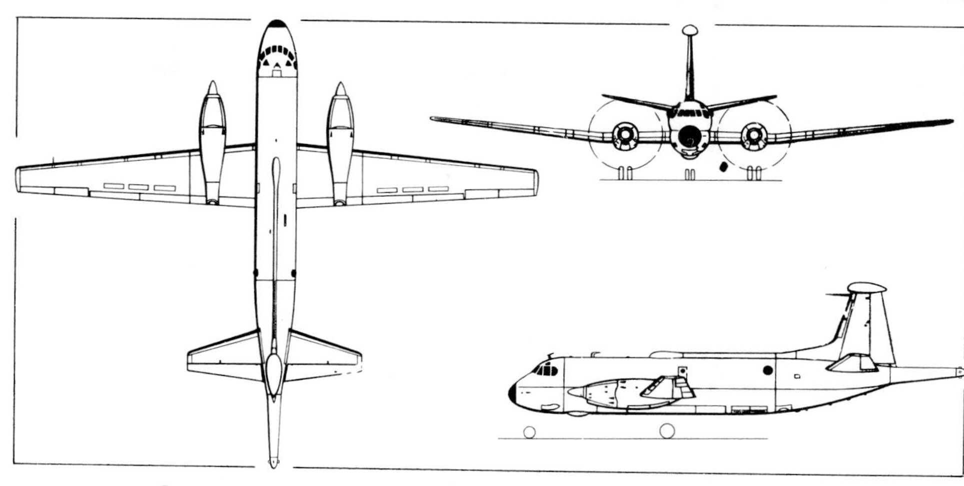
Breguet 1150 Atlantic twin-turboprop maritime patrol aircraft *(Pilot Press)*

All versions of the Mystère 20/Falcon 20 can be modified as follows for specific duties:

Calibration: Two aircraft ordered by the SGAC and one by the Spanish government are used for the calibration of radio navigation aids. The equipment includes a removable console, thus retaining the full passenger-carrying capability of the aircraft.

Airline crew training: Since 20 September 1966, several Mystère 20/Falcons have been used by Air France to train pilots for their jet airliners, with up to five aircraft being used simultaneously. Japan Air Lines also bought three of this version.

Cross-country: Similar to basic aircraft, but with low-pressure tyres for soft-field operation at the same take-off and landing weights. Described in 1968-69 *Jane's*.

Quick-change and cargo: A quick-change kit, consisting of an assembly of nets and supports, keeps the centre aisle free and allows direct access to nine freight compartments. Total usable volume of these compartments is 6·65 m³ (235 cu ft), and transformation from executive configuration to cargo configuration, or vice versa, takes less than one hour. A cargo version of the Falcon is also available and is described separately. For both versions an increase of the maximum zero-fuel weight from 8,900 kg (19,600 lb) to 9,980 kg (22,000 lb) allows an increased payload of up to 3,000 kg (6,615 lb).

Aerial photography: The French Institute Géographique National has a Mystère 20/Falcon fitted with two cameras (Zeiss RMK 610 mm focal length, and Wild RC8, RC9 or RC10) and an intervalometer. This enables the aircraft to be used for high-altitude photography and photogrammetry duties.

Systems trainer: Two aircraft fitted with Mirage III-E combat radar and navigation systems are in service with the French Air Force for training its Mirage pilots. This version, known as the **Falcon ST,** has been sold also to the Libyan Republic.

TYPE: Twin-turbofan executive transport.

WINGS: Cantilever low-wing monoplane. Thickness/chord ratio varies from 10·5 to 8%. Dihedral 2°. Incidence 1° 30′. Sweepback at quarter-chord 30°. All-metal (copper-bearing alloys) fail-safe torsion-box structure with machined stressed skin. Ailerons are each operated by Dassault twin-body actuators, from dual hydraulic systems, and have artificial feel. Non-slotted slats inboard of fence, and slotted slats outboard, on each wing. Hydraulically-actuated airbrakes forward of the hydraulically-actuated two-section single-slotted flaps. Leading-edges anti-iced by engine bleed air.

FUSELAGE: All-metal semi-monocoque structure of circular cross-section, built on fail-safe principles.

TAIL UNIT: Cantilever all-metal structure, with electrically-controlled variable-incidence tailplane mounted halfway up fin. Elevators and rudder each actuated by twin hydraulic servos. No trim tabs.

LANDING GEAR: Retractable tricycle type, by Messier-Hispano, with twin wheels on all three units. Hydraulic retraction, main units inward, nosewheels forward. Oleo-pneumatic shock-absorbers. Goodyear disc brakes and anti-skid units. Normal tyre pressure 9·15 bars (133 lb/sq in) on all units. Low-pressure gear (4·5 bars; 65 lb/sq in) available optionally. Steerable and self-centering nosewheels. Braking parachute standard.

POWER PLANT: Two General Electric CF700-2D-2 turbofan engines (each 19·2 kN; 4,315 lb st) mounted in pods on each side of rear fuselage. Fuel in two integral tanks in wings and two auxiliary tanks aft of rear pressure bulkhead in fuselage, with total capacity of 5,240 litres (1,150 Imp gallons; 1,385 US gallons). Separate fuel system for each engine, with provision for cross-feeding. Single-point pressure refuelling. Emergency refuelling by gravity.

ACCOMMODATION: Crew of two on flight deck, with full dual controls and airline-type instrumentation. Normal seating for eight or ten passengers in individual reclining chairs, with tables between forward pairs of seats and a central 'trench' aisle, or 12-14 passengers at reduced pitch without tables. Toilet at rear. Baggage space and wardrobe on starboard side, immediately aft of flight deck opposite door, and at rear of cabin. Buffet with ice-box, food and liquid storage at front of cabin on port side. Downward-opening door has built-in steps.

SYSTEMS: Duplicated air-conditioning and pressurisation system, supplied with air bled from both engines. Pressure differential 0·57 bars (8·3 lb/sq in). Two independent hydraulic systems, pressure 207 bars (3,000 lb/sq in), with twin engine-driven pumps and emergency electric pump, actuate primary flying controls, flaps, landing gear, wheel brakes, airbrakes and nosewheel steering. 28V DC electrical system with a 9kW 28V DC starter/generator on each engine, one 1500VA and two 750VA 400Hz 118/208V inverters and two 40Ah batteries. Automatic emergency oxygen system. 9kW Microturbo Saphir II APU optional.

ELECTRONICS AND EQUIPMENT: Standard equipment includes duplicated VHF and VOR/glideslope, single ADF and DME, marker beacon receiver, ATC transponder, cockpit audio and duplicated blind-flying instrumentation. Optional equipment includes integrated flight instrument system, weather radar, HF communications radio, autopilot, second ADF and DME, and cabin address system.

Dassault Mystère 20 executive transport (two General Electric CF700-2D-2 turbofan engines) *(Liam Byrne)*

DIMENSIONS, EXTERNAL:
Wing span 16·30 m (53 ft 6 in)
Wing chord (mean) 2·85 m (9 ft 4 in)
Wing aspect ratio 6·4
Length overall 17·15 m (56 ft 3 in)
Length of fuselage 15·55 m (51 ft 0 in)
Height overall 5·32 m (17 ft 5 in)
Tailplane span 6·74 m (22 ft 1 in)
Wheel track 3·69 m (12 ft 1¼ in)
Wheelbase 5·74 m (18 ft 10 in)
Passenger door:
Height 1·52 m (5 ft 0 in)
Width 0·80 m (2 ft 7½ in)
Height to sill 1·09 m (3 ft 7 in)
Emergency exits (each side, over wing):
Height 0·66 m (2 ft 2 in)
Width 0·48 m (1 ft 7 in)

DIMENSIONS, INTERNAL:
Cabin, incl fwd baggage space and rear toilet:
Length 7·08 m (23 ft 2¾ in)
Max width 1·87 m (6 ft 1¾ in)
Max height 1·73 m (5 ft 8 in)
Volume 20·0 m³ (700 cu ft)
Baggage compartment (fwd) 0·70 m³ (24·7 cu ft)
Baggage compartment (aft) 0·37 m³ (13·1 cu ft)

AREAS:
Wings, gross 41·00 m² (440 sq ft)
Horizontal tail surfaces (total) 11·30 m² (121·6 sq ft)
Vertical tail surfaces (total) 7·60 m² (81·8 sq ft)

WEIGHTS:
Weight empty, equipped 7,240 kg (15,970 lb)
Max payload 1,500 kg (3,320 lb)
Max T-O and ramp weight 13,000 kg (28,660 lb)
Max zero-fuel weight 8,900 kg (19,600 lb)
Typical landing weight 8,560 kg (18,870 lb)

PERFORMANCE:
Never-exceed speed at S/L
350 knots (650 km/h; 404 mph) IAS
Never-exceed speed at 7,000 m (23,000 ft)
390 knots (725 km/h; 450 mph) IAS
Max cruising speed at 7,620 m (25,000 ft) at AUW of 9,071 kg (20,000 lb)
465 knots (862 km/h; 536 mph)
Econ cruising speed at 12,200 m (40,000 ft)
405 knots (750 km/h; 466 mph)
Stalling speed 82 knots (152 km/h; 95 mph)
Absolute ceiling 12,800 m (42,000 ft)
Service ceiling, one engine out, at AUW of 8,500 kg (18,700 lb) 7,480 m (24,500 ft)
T-O to 10·7 m (35 ft) at AUW of 12,300 kg (27,130 lb) (full tanks, 8 passengers and baggage)
1,155 m (3,790 ft)
FAR 25 balanced T-O field length, AUW as above
1,450 m (4,750 ft)
FAR 121 landing field length at AUW of 8,550 kg (18,870 lb) (8 passengers, 45 min reserves)
985 m (3,230 ft)
Landing from 15 m (50 ft) 590 m (1,930 ft)
Range with max fuel and 725 kg (1,600 lb) payload at econ cruising speed, with reserves for 45 min cruise 1,930 nm (3,570 km; 2,220 miles)

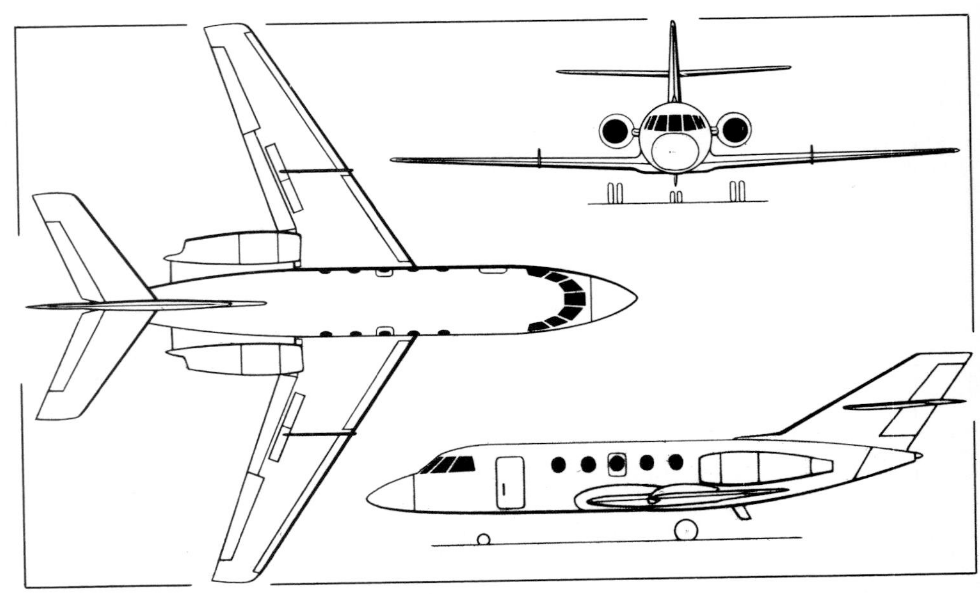
Dassault Mystère 20/Falcon 20 Series F twin-turbofan executive transport *(Pilot Press)*

Installing the freight door on a Dassault Falcon Cargo Jet of Federal Express

DASSAULT FALCON CARGO JET

Under contract from Pan American Business Jets, Little Rock Airmotive converted a Falcon 20 into a specialised cargo aircraft. Known as the Falcon Cargo Jet, the prototype flew for the first time on 28 March 1972. By the Summer of the same year, an operator named Federal Express Corporation, of Little Rock, had three similar aircraft in service and has since expanded its fleet to a total of 33 Falcon D Cargo Jets.

The cargo conversion can be applied to any Falcon 20/Mystère 20 and is offered on the current Series F aircraft.

Basic feature of the conversion is replacement of the standard cabin door by a hydraulically-actuated cargo door 1·88 m wide by 1·44 m high (6 ft 2 in × 4 ft 9 in), forward of the wing on the port side. This door opens upward, with its sill at cabin floor level. The flooring itself is new and offers a completely flat area 7·01 m long by 1·62 m wide (23 ft × 5 ft 4 in). Made of aluminium honeycomb, it can sustain loadings of up to 488 kg/m² (100 lb/sq ft) and affords a vast number of alternative tie-down points for retainer nets and pallets. Floor-mounted rollers are optional.

The Falcon Cargo Jet's Category II solid-state electronics standard includes dual com/nav, dual flight directors, autopilot and weather radar. Specifically, in the case of the Federal Express fleet, the fit comprises RCA com/nav with DME, dual Collins FD-108 flight directors, RCA AVQ-21 radar, Collins AP-105 autopilot, Teledyne angle-of-attack system, Collins ADF and RCA transponder. Also installed on these aircraft is a Fairchild integral electronic weight and balance system, which indicates as a cockpit readout whether or not cargo weight and distribution are within the legal limits. Standard safety provisions include a quick-release cargo restraint system able to withstand 9*g*.

Dimensions, weights and performance are largely unchanged by this conversion scheme. Nominal empty weight is 6,963 kg (15,350 lb) and max zero-fuel weight is increased to 9,980 kg (22,000 lb). Range varies from 1,215 nm (2,250 km; 1,400 miles) with max payload to 1,736 nm (3,215 km; 2,000 miles) with max fuel and a 2,040 kg (4,500 lb) payload. Usable cabin volume is 14·15 m³ (500 cu ft).

DASSAULT FALCON 10

First announced in June 1969, the Falcon 10 is basically a scaled-down version of the Mystère 20/Falcon 20, with similar wing high-lift devices to those of the Falcon F and powered by two small turbofans in the 14·6 kN (3,300 lb st) class. It is designed to fail-safe principles and to comply with US FAR 25 transport category requirements.

A prototype (F-WFAL), with General Electric CJ610 turbojets, made its first flight on 1 December 1970. Flight testing was resumed on 7 May 1971 following modifications to the angles of wing incidence and dihedral and an increase in wing sweepback. The modified aircraft set up a 1,000 km closed-circuit speed record of 502·05 knots (930·4 km/h; 578·13 mph) in FAI Class C1f on 1 June 1971.

A second prototype (F-WTAL), with Garrett AiResearch TFE 731-2 engines, flew for the first time on 15 October 1971, followed by a third aircraft on 14 October 1972, with similar engines. This third Falcon 10 set up a 2,000 km closed-circuit speed record of 494·83 knots (917·02 km/h; 569·809 mph) in Class C1f on 29 May 1973.

The first production Falcon 10 with TFE 731-2 engines made its first flight on 30 April 1973. French certification of this version was granted on 11 September 1973, followed by FAA certification nine days later, allowing deliveries of production aircraft to begin on 1 November. Sixty-five had been delivered by 1 January 1976.

The 32nd Falcon 10 delivered from Mérignac, in April 1975, was the first of two **Falcon 10MER** aircraft for the French Navy, which uses them for radar training of Super Etendard pilots as well as for communications duties. The Navy has an option on three more.

Falcon Jet Corporation, distributor for the western hemisphere, has placed initial orders for 54, with options on a further 106. At 1 January 1976, orders totalled 84, plus 90 options.

The second prototype was adapted for use as a Larzac testbed. It resumed flying, with a Larzac 02 in its starboard nacelle, on 22 May 1973, the TFE 731-2 in its port nacelle being retained. Subsequently, the test engine was replaced by a Larzac 04 as part of the Alpha Jet development programme.

Like the Mystère 20/Falcon 20, the Falcon 10 can be equipped for liaison, executive transport, navigation/attack system training, aerial photography, radio navigation aid calibration and ambulance duties.

TYPE: Twin-turbofan executive transport.

WINGS: Cantilever low-wing monoplane with increased sweepback on inboard leading-edges. All-metal torsion-box structure, with leading-edge slats and double-slotted trailing-edge flaps and plain ailerons. Two-section spoilers above each wing, forward of flaps.

FUSELAGE: All-metal semi-monocoque structure, designed to fail-safe principles.

TAIL UNIT: Cantilever all-metal structure, similar to that of Falcon 20.

LANDING GEAR: Retractable tricycle type, manufactured by Messier-Hispano, with twin wheels on main gear, single wheel on nose gear. Hydraulic retraction, main units inward, nosewheel forward. Oleo-pneumatic shock-absorbers. Low-pressure tyres for soft-field operation.

POWER PLANT: Two Garrett AiResearch TFE 731-2 turbofan engines (each 14·4 kN; 3,230 lb st), mounted in pod on each side of rear fuselage. Fuel in two integral tanks in wings and two feeder tanks aft of rear bulkhead, with total capacity of 3,340 litres (735 Imp gallons; 882 US gallons). Separate fuel system for each engine, with provision for cross-feeding. Pressure refuelling system.

ACCOMMODATION: Crew of two on flight deck, with dual controls and airline-type instrumentation. Provision for third crew member on a jump-seat. Normal seating for four passengers (two individual seats and a three-seat sofa) or for seven passengers, with two individual seats added. Each pair of single seats is separated by a table. Coat compartment on starboard side, immediately aft of flight deck opposite door; rear baggage compartment behind sofa. Galley on left of entrance. Optional front toilet compartment. Downward-opening door with built-in steps.

SYSTEMS: Duplicated air-conditioning and pressurisation systems supplied with air bled from both engines. Pressure differential 0·60 bars (8·8 lb/sq in). Two independent hydraulic systems, each of 207 bars (3,000 lb/sq in) pressure and with twin engine-driven pumps and emergency electric pump, to actuate primary flight controls, flaps, landing gear, wheel brakes, airbrakes, yaw damper and nosewheel steering. 28V DC electrical system with a 9kW DC starter/generator on each engine, three 750VA 400Hz 115V inverters and two 23Ah batteries. Automatic emergency oxygen system.

ELECTRONICS AND EQUIPMENT: Standard equipment includes duplicated VHF and VOR/glideslope, single ADF, marker beacon receiver, ATC transponder, autopilot, intercom system and duplicated blind-flying instrumentation. Optional equipment includes duplicated DME and flight director, second ADF, weather radar and radio altimeter.

DIMENSIONS, EXTERNAL:	
Wing span	13·08 m (42 ft 11 in)
Wing chord (mean)	2·046 m (6 ft 8½ in)
Wing aspect ratio	7·1
Length overall	13·85 m (45 ft 5 in)
Length of fuselage	12·47 m (40 ft 11 in)
Height overall	4·61 m (15 ft 1½ in)
Tailplane span	5·82 m (19 ft 1 in)
Wheel track	2·86 m (9 ft 5 in)
Wheelbase	5·38 m (17 ft 8 in)
Passenger door:	
Height	1·47 m (4 ft 10 in)
Width	0·80 m (2 ft 7 in)
Height to sill	0·884 m (2 ft 10¾ in)
Emergency exit (stbd side, over wing):	
Height	0·914 m (3 ft 0 in)
Width	0·508 m (1 ft 8 in)
DIMENSIONS, INTERNAL:	
Cabin, excl flight deck:	
Length	5·00 m (16 ft 5 in)
Max width	1·46 m (4 ft 9 in)
Max height	1·50 m (4 ft 11 in)
Volume	7·50 m³ (264·6 cu ft)
Baggage compartment volume:	
front (wardrobe)	0·35 m³ (12·35 cu ft)
rear	0·70 m³ (24·7 cu ft)
AREAS:	
Wings, gross	24·1 m² (259 sq ft)
Horizontal tail surfaces (total)	6·75 m² (72·65 sq ft)
Vertical tail surfaces (total)	4·54 m² (48·87 sq ft)
WEIGHTS:	
Weight empty, equipped	4,880 kg (10,760 lb)
Max payload	789 kg (1,740 lb)
Max T-O weight	8,500 kg (18,740 lb)
Max zero-fuel weight	5,850 kg (12,900 lb)
Max landing weight	8,000 kg (17,640 lb)

PERFORMANCE:
Never-exceed speed at S/L 350 knots (648 km/h; 402 mph)
Max operating Mach number 0·87
Max cruising speed 494 knots (915 km/h; 568 mph)
FAR 25 balanced T-O field length with four passengers and fuel for a 1,000 nm (1,850 km; 1,150 mile) stage, 45 min reserves 930 m (3,050 ft)
FAR 25 balanced T-O field length, with four passengers and max fuel 1,250 m (4,100 ft)
FAR 25 landing run, with four passengers and 45 min reserves 630 m (2,070 ft)
Range with four passengers and 45 min reserves 1,918 nm (3,555 km; 2,209 miles)

Dassault Falcon 10 executive transport (two Garrett AiResearch TFE 731-2 turbofan engines)

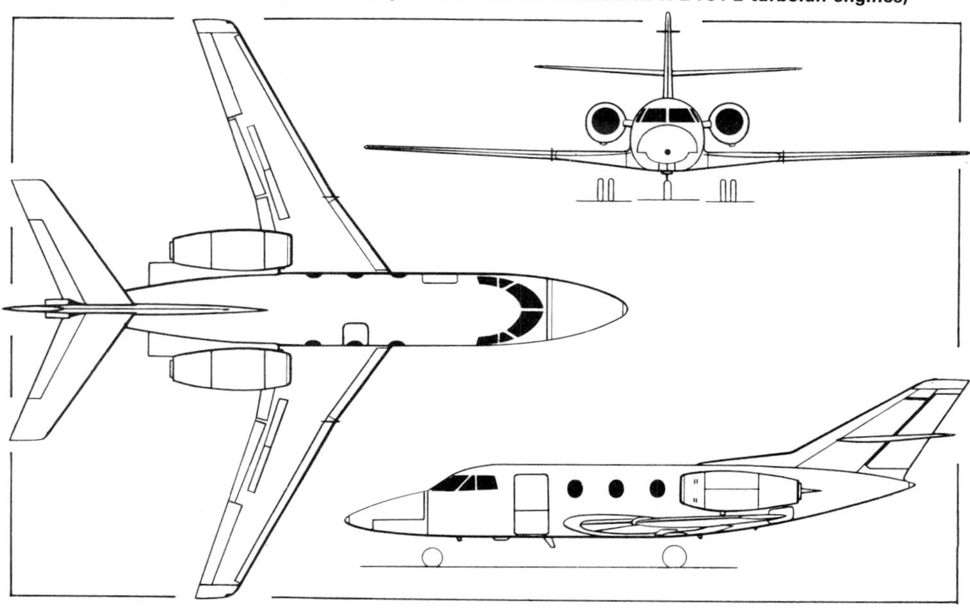
Dassault Falcon 10 four/seven-passenger executive transport *(Pilot Press)*

Dassault Falcon 10MER (Marine Entrainement Radar) aircraft ordered for fleet support, instrument training, Super Etendard systems training and liaison duties with the Aéronavale *(Austin J. Brown)*

DASSAULT-BREGUET FALCON 50

The decision to build a prototype of the extended-range three-engined Falcon 50 was announced at the end of May 1974. Subsequently, it was decided to redesign the wings, benefiting from experience gained with the Mercure. Fitted with advanced high-lift devices, including double-slotted trailing-edge flaps, these wings are well suited to high Mach numbers.

Some components of the Falcon 50 are identical with those of the twin-turbofan Falcon 20, including the front and centre fuselage sections. Basic accommodation is for eight or ten passengers in a high standard of comfort, as in the Falcon 20. A major change is the introduction of a very large fuselage fuel tank aft of the coat-rack and toilet compartment, on the aircraft's CG. Between this additional tankage and the rear pressure bulkhead is a baggage compartment of 2·5 m³ (90 cu ft) capacity. These changes, and others associated with the installation of the centre engine, required lengthening of the fuselage by comparison with the Falcon 20. The vertical tail surfaces are also considerably larger.

The power plant comprises three Garrett AiResearch TFE 731-3 turbofan engines, each rated at 16·5 kN (3,700 lb st) for take-off. Two are pod-mounted on the sides of the rear fuselage; the third, fitted with a thrust reverser, is housed in the fuselage tailcone, with an over-fuselage air intake forward of the base of the fin.

Dassault-Breguet is producing airframes for static and fatigue testing, and hopes to supplement the prototype with the first two production aircraft to speed certification.

The prototype was scheduled to fly in November 1976. Options had been taken out for 49 production Falcon 50s by mid-1976, 43 of them for service in North America. Deliveries are expected to begin in July 1978.

DIMENSIONS, EXTERNAL:	
Wing span	18·86 m (61 ft 10½ in)
Wing aspect ratio	7·73
Length overall	18·43 m (60 ft 5¾ in)
Height overall	5·70 m (18 ft 8½ in)
DIMENSIONS, INTERNAL:	
Cabin: Length	7·24 m (23 ft 9 in)
Height	1·72 m (5 ft 8 in)
Width	1·80 m (5 ft 10¾ in)
AREA:	
Wings, gross	46 m² (495 sq ft)
WEIGHTS:	
Weight empty, equipped	9,000 kg (19,840 lb)
Max fuel	6,950 kg (15,320 lb)
Max T-O weight	16,600 kg (36,600 lb)
Max landing weight	15,810 kg (34,855 lb)
Max zero-fuel weight	10,250 kg (22,600 lb)

Left to right: **Mockup of Falcon 50 with production examples of the Mystère 20/Falcon 20 Series F and Falcon 10**

PERFORMANCE (estimated):
- Max cruising speed: Mach 0·8 (470 knots; 870 km/h; 540 mph)
- T-O run: 1,650 m (5,415 ft)
- Landing run: 1,000 m (3,280 ft)
- Range at Mach 0·73 with four passengers and FAR 121 reserves: 3,000 nm (5,560 km; 3,450 miles)

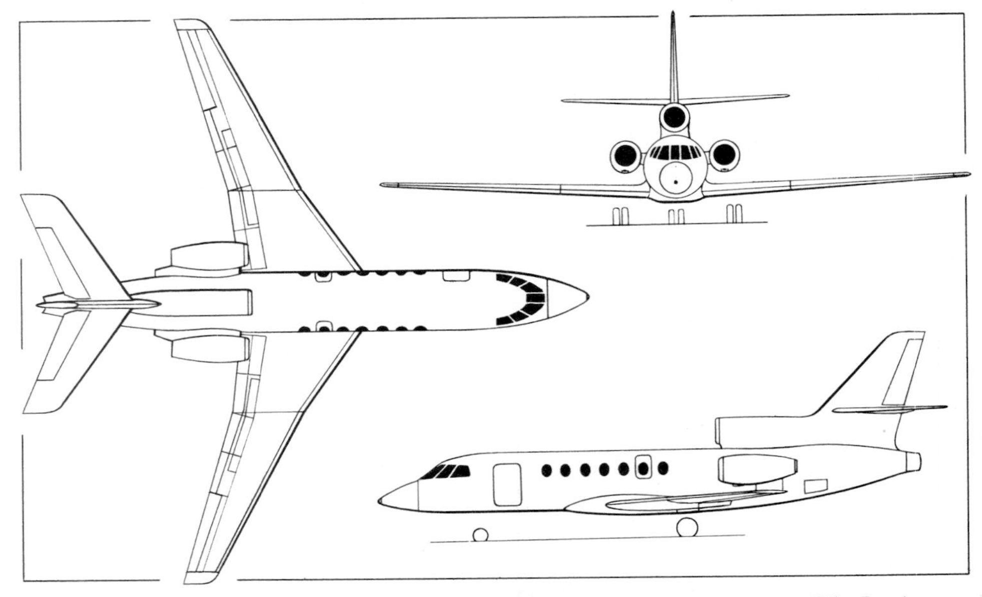

Dassault-Breguet Falcon 50 extended-range three-turbofan executive transport *(Pilot Press)*

DASSAULT MERCURE

The Mercure is a 120/162-seat twin-engined short-haul transport aircraft, optimised for ranges of 100-1,100 nm (185-2,040 km; 115-1,270 miles). Development was started in 1967, and the first of two prototypes (F-WTCC), powered by 66·7 kN (15,000 lb st) Pratt & Whitney JT8D-11 turbofan engines, was flown for the first time on 28 May 1971. The second prototype (F-WTMD), which flew on 7 September 1972, had more powerful JT8D-15 engines. JT8D-15s are also fitted to production Mercures, the first of which flew for the first time on 19 July 1973.

Certification of the Mercure was received on 12 February 1974, by which date a total of 1,225 flying hours had been logged by the first three aircraft. Air Inter ordered 10 Mercures, the first of which was delivered on 15 May 1974. Cat III certification was obtained on 30 September 1974.

The launching programme for the Mercure cost 1,000m Fr (£75m), covering the construction of two prototypes, two airframes for static and fatigue testing, certification and production tooling. Of this sum, the French contribution represented 70 per cent. The remaining amount was shared principally between Aeritalia of Italy (10 per cent), which manufactured the tail unit and fuselage tailcone; CASA of Spain (10 per cent), which manufactured the first and second fuselage sections; SABCA of Belgium (6 per cent), which built the flaps, ailerons, spoilers and air-brakes; and F+W (Emmen) of Switzerland (2 per cent), which was responsible for the engine air intakes and cowling panels.

Under an agreement signed in February 1972, Canadair Ltd of Montreal, Canada, manufactured wing panels, tracks for the wing leading-edges and flaps, and engine nacelle pylons for production Mercures, equivalent to 5·2 per cent of the airframe work.

A full structural description and specification can be found in the 1975-76 *Jane's*.

POWER PLANT: Two 68·9 kN (15,500 lb st) Pratt & Whitney JT8D-15 turbofan engines in underwing pods, fitted with thrust reversers and Dassault-developed noise absorbers. Total fuel capacity of 18,400 litres (4,048 Imp gallons; 4,860 US gallons). Total oil capacity 45 litres (9·9 Imp gallons; 11·9 US gallons).

ACCOMMODATION: Crew of two side by side on flight deck, with two extra optional seats. Typical mixed-class accommodation provides 12 seats four-abreast at 96·5 cm (38 in) pitch and 108 seats six-abreast at 86·5 cm (34 in) pitch. Basic tourist class accommodation provides 150 seats at 81·5 cm (32 in) pitch. High-density layout for up to 162 seats six-abreast at 76·2 cm (30 in) pitch. Six possible locations of toilets and galleys at front and rear, according to layout. Two passenger doors, at front and rear on port side. Aérazur retractable integral stairway built into fuselage below forward passenger door; provision for similar stairway below rear passenger door. Two service doors, at front and rear on starboard side, and two emergency exits over each wing. Cargo/baggage holds beneath cabin floor, one forward and two aft of wings. Forward hold can accommodate 3,500 kg (7,715 lb) or five standard Boeing 727 freight containers; aft hold No. 1 can accommodate 2,800 kg (6,170 lb) or four Boeing 727 containers; aft hold No. 2 can accommodate 1,800 kg (3,965 lb) of baggage.

ELECTRONICS AND EQUIPMENT: To customer's requirements. Basic aircraft designed for all-weather (Cat III) operation.

DIMENSIONS, EXTERNAL:
- Wing span: 30·55 m (100 ft 3 in)
- Wing chord at root: 6·00 m (19 ft 8¼ in)
- Wing chord at tip: 1·74 m (5 ft 8½ in)
- Wing aspect ratio: 8
- Length overall: 34·84 m (114 ft 3½ in)
- Length of fuselage: 34·41 m (112 ft 11 in)
- Height overall: 11·36 m (37 ft 3¼ in)
- Tailplane span: 12·77 m (41 ft 11 in)
- Wheel track: 6·20 m (20 ft 4 in)
- Wheelbase: 12·42 m (40 ft 9 in)

Dassault Mercure short-haul large-capacity transport in the insignia of Air Inter *(Air & General Photographs)*

DIMENSIONS, INTERNAL:
Cabin, excl flight deck:
Length 25·50 m (83 ft 7 in)
Max width 3·66 m (11 ft 11 in)
Max height 2·20 m (7 ft 2¾ in)
Floor area 82·00 m² (882 sq ft)
Volume 162·0 m³ (5,717 cu ft)
Freight hold volume:
forward 15·0 m³ (529 cu ft)
aft No. 1 11·5 m³ (406 cu ft)
aft No. 2 7·5 m³ (265 cu ft)
WEIGHTS AND LOADINGS:
Operating weight, empty 31,800 kg (70,107 lb)
Max payload 16,200 kg (35,715 lb)
Max ramp weight 57,000 kg (125,660 lb)
Max T-O weight 56,500 kg (124,560 lb)
Max landing weight 52,000 kg (114,640 lb)
Max zero-fuel weight 48,000 kg (105,820 lb)
Max wing loading 487 kg/m² (99·75 lb/sq ft)
Max power loading 410 kg/kN (4·02 lb/lb st)
PERFORMANCE:
Max operating speed (VMO/MMO)
380 knots (704 km/h; 437 mph) EAS up to 6,100 m (20,000 ft) and Mach 0·85 above 20,000 ft
Max cruising speed at 6,100 m (20,000 ft)
500 knots (926 km/h; 575 mph)
Max rate of climb at 2,135 m (7,000 ft) at AUW of 45,360 kg (100,000 lb) 1,007 m (3,300 ft)/min
Typical short-haul stage (600 nm; 1,110 km; 690 miles) with 150 passengers and 3,630 kg (8,000 lb) fuel reserves:
FAR 25 T-O distance (S/L, ISA)
2,100 m (6,900 ft)
Flight time 1 hr 22 min
Approach speed 117 knots (217 km/h; 135 mph)
FAR 121 landing distance (S/L, ISA + 14°C)
1,755 m (5,760 ft)
Max range with 150 passengers and 4,100 kg (9,000 lb) fuel reserves 1,125 nm (2,084 km; 1,295 miles)

DASSAULT-BREGUET MERCURE 200

On the basis of experience gained with the Mercure, Dassault-Breguet completed in the first half of 1975 a preliminary study for a higher-capacity version, powered by SNECMA/General Electric CFM56 turbofans, under contract to the French authorities.

At the instigation of the French Transport Ministry, Dassault-Breguet made contact subsequently with McDonnell Douglas Corporation in the USA, to investigate the possibility of joint development of the new aircraft as a successor to both the original Mercure and the DC-9. Seven months of joint work by the two companies led to further design changes, and the resulting proposal was submitted to the French Ministry of Transport on 8 April 1976 as the Mercure 200.

Basic changes by comparison with the original Mercure are as follows:

The power plant consists of two CFM56 high by-pass ratio turbofans. These will each be rated at 97·9 kN (22,000 lb st) for the first two years, but in the event of an engine failure the remaining CFM56 will be able to deliver 106·75 kN (24,000 lb st).

The engine pods are suspended under the wings, instead of being integral with them. This requires lengthening of each main landing gear unit by 270 mm (10¾ in) and the nose unit by 380 mm (1 ft 3 in), with associated redesign of wheel wells. Aircraft ground clearance is improved, and forward location of the engines reduces the possibility of wing damage in the event of a rotor break-up.

Redesign of the rear portion of the wing aerofoil, with trailing-edge camber, provides a more highly accentuated supercritical-type wing without changes forward of the rear spar. Addition of wingtip extensions offers a reduction in induced drag when the wing is working under high lift factors, notably during take-off and high-altitude cruising.

The trailing-edge flaps are redesigned in two segments, taking advantage of recent McDonnell Douglas research into high-lift double-slotted flaps.

As a result of these modifications, it is possible to increase take-off weight and provide increased accommodation by inserting two 'plugs' with a total length of 6 m (19 ft 8¼ in) in the fuselage fore and aft of the wings. With six-abreast seating and centre aisle, this offers maximum accommodation for 186 passengers at a seat pitch of 813 mm (32 in). Normal accommodation would be for 174 passengers at a pitch of 864 mm (34 in), with an alternative mixed class version for 16 first class passengers, four-abreast at a pitch of 965 mm (38 in), and 144 economy class six-abreast at 864 mm (34 in) pitch. In all cases, the normal fore and aft galleys and toilets are retained.

These changes, apart from the engines and nacelles, affect only 23% of the total airframe structure and 5% of the aircraft systems by comparison with the original Mercure. Structure life is calculated to be 40,000 flights or 40,000 flying hours.

DIMENSIONS, EXTERNAL:
Wing span 31·95 m (104 ft 10 in)
Wing aspect ratio 6·6
Length overall 40·93 m (134 ft 3½ in)
Tailplane span 12·77 m (41 ft 10¾ in)
Wheel track 6·60 m (21 ft 7¾ in)
Wheelbase 15·71 m (51 ft 6½ in)
AREA:
Wings, gross 118·7 m² (1,278 sq ft)
WEIGHTS AND LOADING:
Normal fuel load 14,750 kg (32,518 lb)
Max zero-fuel weight 59,000 kg (130,072 lb)
Max ramp weight 70,500 kg (155,425 lb)
*Max T-O weight 70,000 kg (154,323 lb)
Max landing weight 63,000 kg (138,890 lb)
Max wing loading 590 kg/m² (120·8 lb/sq ft)
*Eventual uprating of engines to 106·75 kN (24,000 lb st) will permit increase in T-O weight to 71,000 kg (156,528 lb)
PERFORMANCE (estimated, at max T-O weight):
Max operating speed
380 knots (703 km/h; 437 mph) EAS
Max operating Mach number 0·85
T-O run 2,450 m (8,000 ft)
Range with 174 passengers:
standard fuel 1,526 nm (2,830 km; 1,758 miles)
with increased fuel 1,665 nm (3,085 km; 1,917 miles)

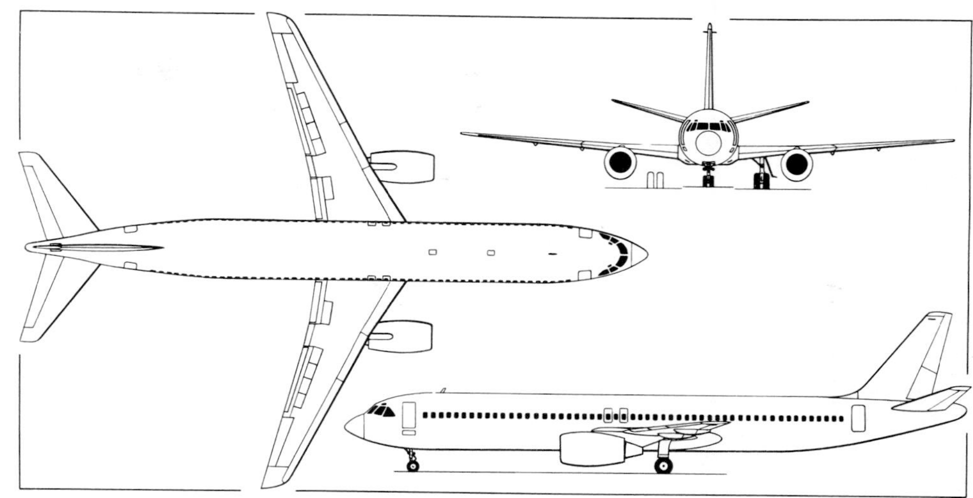
Dassault-Breguet Mercure 200 twin-turbofan short-range large-capacity transport *(Pilot Press)*

FOURNIER
AVIONS FOURNIER

HEAD OFFICE AND WORKS:
Aérodrome d'Athée-Nitray, 37270-Montlouis
Telephone: (47) 50-68-30
DIRECTOR GENERAL: René Fournier

Avions Fournier is producing a two-seat light aircraft, designated RF-6B, of which details follow. A related design by M René Fournier is the RF-6 Sportsman, manufactured by Sportavia (which see) in Germany.

FOURNIER RF-6B CLUB

The prototype of this two-seat aerobatic aircraft, designed by M René Fournier, flew for the first time on 12 March 1974; certification was obtained in April 1975.

Production by Avions Fournier of five pre-series RF-6Bs began during the first half of 1975, and the first of these (F-BVKS) flew for the first time on 4 March 1976. It has a longer nose than the prototype and a more powerful engine. By early 1976, orders for 30 aircraft had been received and it was planned to build up production to a rate of five per month.

The standard RF-6B, as described, has a 74·5 kW (100 hp) engine. A deluxe version with an 89·5 kW (120 hp) engine is under consideration.

Intended primarily for use as an aerobatic or training aircraft, the RF-6B is stressed to +9*g* and −4·5*g*.

TYPE: Two-seat aerobatic, training and sporting aircraft.

WINGS: Cantilever low-wing monoplane. Wing section NACA 23015 at root, NACA 23013 at tip. Dihedral 3° 30′. Incidence 3°. All-wood single-spar structure with plywood and Dacron covering. Frise-type ailerons of wooden construction, Dacron covered. Plain trailing-edge flaps of wooden construction with Dacron covering.

Fournier RF-6B (Rolls-Royce Continental O-200-A engine) *(J. M. G. Gradidge)*

FUSELAGE: All-wood oval structure, plywood covered.

TAIL UNIT: Cantilever structure of wood with Dacron covering. Fixed-incidence tailplane. Trim tab in port elevator.

LANDING GEAR: Non-retractable tricycle type. Oleo-pneumatic shock-absorber in each unit. Steerable nosewheel. Main-wheel tyres size 380-150. Nosewheel tyre size 300 × 100. Hydraulic disc brakes.

POWER PLANT: One 74·5 kW (100 hp) Rolls-Royce Continental O-200-A flat-four engine, driving a Hoffmann two-blade wooden fixed-pitch propeller with spinner. Fuselage fuel tank, immediately aft of firewall, capacity

80 litres (17·6 Imp gallons). Refuelling point on fuselage upper surface, forward of windscreen. Oil capacity 4 litres (0·88 Imp gallons).

ACCOMMODATION: Two seats side by side under transparent canopy, which swings upward and aft for access to cockpit. Dual controls standard. Cockpit heated and ventilated. Baggage space aft of seats.

SYSTEMS: Hydraulic system for brakes only. Electrical power supplied by 12V engine-driven alternator.

ELECTRONICS AND EQUIPMENT: Nav and com radios optional. Blind-flying instrumentation optional.

DIMENSIONS, EXTERNAL:

Wing span	10·50 m (34 ft 5½ in)
Wing chord at root	1·53 m (5 ft 0¼ in)
Wing chord at tip	0·83 m (2 ft 8¾ in)
Length overall	7·00 m (22 ft 11¾ in)
Height overall	2·52 m (8 ft 3 in)
Tailplane span	3·40 m (11 ft 1¾ in)
Propeller diameter	1·75 m (5 ft 9 in)
Propeller ground clearance	0·28 m (11 in)

AREAS:

Wings, gross	13·00 m² (139·9 sq ft)
Ailerons (total)	1·07 m² (11·52 sq ft)
Trailing-edge flaps (total)	1·52 m² (16·36 sq ft)
Rudder	0·74 m² (7·97 sq ft)
Tailplane	1·65 m² (17·76 sq ft)
Elevators (incl tab)	1·00 m² (10·76 sq ft)

WEIGHTS AND LOADINGS:

Weight empty	475 kg (1,047 lb)
Max T-O weight:	
Aerobatic	675 kg (1,488 lb)
Utility	740 kg (1,631 lb)
Max wing loading	57 kg/m² (11·6 lb/sq ft)
Max power loading	9·93 kg/kW (16·31 lb/hp)

PERFORMANCE (at max T-O weight):

Never-exceed speed	162 knots (300 km/h; 186 mph)
Max level speed at S/L	113 knots (210 km/h; 130 mph)
Normal cruising speed at S/L	102 knots (190 km/h; 118 mph)
Stalling speed	46 knots (85 km/h; 53 mph)
Service ceiling	4,000 m (13,125 ft)
T-O to 15 m (50 ft)	290 m (950 ft)
Landing from 15 m (50 ft)	250 m (820 ft)
Landing run, no braking	200 m (655 ft)

INDRAÉRO
INDRAÉRO SA

HEAD OFFICE AND WORKS:
Usine de Vavre, 36200-Argenton-sur-Creuse
Telephone: 04 07-75
Telex: 76 534 Chamco Chateauroux No. 006 1
PRESIDENT-DIRECTOR GENERAL:
Marcel Crepin

Indraéro SA has facilities for the design, construction and overhaul of major aircraft subassemblies, and for other, non-aviation products. Its works cover an area of about 12,500 m² (134,550 sq ft) and it employs about 100 people.

During the 1950s and 1960s it provided assistance in the construction of a number of light aeroplanes designed by M Blanchet and M Jean Chapeau. More recently it acquired exclusive manufacturing rights in the RF8 all-metal two-seat light aircraft designed by M René Fournier. The prototype was described and illustrated in the 1975-76 *Jane's*. There has been no recent news of progress or the start of production.

MUDRY
AVIONS MUDRY ET COMPAGNIE

ADDRESS:
Aérodrome, 27300-Bernay
Telephone: (16-32) 43.07.93
DIRECTOR:
Auguste Mudry

M Auguste Mudry, who is also Director of C.A.A.R.P. (which see), established this company in the works of the former Société Aéronautique Normande (see 1969-70 *Jane's*) at Bernay. Between them the two companies are responsible for production of the CAP 10 and CAP 20 aerobatic light aircraft developed by C.A.A.R.P. from the Piel Emeraude.

Fuselages for the CAP 10, which is described below, are manufactured at Beynes by C.A.A.R.P., final assembly and flight testing being undertaken by Avions Mudry et Cie at Bernay. The CAP 20 is manufactured entirely by C.A.A.R.P., and is described under that company's heading in this section.

CAP 10 two-seat aerobatic light aircraft (Lycoming IO-360-B2F engine)

C.A.A.R.P./MUDRY CAP 10

Developed from the Piel Emeraude two-seat light aircraft (which see), via the prototype C.P. 100 aerobatic version built by C.A.A.R.P., the CAP 10 is intended for use as a training, touring or aerobatic aeroplane. The prototype was flown for the first time in August 1968, and certification of the CAP 10 was granted on 4 September 1970. Construction is to French AIR 2052 (CAR 3) Category A standards for aerobatic flying.

A total of 50 CAP 10s had been completed by early 1974, of which four were delivered to customers in Italy, Germany and Belgium, and 30 to the French Air Force. Some are in service with the Equipe de Voltige Aérienne (EVA) at Salon-de-Provence, and others with the basic flying training school at Clermont-Ferrand-Aulnat. During 1974, ten CAP 10s were sold in France and two were exported, in addition to two which were sent, for demonstration and sales purposes, to Orange County Airport, near New York, USA. Deliveries totalled 67 by early 1976.

TYPE: Two-seat aerobatic light aircraft.

WINGS: Cantilever low-wing monoplane. Wing section NACA 23012. Dihedral 5° from roots. Incidence 0°. No sweepback. All-spruce single-spar torsion-box structure, with trellis ribs, rear auxiliary spar and okoumé plywood covering. Inner section of each wing is rectangular in plan, outer section semi-elliptical. Wooden trailing-edge plain flaps and slotted ailerons.

FUSELAGE: Conventional spruce girder structure, built in two halves and joined by three main frames. Of basically rectangular section with rounded top-decking. Fabric covering. Forward section also has an inner plywood skin for added strength. Engine cowling panels of non-inflammable laminated plastics.

TAIL UNIT: Conventional cantilever structure. All-wood single-spar fin, integral with fuselage, and tailplane. All surfaces plywood-covered except rudder, which is covered with both plywood and fabric. Tailplane incidence adjustable on ground. Trim tab in each elevator.

LANDING GEAR: Non-retractable tailwheel type. Main-wheel legs of light alloy, with ERAM type 9 270 C oleo-pneumatic shock-absorbers. Single wheel on each main unit, tyre size 380 × 150. Solid tailwheel tyre, size 6 × 200. Tailwheel is steerable by rudder linkage but can be disengaged for ground manoeuvring. Hydraulically-actuated main-wheel brakes and parking brake. Streamline fairings on main wheels and legs.

POWER PLANT: One 134 kW (180 hp) Lycoming IO-360-B2F flat-four engine, driving a Hoffmann two-blade fixed-pitch wooden propeller. Fuel in two main tanks in fuselage, one aft of engine fireproof bulkhead and one beneath baggage compartment, with total capacity of 150 litres (33 Imp gallons). Fuel and oil systems modified to permit periods of inverted flying.

ACCOMMODATION: Side-by-side adjustable seats for two persons, with provision for back parachutes, under rearward-sliding moulded transparent canopy. Space for 20 kg (44 lb) of baggage aft of seats in training and touring models.

SYSTEMS: Electrical system includes Delco-Rémy engine-driven alternator and SAFT 12V DC battery.

ELECTRONICS AND EQUIPMENT: CSF 262 12-channel VHF radio fitted.

DIMENSIONS, EXTERNAL:

Wing span	8·06 m (26 ft 5¼ in)
Wing aspect ratio	5·96
Length overall	7·16 m (23 ft 6 in)
Height overall	2·55 m (8 ft 4½ in)
Tailplane span	2·90 m (9 ft 6 in)
Wheel track	2·06 m (6 ft 9 in)

DIMENSION, INTERNAL:

Cabin: Max width	1·054 m (3 ft 5½ in)

AREAS:

Wings, gross	10·85 m² (116·79 sq ft)
Ailerons (total)	0·79 m² (8·50 sq ft)
Vertical tail surfaces (total)	1·32 m² (14·25 sq ft)
Horizontal tail surfaces (total)	1·86 m² (20·0 sq ft)

WEIGHTS (A: Aerobatic, U: Utility):

Weight empty, equipped:	
A, U	540 kg (1,190 lb)
Fuel load:	
A	54 kg (119 lb)
U	108 kg (238 lb)
Max T-O weight:	
A	760 kg (1,675 lb)
U	830 kg (1,829 lb)

PERFORMANCE (at max T-O weight):

Never-exceed speed	183 knots (340 km/h; 211 mph)
Max level speed at S/L	146 knots (270 km/h; 168 mph)
Max cruising speed (75% power)	135 knots (250 km/h; 155 mph)
Stalling speed, flaps up	52 knots (95 km/h; 59·5 mph)
Stalling speed, flaps down	44 knots (80 km/h; 50 mph)
Max rate of climb at S/L	over 360 m (1,180 ft)/min
Service ceiling	5,000 m (16,400 ft)
Max range	647 nm (1,200 km; 745 miles)

REIMS AVIATION
REIMS AVIATION SA

OFFICE AND WORKS:
Reims-Prunay Airport, BP 2745, 51062-Reims Cédex
Telephone: (26) 49.10.88
Telex: REMAVIA 830754
PARIS OFFICE:
18 Quai Alphonse le Gallo, 92100-Boulogne-Billancourt
Telephone: 604.81.36
PRESIDENT DIRECTOR-GENERAL:
Pierre Clostermann
DIRECTOR-GENERAL ADJOINT AND WORKS DIRECTOR: Jean Pichon
FINANCIAL DIRECTOR: Jean Luc Varga
ADMINISTRATIVE DIRECTOR: Armand Blang
PUBLIC RELATIONS: Frédéric Amanou
CHIEF PILOT: Franck Bardou

Under an agreement signed on 16 February 1960, the Cessna Aircraft Company of Wichita, Kansas, USA, acquired a 49% holding in this company, which was then known as Société Nouvelle des Avions Max Holste.

Reims Aviation has the right to manufacture under licence Cessna designs for sale in Europe, Africa and Asia. By 1 January 1976 it had assembled a total of 1,271 Cessna F 150 and 288 FRA 150 two-seat aircraft, 1,425 F 172, 568 FR 172 Reims Rocket and 47 F 182 four-seat aircraft, 140 F 177 RG four-seat aircraft, 3 Cessna 185 utility aircraft, 91 F 337, FT 337 and FT 337P five- or six-seat aircraft, and 49 FTB 337 six-seat STOL aircraft. Nearly 86% of all products are exported.

The 3,000th aircraft completed by Reims Aviation, a Reims Rocket, left the Reims-Prunay works at the beginning of December 1973. A total of 413 aircraft were delivered in the 1975 calendar year.

Reims Aviation is taking part in production of the Aérospatiale N 262/Frégate twin-turboprop light transport, by manufacturing tail units and fuselage tailcones, wheel fairings, flaps, ailerons, wing leading-edges, wingtips, engine nacelles and instrument panels. It is a subcontractor to Dassault-Breguet in the Falcon 10 programme, for which it supplies rear fuselage sections, and in the Super Etendard programme. It is also continuing the overhaul and servicing of M.H.1521 Broussard utility monoplanes.

Reims Aviation employed 508 people on 1 January 1976. Its offices and factory at Reims-Prunay Airport have an area of 23,000 m² (260,500 sq ft).

CESSNA F 150 and FRA 150 AEROBAT

Cessna 150 aircraft assembled under licence by Reims Aviation are designated **F 150.** The first example was flown on 22 February 1966. Production was at the rate of 90 aircraft per year in early 1976.

In addition, Reims Aviation produces an aerobatic version known as the **FRA 150** Aerobat. This is a two-seater with a 97 kW (130 hp) Rolls-Royce Continental engine. Production was at a rate of 30 aircraft per year in early 1976.

A full description of the current Cessna 150 and A150 is given in the US section.

CESSNA F 172 and FR 172/FRB 172 REIMS ROCKET

Cessna 172 aircraft assembled under licence by Reims Aviation are designated **F 172.**

Until 1971, Reims Aviation retained a 145 hp Rolls-Royce Continental engine in the F 172, of which the first example was flown on 4 January 1963, with 805 delivered by the end of 1971. The current F 172 has a 112 kW (150 hp) Lycoming O-320-E2D, like the standard Cessna 172, which is described fully in the US section of this edition. Production in early 1976 was at the rate of 130 aircraft per year.

First displayed at the 1967 Paris Air Show, the **FR 172** Reims Rocket was developed by Reims Aviation from the F 172. The first Rocket was flown in early 1967; the current rate of production is 30 aircraft per year. Reims produces the Rocket exclusively, for worldwide sale.

The Rocket has a 157 kW (210 hp) Rolls-Royce Continental IO-360-H flat-six engine, driving a constant-speed propeller. Wing fuel tank capacity is increased to 197 litres (43·3 Imp gallons), of which 186 litres (41 Imp gallons) are usable. There is also a version with a 64 litre (14·1 Imp gallon) auxiliary fuselage tank, increasing total capacity to 261 litres (57·4 Imp gallons), of which 238 litres (52·3 Imp gallons) are usable.

A number of modifications are embodied in the current model, to improve passenger comfort and performance and increase the useful load. These include conical-camber wingtips, redesigned seats, inertia-reel shoulder harness for front-seat occupants, improved baggage door hatch and a more flexible electrical system. Optional items include vertically-adjustable, fully-articulated front seats, child's foldaway seat in the baggage area, and glider-towing capability. The 0·5 m³ (17·6 cu ft) baggage compartment aft of the rear seats holds a maximum of 90 kg (200 lb).

Also available is a STOL (ADAC) version of the Reims Rocket, designated **FRB 172,** with high-lift trailing-edge flaps and ski landing gear, for operation into high-altitude mountain landing sites.

The following data apply to the standard FR 172 Reims Rocket, except where indicated otherwise:

DIMENSIONS, EXTERNAL:
- Wing span: 10·92 m (35 ft 10 in)
- Length overall: 8·28 m (27 ft 2 in)
- Height overall: 2·68 m (8 ft 9½ in)
- Tailplane span: 3·45 m (11 ft 4 in)
- Wheel track: 2·53 m (8 ft 3½ in)
- Propeller diameter: 1·93 m (6 ft 4 in)

AREA:
- Wings, gross: 16·16 m² (174 sq ft)

WEIGHTS AND LOADINGS:
- Weight empty: 660 kg (1,455 lb)
- Max T-O weight: 1,157 kg (2,550 lb)
- Max wing loading: 71·6 kg/m² (14·7 lb/sq ft)
- Max power loading: 7·37 kg/kW (12·1 lb/hp)

PERFORMANCE (at max T-O weight):
- Max level speed at S/L: 135 knots (251 km/h; 156 mph)
- Max cruising speed (75% power) at 1,675 m (5,500 ft): 129 knots (240 km/h; 149 mph)
- Econ cruising speed at 3,050 m (10,000 ft): 97 knots (179 km/h; 111 mph)
- Stalling speed, power off, flaps up: 53 knots (98 km/h; 61 mph)
- Stalling speed, power off, flaps down: 46 knots (86 km/h; 53 mph)
- Max rate of climb at S/L: 268 m (880 ft)/min
- Service ceiling: 5,180 m (17,000 ft)
- T-O run:
  - FR 172: 226 m (740 ft)
  - FRB 172: 180 m (590 ft)
- T-O to 15 m (50 ft):
  - FR 172: 375 m (1,230 ft)
- Landing from 15 m (50 ft):
  - FR 172: 375 m (1,230 ft)
  - FRB 172: 335 m (1,100 ft)
- Landing run:
  - FR 172: 189 m (620 ft)
  - FRB 172: 135 m (445 ft)
- Range at max cruising speed, no reserve:
  - standard fuel: 555 nm (1,030 km; 640 miles)
  - with aux tank: 751 nm (1,392 km; 865 miles)
- Range at econ cruising speed, no reserve:
  - standard fuel: 729 nm (1,352 km; 840 miles)
  - with aux tank: 981 nm (1,819 km; 1,130 miles)

Reims-Cessna FRA 150 Aerobat (Rolls-Royce Continental O-240-A engine)

Reims-Cessna FR 172 Reims Rocket (Rolls-Royce Continental IO-360-H engine)

Reims-Cessna F 182 (Continental O-470-S engine)

Reims-Cessna F 185 (Continental IO-520-D engine), described under Cessna entry in US section

CESSNA F 177 RG

The Cessna Cardinal RG four-seat light aircraft with retractable landing gear, when assembled by Reims Aviation, has the designation F 177 RG. First example flown on 4 February 1971. Production was at the rate of 15 aircraft per year in early 1976.

CESSNA F 182

The first Reims-assembled F 182 flew for the first time on 2 June 1975. A total of 47 had been delivered by early 1976, and Reims expected to complete 75 in its 1975-76 fiscal year.

A full description of the 182 can be found under the Cessna entry in the US section.

CESSNA F 337, FA 337 and FT 337P PRESSURISED SKYMASTER

In 1969 Reims Aviation began the assembly under licence of the Cessna 337 Super Skymaster six-seat twin-engined light aircraft. Primary structures are supplied by Cessna, and engines by Rolls-Royce; smaller components and equipment are French-built.

There are three current French production versions, each available in standard Skymaster or Skymaster II form, as follows:

F 337. Standard Cessna 337 as built by Reims Aviation. Two 157 kW (210 hp) Rolls-Royce Continental engines. First flown in 1972.

FA 337. Similar to F 337, but with optional STOL (ADAC) modifications, comprising high-lift trailing-edge flaps which reduce T-O run to 200 m (655 ft), T-O to 15 m (50 ft) to 350 m (1,150 ft), landing from 15 m (50 ft) to 340 m (1,115 ft), and landing run to 150 m (495 ft).

FT 337P. Pressurised Skymaster. Similar to F 337 except for having pressurised cabin and 168 kW (225 hp) Continental turbocharged fuel-injection engines.

Production was continuing at the rate of 3 F 337s and 2 Pressurised Skymasters per year in early 1976.

Full descriptions of the current models of the Cessna 337 appear in the US section of this edition. As explained there, the Skymaster II models differ from standard Skymasters in having the extended range fuel system, IFR electronics and other normally optional items installed as standard equipment.

Reims-Cessna FTB 337 G with underwing radar pod for special duties

REIMS-CESSNA FTB 337 G

The airframe of this five/six-seat push-and-pull light twin is basically similar to that of the commercial FA 337 with STOL (ADAC) modifications; but it is powered by two 168 kW (225 hp) Continental TSIO-360-D turbocharged engines. Developed at the request of various government agencies, it is not pressurised but can be equipped for maritime or overland patrol duties, sea or land rescue, or other specialised tasks, with four underwing pylons for containers of food and medicine, dinghies and locator beacons, radar, or equipment to detect oil slicks at sea or forest fires, including an infra-red seeker. The rear of the cabin can be cleared to carry cargo or two stretchers. The aircraft can also be equipped for navigation and IFR training.

By 1 January 1976 Reims Aviation had delivered 49 FTB 337s. It expected to build 20 in the 1975-76 fiscal year and at least 20-30 in 1976-77.

DIMENSIONS, EXTERNAL:
As Cessna Model 337, except:
- Wing span: 12·10 m (39 ft 8½ in)
- Height overall: 2·84 m (9 ft 4 in)

AREA:
- Wings, gross: 18·81 m² (202·5 sq ft)

WEIGHTS AND LOADING:
- Weight empty: 1,454 kg (3,206 lb)
- Max T-O weight: 2,100 kg (4,630 lb)
- Max wing loading: 113 kg/m² (23·2 lb/sq ft)

PERFORMANCE (at max T-O weight):
- Max level speed at S/L: 205 knots (380 km/h; 236 mph)
- Cruising speed (75% power):
  - at 3,000 m (10,000 ft): 185 knots (344 km/h; 214 mph)
  - at 6,000 m (20,000 ft): 200 knots (370 km/h; 230 mph)
- Max rate of climb at S/L: 381 m (1,250 ft)/min
- Rate of climb at 3,000 m (10,000 ft): 346 m (1,135 ft)/min
- Max certificated operating height, on one or two engines: 6,000 m (20,000 ft)
- STOL T-O run: 225 m (738 ft)
- STOL T-O to 15 m (50 ft): 380 m (1,245 ft)
- STOL landing from 15 m (50 ft): 340 m (1,115 ft)
- STOL landing run: 150 m (495 ft)
- Max range:
  - 75% power at 6,000 m (20,000 ft): 955 nm (1,770 km; 1,100 miles)
  - econ power at 3,000 m (10,000 ft): 1,085 nm (2,012 km; 1,250 miles)
  - econ power at 6,000 m (20,000 ft): 1,150 nm (2,132 km; 1,325 miles)

ROBIN
AVIONS PIERRE ROBIN

HEAD OFFICE AND WORKS:
BP 87, 21121-Fontaine-les-Dijon Cédex
Telephone: (80) 35-40-40
Telex: 35-818 Robin

COMMERCIAL MANAGEMENT:
Robin SA
Aérodrome de Toussus-le-Noble, 78530-Buc
Telephone: 956-35-14 and 956-34-62
Telex: 91-708

SERVICE STATIONS:
Dijon-Darois, Paris Toussus-le-Noble, La Rochelle-Laleu, Amiens-Boves

PRESIDENT DIRECTOR GENERAL, AVIONS PIERRE ROBIN:
Pierre Robin

PRESIDENT DIRECTOR GENERAL, ROBIN SA:
Thérèse Robin

GENERAL DIRECTOR:
Ned Allard

COMMERCIAL MANAGER (Paris, Toussus-le-Noble):
Pierre Meynard

SALES SUPERVISOR:
Michel Messud

PRODUCTION DIRECTOR:
J. Lécrivain

TECHNICAL DIRECTOR:
Michel Brandt

PUBLIC RELATIONS (Toussus-le-Noble):
Catherine Denjean

This company was formed in October 1957 as Centre Est Aéronautique to design, manufacture and sell touring aircraft. In 1969 the name of the company was changed to Avions Pierre Robin. Marketing and after-sales service of its products are the responsibility of Robin SA.

The founder, M Pierre Robin, in collaboration with M Jean Delemontez, the engineer responsible for the well-known Jodel series of light aircraft, began by developing for production a three-seat high-performance lightplane of wooden construction. This was the Jodel DR 100 Ambassadeur, which flew on 14 July 1958 and was followed by many refined, enlarged and more powerful types embodying the same basic design features. All of these have been described in earlier editions of *Jane's*.

Since 1973, Avions Pierre Robin has manufactured the DR 400 series of wooden light aircraft, all of which represent highly-refined developments of the company's earlier Jodel designs and were first flown in prototype form in 1972. They are described in detail, together with the company's range of HR 100 and HR 200 all-metal light aircraft, some with retractable landing gear.

Production in 1975 totalled 89 DR 400s, 23 HR 100s and 13 HR 200s. It was hoped to maintain production of 8 DR 400s, 4 HR 100s and 4 HR 200s each month in 1976.

The company's works currently cover an area of about 11,000 m² (118,400 sq ft) and it employs 160 people.

ROBIN DR 400/100 2+2

As its name implies, this smallest aircraft in the current Robin 400 series is a refined development of the earlier DR 220 '2+2' and DR 300/108 '2+2 Tricycle' described in the 1972-73 *Jane's*.

Design of the 400 series, and construction of the first example of each type, began towards the end of 1971. Features common to all six designs include a transparent canopy which slides forward to give access to all seats, and lowered walls on each side of the cabin to provide easier access and improved visibility.

The DR 400/100 2+2 flew for the first time on 24 November 1972 and received SGAC certification on 19 December 1972. Deliveries totalled 42 by February 1975, of which 18 were delivered in 1974.

TYPE: Two/four-seat light aircraft.

WINGS: Cantilever low-wing monoplane. Wing section NACA 23013·5 (modified). Centre-section has constant chord and no dihedral; outer wings have a dihedral of 14°. All-wood one-piece structure, with single box-spar. Leading-edge plywood-covered; polyester-fibre covering overall. Wooden ailerons, covered with polyester-fibre. Aluminium alloy flaps. Ailerons and flaps interchangeable port and starboard. Manually-operated airbrake under spar outboard of landing gear on each side. Picketing ring under each wingtip.

FUSELAGE: Wooden semi-monocoque structure of basic rectangular section, plywood-covered.

TAIL UNIT: Cantilever all-wood structure, covered with polyester-fibre. Sweptback fin and rudder. All-moving one-piece horizontal surface, with tab.

LANDING GEAR: Non-retractable tricycle type, with Jodel-Beaune oleo-pneumatic shock-absorbers and Manu hydraulically-actuated drum brakes. All three wheels and tyres are size 380 × 150, pressure 1·57 bars (22·8 lb/sq in) on nose unit, 1·77 bars (25·6 lb/sq in) on main units. Nosewheel steerable via rudder bar. Fairings over all three legs and wheels. Tailskid with damper. Parking brake.

POWER PLANT: One 74·5 kW (100 hp) Lycoming O-235-H2C flat-four engine, driving a McCauley IA-BCM-7056 two-blade fixed-pitch metal propeller. Fuel tank in fuselage, capacity 110 litres (24 Imp gallons).

ACCOMMODATION: Basic accommodation for two persons side by side, on adjustable seats, in enclosed cabin, with access via forward-sliding jettisonable transparent canopy. Optional bench seat to rear for one or two persons. Max payload 154 kg (340 lb) total on front seats, 111 kg (244 lb) total on rear seat. Dual controls standard. Cabin heated and ventilated. Baggage compartment with internal access.

SYSTEMS AND EQUIPMENT: Standard equipment includes a 12V 50A alternator, 12V 32Ah battery, push-button starter, audible stall warning, and windscreen de-icing. Radio, blind-flying equipment, and navigation, landing and anti-collision lights to customer's requirements.

DIMENSIONS, EXTERNAL:
- Wing span: 8·72 m (28 ft 7¼ in)
- Wing chord, centre-section (constant): 1·71 m (5 ft 7½ in)
- Wing chord at tip: 0·90 m (3 ft 0 in)
- Wing aspect ratio: 5·6
- Length overall: 6·96 m (22 ft 10 in)
- Height overall: 2·23 m (7 ft 3¾ in)
- Tailplane span: 3·20 m (10 ft 6 in)
- Wheel track: 2·60 m (8 ft 6¼ in)
- Wheelbase: 5·20 m (17 ft 0¾ in)
- Propeller diameter: 1·78 m (5 ft 10 in)

DIMENSIONS, INTERNAL:
- Cabin: Length: 1·62 m (5 ft 3¾ in)
  - Max width: 1·10 m (3 ft 7¼ in)
  - Max height: 1·23 m (4 ft 0½ in)
- Baggage space, volume: 0·39 m³ (13·75 cu ft)

AREAS:
- Wings, gross: 13·60 m² (146·39 sq ft)
- Ailerons, total: 1·15 m² (12·38 sq ft)
- Flaps, total: 0·70 m² (7·53 sq ft)
- Fin: 0·61 m² (6·57 sq ft)
- Rudder: 0·63 m² (6·78 sq ft)
- Horizontal tail surfaces, total: 2·88 m² (31·00 sq ft)

WEIGHTS AND LOADINGS:
- Weight empty, equipped: 520 kg (1,146 lb)
- Max T-O and landing weight: 865 kg (1,907 lb)
- Max wing loading: 63·6 kg/m² (13·03 lb/sq ft)
- Max power loading: 11·61 kg/kW (19·07 lb/hp)

PERFORMANCE (at max T-O weight):
- Never-exceed speed: 166 knots (308 km/h; 191 mph)
- Max level speed at S/L: 117 knots (216 km/h; 134 mph)
- Max cruising speed (75% power) at 2,440 m (8,000 ft): 114 knots (211 km/h; 131 mph)
- Econ cruising speed (55% power): 96 knots (178 km/h; 111 mph)
- Stalling speed, flaps down: 44·5 knots (81 km/h; 51 mph)
- Stalling speed, flaps up: 53·5 knots (99 km/h; 61·5 mph)
- Max rate of climb at S/L: 198 m (650 ft)/min
- Service ceiling: 3,800 m (12,460 ft)
- T-O run: 240 m (732 ft)
- T-O to 15 m (50 ft): 510 m (1,675 ft)
- Landing from 15 m (50 ft): 450 m (1,477 ft)
- Landing run: 180 m (590 ft)
- Range with max fuel at max cruising speed, no reserve: 555 nm (1,030 km; 640 miles)

ROBIN DR 400/120 PETIT PRINCE

The prototype of this DR 400 series Petit Prince flew for the first time on 15 May 1972 and received SGAC certification on the 10th of that month, followed by FAA and CAA certification in December 1972. Deliveries totalled 46 by February 1975, of which 5 were delivered during 1974. For 1975, the former 125 hp DR 400/125 model was replaced by the 118 hp DR 400/120, to which the following details apply:

Type: Three/four-seat light training and touring aircraft.

Wings, Fuselage, Tail Unit and Landing Gear: As for DR 400/100 2+2.

Power Plant: One 88 kW (118 hp) Lycoming O-235-L2A flat-four engine, driving a McCauley two-blade fixed-pitch metal propeller, or Hoffmann two-blade fixed-pitch wooden propeller. Fuel tank in fuselage, capacity 110 litres (24 Imp gallons); optional 50 litre (11 Imp gallon) auxiliary tank. Oil capacity 5·7 litres (1·25 Imp gallons).

Accommodation: Seats for three or four persons, in pairs, up to a max weight of 154 kg (340 lb) on front pair and 136 kg (300 lb), including baggage, at rear. Otherwise as for DR 400/100 2+2.

Systems and Equipment: As for DR 400/100 2+2.

Dimensions, external, internal, and Wing Area: As DR 400/100 2+2

Weights and Loadings:
- Weight empty, equipped 530 kg (1,169 lb)
- Max T-O and landing weight 900 kg (1,984 lb)
- Max wing loading 66·2 kg/m² (13·56 lb/sq ft)
- Max power loading 10·23 kg/kW (16·8 lb/hp)

Performance (at max T-O weight):
- Never-exceed speed 166 knots (308 km/h; 191 mph)
- Max level speed at S/L 123 knots (228 km/h; 141 mph)
- Max cruising speed at 2,440 m (8,000 ft) 118 knots (220 km/h; 136 mph)
- Econ cruising speed at 3,650 m (12,000 ft) 110 knots (205 km/h; 127 mph)
- Stalling speed, flaps down 45 knots (83 km/h; 52 mph)
- Stalling speed, flaps up 51 knots (94 km/h; 58·5 mph)
- Max rate of climb at S/L 190 m (623 ft)/min
- Service ceiling 3,500 m (11,500 ft)
- T-O run 360 m (1,180 ft)
- T-O to 15 m (50 ft) 640 m (2,100 ft)
- Range with max fuel at max cruising speed, no reserve 491 nm (910 km; 565 miles)

ROBIN DR 400/140B MAJOR

Type: Four-seat light monoplane.

Wings, Fuselage, Tail Unit, Landing Gear: As for DR 400/100 2+2.

Power Plant: One 119 kW (160 hp) Lycoming O-320-D flat-four engine, driving a Sensenich two-blade metal fixed-pitch propeller. Otherwise as for DR 400/120.

Accommodation: Seating for four persons, on two side-by-side adjustable front seats (max load 154 kg; 340 lb total) and rear bench seat (max load 154 kg; 340 lb total). Forward-sliding transparent canopy gives access to all seats. Up to 62 kg (137 lb) of baggage can be stowed aft of rear seats when four occupants are carried.

Systems and Equipment: As for DR 400/100 2+2.

Dimensions, external:
- As for DR 400/100 2+2, except:
- Propeller diameter 1·83 m (6 ft 0 in)

Dimensions, internal: As for DR 400/100 2+2.

Areas: As for DR 400/100 2+2, except:
- Flaps (total) 0·70 m² (7·53 sq ft)

Weights and Loadings:
- Weight empty, equipped 560 kg (1,235 lb)
- Max T-O and landing weight 1,000 kg (2,205 lb)
- Max wing loading 73·5 kg/m² (15·05 lb/sq ft)
- Max power loading 8·40 kg/kW (13·78 lb/hp)

Performance (at max T-O weight):
- Never-exceed speed 166 knots (308 km/h; 191 mph)
- Max level speed at S/L (60% power) 141 knots (262 km/h; 163 mph)
- Max cruising speed (60% power) at 2,440 m (8,000 ft) 125 knots (232 km/h; 144 mph)
- Econ cruising speed at 3,650 m (12,000 ft) 125 knots (232 km/h; 144 mph)
- Stalling speed, flaps down 47 knots (87 km/h; 54 mph)
- Stalling speed, flaps up 48 knots (89 km/h; 55·5 mph)
- Max rate of climb at S/L 240 m (785 ft)/min
- Service ceiling 4,560 m (14,950 ft)
- T-O run 200 m (655 ft)
- T-O to 15 m (50 ft) 500 m (1,640 ft)
- Landing from 15 m (50 ft) 490 m (1,608 ft)
- Landing run 220 m (722 ft)
- Range with max fuel at max cruising speed, no reserve 491 nm (910 km; 565 miles)

ROBIN DR 400/160 CHEVALIER

This four-seat light aircraft replaced the earlier DR 360 Major 160, described in the 1972-73 *Jane's*. It is generally similar to the DR 400 series aircraft listed earlier, but has a more powerful engine and increased fuel capacity. The first DR 400/160 flew on 29 June 1972. It was awarded SGAC certification on 6 September 1972, and both FAA and CAA certification in December of the same year. Deliveries totalled 43 by February 1975, of which 5 were delivered in 1974.

Type: Four-seat light aircraft.

Wings, Fuselage, Tail Unit, Landing Gear: Generally as for DR 400/100 2+2, but with external baggage door aft of cabin, in top of fuselage on port side.

Power Plant: One 119 kW (160 hp) Lycoming O-320-D flat-four engine, driving a Sensenich two-blade metal fixed-pitch propeller. Fuel tank in fuselage, capacity 110 litres (24 Imp gallons), and two tanks in wing-root leading-edges, giving total capacity of 190 litres (41·75 Imp gallons). Provision for auxiliary tank, raising total capacity to 240 litres (52·75 Imp gallons). Oil capacity 7·55 litres (1·66 Imp gallons).

Accommodation, Systems and Equipment: As for DR 400/140, except baggage capacity 40 kg (88 lb).

Dimensions and Areas:
- As for DR 400/140B, except:
- Baggage door:
  - Height 0·47 m (1 ft 6½ in)
  - Width 0·55 m (1 ft 9½ in)
- Wing area 14·20 m² (152·8 sq ft)

Weights and Loadings:
- Weight empty, equipped 565 kg (1,245 lb)
- Max T-O and landing weight 1,050 kg (2,315 lb)
- Max wing loading 74·2 kg/m² (15·20 lb/sq ft)
- Max power loading 8·82 kg/kW (14·47 lb/hp)

Performance (at max T-O weight):
- Never-exceed speed 166 knots (308 km/h; 191 mph)
- Max level speed at S/L 143 knots (266 km/h; 165 mph)
- Max cruising speed (75% power) at 2,440 m (8,000 ft) 137 knots (255 km/h; 158 mph)
- Econ cruising speed (60% power) at 3,650 m (12,000 ft) 127 knots (236 km/h; 146 mph)
- Stalling speed, flaps down 50 knots (93 km/h; 58 mph)
- Stalling speed, flaps up 56 knots (103 km/h; 64 mph)
- Max rate of climb at S/L 228 m (748 ft)/min
- Service ceiling 4,430 m (14,525 ft)
- T-O run 310 m (1,017 ft)
- T-O to 15 m (50 ft) 620 m (2,034 ft)
- Landing from 15 m (50 ft) 545 m (1,788 ft)
- Landing run 250 m (820 ft)
- Range with max fuel at econ cruising speed, no reserve 863 nm (1,600 km; 994 miles)

ROBIN DR 400/180 RÉGENT

First flown on 27 March 1972, this most powerful, four/five-seat member of the wooden DR 400 series replaced the earlier DR 253 Régent and DR 380 Prince. It received SGAC certification on 10 May 1972, and both FAA and CAA certification in December 1972. Deliveries totalled 74 by February 1975, of which 25 were delivered in 1974.

The DR 400/180 is generally similar to the DR 400/160 Chevalier, except in the following details:

Power Plant: One 134 kW (180 hp) Lycoming O-360-A flat-four engine. Propeller and fuel tankage as for DR 400/160.

Accommodation, Systems and Equipment: Basically as for DR 400/160, but optional seating for three persons on rear bench seat. Baggage capacity 55 kg (121 lb).

Dimensions and Areas:
- As for DR 400/160, except:
- Propeller diameter 1·93 m (6 ft 4 in)

Weights and Loadings:
- Weight empty, equipped 600 kg (1,322 lb)
- Max T-O and landing weight 1,100 kg (2,425 lb)
- Max wing loading 77·7 kg/m² (15·91 lb/sq ft)
- Max power loading 8·21 kg/kW (13·47 lb/hp)

Performance (at max T-O weight):
- Never-exceed speed 166 knots (308 km/h; 191 mph)
- Max level speed at S/L 150 knots (278 km/h; 173 mph)
- Max cruising speed (75% power) at 2,440 m (8,000 ft) 144 knots (267 km/h; 166 mph)
- Econ cruising speed (60% power) at 3,650 m (12,000 ft) 134 knots (249 km/h; 155 mph)
- Stalling speed, flaps down 51·5 knots (95 km/h; 59 mph)
- Stalling speed, flaps up 56·5 knots (105 km/h; 65 mph)
- Max rate of climb at S/L 252 m (825 ft)/min
- Service ceiling 4,720 m (15,475 ft)
- T-O run 315 m (1,035 ft)
- T-O to 15 m (50 ft) 610 m (2,000 ft)
- Landing from 15 m (50 ft) 530 m (1,740 ft)
- Landing run 249 m (817 ft)
- Range with max fuel at econ cruising speed, no reserve 793 nm (1,470 km; 913 miles)

ROBIN DR 400/180R REMORQUEUR

The DR 400/180R is a member of the DR 400 range designed for use as a glider-towing aircraft, although it can also be flown as a normal four-seat tourer. The prototype first flew on 6 November 1972 and received SGAC certification on the 28th of that month. Deliveries totalled 57 by February 1975, of which 29 were delivered in 1974. Details are generally the same as for the DR 400/180 Régent, except for the following items:

Fuselage: No external baggage door.

Power Plant: One 134 kW (180 hp) Lycoming O-360-A flat-four engine, driving (for glider-towing) a Sensenich 76 EM 8S5 058 two-blade propeller. For touring operation a Sensenich 76 EM 8S5 064 propeller of the same diameter is fitted. Fuel capacity as for DR 400/100 2+2.

Dimensions and Areas:
- As for DR 400/140B

Weights and Loadings:
- Weight empty, equipped 560 kg (1,234 lb)
- Max T-O and landing weight 1,000 kg (2,205 lb)
- Max wing loading 73·5 kg/m² (15·05 lb/sq ft)
- Max power loading 7·46 kg/kW (12·25 lb/hp)

Performance (glider tug, at max T-O weight):
- Never-exceed speed 166 knots (308 km/h; 191 mph)

Robin DR 400/140B Major (Lycoming O-320-D engine) *(R. Kunert)*

Robin DR 400/180 Régent (Lycoming O-360-A engine)

Max level speed at S/L (70% power)
124 knots (230 km/h; 143 mph)
Max cruising speed at 2,440 m (8,000 ft)
124 knots (230 km/h; 143 mph)
Econ cruising speed (56% power) at 3,650 m (12,000 ft)
122 knots (226 km/h; 140 mph)
Stalling speed, flaps down
47 knots (87 km/h; 54 mph)
Stalling speed, flaps up
53·5 knots (99 km/h; 61·5 mph)
Max rate of climb at S/L towing Bijave sailplane
210 m (690 ft)/min
Service ceiling 6,000 m (19,675 ft)
T-O run 205 m (673 ft)
T-O to 15 m (50 ft) 400 m (1,313 ft)
Landing from 15 m (50 ft) 470 m (1,542 ft)
Landing run 220 m (722 ft)
Range at econ cruising speed, max fuel, no reserve
444 nm (825 km; 512 miles)

ROBIN HR 100/180

This is a lightweight, simplified version of the HR 100/210 Safari II, with a 134 kW (180 hp) Lycoming O-360 engine. No details are available, except that the prototype flew for the first time in early 1976.

ROBIN HR 100/210 SAFARI II

Some years ago Avions Pierre Robin began studies with a view to the eventual production of an all-metal light aircraft similar in concept to its original range of wooden designs. As a part of this programme, the prototype DR 253 Régent was flown in late 1967 with all-metal wings, and construction of an all-metal prototype aircraft was begun in 1968.

This prototype (F-WPXO), designated HR 100/180 and powered by a 134 kW (180 hp) Lycoming O-360 engine, flew for the first time on 3 April 1969. In 1970 three pre-production aircraft were built, and manufacture of the initial production version, the HR 100/200 with 149 kW (200 hp) Lycoming IO-360-A1D6 engine, began in January 1971. Thirty-one HR 100/200s were built.

A description of the HR 100/200 appeared in the 1972-73 *Jane's*; that which follows applies to the current HR 100/210, first flown on 8 April 1971, and of which production started in July 1972. Deliveries totalled 78 by February 1976.

TYPE: Four-seat light aircraft.

WINGS: Cantilever low-wing monoplane. Wing section NACA 64 series (modified), with max thickness at 45% chord. Constant chord over most of span. Thickness/chord ratio 15%. Dihedral 6° 20′ from roots. No sweepback. All-metal single-spar structure, attached to fuselage by four bolts on each side. Outer skin of flush-riveted Duralinox-Cégédur aluminium alloy. Wingtips, which incorporate navigation and landing lights, are of polyester. All-metal trailing-edge slotted flaps are actuated electrically and have a max setting of 30°. All-metal Frise-type ailerons each have a piano-type hinge on the upper surface; they are controlled by rods and cables.

FUSELAGE: All-metal box-girder load-bearing structure, covered mainly with flush-riveted Duralinox-Cégédur aluminium alloy. Top-decking, between engine firewall and front of cabin, and from rear of cabin to fin, is of polyester; the forward panels are removable to provide easy access to instruments and controls.

TAIL UNIT: Cantilever structure, of similar construction to wings, with slight sweepback on fin and rudder. One-piece single-spar all-moving tailplane, with mass-balance and anti-tab. Rudder and tailplane cable-controlled.

LANDING GEAR: Non-retractable tricycle type, with oleo-pneumatic shock-absorbers. Nosewheel leg is offset to starboard. Single wheel on each unit; all wheels and tyres same size, 420 × 150. Tyre pressures 1·96 bars (28·4 lb/sq in) on nosewheel, 2·16 bars (31·3 lb/sq in) on main units. Streamlined leg and wheel fairings on all units. Hydraulic disc brakes. Parking brake. Tailskid with damper.

POWER PLANT: One 157 kW (210 hp) Continental IO-360-H flat-six engine, driving a McCauley two-blade constant-speed propeller. Fuel in 110 litre (24·2 Imp gallon) flexible tank in each leading-edge; total capacity 220 litres (48·4 Imp gallons). Provision for two auxiliary tanks to double this capacity. Refuelling point above each tank.

ACCOMMODATION: Seating for four persons, in pairs, under jettisonable transparent canopy which slides forward to provide access to all seats. Individual adjustable front seats; bench seat at rear. Max load on each seat 77 kg (170 lb). Space for 50 kg (110 lb) baggage aft of rear seats, accessible internally or by upward-opening external door on port side. Dual controls standard. Cabin ventilated.

SYSTEM: Electrical system includes 12V 60A alternator and 12V 45Ah battery.

ELECTRONICS AND EQUIPMENT: Standard equipment includes push-button starter, audible stall warning indicator, windscreen de-icing and towbar. Optional equipment includes VHF radio, VOR, navigation and landing lights, rotating anti-collision beacon, and full IFR equipment, including autopilot.

DIMENSIONS, EXTERNAL:
Wing span 9·08 m (29 ft 9½ in)
Wing chord (constant) 1·61 m (5 ft 3¼ in)
Length overall 7·45 m (24 ft 5¼ in)
Height overall 2·26 m (7 ft 5 in)
Tailplane span 3·20 m (10 ft 6 in)
Wheel track 3·20 m (10 ft 6 in)
Wheelbase 1·77 m (5 ft 9¾ in)
Propeller diameter 1·87 m (6 ft 1¾ in)

AREA:
Wings, gross 15·2 m² (163·6 sq ft)

WEIGHTS AND LOADINGS:
Weight empty 730 kg (1,609 lb)
Max T-O weight 1,250 kg (2,755 lb)
Max wing loading 82·3 kg/m² (16·85 lb/sq ft)
Max power loading 7·96 kg/kW (13·12 lb/hp)

PERFORMANCE (at max T-O weight):
Never-exceed speed 165 knots (306 km/h; 190 mph)
Max level speed at S/L
148 knots (275 km/h; 171 mph)
Max cruising speed (75% power) at optimum height
140 knots (260 km/h; 162 mph)
Econ cruising speed (65% power)
134 knots (248 km/h; 154 mph)
Stalling speed, flaps down
57·5 knots (106 km/h; 66 mph)
Max rate of climb at S/L 270 m (885 ft)/min
Service ceiling 4,500 m (14,775 ft)
Range with standard fuel, no reserve:
Max cruising speed 734 nm (1,360 km; 845 miles)
Econ cruising speed 795 nm (1,475 km; 916 miles)
Range with auxiliary fuel, no reserve:
Max cruising speed
1,467 nm (2,720 km; 1,690 miles)
Econ cruising speed
1,592 nm (2,950 km; 1,833 miles)

ROBIN HR 100/230TR

This aircraft was identical to the HR 100/285 Tiara, except for having a 175 kW (235 hp) Lycoming O-540-B flat-six engine, driving a Hartzell two-blade variable-pitch propeller. The prototype (F-WVKA) flew for the first time on 7 March 1974 and was known originally as the HR 100/235. The designation was changed to HR 100/230TR in 1975, in which year further work on the project was abandoned.

ROBIN HR 100/250TR

Except for having a different engine, this aircraft is similar to the HR 100/285 Tiara. The prototype (F-WVKA) was produced by converting the 'one-off' HR 100/230TR, by installation of a 186 kW (250 hp) Lycoming IO-540-C4B5 fuel-injection engine, driving a Hartzell two-blade constant-speed propeller of 2·03 m (6 ft 8 in) diameter.

Production began in the second half of 1975, with initial deliveries to the Centre d'Essais en Vol.

WEIGHT:
Max T-O weight 1,400 kg (3,086 lb)

PERFORMANCE (at max T-O weight):
Max level speed at S/L
170 knots (315 km/h; 196 mph)
Max cruising speed at 2,135 m (7,000 ft)
160 knots (297 km/h; 185 mph)
Cruising speed (65% power) at 3,050 m (10,000 ft)
154 knots (285 km/h; 177 mph)
Stalling speed, wheels and flaps up
72 knots (132·5 km/h; 82·5 mph) IAS
Stalling speed, wheels down, 30° flap
62 knots (114 km/h; 71 mph) IAS
Max rate of climb at S/L 324 m (1,065 ft)/min

ROBIN HR 100/285 TIARA

This was the first Robin light aircraft to be built with a retractable landing gear. Design began in September 1971 as a further major development stage beyond the HR 100/210, with new vertical tail surfaces, a redesigned wing structure, and a Teledyne Continental Tiara engine. Construction of the prototype (F-WSQV) was started in April 1972 and the first flight was made on 18 November 1972. Certification was received on 23 July 1974, when the HR 100/285 Tiara entered production.

The prototype had a 320 hp engine, but production HR 100/Tiaras have a 212·5 kW (285 hp) engine. Fifteen have been delivered to the Service de la Formation Aéronautique (SFA) for use at its Montpellier Training Centre.

TYPE: Four/five-seat all-metal light aircraft.

WINGS: Cantilever low-wing monoplane. Wing section NACA 64A515 (modified). Dihedral 6° 18′ from roots. Incidence 4° 41′. No sweepback. Aluminium alloy single-spar structure of constant chord. All-metal Frise-type ailerons and electrically-actuated NACA slotted flaps. No tabs.

FUSELAGE: Aluminium alloy semi-monocoque structure in cabin section. Rear fuselage top-decking and engine cowling are of non-stressed polyester.

TAIL UNIT: Cantilever structure, similar to wings in construction. Sweptback vertical surfaces. One-piece all-moving horizontal surfaces, with automatic anti-tab inboard on each trailing-edge. Trim tab in rudder.

LANDING GEAR: Retractable tricycle type, with single wheel on each unit. Electro-hydraulic retraction. Nosewheel protrudes slightly when retracted, to reduce damage in a wheels-up landing. Oleo-pneumatic shock-absorbers. Main-wheel tyres size 420 × 150 or 6·50-3, pressure 2·17 bars (31·5 lb/sq in). Nosewheel tyre size 330 × 130 or 5·00-5. Hydraulic disc brakes. Parking brake. Tailskid with damper.

POWER PLANT: One 212·5 kW (285 hp) Teledyne Continental Tiara 6-285B flat-six engine, driving a Hoffmann

Robin HR 100/250TR four/five-seat light aircraft (Lycoming IO-540-C4B5 engine)

Robin HR 100/180, a lightweight version of the Safari II

three-blade constant-speed wooden propeller. Four fuel tanks in wings, with total capacity of 440 litres (97 Imp gallons).

ACCOMMODATION: Two persons side by side in individual front bucket seats, with dual controls. Rear bench seat for two or three passengers. Access via forward-sliding jettisonable canopy. Baggage space aft of rear seats, accessible internally or by upward-opening external door on port side.

SYSTEM, ELECTRONICS AND EQUIPMENT: As for HR 100/210.

DIMENSIONS, EXTERNAL:
Wing span 9·08 m (29 ft 9½ in)
Wing chord (constant) 1·675 m (5 ft 6 in)
Wing aspect ratio 5·36
Length overall 7·59 m (24 ft 10¾ in)
Height overall 2·71 m (8 ft 10¾ in)
Tailplane span 3·20 m (10 ft 10 in)
Wheel track 3·225 m (10 ft 7 in)
Wheelbase 2·16 m (7 ft 1 in)
Propeller diameter 2·00 m (6 ft 6¾ in)
Propeller ground clearance 0·34 m (1 ft 1¼ in)

DIMENSIONS, INTERNAL:
Cabin: Length 2·80 m (9 ft 2 in)
Max width 1·115 m (3 ft 8 in)
Max height 1·20 m (3 ft 11¼ in)

AREAS:
Wings, gross 15·2 m² (163·6 sq ft)
Ailerons, total 1·02 m² (10·98 sq ft)
Trailing-edge flaps, total 1·55 m² (16·68 sq ft)
Fin 1·015 m² (10·93 sq ft)
Rudder 0·67 m² (7·21 sq ft)
Tailplane, incl tabs 2·76 m² (29·71 sq ft)

WEIGHTS AND LOADINGS:
Weight empty 840 kg (1,852 lb)
Max T-O weight 1,400 kg (3,086 lb)
Max wing loading 92·1 kg/m² (18·86 lb/sq ft)
Max power loading 6·59 kg/kW (10·82 lb/hp)

PERFORMANCE (at max T-O weight):
Never-exceed speed 194 knots (360 km/h; 223 mph)
Max level speed at S/L 175 knots (325 km/h; 202 mph)
Max cruising speed (75% power) at 2,135 m (7,000 ft) 167 knots (310 km/h; 193 mph)
Cruising speed (65% power) at optimum height 158 knots (293 km/h; 182 mph)
Stalling speed, flaps down 60 knots (110 km/h; 69 mph)
Max rate of climb at S/L 360 m (1,180 ft)/min
Service ceiling 5,700 m (18,700 ft)
T-O run 325 m (1,066 ft)
T-O to 15 m (50 ft) 600 m (1,970 ft)
Landing from 15 m (50 ft) 660 m (2,166 ft)
Landing run 350 m (1,150 ft)
Range with max fuel, no reserve:
Max cruising speed 1,149 nm (2,130 km; 1,323 miles)
Econ cruising speed 1,264 nm (2,344 km; 1,456 miles)

ROBIN HR 100/4+2

This aircraft is designed to carry six persons with 240 litres (52·8 Imp gallons) of fuel and no baggage, or four persons with 400 litres (88 Imp gallons) of fuel and 45 kg (99 lb) of baggage.

The wings, tail unit and landing gear are identical to those of the HR 100/285 Tiara, as is the forward part of the fuselage. The lengthened rear fuselage is entirely different, being a conventional metal structure, with frames and stringers, and without any plastics components. The standard power plant is a 239 kW (320 hp) Teledyne Continental Tiara 6-320 flat-six engine, but the prototype first flew on 23 May 1975 with a 186·5 kW (250 hp) Lycoming, and this will be available optionally in production aircraft. The prototype was being flight tested with a Tiara 6-320 engine in the Spring of 1976.

Projected new versions for the future include one with non-retractable landing gear and a 186·5 kW (250 hp) engine.

DIMENSIONS:
As for HR 100/Tiara, except:
Length overall 7·89 m (25 ft 10¾ in)

WEIGHTS AND LOADINGS:
Weight empty 870 kg (1,918 lb)
Max T-O weight 1,500 kg (3,307 lb)
Max wing loading 99·3 kg/m² (20·3 lb/sq ft)
Max power loading 6·28 kg/kW (10·32 lb/hp)

PERFORMANCE (estimated, at max T-O weight, with 6-320 engine):
Max level speed at S/L 180 knots (333 km/h; 207 mph)
Cruising speed:
75% power at 2,135 m (7,000 ft) 172 knots (318 km/h; 198 mph)
65% power at 3,050 m (10,000 ft) 165 knots (305 km/h; 190 mph)
55% power at 3,050 m (10,000 ft) 151 knots (280 km/h; 174 mph)
Stalling speed, flaps down 60 knots (111 km/h; 69 mph)
Max rate of climb at S/L 360 m (1,180 ft)/min

Robin HR 100/285 Tiara (Teledyne Continental Tiara 6-285B engine)

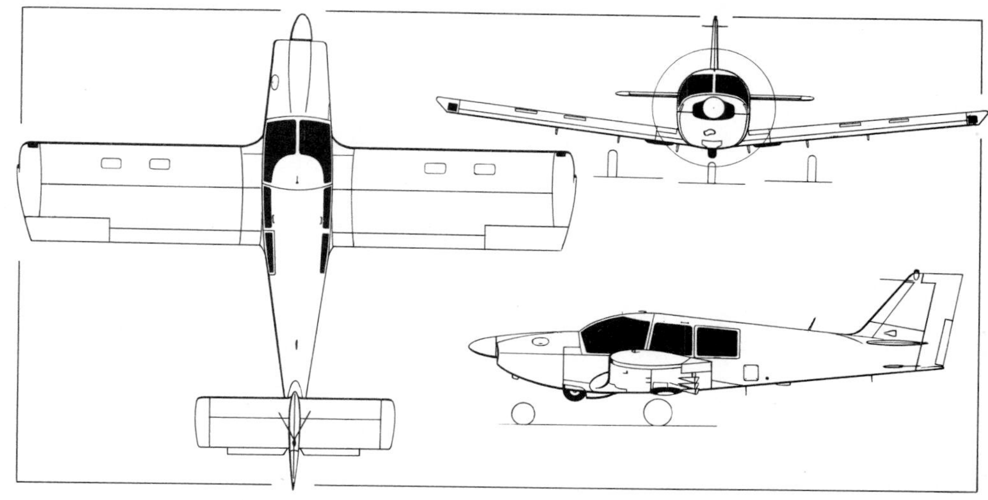
Robin HR 100/4+2 four/six-seat light aircraft *(Pilot Press)*

Robin HR 100/4+2 (Teledyne Continental Tiara 6-320 engine)

Range with six persons and 240 litres fuel, no reserve:
75% power 588 nm (1,090 km; 677 miles)
55% power 690 nm (1,280 km; 795 miles)
Range with four persons, baggage and 400 litres fuel, no reserve:
75% power 980 nm (1,817 km; 1,129 miles)
55% power 1,160 nm (2,150 km; 1,336 miles)
Max range with 440 litres (97 Imp gallons) fuel, no reserve 1,267 nm (2,350 km; 1,460 miles)

ROBIN HR 200

Although this all-metal two-seat aircraft bears an external resemblance to the HR 100/210, it is an entirely separate design of smaller overall dimensions, intended specifically for clubs and flying schools. Design was started in September 1970. Construction of the prototype began in December 1970 and this aircraft flew for the first time on 29 July 1971.

The following versions are available:

HR 200/100S. Basic model in HR 200 series, with minimum equipment and simple cabin arrangement. Generally similar to HR 200/100 Club, but without wheel fairings and with Hoffmann propeller.

HR 200/100 Club. Version with 80·5 kW (108 hp) Lycoming O-235-H2C engine, driving McCauley two-blade fixed-pitch metal propeller. Certificated and in production; 53 delivered by February 1975.

HR 200/120B. Prototype flew with 93 kW (125 hp) Lycoming O-235-J2A engine and was certificated as HR 200/120, but that version is no longer produced. Current production HR 200/120B has an 88 kW (118 hp) Lycoming O-235-L2A, driving a McCauley two-blade fixed-pitch metal propeller. Total of 19 delivered by February 1975.

HR 200/160. Version with 119 kW (160 hp) Lycoming O-320-D engine, driving a Sensenich two-blade fixed-pitch metal propeller. Prototype flying in 1975.

A fully aerobatic version of the HR 200/160 was expected to be tested during 1975.

TYPE: Two-seat light training aircraft.

WINGS: Cantilever low-wing monoplane. Wing section NACA 64A515 (modified). Constant chord. Dihedral 6° 18′ from roots. Incidence 6°. No sweepback. All-metal construction, with I-section spar at 40% chord, aluminium stressed skin and ribs; no stringers. Plain Frise-type ailerons, piano-hinged to upper surface; aerodynamically and statically balanced, and cable-operated. Slotted trailing-edge flaps, piano-hinged to lower surface and actuated electrically.

FUSELAGE: Conventional aluminium stressed-skin structure of basic rectangular section with rounded top-decking.

TAIL UNIT: Cantilever structure, similar to wings in construction. One-piece all-moving horizontal surface of constant chord, with trim and anti-servo tabs. Slightly-swept vertical fin and rudder with aerodynamic balance in upper leading-edge of rudder. Control surfaces aerodynamically and statically balanced, and cable-operated.

LANDING GEAR: Non-retractable tricycle type, with low-pressure oleo-pneumatic shock-absorbers. Single wheel on each unit; all wheels and tyres same size, 380 × 150. Nosewheel on leg offset to starboard, steered by rudder bar. Hydraulic disc brakes and parking brake. Streamlined leg and wheel fairings on all units. Damped tailskid.

POWER PLANT: One Lycoming flat-four engine, driving a two-blade metal fixed-pitch propeller (details under model listings). Single fuel tank in fuselage, capacity 120 litres (26·5 Imp gallons). Auxiliary fuel tanks optional. Refuelling point in port side of fuselage.

ACCOMMODATION: Pilot and passenger side by side, normally on bench seat but with optional individual adjustable seats (standard on HR 200/160). Dual controls standard. Forward-sliding jettisonable Plexiglas canopy. Max load on front seats (total) 154 kg (340 lb). Baggage space at rear of cabin, capacity 25 kg (55 lb) on HR 200/100 and 120B, 30 kg (66 lb) on HR 200/160.

SYSTEMS: Cabin ventilated and heated, with windscreen defrosting standard. Electrical system includes a 12V 32Ah battery, 12V 50A alternator and starter.

ELECTRONICS AND EQUIPMENT: Standard items include push-button starter, audible stall warning, safety belts, front-seat ventilation and heating, windscreen de-icing and towbar. Optional items include radio com/nav equipment, DME, ADF, VOR; two-axis or three-axis autopilot; blind-flying instrumentation; anti-collision, navigation, landing and IFR panel lights; full cabin heating; clock; precision altimeter; heated pitot; and external temperature gauge.

DIMENSIONS, EXTERNAL:
- Wing span: 8·33 m (27 ft 4 in)
- Wing chord (constant): 1·50 m (4 ft 11 in)
- Length overall: 6·64 m (21 ft 9½ in)
- Height overall: 1·94 m (6 ft 4½ in)
- Tailplane span: 2·64 m (8 ft 8 in)
- Wheel track: 2·88 m (9 ft 5½ in)
- Wheelbase: 1·465 m (4 ft 9½ in)
- Propeller diameter: 1·78 m (5 ft 10 in)

DIMENSION, INTERNAL:
- Cabin: Max width: 1·07 m (3 ft 6 in)

AREAS:
- Wings, gross: 12·50 m² (134·5 sq ft)
- Ailerons, total: 1·06 m² (11·41 sq ft)
- Trailing-edge flaps, total: 1·34 m² (14·42 sq ft)
- Elevators, incl tabs: 2·03 m² (21·85 sq ft)

WEIGHTS:
- Weight empty:
  - HR 200/100, 120B: 515 kg (1,135 lb)
  - HR 200/160: 530 kg (1,168 lb)
- Max T-O weight:
  - HR 200/100, 120B: 780 kg (1,719 lb)
  - HR 200/160: 800 kg (1,763 lb)

PERFORMANCE (at max T-O weight):
- Never-exceed speed: 160 knots (296 km/h; 184 mph)
- Max level speed at S/L:
  - HR 200/100: 116 knots (215 km/h; 133 mph)
  - HR 200/120B: 124 knots (231 km/h; 143 mph)
  - HR 200/160: 140 knots (260 km/h; 161 mph)
- Max cruising speed (75% power) at optimum altitude:
  - HR 200/100: 106 knots (197 km/h; 122 mph)
  - HR 200/120B: 122 knots (227 km/h; 141 mph)
  - HR 200/160: 135 knots (250 km/h; 155 mph)
- Stalling speed (full idle):
  - HR 200/100, 120B: 49·5 knots (91·5 km/h; 57 mph)
  - HR 200/160: 50·5 knots (93 km/h; 58 mph)
- Max rate of climb at S/L:
  - HR 200/100: 228 m (748 ft)/min
  - HR 200/120B: 246 m (807 ft)/min
  - HR 200/160: 372 m (1,220 ft)/min
- Service ceiling:
  - HR 200/100: 3,900 m (12,800 ft)
  - HR 200/120B: 4,150 m (13,620 ft)
  - HR 200/160: over 5,000 m (16,400 ft)
- Range at 75% power, standard fuel, no reserve:
  - HR 200/100: 566 nm (1,050 km; 652 miles)
  - HR 200/120B: 563 nm (1,045 km; 649 miles)
  - HR 200/160: 504 nm (935 km; 581 miles)

Robin HR 200/100 Club two-seat light aircraft (Lycoming O-235-H2C engine)

SOCATA

SOCIÉTÉ DE CONSTRUCTION D'AVIONS DE TOURISME ET D'AFFAIRES (Subsidiary of AÉROSPATIALE)

HEAD OFFICE, WORKS AND AFTER-SALES SERVICE:
Aéroport de Tarbes-Ossun-Lourdes, BP 38, 65001-Tarbes Cédex
Telephone: (62) 93-97-30
Telex: SOCATA 520828

SALES:
37 Boulevard de Montmorency, 75781-Paris Cédex 16
Telephone: 524-43-21
Telex: 620059 AIRSPA
Aéroport de Toussus-le-Noble, BP No. 2, 78530-Buc
Telephone: 956-21-00
Telex: 600 836 SOCAERO

CHAIRMAN:
Jean Soissons

This company was formed in 1966, as a subsidiary of Sud-Aviation. It is responsible for production of the various versions of the Rallye three/four-seat light aircraft.

During 1975 Socata delivered 107 Rallyes, 50 per cent of them for export.

Socata also produces the fuselage of the Corvette business aircraft, and components for the Concorde, A300 Airbus, Falcon 10 and 20 business aircraft, and Super Frelon, Puma and Alouette helicopters. It is responsible for overhaul and repair of MS 760 Paris light jet aircraft.

Socata's works cover an area of 47,672 m² (513,135 sq ft), and employed a total of 887 people in February 1976.

SOCATA RALLYE

The Rallye had its origin in a competition organised by the SFATAT in 1958 and was developed originally by the old-established Morane-Saulnier company. The prototype (67 kW; 90 hp MS 880A) Rallye-Club flew on 10 June 1959, and the initial production versions were the MS 880B Rallye-Club and the MS 885 Super Rallye. FAA certification of the design was obtained on 21 November 1961.

Production of the MS 885 (212 built) and MS 890 (8 built) with 108 kW (145 hp) Continental engine, MS 881 (12 built) with 78 kW (105 hp) Potez engine, MS 883 (77 built) with 86 kW (115 hp) Lycoming engine and MS 886 (3 built) with 112 kW (150 hp) Lycoming engine has ended.

The current GT models and Rallye 235 have a changed external appearance and greatly improved cabin comfort, as the result of a modernisation programme undertaken at the beginning of 1972. Features include dual wheel control, with improved aileron efficiency; electrically-controlled flaps offering all positions from 0° to 30°; and a central console grouping the engine controls (throttle, propeller governor and mixture) and the trim and air-conditioning controls in such a way that they are accessible to both pilots. Model GT Rallyes can be fitted with IFR and night flying equipment, following receipt of SGAC approval in June 1973.

The 1976 range of Rallyes is as follows:

Rallye 100 S (= Sport). First flown on 30 March 1973, this version is equipped as a two-seater and, except for the generally similar Rallye 100 ST in two-seat form, is the only aircraft in the Rallye series cleared for spinning. Generally similar to the Rallye 100 T, it first flew on 30 March 1973 and received SGAC certification on 6 April 1973. A total of 54 had been built by 1 January 1976. Ten delivered to French Navy Training School at Lanvéoc-Poulmic in April 1974.

Rallye 100 T (= Tourisme). Basic 3/4-seat model with 74·5 kW (100 hp) Rolls-Royce Continental O-200-A engine and fixed-pitch propeller. Prototype flew on 12 February 1961. Production of a further series, with electrically-controlled wing flaps and more comfortable cabin, was started in 1973. Total of 986 built by 1 January 1976.

Rallye 100 ST. This aircraft may be operated either as a two-seater cleared for spinning, like the Rallye 100 S, or as a three/four-seater with spins prohibited, like the Rallye 100 T. It flew for the first time on 25 September 1974 and received SGAC certification on 4 October 1974. Total of 22 built by 1 January 1976. Five ordered as trainers by the French Navy.

Rallye 125. Four-seat version of the Rallye 100 T with 93 kW (125 hp) Lycoming O-235-F2A engine and fixed-pitch propeller. Prototype flew for the first time on 10 February 1972. Rallye 125 received SGAC certification on 31 May 1972, and 25 had been built by 1 January 1976.

Rallye 150 GT. Four-seat version with 112 kW (150 hp) Lycoming O-320-E2A engine and fixed-pitch propeller, strengthened structure for increased AUW, larger rudder, larger ailerons, fillets of increased size between wing trailing-edges and fuselage, longer nosewheel leg to give increased propeller clearance, enlarged dorsal fin fairing, modified cockpit canopy and a baggage compartment. Streamlined wheel fairings optional. Prototype flew for the first time on 6 February 1964. Total of 303 built by 1 January 1976.

Rallye 150 T. Similar to Rallye 100 T, but with 112 kW (150 hp) Lycoming O-320-E2A engine, enabling four persons to be carried economically over ranges up to 540 nm (1,000 km; 621 miles). First flown on 8 November 1974. SGAC certification received on 26 November 1974. Ten built by 1 January 1976.

Rallye 150 ST. Similar to Rallye 100 ST, but with 112 kW (150 hp) Lycoming O-320-E2A engine. Developed as a high-performance aircraft for flying clubs, and for use by clubs in countries where high temperature or high altitude might impose limitations on lower-powered aircraft. First Rallye 150 ST flew on 10 January 1975 and received SGAC certification in March 1975. Total of 16 built by 1 January 1976.

Rallye 180 GT. Basically similar to 150 GT, but with 134 kW (180 hp) Lycoming O-360-A3A engine and fixed-pitch or constant-speed propeller, giving extra

SOCATA Rallye 235 E four-seat light aircraft (Lycoming O-540-B4B5 engine)

power for duties such as agricultural spraying and dusting, glider and banner towing. Prototype flew for first time on 7 December 1964. French type approval received on 27 April 1965; FAA Type Approval on 23 June 1971. Total of 639 built by 1 January 1976; 100 ordered by the SFA for duty as glider tugs at French gliding centres.

Rallye 220 GT. Generally similar to 180 GT, but with 164 kW (220 hp) Franklin 6A-350-C1 engine and constant-speed propeller. Prototype first flown on 12 May 1967. Received FAA Type Approval 29 April 1968. Total of 250 built by 1 January 1976.

Rallye 235 E. High-performance version, with 175 kW (235 hp) Lycoming O-540-B4B5 engine, driving a Hartzell HC-C2YK-1/8468-6 two-blade constant-speed metal propeller, and with accommodation for four persons. Construction of prototype started in January 1975. First flight made on 1 April 1975. Total of 10 built by 1 January 1976.

Rallye 235 G. Generally similar to Rallye 235 E but equipped for patrol and surveillance missions, with four underwing racks for survival canisters and other stores.

The Rallye 150 GT, 180 GT and 220 GT are authorised for use as ambulance aircraft carrying a pilot, one stretcher patient and medical attendant. They can also be used for glider towing, and some 400 Rallyes are employed in this role, including more than 250 in France. Agricultural spraygear can be fitted and tests have been conducted with various models on ski landing gear.

The following details apply to all versions listed:

TYPE: Two/four-seat light monoplane.

WINGS: Cantilever low-wing monoplane. Wing section NACA 63A416 (modified). Dihedral 7°. Incidence 4°. All-metal single-spar structure. Wide-chord slotted ailerons. Full-span automatic slats. Long-span slotted flaps. Ailerons and flaps have corrugated metal skin. Ground-adjustable aileron tabs. No anti-icing equipment.

TAIL UNIT: Cantilever all-metal structure with corrugated skin on the mass-balanced control surfaces. Fixed-incidence tailplane. One automatic tab and one controllable tab on elevator. One controllable tab on rudder.

LANDING GEAR: Non-retractable tricycle type. ERAM oleo-pneumatic shock-absorbers. Castoring nosewheel. Rallye 235 has Cleveland main wheels with tyres size 6·00-6, pressure 1·8 bars (26·1 lb/sq in); nosewheel tyre size 5·00-4, pressure 1·4 bars (20·3 lb/sq in). Cleveland hydraulic disc brakes. Provision for fitting skis or floats.

POWER PLANT: One flat-four or flat-six engine (details under entries for individual models), driving a two-blade fixed-pitch or constant-speed metal propeller. Fuel in two metal tanks in wings, with total capacity of 96 litres (21 Imp gallons) in Rallye 100, 125 and 150 ST, 170 litres (37·5 Imp gallons) in Rallye 150 T and GT, 220 litres (48 Imp gallons) in Rallye 180 GT and 220, and 270 litres (59·4 Imp gallons) of which 255 litres (56 Imp gallons) are usable in Rallye 235. Refuelling points above wings. Oil capacity 6 litres (1·3 Imp gallons) in Rallye 100 and 125, 8 litres (1·75 Imp gallons) in Rallye 150 and 180 GT, 8·3 litres (1·8 Imp gallons) in Rallye 220, 8·8 litres (1·9 Imp gallons) in Rallye 235.

ACCOMMODATION: Two seats side by side in Rallye 100 S. Other versions have also a bench seat at rear, under large rearward-sliding canopy. Two persons up to a total weight of 110 kg (242 lb) can occupy rear seat of Rallye 100 T. Other versions are full four-seaters. Dual control columns on Rallye 100 and 125. Dual control wheels on Rallye 150, 180, 220 and 235. Individual adjustable front seats and baggage space aft of rear seats (accessible internally) on the GT models and Rallye 235. Heating and ventilation standard.

ELECTRONICS AND EQUIPMENT: The instrument panel is fitted with an anti-glare visor, and is designed to take full radio-navigation equipment to customer's requirements.

DIMENSIONS, EXTERNAL:
- Wing span: 9·74 m (31 ft 11 in)
- Wing chord (constant): 1·30 m (4 ft 3 in)
- Wing aspect ratio: 7·57
- Length overall:
  - 100 T: 6·97 m (22 ft 10½ in)
  - 100 S/ST: 7·045 m (23 ft 1¼ in)
  - 125: 7·16 m (23 ft 5¾ in)
  - 150, 180: 7·24 m (23 ft 9 in)
  - 220: 7·25 m (23 ft 9½ in)
  - 235: 7·28 m (23 ft 10½ in)
- Height overall:
  - 100 T, 125: 2·60 m (8 ft 6¼ in)
  - 100 S/ST, 150, 180, 220, 235: 2·80 m (9 ft 2¼ in)
- Tailplane span: 3·67 m (12 ft 0½ in)
- Wheel track: 2·01 m (6 ft 6½ in)
- Wheelbase: 1·71 m (5 ft 7¼ in)
- Propeller diameter:
  - 235: 1·98 m (6 ft 6 in)

DIMENSIONS, INTERNAL:
- Cabin:
  - Length: 2·25 m (7 ft 4 in)
  - Width: 1·13 m (3 ft 8½ in)

AREAS:
- Wings, gross: 12·28 m² (132 sq ft)
- Ailerons (total): 1·56 m² (16·79 sq ft)
- Trailing-edge flaps (total): 2·40 m² (25·83 sq ft)
- Vertical tail surfaces (total):
  - 100 T, 125: 1·39 m² (14·96 sq ft)
  - 100 S/ST, 150, 180, 220, 235: 1·74 m² (18·73 sq ft)
- Horizontal tail surfaces (total): 3·48 m² (37·50 sq ft)

WEIGHTS AND LOADINGS:
- Weight empty, equipped:
  - 100: 450 kg (992 lb)
  - 125: 510 kg (1,125 lb)
  - 150 GT: 550 kg (1,213 lb)
  - 150 ST: 530 kg (1,168 lb)
  - 150 T: 545 kg (1,202 lb)
  - 180: 570 kg (1,257 lb)
  - 220: 630 kg (1,389 lb)
  - 235: 694 kg (1,530 lb)
- Max T-O and landing weight:
  - 100 S: 750 kg (1,653 lb)
  - 100 T/ST: 770 kg (1,697 lb)
  - 125: 840 kg (1,852 lb)
  - 150 GT: 980 kg (2,160 lb)
  - 150 T: 950 kg (2,094 lb)
  - 150 ST: 870 kg (1,918 lb)
  - 180: 1,050 kg (2,315 lb)
  - 220: 1,100 kg (2,425 lb)
  - 235: 1,170 kg (2,580 lb)
- Max wing loading:
  - 235: 95 kg/m² (19·45 lb/sq ft)
- Max power loading:
  - 235: 6·69 kg/kW (10·98 lb/hp)

PERFORMANCE (at max T-O weight; data for Rallye 235 estimated):
- Never-exceed speed:
  - 100 S: 145 knots (270 km/h; 167 mph)
- Max level speed at S/L:
  - 100, 125: 105 knots (195 km/h; 121 mph)
  - 150 GT: 113 knots (210 km/h; 130 mph)
  - 150 T, ST: 116 knots (215 km/h; 133 mph)
  - 180: 129 knots (240 km/h; 149 mph)
  - 220: 143 knots (266 km/h; 165 mph)
  - 235: 148 knots (275 km/h; 171 mph)
- Cruising speed (75% power) at 1,500 m (5,000 ft):
  - 100 S: 94 knots (175 km/h; 109 mph)
  - 100 T: 92 knots (170 km/h; 105 mph)
  - 125: 97 knots (180 km/h; 112 mph)
  - 150: 108 knots (200 km/h; 124 mph)
  - 180: 121 knots (225 km/h; 140 mph)
  - 220, 235: 132 knots (245 km/h; 152 mph)
- Stalling speed, flaps down:
  - 100: 41 knots (75 km/h; 47 mph)
  - 125: 44 knots (80 km/h; 50 mph)
  - 150 GT: 49 knots (90 km/h; 56 mph)
  - 150 ST: 45 knots (83 km/h; 52 mph)
  - 150 T: 48 knots (88 km/h; 55 mph)
  - 180: 50 knots (92 km/h; 57·5 mph)
  - 220: 52 knots (95 km/h; 59·5 mph)
  - 235: 53 knots (98 km/h; 61 mph)
- Max rate of climb at S/L:
  - 100 S: 177 m (580 ft)/min
  - 100 T: 165 m (541 ft)/min
  - 125: 171 m (561 ft)/min
  - 150 GT: 180 m (590 ft)/min
  - 150 ST: 264 m (866 ft)/min
  - 150 T: 204 m (670 ft)/min
  - 180: 231 m (758 ft)/min
  - 220, 235: 300 m (984 ft)/min
- Service ceiling:
  - 100 S: 3,500 m (11,480 ft)
  - 100 T: 3,200 m (10,500 ft)
  - 125, 150 GT: 2,600 m (8,530 ft)
  - 150 T: 3,000 m (9,840 ft)
  - 150 ST, 220: 4,000 m (13,125 ft)
  - 180: 3,600 m (11,800 ft)
  - 235: 4,500 m (14,760 ft)
- T-O run:
  - 100 S, 220, 235: 120 m (393 ft)
  - 100 T, 150 ST: 130 m (430 ft)
  - 125, 150 GT: 140 m (459 ft)
  - 150 T, 180: 135 m (443 ft)
- T-O to 15 m (50 ft):
  - 235: 305 m (1,000 ft)

Socata Rallye 235 G fitted with survival containers for search and rescue duties

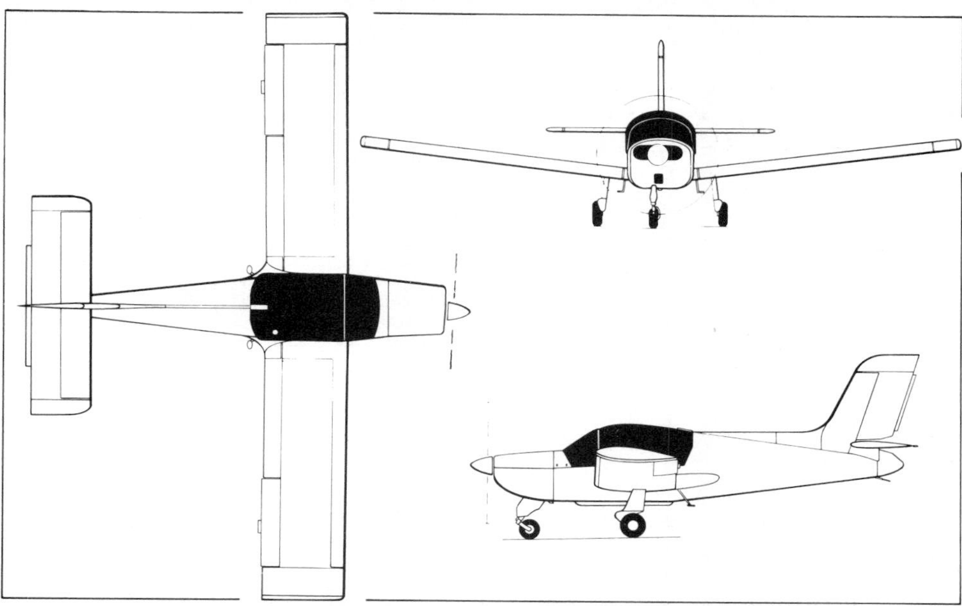

Socata Rallye 235 E high-performance four-seat light aircraft *(Roy J. Grainge)*

Landing run:

100, 220, 235	100 m (328 ft)
150 ST	120 m (394 ft)
125, 150 T and GT, 180	125 m (410 ft)

Range with max fuel:

100 S	405 nm (750 km; 465 miles)
100 T	395 nm (730 km; 455 miles)
125	400 nm (740 km; 460 miles)
150 GT, T	540 nm (1,000 km; 621 miles)
150 ST	324 nm (600 km; 372 miles)
180	702 nm (1,300 km; 807 miles)
220	860 nm (1,600 km; 995 miles)
235	675 nm (1,250 km; 776 miles)

SOCATA ST 10 DIPLOMATE

The ST 10 four-seat light aircraft, designed by Socata, embodies a number of components common to the earlier GY-80 Horizon, from which it differs chiefly in having a longer fuselage and redesigned cabin. The prototype (F-WOFR), known for a time as the Provence, was flown for the first time on 7 November 1967.

Early in 1969, it was modified to flight test refinements that became standard on production aircraft, including a longer fuselage, reduced tailplane span, increased rudder area and a shallow 'keel' beneath the rear fuselage. In this form the Diplomate received type approval by the SGAC on 26 November 1969. Fifty-six were delivered.

Details can be found in the 1975-76 *Jane's*.

WASSMER

SOCIÉTÉ NOUVELLE WASSMER AVIATION

Head Office, Delivery and After-Sales Service:
BP 7, 63501 Aérodrome d'Issoire
Telephone: (73) 89-19-15 and 89-01-54

Works:
Route de Parentignat, 63501-Issoire
Telephone: (73) 89-23-86
Telex: 990 185

Director General:
Roger Liévin

Director of Design Studies:
Daniel Pizzolato

This company was founded in 1905 by M Benjamin Wassmer, under the title Société Wassmer, and in its early days was concerned with overhaul and repair of military aircraft and the manufacture of propellers.

When activities were resumed after the second World War, Wassmer was again concerned initially with repair work, later building the designs of other manufacturers under licence. In 1955 a design department was created; its first product was the Jodel-Wassmer D.120 Paris-Nice, and subsequently more than 300 Jodel aircraft were built by Wassmer.

Today, Wassmer's facilities at Issoire occupy a total area of 25,000 m² (269,100 sq ft), including 6,000 m² (64,600 sq ft) of roofed accommodation, comprising one factory for component manufacture and one for repair work and maintenance of army helicopters at the Aérodrome d'Issoire. It employed 110 people in the Spring of 1976, of whom 11 were engaged on design work.

Current production types include the WA 52 Europa, WA 54 Atlantic and WA 80 Piranha all-plastics light aircraft of Wassmer's own design. In addition, Wassmer is manufacturing and marketing, in partnership with Siren SA, an all-metal derivative of its earlier, wooden WA 4/21 light aircraft, known as the CE.43 Guépard. A jointly-owned company named CERVA(GIE) was formed to co-ordinate this programme, of which details are given under the CERVA heading in this section.

Details of Wassmer's sailplane development and manufacture can be found in the Sailplanes section.

WASSMER WA 52 EUROPA and WA 54 ATLANTIC

First flown on 22 March 1966, the WA 50 was a prototype four-seat light aircraft of which the airframe was made entirely of plastics. Its development was started in 1962, with official support, and the Société du Verre Textile provided considerable help in selecting the most suitable materials for construction.

The airframe was built up of large components moulded in thin layers of glassfibre, reinforced either by stringers or by a double corrugated skin.

Following flight and ground testing of the prototype WA 50 (see 1969-70 *Jane's*), Wassmer developed the following production models:

WA 51 Pacific. Powered by 112 kW (150 hp) Lycoming O-320-E2A engine. Differed from WA 50 prototype in having non-retractable landing gear and modifications to the tail-fin and rear cabin windows. First flown on 17 May 1969. Described in 1974-75 *Jane's*. No longer in production.

WA 52 Europa. Generally similar to WA 51, but powered by a 119 kW (160 hp) Lycoming O-320 engine, driving a variable-pitch propeller, and available with an auxiliary fuel tank.

WA 54 Atlantic. Generally similar to WA 52, but powered by a 134 kW (180 hp) Lycoming O-360 engine. Baggage space increased to 1 m³ (35 cu ft). Revised oleo-pneumatic main landing gear, with forks of the kind fitted to the earlier Wassmer Baladou and new fairings. Improved nosewheel steering. Revised cowling air intake, embodying taxi light. Prototype flew for the first time on 20 February 1973, and production began in June 1973.

A total of 150 Pacifics, Europas and Atlantics had been sold by May 1976.

The following details apply to both the WA 52 and WA 54, except where indicated otherwise:

Type: Four-seat light aircraft.

Wings: Cantilever low-wing monoplane. Wing section NACA 63-418. Incidence 4° at root, 1° at tip. Dihedral 6° 40′. Structure of each wing comprises a one-piece top surface and leading-edge moulding, a bottom skin panel, main front spar, auxiliary rear spar, ten ribs and stringers, all of plastics. Each mechanically-operated aileron is a simple box structure, with corrugated skin, two end ribs and two internal ribs. Three-position mechanically-operated slotted flaps.

Wassmer WA 54 Atlantic four-seat all-plastics light aircraft

Fuselage: Main fuselage shell and integral fin moulded in two halves from glassfibre, with frames and stringers also of glassfibre.

Tail Unit: Cantilever all-plastics structure, with swept vertical surfaces. All-moving one-piece tailplane, with anti-tab each side.

Landing Gear: Non-retractable tricycle type. ERAM oleo-pneumatic main-wheel shock-absorbers. Steerable nosewheel on telescopic shock-strut, similar to that of earlier Wassmer Baladou. Main-wheel brakes, parking brake and wheel fairings standard.

Power Plant: One Lycoming flat-four engine (details under individual model listings), driving a McCauley two-blade variable-pitch metal propeller. Hoffmann three-blade constant-speed composite propeller is to be made available optionally on the WA 54. Integral fuel tank in each sweptforward wing-root leading-edge, with total capacity of 150 litres (33 Imp gallons). Provision for an auxiliary fuel tank, capacity 70 litres (15·5 Imp gallons).

Accommodation: Four seats, in pairs, in enclosed cabin. Front two seats are adjustable. Baggage compartment behind rear seats. Upward-hinged door on each side. Cabin heated and ventilated.

Electronics and Equipment: Electrical equipment includes Delco-Rémy 12V engine starter and 12V 50Ah alternator. VOR and VHF radio standard.

Dimensions, external:

Wing span	9·40 m (30 ft 10 in)
Wing chord at c/l	2·10 m (6 ft 10¾ in)
Wing chord at tip	1·00 m (3 ft 3¼ in)
Wing mean aerodynamic chord	1·375 m (4 ft 6 in)
Wing aspect ratio	7·15
Length overall (WA 54)	7·40 m (24 ft 3½ in)
Height overall	2·26 m (7 ft 5 in)
Tailplane span	3·00 m (9 ft 10 in)
Wheel track	3·00 m (9 ft 10 in)
Wheelbase	1·65 m (5 ft 5 in)
Propeller diameter	1·85 m (6 ft 0¾ in)

Area:

Wings, gross	12·40 m² (40·68 sq ft)

Weights:

Weight empty:	
WA 52	610 kg (1,344 lb)
WA 54	620 kg (1,367 lb)
Max T-O weight:	
WA 52	1,080 kg (2,380 lb)
WA 54	1,130 kg (2,491 lb)

Performance:

Never-exceed speed	193 knots (360 km/h; 223 mph)
Max level speed at S/L:	
WA 54	151 knots (280 km/h; 174 mph)
Cruising speed at 1,675 m (5,500 ft):	
WA 52	135 knots (250 km/h; 155 mph)
WA 54	140 knots (260 km/h; 161 mph)
Max rate of climb at S/L:	
WA 54	360 m (1,180 ft)/min
Range with auxiliary fuel:	
WA 52	755 nm (1,400 km; 870 miles)
WA 54	729 nm (1,350 km; 839 miles)
Endurance with auxiliary fuel:	
WA 54	6 hours

WASSMER WA 80 PIRANHA

First flown in November 1975, the WA 80 Piranha two-seat trainer evolved from the projected WA 70, with T-tail, of which a model was exhibited at the General Aviation Show at Toussus-le-Noble in mid-1974. It is an unusually roomy two-seater, having a fuselage generally similar to that of the four-seat WA 51 Pacific. The basic version is as described in detail. A glider towing variant is planned, with a 112-119 kW (150-160 hp) Lycoming O-320 engine.

The second prototype Piranha was at the CEAT for static testing in the Spring of 1976. Production deliveries were scheduled to begin before the end of the year.

Type: Two-seat light training and general purpose aircraft.

Wings: Cantilever low-wing monoplane. Wing section NACA 63418. Dihedral 6° 40′ from roots. All-plastics (polyester resin) structure produced by vacuum-forming process to give a very precise smooth surface. Sweptforward wing-root leading-edges, embodying stall strips. No sweep on outer wings at quarter-chord. Slightly-inset ailerons, with foam rubber gap seals, are operated through control rods. Manually-operated four-position flaps.

Fuselage: All-plastics (polyester resin) structure, generally similar to that of WA 51/54 series.

Tail Unit: Conventional cantilever configuration, with sweptback vertical surfaces and small dorsal fin. Elevators operated through control rods, rudder by cables. All-plastics (polyester resin) construction. Con-

Wassmer WA 80 Piranha (Rolls-Royce Continental O-200-A engine) *(Flight International)*

trollable trim tab on port elevator.

LANDING GEAR: Non-retractable tricycle type. Main wheels mounted inside each end of a one-piece laminated polyester resin spring. Nosewheel carried on steerable oleo-pneumatic leg. Disc brakes on main wheels. Small tail bumper.

POWER PLANT: One 74·5 kW (100 hp) Rolls-Royce Continental O-200-A flat-four engine, driving a two-blade fixed-pitch EVRA wooden propeller. Fuel tank in fuselage, aft of cockpit, capacity 90 litres (19·75 Imp gallons).

ACCOMMODATION: Two seats side by side, with stick-type dual controls standard. Adjustable seats. Gull-wing door on each side, port one jettisonable. Baggage compartment aft of seats, over fuel tank.

DIMENSIONS, EXTERNAL:

Wing span	9·40 m (30 ft 10 in)
Wing aspect ratio	7·1
Length overall	7·50 m (24 ft 7¼ in)
Height overall	2·10 m (6 ft 10¾ in)
Wheel track	2·10 m (6 ft 10¾ in)

DIMENSIONS, INTERNAL:

Cabin: Length	2·40 m (7 ft 10½ in)
Max height	1·18 m (3 ft 10½ in)
Baggage space	1·2 m³ (42·3 cu ft)

AREA:

Wings, gross	12·40 m² (133·5 sq ft)

WEIGHTS AND LOADINGS:

Weight empty	500 kg (1,102 lb)
Max T-O weight	800 kg (1,763 lb)
Max wing loading	64·51 kg/m² (13·2 lb/sq ft)
Max power loading	10·74 kg/kW (17·63 lb/hp)

PERFORMANCE (at max T-O weight):

Never-exceed speed	163 knots (302 km/h; 187 mph)
Max speed at S/L	129 knots (240 km/h; 149 mph)
Max cruising speed at S/L	102 knots (190 km/h; 118 mph)
Stalling speed, flaps up	51·5 knots (95 km/h; 59 mph)
Stalling speed, flaps down	43·5 knots (80 km/h; 50 mph)
Rate of climb at S/L	210 m (690 ft)/min
Service ceiling	4,000 m (13,125 ft)
T-O run	200 m (655 ft)
T-O to 15 m (50 ft)	300 m (985 ft)
Landing from 15 m (50 ft)	300 m (985 ft)
Landing run	150 m (495 ft)
Range with max fuel (75% power)	378 nm (700 km; 435 miles)

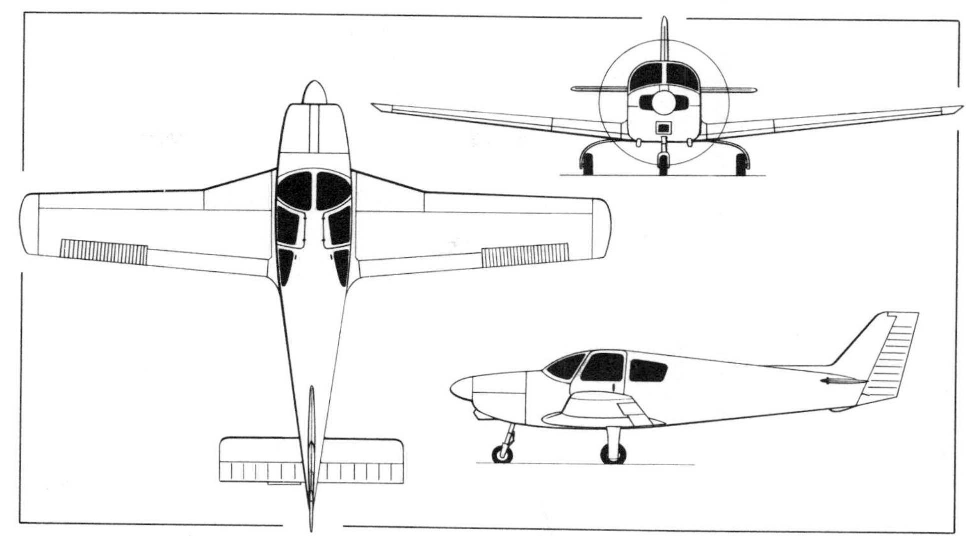

Wassmer WA 80 Piranha two-seat light aircraft *(Pilot Press)*

GERMANY
(FEDERAL REPUBLIC)

AIR-METAL
AIR-METAL FLUGZEUGBAU UND ENTWICKLUNGS—GmbH & Co BETRIEBS-KG

HEAD OFFICE:
D-8058 Erding, Ringstrasse 17
Telephone: (08122) 23 46

DESIGN OFFICE:
D-8000 München 83, Josef-Beiser-Strasse 24
Telephone: (089) 67 20 21/22
Telex: 5-24910 AMFED

WORKS:
D-8300 Landshut, Flugplatz Ellermühle
Telephone: (08765) 256

MANAGING DIRECTOR:
Wolfgang Grabowski

AIR-METAL AM-C 111

Planned production of a prototype STOL transport aircraft, designated AMZ-102T, was reported in the 1971-72 *Jane's*. This project was subsequently abandoned, and Air-Metal began constructing instead the prototype of a more refined utility STOL transport, under the designation AM-C 111 Series 400 SP. In all, eight variants of the AM-C 111 are planned:

100. Basic version with standard fuselage. Quick-change interior and side-loading door optional.

100HD. High-density passenger-carrying version of Series 100.

200. As Series 100, but with pressurised cabin.

200HD. As Series 100HD, but with pressurised cabin.

200SP. Special performance version of Series 200.

300. As Series 100, with added rear loading ramp. Quick-change interior standard.

400. As Series 200, but with added rear loading ramp. Quick-change interior standard.

400SP. Special performance version of Series 400.

Construction of the Series 400SP prototype was well advanced in the Spring of 1976, and negotiations are underway for this version, and the 200SP, to be built in Turkey for the Turkish Air Force.

The description which follows applies in particular to the 400SP:

TYPE: Twin-turboprop STOL transport and utility aircraft.

WINGS: Cantilever high-wing monoplane. Rectangular centre-section of NACA 23015 section. Trapezoidal outer panels of NACA 23015 section at root, NACA 23012 at tip. Incidence 3°. No dihedral. Conventional two-spar structure of light alloy. Detachable leading-edges. Electrically-operated double-slotted Fowler-type trailing-edge flaps of light alloy construction, in two sections on each wing, extending from wing root to aileron. Ailerons, of light alloy construction, are statically and dynamically balanced. Glassfibre-reinforced plastics wingtips. Pneumatic de-icing of wing leading-edges.

FUSELAGE: Semi-monocoque fail-safe structure of light alloy, built in three sections. The forward section comprises the unpressurised nose and pressurised flight deck. Nose can accommodate a radar scanner, and the standard nose cap is a glassfibre honeycomb radome. Nose section aft of the scanner houses the retractable nosewheel, a baggage hold and electronic equipment. The forward pressure bulkhead forms the forward end of the flight deck. Crash wall between flight deck and cabin. Aft pressure bulkhead is a dome of laminated plastics, which is detachable to permit the loading of outsize cargo via the rear loading ramp. The rear fuselage section, embodying the cargo door/ramp, is unpressurised.

TAIL UNIT: Cantilever all-metal structure. Tailplane incidence variable by dual electric actuators. Elevators horn-balanced, with manually-operated trim tab in

Prototype Air-Metal AM-C 111 under construction at Landshut in Spring 1976

each. Rudder horn-balanced, with manually-operated trim tab. Leading-edges of fin and tailplane de-iced pneumatically.

LANDING GEAR: Hydraulically-retractable tricycle type of Menasco design with single wheel on each unit. Nose-wheel unit retracts forward and is enclosed by hinged doors when retracted. Main wheels retract into fairings on each side of fuselage. Oleo-pneumatic shock-absorbers. Emergency extension of landing gear by hand-operated hydraulic pump. Steerable nosewheel with tyre size 9·00-6, pressure 3·45 bars (50 lb/sq in). Main-wheel tyres size 34 × 10·75-16, pressure 4·48 bars (65 lb/sq in). Six-puck hydraulic disc brakes.

POWER PLANT: Two 835 kW (1,120 shp) Pratt & Whitney Aircraft of Canada PT6A-45 turboprop engines, each driving a Hartzell five-blade metal constant-speed reversible-pitch propeller. Fuel contained in integral wing tanks, one in each outer wing and two in centre-section, plus one tank in each engine nacelle; total capacity 2,400 litres (634 US gallons). Auxiliary wing-tip tanks optional, with combined capacity of 696 litres (184 US gallons), providing maximum fuel capacity of 3,096 litres (818 US gallons). Pressure refuelling point in engine nacelle; gravity refuelling points on the upper surface of each wing. Electrical de-icing of propellers, spinners and engine air intakes.

ACCOMMODATION: Pilot and co-pilot on flight deck. Military-type canvas seats for 24 passengers. Quick-change interior kits permit the Series 400SP to be used as a cargo aircraft able to accept bulky cargo, including LD-1 and LD-3 containers, air cargo pallets and military vehicles. An ambulance version can accommodate up to 15 litters, with 5 seats for sitting patients or medical attendants; the ambulance version can be equipped also as an operating theatre for on-board treatment. Other versions cater for the transport of airborne troops, training paratroops, military supply, rescue, air patrol, survey and target towing. Cabin floor is designed for distributed loading of 975 kg/m² (200 lb/sq ft), or 1,220 kg/m² (250 lb/sq ft) between the spar frames. Vehicles may be carried with single wheel loads not exceeding 500 kg (1,102 lb). Tiedown fittings for containers and pallets. Lashing points provided throughout the cabin, in floor, walls and ceiling. Crew door on port side, forward of propeller; passenger door on port side of cabin, aft of wheel fairing; both doors have integral airstairs. Toilet and baggage racks at aft pressure bulkhead. Six double-pane windows on each side of cabin. Two emergency exits on starboard side. Dual controls standard. Electrical windscreen de-icing; electrically-operated windscreen wipers. Accommodation air-conditioned and pressurised.

SYSTEMS: Air-conditioning and pressurisation by engine bleed air, utilising duplicated Hamilton Standard R70/3W bootstrap air-cycle refrigeration systems; pneumatic cabin pressure controller; max pressure differential of 0·29 bars (4·2 lb/sq in). Power plant fire detection and extinguishing system. Hydraulic system, pressure 207 bars (3,000 lb/sq in), supplied by dual engine-driven pumps, either of which can maintain full hydraulic services; handpump for emergency extension of landing gear. Electrical system powered by dual 300A starter/generators; voltage maintained at 28V DC by static voltage regulator. Single-phase AC supply maintained by two 115V 400Hz static transverters. Emergency power from 24V 26Ah battery. Oxygen system fed by storage cylinder, with crew masks stowed in the cockpit console, and cabin drop-out masks for passengers. A 300 litre portable cylinder with smoke mask is available for crew use.

ELECTRONICS AND EQUIPMENT: Electronics comprise RCA AVC-110A VHF transceiver, AVC-111A VHF transceiver, duplicated AVN-220A VHF nav systems, duplicated AVI-203 RMI systems, Primus-40 radar receiver/transmitter, indicator, wave guide and antenna, Primus-10 DME interrogator, Primus-10 DME indicator, AVQ-95 transponder, and associated antennae. Standard equipment includes three-axis autopilot with automatic stabilisation, yaw-damping and turn co-ordination, automatic heading control, automatic climb and descent control, automatic pitch trim and other optional functions; full blind-flying instrumentation for pilot and co-pilot; outside air temperature gauge; annunciator panel; heated pitot heads; and external power socket.

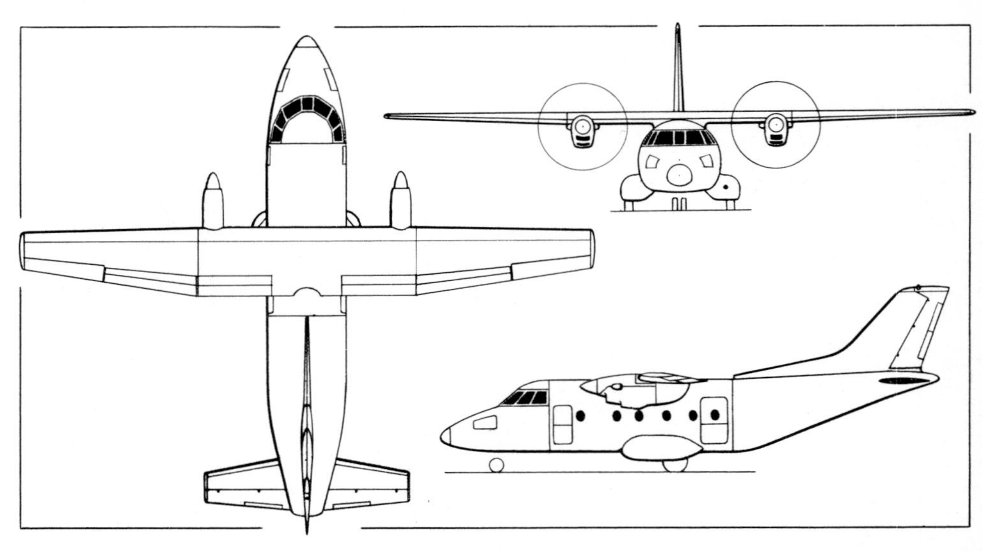

Air-Metal AM-C 111 twin-engined STOL transport *(Pilot Press)*

DIMENSIONS, EXTERNAL:
- Wing span: 19·20 m (63 ft 0 in)
- Wing chord at root: 2·30 m (7 ft 6½ in)
- Wing chord, mean aerodynamic: 1·97 m (6 ft 5½ in)
- Wing aspect ratio: 9·71
- Length overall: 16·81 m (55 ft 2 in)
- Height overall: 6·40 m (21 ft 0 in)
- Tailplane span: 7·80 m (25 ft 7 in)
- Wheel track: 3·26 m (10 ft 8½ in)
- Wheelbase: 5·60 m (18 ft 4½ in)
- Propeller diameter: 2·74 m (9 ft 0 in)
- Passenger door (aft, port):
  - Height: 1·60 m (5 ft 3 in)
  - Width: 0·80 m (2 ft 7½ in)
- Crew door (forward, port):
  - Height: 1·34 m (4 ft 4¾ in)
  - Width: 0·58 m (1 ft 10¾ in)
- Emergency exit (forward, starboard):
  - Height: 1·34 m (4 ft 4¾ in)
  - Width: 0·58 m (1 ft 10¾ in)
- Emergency exit (aft, starboard):
  - Height: 1·28 m (4 ft 2½ in)
  - Width: 0·80 m (2 ft 7½ in)
- Cargo door (aft fuselage):
  - Max width: 2·00 m (6 ft 6¾ in)

DIMENSIONS, INTERNAL:
- Cabin, excl flight deck:
  - Length: 5·94 m (19 ft 6 in)
  - Max width: 2·40 m (7 ft 10½ in)
  - Max height: 1·98 m (6 ft 6 in)
  - Floor area: 11·88 m² (127·9 sq ft)
  - Volume: 23·76 m³ (839·1 cu ft)
- Rear baggage hold volume: 5·20 m³ (183·64 cu ft)
- Nose baggage hold volume: 1·20 m³ (42·38 cu ft)

AREAS:
- Wings, gross: 37·96 m² (408·6 sq ft)
- Ailerons (total): 2·28 m² (24·54 sq ft)
- Trailing-edge flaps (total): 7·14 m² (76·85 sq ft)
- Tailplane: 7·38 m² (79·44 sq ft)
- Elevators (total, incl tabs): 3·62 m² (38·97 sq ft)
- Fin: 3·65 m² (39·29 sq ft)
- Rudder (incl tab): 2·88 m² (31·00 sq ft)

WEIGHTS AND LOADINGS:
- Weight empty: 4,060 kg (8,951 lb)
- Operating weight empty: 4,300 kg (9,480 lb)
- Normal max T-O weight: 8,450 kg (18,629 lb)
- Overload max T-O weight: 9,200 kg (20,282 lb)
- Max wing loading: 242·3 kg/m² (49·63 lb/sq ft)
- Max power loading: 5·51 kg/kW (9·05 lb/shp)

PERFORMANCE (estimated, at normal military T-O weight of 8,450 kg; 18,629 lb):
- Max level speed at 3,050 m (10,000 ft): 228 knots (422 km/h; 262 mph)
- Max cruising speed at 6,100 m (20,000 ft): 217 knots (402 km/h; 250 mph)
- Stalling speed, flaps up: 92 knots (171 km/h; 106 mph)
- Stalling speed, flaps down: 74 knots (138 km/h; 85·5 mph)
- Max rate of climb at S/L: 510 m (1,673 ft)/min
- Max rate of climb at S/L, one engine out: 144 m (472 ft)/min
- Service ceiling: 7,100 m (23,300 ft)
- Service ceiling, one engine out: 3,300 m (10,825 ft)
- T-O run: 750 m (2,460 ft)
- T-O to 15 m (50 ft): 875 m (2,870 ft)
- Landing from 15 m (50 ft), wheel brakes only: 825 m (2,705 ft)
- Landing from 15 m (50 ft), wheel brakes and propeller reversal: 680 m (2,230 ft)
- Landing run, wheel brakes only: 455 m (1,495 ft)
- Landing run, wheel brakes and propeller reversal: 310 m (1,015 ft)
- Range at 80% cruise power, standard fuel, 2,000 kg (4,409 lb) payload and 30 min reserves:
  - at 3,050 m (10,000 ft): 1,080 nm (2,000 km; 1,243 miles)
  - at 4,575 m (15,000 ft): 1,200 nm (2,224 km; 1,382 miles)
  - at 6,100 m (20,000 ft): 1,320 nm (2,444 km; 1,520 miles)
- Range at overload T-O weight of 9,200 kg (20,282 lb), 80% cruise power, standard fuel, 2,700 kg (5,952 lb) payload and 30 min reserves:
  - at 3,050 m (10,000 ft): 1,060 nm (1,963 km; 1,220 miles)
  - at 4,575 m (15,000 ft): 1,190 nm (2,205 km; 1,370 miles)
  - at 6,100 m (20,000 ft): 1,310 nm (2,425 km; 1,507 miles)

DEUTSCHE AIRBUS
DEUTSCHE AIRBUS GmbH

ADDRESS:
D-8000 München 19, Leonrodstrasse 68
Telephone: (089) 1 79 61
Telex: 5215149

CHAIRMAN OF THE BOARD OF DIRECTORS:
Bundesminister a.D. Dr Franz-Josef Strauss

MANAGEMENT:
Dipl Kfm Rolf Siebert
Dipl-Ing Johannes Schäffler

PUBLIC RELATIONS:
Jochen H. Eichen

This company is the German partner in the consortium formed for development and production of the European high-capacity A300 transport aircraft described under the 'Airbus' heading in the International section.

DORNIER
DORNIER GmbH

HEAD OFFICE:
Postfach 1420, 7990 Friedrichshafen/Bodensee
Telephone: Immenstaad (07545) 81+
Telex: 0734372

WORKS:
Research and Development: 7759 Immenstaad/Bodensee (near Friedrichshafen)
Production: Postfach 2160, Trimburgstrasse, 8000 München 66

AIRFIELD AND FLIGHT TEST CENTRE:
8031 Oberpfaffenhofen, near München

BONN OFFICE:
Allianzplatz, 5300 Bonn

BOARD OF DIRECTORS:
Dipl-Ing Claudius Dornier Jr (Chairman)
Dipl-Ing Heinz Boldt
Dipl-Ing Dr Bernhard Schmidt
Dipl-Ing Dr jur Karl-Wilhelm Schäfer
Rainer Hainich

PUBLIC RELATIONS:
Gerhard Patt

Dornier GmbH, formerly Dornier-Metallbauten, was formed in 1922 by the late Professor Claude Dornier (who died on 5 December 1969) as the successor to the 'Do' division of the former Zeppelin Werke, Lindau, GmbH. It has been operated in the form of a Gesellschaft mit beschränkter Haftung since 22 December 1972.

In 1974 the Dornier group employed a total of 7,000 people. Member companies, in addition to Dornier GmbH, include Dornier-Reparaturwerft GmbH, at Oberpfaffenhofen (aircraft servicing and maintenance), Dornier System GmbH of Friedrichshafen (spaceflight, new technologies, electronics, management consultancy and contract research) and Lindauer Dornier GmbH of Lindau, which produces machinery for the textile industry and for the manufacture of plastics foils.

The Do 28 D Skyservant twin-engined STOL transport and utility aircraft continues in production; and technical and marketing efforts are being maintained on the Do 24/72 amphibian flying-boat project. Dornier is also active in development of various types of RPVs and tethered rotor platforms (see RPVs and Targets section).

The company is collaborating with the Dassault-Breguet group in France in developing and producing the Alpha Jet training/light attack aircraft, described in the International section. Under German government contract, it will design and test a supercritical wing on one of the Alpha Jet prototypes. It is also involved in licence production of components for the McDonnell Douglas Phantom fighter, and is collaborating with Pilatus of Switzerland (which see) in marketing the latter company's Turbo-Trainer.

DORNIER Do 28 D-2 SKYSERVANT

Despite its designation, the Skyservant inherited only the basic configuration of the earlier Do 28. The prototype (D-INTL) first flew on 23 February 1966. Type approval for the basic Do 28 D was granted on 24 February 1967, and for the developed Do 28 D-1 on 6 November 1967. FAA certification of the Do 28 D-1 was granted on 19 April 1968. Military type approval of the Do 28 D-1 was granted in January 1970, and of the Do 28 D-2 in late 1971, in accordance with MIL-specification standards. Initial deliveries of the Skyservant were made in the Summer of 1967; sales have reached 200 in 23 countries, including 121 Do 28 D-2s delivered to the Federal German Luftwaffe (101) for general duties and Navy Air Arm (20) for support duties.

On 15 March 1972, F. M. Tuytjens set up six international records in a Do 28 D-1 Skyservant in Class C1e for piston-engined business aircraft in the 3,000-6,000 kg weight category. Those still standing in mid-1976 included an altitude of 8,624 m (28,294 ft) with a 1,000 kg payload; a record payload of 1,000 kg (2,205 lb) carried to a height of 2,000 m (6,562 ft); and time-to-height records of 6 min 6 sec to 3,000 m (9,843 ft), 12 min 2 sec to 6,000 m (19,685 ft), and 44 min 4 sec to 9,000 m (29,528 ft).

The following details apply to the Do 28 D-2, which introduced a number of aerodynamic and detail refinements, including a 192 kg (423 lb) increase in AUW and more extensive standard equipment such as dual controls, dual brake system, directional slaved gyro, cabin heating, 100A alternators and provisions for de-icing system and IFR com/nav antennae installation.

TYPE: Twin-engined STOL transport and utility aircraft.

WINGS: Cantilever high-wing monoplane. Wing section NACA 23018 (modified), with nose slot in the outer half of each wing. Dihedral 1° 30′. Incidence 4°. All-metal box-spar structure. Double-slotted ailerons and flaps have metal structure, partly Eonnex-covered. Balance tabs on ailerons. Pneumatic de-icing optional.

FUSELAGE: Conventional all-metal stressed-skin structure.

TAIL UNIT: Cantilever all-metal structure, with rudder and horizontal surfaces partly Eonnex-covered. All-moving horizontal surface, with combined anti-balance and trim tab. Trim tab in rudder. Pneumatic de-icing optional.

LANDING GEAR: Non-retractable tailwheel type. Dornier oleo-pneumatic shock-absorbers on main units, glassfibre sprung tailwheel unit. Main-wheel tyres size 8·50-10, pressure 3·38 bars (49 lb/sq in). Twin-contact tailwheel tyre size 5·50-4, pressure 2·76 bars (40 lb/sq in). Double-disc hydraulic brakes. Fairings on main legs and wheels standard.

POWER PLANT: Two 283 kW (380 hp) Lycoming IGSO-540-A1E flat-six engines, mounted on stub-wings and each driving a Hartzell three-blade constant-speed propeller. Fuel tanks in engine nacelles, with total usable capacity of 893 litres (196·5 Imp gallons). Refuelling points above nacelles. Provision for two underwing auxiliary fuel tanks, increasing range by approx 50 per cent. Total capacity of separate oil tanks, 33 litres (7·25 Imp gallons).

ACCOMMODATION: Pilot and either co-pilot or passenger side by side on flight deck. Main cabin fitted with up to 12 seats, with aisle, or 13 inward-facing folding seats, or five stretchers and five folding seats, all layouts including toilet and/or baggage compartment and/or darkroom for aerial survey missions aft of cabin. Alternatively, cabin can be stripped for cargo-carrying. Door on each side of flight deck. Emergency exit on starboard side of cabin. Combined two-section passenger and freight door on port side of cabin, at rear.

ELECTRONICS AND EQUIPMENT: IFR instruments and electronics to customer's specifications.

DIMENSIONS, EXTERNAL:

Wing span	15·55 m (51 ft 0¼ in)
Wing chord (constant)	1·90 m (6 ft 2¾ in)
Wing aspect ratio	8·3
Length overall	11·41 m (37 ft 5¼ in)
Height overall	3·90 m (12 ft 9½ in)
Tailplane span	6·61 m (21 ft 8¼ in)
Wheel track	3·52 m (11 ft 6 in)
Wheelbase	8·63 m (28 ft 3¾ in)
Propeller diameter	2·36 m (7 ft 9 in)
Passenger door (port, rear):	
Height	1·34 m (4 ft 4¾ in)
Width	0·65 m (2 ft 1½ in)
Height to sill	0·60 m (1 ft 11½ in)
Freight door (port, rear):	
Height	1·34 m (4 ft 4¾ in)
Width	1·28 m (4 ft 2½ in)

DIMENSIONS, INTERNAL:

Cabin: Max length	3·97 m (13 ft 0½ in)
Max width	1·37 m (4 ft 6 in)
Max height	1·52 m (4 ft 11⅞ in)
Floor area	5·30 m² (57·05 sq ft)
Volume	8·10 m³ (286 cu ft)

AREAS:

Wings, gross	29·00 m² (312·2 sq ft)

Survey version of Dornier Skyservant STOL transport and utility aircraft, with camera windows in cabin undersurface

Dornier Do 28 D-2 Skyservant equipped with auxiliary fuel tanks on wing pylons

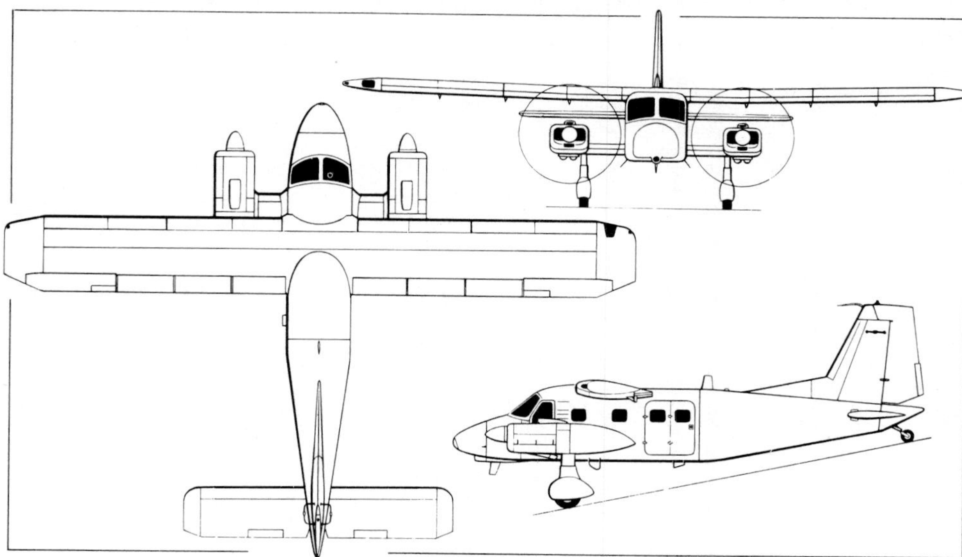

Dornier Do 28 D-2 Skyservant STOL utility transport aircraft *(Pilot Press)*

Ailerons (total)	2·64 m² (28·4 sq ft)
Trailing-edge flaps (total)	4·80 m² (51·6 sq ft)
Fin	4·65 m² (50·0 sq ft)
Rudder, incl tab	1·40 m² (15·1 sq ft)
Tailplane, incl tab	7·65 m² (82·3 sq ft)

WEIGHTS AND LOADINGS:

Weight empty, standard	2,304 kg (5,080 lb)
Max T-O weight	4,015 kg (8,853 lb)
Max ramp weight	4,035 kg (8,900 lb)
Max landing weight	3,650 kg (8,050 lb)
Max wing loading	138 kg/m² (28·36 lb/sq ft)
Max power loading	7·09 kg/kW (11·65 lb/hp)

PERFORMANCE (at max T-O weight):

Max level speed at 3,050 m (10,000 ft)
175 knots (325 km/h; 202 mph)

Cruising speed, 65% power at 3,050 m (10,000 ft)
148 knots (273 km/h; 170 mph)

Econ cruising speed, 50% power at 3,050 m (10,000 ft)
130 knots (241 km/h; 150 mph)

Stalling speed, power off, flaps down
56·5 knots (104 km/h; 65 mph)

Min control speed, power on, flaps down
35 knots (65 km/h; 40 mph)

Max rate of climb at S/L	320 m (1,050 ft)/min
Service ceiling	7,680 m (25,200 ft)
Service ceiling, one engine out	2,620 m (8,600 ft)
T-O run	310 m (1,020 ft)
Landing run	228 m (748 ft)

Range with max fuel
1,590 nm (2,950 km; 1,831 miles)

DORNIER Do 24/72

To meet a need expressed by the Spanish Air Force to replace the Grumman HU-16 Albatross amphibians employed on its sea-air rescue services, Dornier proposed an updated turboprop-powered version of the Do 24T sea-going flying-boat that performed these duties in Spain for many years. Designated Do 24/72, the new design is generally similar to that of the wartime flying-boat except for the change in power plant and provision of retractable tricycle landing gear. An alternative application is that of firefighting, for which the Do 24/72 has a water-bombing capacity of 8,000 litres (1,760 Imp gallons). A decision to begin production is dependent upon sufficient demand and the availability of capital to create a production line.

TYPE: Three-engined amphibian, primarily for sea-air rescue, but suitable for maritime patrol, passenger and cargo transport, ambulance and water bombing roles.

WINGS: Parasol monoplane, with the wing carried on an inverted Vee strut extending from beneath the outboard engine to the stabilising sponson on each side, and on short centre-section struts beneath the centre engine. Conventional all-metal structure. Double-slotted flaps along entire trailing-edge from aileron to aileron. Conventional ailerons, each with trim tab.

HULL: All-metal two-step hull, with sponson on each side to provide lateral stability on the water.

TAIL UNIT: All-metal cantilever structure, with twin end-plate fins and rudders. Elevator in two sections. Trim tab in each elevator and rudder.

LANDING GEAR: Retractable tricycle type, with twin wheels on each unit. Nose unit retracts aft into under-surface of hull, main units into housings at the tips of the stabilising sponsons.

POWER PLANT: Three 1,193 kW (1,600 shp) Lycoming T5321A turboprop engines, mounted in nacelles faired into the leading-edge of the wing.

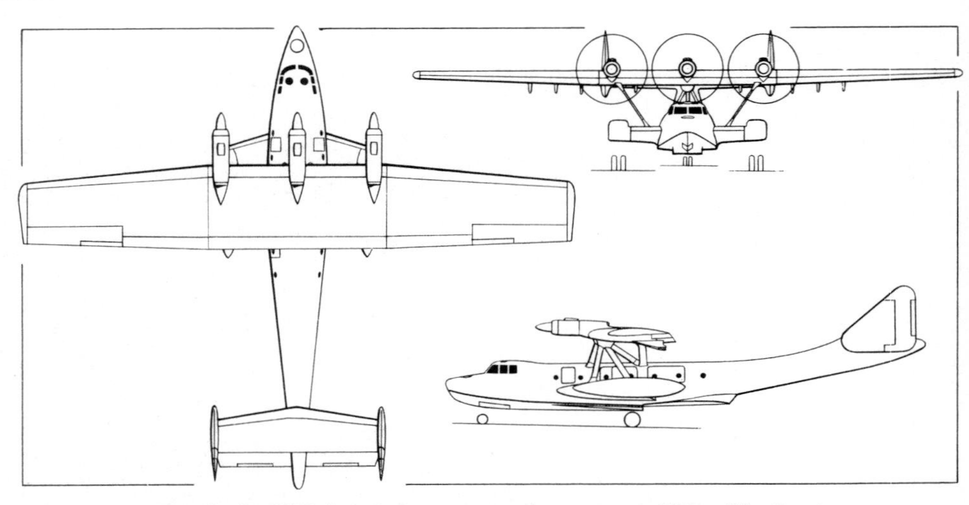

Dornier Do 24/72 three-turboprop general-purpose amphibian *(Pilot Press)*

ACCOMMODATION: Cabin is capable of seating a maximum of 40 persons.

DIMENSIONS, EXTERNAL:

Wing span	28·00 m (91 ft 10¼ in)
Length overall	24·35 m (79 ft 10¾ in)
Height overall	7·30 m (23 ft 11½ in)
Max width of hull	3·00 m (9 ft 10 in)
Width over sponsons	8·00 m (26 ft 3 in)
Wheel track (c/l of main units)	7·00 m (22 ft 11¾ in)

WEIGHTS (estimated):

Weight empty, basic aircraft	10,407 kg (22,943 lb)
Max T-O weight, land or water	18,600 kg (41,005 lb)

PERFORMANCE (estimated):

Max level speed	224 knots (416 km/h; 258 mph)
T-O to 15 m (50 ft) on land	520 m (1,705 ft)
Landing from 15 m (50 ft) on land with reverse thrust	350 m (1,150 ft)
T-O time on water	15 sec

Range with max fuel
1,726 nm (3,200 km; 1,988 miles)

Endurance with max fuel	14 hr

EQUATOR AIRCRAFT
EQUATOR AIRCRAFT GESELLSCHAFT FÜR FLUGZEUGBAU mbH ULM

HEAD OFFICE:
8 München 40, Adalbertstrasse 110
Telephone: (089) 37 24 21
Telex: 5-215531

PRESIDENT: Günther Pöschel

Günther Pöschel, formerly President of Pöschel Aircraft GmbH, designed and built the prototype of a five/six-seat light STOL aircraft, designated P-300 Equator. Powered by a 231 kW (310 hp) Lycoming TIO-541 flat-six engine, this flew for the first time on 8 November 1970, and was last illustrated and described in the 1972-73 *Jane's*.

At that time there was the intention to produce a turboprop-powered version, under the designation P-400 Meridian Turbo-Stol-Amphibian, but lack of capital was responsible for a halt to the development of this aircraft. Herr Pöschel has now formed this new company to continue development and production of the turboprop-powered version, re-designated P-400 Turbo-Equator, as well as production of the piston-engined P-300 Equator.

EQUATOR AIRCRAFT P-400 TURBO-EQUATOR

TYPE: Lightweight STOL amphibian executive aircraft.

WINGS: Cantilever high-wing monoplane. Conventional single-spar structure with outer skin of reinforced glassfibre. Full-span double-slotted trailing-edge flaps.

FUSELAGE: Conventional semi-monocoque structure, with outer skin of reinforced glassfibre, forming a watertight hull.

TAIL UNIT: Cantilever structure with outer skin of reinforced glassfibre. Cruciform tail unit; fin integral with fuselage. Tailplane and elevators mounted approximately two-thirds up the fin, in line with propeller axis. Rudder extends from base of fin to just below tailplane. Trim tab in rudder.

LANDING GEAR: Conventional tricycle-type retractable landing gear for operation from land. When gear is retracted into fuselage, the structure is completely watertight to permit operation from water. Small stabilising sponsons, on each side of fuselage, directly below wings, to give stability when operating on water.

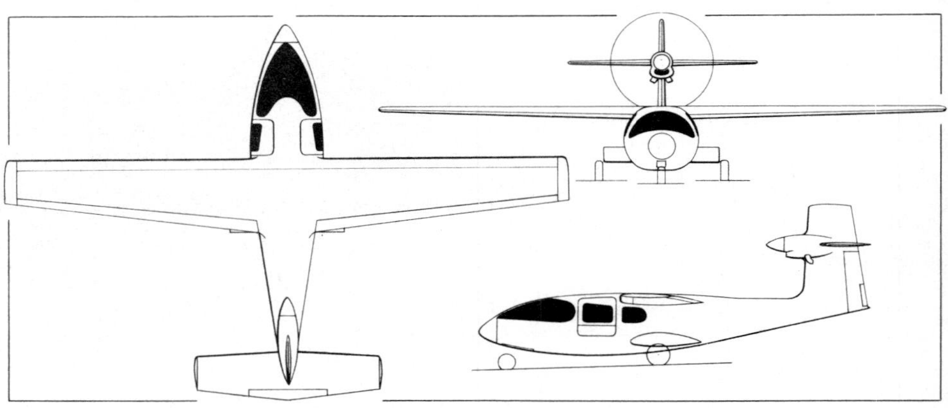

P-400 Turbo-Equator eight-seat turboprop-powered light transport *(Pilot Press)*

POWER PLANT: One 313 kW (420 shp) Allison 250-B17B turboprop engine, driving a three-blade propeller. The small diameter of this engine (only 483 mm; 19 in), allows it to be mounted in the tail unit, with its axis coincident with the intersection of the tailplane and fin.

ACCOMMODATION: Standard seating for pilot and up to seven passengers. Access by means of door on each side of fuselage, forward of wing. Cabin heated and air-conditioned. Dual controls standard.

DIMENSIONS, EXTERNAL:

Wing span	12·40 m (40 ft 8¼ in)
Length overall	8·60 m (28 ft 2¾ in)
Height overall	3·55 m (11 ft 7¾ in)

AREA:

Wings, gross	18·00 m² (193·8 sq ft)

WEIGHTS AND LOADING:

Weight empty	950 kg (2,094 lb)
Max T-O weight	2,000 kg (4,409 lb)
Max wing loading	111·11 kg/m² (22·76 lb/sq ft)

PERFORMANCE (estimated, at max T-O weight):

Max level speed at 2,400 m (7,875 ft)
254 knots (470 km/h; 292 mph)

Max cruising speed at 3,600 m (11,800 ft)
248 knots (460 km/h; 286 mph)

Cruising speed at 7,200 m (23,625 ft)
237 knots (440 km/h; 273 mph)

Landing speed	55 knots (102 km/h; 63·5 mph)
Max rate of climb at S/L	720 m (2,360 ft)/min
Service ceiling	8,500 m (27,875 ft)
T-O run	170 m (560 ft)
T-O to 15 m (50 ft)	245 m (805 ft)
Landing from 15 m (50 ft)	250 m (820 ft)
Landing run	165 m (540 ft)
Landing run, with propeller reversal	50 m (165 ft)

Range with max fuel at 3,600 m (11,800 ft)
755 nm (1,400 km; 870 miles)

HIRTH
WOLF HIRTH GmbH

ADDRESS:
7312 Kirchheim/Teck-Nabern
Telephone: Kirchheim/Teck (070 21) 5 53 77

In 1976, a major share in Wolf Hirth was acquired by MBB, as a result of which the company is devoting its efforts mainly to support of the Bolkow BO 107, 207, 208 and 209 aircraft. It intends to develop a new light aircraft in the near future.

HIRTH ACROSTAR Mk III

Designed for advanced competitive aerobatic flying, the Acrostar uses an entirely symmetrical basic aerofoil section, with variable positive and negative camber, and was conceived originally by the Swiss aerobatic champion, Arnold Wagner. The project was sponsored by Herr Wagner, Herr Josef Hössl (then German aerobatic champion), Herr Walter Wolfrum (a former German champion) and Herr Horst Gehm. Design of the aircraft, which was carried out by Wolf Hirth GmbH under the supervision of Prof Eppler of the Technische Hochschule of Stuttgart, began in the late Summer of 1969. Construction began in December of that year, and the Acrostar Mk II prototype (D-EMKB) was flown for the first time on 16 April 1970, making a successful appearance in the World Aerobatics Competition at Hullavington, England, three months later. It took part subsequently in international meetings in France and Switzerland, gaining first place at its debut in each country.

The first production aircraft (Serial No. 2) was used for type certification tests. Full type certification in the Normal and Aerobatic categories, based on the requirements of FAR 23, was granted by the Luftfahrt Bundesamt.

A new trim system represented the only significant change between the prototype and Mk II production aircraft: two tabs on the inboard ends of the integrated trailing-edge flaps replaced the elevator trim tab of the prototype.

Four Acrostars were completed and delivered by 25 January 1972, and additional deliveries were made in time for the 1972 World Aerobatics Competition.

The subsequent Mk III version introduced more than 50 modifications to reduce weight and improve performance and controllability. These include lightweight control surfaces, improved fairing at the fuselage/wing junction, and changes in the oil system and engine cooling that permit glider towing without danger of engine overheating. The Acrostar is stressed to $\pm 8g$ ($\pm 12g$ ultimate) with a max Aerobatic gross weight of 630 kg (1,389 lb). One example of this Mk III version was manufactured in 1974. Parts for five more Acrostars are in stock. All tooling remains available, and studies for versions with Lycoming or Continental engines are underway.

During 1973 Hirth Acrostars won the Scandinavian Cup, Zadar Cup in Yugoslavia, Swiss International Championships, West German Championships and the Coupe Champion Amberieu in France. In 1974 seven contests were won by Acrostars, the most important being for the Coupe Biancotto.

TYPE: Single-seat advanced aerobatic aircraft.

WINGS: Cantilever low-wing monoplane. Thick, symmetrical aerofoil section, of 20% thickness/chord ratio, designed by Prof Eppler. Single glassfibre main spar, plastics foam reinforced ribs and plywood covering. Slight sweepback on leading-edge. No dihedral or incidence. Inboard trailing-edge flaps and large proportionally-moving ailerons are coupled to the elevator, providing variable camber as a function of stick position. This gives flight characteristics which are equal in both positive and negative manoeuvres. Trim tab on the inboard end of each trailing-edge flap, which can be positioned independently, to provide both pitch and roll trim.

FUSELAGE: Streamlined semi-monocoque structure, of wooden construction except for steel-tube engine mount which is integral with the main landing gear assembly. Fuselage aft of cockpit is detachable for transportation.

Hirth Acrostar Mk II advanced aerobatic aircraft *(Howard Levy)*

TAIL UNIT: Conventional cantilever structure, with single fin and balanced rudder. One-piece elevator. Underfin beneath rear fuselage.

LANDING GEAR: Non-retractable tailwheel type, with optional streamline fairings over main wheels and landing gear struts. Main gear legs built integrally with engine mount. Böhler main units and wheels with dual hydraulic brakes. Steerable tailwheel.

POWER PLANT: One 164 kW (220 hp) Franklin 6A-350-C1 flat-six engine, driving a Hartzell two-blade constant-speed propeller. Standard fuel capacity for aerobatics 50 litres (11 Imp gallons), contained in wing-root leading-edge tank. Optional cruise tank available, capacity 40-60 litres (8·8-13·2 Imp gallons). Oil system specially modified by Hirth to provide lubrication in all aerobatic manoeuvres.

ACCOMMODATION: Single seat under fully-transparent rearward-sliding canopy which incorporates nose-over protection bars. Space aft of seat for 15 kg (33 lb) of baggage.

SYSTEMS: Hydraulic system for brakes only. No electrical system standard, but an external power socket is fitted as standard to permit use of an electric engine starter. Electrical system to include battery, battery stowage, navigation and internal lights optional.

ELECTRONICS AND EQUIPMENT: Radio rack, antenna, microphone and button control and a selection of com radios available to customers' requirements. Optional equipment includes window in cockpit floor with fire protection cover, stopwatch, cylinder head temperature gauge, rate of climb indicator, sideslip indicator, inverted accelerometer, anatomic control column, baggage compartment, glider tow hook and Hoffmann three-blade propeller.

DIMENSIONS, EXTERNAL:
- Wing span: 8·00 m (26 ft 3 in)
- Wing aspect ratio: 6·03
- Length overall: 6·00 m (19 ft 8¼ in)
- Height overall: 2·50 m (8 ft 2½ in)

AREA:
- Wings, gross: 10·60 m² (114·1 sq ft)

WEIGHTS:
- Weight empty, standard equipment: 450 kg (992 lb)
- Weight empty with optional equipment: up to 495 kg (1,091 lb)
- Max T-O weight for aerobatics: 630 kg (1,389 lb)
- Max T-O weight for normal flight: 700 kg (1,543 lb)

PERFORMANCE (at max T-O weight of 700 kg; 1,543 lb, except where indicated):
- Never-exceed speed: 226 knots (420 km/h; 261 mph)
- Max level speed at S/L: 159 knots (295 km/h; 183 mph)
- Max cruising speed, 75% power at 1,830 m (6,000 ft): 157 knots (291 km/h; 181 mph)
- Max manoeuvring speed: 147 knots (273 km/h; 170 mph)
- Econ cruising speed at 3,050 m (10,000 ft): 136 knots (252 km/h; 157 mph)
- Stalling speed: 52 knots (96 km/h; 60 mph)
- Stalling speed inverted: 51 knots (94 km/h; 59 mph)
- Max rate of climb at S/L: 720 m (2,360 ft)/min
- Max rate of climb at S/L, at 630 kg (1,389 lb) gross weight: 900 m (2,950 ft)/min
- T-O run: 70 m (230 ft)
- T-O to 15 m (50 ft): 130 m (425 ft)
- Landing from 15 m (50 ft): 320 m (1,050 ft)
- Landing run: 110 m (360 ft)
- Range with max fuel, no reserves: 465 nm (864 km; 536 miles)

MBB
MESSERSCHMITT-BÖLKOW-BLOHM GmbH

HEAD OFFICE:
Ottobrunn bei München, 8 München 80, Postfach 801220

Telephone: (089) 60001
Telex: 0522279 mbbo

WORKS:
Augsburg, Donauwörth, Hamburg-Finkenwerder, Laupheim, Manching, Munich, Nabern/Teck, Ottobrunn, Schrobenhausen and Stade

PRESIDENT:
Dipl-Ing Ludwig Bölkow

EXECUTIVE VICE-PRESIDENTS:
Dr Johannes Broschwitz
Gunther Horstkotte
Sepp Hort
Peter C. Küchler
Ernst-Georg Pante
Hans Wallner

PUBLIC RELATIONS:
Eduard Roth

EXECUTIVES:
Helmut Langfelder (Aircraft Division)
Ernst-Georg Pantel (Commercial Aircraft Division)
Günther Kuhlo (Dynamics Division)
Peter Schulz (Surface Transport Division)
Johannes Schubert (Space Division)
Kurt Pfleiderer (Helicopter Division)
Kyrill von Gersdorff (Administrative Services Division)

CHAIRMAN OF THE SUPERVISORY BOARD:
Dr Karl-Heinz Sonne

HONORARY CHAIRMAN OF THE SUPERVISORY BOARD:
Prof Dr-Ing E.h. Willy Messerschmitt

In May 1969 the former Messerschmitt-Bölkow GmbH and Hamburger Flugzeugbau GmbH (see 1968-69 *Jane's*) respectively endorsed the merger between their two companies to form a new group known as Messerschmitt-Bölkow-Blohm GmbH. Major shareholder in this company is the Blohm family, which has a 22·05% interest; other shareholders are Prof Dr-Ing E.h. Willy Messerschmitt (16·3%), Dipl-Ing Ludwig Bölkow (13·42%), The Boeing Company and Aérospatiale (8·9% each), Siemens AG (8·35%), the Bavarian State (7·8%), the Bavarian Reconstruction Finance Institute (5·93%) and August Thyssen-Hütte AG (8·35%). The company is the largest aerospace concern in Germany, with a total work force in 1975 of some 20,000 employees, and is affiliated with a number of national and international corporations.

In the military aircraft field, development work on the Panavia Tornado multi-role combat aircraft, as Germany's prime contractor, is the company's main activity. MBB has a 42·5% shareholding in Panavia Aircraft GmbH (which see), together with BAC (42·5%) and Aeritalia (15%). MBB participated substantially in licence production of the F-104G Starfighter, a present standard weapon system of the German air force, and the company's Manching facilities are responsible for overhaul of these aircraft, as well as of the F-4 Phantom. MBB will also take part in the resumed production of the Transall C-160 military transport aircraft.

The main civil aircraft programme involves work on the European A300 Airbus for Airbus Industrie (which see). MBB's share in the development and production of this aircraft is around 30%, which represents 65% of the total German share.

MBB's BO 105 was the world's first twin-turbine lightweight utility helicopter to enter series production. The glassfibre-reinforced plastics technology developed for its rotor blades has found many applications in engineering fields.

MBB weapon systems include the Kormoran air-to-ship stand-off missile, the Jumbo air-to-ground stand-off missile, the Armbrust-300 anti-tank weapon and, in partnership with Aérospatiale of France, the Roland low-level anti-aircraft missile, and the Hot and Milan second-generation anti-tank weapon systems. More than 150,000 Cobra first-generation anti-tank missiles have been produced, including licence production.

MBB has built a number of European satellites, being the prime contractor or system leader for seven of these, and leader of the German portion of the Symphonie contract. Successful satellite launches since 1974 include COS-B and Symphonie 2 in 1975 and Helios B in 1976. MBB is involved in the METEOSAT, OTS and MAROTS satellite programmes, and the Ariane launcher programme; it builds ground stations and payload cones and conducts satellite experiments. MBB plans in future to concentrate more on applications satellites (such as the ARCOMSAT proposal) and, in the manned space field, in the development, construction, testing and integration of individual experiments and complete payloads. Preliminary work on European utilisation of the Spacelab, commissioned by ESA, is a pointer in this direction.

Constantly growing importance is attached by MBB to ground-based forms of transport, and the world's first experimental vehicle using the magnetic cushion effect and the linear induction motor was used by MBB for testing possible components of such a rail system. A 400 km/h (250 mph) unmanned research vehicle was undergoing further testing in early 1976. Another important project for transport operations of the future is the cabin taxi personal rapid transit system being developed by MBB and DEMAG.

MBB BO 105

As a first stage of the development of this light utility helicopter, Bölkow tested a full-size rotor on a ground rig, under German government contract. Design of the BO 105 was started in July 1962 and construction of prototypes began in 1964, under a further government contract.

The rotor system features a rigid titanium hub, with feathering hinges only, and hingeless flexible glassfibre blades. Initial flight tests were made on an Alouette II Astazou helicopter, under a programme conducted jointly with Sud-Aviation (now Aérospatiale) of France.

The first BO 105 prototype was fitted with an existing conventional rotor and two Allison 250-C18 turboshaft engines; subsequent aircraft have had the 'hingeless' rotor. From the Spring of 1970 'droop-snoot' rotor blades of MBB design were introduced.

The second BO 105 flew for the first time on 16 February 1967, also powered by two Allison 250-C18 turboshaft engines. The third prototype, with MAN-Turbo 6022 engines, was flown on 20 December 1967.

The first two pre-production aircraft (the V4 and V5) were completed in the Spring of 1969, and the V4 flew for the first time on 1 May 1969. It was fitted subsequently with two Allison 250-C20 engines, with which it made its first flight on 11 January 1971. Standard production aircraft, designated **BO 105C**, have been fitted with Allison 250-C20 engines since 1973. The version supplied to the UK, with modified equipment, is designated **BO 105D.**

The **BO 105 HGH,** described in the 1975-76 *Jane's,* is an experimental high-speed version which continues in use as a rotor testbed. The **BO 105 S** has increased seating or cargo capacity in a 0·25 m (9·8 in) longer fuselage; and the **BO 105 VBH** is a military communications and observation version of which 12 have been evaluated by the Federal German Ministry of Defence.

German LBA type certification with the Allison 250-C18 power plant was granted in October 1970, and FAA certification in March 1971. LBA and FAA certification with the -C20 power plant were granted in August 1971 and April 1972 respectively. Canadian MoT certification for the BO 105C with -C20 power plant was granted in April 1973, and UK CAA certification for the BO 105D in July 1973. Registro Aeronautico Italiano (RAI) certification was received in March 1974.

The BO 105D has been approved since 1973 by the CAA for commercial single-pilot IFR operation, even in controlled airways.

A total of 250 BO 105s had been delivered (of some 290 ordered) by 1 January 1976, when production was continuing at a rate of six to eight aircraft each month. Orders at that time included 30 for the Netherlands Army, of which 12 were delivered during 1975, with the balance scheduled for delivery in 1976. The first two of an order for the Dutch State Police were due to be delivered in the Summer of 1976. Fourteen have been delivered to Petroleum Helicopters Inc, the largest single civil customer outside Germany. Nearly 50 have been sold by Boeing Vertol Company, which markets the BO 105 in the USA.

The BO 105 is in service also with the Nigerian Air Force for search and rescue; and about 300 BO 105 VBH are expected to be ordered by the German Army, which has selected this type to replace existing Alouette IIs in the liaison and observation roles.

In the Philippines, PADC (which see) is participating in a programme which involves initially the licence assembly of 33 BO 105Cs, following the delivery of five fully-assembled aircraft by MBB.

The Federal German government ordered 20 BO 105s under its 'Katastrophenschutz' scheme to provide speedy assistance after major disasters. These helicopters have already saved at least 1,000 lives in more than 10,000 missions, and rescue versions are also in service in Argentina, Nigeria, the Philippines and Switzerland. The BO 105 can also be fitted with outriggers to carry a total of six Hot anti-tank missiles, three on each side of the cabin, with a stabilised sight above the co-pilot's position.

Type: Five-seat light helicopter.

Rotor System: Four-blade main rotor, with rigid titanium hub and flexible glassfibre-reinforced plastics blades. MBB-designed 'droop-snoot' blades of NACA 23012 asymmetrical section, and having a specially-designed trailing-edge giving improved control in pitching moment. Two-blade semi-rigid tail rotor, with blades of glassfibre-reinforced plastics.

Rotor Drive: Main transmission utilises two stages of spur gears and single stage of bevel gearing. Planetary reduction gear, freewheeling clutch and transmission accessory gear. Tail rotor gearbox on fin. Main rotor/engine rpm ratio 1 : 14·1. Tail rotor/engine rpm ratio 1 : 2·7.

Fuselage: Conventional light alloy semi-monocoque structure of pod and boom type. Glassfibre-reinforced cowling over power plant.

Tail Unit: Horizontal stabiliser of conventional light alloy construction.

Landing Gear: Skids, to which inflatable emergency floats can be attached.

Power Plant: Two 298 kW (400 shp) Allison 250-C20 turboshaft engines. One 570 litre (125 Imp gallon) integral fuel tank under cabin floor. Fuelling point on port side of cabin. Provision for fitting auxiliary tanks in freight compartment. Oil capacity: engine 4 kg (8·8 lb), gearbox: 7 kg (15·4 lb).

Accommodation: Pilot and passenger on individual front seats. Removable dual controls. Bench seat at rear for three persons, removable for cargo and stretcher carrying. Entire rear fuselage aft of seats and under power plant available as freight and baggage space, with access through two clamshell doors at rear. Two standard stretchers can be accommodated in ambulance role. One forward-opening door and one sliding door on each side of cabin.

MBB BO 105 in German Police service (two Allison 250-C20 turboshaft engines) *(Austin J. Brown)*

MBB BO 105 fitted experimentally with six Hot anti-tank missiles and stabilised sight

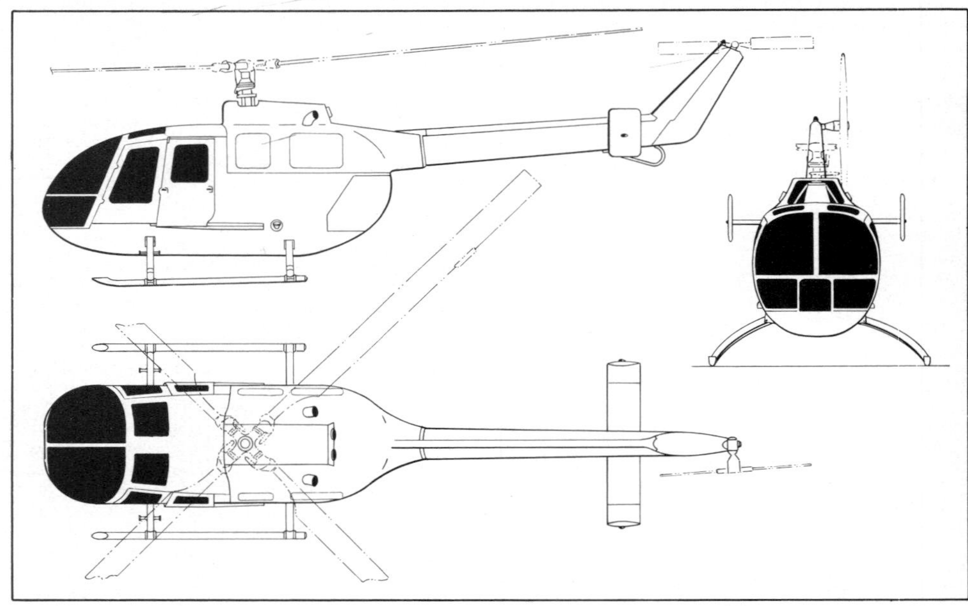

BO 105 five-seat light helicopter (two Allison 250-C20 turboshaft engines) *(Roy J. Grainge)*

Systems: Hydraulic system for powered main rotor controls. 24Ah battery and starter/generator, with provision for external connection.

Electronics and Equipment: A wide range of optional electronics and equipment is available, including search radar, Doppler navigation, emergency flotation gear, auxiliary fuel tanks, cargo hook and two-man rescue winch. A completely equipped ambulance version is available.

Armament (Military versions): Provision for a variety of alternative military loads, including six Hot anti-tank missiles and associated stabilised sight.

Dimensions, external:

Diameter of main rotor	9·82 m (32 ft 2¾ in)
Diameter of tail rotor	1·90 m (6 ft 2¾ in)
Distance between rotor centres	5·95 m (19 ft 6¼ in)
Length, incl main rotor	11·84 m (38 ft 10¾ in)
Length, excl main rotor	8·55 m (28 ft 0½ in)
Height overall	2·98 m (9 ft 9⅜ in)
Skid track	2·50 m (8 ft 2½ in)
Rear loading doors:	
Height	0·64 m (2 ft 1 in)
Width	1·40 m (4 ft 7 in)

DIMENSIONS, INTERNAL:
Cabin, incl cargo compartment:
Length 4·30 m (14 ft 1 in)
Max width 1·40 m (4 ft 7 in)
Max height 1·25 m (4 ft 1 in)
Floor area (cargo compartment) 2·20 m² (23·68 sq ft)
Volume 4·80 m³ (169 cu ft)
Cargo compartment volume 1·50 m³ (53 cu ft)

WEIGHTS AND LOADINGS:
Weight empty 1,120 kg (2,469 lb)
Max T-O weight 2,300 kg (5,070 lb)
Normal disc loading 26·5 kg/m² (5·43 lb/sq ft)
Max disc loading 30·5 kg/m² (6·25 lb/sq ft)

PERFORMANCE (at normal T-O weight, Allison 250-C20 engines):
Never-exceed speed at S/L 145 knots (270 km/h; 167 mph)
Max cruising speed at S/L 125 knots (232 km/h; 144 mph)
Max rate of climb at S/L 420 m (1,378 ft)/min
Rate of climb at S/L, one engine out 46 m (150 ft)/min
Max operating height 5,180 m (17,000 ft)
Hovering ceiling in ground effect 2,715 m (8,900 ft)
Hovering ceiling out of ground effect 1,735 m (5,700 ft)
Range with standard fuel, no reserves:
at S/L 315 nm (585 km; 363 miles)
at 1,525 m (5,000 ft) 355 nm (656 km; 408 miles)
Max range with auxiliary tanks:
at S/L 556 nm (1,030 km; 640 miles)
at 1,525 m (5,000 ft) 626 nm (1,158 km; 720 miles)
Max endurance with auxiliary tanks 6 hr 35 min

MBB BO 106 six/seven-seat twin-turboshaft light helicopter

MBB BO 106

First flown on 25 September 1973, the BO 106 is generally similar to the BO 105 but has the cabin widened by 500 mm (19·7 in) to seat two or three persons in front and four on the rear bench. Its uprated Allison 250-C20B engines each develop a maximum of 313 kW (420 shp) and give 37·3 kW (50 shp) more than the 250-C20 at ISA + 20°C. This gives the BO 106 a performance similar to that of the BO 105, with a cruising speed of 124 knots (230 km/h; 143 mph) and a range of 302 nm (560 km; 348 miles).

An uprated transmission caters for a twin-engine power output of up to 516 kW (692 shp), and single-engine output of 283 kW (380 shp), compared with 474 and 276 kW (636 and 370 shp) respectively for the BO 105 transmission.

The prototype was developed with government aid (60%). A decision on whether to market the type was still pending in early 1976.

MBB BO 107

This enlarged 7/8-seat development of the BO 105 is in the design concept stage. Power plant will be in the 447-485 kW (600-650 shp) range.

MYLIUS

ENTWICKLUNGSGEMEINSCHAFT LEICHTFLUGZEUGE DIPL ING HERMANN MYLIUS

ADDRESS:
8011 Brunnthal-Gudrunsiedlung, Kuckucksweg 6

In addition to heading the light aircraft technical development activities of MBB, Dipl Ing Hermann Mylius develops and builds sporting aircraft privately. One such aircraft was the MHK 101, intended as the first of a family of related designs, which flew for the first time on 22 December 1967 and was manufactured by MBB in developed form as the BO 209 Monsun. The MHK 101 was described in the 1968-69 *Jane's*. Details of the BO 209, of which 102 were built, can be found under the MBB entry in the 1972-73 *Jane's*.

In July 1971, Dipl Ing Mylius began privately the development of a single-seat version of the MHK 101, intended for competitive aerobatics. Construction of a prototype was started in December 1971, and the prototype flew for the first time on 7 July 1973. Known as the MY 102 Tornado, it had accumulated a total of 150 flight hours by the beginning of 1975, and had participated successfully in both the 1973 German Aerobatic Championships and the 1974 Coupe Biancotto.

Mylius MY 102 Tornado prototype single-seat sporting monoplane (Lycoming AIO-360-B1B engine)

MYLIUS MY 102 TORNADO

TYPE: Single-seat sporting and aerobatic aircraft.

WINGS: Cantilever low-wing monoplane. Wing section NACA 64218 at root, NACA 64212 at tip. Dihedral 2° 30′. Incidence 2°. Sweepback 1° 24′ at quarter-chord. Single-spar all-metal structure with glassfibre wingtips. All-metal differentially-operated ailerons. Electrically-operated all-metal plain trailing-edge flaps. By removing three bolts on each side, the wings can be folded back alongside the fuselage to facilitate stowage in a confined space or to permit the aircraft to be towed on ordinary roads behind a car. Control lines to flaps and ailerons disconnect and reconnect automatically during folding and unfolding.

FUSELAGE: Conventional semi-monocoque structure of light alloy, except for engine cowlings which are of glassfibre.

TAIL UNIT: Single all-metal fin and rudder, with sweepback on fin leading-edge and dorsal fairing from base of fin to rear of cockpit canopy. All-moving metal tailplane, with glassfibre tips, has slight taper on leading- and trailing-edges, and is mid-mounted on extreme rear of fuselage, with cable-controlled full-span anti-servo tab.

LANDING GEAR: Tricycle type, with fixed or optionally rearward-retracting nosewheel and non-retractable main wheels. Nosewheel is steerable by means of the rudder pedals, the controls being disconnected automatically during retraction, which is accomplished electrically. Nosewheel can be locked in the down position during flight, if required. Main-gear legs are cantilever steel struts, inclined outward at 45° from fuselage main bulkhead. Cleveland wheels size 5·00-5; Continental tyres size 5·00-5 on nosewheel and 5·50-5 on main wheels. Streamlined fairings on main wheels. Cleveland hydraulically-actuated brakes. Small skid under rear fuselage, which can be fitted with an adapter for transport by road.

POWER PLANT: One 149 kW (200 hp) Lycoming AIO-360-B1B flat-four engine, driving a Hoffmann type HO-V123/180R three-blade metal constant-speed propeller with spinner. Fuel in two tanks in wings, each with capacity of 73·4 litres (19·4 US gallons). Total fuel capacity 146·8 litres (38·8 US gallons). Refuelling point in upper surface of each wing.

ACCOMMODATION: Single seat, beneath rearward-sliding tinted canopy.

SYSTEMS: DC electrical system for flaps, nose gear actuation and engine starting, provided by 40A engine-driven alternator and 12V 33Ah battery. Hydraulic system for brakes only.

ELECTRONICS: Becker AR400 radio.

DIMENSIONS, EXTERNAL:
Wing span 8·10 m (26 ft 6¾ in)
Wing chord at root 1·40 m (4 ft 7 in)
Wing chord at tip 1·00 m (3 ft 3¼ in)
Wing aspect ratio 6·7
Length overall 6·40 m (21 ft 0 in)
Height overall 2·31 m (7 ft 7 in)
Tailplane span 2·80 m (9 ft 2¼ in)
Wheel track 1·94 m (6 ft 4¼ in)
Wheelbase 1·33 m (4 ft 4¼ in)
Propeller diameter 1·80 m (5 ft 11 in)

AREAS:
Wings, gross 9·80 m² (105·5 sq ft)
Ailerons (total) 0·80 m² (8·61 sq ft)
Trailing-edge flaps (total) 0·85 m² (9·15 sq ft)
Fin 0·50 m² (5·38 sq ft)
Rudder 0·52 m² (5·60 sq ft)
Tailplane (incl tab) 1·66 m² (17·87 sq ft)

WEIGHTS AND LOADINGS (A: Normal; B: Utility; C: Aerobatic):
Weight empty 530 kg (1,168 lb)
Max T-O weight:
A 820 kg (1,807 lb)
B 740 kg (1,631 lb)
C 650 kg (1,433 lb)
Max wing loading:
A 83·5 kg/m² (17·1 lb/sq ft)
B 75·0 kg/m² (15·4 lb/sq ft)
C 66·5 kg/m² (13·6 lb/sq ft)
Max power loading:
A 5·50 kg/kW (9·04 lb/hp)
B 4·97 kg/kW (8·16 lb/hp)
C 4·36 kg/kW (7·17 lb/hp)

PERFORMANCE (at max Normal T-O weight):
Never-exceed speed 214 knots (396 km/h; 246 mph)
Max level speed at S/L 174 knots (322 km/h; 200 mph)
Max cruising speed 160 knots (296 km/h; 184 mph)
Stalling speed, flaps down 52 knots (97 km/h; 60 mph)
Max rate of climb at S/L 732 m (2,400 ft)/min
Service ceiling 9,145 m (30,000 ft)
T-O run 100 m (330 ft)
Landing run 180 m (590 ft)
Range with max fuel at AUW of 750 kg (1,653 lb), no reserve 517 nm (960 km; 596 miles)

POLIGRAT—*see 'Master Porter' in Addenda*

RFB
RHEIN-FLUGZEUGBAU GmbH (Subsidiary of VFW-Fokker GmbH)

HEAD OFFICE AND MAIN WORKS:
D-4050 Mönchengladbach, Flugplatz, Postfach 408
Telephone: (0 21 61) 662031
Telex: 08/52506
OTHER WORKS:
D-5050 Porz-Wahn, Flughafen Köln-Bonn, Halle 6; and D-2401 Lübeck-Blankensee, Flugplatz
EXECUTIVE DIRECTORS:
Dipl-Volkswirt Wolfgang Kutscher
Dipl-Ing Alfred Schneider

This company, founded in 1956, has since 1974 held 67% of the stock of Sportavia-Pützer (which see).

RFB is engaged in the development and construction of airframe structural components, with particular reference to wings and fuselages made entirely of glassfibre-reinforced resins. Research and design activities include studies for the Federal German Ministry of Defence.

Current manufacturing programmes include series and individual production of aircraft components and assemblies made of light alloy, steel and glassfibre-reinforced resin, for aircraft in quantity production by other German companies, as well as spare parts and ground equipment. The company is also concerned in shelter and container construction.

Under contract to the German government, RFB is servicing military aircraft, and is providing target-towing flights and other services with special aircraft. It operates a factory-certificated service centre for all types of Piper aircraft and for the Mitsubishi MU-2 utility transport aircraft. General servicing of other types of all-metal aircraft is undertaken, together with the servicing, maintenance, repair and testing of all kinds of flight instruments, engine instruments and navigation and communications electronics.

In the aircraft propulsion field, RFB has been engaged for some years in the development of specialised applications for ducted propellers, leading to the Fantrainer AWI 2 project. Construction of two demonstrators of this military multi-purpose training aircraft was started at the end of 1974. As a variation for civil use, RFB started in 1972 the Fanliner project, the first flight of a prototype taking place in October 1973. In April 1974 it was announced that RFB and Grumman American Aviation (which see) had decided to collaborate in development of this new two-seat light aircraft, which utilises the ducted-fan propulsion system evolved by RFB.

Under the scientific direction of the late Dr A. M. Lippisch, the all-plastics X-113 Am Aerofoil Boat was built and tested; brief details of this can be found in the 1974-75 *Jane's*. Development of a six-seat version of this ground-effect aircraft began in the Autumn of 1974. Designated X-114, this will have a maximum take-off weight of 1,350 kg (2,976 lb) and, being fitted with retractable landing gear, will be able to operate as an amphibian. It is estimated that it will have a cruising speed of 67·5 knots (125 km/h; 77·5 mph).

RFB/Grumman American Fanliner, powered by a Wankel engine driving a ducted fan

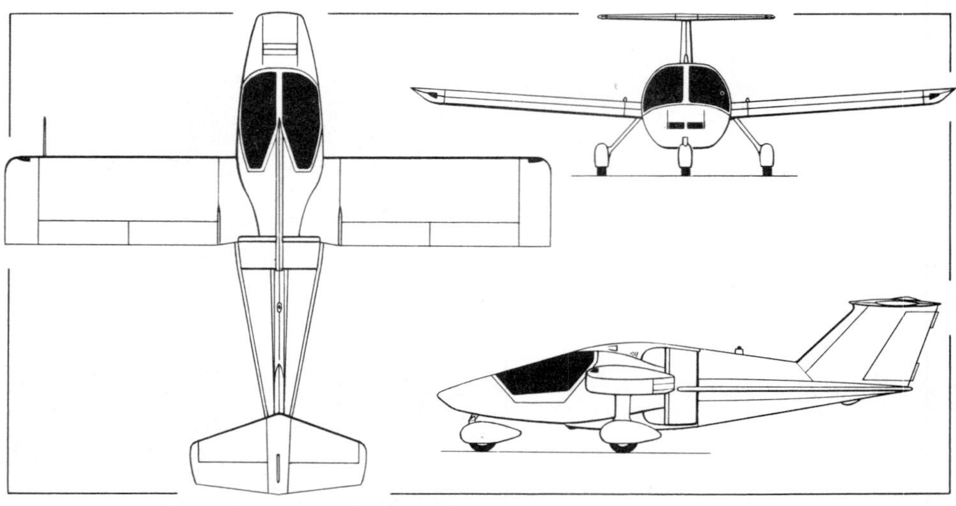
RFB/Grumman American Fanliner two-seat light aircraft *(Pilot Press)*

RFB/GRUMMAN AMERICAN FANLINER

Announced in April 1974, the Fanliner is a two-seat lightweight aircraft developed jointly by RFB of Germany and Grumman American Aviation of the USA. Evolving from RFB's studies of ducted fans, the prototype (D-EJFL) has such a propulsion system and is powered by an NSU Wankel-type engine of 85 kW (114 hp). Airframe construction benefits from adhesive bonding experience with the Grumman American Trainer and Traveler, and a number of components are common between these aircraft and the Fanliner.

Flight testing of the prototype has shown that the ducted fan propulsion system offers a more efficient utilisation of engine power than does a conventional propeller installation. In addition, the rear-mounted engine with central, ducted pusher propeller provides improved visibility, lower cabin noise level, more convenient access for pilot and passenger and, with the propeller shielded by a duct, a reduced ground hazard.

Fanliner flight tests began on 8 October 1973, and both companies are enthusiastic about the results achieved. No decision regarding production had been made by early 1976.

TYPE: Two-seat lightweight experimental aircraft.

WINGS: Cantilever mid-wing monoplane of light alloy construction, similar to those of Grumman American Traveler. Dihedral from roots. No sweep. Constant chord. Ailerons and trailing-edge flaps of light alloy bonded construction. No tabs.

FUSELAGE: Semi-monocoque forward structure of light alloy, comprising nose, cockpit and enclosed engine mounting. Aft fuselage, to carry tail unit, consists of a narrow-section structure continuing the upper and lower lines of the forward fuselage, with a bracing beam extending from the trailing-edge of the wing centre-section to the tail unit on each side.

TAIL UNIT: T-tail of light alloy construction, with swept vertical surfaces. Rudder and elevators of light alloy bonded construction.

LANDING GEAR: Non-retractable tricycle type. Cantilever main-gear legs. Single wheel and speed fairing on each unit.

POWER PLANT: One 85 kW (114 hp) Audi NSU Ro 135 Wankel-type two-chamber rotating-piston engine, driving a pusher propeller, with three plastics blades, mounted within an annular duct.

ACCOMMODATION: Two seats side by side in enclosed cockpit. Individual upward-opening transparent cockpit canopies, hinged on centreline. Dual controls standard. Baggage space aft of seats. Provision for coat locker in nose.

DIMENSIONS, EXTERNAL:

Wing span	7·45 m (24 ft 5¼ in)
Wing aspect ratio	6·0
Length overall	6·10 m (20 ft 0 in)
Height overall	2·03 m (6 ft 8 in)

AREA:

Wings, gross	9·30 m² (100·1 sq ft)

WEIGHTS AND LOADINGS:

Weight empty	520 kg (1,146 lb)
Max T-O weight	750 kg (1,653 lb)
Max wing loading	80·64 kg/m² (16·5 lb/sq ft)
Max power loading	8·82 kg/kW (14·50 lb/hp)

PERFORMANCE (at max T-O weight):

Max level speed at S/L	119 knots (220 km/h; 137 mph)
Max cruising speed at S/L	97 knots (180 km/h; 112 mph)
Max rate of climb at S/L	198 m (650 ft)/min
Range	356 nm (660 km; 410 miles)

RFB FANTRAINER AWI 2

This tandem two-seat training aircraft was first projected in 1970, at which time a model was exhibited at the Hanover Air Show. It utilises a ducted fan propulsion

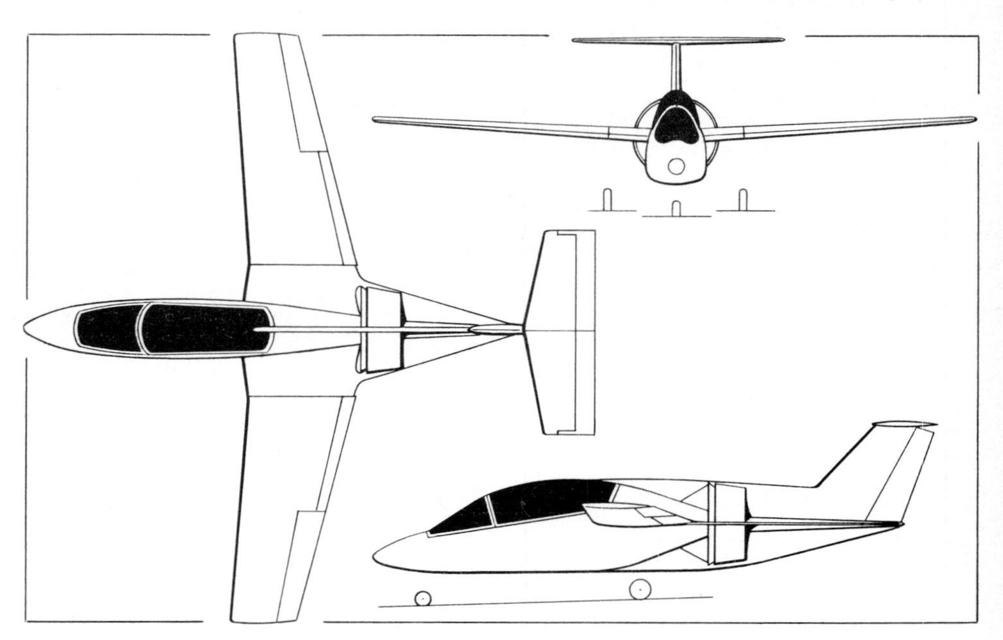
RFB Fantrainer AWI 2 tandem two-seat training aircraft *(Pilot Press)*

system which, in the original concept, comprised a Dowty variable-pitch fan, integral with the rear fuselage and driven by a 224 kW (300 hp) Wankel four-disc rotary engine.

Since the Fantrainer project was initiated, RFB has gained experience of ducted fan propulsion with the Sirius I and II powered sailplanes and the Fanliner two-seat light aircraft.

In March 1975 it was announced that the Federal German Defence Ministry had awarded the company a contract valued at DM 7·5 million to develop and build two Fantrainer prototypes. These will be evaluated as potential replacements for the Piaggio P.149D primary trainers now used by the Luftwaffe. It is suggested that pupils might be able to make direct transition from the Fantrainer to the Alpha Jet. If flight testing proves the practicability of such a scheme, it is possible that about 100 Fantrainers might be put into service from 1979.

The power plant specified for the prototypes comprises two 112 kW (150 hp) Wankel engines, driving a variable-pitch fan through a reduction gearbox, with airbrakes in the fan shroud. Later versions with turboshaft engines of up to 597 kW (800 shp) are under consideration.

Dowty Rotol in the UK announced on 13 January 1976 the receipt of a £30,000 order from RFB for the supply of variable-pitch ducted propulsor fans for the Fantrainer programme.

The general appearance of the AWI 2 Fantrainer can be seen in the accompanying illustration. The first prototype is expected to make its first flight in 1977.

DIMENSIONS, EXTERNAL:

Wing span	9·60 m (31 ft 6 in)
Wing chord at tip	0·55 m (1 ft 9¾ in)
Wing aspect ratio	6·65
Wing sweepforward at 25% chord	6°
Length overall	8·02 m (26 ft 3½ in)
Height overall	2·90 m (9 ft 6 in)

AREA:

Wings, gross	13·90 m² (149·6 sq ft)

WEIGHTS AND LOADING:

Weight empty, equipped	915 kg (2,017 lb)
Max T-O weight	1,350 kg (2,975 lb)
Max wing loading	97 kg/m² (19·9 lb/sq ft)

PERFORMANCE (estimated):

Max cruising speed	172 knots (320 km/h; 199 mph)
Range with max fuel	700 nm (1,300 km; 805 miles)

SPORTAVIA

SPORTAVIA-PÜTZER GmbH u Co KG

HEAD OFFICE AND WORKS:
D-5377 Dahlem-Schmidtheim, Flugplatz Dahlemer Binz

Telephone: (02447) 277/8

Telex: 08 33 602

SALES MANAGER: Alfons Pützer

This company was formed in 1966 by Comte Antoine d'Assche, director of the French company Alpavia SA, and Herr Alfons Pützer, to take over from Alpavia manufacture of the Avion-Planeur series of light aircraft designed by M René Fournier.

In 1969, RFB (which see), a subsidiary of VFW-Fokker GmbH, acquired a 50% holding in Sportavia.

Recently, the company has designed and built two prototypes of a lightweight 2+2-seat sporting aircraft designated RF6.

SPORTAVIA FOURNIER RF6 SPORTSMAN

M René Fournier began design of the RF6 in December 1970. Construction of the first of two prototypes started fourteen months later, and this aircraft made its first flight on 1 March 1973, powered by a 93 kW (125 hp) Lycoming O-235 engine.

A generally similar but somewhat smaller version of this aircraft, powered by a 74·5 kW (100 hp) Rolls-Royce Continental engine and with the designation RF-6B Club, has been built by Avions Fournier in France (which see).

The description which follows applies to the pre-production version of the RF6 (see Addenda):

TYPE: 2+2-seat lightweight sporting aircraft.

WINGS: Cantilever low-wing monoplane. Wing section NACA 23015 at root, NACA 23012 at tip. Dihedral 4°. All-wood single-spar structure, with plywood and fabric covering. Fabric-covered wooden trailing-edge flaps and Frise-type ailerons. No trim tabs.

FUSELAGE: All-wood oval structure, plywood-covered.

TAIL UNIT: Cantilever wooden structure. Fixed-incidence tailplane. Flettner tab in port elevator.

LANDING GEAR: Non-retractable tricycle type. Main units have oleo-pneumatic shock-absorption. Nosewheel carried on strut of E 6150 steel tube, with large free-swivelling fork. Single wheel on each unit. Cleveland single-disc hydraulic brakes. Optional wheel fairings.

POWER PLANT: One 112 kW (150 hp) Lycoming O-320-A2B flat-four engine, driving a Hoffmann HO-V-72 variable-pitch two-blade metal propeller, or HO 14-178-130 constant-speed propeller, both with spinner. Fuel tank in each wing, capacity 45 litres (9·75 Imp gallons). Total fuel capacity 90 litres (19·5 Imp gallons). Refuelling points on wing upper surface. Oil capacity 6·5 litres (1·5 Imp gallons).

ACCOMMODATION: Two front seats side by side under transparent bubble canopy, with two 'occasional' rear seats. Cabin heated and ventilated.

SYSTEMS: Hydraulic system for brakes only. Electrical power provided by 12V 60A engine-driven generator.

ELECTRONICS AND EQUIPMENT: Radio, radio-navigation equipment and full blind-flying instrumentation available to customer's requirements.

DIMENSIONS, EXTERNAL:

Wing span	10·30 m (33 ft 9½ in)
Wing chord at root	1·52 m (5 ft 0 in)
Wing chord at tip	0·83 m (2 ft 8¾ in)
Wing aspect ratio	8·44
Length overall	7·02 m (23 ft 0¼ in)
Height overall	2·36 m (7 ft 8¾ in)
Tailplane span	3·26 m (10 ft 8½ in)
Propeller diameter	1·78 m (5 ft 10 in)
Propeller ground clearance	0·25 m (9¾ in)

AREAS:

Wings, gross	12·57 m² (135·3 sq ft)
Ailerons (total)	1·16 m² (12·49 sq ft)
Trailing-edge flaps (total)	1·54 m² (16·58 sq ft)
Fin	1·05 m² (11·30 sq ft)
Rudder	0·35 m² (3·77 sq ft)
Horizontal surfaces, incl tab	2·64 m² (28·42 sq ft)

Sportavia Fournier RF6 Sportsman (Lycoming O-320-A2B engine)

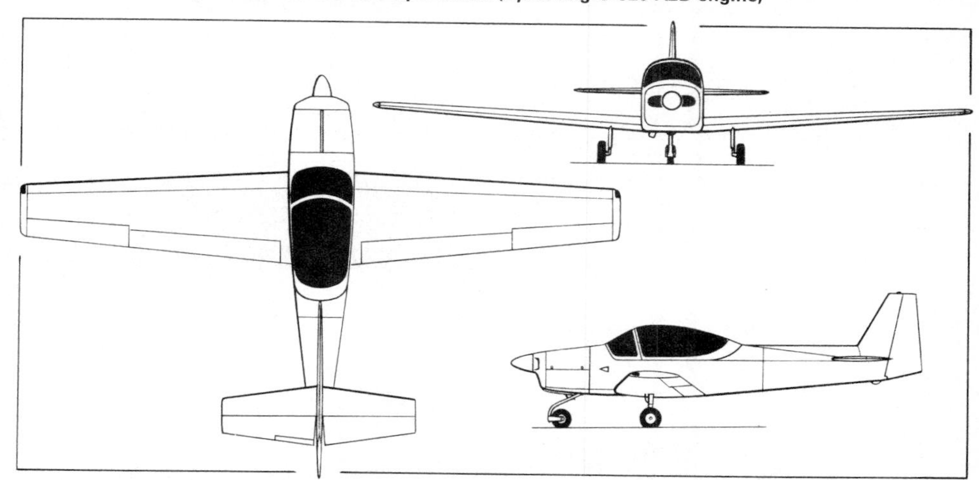

Sportavia Fournier RF6 Sportsman 2 + 2-seat lightweight sporting aircraft *(Pilot Press)*

WEIGHTS AND LOADINGS:

Weight empty	530 kg (1,168 lb)
Max T-O weight	900 kg (1,984 lb)
Max wing loading	71·6 kg/m² (14·7 lb/sq ft)
Max power loading	8·04 kg/kW (13·23 lb/hp)

PERFORMANCE (at max T-O weight):

Max level speed at S/L	135 knots (250 km/h; 155 mph)
Max cruising speed at S/L	108 knots (200 km/h; 124 mph)
Landing speed	41 knots (76 km/h; 47 mph)
Max rate of climb at S/L	228 m (748 ft)/min
Service ceiling	4,000 m (13,125 ft)
T-O and landing run	200 m (656 ft)
T-O to 15 m (50 ft)	480 m (1,575 ft)
Landing from 15 m (50 ft)	360 m (1,180 ft)
Range	453 nm (840 km; 522 miles)

TAIFUN
TAIFUN FLUGZEUGBAU GmbH

Taifun Flugzeugbau planned to put into production an improved version of the Messerschmitt Bf (Me) 108 Taifun, as reported in the 1975-76 *Jane's*. This project has been abandoned.

VFW-FOKKER
VEREINIGTE FLUGTECHNISCHE WERKE-FOKKER GmbH (Subsidiary of ZENTRALGESELLSCHAFT VFW-FOKKER mbH)

HEAD OFFICE:
Hünefeldstrasse 1-5, 2800 Bremen 1 (Postfach 1206)
Telephone: (0421) 538-1
Telex: 245 821

WORKS:
Bremen, Einswarden, Hoykenkamp, Lemwerder, Speyer and Varel

EXECUTIVE DIRECTORS:
Dr Fritz Wenck (Chairman)
Dipl-Volksw A. Ackermann
Prof Dr Rolf Riccius
RA. W. Schaarschmidt
Dipl-Ing Joh. Schäffler

PUBLIC RELATIONS MANAGER: Joseph Grendel

The Vereinigte Flugtechnische Werke GmbH (VFW) was formed at the end of 1963 by a merger of the two Bremen-based aircraft companies of Focke-Wulf GmbH and 'Weser' Flugzeugbau GmbH. They were joined in 1964 by Ernst Heinkel Flugzeugbau GmbH. Details of the history and products of these former companies can be found in earlier editions of *Jane's*.

In 1968-69, VFW acquired 65% of the shares of RFB (Rhein-Flugzeugbau GmbH, which see) but has since become 100% shareholder and has also a 50% holding in Henschel Flugzeugwerke AG of Kassel.

With effect from 1 January 1970, VFW became an equal partner with Fokker of the Netherlands in a company known as Zentralgesellschaft VFW-Fokker mbH, with headquarters in Düsseldorf. The two partners continue to operate independently, as subsidiaries of Zentralgesellschaft VFW-Fokker. The name of the VFW company was changed to VFW-Fokker GmbH as a consequence of the amalgamation.

Current activities of VFW-Fokker include development and production of the VFW 614 twin-turbofan light passenger and freight transport. In addition, VFW is a partner in the Fokker-VFW F28 and Transall C-160 transport programmes.

Other work includes the overhaul and repair of Transall C-160, JetStar, CH-53G and Sea King aircraft, together with modification work on the F-104G.

Through its wholly-owned subsidiary, ERNO Raumfahrttechnik GmbH, VFW-Fokker is participating in the Spacelab programme as prime contractor.

VFW-Fokker is collaborating also in development and production of the Panavia Tornado multi-role combat aircraft and the Airbus A300.

VFW-FOKKER VFW 614

Development of this twin-turbofan short-haul jet transport was undertaken with the financial backing of the Federal German government. Construction of the first of three prototypes started on 1 August 1968, and this aircraft flew for the first time on 14 July 1971.

VFW-Fokker gained LBA certification of the VFW 614 on 23 August 1974 and FAA certification on 4 December 1975.

Two airframes were built for static and dynamic fatigue tests. The former were completed successfully in 1973, the latter in 1975, after 150,000 simulated flight cycles.

By 1 February 1976 VFW-Fokker had received firm orders for 16 aircraft, for Touraine Air Transport (8), Air Alsace (3), Cimber Air (2) and the Federal German government (3). The first production aircraft flew on 28 April 1975 and entered service with Cimber Air in November 1975. Four additional aircraft had been delivered by March 1976.

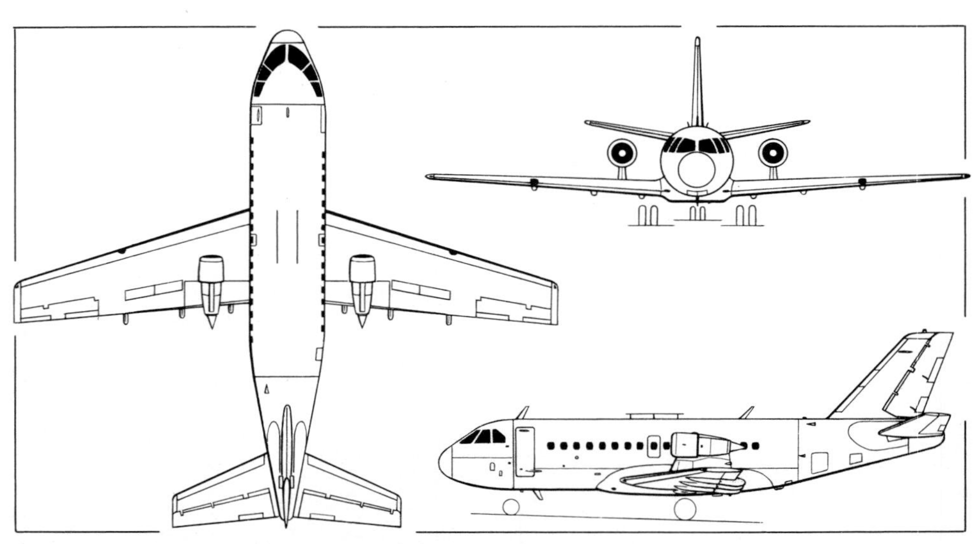

VFW 614 short-haul transport (two Rolls-Royce/SNECMA M45H turbofan engines) *(Pilot Press)*

Manufacture of the VFW 614 is a collaborative venture under the leadership of VFW-Fokker, with participation in the development and production programme by MBB in Germany, Fokker-VFW in the Netherlands, and SABCA and Fairey in Belgium.

TYPE: Twin-turbofan short-haul transport.

WINGS: Cantilever low-wing monoplane. Wing section NACA 63_2A-015 at root, NACA 65_1A-012 at tip. Dihedral 3°. Incidence 3° at root, −1° at tip. Sweepback 15° at quarter-chord. Continuous two-spar fail-safe torsion-box dural structure, consisting of a centre-section integral with the fuselage, and two outer wings. Manually-operated Flettner-type bonded duralumin ailerons, with trim tab of honeycomb construction in port aileron. Hydraulically-operated single-slotted Fowler-type trailing-edge flaps of bonded dural construction. Two split-type flight spoilers and eight ground spoilers, of bonded dural construction. TKS liquid de-icing system.

FUSELAGE: Conventional semi-monocoque fail-safe pressurised structure of circular section, built of high-strength aluminium alloys.

TAIL UNIT: Part-bonded all-swept all-metal structure, with variable-incidence dihedral tailplane. Rudder actuated manually via spring tab with hydraulic boost. Elevators actuated mechanically by two boosters. Elevators may also be actuated manually, and to reduce pilot work load in this mode each elevator has a geared tab. Trim tab in rudder. Electrically-operated tailplane trim, with mechanical backup system. TKS liquid de-icing system for leading-edge of tailplane only.

LANDING GEAR: Hydraulically retractable tricycle type of Dowty Rotol design, with twin wheels on each unit. Steerable nosewheel unit retracts forward. Main units retract inward into fuselage. Oleo-pneumatic telescopic shock-absorbers. Goodyear tyres size 24 × 7·7 on nosewheels; B. F. Goodrich tyres size 34 × 12-12 on main wheels. Tyre pressure 4·69 bars (68 lb/sq in) on nose unit, 5·17 bars (75 lb/sq in) on main units. Goodrich 10 × 5 disc brakes on main units. Messier anti-skid units.

POWER PLANT: Two Rolls-Royce/SNECMA M45H Mk 501 turbofan engines, each rated at a nominal thrust (without losses) of 32·4 kN (7,280 lb), mounted on overwing pylons aft of the wing rear spar. Fuel in integral wing tanks with total capacity of 6,320 litres (1,390 Imp gallons). Single-point pressure refuelling point in starboard outer wing. Provision for overwing gravity refuelling.

ACCOMMODATION: Crew of two on flight deck. Full dual controls and instruments. Standard version provides 40 passenger seats (alternative for 44 passengers) at 810-840 mm (32-33 in) pitch in main cabin, in rows of four with 460 mm (18·1 in) wide centre aisle. Passenger door, with built-in stairs, at front of cabin on port side. Cargo door for cabin baggage compartment at front on starboard side. Catering service door beside pantry at rear of cabin on starboard side. Removable baggage compartment at front of cabin (stbd). Toilet on port side, at rear of cabin. Underfloor baggage holds fore and aft of wing. Cabin pressurised and air-conditioned. Seats for an observer and second cabin attendant optional.

SYSTEMS: Garrett-Normalair double air-cycle air-conditioning system, using engine bleed air. Pressure differential 0·45 bars (6·55 lb/sq in). Two separate hydraulic systems, pressure 207 bars (3,000 lb/sq in), supplied by a pump on each engine and electrically-powered (DC) auxiliary hydraulic pump, with Skydrol 500B fluid, for nosewheel steering, landing gear, flight and ground spoilers, flaps, boosters and brakes. Constant-frequency three-phase 200/115V 400Hz AC electrical system. Two transformer-rectifiers provide 28V DC power, with batteries for emergency supply. Priority oxygen supply for flight crew, with second oxygen system for passengers and cabin staff. TKS liquid de-icing system for wings and tailplane. Garrett-AiResearch GTCP 36-28 APU, mounted in rear fuselage, provides air supply for engine starting and

VFW-Fokker VFW 614 short-haul transport in the insignia of Touraine Air Transport

air-conditioning on ground, and electrical power for pre-flight check of electronics and systems without running engines.

ELECTRONICS AND EQUIPMENT: Standard equipment includes two VOR/ILS, marker, ADF, two compass systems, flight data recorder, autopilot/flight director, weather radar, stall warning system, audio system, flight director indicator, attitude reference system, air data unit and voice recorder. Optional items include an additional ADF, HF communications installation, ATC transponder, DME, radio altimeter, ground proximity warning system, comparator warning system, low weather minima system, digital flight data recorder and altitude alert. Blind-flying instrumentation standard.

DIMENSIONS, EXTERNAL:

Wing span	21·50 m (70 ft 6½ in)
Wing chord at root	4·25 m (13 ft 11¼ in)
Wing chord at tip	1·71 m (5 ft 7¼ in)
Wing aspect ratio	7·22
Length overall	20·60 m (67 ft 7 in)
Length of fuselage	20·15 m (66 ft 1¼ in)
Height overall	7·84 m (25 ft 8 in)
Tailplane span	9·00 m (29 ft 6¼ in)
Wheel track	3·90 m (12 ft 9½ in)
Wheelbase	7·02 m (23 ft 0¼ in)
Passenger door (fwd, port):	
Height	2·10 m (6 ft 10¾ in)
Width	0·75 m (2 ft 5½ in)
Height to sill	1·65 m (5 ft 5 in)
Freight door (fwd, stbd):	
Height	1·17 m (3 ft 10 in)
Width	1·10 m (3 ft 7¼ in)
Height to sill	1·99 m (6 ft 6¼ in)
Catering service door (rear of cabin, stbd):	
Height	1·53 m (5 ft 0¼ in)
Width	0·61 m (2 ft 0 in)
Height to sill	1·96 m (6 ft 5¼ in)
Underfloor baggage door (fwd):	
Height	0·78 m (2 ft 6¾ in)
Width	0·85 m (2 ft 9½ in)
Height to sill	1·28 m (4 ft 2½ in)
Underfloor baggage door (rear):	
Height	0·78 m (2 ft 6¾ in)
Width	0·85 m (2 ft 9½ in)
Height to sill	1·25 m (4 ft 1½ in)

DIMENSIONS, INTERNAL:

Cabin, excl flight deck:	
Length	11·21 m (36 ft 9¼ in)
Max width	2·66 m (8 ft 8¾ in)
Max height	1·92 m (6 ft 3½ in)
Floor area	23·11 m² (248·75 sq ft)
Volume	49·50 m³ (1,748 cu ft)
Baggage compartment volume:	
Front fuselage	2·25 m³ (79·5 cu ft)
Underfloor (fwd)	1·79 m³ (63·2 cu ft)
Underfloor (aft)	1·45 m³ (51·2 cu ft)

AREAS:

Wings, gross	64·00 m² (688·89 sq ft)
Ailerons (total)	3·24 m² (34·87 sq ft)
Trailing-edge flaps (total)	11·60 m² (124·86 sq ft)
Spoilers (total)	5·05 m² (54·36 sq ft)
Fin	5·86 m² (63·08 sq ft)
Rudder	3·59 m² (38·64 sq ft)
Tailplane	13·08 m² (140·79 sq ft)
Elevators	4·92 m² (52·96 sq ft)

WEIGHTS AND LOADING:

Operating weight, empty	12,180 kg (26,850 lb)
Max design zero-fuel weight	16,600 kg (36,600 lb)
Max design taxi weight	20,050 kg (44,200 lb)
Max design T-O and landing weight	19,950 kg (44,000 lb)
Max wing loading	312 kg/m² (64 lb/sq ft)

PERFORMANCE (at max design T-O weight, ISA, except where indicated otherwise):

Max level speed at 6,700 m (21,975 ft), at AUW of 18,200 kg (40,124 lb)
385 knots (713 km/h; 443 mph)

Max cruising speed at 7,620 m (25,000 ft), at AUW of 18,200 kg (40,124 lb)
380 knots (704 km/h; 438 mph)

Stalling speed, flaps up
111·5 knots (207 km/h; 128·5 mph)

Stalling speed, flaps down
85 knots (158 km/h; 98 mph)

Max rate of climb at S/L	945 m (3,100 ft)/min
Max certificated flight altitude	7,620 m (25,000 ft)
Service ceiling, one engine out, at AUW of 18,200 kg (40,124 lb)	4,260 m (13,975 ft)
Runway LCN	16
FAA T-O field length	1,325 m (4,350 ft)
FAA landing field length	1,036 m (3,400 ft)

Range with 40 passengers or 3,625 kg (7,992 lb) payload at 7,620 m (25,000 ft) with Mach 0·63 climb and descent, at 250 knots (463 km/h; 288 mph) CAS with reserves for 150 nm (278 km; 173 mile) alternate and 45 min hold
650 nm (1,204 km; 748 miles)

OPERATIONAL NOISE CHARACTERISTICS (FAR Pt 36):

T-O noise level	90·5 EPNdB
Approach noise level on 3° glideslope	97·1 EPNdB
Sideline noise level	92·2 EPNdB

WILDEN
ING HELMUT WILDEN

ADDRESS: 5201 Kümpeler Hof, Hennef-Sieg
Telephone: 02242-1086

Ing Helmut Wilden has designed and built the prototype of an ultra-lightweight sporting aircraft, which he has designated VoWi 10. It consists essentially of aerofoil surfaces linked together by light alloy tube, resulting in an empty weight of only 70 kg (154 lb) for an aircraft with a wing span of 10·40 m (34 ft 1½ in). Its design originated in 1974, and the prototype, powered by two 6 kW (8 hp) Stihl-Baumsägen single-cylinder two-stroke aircooled engines, flew for the first time on 16 April 1975. Since that time the aircraft has been further developed and is currently powered by a single two-cylinder two-stroke engine. The simplicity of the aircraft's construction makes it possible to market it at a realistic price, with the result that Ing Wilden had orders for more than 50 aircraft in the Spring of 1976. To meet this demand and future orders, he has come to an arrangement under which Start + Flug GmbH is manufacturing the wings and tail unit under subcontract. Final assembly of the complete aircraft, testing and delivery will be made from Bonn-Hangelar airfield.

An earlier design by Ing Wilden, the VoWi 8, was described and illustrated in the 1974-75 *Jane's*.

Wilden VoWi 10 ultra-light aircraft (Lloyd engine) *(Flug Revue Flugwelt)*

WILDEN VoWi 10

TYPE: Ultra-lightweight single-seat sporting aircraft.

WINGS: Wire-braced monoplane. Wing section Göttingen 535. Dihedral 7°. No sweepback. All-moving wing, without ailerons or trailing-edge flaps. Each panel has a leading-edge torsion-box structure of glassfibre and Conticell foam sandwich, with the main part of the wing between the leading-edge and trailing-edge fabric-covered. Wings wire-braced to kingpost above the wing and to landing gear.

FUSELAGE: Light alloy tubular frame, uncovered and consisting of one tube extending from forward of wing to tail unit, a kingpost above the wing, and an A-frame beneath the wing to carry the main landing gear. Tubes extend aft from the extremities of the A-frame to form the mounting for the pilot's seat, and continue behind his back to an attachment on the main fuselage tube at approximately three-quarters of wing chord.

TAIL UNIT: Cantilever Vee-tail, comprising twin fins and rudders, of similar construction to wings.

LANDING GEAR: Non-retractable tailwheel type, with single wheel on each unit. Tailwheel carried on lightweight tube extending below, and attached to, fuselage at leading-edge of fins.

POWER PLANT: Production aircraft have one 15 kW (20 hp) Lloyd two-cylinder two-stroke aircooled engine, mounted at forward end of fuselage tube and driving a two-blade wooden fixed-pitch propeller. Fuel tank contains 10 kg (22 lb) of fuel.

ACCOMMODATION: Single open seat, beneath the wing, mounted on tubular framework extending aft from landing gear and attached to fuselage tube.

DIMENSIONS, EXTERNAL:

Wing span	10·40 m (34 ft 1½ in)
Wing aspect ratio	8·51
Length overall	6·00 m (19 ft 8¼ in)
Height overall	2·92 m (9 ft 7 in)

AREA:

Wings, gross	12·70 m² (136·7 sq ft)

WEIGHTS AND LOADINGS:

Weight empty	70 kg (154 lb)
Max T-O weight	170 kg (375 lb)
Max wing loading	13·4 kg/m² (2·74 lb/sq ft)
Max power loading	11·3 kg/kW (18·75 lb/hp)

PERFORMANCE (at max T-O weight):

Max level speed at S/L	46 knots (85 km/h; 53 mph)
Max cruising speed at S/L	35 knots (65 km/h; 40 mph)
Max rate of climb at S/L	180 m (590 ft)/min
T-O run	40 m (131 ft)
T-O to 15 m (50 ft)	105 m (345 ft)
Landing run	30 m (98 ft)
Max range	108 nm (200 km; 124 miles)

GREECE

HAI
HELLENIC AEROSPACE INDUSTRIES

Steps to establish a national 'aerospace support facility' in Greece were first taken by the Greek government in late 1971, following which various proposals by teams of foreign aerospace companies were evaluated, as recorded in the 1973-74 and subsequent editions of *Jane's*.

A bill approving the establishment of such a facility was passed, by the government then in power, in the Summer of 1975; and on 26 November 1975 contracts worth $120 million were signed in Athens between the Greek government and the four US companies which are to design, build and operate the facility: Lockheed Aircraft Corporation, the Austin Company of New York, Westinghouse Electric Corporation and General Electric.

Lockheed Aircraft International will have overall responsibility for managing the centre, and for all airframe work; personnel will be supplied initially by Lockheed Aircraft Service Co. Austin will be architect and general contractor for the plant. Westinghouse will be responsible for the workshops, and overhaul and repair of electronics, and General Electric for aero-engine overhaul and maintenance.

The new facility, known as Hellenic Aerospace Industries (HAI), will be built on a 71 hectare (175 acre) site adjacent to the present Hellenic Air Force base at Tanagra, about 80 km (50 miles) north of Athens, and will take 2½ years to complete. Hangars, workshops and offices will occupy some 83,613 m² (900,000 sq ft). The facility will provide depot level maintenance for the Hellenic Air Force, maintenance and manufacturing for airlines, and will eventually employ about 3,000 people.

INDIA

CIVIL AVIATION DEPARTMENT

TECHNICAL CENTRE, CIVIL AVIATION DEPARTMENT

HEAD OFFICE:
Civil Aviation Department, R. K. Puram, New Delhi 22

WORKS:
Technical Centre, opposite Safdarjung Airport, New Delhi 110003

Telephone: 611504

DIRECTOR GENERAL: S. Ramamritham

DIRECTOR (R&D): K. B. Ganesan

In addition to its work on the design and development of sailplanes, described in the appropriate section of this edition, the Civil Aviation Department has designed and developed a two/three-seat light aircraft named the Revathi.

REVATHI Mk II

The prototype Revathi Mk I (see 1969-70 *Jane's*) flew for the first time on 13 January 1967, and was type certificated in January 1969.

It was subsequently developed into the Revathi Mk II, with constant-chord metal wings, increased fuel capacity and AUW, and other changes. In this form it flew for the first time on 20 May 1970, and was type certificated on 31 October 1972.

Extensive flight tests, including the successful execution of loops, spins, slow rolls and stall turns, were carried out by the Aircraft and Systems Testing Establishment of the Indian Air Force.

After the incorporation of modifications, the Revathi Mk II resumed flying in April 1974. These modifications included the replacement of the original wooden flaps and ailerons by metal components. An all-metal horizontal tail has been designed and built to replace the former wooden one. A redesigned instrument panel is being developed.

The Revathi Mk II is intended for use by flying clubs, as a basic trainer for ab initio pilot training, including instruction in spinning, night flying, instrument flying and cross-country navigation; it is also suitable for use as a private aircraft. As a trainer, it is designed to meet the Utility category requirements of FAR Pt 23 Appendix A, with pupil and instructor seated side by side. As a private aircraft it meets the Normal category requirements and can seat three people.

The following description applies to the Revathi Mk II prototype with the modifications referred to above:

TYPE: Two/three-seat light aircraft.

WINGS: Cantilever low-wing monoplane. Wing section NACA 23015 at root and up to 60·6% of each half-span; NACA 4412 at tip. Dihedral 5°. Incidence 4°. Two-spar stressed-skin structure. The nose cell and, at root, the rear cell also, are covered with aluminium alloy sheet; the remainder of the wing is fabric-covered. Ailerons and slotted trailing-edge flaps of aluminium alloy.

FUSELAGE: Welded steel tube truss structure, covered with sheet aluminium alloy to rear of cockpit and fabric elsewhere.

TAIL UNIT: Cantilever horizontal surfaces of aluminium alloy; rear cells of elevators fabric-covered. Sweptback vertical surfaces of steel tube construction, covered with fabric. Fin integral with fuselage. Trim tab on rudder and in starboard elevator. Elevator incorporates shielded horn balance; rudder has an unshielded horn balance.

Revathi Mk II two/three-seat light aircraft, developed by the Technical Centre of the Indian Civil Aviation Department

LANDING GEAR: Non-retractable tailwheel type. Rubber rings in tension provide shock-absorption of main units. Dunlop main wheels, size 6·00-6·5, pressure 1·59-2·41 bars (23-35 lb/sq in). Dunlop hydraulic disc brakes. Castoring tailwheel, with solid tyre, carried on leaf springs, coupled flexibly to rudder.

POWER PLANT: One 108 kW (145 hp) Rolls-Royce Continental O-300-C flat-six engine, driving a Sensenich M74DC54 two-blade fixed-pitch metal propeller. Integral fuel tank in each wing, total capacity 148 litres (32·6 Imp gallons). Auxiliary tank, capacity 50 litres (11 Imp gallons), aft of cockpit. Oil capacity 7·5 litres (1·6 Imp gallons).

ACCOMMODATION: Enclosed cabin, seating two persons in front, side by side, and one to the rear. Dual controls standard, including duplicated wheel brake and throttle controls. Jettisonable door on each side. Compartment for baggage.

ELECTRONICS AND EQUIPMENT: Blind- and night-flying instrumentation and Bendix RT-221A-14 380-channel VHF transceiver standard. Bendix RN-222 VHF navigation receiver (with glideslope supplement), IN-224 VOR/ILS and 204A marker receiver optional. Landing light in each wing leading-edge.

DIMENSIONS, EXTERNAL:

Wing span	9·40 m (30 ft 10 in)
Wing chord (constant)	1·50 m (4 ft 11 in)
Wing aspect ratio	6·27
Length overall	7·58 m (24 ft 10 in)
Height overall (tail up)	2·97 m (9 ft 8¾ in)
Tailplane span	2·74 m (9 ft 0 in)
Wheel track	1·96 m (6 ft 5 in)
Wheelbase	5·33 m (17 ft 6 in)
Propeller diameter	1·88 m (6 ft 2 in)
Propeller ground clearance	0·24 m (9½ in)
Cabin doors (each):	
Height	1·22 m (4 ft 0 in)
Width	0·88 m (2 ft 10½ in)

DIMENSIONS, INTERNAL:

Cabin: Length	2·20 m (7 ft 2½ in)
Max width	0·95 m (3 ft 1½ in)
Max height	1·21 m (3 ft 11¾ in)
Baggage compartment volume	0·085 m³ (3 cu ft)

AREAS:

Wings, gross	14·09 m² (151·7 sq ft)
Ailerons (total)	1·16 m² (12·44 sq ft)
Trailing-edge flaps (total)	1·35 m² (14·52 sq ft)
Fin	0·084 m² (8·65 sq ft)
Rudder, incl tab	0·78 m² (8·40 sq ft)
Tailplane	1·22 m² (13·10 sq ft)
Elevators, incl tab	1·29 m² (13·90 sq ft)

WEIGHTS AND LOADINGS:

Weight empty, equipped	616 kg (1,357 lb)
Max T-O and landing weight:	
Normal	962 kg (2,120 lb)
Utility	831 kg (1,832 lb)
Max wing loading:	
Normal	68·2 kg/m² (13·98 lb/sq ft)
Utility	58·95 kg/m² (12·08 lb/sq ft)
Max power loading:	
Normal	8·91 kg/kW (14·62 lb/hp)
Utility	7·69 kg/kW (12·61 lb/hp)

PERFORMANCE (at max T-O weight):

Never-exceed speed	140 knots (260 km/h; 161 mph)
Max level speed at S/L	104 knots (193 km/h; 120 mph)
Max cruising speed (75% power) at 1,830 m (6,000 ft)	91 knots (169 km/h; 105 mph)
Stalling speed, flaps down:	
Normal	51·5 knots (95 km/h; 59 mph)
Utility	48 knots (89 km/h; 55 mph)
Max rate of climb at S/L	182 m (600 ft)/min
Service ceiling	3,505 m (11,500 ft)
T-O run	228 m (750 ft)
Landing run	201 m (660 ft)
Range with wing fuel only, no reserves	347 nm (643 km; 400 miles)
Range with max fuel (incl auxiliary tank), no reserves	434 nm (804 km; 500 miles)

HAL

HINDUSTAN AERONAUTICS LIMITED

ADDRESS:
Indian Express Building, Vidhana Veedhi, PO Box 5150, Bangalore 560 001

Telephone: 75004/5/6, 75034/35/36 and 75053/54

Telex: 043-266 HAL BG

CHAIRMAN:
Air Marshal S. J. Dastur

DIRECTORS:
Shri K. R. Baliga
Shri P. N. Bhalla
Shri S. K. Bhatnagar
Shri P. D. Chopra
Shri S. C. Das
Air Marshal H. C. Dewan
Shri G. C. Katoch
Dr B. D. Nag Chaudhuri
Air Marshal K. Narasimhan
Dr S. R. Valluri

GENERAL MANAGERS:
Bangalore Complex:
Shri Raj Mahindra (Aircraft Division)
Shri K. K. Kirtikar (Engine Division)
Gp Capt Willie Raj (Overhaul Division)
(*appointment awaited*) (Helicopter Division)
Dr S. Ramamurthy (Foundry and Forge Division)
MiG Complex:
Shri Harsimran Singh (Nasik Division)
Shri G. Narasimhan (Koraput Division)
Gp Capt (Retd) R. S. Sivaswamy (Hyderabad Division)
Kanpur Division:
Shri J. Bhandari
Lucknow Division:
Gp Capt (Retd) B. K. Kapur

Hindustan Aeronautics Limited (HAL) was formed on 1 October 1964, amalgamating the former Hindustan Aircraft Ltd (formed 1940) and Aeronautics India Ltd (formed 1963), and has 10 Divisions, five at Bangalore and one each at Nasik, Koraput, Hyderabad, Kanpur and Lucknow, with a total work force of some 40,045 people on 1 January 1976. The company, whose principal customer is the Indian Air Force, is currently manufacturing and overhauling many types of aircraft, including helicopters, and their related aero-engines, electronics, instruments and accessories, including air-to-air missiles.

The Bangalore Complex is engaged in the manufacture of civil and military aircraft and aero-engines, both under licence and of indigenous design. This Complex also has a large organisation undertaking repair and overhaul of airframes, engines, and allied instruments and accessories.

Kanpur Division has been engaged mainly in the manufacture of different versions of the Hawker Siddeley 748 under licence. It has more recently been assigned the responsibility for series production of the HA-31 Mk II Basant agricultural aircraft, designed at the Bangalore Complex.

Nasik, Koraput and Hyderabad form the MiG Complex, which currently undertakes the licence manufacture of Soviet MiG-21 fighters with the collaboration of the USSR.

Lucknow Division, formed in late 1969, is producing aircraft accessories under licence from various manufacturers in the UK, France and the USA, including brake and other hydraulic equipment, flight instruments, air-conditioning, pressurisation and fuel system equipment and ejection seats. On 30 August 1973 the Division began production of landing gear, ejection seats, fuel systems and instruments for the Marut and Kiran. It will eventually undertake the manufacture of gyro instruments for all HAL aircraft, and of accessories for MiG-21 aircraft in Indian service.

BANGALORE COMPLEX

ADDRESS:
Bangalore-17 (Mysore State)

Telephone: 53201 and 50773

The Bangalore Complex of HAL consists essentially of the former Hindustan Aircraft Limited, the activities of which, since its formation in 1940, were described in previous editions of *Jane's*. The Complex is subdivided into an Aircraft Division, Helicopter Division, Engine Division, Overhaul Division, Foundry and Forge Division and Design Bureau.

Bangalore Complex is engaged in developing and building aircraft and aero-engines of its own design, and also manufactures various aircraft and aero-engines under licence. The Engine Division's activities are described in the appropriate section of this edition.

Licence production of the Aérospatiale SA 315B Lama and SA 316B Alouette III helicopters, undertaken initially by the Aircraft Division, is now undertaken by the Helicopter Division, which was officially inaugurated on 19 July 1974.

The Overhaul Division of Bangalore Complex repairs and overhauls Hawker Siddeley (de Havilland) Dove/Devon aircraft, DHC-4 Caribou, Fairchild C-119 Packet transports and English Electric Canberra bombers. Various piston engines and jet engines are overhauled at the Engine Division. The branch factory in Calcutta is continuing to concentrate on the repair and overhaul of DC-3s belonging to the Indian Air Force and non-scheduled operators.

In collaboration with the National Aeronautical Laboratory, the Design Bureau is working on the design requirements of an all-Indian armed light helicopter, with an AUW in the region of 2,500 kg (5,500 lb) and powered by a turboshaft engine of approx 1,044 kW (1,400 shp). It also designed the HPT-32 piston-engined trainer, which has an indigenously-developed 194 kW (260 hp) six-cylinder engine. Other projects in the design stage include an armed version of the Kiran, the Kiran Mk II, which has a derated HAL-built Orpheus 701 turbojet engine; and a trainer version of the Ajeet.

HAL HF-24 MARUT (WIND SPIRIT)

Development of the HF-24 Marut single-seat fighter was started by HAL in 1956, under the design leadership of Dr Kurt Tank, who was responsible for the wartime Focke-Wulf aeroplanes. The first prototype HF-24 Mk I (constructor's number HF-001; Indian Air Force serial number BR-462), powered by two Rolls-Royce Bristol Orpheus 703 turbojet engines, flew for the first time on 17 June 1961. It was followed by the second Mk I prototype (HF-003; BR-463) on 4 October 1962.

The HF-24 is being manufactured to Mk I standard as a ground attack fighter, with Orpheus 703 non-afterburning engines. The first of 18 pre-production Maruts (HF-004; BD-828) flew in March 1963, and a token delivery of two aircraft to the Indian Air Force was made on 10 May 1964. By 1967, 12 more pre-production Mk Is had been handed over to the IAF, the other four being used for test and development programmes. The latter included one aircraft (HF-005; BD-830), designated Mk IA, for early trials in 1966 with an afterburner fitted to its Orpheus 703 engine.

The first series production Mk I (HF-022; BD-844) flew on 15 November 1967, and this version equips Nos. 10, 31 and 220 Squadrons of the Indian Air Force, which used its Maruts successfully, without loss, in the December 1971 war with Pakistan. A total of 116 Maruts had been built by 1 February 1976.

The first of two prototype **Mk IT** tandem two-seat training versions (HF-046; BD-888 and HF-047; BD-889) began its flight tests on 30 April 1970, in the hands of Wg Cdr R. D. Sahni, then chief test pilot of the Bangalore Division of HAL. It was followed by the second Mk IT in March 1971. Differences by comparison with the Mk I are minimal. The internal Matra rocket launcher is removed to make way for the Martin-Baker Mk 84C second seat; full dual controls are fitted, and a wide choice of equipment enables the Mk IT to be used for several advanced training roles, including dual and solo flying; operational training in the all-weather ground attack role; and instrument flying and armament training. Eight Mk ITs had been delivered to the IAF by 1 February 1976.

Development of the Mk IT, and of other versions of the HF-24 (see 1974-75 *Jane's*), has been the responsibility of an all-Indian design team led by Mr S. C. Das since the departure of Dr Tank and his German team in 1967. The search continues for an engine that could give the Marut its intended Mach 2 performance in a future HF-24 Mk III version.

The following description applies to the current production HF-24 Mk I:

TYPE: Single-seat ground attack fighter, stressed to +9·34*g* (limit).

WINGS: Cantilever low-wing monoplane of thin section. Sweepback approx 45° at quarter-chord. All versions have extended-chord leading-edges on outer panels; in addition, overall wing chord is increased on current production aircraft, from HF-048; BD-1192 onwards. (Details of initial version can be found in the 1975-76 *Jane's*.) Conventional torsion-box structure. Hydraulically-actuated ailerons and trailing-edge flaps, with provision for selecting manual control. No de-icing system.

FUSELAGE: Conventional all-metal semi-monocoque structure, narrowed in accordance with area rule in region of wing trailing-edge. Rear fuselage detaches at transport joint for engine removal. Two hydraulically-operated box-type airbrakes on lower fuselage aft of main-wheel wells, opening downward. Engine air intake, with non-adjustable half-cone centrebody, on each side of cockpit.

TAIL UNIT: Cantilever all-metal structure with sweepback on all surfaces. Hydraulically-operated low-set variable-incidence tailplane with electrical trim facility.

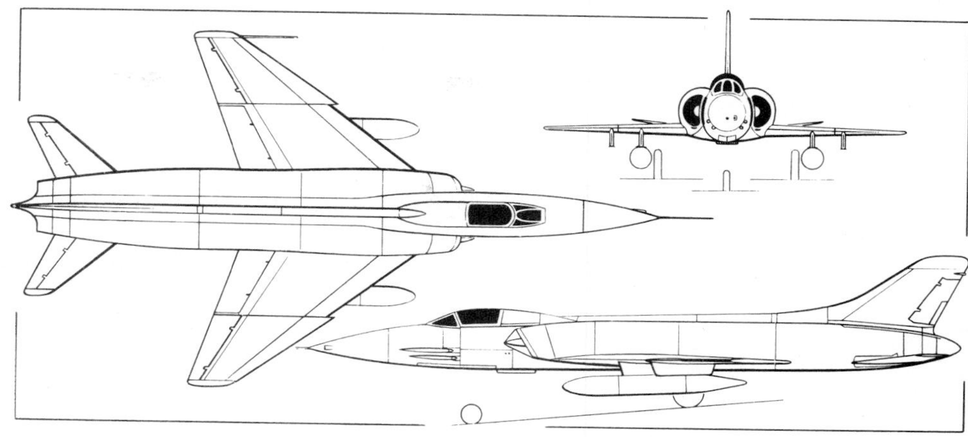

HAL HF-24 Marut Mk I single-seat ground attack fighter *(Pilot Press)*

HAL HF-24 Marut Mk I single-seat fighter, in squadron service with the Indian Air Force

HAL HF-24 Mk IT tandem two-seat training version of the Marut, equipped with underwing fuel tanks

Elevators can be operated either hydraulically or manually. Manually-operated rudder, with trim tab, on early models. Later aircraft, including Mk IT, are fitted with hydraulically-actuated rudder.

LANDING GEAR: Retractable tricycle type, with single Dunlop wheel on each unit, supplied by Dowty Rotol. Hydraulic actuation, nosewheel retracting forward, main units inward into fuselage. Steerable nosewheel. Main-wheel tyres size 29 × 8-15, pressure 6·90 bars (100 lb/sq in). Nosewheel tyre size 19 × 6·25-9, pressure 9·65 bars (140 lb/sq in). Maxaret anti-skid system. No brake cooling. RFD-GQ Type LB-52 Mk 2 ring-slot braking parachute, diameter 3·20 m (10 ft 6 in), located in top of rear fuselage.

POWER PLANT: Two HAL-built Rolls-Royce Bristol Orpheus 703 turbojet engines, each rated at 21·6 kN (4,850 lb st), side by side in rear fuselage. Fuel in main fuselage collector tank, wing centre-section supply tank and two integral wing tanks, with total usable capacity of 2,491 litres (549 Imp gallons). Provision for up to four 454 litre (100 Imp gallon) underwing drop-tanks, and an internal auxiliary tank of 400 litres (88 Imp gallons) capacity in place of the Matra rocket launcher.

ACCOMMODATION: Pilot only, on Martin-Baker Mk 84C zero-altitude ejection seat, under rearward-sliding blister canopy. Windscreen heated by sandwiched gold-film electrode. Side screens and canopy demisted by warm air from air-conditioning system.

SYSTEMS: Air-conditioning system includes two air-cycle heat exchangers and cold air unit. Cockpit is pressurised to differential of 0·24 bars (3·5 lb/sq in) between 7,300 and 12,200 m (24,000 to 40,000 ft). Dowty Rotol hydraulic system, pressure 276 bars (4,000 lb/sq in), supplied by two engine-driven pumps, for all services. Nitrogen system, pressure 207 bars (3,000 lb/sq in), to provide emergency power for landing gear, airbrakes and flaps. 24V DC single-wire earth return electrical system, with two 24V 25Ah batteries and 4Ah emergency supply battery.

ELECTRONICS AND EQUIPMENT: Standard equipment includes DFA 73 D/F, TA and RA Bendix receiver, 12-channel VHF, and Ferranti ISIS (integrated strike and interception system) two-axis rate gyro gunsight.

ARMAMENT: Four 30 mm Aden Mk 2 guns in nose, with 120 rds/gun, and Matra Type 103 retractable pack of 50 SNEB 68 mm air-to-air rockets in lower fuselage aft of nosewheel unit. Attachments for four 1,000 lb bombs, napalm tanks, Type 116 SNEB rocket packs, clusters of T10 air-to-surface rockets, drop-tanks or other stores, under wings.

DIMENSIONS, EXTERNAL:

Wing span	9·00 m (29 ft 6¼ in)
Wing chord at root	4·40 m (14 ft 5¼ in)
Wing chord at tip	1·10 m (3 ft 7¼ in)
Wing aspect ratio	2·90
Length overall	15·87 m (52 ft 0¾ in)
Height overall	3·60 m (11 ft 9¾ in)
Tailplane span	5·104 m (16 ft 9 in)
Wheel track	2·80 m (9 ft 2 in)
Wheelbase	5·555 m (18 ft 2¾ in)

AREAS:

Wings, gross	28·00 m² (301·4 sq ft)
Ailerons (total)	1·254 m² (13·50 sq ft)
Trailing-edge flaps (total)	2·44 m² (26·26 sq ft)
Fin	3·83 m² (41·23 sq ft)
Rudder, incl tab	0·494 m² (5·32 sq ft)
Tailplane	5·596 m² (60·24 sq ft)
Elevators (total)	0·764 m² (8·22 sq ft)

WEIGHTS AND LOADINGS:

Weight empty, equipped:

Mk I with auxiliary ventral tank 6,195 kg (13,658 lb)

Mk IT 6,250 kg (13,778 lb)
T-O weight, 'clean':
Mk I with auxiliary ventral tank 8,951 kg (19,734 lb)
Max T-O weight:
Mk I 10,908 kg (24,048 lb)
Mk IT 10,812 kg (23,836 lb)
Max wing loading:
Mk I 390 kg/m² (79·8 lb/sq ft)
Mk IT 386 kg/m² (79·1 lb/sq ft)
Thrust/weight ratio:
Mk I at T-O weight, 'clean' 0·492

PERFORMANCE:
Max level speed attained at 12,000 m (39,375 ft):
Mk I Mach 1·02
Mk IT Mach 1·00
Max permitted speed at S/L 600 knots (1,112 km/h; 691 mph) IAS
Stalling speed at AUW of 8,951 kg (19,734 lb):
flaps and landing gear up 138 knots (256 km/h; 159 mph)
flaps and landing gear down 133 knots (248 km/h; 154 mph)
Normal landing speed 145 knots (268 km/h; 167 mph)
Time to climb from S/L to 12,200 m (40,000 ft) ('clean' aircraft, ISA + 15°C) 9 min 20 sec
T-O run at S/L 850 m (2,790 ft)
Min ground turning radius 5·22 m (17 ft 2 in)
Radius of action (Mk IT):
at low level 129 nm (238 km; 148 miles)
interception mission at 12,000 m (39,375 ft) 214 nm (396 km; 246 miles)
Ferry range (Mk IT) at 9,145 m (30,000 ft) 780 nm (1,445 km; 898 miles)

HAL HJT-16 Mk I KIRAN (RAY OF LIGHT)

In December 1959, the government of India approved the design and development by HAL of a side-by-side two-seat jet basic trainer designated HJT-16 Mk I, powered by a Rolls-Royce Bristol Viper 11 turbojet.

Detailed design work on the HJT-16 Mk I began in April 1961 under the leadership of Dr V. M. Ghatage. The first prototype flew for the first time on 4 September 1964. It was followed by a second aircraft in August 1965.

A total of 24 pre-production HJT-16 Mk Is were delivered to the Indian Air Force, the initial delivery (of six aircraft) being made in March 1968. The HJT-16 Mk I is now in series production. By 1 February 1976 a total of 94 had been delivered, to meet the requirements of both the Indian Air Force and Indian Navy. Total IAF/IN requirement is for 180 aircraft.

The Kiran is suitable for use in armament training or light attack roles, and the 68th and subsequent Mk Is are fitted with a hardpoint beneath each wing capable of carrying weapons or a drop-tank. Ground firing trials of the HAL gun pods were conducted in 1974. In addition, HAL is developing a Mk II version of the Kiran for these roles, and this is described separately.

TYPE: Two-seat jet basic trainer.

WINGS: Cantilever low-wing monoplane. Wing section NACA 23015 at root, NACA 23012 at tip. Dihedral 4° from roots. Incidence 0° 30′ at root. Conventional all-metal three-spar structure. Frise-type differential ailerons. Hydraulically-actuated trailing-edge split flaps.

FUSELAGE: All-metal semi-monocoque structure of light alloy. Hydraulically-actuated door-type airbrake under centre of fuselage.

TAIL UNIT: Cantilever all-metal structure. Electrically-operated variable-incidence tailplane. Ground-adjustable tab on rudder.

LANDING GEAR: Retractable tricycle type, of HAL manufacture. Hydraulic actuation. Main units retract inward into fuselage; self-centering twin-contact non-steerable nosewheel retracts forward. Oleo-pneumatic shock-absorbers. Main-wheel tyres size 19 × 6·25-9, pressure 6·21 bars (90 lb/sq in). Nosewheel tyre size 15·4 × 4-6, pressure 4·83 bars (70 lb/sq in). Hydraulic brakes, without cooling.

POWER PLANT: One 11·12 kN (2,500 lb st) Rolls-Royce Bristol Viper 11 turbojet engine. Internal fuel in main saddle tanks in fuselage (two 46 Imp gallon), wing centre-section collector tank (62 Imp gallons) and outboard wing tanks (two 48 Imp gallon), with total capacity of 1,137 litres (250 Imp gallons). Provision for two underwing tanks with total capacity of 454 litres (100 Imp gallons). System permits 30 sec of inverted flight.

ACCOMMODATION: Crew of two side by side in air-conditioned and pressurised cockpit, on Martin-Baker Mk H4 HA zero-altitude fully-automatic ejection seats. Clamshell-type canopy. Dual controls and duplicated blind-flying instruments.

SYSTEMS: Air-conditioning system has max pressure differential of 0·12 bars (1·75 lb/sq in). Dowty hydraulic system for landing gear, flaps and airbrake, pressure 207 bars (3,000 lb/sq in). Accumulator for manual emergency system. Electrical system is of 28V DC single-wire earth return type, with two 24V 25Ah batteries. Normalair pressure-demand oxygen system.

ELECTRONICS AND EQUIPMENT: STR 9X/M 10-channel VHF transceiver, AX-3 single-channel VHF standby set and Marconi-Elliott DFA-73 ADF manufactured by BEL-India. Landing light in nose.

ARMAMENT: Hardpoint beneath each wing of 68th and subsequent aircraft, each capable of carrying a 500 lb bomb, an HAL pod containing two 7·62 mm FN machine-guns, a pod containing seven 68 mm SNEB rockets, or a 227 litre (50 Imp gallon) drop-tank.

DIMENSIONS, EXTERNAL:
Wing span 10·70 m (35 ft 1¼ in)
Wing chord at root 2·35 m (7 ft 8½ in)
Wing chord at tip 1·02 m (3 ft 4 in)
Wing aspect ratio 6
Length overall 10·60 m (34 ft 9 in)
Height overall 3·635 m (11 ft 11 in)
Tailplane span 3·90 m (12 ft 9½ in)
Wheel track 2·42 m (7 ft 11 in)
Wheelbase 3·50 m (11 ft 6 in)

AREAS:
Wings, gross 19·00 m² (204·5 sq ft)
Ailerons (total) 1·55 m² (16·68 sq ft)
Flaps (total) 2·34 m² (25·19 sq ft)
Vertical tail surfaces (total) 2·10 m² (22·60 sq ft)
Rudder, incl tab 0·714 m² (7·69 sq ft)
Horizontal tail surfaces (total) 3·72 m² (40·04 sq ft)
Elevators 1·14 m² (12·27 sq ft)

WEIGHTS AND LOADINGS:
Weight empty 2,560 kg (5,644 lb)
Normal T-O weight 3,600 kg (7,936 lb)
Max T-O weight (with two 50 Imp gallon drop-tanks) 4,100 kg (9,039 lb)
Max wing loading 190 kg/m² (38·9 lb/sq ft)
Thrust/weight ratio 0·315

PERFORMANCE (at normal T-O weight):
Max level speed at S/L 375 knots (695 km/h; 432 mph)
Max level speed at 9,150 m (30,000 ft) 371 knots (687·5 km/h; 427 mph)
Max cruising speed 175 knots (324 km/h; 201 mph)
Stalling speed, flaps and landing gear up 81 knots (151 km/h; 94 mph)
Stalling speed, flaps and landing gear down 71 knots (132 km/h; 82 mph)
Ceiling 9,850 m (30,000 ft)
Time to 9,850 m (30,000 ft) 20 min
Min ground turning radius 5·50 m (18 ft 0½ in)
T-O run 442 m (1,450 ft)
Endurance on internal fuel at 230 knots (426 km/h; 265 mph) at 9,150 m (30,000 ft) 1 hr 45 min

HAL HJT-16 Mk I Kiran two-seat jet basic trainer *(Air Portraits)*

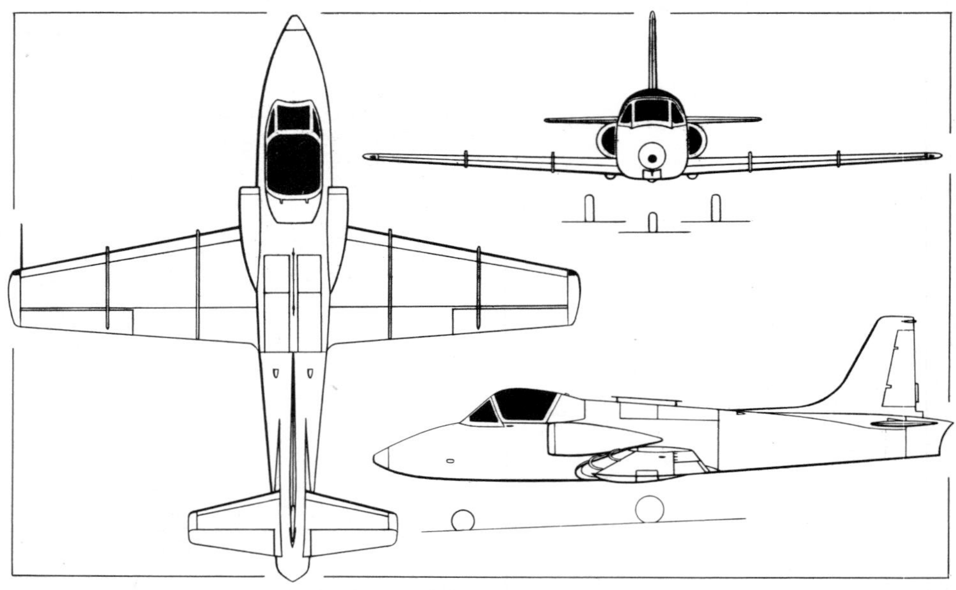
HAL HJT-16 Mk I Kiran side-by-side two-seat jet basic trainer *(Pilot Press)*

HAL KIRAN Mk II

This version of the Kiran, suitable for armament training or counter-insurgency duties, is being developed by HAL. Principal differences include improved weapon-carrying capability and a more powerful engine. The latter is a derated version of the Orpheus 701 turbojet, rated at 15·1 kN (3,400 lb st), which gives the Kiran Mk II improved maximum speed, climb and manoeuvrability. The range, with two underwing drop-tanks fitted as standard, remains the same as for the Mk I. A prototype was expected to fly in October 1976.

ARMAMENT: Two 7·62 mm machine-guns in nose, with 250 rds/gun. Two hardpoints beneath each wing, the inboard ones each capable of carrying a 227·3 litre (50 Imp gallon) drop-tank, a 500 lb high-explosive bomb, a Type 122·1 68 mm rocket pod or a 25 lb practice bomb carrier. Each outboard pylon is capable of carrying a similar weapon load, but is not equipped for a drop-tank.

HAL AJEET (UNCONQUERED)

The Hawker Siddeley Gnat light fighter and fighter-bomber was built under licence by HAL between 1962 and 1974, as described in the 1974-75 and 1975-76 editions of *Jane's*.

The Bangalore Complex completed in early 1974 the design of a Mk II version of the Gnat known as the Ajeet, with improved performance characteristics and equipment, including improved communications and navigation systems; more reliable longitudinal control; and increased combat capability. The last-named characteristic is achieved by a redesigned fuel system, dispensing with the underwing drop-tanks in favour of integral wing tanks, so permitting additional underwing armament to be carried.

The last two Gnat Mk I aircraft were converted as prototypes for the Ajeet, and the first of these was flown during 1975. Flight testing of the hydraulic system and electronics is under way in two other Gnats (IE1071 and 1080), designated Mk IA, and a third is being used for ground testing. Two additional prototypes are being used to test armament and the tailplane ratio changer.

Type: Single-seat lightweight interceptor and ground attack aircraft.

Wings: Cantilever shoulder-wing monoplane. Sweptback wings, of RAE 102 (modified) section. Thickness/chord ratio 8%. Anhedral 5°. Sweepback 40° at quarter-chord. One-piece wing of two-spar thick-skin light alloy construction, fitting into recess in top of fuselage and secured by bolts at four points. Inboard ailerons, powered by Automotive Products hydraulic actuators, droop 20° to serve as flaps when the landing gear is lowered.

Fuselage: Light alloy semi-monocoque structure of pressed frames and extruded stringers.

Tail Unit: Cantilever all-metal structure. One-piece tailplane, operated hydraulically by Hobson actuator with ratio changer. Rear portions of tailplane can be unlocked to perform as elevators.

Landing Gear: Retractable tricycle type, all units retracting rearward into fuselage. Automotive Products hydraulic actuation. Dowty Rotol oleo-pneumatic shock-absorber struts. Wheel well fairings attached to individual landing gear units serve as airbrakes when landing gear is partly lowered, the relative movements of the airbrakes being so adjusted that no change of trim occurs at any speed. A stop on the landing gear airbrake selector prevents gear from being lowered fully when braking only is desired. Main-wheel tyres size 20 × 5·25, pressure 9·3 bars (135 lb/sq in); twin nosewheel tyres, size 17 × 3·25. Toe-operated Dunlop disc brakes. Braking parachute in fairing at base of fin.

Power Plant: One Rolls-Royce Bristol Orpheus 701-01 (BOr.2) non-afterburning turbojet engine, rated at 20 kN (4,500 lb st). Compressed-air starting. Air intakes in sides of fuselage. Seven (plus two optional) fuselage fuel tanks, and integral wing tanks, with common collector tank and one booster pump. Total internal fuel capacity (incl optional fuselage tanks) 1,365 litres (300 Imp gallons). Provision for two 136·5 litre (30 Imp gallon) underwing drop-tanks.

Accommodation: Pilot only, on Martin-Baker GF-4 zero-height/90 knot (167 km/h; 104 mph) lightweight ejection seat, in pressurised cockpit with jettisonable canopy.

Systems: Normalair pressure and temperature control systems, cold air unit and oxygen breathing equipment. Pressure differential 0·28 bars (4·0 lb/sq in). Dowty hydraulic system of 207 bars (3,000 lb/sq in) for aileron, landing gear and tailplane actuation. Dowty electrical system.

Electronics and Equipment: VHF (initially; V/UHF later) transceiver and standby VHF set; ADF; IFF Mk 10; telebriefing unit.

Armament: Two 30 mm Aden cannon in air intake fairings, one on each side of fuselage, with 90 rds/gun. Ferranti Type 195 ISIS gunsight. Four underwing hardpoints on which can be carried two 500 lb bombs, four 18 × 68 mm Arrow rocket pods or two 136·5 litre (30 Imp gallon) drop-tanks.

Dimensions, external:

Wing span	6·73 m (22 ft 1 in)
Wing chord (mean)	1·88 m (6 ft 2 in)
Wing aspect ratio	3·575
Length overall	9·04 m (29 ft 8 in)
Height overall	2·46 m (8 ft 1 in)
Tailplane span	2·84 m (9 ft 4 in)
Wheel track	1·55 m (5 ft 1 in)
Wheelbase	2·36 m (7 ft 9 in)

Weights:

Basic empty weight	2,302 kg (5,074·5 lb)
T-O weight 'clean'	3,539 kg (7,803 lb)
Max T-O weight	4,170 kg (9,195 lb)
Normal landing weight	2,613 kg (5,760 lb)

Performance: (in configurations indicated; A: ISA; B: ISA + 15°C; C: ISA + 30°C):

Max Mach No. at 12,000 m (39,375 ft) at 'clean' T-O weight:

A	0·96
B	0·953
C	0·948

Max level speed at S/L at 'clean' T-O weight:

A	595 knots (1,102 km/h; 685 mph)
B	612 knots (1,134 km/h; 705 mph)
C	622 knots (1,152 km/h; 716 mph)

Time to 12,000 m (39,375 ft) from brakes off, at 'clean' T-O weight:

A	6 min 2 sec
B	7 min 43 sec
C	9 min 33 sec

Service ceiling:

A, B, C	12,000 m (39,375 ft)

Turning performance at 450 knots (834 km/h; 518 mph) IAS at S/L:

A	5·30g
B	5·28g
C	5·00g

T-O run at S/L, zero wind, at T-O weight of 4,136 kg (9,118 lb) with two Arrow rocket pods and two 30 Imp gallon drop-tanks:

A	1,034 m (3,390 ft)
B	1,180 m (3,870 ft)
C	1,376 m (4,515 ft)

Landing run 'clean' at S/L, zero wind, at normal landing weight, no brake 'chute:

A	951 m (3,120 ft)
B	997 m (3,270 ft)
C	1,047 m (3,435 ft)

Landing run 'clean' at S/L, zero wind, at normal landing weight, with brake 'chute:

A	658 m (2,160 ft)
B	695 m (2,280 ft)
C	725 m (2,379 ft)

Combat radius (A, B and C), low level ground attack mission;

with two 500 lb bombs on inboard stations
110 nm (204 km; 127 miles)

with two Arrow rocket pods inboard and two 30 Imp gallon drop-tanks outboard
140 nm (259 km; 161 miles)

with four Arrow rocket pods
104 nm (193 km; 120 miles)

Prototype of the Ajeet lightweight combat aircraft, developed by HAL from the Hawker Siddeley Gnat

HAL AJEET TRAINER

This tandem two-seat trainer version of the Ajeet, which retains the four underwing hardpoints and full combat capability of the single-seater, is evolved in a similar manner to that in which the T.Mk 1 trainer was evolved from the original Gnat fighter. To accommodate the second cockpit, two of the fuselage fuel tanks are deleted, although this can be offset by deleting the Aden cannon to make room for an additional 273 litres (60 Imp gallons) of fuel internally; in this event, provision is made for carrying a 7·62 mm gun pod on each of the inboard wing pylons. Sections 0·70 m (2 ft 3½ in) long are inserted in the fuselage fore and aft of the wings, increasing the overall length of the Ajeet Trainer to 10·44 m (34 ft 3 in).

The Ajeet Trainer is powered by a single Orpheus 701 turbojet engine, and retains the main hydraulic system and powered flying controls of the single-seater. Normal and emergency operation of the landing gear is also similar, but with duplicated controls and a mechanical override facility in the rear cockpit. All instruments, including those for blind flying, are duplicated in the rear cockpit, which will be illuminated, air-conditioned and pressurised similarly to the front cockpit. A gunsight is fitted in the front cockpit only. Both occupants are provided with Martin-Baker GF-4 ejection seats. Electronics and equipment include a multi-channel VHF transceiver (to be replaced eventually by a V/UHF set), ADF and IFF Mk 10.

HAL HPT-32

Currently under development for the Indian Air Force, the HPT-32 is a fully-aerobatic piston-engined basic trainer, with side-by-side seats for instructor and pupil and a third seat at the rear. It can be used for a wide range of ab initio training, including instrument, navigation, night flying and formation flying; for armed patrol; for observation, liaison or sport flying; or for weapon training, light strike duties, supply dropping, search and rescue, reconnaissance, or glider or target towing. The airframe, which is of all-metal construction, is designed to FAR 23, and is expected to have a fatigue life of 6,500 hr. General appearance of the HPT-32 can be seen in the accompanying three-view drawing.

Two prototypes have been ordered, the first of which is expected to fly in early 1977.

Type: Two/three-seat basic trainer.

Wings: Cantilever low-wing monoplane of all-metal construction. Dihedral 5° from roots. Incidence 2° 30′ at root.

Fuselage: All-metal semi-monocoque structure.

Tail Unit: Cantilever all-metal structure, with sweptback vertical surfaces.

Landing Gear: Retractable tricycle type. Nose unit retracts rearward, main units inward into wings.

Power Plant: Prototypes powered by 194 kW (260 hp) Lycoming AEIO-540-D4B5 flat-six engine, driving a Hartzell two-blade constant-speed propeller with spinner. Two integral wing fuel tanks, with total capacity of 227 litres (50 Imp gallons); provision for 136·5 litre (30 Imp gallon) tank in place of rear seat. For production aircraft, an indigenous engine of similar power is under development.

Accommodation: Side-by-side seats for two persons, with third seat to rear, under rearward-sliding jettisonable bubble canopy. Front two seats adjustable in height by 127 mm (5 in); rear seat is not adjustable. Baggage space beside rear seat. Full dual controls, and adjustable rudder pedals, for instructor and pupil.

Armament and Equipment: Four underwing attachments for armament or other stores, up to a total of 255 kg

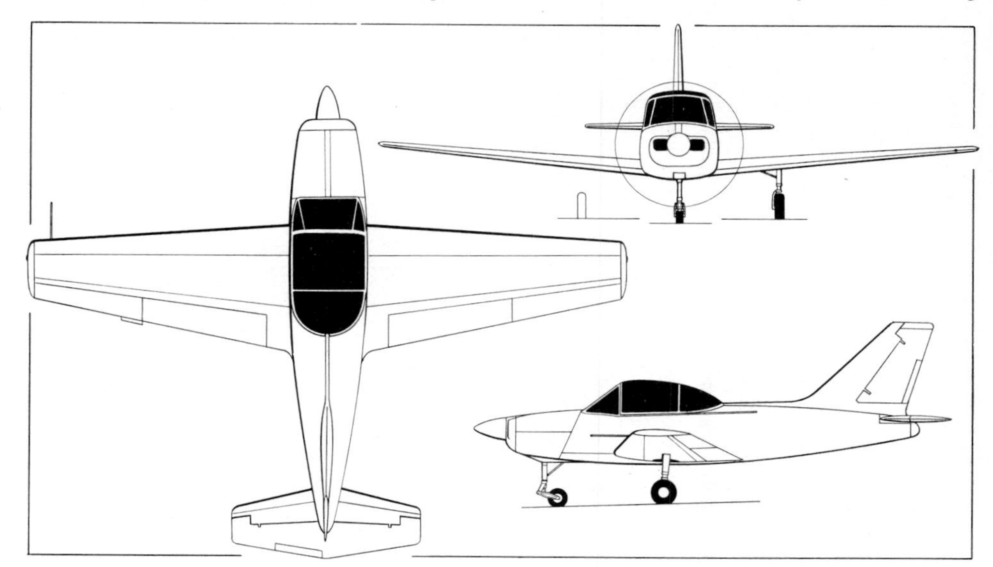

HAL HPT-32 two/three-seat basic trainer *(Pilot Press)*

(562 lb). VHF radio, ADF and marker beacon.

DIMENSIONS, EXTERNAL:
- Wing span 9·50 m (31 ft 2 in)
- Wing aspect ratio 6·0
- Length overall 7·72 m (25 ft 4 in)
- Height overall 2·93 m (9 ft 7¼ in)
- Wheel track 3·30 m (10 ft 10 in)
- Wheelbase 1·71 m (5 ft 7¼ in)

AREA:
- Wings, gross 15·00 m² (161·46 sq ft)

WEIGHTS:
- Weight empty (without electronics) 850 kg (1,874 lb)
- Normal T-O weight 1,200 kg (2,645 lb)
- Max T-O weight 1,500 kg (3,307 lb)

PERFORMANCE (estimated, at 1,200 kg; 2,645 lb normal T-O weight. A: landing gear down; B: landing gear retracted):
- Max level speed at S/L:
  - A 140 knots (260 km/h; 161 mph) EAS
  - B 165 knots (307 km/h; 190 mph) EAS
- Stalling speed, flaps up:
  - A 59·5 knots (110 km/h; 68·5 mph) EAS
  - B 58·5 knots (108 km/h; 67·5 mph) EAS
- Stalling speed, flaps down:
  - A 51 knots (94 km/h; 58·5 mph) EAS
  - B 50 knots (92 km/h; 57·5 mph) EAS
- Max rate of climb at S/L:
  - A 417 m (1,368 ft)/min
  - B 479 m (1,571 ft)/min
- Service ceiling:
  - A 5,750 m (18,865 ft)
  - B 6,500 m (21,325 ft)
- T-O to 15 m (50 ft):
  - A, B 280 m (918 ft)
- Landing from 15 m (50 ft):
  - A, B 390 m (1,280 ft)
- Range at 2,000 m (6,560 ft):
  - A 458 nm (850 km; 528 miles)
  - B 539 nm (1,000 km; 621 miles)
- Endurance at 2,000 m (6,560 ft):
  - A, B (50 Imp gallons fuel) 4 hr 30 min
  - A, B (80 Imp gallons fuel) 7 hr 0 min

HAL Cheetah, Indian-assembled version of the Aérospatiale SA 315B Lama helicopter

HAL-assembled Aérospatiale Alouette III helicopter in Indian military insignia

HAL (AÉROSPATIALE) SA 315B LAMA

Indian name: Cheetah

The Bangalore Complex's Helicopter Division is building the French Aérospatiale SA 315B Lama five-seat general-purpose helicopter (which see) under licence in India, where it is known as the Cheetah.

Initial production is from French-built components. The first Indian-assembled Cheetah was test-flown on 6 October 1972, and a total of 38 had been delivered by 1 February 1976. Deliveries of aircraft manufactured from locally-produced raw materials was scheduled to begin in August 1976.

An agricultural version of the Cheetah is under development. Preliminary spraying trials have been conducted, and further modifications are in progress.

HAL (AÉROSPATIALE) SA 316B ALOUETTE III

Indian name: Chetak

The Bangalore Complex's Helicopter Division is building the French Aérospatiale SA 316B Alouette III under a licence granted in June 1962. The first Indian-assembled Alouette III was flown for the first time on 11 June 1965.

By 1 February 1976, a total of 219 Alouette IIIs had been ordered from HAL, of which 174 had been completed, including three converted to civil standard. A few Alouette IIIs were presented to the Royal Nepal Army in 1974. HAL also supplies Indian-built components for French-built Alouette IIIs.

An armed version, known as the Chetak, is being developed by HAL, carrying four air-to-surface missiles on laterally-mounted booms. Target identification and fire control is via a monocular periscopic sight on the cabin roof. Preliminary firing trials have been carried out successfully.

HAL ARMED LIGHT HELICOPTER (ALH)

The Helicopter Division of HAL has under development a single-engined high-performance armed light helicopter, powered by a Turboméca Astazou XX turboshaft engine mounted on top of the fuselage aft of the rotor head. Two versions are being developed: a standard version for Indian Army/Air Force use, and a variant for the Indian Navy. The former will have a capability for combat missions, communications duties, armed reconnaissance and surveillance, casualty evacuation, crew rescue, external cargo carrying and training. The naval version will be able to perform anti-submarine search and strike, air to surface vessel search and strike, search and rescue, reconnaissance, casualty evacuation, and vertical replenishment duties at sea.

The ALH will have a light alloy semi-monocoque fuselage, with accommodation in the standard version for a crew of two and five passengers or equivalent load. A four-blade semi-rigid single main rotor, with blades made of composite materials, will be fitted, and provision is made for blade folding. A skid-type landing gear will be fitted to the Army/Air Force version; the Navy version will have a non-retractable tricycle landing gear, with a fully-castoring and self-centering nosewheel and a harpoon deck-lock securing system.

Combat equipment on the standard version will include a variety of armament combinations such as miniguns, missiles and rockets; the naval version will have torpedoes, depth charges and missiles, with accompanying electronics appropriate to a given mission.

DIMENSIONS, EXTERNAL:
- Diameter of main rotor 13·00 m (42 ft 8 in)
- Diameter of tail rotor 1·00 m (3 ft 3¼ in)
- Length overall, main rotor turning 14·915 m (48 ft 11¼ in)
- Length, main rotor folded 12·195 m (40 ft 0 in)
- Width, rotors folded 5·60 m (18 ft 4½ in)
- Height overall 4·01 m (13 ft 2 in)
- Height with rotors and tail folded 3·95 m (12 ft 11½ in)
- Cabin door:
  - Height 1·20 m (3 ft 11¼ in)
  - Width 1·10 m (3 ft 7¼ in)

DIMENSIONS, INTERNAL:
- Cabin: Max width 1·35 m (4 ft 5¼ in)
  - Max height 1·35 m (4 ft 5¼ in)

WEIGHTS:
- Weight empty, without equipment:
  - Army version 1,500 kg (3,307 lb)
  - Naval version 1,550 kg (3,417 lb)
- Max T-O weight:
  - Army version 2,500 kg (5,511 lb)
  - Naval version 3,000 kg (6,613 lb)

PERFORMANCE (estimated, at max T-O weight):
- Never-exceed speed:
  - at S/L 186 knots (345 km/h; 214 mph)
  - at 3,050 m (10,000 ft) not less than 162 knots (300 km/h; 186 mph)
  - at 4,375 m (14,350 ft) not less than 140 knots (260 km/h; 161 mph)
- Max continuous cruising speed:
  - at 4,875 m (16,000 ft) 134 knots (250 km/h; 155 mph)
  - at 6,100 m (20,000 ft) 124 knots (230 km/h; 143 mph)
- Normal cruising speed:
  - at 4,875 m (16,000 ft) 129 knots (240 km/h; 149 mph)
  - at 6,100 m (20,000 ft) 100 knots (185 km/h; 115 mph)
- Approach speed:
  - normal 60 knots (110 km/h; 69 mph) IAS
  - precautionary 35·5 knots (65 km/h; 41 mph) IAS
  - autorotative 60 knots (110 km/h; 69 mph) IAS
- Service ceiling 8,000 m (26,250 ft)
- Hovering ceiling out of ground effect 4,875 m (16,000 ft)

KANPUR DIVISION

ADDRESS:
Chakeri, Kanpur
Telephone: HAL PABX 62471-4
Telex: HAL KP 243

When the decision was taken to build the Hawker Siddeley 748 twin-turboprop transport in India, as a replacement for the Dakotas of the Indian Air Force, four hangars at Kanpur were taken over, on 23 January 1960, as the IAF Aircraft Manufacturing Depot. The Depot was incorporated in Aeronautics (India) Ltd in June 1964 and subsequently became the Kanpur Division of Hindustan Aeronautics Ltd.

HAL (HAWKER SIDDELEY) 748

The first Indian-built 748 flew on 1 November 1961, followed by the second one on 13 March 1963. The first four Indian 748s were Srs 1 aircraft, utilising components imported from the UK.

The first Indian-built Srs 2 flew for the first time on 28 January 1964; by January 1973, 39 Srs 2 aircraft had been built and delivered. Most of the airframe components were manufactured by HAL from locally-produced raw materials. The aircraft's 1,570 kW (2,105 ehp) Dart 531 turboprop engines were built by the Bangalore Complex of HAL. Some HF, VHF and other radio equipment was manufactured by Bharat Electronics Ltd of Bangalore.

Indian Airlines placed an initial order for 14 Srs 2 aircraft; the first of these was delivered on 28 June 1967 and the last in March 1970. Deliveries of a further 10 aircraft ordered by the airline, of which seven are believed to have been completed, have been placed in abeyance pending the report of the Dhawan Commission; this was presented in October 1975 and was under consideration by the Indian government in the Spring of 1976.

HAL has also produced the aircraft in several versions for the Indian Air Force, which ordered four Srs 1s and 41 Srs 2s. These include 16 equipped as executive transports, 7 as navigation trainers, 4 as air signals trainers, and 18 as trainers for pilots of multi-engined aircraft.

A prototype HS 748(M) military freighter, developed by Kanpur Division, flew for the first time on 16 February 1972 and successfully completed flight trials with the Indian Air Force. An order to Kanpur Division for 10 more HS 748s was announced in June 1975.

HAL HA-31 Mk II BASANT (SPRING)

Design of this agricultural aircraft began at the Bangal-

Pre-production HAL HA-31 Mk II Basant agricultural aircraft built by HAL's Bangalore Complex *(Alan W. Hall)*

ore Complex of HAL in mid-1968. The prototype, designated HAL-31 Mk I, was powered by a 186 kW (250 hp) Rolls-Royce Continental engine, and was described in the 1971-72 *Jane's*. The aircraft was subsequently completely redesigned as the HA-31 Mk II, with a 298 kW (400 hp) engine, and a prototype of this version flew for the first time on 30 March 1972. A second, pre-production prototype flew in September 1972, and certification was obtained in March 1974. A pre-production batch of 20 Basants were built by the Bangalore Complex, and the first eight of these were handed over to the Indian Ministry of Food and Agriculture on 21 June 1974. Responsibility for series manufacture of the Basant has been assigned to the Kanpur Division, which is to build 100. The first two of these had been completed by February 1976.

The Basant is intended primarily for aerial application of pesticides and fertilisers. It can also be used for aerial survey, fire/patrol duties and cloud seeding.

TYPE: Single-seat agricultural and utility aircraft.

WINGS: Strut-braced low-wing monoplane. Constant-chord wings, of USA 35B section with rounded tips. Thickness/chord ratio 11·6%. Dihedral 5° from roots. No incidence or sweepback. Spars (two) and ribs of light alloy, with fabric covering. Fowler-type trailing-edge flaps and Frise-type ailerons, of similar construction to wings and operated manually. Tab in port aileron. Single inverted-Vee bracing strut and cross-strut on each side, attached to fuselage forward of cockpit.

FUSELAGE: Conventional structure, of welded chrome-molybdenum steel tube with fabric covering in most areas. Structure forward of cockpit designed to absorb impact in the event of a crash. The cockpit structure, the seat and its attachment, seat belt and shoulder harness are all designed to withstand 40*g* loads in the event of a crash.

TAIL UNIT: Conventional alloy structure, with fixed-incidence metal-skinned tailplane. One-piece elevator, with central trim tab. Fabric-covered fin and rudder. Fixed tab in rudder.

LANDING GEAR: Non-retractable tailwheel type, with HAL oleo-pneumatic shock-absorbers on all units. Size 24·7 × 7·5-10 tyres on main wheels, 11·28 × 4-3·5 on tailwheel. Tyre pressure 2·41 bars (35 lb/sq in) on main units. Dunlop hydraulic disc brakes.

POWER PLANT: One 298 kW (400 hp) Lycoming IO-720-C1B flat-eight engine, driving a Hartzell three-blade constant-speed metal propeller. Two inboard fuel tanks (each 91 litres; 20 Imp gallons) and two outboard tanks (each 63·5 litres; 14 Imp gallons) in each wing, and 9 litres (2 Imp gallons) in collector tank. Total fuel capacity 318 litres (70 Imp gallons). Refuelling point on each tank. Oil capacity (nominal) 19·3 litres (4·25 Imp gallons).

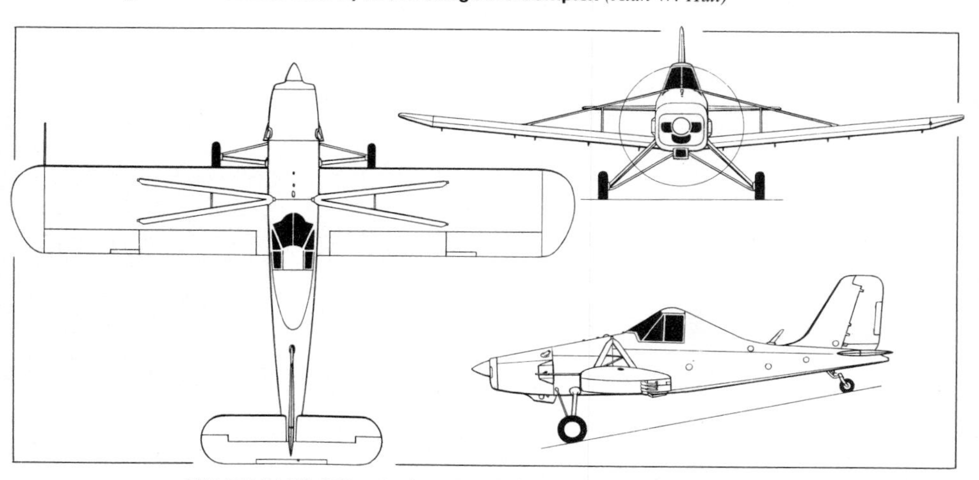
HAL HA-31 Mk II Basant (Lycoming IO-720-C1B engine) *(Pilot Press)*

ACCOMMODATION: Single seat in fully-enclosed cockpit. Forward-hinged door on starboard side; emergency door on port side. Cockpit heated and ventilated.

SYSTEMS: Hydraulic motor for crop spraying. Electrical power, from 24V 50A alternator and 24V 25Ah battery, for engine starting, instruments, VHF radio etc.

ELECTRONICS AND EQUIPMENT: VHF radio optional. Glassfibre hopper, installed between engine firewall and front wall of cockpit enclosure, has 0·93 m³ (33 cu ft) capacity and can carry up to 605 kg (1,333 lb) of pesticide for Normal category operation and up to 907 kg (2,000 lb) in Restricted category.

DIMENSIONS, EXTERNAL:

Wing span	12·00 m (39 ft 4½ in)
Wing chord (constant)	2·00 m (6 ft 6½ in)
Wing aspect ratio	6·17
Length overall (tail down)	9·00 m (29 ft 6¼ in)
Height overall (tail down)	2·55 m (8 ft 4½ in)
Tailplane span	3·86 m (12 ft 8 in)
Wheel track	2·70 m (8 ft 10¼ in)
Wheelbase	6·00 m (19 ft 8¼ in)
Propeller diameter	2·13 m (7 ft 0 in)
Propeller ground clearance	0·25 m (10 in)

AREAS:

Wings, gross	23·34 m² (251·23 sq ft)
Ailerons (total)	2·14 m² (23·03 sq ft)
Trailing-edge flaps (total)	2·618 m² (28·18 sq ft)
Fin	1·06 m² (11·41 sq ft)
Rudder, incl tab	0·86 m² (9·26 sq ft)
Tailplane	2·44 m² (26·26 sq ft)
Elevator, incl tab	1·80 m² (19·37 sq ft)

WEIGHTS AND LOADINGS:

Weight empty	1,200 kg (2,645 lb)
Basic operating weight	1,954 kg (4,300 lb)
Max T-O weight	2,270 kg (5,000 lb)
Max wing loading	97·25 kg/m² (19·92 lb/sq ft)
Max power loading	7·62 kg/kW (12·6 lb/hp)

PERFORMANCE (at basic operating weight):

Never-exceed speed	164 knots (305 km/h; 189 mph)
Max level speed at S/L	121 knots (225 km/h; 140 mph)
Max cruising speed at 2,625 m (8,000 ft)	100 knots (185 km/h; 115 mph)
Econ cruising speed	87 knots (161 km/h; 100 mph)
Stalling speed, flaps up	52 knots (96 km/h; 60 mph)
Stalling speed, flaps down	49 knots (91 km/h; 57 mph)
Max rate of climb at S/L	228 m (750 ft)/min
Service ceiling	3,800 m (12,500 ft)
Min ground turning radius	7·00 m (22 ft 11 in)
T-O run	214 m (700 ft)
T-O to 15 m (50 ft)	365 m (1,200 ft)
Landing from 15 m (50 ft)	305 m (1,000 ft)
Landing run	183 m (600 ft)
Range with max fuel, no payload	348 nm (645 km; 400 miles)
Endurance with max payload, 30 min reserves	1 hr

MiG COMPLEX

The MiG Complex is formed from the Nasik, Koraput and Hyderabad Divisions of HAL, which, under an agreement concluded in 1962, build respectively the airframes, power plants and electronics equipment of MiG-21 fighters under licence from the USSR.

HAL (MIKOYAN) MiG-21

Indian Air Force designations: Type 66-400, 66-600, 74, 76, 77 and 88

The following versions of the MiG-21 (NATO reporting name *Fishbed)* have been supplied to or manufactured in India:

MiG-21F. Six Soviet-built MiG-21F *(Fishbed-C)* day fighters were supplied to the Indian Air Force in early 1963 and were assigned to No. 28 Squadron. Four more were delivered in mid-1964. IAF designation **Type 74**.

MiG-21PF. Two Soviet-built MiG-21PF *(Fishbed-D)* all-weather fighters supplied to IAF (No. 28 Squadron) in mid-1964. Planned import of a further 18 prevented by Indo-Pakistan conflict of September 1965, but later reports suggest that a total of 96 were received from USSR. IAF designation **Type 76**.

MiG-21FL. Soviet export designation of the late-production model MiG-21PF. First version to be manufactured in India, following the supply of some aircraft from Soviet production and an initial order for 60 in 1964. Production by HAL, initially from knock-down components and with Soviet-built Tumansky R-11 engines, began in late 1966; about 100 were assembled in this way, with first deliveries to IAF in 1967. First examples built from raw materials, with 60% indigenous content, handed over to IAF on 19 October 1970, and by December 1971 MiG-21s (all versions) were in service with Nos. 1,4,8, 28, 29, 30, 45 and 47 Squadrons. Indian production of the FL version, completed in 1973, is believed to have totalled 196. IAF designation **Type 77**.

MiG-21M. Current production version with R-11F2S-300 engine. First aircraft handed over to IAF on 14 February 1973, and about 15 delivered by Spring 1974. IAF designation **Type 88**. Fifty Soviet-built MiG-21PFMAs, bought to supplement Indian production, have been placed in service with Nos. 7 and 108 Squadrons. No. 26 Squadron is among those reportedly equipped with MiG-21M.

HAL-built MiG-21M, the first example of which was completed in February 1973, in Indian Air Force insignia

MiG-21MF *(Fishbed-J)*. Improved production version, having more powerful Tumansky R-13-300 turbojet engine and increased fuel. Total of 150 ordered, for delivery at a planned rate of 30 per year.

MiG-21U. Tandem two-seat training versions (NATO reporting name *Mongol)*. Small batch of MiG-21U *Mongol-A* delivered from USSR in 1965, these being given the IAF designation **Type 66-400 Series**. Main version in current service is the *Mongol-B* (with broad-chord fin), the IAF designation of which is **Type 66-600 Series**. Total of 42 received.

A full description of the MiG-21 appears in the USSR section of this edition.

An example of the MiG-21FL, first version of this Soviet aircraft to be manufactured in India

INDONESIA

LIPNUR

LEMBAGA INDUSTRI PENERBANGAN NURTANIO (Department of the Indonesian Air Force, Nurtanio Aircraft Industry)

ADDRESS:
Lanuma Husein Sastranegara (Husein Sastranegara Air Base), Bandung
Telephone: Bandung 56191

DIRECTOR:
Colonel Ir Yuwono

ENGINEERING MANAGER:
Lt Col Daniel D. Suprapto

To honour the service of the late Air Marshal Nurtanio Pringgoadisurjo, who was largely responsible for establishing an aircraft industry in Indonesia, the Department of the Air Force of Indonesia named after him the former Institute for Aero Industry Establishment.

The Institute had come into being in August 1961, as a successor to the Indonesian Air Force Design, Development and Production Depot.

Lipnur has facilities for training aircraft factory workers. Its engineering and production facilities were until 1974 used for the manufacture under licence of the Polish PZL-104 utility aircraft, under the Indonesian name Gelatik.

Lipnur's latest aircraft programme is the LT-200, a modified version of the Pazmany PL-2.

PZL-104 GELATIK 32 (RICE BIRD)

The initial Indonesian production version of the Gelatik, with 225 hp Continental O-470-13A engine, was described in the 1970-71 *Jane's*.

This was followed by the Gelatik 32, fitted with a 230 hp Continental O-470-L or O-470-R engine.

The Gelatik 32 is a licence-built version of the Polish PZL-104 Wilga 32 utility aircraft (see 1974-75 *Jane's*), with detail changes compared with the Polish production model, and was described in the 1975-76 *Jane's*.

A total of 39 Gelatiks of all versions were built, of which the first was flown in 1964; production ended in 1975, but is capable of being reinstated if future requirements make this necessary.

LIPNUR LT-200

The Lipnur LT-200 is a two-seat light aircraft based on the design of the Pazmany PL-2 (see US section). It is intended for use as a military and civil trainer, and for club and private flying.

Construction of the first of two prototypes began in September 1973, and this aircraft (IN-200) flew for the first time on 9 November 1974. In the following month construction began of two modified and improved aircraft, and in 1976 the Lipnur factory planned to complete a pre-production batch of six aircraft. Tooling is under way in anticipation of series production. It is expected that about 30 LT-200s will be ordered initially for the Indonesian Air Force, Civil Flying School and flying clubs, which could be completed within two years from the start of production. Additional production will be subject to market requirements.

The following description applies to the prototype LT-200:

TYPE: Two-seat light aircraft.

WINGS: Cantilever low-wing monoplane. Wing section NACA 63_2615. Dihedral 5°. Incidence 0°. Single-spar structure of 2024S aluminium alloy. Plain ailerons and trailing-edge slotted flaps of similar construction. No tabs.

FUSELAGE: Semi-monocoque structure of 2024S aluminium alloy.

TAIL UNIT: Cantilever structure of 2024S aluminium alloy. Single-spar fin and rudder, with slight sweepback. Trim tab on rudder. Single-spar all-moving tailplane, with servo tab.

LANDING GEAR: Non-retractable tricycle type, with oleo-pneumatic shock-absorber on all three units. Single Goodyear 5·00-5 wheel and size 15 × 5·00-5 tyre on each unit. Tyre pressure 2·07 bars (30 lb/sq in) on main wheels, 1·93 bars (28 lb/sq in) on nosewheel. Goodyear hydraulic disc brakes.

POWER PLANT: One 112 kW (150 hp) Lycoming O-320-E2A flat-four engine, driving a McCauley AGM 7250 two-blade fixed-pitch metal propeller with spinner. All fuel contained in two permanently-attached wingtip tanks, total capacity 94·6 litres (25 US gallons; 20·8 Imp gallons). Refuelling point on top of each tank. Oil capacity 7·7 litres (2 US gallons; 1·7 Imp gallons).

ACCOMMODATION: Side-by-side seats for pilot/instructor and one passenger/pupil under one-piece rearward-sliding canopy. Cabin ventilated. Space for 18 kg (40 lb)

Lipnur LT-200 two-seat light aircraft, based on the Pazmany PL-2

of baggage aft of seats.

SYSTEMS AND EQUIPMENT: 14V 50A electrical system for communications, instruments and lighting. Portable oxygen system. ARC 300 or Narco Com 111 radio, and blind-flying instrumentation, standard.

DIMENSIONS, EXTERNAL:

Wing span over tip-tanks	8·69 m (28 ft 6 in)
Wing chord (constant)	1·27 m (4 ft 2 in)
Wing aspect ratio	6·7
Length overall	5·89 m (19 ft 4 in)
Height overall	2·31 m (7 ft 7 in)
Tailplane span	2·44 m (8 ft 0 in)
Wheel track	2·36 m (7 ft 9 in)
Wheelbase	1·24 m (4 ft 1 in)
Propeller diameter	1·83 m (6 ft 0 in)
Propeller ground clearance	0·22 m (8½ in)

DIMENSION, INTERNAL:

Cabin: Max width	1·02 m (3 ft 4 in)

AREAS:

Wings, gross	10·78 m² (116·0 sq ft)
Ailerons (total)	0·975 m² (10·5 sq ft)
Trailing-edge flaps (total)	1·72 m² (18·5 sq ft)
Fin	0·53 m² (5·7 sq ft)
Rudder	0·44 m² (4·7 sq ft)
Tailplane, incl tab	1·67 m² (18·0 sq ft)

WEIGHTS AND LOADINGS:

Basic weight empty	409 kg (902 lb)
Max payload	172 kg (380 lb)
Max T-O and landing weight	656 kg (1,447 lb)
Max wing loading	56·6 kg/m² (11·6 lb/sq ft)
Max power loading	5·86 kg/kW (9·65 lb/hp)

PERFORMANCE (at max T-O weight):

Never-exceed speed (structural)	161 knots (299 km/h; 186 mph)
Max level speed at S/L	133 knots (246 km/h; 153 mph)
Econ cruising speed at S/L	118 knots (219 km/h; 136 mph)
Stalling speed, flaps down	47 knots (87 km/h; 54 mph)
Max rate of climb at S/L	426 m (1,400 ft)/min
Service ceiling	4,570 m (15,000 ft)
Min ground turning radius	5·64 m (18 ft 6 in)
T-O run	166 m (545 ft)
T-O to 15 m (50 ft)	350 m (1,150 ft)
Landing from 15 m (50 ft)	280 m (920 ft)
Landing run	189 m (620 ft)
Range with max fuel	330 nm (613 km; 381 miles)

INTERNATIONAL PROGRAMMES

AIRBUS
AIRBUS INDUSTRIE

HEAD OFFICE:
Avenue Lucien Servanty, BP No. 33, 31700-Blagnac, France
Telephone: (61) 49 11 44
Telex: AI TO A 530526 F

PARIS OFFICE:
12bis Avenue Bosquet, 75007 Paris, France
Telephone: 551 40 95

AIRFRAME PRIME CONTRACTORS:
Aérospatiale (SNIAS), 37 Boulevard de Montmorency, 75781 Paris-cédex 16, France
Deutsche Airbus GmbH, 8 München 19, Leonrodstrasse 68, Postfach 47, German Federal Republic

CHAIRMAN OF SUPERVISORY BOARD:
Dr Franz-Josef Strauss

PRESIDENT AND CHIEF EXECUTIVE:
Bernard Lathière

EXECUTIVE VICE-PRESIDENT AND CHIEF OPERATING OFFICER:
Roger Beteille

SENIOR VICE-PRESIDENTS:
J. Roeder (Technical)
G. Warde (Customer Services)
D. Krook (Marketing)
F. Kracht (Production)
G. Ville (Finance and Administration)
B. Ziegler (Flight Test)

VICE-PRESIDENT, PUBLIC RELATIONS:
John D. S. Keatinge

Airbus Industrie was set up as a 'Groupement d'Intérêt Economique' to manage the development, manufacture and marketing of the twin-engined large-capacity short/medium-range A300 transport aircraft. It has design leadership and is responsible for the A300 activities of Aérospatiale of France, Deutsche Airbus (MBB and VFW-Fokker) of Germany, Hawker Siddeley Aviation of the UK, Fokker-VFW of the Netherlands and CASA of Spain.

The associated companies raised the necessary funding in the form of repayable loans from the French, German, Dutch and Spanish governments respectively. Hawker Siddeley Aviation is financing separately a part of the development. A levy on sales of the A300 will repay the development loans. In addition, a consortium of major French and German banks was formed to provide finance for sales.

In early 1974 Airbus Industrie moved its headquarters from Paris to new accommodation at Toulouse-Blagnac, where assembly and flight testing of the A300 are undertaken.

Aérospatiale is responsible for manufacturing the entire nose section (including the flight deck), lower centre fuselage and engine pylons, and for final assembly. Deutsche Airbus is responsible for manufacturing the forward fuselage, between the flight deck and wing box, the upper centre fuselage, the rear fuselage and the vertical tail surfaces. Hawker Siddeley has design responsibility for the wings (for which it received the Queen's Award for Technological Achievement in 1976), builds the wing fixed structures, and is working in collaboration with Fokker-VFW, which is building the wing moving surfaces. CASA manufactures the horizontal tail surfaces, fuselage main doors and landing gear doors.

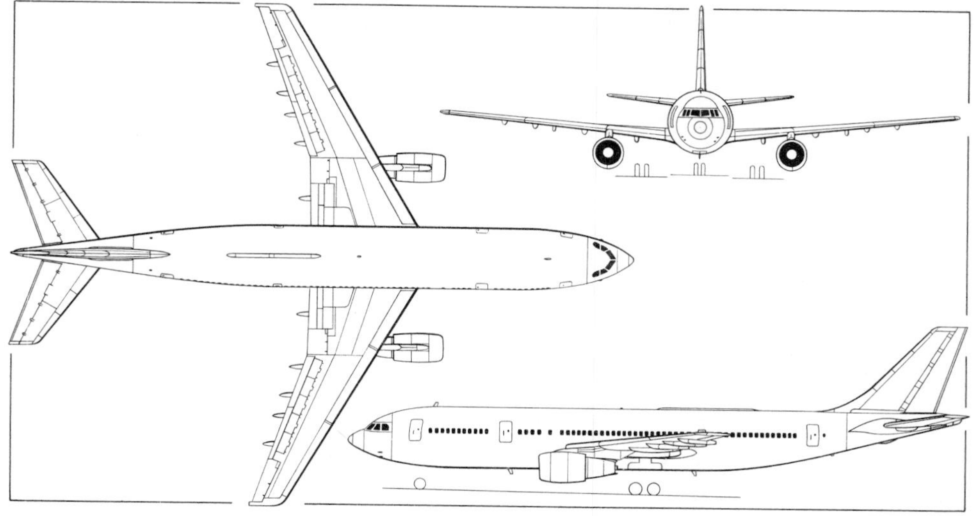

Airbus A300B2 wide-bodied short/medium-range transport *(Pilot Press)*

First B4 version of the twin-turbofan Airbus Industrie A300 for Korean Air Lines

AIRBUS A300

The Airbus A300 is basically a wide-bodied aircraft with underwing pods for two turbofan engines. The early history of the project has appeared in previous editions of *Jane's*.

It is currently being offered with two 227 kN (51,000 lb st) General Electric CF6-50C turbofans, but the underwing location of the power plant enables the A300 to use any advanced technology turbofan engine in the 222·4 kN (50,000 lb st) thrust class. The engines at present offered have considerable development potential and are installed in pods interchangeable with those of the McDonnell Douglas DC-10 Series 30. The CF6-50L of 240·2 kN (54,000 lb st) will be available in 1978.

Six airframes were involved in the certification programme, including one static test specimen and separate components, to cover the complete structure, for fatigue tests.

Construction of the first A300, a B1, began in September 1969. This aircraft (F-WUAB, later F-OCAZ) made its first flight on 28 October 1972, and was followed by the second B1 (F-WUAC) on 5 February 1973. Initial certification covered automatic approach and landing in Category 2 weather conditions. Certification for Category 3A was granted on 30 September 1974. The first automatic landing was made, by the second B1 development aircraft, on 2 May 1973.

The following major versions have been or are being built:

A300B1. Initial version, with 218 kN (49,000 lb st) CF6-50A engines, overall length of 50·97 m (167 ft 2¼ in) and max seating for 302 passengers. Described in detail in 1971-72 *Jane's*. First and second development aircraft built to this specification.

A300B2. Basic production version, to which the following description applies. Third and fourth aircraft (F-WUAD and F-WUAA) are to this configuration, and flew for the first time on 28 June and 20 November 1973 respectively. Next three (ie first three production) aircraft for Air France; the first of these flew for the first time on 15 April 1974 and was delivered on 11 May to Air France, with which it entered service on 23 May 1974 on the Paris-London route. Type certificated by SGAC and LBA on 15 March 1974, and by FAA on 30 May 1974.

A300B2K. Basically as B2, but fitted with the Krueger wing-root leading-edge flaps developed originally for B4, enabling the B2K to operate from 'hot and high' airports.

A300B4. Long-range development of B2, with same commercial capacity but increased design weights and fuel capacity. Krueger flap on each leading-edge wing root to improve take-off performance. Ninth aircraft (F-WLGA)

Airbus Industrie A300B2 wide-bodied transport aircraft (two General Electric CF6-50C turbofan engines) in the insignia of Lufthansa

is the first built to B4 configuration; first flight took place on 26 December 1974. French and German certification granted on 26 March 1975. First delivery made in May 1975 to Germanair, with which it entered service on 1 June 1975.

Possible future versions currently under consideration in 1976 include freighter/convertible versions of the B4 with large cargo doors; a military refuelling tanker version of the B4; the B9 'stretched' version with 240·2 kN (54,000 lb st) engines and six more rows of seats than the B2; the B10 (with Rolls-Royce RB.211, CF6-6 or JT9D engines and up to 214 passengers in a shortened fuselage); and the B11, a short-fuselage (180-200 passengers) extended-span version with four 111·2-133·4 kN (25,000-30,000 lb st) engines and a range of 6,000 nm (11,119 km; 6,909 miles).

To transport large components of the A300 between the various factories involved, Airbus Industrie acquired two Guppy-201 transport aircraft from the USA, which are operated on its behalf by Aéromaritime.

By 1 May 1976, orders and options totalling 57 aircraft had been signed, as follows:

	Orders	*Options*
Air France	6 B2/3 B4	7 B2/B4
Air Inter	3 B2	—
Germanair	2 B4	—
Indian Airlines	3 B2	3 B2
Korean Air Lines	6 B4	—
Lufthansa	4 B2	8 B2
South African Airways	4 B2K	—
Transavia (Holland)	1 B4*	—
Trans European Airways (Belgium)	1 B1*/1 B4	1 B4
Undisclosed	—	4 B2K

**Leased from Airbus Industrie*

A total of 19 A300s had been delivered by 1 May 1976. Production was scheduled to reach the rate of two aircraft per month by the end of 1976. Manufacture of the first 84 production aircraft has been authorised, and long-lead items have been ordered for the next 32.

Type: Large-capacity short/medium-range transport.

Wings: Cantilever mid-wing monoplane. Thickness/chord ratio 10·5%. Sweepback 28° at quarter-chord. Primary two-spar box-type structure, integral with fuselage and incorporating fail-safe principles, built of high-strength aluminium alloy. Third spar across inboard sections. Machined skin with open-sectioned stringers. Each wing has three-section leading-edge slats (no slat cutout over the engine pylon), and three Fowler-type double-slotted flaps on trailing-edge; a Krueger flap on the leading-edge wing root (B2K/B4); an all-speed aileron between inboard flap and outer pair; and a low-speed aileron outboard of the outer pair of flaps. Lift dump facility by combination of three spoilers (outboard) and two airbrakes (inboard) on each wing, forward of outer pair of flaps, plus two additional airbrakes forward of inboard flap. The flaps extend over 84% of each half-span, and increase the wing chord by 25% when fully extended. The datum of the all-speed aileron is deflected downward by up to 10° with flap operation to maintain trailing-edge continuity with deflected flaps. Drive mechanisms for flaps and slats are similar to one another, each powered by twin motors driving ball screwjacks on each surface with built-in protection against asymmetric operation. Two slat positions for take-off and landing. Pre-selection of the airbrake/lift dump lever allows automatic extension of the lift dumpers on touchdown. All flight controls are powered by triplex hydraulic servo-jacks, with no manual reversion. Anti-icing of wing leading-edges, outboard of engine pods, is by hot air bled from engines.

Fuselage: Conventional semi-monocoque structure of circular cross-section, with frames and open Z-section stringers. Built mainly of high-strength aluminium alloy, with steel or titanium for some major components. Skin panels integrally machined in areas of high stress. Honeycomb panels or restricted glassfibre laminates for secondary structures.

Tail Unit: Cantilever all-metal structure, with sweepback on all surfaces. Variable-incidence tailplane and separately-controlled elevators. Tailplane powered by two motors driving a fail-safe ball screwjack. No anti-icing of leading-edges.

Landing Gear: Hydraulically-retractable tricycle type, of Messier-Hispano design, with Messier-Hispano shock-absorbers and wheels. Twin-wheel nose unit retracts forward, main units inward into fuselage. Free-fall extension. Each four-wheel main unit comprises two tandem-mounted bogies, interchangeable left with right, with tyres size 46 × 16·20, pressure 11·6 bars (168 lb/sq in) on B2, 13·3 bars (193 lb/sq in) on B4. Nosewheel tyres size 40 × 14-16, pressure 7·93 bars (115 lb/sq in) on B2, 9·52 bars (138 lb/sq in) on B4. Steering angles 65°/95°. SNECMA (Hispano) hydraulic disc brakes on all main wheels. Duplex anti-skid units fitted, with a third standby hydraulic supply for wheel brakes.

Power Plant (B2 and B4): Two General Electric CF6-50C turbofan engines, each of 227 kN (51,000 lb st), assembled under licence by SNECMA and MTU and mounted in underwing pods, fitted with thrust reversers which are actuated pneumatically by engine bleed air. Nacelles supplied by McDonnell Douglas. Fuel in two integral tanks in each wing, with total usable capacity of 43,000 litres (9,460 Imp gallons) in B2. Total usable capacity 56,600 litres (12,450 Imp gallons) in B4. Two refuelling points standard beneath each wing, outboard of engines.

Accommodation (B2 and B4): Crew of three on flight deck, with provision for two-man operation. Electrical de-icing and demisting of windscreen. Seating for between 220 and 320 passengers in main cabin in six, seven or eight-abreast layout with 787/864 mm (31/34 in) seat pitch. Typical economy class layout has 269 seats, eight abreast with two aisles, at 864 mm (34 in) seat pitch. This layout includes one galley and one toilet forward, with provision for a second one, and one more galley and four toilets aft. Up to 336 passengers can be carried at 762 mm (30 in) seat pitch in single-class high-density layout. Closed hatracks on each side, forming baggage lockers (max capacity 0·062 m³; 2·19 cu ft). Provision for central double-sided rack. Two outward parallel-opening plug-type passenger doors ahead of wing leading-edge on each side, and one on each side at rear. Underfloor baggage/cargo holds fore and aft of wings, with doors on starboard side. The forward hold will accommodate four 2·24 × 3·18 × 1·63 m (88 × 125 × 64 in) pallets or twelve LD3 or IATA A1 containers; the rear hold will accommodate eight LD3 containers each of 4·25 m³ (150 cu ft) capacity. Additional bulk loading of freight provided for in an extreme rear compartment with usable volume of 16·0 m³ (565 cu ft). The latter compartment can be used for the transport of livestock. Entire accommodation is pressurised, including freight, baggage and electronics compartments.

Systems: Air for air-conditioning system can be provided from engines, the APU or a high pressure ground source. Supply is controlled by separate and parallel bootstrap-type units, each of which includes a flow limiting unit, cooler unit, water separator and temperature control unit. In addition, air from each engine passes through a pressure control pre-cooler unit. Distribution in flight deck and three cabin areas, with independent regulation. Two independent automatic systems, with manual override, control the cabin altitude, its rate of change and the differential pressure. Cabin pressure differential for normal operations is 0·57 bars (8·25 lb/sq in). Hydraulic system comprises three fully-independent circuits, operating simultaneously. Fluid used is a fire-resistant phosphate-ester type, working at a pressure of 207 bars (3,000 lb/sq in). The three circuits provide triplex power for primary flying controls; if any circuit fails, full control of the aircraft is retained without any necessity for action by the crew. All three circuits supply the all-speed and low-speed ailerons, rudder and elevator; 'blue' circuit additionally supplies tail trim, spoilers, slats and rudder variable-gear unit; 'green' circuit additionally supplies airbrakes, spoilers, slats, elevator artificial feel units, flaps, steering, wheel brakes and normal landing gear requirements; 'yellow' circuit additionally supplies tail trim, airbrakes, lift dumpers, rudder variable-gear unit, elevator artificial feel unit, flaps, wheelbrakes and steering. Each circuit normally powered by engine-driven self-regulating pumps, one on each engine for the green circuit and one each for the blue and yellow circuits. Dowty Rotol ram-air turbine-driven pump provides standby hydraulic power should both engines become inoperative. Main electrical power is supplied by two Westinghouse three-phase constant-frequency AC generators mounted on the engines. A third identical generator, driven by the APU, can supply power both in flight, to replace a failed engine-driven generator, and on the ground. Supply frequency is 400Hz and voltage is 115/200V. Any one generator can supply sufficient power to operate all equipment and systems necessary for take-off and landing. A conventional generator CSD system is installed, the two units being mounted on opposite sides of the engine gearbox with the CSD driving an aircooled generator at a constant 8,000 rpm. Each generator is rated at 90kVA, with overload ratings of 135kVA for 5 minutes and 180kVA for 5 seconds. The APU generator is driven at constant speed through a gearbox. Three unregulated transformer-rectifier units (TRUs) supply 28V DC power. Three 24V 25Ah nickel-cadmium batteries are used for APU starting and fuel control, engine starter control, standby lights and, by selection, emergency busbar. This busbar and a 115V 400Hz static inverter provide standby power in flight if normal power is unavailable. This system is separated completely from the main system. Hot air protection for engine intakes and slat sections on the wings outboard of the engine. Garrett-AiResearch TSCP 700-5 APU in tailcone, exhausting upward. The installation incorporates APU noise attenuation. Fire protection system is self-contained, and firewall panels protect main structure from an APU fire. The APU can be operated on the ground, in flight up to 10,675 m (35,000 ft), and in icing conditions. Relights are possible up to 7,620 m (25,000 ft). Aircraft is completely independent of ground power sources, since all major services can be operated by the APU.

Electronics and Equipment: Standard communications equipment includes two VHF sets and one Selcal system, plus interphone and passenger address systems. An accident recorder and voice recorder are also installed. Standard navigation equipment includes two VOR, two ILS, two radio altimeters, one marker beacon, two ADF, two DME, two ATC transponders and a weather radar. Most other electronic equipment available to customer's requirements, only those related to the blind landing system (VOR/ILS and radio altimeter) being selected and supplied by the manufacturer. Additional optional equipment includes one or two HF sets, third VHF, second marker beacon, third VOR/ILS, second radar, navigation computer and pictorial display. Both the pilot and co-pilot have an integrated instrument system combining heading and attitude (three SAGEM MGC 10/ARINC 569 are standard in B2, but one or two of these can be replaced by MGC 30/ARINC 571 Mk 1 inertial sensors, which are modular with the MGC 10); SFENA flight director system; and radio information. The SFENA/Smiths/Bodenseewerk automatic flight control system includes a comprehensive range of en-route facilities such as VOR coupling, heading select, height acquire, turbulence, rate of descent (if required) and control wheel steering, in addition to the normal height, speed, pitch-and-roll attitude and heading locks. Dual automatic landing system provides coupled approach and automatic landing facilities suitable for Category 3A operation. The system is designed to allow future extension to Category 3B automatic landing capability.

DIMENSIONS, EXTERNAL (B2 and B4):
Wing span — 44·84 m (147 ft 1 in)
Wing aspect ratio — 7·73
Length overall — 53·62 m (175 ft 11 in)
Length of fuselage — 52·03 m (170 ft 8½ in)
Fuselage max diameter — 5·64 m (18 ft 6 in)
Height overall — 16·53 m (54 ft 2¾ in)
Tailplane span — 16·94 m (55 ft 7 in)
Wheel track — 9·60 m (31 ft 6 in)
Wheelbase (c/l of shock-absorbers) — 18·60 m (61 ft 0 in)
Passengers doors (each):
Height — 1·93 m (6 ft 4 in)
Width — 1·07 m (3 ft 6 in)
Height to sill:
fwd — 4·60 m (15 ft 1 in)
centre — 4·80 m (15 ft 9 in)
rear — 5·50 m (18 ft 0½ in)
Emergency exits (each):
Height — 1·60 m (5 ft 3 in)
Width — 0·61 m (2 ft 0 in)
Height to sill — 4·87 m (15 ft 10 in)
Underfloor cargo door (fwd):
Height — 1·70 m (5 ft 7 in)
Width — 2·44 m (8 ft 0 in)
Height to sill — 2·56 m (8 ft 4¾ in)
Underfloor cargo door (rear):
Height — 1·70 m (5 ft 7 in)
Width — 1·81 m (5 ft 11¼ in)
Height to sill — 2·96 m (9 ft 8½ in)
Underfloor cargo door (extreme rear):
Height — 0·95 m (3 ft 1 in)
Width — 0·95 m (3 ft 1 in)
Height to sill — 3·30 m (10 ft 10 in)

DIMENSIONS, INTERNAL (B2 and B4):
Cabin, excl flight deck:
Length — 39·15 m (128 ft 6 in)
Max width — 5·35 m (17 ft 7 in)
Max height — 2·54 m (8 ft 4 in)
Underfloor cargo hold:
Length:
fwd — 10·60 m (34 ft 9¼ in)
rear — 6·89 m (22 ft 7¼ in)
extreme rear — 3·10 m (10 ft 2 in)
Max height — 1·76 m (5 ft 9¼ in)
Max width — 4·20 m (13 ft 9¼ in)
Underfloor cargo hold volume:
fwd — 75·1 m³ (2,652 cu ft)
rear — 46·8 m³ (1,652 cu ft)
extreme rear — 16·0 m³ (565 cu ft)
Max total volume for bulk loading — 140·0 m³ (4,944 cu ft)

AREAS (B2 and B4):
Wings, gross — 260·0 m² (2,799 sq ft)
Vertical tail surfaces (total) — 45·2 m² (486·5 sq ft)
Horizontal tail surfaces (total) — 69·5 m² (748·1 sq ft)

WEIGHTS AND LOADINGS:
Manufacturer's weight empty:
B2 — 77,062 kg (169,890 lb)
B4 — 78,655 kg (173,404 lb)
Typical operating weight empty:
B2 — 85,900 kg (189,375 lb)
B4 — 88,200 kg (194,445 lb)
Max payload (structural):
B2 — 30,590 kg (67,440 lb)
B4 — 35,050 kg (77,270 lb)
Max T-O weight:
B2 (original certification) — 137,900 kg (304,015 lb)
B2 (current) — 142,000 kg (313,055 lb)
B4 (standard) — 150,000 kg (330,700 lb)
B4 (optional) — 157,500 kg (347,225 lb)
Max ramp weight:
B2 — 142,900 kg (315,040 lb)
B4 — 158,400 kg (349,210 lb)
Max landing weight:
B2 — 130,000 kg (286,600 lb)
B4 — 133,000 kg (293,200 lb)
Max zero-fuel weight:
B2 — 120,500 kg (265,655 lb)
B4 — 122,000 kg (269,000 lb)
Max wing loading:
B2 — 546 kg/m² (111·8 lb/sq ft)
B4 — 606 kg/m² (124·1 lb/sq ft)
Max power loading:
B2 — 312·8 kg/kN (3·07 lb/lb st)
B4 — 330·4 kg/kN (3·24 lb/lb st)

PERFORMANCE (at max T-O weight except where indicated):
Max operating speed (V_{MO}):
B2 at 137,900 kg (304,015 lb) AUW, B4 at 150,000 kg (330,700 lb) — 360 knots (668 km/h; 415 mph) CAS
B4 at optional AUW of 157,500 kg (347,225 lb) — 345 knots (639 km/h; 397 mph) CAS
Max operating Mach number (M_{MO}):
B2, B4 — Mach 0·86
Max cruising speed at 7,620 m (25,000 ft):
B2, B4 — 492 knots (911 km/h; 567 mph) TAS
Typical high-speed cruise at 9,145 m (30,000 ft):
B2, B4 — 495 knots (917 km/h; 570 mph) TAS
Typical long-range cruising speed at 9,450 m (31,000 ft):
B2, B4 — 457 knots (847 km/h; 526 mph) TAS
Approach speed at typical weight:
B2 — 131 knots (243 km/h; 151 mph)
B4 — 132 knots (245 km/h; 152 mph)
Max operating altitude:
B2 — 10,675 m (35,000 ft)
Min ground turning radius (wingtips) — 33·51 m (109 ft 11¼ in)
Runway LCN at max T-O weight:
0·76 m (30 in) radius of rigidity:
B2 — 63
B4 — 72
1·02 m (40 in) radius of rigidity:
B2 — 74
B4 — 85
T-O field length (S/L, ISA + 15°C):
B2 — 1,951 m (6,400 ft)
B4 — 2,500 m (8,200 ft)
Landing field length at typical weight:
B2 — 1,630 m (5,350 ft)
B4 — 1,660 m (5,445 ft)
Range with 269 passengers and baggage:
B2 — 1,800 nm (3,334 km; 2,074 miles)
B4 at 157,500 kg (347,225 lb) AUW — 2,500 nm (4,631 km; 2,878 miles)
Range with max fuel:
B2 — 2,300 nm (4,261 km; 2,648 miles)
B4 at 157,500 kg (347,225 lb) AUW — 3,200 nm (5,930 km; 3,685 miles)

OPERATIONAL NOISE CHARACTERISTICS (FAR Pt 36):
T-O noise level:
B2 — 90 EPNdB
B4 — 92 EPNdB
Approach noise level:
B2, B4 — 101 EPNdB
Sideline noise level:
B2, B4 — 95 EPNdB

ALPHA JET

AIRFRAME PRIME CONTRACTORS:

Avions Marcel Dassault/Breguet Aviation,
BP 32, 92420-Vaucresson, France
Telephone: 970-75-21
Telex: 0734372

Dornier GmbH, Postfach 317, 7990 Friedrichshafen, German Federal Republic
Telephone: (07545) 81
Telex: 0734372

The Alpha Jet 04 prototype, representative of the trainer version for the French Air Force

On 22 July 1969 the French and German governments announced a joint requirement for a new subsonic basic and advanced training aircraft to enter service with the French and German armed forces in the mid-1970s. Each government had a potential requirement for about 200 such aircraft to replace Magister and Lockheed T-33A trainers in service, and two designs were studied during the first half of 1970. These were the Aérospatiale/MBB E 650 Eurotrainer and the Dassault-Breguet/Dornier Alpha Jet.

On 24 July 1970, it was announced that the Alpha Jet design had been selected for development to meet the requirement. The aircraft has been developed also for close air support and battlefield reconnaissance duties, following a change in Luftwaffe requirements.

DASSAULT-BREGUET/DORNIER ALPHA JET

The Dassault-Breguet group of France and Dornier of Germany are jointly developing the Alpha Jet, with Dassault-Breguet as main contractor and Dornier as industrial collaborator, the total work load being shared primarily between the two groups. SABCA and Fairey of Belgium will manufacture nose sections and wing flaps.

On 15 February 1971 the project definition phase of the Alpha Jet was completed, and design work for the development phase was begun in the Autumn of 1971. This received joint Franco-German government approval in late 1972; approval to proceed with the production phase was announced on 26 March 1975, and work on the first production aircraft began towards the end of that year.

Four prototypes were built, the first and third assembled in France and the second and fourth in Germany, plus two airframes for static and fatigue testing.

Flight testing is being carried out predominantly in France, by both French and German pilots, each prototype having made its first few test flights in the country where it was assembled. Prototypes 01 and 02 are being used to finalise systems installations and for flight and performance evaluation; the 03 is representative of the production close support version, and the 04 of the French trainer version. The 01 was completed at St Cloud in mid-June 1973, and the first functional test (of the fuel system) was made on 26 July 1973. This prototype made its first flight, at Istres, on 26 October 1973, followed by the 02 (D-9594) at Oberpfaffenhofen on 9 January 1974, the 03 (40 + 01) at Istres on 6 May 1974 and the 04 (D-9595/F-ZWRX) at Oberpfaffenhofen on 11 October 1974. Flight tests with external stores (on the 02) began on 19 April 1974. By the Spring of 1976 the four prototypes had accumulated more than 1,150 flying hours. The 04 was lost in a crash in the Summer of 1976.

As soon as the flight test programme permits, one Alpha Jet prototype is to be modified by Dornier, under a German Ministry of Defence programme, by fitting an improved transonic wing of supercritical aerofoil section with manoeuvre flaps on the trailing-edge and along the full span of the leading-edge.

French and German production Alpha Jets will have identical airframe, power plant, landing gear and standard equipment; there are assembly lines for production Alpha Jets in each country. The outer wings, tail unit, rear fuselage and cold-flow exhaust are manufactured in Germany; the forward and centre fuselage (with integrated wing centre-section) are manufactured in France. Fuselage nose sections and wing flaps will be manufactured in Belgium by Fairey SA and SABCA. Final assembly of the trainer version takes place in France, and of the close support version in Germany. The power plant prime contractors are Turboméca and SNECMA in France, and MTU and KHD in Germany; and, for the landing gear, Messier-Hispano in France and Liebherr Aero Technik in Germany.

The Federal German government has approved the purchase of 175 Alpha Jets of the close support version; deliveries of these are scheduled to begin in mid-1978. Firm initial contracts were placed in January 1976 for 56 Alpha Jet trainers for France and 84 close support Alpha Jets for Germany. Belgium has chosen the Alpha Jet as its next military trainer, and has ordered 16 initially, with a further 17 to follow.

TYPE: Tandem two-seat basic, low-altitude and advanced jet trainer and close support and battlefield reconnaissance aircraft.

WINGS: Cantilever shoulder-wing monoplane, with 6° anhedral from roots. Thickness/chord ratio 10·2% at root, 8·6% at tip. Sweepback 28° at quarter-chord. All-metal numerically- or chemically-milled structure, consisting of two outer wings bolted to a centre frame. Hydraulically-actuated double-slotted flaps on each

Third Alpha Jet prototype, representative of the production close support/battlefield reconnaissance version for the Federal German Luftwaffe

trailing-edge. Ailerons actuated by double-body hydraulic servo, with trimmable artificial feel system.

FUSELAGE: All-metal semi-monocoque structure, numerically or chemically milled, of basically oval cross-section. Built in three sections: nose (including cockpit), centre-section (including engine air intake trunks and main landing gear housings) and rear (including engine mounts and tail assembly). Electrically-controlled, hydraulically-actuated airbrake on each side of rear upper fuselage, of carbon-fibre-reinforced epoxy resin construction.

TAIL UNIT: Cantilever all-metal type, of similar construction to wings, with 45° sweepback on fin leading-edge and 30° on tailplane leading-edge. Dorsal spine fairing between cockpit and fin. All-flying tailplane, with trimmable and IAS-controlled artificial feel system. Double-body hydraulic servo-actuated rudder, with trimmable artificial feel system.

LANDING GEAR: Forward-retracting tricycle type, of Liebherr/Messier-Hispano design. All units retract hydraulically, main units into underside of engine air intake trunks. Single wheel and low-pressure tyre (approx 3·93 bars; 57 lb/sq in at normal T-O weight) on each unit. Tyre sizes 615 × 255-10 on main units, 380 × 150-4 on nose unit. Steel disc brakes and anti-skid units on main gear. Emergency braking system. Hydraulic nosewheel steering and arrester hook on German version. Nosewheel offset to starboard to permit ground firing from gun pod.

POWER PLANT: Two SNECMA/Turboméca Larzac 04 turbofan engines, each rated at 13·2 kN (2,965 lb st) for production aircraft, mounted on sides of fuselage. Splitter plate in front of each intake. Fuel in two integral tanks in outer wings, one in centre-section and three fuselage tanks. Internal fuel capacity 1,380 litres (303·5 Imp gallons) in French basic trainer version; 1,900 litres (418 Imp gallons) in French low-altitude trainer and German close-support versions. Provision for 310 litre (68·2 Imp gallon) capacity drop-tank on each outer wing pylon. Pressure refuelling standard for all tanks, including drop-tanks. Gravity system optional for fuselage tanks and drop-tanks. Pressure refuelling point near starboard engine air intake. Fuel system incorporates provision for short periods of inverted flying.

ACCOMMODATION: Two persons in tandem, in pressurised cockpit under individual upward-opening canopies. Cabin pressure differential 0·30 bars (4·3 lb/sq in). Rear seat (for instructor in trainer versions) is elevated. Prototypes and French trainer versions fitted with Martin-Baker Mk 4 ejection seats, operable (including ejection through canopy if necessary) at zero height and speeds down to 90 knots (166 km/h; 103 mph). Aircraft for Germany fitted with licence-built (by RFB) Stencel SIIIS zero-zero ejection seats.

SYSTEMS: Cockpit air-conditioning and demisting system. Two independent hydraulic systems, each 207 bars (3,000 lb/sq in), with engine-driven pumps (emergency electric pump on one circuit), for actuating control surfaces, landing gear, brakes, flaps, airbrakes, and (when fitted) nosewheel steering. Pneumatic system, for cockpit pressurisation and air-conditioning, occupants' pressure suits and fuel tank pressurisation, is supplied by compressed air from engines. Main electrical power supplied by two 28V 9kW starter/generators, one on each engine. Circuit includes a 36Ah nickel-cadmium battery and two static inverters for supplying 115V AC current at 400Hz to auxiliary systems. An external ground DC power receptacle is fitted. Hydraulic and electrical systems can be sustained by either engine in the event of the other engine becoming inoperative. Oxygen mask for each occupant, supplied by liquid oxygen converter of 10 litres (2·2 Imp gallons) capacity. Emergency gaseous oxygen bottle for each occupant.

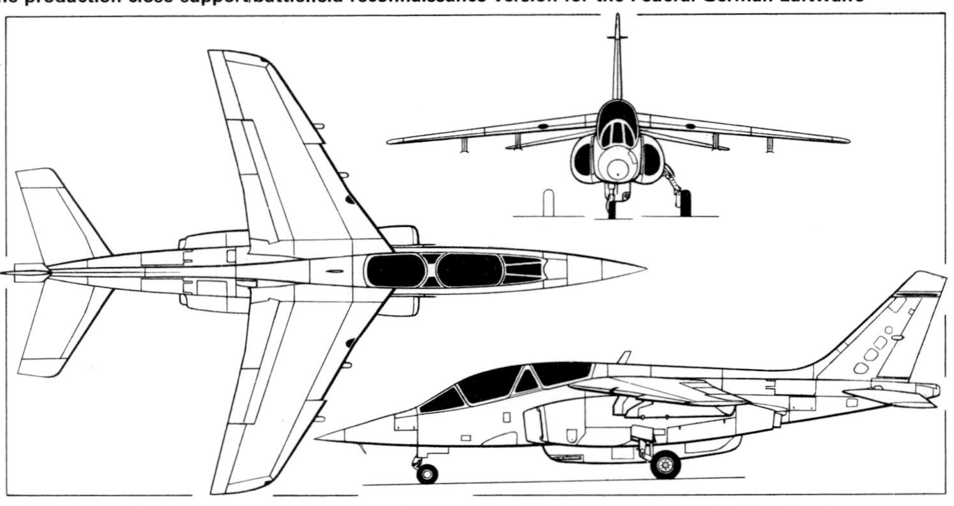

Alpha Jet basic and advanced jet trainer and close support aircraft *(Pilot Press)*

ELECTRONICS AND EQUIPMENT: Dual controls standard. Large electronics bay in rear fuselage, containing most of the radio and navigation equipment. Standard equipment includes VHF and UHF transceivers (optionally, UHF and emergency UHF respectively), IFF/SIF, VOR/ILS and intercom. Landing light in each inboard wing leading-edge. Optional equipment includes Dornier crash recorder, VOR/ILS with marker, Tacan, navigation computer and radio altimeter.

ARMAMENT AND OPERATIONAL EQUIPMENT: For armament training and light close support missions, the Alpha Jet can be equipped with an underfuselage detachable pod containing a 30 mm DEFA or 27 mm Mauser cannon with 150 rds, or a pod with two 12·7 mm machine-guns and 250 rds/gun. Provision also for two hardpoints under each wing, with non-jettisonable pylons, the inner ones each stressed for loads of up to 615 kg (1,356 lb) and the outer ones for up to 285 kg (630 lb). On these can be carried, within the load capacity for each station, pods of up to thirty-six 2·75 in rockets; HE bombs of 50, 125, 250 or 400 kg; 625 lb cluster dispensers; 690 or 825 lb fire bombs; combined launchers for rockets and 360 lb bombs; practice launchers for bombs or rockets; or two drop-tanks (outer pylons only). Provision also for carrying target demonstration devices or an underfuselage reconnaissance pod. Max permissible payload for all five stations is 2,200 kg (4,850 lb). Fire control system for air-to-air or air-to-ground firing, dive bombing and low-level bombing. Firing by trainee pilot (in front seat) is governed by a safety interlock system controlled by the instructor, which energises the forward station trigger circuit and illuminates a fire clearance indicator in the trainee's cockpit. Thompson-CSF weapon sight (Type 902 in French version, Type 903 in German version) and gun camera.

DIMENSIONS, EXTERNAL:
- Wing span: 9·11 m (29 ft 10¾ in)
- Wing aspect ratio: 4·8
- Length overall (excl nose-probe): 12·29 m (40 ft 3¾ in)
- Height overall (at normal T-O weight): 4·19 m (13 ft 9 in)
- Tailplane span: 4·34 m (14 ft 2¾ in)
- Wheel track: 2·71 m (8 ft 10¾ in)
- Wheelbase: 4·716 m (15 ft 5¾ in)

AREAS:
- Wings, gross: 17·50 m² (188·4 sq ft)
- Ailerons (total): 1·04 m² (11·19 sq ft)
- Trailing-edge flaps (total): 2·86 m² (30·78 sq ft)
- Airbrakes (total): 0·74 m² (7·97 sq ft)
- Fin: 2·97 m² (31·97 sq ft)
- Rudder: 0·62 m² (6·67 sq ft)
- Horizontal tail surfaces (total): 3·94 m² (42·41 sq ft)

WEIGHTS AND LOADINGS:
- Weight empty, equipped: 3,150 kg (6,944 lb)
- Operational weight empty: 3,475 kg (7,661 lb)
- Normal T-O weights:
  - trainer, 'clean': 4,540 kg (10,010 lb)
  - weapon training or close support: 6,000 kg (13,227 lb)
- Max T-O weight (exceptional): 7,000 kg (15,432 lb)
- Combat wing loading ('clean'): 230 kg/m² (47·1 lb/sq ft)
- Combat power loading ('clean'): 246 kg/kN (1·5 lb/lb st)

PERFORMANCE (at normal 'clean' T-O weight, except where indicated):
- Max level speed at high altitude: Mach 0·85
- Max level speed at low altitude: 535 knots (991 km/h; 616 mph)
- Approach speed: 120 knots (222 km/h; 138 mph)
- Landing speed at normal landing weight: less than 100 knots (185 km/h; 115 mph)
- Rate of climb at S/L, one engine out, at 4,782 kg (10,542 lb) AUW, in landing configuration: 330 m (1,085 ft)/min
- Time to 12,000 m (39,375 ft): less than 10 min
- Service ceiling: 14,020 m (46,000 ft)
- T-O run: 400 m (1,310 ft)
- Landing run: 520 m (1,705 ft)
- Endurance:
  - low altitude: more than 2 hr
  - high altitude: more than 3 hr
- Radius of action:
  - trainer ('clean'), low altitude: 237 nm (440 km; 273 miles)
- Combat radius (incl 5 min combat) with max external load:
  - ground attack, lo-lo-lo: 220 nm (410 km; 254 miles)
  - ground attack, hi-lo-hi: 339 nm (630 km; 391 miles)
- Ferry range (internal fuel and two 310 litre external tanks): 1,402 nm (2,600 km; 1,615 miles)
- *g* limits (ultimate): +12; −6·4

CONCORDE

CONCORDE SUPERSONIC TRANSPORT

AIRFRAME PRIME CONTRACTORS:

British Aircraft Corporation Ltd, Brooklands Road, Weybridge, Surrey KT13 0RN, England
Telephone: Weybridge (97) 45522

Aérospatiale (SNIAS), 37 Boulevard de Montmorency, 75781 Paris-cédex 16, France
Telephone: 224-84-00

POWER PLANT PRIME CONTRACTORS:

Rolls-Royce (1971) Ltd, PO Box 3, Filton, Bristol, England
Telephone: 0272-693871

Société Nationale d'Etude et de Construction de Moteurs d'Aviation, 150 Boulevard Haussmann, 75361 Paris-cédex 08, France
Telephone: 227-33-94

CONCORDE

Anglo-French negotiations concerning the development of a supersonic transport aircraft culiminated on 29 November 1962 in the signing of two agreements, one between the French and British governments, the other between the manufacturers to whom the project was entrusted. The agreements provided for a fair division of the work, responsibility and development costs among the partners, and covered the manufacture of two Concorde prototypes, followed by two pre-production aircraft and two airframes for static and fatigue testing. The static test programme was completed in September 1973, and this airframe was tested to destruction in June 1974. Fatigue testing is programmed to continue until two aircraft 'lives' (about 48,000 flights) have been attained.

The planned test programme, involving the two prototype flight, two pre-production and first four production Concordes, achieved its target of 5,335 hr flying at the time when the full passenger-carrying certificate of airworthiness was granted by the SGAC and CAA in late 1975. By the time of first deliveries to British Airways and Air France in early 1976, the 10 Concordes to have flown had amassed a total block time of 5,700 hr, of which more than 2,000 hr were supersonic. The total number of flights made up to that time exceeded 2,500.

Initially, delivery positions for 74 Concordes were reserved by 16 airlines, as listed in the 1972-73 *Jane's,* but this option system was abolished on 28 March 1973. The first firm order, for five aircraft, was announced by BOAC on 25 May 1972. It was followed shortly afterwards by announcement of an Air France order for four Concordes and these two contracts were signed on 28 July 1972.

Operationally, development of the new Type 28 thrust reverser nozzle promises a continuing substantial reduction in noise levels, and the engine smoke problem was virtually eliminated with the installation of Olympus 593 Mk 602 engines in the 02 (second pre-production) and subsequent aircraft.

When sonic boom considerations preclude use of the normal climb technique, sufficient power is available to increase the transonic acceleration height to over 12,200 m (40,000 ft). Normally, however, the aircraft will accelerate and climb from 200 knots (370 km/h; 230 mph) CAS at S/L to 400 knots (740 km/h; 460 mph) CAS at 1,500 m (5,000 ft), then climb at a constant CAS of 400 knots (740 km/h; 460 mph) to 11,000 m (36,000 ft) where its speed will be Mach 1·15, climb and accelerate to Mach 1·8 (530 knots; 980 km/h; 610 mph CAS) at 13,800 m (45,300 ft) and continue climbing at this CAS until the cruise Mach number is reached, finally climbing to cruising height at cruising Mach number.

Airframe development and production of the Concorde is undertaken jointly by Aérospatiale and BAC, with two final assembly lines, at Toulouse and Filton respectively. There is no duplication of main production jigs.

Aérospatiale is responsible for development and production of the rear cabin section, wings and wing control surfaces, hydraulic systems, flying controls, navigation systems, radio and air-conditioning system. The automatic flight control system is designed by Marconi-Elliott in the UK and SFENA in France, under contract to Aérospatiale. BAC is responsible for the three forward sections of the fuselage, the rear fuselage and vertical tail surfaces, the engine nacelles and ducting, the electrical system, sound and thermal insulation, oxygen system, fuel system, engine installation, and fire warning and extinguishing systems.

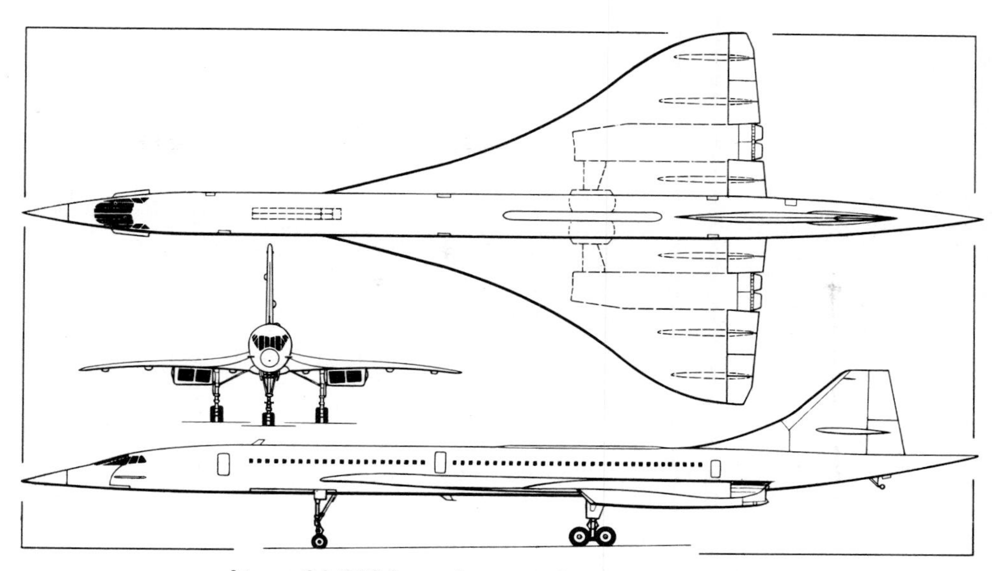

Aérospatiale/BAC Concorde supersonic transport *(Pilot Press)*

The following versions of Concorde have been built:

Concorde 001. First prototype (F-WTSS), assembled by Aérospatiale and first flown on 2 March 1969. Generally similar to later aircraft, but with lower-powered Olympus 593 engines, shorter fuselage (56·24 m; 184 ft 6 in) and design gross weight of 148,000 kg (326,000 lb). Described in previous editions of *Jane's.* Retired on 19 October 1973 to French Air Museum at Le Bourget Airport.

Concorde 002. Second prototype (G-BSST), assembled by BAC and first flown on 9 April 1969. Design gross weight and general configuration as 001. Used for 'hot and high' trials at Jan Smuts Airport, Johannesburg, and for high-temperature certification trials at Torrejon, Spain, in 1973. Has completed its part of the flight test programme. Further details in previous editions of *Jane's.*

Concorde 01. First pre-production aircraft (G-AXDN), assembled in UK and first flown on 17 December 1971. Lengthened fuselage, fully transparent retractable windscreen visor and increased weights. On 7 November 1974, en route from Fairford to Bangor, Maine, made the fastest-ever civil crossing of the North Atlantic to the US mainland, in 2 hr 56 min. Used for tropical icing trials in Nairobi in 1975. Has completed its part of the flight test programme. Further details in previous editions of *Jane's.*

Concorde 02. Second pre-production aircraft (F-WTSA), assembled in France and first flown on 10 January 1973. First Concorde to incorporate Type 28 thrust reverser nozzles; production-standard longer rear fuselage and increased fuel capacity; improvements in wing leading-edge camber (to improve airflow into the engines) and wingtip shape; and modifications to engine exhaust assembly. First to be powered by initial production standard Olympus 593 Mk 602 engines. After its third flight, on 13 January 1973, Concorde 02 made its first fully-automatic landing. Used for cold-weather trials at Fairbanks, Alaska, in 1974. Two return trips between 13-18 June 1974, from Paris to Boston and back. On one of these, on 17 June, Concorde left Boston as a Boeing 747 left Paris for Boston; Concorde flew to Paris, and returned to Boston, before the 747 had reached Boston. Since that time it has been engaged in carbon brake and debris ingestion tests. Details of earlier performance achievements in previous editions of *Jane's.*

Production Concorde. First two production aircraft flown on 6 December 1973 at Toulouse (F-WTSB) and 13 February 1974 at Filton (G-BBDG), each attaining a speed of approx 868 knots (1,610 km/h; 1,000 mph) on its first flight. Second production aircraft used for hot weather flight and ground handling trials in Bahrain, and for a series of Middle East demonstration flights, in August 1974. These were followed by a 3,456 nm (6,405 km; 3,980 mile) flight from Bahrain to Singapore in 3 hr 38 min, on 3 September 1974.

The first two production aircraft flew to Casablanca, one for high-altitude low-temperature air intake tests, the other for take-off and climb performance, systems engineering and noise measurement tests, in October 1974.

Third and fourth production aircraft (F-WTSC and G-BOAC) flew for the first time at Toulouse and Filton on 31 January and 27 February 1975 respectively. These two aircraft were used for the endurance flying programme involving the manufacturers, Air France and British Airways in 830 hr of simulated airline operations to North and South America, the Middle and Far East, and Australia. For this purpose, special category certificates of airworthiness were granted in May and June 1975 by the SGAC and CAA, anticipating the full airworthiness certificates which were granted on 13 October and 5 December 1975. On 1 September 1975, the fourth production Concorde became the first aircraft to make two return crossings of the North Atlantic (London Heathrow-Gander-London) in one day.

The fifth and sixth production aircraft made their first flights on 25 October and 5 November 1975 respectively, and entered service simultaneously on 21 January 1976 between London-Bahrain (British Airways G-BOAA) and Paris-Dakar-Rio de Janeiro (Air France F-BVFA). Simultaneous opening of services to the United States began on 24 May 1976 with flights to Washington's Dulles International Airport from London and Paris. The seventh and eighth production aircraft flew on 6 March and 18 May 1976, and construction of a further eight aircraft is proceeding. British Airways has ordered five production Concordes, and Air France four. Both airlines

Fourth production example of the Aérospatiale/BAC Concorde four-jet supersonic transport aircraft, in British Airways insignia

Fifth production Concorde, used on the first scheduled Air France service from Paris to Rio de Janeiro on 21 January 1976

plan to operate trans-Atlantic services to New York, subject to clearance by the Port of New York Authority. The FAA has already announced, on 4 March 1975, that it has no objection to the Concorde operating into and out of New York, and a 16-month trial period of operation into the United States began with the opening of the service to Washington. Preliminary purchasing agreements have been signed by Iran Air and CAAC of China for three Concordes each.

The following description applies to the production Concorde, except where otherwise indicated:

TYPE: Four-jet supersonic transport.

WINGS: Cantilever low wing of ogival delta planform. Thickness/chord ratio 3% at root, 2·15% from nacelle outboard. Slight anhedral. Continuous camber. Multi-spar torsion-box structure, manufactured mainly from RR.58 (AU2GN) aluminium alloy. Integrally-machined components used for highly loaded members and skin panels. In centre wing, spars are continuous across fuselage, the spars and associated frames being built as single assemblies extending between the engine nacelles. Forward wing sections built as separate components attached to each side of fuselage, spar loads being transferred to cross-members in lower part of main fuselage frames. Three elevons on trailing-edge of each wing, of aluminium alloy honeycomb construction. Each elevon is independently operated by a tandem jack, each half supplied from an independent hydraulic source and controlled by a separate electrical system. Dowty Boulton Paul power control units. Hydraulic artificial feel units protect the aircraft against excessive aerodynamic loads induced by pilot through over-control. Autostabilisation is provided. Autopilot control is by signals fed into normal control circuit. No high-lift devices. Leading-edges ahead of air intakes are de-iced electrically.

FUSELAGE: Mainly-conventional pressurised aluminium alloy semi-monocoque structure of constant cross-section, with unpressurised nose and tail cones. Hoop frames at approx 0·55 m (21·5 in) pitch support integrally-machined panels having closely-pitched longitudinal stringers. Window surrounds in passenger cabin formed of integral skin-stringer panels machined from aluminium alloy planks. Nose is drooped hydraulically to improve forward view during take-off, initial climb, approach and landing. Retractable visor is raised hydraulically to fair in windscreen in cruising flight.

TAIL UNIT: Vertical fin and rudder only. Fin is multi-spar torsion box of similar construction to wings. Two-section aluminium rudder controlled in same way as elevons. No de-icing system.

LANDING GEAR: Hydraulically-retractable tricycle type. Messier-Hispano nose and main units, with Kléber wheels and tyres. Twin-wheel steerable nose unit retracts forward. Four-wheel bogie main units retract inward. Oleo-pneumatic shock-absorbers. Main wheels and tyres size 47 × 15·75-22, pressure 12·9 bars (187 lb/sq in). Nosewheels and tyres size 31 × 10·75-14, pressure 12 bars (174 lb/sq in). Dunlop carbon disc brakes. SNECMA (Hispano) SPAD anti-skid units. Retractable tail bumper.

ENGINE NACELLES: Each consists of hydraulically-controlled variable-area (by ramp) air intake, engine bay and nozzle support structure. Intakes are of RR.58 or AU2GN aluminium alloy with steel leading-edges. The engine bay has an Inconel centre wall with aluminium alloy forward doors and titanium rear doors. The nozzle bay, aft of the rear spar, is of welded Stresskin sandwich panels and heat-resistant nickel alloys. Reverser buckets, which are also used as a secondary nozzle, are actuated by ball-screw jacks driven by compressed air through flexible shafts. Leading-edges of intakes, rear ramp sections and intake auxiliary door are electrically de-iced. Engine nose bullet and inlet guide vanes are de-iced by hot engine bleed air.

POWER PLANT: Four Rolls-Royce/SNECMA Olympus 593 Mk 602 turbojet engines, each rated at 169·3 kN (38,050 lb st) with 17% afterburning, and Type 28 thrust reversers. Fuel system is used also as heat sink and to maintain aircraft trim. All tanks are of integral construction and are in two groups, with total usable capacity of 119,695 litres (26,330 Imp gallons). Main group comprises five tanks in each wing and four tanks in fuselage and maintains CG automatically in cruising flight. Trim tank group (three tanks) comprises two tanks at the front and a tank of 13,150 litres (2,892 Imp gallons) capacity in fuselage beneath tail fin. This group maintains correct relationship between CG and aerodynamic centre of pressure by transferring fuel rearward during acceleration and forward during return to subsonic flight. Four pressure refuelling points in bottom fairing, two forward of each main landing gear unit. Oil capacity 22·75 litres (5 Imp gallons) per engine.

ACCOMMODATION: Pilot and co-pilot side by side on flight deck, with third crew member behind on starboard side. Provision for supernumerary seat behind pilot. Wide variety of four-abreast seating layouts to suit individual requirements of airlines. With all normal toilet and galley service facilities, up to 128 economy class passengers can be carried with 864 mm (34 in) seat pitch. A version with 144 passenger seats at 813 mm (32 in) pitch is available. Toilets at front and centre of cabin. Baggage space under forward cabin and aft of cabin. Passenger doors forward of cabin and amidships on port side, with service doors opposite. Baggage door aft of cabin on starboard side. Emergency exits in rear half of cabin on each side. Two galley areas.

SYSTEMS: Hawker Siddeley Dynamics air-conditioning system, comprising four independent subsystems, with Hamilton Standard heat exchangers. Pressure differential 0·74 bars (10·7 lb/sq in). In each subsystem the air passes through a primary ram-air heat exchanger to an air cycle cold-air unit, and then through secondary air/air and air/fuel heat exchangers. The air is then mixed with hot air and fed to cabins, flight deck, baggage holds, landing gear, equipment and radar bays. Hydraulic services utilise two primary systems and one standby, pressure 275·8 bars (4,000 lb/sq in), each actuated by two engine-driven pumps. Temperature of the Oronite M.2V fluid is limited by heat exchangers. Main systems actuate flying control surfaces, artificial feel units, landing gear, wheel brakes, nosewheel steering, windscreen visor, nosecone droop, engine intake ramps and fuel pumps in rear transfer tank. Electrical system powered by four 60kVA engine-driven constant-speed brushless alternators giving 200/115V AC at 400Hz. Four 150A transformer-rectifiers and two 25Ah batteries provide 28V DC supply.

ELECTRONICS: SFENA/Marconi-Elliott automatic flight control system (AFCS). Litton LTN-72 primary navigation system comprises three identical inertial platforms, each coupled to a digital computer to form three self-contained units, two VOR/ILS systems, one ADF (Marconi-Elliott AD-380 in British Airways aircraft), two DME systems, one marker, two RCA AVQ-X weather radars and two TRT AHV-5 radio altimeters. Plessey flight data recording system in British Airways aircraft. Provision for supplementary system including a long-distance radio fixing system of the Loran C type. Optional equipment includes a second ADF. Basic communications equipment consists of two VHF and two HF transceivers, one Selcal decoder and two ATC transponders (Cossor SSR 2700 in British Airways aircraft). Nose radome by Reinforced Microwave Plastics. Provision for a third VHF transceiver and data link equipment.

DIMENSIONS, EXTERNAL:
- Wing span — 25·56 m (83 ft 10 in)
- Wing aerodynamic reference chord at root — 27·66 m (90 ft 9 in)
- Wing aspect ratio — 1·7
- Length overall — 62·10 m (203 ft 9 in)
- Height overall — 12·19 m (40 ft 0 in)
- Fin aerodynamic reference chord at base — 10·59 m (34 ft 9 in)
- Wheel track — 7·72 m (25 ft 4 in)
- Wheelbase — 18·19 m (59 ft 8¼ in)
- Passenger doors (each):
  - Height — 1·68 m (5 ft 6 in)
  - Width — 0·76 m (2 ft 6 in)
  - Height to sill: fwd — 4·95 m (16 ft 3 in)
  - amidships — 4·74 m (15 ft 7 in)
- Service doors (each):
  - Height — 1·22 m (4 ft 0 in)
  - Width — 0·63 m (2 ft 0·8 in)
  - Height to sill: fwd — 4·95 m (16 ft 3 in)
  - amidships — 4·75 m (15 ft 7 in)
- Baggage hold door (underfloor):
  - Length — 0·99 m (3 ft 3 in)
  - Width — 0·84 m (2 ft 9·2 in)
  - Height to sill — 3·54 m (11 ft 7 in)
- Baggage hold door (rear, stbd):
  - Height — 1·52 m (5 ft 0 in)
  - Width — 0·76 m (2 ft 6 in)
  - Height to sill — 3·94 m (12 ft 11 in)

DIMENSIONS, INTERNAL:
- Cabin:
  - Length, flight deck door to rear pressure bulkhead, incl galley and toilets — 39·32 m (129 ft 0 in)
  - Width — 2·63 m (8 ft 7½ in)
  - Height — 1·96 m (6 ft 5 in)
  - Volume — 238·5 m³ (8,440 cu ft)
- Baggage/freight compartments:
  - underfloor — 6·43 m³ (227 cu ft)
  - rear fuselage (total) — 13·31 m³ (470 cu ft)

AREAS:
- Wings, gross — 358·25 m² (3,856 sq ft)
- Elevons (total) — 32·00 m² (344·44 sq ft)
- Fin (excl dorsal fin) — 33·91 m² (365 sq ft)
- Rudder — 10·40 m² (112 sq ft)

WEIGHTS AND LOADINGS:
- Operating weight empty — 79,265 kg (174,750 lb)
- Typical payload — 11,340 kg (25,000 lb)
- Max payload — 12,740 kg (28,087 lb)
- Max T-O weight — 181,435 kg (400,000 lb)
- Max zero-fuel weight — 92,080 kg (203,000 lb)
- Max landing weight — 111,130 kg (245,000 lb)
- Max wing loading — approx 488 kg/m² (100 lb/sq ft)
- Max power loading — approx 268 kg/kN (2·5 lb/lb st)

WEIGHT AND PERFORMANCE (flight envelope explored by 1 January 1976):
- Altitude — 20,725 m (68,000 ft)
- Airspeed — 565 knots (1,047 km/h; 651 mph) CAS
- Mach number — 2·23
- Minimum airborne speed — 119 knots (221 km/h; 137 mph)
- Incidence — 23·7°
- Weight at T-O — 181,700 kg (400,580 lb)
- Measured crosswind component — 25 knots (46·5 km/h; 29 mph)
- Landing with nose and visor up (simulated)
- CG in T-O and landing — 52-53%
- CG in flight — 52-60%

PERFORMANCE (production version, calculated, at max T-O weight):
- Max cruising speed at 15,635 m (51,300 ft) — Mach 2·02 or 530 knots CAS, whichever is the lesser, equivalent to TAS of 1,176 knots (2,179 km/h; 1,354 mph)
- T-O speed — 237 knots (440 km/h; 273 mph)
- Landing speed — 195 knots (360 km/h; 224 mph)
- Rate of climb at S/L — 1,525 m (5,000 ft)/min
- Service ceiling — approx 18,290 m (60,000 ft)
- Min ground turning radius — 10·60 m (34 ft 9½ in)
- Runway LCN at max T-O weight — 89
- T-O to 10·7 m (35 ft) — 3,230 m (10,600 ft)
- Landing from 10·7 m (35 ft) — 2,444 m (8,020 ft)
- Range with max fuel, FAR reserves and 10,659 kg (23,500 lb) payload — 3,360 nm (6,226 km; 3,869 miles)
- Range with max payload, FAR reserves:
  - at Mach 0·93 at 9,100 m (30,000 ft) — 2,690 nm (4,984 km; 3,097 miles)
  - at Mach 2·02 cruise/climb — 3,315 nm (6,142 km; 3,817 miles)

OPERATIONAL NOISE CHARACTERISTICS (FAR Pt 36):
- T-O noise level — 112 EPNdB
- Approach noise level — 118 EPNdB
- Sideline noise level — 115 EPNdB

EUROPEAN AIRLINER

On 6 September 1974 six European aerospace companies announced the signing of an agreement to work together to meet European airline requirements for the 1980s. Dassault-Breguet joined the group in the latter part of 1975. The companies concerned are BAC and Hawker Siddeley in the UK; Aérospatiale and Dassault-Breguet in France; and Dornier, MBB and VFW-Fokker in Germany.

These companies are already planning a joint response to future airline requirements, with particular emphasis upon those of their national airlines. The co-operation may later be extended to include other European aircraft companies.

HELI-EUROPE
HELI-EUROPE INDUSTRIES LTD

Address:
6 Rue Raffet, 75016-Paris, France

Formation of this new joint company was announced on 31 May 1973 by Aérospatiale and Westland. It is registered in England and has its offices in Paris, and was formed to exploit the existing industrial co-operation between the two companies with regard to future joint designs in the helicopter field. On 16 June 1975 the British and French companies signed, with Agusta of Italy and MBB of Germany, a memorandum of understanding with regard to future joint helicopter development.

PANAVIA
PANAVIA AIRCRAFT GmbH

Head Office:
8 München 86, Postfach 860629, Arabellastrasse 16, German Federal Republic
Telephone: Munich (089) 92171
Telex: 05 29 825

Directors:
Dott R. Bonifacio (Chairman)
H. R. Baxendale
Dott h.c. L. Bölkow (Deputy Chairman)
Dott F. Cereti
F. Forster-Steinberg
A. H. C. Greenwood (Deputy Chairman)
Dott G. Innocenti
F. W. Page
H. Langfelder

Managing Director:
G. Madelung

Deputy Managing Director:
Dr I. A. M. Hall

Functional Directors:
R. P. Beamont (Flight Operations)
O. Friedrich (Systems Engineering, Munich)
Dr I. A. M. Hall (Programme Management)
B. O. Heath (Systems Engineering, Warton)
H. J. Klapperich (Finance and Contracts)
J. K. Quill (Marketing)
Dott R. Sassi
J. A. Thornber (Procurement)

Publicity Manager:
F. Oelwein

Panavia Aircraft GmbH is an international European industrial company formed on 26 March 1969 to design, develop and produce a multi-role combat aircraft (MRCA) for service from the late 1970s with the air forces of the United Kingdom, the Federal Republic of Germany and Italy, and the German Navy. This programme is one of the largest European industrial programmes ever undertaken. The three component companies of Panavia are British Aircraft Corporation, Messerschmitt-Bölkow-Blohm and Aeritalia.

The German, British and Italian governments have set up a joint organisation known as NAMMO (NATO MRCA Management and Production Organisation). This has its executive agency NAMMA (NATO MRCA Management Agency) in Munich in the same building as Panavia.

The project was the subject of a feasibility study, which ended on 1 May 1969, when the project definition phase began. This saw the completion of the detailed design work and costing.

In mid-1970 the governments announced the satisfactory outcome of the definition phase and the beginning of the development phase, which was to lead to the flight of the first of nine prototypes. A further tri-national governmental review of the programme, concluded in March 1973, preceded the production investment phase. The estimated total cost of development work, up to September 1974, was £345 million, of which the UK share was £166 million. In December 1975 the British government announced that the unit cost "is now estimated at £4·96 million, allowing for inflation and fluctuations in exchange rates".

In addition to the MRCA, now named Tornado, Panavia is also undertaking, as a joint private venture, studies of a range of other military aircraft complementary to the MRCA.

PANAVIA TORNADO

The Tornado is a twin-engined two-seat supersonic aircraft capable of fulfilling the agreed operational requirements of its three sponsoring countries. The use of a variable-geometry wing gives it the necessary flexibility to achieve this.

The aircraft is intended to fulfil six major requirements, some of which are shared by more than one of the partners. These are:
(a) Close air support/battlefield interdiction
(b) Interdiction/counter air strike
(c) Air superiority
(d) Interception
(e) Naval strike
(f) Reconnaissance

In addition, a trainer version is being built which will also have an operational capability.

The Royal Air Force is to order 385 Tornados initially. These are due to begin entering service with Strike Command in 1978 and will, in the first instance, replace the Vulcans and Buccaneers of Nos. 9, 12, 15, 16, 35, 44, 50, 101 and 617 Squadrons in the overland strike and reconnaissance roles. Later, the air defence version will succeed the Phantom; and finally the Tornado will replace the Buccaneer for maritime strike tasks. Some two-thirds of the RAF's front-line aircraft will eventually be Tornados, according to the Chief of the Air Staff.

The Luftwaffe is to receive 211 Tornados, to replace the Lockheed F-104G in the battlefield interdiction, counter air and close air support roles. Four wings (Jabos 31, 32, 33 and 34) and one training squadron are to be equipped, starting in 1978. The 113 for MFG1 and 2 of the German Navy will be equipped for strike missions against sea and coastal targets, and for reconnaissance. Approval for the procurement of these 324 aircraft was announced by the Federal German Defence Committee on 19 May 1976.

The Italian Air Force will use its 100 Tornados to replace F-104G and G91R aircraft of the 20°, 102°, 154° and 186° Gruppi in the air superiority, ground attack and reconnaissance roles.

Structural design of the Tornado was completed in August 1972. Nine flying prototypes are being built—four in the UK, three in Germany and two in Italy. Static tests with a complete airframe began in the Spring of 1974.

The 01 first prototype (D-9591), assembled by MBB, made its first flight at Manching, Germany, on 14 August 1974, piloted by Paul Millett, chief test pilot of BAC's Military Aircraft Division. The 02 prototype (XX946) made its first flight, also piloted by Paul Millett, at Warton, Lancashire, on 30 October 1974. The 03 aircraft (XX947), the first to be fitted with dual controls, made its first flight at Warton on 5 August 1975, piloted by David Eagles, Tornado project pilot of BAC's Military Aircraft Division, and was followed by 04 (D-9592) on 2 September 1975 at Manching, flown by MBB's chief test pilot, Hans-Friedrich Rammensee. The 04 was the first aircraft to be fitted with the Tornado's integrated electronics system. The first Italian prototype, the 05 (X-586), flew at Caselle, Turin, on 5 December 1975, piloted by Pietro Trevisan, Aeritalia's Tornado project pilot, and 06 (XX948) followed on 20 December at Warton, with David Eagles at the controls. The 05 and 06 prototypes are being used to evaluate the Tornado as a weapons system when carrying and delivering external stores; 06 was the first of the prototypes to be fitted with the 27 mm Mauser cannon.

In July 1975, prototype 02 undertook in-flight refuelling trials, with an RAF Victor tanker, which were cleared substantially in one flight. In October, 02 cleared the initial external stores carriage programme, carrying large fuel tanks on the inner wing pylons and four 1,000 lb bombs on the underfuselage pylons.

The 07 flew on 30 March 1976, and the 08 on 15 July 1976 (the 470th Tornado flight); the remaining prototype was due to fly later in that year. They will be followed in 1977 by six pre-production aircraft in advance of the main production stream, for which component manufacture has already begun.

Air testing of the RB.199 power plant under a Hawker Siddeley Vulcan testbed (XA903) started on 19 April 1973. The 27 mm Mauser cannon was test-flown in, and fired from, a BAC Lightning fighter. Marshall of Cambridge (see UK section) modified two Buccaneer aircraft to flight test the nav/attack system; and in the USA, the performance of the Texas Instruments radar was assessed in a Convair 240 testbed aircraft prior to installation in the 04 Tornado prototype.

Type: Twin-engined multi-purpose military aircraft.

Wings: Cantilever shoulder-wing monoplane. All-metal wings, of variable geometry, having a sweep of 25° in the fully forward position and 66° when fully swept. Wing carry-through box is of electron-beam-welded titanium alloy; majority of remaining wing structure is of aluminium alloy, with integrally stiffened skin. The wings each pivot hydraulically, on Teflon-plated bearings, from a point in the centre-section just outboard of the fuselage. The root of the outer wing mates with the pivot pin through attachment members made of titanium alloy and fixed to the upper and lower light alloy panels of the outer wing box, and a so-called 'round rib', also of titanium alloy, transmitting the normal aerodynamic force. Sweep actuators are of the ballscrew type, with hydraulic motor drive. In the event of wing sweep failure, the aircraft can land safely with the wings fully swept. High-lift devices on the outer wings include full-span leading-edge slats (three sections on each side), full-span double-slotted trailing-edge flaps (four section on each side), and spoilers (two on upper surface on each side). Spoilers give augmented roll control at unswept and intermediate wing positions at low speed, and also act as lift dumpers after touchdown. All flying control surfaces actuated by electrically-controlled tandem hydraulic jacks. There

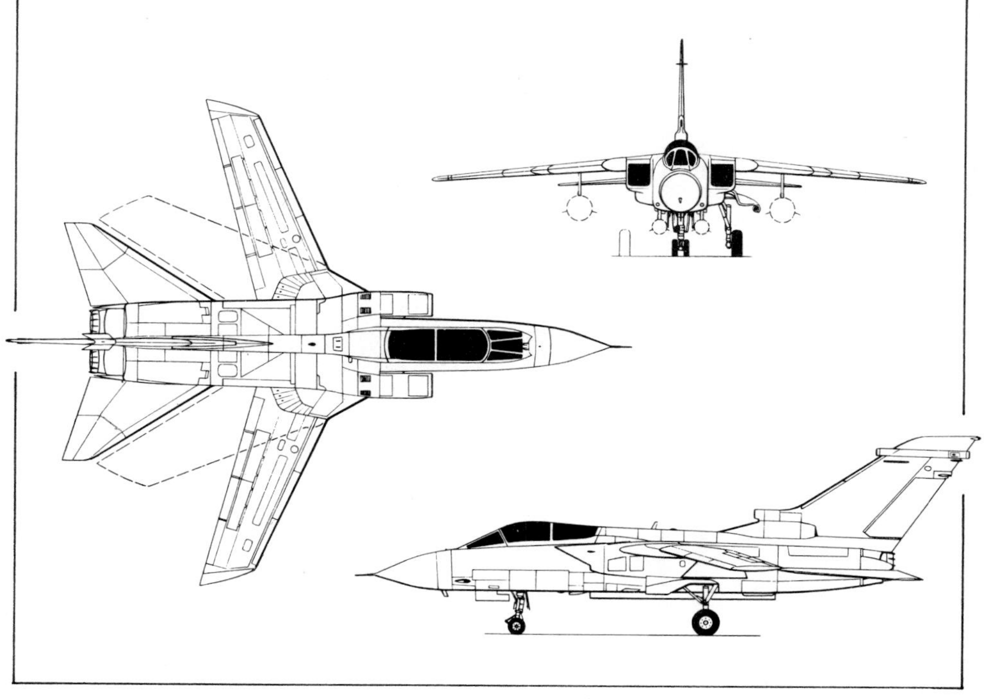

Panavia Tornado multi-role combat aircraft *(Pilot Press)*

Sixth prototype of the Panavia Tornado multi-role combat aircraft, the first to be fitted with Mauser 27 mm rapid-firing cannon

are no ailerons. Entire outer wings, including control surfaces, are Italian-built, Aeritalia having prime responsibility for final assembly and production of these units, assisted by Aermacchi, Aeronavali Venezia, Piaggio, Saca and SIAI-Marchetti as subcontractors. Microtecnica (Italy) is prime subcontractor for the wing sweep system.

Fuselage: Conventional all-metal semi-monocoque structure, mainly of aluminium alloy, built in three main sections. MBB in Germany is prime contractor (with participation by VFW-Fokker for the prototype and pre-production aircraft) for the centre fuselage section, including the engine air intake trunks and wing centre-section box and pivot mechanism. This task includes responsibility for the surface interface between the movable wing and the fixed portion, to ensure both a smooth and slender external contour and proper sealing against aerodynamic pressure over a range of wing sweep positions. The present design uses fibre-reinforced plastics in these areas, and an elastic seal between the outer wings and the fuselage sides. Responsibility for the front fuselage, including both cockpits, and for the rear fuselage, including the engine installation, is undertaken by the Military Aircraft Division of British Aircraft Corporation. Radar-transparent nosecone by AEG-Telefunken, assisted by Aeritalia and BAC, hinges sideways to starboard. Door-type airbrake on each side at top of rear fuselage.

Tail Unit: Cantilever all-metal structure, consisting of single sweptback two-spar fin and rudder, and low-set all-moving horizontal surfaces ('tailerons') which operate together for pitch control and differentially for roll control, assisted by use of the wing spoilers when the wings are not fully swept. Rudder and tailerons actuated by electrically-controlled tandem hydraulic jacks. Passive ECM antenna fairing near top of fin. Ram-air intake for heat exchanger at base of fin. Entire tail unit is the responsibility of BAC.

Landing Gear: Hydraulically-retractable tricycle type, with forward-retracting twin-wheel steerable nose unit. Single-wheel main units retract forward and upward into centre section of fuselage. Development and manufacture of the complete landing gear and associated hydraulics is headed by Dowty Rotol (UK). Dunlop aluminium alloy wheels, brakes, low-pressure tyres (to permit operation from soft, semi-prepared surfaces) and anti-skid units. Main-wheel tyres size 30 × 11·50-14·5, Type VIII (20 ply); nosewheel tyres size 18 × 5·5, Type VIII (12 ply). Runway arrester hook beneath rear of fuselage.

Power Plant: Two Turbo-Union RB.199-34R-2 three-spool turbofan engines, each rated at 37·8 kN (8,500 lb st) dry and 64·5 kN (14,500 lb st) with afterburning, fitted with bucket-type thrust reversers and installed in rear fuselage with downward-opening doors for servicing and engine change. Four large 'blow-out' doors in top of each trunk, above the wedge-shaped two-dimensional intake. Dowty Boulton Paul air intake ramp actuators. All integral fuel in multi-cell Uniroyal self-sealing integral fuselage tanks and/or wing box tanks, all fitted with press-in fuel sampling and water drain plugs, and all refuelled from a single-point NATO connector. Detachable and retractable in-flight refuelling probe can be mounted on starboard side of fuselage, adjacent to cockpit. System also designed to accept a buddy-to-buddy refuelling pack. Provision for drop-tanks of various sizes to be carried beneath outer wings. Dowty Fuel Systems/Lucas/Microtecnica afterburning fuel control system. AEG-Telefunken intake de-icing system.

Accommodation: Crew of two on tandem Martin-Baker Mk 10A ejection seats under Kopperschmidt/AIT one-piece canopy, which is hinged at rear and opens upward. Flat centre windscreen panel and curved side panels, built by Lucas Aerospace, incorporate Sierracote electrically-conductive heating film for de-icing and internal demisting. Seats provide safe escape at zero altitude and at speeds from zero up to 630 knots (1,166 km/h; 725 mph) IAS.

Seventh Tornado prototype, used with 04 to evaluate and develop the electronics systems

Systems: Nordmicro/HSD/Microtecnica air intake control system, and Dowty Boulton Paul/Liebherr Aerotechnik engine intake ramp control actuators. Two separate independent hydraulic systems, one driven by each engine, provide fully duplicated power for wing sweep, flaps, slats, spoilers, airbrakes, landing gear, tailerons and rudder. Main system includes Vickers pump, Dowty accumulators and Teves power pack. Fairey Hydraulics system for actuation of spoilers, rudder and taileron control. Provision for reversion to single-engine drive of both systems, via a mechanical cross-connection between the two engine auxiliary gearboxes, in the event of a single engine failure. In the event of a double engine flameout, an emergency pump in No. 1 system has sufficient duration for re-entry into the engine cold relight boundary. Flying control circuits are protected from loss of fluid due to leaks in other circuits by isolating valves which shut off the utility circuits if the reservoir contents drop below a predetermined safety limit level. Duplicated AC and DC electrical power is provided by two alternators, each driven by its respective engine auxiliary gearbox, to two separate main AC busbars and one essential AC busbar, and through two fan-cooled transformer-rectifier units (TRUs) to two main DC busbars. Lucas/Siemens 40/60kVA 200V 400Hz three-phase constant-frequency AC generating system. Either generator can cope with the full demand of the electrical systems in the event of a single generator failure. If both TRUs fail, an on-board Varta battery supplies the essential DC busbar. In the event of a total loss of power the battery also drives an electro-hydraulic pump which provides power for the primary flying controls. Under normal conditions the battery drives the KHD/Microtecnica/Lucas T312 APU for engine starting, but a DC ground supply is provided to assist starting if required. Plessey emergency power unit (EPU) and power systems controller. Normalair-Garrett precooler and cold-air unit, Marston Excelsior intercooler and Teddington temperature control system. Normalair-Garrett/Draegerwerk/OMI demand-type oxygen system, using a lox converter. KHD accessory drive gearboxes and Rotax/Lucas/Siemens integrated drive generator. Marconi-Elliott flow-metering system. Eichweber fuel gauging system and Flight Refuelling flexible couplings. Graviner fire detection and extinguishing systems. Rotax contactors. Smiths engine speed and temperature indicators.

Electronics and Equipment: Communications equipment includes Plessey PTR 1721 (UK and Italy) or Rohde und Schwarz (Germany) UHF/VHF transceiver; AEG-Telefunken UHF/DF (UK and Germany only); Chelton UHF homer aerial; SIT/Siemens emergency UHF with Rohde und Schwarz switch; BAC HF/SSB aerial tuning unit; Rohde und Schwarz (UK and Germany) or Montedel (Italy) HF/SSB radio; Ultra communications control system; Marconi-Elliott central suppression unit; Epsylon voice recorder; and Chelton communications and landing system aerials.

Primary self-contained nav/attack system includes Texas Instruments multi-mode forward-looking radar (Marconi-Elliott multi-mode airborne interception radar for RAF interceptor version); Ferranti three-axis digital inertial navigation system (DINS) and combined radar display; Decca Type 72 Doppler radar system; Microtecnica air data computer; Litef Spirit 3 16-bit central digital computer; Aeritalia radio/radar altimeter; Smiths/Teldix/OMI electronic head-up display with Davall camera; Ferranti nose-mounted laser ranger and marked target receiver; Marconi-Elliott TV tabular display; Astronautics (USA) bearing distance heading indicator and contour map display. Defensive equipment includes Siemens (Germany) or Cossor SSR-3100 (UK) IFF transponder; Elettronica warning radar; and MSDS/Plessey/Decca passive ECM system.

Flight control system includes a Marconi-Elliott command stability augmentation system (CSAS), incorporating fly-by-wire and autostabilisation; Marconi-Elliott autopilot and flight director (APFD), using two self-monitoring digital computers; Marconi-Elliott triplex transducer unit (TTU), with analogue computing and sensor channels; Marconi-Elliott terrain-following E-scope (TFE); Fairey/Marconi-Elliott quadruplex electro-hydraulic actuator; and Microtecnica air data set. The APFD provides preselected attitude, heading or barometric height hold, heading and track acquisition, and Mach number or airspeed hold with autothrottle. Flight director operates in parallel with, and can be used as backup for, the autopilot. Automatic approach, terrain-following and radio height-holding modes are also available. Other instrumentation includes Smiths horizontal situation indicator, vertical speed indicator and standby altimeter; AEG-Telefunken ADF; Lital standby attitude and heading reference system; SEL (with Setac) or (in UK aircraft) Marconi-Elliott AD2770 (without Setac) Tacan; Cossor CILS 75 ILS; Bodenseewerk attitude direction indicator; and Dornier System flight data recorder.

Overall responsibility for the electronics rests with

the three-nation group Avionica Systems Engineering, combining the activities of EASAMS (UK), ESG (Germany) and SIA (Italy). The electronics systems, while standardised as far as possible, retain the flexibility necessary to perform the various roles required. They provide accurate low- and high-level navigation; precision visual attack on ground targets in blind and poor weather conditions; air-to-ground and air-to-air attack with a wide variety of weapons; manually controlled and automatic attack; and comprehensive on-board checkout and mission data recording; with minimisation of ground support facilities at bases and the front line.

Armament: All Tornados are fitted with two 27 mm Mauser cannon, one in each side of the lower forward fuselage. Other armament varies according to version, with emphasis on the ability to carry a wide range of advanced non-nuclear weapons on three underfuselage attachments and up to four swivelling hardpoints beneath the outer wings. A Marconi-Elliott stores management system is fitted, and Sandall Mace 355 mm (14 in) ejector release units are standard. Initial weapon systems evaluation will include trials in the fourth prototype of a modified Raytheon Sparrow missile, fitted with a British warhead and fuse. The battlefield interdiction version will be capable of dropping defensive 'streuwaffen' (scatter weapons) and of carrying weapons to suit 'hard' or 'soft' targets. The naval and interdictor strike versions will have provision for carrying additional, externally-mounted fuel tanks. The air superiority version will be able to carry a wide range of guided and semi-active homing air-to-air weapons. Among the weapons already specified for, or suitable for carriage by, the Tornado are the Sparrow and Aspide 1A air-to-air missiles; AS.30, Martel, Kormoran and Jumbo air-to-surface missiles; napalm; BL-755 600 lb cluster bombs; and 'smart' or retarded bombs.

Dimensions, external:
- Wing span:
  - fully spread: 13·90 m (45 ft 7¼ in)
  - fully swept: 8·60 m (28 ft 2½ in)
- Length overall: 16·70 m (54 ft 9½ in)
- Height overall: 5·70 m (18 ft 8½ in)

Weights (estimated):
- Weight empty, equipped: 9,980-10,430 kg (22,000-23,000 lb)
- Max T-O weight: 17,240-18,145 kg (38,000-40,000 lb)

Performance:
- Max level speed at 11,000 m (36,000 ft): above 1,146 knots (2,125 km/h; 1,320 mph)
- Max level speed at low altitude: approx 790 knots (1,465 km/h; 910 mph)
- T-O run: 700 m (2,295 ft)
- Landing run: 900 m (2,955 ft)

SEPECAT
SOCIÉTÉ EUROPÉENNE DE PRODUCTION DE L'AVION E.C.A.T.

Airframe Companies:

British Aircraft Corporation Ltd, Brooklands Road, Weybridge, Surrey KT13 0RN, England
Telephone: Weybridge (97) 45522

Avions Marcel Dassault/Breguet Aviation, BP 32, 92420-Vaucresson, France
Telephone: 970-38-50

Directors:
H. R. Baxendale (alternate Chairman)
J. P. Fort (alternate Chairman)
J. Bonnet
F. W. Page
Paul Jaillard
A. H. C. Greenwood
Jeffrey Quill
C. Edelstenne

Management Committee:
Directors as above, plus:
M. Berjon
T. O. Williams
I. R. Yates

Public Relations:
G. B. Hill (BAC)
C. P. Raffin (Dassault-Breguet)

This Anglo-French company was formed in May 1966 by Breguet Aviation and British Aircraft Corporation, the two partners in the design and production of the Jaguar supersonic strike fighter/trainer.

The Jaguar project was initiated by the Defence Ministries of Britain and France on 17 May 1965. The governments of the two countries appointed an official Jaguar Management Committee to look after their interests. SEPECAT is the complementary industrial organisation.

SEPECAT JAGUAR

The Jaguar, which was evolved from the Breguet Br121 project, was designed by Breguet and BAC to meet a common requirement of the French and British air forces laid down in early 1965. This requirement called for a dual-role aircraft, to be used as an advanced and operational trainer and a tactical support aircraft of light weight and high performance, to enter French service in 1972 and with the RAF in 1973. The Jaguar M French naval version (1972-73 *Jane's*) was abandoned in 1973.

The following versions of the Jaguar are in production:

Jaguar A. French single-seat tactical support version. Prototypes (A-03 and A-04) first flown on 29 March and 27 May 1969. Total of 160 ordered.

By 19 May 1976 a total of 60 production Jaguar As had been delivered. The first operational Armée de l'Air Jaguar unit (Esc. 1/7 'Provence') was formed at St Dizier in eastern France on 19 June 1973. The French Air Force plans to equip eight squadrons with Jaguars; of these, Esc.1/7 'Provence', Esc.2/7 'Argonne', Esc. 3/7 'Languedoc' and Esc. 3/11 'Corse', were operational by early 1976; these will be followed by Esc. 1/11 'Roussillon' and Esc. 2/11 'Vosges' by the end of 1976, with the other two squadrons due to equip in 1977.

Jaguar B (RAF designation: Jaguar T.Mk 2). British two-seat operational training version. Prototype B-08 (XW566) first flown on 30 August 1971. Total of 37 ordered by end of 1975, of which 35 had been delivered. First T.Mk 2 delivered to RAF was XX137.

Jaguar E. French two-seat advanced training version. Prototypes (E-01 and E-02) first flown on 8 September 1968 and 11 February 1969. Total of 40 ordered by end of 1975, of which 38 had been delivered. First production Jaguar, designated E-1, flew for the first time on 2 November 1971, and deliveries to the CEAM at Air Base 118, Mont de Marsan, began in May 1972. The first unit to equip with this version was Esc. 1/7 at St Dizier. Deliveries to Esc. 3/11 at Toul began in January 1975.

Jaguar S (RAF designation: Jaguar GR. Mk 1). British single-seat tactical support version, basically similar to A but with an advanced inertial navigation and weapon-aiming system (NAVWASS) controlled by a digital computer. Prototypes S-06 (XW560) and S-07 (XW563) first flown on 12 October 1969 and 12 June 1970. Total of 165 ordered by end of 1975, of which 95 had been delivered. The first production GR. Mk 1 (XX108) flew on 11 October 1972; the first production Jaguar GR.Mk 1 for the RAF (S-4/XX111) was officially handed over at Lossiemouth on 30 May 1973, and was used for ground crew training prior to the formation of the first Jaguar operational conversion team in October. The first for flying training (XX114 and XX115) were delivered on 13 September 1973; aircrew conversion training began in January 1974. The first operational RAF Jaguar unit, No. 54 Squadron, formed at Lossiemouth on 29 March 1974, moving on 9 August to Coltishall, where it became fully operational on 1 January 1975. The second RAF operational unit, No. 6 Squadron, also formed at

SEPECAT Jaguar A single-seat tactical support aircraft of the French Armée de l'Air, with underwing bombs and underfuselage drop-tank

Jaguar S-3, a British single-seater re-engined with Adour Mk 804 turbofans, as specified for the Jaguar International, carrying two Matra 550 Magic air-to-air missiles on overwing pylons

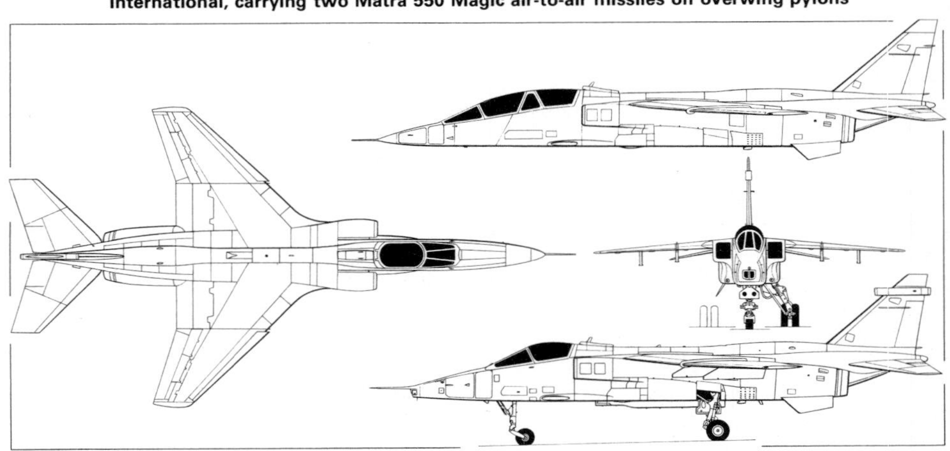

SEPECAT Jaguar S single-seat strike aircraft, with additional side view (top) of Jaguar B two-seat operational training version *(Pilot Press)*

Lossiemouth, moving to Coltishall on 6 November 1974. In all, it is planned to equip eight RAF front-line squadrons with Jaguars: Nos. 2, 14, 17, 20 and 31 with the Second Allied Tactical Air Force in Germany and an eighth squadron, probably in the UK. Four of the squadrons in Germany were operational by early 1976; of these, No. 2 Squadron is a reconnaissance unit.

Jaguar International. Export version, announced on 28 August 1974. Differs little from the standard single-seat Jaguar S except in having more powerful Adour turbofan engines (the RT. 172-26 Adour Mk 804, rated at 23·46 kN; 5,275 lb st dry and 35·59 kN; 8,000 lb st with afterburning) which increase total S/L thrust by nearly 10% for T-O and about 27% at Mach 0·8/0·9. Test flying of these engines, in a converted Jaguar S, took place in 1975, and has confirmed an improved combat performance with substantially enhanced manoeuvrability and acceleration in the low-level speed range. All export Jaguars will be fitted with this mark of engine; other customer options being developed include overwing pylons compatible with Magic or similar dogfight missiles; a multi-purpose nose radar such as the Thomson-CSF Agave; anti-shipping weapons such as Exocet and Kormoran; and night sensors such as low light level TV. Initial orders placed by the Sultan of Oman's Air Force (12) and Ecuadorean Air Force (12). Deliveries to SOAF are due to begin in February 1977; these aircraft will be fitted with a Marconi-Elliott 920ATC NAVWASS computer and provision for carrying Matra R.550 Magic air-to-air missiles on overwing pylons.

Under the terms of the production agreement signed by the British and French Defence Ministries on 9 January 1968, 202 Jaguars are being built for the Royal Air Force and 200 for the French air force. The first formal production contract, placed in the Autumn of 1969, covered 50 Jaguars for France; the second was for 30 for the RAF. Subsequent contracts had brought total Anglo-French orders to 402 by early 1976, of which 220 had been delivered.

Dassault-Breguet factories at Toulouse and Biarritz are responsible for the front and centre fuselage. The Military Aircraft Division of BAC has responsibility for the rear fuselage, air intakes, wings and tail unit. There are final assembly lines for complete aircraft in both Britain and France.

The Jaguar is fully power-controlled in all three axes and is automatically stabilised as a weapons platform by gyros which sense disturbances and feed appropriate correcting data through a computer to the power control assemblies, in addition to the human pilot manoeuvre demands. The power controls are all of duplex tandem arrangement, with both mechanical and electrical servo-valves of the established Fairey platen design.

Training versions are able to operate from conventional runways only 2,000 m (6,560 ft) long, with full provision for safety in the event of an engine failure at the critical point of take-off.

The following description applies to the standard versions of the Jaguar in production for the RAF and the Armée de l'Air:

TYPE: Single-seat tactical support aircraft (Jaguar A, S and International) and two-seat operational or advanced trainer (Jaguar B and E).

WINGS: Cantilever shoulder-wing monoplane. Anhedral 3°. Sweepback 40° at quarter-chord. All-metal two-spar torsion-box structure, the skin of which is machined from solid aluminium alloy, with integral stiffeners. Main portion built as single unit, with three-point attachment to each side of fuselage. Outer panels fitted with slat which also gives effect of extended-chord leading-edge. No conventional ailerons. Lateral control by two-section spoilers, forward of outer flap on each wing, in association (at low speeds) with differential tailplane. Hydraulically-operated (by screwjack) full-span double-slotted trailing-edge flaps. Fairey Hydraulics powered flying controls. Leading-edge slats can be used in combat. Entire wing unit is British-built.

FUSELAGE: All-metal structure, mainly aluminium, built in three main units and making use of sandwich panels and, around the cockpit(s), honeycomb panels. Local use of titanium alloy in engine bay area. The forward and centre fuselage, up to and including the main undercarriage bays, and including cockpit(s), main systems installations, forward fuel tanks and landing gear, are of French construction. The air intakes, and entire fuselage aft of the main-wheel bays, including engine installation, rear fuel tanks and complete tail assembly, are British-built. Two door-type airbrakes under rear fuselage, immediately aft of each main-wheel well. Structure and systems, aft of cockpit(s), are identical for single-seat and two-seat versions.

TAIL UNIT: Cantilever all-metal two-spar structure, covered with aluminium alloy sandwich panels. Rudder and outer panels and trailing-edge of tailplane have honeycomb core. Sweepback at quarter-chord 40° on horizontal, 43° on vertical surfaces. All-moving slab-type tailplane, with 10° of anhedral, the two halves of which can operate differentially to supplement the spoilers. No separate elevators. Fairey Hydraulics powered flying controls. Ventral fins beneath the rear fuselage. Entire tail unit is British-built.

LANDING GEAR: Messier-Hispano retractable tricycle type, all units having Dunlop wheels and low-pressure tyres for rough-field operation. Hydraulic retraction, with oleo-pneumatic shock-absorbers. Forward-retracting main units each have twin wheels, tyre size 615 × 225-10, tyre pressure 5·8 bars (84 lb/sq in). Wheels pivot during retraction to stow horizontally in bottom of fuselage. Single rearward-retracting nose-wheel, with tyre size 550 × 250-6 and pressure of 3·9 bars (57 lb/sq in). Twin landing lights in nosewheel door. Dunlop hydraulic brakes. Anti-skid units and arrester hook standard. Irvin brake parachute of 5·5 m (18 ft 0½ in) diameter housed in fuselage tailcone.

POWER PLANT: Two Rolls-Royce/Turboméca Adour Mk 102 turbofan engines (each rated at 22·75 kN; 5,115 lb st dry and 32·5 kN; 7,305 lb st with afterburning). Lateral-type fixed-geometry air intake on each side of fuselage aft of cockpit. Fuel in six tanks, one in each wing and four in fuselage. Total internal fuel capacity 4,200 litres (924 Imp gallons). Armour protection is provided for critical fuel system components. In the basic tactical sortie the loss of fuel from one tank at the halfway point would not prevent the aircraft from regaining its base. Provision for carrying three auxiliary drop-tanks, each of 1,200 litres (264 Imp gallons) capacity, on fuselage and inboard wing pylons. Jaguar A and S equipped for in-flight refuelling, with a retractable probe forward of the cockpit on the starboard side.

ACCOMMODATION (Jaguar B and E): Crew of two in tandem on Martin-Baker Mk 9 zero-zero ejection seats (Jaguar B) or Mk 4 seats (Jaguar E) giving zero-altitude ejection at speeds down to 90 knots (167 km/h; 104 mph). Individual rearward-hinged canopies. Rear seat is 381 mm (15 in) higher than front seat. Windscreen bulletproof against 7·5 mm rifle fire.

ACCOMMODATION (Jaguar A and S): Enclosed cockpit for pilot, with rearward-hinged canopy and Martin-Baker Mk 9 (Jaguar S) or Mk 4 (Jaguar A) ejection seat as in two-seaters. Bulletproof windscreen, as in two-seat versions.

SYSTEMS: Air-conditioning and pressurisation systems maintain automatically, throughout the flight envelope, comfortable operating conditions for the crew, and also control the temperature in certain equipment bays. Two independent hydraulic systems, powered by two Vickers engine-driven pumps. Hydraulic pressure 207 bars (3,000 lb/sq in). First system (port engine) supplies one channel of each actuator for the flying controls, the hydraulic motors which actuate the flaps and slats, the landing gear retraction and extension system, the brakes and anti-skid units. The second system supplies the other half of each flying control actuator, two further hydraulic motors actuating the slats and flaps, the airbrake and landing gear emergency extension jacks, nosewheel steering system and the wheel brakes. In addition to the duplicated hydraulic power systems, there is an emergency hydraulic power transfer unit. Electrical power provided by two 15kVA AC generators, either of which can sustain functional and operational equipment without load-shedding. DC power provided by two 4kW transformer-rectifiers. Emergency power for essential instruments provided by 15Ah battery and static inverter. De-icing, rain clearance and demisting standard. Liquid oxygen system installed, which also pressurises the pilot's anti-*g* suit.

ELECTRONICS AND OPERATIONAL EQUIPMENT (French versions): Equipment of Jaguar E includes VHF/UHF radio, VOR/ILS and IFF; Tacan with Crouzet Type 90 navigation indicator; SFIM 153-6 twin-gyro inertial platform with two SFIM 810 all-attitude roll and pitch spherical indicators; SFIM 511 directional compass; Jaeger ELDIA air data system with Jaeger altitude indicator; CSF RL 50Pj incidence probe with angle of attack indicator; CSF 121 fire control sighting unit with weapon selector and adaptor for sighting head camera. Except for the use of a SFIM 250-1 twin-gyro platform, and the addition of a vector adder to the navigation indicator, this equipment is repeated in the Jaguar A, which has in addition a panoramic camera, Dassault-built Decca RDN 72 Doppler radar, Crouzet Type 90 navigation computer with target selector, CFTH passive radar warning (ECM) detector, a CSF 31 weapon aiming computer, a Dassault fire control computer for Martel anti-radar missiles and a CSF laser rangefinder. Provision for the addition to these basic installations of such other items as terrain-following radar or sighting equipment for low light level targets.

ELECTRONICS AND OPERATIONAL EQUIPMENT (British versions): Basic equipment of both the Jaguar B and S is similar. It includes a Smiths-built Honeywell radio altimeter, slip indicator, E2B standby compass and autostabilising system, Plessey PTR 377 VHF/UHF radio and Marconi-Elliott HF radio; Cossor CILS 75 ILS; IFF; Tacan; Marconi-Elliott digital/inertial navigation and weapon aiming subsystem (NAVWASS) with MCS 920M digital computer, E3R three-gyro inertial platform, inertial velocity sensor, navigation control unit and projected-map display; Marconi-Elliott air data computer; Smiths electronic head-up display; Smiths FS6 horizontal situation indicator; Sperry C2J gyro amplifier master unit, compass controller and magnetic detector; Plessey weapon control system. Jaguar S fitted with Ferranti laser rangefinder and marked target seeker in modified nose.

ARMAMENT (Jaguar A and S): Two 30 mm cannon (DEFA 553 in Jaguar A, Aden in Jaguar S) in lower fuselage aft of cockpit. One stores attachment point on fuselage centreline and two under each wing. Centreline and inboard wing points can each carry up to 1,000 kg (2,000 lb) of weapons, and the outboard underwing points up to 500 kg (1,000 lb) each. Maximum external stores load 4,535 kg (10,000 lb). Jaguar As in service can carry the AN 52 tactical nuclear weapon. Typical alternative loads include one Martel AS.37 anti-radar missile and two 1,200 litre (264 Imp gallon) drop-tanks; eight 1,000 lb bombs; various combinations of free-fall and retarded bombs, Magic missiles and air-to-surface rockets, including the 68 mm SNEB rocket; a reconnaissance-camera pack; or two drop-tanks. The BAC-designed flush-fitting reconnaissance pod for RAF Jaguars, carried on the fuselage centreline, has optical cameras for horizon-to-horizon coverage and Hawker Siddeley Type 401 infra-red linescan (IRLS) for additional daylight, poor weather and night capability. Cameras are installed in two rotatable drums within the pod, the forward drum (for low/medium altitude missions) containing two Vinten F95 Mk 10 and one F95 Mk 7 oblique cameras. The rear drum can carry alternative modules: two F95 Mk 10 cameras for low-level sorties, or a single F126 for medium altitude reconnaissance. The IRLS package is installed in the rear end of the pod, adjacent to a data conversion unit linked to the NAVWASS digital computer in the aircraft.

Armament (Jaguar B and E): Two 30 mm DEFA 553 cannon in Jaguar E; Jaguar B has single 30 mm Aden cannon on port side. The two-seat versions have similar weapons capability to the tactical models, and can be employed for operational missions as required.

Dimensions, external:
- Wing span: 8·69 m (28 ft 6 in)
- Wing chord at root: 3·58 m (11 ft 9 in)
- Wing chord at tip: 1·13 m (3 ft 8½ in)
- Wing aspect ratio: 3·12
- Length overall:
  - A and S: 16·83 m (55 ft 2½ in)
  - B and E: 17·53 m (57 ft 6¼ in)
- Height overall: 4·89 m (16 ft 0½ in)
- Tailplane span: 4·53 m (14 ft 10¼ in)
- Wheel track: 2·40 m (7 ft 10½ in)
- Wheelbase: 5·69 m (18 ft 8 in)

Areas:
- Wings, gross: 24·00 m² (258·33 sq ft)
- Leading-edge slats (total): 1·05 m² (11·30 sq ft)
- Trailing-edge flaps (total): 4·12 m² (44·35 sq ft)
- Spoilers (total): 0·90 m² (9·67 sq ft)
- Vertical tail surfaces (total): 3·90 m² (42·00 sq ft)
- Horizontal tail surfaces (total): 7·80 m² (83·96 sq ft)

Weights and Loadings:
- Weight empty: 7,000 kg (15,432 lb)
- Normal T-O weight: 11,000 kg (24,000 lb)
- Max T-O weight: 15,500 kg (34,000 lb)
- Max wing loading: 604 kg/m² (126·3 lb/sq ft)
- Max power loading: 238·5 kg/kN (2·33 lb/lb st)

Performance (initial production aircraft):
- Max level speed at S/L: Mach 1·1 (729 knots; 1,350 km/h; 840 mph)
- Max level speed at 11,000 m (36,000 ft): Mach 1·5 (860 knots; 1,593 km/h; 990 mph)
- Landing speed: 115 knots (213 km/h; 132 mph)
- T-O run with typical tactical load: 580 m (1,900 ft)
- T-O to 15 m (50 ft) with typical tactical load: 940 m (3,085 ft)
- Landing from 15 m (50 ft) with typical tactical load: 860 m (2,825 ft)
- Landing run with typical tactical load: 470 m (1,545 ft)
- Typical attack radius, internal fuel only:
  - hi-lo-hi: 440 nm (815 km; 507 miles)
  - lo-lo-lo: 310 nm (575 km; 357 miles)
- Typical attack radius with external fuel:
  - hi-lo-hi: 710 nm (1,315 km; 818 miles)
  - lo-lo-lo: 450 nm (835 km; 518 miles)
- Ferry range with external fuel: 2,270 nm (4,210 km; 2,614 miles)
- *g* limits: +8·6; +12 (ultimate)

TRANSALL

ARBEITSGEMEINSCHAFT TRANSALL

Address:
Hünefeldstrasse 1-5, 28 Bremen 1, German Federal Republic
Telephone: (0421) 5181
Telex: 245821b

The Transall (Transporter Allianz) group was formed in January 1959 by Messerschmitt-Bölkow-Blohm, Aérospatiale and VFW-Fokker, to undertake joint development and production of the C-160 twin-turboprop military transport for the French and German air forces. Others were built for the air forces of South Africa and Turkey.

Production ended originally in 1972, but was expected to be reinstated in 1976 to meet an additional French order.

TRANSALL C-160

The Transall C-160 was developed to meet the specific requirements of the Federal German and French governments for a military transport aircraft capable of carrying troops, casualties, freight, supplies and vehicles, and of operating from semi-prepared surfaces.

Initial production, of the C-160 D (110), C-160 F (60) and C-160 Z (9), was shared between the three participating companies and ended in 1972, as described in earlier editions of *Jane's*.

On 7 May 1976 it was announced that the C-160 was to be reinstated in production, to meet an additional requirement of the French Armée de l'Air. At least 75 are scheduled to be built, including 25 or more for the Armée de l'Air. Work-sharing on this further production batch will be between Aérospatiale (50%) and the two German companies (50%), with a single final assembly line at Toulouse. Aérospatiale will build the wings and undertake final assembly; VFW-Fokker the central fuselage and tail unit; and MBB the cockpit and rear fuselage. The engines, as before, will be manufactured jointly by Rolls-Royce, SNECMA, MTU and FN.

First of the new production aircraft is due to be completed in 1979, when a planned output of two per month should be achieved.

The following details refer to the C-160 F:

Type: Twin-engined turboprop transport.

Wings: Cantilever high-wing monoplane. Dihedral on outer wings 3° 26′. All-metal two-spar structure designed on fail-safe principles. Wing in three sections, comprising a centre-section, which carries the engines, and two outer panels. All-metal ailerons and hydraulically-operated double-slotted flaps. Hydraulically-operated airbrakes (inboard) and spoilers (outboard) forward of flaps on each wing. Electrical anti-icing of leading-edges.

Fuselage: Aluminium alloy (2024-T3) semi-monocoque structure of circular basic section, flattened at the bottom, and designed on fail-safe principles. Underside of upswept rear fuselage lowers to form loading ramp for vehicles.

Tail Unit: Cantilever aluminium alloy (2024-T3) structure. Electrical anti-icing of tailplane leading-edges.

Landing Gear: Retractable tricycle type of Messier design. Hydraulic retraction and hydraulic/pneumatic shock-absorption. Each main unit comprises two pairs of wheels in tandem and is mounted inside a fairing on the side of the fuselage. Wheels can be raised to lower the fuselage for loading. Steerable twin-wheel nose unit. Main wheel tyres size 1,057 mm × 15·30 in; nosewheel tyres size 963 mm × 12·75 in. Tyre pressure (all units) 2·94 bars (42·7 lb/sq in). Messier brakes.

Power Plant: Two 4,549 kW (6,100 ehp) Rolls-Royce Tyne RTy.20 Mk 22 turboprop engines, each driving a Ratier Figeac-built HSD Type 4/8000/6, DB244 Re-15 four-blade constant-speed fully-feathering reversible-pitch propeller. Single-point pressure refuelling. Fuel in four integral wing tanks with total capacity of 16,490 litres (3,625 Imp gallons). Water-methanol capacity 325 litres (71·5 Imp gallons). Oil capacity (total) 68·4 litres (15 Imp gallons).

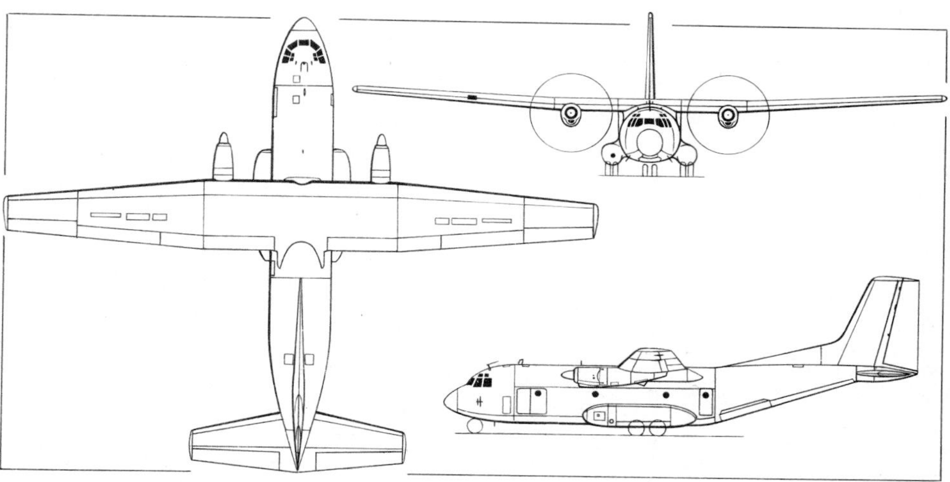

Transall C-160 twin-turboprop general-purpose transport aircraft *(Pilot Press)*

Accommodation: Pressurised accommodation for crew of four, comprising pilot, co-pilot, navigator and flight engineer. Typical payloads include 93 troops or 61-81 fully-equipped paratroops; 62 stretchers and four attendants; armoured vehicles, tanks and tractors not exceeding 16,000 kg (35,274 lb) total weight; one empty five-ton truck and crew; two empty three-ton trucks and crews; or three jeeps with partially-loaded trailers and crews. Flight deck and cargo compartment air-conditioned and pressurised in flight and on the ground. Power-assisted controls. Cargo compartment is provided with a freight door at the front on the port side, a paratroop door on each side immediately aft of the landing gear fairings and a hydraulically-operated rear loading ramp. The floor and all doors are at truckbed height. The floor is provided with lashing points of 5,000 kg (11,023 lb) and 12,000 kg (26,455 lb) capacity, arranged in a 508 mm (20 in) grid, and is stressed to carry large military vehicles. Loads which cannot be driven in can be taken on board rapidly by a winch and system of roller conveyors. Individual loads of up to 8,000 kg (17,637 lb) can be air-dropped.

Systems: Normalair pressurisation and air-conditioning system, differential 0·32 bars (4·59 lb/sq in). Two separate primary hydraulic systems, pressure 207 bars (3,000 lb/sq in), for flying controls, loading ramp, landing gear, wheel brakes, flaps, spoilers, airbrakes, nosewheel steering and other auxiliaries. Two more systems, pressure 175 bars (2,538 lb/sq in), for emergency and ground services, as well as a hand-pump driven emergency system. AC electrical system includes two 60kVA 400-600Hz generators, one 60kVA 400Hz generator and two 9kVA 400Hz generators. 28V DC system and 40Ah battery. AiResearch GTCP-85-160A APU in forward part of port main undercarriage fairing.

Electronics and Equipment: Collins 618M-1A VHF; Magnavox AN/ARC-34C/G UHF; CIT TR-AP-33-A HF; TEAM AS-1227-B PA system; TEAM TF-AP-14-A intercom; Bendix RDR-1DM weather radar; Lorenz AN/ARN-21b Tacan; Cotelec NR-AN16 UHF/DF; Kollsmann A28450 00 002 sextant; Bendix PN 2949-2E-A-78 drift meter; Bendix DRA-12-B/168 Doppler radar; Collins 51 RV-1C VHF nav with glide-slope; Collins 51 Z-4 marker beacon; Telefunken ADF-73F ADF; Sperry C-11 directional gyro; Sperry SYP-820 twin-gyro platform; Teldix PHI-3B-10-3 position and homing indicator; and Siemens STR 700/2 IFF/SIF.

Dimensions, external:
- Wing span: 40·00 m (131 ft 3 in)
- Wing chord at root: 4·84 m (15 ft 10½ in)
- Wing chord at tip: 2·428 m (7 ft 11½ in)
- Wing chord (mean): 4·176 m (13 ft 8½ in)
- Wing aspect ratio: 10
- Length overall: 32·40 m (106 ft 3½ in)
- Height overall: 12·36 m (40 ft 6¾ in)
- Tailplane span: 14·50 m (47 ft 7 in)
- Wheel track: 5·10 m (16 ft 9 in)
- Wheelbase: 10·48 m (34 ft 4½ in)
- Propeller diameter: 5·486 m (18 ft 0 in)
- Propeller ground clearance: 1·30 m (4 ft 3¼ in)
- Distance between propeller centres: 10·90 m (35 ft 9¼ in)
- Paratroop door (each side):
  - Height: 1·90 m (6 ft 2½ in)
  - Width: 0·90 m (3 ft 0 in)
- Cargo door (front, port side):
  - Height: 1·90 m (6 ft 2½ in)
  - Width: 1·98 m (6 ft 6 in)
- Rear loading ramp:
  - Length: 3·70 m (12 ft 1½ in)
  - Width: 3·15 m (10 ft 3½ in)
- Emergency exits:
  - Main hold, fwd, stbd side (one):
    - Height: approx 0·88 m (2 ft 10½ in)
    - Width: approx 0·54 m (1 ft 9¼ in)
  - Flight deck roof (one); roof of main hold, fwd (one); and two in roof of main hold at rear (one each side of dorsal fin):
    - Height: approx 0·52 m (1 ft 8½ in)
    - Width: approx 0·60 m (1 ft 11¾ in)

Dimensions, internal:
- Cabin, excl flight deck and ramp:
  - Length: 13·51 m (44 ft 4 in)
  - Max width: 3·15 m (10 ft 3½ in)

Transall C-160 F (two Rolls-Royce Tyne RTy.20 Mk 22 turboprop engines) of the French Armée de l'Air

Max height 2·98 m (9 ft 8½ in)
Floor area 42·6 m² (458·5 sq ft)
Volume 115·3 m³ (4,072 cu ft)
Cabin, incl ramp:
Length 17·21 m (56 ft 6 in)
Floor area 54·25 m² (584 sq ft)
Volume 139·9 m³ (4,940 cu ft)

AREAS:
Wings, gross 160·10 m² (1,722·7 sq ft)
Ailerons (total) 7·12 m² (76·64 sq ft)
Trailing-edge flaps (total) 34·10 m² (366·92 sq ft)
Spoilers (total) 0·80 m² (8·61 sq ft)
Fin:
excl dorsal fin 29·50 m² (317·5 sq ft)
incl dorsal fin 36·04 m² (388 sq ft)
Rudder 10·55 m² (113·5 sq ft)
Tailplane 40·22 m² (433 sq ft)
Elevators 10·30 m² (111 sq ft)

WEIGHTS AND LOADING:
Weight empty, equipped 28,758 kg (63,400 lb)
Basic operating weight, empty 28,946 kg (63,815 lb)
Max payload 16,000 kg (35,274 lb)
Max T-O weight 51,000 kg (112,435 lb)
Max zero-fuel weight 45,000 kg (99,208 lb)
Max landing weight 47,000 kg (103,617 lb)
Max wing loading 320 kg/m² (65·54 lb/sq ft)

PERFORMANCE (at max T-O weight except where indicated):
Never-exceed speed (4,875-9,145 m; 16,000-30,000 ft) Mach 0·64
Max level speed at 4,875 m (16,000 ft) 320 knots (592 km/h; 368 mph)
Econ cruising speed at 6,100 m (20,000 ft) 245 knots (454 km/h; 282 mph)
Stalling speed:
flaps up 109 knots (203 km/h; 126 mph)
10° flap 100 knots (185 km/h; 115 mph)
20° flap 93 knots (172 km/h; 107 mph)
40° flap 89 knots (166 km/h; 103 mph)
60° flap 82 knots (153 km/h; 95 mph)
Max rate of climb at S/L 396 m (1,300 ft)/min
Rate of climb at S/L, one engine out 91 m (300 ft)/min
Service ceiling 7,770 m (25,500 ft)
Service ceiling, one engine out 4,270 m (14,000 ft)
T-O run, 30° flap 747 m (2,450 ft)
T-O to 15 m (50 ft), 30° flap 900 m (2,950 ft)
Landing from 15 m (50 ft), 40° flap, at max landing weight without propeller reversal 869 m (2,850 ft)
Landing run, normal 550 m (1,800 ft)
Min ground turning radius 28·60 m (93 ft 10 in)
Range, 10% reserves and allowances for 30 min at 4,000 m (13,125 ft):
with 8,000 kg (17,637 lb) payload 2,590 nm (4,800 km; 2,982 miles)
with 16,000 kg (35,274 lb) payload 917 nm (1,700 km; 1,056 miles)
Max ferry range 2,805 nm (5,200 km; 3,230 miles)

VTI/CIAR

PARTICIPANTS:
Vazduhoplovno-Tehnicki Institut, 11132-Zarkovo, Yugoslavia
Centrala Industriala Aeronautica Romana, Bucharest, Romania

VTI/CIAR ORAO (EAGLE)

The Orao is a single-seat twin-jet ground attack fighter under development to meet a joint requirement of the air forces of Romania and Yugoslavia. It is powered by two 17·8 kN (4,000 lb st) Rolls-Royce Viper 623 non-afterburning turbojet engines, and was originally referred to as the 'Jurom' (from *Ju*goslavia-*Rom*ania). The landing gear is of Messier-Hispano design and manufacture. Soko of Yugoslavia is one of the principal companies concerned in the manufacturing programme.

The Orao is believed to have flown for the first time in August 1974, and to have made about 10 flights by the end of that year. On 15 April 1975 it was demonstrated publicly during the Victory Day parade at Batajnica military airfield near Belgrade. The accompanying photograph was taken on that occasion. A second prototype is believed to have been completed in Romania, and a pre-series batch was reportedly under construction in early 1976.

Fairey Hydraulics Ltd of the UK has a contract to supply filters and sampling valves for an undisclosed number of pre-series aircraft, and follow-on contracts are anticipated. Graviner of the UK is supplying the Firewire and BCF fire detection and extinguishing systems. It is anticipated that 200 or more Oraos may be built eventually for the two air forces, including a proportion of two-seat operational trainers.

ARMAMENT: Two 30 mm cannon in lower front fuselage; one underfuselage and four underwing stations for external stores.

DIMENSIONS (estimated):
Wing span 7·50 m (24 ft 7¼ in)
Length overall 13·00 m (42 ft 8 in)
Height overall 3·70 m (12 ft 1¾ in)
Wing area, gross 18·00 m² (193·75 sq ft)

WEIGHTS (estimated):
Weight empty, equipped 4,400 kg (9,700 lb)
Max external stores load 2,000 kg (4,409 lb)
Max T-O weight, with stores 9,000 kg (19,840 lb)

PERFORMANCE (estimated):
Max level speed:
'clean' 594 knots (1,100 km/h; 683 mph)
with external stores 431 knots (800 km/h; 497 mph)

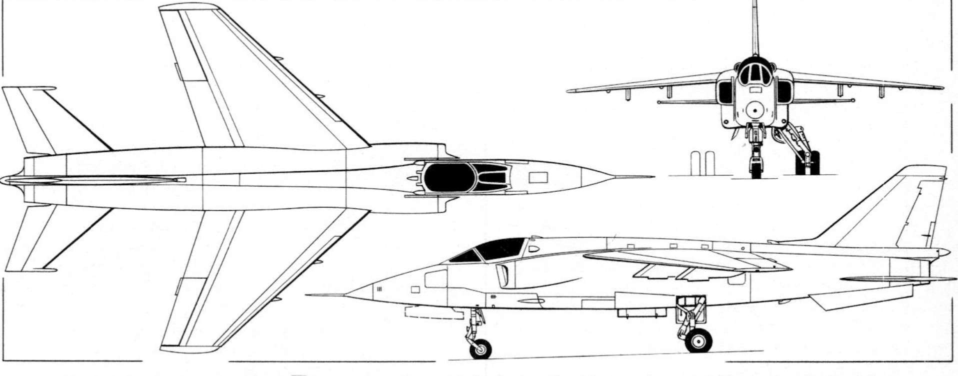

Orao single-seat tactical fighter, developed jointly by the Romanian and Yugoslav industries
(Pilot Press, provisional)

Prototype of the Orao single-seat tactical fighter (two Rolls-Royce Viper turbojet engines)

Max cruising speed:
'clean' 485 knots (900 km/h; 559 mph)
with external stores 378 knots (700 km/h; 435 mph)
Max rate of climb at S/L 5,400 m (17,725 ft)/min
Service ceiling 13,500 m (44,300 ft)
Combat radius with 2,000 kg (4,409 lb) external load:
lo-lo-lo 135 nm (250 km; 155 miles)
hi-lo-hi 243 nm (450 km; 280 miles)

ZENTRALGESELLSCHAFT VFW-FOKKER mbH

ADDRESS:
Gartenstrasse 15, 4000 Düsseldorf, German Federal Republic
Telephone: Düsseldorf 44 941
Telex: 85 84 344 ZGVF-D
BOARD OF MANAGEMENT:
Gerrit Cornelis Klapwijk (Chairman)
Alan R. Buley
Prof Gerhard Eggers
Hubertus Johannes Grobben
Ir Agni Aldo Holle
Dr Martin Lexis
Dr Fritz Wenck
SECRETARY GENERAL:
Dr G. Sadtler
DIRECTORS, VFW-FOKKER GmbH:
See Aircraft section (Germany)
DIRECTORS, FOKKER-VFW BV, AMSTERDAM:
See Aircraft section (Netherlands)
DIRECTORS, ERNO RAUMFAHRTTECHNIK GMBH, BREMEN:
See Spaceflight section (Germany)
DIRECTOR, AVIO-DIEPEN BV, AMSTERDAM:
D. G. de Rooij

In May 1969, NVKNV Fokker, Amsterdam, and Vereinigte Flugtechnische Werke GmbH, Bremen, combined their activities on a parity basis. The effective operating date of the association was made retrospective to 1 January 1969. To accomplish the merger, a new central company named Zentralgesellschaft VFW-Fokker mbH was created, with headquarters in Düsseldorf.

Shareholders of the Zentralgesellschaft are the holding companies NVKNV Fokker, Amsterdam, and VFW-Verwaltungsgesellschaft mbH, Bremen. The Zentralgesellschaft operates through Fokker-VFW BV, Amsterdam, and VFW-Fokker mbH, Bremen, described fully in the Netherlands and German sections respectively. All marketing and product support of civil aircraft of VFW and Fokker design are undertaken by a separate company, Fokker-VFW International BV, with offices at Schiphol-Oost, Netherlands (PO Box 7600).

Companies also operating directly under Zentralgesellschaft VFW-Fokker mbH are ERNO Raumfahrttechnik GmbH, Bremen, the space division; Rhein-Flugzeugbáu GmbH, Mönchengladbach (which see), for light aircraft; and Avio-Diepen BV, Amsterdam, a trading company.

The entire organisation employs about 18,000 people in 15 factories.

The company has, together with Dassault-Breguet of France, a parity interest in the Belgian company SABCA (which see), with factories at Haren and Gosselies.

IRAN

As a major step in a national industrialisation programme, the government of Iran has selected Bell Helicopter Textron (see US section) as its partner in establishing a modern helicopter industry in the country. The selection was made after an international competition under which proposals were considered from a number of major helicopter manufacturers in the United States and Europe; as a result Bell is to establish a new company, Bell Operations Corporation, staffed with experienced personnel and having full responsibility to carry out Bell's role in the programme.

The agreement includes long-term provision for co-production of 400 Model 214A transport helicopters (in addition to those currently being manufactured for the Iranian government at Bell's Fort Worth, Texas, plant); constructing and equipping a major helicopter production facility in Iran; building housing and community support facilities; and an extensive training programme to prepare Iranian personnel to operate and manage the helicopter industry in Iran.

Bell Helicopter Textron will furnish special tooling and production parts, to a total value of more than $125 million, in support of the co-production programme. In addition, the US manufacturers of many important components of the Model 214A (eg, engines, electronics, instruments, hydraulics and special materials) will participate in the programme.

ISRAEL

IAI
ISRAEL AIRCRAFT INDUSTRIES LTD

HEAD OFFICE AND WORKS:
Ben-Gurion International Airport, Lydda (Lod)
Telephone: 03-973111
Telex: Isravia 031114
PRESIDENT:
A. W. Schwimmer
EXECUTIVE VICE-PRESIDENTS:
A. Ben Yoseph
I. Roth
E. P. Wohl (Finance)
SENIOR VICE-PRESIDENT:
S. Yoran (Gen Manager, Bedek Aviation Division)
VICE-PRESIDENTS:
M. Blumkin (General Manager, Engineering Division)
G. Gidor (General Manager, Manufacturing Division)
Z. Yaari (Central Services)
A. Ostrinsky (Personnel and Administration)
S. Mendes (Western Hemisphere Operations)
D. Mozes (Comptroller)
S. Kadmon (Treasurer)
Mrs S. Ron (General Counsel)
COMMERCIAL DIRECTOR, ARAVA:
H. Pearlman
COMMERCIAL DIRECTOR, WESTWIND:
S. Samach
DIRECTOR, SYSTEMS SALES:
N. Rosen
DIRECTOR OF EXTERNAL RELATIONS:
Elkana Galli

This company was established in 1953 as Bedek Aircraft Company. The change of name, to Israel Aircraft Industries, was made on 1 April 1967.

IAI employs about 17,000 people in all its facilities, which occupy a total covered floor area of 300,000 m² (3,229,170 sq ft). It is licensed by, among others, the Israel Civil Aviation Administration, US Federal Aviation Administration, British Civil Aviation Authority and the Israeli Air Force as an approved repair station and maintenance organisation.

Israel Aircraft Industries Ltd is composed of several divisions, plants and subsidiary companies, as follows:

Bedek Aviation is an internationally approved multi-faceted single-site civil and military aircraft service centre. Present programmes include the turnaround inspection, overhaul, repair, retrofitting, outfitting and testing of 30 types of aircraft, including the Boeing 707 and McDonnell Douglas DC-8; 28 types of engine, including the 244·7 kN (55,000 lb st) Pratt & Whitney JT9D; and 6,000 types of components, accessories and systems. Offshore workload includes the supply of total technical support to several international operators. The division holds warranty and/or approved service centre approvals from many of the world's leading component manufacturers.

Bedek has refurbished and resold numerous Boeing 707s. The refurbishment process on a number of these has included conversion from passenger to cargo configuration. The procedure for the structural modification was developed jointly by Boeing and Bedek Aviation.

The Bedek Aviation centre has a total floor area of some 74,322 m² (800,000 sq ft) and employed a staff of about 3,500 in the Spring of 1976.

The **Aircraft Manufacturing Division** manufactures the IAI-designed Arava STOL transport, the 1123 Westwind business jet and the turbofan-powered 1124 Westwind. In addition, the Division is engaged in the manufacture of a vast variety of spares and assemblies for aircraft and jet engines, to meet Israeli Air Force requirements. As a subcontractor to many US and European aircraft manufacturers, the Division produces major aircraft structures, flight control surfaces, cargo loading systems and spares.

The **Engineering Division** is responsible for engineering research, design, development and testing of aerospace systems. It provides engineering support in system analysis, aerodynamics, materials and processing, landing and control systems, and in structural, flight and environmental testing. The Division performed modification and production support for the manufacture of the Magister jet trainer for the IAF; and major structural conversions of the Boeing Model 377, for military applications such as swing-tail freighter and hose-refuelling tanker. The Division designed and developed the Arava STOL transport aircraft, and developed both the 1123 Westwind business jet and a new version known as the 1124 Westwind.

IAI's diversification has led to the creation of a number of subsidiary companies which produce specialised products:

The company's electronics subsidiary, **Elta**, although wholly owned, is fully autonomous. It is specialising in the design, development and production of sophisticated electronic equipment such as airborne, ground and shipborne communications and radars, transceivers and navigational aids, general communications equipment, automatic test systems, and such electronic medical devices as cardiac resuscitation instruments.

MBT Weapons Systems is concerned with advanced electronic research, design and production; it participated in the development of the Gabriel missile system, among others, as well as of an Electronic Warning Fence and an Audible Bomb Release Altimeter.

MBT had a work force of some 1,700 persons in the Spring of 1976.

SHL Servo-Hydraulics Lod designs, develops and manufactures hydraulic and fuel system components, hydraulic flight control servo-systems, landing gears and brake systems.

SHL has about 10,000 m² (107,640 sq ft) of floor space, and employed some 660 persons in the Spring of 1976.

TAMAM Precision Industries manufactures and assembles high-precision electromechanical components and servo-systems for such mechanisms as aerosystems, torque motors and gyroscopes.

PML Precision Mechanisms Ltd includes among its products air-actuated chucks, miniature gears, clutches and brakes.

Orlite Engineering Ltd is a custom-moulder of reinforced plastics, and produces parts for aircraft, cabs and trucks, concrete casting moulds and sheet products.

Turbochrome is specialised in the application of corrosion and abrasion resistant coating by the diffusion process.

The **RAMTA** plant manufactures ground support equipment, stainless steel tanks, the Dabur patrol boat and the RBY armoured vehicle.

IAI NESHER (EAGLE)

Following the French embargo on the delivery of Dassault Mirage 5 fighters to Israel, the decision was taken in Israel to manufacture aircraft of generally similar design to the Mirage. The ultimate outcome of this policy is the IAI Kfir, with a General Electric J79 turbojet instead of the SNECMA Atar fitted to French-built Mirage III/5s. As an interim step, IAI undertook responsibility for manufacturing spares for Mirage III-CJ fighters operated by the Israeli Air Force and for putting into production an aircraft named the Nesher (Eagle). This comprised a locally-built airframe, similar to that of the Mirage III/5, fitted with an Atar 9C afterburning turbojet and Israeli electronics and equipment.

According to a book published in Israel in 1976, under the title *Israel, Army and Defence—a Lexicon,* the prototype Nesher flew for the first time in September 1969. Deliveries began in 1972, and some 40 Neshers are said to have taken part in the October 1973 war.

IAI KFIR (LION CUB)

Following manufacture of the Nesher fighter, powered by an Atar turbojet, IAI developed a more extensively modified and further improved version of the same airframe, powered by a General Electric J79 afterburning turbojet engine. Details were officially made public for the first time on 14 April 1975, when two of the new aircraft, now known as the Kfir, were displayed at Ben-Gurion Airport, Lydda.

The Kfir utilises a basic airframe similar to that of the Dassault Mirage 5, the main changes being a shorter but larger-diameter rear fuselage, to accommodate the J79 engine; an enlarged and flattened undersurface to the forward portion of the fuselage; introduction of a dorsal airscoop, in place of the triangular dorsal fin, to provide cooling air for the afterburner; a strengthened landing gear, with longer-stroke oleos; an elongated nose, extending the overall length to approx 15·3 m (50 ft 2 in); and modified wing leading-edges. Metal Resources Inc of Gardena, California, has an IAI subcontract to manufacture replacement wing components for Israeli Mirages. Several internal changes have also been made, including a redesigned cockpit layout, addition of a considerable amount of Israeli-built electronics equipment, and possibly a slight increase in internal fuel tankage compared with the Mirage 5.

Intended for both air defence and ground attack roles, the Kfir retains the standard Mirage fixed armament of two 30 mm DEFA cannon, and can carry a variety of

Interceptor version of the IAI Kfir single-seat multi-purpose combat aircraft; see Addenda for details of Kfir-C2

external weapons including the Rafael Shafrir air-to-air missile (see Air-Launched Missiles section). It has demonstrated stall-free gun firing throughout the flight envelope.

A description of the Mirage 5 can be found under the Dassault-Breguet heading in the French section of this edition. The following are the main differences of which details had been learned up to the Spring of 1976:

TYPE: Single-seat interceptor and close support aircraft.

WINGS: Bascially as Mirage 5, but with modified leading-edges.

FUSELAGE: Similar to Mirage 5, but with elongated forward section and nosecone (built of locally-developed composites), locating pilot further forward of engine air intakes, and enlarged-diameter rear section with approx 0·61 m (2 ft) shorter tailpipe. Interceptor version has longer and more pointed nose radome than ground attack version. Cross-section of forward fuselage has a wider and flatter undersurface than that of Mirage. Ventral fairing under rear fuselage.

TAIL UNIT: Basically as Mirage 5, but with dorsal fin replaced by triangular-section airscoop to provide cold air for afterburner cooling. Brake parachute in fairing at base of rudder.

LANDING GEAR: Similar to Mirage 5, but with longer-stroke oleos and strengthened to allow for higher operating weights. Main-gear leg fairings shorter than on Mirage; inner portion of each main-leg door is integral with fuselage-mounted wheel door.

POWER PLANT: One General Electric J79 turbojet engine (modified GE-17?), with variable-area nozzle, rated at 52·8 kN (11,870 lb st) dry and 79·62 kN (17,900 lb st) with afterburning. Air intakes enlarged to allow for higher mass flow. Standard internal fuel capacity of the Mirage III is 3,330 litres (732·5 Imp gallons); Kfir may have provision for an additional fuel tank in the fuselage, and probably has an external capability similar to that of the Mirage 5, which can carry up to 4,700 litres (1,034 Imp gallons) of auxiliary fuel in external drop-tanks or 1,000 litres (220 Imp gallons) in combination with 4,000 kg (8,820 lb) of ordnance. Provision for ventral rocket booster motor is apparently retained.

ACCOMMODATION: Pilot only, on Martin-Baker JM.6 zero-zero ejection seat, under rearward-hinged upward-opening canopy. Revised cockpit layout compared with Mirage 5.

ELECTRONICS AND EQUIPMENT: MBT Weapons Systems twin-computer fly-by-wire flight control system. Elta Electronics multi-mode navigation and weapon delivery system. Israeli-built head-up display and gunsight. Aerial under nose for Doppler radar or radar altimeter. Two blade-type antennae also under nose, between radome and nosewheel unit.

ARMAMENT: Two 30 mm DEFA cannon in undersides of engine air intake trunks, as in Mirage 5. Rafael Shafrir dogfight missiles for air-to-air combat. Ground attack version can carry conventional or 'dibber' bombs, rocket pods, or Shrike, Maverick or Hobos air-to-surface missiles.

DIMENSIONS, EXTERNAL:
Similar to Mirage 5 (which see)

WEIGHTS:
Typical combat weight (interceptor), 50% fuel and 2 Shafrir missiles 9,305 kg (20,514 lb)
Max combat T-O weight (ground attack) 14,600 kg (32,188 lb)

PERFORMANCE (estimated):
Max level speed at high altitude over Mach 2·2 (1,260 knots; 2,335 km/h; 1,450 mph)
Max rate of climb at S/L 14,000 m (45,950 ft)/min
Time to 11,000 m (36,100 ft) 1 min 45 sec
Stabilised ceiling (combat configuration) above 15,240 m (50,000 ft)

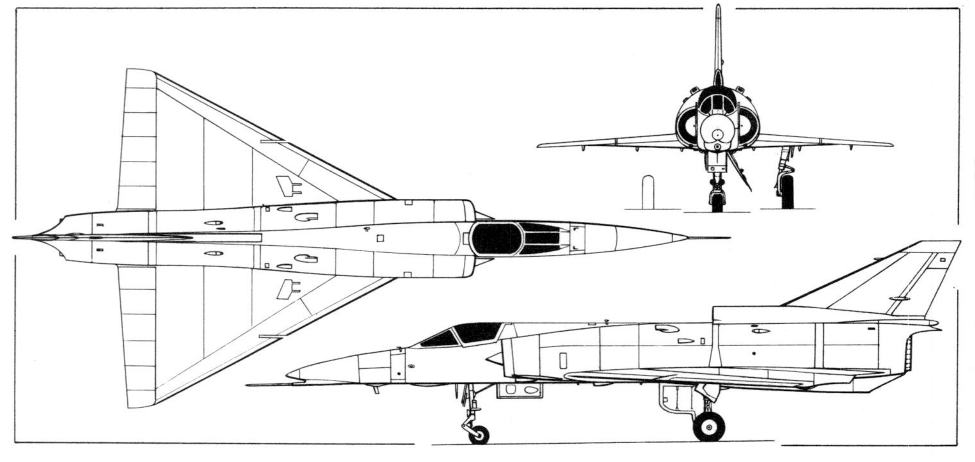

IAI Kfir (General Electric J79 afterburning turbojet engine) *(Pilot Press)*

T-O run at 11,000 kg (24,250 lb) AUW 700 m (2,300 ft)
Landing run at 9,000 kg (19,840 lb) AUW 450 m (1,475 ft)
Combat radius:
interceptor, two 600 litre drop-tanks 200-288 nm (370-535 km; 230-332 miles)
ground attack, lo-lo-lo 351 nm (650 km; 404 miles)
ground attack, hi-lo-hi 700 nm (1,300 km; 807 miles)

IAI-101 and IAI-201 ARAVA

The Arava was designed to fufil the need for a light transport with STOL performance and rough-field landing capabilities.

Design work started in 1966 and construction of a prototype began towards the end of the same year. This airframe was used for structural testing; it was followed by a flying prototype (4X-IAI), which made its first flight on 27 November 1969. A second Arava (4X-IAA) began flight trials on 8 May 1971.

The following versions are available:

IAI-101. Civil transport version, type certificated in April 1972.

IAI-201. Military transport version, based upon the original IAI-101. A prototype (4X-IAB) began its flight tests on 7 March 1972, and this version is now in full production. The standard equipment available for the IAI-201 enables a wide variety of missions to be undertaken.

Prior to the 'Yom Kippur' war in October 1973 a total of 15 military Aravas had been ordered, 14 of them for export. During the conflict three Aravas were lease-operated by the Israeli Air Force.

Sales of the Arava had reached 50 by the beginning of 1976, and by March 1976 a total of 30 of these had been delivered. Customers include the Israeli Air Force, Bolivian Air Force, Ecuadorean Army, Guatemalan Air Force, Honduran Air Force, Mexican Air Force, Nicaraguan Air Force and Salvadorean Air Force. Production of the Arava was at the rate of three per month in the Spring of 1976.

The following description applies to the IAI-201:

TYPE: Twin-turboprop STOL light military transport.

WINGS: Braced high-wing monoplane, with single streamline-section bracing strut each side. Wing section NACA 63(215)A 417. Dihedral 1° 30′. Incidence 0° 27′. No sweepback. Light alloy two-spar torsion-box structure. Frise-type light alloy ailerons. Electrically-operated double-slotted light alloy flaps. Scoop-type light alloy spoilers, for lateral control, above wing at 71% chord. Electrically-actuated trim tab in port aileron.

FUSELAGE: Conventional semi-monocoque light alloy structure of stringers, frames and single-skin panels.

TAIL UNIT: Cantilever light alloy structure, with twin fins and rudders, carried on twin booms extending rearward from engine nacelles. Fixed-incidence tailplane. Geared tab and electrically-actuated trim tab in elevator and geared trim tab in each rudder.

LANDING GEAR: Non-retractable tricycle type, of Electro-Hydraulics manufacture, with single main wheels and steerable nosewheel. Main wheels carried on twin struts, incorporating oleo-pneumatic shock-absorbers. Main wheels size 11·00-12. Nosewheel size 9·00-6. Tyre pressure 3·31 bars (48 lb/sq in) on main units, 2·90 bars (42 lb/sq in) on nose unit. Disc brakes on main units.

POWER PLANT: Two 559 kW (750 shp) Pratt & Whitney Aircraft of Canada PT6A-34 turboprop engines, each driving a Hartzell HC-B3TN three-blade hydraulically-actuated fully-feathering reversible-pitch metal propeller. Electrical de-icing of propellers optional. Two integral fuel tanks in each wing, with total usable capacity of 1,663 litres (366 Imp gallons). Four overwing refuelling points. Optional pressure refuelling point in fuselage/strut fairing. Two cabin-mounted tanks, each of 1,022 litres (225 Imp gallons), are available optionally; these are used for ferrying purposes.

ACCOMMODATION: Crew of one or two on flight deck, with door on starboard side. Main cabin has folding inward-facing metal-framed fabric seats along each side, and can accommodate 20 passengers (IAI-101), 24 fully-equipped troops or 17 paratroops and two dispatchers. Outward-opening door at rear of cabin, opposite which, at floor level, is an emergency exit door/cargo door on the starboard side. Aft section of fuselage is hinged to swing sideways through more than 90° to provide unrestricted access to main cabin. Alternative interior configurations available for ambulance role (twelve stretchers and two sitting patients/medical attendants) or as all-freight transport carrying (typically) a jeep-mounted recoil-less rifle and its four-man crew.

SYSTEMS: Hydraulic system (pressure 172 bars; 2,500 lb/sq in) for brakes and nosewheel steering only. Electrical system includes two 28V 170A DC engine-driven starter/generators, a 28V 40Ah nickel-cadmium battery and two 250VA 115/26V 400Hz static inverters.

Electronics and Equipment: Blind-flying instrumentation standard. Optional equipment includes VHF, VOR/ILS, ADF, marker beacon and PA system.

Armament: Optional 0·50 in Browning machine gun pack on each side of fuselage, above a pylon for a pod containing seven 68 mm rockets. Provision for aft-firing machine-gun. Librascope gunsight.

Dimensions, external:

Wing span	20·96 m (68 ft 9 in)
Wing chord (constant)	2·09 m (6 ft 10½ in)
Wing aspect ratio	10
Length overall	13·03 m (42 ft 9 in)
Length of fuselage pod	9·33 m (30 ft 7¼ in)
Diameter of fuselage	2·50 m (8 ft 2½ in)
Height overall	5·21 m (17 ft 1 in)
Tailplane span (c/l of tailbooms)	5·21 m (17 ft 1 in)
Wheel track	4·01 m (13 ft 2 in)
Wheelbase	4·62 m (15 ft 2 in)
Propeller diameter	2·59 m (8 ft 6 in)
Propeller ground clearance	1·75 m (5 ft 9 in)
Crew door (fwd, stbd):	
Height	0·81 m (2 ft 8 in)
Width	0·46 m (1 ft 6 in)
Passenger door (rear, port):	
Height	1·70 m (5 ft 7 in)
Width	0·89 m (2 ft 11 in)
Cargo drop door (rear, port):	
Height	2·21 m (7 ft 3 in)
Width	0·97 m (3 ft 2 in)
Emergency/baggage door (rear, stbd):	
Height	1·14 m (3 ft 9 in)
Width	0·64 m (2 ft 1¼ in)
Emergency window exits (each):	
Height	0·66 m (2 ft 2 in)
Width	0·48 m (1 ft 6¾ in)

Dimensions, internal:

Cabin, excl flight deck and hinged tailcone:	
Length	3·87 m (12 ft 8¼ in)
Max width	2·33 m (7 ft 8 in)
Max height	1·75 m (5 ft 9 in)
Floor area	7·16 m² (77·1 sq ft)
Volume	13·2 m³ (466·2 cu ft)
Baggage compartment volume	2·60 m³ (91·8 cu ft)
Cargo door volume	3·20 m³ (113 cu ft)

Areas:

Wings, gross	43·68 m² (470·2 sq ft)
Ailerons (total)	1·75 m² (18·84 sq ft)
Trailing-edge flaps (total)	8·80 m² (94·72 sq ft)
Spoilers (total)	0·85 m² (9·2 sq ft)
Fins (total)	4·86 m² (52·31 sq ft)
Rudders (total incl tabs)	3·44 m² (37·03 sq ft)
Tailplane	9·36 m² (100·75 sq ft)
Elevator, incl tabs	2·79 m² (30·03 sq ft)

Weights and Loadings:

Basic operating weight	3,999 kg (8,816 lb)
Max payload	2,351 kg (5,184 lb)
Max T-O and landing weight	6,803 kg (15,000 lb)
Max zero-fuel weight	6,350 kg (14,000 lb)
Max wing loading	153·5 kg/m² (31·44 lb/sq ft)
Max power loading	6·08 kg/kW (10·00 lb/hp)

Performance (at max T-O weight):

Never-exceed speed	215 knots (397 km/h; 247 mph)
Max level speed at 3,050 m (10,000 ft)	176 knots (326 km/h; 203 mph)
Max cruising speed at 3,050 m (10,000 ft)	172 knots (319 km/h; 198 mph)
Econ cruising speed at 3,050 m (10,000 ft)	168 knots (311 km/h; 193 mph)
Stalling speed, 0° flap	75 knots (140 km/h; 87 mph)
Stalling speed, 54° flap	62 knots (115 km/h; 71·5 mph)
Max rate of climb at S/L	393 m (1,290 ft)/min
Rate of climb at S/L, one engine out	55 m (180 ft)/min
Service ceiling	7,620 m (25,000 ft)
Service ceiling, one engine out	2,375 m (7,800 ft)
STOL T-O run	293 m (960 ft)
STOL T-O to 15 m (50 ft)	463 m (1,520 ft)
STOL landing from 15 m (50 ft)	469 m (1,540 ft)
STOL landing run	250 m (820 ft)
Range with max fuel, 45 min reserves	705 nm (1,306 km; 812 miles)
Range with max payload, 45 min reserves	151 nm (280 km; 174 miles)

IAI 1123 WESTWIND

In 1967, Israel Aircraft Industries acquired all production and marketing rights for the Rockwell-Standard Corporation (formerly Aero Commander) Jet Commander business jet transport. The current version is known as the 1123 Westwind business jet, and is certificated in the FAA transport category.

The first of two prototype Jet Commanders (N601J) was flown for the first time in the US on 27 January 1963. The US production line was phased out between 1967 and mid-1969. A total of 150 Commodore Jet/Jet Commander aircraft (US and IAI production) were delivered, in three basic models: the 1121, 1121A and 1121B.

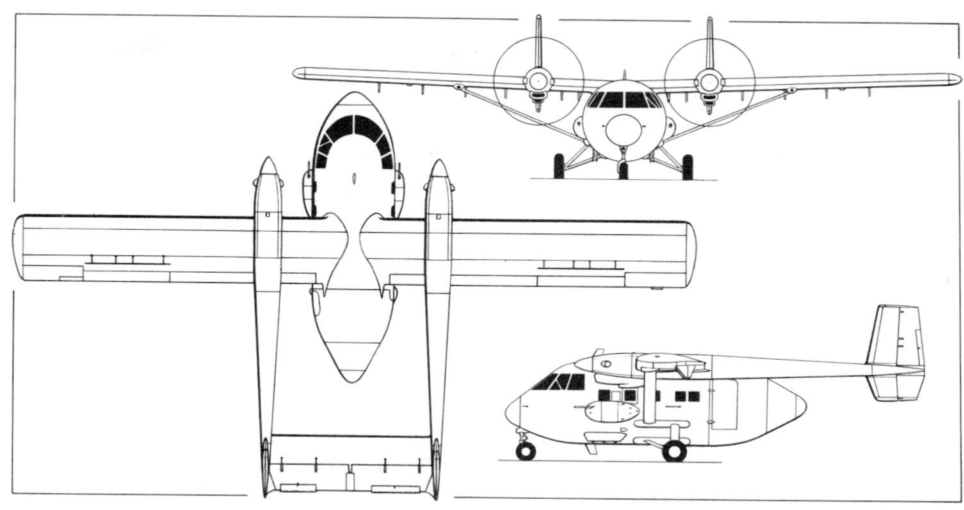

IAI-201 Arava twin-turboprop STOL light military transport *(Pilot Press)*

IAI-201 Arava twin-turboprop military transport of the Salvadorean Air Force

Many modifications and improvements were incorporated by IAI into the original Jet Commander/Commodore Jet 1121, as described in previous editions of *Jane's*.

The 1123 Westwind, to which the following description applies, has a 0·51 m (1 ft 8 in) longer cabin, wingtip auxiliary fuel tanks, more powerful engines, an APU, strengthened landing gear, simplified electrical system, double-slotted flaps, drooped wing leading-edges and two additional cabin windows. A prototype was flown for the first time on 28 September 1970, and was certificated by the Israel CAA and by the FAA at the end of 1971.

Deliveries of the 1123 Westwind have been made to customers in Israel, the USA, Canada, the German Federal Republic, Mexico and Panama. Atlantic Aviation Corporation of Wilmington, Delaware, has sole marketing rights for the aircraft in the USA and Canada.

Production of the 1123 Westwind was scheduled to terminate in mid-1976 with delivery of the 36th aircraft; its place will be taken by the turbofan-powered 1124 Westwind (described separately), of which production has already begun.

Type: Twin-jet business transport.

Wings: Cantilever mid-wing monoplane. Wing section NACA 64A212. Dihedral 2°. Incidence 1° at root, −1° at tip. Sweepback 4° 37′ at quarter-chord. All-metal flush-riveted two-spar fail-safe structure. Manually-operated all-metal ailerons. Electrically-operated all-metal double-slotted trailing-edge Fowler-type flaps and detachable drooped leading-edge. Electrically-operated trim tab in port aileron. Hydraulically-actuated speed brake and two lift dumpers above each wing, forward of flap. All skins chemically milled and fully sealed. All primary control surfaces, including tabs, are fully mass-balanced. Pneumatic anti-icing boots standard.

Fuselage: All-metal semi-monocoque flush-riveted structure with pressurised fail-safe cabin and baggage compartment. Built in two main sections and joined at aft pressure bulkhead. Forward section, except for nosecone, is fully pressurised.

Tail Unit: Cantilever all-metal structure, with 28° sweepback at quarter-chord. Variable-incidence tailplane, actuated electrically. Manually-operated statically-balanced elevators and rudder. Trim tab in rudder. Pneumatic anti-icing boots standard.

Landing Gear: Hydraulically-retractable tricycle type, main wheels retracting outward into wings, twin nosewheels rearward. Oleo-pneumatic shock-absorbers. Single wheels on main units, pressure 10·69 bars (155 lb/sq in). Nose unit steerable and self-centering. Nosewheel tyre pressure 3·45 bars (50 lb/sq in). Goodyear multiple-disc brakes. Fully-modulated anti-skid system has automatic computer/sensor to prevent wheel lock and maintain brake effectiveness. Parking brake fitted.

Power Plant: Two 13·79 kN (3,100 lb st) General Electric CJ610-9 turbojet engines, mounted in pod on each side of rear fuselage. 85% of wing area forms an integral fuel tank, and additional fuel is carried separately in wingtip tanks and rear fuselage tanks. Total usable capacity 5,035 litres (1,107 Imp gallons; 1,330 US gallons), including wingtip tanks. Refuelling points in wingtips and fuselage. Thrust reversers and single-point pressure refuelling available optionally.

Accommodation: Standard seating for two pilots and up to 10 passengers in pressurised cabin. Interior layout to customer's requirements, with galley and toilet standard. Separate pressurised compartment for up to 181 kg (400 lb) of baggage. Passenger door at front on port side; emergency exit on each side, forward of wing. Entire accommodation heated, ventilated and air-conditioned.

Systems: Primary hydraulic system, pressure 138 bars (2,000 lb/sq in), operates through two engine-driven pumps to actuate landing gear, wheel brakes, nosewheel steering, speed brakes and lift dumpers. Electrically-operated emergency system, pressure 69 bars (1,000 lb/sq in), for brakes only. DC electrical system with two 350A 24V engine-driven starter/generators and two 21Ah long-life nickel-cadmium batteries. One main bus for each generator, connected to the central battery bus. A 400Hz 115V AC system is installed, and is powered by two solid-state static inverters, each of 1,000VA capacity, with power fed from the main DC buses. Each inverter is independently capable of supplying the entire AC load if required. Pneumatic system for anti-icing of wing and tail leading-edges only. Warm air bled from engines to prevent icing of air intakes. Windscreen is heated electrically. Microturbo Saphir III APU, installed in tailcone, for ground air-conditioning and ground electrical power supply. Cabin air pressure max differential of 0·60 bars (8·7 lb/sq in) up to 13,725 m (45,000 ft).

Equipment: Full blind-flying instrumentation standard. Radio and electronics to customer's requirements.

DIMENSIONS, EXTERNAL:
Wing span 13·65 m (44 ft 9½ in)
Wing chord at root 3·20 m (10 ft 6 in)
Wing chord at tip 1·07 m (3 ft 6 in)
Wing aspect ratio 6·51
Length overall 15·93 m (52 ft 3 in)
Fuselage: Max width 1·57 m (5 ft 2 in)
Max depth 1·83 m (6 ft 0 in)
Height overall 4·81 m (15 ft 9½ in)
Tailplane span 6·40 m (21 ft 0 in)
Wheel track 3·35 m (11 ft 0 in)
Passenger door:
Height 1·37 m (4 ft 6 in)
Width 0·61 m (2 ft 0 in)
Height to sill 0·51 m (1 ft 8 in)

DIMENSIONS, INTERNAL:
Cabin, excl flight deck:
Length 4·72 m (15 ft 6 in)
Max width 1·45 m (4 ft 9 in)
Max height 1·50 m (4 ft 11 in)
Volume 9·83 m³ (347 cu ft)
Baggage compartment volume 0·71 m³ (25 cu ft)

AREAS:
Wings, gross 28·64 m² (308·26 sq ft)
Ailerons (total) 1·43 m² (15·39 sq ft)
Trailing-edge flaps (total) 3·86 m² (41·58 sq ft)
Fin 4·51 m² (48·60 sq ft)
Rudder, incl tab 0·99 m² (10·69 sq ft)
Tailplane 4·87 m² (52·42 sq ft)
Elevators 1·64 m² (17·66 sq ft)

WEIGHTS AND LOADINGS:
Weight empty, equipped 4,250 kg (9,370 lb)
Typical basic operating weight 5,125 kg (11,300 lb)
Max payload 1,000 kg (2,200 lb)
Max T-O weight 9,389 kg (20,700 lb)
Max ramp weight 9,525 kg (21,000 lb)
Max zero-fuel weight 6,577 kg (14,500 lb)
Max landing weight 8,618 kg (19,000 lb)
Max cabin floor loading 976·5 kg/m² (200 lb/sq ft)
Max wing loading 327·1 kg/m² (67·0 lb/sq ft)
Max power loading 340 kg/kN (3·34 lb/lb st)

PERFORMANCE (at max T-O weight, ISA, except where indicated):
Max operating Mach number at 5,485 m (18,000 ft) and above Mach 0·765
Max level speed at 5,485 m (18,000 ft)
471 knots (872 km/h; 542 mph) TAS
Max operating speed at 5,485 m (18,000 ft)
372 knots (689 km/h; 428 mph) CAS
Max operating speed at S/L
360 knots (668 km/h; 415 mph) CAS
Econ cruising speed at 12,500 m (41,000 ft)
365 knots (676 km/h; 420 mph) TAS
Stalling speed, flaps and landing gear down, at max landing weight
97 knots (180 km/h; 112 mph) CAS
Stalling speed, flaps and landing gear down, at 5,443 kg (12,000 lb) AUW
79 knots (146·5 km/h; 91 mph) CAS
Max rate of climb at S/L 1,231 m (4,040 ft)/min
Rate of climb at S/L, one engine out
335 m (1,100 ft)/min
Service ceiling 13,725 m (45,000 ft)
Service ceiling, one engine out 6,860 m (22,500 ft)
T-O to 10·7 m (35 ft) 1,250 m (4,100 ft)
FAA balanced T-O field length 1,631 m (5,350 ft)
Balanced field length at 8,164 kg (18,000 lb) AUW 1,158 m (3,800 ft)
Landing from 15 m (50 ft) at max landing weight
1,036 m (3,400 ft)
Landing from 15 m (50 ft) at 6,350 kg (14,000 lb) AUW 808 m (2,650 ft)
Range with max fuel, 5 passengers and baggage, 45 min reserves 1,840 nm (3,410 km; 2,120 miles)
Range with max payload, 45 min reserves
1,390 nm (2,575 km; 1,600 miles)
IFR range with 6 passengers after T-O from balanced field length of 1,220 m (4,000 ft)
1,233 nm (2,285 km; 1,420 miles)

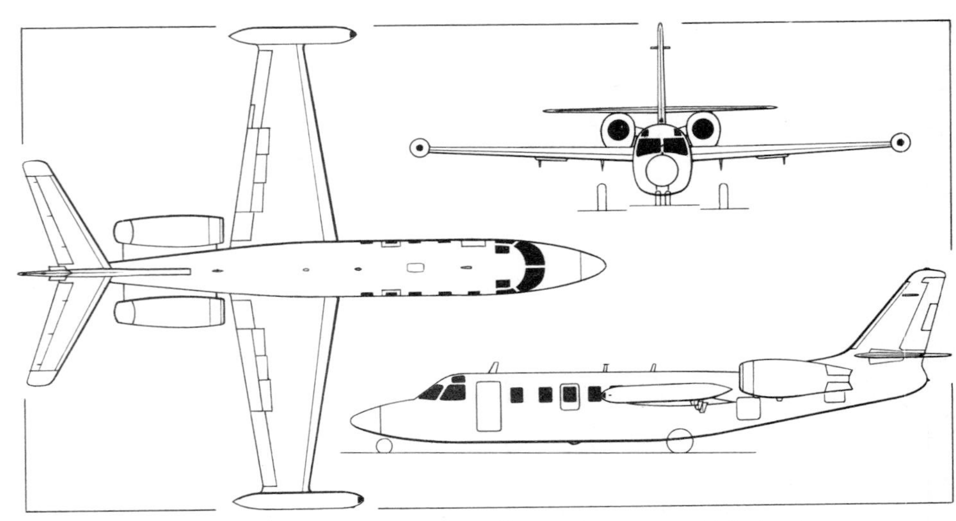

IAI 1124 Westwind twin-turbofan light executive transport *(Pilot Press)*

IAI 1124 Westwind (two Garrett-AiResearch TFE 731-3 turbofan engines)

IAI 1124 WESTWIND

Military designation: 1124N (Coastal Reconnaissance)

This longer-range version of the Westwind was certificated by the FAA in the Spring of 1976, at which time 15 had been ordered by customers in North America.

Two 1123 Westwinds were modified as flight test aircraft and the first of these made its first flight on 21 July 1975. Production was at the rate of two per month in 1976, and is expected to increase to three per month in 1977 following the phasing out of the turbojet-powered 1123 Westwind (which see).

The 1124 has essentially the same airframe as the 1123 Westwind, but is powered by two 16·46 kN (3,700 lb st) Garrett-AiResearch TFE 731-3 turbofan engines, with Grumman thrust reversers as a standard fit.

The following improvements and new systems have been added to the standard equipment list for the 1124 Westwind: single-point pressure refuelling system; new Garrett-AiResearch cabin environmental control system; two 37Ah batteries; new landing gear wheel brakes with a fully-modulated anti-skid system; improved drooped wing leading-edge; and redesigned dorsal fin. Fuel capacity is the same as for the 1123; but runway requirements are substantially reduced. Other changes include a strengthened landing gear and improved electronics, including a Bendix RTR-1200 radar, Collins flight director, autopilot and R/NAV systems.

WEIGHTS:
Basic operating weight (incl two pilots and service load)
5,806 kg (12,800 lb)
Max T-O weight 10,364 kg (22,850 lb)
Max ramp weight 10,432 kg (23,000 lb)
Max landing weight 8,618 kg (19,000 lb)
Max zero-fuel weight 7,257 kg (16,000 lb)

PERFORMANCE (estimated, at max T-O weight):
Max level speed, S/L to 5,900 m (19,400 ft)
471 knots (872 km/h; 542 mph) IAS
Max level speed above 5,900 m (19,400 ft)
Mach 0·765
Service ceiling 13,725 m (45,000 ft)
T-O balanced field length 1,475 m (4,840 ft)
Landing distance at max landing weight
732 m (2,400 ft)
Range with 7 passengers and baggage, IFR reserves
more than 2,400 nm (4,447 km; 2,764 miles)
Max range with 3 passengers and baggage, 45 min reserves
more than 2,500 nm (4,633 km; 2,878 miles)

ITALY

AERITALIA
AERITALIA SpA

HEAD OFFICE:
Piazzale Vincenzo Tecchio 51 (Casella Postale 3065), 80125 Naples
Telephone: (081) 619522, 619721, 619845, 619149, 619703
Telex: N. 71370 (AERIT)

OFFICE OF THE CHAIRMAN:
Via Panama 52, 00198 Rome
Telephone: (06) 8440341
Telex: N. 62395 (AERIT)

BRANCH OFFICE (COMMERCIAL):
Via Panama 52, 00198 Rome
Telephone: (06) 8440341
Telex: N. 62395 (AERIT)

MANAGING DIRECTOR:
Ing Renato Bonifacio

GENERAL MANAGER:
Ing Corrado Innocenti

BOARD OF DIRECTORS:
Dott Ercole Agosta
Dott Romolo Arena
Avv Alberto Boyer
Ing Renato Bonifacio
Ing Luigi di Giorgio
Ing Luigi Galleani d'Agliano
Ing Franco Giura
Ing Ermanno Pedrana
Ing Amilcare Porro
Gen SA (r) Mario Porru Locci
Ing Bruno Ressico
Dott Giovanni Mario Rossignolo
Ing Beppe Sacchi
Dott Fabio Massimo Tafuri

EXECUTIVE DIRECTORS:
Ing Giulio Ciampolini (Combat Aircraft Group)
Ing Fausto Cereti (Transport Aircraft Group)
Ing Stefano Abba (Diversified Activities Group)
Dott Michele Crosio (Personnel)
Ing Edoardo Dello Siesto (Market Research)
Ing Corrado Innocenti (Finance) (interim)
Dott Massimo Rizzo (General Secretary and External Relations)

PRESS SERVICE MANAGER:
Baldassare Catalanotto

PUBLIC RELATIONS:
Luigi Azais

Aeritalia is a joint stock company which was formed on 12 November 1969 by an equal shareholding of Fiat and Finmeccanica-IRI, to combine Fiat's aerospace activities (except those which concern aero-engines) and those of Aerfer and Salmoiraghi belonging to the Finmeccanica group. The company became fully operational under the new title on 1 January 1972. Aeritalia had a total work force, in 1976, of more than 9,500 persons.

Aeritalia's organisation is based upon a centralised general management and three operational groups: Combat Aircraft Group, Transport Aircraft Group, and the Diversified Activities Group. The production centres are located in Turin (Corso Marche, Caselle Nord and Caselle Sud), Milan (Nerviano) and Naples (Pomigliano d'Arco and Capodichino). The Italian Interministerial Committee for Economic Planning has approved Aeritalia's plans for a new industrial plant in the Foggia area.

Aeritalia has a co-operation agreement with Boeing of the USA to develop an advanced commercial transport aircraft. This programme began in the Spring of 1971 and was continuing in 1976.

COMBAT AIRCRAFT GROUP

HEADQUARTERS AND TURIN WORKS:
Corso Marche 41, 00146 Turin
Telephone: (011) 790166, 720072
Telex: N. 21076 (AERITOR)
CASELLE WORKS: Turin Airport CP, 10100 Turin
Telephone: (011) 991362
Telex: N. 21095

The Turin area factories are engaged in the manufacture, assembly and flight testing of the G91Y tactical fighter-bomber and reconnaissance aircraft, F-104S interceptor fighter, and G222 military transport aircraft for the Italian armed forces; in the design and construction of structural components for the Panavia Tornado (see International section); and repair, overhaul and maintenance of test equipment. Other activities include the repair, overhaul and maintenance of F-104G, F-104S and TF-104G aircraft.

TRANSPORT AIRCRAFT GROUP

HEADQUARTERS:
Via Vespucci 9, 80125 Naples
Telephone: (081) 267344
Telex: 71522 (AERVEL)
POMIGLIANO D'ARCO WORKS:
80038 Pomigliano d'Arco, Naples
Telephone: (081) 8841544
Telex: N. 71082 (AERITPOM)
CAPODICHINO WORKS: Via del Riposo alla Doganella, Aeroporto di Capodichino, 80144 Naples
Telephone: (081) 444166
Telex: N. 71356

Aeritalia's principal activities in the Naples area comprise construction of the complete series of fuselage structural panels for the McDonnell Douglas DC-9, and fuselage upper panels and the vertical tail surfaces for the DC-10 commercial airliner; construction of engine support pylons for the Boeing 747; and construction of fuselages for the Aeritalia G222, the Aeritalia (Lockheed) F-104S and the Aeritalia G91Y. Other activities include the repair, overhaul and modification of aircraft of various nations, including Italy and the United States, and the repair and maintenance of Breguet 1150 Atlantic, and G91R, G91T and G91Y aircraft.

FOGGIA BRANCH OFFICE:
Via Giannone 1, 70100 Foggia
Telephone: (0881) 79641
AMENDOLA WORKS:
Under development and construction

DIVERSIFIED ACTIVITIES GROUP

HEADQUARTERS:
Caselle-Turin
Telephone: (011) 991363
Telex: 22083 (AITCEA)
ELECTRONICS:

After earlier work on the concept, design and interface of complex electronics systems and the development of specific equipment for space purposes, Aeritalia diversified its project activity in this field to include the capability to conceive, design and interface such aerospace systems as satellite attitude control and new systems for civil utility purposes, ie vehicle control, automation of airport functions, and simulation and design of production multi-loop controls. Production activity in this area comprises the manufacture, repair and overhaul of sophisticated electronics and space equipment, both for aircraft and space components of Aeritalia's own production and for the international market. See also the Aeritalia entry in the Spaceflight and Research Rockets section.

INSTRUMENTATION:
NERVIANO WORKS: Viale Europa, 20014 Nerviano, Milan
Telephone: (0331) 587330

The Nerviano works manufactures electronics equipment and instrumentation, including systems and instruments for aeronautical, missile and space applications such as accelerometers, altimeters, electrical equipment, gyro compasses, gyroscopic platforms, magnetic compasses, pressure, stress and temperature transducers and load amplifiers.

AERITALIA G91Y

The G91Y is a twin-engined development of the earlier single-engined Fiat G91 (see 1966-67 *Jane's*), based upon the airframe of the G91T version.

Two G91Y prototypes were built, of which the first flew for the first time on 27 December 1966. They were followed by 20 pre-series G91Ys for the Italian Air Force, the first of which was flown in July 1968. All 20 were delivered to the 1° Group of the 8° Wing of the Italian Air Force, based at Cervia.

Delivery of the initial series of 35 production G91Ys to the Italian Air Force began in September 1971, and was completed by mid-1973. Delivery of an additional 10 aircraft was due to be completed by mid-1976. The G91Y is also in service with the 13° Group of the 32° Wing of the Italian Air Force, based at Brindisi.

TYPE: Lightweight single-seat tactical fighter-bomber and reconnaissance aircraft.

WINGS: Cantilever low-wing monoplane. Laminar-flow section. Sweepback at quarter-chord 37° 40′ 38″. Two-spar structure, with milled skin panels and detachable leading-edges. Ailerons with hydraulic servo control. Electrically-actuated slotted trailing-edge flaps. Automatic full-span leading-edge slats.

FUSELAGE: Semi-monocoque structure. Rear fuselage detachable for engine replacement. Two door-type airbrakes under centre-fuselage.

TAIL UNIT: Cantilever structure. Electrically-actuated variable-incidence tailplane. Auxiliary fin beneath each side of rear fuselage.

LANDING GEAR: Retractable tricycle type of Messier-Hispano design. Hydraulic actuation. Main-wheel tyre pressure 3·93 bars (57 lb/sq in). Hydraulic brakes. Brake-chute housed at base of rudder. Arrester hook under rear fuselage.

POWER PLANT: Two General Electric J85-GE-13A turbojet engines (each 12·1 kN; 2,720 lb st dry, 18·15 kN; 4,080 lb st with afterburning), mounted side by side in rear fuselage. Provision for JATO units for assisted take-off. Fuel in main tanks in fuselage and inner wing panels with total capacity of 3,200 litres (703 Imp gallons). Provision for underwing auxiliary tanks.

ACCOMMODATION: Pilot only, on fully-automatic zero-zero ejection seat, under electrically-actuated rearward-hinged jettisonable canopy. Cockpit armoured, pressurised and air-conditioned.

ARMAMENT AND OPERATIONAL EQUIPMENT: Two 30 mm DEFA cannon and cameras in nose. Four underwing attachments for 1,000 lb bombs, 340 kg (750 lb) napalm tanks, four 7×2 in rocket packs, four 28×2 in rocket packs or four 5 in rocket containers. Nav/attack system includes Computing Devices of Canada 5C-15 position and homing indicator, Sperry SYP-820 twin-axis gyro platform, Bendix RDA-12 Doppler radar and AiResearch air data computer, Ferranti ISIS B gyro-gunsight, Smiths electronic head-up display, Honeywell AN/APN-171 radar altimeter and Marconi-Elliott AD 370 ADF.

DIMENSIONS, EXTERNAL:
Wing span 9·01 m (29 ft 6½ in)
Wing chord at root 2·526 m (8 ft 3½ in)
Wing chord at tip 1·274 m (4 ft 2¼ in)
Wing aspect ratio 4·475
Length overall 11·67 m (38 ft 3½ in)
Height overall 4·43 m (14 ft 6 in)
Tailplane span 4·00 m (13 ft 1½ in)
Wheel track 2·94 m (9 ft 8 in)
Wheelbase 3·56 m (11 ft 8 in)

AREAS:
Wings, gross 18·13 m² (195·15 sq ft)
Ailerons (total) 1·742 m² (18·75 sq ft)
Trailing-edge flaps (total) 1·736 m² (18·69 sq ft)
Fin (excl ventral fins) 1·753 m² (18·87 sq ft)
Rudder 0·398 m² (4·28 sq ft)
Horizontal tail surfaces (total) 2·810 m² (30·25 sq ft)

WEIGHTS AND LOADINGS:
Weight empty 3,900 kg (8,598 lb)
Normal T-O weight 7,800 kg (17,196 lb)
Max T-O weight (semi-prepared surface) 7,000 kg (15,432 lb)
Max T-O weight (hard runway) 8,700 kg (19,180 lb)
Max wing loading 480 kg/m² (98·3 lb/sq ft)
Max power loading 239·7 kg/kN (2·35 lb/lb st)

PERFORMANCE (at max T-O weight, except where indicated):
Max level speed at 9,145 m (30,000 ft) Mach 0·95
Max level speed at S/L 600 knots (1,110 km/h; 690 mph)
Stalling speed, flaps down 125 knots (230 km/h; 143 mph)
Max rate of climb at S/L:
with afterburning 5,180 m (17,000 ft)/min
without afterburning 2,134 m (7,000 ft)/min
Time to 12,200 m (40,000 ft):
with afterburning 4 min 30 sec
without afterburning 11 min
Service ceiling 12,500 m (41,000 ft)
Service ceiling, one engine out (with afterburning) 6,000 m (19,685 ft)
*T-O run:
hard runway 1,219 m (4,000 ft)
semi-prepared surface 914 m (3,000 ft)
semi-prepared surface, with JATO 457 m (1,500 ft)
*T-O to 15 m (50 ft):
hard runway 1,829 m (6,000 ft)
semi-prepared surface 1,372 m (4,500 ft)
semi-prepared surface, with JATO 762 m (2,500 ft)
Landing from 15 m (50 ft) 600 m (1,970 ft)
Typical combat radius at S/L 323 nm (600 km; 372 miles)
Ferry range with max fuel 1,890 nm (3,500 km; 2,175 miles)

**See 'Weights' above*

Aeritalia G91Y tactical combat aircraft of the 32° Wing of the Italian Air Force

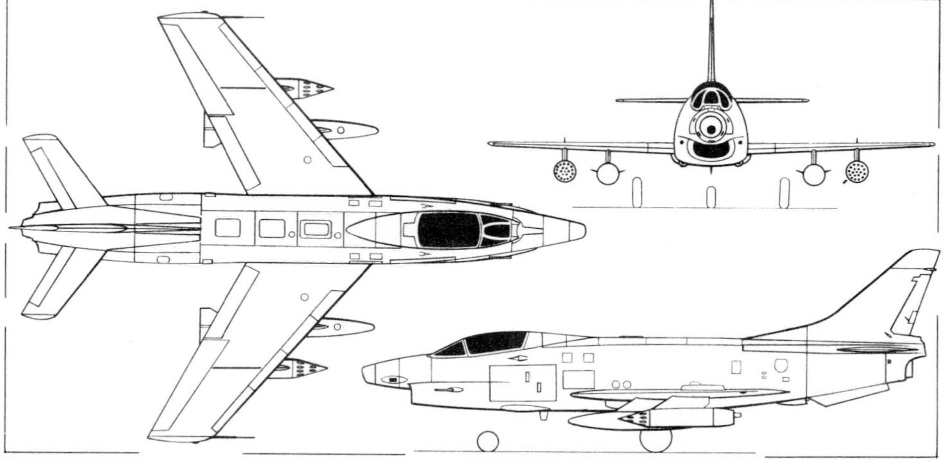
Aeritalia G91Y single-seat twin-engined tactical reconnaissance/fighter-bomber *(Pilot Press)*

AERITALIA G222

The G222 was originally conceived in four separate configurations, three of which were halted at the research project stage. Two prototypes were built of the military transport version, designated G222 TCM, to which the description applies. The first of these flew for the first time on 18 July 1970 and the second on 22 July 1971. The first prototype was handed over to the Italian Air Force in December 1971 for operational evaluation. An additional airframe was completed for static fatigue testing.

The Italian Air Force has ordered 44 production G222s, the first of which flew on 23 December 1975. Deliveries were scheduled to start in 1976.

In December 1974 the Argentine government ordered two G222s, with an option on a third (since taken up). In February 1976 the Dubai government ordered one G222, with an option for a second.

Several major Italian airframe companies are sharing in the construction programme, including Aermacchi (outer wings); Piaggio (wing centre-section); SIAI-Marchetti (tail unit); SACA (miscellaneous airframe components); and CIRSEA (landing gear).

TYPE: Twin-engined general-purpose transport aircraft.

WINGS: Cantilever high-wing monoplane. Thickness/chord ratio 15%. Light alloy three-spar fail-safe structure in three portions, the outer panels having taper on the leading- and trailing-edges and slight dihedral. One-piece centre-section fits in recess in top of fuselage and is secured by bolts at six main points. All-metal ailerons and double-slotted flaps, the latter extending over 60% of the trailing-edge. Spoilers ahead of each outboard flap section. Servo tabs in each aileron. Controls are hydraulically powered.

FUSELAGE: Stressed-skin aluminium alloy fail-safe structure of circular cross-section.

TAIL UNIT: Cantilever aluminium alloy two-spar structure, with sweptback vertical surfaces. Variable-incidence tailplane. Elevators hydraulically powered. Tabs in each elevator. No rudder tabs.

LANDING GEAR: Hydraulically-retractable tricycle type, suitable for use from prepared runways or grass fields. Messier-Hispano design, built under licence by CIRSEA (Nardi-Magnaghi). Twin-wheel nose unit retracts forward, tandem-wheel main units rearward into fairings on sides of fuselage. Oleo-pneumatic shock-absorbers. Gear can be lowered by gravity in emergency, the nose unit being aided by aerodynamic action and the main units by the shock-absorbers, which remain compressed in the retracted position. Oleo pressure in shock-absorbers is adjustable to permit variation in height of the cabin floor from the ground. Low-pressure tubeless tyres on all units, pressure 3·45-3·93 bars (50-57 lb/sq in). Hydraulic multi-disc brakes. No anti-skid units.

POWER PLANT: Two 2,535 kW (3,400 shp) Fiat-built General Electric T64/P4D turboprop engines, each driving a Hamilton Standard 63E60 three-blade variable-pitch metal propeller. Provision in fuselage for eight Aerojet General JATO rockets with total additional thrust of 35·3 kN (7,937 lb) for T-O with extra-heavy loads. Fuel in two outer-wing main tanks, total capacity 6,800 litres (1,495 Imp gallons) and two centre-section auxiliary tanks, total capacity 5,200 litres (1,143 Imp gallons). Total overall fuel capacity 12,000 litres (2,638 Imp gallons).

ACCOMMODATION: Crew of three (two pilots and wireless operator/flight engineer) or four on flight deck. Standard seating for 44 fully-equipped troops or 32 paratroops. Alternative payloads include 36 stretcher patients and eight medical attendants or sitting casualties; or freight. Typical Italian military equipment loads can include two CL-52 light trucks, one CL-52 with a 105 mm L4 howitzer or one-ton trailer, Fiat AR-59 Campagnola reconnaissance vehicle with 106 mm recoil-less gun or 250 kg (550 lb) trailer, or five standard A-22 freight containers. In the ambulance role a second toilet can be installed, and provision can be made to increase the water supply and to install supplementary electrical points and hooks for medical treatment bottles. In the freight role a 1,500 kg (3,306 lb) capacity cargo hoist can be installed, and there is provision for up to 135 cargo tie-down points. Crew door forward of cabin on port side. Doors at front and rear of main cabin on starboard side and at rear on port side. Underside of upswept rear fuselage lowers to form loading ramp, which can be opened in flight for air-drop operations. Provision is made for pressurisation of cabin in production aircraft, but prototypes have air-conditioning only.

SYSTEMS: Starboard main landing gear fairing houses a 113·3 kW (152 hp) Garrett-AiResearch APU for engine starting, hydraulic pump and alternator actuation. Two hydraulic systems, the primary system actuating the flying controls and the secondary system the landing gear, brakes, part of the flying control system and the auxiliaries. Emergency system, fed by APU, can take over from secondary system in flight. Standby hand pump for emergency use to lower landing gear and, on the ground, for propellers and parking brakes.

ELECTRONICS AND EQUIPMENT: Navigation equipment includes inertial PHI system with 12 pre-selectable stations, Doppler, two-axis gyro platform, VOR/ILS, Tacan, radio direction finder, DME, marker beacon, weather radar with secondary navigation capability, radar altimeter and ATC/IFF. Provision for installing head-up display. Communications equipment includes a 3,500-channel UHF/AM radio, a three-channel emergency UHF/AM radio with 1,630 channels, a 930-channel VHF/FM set usable as a direction finder, a 28,000-channel HF/AM SSB CW set, and an intercom acting as mixer and amplifier for all other systems.

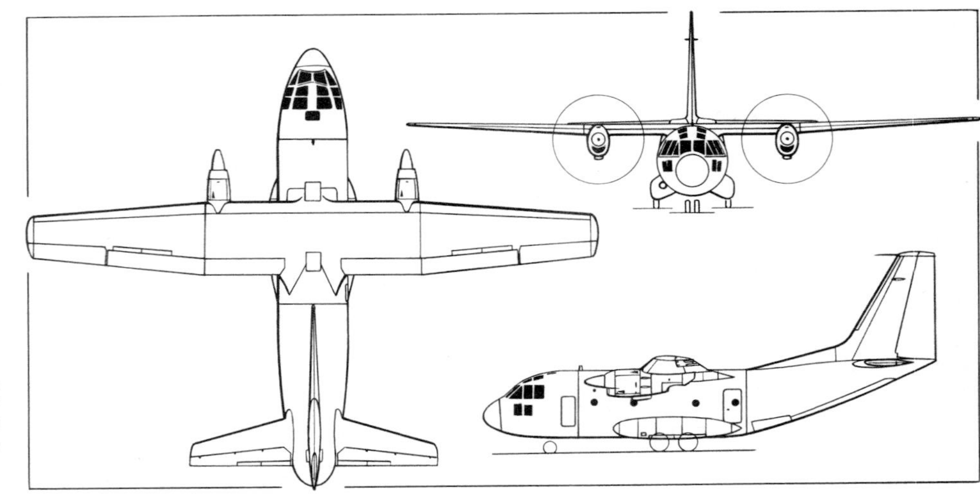

Aeritalia G222 twin-turboprop general-purpose military transport aircraft *(Pilot Press)*

DIMENSIONS, EXTERNAL:

Wing span	28·80 m (94 ft 6 in)
Wing chord at root	3·40 m (11 ft 1¾ in)
Wing chord at tip	1·685 m (5 ft 6¼ in)
Wing aspect ratio	9·15
Length overall	22·70 m (74 ft 5½ in)
Height overall	9·80 m (32 ft 1¾ in)
Fuselage: Max diameter	3·55 m (11 ft 7¾ in)
Tailplane span	12·40 m (40 ft 8¼ in)
Wheel track	3·67 m (12 ft 0½ in)
Wheelbase (to c/l of main units)	6·235 m (20 ft 5½ in)
Propeller diameter	4·42 m (14 ft 6 in)
Distance between propeller centres	9·50 m (31 ft 2 in)

DIMENSIONS, INTERNAL:

Main cabin: Length	8·58 m (28 ft 1¾ in)
Width	2·45 m (8 ft 0½ in)
Height	2·25 m (7 ft 4½ in)
Volume	74·0 m³ (2,613 cu ft)

AREAS:

Wings, gross	82·00 m² (882·6 sq ft)
Ailerons (total)	3·65 m² (39·29 sq ft)
Trailing-edge flaps (total)	18·40 m² (198·06 sq ft)
Spoilers (total)	1·65 m² (17·76 sq ft)
Vertical tail surfaces (total)	19·21 m² (206·67 sq ft)
Horizontal tail surfaces (total)	23·70 m² (255·11 sq ft)

WEIGHTS AND LOADINGS:

Weight empty	14,590 kg (32,165 lb)
Weight empty, equipped	15,400 kg (33,950 lb)
Max payload	9,000 kg (19,840 lb)
Normal T-O weight	24,500 kg (54,013 lb)
Max T-O and landing weight	26,500 kg (58,422 lb)
Max zero-fuel weight	24,400 kg (53,792 lb)
Max wing loading	323 kg/m² (66·2 lb/sq ft)
Max cargo floor loading	750 kg/m² (155 lb/sq ft)
Max power loading	5·23 kg/kW (8·6 lb/shp)

PERFORMANCE (at max T-O weight except where indicated):

Max level speed at 4,575 m (15,000 ft)	291 knots (540 km/h; 336 mph)
Cruising speed at 4,500 m (14,750 ft)	194 knots (360 km/h; 224 mph)
Min speed	84 knots (155 km/h; 96·5 mph)
Time to 4,500 m (14,750 ft)	8 min 35 sec
Max rate of climb at S/L	620 m (2,034 ft)/min

First production example of the Aeritalia G222 twin-turboprop military transport aircraft for the Italian Air Force

Rate of climb at S/L, one engine out	125 m (410 ft)/min
Service ceiling	9,000 m (29,525 ft)
T-O run	525 m (1,720 ft)
T-O to 15 m (50 ft)	825 m (2,707 ft)
Landing from 15 m (50 ft)	720 m (2,362 ft)
Landing run at max landing weight	440 m (1,444 ft)
Accelerate/stop distance	1,200 m (3,937 ft)
Min ground turning radius	20·80 m (68 ft 3 in)
Basic mission range with 5,000 kg (11,025 lb) payload	1,591 nm (2,950 km; 1,833 miles)
Ferry range with max fuel	2,670 nm (4,950 km; 3,075 miles)

AERITALIA (LOCKHEED) F-104S

The first of two Lockheed-built F-104S prototypes flew during December 1966. Aeritalia is building 205 under licence for the Italian Air Force, of which the first was flown on 30 December 1968. Deliveries began in the Spring of 1969. In October 1974 the Turkish Air Force ordered 18 F-104Ss. Delivery of these, which began in December, was completed by May 1975; but in the Spring of 1975 an additional order was placed by the Turkish Air Force for 22 F-104Ss. Delivery of these began in December 1975 and was to be completed by mid-1976. Production of the balance of the Italian Air Force order is continuing, and the 200th F-104S was flown in May 1976.

A programme to modernise the R21G radar system of the F-104S was announced in the Spring of 1974. The improved installation, built under licence by CGE-Fiar in Italy, incorporates a moving-target indication and tracking capability based upon a Rockwell Missile Systems Division moving-target detection processor. CGE-Fiar has itself developed an improved antenna and a number of ECCM (electronic counter-countermeasures) features.

Details of Lockheed production of earlier models of the F-104 Starfighter can be found in the 1972-73 and earlier editions of *Jane's*. The following description applies to the Aeritalia-built F-104S:

TYPE: Single-seat multi-purpose combat aircraft.

WINGS: Cantilever mid-wing monoplane. Bi-convex supersonic wing section with a thickness/chord ratio of 3·36%. Anhedral 10°. No incidence. Sweepback 18° 6′ at quarter-chord. Leading-edge nose radius of 0·41 mm (0·016 in) and razor-sharp trailing-edge. All-metal structure with two main spars, 12 spanwise intermediate channels between spars and top and bottom one-piece skin panels, tapering from thickness of 6·3 mm (0·25 in) at root to 3·2 mm (0·125 in) at tip. Each half-wing measures 2·31 m (7 ft 7 in) from root to tip and is a separate structure cantilevered from five forged frames in fuselage. Full-span electrically-actuated drooping leading-edge. Entire trailing-edge hinged, with inboard sections serving as landing flaps and outboard sections as ailerons. Ailerons are of aluminium, each powered by a servo control system which is irreversible and hydraulically powered, and each actuated by ten small hydraulic cylinders. Trim control is applied to position the aileron relative to the servo control position. An electric actuator positions the aileron trim. Flaps are of aluminium, actuated electrically. Above each flap is the air delivery tube of a boundary layer control system, which ejects air bled from the engine compressor over the entire flap span when the flaps are lowered to the landing position.

FUSELAGE: All-metal monocoque structure. Hydraulically-operated aluminium airbrake on each side of rear fuselage.

TAIL UNIT: T-type cantilever unit with 'all-flying' one-piece horizontal tail surface hinged at mid-chord point at top of the vertical fin and powered by a hydraulic servo. Tailplane has similar profile to wing and is all-metal. Rudder is fully powered by a hydraulic servo. Trim control is applied to position the tailplane relative to the servo control position, by means of an electric actuator. Rudder trim is operated by an electric actuator located in the fin. The rudder itself is trimmed in the same way as the tailplane. Narrow-chord ventral fin on centreline and two smaller lateral fins under fuselage to improve stability.

LANDING GEAR: Retractable tricycle type with Dowty patent liquid-spring shock-absorbers. Hydraulic actuation. Main wheels raised in and forward. Steerable nosewheel retracts forward into fuselage. Main-wheel legs are hinged on oblique axes so that the wheels lie flush within the fuselage when retracted. Main wheels size 26 × 8·0, with Goodrich tyres size 26 × 8·0 type VIII, pressure 11·93 bars (173 lb/sq in). Nosewheel tyre size 18 × 5·5 type VII. Bendix hydraulic disc brakes with Goodyear anti-skid units. Arrester hook under rear of fuselage.

POWER PLANT: One General Electric J79-GE-19 turbojet engine, rated at 52·8 kN (11,870 lb st) dry and 79·62 kN (17,900 lb st) with afterburning. Electrical de-icing elements fitted to air intakes. Most of the aircraft's hydraulic equipment mounted inside large engine bay door under fuselage to facilitate servicing. Internal fuel in five bag-type fuselage tanks with total standard capacity of 3,392 litres (896 US gallons). Provision for external fuel in two 740 litre (195 US gallon) pylon tanks and two 645 litre (170 US gallon) wingtip tanks. Pressure refuelling of all internal and external tanks through single point on upper port fuselage just forward of air intake duct. Gravity fuelling point for internal tanks aft of pressure refuelling point, with individual gravity fuelling of external tanks. In-flight refuelling can be provided through Lockheed-designed probe-drogue system. Probe, mounted below port sill of cockpit, is removable but when installed is non-retractable. Oil capacity 15 litres (4 US gallons).

ACCOMMODATION: Pressurised and air-conditioned cockpit well forward of wings. Canopy hinged to starboard for access. Lockheed Model C-2 ejection seat.

SYSTEMS: Air-conditioning package by AiResearch, using engine bleed air. Pressure differential 0·34 bars (5 lb/sq in). Two completely separate hydraulic systems, using engine-driven pumps operating at 207 bars (3,000 lb/sq in). No. 1 system operates one side of tailplane, rudder and ailerons, also the automatic pitch control actuator and autopilot actuators. No. 2 system operates other half of tailplane, rudder and ailerons, also the landing gear, wheel brakes, airbrakes, nosewheel steering and constant-frequency electrical generator. Emergency ram-air turbine supplies emergency hydraulic pump and 4·5kVA 115/200V electric generator. Electrical system supplied by two engine-driven 20kVA 115/200V variable-frequency (320-520Hz) generators. Constant-speed hydraulic motor drives 2·5kVA 115/200V generator to supply fixed-frequency AC. DC power supplied by two batteries and an inverter.

ELECTRONICS AND EQUIPMENT: Integrated electronics system in which various communications and navigation components may be installed as a series of interconnecting but self-sustaining units which may be varied to provide for different specific missions. Equipment includes autopilot with 'stick steering', which includes modes for pre-selecting and holding altitude, speed, heading and constant rate of turn; multi-purpose NASARR R21G/F15G radar; fixed-reticle gunsight; bombing computer; air data computer; dead reckoning navigation device; Tacan radio air navigation system; provision for data link-time division set and UHF radio; lightweight fully-automatic inertial navigation system; and provision for fitting a camera pod under the fuselage for reconnaissance duties.

ARMAMENT: Nine external attachment points, at wingtips, under wings and under fuselage, for bombs, rocket pods, auxiliary fuel tanks and air-to-air missiles. Normal primary armament consists of two Raytheon AIM-7 Sparrow III air-to-air missiles under wings and permanently-installed M61 20 mm rotary cannon in port underside of fuselage. Provision for two Sidewinders under fuselage and either a Sidewinder or 645 litre (170 US gallon) fuel tank on each wingtip.

Six of the Aeritalia-built Lockheed F-104S combat aircraft supplied to the Turkish Air Force

DIMENSIONS, EXTERNAL:	
Wing span without tip-tanks	6·68 m (21 ft 11 in)
Wing chord (mean)	2·91 m (9 ft 6·6 in)
Wing aspect ratio	2·45
Length overall	16·69 m (54 ft 9 in)
Length of fuselage	15·62 m (51 ft 3 in)
Height overall	4·11 m (13 ft 6 in)
Tailplane span	3·63 m (11 ft 11 in)
Wheel track	2·71 m (8 ft 10¾ in)
Wheelbase	4·59 m (15 ft 0½ in)
AREAS:	
Wings, gross	18·22 m² (196·1 sq ft)
Ailerons (total)	0·85 m² (9·2 sq ft)
Trailing-edge flaps (total)	2·11 m² (22·7 sq ft)
Leading-edge flaps (total)	1·51 m² (16·2 sq ft)
Airbrakes (total)	0·77 m² (8·25 sq ft)
Fin	3·50 m² (37·7 sq ft)
Ventral fin (centreline)	0·55 m² (5·9 sq ft)
Rudder, incl tab	0·51 m² (5·5 sq ft)
Tailplane	4·48 m² (48·2 sq ft)
WEIGHTS AND LOADING:	
Weight empty	6,760 kg (14,900 lb)
T-O weight ('clean')	9,840 kg (21,690 lb)
Max T-O weight	14,060 kg (31,000 lb)
Max wing loading	540 kg/m² (110·7 lb/sq ft)
PERFORMANCE (at 9,840 kg; 21,690 lb AUW):	
Never-exceed speed	Mach 2·2
Max level speed at 11,000 m (36,000 ft)	Mach 2·2 (1,259 knots; 2,330 km/h; 1,450 mph)
Max level speed at S/L	Mach 1·2 (790 knots; 1,464 km/h; 910 mph)
Max cruising speed at 11,000 m (36,000 ft)	530 knots (981 km/h; 610 mph)
Econ cruising speed	Mach 0·85
Service ceiling	17,680 m (58,000 ft)
Zoom altitude	more than 27,400 m (90,000 ft)
Time to accelerate from Mach 0·92 to Mach 2·0	2 min
Time to climb to 10,670 m (35,000 ft)	1 min 20 sec
Time to climb to 17,070 m (56,000 ft)	2 min 40 sec
Radius with max fuel	673 nm (1,247 km; 775 miles)
Ferry range (excl flight refuelling)	1,576 nm (2,920 km; 1,815 miles)

AERITALIA AM.3C

The AM.3C three/four-seat monoplane was developed jointly by Aerfer (now part of Aeritalia) and Aeronautica Macchi for forward air control, observation, liaison, transport of passengers and cargo, casualty evacuation, tactical support of ground forces and similar duties. The first of three prototypes flew for the first time on 12 May 1967.

The AM.3C was produced to meet orders from the South African Air Force (40) and the Rwanda Air Force (3). Deliveries of these were completed by December 1974. Those of the SAAF are known by the South African name **Bosbok**.

A full description and illustration of the AM.3C can be found in the 1975-76 *Jane's*.

AERMACCHI
AERONAUTICA MACCHI SpA

HEAD OFFICE:
Corso Vittorio Emanuele 15, 20122 Milan

OFFICES AND WORKS:
Via Silvestro Sanvito 80, Casella Postale 246, 21100 Varese

Telephone: (0332) 283100

Telex: 38070 Aviomacc

PRESIDENT:
Dott Ing Paolo Foresio

VICE-PRESIDENT AND MANAGING DIRECTOR:
Gen Ing Mario Matacotta

GENERAL MANAGERS:
Dott Ing Ermanno Bazzocchi
Dott Fabrizio Foresio

TECHNICAL DIRECTOR:
Dott Ing Alberto Notari

SALES MANAGER:
Dott Ing Gianni Cattaneo

PRESS AND PUBLIC RELATIONS:
Andrea Artoni

The Macchi company was founded in 1912 in Varese and its first aeroplane was built in 1913. In addition to its former factory area of 36,900 m² (397,200 sq ft), a new plant is under construction on Venegono airfield, and about 13,500 m² (145,310 sq ft) of covered assembly line area has been completed.

Lockheed Aircraft International acquired a substantial minority interest in Aermacchi in December 1959, and Aermacchi subsequently built the Lockheed 60 light utility transport as the Aermacchi-Lockheed AL.60 (see 1972-73 *Jane's*). In current production are various single- and two-seat versions of Aermacchi's own design, the M.B. 326. Two prototypes of the M.B. 339, a new two-seat trainer, were due to fly in 1976. The Italian Air Force has indicated a requirement for 100 M.B. 339s.

In association with Aeritalia, Aermacchi also built the AM.3C military AOP and liaison monoplane.

AERMACCHI M.B. 326

The first prototype of the original Aermacchi M.B. 326 jet trainer flew for the first time on 10 December 1957, powered by a Rolls-Royce Bristol Viper 8 turbojet engine. The more powerful Viper 11 engine is fitted in six production versions of the aircraft built for the air forces of Italy (M.B. 326), Tunisia (M.B. 326B), Ghana (M.B. 326F), Australia (M.B. 326H) and South Africa (M.B. 326M), and for Alitalia (M.B. 326D); and in the M.B. 326E currently being built for the Italian Air Force.

The South African Air Force version, which was built by Atlas (which see), is known as the Impala Mk 1; an improved version of this, the Impala Mk 2, is currently in production.

Other versions with more powerful Viper engines, armament changes and other modifications are as follows:

M.B. 326E. Two-seat advanced trainer for service with the Italian Air Force's Scuola di Volo Basico Iniziale at Lecce. Similar to original M.B. 326 with Viper 11 engine (see 1970-71 *Jane's*), but with the strengthened wings and six underwing hardpoints of the M.B. 326GB. New electronics and equipment include UHF radio, miniaturised Tacan, gyroscopic weapons sight and gun camera. Twelve ordered, of which six are conversions of existing M.B. 326s and six of new construction.

M.B. 326GB. Two-seat dual-control advanced training and counter-insurgency attack version, with airframe modifications and a more powerful Viper turbojet engine. The M.B. 326G prototype flew for the first time in the Spring of 1967, and the similar M.B. 326GB is now in production. Customers include the Argentine Navy (8), and air forces of Zaïre (17) and Zambia (20). In addition, 112 similar aircraft are being assembled in Brazil under licence by EMBRAER (which see) for the Brazilian Air Force, as the **AT-26 Xavante**. A follow-on production batch of 40 Xavantes was authorised during 1975, with options for another 30.

M.B. 326K. Single-seat operational trainer and light ground attack version. Retains most of the structure and systems of the M.B. 326GB, but has a more powerful Viper 632 engine, no second cockpit, additional fuselage fuel tanks, and increased weapon-carrying capacity. Prototype (I-AMKK), with Viper 540 engine, first flown on 22 August 1970; second prototype, with Viper 632 engine, first flown in 1971. In production for Dubai government, which has reportedly ordered three, and South African Air Force, to which four were reported to have been delivered in 1974. In South Africa, the Atlas Impala Mk 2 (which see) is believed to be based on the M.B. 326K, as is Aermacchi's own M.B. 339, described separately.

M.B. 326L. Two-seat advanced trainer version, announced in 1973. Combines the airframe of the single-seat M.B. 326K with the standard two-seat dual control cockpit installation. Dubai government reported to have ordered one.

TYPE: Two-seat basic and advanced trainer (E, GB, GC and L) or single-seat operational trainer (K); all versions have light ground attack capability.

WINGS: Cantilever low/mid-wing monoplane. Wing section NACA 6A series (modified). Thickness/chord ratio 13·7% at root, 12% at tip. Dihedral 2° 55′. Incidence 2° 30′. All-metal two-spar stressed-skin structure in three sections, of which the centre-section is integral with the fuselage. Single fence on each wing at approx two-thirds span. Manually-operated all-metal ailerons and hydraulically-operated slotted flaps. Electrically-actuated balance and trim tab in port aileron. Geared balance tab in starboard aileron. M.B. 326K and L have strengthened structure, hydraulically-operated single-slotted flaps and hydraulically servo-powered ailerons. Automatic flap retraction on 326L above 160 knots (296 km/h; 184 mph).

FUSELAGE: All-metal semi-monocoque structure. Hydraulically-operated airbrake under centre-fuselage.

TAIL UNIT: Cantilever all-metal structure. Electrically-actuated trim tab in rudder and each elevator.

LANDING GEAR (M.B. 326GB): Hydraulically-retractable tricycle type, with oleo-pneumatic shock-absorbers. Nosewheel retracts forward, main units outward into wings. Pirelli main wheels and tyres, size 6·50-10 (8-ply). Steerable and self-centering nosewheel with anti-shimmy device. Dunlop twin-contact nosewheel tyre, size 5-4·5. Hydraulic disc brakes.

LANDING GEAR (M.B. 326K and L): As M.B. 326GB, except for more powerful Dunlop high-capacity hydraulic disc brakes and separate emergency extension system. Tyre pressure (K) at max T-O weight 6·90 bars (100 lb/sq in).

Aermacchi M.B. 326K single-seat training and attack aircraft in Italian Air Force insignia

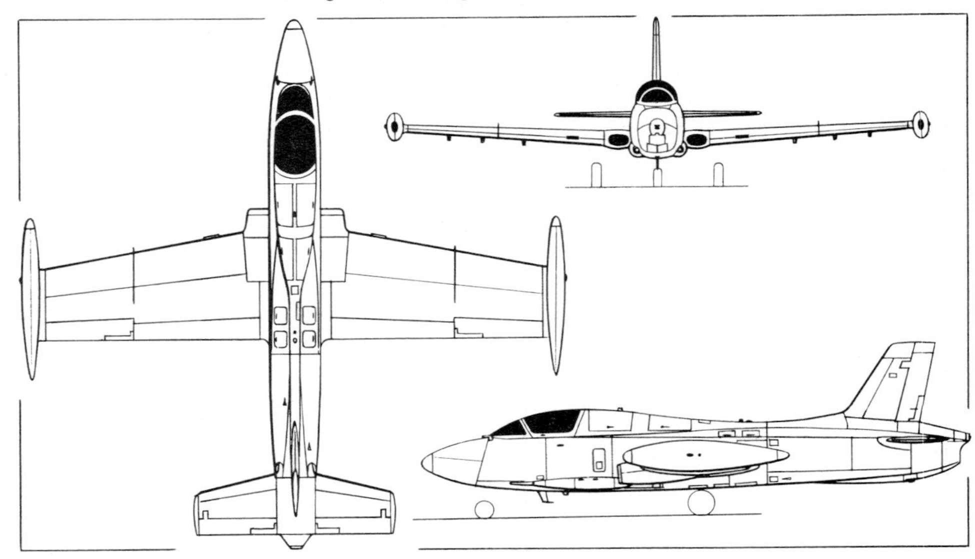

Aermacchi M.B. 326K single-seat training and light ground attack aircraft *(Pilot Press)*

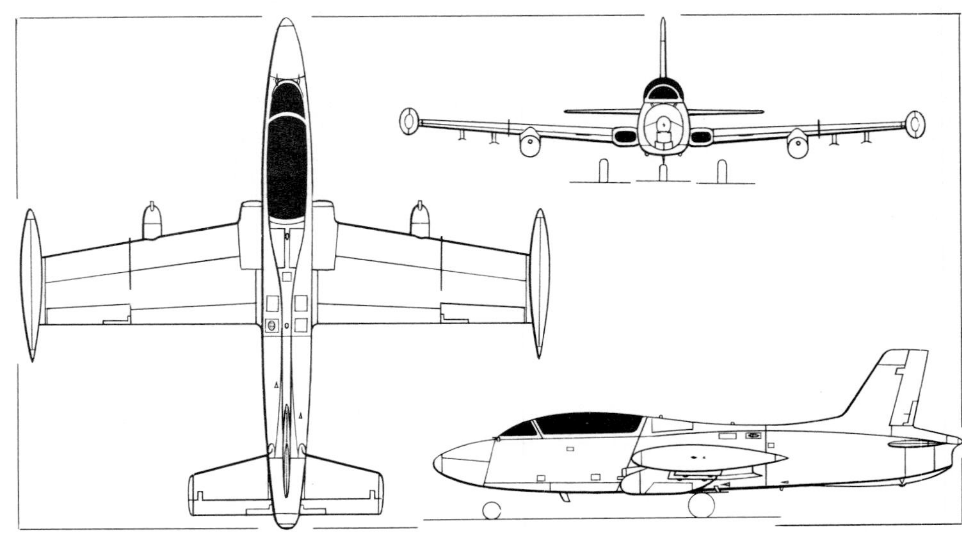

Aermacchi M.B. 326GB two-seat basic training and light tactical attack aircraft *(Pilot Press)*

POWER PLANT: (M.B. 326E): One Rolls-Royce Bristol Viper 11 turbojet engine, rated at 11·12 kN (2,500 lb st).

POWER PLANT (M.B. 326GB): One Rolls-Royce Bristol Viper 20 Mk 540 turbojet engine, rated at 15·17 kN (3,410 lb st). Fuel in flexible rubber main tank in fuselage, capacity 782 litres (172 Imp gallons), and two 305 litre (67 Imp gallon) non-jettisonable wingtip tanks. Total standard fuel capacity 1,392 litres (306 Imp gallons). Provision for two 332 litre (73 Imp gallon) jettisonable underwing tanks, to give total capacity of 2,056 litres (452 Imp gallons). Single-point pressure refuelling receptacle under fuselage. Fuel dump valves permit quick emptying of tip-tanks.

POWER PLANT (M.B. 326K and L): One Rolls-Royce Bristol Viper Mk 632-43 turbojet engine, rated at 18·79 kN (4,000 lb st). M.B. 326L fuel capacity same as for GB. Fuel in M.B. 326K contained in three rubber fuselage tanks and two permanent wingtip tanks, total usable capacity 1,660 litres (366 Imp gallons). Provision to install self-sealing fuselage tanks and reticulated foam anti-explosive filling in all tanks, including those at wingtips. Two underwing stations equipped normally to carry jettisonable auxiliary tanks of up to 340 litres (75 Imp gallons) each. Single-point pressure refuelling receptacle and auxiliary gravity refuelling points.

ACCOMMODATION (M.B. 326E, GB and L): Crew of two in tandem under a one-piece moulded Perspex canopy which hinges sideways to starboard. Pressurised cockpit, differential 0·21 bars (3·0 lb/sq in) in GB, 0·24 bars (3·5 lb/sq in) in L. Dual controls and instruments. Blind-flying screens for pupil. Martin-Baker Mk 04A lightweight ejection seats fitted to M.B. 326GB; Mk 06A zero-zero seats in M.B. 326E and L.

ACCOMMODATION (M.B. 326K): Pilot only, on Martin-Baker Mk 6 zero-zero rocket ejection seat in pressurised and air-conditioned cockpit (differential 0·24 bars; 3·5 lb/sq in). Separately-controlled canopy jettison system provided, but seat is fitted with breakers to permit ejection through canopy in extreme emergency. Canopy hinges sideways to starboard. Provision for armour protection for pilot and other vital areas.

SYSTEMS: Air-conditioning and pressurisation system uses air bled from engine compressor and incorporates turbo-refrigerator unit. Hydraulic system, pressure 172·5 bars (2,500 lb/sq in), for landing gear and doors, flaps, airbrake and wheel brakes. Hydraulic system on M.B. 326K operates through a self-regulating engine-driven pump to provide power for landing gear, flap, airbrake, aileron servo and wheel brake actuation. Independent manually-operated hydraulic system for emergency landing gear extension. DC electrical supply from 30V 9kW starter/generator and two 24V 22Ah batteries. Fixed frequency AC system powered by 750VA main inverter, with 250VA standby unit (two 600VA inverters on M.B. 326K and L). A 6kVA alternator supplies engine air intake anti-icing system and can feed primary electrical system, via transformer-rectifier, in event of DC generator failure.

Electronics and Equipment (M.B. 326GB and L): To customer's specification. Standard configuration includes UHF transmitter-receiver type AN/ARC-51BX with 3,500 channels or 26 preset channels, auxiliary UHF system with 5-channel Collins 718B-8C transceiver, AN/AIC-18 interphone, AN/ARN-52(V) Tacan with USAF type AQU-4/A horizontal situation indicator, and AN/ARN-83 ADF, AN/ARA-50 UHF/DF, Collins 51RV-1 (ARINC 547) VOR/ILS and CPU-76/A flight director computer. Standard AN/APX-72 IFF transponder can be replaced by Bendix-Fiar TRA-62A, AN/APX-68 or AN/APX-77 IFF/SIF system. Cockpit side consoles widened on 326L to provide additional space for equipment.

Electronics and Equipment (M.B. 326K): Navigational and tactical equipment, to customer's specification, can include main (3,500-channel) and standby (5-channel) UHF transceivers or two 680-channel VHF transceivers, Tacan, VOR/ILS and marker beacon, flight director computer with integrated instrumentation, ADF, UHF/DF, navigation computer and Doppler radar. Weapon-sighting equipment may range from a fixed reflector gunsight to a gyroscopic lead-computing sight, with provision to install a laser rangefinder and a bombing computer.

Armament (M.B. 326GB and L, optional): Up to 1,814 kg (4,000 lb) of armament on six underwing attachments. Typical weapon loads include following alternatives: two LAU-3/A packs each containing nineteen 2·75 in FFAR rockets and two packs each containing eight Hispano-Suiza SURA 80 mm rockets; two 12·7 mm gun pods and four packs each containing six SURA 80 mm rockets; one 7·62 mm Minigun, one 12·7 mm gun pod, two Matra 122 rocket packs and two packs each containing six SURA 80 mm rockets; two 500 lb bombs and eight 5 in HVAR rockets; two AS.12 missiles; one 12·7 mm gun pod, one reconnaissance pack containing four Vinten cameras and two 272 kg (600 lb) drop-tanks, or two Matra SA-10 packs each containing a 30 mm Aden gun and 150 rounds. SFOM type 83 fixed gunsight or Ferranti LFS 5/102A gyro-sight. Gun camera in nose.

Armament (M.B. 326K): Standard fixed armament of two 30 mm DEFA electrically-operated cannon in lower front fuselage, with 125 rds/gun. Six underwing pylons, the inboard four stressed to carry up to 454 kg (1,000 lb) each and the outboard pair up to 340 kg (750 lb) each. Max external military load (with reduced fuel) is 1,814 kg (4,000 lb). Each pylon fitted with standard NATO 355 mm (14 in) MA-4A stores rack. Typical loads may include two 750 lb and four 500 lb bombs, four napalm containers, two AS.11 or AS.12 air-to-surface missiles, two machine-gun pods, two Matra 550 air-to-air missiles, six SUU-11A/A 7·62 mm Minigun pods, and various Matra or other launchers for 37 mm, 68 mm, 100 mm, 2·75 in or 5 in rockets. A four-camera tactical reconnaissance pod can be carried on the port inner pylon without affecting the weapon capability of the other five stations.

Dimensions, external:
Wing span over tip-tanks 10·85 m (35 ft 7 in)
Wing chord at root 2·43 m (7 ft 11½ in)
Wing chord at tip 1·40 m (4 ft 7·1 in)
Wing aspect ratio 6·08
Length overall 10·64 m (34 ft 11 in)
Height overall 3·71 m (12 ft 2 in)
Tailplane span 4·08 m (13 ft 4½ in)
Wheel track 2·31 m (7 ft 7 in)
Wheelbase 4·14 m (13 ft 7 in)

Areas:
Wings, gross 19·35 m² (208·3 sq ft)
Ailerons (total) 1·33 m² (14·3 sq ft)
Trailing-edge flaps (total) 2·55 m² (27·5 sq ft)
Fin 1·55 m² (16·7 sq ft)
Rudder, incl tab 0·71 m² (7·6 sq ft)
Tailplane 2·62 m² (28·2 sq ft)
Elevators, incl tabs 0·88 m² (9·5 sq ft)

Weights (M.B. 326E):
Basic operating weight, excl crew 2,618 kg (5,772 lb)
Max T-O weight (full internal fuel and wingtip tanks) 3,593 kg (7,922 lb)
Max T-O weight with armament 4,350 kg (9,600 lb)

Weights and Loadings (M.B. 326GB. T: Trainer; A: Attack):
Basic operating weight, excl crew:
T 2,685 kg (5,920 lb)
*A 2,558 kg (5,640 lb)
Max zero-fuel weight:
T 2,849 kg (6,280 lb)
*A 2,640 kg (5,820 lb)
Max T-O weight (full internal fuel, wingtip and underwing tanks):
T 4,577 kg (10,090 lb)
A, no armament 4,447 kg (9,805 lb)
A, with 769 kg (1,695 lb) armament 5,216 kg (11,500 lb)
Max T-O weight (max armament):
*A, with fuel in fuselage tank only and 1,962 kg (4,325 lb) armament 5,216 kg (11,500 lb)
Max wing loading 269·5 kg/m² (55·2 lb/sq ft)
Max power loading 343·8 kg/kN (3·37 lb/lb st)
* *Without tip-tanks and aft ejection seat*

Weights and Loadings (M.B. 326K):
Weight empty, equipped 3,123 kg (6,885 lb)
T-O weight ('clean') 4,645 kg (10,240 lb)
Typical operational T-O weights:
patrol and visual reconnaissance 5,048 kg (11,130 lb)
photographic reconnaissance 5,111 kg (11,270 lb)
Max T-O weight 5,897 kg (13,000 lb)
Max landing weight 5,443 kg (12,000 lb)
Normal design landing weight 4,535 kg (10,000 lb)
Max wing loading 305 kg/m² (62·4 lb/sq ft)
Max power loading 313·8 kg/kN (3·25 lb/lb st)

Weights and Loadings (M.B. 326L):
Weight empty, equipped 2,964 kg (6,534 lb)
T-O weight, training configuration 4,211 kg (9,285 lb)
Max T-O weight 5,897 kg (13,000 lb)
Max wing loading 305 kg/m² (62·4 lb/sq ft)
Max power loading 313·8 kg/kN (3·25 lb/lb st)

Performance (M.B. 326GB. T: Trainer at typical weight of 3,937 kg (8,680 lb), representing max T-O weight without underwing tanks; AC: Attack version at combat weight of 4,763 kg (10,500 lb); AM: Attack version at max T-O weight):
Never-exceed speed:
T Mach 0·82 (469 knots; 871 km/h; 541 mph EAS)
AC Mach 0·75 (419 knots; 778 km/h; 483 mph EAS)
Max level speed:
T 468 knots (867 km/h; 539 mph)
Max cruising speed:
T 430 knots (797 km/h; 495 mph)
Max rate of climb at S/L:
T 1,844 m (6,050 ft)/min
AC 1,082 m (3,550 ft)/min
AM 945 m (3,100 ft)/min
Time to 3,050 m (10,000 ft):
AC 3 min 10 sec
AM 4 min 0 sec
Time to 6,100 m (20,000 ft):
T 4 min 10 sec
AC 8 min 0 sec
AM 9 min 20 sec
Time to 9,150 m (30,000 ft):
T 7 min 40 sec
AC 15 min 0 sec
AM 18 min 40 sec
Time to 12,200 m (40,000 ft):
T 13 min 5 sec
Service ceiling:
T 14,325 m (47,000 ft)
AC 11,900 m (39,000 ft)
T-O run, ISA:
T 412 m (1,350 ft)
AC 640 m (2,100 ft)
AM 845 m (2,770 ft)
T-O run, ISA + 25°C:
T 506 m (1,660 ft)
AC 805 m (2,640 ft)
AM 1,000 m (3,280 ft)
T-O to 15 m (50 ft), ISA:
T 555 m (1,820 ft)
AC 866 m (2,840 ft)
AM 1,140 m (3,740 ft)
T-O to 15 m (50 ft), ISA + 25°C:
T 704 m (2,310 ft)
AC 1,113 m (3,650 ft)
AM 1,411 m (4,630 ft)
Landing from 15 m (50 ft), ISA
T at landing weight of 3,175 kg (7,000 lb) 631 m (2,070 ft)
AC at landing weight of 4,195 kg (9,250 lb) 802 m (2,630 ft)
Landing from 15 m (50 ft), ISA + 25°C:
T at landing weight of 3,175 kg (7,000 lb) 671 m (2,200 ft)
AC at landing weight of 4,195 kg (9,250 lb) 857 m (2,810 ft)
Range (T, with 113 litres; 25 Imp gallons reserve):
fuselage and tip-tanks 998 nm (1,850 km; 1,150 miles)
fuselage, tip and underwing tanks 1,320 nm (2,445 km; 1,520 miles)
Combat radius (A at max AUW):
max fuel, 769 kg (1,695 lb) armament, 90 kg (200 lb) fuel reserve, out at 6,100 m (20,000 ft), return at 7,620 m (25,000 ft) 350 nm (648 km; 403 miles)
fuselage tank only, 1,814 kg (4,000 lb) armament, 90 kg (200 lb) fuel reserve, cruise at 3,050 m (10,000 ft), five minutes over target 69 nm (130 km; 80 miles)
max fuel, 771 kg (1,700 lb) armament, 90 kg (200 lb) fuel reserve, cruise at 3,050 m (10,000 ft), 1 hr 50 min patrol at 150 m (500 ft) over target 49·5 nm (92 km; 57 miles)

Performance (M.B. 326K. A: aircraft 'clean', at AUW of 4,390 kg; 9,680 lb; B: armed aircraft at 5,443 kg; 12,000 lb AUW):
Max design limit speed at S/L 500 knots (927 km/h; 576 mph) EAS
Max limiting Mach number 0·82
Max level speed at 1,525 m (5,000 ft):
A 480 knots (890 km/h; 553 mph) TAS
Max level speed at 9,150 m (30,000 ft):
B 370 knots (686 km/h; 426 mph) TAS
Stalling speed, flaps up:
A 102 knots (190 km/h; 118 mph) CAS
B 113 knots (211 km/h; 131 mph) CAS
Stalling speed, flaps down:
A 91 knots (169 km/h; 105 mph) CAS
B 102 knots (190 km/h; 118 mph) CAS
Max rate of climb at S/L:
A 1,980 m (6,500 ft)/min
B 1,143 m (3,750 ft)/min
Time to 10,670 m (35,000 ft):
A 9 min 30 sec
B 23 min 0 sec
Runway LCN at max T-O weight 5
T-O run, ISA:
A 411 m (1,350 ft)
B 670 m (2,200 ft)
T-O run, ISA + 20°C:
A 518 m (1,700 ft)
B 815 m (2,675 ft)
T-O to 15 m (50 ft), ISA:
A 572 m (1,875 ft)
B 914 m (3,000 ft)
T-O to 15 m (50 ft), ISA + 20°C:
A 709 m (2,325 ft)
B 1,158 m (3,800 ft)
Max rate of descent at touchdown:
at 4,535 kg (10,000 lb) AUW 3·05 m (10 ft)/sec
at 5,443 kg (12,000 lb) AUW 2·13 m (7 ft)/sec
Typical combat radius:
B (internal fuel and 1,280 kg; 2,822 lb external weapons), lo-lo-lo 145 nm (268 km; 167 miles)
B (reduced fuel and 1,814 kg; 4,000 lb external weapons), lo-lo-lo 70 nm (130 km; 81 miles)
visual reconnaissance with two external fuel tanks 400 nm (740 km; 460 miles)
photo reconnaissance with two auxiliary tanks and camera pod, hi-lo-hi 560 nm (1,036 km; 644 miles)
Max ferry range (two underwing tanks) more than 1,149 nm (2,130 km; 1,323 miles)
g limits +7·33; −3·5

Performance (M.B. 326L at training T-O weight):
Max level speed at S/L 485 knots (898 km/h; 558 mph)
Max level-flight Mach number at 11,000 m (36,000 ft) Mach 0·77
Max rate of climb at S/L 2,134 m (7,000 ft)/min
T-O run 415 m (1,360 ft)
T-O to 15 m (50 ft) 558 m (1,830 ft)

AERMACCHI M.B. 339

In early 1975 the Italian Air Force authorised the manufacture of two flying prototypes of this tandem two-seat trainer/ground attack aircraft, which it plans to order in quantity as a successor to the M.B. 326s currently in service. A ground test airframe will also be built.

The M.B. 339, of which a full-size engineering mockup was completed in 1974, is based essentially upon the airframe and Viper 632 power plant of the M.B. 326K (which see), but has a reshaped forward fuselage, an improved two-seat cockpit, uprated electronics equipment, and other detail changes. Its general appearance can be seen in the accompanying illustrations.

The first of two M.B. 339 prototypes flew on 12 August 1976. Deliveries of production aircraft, expected to number 100, will follow in 1977-78.

Type: Two-seat basic and advanced trainer and ground attack aircraft.

Airframe: Structural design criteria based on MIL-A-008860A; 8g limit load factor in 'clean' configuration. Cockpit designed for 40,000 pressurisation cycles. Service life requirement 10,000 flying hours and 20,000 landings in the training role.

Wings: Cantilever low/mid-wing monoplane. Wing section NACA 64A series. Leading-edge swept back 11° 18′. Sweepback at quarter-chord 8° 29′. All-metal single-spar stressed-skin structure in two panels, bolted to fuselage. Wingtip tanks permanently attached. Single fence on each wing at approx two-thirds span. Servo-powered ailerons embody 'Irving'-type aerodynamic balance provision, and electrically-actuated balance tabs to facilitate reversion to manual operation in the event of servo failure. Hydraulically-operated single-slotted flaps.

FUSELAGE: All-metal semi-monocoque structure. Hydraulically-operated speed brake under centre-fuselage.

TAIL UNIT: Cantilever all-metal structure. Slightly swept-back vertical surfaces. Electrically-actuated trim tab in rudder and each elevator.

LANDING GEAR: Hydraulically-retractable tricycle type, with oleo-pneumatic shock-absorbers. Nosewheel retracts forward, main units outward into the wings. Low-pressure main-wheel tyres size 175 × 254 × 545 10PR; low-pressure nosewheel tyre size 5·00 × 4·5 6PR.

POWER PLANT; One Fiat-built Rolls-Royce Viper Mk 632-43 turbojet engine, rated at 17·79 kN (4,000 lb st). Internal fuel capacity (fuselage and wingtip tanks) 1,390 litres (306 Imp gallons). Provision for two underwing drop-tanks, each of 330 litres (72·5 Imp gallons) capacity.

ACCOMMODATION: Crew of two in tandem, on Martin-Baker Mk F10 zero-zero ejection seats. Elevated rear seat. Two-piece moulded transparent jettisonable canopy, opening sideways to starboard.

ELECTRONICS AND EQUIPMENT: Integrated attitude, heading and flight director system. Two electronics standards: U-version (for countries with UHF radio communications and military ground navigation network) includes Collins AN/ARC-159(V) or Magnavox AN/ARC-164 primary UHF, SIT 301 emergency UHF, Collins IA-102A audio and interphone control, Hoffman ANS 952 Tacan, Collins DF-206 or Marconi-Elliott AD 380C LF/ADF, Bendix AN/APX-100 IFF. V-version (for countries using civilian air traffic control) has two AN/ARC-175(V) (Collins VHF 20B), Collins IA-102A audio, AN/ARN-126 (Collins VIR-31A), VOR/ILS/marker, Collins DME 40, Collins DF-206 or Marconi-Elliott AD 380C LF/ADF, and Collins TDR-90 ATC transponder.

ARMAMENT: Provision for a 7·62 mm GAU-2B/A multi-barrel machine-gun with 1,500 rds in underside of forward fuselage, or a 30 mm DEFA cannon with 120 rds in a flush-fitting underfuselage pod. Ammunition for the GAU-2B/A is stowed internally, in a compartment which can also accept, as alternative packages, photo-reconnaissance equipment, special-mission electronics, a variable-stability system, or baggage. Photographic pod contains four 70 mm Vinten cameras. Provision for Aeritalia or gyroscopic Thomson-CSF gunsight, OMERA-SEGID gun camera and Astronautics head-up display. Up to 1,815 kg (4,000 lb) of external stores can be carried on six underwing hardpoints, the inner four of which are stressed for loads of up to 454 kg (1,000 lb) each and the outer two for up to 340 kg (750 lb) each. Typical loads can include two AS.11 or AS.12 air-to-surface missiles; two Matra 550 air-to-air missiles; two AN/M-3 12·7 mm machine-gun pods on the inner stations, with 350 rds/gun; four 750 lb bombs or napalm containers; six SUU-11A/A 7·62 mm Minigun pods with 1,500 rds/gun; six Matra 155 launchers, each for eighteen 68 mm rockets; six Matra F-2 practice launchers, each for seven 68 mm rockets; six LAU-3/A launchers, each for nineteen 2·75 in rockets; six Simpres LR-25-0 launchers, each for twenty-five 50 mm rockets; six LAU-32 or LAU-59 rocket launchers; or two 330 litre (72·5 Imp gallon) drop-tanks.

DIMENSIONS, EXTERNAL:
Wing span over tip-tanks 10·86 m (35 ft 7½ in)
Length overall 10·97 m (36 ft 0 in)
Height overall 3·74 m (12 ft 3¼ in)
Wheel track 2·48 m (8 ft 1¾ in)
Wheelbase 4·36 m (14 ft 3½ in)

AREA:
Wings, gross 19·30 m² (207·74 sq ft)

WEIGHTS:
Weight empty 3,075 kg (6,780 lb)
T-O weight, 'clean' 4,350 kg (9,590 lb)
Max T-O weight, with external stores 5,895 kg (13,000 lb)

PERFORMANCE (estimated; 'clean' aircraft except where stated):
EAS limit/Mach limit 500 knots (Mach 0·82)
Max level speed at S/L 485 knots (898 km/h; 558 mph)
Max level speed at 9,150 m (30,000 ft) Mach 0·77 (441 knots; 817 km/h; 508 mph)
Max rate of climb at S/L 2,135 m (7,000 ft)/min
Time to 9,150 m (30,000 ft) 7 min
Service ceiling (30·5 m; 100 ft/min rate of climb) 14,630 m (48,000 ft)
T-O run 455 m (1,490 ft)
T-O to, and landing from, 15 m (50 ft) 690 m (2,265 ft)
Landing run 405 m (1,330 ft)
Max range (internal fuel) 950 nm (1,760 km; 1,093 miles)
Max endurance at 9,150 m (30,000 ft) (internal fuel) 2 hr 50 min
Max ferry range with two underwing drop-tanks, 10% reserves 1,140 nm (2,110 km; 1,310 miles)
g limits +8·0; −4·0

Prototype Aermacchi M.B.339 two-seat training and attack aircraft, first flown on 12 August 1976

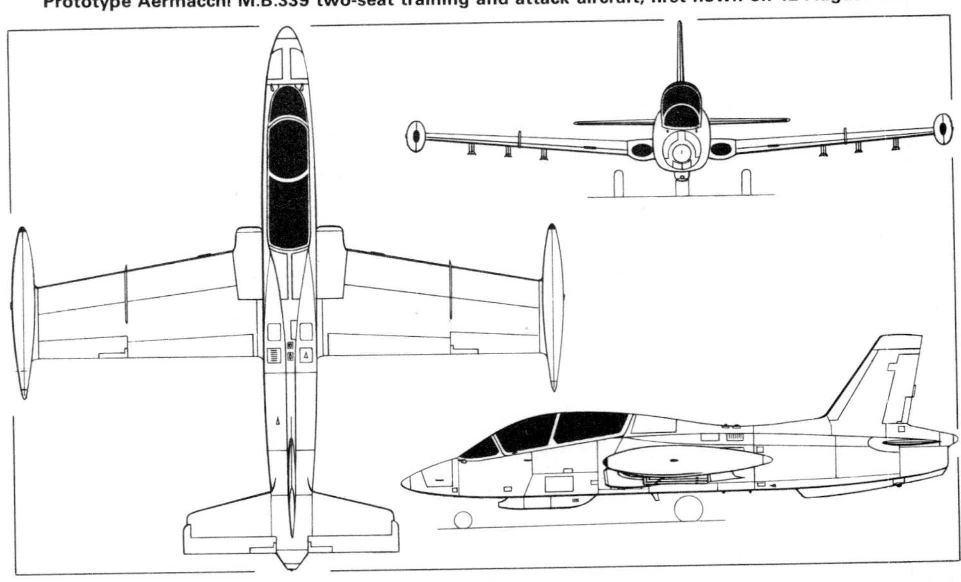

Aermacchi M.B. 339 two-seat jet trainer and light attack aircraft *(Pilot Press)*

AGUSTA
COSTRUZIONI AERONAUTICHE GIOVANNI AGUSTA SpA

HEAD OFFICE AND WORKS:
Casella Postale 193, 21017 Cascina Costa, Gallarate
Telephone: (0331) 220478
Telex: 39569 Agusta

CHAIRMAN AND PRESIDENT:
Conte Corrado Agusta

DEPUTY CHAIRMAN:
Dott E. Marelli

PRESIDENT, CHIEF EXECUTIVE OFFICER:
Dott Ing P. Fascione

GENERAL MANAGERS:
Dott Ing A. Antichi
Dott Ing G. Brazzelli

TECHNICAL DIRECTOR:
Dott Ing L. Passini

PUBLIC RELATIONS:
Giorgio Apostolo

This company was established in 1907 by Giovanni Agusta and built many experimental and production aircraft before the second World War.

In 1952 Agusta acquired a licence to manufacture the Bell Model 47 helicopter and the first Agusta-built Model 47G made its maiden flight on 22 May 1954.

In addition to versions of the Model 47, Agusta is now producing under licence in Italy the Bell Iroquois Models UH-1B and UH-1D/H, as the Agusta-Bell 204B and 205 respectively, the twin-engined Model 212, and the light turbine-powered Model 206 JetRanger helicopter series.

Under licence from Sikorsky, production of SH-3D helicopters began in 1967, and production of the HH-3F (S-61R) started in 1974. Agusta is also engaged, together with Meridionali, SIAI-Marchetti and other Italian companies, in quantity production under licence of the Boeing Vertol CH-47C Chinook helicopter (see SIAI-Marchetti entry).

Details of these aircraft (except the Chinook) are given hereafter, following the description of the A 109A helicopter designed by Agusta.

AGUSTA A 109A

The Agusta A 109A is a high-speed, high-performance twin-engined helicopter. The basic version accommodates a pilot and seven passengers, and has a large baggage compartment in the rear of the fuselage. Alternatively, the A 109A can be adapted for freight-carrying, as an ambulance, or for search and rescue. A military version is described separately.

The first of three Agusta A 109 flying prototypes (NC7101) flew for the first time on 4 August 1971. RAI and FAA certification was announced on 1 June 1975; delivery of production aircraft, designated A 109A, started at the beginning of 1976.

TYPE: Twin-engined general-purpose helicopter.

ROTOR SYSTEM AND DRIVE: Fully-articulated four-blade single main rotor and two-blade semi-rigid delta-hinged tail rotor. Main transmission assembly is housed in fairing above the passenger cabin, driving the main rotor through a coupling gearbox and main reduction gearbox, and the tail rotor through a 90° gearbox. Main rotor blades can be folded back for stowage. Main rotor/engine rpm ratio 1 : 15·62. Tail rotor/engine rpm ratio 1 : 2·88. Rotor brake optional.

FUSELAGE AND TAIL UNIT: Pod and boom type, of aluminium alloy construction, built in four main sections: nose, cockpit, passenger cabin and tailboom. Sweptback vertical fins (above and below fuselage), and non-swept elevators, mounted on rear of tailboom. Tail rotor on port side.

LANDING GEAR: Retractable tricycle type, with single main wheels and self-centering steerable nosewheel. Hydraulic retraction, nosewheel forward, main wheels upward into fuselage. Hydraulic emergency retraction. Brakes on main wheels, locking mechanism on nosewheel. All tyres are of tubeless type, size 360 × 135-6.

POWER PLANT: Two Allison 250-C20B turboshaft engines (each 313 kW; 420 shp for T-O, 287 kW; 385 shp max continuous power, 276 kW; 370 shp max cruise power, derated to 258 kW; 346 shp for twin-engine operation), mounted side by side in upper rear fuselage and separated from passenger cabin and from each other by firewalls. Fuel tank in lower rear fuselage, usable capacity 550 litres (121 Imp gallons). Oil capacity 5·5 litres (1·2 Imp gallons) for each engine and 7·5 litres (1·6 Imp gallons) for transmission.

ACCOMMODATION: Crew of one or two on flight deck, which has a door on each side. Dual controls. Main cabin seats up to six passengers, in two rows of three at 810 mm (32 in) pitch, with large space at rear for baggage. A seventh passenger can be carried in lieu of second crew member. Door to passenger cabin on each side. First row of seats removable to permit use as freight transport. Ambulance version can accommodate two stretchers, one above the other, and two medical attendants, in addition to the pilot.

SYSTEMS: Utility hydraulic system, with emergency accumulators, for landing gear operation, wheel and rotor braking and nosewheel locking. Two separate hydraulic systems provide for dual flight servo-controls. 28V DC electrical system, using two 150A starter/generators, and one 24V 13Ah battery. 115V 400Hz AC power supplied by 250VA static inverter.

ELECTRONICS AND EQUIPMENT: Standard flight instrumentation. Additional instrumentation and equipment to customer's requirements, including provision for VHF-FM, UHF-AM, VOR (with Area Navigation if required), ILS, DME, ADF, etc.

DIMENSIONS, EXTERNAL:
- Diameter of main rotor: 11·00 m (36 ft 1 in)
- Diameter of tail rotor: 2·03 m (6 ft 8 in)
- Length overall, both rotors turning: 13·05 m (42 ft 10 in)
- Length of fuselage: 10·71 m (35 ft 1¾ in)
- Height overall: 3·30 m (10 ft 10 in)
- Elevator span: 2·88 m (9 ft 5½ in)
- Wheel track: 2·45 m (8 ft 0½ in)
- Passenger doors (each):
  - Height: 1·06 m (3 ft 5¾ in)
  - Width: 1·15 m (3 ft 9¼ in)
  - Height to sill: 0·65 m (2 ft 1½ in)

DIMENSIONS, INTERNAL:
- Cabin, excl flight deck:
  - Length: 1·62 m (5 ft 3¾ in)
  - Width: 1·36 m (4 ft 5½ in)
  - Height: 1·28 m (4 ft 2½ in)
  - Volume: 2·82 m³ (100 cu ft)
- Baggage compartment volume: 0·52 m³ (18·4 cu ft)

AREAS:
- Main rotor disc: 95·00 m² (1,022·6 sq ft)
- Tail rotor disc: 3·23 m² (34·75 sq ft)

WEIGHTS AND LOADINGS:
- Weight empty: 1,415 kg (3,120 lb)
- Max T-O weight: 2,450 kg (5,400 lb)
- Max disc loading: 25·8 kg/m² (5·28 lb/sq ft)
- Max power loading: 3·91 kg/kW (6·43 lb/shp)

PERFORMANCE (at 2,450 kg; 5,400 lb AUW. A: ISA, B: ISA + 20°C):
- Never-exceed speed: 168 knots (311 km/h; 193 mph)
- Max cruising speed at max continuous power:
  - A: 144 knots (266 km/h; 165 mph)
- Optimum cruising speed at S/L:
  - A: 125 knots (231 km/h; 143 mph)
- Max rate of climb at S/L:
  - A: 493 m (1,620 ft)/min
- Rate of climb at S/L, one engine out:
  - A: 103 m (340 ft)/min
- Service ceiling:
  - A: 4,968 m (16,300 ft)
- Service ceiling, one engine out:
  - A: 914 m (3,000 ft)
- Hovering ceiling in ground effect:
  - A: 2,987 m (9,800 ft)
  - B: 2,133 m (7,000 ft)
- Hovering ceiling out of ground effect:
  - A: 2,042 m (6,700 ft)
  - B: 1,219 m (4,000 ft)
- Max range at S/L:
  - A: 305 nm (565 km; 351 miles)
- Max endurance at S/L:
  - A: 3 hr 18 min

AGUSTA A 109 MILITARY VERSION

A military version of the A 109A civil helicopter is under development for advanced scout, light observation, anti-tank, utility and electronic warfare duties. Its general configuration, structure and power plant are similar to those of the A 109A (which see).

Five military A 109s, each armed with four TOW air-to-surface missiles, are to be evaluated by the Italian Army.

ACCOMMODATION: Standard seating for a pilot and seven troops. Ambulance version accommodates two stretcher patients and two medical attendants.

ARMAMENT: Basically one 7·62 mm flexibly-mounted machine-gun, with stabilised sight system, and two XM-157 rocket launchers, each with seven 2·75 in rockets. Alternative weapons include Hot or TOW missiles; an electrically-operated 7·62 mm Minigun on flexible mounting, with 1,000 rounds; a fully-automatic MG3 7·62 mm machine-gun with 5,000 rounds; an XM-159C launcher for nineteen 2·75 in rockets, Agusta launcher for seven 81 mm rockets, or 200A-1 launcher for nineteen 2·75 in rockets.

AGUSTA A 129

This is a project for a light anti-tank helicopter,

Agusta A 109A general-purpose helicopter (two Allison 250-C20B turboshaft engines)

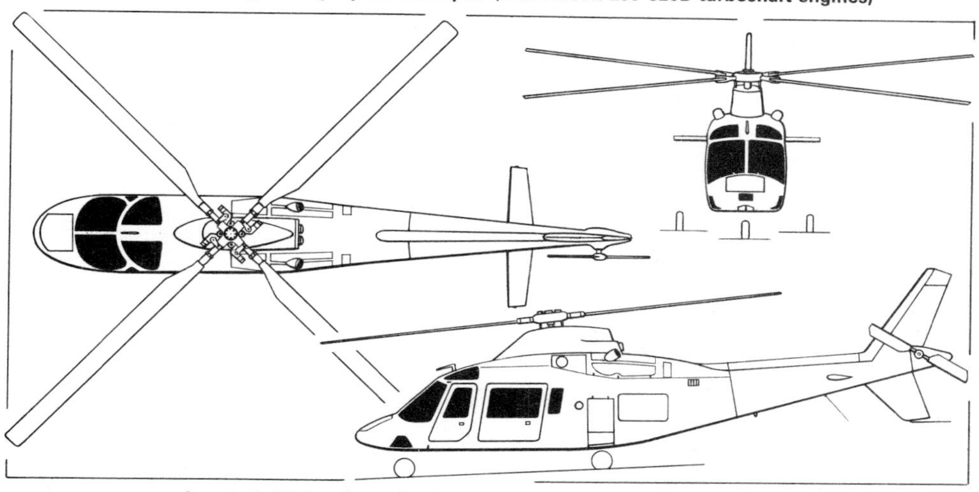
Agusta A 109A twin-engined general-purpose helicopter *(Pilot Press)*

Military version of the Agusta A 109A twin-turboshaft helicopter

Model of the projected Agusta A 129 anti-tank helicopter, developed from the A 109A

developed from the A 109A and utilising most of the dynamic components of the latter aircraft.

A narrow fuselage, 1·00 m (39 in) wide, accommodates the pilot's and co-pilot/gunner's seats in tandem, with the pilot to the rear. Each crew position is fitted with flying controls, instruments and an armoured seat.

Two stub wings are standard, each with two external stores stations. The landing gear can be of either wheel or skid type.

POWER PLANT: Two Allison 250-C30 turboshaft engines, each derated to 335 kW (450 shp) for normal operations. In the event of failure of one engine, the remaining engine can be operated at 410 kW (550 shp) to provide adequate single-engine performance.

ARMAMENT: Hot or TOW anti-tank missiles; air-to-surface rockets; 7·62 mm Minitat machine-gun system. Inner stores stations can each carry 240 kg (529 lb) loads; outer stations can each carry 200 kg (441 lb).

AGUSTA-BELL MODEL 47 SERIES

Agusta has been building the Bell Model 47 helicopter under licence since 1954, and is now the only company still doing so. A total of more than 1,100 had been built in Italy by the end of 1971, and production continues on a limited basis.

Most recent versions built by Agusta have included the standard Models 47G-3B-1, G-3B-2, G-4A, G-5 and J-2A. In addition, Agusta developed from the J-2A two special variants of its own design. These are:

AB 47J-3. Differs from the standard Bell 47J-2A (1966-67 *Jane's*) in having a modified main transmission able to absorb greater power input, ie 201 kW (270 hp) for T-O and a max continuous output of 194 kW (260 hp), in lieu of the normal 194 kW (260 hp) and 164 kW (220 hp) respectively. Performance remains similar to that of the Bell 47J-2A. A special ASW version was evolved for the Italian Navy, for shipboard operation. This has instrumentation for over-sea operation in reduced visibility, and a high-efficiency rotor brake. Its armament comprises one Mk 44 torpedo.

AB 47J-3B-1. High-altitude version, powered by a Lycoming TVO-435-B1A engine rated at 201 kW (270 hp) for T-O and 164 kW (220 hp) for continuous operation. The engine is equipped with an exhaust-driven supercharger, fitted with an automatic control which maintains S/L conditions up to at least 4,300 m (14,000 ft). This version also has a high-inertia rotor and servo-control on both the cyclic and collective pitch control systems.

All Agusta-built versions can be fitted with the full range of optional equipment originally offered by Bell, including pontoons and stretchers.

The following description applies to the AB 47J-3 and J-3B-1:

TYPE: Four-seat general-purpose helicopter.

ROTOR SYSTEM: Two-blade semi-rigid main rotor, with interchangeable blades of bonded all-metal construction; stabilising bar below and at right angles to blades. Conventional swashplate assembly for cyclic and collective pitch control (servo-assisted on 47J-3B-1). Blades do not fold. Brake on main rotor optional. Two-blade all-metal tail rotor.

ROTOR DRIVE: Through centrifugal clutch and two-stage planetary transmission; shaft drive to tail rotor.

FUSELAGE: In three sections: centre, tail and cabin. Rear section of triangular cross-section, supporting the tail rotor assembly and its drive-shaft.

TAIL UNIT: Small synchronised elevator at rear end of fuselage, responding to the fore and aft motion of the cyclic pitch control to improve stability and increase permissible CG travel.

LANDING GEAR: Tubular skid type, with small ground handling wheels and tie-down and towing attachments. Cross-tubes serve as supports for external loads such as stretchers or cargo containers. Provision for inflatable floats to be attached for amphibious operation.

POWER PLANT: One vertically-mounted Lycoming flat-six engine (for details see introductory copy), with clutch, drive-shaft and rotor assembly in an integral unit. Two interconnected saddle-mounted fuel tanks (total capacity 182 litres; 48 US gallons) on CG, with gravity feed.

ACCOMMODATION: Pilot seated centrally in front, with bench seat to rear for three persons, in enclosed cabin with free-blown Plexiglas canopy. Door on each side, with sliding panels in windows. Compartment for 113 kg (250 lb) of baggage at forward end of tailboom. Passenger seats can be removed to permit the installation of two stretchers and a jump-seat for a medical attendant or for the carriage of cargo. With the port-side door removed a trap can be removed to permit the use of a 182 kg (400 lb) capacity internally-mounted electrically-powered hoist for rescue work. Conversion to any of the cabin configurations can be made in minutes without the use of specialised tools.

ELECTRONICS AND EQUIPMENT: Standard equipment includes complete VFR flight and engine instruments, hydraulic boost controls, 28V 50A generator, electric starter, ground handling wheels, heavy-duty battery etc. Optional equipment includes VHF transceiver, night flying equipment, rotating beacons, rotor brake (standard on J-3 ASW version), pontoon landing gear, cabin heater/defroster, stretchers, cargo carriers, dual controls, and agricultural equipment.

DIMENSIONS, EXTERNAL:
- Diameter of main rotor: 11·32 m (37 ft 1½ in)
- Diameter of tail rotor: 1·78 m (5 ft 10 in)
- Main rotor blade chord: 0·28 m (11 in)
- Length overall, main rotor fore and aft: 13·20 m (43 ft 4¾ in)
- Length of fuselage: 9·87 m (32 ft 4¾ in)
- Width, main rotor fore and aft: 2·54 m (8 ft 4 in)
- Height to top of rotor hub: 2·83 m (9 ft 3½ in)
- Skid track: 2·29 m (7 ft 6 in)

AREAS:
- Main rotor blades (each): 1·59 m² (17·14 sq ft)
- Tail rotor blades (each): 0·11 m² (1·20 sq ft)
- Main rotor disc: 100·8 m² (1,085·00 sq ft)
- Tail rotor disc: 2·35 m² (25·31 sq ft)

WEIGHTS:
- Weight empty:
  - J-3: 825 kg (1,819 lb)
  - J-3B-1: 845 kg (1,863 lb)
- Max T-O weight: 1,340 kg (2,950 lb)

PERFORMANCE (47J-3B-1, at max T-O weight):
- Max level speed at S/L: 91 knots (169 km/h; 105 mph)
- Normal cruising speed at 1,525 m (5,000 ft): 75 knots (138 km/h; 86 mph)
- Max rate of climb at S/L: 276 m (905 ft)/min
- Max rate of climb at 4,300 m (14,000 ft): 240 m (785 ft)/min
- Service ceiling: 5,340 m (17,500 ft)
- Hovering ceiling in ground effect: 5,030 m (16,500 ft)
- Hovering ceiling out of ground effect: 3,720 m (12,200 ft)
- Range with max fuel at 1,525 m (5,000 ft), no reserves: 182 nm (338 km; 210 miles)
- Max endurance at 1,525 m (5,000 ft), no reserves: 3 hr 30 min

AGUSTA-BELL 204B and 204AS

The Agusta-Bell 204B is a medium-size utility helicopter, similar to the Bell UH-1B Iroquois; it was in production from 1961 to 1974 for the armed services of Italy, and for military and commercial operators in other countries.

Agusta also designed and built a special ASW version, the Agusta-Bell 204AS, for the Italian and Spanish navies. Further details of these helicopters can be found in the 1975-76 *Jane's*.

AGUSTA-BELL 205 and 205A-1

The Agusta-Bell Model 205 is a multi-purpose utility helicopter, corresponding to the UH-1D/UH-1H versions adopted by the US armed forces. It can be used to transport passengers, equipment and troops, or for casualty evacuation, tactical support, rescue or other missions. Various special installations such as stretchers, floats, snow skids, armament or a rescue hoist can be fitted according to role. The cabin will accommodate a pilot and 14 passengers, and has a clear volume of 6·2 m³ (220 cu ft) when stripped for cargo carrying. The AB 205 is fitted with IFR and night flying instruments, and for normal operation only one pilot is carried.

The AB 205 is in service with the Italian armed forces and has been ordered by those of Iran, Kuwait, Morocco, Saudi Arabia, Spain, Turkey, the United Arab Emirates, Zambia and other countries.

In 1969 Agusta began production of a developed version known as the **AB 205A-1**, certificated for commercial and passenger transport operation. The Agusta-Bell 205A-1 can accommodate up to 14 passengers and has a 0·8 m³ (28·3 cu ft) baggage compartment in the tailboom. A wide variety of equipment can be fitted to fulfil a number of other roles. Power plant is a 1,044 kW (1,400 shp) Lycoming T5313B turboshaft engine, derated to 932 kW (1,250 shp) for take-off.

DIMENSIONS, EXTERNAL:
- Main rotor diameter: 14·63 m (48 ft 0 in)
- Tail rotor diameter: 2·59 m (8 ft 6 in)
- Fuselage length: 12·78 m (41 ft 11 in)
- Width overall: 2·76 m (9 ft 0½ in)
- Height overall: 4·48 m (14 ft 8 in)

WEIGHTS (AB 205):
- Weight empty (standard): 2,177 kg (4,800 lb)
- Normal T-O weight: 3,860 kg (8,500 lb)
- Max T-O weight: 4,310 kg (9,500 lb)

Agusta-Bell 47G-4A three-seat general-purpose helicopter (Lycoming VO-540 engine)

Agusta-Bell 205 multi-purpose helicopter (Lycoming T53-L-13B turboshaft engine)

WEIGHTS (AB 205A-1):
Weight empty (standard) 2,356 kg (5,195 lb)
Max T-O weight (FAA cert):
internal load 4,310 kg (9,500 lb)
external load 4,762 kg (10,500 lb)

PERFORMANCE (AB 205 at AUW of 3,860 kg; 8,500 lb, with T53-L-13 engine):
Max level speed at S/L 120 knots (222 km/h; 138 mph)
Cruising speed 115 knots (212 km/h; 132 mph)
Max rate of climb at S/L 548 m (1,800 ft)/min
Hovering ceiling in ground effect 5,180 m (17,000 ft)
Hovering ceiling out of ground effect 3,350 m (11,000 ft)
Max range, standard tanks, no reserves 312 nm (580 km; 360 miles)
Max endurance, standard tanks, no reserves 3 hr 48 min

PERFORMANCE (AB 205A-1, at max T-O weight with internal load):
Max level speed 120 knots (222 km/h; 138 mph)
Max cruising speed 109 knots (203 km/h; 126 mph)
Max rate of climb at S/L 619 m (2,030 ft)/min
Hovering ceiling in ground effect 3,350 m (11,000 ft)
Hovering ceiling out of ground effect 2,075 m (6,800 ft)
Max range, standard tanks, no reserves 287 nm (532 km; 331 miles)
Max endurance, standard tanks, no reserves 3 hr 18 min

AGUSTA-BELL 206B JETRANGER II

Swedish military designation: HKP 6

The Agusta-Bell 206 JetRanger has been manufactured under licence from Bell since the end of 1967; production of the Model 206A (see 1973-74 *Jane's*) has ended.

Deliveries began in 1972 of the Agusta-Bell 206B JetRanger II, which is now the standard Agusta-built version. The AB 206B has a 298 kW (400 shp) (max) Allison 250-C20 engine, derated to 236 kW (317 shp). The AB 206B has an increased gross weight and an improved performance in 'hot and high' conditions.

By the beginning of 1976 several hundred Agusta-Bell 206-series helicopters had been built and delivered to commercial and military operators, including the armed forces of Iran, Italy, Saudi Arabia, Spain, Sweden and Turkey. The AB 206As for Sweden (Swedish designation HKP 6) have long-leg skid gear and underfuselage weapon racks.

The cabin can accommodate four persons in addition to the pilot; the baggage compartment has a capacity of 113 kg (250 lb) and a usable volume of 0·45 m³ (16 cu ft).

The following details refer to the current Agusta-Bell 206B:

DIMENSIONS, EXTERNAL:
Main rotor diameter 10·16 m (33 ft 4 in)
Length overall, rotors turning 11·94 m (39 ft 2 in)
Length of fuselage 9·50 m (31 ft 2 in)

WEIGHTS:
Weight empty (standard) 682 kg (1,504 lb)
Max T-O weight (internal load) 1,452 kg (3,200 lb)
Max T-O weight (external load) 1,519 kg (3,350 lb)

PERFORMANCE (at AUW of 1,452 kg; 3,200 lb, ISA):
Max level speed at S/L 122 knots (226 km/h; 140 mph)
Cruising speed 116 knots (214 km/h; 133 mph)
Max rate of climb at S/L 414 m (1,358 ft)/min
Hovering ceiling in ground effect 3,450 m (11,325 ft)
Hovering ceiling out of ground effect 1,770 m (5,800 ft)
Max range, standard fuel, no reserves 363 nm (673 km; 418 miles)
Max endurance, standard fuel, no reserves 4 hr 0 min

AGUSTA-BELL 206A-1 and 206B-1

The Agusta-Bell 206A-1 and 206B-1, corresponding to the Bell OH-58A Kiowa (see US section), have been developed expressly for military operation, and are in service with the Italian armed forces and other operators. In general configuration they are similar to the AB 206A and B, with accommodation for a pilot and four passengers. Main differences are a larger-diameter main rotor, and appropriate military modifications, such as local strengthening of the airframe, provision for armament and additional cabin doors. The AB 206A-1 and AB 206B-1 are powered respectively by Allison 250-C18 and 250-C20 engines.

Production of the AB 206B-1 began in 1972.

DIMENSIONS, EXTERNAL:
Main rotor diameter 10·77 m (35 ft 4 in)
Length overall, rotors turning 12·50 m (41 ft 0 in)
Length of fuselage 9·85 m (32 ft 4 in)

WEIGHTS:
Weight empty (standard):
206A-1 682 kg (1,504 lb)
206B-1 698 kg (1,540 lb)
Max T-O weight (internal load):
206A-1 1,360 kg (3,000 lb)
206B-1 1,452 kg (3,200 lb)
Max T-O weight (external load):
206A-1 and B-1 1,519 kg (3,350 lb)

PERFORMANCE (at AUW of 1,360 kg; 3,000 lb, ISA for 206A-1; AUW of 1,452 kg; 3,200 lb, ISA for 206B-1):
Max level speed at S/L:
206A-1 114 knots (211 km/h; 131 mph)
206B-1 120 knots (222 km/h; 138 mph)
Cruising speed:
206A-1 110 knots (204 km/h; 127 mph)
206B-1 112 knots (208 km/h; 129 mph)
Max rate of climb at S/L:
206A-1 475 m (1,560 ft)/min
206B-1 396 m (1,300 ft)/min
Hovering ceiling in ground effect:
206A-1 3,325 m (10,900 ft)
206B-1 3,660 m (12,000 ft)
Hovering ceiling out of ground effect:
206A-1 1,825 m (6,000 ft)
206B-1 2,440 m (8,000 ft)
Max range, standard fuel, no reserves:
206A-1 319 nm (592 km; 368 miles)
206B-1 290 nm (538 km; 334 miles)
Max endurance, standard fuel, no reserves:
206A-1 4 hr 0 min
206B-1 3 hr 30 min

AGUSTA-BELL 206L LONGRANGER

This aircraft is a seven-seat single-engined light helicopter, derived from the AB 206B and powered by a 313 kW (420 shp) Allison 250-C20B turboshaft engine. A Nodamatic transmission suspension system gives virtually vibration-free flight, as described under the Bell Helicopter entry in the US section. Max T-O weight of the AB 206L is 1,814 kg (4,000 lb).

AGUSTA-BELL 212

The Agusta-Bell 212 is a twin-engined utility transport helicopter particularly suited to passenger transport duties. Deliveries began in the second half of 1971, and by the beginning of 1976 more than 40 had been built, with others on order, for Italian and other commercial and military operators. The first IFR flight by an AB 212 fitted with a SFENA Helistab autopilot took place at Gallarate on 5 April 1973.

The general configuration of the AB 212 is similar to that of the AB 205/205A-1, but it embodies considerable modifications to the dynamic components, systems and structure.

Power plant is a Pratt & Whitney Aircraft of Canada PT6T-3 Turbo Twin Pac, derated in the AB 212 to 962 kW (1,290 shp) for T-O and a max continuous rating of 843 kW (1,130 shp). Fuel capacity is 813 litres (215 US gallons; 179 Imp gallons).

Standard accommodation is for a pilot and up to 14 passengers, but the helicopter is readily adaptable to alternative configurations, including a de luxe interior or as a VIP transport.

Optional kits for alternative roles include rescue hoist, cargo hook, auxiliary fuel tanks, stretchers and float landing gear.

An extensively-modified naval version, the AB 212ASW, is being produced by Agusta and is described separately.

DIMENSIONS, EXTERNAL:
Diameter of main rotor 14·63 m (48 ft 0 in)
Diameter of tail rotor 2·59 m (8 ft 6 in)
Length overall, rotors turning 17·40 m (57 ft 1 in)
Fuselage length 14·02 m (46 ft 0 in)
Height to top of cabin roof 2·34 m (7 ft 8 in)
Height overall, tail rotor turning 4·40 m (14 ft 5 in)
Elevator span 2·84 m (9 ft 4 in)
Width over skids 2·64 m (8 ft 8 in)

WEIGHTS:
Weight empty (standard) 2,630 kg (5,800 lb)
Max T-O weight, internal or external load 5,081 kg (11,200 lb)

PERFORMANCE (at AUW of 4,536 kg; 10,000 lb, ISA):
Cruising speed at S/L 110 knots (204 km/h; 127 mph)
Max rate of climb at S/L 567 m (1,860 ft)/min
Service ceiling 5,180 m (17,000 ft)
Hovering ceiling in ground effect 3,960 m (13,000 ft)
Hovering ceiling out of ground effect 3,050 m (10,000 ft)
Max range at 1,525 m (5,000 ft) with standard fuel, no reserves:
on two engines 267 nm (494 km; 307 miles)
on one engine 318 nm (589 km; 366 miles)

AGUSTA-BELL 212ASW

The Agusta-Bell 212ASW helicopter has been developed as a medium-sized twin-engined naval helicopter designed for anti-submarine search and attack missions, and attack missions against surface vessels. It is also suitable for search and rescue and utility roles. It is an extensively modified version of the standard Agusta-Bell 212 (which see), utilising naval operational experience gained with the AB 204AS, and because of its similarity in size to the 204AS can also operate from small ship decks. A prototype has been successfully evaluated, and the AB 212ASW is now being produced and delivered to meet orders from the Italian Navy and from Turkey and other foreign operators.

Apart from some local strengthening and the provision of deck-mooring equipment, the airframe structure remains essentially similar to that of the commercial Model 212 and military UH-1N, described under the Bell entry in the US section. Main differences from the Agusta-Bell 212 are as follows:

TYPE: Twin-engined anti-submarine and anti-surface-vessel helicopter.

POWER PLANT: As AB 212, but with protection against salt water corrosion. Provision for one internal or two external auxiliary fuel tanks.

ACCOMMODATION: Crew of three or four.

SYSTEMS: Standard duplicated hydraulic systems for flight controls, as in AB 212. Either hydraulic system is capable of operating the automatic flight control system. Third, self-contained system for operation of sonar, rescue hoist and other utilities. Electrical system capacity increased to cater for higher power demand; the two standard generators are integrated with a 20kVA alternator.

ELECTRONICS AND EQUIPMENT: Complete instrumentation for day and night sea operation in all weathers. Electronics installed are AN/ARC-150 UHF transceiver, Collins SSB/DSB 718 U-5 HF transceiver, and Agusta AG-03-M intercom, for communications; Marconi-Elliott AD 370B ADF, Hoffman AN/ARN-91 Tacan and Collins AN/ARA-50 homing UHF, for navigation assistance; Aeritalia (Honeywell) AN/APN-171 radar altimeter, Canadian Marconi AN/APN-172(V)2 Doppler radar, Canadian Marconi CMA-708B/ASW navigation computer, and automatic flight control system with General Electric SR-3 gyro platform, Agusta ASE-531A automatic stabilisation equipment and Agusta AATH-547A automatic approach to hover, for automatic navigation; Siemens AN/APX-77 IFF/SIF transponder; MEL ARI-5955 search radar and Motorola SST-119X radar transponder; and Bendix AN/AQS-13B sonar for ASW search.

ARMAMENT AND OPERATIONAL EQUIPMENT: Weapons may consist of two Mk 44 or Mk 46 homing torpedoes, depth charges or two air-to-surface missiles. Provisions for auxiliary installations such as a 270 kg (595 lb) capacity rescue hoist, 2,270 kg (5,000 lb) capacity cargo sling, inflatable emergency pontoons, internal and external auxiliary fuel tanks, according to mission.

ASW MISSION: The basic sensor system employed for the ASW search and attack mission is the AN/AQS-13B variable-depth sonar, with a max operating depth of 137 m (450 ft). The automatic navigation system permits the positioning of the helicopter over any desired 'dip' point of a complex search pattern. The position of the helicopter, computed by the automatic navigation system, is

Agusta-Bell 212ASW anti-submarine and anti-surface-vessel helicopter of the Italian Navy

VIP transport version of the Agusta-Sikorsky SH-3D helicopter

integrated with sonar target information in the radar tactical display where both the surface and the underwater tactical situations can be continuously monitored. Additional navigation and tactical information is provided by accurate UHF direction-finding equipment, from an A/A mode-capable Tacan and from a radar transponder. The automatic flight control system (AFCS) integrates the basic automatic stabilisation equipment with signal output from the radar altimeter, the Doppler radar, sonar cable angle signals and outputs from the dry cable transducer. The effectiveness of this system results in hands-off flight from cruise condition to sonar hover in all weathers and under rough sea conditions. A specially designed cockpit display shows the pilots all flight parameters for each phase of the ASW operation.

The attack mission is carried out with two Mk 44 or Mk 46 homing torpedoes, or with depth charges.

AWW Mission: For this mission the AB 212ASW carries a high-performance long-range search radar, with a very efficient scanner design and installation possessing high discrimination in rough sea conditions. Provisions have also been made to permit incorporation of future radar system developments. The automatic navigation systems and the search radar are integrated to permit a continuously updated picture of the tactical situation. Provisions are also incorporated for the installation of the most advanced ECM systems. The surface attack is performed with air-to-surface wire-guided missiles. In operation, the co-pilot aims and 'flies' the missiles to the target through a gyro-stabilised sight system of the XM-58 type.

Dimensions, external: As AB 212, except:
- Max width:
  - with torpedoes 3·95 m (12 ft 11½ in)
  - with missiles 4·17 m (13 ft 8¼ in)

Weights (A: ASW mission with Mk 46 torpedoes; B: AWW mission with AS.12 missiles; C: search and rescue mission; all at S/L, ISA):
- Weight empty, equipped:
  - A, B, C 3,420 kg (7,540 lb)
- Crew of three:
  - A, B, C 240 kg (529 lb)
- Mission equipment:
  - A (two Mk 46 torpedoes) 500 kg (1,101 lb)
  - B (AS.12 installation and XM-58 sight) 223 kg (491 lb)
  - C (rescue hoist) 40 kg (88 lb)
- Full fuel (normal tanks) 1,035 kg (2,282 lb)
- Auxiliary external tanks 32 kg (70 lb)
- Auxiliary fuel 356 kg (785 lb)
- Mission T-O weight:
  - A 5,079 kg (11,196 lb)
  - B 4,918 kg (10,841 lb)
  - C 4,863 kg (10,720 lb)

Performance (at max T-O weight, ISA):
- Max level speed at S/L 106 knots (196 km/h; 122 mph)
- Max cruising speed with armament 100 knots (185 km/h; 115 mph)
- Max rate of climb at S/L:
  - A 463 m (1,519 ft)/min
  - B 365 m (1,197 ft)/min
- Rate of climb at S/L, one engine out:
  - A 129 m (423 ft)/min
  - B 106 m (348 ft)/min
- Hovering ceiling in ground effect:
  - A 3,810 m (12,500 ft)
- Hovering ceiling out of ground effect:
  - A 1,220 m (4,000 ft)
- Search endurance (A) with 50% at 90 knots (167 km/h; 103·5 mph) cruise and 50% hovering out of ground effect, 10% reserve fuel 3 hr 0 min
- Search range (B) with 10% reserve fuel 323 nm (598 km; 372 miles)
- Endurance (B), no reserves 3 hr 45 min
- Endurance (C) at 90 knots (167 km/h; 103·5 mph) search speed 4 hr 15 min
- Max range with auxiliary tanks, 100 knots (185 km/h; 115 mph) cruise at S/L, 15% reserves 360 nm (667 km; 414 miles)
- Max endurance with auxiliary tanks, no reserves 5 hr 0 min

AGUSTA-BELL 214A

The AB 214A is a medium-sized helicopter configured for various military roles, and is powered by a 2,215 kW (2,970 shp) Lycoming LTC4B-8D turboshaft engine. It can accommodate a pilot and up to 15 passengers.

Weights:
- Weight empty, equipped 3,380 kg (7,450 lb)
- Max T-O weight 7,257 kg (16,000 lb)

Performance (at max internal load T-O weight):
- Max vertical rate of climb at S/L 525 m (1,720 ft)/min
- Hovering ceiling in ground effect 4,875 m (16,000 ft)
- Hovering ceiling out of ground effect 3,960 m (13,000 ft)
- Max range at 1,525 m (5,000 ft) 219 nm (405 km; 252 miles)

AGUSTA-SIKORSKY SH-3D

During 1967, Agusta began the construction under licence of an initial batch of 24 Sikorsky SH-3D anti-submarine helicopters for the Italian Navy. Delivery began in 1969. Additional orders have since been placed, both for the Italian armed forces and for the Imperial Iranian Navy, in various configurations including ASW, VIP transport and rescue. These aircraft are fitted with Teledyne AN/APN-182 Doppler velocity sensor radar and AN/APN-195 search radar.

Full details of the SH-3D can be found under the Sikorsky heading in the US section of this edition.

AGUSTA-SIKORSKY S-61A-4

The AS S-61A-4 is a derivative of the AS SH-3D, and is suitable for a wide range of duties including troop and cargo transport, medical evacuation, and search and rescue. Power plant comprises two 1,118·5 kW (1,500 shp) General Electric T58-GE-5 turboshaft engines. Accommodation is provided for a crew of three and up to 31 fully-equipped troops or 15 stretcher patients.

AGUSTA-SIKORSKY HH-3F (S-61R)

In 1974 Agusta began production of this multi-purpose rescue helicopter. Twenty are being built initially, for the Italian Air Force and foreign operators. Deliveries started in 1976.

Details of the HH-3F can be found under the Sikorsky heading in the US section of this edition.

BREDANARDI

BREDANARDI COSTRUZIONI AERONAUTICHE SpA

Head Office and Works:
Casella Postale 108, 63039 San Benedetto del Tronto, Ascoli Piceno
Telephone: (735) 67358

Rome Office:
Via XXIV Maggio 43/45, 00187 Rome
Telephone: (6) 4660
Telex: 61050

Milan Office:
Aeroporto Forlanini, 20090 Milan
Telephone: (2) 7385251
Telex: 33666

President:
Ing Giovanni Berardi

Managing Director:
Dott Elto Nardi

Director, International Marketing:
Nathaniel R. Hoskot

The Nardi company was established by the four Nardi brothers in 1933. Their first product, the FN-305, flew in 1935, and subsequently more than 600 of these aircraft were built by Nardi and, under licence, by Piaggio. They

Agricultural version of the NH-300C helicopter built under licence by BredaNardi

were followed by other Nardi designs, produced in large numbers.

The Milan factory was rebuilt after the second World War, and the prototype of the company's first post-war aircraft, the FN-333 all-metal amphibian, flew there for the first time on 4 December 1952.

On 15 February 1971 Nardi established the new company BredaNardi in equal partnership with Breda, a member company of the EFIM state-owned financial group, to undertake helicopter production under a manufacturing licence from Hughes Helicopters of the USA. Breda and Nardi each have a 50% interest.

Production of the three-seat Hughes 300C (Italian designation NH-300C) and five-seat Hughes 500C and 500M (Italian designations NH-500C and NH-500M) started at the company's new facility at Porto d'Ascoli, in central Italy, in mid-1976. Variants include the NH-500MC twin-float helicopter for the Italian coastal defence force. Production of the six-seat Hughes 500D (Italian designation NH-500D) is scheduled to start in 1977. Details of the basic Models 300C, 500D and 500M can be found under the Hughes heading in the US section of this edition.

BredaNardi has obtained a licence from Cessna to manufacture the Cessna 150A(M) Aerobat; but it is reported that production of this aircraft is dependent upon a minimum initial order from Italian aero clubs.

BredaNardi has also signed an agreement to help build in Pakistan fifty NH-500C helicopters a year. The helicopters will be produced by Kiyuski International Aviation Company of Campbellpur.

The works at Porto d'Ascoli, where airfield facilities are available, undertake the overhaul, repair and manufacture of components for other types of helicopter and other aircraft.

In addition to aircraft manufacture, Nardi itself produces wheels, brakes, retractable landing gear, hydraulic and electrical aircraft controls, fuel pumps, armament installations and aircraft accessories generally. Among these is the production of landing gear, flaps and other accessories for F-104G and F-104S Starfighters and for the Panavia Tornado programme.

BredaNardi NH-500MC twin-float helicopter of the Italian coastal defence force

GENERAL AVIA
COSTRUZIONI AERONAUTICHE GENERAL AVIA

ADDRESS:
Via Trieste 24, 20096 Pioltello, Milan
Telephone: 9046774
TECHNICAL DIRECTOR:
Dott Ing Stelio Frati
SECRETARY-TREASURER:
Lamberto Frati

Dott Ing Stelio Frati is well known for the many successful light aircraft which, as a freelance designer, he has evolved since 1950.

These have been built in prototype and production series by several Italian manufacturers, and have included such aircraft as the two-seat F.4 and three-seat F.7 Rondone, built by Ambrosini; the Caproni F.5 two-seat light jet aircraft and the twin-engined F.6 Airone built for Pasotti; the Aviamilano F.8 Falco and F.14 Nibbio; the Procaer F.15 Picchio and F.400 Cobra; and the F.250.

The 260 hp developed version of the F.250 is manufactured by SIAI-Marchetti as the SF.260, and is described under that company's heading in this section.

In early 1970, Dott Ing Frati established the General Avia company, of which he is Technical Director, and acquired extensive and well-equipped workshops where, since April 1970, prototypes of his current designs have been built.

The F.20 Pegaso, of which General Avia built two prototypes, was described under the Italair heading in the 1975-76 *Jane's*. General Avia is currently developing prototypes of the F15F, derived from the Procaer F15E Picchio, and the F.600 Canguro.

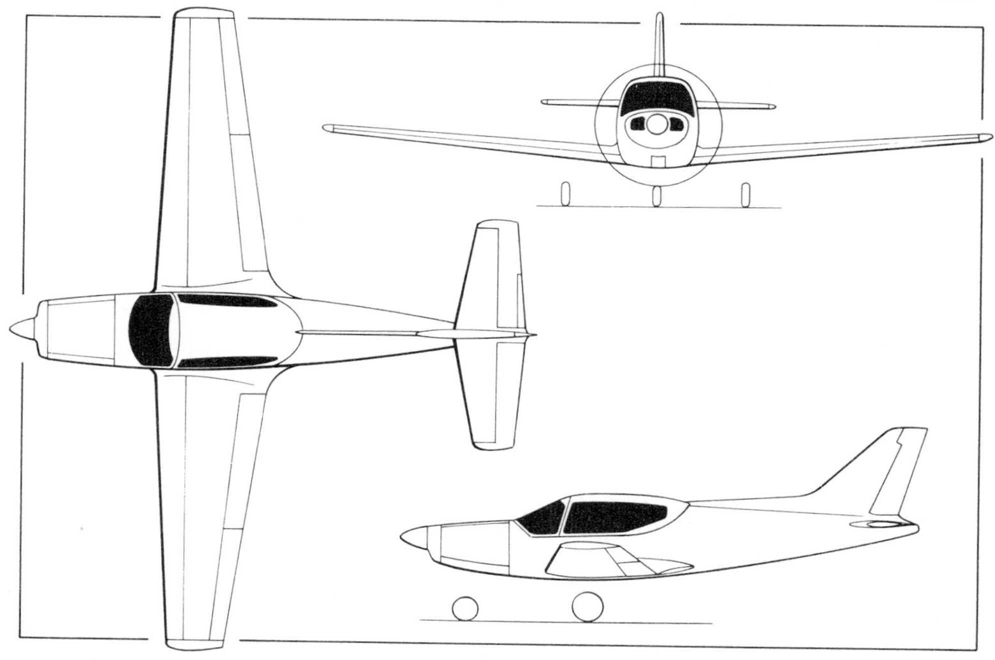

General Avia F15F, derived from the Procaer Picchio *(Michael A. Badrocke)*

GENERAL AVIA (PROCAER) F15F

The F15F is a derivative of the Procaer F15E Picchio (see 1974-75 *Jane's*), designed by Dott Ing Frati. A prototype has been completed, and was scheduled to fly during 1976.

TYPE: Four-seat light monoplane.

WINGS: Cantilever low-wing monoplane. NACA 64-215/64-210 wing sections. Dihedral 6° from roots. Incidence 4°. One-piece metal structure with single main spar, rear spar carrying aileron and flap hinges, and short front spar to carry landing gear loads. All-metal Frise ailerons and electrically-actuated Fowler flaps.

FUSELAGE: All-metal semi-monocoque structure.

TAIL UNIT: Cantilever all-metal structure. Trim tab in starboard elevator.

LANDING GEAR: Retractable tricycle type. Electrical or mechanical retraction. Oleo-pneumatic shock-absorbers. Main wheels size 6·00-6. Steerable nose-wheel size 5·00-5. Hydraulic disc brakes.

POWER PLANT: One 149 kW (200 hp) Lycoming IO-360-A flat-four engine, driving a two-blade constant-speed metal propeller.

ACCOMMODATION: Four persons in pairs in enclosed cabin. Space for 45 kg (100 lb) of baggage behind rear seats. Dual controls. Cabin soundproofed, heated and ventilated.

SYSTEMS AND EQUIPMENT: Two 12V 35Ah batteries, connected in series, provide power for landing gear and flap actuation. 600W engine-driven generator. Blind-flying instruments and radio optional.

DIMENSIONS, EXTERNAL:
Wing chord at root 1·72 m (5 ft 8 in)
Wing chord at tip 0·85 m (2 ft 9 in)
Length overall 7·75 m (25 ft 5¼ in)
Height overall 2·80 m (9 ft 2½ in)
Tailplane span 3·55 m (11 ft 8 in)
Wheel track 2·78 m (9 ft 1¼ in)
Wheelbase 1·73 m (5 ft 8½ in)

WEIGHTS AND LOADINGS:
Weight empty 750 kg (1,653 lb)
Max T-O weight (Utility category) 1,225 kg (2,700 lb)
Max wing loading 91·7 kg/m² (18·8 lb/sq ft)
Max power loading 8·22 kg/kW (13·45 lb/hp)

PERFORMANCE (estimated, at max T-O weight):
Max level speed 167 knots (310 km/h; 193 mph)
Cruising speed 151 knots (280 km/h; 174 mph)
Stalling speed 55·5 knots (102 km/h; 64 mph)
Max rate of climb at S/L 270 m (885 ft)/min
Service ceiling 5,200 m (17,050 ft)
T-O run 260 m (855 ft)
Landing run 200 m (655 ft)
Max range 755 nm (1,400 km; 870 miles)

GENERAL AVIA F.600 CANGURO (KANGAROO)

A prototype of the F.600 Canguro is under construction; the fuselage and wings had been completed by July 1976.

TYPE: Twin-engined freight, ambulance and general utility transport.

WINGS: Cantilever high-wing monoplane. Wing section GAW-1, with 17% thickness/chord ratio. Dihedral 2°. Incidence (constant) 1° 30′. All-metal riveted structure in light alloy, with stressed skin. Centre-section has main spar and two auxiliary spars; outboard of engines, wings have two spars. All-metal ailerons and electrically-operated double-slotted flaps.

FUSELAGE: All-metal semi-monocoque structure, with stressed skin.

TAIL UNIT: Cantilever all-metal stressed-skin structure. Trim tabs in rudder and each elevator.

LANDING GEAR: Non-retractable tricycle type, with rubber shock-absorption.

POWER PLANT: Two 224 kW (300 hp) Lycoming IO-540-K flat-six engines, each driving a Hartzell fully-feathering constant-speed propeller. Fuel in four wing tanks, total capacity 900 litres (198 Imp gallons).

ACCOMMODATION: Crew of one or two. Cabin accommodates up to 10 passengers or paratroops, or four stretcher patients and two medical attendants, or 907 kg (2,000 lb) of freight. Forward door on each side for crew and passengers, and a third, wider door at rear on starboard side for freight loading.

DIMENSIONS, EXTERNAL:
Wing span 13·50 m (44 ft 3½ in)
Wing chord (constant) 1·60 m (5 ft 3 in)
Wing aspect ratio 8·5
Length overall 11·80 m (38 ft 8½ in)
Tailplane span 5·06 m (16 ft 7¼ in)
Rear cargo door width 1·44 m (4 ft 8¾ in)

DIMENSIONS, INTERNAL:
Cabin: Length 4·30 m (14 ft 1¼ in)
Width 1·23 m (4 ft 0½ in)
Height 1·27 m (4 ft 2 in)

AREAS:
Wings, gross 21·50 m² (231·42 sq ft)
Ailerons (total) 1·28 m² (13·78 sq ft)
Trailing-edge flaps (total) 2·32 m² (24·97 sq ft)
Fin 1·46 m² (15·72 sq ft)
Rudder, incl tab 0·90 m² (9·69 sq ft)
Tailplane 3·06 m² (32·94 sq ft)
Elevators (total, incl tabs) 2·50 m² (26·91 sq ft)

WEIGHTS AND LOADINGS:
Weight empty 1,600 kg (3,527 lb)
Max T-O weight 2,800 kg (6,173 lb)
Max wing loading 130 kg/m² (26·63 lb/sq ft)
Max power loading 6·25 kg/kW (10·29 lb/hp)

PERFORMANCE (estimated, at max T-O weight):
Max level speed at S/L
167 knots (310 km/h; 193 mph)
Max cruising speed (75% power)
146 knots (270 km/h; 168 mph)
Econ cruising speed (60% power)
129 knots (240 km/h; 149 mph)
Stalling speed, flaps down
57 knots (105 km/h; 65·5 mph)
Max rate of climb at S/L 402 m (1,319 ft)/min
Rate of climb at S/L, one engine out
114 m (374 ft)/min
Service ceiling 5,300 m (17,400 ft)
Service ceiling, one engine out 1,800 m (5,900 ft)
T-O run 275 m (902 ft)
Landing run 285 m (935 ft)

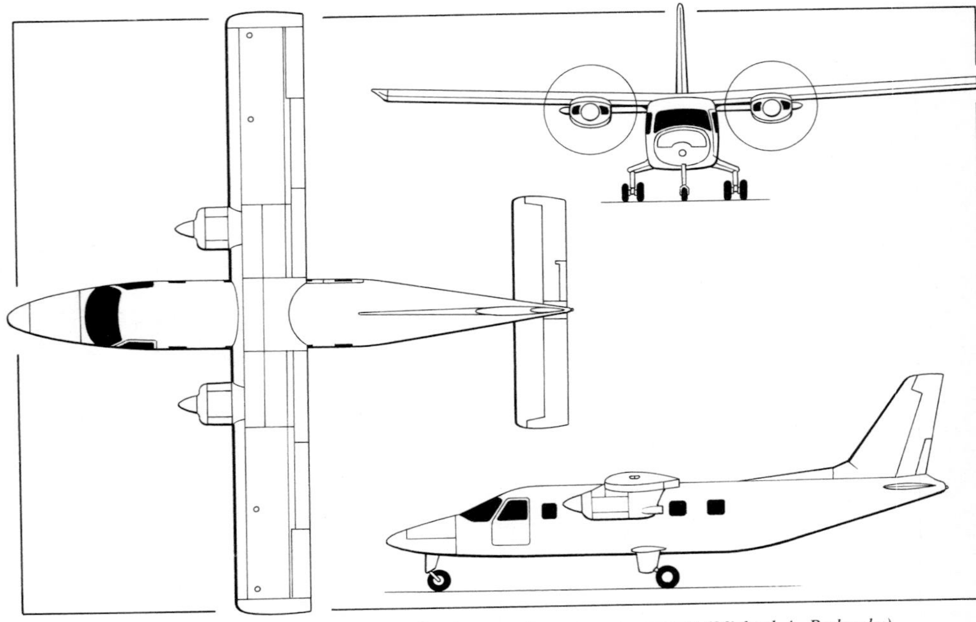

General Avia F.600 Canguro twin-engined general-purpose transport *(Michael A. Badrocke)*

ITALAIR
ITALAIR SpA

ADDRESS:
Via Santa Zita 1-11, 16129 Genoa
Telephone: (010) 589026

PRESIDENT AND MANAGING DIRECTOR:
Mario Del Bianco

DIRECTORS:
Dott Gr Uff Nando Benini
Dott Ing Giorgio Lucarelli
Dott Ing Benito Casinghini
Dott Ferruccio Lanata
Elfo Frignani
Giustino Meneghini

Dott Mazzocchi, a well-known Italian publisher, initiated the formation of Italair primarily to develop, build and market a range of light aircraft and twin-engined executive aircraft.

The first such design is the F.20 Pegaso.

ITALAIR F.20 PEGASO (PEGASUS)

The F.20 Pegaso is a six-seat light business twin. Its design, by General Avia, started in January 1970, and construction began in September 1970.

Two prototypes were built by General Avia, of which the first (I-GEAV) was flown for the first time on 21 October 1971 and the second (I-CBIE) on 11 August 1972. These were acquired subsequently by Italair. RAI Italian certification was granted on 19 November 1974, and FAA certification in the Spring of 1975.

No production of the Pegaso has yet been undertaken. A description of the second (modified) prototype appeared in the 1975-76 *Jane's*.

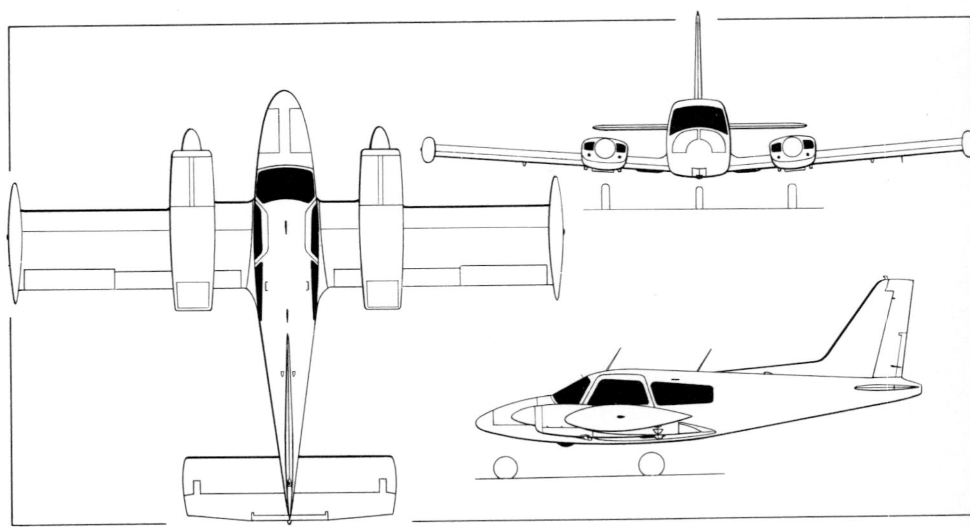

Italair F.20 Pegaso light twin-engined executive aircraft *(Pilot Press)*

MAGNI
VITTORIO MAGNI

ADDRESS:
Via Novara 72, 21017 Samarate (Varese)

Sr Magni built four small rotorcraft, of which the first and second were, respectively, single- and two-seat gyrogliders of the Bensen type. The third aircraft was powered by a Volkswagen engine. Sr Magni's fourth autogyro was displayed at the 1972 Turin Air Show and was described in the 1973-74 *Jane's*.

Sr Magni purchased from Mr Jukka Tervamäki of Finland (which see) the prototype, and all tooling, moulds and production rights, of the latter's JT-5 single-seat autogyro; this was to be marketed in Italy as the MT5.

In addition, Sr Magni has world marketing rights for rotor blades and other components of this aircraft, and rights to sell plans of the aircraft to amateur constructors in French-, Italian- and Spanish-speaking countries. For customers in other countries, plans are supplied by Mr Tervamäki. A description of the JT-5/MT5 appears under the Tervamäki heading in the Homebuilt Aircraft section of this edition.

MERIDIONALI
ELICOTTERI MERIDIONALI SpA

WORKS:
Via Giovanni Agusta 1, Frosinone
Telephone: (0775) 80841/2/3/4
Telex: Elmef 62377

CHAIRMAN AND PRESIDENT:
Conte Corrado Agusta

PRESIDENT, CHIEF EXECUTIVE OFFICER:
Dott Ing P. Fascione

GENERAL MANAGER:
Ing Edi Fieschi

This company was formed with assistance from Agusta (which see) and began to operate in October 1967. Initially, its activities consisted of overhauling helicopters of the Italian armed forces and other organisations, and the manufacture of helicopter components and sub-assemblies. In April 1968 it was announced that an agreement had been concluded with Boeing in the USA, whereby Meridionali acquired rights to the co-production, marketing and servicing of the Boeing Vertol CH-47C Chinook transport helicopter for customers in Italy, Austria, Switzerland and the Middle East. Italian production of the CH-47C is undertaken by SIAI-Marchetti, another member of the Agusta consortium.

Meridionali, whose works occupy a total area of more than 300,000 m² (3,229,170 sq ft), also participates in the manufacturing programme for the Agusta-Bell 206B JetRanger II helicopter (see under Agusta heading in this section).

PARTENAVIA
PARTENAVIA COSTRUZIONI AERONAUTICHE SpA

HEAD OFFICE AND WORKS:
Via Cava, CP 2179, 80026 Casoria (Naples)
Telephone: 596311 (PBX)
Telex: 77199 Partenav

PRESIDENT:
Prof Ing Luigi Pascale

DIRECTORS:
Dott G. Bulgari
Dott G. Fiore
Dott D. Marchiorello
Ing G. Regazzoni

PRODUCTION DIRECTOR:
Ing Nino Pascale

INTERNATIONAL MARKETING:
Ian A. Forbes

This company was founded in 1957 by Prof Ing Luigi Pascale and his brother, Ing Nino Pascale, and has since built a series of light aircraft designed by Prof Ing Pascale.

On 1 March 1974 the company moved from its small factory at Arzano, near Naples, to a 12,000 m² (129,165 sq ft) facility on Capodichino Airport, Naples, where it is now concentrating on production and development of various versions of the P.68 Victor twin-engined light aircraft. Under development also is the P.66C-150 Charlie, a successor to the P.64/P.66 Oscar series.

PARTENAVIA P.64 and P.66 OSCAR SERIES

Design of an improved version of the P.64 Oscar four-seat light aircraft (see 1966-67 *Jane's*) was started in November 1966. The prototype flew in the first half of 1967, and from it were developed the P.64B Oscar-180 (formerly Oscar-B) and Oscar-200, and the P.66B

Oscar-100 and Oscar-150. Production of these aircraft, which were described in the 1975-76 *Jane's,* has now ended. They have been succeeded by the P.66C-150 Charlie, a description of which follows.

PARTENAVIA P.66C-150 CHARLIE

This two/four-seat light aircraft is a progressive development of the P.64B/P.66B Oscar series (see 1975-76 *Jane's*), which it is intended to replace. A prototype had completed certification flight testing by February 1976.

Type: Two/four-seat light monoplane.

Wings: Braced high-wing monoplane with single streamline-section bracing strut each side. Wing section NACA 63 series. Thickness/chord ratio 15%. Dihedral 1° 30'. Incidence at root 1° 40'. No sweepback. Stressed-skin single-spar torsion-box structure of aluminium alloy, with glassfibre-reinforced leading-edges. Ailerons and manually-operated slotted trailing-edge flaps of similar construction to wings.

Fuselage: Forward portion, to rear of cabin, has a welded steel tube basic structure to which are attached light alloy skin panels. Rear fuselage is of conventional light alloy stressed-skin construction.

Tail Unit: Cantilever stressed-skin metal structure with sweptback vertical surfaces. All-moving tailplane in two symmetrical halves joined by steel cross-tube. Anti-balance tab in trailing-edge of tailplane, over 80% of span.

Landing Gear: Non-retractable tricycle type, with steerable nosewheel. Cantilever spring steel main legs. Oleo nosewheel shock-absorber. Cleveland main wheels type 40-28, with Pirelli tyres size 6·00-6. Goodyear nosewheel tyre size 5·00-5. Cleveland type 30-18 hydraulic disc brakes.

Power Plant: One 112 kW (150 hp) Lycoming O-360-A1A flat-four engine driving a Hoffmann HO 23-183.150 two-blade fixed-pitch propeller. Two fuel tanks in wing roots, total usable capacity 132 litres (29 Imp gallons). Refuelling points above wings. Oil capacity 7·5 litres (1·66 Imp gallons).

Accommodation: Enclosed cabin seating two or four persons in pairs; front seats are of the adjustable sliding type. Three forward-hinged doors: one by each front seat and on starboard side at rear. Baggage space aft of rear seats, with separate door on starboard side. Dual controls, heating, ventilation and soundproofing standard.

Electronics and Equipment: Optional items include full IFR instrumentation, Grimes rotating beacon, VHF radio, VOR and ADF.

Dimensions, external:
- Wing span: 9·99 m (32 ft 9¼ in)
- Wing chord (constant): 1·36 m (4 ft 5½ in)
- Wing aspect ratio: 7·45
- Length overall: 7·23 m (23 ft 8¾ in)
- Height overall: 2·77 m (9 ft 1 in)
- Tailplane span: 3·10 m (10 ft 2 in)
- Wheel track: 2·10 m (6 ft 10½ in)
- Propeller diameter: 1·88 m (6 ft 2 in)

Dimensions, internal:
- Cabin: Max width: 1·06 m (3 ft 5¾ in)
- Max height: 1·20 m (3 ft 11¼ in)

Areas:
- Wings, gross: 13·40 m² (144·2 sq ft)
- Ailerons (total): 1·29 m²(13·88 sq ft)
- Trailing-edge flaps (total): 1·71 m² (18·40 sq ft)
- Fin: 0·73 m² (7·86 sq ft)
- Rudder: 0·45 m² (4·84 sq ft)
- Tailplane, incl tab: 2·17 m² (23·36 sq ft)

Weights and Loadings:
- Weight empty: 610 kg (1,345 lb)
- Max T-O weight: 990 kg (2,183 lb)
- Max wing loading: 73·9 kg/m² (15·14 lb/sq ft)
- Max power loading: 8·84 kg/kW (14·55 lb/hp)

Performance (at max T-O weight):
- Max level speed at S/L: 130 knots (241 km/h; 150 mph)
- Max cruising speed (75% power) at 2,150 m (7,000 ft): 118 knots (218 km/h; 135 mph)
- Stalling speed, flaps down: 45 knots (83·5 km/h; 52 mph)
- Max rate of climb at S/L: 275 m (900 ft)/min
- Service ceiling: 4,420 m (14,500 ft)
- T-O run: 245 m (805 ft)
- Landing run: 120 m (395 ft)
- Range with max fuel (65% power): 458 nm (848 km; 527 miles)

PARTENAVIA P.68B VICTOR

Developed from the P.68, which was designed by Prof Ing Luigi Pascale in 1968 and described in the 1975-76 *Jane's,* the P.68B Victor twin-engined light transport has been in production in Partenavia's factory at Naples Airport since the Spring of 1974. By March 1976 more than 50 aircraft had been delivered to operators in Australia, Belgium, Denmark, England, Finland, France, Germany, Holland, Israel, Morocco, Norway, Sweden, the United States and Venezuela. By mid-1976, the annual production capacity had been almost entirely committed, and line positions extended well into 1977.

Partenavia is developing from the P.68B a series of aircraft of the same basic configuration, using many common components. Announced variants include the P.68R (retractable landing gear), P.68T (turbocharged engines) and P.68RT (retractable landing gear and turbocharged engines), which are described separately.

The following description applies to the standard P.68B:

Type: Six-seat light transport and trainer.

Wings: Cantilever high-wing monoplane. Wing section NACA 63 series. Thickness/chord ratio 15%. Dihedral 1°. Incidence 1° 30'. No sweepback. Stressed-skin single-spar torsion-box structure, of aluminium alloy except for area forward of main spar (30% of total area) which is of glassfibre-reinforced plastics. All-metal ailerons and electrically-operated single-slotted trailing-edge flaps. Hoerner wingtips. Trim tab on port aileron.

Fuselage: Conventional all-metal semi-monocoque structure of frames and longerons with four main longerons and stressed-skin covering. Fuselage/wing intersection mainly of glassfibre-reinforced plastics.

Tail Unit: Cantilever stressed-skin metal structure. All-moving tailplane, in two symmetrical halves joined by steel cross-tube and of constant chord except for increase at leading-edge roots. Balance tab in tailplane trailing-edge, over 80% of span. Sweptback fin and rudder, with small dorsal fin. Trim tab in rudder.

Landing Gear: Initial production version has non-retractable tricycle type, with steerable nosewheel. Cantilever spring steel main legs. Oleo-pneumatic shock-absorber on nosewheel. Cleveland main wheels, type 40-96, with Pirelli tyres size 6·00-6. Goodyear nosewheel tyre, size 5·00-5. Cleveland type 30-61 hydraulic disc brakes. Streamlined wheel fairings optional.

Power Plant: Two 149 kW (200 hp) Lycoming IO-360-A1B6 flat-four engines, each driving a Hartzell HC-C2YK-2C/C-7666A-4 two-blade variable-pitch constant-speed fully-feathering propeller. Integral fuel tank in each outer wing, total capacity 410 litres (90 Imp gallons). Two optional long-range tanks, each of 40 litres (8·8 Imp gallons) capacity. Refuelling point above each wing. Oil capacity 15 litres (3·3 Imp gallons).

Accommodation: Seating for six persons in cabin, including pilot, in three rows of two, with space for baggage aft of rear pair. Optional bench seat, accommodating three persons, in place of rear seats. Front seats are of the adjustable sliding type. Access to all seats via large car-type door on port side of cabin. Access to baggage compartment via separate large door on starboard side. Two stretchers can be carried when all passenger seats are removed. Dual controls, cabin heating, ventilation and soundproofing standard.

Systems and Equipment: Optional equipment includes full IFR instrumentation, two 360-channel VHF transceivers, two navigation receivers, VOR/ILS, marker beacon, ADF and DME. Electrical power supplied by two 24V 50A alternators. Bendix FCS 810 autopilot system and Goodrich pneumatic de-icing system optional.

Dimensions, external:
- Wing span: 12·00 m (39 ft 4½ in)
- Wing chord (constant): 1·55 m (5 ft 1 in)
- Wing aspect ratio: 7·74
- Length overall: 9·35 m (30 ft 8 in)
- Height overall: 3·40 m (11 ft 1¾ in)
- Tailplane span: 3·90 m (12 ft 9½ in)
- Wheel track: 2·40 m (7 ft 10½ in)
- Wheelbase: 3·50 m (11 ft 5¾ in)
- Propeller diameter: 1·88 m (6 ft 2 in)
- Baggage door, stbd:
  - Height: 0·80 m (2 ft 7½ in)
  - Width: 0·80 m (2 ft 7½ in)

Dimensions, internal:
- Cabin:
  - Length: 4·00 m (13 ft 1½ in)
  - Max width: 1·16 m (3 ft 9½ in)
  - Max height: 1·21 m (3 ft 11½ in)
- Baggage space: 0·57 m³ (20 cu ft)

Areas:
- Wings, gross: 18·60 m² (200 sq ft)
- Ailerons (total): 1·79 m² (19·27 sq ft)
- Trailing-edge flaps (total): 2·37 m² (25·51 sq ft)
- Fin: 1·59 m² (17·11 sq ft)
- Rudder, incl tab: 0·44 m² (4·74 sq ft)
- Tailplane, incl tab: 4·41 m² (47·47 sq ft)

Weights and Loadings:
- Weight empty: 1,260 kg (2,778 lb)
- Max T-O weight: 1,960 kg (4,321 lb)
- Max landing weight: 1,860 kg (4,100 lb)
- Max wing loading: 105 kg/m² (21·6 lb/sq ft)
- Max power loading: 6·58 kg/kW (10·80 lb/hp)

Performance (at max T-O weight):
- Max level speed at S/L: 173 knots (320 km/h; 199 mph)
- Max cruising speed (75% power) at 1,675 m (5,500 ft): 163 knots (302 km/h; 188 mph)

Partenavia P.68B Victor six-seat light aircraft (two Lycoming IO-360 engines) *(Martin Fricke)*

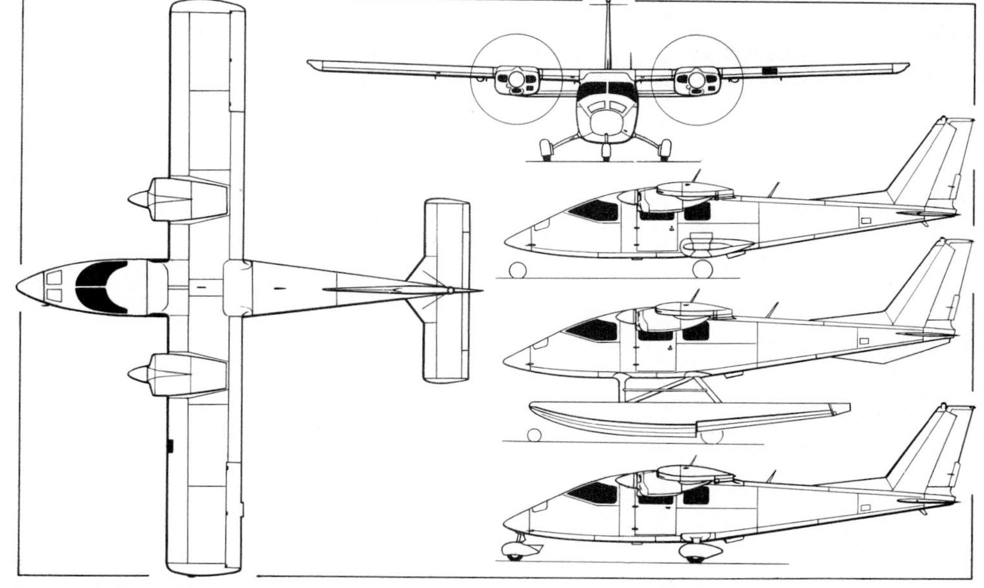

Partenavia P.68B Victor, with additional side views of P.68R with retractable landing gear (top) and P.68 amphibian (centre) *(Pilot Press)*

Cruising speed (65% power) at 2,440 m (8,000 ft)
158 knots (293 km/h; 182 mph)
Cruising speed (55% power) at 3,200 m (10,500 ft)
146 knots (270 km/h; 168 mph)
Stalling speed, flaps up
64 knots (119 km/h; 74 mph)
Stalling speed, flaps down
56 knots (104 km/h; 64·5 mph)
Max rate of climb at S/L 424 m (1,390 ft)/min
Service ceiling 6,100 m (20,000 ft)
Service ceiling, one engine out 2,135 m (7,000 ft)
T-O run 278 m (912 ft)
T-O to 15 m (50 ft) 469 m (1,539 ft)
Landing from 15 m (50 ft) 395 m (1,295 ft)
Landing run 250 m (820 ft)
Max range at max cruising speed
825 nm (1,528 km; 950 miles)
Max range at econ cruising speed
907 nm (1,681 km; 1,045 miles)

PARTENAVIA P.68R

A prototype of this retractable landing gear version of the P.68 was expected to fly for the first time in the Summer of 1976, with certification following later in the year. The nosewheel retracts rearward, the main wheels inward into fairings on the fuselage sides.

PERFORMANCE (estimated, at AUW of 1,960 kg; 4,321 lb):
Max level speed at S/L
186 knots (344 km/h; 214 mph)
Cruising speed:
75% power at 1,675 m (5,500 ft)
174 knots (322 km/h; 200 mph)
65% power at 2,440 m (8,000 ft)
169 knots (313 km/h; 195 mph)
55% power at 3,200 m (10,500 ft)
157 knots (291 km/h; 181 mph)
Max rate of climb at S/L 453 m (1,485 ft)/min
Rate of climb at S/L, one engine out
110 m (361 ft)/min
Service ceiling 6,705 m (22,000 ft)
Service ceiling, one engine out 2,925 m (9,600 ft)
T-O run 278 m (912 ft)
T-O to 15 m (50 ft) 469 m (1,539 ft)
Landing from 15 m (50 ft) 417 m (1,369 ft)
Landing run 262 m (861 ft)
Optimum cruising range, no reserves:
75% power 898 nm (1,664 km; 1,034 miles)
65% power 872 nm (1,615 km; 1,004 miles)
55% power 1,055 nm (1,955 km; 1,215 miles)

PARTENAVIA P.68RT

This projected variant of the P.68B combines the retractable landing gear of the P.68R and the turbo-charged engines of the P.68T.

PERFORMANCE (estimated):
Max level speed at S/L 218 knots (404 km/h; 251 mph)
Cruising speed (75% power) at 6,100 m (20,000 ft)
204 knots (378 km/h; 235 mph)
Cruising speed (65% power) at 5,490 m (18,000 ft)
190 knots (352 km/h; 219 mph)
Stalling speed, flaps up 66 knots (122 km/h; 76 mph)
Stalling speed, flaps down
58 knots (108 km/h; 67 mph)
Max rate of climb at S/L 449 m (1,472 ft)/min
Max rate of climb at S/L, one engine out
106 m (348 ft)/min
Service ceiling 9,450 m (31,000 ft)
Service ceiling, one engine out 4,570 m (15,000 ft)
T-O run 260 m (853 ft)
T-O to 15 m (50 ft) 455 m (1,491 ft)
Landing from 15 m (50 ft) 417 m (1,369 ft)
Landing run 262 m (861 ft)

Prototype Partenavia P.68 Observer, modified from a standard P.68B by Sportavia-Pützer in Germany

PARTENAVIA P.68T

This variant of the P.68B is fitted with two Lycoming TIO-360 turbocharged engines, each rated at 156·5 kW (210 hp) from S/L to 4,260 m (14,000 ft). A prototype, under construction in early 1976, was expected to fly later in the same year.

PERFORMANCE (estimated):
Max level speed at S/L 202 knots (375 km/h; 233 mph)
Cruising speed (75% power) at 6,100 m (20,000 ft)
189 knots (351 km/h; 218 mph)
Cruising speed (65% power) at 5,490 m (18,000 ft)
158 knots (293 km/h; 182 mph)
Stalling speed, flaps up 64 knots (119 km/h; 74 mph)
Stalling speed, flaps down
56 knots (104 km/h; 64·5 mph)
Max rate of climb at S/L 452 m (1,482 ft)/min
Max rate of climb at S/L, one engine out
100 m (329 ft)/min
Service ceiling 9,150 m (30,000 ft)
Service ceiling, one engine out 3,810 m (12,500 ft)
T-O run 244 m (802 ft)
T-O to 15 m (50 ft) 396 m (1,299 ft)
Landing from 15 m (50 ft) 395 m (1,295 ft)
Landing run 250 m (820 ft)

PARTENAVIA P.68 FLOATPLANE/AMPHIBIAN

Development of the P.68 floatplane/amphibious version of the P.68B was undertaken in conjunction with Société d'Etudes et Fabrications Aéronautiques of France, which designed and built the floats. A prototype was undergoing certification trials in early 1976; these were expected to be completed later in the same year.

Apart from the float installation, the basic aircraft is a P.68B (which see). Payload is about 544 kg (1,200 lb), some 160 kg (350 lb) less than that of the P.68B.

PERFORMANCE (estimated):
Max level speed at S/L 148 knots (274 km/h; 170 mph)
Cruising speed (75% power) at 1,675 m (5,500 ft)
140 knots (259 km/h; 161 mph)
Cruising speed (65% power) at 2,745 m (9,000 ft)
136 knots (253 km/h; 157 mph)
Max rate of climb at S/L 335 m (1,100 ft)/min
Service ceiling 4,875 m (16,000 ft)
Service ceiling, one engine out 762 m (2,500 ft)
T-O run 335 m (1,100 ft)

PARTENAVIA P.68 OBSERVER

In collaboration with Sportavia-Pützer, a member of the VFW-Fokker group and Partenavia's West German distributor, a special version of the P.68 known as the Observer is being developed, with a forward and downward visibility for the crew equal to that of a helicopter.

The new Plexiglas nose, cockpit and associated structure were designed by Sportavia-Pützer, and the prototype was constructed at that company's Dahlemer-Binz factory. Production P.68 Observers will be manufactured by Partenavia at Naples.

With its good low-speed handling characteristics, the Observer is considered to be capable of performing many roles allocated normally to helicopters. It is intended particularly for patrol and observation operations.

PARTENAVIA P.70 ALPHA

The Alpha programme has been discontinued. A full description and illustrations of the prototype can be found in the 1975-76 *Jane's*.

PIAGGIO
INDUSTRIE AERONAUTICHE E MECCANICHE RINALDO PIAGGIO SpA

HEAD OFFICE:
Viale Brigata Bisagno 14, Casella Postale 1396, 16121 Genoa
Telephone: 540.521
Telex: 27695 AERPIAG
BRANCH OFFICE:
Via A. Gramsci 34, Rome
WORKS:
Genoa-Sestri P. (Aircraft Division)
Finale Ligure (Aero-Engine Division)
CHAIRMAN:
Ing Armando Piaggio
VICE-PRESIDENT AND CHIEF EXECUTIVE OFFICER:
Dott Lucio Lotti
MANAGING DIRECTOR:
Dott Rinaldo Piaggio
DIRECTOR GENERAL:
Ing Umberto Barnato
TECHNICAL CONSULTANT:
Ing Giovanni P. Casiraghi
TECHNICAL DIRECTOR:
Ing Alessandro Mazzoni
MARKETING DIRECTOR:
Commander G. B. Pizzinato

Piaggio began the construction of aeroplanes in its Genoa-Sestri plant in 1916, and later in the Finale Ligure works. A second large plant covering approx 40,000 m² (430,000 sq ft) was built in 1968 at Genoa-Sestri, on the edge of Genoa's Christopher Columbus Airport, for assembly and overhaul of the P.166 twin-engined aircraft and of the PD-808 light twin-jet utility aircraft.

The present company was formed on 29 February 1964, and has since operated as an independent concern. It employs about 1,300 people.

The company is organised into two production Divisions: Aircraft Division at Genoa-Sestri, and Aero-Engine Division, the activities of which are described in the appropriate section of this edition.

R. PIAGGIO P.166-DL2

The P.166 is a twin-engined light transport aircraft, with standard accommodation for 6/8 passengers. An alternative layout offers seats for 12 passengers, and the aircraft can be used for a variety of different duties.

The P.166 has been produced or is available in several basic versions. Of these, the original P.166 production version (32 built) was described in the 1963-64 *Jane's*; the military P.166M (51 built), the P.166B Portofino (5 built) and P.166C (2 built) in the 1971-72 *Jane's*; and the P.166S search and surveillance version (20 built) in the 1974-75 *Jane's*. Production of these versions has been completed.

The current piston-engined version is the P.166-DL2, which superseded the P.166B as a standard commercial version, with increased fuel capacity in integral wingtip tanks and increased max T-O weight.

A demonstration model, known as the P.166-BL2, flew for the first time on 2 May 1975, after which it was planned to build eight P.166-DL2 production examples. Four of these will be equipped as photogrammetric aircraft (each fitted with two cameras), and all will have a new-design extra-large cabin door and an emergency exit.

The turboprop-powered P.166-DL3 is described separately. The following description applies to the P.166-DL2:

TYPE: Twin-engined light transport.

WINGS: Shoulder gull-wing cantilever monoplane. NACA 230 wing section. Dihedral 21° 30′ on inner portion, 2° 8′ on outer wings. Incidence 2° 43′. Aluminium alloy flush-riveted structure. All-metal slotted ailerons, with geared and trim tab in starboard aileron. All-metal slotted flaps. Rubber-boot de-icing of leading-edges optional.

FUSELAGE: Aluminium alloy flush-riveted semi-monocoque structure.

TAIL UNIT: Cantilever aluminium alloy flush-riveted structure. Geared and trim tabs in elevators; trim tab in rudder.

LANDING GEAR: Retractable tricycle type. Magnaghi oleo-pneumatic shock-absorbers. Hydraulic retraction. Nosewheel retracts rearward, main units outward. Goodyear main wheels with size 25·65 × 8·70-10 tyres, pressure 3·59 bars (52 lb/sq in). Goodyear nosewheel with size 17·5 × 6·30-6 tyre, pressure 2·90 bars (42 lb/sq in). Goodyear or Magnaghi hydraulic brakes.

POWER PLANT: Two 283 kW (380 hp) Lycoming IGSO-540-A1H flat-six engines, each driving a Hartzell type HC-BZ30-2 BL/L10151-8 three-blade feathering constant-speed pusher propeller. Fuel in two internal tanks in outer wings (each 212 litres; 46·6 Imp gallons), and two external wingtip tanks (each 323 litres; 71 Imp gallons). Total fuel capacity 1,070 litres (235·2 Imp gallons). Total oil capacity 34 litres (7·5 Imp gallons).

ACCOMMODATION: Standard seating for pilot and five passengers in individual seats, with toilet and bar at rear. Alternative layouts include an 8-seat executive version, with two rows of three seats in cabin, facing each other; high-density 10-seat version with three individual seats on each side of central aisle in cabin, and curved rear twin seats in place of toilet and bar; cargo version with stripped cabin; ambulance and air survey versions. Main door in centre of cabin on port side. Separate door to flight deck on each side. Emergency exit forward of wing on starboard side. Outside door to baggage compartment aft of cabin on port side. Dual controls standard.

SYSTEMS: Hydraulic system, pressure 127 bars (1,840 lb/sq in), operates landing gear, flaps and brakes. 28V engine-driven DC generator for electrical system.

ELECTRONICS AND EQUIPMENT: Optional equipment includes VHF radio, VOR, ADF, marker beacon receiver, glideslope receiver and Sperry SPL 45 or Collins AP 106 autopilot and horizon gyro unit.

DIMENSIONS, EXTERNAL:

Wing span:	
without tip-tanks	13·51 m (44 ft 4 in)
with tip-tanks	14·69 m (48 ft 2½ in)
Wing chord at root	2·40 m (7 ft 10½ in)
Wing chord at tip	1·15 m (3 ft 9¼ in)
Wing aspect ratio	7·3
Length overall	11·90 m (39 ft 3 in)
Height overall	5·00 m (16 ft 5 in)
Tailplane span	5·10 m (16 ft 9 in)
Wheel track	2·66 m (8 ft 9 in)
Wheelbase	4·71 m (15 ft 5½ in)
Cabin door:	
Height	1·38 m (4 ft 6 in)
Width	1·28 m (4 ft 2 in)

DIMENSIONS, INTERNAL:

Cabin: Length	3·55 m (11 ft 8 in)
Max width	1·57 m (5 ft 2 in)
Max height	1·76 m (5 ft 9 in)
Floor area	5·14 m² (55·3 sq ft)
Volume	6·63 m³ (234·1 cu ft)
Baggage compartment (front)	0·77 m³ (27·2 cu ft)
Baggage compartment (rear)	1·80 m³ (63·6 cu ft)

AREAS:

Wings, gross	26·56 m² (285·9 sq ft)
Ailerons (total)	1·95 m² (21·00 sq ft)
Trailing-edge flaps (total)	2·38 m² (25·60 sq ft)
Fin	1·62 m² (17·44 sq ft)
Rudder, incl tab	1·23 m² (13·24 sq ft)
Tailplane	3·50 m² (37·67 sq ft)
Elevators, incl tabs	1·29 m² (13·88 sq ft)

WEIGHTS AND LOADINGS:

Basic weight empty	2,250 kg (4,960 lb)
Max payload	1,166 kg (2,571 lb)
Max T-O weight	4,100 kg (9,039 lb)
Max zero-fuel weight	3,800 kg (8,377 lb)
Max landing weight	3,800 kg (8,377 lb)
Max wing loading	154·4 kg/m² (31·61 lb/sq ft)
Max power loading	7·24 kg/kW (11·89 lb/hp)

PERFORMANCE (at max T-O weight, except where indicated otherwise):

Never-exceed speed	256 knots (476 km/h; 295 mph)
*Max level speed at 3,450 m (11,300 ft)	214 knots (396 km/h; 246 mph)
*Max cruising speed (75% power) at 4,550 m (15,000 ft)	194 knots (395 km/h; 223 mph)
*Econ cruising speed (45% power) at 4,550 m (15,000 ft)	153 knots (283 km/h; 176 mph)
Stalling speed, flaps and wheels down, at max landing weight	57·5 knots (106 km/h; 66 mph)
Max rate of climb at S/L	430 m (1,415 ft)/min
Service ceiling	8,230 m (27,000 ft)
Service ceiling, one engine out	3,505 m (11,500 ft)
Min ground turning radius	6·00 m (19 ft 8¼ in)
Runway LCN	4·5
T-O run, FAR 23	402 m (1,320 ft)
T-O to 15 m (50 ft), FAR 23	588 m (1,930 ft)
*Landing from 15 m (50 ft), FAR 23	427 m (1,400 ft)
*Landing run, FAR 23	280 m (920 ft)
Range with max fuel, no reserves, at 3,050 m (10,000 ft)	1,300 nm (2,410 km; 1,500 miles)
Range with max payload, no reserves	337 nm (626 km; 388 miles)

*at intermediate flying weight of 3,500 kg (7,715 lb)

R. Piaggio P.166-BL2 demonstration aircraft, predecessor of the production P.166-DL2 *(Martin Fricke)*

PIAGGIO P.166-DL3

This aircraft is generally similar to the P.166-DL2 (which see) but is powered by two Lycoming turboprop engines. The basic structure, flying controls and systems are identical to those of the DL2, except for changes associated with the installation of the turboprop engines. The prototype made its first flight on 3 July 1976.

POWER PLANT: Two Avco Lycoming LTP 101 turboprop engines, each developing 438 kW (587 shp) at take-off, and each driving a Hartzell HC-B3TN-3/T10282-95 three-blade feathering constant-speed pusher propeller. Internal fuel capacity as for P.166-DL2. Two optional underwing tanks (each 284 litres; 62·5 Imp gallons). Total fuel capacity with normal and optional tanks 1,638 litres (360·2 Imp gallons).

DIMENSIONS AND AREAS: As for P.166-DL2

WEIGHTS AND LOADINGS:

Basic weight empty	2,126 kg (4,688 lb)
Max payload	1,306 kg (2,680 lb)
Max T-O weight	4,300 kg (9,480 lb)
Max zero-fuel weight	3,800 kg (8,377 lb)
Max landing weight	3,800 kg (8,377 lb)
Max wing loading	162 kg/m² (33·2 lb/sq ft)
Max power loading	4·91 kg/kW (8·07 lb/hp)

PERFORMANCE (at AUW of 3,855 kg; 8,500 lb except where indicated):

Max level speed at 3,050 m (10,000 ft)	225 knots (417 km/h; 259 mph)
Max cruising speed at 3,050 m (10,000 ft)	218 knots (404 km/h; 250 mph)
Econ cruising speed at 3,050 m (10,000 ft)	162 knots (300 km/h; 186 mph)
Stalling speed, flaps and wheels down, at max landing weight	63 knots (117 km/h; 72 mph)
Max rate of climb at S/L	549 m (1,800 ft)/min
Rate of climb, one engine out	198 m (650 ft)/min
Service ceiling	7,465 m (24,500 ft)
Service ceiling, one engine out	4,267 m (14,000 ft)
T-O run	274 m (900 ft)
T-O to 15 m (50 ft)	500 m (1,640 ft)
Landing from 15 m (50 ft), at AUW of 3,400 kg (7,500 lb) and using reverse thrust	300 m (984 ft)
Range with normal fuel tanks, 30 min reserves	930 nm (1,723 km; 1,070 miles)
Range with normal and optional tanks, 30 min reserves	1,450 nm (2,687 km; 1,667 miles)
Range with max payload, 30 min reserves	400 nm (741 km; 460 miles)

R. PIAGGIO PD-808 526

The PD-808 is a 6/10-seat light jet utility aircraft which was intended for both civil and military use. The Italian Defence Ministry assisted Piaggio by purchasing the two prototypes (powered by Viper Mk 525 engines) and by providing test facilities. The first prototype made its first flight on 29 August 1964.

In mid-1965, the Italian Defence Ministry ordered a number of production PD-808s. These aircraft have more powerful Viper Mk 526 engines and differ from the original prototype in having larger tip-tanks, a longer dorsal fin and forward-sliding nose fairing.

Production of the PD-808 526 has ended. Full details can be found in the 1975-76 *Jane's*, together with data for a version with turbofan engines, the **PD-808 TF**, of which Piaggio has completed design studies.

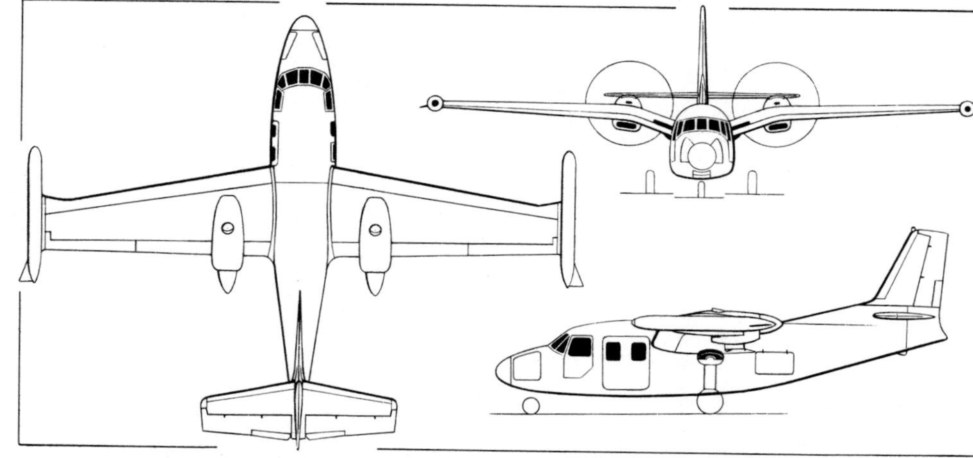

R. Piaggio P.166-DL3 twin-turboprop light transport *(Pilot Press)*

PROCAER

PROGETTI COSTRUZIONI AERONAUTICHE SpA

HEAD OFFICE:
Via Cardinale Ascanio Sforza 85, 20141 Milan
Telephone: 84.93.797

WORKS:
Strada Alzaia Naviglio Pavese 78, Milan

PRESIDENT: Dott Ing Rico Neeff

Production by this company in recent years was concentrated on various versions of the Picchio F15 four-seat light aircraft. The latest Procaer-built version was the all-metal F15E, a description and illustration of which appeared in the 1974-75 *Jane's*.

The F15E prototype (I-PROM) flew for the first time on 21 December 1968, and was certificated by the RAI and FAA on 6 November 1970 and 16 July 1971 respectively. A second prototype, embodying modifications, was undergoing flight tests in 1976 and was the subject of a tender to the Aero Club d'Italia.

A prototype of a developed version, known as the **F15F**, has been built by General Avia (which see) and was due to fly during 1976.

SIAI-MARCHETTI
SIAI-MARCHETTI SOCIETA PER AZIONI

MANAGEMENT AND WORKS:
Via Indipendenza 2, 21018 Sesto Calende (Varese)
Telephone: (0331) 924421
Telex: 39601 Siaiavio
AERODROME: Vergiate (Varese)
ROME OFFICE: Via Barberini 36, 00187 Rome
Telephone: (06) 482811
PRESIDENT:
Dott Ermenegildo Marelli
MANAGING DIRECTOR:
Dott Ing Giovanni Angiola

Founded in 1915, the SIAI-Marchetti company was known originally as Savoia-Marchetti. It has produced a wide range of military and civil landplanes and flying-boats, but is becoming increasingly involved in helicopter manufacture, in association with Agusta and Elicotteri Meridionali. Its current products include civil and military light aircraft of its own design or development, and Boeing Vertol CH-47 helicopters, built under licence. Work on licence-built Sikorsky helicopters was planned to begin in 1975.

Siai-Marchetti is engaged on the overhaul and repair of various types of large aircraft (notably the C-119 and C-130) serving with the Italian Air Force. It also participates in national or multi-national programmes for the Aeritalia G91Y and G222, Aeritalia (Lockheed) F-104S and Panavia Tornado.

SIAI-Marchetti SF.260AMI (SF.260MX series) two/three-seat military trainer for the Italian Air Force

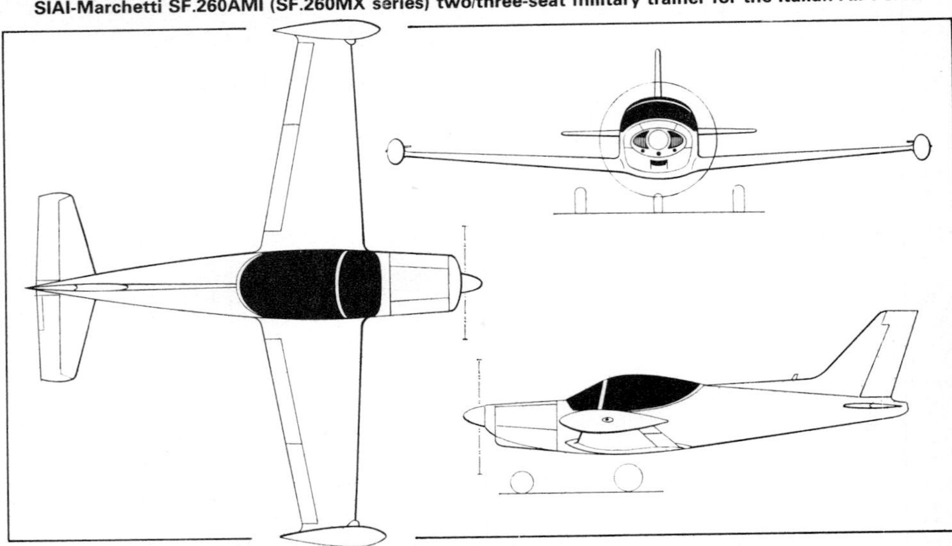

SIAI-Marchetti SF.260MX military trainer (Lycoming O-540-E4A5 engine) *(Roy J. Grainge)*

SIAI-MARCHETTI SF.260MX

The SF.260MX was developed from the basic SF.260 (which see) specifically for military duties, and was first flown on 10 October 1970.

Details of the 138 aircraft ordered by 1975 were given in the 1975-76 *Jane's*. Currently available versions are the SF.260M for training duties, SF.260W for training and tactical support, and the SF.260SW for surveillance duties.

The following details apply to the SF.260M, which incorporates a number of important structural modifications compared with the SF.260. The SF.260W and SF.260SW are described separately.

TYPE: Two/three-seat military training aircraft.

WINGS: Cantilever low-wing monoplane. Wing section NACA 64_1-212 (modified) at root, NACA 64_1-210 at tip. Thickness/chord ratio 13% at root, 10% at tip. Dihedral 6° 20′. Incidence 2° 45′ at root, 0° at tip. No sweepback. Increased wing leading-edge radius compared with basic SF.260, with lower datum line, to improve stall characteristics. All-metal light alloy structure, with single main spar and auxiliary rear spar, built in two portions bolted together at centreline and attached to fuselage by six bolts. Press-formed ribs, with dimpled stiffening holes. Skin, which is butt-jointed and flush-riveted, is stiffened by stringers between main and rear spars. Differentially-operating Frise-type light alloy mass-balanced ailerons (travel 24° up, 13° down), and electrically-actuated light alloy single-slotted flaps (max travel 50°). Flaps and ailerons operated by pushrods and cables. Ground-adjustable tab on each aileron.

FUSELAGE: Semi-monocoque structure of frames and stringers, exclusively of light alloy except for welded steel tube engine mounting, glassfibre front panel of engine cowling, and detachable glassfibre tailcone.

TAIL UNIT: Cantilever light alloy structure, with swept-back vertical surfaces (approx 20 per cent greater in area than those of basic SF.260), fixed-incidence tailplane and one-piece balanced elevator. Two-spar fin and tailplane, bolted to fuselage; single-spar elevator and balanced rudder. Reinforced tail unit/fuselage joints compared with basic SF.260. Rudder (30° travel to left or right) and elevator (travel 24° up, 16° down) operated by cables. Controllable trim tab in starboard half of elevator; ground-adjustable tab on rudder.

LANDING GEAR: Electrically-retractable tricycle type, with mechanical standby for emergency use. Small tail bumper under rear fuselage. Inward-retracting main wheels and rearward-retracting nosewheel each embody Magnaghi oleo-pneumatic shock-absorber (type 2/22028 on main units). Cleveland P/N 3080A main wheels, with size 6·00-6 tube and tyre (6-ply rating), pressure 2·45 bars (35·5 lb/sq in). Cleveland P/N 40-77A nosewheel, with size 5·00-5 tube and tyre (6-ply rating), pressure 1·96 bars (28·4 lb/sq in). Cleveland P/N 3000-500 independent hydraulic single-disc brake on each main wheel. Nosewheel steering (20° to left or right) is operated directly by the rudder pedals, to which it is linked by pushrods. Up-lock secures main gear in retracted position during flight; anti-retraction system prevents main gear from retracting whenever strut is compressed by weight of aircraft. Compared with basic SF.260, the SF.260MX has a reinforced nosewheel drag brace attachment and landing gear retraction supports; increased use of light alloy forgings, instead of welded steel, in certain landing gear structures; and improved retraction locking mechanism.

POWER PLANT: One 194 kW (260 hp) Lycoming O-540-E4A5 flat-six aircooled engine, driving a Hartzell HC-C2YK-1B/8477-8R two-blade constant-speed metal propeller with spinner. Fuel in two internal light alloy wing tanks, each of 49·5 litres (10·9 Imp gallons), and two wingtip tanks, each of 72 litres (15·85 Imp gallons) capacity. Total fuel capacity 243 litres (53·5 Imp gallons), of which 235 litres (51·7 Imp gallons) are usable. Individual refuelling point for each tank. Oil capacity 10 kg (22 lb).

ACCOMMODATION: Side-by-side front seats for instructor and pupil, with third seat centrally at rear. Front seats are individually adjustable fore and aft, and have forward-folding backs and provision for back-type parachute packs. All three seats equipped with aerobatic-type safety belts. Baggage compartment aft of rear seat. One-piece fully-transparent rearward-sliding Plexiglas canopy. Emergency canopy ejection system, instead of the rubber-cord canopy release of the basic SF.260. Steel tube windscreen frame, for protection in the event of an overturn. Cabin is carpeted, air-conditioned, heated and ventilated, and walls are thermally insulated and soundproofed by a glassfibre lining. Dual controls standard.

SYSTEMS: Hydraulic equipment for main-wheel brakes only. No pneumatic system. 24V DC electrical system of single-conductor type, including 24V 70A Prestolite engine-mounted alternator/rectifier and 24V 25Ah Varley battery, for engine starting, flap and landing gear actuation, fuel booster pumps, electronics and lighting. Sealed battery compartment in rear of fuselage on port side. External power receptacle on port side at rear. Connection of an external power source automatically disconnects the battery. Heating system for carburettor air intake. Emergency electrical system for extending landing gear if normal electrical actuation fails; provision for mechanical extension in the event of total electrical failure. Cabin heating, and windscreen de-icing and demisting, by heat exchanger using engine exhaust air. Additional manually-controlled warm-air outlets for general cabin heating.

ELECTRONICS AND EQUIPMENT: Basic instrumentation and military equipment to customer's requirements. Blind-flying instrumentation and communications equipment optional. Landing light in nose, below spinner. Instrument panel can be slid rearward to provide access to rear of instruments. Compared with basic SF.260, the SF.260MX has various improvements to flight controls, engine controls (duplicated propeller and throttle controls), electrical system, radio, and other equipment installations.

DIMENSIONS, EXTERNAL:

Wing span	8·25 m (27 ft 0¾ in)
Wing span over tip-tanks	8·35 m (27 ft 4¾ in)
Wing chord at root	1·60 m (5 ft 3 in)
Wing chord at tip	0·784 m (2 ft 6¾ in)
Wing mean aerodynamic chord	1·325 m (4 ft 4¼ in)
Wing aspect ratio (without tip-tanks)	6·33
Wing taper ratio	2·24
Length overall	7·10 m (23 ft 3½ in)
Length of fuselage	5·10 m (16 ft 8¾ in)
Fuselage: Max width	1·10 m (3 ft 7¼ in)
Max depth	1·042 m (3 ft 5 in)
Height overall	2·41 m (7 ft 11 in)
Tailplane span	3·01 m (9 ft 10½ in)
Wheel track	2·274 m (7 ft 5½ in)
Wheelbase	1·66 m (5 ft 5¼ in)
Propeller diameter	1·93 m (6 ft 4 in)
Min propeller ground clearance	0·20 m (8 in)

DIMENSIONS, INTERNAL:

Cabin: Length	1·66 m (5 ft 5¼ in)
Max width	1·00 m (3 ft 3¼ in)
Height (seat squab to canopy)	0·92 m (3 ft 0¼ in)
Volume	1·50 m³ (53 cu ft)
Baggage compartment volume	0·18 m³ (6·36 cu ft)

AREAS:

Wings, gross	10·10 m² (108·7 sq ft)
Ailerons (total)	0·762 m² (8·20 sq ft)
Trailing-edge flaps (total)	1·18 m² (12·70 sq ft)
Fin	0·76 m² (8·18 sq ft)
Rudder, incl tab	0·60 m² (6·46 sq ft)
Tailplane	1·46 m² (15·70 sq ft)
Elevators, incl tab	0·96 m² (10·30 sq ft)

WEIGHTS AND LOADINGS:

Weight empty, equipped	755 kg (1,664 lb)
Max T-O and landing weight:	
Aerobatic	1,100 kg (2,425 lb)
Utility	1,200 kg (2,645 lb)
Max wing loading	134·6 kg/m² (27·6 lb/sq ft)
Max power loading	6·19 kg/kW (10·17 lb/hp)

PERFORMANCE (at max T-O weight of 1,200 kg; 2,645 lb, except where indicated):
Never-exceed speed 235 knots (436 km/h; 271 mph)
Max level speed at S/L
183 knots (340 km/h; 211 mph)
Max cruising speed (75% power) at 1,500 m (4,925 ft) 161·5 knots (300 km/h; 186 mph)
Stalling speed, flaps up
74 knots (137 km/h; 85·5 mph)

Stalling speed, flaps down, power off
64 knots (118 km/h; 73·5 mph)

Max rate of climb at S/L	456 m (1,496 ft)/min
Time to 1,500 m (4,925 ft)	4 min 0 sec
Time to 2,300 m (7,550 ft)	6 min 50 sec
Time to 3,000 m (9,850 ft)	10 min 0 sec
Service ceiling	5,000 m (16,400 ft)
T-O run at S/L	560 m (1,837 ft)
T-O to 15 m (50 ft) at S/L	775 m (2,543 ft)
Landing from 15 m (50 ft) at S/L	690 m (2,264 ft)
Landing run at S/L	345 m (1,132 ft)
Range with max fuel	805 nm (1,490 km; 925 miles)

g limits:
at max Aerobatic T-O weight +6; −3
at max Utility T-O weight (without external load) +4·4; −2·2

SIAI-MARCHETTI SF.260W

The SF.260W, first flown (I-SJAV) in May 1972, is a developed version of the SF.260MX (which see) combining the structural and technical characteristics of the SF.260M with the ability to carry external loads, up to a maximum of 300 kg (660 lb), on two underwing pylons. The SF.260W can undertake such roles as low-level strike with rockets, anti-tank missiles or machine-guns; forward air control; forward area support, with droppable supply containers; armed reconnaissance; camouflage inspection; or liaison.

The SF.260W also meets the requirements of modern primary flying training, including basic flying training; instrument flying; aerobatics, including deliberate spinning and recovery; night flying; navigation flying; and formation flying.

Sixteen SF.260Ws were ordered by the Philippine Air Force, 12 by the Tunisian Air Force, 10 by the Irish Air Corps and one by the Dubai Police Air Wing.

ARMAMENT: Typical alternative underwing loads when carrying a crew of two include two Matra MAC AAF1 7·62 mm gun pods; two 50 kg bombs; two Matra F2 launchers, each with six 68 mm SNEB 253 rockets; two Simpres AL 8-70 launchers, each with nine 2·75 in FFAR rockets; two Simpres AL 18-50 launchers, each with eighteen 2 in SNIA ARF/8M2 rockets; or two Alkan 500B cartridge throwers for Lacroix 74 mm explosive cartridges, flare cartridges, or F.130 smoke cartridges. As a single-seater, two 120 kg bombs can be carried.

SIAI-MARCHETTI SF.260SW

The SF.260SW is a version of the SF.260MX (which see) developed specially for long-range surveillance, search and rescue and supply missions.

It retains the two underwing pylons of the SF.260W and is equipped with special wingtip tanks incorporating a lightweight radar system on one side and a photo-reconnaissance system on the other side.

Two auxiliary external fuel tanks, each of 83 litres (18·25 Imp gallons) capacity, provide a total fuel capacity of 409 litres (90 Imp gallons) and increase the endurance to 10 hr 55 min.

The Bendix RDR 1400 radar system installed in the nose section of the port tip-tank is of the digital type, and is designed for search and ground mapping as well as for weather avoidance. The basic arrangement consists of a radome housing a stabilised flat plate antenna, aft of which is a compartment containing the transmitter-receiver unit and related accessories. The central compartment comprises a fuel cell with a capacity equal to that of the standard SF.260W wingtip tank. Controls for the radar and camera are mounted at the co-pilot's station, together with the radar display.

SIAI-MARCHETTI SM.1019E

The SM.1019 light STOL aircraft is suitable for observation, light ground attack or utility duties. Its design was started in January 1969, and construction of a prototype began two months later. This aircraft (I-STOL) flew for the first time on 24 May 1969, with an Allison 250-B15C engine, and was granted Normal and Utility category certification by the RAI on 25 October 1969.

A second prototype (I-SJAR), which flew for the first time on 18 February 1971, was designated SM.1019A. It had an improved fuel system, two doors and two instrument panels, and received RAI civil certification in the Normal and Utility categories.

Production began in 1974 of an initial series of 100 military SM.1019EIs for the Aviazione Leggera dell'Esercito (ALE, or Italian Army Light Aviation). Deliveries began in the Summer of 1975, and were scheduled to be completed by the end of 1976.

The following description applies generally to all models of the SM.1019, except where a specific version is indicated:

TYPE: Two-seat STOL light aircraft.

WINGS: High-wing monoplane, braced by single strut on each side. Wing section NACA 2412. Dihedral 2° 8′. Incidence 1° 30′. Washout 3°. Conventional all-metal structure, with detachable tapered outer panels. Metal Frise-type ailerons and electrically-actuated trailing-edge slotted flaps. Trim tab in starboard aileron. Tiedown point at each wingtip.

SIAI-Marchetti SF.260WP light strike aircraft of the Philippine Air Force, with two underwing 7·62 mm gun pods

FUSELAGE: Conventional all-metal stressed-skin structure.

TAIL UNIT: Conventional cantilever all-metal structure, with horizontal surfaces mounted on top of fuselage. Dorsal fin. Fixed-incidence tailplane. Elevators and rudder horn-balanced. Manually-operated mechanically-actuated trim tab in starboard elevator; servo tab in port elevator. Ground-adjustable trim tab on rudder.

LANDING GEAR: Non-retractable tailwheel type, with cantilever leaf-type spring steel main-wheel legs. Goodyear 511960 main wheels, with low-pressure tyres, size 7·00-6, pressure 2·07 bars (30 lb/sq in); Scott 3200A tailwheel, with size 8-3·00 tyre, pressure 2·41 bars (35 lb/sq in). Goodyear independent hydraulic single-disc brakes on main wheels, controllable from either seat. Parking brake. Combined wheel/ski gear, with hydraulic retraction and extension of skis, is optional.

POWER PLANT: One 298 kW (400 shp) Allison 250-B17 turboprop engine, driving a Hartzell HC-B3TF-7/T10173-11R three-blade constant-speed reversible-pitch metal propeller. Fuel in two tanks in each wing, each of 80 litres (17·5 Imp gallons; 21 US gallons) capacity; total capacity 320 litres (70 Imp gallons; 84 US gallons). Refuelling point for each tank on top of wings. Provision for auxiliary underwing tanks. Oil capacity 8 litres (1·75 Imp gallons; 2·1 US gallons).

ACCOMMODATION: Pilot and co-pilot or observer/systems operator seated in tandem in fully-enclosed and extensively-glazed cabin. Two forward-hinged doors on starboard side. Cabin heated, by engine bleed air, and ventilated. Dual controls standard.

SYSTEMS: 28V DC electrical power provided by 30V 150A Lear Siegler P/N230320020 engine-driven starter/generator and 24V 25Ah nickel-cadmium battery. External ground power receptacle. Windscreen defrosting and engine compressor inlet heating standard. Oxygen system optional.

ELECTRONICS AND EQUIPMENT: Choice of VHF/UHF/HF communication systems. VLF/Omega navigation. ADF; IFF; high-performance intercom and compass system. Provision for specialised equipment (VHF/FM, radar warning, Tacan and HLS) to customer's requirements. Twin taxying and landing lights in port outer wing leading-edge. Anti-collision light on top of rudder.

ARMAMENT AND OPERATIONAL EQUIPMENT (SM.1019EI): Two hardpoints beneath each wing for 2·75 in rocket launchers, gun pods, missiles, bombs, auxiliary fuel tanks or a reconnaissance pod. Electronic, photographic and navigation equipment for use as day or night reconnaissance aircraft.

DIMENSIONS, EXTERNAL:

Wing span	10·972 m (36 ft 0 in)
Wing chord at root	1·63 m (5 ft 4¼ in)
Wing chord at tip	1·09 m (3 ft 7 in)
Wing aspect ratio	7·44
Length overall (tail up)	8·52 m (27 ft 11½ in)
Height overall (tail down)	2·86 m (9 ft 4½ in)
Tailplane span	3·42 m (11 ft 2¾ in)
Wheel track	2·29 m (7 ft 6¼ in)
Wheelbase	6·23 m (20 ft 5¼ in)
Propeller diameter	2·29 m (7 ft 6 in)
Propeller ground clearance	0·23 m (9 in)
Cabin doors, each:	
Height	1·06 m (3 ft 5¾ in)
Width	0·60 m (1 ft 11½ in)
Baggage door:	
Height	0·47 m (1 ft 6½ in)
Width	0·53 m (1 ft 9 in)
Height to sill	0·62 m (2 ft 0¼ in)

DIMENSIONS, INTERNAL:

Cabin: Max length	2·00 m (6 ft 6¾ in)
Max width	0·63 m (2 ft 0¾ in)
Max height	1·25 m (4 ft 1¼ in)
Volume	1·10 m³ (38·8 cu ft)
Baggage compartment volume	0·1 m³ (3·5 cu ft)

AREAS:

Wings, gross	16·16 m² (173·95 sq ft)
Ailerons (total)	1·70 m² (18·30 sq ft)
Trailing-edge flaps (total)	1·96 m² (21·10 sq ft)
Fin	0·957 m² (10·30 sq ft)
Rudder	1·295 m² (13·94 sq ft)
Tailplane	1·896 m² (20·41 sq ft)
Elevators (total)	1·584 m² (17·05 sq ft)

WEIGHTS AND LOADINGS (without external stores):

Weight empty, equipped	690 kg (1,521 lb)
Basic weight empty	730 kg (1,609 lb)

T-O weight:
SM.1019A (Utility category), SM.1019EI (training) 1,300 kg (2,866 lb)

Max T-O weight:
SM.1019A (Normal category), SM.1019EI (helicopter escort, reconnaissance) 1,450 kg (3,196 lb)

SIAI-Marchetti SM.1019EI in Italian Army insignia

Wing loading at 1,300 kg (2,866 lb) AUW
80·4 kg/m² (16·5 lb/sq ft)
Max wing loading 89·7 kg/m² (18·4 lb/sq ft)
Power loading at 1,300 kg (2,866 lb) AUW
4·36 kg/kW (7·2 lb/shp)
Max power loading 4·87 kg/kW (8·0 lb/shp)

PERFORMANCE (A: Utility, AUW of 1,300 kg; 2,866 lb. B: helicopter escort, AUW of 1,450 kg; 3,196 lb. C: reconnaissance, AUW of 1,450 kg; 3,196 lb, except where indicated):
Never-exceed speed 169 knots (313 km/h; 194 mph)
Max cruising speed at S/L:
A 160 knots (296 km/h; 184 mph)
B, C 152 knots (281 km/h; 175 mph)
Max cruising speed at 2,500 m (8,200 ft):
A 162 knots (300 km/h; 186 mph)
B, C 154 knots (285 km/h; 177 mph)
Cruising speed (75% power) at 2,500 m (8,200 ft):
A 152 knots (281 km/h; 175 mph)
B, C 145 knots (268 km/h; 167 mph)
Stalling speed, flaps up, with fuel injection:
A 53 knots (98 km/h; 61 mph)
B, C 58 knots (107 km/h; 66·5 mph)
Stalling speed, flaps down, with fuel injection:
A 38 knots (70 km/h; 43·5 mph)
B, C 46 knots (85 km/h; 53 mph)
Max rate of climb at S/L:
A 551 m (1,810 ft)/min
B, C 499 m (1,640 ft)/min
Operational ceiling:
A, B, C 7,620 m (25,000 ft)
T-O run at S/L:
A 112 m (368 ft)
B, C 218 m (716 ft)
T-O to 15 m (50 ft) at S/L:
A 220 m (722 ft)
B, C 361 m (1,185 ft)
Landing from 15 m (50 ft) at S/L:
A 220 m (722 ft)
B, C 281 m (922 ft)
Landing run at S/L:
A 91·5 m (300 ft)
B, C 135 m (443 ft)
Typical operational radius:
B, with two rocket launchers, at AUW of 1,400 kg (3,086 lb) 60 nm (111 km; 69 miles)
Max range at S/L:
A 499 nm (925 km; 575 miles)
B 421 nm (780 km; 485 miles)
C 610 nm (1,130 km; 702 miles)
Max range at 3,000 m (9,845 ft):
A 588 nm (1,090 km; 677 miles)
B 505 nm (935 km; 581 miles)
C 723 nm (1,340 km; 832 miles)
Max range with two external tanks, AUW of 1,400 kg (3,086 lb):
C, at 610 m (2,000 ft)
623 nm (1,154 km; 717 miles)
C, at 2,745 m (9,000 ft)
730 nm (1,352 km; 840 miles)
Max endurance at S/L:
A 5 hr 45 min
B 5 hr 0 min
C 7 hr 20 min
Max endurance at 3,000 m (9,845 ft):
A 6 hr 40 min
B 6 hr 5 min
C, with auxiliary fuel tanks 8 hr 45 min

SIAI-MARCHETTI S.208

First flown on 22 May 1967, the S.208 entered production in the Spring of 1968, and by mid-February 1973 approx 80 had been delivered, to customers in Europe and Africa. Production included 44 of a version designated **S.208M** for the Italian Air Force. These have a jettisonable cabin door, and are used for liaison and training duties.

A version designated **S.208AG** has been developed for general duties, including agricultural and ambulance work and light cargo carrying. This version has a non-retractable landing gear and two doors without a central pillar on the starboard side.

Full details can be found in the 1974-75 *Jane's* of the initial S.208 production version, to which the following details refer:

POWER PLANT: One 194 kW (260 hp) Lycoming O-540-E4A5 flat-six engine, driving a Hartzell two-blade constant-speed metal propeller. Two wing fuel tanks and two auxiliary wingtip tanks, total capacity 446 litres (98 Imp gallons).

ACCOMMODATION: Enclosed cabin seating pilot and up to four passengers. Forward-opening cabin door on starboard side (jettisonable on S.208M; double door on S.208AG). Second door, on port side, available optionally.

DIMENSIONS, EXTERNAL:
Wing span 10·86 m (35 ft 7½ in)
Length overall 8·00 m (26 ft 3 in)
Height overall 2·89 m (9 ft 5¾ in)
Wheel track 3·55 m (11 ft 8 in)
Wheelbase 1·90 m (6 ft 2¾ in)

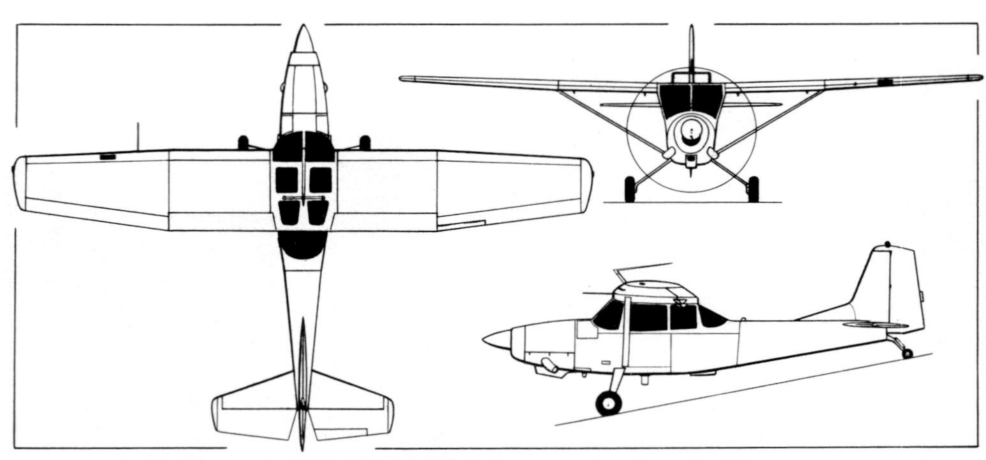

SIAI-Marchetti SM.1019E two-seat STOL light military aircraft *(Pilot Press)*

Propeller diameter 1·88 m (6 ft 2 in)

WEIGHTS AND LOADINGS:
Weight empty, equipped 820 kg (1,807 lb)
Max T-O weight 1,500 kg (3,306 lb)
Max wing loading 93·5 kg/m² (19·2 lb/sq ft)
Max power loading 7·73 kg/kW (12·79 lb/hp)

PERFORMANCE (S.208, at max T-O weight):
Never-exceed speed
175 knots (325 km/h; 202 mph)
Max level speed at S/L
173 knots (320 km/h; 199 mph)
Max cruising speed
162 knots (300 km/h; 187 mph)
Econ cruising speed 140 knots (260 km/h; 162 mph)
Stalling speed 50 knots (92 km/h; 58 mph)
Service ceiling 5,400 m (17,725 ft)
T-O run 340 m (1,115 ft)
T-O to 15 m (50 ft) 500 m (1,640 ft)
Landing from 15 m (50 ft) 400 m (1,310 ft)
Landing run 325 m (1,065 ft)
Range with max internal fuel
647 nm (1,200 km; 746 miles)
Range with max fuel (incl tip-tanks)
1,085 nm (2,000 km; 1,250 miles)

SIAI-MARCHETTI SF.260

Designed by Dott Ing Stelio Frati, the SF.260 is certificated for aerobatic flying.

The prototype, known as the F.250, was built by Aviamilano, with a 186·5 kW (250 hp) Lycoming engine (see 1965-66 *Jane's*), and flew for the first time on 15 July 1964.

The version developed for production was manufactured, initially under licence from Aviamilano, by SIAI-Marchetti, and is designated SF.260. SIAI-Marchetti is now the official holder of the type certificate and of all manufacturing rights in the SF.260. The initial version received FAA Type Approval on 1 April 1966.

Following delivery of the first 50 production aircraft, a second series of 50 civil SF.260s, incorporating many of the structural and aerodynamic improvements of the military SF.260MX, entered production to fulfil orders from Air France, Sabena, Royal Air Maroc and other customers. FAA Type Approval of this version was received on 10 June 1974.

The SF.260 holds two FAI speed records in Class C1b. The first of these is for a speed of 174 knots (322·52 km/h; 200·4 mph) over a 1,000 km closed circuit near Santa Monica, California, on 25 March 1969. The second, set up

S.208M military version of the SIAI-Marchetti S.208, in Italian Air Force insignia

SIAI-Marchetti SF.260 three-seat light aircraft, in Sabena Belgian World Airlines insignia

on 29 March 1969 near Los Angeles, is for a speed of 199·4 knots (369·43 km/h; 229·6 mph) over a 100 km closed circuit. These records are listed under the aircraft's US type name of Waco Meteor.

TYPE: Three-seat cabin monoplane.

WINGS: Cantilever low-wing monoplane. Wing section NACA 64212 at root, NACA 64210 at tip. Dihedral 5°. All-metal single-spar structure in two portions. All-metal Frise-type ailerons and electrically-operated slotted flaps.

FUSELAGE: All-metal semi-monocoque structure, with comparatively thick skin and few stringers.

TAIL UNIT: Cantilever all-metal structure with swept vertical surfaces. Rudder and elevators statically and dynamically balanced. Controllable trim tab in elevator.

LANDING GEAR: Retractable tricycle type. Electrical retraction with manual emergency actuation. Oleo-pneumatic shock-absorbers. Steerable nosewheel with tyre size 5·00-5. Main wheels and tyres size 6·00-6. Cleveland single-disc hydraulic brakes.

POWER PLANT: One 194 kW (260 hp) Lycoming O-540-E4A5 flat-six engine, driving a two-blade Hartzell type HC-C2YK-1B/8477-8R or HC-C2YK-4F/FC 8477-8R constant-speed metal propeller. Fuel in two tanks in wings, total capacity 100 litres (22 Imp gallons), and two on wingtips, total capacity of 140 litres (31 Imp gallons). Overall fuel capacity (four tanks) 240 litres (53 Imp gallons).

ACCOMMODATION: Three seats in enclosed cockpit, two in front, one at rear. Two children with a combined weight not exceeding 113 kg (250 lb) may use rear seat. Rearward-sliding transparent canopy. Baggage compartment capacity 40 kg (88 lb). Cabin soundproofed with glassfibre, heated and ventilated.

EQUIPMENT: Optional equipment includes blind-flying instrumentation, communications radio and oxygen system.

DIMENSIONS, EXTERNAL:

Wing span over tip-tanks	8·25 m (26 ft 11¾ in)
Wing chord at root	1·60 m (5 ft 3 in)
Wing chord at tip	0·78 m (2 ft 6¾ in)
Wing aspect ratio	6·4
Length overall	7·02 m (23 ft 0 in)
Height overall	2·60 m (8 ft 6 in)
Tailplane span	3·00 m (9 ft 9¾ in)
Wheel track	2·26 m (7 ft 5 in)
Wheelbase	1·62 m (5 ft 3½ in)
Propeller diameter	1·93 m (6 ft 4 in)

DIMENSIONS, INTERNAL: As SF.260MX

AREAS:

Wings, gross	10·10 m² (108·5 sq ft)
Ailerons (total)	0·79 m² (8·50 sq ft)
Flaps (total)	1·19 m² (12·7 sq ft)
Fin	0·75 m² (8·05 sq ft)
Rudder	0·50 m² (5·38 sq ft)
Tailplane	1·39 m² (14·9 sq ft)
Elevators, incl tab	0·93 m² (10·0 sq ft)

WEIGHTS AND LOADINGS:

Weight empty, equipped	700 kg (1,543 lb)
Max T-O weight:	
Aerobatic	1,000 kg (2,205 lb)
Utility	1,102 kg (2,430 lb)
Max wing loading	109 kg/m² (22·4 lb/sq ft)
Max power loading	5·91 kg/kW (9·33 lb/hp)

PERFORMANCE (at max Utility T-O weight):

Max level speed at S/L	204 knots (375 km/h; 235 mph)
Max cruising speed at 3,050 m (10,000 ft)	186 knots (345 km/h; 214 mph)
Stalling speed, flaps down	57 knots (104 km/h; 65 mph)
Max rate of climb at S/L	540 m (1,770 ft)/min
Service ceiling	6,500 m (21,370 ft)
T-O run on runway	250 m (820 ft)
T-O run on grass	290 m (950 ft)
T-O to 15 m (50 ft)	425 m (1,390 ft)
Landing from 15 m (50 ft)	490 m (1,610 ft)
Landing run	240 m (790 ft)
Range with max fuel (two persons)	1,107 nm (2,050 km; 1,275 miles)

SIAI-Marchetti-built Boeing Vertol CH-47C helicopter in the insignia of the Imperial Iranian Army Aviation

SIAI-MARCHETTI (BOEING VERTOL) CH-47C

Manufacture of the CH-47C began in the Spring of 1970, to meet orders for 20 (since increased to 42) for the Imperial Iranian Army and Air Force. Twenty-six were ordered by Italian Army Aviation, and funds for these were released in 1973. The CH-47C is now in service with the 11th and 12th Squadrons of the 1st (Antares) Regiment of the Italian Army.

SILVERCRAFT
SILVERCRAFT SpA

HEAD OFFICE AND WORKS:
Strada del Sempione 114, Casella Postale 37, 21018 Sesto Calende (Varese)
Telephone: (0331) 924842 and 923598
MILAN OFFICE:
Via Oglio 12, 20139 Milan
Telephone: (02) 533362

PRESIDENT:
Giovanni Barone Silvestri

PUBLIC RELATIONS MANAGER:
Dott Ing Pier Maria Pellò

Silvercraft was established in early 1962 to develop a light multi-purpose helicopter known as the Silvercraft XY, which flew for the first time in October 1963. This was subsequently developed into the Silvercraft SH-4. The company's Sesto Calende factory is adjacent to the SIAI-Marchetti airfield at Vergiate, and occupies an area of more than 25,000 m² (269,100 sq ft).

SILVERCRAFT SH-4

The SH-4 is a three-seat light helicopter suitable for pilot training, utility, agricultural, survey, police, ambulance, military liaison and observation duties.

The prototype flew for the first time in March 1965, and five pre-production models were completed by the end of 1967. On 4 September 1968 the SH-4 became the first all-Italian helicopter to receive both FAA and RAI certification, and was subsequently certificated also in France by the SGAC.

An initial series of 50 production aircraft was laid down in the Silvercraft works at Sesto Calende, with the co-operation of the Aero-Engine Division of Fiat (which see) which produces mechanical components for the rotor transmission system. These 50 aircraft were fully covered by existing orders, and a follow-on series of 200 SH-4s was planned. Deliveries, to Italian and overseas customers, began in early 1970.

The following versions are currently available:

SH-4. Standard general-purpose version.

SH-4A. Agricultural version. Fitted with 10·00 m (32 ft 9½ in) spraybars, capable of covering a 33 m (108 ft) swath width with liquid chemicals. Max capacity of chemical tanks 200 litres (53 US gallons) up to a max weight of 200 kg (441 lb). Max rate of distribution 91 litres (24 US gallons)/min; chemical tanks can be replenished in 1 minute. Total weight (empty) of spray installation 37 kg (82 lb).

In 1976 a new power plant installation, utilising a 153 kW (205 hp), derated to 127 kW (170 hp), Lycoming LHIO-360-C1A engine, was under development as an alternative to the standard Franklin engine.

Silvercraft SH-4 three-seat helicopter (Franklin 6A-350-D1B engine) *(Ronaldo S. Olive)*

The following details apply to the Franklin-powered models:

TYPE: Three-seat light helicopter.

ROTOR SYSTEM: Two-blade semi-rigid main and tail rotors. Blades constructed of laminated wood, with glassfibre covering and (on main rotor) steel weights at blade tips to augment the inertia of the rotor. Aluminium alloy attachment fittings and steel hubs.

ROTOR DRIVE: Rotors driven through steel shafting. Primary gearbox, consisting of two sets of planetary gears (reduction ratio 1 : 0·89), mounted aft of engine; secondary bevel gearboxes at base of main rotor driveshaft (reduction ratio 1 : 0·164) and in rear of tailboom for main and tail rotors respectively. Rotor design eliminates need for stabilisation bars and dampers, and rotor is controlled directly without hydraulic servo-command system. Main rotor/engine rpm ratio 418 : 2,850. Tail rotor/engine rpm ratio 2,434 : 2,850.

FUSELAGE: Central structure of aluminium alloy, except engine fireproof bulkheads which are of titanium alloy. Semi-monocoque cabin at front and semi-monocoque tailboom are also of aluminium construction. Cabin door frames and window frames of reinforced glassfibre.

TAIL UNIT: Horizontal stabiliser mid-mounted on tailboom. Ventral stabilising fin beneath tip of tailboom.

LANDING GEAR: Tubular skid type, with provision to fit ground manoeuvring wheels or skis. Tailskid at base of ventral stabilising fin, to protect tail rotor. Alternative pontoon gear available for amphibious operation.

POWER PLANT: One 175 kW (235 hp) (derated to 127 kW; 170 hp) Franklin 6A-350-D1B flat-six engine, installed horizontally, offset to port. Fuel in two main tanks, one on each side of base of pylon, with total capacity of 130 litres (28·6 Imp gallons). For short-range missions, one tank may be omitted, the remaining tank containing enough fuel for up to 1 hour's flight. Oil capacity 9 litres (2 Imp gallons).

ACCOMMODATION: Bench seat for pilot and two passengers side by side in enclosed cabin. Forward-hinged, easily removable car-type door on each side. Roof panels of blue tinted Plexiglas. Large baggage compartment. Agricultural version normally flown as single-seater.

SYSTEMS: 12V electrical system includes generator, 37Ah battery and engine starter.

OPTIONAL EQUIPMENT: Dual controls; radio; cabin heating system; navigation, landing and cabin lights. External cargo hook under engine platform for 200 kg (441 lb) slung load. Baggage container on starboard side of engine platform, aft of cabin. Provision in ambulance version for an enclosed stretcher pannier to be mounted externally on brackets on port side of cabin. Agricultural installation of 10·00 m (32 ft 9½ in) sprayboom and twin tanks, one mounted externally each side of cabin and each containing 100 litres (26·5 US gallons) of liquid chemical.

Dimensions, external:
- Diameter of main rotor: 9·03 m (29 ft 7½ in)
- Diameter of tail rotor: 1·39 m (4 ft 6¾ in)
- Length overall, main rotor fore and aft: 10·47 m (34 ft 4¼ in)
- Length of fuselage, incl tailskid: 7·65 m (25 ft 1¼ in)
- Span of horizontal stabiliser: 2·05 m (6 ft 8¾ in)
- Max width of fuselage: 1·54 m (5 ft 0¾ in)
- Height to top of cabin roof: 1·73 m (5 ft 8⅛ in)
- Height overall: 2·98 m (9 ft 9¼ in)
- Width over skids: 1·74 m (5 ft 8½ in)

Dimensions, internal:
- Cabin:
  - Length: 1·47 m (4 ft 9¾ in)
  - Max height: 1·24 m (4 ft 0¾ in)

Areas:
- Main rotor blades (each): 1·17 m² (12·61 sq ft)
- Tail rotor blades (each): 0·09 m² (0·97 sq ft)
- Main rotor disc: 64·04 m² (689·32 sq ft)
- Tail rotor disc: 1·52 m² (16·32 sq ft)

Weights and Loadings:
- Weight empty:
  - SH-4: 518 kg (1,142 lb)
- Max T-O weight (Normal): 862 kg (1,900 lb)
- Max disc loading: 13·36 kg/m² (2·76 lb/sq ft)
- Max power loading: 6·79 kg/kW (11·2 lb/hp)

Performance (at max T-O weight):
- Max level speed at S/L: 87 knots (161 km/h; 100 mph)
- Max cruising speed: 70 knots (130 km/h; 81 mph)
- Econ cruising speed: 63·5 knots (117 km/h; 73 mph)
- Max rate of climb at S/L: 360 m (1,180 ft)/min
- Service ceiling:
  - SH-4: 4,600 m (15,090 ft)
- Hovering ceiling in ground effect:
  - SH-4: 3,000 m (9,845 ft)
- Hovering ceiling out of ground effect:
  - SH-4: 2,400 m (7,875 ft)
- Range with max fuel: 173 nm (320 km; 200 miles)
- Max endurance: 3 hr

JAPAN

CTDC

ZAIDAN HOZIN MINKAN YUSOOKI KAIHATSU KYOKAI (Civil Transport Development Corporation)

Head Office:
Toranomon Daiichi Building, No. 1, Kotohira-cho, Shiba, Minato-ku, Tokyo 105
Telephone: Tokyo (03) 503-3211
Telex: 2222863 NAMC J

Directors:
Kiyoshi Yotsumoto (Chairman)
Reizo Wakasugi (Senior Directing Manager)
Kyoku Hirano (Directing Manager)

The CTDC was established in April 1973 to manage, on behalf of the Japanese government and aerospace industry, the YX programme to design and develop a civil transport aircraft to succeed the NAMC YS-11.

Chairman of the board of directors is the current president of Kawasaki Heavy Industries Ltd; other board members include the chairman of Shin Meiwa; the presidents of Fuji, Mitsubishi, Nippi, Ishikawajima-Harima, All Nippon Airways, Japan Air Lines and Toa Domestic Airlines; and the managing directors of Fuji, Kawasaki and Mitsubishi.

YX PROGRAMME

With substantial support from the Japanese government, the Japanese aerospace industry is planning to develop a civil transport aircraft in collaboration with Boeing and Aeritalia. The aircraft will be in the short/medium-range category, with a passenger capacity of 200/240.

Activity during 1975 was concerned with exploring the feasibility of developing and producing the new transport aircraft for commercial service in the late 1970s or early 1980s. Exploratory work was carried out in collaboration with Boeing, under an agreement signed in April 1973.

A further agreement covering collaboration in the development of the new transport aircraft was to be signed in 1976.

FUJI

FUJI HEAVY INDUSTRIES LTD (Fuji Jukogyo Kabushiki Kaisha)

Head Office:
Subaru Building, 7-2, 1-chome, Nishi-shinjuku, Shinjuku-ku, Tokyo
Telephone: Tokyo (03) 347-2505
Telex: 0-232-2268

Aircraft Factory (Utsunomiya Manufacturing Division):
Utsunomiya City, Tochigi Prefecture
Telephone: Utsunomiya (0286) 58-1111

President:
Eiichi Ohara

Executive Managing Directors:
Nobuhiro Sakata
Sukemitsu Irie

Managing Directors:
Shoji Nagashima
Shigeichi Ota
Iwao Shibuya
Kiyoyuki Kawabata
Yoshishige Suzuki
Hiroshi Yamamoto

General Manager of Aircraft Division:
Atsushi Kasai

Manager of Aircraft Sales Department:
Takaaki Hosaka

Superintendent of Utsunomiya Aircraft Factory:
Saburo Watanabe

Fuji Heavy Industries Ltd was established on 15 July 1953. It is a successor to the Nakajima Aircraft Company, which was established in 1914 and built 30,000 aircraft up to the end of the second World War.

The present Utsunomiya Manufacturing Division (Aircraft and Rolling Stock Factories) occupies a site of 512,070 m² (5,511,870 sq ft) including a floor area of 161,532 m² (1,738,710 sq ft) and in 1975 employed 3,560 people.

Under licence from Cessna, Fuji produced 22 L-19E Bird Dog observation aircraft for the Japan Ground Self-Defence Force.

Under a 1953 licence and technical assistance agreement with Beech Aircraft Corporation, Fuji built the Beechcraft Mentor at Utsunomiya. Several modified versions of the Mentor, designated LM-1 Nikko, LM-2, KM and KM-2, were also built by Fuji. Details of these can be found in earlier editions of *Jane's*. Eight more KM-2s were ordered for the JMSDF in FY 1976.

The modified KM-2B, combining features of the KM-2 and the Beechcraft T-34A, has been ordered by the JASDF as a primary trainer.

Under another major agreement, the company is now producing in Japan the Bell Model 204/205 series of helicopters. The first 204B covered by the agreement arrived in Japan in kit form in May 1962 for assembly by Fuji.

First aircraft designed entirely by Fuji was the T-1 intermediate two-seat jet trainer, described fully in the 1967-68 *Jane's*. It was followed by a four-seat light aeroplane known as the FA-200 Aero Subaru, details of which follow.

Fuji was responsible for construction of the VTOL flying testbed designed by the National Aerospace Laboratory (see 1972-73 *Jane's),* and since 1972 has continued the study of VTOL research aircraft under contract to NAL.

A twin-engined six/eight-seat aircraft, known as the FA-300, is being developed jointly with Rockwell International, and was undergoing flight testing in 1976.

Fuji is currently studying two proposals to meet the JASDF's JT-X requirement for a new jet trainer, acquisition of which is expected to form a part of the 6th defence buildup programme (1982-87). One of these proposals is essentially an improved T-1; the other is a new twin-engined design.

In 1971 Fuji began to produce Firebee I subsonic target drones for the JMSDF, under a technical and licensing agreement with Teledyne Ryan (see RPVs and Targets section).

FUJI FA-200 AERO SUBARU

Fuji began detail design of this light aircraft in 1964 and the prototype flew for the first time on 12 August 1965. Subsequently, refinements were made to the aircraft, and it went into production as the FA-200-160 and FA-200-180. One example was also produced of the FA-203S STOL version, and this was described in the 1970-71 *Jane's*.

Versions in current production are designated as follows:

FA-200-160. Basic four-seat light aircraft, with 119 kW (160 hp) Lycoming engine. Received Japan Civil Aviation Bureau Normal category type certificate as a four-seater on 1 March 1966, Utility category certification as a three-seater on 6 July 1966 and Aerobatic category certification as a two-seater on 29 July 1967. FAA Type Approval in all three categories followed on 26 September 1967.

FA-200-180. Developed version with 134 kW (180 hp) Lycoming engine. Certification by JCAB in Normal (four-seat), Utility (four-seat) and Aerobatic (two-seat) categories was received on 28 February 1968, and FAA Type Approval in all three categories on 25 April 1968.

FA-200-180AO. Version with 134 kW (180 hp) Lycoming engine and fixed-pitch propeller. Certificated by JCAB in Normal (four-seat), Utility (four-seat) and Aerobatic (two-seat) categories on 27 September 1973; and by FAA in all three categories on 1 February 1974.

Production of the FA-200 began in March 1968, and 305 had been completed by 1 February 1976, of which more than 140 were for export.

The following description applies generally to all three current versions, except where a specific version is indicated. 'Subaru' is the Japanese name for the Pleiades group of six stars in the constellation of Taurus, and was chosen to represent the six companies which merged to form Fuji Heavy Industries Ltd.

Type: Four-seat light monoplane.

Wings: Cantilever low-wing monoplane. Dihedral 7°. Incidence 2° 30′. All-metal structure, with single extruded main spar at 42% chord. All-metal riveted Frise-type ailerons and single-slotted flaps. Trim tab on each aileron.

Fuselage: All-metal semi-monocoque structure of frames and stringers.

Tail Unit: Cantilever all-metal structure, with swept vertical surfaces. One-piece tailplane. Trim tab in port elevator. Manually-adjustable tab in rudder.

Landing Gear: Non-retractable tricycle type. Oleo-pneumatic shock-absorbers on all units. Nosewheel steerable. Tube-type 4-ply tyres size 6·00-6 on main wheels, 5·00-5 on nosewheel. Hydraulic disc brakes. Parking brake. Streamlined wheel fairings optional.

Power Plant: One 119 kW (160 hp) Lycoming O-320-D2A (134 kW; 180 hp Lycoming IO-360-B1B in FA-200-180 and 134 kW; 180 hp Lycoming O-360-A5AD in FA-200-180AO) flat-four engine, driv-

Fuji FA-200-180 Aero Subaru four-seat light aircraft (Lycoming IO-360 engine)

ing a McCauley 1C 160/FGM 7656 two-blade fixed-pitch metal propeller in FA-200-160 (McCauley B2D34C53/74E-0 two-blade constant-speed metal propeller in FA-200-180, and McCauley 1A 170/EFA 7658 two-blade fixed-pitch metal propeller in FA-200-180AO). Fuel in two integral tanks in inner wings with total capacity of 204·5 litres (45 Imp gallons). Overwing refuelling point on each tank. Oil capacity 7 litres (1·5 Imp gallons).

Accommodation: Four seats in pairs in enclosed cabin. Individual adjustable front seats, with dual controls. Optional shoulder harness on each of the four seats. Large rearward-sliding canopy, which can be opened in flight. Two tinted roof windows optional. Cabin heating and ventilation, and windscreen defrosting, standard. Baggage compartment in rear of fuselage, capacity 80 kg (176 lb). Baggage shelf aft of rear seats, capacity 20 kg (44 lb).

Electronics and Equipment: Standard electrical equipment includes 12V 50/60A alternator and 12V 38Ah battery. Optional extras include HF and VHF radio, full blind-flying instrumentation, VOR, ADF, ILS, ATC transponder, landing and navigation lights, cabin lights, instrument lights and anti-collision light.

Dimensions, external:	
Wing span	9·42 m (30 ft 11 in)
Wing chord (constant)	1·525 m (5 ft 0 in)
Wing aspect ratio	6·34
Length overall	8·17 m (26 ft 9½ in)
Height overall	2·59 m (8 ft 6 in)
Tailplane span	3·47 m (11 ft 4½in)
Wheel track	2·63 m (8 ft 7½ in)
Wheelbase	1·75 m (5 ft 8¾ in)
Propeller diameter:	
160, 180AO	1·93 m (6 ft 4 in)
180	1·88 m (6 ft 2 in)
Dimensions, internal:	
Cabin: Length	1·74 m (5 ft 8½ in)
Width	1·03 m (3 ft 4½ in)
Areas:	
Wings, gross	14·0 m² (150·7 sq ft)
Ailerons (total)	1·13 m² (12·11 sq ft)
Flaps (total)	1·87 m² (20·13 sq ft)
Fin	1·50 m² (16·11 sq ft)
Rudder, incl tab	0·89 m² (9·58 sq ft)
Tailplane	3·32 m² (35·74 sq ft)
Elevators, incl tab	1·43 m² (15·4 sq ft)
Weights and Loadings (N: Normal; U: Utility; A: Aerobatic category):	
Weight empty:	
160	620 kg (1,366 lb)
180	650 kg (1,433 lb)
180AO	640 kg (1,411 lb)
Max T-O weight:	
N (160)	1,060 kg (2,335 lb)
N (180)	1,150 kg (2,535 lb)
N (180AO)	1,140 kg (2,513 lb)
U (160)	970 kg (2,138 lb)
U (180/180AO)	1,100 kg (2,425 lb)
A (160)	880 kg (1,940 lb)
A (180/180AO)	940 kg (2,072 lb)
Max wing loading:	
N (160)	75·7 kg/m² (15·5 lb/sq ft)
N (180)	82·1 kg/m² (16·8 lb/sq ft)
Max power loading:	
N (160)	8·91 kg/kW (14·59 lb/hp)
N (180)	8·58 kg/kW (14·08 lb/hp)
Performance (N: Normal category; A: Aerobatic category, at max T-O weight):	
Max level speed at S/L:	
N (160)	120 knots (222 km/h; 138 mph)
N (180)	126 knots (233 km/h; 145 mph)
N (180AO)	123 knots (229 km/h; 142 mph)
A (160)	122 knots (225 km/h; 140 mph)
A (180)	128 knots (237 km/h; 147 mph)
A (180AO)	126 knots (233 km/h; 145 mph)
Max cruising speed (75% power):	
N (160) at 1,525 m (5,000 ft)	106 knots (196 km/h; 122 mph)
N (180) at 1,525 m (5,000 ft)	110 knots (204 km/h; 127 mph)
N (180AO) at 1,525 m (5,000 ft)	112 knots (208 km/h; 129 mph)
A (160) at 2,290 m (7,500 ft)	114 knots (211 km/h; 131 mph)
A (180 and 180AO) at 2,290 m (7,500 ft)	119 knots (220 km/h; 137 mph)
Econ cruising speed (55% power):	
N (160) at 1,525 m (5,000 ft)	89 knots (164 km/h; 102 mph)
N (180) at 1,525 m (5,000 ft)	90 knots (167 km/h; 104 mph)
N (180AO) at 1,525 m (5,000 ft)	92 knots (171 km/h; 106 mph)
A (160) at 2,290 m (7,500 ft)	96 knots (177 km/h; 110 mph)
A (180) at 2,290 m (7,500 ft)	100 knots (185 km/h; 115 mph)
A (180AO) at 2,290 m (7,500 ft)	102 knots (190 km/h; 118 mph)

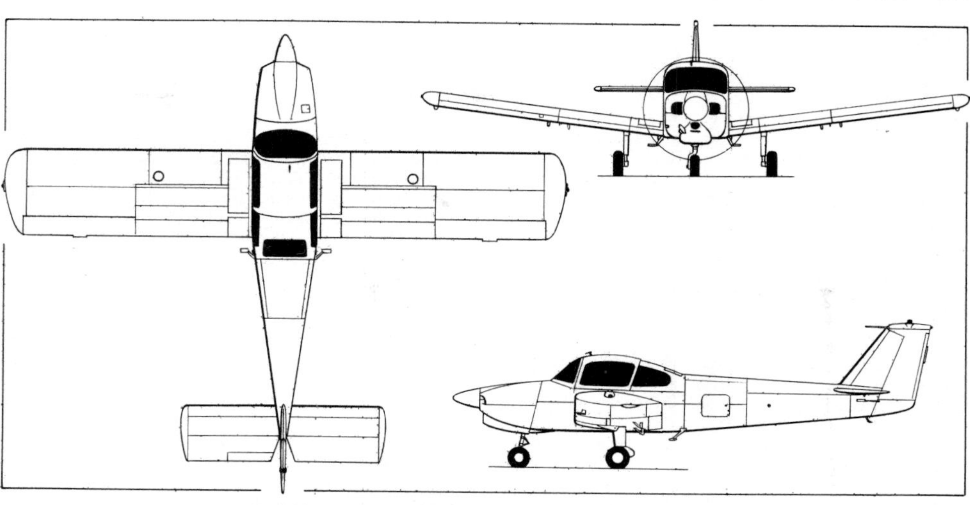

Fuji FA-200-180AO four-seat light aircraft *(Pilot Press)*

Stalling speed, flaps down:	
N (160)	49 knots (90 km/h; 56 mph)
N (180)	53 knots (97 km/h; 60 mph)
N (180AO)	52 knots (95 km/h; 59 mph)
A (160)	45·5 knots (84 km/h; 52·5 mph)
A (180 and 180AO)	47 knots (87 km/h; 54 mph)
Max rate of climb at S/L:	
N (160)	207 m (680 ft)/min
N (180)	232 m (760 ft)/min
N (180AO)	204 m (670 ft)/min
A (160)	302 m (991 ft)/min
A (180)	344 m (1,129 ft)/min
A (180AO)	311 m (1,020 ft)/min
Service ceiling:	
N (160 and 180AO)	3,480 m (11,400 ft)
N (180)	4,175 m (13,700 ft)
A (160)	4,725 m (15,500 ft)
A (180)	5,790 m (19,000 ft)
A (180AO)	5,640 m (18,500 ft)
T-O run:	
A (160)	160 m (525 ft)
A (180 and 180AO)	190 m (623 ft)
T-O to 15 m (50 ft):	
N (160)	465 m (1,530 ft)
N (180 and 180AO)	500 m (1,640 ft)
A (160)	310 m (1,020 ft)
A (180 and 180AO)	305 m (1,000 ft)
Landing from 15 m (50 ft):	
N (160)	340 m (1,115 ft)
N (180 and 180AO)	350 m (1,150 ft)
A (160)	315 m (1,033 ft)
A (180 and 180AO)	325 m (1,066 ft)
Landing run:	
A (160)	115 m (377 ft)
A (180 and 180AO)	125 m (410 ft)
Range with max fuel (55% power at 2,290 m; 7,500 ft, no reserves):	
N (160)	655 nm (1,215 km; 755 miles)
N (180)	725 nm (1,343 km; 835 miles)
N (180AO)	675 nm (1,252 km; 778 miles)
A (160)	816 nm (1,512 km; 940 miles)
A (180)	755 nm (1,400 km; 870 miles)
A (180AO)	748 nm (1,387 km; 862 miles)

FUJI FA-300/ROCKWELL COMMANDER 700

Design and development of the FA-300, following more than two years of market research, began in Japan in the latter half of 1971. It is currently proceeding as a collaborative venture between Fuji and Rockwell International, following the signing of an agreement between the two companies on 28 June 1974.

The FA-300, known in the USA as the **Rockwell Commander 700**, is designed to conform to FAR 23, Amendment 14. Five flying prototypes have been built, the first of which was rolled out on 5 September 1975 and made its first flight at Utsunomiya on 13 November 1975. The second (N9901S), assembled by Rockwell, flew for the first time at Bethany, Oklahoma, on 25 February 1976. A second Japanese prototype, and two more assembled by Rockwell, had flown by the Summer of 1976. Two other airframes are being used for ground testing. Certification by the JCAB and FAA was anticipated by the end of 1976, with deliveries planned to begin in February 1977.

For production aircraft, Fuji will be responsible for the basic structure of all FA-300/Commander 700 aircraft built; Rockwell will be responsible for the assembly, equipment installation and interior furnishing of those intended for sale in the Americas. The FA-300/Commander 700 is a basic six/eight-seat version, from which other models will be developed.Among the latter has been reported a version designated FA-300-Kai/Commander 710 with uprated engines.

The following description applies to the prototypes:

Type: Twin-engined six/eight-seat cabin monoplane.

Wings: Cantilever low-wing monoplane, with exclusive Fuji-developed aerofoil sections. Dihedral 7°. Sealed box-beam structure, forming integral fuel tank. Trim tab in each aileron. Pneumatic de-icing of leading-edges optional.

Fuselage: Conventional semi-monocoque structure, with rather more frames and fewer stringers than comparable types of aircraft. All-metal construction, primarily of 2024 aluminium alloy, with 7075 aluminium alloy for high-stress members.

Tail Unit: Cantilever all-metal structure, with sweptback vertical surfaces and shallow dorsal fin. Fixed-incidence non-swept tailplane, mounted part-way up fin. Balanced elevators and rudder, each with trim tab. Pneumatic de-icing system optional.

Landing Gear: Hydraulically-retractable tricycle type, all units retracting forward. Free-fall emergency extension. Oleo-pneumatic shock-absorbers. Main-wheel tyres size 6·50-8 (8-ply rating); nosewheel tyre size 6·00-6 (6-ply rating).

Power Plant: Two 242 kW (325 hp) Lycoming TIO-540-R2AD turbocharged flat-six engines, each driving a Hartzell three-blade constant-speed fully-feathering metal propeller with spinner. Electrical propeller de-icing system optional. Integral fuel tanks in wings, total capacity 719 litres (190 US gallons; 158 Imp gallons). Oil capacity 11·5 litres (3 US gallons; 2·5 Imp gallons).

Accommodation: Pilot and co-pilot on individual adjustable and reclining seats. Dual controls standard. Pilot's storm window. Heated windscreen and windscreen wiper optional. Seats for four to six persons in pressurised cabin. Forward and aft cabin dividers optional.

Prototype Fuji FA-300/Rockwell Commander 700 six/eight-seat light transport aircraft

Baggage compartments in nose and rear of pressurised cabin. Door with built-in airstair on port side; emergency exit on starboard side. Cabin heated, air-conditioned and pressurised.

SYSTEMS: Air-conditioning and pressurisation system (differential 0·38 bars; 5·5 lb/sq in). Freon-type 16,000 BTU air-conditioner optional. 45,000 BTU capacity combustion heater, with windscreen defroster. Hydraulic system supplied by electro-hydraulic power package. Pressure pumps, driven by each engine, supply air pressure to gyro instruments, cabin door seal, and (when fitted) to wing and tail de-icing systems. Electrical system supplied by two 28V 100A alternators and 24V 25Ah lead-acid battery.

ELECTRONICS AND EQUIPMENT: Installed standard equipment is extensive. Wide range of optional electronics available, including radar, communications, area navigation, autopilot, flight director and radar altimeter. Other optional items include heated windscreen, windscreen wiper, wing and tail pneumatic de-icer boots, ice inspection light, propeller synchroniser, flight hour meter and strobe light.

DIMENSIONS, EXTERNAL:
Wing span 12·94 m (42 ft 5½in)
Length overall 12·00 m (39 ft 4½ in)
Length of fuselage 11·635 m (38 ft 2 in)
Height overall 3·90 m (12 ft 9½ in)
Tailplane span 4·92 m (16 ft 1¾ in)
Wheel track 5·045 m (16 ft 6½ in)
Wheelbase 3·16 m (10 ft 4½ in)
Propeller diameter 2·06 m (6 ft 9 in)
Distance between propeller centres 4·75 m (15 ft 7 in)
Propeller/fuselage clearance 0·57 m (1 ft 10½ in)
Propeller ground clearance 0·30 m (11¾ in)

DIMENSIONS, INTERNAL:
Cabin: Length 5·005 m (16 ft 5 in)
Max width 1·45 m (4 ft 9 in)
Max height 1·45 m (4 ft 9 in)
Baggage volume (nose and rear of cabin, total) 1·50 m³ (53·0 cu ft)

AREAS:
Wings, gross 18·60 m² (200·2 sq ft)
Fin 3·71 m² (39·9 sq ft)
Tailplane 5·15 m² (55·4 sq ft)

WEIGHTS:
Weight empty, standard 1,995 kg (4,400 lb)
Max T-O and landing weight 2,993 kg (6,600 lb)
Max ramp weight 3,011 kg (6,640 lb)

PERFORMANCE (estimated, at max T-O weight except where indicated):
Max level speed at 2,766 kg (6,100 lb) average cruising weight:
full power at 6,100 m (20,000 ft) 231 knots (428 km/h; 266 mph)
Max cruising speed at 2,766 kg (6,100 lb) average cruising weight:
75% power at 7,315 m (24,000 ft) 219 knots (405 km/h; 252 mph)
45% power at 4,570 m (15,000 ft) 154 knots (285 km/h; 177 mph)
Approach speed 90 knots (167 km/h; 104 mph)
Stalling speed, power off, flaps and landing gear up 85·5 knots (158 km/h; 98 mph)
Stalling speed, power off, flaps and landing gear down 70 knots (129 km/h; 80 mph)
Max rate of climb at S/L 445 m (1,460 ft)/min
Rate of climb, one engine out, at 1,525 m (5,000 ft) 78 m (255 ft)/min
Max operating altitude 7,620 m (25,000 ft)
Service ceiling (30·5 m; 100 ft/min climb) 9,265 m (30,400 ft)
Service ceiling, one engine out (15·25 m; 50 ft/min climb) 4,085 m (13,400 ft)
T-O to 15 m (50 ft) 738 m (2,420 ft)
Landing from 15 m (50 ft) 634 m (2,080 ft)
Range at max (75%) cruising power with 606 litres (160 US gallons; 133 Imp gallons) fuel 703 nm (1,303 km; 810 miles)

FUJI KM-2B

The KM-2B is a modification of the original KM-2 development of the Beechcraft T-34A Mentor, described in the 1969-70 *Jane's,* combining the airframe and power plant of the KM-2 with the two-seat cockpit installation of the T-34A. The first KM-2B (JA3725) was flown for the first time on 26 September 1974, and received JCAB category A certification on 28 November 1974.

In 1975 the JASDF announced that it had selected the KM-2B to replace the T-34A in the primary trainer role. Purchase of 62 KM-2Bs is planned, of which the first six were due to be delivered in FY 1976.

TYPE: Two-seat primary trainer.

WINGS: Cantilever low-wing monoplane. Wing section NACA 23016·5 at root, NACA 23012 at tip. Dihedral 6° from roots. Incidence 4° at root. 3° geometric twist. No sweep at quarter-chord. All-metal tapered two-spar structure with stressed skin. Aluminium alloy single-slotted flaps and plain ailerons. Servo tab in each aileron; port tab controllable for trim.

FUSELAGE: Conventional all-metal semi-monocoque structure.

TAIL UNIT: Cantilever all-metal structure. Fixed-incidence tailplane, with elevators. Controllable tab in each elevator; anti-servo tab in rudder.

LANDING GEAR: Electrically retractable tricycle type, with single wheel and oleo-pneumatic shock-absorber on each unit. Nosewheel retracts rearward, main wheels inward into wings. Goodyear wheels and Type III tyres on all units: size 6·50-8 (6 ply), pressure 2·34 bars (34 lb/sq in) on main units; size 5·50-5 (4 ply), pressure 2·76 bars (40 lb/sq in) on nose unit. Goodyear automatically adjustable single-disc brakes on main units.

POWER PLANT: One 254 kW (340 hp) Lycoming IGSO-480-A1A6 flat-six engine, driving a Hartzell HC-A3X20-1E/9333C-3 three-blade constant-speed propeller with spinner. One metal and one bladder-type fuel tank in each wing, total capacity 265 litres (70 US gallons; 58·3 Imp gallons). Refuelling points on top of wings. Oil capacity 11·35 litres (3 US gallons; 2·5 Imp gallons).

ACCOMMODATION: Crew of two in tandem, on adjustable seats in heated and ventilated cabin. Dual controls standard. Framed canopy, with rearward-sliding section over each seat. Space for 45 kg (100 lb) of baggage aft of rear seat.

SYSTEMS AND EQUIPMENT: Blind-flying instrumentation standard. Prototype has one 28V 50A DC generator and one 24V 24Ah battery for electrical power; production aircraft will have a 100A generator and two 250VA static inverters. Standard electronics equipment includes King KTR-900A VHF, King KDF-800 ADF and Collins ICS-356C-4. Production aircraft will have ATC transponder and Tacan.

DIMENSIONS, EXTERNAL:
Wing span 10·00 m (32 ft 9¾in)
Wing chord at root 2·13 m (7 ft 0 in)
Wing chord at tip 1·07 m (3 ft 6 in)
Wing aspect ratio 6·1
Length overall 8·04 m (26 ft 4½ in)
Height overall 3·02 m (9 ft 11 in)
Tailplane span 3·71 m (12 ft 2 in)
Wheel track 2·92 m (9 ft 7 in)
Wheelbase 2·27 m (7 ft 5½ in)
Propeller diameter 2·29 m (7 ft 6 in)
Propeller ground clearance 0·25 m (9¾ in)

DIMENSIONS, INTERNAL:
Cabin:
Max width approx 0·90 m (2 ft 11½ in)
Max height approx 1·30 m (4 ft 3¼ in)

AREAS:
Wings, gross 16·50 m² (177·6 sq ft)
Ailerons (total) 1·07 m² (11·52 sq ft)
Trailing-edge flaps (total) 1·98 m² (21·31 sq ft)
Fin 0·97 m² (10·44 sq ft)
Rudder, incl tab 0·61 m² (6·57 sq ft)
Tailplane 3·46 m² (37·24 sq ft)
Elevators, incl tab 1·39 m² (14·96 sq ft)

WEIGHTS AND LOADINGS:
Weight empty 1,120 kg (2,469 lb)
Max T-O weight 1,510 kg (3,329 lb)
Max wing loading 91·5 kg/m² (18·74 lb/sq ft)
Max power loading 5·94 kg/kW (9·79 lb/hp)

PERFORMANCE (at max T-O weight):
Never-exceed speed 223 knots (413 km/h; 257 mph) EAS

Prototype Fuji KM-2B two-seat primary trainer, developed from the KM-2 and the Beechcraft T-34A

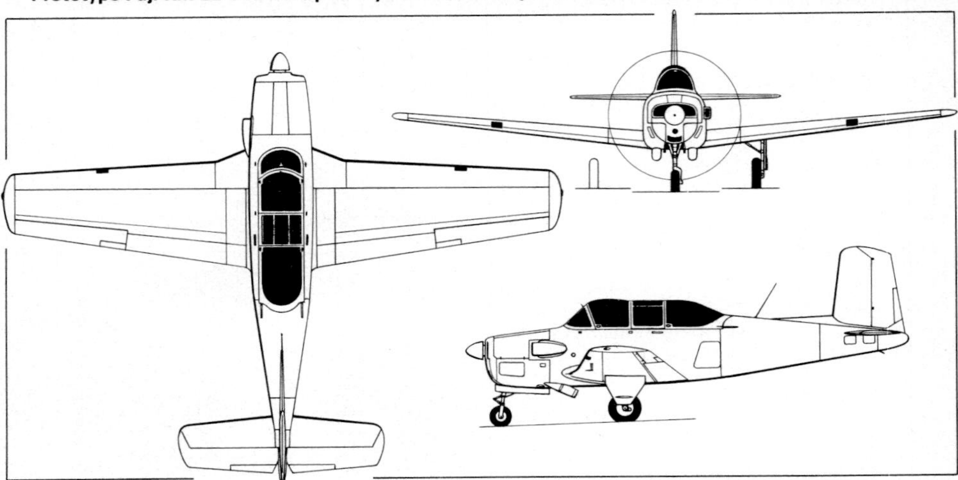
Fuji KM-2B two-seat primary trainer (Lycoming IGSO-480-A1A6 engine) *(Pilot Press)*

Max level speed at 4,875 m (16,000 ft) 203 knots (377 km/h; 234 mph)
Max cruising speed at 2,440 m (8,000 ft) 177 knots (328 km/h; 204 mph)
Econ cruising speed at 2,440 m (8,000 ft) 137 knots (254 km/h; 158 mph)
Stalling speed, flaps up 66 knots (123 km/h; 76 mph)
Stalling speed, flaps down 54 knots (100 km/h; 62 mph)
Max rate of climb at S/L 463 m (1,520 ft)/min
Service ceiling 8,170 m (26,800 ft)
T-O run 265 m (870 ft)
T-O to 15 m (50 ft) 503 m (1,650 ft)
Landing from 15 m (50 ft) 436 m (1,430 ft)
Landing run 238 m (780 ft)
Range with max fuel 560 nm (1,038 km; 645 miles)

FUJI-BELL 204B/204B-2 and 205A-1/UH-1H

Fuji is manufacturing Bell Model 204B and UH-1H helicopters under sub-licence from Mitsui and Co Ltd, Bell's Japanese legal licensee. By 1 February 1976 a total of 43 Fuji-Bell 204Bs had been produced for civilian operators.

Orders for the UH-1B military version of the 204B, from the Japan Ground Self-Defence Force, totalled 90, all of which were delivered by early 1973. Subsequent orders are for the UH-1H version, of which the first example flew for the first time on 17 July 1973; 44 had been ordered by March 1975; a further 10 were to be ordered in FY 1976. The civil 205A-1 will be available to order.

In 1974, six UH-1Bs were evaluated by the JGSDF in an armed light helicopter role, each fitted with two pods containing nineteen 70 mm rockets apiece.

The Fuji-Bell 204B is identical with that built by Bell Helicopter Textron (see US section). It is powered by an 820 kW (1,100 shp) Kawasaki-built Lycoming K5311A turboshaft engine. The Fuji-Bell UH-1H has the same airframe and dynamic components as the Bell-built UH-1H, but has a tractor-type tail rotor and is powered by a 1,044 kW (1,400 shp) Kawasaki-built Lycoming T53-K-13B turboshaft engine.

In October 1973 Fuji developed a higher-powered version of the 204B. Powered by a 1,044 kW (1,400 shp) Lycoming T5313B turboshaft engine, it has the same basic airframe and dynamic components as the 204B, but has a tractor-type tail rotor. The first example of this version, which is designated **204B-2**, was delivered to the Asahi Helicopter Co in early 1974.

The following details apply to the standard Fuji-Bell 204B/204B-2/UH-1H:

Fuji-Bell UH-1H medium helicopter in the insignia of the Japan Ground Self-Defence Force

DIMENSIONS, EXTERNAL:
Diameter of main rotor 16·63 m (48 ft 0 in)
Diameter of tail rotor 2·59 m (8 ft 6 in)
Length overall, tail rotor turning:
204B/B-2 13·61 m (44 ft 7¾ in)
UH-1H 13·67 m (44 ft 10 in)
Length of fuselage:
204B/B-2 12·31 m (40 ft 4¾ in)
UH-1H 12·37 m (40 ft 7 in)
Height overall, tail rotor turning 4·42 m (14 ft 6 in)
Height to top of rotor hub:
204B/B-2 3·77 m (12 ft 4½ in)
UH-1H 3·98 m (13 ft 0¾ in)
Max width over landing skids:
204B/B-2 2·64 m (8 ft 8 in)
UH-1H 2·60 m (8 ft 6½ in)
Tailplane span 2·84 m (9 ft 4 in)

WEIGHTS:
Weight empty:
204B/B-2 2,177 kg (4,800 lb)
UH-1H 2,390 kg (5,270 lb)
Max T-O weight:
204B/B-2 3,855 kg (8,500 lb)
UH-1H 4,309 kg (9,500 lb)

PERFORMANCE (at max T-O weight):
Max level and cruising speed 110 knots (204 km/h; 127 mph)
Max rate of climb at S/L:
204B 463 m (1,520 ft)/min
204B-2 588 m (1,930 ft)/min
UH-1H 488 m (1,600 ft)/min
Service ceiling:
204B 4,480 m (14,700 ft)
204B-2 5,790 m (19,000 ft)
UH-1H 3,840 m (12,600 ft)
Hovering ceiling in ground effect:
204B 2,985 m (9,800 ft)
204B-2 4,635 m (15,200 ft)
UH-1H 4,145 m (13,600 ft)
Hovering ceiling out of ground effect:
204B 1,310 m (4,300 ft)
204B-2 3,200 m (10,500 ft)
UH-1H 335 m (1,100 ft)
Range at S/L:
204B 206 nm (381 km; 237 miles)
204B-2 207 nm (383 km; 238 miles)
UH-1H 252 nm (467 km; 290 miles)

KAWASAKI
KAWASAKI JUKOGYO KABUSHIKI KAISHA (Kawasaki Heavy Industries Ltd)

HEAD OFFICE:
2-16-1 Nakamachi-Dori, Ikuta-ku, Kobe
TOKYO AND AIRCRAFT GROUP OFFICE:
World Trade Center Building, 4-1, Hamamatsu-cho 2-chome, Minato-ku, Tokyo
Telephone: Tokyo (03) 435-2971
Telex: J22672 and J26888
PRESIDENT:
Kiyoshi Yotsumoto
EXECUTIVE VICE-PRESIDENTS:
Riichi Kato
Kenji Hasegawa
WORKS:
Gifu
GENERAL MANAGER, AIRCRAFT GROUP:
Kenji Uchino
ASST GENERAL MANAGER, AIRCRAFT GROUP, AND GENERAL MANAGER, AIRCRAFT SALES DIVISION:
Teruaki Yamada

With effect from 1 April 1969, Kawasaki Aircraft Co Ltd was amalgamated with the Kawasaki Dockyard Co Ltd and the Kawasaki Rolling Stock Mfg Co Ltd, to form a new company known as Kawasaki Heavy Industries Ltd. The Aircraft Division of the former Kawasaki Aircraft Co Ltd, which employs some 4,000 people, continues its activities as the Aircraft Group of this company. Kawasaki has a 34% holding in Nippi (which see).

In addition to extensive overhaul work, Kawasaki has built many US aircraft under licence since 1955.

Between 1959 and 1963 Kawasaki delivered 42 Lockheed P2V-7 (P-2H) Neptune anti-submarine aircraft to the Japan Maritime Self-Defence Force, and six more were delivered in 1965. Kawasaki then developed from the Neptune a new anti-submarine aircraft, designated P-2J, which continues in production.

Design studies are being undertaken for a new ASW aircraft, provisionally designated PX-L, to succeed the P-2J.

Kawasaki began producing the Bell Model 47 helicopter under licence in 1953 and by 1 February 1976 had built 11 Model 47Ds, 15 Model 47Gs, 180 Model 47G-2s and 33 Model 47G-2As; in addition, it built 211 Model KH-4s developed from the Model 47 by its own design staff.

Kawasaki has exclusive rights to manufacture and sell the twin-engined Boeing Vertol 107 Model II helicopter and its own KV-107/IIA development of it. The Hughes Model 369 (500S and 500M) light observation helicopter is also being assembled by Kawasaki under a licence agreement concluded in October 1967. By 1 March 1976 a total of 101 KV-107 helicopters had been delivered to customers in Japan and other countries including Sweden, Thailand and the US; and a total of 134 Hughes 500s to government and commercial operators in Japan.

Kawasaki built main wings and nacelle structures, including the landing gear, for the NAMC YS-11 turboprop transport. It is now prime contractor for the JASDF's C-1 transport; and builds main wing and tail assemblies under subcontract for Japanese-built (see Mitsubishi entry) F-4EJ Phantom II fighters for the JASDF. Cargo and passenger doors for the Lockheed L-1011 TriStar jet transport are manufactured at the Gifu works, and Kawasaki has a Boeing contract to manufacture outboard trailing-edge flaps for the Boeing 747SP transport aircraft.

Kawasaki is engaged in missile development and production; its aero-engine activities are described in the appropriate section of this edition.

KAWASAKI P-2J

JMSDF designation: P-2J

The P-2J was developed by Kawasaki, originally under the designation GK-210, to meet a JMSDF requirement for a new anti-submarine aircraft to replace its P2V-7 Neptunes in service during the 1970s. Design is based very closely upon that of the P2V-7 (P-2H), and began in October 1961. Work on the conversion of a standard P2V-7 as the P-2J prototype began in June 1965, and this aircraft flew for the first time on 21 July 1966.

The first production P-2J was flown on 8 August 1969, and was delivered to the JMSDF on 7 October. The second aircraft was also flown before the end of 1969, and eight current contracts cover the delivery of 76 P-2Js by March 1978. The first 62 P-2Js were delivered to the JMSDF by 31 March 1976. The JMSDF is to purchase an additional 6 of these aircraft, bringing the total to 82.

TYPE: Four-engined anti-submarine and maritime patrol aircraft.

WINGS: Cantilever all-metal mid-wing monoplane, with taper on outer panels. Wing section NACA 2419 (modified) at root, NACA 4410·5 at tip. Dihedral 5° on outer panels. Incidence 3° 30′ at root. No sweepback. Wing designed to give temporary flotation in event of ditching. Centre-section box beam is continuous through fuselage. All-metal ailerons, each incorporating a spring and trim tab. Fowler-type all-metal inboard and outboard trailing-edge flaps. All-metal two-section spoilers in upper surface of outer wing panels, inboard of ailerons. Thermal de-icing of leading-edges.

FUSELAGE: Conventional all-metal unpressurised semi-monocoque structure, basically as P2V-7 (P-2H) but with extra 1·27 m (4 ft 2 in) section inserted between wing leading-edge and cockpit to house improved electronic equipment.

TAIL UNIT: Cantilever all-metal structure, incorporating 'Varicam' (variable camber), a movable trimming surface between the fixed tailplane and each elevator which is operated by hydraulically-driven screwjack. Spring tab and trim tab in rudder, balance tab and spring tab in each elevator. Tail unit has thermal de-icing and is virtually unchanged from P2V-7 except for an increase in rudder area by extending the chord by 0·30 m (1 ft) at the top.

LANDING GEAR: Retractable tricycle type, with single steerable nosewheel and twin-wheel main units. Hydraulic retraction, nosewheel rearward, main wheels forward into inboard engine nacelles. Sumitomo Precision oleo-pneumatic shock-absorbers. Goodyear Type VII tubeless tyres on all units, size 9·9 × 34-14 on nosewheel and 13 × 39-16 on main wheels. Tyre pressures 6·21 bars (90 lb/sq in) on nosewheel, 6·90 bars (100 lb/sq in) on main wheels. Goodyear disc brakes and on/off-type anti-skid units on main units.

POWER PLANT: Two 2,125 kW (2,850 ehp) Japanese-built General Electric T64-IHI-10 turboprop engines, mounted on wing centre-section and each driving a Sumitomo Precision 63E60-19 three-blade variable-pitch metal propeller. Outboard of these, on underwing pylons, are two pod-mounted Ishikawajima-Harima J3-IHI-7C turbojets, each rated at 13·72 kN (3,085 lb st). Fuel in inboard and outboard wing tanks with total capacity of 11,433 litres (2,515 Imp gallons; 3,020 US gallons), plus 1,514 litres (333 Imp gallons; 400 US gallons) in port wingtip tank. For ferry purposes a 2,650 litre (583 Imp gallon; 700 US gallon) auxiliary tank can be installed in the weapons bay. Oil capacity 23·6 litres (5·2 Imp gallons; 6·2 US gallons) for each turboprop and 11 litres (2·4 Imp gallons; 2·9 US gallons) for each turbojet engine.

ACCOMMODATION: Crew of 12, including two pilots on flight deck, seven men in tactical compartment in for-

Kawasaki P-2J patrol aircraft of the Japan Maritime Self-Defence Force (two T64 turboprop engines and two J3-IHI-7C auxiliary turbojets)

ward fuselage and three aft of centre-section wing box beam. Aft of the tactical compartment, in centre fuselage, are an ordnance room, galley and toilet. Crew escape hatches in flight deck, tactical and ordnance compartments. All accommodation heated, ventilated and air-conditioned.

SYSTEMS: Primary hydraulic system, pressure 207 bars (3,000 lb/sq in) for Varicam (tail unit), landing gear and nosewheel steering. Secondary system, pressure 103·5 bars (1,500 lb/sq in) for flaps, spoilers, jet pod doors, main-wheel brakes and propeller braking. Two 40kVA generators provide 115/200V AC power at 400Hz. DC power from three 28V 200A transformer-rectifiers.

ELECTRONICS AND EQUIPMENT (early version): Communications and navigation equipment comprised HIC-3 intercom system, AN/ARC-552 UHF transceiver, HRC-6 VHF transceiver, two HRC-7 HF transceivers with N-CU-58/HRC couplers, HGC-1 teletypewriter with N-CV-55/HRC converter, RRC-15 HF emergency radio set, AN/APN-153(V) Doppler navigation system, AN/AYK-2 navigation computer, N-PT-3 plotter, N-OA-35/HSA tactical plotter, HRN-4 Loran, two HRN-2 ADF, AN/ARA-50 UHF/DF, AN/ARN-52(V) Tacan, AN/APN-171(V) radar altimeter, HRN-3 marker beacon receiver, N-O-41 altitude warning oscillator, and PB-60J autopilot. ASW equipment included AN/APS-80(J) search radar, AN/APA-125A radar display, HLR-1 ECM, AN/ASQ-10A MAD, AN/ASR-3 trail detector, two AN/ARR-52A(V) sonobuoy receivers, AQA-1 sonobuoy indicator, AQA-5 Jezebel recorder, AN/ASA-20B Julie recorder, N-R-86/HRA OTPI, AN/ASA-16 tactical display system, AN/APX-68 SIF transponder, HPX-1 IFF interrogator, N-KY-22/HPX IFF decoder group, AN/ASA-50 ground speed and bearing computer, N-RO-5/HMH BT recorder, HQH-101 ionisation data recorder, and HSA-1 sonobuoy data display system. Searchlight in starboard wingtip pod.

ELECTRONICS AND EQUIPMENT (later, modernised version): Communications and navigation equipment comprises HIC-3 interphone, HRC-110 UHF transceiver, HRC-106 VHF transceiver, two HRC-107 HF transceivers, two N-CU-58/HRC HF antenna couplers, HGC-102 teletypewriter, HGA-101 TTY converter, HSC-1 TTY security unit, RRC-15 emergency radio, AN/APN-187B-N Doppler radar, N-PT-3 navigation plotter, N-OA-35/HSA tactical plotter, HRN-104 Loran, HRN-101 ADF, AN/ARA-50 UHF/DF, HRN-105 Tacan, AN/APN-171-N1 radar altimeter, HRN-106B VOR/ILS/MKR receiver, PB-60J autopilot, MAD manoeuvre programmer and flight director system. ASW equipment comprises HSA-116 integrated data display system and digital data processor (for both navigation and tactical data), AN/APS-80-N search radar, AN/APA-125-N radar indicator, HLR-101 ESM, HSQ-101 MAD, HSA-102 AMC, HSA-103 SAD, AN/ASA-20B Julie recorder, AN/AQA-5-N Jezebel recorder, HQA-101 active sonobuoy indicator, AN/ARR-52A(V) sonobuoy receiver, N-R-86/HRA OTPI, HQH-101 data recorder, AN/APX-68-N SIF transponder, HPX-101 IFF interrogator, N-KY-122/HPX IFF decoder, and HSA-1B sonobuoy data display system. Searchlight in starboard wingtip pod.

DIMENSIONS, EXTERNAL:
- Wing span: 29·78 m (97 ft 8½ in)
- Wing span over tip-tanks: 30·87 m (101 ft 3½ in)
- Wing chord at root: 4·45 m (14 ft 7¼ in)
- Wing chord at tip: 2·22 m (7 ft 3½ in)
- Wing aspect ratio: 10

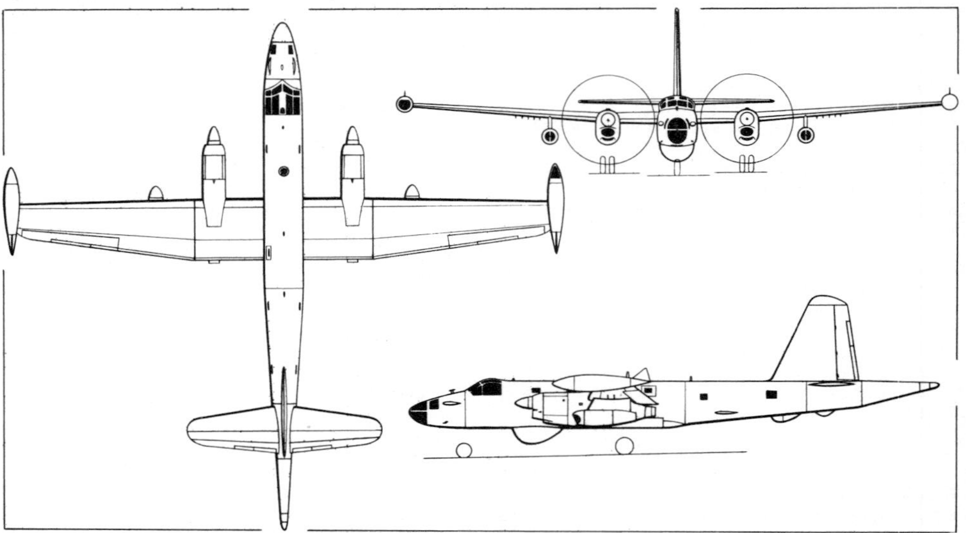

Kawasaki P-2J twin-turboprop development of the Lockheed Neptune *(Pilot Press)*

- Length overall: 29·23 m (95 ft 10¾ in)
- Height overall: 8·93 m (29 ft 3½ in)
- Tailplane span: 10·36 m (34 ft 0 in)
- Wheel track (c/l of shock-absorbers): 7·62 m (25 ft 0 in)
- Wheelbase: 8·84 m (29 ft 0 in)
- Propeller diameter: 4·43 m (14 ft 6¼ in)
- Distance between propeller centres: 7·62 m (25 ft 0 in)

DIMENSIONS, INTERNAL (tactical compartment):
- Cabin: Length: 5·49 m (18 ft 0 in)
  - Max width: 2·03 m (6 ft 8 in)
  - Max height (at c/l): 1·55 m (5 ft 1 in)
  - Floor area: 11·15 m² (120 sq ft)
  - Volume: 14·16 m³ (500 cu ft)

AREAS:
- Wings, gross: 92·9 m² (1,000 sq ft)
- Ailerons (total): 5·87 m² (63·2 sq ft)
- Trailing-edge flaps (total): 16·70 m² (197·6 sq ft)
- Spoilers (total): 1·41 m² (15·2 sq ft)
- Fin: 17·65 m² (190·0 sq ft)
- Rudder, incl tabs: 4·04 m² (43·5 sq ft)
- Tailplane: 21·50 m² (231·0 sq ft)
- Elevators, incl tabs and Varicam: 8·45 m² (91·0 sq ft)

WEIGHTS AND LOADINGS:
- Weight empty: 19,277 kg (42,500 lb)
- Max T-O weight: 34,019 kg (75,000 lb)
- Max zero-fuel weight: 23,087 kg (50,900 lb)
- Max landing weight: 28,122 kg (62,000 lb)
- Max wing loading: 366·2 kg/m² (75·00 lb/sq ft)
- Max power loading: 8·00 kg/kW (13·16 lb/ehp)

PERFORMANCE (at max T-O weight):
- Never-exceed speed: 350 knots (649 km/h; 403 mph)
- Max cruising speed: 217 knots (402 km/h; 250 mph)
- Econ cruising speed at 3,050 m (10,000 ft): 200 knots (370 km/h; 230 mph)
- Stalling speed, flaps down: 90 knots (166 km/h; 103 mph)
- Max rate of climb at S/L: 550 m (1,800 ft)/min
- Service ceiling: 9,150 m (30,000 ft)
- T-O to 15 m (50 ft): 1,100 m (3,600 ft)
- Landing from 15 m (50 ft): 880 m (2,880 ft)
- Range with max fuel: 2,400 nm (4,450 km; 2,765 miles)

KAWASAKI C-1

The C-1 is a medium-sized troop and freight transport designed to meet the JASDF's requirement for a replacement for its fleet of Curtiss C-46 transports. Preliminary design was started by NAMC in 1966, and in 1968 a prototype development contract was awarded. Following the completion of a full-size mockup in March 1968, construction by NAMC began in the following Autumn of two XC-1 flying prototypes and one airframe for static tests. The first flying prototype, assembled at Kawasaki's Gifu factory, made its first flight on 12 November 1970, followed by the second on 16 January 1971. These prototypes were handed over to the Japan Defence Agency on 24 February and 20 March 1971 respectively. Further development and evaluation tests by the JDA were completed in March 1973.

Two pre-production aircraft had been delivered by the end of February 1974. Airframe fatigue testing by the JDA was completed in November 1974.

Eleven production C-1s were ordered in FY 1972, of which the first was delivered in December 1974; all 11 had been delivered by the end of March 1976.

An additional 13 C-1s were ordered during FY 1975; these are due to be delivered during FY 1977, and further production is anticipated.

Prime contractor in the C-1 programme is Kawasaki, which builds the front fuselage and wing centre-section and undertakes final assembly and flight testing. Major subcontractors are Fuji (outer wing panels); Mitsubishi (centre and aft fuselage sections and tail surfaces); Nihon Hikoki (Nippi) (flaps, slats, spoilers, ailerons, engine pylons and pods); and Shin Meiwa (cargo loading equipment).

TYPE: Twin-turbofan medium-range transport.

WINGS: Cantilever high-wing monoplane. Wings have 20° sweepback at quarter-chord, with slightly increased leading-edge sweep inboard of the engine pylons. Thickness/chord ratio 12% at root, 11% at tip. Anhedral 5° 30′ from centre-section. Conventional two-spar fail-safe structure of aluminium alloy, including control surfaces. Two quadruple-slotted flaps on each trailing-edge, with 75° travel. Forward of these, on each wing, are three flight spoilers and a ground spoiler. Drooping leading-edge slats, in four sections, on each wing. Manually-operated aileron outboard of each outer flap. Trim tab in port aileron. Thermal anti-icing of leading-edges, using engine bleed air.

FUSELAGE: Conventional semi-monocoque fail-safe structure of aluminium alloy, with a circular cross-section.

TAIL UNIT: Aluminium alloy cantilever T-tail, with sweepback on all surfaces (30° at fin quarter-chord, 25° at tailplane quarter-chord). Tailplane has 5° anhedral. Variable-incidence tailplane, with elevators. Balance tab in each elevator and anti-balance tab in rudder. Elevators and rudder are each operated by two independent hydraulic actuator systems; the elevators can be operated manually in an emergency. Thermal de-icing of tailplane, using electric heater mat.

LANDING GEAR: Hydraulically-retractable tricycle type, of Sumitomo design. Each main unit has two pairs of wheels in tandem, retracting forward into fairings built on to the sides of the fuselage. Forward-retracting nose unit has twin wheels. Oleo shock-absorbers. Kayaba wheels with Dunlop tyres, which on main units have pressure of 5·17 bars (75 lb/sq in). Kayaba hydraulic brakes (two-rotor on first 10 production aircraft, three-rotor from 11th aircraft onward); Sumisei anti-skid units.

POWER PLANT: Two 64·5 kN (14,500 lb st) Mitsubishi (Pratt & Whitney) JT8D-M-9 turbofan engines, installed in pylon-mounted underwing pods and fitted with thrust reversers. Four integral wing fuel tanks with total capacity of 15,200 litres (3,344 Imp gallons). Single pressure-refuelling point for all tanks, plus overwing gravity refuelling point for each tank.

ACCOMMODATION: Crew of five, comprising pilot, co-pilot, navigator, flight engineer and load supervisor. Escape hatch in flight deck roof on starboard side. Flight deck and main cabin pressurised and air-conditioned. Standard complements are as follows: troops (max) 60, paratroops (max) 45, stretchers 36 plus attendants. As a cargo carrier, loads can include a 2½ ton truck, a 105 mm howitzer, two ¾ ton trucks or three jeeps. Up to three preloaded freight pallets, 2·24 m (7 ft 4 in) wide and 2·74 m (9 ft 0 in) long, can be carried. Floor is stressed for loads of up to 7 kg/cm² (100 lb/sq in). Access to flight deck via downward-opening door, with built-in stairs, on port side of forward fuselage. Paratroop door on each side of fuselage, aft of wing trailing-edge. For air-dropping, the rear-loading ramp/door at the rear of the cabin can be opened in flight to the full cabin cross-section.

SYSTEMS: Pressurisation and air-conditioning systems utilise engine bleed air. APU in front section of starboard landing gear fairing supplies electrical power on ground and in the air, and bleed air on ground. Three independent hydraulic systems, each 207 bars (3,000 lb/sq in). No. 1 system actuates flight controls; No. 2 actuates flight controls, high-lift devices, landing gear, nosewheel steering and brakes; No. 3 system is used for cargo door operation, and provides emergency backup for brakes and high-lift devices. Electrical power supplied by three 40kVA AC generators, two engine-driven and aircooled, and one driven by the APU. 28V DC power is obtained from AC source through a transformer-rectifier. One 24V 30Ah nickel-cadmium battery for emergency DC power.

ELECTRONICS AND EQUIPMENT: Standard equipment includes autopilot, Doppler radar, radio altimeter, HF, VHF and UHF radio, ADF, UHF/DF, marker beacon, VOR/ILS, Tacan, SIF, dual compass system and flight director system. Optional equipment includes Loran and weather radar.

DIMENSIONS, EXTERNAL:
Wing span 30·60 m (100 ft 4¾ in)
Wing chord at root 6·30 m (20 ft 8 in)
Wing chord at tip 2·00 m (6 ft 6¾ in)
Wing aspect ratio 7·8
Length overall 29·00 m (95 ft 1¾ in)
Length of fuselage 26·50 m (86 ft 11¼ in)
Height overall 9·99 m (32 ft 9¼ in)
Tailplane span 11·30 m (37 ft 1 in)
Wheel track 4·40 m (14 ft 5¼ in)
Wheelbase 9·33 m (30 ft 7¼ in)
Rear-loading ramp/door:
Length 2·67 m (8 ft 9¼ in)
Width 2·70 m (8 ft 10¼ in)
Height to sill 1·25 m (4 ft 1¼ in)

DIMENSIONS, INTERNAL:
Cabin: Max length 10·80 m (35 ft 5¼ in)
Max width 3·60 m (11 ft 9¾ in)
Max height 2·55 m (8 ft 4½ in)
Floor area 28·6 m² (308 sq ft)
Volume (excl ramp area) 73·8 m³ (2,606 cu ft)

AREAS:
Wings, gross 120·5 m² (1,297 sq ft)
Ailerons (total) 3·4 m² (36·6 sq ft)
Trailing-edge flaps (total) 22·9 m² (246·5 sq ft)
Spoilers (total) 8·9 m² (95·8 sq ft)
Fin 15·8 m² (170·1 sq ft)
Rudder, incl tabs 6·4 m² (68·9 sq ft)
Tailplane 18·3 m² (197·0 sq ft)
Elevators, incl tabs 6·5 m² (70·0 sq ft)

WEIGHTS AND LOADINGS:
Weight empty 23,320 kg (51,410 lb)
Weight empty, equipped 24,300 kg (53,572 lb)
Normal payload 7,900 kg (17,416 lb)
Max T-O weight 38,700 kg (85,320 lb)
Max wing loading 321·2 kg/m² (65·79 lb/sq ft)
Max power loading 300 kg/kN (2·94 lb/lb st)

PERFORMANCE (at max T-O weight except where indicated):
Max level speed at 7,620 m (25,000 ft) at 35,450 kg (78,150 lb) AUW 435 knots (806 km/h; 501 mph)
Econ cruising speed at 10,670 m (35,000 ft) at 35,450 kg (78,150 lb) AUW 354 knots (657 km/h; 408 mph)
Max rate of climb at S/L 1,065 m (3,500 ft)/min
Service ceiling 11,580 m (38,000 ft)
Service ceiling, one engine out 5,485 m (18,000 ft)
T-O run 640 m (2,100 ft)
T-O to 15 m (50 ft) 910 m (3,000 ft)
Landing from 15 m (50 ft) at 36,860 kg (81,260 lb) weight 823 m (2,700 ft)
Landing run at 36,860 kg (81,260 lb) weight 455 m (1,500 ft)
Range with max fuel and 2,200 kg (4,850 lb) payload 1,810 nm (3,353 km; 2,084 miles)
Range with 7,900 kg (17,416 lb) max payload 700 nm (1,300 km; 807 miles)

KAWASAKI KH-4

JGSDF designation: H-13KH

Developed by Kawasaki from the three-seat Bell Model 47G-3B, the KH-4 four-seat light general-purpose helicopter (201 kW; 270 hp Lycoming TVO-435-D1A flat-six engine) flew for the first time in August 1962 and received JCAB Normal category type approval on 9 November 1962.

A total of 211 KH-4s were built before production ended in mid-1975. Of this total, 158 were delivered to civil operators, 19 to the JGSDF, 28 to Thailand, 5 to Korea and 1 to the Philippines.

A description and illustration of the KH-4 can be found in the 1975-76 *Jane's*.

KAWASAKI KH-7

Kawasaki has completed the design of a new 7/10-seat utility helicopter designated KH-7. To be powered by two Lycoming turboshaft engines in the 590 shp class, it will have a max T-O weight of about 2,700 kg (5,950 lb) and a range of about 300 nm (555 km; 345 miles), and will have both civil and military applications. A mockup has been completed, and negotiations continue with potential foreign collaborators in the programme.

KAWASAKI (BOEING VERTOL) KV-107/II and KV-107/IIA

Swedish Navy designation: HKP 4C

Kawasaki has exclusive rights to manufacture and sell the Boeing Vertol 107 Model II helicopter. The first KV-107 to be produced by Kawasaki under this licence agreement flew for the first time in May 1962.

In 1965, Kawasaki obtained world-wide sales rights in the KV-107 from The Boeing Company's Vertol Division. In November 1965, it was awarded a type certificate for the KV-107 by the FAA.

An improved model, the **KV-107/IIA**, is available in any of the KV-107/II forms, powered by two 1,044 kW (1,400 shp) General Electric CT58-140-1 or Ishikawajima-

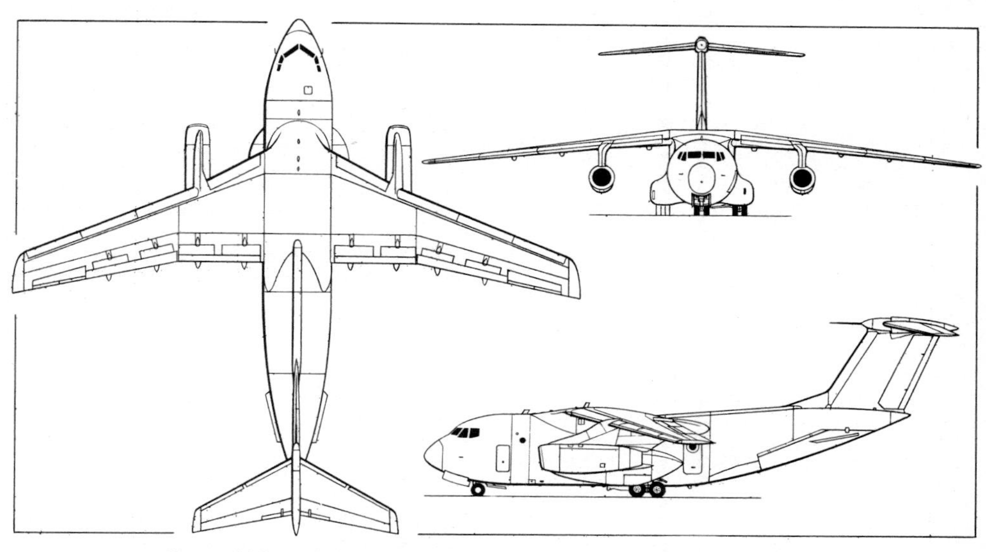

Kawasaki C-1 twin-turbofan medium-range military transport *(Pilot Press)*

First production Kawasaki C-1 medium-range military transport aircraft for the JASDF (two Pratt & Whitney JT8D-9 turbofan engines)

Kawasaki KV-107/IIA-3 tandem-rotor helicopter of the JMSDF's 111th Air Wing performing minesweeping drill

Harima CT58-IHI-140-1 turboshaft engines (max continuous rating 932 kW; 1,250 shp), which give improved performance during VTOL and in 'hot and high' conditions. Fuel capacity 1,323 litres; 350 US gallons (standard), 3,785 litres; 1,000 US gallons (max). A prototype (JA9509) was first flown on 3 April 1968. Type approval granted by JCAB on 26 September 1968 and by FAA on 15 January 1969.

The following versions of the KV-107/II and IIA have been announced:

KV-107/II-1. Basic utility helicopter. None yet built.

KV-107/II-2. Basic airline helicopter. Ten built by 31 March 1968, for Thailand (three), Pan American (two, for operation by New York Airways), New York Airways (one) and Air Lift Inc of Japan (formerly Kanki Airlines, two). Remaining two used as company test aircraft.

KV-107/II-3. Mine countermeasures (MCM) helicopter for JMSDF with extended-range fuel tanks, towing hook and cargo sling. Nine ordered, all of which had been delivered by early 1976, including seven of the **KV-107/IIA-3** version with uprated power. All of these are fitted with minesweeping and retrieval equipment. Each aircraft of the JMSDF's minesweeping unit, the 111th Air Wing, logs an average of approx 40 hr per month minesweeping drill.

KV-107/II-4. Tactical cargo/troop transport for JGSDF, with foldable seats for 26 troops or 15 casualty litters. Strengthened floor for carrying heavy vehicles. Orders for 54 placed, of which 52 had been delivered by early 1976, including one equipped as a VIP transport for Cabinet use. The latest 10 aircraft of the 52 delivered are of the **KV-107/IIA-4** version with uprated power, and four of them are fitted with extended-range fuel tanks.

KV-107/II-5. Long-range search and rescue helicopter for JASDF. Orders for 24 placed, of which 22 had been delivered by early 1976, including seven uprated **KV-107/IIA-5s**. Large additional fuel tank each side of fuselage, making total capacity 3,785 litres (1,000 US gallons). Extensive nav/com equipment, four searchlights, domed observation window and rescue hoist. Seven (plus one additional) basically similar aircraft ordered for Swedish Navy have Kawasaki-built airframes and rotor assemblies but were fitted in Sweden with Rolls-Royce Bristol Gnome H.1200 turboshaft engines and a Decca navigation system. They have a Kawasaki/Boeing automatic flight control system, enabling the aircraft to cruise at preselected altitude and speed, descend at a programmed rate and distance, and come to hover at a preselected altitude. Also provided are automatic climb-out to the cruise mode; standard distance approach; a turns coupler to a preselected heading; altitude sensing, with dual radar altimeters, for added safety in IFR operations; and a vernier flight control to permit critical positioning during rescue hoist operations. Provision for additional features such as a programmed procedural turn for automatic approach to rescue, and automatic approach guided by radio signal from person being rescued. First of these aircraft was delivered to the Swedish Navy on 30 October 1972; all had been delivered by June 1974.

KV-107/II-6. De luxe transport version. None yet built.

KV-107/II-7. De luxe VIP transport with 6-11 seats. One sold to Thailand in 1964.

KV-107/IIA-17. Long-range passenger and cargo transport version for Tokyo Metropolitan Police Department; one delivered in February 1973. Cabin divided into two compartments: front section with 12 passenger seats, rear section capable of accommodating 2,268 kg (5,000 lb) of cargo, six stretcher patients or 12 troops.

The description which follows applies to the commercial KV-107/II-2, except where shown:

TYPE: Twin-engined transport helicopter.

ROTOR SYSTEM: Two three-blade rotors in tandem, rotating in opposite directions. Each blade is made up of a steel D spar to which is bonded a trailing-edge box constructed of aluminium ribs and glassfibre or aluminium skin.

ROTOR DRIVE: Power is transmitted from each engine through individually-overrunning clutches into the aft transmission, which combines the engine outputs, thereby providing a single power output to the interconnecting shaft which enables both rotors to be driven by either engine.

FUSELAGE: Basically square-section semi-monocoque structure built primarily of high-strength bare and Alclad aluminium alloy. Transverse bulkheads and built-up frames support transmission, power plant and landing gear. Loading ramp forms undersurface of upswept rear fuselage on utility and military models. Baggage container replaces ramp on airliner version. Fuselage is sealed to permit operation from water.

LANDING GEAR: Non-retractable tricycle type, with twin wheels on all three units. Oleo-pneumatic shock-absorbers. Tubeless tyres, size 18 × 5·5, pressure 10·34 bars (150 lb/sq in), on all wheels. Disc brakes. Wheel/ski gear optional.

POWER PLANT (KV-107/II): Two 932 kW (1,250 shp) General Electric CT58-110-1 or Ishikawajima-Harima CT58-IHI-110-1 turboshaft engines, mounted side by side at base of rear rotor pylon. Alternatively, two Rolls-Royce Bristol Gnome H.1200 turboshafts (in HKP 4C). Fuel tanks in sponsons, capacity 1,323 litres (350 US gallons). KV-107/IIA has more powerful CT58 engines and provision for increased fuel capacity (see introductory copy).

ACCOMMODATION: Standard accommodation for two pilots, stewardess and 25 passengers in airliner version. Seats in eight rows, in pairs on port side and single seats on starboard side (two pairs at rear of cabin) with central aisle. Airliner fitted with parcel rack and a roll-out baggage container, with capacity of approximately 680 kg (1,500 lb), located in underside of rear fuselage. Ramp of utility model is power-operated on the ground or in flight and can be removed or left open to permit carriage of extra-long cargo.

ELECTRONICS AND EQUIPMENT: Standard equipment includes stability augmentation system (SAS) and automatic speed trim system (AST). Optional equipment includes automatic stabilisation equipment (ASE); automatic flight control system (AFCS); Doppler radar; radio altimeter; HF, VHF and UHF radio; ADF; VOR/ILS; Tacan; compass system and attitude director indicator system; and intercom system.

DIMENSIONS, EXTERNAL:
- Rotor diameter (each): 15·24 m (50 ft 0 in)
- Length overall, blades turning: 25·40 m (83 ft 4 in)
- Length of fuselage: 13·59 m (44 ft 7 in)
- Height to top of rear rotor hub: 5·09 m (16 ft 8½ in)
- Wheel track: 3·92 m (12 ft 10½ in)
- Wheelbase: 7·57 m (24 ft 10 in)
- Passenger door(fwd):
  - Height: 1·60 m (5 ft 3 in)
  - Width: 0·91 m (3 ft 0 in)

DIMENSIONS, INTERNAL:
- Cabin, excl flight deck:
  - Length: 7·37 m (24 ft 2 in)
  - Normal width: 1·83 m (6 ft 0 in)
  - Max height: 1·83 m (6 ft 0 in)
  - Floor area: 13·47 m² (145 sq ft)
  - Volume (usable): 24·5 m³ (865 cu ft)

AREAS:
- Rotor blades (each): 3·48 m² (37·50 sq ft)
- Rotor discs (total): 364·6 m² (3,925 sq ft)

WEIGHTS AND LOADINGS:
- Weight empty, equipped:
  - II-2: 4,868 kg (10,732 lb)
  - IIA-1: 4,589 kg (10,118 lb)
  - IIA-2: 5,250 kg (11,576 lb)
- Max T-O and landing weight: 8,618 kg (19,000 lb) *or* 9,706 kg (21,400 lb)
- Max disc loading: 23·6 kg/m² (4·84 lb/sq ft)
- Max power loading: 4·62 kg/kW (7·6 lb/shp)

PERFORMANCE (A: KV-107/II-2 at 8,618 kg; 19,000 lb AUW. B: KV-107/IIA at same AUW):
- Never-exceed speed:
  - A, B: 146 knots (270 km/h; 168 mph)
- Max speed at S/L, normal rated power:
  - A: 136 knots (253 km/h; 157 mph)
  - B: 137 knots (254 km/h; 158 mph)
- Cruising speed at 1,525 m (5,000 ft):
  - A, B: 130 knots (241 km/h; 150 mph)
- Max rate of climb at S/L:
  - A, normal rated power: 463 m (1,520 ft)/min
  - B: 625 m (2,050 ft)/min
- Max vertical rate of climb at S/L:
  - B: 381 m (1,250 ft)/min
- Service ceiling:
  - A, normal rated power: 4,570 m (15,000 ft)
  - B: 5,180 m (17,000 ft)
- Service ceiling, one engine out:
  - A, military power, yaw, 248 rpm: 107 m (350 ft)
  - B: 1,740 m (5,700 ft)
- Hovering ceiling in ground effect:
  - A: 2,895 m (9,500 ft)
  - B: 3,565 m (11,700 ft)
- Hovering ceiling out of ground effect:
  - A: 1,890 m (6,200 ft)
  - B: 2,680 m (8,800 ft)
- Min landing area (A, B):
  - Length: 38 m (126 ft)
  - Width: 23 m (75 ft)
- T-O to 15 m (50 ft):
  - B: 131 m (430 ft)
- Landing from 15 m (50 ft), one engine out:
  - B: 84 m (275 ft)
- Range:
  - A with 3,000 kg (6,600 lb) payload, 10% reserves: 94 nm (175 km; 109 miles)
  - B with standard fuel: 192 nm (357 km; 222 miles)
  - B with max fuel: 592 nm (1,097 km; 682 miles)

KAWASAKI (HUGHES) 369/MODEL 500 SERIES

JGSDF and JMSDF designation: OH-6J

A total of 101 Model 369HM helicopters, assembled by Kawasaki under licence from the Hughes Helicopters Division of Summa Corporation, had been delivered to the JGSDF (97), JMSDF (3) and civil operators (1) by 1 March 1976. By the same date, a total of 33 Model 369HS helicopters, including four Model 500Cs, had been delivered to civil customers in Japan. Power plants are the 236 kW (317 shp) Allison 250-C18A turboshaft engine, built in Japan by Mitsubishi, for the OH-6J and Model 369HS/500; and the 298 kW (400 shp) Allison 250-C20 turboshaft for the Model 369HS/500C.

DIMENSIONS AND AREAS: As for Hughes Model 369 (see US section)

WEIGHTS AND LOADING:

Weight empty:	
OH-6J, 500	537 kg (1,185 lb)
500C	545 kg (1,203 lb)
Max T-O and landing weight	1,156 kg (2,550 lb)
Design disc loading	22·8 kg/m² (4·68 lb/sq ft)

PERFORMANCE (at max T-O weight, ISA):

Max level and never-exceed speed	130 knots (241 km/h; 150 mph)
Cruising speed	125 knots (232 km/h; 144 mph)
Max rate of climb at S/L	518 m (1,700 ft)/min
Hovering ceiling in ground effect:	
OH-6J, 500	2,500 m (8,200 ft)
500C	3,930 m (12,900 ft)
Hovering ceiling out of ground effect:	
OH-6J, 500	1,615 m (5,300 ft)
500C	2,040 m (6,700 ft)
Range with max fuel:	
OH-6J, 500	307 nm (568 km; 353 miles)
500C	301 nm (557 km; 346 miles)
Max endurance:	
OH-6J, 500	3 hr 36 min
500C	3 hr 24 min

Kawasaki (Hughes) Model 369HM light helicopter of the Japan Ground Self-Defence Force

MITSUBISHI
MITSUBISHI JUKOGYO KABUSHIKI KAISHA (Mitsubishi Heavy Industries Ltd)

HEAD OFFICE:
5-1, Marunouchi 2-Chome, Chiyoda-ku, Tokyo 100
Telephone: Tokyo (03) 212-3111
Telex: J22381 and J22443

NAGOYA AIRCRAFT WORKS:
10, Oye-cho, Minato-ku, Nagoya 455

CHAIRMAN OF BOARD OF DIRECTORS:
Shigeichi Koga

PRESIDENT: Gakuji Moriya

EXECUTIVE VICE-PRESIDENTS:
Kazuo Naito
Masao Kanamori
Okichi Salto

MANAGING DIRECTOR AND GENERAL MANAGER OF AIRCRAFT HEADQUARTERS:
Teruo Tojo

DIRECTOR AND VICE GENERAL MANAGER OF AIRCRAFT HEADQUARTERS:
Kenji Ikeda

MANAGER, AIRCRAFT ADMINISTRATION DEPARTMENT:
Masayoshi Sato

MANAGER, AIRCRAFT DEPARTMENT:
Hiroshi Hamada

MANAGER, AIRCRAFT EQUIPMENT DEPARTMENT:
Seiichi Tsukada

MANAGER, SPACE SYSTEM DEPARTMENT:
Yoshiaki Kato

MANAGER OF MU-2 ADMINISTRATION SECTION:
Kotaro Yoshizawa

GENERAL MANAGER, NAGOYA AIRCRAFT WORKS:
Chushichi Ueda

Mitsubishi began the production of aircraft in the present Oye plant of its Nagoya Engineering Works in 1921, and manufactured a total of 18,000 aircraft of approximately 100 different types during the 24 years prior to the end of the second World War in 1945.

The company was also one of the leading aero-engine manufacturers in Japan, and produced a total of 52,000 engines in the 1,000-2,500 hp range.

The conclusion of the Peace Treaty in 1952 enabled the aircraft industry in Japan to recommence, and in December of that year the company constructed its present Komaki South plant.

This factory, together with Mitsubishi's Oye, Daiko and Komaki North plants, was later separated from the original Nagoya Engineering Works and consolidated as the Nagoya Aircraft Works, with a combined floor area of 247,500 m² (2,750,000 sq ft).

Like other Japanese companies, Mitsubishi restarted with overhaul work for the USAF. Contracts to overhaul F-86 Sabre fighters led, in June 1955, to the selection of Mitsubishi as the company to manufacture 300 F-86F fighters for the Japan Air Self-Defence Force under a licence agreement with North American Aviation Inc.

It subsequently produced a total of 230 Lockheed F-104J and F-104DJ Starfighters, in co-operation with Kawasaki.

In co-operation with Kawasaki, Mitsubishi is producing F-4EJ Phantom tactical fighters for the JASDF, under licence from McDonnell Douglas Corporation. The first two F-4EJs, built by McDonnell Douglas, were delivered to Japan in July 1971. The next eight were assembled from knock-down components, the first of these making its first flight on 12 May 1972. The remaining 118 are being built entirely in Japan. The first JASDF unit to equip with the F-4EJ was the 301st Squadron at Hyakuri, which was formed in August 1972, and a total of 82 F-4EJs had been delivered to the JASDF by the end of March 1976. The second and third F-4EJ units are the 302nd Squadron at Chitose and the 303rd at Komatsu; the 304th and 305th Squadrons were due to form with F-4EJs during 1976. Fourteen US-built RF-4EJs were delivered by mid-1975, and these are in service with the 501st Squadron.

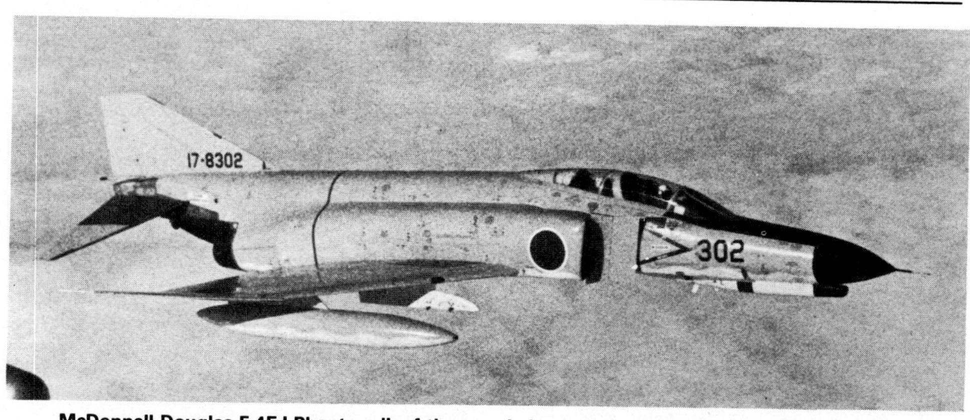

McDonnell Douglas F-4EJ Phantom II, of the type being produced by Mitsubishi for the JASDF

Mitsubishi's overhaul work on Sikorsky S-55 helicopters, started in 1954, led in December 1958 to a licence agreement for the manufacture of this type in the Oye and Komaki plants, and 44 were delivered for civil and military use. Subsequently, Mitsubishi assembled 20 S-58/HSS-1s for the JMSDF (17), Maritime Safety Agency (2) and Asahi Helicopter Ltd (1).

Today, Mitsubishi holds licence agreements to manufacture the Sikorsky S-61A, S-61A-1, S-61B (HSS-2) and S-61B-1 (HSS-2A) helicopters. By 31 March 1976, Mitsubishi had built three S-61As (for the JMSDF, for use in support of the Japanese Antarctic Expedition) and had delivered 67 of an order for 73 S-61Bs (HSS-2s) and S-61B-1s (HSS-2As) to the JMSDF for anti-submarine duties. From 1976, two S-61A-1s for use in support of the Japanese Antarctic Expedition, two S-61A-1s for rescue duties, and 16 S-61B-1s for anti-submarine duties with the JMSDF, are expected to be produced. Mitsubishi had previously delivered 25 S-62As, including 9 to the JASDF, 9 to the JMSDF, 1 to the Maritime Safety Agency, 3 to civil operators, 2 to the Philippines and 1 to Thailand.

Mitsubishi is producing a twin-turboprop utility transport designated MU-2. It built front and centre fuselage sections of the NAMC YS-11 transport and was responsible for final assembly of this aircraft. Mitsubishi is also a subcontractor in the production programme for the Kawasaki C-1 (which see).

In September 1967 the Japan Defence Agency nominated Mitsubishi as prime contractor for development of the T-2 supersonic trainer and FS-T2-KAI close-support combat aircraft for the JASDF, with Fuji, Nippi and Shin Meiwa as principal subcontractors.

Current activities at the Daiko plant include various aero-engine activities; these are described in the appropriate section of this edition.

MITSUBISHI MU-2
JGSDF designation: LR-1

The MU-2 is a twin-turboprop STOL utility transport, the basic design of which was begun in 1960. Prototype construction began in 1962 and the first aircraft was flown on 14 September 1963. By 1 February 1976, total orders for the MU-2 (all versions) had reached 422, including 387 for export and 35 for Japanese customers. Eleven versions have been announced, of which the MU-2A (3 built), MU-2B (34 built), MU-2D (18 built), MU-2E (16 built), MU-2F (95 built) and MU-2G (41 built) have been described in the 1965-66 and subsequent editions of *Jane's*. The five current versions are as follows:

MU-2C. Unpressurised liaison and reconnaissance/support version for JGSDF. First flown on 11 May 1967, and first production aircraft delivered on 30 June 1967. Seven delivered to JGSDF, whose designation is **LR-1.** On first aircraft only, wingtip fuel tanks were replaced by fuselage tank aft of cabin. Remainder have wingtip tanks as standard. One vertical and one swing-type oblique camera in photographic version.

MU-2J. Basically similar to MU-2G (1973-74 *Jane's*), but with more powerful engines. First flown August 1970. JCAB certification February 1971, FAA Type Approval May 1971. Total of 115 delivered by 1 January 1976. Two ordered by JASDF for flight calibration duties, of which one delivered. In production.

MU-2K. Basically similar to MU-2F (1973-74 *Jane's*), but with more powerful AiResearch engines and higher max cruising speed. First flown in August 1971. JCAB certification February 1972, FAA Type Approval May 1972. Total of 82 delivered by 31 March 1976. Six ordered by JASDF for rescue duties, of which five delivered. In production.

MU-2L. Basically similar to MU-2J, but max T-O weight increased to 5,250 kg (11,575 lb), representing a net increase of more than 272 kg (600 lb) in useful load. JCAB certification June 1974, FAA Type Approval July 1974. Total of 35 delivered by 31 January 1976. In production.

MU-2M. Basically similar to MU-2K, but max T-O weight increased to 4,750 kg (10,470 lb). JCAB and FAA certification dates as for MU-2L. Total of 22 delivered by 31 January 1976. In production.

A subsidiary company, Mitsubishi Aircraft International Inc, in San Angelo, Texas, was established on 1 October 1967 for final assembly in the US of semi-finished MU-2s shipped from Japan and for marketing of the aircraft in the western hemisphere.

The following description applies generally to the MU-2J, K, L and M, except where a specific version is indicated:

TYPE: Twin-turboprop utility transport.

WINGS: Cantilever high-wing monoplane. Wing section NACA 64A415 at root, NACA 63A212 (modified) at tip. No dihedral. Incidence 2°. Washout 3°. Sweepback 0° 21′ at quarter-chord. One-piece two-spar all-metal structure with chemically milled skins of 2024 and 7075 aluminium alloy. Spoilers for lateral control, between rear spar and flaps. Electrically-actuated full-span double-slotted Fowler-type flaps of aluminium alloy and plastics construction. Outboard flap section each side incorporates trim aileron. All primary controls manually-operated. Pneumatic de-icing boots.

Fuselage: Circular-section aluminium alloy semi-monocoque structure.

Tail Unit: Cantilever structure of aluminium alloy, except for top of fin, which is of reinforced plastics. Small auxiliary fin beneath each side of rear fuselage on MU-2J and L. Trim tab in rudder and each elevator. Pneumatic de-icing boots.

Landing Gear (MU-2J and L): Retractable tricycle type, with single wheel on each main unit and twin-wheel steerable nose unit. All wheels retract electrically, nosewheel forward, main wheels upward into fairings on fuselage sides. Manual backup system provided. Oleo-pneumatic shock-absorbers. Main-wheel tyres Type III, size 8·50-10 (10-ply). Nosewheel tyres Type III, size 5·00-5 (6-ply). Tyre pressure 2·76-4·76 bars (40-69 lb/sq in) on main units, 3·79 bars (55 lb/sq in) on nose unit. Goodrich single-disc nine-spot hydraulic brakes.

Landing Gear (MU-2K and M): Retractable tricycle type, with single wheel on each main unit and twin-wheel steerable nose unit. All wheels retract electrically into fuselage. Oleo-pneumatic shock-absorbers. Main-wheel tyres Type III 8·50-10 (10-ply). Nose-wheel tyres Type III 5·00-5 (6-ply). Main-wheel tyre pressure 2·76-4·21 bars (40-61 lb/sq in), nosewheel tyre pressure 3·79 bars (55 lb/sq in). Goodrich single-disc nine-spot hydraulic brakes.

Power Plant: Two AiResearch TPE 331-6-251M turboprop engines, each rated at 540 kW (724 ehp) in MU-2J, K and M; and 579 kW (776 ehp) in MU-2L. Hartzell HC-B3TN-5/T10178HB-11 or -11R fully-feathering three-blade reversible-pitch constant-speed propellers. Fuel in two wing leading-edge tanks, total usable capacity 706 litres (186 US gallons; 155 Imp gallons) and two fixed wingtip tanks with total usable capacity of 682 litres (180 US gallons; 150 Imp gallons). Max total fuel capacity of 1,388 litres (366 US gallons; 305 Imp gallons). Oil capacity 11·8 litres (3·1 US gallons; 2·6 Imp gallons).

Accommodation (MU-2J and L): Seats for pilot and co-pilot or passenger on flight deck. Seating in main cabin, on rearward- and forward-facing seats, for 4 to 12 persons. Separate compartment at rear of cabin provides coat locker, toilet and baggage compartment. Door at rear of cabin on port side with built-in steps. Emergency exit door under wing on starboard side.

Accommodation (MU-2K and M): Seats for pilot and co-pilot or passenger on flight deck. Typical seating for five passengers in main cabin, two on individual rearward-facing seats, three on forward-facing rear bench seat. Optional arrangements for up to nine persons, including pilot. Pressurised baggage compartment over main-wheel bays, capacity 100 kg (220 lb). Non-pressurised baggage compartment aft of main-wheel bays, capacity 70 kg (154 lb). Space for coats and small baggage at rear of cabin. Door under wing on port side. Emergency exit door opposite main door.

Systems: Air-cycle pressurisation and air-conditioning system. Differential 0·34 bars (5·0 lb/sq in) in MU-2J and K; 0·41 bars (6·0 lb/sq in) in MU-2L and M. Hydraulic braking system. 28V DC primary electrical system, supplemented by 115V AC system for instruments and electronics. DC power supplied by two 30V 200A generators and two 24V 40Ah nickel-cadmium batteries. Oxygen system standard.

Electronics and Equipment: Blind-flying instrumentation standard. Radio and radar to customers' requirements. Optional equipment includes VOR/LOC, glideslope, ADF and marker beacon receivers; ATC transponder; DME; VHF or other communications systems; compass systems; autopilot; and weather radar.

Dimensions, external:	
Wing span over tip-tanks	11·94 m (39 ft 2 in)
Wing chord (mean)	1·54 m (5 ft 0¾ in)
Wing aspect ratio	7·71
Length overall:	
J, L	12·01 m (39 ft 5 in)
K, M	10·13 m (33 ft 3 in)
Length of fuselage:	
J, L	11·84 m (38 ft 10 in)
K, M	9·98 m (32 ft 9 in)
Height overall:	
J, L	4·17 m (13 ft 8 in)
K, M	3·94 m (12 ft 11 in)
Tailplane span	4·80 m (15 ft 9 in)
Wheel track:	
J, L	2·41 m (7 ft 11 in)
K, M	2·36 m (7 ft 9 in)
Wheelbase:	
J, L	4·39 m (14 ft 5 in)
K, M	4·52 m (14 ft 10 in)
Propeller diameter	2·29 m (7 ft 6 in)
Distance between propeller centres	4·50 m (14 ft 9 in)
Propeller ground clearance:	
J, L	0·79 m (2 ft 7 in)
K, M	0·64 m (2 ft 1 in)
Cabin door:	
Height	1·22 m (4 ft 0 in)
Width	0·76 m (2 ft 5½ in)
Emergency exit door:	
Height	0·72 m (2 ft 4½ in)
Width	0·70 m (2 ft 3½ in)
Dimensions, internal:	
Cabin: Length:	
J, L	5·99 m (19 ft 8 in)
K, M	3·35 m (11 ft 0 in)
Max width:	
J, K, L, M	1·50 m (4 ft 11 in)
Max height:	
J, K, L, M	1·30 m (4 ft 3·2 in)
Baggage compartment:	
J, L	1·08 m³ (38·00 cu ft)
K, M (two, total)	1·22 m³ (43·08 cu ft)
Areas (all versions):	
Wings, gross	16·55 m² (178 sq ft)
Flaps (total)	3·90 m² (42·0 sq ft)
Spoilers (total)	0·54 m² (5·82 sq ft)
Fin	2·85 m² (30·68 sq ft)
Rudder, incl tab	1·17 m² (12·60 sq ft)
Tailplane	4·02 m² (43·26 sq ft)
Elevators, incl tabs	1·39 m² (15·04 sq ft)
Weights and Loadings:	
Weight empty, equipped:	
J	3,084 kg (6,800 lb)
K	2,685 kg (5,920 lb)
L	3,433 kg (7,570 lb)
M	3,113 kg (6,864 lb)
Max ramp weight:	
L	5,273 kg (11,625 lb)
M	4,770 kg (10,520 lb)
Max T-O weight:	
J	4,900 kg (10,800 lb)
K	4,500 kg (9,920 lb)
L	5,250 kg (11,575 lb)
M	4,750 kg (10,470 lb)
Max landing weight:	
J	4,655 kg (10,260 lb)

Mitsubishi MU-2K, rescue version in JASDF insignia *(Norman E. Taylor)*

US-assembled examples of the Mitsubishi MU-2L (foreground) and MU-2M twin-turboprop transport aircraft

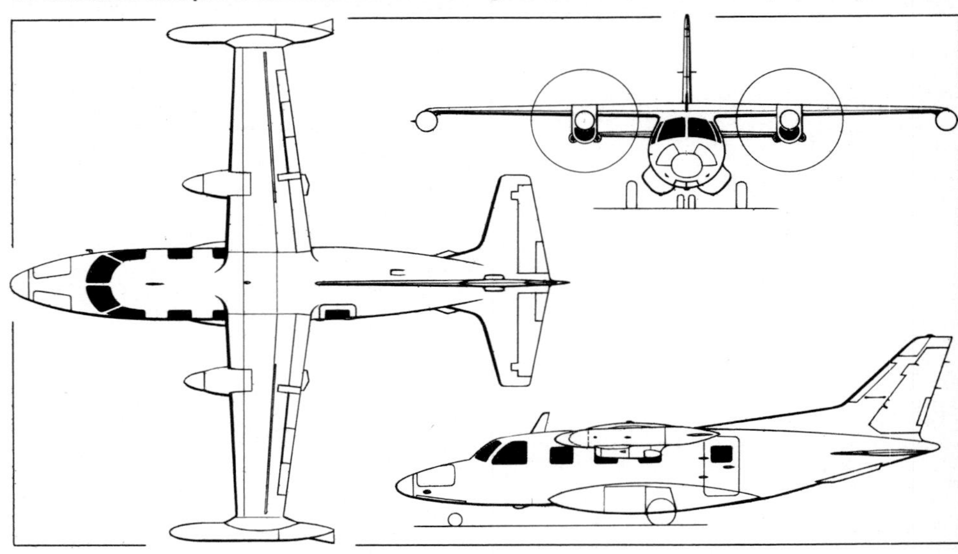
Mitsubishi MU-2L twin-turboprop utility transport *(Pilot Press)*

K	4,280 kg (9,435 lb)
L	5,000 kg (11,025 lb)
M	4,515 kg (9,955 lb)
Max wing loading:	
J	296 kg/m² (60·6 lb/sq ft)
K	272 kg/m² (55·7 lb/sq ft)
L	317·4 kg/m² (65·0 lb/sq ft)
M	287·1 kg/m² (58·8 lb/sq ft)
Max power loading:	
J, L	4·54 kg/kW (7·46 lb/ehp)
K	4·16 kg/kW (6·85 lb/ehp)
M	4·40 kg/kW (7·23 lb/ehp)

PERFORMANCE (at respective AUWs, except where indicated, of 4,175 kg; 9,200 lb (J), 3,960 kg; 8,730 lb (K), 4,695 kg; 10,350 lb (L), 4,195 kg; 9,250 lb (M)):

Never-exceed speed:	
J	330 knots (611 km/h; 380 mph)
K	315 knots (583 km/h; 362 mph)
Max operating speed:	
L, M (all weights)	249 knots (462 km/h; 287 mph) CAS
Max cruising speed at 4,575 m (15,000 ft):	
J	300 knots (556 km/h; 345 mph)
K	315 knots (584 km/h; 363 mph)
L at 4,175 kg (9,200 lb) AUW	295 knots (547 km/h; 340 mph)
M at 3,960 kg (8,730 lb) AUW	317 knots (587 km/h; 365 mph)
Econ cruising speed at 7,620 m (25,000 ft):	
J	265 knots (491 km/h; 305 mph)
K	270 knots (500 km/h; 311 mph)
L at 4,175 kg (9,200 lb) AUW	261 knots (483 km/h; 300 mph)
Econ cruising speed at 8,535 m (28,000 ft):	
M at 3,960 kg (8,730 lb) AUW	269 knots (499 km/h; 310 mph)
Stalling speed, flaps up:	
J	96 knots (178 km/h; 111 mph) CAS
K	95 knots (176 km/h; 109 mph) CAS
L	100 knots (185 km/h; 115 mph) CAS
M	98 knots (181 km/h; 112 mph) CAS
Stalling speed, flaps down:	
J, M	73 knots (135·5 km/h; 84 mph)
K	71 knots (132 km/h; 82 mph)
L	76·5 knots (142 km/h; 88 mph)
Max rate of climb at S/L:	
J	820 m (2,700 ft)/min
K	945 m (3,100 ft)/min
L	802 m (2,630 ft)/min
M	866 m (2,840 ft)/min
Rate of climb at S/L, one engine out:	
J	258 m (845 ft)/min
K	280 m (920 ft)/min
L	206 m (675 ft)/min
M	232 m (760 ft)/min
Service ceiling:	
J	9,390 m (30,800 ft)
K	10,120 m (33,200 ft)
L	9,020 m (29,600 ft)
M	9,815 m (32,200 ft)
Service ceiling, one engine out:	
J	5,700 m (18,700 ft)
K	6,035 m (19,800 ft)
L	4,710 m (15,450 ft)
M	5,485 m (18,000 ft)
T-O to 15 m (50 ft) at max T-O weight:	
J	570 m (1,870 ft)
K	518 m (1,700 ft)
L	661 m (2,170 ft)
M	548 m (1,800 ft)
Landing from 15 m (50 ft):	
J at 3,910 kg (8,620 lb) AUW	509 m (1,670 ft)
K at 3,510 kg (7,740 lb) AUW	333 m (1,090 ft)
L at 4,297 kg (9,475 lb) AUW	573 m (1,880 ft)
M at 3,783 kg (8,340 lb) AUW	488 m (1,600 ft)
Max range with wing tanks and wingtip tanks full, 30 min reserves:	
J at 7,620 m (25,000 ft)	1,350 nm (2,500 km; 1,554 miles)
K at 7,620 m (25,000 ft)	1,460 nm (2,705 km; 1,680 miles)
L at 7,620 m (25,000 ft)	1,259 nm (2,334 km; 1,450 miles)
M at 8,535 m (28,000 ft)	1,460 nm (2,705 km; 1,680 miles)

MITSUBISHI T-2

The T-2, the first supersonic aircraft to be developed by the Japanese aircraft industry, is a twin-engined two-seat jet trainer designed to meet the requirements of the JASDF.

Mitsubishi was selected as prime contractor for the development programme in September 1967. Preliminary and detailed design, under the leadership of Dr Kenji Ikeda, were followed by the completion of a full-size mockup in January 1969, and in March 1970 a development contract for prototype construction was awarded. The first XT-2 prototype (19-5101) flew for the first time on 20 July 1971, and flew supersonically for the first time in level flight (Mach 1·03) during its 30th flight, on 19 November 1971. The first flight of the second prototype (29-5102) followed on 2 December 1971. These two aircraft were delivered to the JASDF in December 1971 and March 1972 respectively for further flight testing. A static test airframe was delivered in March 1971.

Meanwhile, in 1970 two additional development aircraft were ordered, for operational flight testing. These made their first flights on 28 April and 20 July 1972; the flight test programme was completed in March 1974. A fatigue test airframe was delivered in January 1975.

Production orders have been placed for 42 T-2s (18 as advanced trainers, 22 as **T-2A** combat trainers, and two as prototypes for the FS-T2-KAI close-support fighter version, described separately). Twenty of these had been delivered by March 1976, to the 4th Air Wing at Matsushima. The current schedule calls for the remaining 22 T-2s to be delivered by March 1978. They will equip the 21st and 22nd Squadrons of the JASDF.

The fourth national DBP (defence buildup programme) provides for the purchase of 17 more T-2A combat trainers in FY 1976, for delivery by March 1980. Mitsubishi, as prime contractor, is responsible for fuselage construction, final assembly and flight testing of production aircraft. Major programme subcontractors are Fuji (wings and tail unit), Nippi (pylons and launchers) and Shin Meiwa (drop-tanks).

TYPE: Two-seat supersonic jet trainer.

WINGS: Cantilever all-metal shoulder-wing monoplane. Wing section NACA 65 series (modified)). Thickness/chord ratio 4·8%. Anhedral 9° from roots. Sweepback on leading-edges 68° at root, 42° inboard of outer extended-chord panels and 36° on outer panels. Multi-spar torsion box machined from tapered thick panels and constructed mainly of 7075 and 7079 aluminium alloy. Electrically-actuated aluminium honeycomb leading-edge flaps, the outer portions of which have extended chord. Electrically-actuated all-metal single-slotted flaps, with aluminium honeycomb trailing-edges over 70% of each half-span. No conventional ailerons. Lateral control by hydraulically-actuated all-metal two-section slotted spoilers ahead of flaps.

FUSELAGE: Conventional all-metal semi-monocoque structure, mainly of 7075 and 7079 aluminium alloy. Approx 10% of structure, by weight, is of titanium, mostly around engine bays. Two hydraulically-actuated door-type airbrakes under rear fuselage, aft of main-wheel bays.

TAIL UNIT: Cantilever all-metal structure. One-piece hydraulically-actuated all-moving swept tailplane, with 15° anhedral. Inner leading-edges of titanium; outer leading-edges of aluminium. Trailing-edges of aluminium honeycomb construction. Small ventral fin under each side of fuselage at rear. Hydraulically-actuated rudder.

LANDING GEAR: Hydraulically-retractable tricycle type, with pneumatic backup for emergency extension. Main units retract forward into fuselage, nose unit rearward. Single wheel on each unit. Nosewheel steerable through 72°. Oleo-pneumatic shock-absorbers. Hydraulic brakes and Hydro-Aire anti-skid units. Runway arrester hook beneath rear fuselage. Brake parachute in tailcone.

POWER PLANT: Two Rolls-Royce/Turboméca Adour turbofan engines, each rated at 20·95 kN (4,710 lb st) dry and 31·76 kN (7,140 lb st) with afterburning, mounted side by side in centre of fuselage. (Engines licence-built by Ishikawajima-Harima, under designation TF40-IHI-801A.) Fixed-geometry air intake, with auxiliary 'blow-in' intake doors, on each side of fuselage aft of rear cockpit. Fuel in seven fuselage tanks with total capacity of 3,823 litres (841 Imp gallons; 1,010 US gallons). Pressure refuelling point in starboard side of fuselage, forward of main-wheel bay. Provision for carrying up to three 833 litre (183 Imp gallon; 220 US gallon) drop-tanks under wings and fuselage.

ACCOMMODATION: Crew of two in tandem on Weber ES-7J

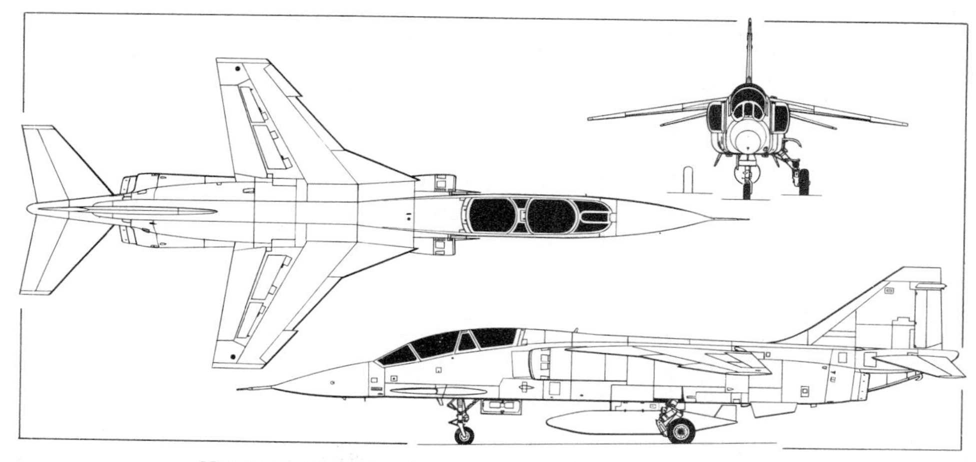

Mitsubishi T-2 tandem two-seat supersonic jet trainer *(Pilot Press)*

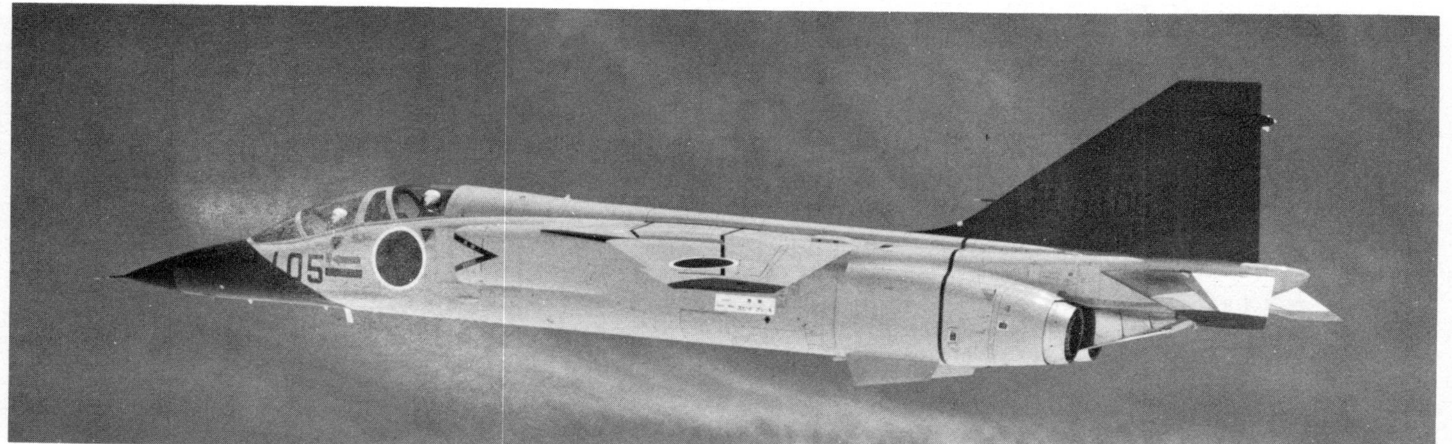

First production example of the Mitsubishi T-2 tandem two-seat supersonic jet trainer built for the JASDF

zero-zero ejection seats in pressurised and air-conditioned cockpits, separated by windscreen. Rear seat elevated 0·28 m (11 in) above front seat. Individual manually-operated rearward-hinged jettisonable canopies. Liquid oxygen equipment.

SYSTEMS: Cockpit air-conditioning system. Two independent hydraulic systems, each 207 bars (3,000 lb/sq in), for flight controls and utilities. Pneumatic bottle for landing gear emergency extension. Primary electrical power from two 12/15kVA AC generators.

ELECTRONICS AND EQUIPMENT: Mitsubishi Electric J/ARC-51 UHF. Nippon Electric J/ARN-53 Tacan and Toyo Communication J/APX-101 SIF/IFF. Mitsubishi Electric J/AWG-11 fire control system in nose. Lear 5010BL attitude and heading reference system.

ARMAMENT (combat trainer version): One Vulcan JM-61 multi-barrel 20 mm cannon in lower fuselage, aft of cockpit on port side. Attachment point on underfuselage centreline and two under each wing for drop-tanks or other stores. Wingtip attachments for air-to-air missiles.

DIMENSIONS, EXTERNAL:
Wing span 7·88 m (25 ft 10¼ in)
Wing aspect ratio 3
Wing taper ratio 3·7
Length overall 17·85 m (58 ft 6¾ in)
Height overall 4·39 m (14 ft 4¾ in)
Wheel track 2·70 m (8 ft 10¼ in)
Wheelbase 5·70 m (18 ft 8½ in)

AREAS:
Wings, gross 21·18 m² (228·0 sq ft)
Vertical tail surfaces (total, excl ventral fins) 5·00 m² (53·82 sq ft)
Horizontal tail surfaces (total) 6·70 m² (72·12 sq ft)

WEIGHTS:
Operational weight empty 6,197 kg (13,662 lb)
Max T-O weight, 'clean' 9,675 kg (21,330 lb)

PERFORMANCE (at max 'clean' T-O weight except where indicated):
Max level speed at 11,000 m (36,000 ft) Mach 1·6
Max rate of climb at S/L 10,670 m (35,000 ft)/min
Service ceiling 15,240 m (50,000 ft)
Required field length 1,525 m (5,000 ft)
Max ferry range with external tanks 1,550 nm (2,870 km; 1,785 miles)

First of two FS-T2-KAI prototypes, shown landing with bombs and drop-tanks on underwing and underfuselage pylons

MITSUBISHI FS-T2-KAI

To replace its North American F-86F Sabres, the JASDF decided to develop a single-seat close-support fighter version of the T-2. While under development, this has the provisional designation FS-T2-KAI; upon entry into service it will be designated Mitsubishi F-1. The first JASDF squadron of F-1s is due to be formed in late 1977; the 4th national defence buildup programme provides for the eventual purchase of 66 of these aircraft, of which the first 18 were ordered in FY 1975 and a further 8 in FY 1976.

The second and third production T-2 trainers (59-5106 and 59-5107) were converted as FS-T2-KAI prototypes, in which form they made their first flights on 7 and 3 June 1975 respectively. These aircraft retain the rear cockpit and canopy of the T-2, but this area is occupied by the fire control system and test equipment instead of a second occupant. Externally, they can be distinguished from the T-2 by the presence of a tubular fairing at the top of the fin, housing a passive warning radar antenna.

These prototypes were delivered to the JASDF Air Proving Wing at Gifu in July and August 1975, for continuation of the flight test and evaluation programme, and the single-seat version was expected to gain JDA type approval by October 1976, with production deliveries to begin just over a year later.

TYPE: Single-seat close-support fighter.

AIRFRAME, POWER PLANT AND SYSTEMS: Generally similar to T-2.

ACCOMMODATION: Generally similar to T-2, but without rear seat.

ELECTRONICS AND EQUIPMENT: To include air-to-air and air-to-ground radar, Ferranti inertial navigation system, radio altimeter, air data computer, Thomson-CSF head-up display, Mitsubishi fire control system and bombing computer, strike camera system and homing and warning system.

ARMAMENT: Single multi-barrel 20 mm cannon, as in T-2. Eight to twelve 500 lb bombs, two or four infra-red air-to-air missiles, two air-to-ship missiles, or rockets, on external attachments. Underwing hardpoints increased to three on each side.

DIMENSIONS AND AREAS:
As for T-2

WEIGHTS:
Operational weight empty 6,288 kg (13,862 lb)
Max T-O weight 13,614 kg (30,013 lb)

PERFORMANCE (estimated):
Max level speed at 11,000 m (36,000 ft) Mach 1·6
Time to 11,000 m (36,000 ft) 2 min
T-O run 1,220 m (4,000 ft)
Combat radius (hi-lo-hi) with eight 500 lb bombs 300 nm (555 km; 345 miles)

NAL
NATIONAL AEROSPACE LABORATORY

ADDRESS:
1880 Jindaiji-machi, Chofu City, Tokyo
Telephone: Musashino (0422) 47-5911

DIRECTOR:
Masao Yamanouchi

HEAD OF V/STOL AIRCRAFT RESEARCH GROUP:
Shun Takeda

The National Aerospace Laboratory (NAL) is a government establishment responsible for research and development in the field of aeronautical and space sciences. Since 1962 it has extended its activity in the field of V/STOL techniques.

NAL VTOL RESEARCH AIRCRAFT

In order to study the problems associated with the hovering, vertical take-off and landing of VTOL aircraft, a flying testbed (FTB) hovering test rig was developed by NAL and constructed by Fuji Heavy Industries Co Ltd. A description and illustration of the aircraft appeared in the 1972-73 *Jane's*.

Utilising the data gained from this vehicle, NAL is to build a further experimental VTOL research vehicle, intended for investigation of automatic flight in the VTOL mode. The specification, as at May 1976, called for a single-seat vehicle having an overall length of 11·54 m (37 ft 10¼ in), a wing span of 10·05 m (32 ft 11¾ in), and a height of 4·32 m (14 ft 2 in); and a power plant of one 13·79 kN (3,100 lb st) General Electric CJ610-9 turbojet engine for horizontal thrust and four 13·04 kN (2,932 lb st) IHI/NAL JR-100V lift-jets installed in tandem in the centre of the fuselage. It is planned to build two prototypes, one for flight testing and one for static test. The flying prototype is scheduled to be completed in 1979 and to begin flight testing in 1980.

NAMC
NIHON KOKUKI SEIZO KABUSHIKI KAISHA (Nihon Aeroplane Manufacturing Co Ltd)

HEAD OFFICE:
Toranomon Daiichi Building, No. 1, Kotohira-cho, Shiba, Minato-Ku, Tokyo 105
Telephone: Tokyo (03) 503-3211
Telex: 222-2863 (NAMC J) and 222-2864 (NAMC)

PRESIDENT:
Yuji Koyama

As detailed in previous editions of *Jane's*, NAMC was responsible for the development and production of the YS-11 twin-turboprop transport aircraft. It continues to be responsible for YS-11 after-sales support.

Full descriptions of the various versions of the YS-11 can be found in the 1967-68 and subsequent editions of *Jane's*.

About 50 of the NAMC work force were transferred to work on the YX programme to develop a successor to the YS-11 (see CTDC entry in this section).

NAMC also built the two XC-1 prototypes of the C-1 military transport for the JASDF, described under the Kawasaki heading in this section.

NIPPI
NIHON HIKOKI KABUSHIKI KAISHA (Japan Aircraft Manufacturing Co Ltd)

HEAD OFFICE AND SUGITA WORKS:
No. 3175 Showa-machi, Kanazawa-ku, Yokohama 236
Telephone: Yokohama (045) 771-1251
Telex: (3822) 267

OTHER WORKS: Atsugi

CHAIRMAN: Masami Takasaki

PRESIDENT: Masao Nagahisa

PUBLIC RELATIONS MANAGER:
Taketoshi Kitamura

Nippi has two works. The Sugita plant, to which the head office was transferred in early 1971, has facilities with a floor area of 45,133 m² (485,805 sq ft) and employs 953 persons. The Atsugi plant, which employs 1,092 persons, has a floor area of 35,759 m² (384,905 sq ft). Kawasaki has a 34% holding in Nippi.

The Atsugi plant is engaged chiefly in the overhaul, repair and maintenance of various types of aircraft and helicopters, including those of the Japan Defence Agency and Maritime Safety Agency, and carrier-based aircraft of the US Navy. The Sugita plant manufactures components and assemblies for the Kawasaki P-2J and C-1, Mitsubishi T-2, F-4EJ and MU-2, and Shin Meiwa PS-1; airframe and dynamic components for the Kawasaki KV-107; dynamic components for the Fuji-Bell UH-1B and Kawasaki-Hughes OH-6J; body structures for Japanese satellites; and tail units for Japanese-built rocket vehicles.

Nippi's latest product is the NP-100A powered sailplane, a description of which can be found in the Sailplanes section.

SHIN MEIWA
SHIN MEIWA INDUSTRY Co Ltd

TOKYO OFFICE:
c/o Shin Ohtemachi Building, 5th Floor, 2-1, 2-chome, Ohtemachi, Chiyoda-ku, Tokyo 100
Telephone: Tokyo (03) 279-3531
Telex: 222 2431 SMICAIR TOK

HEAD OFFICE:
1-5-25, Kosone-Cho, Nishinomiya-Shi, Hyogo-Ken
Telephone: Nishinomiya (0798) 47-0331
Telex: 5644493

WORKS (AIRCRAFT DIVISION):
Konan Plant, 1-1-1 Ogi, Higashinada-ku, Kobe 658
Telephone: (078) 431-4151
Itami Plant, 3-7-1 Minowa, Toyonaka, Osaka 560
Telephone: (068) 54-1151

CHAIRMAN OF THE BOARD:
Toshio Itoh

PRESIDENT:
Yoshio Yagi

SENIOR VICE-PRESIDENT:
Hiroshi Kohno

MANAGING DIRECTOR AND GENERAL MANAGER, AIRCRAFT DIVISION:
Tadao Uno

ASSISTANT GENERAL MANAGER:
Hajime Kawanishi (Managing Director)

Senior Technical Consultant:
Dr Shizuo Kikuhara
Chief Designer:
Dr Koichi Tokuda
Sales Manager and Public Relations (Tokyo Office):
Shigemi Matsui

The former Kawanishi Aircraft Company became Shin Meiwa in 1949 and established itself as a major overhaul centre for Japanese and US military and commercial aircraft.

Shin Meiwa's principal current activities concern the series production of the PS-1 medium-range STOL flying-boat for the JMSDF, an amphibious search and rescue version, the PS-1 Mod, and overhaul work on flying-boats and amphibians.

Shin Meiwa is also engaged in the manufacture of components for other aircraft. In particular, it produces nose and tail components for the Kawasaki P-2J, underwing drop-tanks for the Mitsubishi T-2 supersonic jet trainer and the cargo loading system for the Kawasaki C-1 transport aircraft.

SHIN MEIWA SS-2 and SS-2A

JMSDF designations: PS-1 and PS-1 Mod

After seven years of basic study and research, Shin Meiwa was awarded a contract in January 1966 to develop a new anti-submarine flying-boat for the Japan Maritime Self-Defence Force. As part of this programme, the company first rebuilt a Grumman UF-1 Albatross as a dynamically-similar flying scale model of the new design, under the designation UF-XS. This aircraft was described and illustrated in the 1964-65 *Jane's*.

Company designation for the basic flying-boat is **SS-2**; in ASW configuration this has the JMSDF designation **PS-1**. When adapted for amphibious operation the basic aircraft has the company designation **SS-2A**. The search and rescue version of this amphibian has the company designation **US-1** and JMSDF designation **PS-1 Mod**.

The first PS-1 prototype (5801) flew for the first time on 5 October 1967. It was delivered to the JMSDF on 31 July 1968. The second prototype, which flew on 14 June 1968, was handed over on 30 November 1968. These two aircraft were delivered to the 51st Flight Test Squadron of the JMSDF at Iwakuni. JDA type approval was granted in Autumn 1970.

In addition to the two prototypes, Shin Meiwa had by May 1976 delivered 14 production PS-1s ordered under the 3rd national defence programme. These are in service with the 31st Air Group of the JMSDF. A further seven PS-1s and PS-1 Mods have been ordered. The first PS-1 Mod was delivered on 5 March 1975, the second in July 1975 and the third in early February 1976. At that time all three were based at Iwakuni, where they were undergoing operational testing prior to the formation of a rescue squadron.

Design of the US-1 (PS-1 Mod) began in June 1970; the first example (9071) made its first flight, following a waterborne take-off, on 16 October 1974, and its first flight from a land base on 3 December 1974.

Shin Meiwa is continuing design study of a water bomber version of the aircraft, and funds have been allocated to convert one of the PS-1 prototypes into a water bombing testbed. Shin Meiwa will develop this version in co-operation with the JMSDF and the National Fire Agency, and plans to achieve eventually a 16 ton water load capacity. Initial tests, due to begin in mid-1976, will be made with two fuselage tanks with a total capacity of 8 tons. It is estimated that the testbed aircraft will, in 3·6 hr, be able to deliver a maximum of 185 tonnes (182·1 tons) of water, picked up at 5·2 nm (10 km; 6 miles) from base, to a site 38 nm (70 km; 43·5 miles) away. The proposed production version, with double the capacity, could deliver 660 tonnes (649·6 tons) to a site 54 nm (100 km; 62 miles) from its base before needing to land for refuelling.

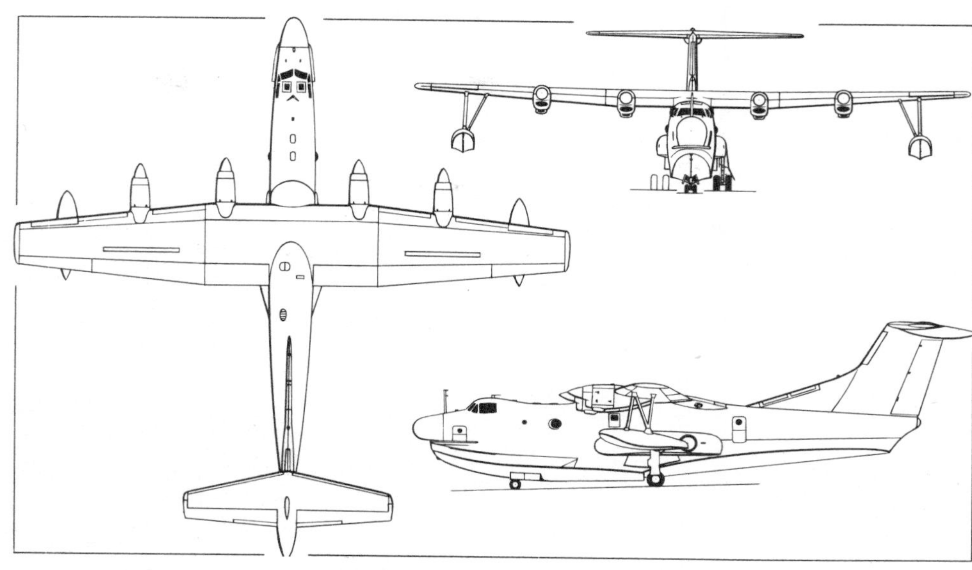

Shin Meiwa US-1 search and rescue amphibian, developed from the PS-1 *(Pilot Press)*

To make possible very low landing and take-off speeds, the PS-1 has both a boundary layer control system and extensive flaps for propeller slipstream deflection. Control and stability in low-speed flight are enhanced by 'blowing' the rudder, flaps and elevators, and by use of an automatic flight control system.

The PS-1 is designed to dip its large sonar deep into the sea during repeated landings and take-offs, and can land on very rough water, in winds of up to 25 knots (47 km/h; 29 mph). Take-offs and landings have been made successfully in seas with wave heights of up to 4 m (13 ft).

The following description applies to both the PS-1 and PS-1 Mod, except where a specific version is indicated:

Type: Four-turboprop STOL anti-submarine flying-boat (PS-1) or air/sea rescue amphibian (PS-1 Mod).

Wings: Cantilever high-wing monoplane. Conventional all-metal two-spar structure with rectangular centre-section and tapered outer panels. High-lift devices include outboard leading-edge slats extending over nearly 17% of the span and large outer and inner blown trailing-edge flaps extending 60° and 80° respectively. Two spoilers are located in front of the outer flap on each wing. Powered ailerons. Leading-edge de-icing boots.

Fuselage: All-metal semi-monocoque hull structure, with high length/beam ratio. Vee-shaped single-step planing bottom, with curved spray suppression strakes along sides of nose and spray suppressor slots in fuselage undersides aft of inboard propeller line. Double-deck interior.

Tail Unit: Cantilever all-metal T-type structure. Large dorsal fin. Tailplane has slats and de-icing boots on leading-edge. Blown rudder and elevators. Tab in each elevator.

Alighting and Beaching Gear (PS-1): Hull; and fixed stabilising floats near wingtips. Retractable tricycle-type beaching gear installed, with aft-retracting single-wheel main gear unit on each side of hull and forward-retracting twin steerable nosewheels, making aircraft independent of ground beaching aids.

Landing Gear (PS-1 Mod): Hull, as PS-1, plus hydraulically-retractable Sumitomo tricycle landing gear with twin wheels on all units. Oleo-pneumatic shock-absorbers. Main units, which retract rearward into fairings on hull sides, have size 40 × 14-22 tyres, pressure 7·79 bars (113 lb/sq in). Nosewheel tyres size 25 × 6·75-18, pressure 20·69 bars (300 lb/sq in). Three-rotor hydraulic disc brakes. No anti-skid units.

Power Plant: Four 2,282 kW (3,060 ehp) Ishikawajima-built General Electric T64-IHI-10 turboprop engines, each driving a Hamilton Standard 63E60-19 three-blade constant-speed reversible-pitch propeller. Additionally, one 1,044 kW (1,400 ehp) Ishikawajima-built General Electric T58-IHI-10 gas turbine (932 kW; 1,250 ehp T58-IHI-10-M1 in PS-1 Mod) is housed in the upper centre portion of the fuselage to provide power for boundary layer control system on rudder, flaps and elevators. Fuel in two bladder-type rear-fuselage tanks and five wing tanks, with total usable capacity of 17,900 litres (3,938 Imp gallons) in PS-1. PS-1 Mod fuel is in wing tanks (10,851 litres; 2,387 Imp gallons) and fuselage tanks (11,649 litres; 2,563 Imp gallons); total capacity 22,500 litres (4,950 Imp gallons). Refuelling point near bow hatch. Oil capacity 101 litres (22·3 Imp gallons).

Accommodation (PS-1): Two pilots and flight engineer on flight deck, which has wide-visibility bulged windows at sides. Aft of this on the upper deck is a tactical compartment, housing two sonar operators, a navigator, MAD operator, radar operator, radio operator and a tactical co-ordinator. Electronic, magnetic and sonic equipment is installed on starboard side, with crew's rest area and bunks on port side. Aft of tactical compartment is the weapons compartment. On the lower deck, from nose to rear, are the electronics compartment, oxygen-bottle bay, main gear bay and two fuel tanks. Door on port side of rear fuselage.

Shin Meiwa PS-1 four-turboprop STOL anti-submarine flying-boat, thirteenth production aircraft for the Japan Maritime Self-Defence Force

First example of the Shin Meiwa PS-1 Mod air/sea rescue amphibian (four Japanese-built General Electric T64-IHI-10 turboprop engines)

Accommodation (PS-1 Mod): Search and rescue version has accommodation for crew of nine and 12 survivors, with 12 stretchers, one auxiliary seat and two observers' seats. Alternatively, up to 36 stretchers can be accommodated. Rescue hatch on port side of fuselage, aft of wing. Transport version can seat up to 69 passengers in mainly four-abreast seating with centre aisle; rear portion of cabin convertible to cargo compartment.

Systems: Cabin air-conditioning system. Two independent hydraulic systems, each 207 bars (3,000 lb/sq in). Air/sea rescue version has oxygen system for all crew and stretcher stations. AiResearch GTCP85-131J APU provides power for starting main engines and shaft power for 40kVA emergency AC generator. BLC system includes a C-2 compressor, driven by T58-IHI-10 gas turbine, which delivers compressed air at a flow of 14 kg (30·9 lb)/sec and pressure of 1·86 bars (27 lb/sq in) for ducting to inner and outer flaps, rudder and elevators. Electrical system includes three-phase 400Hz constant-frequency AC and converted 27V DC. Two 40kVA AC generators, driven by Nos. 2 and 3 main engines. Emergency AC generator driven by APU. Anti-icing, air-conditioning, and fire detection and extinguishing systems are standard on PS-1 Mod.

Electronics and Equipment (PS-1): AN/ARA-50 UHF direction finder, AN/ARN-52 Tacan, HRN-4 Loran, HPN-101B wave height meter, AN/APN-153 Doppler radar, AN/AYK-2 navigation computer, A/A24G-9 TAS transmitter, AN/APS-80 search radar, HGC-102 teletypewriter, AN/APA-125A indicator group, HRN-101 ADF, N-PT-3 dead reckoning plotting board, AN/APX-68 SIF, N-OA-35/HSA tactical plotting group, HLR-1 countermeasure device, AN/ARR-52A sonobuoy receiver, AN/ASQ-10A magnetic anomaly detector, N-MX-143/HSQ error voltage monitor, HQS-101B dipping sonar, AN/ASA-20B recorder group, AN/AQA-5 sonobuoy recorder, AN/AQA-1 sonobuoy indicator group, N-R-86/HRA OTPI (on top position indicator), AN/ASA-16 integrated display system, AN/ASA-50 computer group, HQH-101 sonobuoy data recorder, HSA-1 SDDS, RRC-15 emergency transmitter, HSA-2 automatic magnetic compensator device, N-CU-58/HRC antenna coupler, N-ID-66/HRN BDHI, N-RO-14B BT recorder, HIC-3 interphone and HRC-7 HF.

Electronics and Equipment (PS-1 Mod): HIC-3 interphone, HRC-107 HF, N-CU-58/HRC antenna coupler, HGC-102 teletypewriter, HRC-106 radio, HRC-110 radio, HRN-101 ADF, AN/ARA-50 UHF/DF, HRN-105 Tacan, HRN-104 Loran, HRN-3 marker beacon receiver, AN/APN-171 (N2) radar altimeter, HPN-101B wave height meter, AN/APN-153 Doppler radar, AN/AYK-2 navigation computer, A/A24G-9 TAS transmitter, N-PT-3 dead reckoning plotting board, N-OA-35/HSA tactical plotter group, AN/APS-80N search radar, AN/APA-125N indicator group, AN/APX-68N IFF transponder, RRC-15 emergency transmitter and N-ID-66/HRN BDHI.

Armament and Operational Equipment (PS-1): Weapons bay on upper deck, aft of tactical compartment, in which are stored AQA-3 Jezebel passive long-range acoustic search equipment with 20 sonobuoys and their launchers, Julie active acoustic echo ranging with 30 explosive charges, four 330 lb anti-submarine bombs, and smoke bombs. External armament includes two underwing pods, one between each pair of engine nacelles and each containing two homing torpedoes, and a launcher beneath each wingtip for three 5 in air-to-surface rockets. Searchlight below starboard outer wing.

Operational Equipment (PS-1 Mod): Marker launcher, 10 marine markers, 6 green markers, 2 droppable message cylinders, 10 float lights, pyrotechnic pistol, parachute flares, 2 flare storage boxes, binoculars, 2 rescue equipment kits, 2 droppable life-raft containers, rescue equipment launcher, lifeline pistol, lifeline, 3 lifebuoys, portable speaker, hoist unit, floating mat, lifeboat with outboard motor, camera, and 12 stretchers. Stretchers can be replaced by troop seats.

Dimensions, external:

Wing span	33·14 m (108 ft 8¾ in)
Wing chord at root	5·00 m (16 ft 4¾ in)
Wing chord at tip	2·39 m (7 ft 10 in)
Wing aspect ratio	8
Length overall	33·50 m (109 ft 11 in)
Height overall:	
PS-1	9·715 m (31 ft 10½ in)
PS-1 Mod	9·78 m (32 ft 1 in)
Tailplane span	12·36 m (40 ft 6½ in)
Wheel track:	
PS-1	3·10 m (10 ft 2 in)
PS-1 Mod	3·56 m (11 ft 8¼ in)
Wheelbase:	
PS-1	8·20 m (26 ft 10¾ in)
PS-1 Mod	8·33 m (27 ft 4 in)
Propeller diameter	4·42 m (14 ft 6 in)
Rescue hatch, PS-1 Mod (port side, rear fuselage):	
Height	1·41 m (4 ft 7½ in)
Width	0·79 m (2 ft 7 in)

Areas:

Wings, gross	135·8 m^2 (1,462 sq ft)
Ailerons (total)	6·40 m^2 (68·9 sq ft)
Inner flaps (total)	9·40 m^2 (101·18 sq ft)
Outer flaps (total)	14·20 m^2 (152·85 sq ft)
Leading-edge slats (total)	6·01 m^2 (64·7 sq ft)
Spoilers (total)	2·10 m^2 (22·60 sq ft)
Fin	17·56 m^2 (189 sq ft)
Dorsal fin	6·32 m^2 (68·03 sq ft)
Rudder	7·01 m^2 (75·5 sq ft)
Tailplane	23·05 m^2 (248 sq ft)
Elevators, incl tab	8·78 m^2 (94·5 sq ft)

Weights and Loadings (PS-1):

Weight empty	26,300 kg (58,000 lb)
Normal T-O weight	36,000 kg (79,365 lb)
Max T-O weight	43,000 kg (94,800 lb)
Max wing loading	316·6 kg/m^2 (64·84 lb/sq ft)
Max power loading	4·71 kg/kW (7·74 lb/ehp)

Weights (PS-1 water bomber testbed):

Weight empty	25,200 kg (55,555 lb)
Water payload	8,100 kg (17,857 lb)
Max T-O and landing weight	43,000 kg (94,800 lb)

Weights and Loadings (PS-1 Mod, search and rescue):

Weight empty, equipped	25,500 kg (56,218 lb)
Max oversea operating weight	36,000 kg (79,365 lb)
Max T-O weight on land	45,000 kg (99,200 lb)
Max wing loading	331·4 kg/m^2 (67·9 lb/sq ft)
Max power loading	4·93 kg/kW (8·11 lb/ehp)

Performance (PS-1 at normal T-O weight):

Max level speed at 1,525 m (5,000 ft)	295 knots (547 km/h; 340 mph)
Cruising speed at 1,525 m (5,000 ft):	
4 engines	230 knots (426 km/h; 265 mph)
2 engines	170 knots (315 km/h; 196 mph)
Approach speed	47 knots (87 km/h; 54 mph)
Touchdown speed	41 knots (76 km/h; 47 mph)
Stalling speed	40 knots (75 km/h; 46 mph)
Max rate of climb at S/L	690 m (2,264 ft)/min
Service ceiling	9,000 m (29,500 ft)
Time to 3,050 m (10,000 ft)	5 min
T-O run	250 m (820 ft)
Landing run	180 m (590 ft)
Min ground turning radius	25·00 m (82 ft 0 in)
Normal range	1,169 nm (2,168 km; 1,347 miles)
Max ferry range	2,560 nm (4,744 km; 2,948 miles)
Endurance	15 hr

Performance (PS-1 water bomber testbed, estimated):

Max level speed	265 knots (491 km/h; 305 mph)
Cruising range	1,400 nm (2,593 km; 1,612 miles)

Performance (PS-1 Mod, search and rescue version, at max T-O weight on land, except where indicated):

Max level speed	260 knots (481 km/h; 299 mph)
Cruising speed at 3,050 m (10,000 ft)	230 knots (426 km/h; 265 mph)
Max rate of climb at S/L, AUW of 36,000 kg (79,365 lb)	725 m (2,380 ft)/min
Service ceiling, AUW of 36,000 kg (79,365 lb)	8,535 m (28,000 ft)
T-O to 15 m (50 ft)	660 m (2,165 ft)
Landing from 15 m (50 ft) at 36,000 kg (79,365 lb) AUW	900 m (2,950 ft)
Runway LCN requirement at AUW of 43,000 kg (94,798 lb)	42
Min ground turning radius	21·20 m (69 ft 6½ in)
Radius of search operation at AUW of 45,000 kg (99,200 lb), including 2·3 hr search	900 nm (1,665 km; 1,035 miles)

SHOWA

SHOWA HIKOKI KOGYO KABUSHIKI KAISHA (Showa Aircraft Industry Co Ltd)

Head Office:
Mitsui Building, No. 3, 3-Chome, Nihonbashi-Muromachi, Chuo-ku, Tokyo
Telephone: Tokyo (03) 270-1451
Sales Office:
No. 1, 2-Chome, Nihonbashi-Muromachi, Chuo-ku, Tokyo
Telephone: Tokyo (03) 279-1451
Works:
No. 600, Tanaka-machi, Akishima-shi, Tokyo
President: Teiji Asano

Showa was the first Japanese aircraft manufacturing company to resume post-war operations when it undertook the overhaul and repair of aircraft of the US Air Force.

The company's present activities comprise mainly the manufacture of wingtip floats, tail fin, partition and other doors, torpedo pods and hatches for the Shin Meiwa PS-1 flying-boat; and the supply of aluminium and non-metal honeycomb and honeycomb sandwich panels for aircraft floors and airframe construction. Showa also manufactures a variety of airborne equipment, including galleys, service carts and baggage/cargo containers.

MEXICO

AAMSA
AERONAUTICA AGRICOLA MEXICANA SA

ADDRESS:
171 Oriente No. 398, Colonia Agron, Apartado 26783, Mexico 14, DF

ENQUIRIES TO:
Bill M. Humes, Director of Operations, Rockwell International (General Aviation Division), 5001 North Rockwell Avenue, Bethany, Oklahoma 73008, USA

As the result of an agreement between Rockwell International Corporation of the US (which see) and Industrias Unidas SA of Mexico, this company was formed in 1971 for the purpose of taking over from the former's Commercial Products Group the manufacture of Aero Commander Quail Commander and Sparrow Commander agricultural aircraft. Rockwell International has a 30% holding in the Mexican company.

Aeronautica Agricola Mexicana SA purchased the type design, tooling and all production materials for the Sparrow and Quail Commander agricultural aircraft; it was to build them at a new Industrias Unidas manufacturing complex in Pasteje, Mexico. Rockwell retains the right to market these aircraft in the United States.

Up to the end of 1975, only a dozen of these aircraft had been completed in Mexico, and the Sparrow Commander programme had been terminated. Negotiations in 1975 between Rockwell International and AAMSA resulted in a new programme, which was scheduled to produce 50 Quail Commanders during 1976.

AAMSA (AERO COMMANDER) QUAIL COMMANDER

The Quail Commander is a small agricultural aircraft with a 795 litre (210 US gallon) hopper and low operating cost.

The entire primary structure of the aircraft is coated with Copon, an epoxy resin catalyst paint which is resistant to all known agricultural chemicals.

The Quail Commander can be equipped with any type of dispersal equipment required, ie straight sprayer of either high or low volume, dust dispersal gear, or a quick-change combination dust or spray unit. The units normally offered as optional extras are the Transland Boommaster spray system, Strutmaster spray system with 2 in Simplex pump, invert emulsion spray system, Buckeye bottom loader spray system, Micronair spray system, standard dust spreader, spreader with gate box and agitator, and Transland Swathmaster dry spreader.

TYPE: Single-seat agricultural monoplane.

WINGS: Braced low-wing monoplane. Modified Clark Y wing section. Dihedral 5° 8'. Incidence 0° 20'. Composite structure with spruce spars, metal-covered leading-edge and fabric covering on remainder of wing. Multiple steel tube overwing bracing struts on each side. Hoerner wingtips. Fabric-covered wooden ailerons. Flaps and drooping ailerons.

FUSELAGE: Steel tube structure with fabric covering. Removable side panels.

TAIL UNIT: Wire-braced steel tube structure with fabric covering. Wire deflector from canopy to fin.

LANDING GEAR: Non-retractable tailwheel type. CallAir spring shock-absorbers. Cleveland main wheels, with Goodyear tyres, size 8·50-6 (6-ply). Scott 203 mm (8 in) steerable tailwheel. Cleveland toe-actuated brakes. Wire-cutters on main legs.

POWER PLANT: One 216 kW (290 hp) Lycoming IO-540-G1C5 flat-six engine, driving a McCauley Type 1A200-DFA9045 two-blade fixed-pitch metal propeller. Two-position adjustable-pitch McCauley Type 2D34CT-84HF two-blade metal propeller optional. All fuel in wing tanks, capacity 151 litres (40 US gallons). Oil capacity 11 litres (3 US gallons).

ACCOMMODATION: Single seat in open-sided cockpit aft of hopper. Side doors. Wire-cutters on windscreen. Cabin heater standard.Capacity of standard hopper 795 litres (210 US gallons).

SYSTEMS: Electrical system includes 50A 24V alternator and 35Ah battery.

ELECTRONICS: Narco Mk III Omnigator radio and other electronics available as optional extras.

EQUIPMENT: Optional equipment includes night lighting system, landing light in nose cowl, alternative retractable landing light in wing, rotating beacon and AC full-flow oil filter.

DIMENSIONS, EXTERNAL:

Wing span	10·59 m (34 ft 9 in)
Wing chord, constant	1·59 m (5 ft 2¾ in)
Length overall	7·16 m (23 ft 6 in)
Height overall	2·31 m (7 ft 7 in)
Tailplane span	3·20 m (10 ft 6 in)
Wheel track	2·08 m (6 ft 10 in)
Wheelbase	5·21 m (17 ft 1 in)
Propeller diameter:	
standard	2·29 m (7 ft 6 in)
optional	2·13 m (7 ft 0 in)

AREAS:

Wings, gross	16·90 m² (182 sq ft)
Ailerons (total)	1·99 m² (21·4 sq ft)
Fin	0·80 m² (8·6 sq ft)
Rudder	0·84 m² (9·0 sq ft)
Tailplane	1·47 m² (15·8 sq ft)
Elevators	1·30 m² (14·0 sq ft)

WEIGHTS AND LOADINGS:

Weight empty	726 kg (1,600 lb)
Max payload	726 kg (1,600 lb)
Max T-O weight:	
CAR.3	1,360 kg (3,000 lb)
CAR.8	1,633 kg (3,600 lb)
Max wing loading	80·0 kg/m² (16·4 lb/sq ft)
Max power loading	7·56 kg/kW (12·41 lb/hp)

PERFORMANCE (at CAR.8 max T-O weight, except where indicated):

Max level speed at S/L	104 knots (193 km/h; 120 mph)
Max cruising speed (75% power) at 1,360 kg (3,000 lb) AUW	100 knots (185 km/h; 115 mph)
Normal operating speed	78-87 knots (145-161 km/h; 90-100 mph)
Stalling speed	54 knots (100 km/h; 62 mph)
Stalling speed as usually landed	35 knots (65 km/h; 40 mph)
Max rate of climb at S/L	259 m (850 ft)/min
Service ceiling	4,875 m (16,000 ft)
T-O run	244 m (800 ft)
Landing run at normal landing weight	136 m (447 ft)
Range at 50% power	260 nm (483 km; 300 miles)

ANAHUAC
FABRICA DE AVIONES ANAHUAC SA

ADDRESS:
Calzada Adolfo López Mateos 478, Aeropuerto Internacional, Mexico 9, DF
Telephone: 558-27-57

This company was formed to initiate in Mexico the development of aircraft suited to the particular needs of agricultural aviation in that country, taking its name from the former Aztec valley where Mexico City is now situated. Anahuac's first product was a single-seat agricultural aircraft known as the Tauro 300.

ANAHUAC TAURO 300 (BULL)

Design of the Tauro was begun in January 1967, and the prototype was flown for the first time on 3 December 1968.

The first production Tauro was flown on 5 June 1970, following the award on 8 August 1969 of the Mexican DGAC's approved type certificate No. 1. Seven production aircraft had been completed by the end of 1971, after which studies were made for an improved version of the Tauro to incorporate improvements suggested by early operational use. Mexican government approval was given on 21 December 1972 for financial support to expand the manufacturing programme to meet orders from Mexican customers and for export; but up to the Spring of 1976 it had not been possible to provide this support.

A full description of the initial production version of the Tauro appeared in the 1974-75 *Jane's*.

NETHERLANDS

FOKKER-VFW
FOKKER-VFW BV (Subsidiary of Zentralgesellschaft VFW-Fokker mbH)

HEAD OFFICE AND MAIN FACTORY:
PO Box 7600, Schiphol-Oost (Amsterdam Airport)
Telephone: Amsterdam (020) 5449111
Telex: 12227 SIFO NL

OTHER FACTORIES AND COMPANIES:
Fokker-VFW BV, Drechtsteden Division, with plants at Papendrecht and Dordrecht
Fokker-VFW BV, Avio-Fokker Division, with Ypenburg Works at Ypenburg Air Base, near the Hague; and Woensdrecht Works at Woensdrecht Air Base, near Bergen op Zoom
Fokker-VFW, Hoogeveen Division
Trading Company Avio-Diepen BV

SUPERVISORY BOARD:
H. Buiter
G. C. Klapwijk
Ir A. A. Holle
Prof Dr W. H. J. Reynaerts

MANAGEMENT:
Ir J. Cornelis
H. J. Grobben
Ir A. van Wijlen

MANAGER, MARKETING INFORMATION DEPARTMENT (Press, Publicity and Public Affairs):
Brian H. Railton

Fokker-VFW BV, Netherlands Aircraft Factories, is the Dutch manufacturing company of the Zentralgesellschaft VFW-Fokker mbH (see International section), which was formed when the 50-year-old Royal Netherlands Aircraft Factories Fokker and Vereinigte Flugtechnische Werke GmbH of Germany joined forces on a parity basis in 1969. Marketing and product support of civil airliners produced by the two companies is undertaken by a separate company, Fokker-VFW International BV, whose offices are at PO Box 7600, Schiphol-Oost, The Netherlands.

Fokker-VFW BV forms the entire aircraft industry in the Netherlands, with six plants, in which about 7,000 people are employed. Earlier collaborative ventures, other than those with VFW, included participation in the manufacturing programmes for the Gloster Meteor, Hawker Hunter and Lockheed F-104G, with final assembly lines at Schiphol; and for the Breguet Atlantic and Canadair (Northrop) CF-5/NF-5. Fokker-VFW will have an important share in the forthcoming European manufacturing programme for the General Dynamics F-16 light fighter.

Some 4,300 people are employed at the Schiphol-Oost works, which accommodates the company headquarters and administration together with the main F27 and F28 assembly lines and test flying facilities. Production is continuing of the F27 and F28, each in various versions, and components are being produced for the Airbus A300 (wing moving surfaces). Production of outer wings and struts for the Shorts SD3-30 was transferred from Schiphol-Oost to the Woensdrecht works in 1975. Also at Schiphol are the design offices, research department, numerically-controlled milling department, metal bonding department, electronics division, space division and scientific and administrative computer facilities.

The Drechtsteden plant, formed by the integrated production facilities at Dordrecht, Papendrecht and Hoogeveen, employs some 1,550 people. Most of these are engaged on detail production and component assembly for the F27 and F28, VFW 614 and Airbus A300; other work includes the manufacture of antennae and other specialised products. Several types of Aviobridge airport passenger gangways are also manufactured at Papendrecht.

Avio-Fokker is a Division of Fokker-VFW, employing some 1,300 people, and comprises the former Avio-Diepen plant at Ypenburg Air Base near the Hague and the former Aviolanda plant at Woensdrecht Air Base. At both facilities maintenance, overhaul, repair and modification work is carried out on a wide variety of military and civil aircraft.

Nike missile radomes, and reinforced plastics components for the Friendship, Fellowship, F-104G Starfighter, Airbus A300 and Shorts SD3-30 are manufactured at Ypenburg.

At Woensdrecht the ELMO division is engaged on producing electrical and electronic systems and wire harnesses.

Hoogeveen Division, a facility of the Drechtsteden plant, employs about 200 people in the manufacture of parts for the aerospace industry, radar and telecommunications and other industries. Quantity production of LD3 freight containers is also undertaken in this factory.

FOKKER-VFW F27 FRIENDSHIP

The F27 Friendship is a medium-sized short/medium-range airliner. Two prototypes were built. The first made its first flight on 24 November 1955, and was designed to accommodate 28 passengers in a 22·3 m (73 ft) long fuselage. The second, which flew on 29 January 1957, was representative of Series 100 production aircraft, with Dart 511 engines and 32 seats in a 23·1 m (76 ft) fuselage. Two further airframes were built for static and fatigue testing.

The F27 has been in series production for many years, both by Fokker and by Fairchild Industries in the United States. Deliveries by Fokker began in November 1958. US

Fokker-VFW F27 Mk 600 twin-turboprop military transport in the insignia of the Ghana Air Force

production of the F-27 and FH-227 has ended, a total of 205 having been sold by Fairchild.

The following F27 orders by airlines, air forces and government agencies had been announced by 15 July 1976:

Mk 100 (1967-68 *Jane's;* 83 built, incl 2 corporate; orders listed in 1971-72 *Jane's)*

Mk 200 (114 built, incl 1 corporate); orders as listed in 1973-74 *Jane's*

Mk 300 (1967-68 *Jane's;* 13 built; orders listed in 1971-72 *Jane's)*

Mk 400/600 (174 sold, incl 9 corporate); orders as listed in 1974-75 *Jane's,* plus:

AeroPeru (Mk 600)	2
Air Algérie	3
Indonesian Air Force (Mk 400M)	8
Imperial Iranian government (Mk 400M/600)	3
Somali Airlines (Mk 600)	2
Union of Burma Airways (Mk 600)	1

Mk 500 (62 sold, incl 4 corporate); orders as listed in 1974-75 *Jane's,* plus:

Air Rouergue (France)	1
Ansett (Airlines of New South Wales)	4
Aramco (rough-field version)	4
East-West Airlines (Australia)	2
New Zealand National Airways	2

F-27 (128 built, incl 49 corporate; orders listed in 1971-72 *Jane's)*

FH-227 (77 sold, incl 4 corporate; orders listed in 1973-74 *Jane's)*

By 15 July 1976, total sales were 653 (448 by Fokker-VFW and 205 by Fairchild).

Fokker is standardising currently on the Mks 400, 400M, 500 and 600, but any of the following versions of the F27 are available to order:

F27 Mk 200. Basic airliner or executive model with Dart RDa.7 Mk 532-7R turboprops. First flight 20 September 1959.

F27 Mk 400 Combiplane. Cargo or combined cargo/passenger version of Mk 200. Large cargo door. First flight 6 October 1961.

F27 Mk 400M. Military version, with accommodation for 45 parachute troops, 6,025 kg (13,283 lb) of freight or 24 stretchers and 9 attendants. Large cargo door and enlarged parachuting door on each side. First flight 24 April 1965.

F27 Mk 400M Cartographic version. Aerial survey version with two super-wide-angle cameras, remotely controlled from central navigation station, and navigation sight. Inertial navigation system, with digital readout at navigation station and recorded on each picture. Photography through optical glass window panes. Electrically-operated window doors. First flight 24 August 1973.

F27 Mk 500. Similar to F27 Mk 200, but with lengthened fuselage and large cargo door. The 15 aircraft for the French Ministère des Postes et Télécommunications (Air France) have special para-dropping type large doors on both sides. First flight 15 November 1967.

F27 Mk 500M. Military version, with lengthened fuselage; similar to Mk 400M but with accommodation for 50 paratroops, 6,617 kg (14,588 lb) of freight or 30 stretchers and 6 attendants.

F27 Mk 600. Similar to Mk 200, but with a large cargo door. Does not have the reinforced and watertight flooring of the Combiplane. Can be fitted with quick-change interior, featuring roller tracks and palletised seats and/or cargo pallets. First flight 28 November 1968.

Any of the above models can be fitted, at customer's option, with a Dowty Rotol rough-field landing gear having two-stage oleos with a 100 mm (4 in) increase in stroke, giving increased overall height and propeller ground clearance.

TYPE: Twin-turboprop medium-range airliner.

WINGS: Cantilever high-wing monoplane. Wing section NACA 64-421 at root, 64-415 at tip. Dihedral 2° 30′. Incidence 3° 30′. All-metal riveted and metal-bonded two-spar stressed-skin structure, consisting of centre-section and two detachable outer sections. Detachable honeycomb-core sandwich leading-edges with rubber-boot de-icers. Glassfibre-reinforced plastics trailing-edges. Mechanically-operated single-slotted flaps, divided by engine nacelles. Electrically-operated trim tab in each aileron.

FUSELAGE: All-metal stressed-skin structure, built to fail-safe principles, with cylindrical portions metal bonded and conical parts riveted. Fuselage is pressurised between rear bulkhead of nosewheel compartment and circular pressure bulkhead aft of the baggage compartment. Length of pressurised section 16·16 m (53 ft 0 in), except for Mks 500/500M in which the pressurised section is 17·66 m (57 ft 11 in) long. The slightly flattened fuselage bottom is reinforced by underfloor members.

TAIL UNIT: Cantilever all-metal stressed-skin structure. Fin and tailplane, as well as leading-edges of surfaces, are detachable. Trim tab in each elevator. Pneumatic-boot anti-icing.

LANDING GEAR: Retractable tricycle type. Pneumatic retraction. Dowty oleo-pneumatic shock-absorbers. Twin-wheel main units retract backward into engine nacelles. Single-wheel steerable nose unit retracts forward into non-pressurised nosecone. Main-wheel tyre pressure 5·62 bars (81·5 lb/sq in), nosewheel tyre pressure 3·87 bars (56 lb/sq in). Pneumatic brakes on main wheels, with Dunlop Maxaret automatic anti-skid system. Provision on all currently-available models for Dowty Rotol rough-field landing gear in which, at 19,730 kg (43,500 lb) AUW, the total stroke in the main gear is lengthened from 305 mm (12 in) to 406 mm (16 in), increasing the aircraft's static height and propeller ground clearance by 76 mm (3 in). Low-pressure main-wheel tyres are fitted, pressure 4·2 bars (61 lb/sq in) below 18,143 kg (40,000 lb) AUW and 4·57 bars (66 lb/sq in) at higher operating weights. Nose unit is of levered-suspension type, with tyre pressure of 3·87 bars (56 lb/sq in).

POWER PLANT (all current versions): Two Rolls-Royce Dart Mk 532-7R (RDa.7 rating) turboprop engines, each developing 1,596 kW (2,140 shp) plus 2·34 kN (525 lb st) for take-off. Four-blade Dowty Rotol constant-speed propellers. Integral fuel tanks in outer wings, capacity 5,136 litres (1,130 Imp gallons). Optionally, bag tanks for an additional 2,289 litres (503·5 Imp gallons) may be fitted. Overwing fuelling, but pressure refuelling optional. Provision for carrying two 950 litre (209 Imp gallon) external fuel tanks under wings. Methyl-bromide fire-extinguishing system with flame detectors.

ACCOMMODATION (Mks 200 and 600): Flight compartment seats two pilots side by side, with folding seat for third crew member if required. Main cabin has standard four-abreast seating for 44 passengers (800 mm; 31·5 in seat pitch in Mk 200, 876 mm; 34·5 in in Mk 600); alternative arrangements allow this number to be increased to 48 in Mk 200, 60 in Mk 600. Passenger door at rear of cabin, on port side, with toilet opposite. Standard cargo door at front of Mk 200 on port side; large cargo door in same position on Mk 600, with sill at truck-bed level. Cargo holds forward and aft of main cabin, size dependent on interior arrangement.

ACCOMMODATION (executive and VIP versions): Can be furnished to customer's specification, but a basic layout is available. In this, the cabin is divided into three sections: a conference room with six seats, a rest room with settee and divan, and a lounge with four seats. Toilet, galley, wardrobe, baggage space and seat for attendant in forward fuselage. Second toilet and baggage space at rear.

ACCOMMODATION (Mk 400 Combiplane): Principal features of this version are a large cargo loading door forward of the wings on the port side, with the sill at truck-bed height, and a reinforced cargo floor with tiedown rings. Typical layouts include 40 passengers

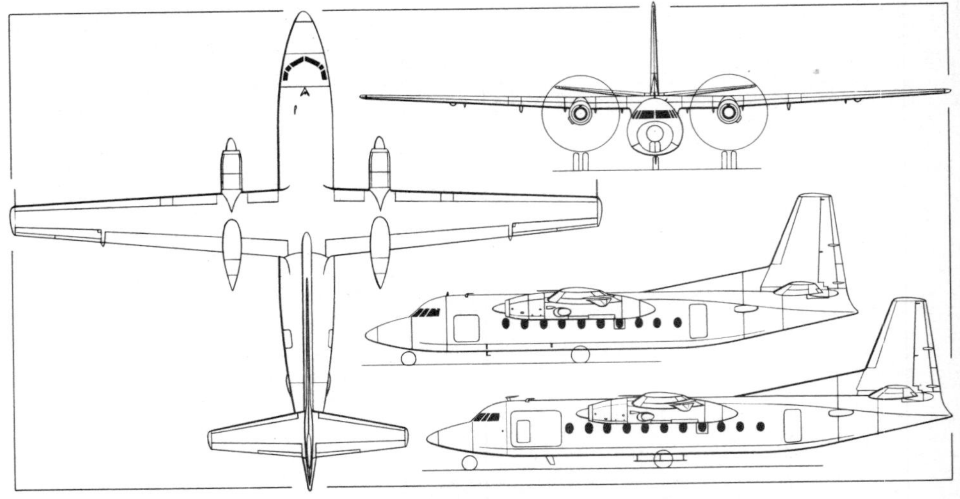

Fokker-VFW F27 Friendship 200, with additional side view (bottom) of Friendship 500 *(Pilot Press)*

four abreast at 900 mm (35·5 in) seat pitch, plus 6·17 m³ (218 cu ft) of cargo space; 28 passengers at same seat pitch in rear of cabin, plus 16·65 m³ (588 cu ft) cargo space; or all-cargo version with 48·90 m³ (1,727 cu ft) of cargo space. Alternative layouts for up to 48 passengers.

ACCOMMODATION (Mk 400M): Folding canvas seats, with safety harnesses, along cabin sides for up to 45 paratroops. Toilet and provision for medical supply box or pantry unit at rear. Ambulance version can accommodate 24 USAF-type stretchers, in eight tiers of three, with seats at front and rear for up to nine medical attendants or sitting casualties. All-cargo version fitted with skid strips, tie-down fittings, protection plates and hinged hatracks. Despatch door on each side of fuselage at rear for dropping supplies and personnel.

ACCOMMODATION (Mk 500): Main cabin has standard seating for 52 passengers four abreast at 895 mm (35·25 in) seat pitch; alternative layouts enable up to 60 passengers to be carried at 720 mm (28·5 in) pitch.

SYSTEMS: Pressurisation and air-conditioning system utilises two Rootes-type engine-driven blowers. Choke heating and air-to-air heat exchanger; optional bootstrap cooling system. Pressure differential 0·29 bars (4·16 lb/sq in) in Mks 400, 500 and 600; 0·38 bars (5·5 lb/sq in) in Mk 200. No hydraulic system. Pneumatic system, pressure 235 bars (3,400 lb/sq in), for landing gear retraction, nosewheel steering and brakes. Emergency pneumatic circuits for landing gear extension and brakes. Primary 28V electrical system supplied by two 375A 28V DC engine-driven generators. Secondary system supplied via two 115V 400Hz AC constant-frequency inverters. Variable-frequency AC power supply, from 120/208V 15kVA engine-driven alternators, for anti-icing and heating. Two 24V 40Ah nickel-cadmium batteries. 1·12 m³ (39·4 cu ft) oxygen system for pilots.

ELECTRONICS AND EQUIPMENT: Standard provisions for VHF and HF transceivers, VHF navigation system (including glideslope), ADF, ILS, marker beacon, dual gyrosyn compass system and intercom system. Provision for weather radar, autopilot etc.

DIMENSIONS, EXTERNAL:
Wing span 29·00 m (95 ft 2 in)
Wing chord at root 3·45 m (11 ft 4 in)
Wing chord at tip 1·40 m (4 ft 7 in)
Wing aspect ratio 12
Length overall:
except Mk 500 23·56 m (77 ft 3½ in)
Mk 500 25·06 m (82 ft 2½ in)
Fuselage: Max width 2·70 m (8 ft 10¼ in)
Max height 2·79 m (9 ft 1¾ in)
Height overall, standard landing gear:
except Mk 500 8·50 m (27 ft 11 in)
Mk 500 8·71 m (28 ft 7¼ in)
Height overall, rough-field landing gear:
except Mk 500 8·59 m (28 ft 2 in)
Tailplane span 9·75 m (32 ft 0 in)
Wheel track (c/l shock-absorbers)
7·20 m (23 ft 7½ in)
Wheelbase:
except Mk 500 8·74 m (28 ft 8 in)
Mk 500 9·74 m (31 ft 11¼ in)
Propeller diameter 3·50 m (11 ft 6 in)
Propeller ground clearance:
standard landing gear:
except Mk 500 0·94 m (3 ft 1 in)
Mk 500 0·99 m (3 ft 3 in)
rough-field landing gear:
except Mk 500 1·02 m (3 ft 4¼ in)
Passenger door (aft, port):
Height 1·65 m (5 ft 5 in)
Width 0·73 m (2 ft 4¾ in)
Height to sill:
except Mk 500 1·22 m (4 ft 0 in)
Mk 500 1·39 m (4 ft 6¾ in)
Service/emergency door (aft, stbd):
Height 1·12 m (3 ft 8 in)
Width 0·74 m (2 ft 5 in)
Height to sill 0·99 m (3 ft 3 in)
Standard cargo door (Mk 200 only):
Height 1·19 m (3 ft 11 in)
Width 1·04 m (3 ft 5 in)
Height to sill 0·99 m (3 ft 3 in)
Large cargo door (Mks 400, 500 and 600):
Height 1·78 m (5 ft 10 in)
Width 2·32 m (7 ft 7½ in)
Height to sill:
except Mk 500 0·99 m (3 ft 3 in)
Mk 500 1·03 m (3 ft 4½ in)
Despatch doors (Mk 400M only, aft, port and stbd, each):
Height 1·65 m (5 ft 5 in)
Width 1·19 m (3 ft 11 in)
Height to sill 1·22 m (4 ft 0 in)

DIMENSIONS, INTERNAL:
Cabin, excl flight deck:
Length:
except Mk 500 14·46 m (47 ft 5 in)
Mk 500 15·96 m (52 ft 4 in)
Max width 2·55 m (8 ft 4½ in)
Max height 2·02 m (6 ft 7½ in)

Mk 200 version of the F27 Friendship in the insignia of Air Anglia

Volume:
except Mk 500 60·5 m³ (2,136 cu ft)
Mk 500 66·8 m³ (2,360 cu ft)
Freight hold (fwd) max:
Mk 200 4·78 m³ (169 cu ft)
Mks 400, 500, 600 5·58 m³ (197 cu ft)
Freight hold (aft) max:
all versions 2·83 m³ (100 cu ft)

AREAS:
Wings, gross 70·00 m² (753·5 sq ft)
Ailerons (total) 3·51 m² (37·80 sq ft)
Trailing-edge flaps (total) 12·72 m² (136·90 sq ft)
Vertical tail surfaces (total) 14·20 m² (153 sq ft)
Horizontal tail surfaces (total) 16·00 m² (172 sq ft)

WEIGHTS AND LOADINGS:
Manufacturer's weight, empty:
Mk 200, 44 seats 10,177 kg (22,436 lb)
Mk 400, 40 seats 10,564 kg (23,290 lb)
Mk 400M 10,596 kg (23,360 lb)
Mk 500, 52-56 seats 10,695 kg (23,578 lb)
Mk 500M 11,034 kg (24,325 lb)
Mk 600, 44 seats 10,336 kg (22,786 lb)
Operating weight, empty:
Mk 200, 44 seats 11,164 kg (24,612 lb)
Mk 400, 40 seats 11,283 kg (24,875 lb)
Mk 400M, all-cargo 10,862 kg (23,947 lb)
Mk 400M, medical evacuation
11,286 kg (24,880 lb)
Mk 400M, paratrooper 11,039 kg (24,336 lb)
Mk 500, 52-56 seats 11,789 kg (25,990 lb)
Mk 500M, all-cargo 11,300 kg (24,912 lb)
Mk 500M, medical evacuation
11,804 kg (26,023 lb)
Mk 500M, paratrooper 11,491 kg (25,332 lb)
Mk 600, 44 seats 11,314 kg (24,943 lb)
Max payload (weight limited):
Mk 200, 44 seats 5,846 kg (12,888 lb)
Mk 400, 40 seats 5,727 kg (12,625 lb)
Mk 400M, all-cargo 6,148 kg (13,553 lb)
Mk 400M, medical evacuation
5,721 kg (12,612 lb)
Mk 400M, paratrooper 5,971 kg (13,164 lb)
Mk 500, 52-56 seats 6,128 kg (13,510 lb)
Mk 500M, all-cargo 6,617 kg (14,588 lb)
Mk 500M, medical evacuation
6,113 kg (13,477 lb)
Mk 500M, paratrooper 6,427 kg (14,168 lb)
Mk 600, 44 seats 5,696 kg (12,557 lb)
Max T-O weight:
all versions 20,410 kg (45,000 lb)
Max landing weight:
Mks 200, 400, 400M and 600 18,600 kg (41,000 lb)
Mks 500 and 500M 19,050 kg (42,000 lb)
Max zero-fuel weight:
Mks 200, 400, 400M and 600 17,010 kg (37,500 lb)
Mks 500 and 500M 17,900 kg (39,500 lb)
Max wing loading:
all versions 291·5 kg/m² (59·7 lb/sq ft)
Max power loading:
all versions 6·39 kg/kW (10·5 lb/shp)

PERFORMANCE (at weights indicated):
Normal cruising speed at 6,100 m (20,000 ft) and AUW of 17,237 kg (38,000 lb):
all versions 259 knots (480 km/h; 298 mph)
Rate of climb at S/L, AUW of 18,143 kg; 40,000 lb):
all civil versions 451 m (1,480 ft)/min
both military versions 494 m (1,620 ft)/min
Service ceiling at AUW of 17,237 kg (38,000 lb):
all civil versions 8,990 m (29,500 ft)
both military versions 9,145 m (30,000 ft)
Service ceiling, one engine out, at AUW of 17,237 kg (38,000 lb):
all civil versions 3,565 m (11,700 ft)
both military versions 4,055 m (13,300 ft)
Runway LCN at max T-O weight, hard runway, standard landing gear 16
Required T-O field length (ICAO-PAMC) at AUW of 18,143 kg (40,000 lb), all civil versions:
S/L, ISA 988 m (3,240 ft)
S/L, ISA +15°C 1,088 m (3,570 ft)
914 m (3,000 ft), ISA 1,210 m (3,970 ft)
Required T-O field length (military) at AUW of 18,143 kg (40,000 lb), both military versions:
S/L, ISA 704 m (2,310 ft)
S/L, ISA +15°C 765 m (2,510 ft)
914 m (3,000 ft), ISA 838 m (2,750 ft)
Required landing field length (ICAO-PAMC) at AUW of 16,329 kg (36,000 lb), all civil versions:
S/L 1,003 m (3,290 ft)
1,525 m (5,000 ft) 1,076 m (3,530 ft)
Required landing field length (military) at AUW of 17,010 kg (37,500 lb), both military versions:
S/L 579 m (1,900 ft)
914 m (3,000 ft) 622 m (2,040 ft)
Range (ISA, zero wind conditions) with FAR 121.645 reserves for alternate, 30 min hold at 3,050 m (10,000 ft) and 10% flight fuel:
Mks 200 and 600, 44 passengers
1,020 nm (1,926 km; 1,197 miles)
Mk 400, 40 passengers
1,025 nm (1,935 km; 1,203 miles)
Mk 500, 52 passengers
935 nm (1,741 km; 1,082 miles)
Military transport range (ISA, zero wind conditions) at max T-O weight, reserves for 30 min hold at S/L and 5% initial fuel:
Mks 400M and 500M, all-cargo, max standard fuel
1,195 nm (2,213 km; 1,375 miles)
Mks 400M and 500M, all-cargo, max possible fuel
2,370 nm (4,389 km; 2,727 miles)
Military combat radius, conditions as above:
Mks 400M and 500M, all-cargo, max standard fuel
625 nm (1,158 km; 719 miles)
Mks 400M and 500M, all-cargo, max possible fuel
1,230 nm (2,278 km; 1,416 miles)
Max endurance at 6,100 m (20,000 ft):
Mk 400M, max standard fuel 7 hr 25 min
Mk 400M, max possible fuel 12 hr 47 min
Mk 500M, max standard fuel 7 hr 14 min
Mk 500M, max possible fuel 12 hr 26 min

OPERATIONAL NOISE CHARACTERISTICS (FAR Pt 36):
T-O noise level 90·6 EPNdB
Approach noise level 100·3 EPNdB
Sideline noise level 92·2 EPNdB

FOKKER-VFW F27 MARITIME

The F27 Maritime is a medium-range maritime patrol version of the Friendship, intended for customers who do not require a more sophisticated long-range patrol aircraft. The basic design was defined in July 1975, and shortly afterwards Fokker-VFW began converting an ex-airline F27 to serve as a prototype/demonstration aircraft. This prototype made its first flight on 25 March 1976.

The F27 Maritime is not intended for anti-submarine duties, but rather for patrol of fishery areas and coastal shipping lanes, surveillance of offshore oil industry operations, search and rescue, environmental control and similar duties. It is operated by a crew of up to seven persons, and the provision of additional fuel capacity in centre-wing bag tanks and wing pylon tanks gives the aircraft an endurance of 10-12 hours, or a range of up to 3,560 nm (4,100 km; 2,547 miles), depending on the mission to be flown. For further details see Addenda.

The extent and complexity of electronic surveillance systems will depend upon individual customer requirements, but a complete range of possible equipment was to be evaluated during the flight test programme in 1976. Basic equipment on board includes a Litton AN/APS-503F search radar, with its scanner mounted in a ventral radome to provide 360° coverage; a Litton LTN-72 long-range inertial navigation system; and a Collins 301E

VHF/UHF D/F system. Two 'bubble' observation windows are provided in the rear fuselage, in which is also installed a marine marker launcher.

Fokker-VFW F27 Maritime coastal patrol aircraft, with ventral radome. Two have been ordered by the Peruvian Navy and one by the Icelandic Coast Guard

FOKKER-VFW F28 FELLOWSHIP

Announced in April 1962, the F28 Fellowship twin-turbofan short-haul transport was developed in collaboration with other European aircraft manufacturers and with the financial support of the Netherlands government. One half of the Dutch share of the development cost was supplied through the Netherlands Aircraft Development Board, the other half through a loan guaranteed by the government.

Under agreements signed in the Summer of 1964, production is undertaken by Fokker-VFW in association with MBB and VFW-Fokker in Germany and Short Bros and Harland in the UK.

Fokker-VFW is responsible for the front fuselage, to a point just aft of the flight deck, the centre fuselage and wing-root fairings. MBB builds the fuselage, from the wing trailing-edge to the rear pressure bulkhead, and the engine nacelles and support stubs. VFW-Fokker is responsible for the rear fuselage and tail unit, and for the cylindrical fuselage section between the wing leading-edge and flight deck. Shorts are responsible for the wings (including the slatted wings for the Mk 6000), and other components, including the main-wheel and nosewheel doors.

First flight of the first prototype F28 (PH-JHG) was made on 9 May 1967, and the second prototype, PH-WEV, flew on 3 August 1967. The third F28 (PH-MOL) flew for the first time on 20 October 1967 and was brought up to production standard in the early Summer of 1968.

The Dutch RLD granted a C of A to the F28 on 24 February 1969, and the first delivery (of the fourth aircraft, to LTU) was made on the same day. The aircraft received FAA Type Approval on 24 March 1969 and German certification on 30 March 1969. RLD certification for operation from unpaved runways was granted in mid-1972. The Mk 1000 was granted FAA-approved noise certification on 31 December 1971.

A total of 115 Fellowships had been ordered by 1 September 1976, as follows:

Mk 1000/1000C (97 ordered, incl 7 Mk 1000C; orders listed in 1975-76 *Jane's,* plus 5 Mk 1000 for Garuda)
Mk 2000 (10 ordered; orders listed in 1975-76 *Jane's)*
Mk 4000
Linjeflyg (Sweden) 8

Six versions have been announced, of which production of the Mks 1000 and 2000 was to end in 1976 in favour of the Mks 3000, 4000 and 6000. The Mk 5000, referred to in the 1975-76 *Jane's,* can be produced to special order. Fokker-VFW also plans a new longer-term development of the F28 with a supercritical wing. This is described separately; descriptions of the current versions follow:

Mk 1000. Initial short-fuselage version, with seating for up to 65 passengers. In service. Being superseded in production in 1976 by Mk 3000. First F28 commercial service was flown by Braathens on 28 March 1969. Available optionally, for all-cargo or mixed passenger/cargo operations, with large freight door at front on port side, aft of passenger door, in which form it is designated **Mk 1000C.**

Mk 2000. Similar to Mk 1000 except for lengthened fuselage, permitting an increase in accommodation for up to 79 passengers in all-tourist layout. F28 first prototype modified to Mk 2000 standard and flown for first time on 28 April 1971. Dutch certification awarded on 30 August 1972. In service. Being superseded in production from mid-1976 by Mk 4000.

Mk 3000. Similar to Mk 4000, but with short fuselage seating up to 65 passengers. Available also in 15-passenger VIP layout, with range of up to 2,200 nm (4,074 km; 2,533 miles).

Mk 4000. High-density long-fuselage version, announced in early 1975, to seat up to 85 passengers at 737 mm (29 in) pitch. Airframe basically as Mk 6000, except for omission of leading-edge slats. Uprated Spey Mk 555-15H power plant. Two additional overwing emergency exits (making a total of four).

Mk 5000. Combines standard 65-seat fuselage of Mk 3000 with the increased-span wings and leading-edge slats of Mk 6000. A non-standard version, but could be produced to special order for customers requiring exceptional short-field performance.

Mk 6000. Long-fuselage version, similar to Mk 2000 except for slatted, long-span wings and improved Spey engines. Prototype, modified from F28 first prototype (previously used for Mk 2000 certification flying) and fitted with modified wings from the second prototype, made its first flight on 27 September 1973. Dutch certification granted on 30 October 1975. Normal max seating capacity 79 passengers, but up to 85 can be carried with reduced fuel load. Two built by 1 January 1976.

The following details apply generally to all versions, except where a specific model is indicated:

Type: Twin-turbofan short-range airliner.

Wings: Cantilever low/mid-wing monoplane. Wing section NACA 0000-X 40Y series with camber varying along span. Thickness/chord ratio up to 14% on inner panels, 10% at tip. Dihedral 2° 30′. Sweepback at quarter-chord 16°. Single-cell two-spar light alloy torsion-box structure, comprising centre-section, integral with fuselage, and two outer panels. Fail-safe construction. Lower skin made of three planks. Taper-rolled top skin. Forged ribs in centre-section, built-up ribs in outer panels. Double-skin leading-edge with ducts for hot-air de-icing. Irreversible hydraulically-operated ailerons. Emergency manual operation of ailerons, through tabs. Hydraulically-operated Fowler double-slotted flaps over 70% of each half-span with electrical emergency extension. Five-section hydraulically-operated lift dumpers in front of flaps on each wing. Trim tab in each aileron. Mks 3000, 4000 and 6000 have extended-span wings, which on the Mk 6000 have full-span hydraulically-operated leading-edge slats.

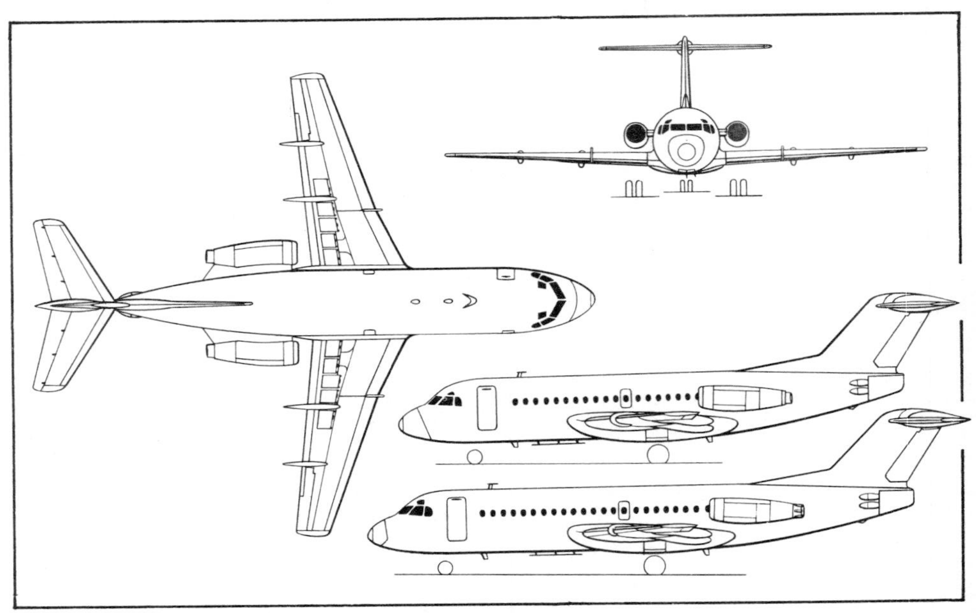
Fokker-VFW F28 Fellowship 1000, with additional side view (bottom) of Fellowship 6000 *(Pilot Press)*

Fuselage: Circular-section semi-monocoque light alloy fail-safe structure, made up of skin panels with Redux-bonded Z-stringers. Bonded doubler plates at door and window cutouts. Quickly-detachable sandwich (metal/end-grain balsa) floor panels. Hydraulically-operated petal airbrakes form aft end of fuselage.

Tail Unit: Cantilever light alloy structure, with hydraulically-actuated variable-incidence T tailplane. Electrical emergency actuation of tailplane. Hydraulically-boosted elevators. Hydraulically-operated rudder with duplicated actuators and emergency manual operation. Honeycomb sandwich skin panels used extensively, in conjunction with multiple spars. Double-skin leading-edges for hot-air de-icing.

Landing Gear: Retractable tricycle type of Dowty-Rotol manufacture, with twin wheels on each unit. Hydraulic retraction, nosewheels forward, main units inward into fuselage. Oleo-pneumatic shock-absorbers. Goodyear wheels, tyres and electronically-controlled braking system. Steerable nosewheel. Main-wheel tyres size 39 × 13, 16-ply rating, pressure 7·03 bars (102 lb/sq in) on Mk 1000, 7·17 bars (104 lb/sq in) on Mk 2000, 7·59 bars (110 lb/sq in) on Mks 3000, 4000 and 6000. Nosewheel tyres size 24·5 × 8·5, 10-ply rating, pressure 5·98 bars (86·5 lb/sq in) on Mk 1000, 5·5 bars (80 lb/sq in) on Mk 2000, and 5·3 bars (77 lb/sq in) on Mk 6000. Low-pressure tyres optional on all units (main-wheel tyre pressure 5·16 bars; 75 lb/sq in on Mks 1000 and 2,000, 5·34 bars; 77·5 lb/sq in on Mks 3000, 4000 and 6000).

Power Plant (Mks 1000 and 2000): Two Rolls-Royce RB.183-2 Spey Mk 555-15 turbofan engines with blade-cooling (each 43·82 kN; 9,850 lb st, flat-rated to 22·5°C), mounted in pod on each side of rear fuselage. No water injection or thrust reversers. Thermal anti-icing for air intakes. For Mks 3000, 4000 and 6000, a Mk 555-15H version of the Spey engine is specified. This has a thrust rating of 44 kN (9,900 lb st), flat-rated to 29·7°C, and has acoustic liners and exhaust silencers. Integral fuel tank in each outer wing panel with total usable capacity of 9,740 litres (2,143 Imp gallons) in Mks 1000/2000/3000/4000; 9,682 litres (2,130 Imp gallons) in Mk 6000. All except Mk 2000 have optional seven bladder-type tank units in wing centre-section with total usable capacity of 3,300 litres (726 Imp gallons). Single refuelling point under starboard wing, near root.

Accommodation: Crew of two side by side on flight deck, with jump-seat for third crew member. Electrically-heated windscreen. Pantry/baggage space immediately aft of flight deck on starboard side, followed by entrance lobby with hydraulically-operated airstair door on port side, service and emergency door on starboard side, and seat for cabin attendant. On Mk 1000C, an optional upward-opening cargo door, to permit all-cargo or all-passenger operation, can be added aft of the passenger airstair door. Additional emergency door on each side of main cabin, over wing (two each side on Mk 4000). Main cabin layout of Mks 1000/3000 can be varied to accommodate 55, 60 or 65 passengers five abreast at

Fokker-VFW F28 Fellowship Mk 1000 short-haul transport, in the insignia of AeroPeru

940, 813/838 or 787 mm (37, 32/33 or 31 in) seat pitch respectively. In Mks 2000/6000, layout can accommodate up to 79 passengers at 787 mm (31 in) seat pitch, and in Mk 4000/6000 to seat 85 passengers at 737 mm (29 in) pitch. Aft of cabin are a wardrobe (port), baggage compartment (port) and toilet compartment (starboard). Underfloor cargo compartments fore and aft of wing, with single door on starboard side of forward hold, with one door on rear hold of each version.

SYSTEMS: AiResearch air-conditioning system, using engine bleed air. Max pressure differential 0·51 bars (7·45 lb/sq in). Two independent hydraulic systems, pressure 207 bars (3,000 lb/sq in). Primary system for flight controls, landing gear, nosewheel steering and brakes, secondary system for duplication of certain essential flight controls. Flying control hydraulic components supplied by Jarry Hydraulics. All-AC electrical system utilises two 20kVA Westinghouse engine-driven generators to supply three-phase constant-frequency 115/200V 400Hz power. One 20Ah battery for starting APU and for emergency power. AiResearch GTCP 36-4A APU, mounted aft of rear pressure bulkhead, for engine starting, ground air-conditioning and ground electrical power, and to drive a third AC generator for standby use on essential services in flight.

ELECTRONICS AND EQUIPMENT: Standard equipment includes VHF transceivers, VHF navigation system (with glideslope), DME, marker beacon, weather radar, ADF, ATC transponder, dual compass system, interphone and public address systems, Smiths SEP6 autopilot, Collins FD 108 flight director, flight guidance caution system, flight data recorder and voice recorder. Thermal bleed air system for wing leading-edges (slats on Mks 5000/6000), tailplane leading-edge and engine air intakes. Stick pusher system on Mk 6000. Optional equipment to customer's requirements, including equipment for operation in Cat. 2 weather minima.

DIMENSIONS, EXTERNAL:
- Wing span:
  - 1000, 2000: 23·58 m (77 ft 4¼ in)
  - 3000, 4000, 6000: 25·07 m (82 ft 3 in)
- Wing chord at root:
  - all versions: 4·80 m (15 ft 9 in)
- Wing chord at tip:
  - 1000, 2000: 1·77 m (5 ft 9¾ in)
- Wing aspect ratio:
  - 1000, 2000: 7·27
- Length overall:
  - 1000, 3000: 27·40 m (89 ft 10¾ in)
  - 2000, 4000, 6000: 29·61 m (97 ft 1¾ in)
- Length of fuselage:
  - 1000, 3000: 24·55 m (80 ft 6½ in)
  - 2000, 4000, 6000: 26·76 m (87 ft 9½ in)
- Fuselage: Max width: 3·30 m (10 ft 10 in)
- Height overall: 8·47 m (27 ft 9½ in)
- Tailplane span: 8·64 m (28 ft 4¼ in)
- Wheel track (c/l of shock-absorbers): 5·04 m (16 ft 6½ in)
- Wheelbase:
  - 1000, 3000: 8·90 m (29 ft 2½ in)
  - 2000, 4000, 6000: 10·35 m (33 ft 11½ in)
- Passenger door (fwd, port):
  - Height: 1·93 m (6 ft 4 in)
  - Width: 0·86 m (2 ft 10 in)
- Service/emergency door (fwd, stbd):
  - Height: 1·27 m (4 ft 2 in)
  - Width: 0·61 m (2 ft 0 in)
- Emergency exits (centre, each):
  - Height: 0·91 m (3 ft 0 in)
  - Width: 0·51 m (1 ft 8 in)
- Freight hold doors (each):
  - Height (fwd, each): 0·90 m (2 ft 11½ in)
  - Height (aft): 0·80 m (2 ft 7½ in)
  - Width (fwd, each): 0·95 m (3 ft 1½ in)
  - Width (aft): 0·89 m (2 ft 11 in)
  - Height to sill (fwd, each): 1·47 m (4 ft 10 in)
  - Height to sill (aft): 1·59 m (5 ft 2½ in)
- Baggage door (rear, port, optional):
  - Height: 0·60 m (1 ft 11½ in)
  - Width: 0·51 m (1 ft 8 in)
- Cargo door (fwd, port, optional):
  - Height: 1·87 m (6 ft 1¾ in)
  - Width: 2·49 m (8 ft 2 in)
  - Height to sill: 2·24 m (7 ft 4¼ in)

DIMENSIONS, INTERNAL:
- Cabin, excl flight deck:
  - Length:
    - 1000, 3000: 13·10 m (43 ft 0 in)
    - 2000, 4000, 6000: 15·31 m (50 ft 3 in)
  - Max length of seating area:
    - 1000, 3000: 10·74 m (35 ft 2¾ in)
    - 2000, 4000, 6000: 12·95 m (42 ft 6¾ in)
  - Max width: 3·10 m (10 ft 2 in)
  - Max height: 2·02 m (6 ft 7¼ in)
  - Floor area:
    - 1000, 3000: 38·4 m^2 (413·3 sq ft)
    - 2000, 4000, 6000: 44·8 m^2 (482·2 sq ft)
  - Volume:
    - 1000, 3000: 71·5 m^3 (2,525 cu ft)
    - 2000, 4000, 6000: 83·0 m^3 (2,931 cu ft)
- Freight hold (underfloor, fwd):
  - 1000, 3000: 6·90 m^3 (245 cu ft)
  - 2000, 4000, 6000: 8·70 m^3 (308 cu ft)
- Freight hold (underfloor, rear):
  - 1000, 3000: 3·80 m^3 (135 cu ft)
  - 2000, 4000, 6000: 4·80 m^3 (169 cu ft)
- Baggage hold (aft of cabin), max:
  - all versions: 2·30 m^3 (81·22 cu ft)

AREAS:
- Wings, gross:
  - 1000, 2000: 76·40 m^2 (822 sq ft)
  - 3000, 4000, 6000: 79·00 m^2 (850 sq ft)
- Ailerons (total): 2·67 m^2 (28·74 sq ft)
- Trailing-edge flaps (total): 14·00 m^2 (150·7 sq ft)
- Fuselage airbrakes (total): 3·62 m^2 (38·97 sq ft)
- Fin (incl dorsal fin): 12·30 m^2 (132·4 sq ft)
- Rudder: 2·30 m^2 (24·76 sq ft)
- Tailplane: 19·50 m^2 (209·9 sq ft)
- Elevators (total): 3·84 m^2 (41·33 sq ft)

WEIGHTS AND LOADINGS:
- Manufacturer's weight empty:
  - 1000, 65 seats: 14,492 kg (31,954 lb)
  - 2000, 79 seats: 14,936 kg (32,929 lb)
  - 6000, 79 seats: 15,638 kg (34,477 lb)
- Operating weight empty:
  - 1000, 65 seats: 16,112 kg (35,521 lb)
  - 2000, 79 seats: 16,711 kg (36,841 lb)
  - 3000, 65 seats: 16,324 kg (35,988 lb)
  - 4000, 85 seats: 16,962 kg (37,394 lb)
  - 6000, 79 seats: 17,381 kg (38,318 lb)
- Max weight-limited payload:
  - 1000: 8,608 kg (18,977 lb)
  - 2000: 8,009 kg (17,656 lb)
  - 3000: 9,076 kg (20,009 lb)
  - 4000: 9,118 kg (20,101 lb)
  - 6000: 8,019 kg (17,678 lb)
- Max T-O weight:
  - 1000, 2000: 29,480 kg (65,000 lb)
  - 3000, 4000: 32,200 kg (70,988 lb)
  - 6000, 79 passengers (normal range): 32,115 kg (70,800 lb)
  - 6000, 79 passengers (extended range): 33,110 kg (72,995 lb)
- Max zero-fuel weight:
  - 1000, 2000: 24,720 kg (54,500 lb)
  - 3000, 6000: 25,400 kg (56,000 lb)
  - 4000: 26,080 kg (57,496 lb)
- Max landing weight:
  - 1000, 2000: 26,760 kg (59,000 lb)
  - 3000, 4000, 6000: 29,030 kg (64,000 lb)
- Max wing loading:
  - 1000, 2000: 386 kg/m^2 (79·1 lb/sq ft)
  - 3000, 4000: 407 kg/m^2 (83·4 lb/sq ft)
  - 6000, 79 passengers (normal range): 406 kg/m^2 (83·3 lb/sq ft)
  - 6000, 79 passengers (extended range): 419 kg/m^2 (85·8 lb/sq ft)
- Max cabin floor loading:
  - all passenger versions: 366 kg/m^2 (75 lb/sq ft)
  - 1000, with large cargo door: 610 kg/m^2 (125 lb/sq ft)
- Max power loading:
  - 1000, 2000: 336 kg/kN (3·3 lb/lb st)
  - 3000, 4000: 367·5 kg/kN (3·6 lb/lb st)
  - 6000, 79 passengers (normal range): 366·4 kg/kN (3·6 lb/lb st)
  - 6000, 79 passengers (extended range): 378 kg/kN (3·7 lb/lb st)

PERFORMANCE (Mks 1000 and 2000 at AUW of 26,760 kg; 59,000 lb, Mks 3000, 4000 and 6000 at AUW of 29,000 kg; 63,934 lb, ISA, except where indicated):
- Never-exceed speed (all versions): 390 knots (723 km/h; 449 mph) EAS or Mach 0·83
- Max permissible operating speed (all versions): 330 knots (611 km/h; 380 mph) EAS or Mach 0·75
- Max cruising speed at 7,000 m (23,000 ft) (all versions): 455 knots (843 km/h; 523 mph) TAS
- Econ cruising speed at 9,150 m (30,000 ft):
  - 1000, 2000: 362 knots (670 km/h; 416 mph) TAS
  - 3000, 4000, 6000: 366 knots (678 km/h; 421 mph) TAS
- Threshold speed at max landing weight:
  - 1000, 2000: 119 knots (220 km/h; 137 mph) EAS
  - 6000: 110 knots (204 km/h; 127 mph) EAS
- Max cruising altitude:
  - all versions: 10,675 m (35,000 ft)
- Min ground turning radius:
  - 1000, 3000: 9·60 m (31 ft 6 in)
  - 2000, 4000, 6000: 10·90 m (35 ft 9 in)
- Runway LCN at max T-O weight (hard runway):
  - 1000, 2000, standard tyres: 27
  - 1000, 2000, low-pressure tyres: 22·5
  - 3000, 4000, 6000, standard tyres: 30
  - 3000, 4000, 6000, low-pressure tyres: 24
- Runway LCN at max T-O weight (flexible runway):
  - 1000, 2000, standard tyres: 21·5
  - 1000, 2000, low-pressure tyres: 18
  - 3000, 4000, 6000, standard tyres: 24
  - 3000, 4000, 6000, low-pressure tyres: 19
- FAR T-O field length at max T-O weight (1000, 2000):
  - S/L: 1,673 m (5,490 ft)
  - S/L, ISA + 10°C: 1,774 m (5,820 ft)
  - S/L, ISA + 15°C: 1,878 m (6,160 ft)
  - 610 m (2,000 ft): 1,820 m (5,970 ft)
  - 915 m (3,000 ft): 1,926 m (6,320 ft)
- FAR T-O field length at max T-O weight (3000, 4000):
  - S/L: 1,740 m (5,710 ft)
  - S/L, ISA + 10°C: 1,800 m (5,905 ft)
  - S/L, ISA + 15°C: 1,880 m (6,168 ft)
  - 610 m (2,000 ft): 1,870 m (6,135 ft)

915 m (3,000 ft) 2,025 m (6,645 ft)
FAR T-O field length at max T-O weight (6000):
S/L 1,310 m (4,300 ft)
S/L, ISA + 10°C 1,370 m (4,495 ft)
S/L, ISA + 15°C 1,545 m (5,070 ft)
610 m (2,000 ft) 1,490 m (4,890 ft)
915 m (3,000 ft) 1,680 m (5,510 ft)
FAR landing field length at max landing weight (1000, 2000):
S/L 1,079 m (3,540 ft)
1,525 m (5,000 ft) 1,222 m (4,010 ft)
FAR landing field length at max landing weight (3000, 4000):
S/L 1,080 m (3,545 ft)
1,525 m (5,000 ft) 1,340 m (4,395 ft)
FAR landing field length at max landing weight (6000):
S/L 890 m (2,920 ft)
1,525 m (5,000 ft) 1,000 m (3,280 ft)
Range, high-speed schedule, FAR 121.645 reserves:
1000, 65 passengers
1,020 nm (1,889 km; 1,174 miles)
2000, 79 passengers 630 nm (1,167 km; 725 miles)
*3000, 65 passengers
975 nm (1,805 km; 1,122 miles)
4000, 85 passengers and 6000, 79 passengers
900 nm (1,667 km; 1,036 miles)
Range, long-range schedule, FAR 121.645 reserves:
1000, 65 passengers
1,130 nm (2,093 km; 1,300 miles)
2000, 79 passengers 700 nm (1,296 km; 806 miles)
*3000, 65 passengers
1,400 nm (2,593 km; 1,611 miles)
4000, 85 passengers
1,000 nm (1,852 km; 1,151 miles)
6000, 79 passengers
1,030 nm (1,908 km; 1,185 miles)
Range (6000) at 33,110 kg (72,995 lb) max T-O weight, FAR 121.645 reserves (centre wing tanks included):
85 passengers, high-speed configuration
880 nm (1,630 km; 1,013 miles)

F28 Fellowship Mk 2000 twin-turbofan airliner in the insignia of Ghana Airways

85 passengers, long-range configuration
980 nm (1,815 km; 1,128 miles)
75 passengers, high-speed configuration
1,050 nm (1,944 km; 1,209 miles)
75 passengers, long-range configuration
1,175 nm (2,176 km; 1,353 miles)
With wing centre-section tanks

Operational Noise Characteristics (FAR Pt 36):
T-O noise level:
1000, 2000 90 EPNdB
3000, 4000 (estimated) 91·5 EPNdB
6000 (estimated) 92·4 EPNdB
Approach noise level:
1000 101·2 EPNdB
2000 101·8 EPNdB
3000, 4000 (estimated) 98 EPNdB
6000 (estimated) 96·6 EPNdB
Sideline noise level:
1000, 2000 99·5 EPNdB
3000, 4000 (estimated) 97·5 EPNdB
6000 (estimated) 98·6 EPNdB

FOKKER-VFW F28-2

As a further development of the F28 Fellowship, Fokker-VFW is studying a version fitted with a new wing, of increased span and supercritical section, and having a further-lengthened fuselage capable of accommodating 95-105 passengers. Some two years' design work is likely to be involved before the detail design of this new version is 'frozen', but among the possibilities being considered are the use of carbon fibre composites in the wing construction and the use of developed versions of the Spey engine, including the so-called 'refanned' Spey.

NEW ZEALAND

AEROSPACE
NEW ZEALAND AEROSPACE INDUSTRIES LIMITED

Head Office and Works:
Hamilton Airport, R.D.2, Hamilton
Telephone: Hamilton 436-144 and 436-069
Telex: NZASIL 21242
General Manager:
G. H. Willetts
Marketing Manager:
G. Bates
Chief Designer:
P. W. C. Monk
Production Manager:
C. R. S. Wood
Quality Control Manager:
N. H. Cribb
Chief Maintenance Engineer:
D. G. Mundell
Company Secretary:
J. D. Linch

Aero Engine Services Ltd and Air Parts (NZ) Ltd (see 1972-73 *Jane's*) amalgamated on 1 April 1973 to form New Zealand Aerospace Industries Ltd. This company has a share capital of $A1·3 million, half of which is held by shareholders of the two constituent companies and the remainder in equal proportions by Air New Zealand and New Zealand National Airways Corporation.

Among the initial tasks of the new company was to integrate production of the (formerly Air Parts) Fletcher agricultural aircraft and the (formerly AESL) Airtrainer CT4, in a new 1,579 m² (17,000 sq ft) facility at the AESL premises at Hamilton Airport.

AEROSPACE AIRTRAINER CT4

In 1967, Victa Ltd in Australia built and flew a prototype four-seat development of the Airtourer two-seat fully-aerobatic all-metal lightplane, known as the Aircruiser. This project was shelved when AESL purchased the Airtourer later that year, but AESL retained first option on the Aircruiser, and in mid-1971 purchased the latter project, including the Victa-built prototype. The project was brought to New Zealand, where AESL decided to manufacture a military trainer based on the original Victa Aircruiser.

The Victa aircraft was stressed only for flying between *g* limits of +3·8 and −1·5; AESL redesigned and restressed the aircraft to make it suitable for aerobatic flying with limits of +6 and −3*g*. The resulting military version, the Airtrainer CT4, differs from the original civil Victa Aircruiser in having a hinged, clear-Perspex cockpit canopy; side-by-side seating for two persons, with an optional third seat at the rear; and stick-type (instead of wheel-type) control columns.

A prototype of the CT4 (ZK-DGY) flew for the first time on 23 February 1972. The Royal Thai Air Force ordered 24 Airtrainers, delivery of which has been completed. Deliveries of 37 to the Royal Australian Air Force have also been completed and production is on schedule of 13 for the Royal New Zealand Air Force. The RNZAF version is known as the **CT4B**. Production of 14 Airtrainers for another customer is proceeding concurrently with those for the RNZAF. Negotiations are continuing with several other air forces, and the production rate has been increased.

Design development of the CT4 is continuing with uprated power plants, and the aircraft has been modified for evaluation in the forward air control role.

Type: Two/three-seat fully-aerobatic light training aircraft.

Wings: Cantilever low-wing monoplane. Wing section NACA 23012 (modified) at root, NACA 4412 (modified) at tip. Dihedral 6° 45′ at chord line. Incidence 3° at root, 0° at tip. Root chord increased by forward sweep of the inboard leading-edges. Single main spar light alloy stressed-skin structure, with glassfibre wingtips which are detachable to permit optional wingtip fuel tanks to be fitted. Single-slotted electrically-actuated flap and aerodynamically-balanced bottom-hinged aileron on each trailing-edge, of light alloy construction with fluted skins. No tabs.

Fuselage: All-metal stressed-skin semi-monocoque structure. Glassfibre engine cowling.

Tail Unit: Cantilever light alloy structure, with some aerodynamic balance. One-piece elevator, statically balanced. Ground-adjustable tab on rudder. Rudder controlled by rod and cable linkage, elevator by rod and mechanical linkage. Electrically-actuated trim control for rudder and elevator.

Landing Gear: Non-retractable tricycle type. Cantilever spring steel main legs. Steerable nosewheel, carried on telescopic strut and oleo shock-absorber. Main units fitted with Dunlop Australia wheels and tubeless tyres size 6·00-6; nosewheel fitted with tubeless tyre size 5·00-5. Tyre pressure 1·59 bars (23 lb/sq in) on main units, 1·10 bars (16 lb/sq in) on nose unit. Dunlop Australia single-disc toe-operated hydraulic brakes, with hand-operated parking lock. Landing gear designed to shear prior to any excess impact loading being transmitted to wing, to minimise structural damage in the event of a crash landing.

Power Plant: One 157 kW (210 hp) Continental IO-360-H flat-six engine standard, driving a Hartzell HC-C2YF-1 two-blade constant-speed metal propeller. Total fuel capacity 204·5 litres (45 Imp gallons). Wingtip tanks, each of 77 litres (17 Imp gallons) capacity, available optionally. 149 kW (200 hp) Lycoming IO-360-B flat-four engine available optionally.

Accommodation: Two seats side by side under hinged, fully-transparent Perspex canopy. Space to rear for optional third seat or 52 kg (115 lb) of baggage or

One of the 24 Airtrainer CT4s built for the Royal Thai Air Force

equipment. Dual controls standard.

DIMENSIONS, EXTERNAL:
Wing span 7·92 m (26 ft 0 in)
Wing span over tip-tanks 8·20 m (26 ft 11 in)
Wing chord at root 2·17 m (7 ft 1¼ in)
Wing chord at tip 0·98 m (3 ft 2½ in)
Wing aspect ratio 5·25
Length overall 7·06 m (23 ft 2 in)
Height overall 2·59 m (8 ft 6 in)
Fuselage: Max width 1·12 m (3 ft 8 in)
Max depth 1·40 m (4 ft 7¼ in)
Tailplane span 3·61 m (11 ft 10 in)
Wheeltrack 2·97 m (9 ft 9 in)
Wheelbase 1·71 m (5 ft 7⅜ in)
Propeller diameter 1·93 m (6 ft 4 in)
Propeller ground clearance 0·43 m (1 ft 5 in)

DIMENSIONS, INTERNAL:
Cabin: Length 2·74 m (9 ft 0 in)
Max width 1·08 m (3 ft 6½ in)
Max height 1·35 m (4 ft 5 in)

AREAS:
Wings, gross 11·98 m² (129·0 sq ft)
Ailerons (total) 0·86 m² (9·24 sq ft)
Flaps (total) 2·10 m² (22·60 sq ft)
Fin 0·60 m² (6·43 sq ft)
Rudder, incl tab 0·58 m² (6·26 sq ft)
Tailplane 1·44 m² (15·50 sq ft)
Elevator 1·26 m² (13·60 sq ft)

WEIGHTS AND LOADINGS:
Basic weight empty 662 kg (1,460 lb)
Weight empty, equipped 675 kg (1,490 lb)
Max T-O weight 1,088 kg (2,400 lb)
Max wing loading 90·8 kg/m² (18·6 lb/sq ft)
Max power loading (IO-360-H) 6·93 kg/kW (11·43 lb/hp)

PERFORMANCE (at T-O weight of 1,066 kg; 2,350 lb, ISA, 210 hp engine):
Never-exceed speed 230 knots (426 km/h; 265 mph)
Max level speed at S/L 155 knots (286 km/h; 178 mph)
Max level speed at 3,050 m (10,000 ft) 142 knots (262 km/h; 163 mph)
Cruising speed at S/L:
75% power 140 knots (259 km/h; 161 mph)
65% power 129 knots (240 km/h; 149 mph)
55% power 118 knots (219 km/h; 136 mph)
Cruising speed at 3,050 m (10,000 ft):
75% power 125 knots (232 km/h; 144 mph)
Stalling speed at S/L:
flaps up 56 knots (103 km/h; 64 mph)
flaps down 46 knots (85·5 km/h; 53 mph)
Stalling speed at 3,050 m (10,000 ft):
flaps up 65·5 knots (121 km/h; 75 mph)
flaps down 51·5 knots (95 km/h; 59 mph)
Max rate of climb at S/L 411 m (1,350 ft)/min
Time to altitude of:
915 m (3,000 ft) 2 min 31 sec
1,525 m (5,000 ft) 4 min 36 sec
3,050 m (10,000 ft) 11 min 40 sec
Service ceiling 5,455 m (17,900 ft)
T-O run 224 m (733 ft)
T-O to 15 m (50 ft) 377 m (1,237 ft)
Landing from 15 m (50 ft) 335 m (1,100 ft)
Landing run 155 m (510 ft)
Max range at S/L at 102·5 knots (190 km/h; 118 mph) 767 nm (1,422 km; 884 miles)
Range with 10% reserves (without tip-tanks):
75% power at S/L 595 nm (1,104 km; 686 miles)
75% power at 1,525 m (5,000 ft) 686 nm (1,271 km; 790 miles)
55% power at S/L 645 nm (1,195 km; 743 miles)
65% power at 1,525 m (5,000 ft) 707 nm (1,311 km; 815 miles)
Max endurance with 10% reserves (without tip-tanks):
75% power at S/L 4 hr 15 min
75% power at 1,525 m (5,000 ft) 5 hr 10 min
55% power at S/L 5 hr 28 min
65% power at 1,525 m (5,000 ft) 5 hr 47 min
g limits +6; −3

AEROSPACE FLETCHER FU-24-950

The FU-24 was developed by the Sargent-Fletcher Company of El Monte, California, initially for agricultural top-dressing work in New Zealand. The prototype flew in July 1954, followed by the first production aircraft five months later. All manufacturing and sales rights for the FU-24 were acquired by Air Parts (NZ) Ltd in 1964, and production was undertaken subsequently in this company's factory at Hamilton Airport, New Zealand.

The initial production series of 100 was delivered to New Zealand operators for top-dressing work. By January 1976, a total of 218 Fletcher aircraft had been produced, including 10 for Iraq and others for customers in Australia, Bangladesh, Pakistan, Thailand, Uruguay and several Pacific Island countries.

The current model is the FU-24-950 agricultural version with 400 hp Lycoming IO-720 engine, to which the following description applies:

TYPE: Agricultural and general-purpose aircraft.

WINGS: Cantilever low-wing monoplane. NACA 4415 wing section. Dihedral (outer wings) 8°. Incidence 2°. All-metal two-spar structure. All-metal plain-hinged ailerons. All-metal slotted flaps.

FUSELAGE: All-metal semi-monocoque structure. Cockpit area stressed for 18*g* impact.

TAIL UNIT: Cantilever all-metal structure. All-movable horizontal tail with anti-servo tab.

LANDING GEAR: Non-retractable tricycle type, with steerable nosewheel. Fletcher air-oil shock-absorber struts. Cleveland wheels and hydraulic disc brakes on main units. Goodyear tyres, size 8·50-6 (6-ply), pressure range 0·76-2·07 bars (11-30 lb/sq in). Wheel fairings optional.

POWER PLANT: One 298 kW (400 hp) Lycoming IO-720-A1A or A1B flat-eight engine, driving a Hartzell HC-C3YR-1R/8475R three-blade constant-speed variable-pitch metal propeller with spinner. Fuel tanks in wing leading-edges; total usable capacity 253 litres (67 US gallons).

ACCOMMODATION (Agricultural models): Enclosed cockpit for pilot and one passenger on side-by-side seats under rearward-sliding canopy. Optional equipment includes large cargo door on port side, additional cargo floor area, small rear door, and dual controls.

ACCOMMODATION (Utility models): Enclosed cabin for pilot and up to seven passengers or equivalent freight. Dual controls optional. Rearward-sliding hood over front two seats. Large passenger/cargo door on port side.

AGRICULTURAL EQUIPMENT: Hopper outlets for spreading of solids (fertiliser, dry ice, poison bait, etc). Transland Swathmaster for top-dressing, seeding and high-volume spraying. Transland Boommaster for liquid spraying with booms, nozzles, fan-driven pump, etc, for low- and high-volume spraying. Micronair spraying equipment with electrically- or fan-driven pump, varied control systems, side-loading valve for liquids, and special adaptor plate for interchangeability of equipment.

GENERAL OPTIONAL EQUIPMENT (all models): Full blind-flying instrumentation with ADF, VHF, VOR and DME. Full dual controls; dual main wheels and brakes, wheel and leg fairings; long-range fuel tanks; cabin heating and air-conditioning systems; metric instrumentation.

DIMENSIONS, EXTERNAL:
Wing span 12·81 m (42 ft 0 in)
Wing chord (constant) 2·13 m (7 ft 0 in)

Aerospace Airtrainer CT4 in Royal Australian Air Force insignia

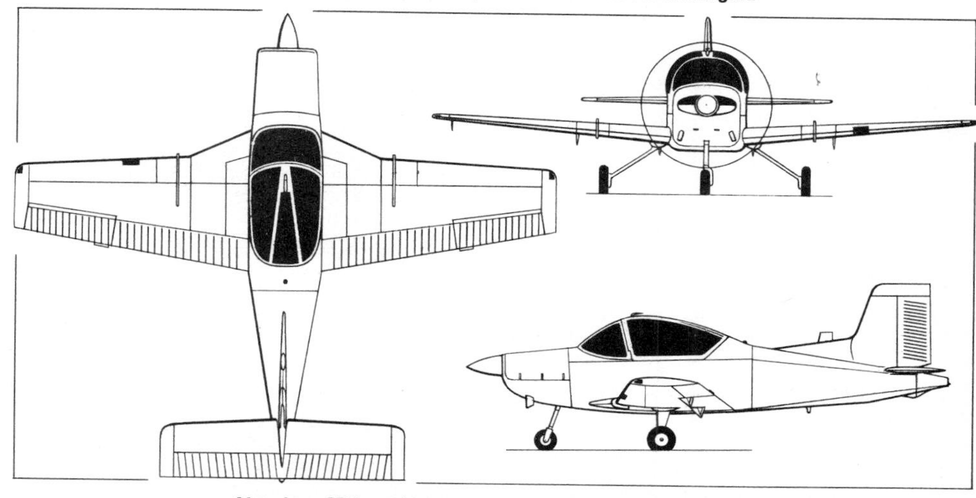

Airtrainer CT4 two/three-seat aerobatic trainer *(Pilot Press)*

Aerospace Fletcher FU-24-950 agricultural aircraft (Lycoming IO-720 engine)

Wing aspect ratio 6
Length overall 9·70 m (31 ft 10 in)
Height overall 3·11 m (10 ft 2½ in)
Tailplane span 4·22 m (13 ft 10 in)
Wheel track 3·71 m (12 ft 2 in)
Wheelbase 2·28 m (7 ft 6 in)
Propeller diameter 2·18 m (7 ft 2 in)
Passenger/cargo door (port, rear):
  Height 0·97 m (3 ft 2 in)
  Width 0·94 m (3 ft 1 in)
Optional small cargo door (rear):
  Height 0·44 m (1 ft 5½ in)
  Width 0·76 m (2 ft 6 in)

DIMENSIONS, INTERNAL:
  Cabin: Length 3·18 m (10 ft 5 in)
    Max width 1·22 m (4 ft 0 in)
    Max height 1·27 m (4 ft 2 in)
    Floor area 3·87 m² (41·7 sq ft)

AREAS:
  Wings, gross 27·31 m² (294 sq ft)
  Ailerons (total) 1·82 m² (19·6 sq ft)
  Trailing-edge flaps (total) 3·16 m² (34·0 sq ft)
  Fin 1·26 m² (13·6 sq ft)
  Rudder 0·64 m² (6·9 sq ft)
  Tailplane 4·00 m² (43·1 sq ft)
  Tailplane tab 0·45 m² (4·9 sq ft)

WEIGHTS AND LOADINGS:
  Weight empty, equipped 1,188 kg (2,620 lb)
  Max payload (agricultural) 1,052 kg (2,320 lb)
  Normal max T-O weight 2,204 kg (4,860 lb)
  Max agricultural T-O weight 2,463 kg (5,430 lb)
  Cabin floor loading 1,885 kg/m² (386 lb/sq ft)
  Normal wing loading 80·6 kg/m² (16·5 lb/sq ft)
  Normal power loading 7·40 kg/kW (12·15 lb/hp)

PERFORMANCE (at Normal max T-O weight):
  Never-exceed speed 143 knots (265 km/h; 165 mph)
  Max level speed at S/L 126 knots (233 km/h; 145 mph)
  Max cruising speed (75% power) 106 knots (196 km/h; 122 mph)
  Stalling speed 49 knots (91 km/h; 57 mph)
  Max rate of climb at S/L 280 m (920 ft)/min
  Service ceiling 4,875 m (16,000 ft)
  T-O run 283 m (930 ft)
  T-O to 15 m (50 ft) 372 m (1,220 ft)
  Landing from 15 m (50 ft) 390 m (1,280 ft)
  Landing run 219 m (720 ft)
  Range with max fuel, 45 min reserves 383 nm (709 km; 441 miles)

FLIGHT ENGINEERS
FLIGHT ENGINEERS LTD

ADDRESS:
PO Box 177, Papakura, Auckland
Telephone: Papakura 89-384
CHAIRMAN OF DIRECTORS:
J. T. Barr
MANAGING DIRECTOR:
M. Curley

Flight Engineers Ltd was formed jointly by Barr Brothers Ltd, an agricultural operating company, and Marine Helicopters Ltd, to provide servicing facilities for the two operating companies. As an extension to this business it was decided to undertake the licence assembly in New Zealand of the Transavia PL-12 Airtruk.

By late May 1973 Flight Engineers Ltd had assembled three of these aircraft, with a fourth nearly complete and a further eight kits to be supplied from Australia. Arrangements progressed subsequently until approximately 80% of the structure was built in New Zealand.

Production was at the rate of one PL-12 a month for a period, but this was not maintained during 1974 due to supply shortages, and production in 1975 was halted due to unfavourable economic conditions. No further news has been received since that time.

PAKISTAN

KIYUSKI INTERNATIONAL

ADDRESS:
Campbellpur

It was reported in early 1976 that this Pakistan company had concluded agreements with Cessna Aircraft Co and Hughes Helicopters of the USA, respectively for licence production in Pakistan of the Cessna T-41D basic training aircraft and the Hughes Model 500 helicopter. An output of 50 of each type per year is planned, initially by assembling imported US-built components. It is understood that BredaNardi of Italy (which see) is also involved in the Pakistani Hughes 500 programme.

PAKISTAN ARMY AVIATION (No. 503 WORKSHOP)

ADDRESS:
Dhamial, near Rawalpindi, Western Punjab

Two types of aircraft are currently being manufactured at the Dhamial base of the Pakistan Army Aviation, which accommodates, besides No. 503 Workshop, the Army Aviation School and three operational PAA squadrons.

The Pakistan Army received about 60 Cessna O-1 Bird Dog observation and liaison aircraft from the USA in the 1950s, and a substantial proportion of these are still in service. Based on its experience of repairing and overhauling these aircraft, and taking advantage of a substantial quantity of spares, the No. 503 Workshop is manufacturing, without a licence, new-production O-1s for the PAA, at an approximate rate of one per month. About 60% of the components of each aircraft are manufactured locally.

The Workshop is, in addition, manufacturing at about the same rate the Aérospatiale Alouette III helicopter, for which the Pakistan Army originally placed an order in 1968. An initial quantity of French-built Alouette IIIs was followed by the supply of knock-down kits for assembly in Pakistan, and domestically-built examples are currently being supplied to all three branches of the Pakistan armed forces.

Above right: **Aérospatiale Alouette III helicopter built in Pakistan** *(John Fricker)*

Right: **Cessna O-1 Bird Dog built at No. 503 Workshop** *(John Fricker)*

PHILIPPINES
(REPUBLIC OF)

PADC
PHILIPPINE AEROSPACE DEVELOPMENT CORPORATION

ADDRESS:
PADC Building, Domestic Terminal Road, Nichols Field, Pasay City 3129
Telephone: 839081 to 839089
Telex: 2440 RPI PH

PRESIDENT:
Roberto H. Lim

EXECUTIVE VICE-PRESIDENT:
Luis M. Mirasol Jr

SENIOR VICE-PRESIDENTS:
Romeo S. David (Aircraft Manufacturing)
Adriano Cruz (Air Transport Services)
Marte U. Iglesias (Maintenance Engineering)

DIRECTORS:
Alfredo Juinio (Chairman)
Juan Ponce Enrile
Vicente Paterno
Leonides Virata
Cesar Virata
Panfilo Domingo
Geronimo Velasco
Roberto H. Lim

PADC is a government corporation established in 1973 by President Ferdinand E. Marcos to promote the development of an aerospace industry in the Philippines.

PADC (MBB) BO 105

In 1974, PADC began an assembly and manufacturing programme for the BO 105C helicopter, under a licence agreement with MBB of Germany. The first BO 105C assembled from knocked-down kits by Rotorcraft Philippines, a PADC subsidiary, was completed in August 1974: eight others were assembled and sold during that year, and a further 10 were completed in 1975. These aircraft are part of a total of 38 involved in the current programme.

PADC (BRITTEN-NORMAN) ISLANDER

PADC has arranged with Britten-Norman (Bembridge) Ltd to produce the Islander transport aircraft in the Philippines, and to develop an amphibious floatplane version of this aircraft.

Phase 1 of this programme began in November 1974 with delivery of the first of six 300 hp Islanders to PADC. The second phase involved 14 unpainted aircraft, delivered to Manila without cabin trim, furnishings and electronics. PADC completed these aircraft at the rate of one a month from February 1975, increasing to two a month by May. By mid-1975 the initial six Islanders and the first few phase 2 aircraft had been delivered, including a small number to the Philippine armed forces for air/sea rescue duties. The second phase was completed by the end of 1975.

The 20 aircraft involved in phase 3 are being assembled by PADC from knocked-down kits supplied by Britten-Norman and shipped to the Philippines. During the first quarter of 1976, one of these aircraft was due for flight testing and seven others were on the assembly line. The final phase will include the manufacture of subassemblies and other aircraft components, using jigs and detailed parts supplied from the United Kingdom.

Of the 60 aircraft to be assembled in phase 4, about 25 will be repurchased by Britten-Norman for sale throughout the world.

The first float-equipped Islander was due to make its first flight at Bembridge during 1976. Once this version is in operation, PADC will become the sole installation centre for the floats in Australasia.

Philippine Aerotransport and Philippine Aerosystems, both PADC subsidiaries, will operate and maintain, respectively, most of the Islander aircraft acquired through the programme.

PADC-assembled Britten-Norman Islander in Philippine Navy insignia

PADC FIXED-WING AIRCRAFT PROTOTYPE

This prototype development project is a joint venture between PADC and the Philippine government's National Science and Development Board (NSDB); the Metals Industry Research and Development Center (MIRDC) is co-operating in the programme.

The aircraft will be an all-metal, externally-braced high-wing monoplane, accommodating four persons including the pilot. It is intended for carrying passengers or cargo, and will be easily convertible into an agricultural crop dusting or seeding aircraft, with the necessary manoeuvrability to carry out such operations over small field areas.

Phase 1 of the programme, which began in October 1975, covered the preliminary design and engineering studies necessary to ensure smooth development; phase 2, which began in January 1976, concerns the detail design, construction and flight testing of a prototype. First flight is planned for mid-1978.

TYPE: Four-seat light utility and agricultural aircraft.

WINGS: High-wing monoplane, braced by a single strut on each side. Wing section NACA 2415 (constant). No anhedral, dihedral or sweepback. Incidence 2°. Trailing-edge flaps and ailerons over virtually entire span. Turned-down wingtips.

FUSELAGE: All-metal pod and boom type.

TAIL UNIT: Cantilever all-metal structure, with slight sweepback on vertical surfaces. Shallow dorsal fin. Balanced rudder and balanced one-piece elevator.

LANDING GEAR: Non-retractable tricycle type. All three wheels same size. Streamline wheel fairing on each unit.

POWER PLANT: One 224 kW (300 hp) Lycoming IO-540-K1B5 flat-six engine, driving a Hartzell constant-speed variable-pitch propeller with spinner. Fuel tank in each wing, combined capacity 189 litres (50 US gallons; 41·5 Imp gallons). Overwing refuelling point above each tank. Oil capacity 11·4 litres (3 US gallons; 2·5 Imp gallons).

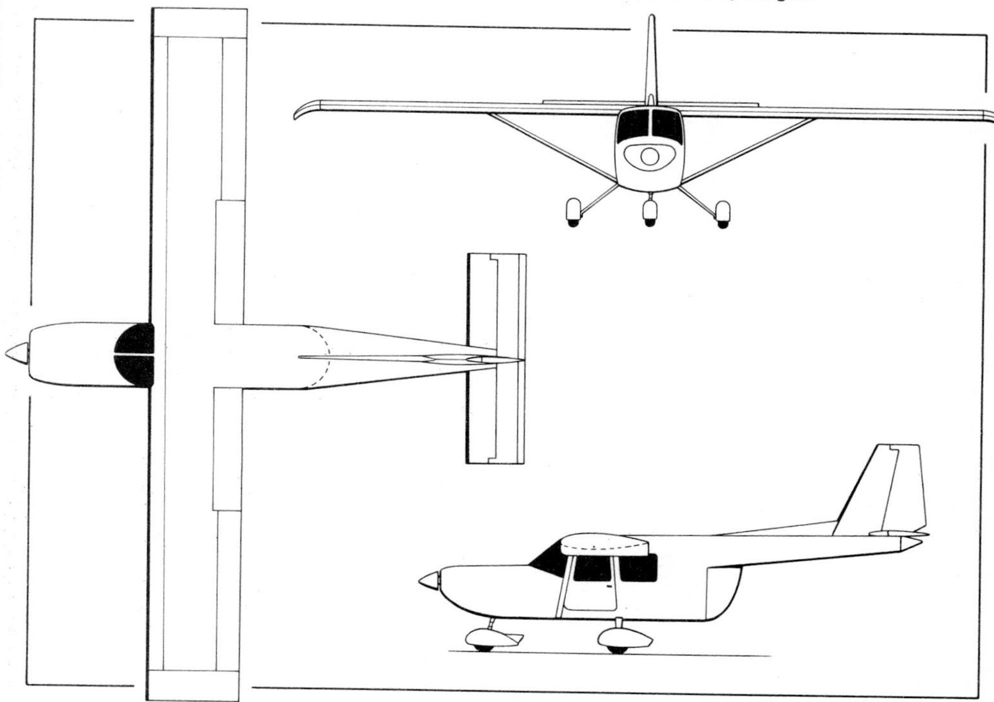

General arrangement of the PADC prototype four-seat utility aircraft *(Michael A. Badrocke)*

ACCOMMODATION: Pilot and up to three passengers, in pairs, in fully-enclosed cabin. Forward-opening car-type door on each side, each with pull-in window for emergency exit. Freight/baggage space aft of rear pair of seats, access to which is via clamshell rear-loading doors. Cabin ventilated.

DIMENSIONS, EXTERNAL:

Wing span	11·66 m (38 ft 3 in)
Wing chord (constant)	1·52 m (5 ft 0 in)
Wing aspect ratio	7·65
Length overall	8·43 m (27 ft 8 in)
Height overall	3·55 m (11 ft 7¾ in)
Tailplane span	3·48 m (11 ft 5 in)
Wheel track	2·40 m (7 ft 10½ in)
Wheelbase	2·20 m (7 ft 2½ in)
Propeller diameter	1·95 m (6 ft 4¾ in)
Propeller ground clearance	0·25 m (10 in)

DIMENSIONS, INTERNAL:

Cabin: Length	2·95 m (9 ft 8 in)
Max width	1·04 m (3 ft 5 in)
Max height	1·22 m (4 ft 0 in)
Floor area	3·07 m² (33·0 sq ft)
Volume	3·82 m³ (135·0 cu ft)
Baggage compartment volume	0·85 m³ (30·0 cu ft)
Freight compartment volume	1·69 m³ (59·6 cu ft)

AREAS:

Wings, gross	17·77 m² (191·3 sq ft)
Ailerons (total)	2·14 m² (23·0 sq ft)
Trailing-edge flaps (total)	1·95 m² (21·0 sq ft)
Fin	1·67 m² (18·0 sq ft)
Rudder	0·84 m² (9·0 sq ft)
Tailplane	3·34 m² (36·0 sq ft)
Elevator	1·51 m² (16·2 sq ft)

WEIGHTS AND LOADINGS (estimated):

Weight empty	992 kg (2,188 lb)
Max T-O weight	1,496 kg (3,300 lb)
Max wing loading	84·2 kg/m² (17·25 lb/sq ft)
Max power loading	3·72 kg/kW (11·0 lb/hp)

PERFORMANCE (estimated, at max T-O weight):

Max level speed at S/L	168·5 knots (312 km/h; 194 mph)
Max cruising speed at S/L	154·5 knots (286 km/h; 178 mph)
Stalling speed, flaps up	55·5 knots (103 km/h; 64 mph)
Stalling speed, flaps down	46·5 knots (86 km/h; 53·5 mph)
Max rate of climb at S/L	372 m (1,220 ft)/min
Service ceiling	5,300 m (17,400 ft)
T-O run	204 m (669 ft)
T-O to 15 m (50 ft)	436 m (1,430 ft)
Landing from 15 m (50 ft)	568 m (1,865 ft)
Landing run	305 m (1,000 ft)
Max range	424 nm (785 km; 488 miles)

PATS/NSDB

PHILIPPINE AERONAUTICS TRAINING SCHOOL/NATIONAL SCIENCE DEVELOPMENT BOARD

As a joint venture, the Philippine Aeronautics Training School and the National Science Development Board are undertaking two fairly simple aeronautical projects designed to utilise local raw materials. These are the Project 7307En light aircraft and the XG-001 two-seat sailplane.

PATS/NSDB PROJECT 7307En

This designation refers to a two-seat low-cost basic training aircraft, powered by a 63·5 kW (85 hp) engine. The airframe is of bamboo, covered with a native Philippine fabric known as ramie tetoron. The engine, which drives a propeller made from Philippine hardwood, and the main landing gear and instruments, are taken from a Piper Cub. Max T-O weight is quoted as approx 600 kg (1,455 lb), and cruising speed as 47·5 knots (88 km/h; 54·5 mph).

PHILIPPINE AIR FORCE

PAF XT-001

First news of this three-seat primary trainer prototype appeared in the late Summer of 1975. Built by the Self-Reliance Development Wing of the Philippine Air Force, reportedly using 'locally designed and manufactured jigs and fixtures', it bears a close and obvious resemblance (apart from the modified wingtips and cockpit) to the SIAI-Marchetti SF.260MP two/three-seat trainer, of which 32 were purchased by the PAF. It is thought likely that the prototype is, in fact, a converted SF.260MP.

No part in the completion of this prototype was taken by Dott Ing Stelio Frati, the designer of the SF.260.

Comparison of data with those for the SF.260MP indicates a slight increase in wing span (though not in gross area), a lower empty weight, and (despite a similar power plant and identical max T-O weight) a slightly reduced performance.

POWER PLANT: One 194 kW (260 hp) Lycoming O-540-E4A5 flat-six engine, driving a two-blade constant-speed propeller with spinner.

DIMENSIONS, EXTERNAL:
- Wing span 8·40 m (27 ft 6¾ in)
- Length overall 7·10 m (23 ft 3½ in)
- Height overall 2·40 m (7 ft 10½ in)

XT-001 prototype primary trainer built by the Self-Reliance Development Wing of the Philippine Air Force

AREA:
- Wings, gross 10·10 m² (108·7 sq ft)

WEIGHTS:
- Weight empty, equipped 720 kg (1,587 lb)
- Max T-O weight 1,200 kg (2,645 lb)

PERFORMANCE (estimated):
- Max cruising speed 140 knots (260 km/h; 162 mph)
- Stalling speed 65 knots (120 km/h; 75 mph)

POLAND

PZL

POLSKIE ZAKLADY LOTNICZE, ZJEDNOCZENIE PRZEMYSLU LOTNICZEGO I SILNIKOWEGO PZL (Polish Aviation Works, Aircraft and Engine Industry Union)

HEAD OFFICE:
ul. Miodowa 5, 00251 Warsaw

Telephone: Warsaw 279985

Telex: 814281

GENERAL MANAGER:
Ing Krzysztof Kuczynski

VICE-DIRECTORS:
Ing Zbigniew Pawlak (Research and Development)
Ing Kazimierz Brejnak (Technical)
Dr Józef Jablonski (Sales)

The Polish aircraft industry is under the central direction of the Zjednoczenie Przemyslu Lotniczego i Silnikowego PZL (Aircraft and Engine Industry Union).

The principal factories concerned with aircraft manufacture are the WSK-Mielec, WSK-Okecie and WSK-Swidnik. Other Polish factories, engaged in the production of sailplanes and aero-engines, are listed in the relevant sections of this edition.

The Polish aircraft industry manufactures the Soviet Antonov An-2 utility biplane under licence, and is responsible for development and production of the Mil Mi-2 turbine-powered helicopter. Several aircraft of Polish origin are also in production.

The export sales of all Polish aviation products are handled by PEZETEL Foreign Trade Enterprise, whose address is given below:

PEZETEL

PEZETEL Foreign Trade Enterprise of Aviation Industry

ADDRESS:
PO Box 371, 00950 Warsaw

Telephone: Warsaw 285071

Telex: 813430

GENERAL MANAGER:
Dr Józef Jablonski

MANAGER OF AVIATION DEPARTMENT:
Stanislas Ferenstein

MANAGER OF PUBLICITY DEPARTMENT:
Ing Janusz Matuszewski

This organisation is the trade representative of the Polish aviation industry for foreign markets.

IL

INSTYTUT LOTNICTWA (Aeronautical Institute)

ADDRESS:
Al. Krakowska 110/114, 02-256 Warsaw-Okecie

Telephone: Warsaw 460993

Telex: 813537

MANAGING DIRECTOR: Ing Zbigniew Pawlak

DEPUTY DIRECTOR: Dipl Ing Jerzy Grzegorzewski

SCIENTIFIC DIRECTOR: Dr Czeslaw Skoczylas

CHIEF OF TECHNICAL INFORMATION DIVISION:
Dipl Ing Andrzej Glass

The Instytut Lotnictwa was founded in 1926. It belongs to the PZL Polish Aviation Works group, under the general management of the PZL Aircraft and Engine Industry Union. The IL is responsible for the control of all research and development work in the Polish aircraft industry. It conducts scientific research, including the investigation of problems associated with low-speed and high-speed aerodynamics, static and fatigue tests, development and testing of aero-engines, flight instruments and other equipment, flight tests, and materials technology. It is also responsible for the construction of experimental aircraft and aero-engines. Descriptions of two of the IL's most recent experimental aircraft, a winged version of the SM-1 helicopter and the Lala-1 'flying laboratory', appeared in the 1973-74 and 1974-75 *Jane's*.

WSK-PZL-MIELEC

WYTWORNIA SPRZETU KOMUNIKACYJNEGO-PZL-MIELEC (Transport Equipment Manufacturing Centre, Mielec)

HEAD OFFICE AND WORKS: ul. Ludowego Wojska Polskiego 3, 39-301 Mielec

Telephone: Mielec 70

Telex: 83214 and 83293

GENERAL MANAGER:
Dipl Ing Tadeusz Ryczaj

Largest and best-equipped aircraft factory in Poland, the WSK at Mielec was engaged mainly on licence production of MiG single-seat jet fighters for several years. These aircraft carry the Polish designation of LiM, meaning Licence MiG.

After completion of the production order for LiM-5s (MiG-17s) for the Polish Air Force, licence production of MiG fighters ceased in Poland in about 1959.

Following a reduction in orders for combat aircraft in 1955, Mielec began production in 1956 of the TS-8 Bies basic trainer, described in the 1962-63 *Jane's*. Four years later, the Soviet-designed An-2 general utility biplane went into production at Mielec. In parallel production with the An-2 at the present time is the TS-11 Iskra jet trainer and light attack aircraft.

There is a design office at the factory for development of original aircraft. Its latest known product is the M-15 agricultural aircraft.

WSK-MIELEC (ANTONOV) An-2

NATO reporting name: *Colt*

The prototype of this large biplane was designed to a specification of the Ministry of Agriculture and Forestry of the USSR and made its first flight on 31 August 1947. It was powered by a 567 kW (760 hp) ASh-21 engine and was known as the SKh-1 (Selskokhozyaistvennyi-1: agricultural-economic-1). This designation was dropped subsequently and in 1948 the design went into production in the USSR as the An-2, with a 746 kW (1,000 hp) ASh-62 engine.

By 1960, more than 5,000 An-2s had been built in the Soviet Union for service with the Soviet armed forces, Aeroflot and other civilian organisations and the various Soviet-built versions have been fully described in previous editions of *Jane's*. Many were exported, to all of the Socialist States, and to Greece, Afghanistan, Mali, Nepal, India and Cuba, and licence rights were granted to China, where the first locally-produced An-2 (Chinese designation Fong Shou No. 2) was completed in December 1957.

Since 1960, apart from a small Soviet-built quantity of a developed version known as the An-2M (see 1971-72 *Jane's*), the continued production of the An-2 has been the responsibility of the Polish WSK factory at Mielec, the original licence arrangement providing for two basic versions: the An-2T transport and An-2R agricultural version. The first 10 Polish-built An-2s were completed in 1960, and WSK-Mielec has since built considerable numbers of this aircraft for domestic use and for export to the USSR, Bulgaria, Czechoslovakia, France, the German Democratic Republic, Hungary, North Korea, Mongolia, the Netherlands, Romania and Yugoslavia. The 5,000th Polish-built An-2 was delivered to the USSR on 3 February 1973. Since beginning An-2 production, WSK-Mielec has made numerous improvements to the airframe of the An-2R, resulting in an increase in TBO from 900 hr in 1961 to 1,500 hr in 1970 and 2,000 hr in 1973.

By the beginning of 1976 more than 6,900 An-2s (all versions) had been built at Mielec, including 3,670 of the An-2R agricultural version. More than 90 per cent of these were for export, chiefly to the USSR. During 1974-75 Polish-operated teams of An-2Rs carried out extensive agricultural operations in Algeria, Egypt, Ethiopia, Hungary, the Sudan and Tunisia.

The Polish-built versions have different designations from those built in the USSR. These are as follows:

An-2P. Passenger version, with seating for 12 adult passengers and two children. Compared with Soviet-built An-2P (see 1971-72 *Jane's*) has improvements in passenger cabin layout and comfort, better soundproofing, a new propeller and spinner, and weight-saving instrumentation and equipment. Entered production in 1968.

An-2PK. Five-seat executive version, having two seats on starboard side and three on port side, with foldaway table between pairs of seats on each side.

An-2P-Photo. Photogrammetry version.

An-2R. Agricultural version, with 1,350 kg (2,976 lb) capacity glassfibre-reinforced epoxy-resin hopper or tank for dry or liquid chemicals. Similar to Soviet-built An-2S. One aircraft converted to IL Lala-1 (see 1973-74 and 1974-75 *Jane's*) in 1971 as testbed for M-15 agricultural aircraft. Modernisation of agricultural equipment under way.

An-2S. Ambulance version, equipped to carry six stretcher patients and their medical attendants.

An-2T. Basic general-purpose version, with accommodation for 12 passengers and baggage or 1,500 kg (3,306 lb) of cargo.

An-2TD. Paratroop transport and training version with six tip-up seats along each side of cabin. Granted French type certificate No. IM-55 (for import licence) in 1972.

An-2TP. Cargo/passenger version, similar to AN-2TD with six tip-up seats along each side of cabin.

An-2M. Twin-float version of An-2T, similar to Soviet-built An-2V. Built in small numbers only.

An-2 Geofiz. Geophysical survey version, developed for the State Prospecting Company in Warsaw.

The following details apply to the WSK-Mielec An-2P:

Type: Single-engined general-purpose biplane.

Wings: Unequal-span single-bay biplane. Wing section RPS 14% (constant). Dihedral, both wings, approx 2° 48′. All-metal two-spar structure, fabric-covered aft of front spar. I-type interplane struts. Differential ailerons and full-span automatic leading-edge slots on upper wings, slotted trailing-edge flaps on both upper and lower wings. Flaps operated electrically, ailerons mechanically by cables and push/pull rods. Electrically-operated trim tab in port aileron.

Fuselage: All-metal stressed-skin semi-monocoque structure of circular section forward of cabin, rectangular in the cabin section and oval in the tail section.

Tail Unit: Braced metal structure. Fin integral with rear fuselage. Fabric-covered tailplane. Elevators and rudder operated mechanically by cables and push/pull rods. Electrically-operated trim tab in rudder and port elevator.

Landing Gear: Non-retractable split-axle type, with long-stroke oleo shock-absorbers. Main wheel tyres size 800 × 260 mm, pressure 2·25 bars (32·7 lb/sq in). Pneumatic shoe brakes on main units. Fully-castoring and self-centering tailwheel with electro-pneumatic lock. For rough-field operation the oleo-pneumatic shock-absorbers can be charged from a compressed air cylinder installed in the rear fuselage. Interchangeable ski landing gear available optionally.

Power Plant: One 746 kW (1,000 hp) Shvetsov ASh-62IR nine-cylinder radial aircooled engine, driving an AW-2 four-blade variable-pitch metal propeller. Six fuel tanks in upper wings, with total capacity of 1,200 litres (264 Imp gallons). Oil capacity 120 litres (26·4 Imp gallons).

Accommodation: Crew of two on flight deck, with access via passenger cabin. Standard accommodation for 12 passengers, in four rows of three with centre aisle. Two foldable seats for children in aisle between first and second rows, and infant's cradle at front of cabin on starboard side. Toilet at rear of cabin on starboard side. Overhead racks for up to 160 kg (352 lb) of baggage, with space for coats and additional 40 kg (88 lb) of baggage between rear pair of seats and toilet. Emergency exit on starboard side at rear. Walls of cabin are lined with glass-wool mats and inner facing of plywood to reduce internal noise level. Cabin floor is carpeted. Cabin heating and starboard windscreen de-icing by engine bleed air; port and centre windscreens are electrically de-iced. Cabin ventilation by ram-air intakes on underside of top wings. Air-conditioning system in An-2R.

Systems: Compressed air cylinder, of 8 litres (490 cu in) capacity, for pneumatic charging of shock-absorbers and operation of tailwheel lock at 49 bars (711 lb/sq in) pressure and operation of main-wheel brakes at 9·80 bars (142 lb/sq in). Contents of cylinder are maintained by AK-50 M engine-driven compressor, with AD-50 automatic relief device to prevent overpressure. DC electrical system is supplied with basic 27V power (and 36V or 115V where required) by an engine-driven generator and a storage battery. CO_2 fire extinguishing system with automatic fire detector.

Electronics and Equipment: Dual controls and blind-flying instrumentation standard. R-842 short wave and R-860 ultra short wave lightweight radio transceivers, RW-UM radio altimeter, ARK-9 radio compass, MRP-56P marker, GIK-1 gyro compass, GPK-48 gyroscopic direction indicator and SPU-7 intercom.

Dimensions, external:

Wing span:	
upper	18·18 m (59 ft 8½ in)
lower	14·24 m (46 ft 8½ in)
Wing chord (constant):	
upper	2·40 m (7 ft 10½ in)
lower	2·00 m (6 ft 6¾ in)
Wing aspect ratio:	
upper	7·57
lower	7·12
Wing gap	2·17 m (7 ft 1½ in)
Length overall:	
tail up	12·74 m (41 ft 9½ in)
tail down	12·40 m (40 ft 8¼ in)
Height overall:	
tail up	6·10 m (20 ft 0 in)
tail down	4·00 m (13 ft 1½ in)
Tailplane span	7·20 m (23 ft 7½ in)
Wheel track	3·45 m (11 ft 3¾ in)
Wheelbase	3·36 m (11 ft 0¼ in)
Propeller diameter	3·60 m (11 ft 9¾ in)
Propeller ground clearance	0·70 m (2 ft 3½ in)
Emergency exit (stbd, rear):	
Height	0·65 m (2 ft 1½ in)
Width	0·51 m (1 ft 8 in)

Areas:

Wings, gross:	
upper	43·6 m² (469 sq ft)
lower	28·0 m² (301 sq ft)
Ailerons (total)	5·90 m² (63·5 sq ft)
Trailing-edge flaps (total)	9·60 m² (103 sq ft)
Fin	5·85 m² (62·97 sq ft)
Rudder, incl tab	2·65 m² (28·52 sq ft)
Tailplane	12·28 m² (132·18 sq ft)
Elevators (total, incl tab)	4·72 m² (50·81 sq ft)

Weights and Loadings:

Weight empty	3,450 kg (7,605 lb)
Max T-O weight	5,500 kg (12,125 lb)
Max landing weight	5,250 kg (11,574 lb)
Max wing loading	76·82 kg/m² (15·7 lb/sq ft)
Max power loading	7·38 kg/kW (12·13 lb/hp)

Performance (at AUW of 5,250 kg; 11,574 lb):

Max level speed at 1,750 m (5,740 ft)	139 knots (258 km/h; 160 mph)
Econ cruising speed	100 knots (185 km/h; 115 mph)
Min flying speed	49 knots (90 km/h; 56 mph)
T-O speed	43 knots (80 km/h; 50 mph)
Landing speed	46 knots (85 km/h; 53 mph)
Max rate of climb at S/L	210 m (689 ft)/min
Service ceiling	4,400 m (14,425 ft)
Time to 4,400 m (14,425 ft)	30 min
T-O run:	
hard runway	150 m (492 ft)
grass	170 m (558 ft)
T-O to 10·7 m (35 ft):	
hard runway	300 m (984 ft)
grass	320 m (1,050 ft)
Landing run:	
hard runway	170 m (558 ft)
grass	185 m (607 ft)
Range at 1,000 m (3,280 ft) with 500 kg (1,102 lb) payload	485 nm (900 km; 560 miles)

WSK-Mielec (Antonov) An-2 general-purpose aircraft, with optional ski landing gear

WSK-MIELEC TS-11 ISKRA (SPARK)

Developed by the OKL under the supervision of Docent Ing T. Soltyk, the TS-11 Iskra two-seat jet trainer was produced as a replacement for the piston-engined TS-8 Bies. The prototype, built at the WSK Warsaw-Okecie, began flight trials on 5 February 1960. Quantity production commenced at the WSK-Mielec in 1962. The formal handing over of the first Iskra for service with the Polish Air Force took place in March 1963. Early production aircraft were powered by a 7·85 kN (1,764 lb st) Type HO-10 Polish-designed axial-flow turbojet engine, which was superseded by the more powerful SO-3.

Several hundred Iskras had been built by 1975, and production continues. A version with underwing armament pods is designated **Iskra 100**; the single-seat version for ground attack duties, which first flew in June 1972, is also in production. The purchase of 50 Iskra 100s by the Indian Air Force was reported in May 1975, and the first of these aircraft were handed over in the following Autumn. The current version, incorporating some modifications introduced in 1975, is designated **Iskra 200**.

The following description applies to the current production versions:

Type: Fully aerobatic two-seat jet primary and basic trainer and single-seat light ground attack aircraft.

Wings: Cantilever mid-wing monoplane. Wing section NACA 64209 at root, NACA 64009 at tip. Sweepback at quarter-chord 7°. Marked dihedral. All-metal torsion box structure with steel main spar and duralumin stressed skin. Hydraulically-servo-assisted ailerons. Two-section double-slotted flaps and airbrakes fitted. One boundary layer fence on each wing.

Fuselage: All-metal semi-monocoque structure of pod and boom type.

Tail Unit: Cantilever all-metal structure. Fin integral with fuselage. Mass- and aerodynamically-balanced elevators and rudder.

Landing Gear: Retractable tricycle type with single

WSK-Mielec TS-11 Iskra, single-seat light ground attack version with underwing rocket and gun pods

Standard two-seat Iskra 100 armament trainer, with underwing practice bombs and rocket pods

wheel on each unit. Nosewheel retracts forward, main wheels inward into wing-root air intake trunks. Hydraulic retraction. Main wheels size 600 × 180, tyre pressure 5·38 bars (78 lb/sq in). Nosewheel size 380 × 150, tyre pressure 3·45 bars (50 lb/sq in). AMG-10 oleo-pneumatic shock-absorbers. Anti-shimmy nosewheel.

Power Plant: One SO-3 turbojet, rated at 9·81 kN (2,205 lb st), mounted in fuselage aft of cockpit section, with nozzle under tailboom. Fuel in two 315 litre (69 Imp gallon) integral wing tanks, one 500 litre (110 Imp gallon) fuselage main tank (700 litre; 154 Imp gallon in single-seater) and one 70 litre (15·5 Imp gallon) fuselage collector tank. Total fuel capacity 1,200 litres (263·5 Imp gallons) in two-seater, 1,400 litres (308 Imp gallons) in single-seater.

Accommodation: Crew of one, or two in tandem, on lightweight ejection seat(s), under a one-piece hydraulically-actuated rearward-hinged jettisonable canopy. Cockpit pressurised and air-conditioned. Rear seat of trainer slightly raised.

Systems: Hydraulic system pressure 138 bars (2,000 lb/sq in). Pneumatic system pressure 118 bars (1,710 lb/sq in). 28V electrical system, with 28Ah battery.

Electronics and Equipment: Trainer has complete dual controls and instrumentation, including blind-flying panels. UHF transceiver, intercom and oxygen equipment standard. Position and homing indicator.

Armament (Iskra 200 and single-seater): Forward-firing 23 mm cannon in nose on starboard side, with gun camera. Four attachments for a variety of underwing stores, including bombs of up to 100 kg (220 lb), rockets and 7·62 mm guns.

Dimensions, external:
- Wing span 10·06 m (33 ft 0 in)
- Wing chord at root 2·254 m (7 ft 4¾ in)
- Wing chord at tip 1·162 m (3 ft 9¾ in)
- Wing aspect ratio 5·71
- Length overall 11·15 m (36 ft 7 in)
- Height overall 3·50 m (11 ft 5½ in)
- Tailplane span 3·84 m (12 ft 7¼ in)
- Wheel track 3·48 m (11 ft 5 in)
- Wheelbase 3·44 m (11 ft 3½ in)

Areas:
- Wings, gross 17·5 m^2 (188·37 sq ft)
- Ailerons (total) 1·50 m^2 (16·15 sq ft)
- Trailing-edge flaps (total) 1·68 m^2 (18·08 sq ft)
- Vertical tail surfaces (total) 2·25 m^2 (24·2 sq ft)
- Horizontal tail surfaces (total) 3·54 m^2 (38·1 sq ft)

Weights and Loadings (trainer):
- Weight empty 2,560 kg (5,644 lb)
- Max T-O weight 3,840 kg (8,465 lb)
- Normal T-O weight 3,800 kg (8,377 lb)
- Max landing weight 3,500 kg (7,716 lb)
- Max wing loading 219 kg/m^2 (44·85 lb/sq ft)
- Max power loading 387·4 kg/kN (3·8 lb/lb st)

Performance (trainer, at normal T-O weight except where indicated):
- Never-exceed speed Mach 0·8 (404 knots; 750 km/h; 466 mph)
- Max level speed at 5,000 m (16,400 ft) 388 knots (720 km/h; 447 mph)
- Normal cruising speed 324 knots (600 km/h; 373 mph)
- Landing speed 92 knots (170 km/h; 106 mph)
- Stalling speed, power off, flaps down 75·5 knots (140 km/h; 87 mph)
- Max rate of climb at S/L 888 m (2,913 ft)/min
- Time to 6,000 m (19,685 ft) 9 min 36 sec
- Service ceiling 11,000 m (36,000 ft)
- T-O run 700 m (2,296 ft)
- T-O to 15 m (50 ft) 1,190 m (3,904 ft)
- Landing from 15 m (50 ft) 1,110 m (3,642 ft)
- Landing run 650 m (2,132 ft)
- Range with max fuel 673 nm (1,250 km; 776 miles)

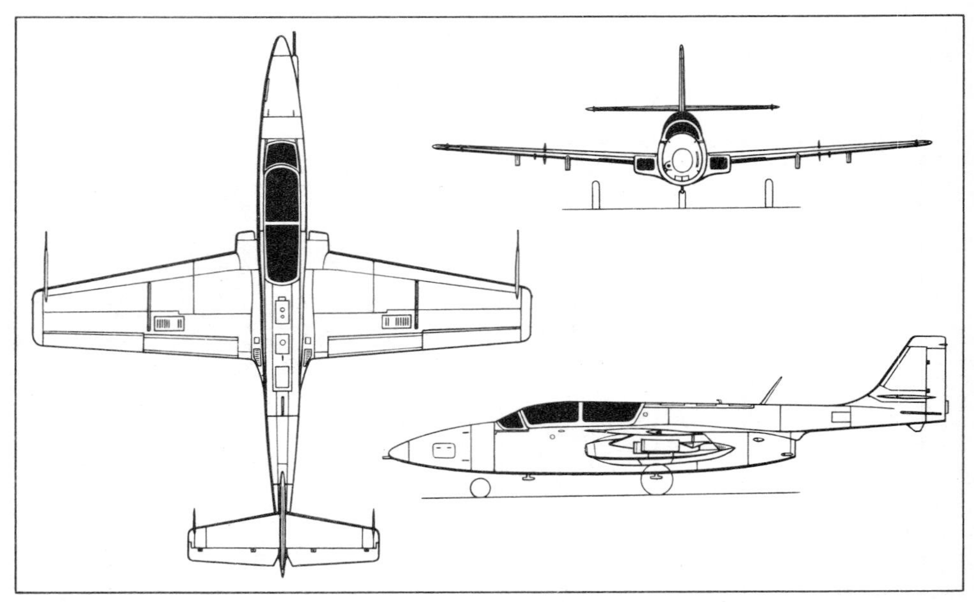

WSK-Mielec TS-11 Iskra 200 jet primary and basic trainer, with underwing weapon attachments *(Pilot Press)*

WSK-MIELEC M-15

On 1 March 1971, an agreement was concluded in Warsaw between the Polish and Soviet governments regarding the development and production of new aviation products, including large and medium-sized agricultural aircraft, light single- and twin-engined helicopters, sailplanes and powered sailplanes. The USSR has not manufactured any specialised agricultural aircraft, apart from a small quantity of An-2Ms, since it transferred production to Poland of the Antonov An-2 in 1960. Consequently, following the 1971 agreement, one subject of discussion between the Polish Ministry of Civil Aviation and the Soviet Ministry of Aircraft Industry was the development of a new, large agricultural aircraft known as the M-15, together with associated agricultural and ground support equipment.

The Soviet government has indicated a requirement for about 3,000 such aircraft, and on 2 December 1971 signed an agreement with the Polish government for large-scale production of the M-15.

Initial design of the M-15 was undertaken by a design bureau at Mielec under Soviet chief consulting engineer R. A. Ismailov and Polish designer K. Gocyla, and staffed by Polish and Soviet specialists. The agricultural equipment for the aircraft is being developed jointly by the Instytut Lotnictwa at Warsaw (which see) and the Soviet Research Institute of Special and Utility Aviation at Krasnodar.

A prototype, designated LLP-M15 (Laboratorium Latajace Prototyp M-15: Flying Laboratory Prototype M-15), was flown on 20 May 1973; the first fully-representative M-15 prototype made its first flight on 9 January 1974. On 2 April 1975 five pre-series M-15s were sent to the USSR for evaluation, and series production (reportedly of an initial batch of 20) began in the same year.

A passenger version has been proposed, in which the agricultural hoppers would be replaced by enlarged between-wings fairings, equipped as passenger cabins with nose baggage compartments.

The following description refers to the agricultural production version:

Type: Three-seat agricultural aircraft.

Wings: Biplane wings, of mainly metal construction and unequal span, built chiefly of aluminium and steel alloys and glassfibre laminates. The upper wing has a constant-chord centre-section and tapered outer panels; the centre-section is faired to the top of the engine pod. The shorter-span lower wings, which house the agricultural dispersal pipes, are of generally similar planform and are joined to the fuselage nacelle at floor level. The entire trailing-edge of the upper wing is hinged, and is made up of hydraulically-operated double-slotted flaps and single-slotted ailerons. There are automatically-operated slats on the leading-edge. In line with each tailboom, and occupying the full depth of the gap between the upper and lower wings, is a narrow streamlined hopper for agricultural chemical, and there is a single outward-sloping bracing strut outboard of each hopper fairing. Trim tab in port aileron.

Fuselage: Central semi-monocoque nacelle, of narrow rectangular section, built of same materials as wings.

Tail Unit: Cantilever metal/glassfibre structure, consisting of twin sweptback endplate fins and rudders, bridged by a high-mounted tailplane and full-span elevators, supported on two slender tailbooms located at approx one-quarter span on the upper wing.

Landing Gear: Non-retractable tricycle type, with single wheel on each unit. Main wheels and tyres size 720 × 360; nosewheel and tyre size 700 × 250. Nosewheel steerable hydraulically, 50° to left or right. Brakes on main wheels.

Power Plant: One 14·7 kN (3,306 lb st) Ivchenko AI-25 turbofan engine, mounted in a pod on top of the fuselage. Five fuel tanks in the upper wing.

Accommodation: Seat for pilot in fully-enclosed cockpit in extreme nose of fuselage. Two seats in cabin, to rear of pilot's compartment, for carrying ground staff during ferry flights. Cockpit air-conditioning by engine compressor bleed air.

Equipment: Full flight and navigation instrumentation, including stall-warning indicator. VFR radio/navigation equipment optional. The two between-wings hoppers have a combined capacity for 2,900 litres (638 Imp gallons) of liquid or 2,200 kg (4,850 lb) of dry (powdered or granulated) chemical. Ivchenko AI-9 APU, normally removed from aircraft during agricultural operations, provides power for engine starting, ground refuelling and filling of hoppers with liquid chemical. Twin landing lights in nose.

Dimensions, external:
- Wing span:
  - upper 22·33 m (73 ft 3¼ in)
  - lower 16·43 m (53 ft 10¾ in)
- Wing mean aerodynamic chord (upper) 1·84 m (6 ft 0½ in)
- Length overall 12·72 m (41 ft 8¾ in)
- Height overall 5·34 m (17 ft 6½ in)
- Wheel track 4·32 m (14 ft 2 in)
- Wheelbase 4·90 m (16 ft 1 in)

Areas:
- Wings (total) 67·50 m^2 (726·6 sq ft)
- Ailerons (total) 9·03 m^2 (97·20 sq ft)
- Trailing-edge flaps (total) 4·99 m^2 (53·71 sq ft)
- Fins (total) 5·53 m^2 (59·52 sq ft)
- Rudders (total) 4·00 m^2 (43·06 sq ft)
- Tailplane 5·92 m^2 (63·72 sq ft)
- Elevators (total) 4·08 m^2 (43·92 sq ft)

Weights and Loading:
- Weight empty 3,090 kg (6,812 lb)
- Max T-O weight 5,650 kg (12,456 lb)
- Max wing loading 84·0 kg/m^2 (17·2 lb/sq ft)

Performance (at max T-O weight):
- Max cruising speed 108 knots (200 km/h; 124 mph)
- Normal operating speed 75·5-89 knots (140-165 km/h; 87-103 mph)
- Stalling speed 58·5 knots (108 km/h; 67·5 mph)
- Max rate of climb at S/L 290 m (950 ft)/min

WSK-Mielec M-15 agricultural biplane (Ivchenko AI-25 turbofan engine)

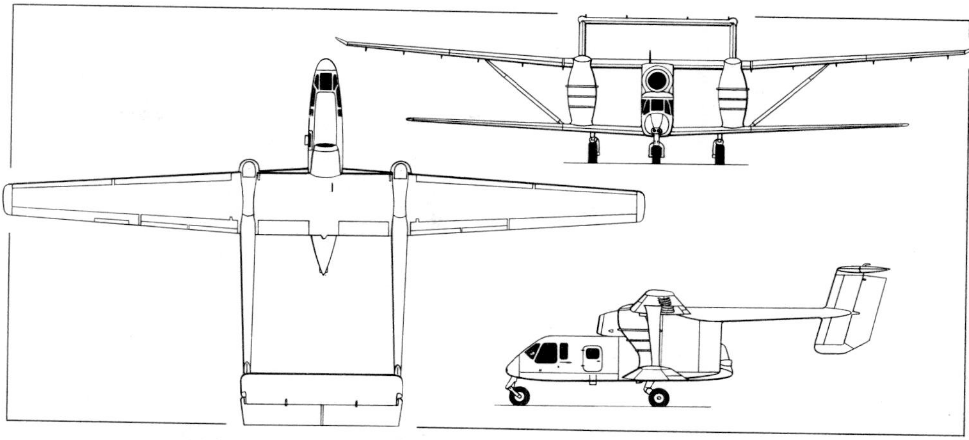

WSK-Mielec M-15 turbofan-engined agricultural aircraft *(Pilot Press)*

T-O run 250 m (820 ft)
Landing run 210 m (690 ft)
Max range at 3,000 m (9,850 ft)
216 nm (400 km; 248 miles)
Swath width 60 m (197 ft)

WSK-MIELEC M-17

A team of students from the Aeronautical Department of Warsaw Technical University, under the leadership of Dipl Ing Edward Marganski, has designed a two-seat all-metal light aircraft known as the M-17 (formerly EM-5A).

The M-17 was designed in 1969-71, as a two-seat training aircraft, and construction began in September 1972 at the WSK-Mielec factory. Two flight test prototypes and a static test airframe are being built. The first flight was reported to have taken place in the Autumn of 1975. The general appearance of the M-17 can be seen in the accompanying illustration.

TYPE: Two/three-seat light aircraft.

WINGS: Cantilever mid-wing monoplane. Wing section NACA 66-215. No anhedral, dihedral or sweepback. Incidence 2°. Constant-chord single-spar all-metal wings. All-metal ailerons and flaps, with beaded alloy skin. No tabs.

FUSELAGE: All-metal semi-monocoque stressed-skin central nacelle and twin tailbooms.

TAIL UNIT: Cantilever all-metal structure, with twin sweptback fins and rudders. Mass-balanced all-moving horizontal surface, with electrically-actuated balance tab in port half. Trim tab on port rudder.

LANDING GEAR: Retractable tricycle type, with electrical actuation. Main units retract inward, nosewheel forward. Nosewheel self-centering. Oleo-pneumatic shock-absorber on each unit. Main wheels from Zlin 526 F, tyre size 420 × 150 mm, pressure 2·94 bars (42·7 lb/sq in). Nosewheel from SZD-12A Mucha 100 sailplane, tyre size 300 × 125 mm, pressure 1·96 bars (28·4 lb/sq in). Aircooled and power- assisted hydraulic brake system from Zlin 526 F.

POWER PLANT: One 134 kW (180 hp) Avia M 137 inverted six-cylinder aircooled in-line engine, with low-pressure injection pump, driving a three-blade controllable-pitch metal pusher propeller. Four integral fuel tanks, two in wing centre-section and one in each outer panel, total capacity 310 litres (68 Imp gallons). Individual refuelling point on top of each tank. Oil capacity 12 litres (2·64 Imp gallons).

ACCOMMODATION: Individual side-by-side seats for two persons in enclosed cabin. Canopy opens forward and upward. Dual controls. Space for 30 kg (66 lb) of baggage, or third seat, aft of front seats. Cabin heated and ventilated.

SYSTEMS: Hydraulic braking system. Electrical system (27·5V 500W AC, and 24V DC battery) for engine starting and landing gear and balance tab actuation. No pneumatic or oxygen system.

ELECTRONICS AND EQUIPMENT: R-860 VHF transceiver, ARK-9 radio compass, gyro-magnetic compass, MRP-66 marker, and gyro horizon.

DIMENSIONS, EXTERNAL:
Wing span 9·20 m (30 ft 2¼ in)
Wing chord (constant) 1·25 m (4 ft 1¼ in)
Wing aspect ratio 7·5
Length overall 6·50 m (21 ft 4 in)
Height overall 1·70 m (5 ft 7 in)
Tailplane span 2·50 m (8 ft 2½ in)
Wheel track 2·70 m (8 ft 10¼ in)
Wheelbase 2·40 m (7 ft 10½ in)
Propeller diameter 1·70 m (5 ft 7 in)
Propeller ground clearance 0·40 m (1 ft 3¾ in)

DIMENSIONS, INTERNAL:
Cabin: Max length 1·60 m (5 ft 3 in)
Max width 1·20 m (3 ft 11¼ in)
Max height 0·72 m (2 ft 4¼ in)

AREAS:
Wings, gross 11·30 m^2 (121·6 sq ft)
Ailerons (total) 0·75 m^2 (8·07 sq ft)
Trailing-edge flaps (total) 0·75 m^2 (8·07 sq ft)
Fins (total) 0·98 m^2 (10·55 sq ft)
Rudders (total, incl tab) 0·54 m^2 (5·81 sq ft)
Tailplane, incl tab 2·00 m^2 (21·53 sq ft)

WEIGHTS AND LOADINGS:
Weight empty 650 kg (1,433 lb)
Max T-O and landing weight:
semi-aerobatic 900 kg (1,984 lb)
non-aerobatic 1,050 kg (2,315 lb)
Max wing loading 76·6 kg/m^2 (15·7 lb/sq ft)
Max power loading 6·72 kg/kW (11·02 lb/hp)

PERFORMANCE (estimated, at 900 kg; 1,984 lb T-O weight except where indicated):
Never-exceed speed 269 knots (500 km/h; 310 mph)
Max level speed at S/L
194 knots (360 km/h; 224 mph)
Max cruising speed at 3,000 m (9,850 ft)
162 knots (300 km/h; 186 mph)
Econ cruising speed at 3,000 m (9,850 ft)
135 knots (250 km/h; 155 mph)
Stalling speed, flaps up 60 knots (110 km/h; 69 mph)
Stalling speed, flaps down
52 knots (95 km/h; 59 mph)
Max rate of climb at S/L 474 m (1,555 ft)/min
Service ceiling 6,500 m (21,325 ft)
T-O and landing run 180 m (590 ft)
T-O to 15 m (50 ft) 320 m (1,050 ft)
Landing from 15 m (50 ft) 290 m (950 ft)
Max range, with 45 min reserves:
with 100 litres (22 Imp gallons) in centre-section tanks and outer tanks empty
323 nm (600 km; 372 miles)
with full fuel at max T-O weight of 1,050 kg (2,315 lb) 1,187 nm (2,200 km; 1,367 miles)

WSK-Mielec M-17 two/three-seat light aircraft nearing completion in 1975

WSK-PZL-OKECIE

WYTWORNIA SPRZETU KOMUNIKACYJNEGO-PZL-OKECIE (Transport Equipment Manufacturing Centre, Okecie)

HEAD OFFICE AND WORKS:
02-256 Warsaw-Okecie, Al. Krakowska 110/114
Telephone: Warsaw 461173

MANAGERS:
Jerzy Malkinski, Eng MSc (General Manager)
Jerzy Milczarek, Eng (Technical)
Roman Kojder, MSc (Sales)

PUBLIC RELATIONS:
Jerzy Pasterski, Eng MSc

The Okecie factory is responsible for light aircraft development and production, and for the design and manufacture of associated agricultural equipment for its own aircraft and for those built at other factories in the Aircraft and Engine Industry Union.

PZL-104 WILGA 35 (THRUSH)

The PZL-104 Wilga is a light general-purpose aircraft intended for a wide variety of general aviation and flying club duties. The original prototype (SP-PAZ), known as the Wilga 1, with a 180 hp WN-6B engine, flew for the first time on 24 April 1962. This aircraft, together with the Wilga 2, C and 3 prototypes, 3A and 3S production versions and other early models, was described in the 1968-69 *Jane's*.

In 1967 the basic design was further modified, with improved cabin comfort, redesigned landing gear and glassfibre tailwheel leg. This version is known as the Wilga 35 (first flight 28 July 1967) when fitted with a 194 kW (260 hp) AI-14R engine, and as the Wilga 32 (first flown 12 September 1967) with a 171·5 kW (230 hp) Continental O-470-K, -L or -R engine and shorter landing gear. Both the Wilga 32 and Wilga 35 received a Polish type certificate on 31 March 1969, having entered production in 1968. A British C of A was granted for the import of the Wilga 32 into the UK.

Production of the Wilga 32 and 32A (described in the 1974-75 *Jane's*) has ended; current production models are the **Wilga 35A** (Aeroclub), **35P** (Passenger/liaison) and **35S** (ambulance). Examples of the Wilga 35A have been sold to customers in Austria, Bulgaria, Czechoslovakia, Egypt, Germany (Democratic Republic), Germany (Federal Republic), Hungary, Romania, Spain, the UK, the USSR and Venezuela.

Details of the Wilga 40 and 43 experimental versions were given in the 1972-73 *Jane's*.

By early 1976 more than 260 Wilgas, of all versions, had been built in Poland. A modified version of the Wilga 32 was produced in Indonesia as the Lipnur Gelatik 32 (which see).

The following description applies to the current version of the Wilga 35A:

TYPE: Single-engined general-purpose monoplane.

WINGS: Cantilever high-wing monoplane. Wing section NACA 2415. Dihedral 1°. All-metal single-spar structure, with leading-edge torsion box and beaded metal skin. Each wing attached to fuselage by three bolts, two at spar and one at forward fitting. All-metal aerodynamically and mass-balanced slotted ailerons, with beaded metal skin. Ailerons can be drooped to supplement flaps during landing. Manually-operated all-metal slotted flaps with beaded metal skin. Fixed metal slat on the leading-edge along the full span of the wing and over the fuselage. Tab on starboard aileron.

FUSELAGE: All-metal semi-monocoque structure in two portions, riveted together. Forward section incorporates main wing spar carry-through structure. Rear section is in the form of a tailcone. Beaded metal skin. Floor in cabin is of sandwich construction, with a paper core, covered with foam rubber.

TAIL UNIT: Braced all-metal structure, with sweptback vertical surfaces. Stressed-skin single-spar tailplane attached to fuselage by a single centre fitting and supported by a single aluminium alloy strut on each side. Stressed-skin two-spar fin structure of semi-monocoque construction. Rudder and one-piece elevator are aerodynamically horn-balanced and mass-balanced, with counterweights in the form of metal slats attached to the front section of the elevator. Controllable trim tab in centre of elevator trailing-edge.

LANDING GEAR: Non-retractable tailwheel type. Semi-cantilever legs, of rocker type, have PZL oleo-pneumatic shock-absorbers. Stomil tubeless low-pressure tyres size 500 × 200 on main wheels. Hydraulic brakes. Steerable tailwheel, size 255 × 110, carried on rocker frame with oleo-pneumatic damper. Retractable metal ski landing gear optional.

Power Plant: One 194 kW (260 hp) Ivchenko AI-14RA nine-cylinder radial aircooled engine, driving a US-122000 two-blade constant-speed wooden propeller. Two removable fuel tanks in each wing, with total capacity of 195 litres (43 Imp gallons). Refuelling point on each side of fuselage, at junction with wing. Oil capacity 16 litres (3·5 Imp gallons). It is intended to power this version with an AI-14RC engine having electrical starting.

Accommodation: Passenger version accommodates four persons in pairs, with adjustable front seats. Baggage compartment aft of seats. Upward-opening door on each side of cabin, jettisonable in emergency. In the parachute training version the starboard door is removed and replaced by a tubular upright with a horizontal strap, and the starboard front seat is rearward-facing. Backrests of the rear seats are removable, and jumps are facilitated by a step on the starboard side and by a parachute hitch. A controllable towing hook can be attached to the tail landing gear permitting the Wilga, in this role, to tow a single glider of up to 650 kg (1,433 lb) weight or two or three gliders with a total combined weight of 1,125 kg (2,480 lb).

Systems: Hydraulic system pressure 39 bars (570 lb/sq in). Engine starting is effected pneumatically by a built-in system of 7 litres (427 cu in) capacity with a pressure of 49 bars (710 lb/sq in); electrical system includes GSK-1500 generator and a 10Ah battery for 24V DC power.

Electronics and Equipment: Standard equipment includes R 860 VHF radio and blind-flying instrumentation.

Dimensions, external:
- Wing span: 11·12 m (36 ft 5¾ in)
- Wing chord (constant): 1·40 m (4 ft 7¼ in)
- Wing aspect ratio: 7·95
- Length overall: 8·10 m (26 ft 6¾ in)
- Height overall: 2·94 m (9 ft 7¾ in)
- Tailplane span: 3·70 m (12 ft 1¾ in)
- Wheel track: 2·85 m (9 ft 4¼ in)
- Wheelbase: 6·70 m (21 ft 11¾ in)
- Propeller diameter: 2·65 m (8 ft 8 in)
- Passenger doors (each):
  - Height: 1·00 m (3 ft 3¼ in)
  - Width: 1·50 m (4 ft 11 in)

Dimensions, internal:
- Cabin: Length: 2·20 m (7 ft 2½ in)
  - Max width: 1·20 m (3 ft 10 in)
  - Max height: 1·50 m (4 ft 11 in)
  - Floor area: 2·20 m² (23·8 sq ft)
  - Volume: 2·40 m³ (85 cu ft)
- Baggage compartment: 0·50 m³ (17·5 cu ft)

Areas:
- Wings, gross: 15·50 m² (166·8 sq ft)
- Ailerons (total): 1·57 m² (16·90 sq ft)
- Trailing-edge flaps (total): 1·97 m² (21·20 sq ft)
- Fin: 0·97 m² (10·44 sq ft)
- Rudder: 0·92 m² (9·90 sq ft)
- Tailplane: 3·16 m² (34·01 sq ft)
- Elevator, incl tab: 1·92 m² (20·67 sq ft)

Weights and Loadings:
- Weight empty, equipped: 850 kg (1,874 lb)
- Max T-O and landing weight: 1,230 kg (2,711 lb)
- Max wing loading: 79·4 kg/m² (16·3 lb/sq ft)
- Max power loading: 6·34 kg/kW (10·43 lb/hp)

Performance (at max T-O weight):
- Never-exceed speed: 150 knots (279 km/h; 173 mph)
- Max level speed: 108 knots (201 km/h; 125 mph)
- Max cruising speed: 104 knots (193 km/h; 120 mph)
- Econ cruising speed: 69·5 knots (128 km/h; 80 mph)
- Stalling speed, power on: 37 knots (68 km/h; 42·5 mph)
- Max rate of climb at S/L: 378 m (1,240 ft)/min
- Service ceiling: 4,580 m (15,025 ft)
- T-O run: 80 m (260 ft)
- T-O to 15 m (50 ft): 186 m (610 ft)
- Landing from 15 m (50 ft): 230 m (755 ft)
- Landing run: 95 m (310 ft)
- Range with max fuel, 30 min reserves: 366 nm (680 km; 422 miles)

Performance as Glider Tug (at max T-O weight):
- Rate of climb at S/L:
  - with 1 glider: 234 m (770 ft)/min
  - with 2 gliders: 204 m (670 ft)/min
  - with 3 gliders: 132 m (435 ft)/min
- Service ceiling:
  - with 1 glider: 4,000 m (13,125 ft)
  - with 2 gliders: 3,900 m (12,800 ft)
  - with 3 gliders: 2,500 m (8,200 ft)
- Time to reach service ceiling:
  - 1 glider: 37 min 12 sec
  - 2 gliders: 43 min 12 sec
  - 3 gliders: 36 min 36 sec
- Range:
  - 1 glider: 344 nm (637 km; 395 miles)
  - 2 gliders: 272 nm (504 km; 313 miles)
  - 3 gliders: 227 nm (420 km; 260 miles)

PZL-104 Wilga 35 four-seat utility aircraft in the insignia of the DOSAAF Soviet flying training organisation

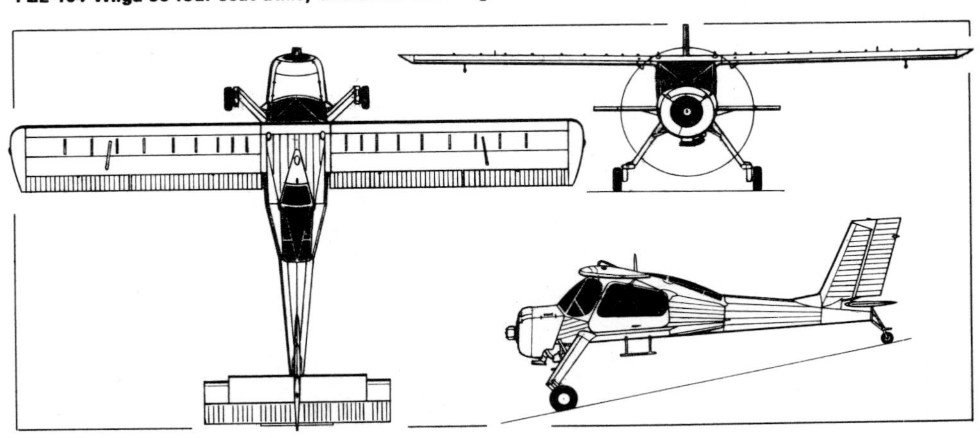

PZL-104 Wilga 35A general-purpose monoplane *(Pilot Press)*

PZL-106 KRUK (RAVEN)

In the early 1960s the WSK-Okecie design team began project studies for an agricultural aircraft to replace the PZL-101 Gawron (Rook), of which WSK-Okecie manufactured more than 330 in the period 1958-73. The initial design, known as the PZL-101M Kruk (Raven), was an extensively redesigned development of the Gawron, with a 194 kW (260 hp) Ivchenko AI-14R radial engine. Progressive refinement led to the current PZL-106 Kruk, with a more powerful engine and the braced low-wing monoplane configuration that has become customary for agricultural aircraft.

The PZL-106 was designed in early 1972 by a team led by Andrzej Frydrychewicz. The first prototype (SP-PAS) was built in seven months and flew for the first time on 17 April 1973, piloted by Jerzy Jedrzejewski and powered by a 298 kW (400 hp) Lycoming engine. Ten days later, it was demonstrated before senior government officials, and has since been followed by a second Lycoming-engined prototype (SP-PBG) and four prototypes fitted with the 447 kW (600 hp) PZL-3S radial engine intended for production aircraft. Production was expected to begin in 1976.

To meet the requirements of potential customers, the production version has been developed to carry a greater chemical load, in a larger hopper. It will be certificated at a max T-O weight of 3,000 kg (6,614 lb), although normal T-O weight with 1,000 kg (2,204 lb) of chemical will be 2,800 kg (6,173 lb). An output of some 600 aircraft for the member countries of the CMEA (Council for Mutual Economic Aid) is anticipated.

The following description applies to the intended production version:

Type: Single-engined agricultural aircraft.

Wings: Braced low-wing monoplane with upward-cambered tips. Clark Y wing section throughout span, except at tips. Thickness/chord ratio 11·7%. Dihedral 4° from roots. Incidence 6° 6'. Sweepback 4° at quarter-chord. All-metal two-spar structure, of constant chord throughout most of span. Metal and synthetic fabric covering. Full-span six-section fixed leading-edge slats. Fabric-covered duralumin slotted ailerons, in three sections on each wing. Production version has no flaps. Streamline-section Vee bracing struts, with jury struts.

Fuselage: Welded steel tube structure, covered with glassfibre-reinforced plastics and synthetic fabric.

Tail Unit: Conventional duralumin structure, with single bracing strut each side. Fixed surfaces metal-covered; rudder and mass-balanced elevators synthetic fabric-covered. Trim tab in starboard elevator.

Landing Gear: Non-retractable tailwheel type, with WSK-Okecie oleo-pneumatic shock-absorber on each unit. Main wheels, with Stomil-Poznan low-pressure tyres (size 800 mm × 260 mm), each carried on side Vee and half-axle. Main-wheel tyre pressure 1·97 bars (29·4 lb/sq in). Pneumatically-operated WSK-Okecie disc brakes. Steerable tailwheel, with Stomil-Poznan tyre size 350 mm × 135 mm, pressure 3·45 bars (51·5 lb/sq in).

Power Plant: One 447 kW (600 hp) PZL-3S seven-cylinder radial aircooled engine, driving a PZL US-129000 four-blade constant-speed glassfibre-reinforced plastics propeller. Fuel in two wing tanks, total capacity 300 litres (66 Imp gallons). Gravity refuelling point on each wing; semi-pressurised refuelling point on starboard side of fuselage. Oil capacity 30 litres (6·6 Imp gallons) max, 7 litres (1·54 Imp gallons) min. Air filter fitted.

Accommodation: Pilot in enclosed, ventilated and air-conditioned cockpit. Second (mechanic's) rearward-facing seat to rear. Combined window/door on each side of cabin with emergency opening. Cockpit fan and heater available optionally. Cockpit area strengthened to resist 40*g* impact.

Systems: Pneumatic system, rated at 49·3 bars (735 lb(sq in), for brakes and agricultural equipment. Electrical power, from 27·5V DC alternator and batteries, for engine starting, pneumatic system control, aircraft lights, instruments and semi-pressurised refuelling.

Equipment: VHF radio. Easily removable non-corrosive (glassfibre-reinforced plastics) hopper/tank, forward of cockpit, can carry 1,000 kg (2,204 lb) of dry or liquid chemical, and has a maximum capacity of 1,400 litres (308 Imp gallons). The hopper has a quick-dump system that can release 1,000 kg of chemical in less than 5 sec. A pneumatically operated intake for the loading of dry chemicals is optional. Distribution system for liquid chemical is powered by a fan-driven centrifugal pump. A precise and reliable dispersal system, with positive on/off action for dry chemicals, gives effective swath widths of 10-35 m (33-115 ft).

Dimensions, external (A: prototypes, B: production version):
- Wing span:
  - A: 13·00 m (42 ft 7¾ in)
  - B: 14·80 m (48 ft 6½ in)
- Wing chord (constant, A, B): 1·90 m (6 ft 2¾ in)
- Wing aspect ratio:
  - A: 6·8
  - B: 7·9

Length overall:
A 8·40 m (27 ft 6¾ in)
B 8·90 m (29 ft 2½ in)
Height overall:
A 2·90 m (9 ft 6¼ in)
B 3·60 m (11 ft 9¾ in)
Tailplane span (A, B) 5·40 m (17 ft 8¾ in)
Wheel track (A, B) 2·80 m (9 ft 2¼ in)
Wheelbase (A, B) 7·40 m (24 ft 3¼ in)
Propeller diameter:
A 2·135 m (7 ft 0 in)
B 2·58 m (8 ft 5½ in)
Propeller ground clearance (A, B) 0·70 m (2 ft 3½ in)
Fuselage doors (each):
Height 0·91 m (2 ft 11¾ in)
Width 1·06 m (3 ft 5¾ in)
Baggage door:
Height 0·70 m (2 ft 3½ in)
Width 0·60 m (1 ft 11¾ in)
Dimensions, internal (approx):
Cabin: Length 1·30 m (4 ft 3¼ in)
Max width 1·20 m (3 ft 11¼ in)
Max height 1·30 m (4 ft 3¼ in)
Floor area 1·00 m² (10·76 sq ft)
Areas:
Wings, gross:
A 24·57 m² (264·5 sq ft)
B 28·40 m² (305·7 sq ft)
Ailerons (total):
A, B 1·68 m² (18·08 sq ft)
Fin:
A 2·81 m² (30·25 sq ft)
B 1·92 m² (20·67 sq ft)
Rudder, incl tab:
A 1·28 m² (13·78 sq ft)
B 1·14 m² (12·27 sq ft)
Tailplane:
A 5·62 m² (60·49 sq ft)
B 7·40 m² (79·65 sq ft)
Elevators, incl tab:
A 3·31 m² (35·63 sq ft)
B 4·30 m² (46·28 sq ft)
Weights and Loadings:
Basic weight empty:
A 1,150 kg (2,535 lb)
B 1,550 kg (3,417 lb)
Max chemical payload:
A 800 kg (1,763 lb)
B 1,000 kg (2,205 lb)
Normal T-O weight:
B 2,800 kg (6,173 lb)
Max T-O weight:
A 2,250 kg (4,960 lb)
B 3,000 kg (6,614 lb)
*Max ramp weight:
B 3,000 kg (6,614 lb)
Max wing loading:
B 105 kg/m² (21·51 lb/sq in)
Max power loading:
B 6·71 kg/kW (11·02 lb/hp)
* *Aircraft stressed for 3,000 kg (6,614 lb) T-O weight in Normal Category, BCAR Section K*
Performance (A: prototypes at 2,250 kg; 4,960 lb AUW; B: production version at 2,800 kg; 6,173 lb, except where indicated):
Never-exceed speed:
B 140 knots (260 km/h; 161 mph)
Max level speed at 2,000 m (6,560 ft):
B 108 knots (200 km/h; 124 mph)
Max cruising speed:
A 86·5 knots (160 km/h; 99·5 mph)
B 97 knots (180 km/h; 112 mph)
Operating speed:
A with 800 kg (1,763 lb) of chemical
65-86·5 knots (120-160 km/h; 74·5-99·5 mph)
B with 1,000 kg (2,205 lb) of chemical
65-86·5 knots (120-160 km/h; 74·5-99·5 mph)
Max rate of climb at S/L, chemical loads as above:
A 180 m (591 ft)/min
B 250 m (820 ft)/min
T-O run:
B at max T-O weight 250 m (820 ft)
Landing run:
B 200 m (656 ft)
Min ground turning radius:
B 10·00 m (32 ft 10 in)
Range, approx:
B 215 nm (400 km; 248 miles)

Lycoming-engined second prototype of the PZL-106 Kruk (Raven) agricultural aircraft *(BIIL)*

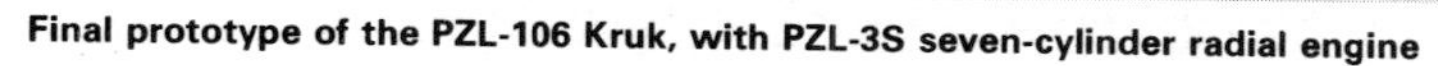

Final prototype of the PZL-106 Kruk, with PZL-3S seven-cylinder radial engine

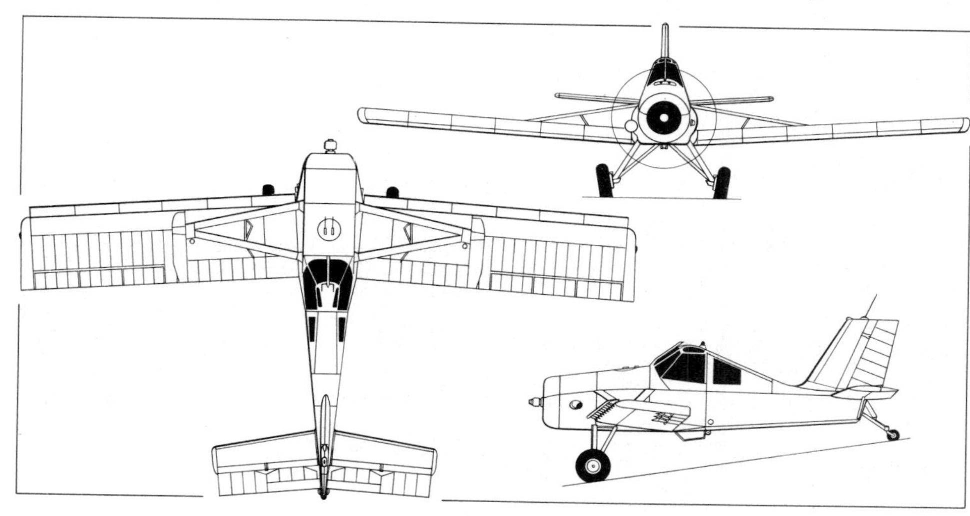

PZL-106A production version of the Kruk (PZL-3S engine) *(Pilot Press)*

WSK-PZL-SWIDNIK

WYTWORNIA SPRZETU KOMUNIKACYJNEGO Im. ZYGMUNTA PULAWSKIEGO-PZL-SWIDNIK (Zygmunt Pulawski's Transport Equipment Manufacturing Centre, Swidnik)

Head Office and Works:
Swidnik k/Lublina
Telephone: Lublin 12061
Telex: 84212 and 84302
General Manager:
Dipl Ing Jozef Lipinski

In 1955, when the manufacture of combat aircraft was drastically reduced in Poland, the WSK at Swidnik began licence production of the Soviet-designed Mi-1 helicopter, which was built under the designation SM-1. A small design office was formed subsequently at the factory to work on variants and developments of the basic SM-1 design and on original projects.

In September 1957, the WSK-Swidnik works was named after the famous pre-war PZL designer Zygmunt Pulawski. Production is concentrated now on various versions of the Soviet-designed Mil Mi-2 turbine-powered helicopter.

The WSK design office is working on a new, lightweight general-purpose helicopter with four/five seats and an AUW of about 1,700 kg (3,750 lb). It will have a max speed of 108-135 knots (200-250 km/h; 125-155 mph), and a range of about 270 nm (500 km; 310 miles). Initially, it will have a single Isotov GTD-350 turboshaft engine of the type that powers the Mi-2, but eventually a developed version of this engine, of 373-522 kW (500-700 shp), is likely to be used.

WSK-SWIDNIK (MIL) Mi-2

NATO reporting name: *Hoplite*

The Mil Mi-2, announced in the Autumn of 1961, was designed in the USSR by the Mikhail L. Mil bureau. It retains the basic configuration of the earlier Mi-1 helicopter, but instead of the latter's single piston engine has two Isotov turboshaft engines mounted side by side above the cabin.

Development of the Mi-2 prototype continued in the USSR until the helicopter had completed its initial State trials programme of flying. Then, in accordance with an agreement signed in January 1964, further development, production and marketing of the Mi-2 were assigned exclusively to the Polish aircraft industry, which had flown its own first example of the Mi-2 in November 1963.

Production by WSK-Swidnik began in 1965, and this factory has since built many hundreds in a variety of versions for both civil and military customers. Among the latter are the air forces of Czechoslovakia, Poland, Romania and the USSR. In recent years it has undertaken a development programme to improve and modernise the original design. The first new version has uprated engines of 335 kW (450 shp) each and is essentially similar to the basic Mi-2. It flew for the first time in 1974. On another version, the metal stabiliser, tail rotor blades and main rotor blades were replaced with similar components made of plastics, intended to simplify production and improve performance. The new rotor blades were designed, manufactured and tested at the WSK-Swidnik transport equipment plant.

The development programme has led to a further version, with enlarged cabin, redesigned landing gear with retractable nose unit, and 335 kW (450 shp) engines as standard. This version, known as the **Mi-2M**, is described separately.

There are several versions of the basic Mi-2, as follows:

(a) Convertible passenger/cargo transport;
(b) Passengers-only version, for six or eight passengers;
(c) Ambulance version (Mi-2R);
(d) Agricultural version;
(e) Search and rescue version, with electrically-operated external hoist;
(f) Freighter version, with external cargo sling;
(g) Pilot training version, designed by WSK-Swidnik;
(h) Photogrammetric version;
(i) Television version (for transmission from the air);
(j) Version with 260 kg (573 lb) capacity hoist;
(k) Version (under development) with inflatable pontoon landing gear.

The following details apply specifically to the basic Mi-2:

TYPE: Twin-turbine general-purpose light helicopter.

ROTOR SYSTEM: Three-blade main rotor fitted with hydraulic blade vibration dampers. All-metal blades of NACA 230-13M section. Flapping, drag and pitch hinges on each blade. Main rotor blades and those of two-blade tail rotor each consist of an extruded duralumin spar with bonded honeycomb trailing-edge pockets. Anti-flutter weights on leading-edges, balancing plates on trailing-edges. Hydraulic boosters for longitudinal, lateral and collective pitch controls. Coil spring counterbalance mechanism in main and tail rotor systems. Pitch-change centrifugal loads on tail rotor carried by ribbon-type steel torsion elements. Rotors do not fold. Electrical blade de-icing system for main and tail rotors. Rotor brake fitted.

ROTOR DRIVE: Main rotor shaft driven via gearbox on each engine; three-stage WR-2 main gearbox, intermediate gearbox and tail rotor gearbox. Main rotor/engine rpm ratio 1 : 24·6; tail rotor/engine rpm ratio 1 : 4·16. Main gearbox provides drive for auxiliary systems and take-off for rotor brake. Freewheel units permit disengagement of a failed engine and also autorotation.

FUSELAGE: Conventional semi-monocoque structure of pod and boom type, made up of three main assemblies: the nose (including cockpit), central section and tailboom. Construction is of sheet duralumin, bonded and spot-welded or riveted to longerons and frames. Main load-bearing joints are of steel alloy.

TAIL UNIT: Variable-incidence horizontal stabiliser controlled by collective-pitch lever.

LANDING GEAR: Non-retractable tricycle type, plus tailskid. Twin-wheel nose unit. Single wheel on each main unit. Oleo-pneumatic shock-absorbers on all units, including tailskid. Main shock-absorbers designed to cope with both normal operating loads and possible ground resonance. Main-wheel tyres size 600 × 180, pressure 4·41 bars (64 lb/sq in). Nosewheel tyres size 300 × 125, pressure 3·45 bars (50 lb/sq in). Pneumatic brakes on main wheels. Metal ski landing gear optional.

POWER PLANT: Two 298 or 335 kW (400 or 450 shp) Polish-built Isotov GTD-350P turboshaft engines, mounted side by side above cabin. Fuel in single rubber tank, capacity 600 litres (131 Imp gallons), under cabin floor. Provision for carrying a 328 litre (52·5 Imp gallon) external tank on each side of cabin. Refuelling point in starboard side of fuselage. Oil capacity 25 litres (5·4 Imp gallons). Engine air intake de-icing by engine bleed air.

ACCOMMODATION: Normal accommodation for one pilot on flight deck (port side). Seats for up to eight passengers in air-conditioned cabin, there being back-to-back bench seats for three persons each, with two optional extra starboard side seats at the rear, one behind the other. All seats are removable for carrying up to 700 kg (1,543 lb) of internal freight. Access to cabin via forward-hinged doors on each side at front of cabin and aft on port side. Pilot's sliding window jettisonable in emergency. Ambulance version has accommodation for four stretchers and a medical attendant or for two stretchers and two sitting casualties. Side-by-side seats and dual controls in pilot training version. Cabin heating, ventilation and air-conditioning standard. Electrical de-icing of windscreen.

OPERATIONAL EQUIPMENT: As an agricultural aircraft, the Mi-2 carries a hopper on each side of the fuselage (total capacity 1,000 litres; 220 Imp gallons) and either a spraybar to the rear of the cabin on each side or a distributor for dry chemicals under each hopper. Swath width covered by the spraying version is 40-45 m (130-150 ft). As a search and rescue aircraft, an electric hoist, capacity 120 kg (264 lb), is fitted. In the freight role an underfuselage hook can be fitted for suspended loads of up to 800 kg (1,763 lb).

SYSTEMS: Cabin heating, by engine bleed air, and ventilation; heat exchangers warm atmospheric air for ventilation system during cold weather. Hydraulic system, pressure 59-78·6 bars (855-1,140 lb/sq in), for cyclic and collective pitch control boosters. Pneumatic system, pressure 49 bars (710 lb/sq in), for main-wheel brakes. AC electrical system, with two STG-3 3kW engine-driven starter/generators and 208V 16kVA three-phase alternator. 24V DC system, with two 28Ah lead-acid batteries.

Agricultural version of the WSK-Swidnik (Mil) Mi-2 twin-turbine light helicopter, with hopper on each side of cabin

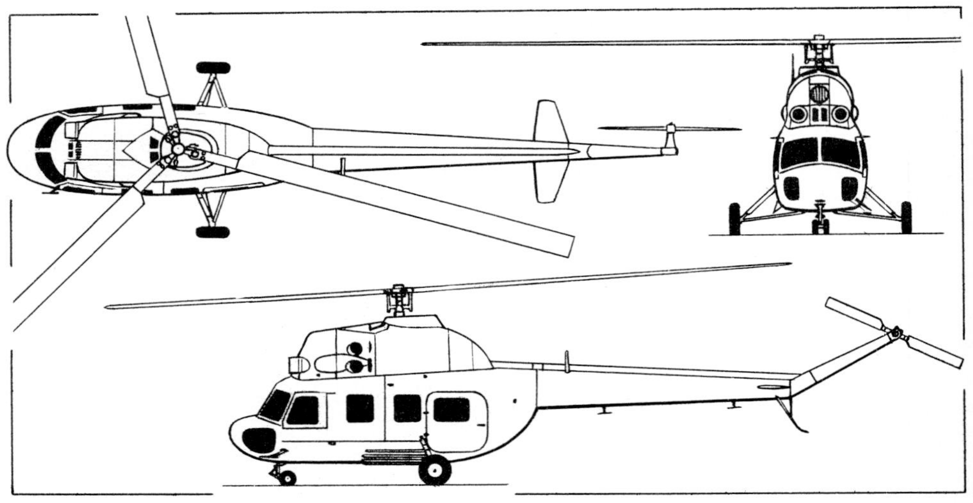

WSK-Swidnik (Mil) Mi-2 twin-turbine general-purpose light helicopter *(Pilot Press)*

ELECTRONICS AND EQUIPMENT: Standard equipment includes two transceivers (medium and short wave), gyro compass, radio compass, radio altimeter, intercom system and blind-flying panel. Electrically-operated wiper for pilot's windscreen. Fire extinguishing system, for engine bays and main gearbox compartment, is generally similar to, but simpler than, the Freon system fitted to the Soviet Mil Mi-8, and can be actuated automatically or manually.

ARMAMENT: A photograph published in the Polish press in 1975 showed an Mi-2 of the Polish Air Force equipped with air-to-surface rocket pods mounted on each side of the cabin.

DIMENSIONS, EXTERNAL:
- Diameter of main rotor: 14·50 m (47 ft 6¾ in)
- Main rotor blade chord (constant, each): 0·40 m (1 ft 3¾ in)
- Diameter of tail rotor: 2·70 m (8 ft 10¼ in)
- Length overall, rotors turning: 17·42 m (57 ft 2 in)
- Length of fuselage: 11·94 m (39 ft 2 in)
- Height to top of rotor hub: 3·75 m (12 ft 3½ in)
- Stabiliser span: 1·85 m (6 ft 0¾ in)
- Wheel track: 3·05 m (10 ft 0 in)
- Wheelbase: 2·71 m (8 ft 10¾ in)
- Tail rotor ground clearance: 1·59 m (5 ft 2¾ in)
- Cabin door (port, rear):
  - Height: 1·065 m (3 ft 5¾ in)
  - Width: 1·115 m (3 ft 8 in)
- Cabin door (stbd, front):
  - Height: 1·11 m (3 ft 7¾ in)
  - Width: 0·75 m (2 ft 5½ in)
- Cabin door (port, front):
  - Height: approx 1·40 m (4 ft 7 in)
  - Width: approx 1·20 m (3 ft 11¼ in)

DIMENSIONS, INTERNAL:
- Cabin:
  - Length, incl flight deck: 4·07 m (13 ft 4¼ in)
  - Length, excl flight deck: 2·27 m (7 ft 5½ in)
  - Mean width: 1·20 m (3 ft 11¼ in)
  - Mean height: 1·40 m (4 ft 7 in)

AREAS:
- Main rotor blades (each): 2·40 m² (25·83 sq ft)
- Tail rotor blades (each): 0·22 m² (2·37 sq ft)
- Main rotor disc: 166·4 m² (1,791·11 sq ft)
- Tail rotor disc: 5·70 m² (61·35 sq ft)
- Horizontal stabiliser: 0·70 m² (7·53 sq ft)

WEIGHTS AND LOADING:
- Basic operating weight, empty:
  - single-pilot versions: 2,365 kg (5,213 lb)
  - dual-control version: 2,424 kg (5,344 lb)
- Max payload, excl pilot, oil and fuel: 800 kg (1,763 lb)
- Normal T-O weight: 3,550 kg (7,826 lb)
- Max T-O weight: 3,700 kg (8,157 lb)
- Max disc loading: 22·4 kg/m² (4·6 lb/sq ft)

PERFORMANCE (at normal T-O weight):
- Max level speed at 500 m (1,640 ft): 113 knots (210 km/h; 130 mph)
- Max cruising speed at 500 m (1,640 ft): 108 knots (200 km/h; 124 mph)
- Econ cruising speed for max range at 500 m (1,640 ft): 102 knots (190 km/h; 118 mph)
- Econ cruising speed for max endurance at 500 m (1,640 ft): 54 knots (100 km/h; 62 mph)
- Max rate of climb at S/L: 270 m (885 ft)/min
- Service ceiling: 4,000 m (13,125 ft)
- Hovering ceiling in ground effect: 2,000 m (6,550 ft)
- Hovering ceiling out of ground effect: 1,000 m (3,275 ft)
- Min landing area: 30 × 30 m (100 × 100 ft)
- Range at 500 m (1,640 ft) with max internal and auxiliary fuel, 30 min reserves: 313 nm (580 km; 360 miles)
- Range at 500 m (1,640 ft) with max payload, 5% fuel reserves: 91 nm (170 km; 105 miles)

WSK-SWIDNIK Mi-2M

Design of this enlarged, modified and more powerful version of the Mi-2 began at WSK-Swidnik in 1968; construction began in the following year. Five development aircraft were built, the first of which made its first flight on 1 July 1974. Production, in passenger/cargo, ambulance and agricultural versions, began in the Spring of 1975.

TYPE: Twin-turbine general-purpose light helicopter.

ROTOR SYSTEM: Generally similar to Mi-2. Three-blade fully articulated single main rotor, with flapping, drag and pitch hinges and anti-torque vibration dampers. Blades of NACA 230-13M constant section, built of aluminium alloy with an extruded spar and bonded honeycomb trailing-edge pockets. Blade spars attached to steel hub by steel grips. Rotor brake fitted. Two-blade semi-rigid tail rotor, with aluminium alloy bonded blades and steel hub. Main rotor blades, tail section and tail rotor blades do not fold.

ROTOR DRIVE: Twin turbines drive through freewheeling units and rotor brake to main gearbox. Steel driveshafts. Tail rotor shaft-driven through intermediate and tail gearboxes. Main and tail rotor/engine rpm ratios as for standard Mi-2.

FUSELAGE: As for Mi-2, but larger and with a completely flat floor.

TAIL UNIT: As for Mi-2.

LANDING GEAR: Tricycle type, with single-wheel non-retractable main units and twin-wheel fully-retractable steerable nose unit. Oleo-pneumatic shock-absorbers on all three units. Main-wheel tyres size 600 × 180 mm, pressure 4·4 bars (64 lb/sq in); nosewheel tyres size 400 × 150 mm, pressure 3·45 bars (50 lb/sq in). Pneumatic brakes on main wheels. Tailskid of reinforced plastics.

POWER PLANT: Two 335 kW (450 shp) Polish-built Isotov GTD-350P turboshaft engines, mounted side by side above cabin. Fuel in integral tank beneath cabin floor, capacity 830 litres (182·5 Imp gallons). Provision for carrying a 238 litre (52·5 Imp gallon) auxiliary tank externally on each side of fuselage. Oil capacity 25 litres (5·4 Imp gallons).

ACCOMMODATION: Normal accommodation for one pilot on flight deck and nine passengers in cabin. Sliding doors on each side to both flight deck and main cabin. Cabin heated and ventilated.

SYSTEMS: Single hydraulic system, pressure 64 bars (925 lb/sq in), for control unit. Pneumatic system, pressure 49 bars (710 lb/sq in), for main-wheel brakes. 24V DC and 208V 400Hz AC electrical systems, including two 3kW starter/generators, one 16kVA alternator and two 28Ah batteries. Electrical anti-icing of main and tail rotor blades. Engine air intake de-icing by engine bleed air. Automatic or manual fire extinguishing system for engines and main gearbox compartment.

ELECTRONICS AND EQUIPMENT: Standard equipment includes two medium and short wave transceivers, gyro compass, radio compass, and radio altimeter. Special equipment includes rescue hoist and engine air intake filters.

DIMENSIONS, EXTERNAL: As for Mi-2, except:
- Distance between rotor centres 8·80 m (28 ft 10½ in)
- Height to top of rotor hub 3·95 m (12 ft 11½ in)
- Wheel track 3·588 m (11 ft 9¼ in)
- Wheelbase 2·469 m (8 ft 1¼ in)

Mi-2M modified version of the Mi-2, with widened 10-seat cabin and retractable nosewheel

Cabin doors (port and stbd, front):
- Height 1·10 m (3 ft 7¼ in)
- Width 0·60 m (1 ft 11¾ in)

Cabin doors (port and stbd, rear):
- Height 1·10 m (3 ft 7¼ in)
- Width 1·10 m (3 ft 7¼ in)

DIMENSIONS, INTERNAL:
Cabin:
- Length, incl flight deck 4·07 m (13 ft 4¼ in)
- Mean width 1·45 m (4 ft 9 in)
- Mean height 1·45 m (4 ft 9 in)
- Floor area 5·60 m² (60·28 sq ft)
- Volume 6·40 m³ (226·0 cu ft)

AREAS: As for Mi-2

WEIGHTS AND LOADINGS:
- Basic operating weight empty 2,365 kg (5,213 lb)
- Max payload 900 kg (1,984 lb)
- Max T-O weight 3,700 kg (8,157 lb)
- Max disc loading 22·4 kg/m² (4·6 lb/sq ft)
- Max power loading 5·51 kg/kW (9·14 lb/shp)

PERFORMANCE: As for Mi-2, except:
- Max rate of climb at S/L 402 m (1,320 ft)/min
- Service ceiling 3,640 m (11,950 ft)
- Hovering ceiling out of ground effect 500 m (1,640 ft)
- Range at 500 m (1,640 ft) with max internal and auxiliary fuel, 30 min reserves 210 nm (390 km; 242 miles)

PORTUGAL

OGMA

OFICINAS GERAIS DE MATERIAL AERONÁUTICO (GENERAL AERONAUTICAL MATERIAL WORKSHOPS)

WORKS:
Alverca do Ribatejo
Telephone: 258/803

DIRECTOR:
Brigadier General Engineer Rui do Carmo da Conceição Espadinha

ASSISTANT DIRECTOR:
Colonel Engineer António Pedro da Silva Gonçalves

OGMA was founded in 1918 and has been in continuous operation since then. It is the Air Force department responsible for maintenance and repair, at depot level, of all aircraft, electronics, engines, structures, ground communications and radar equipment of the Portuguese Air Force.

OGMA has a total floor space of 110,730 m² (1,191,890 sq ft) and a work force of approx 3,300 people.

In addition to major overhaul and IRAN work on aircraft, aero-engines and components for the Portuguese Air Force, OGMA performs maintenance work on Transall C-160 transport aircraft for the Federal German Air Force. Under a contract signed in 1959, it also undertakes IRAN, refurbishing and rehabilitation, periodic inspection and emergency maintenance and crash repair of USAF and US Navy aircraft.

Under an important contract with Aérospatiale of France, OGMA is manufacturing tail structures for the SA 315B Lama and SA 318 Alouette II helicopters, and minor structural components for the SA 330 Puma.

ROMANIA

CIAR

CENTRALA INDUSTRIALA AERONAUTICA ROMANA (Industrial Centre for Romanian Aviation)

HEADQUARTERS: 133 Calea Victoriei, Sector 1, Bucharest

Romania has had a tradition of aviation since the earliest days of flying, dating from the first monoplane built in France in early 1906 by the Romanian engineer Traian Vuia, the original monoplane of Aurel Vlaicu which, on 17 June 1910, became the first nationally-designed aeroplane to be flown in Romania, and the famous aeroplanes designed and built in France and Britain by Henri Coanda in 1910-14.

Since that time the Romanian aircraft industry (IAR) has produced some 80 different types of landplane (of which 70 were Romanian-designed) and three types of seaplane (two being of Romanian design), and has developed and manufactured 39 different types of sailplane. In addition, many other achievements in the fields of theoretical and experimental aerodynamics have been made by teams of Romanian engineers led by Prof Elie Carafoli, Prof Ion Stroiescu, Prof Ion Grosu, Ing Radu Manicatide, Ing Iosif Silimon and others.

Before the second World War the Romanian aircraft industry employed more than 20,000 people, the most important centres being the IAR (Industria Aeronautica Romana) at Brasov, with 8,000 employees, SET at Bucharest, and ICAR, also at Bucharest. The IAR factory, destroyed by bombing in 1944, was rebuilt after the war for the manufacture of tractors, and resumed its aeronautical activities with the IAR-811 training aircraft designed by Dipl Ing Radu Manicatide and flown for the first time in March 1949. Known until 1959 as URMV-3 (aircraft component repair factory 3), the Brasov factory during that period produced more than 200 aircraft of different types (designed by Dipl Ing Manicatide) and more than 20 types of sailplane (designed by Dipl Ing Iosif Silimon). At the same time a number of sailplanes were produced at the Combinatul de Lemn (wood factory) at Reghin by Vladimir Novitzchi; and two repair factories, subordinate to the Ministry of Military Forces, were set up at Medias (ARMV-1) and Bucharest (ARMV-2).

The Orao single-seat fighter, built in partnership by Romania and Yugoslavia and described in the International section *(Front Magazine)*

A major reorganisation took place in 1959, when the URMV-3 at Brasov was dissolved and its staff were divided into two teams. One of these was placed under the leadership of Dipl Ing Manicatide at ARMV-2, which was then renamed IRMA (Intreprinderea de Reparat Material Aeronautic). The other, led by Dipl Ing Silimon, was set up at Ghimbav as a division of IIL-Brasov, to concentrate on sailplane design. Up to 1968 IRMA, which was then responsible to the Ministry of Transport, built more than 140 aircraft for ambulance, training, agricultural and other duties.

The Romanian aeronautical industry was reorganised in 1968, and its activities are now undertaken, within the Ministry of Machine Tool Building Industry and Electrotechnics, by the CIAR (Industrial Centre for Romanian Aviation).

The major activities of CIAR are carried out in two factories: Intreprinderea de Reparat Material Aeronautic (IRMA-Bucuresti) and Intreprinderea de Constructii Aeronautice (ICA-Brasov). Research and development in the aeronautical field are undertaken by the Institute of Fluid Mechanics and Aerospace Construction (IMFCA-Bucuresti).

For the selection and training of specialist workers, there is an Engineering Faculty for Aerospace Construction at the Polytechnic Institute of Bucharest. This Faculty has three sections, devoted to aircraft, propulsion systems and aviation electronics.

The Romanian aircraft industry is collaborating with that of Yugoslavia in the development of the Orao twin-jet (Rolls-Royce Viper 623) fighter and ground attack aircraft to meet a requirement of the air forces of the two countries. All known details of this aircraft can be found in the International section. The Romanian industry also has a long-term agreement with VFW-Fokker of Germany covering manufacture of the VFW 614 twin-turbofan transport aircraft under licence in Romania.

Other current products and activities of the IRMA and ICA factories, which have a combined work force of about 5,000 persons, are as follows:

IRMA-BUCURESTI

INTREPRINDEREA DE REPARAT MATERIAL AERONAUTIC (Aircraft Component Repair Factory)

ADDRESS:
Baneasa Airport, Bucharest

As indicated in the introductory copy for CIAR, IRMA was formed in 1959 from part of the former URMV-3 at Brasov, having also an aircraft design and production centre under the leadership of Dipl Ing Radu Manicatide. It currently specialises in the repair and overhaul of various types of large and small aircraft and aero-engines on behalf of various airlines, including Tarom, the Romanian state airline. It is also the agent and repair centre for Lycoming engines. IRMA was responsible for series production of the IAR-818 ambulance and agricultural aircraft (1965-66 *Jane's*) and, more recently, the IAR-821 and 821B (1973-74 *Jane's*). It was also responsible for building the IAR-822 and 822B prototypes described under the ICA-Brasov heading. In addition, IRMA-Bucuresti is manufacturing under licence the Britten-Norman BN-2A Islander.

IRMA (BRITTEN-NORMAN) ISLANDER

In 1968 it was announced that the Britten-Norman BN-2A Islander (see UK section) was to be manufactured under licence in Romania, and the aircraft is now in production by IRMA.

The current agreements with IRMA cover the production of 315 Islanders, and the first Romanian-built example flew for the first time at Baneasa Airport, Bucharest, on 4 August 1969. A total of 84 had been completed by the end of 1972. A further 30 were built in 1973, with production scheduled to increase to 40 in 1974 and 50 in 1975.

Since September 1972 the production of Islanders has continued in co-operation with the new Fairey Group Britten-Norman company, with which Romania also plans to collaborate in other projects.

Britten-Norman BN-2A Islander twin-engined feederline transport of Air Liberia, built in Romania by IRMA *(Peter J. Bish)*

ICA-BRASOV

INTREPRINDEREA DE CONSTRUCTII AERONAUTICE (Aircraft Construction Factory)

ADDRESS:
Brasov

ICA-Brasov, created in 1968, continues the work begun in 1926 by IAR-Brasov and continued in 1950-59 as URMV-3 Brasov. Today, it manufactures aircraft and sailplanes of its own design, including the IAR-824, of which a description follows; and the IS-28 and IS-29 series of sailplanes, described in the appropriate section of this edition.

In addition, ICA-Brasov undertakes the repair and overhaul of light aircraft; it participates in the manufacturing programme for the Britten-Norman BN-2A Islander; manufactures the Aérospatiale SA 316B Alouette III helicopter under licence in Romania; and is also responsible for building the production-series IAR-822 and -822B agricultural and training aircraft and the prototypes of the IAR-823 training and IAR-826 agricultural aircraft.

In 1970, ICA-Brasov was awarded a Diploma of Honour by the FAI in recognition of its work in the field of aeronautical construction.

IAR-822 and IAR-826

Design of the **IAR-822** was initiated at IMFCA in October 1968, by a team under the leadership of Dipl Ing Radu Manicatide, and the prototype, built by IRMA, flew for the first time in March 1970. Certification of the standard version was granted in October 1971, and of the agricultural version in January 1972.

In 1971 manufacture was begun at ICA-Brasov, which is responsible for series production, of the present series of 200 IAR-822s. The first production aircraft was flown for the first time in August 1971, and by early 1973 a total of 20 had been built. These were of mixed construction; from 1973, the aircraft became available also in all-metal form as the **IAR-826**. The structural description covers both models. The weights and performance figures apply to the IAR-822, but are generally similar for the IAR-826. A developed version, known as the **IAR-827**, is described separately.

Production is both for Romanian use and for export. Applications of the aircraft, in addition to agricultural duties, include highway de-icing, firefighting, aerial survey work, glider towing (up to three gliders), power and pipeline inspection, geological survey, fishery patrol, training, and the transport of up to 700 kg (1,543 lb) of mail or other cargo.

TYPE: Single/two-seat utility aircraft, designed primarily for agricultural operations.

WINGS: Cantilever low-wing monoplane. Wing section NACA 23014. Constant-chord safe-life wings, with 0° dihedral on centre-section and 5° on outer panels. Incidence 5°. Centre-section is a welded chrome-molybdenum steel tube structure, with metal skin. Outer panels, which are attached to centre-section at three points, are of single-spar all-wood construction (Romanian spruce) with birch plywood skin, varnished on the inside and protected externally by a fabric coating (IAR-822); or of all-metal single-spar construction with duralumin skin (IAR-826). Mechanically-actuated single-slotted wood or metal flaps and ailerons. Ground-adjustable tab.

FUSELAGE: Welded chrome-molybdenum steel tube truss-type structure, of basically rectangular section with rounded top-decking. Airtight fabric covering. Fuselage is designed to collapse progressively from the front and is fitted with a steel tube overturn structure to protect the pilot in the event of a crash.

TAIL UNIT: Cantilever all-metal or wood/fabric structure of similar construction to outer wings. Fixed-incidence braced tailplane, of constant chord and NACA 0012 section. One automatic trim tab and one controllable tab in elevators. Ground-adjustable tab on rudder.

LANDING GEAR: Non-retractable tailwheel type, with main

IAR-822 single/two-seat agricultural aircraft (Lycoming IO-540 engine)

units interchangeable left/right. Main units, with fork legs and Vee bracing struts, have oleo-pneumatic shock-struts located in the centre-section. Rubber shock-absorber on tailwheel. Main wheels and tyres size 600 × 180 mm, tyre pressure 3·45 bars (50 lb/sq in); self-centering, fully-steerable tailwheel, size 290 × 110 mm. Hydraulic brakes. Optional ski gear.

POWER PLANT: One 216 kW (290 hp) Lycoming IO-540-G1D5 flat-six engine, driving a Hartzell HC 92 WK-1D-9349-4-6 two-blade constant-speed variable-pitch metal propeller. Fuel tank in each outer wing, total capacity 200 litres (44 Imp gallons). Refuelling point above each wing. Optional auxiliary tank, capacity 125 litres (27·5 Imp gallons), can be mounted in hopper. Oil capacity 11·4 litres (2·5 Imp gallons).

ACCOMMODATION: Single adjustable seat in specially-strengthened enclosed cockpit with steel overturn structure. Upward-hinged door on port side. Cockpit designed to remain substantially undamaged in a low-speed crash. Provision for fitting jump-seat below and behind pilot, to accommodate mechanic or loader. Cabin heated, ventilated and soundproofed.

ELECTRONICS: Standard equipment includes an electrical system for engine starting, pitot heating and flying instruments, a 24V 30Ah battery, and complete instrumentation for agricultural operation and for long-distance flights between operations. Optional equipment includes Bendix RT 221-AE transceiver, position lights, and blind- and night-flying instrumentation.

OPERATIONAL EQUIPMENT: Chemical is carried in epoxy-treated riveted duralumin-sheet hopper in forward fuselage, located at the CG point to avoid trim changes as the load is reduced. The hopper, which is loaded through a rubber-sealed door on the top of the fuselage, has an internal volume of 0·8 m³ (28·25 cu ft) and can accommodate up to 600 litres (132 Imp gallons) of liquid or up to 630 kg (1,389 lb) of dry (powdered or granulated) chemicals. The corners are rounded to avoid dust catchment. Wide range of application equipment available, including a windmill-driven centrifugal pump and spraybars, which can be changed in less than 30 min in the field for a venturi-type solid chemical distributor. Rotary atomisers optional. Entire hopper load can be jettisoned in an emergency.

DIMENSIONS, EXTERNAL:

Wing span	12·80 m (42 ft 0 in)
Wing chord (constant)	2·10 m (6 ft 10¾ in)
Wing aspect ratio	6·3
Wing centre-section span	3·656 m (12 ft 0 in)
Length overall	9·40 m (30 ft 10 in)
Height overall (tail up)	3·10 m (10 ft 2 in)
Height overall (tail down)	2·60 m (8 ft 6½ in)
Wheel track	2·94 m (9 ft 7¾ in)
Wheelbase	5·99 m (19 ft 7¾ in)
Propeller diameter	2·20 m (7 ft 3 in)

AREAS:

Wings, gross	26·00 m² (279·86 sq ft)
Ailerons (total)	2·78 m² (29·92 sq ft)
Flaps (total)	3·64 m² (39·18 sq ft)
Fin	0·67 m² (7·21 sq ft)
Rudder, incl tab	1·46 m² (15·72 sq ft)
Tailplane	2·52 m² (27·12 sq ft)
Elevators, incl tabs	1·98 m² (21·31 sq ft)

WEIGHTS AND LOADINGS:

Weight empty:	
IAR-822	1,080 kg (2,380 lb)
IAR-826	1,120 kg (2,469 lb)
Max T-O weight	1,900 kg (4,188 lb)
Max wing loading	73·0 kg/m² (14·95 lb/sq ft)
Max power loading	8·80 kg/kW (14·44 lb/hp)

PERFORMANCE (IAR-822. S: standard, no external agricultural equipment, at 1,300 kg; 2,866 lb AUW. A: agricultural version, with equipment, at 1,900 kg; 4,188 lb AUW):

Never-exceed speed:	
A	134·5 knots (250 km/h; 155 mph)
Max level speed at S/L:	
S	121 knots (224 km/h; 139 mph)
A	97 knots (180 km/h; 112 mph)
Max cruising speed (75% power) at S/L:	
S	112 knots (208 km/h; 129 mph)
A	89 knots (165 km/h; 103 mph)
Operating speed:	
A	65-86 knots (120-160 km/h; 75-99 mph)
Stalling speed, flaps up:	
S, power on	32·5 knots (60 km/h; 37·5 mph)
A, power on	49 knots (90 km/h; 56 mph)
S, power off	42·5 knots (78 km/h; 48·5 mph)
A, power off	52·5 knots (97 km/h; 60·5 mph)
Stalling speed, 40° flap:	
S, power on	27 knots (50 km/h; 31 mph)
A, power on	40 knots (74 km/h; 46 mph)
S, power off	36·5 knots (67 km/h; 42 mph)
A, power off	48 knots (88 km/h; 55 mph)
Max rate of climb at S/L:	
S	300 m (984 ft)/min
A	210 m (689 ft)/min
Service ceiling:	
S	5,200 m (17,050 ft)
A	3,000 m (9,850 ft)
T-O run, on grass:	
S	80 m (262 ft)
A	180 m (591 ft)
T-O to 15 m (50 ft):	
S	220 m (722 ft)
A	360 m (1,181 ft)
Landing from 15 m (50 ft):	
S	200 m (656 ft)
A	300 m (984 ft)
Landing run:	
S	80 m (262 ft)
A	140 m (459 ft)
Max range, no reserves:	
S	323 nm (600 km; 372 miles)
A	242 nm (450 km; 279 miles)
Max endurance, no reserves:	
S, A	3 hr
Swath width (A):	
sprayer	20-25 m (66-82 ft)
granule duster	12-18 m (39-59 ft)
powder duster	30-40 m (98-131 ft)

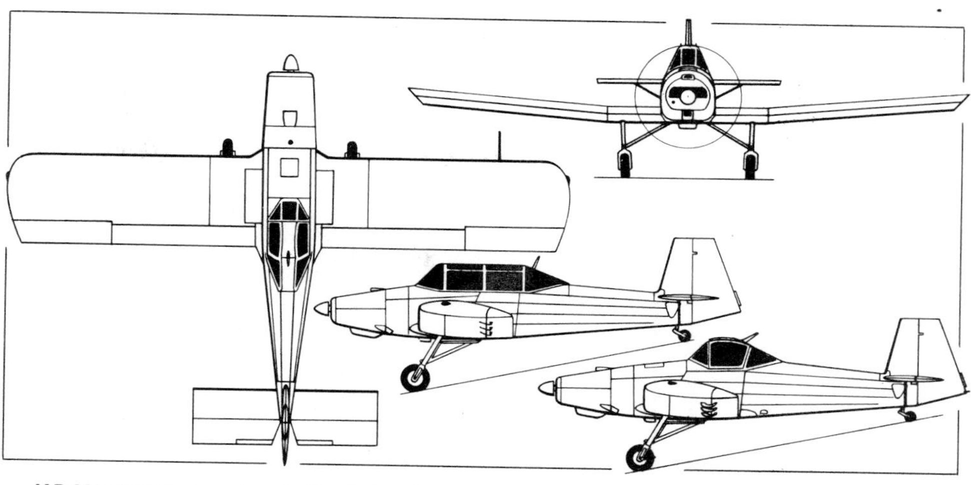

IAR-822 single/two-seat agricultural aircraft, with additional side view (centre) of two-seat IAR-822B *(Pilot Press)*

IAR-826 all-metal version of the IAR-822, with underfuselage dispenser for chemicals

IAR-822B

The IAR-822B is a tandem two-seat version of the IAR-822, design of which began at IMFCA in February 1972. It is intended for training, glider towing and simulated agricultural operations.

Construction of a prototype began at IRMA in May 1972, and this aircraft flew for the first time in December 1972. Production by ICA-Brasov has begun; five aircraft were due to be completed by early 1974, but no news of progress since that time has been received.

The description of the IAR-822 applies also to the IAR-822B, except in the following respects:

TYPE: Two-seat training, glider towing and agricultural training aircraft.

ACCOMMODATION: Enclosed cabin, with fully glazed canopy, seating two persons in tandem. Fuselage structure modified to take second cockpit at front, in place of chemical hopper. Both seats adjustable and can accept parachutes. Dual controls are standard, and aircraft can be fully controlled from either seat. Two upward-hinged doors on port side.

EQUIPMENT: Optionally, to simulate agricultural operations, aircraft can be fitted with dispersal equipment of IAR-822, together with a reduced-capacity hopper (100 litres; 22 Imp gallons) located under front seat. Can also be fitted with release gear for glider towing.

DIMENSIONS, EXTERNAL, AND AREAS:

As for IAR-822, except:	
Length overall	9·55 m (31 ft 4 in)
Wheel track	2·70 m (8 ft 10¼ in)

WEIGHTS AND LOADING:

Weight empty	1,090 kg (2,403 lb)
Max T-O weight	1,450 kg (3,196 lb)
Max wing loading	56·0 kg/m² (11·5 lb/sq ft)

PERFORMANCE (at max T-O weight):

Never-exceed speed	140 knots (260 km/h; 161 mph)
Max level speed at S/L	128 knots (238 km/h; 148 mph)
Max cruising speed (75% power) at S/L	116 knots (215 km/h; 134 mph)
Stalling speed, flaps up:	
power on	35·5 knots (65 km/h; 40·5 mph)
power off	43·5 knots (80 km/h; 50 mph)
Stalling speed, flaps down:	
power on	30 knots (55 km/h; 34·5 mph)
power off	38 knots (70 km/h; 43·5 mph)

Two-seat IAR-822B, developed from the IAR-822 for flying simulated agricultural missions

Max rate of climb at S/L	300 m (984 ft)/min
Service ceiling	5,200 m (17,050 ft)
T-O run	95 m (312 ft)
T-O to 15 m (50 ft)	250 m (820 ft)
Landing from 15 m (50 ft)	220 m (722 ft)
Landing run	90 m (295 ft)
Max range, no reserves	340 nm (630 km; 391 miles)
Max endurance, no reserves	3 hr

IAR-823

The IAR-823 is a two/five-seat training or touring light aircraft, with retractable landing gear. It was designed at IMFCA, work beginning in May 1970, by a team led by Dipl Ing Radu Manicatide. Construction of a prototype began at ICA-Brasov in the Autumn of 1971, and this aircraft made its first flight in July 1973. The first production aircraft flew in 1974.

As a two-seater, the IAR-823 is fully aerobatic and is intended for training duties. With a rear bench seat for up to three more persons it is suitable as an executive, taxi or touring aircraft. Provision is made for two underwing pylons for the carriage of drop-tanks or practice weapons.

TYPE: Two/five-seat cabin monoplane.

WINGS: Cantilever low-wing monoplane. Wing section NACA 23012 (modified). Dihedral 7° from roots. Incidence 3° at root, 1° at tip. Conventional all-metal structure, with single main spar and rear auxiliary spar; three-point attachment to fuselage. Riveted spars, ribs and skin of corrosion-proof aluminium alloy. Leading-edges riveted, and sealed to ribs and main spar to form main torsion box and integral fuel tanks. Electrically-actuated fabric-covered metal single-slotted flaps and fabric-covered Frise-type slotted metal ailerons. Ground-adjustable tab.

FUSELAGE: All-metal semi-monocoque structure. Glassfibre engine cowling.

TAIL UNIT: Cantilever metal structure. Two-spar duralumin-covered fin and tailplane; fabric-covered duralumin horn-balanced rudder and elevators. Electrically-actuated automatic trim tabs in elevators; controllable tab in rudder.

LANDING GEAR: Retractable tricycle type, with steerable nosewheel. Electrical retraction, main units inward, nose unit rearward. Emergency manual actuation. Oleo-pneumatic shock-absorbers. Main-wheel tyres size 6·00-6, pressure 2·93 bars (42·5 lb/sq in). Nosewheel tyre size 355 × 150 mm. Independent hydraulic main-wheel brakes, pedal-controlled from left front seat. Shimmy damper on nose unit. No wheel doors.

POWER PLANT: One 216 kW (290 hp) Lycoming IO-540-G1D5 flat-six engine, driving a Hartzell two-blade constant-speed metal propeller. Fuel in four integral wing tanks, total capacity 360 litres (79 Imp gallons). Provision for two 70 litre (15·4 Imp gallon) drop-tanks on underwing pylons.

ACCOMMODATION: Fully-enclosed cabin, seating two persons side by side on individual adjustable front seats, with removable bench seat at rear for up to three more people. Dual controls standard in training version, optional in other versions. Upward-hinged door (optionally jettisonable) on each side of cabin, which is soundproofed, heated and ventilated. Compartment at rear of cabin for up to 40 kg (88 lb) of baggage. Equipment and layout can be varied for use as air taxi, executive or freight transport, ambulance, liaison or photographic aircraft.

SYSTEMS AND EQUIPMENT: Electrical system, including 50A alternator and 24V 30Ah battery, for engine starting, elevator tab and landing gear actuation, radio communications, landing and navigation lights and cabin and instrument lighting. Other standard equipment includes VFR instrumentation and Bendix RT 221-AE transceiver. Optional equipment, according to mission, includes blind-flying instrumentation and, in civil transport version, marker beacon, nav/com radio, VOR/ILS, ADF and autopilot.

DIMENSIONS, EXTERNAL:

Wing span	10·00 m (32 ft 9¾ in)
Wing chord at c/l	2·00 m (6 ft 6¾ in)
Wing chord at tip	1·00 m (3 ft 3¼ in)
Wing aspect ratio	6·66
Length overall	8·24 m (27 ft 0¼ in)
Height overall	2·52 m (8 ft 3¼ in)
Wheel track	2·48 m (8 ft 1¾ in)
Wheelbase	1·86 m (6 ft 1¼ in)
Propeller diameter	2·23 m (7 ft 4 in)

AREAS:

Wings, gross	15·00 m² (161·5 sq ft)
Ailerons (total)	1·20 m² (12·92 sq ft)
Trailing-edge flaps (total)	1·78 m² (19·16 sq ft)
Horizontal tail surfaces (total)	3·30 m² (35·52 sq ft)
Vertical tail surfaces (total)	1·50 m² (16·15 sq ft)

WEIGHTS AND LOADINGS (A: Aerobatic; U: Utility category):

Weight empty:	
A	900 kg (1,984 lb)
U	910 kg (2,006 lb)
Max T-O weight:	
A	1,190 kg (2,623 lb)
U	1,380 kg (3,042 lb)
Max permissible weight for special missions	1,500 kg (3,307 lb)
Max normal wing loading:	
A	79·0 kg/m² (16·2 lb/sq ft)
U	92·0 kg/m² (18·8 lb/sq ft)
Max normal power loading:	
A	5·51 kg/kW (9·15 lb/hp)

IAR-823 two/five-seat light aircraft (Lycoming IO-540 engine) *(Peter R. March)*

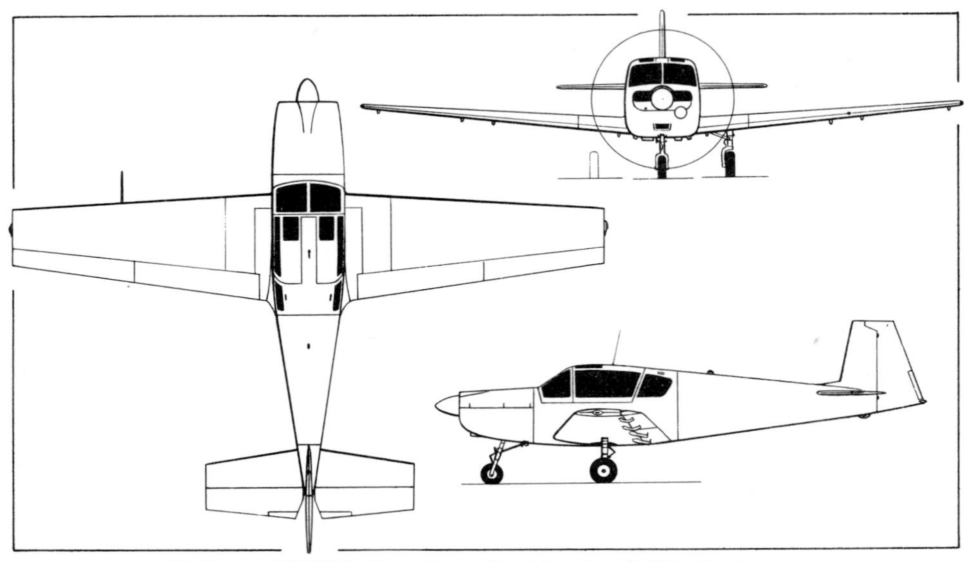

ICA-Brasov IAR-823 light touring and training aircraft *(Pilot Press)*

PERFORMANCE (at 1,400 kg; 3,086 lb AUW except where indicated):

Never-exceed speed (limited)	215 knots (400 km/h; 248 mph)
Max level speed at S/L	167 knots (310 km/h; 192·5 mph)
Max cruising speed (75% power) at 1,750 m (5,750 ft)	162 knots (300 km/h; 186 mph)
Econ cruising speed (60% power) at 3,050 m (10,000 ft)	156 knots (290 km/h; 180 mph)
Stalling speed, flaps up	62·5 knots (115 km/h; 71·5 mph)
Stalling speed, 30° flap	54·5 knots (100 km/h; 62·5 mph)
Max rate of climb at S/L	420 m (1,378 ft)/min
Service ceiling	5,800 m (19,025 ft)
T-O run	230 m (755 ft)
Landing run	200 m (656 ft)
Range, according to mission and payload, 1 hr reserves	431-863 nm (800-1,600 km; 497-994 miles)
Endurance, according to mission and payload	3-6 hr
g limits (at 1,190 kg; 2,623 lb AUW)	+6; −3

IAR-824

Design of the IAR-824, under the leadership of Dipl Ing Iosif Silimon, began at ICA-Brasov in the Winter of 1969 and is based largely upon that of the IS-23A (see 1973-74 *Jane's*). The IS-24 prototype (YR-ISB) flew for the first time on 24 May 1971, and certification was granted on 13 May 1972.

TYPE: Six-seat light multi-purpose aircraft.

WINGS: Cantilever high-wing monoplane. Wing section NACA 64_2-413·5 (constant). No dihedral or sweepback. Two-spar constant-chord all-metal structure of ribs and stringers, with light alloy skin. Automatic leading-edge slats over almost full span. Electrically-operated trailing-edge flaps. Rod- and cable-actuated ailerons, which can be operated differentially or in conjunction with flaps. No tabs.

FUSELAGE: Corrosion-protected aluminium alloy semi-monocoque structure. Engine cowling of metal and glassfibre.

TAIL UNIT: Cantilever metal structure, with variable-incidence tailplane, one-piece elevator and sweptback fin and rudder. Balanced elevator and rudder, with fluted skins. Elevator controlled by rods, tailplane and rudder by cables. Trim tab in rudder.

LANDING GEAR: Non-retractable tricycle type, with hydraulic telescopic shock-absorber on each main unit. Nosewheel steerable by rudder pedals. Main-wheel tyres size 600 × 180; nosewheel tyre size 440 × 130. Hydraulic main-wheel brakes.

POWER PLANT: One 216 kW (290 hp) Lycoming IO-540-C1D5 flat-six engine, driving a Hartzell HC.92.WK.1D-W9350-4·6 two-blade constant-speed metal propeller. Integral fuel tank in wing spar box,

IS-24 prototype of the ICA-Brasov IAR-824 general-purpose monoplane (Lycoming IO-540 engine)

capacity 325 litres (71·5 Imp gallons). Reserves in wing-tip tanks.

ACCOMMODATION: Enclosed, heated, ventilated and soundproofed cabin, accommodating in standard passenger version up to six persons. Passenger seats may be removed for cargo carrying.

DIMENSIONS, EXTERNAL:

Wing span	12·40 m (40 ft 8¼ in)
Wing chord (constant)	1·90 m (6 ft 4 in)
Wing aspect ratio	6·6
Length overall	9·10 m (29 ft 10½ in)
Height overall	3·44 m (11 ft 3½ in)
Wheel track	3·30 m (10 ft 10 in)
Wheelbase	2·30 m (7 ft 6½ in)

AREAS:

Wings, gross	23·60 m² (254 sq ft)
Ailerons (total)	2·40 m² (25·83 sq ft)
Trailing-edge flaps (total)	2·56 m² (27·56 sq ft)
Horizontal tail surfaces (total)	4·00 m² (43·06 sq ft)
Vertical tail surfaces (total)	1·56 m² (16·79 sq ft)

WEIGHTS AND LOADING:

Weight empty	1,240 kg (2,733 lb)
Payload	500 kg (1,102 lb)
Max T-O weight	1,900 kg (4,188 lb)
Max wing loading	81·5 kg/m² (16·7 lb/sq ft)

PERFORMANCE (at max T-O weight):

Max level speed at S/L	110 knots (205 km/h; 127 mph)
Max cruising speed at S/L	97 knots (180 km/h; 112 mph)
Stalling speed	40·5 knots (75 km/h; 47 mph)
Max rate of climb at S/L	180 m (591 ft)/min
Service ceiling	3,000 m (9,850 ft)
T-O run	190 m (623 ft)
Landing run	110 m (361 ft)
Max range	485 nm (900 km; 559 miles)

IAR-827

The IAR-827 is a developed version of the all-metal IAR-826 (which see), and represents a new generation of Romanian agricultural aircraft, having an increased payload, more powerful engine and improved flying and operating characteristics.

TYPE: Single-seat agricultural aircraft.

POWER PLANT: One 298 kW (400 hp) Lycoming IO-720 flat-eight engine, with fuel injection, driving a Hartzell three-blade variable-pitch metal propeller. Total fuel capacity 220 litres (48 Imp gallons).

ICA-Brasov-built Aérospatiale SA 316B Alouette III helicopter *(John Wegg)*

DIMENSIONS, EXTERNAL:

Wing span	14·00 m (45 ft 11¼ in)
Length overall	9·64 m (31 ft 7½ in)
Height overall	2·64 m (8 ft 8 in)
Wheel track	3·60 m (11 ft 9¾ in)

AREA:

Wings, gross	29·00 m² (312·15 sq ft)

WEIGHTS:

Weight empty	1,210 kg (2,667 lb)
Chemical payload for 2 hr mission	900 kg (1,984 lb)
Max payload (FAR 21)	1,000 kg (2,204 lb)
Max T-O weight	2,350 kg (5,180 lb)

PERFORMANCE (estimated. S: standard, no external agricultural equipment, at 1,500 kg; 3,306 lb AUW. A: agricultural version, with equipment, at 2,350 kg; 5,180 lb AUW):

Max level speed:	
S	124 knots (230 km/h; 143 mph)
A	104 knots (193 km/h; 120 mph)
Cruising speed:	
S	114 knots (212 km/h; 132 mph)
A	94 knots (175 km/h; 109 mph)
Stalling speed, flaps down, power off:	
S	40·5 knots (75 km/h; 47 mph)
A	49 knots (90 km/h; 56 mph)
Max rate of climb at S/L:	
S	300 m (984 ft)/min
A	235 m (770 ft)/min
Operational ceiling:	
A	2,000 m (6,560 ft)
Service ceiling:	
S	5,000 m (16,400 ft)
A	3,000 m (9,850 ft)
T-O run:	
A	225 m (738 ft)
T-O to 15 m (50 ft):	
A	450 m (1,476 ft)
Max range:	
S	323 nm (600 km; 372 miles)
Max endurance:	
A	2 hr 30 min

ICA (AÉROSPATIALE) ALOUETTE III

It was announced in 1971 that ICA and Aérospatiale had concluded an agreement for an initial quantity of 50 SA 316B Alouette III helicopters (see French section) to be built in Romania. Production of these is now under way, and Romanian-built components are also being supplied for incorporation in French-built Alouette IIIs.

SOUTH AFRICA

ATLAS

ATLAS AIRCRAFT CORPORATION OF SOUTH AFRICA (PTY) LIMITED

HEAD OFFICE AND WORKS:
PO Box 11, Atlas Road, Kempton Park 1620, Transvaal
Telephone: 973-0111
Telex: J7965

DIRECTORS:
Dr W. J. de Villiers (Chairman)
L. W. Dekker
P. A. Earle
J. F. H. Jagoe
Dr L. B. Knoll
J. S. van Vollenhoven
J. S. Coetzee
F. Nel

GENERAL MANAGER:
G. W. Ward

COMMERCIAL MANAGER:
W. H. Finlay

Atlas has completed manufacture of the Impala Mk 1 (M.B. 326M) jet trainer under licence from Aermacchi. Its present programmes include production of a developed Mk 2 version of the Impala and of the Atlas C4M Kudu, a STOL light transport aircraft which was developed in South Africa. All available details of these programmes follow.

Additionally, Atlas is manufacturing under licence components for the Dassault Mirage F1-CZ and -AZ multi-purpose combat aircraft currently in service with No. 3 Squadron of the South African Air Force. Atlas also undertakes the maintenance and overhaul of SAAF aircraft.

ATLAS IMPALA Mk 2

Manufacture by Atlas of the two-seat Aermacchi M.B. 326M for the South African Air Force has ended. This aircraft has the SAAF designation Impala Mk 1, and has been described in previous editions of *Jane's*. About 150 are reported to have been built.

The Impala Mk 2 is an improved version, based on the single-seat M.B. 326K, which has been developed by Atlas as an advanced trainer for the SAAF and is now in production. Four M.B. 326Ks are reported to have been delivered initially by Aermacchi. No details of the Impala Mk 2 have yet been released officially; a description of the Aermacchi M.B. 326 series can be found in the Italian section of this edition.

First production example of the single-seat Impala Mk 2, based on the Aermacchi M.B. 326K and built by Atlas Aircraft Corporation

ATLAS C4M KUDU

The Kudu is a six/eight-seat light transport aircraft, developed by Atlas in South Africa. It can be converted rapidly from the passenger to the freight role, and vice versa, and can operate from unprepared surfaces. The prototype (ZS-IZF) flew for the first time on 16 February 1974, and civil certification to FAR Pt 23 was granted on 16 June 1975. The third Kudu (ZS-IZG) flew shortly afterwards.

A military prototype flew for the first time on 18 June 1975, and was later handed over to the South African Air Force (SAAF serial number 961) for evaluation.

Kudu is the name of a very agile type of antelope found in southern Africa.

TYPE: Single-engined cabin monoplane.

WINGS: High-wing monoplane, with single bracing strut each side. Wing section NACA 23016 at root, NACA 4412 (modified) at tip. Dihedral 3°. Incidence 4° at root, 0° 27′ at tip. All-metal D-spar torsion-box structure. Electrically-operated Fowler flaps, of all-metal two-spar construction, interchangeable right with left. All-metal piano-hinged ailerons, with inset tabs.

FUSELAGE: All-metal stressed-skin structure, of basically rectangular section.

TAIL UNIT: Cantilever all-metal structure. Electrically-operated variable-incidence tailplane. Servo tab in each elevator.

LANDING GEAR: Non-retractable tailwheel type. Main units each have an independent cantilever leg, and are connected to oleo-pneumatic shock-absorbers mounted below cabin floor level in small underfuselage blister fairings. Main-wheel tyres size 7·00-8, 6-ply rating. Single-disc hydraulic brakes. Tailwheel, mounted in a castoring fork which incorporates a shock-absorber, has a size 5·00-4 6-ply tyre.

POWER PLANT: One 254 kW (340 hp) Piaggio-built Lycoming GSO-480-B1B3 flat-six engine, driving a

Hartzell HC-B3R20-4 three-blade constant-speed metal propeller with spinner. Three removable bag-type fuel tanks in each wing, total capacity 432 litres (95 Imp gallons).

ACCOMMODATION: Enclosed cabin, seating six to eight persons in standard version. Pilot and co-pilot side by side at front, with four individual seats in pairs at rear, or two bench seats each for three persons. Passenger seats can be removed to provide space for up to 560 kg (1,235 lb) of cargo. Heating and ventilation standard. Forward-hinged door on each side for pilot and co-pilot. Main cabin door is on port side, and is in two sections: forward-opening front section for passengers and light cargo, and rearward-opening rear section to supplement this when loading bulky cargo. A sliding door at the rear of the cabin on the starboard side may be opened in flight. A trap-door is provided in the cabin floor for aerial survey purposes.

SYSTEMS: Hydraulic system for main-wheel brakes only. 28V DC electrical system supplied by an engine-driven generator and a 24V 11Ah battery.

ELECTRONICS AND EQUIPMENT: Standard equipment includes instruments, beacon, and cabin, navigation and landing lights. Advanced instrumentation and electronic equipment to customer's specification, including duplicated VHF transceivers, intercom, HF transceiver and ADF.

DIMENSIONS, EXTERNAL:

Wing span	13·005 m (42 ft 8 in)
Wing chord at root	1·73 m (5 ft 8 in)
Wing chord at tip	1·17 m (3 ft 10 in)
Wing aspect ratio	8·07
Length overall (tail up)	9·04 m (29 ft 8 in)
Height overall:	
tail up	4·28 m (14 ft 0½ in)
tail down	2·74 m (9 ft 0 in)
Tailplane span	4·79 m (15 ft 8½ in)
Wheel track (aircraft unladen)	2·68 m (8 ft 9½ in)
Wheelbase	6·61 m (21 ft 8¼ in)
Propeller diameter	2·54 m (8 ft 4 in)
Propeller ground clearance	0·23 m (9 in)
Double doors (port):	
Height	1·04 m (3 ft 5 in)
Width	1·43 m (4 ft 8½ in)
Sliding door (stbd):	
Height	1·04 m (3 ft 5 in)
Width	0·74 m (2 ft 5 in)

DIMENSIONS, INTERNAL:

Cabin: Length, incl flight deck	3·33 m (10 ft 11 in)
Width	1·16 m (3 ft 9½ in)
Height	1·52 m (5 ft 0 in)

AREAS:

Wings, gross	20·91 m² (225·1 sq ft)
Ailerons (total)	2·62 m² (28·19 sq ft)
Trailing-edge flaps (total)	3·77 m² (40·62 sq ft)
Vertical tail surfaces (total)	2·57 m² (27·65 sq ft)
Horizontal tail surfaces (total)	5·48 m² (59·00 sq ft)

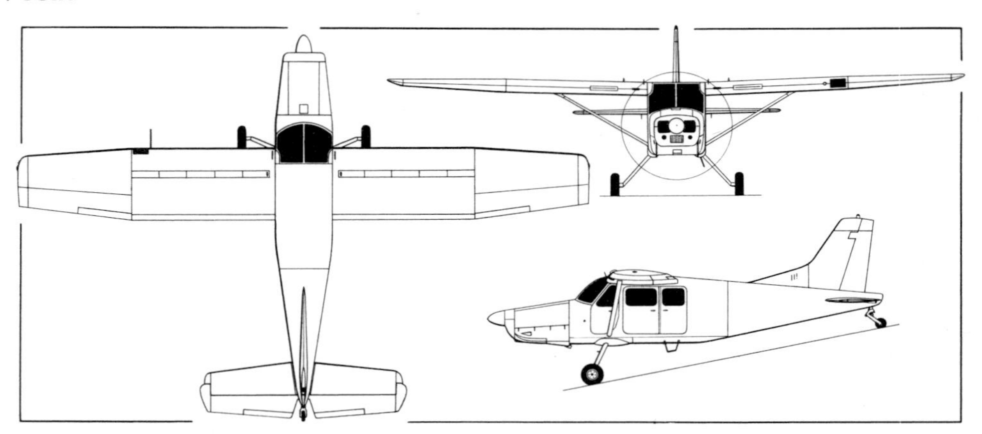

Atlas C4M Kudu six/eight-seat STOL utility light transport *(Michael A. Badrocke)*

Prototype Atlas C4M Kudu light transport (Lycoming GSO-480-B1B3 engine)

WEIGHTS AND LOADING:

Weight empty	1,200 kg (2,646 lb)
Basic operating weight, empty (five passengers plus baggage	1,448 kg (3,194 lb)
Max cargo payload	560 kg (1,235 lb)
Max T-O and landing weight	2,041 kg (4,500 lb)
Max wing loading	97·6 kg/m² (20 lb/sq ft)

PERFORMANCE (at max T-O weight):

Never-exceed speed	160 knots (296 km/h; 184 mph) CAS
Max level speed at 2,440 m (8,000 ft)	140 knots (259 km/h; 161 mph)
Max cruising speed at 3,050 m (10,000 ft)	125 knots (232 km/h; 144 mph)
Econ cruising speed at 3,050 m (10,000 ft)	105 knots (195 km/h; 121 mph)
Stalling speed, flaps up, power on	65 knots (121 km/h; 75 mph)
Stalling speed, flaps down, power on	48 knots (89 km/h; 55·5 mph)
Max rate of climb at S/L	244 m (800 ft)/min
Service ceiling	5,790 m (19,000 ft)
T-O run	215 m (705 ft)
T-O to 15 m (50 ft)	370 m (1,214 ft)
Landing from 15 m (50 ft)	260 m (853 ft)
Landing run	140 m (460 ft)
Range with max fuel, no reserves	700 nm (1,297 km; 806 miles)
Endurance with max fuel, no reserves	7 hr

CSIR

COUNCIL FOR SCIENTIFIC AND INDUSTRIAL RESEARCH (Aeronautics Research Unit, National Mechanical Engineering Research Institute)

ADDRESS:
PO Box 395, Pretoria 0001
Telephone: 74-6011

HEAD OF AERONAUTICS RESEARCH UNIT:
Dr C. G. van Niekerk, DSC(Eng), FRAeS

SENIOR PROJECT LEADERS:
Dr W. J. van der Elst, MRAeS (Low-speed aerodynamics)
M. S. Hunt, FRAeS (Aircraft structure)
A. J. van Wyk, MRAeS (High-speed aerodynamics and Aircraft propulsion)

Initially, the CSIR concentrated on the spin-off from aeronautical research. This work led to the establishment in 1952 of a small aerodynamics division, within a research unit which has since become the National Mechanical Engineering Research Institute. This division grew into what is now the Aeronautics Research Unit, established in 1968.

Current ARU activities include research into lifting rotors, airframe fatigue, synthetic materials, separation of underwing stores, aircraft and missile stability, atmospheric turbulence, and aircraft noise problems.

The ARU designed and is developing a two-seat experimental autogyro, which flew for the first time in late 1972.

CSIR SARA II

The CSIR (ARU) SARA II (South African Research Autogyro) was designed to have a minimum level flight speed of 23·5 knots (43 km/h; 27 mph), a maximum level speed of 86 knots (160 km/h; 99 mph), a rate of climb of 276 m (905 ft)/min and an endurance of 3 hr. The design was started in March 1965, and construction of a prototype began in April 1967. This aircraft (ZS-UGL), after tethered tests from a lorry platform, made its first free flight at Swartkop air force base, near Pretoria, on 30 November 1972, piloted by Capt J. H. Rautenbach.

CSIR SARA II (Lycoming O-360-A engine)

Following initial flight trials various modifications were made, and a description of the prototype in its November 1974 form appeared in the 1975-76 *Jane's*.

This aircraft was severely damaged in a ground handling accident shortly before it was due to appear in the Air Africa International display in 1975. A new, modified experimental aircraft is under construction; details of this had not been released at the time of closing for press.

The following is an abbreviated version of the description of the first prototype given in the 1975-76 *Jane's:*

TYPE: Two-seat experimental autogyro.

ROTOR SYSTEM: Single two-blade teetering rotor. Blades are of constant chord and NACA 8-H-12 section, each attached to hub by a single teeter hinge. Metal trim tab on each trailing-edge, near tip. No rotor brake.

FUSELAGE: Box-type structure of light alloy construction with fairings of glassfibre-reinforced plastics.

TAIL UNIT: Twin fins and rudders, bridged by a fixed-incidence tailplane and supported on twin strut-braced tailbooms.

LANDING GEAR: Non-retractable tricycle type. Shock-absorption by bungee rubber bands and nosewheel oleo leg. Nosewheel steerable and self-centering. Small skid beneath each fin.

POWER PLANT: One 134 kW (180 hp) Lycoming O-360-A flat-four engine, driving a Hartzell two-blade pusher propeller. Power take-off for rotor spin-up. Rubber bag-type fuel tanks in fuselage, capacity 136 litres (30 Imp gallons). Oil capacity 9 litres (2 Imp gallons).

ACCOMMODATION: Crew of two, with dual controls, on side-by-side seats in extensively-glazed cabin.

Forward-opening door, with glazed panels, on each side. Two spaces for baggage above and behind seats.

DIMENSIONS, EXTERNAL:

Rotor diameter	11·13 m (36 ft 6¼ in)
Propeller diameter	1·83 m (6 ft 0 in)
Length of fuselage	4·65 m (15 ft 3 in)
Height to top of rotor hub	2·80 m (9 ft 2¼ in)
Wheel track	2·45 m (8 ft 0½ in)
Wheelbase	2·00 m (6 ft 6¾ in)

AREAS:

Rotor disc	97·20 m² (1,046·25 sq ft)
Fins (total)	1·30 m² (13·99 sq ft)
Rudders (total)	0·60 m² (6·46 sq ft)
Tailplane	1·15 m² (12·38 sq ft)

MEASURED PERFORMANCE (up to mid-January 1975, at T-O weight of 771 kg; 1,700 lb and altitude of 1,465 m; 4,800 ft):

Cruising speed	65 knots (120 km/h; 75 mph)
Min level flight speed	26 knots (48 km/h; 30 mph)
Max rate of climb	140 m (460 ft)/min
T-O run, still air	27·5 m (90 ft)
T-O to 15 m (50 ft)	137 m (450 ft)

NATIONAL DYNAMICS

NATIONAL DYNAMICS (PTY) LTD

ADDRESS:

PO Box 20163, Virginia Airport, Durban North 4016

Telephone: Durban 835350

DIRECTORS:

S. J. Reed (Managing)

S. Reed (Marketing)

Capt J. H. Rautenbach, DFC, DFM (Operations)

This company was formed in 1975 following the acquisition in May of that year by Dr Maitland Reed of the prototype and all rights to the (formerly Patchen) Explorer/Observer four-seat cabin monoplane. Subject to successful certification and the receipt of a series production licence, it is intended to begin production of this aircraft in South Africa.

National Dynamics has also taken over from Air Nova (see 1975-76 *Jane's*) the Falcon aerobatic biplane. In addition, it is engaged in aviation systems manufacture, and in the sale of products of Schweizer Aircraft Corporation of the USA (which see).

NATIONAL DYNAMICS (PATCHEN) EXPLORER/OBSERVER

This aircraft was conceived by the former Thurston Aircraft Corporation (USA) as a landplane version of its TSC-1 Teal amphibian. The original development of both designs was financed by Marvin Patchen Inc, which retained rights to the TSC-2 Explorer when disposing of all rights in the Teal to Schweizer Aircraft Corporation (see US section). Marvin Patchen proposed developing the TSC-2 in two versions, one for civilian use and the other, named the Observer, for law enforcement.

Construction of the prototype Explorer was completed by Aerofab Corporation of Sanford, Maine, on a contract basis, and its initial flight test programme was completed successfully by November 1972. In early February 1973 a market survey was carried out to determine the special requirements of pipeline patrol operators, aerial photographers and law enforcement agencies; simultaneously, work towards certification was continued.

By February 1975 the Explorer prototype had completed 179 flight hours. In May 1975, Marvin Patchen sold the prototype and all rights to the Explorer to Dr Maitland Reed in South Africa where, in early 1976, progress towards certification under FAR Pt 23 was continuing. It is intended to put the aircraft in production once this is obtained.

TYPE: Four-seat cabin monoplane.

WINGS: Cantilever shoulder-wing monoplane. Wing section NACA 4415. Dihedral 1°. Incidence 4°. All-metal D-spar structure. Trailing-edge flaps and ailerons of light alloy construction.

FUSELAGE: All-metal semi-monocoque structure, with glassfibre nose section and cabin top skins. Transparent bubble nose to provide helicopter-like field of view.

TAIL UNIT: Cantilever all-metal T-tail. Trim tabs on elevator and rudder.

LANDING GEAR: Non-retractable tricycle type. Cantilever spring steel main-gear struts. Nose unit has oleo-pneumatic shock-absorber. Main wheels and tyres size 6·00-6; nosewheel and tyre size 5·00-5. Single-disc brakes.

POWER PLANT: One 149 kW (200 hp) Lycoming IO-360-A1A flat-four engine, pylon-mounted above the wing centre-section and driving a Hartzell two-blade metal constant-speed propeller. One 85 litre (22·5 US gallon) fuel tank in each wing leading-edge; total fuel capacity 170 litres (45 US gallons). One optional all-metal fuel tank in fuselage, capacity 94·6 litres (25 US gallons). Total optional fuel capacity 264·6 litres (70 US gallons).

ACCOMMODATION: Pilot and three passengers, seated in pairs, in enclosed cabin. Door on each side of fuselage, sliding fore and aft for cabin access and side photography. Cabin heated and ventilated.

ELECTRONICS AND EQUIPMENT: Panel designed to use Narco Spectrum radio. Optional equipment includes searchlight, stabilised optics slaved to searchlight, siren, PA system, camera mountings, STOL kit and 'quiet' kit.

DIMENSIONS, EXTERNAL:

Wing span	9·75 m (32 ft 0 in)
Wing chord (constant)	1·52 m (5 ft 0 in)
Length overall	7·11 m (23 ft 4 in)
Height overall	2·95 m (9 ft 8 in)
Tailplane span	3·05 m (10 ft 0 in)
Wheel track	2·36 m (7 ft 9 in)
Wheelbase	2·84 m (9 ft 4 in)
Propeller diameter	1·88 m (6 ft 2 in)

Prototype of the National Dynamics (Patchen) Explorer four-seat cabin monoplane in South African markings

AREAS:

Wings, gross	14·59 m² (157 sq ft)
Ailerons (total)	1·19 m² (12·8 sq ft)
Fin	0·99 m² (10·7 sq ft)
Dorsal fin	0·22 m² (2·4 sq ft)
Tailplane	1·77 m² (19·1 sq ft)
Elevators	1·46 m² (15·7 sq ft)

WEIGHTS AND LOADINGS:

Weight empty	621 kg (1,370 lb)
Max T-O and landing weight	998 kg (2,200 lb)
Max wing loading	68·3 kg/m² (14·0 lb/sq ft)
Max power loading	8·20 kg/kW (11·00 lb/hp)

PERFORMANCE (at max T-O weight):

Max cruising speed	111 knots (206 km/h; 128 mph)
*Stalling speed, flaps up, power off	46 knots (85·5 km/h; 53 mph)
*Stalling speed, flaps down, power off	40 knots (74 km/h; 46 mph)
Max rate of climb at S/L	396 m (1,300 ft)/min
T-O to 15 m (50 ft)	271 m (890 ft)
Landing from 15 m (50 ft)	225 m (738 ft)
Range with max fuel at 75% power	770 nm (1,427 km; 887 miles)

**Prototype will not stall power on, with wings level, at any angle of attack*

NATIONAL DYNAMICS (REED) FALCON

As described in the 1975-76 *Jane's*, the Falcon is a redesign of the modified Rooivalk prototype ZS-UDU (1972-73 *Jane's*), designed by Dr Maitland Reed and Capt J. H. Rautenbach, construction of which began in November 1972. It is basically a single-seat aircraft, designed for unlimited aerobatics, but is to be made available also as a two-seater.

The wings of the prototype are the outer panels from the original Rooivalk, fitted with balanced ailerons; the fuselage is shorter than before, but wider.

The following description applies to the single-seat prototype Falcon, which was acquired from the University of Natal in April 1974:

TYPE: Single-seat aerobatic biplane.

WINGS: Strut-braced biplane. Wing section NACA 0012. No anhedral or dihedral. Incidence 0°. Sweepback 10° at quarter-chord. Sitka spruce main spars and ribs, covered with Dacron. Single I bracing strut each side, faired to upper and lower wings; centre-section of upper wing supported by a pair of inverted Vee struts forward of cockpit. Internally-balanced ailerons, of spruce and ply, on both upper and lower wings. No tabs.

FUSELAGE: Welded chrome-molybdenum steel tube (4130) structure, Dacron-covered.

TAIL UNIT: Single fin and rudder, with wire-braced tailplane. Welded steel tube structure, Dacron-covered. Trim tab in each elevator. No rudder tab.

LANDING GEAR: Non-retractable tailwheel type. Vee-type independently-sprung main legs, with rubber-cord shock-absorption. Cleveland main wheels and tyres, size 5·00-5, pressure 2·07 bars (30 lb/sq in). Cleveland hydraulic brakes on main units. Streamline fairings over main wheels.

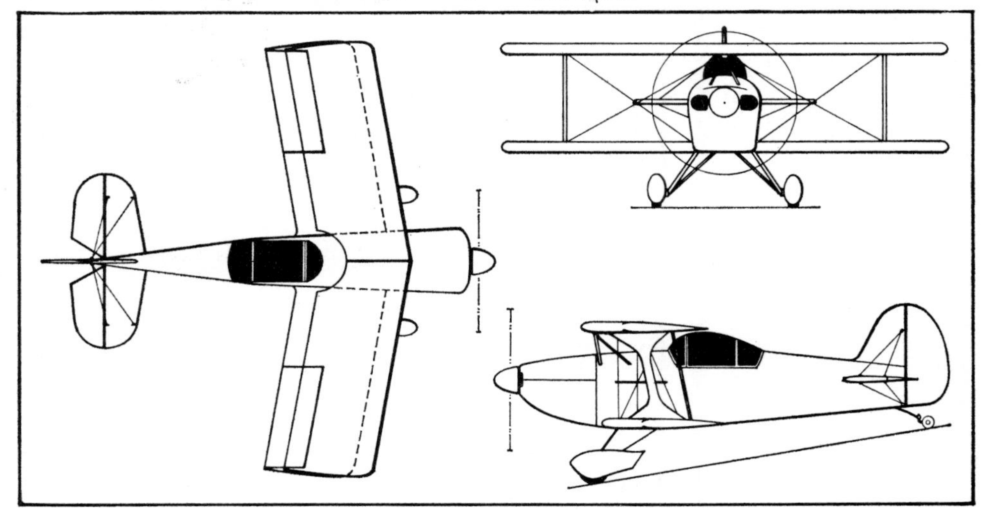

National Dynamics (Reed) Falcon single-seat aerobatic biplane *(Roy J. Grainge)*

POWER PLANT: One 156·5 kW (210 hp) Continental IO-360-C flat-six engine, driving a two-blade constant-speed propeller with spinner. Three fuel tanks in fuselage, total capacity 127 litres (28 Imp gallons), including one 63·6 litre (14 Imp gallon) tank for ferry purposes. Oil capacity 9·1 litres (2 Imp gallons).

ACCOMMODATION: Single cockpit, with fully-transparent rearward-sliding canopy. Design can be adapted to two-seat configuration.

DIMENSIONS, EXTERNAL:

Wing span (upper and lower)	5·87 m (19 ft 3 in)
Wing chord (constant, each)	1·17 m (3 ft 10 in)
Length overall	5·87 m (19 ft 3 in)
Height overall	2·13 m (7 ft 0 in)

AREA:

Wings (total)	12·45 m² (134·0 sq ft)

WEIGHTS:

Weight empty	408 kg (900 lb)
Max T-O weight	657 kg (1,450 lb)

PERFORMANCE (estimated, 156·5 kW; 210 hp engine):

Never-exceed speed	195 knots (362 km/h; 225 mph)
Max level speed	148 knots (274 km/h; 170 mph)
Stalling speed	51·5 knots (95 km/h; 59 mph)
Max rate of climb at S/L	732 m (2,400 ft)/min
Range with max fuel	450 nm (830 km; 515 miles)

VLIÔM

VLIÔM DEVELOPMENT CO (PTY) LTD

ADDRESS:
PO Box 50548, Randburg 2125, Transvaal
Telephone: 706-3197

DIRECTORS:
Dr I. E. Bock
J. N. Bock

VLIÔM

The Vliôm, designed by Dr I. E. Bock and now undergoing development, is a novel form of VTOL vehicle with potential civil applications as a personal aircraft, sports vehicle or commuter transport. The craft is being designed to meet the very high demands of safety, simplicity of control, and economy required for a commuter air vehicle. The name Vliôm is derived from the Afrikaans *Vlieg Hom* (Fly it).

The airframe of the Vliôm consists basically of a shrouded-propeller lift/propulsion system, comprising a vertically-mounted engine and a propeller turning in the horizontal plane, the propeller and craft tilting slightly during forward movement, like a helicopter. The flat shroud-ring around the propeller gives a 60% augmentation in the thrust of the propeller, making possible the use of a small-diameter high-speed propeller. The propeller is fixed-pitch and non-articulated, for reliability and economy of servicing.

The flat shroud-ring has a leading-edge duct which is designed in such a way that it does not affect thrust augmentation during hovering, but controls and neutralises the jet-flap interaction between the propeller and the shroud-ring during forward motion. A sealing lip at the bottom of the shroud-ring, below the propeller tip, makes it possible to increase the propeller-tip/shroud clearance without reducing the shroud's efficiency or increasing the torque demand for yaw stabilisation.

Yaw stabilisation is obtained by louvres placed below the propeller and attached to the shroud-ring. Roll and pitch control are achieved by actuating flaps in the throat of the shroud-ring; such control can also be achieved by differential movement of three sets of equally-spaced louvres placed below the propeller. The flaps in the shroud-ring give full roll control of the propeller during forward movement with the propeller turning essentially in a horizontal plane. The propeller and engine are the only moving parts in the vehicle during steady flight.

Rear parts of the shroud-ring can be opened outward on each side of the craft to form a V-shaped glider wing with the forward part of the shroud-ring; the craft can then descend as a glider in a power failure situation.

The engine and cabin for the crew member(s) are suspended from the ring structure. Landing gear can be varied to suit specific functional requirements.

Design of the Vliôm began in 1967, and powered model tests started in 1971. The present concept has evolved from the results of these tests, and from the fundamental functional needs of a commuter air vehicle. More recent tests, conducted with a 1/10th scale model, have concentrated upon the fundamental aerodynamics and the inherent stability and simplicity of control of the craft. The main work is now concerned primarily with the safety of the craft during the phase immediately after take-off, before it achieves forward speed.

Design studies give the following parameters for a four-seat commuter version, powered by an 89·5 kW (120 hp) lightweight piston engine driving a two-blade fixed-pitch propeller:

DIMENSIONS, EXTERNAL:

Diameter of shroud-ring	4·27 m (14 ft 0 in)
Diameter of propeller	2·44 m (8 ft 0 in)
Height overall	2·44 m (8 ft 0 in)
Height of propeller above ground	2·13 m (7 ft 0 in)
Wheel track	2·44 m (8 ft 0 in)

WEIGHTS (approx):

Weight empty	363 kg (800 lb)
Max T-O weight	726 kg (1,600 lb)

PERFORMANCE (estimated):

Max level speed	104 knots (193 km/h; 120 mph)
Max cruising speed (suburban use)	61 knots (112 km/h; 70 mph)

The 1/10th scale model of the Vliôm photographed from the front, starboard side and above, respectively

SPAIN

AISA

AERONAUTICA INDUSTRIAL SA

HEAD OFFICE:
Plaza de las Cortes 2, Apartado 984, Madrid 14
Telephone: 222 75 80
Telex: 23593 E Madrid

WORKS AND AERODROME:
Cuatro Vientos (Carabanchel Alto), Madrid 25
Telephone: 208 52 40 and 208 96 40

PRESIDENT:
Juan Echevarria

MANAGING DIRECTOR:
Gonzalo Suárez

TECHNICAL MANAGER:
José A. Delgado

AIRCRAFT DESIGN MANAGER:
Juan del Campo

SALES MANAGER:
Rodrígo García

FINANCIAL MANAGER:
Manuel Algarra

This concern has since 1923 been engaged in the manufacture, repair and maintenance of fixed-wing aircraft and helicopters.

During recent years, the AISA design office has been responsible for several liaison, training and sporting aircraft, including the I-11, I-11B, AVD-12 (in collaboration with M Dewoitine) and I-115.

AISA is also engaged in IRAN repair and maintenance of several types of US aircraft, in particular the North American T-6, Beechcraft B55, C90 and F33 Bonanza, and Piper PA-23 and PA-31 aircraft operated by the Spanish Air Force.

Since 1962, AISA has been awarded several US government contracts for IRAN repair work on Sikorsky S-55 (H-19) and S-58 (H-34) helicopters. It is also engaged on the repair and overhaul of Bell 47, 204 and 205, Hughes

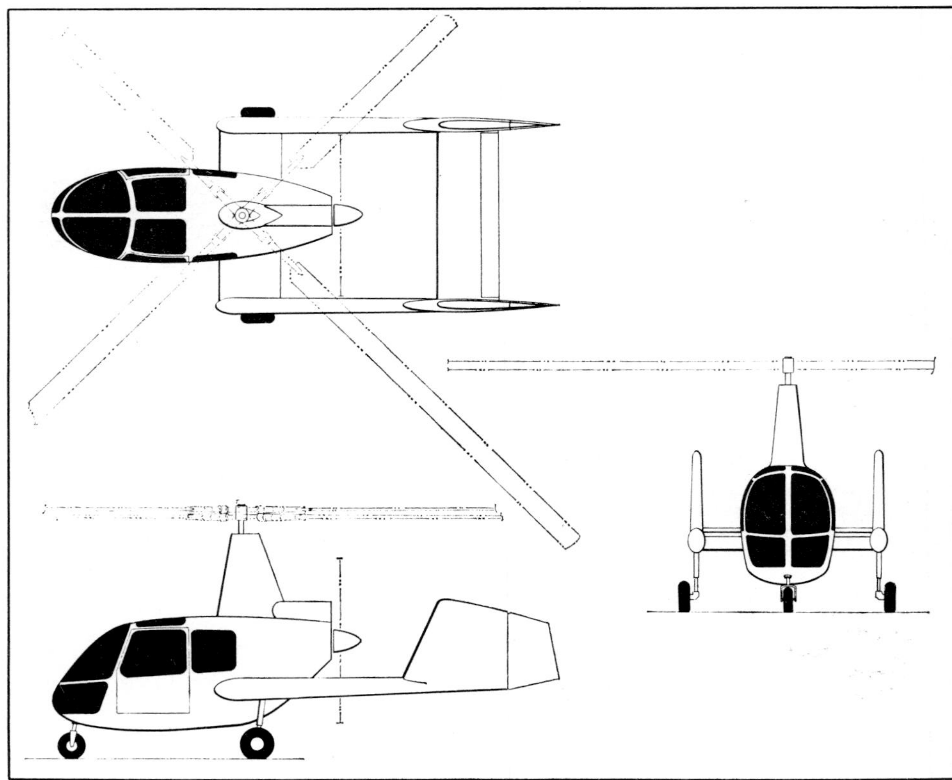

AISA autogyro GN (Lycoming IO-540-K1A5 engine) *(Roy J. Grainge)*

300 and Boeing Vertol CH-47 helicopters of the Spanish Army, Spanish Air Force and civilian operators.

The present facility at Cuatro Vientos has a covered area of 40,000 m² (430,556 sq ft) and employs a work force of some 775 persons.

AISA AUTOGYRO GN

AISA, which in 1927 built some of the earliest Cierva Autogiros, is currently working on a new autogyro, of which design was initiated on 17 June 1970. Construction of the prototype was scheduled to begin in July 1976, with first flight planned for 1977.

The GN, which will have a jump take-off capability, will be powered by a 224 kW (300 hp) Lycoming IO-540-K1A5 flat-six engine, driving a two-blade constant-speed Hartzell pusher propeller. It will have accommodation for a pilot and three passengers, with a single door on each side of the cabin. A four-blade articulated rotor, with rotor brake, will be fitted, and the aircraft will have short-span wings of NACA 0024 section. The pod-shaped fuselage will be an aluminium alloy semi-monocoque structure, with moulded Plexiglas transparencies; the twin tail-booms, tailplane, elevator and twin fins and rudders will also be of all-metal construction. A non-retractable tricycle landing gear will be fitted, with single wheel on each unit.

DIMENSIONS, EXTERNAL:	
Rotor diameter	12·80 m (42 ft 0 in)
Rotor blade chord (each)	0·28 m (11 in)
Wing span	2·60 m (8 ft 6¼ in)
Wing chord	0·85 m (2 ft 9½ in)
Length of fuselage	6·50 m (21 ft 4 in)
Height overall	3·20 m (10 ft 6 in)
Propeller diameter	2·13 m (7 ft 0 in)
Doors (each):	
Height	1·085 m (3 ft 6¾ in)
Width	0·93 m (3 ft 0½ in)
AREA:	
Rotor disc	128·68 m² (1,385·1 sq ft)
WEIGHTS:	
Weight empty	708 kg (1,560 lb)
Max payload	415 kg (915 lb)
Max T-O weight	1,200 kg (2,645 lb)

PERFORMANCE (estimated, at max T-O weight):	
Max level speed at S/L	129 knots (240 km/h; 149 mph)
Cruising speed at S/L	114 knots (212 km/h; 132 mph)
Min level speed	24 knots (44 km/h; 27·5 mph)
Max rate of climb at S/L	390 m (1,280 ft)/min
Landing run (zero wind)	0-5 m (0-16·5 ft)
Range with max fuel	431 nm (800 km; 497 miles)

CASA
CONSTRUCCIONES AERONAUTICAS SA

HEAD OFFICE:
Rey Francisco 4, Apartado 193, Madrid 8
Telephone: 247 25 00
Telex: 27418

WORKS:
Getafe, Seville, San Pablo, Cádiz, Madrid and Ajalvir

HONORARY PRESIDENT:
José Ortiz Echagüe

CHAIRMAN AND PRESIDENT:
Dr Emilio González García

VICE-CHAIRMAN OF THE BOARD:
Dr Eugenio Aguirre Castillo

MANAGING DIRECTOR:
Dr Enrique de Guzmán Ozámiz

GENERAL SECRETARY:
Dr Carlos Marín Jiménez-Ridruejo

DIRECTOR OF MARKETING:
Dr Fernando de Caralt

PUBLIC RELATIONS:
Domingo Balaguer

This company was formed in March 1923 for the primary purpose of producing metal aircraft for the Spanish Air Force. It began by building under licence the Breguet XIX and has since manufactured many other aircraft of foreign design. The most recent of these was the Northrop F-5 fighter.

CASA's own Project Office has designed several aircraft under contract to the Spanish Air Ministry, including the C.212 Aviocar transport, which is currently in production, and the C-101 jet trainer currently in the design and development stage. It also undertakes design and development work for foreign companies and, for example, collaborated in the design of the MBB HFB 320 Hansa light twin-jet executive transport; MBB in turn co-operated in the design of the Aviocar.

Under contract to Dassault-Breguet (which see), CASA is responsible for manufacturing sections of the Mercure transport (forward fuselage up to the wing attachment point) and Falcon 10 light business aircraft (outer wings). CASA is a full member (4·2%) of Airbus Industrie (see International section), and manufactures the horizontal tail surfaces, landing gear doors and forward passenger doors for the Airbus A300 wide-bodied transport aircraft.

CASA undertakes maintenance and modernisation work for the Spanish Air Force and for the US Air Force in Europe. Its principal current activities of this kind concern overhaul and maintenance of McDonnell Douglas F-4 combat aircraft and Bell 47G, 204 and 205 and Sikorsky H-19 helicopters.

In 1972 the former Hispano Aviación SA (see 1972-73 *Jane's*) was merged with CASA, the latter company taking over all of Hispano's offices and other facilities, aircraft production programmes and personnel. In June 1973 ENMASA (Empresa Nacional de Motores de Aviación) was merged into CASA, and now constitutes the CASA División de Motores (see Aero-Engines section). CASA has six factories, employing a total of 7,412 people in early 1976. Including Hispano production, the company had by early 1976 manufactured 3,500 aircraft and overhauled 5,754. CASA has a total covered area in the region of 197,000 m² (2,120,490 sq ft); majority shareholder in the company is the INI (Instituto Nacional de Industria).

CASA C.212 AVIOCAR

Spanish Air Force designation: T.12

The C.212 Aviocar twin-turboprop light utility STOL transport was evolved by CASA to fulfil a variety of military or civil roles, but primarily to replace the mixed fleet of Junkers Ju 52/3m (T.2), Douglas DC-3 (T.3) and CASA-207 Azor (T.7) transport aircraft formerly in service with the Spanish Air Force.

The C.212 is able to fill six main roles—as a 16-seat paratroop transport, military freighter, ambulance, photographic aircraft, crew trainer or 19-seat passenger transport—and has been certificated to joint military and civil standards by the Instituto Nacional de Técnica Aeroespacial (INTA), which was also responsible for the flight test programme. It has a STOL capability that enables it to use unprepared landing strips about 400 m (1,310 ft) in length, and has been optimised for operation in remote areas with a poor infrastructure.

CASA was awarded a contract by the Ministerio del Aire for the development and construction of two flying prototypes of the Aviocar, and one structural test airframe, on 24 September 1968. The first prototype flew for the first time on 26 March 1971, the second on 23 October 1971.

Eight pre-production Aviocars were ordered initially by the Spanish Air Ministry; the first of these made its first flight on 17 November 1972, and all had flown by February 1974.

By early 1976 a total of 88 Aviocars had been sold, and 52 of these had been delivered. Production was then at the rate of four aircraft per month. Two demonstration aircraft are owned by CASA.

In order to promote sales in the Far East, CASA has negotiated the establishment of a C.212 assembly line in Indonesia, as well as full after-sales support in that area.

The following versions have been announced:

C.212A (Spanish Air Force designation T.12B). Military utility transport version, ordered by the air forces of Indonesia (7), Jordan (3), Portugal (20) and Spain (34). First Spanish Air Force squadron is No. 461, based at Gando in the Canary Islands.

C.212B (Spanish Air Force designation TR.12A). Photographic survey version. Six of the eight pre-production Aviocars were completed to this configuration for the Spanish Air Force, and are each equipped with two Wild RC-10 aerial survey cameras. Four others have been ordered, by the Portuguese Air Force.

C.212C. Commercial transport version. Nine ordered by the Indonesian company Pertamina, for operation by Pelita Air Service. First example (PK-PCK) delivered in July 1975. The Spanish and Jordanian Air Forces also have each ordered one C.212C for VIP transport duties.

C.212E. Navigation trainer version. Two of the eight pre-production Aviocars were completed to this configuration for the Spanish Air Force.

TYPE: Twin-turboprop STOL utility transport.

WINGS: Cantilever high-wing monoplane. Wing section NACA 65_3-218. Incidence 2° 30′. No dihedral or sweepback. All-metal light alloy fail-safe structure. All-metal ailerons and double-slotted trailing-edge flaps. Trim tab in port aileron. Rubber-boot de-icing of leading-edges.

FUSELAGE: Semi-monocoque fail-safe structure of light alloy construction.

TAIL UNIT: Cantilever two-spar all-metal structure, with dorsal fairing forward of fin. Tailplane mid-mounted on rear of fuselage. Trim tab in rudder and each elevator. Rubber-boot de-icing of leading-edges.

LANDING GEAR: Non-retractable tricycle type, with single main wheels and single steerable nosewheel. CASA oleo-pneumatic shock-absorbers. Dunlop wheels and tyres, main units size 11·00-12 (8-ply) Type III, nose unit size 8·00-7 Type III. Tyre pressure (all units) 3·10 bars (45 lb/sq in). Dunlop hydraulic disc brakes on main wheels.

POWER PLANT: Two 579 kW (776 ehp) AiResearch TPE 331-5-251C turboprop engines, each driving a Hartzell HC-B4TN-5CL/LT10282HB+4 four-blade constant-speed (Beta-mode on ground) metal propeller (three-blade HC-B3TN-5E in prototype and early pre-series aircraft). Fuel in four outer-wing tanks, with total capacity of 2,100 litres (462 Imp gallons). Oil capacity 6 litres (1·32 Imp gallons) per engine.

ACCOMMODATION: Crew of two on flight deck. For the paratroop role, the main cabin can be fitted with 16 inward-facing seats along the cabin walls, to accommodate 15 paratroops and an instructor/jumpmaster. As an ambulance, the cabin would normally be equipped to carry 12 stretcher patients and 3 sitting casualties, plus medical attendants. As a freighter, the Aviocar can carry up to 2,000 kg (4,410 lb) of cargo in the main cabin, including light vehicles. Photographic version is

Second CASA C.212C Aviocar commercial transport aircraft in the insignia of Pelita Air Service of Indonesia

equipped with two cameras and a darkroom. Aircrew training version accommodation consists of individual desks for an instructor and five pupils, in two rows, fitted with appropriate instrument installations. The civil passenger transport version has standard seating for 19 persons in five rows of three (one to port and two to starboard of centre aisle) at 787 mm (31 in) pitch, plus two rows of two seats. Access to main cabin is via two doors on the port side, one aft of (and providing access to) the flight deck and one aft of the wing trailing-edge. In addition, there is a two-section underfuselage loading ramp/door aft of the main cabin; this door is openable in flight for the discharge of paratroops or cargo, and is fitted with external wheels, to allow the door to remain open during ground manoeuvring. There is an emergency exit door aft of the wing trailing-edge on the starboard side. All versions have a toilet at the forward end of the main cabin on the starboard side, with a baggage compartment opposite on the port side. In the civil transport version, the interior of the rear-loading door can be used for additional baggage stowage.

SYSTEMS: Unpressurised cabin. Hydraulic system, pressure 138 bars (2,000 lb/sq in), operates main-wheel brakes, flaps and nosewheel steering. Electrical system is supplied by two 3kW starter/generators.

ELECTRONICS AND EQUIPMENT: Radio and radar equipment includes Bendix RTA 41B VHF, AN/ARC-34C UHF, VOR/ILS and one ADF. Blind-flying instrumentation standard. Optional equipment includes Tacan, SIF/IFF, Collins 618S-4 HF and a second ADF.

DIMENSIONS, EXTERNAL:
- Wing span: 19·00 m (62 ft 4 in)
- Wing chord at root: 2·50 m (8 ft 2½ in)
- Wing chord at tip: 1·25 m (4 ft 1¼ in)
- Wing aspect ratio: 9
- Length overall: 15·20 m (49 ft 10½ in)
- Height overall: 6·30 m (20 ft 8 in)
- Tailplane span: 7·40 m (24 ft 3¼ in)
- Wheel track: 3·10 m (10 ft 2 in)
- Wheelbase: 5·45 m (17 ft 10½ in)
- Propeller diameter: 2·73 m (8 ft 11½ in)
- Distance between propeller centres: 5·30 m (17 ft 4¾ in)
- Passenger door (port, aft):
  - Max height: 1·58 m (5 ft 2¼ in)
  - Max width: 0·70 m (2 ft 3½ in)
- Crew and servicing door (port, fwd):
  - Max height: 1·10 m (3 ft 7¼ in)
  - Max width: 0·60 m (1 ft 11⅝ in)
- Rear-loading door:
  - Max length: 4·00 m (13 ft 1½ in)
  - Max width: 1·70 m (5 ft 7 in)

DIMENSIONS, INTERNAL:
- Cabin (between flight deck and rear-loading door):
  - Length: 5·00 m (16 ft 4¾ in)
  - Width: 2·00 m (6 ft 6¾ in)
  - Height: 1·70 m (5 ft 7 in)
  - Volume: 17·5 m^3 (618 cu ft)

AREAS:
- Wings, gross: 40·0 m^2 (430·56 sq ft)
- Ailerons (total): 2·45 m^2 (26·37 sq ft)
- Trailing-edge flaps (total): 7·38 m^2 (79·44 sq ft)
- Fin: 4·25 m^2 (45·75 sq ft)
- Rudder, incl tab: 2·02 m^2 (31·74 sq ft)
- Tailplane: 7·36 m^2 (79·22 sq ft)
- Elevators, incl tabs: 3·56 m^2 (38·32 sq ft)

WEIGHTS AND LOADINGS:
- Manufacturer's weight empty: 3,700 kg (8,157 lb)
- Weight empty, equipped: 3,905 kg (8,609 lb)
- Max payload: 2,000 kg (4,410 lb)
- Max T-O weight: 6,300 kg (13,889 lb)
- Max zero-fuel weight: 5,875 kg (12,952 lb)
- Max landing weight: 6,100 kg (13,448 lb)
- Max wing loading: 157·5 kg/m^2 (32·3 lb/sq ft)
- Max power loading: 5·44 kg/kW (9·19 lb/ehp)

PERFORMANCE (at max T-O weight except where indicated):
- Never-exceed speed: 240 knots (445 km/h; 276 mph) EAS
- Max level speed at 3,660 m (12,000 ft): 199 knots (370 km/h; 230 mph)
- Max cruising speed at 3,660 m (12,000 ft): 194 knots (359 km/h; 223 mph)
- Econ cruising speed at 3,660 m (12,000 ft): 170 knots (315 km/h; 196 mph)
- Stalling speed, flaps up, AUW of 6,100 kg (13,448 lb): 72 knots (133 km/h; 83 mph)
- Stalling speed, flaps down: 62 knots (116 km/h; 72 mph)
- Max rate of climb at S/L: 548 m (1,800 ft)/min
- Max rate of climb at S/L, one engine out: 106 m (350 ft)/min
- Service ceiling: 8,140 m (26,700 ft)
- Service ceiling, one engine out: 4,115 m (13,500 ft)
- T-O run: 350 m (1,148 ft)
- T-O to 15 m (50 ft): 484 m (1,588 ft)
- Landing from 15 m (50 ft): 385 m (1,263 ft)
- Landing run: 207 m (679 ft)
- Range at 3,660 m (12,000 ft):
  - with max fuel and 1,045 kg (2,303 lb) payload: 949 nm (1,760 km; 1,093 miles)
  - with max payload: 258 nm (480 km; 298 miles)

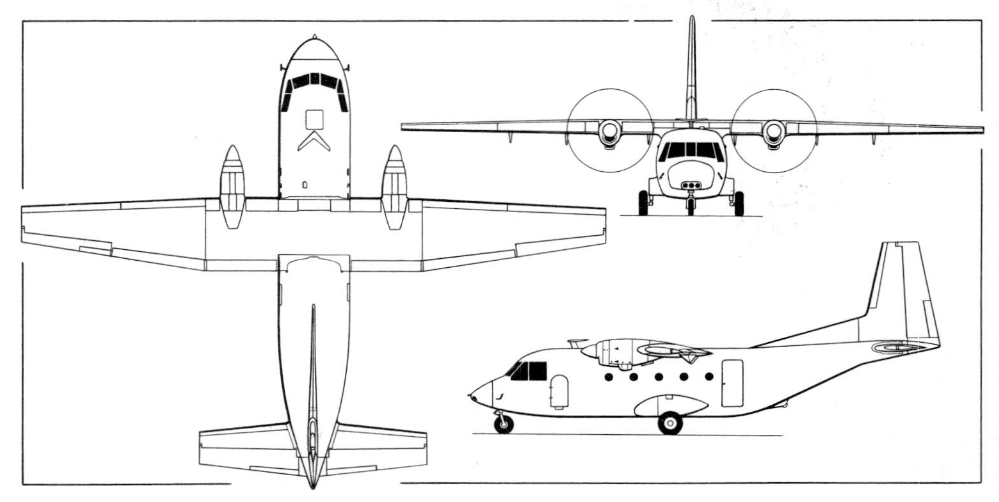

CASA C.212 Aviocar twin-turboprop light transport aircraft *(Pilot Press)*

CASA C-101

On 16 September 1975, CASA and the Spanish Ministerio del Aire signed a development contract for a new basic and advanced military jet trainer aircraft. This has been given the manufacturer's designation C-101. The contract, worth 1,297 million pesetas ($22 million), covers design, development, and the construction of six prototype aircraft of which four are for flight test and two for structural testing. First flight is anticipated in 1977.

To minimise cost and maintenance problems, the C-101 will be built on modular lines, with ample space within the airframe for equipment for any training mission likely to be required in the 1980s. The C-101 will also have the capability of carrying out such additional duties as ground attack, reconnaissance, escort, weapons training, electronic countermeasures (ECM), and photographic missions. Seven external hardpoints will permit the aircraft to carry up to 2,000 kg (4,410 lb) of stores.

The general appearance of the C-101 is shown in the accompanying illustration. Northrop (USA) and MBB (Germany) are to collaborate with CASA in the development programme.

TYPE: Tandem two-seat basic and advanced trainer and light tactical aircraft.

WINGS: Cantilever low-wing monoplane. Wing section NORCASA-15, thickness/chord ratio 15%. Dihedral 5°. Incidence 1°. Sweepback at quarter-chord 1° 53′. All-metal (aluminium alloy) three-spar fail-safe stressed-skin structure. Plain ailerons and slotted trailing-edge flaps, of glassfibre/honeycomb sandwich construction, actuated hydraulically. No tabs.

FUSELAGE: All-metal semi-monocoque fail-safe structure. Hydraulically-operated metal airbrake under centre of fuselage.

TAIL UNIT: Cantilever all-metal structure, with variable-incidence tailplane. Elevators of aluminium honeycomb construction. No tabs.

LANDING GEAR: Hydraulically-retractable tricycle type, with single wheel on each unit. Nose unit retracts forward, main units inward. Oleo-pneumatic shock-absorbers.

POWER PLANT: One Garrett-AiResearch TFE 731-2 or TFE 731-3 non-afterburning turbofan engine (15·57 kN; 3,500 lb st or 16·46 kN; 3,700 lb st), installed in rear fuselage. Lateral intake on each side of fuselage, abreast of second cockpit. Fuel in one 1,125 litre (247 Imp gallon) fuselage bag tank, one 585 litre (129 Imp gallon) integral tank in wing centre-section, and two outer-wing integral tanks each of 345 litres (76 Imp gallons). Total fuel capacity 2,400 litres (528 Imp gallons). Refuelling point in port air intake. Oil capacity 8·5 litres (1·8 Imp gallons).

ACCOMMODATION: Crew of two in tandem, on Martin-Baker Mk 4 ejection seats, under individual canopies which open sideways to starboard. Cockpit pressurised and air-conditioned. Dual controls standard.

SYSTEMS: Hamilton Standard three-wheel bootstrap-type air-conditioning and pressurisation system, differential 0·28 bars (4·1 lb/sq in). Single hydraulic system, pressure 207 bars (3,000 lb/sq in), for landing gear, flaps, airbrake and parking brake. Pneumatic system for air-conditioning, pressurisation and canopy seal. 9kW electrical system. High-pressure oxygen system. De-icing for engine air intakes.

ELECTRONICS AND EQUIPMENT: AN/ARC-164 UHF; Magnavox CA-657 VHF; Bendix AN/ARN-127 VOR; Collins AN/ARN-118 Tacan; Bendix AN/APX-100 IFF/SIF.

DIMENSIONS, EXTERNAL:
- Wing span: 10·60 m (34 ft 9⅜ in)
- Wing chord at root: 2·40 m (7 ft 10½ in)
- Wing chord at tip: 1·40 m (4 ft 7 in)
- Wing aspect ratio: 5·6
- Length overall: 12·25 m (40 ft 2¼ in)
- Height overall: 4·30 m (14 ft 1¼ in)
- Tailplane span: 4·30 m (14 ft 1¼ in)
- Wheel track: 3·20 m (10 ft 6 in)
- Wheelbase: 4·90 m (16 ft 1 in)

AREAS:
- Wings, gross: 20·00 m^2 (215·3 sq ft)
- Ailerons (total): 1·60 m^2 (17·22 sq ft)
- Trailing-edge flaps (total): 2·50 m^2 (26·91 sq ft)
- Fin: 4·70 m^2 (50·59 sq ft)
- Rudder: 1·50 m^2 (16·15 sq ft)
- Tailplane: 4·40 m^2 (47·36 sq ft)
- Elevators: 1·40 m^2 (15·07 sq ft)

WEIGHTS:
- Basic operating weight empty: 2,980 kg (6,570 lb)
- Max T-O and landing weight: 4,700 kg (10,360 lb)

PERFORMANCE (estimated, at 4,600 kg; 10,140 lb AUW):
- Max limiting Mach No.: Mach 0·80
- Max level speed at 7,620 m (25,000 ft): 429 knots (795 km/h; 494 mph) TAS
- Stalling speed, flaps up: 102·5 knots (190 km/h; 118 mph) EAS
- Stalling speed, flaps down: 84 knots (155 km/h; 96·5 mph) EAS

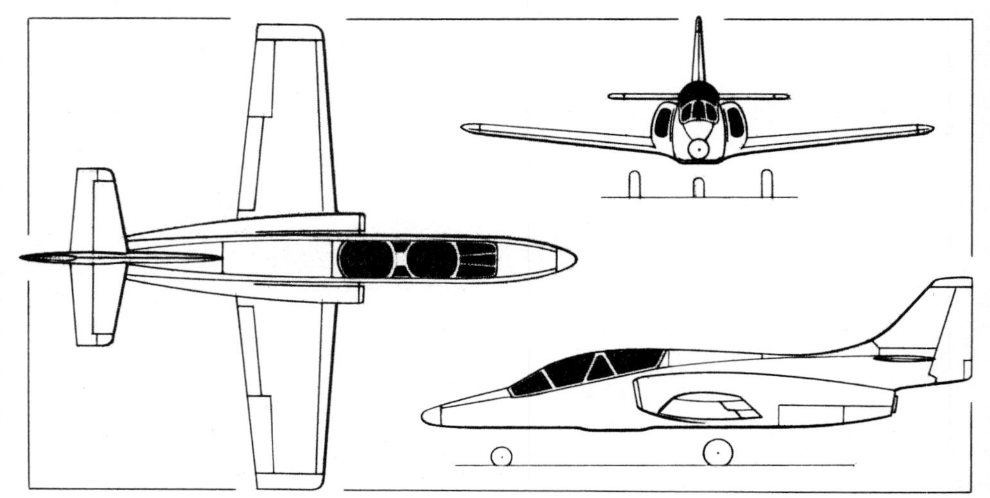

CASA C-101 basic/advanced training aircraft *(Pilot Press)*

Max rate of climb at S/L	1,116 m (3,660 ft)/min
Service ceiling	13,100 m (43,000 ft)
T-O run	675 m (2,215 ft)
T-O to 15 m (50 ft)	900 m (2,950 ft)
Landing from 15 m (50 ft)	660 m (2,165 ft)
Landing run	360 m (1,180 ft)

CASA FLAMINGO

The former Hispano Aviación acquired from MBB in Germany sole rights for further manufacture of the latter's MBB 223 Flamingo light aircraft. The first of 50 Hispano-built Flamingos was flown for the first time on 14 February 1972. Of these, four were delivered to the Spanish Under-Secretary of Civil Aviation, 30 or more to the Syrian Air Force, and three are believed to have been delivered to the Spanish Air Force.

A full description of the Flamingo can be found in the 1974-75 *Jane's.*

SWEDEN

SAAB-SCANIA

SAAB-SCANIA AKTIEBOLAG

HEAD OFFICE:
S-581 88 Linköping
Telephone: Int. +46 13 11 54 00
Telex: 50040 SAABLGS

PRESIDENT:
Curt Mileikowsky

EXECUTIVE VICE-PRESIDENTS:
T. Arnheim
T. Lidmalm

Aerospace Division
Telephone: Int. +46 13 12 90 20
HEAD OF DIVISION: T. Gullstrand
HEAD OF AIRCRAFT SECTOR: H. Schröder
HEAD OF MISSILE AND ELECTRONICS SECTOR:
I. K. Olsson
INFORMATION: Hans G. Andersson

The original Svenska Aeroplan AB was founded at Trollhättan in 1937 for the production of military aircraft. In 1939 this company was amalgamated with the Aircraft Division (ASJA) of the Svenska Järnvägsverkstäderna rolling stock factory in Linköping. Following this merger, Saab moved its head office and engineering departments to Linköping, which has since become the company's main factory.

In 1950, Saab acquired a factory at Jönköping for development and manufacture of airborne equipment. Other post-war expansions include a bombproof underground factory in Linköping, as well as important new production and engineering facilities in Linköping, Jönköping, Trollhättan and Gothenburg. The company's name was changed to Saab Aktiebolag in May 1965.

During 1968 a decision was taken to merge the company with another large Swedish automotive concern, Scania-Vabis, to strengthen the two companies' position in automotive product development, production and export. In that year also, Malmö Flygindustri (MFI) was acquired.

The present Saab-Scania company has more than 37,000 employees, organised in five operating divisions. Of these, more than 5,000 are employed by the Aerospace Division.

Saab-Scania's current aerospace products include the Saab 37 Viggen supersonic multi-purpose STOL combat aircraft, the Saab 35 Draken single-seat all-weather fighter-bomber, the Saab 105 multi-purpose military aircraft and the Safari/Supporter piston-engined trainer and army co-operation aircraft. Since 1949 Saab-Scania has delivered more than 2,000 military jet aircraft to the air forces of four nations. It has also delivered more than 1,500 piston-engined aircraft to military and civil customers around the world. Since 1962, Saab-Scania has also had a dealership for Hughes helicopters in Scandinavia and Finland.

Saab-Scania has greatly expanded its activities in the electronics field. Current production items include computer systems, autopilots, fire control and bombing systems for piloted aircraft, and electronics for guided missiles. A major production programme is the airborne computer for the Saab 37. Space-borne computers, optronic fire control systems and field artillery computer systems are also under development and in production.

Saab-Scania is manufacturing the Saab 05A air-to-surface missile for the Swedish Air Force and a modernised Saab 04E version of the Air Force RB04 homing anti-shipping missile to be carried by the AJ 37 Viggen. Saab 372, a new-generation air-to-air missile, is currently being developed for the JA 37 fighter version of the Viggen.

In addition, Saab-Scania is a member of the MESH space technology consortium which delivered the TD-1A solar research satellite to ESRO. In 1973 ESRO/ESA ordered from MESH the OTS satellite, and subsequently the MAROTS maritime communications satellite.

ATTACK AIRCRAFT SYSTEM 85

The Swedish Air Force has submitted to the Swedish government a classified plan for Attack Aircraft System 85, based on the results of a four-year study for a new generation of attack aircraft. The plan recommends that medium attack units of the SwAF at present being equipped with the Saab AJ 37 Viggen should begin to replace these in the mid-1980s with a modified version of the JA 37 fighter; that a new aircraft should be developed for light attack and training missions, to replace the SK 60 (Saab 105) aircraft currently in service; and that new weapons should be procured, from within Sweden or from abroad, suitable for the missions required of the attack units.

If approved, these proposals would ensure the continuation, until well into the 1990s, of the same basic aircraft type (the Viggen) for fighter, attack and reconnaissance missions; and the maintenance of the domestic capacity for aircraft development and production. During the 1971-75 period, SwKr 15·1 million were allocated for the initial study. Development costs, to be spread over the period 1977-85, are estimated at approx SwKr 2,000 million, the main part of which will be devoted to the new light attack/trainer aircraft. The present studies will continue, and additional material will be submitted to the government, up to 1 January 1977, and a parliamentary decision will be taken during that year.

SAAB 37 VIGGEN (THUNDERBOLT)

Swedish Air Force designations: AJ 37, JA 37, SF 37, SH 37 and SK 37

The Saab 37 Viggen multi-mission combat aircraft is the major component in the System 37 manned weapon system for the Swedish Air Force.

In brief, System 37 comprises the Saab 37 aircraft with power plant, airborne equipment, armament, ammunition and photographic equipment; special ground servicing equipment, including test equipment; and special training equipment, including simulators. Particular attention is paid to the optimum adaptation of System 37 to the SwAF base organisation and air defence control system (STRIL 60).

The Saab 37 is designed as a basic platform which can be readily adapted to fulfil the four primary roles of attack, interception, reconnaissance and training. The aircraft has an advanced aerodynamic configuration, using a foreplane, fitted with flaps, in combination with a main delta wing to confer STOL characteristics.

By employing a Swedish supersonic development of the American Pratt & Whitney JT8D turbofan engine, with a very powerful Swedish-designed afterburner, the Saab 37 can cruise economically and, at the same time, possesses the acceleration and climb performance required for interception duties. The combination of advanced aerodynamic features with this powerful engine, thrust reverser, automatic speed control during landing, and head-up display, enables the aircraft to operate from narrow runways of about 500 m (1,640 ft) length.

The first of seven prototypes of the Saab 37 flew for the first time on 8 February 1967, and by April 1969 all six single-seat prototypes were flying. The seventh Viggen was the prototype for the two-seat SK 37 operational trainer. A number of airframe parts were also completed for static testing.

The following versions have so far been announced:

AJ 37. Single-seat all-weather attack version, with secondary interceptor capability. Initial production version, which began to replace the A 32A Lansen from mid-1971. First production AJ 37 flew on 23 February 1971 and deliveries began on 21 June 1971. First AJ 37 unit was F7 Wing at Såtenäs; by the end of 1975 all three squadrons of F7 were equipped with the AJ 37 (and some SK 37s for training), and deliveries had begun to F15 at Söderhamn.

JA 37. Single-seat interceptor, with Volvo Flygmotor RM8B engine, of improved performance and with secondary capability for attack missions. Preliminary design work began in 1968. Flight testing of selected systems, including the radar, was initiated in early 1973 in a modified Saab 32 Lansen development aircraft. Four modified AJ 37s are being used in the JA 37 development programme. The first of these. for control system tests, flew for the first time in June 1974. The second, for engine tests, made its first flight with an RM8B engine on 27 September 1974; this aircraft was also fitted with a 30 mm Oerlikon long-range cannon, installed in an underbelly

Fifth development aircraft for the JA 37 interceptor version of the Saab Viggen combat aircraft, which first flew in December 1975

Underfuselage mounting for the 30 mm Oerlikon cannon fitted to the JA 37 version of the Viggen

pack, aft of which is a redesigned ventral fin. The third (first flight 22 November 1974) and fourth development aircraft are for electronics and armament system tests respectively. The fifth development aircraft (first flight 15 December 1975) was built from the outset to JA 37 standard. The JA 37 has four elevon hydraulic actuators under each wing, instead of three as on other versions, and a modified, taller tail-fin similar to that of the SK 37.

An initial batch of 30 JA 37s was ordered in September 1974, out of a planned total procurement of 150-200 to re-equip eight or more Draken fighter squadrons of the Swedish Air Force in 1978-85. Production began in late 1974.

SF 37. Single-seat all-weather armed photographic reconnaissance version to replace the S 35E Draken. A production contract was awarded in early 1973. Intended normally for overland reconnaissance, the SF 37 has a modified nose containing cameras and other equipment, permitting reconnaissance at any hour of the day or night, at high or low altitudes and at long distances from its base. First flown on 21 May 1973.

SH 37. Single-seat all-weather maritime reconnaissance version, to replace the S 32C version of the Lansen. Production ordered at same time as the SF 37. Primarily intended to survey, register, and report activities in the neighbourhood of Swedish territory. Can also be used for attack missions. Prototype first flown on 10 December 1973. First production SH 37 delivered on 19 June 1975.

SK 37. Tandem two-seat dual-control training version, in which the rear cockpit takes the place of some electronics and the forward fuselage fuel tank, and is fitted with bulged hood and twin periscopes. Modified, taller tail-fin of increased area. Capable of secondary attack role, with full range of attack armament as in AJ 37. Prototype first flown on 2 July 1970. First production SK 37 delivered in June 1972. In service with F7 Wing at Såtenäs; being delivered also to F15.

Saab 37X. Proposed export version, essentially similar to JA 37.

Initially, 175 aircraft of the AJ 37, SF/SH 37 and SK 37 versions were ordered for the Swedish Air Force; in December 1973 it was announced that five more aircraft (AJ 37s) were to be built within the same overall budget cost. Approx 100 Viggens had been delivered by the beginning of 1976.

The following details refer generally to all versions of the Viggen, except where specific versions are indicated:

Type: Single-seat all-weather multi-purpose combat aircraft and (SK 37) two-seat operational trainer.

Wings: Tandem arrangement of canard foreplane, with trailing-edge flaps, and a rear-mounted delta main wing with two-section hydraulically-actuated powered elevons on each trailing-edge, which can be operated differentially or in unison. Main wing has compound sweep on leading-edge. Outer sections have extended leading-edge. Extensive use of metal-bonded honeycomb panels for wing control surfaces, foreplane flaps and main landing gear doors.

Fuselage: Conventional all-metal semi-monocoque structure, of similar construction to that of Draken, using light metal forgings and heat-resistant plastics bonding. Local use of titanium for engine firewall and other selected areas. Four plate-type airbrakes, one on each side and two below fuselage. Metal-bonded honeycomb construction is used to a large extent. Quick-release handle permits nosecone to be pulled forward on tracks to give access to radar compartment.

Tail Unit: Vertical surfaces only, comprising main fin and powered rudder, supplemented by a small ventral fin. Rudder of metal-bonded honeycomb construction. The main fin can be folded downward to port.

Landing Gear: Retractable tricycle type of Saab origin, built by Motala Verkstad and designed for a max rate of sink of 300 m (985 ft)/min. Power-steerable twin-wheel nose unit retracts forward. Each main unit has two main wheels in tandem and retracts inward into main wing and fuselage. Main oleos shorten during retraction. Nosewheel tyres size 18 × 5·5, pressure 10·7 bars (155 lb/sq in). Main-wheel tyres size 26 × 6·6, pressure 14·8 bars (215 lb/sq in). Goodyear wheels and brakes. Dunlop anti-skid system.

Power Plant (AJ 37, SF/SH 37, SK 37): One Volvo Flygmotor RM8A (supersonic development of the Pratt & Whitney JT8D-22) turbofan engine, fitted with a Swedish-developed afterburner and thrust reverser. This engine is rated at 65·7 kN (14,770 lb st) dry and 115·7 kN (26,015 lb st) with afterburning. Thrust reverser doors are actuated automatically by the compression of the oleo as the nose gear strikes the runway, the thrust being deflected forward via three annular slots in the ejector wall. The ejector is normally kept open at subsonic speeds to reduce fuselage base drag; at supersonic speeds, with the intake closed, the ejector serves as a supersonic nozzle. Fuel is contained in one tank in each wing, a saddle tank over the engine, one tank in each side of the fuselage, and one aft of the cockpit. Electrically-powered pumps deliver fuel to the engine from the central fuselage tank, which is kept filled continuously from the peripheral tanks. Pressure refuelling point beneath starboard wing. Provision for jettisonable external auxiliary tank on underfuselage centreline pylon; this tank is normally a permanent fit on the SK 37.

Power Plant (JA 37, Saab 37X): One Volvo Flygmotor RM8B turbofan engine, rated at 72·1 kN (16,203 lb st) dry and 125 kN (28,108 lb st) with afterburning. Thrust reverser and fuel system details similar to other versions.

Accommodation: Pilot only, on Saab-Scania fully-adjustable rocket-assisted ejection seat beneath rearward-hinged clamshell canopy. Cockpit pressurisation, heating and air-conditioning by engine bleed air, via Delaney Gallay heat exchangers, cooling turbines and water separator. Birdproof windscreen. JA 37 cockpit redesigned and optimised for interceptor mission. SK 37 has twin periscopes, and tandem ejection seats under individual canopies.

Systems: Two independent hydraulic systems, each of 207 bars (3,000 lb/sq in) pressure, each with engine-driven pump; auxiliary electrically-operated standby pump for emergency use. Three-phase AC electrical system supplies 210/115V 400Hz power via a General Electric 60kVA liquid-cooled brushless generator, which also provides 28V DC power via 24V nickel-cadmium batteries and rectifier. Emergency standby power from 6kVA turbogenerator, which is extended automatically into the airstream in the event of a power failure. External power receptacle on port side of fuselage. Graviner fire detection system.

Electronics and Flight Equipment: Altogether, about 50 electronics units, with a total weight of approx 600 kg (1,323 lb), are installed in the Saab 37. Flight equipment includes an automatic speed control system, a Marconi-Elliott (in AJ 37) or Smiths (JA 37) electronic head-up display, AGA aircraft attitude instruments and radio, Phillips (AJ 37) or Garrett-AiResearch (JA 37) air data computer and instruments, L.M. Ericsson radar, Honeywell radar altimeter, Decca Doppler Type 72 navigation equipment, SATT radar warning system, Svenska Radio radar display system and electronic countermeasures, and AIL Tactical Instrument Landing System (TILS), a microwave scanning beam landing guidance system. Most of the electronic equipment in the Viggen is connected to the central digital computer, which is programmed to check out and monitor these systems both on the ground and during flight. The JA 37 has a ram-air intake on the underfuselage centreline, for cooling the electronics compartment.

Armament and Operational Equipment (AJ 37): All armament is carried externally on seven permanent attachment points, three under the fuselage and two under each wing, with standard 750 mm (29·5 in) store ejection racks. Each wing can be fitted with an additional hardpoint if required. Primary armament is the Swedish RB04E air-to-surface homing missile for use against naval targets, or the Saab RB05A air-to-surface missile for use against ground, naval and certain airborne targets. To these can be added pods of Bofors 135 mm air-to-surface rockets, bombs or 30 mm Aden gun pods. The AJ 37 version can be adapted to perform interception missions armed with RB24 (Sidewinder) or RB28 (Falcon) air-to-air missiles. Computations in connection with various phases of an attack, including navigation, target approach and fire control calculations, are handled by a Saab-Scania CK-37 miniaturised digital computer. This computer, which performs 48 specific tasks within the aircraft and is capable of 200,000 calculations per second, also provides data to the head-up display in the cockpit, thus freeing the pilot for concentration on other aspects of a flight. For a typical attack mission, the pilot would feed into the computer the position of the target and flight-path waypoints; the exact time of the attack; details of intended and alternative landing bases; and the type and method of delivery of the weapons to be carried. The computer would then calculate and present to him information regarding engine start and take-off times, navigation and approach to the target (including any deviations from the time schedule), weapon aiming and release, climb-out, return flight path and landing. Continuous monitoring of the flight paths and fuel situation is provided throughout the mission, and the computer can also, when required, release the weapons automatically.

Armament and Operational Equipment (JA 37): Permanent underbelly pack, offset to port side of centreline, containing one 30 mm Oerlikon KCA long-range cannon with a muzzle velocity of 1,050 m (3,445 ft)/sec, a rate of fire of 1,350 rds/min, and a projectile weight of 0·36 kg (0·79 lb). Improved fire control equipment. This gun installation permits retention of the three underfuselage stores attachment points, as in the AJ 37, in addition to the four underwing hardpoints. Advanced target search and acquisition system, based on a high-performance long-range L.M. Ericsson UAP-1023 X-band pulse-Doppler radar which is unaffected by variations of weather and altitude. This radar is not disturbed by ground clutter, and is highly resistant to ECM. Singer-Kearfott SKC-2037 central digital computer and Garrett-AiResearch LD-5 digital air data computer. Singer-Kearfott KT-70L inertial measuring equipment. Honeywell/Saab-Scania SA07 digital automatic flight control system. Weapon system includes provision for long-range homing air-to-air missiles, including the Saab 372 currently under development for the JA 37.

Armament and Operational Equipment (SF 37 and SH 37): Both reconnaissance versions can carry two air-to-air missiles, on the outboard wing stations, for self-

First production Saab SH 37 Viggen sea surveillance and attack aircraft

Ventral reconnaissance pods on a Saab SF 37 Viggen

defence. Equipment in the SF 37 includes a special optical sight, data camera, tape recorder and other registration equipment. The data camera collects and stores on its film co-ordination figures, aircraft position, course, altitude, target location and other data. Four vertical or oblique low-level cameras and two long-range vertical high-altitude cameras and an infra-red camera are installed in the nose, together with the camera sight, an infra-red sensor, and ECM registration equipment. Systems configuration also makes possible the detection of camouflaged targets and horizon-to-horizon (180°) photo coverage. Typical external mission equipment, in addition to air-to-air missiles, includes drop-tanks on the underfuselage stations, and an active or passive ECM pod on each of the inboard underwing pylons.

Internal equipment of the SH 37 includes a nose-mounted surveillance radar similar to that of the AJ 37, a camera for photographing the radar display, ECM registration equipment, and various other registration systems including a data camera and a tape recorder. The inboard and outboard wing pylons can be occupied, respectively, by active or passive ECM pods and air-to-air missiles, as on the SF 37. The underfuselage attachments can carry a drop-tank on the centreline station, a night reconnaissance pod on the port station and a long-range camera pod or a Red Baron night reconnaissance pod on the starboard station.

DIMENSIONS, EXTERNAL:
- Main wing span: 10·60 m (34 ft 9¼ in)
- Main wing aspect ratio: 2·45
- Foreplane span: 5·45 m (17 ft 10½ in)
- Length overall (incl probe):
  - except JA 37, Saab 37X: 16·30 m (53 ft 5¾ in)
  - JA 37, Saab 37X: 16·40 m (53 ft 9¾ in)
- Length of fuselage:
  - except JA 37, Saab 37X: 15·45 m (50 ft 8¼ in)
  - JA 37, Saab 37X: 15·58 m (51 ft 1½ in)
- Height overall:
  - AJ 37: 5·80 m (19 ft 0¼ in)
  - JA 37: 5·90 m (19 ft 4¼ in)
- Height overall, main fin folded: 4·00 m (13 ft 1½ in)
- Wheel track: 4·76 m (15 ft 7½ in)
- Wheelbase (c/l of shock-absorbers):
  - except JA 37, Saab 37X: 5·70 m (18 ft 8½ in)
  - JA 37, Saab 37X: 5·54 m (18 ft 2 in)

AREAS:
- Main wings, gross: 46·00 m² (495·1 sq ft)
- Foreplanes, outside fuselage: 6·20 m² (66·74 sq ft)

WEIGHTS:
- T-O weight:
  - AJ 37: 15,000-20,500 kg (33,070-45,195 lb)
  - JA 37 (normal armament): 17,000 kg (37,478 lb)

PERFORMANCE (AJ 37):
- Max level speed:
  - at high altitude: Mach 2
  - at 100 m (300 ft): above Mach 1·1
- Approach speed: approx 119 knots (220 km/h; 137 mph)
- Time to 10,000 m (32,800 ft) from brakes off, with afterburning: less than 1 min 40 sec
- T-O run: approx 400 m (1,310 ft)
- Landing run: approx 500 m (1,640 ft)
- Required landing field length:
  - conventional landing: 1,000 m (3,280 ft)
  - no-flare landing: 500 m (1,640 ft)
- Tactical radius with external armament:
  - hi-lo-hi: over 540 nm (1,000 km; 620 miles)
  - lo-lo-lo: over 270 nm (500 km; 310 miles)
- g limit: +12 (ultimate)

WEIGHTS AND PERFORMANCE (Saab 37X, as assessed in NATO Steering Committee report, March 1975):
- T-O weight, 'clean', with 2 Sidewinders: 16,783 kg (37,000 lb)
- *External load with max internal fuel: 3,674 kg (8,100 lb)
- Thrust/weight ratio at 16,783 kg (37,000 lb) AUW: 0·78
- Max level speed at 11,000 m (36,000 ft) with 2 Sidewinders: Mach 2·0
- Max rate of climb in 5g turn at low level at Mach 0·7, with 6 Mk 82 bombs: 4,755 m (15,600 ft)/min
- Sustained turn rate at 6,100 m (20,000 ft), with max internal fuel and 2 Sidewinders: 6·3°/sec
- Sustained air turning radius at low level at Mach 0·7, with 6 Mk 82 bombs: 3,353 m (11,000 ft)
- T-O run with 1,814 kg (4,000 lb) external load: 488 m (1,600 ft)
- Landing run with 1,814 kg (4,000 lb) external load, with thrust reversal: 640 m (2,100 ft)
- Radius of action with 6 Mk 82 bombs: 257 nm (476 km; 296 miles)

NATO assessment figure; Saab figure is 6,000 kg (13,227 lb)

A pair of Saab AJ 37 Viggens of the Swedish Air Force's F7 Wing

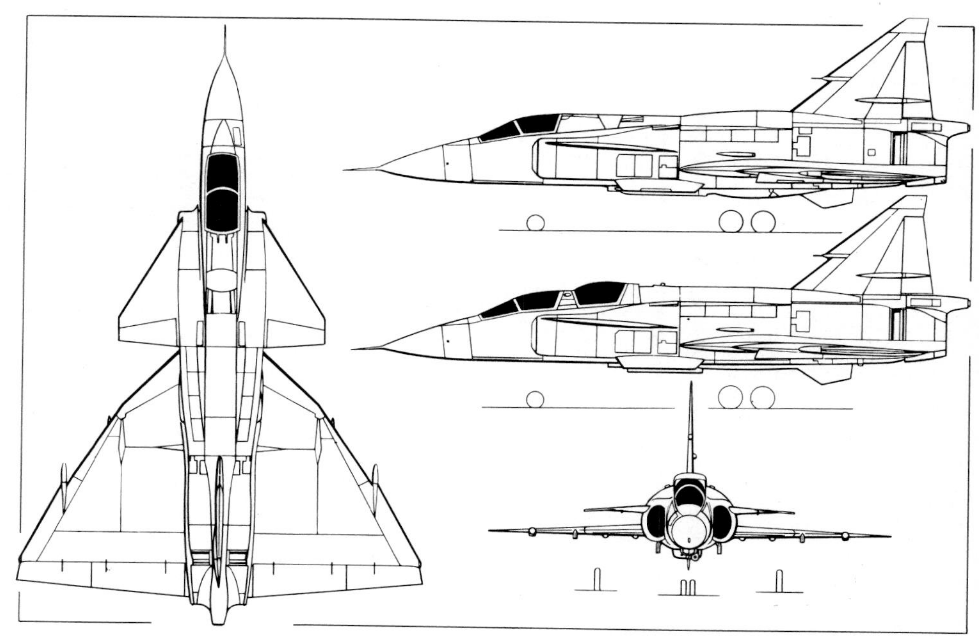
Saab JA 37 Viggen single-seat interceptor, with additional side view (centre) of SK 37 two-seat training version *(Pilot Press)*

SAAB 35 DRAKEN

Swedish Air Force designations: J 35, S 35 and SK 35

The Saab 35 Draken single-seat fighter was originally designed to intercept bombers in the transonic speed range, and carries radar equipment to accomplish this under all weather conditions. It is able also to carry substantial weapon loads for attack duties or cameras for photographic reconnaissance.

The first of three prototypes made its maiden flight on 25 October 1955, and the first version, the J 35A, entered service with the Swedish Air Force at the beginning of 1960. Subsequently the Draken went through several stages of development, and continuously-improved versions for the Swedish Air Force included the J 35B, D and F fighter versions, the SK 35C trainer version and the S 35E reconnaissance version. Production of these models, described in earlier editions of *Jane's*, has been completed.

A new underfuselage photo-reconnaissance pod for the S 35E Drakens of the Swedish Air Force was described in the 1974-75 *Jane's*. The RF-35s of the Royal Danish Air Force are fitted with the Red Baron night reconnaissance pod.

The current version of the Draken is:

Saab 35X. Long-range fighter/attack/reconnaissance version developed for the export market. Externally similar to the J 35F (1972-73 *Jane's*), but has greatly increased attack capability (max external load 4,500 kg; 9,920 lb) and range. For reconnaissance duties, a nose similar to that of the S 35E is fitted. T-O run with nine 1,000 lb bombs is 1,210 m (4,030 ft).

In 1968-69 the Danish Defence Ministry ordered 46 aircraft of this type, designated **Saab 35XD**, for the Royal Danish Air Force. Details of these aircraft, which were delivered in 1970-71, were given in the 1972-73 *Jane's*. A further five two-seat TF-35s were ordered in late 1973.

In April 1970, 12 Drakens (designated **Saab 35S**) were ordered by Finland. These were assembled in Finland by Valmet Oy (which see), for delivery during 1974-75. For familiarisation purposes, six Saab 35B Drakens (designated Saab 35BS) were leased to Finland by the Swedish Air Force.

The total number of Drakens built for Sweden, Denmark and Finland is more than 600.

The following shortened description refers to the basic Saab 35X, of which a full description can be found in the 1975-76 *Jane's:*

TYPE: Single-seat supersonic all-weather fighter, reconnaissance and attack aircraft.

POWER PLANT: One Volvo Flygmotor (Rolls-Royce licence) Avon 300-series engine (Swedish Air Force designation RM6C) with Swedish-developed afterburner. Rating approx 56·9 kN (12,787 lb st) dry and 78·45 kN (17,637 lb st) with afterburner. Internal fuel in integral tanks in inner wings and fuselage bag tanks. Total internal fuel capacity 4,000 litres (880 Imp gallons). Provision for external tanks under fuselage and wings, increasing total capacity to 9,000 litres (1,980 Imp gallons). Additional internal tanks can be fitted in place of guns for ferry purposes. Single-point pressure fuelling system, capacity 840 litres (185 Imp gallons) per minute.

ACCOMMODATION: Pressurised and air-conditioned cockpit, with fully-automatic Saab 73SE-F rocket-assisted ejection seat and GQ parachute system permitting ejection within the normal flight enevelope and down to 54 knots (100 km/h; 62 mph) on the ground. Rearward-hinged canopy.

ARMAMENT: Nine attachment points (each 454 kg; 1,000 lb) for external stores: three under each wing and three under fuselage. Stores can consist of air-to-air missiles and unguided air-to-air rocket pods (19 × 75 mm), 12 × 135 mm Bofors air-to-ground rockets, nine 1,000 lb or fourteen 500 lb bombs, or fuel tanks. Two or four RB24 Sidewinder air-to-air missiles can be carried under wings and fuselage. Two 30 mm Aden cannon (one in each wing) can be replaced by extra internal fuel tanks. With two 1,275 litre (280 Imp gallon) and two 500 litre (110 Imp gallon) drop-tanks, two 1,000 lb or four 500 lb bombs can be carried.

DIMENSIONS, EXTERNAL:

Wing span	9·40 m (30 ft 10 in)
Wing aspect ratio	1·77
Length overall	15·35 m (50 ft 4 in)
Height overall	3·89 m (12 ft 9 in)
Wheel track	2·70 m (8 ft 10½ in)
Wheelbase	4·00 m (13 ft 1 in)

WEIGHTS:

T-O weight 'clean'	11,400 kg (25,130 lb)
T-O weight with two 1,000 lb bombs and two 280 Imp gallon drop-tanks	14,590 kg (32,165 lb)
Max T-O weight	15,000 kg (33,070 lb)
Max overload T-O weight	16,000 kg (35,275 lb)
Normal landing weight	8,800 kg (19,400 lb)

PERFORMANCE (A: AUW of 25,130 lb; B: AUW of 32,165 lb):

Max level speed with afterburning:	
A	Mach 2
B	Mach 1·4
Max rate of climb at S/L with afterburning:	
A	10,500 m (34,450 ft)/min
B	6,900 m (22,650 ft)/min
Time to 11,000 m (36,000 ft) with afterburning:	
A	2 min 36 sec
Time to 15,000 m (49,200 ft) with afterburning:	
A	5 min 0 sec
T-O run with afterburning:	
A	650 m (2,130 ft)
B	1,170 m (3,840 ft)
T-O to 15 m (50 ft) with afterburning:	
A	960 m (3,150 ft)
B	1,550 m (5,080 ft)
Landing run at normal landing weight:	
A and B	530 m (1,740 ft)
Radius of action (hi-lo-hi), internal fuel only:	
A	343 nm (635 km; 395 miles)
Radius of action (hi-lo-hi) with two 1,000 lb bombs and two drop-tanks:	
B	541 nm (1,003 km; 623 miles)
Ferry range with max internal and external fuel	1,754 nm (3,250 km; 2,020 miles)

First Saab 35S Draken assembled by Valmet for the Finnish Air Force

Two-seat Saab 35X (TF-35) of the Royal Danish Air Force

SAAB 105

The Saab 105 twin-jet multi-purpose military aircraft was designed originally for training and ground attack duties, with reconnaissance, target flying and liaison as secondary roles. It normally seats two pilots side by side on ejection seats in a pressurised cabin; four fixed seats can be installed in place of the ejection seats.

The design was developed as a private venture, and the first of two prototypes flew for the first time on 29 June 1963.

Seven versions have been built, of which the SK 60A, B and C versions (150 built) for the Swedish Air Force were described in the 1969-70 *Jane's*; the projected Saab 105XH in the 1971-72 *Jane's*; and the Saab 105XT prototype and Saab 105Ö production version for Austria (40 built) in the 1972-73 *Jane's*.

Development and flight testing of the following version is continuing:

Saab 105G. Current version, developed from Saab 105Ö, with increased armament capability (max external load 2,350 kg; 5,180 lb); more advanced electronics, including a high-precision nav/attack and weapon delivery system; modified wing leading-edges; provision for external fuel tanks on the inboard wing pylons; and increased flap deflection for steeper glideslope. Prototype flown for the first time on 26 May 1972.

The following description refers to the Saab 105G:

TYPE: Multi-purpose twin-jet military aircraft.

WINGS: Cantilever shoulder-wing monoplane. Sweepback 12° 48′ at quarter-chord. Anhedral 6°. Thickness/chord ratio 9·5% at root, 10·9% at tip. One-piece stressed-skin structure with two continous spars. Leading-edge, compared with earlier versions, gives improved manoeuvrability at high Mach numbers. Ailerons, of bonded honeycomb construction, are statically and aerodynamically balanced and have hydraulically-actuated servo assistance, with provision for reversion to manual control in the event of a hydraulic failure. Geared servo tab in each aileron; starboard tab adjustable mechanically for trimming. Hydraulically-operated single-slotted flaps of honeycomb construction. Two small fences on upper surface of each wing.

FUSELAGE: All-metal stressed-skin semi-monocoque structure. Hydraulically-operated perforated airbrakes pivoted in transverse slots in lower fuselage aft of landing gear.

TAIL UNIT: Cantilever all-metal T-tail. Control surfaces of bonded honeycomb construction, statically and aerodynamically balanced. Elevators have hydraulically-actuated servo control, with provision for reversion to manual control in the event of a hydraulic failure. Electrical elevator trimming. Electrically-operated trim tab in rudder. Pneumatic yaw-damper. Geared servo tab in each elevator. Small ventral fin.

LANDING GEAR: Retractable tricycle type. Hydraulic actuation. Main units retract into fuselage, and have provision for gravity extension in an emergency. Forward-retracting hydraulically-steerable nosewheel, with shimmy damper. Oleo-pneumatic shock-absorbers. Hydraulic disc brakes with anti-skid system.

POWER PLANT: Two General Electric J85-GE-17B turbojet engines, each rated at 12·7 kN (2,850 lb st), mounted on sides of fuselage. Engine starting by internal battery. Fuel in two fuselage tanks and two wing tanks. Total internal fuel capacity 2,000 litres (440 Imp gallons). Pressure refuelling point in starboard wingtip. Provision for overwing gravity refuelling. Drop-tanks can be carried on the inboard underwing pylons.

ACCOMMODATION: Two side-by-side ejection seats, though attack missions are normally flown as a single-seater from the left-hand seat. Provision for armour protection in attack role. Birdproof windscreen. Electrically-actuated rearward-hinged canopy of double-curved acrylic glass. Dual controls standard. Alternative provision for four fixed seats for liaison role or as navigation trainer with instructor/pilot and three students.

SYSTEMS: Air-conditioning system includes refrigeration unit. Nominal cabin pressure differential of 0·23 bars (3·4 lb/sq in). Hydraulic system, pressure 207 bars (3,000 lb/sq in), has two pumps (one on each engine) and actuates landing gear, nosewheel steering, brakes, flaps, airbrakes and aileron servo control. System can be operated from one pump only. DC electrical system has two 300A 28V starter/generators and two 22Ah batteries. External power connector installed. AC system provides 3-phase power at 200/115V 400Hz from two 750A converters. *G*-suit connections. Oxygen system, with two 4·5 litre bottles. Fire warning system and fire extinguishers in each engine bay.

ELECTRONICS AND EQUIPMENT: The standard installation, according to mission, includes two VHF, or two UHF, or one VHF and one UHF; one ADF/DME and VOR/ILS with marker beacon, or Tacan, or Decca Doppler 72 radar with Sperry SGP 500 platform, TANS computer and roller map display; Marconi-Elliott air data computer, with probes; Ferranti ISIS F-105/125 gyro sighting head (with or without depressed sight line); Saab BT9R ballistic computer with laser rangefinder; pre-flight and in-flight control box; one transponder.

ARMAMENT AND OPERATIONAL EQUIPMENT: Three attachment points under each wing, the outer points each capable of supporting a 275 kg (606 lb) load, and the other four capable of supporting 450 kg (992 lb) each. Maximum total weapons load, with reduced fuel, is 2,350 kg (5,180 lb); with full internal fuel, up to 1,700 kg (3,748 lb) of external stores can be carried. Wide range of operational loads includes: six 500 lb bombs; four 750 lb bombs and two Sidewinder air-to-air missiles; four 1,000 lb bombs and two Sidewinders; four 500 lb bombs and two 400 litre (106 US gallon; 88 Imp gallon) drop-tanks; two 750 lb bombs, two Sidewinders and two 400 litre tanks; four 500 lb bombs and two 30 mm gun pods; twelve 135 mm rockets; eight 135 mm rockets and two 30 mm gun pods; camera pod (on port outer pylon), two Sidewinders and two 400 litre tanks; two drop-tanks only (liaison mission); or tow reel (port centre pylon), target launcher (port outer pylon) and two 400 litre tanks (target flying mission). For reconnaissance missions, aircraft can be fitted either with a panoramic camera in a special nose housing, which can be used in conjunction with any combination of ground

Saab 105G demonstration aircraft, carrying six bombs on underwing attachments

attack armament, or with a wing-mounted camera pod. This pod contains one forward-looking and four sideways/downward-looking cameras (or a two-camera alternative installation for high-altitude photography); a flashlight pod is available for night photography. Gyro gunsight standard. Specialised equipment available for advanced weapon training, target towing, ECM training, radar warning training, passive and active radar signal augmentation training, and collection of radioactive atmospheric samples.

DIMENSIONS, EXTERNAL:
- Wing span: 9·50 m (31 ft 2 in)
- Length overall: 10·80 m (35 ft 5¼ in)
- Height overall: 2·70 m (8 ft 10 in)
- Wheel track: 2·00 m (6 ft 6¾ in)
- Wheelbase: 3·90 m (12 ft 9½ in)

AREA:
- Wings, gross: 16·3 m² (175 sq ft)

WEIGHTS AND LOADING:
- Weight empty: 3,065 kg (6,757 lb)
- T-O weight:
  - trainer, 'clean': 4,860 kg (10,714 lb)
  - with camera pod, two Sidewinders and two drop-tanks: 5,855 kg (12,908 lb)
  - with six 500 lb bombs: 6,302 kg (13,893 lb)
  - with max external armament: 6,500 kg (14,330 lb)
- Max wing loading: 398 kg/m² (81·5 lb/sq ft)

PERFORMANCE (at weights indicated):
- Never-exceed speed (all weights): Mach 0·86
- Max level speed at S/L:
  - at 4,330 kg (9,546 lb), 'clean': 523 knots (970 km/h; 603 mph)
  - at 5,775 kg (12,731 lb): 475 knots (880 km/h; 547 mph)
- Max level speed at 10,000 m (32,800 ft):
  - at 4,330 kg (9,546 lb), 'clean': 472 knots (875 km/h; 544 mph)
  - at 5,775 kg (12,731 lb): 442 knots (820 km/h; 510 mph)
- Max rate of climb at S/L:
  - at 4,800 kg (10,582 lb): 3,400 m (11,155 ft)/min
  - at 6,302 kg (13,893 lb): 1,820 m (5,971 ft)/min
- Time to 10,000 m (32,800 ft):
  - at 4,800 kg (10,582 lb): 5 min 30 sec
  - at 6,302 kg (13,893 lb): 11 min 50 sec
- Service ceiling: 13,000 m (42,650 ft)
- T-O run:
  - at 4,860 kg (10,714 lb): 410 m (1,345 ft)
  - at 6,302 kg (13,893 lb): 830 m (2,723 ft)
- T-O to 15 m (50 ft):
  - at 4,860 kg (10,714 lb): 700 m (2,297 ft)
  - at 6,302 kg (13,893 lb): 1,270 m (4,167 ft)
- Landing from 15 m (50 ft):
  - at 3,600 kg (7,936 lb): 980 m (3,215 ft)
- Landing run:
  - at 3,600 kg (7,936 lb): 675 m (2,214 ft)
- Typical attack radius, incl 2½ min combat:
  - with six 500 lb bombs:
    - hi-lo-hi, 5% reserves: 375 nm (695 km; 431 miles)
    - lo-lo-lo, 10% reserves: 161 nm (300 km; 186 miles)
  - with four 500 lb bombs and two drop-tanks:
    - hi-lo-hi, 5% reserves: 536 nm (995 km; 618 miles)
    - lo-lo-lo, 10% reserves: 223 nm (415 km; 257 miles)
- Range at 12,000 m (39,375 ft) at 378 knots (700 km/h; 435 mph), 20 min reserves:
  - internal fuel only: 1,068 nm (1,980 km; 1,230 miles)
  - with two drop-tanks: 1,365 nm (2,530 km; 1,572 miles)

SAAB SAFARI

The Safari (formerly Saab-MFI 15) is intended as a basic trainer and utility aircraft. The prototype (SE-301) flew for the first time on 11 July 1969 with a 119 kW (160 hp) engine. Subsequently, its original low-mounted horizontal tail surfaces were replaced by new ones mounted at the top of the fin to prevent interference or damage by snow and debris when operating in Winter from rough airfields. After being re-engined with a 149 kW (200 hp) Lycoming, it resumed flying on 26 February 1971.

The Safari conforms to FAR Pt 23 in the Normal, Utility and Aerobatic categories, and can be adapted to carry up to 300 kg (660 lb) of external stores, such as relief supplies of food or medicines for delivery to disaster areas. Since April 1974 three aircraft have been in use for relief operations in Ethiopia. Up to 7,200 kg (15,873 lb) of grain per aircraft per day has been air-dropped in underwing packages, the aircraft flying at about 59 knots (110 km/h; 68 mph) at heights from 1 to 5 m (3-15 ft) above the ground, each with a 240 kg (529 lb) load. Other typical missions include rescue operations; ambulance role (with internally-stowed stretcher); forest fire or border patrol; road traffic control; and a wide range of basic flying training roles.

A tricycle landing gear is standard, but a tailwheel gear is available optionally, and conversion from one to the other can be accomplished quickly.

Safari of the type used for famine relief in Ethiopia, with droppable underwing loads

A military version, equipped with a weapon delivery system, is known as the Supporter; this is described separately. A pre-series batch of 12 Safari/Supporters was built; of these, two Safaris were delivered to Sierra Leone and five Supporters to Pakistan.

By February 1976 a total of more than 140 Safari/Supporter aircraft had been sold.

TYPE: Two/three-seat light aircraft.

WINGS: Braced shoulder-wing monoplane, with single bracing strut each side. Thickness/chord ratio 10%. Dihedral 1° 30′. All-metal structure, swept forward 5° from roots. Mass-balanced all-metal ailerons. Electrically-operated all-metal plain sealed flaps. Servo tab in starboard aileron.

FUSELAGE: Metal box structure. Glassfibre tailcone, engine cowling panels and wing strut/landing gear attachment fairings.

TAIL UNIT: Cantilever metal T tail comprising swept fin and rudder and one-piece mass-balanced horizontal 'stabilator' with large anti-servo and trimming tab. Glassfibre fin tip. Trim tab in rudder.

LANDING GEAR: Non-retractable tricycle (standard) or tailwheel type. Cantilever composite spring main legs. Goodyear 6·00-6 main wheels and either a 5·00-5 steerable nosewheel or a tailwheel. Cleveland disc brakes on main units. Landes or Finncraft skis, or Edo floats, optional.

POWER PLANT: One 149 kW (200 hp) Lycoming IO-360-A1B6 flat-four engine, driving a Hartzell HC-C2YK-4F/FC7666A-2 two-blade constant-speed metal propeller with spinner. Two integral wing fuel tanks, total capacity 190 litres (41·8 Imp gallons). Oil capacity 7·5 litres (1·6 Imp gallons). From 10-20 sec inverted flight (limited by oil system) permitted.

ACCOMMODATION: Side-by-side adjustable seats, with provision for back-type or seat-type parachutes, for two persons beneath fully-transparent upward-hinged canopy. Dual controls standard. Space aft of front seats for 100 kg (220 lb) of baggage (with external access on port side) or, optionally, a rearward-facing third seat. Upward-hinged door, with window, beneath wing on port side. Cabin heated and ventilated.

SYSTEM: 28V 50A DC electrical system.

ELECTRONICS AND EQUIPMENT: Provision for full blind-flying instrumentation and radio. Six underwing attachments for up to 300 kg (660 lb) of external stores. Landing light in nose.

DIMENSIONS, EXTERNAL:
- Wing span: 8·85 m (29 ft 0½ in)
- Wing chord (outer panels, constant): 1·36 m (4 ft 5½ in)
- Length overall:
  - nosewheel: 7·00 m (22 ft 11½ in)
  - tailwheel: 6·85 m (22 ft 5¾ in)
- Height overall:
  - nosewheel: 2·60 m (8 ft 6½ in)
  - tailwheel (tail down): 1·90 m (6 ft 2¾ in)
- Tailplane span: 2·80 m (9 ft 2¼ in)
- Wheel track:
  - nosewheel: 2·30 m (7 ft 6½ in)
  - tailwheel: 2·025 m (6 ft 7¾ in)
- Wheelbase:
  - nosewheel: 1·59 m (5 ft 2¾ in)
  - tailwheel: 4·75 m (15 ft 7 in)
- Propeller diameter: 1·88 m (6 ft 2 in)
- Cabin door (port):
  - Height: 0·78 m (2 ft 6¾ in)
  - Width: 0·52 m (1 ft 8½ in)

DIMENSIONS, INTERNAL:
- Cabin: Max width: 1·10 m (3 ft 7¼ in)
  - Max height (from seat squab): 1·00 m (3 ft 3¼ in)

AREAS:
- Wings, gross: 11·90 m² (128·1 sq ft)
- Ailerons (total): 0·98 m² (10·55 sq ft)
- Flaps (total): 1·55 m² (16·68 sq ft)
- Fin: 0·77 m² (8·29 sq ft)
- Rudder, incl tab: 0·73 m² (7·86 sq ft)
- Horizontal tail surfaces (total): 2·10 m² (22·6 sq ft)

WEIGHTS:
- Weight empty, equipped: 646 kg (1,424 lb)
- Max T-O weight:
  - Normal: 1,200 kg (2,645 lb)
  - Utility: 1,125 kg (2,480 lb)
  - Aerobatic: 900 kg (1,984 lb)

PERFORMANCE (at max T-O weight, Utility category, nosewheel version):
- Never-exceed speed: 197 knots (365 km/h; 227 mph)
- Max level speed at S/L: 127 knots (236 km/h; 146 mph)
- Cruising speed: 112 knots (208 km/h; 129 mph)
- Stalling speed, flaps down, power off: 58 knots (107 km/h; 67 mph)
- Max rate of climb at S/L: 246 m (807 ft)/min
- Time to 1,830 m (6,000 ft): 9 min 18 sec
- Service ceiling: 4,100 m (13,450 ft)
- T-O run: 210 m (689 ft)
- T-O to 15 m (50 ft): 385 m (1,263 ft)
- Landing from 15 m (50 ft): 390 m (1,280 ft)
- Landing run: 155 m (509 ft)
- Max endurance (65% power) at S/L, 10% reserves: 5 hr 10 min
- *g* limits:
  - Utility: +4·4; −1·76
  - Aerobatic: +6·0; −3·0

SAAB SUPPORTER

Royal Danish Air Force designation: T-17

The basic configuration of the Safari (which see) is retained in the Supporter (formerly Saab-MFI 17), which

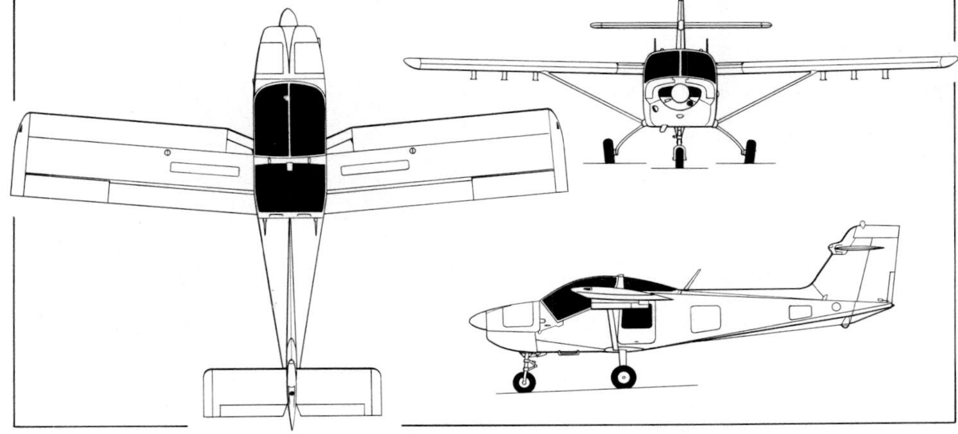

Saab Supporter light army co-operation and training aircraft *(Pilot Press)*

has the added capability to deliver weapons carried on the six underwing stations.

The second Safari was modified to Supporter standard, making its first flight in the new form on 6 July 1972.

The airframe and power plant are the same in each aircraft, but the Supporter can undertake military operations with up to 300 kg (660 lb) of air-to-ground rockets, two pods each housing two machine-guns, or six Bantam wire-guided anti-tank missiles. It is also suitable for use as a military trainer, or for forward air control, reconnaissance, artillery observation, liaison, target towing or other military duties.

Supporters have been ordered by the Pakistan Air Force and Army (45) and Royal Danish Air Force (32). The latter, for delivery by the end of 1976, are for training and observation duties and are designated T-17 in service. Deliveries to Pakistan began in 1974 with five of the 12 pre-production Safari/Supporters, and were to be followed by 40 of the initial production run of 65 aircraft.

AIRFRAME, POWER PLANT AND ACCOMMODATION: As for Safari

ARMAMENT: Six underwing hardpoints, the inner two stressed to carry up to 150 kg (330 lb) each and the outer four up to 100 kg (220 lb) each. Typical loads may include two 7·62 mm machine-gun pods, two Abel pods each with seven 75 mm air-to-surface rockets, four Abel pods each with seven 68 mm rockets, eighteen 75 mm Bofors rockets, or six Bofors Bantam wire-guided anti-tank missiles.

Saab Supporter light multi-purpose military aircraft in Royal Danish Air Force insignia

DIMENSIONS, WEIGHTS AND PERFORMANCE: As for Safari

SWITZERLAND

DÄTWYLER
MAX DÄTWYLER & CO

HEAD OFFICE AND WORKS: Flugplatz, CH-3368 Bleienbach-Langenthal

Telephone: 063 2 06 32
Telex: 68218 mdc ch

Dätwyler & Co is no longer active in the fields of aircraft production and maintenance.

The programme for the Swiss Trainer two-seat basic training aircraft, described in previous editions of *Jane's,* has been abandoned, and the design is now available for purchase.

FFA
FLUG- UND FAHRZEUGWERKE AG ALTENRHEIN

HEAD OFFICE AND WORKS: CH-9422 Staad b/Rorschach
Telephone: (071) 43 01 01
Telex: 77 230

PRESIDENT: Dr C. Caroni

DIRECTOR: Dipl Ing H. Eisenring

WORKS DIRECTOR: O. Wick

CHIEF ENGINEER: Dipl Ing P. Spalinger

This company, known formerly as AG für Dornier Flugzeuge, was originally the Swiss branch of the German Dornier company. It is now an entirely Swiss company.

FFA developed a glassfibre sailplane named Diamant, of which details were given in the Sailplanes section of the 1972-73 *Jane's.* Current activities include production of the AS.202 Bravo light aircraft, and the overhaul, modification and servicing of military and civil aircraft.

The company has about 1,000 employees, approximately half of whom are engaged in its aviation activities.

FFA AS.202 BRAVO

Following an agreement concluded with SIAI-Marchetti of Italy, FFA is engaged in production and development of the AS.202 Bravo light trainer and sporting aircraft.

The first Bravo to fly was a Swiss-assembled AS.202/15 prototype (HB-HEA), which flew for the first time on 7 March 1969. The Italian-built second prototype flew on 7 May 1969. The third aircraft (HB-HEC) made its first flight on 16 June 1969, and the first production aircraft on 22 December 1971. Swiss certification of the AS.202/15 was granted on 15 August 1972; FAA certification was awarded on 16 November 1973.

Two versions are currently available, as follows:

AS.202/15. Two/three-seat initial production version, with 112 kW (150 hp) Lycoming O-320-E2A engine. Optional third seat. In production. The first 25 Bravos are of this version.

AS.202/18A. Two/three-seat aerobatic version with a 134 kW (180 hp) Lycoming engine, Hartzell constant-speed propeller and inverted oil system. First example (HB-HEY) flew for the first time on 22 August 1974.

The following description applies to both current versions, except where a specific model is indicated:

TYPE: Two/three-seat light aircraft.

WINGS: Cantilever low-wing monoplane. Wing section NACA 63_2618 (modified) at centreline, 63_2415 at tip. Thickness/chord ratio 17·63% at root, 15% at tip. Dihedral 5° 43′ from roots. Incidence 3°. Sweepback at quarter-chord 0° 40′. Conventional aluminium single-spar fail-safe structure, with honeycomb laminate skin. Aluminium single-slotted flaps and single-slotted ailerons. Ground-adjustable tab on each aileron.

FUSELAGE: Conventional aluminium semi-monocoque fail-safe structure, with several glassfibre fairings.

TAIL UNIT: Cantilever aluminium single-spar structure with sweptback vertical surfaces. Rudder mass-balanced, with provision for anti-collision beacon. Fixed-incidence tailplane. Two-piece elevator with full-span trim tab on starboard half. Ground-adjustable tab on rudder.

LANDING GEAR: Non-retractable tricycle type, with steerable nosewheel. Rubber-cushioned shock-absorber struts of SIAI-Marchetti design. Main-wheel tyres size 6·00-6; nosewheel tyre size 5·00-5. Tyre pressure (all units) 2·41 bars (35 lb/sq in). Independent hydraulically-operated disc brake on each main wheel.

POWER PLANT (AS.202/15): One 112 kW (150 hp) Lycoming O-320-E2A flat-four engine, driving a McCauley 1C172 MGM two-blade fixed-pitch metal propeller with spinner. Two wing leading-edge fuel tanks with total capacity of 140 litres (30·5 Imp gallons). Refuelling point above each wing. Oil capacity 7·6 litres (1·6 Imp gallons). Additional exhaust muffler available optionally.

POWER PLANT (AS.202/18A): One 134 kW (180 hp) Lycoming AEIO-360-B1F flat-four engine, driving a Hartzell HC-C2YK-1BF/F7666A-2 two-blade constant-speed propeller with spinner. Hoffmann three-blade propeller available optionally. Fuel capacity as for AS.202/15; starboard tank has additional flexible fuel intake for aerobatics. Christen 801 fully-aerobatic oil system, capacity 7·6 litres (1·6 Imp gallons).

ACCOMMODATION: Seats for two persons side by side, in Aerobatic version, under rearward-sliding jettisonable transparent canopy. Space at rear in Utility version for a third seat or 100 kg (220 lb) of baggage. Dual controls, cabin ventilation and heating standard.

SYSTEMS: Hydraulic system for brake actuation. One 12V 60A engine-driven alternator and one 25Ah battery provide electrical power for engine starting, lighting, instruments, communications and navigation installations.

ELECTRONICS AND EQUIPMENT: Provision for VHF radio, VOR, ADF, Nav-O-Matic 200A autopilot, blind-flying instrumentation or other special equipment at customer's option. Clutch-and-release mechanism for glider towing optional.

DIMENSIONS, EXTERNAL:	
Wing span	9·75 m (31 ft 11¾ in)
Wing chord at root	1·88 m (6 ft 2 in)
Wing chord at tip	1·16 m (3 ft 9½ in)
Wing aspect ratio	6·51
Length overall	7·50 m (24 ft 7¼ in)
Length of fuselage	7·15 m (23 ft 5½ in)
Height overall	2·81 m (9 ft 2¾ in)
Tailplane span	3·67 m (12 ft 0½ in)
Wheel track	2·25 m (7 ft 4½ in)
Wheelbase	1·78 m (5 ft 10 in)
Propeller diameter	1·88 m (6 ft 2 in)
Propeller ground clearance	0·31 m (1 ft 0¼ in)
DIMENSIONS, INTERNAL:	
Cabin: Max length	2·15 m (7 ft 0½ in)
Max width	1·02 m (3 ft 4¼ in)
Max height	1·10 m (3 ft 7¼ in)
Floor area	2·15 m² (23·14 sq ft)
AREAS:	
Wings, gross	13·86 m² (149 sq ft)

AS.202/18A version of the FFA Bravo, for a customer in Oman

Ailerons (total) 1·09 m² (11·7 sq ft)
Trailing-edge flaps (total) 1·49 m² (16·04 sq ft)
Fin 0·45 m² (4·84 sq ft)
Rudder, incl tab 0·94 m² (10·12 sq ft)
Tailplane 1·88 m² (20·24 sq ft)
Elevators, incl tab 0·76 m² (8·18 sq ft)

Weights and Loadings:
Weight empty, equipped:
AS.202/15 630 kg (1,388 lb)
AS.202/18A 665 kg (1,466 lb)
Max payload:
AS.202/15, Aerobatic 175 kg (386 lb)
AS.202/15, Utility 270 kg (595 lb)
AS.202/18A, Aerobatic 175 kg (386 lb)
AS.202/18A, Utility 275 kg (606 lb)
Max T-O and landing weight:
AS.202/15, Aerobatic 885 kg (1,951 lb)
AS.202/15, Utility 999 kg (2,202 lb)
AS.202/18A, Aerobatic 950 kg (2,094 lb)
AS.202/18A, Utility 1,050 kg (2,315 lb)
Max wing loading:
AS.202/15 72·2 kg/m² (14·8 lb/sq ft)
AS.202/18A 75·8 kg/m² (15·52 lb/sq ft)
Max power loading:
AS.202/15, Utility 8·92 kg/kW (14·68 lb/hp)
AS.202/18A, Utility 7·84 kg/kW (12·86 lb/hp)

Performance (AS.202/15, Utility version at max T-O weight):
Never-exceed speed 173·5 knots (322 km/h; 200 mph)
Max level speed at S/L 114 knots (211 km/h; 131 mph)
Max cruising speed (75% power) at 2,440 m (8,000 ft) 114 knots (211 km/h; 131 mph)
Econ cruising speed (66% power) at 3,050 m (10,000 ft) 109·5 knots (203 km/h; 126 mph)
Stalling speed, flaps up 59·5 knots (110 km/h; 68·5 mph)
Stalling speed, flaps down 48·5 knots (89 km/h; 55·5 mph)
Max rate of climb at S/L 193 m (633 ft)/min
Service ceiling 4,265 m (14,000 ft)
T-O run 235 m (771 ft)
T-O to 15 m (50 ft) 475 m (1,558 ft)
Landing from 15 m (50 ft) 415 m (1,362 ft)
Landing run 130 m (427 ft)
Range with max fuel, no reserves 480 nm (890 km; 553 miles)

Performance (AS.202/18A, Utility version at max T-O weight):
Never-exceed speed 173·5 knots (322 km/h; 200 mph)
Max level speed at S/L 130 knots (241 km/h; 150 mph)
Max cruising speed (75% power) at 2,440 m (8,000 ft) 122·5 knots (227 km/h; 141 mph)
Econ cruising speed (55% power) at 3,050 m (10,000 ft) 109·5 knots (203 km/h; 126 mph)
Stalling speed, flaps up 60·5 knots (112 km/h; 70 mph)
Stalling speed, flaps down 49·5 knots (91 km/h; 57 mph)
Max rate of climb at S/L 281 m (922 ft)/min
Service ceiling 5,490 m (18,000 ft)
T-O run 210 m (689 ft)
T-O to 15 m (50 ft) 400 m (1,312 ft)
Landing from 15 m (50 ft) 415 m (1,362 ft)
Landing run 130 m (427 ft)
Range with max fuel, no reserves 504 nm (935 km; 581 miles)

PILATUS

PILATUS FLUGZEUGWERKE AG

Head Office and Works:
CH-6370, Stans, near Lucerne
Telephone: (041) 61 14 46
Telex: 78 329

General Manager:
H. Uehlinger

Managers:
Aerospace Division:
D. C. Klöckner
Dr A. Canal (Assistant, Marketing and PR)
W. Damerum (Sales)
K. G. Trautmann (Projects)
O. Masefield (Engineering)
Production:
W. Gubler
Administration:
P. Ebner
Instruments and Vehicles Division:
Dr K. Zimmermann

Pilatus Flugzeugwerke AG was formed in December 1939; details of its early history can be found in previous editions of *Jane's*. It is now a part of the Oerlikon-Bührle Group.

Current Pilatus products are the Turbo-Porter single-engined utility transport and the B4-PC11 sailplane. Its latest venture is the PC-7 Turbo-Trainer.

Pilatus PC-6/B2 STOL utility aircraft with Edo twin-float landing gear

PILATUS PC-6 TURBO-PORTER

The Pilatus PC-6 is a single-engined multi-purpose utility aircraft, with STOL characteristics permitting operation from unprepared strips under harsh environmental and terrain conditions. The aircraft can be converted rapidly from a pure freighter to a passenger transport, and can be adapted for a great number of different missions, including supply dropping, ambulance, aerial survey and photography, parachuting, crop spraying, water bombing and target towing as well as operation from soft ground, snow, glacier or water, and long-range operations.

Design work began in 1957, and the first of five PC-6 piston-engined prototypes made its first flight on 4 May 1959. Swiss certification of the pre-series PC-6, with 253·5 kW (340 hp) Lycoming engine, was received in December 1959, and the entire batch of 20 aircraft was delivered by the Summer of 1961.

Subsequent versions have included the piston-engined PC-6 and PC-6/350 Porters; the PC-6/A, A1, A2, B and C2-H2 Turbo-Porters, with various turboprop power plants. Descriptions of all these can be found in the 1974-75 and earlier editions of *Jane's*.

All aircraft delivered since mid-1966 have a forward-opening door on each side of the cockpit, a large rearward-sliding door on the starboard side of the cabin, and a double door on the port side of the cabin.

Swiss-built piston-engined variants have the name Porter, and turboprop-powered variants are known as Turbo-Porters. In the USA, where the PC-6 is manufactured by Fairchild, it is known simply as the Porter, irrespective of the type of power plant fitted.

The current production version is the **PC-6/B2-H2 Turbo-Porter,** certificated on 30 June 1970 and powered by a 410 kW (550 shp) PT6A-27 turboprop engine. Other versions can be made available on request.

By 1 April 1976, more than 360 PC-6 aircraft, of all models, had been built (including US licence manufacture), and production by Pilatus was continuing at a rate of two to three per month.

Pilatus markets a Q-STOL (Quiet STOL) conversion kit for the B1 and B2 Turbo-Porters fitted with PT6A-20 or -27 engines. This includes a reversal system whereby propeller speed can be altered independently of the engine power setting, and is claimed to reduce the noise level by more than 10 dB for T-O and 20 dB for landing.

The structural description which follows is applicable to the current B2-H2 version. Details of other models can be found in previous editions of *Jane's*. Details of the Turbo-Porter adapted for agricultural duties are given separately.

Type: Single-engined STOL utility transport.

Wings: Braced high-wing monoplane, with single streamline-section bracing strut each side. Wing section NACA 64-514 (constant). Dihedral 1°. Incidence 2°. Single-spar all-metal structure. Entire trailing-edge hinged, inner sections consisting of electrically-operated all-metal double-slotted flaps and outer sections of all-metal single-slotted ailerons. No airbrakes or de-icing equipment. Trim tabs and/or Flettner tabs on ailerons optional; fixed tabs are mandatory if these are not fitted.

Fuselage: All-metal semi-monocoque structure.

Tail Unit: Cantilever all-metal structure. Variable-incidence tailplane. Flettner tabs on elevator.

Landing Gear: Non-retractable tailwheel type. Oleo shock-absorbers of Pilatus design on all units. Steerable/lockable tailwheel. Goodyear Type II main wheels and GA 284 tyres size 24 × 7 or 7·50 × 10 (pressure 2·21 bars; 32 lb/sq in); oversize Goodyear Type III wheels and tyres optional, size 11·0 × 12, pressure 0·88 bars (12·8 lb/sq in). Goodyear tailwheel with size 5·00-4 tyre. Goodyear disc brakes. Alternative Pilatus wheel/ski gear or Edo 58-4580 or 679-4930 floats may be fitted.

Power Plant (PC-6/B2-H2): One 410 kW (550 shp) Pratt & Whitney Aircraft of Canada PT6A-27 turboprop engine, driving a Hartzell HC-B3TN-3D/T-10178 propeller. Standard fuel in integral wing tanks, capacity 480 litres (127 US gallons; 105·5 Imp gallons) normal, 644 litres (170 US gallons; 142 Imp gallons) maximum. Two underwing auxiliary tanks, each of 190 litres (50 US gallons; 42 Imp gallons), available optionally.

Accommodation: Cabin has pilot's seat forward on port side, with one passenger seat alongside, and is normally fitted with six quickly-removable seats, in pairs, to the rear of these for additional passengers. Up to 10 persons can be carried in high-density layout. Floor is level, flush with door sill, and is provided with seat rails. Forward-opening door beside each front seat. Large rearward-sliding door on starboard side of main cabin. Double doors, without central pillar, on port side. Hatch in floor 0·58 × 0·90 m (1 ft 10¾ in × 2 ft 11½ in), openable from inside cabin, for installation of aerial camera or for supply dropping. Hatch in cabin rear wall 0·50 × 0·80 m (1 ft 7 in × 2 ft 7 in) permits stowage of six passenger seats or accommodation of freight items up to 5·0 m (16 ft 5 in) in length. Walls lined with lightweight sound-proofing and heat-insulation material. Adjustable heating and ventilation systems provided. Dual controls optional.

Systems: Cabin heated by bleed air from engine compressor. Scott 8500 oxygen system optional. 200A 30V starter/generator and 24V 34Ah nickel-cadmium battery.

Dimensions, external:
Wing span 15·13 m (49 ft 8 in)
Wing span over navigation lights 15·20 m (49 ft 10½ in)
Wing chord (constant) 1·90 m (6 ft 3 in)
Wing aspect ratio 7·96
Length overall 10·90 m (35 ft 9 in)
Height overall (tail down) 3·20 m (10 ft 6 in)
Elevator span 5·12 m (16 ft 9½ in)
Wheel track 3·00 m (9 ft 10 in)
Wheelbase 7·87 m (25 ft 10 in)
Propeller diameter 2·56 m (8 ft 5 in)
Cabin double door (port) and sliding door (starboard):
Height 1·04 m (3 ft 5 in)
Width 1·58 m (5 ft 2¼ in)

Dimensions, internal:
Cabin, from back of pilot's seat to rear wall:
Length 2·30 m (7 ft 6½ in)
Max width 1·16 m (3 ft 9½ in)
Max height (at front) 1·28 m (4 ft 2½ in)
Height at rear wall 1·18 m (3 ft 10½ in)
Floor area 2·67 m² (28·6 sq ft)
Volume 3·28 m³ (107 cu ft)

Areas:
Wings, gross 28·80 m² (310 sq ft)
Ailerons (total) 3·83 m² (41·2 sq ft)
Flaps (total) 3·76 m² (40·5 sq ft)
Fin 1·70 m² (18·3 sq ft)
Rudder, incl tab 0·96 m² (10·3 sq ft)
Tailplane 4·03 m² (43·4 sq ft)
Elevator, incl tab 4·22 m² (45·4 sq ft)

Weights and Loadings:
Weight empty, equipped 1,215 kg (2,678 lb)
Max T-O and landing weight:
Normal (CAR 3) 2,200 kg (4,850 lb)
Restricted (CAR 8) 2,770 kg (6,100 lb)
Max cabin floor loading 488 kg/m² (100 lb/sq ft)
Max wing loading (Normal) 76·4 kg/m² (15·65 lb/sq ft)
Max power loading (Normal) 5·37 kg/kW (8·82 lb/shp)

Performance (at max T-O weight, Normal category):
Never-exceed speed 151 knots (280 km/h; 174 mph) IAS

Max cruising speed at 3,050 m (10,000 ft)
140 knots (259 km/h; 161 mph) TAS
Econ cruising speed at 3,050 m (10,000 ft)
129 knots (240 km/h; 150 mph) TAS
Stalling speed, power off, flaps up
50 knots (93·5 km/h; 58 mph)
Stalling speed, power off, flaps down
44 knots (82 km/h; 51 mph)
Max rate of climb at S/L 482 m (1,580 ft)/min
Service ceiling 9,150 m (30,025 ft)
T-O run 110 m (360 ft)
T-O to 15 m (50 ft) 235 m (771 ft)
Landing from 15 m (50 ft) 220 m (722 ft)
Landing run 73 m (240 ft)
Max range, no reserves:
internal fuel only 560 nm (1,036 km; 644 miles)
with external fuel 875 nm (1,620 km; 1,007 miles)
Endurance:
internal fuel only 4 hr 20 min
with external fuel 6 hr 45 min
g limits +3·72; −1·50

PILATUS PC-6 TURBO-PORTER (AGRICULTURAL VERSIONS)

The Turbo-Porter can, if required, be equipped for agricultural duties, the necessary equipment being easily removable when not required, to permit the use of the aircraft for other work.

For liquid spraying, a stainless steel tank (capacity 1,330 litres; 292·5 Imp gallons; 351·4 US gallons) is installed behind the two front seats, and 62-nozzle spraybooms are fitted beneath the wings. In this configuration the aircraft can cover a swath width of 45 m (148 ft). An ultra-low-volume system, using four to six atomisers or two to four Micronairs, is also available, permitting increase in swath width up to 400 m (1,310 ft).

For dusting with granulated materials, the lower part of the standard tank can be replaced by a discharge and dispersal door permitting coverage of a swath width of up to 20 m (66 ft). A Transland spreader can be fitted for dust application (swath up to 30 m; 100 ft). Effective swath width of these versions is 13-40 m (43-131 ft), the optimum being approx 20 m (66 ft).

Both versions are fitted with small doors in the fuselage sides, giving access to the tank/hopper for servicing, removal or replenishment, and two single seats or a bench seat for three persons can be installed aft of the tank. Optional items include an engine air intake screen and a loading door for chemical in the top of the fuselage.

Electronics and Equipment: Optional equipment includes Decca Mk 8A navigator, Decca Hi-Fix radio, Decca Doppler 72 radar, gyrosyn CL-11 compass and SR 54A radio altimeter.

Weights (L: liquid spray system; D: dry chemicals system):
Weight empty:
L, D 1,215 kg (2,678 lb)
Agricultural installation:
L 133 kg (293 lb)
D 105 kg (231 lb)
Chemical:
L 1,132 kg (2,497 lb)
D 1,160 kg (2,559 lb)
Fuel, oil and pilot:
L, D 286 kg (630 lb)
Max T-O and landing weight:
L, D 2,770 kg (6,100 lb)

Performance (liquid spray version, PT6A-27 engine, at max T-O weight):
Operating speed
approx 90 knots (167 km/h; 104 mph)
Operating height 6-8 m (20-26 ft)
Spraying duration with full spray tank 6 min

PILATUS PC-7 TURBO-TRAINER

Pilatus announced in the Spring of 1975 details of the PC-7 Turbo-Trainer fully aerobatic two-seat training aircraft, fitted with a 410 kW (550 shp) Pratt & Whitney Aircraft of Canada PT6A-25 turboprop engine.

The PC-7 can be used for basic, transition and aerobatic training, and, with suitable equipment installed, for IFR and tactical training. It meets the requirements of FAR 23 (Aerobatic category) and is also designed to comply with US Air Force military specifications. As a single-seater, it is flown from the front seat. The PC-7T tactical trainer version has six underwing hardpoints, and is capable of carrying up to 1,040 kg (2,293 lb) of underwing stores.

The certification programme was proceeding on schedule in the Summer of 1976. Steps have been taken to implement production, and the first two production aircraft are due for completion in early 1978. Dornier GmbH of Germany will collaborate with Pilatus in marketing the Turbo-Trainer.

Type: Single-engined single/two-seat training aircraft.

Wings: Cantilever low-wing monoplane. Wing section NACA 64_2A series at root, NACA 64_1A series at tip. Thickness/chord ratio 15% at root, 12% at tip. Dihedral 3° on outer panels. Sweepback 1° at quarter-chord. One-piece all-metal single-spar structure, with auxiliary spars and ribs. Constant-chord centre-section and tapered outer panels. Aluminium alloy (2022 or 2024) skin, reinforced by stringers. Some fairings of glassfibre-reinforced plastics. Plain mass-balanced ailerons; split trailing-edge flaps, extending under fuselage. Trim tab in port aileron.

PC-6/B1 Turbo-Porter fitted with Micronair AU-3000 crop-spraying gear

Fuselage: All-metal semi-monocoque structure, with aluminium alloy skin. Some fairings of glassfibre-reinforced plastics.

Tail Unit: Cantilever all-metal structure, of similar construction to wings. Trim tabs in rudder and starboard elevator. All control surfaces mass-balanced.

Landing Gear: Retractable tricycle type. Electrical actuation, with emergency manual extension. Main wheels retract inward, nosewheel rearward. Mechanical/hydraulic shock-absorbers on all units. Castoring nosewheel, with shimmy dampers. Goodyear main wheels and tyres, size 7·00-8, pressure 2·07 bars (30 lb/sq in). Goodyear hydraulic disc brakes on main wheels. Parking brake.

Power Plant: One Pratt & Whitney Aircraft of Canada PT6A-25 turboprop engine, flat-rated at 410 kW (550 shp), driving a Hartzell HC-B3TN-3 three-blade constant-speed reversible-pitch fully-feathering propeller with spinner. Fuel in integral tanks in wings, total usable capacity 470 litres (103 Imp gallons). Fuel system permits up to 30 sec of inverted flight.

Accommodation: Adjustable seats for two persons in tandem, beneath rearward-sliding jettisonable framed Plexiglas canopy. Cockpits ventilated and heated by engine bleed air, which can also be used for windscreen de-icing. Space for 25 kg (55 lb) of baggage aft of seats, with external access.

Systems: Hydraulic system for main-wheel brakes only. No pneumatic system. 28V DC operational (24V nominal) electrical system, incorporating 30V 200A Lear Siegler starter/generator and 24V 34Ah or 40Ah nickel-cadmium battery. Ground power receptacle fitted. 24V 1·4Ah iron-nickel emergency battery optional, for cabin lighting.

Electronics and Equipment: Dual controls standard. Optional equipment includes IFR training shield, to screen rear cockpit; radio, oxygen and tactical training armament to customer's requirements.

Dimensions, external:
Wing span 10·40 m (34 ft 1½ in)
Wing mean aerodynamic chord 1·63 m (5 ft 4·2 in)
Wing mean geometric chord 1·59 m (5 ft 2·6 in)
Wing aspect ratio 6·52
Length overall 9·75 m (32 ft 0 in)
Height overall 3·21 m (10 ft 6½ in)
Propeller diameter 2·36 m (7 ft 9 in)

Area:
Wings, gross 16·60 m² (178·7 sq ft)

Weights and Loadings:
Weight empty, equipped 1,280 kg (2,822 lb)
Normal T-O weight, 'clean' 1,900 kg (4,188 lb)
Max T-O weight, with external stores
2,700 kg (5,952 lb)
Normal landing weight 1,900 kg (4,188 lb)
Normal wing loading, 'clean'
114·5 kg/m² (23·44 lb/sq ft)
Max wing loading 162·7 kg/m² (33·31 lb/sq ft)
Normal power loading, 'clean'
4·63 kg/kW (7·61 lb/shp)
Max power loading 6·59 kg/kW (10·82 lb/shp)

Performance (A: at 1,900 kg; 4,188 lb AUW. B: at 2,700 kg; 5,952 lb AUW):
Never-exceed speed:
A, B 270 knots (500 km/h; 310 mph) EAS
Max structural cruising speed:
A, B 229 knots (425 km/h; 264 mph) EAS
Max cruising speed at 4,000 m (13,125 ft):
A, 100% power
248 knots (460 km/h; 286 mph) TAS
A, 75% power
232 knots (430 km/h; 267 mph) TAS
Max cruising speed at S/L:
A, 75%-100% power
226 knots (420 km/h; 261 mph) TAS
Manoeuvring speed:
A, B 154 knots (285 km/h; 177 mph) EAS
Max speed with flaps and landing gear down:
A, B 118·5 knots (220 km/h; 136 mph) EAS
Stalling speed, flaps up, power off:
A 71 knots (131 km/h; 81·5 mph) EAS

Pilatus PC-7 Turbo-Trainer (Pratt & Whitney Aircraft of Canada PT6A-25 turboprop engine)

B 85 knots (158 km/h; 98 mph) EAS
Stalling speed, flaps down, power off:
A 63 knots (117 km/h; 73 mph) EAS
B 75 knots (139·5 km/h; 86·5 mph) EAS
Normal rate of climb:
A at S/L 630 m (2,065 ft)/min
B at S/L 390 m (1,280 ft)/min
A at 3,050 m (10,000 ft) 588 m (1,930 ft)/min
B at 3,050 m (10,000 ft) 354 m (1,160 ft)/min
A at 6,100 m (20,000 ft) 336 m (1,105 ft)/min
B at 6,100 m (20,000 ft) 84 m (275 ft)/min
Time to 5,000 m (16,400 ft):
A 10 min
B 17 min
Service ceiling:
A 9,500 m (31,175 ft)
B 6,500 m (21,325 ft)
T-O run at S/L, zero wind:
A 185 m (605 ft)
B 280 m (920 ft)
T-O to 15 m (50 ft) at S/L, zero wind:
A 310 m (1,017 ft)
B 600 m (1,970 ft)
Landing from 15 m (50 ft) at S/L, zero wind:
A 490 m (1,607 ft)
Landing run at S/L, zero wind:
A 185 m (605 ft)
Max range:
A, 60% power at 5,000 m (16,400 ft), 5% + 20 min reserves 593 nm (1,100 km; 683 miles)
B, 80% power at 3,050 m (10,000 ft), 5% + 20 min reserves 700 nm (1,300 km; 807 miles)

Pilatus PC-7 Turbo-Trainer, now the subject of a joint programme with Dornier of Germany *(Pilot Press)*

Max endurance:
A 3 hr 38 min
B 4 hr 20 min
g limits (ultimate):
A + 6·0; − 3·0
B + 4·5; − 2·25

SWISS FEDERAL AIRCRAFT FACTORY
EIDGENÖSSISCHES FLUGZEUGWERK—FABRIQUE FÉDÉRALE D'AVIONS—FABBRICA FEDERALE D'AEROPLANI

HEAD OFFICE AND WORKS:
CH-6032 Emmen
Telephone: (041) 59 41 11
Telex: 7 84 80 fwead ch

DIRECTOR:
Lucien Othenin-Girard

DEPUTY DIRECTOR:
Dr Peter Burkhardt

CHIEF DESIGNER:
Heinz Rhomberg

HEAD OF RESEARCH DEPARTMENT:
Heini Kamber

The F+W is the Swiss government's official aircraft establishment for research, development, production, maintenance and modification of military aircraft. F+W was thus prime contractor for licence production of French Aérospatiale Alouette III helicopters and Dassault Mirage III-S/RS fighters for the Swiss Air Force. These and other major aircraft programmes have been described in previous editions of *Jane's*. Recently, the F+W was subcontractor to Hawker Siddeley for final assembly of, and application of Swiss modifications to, 60 refurbished Hunter Mk 58A aircraft and eight two-seat Mk 68s ordered by the Swiss Air Force. The Factory will take part in the assembly programme for 72 Northrop F-5E/F Tiger IIs for the Swiss Air Force.

The F+W employs about 600 people in a works located at Emmen, near Lucerne, covering 35,300 m² (380,000 sq ft). Included are four wind tunnels for speeds of up to Mach 4·5, test cells for piston and turbojet engines with or without afterburners, and a modern data acquisition and processing system.

The F+W is made up of six technical departments. Research Department operates the wind tunnel facilities and performs scientific research and development in the fields of aerodynamics, thermodynamics, systems analysis and flight mechanics. Engineering Department is responsible for the design and development of aircraft, aircraft subsystems and components, and the development of space hardware. Electronics Department undertakes maintenance and modification of the electronics of the Swiss Air Force's aircraft, and provides support for flight evaluations. Quality Assurance Department assures maintenance of high quality and adherence to standards in the production and maintenance of the aircraft and space hardware. The M & P Department maintains up-to-date awareness of new materials and processes, and develops its own process techniques to meet particular requirements. Thus, for example, fabrication of titanium parts and Inconel pressure vessels is undertaken at the F+W. Production Department is competent to undertake the fabrication of aircraft subassemblies and components, as well as final assembly and checkout of complete aircraft.

Among other industrial and commercial activities, the F+W is conducting wind tunnel programmes for foreign aircraft manufacturers, ground transportation developers and users, and for the building industry.

The F+W also offers proprietary products to potential customers, including an electronic audio warntone generator, water separators for aircraft conditioning systems, strain-gauge force measuring scales, and the POHWARO hot water rockets. Such rockets, in sets of six, were used extensively by MBB during 1975 to accelerate its high-speed transportation test vehicle to speeds of up to 191 knots (354 km/h; 220 mph) in just over 4 seconds. The F+W has acquired expertise in detection of aircraft structure fatigue and crack propagation. Fatigue tests on a full-scale Venom airframe with a computer-controlled load installation were concluded successfully in 1975 without a catastrophic failure of the airframe. These test results are now being applied to extend the useful life of Venoms still in Swiss service. Also during 1975, the F+W delivered, installed and put into operation a fatigue test facility for building construction components at the Swiss Federal Institute of Technology in Lausanne.

TAIWAN

AIDC/CAF
AERO INDUSTRY DEVELOPMENT CENTER—CHINESE AIR FORCE

ADDRESS:
PO Box 7173, Taichung, Taiwan 400
Telephone: Taichung 223051 and 223052
Telex: 51140
DIRECTOR: Lieutenant General Y. C. Lee
DEPUTY DIRECTORS:
Major General C. Y. Lee (Manufacturing)
Dr H. M. Hua (Engineering and Research)

The Aero Industry Development Center was established on 1 March 1969 as a successor to the Bureau of Aircraft Industry (BAI), which was established in 1946 in Nanking and moved to Taiwan in 1948.

In October 1968 the Aeronautical Research Laboratory, then a branch of BAI, constructed the first Chinese-built PL-1A (see 1970-71 *Jane's*), which was a slightly modified version of the US Pazmany PL-1 powered by a 93 kW (125 hp) Lycoming O-290-D engine. The PL-1A flew for the first time on 26 October 1968. After flight tests and further modifications, 35 PL-1B Chienshou production models were built by AIDC in 1970, and a further 10 in the Spring of 1972. A third batch, of 10 PL-1Bs, with additional modifications to the engine cowling, nose landing gear, tail unit, flap control and pilot seats, was built and delivered in 1974. These aircraft have been used extensively as primary trainers for CAF air cadets.

AIDC/CAF-built Bell UH-1H Iroquois helicopter for the Chinese Nationalist Army

Under an agreement reached in 1969, the AIDC is producing in Taiwan the Bell UH-1H (Bell Model 205) helicopter under licence for the Chinese Nationalist Army. The AIDC-Bell UH-1H is almost identical with the UH-1H version built by Bell Helicopter Textron (see US section). The original contract was for 50 UH-1Hs, to which in 1972 was added a follow-on order for a further 68 of these aircraft.

The AIDC is engaged also in the licence production of 100 Northrop F-5E Tiger II tactical fighter aircraft (see US section) for the Chinese Nationalist Air Force. The first Chinese-built F-5E (CAF name Chung Cheng) was rolled out on 30 October 1974.

Since 1970, the AIDC has been developing the T-CH-1, a turboprop-powered trainer of its own design with two seats in tandem. It is now designing a twin-turboprop transport designated XC-2.

AIDC T-CH-1

This aircraft is a tandem two-seat trainer, the design of which was started by AIDC in November 1970. Two prototypes were ordered, designated XT-CH-1A and XT-CH-1B; construction began in January 1972.

The first aircraft was completed in September 1973 and was flown for the first time on 23 November 1973, followed by an extensive flight test programme in 1974. A description of the XT-CH-1A appeared in the 1974-75 *Jane's*.

The second prototype (63-3002), which is designated XT-CH-1B, is a modified version able to perform weapon delivery training and counter-insurgency missions. It flew for the first time on 27 November 1974.

The following description applies to the T-CH-1, of which production had begun by early 1976:

TYPE: Turboprop-powered trainer and light ground attack aircraft.

WINGS: Cantilever low-wing monoplane. Wing section NACA 64-2A215 (constant). Dihedral 8° from roots. Incidence 2°. No sweepback. Conventional aluminium alloy stressed-skin structure, with aluminium alloy ailerons and slotted trailing-edge flaps. Link-balance type trim tab in each aileron.

FUSELAGE: Conventional semi-monocoque structure of aluminium alloy.

TAIL UNIT: Cantilever aluminium alloy structure, with fixed-incidence tailplane. Dorsal fin. Link-balance type trim tabs in rudder and each elevator.

LANDING GEAR: Retractable tricycle type. Hydraulic retraction, main wheels inward into wings, nosewheel rearward. Telescopic shock-absorbers. Goodyear brakes. Small tail bumper under rear fuselage.

POWER PLANT: One 1,081 kW (1,450 ehp) Lycoming T53-L-701 turboprop engine, driving a Hamilton Standard 53C51-27 three-blade metal propeller with spinner. Fuel in two tanks in each wing and one in fuselage, with total capacity of 963 litres (255 US gallons; 212 Imp gallons). Oil capacity 30 litres (8 US gallons; 6·6 Imp gallons).

ACCOMMODATION: Crew of two in tandem. Rearward-sliding fully-transparent canopy over each cockpit. Cockpits heated and ventilated.

SYSTEMS: Midland-Ross Corporation heating and ventilating system. 115V 300A system provides AC electrical power at 250VA 400Hz. 28V DC system includes 24V 34Ah battery. Oxygen bottle with volume of 3·5 litres (2,100 cu in).

ELECTRONICS AND EQUIPMENT: Collins AN/ARC-51BX UHF radio and Collins AN/ARN-83 ADF.

DIMENSIONS, EXTERNAL:

Wing span	12·19 m (40 ft 0 in)
Wing chord at root	2·44 m (8 ft 0 in)
Wing chord at tip	1·52 m (5 ft 0 in)
Wing aspect ratio	6
Length overall	10·26 m (33 ft 8 in)
Height overall	3·66 m (12 ft 0 in)
Tailplane span	5·56 m (18 ft 3 in)
Wheel track	3·86 m (12 ft 8 in)
Wheelbase	2·39 m (7 ft 10 in)
Propeller diameter	3·05 m (10 ft 0 in)
Propeller ground clearance	0·74 m (2 ft 5 in)

AREAS:

Wings, gross	25·18 m² (271·0 sq ft)
Ailerons (total)	2·42 m² (26·0 sq ft)
Flaps (total)	5·02 m² (54·0 sq ft)
Fin	1·67 m² (18·0 sq ft)
Rudder, incl tab	1·11 m² (12·0 sq ft)
Elevators, incl tabs	1·81 m² (19·5 sq ft)

WEIGHTS AND LOADINGS:

Weight empty	2,608 kg (5,750 lb)
Max T-O weight	4,173 kg (9,200 lb)
Max landing weight	3,810 kg (8,400 lb)
Max wing loading	166 kg/m² (34·0 lb/sq ft)
Max power loading	2·88 kg/kW (6·34 lb/ehp)

PERFORMANCE (at AUW of 3,447 kg; 7,600 lb):

Never-exceed speed	370 knots (685 km/h; 426 mph)
Max level speed at 4,570 m (15,000 ft)	320 knots (592 km/h; 368 mph)
Max cruising speed at 4,570 m (15,000 ft)	220 knots (407 km/h; 253 mph)
Econ cruising speed at 4,570 m (15,000 ft)	170 knots (315 km/h; 196 mph)
Stalling speed	50 knots (93 km/h; 58 mph)
Max rate of climb at S/L	1,036 m (3,400 ft)/min
Service ceiling	9,755 m (32,000 ft)
T-O run	146 m (480 ft)
T-O to 15 m (50 ft)	244 m (800 ft)
Landing from 15 m (50 ft)	381 m (1,250 ft)
Landing run	183 m (600 ft)
Range with max fuel	1,085 nm (2,010 km; 1,250 miles)

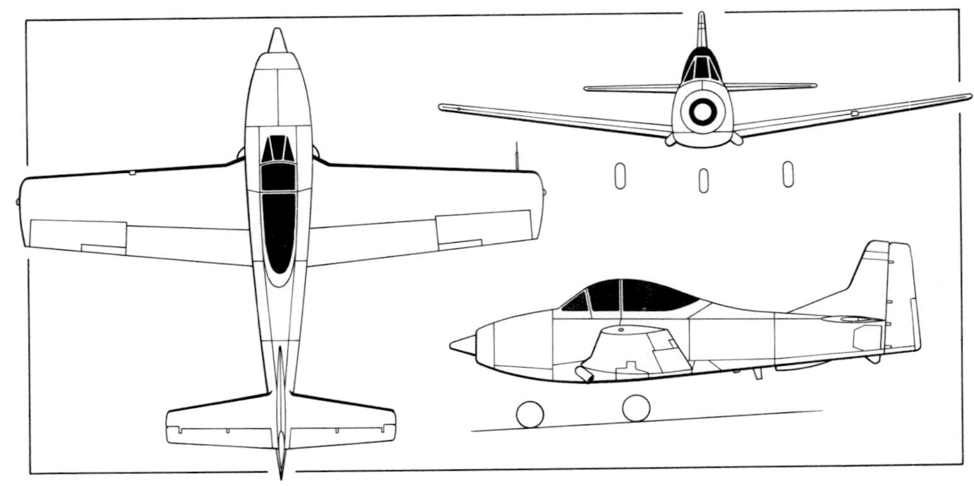

AIDC T-CH-1 tandem two-seat turboprop-powered trainer *(Michael A. Badrocke)*

XT-CH-1B second prototype of this AIDC/CAF-designed tandem two-seat basic trainer

AIDC XC-2

The basic design of the XC-2 twin-turboprop transport, which was started in January 1973, incorporates features of common interest to military and civil operators, including quick-change capability and the ability to operate from short fields and unprepared surfaces. The aircraft, of which a prototype was under construction in mid-1976, will be able to carry up to 38 passengers or 3,855 kg (8,500 lb) of cargo.

TYPE: Twin-turboprop transport aircraft.

WINGS: Cantilever high-wing monoplane. Wing section NACA 65₃-218. Incidence 4°. No dihedral or sweepback at quarter-chord. Light alloy three-spar fail-safe structure, built in three sections: a constant-chord centre-section and tapered outer panels. All-metal manually-operated ailerons and hydraulically-actuated Fowler-type trailing-edge flaps. Servo tab in each aileron.

FUSELAGE: Conventional all-metal semi-monocoque fail-safe structure, of basically rectangular section, upswept at rear to provide clearance for rear loading. Cabin pressurisation optional.

TAIL UNIT: Cantilever aluminium alloy three-spar structure, with sweptback fin and rudder and non-swept horizontal surfaces. Dorsal fin. Horizontal surfaces mounted halfway up fin. Trim and balance tab in rudder and each elevator.

LANDING GEAR: Retractable tricycle type, with hydraulically-steerable twin-wheel nose unit. Single-wheel main units retract into fairings on sides of fuselage.

POWER PLANT: Two 1,082 kW (1,451 ehp) Lycoming T53-L-701A turboprop engines, each driving a Hamilton Standard 53C51-27 three-blade variable-pitch metal propeller with spinner. Fuel in rubber tanks in wings, with combined standard capacity of 3,028 litres (666 Imp gallons; 800 US gallons).

ACCOMMODATION: Crew of three (pilot, co-pilot and flight engineer) on flight deck. Standard seating in main cabin for 38 passengers, four abreast at 787 mm (31 in) pitch. Interior layout has quick-change capability to passenger/cargo or all-cargo configuration. Access to main cabin via forward and rear doors on port side; single door on starboard side; and a two-section loading ramp/door in underside of rear fuselage, aft of main cabin, which is openable in flight for air-drop operations. Provision for toilet, galley and baggage compartment in passenger version.

SYSTEMS: Anti-icing and cabin heating systems standard. Hydraulic system, pressure 207 bars (3,000 lb/sq in), for flaps, landing gear and nosewheel steering. 28V DC primary electrical system, with 300A starter/generator on each engine. Two nickel-cadmium batteries for engine starting and emergency power.

ELECTRONICS AND EQUIPMENT: Communications equipment includes UHF and VHF radios. Navigation equipment includes ADF, Tacan and transponder. Optional equipment includes VOR/ILS and HF.

DIMENSIONS, EXTERNAL:

Wing span	24·90 m (81 ft 8·4 in)
Wing chord (centre-section, constant)	3·05 m (10 ft 0 in)
Wing aspect ratio	9·5
Length overall	19·74 m (64 ft 9 in)
Height overall	7·72 m (25 ft 3·8 in)
Tailplane span	9·12 m (29 ft 10·9 in)
Wheel track	3·86 m (12 ft 7·8 in)
Wheelbase	6·18 m (20 ft 3·4 in)
Propeller diameter	3·05 m (10 ft 0 in)
Propeller ground clearance	0·90 m (2 ft 11·5 in)

DIMENSIONS, INTERNAL:

Cabin, excl flight deck:	
Length	8·095 m (26 ft 6·7 in)

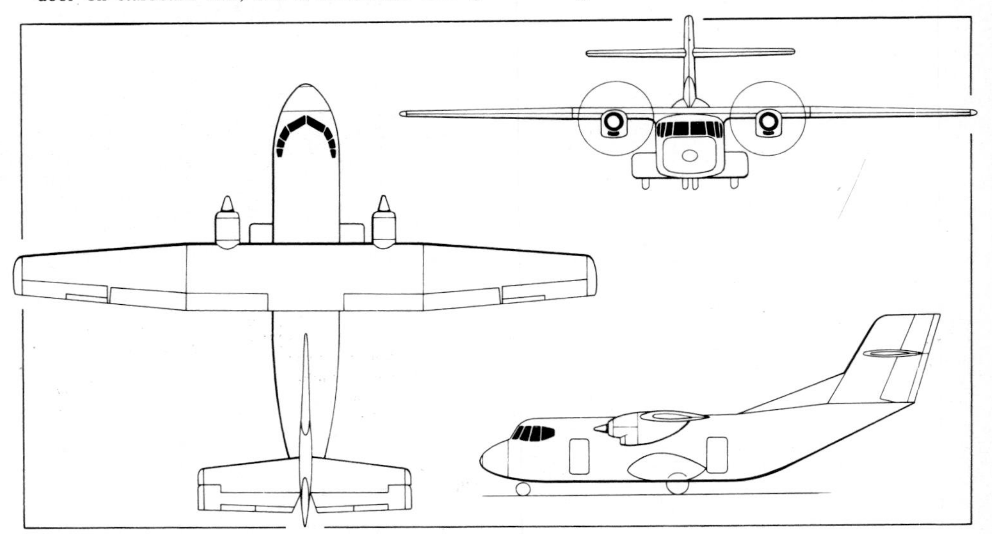

AIDC XC-2 twin-turboprop utility transport aircraft *(Michael A. Badrocke)*

Width	2·57 m (8 ft 5 in)
Height	2·23 m (7 ft 3·7 in)
Volume	45·45 m³ (1,605·0 cu ft)

Areas:

Wings, gross	65·40 m² (704·00 sq ft)
Ailerons (total)	2·12 m² (22·80 sq ft)
Trailing-edge flaps (total)	11·69 m² (125·80 sq ft)
Vertical tail surfaces (total)	11·73 m² (126·30 sq ft)
Horizontal tail surfaces (total)	19·31 m² (207·80 sq ft)

Weights and Loadings:

Weight empty, equipped	5,896 kg (13,000 lb)
Max payload	3,855 kg (8,500 lb)
Max T-O weight	11,340 kg (25,000 lb)
Max landing weight	11,113 kg (24,500 lb)
Max zero-fuel weight	10,120 kg (22,310 lb)
Max wing loading	173·4 kg/m² (35·51 lb/sq ft)
Max power loading	5·24 kg/kW (8·61 lb/ehp)

Performance (estimated, at max T-O weight):

Never-exceed speed	295 knots (546 km/h; 339 mph)
Max level speed at 3,050 m (10,000 ft)	230 knots (426 km/h; 265 mph)
Max cruising speed at 3,050 m (10,000 ft)	220 knots (407 km/h; 253 mph)
Econ cruising speed at 3,050 m (10,000 ft)	175 knots (324 km/h; 201 mph)
Stalling speed, flaps down	66 knots (122·5 km/h; 76 mph)
Max rate of climb at S/L	564 m (1,850 ft)/min
Service ceiling	8,352 m (27,400 ft)
Service ceiling, one engine out	4,572 m (15,000 ft)
T-O run	534 m (1,750 ft)
T-O to 15 m (50 ft)	640 m (2,100 ft)
Landing from 15 m (50 ft)	579 m (1,900 ft)
Landing run	358 m (1,175 ft)
Range with max payload, reserves for 87 nm (161 km; 100 mile) alternate and 45 min hold	310 nm (574 km; 357 miles)
Range with max fuel, 45 min reserves	1,150 nm (2,131 km; 1,324 miles)

AIDC (PAZMANY) PL-1B CHIENSHOU

A description of the PL-1A prototype appeared in the 1970-71 *Jane's*. The PL-1B is the version modified for production, of which 35 were built in 1970 and a further 10 in the Spring of 1972. Significant improvements compared with the PL-1A include a wider cockpit, larger rudder and more powerful engine. A third batch, also of 10 PL-1Bs, was built and delivered in 1974. Additional modifications to the engine cowling, nose landing gear, tail unit, flap control and pilot seats were made in these 10 aircraft

A full description and illustration of the PL-1B appeared in the 1975-76 *Jane's*.

TURKEY

TUSAS

Preliminary steps have been taken by the Turkish government to establish at Kayseri an aircraft manufacturing facility, to be known as TUSAS, financed jointly by the Turkish government (55%) and the Turkish Armed Forces Foundation (45%).

TUSAS was officially established with effect from 11 July 1973 with an initial capital of 300 million Turkish lire; this figure has since been increased to more than 1,000 million lire.

Plans envisage an initial programme for the manufacture in Turkey of some 200 aircraft for the Turkish Air Force. Detailed proposals were considered from major aerospace companies, including BAC and Hawker Siddeley Aviation from the UK, and Lockheed Aircraft Corporation, Northrop Corporation and Vought Corporation from the USA. The evaluation of contenders was co-ordinated by the Turkish Armed Forces Foundation, which in June 1975 concluded an agreement with Air-Metal of Germany (which see) to manufacture under licence that company's AM-C 111 STOL transport aircraft. The versions to be built in Turkey are known as the Series 200SP and 400SP.

UNITED KINGDOM

AJEP
AJEP DEVELOPMENTS

Address:
The Lodge, Marden Hill Farm, nr Hertford, Hertfordshire SG14 2NE
Telephone: Hertford (32) 51936
Managing Director: A. J. E. Perkins

AJEP (WITTMAN) TAILWIND

AJEP is marketing in the UK complete Tailwind aircraft to order. The marketing of plans and kits for construction by amateurs has been discontinued.

The AJEP Tailwind is a much-modified version of the Wittman Tailwind light aircraft (see Homebuilt Aircraft section). The first aircraft constructed by the UK company (G-AYDU), for use as a demonstrator, has been converted to a non-retractable nosewheel landing gear configuration. It is also the first Tailwind fitted with a Hoffmann HO-V 62R variable-pitch propeller. Following successful testing of these modifications, two versions of the AJEP Tailwind are available currently:

Tailwind Series 1. With fixed tailwheel-type landing gear. Available with choice of three alternative engines and variable-pitch or fixed-pitch propeller.

Tailwind Series 2. As Series 1, except for having a fixed tricycle-type landing gear.

A Series 1 aircraft was completed in March 1976 with a Rolls-Royce Continental O-200 engine driving a Hoffmann variable-pitch propeller. It is being used for a series of tests to suggest an optimum fixed-pitch propeller for the Tailwind. A drone version of the Series 2 is under development jointly with the Guided Weapons Division of BAC.

Type: Two-seat lightweight sporting aircraft.

Wings, Fuselage and Tail Unit: As for Wittman Tailwind.

Landing Gear: Series 1 has non-retractable tailwheel type, with Wittman tapered steel cantilever main legs. Cleveland wheels with tyres size 5·00-5. Cleveland disc brakes. Series 2 has non-retractable tricycle gear with single wheel on each unit, size 5·00-5. Nosewheel leg is steerable and has rubber blocks in compression for shock-absorption. Wheel fairings on main wheels of both series and on nosewheel of Series 2.

Power Plant: One 74·5 kW (100 hp) Continental PC60 (GPU) flat-four engine, or 71 kW (95 hp) Rolls-Royce Continental C90 or 74·5 kW (100 hp) O-200-A flat-four engine, driving a Hoffmann HO-V 62R two-blade metal variable-pitch or Hoffmann two-blade metal fixed-pitch propeller with spinner. Fuel tank in fuselage, immediately aft of firewall, capacity 95·5 litres (21 Imp gallons). Oil capacity 6·8 litres (1·5 Imp gallons).

Accommodation: Two seats side by side in enclosed cabin. Space aft of seats for child's optional seat or baggage, max capacity 27 kg (60 lb). Door on each side of cabin, beneath wing.

Dimensions, external:

Wing span	6·86 m (22 ft 6 in)
Wing chord, mean	1·22 m (4 ft 0 in)

AJEP Developments Wittman Tailwind Series 1 with tailwheel landing gear and variable-pitch propeller

AJEP Developments Wittman Tailwind Series 2 with tricycle-type landing gear. Nosewheel fairing and main leg fairings were removed for early test flying

Wing aspect ratio	5·63
Length overall	6·36 m (20 ft 10½ in)
Height overall	1·63 m (5 ft 4 in)
Wheel track:	
Series 1	1·84 m (6 ft 0½ in)
Series 2	1·93 m (6 ft 4 in)
Wheelbase:	
Series 1	4·50 m (14 ft 9 in)
Series 2	1·44 m (4 ft 8¾ in)

Area:

Wings, gross	8·36 m² (90 sq ft)

Weights and Loading:

Weight empty, equipped	377 kg (830 lb)
Max T-O and landing weight	635 kg (1,400 lb)
Max wing loading	75·97 kg/m² (15·56 lb/sq ft)

Performance (at max T-O weight. A: Series 1; B: Series 2 with 74·5 kW; 100 hp engine):

Never-exceed speed:	
A	156 knots (288 km/h; 179 mph)
Max level speed at S/L:	
A	156 knots (288 km/h; 179 mph)
B	152 knots (282 km/h; 175 mph)
Max cruising speed at S/L:	
A	147 knots (272 km/h; 169 mph)
B	143 knots (265 km/h; 165 mph)
Econ cruising speed at S/L:	
A	134 knots (248 km/h; 154 mph)
B	130 knots (241 km/h; 150 mph)
Stalling speed, flaps down:	
A, B	42 knots (78 km/h; 48·5 mph) IAS
Range with max fuel, no reserves:	
A	808 nm (1,496 km; 930 miles)
B	781 nm (1,447 km; 899 miles)
Range with max payload, no reserves:	
A	551 nm (1,020 km; 634 miles)
B	534 nm (990 km; 615 miles)

BAC
BRITISH AIRCRAFT CORPORATION LTD

HEAD OFFICE:
Brooklands Road, Weybridge, Surrey KT13 0RN
Telephone: Weybridge (0932) 45522
Telex: 27111

DIRECTORS:
A. H. C. Greenwood, CBE, JP, CEng, FRAeS (Chairman)
G. E. Knight, CBE (Vice-Chairman)
G. R. Jefferson, CBE, BSc, CEng, MIMechE, FRAeS (Managing Director, Guided Weapons)
F. W. Page, CBE, MA, CEng, FRAeS (Managing Director, Aircraft)
E. G. Barber, BSc, CEng, MRAeS, FIPM (Director of Personnel and Training)
H. R. Baxendale, OBE, ACMA (Deputy Chairman and Managing Director, Military Aircraft Division)
Brian Cookson, LLB (Director of Contracts)
Handel Davies, CB, MSc, CEng, FRAeS, FIAeS, FAIAA (Technical Director)
G. T. Gedge, CBE, CEng, FIProdE (Director of Manufacturing)
T. B. Pritchard, FCA (Financial Director)
J. Ferguson Smith, FCA (Deputy Chairman and Managing Director, Commercial Aircraft Division)

SECRETARY: A. C. Buckley, BSc
TREASURER: J. D. Hanson, LLB, FCA
PUBLICITY MANAGER: C. J. T. Gardner, OBE, MRAeS

Incorporated in 1963, British Aircraft Corporation Ltd was formed to take over the business formerly carried out by its predecessor companies' subsidiaries, Bristol Aircraft Ltd, English Electric Aviation Ltd, Vickers-Armstrongs (Aircraft) Ltd and Hunting Aircraft Ltd. The undertaking is ultimately owned by the General Electric Company Ltd (50%) and Vickers Ltd (50%), and has been rationalised progressively into three Divisions: the Commercial Aircraft Division, the Military Aircraft Division and the Guided Weapons Division, each managed by an individual management company with its own Board of Directors.

The Corporation has the following overseas subsidiaries: British Scandinavian Aviation AB, British Aircraft Corporation (Australia) Pty Ltd and British Aircraft Corporation (USA) Inc; and the following UK subsidiaries: British Aircraft Corporation (Insurance Brokers) Ltd, British Aircraft Corporation (Insurance) Ltd, British Aircraft Corporation (Pension Fund Trustees) Ltd, and British Aircraft Corporation (AT) Ltd. Its associated companies are SEPECAT (formed in May 1966 by BAC and Breguet Aviation to control the development and production of the Jaguar light strike fighter and trainer) and Panavia Aircraft GmbH (formed in March 1969 by BAC, MBB and Aeritalia to foster the development and production of the Tornado multi-role combat aircraft).

COMMERCIAL AIRCRAFT DIVISION

ADDRESS:
Brooklands Road, Weybridge, Surrey KT13 0SF
Telephone: Weybridge (0932) 45522
Telex: 27111
and
Filton House, Filton, Bristol
Telephone: Bristol (0272) 693831
Telex: 44163

DIRECTORS:
F. W. Page, CBE, MA, CEng, FRAeS (Chairman)
J. Ferguson Smith, FCA (Deputy Chairman and Managing Director)
Sir Geoffrey Tuttle, KBE, CB, DFC, FRAeS (Vice-Chairman)
K. Bentley, MA, CEng, MRAeS (Director of Resource Planning)
W. R. Coomber, CEng, FRAeS (Chief Executive, Weybridge)
G. Hanby, FCA, FCWA (Commercial Director)
J. T. Jefferies, CEng, FIProdE (Chief Executive, Filton)
E. E. Marshall, CBE, CEng, FRAeS (Director of Engineering)
H. Smith, FCA (Financial Director)
Dr W. J. Strang, CBE, PhD, BSc, CEng, FRAeS (Technical Director)
E. B. Trubshaw, CBE, MVO, FRAeS (Director of Flight Test)
M. G. Wilde, OBE, BSc, DipAe, CEng, FRAeS (Project Director, Concorde)

SECRETARY: H. T. Fream
PUBLIC RELATIONS MANAGER:
M. Savage

With effect from 1 June 1971 the former Filton and Weybridge Divisions of BAC were replaced by a Commercial Aircraft Division which includes the Filton, Hurn and Weybridge factories and the Fairford flight test centre.

Weybridge, with approximately 5,000 employees, is responsible for the manufacture of all Concorde flight deck and forward fuselage sections, and all rear fuselage, fin and rudder assemblies. In addition it is involved in a variety of subcontract work in support of programmes managed by other BAC Divisions and other aerospace manufacturers, as well as design studies for new subsonic transports.

The factory at Hurn, near Bournemouth, has approximately 2,000 employees and is responsible for BAC One-Eleven final assembly, the production of all Concorde droop nose and visor sections, the assembly of the complete electrical wiring harness for all Concordes and numerous subcontract work programmes. There is also a steady flow of One-Elevens passing through Hurn which are being reconfigured to customers' requirements. Both Weybridge and Hurn are responsible for supporting many hundreds of aircraft in service throughout the world. They include not only One-Elevens, but Viscounts, Vanguards, VC10s, Super VC10s and the Hunting Prince/Pembroke series, all of which have been described in earlier editions of *Jane's*.

Filton, with 5,000 employees (excluding those on guided weapons and space work), is responsible for design and development of the Concorde supersonic transport, in partnership with Aérospatiale of France, and is the location of the British Concorde final assembly centre. Filton is also carrying out a number of subcontract work programmes, and provides after-sales support for former Bristol aircraft which remain in service.

Fairford, in Gloucestershire, is the site of the Division's Flight Test Centre, and is the base from which the British Concorde flight test programme has been controlled.

BAC/AÉROSPATIALE CONCORDE

Details of the Concorde programme can be found in the International section of this edition.

BAC ONE-ELEVEN

Details of the One-Eleven were announced on 9 May 1961, simultaneously with the news that British United Airways had ordered ten. Design and manufacture are shared between three BAC factories, at Weybridge, Filton and Hurn.

Five commercial versions have been produced, and details of the Series 200, 300 and 400 can be found in the 1974-75 *Jane's*. Two versions are available currently, as follows:

Series 475. Combines standard fuselage and accommodation of Series 400 with wings and power plant of Series 500 and a modified landing gear system, using low-pressure tyres, to permit operation from secondary low-strength runways with poorer-grade surfaces. The Srs 400/500 development aircraft (G-ASYD) was converted to serve as prototype and flew for the first time on 27 August 1970. First production Series 475 (G-AYUW) flew for the first time on 5 April 1971. Certification and first production delivery (to Faucett of Peru) in July 1971. The three Srs 475s supplied to the Sultan of Oman's Air Force have a quick-change passenger/cargo interior layout and a 3·05 × 1·85 m (10 ft 0 in × 6 ft 1 in) forward freight door.

Series 500. Derived from Series 300/400, this version incorporates a lengthened fuselage (2·54 m; 100 in fwd of wing, 1·57 m; 62 in aft) which accommodates 97-119 passengers, with a flight crew of two. Wingtip extensions increase span by 1·52 m (5 ft). Take-off performance improved by increased wing area and by installation of two Rolls-Royce Spey Mk 512 DW turbofans, each rated at 55·8 kN (12,550 lb st). Main landing gear strengthened and heavier wing plank stringers used to cater for increased AUW.

Prototype, converted from Srs 400 development aircraft (G-ASYD), flew for the first time on 30 June 1967. First Srs 500 production aircraft (G-AVMH) flew on 7 February 1968. ARB certification 15 August 1968. Deliveries to BEA began on 29 August 1968, and regular services on 17 November 1968.

In addition, the following future versions are being studied:

Series 700/800. Stemming directly from the One-Eleven Series 500, bigger and quieter versions—in the 120-160 seat bracket—are currently being evaluated for the 1980s under these series designations.

Executive and freighter versions of the One-Eleven are available also. More than 30 examples of the former are now in service, notably in the USA and with the Australian and Brazilian governments, with interior and long-range tank conversions being made either by BAC or by specialist contractors. The One-Eleven freighter incorporates a 3·05 m by 1·85 m (10 ft 0 in by 6 ft 1 in) upward-opening hydraulically-powered loading door in the forward fuselage, together with a quickly-removable freight floor overlay and cargo handling system. First to be delivered, in November 1975, was one of three Series 475 freighters operated by the Sultan of Oman's Air Force.

A One-Eleven 'hush kit', comprising an intake duct lining, a by-pass duct lining, an acoustically-lined jetpipe, and a six-chute exhaust silencer, was flown for the first time, on the Srs 475 development aircraft G-ASYD, on 14 June 1974. It is designed to reduce the area within the 90 EPNdB noise contour by approximately 50 per cent, giving a noise footprint equivalent to that of a twin-turboprop aircraft. The effectiveness of the 'hush kit' has been proved and demonstrated by comprehensive testing, performance penalties proving to be lower than estimated, with a 0·75 per cent thrust loss on take-off and a 2 per cent increase in fuel consumption. The installed weight of production kits will be less than 181 kg (400 lb).

Orders for 220 One-Elevens had been received by 1 April 1976, of which 215 had been delivered. Totals of the various series sold by that date are as follows:

Series 200	56
Series 300	9
Series 400	69
Series 475	8
Series 500	78
	220

Operators of the One-Eleven in Spring 1976 were as follows:

Series 200	
Aer Lingus	4
Allegheny	31
BCAL	7
Braniff	1
Dan-Air	2
Executive and military operators	8
Series 300	
Dan-Air	2
Laker Airways	5
Quebecair	2
Series 400	
Austral	4
Bahamasair	3
Bavaria Flug	2
British Airways	7
Dan-Air	3
Gulf Air	4
Liniile Aeriene Romane	2
Quebecair	1
TACA	3
Tarom	5
Executive and military operators	32
Series 475	
Air Malawi	1
Air Pacific	2
Faucett (Peru)	2
Sultan of Oman's Air Force	3
Series 500	
Austral	5
AVIATECA	2
BAC	1
Bavaria Flug	3
BCAL	10
British Airways	18
Dan-Air	5
Germanair	4
LACSA	4
Monarch Airlines	3
PAL	9
Tarom	0†
Transbrasil	6

†5 *due for delivery in* 1977; *to be fitted with hush-kits*

The following description applies to the Series 475 and 500:

TYPE: Twin-engined short/medium-range turbofan transport.

WINGS: Cantilever low-wing monoplane. Modified NACA cambered wing section. Thickness/chord ratio 12½% at root, 11% at tip. Dihedral 2°. Incidence 2° 30′. Sweepback 20° at quarter-chord. All-metal structure of copper-based aluminium alloy, built on fail-safe principles. Three-shear-web torsion box with integrally-machined skin/stringer panels. Ailerons of Redux-bonded light alloy honeycomb, manually operated through servo tabs. Port servo tab used for trimming. Light alloy Fowler flaps hydraulically operated through Hobson actuators. Light alloy spoiler/airbrakes on upper surface of wing, operated hydraulically through Dowty Boulton Paul actuators. Hydraulically-actuated lift dumpers, inboard of spoilers, are standard. Flaps on Series 475 have a glassfibre coating. Thermal de-icing of wing leading-edges with engine bleed air.

FUSELAGE: Conventional circular-section all-metal fail-safe structure with continuous frames and stringers. Skin made from copper-based aluminium alloy.

TAIL UNIT: Cantilever all-metal fail-safe structure, with variable-incidence T tailplane, controlled through duplicated Hobson hydraulic units. Fin integral with rear fuselage. Elevators and rudder actuated hydraulically through Dowty Boulton Paul tandem jacks. Leading-edges of fin and tailplane de-iced by engine bleed air.

LANDING GEAR: Retractable tricycle type, with twin wheels on each unit. Hydraulic retraction, nose unit forward, main units inward. Oleo-pneumatic shock-absorbers manufactured by BAC. Hydraulic nosewheel steering. Dunlop wheels, tubeless tyres and 5-plate heavy-duty hydraulic disc brakes. Hytrol Mk III anti-skid units. Main-wheel tyres size 40 × 12 on Srs 500, pressure 11·03 bars (160 lb/sq in). Dunlop 44 × 16 tyres on Srs 475, pressure 5·72 bars (83 lb/sq in). Nosewheel tyres size 24 × 7·25 on Srs 500, pressure 7·58 bars (110 lb/sq in). Dunlop 24 × 7·7 tyres on Srs 475, pressure 7·24 bars (105 lb/sq in). All tyre pressures are given for aircraft at mid-CG position and operating at max taxi weight.

POWER PLANT: Two Rolls-Royce Spey Mk 512 DW turbofan engines, each rated at 55·8 kN (12,550 lb st), mounted in pod on each side of rear fuselage. Fuel in integral wing tanks of 10,160 litres (2,235 Imp gallons) and centre-section tank of 3,864 litres (850 Imp gallons) capacity; total fuel capacity 14,024 litres (3,085 Imp gallons). Optional 1,591 litre (350 Imp gallon) and 3,182 litre (700 Imp gallon) fuel tanks are available to increase total fuel capacity. Pressure refuelling point in fuselage forward of wing on starboard side. Provision for gravity refuelling. Oil capacity (total engine oil) 13·66 litres (3 Imp gallons) per engine.

ACCOMMODATION (Srs 475): Crew of two on flight deck and up to 89 passengers in main cabin. Single class or mixed class layout, with movable divider bulkhead to permit any first/tourist ratio. Typical mixed class layout has 16 first class (four abreast) and 49 tourist (five abreast) seats. Galley units normally at front on starboard side. Coat space available on port side aft of flight deck. Ventral entrance with hydraulically-operated airstair. Forward passenger door on port side incorporates optional power-operated airstair. Galley service door forward on starboard side. Two baggage and freight holds under floor, fore and aft of wings, with doors on starboard side. Forward freight door on Srs 475s for Sultan of Oman's Air Force. Entire accommodation air-conditioned.

ACCOMMODATION (Srs 500): Crew of two on flight deck and up to 119 passengers in main cabin. Two additional overwing emergency exits, making two on each side. One toilet on each side of cabin at rear. Otherwise generally similar to Srs 475.

SYSTEMS: Fully-duplicated air-conditioning and pressurisation systems with main components by Normalair-Garrett. Air bled from engine compressors through heat exchangers. Max pressure differential 0·52 bars (7·5 lb/sq in). Hydraulic system, pressure 207 bars (3,000 lb/sq in), operates flaps, spoilers, rudder, elevators, tailplane, landing gear, brakes, nosewheel steering, ventral and forward airstairs and windscreen wipers. No pneumatic system. Electrical system utilises two 30kVA Plessey/Westinghouse AC generators, driven by Plessey constant-speed drive and starter units, plus a similar generator mounted on the APU and shaft-driven. AiResearch gas-turbine APU in tailcone to provide ground electrical power, air-conditioning and engine starting, also some system checkout capability. APU is run during take-off to eliminate performance penalty of bleeding engine air for cabin air-conditioning.

ELECTRONICS AND EQUIPMENT: Communications and navigation equipment generally to customers' individual requirements. Typical installation includes dual VHF communications equipment to ARINC 546, dual VHF navigation equipment to ARINC 547A, including glideslope receivers, marker receiver, flight/service interphone system, Marconi AD 370, Bendix DFA 73 or Collins DF 203 ADF, ATC transponder to ARINC 532D, Collins 860 E2 DME, Ekco E 190 or Bendix RDR 1E weather radar. Sperry C9 or CL11 compass systems and Collins FD 108 flight director system (dual) are also installed. The autopilot is the Elliott 2000 Series system and provision is made on the Srs 500 for additional equipment, including automatic throttle control, for low weather minima operation.

DIMENSIONS, EXTERNAL (Srs 475, 500):

Wing span	28·50 m (93 ft 6 in)
Wing chord at root	5·01 m (16 ft 5 in)
Wing chord at tip	1·61 m (5 ft 3½ in)
Wing aspect ratio	8·5
Length overall:	
Srs 475	28·50 m (93 ft 6 in)
Srs 500	32·61 m (107 ft 0 in)
Length of fuselage:	
Srs 475	25·55 m (83 ft 10 in)
Srs 500	29·67 m (97 ft 4 in)
Height overall	7·47 m (24 ft 6 in)
Tailplane span	8·99 m (29 ft 6 in)
Wheel track	4·34 m (14 ft 3 in)
Wheelbase:	
Srs 475	10·08 m (33 ft 1 in)
Srs 500	12·62 m (41 ft 5 in)
Passenger door (fwd, port):	
Height	1·73 m (5 ft 8 in)
Width	0·82 m (2 ft 8 in)
Height to sill	2·13 m (7 ft 0 in)
Ventral entrance:	
Height	1·83 m (6 ft 0 in)
Width	0·66 m (2 ft 2 in)
Height to sill	2·13 m (7 ft 0 in)
Freight door (fwd, starboard):	
Height (projected)	0·79 m (2 ft 7 in)
Width	0·91 m (3 ft 0 in)
Height to sill	1·09 m (3 ft 7 in)
Freight door (rear, starboard):	
Height (projected)	0·66 m (2 ft 2 in)
Width	0·91 m (3 ft 0 in)
Height to sill	1·30 m (4 ft 3 in)
Freight door (fwd, Srs 475 SOAF):	
Height	1·85 m (6 ft 1 in)
Width	3·05 m (10 ft 0 in)
Galley service door (fwd, starboard):	
Height (projected)	1·22 m (4 ft 0 in)
Width	0·69 m (2 ft 3 in)
Height to sill	2·13 m (7 ft 0 in)

DIMENSIONS, INTERNAL (Srs 475):

Cabin, excl flight deck:	
Length	17·31 m (56 ft 10 in)
Max width	3·16 m (10 ft 4 in)
Max height	1·98 m (6 ft 6 in)
Floor area	approx 47·0 m² (506 sq ft)
Freight hold, fwd	10·02 m³ (354 cu ft)
Freight hold, rear	4·42 m³ (156 cu ft)

DIMENSIONS, INTERNAL (Srs 500):

Cabin, excl flight deck:	
Length	21·44 m (70 ft 4 in)
Total floor area	approx 61·78 m² (665 sq ft)
Freight holds (total volume)	19·45 m³ (687 cu ft)

AREAS (Srs 475, 500):

Wings, gross	95·78 m² (1,031 sq ft)
Ailerons (total)	2·86 m² (30·8 sq ft)
Flaps (total)	16·30 m² (175·6 sq ft)
Spoilers (total)	2·30 m² (24·8 sq ft)
Vertical tail surfaces (total)	10·90 m² (117·4 sq ft)
Rudder, incl tab	3·05 m² (32·8 sq ft)
Horizontal tail surfaces (total)	23·90 m² (257·0 sq ft)
Elevators, incl tab	6·55 m² (70·4 sq ft)

WEIGHTS AND LOADINGS:

Operating weight empty:	
Srs 475	23,464 kg (51,731 lb)
Srs 500	24,758 kg (54,582 lb)
Max payload:	
Srs 475	9,647 kg (21,269 lb)
Srs 500	11,983 kg (26,418 lb)
Max T-O weight:	
Srs 475	41,730-44,678 kg (92,000-98,500 lb)
Srs 500	45,200-47,400 kg (99,650-104,500 lb)
Max ramp weight:	
Srs 475	44,905 kg (99,000 lb)
Srs 500	47,625 kg (105,000 lb)
Max landing weight:	
Srs 475	38,100-39,462 kg (84,000-87,000 lb)
Srs 500	39,462 kg (87,000 lb)
Max zero-fuel weight:	
Srs 475	33,112 kg (73,000 lb)
Srs 500	36,741 kg (81,000 lb)
Max wing loading:	
Srs 475	435·5 kg/m² (89·2 lb/sq ft)
Srs 500	472 kg/m² (96·7 lb/sq ft)
Max power loading:	
Srs 475	400·3 kg/kN (3·92 lb/lb st)
Srs 500	424·7 kg/kN (4·16 lb/lb st)

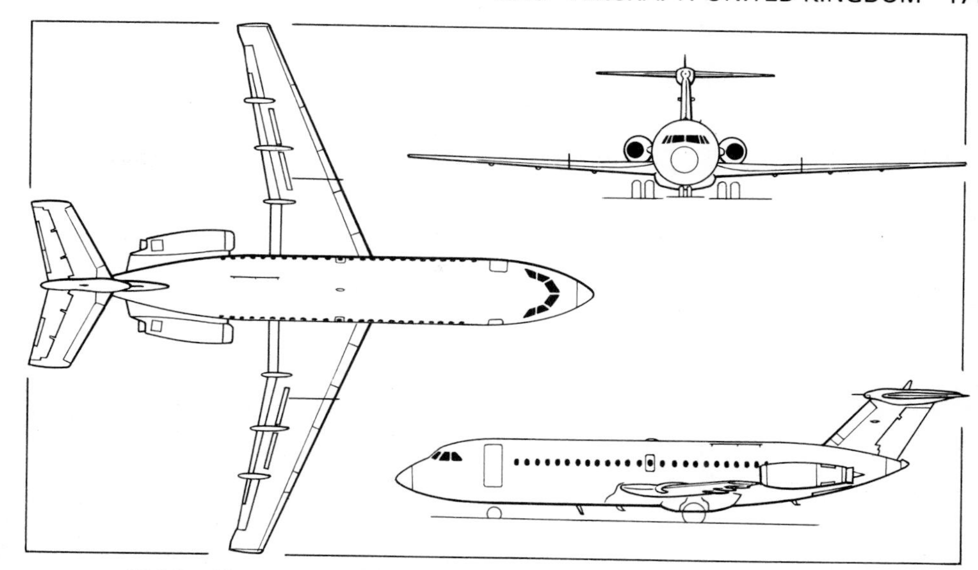

BAC One-Eleven Series 475 twin-turbofan short/medium-range airliner *(Pilot Press)*

BAC One-Eleven Series 475 convertible passenger/cargo transport of the Sultan of Oman's Air Force

BAC One-Eleven Series 500 twin-turbofan transport in the insignia of Philippine Airlines (PAL)

PERFORMANCE (at max T-O weight):
- Never-exceed speed (structural) 410 knots (760 km/h; 472 mph) EAS
- Max level and cruising speed at 6,400 m (21,000 ft) 470 knots (871 km/h; 541 mph) TAS
- Fuel econ cruising speed at 7,620 m (25,000 ft) 400 knots (742 km/h; 461 mph) TAS
- Stalling speed (T-O flap setting):
  - Srs 475 99 knots (184 km/h; 114 mph) EAS
  - Srs 500 105 knots (195 km/h; 121 mph)
- Rate of climb at S/L at 300 knots (555 km/h; 345 mph) EAS:
  - Srs 475 756 m (2,480 ft)/min
  - Srs 500 695 m (2,280 ft)/min
- Max cruising height 10,670 m (35,000 ft)
- Min ground turning radius:
  - Srs 475 17·07 m (56 ft 0 in)
  - Srs 500 17·98 m (59 ft 0 in)
- Runway LCN at max weight, rigid pavement (1: 40):
  - Srs 475 32
  - Srs 500 53
- T-O run at S/L, ISA:
  - Srs 475 1,676 m (5,500 ft)
  - Srs 500 1,981 m (6,500 ft)
- Balanced T-O to 10·7 m (35 ft) at S/L, ISA:
  - Srs 475 1,798 m (5,900 ft)
  - Srs 500 2,225 m (7,300 ft)
- Landing distance (BCAR) at S/L, ISA, at max landing weight:
  - Srs 475 1,439 m (4,720 ft)
- Landing run at S/L, ISA at max landing weight:
  - Srs 475 826 m (2,710 ft)
- Still-air range with max fuel, ISA, with reserves for 200 nm (370 km; 230 mile) diversion and 45 min hold:
  - Srs 475 1,997 nm (3,700 km; 2,300 miles)
  - Srs 500 1,880 nm (3,484 km; 2,165 miles)
- Still-air range with typical capacity payload, ISA, reserves as above:
  - Srs 475 at 44,678 kg (98,500 lb) 1,619 nm (3,000 km; 1,865 miles)
  - Srs 500 at 47,400 kg (104,500 lb) 1,480 nm (2,744 km; 1,705 miles)
  - Srs 475 executive aircraft with additional 3,182 litres (700 Imp gallons) fuel has equivalent range of 2,549 nm (4,725 km; 2,936 miles)

MILITARY AIRCRAFT DIVISION

ADDRESS:
Warton Aerodrome, Preston, Lancashire PR4 1AX
Telephone: Preston (0772) 633333
Telex: 67627

DIRECTORS:
- F. W. Page, CBE, MA, CEng, FRAeS (Chairman)
- H. R. Baxendale, OBE, ACMA (Deputy Chairman and Managing Director)
- A. F. Atkin, OBE, BSc(Hons), DipAe (Hull), CEng, FIMechE, FRAeS (Deputy Managing Director)
- R. P. Beamont, CBE, DSO, DFC, FRAeS (Director of Flight Operations)
- P. Grocock (Commercial Director)
- B. O. Heath, BSc, DIC, CEng, MRAeS (Director of Engineering)
- F. E. Roe, BSc, DIC, CEng, ACGI, FRAeS (Director of Resources and Quality)
- Air Chief Marshal Sir Frederick Rosier, GCB, CBE, DSO (Director and Military Adviser)
- R. H. Sawyer, FCA, FCMA, JDipMA (Financial Director)
- T. O. Williams, MA, CEng, FIMechE, MIEE (Production Director)
- I. R. Yates, BEng, CEng, FRAeS, AMIMechE (Director of Projects)

SPECIAL DIRECTORS:
- R. B. Coles, BSc, CEng, MRAeS (Chief Designer—Systems)
- R. Dickson, MA (Cantab), CEng, FRAeS (Chief of Research)
- S. Gillibrand, MSc, CEng, MRAeS, AMBIM (General Manager, Manufacturing)
- G. M. Hobday, OBE, CEng, MRAeS (Deputy Commercial Director, Product Support)
- W. D. Horsfield, BSc, CEng, MRAeS (Chief Aerodynamicist)
- D. F. McGregor, BSc, CEng, MRAeS (Works Manager, Development)
- J. K. Quill, OBE, AFC (Sales Director, SEPECAT, and Director of Marketing, Panavia)
- W. J. Sarginson (General Manager, Production)
- F. G. Willox, MSc, BSc (Hons), CEng, FRAeS, MIMechE (Tornado Project Manager)

SECRETARY: L. F. Trueman, FCCA

PROJECT MANAGERS:
- R. J. W. Brown (General Manager, Sales)
- M. W. Cara, CEng, MIMechE (Canberra, Lightning, Jet Provost, Strikemaster and Concorde)
- A. Constantine (Jaguar)

PUBLICITY MANAGER:
A. F. Johnston

This Division includes the Preston, Warton and Samlesbury works and has 14,250 employees, including 2,000 in Saudi Arabia. It is responsible for production of the BAC 167, for the refurbishing of Canberra aircraft for overseas customers, after-sales service for operators of this aircraft, and service support of Lightnings and Jet Provosts.

Current programmes also include those for the Panavia Tornado and SEPECAT Jaguar military aircraft, as described in the International section.

BAC JET PROVOST

Under a three-year programme involving 157 aircraft, BAC is installing VOR and DME equipment in Jet Provost T. Mk 3 and T. Mk 5 aircraft in service with the RAF. The refurbished aircraft are designated T. Mk 3A and T. Mk 5A respectively. This programme was scheduled for completion in mid-1976.

BAC 167 STRIKEMASTER

The BAC 167 Strikemaster was developed from the BAC 145 series (see 1972-73 *Jane's*). It has the same airframe, but is powered by a Rolls-Royce Viper Mk 535 turbojet engine (15·2 kN; 3,410 lb st) and has eight underwing hardpoints, enabling it to carry up to 1,360 kg (3,000 lb) of stores. This makes it particularly suitable for counter-insurgency combat operations.

The first BAC 167 (G-27-8) was flown for the first time on 26 October 1967, and the aircraft continues in production. Deliveries of the more recent orders were scheduled to continue until the latter part of 1976.

Orders for the Strikemaster include:

Strikemaster Mk 83 of the Kuwait Air Force, with underwing bombs and drop-tanks

Mk 80. For Royal Saudi Air Force. Twenty-five ordered in 1966. Deliveries began in 1968 and were completed in September 1969.

Mk 80A. Further order, for ten aircraft, for Royal Saudi Air Force.

Mk 81. For South Yemen People's Republic. Four ordered in December 1966; delivery completed in May 1969.

Mk 82. For Sultan of Oman's Air Force. Twelve ordered; delivery completed in December 1969.

Mk 82A. For Sultan of Oman's Air Force. Eight ordered.

Mk 83. For Kuwait Air Force. First order, for six, placed in October 1968; deliveries began in 1969. Order subsequently increased to 12, the last of which was delivered in July 1971.

Mk 84. For Singapore Air Defence Command. Sixteen ordered; delivery completed in September 1970.

Mk 87. For Kenya Air Force. Six ordered.

Mk 88. For Royal New Zealand Air Force. Ten ordered; delivery completed in October 1972. In service with No. 14 Squadron of RNZAF. Six more ordered in Spring 1974. Used also for advanced training.

Mk 89. For Ecuadorean Air Force. Initial order for eight (later increased to 12), of which delivery began in early 1973.

Eight more aircraft (four each for two undisclosed existing operators of Strikemasters) were ordered in Spring 1974. Four of these are reported to be for Ecuador and four for Oman. Eleven additional aircraft have been ordered by Saudi Arabia, and five BAC 145 Strikemasters serve with the Sudan Air Force.

The following description applies primarily to later variants of the Strikemaster, such as the Mk 88, but is also substantially applicable to earlier versions:

TYPE: Two-seat ab initio, basic and advanced trainer, armament trainer and tactical support aircraft.

WINGS: Cantilever low-wing monoplane. Wing section NACA 23015 (modified) at root, NACA 4412 (modified) at tip. Dihedral 6°. Incidence 3° at root, 0° at tip. All-metal structure, with main and subsidiary spars, having three-point attachment to fuselage. Metal-covered ailerons with balance tabs. Hydraulically-operated slotted flaps. Hydraulically-operated air-brakes and lift spoilers on wings at rear spar position ahead of flaps.

FUSELAGE: All-metal semi-monocoque stressed-skin structure, built in three parts, comprising bulkheads, built-up frames and longerons covered with light alloy panels. Hinged nose cap provides access to pressurisation, oxygen, radio and electrical equipment.

TAIL UNIT: Cantilever all-metal structure. One-piece tail-plane, interchangeable elevators, fin and rudder. Fixed surfaces covered with smooth and movable surfaces with fluted alloy skin. Combined trim and balance tab in starboard elevator; balance tabs in port elevator and rudder.

LANDING GEAR: Hydraulically-retractable tricycle type. Main wheels retract inward into wings, nosewheel forward. Dowty oleo-pneumatic shock-absorbers. Dunlop main wheels with tubeless tyres size 21 × 6·75-9, pressure 6·90 bars (100 lb/sq in). Dunlop nosewheel and tubeless tyre size 6·00-4, pressure 6·21 bars (90 lb/sq in). Dunlop hydraulic disc brakes.

POWER PLANT: One Rolls-Royce Bristol Viper Mk 535 turbojet engine (15·2 kN; 3,140 lb st) in fuselage aft of cockpit. Lateral intake on each side of forward fuselage. Internal fuel capacity (one integral tank outboard and three bag tanks inboard in each wing) is 1,227 litres (270 Imp gallons). Refuelling point near each wingtip. Two wingtip fuel tanks, total capacity 436 litres (96 Imp gallons), are a standard fit at all times. All tanks in wings are interconnected. System designed to permit 18 sec of inverted flight. Oil capacity 8 litres (1·75 Imp gallons).

ACCOMMODATION: Two persons side by side in pressurised cabin, on Martin-Baker automatic ejection seats (Mk PB4/1 and PB4/2), suitable for use down to ground level and 90 knots (167 km/h; 104 mph). Power-operated rearward-sliding canopy. Dual controls standard.

SYSTEMS: Pressurisation and air-conditioning system by Normalair and Tiltman Langley, differential 0·21 bars (3 lb/sq in), using engine bleed air. Hydraulic system, pressure 145 bars (2,100 lb/sq in), for landing gear, flaps, airbrakes, lift spoilers and wheel brakes. Engine-driven generator provides 28V DC supply. Three 25Ah batteries. Two inverters supply phased AC to flight instruments and fire warning system. Automatically-controlled gaseous oxygen system for each crew member.

RADIO AND NAVIGATION EQUIPMENT: Varies in different Mks to meet individual customer's requirements. The following radio equipment has been installed in various combinations: ARC 51 BX and ARC 52 UHF; PV 141 UHF homer; D 403 UHF standby; PTR 175 UHF/VHF; Collins 618M VHF; Collins 618 FIA VHF standby; ARI 18120/2 Violet Picture; Sunair ASB 100 and SA 14-RA HF; and PTR 446, SSR 1600 and SSR 2100 IFF. The following navigation equipment has been installed in various combinations: Bendix CNS 220B UHF; Bendix CNS 240B VHF; RCA AVQ-75 DME; AD 370B and ADF 722 ADF; Bendix 221 VOR/ILS; and ARN 84, ARN 52 and ARN 65 Tacan.

ARMAMENT: Two 7·62 mm FN machine-guns, with 550 rds/gun; one in the lower lip of each engine air intake duct. Later variants have SFOM gunsights; GM2L reflector gunsights fitted to some earlier models. Provision for a G90 gun camera and a Smiths camera sight recorder. Four underwing strongpoints for the carriage of external stores. Typical underwing loads include two 341 and two 227 litre (75 and 50 Imp gallon) drop-tanks; four Matra launchers each containing eighteen 68 mm SNEB rockets; four 540 LAU 68 rocket launchers, each with seven rockets; four 540 lb ballistic or retarded bombs, four 250 kg or 500 kg bombs; four PMBR carriers, each with six practice bombs; light-series bomb carriers to carry 8·5, 19 or 25 lb practice bombs; BAC/Vinten five-camera reconnaissance pod; or banks of SURA 80 mm rockets, with four rockets per bank. Other armament, to specific customer requirements, can include napalm tanks, 65 or 125 kg bombs, 2·75 in or 3 in rockets, and 7·62 mm or 20 mm gun packs. Max T-O weight of 5,215 kg (11,500 lb) includes one pilot only, full usable fuel (internal and wingtip tanks) and 1,200 kg (2,650 lb) of external stores. Max possible external stores load 1,360 kg (3,000 lb).

DIMENSIONS, EXTERNAL:
- Wing span over tip-tanks 11·23 m (36 ft 10 in)
- Wing chord at root 2·33 m (7 ft 8 in)

Wing chord at tip	1·27 m (4 ft 2 in)
Wing aspect ratio	5·84
Length overall	10·27 m (33 ft 8½ in)
Height overall	3·34 m (10 ft 11½ in)
Tailplane span	4·11 m (13 ft 6 in)
Wheel track	3·27 m (10 ft 8·9 in)
Wheelbase	2·93 m (9 ft 7·4 in)

AREAS:

Wings, gross	19·85 m² (213·7 sq ft)
Ailerons (total)	1·77 m² (19·06 sq ft)
Trailing-edge flaps (total)	2·30 m² (24·80 sq ft)
Fin	0·86 m² (9·30 sq ft)
Rudder	1·00 m² (10·74 sq ft)
Tailplane	2·18 m² (23·51 sq ft)
Elevators, incl tabs	1·93 m² (20·80 sq ft)

WEIGHTS:
Operating weight empty, equipped, incl crew
2,810 kg (6,195 lb)
Typical T-O weights:
pilot conversion training, 2 crew, full internal fuel 4,219 kg (9,303 lb)
armament training, 2 crew, full internal fuel, practice armament (bombs and racks)
4,808 kg (10,600 lb)
ferry role, 2 crew, full internal fuel plus inboard and outboard drop-tanks
5,213 kg (11,493 lb)
*Max T-O weight 5,215 kg (11,500 lb)
See note under 'Armament' paragraph

PERFORMANCE (at max T-O weight except where indicated):
Never-exceed speed 450 knots (834 km/h; 518 mph)
Max level speed, with 50% fuel, 'clean'
at S/L 391 knots (724 km/h; 450 mph)
at 5,485 m (18,000 ft)
418 knots (774 km/h; 481 mph)
at 6,100 m (20,000 ft)
410 knots (760 km/h; 472 mph)
Stalling speed at 4,309 kg (9,500 lb) AUW:
flaps up 98·5 knots (182 km/h; 113 mph)
flaps down 85·5 knots (158 km/h; 98 mph)
Max rate of climb at S/L (training, full internal fuel) 1,600 m (5,250 ft)/min
Time to height (training, full internal fuel):
to 9,150 m (30,000 ft) 8 min 45 sec
to 12,200 m (40,000 ft) 15 min 30 sec
Service ceiling 12,200 m (40,000 ft)

BAC 167 Strikemaster light attack aircraft *(Pilot Press)*

T-O to 15 m (50 ft):
at 3,579 kg (7,930 lb) AUW (training)
579 m (1,900 ft)
at 5,215 kg (11,500 lb) AUW (combat)
1,067 m (3,500 ft)
Landing from 15 m (50 ft):
at 2,948 kg (6,500 lb) AUW (training)
732 m (2,400 ft)
at 5,103 kg (11,250 lb) AUW (aborted armed sortie) 1,295 m (4,250 ft)
Combat radius (hi-lo-hi), 5 min over target, 10% reserves:
with 1,360 kg (3,000 lb) weapons load
215 nm (397 km; 247 miles)
with 907 kg (2,000 lb) weapons load
355 nm (656 km; 408 miles)
with 454 kg (1,000 lb) weapons load
500 nm (925 km; 575 miles)
Combat radius (lo-lo-lo, at S/L), 5 min over target, 10% reserves:
with 1,360 kg (3,000 lb) weapons load
126 nm (233 km; 145 miles)
with 907 kg (2,000 lb) weapons load
175 nm (323 km; 201 miles)
with 454 kg (1,000 lb) weapons load
240 nm (444 km; 276 miles)
reconnaissance mission
300 nm (555 km; 345 miles)
Range with 91 kg (200 lb) fuel reserves:
at 3,789 kg (8,355 lb) AUW (training)
629 nm (1,166 km; 725 miles)
at 4,558 kg (10,500 lb) AUW (combat)
1,075 nm (1,992 km; 1,238 miles)
at 5,215 kg (11,500 lb) AUW (max T-O)
1,200 nm (2,224 km; 1,382 miles)

BRITISH AEROSPACE

As a first step towards the nationalisation of major sections of the UK aerospace industry, it was announced on 13 April 1976 that six members had been appointed to the Organising Committee, under the Chairmanship of Lord Beswick, as follows:

L. W. Buck, President of the Confederation of Shipbuilding and Engineering Unions; and General Secretary, National Union of Sheet Metal Workers, Coppersmiths and Domestic Heating Engineers
A. H. C. Greenwood, CBE, JP, FRAeS, Chairman, British Aircraft Corporation Ltd
G. R. Jefferson, CBE, Hon BSc (Eng), CEng, MIMechE, FRAeS, Chairman and Managing Director, Guided Weapons Division, British Aircraft Corporation Ltd
Dr A. W. Pearce, CBE, PhD, Chairman, Esso Petroleum Co Ltd
E. G. Rubython, General Manager, Hawker Siddeley Aviation Ltd
J. T. Stamper, MA, CEng, FRAeS, Technical Director, Hawker Siddeley Aviation Ltd.

One further member was appointed on 30 April:
Bernard Friend, lately Chairman and Managing Director of Esso Chemicals UK Ltd

It is intended that Mr Greenwood should be appointed full-time Deputy Chairman of British Aerospace in due course, with Dr Pearce as a part-time member and the others as full-time members. Mr Friend will be Finance Member of British Aerospace.

BRITTEN-NORMAN

BRITTEN-NORMAN (BEMBRIDGE) LTD (Member of the Fairey Group)

HEAD OFFICE:
Bembridge Airport, Bembridge, Isle of Wight PO35 5PR
Telephone: Bembridge 2511/5
Telex: 86277

LONDON SALES OFFICE:
Fairey Britten-Norman Ltd, Cranford Lane, Heston, Middlesex
Telephone: (01) 759-0692
Telex: 886654

DIRECTORS:
D. Lee-Smith (Chairman)
Dr W. G. Watson (Managing Director)
K. W. Mills
P. A. Hatswell
A. A. Brown
D. O. Thurgood (Managing Director, Fairey Britten-Norman Ltd)
P. Desai

CHIEF DESIGNER: A. J. Coombe
PUBLICITY MANAGER: S. R. P. Thomson
WORKS MANAGER: L. Lathwell
COMMERCIAL MANAGER: P. P. Graham
SALES DIRECTOR: P. A. Hatswell

Britten-Norman became a member of the Fairey Group on 31 August 1972, when the assets of Britten-Norman (Bembridge) Limited (see 1972-73 *Jane's*) were acquired by The Fairey Company. The announcement assured the future of the business after some doubt following the original company's financial difficulties which had arisen in 1971.

The Fairey Group has experience in aviation going back to 1915, and is diversified, with marine, nuclear and other interests. One of the largest companies within the Group is Fairey SA at Gosselies, Belgium (which see).

One of Britten-Norman's major problems had been a lack of 'in-house' production facilities, and this was overcome with the formation of a holding company, Fairey Britten-Norman, which looks after the mutual interests of Fairey SA and Britten-Norman (Bembridge) Limited. A new Islander/Trislander production line was established at Gosselies, and Romanian Islander production by IRMA (which see) is also continuing, a further 100 aircraft having been ordered during the past two years.

Final finishing of all aircraft to the specific requirements of individual customers, design, marketing and product support continue at Bembridge.

Britten-Norman announced at the 1974 SBAC Display at Farnborough an agreement with the Philippine Aerospace Development Corporation (which see) for the progressive assembly and manufacture in the Philippines of 100 Islanders in four phases during the period 1975-80.

BRITTEN-NORMAN BN-2A ISLANDER

The Islander is a modern replacement for aircraft in the class of the de Havilland Dragon Rapide, Detail design work began in April 1964 and construction of the prototype (G-ATCT) was started in September of the same year. It flew for the first time on 13 June 1965, powered by two 157 kW (210 hp) Rolls-Royce Continental IO-360-B engines and with wings of 13·72 m (45 ft) span. Subsequently, the prototype was re-engined with more powerful Lycoming O-540 engines, with which it flew for the first time on 17 December 1965. The wing span was also increased by 1·22 m (4 ft) to bring the prototype to production standard.

The production prototype BN-2 Islander (G-ATWU) flew for the first time on 20 August 1966. The Islander received its domestic C of A on 10 August 1967 and an FAA Type Certificate on 19 December 1967.

Deliveries of Islanders began in August 1967, and by mid-June 1976 more than 650 aircraft of the various models had been delivered to operators in 100 countries. Orders totalled more than 750 by mid-1976.

The sale of 100 aircraft to the Philippine Aerospace Development Corporation (PADC) involves also the development of an amphibious floatplane version of the Islander. The first phase of the PADC programme began in 1974 with the delivery of the first of six 224 kW (300 hp) Islanders. Phase two involved 14 unpainted aircraft without cabin trim, furnishing and electronics, which are being finished by PADC. Phase three calls for the assembly by PADC of 20 aircraft supplied in knockdown kit form. The final phase covers 60 aircraft for which subassemblies and certain components will be manufactured in Manila, using jigs and detailed parts supplied from the UK. Of these 60 aircraft, about 25 will be repurchased by Britten-Norman for sale throughout the world.

Initial production aircraft, described in previous editions of *Jane's*, are designated **BN-2.** Those built since 1 June 1969 are known as **BN-2A**, and the following description applies to this version, which continues in production. A military version, known as the **Defender**, a turbine-engined version designated **Turbo Islander** and the three-engined **Trislander** development are described separately.

A further model of the Islander, originally designated **BN-2A-8S**, made its first flight on 22 August 1972 and was announced at Farnborough in September 1972. This version had an extended nose, incorporating 0·79 m³ (28 cu ft) of additional baggage space, which is now available as an option on the BN-2A.

The basic BN-2A Islander is available with a choice of two alternative power plants (see details following) and either standard 14·94 m (49 ft 0 in) span wings or wingtip extensions having raked tips and containing auxiliary fuel tanks. The company has introduced a series of modification kits which are available as standard or optional fits on new production aircraft and which can also be supplied to operators in the field for retrospective fitting to existing aircraft.

The version of the Islander with 224 kW (300 hp) fuel-injection engines was first introduced in 1970, deliveries beginning in November of that year.

A Rajay turbocharging installation was developed in the United States by Jonas Aircraft, the New York based distributors for Britten-Norman in the western hemisphere. The Rajay installation is a bolt-on unit, for manual operation, which can be fitted on to standard 194 kW (260 hp) engines. The superchargers have the effect of increasing the single-engined ceiling to 3,810 m (12,500 ft) and twin-engined ceiling to 7,925 m (26,000 ft). Cruising speed is also increased, from 139 knots (257 km/h; 160 mph) at 2,135 m (7,000 ft) to 146 knots (270 km/h; 168 mph) at 3,050 m (10,000 ft). By the Spring of 1974, ten Islanders fitted with Rajay installations were operating successfully in Mexico.

An amphibious version of the Islander was announced in the early Summer of 1975 and development was continuing in the Summer of 1976. This version has the designation **BN-2A-30** when fitted with the standard Islander wings, or **BN-2A-31** with the extended wingtips. It has twin floats mounted from the landing gear and fuselage, and these incorporate main wheels which retract aft and nose stabilising wheels which retract forward. When the nosewheels are retracted they remain partially exposed at the forward end of the floats to serve as a buffer when docking. Retraction is accomplished by means of an electrically powered hydraulic system.

Maximum T-O weight of this amphibious Islander is calculated at 2,994 kg (6,600 lb) and max landing weight as 2,857 kg (6,300 lb). Total installed weight of the amphibious conversion is 354 kg (780 lb).

The first flight of a prototype BN-2A-30 was scheduled for the Autumn of 1976.

The following description applies to the standard landplane BN-2A, unless otherwise stated:

Type: Twin-engined feederline transport.

Wings: Cantilever high-wing monoplane. NACA 23012 constant wing section. No dihedral. Incidence 2°. No sweepback. Conventional riveted two-spar torsion-box structure in one piece, using L72 aluminium-clad aluminium alloys. Flared-up wingtips of Britten-Norman design. Raked-back extended wingtips optional. Slotted ailerons and single-slotted flaps of metal construction. Flaps operated electrically, ailerons by pushrods and cables. Ground-adjustable tab on starboard aileron. BTR-Goodrich pneumatic de-icing boots optional.

Fuselage: Conventional riveted four-longeron semi-monocoque structure of pressed frames and stringers and metal skin, using L72 aluminium-clad aluminium alloys. Optional 1·15 m (3 ft 9¼ in) nose extension for baggage stowage.

Tail Unit: Cantilever two-spar structure, with pressed ribs and metal skin, using L72 aluminium-clad aluminium alloys. Fixed-incidence tailplane and mass-balanced elevator. Rudder and elevator are actuated by pushrods and cables. Trim tabs in rudder and elevator. Pneumatic de-icing of tailplane and fin optional.

Landing Gear: Non-retractable tricycle type, with twin wheels on each main unit and single steerable nosewheel. Cantilever main legs mounted aft of rear spar. All three legs fitted with Lockheed oleo-pneumatic shock-absorbers. All five wheels and tyres size 16 × 7-7, supplied by Goodyear. Tyre pressure: main 2·41 bars (35 lb/sq in); nose 2·00 bars (29 lb/sq in). Foot-operated aircooled Cleveland hydraulic brakes on main units. Parking brake. Wheel/ski gear available optionally; amphibious float gear under development.

Power Plant: Two Lycoming flat-six engines, each driving a Hartzell HC-C2YK-2B or -2C two-blade metal constant-speed feathering propeller. Standard power plant is the 194 kW (260 hp) O-540-E4C5, but the 224 kW (300 hp) IO-540-K1B5 can be fitted at customer's option. Optional Rajay turbocharging installation on 194 kW (260 hp) engines, to improve high-altitude performance. Integral fuel tank between spars in each wing, outboard of engine. Total fuel capacity (standard) 518 litres (114 Imp gallons; 137 US gallons). With auxiliary tanks in wingtip extensions, total capacity is increased to 741 litres (163 Imp gallons; 196 US gallons). Additional pylon-mounted underwing auxiliary tanks, each of 227 litres (50 Imp gallons; 60 US gallons) capacity, available optionally. Refuelling point in upper surface of wing above each internal tank. Total oil capacity 22·75 litres (5 Imp gallons).

Accommodation: Up to 10 persons, including pilot, on side-by-side front seats and four bench seats. No aisle. Seat backs fold forward. Access to all seats via three forward-opening doors, forward of wing and at rear of cabin on port side and forward of wing on starboard side. Baggage compartment at rear of cabin, with port-side loading door in standard versions. Exit in emergency by removing door windows. Special executive layouts available. Can be operated as freighter, carrying more than a ton of cargo; in this configuration the passenger seats can be stored in the rear baggage bay. In ambulance role, up to three stretchers and two attendants can be accommodated. Other layouts possible, including photographic and geophysical survey, parachutist transport or trainer (with accommodation for up to eight parachutists and a dispatcher), or public health spraying. A 590 litre (130 Imp gallon) chemical tank can be installed in the cabin, supplying liquid to wing-mounted rotary atomiser spray units. Aircraft with this type of equipment have operated successfully in the Far East and with the Desert Locust Control

Britten-Norman BN-2A Islander of Mount Cook Airlines, with wheel/ski landing gear

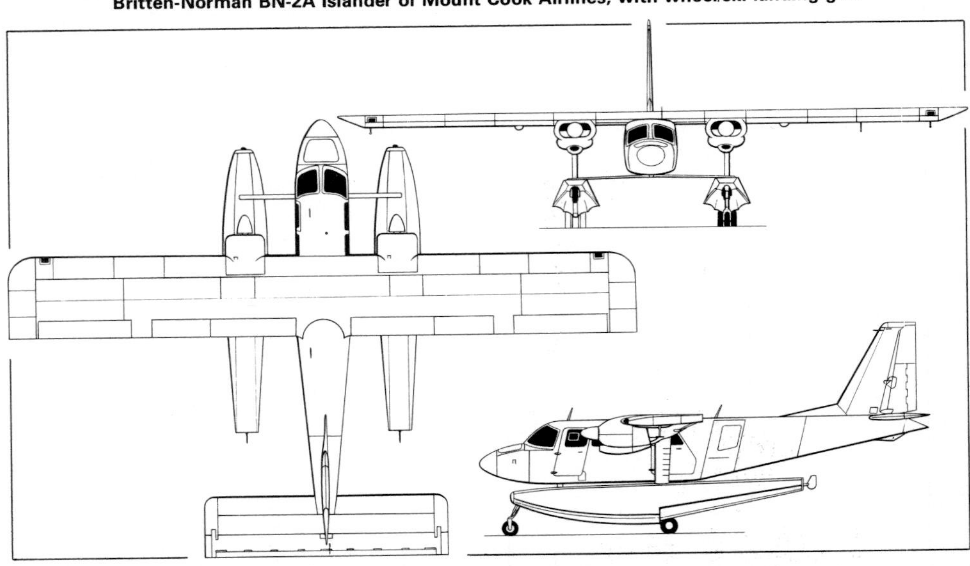

Amphibious version of the Britten-Norman BN-2A Islander *(Pilot Press)*

Organisation at Asmara, Ethiopia.

Systems: Southwind cabin heater standard. 45,000 BTU Stewart Warner combustion unit, with circulating fan, provides hot air for distribution at floor-level outlets and at windscreen demisting slots. Fresh air, boosted by propeller slipstream, is ducted to each seating position for on-ground ventilation. Electrical DC power, for instruments, lighting and radio, from one or two engine-driven 24V 50A self-rectifying alternators and a controller to main busbar and circuit-breaker assembly in nose bay. Emergency busbar with automatic changeover provides a secondary route for essential services. 24V 17Ah heavy-duty lead-acid battery for independent operation. Ground power receptacle provided. Optional electrical de-icing of propellers and windscreen, and pneumatic de-icing of wing and tail unit leading-edges. Intercom system, including second headset, and passenger address system are standard. Oxygen system available optionally for all versions.

Electronics and Equipment: Standard items include blind-flying instrumentation, autopilot, dual flying controls and brake system, and a wide range of VHF or HF communications and navigation equipment.

Dimensions, external:
Wing span:
standard 14·94 m (49 ft 0 in)
with extended tips 16·15 m (53 ft 0 in)
Wing chord (constant) 2·03 m (6 ft 8 in)
Wing aspect ratio:
standard 7·4
with extended tips 7·95
Length overall:
standard 10·86 m (35 ft 7¾ in)
optional nose extension 12·02 m (39 ft 5¼ in)
Fuselage:
Max width 1·21 m (3 ft 11½ in)
Max depth 1·46 m (4 ft 9¾ in)
Height overall 4·18 m (13 ft 8¾ in)
Tailplane span 4·67 m (15 ft 4 in)
Wheel track (c/l of shock-absorbers) 3·61 m (11 ft 10 in)
Wheelbase:
standard 3·99 m (13 ft 1¼ in)
optional nose extension 4·90 m (16 ft 0¾ in)
Propeller diameter 2·03 m (6 ft 8 in)
Cabin door (front, port):
Height 1·10 m (3 ft 7½ in)
Width 0·64 m (2 ft 1¼ in)
Height to sill 0·59 m (1 ft 11¼ in)
Cabin door (front, starboard):
Height 1·10 m (3 ft 7½ in)
Max width 0·86 m (2 ft 10 in)
Height to sill 0·57 m (1 ft 10½ in)
Cabin door (rear, port):
Height 1·09 m (3 ft 7 in)
Width:
top 0·635 m (2 ft 1 in)
bottom 1·19 m (3 ft 11 in)
Height to sill 0·52 m (1 ft 8½ in)
Baggage door (rear, port):
Height 0·69 m (2 ft 3 in)

Dimensions, internal:
Passenger cabin, aft of pilot's seat:
Length 3·05 m (10 ft 0 in)
Max width 1·09 m (3 ft 7 in)
Max height 1·27 m (4 ft 2 in)
Floor area 2·97 m^2 (32 sq ft)
Volume 3·68 m^3 (130 cu ft)
Baggage space aft of passenger cabin:
standard 0·85 m^3 (30 cu ft)
maximum 1·39 m^3 (49 cu ft)
Nose baggage compartment (optional) 0·62 m^3 (22 cu ft)
Freight capacity:
aft of pilot's seat, incl rear cabin baggage space 4·70 m^3 (166 cu ft)
with four bench seats folded into rear cabin baggage space 3·68 m^3 (130 cu ft)

Areas:
Wings, gross:
standard 30·19 m^2 (325·0 sq ft)
with extended tips 31·31 m^2 (337·0 sq ft)
Ailerons (total) 2·38 m^2 (25·6 sq ft)
Flaps (total) 3·62 m^2 (39·0 sq ft)
Fin 3·41 m^2 (36·64 sq ft)
Rudder, incl tab 1·60 m^2 (17·2 sq ft)
Tailplane 6·78 m^2 (73·0 sq ft)
Elevator, incl tabs 3·08 m^2 (33·16 sq ft)

Weights and Loadings (A: standard wings, B: extended wings, C: 194 kW; 260 hp and D: 224 kW; 300 hp engines):
Weight empty, equipped (without electronics):
C 1,627 kg (3,588 lb)
D 1,695 kg (3,738 lb)

Max T-O weight (A, B) 2,993 kg (6,600 lb)
Max zero-fuel weight (BCAR):
A, C, D 2,855 kg (6,300 lb)
B, C, D 2,810 kg (6,200 lb)
Max landing weight (A, B) 2,855 kg (6,300 lb)
Max wing loading:
A 99·1 kg/m² (20·3 lb/sq ft)
B 95·7 kg/m² (19·6 lb/sq ft)
Max floor loading, without cargo panels
586 kg/m² (120 lb/sq ft)
Max power loading:
C 7·71 kg/kW (12·7 lb/hp)
D 6·68 kg/kW (11·0 lb/hp)

PERFORMANCE (at max T-O weight, ISA. C: 194 kW; 260 hp and D: 224 kW; 300 hp engines):
Never-exceed speed:
C, D (standard wings)
177 knots (327 km/h; 203 mph) IAS
C, D (extended wings)
184 knots (340 km/h; 211 mph) IAS
Max level speed at S/L:
C 147 knots (273 km/h; 170 mph)
D 156 knots (290 km/h; 180 mph)
Max cruising speed (75% power) at 2,135 m (7,000 ft):
C 139 knots (257 km/h; 160 mph)
D 147 knots (273 km/h; 170 mph)
Cruising speed (67% power) at 2,750 m (9,000 ft):
C 137 knots (254 km/h; 158 mph)
D 146 knots (270 km/h; 168 mph)
Cruising speed (59% power) at 3,960 m (13,000 ft):
C 133 knots (246 km/h; 153 mph)
D 143 knots (264 km/h; 164 mph)
Stalling speed, flaps up:
C 50 knots (92 km/h; 57 mph) IAS
Stalling speed, flaps down:
C 40 knots (74 km/h; 46 mph)
Max rate of climb at S/L:
C 296 m (970 ft)/min
D 347 m (1,140 ft)/min
Rate of climb at S/L, one engine out:
C 58 m (190 ft)/min
D 61 m (200 ft)/min
Absolute ceiling:
C 4,635 m (15,200 ft)
D 6,020 m (19,750 ft)
Service ceiling:
C 4,025 m (13,200 ft)
Service ceiling, one engine out:
C, standard wings 1,770 m (5,800 ft)
C, extended wings 2,040 m (6,700 ft)
D, standard wings 1,890 m (6,200 ft)
D, extended wings 2,180 m (7,150 ft)
Min ground turning radius 9·45 m (31 ft 0 in)
T-O run at S/L, zero wind, hard runway:
C 169 m (555 ft)
D 201 m (660 ft)
T-O run at 1,525 m (5,000 ft):
D 285 m (936 ft)
T-O to 15 m (50 ft) at S/L, zero wind, hard runway:
C 332 m (1,090 ft)
D 335 m (1,100 ft)
T-O to 15 m (50 ft) at 1,525 m (5,000 ft):
D 475 m (1,560 ft)
Landing from 15 m (50 ft) at S/L, zero wind, hard runway:
C, D 292 m (960 ft)
Landing distance at 1,525 m (5,000 ft):
D 350 m (1,150 ft)
Landing run at 1,525 m (5,000 ft):
D, ISA + 20°C 169 m (555 ft)
Landing run at S/L, zero wind, hard runway:
C, D 137 m (450 ft)
Range at 75% power at 2,135 m (7,000 ft):
C, standard wings 622 nm (1,153 km; 717 miles)
C, extended wings 903 nm (1,673 km; 1,040 miles)
Range at 67% power at 2,750 m (9,000 ft):
C, standard wings 713 nm (1,322 km; 822 miles)
C, extended wings
1,036 nm (1,920 km: 1,193 miles)
Range at 59% power at 3,960 m (13,000 ft):
C, standard wings
755 nm (1,400 km; 870 miles)
C, extended wings
1,096 nm (2,032 km; 1,263 miles)

BRITTEN-NORMAN DEFENDER

The Defender is a variant of the civil Islander which can be adapted for a wide variety of government and military roles such as search and rescue, internal security, long-range patrol, forward air control, troop transport, logistic support and casualty evacuation.

The Defender was first shown at the 1971 Paris Air Show, and later completed an intensive development programme in the UK. The aircraft is available with the same choice of power plant and wing configuration as the current civil version and can be equipped with a wide range of highly sophisticated electronics, including nose-mounted weather radar, providing the aircraft with a marine search capability. Optional equipment includes four NATO standard underwing pylons for a variety of external stores, the inboard pair each carrying up to 317·5 kg (700 lb) and the outboard pair up to 204 kg (450 lb).

Britten-Norman Defender of the Mauretanian Islamic Air Force, carrying 20 SURA rockets on two outboard pylons

Typical underwing loads include twin 7·62 mm machine-guns in pod packs, 250 lb or 500 lb GP bombs, Matra rocket packs, SURA rocket clusters, wire-guided missiles, 5 in reconnaissance flares, anti-personnel grenades, smoke bombs, marker bombs and 227 litre (60 US gallon) drop-tanks.

Internal capacity for passengers, stretcher cases or cargo is the same as that of the civil Islander. Static and air-to-ground firing trials were successfully completed in 1971, during which the forward and beam firing of two pairs of 7·62 mm machine-guns and the forward firing of 68 mm SNEB rocket installations were cleared for operation.

Britten-Norman Defenders/Islanders are in service with the Abu Dhabi Defence Force, Belgian Army, British Army Parachute Association, Ghana Air Force, Guyana Defence Force, Indian Navy, Jamaica Defence Force, Malagasy Air Force, Philippine Navy, Presidential Flight of the Mexican Air Force, Royal Hong Kong Auxiliary Air Force, Panamanian Air Force, Sultan of Oman's Air Force, Mauretanian Islamic Air Force and Rwanda.

The description given for the BN-2A Islander applies also to the Defender, except as follows:

POWER PLANT: Two 224 kW (300 hp) Lycoming IO-540-K1B5 flat-six engines standard.

ELECTRONICS: Typical installation comprises King 360-channel VHF nav/com transceivers with VOR/LOC and VOR/ILS, ADF, marker beacon, KT76 transponder, Sunair ASB 100A HF transceiver, RCA or Bendix radar and Brittain B5 three-axis autopilot.

WEIGHTS AND LOADINGS:
Weight empty 1,682 kg (3,708 lb)
Max T-O weight 2,993 kg (6,600 lb)
Max landing weight 2,855 kg (6,300 lb)
Max wing loading 95·7 kg/m² (19·6 lb/sq ft)
Max power loading 6·68 kg/kW (11·0 lb/hp)

PERFORMANCE (at max T-O weight, ISA. A: no stores on pylons; B: pylons loaded):
Max level speed:
A 153 knots (283 km/h; 176 mph)
B 146 knots (270 km/h; 168 mph)
Cruising speed, 67% power at 3,050 m (10,000 ft):
A 143 knots (265 km/h; 165 mph)
B 136 knots (252 km/h; 157 mph)
Cruising speed, 59% power at 610 m (2,000 ft):
A 128 knots (237 km/h; 147 mph)
B 121 knots (224 km/h; 139 mph)
Stalling speed, flaps down:
A, B 39 knots (73 km/h; 45 mph)
Max rate of climb at S/L:
A 396 m (1,300 ft)/min
B 357 m (1,170 ft)/min
Service ceiling:
A, B 5,180 m (17,000 ft)
Absolute ceiling:
A, B 6,100 m (20,000 ft)
T-O to 15 m (50 ft):
A, B 320 m (1,050 ft)
Landing from 15 m (50 ft):
A, B 303 m (995 ft)
Range with max payload:
A 363 nm (672 km; 418 miles)
B 326 nm (603 km; 375 miles)
Range with standard fuel:
A 1,096 nm (2,027 km; 1,260 miles)
B 1,000 nm (1,850 km; 1,150 miles)
Max range with auxiliary fuel, no reserves, at full mission weight with max endurance power setting
1,497 nm (2,772 km; 1,723 miles)

BRITTEN-NORMAN TURBO ISLANDER

Britten-Norman announced on 29 October 1975 the development of a turboprop-powered version of the Islander. It is generally similar to the piston-engined version of the Islander equipped with the extended baggage nose, except for reinforcement of the wing and fuselage and uprating of the landing gear to cater for a gross weight increase of 318 kg (700 lb). Two versions are available, the **BN-2A-40** with standard wing, and **BN-2A-41** with extended-span wing.

Model of Britten-Norman's Turbo Islander

Powered by two 448 kW (600 shp) Lycoming LTP101 turboprop engines, flat-rated at 298 kW (400 shp) to an altitude of 4,265 m (14,000 ft), preliminary performance estimates indicate that the Turbo Islander should have a cruising speed in excess of 174 knots (322 km/h; 200 mph).

Design of this aircraft began in August 1975, and the first flight of a prototype was scheduled for the Summer of 1976. It is planned to obtain CAA and FAA certification, with initial deliveries of production aircraft during the early months of 1977.

The description of the Islander applies also to the Turbo Islander, except as detailed.

Type: Twin-turboprop business and feederline transport. Available for military general-purpose duties.

Wings: As Islander, except for reinforcement.

Fuselage: As Islander, except for reinforcement. Extended baggage nose standard.

Landing Gear: As Islander, except for uprating to cater for increased gross weight.

Power Plant: Two 448 kW (600 shp) Lycoming LTP101 turboprop engines, flat-rated at 298 kW (400 shp) to 4,265 m (14,000 ft) altitude, each driving a Hartzell three-blade constant-speed fully-feathering propeller. Integral fuel tank between spars in each wing, outboard of engine, and in optional wingtip extensions, with maximum combined capacity of 849 kg (1,872 lb). Refuelling point in upper surface of wing above each internal tank. Additional pylon-mounted underwing auxiliary fuel tanks, each of 227 litres (60 US gallons; 50 Imp gallons) capacity, available optionally.

Armament and Military Equipment: As Defender.

Dimensions, external: As Islander, except:

Length overall	12·02 m (39 ft 5¼ in)
Wheelbase	4·90 m (16 ft 0¾ in)

Weights and Loadings:

Max T-O weight	3,311 kg (7,300 lb)
Max zero-fuel weight	3,084 kg (6,800 lb)
Max landing weight	3,146 kg (6,935 lb)
Max wing loading	109·6 kg/m² (22·45 lb/sq ft)
Max power loading	5·57 kg/kW (9·13 lb/shp)

Performance (estimated, at max T-O weight):

Max cruising speed at 3,050 m (10,000 ft)	more than 191 knots (354 km/h; 220 mph)
Stalling speed, flaps down	50 knots (92·5 km/h; 57·5 mph)
Max rate of climb at S/L	549 m (1,800 ft)/min
Rate of climb at S/L, one engine out	134 m (440 ft)/min
Absolute ceiling	above 9,145 m (30,000 ft)
Absolute ceiling, one engine out	4,360 m (14,300 ft)
T-O to 15 m (50 ft)	355 m (1,165 ft)
Landing from 15 m (50 ft)	320 m (1,050 ft)
Range with max fuel, 45 min reserves	680 nm (1,260 km; 783 miles)

BRITTEN-NORMAN BN-2A Mk III-2 TRISLANDER

In the Autumn of 1970 Britten-Norman introduced an enlarged development of the twin-engined Islander, having a third engine mounted at the rear and a lengthened fuselage seating up to 17 passengers.

The prototype Trislander was produced by converting the second prototype of the twin-engined Islander (G-ATWU), adding a 2·29 m (7 ft 6 in) length of parallel-section fuselage forward of the wing, reinforcing the rear fuselage and fitting a new main landing gear with larger wheels and tyres. The tail unit was modified to act as a mount for the third engine. This aircraft made its first flight on 11 September 1970, appearing in public for the first time at the SBAC Display at Farnborough later the same day. Production aircraft have additional fin area above the rear engine.

The prototype was later dismantled and its fuselage used for structural testing. By the end of 1970 construction had begun of three production aircraft by converting standard Islander airframes from the current production line, and this system has been adopted for all production aircraft, thus maintaining maximum flexibility on what is now a completely integrated Islander/Trislander assembly line. The first production Trislander (G-AYTU) was flown on 6 March 1971, and the first delivery (to Aurigny Air Services in the Channel Islands) was made on 29 June 1971.

ARB certification of the Trislander, granted on 14 May 1971, approved the aircraft for both VFR and IFR operation and for full public transport with one pilot and up to 17 passengers. FAA certification followed on 4 August 1971, to FAR Pt 23 and to the latest air taxi requirements of SFAR Pt 23 and Appendix A of FAR Pt 135. The Appendix A standard is higher than that met by most other commuter aircraft currently offered on world markets, and is achieved primarily because of continued take-off capability and fatigue-free structure.

By mid-1976 orders had been received for more than 50 Trislanders. Of these 35 had been delivered, to customers in the UK, Africa, Australasia, USA, Canada and South America.

Britten-Norman has proposed a military version, the **Trislander M**, able to operate as a 13/17-seat troop carrier, cargo transport or, with underwing auxiliary fuel tanks, in a maritime patrol role.

From early 1975 all Trislanders were equipped with the extended baggage nose of the Islander as standard, and have the designation BN-2A Mk III-2. The original Mk III-1 Trislander, without this feature, is no longer in production.

Type: Three-engined feederline transport.

Wings: Cantilever high-wing monoplane. NACA 23012 constant wing section. No dihedral. Incidence 2°. No sweepback. Conventional riveted two-spar torsion-box structure in one piece, using aluminium -clad aluminium alloys. Increases in skin gauges and spar laminates compared with twin-engined versions. Structure is strictly 'safe-life', but exhibits several fail-safe features and principles. Flared-up wingtips of Britten-Norman design, with raked tips. Slotted ailerons and electrically-operated single-slotted permanently-drooped flaps of metal construction. Ground-adjustable tab in starboard aileron. BTR-Goodrich pneumatic de-icing boots optional.

Fuselage: Conventional riveted four-longeron semi-monocoque structure of pressed frames and stringers and metal skin, using L72 aluminium-clad aluminium alloys. Some reinforcement of fuselage aft of wing to support weight of rear engine. Structure is strictly 'safe-life', but exhibits several fail-safe features and principles.

Tail Unit: Cantilever structure, using L72 aluminium-clad aluminium alloys, with low aspect ratio main fin which also acts as mount for the third engine. Fixed-incidence tailplane (with raked tips) and elevators are similar in construction to those of Islander. Trim tab in rudder. BTR-Goodrich pneumatic de-icing boots for tailplane optional.

Landing Gear: Non-retractable tricycle type, with main-wheel units and single steerable nosewheel. Cantilever main legs mounted aft of rear spar. All five wheels and tyres are Cleveland size 7·00-6. Tyre pressure 3·10 bars (45 lb/sq in) on main units, 2·00 bars (29 lb/sq in) on nose unit. Cleveland foot-operated air-cooled hydraulic disc brakes on main units. Parking brake. No anti-skid units. Fairings fitted to main gear extension tubes below the engine nacelle and above the shock-absorber attachment bolts.

Power Plant: Three 194 kW (260 hp) Lycoming O-540-E4C5 flat-six engines (two mounted on wings and one on vertical tail), each driving a Hartzell HC-C2YK-2G/C8477-4 two-blade constant-speed fully-feathering metal propeller. Fuel in two integral tanks between front and rear wing spars, outboard of the engine nacelles, and two tanks in wingtips. Total fuel capacity 700 litres (154 Imp gallons; 185 US gallons). Overwing refuelling point above each tank. Oil capacity 34 litres (7·5 Imp gallons; 9 US gallons).

Accommodation: Up to 18 persons, including pilot, in pairs on bench seats at approx 787 mm (31 in) pitch. Access to all seats provided by five broad-hinged rearward-opening car-type doors, two on port side and three on starboard side. Baggage compartment at rear of cabin, with external baggage door on port side. Exit in emergency by removing window panels in front four passenger doors. Heating, ventilation and sound insulation standard. Ambulance or VIP interior layouts at customer's option. Dual controls optional.

Systems and Equipment: One Southwind cabin heater fitted as standard. DC electrical system includes two 24V 50A self-rectifying alternators, supplying the instruments, lighting and radio, and a 24V 17Ah battery. No hydraulic or pneumatic systems, except for self-contained hydraulic brakes. Optional equipment includes Bendix M4C or Mitchell Century III autopilot; a wide range of Bendix, King or Narco VHF or HF communications and navigational equipment, including ADF and DME; de-icing systems for propellers (electric), airframe (pneumatic) and windscreen; and second cabin heater, anti-collision strobe beacons, emergency exit beta lights and cargo tiedowns.

Dimensions, external:

Wing span	16·15 m (53 ft 0 in)

Britten-Norman BN-2A Mk III-2 Trislander in the insignia of Trans Jamaican

Wing chord (constant) 2·03 m (6 ft 8 in)
Wing aspect ratio 7·95
Length overall 15·01 m (49 ft 3 in)
Fuselage:
Max width 1·21 m (3 ft 11½ in)
Max depth 1·46 m (4 ft 9¾ in)
Height overall 4·32 m (14 ft 2 in)
Tailplane span 6·48 m (21 ft 3 in)
Wheel track (c/l of shock-absorbers) 3·35 m (11 ft 0 in)
Wheelbase 7·12 m (23 ft 4¼ in)
Propeller diameter 2·03 m (6 ft 8 in)
Propeller ground clearance 0·69 m (2 ft 3 in)
Distance between propeller centres (wing engines) 3·61 m (11 ft 10 in)
Passenger doors (stbd, fwd and centre):
Height 1·10 m (3 ft 7½ in)
Max width 0·89 m (2 ft 10·9 in)
Height to sill 0·57 m (1 ft 10½ in)
Passenger doors (port, fwd and rear):
Height 1·09 m (3 ft 7 in)
Max width 1·21 m (3 ft 11·9 in)
Height to sill 0·57 m (1 ft 10½ in)
Passenger door (stbd, rear):
Height 1·09 m (3 ft 7 in)
Width 0·75 m (2 ft 5½ in)
Baggage compartment door (rear, port):
Height 0·66 m (2 ft 1·95 in)
Width 0·44 m (1 ft 5·2 in)
Nose baggage compartment door (port, optional):
Width 0·79 m (2 ft 7 in)

DIMENSIONS, INTERNAL:
Cabin:
Length, excl flight deck but incl rear baggage compartment 8·24 m (27 ft 0½ in)
Max width 1·09 m (3 ft 7 in)
Max height 1·27 m (4 ft 2 in)
Floor area 7·85 m² (84·45 sq ft)
Volume 9·27 m³ (327·4 cu ft)
Rear baggage compartment volume 0·71 m³ (25·0 cu ft)
Nose baggage compartment volume (optional) 0·62 m³ (22·0 cu ft)

AREAS:
Wings, gross 31·31 m² (337·0 sq ft)
Ailerons (total) 2·38 m² (25·6 sq ft)

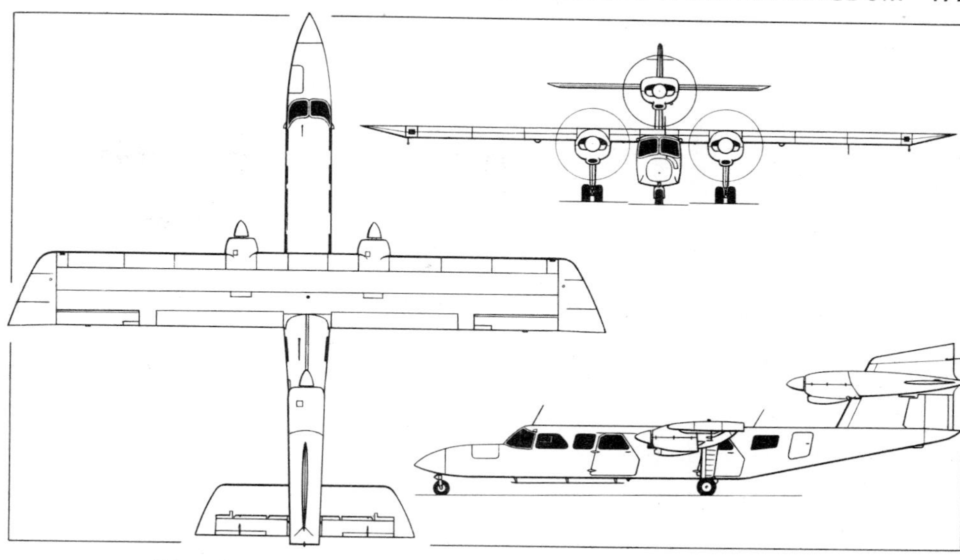

Britten-Norman BN-2A Mk III-2 Trislander feederline transport *(Pilot Press)*

Trailing-edge flaps (total) 3·62 m² (39·0 sq ft)
Fin 5·83 m² (62·7 sq ft)
Rudder, incl tab 1·13 m² (12·2 sq ft)
Tailplane 8·36 m² (90·0 sq ft)
Elevators 2·42 m² (26·0 sq ft)

WEIGHTS AND LOADINGS:
Weight empty, equipped (without electronics) 2,650 kg (5,843 lb)
Max T-O and landing weight 4,536 kg (10,000 lb)
Max wing loading 144·8 kg/m² (29·67 lb/sq ft)
Max power loading 7·79 kg/kW (12·8 lb/hp)

PERFORMANCE (at max T-O weight, ISA):
Max level speed at S/L 156 knots (290 km/h; 180 mph)
Cruising speed (75% power) at 1,980 m (6,500 ft) 144 knots (267 km/h; 166 mph)
Cruising speed (67% power) at 2,470 m (9,000 ft) 139 knots (257 km/h; 160 mph)
Cruising speed (59% power) at 3,960 m (13,000 ft) 135 knots (249 km/h; 155 mph)
Max rate of climb at S/L 298 m (980 ft)/min
Rate of climb at S/L, one engine out 86 m (283 ft)/min
Absolute ceiling 4,450 m (14,600 ft)
Service ceiling 4,010 m (13,150 ft)
Service ceiling, one engine out 2,105 m (6,900 ft)
T-O run at S/L, zero wind, hard runway 393 m (1,290 ft)
T-O to 15 m (50 ft) at S/L, zero wind, hard runway 594 m (1,950 ft)
Landing from 15 m (50 ft) at S/L, zero wind, hard runway 440 m (1,445 ft)
Landing run at S/L, zero wind, hard runway 260 m (852 ft)
Max still-air range at 59% cruising power 868 nm (1,610 km; 1,000 miles)

EKIN—*See 'WHE'*

HAWKER SIDDELEY
HAWKER SIDDELEY GROUP LTD

REGISTERED OFFICE:
18 St James's Square, London SW1Y 4LJ
Telephone: 01-930-6177
Telex: 919011

DIRECTORS:
Sir Arnold Hall, FRS (Chairman and Managing Director)
Sir John Lidbury, FRAeS (Vice-Chairman and Deputy Managing Director)
Air Chief Marshal Sir Harry Broadhurst, GCB, KBE, DSO, DFC, AFC, RAF (Retd)
The Lord Greenhill of Harrow, GCMG, OBE
R. R. Kenderdine, CEng, FIPE
A. J. Laurence, FCA (Finance)
Sir Joseph Lockwood
C. D. MacQuaide, FCA
M. Parkinson, MA
The Rt Hon Lord Shawcross, GBE, QC
Sir Thomas Sopwith, CBE, Hon FRAeS (Founder President)
F. H. Wood, CEng, FIMechE

SECRETARY: C. B. White, MA

Hawker Siddeley Group, with over 80,000 employees, is one of the largest industrial organisations in the world; of this total approximately 46,000 are employed in the Group's aerospace activities. Sales amount to more than £550 million per year; capital employed is more than £180 million. Hawker Siddeley Aviation and Hawker Siddeley Dynamics together form one of the largest aerospace manufacturing organisations in Europe. Hawker Siddeley Diesels is a leading manufacturer of diesel engines; its products are in use throughout the world. Hawker Siddeley Electric is a major supplier of power generation and distribution plant, and manufactures a comprehensive range of electrical equipment. High Duty Alloys Ltd produces many aviation components, including Hiduminium 58 (French designation AU2GN) aluminium-alloy components for the Anglo-French Concorde supersonic transport.

Overseas subsidiaries include Hawker de Havilland Australia Pty Ltd, and Hawker Siddeley Canada Ltd.

HAWKER SIDDELEY AVIATION LTD

HEAD OFFICE:
Richmond Road, Kingston upon Thames, Surrey KT2 5QS
Telephone: 01-546-7741
Telex: 23726

DIRECTORS:
Sir Arnold Hall, FRS (Chairman)
Sir John Lidbury, FRAeS (Deputy Chairman and Managing Director)
Air Chief Marshal Sir Harry Broadhurst, GCB, KBE, DSO, DFC, AFC, RAF (Retd) (Deputy Managing Director)
Air Chief Marshal Sir Peter Fletcher, KCB, OBE, DFC, AFC, RAF(Retd)
J. L. Glasscock, BA, FCIS, JP (Director and General Manager, Kingston)
A. J. Laurence, FCA
E. G. Rubython (Director and General Manager)
J. L. Thorne (Director and General Manager, Hatfield)
J. T. Stamper, MA, CEng, FRAeS (Technical Director)
P. Jefferson, CEng, MRAeS, MIMechE (Production Director)
G. W. Carr, FCIS, MRAeS (Director and General Manager, Manchester)
J. P. Smith, CEng, FRAeS (Director and Chief Engineer, Civil)
B. P. Laight, OBE, CEng, MSc, MIMechE, FRAeS (Director and Chief Project Engineer, Military)
E. F. T. Jenkins, BCom (Commercial Director)
A. S. Watson (Marketing Director)
L. G. Wilgoss (Financial Director)
J. Garston, OBE, CEng, MRAeS (General Manager, Chester)

EXECUTIVE DIRECTORS:
R. G. Adolphus, BSc, CEng, MRAeS (Production, Kingston)
D. W. Allen (Finance, Brough)
H. R. Beattie, BSc(Eng), CEng, MRAeS, AMBIM (Commercial, Manchester)
C. F. Bethwaite, BSc (Hons), CEng, FRAeS (Hatfield)
R. D. Boot, BSc(Eng), FRAeS (Chief Engineer, Brough)
M. J. Brennan, BSc, FIMechE, FRAeS (Special Projects)
N. V. Barber, BA, MSc (General Manager, Brough)
B. J. Champion, CA(SA) (Finance, Chester)
C. M. Chandler, ACWA (Commercial, Kingston)
D. F. Corbett, AMBIM (Production, Hamble)
R. A. Courtman, MIProdE (Manufacturing, Head Office)
J. Cunningham, CBE, DSO, DFC, DL, FRAeS (Chief Test Pilot, Hatfield)
P. Edwards (Production, Chester)
K. Essex-Crosby, CEng, FRAeS (Deputy Chief Engineer, Brough)
J. W. Fozard, DCAe, BSc(Hons), CEng, FRAeS (Deputy Chief Engineer, Kingston)
R. M. Gilbert (Accounting, Head Office)
M. J. Goldsmith, DIC, CEng, MRAeS (Airbus, Hatfield)
H. T. Healy, ACMA (Finance, Manchester)
R. S. Hooper, DCAe, DAe, CEng, MIMechE, MRAeS (Chief Engineer, Kingston)
J. Hosie, FIWM (General Manager, Bitteswell)
E. C. T. Humberstone, FInstPS (Purchasing, Head Office)
J. A. Johnstone, OBE, CEng, FRAeS, AMSLAET (Marketing, Hatfield)
W. Lambert, FCA (Finance, Hatfield)
A. E. Lane (General Manager, Hamble)
J. McGregor Smith, ACWA (Finance, Manchester)
R. Meakins (Commercial, Chester)
H. Mitchell (Legal, Head Office)
F. Murphy, OBE, DFC, MRAeS (Military Sales, Head Office)
D. R. Newman, CEng, FRAeS (Assistant Chief Engineer, Hatfield)
P. R. Owen, CEng, FRAeS (Deputy Chief Engineer, Hatfield)
J. E. Perry, BSc, CEng (Production, Manchester)
A. E. Rowland (Production Planning, Head Office)
J. B. Scott-Wilson, MA, MRAeS (Deputy Chief Engineer, Manchester)
A. F. Smith, BA (Contracts)
A. C. Spencer (Production, Brough)
K. B. Tennant, FCMA (Finance, Hamble)
B. F. W. Tull, CBE (Marketing, Manchester)

J. F. White, FCA, MBIM, MRAeS (Finance, Kingston)
G. A. Whitehead, CBE, FIMechE, CEng, FRAeS, AMCT (Chief Engineer, Manchester)
G. R. Wilkinson (Production, Hatfield)
J. C. Wimpenny, CEng, FRAeS (Research, Head Office)

Secretary: R. D. Smith Wright, FCA

Hawker Siddeley Aviation Ltd is responsible for all aircraft design, development, production and supply activities of the Hawker Siddeley Group.

The Aviation Head Office at Kingston upon Thames administers company policy and co-ordinates the activities of the eleven establishments in the United Kingdom. At present three civil aircraft types, the HS 125 business jet, the HS 748 turboprop transport and the Trident short/medium-range jet airliner; and three military aircraft types, the Harrier V/STOL strike fighter, the Buccaneer low-level strike/reconnaissance aircraft and the Nimrod turbofan-powered maritime reconnaissance aircraft, are in production. In addition, the refurbishing of Hunter fighters for foreign governments continues.

Under development at Kingston are the Hawker Siddeley Hawk multi-purpose jet trainer for the Royal Air Force and a maritime version of the Harrier for the Royal Navy.

Conversion of 27 Handley Page Victors to K. Mk 2 tanker configuration (reportedly for Nos. 55, 57 and 214 Squadrons) is continuing. The first Victor K.2 was delivered to No. 232 OCU at Marham, Norfolk, on 7 May 1974.

Two separate agreements have been made between Hawker Siddeley Aviation and McDonnell Douglas Corporation in the USA. In these, the American company has responsibility for support and any licence production of the Harrier in the United States, and Hawker Siddeley, as weapon system sister design company, has responsibility for in-service support and modification of the McDonnell Douglas Phantoms serving with the Royal Navy and Royal Air Force.

Hawker-Siddeley Aviation is also working in close conjunction with Airbus Industrie, the European company developing and manufacturing the A300 high-capacity short-haul airliner. Under the participation agreement, Hawker Siddeley has responsibility for design of the wing and manufacture of the wing main structure, and is collaborating with Fokker-VFW in the Netherlands, which manufactures the wing moving surfaces.

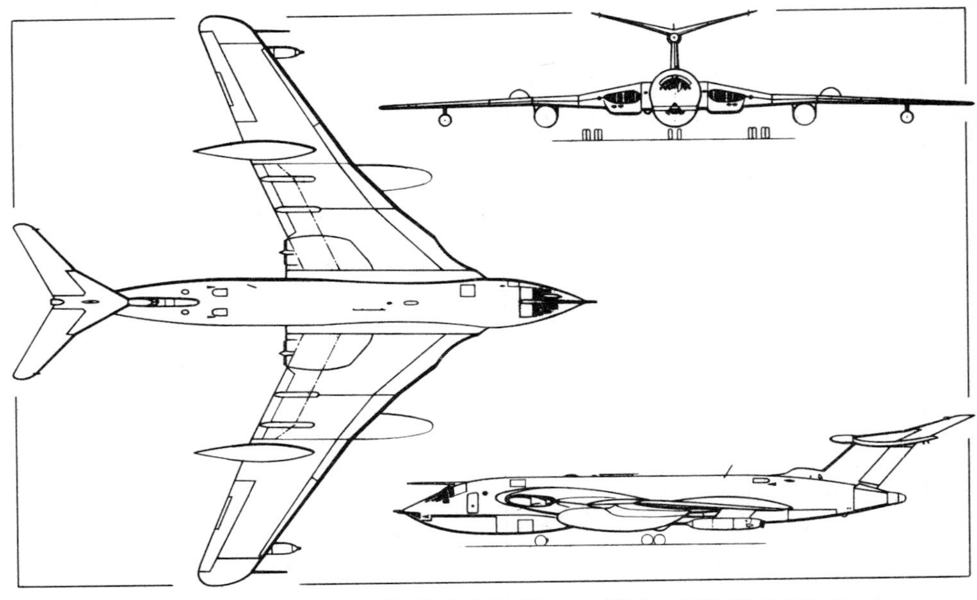

Hawker Siddeley conversion to K. Mk 2 of the Victor B. Mk 2 and SR. Mk 2 *(Pilot Press)*

HAWKER SIDDELEY 125 SERIES 600

The Hawker Siddeley (formerly de Havilland) 125 is a twin-jet business aircraft which is also suitable for use by armed forces in the communications role, as a troop carrier, as an ambulance aircraft, for airways inspection, and as an economical trainer for pilots, navigators and specialised radio and radar operators. All Series of HS 125s can operate from unpaved runways without modification.

The HS 125 was developed as a private venture, and the first of two prototypes flew for the first time on 13 August 1962. Deliveries to customers began in September 1964.

By 21 June 1976 a total of 353 HS 125s had been sold, more than 80 per cent of them for export, including 183 in North America. One British company operates 11 HS 125s, and 12 other operators have fleets of two or more aircraft. The Series 2 navigation trainer version serves as the **Dominie T. Mk 1** with the RAF, whose No. 32 Squadron also operates, in the communications role, four HS 125 Series 400 under the designation **CC. Mk 1** and two Series 600 under the designation **CC.Mk 2**. The HS 125 has also been supplied in the communications role to the air forces of Brazil (eleven), Ghana (one), Malaysia (two), South Africa (seven, known in SAAF service as the **Mercurius**), and to the Argentine Navy (one). Qantas purchased two HS 125 Series 3s for pilot training. Aircraft supplied to the Australian Department of Civil Aviation, the Brazilian government and the South African Department of Civil Aviation are extensively equipped for airways inspection and calibration of radio aids.

To market the HS 125 in North America a new company, Hawker Siddeley Aviation Incorporated, began operation on 1 October 1975. From that date, all such aircraft sold in the USA, Canada and Mexico have the common designation HS 125. Innotech Aviation Ltd has been appointed sole distributor for Canada; Atlantic Aviation, AiResearch Aviation and K.C. Aviation operate approved service centres. Equipping and furnishing of aircraft to the requirements of individual customers can be undertaken by these companies, or by other FAA-approved organisations.

Production of the Hawker Siddeley 125 Series 1 (8 built), 1A (64 built), 1B (13 built), 2 (RAF Dominie T. Mk 1, 20 built), 3 (2 built), 3A (12 built), 3B (15 built), 3A-R and 3A-RA (20 built), 3B-RA (16 built), 400A (69 built) and 400B (47 built) has ended, and these versions have been described in previous editions of *Jane's*. The current versions are:

HS 125 Srs 600A and 600B. Larger, faster development of Srs 400, with 20 per cent more payload, for North American markets (Srs 600A) and the rest of the world (Srs 600B). Changes compared with Srs 400 include more powerful Viper 601 engines, strengthened wings with modified control surfaces, lengthened fuselage (seating a maximum of 14 passengers), taller main fin and extended ventral fin, additional fuel tank in extended dorsal fin, deletion of cockpit canopy fairing, and other detail improvements. First of two development aircraft (G-AYBH) flew for the first time on 21 January 1971, and second (G-AZHS) on 25 November 1971. Certificated by ARB (Special category) on 4 August 1971, and by FAA (600A) on 17 August 1972. In production, with first deliveries in early 1973. Orders up to 1 February 1976 totalled 56, including two (XX507 and XX508) as CC. Mk 2s for No. 32 Squadron, RAF, the remainder being for civilian customers throughout the world.

HS 125 Srs 700. Development of the Srs 600, with the Rolls-Royce Viper 601 turbojet engines replaced by AiResearch turbofan engines, airframe refinements to reduce drag, and equipment changes. Described separately.

The following description applies specifically to the Series 600 version:

Type: Twin-jet business transport aircraft.

Wings: Cantilever low-wing monoplane. Thickness/chord ratio 14% at root, 11% at tip. Dihedral 2°. Incidence 2° 6′ at root, −0° 24′ at tip. Sweepback 20° at quarter-chord. Wings built in one piece and dished to pass under fuselage, to which they are attached by four vertical links, a side link and a drag spigot. All-metal two-spar fail-safe structure, with partial centre spar of approx two-thirds span, sealed to form integral fuel tankage which is divided into two compartments by centreline rib. Skins are single-piece units on each of the upper and lower semi-spans. Detachable leading-edges. Fence on each upper surface at approx two-thirds span. Mass-balanced ailerons, operated manually by cable linkage. Trim tab and geared tab in port aileron, two geared tabs in starboard aileron. Aileron fences to improve lateral stability. Large, four-position double-slotted flaps (45° travel compared with 50° on Srs 400), actuated hydraulically via a screwjack on each flap. Mechanically-operated hydraulic cutout prevents asymmetric operation of the flaps. Flat-plate spoilers above and below each wing, forming part of flap shrouds, provide lift-dumping facility during landing, and have interconnected controls to prevent asymmetric operation. TKS liquid system, using porous stainless steel leading-edge panels, for de-icing or anti-icing.

Fuselage: All-metal semi-monocoque fail-safe structure, making extensive use of Redux bonding. Constant circular cross-section over much of its length. Compared with Srs 400, the Srs 600 has an extra 0·61 m (2 ft 0 in) cabin section added forward of the wings, and 12 cabin windows instead of 10; the nose radome is redesigned and is 152 mm (6 in) longer.

Tail Unit: Cantilever all-metal structure, with fixed-incidence tailplane mounted on fin. Small fairings on tailplane undersurface to eliminate turbulence around elevator hinge cutouts. Triangular ventral fin, and extended dorsal fin. Control surfaces operated manually via cable linkage. Tabs in rudder and each elevator. TKS liquid de-icing or anti-icing of fin and tailplane leading-edges.

Landing Gear: Retractable tricycle type, with twin wheels on each unit. Hydraulic retraction on all units; nosewheels forward, main wheels inward, into wings. Oleo-pneumatic shock-absorbers. Fully-castoring nose unit, steerable 45° to left or right. Dunlop main wheels and 10-ply tyres, size 23 × 7-12, pressure 8·27 bars (120 lb/sq in). Dunlop nosewheels and 6-ply tyres, size 18 × 4·25-10, pressure 5·17 bars (75 lb/sq in). Dunlop double-disc hydraulic brakes with Maxaret anti-skid units on all main wheels.

Power Plant: Two Rolls-Royce Bristol Viper 601-22 turbojet engines, each rated at 16·7 kN (3,750 lb st), pod-mounted on sides of rear fuselage. Hot-air anti-icing of intake lips and bullets. Integral fuel tanks in wings, with total capacity of 4,673 litres (1,028 Imp gallons). Overwing refuelling point near each wingtip. Rear underfuselage tank of 509 litres (112 Imp gallons) capacity, with refuelling point on starboard side, and 231 litre (51 Imp gallon) dorsal fin tank, raising overall total capacity to 5,414 litres (1,191 Imp gallons; 1,430

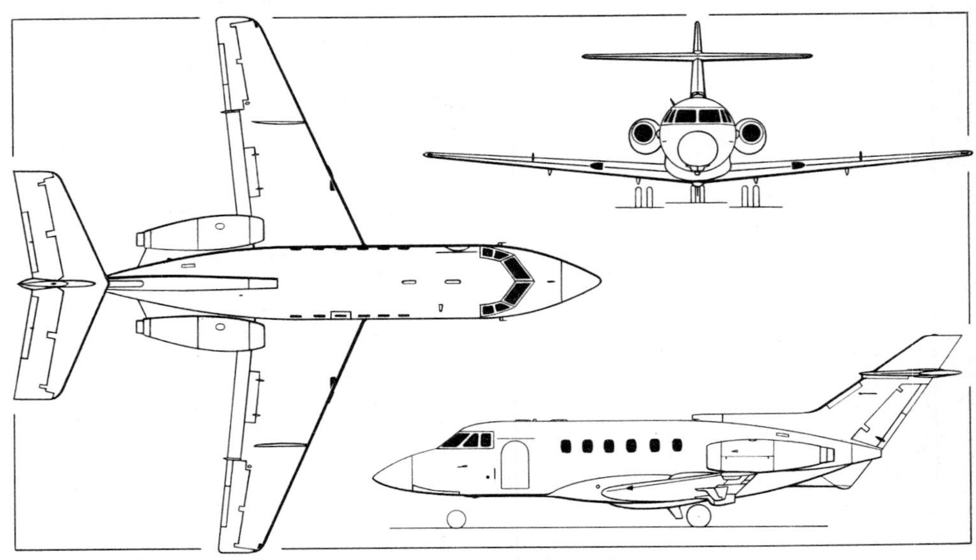

Hawker Siddeley 125 Srs 600 light twin-jet business transport *(Pilot Press)*

Hawker Siddeley 125 Series 600 twin-jet business transport (two Rolls-Royce Bristol Viper 601-22 turbojet engines)

US gallons) of which 5,368 litres (1,181 Imp gallons; 1,418 US gallons) are usable. Self-contained engine re-oiling system, capacity 15·5 litres (27 Imp pints).

ACCOMMODATION: Crew of two on flight deck, which is fully soundproofed, insulated and air-conditioned. Dual controls standard. Seat provided for third crew member. Standard executive layout has seating for eight passengers, with forward baggage compartment, refreshment bar and coat compartment (forward) and toilet (aft). Compared with Srs 400, there are smoother-line roof panels, with individual recessed lights and air louvres. Cabin restyling offers the operator a choice of interchangeable furnishing units to suit individual requirements. The new, wider seats, which on the Srs 600A swivel through 180°, are adjustable fore and aft and sideways, have adjustable lumbar support, and can be reclined hydraulically up to 40°. Typical executive furnishing includes a couch for three persons and five individual seats, foldaway conference table and individual foldaway wall tables. Alternative high-density layout is available, seating up to 14 passengers. Outward-opening door at front on port side, with integral airstairs. Emergency exit over wing on starboard side. Windscreen demisting by engine bleed air; electrical windscreen anti-icing, with methanol spray backup.

SYSTEMS: AiResearch air-conditioning and pressurisation system. Max cabin differential 0·58 bars (8·35 lb/sq in), maintaining S/L cabin pressure up to 6,550 m (21,500 ft). Oxygen system standard, with dropout masks for passengers. Hydraulic system, pressure 159-207 bars (2,300-3,000 lb/sq in), for operation of landing gear, main-wheel doors, flaps, spoilers, nosewheel steering, main-wheel brakes and anti-skid units. Two accumulators provide emergency hydraulic power for wheel brakes in case of a main system failure. Independent auxiliary system for lowering landing gear and flaps in the event of a main system failure. DC electrical system utilises two 300A 9kW engine-driven starter/generators and two 24V 25Ah batteries. A 24V 3·5Ah battery provides separate power for igniter and starter control circuits. AC electrical system includes two 115V 2·5kVA 400Hz three-phase rotary inverters and one 250VA solid-state standby inverter for electronics, and one engine-driven 115V 3kVA frequency-wild alternator for windscreen anti-icing. Ground power receptacle on starboard side at rear of fuselage for 28V external DC supply. AiResearch GTCP-30-92 auxiliary power unit is standard on Srs 600B; Solar T-62T-39 is optional on Srs 600A. Engine ice protection system supplied by engine bleed air. Graviner triple FD Firewire fire warning system and two BCF engine fire extinguishers.

ELECTRONICS AND EQUIPMENT: Standard equipment includes full blind-flying instrumentation, complete ice protection system, stick-shaker stall warning, and electrically-heated rudder auto-bias to apply corrective rudder during asymmetric engine power conditions. A spring and *g* weight are included in the elevator circuit to reduce variations in stick force to a minimum over a wide CG range. Compared with the Srs 400, the layout of flight deck instrumentation has been completely redesigned, all systems (including the electrical and ice protection systems) have been refined, and a new central warning system is incorporated. A combined slot/stereo tape unit and FM/AM self-seeking radio are fitted as standard in Srs 600B, together with storage for additional tape cartridges, magazines and stationery. Comprehensive electronics, available to customer's requirements, include an automatic flight system comprising autopilot (typically, Sperry SP40C or Bendix PB60 for Srs 600A, Collins AP104 for Srs 600A and 600B), flight director and compass; dual VHF nav/com; HF com; dual ADF; marker; ATC transponder; DME; and weather radar. Doppler, Decca Navigator, flight data recorder, and passenger address system may also be installed. Equipment for ICAO Category 2 low weather minima operation will be available as an option. A feature console is provided for fitting customer-specified optional items such as digital readouts and a telephone.

DIMENSIONS, EXTERNAL:
- Wing span: 14·33 m (47 ft 0 in)
- Wing chord (mean): 2·29 m (7 ft 6¼ in)
- Wing aspect ratio: 6·25
- Length overall: 15·39 m (50 ft 6 in)
- Height overall: 5·26 m (17 ft 3 in)
- Fuselage: Max diameter: 1·93 m (6 ft 4 in)
- Tailplane span: 6·10 m (20 ft 0 in)
- Wheel track (c/l of shock-absorbers): 2·79 m (9 ft 2 in)
- Wheelbase: 6·34 m (20 ft 9½ in)
- Passenger door (fwd, port):
  - Height: 1·30 m (4 ft 3 in)
  - Width: 0·69 m (2 ft 3 in)
  - Height to sill: 1·07 m (3 ft 6 in)
- Emergency exit (overwing, stbd):
  - Height: 0·91 m (3 ft 0 in)
  - Width: 0·51 m (1 ft 8 in)

DIMENSIONS, INTERNAL:
- Cabin (excl flight deck):
  - Length: 6·50 m (21 ft 4 in)
  - Max width: 1·80 m (5 ft 11 in)
  - Max height: 1·75 m (5 ft 9 in)
  - Floor area: 5·11 m² (55·0 sq ft)
  - Volume: 17·8 m³ (628·0 cu ft)
- Baggage compartment: 0·84 m³ (29·6 cu ft)

AREAS:
- Wings, gross: 32·8 m² (353·0 sq ft)
- Ailerons (total): 2·76 m² (29·76 sq ft)
- Trailing-edge flaps (total): 5·21 m² (56·06 sq ft)
- Fin, incl dorsal fin: 5·31 m² (57·15 sq ft)
- Ventral fin: 0·61 m² (6·61 sq ft)
- Horizontal tail surfaces (total): 9·29 m² (100 sq ft)

WEIGHTS AND LOADINGS:
- Weight empty: 5,683 kg (12,530 lb)
- Typical operating weight, empty: 6,148 kg (13,555 lb)
- Max payload: 907 kg (2,000 lb)
- Max T-O and ramp weight: 11,340 kg (25,000 lb)
- Max zero-fuel weight: 7,053 kg (15,550 lb)
- Max landing weight: 9,979 kg (22,000 lb)
- Max wing loading: 346 kg/m² (70·8 lb/sq ft)
- Max power loading: 339·5 kg/kN (3·33 lb/lb st)

PERFORMANCE (initial certification, at max T-O weight except where indicated):
- Never-exceed speed: 375 knots (695 km/h; 432 mph) IAS
- Max design Mach number in dive: 0·85
- Max operating speed:
  - fuselage fuel tanks empty: 320 knots (592 km/h; 368 mph) IAS
  - fuel in fuselage fuel tanks: 280 knots (519 km/h; 322 mph) IAS
- Max operating Mach number: 0·78
- Max cruising speed at 8,534 m (28,000 ft): 454 knots (840 km/h; 522 mph)
- Econ cruising speed at 11,890 m (39,000 ft): 403 knots (747 km/h; 464 mph)
- Rough-air speed: 230 knots (426 km/h; 265 mph) IAS
- Landing gear operation speed: 220 knots (407 km/h; 253 mph) IAS
- Flap operating speed:
  - T-O: 220 knots (407 km/h; 253 mph) IAS
  - approach: 175 knots (324 km/h; 201·5 mph) IAS
  - landing: 160 knots (296·5 km/h; 184 mph) IAS
- Stalling speed, flaps down: 83 knots (155 km/h; 96 mph) EAS
- Max rate of climb at S/L: 1,493 m (4,900 ft)/min
- Rate of climb at S/L, one engine out: 420 m (1,380 ft)/min
- Service ceiling: 12,500 m (41,000 ft)
- T-O run: 1,341 m (4,400 ft)
- T-O balanced field length: 1,631 m (5,350 ft)
- Landing from 15 m (50 ft) at typical landing weight, unfactored: 649 m (2,130 ft)
- Landing run (scheduled performance):
  - Srs 600A at typical landing weight: 1,036 m (3,400 ft)
  - Srs 600A at max landing weight: 1,295 m (4,250 ft)
  - Srs 600B at 7,167 kg (15,800 lb) landing weight: 1,137 m (3,730 ft)
- Min ground turning radius (inside wheel): 4·70 m (15 ft 5 in)
- Runway LCN requirement at max T-O weight: 10
- Typical range with 454 kg (1,000 lb) payload, 45 min reserves plus allowances for T-O, approach, landing and taxying: 1,650 nm (3,057 km; 1,900 miles)
- Range with max fuel and max payload, reserves as above: 1,560 nm (2,891 km; 1,796 miles)

HAWKER SIDDELEY 125 SERIES 700

This turbofan version of the HS 125 was introduced by Hawker Siddeley in 1976. The prototype (G-BFAN) was produced by conversion of a Series 600 airframe and flew for the first time on 28 June 1976. Similar conversions of existing turbojet-powered HS 125s will be offered; but new-production Series 700 aircraft will embody many refinements in addition to the change of power plant. As in the case of earlier versions, the intended market will be indicated by a suffix letter: the **Series 700A** will be for the North American market, **Series 700B** for the rest of the world.

Use of Garrett-AiResearch TFE 731-3-1H turbofan engines gives an improved specific fuel consumption by comparison with that of the turbojets in earlier versions of the HS 125. The Series 700 also meets all existing and proposed international noise regulations.

Improvements to the airframe, to reduce drag and enhance its appearance, include redesign of the wing keel skid, use of countersunk rivets instead of mushroom-head types in the flap bottom skin, replacement of the lower airbrake leading-edge castellations by internal castellations, replacement of the mushroom-head bolts and rivets in the inner tank doors and aileron trailing-edges by countersunk types, redesign of the ventral fin and adjacent fairings in glassfibre and enlargement of the area of the ventral fin to improve directional stability, deletion of the

NACA cooling air intake introduced in the nose of the Series 600, addition of fairings over the windscreen wiper blades and two ADF loop aerials, and use of Harper radius countersunk rivets instead of mushroom-head rivets in the fuselage and tail unit.

New interior equipment and furnishings include the use of figured walnut veneer on cabin tables and toilet consoles, leather trim, provision of a Blaupunkt Bamberg combined radio/cassette stereo player and recorder, a luxury toilet compartment, digital cabin clock, slide-out portable bar box, full harness on sideways-facing seats, improved life-jacket stowage under seats, improved plug-in meal tray for divan occupants, and a new range of interior colour schemes.

The first flight of a production HS 125 Series 700 was scheduled for December 1976, with UK certification anticipated by the end of April 1977. The first sale had been made before completion of the prototype.

The description of the Hawker Siddeley 125 Series 600 applies also to the Series 700, except as follows:

TYPE: Twin-turbofan business transport aircraft.

WINGS, FUSELAGE, TAIL UNIT, LANDING GEAR: Basically as for Series 600, with refinements listed in introductory copy.

POWER PLANT: Two 16·46 kN (3,700 lb st) Garrett-AiResearch TFE 731-3-1H turbofan engines, in rear-fuselage pods designed and manufactured by Grumman Aerospace. Engine intake anti-icing by engine bleed air. Fuel system as for Series 600, except for minor detail changes in the overwing refuelling arrangements and the introduction of fuel computers which perform governing, limiting and scheduling functions in response to throttle lever and other engine inputs. Oil capacity 5·7 litres (1·5 US gallons) per engine.

ACCOMMODATION: Basically as for Series 600.

SYSTEMS: As for Srs 600, except new variable-displacement hydraulic pumps; dual independent engine tappings of low-pressure air for rudder auto-bias; and changes to electrical system which include new Lear Siegler starter/generators, static inverters for AC supply, new Lucas 120V 4·4VA alternator to power the windscreen heating system, with a second alternator optional, increased-life nickel-cadmium batteries, and system changes to allow the use of a thermal/cold start switch to put batteries in series for a cold start and cross engine starting from the first generator on line.

ELECTRONICS AND EQUIPMENT: Standard electronics include dual Collins VHF-20A com transceivers, dual Collins 51RV-2B VHF nav receivers, dual Collins DF-206 ADF, Collins 51Z-4 marker beacon receiver, dual Collins MC-103 compass, Collins DME-40 DME, Marconi AD1540 audio control and passenger address system, RCA AVQ-21 weather radar, RCA AVQ-95 ATC transponder and Blaupunkt Bamberg stereo tape and AM/FM radio. Provisions for Collins 718U-5 HF com transceiver and second transponder and DME. Collins APS-80 autopilot and FGS-80 flight director system standard, to provide altitude hold, altitude pre-select, airspeed hold, Mach number hold, vertical speed hold, aircraft heading, VOR/LOC, ILS approach and pitch with electric trim.

DIMENSIONS, EXTERNAL:
As for Srs 600, except:
Length overall 15·46 m (50 ft 8½ in)
Height overall 5·36 m (17 ft 7 in)

DIMENSIONS, INTERNAL, AND AREAS:
As for Srs 600

WEIGHTS AND LOADINGS:
Weight empty 5,747 kg (12,670 lb)
Typical operating weight, empty 6,212 kg (13,695 lb)
Max payload 1,068 kg (2,355 lb)
Max T-O weight 10,977 kg (24,200 lb)
Max ramp weight 11,113 kg (24,500 lb)
Max zero-fuel weight 7,280 kg (16,050 lb)
Max landing weight 9,979 kg (22,000 lb)
Max wing loading 334·9 kg/m² (68·6 lb/sq ft)
Max power loading 333·4 kg/kN (3·27 lb/lb st)

PERFORMANCE (estimated, at max T-O weight unless indicated otherwise):
Never-exceed speed Mach 0·85
Max level speed at S/L 320 knots (592 km/h; 368 mph) IAS
Max cruising speed at 8,380 m (27,500 ft) 436 knots (808 km/h; 502 mph) TAS
Econ cruising speed at 11,275-12,500 m (37,000-41,000 ft) 403 knots (747 km/h; 464 mph) TAS
Stalling speed, flaps down 83 knots (155 km/h; 96 mph) EAS
Service ceiling 12,500 m (41,000 ft)
T-O run 1,275 m (4,180 ft)
T-O to 10·7 m (35 ft), unfactored 1,487 m (4,880 ft)
T-O balanced field length 1,890 m (6,200 ft)
Landing from 15 m (50 ft) at landing weight of 7,166 kg (15,800 lb), unfactored 619 m (2,030 ft)
Landing run at 6,803 kg (15,000 lb) landing weight 1,231 m (4,040 ft)
Range with max fuel, allowances for T-O, approach, landing and taxying, 45 min reserves 2,330 nm (4,318 km; 2,683 miles)
Range with max payload, allowances as above, 45 min reserves 1,920 nm (3,556 km; 2,210 miles)

Prototype of the Series 700 turbofan-engined version of the Hawker Siddeley 125

OPERATIONAL NOISE CHARACTERISTICS (FAR Pt 36):
T-O noise level 86·6 EPNdB
Approach noise level 95·3 EPNdB
Sideline noise level 88·9 EPNdB

HAWKER SIDDELEY 146

Work on this project for a short-range transport aircraft has continued on a low-budget basis during 1975-76, but there is no intention of completing a prototype at this stage. Details of the project can be found in the 1974-75 *Jane's*.

HAWKER SIDDELEY 748 SERIES 2A

Design of the Hawker Siddeley (originally Avro) 748 short/medium-range turboprop airliner started in January 1959. The first prototype flew on 24 June 1960, followed by a second on 10 April 1961. UK production of the Series 1 (18 built) and 2 (including two Andover CC.Mk 2s for The Queen's Flight and four for Air Support Command), described in previous editions of *Jane's,* has been completed.

The Series 2A, which superseded the Series 2 in production from mid-1967, differs only in having 1,700 kW (2,280 ehp) Dart RDa.7 Mk 532-2L or -2S turboprop engines (now designated Mk 534-2 and Mk 535-2 respectively), giving improved performance. The Mk 535-2 is standard in current production aircraft. Production continuing in 1976. Sales (including 31 Andover C.Mk 1s for the RAF: see 1968-69 *Jane's*) totalled 312 by 20 February 1976, including 259 for export. Of these 286 had been delivered. Nine aircraft have been sold with Dart RDa.8 engines, seven being supplied to Bundesanstalt für Flugsicherung (Germany) with calibration equipment for radio navigational aids and two to the Royal Australian Navy with navigational and electronic training equipment.

The HS 748 is the subject of a manufacturing agreement with the Indian government, and 79 aircraft (included in above totals) are being assembled from British-built components by Hindustan Aeronautics Ltd. Of these 24 are for Indian Airlines and 55 for the Indian Air Force.

HS 748 Civil Transport

The Series 2A is available optionally with a large rear freight door which has an opening of 2·67 m by 1·72 m (8 ft 9 in × 5 ft 7¾ in). The first aircraft with a large freight door flew for the first time on 31 December 1971 and underwent extensive air and ground trials, including the dropping of parachutists and supplies. Civil HS 748s with a large freight door are in service with, or on order for, Guyana Airways and a mining company in Tanzania.

HS 748 Military Transport

The military transport version of the Series 2A has the large rear freight door that is available for the civil transport, and has, in addition, a strengthened cabin floor capable of supporting an overall floor loading of 976·5 kg/m² (200 lb/sq ft), and fixed fittings to undertake a wide range of military roles. Optional military overload take-off and landing weights give improved payload/range capabilities. A total of 24 HS 748 military transports are in service with, or on order for, the Belgian Air Force, Brazilian Air Force, Ecuadorean Air Force, Indian Air Force, Nepal Royal Flight and an undisclosed air force.

HS 748 Coastguarder

Latest variant in the HS 748 series, for search and rescue and maritime surveillance roles. Primary sensor is MAREC search radar with large belly-mounted antenna giving full 360° coverage. Large 0·43 m (1 ft 5 in) display with PPI, ground stabilised, offset mode, sector scan, and variable scale range markers. Additional information to tactical navigator's station from Decca Doppler/TANS computer and Omega VLF nav system. Launch chute for flares and rescue dinghies. Long-range inner wing fuel tanks standard, to provide total fuel capacity of 9,956 litres (2,190 Imp gallons). Wide range of optional equipment available. Prototype scheduled to fly in early 1977.

TYPE: Twin-engined passenger or freight transport.

WINGS: Cantilever low-wing monoplane. Wing section NACA 23018 at root, NACA 4412 at tip. Dihedral 7°. Incidence 3°. Sweepback 2° 54′ at quarter-chord. All-metal two-spar fail-safe structure. No cutouts in spars for engines or landing gear. All-metal set-back hinge, shielded horn-balance, manually-operated ailerons and electrically-actuated Fowler flaps. Geared tab in each aileron. Trim tab in starboard aileron. Pneumatic leading-edge de-icing boots.

FUSELAGE: All-metal semi-monocoque riveted fail-safe structure, of circular section.

TAIL UNIT: Cantilever all-metal structure. Fixed-incidence tailplane. Manually-operated controls. Trim tabs in elevators and rudder. Spring tab in rudder.

LANDING GEAR: Retractable tricycle type, with hydraulically-steerable nose unit. All wheels retract forward hydraulically. Main wheels retract into bottom of engine nacelles forward of front wing spar. Dowty Rotol shock-absorbers. Twin wheels, with Dunlop tyres, on all units. Main wheels size 32 × 10·75-14. Nosewheels 25·65 × 8·5-10. Standard tyre pressures: main wheels 5·03 bars (73 lb/sq in); nosewheels 3·79 bars (55 lb/sq in). Minimum tyre pressures: main wheels 4·48 bars (65 lb/sq in); nosewheels 3·45 bars (50 lb/sq in). Dunlop disc brakes with Maxaret anti-skid units. No brake cooling.

POWER PLANT: Two 1,700 kW (2,280 ehp) Rolls-Royce Dart RDa.7 Mk 534-2 or Mk 535-2 turboprop engines, each driving a Dowty Rotol four-blade constant-speed fully-feathering propeller. Fuel in two integral wing tanks, with total capacity of 6,550 litres (1,440 Imp gallons). Underwing pressure refuelling and overwing gravity refuelling. Oil capacity 14·2 litres (25 Imp pints) per engine.

ACCOMMODATION: Crew of two on flight deck, and cabin

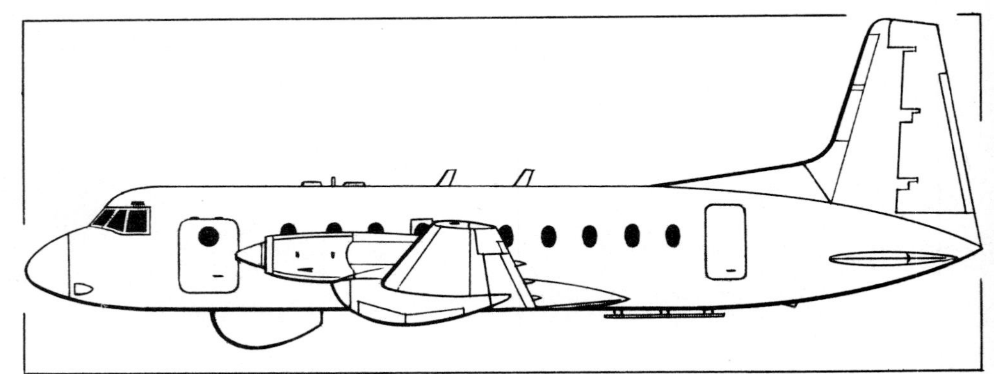

Side elevation of the Hawker Siddeley 748 Coastguarder *(Pilot Press)*

Hawker Siddeley 748 Series 2A with optional large rear freight door

attendant. Normal accommodation for 40-58 passengers in paired seats on each side of central gangway. Baggage compartment forward of cabin, with provision for steward's seat. Galley, toilet and baggage compartment aft of cabin. Forward baggage compartment and steward's seat can be replaced by freight hold with moving partition between hold and passenger cabin. Main passenger door, on port side at rear, with smaller door on starboard side to serve as baggage door and emergency exit. Crew and freight door on port side at front. Hydraulically-operated stairs.

ACCOMMODATION (military transport): Up to 58 troops in airline type seats. Provision for forward and aft baggage compartments and hydraulically-operated airstairs. In paratroop role up to 48 paratroops and despatchers can be accommodated on sidewall folding seats with safety harness. Dropping by static line or free fall. For casualty evacuation up to 24 stretchers and nine nursing staff can be carried, with provision for medical supplies and equipment. For supply dropping a guided roller conveyor system allows twelve 340 kg (750 lb) or seven 680 kg (1,500 lb) loads to be dropped within six seconds. Capacity for almost 6 tonnes (13,227 lb) freight. Large cargo door will accept items up to 1·52 m × 1·52 m × 3·00 m (5 ft × 5 ft × 9 ft 10 in) or small diameter pipes over 12 m (39 ft 4 in) in length. On-board freight hoist and palletised freight system available. Quickly-removable VIP cabin available, and a variety of VIP layouts, with separate toilet, telephone and wide range of options.

SYSTEMS: Normalair pressurisation and air-conditioning system, giving equivalent altitude of 2,440 m (8,000 ft) at 7,620 m (25,000 ft). Pressure differential 0·38 bars (5·5 lb/sq in). Hydraulic system, pressure 172 bars (2,500 lb/sq in), for landing gear retraction, nosewheel steering, brakes and propeller brakes. No pneumatic system. One 9kW 28V DC generator and one 22kVA alternator on each engine. Two 1,800VA inverters.

ELECTRONICS AND EQUIPMENT: Collins or Bendix solid-state radio and radar. Blind-flying instrumentation and weather radar. Smiths SEP.2E autopilot. Provision for flight director system and flight data recorder.

DIMENSIONS, EXTERNAL:
- Wing span: 30·02 m (98 ft 6 in)
- Wing chord at root: 3·49 m (11 ft 5¼ in)
- Wing chord at tip: 1·34 m (4 ft 5 in)
- Wing aspect ratio: 11·967
- Length overall: 20·42 m (67 ft 0 in)
- Fuselage:
  - Max diameter: 2·67 m (8 ft 9 in)
- Height overall: 7·57 m (24 ft 10 in)
- Tailplane span: 10·97 m (36 ft 0 in)
- Wheel track: 7·54 m (24 ft 9 in)
- Wheelbase: 6·30 m (20 ft 8 in)
- Propeller diameter: 3·66 m (12 ft 0 in)
- Propeller ground clearance: 0·61 m (2 ft 0 in)
- Passenger door (port, rear):
  - Height: 1·57 m (5 ft 2 in)
  - Width: 0·76 m (2 ft 6 in)
  - Height to sill: 1·84 m (6 ft 0½ in)
- Freight and baggage door (fwd):
  - Height: 1·37 m (4 ft 6 in)
  - Width: 1·22 m (4 ft 0 in)
  - Height to sill: 1·84 m (6 ft 0½ in)
- Baggage door (rear, stbd):
  - Height: 1·24 m (4 ft 1 in)
  - Width: 0·64 m (2 ft 1 in)
  - Height to sill: 1·84 m (6 ft 0½ in)
- Optional freight door (rear, port):
  - Height: 1·72 m (5 ft 7¾ in)
  - Width: 2·67 m (8 ft 9 in)

DIMENSIONS, INTERNAL:
- Cabin, excl flight deck:
  - Length: 14·17 m (46 ft 6 in)
  - Max width: 2·46 m (8 ft 1 in)
  - Max height: 1·92 m (6 ft 3½ in)
  - Floor area: 27·5 m² (296 sq ft)
  - Volume: 56·35 m³ (1,990 cu ft)
- Max total freight holds: 9·54 m³ (337 cu ft)

AREAS:
- Wings, gross: 75·35 m² (810·75 sq ft)
- Ailerons (total): 3·98 m² (42·90 sq ft)
- Trailing-edge flaps (total): 14·83 m² (159·80 sq ft)
- Fin: 9·81 m² (105·64 sq ft)
- Rudder, incl tabs: 3·66 m² (39·36 sq ft)
- Tailplane: 17·55 m² (188·9 sq ft)
- Elevators, incl tabs: 5·03 m² (54·10 sq ft)

WEIGHTS AND LOADINGS (A: standard aircraft; B: military transport):
- Basic operating weight, incl crew:
  - A: 12,247 kg (27,000 lb)
  - B: 11,574 kg (25,516 lb)
- Max payload:
  - A: 5,216 kg (11,500 lb)
  - B: 5,890 kg (12,985 lb)
  - B, optional overload: 7,930 kg (17,482 lb)
- Max T-O weight:
  - A, B: 21,092 kg (46,500 lb)
  - B, optional overload: 23,133 kg (51,000 lb)
- Max zero-fuel weight:
  - A, B: 17,460 kg (38,500 lb)
  - B, optional overload: 19,504 kg (43,000 lb)
- Max landing weight:
  - A, B: 19,504 kg (43,000 lb)
  - B, optional overload: 21,546 kg (47,500 lb)
- Max wing loading:
  - A: 279·8 kg/m² (57·3 lb/sq ft)
- Max power loading:
  - A: 6·20 kg/kW (10·2 lb/ehp)

PERFORMANCE (A: standard aircraft at max T-O weight unless otherwise indicated; B: military transport at

Hawker Siddeley 748 Srs 2A twin-turboprop transport aircraft operated by British Airways

normal max T-O weight with 20% fuel reserves; C: military transport at optional overload T-O weight with 20% fuel reserves):

Cruising speed:
A, at 17,236 kg (38,000 lb) 244 knots (452 km/h; 281 mph)

Max rate of climb at S/L:
A, at 17,236 kg (38,000 lb) 433 m (1,420 ft)/min

Service ceiling:
A 7,620 m (25,000 ft)

Min ground turning radius:
A, B, C 11·82 m (39 ft)

Runway LCN:
A 9 to 18

T-O run:
A (BCAR) 1,225 m (4,020 ft)
B 756 m (2,480 ft)
C 945 m (3,100 ft)

Balanced field length:
A (BCAR) 1,640 m (5,380 ft)
A (BCAR 650 nm; 1,203 km; 748 mile sector, 40 passengers and reserves for 200 nm; 370 km; 230 miles plus 45 min hold) 884 m (2,900 ft)

T-O to 15 m (50 ft):
B 927 m (3,040 ft)
C 1,158 m (3,800 ft)

Landing field length:
A (BCAR) 1,033 m (3,390 ft)

Landing from 15 m (50 ft):
B 567 m (1,860 ft)
C 625 m (2,050 ft)

Landing run:
B 347 m (1,140 ft)
C 387 m (1,270 ft)

Range with max payload:
A, with reserves for 200 nm (370 km; 230 miles) plus 45 min hold 735 nm (1,361 km; 846 miles)
B 960 nm (1,778 km; 1,105 miles)
C 840 nm (1,556 km; 967 miles)

Range with max fuel:
A, with reserves for 200 nm (370 km; 230 miles) plus 45 min hold 1,720 nm (3,186 km; 1,980 miles)
B, with 4,292 kg (9,462 lb) payload 1,430 nm (2,649 km; 1,646 miles)
C, with 6,334 kg (13,964 lb) payload 1,280 nm (2,372 km; 1,474 miles)

Radius of action:
B, supply drop mission with 12 × 340 kg (750 lb) containers 625 nm (1,158 km; 720 miles)

Operational Noise Characteristics (FAR Pt 36):
T-O noise level 92·5 EPNdB
Approach noise level 103·8 EPNdB
Sideline noise level 96·3 EPNdB

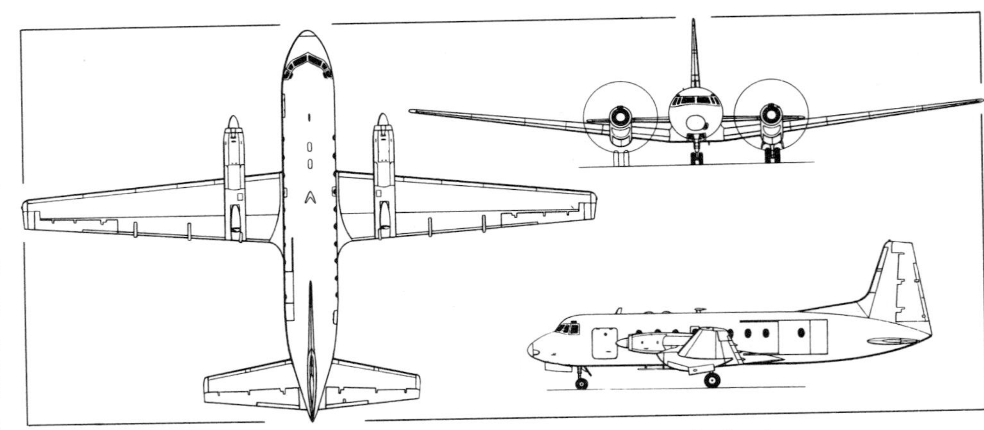

Hawker Siddeley 748 Srs 2A with freight door *(Pilot Press)*

HAWKER SIDDELEY TRIDENT

The Hawker Siddeley (originally de Havilland D.H.121) Trident was ordered into production initially to meet BEA's requirements for a short-haul 520 knot (965 km/h; 600 mph) airliner for service from 1963-64 onwards. Design was started in 1957 and construction of the first airframe began on 29 July 1959. The first Trident (G-ARPA), a production aircraft for BEA, flew for the first time on 9 January 1962.

Five versions of the Trident have been built, of which the Trident 1 and 1E were fully described in previous editions of *Jane's*. Current versions are as follows:

Trident 2E. Developed version; 15 ordered by BEA in August 1965, with accommodation for up to 115 passengers. Two others ordered subsequently by Cyprus Airways, and 33 by CAAC, the national airline of the Chinese People's Republic. Overall length unchanged. Fuel capacity and max T-O weight increased, and take-off performance improved by use of more powerful Spey engines. Leading-edge slats, as in Trident 1E, and increased wing span. Low-drag (Küchemann) wingtips. Some strengthening of undercarriage, and of wing and fuselage by use of thicker panels. The first Trident 2E (G-AVFA) flew for the first time on 27 July 1967, and the first for BEA (G-AVFC) was delivered on 15 February 1968. BEA (British Airways) aircraft, known as Trident Twos, began scheduled services on 18 April 1968.

Trident 2Es were delivered with autoland installed at triplex level and were the first airliners in the world with complete all-weather-operation instrumentation of this kind. ARB certification for Trident autolands in Category 2 weather (100 ft decision height, 400 m RVR) was received in February 1969. Full Category 2 operation was first performed by BEA Tridents in December 1970. AiResearch APU, similar to that in Trident 3B, being fitted retrospectively to all British Airways Trident 2Es. Category 3B certification was announced in mid-1975.

Trident 3B. High-capacity short-haul development of Trident 1E, with fuselage lengthened by 5·00 m (16 ft 5 in) to accommodate from 128 to 180 passengers. Wing span as Trident 2E, but wing area, wing incidence and flap span increased. Powered by same mark of Spey turbofan as Trident 2E, but with Rolls-Royce RB.162-86 turbojet in tail for improved T-O performance. First flight, on 11 December 1969, was made by G-AWYZ without an operational RB.162 engine fitted. The RB.162 was fitted in February 1970, the first flight with this engine operating being made on 22 March 1970. BEA ordered 26, with options on 10 more, and the first of these entered service on 1 April 1971. Trident 3B autoland operation down to 12 ft decision height 270 m RVR, and take-offs in RVR of 90 m (full Category 3A conditions) were certificated by the ARB (now the CAA) in December 1971, and in May 1972 the Authority approved the start of such operations by BEA. All these Tridents, now in the British Airways fleet, were fitted retrospectively to this standard by the Winter of 1972-73. Category 3B certification was announced in mid-1975.

Super Trident 3B. Announced in late 1972, following an order for two of this version by CAAC (Civil Aviation Administration of China). Externally identical to British Airways Trident 3B; major differences are an increase in passenger seating capacity to 152, and increases in fuel capacity, max T-O and max zero-fuel weights, and an effective range increase of 373 nm (692 km; 430 miles). The additional 1,727 litres (380 Imp gallons) of fuel is carried in the wing centre-section tank, and increases the total usable fuel capacity to 27,275 litres (6,000 Imp gallons). First example flew for the first time on 9 July 1975; both aircraft delivered by the end of that year.

The following details apply to the Trident 2E and 3B:

Type: Short/medium-range turbofan-engined airliner.

Wings: Cantilever low-wing monoplane. Wing sections designed with high critical drag rise Mach number for economical operation at ultimate subsonic cruising speeds. Mean thickness/chord ratio approx 9·8%. Dihedral 3°. Incidence 6° 30′ at root, 1° 30′ at tip. Sweepback at quarter-chord 35°. Main wing is continuous from wingtip to wingtip, and comprises a six-cell centre-section box extending across the fuselage, a two-cell box from the wing root out to 40% of the semi-span, and from there a single-cell box to the wingtip. The entire wing box is subdivided to form integral fuel tanks. Skins and stringers are of aluminium alloy, as are the leading-edge and the trailing-edge flaps. Extensive use is made of Reduxing between skins and stringers. Structure is fail-safe, except for slat and flap tracks which are safe-life components tested to at least six times the aircraft life. Conventional all-metal ailerons actuated by triplexed power control system without manual reversion. Three independent hydraulic systems work continuously in parallel and power three separate jacks of Fairey manufacture at each primary flying control surface. Two all-metal double-slotted trailing-edge flaps on each wing. Krueger leading-edge flap at each wing-root. All flaps operated by screwjacks and hydraulic motors of Hobson manufacture. One all-metal spoiler on 2E, two on 3B, forward of outer flap on each wing, act also as airbrakes/lift dumpers. Lift dumpers forward of inner flaps. No trim tabs. Both 2E and 3B have full-span leading-edge slats, in four sections per wing, operated by screwjacks and extending on curved titanium tracks. Thermal anti-icing system.

Fuselage: Consists of a pressure shell extending back to the engines and a rear fuselage carrying the engines and tail unit. Semi-monocoque fail-safe structure of aluminium/copper alloys, using Redux bonding to attach stringers to skin throughout the pressure cell. Unpressurised cutouts for nose and main landing gear and wing centre-section.

Hawker Siddeley Super Trident 3B for delivery to CAAC of China

Tail Unit: Cantilever all-metal T-tail. All-moving tailplane with geared slotted flap on trailing-edge to assist in providing high negative lift coefficient for take-off and landing. No trim tabs. Power control system as for ailerons. Thermal anti-icing of leading-edges.

Landing Gear: Retractable tricycle type. Hydraulic retraction. Hawker Siddeley (main units) and Automotive Products (nose) oleo-pneumatic shock-absorbers. Each main unit consists of two twin-tyred wheels mounted on a common axle: during retraction the leg twists through nearly 90° and lengthens by 152 mm (6 in), enabling wheels to stow within the circular cross-section of the fuselage. Nose unit has twin wheels and is offset 0·61 cm (2 ft 0 in) to port, retracting transversely. Dunlop wheels, tyres and multi-plate disc brakes, with Maxaret anti-skid units. Trident 3B has main-wheel tyres size 36 × 10, pressure 11·38 bars (165 lb/sq in), and nosewheel tyres size 29 × 8, pressure 8·55 bars (124 lb/sq in).

Power Plant: Three Rolls-Royce Spey RB.163-25 Mk 512-5W turbofan engines (two in pods, one on each side of rear fuselage; one inside rear fuselage), each rated at 53·2 kN (11,960 lb st). Additionally, Trident 3B has a 23·35 kN (5,250 lb st) Rolls-Royce RB.162-86 turbojet installed in tail, below the rudder, to boost T-O and climb-out. Five integral fuel tanks, four in wings and one in centre-section. Total usable fuel capacity: Trident 2E, 29,094 litres (6,400 Imp gallons); Trident 3B, 25,548 litres (5,620 Imp gallons). For operators requiring greater range/payload capability, Trident 3B has provision for an additional wing centre-section tank of 1,727 litres (380 Imp gallons) capacity. This tank is fitted in the Super 3B. One pressure refuelling point under each wing. Oil capacity 13·5 litres (3 Imp gallons) per engine.

Accommodation (Trident 2E): Crew of three on flight deck. Mixed-class version has galley (stbd) and toilet (port) at front, then a 12-seat first class compartment, with seats in pairs on each side of central aisle, two galleys (port), 79-seat tourist class cabin with three-seat units on each side of aisle, and two toilets at rear. British Airways' all-tourist version has 97 seats, six-abreast, with galley and toilet at front and two toilets at rear. Provision can be made for 132 passengers in high-density seating arrangement. Two inward-opening plug-type passenger doors, at front and centre of cabin on port side, with provision for built-in airstairs. Doors for crew and servicing at front and amidships on starboard side. Large underfloor baggage holds forward and aft of wing. All crew and passenger accommodation air-conditioned. Provision for air-conditioning forward part of forward baggage hold for animals.

Accommodation (Trident 3B): Basically as for Trident 2E, with four-abreast first class seats at 965 mm (38 in) pitch and six-abreast tourist seating at 787 mm (31 in) pitch. Mixed class version has toilet (port) and two galleys (port and stbd) at front, 14-seat first class cabin, and 122-seat tourist cabin, with two galleys and two toilets (one each port and stbd) at rear. All-tourist version has 152 seats at 762 mm (30 in) pitch, no coat stowage and only one galley (stbd) instead of two at rear, but is otherwise similar. High-density versions can have up to 180 seats (with seven-abreast seating in centre fuselage) at 711 mm (28 in) pitch and no rear galley, but are otherwise similar to the 152-seat version.

Systems: Hawker Siddeley Dynamics air-conditioning and pressurisation system, differential 0·57 bars (8·25 lb/sq in). Two independent supplies, each capable of maintaining full cabin pressurisation, with emergency ram-air system for use below 2,440 m (8,000 ft). Three independent hydraulic systems operating all flying controls, landing gear, nosewheel steering, brakes and windscreen wipers. Each system powered by separate engine-driven pump, operating continuously in parallel at 207 bars (3,000 lb/sq in), using Skydrol fluid. Backup hydraulic power supplied by two electrically-driven pumps, and emergency power from dropout air turbine, capable of feeding any one system. Pneumatic system for toilet flushing, forward water system, stall recovery system and for pressurising hydraulic reservoirs. Electrical system comprises three separate channels, supplied by three 27·5kVA brushless generators. Emergency 30 min AC and DC supply available from 24V battery. AiResearch GTCP 85C APU for engine starting and cabin air-conditioning, driving generator to provide 40kVA of electrical power from which hydraulic systems can also be actuated through standby pumps.

Electronics and Equipment: To customer's specification. Provision for duplicated VOR/ILS, including a third localiser for three-channel automatic landing guidance; integration of navigational aids with flight system, providing coupling facilities for all flight modes except take-off; duplicated ADF, VHF and HF with selective calling; C- or X-band weather radar; triplicated radio altimeters for automatic landing; Doppler; DME and transponder.

Dimensions, external:
Wing span (2E, 3B) 29·87 m (98 ft 0 in)
Wing geometric mean chord:
3B 4·65 m (15 ft 2¾ in)
Wing aspect ratio:
3B 6·43
Length of fuselage:
3B 36·55 m (119 ft 11 in)
Length overall:
2E 34·97 m (114 ft 9 in)
3B 39·98 m (131 ft 2 in)
Height overall:
2E 8·23 m (27 ft 0 in)
3B 8·61 m (28 ft 3 in)
Tailplane span 10·44 m (34 ft 3 in)
Wheel track (c/l of shock-absorbers)
5·83 m (19 ft 1¼ in)
Wheelbase:
2E 13·41 m (44 ft 0 in)
3B 16·01 m (52 ft 6½ in)
Passenger doors (both):
Height 1·78 m (5 ft 10 in)
Width 0·71 m (2 ft 4 in)
Min height to sill 2·87 m (9 ft 5 in)
Max height to sill 3·12 m (10 ft 3 in)
Crew and service doors (fwd stbd on 2E; fwd, centre and rear stbd on 3B. Optional fourth door rear port side on high-density 3B):
Height 1·22 m (4 ft 0¼ in)
Width 0·61 m (2 ft 0 in)
Height to sill approx 2·74 m (9 ft 0 in)
Emergency exits (above centre-section, port and stbd):
Height 1·03 m (3 ft 4 in)
Width 0·51 m (1 ft 8 in)
Baggage hold doors (fwd, stbd):
Height (vertical) 0·89 m (2 ft 11 in)
Width 1·22 m (4 ft 0 in)
Height to sill 1·37 m (4 ft 6 in)
Baggage hold door (rear, port):
Mean height (vertical) 0·81 m (2 ft 8 in)
Width 0·89 m (2 ft 11 in)
Height to sill 1·37 m (4 ft 6 in)

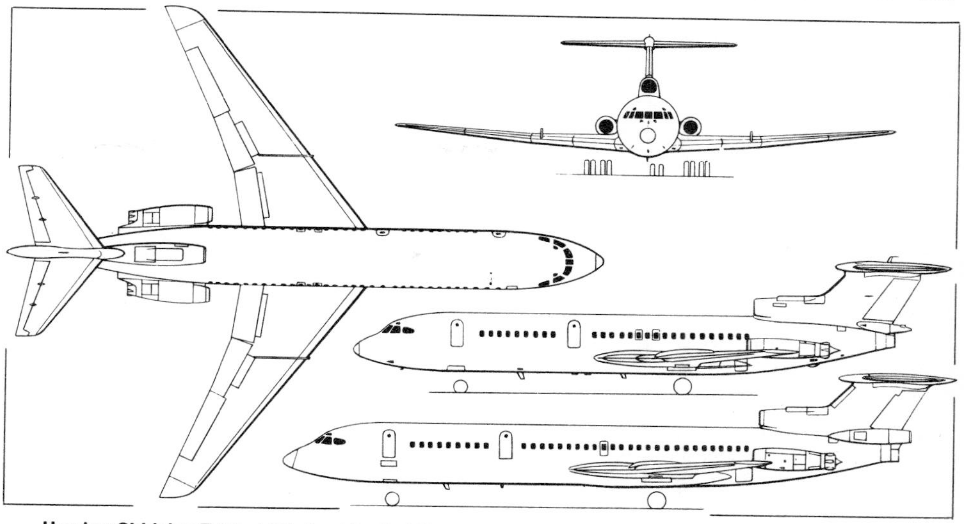

Dimensions, internal:
Cabin, excl flight deck:
Length:
2E 20·46 m (67 ft 1½ in)
3B 25·43 m (83 ft 5 in)
Max width 3·44 m (11 ft 3½ in)
Max height:
2E 2·02 m (6 ft 7½ in)
3B 2·03 m (6 ft 8 in)
Floor area:
2E 65·77 m^2 (708 sq ft)
3B 96·9 m^2 (1,043 sq ft)
Volume:
2E 125·7 m^3 (4,440 cu ft)
3B 158·57 m^3 (5,600 cu ft)
Freight hold (fwd):
2E 13·88 m^3 (490 cu ft)
3B 17·92 m^3 (633 cu ft)
Freight hold (rear):
2E 7·65 m^3 (270 cu ft)
3B 13·51 m^3 (477 cu ft)

Areas:
Wings, gross:
2E 135·82 m^2 (1,462 sq ft)
3B 138·7 m^2 (1,493 sq ft)
Ailerons (total) 4·89 m^2 (52·5 sq ft)
Trailing-edge flaps (total):
3B 27·12 m^2 (291·9 sq ft)
Spoilers 1·42 m^2 (15·3 sq ft)
Fin 18·76 m^2 (202 sq ft)
Rudder 4·84 m^2 (52·1 sq ft)
Tailplane 28·80 m^2 (310 sq ft)

Weights and Loadings:
Operating weight, empty:
2E 33,203 kg (73,200 lb)
3B (152-seat) 37,090 kg (81,778 lb)
Super 3B (152-seat) 37,259 kg (82,143 lb)
Max payload:
2E 12,156 kg (26,800 lb)
3B (152-seat) 15,296 kg (33,722 lb)
Super 3B (152-seat) 16,037 kg (35,357 lb)
Max T-O weight:
2E 65,315 kg (144,000 lb)
3B 68,040 kg (150,000 lb)
Super 3B 71,667 kg (158,000 lb)
Max ramp weight:
3B 68,267 kg (150,500 lb)
Max zero-fuel weight:
2E 45,359 kg (100,000 lb)
3B 52,395 kg (115,500 lb)
Super 3B 53,296 kg (117,500 lb)
Max landing weight:
2E 51,261 kg (113,000 lb)
3B 58,285 kg (128,500 lb)
Super 3B 58,965 kg (130,000 lb)
Max wing loading:
3B 490·7 kg/m^2 (100·5 lb/sq ft)
Max power loading:
3B 371·9 kg/kN (3·65 lb/lb st)

Performance (2E at max T-O weight):
Never-exceed speed (design limit) Mach 0·95
Typical high-speed cruise at 8,230 m (27,000 ft)
Mach 0·88 (525 knots; 972 km/h; 605 mph)
Econ cruising speed at 9,150 m (30,000 ft)
Mach 0·88 (518 knots; 959 km/h; 596 mph)
T-O field length for 868 nm (1,610 km; 1,000 mile) stage, with 9,697 kg (21,378 lb) payload
1,950 m (6,400 ft)
Minimum ground turning radius 15·85 m (52 ft 0 in)
Runway LCN requirement at max weight 58
Range with max fuel* and 7,493 kg (16,520 lb) payload 2,171 nm (4,025 km; 2,500 miles)
Range with typical space-limited payload* of 9,679 kg (21,378 lb) 2,140 nm (3,965 km; 2,464 miles)

Performance (Super 3B, at max T-O weight except where stated):
Never-exceed speed (design limit) Mach 0·95
Max cruising speed at 8,625 m (28,300 ft)
522 knots (967 km/h; 601 mph)
Typical high-speed cruise at 7,620 m (25,000 ft)
505 knots (936 km/h; 581 mph)
Econ cruising speed at 8,800-10,000 m (29,000-33,000 ft) 463 knots (858 km/h; 533 mph)
Stalling speed (at max landing weight, flaps down)
112 knots (208 km/h; 129 mph) EAS
T-O to 10·7 m (35 ft) 2,715 m (8,900 ft)
Landing from 9 m (30 ft) at max landing weight
1,730 m (5,680 ft)
Minimum ground turning radius 18·60 m (61 ft 0 in)
Runway LCN requirement at max weight 66
Range with max fuel* and 12,791 kg (28,200 lb) payload 2,050 nm (3,798 km; 2,360 miles)
Range with max payload*
1,550 nm (2,872 km; 1,785 miles)

**Reserves for 217 nm (450 km; 250 mile) diversion, 45 min hold at 4,570 m (15,000 ft), final reserve, 4·5% en route allowance and allowances for taxi prior to take-off, circuit, approach and land at destination, and taxi after landing*

Operational Noise Characteristics (FAR Pt 36: 2E, estimated):
T-O noise level 109 EPNdB
Approach noise level 109·5 EPNdB
Sideline noise level 106 EPNdB

Operational Noise Characteristics (FAR Pt 36: 3B, estimated):
T-O noise level 105 EPNdB
Approach noise level 110·5 EPNdB
Sideline noise level 108 EPNdB

Hawker Siddeley Trident 2E short-haul airliner, with additional side view (bottom) of Trident 3B *(Pilot Press)*

HAWKER SIDDELEY HAWK

After examining designs submitted by BAC and Hawker Siddeley to meet an RAF requirement for a basic and advanced jet trainer, the Ministry of Defence announced in October 1971 that the Hawker Siddeley 1182 had been selected to meet this requirement. Selection of a non-afterburning version of the Rolls-Royce/Turboméca Adour to power the aircraft was announced on 2 March 1972, and later in the same month the Ministry of Defence confirmed an order for 176 HS 1182s, which were given the RAF name of Hawk. These will consist of one pre-production aircraft (XX154), which first flew on 21 August 1974, and 175 production Hawks, of which deliveries were scheduled to begin in October 1976. There are no separate prototypes; instead, the first five production aircraft are allocated to the development programme. Ten Hawks had flown by the Autumn of 1976.

The Hawk has been designed to be fully aerobatic (it is stressed to +8 and −4*g*) and to have a fatigue life of 6,000 hours. It will eventually replace the Jet Provost, Gnat Trainer and Hunter in RAF service for pre-wings and advanced flying training, and for radio, navigation and weapons training. The basic design is capable of development for other operational roles, and studies of a number of variants have been made.

TYPE: Two-seat basic and advanced jet trainer, with capability for close support role.

WINGS: Cantilever low-wing monoplane. Thickness/chord ratio 10·9% at root, 9% at tip. Dihedral 2°. Sweepback 26° on leading-edge, 21° 30′ at quarter-chord. One-piece wing, with six-bolt attachment to fuselage, employing a machined spars-and-skin torsion box, the greater part of which forms an integral fuel tank. Hydraulically-operated double-slotted flaps and ailerons, the latter operated by Automotive Products tandem actuators.

FUSELAGE: Conventional all-metal structure of frames and stringers, cut out to accept the one-piece wing. Large airbrake under rear of fuselage, aft of wing.

TAIL UNIT: Cantilever all-metal structure, with sweepback on all surfaces. One-piece all-moving power-operated anhedral tailplane, with Automotive Products tandem hydraulic actuators. Mechanically-operated rudder, with electrically-actuated trim tab.

LANDING GEAR: Wide-track retractable tricycle type, with single wheel on each unit. Hydraulic actuation, using Automotive Products jacks. Main units retract inward into wing, ahead of front spar; nosewheel retracts forward. Main wheels and tyres size 6·50-10, pressure 9·86 bars (143 lb/sq in). Nosewheel and tyre size 4·4-16, pressure 8·27 bars (120 lb/sq in). Tail bumper fairing under rear fuselage. Anti-skid wheel brakes.

POWER PLANT: One Rolls-Royce/Turboméca RT.172-06-11 Adour 151 non-afterburning turbofan engine, rated at 23·75 kN (5,340 lb st). Air intake on each side of fuselage, forward of wing leading-edge. Engine starting by integral gas turbine starter. Fuel in one fuselage bag tank (822·5 litres; 181 Imp gallons) and integral wing tank (836·5 litres; 184 Imp gallons); total fuel capacity 1,659 litres (365 Imp gallons). Pressure refuelling point near front of port engine air intake trunk. Provision for carrying one 454 litre (100 Imp gallon) drop-tank on each inboard underwing pylon.

ACCOMMODATION: Crew of two in tandem under one-piece fully-transparent sideways-opening canopy. Fixed front windscreen and separate internal windscreen in front of rear cockpit. Rear seat elevated. Martin-Baker Mk 10B zero-zero rocket-assisted ejection seats, with MDC (miniature detonation cord) system to break canopy before seats eject. The MDC can also be operated from outside the cockpit in case of a ground emergency. Dual controls standard. Entire accommodation pressurised, heated and air-conditioned.

SYSTEMS: Hawker Siddeley Dynamics cockpit air-conditioning and pressurisation systems, using engine bleed air. Duplicated hydraulic systems, each 207 bars (3,000 lb/sq in), for actuation of control jacks, flaps, airbrake, landing gear and anti-skid wheel brakes. Compressed nitrogen accumulators provide emergency power for flaps and landing gear. Hydraulic accumulator for emergency operation of wheel brakes. No pneumatic system. DC electrical power from single brushless generator, with two static inverters to provide AC power and two batteries for standby power. Gaseous oxygen system for crew. Pop-up Dowty Rotol ram-air turbine in upper rear fuselage provides emergency power for flying controls in the event of an engine or No. 2 pump failure.

ELECTRONICS AND EQUIPMENT: Flight instrumentation includes Ferranti gyros and inverter, two Sperry Gyroscope RAI-4 4 in remote attitude indicators and a magnetic detector unit, and Louis Newmark compass system. Radio and navigation equipment includes Sylvania UHF and VHF, Cossor CAT.7000 Tacan, Cossor ILS with CILS.75/76 localiser/glideslope receiver and marker receiver, and IFF/SSR.

ARMAMENT AND OPERATIONAL EQUIPMENT: Ferranti F.195 weapon sight and camera recorder in each cockpit. Trainer version has underfuselage centreline-mounted 30 mm Aden gun and ammunition pack, and two inboard underwing points each capable of carrying a 454 kg (1,000 lb) stores load. Typical underwing armament training loads include two Matra 155 launchers, each with eighteen 2·75 in air-to-surface rockets, or two clusters of four practice bombs. Provision for two outboard underwing pylons, and a pylon in place of the ventral gun pack, also each capable of a 1,000 lb load (2,540 kg; 5,600 lb total external stores load), for close support role. In RAF training roles the normal max external load will probably be about 680 kg (1,500 lb).

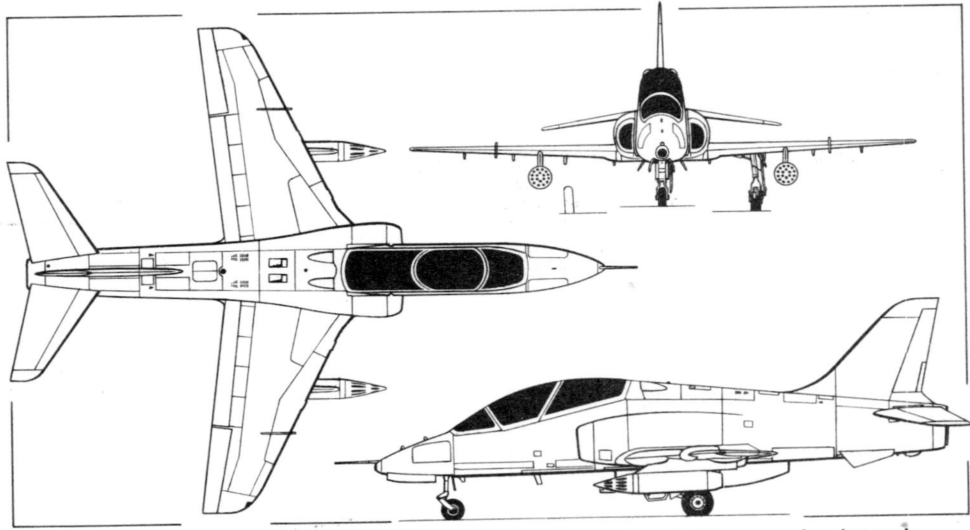

Hawker Siddeley Hawk two-seat advanced training aircraft, equipped with gun and rocket pods
(Pilot Press)

Hawker Siddeley Hawk two-seat basic and advanced jet trainer

Third Hawker Siddeley Hawk to fly, camouflaged and carrying four underwing rocket pods

DIMENSIONS, EXTERNAL:	
Wing span	9·39 m (30 ft 9¾ in)
Wing chord at root	2·65 m (8 ft 8¼ in)
Wing chord at tip	0·90 m (2 ft 11½ in)
Wing aspect ratio	5·284
Length overall, excl probe	11·17 m (36 ft 7¾ in)
Height overall	4·09 m (13 ft 5 in)
Tailplane span	4·39 m (14 ft 4¾ in)
Wheel track	3·34 m (10 ft 11½ in)
AREAS:	
Wings, gross	16·69 m² (179·6 sq ft)
Ailerons (total)	1·05 m² (11·30 sq ft)
Trailing-edge flaps (total)	2·50 m² (26·91 sq ft)
Airbrake	0·53 m² (5·70 sq ft)
Fin	2·51 m² (27·02 sq ft)
Rudder, incl tab	0·58 m² (6·24 sq ft)
Tailplane	4·33 m² (46·61 sq ft)
WEIGHTS:	
Weight empty	3,379 kg (7,450 lb)
T-O weight:	
trainer, 'clean'	5,035 kg (11,100 lb)
trainer, armed	5,443 kg (12,000 lb)
Max T-O weight	7,375 kg (16,260 lb)
Max landing weight	4,649 kg (10,250 lb)
PERFORMANCE:	
Max level speed	562 knots (1,041 km/h; 647 mph) IAS
Max Mach number	1·16
Time to 9,145 m (30,000 ft)	6 min 20 sec
Service ceiling	14,630 m (48,000 ft)
Endurance (trainer, 'clean')	approx 2 hr 0 min
Ferry range with two 455 litre (100 Imp gallon) drop-tanks	1,700 nm (3,150 km; 1,957 miles)

HAWKER SIDDELEY HARRIER

RAF designations: Harrier GR.Mk 1, 1A and 3; and T. Mk 2, 2A and 4

USMC designations: AV-8A and TAV-8A

The Harrier is the western world's only operational fixed-wing V/STOL strike fighter. Developed from six years of operating experience with the P.1127/Kestrel series of aircraft (see 1968-69 *Jane's*), the Harrier is an integrated V/STOL weapon system, incorporating the

Ferranti FE 541 inertial navigation and attack system and Smiths head-up display. The first of six single-seat prototypes (XV276) flew for the first time on 31 August 1966. The major current production version is the single-seat AV-8A (Harrier Mk 50) for the US Marine Corps.

By February 1976, some 230 aircraft of the Harrier family had been built, and had made more than 400,000 lift-offs and landings from a variety of surfaces such as grass, tarmac, concrete, dirt and gravel strips, snow- and ice-covered runways. They had also operated from the decks of 24 ships, including US, Argentinian, Spanish, Indian, French and British aircraft carriers, Italian and British cruisers, and US amphibious support ships. The aircraft had been flown by more than 300 Air Force, Navy, Marine and Army pilots from the UK, the USA and the German Federal Republic.

The following versions of the Harrier have been built:

Harrier GR. Mk 1, 1A and 3. Single-seat close-support and tactical reconnaissance versions, in quantity production for the Royal Air Force. First of 77 production aircraft ordered initially (XV738) flew on 28 December 1967. Entered service with the Harrier OCU, No. 233 Squadron, at RAF Wittering, on 1 April 1969. Delivered to No. 1 Squadron at Wittering and Nos. 3, 4 and 20 in Germany. Harriers of No. 1 Squadron carried out operational trials on board HMS *Ark Royal* in May 1971.

A Harrier GR. Mk 1A, piloted by Sqn Ldr T. L. Lecky-Thompson, set up two international time-to-height records after VTO, in Class H for jet-lift aircraft, on 5 January 1971. The aircraft, after a vertical take-off, reached 9,000 m (29,528 ft) in 1 min 44·7 seconds and 12,000 ft (39,370 m) in 2 min 22·7 seconds. The same RAF pilot also set up a Class H altitude record of 14,040 m (46,063 ft) in a Harrier GR. Mk 1A on 2 January 1971.

The Harrier has, in non-record-attempt flights, been flown to altitudes in excess of 15,240 m (50,000 ft).

Harrier GR. Mk 1 aircraft were fitted initially with 84·5 kN (19,000 lb st) Pegasus 101 engines. When retrofitted subsequently with the 89·0 kN (20,000 lb st) Pegasus 102 engine they were redesignated GR.Mk 1A. Aircraft now in service are equipped with the more powerful Pegasus 103 engine and are designated GR. Mk 3. A further 15 Harriers, comprising 12 GR.Mk 3s and 3 T.Mk 4s, were ordered for the RAF in March 1973.

A Harrier GR.Mk 1 (XV742) was allocated temporarily the civil registration G-VSTO for use as an overseas demonstrator.

Harrier T. Mk 2, 2A and 4. Two-seat versions, retaining the full combat capability of the single-seater in terms of equipment fit and weapon carriage. There is a large degree of commonality in structure and system components, ground support equipment and flight and ground crew training. Differences include a new, longer nose section forward of the wing leading-edge, with two cockpits in tandem; a tailcone approx 1·83 m (6 ft) longer than that of the single-seat model; and enlarged fin surfaces. The two-seat Harrier may be used operationally with the rear seat and compensating tail ballast removed, thus minimising the weight penalty over its single-seat counterpart. First development aircraft (XW174) flew on 24 April 1969, followed by the second on 14 July 1969 and the first production aircraft (XW264) on 3 October 1969. Current orders are for 21 of this version (including two development aircraft), and the first of these entered RAF service in July 1970.

The Harrier T. Mk 2, like the GR. Mk 1, was powered originally by the Pegasus 101 engine. The designations T. Mk 2A and T. Mk 4 apply to aircraft retrofitted with, respectively, the Pegasus 102 and 103.

Sea Harrier FRS.Mk 1. Developed version for Royal Navy; described separately.

Harrier Mk 50 (USMC designation AV-8A). Single-seat close-support and tactical reconnaissance version for the US Marine Corps. Dimensionally the same as GR. Mk 3, but with modifications to customer's specification, including provision for the carriage of Sidewinder missiles. Initial quantity of 12 ordered in 1969. Subsequent firm orders brought this total to 102. The last fiscal order for USMC aircraft included eight Harrier **Mk 54s** with Pegasus 103 engines (a two-seat version designated **TAV-8A**) for operational training; the first of these was delivered in September 1975.

The first AV-8A was delivered to the USA on 26 January 1971. The first 10 AV-8As had Pegasus 102 engines; the next 92 aircraft are powered by Pegasus 103s, which have also been fitted retrospectively to the earlier aircraft. McDonnell Douglas has licence rights to manufacture "any significant numbers" ordered if the US government decides to build in the USA.

The AV-8As equip three USMC combat squadrons: VMA 513, VMA 542 and VMA 231 at Cherry Point, North Carolina. A training squadron, VMA(T) 203, is also based at Cherry Point. During 1974-75, VMA 513 operated from Japan and a detachment was operational from LPH9 USS *Guam.* Six AV-8As and two TAV-8As have been ordered, through the USA, for the Spanish Navy, by whom they will be known as **Matadors.**

Harrier Mk 52. One aircraft built as a demonstrator using HSA and equipment suppliers' private funding. It is similar to the Harrier T. Mk 4, and is fitted with a Pegasus 103 engine; in recognition of its status as the first civil-registered jet V/STOL aircraft in the UK, it has been granted the civil registration G-VTOL. First flight was made on 16 September 1971, with a Pegasus 102 fitted initially.

An Advanced Harrier study was completed in December 1973 by Hawker Siddeley, Rolls-Royce, McDonnell Douglas and Pratt & Whitney. This was funded jointly by the UK and US governments on behalf of the RAF, RN, USMC and USN, but in March 1975 the UK Secretary of State for Defence, Mr Roy Mason, stated that "there is not enough common ground on the Advanced Harrier for us to join in the programme with the US". The US proposals for advanced versions of the Harrier are, therefore, described under the McDonnell Douglas heading in this edition.

The following details apply generally to the Harrier GR. Mk 3 and T. Mk 4, except where a specific version is indicated:

Type: V/STOL close-support and reconnaissance aircraft.

Wings: Cantilever shoulder-wing monoplane. Aerofoil section of HSA design. Thickness/chord ratio 10% at root, 5% at tip. Anhedral 12°. Incidence 1° 45′. Sweepback at quarter-chord 34°. One-piece aluminium alloy three-spar safe-life structure with integrally-machined skins, manufactured by Brough factory of HSA, with six-point attachment to fuselage. Plain ailerons and flaps, of bonded aluminium alloy honeycomb construction. Ailerons irreversibly operated by Fairey tandem hydraulic jacks. Jet reaction control valve built into front of each outrigger wheel fairing. Entire wing unit removable to provide access to engine. For ferry missions, the normal 'combat' wingtips can be replaced by bolt-on extended tips to increase ferry range.

Fuselage: Conventional semi-monocoque safe-life structure of frames and stringers, mainly of aluminium alloy, but with titanium skins at rear and some titanium adjacent to engine and in other special areas. Access to power plant through top of fuselage, ahead of wing. Jet reaction control valves in nose and in extended tailcone. Large forward-hinged airbrake under fuselage, aft of main-wheel well.

Tail Unit: One-piece variable-incidence tailplane, with 15° of anhedral, irreversibly operated by Fairey tandem hydraulic jack. Rudder and trailing-edge of tailplane are of bonded aluminium honeycomb construction. Rudder is operated manually. Trim tab in rudder. Ventral fin under rear fuselage. Fin tip carries suppressed VHF aerial.

Landing Gear: Retractable bicycle type of Dowty Rotol manufacture, permitting operation from rough unprepared surfaces of CBR as low as 3% to 5%. Hydraulic actuation, with nitrogen bottle for emergency extension of landing gear. Single steerable nosewheel retracts forward, twin coupled mainwheels rearward, into fuselage. Small outrigger units retract rearward into fairings slightly inboard of wingtips. Nosewheel leg is of levered-suspension Liquid Spring type. Dowty Rotol telescopic oleo-pneumatic main and outrigger gear. Dunlop wheels and tyres, size 26·00 × 8·75-11 (nose unit), 27·00 × 7·74-13 (main units) and 13·50 × 6·4 (outriggers). GR. Mk 1, 1A and 3 pressure 6·21 bars (90 lb/sq in) on nose and main units, 6·55 bars (95 lb/sq in) on outriggers. T. Mk 2, 2A and 4 tyre pressures 6·90 bars (100 lb/sq in) on nose unit, 6·55 bars (95 lb/sq in) on main and outrigger units. Dunlop multi-disc structural carbon brakes and Dunlop-Hytrol adaptive anti-skid system.

Power Plant: One Rolls-Royce Bristol Pegasus Mk 103 vectored-thrust turbofan engine (95·6 kN; 21,500 lb st), with four exhaust nozzles of the two-vane cascade type, rotatable through 98° from fully-aft position. Engine bleed air from HP compressor used for jet reaction control system and to power duplicated air motor for nozzle actuation. The low-drag intake cowls, with outward-cambered lips, each have 8 automatic suction relief doors aft of the leading-edge to improve intake efficiency by providing extra air for the engine at low forward or zero speeds. Fuel in five integral tanks in fuselage and two in wings, with total capacity of approx 2,865 litres (630 Imp gallons). This can be supplemented by two 455 litre (100 Imp gallon) jettisonable combat tanks or two 1,500 litre (330 Imp gallon)

Hawker Siddeley Harrier GR. Mk 3 with laser rangefinder in modified nose

Hawker Siddeley AV-8A Harriers in service with US Marine Corps Squadron VMA 231

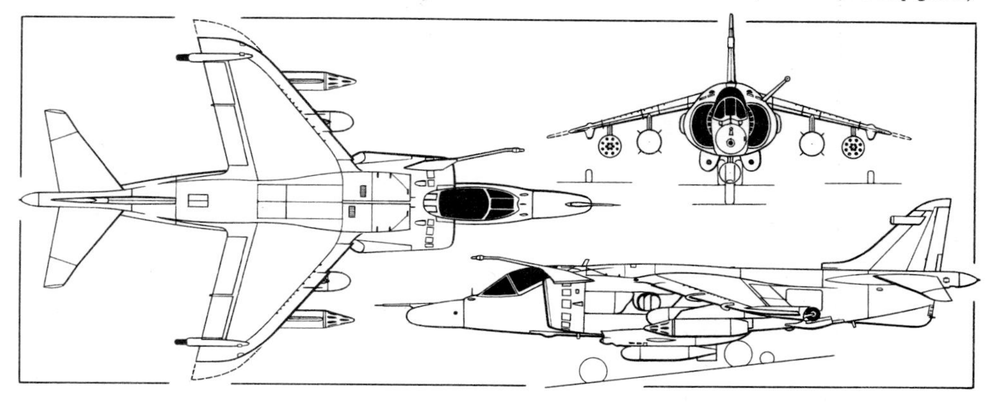

Hawker Siddeley Harrier GR. Mk 3 single-seat V/STOL close-support and reconnaissance aircraft with laser rangefinder in nose. This is being fitted retrospectively to all GR. Mk 3s *(Pilot Press)*

ferry tanks on the inboard wing pylons. Ground refuelling point in port rear nozzle fairing. Provision for in-flight refuelling probe above the port intake cowl.

Accommodation: Crew of one (Mk 3) or two (Mk 4) on Martin-Baker Type 9A Mk 2 zero-zero rocket ejection seats which operate through the miniature detonating cord equipped canopy of the pressurised, heated and air-conditioned cockpit. AV-8A Harriers of the USMC will be retrofitted with Stencel SIIIS-3 ejection seats. Manually-operated canopy, rearward-sliding on single-seat, sideways-opening (to starboard) on two-seat versions. Birdproof windscreen, with hydraulically-actuated wiper. Windscreen de-icing.

Systems: Three-axis limited-authority autostabiliser for V/STOL flight. Pressurisation system of HSA design, with Normalair-Garrett and Marston major components; max pressure differential 0·24 bars (3·5 lb/sq in). Duplicated hydraulic systems, each of 207 bars (3,000 lb/sq in), actuate Fairey flying control and general services and include a retractable ram-air turbine inside top of rear fuselage, driving a small hydraulic pump for emergency power. AC electrical system with transformer-rectifiers to provide required DC supply. One 12kVA Lucas alternator. Two 28V 25Ah batteries, one of which energises a 24V motor to start Lucas gas-turbine starter/APU. This unit drives a 6kVA auxiliary alternator for ground readiness servicing and standby. Normalair-Garrett liquid oxygen system of 5 litres (1 Imp gallon) capacity. Bootstrap-type cooling unit for equipment bay, with intake at base of dorsal fin.

Electronics and Equipment: Plessey U/VHF, Ultra standby UHF, Hoffman Tacan and Cossor IFF, Ferranti FE 541 inertial navigation and attack system (INAS), with Sperry C2G compass, Smiths electronic head-up display of flight information and Smiths air data computer. INAS can be aligned equally well at sea or on land. The weapon aiming computer provides a general solution for manual or automatic release of free-fall and retarded bombs, and for the aiming of rockets and guns, in dive and straight-pass attacks over a wide range of flight conditions and very considerable freedom of manoeuvre in elevation. Communication equipment ranges through VHF in the 100-156MHz band to UHF in the 220-400MHz band.

Armament and Operational Equipment: Optically-flat panel in nose, on port side, for F.95 oblique camera, which is carried as standard. A cockpit voice recorder with in-flight playback facility supplements the reconnaissance cameras, and facilitates rapid briefing and mission evaluation. No built-in armament. Combat load is carried on four underwing and one underfuselage pylons, all with ML ejector release units. The inboard wing points and the fuselage point are stressed for loads of up to 910 kg (2,000 lb) each, and the outboard underwing pair for loads of up to 295 kg (650 lb) each; the two strake fairings under the fuselage can each be replaced by a 30 mm Aden gun pod and ammunition. At present, the Harrier is cleared for operations with a maximum external load exceeding 2,270 kg (5,000 lb), but has flown with a weapon load of 3,630 kg (8,000 lb). The Harrier is able to carry 30 mm guns, bombs, rockets and flares of UK and US designs, and in addition to its fixed reconnaissance camera can also carry a five-camera reconnaissance pod on the underfuselage pylon. A typical combat load comprises a pair of 30 mm Aden gun pods, a 1,000 lb bomb on the underfuselage pylon, a 1,000 lb bomb on each of the inboard underwing pylons, and a Matra 155 launcher with 19 × 68 mm SNEB rockets on each outboard underwing pylon. A Sidewinder installation is provided in the AV-8A version, to give the aircraft an effective air-to-air capability in conjunction with the two 30 mm Aden guns.

Dimensions, external:

Wing span:	
combat	7·70 m (25 ft 3 in)
ferry	9·04 m (29 ft 8 in)
Wing chord at root	3·56 m (11 ft 8 in)
Wing chord at tip	1·26 m (4 ft 1½ in)
Wing aspect ratio:	
combat	3·175
ferry	4·08
Length overall:	
single-seat	13·87 m (45 ft 6 in)
two-seat	17·00 m (55 ft 9½ in)
Height overall:	
single-seat	approx 3·43 m (11 ft 3 in)
two-seat	approx 4·17 m (13 ft 8 in)
Tailplane span	4·24 m (13 ft 11 in)
Outrigger wheel track	6·76 m (22 ft 2 in)
Wheelbase, nosewheel to main wheels	approx 3·45 m (11 ft 4 in)

Areas:

Wings, gross:	
combat	18·68 m² (201·1 sq ft)
ferry	20·1 m² (216 sq ft)
Ailerons (total)	0·98 m² (10·5 sq ft)
Trailing-edge flaps (total)	1·29 m² (13·9 sq ft)
Fin (excl ventral fin):	
single-seat	2·40 m² (25·8 sq ft)
two-seat	3·57 m² (38·4 sq ft)
Rudder, incl tab	0·49 m² (5·3 sq ft)
Tailplane	4·41 m² (47·5 sq ft)

Hawker Siddeley Harrier two-seat combat trainer of the RAF's No. 233 Squadron

Weights and Loading:

Basic operating weight, empty, with crew:	
GR.Mk 1 and Mk 50	5,533 kg (12,200 lb)
T.Mk 2 (solo for combat)	5,896 kg (13,000 lb)
T.Mk 2 (dual)	6,168 kg (13,600 lb)
Max T-O weight (single-seat)	over 11,340 kg (25,000 lb)
Max wing loading (single-seat)	610 kg/m² (125 lb/sq ft)

Performance:

Speed at low altitude	over 640 knots (1,186 km/h; 737 mph) EAS
Mach number (in a dive)	approaching 1·3
Ceiling	more than 15,240 m (50,000 ft)
Endurance with one in-flight refuelling	more than 7 hr
Range with one in-flight refuelling	more than 3,000 nm (5,560 km; 3,455 miles)

HAWKER SIDDELEY SEA HARRIER

On 15 May 1975, the British government announced its decision to proceed with full development of a maritime version of the Harrier, subsequently designated Sea Harrier. The initial requirement is for 25 aircraft, primarily to equip the Royal Navy's new 'Invincible' class of through-deck cruisers from 1979. Of this total, 21 will be **Sea Harrier FRS. Mk 1** combat aircraft, three combat-standard development aircraft (available for operational use if required) and one two-seat trainer. The intended complement of each 'Invincible' class cruiser is nine Sea King helicopters and five Sea Harriers. Sea Harriers are also expected to serve on board the aircraft carrier *Hermes,* which is to be converted for anti-submarine duties.

The first Sea Harrier will fly in mid-1977 and will be a production aircraft, there being no prototype stage. Major changes compared with the Harriers in current service with the Royal Air Force and US Marine Corps will comprise a raised cockpit, revised operational electronics, and installation of multi-mode Ferranti radar in a redesigned nose that will fold to port for carrier stowage. Known by the name Blue Fox, this radar has been under development since March 1973, when the Electronic Systems Department of Ferranti was awarded a study and preliminary development contract. It is a derivative of the frequency-agile Seaspray radar fitted in the Lynx helicopter, but embodies changes to suit its different role, with air-to-air intercept and air-to-surface modes of operation. Two specially modified Hawker Hunter T. Mk 8s, redesignated T.Mk 8M, have been fitted with nose-mounted Blue Fox radars. Intended to speed the development of this radar, they will be used subsequently for radar training. Equipment of the Sea Harrier will include ECM in a container near the tip of the tail-fin and underwing attachments for air-to-air missiles of the Sidewinder type.

The Royal Navy's Sea Harrier FRS. Mk 1 will have a Rolls-Royce Pegasus 104 vectored-thrust turbofan engine. This will give 95·6 kN (21,500 lb st), like the Pegasus 103s fitted to current RAF Harriers. The two variants will differ little in design, except that the Pegasus 104 will incorporate additional anti-corrosion features and will generate greater electrical power.

Harriers have already accumulated thousands of take-offs and landings at sea, from a total of 24 different ships of eight naval services, in a wide range of weather, sea and climatic conditions. These operations have proved that no changes are needed to the aircraft's V/STOL design features to permit routine deployment at sea.

Estimated weights, loadings and performance figures are not yet available for the Sea Harrier. It is expected that the Navy's FRS.Mk 1 will operate at approximately the same weights as the GR.Mk 3, and will be capable of lifting a full military payload with a 152 m (500 ft) deck run into an overdeck wind of 30 knots (55·5 km/h; 34·5 mph).

The description of the GR.Mk 3 applies also to the FRS.Mk 1, except as follows:

Type: V/STOL fighter, reconnaissance and strike aircraft.

Power Plant: As GR.Mk 3, except one Rolls-Royce Bristol Pegasus Mk 104 vectored-thrust turbofan engine of 95·6 kN (21,500 lb st).

Accommodation: As GR.Mk 3, except for provision of Martin-Baker Type 10 ejection seat.

Systems: As GR.Mk 3, except autopilot function on Fairey Hydraulics, giving throughput to aileron power controls as well as to three-axis autostats. Pressurisation system of HSA design with major components from Normalair-Garrett and Delaney Gallay. British Oxygen liquid oxygen system of 5 litres (1 Imp gallon) capacity. Lucas Mk 2 GTS/APU.

Electronics and Equipment: Ferranti multi-mode Blue Fox nose-mounted radar. Smiths electronic head-up display and digital weapon aiming computer. Decca Doppler. Ferranti digital reference platform and navigation computer.

Armament and Operational Equipment: As GR.Mk 3, except for addition of Sidewinder installation similar to that of AV-8A, and provision for two air-to-surface missiles, as yet unspecified.

Dimensions, external: As GR.Mk 3 except:

Wing span	7·70 m (25 ft 3¼ in)
Length overall	14·50 m (47 ft 7 in)
Length overall, nose folded	12·88 m (42 ft 3 in)
Height overall	3·71 m (12 ft 2 in)

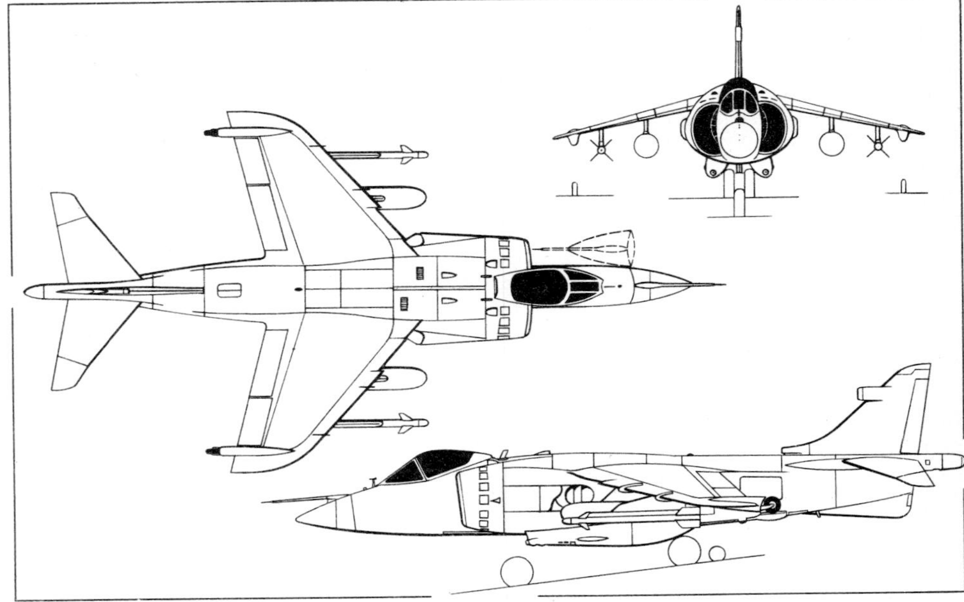

Hawker Siddeley Sea Harrier FRS. Mk 1 for the Royal Navy *(Pilot Press)*

HAWKER SIDDELEY BUCCANEER

The Hawker Siddeley (originally Blackburn) Buccaneer strike aircraft flew for the first time on 30 April 1958, and was produced initially for the Royal Navy (20 development aircraft, 40 S.Mk 1 and 84 S.Mk 2) and the South African Air Force (16 S.Mk 50). Descriptions of these versions were given in the 1970-71 *Jane's*.

Most Royal Navy S.Mk 2s were later transferred to the RAF, the first four being delivered to No. 12 Squadron at RAF Honington on 1 October 1969. Those operated by the RAF are designated **S.Mk 2A** (without Martel capability) and **S.Mk 2B** (with Martels). Other airframe and equipment differences exist between these models, but the capability to carry Martel air-to-ground missiles is the fundamental definition of aircraft standard. The RAF, in addition to the ex-RN aircraft, ordered 43 new-production S.Mk 2Bs, the first of which flew on 8 January 1970. Delivery of these was due to be completed in 1976. All Mk 2As are being brought up to Mk 2B standard. First RAF units to be completely equipped with the Buccaneer S.Mk 2B were Nos. 15 and 16 Squadrons based at Laarbruch in Germany.

Buccaneers remaining in Royal Navy service are now designated **S.Mk 2C** without, and **S.Mk 2D** with, Martel capability.

Three S.Mk 2Bs (XW986-988) were specially built for the Royal Aircraft Establishment, and are being used for development trials of various weapons. Two other Buccaneers were modified by Marshall of Cambridge (which see) as a part of the Panavia Tornado development programme.

The following details apply to the Buccaneer S.Mk 2A/2B:

TYPE: Two-seat strike and reconnaissance aircraft.

WINGS: Cantilever mid-wing monoplane. Sweepback at quarter-chord: 40° at root, decreasing first to 38° 36′ and then to 30° 12′. Thin section. No dihedral. Incidence 2° 30′. Structure is of all-metal multi-spar design with integrally-stiffened thick skins machined from the solid. Inner wings each have an aluminium alloy auxiliary spar and two steel main spars which are bolted to three spar rings in centre fuselage. Outer wings have two aluminium alloy spars. Electrically-actuated ailerons, powered by Dowty Boulton Paul duplicated tandem actuators, can be drooped in conjunction with the inboard flaps to provide a full-span trailing-edge flap system. No trim tabs. Resin-bonded glassfibre tips on wings and ailerons. Super-circulation boundary layer control, with air outlet slots near leading-edges and forward of the drooping ailerons and plain flaps. This system also provides thermal de-icing of the engines and intakes; use of the boundary layer system supplies sufficient heat to de-ice the wing and tailplane leading-edges under most operational conditions. Outer wings fold upward hydraulically for stowage.

FUSELAGE: All-metal semi-monocoque structure, bulged at rear end in conformity with area rule. Built in three main sections, comprising cockpit, centre fuselage and rear fuselage, plus nosecone and tailcone. Upper section of centre fuselage contains the fuel tanks, lower section contains the weapons bay. Engine and jetpipe firewalls and heat shields are titanium. Equipment bay in rear fuselage has strengthened floor to absorb stresses when arrester hook is used and transfer them to main structure. Tailcone is made up of two petal-type airbrakes, hydraulically actuated to hinge sideways into the airstream; these can be opened fully or to any intermediate position. For stowage the resin-bonded glasscloth nosecone hinges sideways to port and the airbrakes are fully opened.

TAIL UNIT: Cantilever all-metal T-tail. Large dorsal fin faired into fuselage dorsal fairing. All-moving tailplane attached to tip of fin, which is pivoted to move with it. Electrically-actuated tailplane trim flap is used only when ailerons are deflected. Flying control surfaces powered by Dowty Boulton Paul duplicated tandem actuators. Super-circulation boundary layer control system, with air outlet slots in underskin of tailplane, just aft of leading-edge.

LANDING GEAR: Retractable tricycle type of Dowty manufacture. Hydraulic retraction, main wheels inward into jetpipe nacelles, nosewheel rearward into front fuselage. Oleo shock-absorbers and single wheels on all units. Goodyear or Dunlop main wheels and tubeless tyres, size 35 × 10, pressure 15·86-17·93 bars (230-260 lb/sq in). Goodyear or Dunlop nosewheel and tyre size 24 × 6·6, pressure 20-20·34 bars (290-295 lb/sq in). Hydraulically steerable nosewheel. Goodyear or Dunlop double-disc hydraulic brakes, with anti-skid system. Sting-type arrester hook under rear fuselage.

POWER PLANT: Two 49·4 kN (11,100 lb st) Rolls-Royce RB.168-1A Spey Mk 101 turbofan engines, housed in nacelle on each side of the fuselage. Standard internal fuel in eight integral tanks in upper part of centre fuselage, total capacity 7,092 litres (1,560 Imp gallons), with provision for cross-feed of all fuel to either engine. In addition, a 1,932 litre (425 Imp gallon) bomb-door fuel tank can be fitted, without detriment to the aircraft's bomb-carrying capability. Provision for additional 2,000 litre (440 Imp gallon) auxiliary tank in weapons bay, and/or two 1,136 or 1,955 litre (250 or 430 Imp gallon) underwing drop-tanks on the inboard pylons. Detachable flight refuelling probe standard. In the tanker role (max capacity 12,797 litres; 2,815 Imp gallons) the inboard starboard pylon is occupied by a 636 litre (140 Imp gallon) Mk 20B or 20C refuelling pod fed continuously from the main fuel system.

ACCOMMODATION: Crew of two in tandem on Martin-Baker zero-zero ejection seats in pressurised cockpit under single electrically-actuated rearward-sliding blown Perspex canopy. Canopy can be jettisoned separately, if necessary, by explosive charge. Windscreen anti-icing by gold film electrical heating system.

SYSTEMS: Liquid oxygen breathing system. Normalair pressurisation and air-conditioning system. Main hydraulic system pressure 276 bars (4,000 lb/sq in); secondary system, for flying controls, pressure 227·5 bars (3,300 lb/sq in). Two 30kVA alternators, one driven by each engine, provide 200V 400Hz three-phase AC electrical power. For certain equipment this is phased through a 115V 400Hz transformer. Two 4·5kW rectifiers supply a 28V battery to provide DC power for certain other systems. Emergency battery provides 20 min of power in the event of failure of main generating system.

ELECTRONICS, ARMAMENT AND OPERATIONAL EQUIPMENT: Standard equipment includes single-sideband HF and UHF/VHF communications equipment with centralised audio selection and telebriefing, air data system, Doppler radar navigation system, master reference gyro, search and fire control radar incorporating terrain warning, and strike sighting and computing system. The rotating weapons bay door can carry four 1,000 lb HE Mk 10 bombs, a 2,000 litre (440 Imp gallon) fuel tank, or a reconnaissance pack containing one vertical F97 night camera and six F95 day cameras (three vertical, two oblique and one forward) with low or high altitude 102 mm or 305 mm (4 in or 12 in) lenses. Other possible reconnaissance equipment includes linescan, electronic flash gear and different camera arrangements. Each of the four wing pylon stations can be adapted to carry a wide variety of external stores. Typical loads for any one pylon include one 1,000 lb HE Mk N1 or Mk 10 bomb; two 500 lb or 540 lb bombs on tandem carriers; one 18-tube 68 mm rocket pod; one 36-tube 2 in rocket pod; 3 in rockets; or an HSD/Matra Martel air-to-surface missile (maximum 3 missiles and a Martel systems pod). Each pylon is also suitable for carrying three 1,000 lb stores on triple release ejection units, or six 500 lb stores on multiple ejection release units, with only small restrictions on the flight envelope. In addition to a Mk 20 in-flight refuelling pod, when operating in the tanker role, an airborne low-pressure starter pod can be carried on an inner pylon; 1,136 litre (250 Imp gallon) or 1,955 litre (430 Imp gallon) drop-tanks can also be carried on these positions. Maximum internal and external stores load is 7,257 kg (16,000 lb).

One of three Hawker Siddeley Buccaneers for RAE weapons trials. Yellow, green and white paint scheme aids tracking cameras

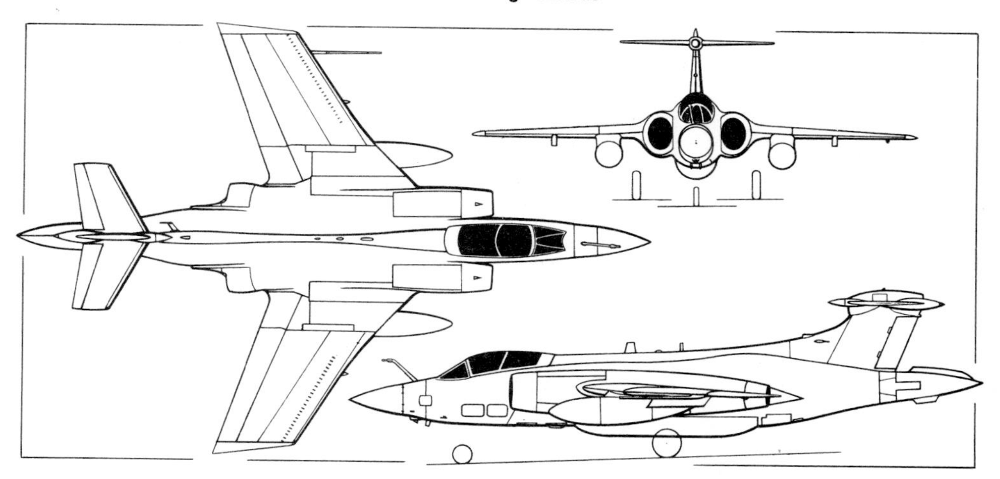

Hawker Siddeley Buccaneer S. Mk 2B twin-engined strike aircraft *(Pilot Press)*

DIMENSIONS, EXTERNAL:

Wing span	13·41 m (44 ft 0 in)
Wing span (folded)	6·07 m (19 ft 11 in)
Wing span (folded, over tank fairings)	6·22 m (20 ft 5 in)
Wing chord at root	4·14 m (13 ft 7 in)
Wing chord at tip	2·44 m (8 ft 0 in)
Wing chord (mean)	3·65 m (11 ft 11½ in)
Wing aspect ratio	3·55
Length overall	19·33 m (63 ft 5 in)
Length folded	15·79 m (51 ft 10 in)
Height overall	4·95 m (16 ft 3 in)
Height folded	5·08 m (16 ft 8 in)
Tailplane span	4·34 m (14 ft 3 in)
Wheel track	3·62 m (11 ft 10½ in)
Wheelbase	6·30 m (20 ft 8 in)

AREAS:

Wings, gross	47·82 m² (514·70 sq ft)
Ailerons (total)	5·09 m² (54·80 sq ft)
Trailing-edge flaps (total)	2·16 m² (23·30 sq ft)
Fin	6·37 m² (68·60 sq ft)
Rudder	1·00 m² (10·74 sq ft)
Tailplane, gross	7·02 m² (75·52 sq ft)
Tailplane trim flap	2·06 m² (22·20 sq ft)

WEIGHTS:

Typical take-off weights	20,865 kg (46,000 lb) to 25,400 kg (56,000 lb)
Max T-O weight	28,123 kg (62,000 lb)
Typical landing weight	15,876 kg (35,000 lb)

PERFORMANCE:

Max design level speed at 61 m (200 ft)	Mach 0·85 (560 knots; 1,038 km/h; 645 mph)
T-O run at S/L, ISA:	
at 20,865 kg (46,000 lb) AUW	720 m (2,360 ft)
at 25,400 kg (56,000 lb) AUW	1,160 m (3,800 ft)
Landing run at 15,876 kg (35,000 lb) landing weight, S/L, ISA	960 m (3,150 ft)
Typical strike range	2,000 nm (3,700 km; 2,300 miles)
Endurance with two in-flight refuellings	9 hr

HAWKER SIDDELEY NIMROD

RAF designations: Nimrod MR. Mks 1 and 2 and R. Mk 1

The Nimrod was evolved to replace the Shackleton maritime reconnaissance aircraft of RAF Strike Command, with which it is scheduled to serve until well into the 1990s. Design of the Nimrod, as the Hawker Siddeley 801, began in June 1964, and government authority to proceed was announced in June 1965.

Based substantially upon the airframe of the Hawker Siddeley (de Havilland) Comet 4C, the Nimrod is a new-production aircraft with a 1·98 m (6 ft 6 in) shorter, modified pressurised fuselage; a new, unpressurised, underslung pannier for operational equipment and weapons; and Rolls-Royce Spey turbofan engines (instead of the Avon turbojets of the Comet), with wider air intakes to allow for the greater mass flow. Other external changes include enlarged flight deck main windows and 'eyebrow' windows; ESM and MAD equipment, in glassfibre fairings on top of the fin and in the tailboom respectively; and a searchlight in the starboard wing external fuel tank. The search radar is housed in a streamlined glassfibre fairing which forms the nose section of the unpressurised lower fuselage.

The Nimrod was designed to combine the advantages of high-altitude, fast transit speed with low wing loading and good low-speed manoeuvring capabilities when operating in its primary roles of anti-submarine warfare, surveillance and anti-shipping strike. When required, two of the four Spey engines can be shut down to extend endurance, and the aircraft can cruise and climb on only one engine. A wide range of weapons can be carried in the 14·78 m (48 ft 6 in) long bomb bay, and large numbers of sonobuoys and markers can be carried and released from the pressurised rear fuselage area.

In addition to its surveillance and ASW roles, the Nimrod can be used for day and night photography, and has a stand-off surface missile capability. It can carry 16 additional personnel in the self-support role, or 45 persons after removal of some equipment in the aft section of the fuselage.

Two prototypes were built, both utilising existing Comet 4C airframes. The first of these (XV148), fitted with Spey engines, flew for the first time on 23 May 1967 and was used for aerodynamic testing. The second (XV147) retained its original Avon engines, was first flown on 31 July 1967, and was used for development of the nav/tac system and special maritime equipment.

The following production versions have been announced:

Nimrod MR. Mk 1. Initial production version, to which the detailed description applies. Thirty-eight ordered initially (XV226-263), the first of which was flown on 28 June 1968. Deliveries began on 2 October 1969 and were completed in August 1972. The MR. Mk 1 was delivered initially to No. 236 OCU, RAF Strike Command, at St Mawgan, Cornwall, and is now in service with No. 42 Squadron, also at St Mawgan; Nos. 120, 201 and 206 Squadrons at Kinloss, Scotland; and No. 203 Squadron of the Near East Air Force, based at Luqa, Malta. An order for eight additional Nimrods was announced in January 1972 and delivery of these began in 1975. Only the first four will now be delivered to Mk 1 standard.

Nimrod R. Mk 1. Designation of three aircraft (additional to the 46 MR. Mk 1s ordered for RAF Strike Command) delivered in 1971 to No. 51 Squadron at Wyton, Huntingdonshire. These aircraft (XW664-666) are replacements for Comet 2s; they are said to be employed for electronic reconnaissance and to monitor hostile radio and radar transmissions, although official statements have referred only to radio/radar calibration duties connected with RAF equipment. They can be identified by the absence of an MAD tailboom.

Nimrod MR. Mk 2. The RAF's Nimrod MR. Mk 1 fleet is being refitted with new communications equipment, and advanced tactical sensor and navigation systems, this programme beginning in 1975. Re-delivery of completely refitted aircraft is scheduled to take place during 1978-80, and the aircraft will then be redesignated MR. Mk 2. Equipment in this version will include an advanced search radar, offering greater range and sensitivity coupled with a higher data processing rate; and a new acoustic processing system, being developed by Marconi-Elliott Avionics Systems, which is intended to be compatible with a wide range of existing and projected sonobuoys. An export model of the Nimrod, equipped to MR. Mk 2 standards, is offered with uprated Spey engines, improved APU and brake cooling.

In addition to the above versions, the RAF has under active consideration an AEW version of the Nimrod, and a description of this is given separately.

Ample space and power is available in the basic Nimrod design to accept additional or alternative sensors such as sideways-looking radar, forward-looking infra-red, infra-red linescan, low light level TV, digital processing of intercepted ESM signals and other new developments. Other roles for which it is suitable include airborne warning and control (AWACS); oversea or overland long-range search and rescue; sea control and fishery protection; emergency personnel transport; and in-flight refuelling tanker.

TYPE: Four-turbofan maritime patrol aircraft.

WINGS: Cantilever low/mid-wing monoplane, of metal construction. Sweepback 20° at quarter-chord. All-metal two-spar structure, comprising a centre-section, two stub-wings and two outer panels. Extensive use of Redux metal-to-metal bonding. All-metal ailerons, operated through duplicated hydraulic and mechanical units. Trim tab in each aileron. Plain flaps outboard of engines, operated hydraulically. Hot-air anti-icing system.

FUSELAGE: All-metal semi-monocoque structure. The circular-section cabin space is fully pressurised. Below this is an unpressurised pannier housing the bomb bay, radome and additional space for operational equipment. Segments of this pannier are free to move relative to each other, so that structural loads in the weapons bay are not transmitted to the pressure-cell. A glassfibre nose radome and tailboom are provided.

TAIL UNIT: Cantilever all-metal structure. Rudder and elevators operated through duplicated hydraulic and mechanical units. A glassfibre pod on top of the fin houses ESM equipment. Trim tab in each elevator. Hot-air anti-icing system.

LANDING GEAR: Retractable tricycle type, similar to Comet 4C but with strengthened main leg and axle beams, stronger wheels and hydraulic brakes of increased capacity. Four-wheel tandem-bogie main units, with size 36 × 10-18 Dunlop tyres, pressure 12·76 bars (185 lb/sq in). Twin-wheel nose unit, with size 30 × 9-15 Dunlop tyres, pressure 6·21 bars (90 lb/sq in).

POWER PLANT: Four Rolls-Royce RB. 168-20 Spey Mk 250 turbofan engines, each rated at 54 kN (12,140 lb st). Reverse thrust fitted on two outer engines. Fuel in fuselage keel tanks, integral wing tanks, and permanent external tank on each wing leading-edge, with total capacity of 48,780 litres (10,730 Imp gallons), equivalent to a fuel weight of 38,940 kg (85,840 lb). Provision for up to six removable tanks to be carried in the weapons bay, increasing max fuel weight to 45,785 kg (100,940 lb) and max overload T-O weight of aircraft to 87,090 kg (192,000 lb). Existing Spey engines will be modified to Mk 806 standard in MR. Mk 2, providing 4% greater thrust in high temperature operating conditions.

ACCOMMODATION: Normal crew of 12, comprising pilot, co-pilot, and flight engineer on flight deck; routine navigator, tactical navigator, radio operator, radar operator, two sonics systems operators, ESM/MAD operator, and two observers/stores loaders in main (pressurised) cabin, which is fitted out as a tactical compartment. In this compartment, from front to rear, are a toilet on the port side; stations for the two navigators (stbd), radio and radar operators (port), and sonics systems operators (stbd) in the forward section; ESM/MAD operator's station, galley, four-seat dining area, rest quarters and sonobuoy stowage in the middle section; and buoy and marker launch area in the rear section. Three hemispherical observation windows forward of wings (one port, two stbd), giving 180° field of view. Two normal doors, emergency door, and four overwing emergency exits. Weapons bay can be utilised for additional fuel tanks (see under 'Power Plant') or for the carriage of freight. Provision is made for a trooping role, in which configuration 45 passengers can be accommodated if some rear-fuselage equipment is removed.

SYSTEMS: Air-conditioning by engine bleed air; Smith-Kollsman pressurisation system, with additional Normalair-Garrett conditioning pack on Mk 2 aircraft, max differential 0·603 bars (8·75 lb/sq in). Anti-icing and bomb-bay heating by engine bleed air. Automotive Products hydraulic system, pressure 172 bars (2,500 lb/sq in), for flying control power units, landing gear shock-absorbers, steering and door jacks, weapons bay door jacks, camera aperture door jacks, and self-sealing couplings for water charging, ground test, engine bay and ancillary services. Lucas APU provides high-pressure air for engine starting. Electrical system utilises four 60kVA engine-driven alternators, with English Electric constant-speed drives, to provide 200V 400Hz three-phase AC supply. Secondary AC comes from two 115V three-phase static transformers, with duplicate 115/26V two-phase static transformers which also feed a 1kVA frequency changer providing a 115V 1,600Hz single-phase supply for radar equipment. Emergency supplies for flight instruments are provided by a 115V single-phase static inverter. DC supply is by four 28V transformer-rectifier units backed up by two nickel-cadmium batteries.

ELECTRONICS AND EQUIPMENT (MR. Mk 1): Routine navigation by Decca Doppler Type 67M/Marconi-Elliott E3 heading reference system, with reversionary heading from a Sperry GM7 duplicated gyro compass system, operating in conjunction with a Ferranti routine dynamic display. Tactical navigation, and stores selection and release, by Marconi-Elliott nav/attack system utilising an 8K Marconi-Elliott 920B digital computer. Tactical display station provides continually-updated information about aircraft position, with present and past track, sonobuoy positions, range circles from sonobuoys, ESM bearings, MAD marks, radar contacts and visual bearings. Course information can be displayed automatically to the pilots on the flight director system; alternatively, the computer can be coupled to the autopilot to allow the tactical navigator to direct the aircraft to a predicted target interception, weapon release point, or any other point on the tactical display. ASW equipment includes Sonics 1C sonar and a new long-range sonar system; EMI ASV-21D air-to-surface-vessel detection radar in nose; Autolycus ionisation detector; Thomson-CSF ESM (electronic support measures) equipment in pod on top of fin; and Emerson Electronics ASQ-10A MAD (magnetic anomaly detector) in extended tailboom. Strong Electric 70 million candlepower searchlight at front of starboard external wing fuel tank. Aeronautical and General Instruments F.126 and F.135 cameras for day and night photography respectively, the latter having Chicago Aero Industries electronic flash equipment. Smiths SFS.6 automatic flight control system, embodying SEP.6 three-axis autopilot, integrated with the navigation and tactical system. Twin Plessey PTR 175 UHF/VHF, and Marconi-Elliott AD 470 HF, communications transceivers; twin Marconi-Elliott AD 260 VOR/ILS; Hoffman ARN 72 Tacan; Decca Loran C/A; Marconi-Elliott AD 360 ADF; Honeywell AN/APN-171(V) radar altimeter. Yaw damper and Mach trim standard.

ELECTRONICS AND EQUIPMENT (MR. Mk 2): New and more flexible operational system, using three separate processors for tactical navigation, radar and acoustics. Marconi-Elliott central tactical system, based on a 920 ATC computer with a greater storage capacity than that of MR. Mk 1, to provide improved computing and

Hawker Siddeley Nimrod MR. Mk 1 four-turbofan maritime patrol aircraft of RAF Strike Command

display facilities and, in conjunction with a Ferranti inertial navigation system, improved navigation capabilities. EMI Searchwater long-range air-to-surface-vessel radar, with its own data processing sub-system incorporating a Ferranti FM 1600D digital computer. This system presents a clutter-free picture, can detect and classify surface vessels, submarine snorts and periscopes at extreme ranges, can track several targets simultaneously, and is designed to operate in spite of countermeasures. AQS 901 acoustics processing and display system, based on twin Marconi-Elliott 920 ATC computers, will be compatible with a wide range of passive and active sonobuoys, either in existence or under development. Communications are being improved by the installation of twin Marconi-Elliott HF transceivers (instead of the original single AD 470), and a radio teletype and encryption system.

ARMAMENT (MR. Mk 1): 14·78 m (48 ft 6 in) long weapons bay, with two pairs of doors, in unpressurised lower fuselage pannier, able to carry up to six lateral rows of ASW weapons, accommodating up to nine torpedoes as well as depth charges, or varying numbers of different-sized mines or bombs. Alternatively, to give greater range and endurance, up to six auxiliary fuel tanks can be fitted in the weapons bay, or a combination of fuel tanks and weapons can be carried. To ensure weapon serviceability, the weapons bay is heated when the ambient temperature falls below +5°C. Bay approx 9·14 m (30 ft) long in rear pressurised part of fuselage for storing and launching of active and passive sonobuoys and marine markers. Two rotary launchers, each capable of holding six size A sonobuoys, are used when the cabin is unpressurised; two single-barrel launchers are used when the aircraft is pressurised. A hard-point is provided beneath each wing, just outboard of the main-wheel doors, on which can be carried two pylon-mounted pairs of AS.12 or other air-to-surface missiles, rocket or cannon pods, or mines, according to mission requirements.

DIMENSIONS, EXTERNAL:

Wing span	35·00 m (114 ft 10 in)
Wing chord at root	9·00 m (29 ft 6 in)
Wing chord at tip	2·06 m (6 ft 9 in)
Wing aspect ratio	6·2
Length overall	38·63 m (126 ft 9 in)
Height overall	9·08 m (29 ft 8½ in)
Tailplane span	14·51 m (47 ft 7¼ in)
Wheel track	8·60 m (28 ft 2½ in)
Wheelbase	14·24 m (46 ft 8½ in)

DIMENSIONS, INTERNAL:

Cabin (incl flight deck, navigation and ordnance areas, galley and toilet):

Length	26·82 m (88 ft 0 in)
Max width	2·95 m (9 ft 8 in)
Max height	2·08 m (6 ft 10 in)
Volume	124·14 m³ (4,384 cu ft)

AREAS:

Wings, gross	197·0 m² (2,121 sq ft)
Ailerons (total)	5·63 m² (60·6 sq ft)
Trailing-edge flaps (total)	23·37 m² (251·6 sq ft)
Fin and rudder (above tailplane centreline)	10·96 m² (118 sq ft)
Dorsal fin	5·67 m² (61 sq ft)
Tailplane	40·41 m² (435 sq ft)
Elevators (incl tabs)	12·57 m² (135·3 sq ft)

WEIGHTS (MR. Mk 1):

Typical weight empty	39,000 kg (86,000 lb)
Max disposable payload	6,120 kg (13,500 lb)
Normal max T-O weight	80,510 kg (177,500 lb)
Max overload T-O weight	87,090 kg (192,000 lb)
Typical landing weight	54,430 kg (120,000 lb)

PERFORMANCE (MR. Mk 1):

Max operational necessity speed, ISA + 20°C	500 knots (926 km/h; 575 mph)
Max transit speed, ISA + 20°C	475 knots (880 km/h; 547 mph)
Econ transit speed, ISA + 20°C	425 knots (787 km/h; 490 mph)
Typical low-level patrol speed (two engines)	200 knots (370 km/h; 230 mph)
Operating height range	S/L to 12,800 m (42,000 ft)
Min ground turning radius	27·1 m (89 ft 0 in)
Runway LCN at T-O weight of 82,550 kg (182,000 lb)	50
T-O run at 80,510 kg (177,500 lb) AUW, ISA at S/L	1,463 m (4,800 ft)
Unfactored landing distance at 54,430 kg (120,000 lb) landing weight, ISA at S/L	1,615 m (5,300 ft)
Typical ferry range	4,500-5,000 nm (8,340-9,265 km; 5,180-5,755 miles)
Typical endurance	12 hr

HAWKER SIDDELEY NIMROD AEW

Hawker Siddeley Aviation has designed and proposed the construction of an airborne early warning (AEW) version of the Nimrod which is intended specifically for European defence. It has been made possible by the development by Marconi-Elliott Avionics of a new radar system which, in addition to an essential maritime capability, satisfies also the air defence requirements of central Europe. The aircraft could provide, at long range and at low or high altitude, detection, tracking and classification of aircraft, missiles and ships; interceptor control; direction of strike aircraft; air defence; air traffic control; and search and rescue facilities.

Designed specifically for installation in this modified version of the maritime reconnaissance Nimrod, the radar requires modification to the nose and tail to permit installation of the newly developed and identically-shaped scanners in fore and aft positions. The aircraft's performance is likely to be affected only marginally by the structural changes and a reduction in lateral stability is compensated by a 0·91 m (3 ft 0 in) increase in fin height.

Mounting the scanners at the extremities of the airframe ensures good all-round coverage, and they do not suffer from airframe obscuration effects. Designed for very low sidelobe level, they are synchronised and each sweeps through 180° in azimuth, the IFF interrogator using the same scanners to aid correlation of IFF and radar returns. With automatic roll- and pitch-stabilisation by dual INS, which compensates for structural flexing, these scanners are able to overcome the cyclic error which is present in other systems.

The associated radar is a pulsed Doppler system that, in addition to the detection of aircraft, has a ship surveillance capability. The rate at which pulses are transmitted can be varied to provide maximum detection in differing terrain conditions or sea states. The system has also highly sophisticated anti-jamming features to cope with the growing efficiency of electronic countermeasures.

The radar passes target plots in terms of range, azimuth, radial velocity and altitude to the advanced digital data handling system; this is based on an airborne computer that controls the flow of data from the scanners and correlates track information between the AEW aircraft and a surface control station. A total of six operator consoles is planned. Each has a tactical situation display, showing the tracks selected by the operator, and a tabular display for the selective presentation of detailed track and control information. Much of the data control is fully automatic; thus, association of radar, IFF and ESM, track initiation, tracking and data storage require no action from the operator. Control of the data handling system is achieved by rolling ball and functionally arranged keyboards, the operator interfacing with the system to carry out system control, track classification, fighter control and data link management.

High standards of communications and navigation are essential to complement the advanced radar and data

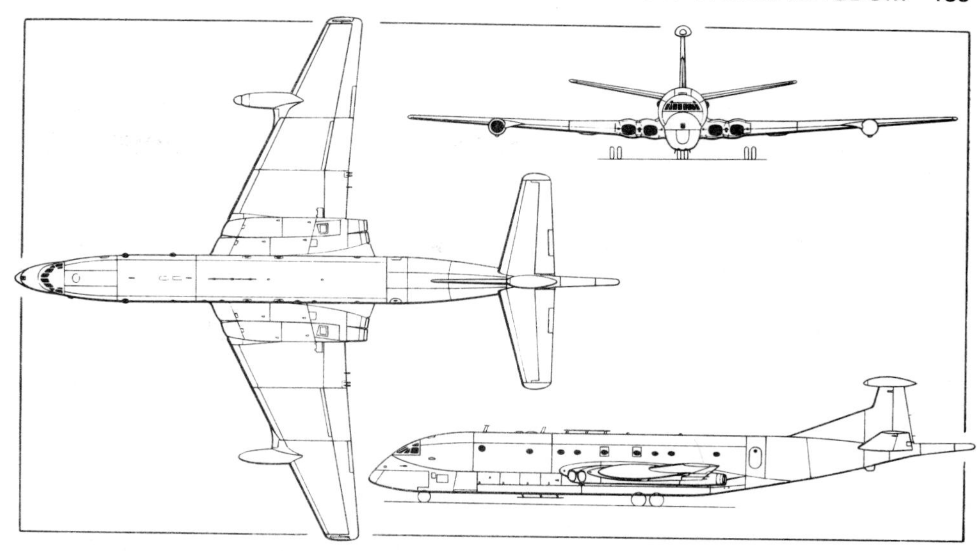

Hawker Siddeley Nimrod MR. Mk 1 four-turbofan maritime patrol aircraft *(Pilot Press)*

Hawker Siddeley Nimrod R. Mk 1 of No. 51 Squadron, RAF. Note the modified tailcone in place of the MAD boom and the revised contours of the port wing leading-edge pod *(Coventry Aviation Society)*

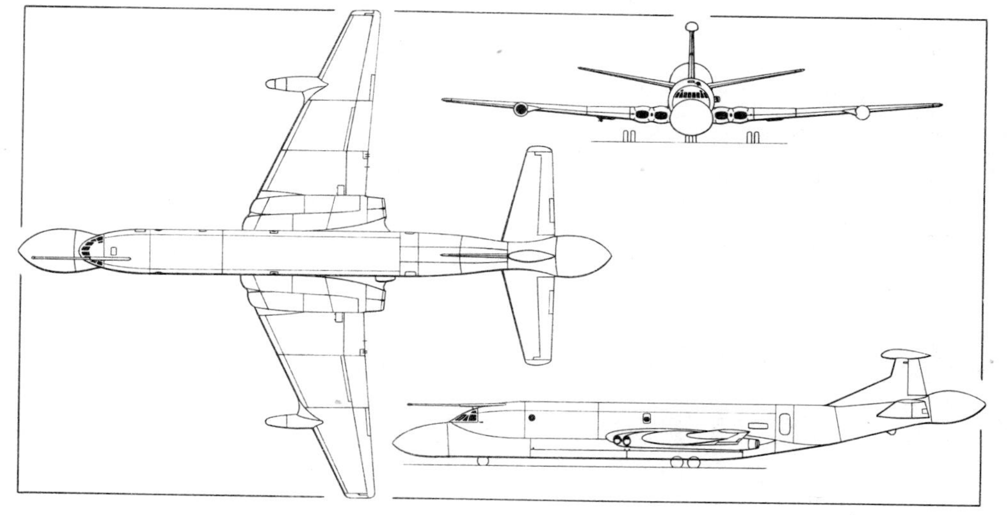

The projected AEW version of the Hawker Siddeley Nimrod *(Pilot Press)*

handling system. For communications the AEW Nimrod will carry tactical UHF transceivers, SIMOP HF transceivers, pilot's U/VHF, RATT, secure voice com, LF receiver and data links. Primary navigation electronics will consist of dual inertial navigation systems (INS). The secondary navigation system will include a gyro magnetic compass, air data computer, twin VOR/ILS, ADF, Tacan, autopilot and a flight director. ESM equipment is housed in the pod at the top of the tail fin and in the two pods on the wing leading-edges.

The general appearance of the AEW Nimrod is shown in the accompanying illustration. Features of special significance for this role are the spacious cabin for electronics and crew, high transit speed and sound low-speed characteristics.

DIMENSIONS, EXTERNAL:
- Wing span: 35·00 m (114 ft 10 in)
- Length overall: 41·37 m (135 ft 8¾ in)
- Height overall: 10·00 m (32 ft 9¾ in)

PERFORMANCE:
- Endurance: in excess of 10 hr

LOCKSPEISER
LOCKSPEISER AIRCRAFT LTD

REGISTERED OFFICE:
652 Grand Buildings, Trafalgar Square, London WC2 5HN
Telephone: 01-839 2777

MANAGING DIRECTOR:
David Lockspeiser, MRAeS, CEng

COMPANY SECRETARY:
Christopher E. Bean, FCA

LOCKSPEISER LDA-01

Mr David Lockspeiser designed a utility aeroplane known as the LDA, or Land Development Aircraft, the production version of which is intended for operation as a passenger, freight or vehicle transport, as an agricultural, ambulance, survey or firefighting aircraft, or for other duties. In a military version, for use as a light troop transport or battlefield support aircraft, the initials stand for Light Defence Aircraft.

The basic concept of the LDA is that of an 'aerial Land-Rover', offering a wide variety of applications, low initial cost and economy of operation, and capable of being easily assembled, inspected and repaired. Many of the major components are interchangeable, and the aircraft can carry a complete set of its own spares, including wings. A primary design consideration was ease of construction for licensed manufacture and assembly.

A 70% scale prototype, registered G-AVOR and known as the LDA-01, was flown for the first time by Mr Lockspeiser on 24 August 1971. Powered at that time by a 63 kW (85 hp) Continental engine, it took off in less than 91 m (300 ft). A description of it with this engine appeared in the 1972-73 *Jane's*.

A feature of the LDA design is a flush-fitting removable ventral container which serves as an interchangeable 'mission pack' and facilitates the quick conversion of the aircraft from one role to another. The landing gear is designed to permit easy manoeuvring of the aircraft on the ground, to pick up a pre-loaded container.

Development flying with the LDA-01 has defined the proposed full-size LDA-1 as a 12·04 m (39 ft 6 in) span aircraft, with 7·08 m³ (250 cu ft) of usable cargo space, a disposable load of 907 kg (2,000 lb) and a max T-O weight of 1,814 kg (4,000 lb). As an alternative to the removable mission pack of the LDA-01, the LDA-1 may have a side-loading double door on the port side and be fitted with a conventional tricycle landing gear.

The description which follows applies to the LDA-01 prototype in its current form:

TYPE: Single-engined general utility aeroplane.

WINGS: Strut-braced main wings at rear and cantilever foreplane at front. Main wings and foreplane are of constant NACA 23012 section and constant chord. Dihedral 3° on main wings; 0° on foreplane. Main wing incidence 0°, foreplane 3° (adjustable on ground). Conventional all-metal alloy construction, with parallel main and rear spars and pop-riveted stressed-skin covering. Built in three basically identical and interchangeable units, two forming the main wings and the third being used as the foreplane. Each panel has four strongpoints at the centre. These serve as attachment points to the fuselage when the panel is positioned as a foreplane; when it is positioned as a port or starboard mainplane they serve as fin-post attachments or as lift-strut and picketing points. They can also be located on a 'luggage rack' under the fuselage when a panel is carried as a spare by an aircraft of the same type. Main wings have trailing-edge flaps inboard and ailerons outboard; in addition to their normal function these are operated in unison to perform the function of an elevator. The foreplane is fitted with a screwjack-operated flap which, in addition to its conventional function, also doubles as a pitch trimmer. This system of control, as distinct from one employing an elevator on the foreplane, gives greater safety at the stall, the foreplane being designed to stall before the main wings. Main wings fitted with fences to contain vortex disturbance from the foreplane tips.

Lockspeiser LDA-01 prototype Land Development or Light Defence Aircraft

FUSELAGE: Conventional box-shaped structure, consisting of a space frame of 19 mm (¾ in) square 22 gauge T.35 steel, welded on a flat jig and covered with an easily removable fabric bag. Nosecone and cowling panels are of glassfibre. Ventral detachable payload container, which fits flush with the basic structure, is of welded steel and light alloy.

TAIL UNIT: Twin wire-braced fins and twin rudders, above and below main wings, of welded steel tube construction with fabric covering.

LANDING GEAR: Non-retractable tricycle type, with cantilever main-gear legs at rear, inclined forwards. Rubber shock-absorbers. Goodyear single or twin nosewheel(s) and tyre(s), size 5·00-5·5, pressure 1·035 bars (15 lb/sq in), steerable from rudder bar. Ackerman steering of nose leg(s). Goodyear 6·00-6 main wheels and tyres, pressure 2·07 bars (30 lb/sq in). Goodyear hydraulic brakes on main wheels.

POWER PLANT: One 119 kW (160 hp) Lycoming O-320-D1A flat-four engine, at rear of fuselage, driving a Hoffmann HO-V-72 two-blade constant-speed metal pusher propeller with spinner. Two fuel tanks in fuselage, one forward and one aft of cargo bay, each of 69 litres (15·2 Imp gallons) capacity. Refuelling points on starboard side.

ACCOMMODATION: Pilot only, in enclosed cabin. Sideways-opening canopy, hinged on port side. Removable payload container in lower centre of fuselage. Production LDA-1 will have optional side-loading double doors, 1·98 m (6 ft 6 in) wide and 1·22 m (4 ft 0 in) in height, instead of the removable container, and will also have access via the roof so that conventional loaders can be used when the aircraft is employed in an agricultural role. Proposed Light Defence Aircraft military version capable of carrying six soldiers and their equipment, or of being fitted with anti-tank missiles or machine-gun pods.

SYSTEMS AND EQUIPMENT: 12V electrical system. Rabat Type 35 battery. Bendix VHF and VOR nav/com system.

DIMENSIONS, EXTERNAL:
- Main wing span: 9·27 m (30 ft 5 in)
- Foreplane span: 4·39 m (14 ft 5 in)
- Main wing chord, constant: 1·14 m (3 ft 9 in)
- Foreplane chord, constant: 1·14 m (3 ft 9 in)
- Main wing aspect ratio: 8·2
- Foreplane aspect ratio: 5·2
- Length overall: 7·14 m (23 ft 5¼ in)
- Fuselage: Max width: 0·91 m (3 ft 0 in)
  - Max depth: 1·07 m (3 ft 6 in)
- Height overall: 2·77 m (9 ft 1 in)
- Wheel track: 2·08 m (6 ft 10 in)
- Wheelbase: 3·91 m (12 ft 10 in)
- Propeller diameter: 1·80 m (5 ft 11 in)
- Fuselage floor/ground clearance: 0·57 m (1 ft 10½ in)
- Removable payload container:
  - Length: 1·98 m (6 ft 6 in)
  - Width: 0·91 m (3 ft 0 in)
  - Depth: 0·38 m (1 ft 3 in)

DIMENSION, INTERNAL:
- Centre fuselage: total internal volume: 1·7 m³ (60 cu ft)

AREAS:
- Main wings, gross: 10·46 m² (112·55 sq ft)
- Foreplane, gross: 3·74 m² (40·25 sq ft)
- Mainplane flaps (total): 0·93 m² (10·0 sq ft)
- Ailerons (total): 1·49 m² (16·0 sq ft)
- Foreplane flaps (total): 0·93 m² (10·0 sq ft)
- Fins (total): 1·58 m² (17·0 sq ft)
- Rudders (total): 1·07 m² (11·5 sq ft)

WEIGHTS:
- Basic weight empty: 561 kg (1,236 lb)
- Operating weight empty: 635 kg (1,401 lb)
- Normal T-O weight: 733 kg (1,617 lb)
- Design max T-O weight: 771 kg (1,700 lb)

PERFORMANCE:
- Cruising speed: 92 knots (170 km/h; 106 mph)
- Optimum climbing speed: 68 knots (126 km/h; 78 mph)
- Stalling speed: 42 knots (78 km/h; 49 mph)
- T-O run, flaps up: 183 m (600 ft)
- Landing run, flaps up: 116 m (380 ft)
- Range: 260 nm (481 km; 299 miles)

MARSHALL
MARSHALL OF CAMBRIDGE (ENGINEERING) LTD (Aircraft Division)

HEAD OFFICE AND WORKS:
Airport Works, Cambridge CB5 8RX
Telephone: Cambridge (0223) 61133
Telex: 81208

MANAGING DIRECTOR:
Sir Arthur Marshall, OBE

COMMERCIAL DIRECTOR:
R. D. Horsbrough

CHIEF ENGINEER:
R. O. Gates

SALES MANAGER:
Norman Sellars

The Aircraft Division of this company (known as Marshalls Flying School Ltd until 1962) has specialised for many years in the modification, overhaul and repair of military and commercial aircraft, including the design and installation of interior furnishing for executive transports.

The company's design department is both CAA and MoD(PE) approved. As an approved service and repair centre for the Lockheed Hercules, Grumman Gulfstream and Cessna Citation aircraft, Marshall of Cambridge also has FAA approval covering most types of American aircraft. The company's conversion, modification and overhaul facilities, which include some of the largest heated hangars in England, with workshop support to full aircraft factory standard, have enabled it to undertake numerous major programmes of work on Viscounts, Britannias, Comets, VC10s, Canberras and a vast number of other civil and military aircraft.

In 1966, Marshall of Cambridge was appointed the designated centre for the Royal Air Force Hercules C. Mk 1 transport aircraft, being responsible for controlling all

Hercules W. Mk 2, modified by Marshall of Cambridge from a C. Mk 1 for the RAF Meteorological Research Flight

technical data, special modifications and development, together with the preparation of these aircraft and painting before delivery to the Service. In 1973, Marshall completed the conversion of an RAF Hercules C. Mk 1 to W. Mk 2 configuration.

During 1966, Marshall of Cambridge was selected to design and manufacture the variable-geometry nose and visor for the pre-production Concorde aircraft, and to design and manufacture the Concorde flight deck and associated electrics, and ground equipment.

Additions to the factory now include a separate hangar for specialised painting of the largest aircraft, and a sculpture milling shop for manufacture of major aircraft components.

Marshall has converted two Hawker Siddeley Buccaneer Mk 2s for use as trials aircraft in connection with the Panavia Tornado development programme.

MARSHALL (LOCKHEED) HERCULES W. Mk 2 CONVERSION

The Hercules W. Mk 2 is a long-range meteorological aircraft, adapted by Marshall from a Lockheed Hercules C. Mk 1 (XV208) procured by the Ministry of Defence for the RAF's Meteorological Research Flight at the RAE, Farnborough, Hampshire, to replace a Vickers Varsity used by the Flight.

The W. Mk 2 flew for the first time on 21 March 1973, and entered service on 3 January 1974. Among its early tasks was participation in the multi-national Project GATE in 1974.

The outward appearance of the W. Mk 2 is shown in the accompanying photograph. Extensive modification to the nose of the aircraft, to incorporate a 5·49 m (18 ft) long nose boom, necessitated mounting the Ekco 280 weather radar scanner in a pod above the flight deck. Instrumentation pods can also be fitted on the wings, outboard of the engine nacelles.

Full details of the equipment and other changes made in the W. Mk 2 were given in the 1974-75 *Jane's*.

The dimensions, weights and performance of the Hercules C. Mk 1 (Lockheed C-130K), as given in the US section of this edition, apply generally also to the W. Mk 2, except in the following respects:

DIMENSIONS, EXTERNAL:

Length overall, incl boom	36·58 m (120 ft 0 in)
Height overall	11·71 m (38 ft 5 in)

WEIGHTS:

Weight empty	32,059 kg (70,678 lb)
Weight empty, equipped	37,149 kg (81,900 lb)
Max normal T-O weight	70,310 kg (155,000 lb)
Max zero-fuel weight	58,422 kg (128,800 lb)
Max landing weight	58,970 kg (130,000 lb)

MARSHALL (HAWKER SIDDELEY) BUCCANEER CONVERSION

Marshall undertook in 1974 the conversion of two Hawker Siddeley Buccaneer S. Mk 2 aircraft (XT272 and XT285) for 'hack' trials work in connection with the development of electronics systems for the Panavia Tornado. This conversion involved aircraft design, airworthiness, electronics installation, monitoring displays, instrumentation and the provision of required environmental systems for the aircraft and equipment.

Distinguishing features include an extended nose with drooped radome, extending the overall length to 20·42 m (67 ft 0 in); an underfuselage laser unit fairing; a camera-window shutter fairing on the bomb-bay door; and an underwing camera pod.

WEIGHTS:

Basic weight empty	15,200 kg (33,509 lb)
Normal T-O weight	20,952 kg (46,193 lb)
Max T-O weight	22,978 kg (50,657 lb)

First of two Buccaneer S. Mk 2s converted by Marshall as electronics testbeds for the Panavia Tornado

SCOTTISH AVIATION

SCOTTISH AVIATION LTD (Member company of the Laird Group)

HEAD OFFICE AND WORKS:
Prestwick International Airport, Ayrshire KA9 2RW
Telephone: Prestwick (0292) 79888
Telex: 77432
OTHER WORKS:
Cumnock
LONDON OFFICE:
60 Buckingham Palace Road, London SW1W 0RR
Telephone: (01) 730-5187
DIRECTORS:
J. A. Gardiner (Chairman)
T. D. M. Robertson, CBE (Deputy Chairman)
H. W. Laughland (Managing Director)
W. L. Denness (Programme Director)
D. McConnell (Commercial)
G. S. Nelson
Dr W. G. Watson (Technical)
J. R. Woods (Works)
E. A. S. Porter (Financial)
SECRETARY: G. T. Lyon
WORKS ACCOUNTANT: J. Baird
MARKETING MANAGER: R. L. Porteous

Scottish Aviation Ltd was formed in 1935 to provide opportunities for employment in the various branches of aviation in Scotland. In doing so, the company developed Prestwick International Airport and on it established an aircraft design and manufacturing industry.

The company's five-seat SA-1 Prestwick Pioneer first flew in 1950 and was followed by the 16-seat SA-2 Twin Pioneer in 1955. A total of 150 aircraft of these types were built. Details can be found in earlier editions of *Jane's*.

Scottish Aviation's activities are currently concentrated on five main programmes. These are: maintenance and modification of CF-104 Starfighter aircraft for the Canadian Armed Forces; production of components for, and overhaul of, Rolls-Royce piston and jet engines; manufacture of major airframe components, including major fuselage sections for the Lockheed C-130 Hercules and doors for the Lockheed TriStar; production and development of Bulldog training aircraft; and production and development of the Jetstream light transport and aircrew training aircraft.

The company's design facilities are CAA- and AQD-approved, and maintenance services are covered by CAA, AQD and FAA approvals.

A separate division of the company, Scottish Air Engine Services, also based at Prestwick Airport, undertakes the overhaul of Pratt & Whitney, Avco Lycoming and Teledyne Continental piston engines.

The Bulldog Series 120 and Jetstream Series 200 are now in full production. In addition, Scottish Aviation offers product support facilities for the Beagle Pup, B.206 and Basset.

Two of the Scottish Aviation Bulldog Model 122 two-seat trainers for the Ghana Air Force

Scottish Aviation Bulldog Model 124 development aircraft with Matra 68 mm rocket pods underwing

SCOTTISH AVIATION SA-3-120 BULLDOG SERIES 120

The Bulldog originated in 1968 as a military trainer version of the Beagle Pup. It differs substantially, however, from the Pup in having a fully-transparent canopy, increased wing span, and strengthened construction to allow full aerobatic operation.

First flight of the Beagle-built prototype (G-AXEH) was made on 19 May 1969. A second prototype (G-AXIG), completed by Scottish Aviation, was flown on 14 February 1971, and a third airframe was completed for static and fatigue tests.

All versions ordered so far are basically similar, except for the equipment fitted. The first production Bulldog, completed by Scottish Aviation, flew for the first time on 22 June 1971 and received full ARB certification on 30 June 1971. The first 98 production Bulldogs were of the Series 100 version, described in the 1972-73 *Jane's*. The second prototype was refurbished, issued with a Normal category C of A, and delivered to a private owner under the designation Model 104.

Production continued with the Series 120, which was awarded full CAA certification on 12 February 1973. A version with retractable landing gear, the Bulldog Series 200/Bullfinch, is described separately.

By mid-1976 orders for the Bulldog Series 120 were as follows:

Model 121. For Royal Air Force, by whom it is designated **T. Mk 1.** Total of 130 ordered, of which the first (XX513) flew for the first time on 30 January 1973 and was delivered to the A & AEE at Boscombe Down on 20 February 1973. By February 1976, all had been delivered, and were in service with No. 2 FTS at Leeming, the CFS at Little Rissington, and University Air Squadrons.

Model 122. For Ghana Air Force. Six ordered initially, delivery of which was completed in September 1973. Further seven ordered in December 1974, delivery of which was completed by February 1976.

Model 123. For Nigerian Air Force. Twenty ordered. Delivery completed December 1974.

Model 124. One aircraft (G-ASAL) used as company demonstration aircraft.

Model 125. For Jordanian Royal Academy of Aeronautics. Five ordered initially, delivery of which was completed in July 1974. Further three ordered in January 1975, delivery of which was completed in May 1975. An additional order for five aircraft, making a total of 13, was placed in January 1976; delivery of these was completed in March 1976.

Model 126. For Lebanese Air Force. Six ordered. Delivery completed in October 1975.

The following description applies to the Bulldog Series 120:

Type: Two/three-seat primary trainer.

Wings: Cantilever low-wing monoplane. Wing section NACA 63_2615. Dihedral 6° 30′. Incidence 1° 9′ at root. Conventional single-spar two-cell riveted stressed-skin structure of light alloy. Electrically-operated slotted trailing-edge flaps and slotted ailerons of similar construction. Fixed tab in starboard aileron.

Fuselage: Conventional light alloy stressed-skin semi-monocoque structure.

Tail Unit: Cantilever two-spar light alloy stressed-skin structure. Fixed-incidence tailplane. Full-span trim tab in starboard elevator. Manually-operated trim tab in rudder. Fixed ventral fin.

Landing Gear: Non-retractable tricycle type, with single wheel on each unit. Steerable nosewheel with Automotive Products oleo-pneumatic shock-absorber and Goodyear wheel and tyre, size 5·00-5, pressure 2·76 bars (40 lb/sq in). Main units have Automotive Products oleo-pneumatic shock-absorbers and Goodyear wheels and tyres, size 6·00-6, pressure 2·07 bars (30 lb/sq in). Goodyear hydraulic disc brakes on main wheels. Optional ski landing gear.

Power Plant: One 149 kW (200 hp) Lycoming IO-360-A1B6 flat-four engine, driving a Hartzell HC-C2YK-4F/FC7666A-2 two-blade constant-speed metal propeller with spinner. Lycoming AEIO-360-A1B6 engine available optionally. Four removable metal fuel tanks, two in each wing, with total usable capacity of 145·5 litres (32 Imp gallons). Refuelling point on top of each wing. Oil capacity 7·6 litres (1·67 Imp gallons).

Accommodation: Enclosed cabin seating pilot and co-pilot or trainee side by side with dual controls, with space at rear for observer's seat or up to 54 kg (120 lb) of baggage. Rearward-sliding jettisonable transparent canopy. Cabin heated and ventilated.

Systems: Heat exchanger for cabin heating. Hydraulic system, pressure 40 bars (580 lb/sq in), for main-wheel brakes only. Vacuum-type pneumatic system available optionally. 24V DC power from engine-driven alternator and 24V 25Ah storage battery. No oxygen or de-icing systems.

Electronics and Equipment: Radio to individual customer's requirements; panel can accommodate dual VHF and navaids. Blind-flying instrumentation standard. Glider towing attachment optional.

Armament: Standard aircraft is unarmed, but has provision for installation of four underwing hardpoints to which can be attached various weapon loads if required. Maximum underwing load 290 kg (640 lb).

Dimensions, external:

Wing span	10·06 m (33 ft 0 in)
Wing chord at root	1·51 m (4 ft 11¼ in)
Wing chord at tip	0·86 m (2 ft 9¾ in)
Wing aspect ratio	8·4
Length overall	7·09 m (23 ft 3 in)
Height overall	2·28 m (7 ft 5¾ in)
Tailplane span	3·35 m (11 ft 0 in)
Wheel track	2·03 m (6 ft 8 in)
Wheelbase	1·40 m (4 ft 7 in)
Propeller diameter	1·88 m (6 ft 2 in)
Propeller ground clearance	0·26 m (10¼ in)

Dimensions, internal:

Cabin: Length	2·11 m (6 ft 11 in)
Max width	1·14 m (3 ft 9 in)
Max height	1·02 m (3 ft 4 in)

Areas:

Wings, gross	12·02 m² (129·4 sq ft)
Ailerons (total)	0·87 m² (9·4 sq ft)
Trailing-edge flaps (total)	1·30 m² (13·95 sq ft)
Vertical tail surfaces (total)	2·11 m² (22·72 sq ft)
Horizontal tail surfaces (total)	2·55 m² (27·50 sq ft)

Weights and Loadings:

Basic weight empty	669 kg (1,475 lb)
Max T-O weight:	
normal and semi-aerobatic	1,066 kg (2,350 lb)
fully aerobatic	1,015 kg (2,238 lb)
Max wing loading	88·6 kg/m² (18·15 lb/sq ft)
Max power loading	7·15 kg/kW (11·75 lb/hp)

Performance (at max T-O weight):

Never-exceed speed (structural)	210 knots (389 km/h; 241 mph)
Max level speed at S/L	130 knots (241 km/h; 150 mph)
Max cruising speed at 1,220 m (4,000 ft)	120 knots (222 km/h; 138 mph)
Econ cruising speed at 1,220 m (4,000 ft)	105 knots (194 km/h; 121 mph)
Stalling speed, flaps down	54 knots (100 km/h; 62 mph) EAS
Max rate of climb at S/L	315 m (1,034 ft)/min
Service ceiling	4,875 m (16,000 ft)
Min ground turning radius	9·75 m (32 ft 0 in)
T-O run	274 m (900 ft)
T-O to 15 m (50 ft)	427 m (1,400 ft)
Landing from 15 m (50 ft)	363 m (1,190 ft)
Landing run	153 m (500 ft)
Range with max fuel, 55% power, no reserves	540 nm (1,000 km; 621 miles)
g limits:	
semi-aerobatic	+4·4; −1·8
fully aerobatic	+6; −3

SCOTTISH AVIATION SA-3-200 BULLDOG SERIES 200 and BULLFINCH

In the Autumn of 1974, Scottish Aviation announced that a further version of the Bulldog was under development. This is the Series 200, in which the major differences are a fully-retractable landing gear and provision for an optional fourth seat. Other improvements, compared with the Series 120, include a longer and cleaner engine cowling; deepened and repositioned firewall, giving more space for electronics and instrumentation, with easier access; a higher-mounted tailplane; a plug-type cockpit canopy of revised contours; and increased aerobatic and non-aerobatic weights. In addition to basic, aerobatic and weapons training roles, the Series 200 is suitable for military observation, liaison, reconnaissance, forward air control, light strike and supply dropping duties.

In civilian form, the aircraft is known as the **Bullfinch.** A prototype of this version flew for the first time on 20 August 1976. Deliveries of production aircraft are scheduled to begin in 1977.

The description of the Bulldog Series 120 applies generally to the Series 200/Bullfinch, except in the following respects:

TYPE: Two/four-seat light aircraft.

LANDING GEAR: Tricycle type, generally similar to that of Series 120, but fully retractable.

POWER PLANT: One 149 kW (200 hp) Lycoming AEIO-360-A1B6 flat-four engine and two-blade constant-speed Hartzell propeller with spinner. Four metal wing fuel tanks, as in Series 120, with combined capacity of 145·5 litres (32 Imp gallons).

ACCOMMODATION: As Series 120, but provision for up to four seats in cabin.

ARMAMENT (Bulldog): Standard aircraft is unarmed, but has provision for installation of four underwing hardpoints to which various weapon loads can be attached if required. Maximum underwing load 290 kg (640 lb).

DIMENSIONS, EXTERNAL:

As Series 120, except:

Wing span	10·29 m (33 ft 9·3 in)
Length overall	7·59 m (24 ft 11 in)
Height overall	2·54 m (8 ft 4 in)
Wheel track	2·29 m (7 ft 6 in)
Wheelbase	1·75 m (5 ft 9 in)

WEIGHTS:

Typical operating weight empty	821 kg (1,810 lb)
Max aerobatic T-O weight	1,045 kg (2,304 lb)
Max T-O weight	1,179 kg (2,601 lb)

PERFORMANCE (estimated, at 1,045 kg; 2,304 lb AUW, ISA):

Max level speed at S/L	150 knots (278 km/h; 173 mph)
Max cruising speed at 1,220 m (4,000 ft)	141 knots (261 km/h; 162 mph)
Max rate of climb at S/L	353 m (1,160 ft)/min
Service ceiling	5,640 m (18,500 ft)
T-O to 15 m (50 ft) at S/L	390 m (1,280 ft)
Landing from 15 m (50 ft) at S/L	377 m (1,238 ft)
Range (55% power) with max fuel	540 nm (1,000 km; 621 miles)
Max endurance	5 hr

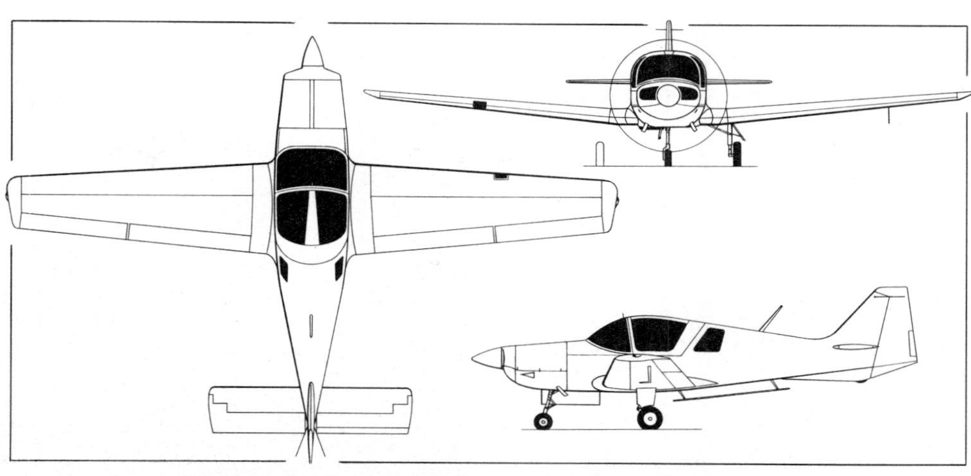

Scottish Aviation Bullfinch, with retractable landing gear *(Pilot Press)*

Prototype of the Scottish Aviation Bullfinch two/four-seat light aircraft

SCOTTISH AVIATION JETSTREAM SERIES 200

RAF designation: Jetstream T. Mk 1

The original H.P. 137 Jetstream was designed and developed between 1966 and 1970 by Handley Page Ltd, and was described in *Jane's* at that time. A number of Handley Page-built Jetstream Mk 1s are in service with operators in Canada, France, the UK, the USA and Zaïre. Some Mk 1s have been converted to Series 200 standard. This model originated with Handley Page, was developed subsequently by Jetstream Aircraft Ltd (see 1972-73 *Jane's*), and is now available from Scottish Aviation. A full UK type certificate in the transport category (passenger), for operations in performance group C, was awarded on 22 November 1972.

The major production commitment so far was for 26 military Series 200s (Model 201) for the Royal Air Force, ordered in February 1972. The first of these (XX 475) flew for the first time on 13 April 1973, eight months after construction began, and was delivered to the A & AEE, Boscombe Down, in July 1973. The RAF aircraft, which are designated Jetstream T. Mk 1, are generally similar to the civil Series 200 except for having Astazou XVI D engines, 'eyebrow' windows above the flight deck, and different instrumentation and electronics installations. They were intended to be used as trainers for pilots of multi-engined aircraft, superseding the Vickers Varsity in this role. The third production Model 201 was delivered to the Central Flying School at Little Rissington and on 12 December 1973 to No. 5 FTS. All 26 aircraft had been delivered to the RAF by the Spring of 1976, but were put temporarily in store at St Athan, pending a decision on future requirements. It was announced in October 1976, that eight would be used by the RAF in the multi-engine pilot training role; twelve are being converted for observer training with the Royal Navy; five are to be held in reserve.

TYPE: Twin-turboprop light transport and aircrew trainer.

WINGS: Cantilever low-wing monoplane. Wing section NACA 63A418 at root, NACA 63A412 at tip. Dihedral 7° from roots. Incidence 2° at root, 0° at tip. Sweepback 0° 34′ at quarter-chord. Aluminium alloy fail-safe structure. Aluminium alloy manually-operated Frise-type ailerons. Hydraulically-operated aluminium alloy double-slotted flaps, with glassfibre slat. No slots or leading-edge flaps. Trim tab in each aileron. Goodrich pneumatic rubber-boot de-icing system for leading-edges.

FUSELAGE: Conventional aluminium alloy semi-monocoque fail-safe structure, with chemically-milled skin panels. Fully pressurised.

TAIL UNIT: Cantilever two-spar aluminium alloy structure. Fixed-incidence tailplane. Manually-operated control surfaces. Trim tabs in rudder and each elevator. Goodrich pneumatic rubber-boot de-icing system for leading-edges.

LANDING GEAR: Retractable tricycle type, with nosewheel steering. Hydraulic retraction, main wheels inward into wings, twin nosewheels forward. Electro-Hydraulics oleo-pneumatic shock-absorbers. Dunlop wheels, tyres and disc brakes on all units. Main-wheel tyres size 28 × 9·00-12, pressure 2·34 bars (34 lb/sq in). Nosewheel tyres size 6·00-6, pressure 3·93 bars (57 lb/sq in). No brake cooling. Dunlop anti-skid units.

POWER PLANT: Two 743 kW (996 ehp) Turboméca Astazou XVI C2 turboprop engines (Astazou XVI D in Model 201), each driving a Hamilton Standard Type 23LF-371 three-blade variable- and reversible-pitch fully-feathering metal propeller. Fuel in integral tank in each wing, total capacity 1,745 litres (384 Imp gallons; 461 US gallons). Refuelling point on top of each outer wing. Oil capacity 9·5 litres (2·1 Imp gallons; 2·5 US gallons) pe. engine. Hot-air de-icing of engine air intakes, electrical de-icing of propellers and spinners.

ACCOMMODATION: Two seats side by side on flight deck, with provision for dual controls, though aircraft can be approved (subject to local regulations) for single-pilot operation. Main cabin can be furnished in executive layout for up to 12 passengers, with individual swivel seats and settees and full galley and toilet facilities; or in airliner layout, for up to 16 passengers, with toilet but

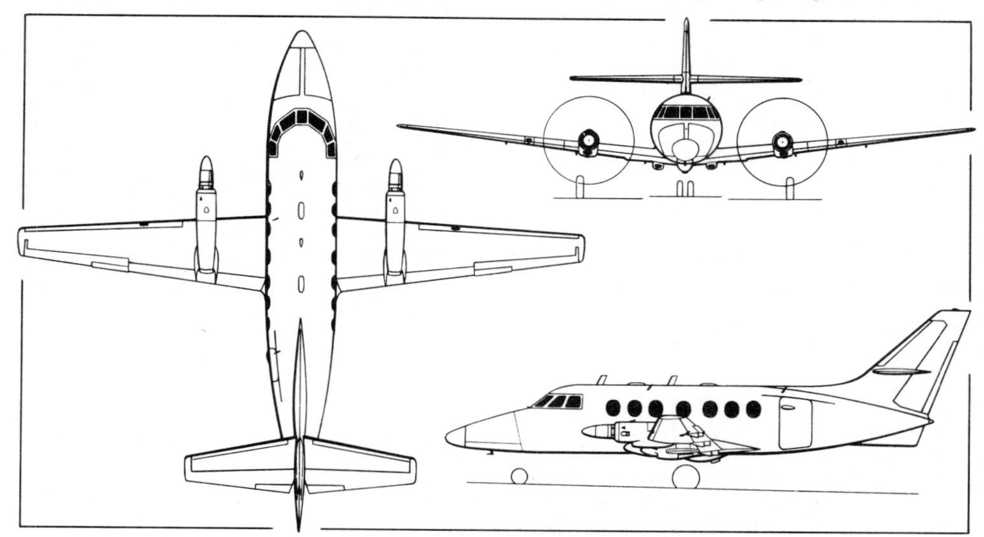

Scottish Aviation Jetstream Srs 200 twin-turboprop light transport *(Pilot Press)*

no galley. RAF T. Mk 1 accommodation includes two pilot seats, four passenger seats, and toilet. Universal seat rails fitted. Downward-opening passenger door, with integral stairs, at rear of cabin on port side. Emergency exit over wing on starboard side. Baggage compartment in rear of cabin, aft of main door. Entire accommodation pressurised, heated, ventilated and air-conditioned. Windscreen de-iced electrically.

SYSTEMS: AiResearch dual air cycle air-conditioning system, using engine bleed air. Cabin pressure control set to 0·38 bars (5·5 lb/sq in). Duplicated hydraulic systems, each of 138 bars (2,000 lb/sq in) pressure, for actuation of flaps, landing gear, brakes and nosewheel steering. Electrical system includes two 3kW 30V DC starter/generators, two 9kVA 200V AC 400Hz alternators, all of Plessey manufacture, and two 24V 25Ah Dagenite batteries. Piped oxygen system, with optional dropout masks.

ELECTRONICS AND EQUIPMENT: All instruments and electronics to customer's specification. Equipment in RAF T. Mk 1 includes Sperry STARS flight director system, Decca VHF nav/com system, Collins 51-series VOR/ILS with marker beacon, Marconi-Elliott AD 370B ADF, Decca DME, Bendix M.4C autopilot, Cossor 1520 transponder and S. G. Brown intercom.

Jetstream executive transport aircraft operated by the Decca Navigator Company *(Peter J. Bish)*

DIMENSIONS, EXTERNAL:

Wing span	15·85 m (52 ft 0 in)
Wing chord at root	2·19 m (7 ft 2½ in)
Wing chord at tip	0·80 m (2 ft 7¼ in)
Wing aspect ratio	10
Length overall	14·37 m (47 ft 1½ in)
Length of fuselage	13·40 m (43 ft 11½ in)
Height overall	5·32 m (17 ft 5½ in)
Fuselage: Max diameter	1·98 m (6 ft 6 in)
Tailplane span	6·60 m (21 ft 8 in)
Wheel track	5·94 m (19 ft 6 in)
Wheelbase	4·60 m (15 ft 1 in)
Propeller diameter	2·59 m (8 ft 6 in)
Passenger door: Height	1·42 m (4 ft 8 in)
Width	0·86 m (2 ft 10 in)
Emergency exit: Height	0·91 m (3 ft 0 in)
Width	0·56 m (1 ft 10 in)

DIMENSIONS, INTERNAL:

Cabin, excl flight deck: Length	7·32 m (24 ft 0 in)
Max width	1·85 m (6 ft 1 in)
Max height	1·80 m (5 ft 11 in)
Floor area	8·35 m² (90 sq ft)
Volume	18·05 m³ (638 cu ft)
Baggage compartment volume (according to layout)	1·94-2·53 m³ (68·5-89·5 cu ft)

AREAS:

Wings, gross	25·08 m² (270 sq ft)
Ailerons, aft of hinge line (total)	1·52 m² (16·4 sq ft)
Trailing-edge flaps (total)	3·25 m² (35·0 sq ft)
Vertical tail surfaces (total)	7·72 m² (83·1 sq ft)
Horizontal tail surfaces (total)	7·80 m² (84·0 sq ft)

WEIGHTS AND LOADINGS:

Manufacturer's weight empty	3,485 kg (7,683 lb)
Max payload	1,730 kg (3,814 lb)
Max T-O and landing weight	5,700 kg (12,566 lb)
Max ramp weight	6,020 kg (13,272 lb)
Max zero-fuel weight	5,556 kg (12,250 lb)
Max wing loading	226 kg/m² (46·3 lb/sq ft)
Max power loading	3·84 kg/kW (6·3 lb/ehp)

PERFORMANCE (at max T-O weight, ISA):

Never-exceed speed (structural)	300 knots (555 km/h; 345 mph)
Max level and cruising speed at 3,050 m (10,000 ft)	245 knots (454 km/h; 282 mph)
Econ cruising speed at 4,575 m (15,000 ft)	234 knots (433 km/h; 269 mph)
Stalling speed, flaps down	76 knots (141 km/h; 87·5 mph)
Max rate of climb at S/L	762 m (2,500 ft)/min
Rate of climb at S/L, one engine out	182 m (600 ft)/min
Service ceiling	7,620 m (25,000 ft)
Service ceiling, one engine out	3,050 m (10,000 ft)
Min ground turning radius	12·52 m (41 ft 1 in)
T-O run	579 m (1,900 ft)
T-O to 15 m (50 ft)	762 m (2,500 ft)
Landing from 15 m (50 ft)	702 m (2,310 ft)
Range with max fuel, reserves for 45 min hold and 5% total fuel	1,200 nm (2,224 km; 1,380 miles)

SHORTS

SHORT BROTHERS & HARLAND LTD

HEAD OFFICE, WORKS AND AERODROME:
PO Box 241, Airport Road, Belfast BT3 9DZ, Northern Ireland
Telephone: 0232-58444
Telex: 74688

OTHER FACTORIES:
Newtownards, Castlereagh, Belfast (2)

LONDON OFFICE:
Berkeley Square House, Berkeley Square, W1X 5LB

CHAIRMAN:
Sir George Leitch, KCB

MANAGING DIRECTOR:
P. F. Foreman, CBE

DIRECTORS:
F. F. H. Charlton
D. W. G. L. Haviland, CB
Dr Llewellyn Smith, CBE
H. E. Trevan-Hawke

SECRETARY:
Gordon Bruce, MA

EXECUTIVE COMMERCIAL DIRECTOR:
D. N. B. McCandless, OBE

EXECUTIVE DIRECTOR, ENGINEERING:
T. D. R. Carroll

EXECUTIVE FINANCIAL DIRECTOR:
N. L. Galloway

GENERAL MANAGER (AIRCRAFT AND AEROSTRUCTURES):
A. F. C. Roberts, OBE

MANAGER, AEROSTRUCTURES DIVISION:
M. I. Wild

GENERAL MANAGER, MISSILE SYSTEMS DIVISION:
K. W. Tyson

MANUFACTURING GENERAL MANAGER:
B. Carlin

MANAGER, FLYING SERVICES DIVISION:
Wg Cdr T. C. Chambers, AFC

CHIEF TEST PILOT:
D. B. Wright

PUBLICITY MANAGER:
G. H. Edgar

The original firm of Short Brothers was established at Battersea in 1903, to manufacture balloons. In 1909 it was moved to Leysdown, becoming the first manufacturer of aeroplanes in the UK when it received a contract to build six Wright biplanes. The company later transferred its works to Eastchurch and then to Rochester.

In June 1936 Short Brothers, in collaboration with Harland & Wolff Ltd, formed a new company known as Short & Harland Ltd to build aircraft in Belfast, and in 1947 activities were concentrated in Belfast under the name Short Bros & Harland Ltd.

In 1954 the Bristol Aeroplane Co Ltd acquired a financial interest and the company is now owned by the British government, Rolls-Royce and Harland & Wolff, with the government holding a 69½% controlling interest. In 1973 the company received the Queen's Award to Industry for the seventh year in succession. It had 6,000 employees in February 1976.

The company's current products include the SD3-30 30-seat commuter airliner, which made its first flight in August 1974 and was about to enter airline service in mid-1976. Also in production are the Skyvan and Skyliner turboprop STOL light transports, which are in use throughout the world for passenger, freight, survey, military and miscellaneous operations.

Internationally, Shorts is collaborating as risk-sharing partner with Fokker-VFW, MBB and VFW-Fokker in production of the F28 Fellowship transport, with responsibility for the wings; and holds contracts to produce ailerons, spoilers, wingtips, landing gear doors, galley doors, environmental control system doors and tail unit rib assemblies for the Lockheed L-1011 TriStar, and landing gear doors for the Boeing 747. During 1967, Shorts began the design and manufacture of pods for Rolls-Royce jet engines, and is responsible for podding Rolls-Royce RB.211 turbofan engines for the TriStar. Deliveries of these direct to Lockheed at Palmdale began in the early Summer of 1970. The company is also podding RB.211 engines for the Boeing 747s of British Airways.

To cope with its involvement in the TriStar programme, Shorts installed some of Europe's most advanced facilities for the hot-forming of titanium and the manipulation of high-temperature creep-resistant alloys. It is producing M45H turbofan pods for the VFW-Fokker VFW 614 and is conducting advanced research into jet-engine noise reduction and metal bonding. Shorts is also quality-approved subcontractor to many major US and UK aerospace companies. Conversely, production of the wings for the Shorts SD3-30 is being undertaken jointly by the British Aircraft Corporation and Fokker-VFW, and production of the SD3-30 landing gear by Menasco in Canada.

In addition to its activities in the field of piloted aircraft, Shorts is engaged on missile development and production, production of supersonic target drones, development of the Skyspy remote control aerial surveillance vehicle, and development, to MoD contract, of the MATS-B target drone.

The company's Flying Services Division operates maintenance units and airfields for various civil and military organisations, and flies and maintains aircraft and target drones for the Ministry of Defence. This includes operation of the Llanbedr target aircraft base, and the target service, supply and recovery flight at the Woomera range in Australia.

SHORTS SC.7 SKYVAN

Design of the SC.7 Skyvan was started as a private venture in 1959, and construction of the first prototype began in 1960. This aircraft (G-ASCN) flew for the first time on 17 January 1963, with two 290 kW (390 hp) Continental GTSIO-520 piston engines, and completed its flight trials by mid-1963. It was then re-engined with 388 kW (520 shp) Astazou II turboprops and first flew in its new form on 2 October 1963.

The following versions of the Skyvan have been built:

Skyvan Srs 1 and Srs 1A. Designation of first prototype, with Continental engines (Srs 1) and later with Astazou IIs (Srs 1A).

Skyvan Srs 2. Three development aircraft (G-ASCO/ASZI/ASZJ) and 16 initial production aircraft, with 544 kW (730 ehp) Astazou XII turboprop engines. First flown on 29 October 1965. Several subsequently re-engined to Srs 3 standard. Descriptions in 1968-69 and 1970-71 *Jane's*.

Skyvan Srs 3. Current civil version, which superseded Srs 2 in 1968. First Srs 3 to fly was the second development aircraft, G-ASZI, which had been equipped originally with Astazous. The first flight with AiResearch engines was made on 15 December 1967, and a second aircraft (G-ASZJ) re-engined with TPE 331s flew on 20 January 1968. Total of 56 ordered by 15 June 1976, the most recent customers being the governments of Mexico (5) and Venezuela (3).

Skyvan Srs 3A. Introduced in September 1970, this version complies with British Civil Airworthiness Requirements, Passenger Transport Category, in Performance Group A and has a max T-O weight of 6,215 kg (13,700 lb) and max landing weight of 6,075 kg (13,400 lb). One ordered.

Skyvan Srs 3M. Military version of Srs 3, modified internally to accept optional equipment for typical military missions. Prototype (G-AXPT) flew for the first time in early 1970. Suitable for paratrooping and supply dropping, assault landing, troop transport, casualty evacuation, staff transport, and vehicle or ordnance transport. Initial order for two, for Austrian Air Force, placed in February 1969.

A total of 50 had been ordered by 15 June 1976, for 13 armed services, including the Argentine Naval Prefectura

Shorts Skyvan Srs 3M twin-turboprop military transport aircraft in the insignia of the Ghana Air Force

(5), Austrian Air Force (2), Ecuador Army Air Force (1), Ghana Air Force (6), Indonesian Air Force (3), Mauretanian Air Force (2), Royal Nepalese Army (2), No. 2 Squadron of the Sultan of Oman's Air Force (16), Singapore Air Defence Command (6), Royal Thai Police (3), Yemen Arab Republic Air Force (2), and undisclosed customers (2). Three of the Singapore aircraft are equipped for search and rescue duties. Those of the Indonesian Air Force are equipped to civil standard and operate social services on behalf of the Ministry of the Interior. One of the Ghanaian aircraft has a Skyliner layout.

Skyliner. All-passenger version, described separately. Total of nine sold.

Total orders for Series 3/3A/3M Skyvans and Skyliners had reached 115 by 15 June 1976. In February 1970 the Skyvan became the first aircraft to be certificated under the British Air Registration Board's new Civil Airworthiness Requirements for STOL operations.

The following description applies to the standard civil Srs 3 and military Srs 3M in current production:

TYPE: Light civil or military STOL utility transport.

WINGS: Braced high-wing monoplane. Wing section NACA 63A series (modified). Thickness/chord ratio 14%. Dihedral 2° 2′. Incidence 2° 30′. Light alloy structure consisting of a two-cell box with wing skins made up of a uniform outer sheet bonded to a corrugated inner sheet. All-metal single-slotted ailerons. Geared tabs on port and starboard ailerons, with manual trim on starboard aileron. All-metal single-slotted flaps. Provision for sintered leading-edge de-icing system.

FUSELAGE: Light alloy structure. Nose and crew cabin section is of conventional skin/stringer design. Elsewhere, the fuselage structure consists of double-skin panels (flat outer sheets bonded to inner corrugated sheets), stabilised by frames.

TAIL UNIT: Cantilever all-metal two-spar structure, with twin fins and rudders. Fixed-incidence tailplane. Geared trim tabs in outer elevators and rudders. Provision for sintered leading-edge de-icing system.

LANDING GEAR: Non-retractable tricycle type. Single wheel on each unit. Steerable nosewheel. Main units carried on short sponsons. Electro-Hydraulics oleo-pneumatic shock-absorbers. Main-wheel tyres size 11·00-12, nosewheel tyre size 7·50-10. Tyre pressure (all units) 2·76 bars (40 lb/sq in). Hydraulically-operated disc brakes, with differential braking for steering. Provision for fitting skis and low-pressure tyres.

POWER PLANT: Two 533 kW (715 shp) Garrett-AiResearch TPE 331-201 turboprop engines, each driving a Hartzell HC-B3TN-5/T10282H three-blade (optionally four-blade) variable-pitch propeller. Fuel in four tanks in pairs on top of fuselage between wing roots, each pair consisting of one tank of 182 litres (40 Imp gallons) capacity and one of 484 litres (106·5 Imp gallons) capacity. Total fuel capacity of 1,332 litres (293 Imp gallons). Provision for increase in total fuel capacity to 1,773 litres (390 Imp gallons) by installing four specially-designed tanks in spaces between fuselage frames on each side, beneath main fuel tank. Oil capacity 7·73 litres (1·7 Imp gallons).

ACCOMMODATION: Crew of one, with provision for two. Accommodation (Srs 3) for up to 19 passengers, or 12 stretcher patients and attendants, or 2,085 kg (4,600 lb) of freight, vehicles or agricultural equipment. Srs 3M can accommodate 22 equipped troops; 16 paratroops and a despatcher; 12 stretcher cases and two medical attendants, or 2,358 kg (5,200 lb) of freight. It carries its own lightweight vehicle loading ramps and has a one-piece door which leaves the fuselage threshold entirely clear of appendages. Executive version provides luxury accommodation and equipment for nine passengers. Full-width rear loading door, and forward door on each side of crew compartment. Rear door can be opened in flight to permit the parachuting of loads up to 1·37 m (4 ft 6 in) in height. Cockpit and cabin heated by engine bleed air mixed with fresh air from intake in nose. Cabin unpressurised. Some aircraft fitted with Rolamat cargo loading equipment.

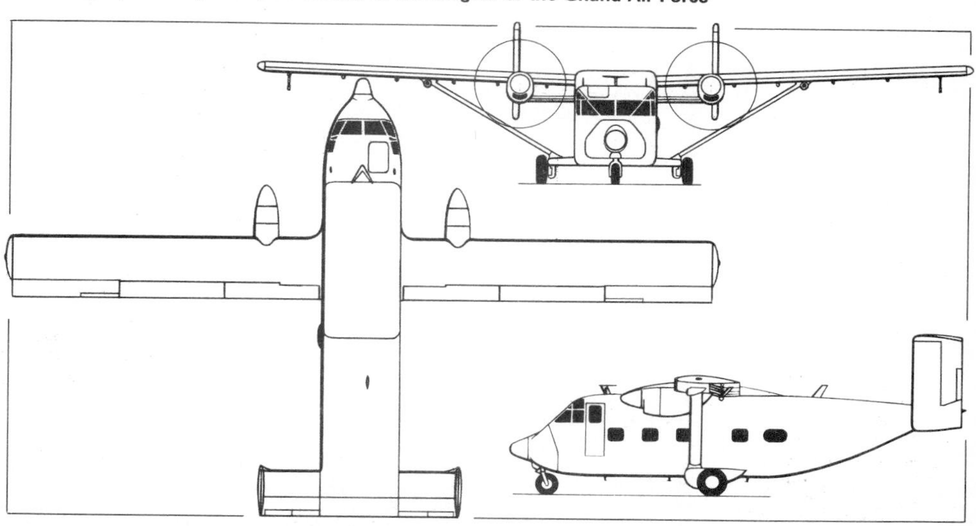

Shorts Skyvan Series 3M military transport with nose radome *(Pilot Press)*

SYSTEMS: Hydraulic system, pressure 172 bars (2,500 lb/sq in), operates flaps, wheel brakes and nosewheel steering. No pneumatic system. Electrical system utilises two busbars, operating independently, each connected to a 28V 125A DC starter/generator, a battery and a 115V 400Hz static inverter. General services are 28V DC; some radio and instruments 115V AC.

ELECTRONICS AND EQUIPMENT: Radio optional. Typical installation for operations in Europe and USA consists of duplicated VHF, duplicated VOR/ILS, marker beacon and ADF. Provision for HF, DME, transponder, Bendix M4C autopilot and weather radar. Blind-flying instrumentation standard.

EQUIPMENT (Srs 3M): Port-side blister window for an air despatcher; two anchor cables for parachute static lines; a guard rail beneath the tail to prevent control surface fouling by the static lines; inward-facing paratroop seats with safety nets; parachute signal light; mounts for NATO-type stretchers; and roller conveyors for easy loading and paradropping of pallet-mounted supplies.

DIMENSIONS, EXTERNAL:

Wing span	19·79 m (64 ft 11 in)
Wing chord (constant)	1·78 m (5 ft 10 in)
Wing aspect ratio	11
Length overall:	
3	12·21 m (40 ft 1 in)
3M, with radome	12·60 m (41 ft 4 in)
Height overall	4·60 m (15 ft 1 in)
Tailplane span	5·28 m (17 ft 4 in)
Wheel track	4·21 m (13 ft 10 in)
Wheelbase	4·52 m (14 ft 10 in)
*Propeller diameter	2·59 m (8 ft 6 in)
Propeller ground clearance	1·52 m (5 ft 0 in)
Crew and passenger doors (fwd, port and stbd):	
Height	1·52 m (5 ft 0 in)
Width	0·51 m (1 ft 8 in)
Height to sill	1·14 m (3 ft 9 in)
Rear loading door:	
Height	1·98 m (6 ft 6 in)
Width	1·96 m (6 ft 5 in)
Height to sill	0·74 m (2 ft 5 in)

**Optional four-blade propellers of 2·51 m (8 ft 3 in) diameter*

DIMENSIONS, INTERNAL:

Cabin, excl flight deck:	
Length	5·67 m (18 ft 7 in)
Max width	1·98 m (6 ft 6 in)
Max height	1·98 m (6 ft 6 in)
Floor area	11·15 m² (120 sq ft)
Volume	22·09 m³ (780 cu ft)

AREAS:

Wings, gross	34·65 m² (373 sq ft)
Ailerons (total)	3·00 m² (32·3 sq ft)
Trailing-edge flaps (total)	5·86 m² (63·1 sq ft)
Fins	7·62 m² (82·0 sq ft)
Rudders, incl tabs	2·41 m² (25·9 sq ft)
Tailplane	7·53 m² (81·0 sq ft)
Elevators, incl tabs	3·62 m² (39·0 sq ft)

WEIGHTS AND LOADINGS (with 1,332 litres; 293 Imp gallons of fuel):

Basic operating weight:	
3	3,331 kg (7,344 lb)
3M	3,356 kg (7,400 lb)
Typical operating weight as freighter:	
3	3,447 kg (7,600 lb)
3M	3,456 kg (7,620 lb)
Typical operating weight with passengers or troops:	
3	3,674 kg (8,100 lb)
3M	3,778 kg (8,330 lb)
Max payload for normal T-O weight:	
3	2,086 kg (4,600 lb)
3M	2,358 kg (5,200 lb)

Max payload for overload T-O weight:
3M 2,721 kg (6,000 lb)
Max T-O weight:
3, normal 5,670 kg (12,500 lb)
3M, normal 6,214 kg (13,700 lb)
3M, overload 6,577 kg (14,500 lb)
Max landing weight:
3 5,670 kg (12,500 lb)
3M 6,123 kg (13,500 lb)
Max wing loading:
3 163·6 kg/m^2 (33·5 lb/sq ft)
3M 179·1 kg/m^2 (36·7 lb/sq ft)
Max power loading:
3 5·32 kg/kW (8·74 lb/shp)
3M 6·17 kg/kW (9·58 lb/shp)

PERFORMANCE (at max T-O weight, with 1,332 litres; 293 Imp gallons of fuel):
Never-exceed speed
217 knots (402 km/h; 250 mph) EAS
Max cruising speed at 3,050 m (10,000 ft):
max continuous power
176 knots (327 km/h; 203 mph)
cruise power 169 knots (314 km/h; 195 mph)
Econ cruising speed at 3,050 m (10,000 ft)
150 knots (278 km/h; 173 mph)
Stalling speed, flaps down:
3 60 knots (111 km/h; 69 mph) EAS
3M 62 knots (115 km/h; 71 mph) EAS
Max rate of climb at S/L:
3 500 m (1,640 ft)/min
3M 466 m (1,530 ft)/min
Service ceiling (30 m; 100 ft/min climb):
3 6,858 m (22,500 ft)
3M 6,705 m (22,000 ft)
Service ceiling, one engine out (15 m; 50 ft/min climb):
3 3,810 m (12,500 ft)
3M 2,895 m (9,500 ft)
Min ground turning radius 3·76 m (12 ft 4 in)
Runway LCN at AUW of 5,670 kg (12,500 lb):
standard tyres 3·5
low-pressure tyres 3·0
T-O run, STOL, unfactored:
3 213 m (700 ft)
3M 238 m (780 ft)
T-O run (normal):
3 (BCAR) 512 m (1,680 ft)
T-O to 15 m (50 ft), STOL, unfactored:
3 320 m (1,050 ft)
3M 384 m (1,260 ft)
T-O to 15 m (50 ft):
3 (BCAR, normal) 610 m (2,000 ft)
3 (BCAR, STOL) 482 m (1,580 ft)
3 (FAR Pt 23) 488 m (1,600 ft)
Landing from 15 m (50 ft):
3 (BCAR, normal) 622 m (2,040 ft)
3 (BCAR, STOL) 567 m (1,860 ft)
3 (FAR Pt 23) 451 m (1,480 ft)
3M (STOL, unfactored) 425 m (1,395 ft)
Landing from 9 m (30 ft):
3 (STOL, unfactored) 351 m (1,150 ft)
3 (BCAR, STOL) 500 m (1,640 ft)
Landing run:
3M (STOL, unfactored) 212 m (695 ft)
Range at long-range cruising speed, 45 min reserves:
3 600 nm (1,115 km; 694 miles)
3M 580 nm (1,075 km; 670 miles)
Range (typical freighter) at long-range cruising speed, 45 min reserves:
3 with 1,814 kg (4,000 lb) payload
162 nm (300 km; 187 miles)
3M with 2,268 kg (5,000 lb) payload
208 nm (386 km; 240 miles)

SHORTS SKYLINER SERIES 1

Concurrent with certification of the Skyvan Series 3A to Performance Group A standards, Shorts unveiled at the 1970 Farnborough Air Show an all-passenger version known as the Skyliner. This takes full advantage of the internal capacity of the Skyvan, and the increase in max take-off weight, to provide a spacious all-passenger cabin, furnished to de luxe standard, including the provision of overhead baggage lockers. This version can be equipped for up to 19 passengers; optional features include a galley and an airstair passenger door.

The principal structural alterations are the provision of a large door in each side of the fuselage at the rear, for ease of passenger boarding. The large cargo-loading door of the basic Skyvan can be replaced by a smaller door which allows access to a large rear baggage compartment.

Up to April 1976 Skyliners had been ordered by British Airways—Scottish (two), Gulf Air (one), Airexecutive Norway (one), Ednasa Hong Kong (three) and Yokohama Air (one).

A special VIP version, the Skyliner Executive, was announced concurrently with the commercial passenger version; one has since been delivered to the Royal Air Flight of Nepal, and one of the Ghana Air Force's aircraft has a Skyliner layout.

External dimensions and most systems of the Skyliner are identical to those of the Skyvan Series 3.

The following are the salient features of the Skyliner:

ACCOMMODATION: Crew of two, with provision for steward or stewardess. Accommodation for up to 19 passengers. Standard optional equipment includes a toilet with washing facilities and a small galley for buffet service.

DIMENSIONS, EXTERNAL: As Skyvan Srs 3 except:
Passenger doors (rear, port and stbd):
Height 1·45 m (4 ft 9 in)
Width 0·76 m (2 ft 6 in)

DIMENSIONS, INTERNAL:
Cabin, excl flight deck and baggage compartment:
Max length 6·10 m (20 ft 0 in)
Max width 1·98 m (6 ft 6 in)
Max height 1·98 m (6 ft 6 in)
Volume 23·5 m^3 (830 cu ft)
Baggage compartment (basic aircraft, excl overhead baggage lockers in cabin) 4·20 m^3 (148 cu ft)

WEIGHTS:
Typical operating weight 4,055 kg (8,940 lb)
Max T-O weight 6,210 kg (13,700 lb)
Max landing weight 6,070 kg (13,400 lb)

PERFORMANCE (Group A, ISA at S/L):
T-O to 10·7 m (35 ft) 1,020 m (3,350 ft)
Landing from 15 m (50 ft) 1,010 m (3,320 ft)
Range at high-speed cruise, reserves for 50 nm (92 km; 57 miles) diversion and 45 min hold
170 nm (315 km; 196 miles)

SHORTS SD3-30

The SD3-30 is a 30-passenger twin-turboprop transport aircraft designed primarily for commuter and regional air service operators whose current 18/20-seat aircraft require replacement by larger aircraft.

Design of the SD3-30 is derived from that of the Skyvan STOL utility transport, and it retains many of the latter type's well-proven characteristics, including the large cabin cross-section. The same fail-safe concept and design philosophy is employed in the structural components. The cabin, including the toilet and galley compartments, is 3·78 m (12 ft 5 in) longer than that of the Skyvan Srs 3.

Two protoypes and the first production aircraft were used for the development programme. The first prototype (G-BSBH) flew for the first time on 22 August 1974. Eight days earlier, the first order for the SD3-30 was placed by Command Airways of Poughkeepsie, New York, for three aircraft; this was followed by an order from Time Air of Canada for three. First to enter service, on 24 August 1976, was a Time Air SD3-30.

CAA certification to full Transport Category requirements was granted on 18 February 1976, and was followed on 18 June 1976 by US FAR Pt 25 and Pt 36 approval. Initial deliveries began later the same month. The SD3-30 will conform with CAB Pt 298 (US) and meets the noise requirements of FAR Pt 36 by a substantial margin. Unrestricted maximum-weight operation is achievable at S/L ambient temperatures up to ISA + 19°C (FAR) or ISA + 20°C (BCAR).

A military version, the **SD3-M**, has also been announced. This will be capable of a variety of roles, including the tactical transportation of troops, cargo and vehicles, paratrooping, supply dropping, casualty evacuation and search and rescue, and will be able to carry up to 32 troops or 3,630 kg (8,000 lb) of cargo.

The following description applies to both the SD3-30 and the SD3-M, except where a specific version is indicated:

TYPE: Twin-turboprop civil and military transport aircraft.

WINGS: Braced high-wing monoplane, of all-metal fail-safe construction, built in three sections. Wing sections NACA 63A series (modified). Thickness/chord ratio 18% at root, 14% on outer panels. Dihedral 3° on outer panels. Centre-section, integral with top of centre-fuselage, has taper on leading- and trailing-edges, and is a two-spar single-cell box structure of light alloy with conventional skin and stringers on the undersurface. The strut-braced outer panels, which are pin-jointed to the centre-section, are reinforced Skyvan constant-chord units, built of light alloy, and each consists of a two-cell box; they, and the centre-section upper surface, have wing skins made up of a smooth outer skin bonded to a corrugated inner skin. All-metal single-slotted ailerons. Geared tabs in port and starboard ailerons, with manual trim on starboard aileron. All-metal single-slotted flaps, each in three sections. Primary control surfaces are rod-actuated.

FUSELAGE: Light alloy structure, built in two main portions: nose (including flight deck, nosewheel bay and forward baggage compartment); and the centre (including main wing spar attachment frames and lower transverse beams which carry the main landing gear and associated fairings) and rear portion (including aft baggage compartment, optional rear-loading door and tail unit attachment frames). The nose and rear underfuselage are of conventional skin/stringer design. The remainder is composed of a smooth outer skin bonded to a corrugated inner skin and stabilised by frames.

TAIL UNIT: Cantilever all-metal two-spar structure with twin fins and rudders, basically similar to that of the Skyvan. Fixed-incidence tailplane, with reinforced leading-edge. Full-span elevator, aerodynamically balanced by set-back hinges. Rudders each have an unshielded horn aerodynamic balance. Primary control surfaces are rod-actuated. Geared trim tabs in elevator and rudders.

LANDING GEAR: Menasco retractable tricycle type, with single wheel on each unit. Main units carried on short sponsons, into which the wheels retract hydraulically. Oleo-pneumatic shock-absorbers. Nosewheel is steerable. Normal tyre pressures: main units 5·17 bars (75 lb/sq in), nose unit 4·00 bars (58 lb/sq in). Special requirements for rough-field operation have been catered for in the design.

POWER PLANT: Two 835 kW (1,120 shp) (max continuous rating 761 kW; 1,020 shp) Pratt & Whitney Aircraft of Canada PT6A-45A turboprop engines, each driving a Hartzell five-blade low-speed propeller. Fuel in main tanks in wing centre-section/fuselage fairing, total capacity 2,182 litres (480 Imp gallons). Normal cross-feed provisions to allow for pump failure. Provision to increase total fuel capacity for special requirements.

ACCOMMODATION (SD3-30): Crew of two on flight deck, plus cabin attendant. Dual controls standard. Standard seating for 30 passengers, in ten rows of three at 762 mm (30 in) pitch, with wide aisle. Seat rails fitted to facilitate changes in configuration. Galley, toilet and cabin attendant's seat at rear. Large overhead baggage lockers. Entire accommodation soundproofed and air-conditioned. Baggage compartments in nose (1·27 m^3; 45 cu ft) and to rear of cabin (2·83 m^3; 100 cu ft), each with external access and capable of holding a combined total of 454 kg (1,000 lb) of baggage. Passenger door is at rear of cabin on port side. Passenger version has two emergency exits on the starboard side, one on the port side and one in the flight deck roof. Mixed-traffic version has full access to these emergency exits. For mixed passenger/freight operation a bulkhead divides the cabin into a rear passenger area (typically for 18 persons) and a forward cargo compartment, the latter being loaded through a large port-side door, capable of admitting ATA 'D' type containers. In all-cargo configuration the cabin can accommodate up to seven 'D' type containers, with ample space around them for additional freight. Cabin floor is flat throughout its length, and is designed to support loadings of 181 kg (400 lb) per foot run at 610·3 kg/m^2 (125 lb/sq ft). Locally-reinforced areas of higher strength are also provided. Seat rails can be used as cargo lashing points. Freight loading is facilitated by the low-level cabin floor.

ACCOMMODATION (SD3-M): Generally similar to SD3-30, but with large rear-loading door. Cabin capable of accommodating up to 32 fully-equipped troops, or 30

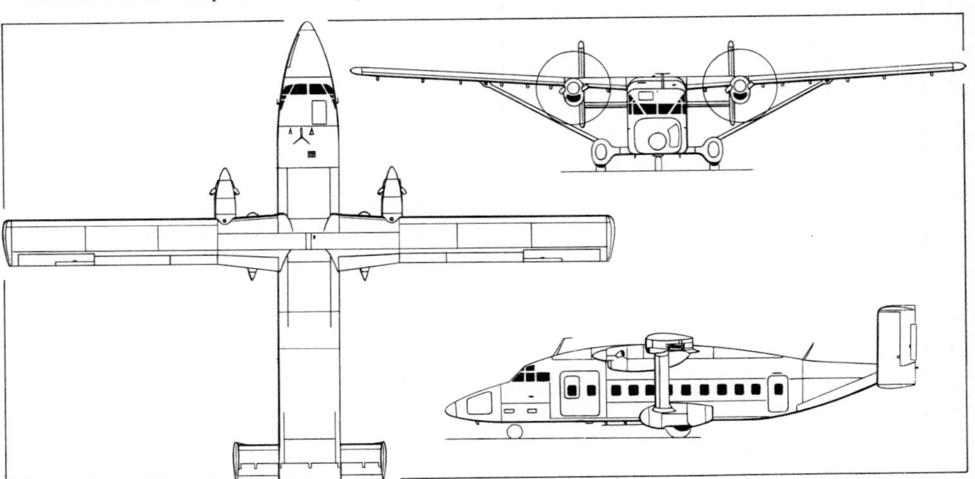

Shorts SD3-30 twin-turboprop civil and military transport *(Pilot Press)*

Shorts SD3-30 in the insignia of the US operator Command Airways, first customer to order this aircraft

fully-armed paratroops and a despatcher, when used for personnel transport. Freighter version can carry up to 3,630 kg (8,000 lb) of cargo, and more than 3,175 kg (7,000 lb) of supplies can be air-dropped. With load spreaders, the SD3-M can also be operated as a vehicle transport, carrying Land-Rovers or similar vehicles; for casualty evacuation, carrying 18 stretchers and three medical attendants; for search and rescue, with up to 10 hr endurance; as a VIP STOL transport; and for aerial survey, border and coastal patrol.

Systems: Hydraulic system of 207 bars (3,000 lb/sq in), supplied by engine-driven pumps, operates landing gear, nosewheel steering, flaps and brakes (at lower pressure) and includes emergency accumulators. Main electrical system, for general services, is 28V DC and is of the split busbar type with cross-coupling for essential services. Lucas 28V 250A DC starter/generator for engine starting and aircraft services, with separate 1·5kW 200V AC output for windscreen anti-icing and demisting. Special AC sources of 115V and 26V available at 400Hz for certain instruments. De-icing/anti-icing system for wing and tail leading-edges; inertial anti-icing system for engine intake ducts; electric mat de-icing for inlet lips and propellers. Full air-conditioning system.

Electronics and Equipment: Anti-icing system, air-conditioning system and passenger safety equipment standard. Wide range of radio and navigation equipment available to customer's requirements. Typical standard equipment would comprise duplicated VHF communications and navigation systems, two glide-slope/markers, two ILS repeaters, two radio magnetic indicators, one ADF, one transponder, one DME, PA system, flight data recorder, voice recorder and weather radar.

Dimensions, external:
- Wing span 22·76 m (74 ft 8 in)
- Wing chord (standard mean) 1·85 m (6 ft 0·7 in)
- Length overall 17·69 m (58 ft 0½ in)
- Height overall 4·95 m (16 ft 3 in)
- Propeller diameter 2·82 m (9 ft 3 in)
- Propeller ground clearance 1·83 m (6 ft 0 in)
- Cabin floor: height above ground 0·94 m (3 ft 1 in)
- Passenger door (port, rear):
  - Height 1·57 m (5 ft 2 in)
  - Width 0·71 m (2 ft 4 in)
- Forward cargo door (port):
  - Height 1·68 m (5 ft 6 in)
  - Width 1·42 m (4 ft 8 in)

Dimensions, internal:
- Cabin: Max length, incl toilet 9·47 m (31 ft 1 in)
  - Max width 1·98 m (6 ft 6 in)
  - Max height 1·98 m (6 ft 6 in)
  - Volume (all-cargo) 34·83 m³ (1,230 cu ft)
- Baggage compartments volume (total usable) 4·11 m³ (145 cu ft)

Areas:
- Wings, gross 42·1 m² (453·0 sq ft)
- Ailerons (total, aft of hinges) 2·55 m² (27·5 sq ft)
- Trailing-edge flaps (total) 7·74 m² (83·3 sq ft)
- Fins (total) 8·65 m² (93·1 sq ft)
- Rudders (total, aft of hinges) 2·24 m² (24·1 sq ft)
- Tailplane 7·77 m² (83·6 sq ft)
- Elevator (total, aft of hinges) 2·54 m² (27·3 sq ft)

Weights:
- Weight empty, equipped (incl crew of three):
  - 3-30 for 30 passengers 6,455 kg (14,230 lb)
- Fuel:
  - standard tanks 1,741 kg (3,840 lb)
  - with additional capacity 2,286 kg (5,040 lb)
- Max payload for normal max T-O weight:
  - 3-30 with 30 passengers and baggage 2,694 kg (5,940 lb)
  - 3-30 freighter 3,400 kg (7,500 lb)
  - 3-M 3,630 kg (8,000 lb)
- Payload for max range, standard tanks:
  - 3-M, normal max T-O weight 2,215 kg (4,885 lb)
  - 3-M, max T-O weight (operational necessity) 3,123 kg (6,885 lb)
- Max T-O weight:
  - 3-30 9,979 kg (22,000 lb)
  - 3-M, normal 9,979 kg (22,000 lb)
  - 3-M (operational necessity) 10,886 kg (24,000 lb)
- Max landing weight:
  - all versions 9,843 kg (21,700 lb)
- Max wing loading 237·14 kg /m² (48·57 lb/sq ft)
- Max power loading 5·975 kg/kW (9·82 lb/shp)

Performance (estimated, at normal max T-O weight, ISA at S/L, except where indicated):
- Max cruising speed at 3,050 m (10,000 ft), AUW of 9,072 kg (20,000 lb):
  - 3-30, 3-M 198 knots (367 km/h; 228 mph)
- Econ cruising speed at 3,050 m (10,000 ft), AUW of 9,072 kg (20,000 lb):
  - 3-30, 3-M 160 knots (296 km/h; 184 mph)
- Stalling speed, flaps and landing gear up:
  - 3-30, 3-M 92 knots (171 km/h; 106 mph)
- Stalling speed at max landing weight, flaps and landing gear down:
  - 3-30, 3-M 74 knots (137 km/h; 85 mph)
- Max rate of climb at S/L:
  - 3-30 369 m (1,210 ft)/min
- Service ceiling, one engine out, AUW of 8,618 kg (19,000 lb):
  - 3-30, 3-M 4,450 m (14,600 ft)
- T-O run:
  - 3-M (STOL) 355 m (1,165 ft)
- T-O distance (FAR Pt 25 and BCAR Gp A):
  - 3-30 1,183 m (3,880 ft)
  - 3-30, ISA + 15°C 1,405 m (4,610 ft)
- T-O to 15 m (50 ft):
  - 3-M (STOL) 552 m (1,810 ft)
- Landing distance, AUW of 8,618 kg (19,000 lb):
  - 3-30, BCAR 1,058 m (3,470 ft)
  - 3-30, FAR 939 m (3,080 ft)
  - 3-M (STOL) 475 m (1,560 ft)
- Landing run, AUW of 8,618 kg (19,000 lb):
  - 3-M (STOL minimum) 226 m (740 ft)
- Runway LCN at max T-O weight 10·1
- Range with max payload, cruising at 3,050 m (10,000 ft), no reserves:
  - 3-30 (passenger) 435 nm (805 km; 500 miles)
- Range at 3,050 m (10,000 ft) with alternative payloads, no reserves:
  - 3-30 (passenger) with 1,815 kg (4,000 lb) payload 903 nm (1,673 km; 1,040 miles)
  - 3-30 (freighter) with 2,495 kg (5,500 lb) payload 846 nm (1,569 km; 975 miles)
  - 3-M with 3,630 kg (8,000 lb) payload 125 nm (231 km; 144 miles)
  - 3-M with 3,630 kg (8,000 lb) payload at max operational necessity T-O weight 560 nm (1,038 km; 645 miles)
- Max range with additional fuel:
  - 3-M 1,170 nm (2,167 km; 1,347 miles)
- Max endurance:
  - 3-M, standard tanks 7 hr 30 min
  - 3-M, additional fuel 10 hr 0 min

WALLIS

WALLIS AUTOGYROS LTD

Head Office:
Reymerston Hall, Norfolk NR9 4QY
Telephone: Mattishall 418

Directors:
Wg Cdr K. H. Wallis, CEng, MRAeS
P. M. Wallis

By adopting a completely new design approach to the mechanical details of the single-seat ultra-light autogyro, Wg Cdr Wallis has produced a much-refined aircraft which can be flown quite safely 'hands and feet off'. The Wallis prototype (G-ARRT, flown for first time in August 1961) introduced many patented features, including a rotor head with offset gimbal system to provide hands and feet off stability and to eliminate pitch-up and 'tuck-under' hazards; a high-speed flexible rotor spin-up shaft with positive disengagement during flight; an automatic system of controlling rotor drive on take-off which allows power to be applied until the last moment; centrifugal stops to control rotor blade teetering; and a novel safe starting arrangement.

All Wallis autogyros are built for special purposes, and are not on public sale.

WALLIS WA-116 and WA-116-T

The WA-116 represents the original Wallis design, of which the prototype (G-ARRT) flew for the first time on 2 August 1961, powered by a 54 kW (72 hp) modified McCulloch 4318 piston engine. Four more WA-116s were built by Beagle and five by Wg Cdr Wallis, as described in the 1973-74 *Jane's*. The last of these was later dismantled for construction of G-AXAS, a tandem two-seat version which was designated WA-116-T and flew for the first time on 3 April 1969.

The WA-116 currently holds the height record for autogyros in Classes E3 and E3a, at 4,639 m (15,220 ft), set on 11 May 1968 by the prototype with its original 54 kW (72 hp) McCulloch engine; and the Class E3 and E3a records for speed in a straight line (96·589 knots; 179·000 km/h; 111·225 mph), set on 12 May 1969 by the same aircraft re-engined with a 67 kW (90 hp) McCulloch.

The WA-116 has remained potentially one of the most promising of the Wallis autogyro designs, particularly following the refitting of G-ASDY in 1971 with a 44 kW (60 hp) Franklin 2A-120-A engine. It is currently fitted with a Franklin 2A-120-B engine, and suffix designations are used to indicate the type of engine fitted to various Wallis autogyros. Thus the two-seat WA-116-T is known as the **WA-116-T/Mc**, and the Franklin-engined aircraft as **WA-116/F**.

The WA-116/F has undergone progressive refinements which include, on the latest Franklin conversions, a four-blade propeller. The former WA-116/F hack G-ASDY has been brought up to 'working WA-116F' standard. A particularly successful conversion to WA-116/F has been that of the WA-116/Mc previously registered in Ceylon as 4R-ACK. This resumed its former British registration G-ATHM and has undergone more extensive conversion than its predecessors, mainly to increase fuel capacity and pilot comfort to fit it for special long-range flights.

The rebuilt G-ATHM has a 50 litre (11 Imp gallon) internal fuel tank, and began test flying, for range, in April 1974. Fitted also with a 36 litre (8 Imp gallon) jettisonable long-range ventral tank, it set up in July 1974 (subject to confirmation) new Class E3 and E3a world records, for non-stop distance in a 100 km closed circuit, of 361 nm (670 km; 416 miles). Additionally, this flight set (also subject to confirmation) new 100 km and 500 km closed-circuit speed records of 70·51 knots (130·67 km/h; 81·19 mph) and 68·07 knots (126·14 km/h; 78·38 mph) respectively. A 91 litre (20 Imp gallon) ventral tank was next fitted, and on 28 September 1975, with this tank contain-

Left: Wg Cdr Wallis with the WA-116/F, holder (subject to confirmation) of the world 100 km closed-circuit speed record for autogyros; right, the WA-117/R-R with Rolls-Royce Continental engine and multi-band photographic equipment pack

ing some 70 litres (15·5 Imp gallons), Wg Cdr Wallis made a nonstop flight from Lydd, Kent, to Wick, Caithness. Subject to confirmation, this flight qualifies for Class E3 and E3a records for distance in a straight line of 471·785 nm (874·315 km; 543·273 miles). In fact, the actual distance flown, to avoid airfield zones and other hazardous areas, was in the order of 521 nm (966 km; 600 miles), flown in 6 hr 25 min at an average speed of approx 81 knots (150 km/h; 93 mph). The ventral tank was not jettisoned after being emptied, and the WA-116/F landed with sufficient fuel remaining for a further 65 nm (121 km; 75 miles). In mid-1976, FAI ratification of the 1974 and 1975 records was still awaited; following this ratification the WA-116 will hold 13 international autogyro records for speed, height and distance; and Wg Cdr Wallis reports that it has not yet been fully exerted.

By early 1975 the WA-116-T/Mc had made more than 130 flights in a programme of multi-spectral experiments by Plessey Radar on behalf of the Home Office. The programme, intended to evaluate this method of detecting illicit graves (eg of murder victims) from the air, entails taking detailed photographs from about 30 m (100 ft) directly above the suspected sites.

WEIGHTS (WA-116/F):
- Weight empty — 143 kg (316 lb)
- Max T-O weight — 317·5 kg (700 lb)

PERFORMANCE (WA-116/F):
- Max level speed — not fully explored
- Cruising speed without long-range tank — 87 knots (161 km/h; 100 mph)
- Max rate of climb at S/L — 305 m (1,000 ft)/min
- Max range with long-range tank (estimated) — 651 nm (1,207 km; 750 miles)

WALLIS WA-117/R-R

Started in 1964, the WA-117 was intended to combine proven features of the WA-116 airframe with a fully-certificated engine, the 74 kW (100 hp) Rolls-Royce Continental O-200-B. An experimental test vehicle (G-ATCV) flew for the first time on 24 March 1965; this was later dismantled for the construction of a true WA-117 prototype (G-AVJV), which made its first flight on 28 May 1967. This aircraft took part in the Loch Ness investigations in 1970 and, as described in the 1972-73 *Jane's*, was evaluated as a carrying vehicle for HSD Linescan 212 infra-red sensor equipment. A third WA-117 (G-AXAR) was sold to Airmark Ltd for a test and certification programme prior to intended marketing by Airmark. Built by Wg Cdr Wallis's cousin, it had a larger cockpit than the other WA-117s, and a larger (63·6 litre; 14 Imp gallon) fuel tank; but it was lost in a crash at the 1970 SBAC Display at Farnborough. The investigation into this accident found that the aircraft behaves normally within a reasonable flight envelope. However, because of its ability to accelerate very quickly beyond what might be considered a normal flight envelope, changes were introduced to make the aircraft less sensitive to mishandling at high speeds.

The G-AVJV prototype, which with its special silencers and special four-blade 'quiet' propeller is one of the quietest powered aircraft of any kind yet built, continues to be a particularly reliable working aircraft, especially for carrying still, ciné and multi-spectral cameras. Recent work for the Department of the Environment, carrying a multi-spectral four-camera pack, was very successful, and it has since been fitted with new multi-band photographic equipment and a radar altimeter, for continued work of this nature. A new programme on behalf of Plessey Radar began in the Spring of 1976.

WEIGHT:
- Max T-O weight — approx 317·5 kg (700 lb)

PERFORMANCE:
- Max level speed — 104 knots (193 km/h; 120 mph)
- Cruising speed — 78 knots (145 km/h; 90 mph)
- Max rate of climb — approx 305 m (1,000 ft)/min

WALLIS WA-118/M METEORITE

Design of the WA-118 Meteorite (G-ATPW) was started in April 1965. Construction began in October 1965 and it flew for the first time on 6 May 1966.

The 89 kW (120 hp) supercharged Italian Meteor Alfa 1 engine was brought up to the then-current modification standards during 1969-70.

The aircraft, intended for speeds of up to 174 knots (322 km/h; 200 mph), was also fitted with a bubble canopy, reclining cockpit and other modifications and was rebuilt as G-AVJW, making its first flight in this new form on 9 August 1969. It is intended for a long-term test programme, and is currently undergoing a complete rebuild to bring it into line with the latest features of the Wallis range.

WALLIS WA-120/R-R

Construction of the WA-120 began in early 1970, under the original designation WA-117-S. It subsequently developed into more than a re-engined version of the WA-117, so justifying the use of a new designation.

The WA-120 (G-AYVO) is powered by a 97 kW (130 hp) Rolls-Royce Continental O-240-A flat-four engine and cruises at a fuel consumption of 15·9 litres (3·5 Imp gallons)/hr. It flew for the first time on 13 June 1971. Flight operation continues; the original horizontal tail surface was removed after tests in varying climatic conditions.

The WA-120 has a forward-sliding transparent cockpit canopy, and can be flown at speeds of up to 60 knots (111 km/h; 69 mph) with this canopy partly open.

WALLIS WA-121

The WA-121, currently in the flight development stage, is the smallest and lightest Wallis autogyro to date. At present, three versions are envisaged: a high-speed version **(WA-121/Mc)** with a Wallis-McCulloch engine of about 74 kW (100 hp); a cross-country version **(WA-121/F)** with a 44·5 kW (60 hp) Franklin 2A-120-B engine; and a high-altitude version **(WA-121/M Meteorite 2)** with a supercharged 89 kW (120 hp) Meteor Alfa 1 radial two-stroke engine and transistorised ignition.

The prototype (G-BAHH) has a high-mounted tailplane and an open cockpit, and made its first flight on 28 December 1972. With the McCulloch engine, it has already exceeded unofficially the speed and altitude records set up by the WA-116 prototype G-ARRT. It employs a number of improvements in control system design, resulting in greater stability at speed, better head resistance and greater pilot comfort. Special features in the rotor head suspension, originally incorporated in the WA-117 prototype G-AVJV, are incorporated also in the WA-120 and WA-121.

The next version to be built will probably be the WA-121/F, but this is currently awaiting the completion of other projects and the evaluation of further experience to be gained with the WA-116/F.

WALLIS WALLBRO MONOPLANE REPLICA

In 1909 the father and uncle of Wg Cdr Wallis, H. S. and P. V. Wallis, formed the Wallis Aeroplane Company at 12 St Barnabas Road, Cambridge, to construct a single-seat tractor aeroplane known as the Wallbro Monoplane. It had a wing span of 9·14 m (30 ft 0 in), a length of 7·62 m (25 ft 0 in), and was powered by an 18·6 kW (25 hp at 1,500 rpm) J.A.P. four-cylinder Vee-type engine driving an A. V. Roe two-blade ground-adjustable tractor propeller of 1·98 m (6 ft 6 in) diameter. Weight, with 6·8 litres (1·5 Imp gallons) of fuel and 0·23 litres (0·5 Imp gallons) of oil, is said to have been 150 kg (330 lb), but may have been higher. Completed and on public display by Whitsun 1910, the aircraft made a number of hop-flights in a field near Cambridge during the Summer of that year, but was

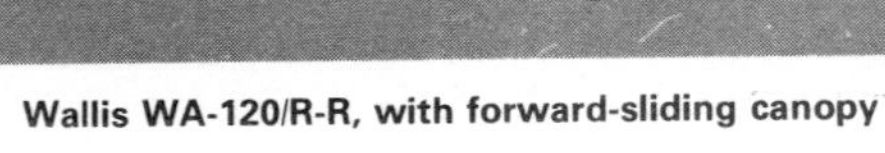
Wallis WA-120/R-R, with forward-sliding canopy

The McCulloch-engined WA-121 high-speed autogyro

later destroyed in its shed by a storm and was never rebuilt.

The most notable feature of the monoplane was its construction from 20 gauge (25·4 mm; 1 in diameter) seamless drawn steel tubing, used by the Wallis brothers because of their experience of its use in the manufacture of racing motorcycles. This was possibly the first use of steel tube as the main structural material of an aeroplane. The Wallbro Monoplane was fitted with conventional ailerons, and had a wheel-type roll control system, mounted on the fore-and-aft pitch lever. The wing ribs were of wood.

The sons of the original constructors, Wg Cdr Kenneth Wallis and his cousin, Geoffrey V. Wallis, were in early 1976 constructing a flying replica of the Wallbro Monoplane. Apart from the necessity to fit a modern engine (only three of the original type of J.A.P. engines are believed to have been built), the design of the 1909 aircraft is being followed meticulously, and the steel tube fuselage frame had been completed by the Spring of 1976.

WESTLAND

WESTLAND AIRCRAFT LTD

HEAD OFFICE, WORKS AND AERODROME:
Yeovil, Somerset BA20 2YB

LONDON OFFICE:
8 The Sanctuary, Westminster, London SW1P 3JU

CHAIRMAN:
Sir David Collins, CBE, CEng, FIMechE, FIProdE, FRAeS

VICE-CHAIRMAN AND CHIEF EXECUTIVE:
B. D. Blackwell, MA, BSc(Eng), CEng, FIMechE, FRAeS, FBIM

VICE-CHAIRMEN:
Walter Oppenheimer, FCA
G. S. Hislop, PhD, BSc, ARCST, CEng, FIMechE, FRAeS, FRSA

DIRECTORS:
The Rt Hon Lord Aberconway
F. E. J. Hallett
Sir Christopher Hartley, KCB, CBE, DFC, AFC, BA
Sir Ronald Melville, KCB
Sir Eric Mensforth, CBE, DSc, CEng
W. T. C. Miller, MA, CEng, MIMechE, MInstM
J. Speechley, OBE, MSc, CEng, FRAeS
S. W. Wiltshire

SECRETARY:
M. Barnes, FCA

PUBLIC RELATIONS EXECUTIVE:
John Teague, CEng, MRAeS, MInstM, MIPR

Westland Aircraft Ltd was formed in July 1935, to take over the aircraft branch of Petters Ltd, previously known as the Westland Aircraft Works, which had been engaged in aircraft design and construction since 1915.

Westland entered the helicopter industry in 1947 by acquiring the licence to build the Sikorsky S-51, of which it produced 133 under the name Westland Dragonfly. This technical association with Sikorsky has continued, and it was decided subsequently to concentrate the company's resources on the design, development and construction of helicopters.

In 1959, Westland acquired Saunders-Roe Ltd. In 1960, it further acquired the Helicopter Division of Bristol Aircraft Ltd and Fairey Aviation Ltd, as a result of which it is the only major organisation engaged in helicopter design and manufacture in the United Kingdom.

Since 1 October 1966, the company's helicopter business has been conducted through a wholly-owned company named Westland Helicopters Ltd.

Through the British Hovercraft Corporation Ltd, Westland is continuing development of the Hovercraft type of vehicle pioneered by Saunders-Roe.

One of Westland's subsidiary companies, Normalair-Garrett Ltd, specialises in the design, development and production of aircraft pressure control, air-conditioning, oxygen breathing and hydraulic systems. Able to supply complete aircraft installations, Normalair-Garrett Ltd is recognised as the foremost European authority in this field. Most British pressurised aircraft, civil and military, use Normalair-Garrett equipment, as do the Panavia Tornado and many aircraft of foreign design. In addition, this company produces data loggers, trace readers and hydraulic equipment for aircraft flying controls.

WESTLAND HELICOPTERS LTD

HEAD OFFICE, WORKS AND AERODROME:
Yeovil, Somerset BA20 2YB
Telephone: Yeovil (0935) 5222
Telex: 46277

CHAIRMAN:
Sir David Collins, CBE, CEng, FIMechE, FIProdE, FRAeS

DEPUTY CHAIRMAN:
B. D. Blackwell, MA, BSc(Eng), CEng, FIMechE, FRAeS, FBIM

MANAGING DIRECTOR:
J. Speechley, OBE, MSc, CEng, FRAeS

DEPUTY MANAGING DIRECTOR:
A. N. Street, MRAeS

ACCOUNTING DIRECTOR AND SECRETARY:
A. R. B. Hobbs, FCCA

FINANCE DIRECTOR:
Malcolm Jones, BSc(Econ), FCMA, AMBIM

RESEARCH DIRECTOR:
J. P. Jones, PhD, BSc(Eng), CEng, FRAeS

COMMERCIAL DIRECTOR:
A. V. N. Reed, BSc(Tech), CEng, FRAeS, AMBIM

WORKS DIRECTOR:
W. Baxter, BEng, CEng, MIMechE, MIProdE

TECHNICAL DIRECTOR:
V. A. B. Rogers, MSc, CEng, FRAeS, FIMechE

DIRECTOR:
Walter Oppenheimer, FCA

SECRETARY:
M. S. Double, MA, FCA

YEOVIL DIVISION

TECHNICAL DIRECTOR:
V. A. B. Rogers, MSc, CEng, FRAeS, FIMechE

GENERAL WORKS MANAGER:
C. F. Read

Helicopters in current production at the Yeovil headquarters of Westland Helicopters are the Sea King, Commando, Gazelle and Lynx. The Gazelle and Lynx, together with the Puma, form part of the Anglo-French helicopter co-operation programme.

Gazelle helicopters are in production for both the British and French armed forces. Production of the Lynx began in 1975, and the first production aircraft, a naval version, began its acceptance trials in early February 1976. Forty Pumas were delivered to the RAF, and Westland is now actively involved with the production of component sets for the Pumas built by Aérospatiale.

WESTLAND SEA KING

The Sea King was developed originally by Westland to meet the Royal Navy's requirement for an advanced anti-submarine helicopter with prolonged endurance. It can also undertake a number of secondary roles, such as search and rescue, tactical troop transport, casualty evacuation, cargo carrying and long-range self-ferry. A land-based general-purpose version, known as the Commando, is described separately.

The Sea King development programme stemmed from a licence agreement for the S-61 helicopter concluded originally with Sikorsky in 1959. This permitted Westland to utilise the basic airframe and rotor system of the Sikorsky SH-3D, of which a description can be found in the US section. Considerable changes were made in the power plant and in specialised equipment, to meet British requirements.

The fuselage is essentially similar to that of the basic Sikorsky aircraft, with a watertight hull which allows water landing in an emergency. The retractable main landing gear is housed in sponsons braced to the fuselage by fixed struts. To improve the lateral stability and flotation capability of the helicopter with the rotor stopped, inflatable buoyancy bags are fitted to the outside of each sponson.

The following versions of the Sea King had been announced up to mid-1976:

Sea King HAS. Mk 1. ASW version for Royal Navy, ordered in 1967. First production HAS. Mk 1 (XV642) flown for the first time on 7 May 1969. Total of 56 built, delivery of which was completed in May 1972. In service with Nos. 814, 819, 820, 824 and 826 Squadrons. Described in previous editions of *Jane's*.

Sea King HAS. Mk 2. Uprated version for ASW and SAR duties with the Royal Navy. Thirteen ordered; deliveries to begin in 1976.

Sea King HAR. Mk 3. Uprated version for SAR duties with the Royal Air Force. Fifteen ordered, deliveries of which are scheduled to begin in Autumn 1977.

Sea King Mk 41. Search and rescue version for Federal German Navy. First example (89 + 50) flown for the first time on 6 March 1972. Twenty-two ordered, of which production and delivery were completed in 1974. First unit to equip with these aircraft was MFG.5, based at Kiel-Holtenau.

Sea King Mk 42. ASW version for Indian Navy. Original order for six, which are in service with No. 330 Squadron. Delivery of a further six was completed in 1974, and these are in service with No. 336 Squadron.

Sea King Mk 43. SAR version for Norwegian Air Force. Ten ordered, all of which were delivered in 1972. In service with No. 330 Squadron at Bodo.

Sea King Mk 45. ASW version for Pakistan Navy. Six ordered, delivery of which was completed during 1975.

Sea King Mk 48. SAR version for Belgian Air Force. Five ordered, for delivery in 1976.

Sea King Mk 50. Version, developed from Mk 1, for No. 817 Squadron of the Royal Australian Navy, which ordered 10. First flight 30 June 1974. Production includes offset manufacture in Australia to 30% of the contract value. Deliveries began in the Autumn of 1974. The Mk 50 was the first fully-uprated version of the Sea King to fly. It is capable of operation in the roles of anti-submarine search and strike, vertical replenishment, tactical troop lift, search and rescue, casualty evacuation, and self-ferry.

In addition, six Sea Kings are included in the order for 30 helicopters placed by Saudi Arabia on behalf of the Egyptian government, bringing to 189 the total number of Sea King and Commando aircraft ordered up to mid-1976.

A description of the Sea King Mk 1, which is applicable also to most export versions except the Mk 50, has appeared in previous editions of *Jane's*. Current production Sea Kings (Mks 2/3 standard) have uprated Gnome engines and transmission, a six-blade tail rotor, increased max T-O weight and other detail improvements; the following description applies to the current version. Most of the improvements are incorporated also in the Mk 50s built for Australia, and in certain other export versions.

POWER PLANT (all current versions): Two 1,238 kW (1,660 shp) (max contingency rating) Rolls-Royce Gnome H.1400-1 turboshaft engines, mounted side by side above cabin. Fuel in underfloor bag tanks, total capacity (SAR versions) 3,636 litres (800 Imp gallons). Internal auxiliary tank may be fitted for long-range ferry purposes. Pressure refuelling point on starboard side, two gravity points on port side.

ACCOMMODATION: Crew of four in ASW role; accommodation for up to 22 survivors in SAR role. Two-section airstair door at front on port side, cargo door at rear on starboard side. Entire accommodation heated and ventilated. Cockpit doors and windows, and two windows each side of cabin, can be jettisoned in an emergency.

SYSTEMS: Three main hydraulic systems. Primary and auxiliary systems operate main rotor control. Utility system (207 bars; 3,000 lb/sq in) for main landing gear, sonar and rescue winches and blade folding. Pressure for windscreen wipers 86 bars (1,250 lb/sq in). Electrical system includes two 20kVA 200V three-phase 400Hz engine-driven generators, a 26V single-phase AC supply fed from the aircraft's 22Ah nickel-cadmium battery

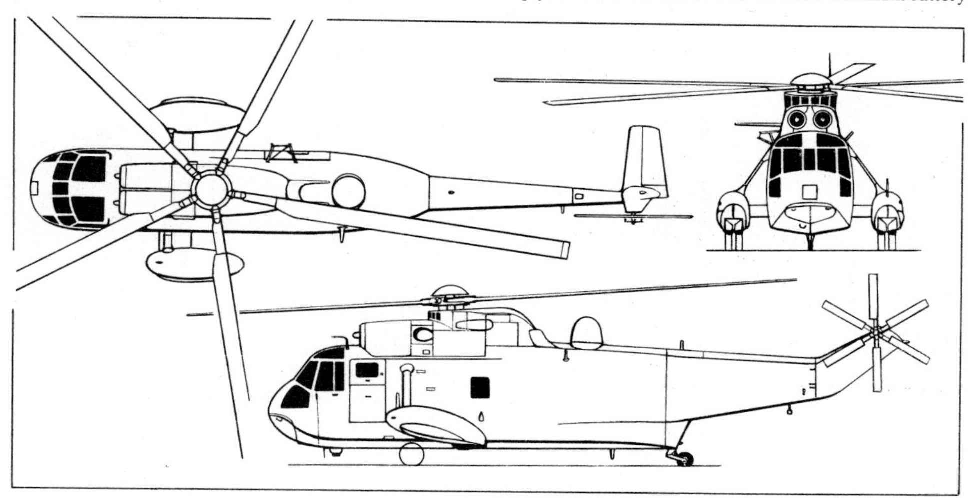

Westland Sea King HAS.Mk 1 anti-submarine helicopter *(Pilot Press)*

Second of the five Westland Sea King Mk 48 helicopters ordered by the Belgian Air Force for search and rescue duties

through an inverter, and DC power provided as a secondary system from two 200A transformer-rectifier units.

OPERATIONAL EQUIPMENT (ASW models): As equipped for this role, the Sea King is a fully-integrated all-weather hunter-killer weapon system, capable of operating independently of surface vessels, and has the following equipment and weapons to achieve this task: Plessey Type 195 dipping sonar, Bendix AN/AQS-13B dipping sonar, Marconi-Elliott AD 580 Doppler navigation system, AW 391 search radar in dorsal radome, transponder beneath rear fuselage, Honeywell AN/APN 171 radio altimeter, Sperry GM7B Gyrosyn compass system, Louis Newmark Mk 31 automatic flight control system, two No. 4 marine markers, four No. 2 Mk 2 smoke floats, up to four Mk 44 homing torpedoes, or four Mk 11 depth charges or one Clevite simulator. Observer/navigator has tactical display on which sonar contacts are integrated with search radar and navigational information. Radio equipment comprises Plessey PTR 377 UHF/VHF and homer, Ultra D 403M standby UHF, Collins 618-T3 HF radio, Ultra UA 60M intercom, Telebrief system and IFF provisions. For secondary role a mounting is provided on the aft frame of the starboard door for a general-purpose machine-gun. The Mk 31 AFCS provides radio altitude displays for both pilots; artificial horizon displays; three-axis stabilisation in pilot-controlled manoeuvres; attitude hold, heading hold and height hold in cruising flight; controlled transition manoeuvres to and from the hover; automatic height control and plan position control in the hover; and an auxiliary trim facility.

OPERATIONAL EQUIPMENT: A wide range of radio and navigation equipment may be installed, including VHF/UHF communications, VHF/UHF homing, radio compass, Doppler navigation system, radio altimeter, VOR/ILS, radar and transponder, of Collins, Plessey, Honeywell and Marconi-Elliott manufacture. A Sperry compass system and a Louis Newmark automatic flight control system are also installed. Sea Kings equipped for search and rescue have in addition a Breeze BL 10300 variable-speed hydraulic rescue hoist of 272 kg (600 lb) capacity mounted above the starboard-side cargo door. Automatic main rotor blade folding and spreading is standard with this version, and for shipboard operation the tail pylon can also be folded. With search radar fitted, a total of 18 survivors and medical staff can be carried; this total can be increased to 22 if the search radar is omitted. In the casualty evacuation role, the Sea King can accommodate up to 9 stretchers and two medical attendants, or intermediate combinations of seats and stretchers; a typical layout might provide for 14 seats and two stretchers. In the troop transport role, the Sea King can accommodate 22 troops, with the majority of seats at 419 mm (16·5 in) pitch, and can carry this load over a range of 300 nm (555 km; 345 miles) under ISA sea level conditions. As a cargo transport, the aircraft has an internal capacity of 2,720 kg (6,000 lb) or a max external load capacity of 3,630 kg (8,000 lb) when a low-response sling is fitted.

DIMENSIONS, EXTERNAL:
- Diameter of main rotor: 18·90 m (62 ft 0 in)
- Diameter of tail rotor: 3·16 m (10 ft 4 in)
- Length overall (rotors turning): 22·15 m (72 ft 8 in)
- Length of fuselage: 17·01 m (55 ft 9¾ in)
- Length overall (main rotor folded): 17·42 m (57 ft 2 in)
- Length overall (rotors and tail folded): 14·40 m (47 ft 3 in)
- Height overall (rotors turning): 5·13 m (16 ft 10 in)
- Height overall (rotors spread and stationary): 4·85 m (15 ft 11 in)
- Height to top of rotor hub: 4·72 m (15 ft 6 in)
- Width overall (rotors folded):
  - with flotation bags: 4·98 m (16 ft 4 in)
  - without flotation bags: 4·77 m (15 ft 8 in)
- Wheel track (c/l of shock-absorbers): 3·96 m (13 ft 0 in)
- Cabin door (port):
  - Height: 1·68 m (5 ft 6 in)
  - Width: 0·91 m (3 ft 0 in)
- Cargo door (stbd):
  - Height: 1·52 m (5 ft 0 in)
  - Width: 1·73 m (5 ft 8 in)
  - Height to sill: 1·14 m (3 ft 9 in)

DIMENSIONS, INTERNAL:
- Cabin: Length:
  - ASW: 5·87 m (19 ft 3 in)
  - SAR: 7·59 m (24 ft 11 in)
  - Max width: 1·98 m (6 ft 6 in)
  - Max height: 1·92 m (6 ft 3½ in)

WEIGHTS (A: anti-submarine, B: SAR, C: troop transport, D: casualty evacuation, E: internal cargo):
- Basic weight:
  - A, B, C, D, E: 6,084 kg (13,413 lb)
  - Mk 50: 5,443 kg (12,000 lb)
- Weight, equipped:
  - A: 6,201 kg (13,672 lb)
  - B, C: 5,613 kg (12,376 lb)
  - D: 5,797 kg (12,781 lb)
  - E: 5,558 kg (12,253 lb)
- Max T-O weight:
  - all current versions: 9,525 kg (21,000 lb)

PERFORMANCE (at max T-O weight, all current versions):
- Cruising speed at S/L: 112 knots (208 km/h; 129 mph)
- Max rate of climb at S/L: 616 m (2,020 ft)/min
- Max vertical rate of climb at S/L: 119 m (390 ft)/min
- Service ceiling, one engine out: 1,220 m (4,000 ft)
- Hovering ceiling in ground effect: 1,525 m (5,000 ft)
- Hovering ceiling out of ground effect: 975 m (3,200 ft)
- Range with max standard fuel: 664 nm (1,230 km; 764 miles)
- Ferry range with max standard and auxiliary fuel: 814 nm (1,507 km; 937 miles)

WESTLAND COMMANDO

First announced by Westland on 14 July 1971, the Commando is a tactical helicopter based on the Sea King.

The payload/range performance and endurance capabilities of the Sea King have been optimised in the design of the Commando, which is intended to operate with maximum efficiency in the primary roles of tactical troop transport, logistic support and cargo transport, and casualty evacuation. In addition, the Commando can operate effectively in the secondary roles of air-to-surface strike and search and rescue.

The first five Commandos, part of a larger order placed on behalf of Egypt by the Saudi Arabian government, were of the **Mk 1** version, a minimally-modified version able to transport up to 21 troops. The first two Commando Mk 1s were flown for the first time on 12 and 13 September 1973, and were delivered to Egypt in January/February 1974. All five have now been delivered.

The major production version, to which the following description applies, is the **Commando Mk 2,** which flew for the first time (G-17-12) on 16 January 1975. The Saudi Arabian order for 31 Sea King/Commando helicopters includes a number of Mk 2s ordered on behalf of the Egyptian government. In addition, four Mk 2s were ordered by the Qatar Emiri Air Force, for late 1975/early 1976 delivery.

TYPE: Twin-turboshaft tactical military helicopter.

ROTOR SYSTEM: Five-blade single main rotor and six-blade tail rotor. Main rotor blades, of NACA 0012 section, attached to hub by multiple bolted joint. Blade construction consists of a light alloy extruded spar, with light alloy trailing-edge pockets. Tail rotor blades are of similar construction. Rotor brake fitted. Tail section folds for stowage; main rotor blades do not.

ROTOR DRIVE: Twin input four-stage reduction main gear-

Westland Commando Mk 2 tactical military helicopter, developed from the Sea King

box, with single bevel intermediate and tail gearboxes. Main rotor/engine rpm ratio 93·43; tail rotor/engine rpm ratio: 15·26.

FUSELAGE: Light alloy stressed-skin structure, unpressurised.

TAIL UNIT: Similar to Sea King, with starboard-side half-tailplane at top of tail rotor pylon.

LANDING GEAR: Non-retractable tailwheel type, with twin-wheel main units. Oleo-pneumatic shock-absorbers. Main-wheel tyres size 6·50-10, tailwheel tyre size 6·00-6.

POWER PLANT: As for current versions of Sea King (which see).

ACCOMMODATION: Crew of two on flight deck. Seats along cabin sides, and single jump seat, for total of 28 troops. Two-piece airstair door at front on port side, cargo door at rear on starboard side. Entire accommodation heated and ventilated. Cockpit doors and windows, and two windows each side of main cabin, are jettisonable in an emergency.

SYSTEMS: Primary and secondary hydraulic systems for flight controls. No pneumatic system. Electrical system includes two 20kVA alternators.

ELECTRONICS AND EQUIPMENT: Blind-flying instrumentation standard. Wide range of radio, radar and navigation equipment available to customer's requirements. Cargo sling and rescue hoist optional.

ARMAMENT: Wide range of guns, missiles, etc may be carried, according to customer's requirements.

DIMENSIONS, EXTERNAL:

Diameter of main rotor	18·90 m (62 ft 0 in)
Diameter of tail rotor	3·16 m (10 ft 4 in)
Distance between rotor centres	11·10 m (36 ft 5 in)
Main rotor blade chord	0·46 m (1 ft 6¼ in)
Length overall (rotors turning)	22·15 m (72 ft 8 in)
Length of fuselage	17·02 m (55 ft 10 in)
Height overall (rotors turning)	5·13 m (16 ft 10 in)
Height to top of rotor hub	4·72 m (15 ft 6 in)
Wheel track (c/l of shock-absorbers)	3·96 m (13 ft 0 in)
Wheelbase	7·21 m (23 ft 8 in)
Passenger door (fwd, port):	
Height	1·68 m (5 ft 6 in)
Width	0·91 m (3 ft 0 in)
Cargo door (aft, stbd):	
Height	1·52 m (5 ft 0 in)
Width	1·73 m (5 ft 8 in)

DIMENSIONS, INTERNAL:

As Sea King (SAR version)

AREAS:

Main rotor disc	280·5 m² (3,019 sq ft)
Tail rotor disc	7·79 m² (83·86 sq ft)
Main rotor blades (each)	4·14 m² (44·54 sq ft)
Tail rotor blades (each)	0·23 m² (2·46 sq ft)
Tailplane	1·80 m² (19·40 sq ft)

WEIGHTS:

Operating weight empty (troop transport, 2 crew)	5,700 kg (12,566 lb)
Max T-O weight	9,525 kg (21,000 lb)

PERFORMANCE (at max T-O weight): As given for Sea King, plus:

Range with max payload (28 troops), reserves for 30 min stand-off and T-O
240 nm (445 km; 276 miles)

WESTLAND/AÉROSPATIALE LYNX

The Lynx is one of three types of aircraft (Puma, Gazelle and Lynx) covered by the Anglo-French helicopter agreement first proposed in February 1967 and confirmed on 2 April 1968. On 1 December 1972 a long-term agreement was signed between Westland Helicopters and Aérospatiale to formalise and strengthen the existing collaboration programme. Westland has design leadership in the Lynx, which is a medium-sized helicopter intended to fulfil general-purpose, naval and civil transport roles. It is the first British aircraft to be designed entirely on a metric basis.

The first of 13 Lynx prototypes (XW835) flew for the first time on 21 March 1971 and was followed by XW837, the third prototype (second Lynx to fly), on 28 September 1971. Initial icing trials, with XW837, were carried out in Denmark in early 1975. Preliminary hot weather trials took place in North Africa during July-August 1975, and operational trials in snow and icing conditions took place in Norway and Denmark in early 1976. The fourth Lynx was the first to have the monobloc rotor head designed for production aircraft.

On 20 and 22 June 1972 respectively the fifth Lynx (XX153), piloted by Roy Moxam, set up Class E1e international speed records of 173·61 knots (321·74 km/h; 199·92 mph) over a 15/25 km straight course and 171·868 knots (318·504 km/h; 197·909 mph) over a 100 km closed circuit. During the flight test programme the Lynx was rolled at more than 100° per second, dived at 200 knots (370 km/h; 230 mph), and flown backwards at 70 knots (130 km/h; 80 mph).

Details of the subsequent development aircraft can be found in the 1975-76 *Jane's*. By the beginning of 1976 these 13 aircraft had completed a total of 2,800 hr of test flying. One example of the Army version and two of the naval version began final type testing and certification

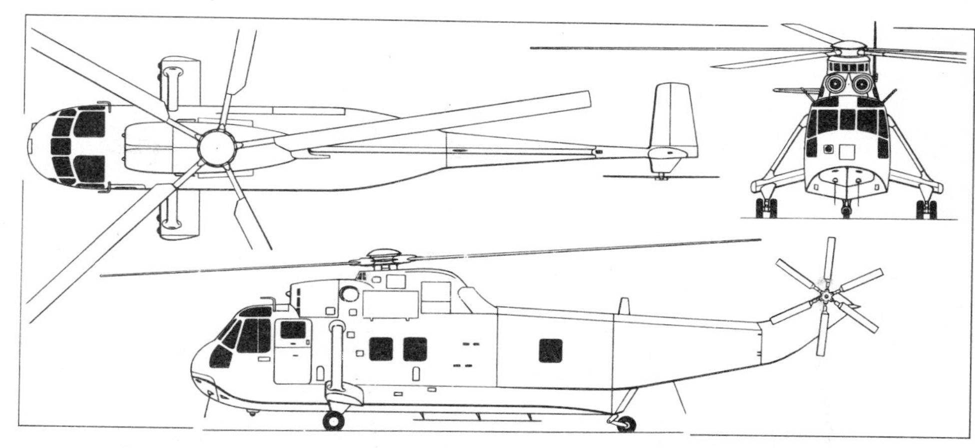

Westland Commando Mk 2 twin-turboshaft tactical military helicopter *(Pilot Press)*

trials at the A & AEE, Boscombe Down, in mid-1975. The first production aircraft, a naval version, began its acceptance trials in early February 1976, and was to be followed shortly afterwards by the first production Army Lynx. A Lynx Intensive Flying Trials Unit was due to form towards the end of 1976.

The following versions of the Lynx have been announced:

Lynx AH. Mk 1. General-purpose and utility version for the British Army. Development aircraft XX153 and XX907 (first flights 12 April 1972 and 20 May 1973). Capable of operation on tactical troop transport, logistic support, armed escort of troop-carrying helicopters, anti-tank strike, search and rescue, casualty evacuation, reconnaissance and command post duties. Total of 63 ordered by early 1976.

Lynx HAS. Mk 2. Version for Royal Navy, for advanced shipborne anti-submarine and other duties. Total of 30 ordered by early 1976. First production aircraft flown on 10 February 1976. Ferranti Seaspray search and tracking radar in modified nose. Capable of operation on anti-submarine classification and strike, air to surface vessel search and strike, search and rescue, reconnaissance, troop transport, fire support, communication and fleet liaison, and vertical replenishment duties. Two development aircraft originally (XX469 and XX510, first flown 25 May 1972 and 6 March 1973). The former was lost in an accident and was replaced by XZ166 (first flown 5 March 1975). Sea trials on board the helicopter support ship RFA *Engadine,* by XX510, began on 29 June 1973; air launches of dummy Sea Skua weapons have been made. Further trials on board the *Tourville* were completed under operational conditions at sea in the Spring of 1974, and a brief trial on board HMS *Sheffield,* by XX510, towards the end of 1975.

Lynx (French Navy). Navy version, generally similar to British HAS. Mk 2 but with more advanced target detection equipment. Two development aircraft (XX904 and XX911, first flights 6 July and 18 September 1973), both of which were handed over to Aérospatiale in the Autumn of 1973 for equipment to Aéronavale standard and continuation of testing in France. Total of 18 ordered by early 1976, with a further eight on option.

Sea Lynx. Westland contender for the US Navy's LAMPS (Light Airborne Multi-Purpose System) requirement for a successor to the Kaman SH-2 Seasprite.

The first Lynx production order was placed by the Ministry of Defence in May 1973; confirmation of this contract, which covers 119 aircraft for the British and French Navies and the British Army, was announced in February 1974. In addition, 16 naval Lynx have been ordered by the Royal Netherlands Navy, of which 10 are at the increased AUW of 4,763 kg (10,500 lb). Nine naval Lynx are on order for the Brazilian Navy. On 30 July 1973 Ferranti announced receipt of a contract for 100 Seaspray radars for installation in the British naval version. Lynx production is shared in the ratio of 70% by Westland to 30% by Aérospatiale.

The first prototype Lynx has been re-engined with two Pratt & Whitney Aircraft of Canada PT6B-34 turboshaft engines. Registered G-BEAD, it flew for the first time with these engines on 19 July 1976.

The following description applies to the basic military general-purpose and naval versions with the standard Gem power plant, except where a specific version is indicated:

TYPE: Twin-engined multi-purpose helicopter.

ROTOR SYSTEM: Single four-blade semi-rigid main rotor and four-blade tail rotor. The main rotor blades, which are interchangeable, are of cambered aerofoil section and embody mass taper. Each blade consists of a two-piece, two-channel stainless steel D-shaped box-spar, to which is bonded a glassfibre-reinforced plastics rear skin stabilised by a Nomex plastics honeycomb core. Blade tips are of moulded glassfibre-reinforced plastics, with a stainless steel anti-erosion sheath forward of the 50% chord line. Each blade is attached to the main rotor hub by titanium root attachment plates and a flexible arm; the inboard portion of each arm accommodates most of the flapping movement of each blade, while the outer portion provides freedom in the lag plane. The rotor

Second Lynx HAS. Mk 2 development aircraft, undergoing deck landing trials in Royal Navy insignia

hub and inboard portions of the flexible arms are built as a complete unit, in the form of a titanium monobloc forging. A feathering hinge, comprising double needle bearings, is incorporated between the inboard and outboard flexible arms. The feathering hinge bearings are relieved of centrifugal loading by a flexible torsion bar which joints the inboard and outboard section of each arm. A two-pin jaw for blade attachment and manual blade folding is provided. Each of the tail rotor blades has a light alloy spar, machined integrally with the root attachment, which forms the nose portion of the aerofoil section and has a flush-fitting stainless steel sheath on the leading-edge. The rear section of each blade is of similar construction to that of the main rotor blades. The tail rotor hub has conventional flapping and feathering hinges, and incorporates torsionally flexible tiebars which carry the centrifugal loads inboard to the flapping hinges. Tail rotor blades are replaceable in matched pairs, and each blade is attached to the hub by the outboard tiebar pin and a six-bolt root-end flanged joint. Main rotor blades of both versions can be folded, and tail rotor pylon of naval version can be folded and spread manually, to reduce overall length for stowage.

Rotor Drive: Transmission consists of three interconnected gearboxes, transmitting power to the main and tail rotors. The engines are mounted from extensions of the gearbox casing through gimbal and flexible couplings which permit a degree of angular misalignment. The drives are taken from the front of the engines into the main gearbox, which is mounted above the cabin forward of the engines. This gearbox interconnects the two engines, with the speed reduction being carried out in two stages. The first stage uses an involute-form spiral bevel pinion and gear. The second stage comprises a conformal pinion meshing with a gear fixed directly to the main rotor drive-shaft. In flight, the accessory gears, which are all at the front of the main gearbox, are driven by one of the two through shafts from the first-stage reduction gears. For system checking on the ground without the rotor turning, the accessories can be driven by the port engine via a through shaft, a lockout freewheel unit being selected manually to isolate the main rotor transmission from the port engine input drive. Freewheel units are mounted in each engine gearbox shaft, and also within the accessory drive chain of gears. Rotor head controls are actuated by three identical tandem servojacks, trunnion-mounted from the main rotor gearbox and powered by two independent hydraulic systems. The collective jack is mounted centrally on the forward end of the main gearbox, with the cyclic jacks positioned at 45° on each side. Duplex autostabiliser actuators are integral with each jack. Cyclic and collective inputs from the three control jacks are translated to the lower bearing housing of a four-arm spider which is located within, and rotates with, the main rotor shaft. The spider is mounted universally within a splined section of the main shaft, above its bearing housing, and is linked to the blade pitch-change levers by four adjustable-length track rods. Rod and lever control runs are employed on both the cyclic and collective systems, and are carried within protective ducts below the cockpit floor, up to cabin roof level on both sides of the aircraft, and finally to the rotor head. Yaw control runs are initially by rod and lever, and then to cables which transmit pedal movements along the tailboom to the tail rotor control jack, which in turn effects blade pitch changes. Spring feel units and electric trim motors for the cyclic control channels are installed below the cockpit floor. Yaw control pedals are adjustable separately over a wide range. Control system incorporates a simple stability augmentation system, which acts in a single channel to provide improved stability in pitch. Provision is made for in-flight blade tracking. Each engine embodies an independent control system which provides full-authority rotor speed governing, pilot control being limited to selection of the desired rotor speed range. In the event of an engine failure, this system will restore power up to single-engine maximum contingency rating to maintain power turbine/rotor governed speed within the prescribed limits. A single, centrally-mounted rotor speed select lever, with a limited authority, sets the datum of the power turbine/rotor speed governing system. This system meters fuel to maintain the selected speed throughout the flight condition range. A fine-adjustment trimming control is provided to facilitate accurate matching of each engine. On the naval versions, the main rotor can provide negative thrust to increase stability on deck after touchdown. Tail rotor drive is taken from the main ring gear. A hydraulically-operated rotor brake is mounted on the main gearbox at the tail rotor drive-shaft coupling, the shaft continuing aft to the single-stage, bevel reduction type intermediate and tail rotor gearboxes. Pitch variation of the tail rotor blades is controlled by a spider, actuated by hydraulic jack via a pushrod which extends through the centre of the tail rotor gearbox.

Fuselage and Tail Unit: Conventional semi-monocoque pod and boom structure, mainly of light alloy, including a cantilever floor structure with unobstructed surface. Glassfibre components used for access panels, doors and fairings. The forward fuselage is free from bulkheads, giving an unrestricted field of view. Single large window in each of the main cabin sliding doors. Provision for internally-mounted defensive armament, and for universal flange mountings on each side of the exterior to carry weapons or other stores. Tailboom is a light alloy monocoque structure bearing the sweptback vertical fin/tail rotor pylon, which has a half-tailplane near the tip on the starboard side. Tailplane leading- and trailing-edges, and bullet fairing over tail rotor gearbox, are of glassfibre.

Landing Gear (general-purpose version): Non-retractable tubular skid type. Provision for a pair of adjustable ground handling wheels on rear of each skid. Flotation gear optional.

Landing Gear (naval versions): Non-retractable oleo-pneumatic tricycle type. Single-wheel main units, mounted on sponsons near rear of main fuselage, are fixed at 27° toe-out for deck landing, and can be manually turned into line and locked fore and aft for movement of aircraft into and out of ship's hangar. Twin-wheel nose unit can be castored hydraulically through 90° by the pilot. Designed for high shock-absorption to facilitate take-off from, and landing on, small decks under severe sea and weather conditions. Sprag brakes (wheel locks) fitted to each wheel prevent rotation on landing or inadvertent deck roll. These locks are disengaged hydraulically and will automatically re-engage in the event of hydraulic failure. Friction brakes may be fitted for shore use. Flotation gear, and hydraulically-actuated harpoon deck-lock securing system, optional.

Power Plant: Two Rolls-Royce BS.360-07-26 Gem turboshaft engines. Each has a max continuous rating of 559 kW (750 shp), a take-off and inter-contingency rating of 619 kW (830 shp), and a max contingency rating (2½ min) of 671 kW (900 shp). Engines mounted side by side on top of the fuselage upper decking, aft of the main rotor shaft and gearbox, and separated from fuselage, transmission area and each other by firewalls. Engine air intakes de-iced electrically. Fuel in five crashproof bag-type tanks, all within the fuselage structure, comprising two main tanks each of 204 kg (450 lb) capacity, two side-by-side collector tanks each of 93 kg (204·5 lb) capacity, and a 148 kg (326 lb) capacity underfloor tank at the forward end of the cabin. Total fuel capacity 733 kg (1,616 lb). Cross-feed system allows fuel to be supplied from both collector tanks to one engine or from one tank to both engines. If required, ferry range can be increased by installing in rear of cabin two metal auxiliary tanks with a combined capacity of 726 kg (1,600 lb). Single-point pressure refuelling (3·79 bars; 55 lb/sq in max) and defuelling; two points for gravity refuelling. A removable 114 litres (25 Imp gallons)/min pressure refuelling/defuelling pack can be fitted in the cabin which, with port engine running, can be used to refuel aircraft from dump stocks on ground or containers suspended from hoist. It is also possible to raise fuel about 5 m (15 ft) while the aircraft is hovering. Fuel jettison capability for main and forward tanks. Provision for self-sealing of both collector tanks (except in Royal Navy version) to provide protection against small-arms fire. Engine oil tank capacity 6·8 litres (1·5 Imp gallons). Main rotor gearbox oil capacity 18 litres (4 Imp gallons). Engine access doors manufactured by Hawker de Havilland Australia.

Accommodation: Pilot and co-pilot or observer on side-by-side seats which can accommodate back-type dinghies and are adjustable fore and aft and for height. Inertia-reel shoulder harness for pilot and co-pilot. Dual controls optional. Additional crew members (eg, gunner, hoist operator) according to role. Individual forward-hinged cockpit door and large rearward-sliding cabin door on each side; all four doors jettisonable. Main cabin doors manufactured by Hawker de Havilland Australia. Cockpit accessible from cabin area. Maximum high-density layout (general-purpose version) for one pilot and 10 armed troops or paratroops, on lightweight bench seats in soundproof cabin. Alternative VIP layouts for four to seven passengers, with additional cabin soundproofing. Seats can be removed quickly to permit the carriage of up to 907 kg (2,000 lb) of freight internally. Tie-down rings are provided at approx 508 mm (20 in) intervals on main cabin floor, which is stressed for loads of up to 976 kg/m² (200 lb/sq ft). Alternatively, loads of up to 1,360 kg (3,000 lb) can be carried externally on freight hook mounted below the cabin floor and fitted, in naval version, with electrically-operated emergency release system. In the casualty evacuation role, with a crew of two, the Lynx can accommodate three standard stretchers and a medical attendant; electrically-heated casualty bags can be provided. Both versions have secondary capability for search and rescue (up to nine survivors) and other roles (see introductory copy and 'Equipment' paragraphs).

Systems: Two independent hydraulic systems in all versions, pressure 141 bars (2,050 lb/sq in). Pumps powered by accessory drive from main rotor gearbox, enabling full power to be drawn from both main systems in event of engine failure. If either No. 1 or No. 2 main systems fails, the other maintains adequate flying control. No. 1 system, additionally, actuates tail rotor yaw control and rotor brake. Tail rotor operation reverts to mechanical control if No. 1 system fails. A third hydraulic system, at the same pressure, is provided in the naval version when sonar equipment, MAD or a hydraulic winch system are installed. When this third hydraulic system, at the same pressure, is provided in the operated by this system. No pneumatic system. 28V DC electrical power supplied by two 6kW engine-driven starter/generators and an alternator. Engines can also be started from external 28V DC power source. 24V 23Ah (optionally 40Ah) nickel-cadmium battery fitted for essential services and emergency engine starting. 200V three-phase AC power available at 400Hz from two 15kVA transmission-driven alternators. AC and DC external ground power sockets on starboard side of fuselage. Graviner Triple FD engine fire detection system; two separate fire suppression systems fitted, but interconnected to permit contents of both bottles to be directed to one engine if neccessary. All versions fitted with centralised standard warning system which provides visual and audio warnings of major emergencies, visual warnings for secondary failure, and visual indications of an advisory nature. Optional cabin heating and ventilation system, using mixing unit combining engine bleed air with outside air. Optional supplementary cockpit heating system. Electrical anti-icing and demisting of windscreen, and electrically-operated windscreen wipers, standard; windscreen washing system optional.

Electronics and Flight Equipment: Main equipment bays are in nose (under upward-hinged door) and at rear of cabin. All versions equipped as standard with navigation, cabin and cockpit lights; adjustable landing light under nose; anti-collision beacon; first aid kit(s); and hand-type fire extinguishers for cabin. Optional equipment common to all roles (general-purpose and RN versions) includes simplex four-axis cross-country autopilot system; Plessey PTR 377 UHF/VHF transceiver with homing; Ultra D 403M standby UHF; S.G. Brown three-position crew intercom. Optional role equipment or installations for both versions include Marconi-Elliott automatic flight control system (AFCS); AN/ARC-44 VHF(FM); Collins 718 UA HF; VOR/ILS; DME; AN/ARN-52 Tacan (general-purpose version only); X-band transponder (naval version only); Sperry C2J or GM9B Gyrosyn compass system; Sperry E2C standby compass; Sperry HAI-5 heading and attitude indicator (Royal Naval version); Louis Newmark 8462 vertical gyro; Plessey PTR 446 IFF transponder; AD 360 radio compass (general-purpose version only); Honeywell AN/APN-198 radar

Westland/Aérospatiale Lynx, first prototype, re-engined with two Pratt & Whitney Aircraft of Canada PT6B-34 turboshaft engines

altimeter; Decca Tactical Air Navigation System (TANS) with Decca Type 71 Doppler radar; Decca Mk 19 flight log; and vortex-type sand filter for engine air intakes. Additional AFCS units in general-purpose version permit automatic turns and radio height hold; in naval version, when sonar is fitted, these units are extended to provide automatic transition to the hover and automatic Doppler hold in the hover. Other optional equipment (both versions) includes signal pistol and cartridges, Aldis lamp and stowage.

Armament and Operational Equipment: For armed escort, anti-tank or air-to-surface strike missions, general-purpose version can be equipped with one 20 mm AME 621 or similar cannon, with 1,500 rds, or a pintle-mounted 7·62 mm GEC Minigun inside cabin; or a Minigun beneath cabin, in Emerson Minitat installation, with 3,000 rds. External pylon can be fitted on each side of cabin for a variety of stores, including two Minigun or other self-contained gun pods; two pods of fourteen and two of seven 2 in rockets; pods of 68 mm SNEB rockets; or up to six BAC Hawkswing or Aérospatiale AS.11, or eight Aérospatiale/MBB Hot or Hughes TOW, or similar air-to-surface missiles. An additional six or eight missiles can be carried in cabin, for rearming in forward areas, and Avimo-Ferranti 530 lightweight stabilised sight is fitted for target detection and missile direction. The Lynx can also transport mobile anti-tank teams of three gunners with missiles and launchers. For search and rescue role, with a crew of three, both versions can be fitted with a waterproof floor, four 5 in flares (three in naval version), and a 272 kg (600 lb) capacity electrically-operated 'clip-on' hoist in starboard side of cabin. Alternative option of hydraulically-operated hoist in naval version when third hydraulic system is installed. Hoist, which can lift a load through 76 m (250 ft) at 30·5 m (100 ft)/min, can be swung back into cabin when not in use, permitting sliding door to be closed. General-purpose version can also be equipped for several other duties, including firefighting and crash rescue, reconnaissance, military command post, liaison, customs and border control, and pilot and operational training. Optional equipment, according to role, can include lightweight sighting system with alternative target magnification, vertical and/or oblique cameras, up to six 5 in flares for night operation, low light level TV, infra-red linescan, searchlight, and specialised communications equipment. Naval version can carry out a number of these roles, but has specialised equipment for its primary duties. For ASW role, this includes two Mk 44 or Mk 46 homing torpedoes, one each on an external pylon on each side of fuselage, and six marine markers; or two Mk 11 depth charges. Detection of submarine can either be carried out by parent ship (in which case the Lynx carries retractable classification and localisation equipment), or the Lynx can itself be equipped for this function, with Alcatel D.U.A.V.4 lightweight dunking sonar, and hydraulically-powered winch and cable hover mode facilities within the AFCS. Ferranti Seaspray lightweight search and tracking radar, for detecting small surface targets in low visibility/high sea conditions. Armament includes BAC CL834 Sea Skua semi-active homing missiles for attacking light surface craft; alternatively, four AS.12 or similar wire-guided missiles can be employed in conjunction with AF 530 or APX-334 lightweight stabilised optical sighting system.

Dimensions, external (A: general-purpose version; N: naval version):
- Diameter of main rotor (A, N) 12·802 m (42 ft 0 in)
- Diameter of tail rotor (A, N) 2·21 m (7 ft 3 in)
- Main rotor blade chord (A, N, constant, each) 0·39 m (1 ft 3·4 in)
- Tail rotor blade chord (A, N, constant, each) 180 mm (7·1 in)
- Length overall, both rotors turning (A, N) 15·163 m (49 ft 9 in)
- Length overall:
  - A, main rotor blades folded 13·165 m (43 ft 2·3 in)
  - N, main rotor blades and tail folded 10·618 m (34 ft 10 in)
- Length of fuselage, nose to tail rotor centre:
  - A 12·06 m (39 ft 6·8 in)
  - N 11·92 m (39 ft 1·3 in)
- Width overall, main rotor blades folded:
  - A, N 2·94 m (9 ft 7·75 in)
- Height overall, both rotors turning:
  - A 3·66 m (12 ft 0 in)
  - N 3·60 m (11 ft 9¾ in)
- Height overall, both rotors stopped:
  - A 3·504 m (11 ft 6 in)
  - N 3·365 m (11 ft 0·5 in)
- Height to top of rotor hub:
  - A 2·964 m (9 ft 8·7 in)
- Height overall, main rotor blades and tail folded:
  - N 3·20 m (10 ft 6 in)
- Tail rotor ground clearance:
  - A 1·41 m (4 ft 7·5 in)
  - N 1·38 m (4 ft 6·3 in)
- Tailplane half-span (from fuselage c/l):
  - A, N 1·776 m (5 ft 9·9 in)
- Skid track (A) 2·032 m (6 ft 8 in)
- Wheel track (N) 2·778 m (9 ft 1·4 in)
- Wheelbase (N) 3·014 m (9 ft 10·7 in)
- Cabin door openings (A, N, each):
  - Mean width 1·372 m (4 ft 6 in)
  - Height 1·194 m (3 ft 11 in)

Dimensions, internal:
- Cabin, from back of pilots' seats:
  - Min length 2·057 m (6 ft 9 in)
  - Max width 1·778 m (5 ft 10 in)
  - Width at rear 1·409 m (4 ft 7·5 in)
  - Max internal floor width 1·715 m (5 ft 7·5 in)
  - Max height 1·422 m (4 ft 8 in)
  - Floor area 3·72 m² (40·04 sq ft)
  - Volume 5·21 m³ (184 cu ft)

Weights (A: general-purpose version, N: naval version):
- Manufacturer's bare weight:
  - A 2,519 kg (5,553 lb)
  - N 2,661 kg (5,866 lb)
- Manufacturer's basic weight:
  - A 2,599 kg (5,730 lb)
  - N 2,762 kg (6,089 lb)
- Operating weight empty, equipped:
  - A, troop transport (pilot and 10 troops) 2,727 kg (6,012 lb)
  - A, anti-tank strike (incl weapon pylons, firing equipment and sight) 3,012 kg (6,640 lb)

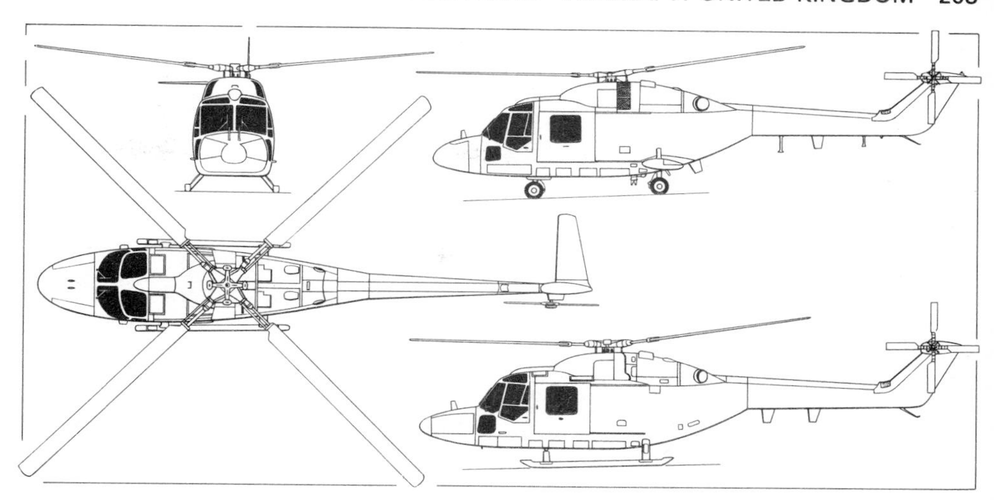

Westland Lynx AH. Mk 1 light general-purpose helicopter, with additional side view (top) of Lynx HAS. Mk 2
(Pilot Press)

First production Lynx HAS. Mk 2 for the Royal Navy

The 150th Westland-built Gazelle, an AH. Mk 1 delivered in June 1976

A, search and rescue (crew of three)
3,904 kg (6,402 lb)
N, anti-submarine strike 3,083 kg (6,797 lb)
N, reconnaissance (crew of two)
3,064 kg (6,755 lb)
N, anti-submarine classification and strike
3,174 kg (6,997 lb)
N, air to surface vessel search and strike (crew of two and two Sea Skuas) 3,143 kg (6,929 lb)
N, search and rescue (crew of three)
3,191 kg (7,035 lb)
N, dunking sonar search and strike
3,391 kg (7,476 lb)
Normal max T-O weight:
A, N 4,309 kg (9,500 lb)
Max T-O weight 4,763 kg (10,500 lb)

PERFORMANCE: (at normal max T-O weight at S/L, ISA, except where indicated. A: general-purpose version; N: naval version):
Never-exceed speed at 3,628 kg (8,000 lb) AUW:
A, N 180 knots (333 km/h; 207 mph)
Max continuous cruising speed:
A, N 148 knots (273 km/h; 170 mph)
A, N (ISA + 20°C)
138 knots (256 km/h; 159 mph)
Max continuous cruising speed (1 hr), one engine out:
A 129 knots (240 km/h; 149 mph)
N 115 knots (213 km/h; 132 mph)
A (ISA + 20°C) 109 knots (203 km/h; 126 mph)
N (ISA + 20°C) 95 knots (175 km/h; 109 mph)
Speed for max endurance:
A, N (ISA and ISA + 20°C)
70 knots (130 km/h; 81 mph)
Min flying speed (max contingency rating, one engine out):
A, N 24 knots (44·5 km/h; 28 mph)
A, N (ISA + 20°C) 34 knots (63 km/h; 39·5 mph)
Max forward rate of climb:
A 670 m (2,198 ft)/min
N 665 m (2,182 ft)/min
A, N (ISA + 20°C) 550 m (1,805 ft)/min
Max forward rate of climb (1 hr power), one engine out:
A, N 210 m (689 ft)/min
A, N (ISA + 20°C) 88 m (289 ft)/min
Max vertical rate of climb:
A, N 340 m (1,115 ft)/min
A, N (ISA + 20°C) 310 m (1,017 ft)/min
Hovering ceiling out of ground effect:
A, N 2,900 m (9,515 ft)
Typical range, with reserves:
A, troop transport 300 nm (556 km; 345 miles)
A, search and rescue 150 nm (278 km; 173 miles)
Radius of action, out and back at max sustained speed, max hover weight 4,309 kg (9,500 lb), allowances for T-O and landing, 15 min loiter in search area, 2 min hover for each survivor, and reserves for 20 min loiter at end of mission:
N, search and rescue (crew of 3 and 2 survivors)
130 nm (241 km; 150 miles)
N, search and rescue (crew of 3 and 8 survivors)
115 nm (213 km; 132 miles)
Time on station at 50 nm (93 km; 58 miles) radius, out and back at max sustained speed, with 2 torpedoes and 6 marine markers, allowances for T-O and landing and reserves for 20 min loiter at end of mission:
N, anti-submarine classification and strike, loiter speed on station 1 hr 43 min
N, anti-submarine strike, loiter on station
1 hr 59 min
N, dunking sonar search and strike, 50% loiter speed and 50% hover on station 50 min
Time on station at 50 nm (93 km; 58 miles) radius, out and back at max sustained speed, with crew of 2 and 2 Sea Skuas, allowances and reserves as above:
N, air to surface vessel strike, en-route radar search and loiter speed on station 2 hr 0 min
Max range:
A 365 nm (676 km; 420 miles)
N 340 nm (629 km; 391 miles)
A (ISA + 20°C) 368 nm (681 km; 423 miles)
N (ISA + 20°C) 340 nm (629 km; 391 miles)
Max endurance:
A, N (ISA + 20°C) 3 hr 26 min
Max ferry range with auxiliary cabin tanks:
A 720 nm (1,334 km; 829 miles)
N 606 nm (1,123 km; 698 miles)

WESTLAND 606

Westland 606 is the designation of the planned civil versions of the Lynx military helicopter, first details of which were announced in late August 1974. A full-scale mockup has been built, and details of the projected variants of the aircraft were given in the 1975-76 *Jane's*. Owing to heavy commitments in its military helicopter programmes, Westland has decided to defer further work on the 606.

WESTLAND/AÉROSPATIALE SA 330 PUMA

Production of the Puma for the British and French armed forces, which began in 1968, is shared between Westland and Aérospatiale. Following the completion of 40 Puma HC.Mk 1s for the RAF in 1972 (see 1973-74 *Jane's*), Westland is building Puma component sets for the French production line.

The Puma is described fully under the Aérospatiale heading in the French section.

WESTLAND/AÉROSPATIALE SA 341 GAZELLE

The Gazelle, described fully under the Aérospatiale heading in the French section, is in joint production in Britain and France under the same Anglo-French agreement as the Puma. The Gazelle has been ordered by the British Army (138 **AH.Mk 1**), Royal Navy (30 **HT.Mk 2**) and Royal Air Force (14 **HT.Mk 3**). Total worldwide sales of the Gazelle had reached more than 700 by the Spring of 1976, including 42 for Egypt.

The first production Gazelle was flown on 6 August 1971, and the first HT.Mk 2 (XW845) on 6 July 1972. The first civil Gazelle (G-BAGJ) was delivered to Point-to-Point Helicopters Ltd, on 27 March 1973, and the first HT. Mk 3 to the RAF Central Flying School on 16 July 1973. The AH. Mk 1 entered service on 6 July 1974, with No. 660 Squadron of the Army Air Corps at Soest, Germany; and the HT. Mk 2 with No. 705 Squadron at RNAS Culdrose on 10 December 1974. Deliveries to the British services and to civil customers totalled 123 by 1 February 1976.

Westland/Aérospatiale Gazelle HT. Mk 3 of Royal Air Force Training Command

WHE

W. H. EKIN (ENGINEERING) CO LTD

ADDRESS:
59 Mill Road, Crumlin, Co Antrim BT29 4XL, Northern Ireland
Telephone: Crumlin 52222
DIRECTORS:
William H. Ekin, BSc(Hons), CEng, MIMechE, DMS
M. J. H. Ekin

This company was formed in March 1969 to undertake the production of six McCandless Mk IV Gyroplanes (see 1972-73 *Jane's*), and the first of these made its first flight in February 1972. For various reasons, an extensive redesign of the McCandless Mk IV was embarked upon by Ekin in Autumn 1971 and a new prototype (G-AXXN) flew for the first time on 1 February 1973. The modified aircraft is now known as the WHE Airbuggy. The first production Airbuggy (G-AXYX) was delivered to an English customer in October 1975, and a second aircraft was ordered in January 1976 for delivery during the following Summer. It was hoped to complete two additional aircraft in 1976.

A change is being made from Bensen-type rotor blades, as fitted to the prototype, to Rotordyne bonded metal blades which offer a substantial improvement in performance. Design and manufacture of a radio-telemetry unit for attachment to the rotor hub was under way in 1975. This was intended to transmit strain gauge data to measure rotor blade vibration and stress levels as the aircraft's AUW was increased progressively.

WHE AIRBUGGY

First production example of the WHE Airbuggy single-seat autogyro

TYPE: Single-seat light autogyro.

ROTOR SYSTEM AND DRIVE: Two-blade semi-rigid teetering rotor with an offset gimbal head, through the centre of which runs the rotor spin-up drive. Blades are secured to hub by bolts, and are of Rotordyne bonded metal construction. Rotor spin-up effected via Vee-belt drive to 9·667 : 1 worm reduction gearbox and universal and sliding joints. No rotor brake or blade folding.

FUSELAGE: Space-frame of T35 and T45 steel tube, assembled by sifbronze welding.

TAIL UNIT: Fin, rudder and tailplane formed from plywood sandwich. Ground-adjustable trim tab on rudder. Endplate auxiliary fins on tailplane of prototype.

LANDING GEAR: Non-retractable tricycle type. All three wheels have rubber bungee for shock-absorption. Nosewheel steerable. Main wheels with tyres sized from 12 × 2·5 to 13·5 × 5 and nosewheel from 12 × 2·5 to 12 × 3·8 according to surface from which aircraft operates. Internal expanding drum brakes.

POWER PLANT: One 56 kW (75 hp) 1,600-1,800 cc modified Volkswagen engine, driving a Hoffmann two-blade pusher propeller via a 1·5 : 1 reduction drive.

Drive transmitted by 10 Vee belts. Fuel capacity 29·5 litres (6·5 Imp gallons), in gravity tank mounted above engine, with refuelling point on top of tank. Oil capacity 2·5 litres (4·5 Imp pints).

ACCOMMODATION: Single seat in open cockpit behind large windscreen. Door on starboard side.

SYSTEM: A 12V 25Ah ground rechargeable battery is installed to power an electric engine starter motor and to supply instruments and radio.

ELECTRONICS: Prototype has Parkair Nipper 24-channel com transceiver.

DIMENSIONS, EXTERNAL:
- Diameter of rotor 6·63 m (21 ft 9 in)
- Rotor blade chord 0·18 m (7 in)
- Propeller diameter 1·45 m (4 ft 9 in)
- Length overall 3·51 m (11 ft 6 in)
- Height overall 2·21 m (7 ft 3 in)
- Wheel track 1·63 m-1·68 m (5 ft 4 in-5 ft 6 in)
- Wheelbase 1·32 m (4 ft 4 in)

AREAS:
- Rotor blades (each) 0·52 m² (5·6 sq ft)
- Rotor disc 37·63 m² (405 sq ft)
- Fin plus endplates (prototype) 0·73 m² (7·9 sq ft)
- Rudder 0·49 m² (5·3 sq ft)

WEIGHTS AND LOADINGS:
- Weight empty 161 kg (355 lb)
- Max T-O and landing weight 295 kg (650 lb)
- Max disc loading 8·5 kg/m² (1·75 lb/sq ft)
- Max power loading 5·27 kg/kW (8·7 lb/hp)

PERFORMANCE (at max T-O weight):
- Never-exceed speed 69 knots (128 km/h; 80 mph)
- Max level speed at S/L 69 knots (128 km/h; 80 mph)
- Max cruising speed 61 knots (113 km/h; 70 mph)
- Econ cruising speed 52 knots (97 km/h; 60 mph)
- Max rate of climb at S/L 305 m (1,000 ft)/min
- T-O run 46-92 m (150-300 ft)
- Landing run, still air 9 m (30 ft)
- Landing run in 13 knot (24 km/h; 15 mph) wind Nil
- Range with max fuel, no allowances 121 nm (225 km; 140 miles)
- Range with max payload, no allowances 86 nm (161 km; 100 miles)

UNITED STATES OF AMERICA

AEREON

AEREON CORPORATION

ADDRESS:
1 Palmer Square, Princeton, New Jersey 08540
Telephone: (609) 921-2131

PRESIDENT AND TREASURER:
William McE. Miller Jr

AEREON 26

Aereon Corporation has built and flown a proof-of-concept research vehicle which is described as having the hull geometry of a lifting-body airship. Designated Aereon 26, its design began in August 1967; construction of the prototype started in the following month. The first flight in ground effect was made on 6 September 1970, and the first airfield circuits were recorded on 6 March 1971. Only limited details of its construction and performance have been made available:

TYPE: Lifting-body research vehicle.

WING/BODY: Basic structure of welded aluminium tube, fabric-covered except for the elevons and fixed centre-section of the aft body trailing-edge. Each elevon, of light alloy construction, has a trim tab extending for the full span of its trailing-edge. Thickness/chord ratio 1 : 4·5.

TAIL UNIT: Twin fins, rudders and anhedral surfaces. Construction similar to that of the wing/body, but with light alloy skins.

LANDING GEAR: Non-retractable tricycle type. Main units have side Vees and half-axles. Rubber bungee shock-absorption.

Aereon 26 lifting-body airship research vehicle

POWER PLANT: One 67 kW (90 hp) McCulloch flat-four two-stroke engine, mounted centrally on a pylon structure adjacent to the trailing-edge of the wing/body. Sensenich hand-carved four-blade wooden fixed-pitch pusher propeller. Fuel tank mounted within the wing/body.

ACCOMMODATION: Single seat beneath jettisonable bubble canopy. Access via clamshell door, hinged at top, on starboard side immediately beneath canopy seal.

DIMENSIONS, EXTERNAL:
- Span, over anhedral surfaces 6·80 m (22 ft 3¾ in)
- Aspect ratio 1·23
- Length overall, excl data boom 8·38 m (27 ft 6 in)
- Height overall 2·28 m (7 ft 3¾ in)

WEIGHT:
- Basic operating weight 544 kg (1,200 lb)

PERFORMANCE:
- Never-exceed speed 110 knots (204 km/h; 127 mph)
- T-O run 244 m (800 ft)

AEROSPACE GENERAL

AEROSPACE GENERAL COMPANY

ADDRESS:
Route 1, Box 208, Odessa, Texas 79763
Telephone: (915) 332-8233

PRESIDENT:
Gilbert Magill

This company was formed by Mr Gilbert Magill who, as former President of Rotor-Craft Corporation, was responsible for the RH-1 Pinwheel ultra-light one-man helicopter described and illustrated in the 1960-61 *Jane's*. The same basic concept, including the use of rotor-tip rockets for propulsion, is embodied in a new powered pilot rescue vehicle known as the Mini-Copter. Three prototypes were ordered by the US Navy for evaluation, and development, under contract to the Naval Air Development Center at Warminster, Pennsylvania, was continuing in early 1976.

Aerospace General also has under development the prototype of a civil version of the Mini-Copter; it is planned to obtain FAA certification of this aircraft and put it into production.

AEROSPACE GENERAL MINI-COPTER

Military designation: MDV-1 (Mini-Copter Demonstration Vehicle-1)

Unlike the Bell and Kaman pilot self-rescue systems described in earlier issues of *Jane's*, the Mini-Copter is intended for air-dropping to a pilot who has been forced down behind enemy lines or in terrain unsuited to conventional rescue craft. It is a miniature helicopter which is being evaluated in three configurations:

Configuration 1. The most basic version, with fuel tanks and control/rotor unit strapped to the pilot. No landing gear.

Configuration 2. The simplest true aircraft form, consisting of a welded steel tube structure carrying the control/rotor unit and fuel tanks, and providing a seat for the pilot. Skid landing gear.

Configuration 3. Generally similar to the second configuration, but with the addition of a 67 kW (90 hp) McCulloch flat-four engine driving a pusher propeller.

Aerospace General Mini-Copter in its most basic Configuration 1

Aerospace General Mini-Copter in Configuration 2, with the addition of simple skid landing gear

Aerospace General Mini-Copter in Configuration 3, with addition of a McCulloch flat-four engine

This model takes off as a helicopter, using its blade-tip power units; at a forward speed of 26 knots (48 km/h; 30 mph) the tip units are turned off, and the Mini-Copter then flies as an autogyro.

In the first configuration, the aircraft is easily folded and stowed into a US Navy CTU/1A aerial delivery container for air-dropping by parachute. Only 1·52 m (5 ft 0 in) in length and 0·53 m (1 ft 9 in) in diameter, the container accommodates the Mini-Copter without difficulty. The two blades of the 5·49 m (18 ft 0 in) diameter rotor each fold at the hub and at half their length.

Basic power plant of all three configurations comprises two small rocket motors which are faired into the blade tips of the main rotor. These contain a silver-plated screen catalyst bed through which liquid 90% hydrogen peroxide is passed. This causes a chemical reaction which decomposes the peroxide into water and free oxygen at a temperature of 1,340°F (727°C), releasing superheated steam through a small nozzle to provide the necessary reaction. Each rocket motor develops about 0·187 kN (42 lb st); their combined weight is little more than 0·45 kg (1 lb).

Design of this aircraft originated in 1972, and the first flight of a prototype was recorded on 31 March 1973.

Type: Basic one-man helicopter.

Rotor System: Two-blade main rotor. Blade section 8H-12. Blades attached to rotor hub by simple flapping hinges. Each main blade is a light alloy extrusion. Single-blade counterbalanced tail rotor of rubber is used for directional control, the tip-driven main rotor being without torque. Main rotor rpm 750; tail rotor rpm 1,500. Rotor brake.

Fuselage: Basic structure, common to all configurations, is a welded tubular A-frame of 4130 steel. A change to light alloy is proposed for production versions, to decrease structure weight and increase payload.

Tail Unit: Fixed Vee tail surface carried on foldable steel tube extension from top of fuselage A-frame, which also carries tail rotor on starboard side.

Landing Gear: None in Configuration 1. Welded steel tube structure with steel tube skids in Configurations 2 and 3. Oleo-pneumatic shock-absorber in each leg. Solid wheel, 0·13 m (5 in) diameter, at aft end of each skid for ground handling only.

Power Plant: Two 0·187 kN (42 lb st) Aerospace General rocket motors, one in tip of each main rotor blade. Configuration 3 has also one 67 kW (90 hp) McCulloch flat-four engine, mounted in landing gear frame beneath pilot's seat and driving a two-blade wooden fixed-pitch pusher propeller. Fuel for rocket motors, in Configurations 1 and 2, carried in two 26·5 litre (7 US gallon) tanks, mounted on each side of basic structure. In Configuration 3 these tanks are replaced by two 38 litre (10 US gallon) tanks, carrying gasoline fuel for the McCulloch engine. In Configuration 3, fuel for the rocket motors is contained in a 15·1 litre (4 US gallon) tank mounted at the top of the A-frame, this being sufficient for six take-offs.

Accommodation: Seat for pilot in Configurations 2 and 3. Suspended control column, operated by one hand, is moved in the direction in which the pilot wishes to fly.

Electronics and Equipment: Communication by Radio Shack walkie-talkie. Cargo sling optional.

Dimensions, external:
- Diameter of main rotor: 5·49 m (18 ft 0 in)
- Main rotor blade chord: 0·15 m (6 in)
- Diameter of tail rotor: 0·76 m (2 ft 6 in)
- Propeller diameter (Conf. 3): 1·32 m (4 ft 4 in)
- Length overall: 2·44 m (8 ft 0 in)
- Width, rotors folded: 1·22 m (4 ft 0 in)
- Height to top of rotor hub: 2·13 m (7 ft 0 in)

Areas:
- Main rotor blades (each): 0·37 m² (4 sq ft)
- Main rotor disc: 23·60 m² (254 sq ft)
- Tail rotor disc: 0·46 m² (5 sq ft)

Weights (A: Configuration 1; B: Configuration 2; C: Configuration 3):
- Weight empty:
  - A: 59 kg (130 lb)
  - B: 75 kg (165 lb)
  - C: 125 kg (275 lb)
- Max T-O weight:
  - A, B: 249 kg (550 lb)
  - C: 295 kg (650 lb)

Performance (at max T-O weight. A: Configuration 1; B: Configuration 2; C: Configuration 3):
- Max level speed:
  - A: 95·5 knots (177 km/h; 110 mph)
  - B: 91 knots (169 km/h; 105 mph)
  - C with rocket motors plus petrol engine: 122 knots (225 km/h; 140 mph)
  - C with petrol engine only: 82·5 knots (153 km/h; 95 mph)
- Cruising speed:
  - A: 78 knots (145 km/h; 90 mph)
  - B: 74 knots (137 km/h; 85 mph)
  - C with petrol engine only: 74 knots (137 km/h; 85 mph)
- Max rate of climb at S/L:
  - A, B: more than 610 m (2,000 ft)/min
  - C with rocket motors plus petrol engine: more than 762 m (2,500 ft)/min
  - C with petrol engine only: 274 m (900 ft)/min
- Hovering ceiling out of ground effect:
  - A, B: more than 6,100 m (20,000 ft)
- Service ceiling:
  - C with rocket motors plus petrol engine: more than 5,485 m (18,000 ft)
  - C with petrol engine only: 3,960 m (13,000 ft)
- Range, with max fuel, 10% reserves:
  - A, B: 17·5 nm (32 km; 20 miles)
  - C: 217 nm (402 km; 250 miles)

AHRENS

AHRENS AIRCRAFT CORPORATION

Address:
2800 Teal Club Road, Oxnard, California 93030
Telephone: (805) 985-2000
Telex: 65-9240

Vice-President: Kim K. Ahrens (Engineering)

Ahrens Aircraft Corporation initiated the design of a four-engined passenger/cargo transport in January 1975; construction of a prototype was begun in August of that year. The first flight of this aircraft was scheduled for the Summer of 1976. It is intended to gain certification under FAR Part 25.

AHRENS AR 404

Type: Passenger/cargo transport.

Wings: Cantilever high-wing monoplane. Wing section NACA 64_3-618. Dihedral 0°. Incidence 0°. Fail-safe structure of conventional light alloy construction, with light alloy skins. Single-slotted trailing-edge flaps of light alloy construction. Plain ailerons of light alloy construction. Trim tab in each aileron.

Fuselage: Semi-monocoque fail-safe structure of light alloy.

Tail Unit: Cantilever structure of light alloy. Servo trim tab in elevator. Trim tab in rudder.

Landing Gear: Tricycle type, with non-retractable main units and electrically-retractable nosewheel unit. Nosewheel retracts forward. Wide track main landing gear has two wheels in tandem on each side, each carried on an individual Vee-shape arm, with rubber-in-compression shock-absorption. Twin nosewheels carried on oleo-pneumatic shock-absorber strut. Cleveland wheels and disc brakes. Main-wheel tyres size 7·00-8, nosewheel tyres size 6·00-6.

Power Plant: Four 313 kW (420 shp) Allison 250-B17B turboprop engines, each driving a Hartzell three-blade metal constant-speed and fully-reversible propeller. Four wing fuel tanks with combined total capacity of 2,082 litres (550 US gallons). Propeller blades de-iced electrically.

Accommodation: Crew of two side by side on flight deck. Accommodation for up to 26 passengers or cargo. Door to flight deck on starboard side. Passenger door aft of wing on starboard side. Split cargo door in aft fuselage, lower half serving also as loading ramp. Accommodation air-conditioned and heated.

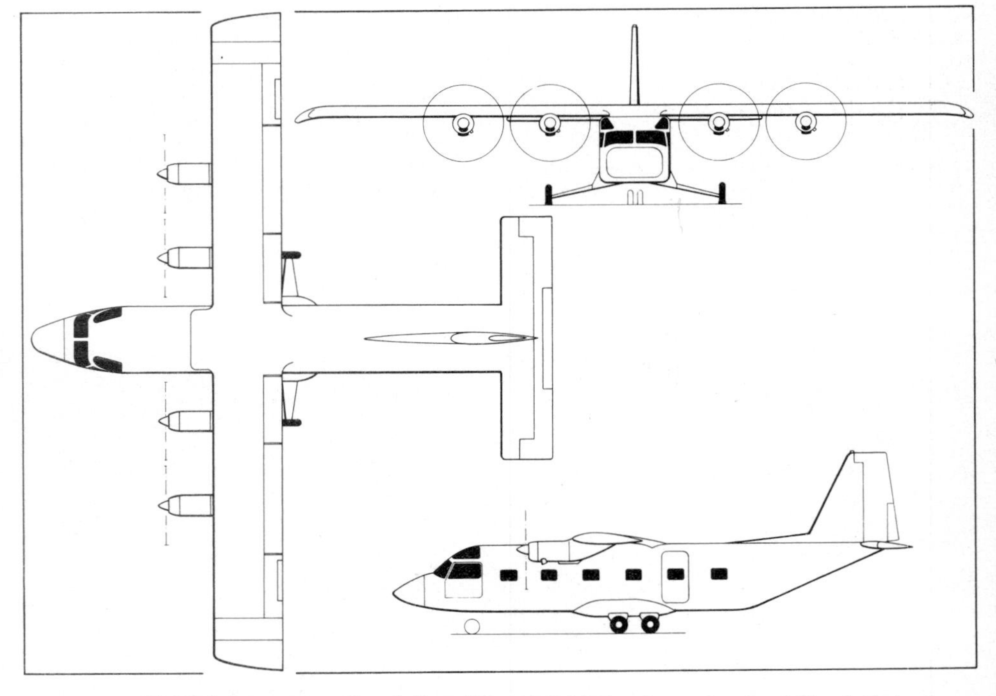

Ahrens AR 404 light transport aircraft (four Allison 250-B17B turboprop engines) *(Roy J. Grainge)*

Systems: Hydraulic system for brakes only. Talley air-conditioning system. Janitrol heating system. 28V DC electrical system.

Electronics and Equipment: Collins nav-com transceivers and blind-flying instrumentation standard.

Dimensions, external:
- Wing span: 19·81 m (65 ft 0 in)
- Wing chord, constant: 1·98 m (6 ft 6 in)
- Wing aspect ratio: 10·0
- Length overall: 14·73 m (48 ft 4 in)
- Height overall: 5·33 m (17 ft 6 in)
- Tailplane span: 7·32 m (24 ft 0 in)
- Wheel track: 5·18 m (17 ft 0 in)
- Propeller diameter: 2·29 m (7 ft 6 in)
- Propeller ground clearance: 1·45 m (4 ft 9 in)

Cabin doors (each):
Height 1·52 m (5 ft 0 in)
Width 0·76 m (2 ft 6 in)
Rear cargo doors (upper and lower):
Width 1·63 m (5 ft 4 in)
DIMENSIONS, INTERNAL:
Cabin:
Length 7·32 m (24 ft 0 in)
Max width 1·85 m (6 ft 1 in)
Max height 1·85 m (6 ft 1 in)
Floor area 13·6 m² (146 sq ft)
Volume 24·64 m³ (870 cu ft)

AREAS:
Wings, gross 39·2 m² (422 sq ft)
Ailerons (total, incl tabs) 3·99 m² (43 sq ft)
Trailing-edge flaps (total) 5·95 m² (64 sq ft)
Vertical tail surfaces (incl tab) 4·83 m² (52 sq ft)
Horizontal tail surfaces (incl tab) 11·15 m² (120 sq ft)
WEIGHTS AND LOADINGS (estimated):
Weight empty 2,812 kg (6,200 lb)
Weight empty, equipped 3,084 kg (6,800 lb)
Max T-O and landing weight 7,348 kg (16,200 lb)
Max wing loading 185·5 kg/m² (38 lb/sq ft)
Max power loading 5·87 kg/kW (9·7 lb/shp)

PERFORMANCE (estimated, at max T-O weight):
Max level speed at 1,525 m (5,000 ft)
182 knots (338 km/h; 210 mph)
Max cruising speed at 1,525 m (5,000 ft)
152 knots (282 km/h; 175 mph)
Econ cruising speed at 1,525 m (5,000 ft)
139 knots (257 km/h; 160 mph)
Stalling speed, flaps down
83 knots (153 km/h; 95 mph)
Max rate of climb at S/L 366 m (1,200 ft)/min
Service ceiling 8,230 m (27,000 ft)
Range with max fuel
1,200 nm (2,220 km; 1,380 miles)

AJI
AMERICAN JET INDUSTRIES INC

HEAD OFFICE AND WORKS:
7701 Woodley Avenue, Van Nuys, California 91406
Telephone: (213) 988-9900
Telex: 662461
PRESIDENT: Allen E. Paulson
VICE-PRESIDENTS:
Robert H. Cooper (Marketing)
Robert W. Lillibridge (Engineering and Product Development)
Roy E. Marquardt (Asst to the President)
DIRECTOR OF AIRCRAFT SALES:
William H. Boone

In 1974 American Jet Industries (AJI) completed an expansion programme. All of its production facilities are now located on a 38-acre site at Van Nuys Airport, California, occupying more than 46,450 m² (500,000 sq ft) of offices and hangars used formerly by Lockheed Aircraft Corporation for the Cheyenne helicopter programme.

Founded in 1951, AJI specialises in the modification and repair of all types of executive and airline transport aircraft. The company is also converting Convair and Lockheed turboprop transport aircraft to all-cargo or passenger/cargo configuration; installing cargo doors and floors in Fairchild Hiller FH-227 aircraft; and converting Cessna Model 401, 402 and 414 aircraft to have turboprop power plants.

In order to speed the transport of damaged aircraft to the works at Van Nuys Airport, AJI announced on 6 March 1974 that it had purchased from Aero Spacelines Inc the latter company's Pregnant Guppy and Mini-Guppy aircraft. The Mini-Guppy has been used extensively on a lease basis for the transport of outsize cargo, but the Pregnant Guppy is no longer in service.

In 1974 AJI modified six Fairchild FH-227s to cargo/passenger configuration, including major overhaul of each aircraft. Five FH-227s are being completely overhauled and refurbished for Pan Adria of Yugoslavia; and four Convair 640 aircraft are being converted to all-cargo configuration for Zantop Air Transport Inc.

A new palletised cargo system designed specifically for the Lockheed Electra includes a 2·06 m (6 ft 9 in) by 3·61 m (11 ft 10 in) hydraulically-operated outward-opening forward cargo door, emergency access, pressurised cargo interior, smoke detection system, floor strengthened to allow a loading of 1,465 kg/m² (300 lb/sq ft) and provision of a pallet loading system with 3,175 kg (7,000 lb) cargo capacity per pallet. In this new configuration, the Lockheed Electra is able to carry a 15,875 kg (35,000 lb) payload over a range of 1,700 nm (3,149 km; 1,957 miles) at a cruising speed of 350 knots (649 km/h; 403 mph). During 1975 Electra conversions were delivered to Fairbanks Air Service in Alaska and Aerocosta of Bogotá, Colombia. Two more Electras are undergoing conversion.

In October 1975 AJI announced plans to build a new general aviation STOL aircraft, designated Hustler Model 400.

AJI TURBO STAR 402

In November 1969, American Jet Industries began conversion of a standard Cessna 402 to turbine power. This involved removal of its Continental piston engines and their replacement by two 298 kW (400 shp) Allison 250-B17 turboprop engines, each driving a Hartzell Type HCa3VF-7 three-blade metal constant-speed fully-feathering reversible propeller with Beta control.

The conversion offered an overall saving of 229 kg (505 lb) in terms of empty weight, giving increased performance, range and payload by comparison with the standard Cessna 402.

The first flight of the Turbo Star 402, as AJI named the conversion, was made on 10 June 1970. Since that time a number of additional modifications have been introduced. Gross weight has been increased to 2,959 kg (6,525 lb), an automatic propeller feathering system has been added, minimum control speed has been reduced by 12·6%, and fuel capacity increased from 477 litres (126 US gallons) to 757 litres (200 US gallons). Recertification of the Turbo Star 402 in this form, in January 1974, was announced by the company on 25 March 1974, and the prototype has since been acquired by Winship Air Service in Alaska. The first two production models entered operation with Scenic Airlines of Las Vegas, Nevada, and a third aircraft was delivered to this operator in 1975.

American Jet Industries Turbo Star 402, a turbine-powered conversion of the Cessna Model 402

The following details apply to the Turbo Star 402 as now in production:
WEIGHTS AND LOADINGS:
Weight empty 1,458 kg (3,214 lb)
Max T-O weight 2,959 kg (6,525 lb)
Max zero-fuel weight 2,857 kg (6,300 lb)
Max landing weight 2,812 kg (6,200 lb)
Max wing loading 163·6 kg/m² (33·5 lb/sq ft)
Max power loading 4·96 kg/kW (8·16 lb/shp)
PERFORMANCE (at max T-O weight):
Max level speed at 2,895 m (9,500 ft)
233 knots (431 km/h; 268 mph)
Max cruising speed at 3,660 m (12,000 ft)
223 knots (414 km/h; 257 mph)
Econ cruising speed at 6,100 m (20,000 ft)
208 knots (386 km/h; 240 mph)
Approach speed
90·5 knots (167 km/h; 104 mph) CAS
Min control speed, one engine out
77·5 knots (144 km/h; 89 mph) CAS
Max rate of climb at S/L 639 m (2,095 ft)/min
Rate of climb, one engine out 148 m (485 ft)/min
Operational ceiling 7,620 m (25,000 ft)
Service ceiling, one engine out 3,870 m (12,700 ft)
T-O run 254 m (832 ft)
T-O to 15 m (50 ft) 440 m (1,445 ft)
Accelerate/stop distance 605 m (1,984 ft)
Balanced field length 605 m (1,984 ft)
Landing from 15 m (50 ft), with propeller reversal, at max landing weight 336 m (1,104 ft)
Landing run, with propeller reversal, at max landing weight 137 m (450 ft)
Range at max cruising speed, no reserves
1,149 nm (2,130 km; 1,324 miles)
Max range at econ cruising speed, no reserves
1,397 nm (2,589 km; 1,609 miles)

AJI TURBO STAR PRESSURISED 414

The prototype of American Jet Industries' conversion of a Cessna Model 414, designated AJI Turbo Star Pressurised 414, made its first flight in mid-1974. Certification was scheduled for the Spring of 1976.

The conversion is generally similar to that developed for the Cessna Model 402, but differs by introducing an uprated version of the Allison turboprop engine, the 250-B17B which develops 313 kW (420 shp) for take-off, and a newly developed hydraulically-driven 'fail-safe' pressurisation system offering significant weight reduction. It is anticipated that the conversion will allow a Cessna Model 414 to equal the performance of the more expensive Cessna 421B Golden Eagle.

Preliminary specification details of the Turbo Star Pressurised 414 are as follows:
WEIGHTS AND LOADINGS:
Weight empty 1,590 kg (3,505 lb)
Max T-O weight 2,959 kg (6,525 lb)
Max zero-fuel weight 2,857 kg (6,300 lb)
Max landing weight 2,812 kg (6,200 lb)
Max wing loading 163·6 kg/m² (33·5 lb/sq ft)
Max power loading 4·73 kg/kW (7·77 lb/shp)
PERFORMANCE (estimated, at max T-O weight):
Max level speed at 3,050 m (10,000 ft)
238 knots (441 km/h; 274 mph)
Max cruising speed at 4,875 m (16,000 ft)
226 knots (418 km/h; 260 mph)
Econ cruising speed at 6,100 m (20,000 ft)
208 knots (386 km/h; 240 mph)
Approach speed
90·5 knots (167 km/h; 104 mph) CAS
Min control speed, one engine out
77·5 knots (144 km/h; 89 mph) CAS
Max rate of climb at S/L 664 m (2,180 ft)/min
Rate of climb at S/L, one engine out
155 m (510 ft)/min
Operational ceiling 7,620 m (25,000 ft)
Service ceiling, one engine out 4,205 m (13,800 ft)
T-O run 241 m (792 ft)
T-O to 15 m (50 ft) 426 m (1,396 ft)
Accelerate/stop distance 591 m (1,940 ft)
Balanced field length 591 m (1,940 ft)
Landing from 15 m (50 ft), with propeller reversal, at max landing weight 336 m (1,104 ft)
Landing run, with propeller reversal, at max landing weight 137 m (450 ft)
Range at max cruising speed, no reserves
1,112 nm (2,061 km; 1,281 miles)
Max range at econ cruising speed, no reserves
1,327 nm (2,459 km; 1,528 miles)

AJI HUSTLER MODEL 400

Plans to build a new general aviation STOL aircraft were announced by American Jet Industries on 24 October 1975. STOL characteristics for the new aircraft—to be known as the Hustler Model 400—stem from the use of a supercritical wing with full-span Fowler trailing-edge flaps and spoilers, instead of ailerons, for lateral control. Power plant is unusual, comprising a turboprop engine installed conventionally in the fuselage nose, with a small turbojet standby engine mounted in the aft fuselage. The turbojet is intended to offer additional safety, by enabling the Hustler to maintain 148 knots (274 km/h; 170 mph) IAS at an altitude of 4,570 m (15,000 ft) with the propeller of the nose engine feathered, and is available in emergency at take-off, being started automatically by a torque-sensing device on the turboprop engine. In situations when it is realised that additional power might be needed at short notice, the pilot will have the option of taking off with the turbojet engine idling. Additionally, this engine can be used to boost the aircraft's maximum cruising speed by some ten per cent, but range would suffer considerably because of the higher rate of fuel consumption. This latter factor is in opposition to the basic design concept, for the primary aim was to produce a fast and economical business/utility aircraft. To this end the cabin is pressurised, to permit cruising altitudes of up to 10,670 m (35,000 ft), and major efforts have been made to develop a well streamlined, low-drag airframe with a power plant that will be economical in operation.

The initial detail design provided small air intake scoops for the aft engine on each side of the lower aft fuselage, but later design analysis has shown the desirability of resiting the air intake at the base of the fin to ensure that water and debris thrown up by the landing gear will not be ingested into the turbojet engine.

American Jet Industries has built a full-scale mockup of the Hustler, and has started the fabrication of components for a static test vehicle and two flight test/certification aircraft, with the first flight of a prototype scheduled for mid-1976. The company anticipates that the flight test programme and FAA certification will occupy approximately 5 months, with initial deliveries of production aircraft being made in late 1976 or early 1977.

Other specialised versions are being studied, including a two-seat basic trainer for the US Navy and a high-altitude photographic survey aircraft with a wing area approaching double that of the basic Hustler, permitting operations at altitudes up to 15,240 m (50,000 ft).

TYPE: Seven-seat business/utility aircraft.

WINGS: Cantilever mid-wing monoplane. Supercritical wing section GAW Mod 4. Thickness/chord ratio 12·5%. Dihedral 2°. Incidence 0°. Sweepback at quarter-chord 15°. Conventional light alloy two-spar structure with ribs, stringers and chemically-milled skins, flush riveted. Full-span single-slotted light alloy Fowler trailing-edge flaps. Dual spoilers of light alloy construction on the upper surface of each wing, forward of the trailing-edge flaps. Electrical de-icing system for wing leading-edges.

FUSELAGE: Semi-monocoque light alloy fail-safe structure of circular cross-section.

TAIL UNIT: Cantilever light alloy structure with swept surfaces and all-moving tailplane. Anti-servo and trim tab in tailplane, trim tab in rudder. Electrical de-icing system for leading-edges of fin and tailplane.

LANDING GEAR: Electrically-retractable tricycle type. Main units retract inward, nose unit aft. Oleo-pneumatic shock-absorbers. Single Goodyear wheel and tyre on each unit. Main wheels and tyres size 22 × 9·00, pressure 3·10 bars (45 lb/sq in); nosewheel size 17·50 × 6·00, pressure 2·41 bars (35 lb/sq in). Goodyear toe-operated hydraulic brakes.

POWER PLANT: One 634 kW (850 shp) Pratt & Whitney Aircraft of Canada PT6A-41 turboprop engine, driving a Hartzell four-blade metal constant-speed reversible propeller with Beta control, and standby power plant comprising one Teledyne CAE 372-2 turbojet engine rated at 2·85 kN (640 lb st). Integral fuel tanks in wings with combined capacity of 795 litres (210 US gallons). Wingtip tanks, each with capacity of 150·5 litres (40 US gallons) optional. Max fuel capacity 1,096 litres (290 US gallons). Refuelling points on upper surface of each wing. Oil capacity 9·5 litres (2·5 US gallons).

ACCOMMODATION: Seven seats, for pilot and six passengers, or crew of two with five passengers. Two seats side-by-side in cockpit, separated from cabin by radio rack on starboard side and small galley on port side. Two aft-facing and two forward-facing seats side by side in cabin, with a third combined seat/toilet on the starboard side at rear of cabin. Baggage space at rear of cabin. Two-part door on port side, aft of wing, with airstairs in lower section. Hinged emergency exit on starboard side of cabin, between aft-and forward-facing seats. Accommodation heated, ventilated, air-conditioned and pressurised.

SYSTEMS: AiResearch air cycle heating and cooling with pressurisation at a differential of 0·51 bars (7·4 lb/sq in). Hydraulic system for brakes only. Electrical system at 24V DC supplied by starter/generator. Nickel-cadmium storage battery. Oxygen system for emergency use.

ELECTRONICS AND EQUIPMENT: A range of standard electronics for communication and navigation is available. Blind-flying instrumentation standard. Weather radar, optional, can be mounted in nose of optional port wing-tip tank.

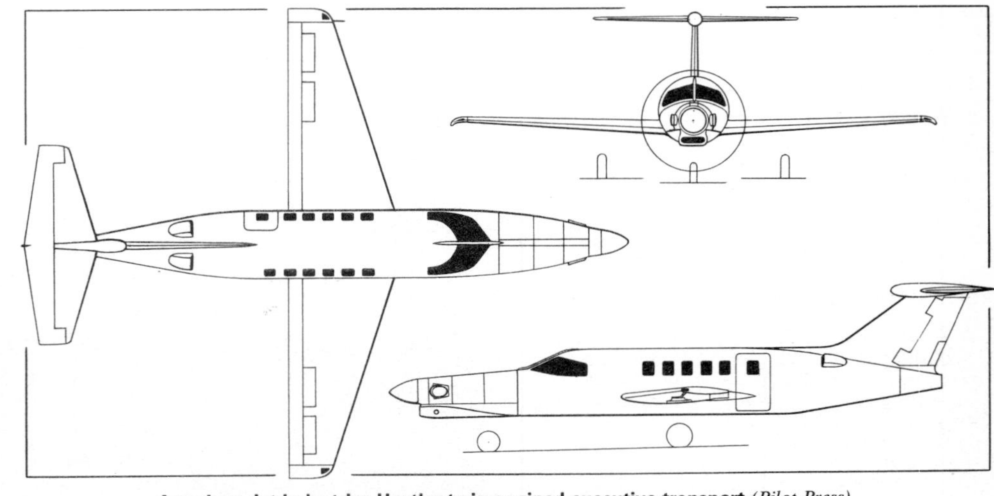

American Jet Industries Hustler twin-engined executive transport *(Pilot Press)*

DIMENSIONS, EXTERNAL:

Wing span	8·53 m (28 ft 0 in)
Wing span over tip-tanks	8·84 m (29 ft 0 in)
Wing chord at root	2·03 m (6 ft 8 in)
Wing chord at tip	1·02 m (3 ft 4 in)
Length overall	10·61 m (34 ft 9¾ in)
Height overall	3·00 m (9 ft 10 in)
Tailplane span	3·66 m (12 ft 0 in)
Wheel track	3·68 m (12 ft 0¾ in)
Wheelbase	3·68 m (12 ft 0¾ in)
Propeller diameter	2·03 m (6 ft 8 in)
Propeller ground clearance	0·32 m (1 ft 0½ in)
Passenger door:	
Height	1·17 m (3 ft 10 in)
Width	0·71 m (2 ft 4 in)
Emergency exit:	
Height	0·86 m (2 ft 10 in)
Width	0·71 m (2 ft 4 in)

DIMENSIONS, INTERNAL:

Cabin, aft of firewall:	
Length	5·18 m (17 ft 0 in)
Max width	1·32 m (4 ft 4 in)
Max height	1·27 m (4 ft 2 in)

AREAS:

Wings, gross	16·62 m² (146·6 sq ft)
Trailing-edge flaps (total)	2·75 m² (29·56 sq ft)
Spoilers (total)	0·91 m² (9·78 sq ft)
Vertical tail surfaces, incl tab	3·02 m² (32·5 sq ft)
Tailplane, incl tabs	3·85 m² (41·39 sq ft)

WEIGHTS AND LOADINGS:

Weight empty	1,588 kg (3,500 lb)
Max T-O weight	2,722 kg (6,000 lb)
Max zero-fuel weight	2,495 kg (5,500 lb)
Max landing weight	2,586 kg (5,700 lb)
Max wing loading	199·8 kg/m² (40·92 lb/sq ft)
Max power loading	4·29 kg/kW (7·06 lb/shp)

PERFORMANCE (at max T-O weight, ISA):

Max level speed and max cruising speed at 6,100 m (20,000 ft) 347 knots (644 km/h; 400 mph)

Econ cruising speed at 10,670 m (35,000 ft) 254 knots (470 km/h; 292 mph)

Stalling speed, flaps down 59 knots (109 km/h; 68 mph)

Max rate of climb at S/L 1,067 m (3,500 ft)/min

Service ceiling, forward engine only 10,670 m (35,000 ft)

Service ceiling, aft engine only 4,570 m (15,000 ft)

T-O run 152 m (500 ft)

T-O to 15 m (50 ft) 274 m (900 ft)

Landing from 15 m (50 ft), with Beta control 305 m (1,000 ft)

Landing run 168 m (550 ft)

Accelerate/stop distance 366 m (1,200 ft)

Range with max fuel, no reserves 2,084 nm (3,862 km; 2,400 miles)

Range with max payload, no reserves 1,911 nm (3,541 km; 2,200 miles)

AMR—*See Addenda*

ARCTIC

ARCTIC AIRCRAFT COMPANY

ADDRESS:

PO Box 6-141, Anchorage, Alaska 99502

Telephone: (907) 277-4918 and (907) 344-2098

SECRETARY: C. J. Diehl

Arctic Aircraft is constructing and marketing an improved version of the Interstate S1B2, first flown more than 30 years ago, as the Arctic Tern.

ARCTIC AIRCRAFT ARCTIC TERN

TYPE: Two-seat sporting and general utility aircraft.

WINGS: High-wing monoplane, with Vee bracing struts each side and auxiliary struts. Wing section NACA 23012. Composite structure with Sitka spruce spars, light alloy ribs and Dacron covering. Hoerner-type wingtips of glassfibre. Fowler-type single-slotted trailing-edge flaps. Plain inset ailerons.

FUSELAGE: Welded structure of 4130 chrome-molybdenum steel tube, Dacron covered. Two-piece engine cowling of glassfibre.

TAIL UNIT: Wire-braced structure of welded 4130 steel tube with Dacron covering. Trim tab in elevator.

LANDING GEAR: Non-retractable tailwheel type. Main wheels carried in two side Vees and half-axles hinged to fuselage. Shock-absorption by rubber bungee. Cleveland main wheels with US Uniroyal tyres size 8·50 × 6. Maule tailwheel, diameter 0·203 m (8 in). Scott toe-operated brakes. Parking brake. Edo 2000 floats or 2500 skis optional.

POWER PLANT: One 112 kW (150 hp) Lycoming O-320 flat-four engine, driving a McCauley two-blade metal fixed-pitch propeller. One fuel tank in each wing, total capacity 151 litres (40 US gallons). Belly-mounted auxiliary fuel tank optional.

ACCOMMODATION: Two seats in tandem; rear seat removable to provide additional space for cargo. Cabin door on starboard side, beneath wing. Cabin step. Baggage space in rear fuselage, with external door on starboard side. Tinted windows and cabin skylight. Safety belts and fittings standard. Dual controls standard. Cabin soundproofed, heated and ventilated. Windscreen de-icing by hot air. Cabin floor carpeted.

SYSTEM: Electrical system includes 55A engine-driven alternator, 12V DC storage battery, engine starter and navigation lights.

ELECTRONICS AND EQUIPMENT: A range of radios is available. Standard items include epoxy priming, three-colour paint scheme, ground manoeuvring handles and engine quick oil drain. Optional items include sheet metal or vinyl interior, belly-mounted cargo pack, lumber rack, landing lights, instrument lights, cabin dome lights, Alcor exhaust gas analyser, folding front seat, shoulder harness and salt water corrosion proofing.

DIMENSIONS, INTERNAL:

Cabin volume	1·38 m³ (48·7 cu ft)
Baggage area volume	0·84 m³ (29·63 cu ft)

WEIGHT:

Max T-O weight	862 kg (1,900 lb)

PERFORMANCE:

Never-exceed speed 152 knots (282 km/h; 175·5 mph)

Max cruising speed at S/L, 75% power 102 knots (188 km/h; 117 mph)

Stalling speed, flaps down 34 knots (63 km/h; 39 mph)

Max rate of climb at S/L, at max T-O weight 389 m (1,275 ft)/min

Absolute ceiling 6,400 m (21,000 ft)

T-O run at max T-O weight 99 m (325 ft)

T-O to 15 m (50 ft) at max T-O weight 152 m (500 ft)

Range with max standard fuel, zero wind 565 nm (1,045 km; 650 miles)

AVIATION SPECIALTIES—*See 'Helitec Corporation'*

BEDE
BEDE AIRCRAFT INC

HEAD OFFICE:
Newton Municipal Airport, PO Box 706, Newton, Kansas 67114

Telephone: (316) 283-8870

PRESIDENT: James R. Bede

Most of the aircraft designed and developed by this company are intended for amateur construction, and are described in the Homebuilts section. There are, however, factory-built versions of the piston-engined BD-5 and jet-powered BD-5J single-seat sporting aircraft, as follows:

BEDE BD-5D

The interest aroused by the original BD-5A/B proposals was such that Bede Aircraft received many requests from those who wished to buy a production BD-5, as opposed to building the aircraft themselves.

To meet this demand a production line was established, and the resulting version of what is essentially a BD-5G with the 52 kW (70 hp) Xenoah engine has the designation BD-5D. It is to be certificated in the Utility category under FAR Part 23. Due to delays in obtaining engines, deliveries of the first production aircraft were scheduled for late 1975.

The description given for the BD-5G applies also to the BD-5D, except that the latter, being a production aircraft, has certain standard equipment. Optional items include advanced instrumentation, with communications and navigation radios to suit individual requirements. An unusual option is an enclosed trailer, in effect a portable hangar for the aircraft.

Specification and performance of the BD-5D are the same as for the BD-5G with 52 kW (70 hp) engine (see Homebuilts section).

BEDE BD-5JP

This is a factory-built version of the jet-powered Bede BD-5J single-seat sporting aircraft, of which details can be found in the Homebuilts section. The two versions are identical in all significant respects. Certification of the BD-5JP was expected by December 1976. Orders totalled 65 in mid-1976.

BEECHCRAFT
BEECH AIRCRAFT CORPORATION

HEAD OFFICE AND MAIN WORKS:
Wichita, Kansas 67201

Telephone: (316) 689-7111

BRANCH DIVISIONS:
Liberal, Kansas; Salina, Kansas; and Boulder, Colorado

CHAIRMAN OF THE BOARD:
Mrs O. A. (Walter H.) Beech

PRESIDENT:
Frank E. Hedrick

GROUP VICE-PRESIDENTS:
E. C. Burns
R. H. McGregor

SENIOR VICE-PRESIDENTS:
James N. Lew (Engineering)
J. A. Elliott (Treasurer)
M. G. Neuburger (International Relations)

VICE-PRESIDENTS:
Seymour Colman
Harold W. Deets (Materiel)
Glen Ehling (Manufacturing)
Leddy L. Greever (Corporate Director)
Dwight C. Hornberger (International Marketing)
J. E. Isaacs (Industrial Relations)
E. C. Nikkel (Aerospace Programmes)
John A. Pike (Production)
C. A. Rembleske (Aircraft Engineering)
Austin Rising (Corporate Planning and Distribution Development)
Darrell L. Schneider (Government Relations)

SECRETARY:
L. Winters

ASSISTANT SECRETARY:
I. Alumbaugh

ASSISTANT TREASURER:
C. W. Dieker

CONTROLLER, COST MANAGEMENT:
L. R. Damon

DIRECTOR, ADVERTISING AND SALES PROMOTION:
R. James Yarnell

DIRECTOR, PUBLIC RELATIONS:
Bill Robinson

Founded jointly in 1932 by Mrs Olive Ann Beech and the late Walter H. Beech, pioneer designer and builder of light aeroplanes in the United States, the Beech Aircraft Corporation is currently engaged in the production of civil and military aircraft, missile targets, aircraft and missile components and cryogenic equipment for spacecraft.

Deliveries by Beech in 1975 were made up of 3 Beechcraft B99 Airliners, 177 King Airs, 11 Queen Airs, 51 Dukes, 320 Barons, 397 Bonanzas, 82 Sierras, 144 Sundowners, 44 Sports and 2 Beechcraft Hawker BH 125 jets. By the end of 1975 Beech had delivered well over 1,600 pressurised aircraft since introducing the King Air 90 in 1964, a record exceeding that of any other general aviation manufacturer. Total production of Beechcraft aeroplanes was approaching 38,000 in mid-1976.

First deliveries of the pressurised Beechcraft C-12A twin-turboprop transport were made in July 1975 to both the US Army and Air Force. In late 1975 Beech announced production plans for the twin-engine T-tail Beechcraft Model 76, with first deliveries scheduled for the Autumn of 1977.

During 1975 four Beech programmes were moved from research and development to production status. These comprise the T-34C Turbo Mentor trainer for the US Navy, and for the Air Force of Morocco; MQM-107 (VSTT) variable-speed recoverable missile target for the US Army; 'Sea Skipper' version of the AQM-37A missile target for the US Navy; and the pressurised Beechcraft Baron 58P twin-engined commercial aircraft.

The first flight of an experimental two-seat trainer, the Beechcraft PD 285, was made during 1975. New features include a GAW-1 wing section and extensive use of honeycomb construction in the fuselage. In early 1976 the flight test programme was continuing with a T-tail installation.

In March 1975 a prototype fanjet version of the Beechcraft Super King Air, developed by Beech following a detailed study of the business jet market, made its first flight. The company reports that, while the aircraft has performed well and met all objectives, evaluation of several configurations would be made before arriving at a production decision.

Production continues under a succession of contracts awarded by Bell Helicopter Textron since 1968 for manufacture of airframes for the turbine-powered Bell JetRanger helicopter. The contracts are assigned to Beech Aircraft's Wichita Plant III, which is responsible for fuselage, skid gear, tailboom, spar, stabiliser and two rear fairing assemblies. Initial deliveries of JetRanger airframes began in December 1968, with first deliveries of Bell Light Observation Helicopter (OH-58) airframes in the preceding month. Units are shipped to Fort Worth, where Bell completes assembly.

In early 1976, Beech announced receipt of a $9·6 million follow-on contract for continued production of airframes for the commercial JetRanger II until December 1977. With this new contract, Beech announced that deliveries and orders on hand for ship sets of airframes and assemblies for the JetRanger totalled more than 2,100, with a contract value, including spares, of more than $60·7 million.

Beech production of subassemblies for the McDonnell Douglas F-4 Phantom II fighter has entered its fourteenth consecutive year.

In September 1975, Beech Aircraft and Hawker Siddeley Aviation Ltd of Kingston upon Thames, England, terminated by mutual consent a five-year agreement under which a Beech Aircraft subsidiary, Beechcraft Hawker Corporation, had been formed to market the Beechcraft Hawker 125-400 and 125-600 (Hawker Siddeley 125 Srs 400 and 600) in North America. Between 1970 and 1975 a total of 64 BH 125-400/600s had been delivered by Beechcraft Hawker.

Beech Aircraft occupies 234,976 m² (2,529,266 sq ft) of plant area at its four major facilities in Wichita, Liberal and Salina, Kansas, and Boulder, Colorado.

The Salina division supplies all wings used in Wichita production and is responsible for manufacture and final assembly of the six-seat Beechcraft Duke and the new pressurised Beechcraft Baron 58P.

All assembly, flight testing and delivery of the Beechcraft Sierra, Sundowner and Sport are carried out at the Liberal Division.

In an expanded marketing programme in the light aircraft field, Beech began franchising Beech Aero Centers in 1972 to ensure optimum sales and service of the Sierra, Sundowner and Sport. By March 1976, 112 of these Centers had been appointed; during 1975 the 3,000th Beech aircraft to be supplied to these Centers was delivered from the Liberal Division.

Work at Boulder involves space vehicle or missile applications, and included design, development, final assembly and testing of the cryogenic gas storage system for NASA's Apollo and Skylab spacecraft. This same system was utilised in the US-USSR Apollo-Soyuz Test Programme carried out in July 1975. Boulder engineers have developed for NASA cryogenic tanks with the capacity of supporting space missions of as long as six months. These tanks can store 50 times as much oxygen and 120 times as much hydrogen as those used in the Apollo programme.

In January 1974 Beech was awarded a subcontract to produce the power reactant storage assembly for NASA's Space Shuttle orbiter. The assembly includes two liquid oxygen and two liquid hydrogen tanks to supply the orbiter's fuel cells and environmental control/life support system. Design, development, test and production will be carried out at the Boulder division, with deliveries scheduled to continue until the end of 1979. During 1975-76 Beech received three additional contracts to develop Space Shuttle ground support equipment systems, which will be used at both the Kennedy Space Center and Vandenberg AFB launch sites. Total Beech contracts on the Space Shuttle programme have reached $11·8 million.

As a direct result of its work for the US space programme, Beech has developed a cryogenic automotive fuel system which uses liquefied natural gas as a fuel source and is designed to utilise liquid hydrogen ultimately, one of the world's most abundant fuel resources. The Beech system has undergone extensive laboratory testing and field testing, in private and commercial road vehicles. Both liquid natural gas and liquid hydrogen applications have been demonstrated as feasible and economically and environmentally acceptable. Upon successful completion of prototype system testing, Beech reported that the programme beyond the development stage would be deferred until such time as cryogenic fuels became more readily available.

Boulder also produces aircraft assemblies for other Beech divisions and the AQM-37A and MQM-107 (VSTT) missile target systems for the US Army (see RPVs and Targets section).

In 1975 Beech was selected by the California Institute of Technology's Jet Propulsion Laboratory (JPL) to support the study phase of a NASA research programme using in-flight injection of hydrogen into aircraft engines as a means of reducing fuel consumption. The programme involves study of the feasibility of applying a catalytic hydrogen generator, developed by JPL, to turbocharged piston engines. Installed in the engines' air induction system, the unit generates hydrogen spontaneously from standard aviation gasoline. By early 1976 the study phase of the three-part NASA programme had been completed and the laboratory testing phase was under way.

Wholly-owned subsidiaries of the parent company include Beech Acceptance Corporation Inc, which is engaged in business aircraft retail finance and leasing; Beechcraft AG, which has its headquarters in Zurich, Switzerland, and supports in Europe the sales, liaison and other activities of the parent company; Travel Air Insurance Company Ltd, a Bermuda-based company organised during 1972 to provide aircraft liability insurance; Beech Holdings Inc, which provides marketing support to the parent company; Beech International Sales Corporation, Wichita, through which all Beech export sales are made; and the following product sales outlets: Houston-Beechcraft Inc, Houston, Texas; Denver-Beechcraft Inc, Denver, Colorado; United Beechcraft Inc, Wichita, Kansas; Beechcraft West Hayward, Van Nuys and Fresno, California; Mission Beechcraft, Santa Ana, California; Indiana Beechcraft Inc, Indianapolis, Indiana; and Beechcraft East Inc, Farmingdale, New York.

BEECHCRAFT TURBO MENTOR

US Navy designation: T-34C

In March 1953 the USAF selected the Beechcraft Model 45 as its new primary trainer and, under the designation T-34A Mentor, a total of 450 were eventually acquired. Power plant consisted of a 168 kW (225 hp) Continental O-470-13 flat-six engine.

Just over a year after the Air Force adopted the Beech Model 45 as its primary trainer, the US Navy reached a similar decision, and a total of 423 T-34B Mentors were built for that service.

Experience in both the USAF and USN showed the Mentor to be a rugged and reliable aircraft, and in 1973 Beech received a USN R & D contract to modify two T-34Bs to see whether the type could be upgraded for a continuing training role. This involved the installation of a turboprop engine and the latest electronics equipment, the primary object being to let student pilots have experience of operating turbine-powered aircraft from the beginning of their flight training.

The power plant selected was the PT6A-25, which has a torque limiter in this application to restrict engine output to 298 kW (400 shp). This not only ensures long engine life, but also provides constant performance over a wide range of temperature and altitude.

Design of the modifications to update the aircraft began in March 1973, and conversion of two T-34Bs (140784 and 140861) started in May 1973. Designated YT-34C, the first of these aircraft flew for the first time on 21 September 1973, and the test programme continued throughout 1974.

By comparison with the original Mentor, the YT-34C had a 454 kg (1,000 lb) increase in gross weight, which required structural modifications to strengthen the fuselage and tail unit. Additional strength for other assemblies and components was achieved largely by adopting off-the-shelf parts from other Beech aircraft.

By February 1976 the two prototype YT-34Cs had flown more than 800 test hours, including nearly 300 hours of evaluation by the Navy.

Beech has received $43·6 million USN contracts for the first 116 new-production T-34Cs, of an anticipated total requirement of nearly 400, with first deliveries scheduled

for the early Autumn of 1976. A contract valued at $5·5 million for the supply of 12 T-34Cs to the air force of Morocco was received during 1975.

Production Turbo Mentors will incorporate improvements developed during the flight test programme; structural strength is to be further increased to permit high limit speeds and a fatigue life of 16,000 hours in a primary flight training role.

TYPE: Two-seat turbine-powered primary training aircraft.

WINGS: Cantilever low-wing monoplane. Wing section NACA 23016·5 (modified) at root, NACA 23012 at tip. Dihedral 7°. Incidence 4° at root, 1° at tip. No sweepback. Conventional box beam structure of light alloy. Ailerons of light alloy construction. Single-slotted trailing-edge flaps of light alloy. Manually operated trim/servo tab in port aileron.

FUSELAGE: Semi-monocoque light alloy structure.

TAIL UNIT: Cantilever structure of light alloy. Fixed-incidence tailplane. Manually-operated trim tabs in elevators and rudder. Twin ventral fins under rear fuselage.

LANDING GEAR: Electrically-retractable tricycle type. Main units retract inward, nosewheel aft. Beech oleo-pneumatic shock-absorbers. Single wheel on each unit. Main wheels and tyres size 6·50-8. Nosewheel and tyre size 5·00-5. Goodyear multiple-disc hydraulic brakes.

POWER PLANT: One 533 kW (715 shp) Pratt & Whitney Aircraft of Canada PT6A-25 turboprop engine, torque limited to 298 kW (400 shp), driving a Hartzell three-blade metal constant-speed fully-feathering propeller. Two bladder-type fuel cells in each wing, with a combined usable capacity of 378·5 litres (125 US gallons). Oil capacity 13·2 litres (3·5 US gallons).

ACCOMMODATION: Instructor and pupil in tandem beneath rearward-sliding cockpit canopy. Cockpit ventilated, and heated by engine bleed air. Dual controls standard.

SYSTEMS: Hydraulic system for brakes only. Pneumatic system for emergency opening of cockpit canopy. Diluter demand gaseous oxygen system, pressure 103·5 bars (1,500 lb/sq in). Electrical power supplied by 200A starter/generator. Air-conditioning system planned for production aircraft but not installed in prototypes.

ELECTRONICS AND EQUIPMENT: Blind-flying instrumentation standard. Electrically-heated pitot and angle of attack indicator. UHF com, Omni, DME, LF/DF and transponder. Intercom. Fluxgate compass system.

ARMAMENT: An armament system similar to that of the Model PD 249 Pave Coin Bonanza, detailed in the 1973-74 *Jane's*, could be provided.

DIMENSIONS, EXTERNAL:
- Wing span — 10·16 m (33 ft 3⅞ in)
- Wing chord at root — 2·55 m (8 ft 4½ in)
- Wing chord at tip — 1·05 m (3 ft 5¼ in)
- Wing aspect ratio — 6·22
- Length overall — 8·75 m (28 ft 8½ in)
- Height overall — 3·02 m (9 ft 10⅞ in)
- Tailplane span — 3·71 m (12 ft 2⅛ in)
- Wheel track — 2·91 m (9 ft 6½ in)
- Wheelbase — 2·41 m (7 ft 11 in)
- Propeller diameter — 2·29 m (7 ft 6 in)
- Propeller ground clearance — 0·45 m (1 ft 5¾ in)

DIMENSIONS, INTERNAL:
- Cabin: Length — 2·74 m (9 ft 0 in)
  - Max width — 0·86 m (2 ft 10 in)
  - Max height — 1·22 m (4 ft 0 in)

AREAS:
- Wings, gross — 16·71 m² (179·9 sq ft)
- Ailerons (total) — 1·06 m² (11·4 sq ft)
- Trailing-edge flaps (total) — 1·98 m² (21·3 sq ft)
- Fin — 1·31 m² (14·1 sq ft)
- Rudder, incl tab — 0·76 m² (8·16 sq ft)
- Tailplane — 2·95 m² (31·8 sq ft)
- Elevators, incl tabs — 1·50 m² (16·2 sq ft)

WEIGHTS AND LOADING:
- Weight empty — 1,193 kg (2,630 lb)
- Max T-O and landing weight — 1,938 kg (4,274 lb)
- Max wing loading — 108·3 kg/m² (22·2 lb/sq ft)

PERFORMANCE (preliminary results at max T-O weight):
- Never-exceed speed — 250 knots (463·5 km/h; 288 mph)
- Max level speed at 5,335 m (17,500 ft) — 223 knots (414 km/h; 257 mph)
- Max cruising speed at 5,335 m (17,500 ft) — 214 knots (397 km/h; 247 mph)
- Stalling speed, flaps up — 55 knots (102 km/h; 63·3 mph) CAS
- Max rate of climb at 3,050 m (10,000 ft) — 388 m (1,275 ft)/min
- Service ceiling — over 9,145 m (30,000 ft)
- Range at 6,100 m (20,000 ft) — 650 nm (1,205 km; 749 miles)

BEECHCRAFT PD 285 (MODEL 77)

Beech Aircraft first announced in late 1974 that the company was building the prototype of a single-engined trainer under the designation PD 285, and this flew for the first time on 6 February 1975. Since that time it has been used as a testbed aircraft, and a brief description of the prototype in its original configuration appeared in the 1975-76 *Jane's*.

Beechcraft YT-34C Turbo Mentor primary trainer (PT6A-25 turboprop engine)

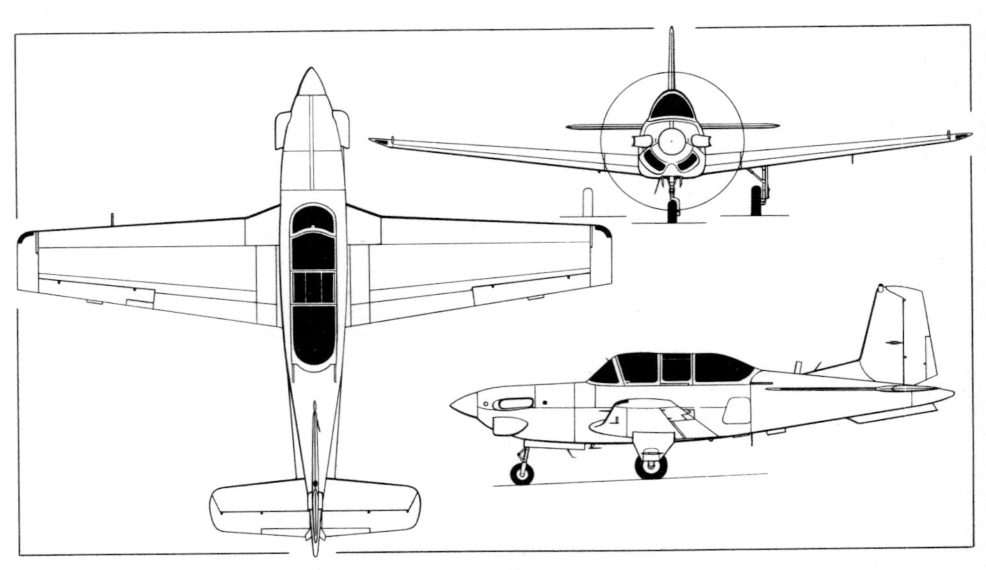

Beechcraft T-34C Turbo Mentor turboprop-powered primary training aircraft *(Pilot Press)*

Beechcraft PD 285, prototype for the Model 77 lightweight training aircraft

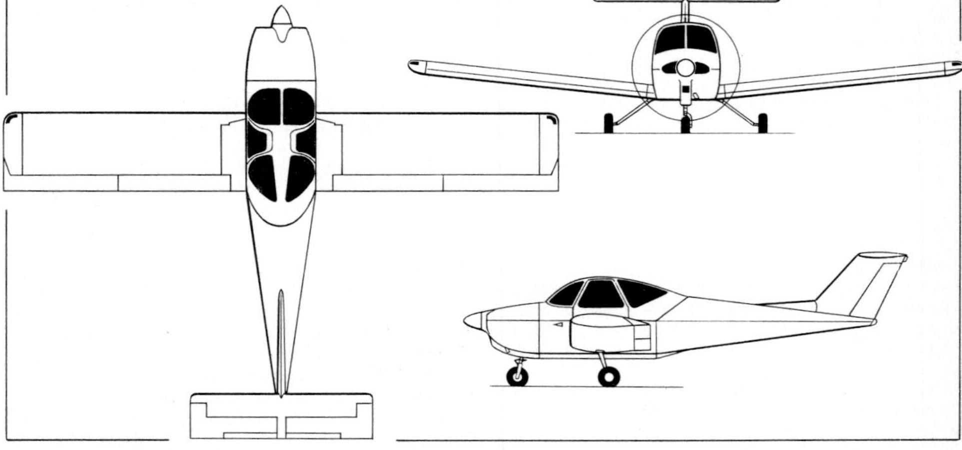

Beechcraft PD 285 experimental light aircraft in latest form with T-tail *(Pilot Press)*

The PD 285 is flying currently with a T-tail, which is expected to confer certain aerodynamic improvements. Intended primarily as a low-cost trainer for Beech Aero Centers, it utilises new construction techniques to reduce manufacturing costs. It has a cantilever low wing of GAW-1 aerofoil section, which resulted from NASA/Beech research into supercritical aerofoils with high-lift characteristics. The aircraft has side by side seats for instructor and pupil in a spacious cabin with excellent view. A door on each side of the cabin provides easy access.

The PD 285 will enter production as the Beechcraft **Model 77**, with an 85·5 kW (115 hp) Lycoming O-235 engine and max T-O weight of 748 kg (1,650 lb). Deliveries will begin in 1978.

BEECHCRAFT SIERRA 200, SUNDOWNER 180 and SPORT 150

In December 1971, Beech introduced a new light aircraft marketing programme centred around three models, which were given individual exterior paint schemes and renamed from their previous Musketeer designations (see 1971-72 *Jane's*).

In 1974 these designations were changed again to indicate the engine horsepower rating. Current names are Beechcraft Sierra 200 (formerly Model A24R Musketeer Super R), Sundowner 180 (Model C23, formerly Musketeer Custom), and Sport 150 (Model B19, formerly Musketeer Sport). The fourth aircraft in the former Musketeer line, the Super, was discontinued at the end of 1971 after a total of 368 had been built.

The Sierra 200 was recertificated in 1974 and redesignated Model B24R, due to the installation of a new engine, improved cowling and redesign of control features. The Sierra 200 has also elliptical entry steps, a firewall-mounted oil cooler, added airscoop on the engine cowling, weighted engine crankshaft and a Hartzell propeller.

The three current aircraft have a cabin door on the port side of the fuselage, the Sport 150 thus becoming the only US low-wing trainer with cabin doors on each side. Standard equipment includes a new quadrant for the engine controls, new low-profile instrument panel to improve visibility, padded glareshield, safety contour door handles and improved door-latching system, automotive-type inner door-latch handles, inertia-reel shoulder restraint belts for all seats, and restyled Bonanza-type interiors. The Sierra 200 has a 0·61 × 0·91 m (24 × 36 in) rear door on the port side of the fuselage for easier passenger use or cargo loading.

Details of the current models are as follows:

Sundowner 180. Basic four-seat version with 134 kW (180 hp) Lycoming O-360-A4K engine, driving a Sensenich Type 76EM8S5-0-60 two-blade fixed-pitch metal propeller, and non-retractable landing gear. Aerobatic version is approved for rolls, Immelmann turns, loops, spins, chandelles and other manoeuvres, carrying two persons. Three windows standard on each side of cabin.

Sport 150. Two/four-seat sporting and training version with 112 kW (150 hp) Lycoming O-320-E3D engine, driving a Sensenich 74DM6S5-0-54 two-blade fixed-pitch metal propeller, and non-retractable landing gear. Aerobatic version is approved for rolls, Immelmann turns, loops, spins, chandelles and other manoeuvres, carrying two persons. Two windows standard on each side of cabin.

Sierra 200. Generally similar to the Sundowner but with accommodation for four to six persons, a 149 kW (200 hp) Lycoming IO-360-A1B6 engine, driving a Hartzell Type HC-M2YR-1BF/F7666A-2R two-blade metal constant-speed propeller, and retractable tricycle landing gear. Electrically-actuated hydraulic system based on a self-contained unit in the rear fuselage, comprising electrically-driven hydraulic pump, fluid reservoir and valves. An emergency valve, sited adjacent to the pilot's feet, allows selection of the landing gear to free-fall within three seconds. Main wheels retract outward into wings; nosewheel turns through 90° as it retracts rearwards. Four windows standard on each side of cabin.

Factory-installed optional equipment packages are as follows:

Weekender. Includes sun visors; lighting group comprising rotating beacon, navigation, cabin dome, overhead instrument and map lights; cabin boarding steps; dual controls and pedal-operated brakes for co-pilot; adding 6·8 kg (15 lb) to basic empty weight.

Holiday. As above, plus wing-mounted landing light; 35Ah battery; instrument group comprising 3 in horizon and directional gyros with vacuum system, turn co-ordinator, rate-of-climb indicator, 8-day clock, and outside air temperature gauge; adding 15·4 kg (34 lb) to basic empty weight.

Professional. As above, plus wing-mounted taxi light; heated pitot tube; electrically-operated tailplane trim and control wheel switch; tinted windscreen and windows; true airspeed indicator; two headrests and instrument post lights; adding 19·5 kg (43 lb) to basic empty weight.

Factory-installed optional electronics packages are as follows (a cabin speaker, microphone with jack, antennae and wiring are common to all eight):

No. 1. Narco Com 10A 360-channel communications transceiver with Nav 10 200-channel navigation frequency selector, and VOR/LOC indicator; adding 4·5 kg (10 lb) to basic empty weight.

No. 2. Narco Com 11B 720-channel transceiver, Narco Nav 11 200-channel navigation receiver/converter and VOR/LOC indicator, Narco AT-50A transponder, Narco ADF-140 ADF system with remote indicator; adding 9·5 kg (21 lb) to basic empty weight.

No. 3. As No. 1 plus No. 2, with Narco CP-125 audio panel, amplifier and 3-light marker beacon display; adding 13·2 kg (29 lb) to basic empty weight.

No. 4. Dual Narco Com 11B 720-channel transceivers, Narco Nav 11 200-channel navigation receiver/converter and VOR/LOC indicator, Narco Nav 12 200-channel navigation receiver/converter and VOR/ILS indicator, Narco UGR-3 glideslope receiver, Narco MBT-12R marker beacon receiver, Narco CP-125 audio panel, amplifier and 3-light marker beacon display, Narco AT-50A transponder, and Narco ADF-140 ADF system with remote indicator; adding 16·3 kg (36 lb) to basic empty weight.

No. 5. King KX-145 nav/com (720-channel communications and 200-channel navigation) with KI-205 VOR/LOC converter-indicator; adding 4·1 kg (9 lb) to basic empty weight.

No. 6. King KX-170B nav/com (720-channel communications and 200-channel navigation) with KI-201C VOR/LOC converter-indicator, King KT-78 transponder, King KR-86 ADF with indicator; adding 11·8 kg (26 lb) to basic empty weight.

No. 7. Dual King KX-170B nav/com (720-channel communications and 200-channel navigation) with dual KI-201C VOR/LOC converter-indicators, King KMA-20 audio panel, amplifier and 3-light marker beacon receiver, King KT-78 transponder and King KR-86 ADF; adding 18 kg (40 lb) to basic empty weight.

No. 8. Dual Collins VHF-251 720-channel com transceiver, dual Collins VIR-351 200-channel nav receiver/converter with IND-350 VOR/LOC indicators, Collins AMR-350 audio control panel, amplifier and 3-light marker beacon receiver, Collins TDR-950 transponder, Collins ADF-650 with IND-650 indicator and Collins combined loop/sense antenna; adding 11·3 kg (25 lb) to basic empty weight.

Production is centred in Beech's Liberal, Kansas, plant. A total of 3,281 Musketeers, Sundowners, Sports and Sierras had been delivered by 1 January 1976. They included 20 aircraft supplied to the Mexican government for military training, 25 for the Canadian Armed Forces, 21 for the Indonesian Department of Transportation, Communication and Tourism for its primary training programme, and 19 for the University of Illinois Institute of Aviation.

The following details apply to all three current models:

TYPE: Two, four or six-seat cabin monoplane.

WINGS: Cantilever low-wing monoplane. Wing section NACA 63_2A415. Dihedral 6° 30′. Incidence 3° at root, 1° at tip. Single extruded main spar at 50% chord. Aluminium skin and stringers are bonded to honeycomb Trussgrid ribs on forward 50% of wing; rear 50% of wing is riveted. Slotted all-metal riveted ailerons and mechanically-controlled (optionally electrically-actuated) flaps have corrugated skin. No trim tabs. Plastics wingtips.

FUSELAGE: Cabin section has basic keel formed by floor and lower skin, with rolled skin side panels, stringers, a minimum number of bulkheads and structural top. Conventional semi-monocoque rear fuselage.

Beechcraft Sierra 200 (top), Sundowner 180 (centre) and Sport 150 light aircraft

TAIL UNIT: Cantilever all-metal structure, with swept vertical surfaces. One-piece all-moving horizontal surface with full-span anti-servo tab. Optional electric tailplane trim. Rudder and aileron controls interconnected for easy cross-country flying.

LANDING GEAR (Sundowner and Sport): Non-retractable tricycle type. Beech rubber-disc shock-absorbers. Tube-type tyres size 15 × 6·00-6, pressure 2·76 bars (40 lb/sq in). Optional size 17·5 × 6·00-6 tyres, pressure 1·52 bars (22 lb/sq in). Cleveland disc-type hydraulic brakes with toe-operated control. Steerable nosewheel. Parking brake.

LANDING GEAR (Sierra): Tricycle type, with electrically-actuated hydraulic retraction. Main units retract outwards and upwards into wing wells; nosewheel unit turns through 90° and retracts rearward to fold flat into a fairing behind the nosewheel. Beech rubber-disc shock-absorbers. Main wheels fitted with tube-type tyres size 17·5 × 6·00-6, pressure 2·21 bars (32 lb/sq in). Nosewheel tyre size 14·2 × 5·00-5, pressure 2·41 bars (35 lb/sq in). Cleveland hydraulic disc brakes with toe-operated control. Parking brake.

POWER PLANT: One flat-four engine (details given under model listings). Two fuel tanks in inboard wing leading-edges, with usable capacity of 196·8 litres (52 US gallons). Refuelling points above tanks. Oil capacity 7·5 litres (2 US gallons).

ACCOMMODATION (Sundowner and Sierra): Pilot and three or five passengers (Sierra); pilot and three passengers (Sundowner); in pairs, in enclosed cabin with door on each side. Compartment for 122 kg (270 lb) baggage, with external door on port side. In-flight adjustable seats, pilot's storm window, windscreen defroster, instrument panel glareshield, air vents, map stowage, wall-to-wall carpeting. Optional aerobatic kit for Sundowner includes *g* meter and quick-release door.

ACCOMMODATION (Sport): Generally as for other versions; pilot and up to three passengers in pairs. Optional aerobatic kit includes *g* meter and quick-release door.

SYSTEMS: Electrical system supplied by 60A alternator, 12V 25Ah battery. 35Ah battery optional. Hydraulic system for brakes only, except on Sierra which has electro-hydraulic retraction system for landing gear. Vacuum system for instruments optional.

ELECTRONICS AND EQUIPMENT: Standard equipment includes stall warning system, ventilation, heating and defrosting system; towbar; tie-down rings; control locks and pitot cover. Optional equipment and electronics as listed earlier. Additional optional equipment includes rear cabin 'family seat', headrests, cabin fire extinguisher, Hobbs hour meter, true airspeed indicator, alternate static source, outside air temperature gauge, wing-mounted taxi light, tinted windscreen and windows, mixture indicator, strobe lights, instrument post lights, external power socket, acrylic enamel paint, internal corrosion proofing, heavy-duty tyres (except Sierra) and electrically-operated flaps.

DIMENSIONS, EXTERNAL:

Wing span	9·98 m (32 ft 9 in)
Wing chord (constant)	1·34 m (4 ft 4¾ in)
Wing aspect ratio	7·5
Length overall:	
Sport, Sierra	7·836 m (25 ft 8½ in)
Sundowner	7·84 m (25 ft 8¾ in)
Height overall:	
Sport, Sundowner	2·50 m (8 ft 2½ in)

Sierra 2·57 m (8 ft 5 in)
Tailplane span 3·25 m (10 ft 8 in)
Wheel track:
Sundowner, Sport 3·61 m (11 ft 10 in)
Sierra 3·86 m (12 ft 8 in)
Wheelbase:
Sundowner, Sport 1·93 m (6 ft 4 in)
Sierra 1·83 m (6 ft 0¼ in)
Propeller diameter:
Sport, Sierra 1·88 m (6 ft 2 in)
Sundowner 1·91 m (6 ft 3 in)
Propeller ground clearance:
Sundowner 0·34 m (1 ft 1½ in)
Sport 0·37 m (1 ft 2½ in)
Sierra 0·38 m (1 ft 3 in)
Cabin doors:
Height 0·97 m (3 ft 2 in)
Width 1·03 m (3 ft 4 in)
Baggage compartment door:
Sundowner:
Height 0·47 m (1 ft 6½ in)
Width 0·60 m (1 ft 11¾ in)
Sierra:
Height 0·91 m (3 ft 0 in)
Width 0·61 m (2 ft 0 in)

DIMENSIONS, INTERNAL:
Cabin, aft of instrument panel:
Length:
Sundowner, Sierra 2·41 m (7 ft 11 in)
Sport 1·80 m (5 ft 11 in)
Max width 1·18 m (3 ft 8 in)
Max height 1·22 m (4 ft 0½ in)
Floor area:
Sundowner, Sierra 2·4 m² (25·84 sq ft)
Volume:
Sundowner, Sierra 2·92 m³ (103·2 cu ft)
Baggage compartment:
Sundowner, Sierra 0·55 m³ (19·5 cu ft)
Sport 0·82 m³ (28·8 cu ft)

AREAS:
Wings, gross 13·57 m² (146 sq ft)
Ailerons (total) 0·92 m² (9·9 sq ft)
Flaps (total) 1·74 m² (18·7 sq ft)
Fin 0·99 m² (10·61 sq ft)
Rudder 0·43 m² (4·62 sq ft)
Tailplane, incl anti-servo tab 2·51 m² (27·0 sq ft)

WEIGHTS AND LOADINGS:
Weight empty (incl oil and unusable fuel):
Sundowner 680 kg (1,500 lb)
Sport 650 kg (1,433 lb)
Sierra 776 kg (1,711 lb)
T-O weight, Utility category:
Sundowner, Sport 920 kg (2,030 lb)
Max T-O weight:
Sundowner 1,111 kg (2,450 lb)
Sport 975 kg (2,150 lb)
Sierra 1,247 kg (2,750 lb)
Max wing loading:
Sundowner 81·9 kg/m² (16·78 lb/sq ft)
Sport 71·9 kg/m² (14·73 lb/sq ft)
Sierra 91·9 kg/m² (18·84 lb/sq ft)
Max power loading:
Sundowner 8·29 kg/kW (13·61 lb/hp)
Sport 8·71 kg/kW (14·33 lb/hp)
Sierra 8·37 kg/kW (13·75 lb/hp)

PERFORMANCE (at max T-O weight):
Max level speed at S/L:
Sundowner 120 knots (222 km/h; 138 mph)
Sport 110 knots (204 km/h; 127 mph)
Sierra 140 knots (259 km/h; 161 mph)
Max cruising speed:
Sundowner 118 knots (219 km/h; 136 mph)
Sport 109 knots (201 km/h; 125 mph)
Sierra 131 knots (243 km/h; 151 mph)
Econ cruising speed:
Sundowner 93 knots (172 km/h; 107 mph)
Sport 85 knots (158 km/h; 98 mph)
Sierra 111 knots (206 km/h; 128 mph)
Stalling speed, flaps down, power off:
Sundowner 52 knots (95 km/h; 59 mph) IAS
Sport 49·5 knots (92 km/h; 57 mph) IAS
Sierra 55 knots (102 km/h; 63 mph) IAS
Max rate of climb at S/L:
Sundowner 241 m (792 ft)/min
Sport 207 m (680 ft)/min
Sierra 272 m (891 ft)/min
Service ceiling:
Sundowner 3,840 m (12,600 ft)
Sport 3,550 m (11,650 ft)
Sierra 4,370 m (14,342 ft)
Absolute ceiling:
Sundowner 4,390 m (14,400 ft)
Sport 4,160 m (13,650 ft)
Sierra 4,618 m (15,150 ft)
Min ground turning radius 6·81 m (22 ft 4 in)
T-O run:
Sundowner 344 m (1,130 ft)
Sport 314 m (1,030 ft)
Sierra 356 m (1,169 ft)
T-O to 15 m (50 ft):
Sundowner 596 m (1,955 ft)
Sport 498 m (1,635 ft)
Sierra 550 m (1,804 ft)
Landing from 15 m (50 ft):
Sundowner 452 m (1,484 ft)
Sport 516 m (1,693 ft)
Sierra 463 m (1,519 ft)
Landing run:
Sundowner 214 m (703 ft)
Sport 251 m (824 ft)
Sierra 254 m (803 ft)
Range at 75% power with max fuel, allowances for warm-up, T-O, climb and 45 min fuel reserves:
Sundowner 459 nm (851 km; 529 miles)
Sport 560 nm (1,038 km; 645 miles)
Sierra 561 nm (1,040 km; 646 miles)
Max cruising range:
Sundowner 508 nm (941 km; 585 miles)
Sport 578 nm (1,072 km; 666 miles)
Sierra 592 nm (1,097 km; 682 miles)

BEECHCRAFT BONANZA MODEL V35B

The prototype Bonanza flew for the first time on 22 December 1945 and the type went into production in 1947. Beech had delivered a total of 9,823 V-tail Bonanzas by 1 January 1976, in which year the Bonanza Model 35 series entered its 29th consecutive year of production.

The current version, designated Model V35B, is described in detail.

The Bonanza Model A36 utility aircraft, and Model F33 series with conventional tail unit, are described separately.

The Bonanzas for 1976 are equipped with a dual-duct fresh air system to increase cabin airflow, and safety features which include a three-light strobe system and single diagonal strap shoulder harness with inertia reel for all occupants as standard equipment. Five optional factory-installed IFR electronics packages include dual communication, navigation, marker beacon, glideslope, DME, and transponder. Three packages meet FAA Technical Standard Order (TSO). Beech was in 1972 the first general aviation manufacturer to acquire IFR approval of factory installation of area navigation equipment on production aircraft with this equipment.

In 1976 Beech was offering five 'Super Utility' packages of optional equipment, as follows:

Package No. 1. Mitchell Century I autopilot with VOR/LOC tracker; 3 in directional gyro and gyro horizon with pressure system; alternate static source; 280 litre (74 US gallon) extended-range fuel tanks; heated pitot tube; super soundproofing; large cargo door; adding 22·7 kg (50 lb) to basic empty weight.

Package No. 2. As above, less Century I autopilot and gyros, plus Mitchell Century III autopilot, which includes roll, pitch and heading control, altitude hold, pitch trim, radio and glideslope couplers and Mitchell 3 in directional gyro and gyro horizon. This installation requires an optional glideslope receiver. Adding 30·4 kg (67 lb) to basic empty weight.

Package No. 3. As package No. 2, but replaces Mitchell Century III by Mitchell Century IV, this including Edo-Aire Mitchell DG-360 directional gyro in place of Mitchell directional gyro. Adding 35 kg (77 lb) to basic empty weight.

Package No. 4. As package No. 1, less Century I autopilot and gyros, plus Bendix FCS-810 autopilot, which includes roll, pitch and heading control, altitude hold, pitch trim, radio and glideslope couplers and Bendix 3 in directional gyro and gyro horizon. This installation requires an optional glideslope receiver. Adding 36 kg (79 lb) to basic empty weight.

Package No. 5. As package No. 1, less Century I autopilot and gyros, plus King KFC-200 automatic flight control system. Includes King KCS-55A electric slaved pictorial navigation system and King air-driven double cue steering horizon. This installation requires King nav receiver, omni converter and glideslope receiver; adding 38·5 kg (85 lb) to basic empty weight.

Other optional extras available on the Bonanza include the Beech-designed 'Magic Hand'. Designed to eliminate the possibility of wheels-up landing or inadvertent retraction of the landing gear on the ground, it lowers the gear automatically on approach when the engine manifold pressure falls below approximately 508 mm (20 in) and airspeed has been reduced to 104 knots (193 km/h; 120 mph). On take-off, it keeps the gear down until the aircraft is airborne and has accelerated to 78 knots (145 km/h; 90 mph) IAS. The system can be switched off by the pilot at will.

An optional 12,000 BTU refrigeration-type air-conditioning system was introduced during 1975, comprising an evaporator located beneath the pilot's seat, condenser mounted on the lower fuselage and an engine-mounted compressor. Air outlets are located on the centre console, and a two-speed blower is provided for air distribution.

TYPE: Four/six-seat light cabin monoplane.

WINGS: Cantilever low-wing monoplane. Wing section Beech modified NACA 23016·5 at root, modified NACA 23012 at tip. Dihedral 6°. Incidence 4° at root, 1° at tip. Sweepback 0° at quarter-chord. Each wing is a two-spar semi-monocoque box-beam of conventional aluminium alloy construction. Symmetrical-section ailerons and single-slotted flaps of aluminium alloy construction. Ground-adjustable trim tab in each aileron.

FUSELAGE: Conventional aluminium alloy semi-monocoque structure. Hat-section longerons and channel-type keels extend forward from cabin section, making the support structure for the engine and nose-wheel an integral part of the fuselage.

TAIL UNIT: Cantilever V tail, with tailplane and elevators set at 33° dihedral angle. Semi-monocoque construction. Fixed surfaces have aluminium alloy structure and skin. Control surfaces, aft of the light alloy spar, are primarily of magnesium alloy, with large controllable trim tab in each. Tail surfaces are interchangeable port and starboard, except for tabs and actuator horns. Electrically-operated elevator trimming optional.

LANDING GEAR: Electrically-retractable tricycle type, with steerable nosewheel. Main wheels retract inward into wings, nosewheel aft. Beech oleo-pneumatic shock-absorbers on all units. Cleveland main wheels, size 6·00-6, and tyres, size 7·00-6, pressure 2·07 bars (30 lb/sq in). Cleveland nosewheel and tyre, size 5·00-5, pressure 2·76 bars (40 lb/sq in). Cleveland ring-disc hydraulic brakes. Parking brake.

POWER PLANT: One 212·5 kW (285 hp) Continental IO-520-BA flat-six engine, driving a McCauley two-blade metal constant-speed propeller. Three-blade McCauley propeller optional. Manually-adjustable engine cowl flaps. Two standard fuel tanks in wing leading-edges, with total usable capacity of 166·5 litres (44 US gallons). Optionally, these can be replaced by tanks with total usable capacity of 280 litres (74 US gallons). Refuelling points above tanks. Optional wingtip tanks, total capacity 151 litres (40 US gallons). Oil capacity 11·5 litres (3 US gallons).

Beechcraft's V-tailed Bonanza Model V35B (Continental IO-520-BA engine)

ACCOMMODATION: Enclosed cabin seating four, five or six persons on individual seats. Centre windows open for ventilation on ground and have release pins to permit their use as emergency exits. Pilot's storm window, port side. Cabin structure reinforced for protection in turn-over. Space for up to 122·5 kg (270 lb) of baggage aft of seats. Passenger door and baggage access door both on starboard side. Cabin heated and ventilated.

SYSTEMS: Electrical system supplied by 70A alternator, 12V 35Ah battery. Hydraulic system for brakes only. Pneumatic system for instrument gyros. Refrigeration-type air-conditioning system optional.

ELECTRONICS AND EQUIPMENT: Standard equipment includes heating, ventilation and defrosting system; electric turn co-ordinator; 8-day clock; outside air temperature gauge; flap position indicator; stall warning system; electroluminescent sub-panel lighting; landing light; taxi light; navigation lights; cabin dome and instrument lights; passengers' reading lights; ultra-violet-proof windscreen and windows; wall-to-wall carpeting; glove compartment; sun visors; pilot's foul weather window; armrests; headrest; inertia reel shoulder harness for all occupants; assist straps; utility shelf; towbar; pitot tube cover; control lock; and winterisation kit. Standard electronics comprise King KX-170B 720-channel nav/com system with KI-201C VOR/LOC converter-indicator, microphone, headset, cabin speaker, Beechcraft B11-1 nav/com/GS antenna and emergency locator transmitter. Optional items include nav/com equipment by Bendix, King, Narco, Sunair, Collins and Pantronics, including marker beacon receiver, ADF, DME, VOR/LOC transmitter-receiver and glideslope receiver, full oxygen system, super soundproofing, co-pilot's toe-operated brakes, internally lighted instruments, instrument post lights, exhaust temperature gauge, control wheel map light, alternate static source, heated pitot tube, standby generator system, external power socket, cabin door courtesy light, electric elevator trim, dual control wheels, rotating beacons, strobe lights, fifth and sixth oxygen outlets, reclining seat adjusters for 2nd, 3rd and 4th seats, headrests and fifth and sixth seats.

DIMENSIONS, EXTERNAL:

Wing span	10·21 m (33 ft 6 in)
Wing chord at root	2·13 m (7 ft 0 in)
Wing chord at tip	1·07 m (3 ft 6 in)
Wing aspect ratio	6·2
Length overall	8·05 m (26 ft 5 in)
Height overall	2·31 m (7 ft 7 in)
Tailplane span	3·10 m (10 ft 2 in)
Wheel track	2·92 m (9 ft 7 in)
Wheelbase	2·13 m (7 ft 0 in)
Propeller diameter:	
Two-blade	2·13 m (7 ft 0 in)
Three-blade	2·03 m (6 ft 8 in)
Passenger door:	
Height	0·91 m (3 ft 0 in)
Width	0·94 m (3 ft 1 in)
Baggage compartment door:	
Height	0·47 m (1 ft 6½ in)
Width	0·57 m (1 ft 10½ in)

DIMENSIONS, INTERNAL:

Cabin, aft of firewall:	
Length	3·07 m (10 ft 1 in)
Max width	1·07 m (3 ft 6 in)
Max height	1·27 m (4 ft 2 in)
Volume	3·31 m³ (117 cu ft)
Baggage space	0·99 m³ (35 cu ft)

AREAS:

Wings, gross	16·80 m² (181 sq ft)
Ailerons (total)	1·06 m² (11·4 sq ft)
Trailing-edge flaps (total)	1·98 m² (21·3 sq ft)
Fixed tail surfaces	2·20 m² (23·8 sq ft)
Movable tail surfaces, incl tabs	1·34 m² (14·4 sq ft)

WEIGHTS AND LOADINGS:

Weight empty, equipped	930 kg (2,051 lb)
Max T-O and landing weight	1,542 kg (3,400 lb)
Max wing loading	91·8 kg/m² (18·80 lb/sq ft)
Max power loading	7·26 kg/kW (11·96 lb/hp)

PERFORMANCE (at max T-O weight):

Max level speed at S/L	182 knots (338 km/h; 210 mph)
Max cruising speed (75% power) at 1,980 m (6,500 ft)	176 knots (327 km/h; 203 mph)
Econ cruising speed (45% power) at 3,660 m (12,000 ft)	142 knots (264 km/h; 164 mph)
Stalling speed, wheels and flaps up	65 knots (119 km/h; 74 mph)
Stalling speed, wheels and flaps down	55 knots (102 km/h; 63 mph)
Max rate of climb at S/L	346 m (1,136 ft)/min
Service ceiling	5,335 m (17,500 ft)
Absolute ceiling	5,850 m (19,200 ft)
Min ground turning radius	3·72 m (12 ft 2½ in)
T-O run	340 m (1,115 ft)
T-O to 15 m (50 ft)	570 m (1,870 ft)
Landing from 15 m (50 ft)	459 m (1,505 ft)
Landing run	243 m (797 ft)
Range, 45% power at 3,660 m (12,000 ft) with max fuel and allowances for warm-up, T-O, climb and 45 min reserves	874 nm (1,620 km ; 1,007 miles)

Beechcraft Model F33C Bonanzas in the insignia of the Mexican Air Force

BEECHCRAFT BONANZA MODEL F33A/C

The **F33A** version of the Bonanza is a four/six-seat single-engined executive aircraft, similar in general configuration to the Bonanza Model V35B, but distinguished by a conventional tail unit with sweptback vertical surfaces. The prototype flew for the first time on 14 September 1959, and the production models were known as Debonairs until 1967.

The **F33C** is generally similar to the A, but approved for both aerobatic and utility operation.

The Model G33 was discontinued in early 1973 after 49 had been produced.

A total of 2,007 Model 33s had been built by 1 January 1976. Twenty-one were bought for pilot training by Lufthansa in Germany, and Pacific Southwest Airlines acquired ten G33s for airline crew training. Deliveries of F33As and aerobatic F33Cs made during 1975 to equip foreign air forces were as follows: Imperial Air Force of Iran, 16 F33Cs; Mexican Air Force, 5 F33Cs; Spanish Air Ministry, 12 F33As and 4 F33Cs, part of an order for 18 of the aerobatic Bonanzas.

Optional extras include the 'Magic Hand' automatic landing gear control system, air-conditioning system and other items described under the Model V35B Bonanza entry.

TYPE: Four/six-seat cabin monoplane.

WINGS: As for V35B Bonanza.

FUSELAGE: As for Bonanza V35B series.

TAIL UNIT: Conventional cantilever all-metal stressed-skin structure, primarily of aluminium alloy but with corrugated magnesium skin on elevators. Large trim tab in each elevator. Fixed tab in rudder.

LANDING GEAR: As for Bonanza V35B series. Main wheels size 6·00-6, with tyres size 7·00-6, pressure 2·07 bars (30 lb/sq in); nosewheel size 5·00-5, tyre pressure 2·76 bars (40 lb/sq in).

POWER PLANT: One 212·5 kW (285 hp) Continental IO-520-BA flat-six engine, driving a McCauley two-blade metal constant-speed propeller. Fuel tanks and capacity as for V35B. Refuelling points on wing upper surface. Oil capacity 11·5 litres (3 US gallons).

ACCOMMODATION: Enclosed cabin with four individual seats in pairs as standard, plus optional forward-facing fifth and sixth seats. Baggage compartment and hat shelf aft of seats. Passenger door and baggage compartment door on starboard side. Heater standard. Large cargo door, on starboard side of fuselage, optional.

SYSTEMS: Electrical system supplied by 70A alternator. 12V 35Ah battery. Hydraulic system for brakes only. Pneumatic system for instrument gyros. Refrigeration-type air-conditioning system optional.

ELECTRONICS AND EQUIPMENT: As for V35B.

DIMENSIONS, EXTERNAL: As for V35B except:

Length overall	8·13 m (26 ft 8 in)
Height overall	2·51 m (8 ft 3 in)
Tailplane span	3·71 m (12 ft 2 in)
Propeller diameter	2·13 m (7 ft 0 in)

DIMENSIONS, INTERNAL: As for V35B

AREAS: As for V35B except:

Fin	0·85 m² (9·1 sq ft)
Rudder, incl tab	0·43 m² (4·6 sq ft)
Tailplane	1·75 m² (18·82 sq ft)
Elevators, incl tabs	1·24 m² (13·36 sq ft)

WEIGHTS AND LOADINGS:

Weight empty	942 kg (2,076 lb)
Max T-O and landing weight	1,542 kg (3,400 lb)
Max wing loading	91·8 kg/m² (18·8 lb/sq ft)
Max power loading	7·26 kg/kW (11·96 lb/hp)

PERFORMANCE:

Max level speed at S/L	181 knots (335 km/h; 208 mph)
Max cruising speed, 75% power at 1,980 m (6,500 ft)	174 knots (322 km/h; 200 mph)
Econ cruising speed, 45% power at 3,660 m (12,000 ft)	136 knots (253 km/h; 157 mph)
Stalling speed, wheels and flaps up	64·5 knots (119 km/h; 74 mph)
Stalling speed, wheels and flaps down	55 knots (101·5 km/h; 63 mph)
Max rate of climb at S/L	346 m (1,136 ft)/min
Service ceiling	5,335 m (17,500 ft)
Absolute ceiling	5,850 m (19,200 ft)
Min ground turning radius	3·72 m (12 ft 2½ in)
T-O run	340 m (1,115 ft)
T-O to 15 m (50 ft)	570 m (1,870 ft)
Landing from 15 m (50 ft)	459 m (1,505 ft)
Landing run	243 m (797 ft)
Range at econ cruising speed with max fuel, allowances for warm-up, T-O, climb and 45 min reserves	847 nm (1,570 km; 976 miles)

BEECHCRAFT BONANZA MODEL A36

This version of the Bonanza, introduced in mid-1968, is a full six-seat utility aircraft developed from the Bonanza Model V35B. It is generally similar to the V35B, but has a conventional tail unit with sweptback vertical surfaces, similar to that of the F33 series of Bonanzas. In addition, the A36 has large double doors on the starboard side of the fuselage aft of the wing root, to facilitate loading and unloading of bulky cargo when used in a utility role. The cabin volume is increased by 0·34 m³ (12 cu ft) compared with the V35B.

Like all Bonanzas, the Model A36 is licensed in the FAA Utility category at full gross weight, with no limitation of performance. A total of 819 civil versions of the Model 36/A36 had been built by 1 January 1976.

The current version of the Bonanza Model A36 introduced the improvements noted for the Bonanza V35B as well as options which include a club-seating interior layout with rear-facing third and fourth seats, executive writing desk, headrests for third and fourth seats, reading lights and fresh air outlets for fifth and sixth seats. Optional extras include the 'Magic Hand' automatic landing gear control system, refrigeration-type air-conditioning system and all other items mentioned under the Model V35B Bonanza entry, except for the large cargo door.

TYPE: Four/six-seat utility light cabin monoplane.

WINGS: As for Model V35B.

FUSELAGE: As for Model V35B but lengthened by 0·254 m (10 in).

TAIL UNIT: Conventional cantilever all-metal stressed-skin structure, primarily of aluminium alloy but with corrugated magnesium skin on elevators. Large trim tab in each elevator. Fixed tab in rudder.

LANDING GEAR: Electrically-retractable tricycle type, similar to that of Baron. Main units retract inward into wings, nosewheel rearward. Beech oleo-pneumatic shock-absorbers. Steerable nosewheel. Cleveland main wheels, size 6·00-6, with tyres size 7·00-6, pressure 2·07 bars (30 lb/sq in). Nosewheel tyre size 5·00-5, pressure 2·76 bars (40 lb/sq in). Cleveland ring-disc hydraulic brakes.

POWER PLANT: As for Model V35B.

ACCOMMODATION: Enclosed cabin seating four,five or six persons on individual seats. Two rear removable seats and two folding seats permit rapid conversion to utility configuration. Optional club-seating layout with rear-facing third and fourth seats, executive writing desk, headrests for third and fourth seats, reading lights and fresh air outlets for fifth and sixth seats. Double doors of bonded aluminium honeycomb construction on starboard side facilitate loading of cargo. As an air ambulance, one stretcher can be accommodated with ample room for a medical attendant and/or other passengers.

Extra windows provide improved view for passengers. Stowage for 181 kg (400 lb) of baggage.

SYSTEMS: Electrical system supplied by 70A alternator, 12V 35Ah battery. Hydraulic system for brakes only. Pneumatic system for instrument gyros. Refrigeration-type air-conditioning system optional.

ELECTRONICS AND EQUIPMENT: Standard equipment includes King KX-170B 720-channel nav/com, with KI-201C VOR/LOC Omni converter/indicator and Beechcraft antenna, but a wide range of optional electronics equipment is available. Optional items of equipment are as detailed for the V35B Bonanza, except as noted.

DIMENSIONS, EXTERNAL: As for V35B except:

Length overall	8·38 m (27 ft 6 in)
Height overall	2·57 m (8 ft 5 in)
Wheelbase	2·39 m (7 ft 10¼ in)
Rear passenger/cargo door:	
Height	1·02 m (3 ft 4 in)
Width	1·14 m (3 ft 9 in)

DIMENSIONS, INTERNAL:

Cabin, aft of firewall:	
Length	3·33 m (10 ft 11 in)
Max width	1·07 m (3 ft 6 in)
Max height	1·27 m (4 ft 2 in)
Volume	3·65 m³ (129 cu ft)
Baggage compartment	1·13 m³ (40 cu ft)

AREA:

Wings, gross	16·8 m² (181 sq ft)

WEIGHTS AND LOADINGS:

Weight empty, equipped	957 kg (2,111 lb)
Max T-O weight	1,633 kg (3,600 lb)
Max wing loading	97·2 kg/m² (19·9 lb/sq ft)
Max power loading	7·68 kg/kW (12·6 lb/hp)

PERFORMANCE:

Max level speed at S/L 178 knots (323 km/h; 204 mph)

Max cruising speed (75% power) at 1,830 m (6,000 ft) 170 knots (315 km/h; 196 mph)

Cruising speed (65% power) at 3,050 m (10,000 ft) 162 knots (301 km/h; 187 mph)

Cruising speed (55% power) at 3,660 m (12,000 ft) 148 knots (274 km/h; 170 mph)

Stalling speed, power off, wheels and flaps up 63 knots (116 km/h; 72 mph)

Stalling speed, power off, wheels down and 30° flap 52·5 knots (97 km/h; 60 mph)

Max rate of climb at S/L	309 m (1,015 ft)/min
Service ceiling	4,875 m (16,000 ft)
Absolute ceiling	5,425 m (17,800 ft)
Min ground turning radius	4·20 m (13 ft 9½ in)
T-O run	383 m (1,257 ft)
T-O to 15 m (50 ft)	660 m (2,167 ft)
Landing from 15 m (50 ft)	480 m (1,575 ft)
Landing run	254 m (833 ft)

Range with max fuel, with allowances for warm-up, T-O, climb and 45 min fuel reserves:

55% power at 3,050 m (10,000 ft) 769 nm (1,425 km; 886 miles)

65% power at 3,050 m (10,000 ft) 749 nm (1,388 km; 863 miles)

75% power at 1,980 m (6,500 ft) 694 nm (1,287 km; 800 miles)

Beechcraft Bonanza Model A36 (Continental IO-520-BA engine)

BEECHCRAFT BARON MODEL 95-B55

US Army designation: T-42A Cochise

The original Baron Model 95-55 was a four/five-seat cabin monoplane developed from the earlier Travel Air but with more power, better all-weather capability and airframe refinements that included a swept tail-fin. It first flew in prototype form on 29 February 1960 and was licensed in the FAA Normal category in November 1960. The Baron Model 95-B55 was similarly licensed in September 1963.

The current Barons are optional four-, five- or six-seaters, with interior features as described for the Bonanza.

In February 1965 the US Army selected the Model 95-B55 as winner of its competition for a twin-engined fixed-wing instrument trainer. Beech identified the military trainer as the Model 95-B55B, and this received FAA Type Approval in the Normal and Utility categories in August 1964. The US Army ordered 65, which were delivered under the designation T-42A. During 1971 Beech delivered five more T-42As to the US Army, for service with the army of Turkey, under the Military Assistance Programme. Export deliveries of the standard Model 95-B55 have also been made, including 15 for the Spanish Air Minstry and six for the Civil Air Bureau of Japan. These aircraft are used as instrument trainers.

A total of 1,996 civil and military Barons of the Model 95-55 series had been delivered by 1 January 1976.

TYPE: Four/six-seat cabin monoplane.

WINGS: Cantilever low-wing monoplane. Wing section NACA 23016·5 at root, NACA 23010·5 at tip. Dihedral 6°. Incidence 4° at root, 0° at tip. No sweepback. Each wing is a two-spar semi-monocoque box beam of conventional aluminium alloy construction. Symmetrical-section ailerons of light alloy construction, with beaded skins. Electrically-operated single-slotted light alloy trailing-edge flaps, with beaded skins. Manually-operated trim tab in port aileron. Pneumatic rubber de-icing boots optional.

FUSELAGE: Semi-monocoque aluminium alloy structure. Hat-section longerons and channel-type keels extend forward from the cabin section, making the support structure for the forward nose section and nosewheel gear an integral part of the fuselage.

TAIL UNIT: Cantilever all-metal structure. Elevators have smooth magnesium alloy skins. Manually-operated trim tab in each elevator and in rudder. Pneumatic rubber de-icing boots optional.

LANDING GEAR: Electrically-retractable tricycle type. Main units retract inward into wings, nosewheel aft. Beech oleo-pneumatic shock-absorbers on all units. Steerable nosewheel with shimmy damper. Cleveland wheels, with 6·50-8 main-wheel tyres, pressure 3·45-3·72 bars (50-54 lb/sq in). Nosewheel tyre size 5·00-5, pressure 3·31-3·59 bars (48-52 lb/sq in). Cleveland ring-disc hydraulic brakes. Parking brake.

POWER PLANT: Two 194 kW (260 hp) Continental IO-470-L flat-six engines, each driving a Hartzell two-blade constant-speed fully-feathering propeller. Optional Hartzell three-blade propellers. Manually-operated cowl flaps. Standard fuel system comprises two interconnected tanks in each wing leading-edge, with total usable capacity of 378 litres (100 US gallons). Optional interconnected fuel tanks may be added in each wing to provide a total usable capacity of 514 litres (136 US gallons). Single refuelling point in each wing for the standard or optional fuel systems. Optional fuel system includes a mechanical sight gauge in each wing leading-edge to give partial fuelling information. Oil capacity 23 litres (6 US gallons). Propeller de-icing optional.

ACCOMMODATION: Standard model has four individual seats in pairs in enclosed cabin, with door on starboard side. Single diagonal strap shoulder harness with inertia reel standard. Optional wider door for cargo. Folding airline-style fifth and sixth seats optional, complete with shoulder harness and inertia reel. Baggage compartments aft of cabin and in nose, both with external doors and with capacity of 181 kg (400 lb) and 136 kg (300 lb) respectively. An extended rear compartment providing for an additional 54 kg (120 lb) of baggage is optional. Pilot's storm window, port side. Openable windows adjacent to the third and fourth seats are used for ground ventilation and as emergency exits. Cabin heated and ventilated. Windscreen defrosting standard. Alcohol de-icing for port side of windscreen optional.

SYSTEMS: Cabin heated by Janitrol 50,000 BTU heater, which serves also for windscreen defrosting. Oxygen system of 1·39 m³ (49 cu ft) or 1·87 m³ (66 cu ft) capacity optional. Electrical system includes two 24V 25A generators. One 24V 17Ah battery. Two 24V 50A engine-driven alternators and/or two 12V 25Ah batteries optional. Hydraulic system for brakes only. Pneumatic vacuum system for instrument gyros and optional wing and tail unit de-icing system.

ELECTRONICS AND EQUIPMENT: Standard equipment includes blind-flying instrumentation; electric turn co-ordinator; outside air temperature gauge; 8-day clock; vacuum gauge; flap position indicator; heated pitot tube; heated fuel vents; ultra-violet-proof windscreen and windows; soundproofing; wall-to-wall carpeting; glove compartment; sun visors; in-flight storage pockets; armrests; pilot's seat headrest; armrests on all seats; two landing lights; navigation lights; cabin dome, instrument and map lights; two cabin reading lights for each passenger; cabin door courtesy light; towbar; control locks; and winterisation kit. Standard electronics comprise 720-channel communications transceiver; 200-channel navigation receiver with VOR/LOC converter-indicator; emergency locator transmitter; microphone; headset; cabin speaker and nav/com GS antenna. Optional electronics include a wide range of nav/coms, ADF, DME, glideslope and marker beacon installations by Bendix, Collins, King, Narco and RCA. Bendix, King or Edo Aire Mitchell autopilot systems, and Bendix, King or RCA radar systems are also optional. Optional equipment includes rotating beacons, dual control wheels, executive desk, oxygen installation, engine and flight hour recorders, exhaust temperature gauge, internally lighted instruments, instrument post lights, control-wheel chronograph, three-light strobe system, nosewheel taxi light, wing ice lights, two 50A belt-driven alternators with alternator failure lights, co-pilot's control wheel map light, heated stall warning vane, reading light and air outlets for fifth and sixth seats, lightweight automatic de-icer installation, alternate static source, Hartzell three-blade propellers, propeller unfeathering accumulators, propeller synchroniser, co-pilot brakes.

Beechcraft Baron Model B55 (two Continental IO-470-L engines)

external power socket, super soundproofing, seat headrests and extended baggage compartment.

Dimensions, external:

Wing span 11·53 m (37 ft 10 in)
Wing chord at root 2·13 m (7 ft 0 in)
Wing chord at tip 0·90 m (2 ft 11·6 in)
Wing aspect ratio 7·16
Length overall 8·53 m (28 ft 0 in)
Height overall 2·92 m (9 ft 7 in)
Tailplane span 4·19 m (13 ft 9 in)
Wheel track 2·92 m (9 ft 7 in)
Wheelbase 2·13 m (7 ft 0 in)
Propeller diameter:
two-blade 1·98 m (6 ft 6 in)
three-blade 1·93 m (6 ft 4 in)
Passenger door:
Height 0·91 m (3 ft 0 in)
Width 0·94 m (3 ft 1 in)
Height to step 0·33 m (1 ft 1 in)
Baggage door (fwd):
Height 0·56 m (1 ft 10 in)
Width 0·64 m (2 ft 1 in)
Baggage door (rear):
Standard:
Height 0·57 m (1 ft 10½ in)
Width 0·47 m (1 ft 6½ in)
Height to sill 0·71 m (2 ft 4 in)
Optional:
Height 0·57 m (1 ft 10½ in)
Width 0·97 m (3 ft 2 in)

Dimensions, internal:

Cabin: Length 3·07 m (10 ft 1 in)
Max width 1·07 m (3 ft 6 in)
Max height 1·27 m (4 ft 2 in)
Baggage compartment (fwd) 0·34 m^3 (12 cu ft)
Baggage compartment (rear) 0·99 m^3 (35 cu ft)

Areas:

Wings, gross 18·50 m^2 (199·2 sq ft)
Ailerons (total) 1·06 m^2 (11·40 sq ft)
Trailing-edge flaps (total) 2·39 m^2 (25·70 sq ft)
Fin 1·02 m^2 (11·00 sq ft)
Rudder, incl tab 1·08 m^2 (11·60 sq ft)
Tailplane 4·46 m^2 (48·06 sq ft)
Elevators, incl tabs 1·51 m^2 (16·20 sq ft)

Weights and Loadings:

Weight empty 1,431 kg (3,156 lb)
Max T-O and landing weight 2,313 kg (5,100 lb)
Max ramp weight 2,322 kg (5,121 lb)
Max wing loading 120·5 kg/m^2 (25·6 lb/sq ft)
Max power loading 5·96 kg/kW (9·8 lb/hp)

Performance (at max T-O weight):

Max level speed at S/L 205 knots (380 km/h; 236 mph)
Max cruising speed, 75% power at 2,135 m (7,000 ft) 195 knots (362 km/h; 225 mph)
Cruising speed, 55% power at 3,660 m (12,000 ft) 180 knots (333 km/h; 207 mph)
Stalling speed (power off, wheels and flaps down) 73 knots (135·5 km/h; 84 mph) IAS
Max rate of climb at S/L 510 m (1,670 ft)/min
Rate of climb at S/L, one engine out 97 m (318 ft)/min
Service ceiling 6,000 m (19,700 ft)
Service ceiling, one engine out 2,135 m (7,000 ft)
Min ground turning radius 3·71 m (12 ft 2 in)
Runway LCN 2
T-O run 409 m (1,340 ft)
T-O to 15 m (50 ft) 511 m (1,675 ft)
Landing from 15 m (50 ft) 561 m (1,840 ft)
Landing run 287 m (940 ft)
Range, 65% power at 3,200 m (10,500 ft), with 514 litres (136 US gallons) fuel, and with allowances for warm-up, taxi, T-O, climb to altitude and 45 min reserves 942 nm (1,746 km; 1,085 miles)

BEECHCRAFT BARON MODEL E55

The Baron E55 had its origin in the Baron 95-B55 when the Model 95-C55 was added to the Baron series of twin-engined aircraft in August 1965. The 95-C55 had Continental IO-520-C engines, a pneumatic vacuum system for instrument gyros and the optional wing and tail unit de-icing system, two 24V 50A engine-driven alternators, increased tailplane span, swept vertical surfaces and an extended nose baggage compartment. It was followed by the D55 in October 1967, this model introducing a pneumatic pressure system in place of the pneumatic vacuum system. The subsequent Model E55, which has an improved interior and systems accessory refinements, was licensed in the FAA Normal category on 12 November 1969.

Beech had delivered a total of 1,063 of this Baron series by 1 January 1976.

Type: Four/six-seat cabin monoplane.

Wings: As for Model 95-B55.

Fuselage: As for Model 95-B55, except nose extended by 0·305 m (1 ft 0 in).

Tail Unit: As for Model 95-B55, except tailplane span increased.

Landing Gear: As for Model 95-B55, except main-wheel tyre pressure 3·59-3·86 bars (52-56 lb/sq in); nosewheel tyre pressure 3·79-4·14 bars (55-60 lb/sq in).

Power Plant: Two 212·5 kW (285 hp) Continental IO-520-C flat-six engines, each driving a Hartzell two-blade metal constant-speed and fully-feathering propeller. Hartzell three-blade propellers optional. Fuel system as for Model 95-B55, except optional maximum capacity 628 litres (166 US gallons). Oil capacity 23 litres (6 US gallons). Full-flow oil filters standard; propeller de-icing optional.

Accommodation: As for Model 95-B55, except that extended rear compartment, providing for an additional 54 kg (120 lb) of baggage, is standard.

Systems: As for Model 95-B55, except pneumatic pressure system for instrument gyros and optional wing and tail unit de-icing system. Electrical system supplied by two 24V 50A engine-driven alternators, with failure lights, and one 24V 12Ah battery.

Electronics and Equipment: Equipment as for Model 95-B55, except that instrument air pressure and flap position indicators are standard; electrical operation of trim tabs optional. Electronics as for Model 95-B55, except that an ADF receiver is standard.

Dimensions, external:

As for Model 95-B55, except:
Length overall 8·84 m (29 ft 0 in)
Height overall 2·79 m (9 ft 2 in)
Tailplane span 4·85 m (15 ft 11 in)
Wheelbase 2·46 m (8 ft 1 in)

Dimensions, internal:

As for Model 95-B55, except:
Cabin: Length (incl extended rear baggage compartment) 3·58 m (11 ft 9 in)
Baggage compartment (fwd) 0·51 m^3 (18 cu ft)
Extended rear baggage compartment 1·27 m^3 (45 cu ft)

Areas:

As for Model 95-B55, except:
Tailplane 4·95 m^2 (53·30 sq ft)
Elevators, incl tabs 1·84 m^2 (19·80 sq ft)

Weights and Loadings:

Weight empty 1,447 kg (3,191 lb)
Max T-O and landing weight 2,405 kg (5,300 lb)
Max ramp weight 2,415 kg (5,324 lb)
Max wing loading 130·0 kg/m^2 (26·6 lb/sq ft)
Max power loading 5·66 kg/kW (9·3 lb/hp)

Performance (at max T-O weight):

Max level speed at S/L 210 knots (390 km/h; 242 mph)
Max cruising speed, 75% power at 2,135 m (7,000 ft) 200 knots (370 km/h; 230 mph)
Econ cruising speed, 55% power at 3,660 m (12,000 ft) 181 knots (335 km/h; 208 mph)
Stalling speed (power off, wheels and flaps down) 72 knots (134 km/h; 83 mph) IAS
Max rate of climb at S/L 510 m (1,670 ft)/min
Rate of climb at S/L, one engine out 102 m (335 ft)/min
Service ceiling 6,370 m (20,900 ft)
Service ceiling, one engine out 2,165 m (7,100 ft)
Absolute ceiling 6,797 m (22,300 ft)
Absolute ceiling, one engine out 2,530 m (8,300 ft)
Min ground turning radius 4·27 m (14 ft 0 in)
Runway LCN 2
T-O run 408 m (1,337 ft)
T-O to 15 m (50 ft) 497 m (1,631 ft)
Landing from 15 m (50 ft) 548 m (1,798 ft)
Landing run 277 m (908 ft)
Range, 65% power at 3,350 m (11,000 ft) with 514 litres (136 US gallons) fuel, allowances for warm-up, taxi, T-O, climb to altitude and 45 min reserves 833 nm (1,545 km; 960 miles)
Range, conditions and allowances as above, but with 628 litres (166 US gallons) fuel 1,052 nm (1,950 km; 1,212 miles)

BEECHCRAFT BARON MODEL 58

In late 1969 Beech introduced a new version of the Baron, designated Model 58. Developed from the Baron D55, it differed by having the forward cabin section extended by 0·254 m (10 in), allowing the windscreen, passenger door, instrument panel and front seats to be moved forward and so provide a more spacious cabin. This change was made without affecting the wing main spar location, but the wheelbase was extended by moving the nosewheel forward, to improve ground handling. New features included double passenger/cargo doors on the starboard side of the cabin, extended propeller hubs, redesigned engine nacelles to improve cooling, and a fourth window on each side of the cabin. The Model 58 Baron was licensed by the FAA in the Normal category on 19 November 1969.

Beech had delivered more than 700 of this Baron series (including Baron 58Ps and 58TCs) by 1 May 1976.

Type: Four/six-seat cabin monoplane.

Wings: As for Model 95-B55.

Fuselage: As for Model E55, except forward cabin section extended by 0·254 m (10 in).

Tail Unit: As for Model E55.

Landing Gear: As for Model E55, except wheelbase extended by 0·254 m (10 in).

Power Plant: As for Model E55, except that the standard and optional Hartzell propellers have extended hubs, and the engine nacelles are lengthened to accommodate these. The standard fuel system has a usable capacity of 514 litres (136 US gallons), with optional usable capacity of 628 litres (166 US gallons). Optional 'wet wingtip' installation also available, increasing usable capacity to 734 litres (194 US gallons).

Accommodation: As for Model E55, except that folding fifth and sixth seats, or club seating comprising folding fifth and sixth seats and aft-facing third and fourth seats, are optional. Double passenger/cargo doors on starboard side of cabin provide access to space for 181 kg (400 lb) of baggage or cargo behind the third and fourth seats.

Systems: As for Model E55.

Electronics and Equipment: Equipment as for Model E55, except heated stall warning vane and double passenger/cargo doors with door-ajar warning light standard. Hartzell three-blade propellers with extended hubs, and club seating arrangement, as described above, optional. Standard and optional electronics as for Model E55.

Dimensions, external, and Areas:

As for Model E55, except:
Length overall 9·09 m (29 ft 10 in)
Height overall 2·90 m (9 ft 6 in)
Wheelbase 2·72 m (8 ft 11 in)
Rear passenger/cargo doors:
Max height 1·02 m (3 ft 4 in)
Width 1·14 m (3 ft 9 in)
Emergency exit window:
Height 0·53 m (1 ft 9 in)
Width 0·61 m (2 ft 0 in)

Dimensions, internal:

As for Model E55, except:
Cabin, incl rear baggage area:
Length 3·84 m (12 ft 7 in)
Floor area 3·72 m^2 (40 sq ft)
Volume 3·85 m^3 (135·9 cu ft)

Weights and Loadings:

Weight empty 1,490 kg (3,286 lb)
Max T-O and landing weight 2,449 kg (5,400 lb)
Max wing loading 132·3 kg/m^2 (27·1 lb/sq ft)
Max power loading 5·76 kg/kW (9·5 lb/hp)

Performance (at max T-O weight):

Never-exceed speed 223 knots (414 km/h; 257 mph)
Max level speed at S/L 210 knots (390 km/h; 242 mph)
Max cruising speed, 75% power at 2,135 m (7,000 ft) 200 knots (370 km/h; 230 mph)
Cruising speed, 65% power at 3,350 m (11,000 ft) 195 knots (362 km/h; 225 mph)
Cruising speed, 55% power at 3,660 m (12,000 ft) 180 knots (333 km/h; 207 mph)
Stalling speed, power off, landing gear and flaps down 74 knots (137 km/h; 85 mph) IAS
Max rate of climb at S/L 516 m (1,694 ft)/min
Rate of climb at S/L, one engine out 116 m (382 ft)/min
Service ceiling 5,425 m (17,800 ft)
Service ceiling, one engine out 2,180 m (7,150 ft)
Min ground turning radius 4·57 m (15 ft 0 in)
Runway LCN 2
T-O run 428 m (1,403 ft)
T-O to 15 m (50 ft) 520 m (1,706 ft)
Landing from 15 m (50 ft) 631 m (2,070 ft)
Landing run 318 m (1,044 ft)
Cruising range, 65% power at 3,350 m (11,000 ft) with 628 litres (166 US gallons) fuel and allowances for warm-up, taxi, T-O, climb to altitude and 45 min reserves 1,052 nm (1,950 km; 1,212 miles)

BEECHCRAFT BARON MODEL 58P

Design of this pressurised version of the Model 58 Baron started in June 1972; the first flight of the prototype was made in August 1973. The first production aircraft was flown in late 1974 and certification under FAR Part 23 was received in early 1975. The Model 58P is powered by two 231 kW (310 hp) Continental TSIO-520-L engines. Its pressurisation system provides a 2,440 m (8,000 ft) cabin environment to an altitude of 5,500 m (18,000 ft).

Deliveries of production aircraft began in late 1975, and a total of 21 Baron 58Ps had been delivered by 1 January 1976.

Type: Four/six-seat cabin monoplane.

Wings: As for Model 95-B55.

Fuselage: As for Model 58, except structural reinforcement to cater for pressurisation.

Tail Unit: As for Model 58.

Landing Gear: As for Model 58, except Cleveland main wheels with tyres size 19·5 × 6·75-6, 10-ply rating, pressure 5·52 bars (80 lb/sq in). Cleveland nosewheel with tyre size 15× 5-5, 6-ply rating, pressure 4 bars (58 lb/sq in). Cleveland aircooled single-ring hydraulic disc brakes.

Power Plant: Two 231 kW (310 hp) Continental TSIO-520-L turbocharged flat-six engines, each driving a Hartzell three-blade metal constant-speed and fully-feathering propeller. Integral fuel tanks in wings, with standard capacity of 651 litres (172 US gallons) of which 401 litres (166 US gallons) are usable. Optional maximum capacity of 742 litres (196 US gallons) of which 719 litres (190 US gallons) are usable. Refuelling points in outboard leading-edge of wings and, for optional maximum fuel, in wingtips. Oil capacity 22·7 litres (6 US gallons). Electrical anti-icing for propellers optional.

ACCOMMODATION: Standard accommodation has four individual seats in pairs, facing forward, with shoulder harness and inertia reel. Fifth and sixth seats optional, as is club layout. Doors on starboard side, adjacent to co-pilot, and at trailing-edge of wing on port side. Baggage space in aft cabin and in fuselage nose, with door on starboard side of nose. Openable storm window for pilot on port side. Cabin heated, air-conditioned and pressurised. Windscreen defrosting by hot air. Alcohol anti-icing system for port side of windscreen optional.

SYSTEMS: AiResearch pressurisation with max differential of 0·26 bars (3·7 lb/sq in). Beechcraft 16,000 BTU air-conditioning. Janitrol 35,000 BTU heater. Engine-driven compressors supply air for flight instruments, pressurisation control and optional pneumatic de-icing boots. Electrical system powered by two 24V 50A alternators, with two 12V 24Ah storage batteries. Two 24V 100A alternators optional. Hydraulic system for brakes only. Oxygen system of 0·42 m³ (15 cu ft) optional.

ELECTRONICS AND EQUIPMENT: Standard electronics package comprises King KX-170B nav/com (720-channel com transceiver and 200-channel nav receiver) with KI-201C VOR/LOC converter-indicator, KR-85 ADF with KI-225 indicator, microphone, headset, cabin speaker and B12 nav and B6 com antennae. Optional electronics by Bendix, Collins, King, Narco and RCA. Standard equipment as for Model E55, plus heated stall warning vane, dual rotating beacons, step light, door-ajar warning lights, nose baggage compartment light, emergency locator transmitter and exterior acrylic enamel paint. Optional equipment includes refrigeration-type air-conditioner, co-pilot brakes, dual control wheels, control wheel clock, executive writing desk, cabin fire extinguisher, engine and flight hour recorders, strobe lights, wing ice lights, instrument post lights, internally illuminated instruments, propeller synchroniser and unfeathering accumulators, static wicks, elevator electric trim control and a variety of interior furnishings.

DIMENSIONS, EXTERNAL:

Wing span	11·53 m (37 ft 10 in)
Wing chord at root	2·62 m (8 ft 7 in)
Wing chord at tip	1·07 m (3 ft 6 in)
Length overall	9·09 m (29 ft 10 in)
Height overall	2·79 m (9 ft 2 in)
Tailplane span	4·85 m (15 ft 11 in)
Wheel track	2·92 m (9 ft 7 in)
Wheelbase	2·72 m (8 ft 11 in)
Propeller diameter	2·03 m (6 ft 8 in)
Propeller ground clearance	0·25 m (9¾ in)
Passenger door (starboard, fwd):	
Height	0·91 m (3 ft 0 in)
Width	0·94 m (3 ft 1 in)
Height to sill	0·51 m (1 ft 8 in)
Passenger door (port, aft):	
Height	0·99 m (3 ft 3 in)
Width	0·56 m (1 ft 10 in)
Height to sill	0·79 m (2 ft 7 in)
Baggage door (nose, stbd):	
Height	0·38 m (1 ft 3 in)
Width	0·64 m (2 ft 1 in)

DIMENSIONS, INTERNAL:

Cabin: Length (incl rear baggage compartment)	3·84 m (12 ft 7 in)
Max width	1·07 m (3 ft 6 in)
Max height	1·27 m (4 ft 2 in)
Floor area	3·72 m² (40 sq ft)
Volume	3·885 m³ (137·2 cu ft)
Baggage compartment (nose)	0·51 m³ (18 cu ft)
Rear baggage compartment	0·28 m³ (10 cu ft)

AREAS: As Model 95-B55 except:

Tailplane	4·95 m² (53·30 sq ft)
Elevators, incl tabs	1·84 m² (19·80 sq ft)

WEIGHTS AND LOADINGS:

Weight empty, equipped	1,808 kg (3,985 lb)
Max T-O weight	2,785 kg (6,140 lb)
Max zero-fuel weight	2,585 kg (5,700 lb)
Max landing weight	2,767 kg (6,100 lb)
Max wing loading	158·2 kg/m² (32·4 lb/sq ft)
Max power loading	6·03 kg/kW (9·90 lb/hp)

PERFORMANCE (at max T-O weight):

Never-exceed speed	235 knots (436 km/h; 271 mph) IAS
Max level speed at 4,875 m (16,000 ft)	196 knots (363 km/h; 226 mph) IAS
Max cruising speed at 4,875 m (16,000 ft)	196 knots (363 km/h; 226 mph) IAS
Econ cruising speed at 4,875 m (16,000 ft)	168 knots (311 km/h; 193 mph) IAS
Stalling speed, flaps down	76 knots (141 km/h; 87·5 mph) IAS
Max rate of climb at S/L	434 m (1,424 ft)/min
Rate of climb at S/L, one engine out	62 m (205 ft)/min
Service ceiling	7,620 m (25,000 ft)
Service ceiling, one engine out	4,030 m (13,220 ft)
T-O run	504 m (1,654 ft)
T-O to 15 m (50 ft)	842 m (2,761 ft)
Landing from 15 m (50 ft)	761 m (2,498 ft)
Landing run	448 m (1,471 ft)

Range with max optional fuel, allowances for start, taxi, T-O, climb and 45 min reserves:

max cruising power at 7,620 m (25,000 ft)	937 nm (1,736 km; 1,079 miles)
econ cruising power at 7,620 m (25,000 ft)	1,126 nm (2,086 km; 1,296 miles)

Beechcraft Baron 58P pressurised four/six-seat cabin monoplane

BEECHCRAFT BARON MODEL 58TC

This turbocharged version of the Baron Model 58 is generally similar to the Model 58P, with the same power plant, but is unpressurised, with detail differences in the airframe and equipment. The design originated in July 1974, and construction of a prototype to production aircraft standard began in February 1975. The first flight of this aircraft was made on 31 October 1975 and FAA certification, in the Normal category, was granted on 23 January 1976. Deliveries began in June 1976.

TYPE: Four/six-seat cabin monoplane.

WINGS: Generally similar to Model 58P, except wing chord reduced.

FUSELAGE: Generally similar to Model 58P, except structural reinforcement for pressurisation has not been carried out.

TAIL UNIT: Generally similar to Model 58P, except area of tailplane increased and that of the elevators reduced.

LANDING GEAR: Generally similar to Model 58P, except main wheels and tyres by Cleveland or Goodrich, size 19·85 × 6·90-9·63, pressure 5·24-5·65 bars (76-82 lb/sq in). Cleveland nosewheel with tyre size 14·20 × 4·95-6·5, pressure 3·17-3·45 bars (46-50 lb/sq in). Cleveland or Goodrich disc brakes.

POWER PLANT: As Model 58P.

ACCOMMODATION: Generally similar to that of Model 58P, except no pressurisation. Port aft door deleted and replaced by double utility doors on starboard side, aft of wing. Centre cabin windows on each side serve also as emergency exits. Cabin heated and ventilated; air-conditioning optional.

SYSTEMS: Beech Freon air-conditioning system optional. Hydraulic system for brakes and propeller unfeathering only. Electrical power from two 24V 50A alternators, with 24V 17Ah storage battery. Two 100A alternators and a 24V 25Ah storage battery optional. Oxygen system to supply crew and passengers.

ELECTRONICS AND EQUIPMENT: King electronics standard, including nav/com, DME, ADF and transponder. Encoding altimeter standard. RCA radar optional. Standard equipment includes blind-flying instrumentation, heated pitot and fuel vents and alternate static source. A wide range of optional electronics and equipment available.

DIMENSIONS, EXTERNAL: As for Model 58P except:

Wing chord at root	2·24 m (7 ft 4 in)
Wing chord at tip	0·90 m (2 ft 11½ in)
Wing aspect ratio	7·59
Propeller diameter	1·98 m (6 ft 6 in)
Propeller ground clearance	0·28 m (10¾ in)
Utility double door (starboard, aft):	
Height	1·02 m (3 ft 4 in)
Width	1·14 m (3 ft 9 in)
Baggage door (nose, starboard):	
Height	0·56 m (1 ft 10 in)
Width	0·64 m (2 ft 1 in)
Emergency exits (port and stbd):	
Height	0·53 m (1 ft 9 in)
Width	0·61 m (2 ft 0 in)

DIMENSIONS, INTERNAL: As for Model 58P

AREAS:

Wings, gross	17·48 m² (188·13 sq ft)
Ailerons (total, incl tabs)	1·08 m² (11·58 sq ft)
Trailing-edge flaps (total)	1·98 m² (21·3 sq ft)
Fin	1·46 m² (15·67 sq ft)
Rudder, incl tab	0·81 m² (8·75 sq ft)
Tailplane	5·11 m² (55·05 sq ft)
Elevators, incl tabs	1·65 m² (17·8 sq ft)

WEIGHTS AND LOADINGS:

Weight empty, equipped	1,715 kg (3,780 lb)
Max T-O and landing weight	2,767 kg (6,100 lb)
Max ramp weight	2,785 kg (6,140 lb)
Max zero-fuel weight	2,585 kg (5,700 lb)
Max wing loading	158·2 kg/m² (32·4 lb/sq ft)
Max power loading	5·99 kg/kW (9·84 lb/hp)

PERFORMANCE:
See Addenda

BEECHCRAFT DUKE B60

Design work on the original version of this 4/6-seat pressurised and turbocharged light twin-engined transport started in early 1965. Construction of the prototype began in January 1966, and the first flight was made on 29 December 1966. FAA Type Approval was granted on 1 February 1968.

The current version of the B60 has an AiResearch Lexan pressurisation system, with a mini controller that allows selection of cabin altitude prior to take-off or landing. This system can also change the aircraft cabin altitude at any desired rate from 15-610 m (50-2,000 ft)/min.

A total of 367 Dukes had been produced by 1 January 1976.

TYPE: Four/six-seat cabin monoplane.

WINGS: Cantilever low-wing monoplane. Wing section NACA 23016·5 at root, NACA 23010·5 at tip. Thickness/chord ratio 13·7% at root, 10·5% at tip. Dihedral 6°. Incidence 4° at root, 0° at tip. Each wing is a two-spar semi-monocoque box beam of conventional aluminium alloy construction. Overhang-balance ailerons constructed of aluminium alloy. Conventional hinged trim tab in port aileron. Electrically-operated single-slotted aluminium alloy flaps. Pneumatic rubber de-icing boots optional.

FUSELAGE: Semi-monocoque aluminium alloy structure. Heavy-gauge chemically-milled aluminium alloy skins.

TAIL UNIT: Cantilever all-metal structure. Aluminium spars and end ribs; magnesium alloy skins reinforced with metal bonded honeycomb stiffeners running chordwise. Dorsal fin. Swept vertical and horizontal surfaces. Tailplane dihedral 10°. Trim tabs in rudder and port elevator. Pneumatic rubber de-icing boots optional.

LANDING GEAR: Electrically-retractable tricycle type. Main units retract inward, nosewheel aft; all three units have fairing doors. Beechcraft oleo-pneumatic shock-absorbers. Goodrich main wheels and tyres size 19·50 × 6·75-8 10-ply rating, pressure 5·52 bars (80 lb/sq in). Goodyear steerable nosewheel with shimmy damper, tyre size 15 × 6·00-6, pressure 3·45 bars (50 lb/sq in). Goodrich single-disc hydraulic brakes. Parking brake.

POWER PLANT: Two 283 kW (380 hp) Lycoming TIO-541-E1C4 turbocharged flat-six engines, each driving a Hartzell three-blade metal constant-speed and fully-feathering propeller. Propeller unfeathering accumulators and electric anti-icing optional. Electrically-operated engine cowl flaps. Two interconnected fuel cells in each wing containing 269 litres (71 US gallons); total usable capacity 538 litres (142 US gallons). Optionally, four interconnected fuel cells in each wing containing 390 litres (103 US gallons); total usable capacity 780 litres (206 US gallons); or five interconnected fuel cells in each wing containing 439 litres (116 US gallons), with total usable capacity of 878 litres (232 US gallons). Refuelling points in leading-edge of each wing, near wingtip. Oil capacity 24·5 litres (6·5 US gallons). Electrical propeller anti-icing optional.

Accommodation: Standard model has four individual seats in pairs, with centre aisle, in enclosed cabin. Door, hinged at forward edge, on port side at rear of cabin. Baggage hold in the nose, capacity 0·91 m^3 (32 cu ft), with external access door on port side of nose. Additional stowage for 0·80 m^3 (28·25 cu ft) of baggage at rear of cabin. Optional extras include fifth and sixth seats, rearward-facing third and fourth seats, headrests, curtain separating passenger and pilot seating, writing desks, refreshment cabinets, toilet, windscreen electrical de-icing and cabin fire extinguishers.

Systems: Cabin pressurisation system, differential 0·32 bars (4·6 lb/sq in), supplied by engine turbocharger bleed air, maintains cabin altitude equivalent to 3,050 m (10,000 ft) at 7,560 m (24,800 ft). Optional engine-driven vapour-cycle air-conditioning system of 14,000 BTU and combustion heater of 45,000 BTU. Automatic altitude controller, for cabin pressurisation system standard. Oxygen system optional, with 0·31 m^3 (11 cu ft), 0·62 m^3 (22 cu ft) or 1·39 m^3 (49 cu ft) bottle. Hydraulic system for brakes only. Pneumatic system for pressure-operated instruments and de-icing boots only. 24V 125A generators standard; 24V 13Ah aircooled nickel-cadmium battery.

Electronics and Equipment: Standard equipment includes full-flow oil filters, alternate static source, blind-flying instrumentation, electric turn co-ordinator, outside air temperature gauge, eight-day clock, heated pitot tube, external power socket, fuel vent anti-icer, heated stall warning device, dual landing lights, navigation lights, rotating beacons, taxi light, map light, passengers' reading lights, cabin dome light, instrument post lights, instrument floodlights, entrance light, nose baggage compartment light, ventilation installation with provisions for air-conditioning, tinted cabin side windows, super soundproofing, wall-to-wall carpeting, sun visors, cabin storage pockets, foul weather window for pilot, armrests, headrests, baggage straps, retracting exterior step, and aircraft towbar. Optional equipment includes de luxe instrument panel with duplicated blind-flying instrumentation for co-pilot, instantaneous vertical speed indicator, tachometer with synchroscope, flight and engine hour recorders, pilot's control wheel chronometer, co-pilot's control wheel map light, co-pilot toe brakes, wing ice lights, internally lighted instruments, propeller synchroniser, urethane paint, pilot's relief tube and strobe lights. Standard electronics comprise RCA AVC-110A main VHF transceiver with integral control and B3 com antenna; RCA AVN-210A manual Omni No. 1 with VOR/ILS converter-indicator, B12 nav antenna, B16 marker antenna and A-326A GS antenna; Collins 356C-4 isolation amplifier; Collins 356F-3 speaker amplifier; marker beacon; glideslope; Beech metal radio panel; static wicks; single audio switch panel; microphone key button in pilot's control wheel; white lighting; microphone, headset and cockpit speaker; electronics master switch; annunciator panel; and emergency locator transmitter. Optional electronics include an extensive range of RCA, King, Collins, Bendix and Beech equipment.

Dimensions, external:
- Wing span: 11·97 m (39 ft 3¼ in)
- Wing chord at fuselage c/l: 2·80 m (9 ft 2⅛ in)
- Wing chord at tip: 0·90 m (2 ft 11⅝ in)
- Wing aspect ratio: 7·243
- Length overall: 10·31 m (33 ft 10 in)
- Height overall: 3·76 m (12 ft 4 in)
- Tailplane span: 5·18 m (17 ft 0 in)
- Wheel track: 3·36 m (11 ft 0¼ in)
- Wheelbase: 2·81 m (9 ft 2½ in)
- Propeller diameter: 1·88 m (6 ft 2 in)
- Passenger door:
  - Height: 1·21 m (3 ft 11½ in)
  - Width: 0·67 m (2 ft 2½ in)
  - Height to sill: 0·81 m (2 ft 8 in)
- Baggage compartment door:
  - Height: 0·60 m (1 ft 11½ in)
  - Width: 0·95 m (3 ft 1½ in)
  - Height to sill: 0·95 m (3 ft 1½ in)

Dimensions, internal:
- Cabin:
  - Length: 3·61 m (11 ft 10 in)
  - Max width: 1·27 m (4 ft 2 in)
  - Max height: 1·32 m (4 ft 4 in)
  - Floor area: 3·36 m^2 (36·2 sq ft)
  - Volume: 4·80 m^3 (169·6 cu ft)

Areas:
- Wings, gross: 19·78 m^2 (212·9 sq ft)
- Ailerons (total): 1·06 m^2 (11·4 sq ft)
- Trailing-edge flaps (total): 2·76 m^2 (29·7 sq ft)
- Fin: 1·52 m^2 (16·38 sq ft)
- Rudder, incl tab: 1·15 m^2 (12·4 sq ft)
- Tailplane: 4·24 m^2 (45·6 sq ft)
- Elevators, incl tab: 1·52 m^2 (16·4 sq ft)

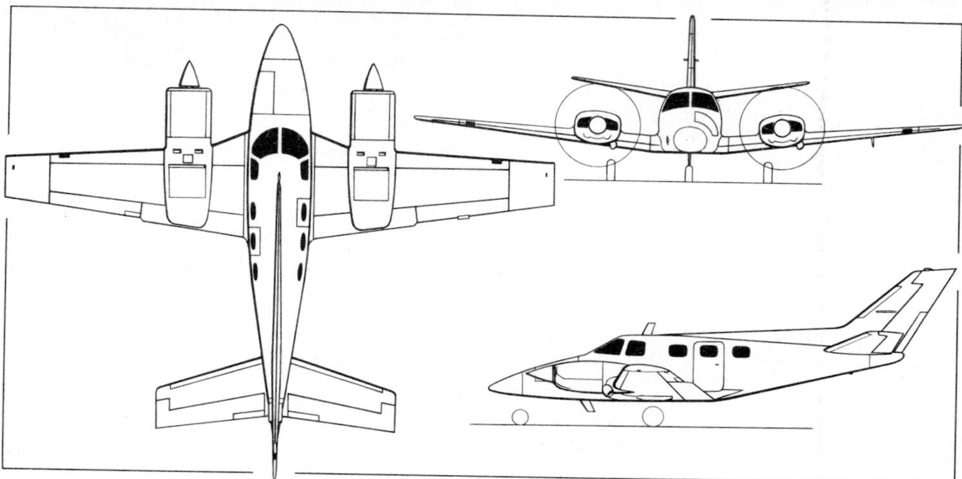

Photograph and three-view drawing *(Pilot Press)* **of the Beechcraft Duke B60 four/six-seat pressurised transport (two Lycoming TIO-541-E1C4 engines)**

Weights and Loadings:
- Weight empty, equipped: 1,939 kg (4,275 lb)
- Max T-O and landing weight: 3,073 kg (6,775 lb)
- Max ramp weight: 3,093 kg (6,819 lb)
- Max wing loading: 155·3 kg/m^2 (31·8 lb/sq ft)
- Max power loading: 5·43 kg/kW (8·9 lb/hp)

Performance (at max T-O weight):
- Never-exceed speed: 235 knots (434·5 km/h; 270 mph) IAS
- Max level speed at 7,010 m (23,000 ft): 248 knots (460 km/h; 286 mph)
- Max cruising speed:
  - 79% power at 7,620 m (25,000 ft): 241 knots (447 km/h; 278 mph)
  - 79% power at 6,100 m (20,000 ft): 230 knots (426 km/h; 265 mph)
  - 79% power at 4,570 m (15,000 ft): 219 knots (406 km/h; 252 mph)
- Cruising speed:
  - 75% power at 7,620 m (25,000 ft): 236 knots (438 km/h; 272 mph)
  - 75% power at 6,100 m (20,000 ft): 225 knots (417 km/h; 259 mph)
  - 75% power at 4,570 m (15,000 ft): 215 knots (398 km/h; 247 mph)
  - 65% power at 7,620 m (25,000 ft): 222 knots (410 km/h; 255 mph)
  - 65% power at 6,100 m (20,000 ft): 211 knots (391 km/h; 243 mph)
  - 65% power at 4,570 m (15,000 ft): 202 knots (373 km/h; 232 mph)
  - 45% power at 7,620 m (25,000 ft): 182 knots (338 km/h; 210 mph)
  - 45% power at 6,100 m (20,000 ft): 178 knots (330 km/h; 205 mph)
  - 45% power at 4,570 m (15,000 ft): 171 knots (317 km/h; 197 mph)
- Stalling speed, wheels and flaps up, power off: 81 knots (150·5 km/h; 93·5 mph)
- Stalling speed, wheels and flaps down, power off: 73 knots (135 km/h; 84 mph)
- Max rate of climb at S/L: 488 m (1,601 ft)/min
- Rate of climb at S/L, one engine out: 94 m (307 ft)/min
- Service ceiling: 9,145 m (30,000 ft)
- Service ceiling, one engine out: 4,600 m (15,100 ft)
- Runway LCN: 4
- T-O run: 611 m (2,006 ft)
- T-O to 15 m (50 ft): 800 m (2,626 ft)
- Landing from 15 m (50 ft): 934 m (3,065 ft)
- Landing run: 402 m (1,318 ft)
- Range, with max optional fuel and allowances for warm-up, taxi, take-off, climb to altitude and 45 min fuel reserves at 45% power, ISA:
  - 75% power at 7,620 m (25,000 ft): 1,005 nm (1,862 km; 1,157 miles)
  - 75% power at 6,100 m (20,000 ft): 979 nm (1,813 km; 1,127 miles)
  - 75% power at 4,570 m (15,000 ft): 956 nm (1,772 km; 1,101 miles)
  - 65% power at 7,620 m (25,000 ft): 1,118 nm (2,071 km; 1,287 miles)
  - 65% power at 6,100 m (20,000 ft): 1,092 nm (2,022 km; 1,257 miles)
  - 65% power at 4,570 m (15,000 ft): 1,068 nm (1,979 km; 1,230 miles)
  - 45% power at 7,010 m (23,000 ft): 1,223 nm (2,266 km; 1,408 miles)
  - 45% power at 6,100 m (20,000 ft): 1,227 nm (2,274 km; 1,413 miles)
  - 45% power at 4,570 m (15,000 ft): 1,211 nm (2,243 km; 1,394 miles)

BEECHCRAFT QUEEN AIR B80 and QUEEN AIRLINER B80

The prototype of the original Queen Air 80, which introduced more powerful engines than those of the A65, flew for the first time on 22 June 1961 and received its FAA Type Certificate on 20 February 1962. It was followed in January 1964 by the Queen Air A80, with increased wing span and AUW, new interior styling, increased fuel capacity and redesigned nose compartment, giving more space for radio. The A80 was followed in turn by the improved B80 and eleven-seat Queen Airliner B80, to which the details apply.

The B80 has two optional equipment packages for factory installation:

Executive Package. Comprising quickly removable forward partition with magazine rack and accordion door separating cockpit from cabin; quickly removable aft cabin partition and door; private lavatory, aft installation with relief tube; four executive cabin chairs with inboard folding armrests, or two such chairs and one four-seat couch (exchange for five commuter chairs with fixed armrests); and aft coat hanger rod installation; adding 19 kg (42 lb) to basic empty weight.

Airliner Package. Comprising seven floor-mounted chairs with inboard folding armrests and removable upholstery, two aft folding chairs with removable upholstery, extended seat tracks, four additional fresh air outlets and reading lights (exchange for five standard chairs); extended aft baggage compartment with baggage restraints; aft baggage door with safety light; fourth cabin window on starboard side; 39Ah nickel-cadmium battery (exchange for 13Ah battery); metal map case under pilot's seat; cockpit fire extinguisher; cockpit separation half-curtain; second outboard fuel filler caps with dipstick (one each side); coat hanger bar; and upholstery for nose baggage compartment; adding 77 kg (170 lb) to basic empty weight.

By 1 January 1976, Beech had built a total of 501 Queen Air 80s, A80s, B80s and Queen Airliner B80s, of which 43 per cent had been exported.

TYPE: Six/eleven-seat business aircraft, commuter airliner and utility aircraft.

WINGS: Cantilever low-wing monoplane. Wing section NACA 23020 at root, NACA 23012 outboard of joint between outer panel and wingtip. Dihedral 7°. Incidence 3° 55′ at root, 0° 1′ at tip. Two-spar all-metal structure of aluminium alloy. All-metal ailerons of magnesium. Trim tab in port aileron. Single-slotted aluminium alloy flaps. Pneumatic rubber de-icing boots optional.

FUSELAGE: Aluminium alloy semi-monocoque structure.

TAIL UNIT: Cantilever all-metal structure of aluminium alloy, with sweptback vertical surfaces. Tailplane dihedral 7°. Trim tabs in rudder and elevators. Pneumatic rubber de-icing boots optional.

LANDING GEAR: Electrically-retractable tricycle type. Main units retract forward, nosewheel aft. Beechcraft oleo-pneumatic shock-absorbers. Goodyear main wheels and tyres, size 8·50-10, 8-ply rating, pressure 3·31 bars (48 lb/sq in). Goodyear steerable nosewheel with tyre size 6·50-10, 6-ply rating, pressure 2·90 bars (42 lb/sq in). Goodyear heat sink and aircooled single-disc hydraulic brakes. Parking brakes.

POWER PLANT: Two 283 kW (380 hp) Lycoming IGSO-540-A1D supercharged flat-six engines, each driving a Hartzell three-blade fully-feathering constant-speed propeller. Propeller synchrophasers optional. Fuel in two inboard wing tanks, each with capacity of 166 litres (44 US gallons) and two outboard tanks, each 238·5 litres (63 US gallons). Total standard fuel capacity 809 litres (214 US gallons). Provision for two optional auxiliary tanks in wings to bring total capacity to 1,000 litres (264 US gallons). Refuelling points above wings. Oil capacity (total) 30 litres (8 US gallons).

ACCOMMODATION: Crew of one or two on flight deck and four to nine passengers in cabin. Basic layout has five commuter passenger seats; executive layout has four lounge chairs, fore and aft partitions and lavatory; airliner layout has seven commuter and two folding passenger seats, extended aft baggage compartment and door, map case, cabin fire extinguisher and fourth cabin window. Door on port side of cabin at rear; optionally, double-width cargo doors. Optional toilet and baggage compartment opposite door, capacity 160 kg (350 lb). Other optional items include sofa, tables, refreshment cabinets and external cargo pod.

SYSTEMS: Optional electrically-driven vapour-cycle air-conditioning system with combustion heater. Standard model has 100,000 BTU heater and ventilating system. Hydraulic system for brakes only. Two 28V 150A DC engine-driven generators and 24V 13Ah nickel-cadmium battery for electrical system. 39Ah nickel-cadmium battery optional. Oxygen system of 1·81 m³ (64 cu ft) capacity optional.

ELECTRONICS AND EQUIPMENT: Standard electronics comprise Narco Com 11B 720-channel transceiver with Nav 12 300-channel nav receiver/converter, VOR/ILS indicator, B3 com antenna and B17 nav/GS antenna; Narco Com 11B 720-channel transceiver with Nav 11 200-channel nav receiver/converter, VOR/LOC indicator and B3 com antenna; Collins 356F-3 speaker amplifier and Collins 356C-4 isolation amplifier; Bendix T-12D ADF with 551RL indicator, voice range filter and sense antenna; Narco MBT-24R marker beacon with B16 marker antenna and panel-mounted marker lights; Narco UGR-2A glideslope with B35 glideslope antenna; Beech metal radio panel, accessories and static wicks; edge-lighted audio switch panel; white lighting; dual microphones, headsets and single cockpit speaker; emergency locator transmitter; and electronics master switch. Standard equipment includes dual controls, blind-flying instrumentation, outside air temperature gauge, 8-day clock, flap position indicator, dynamic brake on landing gear, landing gear warning system, heated stall warning system, dual heated pitot heads, cabin door 'unlocked' warning light, external power socket, two landing lights, navigation lights, dual rotating beacons, cabin door and indirect overhead lighting, reading lights for each passenger, map light, aft compartment dome light, primary and secondary instrument light systems, windscreen defroster, dual storm windows, sun visors, map pockets, four-way adjustable pilot and co-pilot seats, 'No smoking—Fasten seat belt' sign, carpeted floor, provisions for removable cabin partitions and toilet, window curtains, emergency exit, coat rack, towbar, control lock assembly and heated fuel vents. A wide range of electronics and equipment is available to customer's requirements, including Collins, King, Bendix, Sunair, Narco or RCA radar and radio packages; King RNav; RCA weather-avoidance radar; Beech or Mitchell autopilot; de luxe instrument panel carrying duplicate blind-flying instruments, exhaust temperature gauge, flight hour recorder; control-wheel chronograph and map light installation, nosewheel taxi light, three-light strobe system, wing ice lights, alcohol or electrical propeller and windscreen anti-icing, dual windscreen wipers, propeller unfeathering accumulators, nose compartment light, super soundproofing, external power socket, cabin and cockpit fire extinguishers, cargo door, and photographic installation.

Beechcraft Queen Air B80 (two Lycoming IGSO-540-A1D engines)

DIMENSIONS, EXTERNAL:
- Wing span: 15·32 m (50 ft 3 in)
- Wing chord at root: 2·15 m (7 ft 0½ in)
- Wing chord at tip: 1·07 m (3 ft 6 in)
- Wing aspect ratio: 7·51
- Length overall: 10·82 m (35 ft 6 in)
- Height overall: 4·33 m (14 ft 2½ in)
- Tailplane span: 5·25 m (17 ft 2¾ in)
- Wheel track: 3·89 m (12 ft 9 in)
- Wheelbase: 3·75 m (12 ft 3½ in)
- Propeller diameter: 2·36 m (7 ft 9 in)
- Standard passenger door:
  - Height: 1·31 m (4 ft 3¾ in)
  - Width: 0·69 m (2 ft 3 in)
  - Height to sill: 1·17 m (3 ft 10 in)
- Optional cargo door:
  - Height: 1·31 m (4 ft 3¾ in)
  - Width: 1·37 m (4 ft 6 in)
  - Height to sill: 1·17 m (3 ft 10 in)

DIMENSIONS, INTERNAL:
- Cabin, incl flight deck and baggage area:
  - Length: 6·97 m (19 ft 7 in)
  - Max width: 1·37 m (4 ft 6 in)
  - Max height: 1·45 m (4 ft 9 in)
  - Volume: 9·5 m³ (335·4 cu ft)
- Nose baggage compartment: 0·68 m³ (24 cu ft)
- Standard aft baggage compartment: 1·50 m³ (53 cu ft)
- Optional extension to aft baggage compartment: 0·48 m³ (17 cu ft)

AREAS:
- Wings, gross: 27·3 m² (293·9 sq ft)
- Ailerons (total): 1·29 m² (13·90 sq ft)
- Trailing-edge flaps (total): 2·72 m² (29·30 sq ft)
- Fin: 2·20 m² (23·67 sq ft)
- Rudder, incl tab: 1·30 m² (14·00 sq ft)
- Tailplane: 4·39 m² (47·25 sq ft)
- Elevators, incl tabs: 1·66 m² (17·87 sq ft)

WEIGHTS AND LOADINGS:
- Weight empty, equipped: 2,393 kg (5,277 lb)
- Max T-O and landing weight: 3,992 kg (8,800 lb)
- Max ramp weight: 4,016 kg (8,855 lb)
- Max wing loading: 146·0 kg/m² (29·9 lb/sq ft)
- Max power loading: 7·05 kg/kW (11·6 lb/hp)

PERFORMANCE (at max T-O weight, except where indicated):
- Max level speed at 3,500 m (11,500 ft) at average AUW: 215 knots (400 km/h; 248 mph)
- Cruising speed, 70% power at 4,570 m (15,000 ft) at average AUW: 195 knots (362 km/h; 225 mph)
- Econ cruising speed, 45% power at 4,570 m (15,000 ft) at average AUW: 159 knots (294 km/h; 183 mph)
- Stalling speed, wheels and flaps up: 85 knots (157 km/h; 97 mph) IAS
- Stalling speed, wheels and flaps down: 71 knots (131 km/h; 81 mph) IAS
- Max rate of climb at S/L: 388 m (1,275 ft)/min
- Rate of climb at S/L, one engine out: 64 m (210 ft)/min
- Service ceiling: 8,168 m (26,800 ft)
- Service ceiling, one engine out: 3,596 m (11,800 ft)
- Min ground turning radius: 11·58 m (38 ft 0 in)
- Runway LCN: 3·5
- T-O run: 612 m (2,007 ft)
- T-O to 15 m (50 ft): 779 m (2,556 ft)
- Landing from 15 m (50 ft) at max landing weight: 784 m (2,572 ft)
- Landing run at max landing weight: 494 m (1,620 ft)
- Range with max optional fuel, allowances for warm-up, taxi, T-O and climb to altitude, with 45 min reserves:
  - 70% power at 4,570 m (15,000 ft): 957 nm (1,773 km; 1,102 miles)
  - 65% power at 5,180 m (17,000 ft): 1,076 nm (1,994 km; 1,239 miles)
  - 45% power at 4,570 m (15,000 ft): 1,317 nm (2,441 km; 1,517 miles)

BEECHCRAFT KING AIR MODEL C90

USAF designation: VC-6B
US Navy designation: T-44A

The King Air C90 is a pressurised 6/10-seat twin-turboprop business aircraft which superseded the original Models 90, A90 and B90 King Air.

The 1976 King Air C90 is powered by Pratt & Whitney Aircraft of Canada PT6A-21 (previously PT6A-20) turboprop engines, which provide improved performance over a wide range of altitudes and temperatures. Increases in take-off and climb power offer improvements in high altitude and hot weather operation and, since these engines also run cooler, an increase in useful life and lower overhaul costs are anticipated.

Introduced in September 1970, the C90 King Air utilises the more advanced cabin pressurisation and heating system of the King Air 100. This comprises a dual engine bleed air system for cabin pressurisation, with a max differential of 0·32 bars (4·6 lb/sq in).

The 1976 King Air C90 has as standard dual bleed air pressurisation and cabin heating, including supplementary electric heating, air-conditioning, super soundproofing, a full anti-icing system, four cabin seats in club arrangement, forward cabin partition with curtain, aft starboard cabin partition with curtain, polarised cabin windows, exterior urethane paint scheme and a comprehensive electronics package which includes dual nav/com, transponder, DME, ADF, marker beacon, glideslope and dual blind-flying instrumentation.

A total of 836 King Air 90s had been delivered by 1 January 1976, including one for the USAF's 1254th Special Air Missions Squadron at Andrews AFB, Maryland, for VIP transport duties under the designation VC-6B. In 1975 Beech delivered four King Air C90s to the Spanish Air Ministry's Civil Aviation School for instrument training and liaison. The US Navy in 1976 ordered 15 King Air 90s, with options on 56 more, as **T-44A** advanced trainers to meet its VTAM (X) requirement. Deliveries of the first 15 aircraft, which will replace obsolete TS-2As and TS-2Bs, are scheduled for completion by October 1977. They will differ from other King Air 90s by having 559 kW (750 shp) PT6A-34B engines, flat rated to 410 kW (550 shp).

TYPE: Six/ten-seat twin-turboprop business aircraft.

WINGS: Cantilever low-wing monoplane. Wing section NACA 23014·1 (modified) at root, NACA 23016·22 (modified) at outer end of centre-section, NACA 23012 at tip. Dihedral 7°. Incidence 4° 48′ at root, 0° at tip. No sweepback at quarter-chord. Two-spar aluminium alloy structure. All-metal ailerons of magnesium, with adjustable trim tab on port aileron. Single-slotted aluminium alloy flaps. Automatic pneumatic de-icing boots on leading-edges standard.

FUSELAGE: Aluminium alloy semi-monocoque structure.

TAIL UNIT: Cantilever all-metal structure with sweptback vertical surfaces. Fixed-incidence tailplane, with 7° dihedral. Trim tabs in rudder and each elevator. Automatic pneumatic de-icing boots on leading-edges of fin and tailplane.

LANDING GEAR: Electrically-retractable tricycle type. Nosewheel retracts rearward, main wheels forward into engine nacelles. Main wheels protrude slightly beneath nacelles when retracted, for safety in a wheels-up emergency landing. Steerable nosewheel with shimmy damper. Beech oleo-pneumatic shock-absorbers. B.F. Goodrich main wheels with tyres size 8·50-10, pressure 3·79 bars (55 lb/sq in). B.F. Goodrich nosewheel with tyre size 6·50-10, pressure 3·59 bars (52 lb/sq in). Goodrich heat-sink and aircooled multi-disc hydraulic brakes. Parking brakes.

POWER PLANT: Two 410 kW (550 ehp) Pratt & Whitney Aircraft of Canada PT6A-21 turboprop engines, each driving a Hartzell three-blade constant-speed fully-feathering propeller. Propeller electro-thermal anti-icing standard. Fuel in two tanks in engine nacelles, each with capacity of 231 litres (61 US gallons), and bladder type auxiliary tanks in outer wings, each with capacity of 496 litres (131 US gallons). Total fuel capacity 1,454 litres (384 US gallons). Refuelling points in top of each engine nacelle and in wing leading-edge outboard of each nacelle. Oil capacity 13·2 litres (3·5 US gallons). Engine anti-icing system standard.

ACCOMMODATION: Two seats side by side in cockpit with dual controls standard. Normally, four reclining seats are provided in the main cabin, in pairs facing each other fore and aft. Standard furnishings include cabin forward partition, with fore and aft partition curtain, hinged nose baggage compartment door and coat rod. Optional arrangements seat up to eight persons, some with two- or three-place couch and refreshment cabinets. Baggage racks at rear of cabin on starboard side, with optional toilet on port side. Door on port side aft of wing, with built-in airstairs. Emergency exit on starboard side of cabin. Entire accommodation pressurised and air-conditioned. Electrically-heated windscreen standard.

SYSTEMS: Pressurisation by dual engine bleed air system with maximum pressure differential of 0·32 bars (4·6 lb/sq in). Cabin heated by 45,000 BTU dual engine bleed air system and auxiliary electrical heating system. Electrical system utilises two 28V 250A starter/generators, 24V 45Ah aircooled nickel-cadmium battery with failure detector. Complete de-icing and anti-icing equipment. Oxygen system, 0·62 m³ (22 cu ft), 1·39 m³ (49 cu ft) or 1·81 m³ (64 cu ft) capacity, optional. Vacuum system for flight instruments.

ELECTRONICS AND EQUIPMENT: Standard electronics package comprises dual King KX-175B VHF transceivers with KA-39 power adapters and B3 antennae; KN-77 Omni No. 1 VOR/ILS converter, Collins 331A-3G indicator; King KN-73 glideslope with B17 antenna and B35 glideslope antenna; KN-77 Omni No. 2 VOR/ILS converter and KNI-520 indicator with B17 antenna; KR-85 ADF with KI-225 indicator and sense antenna; KT-76 transponder with B18 antenna; KN-65A DME with KI-266 indicator, Nav 1-Nav 2 switching, DME hold and B18 antenna; KMA-20 audio system less lights; KA-40 marker lights and B16 antenna; Collins PN-101 pilot's compass system; Standard Electric gyro horizon for pilot; co-pilot's 3 in CF gyro horizon and directional gyro; dual 125VA Flite-Tronics PC-14B inverters with failure light; dual flight instrumentation; sectional instrument panel; white lighting; radio accessories, static wicks and Beech metal radio panel; microphone key button in pilot and co-pilot control wheels; dual microphones, headsets and cockpit speakers; and electronics master switch. A wide range of optional electronics, including flight director, autopilot and weather radar, available as individual items or in standard packages, can be supplied to customer's requirements. Standard equipment includes heated stall warning system, dual heated pitot heads, external power socket, wing ice lights, two landing lights, taxi light, navigation lights, dual rotating beacons, dual map lights, indirect cabin lighting, reading light, cabin door light, aft compartment dome light, primary and secondary instrument light systems, white cockpit lighting, fresh air outlets, double-glazed windows, curtains and shades, soundproofing, carpeted floor, 'No smoking—Fasten seat belt' sign, windscreen defroster, engine anti-icing system with inertial separators, dual storm windows, sun visors, map pockets, windscreen wipers, heated fuel vents, cabin coat rack, pictorial navigation indicator, electric directional gyro, outside air temperature gauge, vacuum gauge, oxygen pressure and de-icing pressure gauges, cabin rate of climb indicator, cabin altitude and differential pressure indicators, edge-lighted control panels, emergency locator transmitter, towbar and control lock assembly. Optional equipment includes flight hour recorder, chronograph in pilot's control wheel, 24-hour GMT clock in co-pilot's control wheel, reversible-pitch propellers with synchrophaser or automatic propeller feathering system with reversible-pitch propellers and synchrophasers, three-light strobe system, engine fire detection and fire extinguishing systems, electrically-operated elevator trim, cabin fire extinguisher, low profile toilet installation, instantaneous vertical speed indicator and cabin door step lights.

DIMENSIONS, EXTERNAL:

Wing span	15·32 m (50 ft 3 in)
Wing chord at root	2·15 m (7 ft 0½ in)
Wing chord at tip	1·07 m (3 ft 6 in)
Wing aspect ratio	8·57
Length overall	10·82 m (35 ft 6 in)
Height overall	4·33 m (14 ft 2½ in)
Tailplane span	5·25 m (17 ft 2½ in)
Wheel track	3·89 m (12 ft 9 in)
Wheelbase	3·75 m (12 ft 3½ in)
Propeller diameter	2·36 m (7 ft 9 in)
Passenger door:	
Height	1·31 m (4 ft 3¾ in)
Width	0·69 m (2 ft 3 in)
Height to sill	1·17 m (3 ft 10 in)

DIMENSIONS, INTERNAL:

Total pressurised length	5·43 m (17 ft 10 in)
Cabin:	
Length	3·86 m (12 ft 8 in)
Max width	1·37 m (4 ft 6 in)
Max height	1·45 m (4 ft 9 in)
Floor area	6·50 m² (70 sq ft)
Volume	8·89 m³ (314 cu ft)
Baggage compartment, aft	1·51 m³ (53·5 cu ft)

AREAS:

Wings, gross	27·31 m² (293·94 sq ft)
Ailerons (total)	1·29 m² (13·90 sq ft)
Trailing-edge flaps (total)	2·72 m² (29·30 sq ft)
Fin	2·20 m² (23·67 sq ft)
Rudder, incl tab	1·30 m² (14·00 sq ft)
Tailplane	4·39 m² (47·25 sq ft)
Elevators, incl tabs	1·66 m² (17·87 sq ft)

WEIGHTS AND LOADINGS:

Weight empty	2,558 kg (5,640 lb)
Max T-O weight	4,377 kg (9,650 lb)
Max ramp weight	4,402 kg (9,705 lb)
Max landing weight	4,159 kg (9,168 lb)
Max wing loading	160·1 kg/m² (32·8 lb/sq ft)
Max power loading	5·34 kg/kW (8·8 lb/ehp)

PERFORMANCE (at max T-O weight, except where indicated):

Max cruising speed at 3,660 m (12,000 ft)
222 knots (412 km/h; 256 mph)

Max cruising speed at 4,880 m (16,000 ft) at AUW of 3,794 kg (8,365 lb)
219 knots (406 km/h; 252 mph)

Max cruising speed at 6,400 m (21,000 ft) at AUW of 3,794 kg (8,365 lb)
216 knots (401 km/h; 249 mph)

Stalling speed, wheels and flaps up, power off
89 knots (164 km/h; 102 mph) IAS

Stalling speed, wheels and flaps down, power off
76 knots (140 km/h; 87 mph) IAS

Max rate of climb at S/L 596 m (1,955 ft)/min

Rate of climb at S/L, one engine out
164 m (539 ft)/min

Service ceiling	8,565 m (28,100 ft)
Service ceiling, one engine out	4,587 m (15,050 ft)
Min ground turning radius	11·58 m (38 ft 0 in)
Runway LCN	4
T-O to 15 m (50 ft)	689 m (2,261 ft)
Accelerate/stop distance	1,066 m (3,498 ft)

Landing from 15 m (50 ft) without propeller reversal at AUW of 4,159 kg (9,168 lb) 613 m (2,010 ft)

Landing run, without propeller reversal, at AUW of 4,159 kg (9,168 lb) 328 m (1,075 ft)

Range with max fuel at max cruising speed, incl allowance for starting, taxi, take-off, climb, descent and 45 min reserves at max range power, ISA, at:

6,400 m (21,000 ft)
1,202 nm (2,227 km; 1,384 miles)

4,875 m (16,000 ft)
1,057 nm (1,959 km; 1,217 miles)

3,660 m (12,000 ft)
957 nm (1,773 km; 1,102 miles)

Max range at econ cruising power, allowances as above at:

6,400 m (21,000 ft)
1,281 nm (2,374 km; 1,475 miles)

4,875 m (16,000 ft)
1,159 nm (2,147 km; 1,334 miles)

3,660 m (12,000 ft)
1,065 nm (1,973 km; 1,226 miles)

Beechcraft King Air C90 six/ten-seat twin-turboprop business aircraft

BEECHCRAFT KING AIR E90

On 1 May 1972 Beech announced an addition to the King Air range of business aircraft. Designated King Air E90, this combines the airframe of the King Air C90 with the 507 kW (680 ehp) Pratt & Whitney Aircraft of Canada PT6A-28 turboprop engines that power the King Air A100, each flat rated to 410 kW (550 ehp).

The description of the King Air C90 in this edition applies also to the King Air E90, except as follows:

LANDING GEAR: As King Air C90, except main-wheel tyre pressure 3·93 bars (57 lb/sq in).

POWER PLANT: Two 507 kW (680 ehp) Pratt & Whitney Aircraft of Canada PT6A-28 turboprop engines, flat rated to 410 kW (550 ehp), each driving a Hartzell three-blade metal fully-feathering and reversible constant-speed propeller. Standard fuel capacity 1,794 litres (474 US gallons).

Beechcraft King Air E90 (two Pratt & Whitney Aircraft of Canada PT6A-28 turboprop engines)

Electronics and Equipment: Standard electronics include RCA AVC-111A main VHF transceiver with B3 antenna; RCA AVC-110A standby VHF transceiver with B3 antenna; RCA AVN-220A manual omni No. 1 with glideslope, marker beacon receiver, Collins 331A-3G indicator, single marker light, B17 nav antenna, B16 marker antenna and B35 glideslope antenna; RCA AVN-221A manual omni No. 2 on B17 antenna; King KR-85 ADF with KI-225 indicator, voice range filter and sense antenna; RCA AVQ-47 radar with 12 in phased array antenna and standard scope; Bendix TPR-660 transponder with B18 antenna; King KN-65A DME with KI-266 indicator, KA-43 converter, Nav 1 and Nav 2 switching, DME hold and B18 antenna, Collins 356C-4 isolation amplifier and 356F-3 speaker amplifier with single set audio switches in edge-lighted panel; Collins PN-101 compass system; Standard Electric gyro horizon; 3 in CF gyro horizon and directional gyro for co-pilot; dual 125VA Flite-Tronics PC-14B inverters with failure light; white lighting; radio accessories; static wicks; Beech metal radio panel; dual flight instrumentation; sectional instrument panel; microphone button in pilot and co-pilot control wheels; dual microphones, headsets and cockpit speakers; electronics master switch; and emergency locator transmitter. Standard equipment includes engine-driven fuel boost pumps, vertically-arranged engine instruments, polarised cabin windows, alternate static source and items as detailed for King Air C90. Optional equipment includes the items detailed for the King Air C90, plus an extensive range of optional electronics.

Weights and Loadings:

Weight empty	2,670 kg (5,886 lb)
Max T-O weight	4,581 kg (10,100 lb)
Max ramp weight	4,608 kg (10,160 lb)
Max landing weight	4,400 kg (9,700 lb)
Max wing loading	168·0 kg/m² (34·4 lb/sq ft)
Max power loading	4·52 kg/kW (7·43 lb/ehp)

Performance (at max T-O weight, except where indicated):

Max cruising speed at 3,660 m (12,000 ft)	249 knots (462 km/h; 287 mph)
Cruising speed at max recommended cruise power:	
at 4,875 m (16,000 ft)	247 knots (459 km/h; 285 mph)
at 6,400 m (21,000 ft)	245 knots (454 km/h; 282 mph)
Cruising speed for max range	197 knots (365 km/h; 227 mph)
Stalling speed, power off, wheels and flaps up	86 knots (159 km/h; 99 mph) IAS
Stalling speed, power off, wheels and flaps down	77 knots (143 km/h; 89 mph) IAS
Max rate of climb at S/L	570 m (1,870 ft)/min
Rate of climb at S/L, one engine out	143 m (470 ft)/min
Service ceiling	8,419 m (27,620 ft)
Service ceiling, at 3,629 kg (8,000 lb) AUW	9,421 m (30,910 ft)
Service ceiling, one engine out	4,386 m (14,390 ft)
Service ceiling, one engine out, at 3,629 kg (8,000 lb) AUW	6,218 m (20,400 ft)
Min ground turning radius	11·58 m (38 ft 0 in)
Runway LCN	4·5
T-O run	473 m (1,553 ft)
T-O to 15 m (50 ft)	617 m (2,024 ft)
Landing distance, 5° approach angle, full flap, at max landing weight:	
landing from 15 m (50 ft)	643 m (2,110 ft)
landing run	314 m (1,030 ft)
Accelerate/stop distance, incl 2 sec failure recognition time	1,139 m (3,736 ft)
Cruising range at max recommended cruise power:	
at 4,875 m (16,000 ft)	1,125 nm (2,084 km; 1,295 miles)
at 6,400 m (21,000 ft)	1,309 nm (2,425 km; 1,507 miles)
Cruising range at max range power:	
at 4,875 m (16,000 ft)	1,480 nm (2,742 km; 1,704 miles)
at 6,400 m (21,000 ft)	1,625 nm (3,011 km; 1,871 miles)
Range at max T-O weight, max recommended power at 6,400 m (21,000 ft), 45 min reserves, five occupants, 117 kg (258 lb) baggage and 1,440 kg (3,176 lb) fuel before engine start	1,309 nm (2,425 km; 1,507 miles)

BEECHCRAFT B99 AIRLINER

The Beechcraft B99 Airliner is an unpressurised high-performance 17-seat twin-turboprop airliner designed specifically for the scheduled airline and air taxi market. The prototype of the original Model 99 flew for the first time in July 1966 and the first delivery of a production aircraft was made on 2 May 1968.

By 1 January 1976 a total of 164 of these aircraft had been delivered to 64 operators.

Installation of an optional forward-hinged cargo door forward of the standard airstair door permits the Airliner to be used for all-cargo or combined cargo/passenger operations, with a movable bulkhead separating freight and passengers in the latter configuration.

Two versions are currently available:

B99 Airliner. Standard model with gross weight of 4,944 kg (10,900 lb) and powered by two 507 kW (680 ehp) Pratt & Whitney Aircraft of Canada PT6A-28 turboprop engines, as described.

B99 Executive. Basically the same as the standard model, but offering optional seating arrangements for 8 to 17 persons and various corporate interiors.

Earlier versions were the 99 Airliner, with two 410 kW (550 ehp) PT6A-20 turboprop engines, and the 99A Airliner with two 507 kW (680 ehp) PT6A-27 turboprop engines, flat rated to 410 kW (550 ehp).

Nine examples of the latter aircraft were supplied to the Chilean Air Force, for search and rescue and navigation training.

Type: Twin-turboprop light passenger, freight or executive transport.

Wings: Cantilever low-wing monoplane. Wing section NACA 23018 at root, NACA 23016·5 at centre-section joint with outer panel, NACA 23012 at tip. Dihedral 7°. Incidence 4° 48′ at root, 0° at tip. Two-spar all-metal aluminium alloy structure. All-metal ailerons of magnesium. Trim tab in port aileron. Single-slotted aluminium alloy flaps. Optional automatic pneumatic de-icing boots.

Fuselage: All-metal semi-monocoque structure.

Tail Unit: Cantilever all-metal structure, with sweptback vertical surfaces and a ventral stabilising fin. Variable-incidence tailplane. Trim tab in rudder. Pneumatic de-icing boots optional.

Landing Gear: Retractable tricycle type with single steerable nosewheel and twin wheels on each main unit. Electrical retraction, nosewheel rearward, main units forward into engine nacelles. Hydraulic retraction system optional. Beech oleo-pneumatic shock-absorbers. Goodrich wheels and tyres. Main-wheel tyres size 18 × 5·5, pressure 6·34-6·62 bars (92-96 lb/sq in). Nosewheel tyre size 6·50-10, pressure 3·45-3·79 bars (50-55 lb/sq in). B.F. Goodrich heat-sink and aircooled single-disc hydraulic brakes. Parking brake. Shimmy damper on nosewheel.

Power Plant: Two 507 kW (680 ehp) Pratt & Whitney Aircraft of Canada PT6A-28 turboprop engines, each driving a Hartzell three-blade fully-feathering and reversible-pitch constant-speed propeller. Automatic feathering system standard. Rubber fuel tanks in wings, with total capacity of 1,393 litres (368 US gallons).

Accommodation: Crew of two side by side on flight deck, with full dual controls and blind-flying instrumentation. Half-curtain or bulkhead between flight deck and cabin. Standard version has 15 removable high-density cabin chairs, two-abreast with centre aisle (single chair opposite door). Executive version has six standard seats in cabin, the two forward seats facing rearwards. Baggage space aft of rear seats, with external door. Nose baggage compartment with two external doors. An optional underfuselage baggage/cargo pod with a volume of 1·01 m³ (35·5 cu ft) and structural capacity of 363 kg (800 lb) is available, and this does not affect speed appreciably. Airstair door on port side of cabin at rear. Optional forward-hinged cargo door forward of passenger door, to give wide unobstructed opening for cargo loading. Emergency exit on each side at forward end of cabin. A wide selection of corporate interiors and removable chemical or electrical flushing toilet optional.

Systems: Automatic 100,000 BTU heating system and high-capacity ventilation system, with individual fresh air outlets, standard. Optional 24,000 BTU air-conditioning system. Hydraulic system for brakes only, except when optional hydraulically-operated landing gear retraction system is installed. 28V DC electrical system, with two 200A generators, 40Ah nickel-cadmium battery with failure detector, and dual solid-state inverters.

Electronics and Equipment: Standard electronics (domestic aircraft) include dual 360-channel VHF transceivers; dual 200-channel nav receivers, with VOR/ILS indicators; ADF with voice range filter; three-light marker beacon; 40-channel glideslope receiver; 200-channel DME with Nav 1 and Nav 2 switching; transponder; pilot-to-cabin paging, with four speakers; dual 120VA inverters with failure light; pilot's electric gyro horizon; dual microphones, headsets and cockpit speakers; microphone key buttons in pilot's and co-pilot's control wheels; and electronics master switch. The standard export electronics package includes a second ADF and a 10-channel HF transceiver in lieu of DME and transponder. Standard equipment includes electric propeller anti-icing, landing and taxi lights, wing ice lights, cabin instrument and map lights, dual rotating beacons, and fire detection system. Optional equipment includes high-pressure oxygen system, air-conditioning system with Freon compressor, engine fire extinguishing system, weather radar, propeller synchronisers, high-intensity anti-collision lights, autopilot, electrical windscreen anti-icing system and auxiliary engine bleed air heater.

Dimensions, external:

Wing span	14·00 m (45 ft 10½ in)
Wing chord at root	2·15 m (7 ft 0½ in)
Wing chord at tip	1·07 m (3 ft 6 in)
Wing aspect ratio	7·51
Length overall	13·58 m (44 ft 6¾ in)
Height overall	4·38 m (14 ft 4¼ in)
Tailplane span	6·82 m (22 ft 4½ in)
Wheel track	3·96 m (13 ft 0 in)
Wheelbase	5·48 m (17 ft 11¾ in)
Propeller diameter	2·37 m (7 ft 9½ in)
Propeller ground clearance	0·34 m (1 ft 1½ in)
Passenger door:	
Height	1·31 m (4 ft 3½ in)
Width	0·69 m (2 ft 3 in)
Cargo double-door (optional):	
Height	1·31 m (4 ft 3½ in)
Width	1·36 m (4 ft 5½ in)

Dimensions, internal:

Cabin, incl flight deck:	
Length	7·72 m (25 ft 4 in)
Max width	1·40 m (4 ft 7 in)
Max height	1·45 m (4 ft 9 in)
Volume	12·0 m³ (423·6 cu ft)
Baggage space (nose) volume	1·24 m³ (43·9 cu ft)
Baggage space (rear) volume	0·60 m³ (21·1 cu ft)

Weights and Loadings (A: 99 Airliner, B: 99A Airliner, C: B99 Airliner):

Weight empty, equipped:	
A	2,595 kg (5,722 lb)
B	2,607 kg (5,749 lb)
C	2,620 kg (5,777 lb)
Max T-O weight:	
A, B	4,717 kg (10,400 lb)
C	4,944 kg (10,900 lb)
Max design taxi weight:	
C	4,969 kg (10,955 lb)
Max wing loading:	
C	190·3 kg/m² (38·97 lb/sq ft)
Max power loading:	
C	4·88 kg/kW (8·02 lb/ehp)

Performance (A: 99 Airliner, B: 99A Airliner, C: B99 Airliner, at max T-O weight unless stated otherwise):

Max cruising speed at AUW of 4,309 kg (9,500 lb):

A at 2,440 m (8,000 ft) 221 knots (409 km/h; 254 mph)

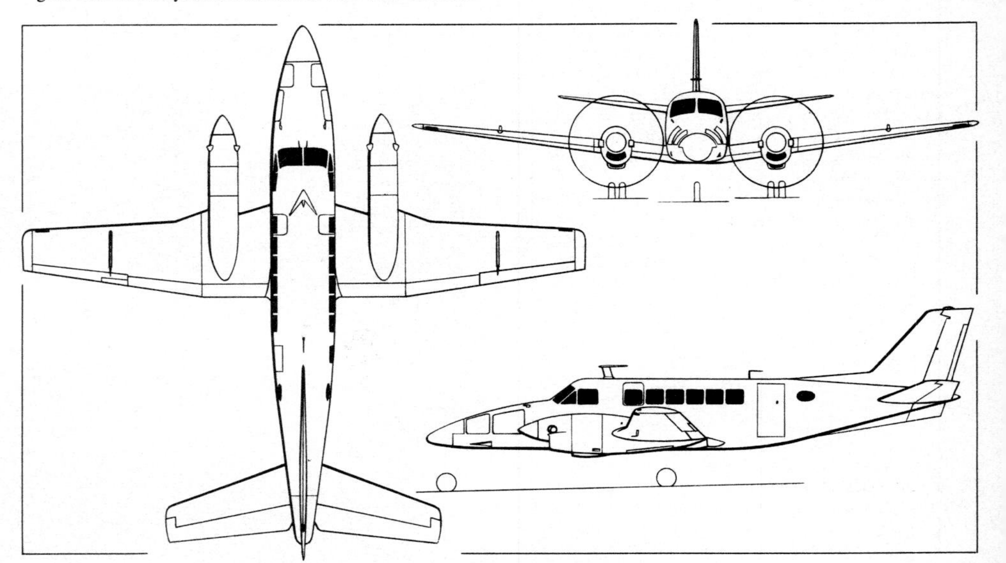

Beechcraft B99 Airliner 17-seat light transport *(Pilot Press)*

Beechcraft B99 Airliner 17-seat light transport in the insignia of Air Champagne Ardenne *(Dr Alan Beaumont)*

A at 3,660 m (12,000 ft)
219 knots (406 km/h; 252 mph)
B at 2,440 m (8,000 ft)
243 knots (451 km/h; 280 mph)
B at 3,660 m (12,000 ft)
247 knots (457 km/h; 284 mph)
C at 2,440 m (8,000 ft)
243 knots (451 km/h; 280 mph)
C at 3,660 m (12,000 ft)
247 knots (459 km/h; 285 mph)
Max rate of climb at S/L:
A, B 518 m (1,700 ft)/min
C 637 m (2,090 ft)/min
Rate of climb at S/L, one engine out:
C 171 m (561 ft)/min
Service ceiling:
A 7,210 m (23,650 ft)
B 7,986 m (26,200 ft)
C 8,020 m (26,313 ft)
Service ceiling, one engine out:
A 2,470 m (8,100 ft)
B 4,025 m (13,200 ft)
C 3,960 m (13,000 ft)
Min ground turning radius 12·2 m (40 ft 0 in)
Runway LCN 4·5
T-O run:
A, B 527 m (1,728 ft)
C 506 m (1,660 ft)
Accelerate/stop distance:
A, B 1,137 m (3,731 ft)
C 1,120 m (3,674 ft)
Landing from 15 m (50 ft), without propeller reversal:
A, B 817 m (2,680 ft)
C 851 m (2,793 ft)
Range at max cruising speed, at 2,440 m (8,000 ft), with max fuel, 45 min reserves:
A 788 nm (1,459 km; 907 miles)
B 765 nm (1,417 km; 881 miles)
C 722 nm (1,339 km; 832 miles)
Range at max cruising speed, at 2,440 m (8,000 ft), with 17 occupants and 218 kg (480 lb) baggage, 45 min reserves:
A 367 nm (679 km; 422 miles)
B 333 nm (616 km; 383 miles)
C 461 nm (853 km; 530 miles)

BEECHCRAFT KING AIR 100 and A100

US Army designation: U-21F

Beech Aircraft announced on 26 May 1969 the addition of a new version of the King Air to its fleet of corporate transport aircraft. Designated King Air 100, this is a pressurised transport with increased internal capacity and more powerful engines, enabling it to carry a useful load of more than two short tons. By comparison with the King Air 90 series, it has a fuselage 1·27 m (4 ft 2 in) longer, reduced wing span, larger rudder and elevator and twin-wheel main landing gear. It is available in a variety of interior configurations, seating six to eight in executive versions, or up to 13 in high-density arrangement, plus crew of two.

The King Air 100 has been approved for Category 2 landing minima by the FAA. Initial deliveries were made in August 1969, following FAA certification. A total of 222 King Air 100s had been delivered by 1 January 1976.

First deliveries of the advanced Model A100, comprising five U-21Fs for the Department of the Army, began in October 1971. Supplied under a $2·5 million contract, they represent the first pressurised aircraft in the Army's inventory.

Export deliveries of the King Air A100 made during 1975 included one to Canada, one to Chile, one to Jamaica and two to Saudi Arabia. Two aircraft equipped with a Beech-developed UNACE package (Universal Aircraft Com/Nav Evaluation) were delivered to Belgium and Indonesia. UNACE-configured aircraft, which provide an economical means of inspecting and calibrating aviation navigation aids, are operating also in Algeria, Canada, Malaysia, Mexico and the USA. Beech is able to modify King Airs for aerial photography, and deliveries of camera-equipped aircraft have been made to Canada, Chile, Jamaica, Saudi Arabia and Thailand, as well as to various US organisations.

TYPE: Twin-turboprop light passenger, freight or executive transport.

WINGS: As for Beechcraft 99 Airliner. Pneumatic de-icing boots standard.

FUSELAGE: As for King Air C90, except length extended by 0·76 m (2 ft 6 in) forward of the wing, and 0·51 m (1 ft 8 in) aft of the wing.

TAIL UNIT: Cantilever all-metal structure with swept vertical surfaces and a ventral stabilising fin. Trim tab in rudder. Electrically-operated adjustment of tailplane incidence. Pneumatic de-icing boots standard.

LANDING GEAR: As for Beechcraft 99 Airliner. Dual main wheels and tubeless tyres size 18 × 5·5, pressure 7·10 bars (103 lb/sq in). Nosewheel with tubeless tyre size 6·50 × 10, pressure 3·93 bars (57 lb/sq in). B. F. Goodrich heat sink and aircooled single-disc hydraulic brakes.

POWER PLANT: Two 507 kW (680 ehp) Pratt & Whitney Aircraft of Canada PT6A-28 turboprop engines, each driving a Hartzell three-blade fully-feathering and reversible-pitch constant-speed propeller, or four-blade propeller on the King Air A100. Rubber fuel cells in wings, with total capacity of 1,469 litres (388 US gallons) on King Air 100; A100 has two additional fuel cells in each outer wing panel, providing a total fuel capacity of 1,779 litres (470 US gallons). Automatic fuel heating systems; inertial engine inlet de-icing system; engine inlet lips de-iced by electro-thermally heated boots. Goodrich electrical propeller anti-icing system.

ACCOMMODATION: Crew of two side by side on flight deck, with full dual controls and instruments. Easily removable partition with sliding door between flight deck and cabin. Six fully-adjustable individual cabin chairs standard, with removable headrests, with a variety of alternative layouts, for up to 13 passengers in commuter role. Polarised cabin windows. Dual storm windows. Fully-carpeted floor. External access door to forward radio compartment. Aft fuselage maintenance access door. Plug-type emergency exit at forward end of cabin on starboard side. Passenger door at rear of cabin on port side, with integral airstair. Easily removable aft cabin partition with sliding doors. Lavatory installation and stowage for up to 186 kg (410 lb) baggage in aft fuselage. Other standard cabin equipment includes reading lights and fresh air outlets for all passengers, cabin coat rack and dual 'No smoking—Fasten seat belt' signs. Electro-thermally heated windscreen, hot air windscreen defroster and windscreen wipers standard. Optional equipment includes cabin fire extinguisher, additional cabin window, flush toilet and a variety of interior cabinets.

SYSTEMS: Cabin pressurisation by dual engine bleed air with a maximum differential of 0·32 bars (4·6 lb/sq in). Cabin heated by 45,000 BTU dual engine bleed air system and auxiliary 27,000 BTU electrical heating system. Oxygen system for flight deck and 0·62 m³ (22 cu ft) oxygen system for cabin standard. Cabin oxygen system of 1·39 m³ (49·2 cu ft) optional. Dual vacuum system for instruments. Hydraulic system for brakes only. Pneumatic system for wing and tail unit de-icing only. Two 250A starter/generators. Aircooled nickel-cadmium 28V 45Ah battery with failure detector. Engine fire detection system.

EQUIPMENT: Standard equipment includes stall warning system, heated nose-mounted pitot heads, heated fuel vents, full blind-flying instrumentation, wing ice lights, dual nose-mounted landing lights, nose gear taxi light, adjustable cabin reading lights, cabin door, baggage compartment, aisle and map lights, blue-white cockpit lighting, primary and secondary instrument lights, rheostat-controlled white cockpit lighting, navigation lights, upper and lower rotating beacons, map pockets, dual adjustable sun visors and polarised cabin windows to reduce external light intensity and glare, super soundproofing and towbar. Optional items include a variety of interior layouts and seating, engine fire extinguisher system, propeller synchrophaser, smoke detection system for radio compartment, flight hour meter, safe-flight-speed control indicator, three-light strobe system, and automatic propeller feathering. Oversize dual main wheels and tyres are available to replace the standard dual main wheels and tyres.

ELECTRONICS: Standard installation of A100 comprises King KTR-905 main VHF transceiver with dual Gables controls and B3 antenna; King KTR-905 standby transceiver with Gables control and B3 antenna; King KNR-630 automatic omni No. 1 with Collins 331A-3G indicator, Gables control and B17 nav antenna; King KNR-630 automatic omni No. 2 with Collins 331H-3G indicator and Gables control; dual omni range filters; Collins dual 356F-3 audio amplifiers with dual 356C-4 isolation amplifiers with dual audio switches; King KMR-675 marker beacon receiver with dual marker lights and B16 antenna; King KDF-805 ADF less indicator, with KFS-580B control, voice-range filter and Beech flush sense antenna; dual glideslope receivers included in No. 1 and No. 2 KNR-630 with B35 glideslope antenna. RCA AVQ-47 radar with 12 in antenna and standard scope; Sperry C-14-43 compass system with pilot's servo amplifier; Collins 332C-10 RMI with VOR-1/ADF and VOR-2/ADF; King KXP-755 transponder with Gables control and B18 antenna; King KDM-705A DME with KDI-571 indicator, Nav 1-Nav 2 switching and DME hold and B18 antenna; dual 600 VA Flite-Tronics PC-17 inverters with failure light; sectional instrument panel; dual flight instrumentation; Standard Electric pilot's gyro horizon; co-pilot's 3 in CF gyro horizon and directional gyro; Beech edge-lighted radio panel, radio accessories, microphone key button in pilot's and co-pilot's control wheels, static wicks and white lighting; dual microphones, headsets and cabin speakers; and electronics master switch. A wide range of optional electronics by Bendix, Collins, King, RCA and Sunair are available to customers requirements.

DIMENSIONS, EXTERNAL:
Wing span 13·98 m (45 ft 10½ in)
Wing chord at root 2·15 m (7 ft 0½ in)
Wing chord at tip 1·07 m (3 ft 6 in)
Wing aspect ratio 7·51
Length overall 12·18 m (39 ft 11⅜ in)
Height overall 4·68 m (15 ft 4¼ in)
Tailplane span 6·81 m (22 ft 4½ in)
Wheel track 3·96 m (13 ft 0 in)
Wheelbase 4·55 m (14 ft 11 in)
Propeller diameter:
three-blade 2·37 m (7 ft 9½ in)
four-blade 2·29 m (7 ft 6 in)
Propeller ground clearance:
four-blade 0·34 m (1 ft 1½ in)

DIMENSIONS, INTERNAL:
Cabin: Length (excl flight deck) 5·08 m (16 ft 8 in)
Max width 1·37 m (4 ft 6 in)
Max height 1·45 m (4 ft 9 in)

Beechcraft King Air A100 eight/fifteen-seat pressurised transport (two Pratt & Whitney Aircraft of Canada PT6A-28 turboprop engines)

Volume, electronics compartment in nose
0·45 m³ (16 cu ft)
Volume, aft baggage compartment
1·51 m³ (53·5 cu ft)

WEIGHTS AND LOADINGS (A: King Air 100, B: King Air A100):
Weight empty:
A 2,905 kg (6,405 lb)
B 3,065 kg (6,759 lb)
Max T-O weight:
A 4,808 kg (10,600 lb)
B 5,216 kg (11,500 lb)
Max ramp weight:
A 4,838 kg (10,668 lb)
B 5,247 kg (11,568 lb)
Max zero-fuel weight:
B 4,354 kg (9,600 lb)
Max landing weight:
B 5,084 kg (11,210 lb)
Max wing loading:
B 199 kg/m² (40·8 lb/sq ft)
Max power loading:
B 5·14 kg/kW (8·46 lb/ehp)

PERFORMANCE (A: King Air 100 at max T-O weight of 4,808 kg: 10,600 lb; B: King Air A100 at max T-O weight of 5,216 kg: 11,500 lb, unless otherwise quoted):
Max cruising speed:
A, at 4,309 kg (9,500 lb) AUW:
at 6,400 m (21,000 ft)
239 knots (443 km/h; 275 mph)
at 4,875 m (16,000 ft)
245 knots 454 km/h; 282 mph)
at 3,050 m (10,000 ft)
248 knots (459 km/h; 285 mph)
B, at 4,762 kg (10,500 lb) AUW:
at 6,400 m (21,000 ft)
235 knots (436 km/h; 271 mph)
at 4,875 m (16,000 ft)
243 knots (450 km/h; 280 mph)
at 3,050 m (10,000 ft)
248 knots (459 km/h; 285 mph)
Cruising speed, low cruise power:
A, at 4,309 kg (9,500 lb) AUW:
at 6,400 m (21,000 ft)
226 knots (418 km/h; 260 mph)
at 4,875 m (16,000 ft)
234 knots (433 km/h; 269 mph)
at 3,660 m (12,000 ft)
237 knots (439 km/h; 273 mph)
B, at 4,762 kg (10,500 lb) AUW:
at 6,400 m (21,000 ft)
221 knots (409 km/h; 254 mph)
at 4,875 m (16,000 ft)
231 knots (428 km/h; 266 mph)
at 3,050 m (10,000 ft)
236 knots (438 km/h; 272 mph)
Stalling speed, power off, wheels and flaps up:
A 92 knots (170·5 km/h; 106 mph)
B 90 knots (167·5 km/h; 104 mph)
Stalling speed, power off, wheels and flaps down:
A 76 knots (141 km/h; 87·5 mph)
B 75 knots (139 km/h; 86·5 mph)
Max rate of climb at S/L:
A 671 m (2,200 ft)/min
B 598 m (1,963 ft)/min
Rate of climb at S/L, one engine out:
A 185 m (608 ft)/min
B 138 m (452 ft)/min
Service ceiling:
A 7,895 m (25,900 ft)
B 7,575 m (24,850 ft)
Service ceiling one engine out:
A 3,595 m (11,800 ft)
B 2,835 m (9,300 ft)
Min ground turning radius 12·2 m (40 ft 0 in)
Runway LCN 4·5
T-O run:
A (using flaps) 443 m (1,452 ft)
B (no flaps) 628 m (2,060 ft)
B (30% flap) 565 m (1,855 ft)
T-O to 15 m (50 ft):
A (using flaps) 527 m (1,729 ft)
B (no flaps) 989 m (3,245 ft)
B (30% flap) 817 m (2,681 ft)
Landing from 15 m (50 ft) at max landing weight, without propeller reversal:
A 652 m (2,138 ft)
B 685 m (2,246 ft)
Landing run at max landing weight, without propeller reversal:
A 378 m (1,240 ft)
B 397 m (1,302 ft)
*Accelerate/stop distance, flaps up:
B 1,303 m (4,275 ft)
*Accelerate/stop distance, 30% flap:
B 1,182 m (3,877 ft)
Range at high cruise power, A with 1,415 litres (374 US gallons) fuel, B with 1,779 litres (470 US gallons) fuel, incl allowances for starting, taxi, take-off, climb, descent and 45 min reserves:
at 6,400 m (21,000 ft):
A 945 nm (1,752 km; 1,089 miles)
B 1,201 nm (2,227 km; 1,384 miles)
at 4,875 m (16,000 ft):
A 830 nm (1,538 km; 956 miles)
B 1,064 nm (1,971 km; 1,225 miles)
at 3,050 m (10,000 ft):
A 706 nm (1,308 km; 813 miles)
B 900 nm (1,667 km; 1,036 miles)
Range at low cruise power, fuel and allowances as above:
at 6,400 m (21,000 ft):
A 1,005 nm (1,863 km; 1,158 miles)
B 1,287 nm (2,385 km; 1,482 miles)
at 4,875 m (16,000 ft):
A 895 nm (1,659 km; 1,031 miles)
B 1,149 nm (2,129 km; 1,323 miles)
at 3,050 m (10,000 ft):
A 762 nm (1,413 km; 878 miles)
B 982 nm (1,818 km; 1,130 miles)

**Includes allowance for failure recognition*

BEECHCRAFT KING AIR B100

On 20 March 1975 Beech recorded the first flight of a new version of the King Air. Designated King Air B100, it is generally similar to its predecessors, except for the installation of two 533 kW (715 shp) AiResearch TPE 331-6-252B turboprop engines, giving improved performance.

The description of the King Air A100 applies also to the B100, except as follows:

TYPE: Twin-turboprop light passenger, freight or executive transport.

WINGS, FUSELAGE, TAIL UNIT, LANDING GEAR: As for Model A100.

POWER PLANT: Two 533 kW (715 shp) Garrett-AiResearch TPE 331-6-252B turboprop engines, each driving a four-blade metal constant-speed and fully-feathering propeller. Fuel system and anti-icing system as for Model A100.

ACCOMMODATION, SYSTEMS, ELECTRONICS AND EQUIPMENT: As for Model A100, except for slight variations in ancillary equipment associated directly with the power plant.

DIMENSIONS, EXTERNAL: As for Model A100, except:
Propeller ground clearance 0·39 m (1 ft 3½ in)

DIMENSIONS, INTERNAL, AND AREAS: As for Model A100

WEIGHTS AND LOADINGS:
Weight empty, equipped 3,232 kg (7,127 lb)
Max T-O weight 5,352 kg (11,800 lb)
Max ramp weight 5,386 kg (11,875 lb)
Max zero-fuel weight 4,354 kg (9,600 lb)
Max landing weight 5,080 kg (11,200 lb)
Max wing loading 206·0 kg/m² (42·2 lb/sq ft)
Max power loading 5·02 kg/kW (8·25 lb/shp)

PERFORMANCE (at max T-O weight):
Max level speed and max cruising speed at 3,660 m (12,000 ft) 268 knots (497 km/h; 309 mph) TAS
Econ cruising speed at 6,100 m (20,000 ft)
262 knots (486 km/h; 302 mph) TAS
Stalling speed, flaps up
93 knots (172·5 km/h; 107 mph) IAS
Stalling speed, flaps down
83 knots (154·5 km/h; 96 mph) IAS
Max rate of climb at S/L 652 m (2,139 ft)/min
Rate of climb at S/L, one engine out
152 m (501 ft)/min
Service ceiling 8,870 m (29,100 ft)
Service ceiling, one engine out 4,617 m (15,150 ft)
T-O run 579 m (1,898 ft)
T-O run, 30% flap 535 m (1,755 ft)
T-O to 15 m (50 ft) 899 m (2,951 ft)
T-O to 15 m (50 ft), 30% flap 821 m (2,694 ft)
Landing from 15 m (50 ft) 1,040 m (3,413 ft)
Landing from 15 m (50 ft) with propeller reversal
817 m (2,682 ft)
Landing run 503 m (1,651 ft)
Landing run, with propeller reversal 514 m (1,686 ft)
Range with 1,779 litres (470 US gallons) usable fuel, with allowances for start, taxi, T-O, climb, descent and 45 min reserves at max range power, ISA:
at max cruising power at
3,660 m (12,000 ft)
1,003 nm (1,857 km; 1,155 miles)
4,875 m (16,000 ft)
1,115 nm (2,066 km; 1,284 miles)
6,100 m (20,000 ft)
1,244 nm (2,304 km; 1,432 miles)
at econ cruising power at:
3,660 m (12,000 ft)
1,108 nm (2,053 km; 1,276 miles)
4,875 m (16,000 ft)
1,205 nm (2,232 km; 1,387 miles)
6,100 m (20,000 ft)
1,304 nm (2,415 km; 1,501 miles)

BEECHCRAFT SUPER KING AIR 200

US military designations: C-12 and RU-21J

Design of the Super King Air 200 began in October 1970, construction of the first prototype and first pre-production aircraft starting simultaneously a year later. The first prototype, serial BB-1, flew for the first time on 27 October 1972, followed by the second aircraft, BB-2, on 15 December 1972. While the flight tests and testing of a static fuselage were under way, construction of the first production aircraft began in June 1973. FAA certification under FAR Part 23 was awarded on 14 December 1973, the aircraft satisfying also the icing requirements of FAR Part 25.

By comparison with the King Air 100, the Super King Air 200 has increased wing span, basically the same fuselage, a new T-tail, more powerful engines, additional fuel capacity, increased cabin pressurisation and a higher gross weight. By 1 January 1976 Beech had delivered 100 Super King Airs to commercial and private operators and 19 military **C-12As** to the USAF and US Army. In addition, during 1974 the US Army added three Super King Airs to its fleet of special mission aircraft, with the designation **RU-21J**. Under an R and D contract, Beech had modified these antenna-laden aircraft for the US Army's Cefly Lancer programme. They are approved for take-off at a special AUW of 6,804 kg (15,000 lb).

In August 1974, Beech received a first contract from the US Army to build and support 34 modified versions of the Super King Air. Of these, the US Army was to receive 20 and the USAF 14. Options in the contract permit both services to purchase additional aircraft, and also to procure worldwide aircraft servicing, including on-site personnel, facilities for inspection and maintenance, and for stocking spare parts at strategic bases.

In August 1975, both the USAF and US Army exercised options to their existing contracts for the C-12A, the Air Force adding 20 and the Army 16. Contract value of the total of 70 C-12As under production contract is $45 million. Deliveries of these military Super King Airs will extend into October 1977.

The C-12As are described as 'standard off-the-shelf Super King Air types, modified slightly to meet military flight requirements and to orient the control systems for two-pilot operation which is standard military practice'. Accommodation is provided for eight passengers, plus two pilots, with easy conversion to cargo missions. The large baggage area has provisions for storing survival gear.

Worldwide deployment of the C-12As began in July 1975. Standard power plant comprises two 559 kW (750 shp) Pratt & Whitney Aircraft of Canada PT6A-38 turboprop engines, each driving a Hartzell three-blade fully-feathering and reversible-pitch constant-speed propeller. Usable fuel capacity is 1,318 litres (348 US gallons). In other respects, the details given for the Super King Air apply generally to the C-12A.

In early 1975 the FAA awarded Beech a $1·6 million contract, under which the company will modify a Super King Air to accommodate FAA-developed electronic equipment for evaluating ground-based navigation aids, including en route facilities, instrument landing systems, low and medium frequency radio stations, approach lights and radar facilities. The aircraft is to be stationed in Hawaii.

Two Super King Airs with camera installations have been ordered for France's Institut Géographique National. These have detachable wingtip fuel tanks, increasing total usable capacity from 2,059 to 2,456 litres (544 to 649 US gallons) and adding approx 1 hr to max endurance. A high flotation main landing gear is fitted to this version, permitting the max T-O and landing weights to be increased to 6,350 kg (14,000 lb) and 6,123 kg (13,500 lb) respectively.

TYPE: Twin-turboprop passenger or executive light transport.

WINGS: Cantilever low-wing monoplane. Wing section NACA 23018·5 (modified) at root, NACA 23011·3 at tip. Dihedral 6°. Incidence 3°48′ at root, −1° 7′ at tip. No sweepback at quarter-chord. Two-spar light alloy structure. Conventional ailerons of light alloy construction, with trim tab in port aileron. Single-slotted trailing-edge flaps of light alloy construction. Pneumatic de-icing boots standard.

FUSELAGE: Light alloy semi-monocoque structure of safe-life design.

TAIL UNIT: Conventional cantilever T-tail structure of light alloy with swept vertical and horizontal surfaces. Fixed-incidence tailplane. Trim tab in each elevator. Anti-servo tab in rudder. Pneumatic de-icing boots standard, on leading-edge of tailplane only.

LANDING GEAR: Electrically-retractable tricycle type, with twin wheels on each main unit. Single wheel on steerable nose unit, with shimmy damper. Main units retract forward, nosewheel aft. Beech oleo-pneumatic shock-absorbers. Goodrich main wheels and tyres size 18 × 5·5, pressure 7·03 bars (102 lb/sq in). Goodrich nosewheel size 6·50 × 10, with tyre size 22 × 6·75-10, pressure 3·93 bars (57 lb/sq in). Goodrich hydraulic multiple-disc brakes. Parking brake.

POWER PLANT: Two 634 kW (850 shp) Pratt & Whitney Aircraft of Canada PT6A-41 turboprop engines, each driving a three-blade metal constant-speed fully-feathering and reversible propeller; PT6A-38 engines in C-12A, as noted in introductory copy. Bladder type fuel cells in each wing, with main system capacity of 1,461 litres (386 US gallons) and auxiliary system capacity of 598 litres (158 US gallons). Total fuel capacity 2,059 litres (544 US gallons). Two refuelling points in upper surface of each wing. Oil capacity 29·5 litres (7·8 US gallons). Anti-icing of engine air intakes by hot air from engine exhaust is standard. Electro-thermal anti-icing for propellers. Wingtip tanks optional.

ACCOMMODATION: Pilot only, or crew of two side by side, on flight deck, with full dual controls and instruments as standard. Six cabin seats standard, with alternative layouts for a maximum of 13 passengers in cabin and 14th beside pilot. Partition with sliding door between cabin and flight deck, and partition at rear of cabin. Door at rear of cabin on port side, with integral airstair. Inward-opening emergency exit on starboard side over wing. Lavatory and stowage for up to 186 kg (410 lb) baggage in aft fuselage. Maintenance access door in rear fuselage; radio compartment access door in nose. Standard equipment includes reading lights and fresh air outlets for all passengers, triple cabin windows with polarised glare control, fully-carpeted floor, 'No smoking—Fasten seat belt' sign, cabin coat rack, fluorescent cabin lighting, aisle and door courtesy lights. Electrically-heated windscreens, hot air windscreen defroster, dual storm windows, sun visors, map pockets and windscreen wipers. Cabin is air-conditioned and pressurised, and can be provided with optional radiant heat panels.

SYSTEMS: Cabin pressurisation by engine bleed air, with a maximum differential of 0·41 bars (6·0 lb/sq in). Cabin air-conditioner of 31,000 BTU capacity. Auxiliary cabin heating by radiant panels optional. Oxygen system for flight deck, and 625 litre (22 cu ft) oxygen system for cabin, standard; system of 1,390 litres (49 cu ft) or 1,810 litres (64 cu ft) optional. Dual vacuum system for instruments. Hydraulic system for brakes only. Pneumatic system for wing and tailplane de-icing. Electrical system has two 250A 28V starter/generators and 24V 34Ah aircooled nickel-cadmium battery with failure detector. AC power provided by dual 600VA inverters.

ELECTRONICS: Standard King Gold Crown electronics package includes King KTR-900A main and standby VHF transceivers with Gables controls and B3 antennae; King KNR-600A manual omni No. 1 with Collins 331A-3G indicator, Gables control and B17 nav antenna; King KNR-660A automatic omni No. 2 with Collins 331H-3G indicator and Gables control; dual omni range filters; Collins dual 356F-3 audio amplifiers with dual 356C-4 isolation amplifiers with dual audio switches; King KDF-800 ADF less indicator, with KFS-580 control, voice range filter and Beech flush sense antenna; King KGM-691 No. 1 glideslope and marker beacon receiver with dual marker lights, B16 antenna and B35 glideslope antenna; King KGS-681 No. 2 glideslope receiver; RCA AVQ-47 radar with 12 in phased array antenna and standard scope; Sperry C-14-23 compass system with servo amplifier; AAR 2105D-B-6 RMI with ADF on single needle and VOR 2 on double needle; King KXP-750A transponder with Gables control and B18 antenna; King KN-65 DME with KI-265 indicator, KA-43 converter, Nav 1 and Nav 2 switching and B18 antenna; dual 600VA Flite-Tronics PC-17 inverters with failure light; sectional instrument panel; dual flight instrumentation; standard electric gyro horizon; co-pilot's 3 in CF gyro horizon and directional gyro; Beech edge-lighted radio panel, radio accessories, microphone key button in pilot's and co-pilot's control wheels, static wicks and white lighting; dual microphones, headsets and cockpit speakers; emergency locator transmitter; and electronics master switch. An extensive range of optional electronics by ARC, Collins, King, RCA and Sperry is available to customer's requirements.

EQUIPMENT: Standard equipment includes an automatic fuel heater system, engine fire detection system, propeller synchroscope indicator, max permissible airspeed indicator, outside air temperature gauge, eight-day clock, chronograph clock, cabin rate of climb indicator, cabin altitude and differential pressure indicator, annunciator panel, heated stall warning system, dual heated pitot heads, external power socket, heated fuel vents, wing ice lights, dual landing lights, taxi light, rotating beacons, dual map lights, cabin dome lights, primary and secondary instrument lighting systems, blue-white flight deck lights, internal corrosion-proofing, super soundproofing, low profile glareshield,

Beechcraft Super King Air 200 (two Pratt & Whitney Aircraft of Canada PT6A-41 turboprop engines). Wingtip tanks are now optional for the Super King Air

urethane paint, towbar, and control lock assembly. Optional equipment includes instantaneous vertical speed indicator, flight hour recorder, propeller synchrophaser, oversize wheels and tyres, three-light strobe system, electrically-powered elevator trim, two- and four-seat couches with and without stowage drawer, folding armrests for couch, sixth starboard cabin window, forward-facing instead of standard toilet, Sherwood flushing toilet, cabin fire extinguisher, aft baggage partition, approach plate/map cases, cabin table, chair covers, entrance door step lights, fin illumination light, a variety of cabinets and a range of galley equipment.

DIMENSIONS, EXTERNAL:
- Wing span: 16·61 m (54 ft 6 in)
- Wing chord at root: 2·18 m (7 ft 1¾ in)
- Wing chord at tip: 0·90 m (2 ft 11⅝ in)
- Wing aspect ratio: 9·8
- Length overall: 13·36 m (43 ft 10 in)
- Height overall: 4·52 m (14 ft 10 in)
- Tailplane span: 5·61 m (18 ft 5 in)
- Wheel track: 5·23 m (17 ft 2 in)
- Wheelbase: 4·56 m (14 ft 11½ in)
- Propeller diameter: 2·50 m (8 ft 2½ in)
- Propeller ground clearance: 0·37 m (1 ft 2½ in)
- Distance between propeller centres: 5·23 m (17 ft 2 in)
- Passenger door:
  - Height: 1·31 m (4 ft 3½ in)
  - Width: 0·68 m (2 ft 2¾ in)
  - Height to sill: 1·17 m (3 ft 10 in)
- Nose electronics service doors (port and stbd):
  - Max height: 0·57 m (1 ft 10½ in)
  - Width: 0·63 m (2 ft 1 in)
  - Height to sill: 1·37 m (4 ft 6 in)
- Emergency exit door (stbd):
  - Height: 0·66 m (2 ft 2 in)
  - Width: 0·50 m (1 ft 7¾ in)

DIMENSIONS, INTERNAL:
- Cabin (from forward to aft pressure bulkhead):
  - Length (Super King Air): 6·71 m (22 ft 0 in)
  - Length (C-12A): 5·00 m (16 ft 5 in)
  - Max width: 1·37 m (4 ft 6 in)
  - Max height: 1·45 m (4 ft 9 in)
  - Floor area: 7·80 m² (84 sq ft)
  - Volume: 11·10 m³ (392 cu ft)
- Baggage hold, rear of cabin:
  - Volume: 1·53 m³ (54 cu ft)

AREAS:
- Wings, gross: 28·15 m² (303 sq ft)
- Ailerons (total): 1·67 m² (18·0 sq ft)
- Trailing-edge flaps (total): 4·17 m² (44·9 sq ft)
- Fin: 3·46 m² (37·2 sq ft)
- Rudder, incl tab: 1·40 m² (15·1 sq ft)
- Tailplane: 4·52 m² (48·7 sq ft)
- Elevators, incl tabs: 1·79 m² (19·3 sq ft)

WEIGHTS AND LOADINGS:
- Weight empty: 3,318 kg (7,315 lb)
- Max T-O and landing weight: 5,670 kg (12,500 lb)
- Max ramp weight: 5,710 kg (12,590 lb)
- Max zero-fuel weight: 4,717 kg (10,400 lb)
- Max wing loading: 201·6 kg/m² (41·3 lb/sq ft)
- Max power loading: 4·47 kg/kW (7·4 lb/shp)

WEIGHTS AND LOADINGS (C-12A): As Super King Air, except:
- Basic empty weight: 3,569 kg (7,869 lb)
- Max T-O and landing weight: 5,670 kg (12,500 lb)
- Max ramp weight: 5,708 kg (12,585 lb)
- Max wing loading: 201·4 kg/m² (41·25 lb/sq ft)
- Max power loading: 4·47 kg/kW (8·33 lb/shp)

PERFORMANCE (at max T-O weight, ISA, unless specified):
- Never-exceed speed: Mach 0·483 (270 knots; 499 km/h; 310 mph) CAS
- Max level speed, average cruise weight at 4,570 m (15,000 ft): 289 knots (536 km/h; 333 mph)
- Max cruising speed, average cruise weight at 7,620 m (25,000 ft): 278 knots (515 km/h; 320 mph)
- Econ cruising speed, average cruise weight at 7,620 m (25,000 ft): 272 knots (503 km/h; 313 mph)
- Stalling, speed, flaps up: 99 knots (183 km/h; 114 mph) CAS
- Stalling speed, flaps down: 75 knots (139·5 km/h; 86·5 mph) CAS
- Max rate of climb at S/L: 747 m (2,450 ft)/min
- Rate of climb at S/L, one engine out: 226 m (740 ft)/min
- Service ceiling: above 9,450 m (31,000 ft)
- Service ceiling, one engine out: 5,835 m (19,150 ft)
- T-O run: 592 m (1,942 ft)
- T-O run, 40% flap: 566 m (1,856 ft)
- T-O to 15 m (50 ft), flaps up: 1,020 m (3,345 ft)
- T-O to 15 m (50 ft), 40% flap: 786 m (2,579 ft)
- Landing from 15 m (50 ft), full flap, without propeller reversal: 867 m (2,845 ft)
- Landing from 15 m (50 ft) with propeller reversal: 632 m (2,074 ft)
- Landing run, full flap, without propeller reversal: 536 m (1,760 ft)
- Landing run with propeller reversal: 341 m (1,120 ft)
- Range with 2,059 litres (544 US gallons) usable fuel, with allowances for start, taxi, climb, descent and 45 min reserves at max range power, ISA:
  - max cruising power at:
    - 5,485 m (18,000 ft): 1,190 nm (2,204 km; 1,370 miles)
    - 7,620 m (25,000 ft): 1,485 nm (2,752 km; 1,710 miles)
    - 9,450 m (31,000 ft): 1,757 nm (3,255 km; 2,023 miles)
  - econ cruising power at:
    - 5,485 m (18,000 ft): 1,487 nm (2,755 km; 1,712 miles)
    - 7,620 m (25,000 ft): 1,737 nm (3,218 km; 2,000 miles)
    - 9,450 m (31,000 ft): 1,887 nm (3,497 km; 2,173 miles)

PERFORMANCE (C-12A at max T-O weight):
- Max level speed at 4,267 m (14,000 ft): 261 knots (484 km/h; 301 mph)
- Max cruising speed at 9,140 m (30,000 ft): 227 knots (421 km/h; 262 mph)
- Service ceiling: 9,420 m (30,900 ft)
- Service ceiling, one engine out: 5,425 m (17,800 ft)
- T-O to 15 m (50 ft): 869 m (2,850 ft)
- Landing from 15 m (50 ft): 766 m (2,514 ft)
- Range at max cruising speed: 1,584 nm (2,935 km; 1,824 miles)

RU-21J version of the Beechcraft Super King Air procured for the US Army's Cefly Lancer programme

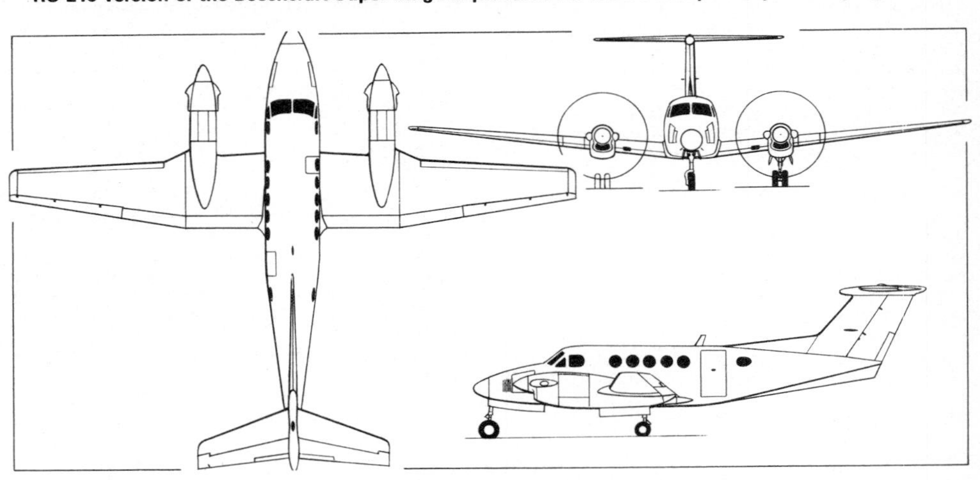
Beechcraft Super King Air 200 twin-turboprop transport *(Pilot Press)*

BEECHCRAFT FAN JET 400

Fan Jet 400 is the name given by Beech to a prototype Super King Air 200 which, since mid-1975, has been undergoing evaluation with a trial installation of two Pratt & Whitney Aircraft of Canada JT15D turbofan engines. Other configurations are to be studied and evaluated before any decision is made regarding a certification or production commitment.

BEECHCRAFT MODEL 76

The Model 76, a four-seat twin-engined light aircraft, is currently under development for certification. A testbed

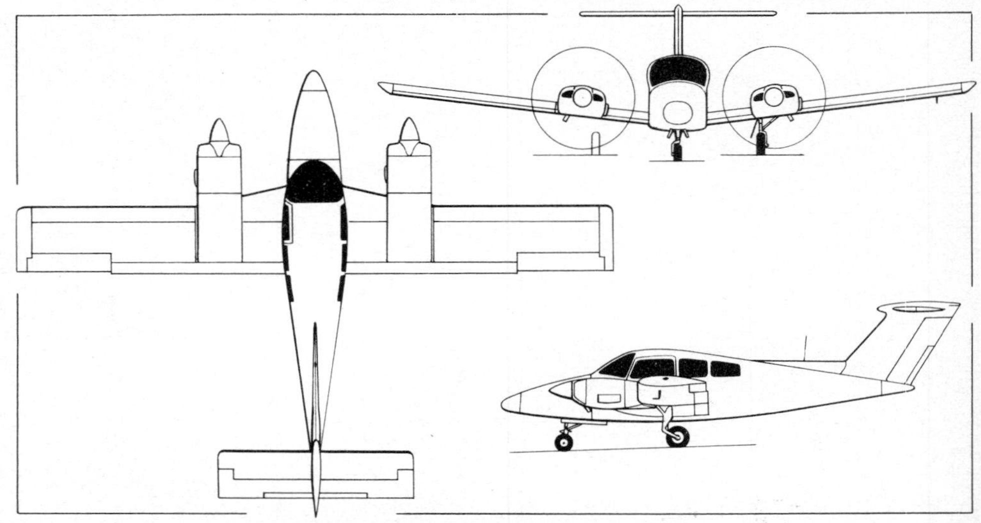
Beechcraft 76 prototype four-seat light twin *(Pilot Press)*

version of this aircraft, designated PD 289, has been undergoing a comprehensive flight test programme since September 1974.

The Model 76 is planned for use by Beech Aero Centers, and is designed for the personal light twin, light charter and multi-engine flight trainer markets. Emphasis has, therefore, been placed on good low speed flight and single-engine handling characteristics. It has a stalling speed of under 61 knots (113·5 km/h; 70·5 mph), a maximum range of more than 695 nm (1,287 km; 800 miles) and a full-fuel payload of 390 kg (860 lb). It is powered by two 134 kW (180 hp) Lycoming O-360 series flat-four engines, providing a maximum cruising speed of 160 knots (297 km/h; 185 mph).

Certification of the Model 76 will be under the Normal category for day and night VFR and IFR. Delivery of the first production aircraft is scheduled for late 1977.

Beechcraft PD 289, prototype for the Model 76 (two Lycoming O-360 series engines)

BELL

BELL AEROSPACE COMPANY DIVISION OF TEXTRON INC

Head Office and Works:
Buffalo, New York 14240
Telephone: (716) 297-1000

President:
William G. Gisel

Executive Vice-Presidents:
Lawrence P. Mordaunt (Niagara Frontier Operations)
Norton C. Willcox (Administration)

Vice-Presidents:
Dr Clifford F. Berninger (Research and Engineering)
John R. Clark Jr (Eastern Region)
John F. Gill (Product Assurance)
Adolph Kastelowitz (Manufacturing)
John J. Kelly (Vice-President and General Manager, New Orleans Operations)
John H. Pamperin (General Manager, Dalmo Victor Operations)
Joseph R. Piselli (Marketing)
Delmar E. Wilson (Western Region)

Bell Aerospace is active in aircraft, missile, propulsion and electronics systems development and advanced technology for aerospace and defence programmes. Its research and development programmes include air cushion vehicles, air cushion landing systems and high-energy lasers. It is producing an advanced upper-stage propulsion system for the USAF's Minuteman III ICBM. Designated Post Boost Propulsion System (PBPS), this has attained a high record of reliability since production began in 1968.

Production of liquid-propellant rocket engines for Lockheed's Agena satellite programme continues and is described in the Aero-Engines section.

Major propulsion systems and components research and development programmes cover the investigation of high-energy and powdered propellants, advanced engine cooling techniques, and new positive expulsion propellant storage and delivery systems.

As an outgrowth of its AN/SPN-42 Automatic Carrier Landing System (ACLS), now in operation on board US Navy aircraft carriers, Bell is updating land-based versions at US Naval air stations where the systems are being used to train pilots and operators in the use of the carrier-based system. In 1972 the company's land-based AN/SPN-42T2 at Cecil Field NAS was certificated for Mode 1 landings on all runways. In 1973, Bell completed a Naval Electronic Systems Command contract for the installation of a system at the Whidbey Island NAS near Seattle, Washington. Both the land- and carrier-based systems permit fully-automatic hands-off landings, and can handle up to 120 aircraft an hour.

In 1975, Bell, teamed with the Bendix Corporation, entered Phase III of the FAA competition for development of an automatic landing system for civilian airports. Through a series of phases, the FAA will eventually select one of the competing teams of companies for production of the Microwave Landing System.

Other electronics work is concerned with precision inertial equipment, transoceanic satellite-relay air traffic control systems, and airborne target location and fire control systems.

In late 1973 Dalmo Victor became an operating unit of Bell Aerospace. With headquarters near San Francisco, this company produces electro-magnetic defence systems, aerospace antennae and electro-optical equipment; its Oregon Technical Products unit, located in Grants Pass, Oregon, manufactures a variety of electro-mechanical products.

de Havilland Canada XC-8A research aircraft taking off with inflated ACLS cushion

BELL AIR CUSHION LANDING SYSTEM (ACLS)

Bell Aerospace began development of an air cushion landing system (ACLS) as a company-funded research project in December 1963. In 1966 it received a $99,000 contract from the USAF Flight Dynamics Laboratory for wind-tunnel testing of the project. Subsequent Air Force contracts included a $99,500 feasibility study in 1966, a $98,700 model test programme in 1967 and a $66,300 flight test programme in 1968.

The initial intention was to determine the best form of ACLS for cargo transports, and the flight test programme was carried out with a modified Lake LA-4 four-seat amphibian (see 1970-71 *Jane's*). Bell also studied the feasibility of using an ACLS on Space Shuttle vehicles, under NASA contract (details in 1972-73 *Jane's*).

Current activities are funded under a joint United States/Canadian programme aimed at adapting the ACLS for military transport aircraft. This would allow such aircraft to operate from a variety of surfaces, including rough fields, soft soils, swamps, water, ice and snow. A contract for the first phase, covering programme definition and air cushion trunk fabrication, was awarded to Bell by the USAF Flight Dynamics Laboratory in November 1970.

It was decided to use a de Havilland Canada XC-8A Buffalo STOL military transport aircraft, loaned by the Canadian Department of National Defence, as the testbed for this programme.

The ACLS is based on the ground effect principle that employs a layer of air instead of wheels as an aircraft's ground contacting medium. The system's trunk, a large inner-tube-like arrangement, encircles the underside of the fuselage. Upon inflation, the trunk provides an air duct and seal for the air cushion.

The cushion air pressure is provided by an on-board auxiliary compressor, and the underside of the rubberised trunk is perforated with hundreds of vent holes through which the air is allowed to escape to form the air cushion.

The ACLS trunk is made of rubber and nylon and is approximately 9·75 m (32 ft) long and 4·27 m (14 ft) wide when inflated. An on-board beta control enables the pilot very quickly to alter or reverse propeller pitch; the underwing float/skids are used to maintain roll stability during operations on water or a solid surface. The conventional wheeled landing gear of the Buffalo is retained in the XC-8A, and transition between this and the inflated trunk is effected prior to an ACLS take-off or landing. This can be done either by using a special raised ground ramp or by inflating the trunk in flight after a conventional wheeled take-off.

Because the ACLS distributes the weight of an aircraft over a considerably larger area than conventional wheeled systems, and itself exerts a ground pressure of less than 0·20 kg/cm² (3 lb/sq in), the concept permits operations on surfaces with very low bearing strengths.

The accompanying photograph shows clearly the balancer floats for operation from water and, beneath them, the sprung skids for operation from land.

Pratt & Whitney Aircraft of Canada was made responsible for development and flight qualification of the auxiliary power system, and de Havilland Aircraft of Canada modified the XC-8A to take the ACLS installation. The Canadian government funded the work of these two companies.

The first ACLS take-off by the XC-8A was made on 31 March 1975, and the first ACLS landing was recorded on 11 April 1975. The flights took place at Wright-Patterson AFB, Ohio, and further take-offs and landings from progressively more difficult surfaces have been made since then. The aircraft was scheduled to be transferred to Cold Lake, Alberta, Canada in early 1976 for cold-weather trials, with the flight test programme planned to continue throughout 1976.

BELL HELICOPTER TEXTRON

Head Office:
PO Box 482, Fort Worth, Texas 76101
Telephone: (817) 280-2011

President:
James F. Atkins

Senior Vice-President:
Hans Weichsel Jr

Vice-Presidents:
M. R. Barcellona
Edwin L. Farmer (Finance)
John Finn (Industrial Relations)
James C. Fuller (Public Relations)
Leonard M. Horner (Operations)
William L. Humphrey (General Manager, Amarillo Facility)
Robert R. Lynn (Research & Engineering)
Joseph Mashman (Special Projects)
Warren T. Rockwell (Washington Operations)
Dwayne K. Jose (Commercial Marketing)
Clifford J. Kalista (Advanced Attack Helicopter)
Frank M. Sylvester (International Marketing)
Ted R. Treff (Treasurer)

Reflecting the fact that Bell Helicopter Company was the largest operating division of Textron Inc, the company's name was changed to Bell Helicopter Textron on 1 January 1976.

Production at Fort Worth is concerned primarily with military and commercial single- and twin-engined versions of the turbine-powered UH-1 Iroquois, the AH-1 HueyCobra armed helicopter developed from the UH-1, and military and commercial versions of the Model 206 JetRanger. The Bell 47, in continuous production in the

USA for more than 25 years, after receiving the first helicopter Approved Type Certificate from the CAA on 8 March 1946, is no longer in production by Bell. Versions of the Model 47 continue in production, however, by Agusta in Italy (which see).

Versions of the UH-1 are built under licence by Agusta in Italy and Fuji in Japan (which see). Bell also has licence agreements with the Republic of China, covering co-production of Model 205 general-purpose helicopters, and with the government of Australia, covering the production of Model 206B-1 Kiowas for the Australian Army. Prime contractor in Australia is the Commonwealth Aircraft Corporation (which see).

Since 1958, when Bell's Model XV-3 tilt-rotor research aircraft achieved the first full in-flight conversion by a machine of this configuration, Bell engineers have continued research in this field and have completed recent US Army/USAF/NASA contracts to investigate proprotor and folding proprotor technology. The contracts included manufacture and wind tunnel testing of examples of both types of rotor.

A full-scale folding proprotor of 7·62 m (25 ft) diameter was built. During 1971 it completed whirl, folding and shake tests. In early 1972 power testing in the NASA wind tunnel at Ames Research Center was carried out, during which the complete stop/fold sequence and blade folding at up to 175 knots (325 km/h; 202 mph) were achieved. Stability was excellent to a speed of more than 195 knots (362 km/h; 225 mph), the maximum attainable wind tunnel speed, and all test requirements were met or exceeded.

Towards the end of 1972, Bell and one other company received contracts from NASA and the US Army for the design of a tilt-rotor VTOL research vehicle. In May 1973 Bell announced that its Model 301 proposal had been selected for development. Two examples are being built, and have been allocated the US Army designation XV-15.

Also under investigation, and already tested extensively, is a method of varying a rotor's diameter in flight. Maximum diameter would provide lift for vertical take-off and climb, with minimum diameter being used when the rotor was serving as a conventional tractor propeller, thus increasing cruise efficiency.

During 1972 Bell achieved a major breakthrough in the elimination of vibration in helicopters with what is known as the nodalisation concept, flight test data and analytical results suggesting that 70 to 90 per cent vibration isolation was practicable. This concept is based on the scientific fact that any beam subjected to vertical vibratory forces, such as those induced by a rotor, will develop flexing to produce a wave form. Points of no relative motion, called the nodal points, appear at equal distances from the centre of the induced wave form, and it is at these points that Bell suspends the helicopter fuselage. Since the nodes have no relative motion, the fuselage becomes virtually free from rotor-induced vibration. Flight tests of a Model 206 JetRanger with its fuselage suspended from a nodalised beam were so convincing that Bell decided to utilise this 'Noda-Matic' technique on new production helicopters, beginning with the Model 206L LongRanger and Model 214.

Bell Helicopter Textron is responsible for management of Bell Operations Corporation, newly formed to co-operate with the government of Iran in establishing a helicopter manufacturing industry in that country. Further details of this programme can be found under the entry for Iran.

Approximately 9,600 people were employed by Bell at the beginning of 1976. The company has produced more than 21,500 helicopters.

BELL MODEL 205

US military designations: UH-1D/H and HH-1H Iroquois

Canadian military designation: CH-118 Iroquois

Although basically similar to the earlier Model 204 (see 1971-72 *Jane's*), the Model 205 introduced a longer fuselage, increased cabin space to accommodate a much larger number of passengers, and other changes. The following military versions have been built:

UH-1D. This US Army version of the Model 205 Iroquois has an 820 kW (1,100 shp) Lycoming T53-L-11 turboshaft, 14·63 m (48 ft) rotor, normal fuel capacity of 832 litres (220 US gallons) and overload capacity of 1,968 litres (520 US gallons). Relocation of the fuel cells increases cabin space to 6·23 m³ (220 cu ft), providing sufficient room for a pilot and twelve troops, or six litters and a medical attendant, or 1,815 kg (4,000 lb) of freight. A contract for a service test batch of seven YUH-1Ds was announced in July 1960 and was followed by further very large production orders from the US Army and from many other nations of the non-Communist world. First YUH-1D flew on 16 August 1961 and delivery to US Army field units began on 9 August 1963, when the second and third production UH-1Ds went to the 11th Air Assault Division at Fort Benning, Georgia. The UH-1D was superseded in production for the US Army by the UH-1H, but 352 UH-1Ds were built subsequently under licence in Germany for the German Army and Air Force. Prime contractor was Dornier.

UH-1H. Following replacement of the original T53-L-11 turboshaft by the 1,044 kW (1,400 shp) T53-L-13, the version of the Model 205 currently in production by Bell for the US Army is designated UH-1H. Deliveries of an initial series of 319 aircraft for the US Army began in September 1967. Subsequent orders included 300 more for the Army in January 1971, and nine for the RNZAF.

Additional orders for a total of 560 UH-IHs were placed in 1971-73. An add-on contract for 54 more UH-1Hs, valued at $11·9 million, was awarded in September 1974, with delivery extending into 1976.

Under a licensing agreement concluded in 1969, the Republic of China is producing UH-1Hs for the Nationalist Chinese Army, with much of the manufacturing and assembly process being carried out at Taichung, Taiwan. The initial production programme was for 50 helicopters, already delivered. Subsequent orders have increased the total procurement to 118.

CH-118. Similar to UH-1H, for Mobile Command, Canadian Armed Forces. First of ten delivered on 6 March 1968. Originally designated CUH-1H.

HH-1H. It was announced on 4 November 1970 that a fixed price contract worth more than $9·5 million had been received from the USAF for 30 HH-1H aircraft (generally similar to the UH-1H) for use as local base rescue helicopters. Deliveries were completed during 1973.

The commercial Model 205A-1 is described separately.

The following details refer specifically to the military UH-1H:

TYPE: Single-rotor general-purpose helicopter.

ROTOR SYSTEM: Two-blade all-metal semi-rigid main rotor with interchangeable blades, built up of extruded aluminium spars and laminates. Stabilising bar above and at right angles to main rotor blades. Underslung feathering axis hub. Two-blade all-metal tail rotor of honeycomb construction. Blades do not fold.

ROTOR DRIVE: Shaft-drive to both main and tail rotors. Transmission rating 820 kW (1,100 shp). Main rotor rpm 294-324.

FUSELAGE: Conventional all-metal semi-monocoque structure.

TAIL SURFACE: Small synchronised elevator on rear fuselage is connected to the cyclic control to increase allowable CG travel.

LANDING GEAR: Tubular skid type. Lock-on ground handling wheels and inflated nylon float-bags available.

POWER PLANT: One 1,044 kW (1,400 shp) Lycoming T53-L-13 turboshaft mounted aft of the transmission on top of the fuselage and enclosed in cowlings. Five interconnected rubber fuel cells, total capacity 832 litres (220 US gallons). Overload fuel capacity of 520 US gallons obtained by installation of kit comprising two

Bell UH-1H Iroquois of the Brazilian Air Force *(Ronaldo S. Olive)*

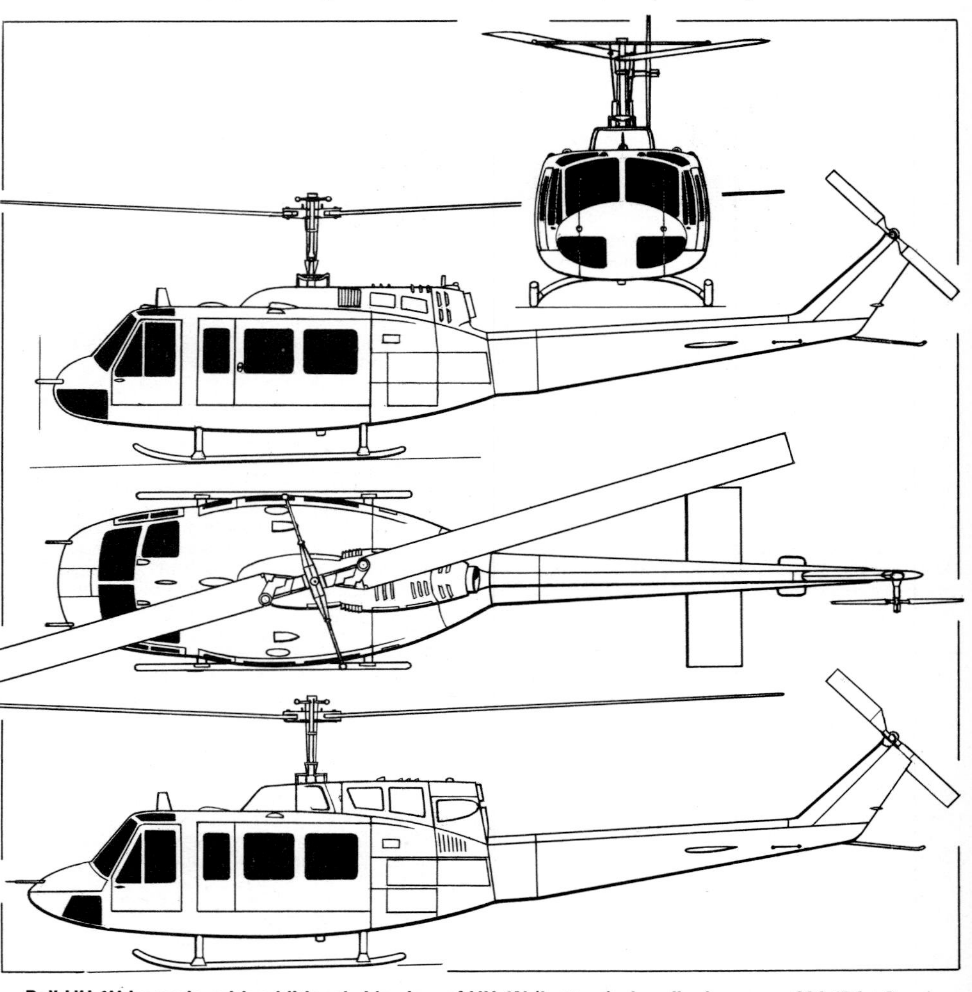

Bell UH-1H Iroquois, with additional side view of UH-1N (bottom), described on page 231 *(Pilot Press)*

568 litre (150 US gallon) internal auxiliary fuel tanks interconnected with the basic fuel system.

ACCOMMODATION: Cabin space of 6·23 m³ (220 cu ft) provides sufficient room for pilot and 11-14 troops, or six litters and a medical attendant, or 1,759 kg (3,880 lb) of freight. Crew doors open forward and are jettisonable. Two doors on each side of cargo compartment; front door is hinged to open forward and is removable, rear door slides aft. Forced air ventilation system.

EQUIPMENT: Bleed air heater and defroster, comprehensive range of engine and flight instruments, power plant fire detection system, 30V 300A DC starter/generator, navigation, landing and anti-collision lights, controllable searchlight, hydraulically-boosted controls. Optional equipment includes external cargo hook, auxiliary fuel tanks, rescue hoist, 150,000 BTU muff heater.

ELECTRONICS: FM, UHF, VHF radio sets, IFF transponder, Gyromatic compass system, direction finder set, VOR receiver and intercom standard. Optional nav/com systems.

DIMENSIONS, EXTERNAL:
- Diameter of main rotor: 14·63 m (48 ft 0 in)
- Main rotor blade chord: 0·53 m (1 ft 9 in)
- Diameter of tail rotor: 2·59 m (8 ft 6 in)
- Tail rotor blade chord: 0·213 m (8·4 in)
- Length overall (main rotor fore and aft): 17·40 m (57 ft 1 in)
- Length of fuselage: 12·77 m (41 ft 10¾ in)
- Height overall: 4·42 m (14 ft 6 in)

AREAS:
- Main rotor disc: 168·06 m² (1,809 sq ft)
- Tail rotor disc: 5·27 m² (56·7 sq ft)

WEIGHTS AND LOADINGS:
- Weight empty: 2,116 kg (4,667 lb)
- Basic operating weight (troop carrier mission): 2,520 kg (5,557 lb)
- Mission weight: 4,100 kg (9,039 lb)
- Max T-O and landing weight: 4,309 kg (9,500 lb)
- Max zero-fuel weight: 3,660 kg (8,070 lb)
- Max disc loading: 25·6 kg/m² (5·25 lb/sq ft)
- Max power loading: 4·13 kg/kW (8·63 lb/shp)

PERFORMANCE (at max T-O weight):
- Never-exceed speed: 110 knots (204 km/h; 127 mph)
- Max level and cruising speed: 110 knots (204 km/h; 127 mph)
- Econ cruising speed at 1,735 m (5,700 ft): 110 knots (204 km/h; 127 mph)
- Max rate of climb at S/L: 488 m (1,600 ft)/min
- Service ceiling: 3,840 m (12,600 ft)
- Hovering ceiling in ground effect: 4,145 m (13,600 ft)
- Hovering ceiling out of ground effect: 335 m (1,100 ft)
- Range with max fuel, no allowances, no reserves, at S/L: 276 nm (511 km; 318 miles)

BELL MODEL 205A-1

The Model 205A-1 is a fifteen-seat commercial utility helicopter developed from the UH-1H, with 1,044 kW (1,400 shp) Lycoming T5313A turboshaft, derated to 932 kW (1,250 shp) for take-off. It is designed for rapid conversion for alternative air freight, flying crane, ambulance, rescue and executive roles. Total cargo capacity is 7·02 m³ (248 cu ft) including baggage space in tailboom, with 2·34 m (7 ft 8 in) by 1·24 m (4 ft 1 in) door openings on each side of the cabin to facilitate loading of bulky freight. External load capacity in flying crane role is 2,268 kg (5,000 lb). The ambulance version can accommodate six litter patients and one or two medical attendants.

Normal fuel capacity is 814 litres (215 US gallons); optional capacity is 1,495 litres (395 US gallons).

The Model 205A-1 is produced under licence in Italy by Agusta (which see) as the AB 205A-1.

The description of the Bell UH-1H applies also to the Model 205A-1, except for the following details:

TYPE: Fifteen-seat commercial utility helicopter.

ELECTRONICS AND EQUIPMENT: Standard equipment includes vertical gyro system, 5 in gyro attitude indicator, gyro compass, master caution panel, bleed air heater, force trim hydraulic boost controls, soundproof headliner, dual windscreen wipers, cabin and engine fire extinguishers, map case and retractable passenger boarding steps. Optional items include dual controls, float landing gear, rotor brake, external cargo suspension, rescue hoist, auxiliary fuel tanks, litter installations, high-output cabin heater, protective covers and customised interiors. Standard electronics comprise 360-channel VHF transceiver and intercom system. An extensive range of optional nav/com systems is available.

DIMENSIONS, EXTERNAL:
- Length of fuselage: 12·65 m (41 ft 6 in)
- Height overall: 4·39 m (14 ft 4¾ in)

WEIGHTS:
- Weight empty, equipped: 2,357 kg (5,197 lb)
- Normal T-O weight: 4,309 kg (9,500 lb)
- Max T-O weight, external load: 4,763 kg (10,500 lb)

PERFORMANCE (at normal T-O weight):
- Max level speed from S/L to 915 m (3,000 ft): 110 knots (204 km/h; 127 mph)
- Max cruising speed at S/L: 110 knots (204 km/h; 127 mph)
- Max cruising speed at 2,440 m (8,000 ft): 96 knots (179 km/h; 111 mph)
- Max rate of climb at S/L: 512 m (1,680 ft)/min
- Max vertical rate of climb at S/L: 259 m (850 ft)/min
- Service ceiling: 4,480 m (14,700 ft)
- Hovering ceiling in ground effect: 3,170 m (10,400 ft)
- Hovering ceiling out of ground effect: 1,830 m (6,000 ft)
- Range at S/L, at max cruising speed: 270 nm (500 km; 311 miles)
- Range at 2,440 m (8,000 ft) at max cruising speed, no reserves: 298 nm (553 km; 344 miles)

BELL MODEL 206B JETRANGER II

In the Spring of 1971, Bell began delivery of the more powerful Model 206B JetRanger II, and this subsequently replaced in production the original Model 206A JetRanger. Military 206B-1 Kiowas assembled in Australia are to Model 206B standard.

Power plant of the Model 206B JetRanger II is the 298 kW (400 shp) Allison 250-C20 turboshaft engine, which Bell was able to install with minimal modification of the basic airframe to meet requests for higher performance under hot-day/high-altitude conditions. This power plant increases power-limited airspeeds by 4·3 knots (8 km/h; 5 mph) at S/L ISA, and by as much as 25 knots (46·5 km/h; 29 mph) at 3,050 m (10,000 ft) in a 35°C ambient temperature. Hovering weights are increased by approximately 181 kg (400 lb) at the same altitude, and hovering ceilings by approximately 1,220 m (4,000 ft) at the same gross weights.

Bell is producing an airframe modification kit to convert Model 206As to JetRanger II standard and Allison made available in late 1971 a conversion kit to modify 250-C18 engines to 250-C20 standard.

In late 1972, Bell announced that it was flying a demonstration Model 206 JetRanger in which the fuselage was suspended from a nodalised beam. Details of this 'Noda-Matic' concept are given in this company's introductory copy.

Under a five-year programme, covered by contracts valued at more than $75 million, Beech Aircraft produced airframes for both the commercial and military versions of the JetRanger and JetRanger II, the first airframe being delivered to Bell on 1 March 1968. A follow-on contract, valued at $8·6 million, was awarded to Beech in December 1972, extending production to August 1974, and additional follow-on contracts were announced in 1973 and 1976. The work involves manufacture of the fuselage, skid gear, tailboom, spar, stabiliser and two rear fairing assemblies.

Bell announced in January 1976 a programme, being actively pursued in conjunction with the Collins Radio Group, to develop an IFR system for the Model 206B, which could be applicable also to the Model 206L Long-Ranger. The intention was to achieve single-pilot IFR certification for the Models 206B and 206L during the first half of 1976, based on use of an autopilot designated Collins AP-107H. This rate-controlled autopilot provides full control authority in the pitch and roll axes. In addition to basic attitude control, it provides automatic navigation in selected heading, VOR capture and track, ILS capture and track, back course localiser capture and track, plus altitude hold and optional airspeed hold modes. A climb-out and go-around mode establishes automatically a straight-ahead pitch-up attitude. It was intended, when certification had been gained, to offer the IFR package also as a retrofit kit.

By January 1976, Bell and its licensees had manufactured more than 5,000 of the Model 206 series for military and commercial customers. Largest commercial operator is Petroleum Helicopters Inc, which has acquired a total of more than 135 JetRangers and JetRanger IIs. Military operators of the JetRanger II include the Brazilian Navy, with eighteen.

TYPE: Turbine-powered general-purpose light helicopter.

ROTOR SYSTEM: Two-blade semi-rigid see-saw type main rotor, employing pre-coning and underslinging to ensure smooth operation. Blades are of standard Bell 'droop-snoot' section. They have a D-shape aluminium spar, bonded aluminium alloy skin, honeycomb core and a trailing-edge extension. Each blade is connected to the hub by means of a grip, pitch-change bearings and a tension-torsion strap assembly. Two tail rotor blades have bonded aluminium skin but no core. Main rotor blades do not fold, but modification to permit manual folding is possible. Rotor brake available as optional kit.

ROTOR DRIVE: Rotors driven through tubular steel alloy shafts with spliced couplings. Initial drive from engine through 90° spiral bevel gear to single-stage planetary main gearbox. Shaft to tail rotor single-stage bevel gearbox. Freewheeling unit ensures that main rotor continues to drive tail rotor when engine is disengaged. Main rotor/engine rpm ratio 1 : 15; main rotor rpm 374-394. Tail rotor/engine rpm ratio 1 : 2·3.

FUSELAGE: Forward cabin section is made up of two aluminium alloy beams and 25 mm (1 in) thick aluminium honeycomb sandwich. Rotor, transmission and engine are supported by upper longitudinal beams. Upper and lower structures are interconnected by three fuselage bulkheads and a centrepost to form an integrated structure. Intermediate section is of aluminium alloy semi-monocoque construction. Aluminium monocoque tailboom.

TAIL UNIT: Fixed stabiliser of aluminium monocoque construction, with inverted aerofoil section. Fixed vertical tail-fin in sweptback upper and ventral sections, made of aluminium honeycomb with aluminium alloy skin.

LANDING GEAR: Aluminium alloy tubular skids bolted to extruded cross-tubes. Tubular steel skid on ventral fin to protect tail rotor in tail-down landing. Special high skid gear (0·254 m; 10 in greater ground clearance) available for use in areas with high brush. Inflated bag-type pontoons or stowed floats capable of in-flight inflation available as optional kits.

POWER PLANT: One 298 kW (400 shp) Allison 250-C20 turboshaft engine. Fuel tank below and behind rear passenger seat, capacity 288 litres (76 US gallons). Refuelling point on starboard side of fuselage, aft of cabin. Oil capacity 5·2 litres (5·5 US quarts).

ACCOMMODATION: Two seats side by side in front and rear bench seat for three persons. Two forward-hinged doors on each side, made of formed aluminium alloy with transparent panels. Baggage compartment aft of rear seats, capacity 113 kg (250 lb), with external door on port side.

SYSTEMS: Hydraulic system, pressure 41·5 bars (600 lb/sq in), for cyclic, collective and directional controls. Electrical supply from 150A starter/generator. One 24V 13Ah nickel-cadmium battery.

ELECTRONICS: Full range of electronics available in form of optional kits, including VHF communications and omni navigation kit, glideslope kit, ADF, DME, marker beacon, transponder and intercom and speaker system.

EQUIPMENT: Standard equipment includes night lighting equipment, dynamic flapping restraints, door locks, fire extinguishers and first aid kit. Optional items include dual controls, custom seating, external cargo sling with 545 kg (1,200 lb) capacity, heater, high-intensity night lights, turn and slip indicator, clock, engine oil vent, fire detection system, engine fire extinguisher, fairing kit, camera access door, engine hour meter, internal litter kit and stability and control augmentation system.

DIMENSIONS, EXTERNAL:
- Diameter of main rotor: 10·16 m (33 ft 4 in)
- Main rotor blade chord: 0·33 m (1 ft 1 in)
- Diameter of tail rotor: 1·57 m (5 ft 2 in)
- Distance between rotor centres: 5·96 m (19 ft 6½ in)
- Length overall, blades turning: 11·82 m (38 ft 9½ in)
- Length of fuselage: 9·50 m (31 ft 2 in)

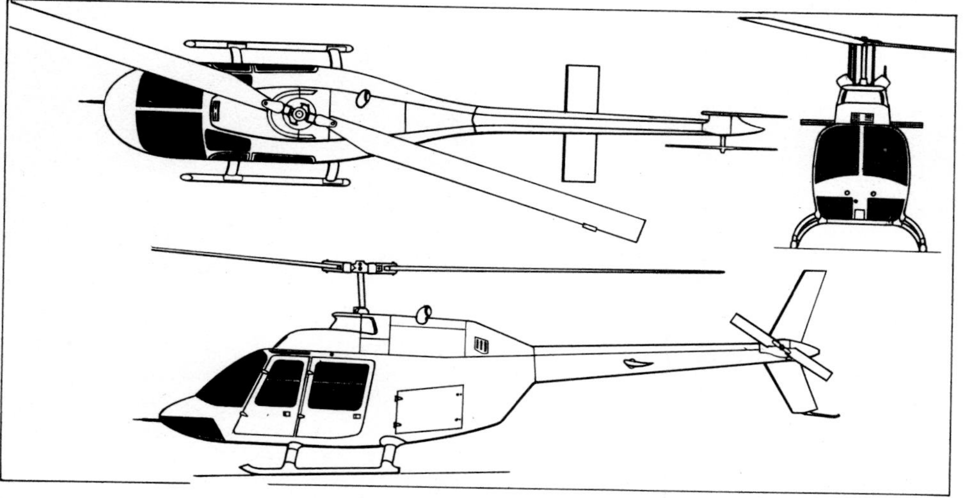

Bell 206B JetRanger II light utility helicopter *(Pilot Press)*

Height overall 2·91 m (9 ft 6½ in)
Stabiliser span 1·96 m (6 ft 5¼ in)
Width over skids 1·92 m (6 ft 3½ in)

DIMENSIONS, INTERNAL:
Cabin: Length 2·13 m (7 ft 0 in)
Max width 1·27 m (4 ft 2 in)
Max height 1·28 m (4 ft 3 in)
Baggage compartment 0·45 m³ (16 cu ft)

AREAS:
Main rotor blades (total) 3·35 m² (36·1 sq ft)
Tail rotor blades (total) 0·21 m² (2·26 sq ft)
Main rotor disc 81·1 m² (873 sq ft)
Tail rotor disc 1·95 m² (20·97 sq ft)
Stabiliser 0·90 m² (9·65 sq ft)

WEIGHTS:
FAA empty weight 660 kg (1,455 lb)
Max T-O weight 1,451 kg (3,200 lb)

PERFORMANCE (at max T-O weight):
Max level speed at S/L 122 knots (225 km/h; 140 mph)
Max level speed at 1,525 m (5,000 ft) 116 knots (216 km/h; 134 mph)
Max cruising speed at S/L 118 knots (219 km/h; 136 mph)
Max and econ cruising speed at 1,525 m (5,000 ft) 120 knots (222 km/h; 138 mph)
Max rate of climb at S/L 384 m (1,260 ft)/min
Vertical rate of climb at S/L 85 m (280 ft)/min
Service ceiling over 6,100 m (20,000 ft)
Hovering ceiling in ground effect 3,445 m (11,300 ft)
Hovering ceiling out of ground effect 1,770 m (5,800 ft)
Range with max fuel and max payload at S/L, no reserves 299 nm (555 km; 345 miles)
Range with max fuel and max payload at 1,525 m (5,000 ft), no reserves 337 nm (624 km; 388 miles)

BELL KIOWA

US Army designation: OH-58
Canadian military designation: CH-136

On 8 March 1968 the US Army named Bell as winner of its reopened light observation helicopter competition, and awarded the company the first increment of a planned total order for 2,200 **OH-58A** Kiowa aircraft generally similar to the Model 206A.

The first OH-58A was delivered to the US Army on 23 May 1969 and deployment in Vietnam began in the early Autumn of 1969.

On 1 May 1970 it was announced that 74 COH-58As had been ordered for the Canadian Armed Forces. The first of these was handed over officially at Uplands Airport, Ottawa, on 16 December 1971, and deliveries (from US Army production) were completed in October 1972. In January 1973 the US Army ordered an additional 74 OH-58As, but these represented replacements for the COH-58As delivered to the Canadian Armed Forces and the total US Army order remained at 2,200 aircraft; the delivery of these was completed by the end of 1973. The Canadian aircraft are now designated CH-136.

In early 1971 it was announced that Bell Helicopter Company and the Australian government had entered upon a co-production programme under which 75 Model 206B-1 Kiowa military light observation helicopters (similar to the OH-58A) would be delivered over an eight-year period. This was amended to a total of 56 units when the Australian government reduced defence expenditure during 1974. The initial 12 206B-1s were built by Bell, and the first of these was handed over officially at Eagle Farm Airport, Brisbane, on 22 November 1971. Commonwealth Aircraft Corporation (which see) is prime Australian licensee, with responsibility for final assembly of the remainder. Only the engines and electronics are being supplied from US sources.

Under a US Army development qualification contract placed on 30 June 1976, Bell is converting three OH-58As to improved **OH-58C** standard.

Major difference between the OH-58A Kiowa and JetRanger concerns the main rotor, that of the Kiowa having an increased diameter. There are also differences in the internal layout and electronics.

TYPE: Turbine-powered light observation helicopter.

ROTOR SYSTEM: Two-blade semi-rigid see-saw type main rotor, employing pre-coning and underslinging to ensure smooth operation. Blades of standard Bell 'droop-snoot' section, with D-shape aluminium spar, bonded light alloy skin, honeycomb core and trailing-edge extension. Each blade is connected to the hub by means of a grip, pitch-change bearings and a tension-torsion strap assembly. The two tail rotor blades have bonded aluminium skin but no core. Main rotor blades do not fold, but modification to permit manual folding is possible. Rotor brake available as optional kit.

ROTOR DRIVE: Rotors driven through tubular steel alloy shafts with spliced couplings. Initial drive from engine through 90° spiral bevel gear to single-stage planetary main gearbox. Shaft to tail rotor single-stage bevel gearbox. Freewheeling unit ensures that main rotor continues to drive tail rotor when engine is disengaged. Main rotor/engine rpm ratio 1 : 17·44; main rotor rpm 354. Tail rotor/engine rpm ratio 1 : 2·353.

FUSELAGE: Forward cabin section is made up of two aluminium alloy beams and 25 mm (1 in) thick aluminium honeycomb sandwich. Rotor, transmission and engine are supported by upper longitudinal beams. Upper and lower structures are interconnected by three fuselage bulkheads and a centrepost to form an integrated structure. Intermediate section is of light alloy semi-monocoque construction. Aluminium monocoque tailboom.

Bell Model 206B JetRanger II light helicopter (Allison 250-C20 turboshaft engine)

TAIL UNIT: Fixed stabiliser of aluminium monocoque construction, with inverted aerofoil section. Fixed vertical fin in sweptback upper and ventral sections, constructed of aluminium honeycomb with light alloy skins.

LANDING GEAR: Light alloy tubular skids bolted to extruded cross-tubes. Tubular steel skid on ventral fin to protect tail rotor in tail-down landing. Special high skid gear available, with 0·254 m (10 in) greater ground clearance, for use in areas with high brush. Inflated bag-type pontoons, or stowed floats capable of in-flight inflation, available as optional kits.

POWER PLANT: One 236·4 kW (317 shp) Allison T63-A-700 turboshaft engine. Fuel tank below and behind aft passenger seat, total usable capacity 276 litres (73 US gallons). Refuelling point on starboard side of fuselage, aft of cabin. Oil capacity 5·6 litres (1·5 US gallons).

ACCOMMODATION: Forward crew compartment seats pilot and co-pilot/observer side by side. Entrance to this compartment is provided by single door on each side of fuselage. The cargo/passenger compartment, which has its own access doors, one on each side, provides approximately 1·13 m³ (40 cu ft) of cargo space, or provision for two passengers by installation of two seat cushions, seat belts and shoulder harnesses.

SYSTEMS: Hydraulic system, pressure 41·5 bars (600 lb/sq in) for cyclic and collective controls. Electrical supply from 150A starter/generator. One 24V 13Ah nickel-cadmium battery.

ELECTRONICS: C-6533/ARC intercommunication subsystem, AN/ARC-114 VHF-FM, AN/ARC-115 VHF-AM, AN/ARC-116 UHF-AM, AN/ARN-89 ADF, AN/ASN-43 gyro magnetic compass, AN/APX-72 transponder, TSEC/KY-28 communications security set, C-8157/ARC control indication, MT-3802/ARC mounting, TS-1843/APX transponder test set and mounting, KIT-1A/TSEC computer and mounting, and duplicate AN/ARC-114.

ARMAMENT: Standard equipment is the XM-27 armament kit, utilising the 7·62 mm Minigun.

DIMENSIONS, EXTERNAL: As JetRanger II, except:
Diameter of main rotor 10·77 m (35 ft 4 in)
Length overall, blades turning 12·49 m (40 ft 11¾ in)
Length of fuselage 9·93 m (32 ft 7 in)

AREAS: As JetRanger II, except:
Main rotor blades (total) 3·55 m² (38·26 sq ft)
Main rotor disc 90·93 m² (978·8 sq ft)

WEIGHTS AND LOADING:
Weight empty 664 kg (1,464 lb)
Operating weight 1,049 kg (2,313 lb)
Max T-O and landing weight 1,360 kg (3,000 lb)
Max zero-fuel weight 1,145 kg (2,525 lb)
Max disc loading 14·9 kg/m² (30·7 lb/sq ft)

PERFORMANCE (at observation mission gross weight of 1,255 kg; 2,768 lb, ISA, except where indicated):
Never-exceed speed at S/L 120 knots (222 km/h; 138 mph)
Cruising speed for max range 102 knots (188 km/h; 117 mph)
Loiter speed for max endurance 49 knots (90·5 km/h; 56 mph)
Max rate of climb at S/L 543 m (1,780 ft)/min
Service ceiling 5,760 m (18,900 ft)
Hovering ceiling in ground effect 4,145 m (13,600 ft)
Hovering ceiling out of ground effect 2,682 m (8,800 ft)
Hovering ceiling out of ground effect (armed scout mission at 1,360 kg; 3,000 lb) 1,828 m (6,000 ft)
Max range at S/L, 10% reserves 259 nm (481 km; 299 miles)
Max range at S/L, armed scout mission at 1,360 kg (3,000 lb), no reserves 264 nm (490 km; 305 miles)
Endurance at S/L, no reserves 3 hr 30 min

BELL MODEL 206L LONGRANGER

First announced on 25 September 1973, Bell's LongRanger is intended to satisfy a requirement for a turbine-powered general-purpose light helicopter in a size and performance range between the five-seat JetRanger II and 15-seat Model 205A-1.

Developed from the JetRanger II, it has a fuselage

Bell OH-58A Kiowa turbine-powered light observation helicopter in US Army service *(Norman E. Taylor)*

which is 0·64 m (2 ft 1 in) longer, an Allison 250-C20B engine with a take-off rating of 313 kW (420 shp) and continuous rating of 276 kW (370 shp), new rotor, and transmission system rated at 319 kW (428 shp). It incorporates Bell's new Noda-Matic cabin suspension system, described in the introductory copy to this company's entry.

An increase of 83·3 litres (22 US gallons) in fuel capacity, to a total of 371 litres (98 US gallons), extends range by over 39 nm (72 km; 45 miles) at a maximum take-off weight of 1,769 kg (3,900 lb) by comparison with the JetRanger II.

The company's latest developments in transmission technology provide a power rating increase of more than one-third over the earlier light-turbine transmission, while adding only 3·6 kg (8 lb) to component weight.

The Noda-Matic transmission suspension system not only gives a substantial reduction in rotor-induced vibration, particularly noticeable in high-speed cruise and manoeuvring conditions, but also, through the use of elastomerics, isolates structure-borne noise from the cabin environment. This results in a standard of comfort comparable with that of turboprop-powered fixed-wing aircraft.

With a cabin volume of 2·35 m³ (83 cu ft), compared with the 1·39 m³ (49 cu ft) of the JetRanger II, utility is enhanced by innovations that allow the maximum use of this space. For example, the port forward passenger seat has a folding back to allow loading of a container measuring 2·44 × 0·91 × 0·30 m (8 × 3 × 1 ft), making possible the carriage of such items as survey equipment, skis, and long components that cannot be accommodated in any other light helicopter. Double doors on the port side of the cabin provide an opening 1·52 m (5 ft 0 in) in width, for easy straight-in loading of litter patients or utility cargo; in an ambulance or rescue role two litter patients and two ambulatory patients/attendants may be carried. With a crew of two, the standard cabin layout accommodates five passengers in two canted aft-facing seats and three forward-facing seats. An optional executive cabin layout has four individual passenger seats.

Detail improvements include a redesigned instrument panel, pedestal and glareshield, to give the pilot improved view over the nose and through the lower forward windows. To simplify maintenance, the LongRanger's forward upper engine cowling is hinged and the tail rotor gearbox has an easy-access cover.

A prototype (N206L) first flew on 11 September 1974, and production deliveries began in October 1975. Optional kits include emergency flotation gear, a 907 kg (2,000 lb) cargo hook, and an engine bleed air environmental control unit.

DIMENSIONS, EXTERNAL:
Diameter of main rotor 11·28 m (37 ft 0 in)
Main rotor rpm 394

WEIGHTS:
Weight empty, standard configuration 890 kg (1,962 lb)
Max T-O weight 1,814 kg (4,000 lb)

PERFORMANCE (ISA at max T-O weight):
Max level speed at S/L 130 knots (241 km/h; 150 mph)
Cruising speed at S/L 110 knots (204 km/h; 127 mph)
Service ceiling at max cruise power 3,930 m (12,900 ft)
Hovering ceiling in ground effect 1,950 m (6,400 ft)
Hovering ceiling out of ground effect 366 m (1,200 ft)
Range at S/L 300 nm (555 km; 345 miles)
Range at 1,525 m (5,000 ft) 335 nm (620 km; 385 miles)

BELL MODEL 209 HUEYCOBRA (single-engined)

US Army designations: AH-1G, AH-1Q and AH-1S
Spanish Navy designation: Z-16

First flown on 7 September 1965, six months after its development was started, the Model 209 HueyCobra is a development of the UH-1B/C Iroquois intended specifically for armed helicopter missions. It combines the basic transmission and rotor system and (in its initial form) the power plant of the UH-1C with a new streamlined fuselage designed for maximum speed, armament load and crew efficiency.

The prototype was sent to Edwards AFB for US Army evaluation in December 1965. On 11 March 1966, the Army announced its intention to order the HueyCobra into production.

Versions announced so far are as follows:

AH-1G. Original version for US Army, powered by 1,400 shp (1,044 kW; derated to 820 kW; 1,100 shp) Lycoming T53-L-13 turboshaft, driving a two-blade wide-chord Model 540 'door-hinge' rotor of the kind fitted to the UH-1C. Main rotor rpm is 294-324. A development contract for two pre-production prototypes was placed on 4 April 1966, followed on 13 April by an initial contract for 110 production aircraft plus long lead-time spares. Subsequent contracts raised the total on order to 838 by October 1968. On 30 January 1970 the US Army ordered a further 170 AH-1Gs for delivery between July 1971 and August 1972, followed by a further order

Bell Model 206L LongRanger seven-seat general-purpose light helicopter *(Brian M. Service)*

for 70 in mid-1971. Twenty have been ordered by the Spanish Navy for anti-shipping strike duties.

Deliveries of the original production series began in June 1967 and operational deployment to Vietnam began in the early Autumn of 1967. The US Marine Corps acquired 38 AH-1Gs during 1969, for transition training and initial deployment, pending delivery of the AH-1J; these are included in the above totals.

AH-1J. Twin-turbine version for US Marine Corps. Described separately.

AH-1Q. Anti-armour version of the AH-1G HueyCobra, equipped to fire Hughes TOW anti-tank missiles. The first of eight pre-production examples was delivered to the US Army in early1973, when it was anticipated that half of the US Army's inventory of approximately 600 AH-1Gs would be modified to AH-1Q standard. A $59·2 million contract awarded in January 1974 covered the conversion of 101 aircraft; a $54·2 million contract was received subsequently for the modification of an additional 189 AH-1Gs to AH-1Q configuration. This involved the installation of eight TOW missile containers, disposed as two two-round pods on each of the outboard underwing pylons, plus a helmet sight subsystem produced by the Univac Division of Sperry Rand. The inboard wing weapon pylons were still available for other stores. TOW/Cobras demonstrated their anti-armour proficiency in Vietnam in 1972, but proved unable to carry a full weapon load in that environment, leading to development of the AH-1S. Production deliveries began on 10 June 1975. Only 92 of the planned 290 conversions remain to AH-1Q standard; others have been converted to AH-1S.

AH-1R. As AH-1G, with uprated power plant (1,360 kW; 1,825 shp T53-L-703) but without TOW missile installation.

AH-1S. Advanced version of AH-1G equipped to fire TOW missiles, with upgraded power plant (1,360 kW; 1,825 shp T53-L-703), gearbox and transmission. US Army planning calls for all but 92 of the AH-1Gs converted to AH-1Q TOW configuration to be modified in the field to AH-1S standard. The remaining AH-1Gs not yet converted are being produced directly to AH-1S standard. In 1975-76 the US Army awarded Bell $53 million contracts covering the production of 66 new examples of the AH-1S, additional to the 198 converted aircraft, plus $3·3 million for cockpit and engineering changes, including incorporation of the Bell-developed flat plate canopy. Deliveries are to take place in 1977-78.

AH-1T. Improved version of the AH-1J SeaCobra for the US Marine Corps. Described separately.

Relatively small, the HueyCobra has a low silhouette and narrow profile, with a fuselage width of only 0·965 m (38 in). These features make it easy to conceal with small camouflage nets or to move under cover of trees. Tandem seating for the crew of two provides maximum field of view for the pilot and forward gunner. The skid landing gear is non-retractable. Stub-wings carry armament and help to offload the rotor in cruising flight.

Emerson Electric designed and developed for the initial production version of the AH-1G HueyCobra the TAT-102A tactical armament turret, which was faired into the front fuselage undersurface and housed a GAU-2B/A (formerly XM-134) Minigun six-barrel 7·62 mm machine-gun, with 8,000 rounds. This turret was superseded on the AH-1G by the XM-28 subsystem, mounting either two Miniguns, with 4,000 rounds each; two XM-129 (similar to the XM-75) 40 mm grenade launchers, each with 300 rounds; or one Minigun and one XM-129. Two rates of fire are provided for the TAT-102A and XM-28 Miniguns, namely 1,600 and 4,000 rounds per minute. The lower rate is for searching or registry fire; the higher rate is used for attack, the rate of fire being controlled by the gunner's trigger. The XM-129 fires at a single rate of 400 rds/min.

Four external stores attachments were provided under the stub-wings of the AH-1G, to accommodate a total of seventy-six 2·75 in rockets in four XM-159 packs, twenty-eight similar rockets in four XM-157 packs, or two XM-18E1 Minigun pods.

In late 1969 an XM -35 20 mm cannon kit was added to the weapons available for the AH-1G, and an initial batch of six aircraft equipped with the XM-35 was delivered to the US Army in December 1969. A total of 350 kits was ordered subsequently by the Army.

Designed jointly by Bell and General Electric, the XM-35 armament subsystem consists of a six-barrel 20 mm automatic cannon, two ammunition boxes and certain structural and electrical modifications. Mounted on the inboard stores attachment of the port stub-wing, the XM-35 has a firing rate of 750 rounds per minute. Two ammunition boxes faired flush to the fuselage below the stub wings accommodate 1,000 rounds. Total installed weight of the system is 531 kg (1,172 lb).

In normal operation, the co-pilot/gunner of the AH-1G controls and fires the turret armament, using a hand-held pantograph-mounted sight to which the turret is slaved. The gunner can fire throughout a field of 230° (115° each side of the aircraft centreline) and can depress the turreted weapons 50° and elevate them 25°. Velocity jump compensation automatically computes the lead angle with respect to the relative motion of aircraft and target. In addition, the gunner has the capability of firing the wing stores.

The pilot can fire the turreted weapons only in the stowed position, dead ahead. The turret returns to the stowed position automatically when the gunner releases his grip on the slewing switch.

The pilot normally fires the wing stores, utilising the XM-73 adjustable rocket sight. Rockets are fired in pairs, made up of one rocket from each opposing wing station. Any desired number of pairs from one to nineteen can be

This AH-1Q TOW/Cobra has Bell-developed flat plate canopy, standard on AH-1S

preselected on the cockpit-mounted intervalometer. The inboard wing points are equipped to fire either the XM-18 or XM-18E1 Minigun pod. All wing stores are symmetrically or totally jettisonable.

The crew of the HueyCobra are protected by seats and side panels made of NOROC armour, manufactured by the Norton Company. Other panels protect vital areas of the aircraft.

On missions of 50 nm (92 km; 57 miles) radius the HueyCobra can reach the target area in half the time taken by a UH-1B and operate in the target area for three times as long. During flight tests it has been dived at a speed of 214 knots (397 km/h; 246 mph). Normal fuel capacity is 1,014 litres (268 US gallons).

In November 1970 three AH-1G HueyCobras equipped with prototypes of a day/night fire control system known as SMASH (South-east Asia Multi-sensor Armament System for HueyCobra) were delivered to the US Army for tests. Major elements of the system included a nose-mounted Sighting System Passive Infra-red (SSPI) sensor developed by Aerojet Electro-Systems Company, a Moving Target Indicator (MTI) radar (which is an AN/APQ-137B high-resolution radar system developed by Emerson Electric) mounted on the starboard wing station, pilot and co-pilot/gunner displays and consoles, and an Interface Control Unit (ICU) that electronically 'married' the fire control subsystem sensing, control and display units to the aircraft's armament.

DIMENSIONS, EXTERNAL:
- Diameter of main rotor: 13·41 m (44 ft 0 in)
- Main rotor blade chord: 0·69 m (2 ft 3 in)
- Diameter of tail rotor: 2·59 m (8 ft 6 in)
- Tail rotor blade chord: 0·21 m (8·4 in)
- Wing span: 3·15 m (10 ft 4 in)
- Length overall (main rotor fore and aft): 16·14 m (52 ft 11½ in)
- Length of fuselage: 13·54 m (44 ft 5 in)
- Height overall: 4·10 m (13 ft 5½ in)
- Width over skids: 2·13 m (7 ft 0 in)

AREAS:
- Main rotor disc: 141·2 m² (1,520·4 sq ft)
- Tail rotor disc: 5·27 m² (56·8 sq ft)

WEIGHTS (AH-1G):
- Operating weight: 2,754 kg (6,073 lb)
- Mission weight: 4,266 kg (9,407 lb)
- Max T-O and landing weight: 4,309 kg (9,500 lb)

PERFORMANCE (AH-1G, at max T-O weight):
- Never-exceed speed: 190 knots (352 km/h; 219 mph)
- Max level speed: 190 knots (352 km/h; 219 mph)
- Max rate of climb at S/L: 375 m (1,230 ft)/min
- Service ceiling: 3,475 m (11,400 ft)
- Hovering ceiling in ground effect: 3,015 m (9,900 ft)
- Max range at S/L, max fuel, 8% reserves: 310 nm (574 km; 357 miles)

BELL MODEL 209 HUEYCOBRA (twin-engined)

US military designation: AH-1J SeaCobra

This is a modified version of the Bell AH-1G, initially for the US Marine Corps. A batch of 49 AH-1Js was ordered for the USMC (by which they are known as SeaCobras) in May 1968, and a pre-production aircraft was displayed to representatives of the US armed forces at Enless, Texas, on 14 October 1969.

Delivery of the Marine AH-1Js began in mid-1970 and was completed in the following year. Twenty more were ordered in early 1973, for delivery to the USMC between April 1974 and February 1975. On 5 June 1974, Bell received a contract, administered by the US Army Aviation Systems Command, to modify the last two of these SeaCobras to have uprated components for increased payload and performance. A contract for 10 production examples of this improved version was announced in June 1975, and it has since been given the USMC designation **AH-1T** (described separately).

Bell announced on 22 December 1972 the receipt of a further order for 202 AH-1Js from the US Army, with an initial funding of $38·5 million. They are being acquired by Iran through the US government, and are similar to the aircraft delivered to the US Marine Corps. Deliveries to Iran began in 1974.

The standard AH-1J differs from the single-engined AH-1G in having a 1,342 kW (1,800 shp) Pratt & Whitney Aircraft of Canada T400-CP-400 coupled free-turbine turboshaft power plant (a military version of the PT6T-3 Turbo Twin Pac power plant as described for the Model 212/UH-1N). Engine and transmission are flat-rated for 820 kW (1,100 shp) continuous output, with increase to 932 kW (1,250 shp) for take-off or 5 min emergency power. To cater for the increased power, the tail rotor pylon has been strengthened and the tail rotor blade chord increased.

An electrically-driven 20 mm turret system, developed by the General Electric Company, is faired into the forward lower fuselage, and houses an XM-197 three-barrel weapon, which is a lightweight version of the General Electric M-61 cannon. The firing rate is 750 rounds per minute, but a 16-round burst limiter is incorporated in the firing switch. The gun has a tracking capability of 220° in azimuth, 50° depression and 18° elevation, and can be slewed at a rate of 80° per second. A barrel length of 1·52 m (5 ft) makes it imperative that the XM-197 is centralised before wing stores are fired. An ammunition container of 750-round capacity is located in the fuselage directly aft of the turret. Four external stores attachment points under the stub-wings can accommodate various loads, including XM-18E1 7·62 mm Minigun pods as well as 2·75 in folding-fin rockets in either seven-tube (XM-157) or 19-tube (XM-159) packs.

The USMC SeaCobra also differs from the AH-1G in having Marine electronics.

DIMENSIONS, EXTERNAL:
- As for AH-1G except:
- Tail rotor blade chord: 0·292 m (11½ in)
- Length overall (main rotor fore and aft): 16·26 m (53 ft 4 in)
- Length of fuselage: 13·59 m (44 ft 7 in)
- Width of fuselage: 0·98 m (3 ft 2½ in)
- Height overall: 4·15 m (13 ft 8 in)

WEIGHTS:
- Operating weight, incl 181 kg (400 lb) crew, fluids, electronics and armour: 3,294 kg (7,261 lb)
- Basic combat weight: 4,523 kg (9,972 lb)
- Max T-O and landing weight: 4,535 kg (10,000 lb)

PERFORMANCE (at max T-O weight):
- Never-exceed speed: 180 knots (333 km/h; 207 mph)
- Max level speed: 180 knots (333 km/h; 207 mph)
- Max rate of climb at S/L: 332 m (1,090 ft)/min
- Service ceiling: 3,215 m (10,550 ft)
- Hovering ceiling in ground effect: 3,794 m (12,450 ft)
- Max range, no reserves: 311 nm (577 km; 359 miles)

Bell AH-1J SeaCobra of the Imperial Iranian Army Aviation service

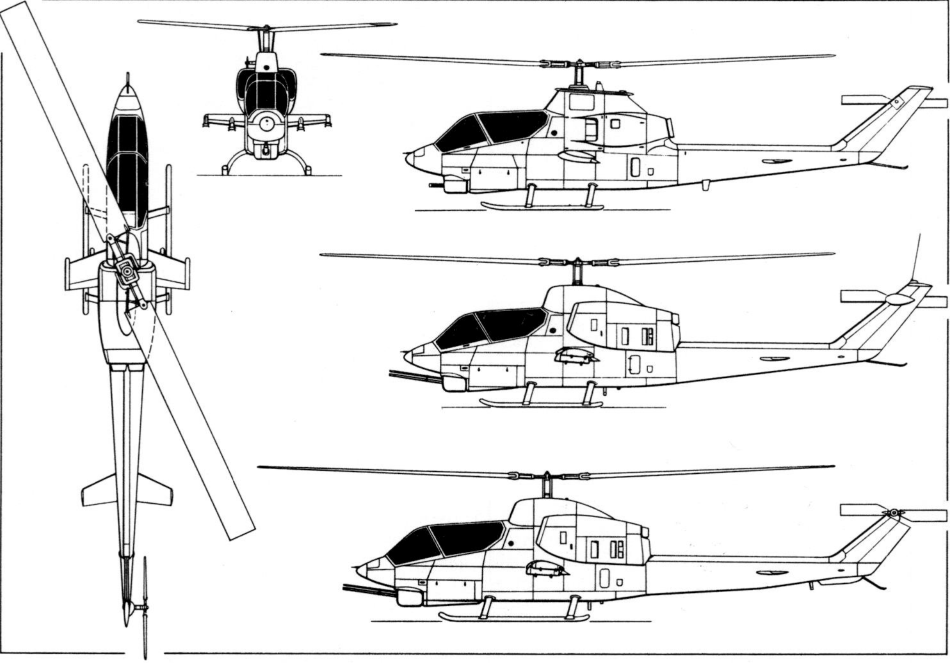
Bell AH-1T SeaCobra, with additional side views of AH-1J (centre) and AH-1G (top) *(Pilot Press)*

BELL MODEL 209 IMPROVED SEACOBRA

US military designation: AH-1T SeaCobra

The improved version of the AH-1J SeaCobra, of which ten production examples were ordered for the US Marine Corps in June 1975, has been given the designation AH-1T. It incorporates features of the AH-1J airframe, but embodies the dynamic system of the Model 214, technology developed for the Model 309 KingCobra and an upgraded engine, resulting in significant load and performance increases by comparison with the AH-1J.

ROTOR SYSTEM: Similar to that developed for the Model 214 Huey Plus (to take advantage of a very considerable increase in power), which in turn was developed from the Model 540 rotor of the 'door hinge' type, first introduced on the UH-1C. The strengthened rotor hub incorporates elastomeric and Teflon-faced bearings, the rotor blade chord is increased, and the blades have swept tips which reduce noise and improve high-speed performance. The tail rotor is similar to that developed for the Model 214, with increased diameter and blade chord.

ROTOR DRIVE: Uprated transmission system of the type

Bell AH-1T, improved version of the Model 209 SeaCobra for the US Marine Corps

developed originally for the HueyTug and having a rating of 1,529 kW (2,050 shp).

FUSELAGE: As for the AH-1J, but lengthened by the insertion of a 0·305 m (1 ft 0 in) fuselage splice to accommodate tankage for an extra 181·5 kg (400 lb) of fuel.

TAIL UNIT: As for the AH-1J, but tailboom lengthened by 0·79 m (2 ft 7 in).

POWER PLANT: One 1,469 kW (1,970 shp) Pratt & Whitney Aircraft of Canada T400-WV-402 coupled free-turbine turboshaft engine.

DIMENSIONS, EXTERNAL:
- Diameter of main rotor: 14·63 m (48 ft 0 in)
- Main rotor blade chord: 0·84 m (2 ft 9 in)
- Diameter of tail rotor: 2·96 m (9 ft 8½ in)
- Tail rotor blade chord: 0·305 m (1 ft 0 in)
- Length overall (main rotor fore and aft): 17·35 m (56 ft 11 in)
- Length of fuselage: 14·68 m (48 ft 2 in)

WEIGHTS:
- Weight empty: 3,635 kg (8,014 lb)
- Operating weight: 3,904 kg (8,608 lb)
- Max T-O weight: 6,350 kg (14,000 lb)
- Max useful load (fuel and disposable ordnance): 2,445 kg (5,392 lb)

BELL MODEL 212 TWIN TWO-TWELVE

US military designation: UH-1N

Canadian military designation: CH-135

Bell announced on 1 May 1968 that the Canadian government had approved development of a twin-engined UH-1 helicopter to be powered by a Pratt & Whitney Aircraft of Canada PT6T power plant. Subsequently, the Canadian government ordered 50 of these aircraft (designated CUH-1N) for the Canadian Armed Forces, with options on 20 more. Simultaneously, orders totalling 141 aircraft for the United States services were announced, comprising 79 for the USAF, 40 for the USN and 22 for the USMC, all having the designation **UH-1N**. Subsequent orders cover the delivery of 117 more UH-1Ns to the US Navy and Marine Corps in 1973-78.

Initial deliveries for the USAF began in 1970, and the first CUH-1N for the Canadian Armed Forces was handed over officially at Uplands Airport, Ottawa, on 3 May 1971; the Canadian order was completed one year later. Deliveries to the USN and USMC began during 1971. Canadian aircraft are now designated CH-135.

A commercial version, known as the Twin Two-Twelve, is also in full-scale production. This received FAA type certification in October 1970, and on 30 June 1971 the Two-Twelve was granted FAA Transport Type Category A certification.

More than 250 commercial Model 212s have been delivered.

The Model 212/UH-1N utilises a Bell 205A/UH-1H airframe, and civil and military versions have basically the same configuration, but differ in mission kits and electronics. They each accommodate a pilot and 14 passengers or, in cargo configuration, provide 6·23 m³ (220 cu ft) of internal capacity. The Model 212 has the capability of carrying an external load of 2,268 kg (5,000 lb), and the military UH-1N a load of 1,534 kg (3,383 lb).

Power plant is a Pratt & Whitney Aircraft of Canada PT6T-3 Turbo Twin Pac, which consists of two PT6 turboshaft engines coupled to a combining gearbox with a single output shaft. Producing 1,342 kW (1,800 shp), the Twin Pac is flat rated to 932 kW (1,250 shp) for T-O and 820 kW (1,100 shp) for continuous operation. In the event of an engine failure, the remaining engine is capable of delivering 671 kW (900 shp) for 30 minutes or 570·5 kW (765 shp) continuously, which is adequate to maintain cruise performance at maximum gross weight.

On 6 March 1972 a UH-1N helicopter of the US Navy Antarctic Development Squadron Six (VXE-6) stationed at McMurdo carried parachute rigger Hendrick V. Gorick to a height of 6,248 m (20,500 ft), from which altitude he jumped to set a new parachute jump record for the Antarctic continent.

Bell announced in January 1973 that two Twin Two-Twelves had been modified and flown in a programme to gain IFR certification from the UK's CAA and America's FAA. Conversion of the Model 212 from VFR to IFR configuration requires a new electronics package, new instrument panel and aircraft stabilisation controls. The Model 212 has also qualified for IFR certification by the Norwegian DCA and the Canadian MoT.

The new electronics include dual King KTR-900A com transceivers; dual King KNR-660A VOR/LOC/RMI receivers; King KDF-800 ADF; King KMD-700A DME; King KXP-750A transponder; King KGM-690 marker beacon/glideslope receiver; and dual Sperry Tarsyn-444 three-axis gyro units. The new panel has blue-white lighted instruments, and the pilot's and co-pilot's 5 in attitude director indicator and horizontal situation indicator are as used in Astronautics Corporation's Model 11300 flight director system. Installation of a mechanical Stability Control Augmentation System (SCAS), an automatic flight control system, separation of the two DC generator circuits and provision of a triple-redundant AC power system completes the IFR installation.

Bell announced simultaneously with the above that Helicopter Service AS of Oslo, Norway, had ordered six Model 212s equipped for IFR, which were to be used in support of offshore oil operations in the North Sea. Delivery was completed in 1973, and the first of four IFR 212s ordered by the Japan Maritime Safety Agency was delivered in November 1973. By the end of that year a total of 20 IFR-configured 212s had been delivered and about 50% of the 212s sold in 1974 were of IFR configuration.

Bell Helicopter Textron's largest single order for non-military helicopters was received from the government of Peru in October 1973. Valued at more than $11·5 million, it covered the purchase of 14 Model 212s, plus spares and technical services. The helicopters, assigned to the Peruvian Air Force Group 3, are to be used on commercial contracts in support of oil exploration, drilling and production operations. They supplement three Model 212s purchased earlier by Peru.

The description given for the Bell UH-1H applies generally to the Model 212, but the twin-engined power plant makes considerable changes to the specification, which is detailed below for both the Models 212 and UH-1N:

DIMENSIONS, EXTERNAL:
- Diameter of main rotor (with tracking tips): 14·69 m (48 ft 2¼ in)
- Main rotor blade chord: 0·72 m (2 ft 4⅜ in)
- Diameter of tail rotor: 2·59 m (8 ft 6 in)
- Tail rotor blade chord: 0·292 m (11½ in)
- Length overall (main rotor fore and aft): 17·46 m (57 ft 3¼ in)
- Length of fuselage: 12·92 m (42 ft 4¾ in)
- Height overall: 4·39 m (14 ft 4¾ in)

WEIGHTS (A: Model 212, B: UH-1N):
- FAA empty weight plus usable oil:
  - A: 2,517 kg (5,549 lb)
- Max T-O weight and mission weight:
  - A: 5,080 kg (11,200 lb)
  - B: 4,762 kg (10,500 lb)

PERFORMANCE (at max T-O weight):
- Never-exceed speed at S/L:
  - A, B: 109 knots (203 km/h; 126 mph)
- Max level speed at S/L:
  - A, B: 109 knots (203 km/h; 126 mph)
- Max rate of climb at S/L:
  - A, B: 532 m (1,745 ft)/min
- Service ceiling:
  - A: 5,305 m (17,400 ft)

Bell Model 212 general-purpose helicopter in service with the Royal Brunei Regiment

Bell UH-1N of Antarctic Development Squadron Six (VXE-6), US Navy

IFR-equipped version of the Bell Model 212 general-purpose helicopter *(Howard Levy)*

B	4,570 m (15,000 ft)
Hovering ceiling in ground effect:	
A	4,480 m (14,700 ft)
B	3,930 m (12,900 ft)
Hovering ceiling out of ground effect:	
A	3,230 m (10,600 ft)
B	1,495 m (4,900 ft)
Max range at S/L, no reserves:	
A	237 nm (439 km; 273 miles)
B	216 nm (400 km; 248 miles)

BELL MODEL 214A/C

On 22 December 1972, Bell announced that it had received an order for 287 advanced **Model 214A** 16-seat utility helicopters from the US Army. With an initial funding of $63 million, these are being acquired by Iran through the US government. Present plans envisage co-production of 400 more Model 214As, with Iran's new helicopter industry. Iranian name for the Model 214A is **Isfahan.**

Developed from the Model 214 Huey Plus, which was described in the 1974-75 *Jane's,* the 214A is powered by a 2,185 kW (2,930 shp) Lycoming LTC4B-8D turboshaft engine, an improved version of the T55-L-7C that was fitted in the original Model 214A demonstrator when it went to Iran. It has the 1,529 kW (2,050 shp) transmission and rotor drive systems developed for the KingCobra experimental gunship helicopter, described in the 1973-74 *Jane's,* and embodies Bell's Noda-Matic nodalised beam concept to minimise vibration.

At a max T-O weight of 5,896 kg (13,000 lb), the Model 214A has a cruising speed of 130 knots (241 km/h; 150 mph) and a range of 260 nm (481 km; 299 miles). It has an external load max T-O weight of 6,803 kg (15,000 lb).

The first Model 214A for Iran flew for the first time on 13 March 1974. Deliveries began on 26 April 1975, and were scheduled to build up to a rate of ten a month within a year.

On 29 April 1975, three days after delivery of the first production Model 214A to Iran, this aircraft was used to set up five records for altitude and time-to-height in Class E1e (3,000 to 4,500 kg). The pilot in each case was Maj Gen Manouchehr Khosrodad, commanding general of the Imperial Iranian Army Aviation. Co-pilot was Clem Bailey, Bell's assistant chief production test pilot. The records are for a maximum altitude of 9,071 m (29,760 ft), maximum sustained altitude in horizontal flight, also of 9,071 m, and times of 1 min 58 sec to 3,000 m, 5 min 13·2 sec to 6,000 m, and 15 min 5 sec to 9,000 m.

In February 1976, the Iranian government ordered 39 **Model 214Cs,** similar to the 214A but equipped for search and rescue.

BELL MODEL 214B BIGLIFTER

Bell announced on 4 January 1974 that a commercial version of the Model 214A, known as the Model 214B, would become available in 1975, emphasising that it would have a lift capability better than any existing commercial helicopter in the medium category. FAA certification was received on 27 January 1976.

Powered by a 2,185 kW (2,930 shp) Lycoming T5508D turboshaft engine, the Bell 214B BigLifter has the same rotor drive and transmission system as the Model 214A. The engine is flat-rated at a maximum of 1,678 kW (2,250 shp) and the transmission is rated at 1,529 kW (2,050 shp) for take-off, with a maximum continuous power output of 1,379·5 kW (1,850 shp). The main rotor has a Wortmann blade section, raked tips, and an advanced rotor hub with elastomeric bearings on the flapping axis. The tail rotor also has raked tips, and a hub that requires no lubrication. Other features include an automatic flight control system, with stability augmentation and attitude retention; nodalised suspension; separate dual hydraulic systems; and a large engine deck that serves also as a maintenance platform. Differences by comparison with the military Model 214A include the addition of an engine fire extinguishing system, push-out escape windows in the cargo doors, and commercial electronics.

It has been stated that the Model 214B can cruise at 130 knots (241 km/h; 150 mph) with an internal load of 1,814 kg (4,000 lb). Alternative loads can include 14 men, nearly four US tons of chemicals in an agricultural role, or 3,025 litres (800 US gallons) of water or suppressant in a firefighting role. It is able to haul an external load of 3,175 kg (7,000 lb) on its cargo hook.

Production in mid-1976 was at the rate of two aircraft each month.

DIMENSIONS, EXTERNAL:	
Main rotor diameter	15·24 m (50 ft 0 in)
Main rotor blade chord	0·84 m (2 ft 9 in)
Tail rotor diameter	2·95 m (9 ft 8 in)
Tail rotor blade chord	0·305 m (1 ft 0 in)
WEIGHT:	
Max T-O weight	7,257 kg (16,000 lb)

BELL MODEL 222

In April 1974, Bell announced its intention of developing the Model 222, described as the first commercial light twin-engined helicopter to be built in the USA. The first of three prototypes flew for the first time on 13 August 1976; deliveries of production aircraft are planned to begin in 1978.

Bell Model 214A Isfahan in Iranian Army camouflage (Lycoming LTC4B-8D turboshaft engine)

Bell Model 214B BigLifter utility helicopter (Lycoming T5508D turboshaft engine)

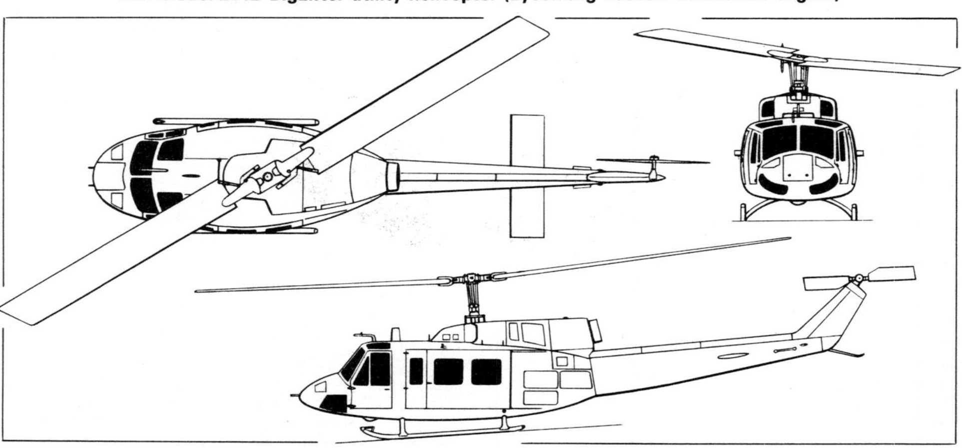
Bell Model 214B medium-size commercial heavy-lift helicopter *(Pilot Press)*

Prototype Bell Model 222 (two Avco Lycoming LTS 101-650C turboshaft engines)

Designed to meet FAR Part 29 Transport Category requirements, the general appearance of the Model 222 is shown in the accompanying illustration. Before taking a development decision, Bell displayed a full-scale concept mockup, designated D306, at the annual convention of the Helicopter Association of America, in San Diego, in January 1974. The response of potential operators encouraged the development go-ahead, and customer suggestions were embodied in the definitive Model 222 design. In particular, the lower glazing of the flight deck was revised to provide increased visibility for rooftop landings, and the cabin was both lengthened and widened at the rear to give more spacious accommodation in high-density passenger-carrying configuration. There is now 3·51 m³ (124 cu ft) of cabin space and another 1·19 m³ (42 cu ft) in the baggage compartment.

The main cabin is 3·81 m (12 ft 6 in) long from the rear of the baggage compartment to the rear of the flight deck, to which the passengers have full access. Alternative configurations range from six-person super executive to ten-person high-density, with provision for rapid conversion from passenger to utility arrangements. As an ambulance the Model 222 will accommodate two NATO/Stokes litters, two attendants and two crew.

The Model 222 is powered by two 447·5 kW (600 shp) Avco Lycoming LTS 101-650C turboshaft engines, mounted in a streamlined housing above the cabin and aft of the rotor pylon. The airframe incorporates Bell's focused pylon and nodalisation for smoothness, and isolates structure-borne vibration by mounting the pylon and engines in elastomerics. The fully-retractable landing gear has individually braked main wheels and a fully swivelling nosewheel incorporating a locking device. Bell is developing a range of accessory kits for the Model 222, including emergency flotation gear, high skid landing gear and comprehensive electronics packages.

Safety features include dual electrical, fuel and hydraulic systems, fail-safe structures and redundancy throughout the systems. The completely dry main rotor hub has conical elastomerics. When equipped with the appropriate avionics, the Model 222 is intended to qualify for IFR certification.

DIMENSIONS, EXTERNAL:
- Diameter of main rotor: 11·87 m (38 ft 11⅜ in)
- Main rotor blade chord: 0·66 m (2 ft 2 in)
- Diameter of tail rotor: 1·98 m (6 ft 6 in)
- Tail rotor blade chord: 0·254 m (10 in)
- Length overall (rotors fore and aft): 14·41 m (47 ft 3¼ in)
- Length of fuselage: 12·12 m (39 ft 9 in)
- Height to top of rotor hub: 3·39 m (11 ft 1½ in)
- Tailplane span: 1·73 m (5 ft 8 in)
- Wheel track: 2·77 m (9 ft 1 in)
- Wheelbase: 3·59 m (11 ft 9½ in)

WEIGHTS:
- Weight empty, equipped: 1,830 kg (4,035 lb)
- Max T-O weight: 3,039 kg (6,700 lb)

PERFORMANCE (estimated):
- Speed at max continuous power: 156 knots (290 km/h; 180 mph)
- Normal cruising speed: over 130 knots (240 km/h; 150 mph)
- Service ceiling, one engine out: over 3,050 m (10,000 ft)
- Range, with 20 min reserves: 370 nm (685 km; 425 miles)

BELL MODEL 301

US Army designation: XV-15

Bell Helicopter announced in May 1973 that it had been chosen by NASA and the US Army to build and test two twin-engined tilt-rotor research aircraft. Estimated cost of the four-year programme is $29·2 million.

The company has been working on tilt-rotor technology since the mid-1950s, proving the concept feasible with its XV-3 prototype, described in the 1962-63 *Jane's*. Since that time development of tilt-rotor systems has progressed steadily, leading to the Model 301 which Bell proposed to meet the NASA/Army requirement. The two research aircraft being built have been allocated the official designation XV-15. They have fuselages and tail units built under subcontract by Rockwell International's Tulsa Division.

As shown in the accompanying three-view drawing of the XV-15, the airframe structure is basically that of a conventional fixed-wing aircraft. However, the wingtip-mounted engines and rotors can be swivelled to a vertical position for VTOL operations, and are then moved forward gradually to provide transition to cruising flight at speeds in excess of 300 knots (556 km/h; 345 mph).

Power plant comprises two uprated examples of the Lycoming T53 turboshaft engine, each with a contingency rating of 1,342 kW (1,800 shp). Qualified to operate in the vertical mode, they have been redesignated LTC1K-4K. The three-blade rotors are stiff in plane and gimballed, with an elastomeric hub spring to increase control power and damping. The blades are a high-twist design, suitable for both helicopter and high-speed aircraft flight modes. Interconnect drive-shafts and redundant tilting mechanisms permit single-engine operation and fail-safe tilt capability.

The XV-15 is fitted with a stability and control augmentation system to improve the handling qualities and enhance pilot efficiency. Ejection seats will be installed as a safety feature during flight trials.

Future commercial and military aircraft which might be derived from the XV 15 would have a wing span of about 10·67 m (35 ft) and fuselage length of 12·50 m (41 ft). They would carry 15 troops in military service or 12 passengers as civil transports.

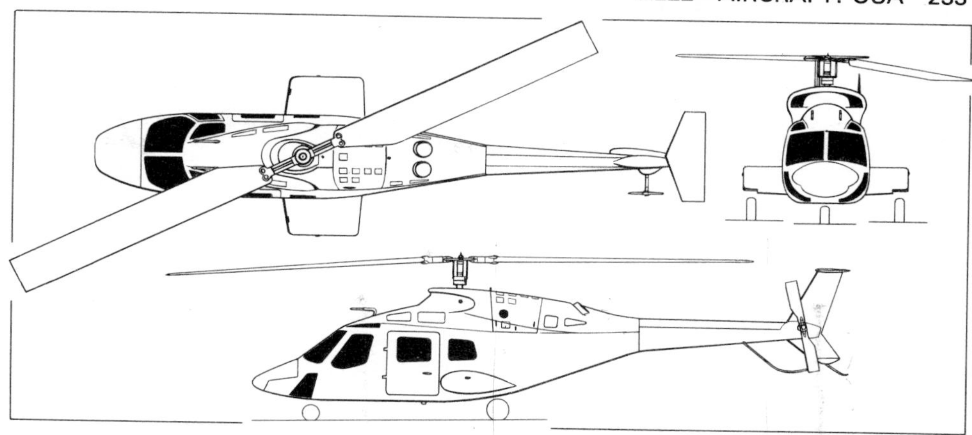

Bell Model 222 light twin-engined commercial helicopter *(Pilot Press)*

First Bell XV-15, before installation of its two three-blade rotors on the tilting engines

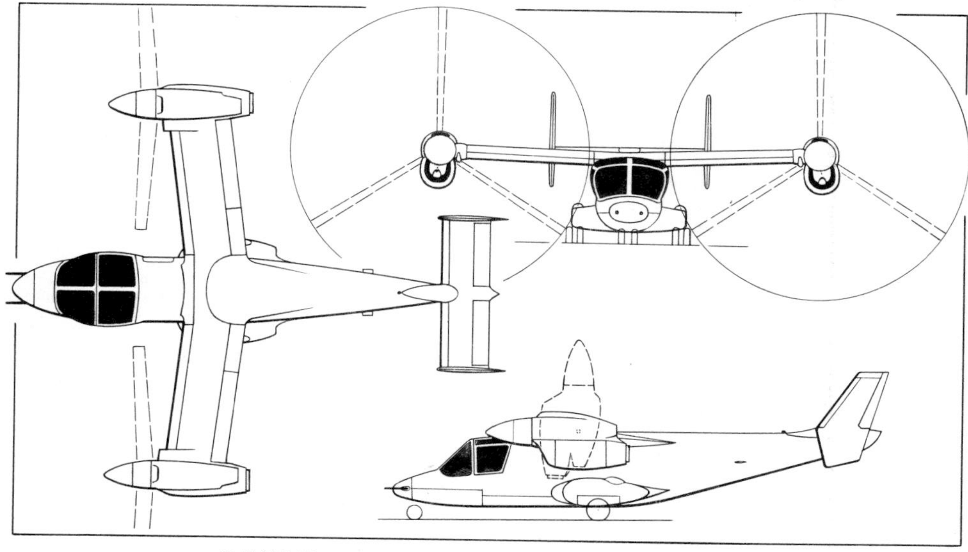

Bell XV-15 prototype tilt-rotor research aircraft *(Pilot Press)*

The programme is being funded and managed jointly by the NASA Ames Research Center and the US Army's Air Mobility Research and Development Laboratory. The two XV-15s, when available, will be used in a research programme to prove the concept, explore the limits of the operational flight envelope and assess its application to military and civil transport needs.

The first XV-15 will be tested in the 12 m × 24 m (40 ft × 80 ft) wind tunnel at NASA's Ames Research Center prior to the first flight, by the second aircraft, which is scheduled for 1977.

DIMENSIONS, EXTERNAL:
- Rotor diameter: 7·62 m (25 ft 0 in)
- Rotor blade chord: 0·36 m (1 ft 2 in)
- Wing span, between rotor axes: 9·80 m (32 ft 2 in)
- Wing chord (constant): 1·60 m (5 ft 3 in)
- Wing aspect ratio: 6·12
- Width overall, rotors turning: 17·44 m (57 ft 2⅜ in)
- Length overall, excl instrument boom: 12·83 m (42 ft 1¼ in)
- Height overall, rotors in hover mode: 4·66 m (15 ft 3½ in)
- Tailplane span: 3·91 m (12 ft 10 in)

AREAS:
- Rotor disc (each): 45·60 m² (490·87 sq ft)
- Wings, gross: 15·70 m² (169 sq ft)
- Trailing-edge flaps (total): 1·02 m² (11 sq ft)
- Flaperons (total): 1·88 m² (20·2 sq ft)
- Fins (total): 3·99 m² (43·0 sq ft)
- Rudders (total): 0·70 m² (7·5 sq ft)
- Tailplane: 3·46 m² (37·25 sq ft)
- Elevators (total): 1·21 m² (13·0 sq ft)

WEIGHTS AND LOADINGS (estimated):
- Weight empty: 4,341 kg (9,570 lb)
- Design VTOL gross weight: 5,897 kg (13,000 lb)
- Max VTOL gross weight: 6,804 kg (15,000 lb)
- Rotor disc loading at 5,897 kg (13,000 lb): 64·66 kg/m² (13·24 lb/sq ft)
- Wing loading at 5,897 kg (13,000 lb): 375·6 kg/m² (76·92 lb/sq ft)

PERFORMANCE (estimated, at design VTOL gross weight, unless stated otherwise):
Max diving speed at 3,810 m (12,500 ft)
360 knots (666 km/h; 414 mph)
Max cruising speed at 5,180 m (17,000 ft)
330 knots (612 km/h; 380 mph)
Max speed in helicopter mode, mast angle 75°
120 knots (222 km/h; 138 mph)
Min control speed with rotors operating as tractor propellers 120 knots (222 km/h; 138 mph)
Endurance in hover mode at S/L, ISA, 10% reserve fuel 1 hr
Endurance, cruising at 200 knots (370 km/h; 230 mph) at 6,100 m (20,000 ft) 2 hr

BELL MODEL 409

US Army designation: YAH-63

The Bell YAH-63 is one of two contenders in the US Army's AAH (Advanced Attack Helicopter) competition, for which RFPs (Requests for Proposals) were issued in November 1972. Initial submissions were received from Bell Helicopter, Boeing Vertol, Hughes, Lockheed and Sikorsky, and on 22 June 1973 it was announced that the Bell and Hughes designs had been selected for development. These are designated, respectively, YAH-63 and YAH-64.

The Bell contract, valued at $44·7 million, covers the construction of two flight test prototypes and a ground test vehicle. A static test airframe has also been completed.

The ground test vehicle was rolled out on 31 January 1975, and on 19 April 1975 began a programme totalling more than 100 hours which included 50 hours of ground running, and vibration, proof-load and other tests. The first flying prototype (22246) made its first flight at Arlington, Texas, on 1 October 1975, and was followed by the second on 21 December 1975. Both aircraft, and the two Hughes prototypes, were delivered to the US Army during 1976 for a fly-off competition lasting approximately four months, and announcement of the winning design was expected by November 1976. This will mark the end of Phase 1 of the AAH programme.

Phase 2 will involve fitting the winning prototypes with advanced electronics, electro-optical equipment and weapon fire control systems, for further evaluation; continued development of the airframe; and the manufacture of three more aircraft. The US Army has stated a requirement for 472 AAHs; production, if approved, would begin in the late 1970s.

The following description applies to the YAH-63 prototypes:

TYPE: Prototype armed helicopter.

ROTOR SYSTEM: Two-blade semi-rigid teetering main rotor and two-blade tail rotor. Main rotor blades are of wide chord and constant Wortmann FX-69-H-083 section throughout, with raked tips. Each blade has a leading-edge, twin spars and forward skin of stainless steel, with an aluminium honeycomb filling between the spars; and a non-structural rear portion with a non-corroding Nomex core and glassfibre skin. The blades are attached to the hub by flapping axis moment springs which provide control power in zero *g* manoeuvres to give instant fuselage response to cyclic control without risk of control reversal. These springs also eliminate the need for blade tiedown, and permit starting in winds of up to 60 knots (111 km/h; 69 mph). The blades, which can be folded manually, are tested to retain their structural integrity after a chord-line hit by a 23 mm shell. The main rotor hub incorporates elastomeric bearings, to accommodate all flapping and feathering motions and to ease maintenance requirements. A two-position rotor mast is fitted: this is in the extended position for flight and for weapons 'super elevation' clearance, but can be retracted manually to reduce the aircraft's profile for air transportation by C-141 or C-5A. The rotor pylon suspension system incorporates nodalised dynamic beams to reduce crew fatigue, extend airframe component and subsystems life, and provide a more stable gun platform. The tail rotor, located on the port side, has wide-chord, high-thrust, stainless steel blades, a flex-beam hub, and redundant pitch-change controls.

ROTOR DRIVE: Main transmission is driven directly by the two engines via a 'flat-pack' of laterally-disposed herringbone and spiral bevel gears and a collector gear, without the need for intermediate or reduction gearboxes. The rotating controls are ballistically tolerant, with pitch links and clevis arms tested to continue operating safely after a 12·7 mm hit. All fixed controls are redundant and well separated. Tail rotor is driven by an externally-mounted driveshaft (also 12·7 mm survivable) via a single tail rotor gearbox.

WINGS: Cantilever mid-wing monoplane, of low aspect ratio, mounted aft of cockpit. Two hardpoints under each wing for the carriage of ordnance and/or drop-tanks.

FUSELAGE: Of low-profile gunship configuration. Forward portion, of conventional semi-monocoque construction with a minimum of compound curvatures, forms the major load-bearing structure. Circular-section tailboom, with riveted skin, is survivable against hits from 23 mm weapons.

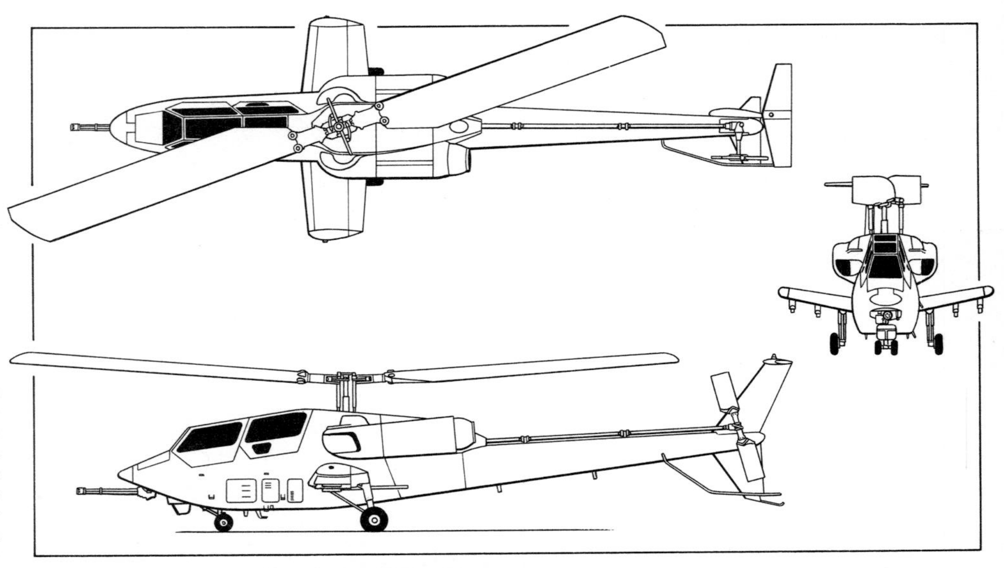

Bell YAH-63 advanced attack helicopter (two General Electric T700-GE-700 turboshaft engines) *(Michael A. Badrocke)*

Bell YAH-63 advanced attack helicopter prototype

TAIL UNITS: Of all-swept 'I' configuration, comprising main and ventral fins, fixed-incidence tailplane and a smaller, lower horizontal surface, also fixed. Tail surfaces are removable for transportation.

LANDING GEAR: High-flotation, non-retractable type, with single wheel on each main unit and twin nosewheels. Main legs fold rearward to reduce profile for storage or transportation. Gear designed to absorb sink rate of up to 9·75 m (32 ft)/sec.

POWER PLANT: Two 1,145 kW (1,536 shp) General Electric T700-GE-700 turboshaft engines, mounted one on each side of fuselage aft of stub-wings. Fuel in two fuselage tanks aft of cockpit and below engine air intakes. All fuel tanks are crash-resistant, self-sealing, and incorporate internal and external void-filling foam.

ACCOMMODATION: Pilot and co-pilot/gunner in tandem, under four-plane flat-surface glint-reducing canopy. Both crew members sit in armoured (12·7 mm resistant) bucket seats, and are separated by a 23 mm resistant transparent plastics screen; pilot occupies front seat, co-pilot/gunner the elevated rear seat. Access to both cockpits is from port side, via a one-piece upward-opening framed transparency to which the side-panel armour is attached. A redundant ballistic canopy-jettison system permits emergency egress from either side of the aircraft. Armour protection panels in sides and floor of cockpit. Barrier of approx 102 mm (4 in) thick aluminium/glassfibre-reinforced plastics sandwich between cockpit and ammunition compartment.

SYSTEMS AND EQUIPMENT: Hydraulic boost system for main rotor. Large electronics compartment in forward fuselage, below cockpits, with external access via three doors on each side. Accessory gearbox, mounted between engines, is driven by the main transmission or by an integral APU. Navigation equipment includes Loran C/D.

ARMAMENT AND OPERATIONAL EQUIPMENT: Fixed armament consists of a General Electric XM-188 three-barrel 30 mm cannon, mounted in a turret under the extreme nose. This gun, which is aimed and fired by the pilot, has a normal rate of fire of 600 rds/min (200 rds each barrel), but this can easily be doubled or tripled if required. The chin-mounted stabilised sight is aft of the gun turret and incorporates a night vision FLIR (forward-looking infra-red), optics and laser. Both crew stations are equipped with non-mechanical helmet sights, and the pilot also has a direct-fire sight. Space, weight and power provisions are made for a pilot's night vision system (PNVS). The linkless ammunition for the XM-188 gun (800-1,200 rds) is carried in a container outside the primary airframe structure, in the fuselage floor amidships, and is stored pointing downward so that it would explode away from the crew compartment in the event of a hit in the container. Although designed to survive such a hit, the ammunition container can be jettisoned if required. The co-pilot/gunner is responsible for the air-launched weapons carried on the four underwing stations. These can comprise up to sixteen TOW anti-tank missiles or seventy-six 2·75 in folding-fin rockets in their launchers, or combinations of both weapons, or up to four drop-tanks. The TOW missiles are guided by stabilised telescopic sight by day, and by an infra-red vision system at night.

DIMENSIONS, EXTERNAL:
Diameter of main rotor 15·54 m (51 ft 0 in)
Diameter of tail rotor 2·90 m (9 ft 6 in)
Main rotor blade chord (constant, each) 1·08 m (3 ft 6·6 in)
Tail rotor blade chord (constant, each) 0·43 m (1 ft 5 in)
Wing span 5·24 m (17 ft 2·4 in)
Length overall, tail rotor turning 16·00 m (52 ft 5·85 in)
Length overall, both rotors turning 18·51 m (60 ft 8·85 in)
Height to top of cabin 3·01 m (9 ft 10·6 in)
Height to rotor hub 3·73 m (12 ft 2·72 in)

WEIGHT:
Design mission T-O weight approx 6,805 kg (15,000 lb)

PERFORMANCE (estimated, with 8 TOWs and 800 rds of 30 mm ammunition, at 1,220 m; 4,000 ft and at 35°C):
Sustained cruising speed 145-175 knots (269-325 km/h; 167-202 mph)
Vertical rate of climb at 95% power more than 152 m (500 ft)/min
Hovering ceiling out of ground effect 1,980 m (6,500 ft)
Endurance 1 hr 54 min

PERFORMANCE (envelope explored to 18 November 1975):
Total flight time 27 hr
Max T-O weight 7,320 kg (15,940 lb)
Forward speed 142 knots (263 km/h; 163·5 mph)
Sideways speed 35 knots (65 km/h; 40 mph)
Backward speed 20 knots (37 km/h; 23 mph)
High-speed taxi 40 knots (74 km/h; 46 mph)
Altitude 1,220 m (4,000 ft)
Continuous run 1 hr 30 min

BELLANCA

BELLANCA AIRCRAFT CORPORATION

HEAD OFFICE AND WORKS:
Box 624, Municipal Airport, Alexandria, Minnesota 56308
Telephone: (612) 762-1501
CHAIRMAN: J. K. Downer
VICE-CHAIRMAN: James M. Miller
PRESIDENT: Warren A. Wilbur III
VICE-PRESIDENT AND GENERAL MANAGER:
J. M. Keating
VICE-PRESIDENTS:
James L. Brown
Robert E. DePalma
John J. McCarten
Mrs T. E. Mitchell
VICE-PRESIDENT/TREASURER:
C. A. Duray
GENERAL SALES MANAGER: F. O. Thompson

Known originally as International Aircraft Manufacturing Inc (Inter-air), Bellanca Sales Company (a subsidiary of Miller Flying Service) acquired the assets of Champion Aircraft Corporation on 30 September 1970. Following the merger, the name Bellanca Aircraft Corporation was adopted, and Bellanca now markets both its own products and those of Champion Aircraft.

In addition to continued production of the four-seat Viking series, Bellanca markets the two-seat Citabria and an advanced aerobatic aircraft named the Decathlon. Instead of the 112 kW (150 hp) Model 7GCBC Scout utility aircraft, described in the 1974-75 *Jane's,* a more powerful version, designated Model 8GCBC, is now in production. Further work on the Bellanca Trainer (1974-75 *Jane's*) has been postponed.

BELLANCA VIKING SERIES

There are three current aircraft in the Viking series, developed from the earlier Bellanca 260C and Standard Viking 300 (see 1971-72 *Jane's*), as follows:

Model 17-30A Super Viking 300A. Powered by a 224 kW (300 hp) Continental IO-520-K flat-six engine, driving a McCauley two- or three-blade metal constant-speed propeller.

Model 17-31A Super Viking 300A. This is identical to the foregoing version except for the installation of a 224 kW (300 hp) Lycoming IO-540-K1E5 engine, driving a Hartzell three-blade constant-speed propeller. Total of 695 Super Vikings (both models) delivered by end of 1975.

Model 17-31ATC Turbo Viking 300A. Powered by a 224 kW (300 hp) Lycoming IO-540-K1E5 engine with two Rajay turbochargers. Hartzell three-blade constant-speed propeller. Total of 120 Turbo Vikings delivered by end of 1975.

By the end of 1975 a total of 1,440 Vikings of all models had been built.

TYPE: Four-seat light business aircraft.

WINGS: Cantilever low-wing monoplane. Bellanca B wing section. Dihedral 4° 30′. Incidence 0° at root, −3° at tip. Structure consists of two laminated Sitka spruce spars, mahogany plywood and spruce ribs and mahogany plywood skin, covered with Dacron. Ailerons and electrically-actuated flaps are Dacron-covered wooden structures.

FUSELAGE: Welded 4130 steel tube structure, covered with Dacron. Two-piece glassfibre engine cowling, suspended from firewall.

TAIL UNIT: Strut-braced welded 4130 steel tube structure, covered with Dacron. Sweptback vertical surfaces. Trim tab in port elevator.

LANDING GEAR: Tricycle type, with Auto-Axion electro-hydraulic retraction, which lowers gear automatically during approach if pilot omits to do so, and prevents accidental retraction on ground. Manual emergency extension. Nosewheel protrudes slightly when retracted to reduce damage in a wheels-up landing. Nosewheel retracts rearward, main wheels forward into underwing fairings, optionally enclosed by doors. Spring-air-oil shock-absorbers. Main-wheel tyres size 6·00-6 6-ply. Steerable nosewheel. Goodyear type 2-747 hydraulic disc brakes.

POWER PLANT: One flat-six engine (details given under model descriptions). Two fuel tanks in wings and one in fuselage, aft of cabin, with total usable capacity of 227 litres (60 US gallons). Optional auxiliary fuel tanks in fuselage, increasing max usable capacity to 283 litres (75 US gallons). Refuelling points above each wing and on starboard side of fuselage. Oil capacity 11·5 litres (3 US gallons).

ACCOMMODATION: Four seats in pairs in enclosed cabin. Dual controls standard, with brakes on port side only. Moulded glassfibre door on starboard side of cabin. Tinted glass. Baggage space, capacity 84 kg (186 lb), aft of rear seats, with glassfibre external door and in-flight access. Provision for tube for carrying skis, max weight 9 kg (20 lb). Heating and ventilation standard.

SYSTEM: 12V electrical system, with Prestolite 60A alternator, solid-state regulator and 33Ah battery. Landing, taxi and navigation lights standard.

ELECTRONICS AND EQUIPMENT: Standard equipment includes artificial horizon, directional gyro, electric turn co-ordinator, rate of climb indicator, vacuum gauge, 8-day clock, outside air temperature gauge, tinted glass, sun visor, towbar, stall warning device, soundproofing, custom interior, tie-down straps. With factory-installed radio equipment the following additional equipment is standard: Narco omni antenna, Electro Voice microphone, power cable, Narco VP-10 broad-band transmitting antenna and microphone jacks. Mitchell Century I, II or III autopilot optional, with optional accessories which include radio tracker, radio coupler and automatic trim for Century II or III, glideslope coupler for Century III, electric trim and switch kits. Optional radio and navigation equipment includes Bendix, King and Narco VHF transceivers, transponders and marker beacon receivers, Bendix, King, Kett and Narco ADF radio receivers, King and Narco DME and Narco course line computer. Miscellaneous optional equipment includes Alcor engine analyser and EGT meter, heated pitot tube, anxiliary power source, stereo tape player, oxygen system, true air speed indicator, alternative paint schemes, 3 in gyros, strobes, quick oil drain, alternative static source, shoulder harness for front seat, dual brakes and map pockets.

Bellanca Super Viking 300A four-seat light aircraft

DIMENSIONS, EXTERNAL:
- Wing span: 10·41 m (34 ft 2 in)
- Length overall: 8·02 m (26 ft 4 in)
- Height overall: 2·24 m (7 ft 4 in)
- Tailplane span: 3·71 m (12 ft 2 in)
- Wheel track: 2·74 m (9 ft 0 in)
- Wheelbase: 2·24 m (7 ft 4 in)
- Propeller diameter: 2·03 m (6 ft 8 in)
- Cabin door:
  - Height: 0·95 m (3 ft 1½ in)
  - Max width: 0·88 m (2 ft 10½ in)
- Baggage compartment door:
  - Height: 0·61 m (2 ft 0 in)
  - Width: 0·51 m (1 ft 8¼ in)

DIMENSIONS, INTERNAL:
- Cabin: Length, firewall to rear wall: 3·10 m (10 ft 2 in)
  - Max width: 1·09 m (3 ft 7 in)
  - Max height: 1·19 m (3 ft 11 in)
- Baggage compartment volume: 0·34 m³ (12·08 cu ft)

AREAS:
- Wings, gross: 15·00 m² (161·5 sq ft)
- Ailerons (total): 1·09 m² (11·77 sq ft)
- Trailing-edge flaps (total): 1·50 m² (16·16 sq ft)

WEIGHTS (A: IO-520, B: IO-540, C: turbocharged IO-540):
- Weight empty:
  - A: 1,005 kg (2,217 lb)
  - B: 1,014 kg (2,236 lb)
  - C: 1,076 kg (2,372 lb)
- Max T-O weight: 1,508 kg (3,325 lb)

PERFORMANCE (at max T-O weight, A: IO-520, B: IO-540, C: turbocharged IO-540):
- Never-exceed speed:
  - A, B, C: 196 knots (363 km/h; 226 mph) IAS
- Max cruising speed (75% power):
  - A: 163 knots (302 km/h; 188 mph) TAS
  - B: 165 knots (305 km/h; 190 mph) TAS
  - C at 7,315 m (24,000 ft): 193 knots (357 km/h; 222 mph) TAS
- Cruising speed (65% power):
  - A: 157 knots (291 km/h; 181 mph) TAS
  - B: 161 knots (298 km/h; 185 mph) TAS
  - C at 7,315 m (24,000 ft): 175 knots (325 km/h; 202 mph) TAS
- Stalling speed, wheels and flaps down:
  - A, B, C: 61 knots (113 km/h; 70 mph) CAS
- Max rate of climb at S/L:
  - A, B, C: 356 m (1,170 ft)/min
- Service ceiling:
  - A: 5,180 m (17,000 ft)
  - B: 5,550 m (18,200 ft)
- Certificated ceiling:
  - C: 7,315 m (24,000 ft)
- T-O to 15 m (50 ft):
  - A, B, C: 433 m (1,420 ft)
- Landing from 15 m (50 ft):
  - A, B, C: 409 m (1,340 ft)
- Max range, standard fuel:
  - A: 638 nm (1,181 km; 734 miles)
  - B: 655 nm (1,215 km; 755 miles)
  - C: 716 nm (1,327 km; 825 miles)
- Max range, max fuel:
  - A: 803 nm (1,485 km; 923 miles)
  - B: 816 nm (1,512 km; 940 miles)
  - C: 894 nm (1,657 km; 1,030 miles)

CHAMPION (BELLANCA) MODEL 7ECA/7GCAA/7KCAB CITABRIA

The Citabria ('airbatic' spelled backwards) represents Bellanca's advanced development of the Model 7 Champion airframe. There are three current versions, as follows:

Model 7ECA. Basic version, with 85·5 kW (115 hp) Lycoming O-235-C1 engine and standard wings. Design and prototype construction started on 1 January 1964. Prototype flew for the first time on 1 May 1964 and first production model on 18 August 1964. FAA certification received 5 August 1964. During 1969 the Model 7ECA received FAA certification for operation on Edo floats.

Model 7GCAA. Generally similar to Model 7ECA but with 112 kW (150 hp) Lycoming O-320-A2D engine. Design started 15 February 1965. Construction of prototype began on 1 May 1965, and it flew on 30 May, followed by the first production model on 20 July 1965. FAA certification received 30 July 1965.

Model 7KCAB. Generally similar to Model 7ECA but with 112 kW (150 hp) Lycoming AEIO-320-E2B engine. Currently certificated with a special fuel and oil system for prolonged inverted flying.

TYPE: Two-seat light cabin monoplane.

WINGS: Braced high-wing monoplane. NACA 4412 wing section. Dihedral 2°. Incidence 1°. Two spruce spars, aluminium ribs, Dacron covering. Steel tube Vee bracing struts. Single-spar Dacron-covered metal ailerons. Glassfibre-reinforced polyester wingtips.

FUSELAGE: Welded chrome-molybdenum steel tube structure, covered with Dacron.

TAIL UNIT: Wire-braced welded steel tube structure, with Dacron covering. Fixed-incidence tailplane. Counter-balanced elevators. Controllable trim tab in elevator.

LANDING GEAR: Non-retractable tailwheel type. All aircraft produced since January 1968 have cantilever spring steel main gear with 6·00-6 wheels and tyres as standard. Tyre pressure 1·65 bars (24 lb/sq in). Cleveland disc brakes. Wheel fairings optional. Pee Kay 1800 or Edo floats, and Federal A-200-A skis, available on Model 7ECA.

POWER PLANT: One flat-four engine, as described under individual model listings. McCauley two-blade fixed-pitch metal propeller: type 1C90ALM on 85·5 kW (115 hp) model, and type 1C172AGM on 112 kW (150 hp) models. Two aluminium fuel tanks in wings, total capacity 136 litres (36 US gallons), of which 132 litres (35 US gallons) are usable. Refuelling points above tanks. Oil capacity 5·75 litres (1·5 US gallons) on version with 85·5 kW (115 hp) engine, 7·5 litres (2 US gallons) on 112 kW (150 hp versions).

ACCOMMODATION: Enclosed cabin seating two persons in tandem. Dual controls. Heater standard. Quick-jettison door on starboard side. Space for 45 kg (100 lb) baggage.

SYSTEMS: Hydraulic system for brakes only. Electrical system powered by engine-driven generator.

ELECTRONICS AND EQUIPMENT: Wide range of King and Narco radio equipment optional, including omni, ILS and ADF. Blind-flying instrumentation optional. Standard equipment includes landing and navigation lights.

DIMENSIONS, EXTERNAL:
- Wing span: 10·19 m (33 ft 5 in)
- Wing chord (constant): 1·52 m (5 ft 0 in)
- Wing aspect ratio: 6·72
- Length overall: 6·91 m (22 ft 8 in)

Height overall 2·36 m (7 ft 9 in)
Wheel track 1·93 m (6 ft 4 in)
Wheelbase 4·90 m (16 ft 1 in)
Cabin door: Height 0·94 m (3 ft 1 in)
Width 0·94 m (3 ft 1 in)
Height to sill 0·44 m (1 ft 5½ in)

AREAS:
Wings, gross 15·33 m^2 (165 sq ft)
Ailerons (total) 1·53 m^2 (16·5 sq ft)
Fin 0·65 m^2 (7·02 sq ft)
Rudder 0·63 m^2 (6·83 sq ft)
Tailplane 1·14 m^2 (12·25 sq ft)
Elevators, incl tab 1·35 m^2 (14·58 sq ft)

WEIGHTS AND LOADINGS:
Weight empty, equipped:
7ECA 481 kg (1,060 lb)
7GCAA 503 kg (1,109 lb)
7KCAB 517 kg (1,140 lb)
Max T-O and landing weight:
landplanes 748 kg (1,650 lb)
Max wing loading 48·8 kg/m^2 (10 lb/sq ft)
Max power loading:
7ECA 8·72 kg/kW (14·35 lb/hp)
7GCAA, 7KCAB 6·68 kg/kW (11·0 lb/hp)

PERFORMANCE (at max T-O weight):
Never-exceed speed 140 knots (261 km/h; 162 mph)
Max level speed at S/L:
7ECA 102 knots (188 km/h; 117 mph)
7ECA seaplane 75 knots (138 km/h; 86 mph)
7GCAA 113 knots (209 km/h; 130 mph)
7KCAB 113 knots (209 km/h; 130 mph)
Max cruising speed (75% power) at optimum height:
7ECA 108 knots (200 km/h; 124 mph)
7GCAA, 7KCAB 112 knots (207 km/h; 129 mph)
Cruising speed (65% power):
7ECA 102 knots (189 km/h; 117 mph)
7GCAA, 7KCAB 107 knots (198 km/h; 123 mph)
Stalling speed:
all models 44·5 knots (82 km/h; 51 mph)
Max rate of climb at S/L:
7ECA 221 m (725 ft)/min
7ECA seaplane 157 m (515 ft)/min
7GCAA, 7KCAB 341 m (1,120 ft)/min
Service ceiling:
7ECA 3,660 m (12,000 ft)
7GCAA, 7KCAB 5,180 m (17,000 ft)
T-O run:
7ECA 139 m (455 ft)
7GCAA, 7KCAB 116 m (382 ft)
T-O to 15 m (50 ft):
7ECA 273 m (895 ft)
7GCAA, 7KCAB 202 m (663 ft)
Landing from 15 m (50 ft):
7ECA 236 m (775 ft)
7GCAA, 7KCAB 230 m (755 ft)
Landing run:
all models 121 m (400 ft)
Max range:
7ECA 602 nm (1,116 km; 694 miles)
7GCAA, 7KCAB 486 nm (900 km; 559 miles)
g limits +5; −2

Champion Citabria two-seat aerobatic light aircraft

Bellanca Model 8GCBC Scout two-seat light aircraft, equipped with Sorensen spraygear
(Dr Alan Beaumont)

BELLANCA MODEL 8GCBC SCOUT

This version of the Scout, with a 134 kW (180 hp) Lycoming engine, received type approval on 30 April 1974, and went into immediate production to meet large orders. By the end of 1975 a total of 197 8GCBC Scouts had been produced.

The description of the basic Citabria applies also to the Scout, except for the following details:

WINGS: As for Citabria, except increased wing span and provision of fabric-covered trailing-edge 27° high-lift flaps of metal construction. Dihedral reduced to 1°. Hoerner wingtips.

FUSELAGE: Removable metal skin panels on undersurface.

TAIL UNIT: As for Citabria, but areas increased.

LANDING GEAR: Non-retractable tailwheel type. Cantilever spring steel main gear with wheels and tyres size 7·00-6. Heavy-duty Scott tailwheel. Ski installations approved are Fluidyne 2000 and Airgas Landis 2000A; approved floats are Edo 2000, Pee Kay 2000 and Canadian Aircraft Products 2000A. Hydraulic disc brakes. Parking brake.

POWER PLANT: One 134 kW (180 hp) Lycoming O-360-C2E flat-four engine, driving a McCauley two-blade metal fixed-pitch propeller. Two fuel tanks in wings with a total capacity of 136 litres (36 US gallons), of which 132 litres (35 US gallons) are usable. Refuelling points above tanks. Oil capacity 7·5 litres (2 US gallons).

ACCOMMODATION: Enclosed cabin seating two persons in tandem.

EQUIPMENT: Standard equipment includes fire extinguisher, crash locator beacon, landing light, navigation lights, de luxe interior trim, two-colour external paint scheme and seaplane corrosion proofing. Optional items include a 340 litre (90 US gallon) Sorensen belly tank and underwing spraybooms, glider towing hook, and ski and float installations as detailed under landing gear.

DIMENSIONS, EXTERNAL:
Wing span 11·02 m (36 ft 2 in)
Wing chord (constant) 1·52 m (5 ft 0 in)
Length overall 6·93 m (22 ft 9 in)
Height overall 2·24 m (7 ft 4 in)
Tailplane span 3·10 m (10 ft 2¼ in)
Wheel track 2·10 m (6 ft 10½ in)
Wheelbase, tail up 5·02 m (16 ft 5½ in)
Propeller diameter 2·03 m (6 ft 8 in)

AREAS:
Wings, gross 16·7 m^2 (180 sq ft)
Vertical tail surfaces 1·53 m^2 (16·5 sq ft)
Horizontal tail surfaces 2·42 m^2 (26 sq ft)

WEIGHTS:
Weight empty 597 kg (1,315 lb)
Normal category payload 379 kg (835 lb)
Restricted category payload 517 kg (1,140 lb)
Max T-O weight:
Normal 975 kg (2,150 lb)
Restricted 1,179 kg (2,600 lb)

PERFORMANCE (at Normal category max T-O weight):
Never-exceed speed
140 knots (260 km/h; 162 mph) CAS
Max level speed at S/L
117 knots (217 km/h; 135 mph) TAS
Cruising speed at 75% power
107 knots (198 km/h; 123 mph) TAS
Stalling speed, flaps down, power off
45 knots (84 km/h; 52 mph)
Max rate of climb at S/L 329 m (1,080 ft)/min
T-O run 156 m (510 ft)

CHAMPION (BELLANCA) MODEL 8KCAB DECATHLON

The Champion Model 8KCAB Decathlon is an aerobatic competition aircraft designed for loads of +6*g* and −5*g*, and has been arbitrarily cleared for two minutes of inverted flight, although the aircraft has been flown inverted in excess of four minutes without loss of oil or oil pressure. FAA certification under FAR 23, for both Normal and Aerobatic categories, was granted on 16 October 1970.

Following sale of the first hand-built production batch of 14 aircraft (first one delivered on 24 February 1971), Bellanca decided to begin full-scale production with the 15th aircraft, thus making the Decathlon the only unlimited aerobatic competition aircraft in production in the USA. By the end of 1975 a total of 230 Decathlons had been produced.

TYPE: Two-seat light cabin monoplane.

WINGS: Braced high-wing monoplane. Wing section NACA 1412 modified. Dihedral 1°. Incidence 1° 30′. Size of front spar substantially increased and ribs stronger and more closely-spaced by comparison with Citabria. Trusses added between front and rear spars in aileron area. Aluminium (front) and steel tube (rear) Vee bracing struts each side, of enlarged section.

FUSELAGE: Welded steel tube structure, fabric covered.

TAIL UNIT: Wire-braced welded steel tube structure. Fixed-incidence tailplane. Trim tab in elevator.

LANDING GEAR: Non-retractable tailwheel type. Cantilever spring steel main gear. Fairings on main wheels.

POWER PLANT: One 112 kW (150 hp) Lycoming AEIO-320-E1B flat-four engine, driving a Hartzell HC-C2YL-4/C7663-4 two-blade metal counterweighted constant-speed propeller. Two wing fuel tanks, with total usable capacity of 151·4 litres (40 US gallons). Oil capacity 7·5 litres (2 US gallons).

Accommodation: Enclosed cabin seating two persons in tandem. Quick-jettison door on starboard side. Space for 45 kg (100 lb) baggage.

Dimensions, external:

Wing span	9·75 m (32 ft 0 in)
Wing chord (constant)	1·63 m (5 ft 4 in)
Length overall	6·98 m (22 ft 10¾ in)
Height overall	2·36 m (7 ft 9 in)
Wheel track	1·93 m (6 ft 4 in)
Wheelbase	4·98 m (16 ft 4 in)

Areas:

Wings, gross	15·79 m² (170 sq ft)
Ailerons (total)	1·92 m² (20·68 sq ft)
Fin	0·65 m² (7·02 sq ft)
Rudder	0·63 m² (6·83 sq ft)
Tailplane	1·14 m² (12·25 sq ft)
Elevators, incl tab	1·35 m² (14·58 sq ft)

Weights and Loadings:

Weight empty	578 kg (1,275 lb)
Max T-O weight	815 kg (1,800 lb)
Max wing loading	51·75 kg/m² (10·6 lb/sq ft)
Max power loading	7·28 kg/kW (12·0 lb/hp)

Performance (at max T-O weight):

Never-exceed speed	156 knots (289 km/h; 180 mph)
Max level speed	123 knots (228 km/h; 142 mph)
Max cruising speed, 75% power at 2,440 m (8,000 ft)	117 knots (217 km/h; 135 mph)
Stalling speed	46 knots (86 km/h; 53 mph)
Max rate of climb at S/L	312 m (1,025 ft)/min
Service ceiling	4,875 m (16,000 ft)
Max range	478 nm (885 km; 550 miles)

Champion Decathlon two-seat light aircraft, in inverted flight

BELLANCA

BELLANCA AIRCRAFT ENGINEERING INC

Head Office and Works:
PO Box 70, Scott Depot, nr Charleston, West Virginia 25560

Telephone: (304) 755-4354

President: August T. Bellanca

Managing Director:
Henry E. Payne

The original Bellanca Aircraft Corporation of New Castle, Delaware, merged with companies not engaged in aircraft manufacture and lost its identity in 1959. The present company, formed by Mr August Bellanca and his father, the late G. M. Bellanca, bought all of the original Bellanca aircraft designs with the exception of the Model 14-19.

Following extensive research into the use of glassfibre composite materials for airframe construction, Bellanca Aircraft Engineering began the testing of full-scale structures, embodying a variety of different design and fabrication techniques. This led, in 1963, to the decision to construct a prototype, designated Model 19-25 Skyrocket, constructed of high-strength glassfibre-epoxy laminates. Complete aerodynamic testing in a 2·13 m by 3·05 m (7 ft by 10 ft) wind tunnel was completed successfully in 1967.

On 1 December 1971 the company was reorganised and acquired corporate offices and production plant at Scott Depot, West Virginia. Construction of the Skyrocket prototype was well advanced by February 1973, but the decision to install a 324·5 kW (435 hp) Continental engine, instead of the 298 kW (400 hp) Lycoming IO-720-A1A as specified originally, delayed completion of the prototype and it was not until March 1975 that the first flight was recorded.

Prototype of the Bellanca Model 19-25 Skyrocket II six-seat light aircraft

BELLANCA MODEL 19-25 SKYROCKET II

Research and design of this aircraft was initiated in 1956 by the late G. M. Bellanca and his son, August T. Bellanca. The early decision to fabricate this aircraft from glassfibre composites resulted from much research and testing. Basically, the advantages are that smooth aerodynamic surfaces are obtained, together with high strength. The materials used in the Model 19-25 Skyrocket have a higher strength-to-weight ratio than aluminium, with better durability and fatigue resistance.

Since the first flight in March 1975, the Skyrocket II prototype has demonstrated its capabilities by setting record speeds of 245·80 knots (455·23 km/h; 282·87 mph) over a 100 km closed circuit, 257·64 knots (477·15 km/h; 296·49 mph) over 500 km, and 261·67 knots (484·62 km/h; 301·13 mph) over 1,000 km in the 1,000-1,750 kg weight class; and a speed of 272·85 knots (505·31 km/h; 313·99 mph) over 1,000 km in the 1,750-3,000 kg class. Orders had been received by 1 February 1976 for a total of 70 aircraft.

Type: Six-seat light cabin monoplane.

Wings: Cantilever low-wing monoplane. NACA $63_{2}215$ laminar-flow wing section. Dihedral 2°. Incidence 2°. Composite structure of glassfibre epoxy laminate and light alloy honeycomb sandwich, made in two half-shells, with two glassfibre/carbon spars. Plain ailerons and trailing-edge flaps of similar construction. Anti-icing of wing leading-edges by hot air.

Fuselage: Semi-monocoque fail-safe pressurised structure of glassfibre epoxy laminate. Airbrake in fuselage undersurface. Glassfibre stringers. Fuselage is moulded in two halves, each with an integral wing root, vertical fin half and tailplane root.

Tail Unit: Cantilever structure of glassfibre epoxy laminate with light alloy honeycomb sandwich. Electrically-adjustable variable-incidence tailplane. No trim tabs. Anti-icing of leading-edges by hot air.

Landing Gear: Hydraulically-retractable tricycle type. Nosewheel retracts aft, main units inward. Ozone oleo-pneumatic shock-absorbers with single wheel on each unit. Goodyear main wheels and tyres size 15 × 6·00-6.

Power Plant: One 324 kW (435 hp) Continental GTSIO-520-F flat-six engine, flat rated to an altitude of 5,790 m (19,000 ft), driving a Hartzell three-blade metal constant-speed propeller. One integral fuel tank in each wing with a combined capacity of 605·5 litres (160 US gallons) of which 568 litres (150 US gallons) is usable. Refuelling points at wingtips. Oil capacity 14·2 litres (3·75 US gallons).

Accommodation: Pilot and five passengers in enclosed cabin. Door on starboard side. Stowage for 90·7 kg (200 lb) of baggage.

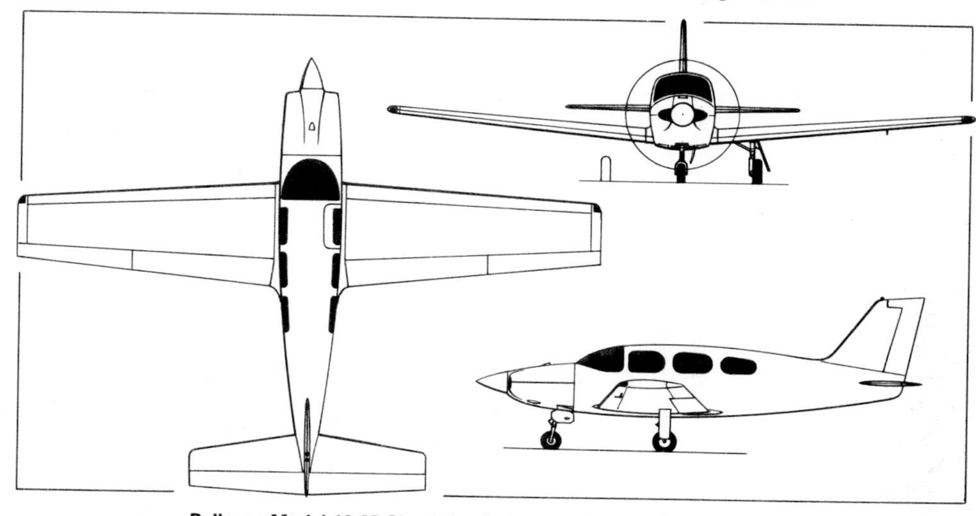

Bellanca Model 19-25 Skyrocket II six-seat light aircraft *(Pilot Press)*

Dimensions, external:

Wing span	10·67 m (35 ft 0 in)
Wing aspect ratio	6·7
Length overall	8·81 m (28 ft 11 in)
Height overall	2·82 m (9 ft 3 in)
Propeller diameter	2·08 m (6 ft 10 in)

Area:

Wings, gross	16·96 m² (182·6 sq ft)

Weights and Loadings:

Weight empty	1,043 kg (2,300 lb)
Max T-O weight	1,860 kg (4,100 lb)
Max wing loading	109·37 kg/m² (22·4 lb/sq ft)
Max power loading	5·74 kg/kW (9·43 lb/hp)

Performance (at max T-O weight):
Max level speed at 8,840 m (29,000 ft) 287 knots (532 km/h; 331 mph)
Max cruising speed, 75% power at 7,300 m (24,000 ft) 262 knots (486 km/h; 302 mph)
Cruising speed, 65% power at 4,570 m (15,000 ft) 221 knots (410 km/h; 255 mph)
Stalling speed, flaps down 56·5 knots (105 km/h; 65 mph)
Max rate of climb at S/L 579 m (1,900 ft)/min
Service ceiling above 9,145 m (30,000 ft)
T-O run 274 m (900 ft)
T-O to 15 m (50 ft) 546 m (1,790 ft)
Landing from 15 m (50 ft) 507 m (1,665 ft)
Landing run 241 m (790 ft)
Range, 75% power, at 4,570 m (15,000 ft) more than 1,055 nm (1,955 km; 1,215 miles)
Range, 65% power, at 4,570 m (15,000 ft) more than 1,272 nm (2,357 km; 1,465 miles)

BOEING
THE BOEING COMPANY

Head Office:
PO Box 3707, Seattle, Washington 98124
Established: July 1916
Chairman of the Board and Chief Executive Officer:
T. A. Wilson
President:
Malcolm T. Stamper
Senior Vice-Presidents:
H. W. Haynes (Executive Vice-President, Chief Financial Officer)
W. N. Maulden (Corporate Operations)
J. E. Prince (Administration)
Vice-Presidents:
R. E. Bateman (General Manager, Boeing Marine Systems)
W. L. Hamilton (International Business)
H. K. Hebeler (General Manager, Boeing Engineering and Construction Division)
V. F. Knutzen (Controller)
S. M. Little (Industrial and Public Relations)
H. W. Neffner (Contracts)
Other Vice-Presidents are listed under Divisions
Treasurer: J. B. L. Pierce
Public Relations and Advertising Director:
R. P. Bush

Boeing Commercial Airplane Company:

Address:
PO Box 707, Renton, Washington 98055
President:
E. H. Boullioun
Vice-Presidents:
W. W. Buckley (General Manager, 707/727/737 Division)
K. F. Holtby (General Manager, 747 Division)
J. E. Steiner (Technology and New Programme Development)
J. F. Sutter (Programme Operations)
D. D. Thornton (Finance, Contracts and International Operations)
R. W. Welch (Executive Vice-President)
D. D. Whitford (General Manager, Fabrication Division)
C. F. Wilde (Sales and Marketing)
H. W. Withington (Engineering)
B. S. Wygle (Customer Support)

Boeing Aerospace Company:

Address:
Kent, Washington
President:
O. C. Boileau
Vice-Presidents:
L. D. Alford (General Manager, Missiles and Space Group)
D. A. Cole (General Manager, Naval Systems Division)
D. E. Graves (Military Applications)
W. T. Hamilton (Manager, New Military Transport Programmes)
H. E. Hurst (Prototype Airplane Operations, Aeronautical and Information Systems Division)
J. C. Maxwell (General Manager, Military Airplane Development)
B. T. Plymale (Product Development Manager)
R. W. Taylor (General Manager, Military Systems Group)

Wichita Division:

Address:
3801 South Oliver, Wichita, Kansas 67210
Vice-President and General Manager:
O. H. Smith

Boeing Vertol Company:

Address:
Boeing Centre, PO Box 16858, Philadelphia, Pa 19142
President:
H. N. Stuverude
Executive Vice-President:
G. D. Nible
Vice-President and Asst General Manager:
C. W. Ellis

In May 1961 The Boeing Airplane Company changed its proprietary name to The Boeing Company as a recognition of the company's diversified interests. The change did not imply any decreased interest in the design and manufacture of aircraft.

On 19 December 1972 it was announced that three of the operating organisations of The Boeing Company had been designated as companies, comprising Boeing Commercial Airplane Company, Renton, Washington; Boeing Aerospace Company, Kent, Washington; and Boeing Vertol Company, Philadelphia, Pennsylvania.

The Wichita Division at Wichita, Kansas, continues modification programmes, 737 and 747 parts fabrication, research, programmes on military aircraft currently in use with the armed forces (B-52 and KC-135) and other support functions for the company. A new factory with an area of 75,000 m² (807,300 sq ft) was opened on 9 November 1971 at St James-Assiniboia Airport, near Winnipeg, to produce 747 components.

BOEING COMMERCIAL AIRPLANE COMPANY

The Boeing Commercial Airplane Company, with headquarters at the company's Renton, Washington, facility just south of Seattle, has three divisions. The 707/727/737 Division and the 747 Division continue to manufacture aircraft of those series; the Fabrication Division handles central fabrication services from its Auburn, Washington, plant, and other services such as CAG facilities. The Engineering Organisation is responsible for such functions as technology, quality control and flight operations.

The 747 Division is at Everett, Washington, 30 miles north of Seattle; and the 707/727/737 Engineering Organisations are based at Renton.

Boeing delivered its 2,500th commercial jet transport, an Advanced 737 for Transavia of the Netherlands, on 17 May 1974. Including military derivatives, deliveries during 1975 totalled 8 Model 707s, 91 Model 727s, 51 Model 737s and 21 Model 747s.

BOEING MODEL 707

US Air Force designation: VC-137

The prototype of the Boeing Model 707 was the first jet transport designed as such to be completed and flown in the United States. It made its first flight on 15 July 1954.

Designated Model 367-80, it was built as a private venture and was used to demonstrate the potential of commercial and military developments of the design for a period of more than 15 years. During its early test programme, it was fitted with a flight refuelling boom, to prove the capability of this type of aircraft for refuelling present and future jet bombers, fighters and reconnaissance aircraft at or near their operational altitudes and speeds. As a result, a developed version was ordered in large numbers for the USAF under the designation KC-135.

Boeing announced on 25 April 1972 that the Model 367-80, widely known as the Dash-Eighty, was to be given to the Smithsonian Institution.

On 13 July 1955 Boeing was given clearance by the USAF to build commercial developments of the prototype concurrently with the production of military KC-135 tanker-transports. These transport aircraft have the basic designations of Boeing 707 and 720, but were made available in many versions, of which a total of 919 had been sold and 905 delivered by 31 August 1976. These totals include five specially-equipped aircraft delivered to the USAF under the designations VC-137A (now VC-137B) and VC-137C, and two AWACS (Airborne Warning And Control System) aircraft which were used, under the designation EC-137D, for competitive trials of downward-looking radars. They are now being used in the next phases of the E-3A programme (which see).

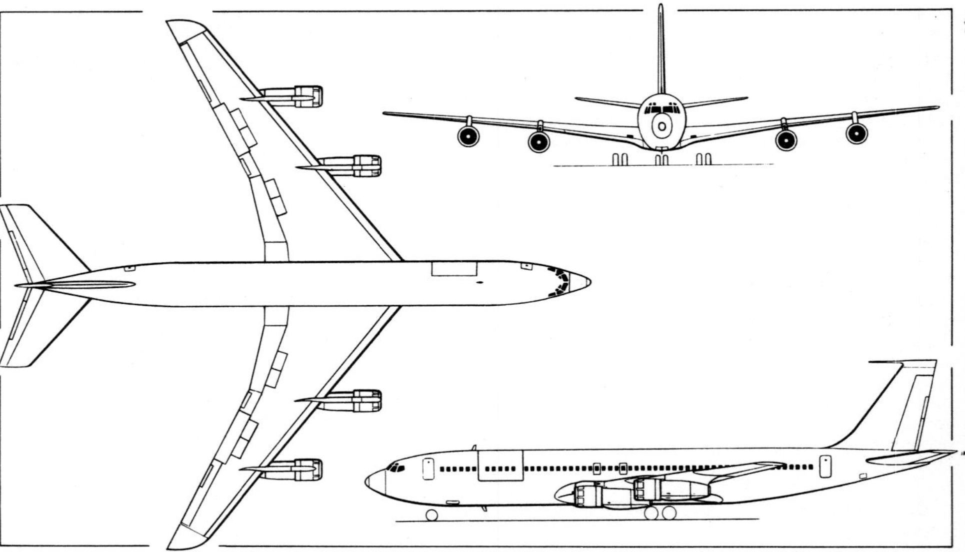
Boeing 707-320C four-turbofan passenger/cargo transport aircraft *(Pilot Press)*

The only version of the 707 available in 1976 is as follows:

707-320C Convertible. Certificated to carry up to 219 passengers, this version can also operate mixed passenger/cargo or all-cargo services. Loading is through a 2·34 m × 3·40 m (92 in × 134 in) forward cargo door, with cargo on pallets or in containers. A Boeing-developed cargo handling system is installed on seven rows of seat tracks in the floor. Upper-deck cargo space comprises 161·21 m³ (5,693 cu ft) palletised and lower deck 48·14 m³ (1,700 cu ft) bulk. The cargo system can carry thirteen 2·24 m × 3·18 m (88 in × 125 in) or 2·24 m × 2·74 m (88 in × 108 in) 'A' type containers. There is a crew rest area aft of the flight deck. Received FAA Type Approval on 30 April 1963, and first entered service with Pan American in June 1963. Five were delivered to the the Canadian Armed Forces during 1970-71 to serve as troop and staff transports and military cargo carriers. Two of these are equipped as flight refuelling tankers, utilising wingtip pods containing a hinged boom and trailing hose and drogue manufactured by Beech.

Type: Four-turbofan airliner.

Wings: Cantilever low-wing monoplane. Dihedral 7°. Incidence 2°. Sweepback at quarter-chord 35°. All-metal two-spar fail-safe structure. Centre-section continuous through fuselage. Normal outboard aileron and small inboard aileron on each wing, built of aluminium honeycomb panels. Two Fowler flaps and one fillet flap of aluminium alloy on each wing. Full-span leading-edge flaps. Four hydraulically-operated aluminium alloy spoilers on each wing, forward of flaps. Primary flying controls are aerodynamically balanced and manually operated through spring tabs. Lateral control at low speeds by all four ailerons, supplemented by spoilers which are interconnected with the ailerons. Lateral control at high speeds by inboard ailerons and spoilers only. Operation of flaps adjusts linkage between inboard and outboard ailerons to permit outboard operation with extended flaps. Spoilers may also be used symmetrically as speed brakes. Thermal anti-icing of wing leading-edges.

Fuselage: All-metal semi-monocoque fail-safe structure with cross-section made up of two circular arcs of different radii, the larger above, faired into smooth-contoured ellipse.

Tail Unit: Cantilever all-metal structure. Anti-balance

Boeing Model 707-366C (four Pratt & Whitney JT3D-7 turbofan engines) in the insignia of EgyptAir

tab and trim tab in rudder. Trim and control tabs in each elevator. Electrically and manually operated variable-incidence tailplane. Powered rudder.

LANDING GEAR: Hydraulically-retractable tricycle type. Main units are four-wheel bogies which retract inward into underside of thickened wing-root and fuselage. Dual nosewheel unit retracts forward into fuselage. Landing gear doors close when legs fully extended. Gear can be extended in flight to give maximum rate of descent of 4,570 m/min (15,000 ft/min) when used in conjunction with spoilers. Boeing oleo-pneumatic shock-absorbers. Main wheels and tyres size 46 × 16. Nosewheels and tyres size 39 × 13. Tyre pressures: main wheels 12·41 bars (180 lb/sq in), nosewheels 7·93 bars (115 lb/sq in). Multi-disc brakes by Goodyear. Hydro-Aire flywheel detector type anti-skid units.

POWER PLANT: Four Pratt & Whitney JT3D-7 turbofan engines, each developing 84·5 kN (19,000 lb st), in pods under wings. Fuel in four main, two reserve and one centre main integral wing tanks with total capacity of 90,299 litres (23,855 US gallons). Provision for both pressure and gravity refuelling. Total oil capacity 114 litres (30 US gallons).

ACCOMMODATION: Max accommodation for up to 219 passengers. Typical arrangement has 14 first class seats, a 4-seat lounge and 133 coach class seats, with four galleys and five toilets. There are two passenger doors, forward and aft on port side. Galley servicing doors forward and aft on starboard side. Baggage compartments fore and aft of wing in lower segment of fuselage below cabin floor. Entire accommodation, including baggage compartments, air-conditioned and pressurised.

SYSTEMS: Air-cycle air-conditioning and pressurisation system, using three AiResearch engine-driven turbo-compressors. Pressure differential 0·59 bars (8·6 lb/sq in). Hydraulic system, pressure 207 bars (3,000 lb/sq in), for landing gear retraction, nosewheel steering, brakes, flaps, flying controls and spoilers. Electrical system includes four 30kVA or 40kVA 115/200V 3-phase 400Hz AC alternators and four 75A transformer-rectifiers giving 28V DC. APU optional.

ELECTRONICS AND EQUIPMENT: To customer's specification.

DIMENSIONS, EXTERNAL:
Wing span 44·42 m (145 ft 9 in)
Wing chord at root 10·33 m (33 ft 10·7 in)
Wing chord at tip 2·84 m (9 ft 4 in)
Wing aspect ratio 7·056
Length overall 46·61 m (152 ft 11 in)
Length of fuselage 44·35 m (145 ft 6 in)
Width of fuselage 3·76 m (12 ft 4 in)
Height overall 12·93 m (42 ft 5 in)
Tailplane span 13·95 m (45 ft 9 in)
Wheel track 6·73 m (22 ft 1 in)
Wheelbase 17·98 m (59 ft 0 in)
Passenger doors (each):
Height 1·83 m (6 ft 0 in)
Width 0·86 m (2 ft 10 in)
Height to sill:
fwd 3·25 m (10 ft 8 in)
aft 3·20 m (10 ft 6 in)
Cargo door:
Height 2·34 m (7 ft 8 in)
Width 3·40 m (11 ft 2 in)
Height to sill 3·20 m (10 ft 6 in)
Forward baggage compartment door:
Height 1·27 m (4 ft 2 in)
Width 1·22 m (4 ft 0 in)
Height to sill 1·55 m (5 ft 1 in)
Rear baggage compartment door (fwd):
Height 1·24 m (4 ft 1 in)
Width 1·22 m (4 ft 0 in)
Height to sill 1·47 m (4 ft 10 in)
Rear baggage compartment door (aft):
Height 0·89 m (2 ft 11 in)
Width 0·76 m (2 ft 6 in)
Height to sill 1·93 m (6 ft 4 in)

DIMENSIONS, INTERNAL:
Cabin, excl flight deck:
Length 33·93 m (111 ft 4 in)
Max width 3·55 m (11 ft 8 in)
Max height 2·34 m (7 ft 8 in)
Floor area 106·18 m² (1,143 sq ft)
Volume 228·6 m³ (8,074 cu ft)
Baggage compartment (fwd) 23·65 m³ (835 cu ft)
Baggage compartment (rear) 24·50 m³ (865 cu ft)

AREAS:
Wings, gross 283·4 m² (3,050 sq ft)
Ailerons (total) 11·24 m² (121 sq ft)
Trailing-edge flaps (total) 44·22 m² (476 sq ft)
Leading-edge flaps 14·31 m² (154 sq ft)
Fin 30·47 m² (328 sq ft)
Rudder, incl tabs 9·48 m² (102 sq ft)
Tailplane 58·06 m² (625 sq ft)
Elevators, incl tabs 14·03 m² (151 sq ft)

WEIGHTS AND LOADINGS:
Basic operating weight, empty:
Passenger 66,768 kg (147,200 lb)
Cargo 64,000 kg (141,100 lb)
Max payload:
Cargo 40,324 kg (88,900 lb)
Max T-O weight 151,315 kg (333,600 lb)
Max ramp weight 152,405 kg (336,000 lb)
Max zero-fuel weight 104,330 kg (230,000 lb)
Max landing weight 112,037 kg (247,000 lb)
Max wing loading 537·1 kg/m² (110·0 lb/sq ft)
Max power loading 448 kg/kN (4·39 lb/lb st)

PERFORMANCE (at average cruising weight, unless indicated otherwise):
Never-exceed speed Mach 0·95
Max level speed 545 knots (1,010 km/h; 627 mph)
Max cruising speed at 7,620 m (25,000 ft) 525 knots (973 km/h; 605 mph)
Econ cruising speed 478 knots (886 km/h; 550 mph)
Stalling speed (flaps down, at max landing weight) 105 knots (195 km/h; 121 mph)
Max rate of climb at S/L 1,219 m (4,000 ft)/min
Service ceiling 11,885 m (39,000 ft)
CAR T-O to 10·7 m (35 ft) 3,054 m (10,020 ft)
CAR landing from 15 m (50 ft) 1,095 m (6,250 ft)
Landing run 785 m (2,575 ft)
Range with max fuel, allowances for climb and descent, no reserves 6,493 nm (12,030 km; 7,475 miles)
Range with max payload, allowances for climb and descent, no reserves 3,735 nm (6,920 km; 4,300 miles)

BOEING MODEL 727

On 5 December 1960, Boeing announced its intention to produce a short/medium-range transport designated Boeing 727. Design work had been under way since June 1959 and component manufacture had been started in October 1960.

A major innovation, compared with the company's earlier designs, was the choice of a rear-engined layout. In other respects the 727 bears a resemblance to the 707 and 720 series. It has an identical upper fuselage section and many parts and systems are interchangeable between the three types.

The 727-100, 727-100C, 727-100QC and 727-100 Business Jet versions of the Model 727 are no longer in production; details of these can be found in the 1973-74 Jane's. Versions which remain in production are as follows:

727-200. Lengthened version announced on 5 August 1965 with basic accommodation for 163 passengers and maximum capacity of 189 passengers. Fuselage extended by 3·05 m (10 ft) both forward and aft of main undercarriage wheel well. Structural modification corresponding to higher loads. Revised centre engine air intake. Three JT8D-9 turbofans, each flat rated at 64·5 kN (14,500 lb st) to 29°C, are standard. Optionally, JT8D-11s rated at 66·7 kN (15,000 lb st), JT8D-15s rated at 68·9 kN (15,500 lb st), orJT8D-17s rated at 71·2 kN (16,000 lb st) can be fitted. Construction of the first 727-200 began in September 1966 and the first flight was made on 27 July 1967. FAA certification was awarded on 30 November 1967.

Advanced 727-200. On 12 May 1971, Boeing announced it was offering the Advanced 727-200 at 86,635 kg (191,000 lb) gross ramp weight and with sound suppression characteristics that would make this version quieter than any other commercial jet transport then in use; initial deliveries of this version began in June 1972. Increased fuel capacity gives a range capability at least 694 nm (1,287 km; 800 miles) greater than that of the earlier 727-200. The interior features the 'Superjet-look'.

A later version, first ordered by Sterling Airways in May 1972, has Pratt & Whitney JT8D-15 engines, ramp weight of 94,350 kg (208,000 lb) and more fuel.

A total of 1,305 Model 727s had been sold by 31 August 1976, of which 1,213 had been delivered. The 727 is the only commercial transport aircraft to have exceeded a sales figure of one thousand.

TYPE: Three-turbofan airliner.

WINGS: Cantilever low-wing monoplane. Special Boeing aerofoil sections. Thickness/chord ratio from 9% to 13%. Dihedral 3°. Incidence 2°. Sweepback at quarter-chord 32°. Primary structure is a two-spar aluminium alloy box with conventional ribs. Upper and lower surfaces are of riveted skin-stringer construction. There are no chordwise splices in the primary structure from the fuselage to the wingtip. Advanced 727-200s at gross weight options have modified stringers and in-spar webs, as well as upper and lower surface wing skins of increased gauge. Structure is fail-safe. Hydraulically-powered aluminium ailerons, in inboard (high speed) and outboard (low speed) units, operate in conjunction with flight spoilers. Triple-slotted trailing-edge flaps constructed primarily of aluminium and aluminium honeycomb. Four aluminium leading-edge slats on outer two-thirds of wing. Three Krueger leading-edge flaps on inboard third of wing, made from magnesium or aluminium castings. Seven aluminium (plus some magnesium) spoilers on each wing, consisting of five flight spoilers outboard and two ground spoilers inboard. Spoilers function also as airbrakes. Balance tab in each outboard aileron; control tab in each inboard aileron. Controls are hydraulically-powered dual systems with automatic reversion to manual control. Actuators manufactured primarily by Weston, National Waterlift and Bertea. Thermal anti-icing of wing leading-edges by engine bleed air.

FUSELAGE: Semi-monocoque fail-safe structure, with aluminium alloy skin reinforced by circumferential frames and longitudinal stringers.

TAIL UNIT: Cantilever structure, built primarily of aluminium alloys, with tailplane mounted near tip of fin. Dual-powered variable-incidence tailplane, with direct manual reversion. Hydraulically-powered dual elevator

Boeing Model 727-200 (three Pratt & Whitney JT8D-15 turbofan engines) in the insignia of Air Jamaica

control system with control tab manual reversion. Hydraulically-powered rudders, utilising two main systems with backup third system for lower rudder. Anti-balance tabs; rudder trim by displacing system neutral.

LANDING GEAR: Hydraulically-retractable tricycle type, with twin wheels on all three units. Nosewheels retract forward, main gear inward into fuselage. Boeing oleo-pneumatic shock-absorbers. B.F. Goodrich nose-gear wheels, tyres and brakes are standard on all models. Goodrich and Bendix are both approved suppliers of main-gear wheels, tyres and brakes for all model 727s. Nosewheels and tyres are size 32 × 11·5 Type VIII. Main-gear wheels and tyres size 49 × 17 Type VII are standard on all models, with size 50 × 21 optional on the 727-200. Increased gross weight versions of 83,000 kg (183,000 lb) and over have 50 × 21 main-wheel tyres as standard equipment.

POWER PLANT: Three Pratt & Whitney JT8D turbofan engines (details under individual model listings) with thrust reversers. Each has individual fuel system fed from integral tanks in wings, but all three tanks are interconnected. Optional fuselage fuel tanks can be installed, displacing forward and/or aft cargo compartment volume. Standard total fuel capacity 30,623 litres (8,090 US gallons). Modular design bladder cell tanks with dual fuel barrier can be installed to contain up to approximately 9,387 litres (2,480 US gallons). Single pressure fuelling point, rated at 2,271 litres (600 US gallons)/min, near wing leading-edge on underside of starboard wing at mid-span. Total usable oil capacity 45·5 litres (12 US gallons).

ACCOMMODATION: Crew of three on flight deck. Basic accommodation for 163 passengers, six abreast. Max capacity 189 passengers. Two galleys forward and/or aft. One toilet forward and two aft. Other layouts to customer's specification. A 'Superjet-look' passenger interior design is standard. The wide-body effect is achieved (without any changes in cross-section dimensions) by lighting and architectural redesign. Retrofit kits for the 'Superjet-look' are offered. Entry via hydraulically-operated integral aft stairway under centre engine and door at front on port side with optional Weber Aircraft electrically-operated airstairs. Two Type III emergency exits in mid-cabin on each side and aft service door on each side. The starboard forward service door is opposite the port forward passenger door. Two heated and pressurised baggage and freight compartments under floor, forward and aft of main landing gear bay. Each compartment has one outward-opening cargo door; a second cargo door is optional for the aft compartment.

SYSTEMS: AiResearch air-conditioning and pressurisation system, using engine bleed air combined with air-cycle refrigeration. Pressure differential 0·59 bars (8·6 lb/sq in). Three independent 207 bar (3,000 lb/sq in) hydraulic systems, utilising Boeing Material Specification BMS 3-11 hydraulic fluid, provide power for flying controls, landing gear and aft airstairs. Electrical system includes three 40kVA 400Hz constant-frequency AC generators, three 50A transformer-rectifier units, one 22Ah battery. AiResearch APU provides electrical power and compressed air for engine starting, air-conditioning and electrical systems on ground.

ELECTRONICS AND EQUIPMENT: Standard equipment includes dual Collins 618M-3 VHF com installations, Motorola NA-134D2 Selcal, flight and service attendants' interphone, Sundstrand passenger address system, ARINC 542 flight recorder and remote encoder, Collins 642C1 voice recorder, dual Collins 51RV-2B VHF nav systems, Collins 51Y-7 ADF, dual King KDM-7000 DME, dual Bendix TRA-63A ATC, Collins 51Z-4 marker beacon, Bendix RDR-1E X-band weather radar, Sperry SP-50 single-channel Mod B1k V autopilot, dual yaw dampers, dual vertical gyros, dual Sperry C-9D compasses, Bendix ALA-51A radio altimeter, instrument comparison and warning system, ARINC 545 air data system, variable instrument switching, dual FD-108 flight directors with glideslope gain programming, and Sundstrand ground proximity warning system. Optional equipment includes dual Bendix ARINC 540 Doppler, Collins FD-110 or Bendix 2-15 flight directors, Sperry C-11 compass system, HF com, Bendix VHF, third VHF, RCA weather radar, dual weather radar systems, Bendix DFA-73A-1 ADF, dual ADF, autothrottles, speed command, dual autopilot channels, roll monitor and flare coupler.

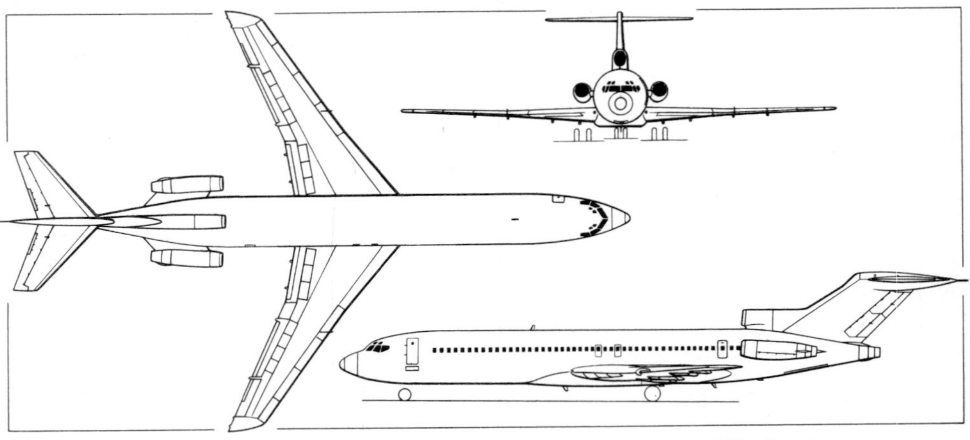

Boeing 727-200 three-turbofan short/medium-range transport *(Pilot Press)*

DIMENSIONS, EXTERNAL:

Wing span	32·92 m (108 ft 0 in)
Wing chord at root	7·70 m (25 ft 3 in)
Wing chord at tip	2·34 m (7 ft 8 in)
Wing aspect ratio	7·2
Length overall	46·69 m (153 ft 2 in)
Length of fuselage	41·51 m (136 ft 2 in)
Height overall	10·36 m (34 ft 0 in)
Tailplane span	10·90 m (35 ft 9 in)
Wheel track	5·72 m (18 ft 9 in)
Wheelbase	19·28 m (63 ft 3 in)
Passenger door (ventral):	
Height	1·93 m (6 ft 4 in)
Width	0·81 m (2 ft 8 in)
Passenger door (fwd):	
Height	1·83 m (6 ft 0 in)
Width	0·86 m (2 ft 10 in)
Height to sill	2·67 m (8 ft 9 in)
Service door (each):	
Height	1·52 m (5 ft 0 in)
Width	0·76 m (2 ft 6 in)
Baggage hold door (fwd):	
Height	1·07 m (3 ft 6 in)
Width	1·37 m (4 ft 6 in)
Baggage hold door (aft):	
Height	1·12 m (3 ft 8 in)
Width	1·37 m (4 ft 6 in)

DIMENSIONS, INTERNAL:

Cabin (aft of flight deck to rear pressure bulkhead):	
Length	28·24 m (92 ft 8 in)
Max width	3·55 m (11 ft 8 in)
Max height	2·11 m (6 ft 11 in)
Floor area	91·05 m² (980 sq ft)
Volume	188·4 m³ (6,652 cu ft)
Baggage hold (fwd)	20·1 m³ (710 cu ft)
Baggage hold (aft):	
standard	23·1 m³ (815 cu ft)
with optional 2nd door	21·1 m³ (745 cu ft)

AREAS:

Wings, gross	157·9 m² (1,700 sq ft)
Ailerons (total)	5·30 m² (57 sq ft)
Trailing-edge flaps, retracted (total)	26·10 m² (281 sq ft)
Trailing-edge flaps, extended (total)	36·04 m² (388 sq ft)
Flight spoilers (total)	7·41 m² (79·8 sq ft)
Fin	33·07 m² (356 sq ft)
Rudder, incl tabs	6·13 m² (66 sq ft)
Tailplane	34·93 m² (376 sq ft)
Elevators, incl tabs	8·83 m² (95 sq ft)

WEIGHTS AND LOADINGS (A: AUW of 78,470 kg (173,000 lb), B: brake release weight of 83,820 kg (184,800 lb), C: brake release weight of 86,405 kg (190,500 lb), D: brake release weight of 95,027 kg (209,500 lb)):

Operating weight empty (basic specification):	
727-200 (A)	44,565 kg (98,250 lb)
727-200 (B)	45,086 kg (99,398 lb)
727-200 (C)	46,400 kg (102,300 lb)
Operating weight empty (typical airline):	
727-200 (A, B, C)	45,360 kg (100,000 lb)
Max payload (structural, based on airline operating weight empty):	
727-200 (A, B)	16,329 kg (36,000 lb) or 17,236 kg (38,000 lb)
727-200 (C)	18,915 kg (41,700 lb)
Max T-O weight:	
727-200 (A)	78,015 kg (172,000 lb)
727-200 (B)	83,820 kg (184,800 lb)
727-200 (C)	86,405 kg (190,500 lb)
727-200 (D)	95,027 kg (209,500 lb)
Max ramp weight:	
727-200 (A)	78,470 kg (173,000 lb)
727-200 (B)	84,275 kg (185,800 lb)
727-200 (C)	86,635 kg (191,000 lb)
727-200 (D)	95,254 kg (210,000 lb)
Max zero-fuel weight:	
727-200 (A)	61,690 kg (136,000 lb) or 62,595 kg (138,000 lb)
727-200 (B)	62,595 kg (138,000 lb)
727-200 (C)	63,500 kg (140,000 lb)
727-200 (D)	65,315 kg (144,000 lb)
Max landing weight:	
727-200 (A)	68,035 kg (150,000 lb) or 70,080 kg (154,500 lb) or 72,575 kg (160,000 lb)

727-200 (B, C) 70,080 kg (154,500 lb) or 72,575 kg (160,000 lb)
727-200 (D) 72,575 kg (160,000 lb)
Max wing loading:
727-200 (A) 494·1 kg/m² (101·2 lb/sq ft)
727-200 (B) 530·7 kg/m² (108·7 lb/sq ft)
727-200 (C) 544·4 kg/m² (111·5 lb/sq ft)
727-200 (D) 595·7 kg/m² (122 lb/sq ft)
Max power loading:
727-200 (A) 367 kg/kN (3·9 lb/lb st)
727-200 (B) 392 kg/kN (4·2 lb/lb st)
727-200 (C) 404·5 kg/kN (4·1 lb/lb st)
727-200 (D) 445 kg/kN (4·4 lb/lb st)

PERFORMANCE (727-200 (B) at brake release weight of 83,820 kg (184,800 lb), 727-200 (C) at brake release weight of 86,405 kg (190,500 lb), and 727-200 (D) at brake release weight of 95,027 kg (209,500 lb), except where indicated):
Max operating speed Mach 0·90
Max level speed:
727-200 (B) at 6,585 m (21,600 ft) 549 knots (1,017 km/h; 632 mph)
727-200 (C, D) at 6,250 m (20,500 ft) 539 knots (999 km/h; 621 mph)
Max cruising speed:
727-200 (B) at 6,705 m (22,000 ft) 514 knots (953 km/h; 592 mph)
727-200 (C,D) at 7,530 m (24,700 ft) 520 knots (964 km/h; 599 mph)
Econ cruising speed at 9,145 m (30,000 ft) 495 knots (917 km/h; 570 mph)
Stalling speed at S/L, flaps up:
727-200 (B) 171 knots (317 km/h; 197 mph)
Stalling speed at S/L, flaps down:
at 72,575 kg (160,000 lb) 106 knots (197 km/h; 122 mph)
Max rate of climb at S/L:
727-200 (B) 762 m (2,500 ft)/min
727-200 (C) 793 m (2,600 ft)/min
Service ceiling:
727-200 (B) 10,060 m (33,000 ft)
727-200 (C) 10,210 m (33,500 ft)
Min ground turning radius (Advanced 727-200) 24·49 m (80 ft 4 in)
Runway LCN (Advanced 727-200) at max weight of 86,635 kg (191,000 lb), optimum tyre pressure and 0·51 m (20 in) flexible pavement:
50 × 21 tyres 74
T-O run:
727-200 (B) 2,515 m (8,250 ft)
727-200 (C) 2,301 m (7,550 ft)
CAR T-O distance to 10·7 m (35 ft):
727-200 (B) 2,847 m (9,340 ft)
727-200 (C) 2,591 m (8,500 ft)
CAR landing distance from 15 m (50 ft):
at 72,575 kg (160,000 lb) 1,430 m (4,690 ft)
Landing run:
at 72,575 kg (160,000 lb) 853 m (2,800 ft)
Range at long-range cruising speed, with fuel and load as specified, ATA domestic reserves:
727-200 (B) with 33,878 litres (8,950 US gallons) fuel, brake release weight of 83,550 kg (184,200 lb) and payload of 11,521 kg (25,400 lb) 2,300 nm (4,260 km; 2,645 miles)
727-200 (C) with 37,000 litres (9,775 US gallons) fuel, brake release weight of 86,405 kg (190,500 lb) and payload of 11,974 kg (26,400 lb) 2,475 nm (4,585 km; 2,850 miles)
Range with max payload, at long-range cruising speed, with load specified, ATA domestic reserves:
727-200 (B) at brake release weight of 83,550 kg (184,200 lb) 1,450 nm (2,685 km; 1,670 miles)
727-200 (C) at brake release weight of 86,405 kg (190,500 lb) 1,605 nm (2,970 km; 1,845 miles)
727-200 (D) at brake release weight of 95,027 kg (209,500 lb) approx 2,500 nm (4,635 km; 2,880 miles)

OPERATIONAL NOISE CHARACTERISTICS (Advanced 727-200, with JT8D-15 engines, FAR Pt 36):
T-O noise level at brake release weight of 86,405 kg (190,500 lb) 100 EPNdB
Approach noise level at 70,080 kg (154,500 lb) landing weight and 30° flap 100·4 EPNdB
Sideline noise level 102·2 EPNdB

BOEING MODEL 737

US Air Force designation: T-43A

The decision to build this short-range transport was announced by Boeing on 19 February 1965. Simultaneously, a first order for 21 aircraft was placed by Lufthansa.

The original Model 737 was designed to utilise many components and assemblies already in production for the Boeing 727. Design began on 11 May 1964, and the first Model 737 flew on 9 April 1967. Deliveries began before the end of 1967, following FAA certification on 15 December. Sales of the 737 totalled 489 on 31 August 1976, of which 469 had been delivered. These totals include 19 Model 737-200s modified as T-43A navigation trainers for the USAF (see 1975-76 *Jane's*).

Details of the early production versions of the Model 737, and of subsequent design development, can be found in the 1974-75 *Jane's*. Versions currently available are as follows:

Advanced 737-200. Current standard model, with max ramp weight of 52,605 kg (116,000 lb) and max T-O weight of 52,390 kg (115,500 lb), JT8D-9A engines as standard (JT8D-15 or JT8D-17 engines optional) and basic fuel capacity of 19,547 litres (5,164 US gallons). It accommodates 115 passengers and baggage, with 86 cm (34 in) pitch seating, or up to 130 passengers in 74 cm (29 in) pitch seating with no reduction in cabin facilities. A gross weight option with a max ramp weight of 53,295 kg (117,500 lb) and max T-O weight of 53,070 kg (117,000 lb) is available.

Avanced 737-200C/QC. Standard convertible passenger/cargo model with strengthened fuselage and floor, and a large two-position upper-deck cargo door size 2·18 m × 3·40 m (7 ft 2 in × 11 ft 2 in). The quick-change (QC) feature allows more rapid conversion by using palletised passenger seating and other special interior furnishings. A gross weight option with a max ramp weight of 53,295 kg (117,500 lb) and max T-O weight of 53,070 kg (117,000 lb) is available.

Advanced 737-200 Business Jet. Same as standard Advanced 737-200, except interiors are adapted to special business and executive luxury requirements. Additional fuel capacity offered by installation of fuel cells in lower cargo compartments. With max fuel this model can carry 15 passengers up to 3,350 nm (6,207 km; 3,857 miles).

Advanced 737-200 Long Range. Higher gross weight version of the Advanced 737-200 for longer-range use. Maximum T-O weight is 58,740 kg (129,500 lb) with JT8D-15, -17 or -17R engines and a fuel capacity of 23,598 litres (6,234 US gallons). The additional capacity for a 600 nm (1,112 km; 691 mile) increase in range capability is provided by a 4,051 litre (1,070 US gallon) fuel tank installed in the aft lower cargo compartment. Aircraft is identical to the current Advanced 737-200 except for the auxiliary fuel tank, new wheels, tyres and brakes, and strengthened landing gear and wing structure. Sectors of more than 2,400 nm (4,446 km; 2,763 miles) can be served with a 130-passenger payload and typical fuel reserves.

An FAA-certificated kit is available which enables the Model 737 to operate from unpaved or gravel runways. The kit includes a vortex dissipator for each engine, consisting of a short hollow boom that protrudes from under each engine's forward edge. The boom is capped by a plug with downward-facing orifices. Pressurised engine bleed air forced through these orifices destroys any ground-level vortex and prevents small pieces of gravel being ingested by the engines. Other items include a gravel deflection 'ski' on the nosewheel, deflectors between the main landing gear wheels, protective shields over hydraulic tubing and speed brake cable on the main gear strut, glassfibre reinforcement of lower inboard flap surfaces, application of Teflon-base paint to fuselage and wing undersurfaces and provision of more robust DME, ATC and VHF antennae.

TYPE: Twin-turbofan short-range transport.

WINGS: Cantilever low-wing monoplane. Special Boeing wing sections. Average thickness/chord ratio 12·89%. Dihedral 6°. Incidence 1° at root. Sweepback at quarter-chord 25°. Aluminium alloy dual-path fail-safe two-spar structure. Ailerons of aluminium honeycomb construction. Boeing-developed triple-slotted trailing-edge flaps, all of aluminium with trailing-edges of aluminium honeycomb. Aluminium alloy Krueger flaps on leading-edge, inboard of nacelles. Three leading-edge slats of aluminium alloy with aluminium honeycomb trailing-edge on each wing from engine to wingtip. Two-section aluminium honeycomb flight spoilers on each outer wing serve both as airbrakes in the air and for lateral control, in association with ailerons. Two-section aluminium honeycomb ground spoilers on each wing, inboard of engine, are used only during landing. Ailerons are hydraulically powered by two hydraulic systems with manual reversion. Trailing-edge flaps are hydraulically powered, with electrical backup. Leading-edge slats and Krueger flaps are symmetrically powered by the two hydraulic systems. Flight spoilers are symmetrically powered by the two main individual hydraulic systems. Engine bleed air for anti-icing supplied to engine nose cowls and all wing leading-edge slats.

FUSELAGE: Aluminium alloy semi-monocoque fail-safe structure.

TAIL UNIT: Cantilever aluminium alloy multi-spar structure. Variable-incidence tailplane. Elevator has dual hydraulic power, with manual reversion. Rudder is powered by a dual actuator from two main hydraulic systems, with a standby hydraulic actuator and system. Tailplane trim has dual electric drive motors, with manual backup. Elevator control tabs for manual reversion are locked out during hydraulic actuation.

LANDING GEAR: Hydraulically-retractable tricycle type, with free-fall extension. Nosewheels retract forward, main units inward. No main gear doors: wheels form wheel-well seal. Twin wheels on each main and nose unit. Boeing oleo-pneumatic shock-absorbers. Main wheels and tyres size 40 × 14 (low-pressure 40 × 18-17 tyres, or 40 × 14-21 cantilever tyres with heavy-duty wheel brakes, are available optionally). Nosewheels and tyres size 24 × 7·7 (low-pressure 24·5 × 8·5 tyres available optionally). Bendix multi-disc brakes. Hydro-Aire Mk III anti-skid units and automatic brakes standard.

POWER PLANT: Two Pratt & Whitney JT8D turbofan engines (details under individual model listings), in underwing pods. High-performance target-type thrust reversers installed on all aircraft delivered since February 1969, in place of a thrust reverser of earlier design. Advanced models have standard fuel capacity of up to 19,547 litres (5,164 US gallons), with integral fuel cells in wing centre-section as well as two integral wing tanks. Long-range version has auxiliary fuel tank in aft lower cargo compartment, giving max fuel capacity of 23,598 litres (6,234 US gallons). Single-point pressure refuelling through leading-edge of starboard wing. Fuelling rate 1,135 litres (300 US gallons)/min. Auxiliary overwing fuelling points. Total oil capacity 41·5 litres (11 US gallons).

ACCOMMODATION: Crew of two side by side on flight deck. Details of passenger accommodation given under individual model descriptions. Passenger versions are equipped with forward airstair; an aft airstair is optional. Convertible passenger/cargo versions have the aft airstair as standard and forward airstair optional. One plug-type door at each corner of cabin, with passenger doors on port side and service doors on starboard side. Overwing escape hatches on each side. Basic passenger cabin has one lavatory and one galley at each end. Provision for a large variety of interior arrangements. Freight holds forward and aft of wing, under floor.

SYSTEMS: Air-conditioning and pressurisation system utilises engine bleed air. Max differential 0·52 bars (7·5 lb/sq in). Two independent hydraulic systems, using fire-resistant hydraulic fluid, for flying controls, flaps, slats, landing gear, nosewheel steering and brakes; pressure 207 bars (3,000 lb/sq in). No pneumatic system. Electrical supply provided by engine-driven generators. AiResearch APU for air supply and electrical power in flight and on the ground, as well as engine starting.

ELECTRONICS AND EQUIPMENT: Equipment to satisfy FAA Category II low weather minimum criteria is standard. Autopilot, specially designed for ILS localiser and glideslope control, with control wheel steering.

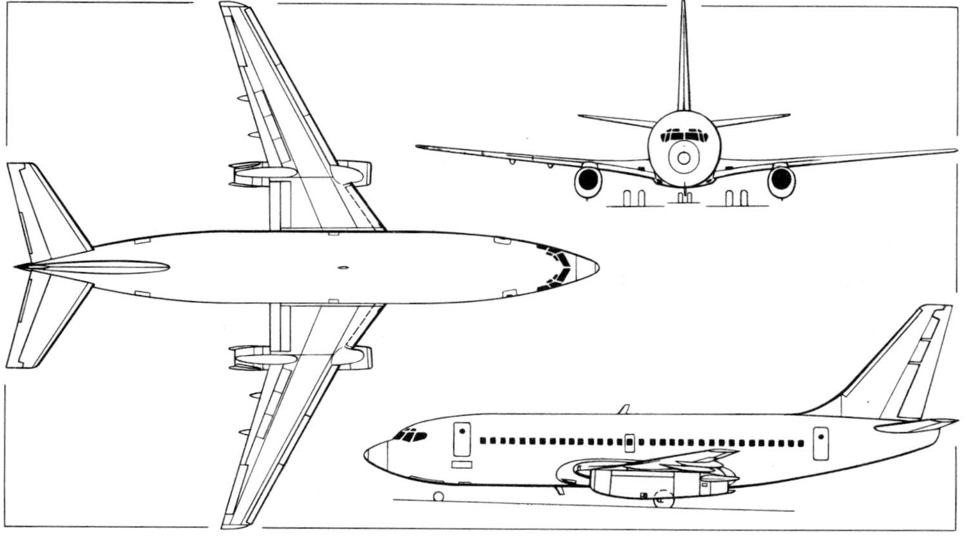

Boeing 737-200 twin-turbofan short-range transport *(Pilot Press)*

Boeing Model 737-200C (two Pratt & Whitney JT8D-17 turbofan engines) in the insignia of Air Nauru

DIMENSIONS, EXTERNAL:
Wing span 28·35 m (93 ft 0 in)
Wing chord at root 4·71 m (15 ft 5·6 in)
Wing chord at tip 1·60 m (5 ft 3 in)
Wing aspect ratio 8·83
Length overall 30·48 m (100 ft 0 in)
Length of fuselage 29·54 m (96 ft 11 in)
Height overall 11·28 m (37 ft 0 in)
Tailplane span 10·97 m (36 ft 0 in)
Wheel track 5·23 m (17 ft 2 in)
Wheelbase 11·38 m (37 ft 4 in)
Main passenger door (port, front):
Height 1·83 m (6 ft 0 in)
Width 0·86 m (2 ft 10 in)
Height to sill 2·62 m (8 ft 7 in)
Passenger door (port, rear):
Height 1·83 m (6 ft 0 in)
Width 0·76 m (2 ft 6 in)
Width with airstair 0·86 m (2 ft 10 in)
Height to sill 2·72 m (8 ft 11 in)
Galley service door (stbd, front):
Height 1·65 m (5 ft 5 in)
Width 0·76 m (2 ft 6 in)
Height to sill 2·62 m (8 ft 7 in)
Service door (stbd, rear):
Height 1·65 m (5 ft 5 in)
Width 0·76 m (2 ft 6 in)
Height to sill 2·72 m (8 ft 11 in)
Freight hold door (stbd, fwd):
Height 1·30 m (4 ft 3 in)
Width 1·22 m (4 ft 0 in)
Height to sill 1·30 m (4 ft 3 in)
Freight hold door (stbd, rear):
Height 1·22 m (4 ft 0 in)
Width 1·22 m (4 ft 0 in)
Height to sill 1·45 m (4 ft 9 in)

DIMENSIONS, INTERNAL:
Cabin, incl galley and toilet:
Length 20·88 m (68 ft 6 in)
Max width 3·52 m (11 ft 6½ in)
Max height 2·18 m (7 ft 2 in)
Floor area 63·8 m² (687 sq ft)
Volume 131·28 m³ (4,636 cu ft)
Freight hold (fwd) volume 10·48 m³ (370 cu ft)
Freight hold (rear) volume 14·30 m³ (505 cu ft)

AREA:
Wings, gross 91·05 m² (980 sq ft)

WEIGHTS AND LOADINGS (standard aircraft at brake release weight of 52,390 kg; 115,500 lb except where indicated):
Operating weight empty:
737-200 27,397 kg (60,400 lb)
737-200C all passenger 28,803 kg (63,500 lb)
737-200C all cargo 27,215 kg (60,000 lb)
737-200QC all passenger 27,691 kg (61,050 lb)
737-200QC all cargo 29,030 kg (64,000 lb)
Max payload:
737-200 15,694 kg (34,600 lb)
737-200C all passenger 14,288 kg (31,500 lb)
737-200C all cargo 15,875 kg (35,000 lb)
737-200QC all passenger 12,972 kg (28,600 lb)
737-200QC all cargo 15,399 kg (33,950 lb)
737 Business Jet 2,267 kg (5,000 lb)
Max T-O weight:
All models 52,390 kg (115,500 lb)
Optional 53,070 kg (117,000 lb)
Max ramp weight:
All models 52,615 kg (116,000 lb)
Optional 53,297 kg (117,500 lb)
Max zero-fuel weight:
All models 43,091 kg (95,000 lb)
Optional for 200C 43,771 kg (96,500 lb)
Max landing weight:
All models 46,720 kg (103,000 lb)
Optional 47,627 kg (105,000 lb)
Max wing loading:
All models 582·96 kg/m² (119·4 lb/sq ft)
Max power loading (JT8D-17):
All models 368 kg/kN (3·61 lb/lb st)

WEIGHTS AND LOADINGS (at brake release weight of 58,740 kg; 129,500 lb):
Operating weight empty 29,528 kg (65,100 lb)
Max payload 13,562 kg (29,900 lb)
Max T-O weight 58,740 kg (129,500 lb)
Max ramp weight 58,967 kg (130,000 lb)
Max zero-fuel weight 43,091 kg (95,000 lb)
Max landing weight 47,627 kg (105,000 lb)
Max wing loading 644·5 kg/m² (132 lb/sq ft)
Max power loading (JT8D-17) 408 kg/kN (4 lb/lb st)

PERFORMANCE (ISA, with JT8D-9 engines):
Max operating speed, all models Mach 0·84
Max level speed, all models, at 7,165 m (23,500 ft) 509 knots (943 km/h; 586 mph)
Max cruising speed, 737-200 at an average cruise weight of 40,823 kg (90,000 lb) at 6,890 m (22,600 ft) 500 knots (927 km/h; 576 mph)
Econ cruising speed at 9,145 m (30,000 ft) Mach 0·73
Stalling speed, flaps down, at max landing weight 99 knots (184 km/h; 114 mph)
Rate of climb at S/L, all models, at 45,355 kg (100,000 lb) AUW 1,280 m (4,200 ft)/min
Runway LCN (Advanced 737-200 at max taxi weight of 52,615 kg; 116,000 lb, optimum tyre pressure and 20 in flexible pavement):
40 × 14-16 tyres 51
40 × 14-21 tyres 51
40 × 18-17 tyres 36
FAR T-O distance to 10·7 m (35 ft), 737-200 at 49,435 kg (109,000 lb) AUW:
JT8D-9 engines 1,935 m (6,350 ft)
JT8D-17 engines 1,524 m (5,000 ft)
FAR landing distance from 15 m (50 ft), 737-200 at max landing weight 1,250 m (4,100 ft)
Min ground turning radius 17·58 m (57 ft 8 in)
Range with max fuel, cruising at 9,145 m (30,000 ft), including reserves for 174 nm (321 km; 200 miles) flight to alternate airport and 45 min continued cruise, 737-200 at 52,615 kg (116,000 lb) taxi weight with 107 passengers 2,200 nm (4,075 km; 2,530 miles)
Range with max payload, conditions as above, 737-200 with 115 passengers 2,060 nm (3,815 km; 2,370 miles)

OPERATIONAL NOISE CHARACTERISTICS (Advanced 737-200 with JT8D-9 engines and nacelle acoustic treatment, FAR Pt 36):
T-O noise level at 52,390 kg (115,500 lb) brake release weight 95·3 EPNdB
Sideline noise level at 52,390 kg (115,500 lb) brake release weight 100·6 EPNdB
Approach noise level at 46,720 kg (103,000 lb) max landing weight 101·1 EPNdB

BOEING MODEL 747

USAF designation: E-4

First details of this very large commercial transport were announced on 13 April 1966, simultaneously with the news that Pan American World Airways had placed a $525 million contract for 25 Boeing 747s, including spares.

There was no prototype. The first 747 made its first flight on 9 February 1969 and FAA certification was granted on 30 December 1969. The first 747 to be delivered was received by Pan American in late 1969, and this company inaugurated commercial service with the type on its New York/London route on 22 January 1970. Orders for differing versions of the 747 totalled 310 by 31 August 1976, by which date 288 had been delivered. By 1 April 1976 more than 111 million passengers had been carried in 747s, flight hours had exceeded 3·4 million, and the airliners had flown more than 1,582 million nm (2,930 million km; 1,821 million miles).

Versions of the Boeing 747 are available as follows:

747-100. Two versions, each of which seats up to 500 passengers. Typical configuration accommodates 48 first class and 337 tourist passengers. Versions with max taxi weights of 323,400 kg (713,000 lb) and 334,750 kg (738,000 lb) are available.

747SP. Lighter-weight derivative of the 747-100, designed for longer range/lower density routes. Described separately.

747SR. This short-range version of the 747-100 embodies structural changes required for high take-off and landing cycles. The purchase of four 747SRs by Japan Air Lines was announced on 30 October 1972, and these aircraft have max taxi weights of 273,515 kg (603,000 lb) and 237,225 kg (523,000 lb). The first delivery of this version was made during September 1973.

747-200B. Passenger version, with same accommodation as 747-100. Basic version had max T-O weight of 351,530 kg (775,000 lb) and increased fuel capacity. Available now with 216 kN (48,570 lb st) JT9D-7A engines and max T-O weight of 356,070 kg (785,000 lb); 213·5 kN (48,000 lb st) dry or 225·5 kN (50,000 lb st) wet JT9D-7F, 235·75 kN (53,000 lb st) JT9D-70A, and 233·5 kN (52,500 lb st) General Electric CF6-50E engines and max T-O weight of 362,870 kg (800,000 lb).

747-200C. Version of 747-200B which can be converted from all-passenger to all-cargo, or a combination of both. The first 747-200C was delivered to World Airways in April 1973. Max T-O weight of 362,870 kg (800,000 lb) with JT9D-7F engines, or 371,945 kg (820,000 lb) with JT9D-70A or CF6-50E engines.

747-200F. Freighter version, capable of delivering 90,720 kg (200,000 lb) of palletised cargo over a range of 3,744 nm (6,940 km; 4,312 miles). Certification of the first 747-200F, which is described separately, was awarded by the FAA on 7 March 1972. T-O weights as detailed for 747-200C.

747-200B (RB.211). Version with 223 kN (50,100 lb st) Rolls-Royce RB.211-524B turbofans and 362,870 kg (800,000 lb) gross weight. First ordered by British Airways in June 1975; deliveries due in 1977. Wings stressed for eventual operation, when 236 kN (53,000 lb st) version of RB.211 becomes available, at a max T-O weight of 371,945 kg (820,000 lb).

The Advanced Airborne Command Post version of the 747, developed for the USAF as the Boeing **E-4**, is described separately under the Boeing Aerospace heading.

Boeing Model 747-206B in the insignia of KLM Royal Dutch Airlines (four General Electric CF6-50E turbofan engines)

Orders announced by 27 August 1976 for various versions of the Model 747 were as follows:

Customer	Number	Model
Aerolineas Argentinas	1	747-287B
Air Canada	5	747-133
Air Canada	1	747-233B
Air France	16	747-128
Air France (GE engines)	1	747-228C
Air France	2	747-228F
Air-India	5	747-237B
Alitalia	2	747-143
Alitalia	3	747-243B
American Airlines	16	747-123
Braniff	1	747-127
British Airways	18	747-136
British Airways (Rolls-Royce engines)	6	747-236B
China Air	1	747SP
Condor	2	747-230B
Continental	4	747-124
CP Air	4	747-217B
Delta Air Lines	5	747-132
Eastern Air Lines	4	747-131
El Al	3	747-258B
El Al	1	747-258C
Iberia	2	747-156
Iberia	1	747-256B
Iran Air	2	747-286B
Iran Air	3	747SP-86
Iraqi Airways	2	747-270C
Irish International	2	747-148
Japan Air Lines	8	747-146
Japan Air Lines	12	747-246B
Japan Air Lines	7	747SR-46
Japan Air Lines	1	747-246F
KLM	9	747-206B
Korean Air Lines	2	747-2B5B
Lufthansa	3	747-130
Lufthansa	2	747-230B
Lufthansa	1	747-230F
Lufthansa (GE engines)	2	747-230B
Middle East Airlines	3	747-2B4B
National Airlines	2	747-135
Northwest Orient	10	747-151
Northwest Orient	5	747-251B
Northwest Orient	4	747-251F
Olympic Airways	2	747-284B
Pan American	33	747-121
Pan American	5	747SP-21
Qantas	15	747-238B
Sabena	2	747-129
SAS	2	747-283B
Seaboard World	2	747-245F
Singapore	5	747-212B
South African Airways	5	747-244B
South African Airways	6	747SP-44
Swissair	2	747-257B
Syrian Arab	2	747SP-94
TAP Portugal	4	747-282B
TWA	15	747-131
United Air Lines	18	747-122
USAF	4	747-E4A
Wardair	1	747-1D1
World Airways	3	747-273C

The following details apply specifically to the basic Model 747 passenger airliner:

TYPE: Four-turbofan heavy commercial transport.

WINGS: Cantilever low-wing monoplane. Special Boeing wing sections. Thickness/chord ratio 13·44% inboard, 7·8% at mid-span, 8% outboard. Dihedral 7°. Incidence 2°. Sweepback 37° 30′ at quarter-chord. Aluminium alloy dual-path fail-safe structure. Low-speed outboard ailerons; high-speed inboard ailerons. Triple-slotted trailing-edge flaps. Six aluminium honeycomb spoilers on each wing, comprising four flight spoilers outboard and two ground spoilers inboard. Ten variable-camber leading-edge flaps outboard and three-section Krueger flaps inboard on each wing leading-edge. All controls fully powered.

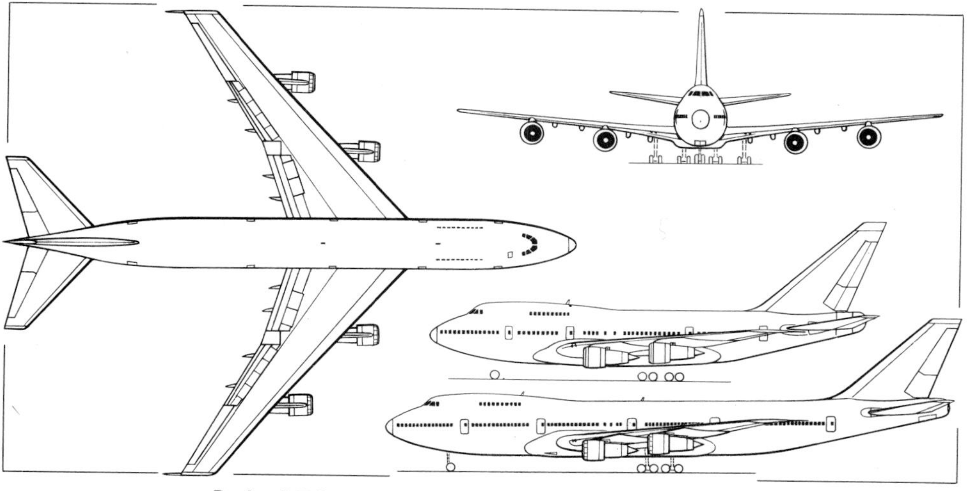

Boeing 747 four-turbofan heavy transport aircraft *(Pilot Press)*

FUSELAGE: Conventional semi-monocoque structure, consisting of aluminium alloy skin, longitudinal stiffeners and circumferential frames. Structure is of fail-safe design, utilising riveting, bolting and structural bonding.

TAIL UNIT: Cantilever aluminium alloy dual-path fail-safe structure. Variable-incidence tailplane. No trim tabs. All controls fully powered.

LANDING GEAR: Hydraulically-retractable tricycle type. Twin-wheel nose unit retracts forward. Main gear comprises four four-wheel bogies: two, mounted side by side under fuselage at wing trailing-edge, retract forward; two mounted under wings retract inward. Cleveland Pneumatic oleo-pneumatic shock-absorbers. All 18 wheels and tubeless tyres of model 747-100 are size 46 × 16 Type VII. Tyre pressure: main wheels 14·07 bars (204 lb/sq in), nosewheels 11·38 bars (165 lb/sq in). Main wheels and tyres size 49 × 17 on 747-200B model, pressure 12·76 bars (185 lb/sq in). Disc brakes on all main wheels, with individually-controlled anti-skid units.

POWER PLANT: Four Pratt & Whitney JT9D turbofan engines in pods pylon-mounted on wing leading-edges. All current versions of the 747 are structurally capable of accepting JT9D-3, JT9D-3W, JT9D-7, JT9D-7A, JT9D-7AW, JT9D-7F or JT9D-7W engines and have QEC capability. The JT9D-70A can be installed on the -200 models. Corresponding ratings of the above engines are: 193·5; 200; 202·5; 209; 216; 213·5 dry, 222·5 wet; 209; and 235·75 kN (43,500; 45,000; 45,500; 46,950; 48,570; 48,000 dry, 50,000 wet; 47,000; and 53,000 lb st). The General Electric CF6-50E turbofan engine, rated at 233·5 kN (52,500 lb st), and Rolls-Royce RB.211-524B turbofan engine, rated at 223 kN (50,100 lb st), can be installed in the 747-200 models. Fuel in seven integral tanks. Capacity of centre wing tank varies according to version: 747-100 48,790 litres (12,890 US gallons); 747-200B and 747-200F 63,139 litres (16,680 US gallons). Remaining tanks common to all versions: two inboard main tanks, each 46,560 litres (12,300 US gallons); two outboard main tanks, each 16,731 litres (4,420 US gallons); two outboard reserve tanks, each 1,892 litres (500 US gallons). Total capacity 747-100 178,702 litres (47,210 US gallons); 747-200B and 747-200F 194,680 litres (51,430 US gallons). Refuelling point on each wing between inboard and outboard engines. Total usable oil capacity 19 litres (5 US gallons).

ACCOMMODATION: Normal operating crew of three, on flight deck above level of main deck. Observer station and provision for second observer station are provided. Basic accommodation for 385 passengers, made up of 48 first class, which includes a 16-passenger upper deck lounge, and 337 economy class. Alternative layouts accommodate 447 economy class passengers in nine-abreast seating or 500 ten-abreast. All versions have two aisles. Five passenger doors on each side, of which two forward of wing on each side are normally used. Freight holds under floor, forward and aft of wing, with doors on starboard side. One door on forward hold, two on rear hold. Aircraft is designed for fully-mechanical loading of baggage and freight. An optional side cargo door is available for both passenger and freighter versions of the Model 747. Installed aft of door 4 on the port side of the fuselage, it provides a clear opening 3·40 m (11 ft 2 in) wide and 3·05 m (10 ft 0 in) in height. This door makes it possible to carry main-deck cargo on passenger versions. Addition of this door to the freighter allows loads up to 3·05 m (10 ft) in height to be accommodated aft of the flight deck, and also makes possible simultaneous nose and side cargo handling.

SYSTEMS: Air-cycle air-conditioning system. Pressure differential 0·61 bars (8·9 lb/sq in). Electrical supply from four aircooled 60kVA generators mounted one on each engine. Two 60kVA generators (supplemental cooling allows 90kVA each) mounted on APU for ground operation and to supply primary electrical power when engine-mounted generators are not operating. Three-phase 400Hz constant-frequency AC generators, 115/200V output. 28V DC power obtained from transformer-rectifier units. 24V 30Ah nickel-cadmium battery for selected ground functions and as in-flight backup. Gas-turbine APU for pneumatic and electrical supplies.

ELECTRONICS AND EQUIPMENT: Standard electronics include two ARINC 566 VHF communications systems, two ARINC 533A HF communications systems, two ARINC 531 Selcal, two ARINC 547 VOR/ILS navigation systems, two ARINC 550 ADF, marker beacon, two ARINC 568 DME, two ARINC 532D ATC, two ARINC 552 low-range radio altimeters, two ARINC 564 weather radar units, two ARINC 561 inertial navigation systems, two heading reference systems, ARINC 412 interphone, passenger address system, passenger entertainment system, ARINC 573 flight recorder, ARINC 557 cockpit voice recorder, integrated electronic flight control system to provide automatic stabilisation, path control and pilot assist functions for category II and III landing conditions, two ARINC 565 central air data systems, stall warning system, central instrument warning system, attitude and navigation instrumentation, and standby attitude indication.

DIMENSIONS, EXTERNAL:

Wing span	59·64 m (195 ft 8 in)
Wing chord at root	16·56 m (54 ft 4 in)
Wing chord at tip	4·06 m (13 ft 4 in)
Wing aspect ratio	6·96
Length overall	70·51 m (231 ft 4 in)
Length of fuselage	68·63 m (225 ft 2 in)

Height overall 19·33 m (63 ft 5 in)
Tailplane span 22·17 m (72 ft 9 in)
Wheel track 11·00 m (36 ft 1 in)
Wheelbase 25·60 m (84 ft 0 in)
Passenger doors (ten, each):
Height 1·93 m (6 ft 4 in)
Width 1·07 m (3 ft 6 in)
Height to sill approx 4·88 m (16 ft 0 in)
Baggage door (front hold):
Height 1·73 m (5 ft 8 in)
Width 2·64 m (8 ft 8 in)
Height to sill approx 2·64 m (8 ft 8 in)
Baggage door (forward door, aft hold):
Height 1·73 m (5 ft 8 in)
Width 2·64 m (8 ft 8 in)
Height to sill approx 2·69 m (8 ft 10 in)
Bulk loading door (rear door on aft hold):
Height 1·22 m (4 ft 0 in)
Width 1·12 m (3 ft 8 in)
Height to sill approx 2·90 m (9 ft 6 in)
Optional cargo door (port):
Height 3·05 m (10 ft 0 in)
Width 3·40 m (11 ft 2 in)

Dimensions, internal:
Cabin, incl toilets and galleys:
Length 57·00 m (187 ft 0 in)
Max width 6·13 m (20 ft 1½ in)
Max height 2·54 m (8 ft 4 in)
Floor area, passenger deck 327·9 m^2 (3,529 sq ft)
Volume, passenger deck 789 m^3 (27,860 cu ft)
Baggage hold (fwd, containerised) volume 78·4 m^3 (2,768 cu ft)
Baggage hold (aft, containerised) volume 68·6 m^3 (2,422 cu ft)
Bulk volume 28·3 m^3 (1,000 cu ft)

Areas:
Wings, reference area 511 m^2 (5,500 sq ft)
Ailerons (total) 20·6 m^2 (222 sq ft)
Trailing-edge flaps (total) 78·7 m^2 (847 sq ft)
Leading-edge flaps (total) 48·1 m^2 (518 sq ft)
Spoilers (total) 30·8 m^2 (331 sq ft)
Fin 77·1 m^2 (830 sq ft)
Rudder 22·9 m^2 (247 sq ft)
Tailplane 136·6 m^2 (1,470 sq ft)
Elevators 32·5 m^2 (350 sq ft)

Weights (The following suffixes are used to denote engine installations: (V) RB.211-524B, (W) JT9D-7A or 7AW, (X) JT9D-7FW, (Y) CF6-50E, (Z) JT9D-70A):
Operating weight, empty:
747-100 (W) 161,885 kg (356,900 lb) or 161,975 kg (357,100 lb)
747SR (W) 156,490 kg (345,000 lb)
747-200B (W) 165,925 kg (365,800 lb)
747-200B (X) 166,015 kg (366,000 lb)
747-200B (Y) 167,785 kg (369,900 lb)
747-200B (Z) 169,600 kg (373,900 lb)
747-200B (V) 170,145 kg (375,100 lb)
747-200C, all passenger:
(W) 171,365 kg (377,800 lb)
(X) 171,460 kg (378,000 lb)
(Y) 173,660 kg (382,850 lb)
(Z) 175,470 kg (386,850 lb)
(V) 176,015 kg (388,050 lb)
747-200C, all cargo:
(W) 160,980 kg (354,900 lb)
(X) 161,070 kg (355,100 lb)
(Y) 163,270 kg (359,950 lb)
(Z) 165,085 kg (363,950 lb)
(V) 165,630 kg (365,150 lb)
Max payload:
747-100 (W) 76,930 kg (169,600 lb) or 76,840 kg (169,400 lb)
747SR (W) 58,965 kg (130,000 lb)
747-200B:
(W) 72,890 kg (160,700 lb)
(X) 72,800 kg (160,500 lb)
(Y) 71,035 kg (156,600 lb)
(Z) 69,220 kg (152,600 lb)
(V) 68,675 kg (151,400 lb)
747-200C, all passenger:
(W) 96,250 kg (212,200 lb)
(X) 96,160 kg (212,000 lb)
(Y) 93,960 kg (207,150 lb)
(Z) 92,145 kg (203,150 lb)
(V) 91,600 kg (201,950 lb)
747-200C, all cargo:
(W) 106,640 kg (235,100 lb)
(X) 106,550 kg (234,900 lb)
(Y) 104,350 kg (230,050 lb)
(Z) 102,535 kg (226,050 lb)
(V) 101,990 kg (224,850 lb)
Max T-O weight:
747-100 (W) 322,050 kg (710,000 lb) or 332,480 kg (733,000 lb)
747SR (W) 235,865 kg (520,000 lb) or 272,155 kg (600,000 lb)
747-200B (W, X) 351,530 kg (775,000 lb)
747-200B (V, Y, Z) 362,870 kg (800,000 lb)
747-200C, all passenger and all cargo (X) 351,530 kg (775,000 lb) or 365,140 kg (805,000 lb)
747-200C, all passenger and all cargo:
(V, Y, Z) 362,870 kg (800,000 lb)
(Y, Z) or 371,945 kg (820,000 lb)
Max ramp weight:
747-100 (W) 323,400 kg (713,000 lb) or 334,750 kg (738,000 lb)
747SR (W) 237,225 kg (523,000 lb) or 273,515 kg (603,000 lb)
747-200B (W, X) 352,895 kg (778,000 lb)
747-200B (V, Y, Z) 366,500 kg (808,000 lb)
747-200C, all passenger and all cargo (V, X) 352,895 kg (778,000 lb) or 366,500 kg (808,000 lb)
747-200C, all passenger and all cargo (Y, Z) 366,500 kg (808,000 lb) or 373,305 kg (823,000 lb)
Max zero-fuel weight:
747-100 (W), 747-200B (V, X, Y, Z) 238,815 kg (526,500 lb)
747SR (W) 215,455 kg (475,000 lb)
747-200C, all passenger and all cargo (V, X, Y, Z) 267,620 kg (590,000 lb)
Max landing weight:
747-100 (W), 747-200B (V, X, Y, Z) 255,825 kg (564,000 lb)
747SR (W) 229,060 kg (505,000 lb)
747-200C, all passenger and all cargo (V, X, Y, Z) 285,760 kg (630,000 lb)

Performance (at max T-O weight except where indicated):
Max level speed:
747-100 at 9,150 m (30,000 ft) at AUW of 272,155 kg (600,000 lb) 517 knots (958 km/h; 595 mph)
747-200B at 9,150 m (30,000 ft) at AUW of 272,155 kg (600,000 lb) 528 knots (978 km/h; 608 mph)
Cruise ceiling, all versions 13,715 m (45,000 ft)
Min ground turning radius 22·86 m (75 ft 0 in)
Runway LCN (A: 323,400 kg; 713,000 lb, B: 334,750 kg; 738,000 lb, C: 352,895 kg; 778,000 lb max taxi weight, on h=0·51 m; 20 in flexible pavement):
A 79
B 81
C 83
Runway LCN (weights as above, on l=1·02 m; 40 in rigid pavement):
A 87
B 88
C 90
FAR T-O distance to 10·7 m (35 ft) at S/L, ISA, 747-100 at T-O weight of 332,480 kg (733,000 lb), 747-200B at 356,070 kg (785,000 lb):
747-100 2,880 m (9,450 ft)
747-200B 3,200 m (10,500 ft)
FAR landing field length, at max landing weight:
747-100, 747-200B 1,880 m (6,170 ft)
Range (long-range cruise, FAR 121.645 reserves):
747-100 at 332,480 kg (733,000 lb), with 385 passengers and baggage 4,930 nm (9,136 km; 5,677 miles)
747-200B at 356,070 kg (785,000 lb), with 385 passengers and baggage 5,400 nm (10,005 km; 6,218 miles)
Ferry range (long-range cruise, FAR 121.645 reserves):
747-200B 6,400 nm (11,860 km; 7,370 miles)

Operational Noise Characteristics (A: JT9D-7A engines at brake release weight (BRW) of 332,480 kg; 733,000 lb. B: JT9D-7A (wet) engines at BRW of 356,070 kg; 785,000 lb, FAR Pt 36):
T-O noise level:
A, B, C 107 EPNdB
Approach noise level:
A 107 EPNdB
B 106 EPNdB
Sideline noise level:
A 99 EPNdB
B 98 EPNdB

BOEING MODEL 747SP

The Boeing Company announced on 3 September 1973 that it intended to proceed 'incrementally' with development of a lower-weight longer-range version of the basic Model 747, for use on lower-density routes. A week later came the news that Pan American had placed an order for 10 747SP (Special Performance) aircraft, with an option on 15 more. Subsequently, the Pan American order was reduced to five, but orders have been received from China Air (1), Iran Air (3), South African Airways (6) and Syrian Arab (2).

Retaining a 90 per cent commonality of components with the standard Model 747, the major change is a reduction in overall length of 14·30 m (47 ft 7 in). Construction of the first production aircraft began in April 1974, with rollout on 19 May 1975, first flight on 4 July 1975, and FAA certification on 4 February 1976.

On 23-24 March 1976, taking off at a gross weight of 323,547 kg (713,300 lb) with 50 passengers, the first 747SP for South African Airways made a delivery flight from Paine Field, Washington, to Cape Town of 8,936 nm (16,560 km; 10,290 miles), a world record for nonstop distance flown by a commercial aircraft. The aircraft landed with fuel remaining for a further 2 hr 27 min flying.

The description of the basic Model 747 applies also to the 747SP, except for the following details:

Wings: As Model 747, except that trailing-edge flaps are of single-slotted variable pivot type, and wing structural materials are of reduced gauge. Large flap track fairings replaced by small link fairings. New wing/body fairings and leading-edge fillets.

Fuselage: As Model 747, except length reduced.

Tail Unit: Similar to 747, but tailplane span increased by 3·05 m (10 ft). Two-segment elevators. Height of fin increased by 1·52 m (5 ft 0 in). Double-hinged rudder.

Landing Gear: As Model 747, except structural weight reduced. Main-wheel tyres size 16-46, pressure 12·76 bars (185 lb/sq in). Nosewheel tyres size 17-49, pressure 15·86 bars (230 lb/sq in). Modified 747-100 steel brakes by Bendix.

Power Plant: Four Pratt & Whitney JT9D-7A or -7F turbofan engines, each of 209 kN (46,950 lb) or 213·5 kN (48,000 lb st) respectively; or four General Electric CF6-50E turbofan engines, each of 233·5 kN (52,500 lb st). Fuel system, fuel capacity and oil capacity as for Model 747-100, except Model 747SP has an additional 5,943 litres (1,570 US gallons) reserve fuel.

Accommodation: Normal operating crew of three on flight deck above level of main deck. Observer station and provision for second observer station are provided. Basic accommodation for 288 passengers on main deck, with 28 first class seats in forward area and ten-abreast seating throughout the major part of the main cabin. Seating for 16/32 passengers in upper-deck first class lounge optional, giving total optional capacity of 305/321 passengers. Max high-density accommodation for 360 passengers. Four doors on each side, two forward and two aft of the wing. Crew door on starboard side giving access to upper deck. Freight holds under floor, forward and aft of wing box, each with one door on starboard side.

Systems, Electronics and Equipment: As for 747.

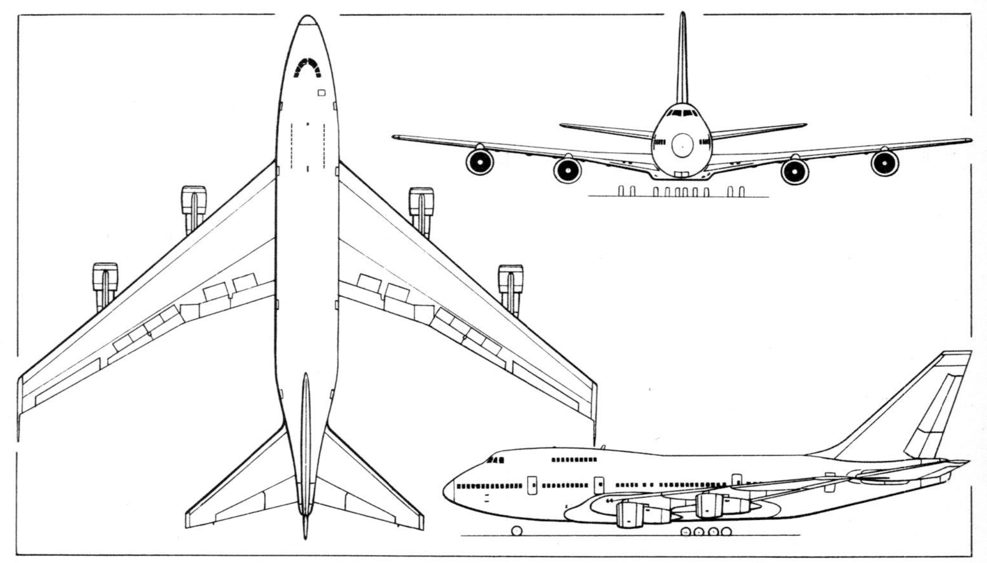

Boeing 747SP short-fuselage long-range version of the 747 *(Pilot Press)*

Boeing 747SP-21 reduced-capacity long-range airliner in the insignia of Pan American

DIMENSIONS, EXTERNAL: As for Model 747 except:
- Length overall: 56·31 m (184 ft 9 in)
- Height overall: 19·94 m (65 ft 5 in)
- Tailplane span: 25·22 m (82 ft 9 in)
- Wheelbase: 20·52 m (67 ft 4 in)

DIMENSIONS, INTERNAL:
- Cabin, incl toilets and galleys:
  - Length: 42·27 m (138 ft 8 in)
  - Max width: 6·13 m (20 ft 1½ in)
  - Max height: 2·54 m (8 ft 4 in)
  - Floor area, passenger deck: 253·2 m² (2,725 sq ft)
  - Volume, passenger deck: 613·34 m³ (21,660 cu ft)
- Baggage hold volume (fwd): 48·99 m³ (1,730 cu ft)
- Baggage hold volume (aft, containerised): 48·99 m³ (1,730 cu ft)
- Bulk compartment volume (aft): 11·33 m³ (400 cu ft)

AREAS: As for Model 747 except:
- Ailerons (total): 20·37 m² (219·3 sq ft)
- Trailing-edge flaps (total): 78·78 m² (848 sq ft)
- Fin: 82·22 m² (885 sq ft)
- Tailplane: 142·51 m² (1,534 sq ft)

WEIGHTS:
- Operating weight empty: 141,935 kg (312,920 lb)
- Max payload: 44,034 kg (97,080 lb)
- Max T-O weight: 299,370 kg (660,000 lb)
- Max ramp weight: 300,730 kg (663,000 lb)
- Max zero-fuel weight: 185,970 kg (410,000 lb)
- Max landing weight: 204,115 kg (450,000 lb)

PERFORMANCE (at max T-O weight, except where indicated):
- Never-exceed speed: Mach 0·92
- Max level speed, AUW of 226,795 kg (500,000 lb) at 9,145 m (30,000 ft): 529 knots (980 km/h; 609 mph)
- Service ceiling: 13,745 m (45,100 ft)
- Min ground turning radius over outer wingtip: 22·25 m (73 ft 0 in)
- Runway LCN (at max ramp weight, h = 0·51 m; 20 in flexible pavement): 73
- Runway LCN (at max ramp weight, l = 1·02 m; 40 in rigid pavement): 74
- FAR T-O distance to 10·7 m (35 ft) at S/L, ISA: 2,165 m (7,100 ft)
- FAR landing field length, at max landing weight: 1,705 m (5,600 ft)
- Range (long-range step cruise, FAR 121.645 reserves) with 305 passengers and baggage: 5,712 nm (10,586 km; 6,578 miles)
- Ferry range (long-range step cruise, FAR 121.645 reserves): 7,253 nm (13,441 km; 8,352 miles)

BOEING MODEL 747-200F FREIGHTER

The Boeing Model 747-200F is a freighter version of the standard Model 747-200, capable of delivering 90,720 kg (200,000 lb) of containerised or palletised cargo over a range of 3,744 nm (6,940 km; 4,312 miles).

Certification of the first 747-200F was awarded on 7 March 1972 and this aircraft was delivered to Lufthansa later in the same month.

To ensure maximum utilisation, the 747-200F has a special loading system that enables two men to handle and stow the maximum load of up to 115,500 kg (254,640 lb) in 30 min. This system comprises rollers, castors, rails and drive wheels which are powered from the aircraft's electrical system, utilising the APU, or from an external power source.

As cargo reaches the nose-door sill area it is steered by the loading manager from a master control station. Loads are propelled by the drive wheels, under control from a series of local control stations, along roller tracks to their assigned positions, where they are locked in place.

An on-board computer-controlled weight and balance system measures, computes and displays all necessary loading data via a series of transducers mounted inside the landing gear axles, providing the operator with very precise CG location and weight readings.

If, as a result of incorrect loading, the CG location is outside a pre-set limit, a visible internal and audible external alarm is actuated. Simultaneously, electrical power to the deck cargo-handling system is isolated, halting the loading operation.

The 747-200F can carry up to 29 containers measuring 3·05 m × 2·44 m × 2·44 m (10 ft long, 8 ft high and 8 ft wide), plus 30 lower-lobe containers, each of 4·90 m³ (173 cu ft) capacity, and 22·65 m³ (800 cu ft) of bulk cargo. The main deck can accommodate ANSI/ISO containers of up to 12·19 m (40 ft) in length, and many combinations of pallets and igloos. The lower hold can accommodate combinations of IATA-A1 or -A2, and ATA LD-1 or -3 half-width containers, full-width or main-deck baggage containers, as well as many combinations of pallets and igloos.

The nose loading door, which is hinged just below the flight deck to allow it to swing forward and upward, gives clear access to the main deck to facilitate the handling of long or large loads.

The description of the Model 747-200B applies also to the Model 747-200F except as follows:

TYPE: Four-turbofan heavy commercial freighter.

FUSELAGE: As for Model 747-200B, except nose cargo loading door with max width of 3·45 m (11 ft 4 in) and max height of 2·49 m (8 ft 2 in), which is hinged at the top and opens forward and upward.

ACCOMMODATION: Normal operating crew of three on flight deck. Nose cargo loading door, hinged at top. Lower lobe cargo doors, on starboard side, one forward and one aft of wing. Bulk compartment cargo door, on starboard side, aft of lower lobe cargo door. Two doors for crew on port side of aircraft. Aircraft is designed for fully-mechanical loading of freight.

DIMENSIONS, EXTERNAL: As for Model 747-200B except:
- Crew doors (two, each):
  - Height: 1·93 m (6 ft 4 in)
  - Width: 1·07 m (3 ft 6 in)
  - Height to sill: approx 4·88 m (16 ft 0 in)
- Nose cargo loading door:
  - Height: 2·49 m (8 ft 2 in)
  - Width at top (min): 2·64 m (8 ft 8 in)
  - Height to sill: approx 4·90 m (16 ft 1 in)

DIMENSIONS, INTERNAL:
- Main cargo deck:
  - Height: 2·54 m (8 ft 4 in)
  - Max width: 5·92 m (19 ft 5 in)
- Lower lobe:
  - Width at floor level: 3·18 m (10 ft 5 in)
- Total cargo volume: 670·83 m³ (23,690 cu ft)

AREAS: As for Model 747-200B

WEIGHTS (The following suffixes are used to denote engine installations: (V) RB.211-524B, (W) JT9D-7A or JT9D-7AW, (X) JT9D-7FW, (Y) CF6-50E, (Z) JT9D-70A):
- Operating weight empty:
  - W: 152,860 kg (337,000 lb)
  - X: 152,950 kg (337,200 lb)
  - Y: 155,150 kg (342,050 lb)
  - Z: 156,965 kg (346,050 lb)
  - V: 157,510 kg (347,250 lb)
- Max T-O weight:
  - X: 351,530 kg (775,000 lb)
  - Y, Z: 362,870 kg (800,000 lb) or 371,945 kg (820,000 lb)
- Max ramp weight:
  - X: 352,895 kg (778,000 lb)
  - Y, Z: 366,500 kg (808,000 lb) or 373,305 kg (823,000 lb)
- Max zero-fuel weight:
  - X, Y, Z: 267,615 kg (590,000 lb)
- Max landing weight:
  - X, Y, Z: 285,760 kg (630,000 lb)

PERFORMANCE (at max T-O weight except where indicated):
- Max level speed at AUW of 272,155 kg (600,000 lb), at

Nose-loading of a Boeing Model 747-246F freighter being demonstrated

Boeing Model 747-251F freighter in the insignia of Northwest Orient

9,145 m (30,000 ft) 528 knots (978 km/h; 608 mph)
Cruise ceiling 13,715 m (45,000 ft)
Min ground turning radius 22·86 m (75 ft 0 in)
FAR T-O distance to 10·7 m (35 ft) at S/L, ISA 3,322 m (10,900 ft)
FAR landing field length, at max landing weight 2,216 m (7,270 ft)
Range, long-range cruise, FAR 121.645 reserves, with 116,962 kg (257,858 lb) payload 2,501 nm (4,630 km; 2,880 miles)
Ferry range with max fuel, long-range cruise, FAR 121.645 reserves 6,903 nm (12,790 km; 7,949 miles)

BOEING 747-123 SPACE SHUTTLE ORBITER CARRIER (NASA 905)

Boeing is modifying the Model 747-123 that NASA acquired from American Airlines in August 1974. The modifications will enable the 747 to carry a Space Shuttle orbiter aircraft 'piggy-back' fashion, as shown in the illustration under the NASA entry in this section. The fairing shown covering the rocket chambers of the orbiter will be fitted only during ferry flights, to reduce drag. The orbiter will also be launched from the 747 during its test programme.

Modification of the 747-123, now designated NASA 905, was scheduled for completion in late 1976.

BOEING AEROSPACE COMPANY

The Boeing Aerospace Company has its headquarters at the company's space centre at Kent, Washington, some 12 miles south of Seattle. It consists of Aerospace Operations, Product Development, Research and Engineering Division, Field Operations and Support Division, Missiles and Space Group, Army Systems Division, Information Systems Division and Military Airplane Development. Responsible for much of Boeing's military, space and diversification efforts, it has a labour force of approximately 17,000. Among its principal current activities are the SRAM missile programme, Minuteman modernisation, advanced surface transportation programmes, military applications of commercial transports, and space projects.

BOEING AWACS

USAF designations: EC-137D and E-3A

The E-3A AWACS (Airborne Warning And Control System) aircraft being developed and produced for USAF service in the late 1970s is equipped with extensive sensing, communications, display and navigational devices.

In concept, an AWACS offers the potential of long-range high- or low-level surveillance of all air vehicles, manned or unmanned, in all weathers and above all kinds of terrain. Its data storage and processing capability would provide real-time assessment of enemy action, and of the status and position of friendly resources. By centralising the co-ordination of complex, diverse and simultaneous air operations, such an aircraft would be able to command and control the total air effort: strike, air superiority, support, airlift, reconnaissance and interdiction.

The primary use of such an aircraft, as deployed by Aerospace Defence Command, will be as a survivable early-warning airborne command and control centre for identification, surveillance and tracking of airborne enemy forces, and for the command and control of NORAD (North American Air Defense) forces. Similar aircraft, operated by Tactical Air Command, will be used as airborne command and control centres for quick-reaction deployment and tactical operations.

Boeing's Aerospace Group was one of two competitors for the AWAC system (the other being McDonnell Douglas), and was awarded an initial contract as prime contractor and systems integrator for the programme on 23 July 1970. Boeing's submission was based on the airframe of the Model 707-320B commercial jet transport. In Phase 1 of the development programme, two of these aircraft, with the prototype designation EC-137D, were modified initially for comparative trials with prototype downward-looking radars designed, respectively, by Hughes Aircraft Company and Westinghouse Electric Corporation.

The first flight by one of these aircraft was made on 9 February 1972. After more than five months of radar test flights, during which each radar accumulated over 290 hours of airborne operating time, Boeing completed its evaluation, and the Westinghouse radar was selected on 5 October 1972. Following successful completion of the radar competition, additional data processing equipment and two tracking displays were installed in the Westinghouse-equipment test aircraft, and a new series of flight tests was conducted to demonstrate the ability of the radar and data processor to detect and maintain continuous tracking of airborne targets. In addition, the capability of the system to maintain several simultaneous tracks was evaluated. These tests also proved successful, and were completed by 6 November 1972.

On 26 January 1973, the USAF announced that, following satisfactory completion of Phase 1, approval had been given for full-scale development of the AWACS aircraft under Phase 2 of the programme. To reduce costs, two major changes were made from the original Phase 2 proposal. The previously planned power plant of eight General Electric TF34-GE-2 turbofan engines was superseded by four Pratt & Whitney TF33-P-7 turbofans, each of 93·4 kN (21,000 lb st); and only three test aircraft were ordered instead of the six originally envisaged.

Phase 2 of the development programme involved systems integration demonstration, and initial operational test and evaluation. Additional subsystems were installed in one of the two existing EC-137D test aircraft, so that it could demonstrate full AWACS capability. This aircraft completed its test programme in mid-1975, after logging more than 1,600 flight hours. It was being reworked in early 1976 for delivery as the sixth production E-3A.

Following successful demonstration of the full AWAC system, it is intended that the three development/operational test aircraft shall be refurbished and will enter the operational inventory. This will be done under Phase 3 which also covers the manufacture of production aircraft, of which 34 are planned to be built. The first two full-scale E-3A development aircraft made their first flights, without AWACS electronics installed, in February and July 1975. The first production contract, for six aircraft, was announced in the Spring of 1975.

On 12 August 1975 one of the pre-production E-3As began a series of aircraft performance tests. These involved flutter, engine performance, flight loads, aerial refuelling and handling qualities, and were scheduled to continue into the Summer of 1976. It was planned to devote approximately 400 flight hours to these tests, the major portion of which were to be flown in the Pacific northwest area. On 31 October 1975 the first E-3A with production electronics installed began engineering test and evaluation, manned by a mixed Boeing and USAF crew. The tests were a preliminary to formal qualification

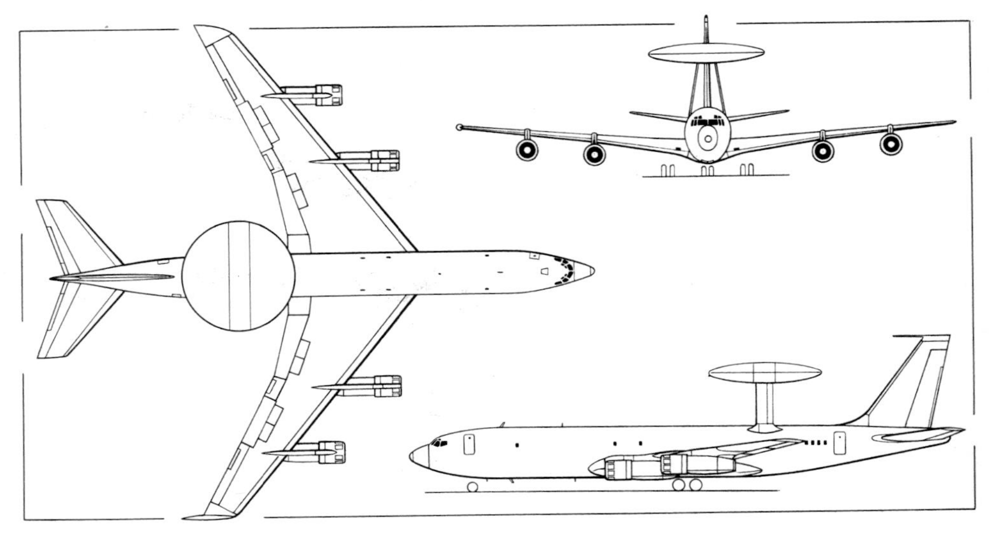

Boeing E-3A airborne warning and control system aircraft *(Pilot Press)*

Pre-production standard Boeing E-3A Airborne Warning and Control System (AWACS) aircraft for the USAF

testing, which was carried out during 1976. The first production delivery to the US Air Force was scheduled for November 1976.

The existing Boeing 707-320 requires relatively minor adaptation to accommodate the AWAC system. External changes include the rotodome assembly, which is mounted on two large struts rooted into the fuselage structure aft of the wing, new engine pylon fairings, specially located windows, doors and hatches, and provisions for in-flight refuelling. Essential antennae can be installed within the wings, fin, tailplane and fuselage, and internal changes require floor reinforcement, provision of crew compartments, and revised cooling and wiring systems.

TYPE: Airborne early-warning and command post aircraft.

WINGS, FUSELAGE, TAIL UNIT AND LANDING GEAR: Basically as Boeing 707-320B, with strengthened fuselage structure and installation of rotodome.

POWER PLANT: Prototypes retained their existing power plants during Phase 1. Pre-production and production aircraft are powered by four Pratt & Whitney TF33-P-7 turbofan engines, redesignated TF33-PW-100/100A in their AWACS-modified configuration. Each rated at 93·4 kN (21,000 lb st), they are mounted in pods beneath the wings.

ACCOMMODATION: Basic operational crew of 17 includes a flight crew complement of four plus thirteen AWACS specialists, though this latter number can vary for tactical and defence missions. Aft of flight deck, from front to rear of fuselage, are communications, data processing and other equipment bays; multi-purpose consoles; communications, navigation and identification equipment; and crew rest area.

SYSTEMS: A liquid cooling system provides protection for the radar transmitter. An air-cycle pack system and a closed-loop ram-cooled environmental control system ensure a suitable environment for crew and electronics equipment. Electrical power generation has a 600kVA capability. The distribution centre for mission equipment power and remote electronics is located in the lower forward cargo compartment. The aft cargo compartment houses the radar transmitter and an APU. External sockets allow intake of power when the aircraft is on the ground. Two separate and independent hydraulic systems power flight-essential and mission-essential equipment, but either system has the capability of satisfying the requirements of both equipment groups in an emergency.

ELECTRONICS AND EQUIPMENT: Prominent above the fuselage is the elliptical cross-section rotodome which is 9·14 m (30 ft) in diameter and 1·83 m (6 ft) in depth. It comprises four essential elements: a strut-mounted turntable, supporting the rotary joint assembly to which are attached sliprings for electrical and waveguide continuity between rotodome and fuselage; a structural centre section of aluminium skin and stiffener construction, which supports the surveillance radar and IFF/TADIL C antennae, radomes, auxiliary equipment for radar operation and environmental control of the rotodome interior; liquid cooling of the radar antenna; and two radomes constructed of multi-layer glassfibre sandwich material, one for the surveillance radar and one for the IFF/TADIL C array. For surveillance operations the rotodome is hydraulically driven at 6 rpm, but during non-operational flights it is rotated at only ¼ rpm, to keep the bearings lubricated. The Westinghouse radar operates in the S band; by use of pulse Doppler technology, with a high pulse repetition frequency, this radar features long range and accuracy in addition to a normal downlook capability. Its antenna, spanning about 7·32 m (24 ft), and 1·52 m (5 ft) deep, scans mechanically in azimuth, and electronically from ground level up into the stratosphere. Heart of the data processing is an IBM 4 Pi CC-1 high-speed computer, the entire group consisting of arithmetic control units, input/output units, main storage units, peripheral control units, mass memory drums, magnetic tape transports, punched tape reader, line printer, and an operator's control panel. Processing speed is in the order of 740,000 operations/sec; input/output data rate has a maximum of 710,000 words/sec; main memory size is 114,688 words (expandable to 180,224), and mass memory size 802,816 words (expandable to 1,204,224). An interface adapter unit developed by Boeing is the key integrating element interconnecting functional data between AWACS avionics subsystems, data processing group, radar, communications, navigation/guidance, display, azimuth and identification. Data display and control is provided by Hazeltine Corporation multi-purpose consoles (MPC) and auxiliary display units (ADU); in present configuration each AWACS aircraft carries nine MPCs and two ADUs. Navigation/guidance relies upon three principal sources of information: Delco AN/ASN-119 (Carousel IV) inertial platform; Northrop AN/ARN-120 Omega navigation; and a Ryan AN/APN-213 Doppler velocity sensor. Communications equipment, supplied by Collins Radio, Electronic Communications Inc, and Hughes Aircraft, provides HF, VHF and UHF communication channels by means of which information can be transmitted or received in clear or secure mode, in voice or digital form. Identification is based on an AN/APX-103 interrogator set being developed by Cutler-Hammer's AIL Division. It is the first airborne IFF interrogator set to offer complete AIMS Mk X SIF air traffic control and Mk XII military identification friend or foe (IFF) in a single integrated system. Simultaneous Mk X and Mk XII multi-target and multi-mode operations will allow the operator to obtain instantaneously the range, azimuth and elevation, code identification and IFF status of all targets within radar range.

BOEING 707-LRPA

To meet a Canadian government requirement for a long-range maritime patrol aircraft, Boeing offered a specially-developed version of the Model 707-320C. Details of this project can be found in the 1975-76 *Jane's*.

BOEING B-52G/H STRATOFORTRESS

Under a $212 million programme, Boeing B-52G and H Stratofortresses of the USAF's Strategic Air Command have been equipped with an AN/ASQ-151 Electro-optical Viewing System (EVS) to improve low-level flight capability.

Clearly seen in the accompanying photograph are the two steerable chin turrets which house the new sensors. That on the starboard side contains a Hughes Aircraft AAQ-6 forward-looking infra-red scanner (FLIR), while the port turret houses a Westinghouse AVQ-22 low-light-level TV camera.

In the first six months of operational service the reliability of the equipment exceeded the specification issued by the Air Force Logistics Command's Oklahoma City Logistics Center, the first seven B-52s to be equipped with EVS recording a mean time between failure of 37·4 hours. This represented only 13 equipment failures in an accumulated 486 flight hours, and demonstrated that the 1,110-hour production reliability test that preceded the first installation served its purpose adequately. In this test, a facsimile of the EVS was subjected to temperature, humidity and vibration conditions that duplicated those of the B-52.

More than 270 EVS kits were to be produced, with the last scheduled for delivery in the first quarter of 1976.

Boeing B-52H fitted with AN/ASQ-151 Electro-optical Viewing System

BOEING T-43A

Experience gained during the war in Vietnam alerted the USAF to the need to increase its supply of trained navigators. To meet this requirement and to improve the standard of training, it was decided to replace the existing fleet of Convair T-29 piston-engined trainers by more modern aircraft of greater capacity.

Boeing's Model 737-200 was selected as the most suitable off-the-shelf basic aircraft for this role, and 19 of these, modified to meet the military requirement, were delivered with the designation T-43A under an $81·7 million contract. Details can be found in the 1975-76 *Jane's*.

BOEING ADVANCED AIRBORNE COMMAND POST

USAF designation: E-4

On 28 February 1973 the USAF's Electronic Systems Division announced from its headquarters at Hanscom

Boeing E-4B advanced airborne command post operated by the USAF Strategic Air Command

Field, Bedford, Massachusetts, that it had awarded The Boeing Company a $59 million fixed-price contract for the supply of two Model 747Bs to be adapted as **E-4A** airborne command posts under the 481B Advanced Airborne Command Post (AABNCP) programme. A contract valued at more than $27·2 million was awarded, in July 1973, for a third aircraft; in December 1973 the fourth aircraft was contracted at $39 million.

The third and fourth aircraft differ from the first two in having General Electric F103-GE-100 turbofan engines, each rated at 233·5 kN (52,500 lb st), instead of the JT9Ds fitted normally to aircraft of the 747 series. The fourth and subsequent aircraft are fitted with more advanced equipment (see below) and are designated **E-4B**.

On 15 January 1976 it was stated that the total planned force was six E-4Bs, with the programme to equip them scheduled for completion in 1983. Estimated costs for research, development, test and evaluation to the end of 1981 are $353·2 million, with an additional $499·5 million for procurement and $28·1 million for military construction of support facilities.

The E-4Bs are to replace EC-135 Airborne Command Posts of the National Military Command System and Strategic Air Command, which are military variants of the Model 707. E-Systems Inc of Greenville, Texas, won a contract to install interim equipment in the first three E-4As. This involved transfer and integration of equipment removed from EC-135s. In this condition the aircraft can be used with an expanded battle staff, allowing a more flexible response capability than was possible with the older aircraft.

The first of the operational E-4As was delivered to Andrews AFB, Maryland, in December 1974. The second and third, also consigned to Andrews AFB, were received in May and September 1975.

In early 1974, Boeing and a team comprising Computer Sciences Corp, of Falls Church, Virginia; Electrospace Systems Inc of Richardson, Texas; and E-Systems Inc, won the contract to design and install the advanced command post equipment in the remainder of the fleet and, eventually, to replace the equipment in the first three aircraft.

The USAF took delivery of its first E-4B testbed aircraft in August 1975. Equipped for flight refuelling, this is the first to be equipped with the advanced command, control and communications equipment, scheduled for completion in late 1979.

The first E-4B was test-flown in the Spring of 1976 with a new 1,200kVA electrical system (two 150kVA generators on each engine), designed to support the advanced electronics equipment to be added later. The latter will include a wide variety of radio communications equipment, including a new LF/VLF system employing a trailing-wire antenna that is towed behind the aircraft in flight.

During the interim period, the three E-4As in service are able to operate as National Emergency Airborne Command Posts (NEACPs), and are providing operational experience that will be invaluable in finalising the design of the equipment that will be installed in the E-4Bs. Subsequently, the E-4As will be equipped to full E-4B standard, and were scheduled to have flight refuelling equipment installed by July 1976.

The E-4 is designed for long-endurance missions, its 429·2 m² (4,620 sq ft) of floor space accommodating almost three times the payload of the EC-135 which it replaces. The main deck of the aircraft is divided into six areas: the National Command Authorities' (NCA) work area, conference room, briefing room, battle staff work area, communications control centre and rest area. The flight deck accommodates the flight crew, navigation station and flight crew rest area. Lobe areas, beneath the main deck, house a technical control facility and a limited on-board maintenance storage area.

The E-4B when it enters service will include accommodation for a larger battle staff than that carried by the E-4A; an air-conditioning system of 226·5 m³ (8,000 cu ft)/min capacity to cool electronic components; nuclear thermal shielding; acoustic controls; an improved technical control facility; a 1,200kVA generator system; super high frequency (SHF) and low frequency/very low frequency (LF/VLF) communications equipment.

Strategic Air Command (SAC) is now sole operational manager of the AABNCP force. Transfer of operational responsibility from Headquarters Command USAF to SAC began in October 1975 and became effective as from 1 November 1975. It is anticipated that the main operating base for the E-4 fleet will be transferred eventually from Andrews AFB to Offutt AFB, Nebraska.

BOEING YC-14 (AMST)

Looking ahead for potential replacements for its fleet of Lockheed C-130 Hercules transport aircraft, the USAF issued requests for proposals to nine US aerospace companies in early 1972. Responses were received from Bell Aerospace, Boeing, Fairchild Industries, a combined Lockheed-Georgia/North American Rockwell team and McDonnell Douglas. From these proposals, those of Boeing and McDonnell Douglas (which see) were selected, and on 10 November 1972 these two companies were each awarded a contract to develop, construct and flight test two aircraft to compete in a prototype fly-off competition.

The advanced medium short take-off and landing transport (AMST) programme is under the management of the Prototype Program Office of the USAF Systems Command's Aeronautical Systems Division, Wright-Patterson AFB, Ohio. Boeing's entry, the two prototypes of which have been allocated the USAF designation YC-14, is being built under a $105·9 million contract. Its Phase 1 requirement, which had a 90-day completion period, demanded the submission of additional design/performance analysis. Both companies completed this stage of the contract in just over a month, and this enabled the USAF to give a go-ahead for Phase 2 of the contract some 30 days ahead of schedule. Phase 2 covered a 45-month period, during which each company was to build and fly two prototypes, emphasis being placed on performance and cost goals rather than rigid adherence to specification requirements.

The first flight of Boeing's first YC-14 prototype was scheduled for September 1975, that of the second aircraft about two months later. After the programme originated, however, Congress set a limit of $25 million on the YC-14 and YC-15 in the FY 1974 budget, instead of the figure of $65·2 million which the USAF had requested. This meant that, although the programme was able to continue, the

First prototype of the Boeing YC-14 advanced medium STOL transport (two General Electric CF6-50D turbofan engines)

Roll-out photograph of the first Boeing YC-14 advanced medium STOL transport prototype for the US Air Force

first flight date for the first prototype had to be deferred to mid-1976. The YC-14 was flown for the first time on 6 August 1976. When all four competing aircraft become available it is anticipated that evaluation may last for about a year.

A significant feature of the Boeing YC-14 is the use of a relatively small supercritical wing, with an overwing installation of the power plant. Benefits accruing from this layout include the presentation of a low infra-red signature to ground-based detectors; an uncluttered underwing surface, simplifying the carriage of external stores, including RPVs; efficient thrust reversal; and a reduced noise footprint. Significant improvement of cargo compartment loading efficiency will result from the adoption of the wide-body fuselage concept, which is now a familiar feature of civil air transports.

Type: Advanced military STOL transport.

Wings: Cantilever shoulder-wing monoplane. Comparatively small supercritical wing of tapered planform, incorporating advanced concepts to enhance STOL capability. Wing upper-surface blowing concept requires the engines to be mounted above and forward of the wing, so that they exhaust over the wing upper surface. Wide-span leading-edge and Coanda-type trailing-edge flaps will, when extended, induce the high-speed airflow from the engines to cling to the surface of the wing/flap system and direct it downward, generating powered lift. Boeing claims that wind tunnel tests of the system have shown it to be superior to other powered lift concepts, such as externally blown flaps or vectored thrust.

Fuselage: Conventional semi-monocoque all-metal structure.

Tail Unit: Cantilever all-metal structure with high T-tail. Double-hinged rudder and elevators.

Landing Gear: Retractable tricycle type. Twin wheels on nose unit. Each main unit is of the four-post levered type, with twin wheels in tandem. Main wheels and nosewheels have tyres size 40 × 18-16.

Power Plant: Two General Electric CF6-50D two-shaft high bypass ratio turbofan engines, each with a max rating of approx 227 kN (51,000 lb st). Mounting of the engines above and forward of the wing is expected to offer significant noise reduction. Internal fuel load 30,118 kg (66,400 lb).

Accommodation: Able to carry 150 troops, or approximately 12,247 kg (27,000 lb) cargo in STOL operations or 36,740 kg (81,000 lb) in conventional operation. Passenger doors on each side of fuselage. Cargo loading ramp in undersurface of rear fuselage. Undersurface of fuselage retracts upward inside fuselage aft of ramp. Digital flight controls are triple-redundant and fail-operational.

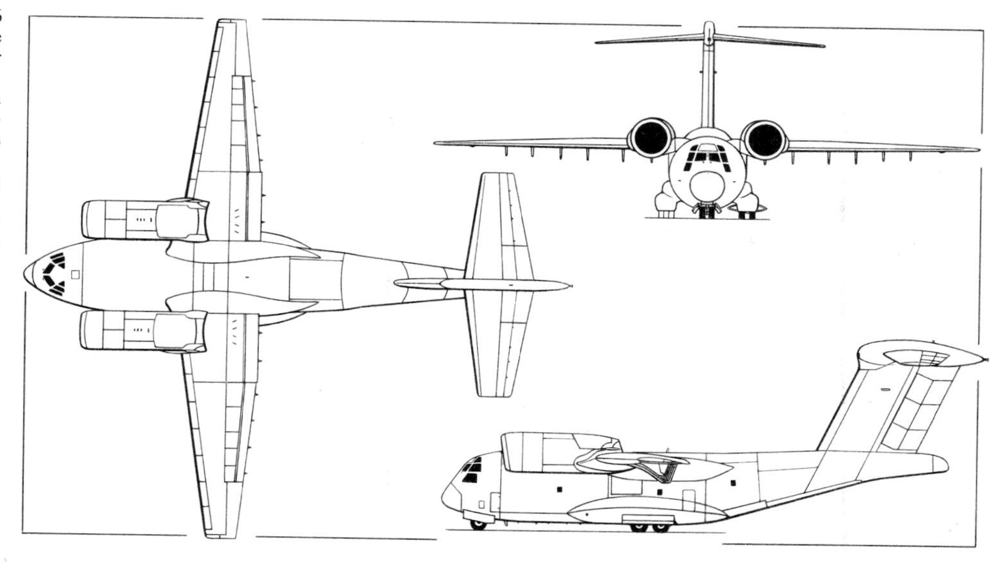

Boeing YC-14 advanced medium STOL transport *(Pilot Press, provisional)*

Dimensions, external:
- Wing span: 39·32 m (129 ft 0 in)
- Length overall: 40·13 m (131 ft 8 in)
- Height overall: 14·73 m (48 ft 4 in)
- Fuselage: Max diameter: 5·44 m (17 ft 10 in)
- Tailplane span: 16·68 m (54 ft 8½ in)
- Wheel track: 5·66 m (18 ft 7 in)
- Wheelbase: 12·50 m (41 ft 0 in)

Dimensions, internal:
- Cargo compartment:
  - Length, incl ramp: 18·66 m (61 ft 2½ in)
  - Length, excl ramp: 14·43 m (47 ft 4 in)
  - Width: 3·50-3·55 m (11 ft 6 in to 11 ft 8 in)
  - Height: 3·40-3·66 m (11 ft 2 in to 12 ft 0 in)

Area:
- Wings, gross: 163·7 m² (1,762 sq ft)

Weights and Loading (estimated):
- Operating weight empty: 53,297 kg (117,500 lb)
- Max STOL T-O weight: approx 77,112 kg (170,000 lb)
- Max T-O weight (2·5g load factor): 107,500 kg (237,000 lb)
- Landing weight (STOL): 72,575 kg (160,000 lb)
- Wing loading (STOL): 444·3 kg/m² (91 lb/sq ft)

Performance (at STOL T-O weight, estimated):
- Max level speed at S/L: 350 knots (649 km/h; 403 mph)
- Max level speed at altitude: 438 knots (811 km/h; 504 mph)
- Long-range cruising speed: 390 knots (723 km/h; 449 mph)
- Approach speed: 86 knots (159 km/h; 99 mph)
- Rate of climb at S/L at 72,575 kg (160,000 lb) AUW: 1,935 m (6,350 ft)/min
- Service ceiling: 13,716 m (45,000 ft)
- T-O run, S/L at 15°C: 305 m (1,000 ft)
- T-O field length, S/L at 15°C: 527 m (1,730 ft)
- Landing field length, idle reverse, S/L at 15°C: 556 m (1,825 ft)
- Landing run, S/L at 15°C: 360 m (1,180 ft)
- Operating radius: 400 nm (740 km; 460 miles)
- Ferry range: 2,770 nm (5,133 km; 3,190 miles)

BOEING VERTOL COMPANY

Boeing Vertol Company produces the CH-47 Chinook helicopter for the US Army. Research and development work on heavy lift helicopters resulted in the company being awarded a contract worth an estimated $76 million to conduct the first phase of the development of a heavy lift helicopter. The company has also demonstrated the BO 105 light helicopter to the US Navy. Boeing has marketing rights for this aircraft in the United States and other parts of the western hemisphere, and has an option to manufacture it in the USA. The BO 105 was developed and is manufactured in Germany by Messerschmitt-Bölkow-Blohm GmbH (which see), the largest German aerospace company, in which Boeing has a holding of just under 10%.

BOEING VERTOL MODELS 114 and 234

US Army designation: CH-47 Chinook

Canadian Armed Forces designation: CH-147

Development of the CH-47 Chinook series of helicopters began in 1956, when the Department of the Army announced its intention to replace its piston-engined transport helicopters with a new generation of turbine-powered helicopters. As a result of a systems capability analysis by a joint Army/Air Force Selection Board, the Boeing Vertol company was awarded an initial contract for five YCH-47As (formerly YHC-1B) by the US Army in June 1959. The first YCH-47A was completed on 28 April 1961. The first hovering flight was made on 21 September 1961. Since then, the effectiveness of the CH-47 has been increased by successive product improvement programmes. A total of 762 Chinooks had been delivered (US Army 699, Australia 12, Iran 26, Italy 10, Spain 7 and Canada 8) by December 1975.

The CH-47 was designed to meet the US Army's requirement for an all-weather medium transport helicopter and, depending upon the series model, is capable of transporting specified payloads under severe combinations of altitude and temperature conditions. The primary mission radius criterion established by the US Army is 100 nm (185 km; 115 miles). The primary mission take-off gross weight is based on the capability of hovering out of ground effect at 1,830 m/35°C (6,000 ft/95°F). The CH-47C has demonstrated its ability to hover out of ground effect with a useful load of 11,453 kg (25,250 lb) at sea level under standard atmospheric conditions.

Three versions of the Chinook have been produced:

CH-47A. Initial production version, powered by two 1,640 kW (2,200 shp) Lycoming T55-L-5 or 1,976 kW (2,650 shp) T55-L-7 turboshaft engines. Operation of the CH-47A by the Vietnamese Air Force (VNAF) began in 1971. Production completed.

CH-47B. Developed version with 2,125 kW (2,850 shp) T55-L-7C turboshaft engines, redesigned rotor blades with cambered leading-edge, blunted rear rotor pylon, and strakes along rear ramp and fuselage for improved flying qualities. First of two prototypes flew for the first time in early October 1966. Deliveries began on 10 May 1967. Production completed.

CH-47C. The current Model 234 achieves increased performance from a combination of strengthened transmissions, two 2,796 kW (3,750 shp) T55-L-11C engines and increased integral fuel capacity totalling 4,137 litres (1,093 US gallons). First flight of the original model CH-47C was made on 14 October 1967, and deliveries of production aircraft began in the Spring of 1968. They were first deployed in Vietnam in September 1968.

Eight CH-47Cs were ordered by Canada in August 1973, and deliveries of these began in September 1974. Designated **CH-147,** these aircraft are to Model 234 standard, with T55-L-11C engines, ISIS, CWFS, forward door rescue hoist, ferry range tank kit, up to 44 troop seats, advanced flight control system, rear ramp with water dam, 12,700 kg (28,000 lb) cargo hook, T-O weight of 22,680 kg (50,000 lb), and weight for water operations of 16,330 kg (36,000 lb) normal or 20,865 kg (46,000 lb) emergency.

Following extensive development work on a Crashworthy Fuel System (CWFS), and an Integral Spar Inspection System (ISIS), these safety features were made available during 1973. Incorporation of the CWFS on US Army CH-47Cs is being accomplished by retrofit kits, delivery of which began in March 1973. All Chinooks delivered to Australia and Canada have this system which provides a total fuel capacity of 3,944 litres (1,042 US gallons).

Rotor blades with ISIS, as well as improved corrosion protection, have been approved for use on all CH-47s; all new blades for initial installation, or supplied as spares, embody these improvements. By the end of 1975, more than 3,700 CH-47 rotor blades had been modified to embody ISIS. Blades in service or already delivered can have ISIS incorporated by means of a kit, deliveries of which began in early 1974; improved corrosion protection can be applied at the first blade overhaul.

CH-47 helicopters are in service in many places, including Alaska, Australia, Germany, Hawaii, Iran, Italy, Korea, Spain and Thailand as well as at numerous US National Guard and US Army installations within the continental United States.

Details of the CH-47A and -47B can be found in the 1974-75 *Jane's.* Those which follow apply specifically to the current CH-47C/Model 234, to which standard all surviving US Army CH-47s are being updated:

TYPE: Twin-engined medium transport helicopter.

ROTOR SYSTEM: Two three-blade rotors, rotating in opposite directions and driven through interconnecting shafts which enable both rotors to be driven by either engine. Rotor blades, of a modified NACA 0012 section, have cambered leading-edge, a strengthened steel spar structure and honeycomb-filled trailing-edge boxes. Two blades of each rotor can be folded manually. Rotor heads are fully articulated, with pitch, flapping and drag hinges. All bearings are submerged completely in oil.

ROTOR DRIVE: Power is transmitted from each engine through individual overrunning clutches, into the combiner transmission, thereby providing a single power output to the interconnecting shafts. Rotor/engine rpm ratio 64 : 1.

Boeing Vertol CH-47C Chinook twin-engined medium transport helicopter

FUSELAGE: Square-section all-metal semi-monocoque structure. Loading ramp forms undersurface of upswept rear fuselage. Fairing pods along bottom of each side are made of metal honeycomb sandwich and are sealed and compartmented, as is the underfloor section of the fuselage, for buoyancy during operation from water.

LANDING GEAR: Menasco non-retractable quadricycle type, with twin wheels on each forward unit and single wheels on each rear unit. Oleo-pneumatic shock-absorbers on all units. Rear units fully castoring and steerable; power steering installed on starboard rear unit. All wheels are government-furnished size 24 × 7·7-VII, with tyres size 8·50-10-III, pressure 4·62 bars (67 lb/sq in). Two single-disc hydraulic brakes. Provision for fitting detachable wheel-skis.

POWER PLANT: Two 2,796 kW (3,750 shp) Lycoming T55-L-11C turboshaft engines, mounted on each side of rear rotor pylon. Combined transmission rating 5,369 kW (7,200 shp); max single-engine transmission limit 3,430 kW (4,600 shp). Self-sealing fuel tanks in external pods on sides of fuselage. Total fuel capacity is 4,137 litres (1,093 US gallons), or 3,944 litres (1,042 US gallons) when equipped with Crashworthy Fuel System. Refuelling points above tanks. Total oil capacity 14 litres (3·7 US gallons).

ACCOMMODATION: Two pilots on flight deck, with dual controls. Jump seat is provided for crew chief or combat commander. Jettisonable door on each side of flight deck. Depending on seating arrangement, 33 to 44 troops can be accommodated in main cabin, or 24 litters plus two attendants, or vehicles and freight. Typical loads include a complete artillery section with crew and ammunition. All components of the Pershing missile system are transportable by Chinooks. Extruded magnesium floor designed for distributed load of 1,465 kg/m² (300 lb/sq ft) and concentrated load of 1,136 kg (2,500 lb) per wheel in tread portion. Floor contains eighty-three 2,270 kg (5,000 lb) tie-down fittings and eight 4,540 kg (10,000 lb) fittings. Rear loading ramp can be left completely or partially open, or can be removed to permit transport of extra-long cargo and in-flight parachute or free-drop delivery of cargo and equipment. Main cabin door, at front on starboard side, comprises upper hinged section which can be opened in flight and lower section with integral steps. Lower section is jettisonable. Up to 12,700 kg (28,000 lb) can be carried on external cargo hook.

SYSTEMS: Cabin heated by 200,000 BTU heater-blower. Hydraulic system provides pressures of 207 bars (3,000 lb/sq in) for flying controls, and 276 bars (4,000 lb/sq in) for engine starting. Electrical system includes two 20kVA alternators driven by transmission drive system. Solar T62 APU runs accessory gear drive, thereby operating all hydraulic and electrical systems.

ELECTRONICS AND EQUIPMENT: All government furnished, including UHF communications and FM liaison sets, transponder, intercom, omni-receiver, ADF and marker beacon receiver. Blind-flying instrumentation standard. Special equipment includes dual electro-hydraulic stability augmentation system, automa-

CH-47C CHINOOK WEIGHTS AND PERFORMANCE

	Condition 1	Condition 2	Condition 3	Condition 4
Weight empty	9,736 kg (21,464 lb)	9,736 kg (21,464 lb)	9,812 kg (21,633 lb)	9,599 kg (21,162 lb)
Payload	5,284 kg (11,650 lb)	2,903 kg (6,400 lb)	9,843 kg (21,700 lb)	—
T-O weight	17,463 kg (38,500 lb)	14,968 kg (33,000 lb)	20,593 kg (45,400 lb)	20,865 kg (46,000 lb)
Max speed, S/L, ISA at normal rated power	155 knots (286 km/h; 178 mph)	164 knots (304 km/h; 189 mph)	127 knots (235 km/h; 146 mph)	—
Average cruise speed	139 knots (257 km/h; 160 mph)	137 knots (254 km/h; 158 mph)	114 knots (211 km/h; 131 mph)	133 knots (246 km/h; 153 mph)
Max rate of climb, S/L, ISA at normal rated power	649 m (2,130 ft)/min	878 m (2,880 ft)/min	421 m (1,380 ft)/min	402 m (1,320 ft)/min
Service ceiling, ISA, normal rated power	3,290 m (10,800 ft)	4,570 m (15,000 ft)	2,560 m (8,400 ft)	2,440 m (8,000 ft)
Hovering ceiling out of ground effect, ISA, max power	2,805 m (9,200 ft)	4,145 m (13,600 ft)	Sea level	—
Mission radius	100 nm (185 km; 115 miles)	100 nm (185 km; 115 miles)	20 nm (37 km; 23 miles)	—
Ferry range	—	—	—	1,156 nm (2,142 km; 1,331 miles)

Condition 1 Criteria: Take-off gross weight equals gross weight to hover out of ground effect at 1,830 m/35°C (6,000 ft/95°F). Radius of action 100 nm (185 km; 115 miles). Fuel reserves 10%. Payload carried internally.

Condition 2 Criteria: Take-off gross weight equals design gross weight. Radius of action 100 nm (185 km; 115 miles). Fuel reserves 10%. Payload carried internally.

Condition 3 Criteria: Take-off gross weight equals gross weight to hover out of ground effect at S/L ISA. Radius of action 20 nm (37 km; 23 miles). Fuel reserves 10%. Payload carried externally. Except for the mission average cruise speed, all other performance is predicated on internal loading of cargo.

Condition 4 Criteria: Take-off gross weight represents alternative design gross weight. Max ferry range (integral and internal auxiliary fuel only), cruise at optimum altitude and standard temperature, no payload, 10% fuel reserves.

Boeing Vertol YUH-61A UTTAS twin-engined military transport helicopter

tic/manual speed trim system, hydraulically-powered winch for rescue and cargo handling purposes, cargo and rescue hatch in floor, external cargo hook of 9,072 kg (20,000 lb) capacity, integral work stands and steps for maintenance, rearview mirror, provisions for paratroops' static lines and for maintenance davits for removal of major components.

DIMENSIONS, EXTERNAL:

Diameter of rotors (each)	18·29 m (60 ft 0 in)
Main rotor blade chord	0·64 m (2 ft 1¼ in)
Distance between rotor centres	11·94 m (39 ft 2 in)
Length overall, rotors turning	30·18 m (99 ft 0 in)
Length of fuselage	15·54 m (51 ft 0 in)
Width, rotors folded	3·78 m (12 ft 5 in)
Height to top of rear rotor hub	5·68 m (18 ft 7·8 in)
Wheel track (c/l of shock-absorbers)	3·20 m (10 ft 6 in)
Wheelbase	6·86 m (22 ft 6 in)
Passenger door (fwd, stbd):	
Height	1·68 m (5 ft 6 in)
Width	0·91 m (3 ft 0 in)
Height to sill	1·09 m (3 ft 7 in)
Rear loading ramp entrance:	
Height	1·98 m (6 ft 6 in)
Width	2·31 m (7 ft 7 in)
Height to sill	0·79 m (2 ft 7 in)

DIMENSIONS, INTERNAL:

Cabin, excl flight deck:	
Length	9·20 m (30 ft 2 in)
Width (mean)	2·29 m (7 ft 6 in)
Width at floor	2·51 m (8 ft 3 in)
Height	1·98 m (6 ft 6 in)
Floor area	21·0 m² (226 sq ft)
Usable volume	41·7 m³ (1,474 cu ft)

AREAS:

Rotor blades (each)	5·86 m² (63·1 sq ft)
Main rotor discs (total)	525·3 m² (5,655 sq ft)

WEIGHTS AND PERFORMANCE:

See table on previous page

BOEING VERTOL YUH-61A

The US Department of Defense announced on 30 August 1972 that it had awarded a $91 million contract to the Boeing Vertol Company to design, build and test three prototype helicopters under the US Army's Utility Tactical Transport Aircraft System (UTTAS) programme.

Each of these aircraft, designated YUH-61A, is a twin-engined single-rotor helicopter, able to carry 12-20 troops and a crew of three. It is intended for use as a troop transport or for medical evacuation and logistics duties. UTTAS aircraft are scheduled to replace the US Army's UH-1H Iroquois in the assault transport role in the late 1970s.

The development programme is to emphasise reliability and maintainability features, in order to keep life-cycle costs to a minimum. Stringent production cost goals have been established and have been made an incentive feature of the development contract.

The YUH-61A incorporates many advances in helicopter technology which should contribute to a substantial reduction in costs, while at the same time improving safety, reliability, performance and flight characteristics. As an example, the use of a simplified hingeless rotor of composite materials has brought a significant reduction in working parts by comparison with previous designs, as well as resulting in superior aircraft stability and safety. Built-in work platforms, direct access to all major components and modularised design offer improved maintainability and should increase the productivity of the aircraft.

The power plant comprises two 1,118 kW (1,500 shp) General Electric T700-GE-700 turboshaft engines. All transmission components are supplied by Litton Systems, Power Transmission Division. Maximum usable fuel totals 1,038 kg (2,288 lb).

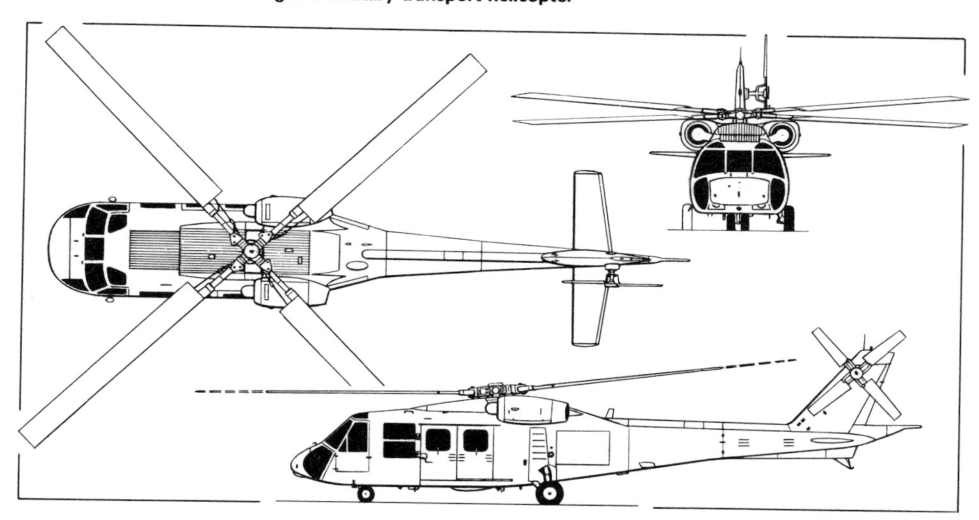

Boeing Vertol YUH-61A Utility Tactical Transport Aircraft System (UTTAS) *(Pilot Press)*

First flight of the prototype YUH-61A was made on 29 November 1974, of the second on 19 February 1975 and the third on 28 May 1975. Following an evaluation competition between UTTAS prototypes developed by Boeing Vertol and Sikorsky Aircraft, it is expected that a production contract will be awarded to one of these companies. A modified version of the YUH-61A is under consideration to meet the US Navy's LAMPS III requirement.

DIMENSIONS, EXTERNAL:

Diameter of main rotro	14·94 m (49 ft 0 in)
Diameter of tail rotor	3·44 m (11 ft 3·6 in)
Length overall, rotors turning	18·50 m (60 ft 8·4 in)
Length of fuselage	16·00 m (52 ft 6 in)
Height to top of rotor hub	2·92 m (9 ft 7 in)
Height over static tail rotor	4·72 m (15 ft 6 in)
Wheel track (outside of wheels)	2·58 m (8 ft 5½ in)

DIMENSIONS, INTERNAL:

Cabin: Length	3·86 m (12 ft 8 in)
Max width	2·18 m (7 ft 2 in)
Max height	1·37 m (4 ft 6 in)
Volume	11·67 m³ (412 cu ft)

WEIGHTS:

Weight empty	4,422 kg (9,750 lb)
Max payload	2,687 kg (5,924 lb)
Mission T-O weight	6,875 kg (15,157 lb)
Max T-O weight	8,935 kg (19,700 lb)

PERFORMANCE (mission gross weight, 35°C day):

Max level speed at 1,220 m (4,000 ft)	155 knots (286 km/h; 178 mph)
Cruising speed at 1,220 m (4,000 ft)	145 knots (268 km/h; 167 mph)
Vertical rate of climb at 1,220 m (4,000 ft), 95% rated power	202 m (664 ft)/min
Service ceiling, one engine out	2,173 m (7,130 ft)
Hovering ceiling out of ground effect	1,966 m (6,450 ft)
Range at cruising speed at 1,220 m (4,000 ft), 30 min reserves	321 nm (595 km; 370 miles)
Endurance with max fuel, cruise speed at 1,220 m (4,000 ft), 30 min reserves	2 hr 48 min

BOEING VERTOL MODEL 179

Boeing Vertol's Model 179 is a commercial derivative of the YUH-61A helicopter designed to meet the US Army's UTTAS requirements.

A 14/20-passenger twin-turbine single-rotor helicopter, the Model 179 has full IFR capability. Advanced technology features developed for the YUH-61A ensure that the Model 179 has high standards of reliability and safety, coupled with low vibration and noise levels, and

Prototype of the Boeing Vertol Model 179, a commercial derivative of the YUH-61A *(Howard Levy)*

substantial reductions in operating costs for a helicopter in this weight category.

A prototype was first flown on 5 August 1975. Petroleum Helicopters Inc has ordered 28 Model 179s, for delivery from 1978.

DIMENSIONS, EXTERNAL AND INTERNAL: As YUH-61A except:
- Length overall, rotors turning 18·14 m (59 ft 6 in)
- Height over static tail rotor 4·63 m (15 ft 2·4 in)

WEIGHTS:
- Weight empty 4,264 kg (9,400 lb)
- Design gross weight 7,756 kg (17,100 lb)
- Max T-O weight 8,482 kg (18,700 lb)

PERFORMANCE (at 7,756 kg; 17,100 lb AUW, ISA):
- Max cruising speed at S/L 156 knots (289 km/h; 180 mph)
- Cruising speed at S/L (99% best range) 134 knots (248 km/h; 154 mph)
- Hovering ceiling out of ground effect 1,722 m (5,650 ft)
- Range at S/L, no reserves 473 nm (876 km; 545 miles)
- Range at 1,525 m (5,000 ft), no reserves 520 nm (963 km; 598 miles)
- Endurance at S/L, no reserves 4 hr 40 min

Boeing Vertol BO 105 Executaire, a lengthened-fuselage version of the West German light helicopter

BOEING VERTOL BO 105 EXECUTAIRE

Boeing Vertol has held US marketing rights for the BO 105 since 1972, and exclusive rights to build the aircraft under licence from MBB for US and other western hemisphere markets. It announced on 31 March 1975 that a modified version of the BO 105, known as the Executaire, had flown for the first time on 18 March 1975. Conversion was done by Carson Helicopters Inc.

The Executaire is aimed specifically at the US market for executive transport helicopters. The primary modification involves lengthening the aft passenger compartment by 0·25 m (10 in), to give more leg room for passengers, as well as to provide space for improved cabin accessories and better temperature control and noise levels. Additionally, the rear sliding doors are replaced by hinged doors, and an extra window has been added each side, adjacent to the rear passengers' seats. In a utility interior, an additional passenger seat can be installed to provide accommodation for six persons, including the pilot.

The Executaire modification adds about 16 kg (35 lb) to the empty weight of the basic BO 105, which is described under the MBB entry in the German section.

BRANTLY-HYNES

BRANTLY-HYNES HELICOPTER INC

HEAD OFFICE AND WORKS:
PO Box 1046, Frederick, Oklahoma 73542
Telephone: (405) 335-2256

PRESIDENT: Michael K. Hynes

This company, formed on 1 January 1975, replaced Brantly Operators Inc which acquired all rights in Brantly helicopters in late 1970. Mr M. K. Hynes acquired also ownership of the Type Certificates for the Brantly B-2, B-2A, B-2B and Model 305.

Brantly-Hynes has put the two-seat B-2B and five-seat Model 305 back into production.

BRANTLY-HYNES MODEL B-2B

TYPE: Two-seat light helicopter.

ROTOR SYSTEM: Three-blade main rotor. Articulated inboard flapping hinges offset 0·07 m (2·67 in) from hub, and coincident flap and lag hinges offset 1·31 m (4 ft 3¾ in) from hub. Symmetrical blade section with 29% thickness ratio on inboard portion; NACA 0012 section outboard of hinge. Inboard portion of each blade is rigid, built around a steel spar blade. Outboard portion is flexible, with an extruded aluminium leading-edge spar and polyurethane core; aluminium skin is bonded to core and riveted to spar. Blades are attached to hub by flapping links and do not fold. A rotor brake is standard equipment. Two-blade all-metal anti-torque tail rotor.

ROTOR DRIVE: Through centrifugal clutch and planetary reduction gears. Bevel gear take-off from main transmission with flexible coupling to tail rotor drive-shaft. Main rotor/engine rpm ratio 1 : 6·158. Tail rotor/engine rpm ratio 1 : 1.

FUSELAGE: Stressed-skin all-metal structure with conical tail section. Tail rotor on swept-up boom extension.

LANDING GEAR: Alternative skid, wheel or float gear. Skid type has small retractable wheels for ground handling, fixed tailskid and four shock-absorbers with rubber in compression. Tyres size 10 × 3½, pressure 4·14 bars (60 lb/sq in). Alternative non-retractable tricycle landing gear has oleo-pneumatic shock-absorbers for all units, with single wheels on the main units and twin nosewheels. Inflatable pontoons, which attach to the standard skids, are available to permit operation from water.

POWER PLANT: One 134 kW (180 hp) Lycoming IVO-360-A1A flat-four engine, mounted vertically, with induction cooling system. Rubber bag-type fuel tank under engine, capacity 117 litres (31 US gallons). Refuelling point on port side of fuselage. Oil capacity 5·7 litres (1·5 US gallons).

ACCOMMODATION: Totally-enclosed circular-section cabin for two persons seated side by side. Forward-hinged door on each side. Dual controls, cabin heater and demisting fan standard. Compartment for 22·7 kg (50 lb) baggage in forward end of tail section.

ELECTRONICS AND EQUIPMENT: Provision for Narco Mk 12, King KY-90 or King KX-150A radio. Twin landing lights in nose.

DIMENSIONS, EXTERNAL:
- Diameter of main rotor 7·24 m (23 ft 9 in)
- Main rotor blade chord:
  - inboard 0·225 m (8·85 in)
  - outboard 0·203 m (8·0 in)
- Diameter of tail rotor 1·29 m (4 ft 3 in)
- Length overall 6·62 m (21 ft 9 in)
- Height overall 2·06 m (6 ft 9 in)
- Skid track 1·73 m (5 ft 8¼ in)
- Passenger doors (each):
  - Height 0·79 m (2 ft 7 in)
  - Width 0·86 m (2 ft 9¾ in)
- Baggage compartment door:
  - Mean height 0·25 m (9¾ in)
  - Length 0·55 m (1 ft 9¾ in)

DIMENSIONS, INTERNAL:
- Max width of cabin 1·27 m (4 ft 2 in)
- Baggage compartment 0·17 m³ (6 cu ft)

AREAS:
- Main rotor blades (each) 0·69 m² (7·42 sq ft)
- Main rotor disc 41·06 m² (442 sq ft)
- Tail rotor disc 1·21 m² (13 sq ft)

WEIGHTS AND LOADINGS:
- Weight empty with skids 463 kg (1,020 lb)
- Weight empty with floats 481 kg (1,060 lb)
- Max T-O weight 757 kg (1,670 lb)
- Max disc loading 18·4 kg/m² (3·77 lb/sq ft)
- Max power loading 5·65 kg/kW (9·27 lb/hp)

PERFORMANCE (at max T-O weight):
- Max level speed at S/L 87 knots (161 km/h; 100 mph)
- Max cruising speed (75% power) 78 knots (145 km/h; 90 mph)
- Max rate of climb at S/L 580 m (1,900 ft)/min
- Service ceiling 3,290 m (10,800 ft)
- Hovering ceiling in ground effect 2,040 m (6,700 ft)
- Range with max fuel, with reserves 217 nm (400 km; 250 miles)

BRANTLY-HYNES MODEL 305

The Model 305 is a five-seat helicopter of similar configuration to the Model B-2B, but larger in every respect. The prototype of the original Model 305 flew for the first time in January 1964, and FAA Type Approval was received on 29 July 1965.

TYPE: Five-seat light helicopter.

ROTOR SYSTEM: Three-blade main rotor. Articulated inboard flapping hinges, offset 0·09 m (3·625 in) from hub, and coincident flap and lag hinges outboard. Inboard portion of each blade is rigid, built around a steel spar blade. All-metal outboard portion has a D-spar and is foam-filled. Two-blade all-metal tail rotor. Each blade has a forged aluminium leading-edge spar, ribs and riveted aluminium skin. Main rotor blades do not fold. Rotor brake is standard.

ROTOR DRIVE: Main rotor shaft-driven through centrifugal clutch and planetary reduction gears. Bevel gear take-off from main transmission, with flexible coupling, through tail rotor drive-shaft and intermediate gearbox to tail gearbox. Main rotor/engine rpm ratio 1 : 6·666. Tail rotor/engine rpm ratio 1 : 0·998.

FUSELAGE: Stressed-skin all-metal structure, with conical tail section. Tail rotor carried on swept-up boom extension.

TAIL UNIT: Small variable-incidence horizontal stabiliser of all-metal stressed-skin construction.

LANDING GEAR: Alternative skid, wheel or float gear. Skid landing gear has four oleo struts, two on each side, and

Brantly-Hynes Model B-2B light helicopter (Lycoming IVO-360-A1A engine)

small retractable ground handling wheels. The wheel gear has two main wheels and twin nosewheels, all on oleo-pneumatic shock-absorbers. Goodyear main wheels and tyres size 6·00-6, pressure 2·07 bars (30 lb/sq in). Goodyear nosewheels and tyres size 5·00-5, pressure 1·93 bars (28 lb/sq in). Goodyear single-disc hydraulic brakes on main wheels.

POWER PLANT: One 227·5 kW (305 hp) Lycoming IVO-540-A1A flat-six engine. One rubber-cell fuel tank under engine, capacity 163 litres (43 US gallons). Refuelling point in port side of fuselage. Oil capacity 9·5 litres (2·5 US gallons).

ACCOMMODATION: Two individual seats side by side with dual controls. Rear bench seat for three persons. Door on each side. Rear compartment for 113 kg (250 lb) of baggage, with downward-hinged door on starboard side.

ELECTRONICS AND EQUIPMENT: King or Narco radio, to customer's specification. Blind-flying instrumentation is available, but helicopter is not certificated for instrument flight.

DIMENSIONS, EXTERNAL:

Diameter of main rotor	8·74 m (28 ft 8 in)
Main rotor blade chord (constant)	0·254 m (10 in)
Diameter of tail rotor	1·30 m (4 ft 3 in)
Length overall, rotor turning	10·03 m (32 ft 11 in)
Length of fuselage	7·44 m (24 ft 5 in)
Height overall	2·44 m (8 ft 0⅛ in)
Wheel track	2·10 m (6 ft 10¾ in)
Wheelbase	2·15 m (7 ft 0½ in)
Passenger doors (each):	
Height	0·82 m (2 ft 8⅛ in)
Width	1·02 m (3 ft 3⅞ in)
Baggage compartment door:	
Mean height	0·30 m (1 ft 0¼ in)
Width	0·69 m (2 ft 3 in)

DIMENSIONS, INTERNAL:

Cabin: Length	2·30 m (7 ft 6½ in)
Max width	1·39 m (4 ft 6¾ in)
Max height	1·22 m (4 ft 0½ in)
Baggage compartment	0·47 m³ (16·7 cu ft)

AREAS:

Main rotor blades (each)	1·09 m² (11·79 sq ft)
Tail rotor blades (each)	0·05 m² (0·50 sq ft)
Main rotor disc	3·33 m² (35·8 sq ft)
Tail rotor disc	1·32 m² (14·18 sq ft)

Brantly-Hynes Model 305 five-seat light helicopter (Lycoming IVO-540-A1A engine)

WEIGHTS AND LOADINGS:

Weight empty	817 kg (1,800 lb)
Max T-O and landing weight	1,315 kg (2,900 lb)
Max zero-fuel weight	1,224 kg (2,700 lb)
Max disc loading	22·7 kg/m² (4·65 lb/sq ft)
Max power loading	5·78 kg/kW (9·84 lb/hp)

PERFORMANCE (at max T-O weight):

Max level speed at S/L	104 knots (193 km/h; 120 mph)
Max cruising speed at S/L	96 knots (177 km/h; 110 mph)
Max rate of climb at S/L	297 m (975 ft)/min
Service ceiling	3,660 m (12,000 ft)
Hovering ceiling in ground effect	1,245 m (4,080 ft)
Range with max fuel and max payload, with 15 min reserves	191 nm (354 km; 220 miles)

CAMAIR

CAMAIR AIRCRAFT CORPORATION

HEAD OFFICE:
PO Box 231, Remsenburg, Long Island, New York 11960
Telephone: (516) 325-0120
PRESIDENT:
Fred Garcia Jr

CAMAIR TWIN NAVION

The Camair Twin Navion is a twin-engined version of the North American/Ryan Navion light aircraft. It embodies structural modifications to cater for the increased power and weight, together with design and aerodynamic refinements which provide improved performance, comfort and styling. Versions are as follows:

CTN-A. Prototype only. Built and flown in 1953 with two 168 kW (225 hp) Continental engines.

CTN-B. Powered by two 179 kW (240 hp) Continental O-470-B engines. First flown in early 1954. Total of 28 delivered in 1955-59.

CTN-C. First flown in 1960, with two 194 kW (260 hp) Continental IO-470-D engines. A number of Model Bs have been converted to Model C configuration.

CTN-D. The prototype of this version was powered originally by two 194 kW (260 hp) Continental TSIO-470-B engines, and had increased fuel capacity and gross weight. It has since had two 224 kW (300 hp) Continental IO-520 engines installed.

Production of new aircraft has been suspended; but Camair is able to modify B and C models to the latest model D configuration, and continues to supply spares to Twin Navion owners.

The following details apply specifically to the CTN-D:

TYPE: Twin-engined four-seat cabin monoplane.

WINGS: Cantilever low-wing monoplane. Wing section NACA 4415R at root, NACA 6410R at tip. Dihedral 7° 30′. Incidence 2° at root, −1° at tip. All-metal structure. All-metal mass-balanced ailerons and hydraulically-operated flaps.

FUSELAGE: All-metal semi-monocoque structure.

TAIL UNIT: Cantilever all-metal structure. Controllable trim tabs on rudder and elevators. Optional electrically-powered elevator trim control.

LANDING GEAR: Hydraulically-retractable tricycle type. Main units retract inward, nose unit aft. Oleo-pneumatic shock-absorbers. Main-wheel tyres size 7·00-8, Type III. Steerable nosewheel, tyre size 6·00-6, Type III. Hydraulic disc brakes.

POWER PLANT: Two 224 kW (300 hp) Continental IO-520 flat-six engines, each driving a Hartzell three-blade constant-speed fully-feathering propeller. Two 75·5 litre (20 US gallon) aluminium alloy fuel tanks in the wing roots, two 132 litre (35 US gallon) wingtip tanks and two 132 litre (35 US gallon) rubber fuel cells located in the aft overwing engine nacelles. Total standard fuel capacity 679 litres (180 US gallons). Optional auxiliary fuselage fuel tank containing 75·5 litres (20 US gallons). Oil capacity 11·5 litres (3 US gallons) per engine.

Camair Twin Navion CTN-D (two Continental IO-520 engines)

ACCOMMODATION: Enclosed cabin seating pilot and co-pilot on individually adjustable and reclining front seats and two passengers on individual rear seats. Rearward-sliding canopy with transverse web which seals the baggage compartment and provides a shelf when the canopy is closed. Dual controls. Cabin soundproofing, air-conditioning and heating. Windscreen demister. Baggage compartment aft of cabin, with outside access, for 82 kg (180 lb); and forward compartment for equipment and baggage in nose, capacity 45 kg (100 lb). Total baggage capacity 127 kg (280 lb).

SYSTEMS: Hydraulic system supplied by two engine-driven pumps to operate flaps, landing gear and brakes. Dual engine-driven vacuum pumps. Dual engine-driven alternators power a 24V electrical system, which has also an external power socket. Oxygen system optional.

ELECTRONICS AND EQUIPMENT: Optional items include complete dual radio, full IFR instrumentation and navigation equipment, plus integrated flight system, weather radar and autopilot. Standard equipment includes dual retractable taxi lights, dual landing lights, triple strobe lights and rotating beacon. Alcohol propeller de-icing system optional.

DIMENSIONS, EXTERNAL:

Wing span	10·57 m (34 ft 8 in)
Wing chord at root	2·20 m (7 ft 2½ in)
Wing chord at tip	1·19 m (3 ft 11 in)
Wing aspect ratio	6·04
Length overall	8·53 m (28 ft 0 in)
Height overall	3·25 m (10 ft 8 in)
Wheel track	2·65 m (8 ft 8½ in)
Wheelbase	2·35 m (7 ft 8½ in)

AREAS:

Wings, gross	17·13 m² (184·34 sq ft)
Ailerons (total)	0·96 m² (10·32 sq ft)
Trailing-edge flaps (total)	2·72 m² (29·23 sq ft)
Fin	1·69 m² (18·20 sq ft)
Rudder	0·76 m² (8·20 sq ft)
Tailplane	4·0 m² (43·05 sq ft)
Elevators	1·31 m² (14·10 sq ft)

WEIGHTS AND LOADINGS:

Weight empty	1,360 kg (3,000 lb)
Max T-O weight	2,041 kg (4,500 lb)
Max landing weight	1,960 kg (4,323 lb)
Max wing loading	119·1 kg/m² (24·4 lb/sq ft)
Max power loading	4·56 kg/kW (7·5 lb/hp)

PERFORMANCE (at AUW of 1,590 kg; 3,500 lb):

Max level speed at S/L	187 knots (346 km/h; 215 mph)
Max cruising speed, 75% power at 1,980 m (6,500 ft)	174 knots (322 km/h; 200 mph)
Stalling speed	52·5 knots (97 km/h; 60 mph)
Max rate of climb at S/L	610 m (2,000 ft)/min
Rate of climb at S/L, one engine out	152 m (500 ft)/min
Service ceiling	6,705 m (22,000 ft)
Service ceiling, one engine out	3,050 m (10,000 ft)
T-O run	122 m (400 ft)
Landing run	183 m (600 ft)

CESSNA

CESSNA AIRCRAFT COMPANY

HEAD OFFICE AND WORKS:
Wichita, Kansas 67201
Telephone: (316) 685-9111
CHAIRMAN OF THE BOARD AND CHIEF EXECUTIVE OFFICER:
Russell W. Meyer, Jr
PRESIDENT:
Malcolm S. Harned
SENIOR VICE-PRESIDENTS:
R. L. Lair (Commercial Aircraft Marketing)
R. P. Bauer (Treasurer and Controller)
VICE-PRESIDENTS:
William A. Boettger (Pawnee Division)
Pierre Clostermann (President, Reims Aviation)
Robert D. Dickerson (Wallace Division)
John W. Dussault (McCauley Accessory Division)
Shelby Law (Fluid Power Division)
Homer G. Nester (Controller and Assistant Treasurer)
Derek Vaughan (Commercial Jet Marketing Division)
Thaine L. Woolsey (Fluid Power Division)
Bill Worford (Personnel Relations)
Lee Zuker (Aircraft Radio and Control Division)
SECRETARY: Vincent E. Moore

Cessna Aircraft Company was founded by the late Clyde V. Cessna, a pioneer in US aviation in 1911, and was incorporated on 7 September 1927.

By the beginning of January 1976 the company had produced a total of 126,246 aircraft, of which 12,191 were for military use. Pawnee Division delivered its 100,000th single-engined aeroplane, a Skyhawk II, on 21 July 1975.

Cessna has four plants in Wichita engaged on production of commercial and military aircraft, and The Fluid Power Division in Hutchinson, Kansas, which manufactures fluid power systems.

Subsidiary companies owned by Cessna are Aircraft Radio and Control Division at Boonton, New Jersey, the McCauley Accessories Division of Dayton, Ohio, Cessna Fluid Power Ltd of Glenrothes, Fife, Scotland, Cessna Finance Corporation and Cessna International Finance Corporation in Wichita. It has a 49% interest in Reims Aviation of France.

In early 1976 Cessna had in production 58 types of commercial aircraft. In addition, it is continuing to produce the T-37B twin-engined jet trainer and A-37 strike aircraft for the USAF and the T-37C for the US Military Assistance Programme.

During 1975 Cessna commercial sales totalled 7,673 aircraft, including units assembled in France by Reims Aviation (which see).

Military subcontract programmes include manufacture of assemblies, including missile ejection racks, wing tank and missile pylons, for the McDonnell Douglas F-4 Phantom II, and crew door subassemblies for Bell helicopters.

CESSNA MODEL 150

The prototype of the Model 150 flew for the first time in September 1957, and Cessna re-entered the two-seat light aircraft market by putting it into production in August 1958. By 1 January 1976 a total of 21,771 Model 150s had been delivered, including aircraft built in France by Reims Aviation as F-150s.

The current American-built Model 150 is available in standard, Commuter, Commuter II and Aerobat (described separately) versions. The Commuter II has the same equipment as the Commuter, plus a second Cessna 300 nav/com, providing a total 720-channel com and 200-channel nav and VOR/LOC indicator, Cessna 300 transponder, true airspeed indicator, emergency locator transmitter and external power socket.

The 1976 versions of the Model 150 have a number of improvements as standard, including a new instrument panel layout to improve readability and provide more space for electronic equipment, an airspeed indicator with the primary reading in knots, a semi solid-state voltage regulator, circuit breakers replacing fuses, and new interior and exterior styling. New options include a slim-line microphone, vertically-adjustable seats for pilot and co-pilot and an anti-precipitation static kit.

The original Model 150 received FAA Type Approval on 10 July 1958.

TYPE: Two-seat cabin monoplane.

WINGS: Braced high-wing monoplane. Wing section NACA 2412 (tips symmetrical). Dihedral 1°. Incidence 1° at root, 0° at tip. All-metal structure, with conical-camber glassfibre tips on Commuter and Commuter II (optional on standard model). Modified Frise all-metal ailerons. Electrically-actuated NACA single-slotted all-metal flaps.

FUSELAGE: All-metal semi-monocoque structure.

TAIL UNIT: Cantilever all-metal structure, with sweptback vertical surfaces. Trim tab in starboard elevator. Ground-adjustable rudder trim tab.

LANDING GEAR: Non-retractable tricycle type. Land-O-Matic cantilever main legs, each comprising a one-piece machined conically-tapered spring steel tube. Steerable nosewheel on oleo-pneumatic shock-absorber strut. Size 6·00-6 wheels, with nylon tube-type tyres, on main wheels; size 5·00-5 nosewheel, with nylon tube-type tyre. Tyre pressure 2·07 bars (30 lb/sq in). Toe-operated single-disc hydraulic brakes. Optional wheel fairings for all three units (standard on Commuter and Commuter II). Parking brake.

Cessna Model 150 (Continental O-200-A engine)

POWER PLANT: One 74·5 kW (100 hp) Continental O-200-A flat-four engine, driving a McCauley two-blade metal fixed-pitch propeller. Two all-metal fuel tanks in wings. Total standard fuel capacity 98 litres (26 US gallons), of which 85 litres (22·5 US gallons) are usable. Optional long-range tanks increase total capacity to 143·8 litres (38 US gallons), of which 132·5 litres (35 US gallons) are usable. Oil capacity 5·7 litres (1·5 US gallons).

ACCOMMODATION: Enclosed cabin seating two side by side. Vertically-adjustable seats for pilot and co-pilot; inertia reel shoulder harness and dual controls optional on standard version. Baggage compartment behind seats, backs of which hinge forward. Baggage capacity 54 kg (120 lb). Alternatively, 'family seat' can be fitted in baggage space, for two children not exceeding 54 kg (120 lb) total weight. Door, with opening window, on each side. Heating and ventilation standard. Windscreen defroster standard. Optional overhead skylights.

SYSTEMS: Hydraulic system for brakes only. Electrical power supplied by 12V 60A alternator and 12V battery.

ELECTRONICS AND EQUIPMENT: Optional equipment includes Cessna 300 Series nav/com with 360-channel com and 160-channel nav with remote VOR indicator, 300 Series transceiver with 360 com channels, 300 Series nav/com with 360-channel com, 200-channel nav with remote VOR/LOC or VOR/ILS indicator, Series 300 ADF, marker beacon with three lights and aural signal, transponder with 4096 code capability and slim-line microphone; blind-flying instrumentation (standard on Commuter and Commuter II); rate of climb indicator, turn co-ordinator indicator, outside air temperature gauge, rearview mirror, sun visors, cowl-mounted landing light and omni-flash beacon (all standard on Commuter and Commuter II). Standard equipment includes a stall warning indicator, control locks, cabin dome light, variable intensity instrument panel red floodlights, windscreen defroster, navigation lights, map compartment, baggage retaining net, and safety belts. Optional extras include a winterisation kit, anti-precipitation static kit, control-wheel mounted map light, electric clock and sensitive altimeter (standard for Commuter and Commuter II), directional and horizon gyros with vacuum (standard for Commuter and Commuter II), true airspeed and turn and bank indicators, directional gyro with movable heading index, flight hour recorder, cabin fire extinguisher, emergency locator transmitter (standard on Commuter II), overhead skylights, full-flow oil filter, internal corrosion proofing, heated pitot (standard on Commuter and Commuter II), glider tow hook, tinted windows, cowl-mounted landing lights, conical-camber wingtips and omni-flash beacon (both standard on Commuter and Commuter II), white strobe lights, advanced-design dry vacuum pump, ground service socket for external battery connection (standard on Commuter II), handle and step for easier refuelling, and a quick-drain oil system.

DIMENSIONS, EXTERNAL:

Wing span:	
Standard	9·97 m (32 ft 8½ in)
Commuter, Commuter II	10·11 m (33 ft 2 in)
Wing chord at root	1·63 m (5 ft 4 in)
Wing chord at tip	1·12 m (3 ft 8½ in)
Wing aspect ratio	6·7
Length overall	7·29 m (23 ft 11 in)
Height overall	2·59 m (8 ft 6 in)
Tailplane span	3·05 m (10 ft 0 in)
Wheel track	2·32 m (7 ft 7¼ in)
Wheelbase	1·47 m (4 ft 10 in)
Propeller diameter	1·75 m (5 ft 9 in)
Passenger doors (each):	
Width	0·86 m (2 ft 10 in)

AREAS:

Wings, gross:	
Standard	14·59 m² (157 sq ft)
Commuter, Commuter II	14·83 m² (159·58 sq ft)
Ailerons (total)	1·66 m² (17·88 sq ft)
Trailing-edge flaps (total)	1·72 m² (18·56 sq ft)
Fin	0·83 m² (8·94 sq ft)
Rudder	0·65 m² (6·98 sq ft)
Tailplane	1·58 m² (17·06 sq ft)
Elevators, incl tab	1·06 m² (11·46 sq ft)

WEIGHTS AND LOADINGS:

Weight empty, equipped, standard tanks:	
Standard	454 kg (1,000 lb)
Commuter	501 kg (1,104 lb)
Commuter II	509 kg (1,122 lb)
Max T-O weight	726 kg (1,600 lb)
Max wing loading:	
Standard	49·8 kg/m² (10·2 lb/sq ft)
Commuter, Commuter II	48·9 kg/m² (10·0 lb/sq ft)
Max power loading	9·74 kg/kW (16·0 lb/hp)

PERFORMANCE (all models, at max T-O weight):

Never-exceed speed	140 knots (261 km/h; 162 mph)
*Max level speed at S/L	109 knots (201 km/h; 125 mph)
*Max cruising speed (75% power) at 2,135 m (7,000 ft)	106 knots (196 km/h; 122 mph)
*Econ cruising speed at 3,050 m (10,000 ft)	82 knots (153 km/h; 95 mph)
Stalling speed, flaps up, power off	48 knots (89 km/h; 55 mph)
Stalling speed, flaps down, power off	42 knots (78 km/h; 48 mph)
Max rate of climb at S/L	204 m (670 ft)/min
Service ceiling	4,265 m (14,000 ft)
T-O run	224 m (735 ft)
T-O to 15 m (50 ft)	422 m (1,385 ft)
Landing from 15 m (50 ft)	328 m (1,075 ft)
Landing run	136 m (445 ft)
Range, recommended lean mixture with allowance for start, taxi, T-O, climb and 45 min reserves at 45% power:	
Standard fuel, 75% power at 2,135 m (7,000 ft)	340 nm (629 km; 391 miles)
Max fuel, 75% power at 2,135 m (7,000 ft)	580 nm (1,075 km; 668 miles)
Standard fuel, econ cruising power at 3,050 m (10,000 ft)	420 nm (777 km; 483 miles)
Max fuel, econ cruising power at 3,050 m (10,000 ft)	735 nm (1,361 km; 846 miles)

**With wheel speed fairings which increase speeds by approximately 1·75 knots (3·2 km/h; 2 mph)*

CESSNA MODEL A150 AEROBAT

Introduced in 1970, the Model A150 Aerobat was designed to combine the economy and versatility of the standard Model 150 with aerobatic capability. Structural changes allow the Aerobat to perform 'unusual attitude' manoeuvres and it is licensed in the Aerobatic category for load factors of +6*g* and −3*g* at full gross weight, permitting the performance of barrel and aileron rolls, snap rolls, loops, Immelmann turns, Cuban eights, spins, vertical reversements, lazy eights and chandelles.

Equipment of the Aerobat differs only slightly from the standard aircraft. Quick-release cabin doors, removable seat cushions and backs, quick-release lap belts, and shoulder harnesses are standard, as are two tinted skylights which offer extra visibility; a ground-adjustable rudder trim tab is fitted and distinct external styling provides immediate recognition of the A150's aerobatic role.

The 1976 version of the Aerobat has the improvements detailed for the standard Model 150. Optional equipment includes an accelerometer, 3 in lightweight non-tumbling gyros, conical-camber glassfibre wingtips, steps and handles to simplify refuelling, and a quick-drain oil valve.

Structural changes have increased the empty weight slightly, by comparison with the standard Model 150.

Customers include the Ecuadorean Air Force, which took delivery of 24 Aerobats during 1974-75.

DIMENSIONS AND AREAS:
As for standard Model 150

WEIGHTS AND LOADINGS:
Weight empty, equipped, standard tanks 488 kg (1,076 lb)
Max T-O weight 726 kg (1,600 lb)
Max wing loading 49·8 kg/m² (10·2 lb/sq ft)
Max power loading 9·74 kg/kW (16·0 lb/hp)

PERFORMANCE (at max T-O weight):
*Max level speed at S/L 108 knots (200 km/h; 124 mph)
*Max cruising speed (75% power) at 2,135 m (7,000 ft) 105 knots (195 km/h; 121 mph)
*Econ cruising speed at 3,050 m (10,000 ft) 82·5 knots (151 km/h; 94 mph)
Stalling speed, flaps up, power off 49 knots (91 km/h; 56·5 mph)
Stalling speed, flaps down, power off 44 knots (82 km/h; 51 mph)
Max rate of climb at S/L 204 m (670 ft)/min
Service ceiling 4,265 m (14,000 ft)
T-O run 224 m (735 ft)
T-O to 15 m (50 ft) 422 m (1,385 ft)
Landing from 15 m (50 ft) 328 m (1,075 ft)
Landing run 136 m (445 ft)
Range, recommended lean mixture with allowance for start, taxi, T-O, climb and 45 min reserves at 45% power:
Standard fuel, 75% power at 2,135 m (7,000 ft) 335 nm (621 km; 386 miles)
Max fuel, 75% power at 2,135 m (7,000 ft) 570 nm (1,055 km; 656 miles)
Standard fuel, econ cruising power at 3,050 m (10,000 ft) 415 nm (769 km; 478 miles)
Max fuel, econ cruising power at 3,050 m (10,000 ft) 725 nm (1,343 km; 835 miles)

**With optional wheel speed fairings which increase speeds by approximately 1·75 knots (3·2 km/h; 2 mph)*

CESSNA SKYHAWK

USAF designation: T-41A Mescalero

With the standard Model 172 (1975-76 *Jane's*) discontinued for 1976, only two versions are now available:

Skyhawk. Basic 1976 version, introducing as standard improvements a redesigned instrument panel; an airspeed indicator with the primary scale in knots; improved cabin soundproofing; new cabin door stop, heater vent outlets, chrome seat adjustment handles and seat-belt retainers; redesigned baggage door hinge; rounded fin and rudder tips; improved lightweight wiring installations and a semi solid-state voltage regulator.

Skyhawk II. As Skyhawk, but including as standard a 300 Series nav/com with 360-channel com and 160-channel nav, dual controls, true airspeed indicator, navigation light detectors, heated pitot, courtesy lights, omni-flash beacon, alternate static source and emergency locator transmitter.

The Skyhawk is certificated for operation as a floatplane, and can be fitted with skis. A version designated F-172 is produced in France by Reims Aviation.

On 31 July 1964, the USAF ordered 170 earlier-type Model 172s, under the designation **T-41A,** for delivery between September 1964 and July 1965. USAF student pilots complete about 30 hours of basic training on the T-41A before passing on to the T-37B jet primary trainer. Eight T-41As have been bought by the Ecuadorean Air Force, five by the Honduran Air Force and 26 by the Peruvian government. The USAF ordered more in July 1967 and a total of 237 had been built by December 1973. The more powerful T-41B/C/D (R172E) are described separately.

A total of 24,822 aircraft in the Model 172/Skyhawk series had been built by 1 January 1976, including 1,234 F-172s built in France.

TYPE: Four-seat cabin monoplane.

WINGS: Braced high-wing monoplane. NACA 2412 wing section. Dihedral 1° 44′. Incidence 1° 30′ at root, −1° 30′ at tip. All-metal structure, except for conical-camber glassfibre wingtips. Single bracing strut on each side. Modified Frise all-metal ailerons. Electrically-controlled NACA all-metal single-slotted flaps inboard of ailerons.

FUSELAGE: All-metal semi-monocoque structure.

TAIL UNIT: Cantilever all-metal structure. Sweepback on fin 35° at quarter-chord. Trim tab in starboard elevator. Ground-adjustable trim tab in rudder.

LANDING GEAR: Non-retractable tricycle type. Cessna Land-O-Matic cantilever main legs, each comprising a one-piece machined conically-tapered spring steel tube. Nosewheel is carried on an oleo-pneumatic shock-strut and is steerable with rudder up to 10° and controllable up to 30° on either side. Cessna main wheels size 6·00-6 and nosewheel size 5·00-5 (optionally 6·00-6), with nylon cord tube-type tyres. Tyre pressure: main wheels 1·59 bars (23 lb/sq in), nosewheel 1·79 bars (26 lb/sq in). Hydraulic disc brakes. Optional wheel fairings. Alternative float and ski gear.

POWER PLANT: One 112 kW (150 hp) Lycoming O-320-E2D flat-four engine, driving a two-blade fixed-pitch metal propeller. One fuel tank in each wing, total capacity 159 litres (42 US gallons). Usable fuel 143·8 litres (38 US gallons). Provision for long-range tanks, giving total capacity of 197 litres (52 US gallons), of which 182 litres (48 US gallons) are usable. Oil capacity 7·5 litres (2 US gallons).

Cessna Skyhawk II four-seat cabin monoplane (Lycoming O-320-E2D engine)

ACCOMMODATION: Cabin seats four in two pairs, with optional fully-articulating front seats. Baggage space aft of rear seats, capacity 54 kg (120 lb). An optional fold-away seat can be fitted in baggage space, for one or two children not exceeding 54 kg (120 lb) total weight. Door on each side of cabin, giving access to all seats, simplifies loading if rear seats are removed and cabin used for freight. Pilot's window opens; co-pilot's opening side window and dual controls optional on Skyhawk; dual controls standard on Skyhawk II. Baggage door on port side. Combined heating and ventilation system. Glassfibre soundproofing. Optional overhead skylights.

SYSTEM: Electrical system includes a 60A 12V alternator, automatic alternator cutout, electric engine starter and 12V battery.

ELECTRONICS AND EQUIPMENT: True airspeed indicator, courtesy lights, emergency locator transmitter, alternate static source, navigation light detectors, heated pitot and omni-flash beacon standard on Skyhawk II, optional on Skyhawk. Optional extras for both models include Cessna Series 300 720-channel transceiver, 720-channel nav/com with remote VOR indicator, 720-channel nav/com with remote VOR/LOC indicator or VOR/ILS indicator, ADF, marker beacon with three lights and aural signal, transponder with 4096 code capability, DME, 10-channel HF transceiver, Nav-O-Matic autopilot with heading control plus VOR, and Series 400 glideslope receiver, boom microphone with control-wheel switch, control-wheel map light, sensitive altimeter, directional gyro with movable heading index, electric clock, outside air temperature gauge, landing light, rate of climb indicator, turn co-ordinator, map and instrument panel light, carburettor air temperature gauge, turn and bank indicator, horizon and directional gyros with vacuum system, sun visors, towbar, flight hour recorder, cabin fire extinguisher, headrests, rear-view mirror, child's foldaway seat, rear seats with individual reclining backs, front seats with articulating recline and vertical adjustment, utility shelf, safety belts for third and fourth seats, inertia reel shoulder harnesses, anti-precipitation static kit, overhead skylights, portable stretcher, rear-seat ventilation system, hinged window on starboard side, full-flow oil filter, engine primer system, wing-strut and fuselage steps and handles for easy refuelling, quick-drain oil valve, internal corrosion proofing, floatplane kit, external power socket, glider tow hook, beacon, dual cowl-mounted landing lights, wingtip strobe lights, tailplane abrasion boots, tinted windows and winterisation kit.

DIMENSIONS, EXTERNAL: (L: landplane; F: floatplane):
Wing span 10·92 m (35 ft 10 in)
Wing chord at root 1·63 m (5 ft 4 in)
Wing chord at tip 1·12 m (3 ft 8½ in)
Wing aspect ratio 7·52
Length overall: L 8·20 m (26 ft 11 in)
F 8·23 m (27 ft 0 in)
Height overall: L 2·68 m (8 ft 9½ in)
F 3·02 m (9 ft 11 in)
Tailplane span 3·45 m (11 ft 4 in)
Wheel track: L 2·53 m (8 ft 3½ in)
Wheelbase: L 1·63 m (5 ft 4 in)
Propeller diameter:
L 1·91 m (6 ft 3 in)
F 2·03 m (6 ft 8 in)
Passenger doors (each):
Height 1·01 m (3 ft 3¾ in)
Width 0·89 m (2 ft 11 in)

AREAS:
Wings, gross 16·17 m² (174 sq ft)
Ailerons (total) 1·70 m² (18·3 sq ft)
Trailing-edge flaps (total) 1·97 m² (21·20 sq ft)
Fin 1·04 m² (11·24 sq ft)
Rudder 0·69 m² (7·43 sq ft)
Tailplane 2·00 m² (21·56 sq ft)
Elevators, incl tab 1·35 m² (14·53 sq ft)

WEIGHTS AND LOADINGS (Skyhawk landplane: L; floatplane: F):
Weight empty, equipped:
L 618 kg (1,363 lb)
F 707 kg (1,558 lb)
Skyhawk II 654 kg (1,441 lb)
Max T-O weight:
L 1,043 kg (2,300 lb)
F 1,007 kg (2,220 lb)
Skyhawk II 1,043 kg (2,300 lb)
Max wing loading:
L 64·4 kg/m² (13·2 lb/sq ft)
F 62·0 kg/m² (12·7 lb/sq ft)
Skyhawk II 64·4 kg/m² (13·2 lb/sq ft)
Max power loading:
L 9·31 kg/kW (15·3 lb/hp)
F 8·99 kg/kW (14·8 lb/hp)

PERFORMANCE (L: Skyhawk and Skyhawk II landplane; F: floatplane, at max T-O weight):
Never-exceed speed:
L 151 knots (280 km/h; 174 mph)
Max level speed at S/L:
L 125 knots (232 km/h; 144 mph)
F 98 knots (182 km/h; 113 mph)
Max cruising speed (75% power):
L, at 2,440 m (8,000 ft) 120 knots (222 km/h; 138 mph)
F, at 2,285 m (7,500 ft) 97 knots (180 km/h; 112 mph)
Econ cruising speed at 3,050 m (10,000 ft):
L 101 knots (187 km/h; 116 mph)
F 86 knots (159 km/h; 99 mph)
Stalling speed, flaps up:
L 50 knots (92 km/h; 57 mph) CAS
F 48 knots (88·5 km/h; 55 mph) CAS
Stalling speed, flaps down:
L 43 knots (79 km/h; 49 mph)
F 44 knots (82 km/h; 51 mph)
Max rate of climb at S/L:
L 196 m (645 ft)/min
F 218 m (715 ft)/min
Service ceiling:
L 3,995 m (13,100 ft)
F 3,660 m (12,000 ft)
T-O run:
L 264 m (865 ft)
F 494 m (1,620 ft)
T-O to 15 m (50 ft):
L 465 m (1,525 ft)
F 729 m (2,390 ft)
Landing from 15 m (50 ft):
L 381 m (1,250 ft)
F 410 m (1,345 ft)
Landing run:
L 158 m (520 ft)
F 180 m (590 ft)
Range, at recommended lean mixture, with allowances for engine start, taxi, T-O, climb and 45 min reserves at 45% power:
Standard fuel at 2,440 m (8,000 ft):
L 450 nm (834 km; 518 miles)
F 365 nm (676 km; 420 miles)
Max fuel at 2,440 m (8,000 ft):
L 595 nm (1,102 km; 685 miles)
F 480 nm (890 km; 553 miles)
Standard fuel at 3,050 m (10,000 ft)
L 480 nm (890 km; 553 miles)
F 385 nm (713 km; 443 miles)
Max fuel at 3,050 m (10,000 ft):
L 640 nm (1,186 km; 737 miles)
F 510 nm (944 km; 587 miles)

CESSNA MODEL R172E

US Army designation: T-41B Mescalero
US Air Force designations: T-41C/D Mescalero

The Cessna Model R172E is a more powerful version of the original Model 172. Its design was started in late 1963, and a prototype was then built, with a 134 kW (180 hp) Continental O-360 engine. Type Approval was received in 1964, but the original power plant was replaced in the production Model R172E by a fuel-injection IO-360 engine.

In August 1966, the US Army ordered 255 aircraft of this type, under the designation **T-41B**, for training and installation support duties. Delivery of these was completed in March 1967.

In October 1967, the US Air Force ordered 45 similar aircraft, with fixed-pitch propellers, under the designation **T-41C**, for cadet flight training at the USAF Academy in Colorado. A total of 52 had been produced by 1 February 1976. Thirty **T-41Ds**, with constant-speed propellers and 28V electrical systems, were ordered initially for the Colombian Air Force, and deliveries of this version to all operators totalled 238 by 1 February 1976.

In addition, a version known as the Reims Rocket is being produced by Reims Aviation in France (which see). A total of 551 Rockets had been delivered by 1 January 1976.

The description of the Skyhawk applies also to the R172E, except for the following details:

Power Plant: One 156·5 kW (210 hp) Continental IO-360-D flat-six engine, driving a McCauley 2A34-C209/78CCA-2 constant-speed propeller. Two fuel tanks in wings with total capacity of 197 litres (52 US gallons), of which 174 litres (46 US gallons) are usable. Provision for long-range tanks, giving total usable capacity of 238 litres (63 US gallons). Refuelling points above wing. Oil capacity 9·5 litres (2·5 US gallons).

Accommodation: Basically as for Skyhawk. T-41B has special crew seatbacks and shoulder harness, with forward-hinged door on each side of cabin by crew seats. Baggage capacity 90·5 kg (200 lb).

Electronics and Equipment: The T-41B, C and D have variations in their electronics and other equipment consistent with their military roles.

Dimensions, external:
Propeller diameter 1·93 m (6 ft 4 in)

Weights and Loadings:
Weight empty, equipped 637 kg (1,405 lb)
Max T-O and landing weight 1,156 kg (2,550 lb)
Max wing loading 71·3 kg/m² (14·6 lb/sq ft)
Max power loading 7·39 kg/kW (12·1 lb/hp)

Performance (at max T-O weight):
Never-exceed speed 158 knots (293 km/h; 182 mph)
Max level speed at S/L 133 knots (246 km/h; 153 mph)
Max cruising speed at 1,675 m (5,500 ft) 126 knots (233 km/h; 145 mph)
Econ cruising speed at 3,050 m (10,000 ft) 91 knots (169 km/h; 105 mph)
Stalling speed, flaps up 55·6 knots (103 km/h; 64 mph)
Stalling speed, flaps down 46 knots (85 km/h; 53 mph)
Max rate of climb at S/L 268 m (880 ft)/min
Service ceiling 5,180 m (17,000 ft)
T-O run 226 m (740 ft)
T-O to 15 m (50 ft) 375 m (1,230 ft)
Landing from 15 m (50 ft) 387 m (1,270 ft)
Landing run 189 m (620 ft)
Range with max fuel at econ cruising speed at 3,050 m (10,000 ft) 877 nm (1,625 km; 1,010 miles)

CESSNA CARDINAL

On 30 September 1967, Cessna introduced its Model 177, a single-engined four-seat aircraft with a cantilever wing, then powered by a 112 kW (150 hp) Lycoming engine, and intended as a luxury addition to its range of single-engined two- and four-seat models. Increased engine power was provided subsequently by installation of the 134 kW (180 hp) Lycoming O-360 engine as standard.

Five versions of this aircraft became available by late 1970, of which the Model 177 was the basic standard version. It has been discontinued for 1976, leaving four commercial versions currently available:

Cardinal. Standard 1976 version, introducing as standard a redesigned instrument p .nel, an airspeed indicator with the primary scale in knots, rigid vent windows, flush-mounted electronics cooling scoop, better cabin ventilation, chrome seat adjustment handles and seat belt retainers, and a semi solid-state voltage regulator.

Cardinal II. As Cardinal, plus dual controls, true airspeed indicator, Cessna Series 300 nav/com with 360-channel com and 160-channel nav with VOR indicator, emergency locator transmitter, alternate static source, heated pitot and courtesy lights as standard.

Cardinal RG and RG II. Versions with retractable landing gear, described separately.

A total of 3,261 Model 177/Cardinals had been built by 1 January 1976.

Type: Four-seat cabin monoplane.

Wings: Cantilever high-wing monoplane. Wing section modified NACA 2400 series. Dihedral 1° 30′. Incidence 3° 30′ at root, 0° 30′ at tip. All-metal structure except for conical-camber glassfibre wingtips. Modified Frise all-metal ailerons. Electrically-operated wide-span all-metal slotted flaps.

Cessna Cardinal II with non-retractable landing gear (Lycoming O-360-A1F6 engine)

Fuselage: All-metal semi-monocoque structure of low profile.

Tail Unit: Cantilever all-metal structure. Sweepback on fin 35° at quarter-chord. All-moving tailplane, with large controllable trim tab. Controllable rudder trim tab.

Landing Gear: Non-retractable tricycle type. Improved Cessna Land-O-Matic cantilever main legs, each comprising a one-piece machined conically-tapered steel tube. Nosewheel is carried on a short-stroke oleo-pneumatic shock-strut, with hydraulic damper, and is steerable with rudder up to 12° each side and controllable up to 45° on each side. Cessna main wheels size 6·00-6 and nosewheel size 5·00-5, with nylon cord tube-type tyres. Tyre pressure 2·07 bars (30 lb/sq in). Single-caliper hydraulic disc brakes. Parking brake locks both main wheels. Wheel fairings optional.

Power Plant: One 134 kW (180 hp) Lycoming O-360-A1F6 flat-four engine, driving a two-blade constant-speed metal propeller. Pointed aluminium spinner. Fuel is carried in a 94·5 litre (25 US gallon) integral fuel tank in each wing, vented at the wingtip. Total usable fuel 185 litres (49 US gallons). Optional fuel system of 230 litres (61 US gallons) capacity, of which 227 litres (60 US gallons) are usable. Refuelling point on top of each wing. Oil capacity 7·5 litres (2 US gallons).

Accommodation: Cabin seats four in two pairs. Optional seat for two children aft of rear seats. Baggage compartment in rear fuselage, capacity 54 kg (120 lb), with large forward-hinged external access door in port side of fuselage. Forward-hinged door on each side of cabin, forward of main landing gear. Combined heating and ventilation system. Glassfibre soundproofing. Dual controls standard on Cardinal II.

Systems: Electrical supply from 12V 60A alternator. 12V 25Ah battery. 12V 33Ah battery optional.

Electronics and Equipment: True airspeed indicator, emergency locator transmitter, alternate static source, heated pitot, courtesy lights and Cessna Series 300 nav/com with 360-channel com and 160-channel nav, with VOR indicator standard on Cardinal II, optional on Cardinal. Optional equipment for both models includes Cessna Series 300 360-channel transceiver, 360-channel nav/com with remote VOR/LOC indicator or VOR/ILS indicator, ADF, marker beacon with three lights and aural signal, transponder with 4096 code capability, DME, 10-channel HF transceiver, Nav-O-Matic autopilot with heading control plus VOR, Series 400 glideslope receiver, boom microphone with control wheel switch, control wheel map light, sensitive altimeter, directional gyro with movable heading index, carburettor air temperature gauge, economy mixture indicator, turn and bank indicator, flight hour recorder, cabin fire extinguisher, headrests, electric clock, outside air temperature gauge, landing light, rate-of-climb indicator, turn co-ordinator indicator, sun visors, horizon and directional gyros with vacuum system, tailcone lift handles, towbar, rearview mirror, child's seat, individual articulating and vertically adjustable front seats, individual reclining rear seats, safety belts for third and fourth seats, stretcher, rear seat ventilation system, internal corrosion proofing, navigation light detectors, external power socket, engine primer system, quick-drain oil valve, cowl-mounted landing lights, wingtip strobe lights, tinted windows, wing leveller system and engine winterisation kit. Standard equipment includes omni-flash beacon, windscreen defroster, stall warning indicator, control locks, full-flow oil filter and navigation lights.

Dimensions, external:
Wing span 10·82 m (35 ft 6 in)
Wing chord at root 1·68 m (5 ft 6 in)
Wing chord at tip 1·22 m (4 ft 0 in)
Wing aspect ratio 7·31
Length overall 8·31 m (27 ft 3 in)
Height overall 2·62 m (8 ft 7 in)
Tailplane span 3·61 m (11 ft 10 in)
Wheel track 2·53 m (8 ft 3½ in)
Wheelbase 1·94 m (6 ft 4½ in)
Propeller diameter 1·93 m (6 ft 4 in)
Passenger doors (each):
Height 1·12 m (3 ft 8 in)
Width 1·22 m (4 ft 0 in)
Height to sill 0·69 m (2 ft 3 in)

Dimensions, internal:
Cabin:
Length 4·46 m (14 ft 7½ in)
Max width 1·22 m (4 ft 0 in)
Max height 1·13 m (3 ft 8½ in)

Areas:
Wings, gross 16·2 m² (174·0 sq ft)
Ailerons (total) 1·75 m² (18·86 sq ft)
Trailing-edge flaps (total) 2·74 m² (29·50 sq ft)
Fin 1·02 m² (11·02 sq ft)
Rudder 0·60 m² (6·41 sq ft)
Tailplane, incl tab 3·25 m² (35·01 sq ft)

Weights and Loadings:
Weight empty:
Cardinal 695 kg (1,533 lb)
Cardinal II 707 kg (1,560 lb)
Max T-O weight:
Cardinal and Cardinal II 1,134 kg (2,500 lb)
Max wing loading:
Cardinal and Cardinal II 70·3 kg/m² (14·4 lb/sq ft)
Max power loading:
Cardinal and Cardinal II 8·46 kg/kW (13·9 lb/hp)

Performance (Cardinal and Cardinal II at max T-O weight):
Max level speed at S/L 139 knots (257 km/h; 160 mph)
Max cruising speed, 75% power at 3,050 m (10,000 ft) 130 knots (241 km/h; 150 mph)
Stalling speed, flaps up 52 knots (96·5 km/h; 60 mph) IAS
Stalling speed, flaps down 40 knots (74 km/h; 46 mph) IAS
Max rate of climb at S/L 256 m (840 ft)/min
Service ceiling 4,450 m (14,600 ft)
T-O run 229 m (750 ft)
T-O to 15 m (50 ft) 427 m (1,400 ft)
Landing from 15 m (50 ft) 372 m (1,220 ft)
Landing run 183 m (600 ft)
Range, recommended lean mixture with allowances for start, taxi, T-O, climb and 45 min reserves at 45% power:
Max cruising speed at 3,050 m (10,000 ft) with standard fuel 535 nm (991 km; 616 miles)
Max cruising speed at 3,050 m (10,000 ft) with max fuel 675 nm (1,250 km; 777 miles)
Econ cruising speed at 3,050 m (10,000 ft) with standard fuel 615 nm (1,139 km; 708 miles)
Econ cruising speed at 3,050 m (10,000 ft) with max fuel 780 nm (1,445 km; 898 miles)

CESSNA CARDINAL RG and RG II

On 3 December 1970 Cessna announced a further version of its Cardinal single-engined four-seat cabin monoplane, with hydraulically-retractable tricycle-type landing gear, a more powerful fuel-injection engine and a number of different standard and optional items.

The landing gear is retracted by a simplified self-contained hydraulic system, with an electrically-powered hydraulic pump that provides a maximum system pressure of 103·5 bars (1,500 lb/sq in). The hand pump, for emergency retraction or extension of the gear, is designed to eliminate the need for complex sequencing valves in the hydraulic power pack. When the landing gear is retracted the nose unit is faired by wheel doors; the main gear is retained flush with the fuselage and has no wheel doors.

Two versions of the Cardinal RG are available for 1976:

Cardinal RG. Standard retractable-gear version, introducing the same standard and optional improvements as the fixed-gear Cardinal, plus new standard features which include improved wheels and brakes, repositioned fuel sump drain, better cabin heating and ventilation, redesigned landing gear and a single audio source with a dual voice coil which provides for normal com usage, and a second coil for landing gear/stall warning signal.

Cardinal RG II. Specially-equipped version of the Cardinal RG which includes as standard Cessna Series 300 nav/com with 720-channel com and 200-channel nav with remote VOR/LOC indicator, ADF, transponder, 200A Nav-O-Matic autopilot with VOR/LOC track and intercept functions, horizon and directional gyros, true airspeed indicator, dual controls, navigation light detectors, external power socket, heated pitot, courtesy lights and emergency locator transmitter. Nav-Pac IFR option adds a second 300 nav/com with VOR/ILS, 400 glideslope and 400 marker beacon.

By 1 January 1976 a total of 886 Cardinal RGs had been delivered, the total including 134 produced by Reims Aviation in France.

Cessna Cardinal RG II, a retractable-gear version of the four-seat Cardinal

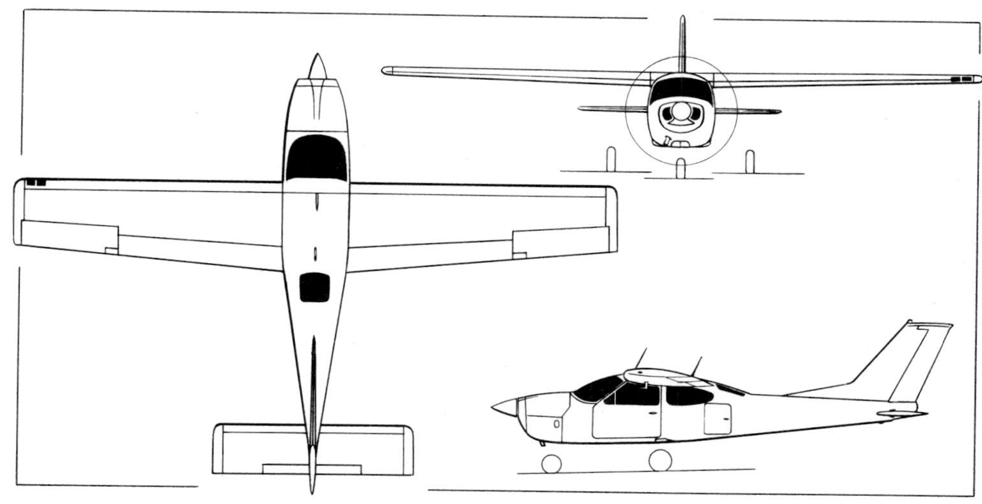

Cessna Cardinal RG four-seat light aircraft with retractable landing gear *(Pilot Press)*

TYPE: Four-seat cabin monoplane.

WINGS: Cantilever high-wing monoplane. Wing section NACA 64A215 at root, NACA 64A412 at tip. Dihedral 1° 30′. Incidence 4° 7′ at root, 0° 43′ at tip. All-metal structure except for glassfibre wingtips. All-metal ailerons. Electrically-operated all-metal trailing-edge flaps.

FUSELAGE: All-metal semi-monocoque structure of low profile.

TAIL UNIT: Cantilever all-metal structure with swept vertical surfaces. All-moving tailplane with large controllable trim tab.

LANDING GEAR: Hydraulically-retractable tricycle type. Tubular spring steel main gear struts, retracting rearward into fuselage. Nosewheel, which retracts rearward, is carried on a short-stroke oleo-pneumatic shock-strut with hydraulic damper, and is steerable. Nosewheel enclosed by doors when retracted. Hydraulic brakes. Parking brake.

POWER PLANT: One 149 kW (200 hp) Lycoming IO-360-A1B6D flat-four engine, driving a two-blade constant-speed metal propeller. Pointed metal spinner. One 115 litre (30·5 US gallon) integral fuel tank in each wing. Total fuel capacity 230 litres (61 US gallons), of which 227 litres (60 US gallons) are usable. Refuelling point in top of each wing. Oil capacity 7·5 litres (2 US gallons).

ACCOMMODATION: Cabin seats four in two pairs. Forward-hinged door on each side of cabin, forward of main landing gear. Cabin heated and ventilated. Baggage compartment in rear fuselage, capacity 54 kg (120 lb), with large forward-hinged external access door on port side of fuselage.

SYSTEMS: Electrical system powered by a 14V 60A alternator and 12V battery for operation of wing flaps, hydraulic motor, electronics and lighting. Hydraulic system for landing gear retraction and brakes.

ELECTRONICS AND EQUIPMENT: Optional items are as detailed for the Cardinal, plus Cessna Series 400 720-channel transceiver, 720-channel nav/com with remote VOR/LOC indicator or VOR/ILS indicator, ADF with digital tuning, transponder with 4096 code capability. Standard equipment includes inertia safety belts for pilot and co-pilot and omni-flash beacon at tip of fin.

DIMENSIONS, EXTERNAL: As for Cardinal, except:
- Wheel track — 2·39 m (7 ft 10 in)
- Propeller diameter — 1·98 m (6 ft 6 in)

WEIGHTS AND LOADINGS:
- Weight empty:
  - RG — 774 kg (1,707 lb)
  - RG II — 802 kg (1,768 lb)
- Max T-O weight — 1,270 kg (2,800 lb)
- Max wing loading — 78·6 kg/m² (16·1 lb/sq ft)
- Max power loading — 8·52 kg/kW (14·0 lb/hp)

PERFORMANCE (at max T-O weight):
- Max level speed at S/L — 156 knots (290 km/h; 180 mph)
- Max cruising speed (75% power) at 2,135 m (7,000 ft) — 148 knots (273 km/h; 170 mph)
- Econ cruising speed at 3,050 m (10,000 ft) — 121 knots (223 km/h; 139 mph)
- Stalling speed, flaps up 57 knots (106 km/h; 66 mph)
- Stalling speed, flaps down — 50 knots (92 km/h; 57 mph)
- Max rate of climb at S/L — 282 m (925 ft)/min
- Service ceiling — 5,210 m (17,100 ft)
- T-O run — 271 m (890 ft)
- T-O to 15 m (50 ft) — 483 m (1,585 ft)
- Landing from 15 m (50 ft) — 411 m (1,350 ft)
- Landing run — 223 m (730 ft)
- Range, recommended lean mixture with allowances for start, taxi, T-O, climb and 45 min reserves at 45% power:
  - Max cruising speed at 2,135 m (7,000 ft) with 227 litres (60 US gallons) usable fuel — 715 nm (1,324 km; 823 miles)
  - Econ cruising speed at 3,050 m (10,000 ft) with 227 litres (60 US gallons) usable fuel — 895 nm (1,657 km; 1,030 miles)

CESSNA MODEL 180 SKYWAGON

The Model 180 Skywagon has a typical Cessna braced high-wing monoplane layout, but with a tailwheel type of landing gear.

The 1976 version of the Model 180 introduces as standard a redesigned instrument panel; an airspeed indicator with the primary scale in knots; improved heater vents, passenger seating and electronics cooling; a new three-way fuel selector valve; and a semi solid-state voltage regulator. New options include retractable tailcone lift handles, refuelling assist handles, tailwheel jack pad and an anti-precipitation static kit. The flaps can be extended at higher speeds than on previous models.

A total of 5,648 Model 180s had been built by 1 January 1976.

TYPE: One/six-seat cabin monoplane.

WINGS: Generally similar in construction to those of Skyhawk. Dihedral 1° 44′.

FUSELAGE: All-metal semi-monocoque structure. Identical to fuselage of Cessna 185, except for firewall and mounting brackets for dorsal fin.

TAIL UNIT: Unswept cantilever all-metal structure with adjustable-incidence tailplane. Normally no trim tabs; but manually-operated rudder trim is optionally available.

LANDING GEAR: Non-retractable tailwheel type. Cessna cantilever spring steel main legs. Tailwheel has tapered tubular spring. Main wheels and nylon tube-type tyres size 6·00-6 (optionally 8·00-6). Scott tailwheel size 8·00 × 2·80. Tyre pressure, main wheels 2·07 bars (30 lb/sq in), tailwheel 3·79-4·48 bars (55-65 lb/sq in) according to load. Hydraulic disc brakes. Parking brake. Alternative Edo Model 628-296 floats, snow ski or amphibian gear.

POWER PLANT: One 171·5 kW (230 hp) Continental O-470-S flat-six engine, driving a McCauley 2A34C203/90DA-8 constant-speed metal propeller. Two fuel tanks in wings, with total standard capacity of 246 litres (65 US gallons) and usable capacity of 227 litres (60 US gallons). Optional long-range tanks with total capacity of 318 litres (84 US gallons), of which 299 litres (79 US gallons) are usable. Oil capacity 11·5 litres (3 US gallons).

ACCOMMODATION: Standard seating is for a pilot only, with a choice of three optional arrangements. Maximum seating is for six persons in three pairs, without baggage space. With fewer seats there is space at rear of cabin for up to 181 kg (400 lb) of baggage. Door on each side of cabin, plus optional cargo door and baggage compartment door on port side. Starboard door has quick-release hinge pins so that it can be removed when loading bulky cargo. Fifth and sixth passenger seats, attached to aft wall of cabin, can be folded when space is required for cargo. Hinged window each side. Instrument lighting controls are transistorised. Heating and ventilation standard. Fully-articulating seats for pilot and co-pilot, child's foldaway seat for the rear cabin and safety belts for rear-seat passengers are available optionally. Dual controls optional.

SYSTEMS: Hydraulic system for brakes only. Electrical system powered by 14V 60A alternator. 12V 33Ah battery. Oxygen system, 1·36 m³ (48 cu ft) capacity, optional.

ELECTRONICS AND EQUIPMENT: Optional electronics include Cessna Series 300 360-channel com transceiver, 360-channel nav/com with 160-channel nav and remote VOR indicator, 720-channel com and 200-channel nav with remote VOR/LOC indicator or VOR/ILS indicator, ADF with digital tuning, marker beacon with three lights and aural signal, transponder with 4096 code capability, DME, Nav-O-Matic single-axis autopilot with heading control and VOR intercept and track; or Series 400 720-channel com transceiver, 720-channel nav/com with remote VOR/LOC or VOR/ILS indicator, transponder with 4096 code capability, glideslope receiver and ADF with digital tuning. Standard equipment includes audible stall warning indicator, instrument panel red floodlights, control locks, windscreen defroster, cabin dome light, landing and taxi lights, and baggage restraint net. Optional equipment includes blind-flying instrumentation, boom microphone, electric clock, control wheel with map light and microphone switch, carburettor air temperature gauge, outside air temperature gauge, true airspeed indicator, economy mixture indicator, rate of climb indicator, turn co-ordinator, turn and bank indicator, instrument panel post lights, rudder pedal extensions, flight hour recorder, pilot and co-pilot headrests, twin beverage pack, cargo tie-down fittings, de luxe interior, inertia-reel shoulder harness, bubble windows, amphibian kit, floatplane kit, jack pad, strobe light, tailwheel lock, ski axles, ski provisions, map and auxiliary instrument light, courtesy lights, co-pilot's seat, stretcher installation, sun visors, emergency locator transmitter, tinted windows, heated pitot, non-congealing oil cooler, oil dilution system, engine winterisation kit, internal corrosion proofing, navigation light detectors, external power socket, omni-flash beacon, quick drain oil valve, overall paint scheme, photographic provisions, agricultural sprayer system, tailplane abrasion boots, anti-precipitation static kit, tailcone lift handles, tailwheel jack pad, refuelling assist handles, alternate static source and cabin fire extinguisher.

DIMENSIONS, EXTERNAL:
- Wing span: 10·92 m (35 ft 10 in)
- Wing chord at root: 1·63 m (5 ft 4 in)
- Wing chord at tip: 1·11 m (3 ft 7¾ in)
- Wing aspect ratio: 7·52
- Length overall:
  - Landplane, skiplane: 7·81 m (25 ft 7½ in)
  - Floatplane: 8·23 m (27 ft 0 in)
  - Amphibian: 8·38 m (27 ft 6 in)
- Height overall:
  - Landplane, skiplane: 2·36 m (7 ft 9 in)
  - Floatplane: 3·71 m (12 ft 2 in)
  - Amphibian: 3·86 m (12 ft 8 in)
- Tailplane span: 3·30 m (10 ft 10 in)
- Wheel track, landplane: 2·33 m (7 ft 8 in)
- Propeller diameter:
  - Landplane, skiplane: 2·08 m (6 ft 10 in)
  - Floatplane, amphibian: 2·24 m (7 ft 4 in)
- Passenger doors (each):
  - Height: 1·01 m (3 ft 3¾ in)
  - Width: 0·89 m (2 ft 11 in)

AREAS:
- Wings, gross: 16·16 m² (174 sq ft)
- Ailerons (total): 1·70 m² (18·3 sq ft)
- Trailing-edge flaps (total): 1·97 m² (21·23 sq ft)
- Fin: 0·84 m² (9·01 sq ft)
- Dorsal fin: 0·19 m² (2·04 sq ft)
- Rudder: 0·68 m² (7·29 sq ft)
- Tailplane: 1·94 m² (20·94 sq ft)
- Elevators: 1·40 m² (15·13 sq ft)

WEIGHTS AND LOADINGS:
- Weight empty, equipped:
  - Landplane: 733 kg (1,617 lb)
  - Floatplane: 873 kg (1,924 lb)
  - Skiplane: 798 kg (1,759 lb)
  - Amphibian: 988 kg (2,179 lb)
- Max T-O weight:
  - Landplane, skiplane: 1,270 kg (2,800 lb)
  - Floatplane, amphibian: 1,338 kg (2,950 lb)
- Max wing loading:
  - Landplane, skiplane: 78·6 kg/m² (16·1 lb/sq ft)
  - Floatplane, amphibian: 83·0 kg/m² (17·0 lb/sq ft)
- Max power loading:
  - Landplane, skiplane: 7·41 kg/kW (12·2 lb/hp)
  - Floatplane, amphibian: 7·80 kg/kW (12·8 lb/hp)

PERFORMANCE (at max T-O weight):
- Never-exceed speed:
  - Landplane: 167 knots (309 km/h; 192 mph)
- Max level speed at S/L:
  - Landplane: 148 knots (274 km/h; 170 mph)
  - Floatplane, amphibian, skiplane: 129 knots (240 km/h; 149 mph)
- Max cruising speed (75% power) at 1,980 m (6,500 ft):
  - Landplane: 141 knots (261 km/h; 162 mph)
  - Floatplane, amphibian: 127 knots (235 km/h; 146 mph)
  - Skiplane: 125 knots (232 km/h; 144 mph)
- Econ cruising speed at 3,050 m (10,000 ft):
  - Landplane: 105 knots (195 km/h; 121 mph)
  - Floatplane, amphibian: 99 knots (183 km/h; 114 mph)
  - Skiplane: 88 knots (162 km/h; 101 mph)
- Stalling speed, flaps up, power off:
  - All versions: 53 knots (98·5 km/h; 61 mph) CAS
- Stalling speed, flaps down, power off:
  - All versions: 48 knots (88·5 km/h; 55 mph) CAS
- Max rate of climb at S/L:
  - Landplane: 332 m (1,090 ft)/min
  - Floatplane, amphibian: 302 m (990 ft)/min
  - Skiplane: 271 m (890 ft)/min
- Service ceiling:
  - Landplane: 5,975 m (19,600 ft)
  - Floatplane, amphibian: 4,877 m (16,000 ft)
  - Skiplane: 4,755 m (15,600 ft)
- T-O run:
  - Landplane: 190 m (625 ft)
  - Floatplane: 390 m (1,280 ft)
  - Amphibian, on land: 415 m (1,360 ft)
  - Amphibian, on water: 390 m (1,280 ft)
- T-O to 15m (50 ft):
  - Landplane: 367 m (1,205 ft)
  - Floatplane: 631 m (2,070 ft)
  - Amphibian, on land: 666 m (2,185 ft)
  - Amphibian, on water: 631 m (2,070 ft)
- Landing from 15 m (50 ft):
  - Landplane: 416 m (1,365 ft)
  - Floatplane: 524 m (1,720 ft)
  - Amphibian, on land: 454 m (1,490 ft)
  - Amphibian, on water: 524 m (1,720 ft)
- Landing run:
  - Landplane: 146 m (480 ft)
  - Floatplane: 224 m (735 ft)
  - Amphibian, on land: 312 m (1,025 ft)
  - Amphibian, on water: 224 m (735 ft)
- Range, at recommended lean mixture with allowances for start, taxi, T-O, climb and 45 min reserves at 45% power:

Cessna Model 180 Skywagon one/six-seat cabin monoplane (Continental O-470-S engine)

- Standard fuel, max cruising speed at 1,980 m (6,500 ft):
  - Landplane: 470 nm (870 km; 541 miles)
  - Floatplane, amphibian: 420 nm (777 km; 483 miles)
  - Skiplane: 410 nm (759 km; 472 miles)
- Max fuel, max cruising speed at 1,980 m (6,500 ft):
  - Landplane: 660 nm (1,223 km; 760 miles)
  - Floatplane, amphibian: 595 nm (1,102 km; 685 miles)
  - Skiplane: 580 nm (1,075 km; 668 miles)
- Standard fuel, econ cruising speed at 3,050 m (10,000 ft):
  - Landplane: 580 nm (1,075 km; 668 miles)
  - Floatplane, amphibian: 550 nm (1,019 km; 633 miles)
  - Skiplane: 460 nm (851 km; 529 miles)
- Max fuel, econ cruising speed at 3,050 m (10,000 ft):
  - Landplane: 825 nm (1,529 km; 950 miles)
  - Floatplane, amphibian: 790 nm (1,464 km; 910 miles)
  - Skiplane: 655 nm (1,213 km; 754 miles)

CESSNA SKYLANE

Three variants of this aircraft were available, of which the Model 182 was the basic standard version. It has been discontinued for 1976, leaving two commercial versions currently available:

Skylane. Standard version.

Skylane II. As Skylane, plus factory-installed electronics package which includes Cessna Series 300 nav/com with 720-channel com and 200-channel nav with remote VOR/LOC, ADF, transponder and 200A Nav-O-Matic autopilot with VOR/LOC track and intercept. It is available optionally with a Nav Pac, which adds a second 300 nav/com plus VOR/ILS, Series 400 glideslope and Series 300 marker beacon. Standard equipment includes horizon and pictorial directional gyros, true airspeed indicator, dual controls, navigation light detectors, external power socket, heated pitot, courtesy lights and emergency locator transmitter.

The 1976 versions of the Skylane have a number of improvements as standard, including a redesigned instrument panel, new airspeed indicator with primary scale in knots, recontoured glareshield, flush-mounted air cooling scoop for the electronics, a semi solid-state voltage regulator and, to improve performance, redesigned landing-gear-to-fuselage and wing root fairings, bonded fuel tank covers, redesigned fin and rudder tip fairings, and new wheels and brakes. New options include an air vent for rear-seat passengers, electrically controlled elevator trim, anti-precipitation static kit, and openable starboard window.

A version designated F182 Skylane is being introduced to the European market in 1975 by Reims Aviation in France (which see).

A total of 15,452 Model 182/Skylanes had been built by 1 January 1976.

TYPE: Four-seat cabin monoplane.

WINGS: Braced high-wing monoplane. Wing section NACA 2412, modified. Incidence at root 0° 47′, at tip −2° 50′. Dihedral 1° 44′. Wing structure similar to Skyhawk, except metal-to-metal bonded leading-edge.

FUSELAGE: All-metal semi-monocoque structure.

TAIL UNIT: Cantilever all-metal structure with swept fin and rudder. Trim tab in starboard elevator. Electrically-operated elevator trim optional.

LANDING GEAR: Non-retractable tricycle type. Land-O-Matic cantilever main legs, each comprising a one-piece machined conically-tapered spring steel tube. Steerable nosewheel with oleo-pneumatic shock-absorption. Cessna main wheels and tyres size 6·00-6, pressure 2·90 bars (42 lb/sq in). Cessna nosewheel and tyre size 5·00-5, pressure 3·38 bars (49 lb/sq in). Cessna hydraulic disc brakes. Parking brake. Optional wheel fairings.

POWER PLANT: Similar to that of Model 180, except for McCauley propeller type 2A34C203/90DCA-8.

ACCOMMODATION: Generally similar to Skyhawk, with standard seating for four; four seat-belts and two shoulder harnesses standard. Optional child's seat. Baggage space aft of rear seats and hatshelf with total capacity of 91 kg (200 lb), with external baggage door. Cargo tiedown net standard. Front seat inertia-reel shoulder harness, rear seat shoulder harness, leather seating, air vent for rear-seat passengers, and openable starboard window optional. Dual controls optional on Skylane.

SYSTEMS: Electrical system powered by 60A 14V engine-driven alternator. 12V battery. Hydraulic system for brakes only. Vacuum system optional. Oxygen system of 1·36 m³ (48 cu ft) capacity optional.

Cessna Skylane II four-seat cabin monoplane, which has a factory-installed electronics package

ELECTRONICS AND EQUIPMENT: Optional electronics include Cessna 200 Series 200A Nav-O-Matic autopilot, 300 Series 360-channel com transceiver, 720-channel nav/com with remote VOR/LOC or VOR/ILS indicator, ADF with digital tuning, marker beacon with three lights and aural signal, transponder with 4096 code capability, DME, 10-channel HF transceiver, 300A Nav-O-Matic single-axis autopilot with heading control plus VOR intercept and track, 400 Series glideslope receiver, ADF with digital tuning and transponder with 4096 code capability. Standard equipment includes audible stall warning device, variable-intensity instrument panel red floodlights, pedestal lights, control locks, armrests, windscreen defrosters, cabin dome light, baggage restraint net, adjustable cabin ventilators, tinted windscreen and windows, landing, taxi and navigation lights and cabin steps. Optional equipment includes blind-flying instrumentation, sensitive altimeter, electric clock, outside air temperature gauge, turn co-ordinator indicator, rate of climb indicator, control wheel with map light and microphone switch, carburettor air temperature gauge, economy mixture indicator, instrument post lights, flight hour recorder, rear window curtain, sun visors, cabin fire extinguisher, headrests, rearview mirror, inertia-reel shoulder harness for front seats, shoulder harness for rear seats, leather seating, child's seat, skylights, stretcher installation, utility shelf, non-congealing oil cooler, full-flow oil filter, quick drain oil valve, engine winterisation kit, engine priming system, overall paint scheme, towbar, internal corrosion proofing, heated stall warning transmitter, glider tow hook, omni-flash beacon, wingtip strobe lights, tailplane abrasion boots and tailcone lift handles. Horizon and pictorial directional gyros, true airspeed indicator, dual controls, navigation light detectors, external power socket, heated pitot, courtesy lights and emergency locator transmitter are standard for Skylane II, optional for Skylane.

DIMENSIONS, EXTERNAL:
- Wing span: 10·92 m (35 ft 10 in)
- Wing chord at root: 1·63 m (5 ft 4 in)
- Wing chord at tip: 1·09 m (3 ft 7 in)
- Length overall: 8·55 m (28 ft 0¾ in)
- Height overall: 2·79 m (9 ft 1¾ in)
- Tailplane span: 3·55 m (11 ft 8 in)
- Wheel track: 2·77 m (9 ft 1 in)
- Wheelbase: 1·69 m (5 ft 6½ in)
- Propeller diameter: 2·08 m (6 ft 10 in)
- Passenger doors (each):
  - Height: 1·02 m (3 ft 4¼ in)
  - Width: 0·90 m (2 ft 11¼ in)

AREAS:
- Wings, gross: 16·16 m^2 (174 sq ft)
- Ailerons (total): 1·70 m^2 (18·3 sq ft)
- Trailing-edge flaps (total): 1·97 m^2 (21·20 sq ft)
- Fin: 1·08 m^2 (11·62 sq ft)
- Rudder: 0·65 m^2 (6·95 sq ft)
- Tailplane: 2·13 m^2 (22·96 sq ft)
- Elevators: 1·47 m^2 (15·85 sq ft)

WEIGHTS AND LOADINGS:
- Weight empty, equipped:
  - Skylane: 777 kg (1,714 lb)
  - Skylane II: 805 kg (1,776 lb)
- Max T-O weight: 1,338 kg (2,950 lb)
- Max wing loading: 82·5 kg/m^2 (16·9 lb/sq ft)
- Max power loading: 7·80 kg/kW (12·8 lb/hp)

PERFORMANCE (at max T-O weight):
- Max level speed at S/L: 148 knots (273 km/h; 170 mph)
- Max cruising speed, 75% power at 2,135 m (7,000 ft): 144 knots (267 km/h; 166 mph)
- Stalling speed, flaps up: 56 knots (103 km/h; 64 mph)
- Stalling speed, flaps down: 50 knots (92 km/h; 57 mph)
- Max rate of climb at S/L: 271 m (890 ft)/min
- Service ceiling: 5,395 m (17,700 ft)
- T-O run: 215 m (705 ft)
- T-O to 15 m (50 ft): 411 m (1,350 ft)
- Landing from 15 m (50 ft): 411 m (1,350 ft)
- Landing run: 180 m (590 ft)
- Range, recommended lean mixture, with allowances for start, taxi, T-O, climb and 45 min reserves at 45% power:
  - Standard fuel, 75% power at 1,980 m (6,500 ft): 475 nm (880 km; 547 miles)
  - Max fuel, 75% power at 1,980 m (6,500 ft): 670 nm (1,241 km; 771 miles)
  - Standard fuel, econ cruising speed at 3,050 m (10,000 ft): 565 nm (1,046 km; 650 miles)
  - Max fuel, econ cruising speed at 3,050 m (10,000 ft): 810 nm (1,500 km; 932 miles)

CESSNA MODEL 185 SKYWAGON

US military designation: U-17

The prototype of the Model 185 Skywagon flew for the first time in July 1960 and the first production model was completed in March 1961. It is generally similar to the Model 180 Skywagon, except for installation of a 224 kW (300 hp) Continental IO-520 engine.

The Model 185 Skywagon can be fitted with Edo 628-2960 floats, or Edo Model 597 amphibious floats, or Fli-Lite skis, and is suitable for agricultural duties, using Quickly-removable Sorensen spraygear. It can carry under its fuselage a detachable glassfibre Cargo-Pack, more than 2·75 m long and 0·79 m wide (9 ft × 2 ft 7 in), with a volume of 0·61 m^3 (21·5 cu ft) and capacity of 136 kg (300 lb). The Pack incorporates loading doors on the side and at the rear.

Cessna Model 185 Skywagon one/six-seat cabin monoplane (Continental IO-520-D engine)

The 1976 version has the same improvements as those detailed for the Model 180.

Cessna has received important contracts to supply U-17A(B Skywagons to the US Air Force for delivery to overseas countries, under the US Military Assistance Programme.

A total of 2,831 Model 185 Skywagons, including U-17A/B Skywagons to the US Air Force for delivery to

TYPE: One/six-seat cabin monoplane.

WINGS AND FUSELAGE: Similar to Model 180.

TAIL UNIT: Same as for Model 180, except for fin of increased area and manually-operated rudder trim as standard equipment.

LANDING GEAR: Similar to Model 180, except for tyre pressures: main wheels (6·00-6) 2·41 bars (35 lb/sq in), main wheels (8·00-6) 1·72 bars (25 lb/sq in), tailwheel 3·79-4·83 bars (55-70 lb/sq in) depending on load. Manual tailwheel lock standard. Optional amphibian, float or ski gear.

POWER PLANT: One 224 kW (300 hp) Continental IO-520-D flat-six engine, driving a McCauley constant-speed metal propeller. Fuel in two tanks in wings, total capacity 246 litres (65 US gallons), of which 235 litres (62 US gallons) are usable. Extended-range tanks available as optional equipment in place of standard tanks, total capacity 318 litres (84 US gallons), of which 306·5 litres (81 US gallons) are usable. Oil capacity 11·4 litres (3 US gallons).

ACCOMMODATION, ELECTRONICS AND EQUIPMENT: Same as for Model 180, except omni-flash beacon and manual tailwheel lock standard.

DIMENSIONS:
- Same as for Model 180, except:
- Propeller diameter:
  - Landplane: 2·08 m (6 ft 10 in)
  - Floatplane, amphibian, skiplane: 2·18 m (7 ft 2 in)

AREAS:
- Same as for Model 180, except:
  - Fin: 1·29 m^2 (13·86 sq ft)

WEIGHTS AND LOADINGS:
- Weight empty, equipped:
  - Landplane: 725 kg (1,600 lb)
  - Floatplane: 866 kg (1,910 lb)
  - Amphibian: 982 kg (2,165 lb)
  - Skiplane: 791 kg (1,745 lb)
- Max T-O weight:
  - Landplane, skiplane: 1,519 kg (3,350 lb)
  - Floatplane: 1,506 kg (3,320 lb)
  - Amphibian, land take-off: 1,481 kg (3,265 lb)
  - Amphibian, water take-off: 1,406 kg (3;100 lb)
- Max wing loading:
  - Landplane, skiplane: 94·2 kg/m^2 (19·3 lb/sq ft)
  - Floatplane: 93·3 kg/m^2 (19·1 lb/sq ft)
  - Amphibian: 91·8 kg/m^2 (18·8 lb/sq ft)
- Max power loading:
  - Landplane, skiplane: 6·78 kg/kW (11·2 lb/hp)
  - Floatplane: 6·72 kg/kW (11·1 lb/hp)
  - Amphibian: 6·61 kg/kW (10·9 lb/hp)

PERFORMANCE (at max T-O weight):
- Never-exceed speed:
  - Landplane: 182 knots (338 km/h; 210 mph)
- Max level speed at S/L:
  - Landplane: 155 knots (286 km/h; 178 mph)
  - Floatplane: 141 knots (261 km/h; 162 mph)
  - Amphibian: 135 knots (251 km/h; 156 mph)
  - Skiplane: 136 knots (252 km/h; 157 mph)
- Max cruising speed (75% power) at 2,285 m (7,500 ft):
  - Landplane: 147 knots (272 km/h; 169 mph)
  - Floatplane: 135 knots (251 km/h; 156 mph)
  - Amphibian: 129 knots (240 km/h; 149 mph)
  - Skiplane: 133 knots (246 km/h; 153 mph)
- Econ cruising speed at 3,050 m (10,000 ft):
  - Landplane: 112 knots (208 km/h; 129 mph)
  - Floatplane: 94 knots (174 km/h; 108 mph)
  - Amphibian: 88 knots (162 km/h; 101 mph)
  - Skiplane: 109 knots (203 km/h; 126 mph)
- Stalling speed, flaps up, power off:
  - Landplane, skiplane, floatplane: 56·5 knots (105 km/h; 65 mph)
  - Amphibian: 55 knots (102 km/h; 63 mph)
- Stalling speed, flaps down, power off:
  - Landplane, skiplane: 49 knots (90·5 km/h; 56 mph)
  - Amphibian: 51 knots (94 km/h; 58 mph)
  - Floatplane: 52·5 knots (97 km/h; 60 mph)
- Max rate of climb at S/L:
  - Landplane: 308 m (1,010 ft)/min
  - Floatplane: 293 m (960 ft)/min
  - Amphibian: 296 m (970 ft)/min
- Service ceiling:
  - Landplane: 5,229 m (17,150 ft)
  - Floatplane: 5,000 m (16,400 ft)
  - Amphibian: 4,663 m (15,300 ft)
- T-O run:
  - Landplane: 235 m (770 ft)
  - Floatplane: 337 m (1,105 ft)
  - Amphibian, on land: 204 m (670 ft)
  - Amphibian, on water: 270 m (885 ft)
- T-O to 15 m (50 ft):
  - Landplane: 416 m (1,365 ft)
  - Floatplane: 530 m (1,740 ft)
  - Amphibian, on land: 389 m (1,275 ft)
  - Amphibian, on water: 436 m (1,430 ft)
- Landing from 15 m (50 ft):
  - Landplane: 427 m (1,400 ft)
  - Floatplane: 466 m (1,530 ft)
  - Amphibian, on land: 378 m (1,240 ft)
  - Amphibian, on water: 450 m (1,480 ft)
- Landing run:
  - Landplane: 146 m (480 ft)
  - Floatplane: 195 m (640 ft)
  - Amphibian, on land: 238 m (780 ft)
  - Amphibian, on water: 183 m (600 ft)
- Range at econ cruising speed at 3,050 m (10,000 ft), long-range tanks, no reserves:
  - Landplane: 898 nm (1,665 km; 1,035 miles)
  - Floatplane: 807 nm (1,496 km; 930 miles)
  - Amphibian: 755 nm (1,400 km; 870 miles)
  - Skiplane: 777 nm (1,440 km; 895 miles)
- Range at max cruising speed, standard tanks, no reserves (amphibian and floatplane only 223 litres; 59 US gallons usable):
  - Landplane: 573 nm (1,062 km; 660 miles)
  - Amphibian: 482 nm (893 km; 555 miles)
  - Floatplane: 503 nm (933 km; 580 miles)
  - Skiplane: 516 nm (957 km; 595 miles)
- Range at max cruising speed, long-range tanks, no reserves:
  - Landplane: 720 nm (1,335 km; 830 miles)
  - Floatplane: 664 nm (1,231 km; 765 miles)
  - Amphibian: 638 nm (1,182 km; 735 miles)
  - Skiplane: 651 nm (1,207 km; 750 miles)

CESSNA AGWAGON and AGTRUCK

On 8 December 1971, Cessna announced the introduction of four new agricultural aircraft, three of them based on the earlier AGwagon low-wing monoplane. Of these, the AGpickup has been discontinued for 1976.The high-wing AGcarryall is described separately.

The current AGwagon and AGtruck are of all-metal

construction and have special corrosion proofing, heavy-duty spring steel Land-O-Matic landing gear and Cessna's Camber-Lift wing to provide better control during low-speed operations. Wing fences are used to smooth airflow over the wing. Special attention has been paid to safety features, and these include ensolite padding on the upper instrument panel, urethane padding on tubular structures in the cabin area and around doors, safe flush switch and control locations and quick-release door hinges. Other standard features include wide wing walks, large hopper loading doors, and fresh-air scoops that slightly pressurise the cockpit and tailcone to prevent the ingress of dust and fumes.

Optional equipment includes a special night operations package to provide brilliant illumination for night operations. This comprises a 100A 24V alternator, taxi/landing lights, instrument panel lights, overhead floodlight, two 600W retractable spray lights, lighting angle control for spray lights, wingtip turning lights, hopper quantity light and a control stick grip incorporating light switches.

The 1976 versions of these agricultural aircraft introduce a number of improvements as standard, including a more efficient dispersal system, quick-release emergency system, and improved exhaust system and hopper door. New options include an electric spray control valve, leading-edge repair kit, wheel fenders and tailcone jack point.

Differences between the two models are as follows:

AGwagon. Basic model, powered by a 224 kW (300 hp) Continental IO-520-D flat-six engine, driving a constant-speed propeller. Standard equipment includes a 757 litre (200 US gallon) hopper, a liquid and dry material dispersal control system, cockpit canopy with all-round vision, tailplane abrasion boots, oversize 8·00-8 × 22 main-wheel tyres, 10 in tailwheel tyre, wire cutters, cable deflector, tailcone lift handles, hopper side-loading system on port side, navigation lights, pilot's four-way adjustable seat, control stick lock, quick oil drain, auxiliary fuel pump, steerable tailwheel and remote fuel strainer drain control.

AGtruck. As AGwagon, except for 1,060 litre (280 US gallon) hopper. Additional standard equipment includes a 22-nozzle engine-driven hydraulic spray system with manually-controlled spray valve and gatebox without agitator, wing fuel tanks, extended conical-cambered wingtips, automatic inertia reel for the safety belt system, sensitive altimeter, pilot's foul weather windows, pilot's four-way adjustable seat, strobe lights, instrument panel lights, landing and taxi lights, three-colour exterior styling, oversize 10 in main and tailwheel tyres and overall polyurethane paint scheme.

By 1 January 1976 deliveries totalled 1,411 AGwagons and 937 AGtrucks. A total of 53 AGpickups had been delivered when production ended.

Type: Single-seat agricultural monoplane.

Wings: Braced low-wing monoplane, with single streamline-section bracing strut each side. Wing section NACA 2412, modified. Dihedral 9°. Incidence 1° 30′ at root, −1° 30′ at tip. All-metal structure with NACA all-metal single-slotted flaps inboard of Frise all-metal ailerons. Aileron leading-edge gaps sealed. Wing fences immediately outboard of bracing strut attachment points. Conical-cambered wingtips, extended on AGtruck.

Fuselage: Rectangular-section welded steel tube structure with removable metal skin panels forward of cabin. All-metal semi-monocoque rear fuselage.

Tail Unit: Cantilever all-metal structure. Fixed-incidence tailplane. Trim tab in starboard elevator.

Landing Gear: Non-retractable tailwheel type. Land-O-Matic cantilever main legs of heavy-duty spring steel. Tapered tubular tailwheel spring shock-absorber. Main wheels and tyres size 8·00-8 × 22 on AGwagon, with oversize 10 in mainwheel tyres on AGtruck. AGwagon has 10 in tailwheel tyre and AGtruck an oversize 10 in tailwheel tyre. Hydraulic disc brakes and parking brake. Wheel fenders optional.

Power Plant: One 224 kW (300 hp) Continental IO-520-D flat-six engine, driving a McCauley two-blade metal constant-speed propeller. Metal fuel tank aft of firewall in AGwagon, capacity 140 litres (37 US gallons), of which 138 litres (36·5 US gallons) are usable or, optionally on AGwagon and standard for AGtruck, two 106 litre (28 US gallon) wing fuel tanks, total capacity 212 litres (56 US gallons) of which 197 litres (52 US gallons) are usable. Oil capacity 11·4 litres (3 US gallons).

Accommodation: Pilot only, on vertically and longitudinally adjustable seat, in enclosed cabin. Steel overturn structure. Combined window and door on each side, hinged at bottom. Heating and ventilation standard.

Systems: Electrical system has a 60A 12V alternator and 12V 24Ah battery as standard. A 60A 24V or a 100A 24V alternator is available optionally.

Electronics and Equipment: Standard equipment is as detailed in model listings. Optional equipment includes fan-driven or engine-driven hydraulic spray systems; 22, 44 or 64 nozzle spraybooms; two spreader systems for either medium or high-volume applications; electric spray control valve; wing leading-edge repair kit; and a tailcone jack point.

Cessna AGwagon agricultural aircraft (Continental IO-520-D engine)

Dimensions, external:
- Wing span: 12·70 m (41 ft 8 in)
- Wing chord at root: 1·63 m (5 ft 4 in)
- Wing chord at tip: 1·12 m (3 ft 8 in)
- Length overall: 8·00 m (26 ft 3 in)
- Height overall: 2·35 m (7 ft 8½ in)
- Tailplane span: 3·30 m (10 ft 10 in)
- Wheel track: 2·24 m (7 ft 4⅜ in)
- Propeller diameter:
  - standard: 2·08 m (6 ft 10 in)
  - optional: 2·18 m (7 ft 2 in)

Areas (A: AGwagon; B: AGtruck):
- Wings, gross:
  - A: 18·77 m² (202 sq ft)
  - B: 19·05 m² (205 sq ft)

Weights and Loadings (A: AGwagon; B: AGtruck):
- Weight empty, approx, with no dispersal equipment installed:
  - A: 900 kg (1,985 lb)
  - B: 934 kg (2,059 lb)
- Weight empty, with liquid dispersal system gatebox and engine-driven hydraulic pump:
  - A: 970 kg (2,140 lb)
  - B: 1,004 kg (2,214 lb)
- T-O weight, Normal category: 1,496 kg (3,300 lb)
- Max T-O weight, Restricted category:
  - A: 1,814 kg (4,000 lb)
  - B: 1,905 kg (4,200 lb)
- Wing loading, Normal category:
  - A: 79·6 kg/m² (16·3 lb/sq ft)
  - B: 78·6 kg/m² (16·1 lb/sq ft)
- Power loading, Normal category: 6·68 kg/kW (11·0 lb/hp)

Performance (AGwagon and AGtruck, at 1,496 kg; 3,300 lb AUW. A: without dispersal equipment; B: with liquid dispersal equipment):
- Max level speed at S/L:
  - A: 131 knots (243 km/h; 151 mph)
  - B: 105 knots (195 km/h; 121 mph)
- Max cruising speed, 75% power at 1,980 m (6,500 ft):
  - A: 122 knots (225 km/h; 140 mph)
  - B: 98 knots (182 km/h; 113 mph)
- Stalling speed, flaps up:
  - A, B: 53 knots (98·5 km/h; 61 mph) CAS
- Stalling speed, flaps down:
  - A, B: 49·5 knots (92 km/h; 57 mph) CAS
- Max rate of climb at S/L:
  - A: 287 m (940 ft)/min
  - B: 210 m (690 ft)/min
- Service ceiling:
  - A: 4,785 m (15,700 ft)
  - B: 3,383 m (11,100 ft)
- T-O run:
  - A: 186 m (610 ft)
  - B: 207 m (680 ft)
- T-O to 15 m (50 ft):
  - A: 296 m (970 ft)
  - B: 332 m (1,090 ft)
- Landing from 15 m (50 ft):
  - A, B: 386 m (1,265 ft)
- Landing run:
  - A, B: 128 m (420 ft)
- Range, recommended lean mixture with allowances for start, taxi, T-O, climb and 45 min reserves at 45% power:
  - Max cruising speed at 1,980 m (6,500 ft) with 138 litres (36·5 US gallons) usable fuel:
    - A: 200 nm (370 km; 230 miles)
    - B: 156 nm (290 km; 180 miles)
  - Max cruising speed at 1,980 m (6,500 ft) with 197 litres (52 US gallons) usable fuel:
    - A: 321 nm (595 km; 370 miles)
    - B: 256 nm (475 km; 295 miles)

CESSNA AGCARRYALL

First announced by Cessna on 8 December 1971, the AGcarryall represented a new multi-purpose concept in this specialised category of aircraft. It is intended for use as a demonstrator of spraying techniques, as a runabout for moving people, equipment or cargo when operating in the field, as a backup aircraft for peak seasonal workloads, as an agricultural pilot trainer, and for use by the farmer who requires spraying capability plus transportation.

Based upon the Model 185, the AGcarryall has two seats as standard, and optional seating for four additional passengers, and is provided with spraybooms and a 571 litre (151 US gallon) chemical tank.

The 1976 version introduces as standard a number of improvements, including an airspeed indicator with the primary scale in knots; redesigned audio selector box; polycarbonate heating ducts; standard fuel selector valve; a new voltage regulator; and new interior and exterior styling.

A total of 62 AGcarryalls had been delivered by 1 January 1976.

Type: One/six-seat agricultural utility monoplane.

Wings: Generally similar to Model 185. Provision for attachment of streamline-section Vee struts on under-

Cessna AGcarryall utility aircraft (Continental IO-520-D engine)

surface of each wing to support outer end of sprayboom.

FUSELAGE AND TAIL UNIT: As for Model 185.

LANDING GEAR: Non-retractable tailwheel type. Land-O-Matic cantilever main legs of heavy-duty spring steel. Tapered tubular tailwheel spring shock-absorber. Hydraulic disc brakes. Parking brake. Wire cutters on main legs.

POWER PLANT: As for Model 185.

ACCOMMODATION: Standard seating is for a pilot and passenger, side by side, on four-way adjustable seats. Optional seating for four additional passengers, in two pairs. Door on each side of cabin with quick-release hinges. Extended baggage floor. Cabin heated and ventilated. Wire cutters on windscreen.

SYSTEM: Electrical system with 60A 12V engine-driven alternator and 12V 24Ah battery standard.

EQUIPMENT: Standard equipment includes corrosion proofing, windscreen defrosting system, remote fuel strainer drain control, interior lights, landing and taxi lights, navigation lights, aft cabin baggage net, cable deflector, two-colour external paint scheme, and wind-driven spray system with associated 30-nozzle boom, liquid material controls, underfuselage chemical tank with capacity of 571 litres (151 US gallons), omni-flash beacon, overhead floodlight, stowable rudder pedals, safety belts for pilot and co-pilot. Optional equipment includes inertia-reel shoulder harness for pilot and co-pilot, ground assist handles, tailcone jack point, improved static discharge system and seating for three to six people.

DIMENSIONS, EXTERNAL: As for Model 185

AREAS: As for Model 185

WEIGHTS AND LOADINGS:

Weight empty	863 kg (1,902 lb)
Max T-O weight	1,519 kg (3,350 lb)
Max wing loading	94·2 kg/m² (19·3 lb/sq ft)
Max power loading	6·78 kg/kW (11·2 lb/hp)

PERFORMANCE (at max T-O weight):

*Max level speed at S/L	129 knots (238 km/h; 148 mph)
*Max cruising speed at 762 m (2,500 ft)	117 knots (217 km/h; 135 mph)
*Max cruising speed at 2,285 m (7,500 ft)	122 knots (225 km/h; 140 mph)
Stalling speed, flaps up	57 knots (105 km/h; 65 mph) CAS
Stalling speed, flaps down	49 knots (91 km/h; 56 mph) CAS
Max rate of climb at S/L	258 m (845 ft)/min
Service ceiling	4,085 m (13,400 ft)
T-O run	270 m (885 ft)
T-O to 15 m (50 ft)	442 m (1,450 ft)
Landing from 15 m (50 ft)	427 m (1,400 ft)
Landing run	146 m (480 ft)

Range, recommended lean mixture with allowances for start, taxi, T-O, climb and 45 min reserves at 45% power:

Max cruising speed at 2,285 m (7,500 ft) with standard fuel	343 nm (635 km; 395 miles)
Max cruising speed at 2,285 m (7,500 ft) with max fuel	490 nm (909 km; 565 miles)

With spraybooms removed, max level speed and cruising speed are increased by 8·7 knots (16 km/h; 10 mph)

CESSNA STATIONAIR

Cessna re-named the former U206 Skywagon and TU206 Turbo-Skywagon as the Stationair and Turbo-Stationair respectively, to emphasise the considerable differences between these six-seat cargo/utility aircraft and the Model 185 Skywagon. In particular, they have swept vertical tail surfaces, a tricycle landing gear, a tailplane of greater span, wide-span flaps, and double cargo doors on the starboard side of the fuselage which permit the easy loading and unloading of a crate more than 1·22 m long, 0·91 m wide and 0·91 m deep (4 ft × 3 ft × 3 ft).

The two basic versions of the Stationair are as follows:

Stationair. Standard cargo utility model with 224 kW (300 hp) Continental IO-520-F engine and double loading doors, as described in detail.

Turbo-Stationair. Similar to the Stationair but with 212·5 kW (285 hp) Continental TSIO-520-C turbocharged engine in modified cowling and provided with a manifold pressure relief valve to prevent overboost.

A utility version of the Stationair is also available, with a single seat for the pilot as standard, vinyl floor covering, two-colour paint scheme and no wheel fairings. Up to five passenger seats can be supplied optionally.

The 1976 models of the two basic Stationair versions introduce a number of improvements as standard, including a redesigned instrument panel, an airspeed indicator with the primary scale in knots, aerodynamic refinements to the fin and rudder tips, bonded fuel tank covers, recontoured glareshield, flush-mounted airscoop for electronics cooling, repositioned microphone and headphone jacks, new muffler components and new interior and exterior styling.

A total of 3,724 Model 206 Skywagons and Stationairs had been built by 1 January 1976, including 643 de luxe Super Skylanes of similar basic design.

TYPE: Single-engined cargo/utility aircraft.

WINGS: Braced high-wing monoplane. Single streamlined-section bracing strut each side. Wing section NACA 2412, modified. Dihedral 1° 44′. Incidence 1° 30′ at root, —1° 30′ at tip. All-metal structure. Glassfibre conical camber tips. Modified Frise-type wide-chord ailerons. Electrically-operated long-span NACA single-slotted flaps. No tabs.

FUSELAGE: Conventional all-metal semi-monocoque structure.

TAIL UNIT: Cantilever all-metal structure, with sweptback vertical surfaces. Large trim tab in starboard elevator. Electrical operation of trim tab optional.

LANDING GEAR: Non-retractable tricycle type. Cessna Land-O-Matic cantilever spring steel main legs. Steerable nosewheel with oleo-pneumatic shock-absorbers. Cessna wheels, tubeless tyres and hydraulic disc brakes. Parking brake. Main wheels and tyres size 6·00-6, pressure 2·90 bars (42 lb/sq in). Nosewheel and tyre size 5·00-5, pressure 3·10 bars (45 lb/sq in). Mainwheel tyres size 8·00-6, nosewheel tyre size 6·00-6 and oversize wheel fairings optional. Available with floats and hydraulically-operated wheel-skis.

POWER PLANT: One Continental flat-six engine (details given under model listings), driving a McCauley three-blade metal constant-speed propeller type D2A32C90/82NC-2 (Stationair) and D3A-32C88/82NC-2 (Turbo-Stationair). Two fuel cells in wings, with total standard capacity of 246 litres (65 US gallons), of which 238·5 litres (63 US gallons) are usable. Optional capacity of 318 litres (84 US gallons), of which 302·8 litres (80 US gallons) are usable. Oil capacity 11·4 litres (3 US gallons).

ACCOMMODATION: Standard seating for pilot, co-pilot and up to four passengers, front seats with inertia safety belts. Pilot's door on port side. Large double cargo doors on starboard side; forward door hinged to open forward, rear door hinged to open rearward. Aircraft can be flown with cargo doors removed for photography, air dropping of supplies or parachuting. Openable starboard window optional. Fully articulating seats for pilot and co-pilot and safety harness for four rear seats optional. Cabin heated and ventilated.

SYSTEMS: Electrical system powered by an engine-driven 60A 14V alternator. 12V 33Ah battery. 28V electrical system optional. Hydraulic system for brakes and optional wheel-skis. Oxygen system of 2·10 m³ (74 cu ft) capacity standard on Turbo-Stationair; 1·36 m³ (48 cu ft) system optional for Stationair. Vacuum system optional.

ELECTRONICS AND EQUIPMENT: Optional electronics as detailed for the Skylane, plus Series 400 Nav-O-Matic two-axis autopilot with heading control, VOR intercept and track and altitude control, plus the optional electronics detailed for the Skylane II. Standard equipment as for the Skylane, plus sensitive altimeter, electric clock, turn co-ordinator indicator, outside air temperature gauge, glareshield, overall paint scheme, and sun visors. Optional equipment, less the above items, is as detailed for the Skylane, plus ambulance kits, casket kit, photographic provisions and skydiving kit. The child's seat and skylights are not available for the Stationair. The Turbo-Stationair has an overboost control valve, absolute pressure controller, pressurised fuel system, turbine access door, pilot's all-purpose control wheel, non-congealing oil cooler, full flow oil filter and alternate static source as standard.

DIMENSIONS, EXTERNAL (L: landplane; F: floatplane; S: skiplane):

Wing span	10·92 m (35 ft 10 in)
Wing chord at root	1·63 m (5 ft 4 in)
Wing chord at tip	1·09 m (3 ft 7 in)
Wing aspect ratio	7·63
Length overall:	
Stationair: L, S	8·53 m (28 ft 0 in)
Stationair: F	8·67 m (28 ft 5½ in)
Turbo-Stationair	8·61 m (28 ft 3 in)
Height overall:	
L, S	2·93 m (9 ft 7½ in)
F	4·25 m (13 ft 11½ in)
Tailplane span	3·96 m (13 ft 0 in)
Wheel track: L	2·48 m (8 ft 1¾ in)
Propeller diameter:	
L	2·03 m (6 ft 8 in)
F	2·18 m (7 ft 2 in)
Pilot's door (port):	
Height, mean	1·03 m (3 ft 4 in)
Cargo double door (stbd):	
Height	0·98 m (3 ft 2½ in)
Width	1·13 m (3 ft 8½ in)
Height to sill	0·64 m (2 ft 1 in)

DIMENSIONS, INTERNAL:

Cabin: Length	3·66 m (12 ft 0 in)
Max width	1·12 m (3 ft 8 in)
Max height	1·26 m (4 ft 1½ in)
Volume available for payload	2·87 m³ (101·2 cu ft)

AREAS:

Wings, gross	16·17 m² (174·0 sq ft)
Ailerons (total)	1·60 m² (17·32 sq ft)
Trailing-edge flaps (total)	2·63 m² (28·35 sq ft)
Fin	1·08 m² (11·62 sq ft)
Rudder, incl tab	0·65 m² (6·95 sq ft)
Tailplane	2·31 m² (24·84 sq ft)
Elevators, incl tab	1·86 m² (20·08 sq ft)

WEIGHTS AND LOADINGS (L: landplane; F: floatplane; S: skiplane):

Weight empty, one seat only:	
Stationair: L	810 kg (1,785 lb)
F	968 kg (2,135 lb)
S	923 kg (2,035 lb)
Turbo-Stationair: L	844 kg (1,860 lb)
S	971 kg (2,140 lb)
Max T-O and landing weight:	
Stationair: L	1,633 kg (3,600 lb)
F	1,588 kg (3,500 lb)
S	1,496 kg (3,300 lb)
Turbo-Stationair: L	1,633 kg (3,600 lb)
S	1,496 kg (3,300 lb)
Max wing loading:	
Stationair: L	101 kg/m² (20·7 lb/sq ft)
F	98 kg/m² (20·1 lb/sq ft)
S	93 kg/m² (19·0 lb/sq ft)
Turbo-Stationair:	
L	101 kg/m² (20·7 lb/sq ft)
S	93 kg/m² (19·0 lb/sq ft)
Max power loading:	
Stationair: L	7·29 kg/kW (12·0 lb/hp)
F	7·09 kg/kW (11·7 lb/hp)
S	6·68 kg/kW (11·0 lb/hp)
Turbo-Stationair: L	7·68 kg/kW (12·6 lb/hp)
S	7·04 kg/kW (11·6 lb/hp)

PERFORMANCE (L: landplane; F: floatplane; S: skiplane):

Max level speed:	
Stationair at S/L:	
L	156 knots (290 km/h; 180 mph)
F	136 knots (252 km/h; 157 mph)
S	121 knots (224 km/h; 139 mph)
Turbo-Stationair at 5,790 m (19,000 ft):	
L	176 knots (326 km/h; 203 mph)
S	145 knots (269 km/h; 167 mph)
Max cruising speed (75% power):	
Stationair at 1,980 m (6,500 ft):	
L	147 knots (272 km/h; 169 mph)
F	130 knots (241 km/h; 150 mph)
S	118 knots (219 km/h; 136 mph)
Turbo-Stationair at 7,320 m (24,000 ft):	
L	167 knots (309 km/h; 192 mph)

Cessna Stationair one/six-seat cargo/utility aircraft

S 131 knots (243 km/h; 151 mph)
Stalling speed, flaps up, power off:
Stationair: L 62 knots (115 km/h; 71·5 mph) CAS
F 61 knots (113·5 km/h; 70·5 mph) CAS
S 59 knots (110 km/h; 68 mph) CAS
Turbo-Stationair:
L 62 knots (115 km/h; 71·5 mph) CAS
S 59 knots (110 km/h; 68 mph) CAS
Stalling speed, flaps down, power off:
Stationair: L 54 knots (101 km/h; 62·5 mph) CAS
F 53 knots (98·5 km/h; 61 mph) CAS
S 52 knots (96·5 km/h; 60 mph) CAS
Turbo-Stationair:
L 54 knots (101 km/h; 62·5 mph) CAS
S 52 knots (96·5 km/h; 60 mph) CAS
Max rate of climb at S/L:
Stationair: L 280 m (920 ft)/min
F 260 m (855 ft)/min
S 244 m (800 ft)/min
Turbo-Stationair: L 314 m (1,030 ft)/min
S 280 m (920 ft)/min
Service ceiling:
Stationair: L 4,511 m (14,800 ft)
F 4,237 m (13,900 ft)
S 3,505 m (11,500 ft)
Turbo-Stationair: L 8,020 m (26,300 ft)
S 7,163 m (23,500 ft)
T-O run:
Stationair:
L 274 m (900 ft)
F 440 m (1,445 ft)
Turbo-Stationair:
L 277 m (910 ft)
T-O to 15 m (50 ft):
Stationair:
L 543 m (1,780 ft)
F 754 m (2,475 ft)
Turbo-Stationair:
L 552 m (1,810 ft)
Landing from 15 m (50 ft):
Stationair:
L 425 m (1,395 ft)
F 479 m (1,570 ft)
Turbo-Stationair:
L 425 m (1,395 ft)
Landing run:
Stationair:
L 224 m (735 ft)
F 212 m (695 ft)
Range, Stationair, recommended lean mixture with allowances for start, taxi, T-O, climb and 45 min reserves at 45% power:
Max cruising speed at 1,980 m (6,500 ft) with standard fuel:
L 450 nm (833 km; 518 miles)
F 400 nm (740 km; 460 miles)
S 365 nm (676 km; 420 miles)
Max cruising speed at 1,980 m (6,500 ft) with max fuel:
L 610 nm (1,130 km; 702 miles)
F 545 nm (1,009 km; 627 miles)
S 490 nm (907 km; 564 miles)
Econ cruising speed at 3,050 m (10,000 ft) with standard fuel:
L 555 nm (1,028 km; 639 miles)
F 500 nm (927 km; 576 miles)
S 420 nm (777 km; 483 miles)
Econ cruising speed at 3,050 m (10,000 ft) with max fuel:
L 755 nm (1,398 km; 869 miles)
F 685 nm (1,268 km; 788 miles)
S 575 nm (1,065 km; 662 miles)
Range, Turbo-Stationair, recommended lean mixture with allowances for start, taxi, T-O, climb and 45 min reserves at 45% power:
Max cruising speed at 6,100 m (20,000 ft) with standard fuel:
L 435 nm (806 km; 501 miles)
S 350 nm (648 km; 403 miles)
Max cruising speed at 6,100 m (20,000 ft) with max fuel:
L 605 nm (1,120 km; 696 miles)
S 485 nm (898 km; 558 miles)
Econ cruising speed at 6,100 m (20,000 ft) with standard fuel:
L 495 nm (917 km; 570 miles)
S 370 nm (685 km; 426 miles)
Econ cruising speed at 6,100 m (20,000 ft) with max fuel:
L 690 nm (1,278 km; 794 miles)
S 520 nm (964 km; 599 miles)

CESSNA SKYWAGON 207 and TURBO-SKYWAGON T207

On 19 February 1969 Cessna announced two new seven-seat versions of its Skywagon utility aircraft. Generally similar to the earlier Model 206 Super Skywagon, the new Skywagon had been 'stretched' to provide improved load-carrying ability while retaining the single engine and operating economy of the Model 206.

In addition to the longer fuselage, new features included a door for the co-pilot or passenger on the starboard side at the front of the cabin, and a separate baggage compartment forward of the cabin, accessible through an external door, also on the starboard side of the fuselage.

Design of this model started in November 1967 and the prototype flew for the first time on 11 May 1968. The first production aircraft, a Model 207, was completed on 13 December 1968 and made its first flight on 3 January 1969, followed three days later by the first flight of a T207 Turbo-Skywagon. Both models received FAA certification on 31 December 1968. A total of 319 Model 207s had been delivered by 1 January 1976.

The 1976 versions of the Model 207 introduce several improvements as standard, including aerodynamic refinements to fin and rudder tip fairings, bonded fuel tank covers, improved flap follow-up system, airspeed indicator with primary scale in knots, improved glareshield, new muffler components and redesigned standard and oversize wheel fairings.

There are two current versions, as follows:

Skywagon 207. Standard passenger/cargo utility model with 224 kW (300 hp) Continental IO-520-F engine.

Turbo-Skywagon T207. Generally similar to Skywagon 207, but with 224 kW (300 hp) Continental TSIO-520-G turbocharged engine, driving a McCauley D2A34C78/90AT-8·5 two-blade metal constant-speed propeller. Three-blade propeller optional. Absolute pressure controller, pressurised fuel system, non-congealing oil cooler, full-flow oil filter, overboost control valve, alternate static source and oxygen system standard.

The following description applies to the Skywagon 207, except where stated otherwise:

TYPE: Single-engined utility aircraft.

WINGS: Braced high-wing monoplane. Single streamline-section bracing strut each side. Wing section NACA 2412 from root to just inboard of tip; wingtip is symmetrical. Dihedral 1° 44′. Incidence 1° 30′ at root, —1° 30′ at tip. All-metal structure. Glassfibre conical-camber tips. Modified Frise-type all-metal wide-chord ailerons. Electrically-operated long-span NACA single-slotted all-metal flaps. No trim tabs.

FUSELAGE: Conventional all-metal semi-monocoque structure.

TAIL UNIT: Cantilever all-metal structure, with sweptback vertical surfaces. Tailplane fixed with —3° incidence. Large trim tab in starboard elevator. Electrical operation of trim tab optional. Rudder trimmed by adjustment of bungee.

LANDING GEAR: Non-retractable tricycle type. Improved Cessna Land-O-Matic cantilever main legs of one-piece tapered steel tube. Steerable nosewheel with Cessna oleo-pneumatic shock-absorber and hydraulic shimmy damper. Cessna wheels, tubeless tyres and hydraulic disc brakes. Main wheels and tyres size 6·00-6, pressure 3·79 bars (55 lb/sq in). Nosewheel and tyre size 5·00-5, pressure 3·38 bars (49 lb/sq in). Optional 8·00-6 main-wheel tyres, pressure 2·41 bars (35 lb/sq in), nosewheel tyre size 6·00-6, pressure 2·00 bars (29 lb/sq in). Wheel fairings standard; oversize wheel fairings optional.

POWER PLANT: One 224 kW (300 hp) Continental IO-520-F flat-six engine, driving a McCauley D2A34C58/90AT-8 two-blade metal constant-speed propeller. A bladder-type fuel tank, capacity 123 litres (32·5 US gallons), is located in the inboard section of each wing. Total fuel capacity 246 litres (65 US gallons), of which 220 litres (58 US gallons) are usable. Optional tankage increases capacity to 159 litres (42 US gallons) in each wing, giving a total capacity of 318 litres (84 US gallons), of which 292 litres (77 US gallons) are usable. Refuelling points in upper surface of each wing. Oil capacity 11·4 litres (3 US gallons).

ACCOMMODATION: Pilot's seat only standard. Optional individual seats for up to seven persons, arranged in three pairs, two-abreast, with a single seat at the rear of cabin. Pilot's door on port side, co-pilot's door on starboard side at front. Large double cargo doors on starboard side at rear of cabin; forward door hinged to open forward, rear door hinged to open rearward. Aircraft can be flown with cargo doors removed for photography, air dropping of supplies or parachuting; optional equipment includes a spoiler for use when the aircraft is flown in this configuration. Openable window, port side; openable window starboard side optional. Separate baggage compartment, forward of cabin, capacity 54 kg (120 lb), accessible through top-hinged door on starboard side. External glassfibre cargo pack, capacity 136 kg (300 lb), carried beneath the fuselage, is available as an optional extra.

SYSTEMS: Hydraulic system for brakes. Electrical system powered by a 14V 60A engine-driven alternator. 12V 33Ah battery. 24V electrical system optional. Oxygen system of 2·15 m³ (76 cu ft) capacity standard on T207; 1·36 m³ (48 cu ft) system optional on 207.

ELECTRONICS AND EQUIPMENT: As described for the Model 180, but with electric clock, sensitive altimeter, outside air temperature gauge, flap position indicator, rate of

Cessna Skywagon 207 one/seven-seat utility aircraft (Continental IO-520-F engine)

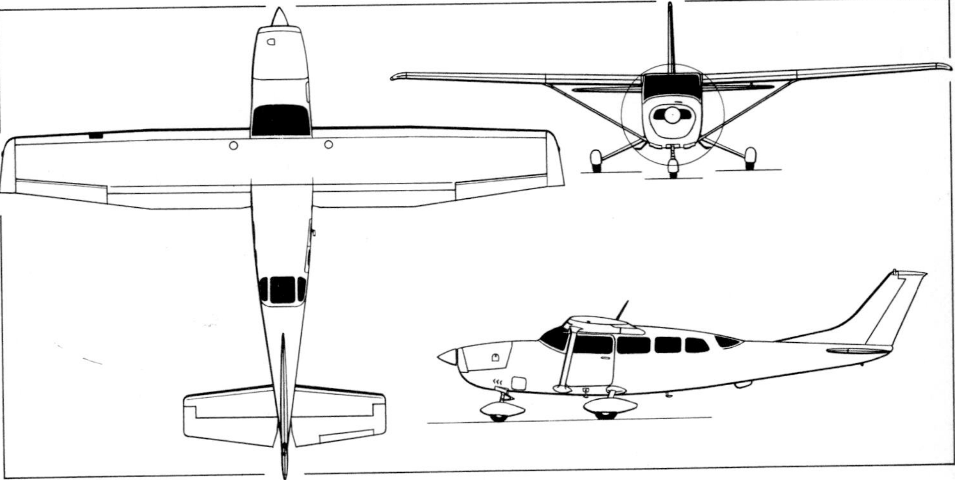

Cessna Skywagon 207 utility aircraft *(Pilot Press)*

climb indicator, turn co-ordinator, elevator and rudder trim controls, cabin steps, electro-luminescent lights for switch and comfort control panels, instrument panel glareshield light, triple dome lights, sun visors, tinted windscreen and windows, emergency locator transmitter, baggage tie-down rings, dual-beam landing lights and towbar standard. Additional optional items include Series 400 transponder with 4096 code capability, instrument post lights, ambulance kit comprising stretcher, oxygen and attendant's seat, heated stall warning transmitter and pitot, centre armrests, rearview mirror, articulating and vertically adjustable front seats, anti-precipitation static kit, inertia-reel shoulder harness for pilot and co-pilot seats, flush com antenna, openable starboard window, Cessna Series 400 electronics, shoulder harness for five passenger seats, glider tow hook and skydiving kit.

DIMENSIONS, EXTERNAL:
Wing span 10·92 m (35 ft 10 in)
Wing chord at root 1·63 m (5 ft 4 in)
Wing chord at tip 1·12 m (3 ft 8 in)
Wing aspect ratio 7·46
Length overall 9·68 m (31 ft 9 in)
Height overall 2·91 m (9 ft 6½ in)
Tailplane span 3·96 m (13 ft 0 in)
Wheel track 3·05 m (10 ft 0 in)
Wheelbase 2·11 m (6 ft 11¼ in)
Propeller diameter:
Two-blade:
207 2·08 m (6 ft 10 in)
T207 2·07 m (6 ft 9½ in)
Three-blade 2·03 m (6 ft 8 in)
Forward cabin doors (each):
Height 1·05 m (3 ft 5½ in)
Width 0·89 m (2 ft 11½ in)
Height to sill 0·71 m (2 ft 4 in)
Cargo double doors (stbd):
Height 0·97 m (3 ft 2 in)
Width 1·13 m (3 ft 8½ in)
Height to sill 0·76 m (2 ft 6 in)
Baggage door (stbd):
Height 0·61 m (2 ft 0 in)
Width 0·34 m (1 ft 1½ in)
Height to sill 1·02 m (3 ft 4 in)

DIMENSIONS, INTERNAL:
Cabin:
Length 4·27 m (14 ft 0 in)
Max width 1·13 m (3 ft 8½ in)
Max height 1·24 m (4 ft 1 in)
Floor area 4·38 m^2 (47·1 sq ft)
Volume 4·40 m^3 (155·5 cu ft)
Forward baggage compartment:
Length 0·43 m (1 ft 5 in)
Max width 1·05 m (3 ft 5½ in)
Max height 0·69 m (2 ft 3 in)
Floor area 0·46 m^2 (4·9 sq ft)
Volume 0·27 m^3 (9·5 cu ft)
Underfuselage cargo pack 0·34 m^3 (12·0 cu ft)

AREAS:
Wings, gross 16·17 m^2 (174·0 sq ft)
Ailerons (total) 1·60 m^2 (17·32 sq ft)
Trailing-edge flaps (total) 2·66 m^2 (26·60 sq ft)
Fin 0·84 m^2 (9·04 sq ft)
Rudder 0·65 m^2 (6·95 sq ft)
Tailplane 2·31 m^2 (24·84 sq ft)
Elevators, incl tab 1·86 m^2 (20·08 sq ft)

WEIGHTS AND LOADINGS:
Weight empty, one seat only:
207 891 kg (1,964 lb)
T207 935 kg (2,062 lb)
Max T-O and landing weight:
207 and T207 1,724 kg (3,800 lb)
Max wing loading:
207 and T207 106·4 kg/m^2 (21·8 lb/sq ft)
Max power loading:
207 and T207 7·70 kg/kW (12·7 lb/hp)

PERFORMANCE (at max T-O weight and with optional wheel fairings, which increase speed by 3-4 knots; 5·5-7·5 km/h; 3·5-4·5 mph):
Never-exceed speed:
207 and T207 182 knots (338 km/h; 210 mph)
Max level speed:
207 at S/L 150 knots (278 km/h; 173 mph)
T207 at 5,180 m (17,000 ft) 170 knots (315 km/h; 196 mph)
Normal cruising speed (75% power):
207 at 1,980 m (6,500 ft) 143 knots (266 km/h; 165 mph)
T207 at 3,050 m (10,000 ft) 147 knots (272 km/h; 169 mph)
T207 at 6,100 m (20,000 ft) 159 knots (295 km/h; 183 mph)
Econ cruising speed:
207 at 3,050 m (10,000 ft) 127 knots (235 km/h; 146 mph)
T207 at 6,100 m (20,000 ft) 137 knots (254 km/h; 158 mph)
Stalling speed, flaps up, power off:
207 and T207 66 knots (123 km/h; 76 mph)
Stalling speed, 30° flap, power off:
207 and T207 58 knots (108 km/h; 67 mph)
Max rate of climb at S/L:
207 247 m (810 ft)/min
T207 270 m (885 ft)/min
Service ceiling:
207 4,054 m (13,300 ft)
T207 7,376 m (24,200 ft)
T-O run:
207 and T207 335 m (1,100 ft)
T-O to 15 m (50 ft):
207 and T207 600 m (1,970 ft)
Landing from 15 m (50 ft):
207 and T207 457 m (1,500 ft)
Landing run:
207 and T207 233 m (765 ft)
Range, recommended lean mixture, allowances for start, taxi, T-O, climb and 45 min reserves at 45% power:
207 at max cruising speed at 1,980 m (6,500 ft) with standard fuel 390 nm (722 km; 449 miles)
207 at max cruising speed at 1,980 m (6,500 ft) with max fuel 565 nm (1,046 km; 650 miles)
T207 at max cruising speed at 6,100 m (20,000 ft) with standard fuel 340 nm (629 km; 391 miles)
T207 at max cruising speed at 3,050 m (10,000 ft) with standard fuel 365 nm (676 km; 420 miles)
T207 at max cruising speed at 6,100 m (20,000 ft) with max fuel 520 nm (964 km; 599 miles)
T207 at max cruising speed at 3,050 m (10,000 ft) with max fuel 535 nm (991 km; 616 miles)
207 at econ cruising speed at 3,050 m (10,000 ft) with standard fuel 470 nm (870 km; 541 miles)
207 at econ cruising speed at 3,050 m (10,000 ft) with max fuel 690 nm (1,278 km; 794 miles)
T207 at econ cruising speed at 6,100 m (20,000 ft) with standard fuel 375 nm (695 km; 432 miles)
T207 at econ cruising speed at 3,050 m (10,000 ft) with standard fuel 445 nm (824 km; 512 miles)
T207 at econ cruising speed at 6,100 m (20,000 ft) with max fuel 595 nm (1,102 km; 685 miles)
T207 at econ cruising speed at 3,050 m (10,000 ft) with max fuel 655 nm (1,213 km; 754 miles)

CESSNA MODEL 210 CENTURION and CENTURION II

The original prototype Model 210, which flew in January 1957, followed the general formula of the Cessna series of all-metal high-wing monoplanes, but was the first to have a retractable tricycle landing gear.

Later versions of the Model 210 have a fully-cantilever wing, eliminating the bracing struts used on earlier models. Their design was started on 24 October 1964 and construction of a prototype began on 29 November 1964. The first T210 with the new wing flew on 18 June 1965.

On 3 December 1970 Cessna announced the introduction of two new versions of the Model 210 to be known as Centurion II and Turbo-Centurion II. These differ from the Centurion and Turbo-Centurion by having as standard equipment a factory-installed IFR electronics package which offers a cost saving of 19% on electronics equipment, plus a gyro panel, dual controls, articulating front seats and all-purpose control wheel.

Improvements in the 1976 versions include as standard a redesigned instrument panel cover, glareshield, fin and rudder tip fairings; all-electric engine instruments and fuel gauges, airspeed indicator with the primary scale in knots, audio panel, audio speaker, muffler components, bonded baggage door and (on Turbo-Centurions only) new cooling air exits; improved elevator control system, landing gear and (on Centurions only) better accessibility to the starboard side of the engine by changes to the air induction and filter system. New options include electric windscreen anti-icing, openable starboard window, anti-precipitation static kit and Cessna Series 400 electronics.

The four current production versions of the Centurion are as follows:

210 Centurion. Standard model, with 224 kW (300 hp) Continental IO-520-L flat-six engine, driving a McCauley D3A32C88/82NC-2 three-blade metal constant-speed propeller.

T210 Turbo-Centurion. Generally similar to Centurion, but powered by a 212·5 kW (285 hp) Continental TSIO-520-H turbocharged engine, driving a McCauley D3A32C88/82NC-2 three-blade metal constant-speed propeller. Absolute pressure controller, full-flow oil filter, pressurised fuel system, oxygen system, non-congealing oil cooler and overboost control valve standard.

The Turbo-Centurion version holds an international altitude record for aircraft of this class with a height of 12,906·5 m (42,344 ft).

210 Centurion II. Identical to Centurion but with a 720-channel Cessna Series 300 nav/com, ADF, transponder, all-purpose control wheel, instrument post lights, true airspeed indicator, horizon and directional gyro with vacuum system and suction gauge, economy mixture indicator, dual controls, reclining and vertically-adjustable co-pilot's seat, emergency locator transmitter, ground power socket, navigation light detectors, heating system for pitot and stall warning transmitter, omni-flash beacon, two courtesy lights and alternate static source as standard. Nav-O-Matic 200A autopilot optional. A Nav Pac is also optional, this including a second Series 300 nav/com, with Series 400 glideslope and marker beacon.

T210 Turbo-Centurion II. Identical to Turbo-Centurion but with additional standard equipment as detailed for Centurion II. The same optional electronics are available also.

The original versions received FAA Type Approval on 23 August 1966. A total of 4,557 Model 210s had been delivered by 1 January 1976.

TYPE: Six-seat cabin monoplane.

WINGS: Cantilever high-wing monoplane. Wing section NACA 64$_2$A215 at root, NACA 64$_1$A412 (A=0·5) at tip. Dihedral 1° 30'. Incidence 1° 30' at root, −1° 30' at tip. All-metal structure, except for glassfibre conical-camber tips. All-metal Frise-type ailerons. Electrically-actuated all-metal Fowler-type flaps. Ground-adjustable tab in each aileron. Pneumatic de-icing system optional.

FUSELAGE: All-metal semi-monocoque structure.

TAIL UNIT: Cantilever all-metal structure with 36° sweepback on fin. Fixed-incidence tailplane. Controllable trim tabs in rudder and starboard elevator. Electric operation of elevator tab optional. Pneumatic de-icing system optional.

LANDING GEAR: Hydraulically-retractable tricycle type with single wheel on each unit. Nose unit retracts forward, main units aft and inward. Wheel doors close when wheels are up or down. Chrome vanadium tapered steel tube main legs. Steerable nosewheel with oleo-pneumatic shock-absorber. Cessna main wheels and tube-type tyres, size 6·00-6, pressure 2·90 bars (42 lb/sq in). Cessna nosewheel and tyre, size 5·00-5, pressure 3·10 bars (45 lb/sq in). Cessna hydraulic disc brakes. Parking brake.

POWER PLANT: One flat-six engine, as described under model listings. Electrical de-icing system for propeller optional. Integral fuel tanks in wings, with max total capacity of 340 litres (90 US gallons). Refuelling points above wing. Oil capacity 9·5 litres (2·5 US gallons) in 210, 10·5 litres (2·75 US gallons) in T210.

ACCOMMODATION: Six persons in pairs in enclosed cabin. Front two seats of fully-articulating type on Centurion II and Turbo-Centurion II (pilot's seat only on other versions). Fifth and sixth seats have folding backs to accommodate articles up to 2·01 m (6 ft 7 in) long. Openable window on port side standard; optional for starboard side. Dual controls standard on Centurion II and Turbo-Centurion II (optional on other models). Forward-hinged door on each side of cabin. Baggage space aft of rear seats, capacity 136 kg (300 lb), with outside door on port side. Combined heating and ventilation system. Windscreen electric anti-icing optional.

SYSTEMS: Integral hydraulic-electric unit for landing gear retraction. Hydraulic system for brakes. Electrical power supplied by 24V 60A engine-driven alternator. 24V battery. Oxygen system standard on Turbo-Centurion, optional for Centurion.

ELECTRONICS AND EQUIPMENT: Optional electronics as for Stationair, except that Series 300 or 400 integrated

Cessna Turbo-Centurion II one/six-seat cabin monoplane (Continental TSIO-520-H engine)

flight control system is available when the Series 200A, 300A, or 400 Nav-O-Matic autopilot is replaced by the Series 400A two-axis autopilot, which has automatic pitch trim and an optional ILS coupler. Series 400 electronics are also available as options. Standard equipment includes sensitive altimeter, electric clock, outside air temperature gauge, audible landing gear and stall warning indicators, turn co-ordinator indicator, electroluminescent lights for switch and comfort control panels, glareshield lights, variable-intensity instrument panel red floodlights, control locks, armrests, windscreen defroster, dome and map lights, baggage restraint net, sun visors, adjustable cabin air ventilation, tinted windscreen and windows, landing lights, taxi light, navigation lights, three-blade propeller, rate of climb indicator, quick fuel drains and sampler cup, overall paint scheme, cabin steps and towbar. Optional equipment includes an alternate static source, a gyro panel, fully-articulating co-pilot seat, all-purpose control wheel, true airspeed indicator, economy mixture indicator, instrument post lights, emergency locator transmitter, external power socket, navigation light detectors, pitot and stall warning heating system, omni-flash beacon and two courtesy lights for Centurion and Turbo-Centurion; plus control wheel map light, turn and bank indicator, boom microphone, flight hour recorder, cabin fire extinguisher, headrests, rearview mirror, stretcher installation, ice detector light, engine priming system, internal corrosion proofing, glider tow hook, wingtip-mounted strobe lights, tailplane abrasion boots, static dischargers and electric elevator trim system. Optional for the Centurion and Centurion II only are a full-flow oil filter, non-congealing oil cooler, and engine winterisation kit.

DIMENSIONS, EXTERNAL:
- Wing span: 11·20 m (36 ft 9 in)
- Wing chord at root: 1·68 m (5 ft 6 in)
- Wing chord at tip: 1·22 m (4 ft 0 in)
- Wing aspect ratio: 7·66
- Length overall: 8·58 m (28 ft 1¾ in)
- Height overall: 2·87 m (9 ft 5 in)
- Tailplane span: 3·96 m (13 ft 0 in)
- Wheel track: 2·64 m (8 ft 8 in)
- Wheelbase: 1·75 m (5 ft 9 in)
- Propeller diameter: 2·03 m (6 ft 8 in)
- Passenger doors (each):
  - Height: 1·02 m (3 ft 4¼ in)
  - Width: 0·90 m (2 ft 11¼ in)
  - Height to sill: 0·91 m (3 ft 0 in)
- Baggage compartment door:
  - Height: 0·57 m (1 ft 10½ in)
  - Width: 0·74 m (2 ft 5 in)

DIMENSIONS, INTERNAL:
- Cabin:
  - Length: 3·50 m (11 ft 6 in)
  - Max width: 1·08 m (3 ft 6½ in)
  - Max height: 1·23 m (4 ft 0½ in)
  - Floor area: 2·69 m² (29·0 sq ft)
  - Volume: 3·96 m³ (139·9 cu ft)
- Baggage space: 0·46 m³ (16·25 cu ft)

AREAS:
- Wings, gross: 16·25 m² (175 sq ft)
- Ailerons (total): 1·75 m² (18·86 sq ft)
- Trailing-edge flaps (total): 2·74 m² (29·50 sq ft)
- Fin, incl dorsal fin: 0·95 m² (10·26 sq ft)
- Rudder, incl tab: 0·65 m² (6·95 sq ft)
- Tailplane: 1·73 m² (18·57 sq ft)
- Elevators, incl tab: 1·87 m² (20·08 sq ft)

WEIGHTS AND LOADINGS:
- Weight empty:
  - Centurion: 984 kg (2,170 lb)
  - Centurion II: 1,018 kg (2,244 lb)
  - Turbo-Centurion: 1,040 kg (2,293 lb)
  - Turbo-Centurion II: 1,073 kg (2,366 lb)
- Max T-O and landing weight:
  - All versions: 1,723 kg (3,800 lb)
- Max wing loading:
  - All versions: 106 kg/m² (21·7 lb/sq ft)
- Max power loading:
  - Centurion, Centurion II: 7·69 kg/kW (12·7 lb/hp)
  - Turbo-Centurion, Turbo-Centurion II: 8·11 kg/kW (13·3 lb/hp)

PERFORMANCE (at max T-O weight):
- Max level speed:
  - Centurion, Centurion II at S/L: 175 knots (325 km/h; 202 mph)
  - Turbo-Centurion, Turbo-Centurion II at 5,800 m (19,000 ft): 205 knots (380 km/h; 236 mph)
- Max cruising speed:
  - Centurion, Centurion II, 75% power at 1,980 m (6,500 ft): 171 knots (317 km/h; 197 mph)
  - Turbo-Centurion, Turbo-Centurion II, 75% power at 7,315 m (24,000 ft): 202 knots (375 km/h; 233 mph)
- Stalling speed, flaps up, power off:
  - All versions: 65 knots (121 km/h; 75 mph) CAS
- Stalling speed, flaps down, power off:
  - All versions: 56 knots (104 km/h; 64·5 mph) CAS
- Max rate of climb at S/L:
  - Centurion, Centurion II: 262 m (860 ft)/min
  - Turbo-Centurion, Turbo-Centurion II: 283 m (930 ft)/min
- Service ceiling:
  - Centurion, Centurion II: 4,725 m (15,500 ft)
  - Turbo-Centurion, Turbo-Centurion II: 8,685 m (28,500 ft)
- T-O run:
  - Centurion, Centurion II: 381 m (1,250 ft)
  - Turbo-Centurion, Turbo-Centurion II: 357 m (1,170 ft)
- T-O to 15 m (50 ft):
  - All versions: 619 m (2,030 ft)
- Landing from 15 m (50 ft):
  - All versions: 457 m (1,500 ft)
- Landing run:
  - All versions: 233 m (765 ft)
- Range, recommended lean mixture with allowances for start, taxi, T-O, climb and 45 min reserves at 45% power:
  - Centurion, Centurion II at max cruising speed at 1,980 m (6,500 ft) with 242 kg (534 lb) usable fuel: 855 nm (1,583 km; 984 miles)
  - Centurion, Centurion II at econ cruising speed at 3,050 m (10,000 ft) with 242 kg (534 lb) usable fuel: 1,060 nm (1,963 km; 1,220 miles)
  - Turbo-Centurion, Turbo-Centurion II at max cruising speed at 6,100 m (20,000 ft) with 242 kg (534 lb) usable fuel: 860 nm (1,593 km; 990 miles)
  - Turbo-Centurion, Turbo-Centurion II at max cruising speed at 3,050 m (10,000 ft) with 242 kg (534 lb) usable fuel: 830 nm (1,537 km; 955 miles)
  - Turbo-Centurion, Turbo-Centurion II at econ cruising speed at 6,100 m (20,000 ft) with 242 kg (534 lb) usable fuel: 1,005 nm (1,862 km; 1,157 miles)
  - Turbo-Centurion, Turbo-Centurion II at econ cruising speed at 3,050 m (10,000 ft) with 242 kg (534 lb) usable fuel: 1,020 nm (1,889 km; 1,174 miles)

CESSNA MODEL 310 and 310 II

USAF designation: U-3

The Model 310 is a twin-engined five/six-seat cabin monoplane, the prototype of which flew on 3 January 1953. It went into production in 1954. The Turbo-System Model 310 was added in late 1968, and the first production model was delivered in December 1968. On 21 December 1973 Cessna announced two new versions of the Model 310, known as the 310 II and the Turbo 310 II, which have factory installed IFR electronics plus other comfort and convenience features as standard. A total of 4,268 examples of the Model 310 had been completed by 1 January 1976.

There are four current versions of the Model 310, as follows:

310. Standard model, as described in detail, powered by two 212·5 kW (285 hp) Continental IO-520-M flat-six engines, driving McCauley three-blade metal fully-feathering constant-speed propellers.

Turbo-System T310. Similar to 310, but with two 212·5 kW (285 hp) Continental TSIO-520-B turbocharged engines, with automatic propeller synchronisation, full-flow oil filters, absolute and pressure ratio controllers, overboost control valves and engine cowl flaps as standard.

310 II. Identical to 310, but having as standard equipment dual 300 Series nav/com with 720-channel com, 200-channel nav, VOR/LOC and VOR/ILS indicators; ADF with digital tuning; 400 Series glideslope receiver; marker beacon; transponder; 400B Nav-O-Matic autopilot with approach coupler; associated antennae; six individual seats; dual controls; starboard landing light; taxi light; rotating beacon; outside air temperature gauge; economy mixture indicator, locator beacon, nosewheel fender, static dischargers; external power socket; and large baggage door.

Turbo-System T310 II. Identical to T310, but with the additional standard equipment as detailed for the 310 II.

The 1976 versions of the Model 310 introduce a number of improvements as standard, including crew seat belts attached to seats, 5th and 6th passenger seats moved 100 mm (4 in) aft to increase legroom, improved door sealing, hydraulic parking brake, increased brake capacity, low fuel warning light, and new wastegate access panel and wing induction inlet on Turbo 310. New interior and exterior styling is introduced on all versions.

The Cessna 310 has been in service with the USAF since 1957, when it won a competition for a light twin-engined administrative liaison and cargo aircraft. Initial orders for a total of 160 'off-the-shelf', under the designation U-3A (formerly L-27A), were followed by a contract for 36 later models, designated U-3B, which were delivered between December 1960 and June 1961. Other military operators include the Zaïre Air Force, which took delivery of 15 Model 310s in 1975.

TYPE: Twin-engined five- or six-seat monoplane.

WINGS: Cantilever low-wing monoplane. Wing section NACA 23018 at centreline, NACA 23009 at tip. Dihedral 5°. Incidence 2° 30′ at root, −0° 30′ at tip. All-metal structure. Electrically-operated split flaps. Trim tab in port aileron. Pneumatic de-icing system optional.

FUSELAGE: All-metal semi-monocoque structure.

TAIL UNIT: Cantilever all-metal structure, with 40° sweepback on fin at quarter-chord. Small ventral fin. Trim tabs in rudder and starboard elevator. Electrically-operated elevator trim optional. Pneumatic de-icing system optional.

LANDING GEAR: Retractable tricycle type. Electro-mechanical retraction. Cessna oleo shock-absorber struts. Nosewheel steerable to 15° and castoring from 15° to 55° each side. Main wheels size 6·50-10, tyre pressure 4·14 bars (60 lb/sq in). Nosewheel size 6·00-6, tyre pressure 2·76 bars (40 lb/sq in). Goodyear single-disc hydraulic brakes. Hydraulic parking brake.

POWER PLANT: Two flat-six engines, as described under individual model listings, driving three-blade propellers. Automatic propeller unfeathering system and propeller de-icing optional; automatic propeller synchroniser standard for T310 and T310 II, optional for 310 and 310 II. Standard fuel in two permanently attached canted wingtip tanks, each holding 193 litres (51 US gallons), of which 189 litres (50 US gallons) are usable. Cross-feed fuel system. Optional fuel in two 77·5 litre (20·5 US gallon) rubber fuel cells installed between the wing spars outboard of each engine nacelle, two 43·5 litre (11·5 US gallon) rubber fuel cells further outboard in each wing, and two 77·5 litre (20·5 US gallon) wing locker fuel tanks, providing a maximum fuel capacity of 783 litres (207 US gallons), of which 768 litres (203 US gallons) are usable. Oil capacity: 310, 22·7 litres (6 US gallons); T310, 24·6 litres (6·5 US gallons).

ACCOMMODATION: Cabin normally seats five, two in front and three on cross-bench behind. Four alternative seating arrangements are available, with up to six individual seats in pairs, all of which can tilt and have fore and aft adjustment, individual air vents, reading lights and magazine pockets. Dual controls optional. Inertia seat-belts for two front seats (optional for rear seats). Pilot's storm window, port side. Cabin windows are double-glazed to reduce noise level. Large door on starboard side giving access to all seats. Cargo door, 1·02 m (3 ft 4 in) wide, for loading of bulky items, standard on 310 II and T310 II, optional on 310 and T310. Baggage compartment at rear of cabin, capacity 163 kg (360 lb), with internal and external access; locker for a further 54·5 kg (120 lb) of baggage in the rear of each engine nacelle; and baggage compartment in extended nose with capacity of 158 kg (350 lb). Total baggage capacity 430 kg (950 lb). Optional cabin accessories include writing desk, window curtains, electrical adjustment of pilot and co-pilot seats, all-leather seats, oxygen system and photographic survey provisions. Windscreen defrosting standard; windscreen alcohol de-icing system optional.

Cessna Model 310 five/six-seat cabin monoplane (two Continental IO-520-M engines)

SYSTEMS: Electrical system powered by two 50A 28V engine-driven alternators and 24V 25Ah battery. 100A alternators optional. Oxygen system of 2·17 m³ (76·6 cu ft) or 1·37 m³ (48·3 cu ft) capacity optional; an automatic altitude compensating regulator is standard with this installation. Janitrol 45,000 BTU thermostatically-controlled blower-type heater for cabin heating and windscreen defrosting. Cabin air-conditioning system rated at 12,000 BTU optional. Vacuum system supplied by two engine-driven pumps with adequate capacity to cater for the pneumatic de-icing boots and flight instruments. Hydraulic system for brakes only.

ELECTRONICS AND EQUIPMENT: Optional electronics include Series 300 nav/com transceiver with 720-channel com and 200-channel nav with remote VOR/LOC or VOR/ILS indicator, ADF with digital tuning, 10-channel HF and flight director system; Series 400 nav/com transceiver with 720-channel com and 200-channel nav with remote VOR/LOC or VOR/ILS indicator, 40-channel glideslope, ADF with digital tuning and BFO, transponder with 4096 code capability, encoding altimeter, Nav-O-Matic 400A two-axis autopilot and integrated flight control system with optional RMI or HSI; or Series 800 720-channel com transceiver, 200-channel nav receiver with remote VOR/ILS indicator, 40-channel glideslope/marker beacon receiver, ADF with digital tuning and BFO, RMI, DME, transponder and integrated flight control system. Additional electronics options include PN-101 pictorial navigation system, X-band weather radar, KNC-610 area nav, AVQ-75 DME, KN-65 DME, radar altimeter, locator beacon, yaw damper, boom microphone and headset. Standard equipment includes sensitive altimeter, electric clock, blind-flying instrumentation, audible landing gear and stall warning indicators, heater overheat light, variable-intensity emergency floodlight, map light, alternator failure lights, instrument post lights, control locks, armrests, reading lights, pilot and co-pilot safety belts, hat shelf, super soundproofing, cabin radio speaker, baggage straps, sun visors, adjustable cabin air ventilators, emergency exit window, aft omni-vision window, tinted dual-pane windows, full-flow oil coolers, navigation light detectors, nosewheel fender, heated pitot, heated fuel vents and stall-warning transmitter, retractable landing light, navigation lights, quick-drain fuel valves, overall paint scheme, retractable cabin step and towbar. Optional equipment includes GMT clock, relief tube, flight hour recorder, co-pilot's blind-flying instrumentation, angle of attack indicator, true airspeed indicator, economy mixture indicator, outside air temperature gauge, instantaneous rate of climb indicator, electro-luminescent panel lighting, turn co-ordinator, cabin curtain, rear window curtains, writing desk, engine fire detection and extinguishing system, cabin fire extinguisher, nacelle and nose baggage compartment courtesy lights, rudder pedal locks, electrically-adjustable pilot and co-pilot seats, all-leather seats, emergency locator transmitter, external power socket, synchronous tachometer, internal corrosion proofing, fuselage ice protection plates, ice detection light, rotating beacon, second retractable landing light, wing walk and cabin step lights, three-light strobe system, taxi light, polyurethane paint, photographic provisions, pilot's boom microphone, combination boom microphone/headset, carpet for nose baggage area, rear-view mirror, inertia-reel shoulder harness for pilot and co-pilot, eight-track stereo with four cabin speakers, stereo headsets, anti-collision light, heated dual static source, static dischargers and radome nose. Additional optional items for the Model 310 and 310 II include automatic propeller synchroniser and partial oxygen system plumbing.

DIMENSIONS, EXTERNAL:

Wing span	11·25 m (36 ft 11 in)
Wing chord at root	1·72 m (5 ft 7½ in)
Wing chord at tip	1·16 m (3 ft 9½ in)
Wing aspect ratio	7·3
Length overall	9·73 m (31 ft 11½ in)
Height overall	3·25 m (10 ft 8 in)
Tailplane span	5·18 m (17 ft 0 in)
Wheel track	3·59 m (11 ft 9½ in)
Wheelbase	2·80 m (9 ft 2¼ in)
Propeller diameter:	
310, 310 II	1·94 m (6 ft 4½ in)
T310, T310 II	1·98 m (6 ft 6 in)

DIMENSIONS, INTERNAL:

Baggage compartment (cabin)	1·26 m³ (44·6 cu ft)
Baggage compartments (nacelles, total)	0·52 m³ (18·5 cu ft)
Baggage compartment (nose)	0·59 m³ (21 cu ft)

AREAS:

Wings, gross	16·63 m² (179 sq ft)
Ailerons (total)	1·06 m² (11·44 sq ft)
Trailing-edge flaps (total)	2·13 m² (22·90 sq ft)
Fin	1·33 m² (14·30 sq ft)
Rudder	1·09 m² (11·76 sq ft)
Tailplane	2·99 m² (32·15 sq ft)
Elevators	2·05 m² (22·10 sq ft)

WEIGHTS AND LOADINGS:

Weight empty:	
310	1,514 kg (3,337 lb)
310 II	1,623 kg (3,578 lb)
T310	1,566 kg (3,453 lb)
T310 II	1,675 kg (3,694 lb)
Max T-O weight:	
All versions	2,494 kg (5,500 lb)
Max landing weight:	
All versions	2,449 kg (5,400 lb)
Max wing loading:	
All versions	150 kg/m² (30·73 lb/sq ft)
Max power loading:	
All versions	5·87 kg/kW (9·65 lb/hp)

PERFORMANCE (at max T-O weight, except where indicated):

Max level speed:
- 310 at S/L: 207 knots (383 km/h; 238 mph)
- T310 at 4,875 m (16,000 ft): 236 knots (438 km/h; 272 mph)

Max cruising speed:
- 310, 75% power at 2,285 m (7,500 ft): 194 knots (359 km/h; 223 mph)
- T310, 75% power at 6,100 m (20,000 ft): 222 knots (412 km/h; 256 mph)

Econ cruising speed with max fuel:
- 310 at 3,050 m (10,000 ft): 144 knots (267 km/h; 166 mph)
- T310 at 6,100 m (20,000 ft): 175 knots (325 km/h; 202 mph)

Min control speed (VMC):	
All versions	81 knots (150 km/h; 93 mph)
Stalling speed:	
All versions	67 knots (124 km/h; 77 mph)
Max rate of climb at S/L:	
310	507 m (1,662 ft)/min
T310	518 m (1,700 ft)/min
Rate of climb at S/L, one engine out:	
310	113 m (370 ft)/min
T310	119 m (390 ft)/min
Service ceiling:	
310	6,020 m (19,750 ft)
T310	8,350 m (27,400 ft)
Service ceiling, one engine out:	
310	2,255 m (7,400 ft)
T310	5,245 m (17,200 ft)
T-O run:	
310	407 m (1,335 ft)
T310	398 m (1,306 ft)
T-O to 15 m (50 ft):	
310	518 m (1,700 ft)
T310	507 m (1,662 ft)

Landing from 15 m (50 ft):
- All versions, at 2,449 kg (5,400 lb): 546 m (1,790 ft)

Landing run:
- All versions, at 2,449 kg (5,400 lb): 195 m (640 ft)

Range, recommended lean mixture, allowances for start, taxi, T-O, climb and 45 min reserves at 45% power:
- 310, 310 II, max cruising speed at 2,285 m (7,500 ft) with 272 kg (600 lb) usable fuel: 494 nm (916 km; 569 miles)
- 310, 310 II, as above with 552 kg (1,218 lb) usable fuel: 1,132 nm (2,097 km; 1,303 miles)
- 310, 310 II, econ cruising speed at 3,050 m (10,000 ft) with 272 kg (600 lb) usable fuel: 616 nm (1,141 km; 709 miles)
- 310, 310 II, as immediately above with 552 kg (1,218 lb) usable fuel: 1,511 nm (2,800 km; 1,740 miles)
- T310, T310 II, max cruising speed at 6,100 m (20,000 ft) with 272 kg (600 lb) usable fuel: 519 nm (961 km; 597 miles)
- T310, T310 II, as immediately above with 552 kg (1,218 lb) usable fuel: 1,242 nm (2,301 km; 1,430 miles)
- T310, T310 II, econ cruising speed at 6,100 m (20,000 ft) with 272 kg (600 lb) usable fuel: 571 nm (1,057 km; 657 miles)
- T310, T310 II, as immediately above with 552 kg (1,218 lb) usable fuel: 1,440 nm (2,668 km; 1,658 miles)

CESSNA MODEL 318

USAF designation: T-37

The T-37 was the first jet trainer designed as such from the start to be used by the USAF. The first of two prototype XT-37s made its first flight on 12 October 1954, and the first of an evaluation batch of 11 T-37As flew on 27 September 1955.

A total of 1,261 T-37s had been delivered by 27 October 1975, with production continuing. In addition to aircraft supplied to the USAF, there have been substantial deliveries to foreign governments by direct purchase, or through the Military Assistance Programme.

Three versions have been built in quantity:

T-37A. Initial production version with Continental J69-T-9 turbojets, 4·1 kN (920 lb st). 534 built. Converted to T-37B standard by retrospective modification.

T-37B. Two Continental J69-T-25 turbojets, 4·56 kN (1,025 lb st). New Omni navigational equipment, UHF radio and instrument panel. First T-37B was accepted into service with the USAF in November 1959. The T-37B has also been supplied to the Royal Thai Air Force and the Cambodian, Chilean and Pakistan Air Forces. Forty-seven ordered by the Federal German government to train Luftwaffe pilots at Sheppard AFB in Texas.

Equipment can be added to the T-37B to enable it to perform military surveillance and low-level attack duties, in addition to training. Range can be extended by two 245 litre (65 US gallon) wingtip fuel tanks. Two armed T-37Bs were evaluated at the USAF Special Air Warfare Center, and were followed by two prototypes of the more powerful and more heavily armed YAT-37D (see entry on A-37). Thirty-nine T-37Bs were later converted to A-37A standard. To replace these and to meet further requirements, the USAF placed further contracts for the T-37B in 1967 and again in 1968, bringing the total ordered to 447.

T-37C. Basically similar to T-37B, but with provision for both armament and wingtip fuel tanks. Initial order for 34 placed by USAF for supply to foreign countries under Military Assistance Programme. Portugal received 30, of which 18 were supplied under this Programme, Peru had 15 and others were supplied to Brazil, Chile, Colombia, Greece, Pakistan, South Korea, Thailand and Turkey. In production. Total orders exceeded 250 on 27 October 1975.

Following 133 bird strikes encountered in 1965-70, the USAF ordered the original windscreens for more than 800 T-37s to be replaced with a birdproof type. Developed by Cessna, using General Electric-developed Lexan polycarbonate plastics, they are sandwiched with thin acrylic material on each side. These screens are 127 mm (0·5 in) thick, weigh about 19 kg (42 lb) and can resist the impact of a 1·8 kg (4 lb) bird at a speed of 250 knots (463 km/h; 288 mph).

The following details refer to the T-37B:

TYPE: Two-seat primary trainer.

WINGS: Cantilever low-wing monoplane. Wing section NACA 2418 at root, NACA 2412 at tip. Dihedral 3°. Incidence at root 3° 30′. Two-spar aluminium alloy structure. Hydraulically-operated all-metal high-lift slotted flaps inboard of ailerons.

FUSELAGE: All-metal semi-monocoque structure. Hydraulically-actuated speed brake below forward part of fuselage in region of cockpit.

TAIL UNIT: Cantilever all-metal structure. Fin integral with fuselage. Tailplane mounted one-third of way up fin. Movable surfaces all have electrically-operated trim tabs.

LANDING GEAR: Hydraulically-retractable tricycle type. Bendix oleo-pneumatic shock-absorbers. Steerable nosewheel. Tyres by General Tire and Rubber Co. Main-wheel tyres size 20 × 4·4. Nosewheel tyre size 16 × 4·4. General Tire and Rubber Co multiple-disc hydraulic brakes.

Cessna T-37C two-seat primary jet trainer of the Portuguese Air Force *(Denis Hughes)*

POWER PLANT: Two Continental J69-T-25 turbojet engines (each 4·56 kN; 1,025 lb st). Six rubber-cell interconnected fuel tanks in each wing, feeding main tank in fuselage aft of cockpit. Total usable fuel capacity 1,170 litres (309 US gallons). Automatic fuel transfer by engine-driven pumps and a submerged booster pump. Provision for two 245 litre (65 US gallon) wing-tip fuel tanks on T-37C only. Oil capacity 11·8 litres (3·12 US gallons).

ACCOMMODATION: Enclosed cockpit seating two side by side with dual controls. Ejection seats and jettisonable clamshell type canopy. Standardised cockpit layout, with flaps, speed brakes, trim tab, radio controls, etc, positioned and operated as in standard USAF combat aircraft.

ELECTRONICS: Standard USAF UHF radio; Collins VHF navigation equipment and IFF.

ARMAMENT AND EQUIPMENT (T-37C only): Provision for two 250 lb bombs or four Sidewinder missiles. Associated equipment includes K14C computing gun-sight and AN-N6 16 mm gun camera. For reconnaissance duties, KA-20 or KB-10A cameras, or HC217 cartographic camera, can be mounted in fuselage.

DIMENSIONS, EXTERNAL:
Wing span 10·30 m (33 ft 9·3 in)
Wing chord (mean) 1·70 m (5 ft 7 in)
Wing chord at tip 1·37 m (4 ft 6 in)
Wing aspect ratio 6·2
Length overall 8·92 m (29 ft 3 in)
Height overall 2·80 m (9 ft 2·3 in)
Tailplane span 4·25 m (13 ft 11½ in)
Wheel track 4·28 m (14 ft 0½ in)
Wheelbase 2·36 m (7 ft 9 in)

AREAS:
Wings, gross 17·09 m² (183·9 sq ft)
Ailerons (total) 1·05 m² (11·30 sq ft)
Trailing-edge flaps (total) 1·40 m² (15·10 sq ft)
Fin 1·07 m² (11·54 sq ft)
Rudder, incl tab 0·58 m² (6·24 sq ft)
Tailplane 3·25 m² (34·93 sq ft)
Elevators, incl tabs 1·09 m² (11·76 sq ft)

WEIGHTS AND LOADINGS (A: T-37B; B: T-37C):
Max T-O weight:
A 2,993 kg (6,600 lb)
B 3,402 kg (7,500 lb)
Max wing loading:
A 175·3 kg/m² (35·9 lb/sq ft)
B 199·2 kg/m² (40·8 lb/sq ft)
Max power loading:
A 328·2 kg/kN (3·21 lb/lb st)
B 373 kg/kN (3·65 lb/lb st)

PERFORMANCE (at max T-O weight except as noted. A: T-37B; B: T-37C):
Max level speed:
A, at 7,620 m (25,000 ft) 370 knots (685 km/h; 426 mph)
B, at 7,620 m (25,000 ft) 349 knots (647 km/h; 402 mph)
Normal cruising speed:
A, at 7,620 m (25,000 ft) 330 knots (612 km/h; 380 mph)
B, at 7,620 m (25,000 ft) 310 knots (574 km/h; 357 mph)
Stalling speed:
A 72 knots (134 km/h; 83 mph)
B 77 knots (143 km/h; 89 mph)
Max rate of climb at S/L:
A 920 m (3,020 ft)/min
B 728 m (2,390 ft)/min
Service ceiling:
A 10,700 m (35,100 ft)
B 9,115 m (29,900 ft)
Service ceiling, one engine out:
A 6,125 m (20,100 ft)
B 4,115 m (13,500 ft)
T-O to 15 m (50 ft):
A 625 m (2,050 ft)
B 838 m (2,750 ft)
Landing from 15 m (50 ft):
A 823 m (2,700 ft)
B 1,036 m (3,400 ft)
Range with MIL-C-5011A reserves at 7,620 m (25,000 ft):
A 472 nm (874 km; 543 miles)
B 703 nm (1,302 km; 809 miles)
Range at normal rated power, 5% reserves at 7,620 m (25,000 ft):
A 525 nm (972 km; 604 miles)
B 738 nm (1,367 km; 850 miles)
Max range, 5% reserves at 7,620 m (25,000 ft):
A 576 nm (1,067 km; 663 miles)
B 819 nm (1,517 km; 943 miles)

Cessna A-37B Dragonfly light strike aircraft (two General Electric J85-GE-17A turbojets)

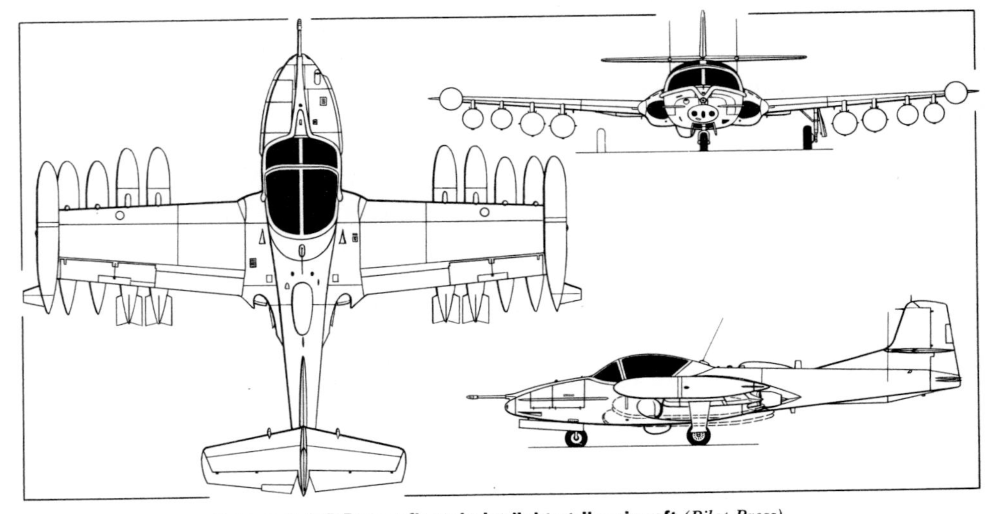

Cessna A-37B Dragonfly twin-jet light strike aircraft *(Pilot Press)*

CESSNA MODEL 318E DRAGONFLY

USAF designation: A-37

The A-37 is a development of the T-37 trainer, intended for armed counter-insurgency (COIN) operations from short unimproved airstrips. Two YAT-37D prototypes were produced initially, for evaluation by the USAF, by modifying existing T-37 airframes. The first of these flew for the first time on 22 October 1963, powered by two 10·68 kN (2,400 lb st) General Electric J85-GE-5 turbojets. There are two production versions, as follows:

A-37A (Model 318D). First 39 aircraft, converted from T-37B trainers. Withdrawn from service in 1974. Details in 1974-75 *Jane's*.

A-37B (Model 318E). Production version, of which design began in January 1967. Construction of prototype started in following month and it flew for the first time in September 1967. A-37B has two General Electric J85-GE-17A turbojets, giving more than double the take-off power available for the T-37, permitting an almost-doubled take-off weight. A total of 511 had been delivered by 13 February 1976. In addition to USAF, operators include the air forces of Chile (16), Guatemala (8), Honduras (6) and Peru (24).

The following details apply to the A-37B:

TYPE: Two-seat light strike aircraft.

WINGS: Cantilever low-wing monoplane. Wing section NACA 2418 (modified) at root, NACA 2412 (modified) at tip. Dihedral 3°. Incidence 3° 38′ at root, 1° at tip. No sweep at 22½% chord. Two-spar aluminium alloy structure. Conventional all-metal ailerons, with forward skin of aluminium alloy and aft skin of magnesium alloy. Electrically-operated trim tab in port aileron with force-sensitive boost tabs in both ailerons, plus hydraulically-operated slot-lip ailerons forward of the flap on the outboard two-thirds of flap span. Hydraulically-operated all-metal slotted flaps of NACA 2h type. No de-icing equipment.

FUSELAGE: All-metal semi-monocoque structure. Hydraulically-operated speed brake, measuring 1·14 m (3 ft 9 in) by 0·30 m (1 ft 0 in), below forward fuselage immediately aft of nosewheel well. Mountings for removable probe for in-flight refuelling on upper fuselage in front of cockpit.

TAIL UNIT: Cantilever all-metal structure. Fin integral with fuselage. Fixed-incidence tailplane mounted one-third of way up fin. Electrically-operated trim tabs in port elevator and rudder. No de-icing equipment.

LANDING GEAR: Retractable tricycle type. Cessna oleo-pneumatic shock-absorber struts on all three units. Hydraulic actuation, main wheels retracting inward, nosewheel forward. Steerable nosewheel. Goodyear tyres and single-disc brakes. Main-wheel tyres size 7·00-8 (14PR). Nosewheel tyre size 6·00-6 (6PR). Tyre pressure: main wheels 7·58 bars (110 lb/sq in), nosewheel 2·55 bars (37 lb/sq in).

POWER PLANT: Two General Electric J85-GE-17A turbojet engines, each rated at 12·7 kN (2,850 lb st). Fuel tank in each wing, each with capacity of 428 litres (113 US gallons); two non-jettisonable tip-tanks, each of 360 litres (95 US gallons) capacity; sump tank in fuselage, aft of cockpit, capacity 344 litres (91 US gallons). Total standard usable fuel capacity 1,920 litres (507 US gallons). Single-point refuelling through in-flight refuelling probe, with adaptor. Alternative refuelling through flush gravity filler cap in each wing and each tip-tank. Four 378 litre (100 US gallon) auxiliary tanks can be carried on underwing pylons. Provision for in-flight refuelling through nose-probe. Total oil capacity 9 litres (2·25 US gallons).

ACCOMMODATION: Enclosed cockpit seating two side by side, with dual controls, dual throttles, full flight instrument panel on port side, partial panel on starboard side, engine instruments in between. Full blind-flying instrumentation. Jettisonable canopy hinged to open upward and rearward. Standardised cockpit layout as in standard USAF combat aircraft. Cockpit air-conditioned but not pressurised. Flak-curtains of layered nylon are installed around the cockpit. Windscreen defrosted by engine bleed air. A polycarbonate bird-resistant windscreen is available optionally.

SYSTEMS: AiResearch air-conditioning system of expansion turbine type, driven by engine bleed air. Hydraulic system, pressure 103·5 bars (1,500 lb/sq in), operates landing gear, main landing gear doors, flaps, thrust attenuator, nosewheel steering system, speedbrake, stall spoiler, inlet screen. Pneumatic system, pressure 138 bars (2,000 lb/sq in), utilises nitrogen-filled 819 cc (50 cu in) air bottle for emergency landing gear extension. Electrical system includes two 28V DC 300A starter/generators, two 24V nickel-cadmium batteries, and provisions for external power source. One main inverter (2,500VA 3-phase 115V 400Hz), and one standby inverter (750VA 3-phase 115V 400Hz), to provide AC power.

ELECTRONICS: Radio and radar installations include UHF communications (AN/ARC-109A, ARC-151 and ARC-164), FM communications (FM-622A), Tacan (AN/ARN-65), ADF (AN/ARN-83), IFF (AN/APX-72), direction finder (AN/ARA-50), VHF communications (VHF-20B), VOR/LOC, glideslope, marker beacon (VIR-31A) and interphone (AIC-18).

ARMAMENT AND OPERATIONAL EQUIPMENT: GAU-2B/A 7·62 mm Minigun installed in forward fuselage. Each wing has four pylon stations, the two inner ones carrying 394 kg (870 lb) each, the intermediate one 272 kg (600 lb) and the outer one 227 kg (500 lb). The following weapons, in various combinations, can be carried on these underwing pylons: SUU-20 bomb and rocket pod, MK-81 or MK-82 bomb, BLU-32/B fire bomb, SUU-11/A gun pod, CBU-24/B or CBU-25/A dispenser and bomb, M-117 demolition bomb, LAU-3/A rocket pod, CBU-12/A, CBU-14/A or CBU-22/A dispenser and bomb, BLU-1C/B fire bomb, LAU-32/A or LAU-59/A rocket pod, CBU-19/A canister cluster and SUU-25/A flare launcher. Associated equipment includes an armament control panel, Chicago Aerial Industries CA-503 non-computing gunsight, KS-27C gun camera and KB-18A strike camera.

DIMENSIONS, EXTERNAL:
- Wing span over tip-tanks: 10·93 m (35 ft 10½ in)
- Wing chord at root: 2·01 m (6 ft 7·15 in)
- Wing chord at tip: 1·37 m (4 ft 6 in)
- Wing aspect ratio: 6·2
- Length overall, excl refuelling probe: 8·62 m (28 ft 3¼ in)
- Height overall: 2·70 m (8 ft 10½ in)
- Tailplane span: 4·25 m (13 ft 11½ in)
- Wheel track: 4·28 m (14 ft 0½ in)
- Wheelbase: 2·39 m (7 ft 10 in)

AREAS:
- Wings, gross: 17·09 m² (183·9 sq ft)
- Ailerons (total): 1·05 m² (11·30 sq ft)
- Trailing-edge flaps (total): 1·40 m² (15·10 sq ft)
- Fin: 1·07 m² (11·54 sq ft)
- Rudder, incl tab: 0·58 m² (6·24 sq ft)
- Tailplane: 3·25 m² (34·93 sq ft)
- Elevators, incl tab: 1·09 m² (11·76 sq ft)

WEIGHTS AND LOADINGS:
- Weight empty, equipped: 2,817 kg (6,211 lb)
- Max T-O and landing weight: 6,350 kg (14,000 lb)
- Normal landing weight: 3,175 kg (7,000 lb)
- Max zero-fuel weight: 4,858 kg (10,710 lb)
- Max wing loading: 319·3 kg/m² (65·4 lb/sq ft)
- Max power loading: 250 kg/kN (2·1 lb/lb st)

PERFORMANCE (at max T/O weight, except as detailed otherwise):
- Never-exceed speed (Mach limitation): 455 knots (843 km/h; 524 mph)
- Max level speed at 4,875 m (16,000 ft): 440 knots (816 km/h; 507 mph)
- Max cruising speed at 7,620 m (25,000 ft): 425 knots (787 km/h; 489 mph)
- Stalling speed at max landing weight, wheels and flaps down: 98·5 knots (182 km/h; 113 mph)
- Stalling speed at normal landing weight, wheels and flaps down: 75 knots (139 km/h; 86·5 mph)
- Max rate of climb at S/L: 2,130 m (6,990 ft)/min
- Service ceiling: 12,730 m (41,765 ft)
- Service ceiling, one engine out: 7,620 m (25,000 ft)
- T-O run: 531 m (1,740 ft)
- T-O to 15 m (50 ft): 792 m (2,596 ft)
- Landing from 15 m (50 ft) at max landing weight: 2,012 m (6,600 ft)
- Landing run at max landing weight: 1,265 m (4,150 ft)
- Landing run at normal landing weight: 521 m (1,710 ft)
- Range with max fuel, including four 378 litre (100 US gallon) drop-tanks, at 7,620 m (25,000 ft) with reserves: 878 nm (1,628 km; 1,012 miles)
- Range with max payload, including 1,860 kg (4,100 lb) ordnance: 399 nm (740 km; 460 miles)

CESSNA MODEL 337 SKYMASTER and SKYMASTER II

USAF designation: O-2

This unorthodox all-metal 4/6-seat business aircraft resulted from several years of study by Cessna aimed at producing a twin-engined aeroplane that would be simple to fly, low in cost, safe and comfortable, while offering all the traditional advantages of two engines. Construction of a full-scale mockup was started in February 1960 and completed two months later. The prototype flew for the first time on 28 February 1961, followed by the first production model in August 1962. FAA Type Approval was received on 22 May 1962 and deliveries of the original Model 336 Skymaster, with non-retractable landing gear, began, in May 1963.

A total of 195 Model 336 Skymasters had been built by January 1965. In the following month, this version was superseded by the Model 337 Skymaster, with increased wing incidence, retractable landing gear, and other changes, making it virtually a new aeroplane. A total of 1,787 Model 336/337 Skymasters had been built by 1 January 1976, and an additional 56 Reims Skymasters have been built by Reims Aviation in France.

Two commercial versions of the Model 337 Skymaster are available for 1976:

Skymaster. Basic version, to which the detailed description applies.

Skymaster II. Generally similar to the Skymaster, but including the following equipment as standard: directional and horizon gyros with associated vacuum system and suction gauge, true airspeed indicator, dual controls, extended range fuel system, ground power socket, heated pitot and stall warning, emergency locator transmitter, alternate static source, Series 300 nav/com with 720-channel com and 200-channel nav with VOR/ILS, Series 300 nav/com with 720-channel com and 200-channel nav with VOR/LOC, ADF and transponder, Series 400 40-channel glideslope, marker beacon, Nav-O-Matic autopilot (400A Nav-O-Matic or Series 300 IFCS offered as alternative exchanges).

Cessna Model 337 Skymaster II four/six-seat twin-engined aircraft

The 1976 versions of the commercial Skymaster introduce as standard a number of improvements, including a new parking brake valve, airspeed indicator with the primary scale in knots, single source audio speaker, additional soundproofing, standardised fuel system with cross-feed provisions; improved electronics cooling, wing-to-fuselage fairings, glareshield and defroster system; relocation of propeller synchrophaser system; redesigned power pack priority valve; optional centre seat which can be moved fore and aft the entire length of the cabin to ease access; new interior and exterior styling.

In addition, two military versions have been delivered to the USAF, as follows:

O-2A. Equipped for forward air controller missions, including visual reconnaissance, target identification, target marking, ground-to-air co-ordination and damage assessment. Dual controls standard. Four underwing pylons for external stores, including rockets, flares or other light ordnance, such as a 7·62 mm Minigun pack. Modified 60A electrical system to support special electronics systems, including UHF, VHF, FM, ADF, Tacan and APX transponder.

Initial contract, dated 29 December 1966, called for 145 O-2As; a follow-on contract awarded in June 1967 brought the total on order to 192, all of which had been delivered by early 1968. A further contract was announced on 26 June 1968, for 45 O-2As, together with modification services and spares, and this was amended in September 1968 to increase the quantity to 154 aircraft. The additional 109 O-2As have lightweight electronics. In early 1970 Cessna delivered 12 O-2As to the Imperial Iranian Air Force for training, liaison and observation duties.

O-2B. Generally similar to the commercial version, but equipped for psychological warfare missions. Advanced communications system and high-power air-to-ground broadcasting system, supplied by University Sound division of LTV Ling Altec and utilising three 600W amplifiers with highly directional speakers. Manual dispenser fitted, for leaflet dropping. Initial contract for 31 placed on 29 December 1966, and the programme was initiated by the repurchase of 31 commercial aircraft, six of which were used for pilot training at Eglin AFB, Florida. First O-2B accepted by USAF on 31 March 1967 and was assigned to Vietnam. A combined total of 510 O-2As and O-2Bs were delivered by December 1970.

The following details apply to the standard commercial Skymasters:

TYPE: Tandem-engined cabin monoplane.

WINGS: Braced high-wing monoplane, with single streamlined bracing strut each side. Wing section NACA 2412 at root, NACA 2409 at tip. Dihedral 3°. Incidence 4° 30′ at root, 2° 30′ at tip. Conventional all-metal two-spar structure. Conical-camber glassfibre wingtips. All-metal Frise ailerons. Electrically-operated all-metal single-slotted flaps. Ground-adjustable tab in port aileron. Pneumatic de-icing system optional.

FUSELAGE: Conventional all-metal semi-monocoque structure.

TAIL UNIT: Cantilever all-metal structure with twin fins and rudders, carried on two slim metal booms. Trim tab in elevator, with optional electric actuation. Optional pneumatic de-icing system.

LANDING GEAR: Hydraulically-retractable tricycle type. Cantilever spring steel main legs. Steerable nosewheel with oleo-pneumatic shock-absorber. Main wheels and tyres size 6·00-6. Nosewheel and tyre size 5·00-5. Main wheel tyre pressure 3·10 bars (45 lb/sq in). Hydraulic disc brakes. Parking brake. Oversize wheels and heavy-duty brakes optional.

POWER PLANT: Two 156·5 kW (210 hp) Continental IO-360-C flat-six engines, each driving a McCauley two-blade fully-feathering constant-speed metal propeller. Electrically-operated cowl flaps. Propeller de-icing optional for forward propeller. Fuel in two main tanks in each outer wing, with total usable capacity of 348 litres (92 US gallons); two additional tanks in inner wings, with total usable capacity of 212 litres (56 US gallons), provide optional long-range system. Total usable capacity with optional tanks 560 litres (148 US gallons). Refuelling points above wings. Oil capacity 15 litres (4 US gallons).

ACCOMMODATION: Standard accommodation for pilot and co-pilot on fully-articulating individual seats, with rear bench seat for two passengers. Dual controls optional. Alternative arrangements include individual seats for fifth and sixth passengers. Optional cabin equipment includes fully-articulating individual seats for passengers and matching headrests. Optional centre seat can be moved fore and aft entire cabin length to ease access. Space for 165 kg (365 lb) of baggage in four-seat version. Airstair door on starboard side. Cabin is heated, ventilated and soundproofed. Adjustable air vents and reading lights available to each passenger. Provision for carrying glassfibre cargo pack, with capacity of 136 kg (300 lb), under fuselage; this reduces cruising speed by only 2·6 knots (5 km/h; 3 mph).

SYSTEMS: Electrical system supplied by two 38A 28V engine-driven alternators. 24V battery. Hydraulic system for landing gear retraction and brakes.

ELECTRONICS AND EQUIPMENT: Standard equipment includes sensitive altimeter, airspeed indicator, rate of climb indicator, electric clock, outside air temperature gauge, audible stall warning device, engine synchronisation indicator, turn co-ordinator indicator, elevator and aileron control locks, windscreen defroster, dome light, map light, reading lights, baggage net, sun visors, hinged window starboard side, all-weather window, tinted windscreen and windows, omni-flash beacon, taxi light, navigation light detectors, anti-precipitation static kit, retractable tie-down rings, towbar and quick drain fuel tank valves. Optional electronics include Cessna Series 300 nav/com with 720-channel com and 200-channel nav with remote VOR/LOC or VOR/ILS indicator, ADF with digital tuning, marker beacon with three lights and aural signal, DME, 10-channel HF transceiver, and Nav-O-Matic single-axis autopilot and integrated flight control system; or Series 400 nav/com with 720-channel com and 200-channel nav with VOR/LOC or VOR/ILS indicator, 40-channel glideslope receiver, ADF with digital tuning and BFO, transponder with 4096 code capability, Nav-O-Matic 400 or 400A two-axis autopilot and integrated flight control system. Optional equipment includes all-purpose control wheel with provision for map light, boom microphone switch, pitch trim switch, autopilot/electric trim disengage switch, blind-flying instrumentation, economy mixture indicator, true airspeed indicator, instrument post lights, approach plate holder, flight hour recorder, alternate static source, cabin fire extinguisher, baggage net, wall-mounted table, safety belts for 3rd, 4th, 5th and 6th seats, oxygen system, portable stretcher, cargo tie-down installation, full-flow oil filters, oil dilution system, external power socket, propeller synchrophaser, winterisation kit, internal corrosion proofing, ice detection system, white strobe lights, photographic provisions, static wicks, windscreen anti-icing panel, pitot heating system, flush glideslope antenna, cargo pack, and emergency exit window for port side.

DIMENSIONS, EXTERNAL:
- Wing span: 11·63 m (38 ft 2 in)
- Wing chord at root: 1·83 m (6 ft 0 in)
- Wing chord at tip: 1·22 m (4 ft 0 in)
- Wing aspect ratio: 7·18
- Length overall: 9·07 m (29 ft 9 in)
- Height overall: 2·79 m (9 ft 2 in)
- Tailplane span: 3·06 m (10 ft 0⅜ in)
- Wheel track: 2·49 m (8 ft 2 in)
- Wheelbase: 2·39 m (7 ft 10 in)

Propeller diameter:
- Front: 1·98 m (6 ft 6 in)
- Rear: 1·93 m (6 ft 4 in)

Passenger door:
- Height: 1·17 m (3 ft 10 in)
- Width: 0·91 m (3 ft 0 in)

DIMENSIONS, INTERNAL:
- Cabin: Length: 3·02 m (9 ft 11 in)
- Max width: 1·12 m (3 ft 8¼ in)
- Max height: 1·30 m (4 ft 3¼ in)
- Volume: 3·62 m^3 (128 cu ft)
- Baggage space: 0·50 m^3 (17 cu ft)

AREAS:
- Wings, gross: 18·81 m^2 (202·5 sq ft)
- Ailerons (total): 1·43 m^2 (15·44 sq ft)
- Trailing-edge flaps (total): 3·43 m^2 (36·88 sq ft)
- Fins (total): 2·85 m^2 (30·68 sq ft)
- Rudders (total): 0·99 m^2 (10·70 sq ft)
- Tailplane: 3·05 m^2 (32·82 sq ft)

WEIGHTS AND LOADINGS:
- Weight empty:
  - Skymaster: 1,229 kg (2,710 lb)
  - Skymaster II: 1,288 kg (2,840 lb)
- Max T-O weight: 2,100 kg (4,630 lb)
- Max landing weight: 1,995 kg (4,400 lb)
- Max wing loading: 112 kg/m^2 (22·9 lb/sq ft)
- Max power loading: 6·71 kg/kW (11·0 lb/hp)

PERFORMANCE (at max T-O weight):
- Max level speed at S/L: 179 knots (332 km/h; 206 mph)
- Max cruising speed, 75% power at 1,675 m (5,500 ft): 169 knots (314 km/h; 195 mph)
- Econ cruising speed at 3,050 m (10,000 ft): 128 knots (237 km/h; 147 mph)
- Stalling speed, flaps up, power off: 70 knots (129 km/h; 80 mph) CAS
- Stalling speed, flaps down, power off: 61 knots (113 km/h; 70 mph) CAS
- Max rate of climb at S/L: 335 m (1,100 ft)/min
- Rate of climb at S/L, front engine only: 82 m (270 ft)/min
- Rate of climb at S/L, rear engine only: 98 m (320 ft)/min
- Service ceiling: 5,490 m (18,000 ft)
- Service ceiling, front engine only: 1,860 m (6,100 ft)
- Service ceiling, rear engine only: 2,160 m (7,100 ft)
- T-O run: 305 m (1,000 ft)
- T-O to 15 m (50 ft): 510 m (1,675 ft)
- Landing from 15 m (50 ft): 503 m (1,650 ft)
- Landing run: 213 m (700 ft)
- Range, recommended lean mixture with allowances for start, taxi, T-O, climb and 45 min reserves at 45% power:
  - Max cruising speed at 1,675 m (5,500 ft) with 239 kg (528 lb) usable fuel: 545 nm (1,009 km; 627 miles)
  - Max cruising speed at 1,675 m (5,500 ft) with 403 kg (888 lb) usable fuel: 990 nm (1,834 km; 1,140 miles)
  - Econ cruising speed at 3,050 m (10,000 ft) with 239 kg (528 lb) usable fuel: 670 nm (1,241 km; 771 miles)
  - Econ cruising speed at 3,050 m (10,000 ft) with 403 kg (888 lb) usable fuel: 1,235 nm (2,288 km; 1,422 miles)

Model P337 pressurised version of the Cessna Skymaster

CESSNA MODEL P337 PRESSURISED SKYMASTER and PRESSURISED SKYMASTER II

On 8 December 1971 Cessna introduced a pressurised version of the Skymaster. Design and construction of the prototype began in January 1971, and the first prototype made its initial flight on 23 July 1971. Construction of pre-production and production aircraft began simultaneously in May 1971, and FAA certification was granted on 2 February 1972. Deliveries began in May 1972, and Cessna claims the P337 to be the world's cheapest twin-engined pressurised aircraft.

The pressurised version is distinguished easily from the standard Skymaster in having four, instead of three, windows on each side of the cabin. Pressurisation is provided from the turbocharged engines, either of which can maintain full pressurisation and air-conditioning. With a maximum differential of 0·23 bars (3·35 lb/sq in), a cabin altitude of 3,050 m (10,000 ft) can be maintained to 6,100 m (20,000 ft). Pilot setting of departure and landing field altitudes on the pressurisation controls is all that is necessary for the system to begin automatic operation.

Two versions of the Model P337 are available for 1976:

Pressurised Skymaster. Basic version, to which the detailed description applies.

Pressurised Skymaster II. Pressurised version of Skymaster II, to which it is identical except for pressurisation installation; a Series 400A Nav-O-Matic autopilot is standard, with Series 300 IFCS offered in exchange; a Series 400 transponder replaces the Series 300 transponder of the Skymaster II.

The 1976 versions of the P337 introduce the improvements detailed for both versions of the Model 337. A total of 208 Model P337 Skymasters had been built by 1 January 1976, plus an additional 19 Reims Pressurised Skymasters built by Reims Aviation in France.

TYPE: Tandem-engined pressurised cabin monoplane.

WINGS: As Model 337 Skymaster.

FUSELAGE: Conventional all-metal semi-monocoque structure, with fail-safe structure in the pressurised section extending between the two engine bulkheads, but excluding aft lower area below cabin floor.

TAIL UNIT: As Model 337, except areas of horizontal surfaces changed.

LANDING GEAR: Hydraulically-retractable tricycle type, main units retracting aft, nosewheel forward. Cantilever spring steel main units. Steerable nosewheel with oleo-pneumatic shock-absorber. Main-wheel tyres size 6·00-6, pressure 3·79 bars (55 lb/sq in). Nosewheel tyre size 15·00 × 6·00-6, pressure 2·90 bars (42 lb/sq in). Cleveland hydraulic disc brakes. Parking brake. Heavy-duty wheels, brakes and tyres optional.

POWER PLANT: Two 168 kW (225 hp) Continental TSIO-360-C turbocharged flat-six engines, each driving a McCauley two-blade metal constant-speed fully-feathering propeller. Propeller de-icing optional for forward propeller. Four interconnected fuel tanks in each wing with a combined capacity of 285 litres (75·3 US gallons). Total fuel capacity 570 litres (150·6 US gallons). Refuelling points in wing upper surfaces. Oil capacity 17 litres (4·5 US gallons).

ACCOMMODATION: Standard accommodation for pilot and co-pilot on fully articulating individual seats, with rear bench seat for two passengers. A third passenger seat at rear of cabin is optional. Space for 165 kg (365 lb) of baggage in four-seat version. Bench seat folds to port to provide easy access to baggage area. Two-section door on starboard side, lower half opening downward and incorporating airstairs. Upper half opens upward. Cabin is heated and ventilated. Double-pane windows. Individual adjustable air ventilators and reading lights for passengers. Windscreen defrosting standard. Windscreen de-icing optional.

SYSTEMS: Electrical system powered by two 28V 38A engine-driven self-rectifying alternators. 24V battery. Electrically-driven hydraulic pump for landing gear retraction. Vacuum system optional for blind-flying instrumentation. Oxygen system optional. Cabin pressurised by engine bleed air, max differential 0·23 bars (3·35 lb/sq in). Cabin heated by 25,000 BTU gasoline heater and/or hot air from the compression section of the pressurisation system.

ELECTRONICS AND EQUIPMENT: Optional electronics are as detailed for the Model 337 Skymaster. Standard equipment is the same as for the Skymaster, plus a manual cabin altitude control, cabin rate-of-climb, cabin differential pressure and cabin altitude gauges, propeller synchrophaser and altitude warning light. Optional equipment includes a true airspeed computer, emergency locator beacon, solid-state oxygen system, engine priming system, cargo pack, and automatic propeller unfeathering system, in addition to the options detailed for the Skymaster.

DIMENSIONS, EXTERNAL:
- As for Model 337 except:
- Length overall: 9·09 m (29 ft 10 in)
- Propeller ground clearance:
  - Front: 0·23 m (9 in)
  - Rear: 0·51 m (1 ft 8 in)
- Passenger door:
  - Height: 1·15 m (3 ft 9¼ in)
  - Width: 0·90 m (2 ft 11¼ in)
  - Height to sill: 0·56 m (1 ft 10 in)

DIMENSIONS, INTERNAL:
- Cabin: Length: 3·02 m (9 ft 11 in)
- Max width: 1·11 m (3 ft 7¾ in)
- Max height: 1·29 m (4 ft 2¾ in)
- Floor area: 2·10 m^2 (22·6 sq ft)
- Volume: 3·62 m^3 (128 cu ft)
- Baggage space: 0·50 m^3 (17 cu ft)

AREAS:
- As for Model 337 except:
- Tailplane: 3·05 m^2 (32·82 sq ft)
- Elevator, incl tab: 1·18 m^2 (12·72 sq ft)

WEIGHTS AND LOADINGS:
- Weight empty:
  - Pressurised Skymaster: 1,349 kg (2,975 lb)
  - Pressurised Skymaster II: 1,401 kg (3,090 lb)
- Max T-O weight: 2,131 kg (4,700 lb)
- Max landing weight: 2,025 kg (4,465 lb)
- Max wing loading: 113·3 kg/m^2 (23·2 lb/sq ft)
- Max power loading: 6·34 kg/kW (10·4 lb/hp)

PERFORMANCE (at max T-O weight):
- Max level speed at 6,100 m (20,000 ft): 217 knots (402 km/h; 250 mph)
- Max cruising speed at 6,100 m (20,000 ft): 205 knots (380 km/h; 236 mph)
- Max cruising speed, 75% power at 3,050 m (10,000 ft): 186 knots (344 km/h; 214 mph)
- Stalling speed, flaps and wheels up, power off: 70 knots (130 km/h; 81 mph) CAS
- Stalling speed, flaps and wheels down, power off: 62 knots (114·5 km/h; 71 mph) CAS
- Max rate of climb at S/L: 381 m (1,250 ft)/min
- Rate of climb at S/L, one engine out: 114 m (375 ft)/min
- Max operating altitude: 6,100 m (20,000 ft)
- Service ceiling, one engine out: 5,700 m (18,700 ft)
- Min ground turning radius: 2·54 m (8 ft 4 in)
- T-O run: 288 m (945 ft)
- T-O to 15 m (50 ft): 457 m (1,500 ft)
- Landing from 15 m (50 ft) at max landing weight: 511 m (1,675 ft)
- Landing run at max landing weight: 242 m (795 ft)
- Range, recommended lean mixture, with allowances for start, taxi, T-O, climb and 45 min reserves at 45% power:
  - Max cruising speed at 6,100 m (20,000 ft) with 403 kg (888 lb) usable fuel: 98p nm (1,815 km; 1,128 miles)
  - Max cruising speed at 3,050 m (10,000 ft) with 403 kg (888 lb) usable fuel: 915 nm (1,694 km; 1,053 miles)
  - Econ cruising speed at 6,100 m (20,000 ft) with 403 kg (888 lb) usable fuel: 1,140 nm (2,111 km; 1,312 miles)
  - Econ cruising speed at 3,050 m (10,000 ft) with 403 kg (888 lb) usable fuel: 1,110 nm (2,056 km; 1,278 miles)

CESSNA MODEL 340A and 340A II

Cessna announced on 8 December 1971 the introduction of a pressurised twin-engined business aircraft designated Model 340. Developed from the Model 310, it had a wing and landing gear generally similar to those of the Model 414, a pressurised fuselage of fail-safe design, a tail unit similar to that of the Model 310 and 212·5 kW (285 hp) Continental TSIO-520-K engines. The Model 340 II followed, with factory installed electronics as standard.

Improved versions of these aircraft available in 1976 are as follows:

340A. Standard model, as described in detail.

340A II. Identical to Model 340A, but with dual Series 300 nav/coms with 720-channel com and 200-channel nav and VOR/LOC and VOR/ILS indicators, ADF, DME, marker beacon, and Series 400 glideslope, transponder and 400A two-axis Nav-O-Matic autopilot as standard. Other standard equipment includes dual controls, external power socket, starboard landing light, taxi light, outside air temperature gauge and all necessary antennae for on-board electronics.

Both models have as standard a number of improvements, including more powerful engines; reduced-diameter three-blade propellers, more efficient air-conditioning system, low fuel warning light, propeller synchrophaser, restyled cabin seats and repositioned oxygen masks, to ease movement within the cabin, and improved interior styling.

A total of 359 Model 340s had been delivered by 1 January 1976.

TYPE: Six-seat pressurised business aircraft.

WINGS: Cantilever low-wing monoplane, with 'Stabila-tip' fixed wingtip fuel tanks. Wing section NACA 23018 (modified) at aircraft centreline, NACA 23015 (modified) at centre-section/outer wing junction, NACA 23009 (modified) at tip. Dihedral 5° on outer panels. Incidence 2° 30′ at root, −0° 30′ at tip. All-metal two-spar structure. All-metal ailerons of single-spar construction; controllable trim tab in starboard aileron. Electrically-actuated all-metal split trailing-edge flaps, of single-spar construction with lower skin, comprising an inboard and outboard panel on each wing. Optional pneumatic de-icing system.

FUSELAGE: All-metal semi-monocoque structure. The pressurised cabin section, extending from station 100·00 aft to station 252·00, is of fail-safe construction. All openings are reinfoced with doublers and frame members, and longitudinal continuity is provided by lightweight extruded T-section stringers.

TAIL UNIT: Cantilever all-metal structure with swept vertical surfaces. Fixed-incidence tailplane of conventional two-spar construction. Elevators of single-spar construction, with controllable trim tab in starboard elevator. Rudder, built up on a formed channel spar and transverse ribs, has a controllable trim tab. Optional pneumatic de-icing system.

LANDING GEAR: Retractable tricycle type, with single wheel on each unit. Electro-mechanical retraction, main units inward into wings and faired by doors when retracted, nose unit rearward into the fuselage nose and faired by two doors when retracted. Mechanically-operated emergency gear extension system. Cessna oleo-pneumatic shock-absorbers. Steerable nosewheel with shimmy damper and self-centering device. Mainwheel tyres size 6·50-10 (8-ply); nosewheel tyre size 6·00-6 (6-ply). Single-disc hydraulic brakes. Parking brake.

POWER PLANT: Two 231 kW (310 hp) Continental TSIO-520-N flat-six turbocharged fuel-injection engines, each driving a McCauley three-blade metal constant-speed and fully-feathering propeller. Fuel system, max usable capacity 768 litres (203 US gallons), as described for Model 310. Manifold pressure relief valves to prevent engines from overboosting are standard equipment. Oil capacity 23·7 litres (6·25 US gallons).

ACCOMMODATION: Standard seating for pilot and co-pilot on tilting and individually adjustable seats. Individual seats for four passengers, two forward-facing on the port side, one aft-facing and one forward-facing on starboard side. Door, on port side aft of wing, is two-piece type with built-in airstairs in bottom portion. Plug-type emergency escape hatch on starboard side of cabin, over wing. Foul-weather window for pilot. Baggage accommodated in nose compartment with external access doors, capacity 159 kg (350 lb), two wing lockers, capacity 54·5 kg (120 lb) each, and in rear cabin area, capacity 154 kg (340 lb). Total baggage capacity 422 kg (930 lb). Cabin pressurised, heated and ventilated. Air-conditioning optional. Windscreen defroster standard; windscreen de-icing optional.

SYSTEMS: Electrical system powered by two 28V 50A engine-driven alternators and 24V 25Ah battery. Vacuum system supplied by two engine-driven pumps. Hydraulic system for brakes only. Cabin pressurised by engine bleed air, max differential 0·29 bars (4·2 lb/sq in). Cabin heated by Stewart Warner 45,000 BTU gasoline heater. Lightweight air-conditioning system optional.

ELECTRONICS AND EQUIPMENT: Standard equipment of Model 340A includes sensitive altimeter, rate of climb indicator, cabin altitude and differential pressure indicator, Accru-Measure fuel gauging system, blind-flying instrumentation, clock, audible stall warning device, variable intensity floodlights and instrument post lights, aileron and elevator control lock, courtesy lights, individual reading lights, safety belts for pilot and co-pilot, super soundproofing, cabin radio speaker, sun visors, full-flow oil filters, quick drain fuel valves and sampler cup, propeller synchroniser, navigation lights with flasher unit, heating system for fuel vents, pitot and stall warning device, retractable landing light in port wing, heater overheat indicator light, 'Not Locked' light for cabin door, all-over paint scheme, two rotating beacons, and towbar. Optional for the Model 340A but standard for the 340A II are the items detailed in the model listing. Optional items common to both versions include turn co-ordinator, economy mixture and instantaneous rate of climb indicators, flight hour recorder, true airspeed indicator, angle of attack indicator, electric elevator trim control, inertia-reel shoulder harness for pilot and co-pilot, stereo system, rudder pedal lock, cabin writing desk, window curtains, flight deck divider curtain, 0·31 m³ (11·0 cu ft) or 2·17 m³ (76·6 cu ft) oxygen system, cabin fire extinguisher, boom microphone, propeller unfeathering system, 100A alternators, engine fire detection and extinguishing system, GMT clock, blind-flying instrumentation for co-pilot, baggage courtesy lights, all-leather seats, emergency locator transmitter, tinted double-pane cabin windows, heated dual static source, polyurethane paint, nosewheel fender, ice detection lights, taxi light, white strobe lights, propeller de-icing system, windscreen alcohol de-icing system, fuselage ice impact panels, radome nose, internal corrosion proofing, static wicks, and dual pitot system. Optional electronics for the Model 340A are as detailed for the Model 310.

Cessna Model 340A six-seat pressurised business aircraft

DIMENSIONS, EXTERNAL:
- Wing span: 11·62 m (38 ft 1·3 in)
- Wing chord, mean aerodynamic: 1·57 m (5 ft 1·68 in)
- Wing aspect ratio: 7·2
- Length overall: 10·46 m (34 ft 4 in)
- Height overall: 3·84 m (12 ft 7 in)
- Tailplane span: 5·18 m (17 ft 0 in)
- Wheel track: 3·93 m (12 ft 10·7 in)
- Wheelbase: 3·12 m (10 ft 2·7 in)
- Propeller diameter: 1·94 m (6 ft 4½ in)
- Passenger door:
  - Height: 1·18 m (3 ft 10½ in)
  - Width: 0·53 m (1 ft 9 in)
- Emergency hatch:
  - Height: 0·48 m (1 ft 7 in)
  - Width: 0·66 m (2 ft 2 in)

DIMENSIONS, INTERNAL:
- Cabin:
  - Length, incl baggage compartment: 3·86 m (12 ft 8 in)
  - Max width: 1·18 m (3 ft 10½ in)
  - Max height: 1·24 m (4 ft 1 in)
  - Volume (total): 4·6 m³ (162·4 cu ft)
- Baggage space:
  - Cabin: 0·52 m³ (18·5 cu ft)
  - Nose: 0·44 m³ (15·5 cu ft)
  - Engine nacelles (each): 0·13 m³ (4·625 cu ft)

AREAS:
- Wings, gross: 17·09 m² (184 sq ft)
- Ailerons (total): 1·06 m² (11·44 sq ft)
- Trailing-edge flaps (total): 2·14 m² (23·06 sq ft)
- Fin: 1·51 m² (16·20 sq ft)
- Rudder, incl tab: 1·09 m² (11·76 sq ft)
- Tailplane: 2·99 m² (32·15 sq ft)
- Elevators, incl tab: 1·97 m² (21·25 sq ft)

WEIGHTS AND LOADINGS:
- Weight empty:
  - 340A: 1,754 kg (3,868 lb)
  - 340A II: 1,858 kg (4,096 lb)
- Max T-O and landing weight: 2,717 kg (5,990 lb)
- Max wing loading: 158·9 kg/m² (32·55 lb/sq ft)
- Max power loading: 5·88 kg/kW (9·66 lb/hp)

PERFORMANCE (at max T/O weight):
- Max level speed at 6,100 m (20,000 ft): 242 knots (448 km/h; 278 mph)
- Max cruising speed, 77·5% power:
  - at 7,470 m (24,500 ft): 228 knots (422 km/h; 262 mph)
  - at 3,050 m (10,000 ft): 199 knots (368 km/h; 229 mph)
- Econ cruising speed:
  - at 7,620 m (25,000 ft) with 272 kg (600 lb) usable fuel: 186 knots (344 km/h; 214 mph)
  - at 7,620 m (25,000 ft) with 552 kg (1,218 lb) usable fuel: 183 knots (340 km/h; 211 mph)
- Max rate of climb at S/L: 503 m (1,650 ft)/min
- Rate of climb at S/L, one engine out: 96 m (315 ft)/min
- Service ceiling: 9,085 m (29,800 ft)
- Service ceiling, one engine out: 4,815 m (15,800 ft)
- T-O run: 492 m (1,615 ft)
- T-O to 15 m (50 ft): 680 m (2,230 ft)
- Landing from 15 m (50 ft): 564 m (1,850 ft)
- Landing run: 232 m (760 ft)
- Range, recommended lean mixture with allowances for start, taxi, T-O, climb, descent and 45 min reserves at 45% power:
  - Max cruising speed at 3,050 m (10,000 ft) with 552 kg (1,218 lb) usable fuel: 989 nm (1,833 km; 1,139 miles)
  - Max cruising speed at 7,470 m (24,500 ft) with 552 kg (1,218 lb) usable fuel: 1,116 nm (2,068 km; 1,285 miles)
  - Econ cruising speed at 3,050 m (10,000 ft) with 552 kg (1,218 lb) usable fuel: 1,244 nm (2,304 km; 1,432 miles)
  - Econ cruising speed at 7,620 m (25,000 ft) with 552 kg (1,218 lb) usable fuel: 1,286 nm (2,383 km; 1,481 miles)

CESSNA MODEL 402 and 402 II

The original Model 402 was announced simultaneously with the Model 401, with a similar airframe and power plant. It was intended for the third-level airline market, with a convertible cabin and reinforced cabin floor of bonded crushed honeycomb construction, enabling it to be changed quickly from a ten-seat commuter to a light cargo transport.

On 8 December 1971 Cessna announced an extension to the Model 400 series, renaming the original Model 402 as the Model 402 Utililiner and introducing a version designated Model 402 Businessliner. On 29 October 1975 Mk II versions of both aircraft were made available for 1976, each including a package of factory-installed equipment and electronics as standard. Four versions of the Model 402 are therefore offered in 1976:

Model 402 Utililiner. Basic version, as described in detail.

Model 402 Businessliner. As basic version, except six/eight seats and optional side-hinged door to provide a total loading door width of 1·02 m (3 ft 4 in). Other options include folding business desks, stereo equipment, refreshment centre and cabin dividers.

Model 402 II Utililiner. As basic version, plus the following factory-installed equipment and electronics as standard: dual controls, economy mixture indicator, starboard landing light, locator beacon, nosewheel fender, external power socket, static dischargers, dual Cessna Series 300 nav/com with 720-channel com, 200-channel nav and VOR/ILS and VOR/LOC indicators, Series 300 ADF, Series 400 marker beacon, glideslope and transponder, Series 400B Nav-O-Matic autopilot, and basic electronics kit comprising antennae, electronics cooling kit and audio system.

Model 402 II Businessliner. As Model 402 Businessliner, plus standard factory-installed equipment and electronics detailed for Model 402 II Utililiner.

All 1976 versions of the Model 402 introduce a number of improvements as standard, including improved air-conditioning and additional heater outlets, resiting of oxygen masks and headset receptacles, new low fuel warning light, larger access panel to wastegate, laminated glassfibre wing induction inlet, new escape hatch seal and latching system to make it easily openable for on-ground ventilation, and new interior and exterior styling. New options provide for installation of the battery in the nose baggage area and provision of high-accuracy EGT.

The same prototype served for Models 401 and 402, and the FAA Type Certificate, awarded on 20 September 1966, also covered both types. A total of 816 Model 402s had been built by 1 January 1976. Twelve Model 402s were delivered in 1975 to the Royal Malaysian Air Force, which uses ten of them for multi-engine training and the other two for photographic and liaison missions.

American Jet Industries (which see) has developed a turboprop conversion of the Model 402, known as the Turbo Star 402.

TYPE: Ten-seat (optional nine-seat) convertible passenger/freight transport (Utililiner) or six/eight-seat business aircraft (Businessliner).

WINGS: Cantilever low-wing monoplane, with 'Stabila-tip' fixed wingtip fuel tanks. Wing section NACA 23018 at aircraft centreline, NACA 23015 at centre-section/outer wing junction, NACA 23009 at tip. Dihedral 5° on outer panels. Incidence 2° at root, —0° 30′ at tip. All-metal two-spar structure. All-metal ailerons and electrically-actuated split flaps. Trim tab in port aileron. Optional pneumatic de-icing system.

FUSELAGE: All-metal semi-monocoque structure.

TAIL UNIT: Cantilever all-metal structure, with 40° sweepback on fin at quarter-chord. Fixed-incidence tailplane. Trim tabs in rudder and starboard elevator. Electric operation of trim tabs optional. Optional Goodrich pneumatic de-icing system.

LANDING GEAR: Retractable tricycle type, with single wheel on each unit. Electro-mechanical retraction, main units inward into wings, nose unit rearward. Cessna oleo-pneumatic shock-absorbers. Cleveland Aircraft Products wheels, with Cessna tyres size 6·50-10 on main wheels, size 6·00-6 on nosewheel. Tyre pressures: main, 4·27 bars (62 lb/sq in); nose, 2·76 bars (40 lb/sq in). Cleveland single-disc hydraulic brakes. Parking brake.

POWER PLANT: Two 224 kW (300 hp) Continental TSIO-520-E flat-six engines, each driving a McCauley three-blade metal constant-speed fully-feathering propeller. Propeller synchronisation, automatic unfeathering and electric de-icing systems optional. Fuel system with max usable capacity of 768 litres (203 US gallons), as described for the Model 310. Oil capacity 24·6 litres (6·5 US gallons). Manifold pressure relief valves to prevent engine overboosting are standard.

ACCOMMODATION: Two seats side by side in pilot's compartment. Dual controls standard on Model 402 II versions, optional for Model 402 versions. The Utililiner cabin has four individual seats in pairs and two double seats. Passenger seats are 'Enviro-form' moulded honeycomb seats, glassfibre reinforced. Businessliner has four individual seats as standard, two additional seats optional, in the main cabin. Passenger reading lights standard on Businessliner, optional on Utililiner. Door with built-in airstair on port side of cabin at centre. Storm windows for pilot and co-pilot. Tinted cabin windows. An emergency escape hatch is provided on the starboard side of the cabin. Optional cargo door and crew access door available. Baggage area, at rear of cabin, with capacity of up to 226 kg (500 lb). Nose baggage compartment, with optional carpeting, is accessible from either side and has capacity of 159 kg (350 lb). Articles up to 1·96 m (6 ft 5 in) in length may be carried in the nose compartment. Electronics or baggage compartment in nose, separate from forward baggage compartment, and accessible through an 'over the top' 180° cam-locked door, has a capacity of 113 kg (250 lb). Optional side access door. Wing lockers, at rear of each engine nacelle, each have capacity of 54 kg (120 lb). Total baggage capacity 606 kg (1,340 lb), if no electronics are carried in the forward nose compartment. Cabin heated and ventilated. Windscreen defrosting standard. Electric anti-icing of pilot's window or alcohol anti-icing of pilot's and co-pilot's windows optional.

SYSTEMS: Electrical system powered by two 24V 50A alternators. 24V 25Ah battery. Battery can be sited optionally in nose baggage area. 100A alternators optional. Hydraulic system for brakes only. Vacuum system provided by two engine-driven pumps. Oxygen system of 1·25 m³ (44 cu ft) or 3·25 m³ (114·9 cu ft) capacity optional. Air-conditioning system optional. Heating and ventilation system with 45,000 BTU gasoline heater standard.

ELECTRONICS AND EQUIPMENT: Optional electronics as detailed for Model 310, plus radio telephone and CCC CIR-10 emergency locator transmitter. Standard equipment includes sensitive altimeter, electric clock, variable intensity floodlight, outside air temperature gauge, full blind-flying instrumentation, audible stall warning and landing gear indicators, cabin door 'Not Locked' light, map light, heater overheat warning light, alternator failure lights, variable intensity instrument post lights, aileron and elevator control lock, sun visors, armrests, pilot and co-pilot safety belt system, super soundproofing, cabin radio speaker, adjustable cabin air ventilators, navigation light detectors, courtesy lights, retractable landing light, navigation lights, two rotating beacons, all-over paint scheme and towbar. Optional equipment includes GMT clock, high efficiency GMT gauges, inertial shoulder restraint system for pilot and co-pilot, co-pilot's blind-flying instrumentation, economy mixture indicator, instantaneous rate of climb indicator, true airspeed indicator, rudder pedal locks, flight hour recorder, turn co-ordinator, cabin fire extinguisher, Utililiner or Businessliner interiors (including flight deck divider curtains, window curtains, headrests, reading lights, 'Seat Belt' and 'No Smoking' signs and various arrangements of seats, tables, refreshment units and toilets), heavy-duty brakes, internal corrosion proofing, external power socket, ice detection light, second retractable landing light, taxi light, three-light strobe system, nosewheel fender, polyurethane paint, photographic provisions, dual heated static source and static dischargers.

DIMENSIONS, EXTERNAL:

Wing span over tip-tanks	12·15 m (39 ft 10¼ in)
Wing chord at root	1·71 m (5 ft 7½ in)
Wing chord at tip	1·16 m (3 ft 9½ in)
Wing aspect ratio	7·5
Length overall	11·0 m (36 ft 1 in)
Height overall	3·56 m (11 ft 8 in)
Tailplane span	5·18 m (17 ft 0 in)
Wheel track	4·50 m (14 ft 9 in)
Wheelbase	3·19 m (10 ft 5½ in)
Propeller diameter	1·94 m (6 ft 4½ in)
Passenger door (standard):	
Height	1·21 m (3 ft 11½ in)
Width	0·58 m (1 ft 11 in)
Height to sill	1·21 m (3 ft 11½ in)
Cargo door (optional):	
Height	1·21 m (3 ft 11½ in)
Width	1·00 m (3 ft 3½ in)
Height to sill	1·21 m (3 ft 11½ in)
Nose baggage doors (each):	
Height	0·51 m (1 ft 8 in)
Width	0·80 m (2 ft 7½ in)
Nacelle baggage doors (each):	
Height	0·30 m (1 ft 0 in)
Width	0·62 m (2 ft 0½ in)

Cessna Model 402 Businessliner six/eight-seat business aircraft

DIMENSIONS, INTERNAL:

Cabin:	
Length	4·83 m (15 ft 10 in)
Max width	1·42 m (4 ft 8 in)
Max height	1·30 m (4 ft 3 in)
Volume	6·30 m³ (222·4 cu ft)

AREAS:

Wings, gross	18·18 m² (195·72 sq ft)
Ailerons (total)	1·06 m² (11·44 sq ft)
Trailing-edge flaps (total)	2·13 m² (22·90 sq ft)
Fin	3·52 m² (37·89 sq ft)
Rudder, incl tab	1·65 m² (17·77 sq ft)
Tailplane	5·64 m² (60·70 sq ft)
Elevators, incl tab	1·64 m² (17·63 sq ft)

WEIGHTS AND LOADINGS:

Weight empty:	
Businessliner	1,752 kg (3,864 lb)
Utililiner	1,767 kg (3,896 lb)
Businessliner II	1,811 kg (3,993 lb)
Utililiner II	1,825 kg (4,025 lb)
Max T-O weight	2,857 kg (6,300 lb)
Max landing weight	2,812 kg (6,200 lb)
Max wing loading	157·2 kg/m² (32·2 lb/sq ft)
Max power loading	6·38 kg/kW (10·5 lb/hp)

PERFORMANCE (at max T-O weight, except where indicated):

Max level speed at 4,875 m (16,000 ft) — 227 knots (420 km/h; 261 mph)

Max cruising speed, 75% power at 3,050 m (10,000 ft) — 190 knots (352 km/h; 219 mph)

Max cruising speed, 75% power at 6,100 m (20,000 ft) — 208 knots (386 km/h; 240 mph)

Max rate of climb at S/L	491 m (1,610 ft)/min
Rate of climb at S/L, one engine out	69 m (225 ft)/min
Service ceiling	7,980 m (26,180 ft)
Service ceiling, one engine out	3,450 m (11,320 ft)
T-O run	517 m (1,695 ft)
T-O to 15 m (50 ft)	677 m (2,220 ft)
Landing from 15 m (50 ft) at max landing weight	538 m (1,765 ft)
Landing run at max landing weight	237 m (777 ft)

Range, recommended lean mixture, with allowances for start, taxi, T-O, climb, descent and 45 min reserves at 45% power:

Max cruising speed at 3,050 m (10,000 ft) with 552 kg (1,218 lb) usable fuel — 1,004 nm (1,860 km; 1,156 miles)

Max cruising speed at 6,100 m (20,000 ft) with 552 kg (1,218 lb) usable fuel — 1,080 nm (2,000 km; 1,243 miles)

Econ cruising speed at 3,050 m (10,000 ft) with 552 kg (1,218 lb) usable fuel — 1,186 nm (2,197 km; 1,365 miles)

Econ cruising speed at 6,100 m (20,000 ft) with 552 kg (1,218 lb) usable fuel — 1,231 nm (2,280 km; 1,417 miles)

CESSNA MODEL 414

Cessna introduced the pressurised twin-engined Model 414 on 10 December 1969 as a 'step-up' aircraft for owners of Cessna or other light unpressurised twins. It combines the basic fuselage and tail unit of the Model 421 with the wing of the Model 402 and has 231 kW (310 hp) turbocharged Continental engines. Flush intakes in the engine cowlings provide improved air cooling of the engine installation, and cabin heating and pressurisation are provided by a Garrett-AiResearch engine bleed air system and a Garrett miniaturised cabin pressure control system. Either engine is able to maintain full pressurisation down to 60% power. A radiant heating system circulates heated air beneath the cabin floor and up the side walls, and an optional 45,000 BTU heater is available to provide heating on the ground or for use during extremely low temperatures. The aircraft is equipped with Cessna's Accru-Measure fuel monitoring system which provides a linear readout in both pounds and gallons to an accuracy of plus or minus 3 per cent.

A choice of eight seating layouts provides accommodation for up to seven persons, including crew, and seats incorporate armrests, tapered backs and headrests.

It was announced by Cessna on 29 October 1975 that a Mk II version of the Model 414 would be available for 1976, with a factory-installed package of equipment and electronics. Two versions of the Model 414 are, therefore, currently available:

414. Standard version, as described in detail.

414 II. As Model 414, but with the following electronics and equipment installed as standard: dual Cessna Series 400 nav/com, each with 720-channel com and 200-channel nav, with VOR/ILS and VOR/LOC indicators, Series 400 ADF, glideslope, marker beacon, transponder, DME, 400B Nav-O-Matic autopilot with associated antennae, electronics cooling kit and audio system, economy mixture indicator, co-pilot's flight instruments, auxiliary wing tanks with capacity of 238 litres (63 US gallons), starboard landing light, strobe lights, taxi light, variable cabin pressure control system, locator beacon, nosewheel fender, external power socket and set of static dischargers.

The prototype of the Model 414 flew for the first time on 1 November 1968 and FAA certification was granted on 18 August 1969. A total of 400 had been built by 1 January 1976.

The 1976 aircraft have as standard a number of improvements, including a propeller synchrophaser, increased air-conditioning cooling capacity, repositioning of oxygen masks and headsets, enlarged wastegate access panel, a low-fuel warning light, high-accuracy EGT reading and new interior and exterior styling.

TYPE: Six/seven-seat pressurised light transport.

WINGS: Cantilever low-wing monoplane, with 'Stabila-tip' fixed wingtip fuel tanks. Wing section NACA 23018 (modified) at aircraft centreline, NACA 23015 (modified) at centre-section/outer wing junction, NACA 23009 (modified) at tip. Dihedral 5° on outer panels. Incidence 2° at root. All-metal two-spar structure with stamped ribs and surface skins reinforced with spanwise stringers. All-metal ailerons and electrically-actuated split flaps. Trim tab in starboard aileron. Optional pneumatic de-icing system.

FUSELAGE: Conventional all-metal semi-monocoque structure, with fail-safe structure in the pressurised section.

TAIL UNIT: Cantilever all-metal structure, with sweptback vertical surfaces. Fixed-incidence tailplane. Trim tabs in rudder and starboard elevator. Optional pneumatic de-icing system.

LANDING GEAR: Retractable tricycle type. Electro-mechanical retraction, main units inward into wings, nosewheel unit rearward. Manual system for emergency retraction or extension. Oleo-pneumatic shock-absorbers. Steerable nosewheel. Magnesium wheels. Main-wheel tyres size 6·50-10 (8-ply), nosewheel tyre size 6·00-6 (6-ply). Goodyear single-disc hydraulic brakes. Parking brakes.

POWER PLANT: Two 231 kW (310 hp) Continental TSIO-520-N flat-six turbocharged engines, each driving a

McCauley 3AF32C93M/82NC-5·5 metal three-blade constant-speed fully-feathering propeller. Unfeathering pressure accumulator and electrical blade de-icing system optional. Fuel system with max usable capacity of 768 litres (203 US gallons), as described for the Model 310. Oil capacity 24·6 litres (6·5 US gallons).

ACCOMMODATION: Two seats side by side in pilot's compartment. Optional curtain, or solid divider with curtain, to separate pilot's compartment from main cabin. Standard seating arrangement for four forward-facing passenger seats. Optional arrangements provide for front passenger seats to face aft and a forward-facing seventh seat. Individual consoles each include reading light and ventilator. Optional items include executive writing desk, tables, hat shelf, stereo equipment, electrically-adjustable pilot's and co-pilot's seats, refreshment and Thermos units, fore and aft cabin dividers, electric shaver converter, all-leather seats, passenger instrument console (clock, true airspeed indicator and altimeter) and intercom. Door is two-piece type with built-in airstairs in bottom portion, on port side of cabin at rear. Plug-type emergency escape hatch on starboard side of cabin. Double-pane cabin windows. Foul-weather windows for pilot and co-pilot, on each side of fuselage. Electrically de-iced windscreen optional. Baggage accommodated in nose compartment with external access doors, capacity 159 kg (350 lb), two wing lockers, capacity 54·5 kg (120 lb) each, and in rear cabin area, capacity 226 kg (500 lb). Total baggage capacity 494 kg (1,090 lb).

SYSTEMS: Cabin pressurisation system, max differential 0·29 bars (4·2 lb/sq in). Electrical system powered by two engine-driven 28V 50A alternators. 24V 25Ah battery. 28V 100A alternators optional. Hydraulic system for brakes only. Vacuum system for blind-flying instrumentation and optional wing and tail unit de-icing system. Oxygen system of 3·25 m³ (114·9 cu ft) capacity, or emergency oxygen system of 0·31 m³ (11·0 cu ft) capacity optional. Air conditioning system optional.

ELECTRONICS AND EQUIPMENT: Standard equipment includes sensitive altimeter, electric clock, dual controls, windscreen defroster, outside air temperature gauge, blind-flying instrumentation, audible stall warning device, instrument post lights, alternator failure lights, aileron and elevator control lock, aircraft systems monitoring device, heater overheat light, cabin door 'Not Locked' light, armrests, aft cabin light, adjustable cabin air ventilators, non-congealing oil coolers, quick-drain fuel valves, heated stall warning transmitter, pitot and fuel vents, navigation light detectors, retractable landing light, overall paint scheme, propeller synchronisers, window curtains, courtesy lights, reading lights, super soundproofing, sun visors, full-flow oil filters, navigation lights, rotating beacons and tow-bar. Optional electronics for Model 414 include Cessna Series 400 720-channel nav/com with 200-channel nav and remote VOR/LOC or VOR/ILS indicator, 40-channel glideslope, ADF with digital tuning and BFO, transponder with 4096 code capability, 400B Nav-O-Matic two-axis autopilot. Optional electronics for both Model 414 and 414 II include integrated flight director system with optional RMI; Series 800 720-channel com transceiver, 200-channel nav receiver with remote VOR/ILS indicator, 40-channel marker beacon/glide-slope receiver, transponder with 4096 code capability, ADF with digital tuning and BFO, RMI, DME and integrated flight control system; PN-101 pictorial navigation system, X-band weather radar, single sideband HF transceiver and aft cabin intercom system. Optional equipment for Model 414 includes blind-flying instrumentation for co-pilot, nosewheel fender, economy mixture indicator, external power socket, locator beacon, starboard landing light, strobe lights, taxi light, static dischargers, variable cabin pressure control system. Optional equipment for both Model 414 and 414 II includes electric elevator trim, GMT clock, polyurethane paint, true airspeed indicator, instantaneous rate of climb indicators, rudder pedal lock, boom microphone, flight hour recorder, turn co-ordinator, electric or alcohol windscreen anti-icing, cabin fire extinguisher, 'Fasten seat belt' and 'Oxygen' signs, toilet with privacy curtain, flight deck/cabin divider or curtain, executive table, refreshment centre, 8-track stereo installation, ventilating fan system, tinted windows, internal corrosion proofing, fuselage ice impact panels, ice detection lights, dual pitot system, radome nose, engine nacelle fire detection and extinguishing system and heavy-duty brakes.

DIMENSIONS, EXTERNAL:

Wing span over tip-tanks	12·17 m (39 ft 11 in)
Wing chord, c/l to nacelles (constant)	1·71 m (5 ft 7½ in)
Wing chord at tip	1·16 m (3 ft 9½ in)
Wing aspect ratio	7·5
Length overall	10·29 m (33 ft 9 in)
Height overall	3·55 m (11 ft 8 in)
Tailplane span	5·18 m (17 ft 0 in)
Wheel track	4·48 m (14 ft 8¼ in)
Wheelbase	3·19 m (10 ft 5¾ in)
Propeller diameter	1·94 m (6 ft 4½ in)

Cessna Model 414 six/seven-seat pressurised light transport (two Continental TSIO-520-N engines)

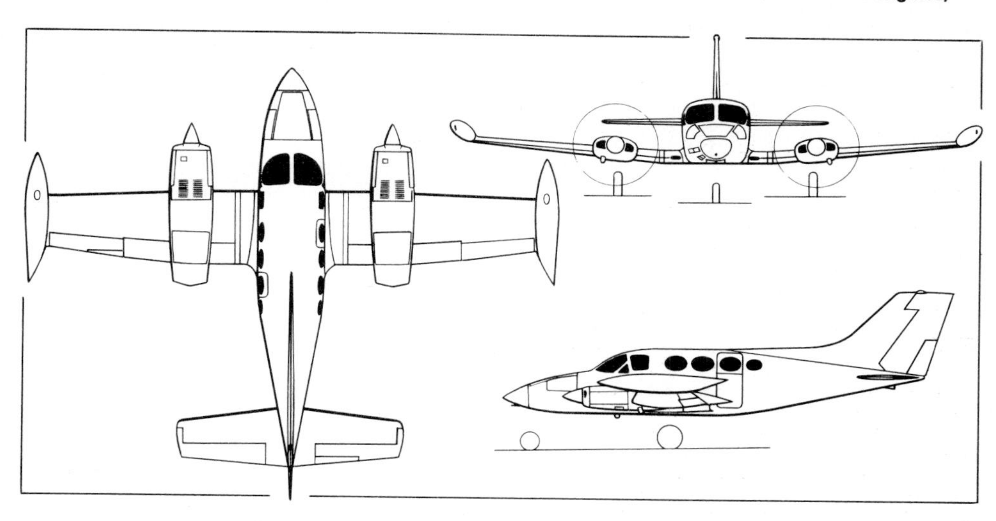

Cessna Model 414 pressurised light transport *(Pilot Press)*

Passenger door:	
Height	1·21 m (3 ft 11½ in)
Width	0·58 m (1 ft 11 in)
Height to sill	1·21 m (3 ft 11½ in)

DIMENSIONS, INTERNAL:

Cabin:	
Length	4·42 m (14 ft 6 in)
Max width	1·40 m (4 ft 7 in)
Max height	1·29 m (4 ft 3 in)
Volume	6·11 m³ (215·6 cu ft)

AREAS:

Wings, gross	18·18 m² (195·72 sq ft)
Ailerons (total)	1·06 m² (11·44 sq ft)
Trailing-edge flaps (total)	2·13 m² (22·90 sq ft)
Fin	3·52 m² (37·89 sq ft)
Rudder, incl tab	1·65 m² (17·77 sq ft)
Tailplane	5·64 m² (60·70 sq ft)
Elevators, incl tab	1·64 m² (17·63 sq ft)

WEIGHTS AND LOADINGS:

Weight empty:	
414	1,871 kg (4,126 lb)
414 II	1,979 kg (4,363 lb)
Max T-O weight	2,880 kg (6,350 lb)
Max landing weight	2,812 kg (6,200 lb)
Max wing loading	158·2 kg/m² (32·4 lb/sq ft)
Max power loading	6·23 kg/kW (10·24 lb/hp)

PERFORMANCE (at max T-O weight, except where indicated):

Max level speed at S/L
197 knots (365 km/h; 227 mph)

Max level speed at 6,100 m (20,000 ft)
236 knots (438 km/h; 272 mph)

Cruising speed, 77·5% power at 7,470 m (24,500 ft)
221 knots (409 km/h; 254 mph)

Cruising speed, 77·5% power at 3,050 m (10,000 ft)
194 knots (359 km/h; 223 mph)

Econ cruising speed at 7,620 m (25,000 ft)
190 knots (352 km/h, 218 mph)

Max rate of climb at S/L 482 m (1,580 ft)/min

Rate of climb at S/L, one engine out
73 m (240 ft)/min

Service ceiling	9,175 m (30,100 ft)
Service ceiling, one engine out	3,460 m (11,350 ft)
T-O run	517 m (1,695 ft)
T-O to 15 m (50 ft)	716 m (2,350 ft)

Landing from 15 m (50 ft) at max landing weight
568 m (1,865 ft)

Landing run at max landing weight 245 m (805 ft)

Range, recommended lean mixture with allowances for start, taxi, T-O, climb, descent and 45 min reserves at 45% power:

Max cruising speed, 77·5% power at 3,050 m (10,000 ft) with 552 kg (1,218 lb) fuel
1,007 nm (1,865 km; 1,159 miles)

Max cruising speed, 77·5% power at 7,470 m (24,500 ft) with 552 kg (1,218 lb) fuel
1,134 nm (2,102 km; 1,306 miles)

Econ cruising speed at 3,050 m (10,000 ft) with 552 kg (1,218 lb) fuel
1,301 nm (2,410 km; 1,498 miles)

Econ cruising speed at 7,620 m (25,000 ft) with 552 kg (1,218 lb) fuel
1,300 nm (2,409 km; 1,497 miles)

CESSNA MODEL 421

On 28 October 1965, Cessna announced a pressurised twin-engined business aircraft designated Model 421, the prototype of which had flown for the first time on 14 October 1965. FAA type approval was received on 1 May 1967 and deliveries began in the same month.

Two developed versions of the Model 421 were produced subsequently as the 421B Golden Eagle and 421B Executive Commuter, remaining in production until replaced by the Model 421C in 1976. Four versions of the 421C are available:

Model 421C Golden Eagle. Announced on 29 October 1975, this improved version of the Model 421B embodies a number of important changes. Most important is a bonded wet wing structure outboard of the engine nacelles, without the former distinctive wingtip tanks, which permits the introduction of a simplified fuel management system. Vertical fin and rudder area has been increased and a hydraulically-operated landing gear introduced; a new turbocharger on each engine increases critical altitude to 6,100 m (20,000 ft). Other improvements include an air-conditioning system with increased cooling capacity and wide-blade advanced technology propellers to increase performance and reduce noise levels. A new engine exhaust system further improves power plant efficiency, and there are changes in the cooling system to reduce drag and improve cooling. Power plant fire detection and extinguishing system is now located in the wing; strobe lights, electric elevator trim and improved static dischargers are standard. A low-fuel warning system is optional.

Model 421C II Golden Eagle. As above, plus a standard factory-installed electronics package and 11 additional items of equipment as standard.

Cessna Model 421C II Golden Eagle pressurised transport (two Continental GTSIO-520-L turbocharged engines)

Model 421C Executive Commuter. This is a ten-seat version of the Golden Eagle, designed specifically for the commuter airline, commercial and corporate flying markets. It differs primarily in having lightweight, easily-removable seating to provide alternative passenger/cargo configurations.

Model 421C II Executive Commuter. As above, plus the electronics and equipment packages available on the 421C II Golden Eagle.

A total of 1,055 Model 421s had been delivered by 1 January 1976.

The description which follows applies to the Model 421C Golden Eagle:

TYPE: Six/eight-seat pressurised light transport.

WINGS: Cantilever low-wing monoplane. Wing section NACA 23018 (modified) at root, NACA 23015 (modified) at centre-section/outer wing junction, NACA 23009 (modified) at tip. Dihedral 5° on outer panels. Incidence 2° 30′ at root, −0° 30′ at tip. All-metal two-spar structure. Outer wing panels of bonded construction. All-metal ailerons and electrically-actuated split flaps. Trim tab in port aileron. Optional pneumatic de-icing system.

FUSELAGE: As for Model 414.

TAIL UNIT: As for Model 414, except area of fin and rudder increased.

LANDING GEAR: Hydraulically-retractable tricycle type, main units retracting inward, nosewheel unit aft. Emergency extension by means of a 138 bar (2,000 lb/sq in) rechargeable nitrogen bottle. Oleo-pneumatic shock-absorbers. Steerable nosewheel. Main wheel tyres size 6·50-10 (8 ply), nosewheel tyre 6·00-6 (6-ply). Goodyear single-disc hydraulic brakes. Parking brake.

POWER PLANT: Two 280 kW (375 shp) Continental GTSIO-520-L flat-six geared and turbocharged engines, each driving a McCauley three-blade metal fully-feathering constant-speed propeller. Standard usable fuel capacity is 806 litres (213 US gallons) of which 780 litres (206 US gallons) are usable, contained in wet wing. Optional wing locker tanks provide a maximum usable capacity of 991 litres (262 US gallons). Oil capacity 24·6 litres (6·5 US gallons).

ACCOMMODATION Generally the same as for Model 414, except passenger cabin will accommodate up to six passengers; seats have tapered backs and headrests. The nose compartment can contain a total of 272 kg (600 lb) of baggage and electronics, and two wing lockers an additional 91 kg (200 lb) each, plus 226 kg (500 lb) in the rear cabin area, making a total capacity of 680 kg (1,500 lb).

SYSTEMS, ELECTRONICS AND EQUIPMENT: Generally as for Model 414, except cabin pressurisation system max differential 0·34 bars (5·0 lb/sq in). Hydraulic system for landing gear operation supplied by dual engine-driven pumps, pressure 103·5 bars (1,500 lb/sq in). Rechargeable nitrogen bottle for emergency extension of landing gear, pressure 138 bars (2,000 lb/sq in).

DIMENSIONS, EXTERNAL: As for Model 414 except:
- Wing span 12·53 m (41 ft 1·5 in)
- Wing chord at root 1·77 m (5 ft 9·86 in)
- Wing chord at tip 1·14 m (3 ft 8·66 in)
- Length overall 11·09 m (36 ft 4·6 in)
- Height overall 3·49 m (11 ft 5·4 in)
- Tailplane span 5·18 m (17 ft 0 in)
- Wheel track 5·48 m (17 ft 11·65 in)
- Wheelbase 3·18 m (10 ft 5·28 in)

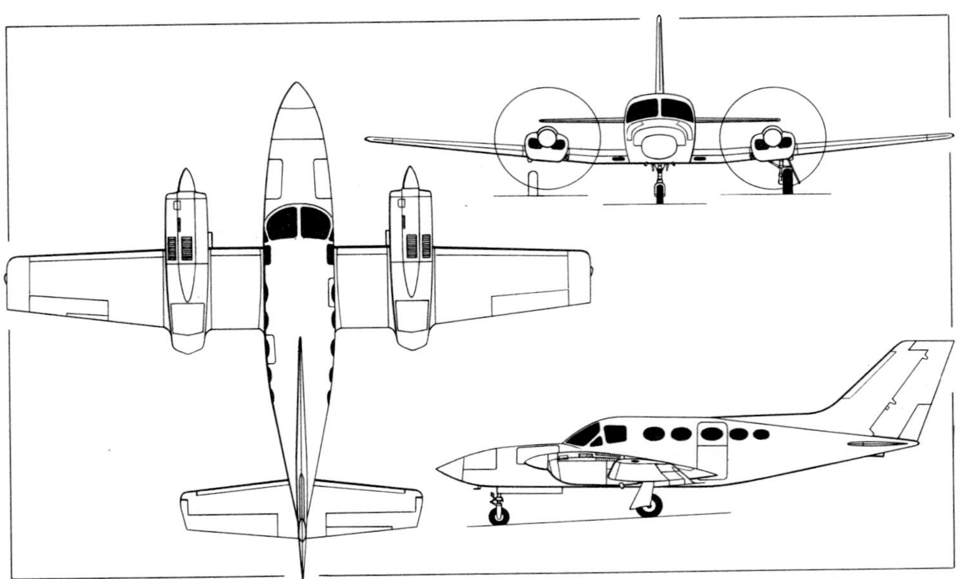

Cessna Model 421C Golden Eagle light transport aircraft *(Pilot Press)*

AREA:
- Wings, gross 19·97 m² (215 sq ft)

WEIGHTS AND LOADINGS:
- Weight empty:
  - 421C Golden Eagle 2,041 kg (4,501 lb)
  - 421C II Golden Eagle 2,145 kg (4,729 lb)
- Max T-O weight 3,379 kg (7,450 lb)
- Max landing weight 3,265 kg (7,200 lb)
- Max wing loading 169·4 kg/m² (34·7 lb/sq ft)
- Max power loading 6·03 kg/kW (9·9 lb/hp)

PERFORMANCE (at max T-O weight):
- Max level speed at 6,100 m (20,000 ft) 256 knots (475 km/h; 295 mph)
- Max cruising speed, 75% power at 7,620 m (25,000 ft) 240 knots (444 km/h; 276 mph)
- Max rate of climb at S/L 591 m (1,940 ft)/min
- Rate of climb at S/L, one engine out 107 m (350 ft)/min
- Service ceiling 9,205 m (30,200 ft)
- Service ceiling, one engine out 4,540 m (14,900 ft)
- T-O run 544 m (1,786 ft)
- T-O to 15 m (50 ft) 708 m (2,323 ft)
- Landing from 15 m (50 ft) 699 m (2,293 ft)
- Landing run 219 m (720 ft)
- Range, recommended lean mixture, with allowances for start, taxi, T-O, climb, descent and 45 min reserves at 45% power:
  - Max cruising speed, 75% power at 3,050 m (10,000 ft) with 713 kg (1,572 lb) fuel 1,104 nm (2,045 km; 1,271 miles)
  - Max cruising speed, 75% power at 7,620 m (25,000 ft) with 713 kg (1,572 lb) fuel 1,251 nm (2,317 km; 1,440 miles)
  - Econ cruising speed at 3,050 m (10,000 ft) with 713 kg (1,572 lb) fuel 1,464 nm (2,711 km; 1,684 miles)
  - Econ cruising speed at 7,620 m (25,000 ft) with 713 kg (1,572 lb) fuel 1,487 nm (2,755 km; 1,712 miles)

CESSNA TITAN

On 16 July 1975 Cessna Aircraft Company announced that it was developing a new twin-engined business/commuter/cargo aircraft, designated Model 404 Titan. The model number was deleted subsequently, and the aircraft is known currently as the Cessna Titan. It was designed to carry a nominal 1,587 kg (3,500 lb) useful load out of a 771 m (2,530 ft) airstrip. A prototype flew for the first time on 26 February 1975, and it was hoped to obtain certification in time for deliveries to begin in 1976.

The Titan offers an increase of more than 30% in ton-miles per gallon by comparison with the Cessna 402. Its wing and landing gear are as described for the Cessna Model 441. The power plant comprises two 280 kW (375 hp) turbocharged engines, driving propellers of advanced aerofoil section to offer new standards of propulsion efficiency. Standard fuel capacity is 1,287 litres (340 US gallons).

The Titan's cabin, which is almost 5·79 m (19 ft 0 in) long, is designed for rapid conversion to satisfy cargo, commuter and executive transport roles. Optional large double cargo doors permit the loading of bulky cargo, including D-size airline-type cargo packs.

The Cessna Titan is available in two versions:

Titan Ambassador. Configured and equipped for passenger carrying. Alternative layout, with a luxury executive interior seating six persons.

Titan Courier. Utility version for passenger/cargo role;

can accommodate up to ten passengers with seating at 89 cm (35 in) pitch.

DIMENSIONS:
Not available

WEIGHTS:

Weight empty	2,177 kg (4,800 lb)
Max T-O weight	3,765 kg (8,300 lb)
Max landing weight	3,674 kg (8,100 lb)

PERFORMANCE (at max T-O weight, except where indicated):
Max level speed at 6,100 m (20,000 ft)
243 knots (450 km/h; 280 mph)
Max level speed at 3,050 m (10,000 ft)
215 knots (399 km/h; 248 mph)
Max cruising speed at 6,100 m (20,000 ft)
214 knots (396 km/h; 246 mph)
Max cruising speed at 3,050 m (10,000 ft)
194 knots (359 km/h; 223 mph)
T-O to 15 m (50 ft) 771 m (2,530 ft)
Landing from 15 m (50 ft) at max landing weight
683 m (2,240 ft)
Range, max cruising speed at 6,100 m (20,000 ft)
approx 1,650 nm (3,057 km; 1,900 miles)
Range, max cruising speed at 3,050 m (10,000 ft)
approx 1,500 nm (2,779 km; 1,727 miles)

CESSNA MODEL 441 CONQUEST

Cessna announced on 15 November 1974 that it was developing a twin-turboprop business aircraft designated Model 441, with initial deliveries scheduled for 1977. This type is designed to slot into the market gap between existing twin piston-engined aircraft and turbofan-powered business aircraft.

The Model 441 is powered by two Garrett-AiResearch TPE 331-8-401 turboprop engines, which have been developed specially to meet the high-altitude high-speed requirements set by Cessna for this aircraft. Its high performance stems in part from use of a new high aspect ratio bonded wing, and from the high-strength trailing-link-type hydraulically retractable tricycle landing gear.

The prototype of the Model 441 flew for the first time on 26 August 1975. It has since been named Conquest.

TYPE: Eight/ten-seat pressurised executive transport.

WINGS: Cantilever low-wing monoplane, with constant-chord centre-section and tapered outer panels. Wing section NACA 23018 on centre-section, NACA 23012 at tip. Dihedral 3° 30′ on constant-chord section, 4° 55′ on outer panels. Incidence 2° at root, −1° at tip. All-metal structure with three-spar centre-section. Bonded construction. Large hydraulically-operated Fowler-type flaps of rigid honeycomb construction. Tab in port aileron.

TAIL UNIT: Cantilever structure with sweptback vertical surfaces. Dihedral 12° on horizontal surfaces. Large tab in each elevator and rudder.

LANDING GEAR: Retractable tricycle type, with single wheel on each unit. Hydraulic actuation, with retraction time of less than 5 sec. Main units retract inward into wings, nose unit rearward. All legs of articulated (trailing-link) type. Steerable nosewheel. Tyres size 7·75-10 on main wheels, 6·00-6 on nosewheel.

POWER PLANT: Two Garrett-AiResearch TPE 331-8-401 turboprop engines, each flat rated at 462 kW (620 shp) to 4,875 m (16,000 ft). Hartzell constant-speed fully-feathering and reversible-pitch three-blade propellers. Total usable fuel capacity 1,703 litres (450 US gallons).

ACCOMMODATION: Seats for eight to ten persons, including pilot, in pressurised and air-conditioned cabin. Door aft of wing on port side, with upward-hinged top portion and downward-hinged lower portion with integral airstairs. Emergency exit over wing on starboard side. Baggage door on each side of nose. Optional items include aft cabin divider, refreshment centre, toilet, writing tables and stereo system.

SYSTEMS: Pressurisation system max differential 0·43 bars (6·3 lb/sq in). Hydraulic system for operation of flaps, landing gear and brakes, pressure 103·5 bars (1,500 lb/sq in). Electronic fuel control system.

DIMENSIONS, EXTERNAL:

Wing span	14·12 m (46 ft 4 in)
Wing chord at root	1·78 m (5 ft 10 in)
Wing chord at tip	1·23 m (4 ft 0¼ in)
Wing aspect ratio	8·7
Length overall	11·89 m (39 ft 0¼ in)
Height overall	3·99 m (13 ft 1¼ in)
Tailplane span	5·81 m (19 ft 1 in)
Wheel track	4·28 m (14 ft 0¾ in)
Wheelbase	3·77 m (12 ft 4½ in)

DIMENSIONS, INTERNAL:

Cabin:	
Length	5·71 m (18 ft 9 in)
Max width	1·40 m (4 ft 7 in)
Max height	1·29 m (4 ft 3 in)

AREAS:

Wings, gross	22·48 m² (242 sq ft)
Vertical tail surfaces	4·05 m² (43·6 sq ft)
Horizontal tail surfaces	5·89 m² (63·38 sq ft)

WEIGHTS:

Weight empty	2,288 kg (5,045 lb)
Max ramp weight	4,343 kg (9,575 lb)
Max T-O weight	4,309 kg (9,500 lb)
Max landing weight	4,218 kg (9,300 lb)

PERFORMANCE (estimated, at max T-O weight, except where indicated):
Max level speed at 4,875 m (16,000 ft)
282 knots (523 km/h; 325 mph)
Max cruising speed at 5,180 m (17,000 ft)
280 knots (519 km/h; 322 mph)
Max rate of climb at S/L 733 m (2,405 ft)/min
Rate of climb at S/L, one engine out
213 m (700 ft)/min
Service ceiling 10,120 m (33,200 ft)
Service ceiling, one engine out 5,600 m (18,350 ft)
T-O to 15 m (50 ft) 748 m (2,455 ft)
Landing from 15 m (50 ft) at max landing weight
739 m (2,425 ft)
Range with max payload at max cruise power, with allowances for starting, taxying, take-off, climb and 45 min reserves:
at 5,180 m (17,000 ft)
755 nm (1,398 km; 869 miles)

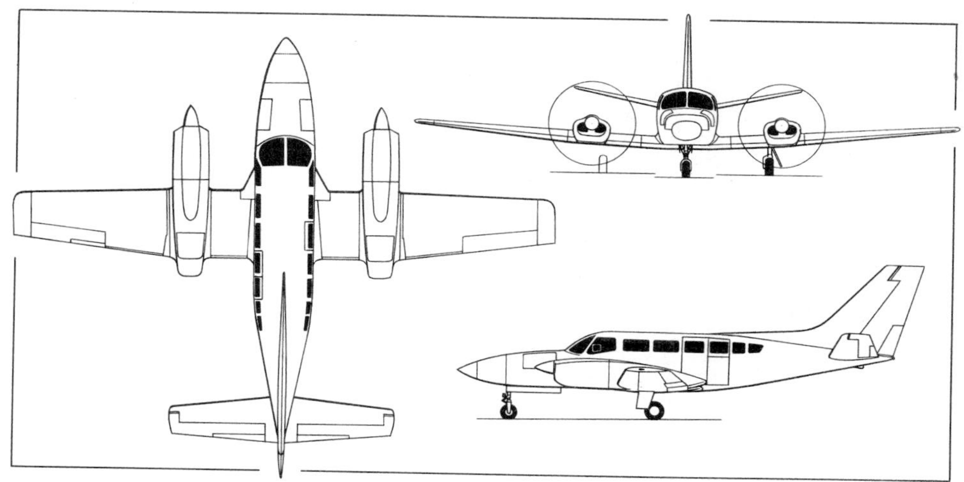

Photograph and three-view drawing *(Pilot Press)* of the Cessna Titan business/commuter transport

Cessna Model 441 Conquest eight/ten-seat pressurised executive transport (two Garrett-AiResearch TPE 331-8-401 turboprop engines)

at 7,620 m (25,000 ft)
940 nm (1,741 km; 1,082 miles)
at 10,060 m (33,000 ft)
1,160 nm (2,148 km; 1,335 miles)
Range with max fuel and 5 people at max cruise power, allowances as above:
at 5,180 m (17,000 ft)
1,160 nm (2,148 km; 1,335 miles)
at 7,620 m (25,000 ft)
1,460 nm (2,704 km; 1,680 miles)
at 10,060 m (33,000 ft)
1,830 nm (3,390 km; 2,106 miles)

CESSNA CITATION 500 SERIES

On 7 October 1968 Cessna announced that it was developing a new eight-seat pressurised executive turbofan aircraft named Fanjet 500, which would be able to operate from most airfields used by light and medium twin-engined aircraft.

After the first flight of the prototype, on 15 September 1969, it was announced that the aircraft's name had been changed to Citation. Subsequently, the gross weight was increased from 4,309 kg (9,500 lb) to 4,695 kg (10,350 lb) and several other changes were made. These included a lengthened front fuselage, movement of the engine nacelles further aft, larger vertical tail, and resiting of, and introduction of dihedral on, the tailplane.

The second Citation flew on 23 January 1970, and by mid-February 1971 the two prototypes had accumulated almost 800 hours of flight time in nearly 600 flights. Two further airframes were built for cyclic and static testing, and by 11 February 1971 all major structural tests had been accepted by the FAA.

On 1 July 1971 Cessna announced that the first production Citation 0001 (N502CC) had recently made its first flight. Final FAA certification under FAR Part 25 was awarded on 9 September 1971.

The Citation is designed to fly from runways as little as 762 m (2,500 ft) in length, and is able also to fly into and out of many unpaved airfields which are not suitable for other commercial jet aircraft. Official tests have shown that the Citation has noise levels at take-off, sideline and approach which are at least 15 EPNdB below the allowable values specified by the FAA's FAR Part 36 noise certification requirements. The Citation is offered on a direct company-to-customer basis in the basic standard configuration or as a complete business aircraft package, including factory-installed interior and electronics, ground and flight training, and one year of computerised maintenance service. Cessna states that factory installation of interior and electronics allows greater payload and also ensures that proper attention is given to the weight distribution of installed equipment.

In February 1972 the Citation was certificated at a maximum T-O weight of 4,921 kg (10,850 lb), and it was subsequently announced, on 30 June 1972, that beginning with production aircraft No. 71, the certificated max ramp weight would be 5,284 kg (11,650 lb), with a max T-O weight of 5,215 kg (11,500 lb). Certification at this max T-O weight was granted on 17 January 1973, since when Cessna has made available a modification kit to provide the new gross weight capability for production aircraft prior to No. 71.

Certification in other countries includes a West German LBA type certificate issued on 21 June 1972, French SGAC type certificate issued on 27 March 1974 and UK FA13 type certificate issued on 30 August 1974. The Citation has also satisfied the requirements for registration in Australia, Austria, Belgium, Canada, Denmark, Japan, Italy, the Netherlands, Spain, Sweden, Switzerland, Yugoslavia and Zambia.

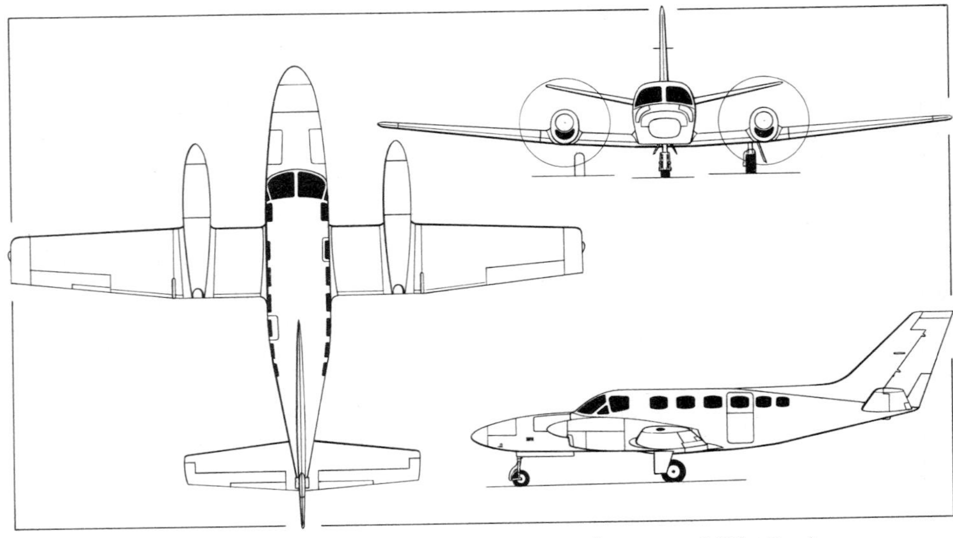

Cessna Model 441 Conquest eight/ten-seat executive transport *(Pilot Press)*

An increase in take-off gross weight to a maximum of 5,375 kg (11,850 lb) and the use of optional Rohr Industries thrust reversers received FAA certification in February 1976. A total of 300 Citations had been delivered by 29 January 1976. They include special 11-seat versions and airways check aircraft.

TYPE: Twin-turbofan executive transport.

WINGS: Cantilever low-wing monoplane without sweepback. Wing section at c/l NACA 23014 (modified), at wing station 247·95 NACA 23012. Incidence at c/l 2° 30′, at wing station 247·95 —0° 30′. Dihedral 4°. All-metal fail-safe structure with two primary spars, an auxiliary spar, three fuselage attachment points and conventional ribs and stringers. Manually-operated ailerons, with manual trim on port aileron. Electrically-operated single-slotted trailing-edge flaps. Hydraulically-operated aerodynamic speed brakes.

FUSELAGE: All-metal pressurised structure of circular section. Fail-safe design, providing multiple load paths.

TAIL UNIT: Cantilever all-metal structure. Horizontal surfaces have dihedral of 9°. Large dorsal fin and smaller ventral fin. Manually-operated control surfaces. Electric elevator trim with manual override; manual rudder trim.

LANDING GEAR: Hydraulically-retractable tricycle type with single wheel on each unit. Main units retract inward into the wing, nose gear forward into fuselage nose. Free-fall and pneumatic emergency extension systems. Goodyear main wheels and tyres of 559 mm (22 in) diameter, pressure 6·21 bars (90 lb/sq in) on aircraft Serial Nos. 0001-0070; aircraft subsequent to Serial No. 0070 have 22 in diameter main-wheel tyres, pressure 6·90 bars (100 lb/sq in). Steerable nosewheel with Goodyear wheel and tyre of 457 mm (18 in) diameter, pressure 8·27 bars (120 lb/sq in). Goodyear hydraulic brakes. Parking brake and pneumatic emergency brake system. Skid warning system optional.

POWER PLANT: Two Pratt & Whitney JT15D-1 turbofan engines, each rated at 9·77 kN (2,200 lb st) for take-off, mounted in pod on each side of rear fuselage. Rohr thrust reversers optional. Integral fuel tanks in wings, with capacity of 1,720 kg (3,793 lb).

ACCOMMODATION: Crew of two on separate flight deck. Fully-carpeted main cabin equipped with two individual forward-facing seats aft, one single forward-facing seat centre port, one single aft-facing seat centre starboard and a fifth aft-facing corner lounge chair at front of cabin on starboard side, all with headrests. Toilet compartment and main baggage area at rear of cabin. Refreshment unit at front of cabin. Second baggage area in nose. Cabin is pressurised, heated and air-conditioned. Individual reading lights and air inlets for each passenger. Drop-out constant-flow oxygen system for emergency use. Plug-type door with integral airstair at front on port side and one emergency exit on starboard side. Doors on each side of nose baggage compartment. Tinted windows, each with curtains. Optional eight-seat layout for crew of two and six passengers, executive table, flush toilet replacing standard toilet, electric razor socket and 110V converter and choice of interior trims.

SYSTEMS: Pressurisation system supplied with engine bleed air, max pressure differential 0·59 bars (8·5 lb/sq in). Hydraulic system, pressure 103·5 bars (1,500 lb/sq in), with two pumps to operate landing gear and speed brakes. Electrical system supplied by two 400A 28V DC starter/generators, with two 600VA inverters and 24V 39Ah nickel-cadmium battery. Oxygen system of 0·62 m³ (22 cu ft) capacity includes two crew demand masks and five drop-out constant-flow masks for passengers.

ELECTRONICS AND EQUIPMENT: Standard electronics

Cessna Citation 500 Series seven/eight-seat twin-turbofan executive transport (two Pratt & Whitney JT15D-1 turbofan engines)

equipment included in the fully-equipped standard Citation (up to construction number 275) comprises Bendix FGS-70 autopilot/flight director; King KDF-800 ADF; dual RCA AVC-110A VHF transceivers; dual RCA AVN-220A nav receivers, VOR, localiser, glideslope and marker beacon; dual RCA AVI-200 RMI; dual Avtech audio amplifiers; Bendix CB-70 compass system; Collins PN-101 HSI and compass system; RCA AVQ-21 radar; RCA AVQ-85 DME; RCA AVQ-95 transponder; Intercontinental Dynamics altitude alerting and reporting; and all related antennae and equipment. Standard Category II electronics package on aircraft subsequent to construction number 275 comprises Sperry SPZ 5200 flight control system with choice of single or double-cue command bars, including Sperry 500 autopilot, Sperry altimeter with altitude alerting and reporting functions, complete vertical navigation capability, air data computer, Sperry Model 600 (port)/Model 044 (starboard) horizontal situation indicator, Sperry ADI Model 300 or Model 600 command and control computer and autopilot servos, Bendix RDR 1200 continuous vision weather radar, dual Collins VHF-20 com transceivers, dual Collins VIR-30 nav receivers, Collins TDR-90 transponder, DME-40, 332-CIO radio magnetic indicator, and ARC-846A ADF. Provision for advanced instrumentation and electronics to customer's specification. Standard equipment includes automatic engine start system; engine fire warning and extinguishing system; inlet anti-icing; birdproof windscreen with de-fog system; windscreen anti-icing, with standby alcohol de-icing system; gust locks; stall warning system; two anti-collision beacons; entry light; emergency exit lights; storm lights; instrument standby lights; tailcone compartment light; 'No Smoking, Fasten Seat Belts' sign; wing ice, taxi, navigation and landing lights; external power receptacle; flight deck sunshades; map lights; internally lighted instruments; audible high Mach/airspeed warning; audible landing gear warning; generator load ammeters; standby magnetic compass; foul weather window; individual life vests; emergency battery pack; low fuel level and battery temperature warning lights; baggage tie-down kit; cabin fire extinguisher; and standard blind-flying instruments. Optional items include high-capacity oxygen system, surface de-icing system, ice detection system, anti-skid warning system, strobe lights, angle of attack indicator, engine fan synchroniser, navigation chart case, hatrack/storage shelf, nose baggage compartment light, electric razor light, refreshment cabinets, storage drawers, executive tables and flush toilets.

DIMENSIONS, EXTERNAL:
Wing span 13·39 m (43 ft 11 in)
Wing aspect ratio 6·6
Length overall 13·26 m (43 ft 6 in)
Height overall 4·36 m (14 ft 3¾ in)
Tailplane span 5·74 m (18 ft 10 in)
Wheel track 3·84 m (12 ft 7 in)
Wheelbase 4·78 m (15 ft 8¼ in)
Cabin door (port):
Height 1·29 m (4 ft 2¾ in)
Width 0·60 m (1 ft 11½ in)
Emergency exit (starboard):
Height 0·95 m (3 ft 1¼ in)
Width 0·56 m (1 ft 10 in)

DIMENSIONS, INTERNAL:
Cabin:
Length, front to rear bulkhead 5·33 m (17 ft 6 in)
Max width 1·50 m (4 ft 11 in)
Max height 1·32 m (4 ft 4 in)
Baggage space:
Cabin 1·22 m³ (43 cu ft)
Nose 0·48 m³ (17 cu ft)

AREAS:
Wings, gross 24·2 m² (260 sq ft)
Horizontal tail surfaces 6·56 m² (70·6 sq ft)
Vertical tail surfaces 4·73 m² (50·9 sq ft)

WEIGHTS AND LOADINGS (from aircraft No. 275 onward):
Weight empty (incl electronics) 2,927 kg (6,454 lb)
Max T-O weight 5,375 kg (11,850 lb)
Max landing weight 4,989 kg (11,000 lb)
Max zero-fuel weight 3,810 kg (8,400 lb)

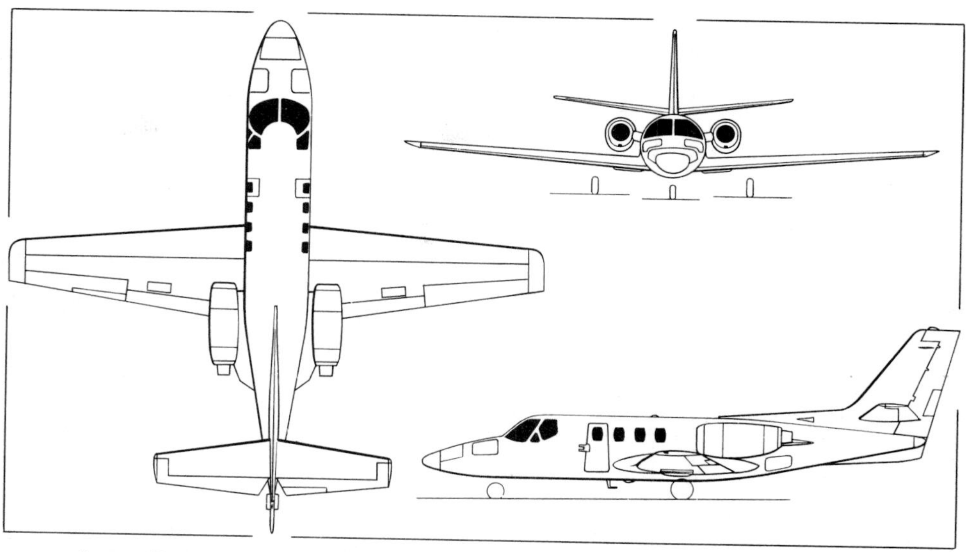

Cessna Citation 500 Series twin-turbofan seven/eight-seat executive transport *(Pilot Press)*

Optional max zero-fuel weights:
4,309 kg (9,500 lb) or 4,762 kg (10,500 lb)
Wing loading 215·8 kg/m² (44·2 lb/sq ft)
Power loading 275 kg/kN (2·69 lb/lb st)

PERFORMANCE (at max T/O weight, except where indicated):
Max operating speed, S/L to 4,265 m (14,000 ft)
260 knots (481 km/h; 299 mph) CAS
Max operating speed, 4,265 to 7,925 m (14,000 to 26,000 ft) 287 knots (531 km/h; 330 mph) CAS
Max operating speed above 7,925 m (26,000 ft)
Mach 0·7
Cruising speed, ±3%, at max cruise thrust, ISA at 7,560 m (24,800 ft) 348 knots (644 km/h; 400 mph) TAS
Flap extension speed, at 15° extension
200 knots (370 km/h; 230 mph) CAS
Flap extension speed, at 40° extension
174 knots (322 km/h; 200 mph) CAS
Landing gear operating and extended speed
174 knots (322 km/h; 200 mph) CAS
Stalling speed, in landing configuration at max landing weight 84·5 knots (157 km/h; 97 mph)
Max certificated altitude 12,495 m (41,000 ft)
Min ground turning radius 17·34 m (56 ft 10¾ in)
Balanced field length, FAR 25, ±5% at S/L, ISA
998 m (3,275 ft)
Landing field length, FAR 25, ±5% at S/L, ISA at landing weight of 3,674 kg (8,100 lb)
570 m (1,870 ft)
Landing field length from 15 m (50 ft), FAR 25, max landing weight at S/L 701 m (2,300 ft)
Range with 621 kg (1,369 lb) payload, max usable fuel and oil, at 12,495 m (41,000 ft), with allowance for T-O, climb, descent and 45 min reserves, based on empty weight of 2,927 kg (6,454 lb), incl electronics
1,307 nm (2,422 km; 1,505 miles)

OPERATIONAL NOISE CHARACTERISTICS (FAR Pt 36):
T-O noise level 78 EPNdB
Approach noise level 88 EPNdB
Sideline noise level 87 EPNdB

CESSNA MODEL 700 CITATION III

An accompanying artist's impression depicts the Model 700 Citation III, development of which was announced by Cessna on 30 October 1974. It will be powered by three Garrett-AiResearch TFE 731-3 turbofan engines, and will have a design range of about 3,000 nm (5,550 km; 3,450 miles) at airline speeds and operating heights. Deliveries are scheduled to begin in 1978.

Artist's impression of Cessna Model 700 Citation III three-engined long-range business aircraft

CONTINENTAL COPTERS
CONTINENTAL COPTERS INC

PO Box 13284, Cardinal Road, Fort Worth, Texas 76118
Telephone: (817) BU1-2330
PRESIDENT: John L. Scott

Continental Copters has developed and is producing a series of specialised single-seat agricultural conversions of various versions of the Bell Model 47 helicopter, under the name El Tomcat. Design work on the original conversion began in 1959 and the prototype El Tomcat Mk II flew in April of that year, receiving an FAA Supplemental Type Certificate shortly afterwards.

The prototype of the improved El Tomcat Mk III flew for the first time in April 1965. Further refinement of the design produced the El Tomcat Mk IIIA in January 1966; details and a picture of this version can be found in the 1966-67 *Jane's*. It was superseded in 1967 by the El Tomcat Mk IIIB, which introduced a number of improvements and was described in the 1968-69 *Jane's*.

In 1968 Continental Copters produced the El Tomcat Mk IIIC, an improved version of the IIIB with cleaner nose profile, wraparound side windows in the roof of the cabin for rear-quarter visibility, and refuelling capability from either side of the aircraft. Also in 1968 the company delivered its first El Tomcat Mk V, generally similar to the IIIC but with a change in power plant.

The Mk V was succeeded subsequently by the Mk V-A, based on the Model 47G-2, with a 194 kW (260 hp) Lycoming VO-435-A1F engine and a 24V electrical system; and the Mk V-B, with a 198 kW (265 hp) Lycoming VO-435-B1A and 24V electrical system. Subsequently, production of both the Mk IIIC and MK V-B was terminated in favour of the Mk V-A.

Next came the El Tomcat Mk VI, based on the Bell Model 47G-3B with turbocharged engine. Variants are the Mks VI-A and VI-B.

During 1975, seven helicopters were completed, and by the beginning of April 1976 the company had delivered a total of 63 Tomcats. These are in service with customers in the USA, Hawaii, Panama, Portugal, Puerto Rico and Turkey, and a Mk VI-B Tomcat was scheduled for delivery to Switzerland in mid-1976.

In early 1975 the company began work on modifications to develop a special-purpose agricultural aircraft from the airframe of the Bell JetRanger. Progress has, however, been delayed by current economic conditions.

In addition, Continental Copters has for some years been producing passenger helicopters conforming to the Bell 47G and G-2 types. These are assembled from spare and/or surplus parts and are listed in the FAA Helicopter Specification H-1. Other helicopters, not included in this list, have been delivered to Latin America.

CONTINENTAL COPTERS EL TOMCAT Mk V-A

Each version of the El Tomcat is basically a Bell Model 47G-2 helicopter which has been converted into a specialised single-seat agricultural aircraft. Payload is increased by deletion of unnecessary structure and equipment. In particular, the original cabin is replaced by a simple functional cab for the pilot. This was improved during 1974 by the development of a strengthened 'cage' for the pilot, consisting of a rigid forward frame to the cockpit, constructed from 254 mm (1 in) square steel tube. Additionally, a wire deflector installation has been developed for this aircraft, consisting of a steel tube above and below the nose, mounted on the aircraft centreline. This would deflect any wires or cables to sharpened jaws mounted on the steel tubes.

Power plant consists of a 194 kW (260 hp) Lycoming VO-435-A1F engine. Standard Bell Model 47D-1 fuel system with a capacity of 109 litres (29 US gallons). Oil capacity 11 litres (3 US gallons). A 24V electrical system is standard.

In the current El Tomcat Mk V-A, which was awarded a full Type Certificate by the FAA in May 1973, the windscreen has been further reduced in area and moved closer to the pilot compared with earlier versions: in addition, the new strengthened forward frame serves as the windscreen mounting. The glassfibre nose has been modified to ensure easy accessibility to all instruments, the battery and other equipment. It provides a flush mounting for two 600W landing lights which are controllable in elevation by the pilot during flight, landing-light switches being mounted on the collective stick, immediately below the throttle. The cabin roof is of glassfibre, lower than on earlier versions of El Tomcat, and incorporates wraparound side windows for rear-quarter view. An FAA-approved folding jump-seat has been developed to permit carriage of a flagman to distant work sites when large fields are being sprayed. Standard equipment includes pilot's shoulder harness.

El Tomcat has a revised control system. The collective control has been altered to conform to standard collective geometry, but Continental retains ball bearings in the collective jack shaft, instead of brass bushings, to provide smoother operation. A Harley Davidson throttle control is fitted. The flying controls are hydraulically-boosted.

The chemical hoppers take the form of two streamlined blister tanks which fit flush against the sides of the fuselage immediately aft of the cabin. A fan-driven pump is mounted adjacent to each tank, aft of the spraybar supports. A filtered ventilation system for the cockpit minimises toxic spray ingress during spraying operations. Types of spraygear fitted include the Bell Agmaster, Simplex Lo Profile and special designs developed by customers.

Apart from the changes noted, the basic structural description of the standard Bell Model 47 (which can be found under the Agusta entry in the Italian section) applies also to El Tomcat.

DIMENSIONS:
As for standard Bell Model 47G-2

WEIGHTS:
Weight empty, less specialised equipment 623 kg (1,375 lb)
Max T-O weight 1,111 kg (2,450 lb)

PERFORMANCE:
As for standard Bell Model 47G-2 except:
Range (with fuel reserve for 30 min) 86 nm (160 km; 100 miles)

Continental Copters El Tomcat Mk V-A, showing cable deflector system

Continental Copters El Tomcat Mk VI-B for export to Switzerland

CONTINENTAL COPTERS EL TOMCAT Mk VI-B

The original El Tomcat Mk VI was a conversion of the Bell Model 47G-3B, with a turbocharged Lycoming TVO-435 engine, developing 201 kW (270 hp), and Bell 47-110-250-23 main rotor blades. The increased length of these blades made it necessary to extend the basic Mk V-A centre frame both fore and aft. The extension aft (as on later production Bell 47s) was needed to provide adequate clearance between main and tail rotor blades, and extension forward to compensate for the resulting rearward movement of the CG.

As in the Mk V-A, the battery was accommodated in the nose of the cab. In addition, the engine oil tank was placed in the nose; an airscoop on top of and an outlet on the bottom of the nose were installed to allow ram air to carry away the heat radiated by the tank. The standard El Tomcat in-flight-adjustable landing lights were installed.

Only this prototype of the El Tomcat Mk VI was completed; it was supplied to an operator in the Rocky Mountain area for spraying operations at high altitude. FAA certification was not obtained prior to delivery.

The prototype was claimed to be very stable, and flight tests demonstrated its ability to hover directly into the wind with the pilot's hands and feet off the controls.

A prototype of the Mk VI-B, which has a component similarity to the Bell Model 47G-5, was subsequently completed and flown, and was awarded an FAA Supplemental Type Certificate in the Standard category; this is now the production version. The FAA required installation of a compass and altimeter to meet certification standards, and Continental Copters anticipated that instrument panel and windscreen changes would be required on subsequent aircraft to facilitate installation of these instruments. This has been catered for when carrying out the pilot's 'cage' modification as described for the Mk V-A.

The hydraulic reservoir/regulator unit for the Simplex Lo-Profile sprayer system fitted to the Mk VI-B has been mounted forward on the port side of the centre frame instead of in the usual rear position. This enables the pilot to see the hydraulic pressure gauge easily and also offsets the normal aft CG condition of the El Tomcat.

Empty weight of the El Tomcat Mk VI-B is 674 kg (1,487 lb) and max T-O weight 1,293 kg (2,850 lb). Maximum permissible speed has been limited to 61 knots (113 km/h; 70 mph) to minimise the airspeed calibration test. In other respects the flight envelope of the Mk VI-B is identical to that of the Bell 47G-5. Three examples of the Mk VI-B have been exported to Portugal.

Conversion kits for the Mk VI-B installation are available, and Continental Copters states that by using the data gained from this programme Supplemental Type Certification of conversions of Bell Model 47G-2A, -2B and -4 helicopters can be carried out with a minimum of FAA testing.

DOMINION

DOMINION AIRCRAFT CORPORATION LTD

ADDRESS:
1005 West Perimeter Road, Renton, Washington 98055
Telephone: (206) 228-3536

PRESIDENT:
Lawrence Matanski

DOMINION SKYTRADER 800

The Skytrader 800 is a twin-engined STOL transport and general-purpose aircraft. Much of the design was carried out by former members of The Boeing Company's staff at Renton, Washington, USA, and construction of wing components and subassemblies of the prototype began at Renton in the Autumn of 1972.

The first flight of the Skytrader 800 was made in the Spring of 1975.

Two Skytraders have been ordered by the Macmillan-

Prototype of the Dominion Skytrader 800 STOL general-purpose transport (two Lycoming IO-720-B1A engines) *(Pilot Press)*

Bloedel Forest Products company.

TYPE: Twin-engined STOL general-purpose transport aircraft.

WINGS: High-wing monoplane, with single bracing strut on each side. Constant-chord wings, with electrically-operated leading-edge slats. Full-span ailerons and flaps on trailing-edge. Trim tab in port aileron. Anti-icing system optional.

FUSELAGE: Conventional structure, with rectangular cabin section and upswept rear end.

TAIL UNIT: Cantilever structure, with horizontal surfaces of constant chord. Dorsal fin. Trim tabs in rudder and each elevator.

LANDING GEAR: Non-retractable tricycle type, with steel legs and rubber shock-absorbers on main units, oleo shock-absorber on nose unit. Goodyear wheels, tyres and disc brakes. Parking brake standard. Provision for alternative twin-float amphibian landing gear, oversize low-pressure tyres, or wheel-ski landing gear, at customer's option. Amphibian gear has retractable 0·38 m (15 in) nosewheel and 0·635 m (25 in) main wheel on each float.

POWER PLANT: Two 298 kW (400 hp) Lycoming IO-720-B1A flat-eight engines in prototype, each driving a three-blade Hartzell constant-speed fully-feathering metal propeller (reversible pitch optional). Provision for fitting 354 kW (475 hp) Lycoming TIO-720-C engines in production aircraft. Internal fuel capacity (outer wing tanks) 605 litres (160 US gallons) normal, 909 litres (240 US gallons) with auxiliary tanks in wing roots. Provision for a further 628 litres (166 US gallons) in two underwing drop-tanks, raising total capacity to 1,537 litres (406 US gallons). Provision for JATO at customer's option.

ACCOMMODATION: Crew of two at front of cabin. Various internal layouts, including six-seat executive, 12-seat passenger transport, or all-freight. Executive layout includes toilet and wardrobe/baggage space. Passenger version has baggage space at rear but no toilet. Fold-away seats can be installed for quick-change passenger/freight conversion. Can be equipped as water bomber, with 1,324 litre (350 US gallon) tank installed by the rear-loading doors. Cabin heated and ventilated. Access to flight deck by forward-hinged door on each side, and to main cabin via port-side double doors, starboard-side single door, and rear-loading doors in underside of fuselage. Airstair door on port side optional.

SYSTEMS AND EQUIPMENT: Electrical system includes two 70A 28V alternators. 45,000 BTU cabin heater. Optional equipment includes anti-icing system, windscreen wipers, weather radar, autopilot, oxygen equipment, engine fire extinguishers, cargo tie-downs and baggage rack, water-bombing installation and external power receptacle.

Dominion Skytrader 800 STOL general-purpose transport *(Pilot Press)*

DIMENSIONS, EXTERNAL:	
Wing span	16·76 m (55 ft 0 in)
Wing aspect ratio	7·9
Length overall	12·50 m (41 ft 0 in)
Height overall	5·76 m (18 ft 10¾ in)
Single passenger door (stbd):	
Width	0·61 m (2 ft 0 in)
Height	1·42 m (4 ft 8 in)
Double doors (port):	
Width	1·22 m (4 ft 0 in)
Height	1·42 m (4 ft 8 in)
Rear-loading cargo doors:	
Width at top	0·97 m (3 ft 2 in)
Width at bottom	1·27 m (4 ft 2 in)
Length	2·03 m (6 ft 8 in)
DIMENSIONS, INTERNAL:	
Cabin:	
Max length, incl flight deck	5·18 m (17 ft 0 in)
Max width	1·27 m (4 ft 2 in)
Max height	1·65 m (5 ft 5 in)
Total volume	10·65 m³ (376 cu ft)
Cargo volume	8·72 m³ (308 cu ft)
AREAS:	
Wings, gross	35·77 m² (385 sq ft)
Ailerons (total)	3·06 m² (32·96 sq ft)
Trailing-edge flaps (total)	5·34 m² (57·48 sq ft)
Fin	2·72 m² (29·33 sq ft)
Rudder, incl tab	2·01 m² (21·67 sq ft)
Tailplane	9·29 m² (100·00 sq ft)
Elevators, incl tabs	4·31 m² (46·35 sq ft)
WEIGHTS AND LOADINGS:	
Weight empty	2,245 kg (4,950 lb)
Max T-O weight	3,855 kg (8,500 lb)
Max wing loading	4·5 kg/m² (22 lb/sq ft)
Max power loading	6·47 kg/kW (10·6 lb/hp)

PERFORMANCE (estimated, at max T-O weight):	
Max level speed	182 knots (338 km/h; 210 mph)
Max cruising speed (75% power) at 3,050 m (10,000 ft)	153 knots (285 km/h; 177 mph)
Cruising speed (55% power) at 762 m (2,500 ft)	130 knots (241 km/h; 150 mph)
Min control speed, one engine out	54 knots (100 km/h; 62 mph)
Stalling speed, flaps down, slats extended, power off	52 knots (97 km/h; 60 mph)
Min speed at which fully manoeuvrable, slats extended, power on	45 knots (84 km/h; 52 mph)
Max rate of climb at S/L	487 m (1,600 ft)/min
Rate of climb at S/L, one engine out	128 m (420 ft)/min
*Service ceiling	5,335 m (17,500 ft)
*Service ceiling, one engine out	2,135 m (7,000 ft)
T-O run	119 m (390 ft)
T-O to 15 m (50 ft)	272 m (890 ft)
Landing from 15 m (50 ft)	223 m (730 ft)
Landing run	95 m (310 ft)
Range with max internal fuel (75% power)	805 nm (1,495 km; 930 miles)
Range with max internal fuel (55% power)	1,240 nm (2,300 km; 1,430 miles)
Range with max internal and external fuel (55% power)	2,125 nm (3,940 km; 2,450 miles)

**without turbocharging*

ECTOR

ECTOR AIRCRAFT COMPANY

ADDRESS:
414 East Hillmont Road, Odessa, Texas 79762
Telephone: (915) 362-1841
PRESIDENT: Alvin H. Parker

ECTOR L-19 MOUNTAINEER and SUPER MOUNTAINEER

The Ector Aircraft Company has in production a civil version of the Cessna L-19 Bird Dog (last described in the 1964-65 *Jane's*), to which it has given the name Mountaineer. This is available in two models:

Mountaineer. Standard model, with 159 kW (213 hp) Continental O-470-11 engine and fixed-pitch propeller. Continental O-470-11-13 (cruising speed increased to 108 knots; 201 km/h; 125 mph at 55% power) or O-470-11-13-15 engine and constant-speed propeller (104 knots; 193 km/h; 120 mph at 57% power) optional.

Super Mountaineer. More powerful version with a 179 kW (240 hp) Lycoming O-540-A4B5 engine, driving a Hartzell Type HC-C2YK-1B two-blade metal constant-speed propeller, and with additional equipment as standard.

Generally similar to the original Cessna L-19, Ector's Mountaineers are rebuilt completely from new off-the-shelf or serviceable components. The entire airframe is corrosion-proofed with zinc chromate before assembly, mounting brackets for floats are built into the basic airframe and all four side windows can be opened in flight.

The rear seat is removable to permit the carriage of cargo.

The Mountaineer is in service with various organisations as a glider tug and for patrol and general-purpose duties; it is also in demand as a sporting aircraft.

A total of more than 30 Mountaineers had been produced by 1 January 1976.

TYPE: Two-seat lightweight cabin monoplane.

WINGS: Braced high-wing monoplane. Single streamline-section bracing strut each side. Wing section NACA 2412. Dihedral 2° 8′. Incidence 1° 30′ at root, −1° 30′ at tip. All-metal single-spar structure, with metal skin. Frise-type all-metal ailerons. Fowler all-metal trailing-edge flaps. No tabs.

FUSELAGE: Conventional all-metal semi-monocoque structure.

TAIL UNIT: Cantilever all-metal structure. Trim tab in elevator. Small auxiliary fins are attached to tailplane tips of floatplane version.

LANDING GEAR: Non-retractable tailwheel type. Cantilever spring steel main legs. Goodyear main wheels with tyres size 6·00-6. Scott steerable tailwheel. Single-disc hydraulic brakes. Floats, skis or tandem landing gear for rough terrain optional.

POWER PLANT: One Continental or Lycoming flat-six engine as detailed in model listings, driving a two-blade fixed-pitch or constant-speed propeller. One fuel tank in each wing root, total capacity 151·4 litres (40 US gallons). Optionally, fuel cells in each wing with a total capacity of 246 litres (65 US gallons). Refuelling points on wing upper surface. Oil capacity 9·4 litres (2·5 US gallons).

ACCOMMODATION: Two seats in tandem in enclosed cabin with 360° field of vision. Door on starboard side. All four cabin side windows can be opened fully. Six skylights in roof. Space for baggage behind rear seats. With rear seat removed, 0·85 m³ (30 cu ft) of space is available for freight. Cabin heated and ventilated.

SYSTEMS: Hydraulic system for brakes only. Electrical system powered by 24V 50A engine-driven generator. Super Mountaineer has a 12V electrical system.

ELECTRONICS AND EQUIPMENT: Radio equipment available to customer's requirements. Navigation and landing lights, and heated pitot, standard. External power socket optional. Stall warning indicator, Hobbs hour meter and Whelen three-light strobe system standard on Super Mountaineer. Wing racks optional.

Ector Super Mountaineer, a civil version of the Cessna L-19 Bird Dog

DIMENSIONS, EXTERNAL:

Wing span	10·97 m (36 ft 0 in)
Wing chord (root)	1·63 m (5 ft 4 in)
Wing chord (tip)	1·09 m (3 ft 7 in)
Wing aspect ratio	7·35
Length overall	7·86 m (25 ft 9½ in)
Height overall	2·29 m (7 ft 6 in)
Tailplane span	3·21 m (10 ft 6½ in)
Propeller diameter	2·29 m (7 ft 6 in)
Propeller ground clearance	0·23 m (9 in)
Door:	
Height	0·64 m (2 ft 1 in)
Width	0·81 m (2 ft 8 in)
Height to sill	1·12 m (3 ft 8 in)

WEIGHTS (standard Mountaineer):

Weight empty, equipped	658 kg (1,450 lb)
Max T-O weight	1,043 kg (2,300 lb)

PERFORMANCE (A: Mountaineer with O-470-11 engine and fixed-pitch propeller; B: Super Mountaineer):

Max level speed at S/L:	
A	87 knots (161 km/h; 100 mph) TAS
B	112 knots (208 km/h; 129 mph) IAS
Max cruising speed:	
B at 2,135 m (7,000 ft)	139 knots (257 km/h; 160 mph) TAS
Econ cruising speed:	
B at 2,895 m (9,500 ft)	109 knots (201 km/h; 125 mph) IAS
Stalling speed, flaps down:	
B	45·5 knots (84 km/h; 52 mph)
Max rate of climb at S/L:	
A	366 m (1,200 ft)/min
B	549 m (1,800 ft)/min
Service ceiling:	
A	6,980 m (22,900 ft)
T-O run:	
A	122 m (400 ft)
Landing run:	
A	98 m (320 ft)
Range, max fuel, no reserves:	
A	651 nm (1,207 km; 750 miles)

EMAIR

EMAIR (a division of Emroth Co)

ADDRESS:

Hangar 38, Industrial Airpark, Harlingen, Texas 78550

Telephone: (512) 425-6363

PRESIDENT: George A. Roth

Under contract to Murrayair Ltd of Hawaii, Air New Zealand engineers began in September 1968 the construction of a single-seat agricultural aircraft, which first flew on 27 July 1969. Subsequently it was dismantled and shipped to Honolulu to complete its trials for FAA certification. FAA Type Approval in the Restricted category was granted on 14 April 1970 at an AUW of 2,834 kg (6,250 lb).

The aircraft is now known as the Emair MA-1 and is in production at Harlingen, Texas.

EMAIR MA-1

The MA-1 was designed specifically for agricultural use. For ease of maintenance and repair the fuselage comprises four bolt-together sections, covered by removable glassfibre side panels. Greater than average gap and stagger of the biplane wings improves pilot view and ensures easy access to the hopper; and the wide-track main landing gear ensures adequate stability for operations from rough fields.

The cockpit, which is fully enclosed, has a bench seat for the pilot and can accommodate also a loader/mechanic.

The hopper, which forms an integral part of the front fuselage, has two outlets, each with an adapter. For granular dispersal a stainless steel rotating gate is attached to each of these adapters; for spray distribution one adapter carries the wind-driven spraypump and control valve, the other carries the bottom-loading connector and liquid jettison door. Spraybooms are mounted in a low-drag area at the trailing-edge of the lower wings.

The initial batch of six MA-1s was operational by mid-1973. A total of 25 aircraft had been completed by 1 January 1976, when the scheduled production programme had ended and aircraft were being built only on receipt of a firm order.

Flight testing of a version designated **MA-1B** began on 1 August 1975. It is essentially the same as the MA-1 except for the installation of an 895 kW (1,200 hp) Wright R-1820 engine, derated to 671 kW (900 hp). This is claimed to make the MA-1B the most powerful agricultural aircraft in current production in the west. Max T-O weight is unchanged, but the increased power permits the carriage of greater payloads, especially under high-altitude conditions and from short or muddy fields. The geared engine drives a slower-turning larger-diameter propeller with broader blades, resulting in better propulsive efficiency and lower noise levels. Deliveries were scheduled to begin in the Spring of 1976, following FAA certification, and production of two aircraft per month was planned for the remainder of 1976.

Emair MA-1 heavy-duty agricultural aircraft (Pratt & Whitney R-1340-AN1 engine)

TYPE: Single-seat agricultural biplane.

WINGS: Strut-braced biplane with forward-staggered wings of unequal span. NACA 4412 (modified) wing section. Dihedral 1° 30′ on upper wing, 3° on lower wings. Incidence 4° 30′ on upper wing, 4° on lower wings. No sweepback. Stagger 31°. Upper wing carried on streamline steel tube struts. Conventional two-spar structure. Centre-section of upper wing is of aluminium construction. Outer panels of top wing, and both lower wing panels, have spruce laminated spars and ribs, with duralumin channel-section compression struts and steel tie-rod internal bracing, and fabric covering. Ailerons, of aluminium construction with fabric covering, on both upper and lower wings, linked by struts. No trim tabs or slats. Butt fittings and inboard compression members of the lower wings are of stainless steel. Glassfibre wingtips.

FUSELAGE: Rectangular welded chrome-molybdenum steel tube framework in four separate bolt-together sections, with glassfibre side panels.

TAIL UNIT: Conventional single fin and rudder, of welded chrome-molybdenum steel tube construction with fabric covering. Wire-braced fixed-incidence tailplane and fin. Trim tab in each elevator.

LANDING GEAR: Non-retractable tailwheel type. Oleo-pneumatic shock-absorbers on main and nosewheel units. Main legs enclosed in streamlined glassfibre fairings. Main wheels fitted with 8-ply nylon tyres size 27 × 9-14, pressure 2·41 bars (35 lb/sq in). Steerable tailwheel with solid tyre size 12 × 4-6. Hayes shoe-type hydraulic main-wheel brakes. Parking brakes.

POWER PLANT: One 447 kW (600 hp) Pratt & Whitney R-1340-AN1 Wasp radial engine (MA-1) or 895 kW (1,200 hp) Wright R-1820 radial engine, derated to 671 kW (900 hp) (MA-1B). The Pratt & Whitney engine drives a Hamilton Standard 12.D.40 two-blade adjustable-pitch constant-speed metal propeller. Fuel tank in upper wing centre-section, capacity 408 litres (108 US gallons). Refuelling point above upper wing. Oil capacity 30·5 litres (8 US gallons). Cowling for engine optional.

ACCOMMODATION: Pilot on bench-type seat which provides sufficient room for carriage of a loader/mechanic on short-duration flights. Fully enclosed ventilated cockpit in centre of fuselage, with reinforced overturn structure aft of cockpit. Sliding side-screens of cockpit canopy retract into cockpit wall.

OPERATIONAL EQUIPMENT: Between the cockpit and the

engine is mounted a glassfibre hopper for dust or liquid, with a capacity of 1·77 m³ (62·5 cu ft) or 1,703 litres (450 US gallons), the largest of its kind ever fitted to a single-engined aircraft. It is built around the fuselage tubing, which in this area is of stainless steel to resist chemical corrosion. A window in the rear wall of the hopper allows in-flight inspection of the contents, and sight gauges are provided on each side of the hopper.

DIMENSIONS, EXTERNAL:

Wing span (upper)	12·70 m (41 ft 8 in)
Wing span (lower)	10·67 m (35 ft 0 in)
Wing chord (both, constant)	1·60 m (5 ft 3 in)
Wing aspect ratio (upper)	7·9
Wing aspect ratio (lower)	6·7
Wing stagger	1·19 m (3 ft 11 in)
Wing gap	1·98 m (6 ft 6 in)
Length overall (tail up)	8·74 m (28 ft 8 in)
Height overall (tail down)	3·58 m (11 ft 9 in)
Propeller diameter	2·74 m (9 ft 0 in)
Propeller ground clearance (tail up)	0·36 m (1 ft 2 in)
Tailplane span	3·81 m (12 ft 6 in)
Wheel track	2·64 m (8 ft 8 in)
Wheelbase	6·71 m (22 ft 0 in)
Hopper opening, above fuselage:	
Length	0·61 m (2 ft 0 in)
Width	0·46 m (1 ft 6 in)

AREAS:

Wings, gross	37·16 m² (400 sq ft)
Ailerons (total)	5·13 m² (55·2 sq ft)
Fin	0·29 m² (3·14 sq ft)
Rudder	1·10 m² (11·83 sq ft)
Tailplane	1·97 m² (21·16 sq ft)
Elevators, incl tabs	1·31 m² (14·14 sq ft)

WEIGHTS AND LOADINGS (MA-1):

Weight empty	1,699 kg (3,746 lb)
Hopper load	1,360 kg (3,000 lb)
Structural design T-O and landing weight	2,834 kg (6,250 lb)
Max T-O weight (agricultural)	3,175 kg (7,000 lb)
Max wing loading	76·65 kg/m² (15·7 lb/sq ft)
Max power loading	6·34 kg/kW (10·4 lb/hp)

PERFORMANCE (MA-1):

Never-exceed speed	128 knots (238 km/h; 148 mph) CAS
Max cruising and manoeuvring speed	102 knots (188 km/h; 117 mph) CAS
Stalling speed (loaded)	51·5 knots (95 km/h; 59 mph)
Min ground turning radius	9·14 m (30 ft)

ENSTROM

THE ENSTROM HELICOPTER CORPORATION

HEAD OFFICE AND WORKS:
PO Box 277, Menominee County Airport, Menominee, Michigan 49858
Telephone: (906) 863-9971
Telex: 26-3451

PRESIDENT AND CHAIRMAN OF THE BOARD:
F. Lee Bailey

SENIOR VICE-PRESIDENT:
Paul L. Shultz

EXECUTIVE VICE-PRESIDENT:
H. F. Moseley

VICE-PRESIDENTS:
D. E. Brandt (Engineering)
M. Hathaway (Finance)
E. L. Medina (Administration)

In its original form, as the R. J. Enstrom Corporation, this company was formed in 1959 to develop an experimental light helicopter built by Rudolph J. Enstrom. This helicopter flew for the first time on 12 November 1960. There followed a design and development programme on a new helicopter, designated F-28, the first of which flew for the first time in May 1962. A limited number of F-28s were built, and were followed by the improved Model F-28A in 1968.

The company was acquired by the Purex Corporation in October 1968 and was operated for a time as part of the Pacific Airmotive Aerospace group. Under this ownership, a turbocharged F-28B version was developed, as well as a Model T-28 turbine-powered version.

The activities of this group ended in February 1970; but with transfer of ownership of the Purex interest to F. Lee Bailey in January 1971 the present company resumed manufacture of the Model F-28, which is being marketed throughout the world.

ENSTROM MODELS F-28 and 280 SHARK

During 1973 an advanced version of the basic Model F-28A known as the **Model 280 Shark** was developed. It is generally similar to the Model F-28A, except for the cabin area which has improved aerodynamic contours, and tail stabilising surfaces that include a small dorsal fin, larger ventral fin with skid to protect the tail rotor in a tail-down landing, small fixed horizontal surfaces and increased fuel capacity. Certification of this version with the standard 153 kW (205 hp) Lycoming HIO-360 fuel-injection engine was gained in September 1974. Turbocharged versions of both aircraft have also been developed, under the designations F-28C and 280C respectively, and received FAA certification on 8 December 1975.

There are, therefore, four versions available currently:

F-28A. Basic version, as described, of which 338 had been built by 1 March 1976.

Model 280 Shark. Improved version, with increased standard fuel capacity. A total of 31 Sharks had been delivered by 1 March 1976.

F-28C. Generally similar to F-28A, except for installation of 153 kW (205 hp) Lycoming HIO-360-E1AD engine with Rajay 301 E-10-2 turbocharger, and standard fuel capacity of 151·4 litres (40 US gallons).

Model 280C. As Model 280 Shark, except for installation of 153 kW (205 hp) Lycoming HIO-360-E1AD engine with Rajay 301 E-10-2 turbocharger.

In addition, a turboshaft-powered version known as the **Spitfire Mk 1** is being offered by Spitfire Helicopters Inc. Further details of this version can be found in the Addenda.

The following description applies basically to the current F-28A:

TYPE: Three-seat light helicopter.

ROTOR SYSTEM: Fully-articulated metal three-blade main rotor. Blades are of bonded light alloy construction, each attached to rotor hub by retention pin and drag link. Blade section NACA 00135. Two-blade teetering tail rotor, with blades of bonded light alloy construction. Tail rotor of F-28A and Model 280 is on starboard side; that of F-28C and 280C is on port side and rotates in opposite direction. Blades do not fold. No rotor brake.

Enstrom Model 280C Shark, with a Lycoming HIO-360-E1D turbocharged engine *(Howard Levy)*

ROTOR DRIVE: Poly Vee-belt drive system. Right-angle drive reduction gearbox. Main rotor/engine rpm ratio 1 : 8·78; tail rotor/engine rpm ratio 1 : 1·226 (F-28A and Model 280). Main rotor/engine rpm ratio 1 : 7·154; tail rotor/engine rpm ratio 1 : 1·156 (F-28C and Model 280C).

FUSELAGE: Glassfibre and light alloy cab structure, with welded steel tube centre-section. Semi-monocoque aluminium tailcone structure.

LANDING GEAR: Skids carried on Enstrom oleo-pneumatic shock-absorbers. Air Cruiser inflatable floats available optionally.

POWER PLANT: One 153 kW (205 hp) Lycoming HIO-360-C1B flat-four engine. Other versions as detailed in model listings. Two fuel tanks, each of 56·5 litres (15 US gallons) or, optionally, 75·7 litres (20 US gallons). Total fuel capacity 113 litres (30 US gallons) or 151·4 litres (40 US gallons). Standard capacity of 151·4 litres (40 US gallons) on F-28C, 280 and 280C. Oil capacity 7·5 litres (2 US gallons).

ACCOMMODATION: Pilot and two passengers, side by side on bench seat, centre seat of which is removable. Fully-transparent removable door on each side of cabin. Baggage space aft of engine compartment, with external access door. Cabin heated and ventilated.

SYSTEMS: Electrical power provided by 12V 70A engine-driven alternator.

ELECTRONICS AND EQUIPMENT: Nav/com equipment available to customer's requirements. Cargo hook, floats, spraygear and litters optional.

DIMENSIONS, EXTERNAL:

Diameter of main rotor	9·75 m (32 ft 0 in)
Diameter of tail rotor	1·42 m (4 ft 8 in)
Distance between rotor centres	5·56 m (18 ft 3 in)
Main rotor blade chord	0·241 m (9½ in)
Length overall	8·94 m (29 ft 4 in)
Height to top of rotor hub	2·79 m (9 ft 2 in)
Skid track	2·36 m (7 ft 9 in)
Cabin doors (each, F-28A, Model 280):	
Height	1·09 m (3 ft 7 in)
Width	1·02 m (3 ft 4 in)
Height to sill	0·61 m (2 ft 0 in)
Cabin doors (each, F-28C, Model 280C):	
Height	1·04 m (3 ft 5 in)
Width	0·84 m (2 ft 9 in)
Height to sill	0·64 m (2 ft 1 in)
Baggage door:	
Height	0·55 m (1 ft 9½ in)
Width	0·39 m (1 ft 3½ in)
Height to sill	0·86 m (2 ft 10 in)

DIMENSIONS, INTERNAL:

Max width of cabin:	
F-28A, Model 280	1·55 m (5 ft 1 in)
F-28C, Model 280C	1·47 m (4 ft 10 in)
Volume of baggage hold	0·20 m³ (7 cu ft)

AREAS:

Main rotor disc	74·69 m² (804 sq ft)
Tail rotor disc	1·58 m² (17·06 sq ft)

WEIGHTS AND LOADINGS (A: F-28A, Model 280; B: F-28C, Model 280C):

Weight empty:	
A	657 kg (1,450 lb)
B	678 kg (1,495 lb)
Max T-O weight:	
A	975 kg (2,150 lb)
B	998 kg (2,200 lb)
Max disc loading:	
A	13·04 kg/m² (2·67 lb/sq ft)
B	13·38 kg/m² (2·74 lb/sq ft)
Max power loading:	
A	6·37 kg/kW (10·49 lb/hp)
B	6·52 kg/kW (10·73 lb/hp)

PERFORMANCE: (at max T-O weight, A: F-28A, Model 280; B: F-28C; C: Model 280C):

Never-exceed speed at S/L:	
A, B	97 knots (180 km/h; 112 mph) IAS
C	102 knots (188 km/h; 117 mph) IAS
Max cruising speed at S/L:	
A, B, C	87 knots (161 km/h: 100 mph)
Max rate of climb at S/L:	
A	290 m (950 ft)/min
B, C	396 m (1,300 ft)/min
Service ceiling:	
A	3,660 m (12,000 ft)
B, C	5,485 m (18,000 ft)
Hovering ceiling out of ground effect:	
A	1,035 m (3,400 ft)
B, C	2,530 m (8,300 ft)
Hovering ceiling in ground effect:	
A	1,705 m (5,600 ft)
B, C	3,475 m (11,400 ft)
Range with max fuel:	
A, B, C	205 nm (381 km; 237 miles)

EXCALIBUR
EXCALIBUR AVIATION COMPANY

ADDRESS:
PO. Box 32007, San Antonio, Texas 78216
Telephone: (512) 927-6201 or (512) 696-4221
PRESIDENT:
William C. Hickey

Excalibur Aviation, which for some time was responsible for production of the improved versions of the Beechcraft Queen Air and Twin-Bonanza marketed by Swearingen Aircraft, acquired all rights of this conversion programme on 1 October 1970, and is continuing to produce these aircraft at Stinson Municipal Airport, San Antonio, Texas. The only change brought about by the new ownership is use of the name Queenaire 800 for the former Swearingen 800.

EXCALIBUR QUEENAIRE 800 and 8800

The Excalibur modification of Queen Air 65s and 80s includes installation of two 298 kW (400 hp) Lycoming IO-720-A1B eight-cylinder engines, each driving a Hartzell three-blade metal constant-speed and fully-feathering propeller; new engine mountings; new exhaust system; new low-drag engine nacelles; new (or zero-time overhauled and certified) accessories; and Excalibur fully-enclosed wheel-well doors. Modifications of the Beechcraft Queen Air 65, A65 and 80 of all serial numbers are designated Queenaire 800; similar modifications to the Queen Air A80 and B80 of all serial numbers have the designation Queenaire 8800.

Weight and performance data are given after the Excalibur and Excalibur 800 description which follows.

EXCALIBUR EXCALIBUR and EXCALIBUR 800

The Excalibur modification of all Beechcraft Twin-Bonanzas except the 1954 B50, and including the installation of new 283 kW (380 hp) Lycoming IGSO-540-A1A engines, is designated as the Excalibur. The modification of all Beechcraft Twin-Bonanzas except the early B50 and C50, and including the installation of new 298 kW (400 hp) Lycoming IO-720-A1B engines, has the designation Excalibur 800.

As with the Queenaire conversions, the new engines (with Hartzell constant-speed and fully-feathering propellers) are installed in low-drag nacelles, and the other modifications as detailed for Queenaires are also carried out. In addition, if the Twin-Bonanza is one which had a fuel capacity of 681 litres (180 US gallons) as standard, the outer wing panels are removed from the aircraft and converted to receive new Beechcraft fuel cells to provide a total fuel capacity of 870 litres (230 US gallons). Depending upon the particular model, several changes may be necessary to cater for the increase in maximum take-off weight. These include, if and as necessary, installation of new nose gear casting, main gear trunnions, fuselage doubler on port side below the pilot's side windows, and structural reinforcement of the wing rear spar in the centre-section area (Excalibur 800 only).

Weights and performance details of the Excalibur conversions which follow are identified as A: Excalibur; B: Excalibur 800; C: Queenaire 800; and D: Queenaire 8800.

WEIGHTS:
Weight empty, equipped (average):
A, B — 2,313 kg (5,100 lb)
C — 2,449 kg (5,400 lb)
D — 2,631 kg (5,800 lb)
Max T-O weight:
A — 3,311 kg (7,300 lb)
B — 3,447 kg (7,600 lb)
C — 3,628 kg (8,000 lb)
D — 3,991 kg (8,800 lb)

Typical Excalibur Queenaire 800/8800 conversion (two Lycoming IO-720-A1B engines)

Excalibur Excalibur conversion of Beechcraft Twin-Bonanza (two Lycoming IGSO-540-A1A engines)

Max landing weight:
A — 3,175 kg (7,000 lb)
B — 3,284 kg (7,240 lb)
C — 3,447 kg (7,600 lb)
D — 3,792 kg (8,360 lb)

PERFORMANCE (at max T-O weight):
Cruising speed, 75% power:
A at 4,115 m (13,500 ft) — 226 knots (418 km/h; 260 mph)
B at 2,530 m (8,300 ft) — 204 knots (378 km/h; 235 mph)
C, D at 2,530 m (8,300 ft) — 201 knots (372 km/h; 231 mph)
Cruising speed, 65% power:
A at 5,180 m (17,000 ft) — 223 knots (414 km/h; 257 mph)
B at 3,050 m (10,000 ft) — 198 knots (367 km/h; 228 mph)
C, D at 3,050 m (10,000 ft) — 195 knots (362 km/h; 225 mph)
Cruising speed, 45% power at 3,050 m (10,000 ft):
A — 187 knots (346 km/h; 215 mph)
B — 174 knots (322 km/h; 200 mph)
C, D — 172 knots (319 km/h; 198 mph)
Stalling speed, gear and flaps up:
A — 78·5 knots (145 km/h; 90 mph)
B, C — 80 knots (148 km/h; 92 mph)
D — 86 knots (160 km/h; 99 mph)
Stalling speed, gear and flaps down:
A, B — 71·5 knots (132 km/h; 82 mph)
C — 68 knots (126 km/h; 78 mph)
D — 70 knots (129 km/h; 80 mph)
Max rate of climb at S/L:
A — 579 m (1,900 ft)/min
B — 570 m (1,870 ft)/min
C — 468 m (1,535 ft)/min
D — 454 m (1,490 ft)/min
Rate of climb at S/L, one engine out:
A — 104 m (340 ft)/min
B — 134 m (440 ft)/min
C — 110 m (360 ft)/min
D — 76 m (250 ft)/min
Service ceiling:
A — 9,145 m (30,000 ft)
B, C — 6,765 m (22,200 ft)
D — 6,100 m (20,000 ft)
Service ceiling, one engine out:
A — 5,335 m (17,500 ft)
B, C — 3,595 m (11,800 ft)
D — 3,110 m (10,200 ft)
Max range at 3,050 m (10,000 ft), A with 30 min reserves, B, C, D with 113·6 litres (30 US gallons) reserve:
A — 2,032 nm (3,765 km; 2,340 miles)
B — 1,335 nm (2,475 km; 1,538 miles)
C — 1,322 nm (2,451 km; 1,523 miles)
D — 1,547 nm (2,867 km; 1,782 miles)

FAIRCHILD INDUSTRIES
FAIRCHILD INDUSTRIES INC

EXECUTIVE OFFICE:
Germantown, Maryland 20767
Telephone: (301) 428-6000

PRESIDENT AND CHIEF EXECUTIVE OFFICER:
Edward G. Uhl
EXECUTIVE VICE-PRESIDENTS:
Charles Collis
John F. Dealy
VICE-PRESIDENTS:
Ralph Bonafede
Wernher von Braun
Emanuel Fthenakis
Norman Grossman
John P. Healey
Thomas Turner
TREASURER: Stuart C. Dew
COMPTROLLER: George S. Attridge
SECRETARY: Richard R. Molleur
DIRECTOR, PUBLIC RELATIONS:
James A. Crandall

Fairchild Republic Company
DIVISIONAL OFFICE AND WORKS:
Farmingdale, Long Island, New York 11735
PRESIDENT: Norman Grossman
HAGERSTOWN, MARYLAND 21740 FACILITY:
GENERAL MANAGER: Leo LaBell

Fairchild Aircraft Service Division
DIVISIONAL OFFICE AND WORKS:
Crestview, Florida 32536
GENERAL MANAGER: R. C. Woods
ST AUGUSTINE, FLORIDA FACILITY:
St Augustine, Florida 32084
FACILITY MANAGER: A. J. Cammareri

Fairchild Space and Electronics Company
Germantown, Maryland 20767
PRESIDENT: Harry Dornbrand

Swearingen Aviation Corporation
San Antonio, Texas 78216
PRESIDENT: Constantine Stathis

Fairchild Stratos Division
Manhattan Beach, California 90266
GENERAL MANAGER: William R. Beckert

Fairchild Industrial Products Division
Winston-Salem, North Carolina 27107
GENERAL MANAGER: Richard G. Orr

Fairchild Burns Company
Winston-Salem, North Carolina 27107
GENERAL MANAGER: R. C. Woods

American Satellite Corporation
Germantown, Maryland 20769
PRESIDENT: Emanuel Fthenakis

S. J. Industries Inc
Alexandria, Virginia 22304
PRESIDENT: R. F. Julius

Fairchild KLIF Inc
Radio Station KLIF, Dallas, Texas 75201
GENERAL MANAGER: Alan Henry
STATION MANAGER: Klee C. Dobra

Fairchild Minnesota Inc
Radio Stations WYOO (AM and FM), Richfield, Minnesota 55423
GENERAL MANAGER: Michael R. Sigelman

Fairchild Industries is a diversified aerospace company with interests in communications, satellite systems, radio broadcasting, industrial products and land development.

The Fairchild Republic Company combines all of Fairchild's military aircraft research, design, manufacture and repair and overhaul facilities and personnel into one organisation. Facilities are located at Farmingdale, Long Island, New York; and Hagerstown, Maryland.

Fairchild Republic Company is proceeding with full-scale development and production of the A-10 close-support aircraft for the USAF. It is also manufacturing under subcontract aft fuselages, fin and tailcone assemblies, rudders, tailplanes and engine access doors for the McDonnell Douglas F/RF-4 Phantom II tactical fighter as well as the twin vertical fins for the Grumman F-14A Tomcat fighter. The company is a major subcontractor in the Boeing 747 programme, manufacturing ailerons, spoilers, and wing trailing-edge and leading-edge flaps. This work is performed at the Farmingdale and Hagerstown facilities.

Fairchild Republic Company is also a major subcontractor to Rockwell International's Space Division, and is responsible for the vertical tail of the Space Shuttle.

Other programmes in which the company was involved included the development and construction of radar site monitoring equipment for the Safeguard anti-ballistic missile system (now deactivated).

Wing structures for Swearingen Aviation Corporation's Merlin and Metro series are manufactured at Hagerstown. This facility also provides support for the F-27 and FH-227 twin-turboprop transport aircraft that were manufactured under licence from Fokker-VFW, as well as the F28 Fellowship and the FH-1100 five-passenger turbine-powered helicopter. The Porter STOL aircraft, manufactured under licence from Pilatus of Switzerland, as well as a military version of that aircraft known as the Peacemaker, are built at Hagerstown.

The Florida-based Fairchild Aircraft Service Division is engaged in the maintenance, repair, modification and support of a wide variety of aircraft. It also modifies North American T-28B aircraft to T-28D-10 fighter-bomber configuration under USAF contract, and, under subcontract to Sperry, is converting Convair F-102As to PQM-102 drone aircraft configuration.

The Fairchild Space and Electronics Company directs the company's efforts in the design, development and manufacture of spacecraft, spacecraft subsystems, rocket payload projects, electronic systems, and letter and parcel handling equipment for the US Postal Service. It was responsible for design and manufacture of NASA's Applications Technology Satellite ATS-6.

Fairchild Space and Electronics Company provides design, analysis, fabrication, testing and launch support services as a prime contractor and subcontractor to NASA, the Department of Defense, Atomic Energy Commission, Comsat/Intelsat and international organisations. As part of this effort the company has conducted a number of advanced spacecraft studies, including nuclear-powered multi-mission spacecraft, and advanced concepts for communications and applications satellites. It is also a major supplier of deployable/retractable tubular structures and their mechanisms, of the type used as antennae and gravity stabilisation booms for spacecraft. Other spacecraft subsystems include thermal control louvres, heat pipes, solar array panels complete with solar cells, and components for such spacecraft as the OAO, Nimbus, ERTS, IMP, LES series, Skylab, VO'75 and sounding rockets.

The Division is active in the design and manufacture of search and meteorological/weather radar systems, and in intelligence data acquisition and management. Its auxiliary data annotation systems (ADAS) are in use in RF-4, RF-101, P-3 and OV-1C and -1D reconnaissance aircraft.

An integrated armament control or stores management system developed by the Division provides F-111D pilots with an inventory control and monitoring unit by integrating, on only two cockpit panels, the displays and controls of all the various weapons and stores. A similar system is being provided for F-14A aircraft.

The Fairchild Stratos Division specialises in the development and manufacture of aerospace and commercial aircraft subsystems, accessories and components. It is now manufacturing valves and other components, including the ammonia boiler system, for the Space Shuttle.

Fairchild AU-23A Peacemaker, a counter-insurgency version of the Pilatus Porter

Commercial products manufactured by this Division include lower lobe galley modules and food service carts for Boeing 747 and McDonnell Douglas DC-10 transports. Fairchild Stratos also supplies the complete food, beverage and waste cart system for the Lockheed L-1011 TriStar. The Division designs and manufactures an extensive line of military and commercial ground and airborne air-conditioning systems, together with air-turbine secondary power systems, subsystems and associated components.

Traditionally, this Division is a major producer of military airborne stores dispensing systems, for both lethal and non-lethal stores such as underwater sound sources, sonobuoys, flares, small bombs, and psychological warfare items for all US military services.

Fairchild Burns Company is one of the world's principal suppliers of commercial aircraft seating.

American Satellite Corporation, a subsidiary of Fairchild Industries, is developing a US domestic communications satellite system under authority from the Federal Communications Commission. Earth stations for the nationwide system are located at sites near New York City, San Francisco, Dallas and Los Angeles; they began commercial service in July 1974.

Swearingen Aviation Corporation, a subsidiary in which Fairchild Industries has a ninety per cent holding, manufactures the Merlin series of turboprop-powered executive aircraft and the Metro 19-passenger aircraft for commuter airlines.

The Corporation has also the following subsidiaries: Fairchild Arms International Ltd, Germantown, Maryland; Fairchild Aviation (Asia) Ltd, Bangkok, Thailand; Fairchild Aviation (Holland) NV, Amsterdam, The Netherlands; and Fairchild-Germantown Development Co Inc, Germantown, Maryland, a participant in the new Century XXI office centre surrounding the company's corporate headquarters.

FAIRCHILD INDUSTRIES (PILATUS) PORTER and PEACEMAKER

USAF designation: AU-23A

Under licence from Pilatus Flugzeugwerke AG of Switzerland, Fairchild Industries produces Porter STOL utility aircraft for agencies of the US government. These aircraft are available with a 507 kW (680 shp) Pratt & Whitney Aircraft of Canada PT6A-27 turboprop engine, flat rated to 410 kW (550 shp). Flat rating is adopted on this engine in order to obtain improved hot-day and/or high-altitude performance.

The first production Fairchild Industries Porter (with PT6A-20 engine) was rolled out on 3 June 1966.

A militarised version of the Porter, known as the Peacemaker, has been developed for counter-insurgency operations, including transport, light armed and photographic reconnaissance, leaflet dropping and loudspeaker broadcasting. This version has an underfuselage hardpoint capable of carrying a 272 kg (600 lb) store, and four underwing hardpoints, of which the inboard pair can carry 261 kg (575 lb) each, and the outboard pair 159 kg (350 lb) each. However, total external load on each wing may not exceed 318 kg (700 lb).

Powered by a 485 kW (650 shp) Garrett TPE 331-1-101F engine, the Peacemaker has a complete military nav/com system, including VHF, UHF and FM electronics, an armament control system, dual controls, and high-flotation tyres for rough-field operation.

Alternative installations include a side-firing XM-197 20 mm cannon, which has a rate of fire of up to 700 rds/min, or 7·62 mm Miniguns with firing rates of 2,000 or 4,0000 rds/min. The underwing weapon stations will accept various combinations of rockets, bombs, canisters, napalm, smoke grenades and flares.

In early 1970, Fairchild demonstrated this version of the Peacemaker, by successful firing of the XM-197 side-firing flexibly-mounted manually-operated 20 mm cannon. Fifteen were acquired by the USAF under the designation AU-23A for evaluation under its Credible Chase programme, in competition with the Helio AU-24A. One of the AU-23As was lost, thirteen were supplied to Thailand and one was retained by the USAF. This last aircraft was scheduled for transfer to the Royal Thai Air Force during 1976. In the latter months of 1975 Fairchild Industries began delivering an additional 20 Peacemakers purchased by the Royal Thai Air Force through the USAF Foreign Military Sales Program. The balance of these 20 Peacemakers was scheduled for delivery during 1976.

Full details of the basic Pilatus Porter can be found in the Swiss section of this edition; available details of the Peacemaker specification follow:

DIMENSIONS, EXTERNAL:

Wing span	15·14 m (49 ft 8 in)
Length overall	11·23 m (36 ft 10 in)
Height overall	3·73 m (12 ft 3 in)

DIMENSIONS, INTERNAL:

Cabin: Length	3·70 m (12 ft 1½ in)
Max width	1·16 m (3 ft 9½ in)
Max height	1·28 m (4 ft 2½ in)

WEIGHT AND LOADINGS:

Max T-O and landing weight	2,767 kg (6,100 lb)
Max wing loading	96·1 kg/m² (19·68 lb/sq ft)
Max power loading	5·71 kg/kW (9·38 lb/shp)

PERFORMANCE (at max T-O weight):

Max level speed	151 knots (280 km/h; 174 mph)
Cruising speed	142 knots (262 km/h; 163 mph)
Slow flight speed	56·5 knots (105 km/h; 65 mph)
Max rate of climb at S/L	457 m (1,500 ft)/min
Service ceiling	6,950 m (22,800 ft)
T-O run	155 m (510 ft)
Landing run	90 m (295 ft)
Max range	485 nm (898 km; 558 miles)

FAIRCHILD REPUBLIC COMPANY

Farmingdale, Long Island, New York 11735

PRESIDENT:

Dr Norman Grossman

Founded on 17 February 1931, as the Seversky Aircraft Company, Republic operated as Republic Aviation Corporation from 1939 until September 1965, when it became a division of Fairchild Hiller Corporation, now Fairchild Industries Inc.

In early 1973 the Fairchild Republic Company was named the winner of the A-X competition for the development of a new close-support aircraft for the USAF. Designated A-10A, it was the first US aircraft designed specifically for this role. Other major contracts cover the manufacture of assemblies for the McDonnell Douglas F/RF-4 Phantom II fighter and the Boeing 747. For the 747, Republic is manufacturing all the wing control surfaces, including ailerons, spoilers, leading-edge flaps and trailing-edge flaps.

Fairchild Republic Company is also responsible for the vertical tail assembly for the Rockwell International Space Shuttle orbiter vehicle.

FAIRCHILD REPUBLIC A-10

USAF designation: A-10A

On 18 December 1970, Fairchild Republic and Northrop were selected as the two companies that were each to build two prototypes for evaluation under the USAF's A-X programme, initiated in 1967, for a close-support aircraft. The first Fairchild Republic prototype (71-1369), designated YA-10A, flew for the first time on 10 May 1972, followed by the second prototype (71-1370) on 21 July 1972. USAF flight evaluation in competition with Northrop's YA-9A began on 10 October and was completed on 9 December 1972. On 18 January 1973 it was announced that the A-10A had been selected as the winner, and the company later received a contract for ten pre-production aircraft (six for development test and evaluation and four for initial operational test and evaluation). The six DT and E aircraft were reduced later to two, the first of which first flew on 15 February 1975. Static and fatigue test airframes were also completed.

These six pre-production aircraft have production flight control and landing gear systems, an increase in wing area of 0·23 m² (2·5 sq ft), improved windscreen, streamlined ordnance pylons and pods, and production TF34-GE-100 engines instead of the YTF34s which powered the prototypes. The first aircraft is being used to study flutter, stability and control, airloads, and general handling characteristics. The remaining five aircraft are for use by the USAF for stores certification, and for systems, performance and climatic testing. Only the first and fourth of these aircraft are not fitted with the GAU-8/A gun system.

An initial order for 22 production A-10As for the USAF was confirmed on 20 December 1974, with the allocation of $99 million in the FY 1975 budget, and delivery of these, to the 333rd Tactical Fighter Training Squadron at Davis-Monthan AFB, Arizona, began in the Spring of 1976, following the first flight by a production A-10A (75-00258) on 21 October 1975. By the Summer of 1976, funds had been approved for the procurement of 73 more A-10As, and FY 1977 funding for a further 100 had been requested. The USAF has indicated an eventual requirement for 733 A-10 aircraft.

The following description applies to the production A-10A:

TYPE: Single-seat close-support aircraft.

WINGS: Cantilever low-wing monoplane, with wide-chord, deep aerofoil section (NACA 6716 on centre-section and at start of outer panel, NACA 6713 at tip) to provide low wing loading. Incidence −1°. Dihedral 7° on outer panels. Aluminium alloy three-spar structure, consisting of one-piece constant-chord centre-section and tapered outer panels with integrally stiffened skins and drooped (cambered) wingtips. Outer panel leading-edges and core of trailing-edges are of honeycomb sandwich. Four-point attachment of wings to fuselage, at front and rear spars. Two-segment, three-position trailing-edge slotted flaps, interchangeable right with left. Wide-span ailerons, made up of dual upper and lower surfaces that separate to serve as airbrakes. Flaps, airbrakes and ailerons actuated hydraulically. Ailerons pilot-controlled by servo tab during manual reversion. Small leading-edge slat inboard of each main-wheel fairing. Redundant and armour-protected flight control system.

FUSELAGE: Semi-monocoque structure of aluminium alloy (chiefly 2024 and 7075), with four main longerons, multiple frames, and lap-jointed and riveted skins. Built in front, centre and aft portions. Single-curvature components aft of nose portion, interchangeable right with left. Centre portion incorporates wing box carry-through structure.

TAIL UNIT: Cantilever aluminium alloy structure, with twin fins and interchangeable rudders mounted at the tips of constant-chord tailplane. Interchangeable elevators, each with an electrically-operated trim tab. Rudders and elevators actuated hydraulically. Redundant and armour-protected flight control system.

LANDING GEAR: Menasco retractable tricycle type with single wheel on each unit. All units retract forward, and have provision for emergency gravity extension. Interchangeable mainwheel units retract into non-structural pod fairings attached to the lower surface of the wings. When fully retracted approximately half of each wheel protrudes from the fairing. Steerable nosewheel is offset to starboard to clear firing barrel of gun. Main wheels size 36 × 11, Type VII; nosewheel size 24 × 7·7-10, Type VII.

POWER PLANT: Two General Electric TF34-GE-100 high bypass ratio turbofan engines, each rated at 40·3 kN (9,065 lb st), enclosed in separate pods, each pylon-mounted to the upper rear fuselage at a point approximately midway between the wing trailing-edges and the tailplane leading-edges. Fuel is contained in two tear-resistant and self-sealing cells in the fuselage, and two smaller, adjacent integral cells in the wing centre-section. Maximum internal fuel capacity 4,853 kg (10,700 lb). All fuel cells are internally filled with reticulated foam, and all fuel systems pipework is contained within the cells except for the feeds to the engines, which have self-sealing covers. Three 2,271 litre (600 US gallon) jettisonable auxiliary tanks can be carried on underwing and fuselage centreline pylons. Provision for in-flight refuelling using universal aerial refuelling receptacle slipway installation (UARRSI).

ACCOMMODATION: Single-seat enclosed cockpit, well forward of wings, with large transparent bubble canopy to provide all-round vision. Bulletproof windscreen. Canopy is hinged at rear and opens upward. Pilot's ejection seat operable at speeds from 450 knots (834 km/h; 518 mph) down to zero speed at zero height. Entire cockpit area is protected by an armoured 'bath-tub' structure of titanium, capable of withstanding projectiles of up to 23 mm calibre. Basic design work for a dual-control two-seat version has been completed.

SYSTEMS: Redundant control system incorporates two 207 bar (3,000 lb/sq in) primary hydraulic flight control systems, each powered by an engine-driven pump, and a manual backup. Hydraulic systems actuate flaps, flying control surfaces, landing gear, brakes and nosewheel steering. Two independent hydraulic motors, either of which is sufficient to sustain half-rate firing, supply drive for 30 mm gun barrel rotation. Electrical system includes two 30/40kVA 115/200V AC engine-driven generators and a standby battery and inverter. Auxiliary power unit. Environmental control system, using engine bleed air for cockpit pressurisation and air-conditioning, pressurisation of pilot's *g* suit, windscreen anti-icing and rain clearance, fuel transfer, gun compartment purging, and other services.

ELECTRONICS AND EQUIPMENT: Head-up display giving airspeed, altitude, and dive angle; weapons delivery package with dual reticle optical sight for use in conjunction with underfuselage Pave Penny laser target seeker pod; target penetration aids; associated equipment for Maverick and other missile systems; IFF/SIF (AIMS); UHF/AM; VHF/AM; VHF/FM; Tacan; UHF/DF; VOR/ILS; X-band transponder; all-altitude heading and attitude reference system (HARS); radar homing and warning (RHAW); secure voice communications; active or passive electronic countermeasures (ECM); armament control panel; and gun camera. Space provisions for HF/SSB, ILS/FDS, Loran C/D, ECM pod, chaff/flare dispenser and other 'growth' electronics and equipment.

Fairchild Republic A-10A single-seat close-support aircraft *(J. M. G. Gradidge)*

ARMAMENT: General Electric GAU-8/A Avenger 30 mm seven-barrel cannon, mounted in nose with 2° depression and offset slightly to port so that as the barrels rotate the firing barrel is always on the aircraft's centreline. Gun and handling system for the linkless ammunition are mechanically synchronised and driven by two motors fed from the aircraft's hydraulic system. The single-drum magazine has a capacity of 1,350 rounds, and has a dual firing rate of either 2,100 or 4,200 rds/min. Four stores pylons under each wing (one inboard and three outboard of each main-wheel fairing), and three under fuselage, for max external load of 7,257 kg (16,000 lb). External load with full internal fuel is 5,482 kg (12,086 lb). The centreline pylon and the two flanking fuselage pylons cannot be occupied simultaneously. The centreline pylon has a capacity of 2,268 kg (5,000 lb); the two fuselage outer pylons and two centre-section underwing pylons 1,587 kg (3,500 lb) each; the two innermost outer-wing pylons 1,134 kg (2,500 lb) each; and the four outermost wing pylons 453 kg (1,000 lb) each. These allow carriage of a wide range of stores, including twenty-eight 500 lb Mk-82 LDGP or Mk-82 retarded bombs; six 2,000 lb Mk-84 general-purpose bombs; eight BLU-1 or BLU-27/B incendiary bombs; four SUU-25 flare launchers; twenty Rockeye II cluster bombs, sixteen CBU-52/71, ten CBU-38, or sixteen CBU-70 dispenser weapons; six AGM-65A Maverick missiles; Mk-82 and Mk-84 laser-guided bombs; Mk-84 electro-optically-guided bombs; two SUU-23 gun pods; chaff or other jammer pods; or up to three drop-tanks.

DIMENSIONS, EXTERNAL:

Wing span	17·53 m (57 ft 6 in)
Wing chord at root	3·04 m (9 ft 11½ in)
Wing chord (mean)	2·73 m (8 ft 11·32 in)
Wing chord at tip	1·99 m (6 ft 6·4 in)
Wing aspect ratio	6·54
Length overall	16·26 m (53 ft 4 in)
Height overall	4·47 m (14 ft 8 in)
Tailplane span	5·74 m (18 ft 10 in)
Wheel track	5·25 m (17 ft 2½ in)
Wheelbase	5·40 m (17 ft 8¾ in)

Firing trials of the 30 mm GAU-8/A gun in the nose of a Fairchild Republic A-10A

AREAS:

Wings, gross	47·01 m² (506·0 sq ft)
Ailerons (total, incl tabs)	4·42 m² (47·54 sq ft)
Trailing-edge flaps (total)	7·99 m² (85·99 sq ft)
Leading-edge slats (total)	0·98 m² (10·56 sq ft)
Airbrakes (total)	8·06 m² (86·78 sq ft)
Fins (total)	7·80 m² (83·96 sq ft)
Rudders (total)	2·18 m² (23·50 sq ft)
Tailplane	8·31 m² (89·40 sq ft)
Elevators (total, incl tabs)	2·69 m² (29·00 sq ft)

WEIGHTS AND LOADINGS:

Manufacturer's empty weight	9,176 kg (20,231 lb)
Basic equipped weight, 'clean'	10,515 kg (23,181 lb)
Operating weight empty	10,977 kg (24,200 lb)
*Basic design weight, equipped	13,628 kg (30,044 lb)
**Forward airstrip weight	13,976 kg (30,813 lb)
Max T-O weight	21,500 kg (47,400 lb)
Max wing loading	449·88 kg/m² (92·14 lb/sq ft)
Max power loading	262·4 kg/kN (2·57 lb/lb st)
Thrust/weight ratio	0·6

**incl six 500 lb bombs, 750 rds of ammunition, and fuel for 300 nm (555 km; 345 miles) with 20 min reserves*
***with four Mk-82 bombs*

PERFORMANCE: (at max T-O weight except where indicated):
Never-exceed speed 450 knots (834 km/h; 518 mph)
Max combat speed at S/L, 'clean' 390 knots (722 km/h; 449 mph)
Combat speed at 1,525 m (5,000 ft), with six Mk-82 bombs 385 knots (713 km/h; 443 mph)
Cruising speed at S/L 300 knots (555 km/h; 345 mph)
Stabilised 45° dive speed below 2,440 m (8,000 ft), AUW of 15,932 kg (35,125 lb) 260 knots (481 km/h; 299 mph)

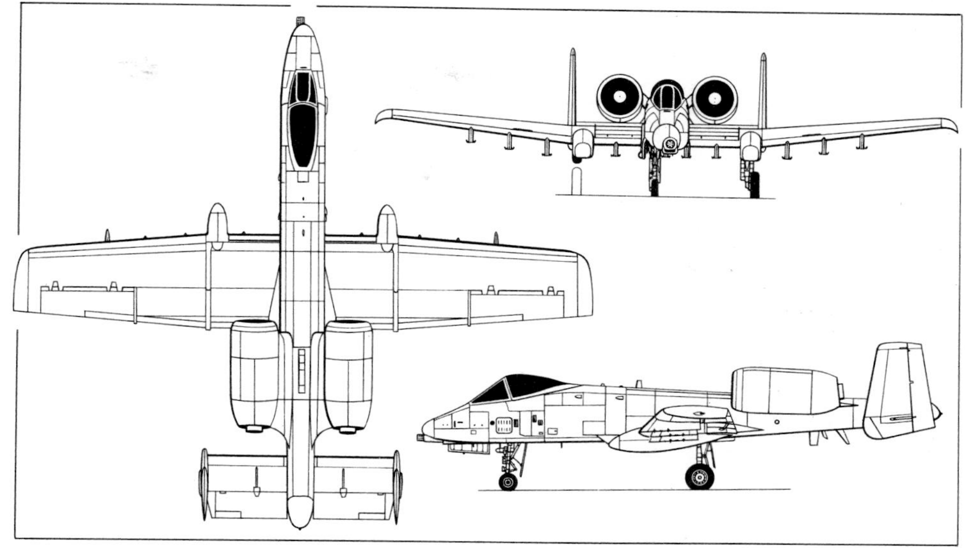

Fairchild Republic A-10A single-seat twin-engined close-support aircraft *(Pilot Press)*

Max rate of climb at S/L at basic design weight 1,828 m (6,000 ft)/min
T-O distance:
at max T-O weight 1,372 m (4,500 ft)
at forward airstrip weight 366 m (1,200 ft)
Landing distance:
at max T-O weight 762 m (2,500 ft)
at forward airstrip weight 366 m (1,200 ft)
Operational radius:
close air support and escort, 2 hr loiter, 20 min reserves 250 nm (463 km; 288 miles)
reconnaissance 400 nm (740 km; 460 miles)
deep strike 540 nm (1,000 km; 620 miles)
Ferry range, zero wind 2,300 nm (4,200 km; 2,647 miles)

FRAKES

FRAKES AVIATION INC

ADDRESS: Route 3, PO Box 229-B, Cleburne, Texas 76031

FRAKES/GRUMMAN TURBO-CAT

Frakes Aviation has modified a Grumman Ag-Cat by removing the standard Pratt & Whitney radial aircooled engine and installing in its place a 559 kW (750 shp) Pratt & Whitney Aircraft of Canada PT6A-34 turboprop engine, driving a Hartzell three-blade metal propeller 2·59 m (8 ft 6 in) in diameter. Drag is reduced significantly by reason of the fact that the small and lightweight turboprop is mounted well forward of the firewall and enclosed in a slender two-piece glassfibre cowling. The PT6A-34 offers economies in maintenance costs and fuel consumption; experience and evaluation by Air Rice Inc of Katy, Texas, may indicate that a lower-output version of the PT6A would be equally suitable for installation in the Ag-Cat, so offering still greater economy.

An extreme shortage of components for the ageing Pratt & Whitney R-1340 radial engine has prompted this, and other, experimental conversions of agricultural aircraft; higher initial cost is likely to be more than offset by greater reliability, fuel economies, reduced maintenance costs, increased TBO of the engine and the lower drag of the small-diameter turbine by comparison with the original radial engine.

GATES LEARJET

GATES LEARJET CORPORATION

CORPORATE OFFICES, AIRCRAFT DIVISION:
Mid-Continent Airport, PO Box 1280, Wichita, Kansas 67201
Telephone: (316) 722-5640
Telex: 417441
JET ELECTRONICS AND TECHNOLOGY INC:
5353 52nd Street, Grand Rapids, Michigan 49508
FIXED BASE OPERATIONS:
Hangar 7, Stapleton International Airport, Denver, Colorado 80207
CHAIRMAN OF THE BOARD:
Charles C. Gates
PRESIDENT: Harry B. Combs
EXECUTIVE VICE-PRESIDENTS:
A. J. Brizzolara (J.E.T.)
W. M. Conlin (Aircraft Division)
L.A. Ulrich (Fixed Base Operations)
VICE-PRESIDENTS:
L. C. Barry (Industrial Relations)
R. E. Cloughley (Marketing)
C. E. Dyas (Marketing Support)
A. F. Green (Manufacturing)
J. R. Greenwood (Corporate Affairs)
D. J. Grommesh (Engineering)
S. Kvassay (International Marketing)
E. C. Mandenberg (Materiel)
R. C. Scott (Operations)
R. E. Wolin (Domestic Marketing and Governmental Relations)
TREASURER: W. H. Webster
SECRETARY: R. C. Troll
DIRECTOR, PUBLIC RELATIONS: A. K. Higdon

Founded in 1960 by William P. Lear Sr, this company was known originally as the Swiss American Aviation Corporation, which was formed to manufacture a high-speed twin-jet executive aircraft known as the Learjet 23 (formerly SAAC-23). Most of the tooling for production of this aircraft was completed in Europe and then, in 1962, all company activities were relocated at Wichita, Kansas; shortly afterwards the company became known as Lear Jet Corporation. In 1967 all of Mr Lear's interests in the company (approximately 60 per cent) were acquired by The Gates Rubber Company of Denver, Colorado, and in January 1970 the company name was changed to Gates Learjet Corporation.

On 28 October 1975 Gates Learjet announced improved models of its Learjets, under the designation Century III series. These include a new Model 24E which has a limited maximum T-O weight and requires a balanced field length of only 853 m (2,800 ft).

All Century III Learjets incorporate a cambered wing and other changes to reduce stall and approach speeds and balanced field length. This wing has a new contour, extending from the leading-edge aft to the second wing spar, and, coupled with other advances including an improved stall warning/prevention system, improves lateral stability and handling qualities in the slower flight regimes. Performance increases from Century III improvements include approach speed reductions of 16-18 knots (30-34 km/h; 18·5-21 mph), and increases in cruise range of as much as seven per cent. Century III improvements are available as a factory retrofit modification for earlier Learjet 24, 25, 35 and 36 models, and can be completed in approximately one week.

During 1975 a total of 79 Learjets was delivered (45 to customers in the US, the balance for export), making this the eleventh consecutive year in which Gates Learjet has led in cumulative deliveries of business jet aircraft. In May 1976 the company delivered the 600th Learjet, a Model 35, to Dart Industries of Los Angeles, California.

Learjets are produced at the company's main facility at Wichita, Kansas. A new maintenance/modification centre at Tucson, Arizona, was scheduled for opening in 1976. Other fixed-base operations, under the name of Combs-Gates, are located at Denver, Colorado; Indianapolis, Indiana; and Palm Springs, California.

In January 1976 employees at the company's Aircraft Division facilities in Wichita numbered approximately 2,300. Combined operations employment, including a wholly-owned subsidiary, Jet Electronics and Technology Inc, totals 2,800. Facilities include 44,593 m² (480,000 sq ft) on a 64-acre site at Wichita and 86,864 m² (935,000 sq ft) in combined operations throughout the USA.

GATES LEARJET 24E and 24F

The prototype Lear Jet twin-jet executive transport flew for the first time on 7 October 1963 and deliveries of production Learjet 23 aircraft began on 13 October 1964. After a total of 104 of this version had been delivered, it was superseded by the Learjet 24, which was certificated under Federal Air Regulations Part 25 (formerly CAR 4B), as have been all subsequent models produced by the company. Deliveries of the Learjet 24 began in March 1966, and a total of 80 were built. This was replaced by a developed version with more powerful engines, known as the Learjet 24B, which received FAA certification on 17 December 1968, and then by the 24D (see 1975-76 *Jane's*).

The current Learjet 24E and 24F replaced the Learjet 24D and 24D/A in 1976. These latter aircraft were the first to feature a tail unit design in which the non-structural bullet at the junction of the tailplane and fin was deleted. The Learjet 24E and 24F are basically similar, differing primarily in standard fuel capacity; the 24E is limited to a max T-O weight of 5,850 kg (12,900 lb) to offer reduced runway requirements.

TYPE: Twin-jet light executive transport.

WINGS: Cantilever low-wing monoplane. Wing section NACA 64A 109. Dihedral 2° 30′. Incidence 1°. Sweepback 13° at quarter-chord. All-metal eight-spar structure with chemically-milled alloy skins. Manually-operated, aerodynamically-balanced all-metal ailerons. Hydraulically-actuated all-metal single-slotted flaps. Hydraulically-actuated all-metal spoilers mounted on trailing-edge ahead of flaps. Trim tab in port aileron. Balance tab in each aileron. Anti-icing by engine bleed air ducted into leading-edges.

FUSELAGE: All-metal flush-riveted semi-monocoque fail-safe structure.

TAIL UNIT: Cantilever all-metal sweptback structure, with electrically-actuated variable-incidence T-tailplane and small ventral fin. Conventional manually-operated control surfaces. Trim tab in rudder. Electrically-heated thermal de-icing of tailplane leading-edge.

LANDING GEAR: Retractable tricycle type, with twin wheels on each main unit and single steerable nosewheel. Hydraulic actuation, with backup pneumatic extension. Oleo-pneumatic shock-absorbers. Main wheels fitted with Goodyear 18 × 5·50 10-ply tyres, pressure 7·93 bars (115 lb/sq in). Nosewheel fitted with Goodyear Dual Chine tyre size 18 × 4·40 10-ply rating, pressure 7·24 bars (105 lb/sq in).Goodyear multiple-disc hydraulic brakes. Anti-skid units.

POWER PLANT: Two General Electric CJ610-6 turbojet engines (each rated at 13·1 kN; 2,950 lb st) mounted in pod on each side of fuselage aft of wings. Fuel in integral wing and wingtip tanks with a total standard fuel capacity of 2,706 litres (715 US gallons) in 24E and 3,180 litres (840 US gallons) in 24F. Oil capacity 3·75 litres (1 US gallon) per engine. Engine inlet anti-icing by bleed air.

ACCOMMODATION: Two seats side by side on flight deck, with dual controls. Up to six passengers in cabin, with one on inward-facing bench seat on starboard side at front, then two on forward or aft-facing armchairs with centre aisle, and three on forward-facing couch. Toilet and stowage space under front inward-facing seat, which can be screened from remainder of cabin by curtain. Refreshment cabinet opposite this seat. Baggage compartment aft of cabin. With back of rear bench seat folded down, baggage compartment and rear of cabin can be used to carry cargo or stretchers. Table at rear. In full cargo version, the rearward-facing armchair seats are also removed. Two-piece door, with upward-hinged portion and downward-hinged portion with integral steps, on port side of cabin at front. Emergency exit on starboard side. Cargo door optional. Windscreen anti-icing by engine bleed air with liquid methyl alcohol backup.

SYSTEMS: Air-conditioning by Freon R12 vapour cycle system, supplemented by ram-air heat exchanger during pressurised flight. Cabin pressurisation, by engine bleed air, has max differential of 0·63 bars (9·2 lb/sq in). Electrical system powered by dual 400A 30V DC starter/generators with AC power supplied from dual 1,000 VA solid-state inverters. Dual 24V 22Ah nickel-cadmium batteries. Emergency battery pack optional. Dual engine-driven hydraulic pumps, each capable of maintaining full system pressure of 103·5 bars (1,500 lb/sq in). Auxiliary electrically-driven hydraulic pump. Pneumatic system at pressure of 124-207 bars (1,800-3,000 lb/sq in) for emergency extension of landing gear and operation of wheel brakes. Engine fire detection and extinguishing system. Oxygen system for emergency use has crew demand masks and passenger dropout masks. Alcohol anti-icing system for radome.

ELECTRONICS AND EQUIPMENT: Complete nav/com systems, to full airline standard, are available to customer's requirements, comprising Collins, Bendix, Sperry or export equipment. All have Learjet autopilot as standard. A typical standard electronics package for the Learjet 24E includes dual Collins VHF-20 com, dual Collins VIR-30 nav, dual marker lamps, dual Avtech audio heads, Collins DME-40, Collins TDR-90 transponder, RCA Primus 35 radar, Collins DF-206 ADF, dual Collins 332C-10 RMI, collins FD112V flight director, Collins/J.E.T. PN101/5-4000 flight indicator (co-pilot), dual J. E. T. VG-206 vertical gyros, dual J.E.T. DN101D directional gyros, and IDC System II alerting and reporting. Standard equipment includes birdproof windscreen; heated pitot tubes and static ports; ice detector lights; anti-collision beacons; strobe, landing/taxi, navigation, cabin dome, map, instrument panel flood, baggage compartment, cabin reading, cabin entry, depressurisation warning and engine fire warning lights; Mach warning and stall warning system; dual clocks; lightning protection; fire axe and cabin fire extinguisher; stereo system and control locks.

DIMENSIONS, EXTERNAL:
- Span over tip-tanks: 10·84 m (35 ft 7 in)
- Wing chord at root: 2·74 m (9 ft 0 in)
- Wing chord at tip: 1·40 m (4 ft 7 in)
- Wing aspect ratio: 5·01
- Length overall: 13·18 m (43 ft 3 in)
- Length of fuselage: 12·50 m (41 ft 0 in)
- Height overall: 3·73 m (12 ft 3 in)
- Tailplane span: 4·47 m (14 ft 8 in)
- Wheel track (c/l shock-absorbers): 2·51 m (8 ft 3 in)
- Wheelbase: 4·93 m (16 ft 2 in)
- Cabin door:
  - Height: 1·57 m (5 ft 2 in)
  - Standard width: 0·61 m (2 ft 0 in)
  - Optional width: 0·91 m (3 ft 0 in)
- Emergency exit:
  - Height: 0·71 m (2 ft 4 in)
  - Width: 0·48 m (1 ft 7 in)

DIMENSIONS, INTERNAL:
- Cabin, between pressure bulkheads:
  - Length: 5·28 m (17 ft 4 in)
  - Max width: 1·50 m (4 ft 11 in)
  - Max height: 1·32 m (4 ft 4 in)
  - Volume, incl baggage compartment: 6·97 m³ (246·0 cu ft)
- Baggage compartment: 1·13 m³ (40·0 cu ft)

AREAS:
- Wings, gross: 21·53 m² (231·77 sq ft)
- Ailerons (total): 1·08 m² (11·70 sq ft)
- Trailing-edge flaps (total): 3·42 m² (36·85 sq ft)
- Spoilers: 0·66 m² (7·05 sq ft)
- Fin: 3·47 m² (37·37 sq ft)
- Rudder, incl tab: 0·67 m² (7·18 sq ft)
- Tailplane: 5·02 m² (54·00 sq ft)
- Elevators: 1·31 m² (14·13 sq ft)

WEIGHTS AND LOADINGS (A: Learjet 24E; B: Learjet 24F):
- Weight empty, equipped:
  - A: 3,186 kg (7,025 lb)
  - B: 3,234 kg (7,130 lb)
- Operating weight empty:
  - A: 3,368 kg (7,425 lb)
  - B: 3,415 kg (7,530 lb)
- Max payload:
  - A: 1,803 kg (3,975 lb)
  - B: 1,755 kg (3,870 lb)
- Max T-O weight:
  - A: 5,850 kg (12,900 lb)
  - B: 6,123 kg (13,500 lb)
- Max ramp weight:
  - A: 5,987 kg (13,200 lb)
  - B: 6,259 kg (13,800 lb)

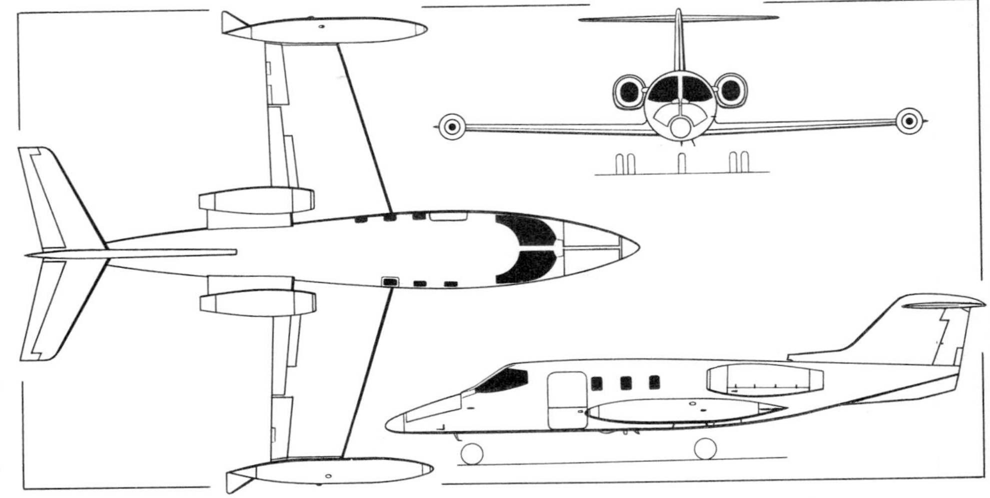

Gates Learjet 24 Series twin-jet light executive transport *(Pilot Press)*

- Max zero-fuel weight:
  - A, B: 5,171 kg (11,400 lb)
- Max landing weight:
  - A, B: 5,388 kg (11,880 lb)
- Max wing loading:
  - A: 271·75 kg/m² (55·66 lb/sq ft)
  - B: 284·4 kg/m² (58·25 lb/sq ft)
- Max power loading:
  - A: 223·3 kg/kN (2·19 lb/lb st)
  - B: 233·7 kg/kN (2·29 lb/lb st)

PERFORMANCE (at max T-O weight, unless stated otherwise, A: Learjet 24E; B: Learjet 24F):
- Never-exceed speed:
  - A, B: Mach 0·86
- Max operating speed at 9,450 m (31,000 ft):
  - A, B: 473 knots (877 km/h; 545 mph)
- Max operating speed at 13,715 m (45,000 ft):
  - A, B: 464 knots (859 km/h; 534 mph)
- Econ cruising speed at 13,715 m (45,000 ft):
  - A, B: 418 knots (774 km/h; 481 mph)
- Stalling speed, 'clean':
  - A, B: 86 knots (160 km/h; 99 mph) IAS
- Stalling speed at max landing weight, wheels and flaps down:
  - A, B: 87 knots (161 km/h; 100 mph) IAS
- Max rate of climb at S/L:
  - A: 2,200 m (7,220 ft)/min
  - B: 2,073 m (6,800 ft)/min
- Service ceiling:
  - A, B: 13,715 m (45,000 ft)
- Service ceiling, one engine out:
  - A: 8,685 m (28,500 ft)
  - B: 8,230 m (27,000 ft)
- Min ground turning radius:
  - A, B: 10·46 m (34 ft 4 in)
- T-O run:
  - A: 626 m (2,055 ft)
  - B: 703 m (2,305 ft)
- T-O to 10·7 m (35 ft), FAA balanced field length:
  - A: 853 m (2,800 ft)
  - B: 975 m (3,200 ft)
- Landing from 15 m (50 ft), at typical landing weight:
  - A, B: 747 m (2,450 ft)
- Landing run, at max landing weight:
  - A, B: 419 m (1,375 ft)
- Range with 4 passengers, max fuel and 45 min reserves:
  - A: 1,265 nm (2,343 km; 1,456 miles)
  - B: 1,472 nm (2,728 km; 1,695 miles)

OPERATIONAL NOISE CHARACTERISTICS (FAR Pt 36):
- T-O noise level: 94 EPNdB
- Approach noise level: 92 EPNdB
- Sideline noise level: 103 EPNdB

GATES LEARJET 25D

First flown on 12 August 1966 as the Learjet 25, this version is 1·27 m (4 ft 2 in) longer than the series 24 aircraft, and will accommodate eight passengers and a crew of two. FAA certification in the air transport category (FAR 25) was obtained on 10 October 1967 and the initial delivery was made in November 1967. British CAA certification was received on 26 June 1974.

The 1976 version has the designation Learjet 25D, and introduces the same Century III improvements as the Learjet 24 series. In addition, 8° flap settings for take-off are approved for both Series 25 aircraft, permitting an increase of more than 907 kg (2,000 lb) in max T-O weight. Factory-installed thrust reversers for the General Electric CJ610-6 turbojet engines are optional.

Two Learjet 25Bs supplied to the Peruvian Air Force in 1974 are each fitted with an underbelly pack containing two Wild RC-10 aerial survey cameras. An accompanying illustration shows a civil-registered Learjet 25C of Cruzeiro do Sul, Brazil, with a similar installation, which leaves the cabin interior quickly convertible for passenger or cargo transport duties.

The description of the Learjet 24E/F applies also to the Model 25D, except in the following details:

DIMENSIONS, EXTERNAL:
- As for Learjet 24E/F except:
  - Length overall: 14·50 m (47 ft 7 in)
  - Wheelbase: 5·84 m (19 ft 2 in)

Gates Learjet 25C with twin camera installation in specially modified fuselage

DIMENSIONS, INTERNAL:
As for Learjet 24E/F except:
Cabin, between pressure bulkheads:
Length 6·27 m (20 ft 7 in)
Volume, incl baggage compartment 8·47 m³ (299 cu ft)

WEIGHTS AND LOADINGS:

Weight empty, equipped	3,465 kg (7,640 lb)
Operating weight empty	3,647 kg (8,040 lb)
Max payload	1,524 kg (3,360 lb)
Max T-O weight	6,804 kg (15,000 lb)
Max ramp weight	7,030 kg (15,500 lb)
Max landing weight	6,033 kg (13,300 lb)
Max wing loading	315·9 kg/m² (64·7 lb/sq ft)
Max power loading	259·7 kg/kN (2·54 lb/lb st)

PERFORMANCE (at max T-O weight, unless stated otherwise):
As for Learjet 24F, except:
Max cruising speed at 12,500 m (41,000 ft) Mach 0·81 (464 knots; 859 km/h; 534 mph)
Stalling speed, wheels and flaps down, at max landing weight 91 knots (169 km/h; 105 mph) IAS
Max rate of climb at S/L 1,844 m (6,050 ft)/min
Max rate of climb at S/L, one engine out 533 m (1,750 ft)/min
Service ceiling, one engine out 7,165 m (23,500 ft)
Min ground turning radius 11·43 m (37 ft 6 in)
T-O run 843 m (2,765 ft)
T-O to 10·7 m (35 ft), FAA balanced field length 1,204 m (3,950 ft)
Landing from 15 m (50 ft), at typical landing weight 799 m (2,620 ft)
Landing run at max landing weight 448 m (1,470 ft)
Range with 4 passengers, max fuel and 45 min reserves 1,535 nm (2,843 km; 1,767 miles)

OPERATIONAL NOISE CHARACTERISTICS:
As for Learjet 24E/F

GATES LEARJET 25F

The first models of this longer-range version of the basic Learjet 25 entered production in 1970. With the addition of a 772 litre (204 US gallon) fuselage fuel tank, the current Learjet 25F has a non-stop range in excess of 1,826 nm (3,383 km; 2,102 miles), plus fuel reserves. The cabin of this version is optionally convertible from a four- or six-seat configuration to a two-bed sleeper compartment. It is otherwise the same as the Learjet 25D, and the descriptions of that aircraft and of the Learjet 24E/F apply also to the 25F except as follows:

DIMENSIONS, INTERNAL:
Volume, incl baggage compartment 6·97 m³ (246 cu ft)

WEIGHTS:

Weight empty, equipped	3,436 kg (7,575 lb)
Operating weight empty	3,617 kg (7,975 lb)
Max payload	1,553 kg (3,425 lb)

PERFORMANCE (at max T-O weight):
Long-range cruising speed at 12,500 m (41,000 ft) Mach 0·73 (418 knots; 774 km/h; 481 mph)
Max certificated operating altitude 13,715 m (45,000 ft)
Range with 4 passengers, max fuel, 45 min reserves 1,652 nm (3,060 km; 1,902 miles)

GATES LEARJET 35A and 36A

It was announced at the Paris Air Show in May 1973 that the company intended to produce two new business jets, identified as the Learjet 35 Transcontinental and Learjet 36 Intercontinental, each to be powered by two 15·6 kN (3,500 lb st) AiResearch TFE 731-2 turbofan engines. A prototype for the Learjet 35 and 36 (known originally as the Learjet Model 26) made its first flight with TFE 731-2 engines on 4 January 1973.

Generally similar in basic configuration to the other members of the Learjet family of aircraft, the Learjet 35 and 36 are slightly larger in size than the Learjet 25B, previously the largest of those in production. The two new aircraft were almost identical, differing in fuel capacity and accommodation. FAA certification was awarded in July 1974, and customer deliveries began later that year.

The 1976 versions of these aircraft have the designations Learjet 35A and 36A, and introduce as standard the Century III improvements and engine synchronisers. New options for both models include thrust reversers, high-energy brakes and a closed-circuit television projection system.

The description of the Learjet 24E/F applies also to the Learjet 35A and 36A, except in the following details:

TYPE: Twin-turbofan light executive transport.

WINGS: As for Learjet 24E/F, except span increased.

FUSELAGE: As for Learjet 24E/F, except length increased.

TAIL UNIT AND LANDING GEAR: As for Learjet 24E/F.

POWER PLANT: Two AiResearch TFE 731-2 turbofan engines, each rated at 15·6 kN (3,500 lb st), mounted in pod on each side of rear fuselage. Fuel in integral wing and wingtip tanks and a fuselage tank, with a combined usable capacity (Learjet 35A) of 3,524 litres (931 US gallons). Learjet 36A has a larger fuselage tank, giving a combined usable total of 4,201 litres (1,110 US gallons). Refuelling point on upper surface of each wingtip tank. Engine nacelle leading-edges anti-iced by engine bleed air. Thrust reversers optional.

ACCOMMODATION: Crew of two on flight deck, with dual controls. Up to eight passengers in Learjet 35A; one on inward-facing bench seat on starboard side at front, then two pairs of forward-facing armchairs with centre aisle, and three on forward-facing couch at rear of cabin. Learjet 36A can accommodate up to six passengers, one pair of forward-facing armchairs being removed. Toilet and stowage space under front inward-facing seat which can be screened from remainder of cabin. Refreshment cabinet opposite this seat, aft of passenger door. Baggage compartment with capacity of 226 kg (500 lb) aft of cabin. Two-piece clamshell door at forward end of cabin on port side, with integral steps built into lower half. Emergency exit on starboard side of cabin.

SYSTEMS: Environmental control system comprises cabin pressurisation, ventilation, heating and cooling. Heating and pressurisation are provided by engine bleed air, with a maximum pressure differential of 0·61 bars (8·9 lb/sq in), maintaining a cabin altitude of 2,440 m (8,000 ft) to an actual altitude of 13,715 m (45,000 ft). Freon R12 vapour cycle cooling system supplemented by a ram-air heat exchanger. Anti-icing system includes distribution of engine bleed air for wing, tailplane and engine nacelle leading-edges and windscreen; electrical heating of pitot heads, stall warning vanes and static ports; and alcohol spray on windscreen and nose radome. Hydraulic system supplied by two engine-driven pumps, each pump capable of maintaining alone the full system pressure of 103·5 bars (1,500 lb/sq in), for operation of landing gear, brakes, flaps and spoilers. Electrically-driven hydraulic pump for emergency operation of all hydraulic services. Pneumatic system of 124 to 207 bars (1,800 to 3,000 lb/sq in) pressure for emergency extension of landing gear and operation of brakes. Electrical system powered by two 30V 400A brushless generators, two 1kVA solid-state inverters to provide AC power, and two 24V 22Ah nickel-cadmium batteries. Oxygen system for emergency use with crew demand masks and dropout masks for each passenger.

ELECTRONICS: A standard electronics package is available, comprising two Collins VHF-20 transceivers, two Collins VIR-30 VOR/ILS/GS/MB with Dorne-Margolin antenna, Avtech 1250 and 1251 compression audio systems, RCA AVQ-85 DME, Wilcox 1014A transponder, RCA AVQ-21 weather radar, Collins DF-206 ADF, two Collins 332C-10 RMI with single or dual switching, Collins FD-108Y flight director indicator

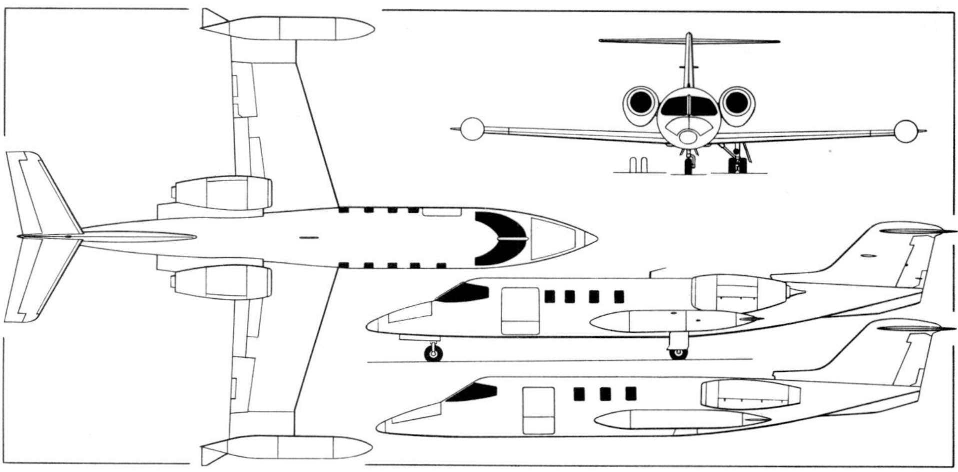
Gates Learjet 35A with optional cargo door. Additional side view (bottom) shows Learjet 25D *(Pilot Press)*

Gates Learjet 35A twin-turbofan light executive transport in Century III styling (two AiResearch TFE 731-2 engines)

with altitude hold, Collins/J.E.T. PN-101/5-4000 co-pilot's flight indicators, two J.E.T. VG-206 vertical gyros, two J.E.T. DN-101D directional gyros, IDC System 2; and J.E.T. PS-823B/AI-804 standby power pack and attitude gyro. Alternatively to this pre-engineered optional package, customers may select electronics equipment to meet specific requirements.

EQUIPMENT: Standard equipment includes engine fire extinguishers, jacking pads, internal corrosion-proofing, soundproofing, ice detector lights, battery overheat warning system, anti-collision beacons, landing/taxi lights, navigation lights, map lights, instrument panel and floodlights, baggage compartment lights, cabin reading lights, cabin entry light, engine fire warning lights, angle of attack indicator, cabin rate of climb indicator, two clocks, wing and tailplane temperature indicators, alternate static source, lightning protection, fire axe, crew and passenger flotation jackets, cabin fire extinguisher, external power source and control locks.

DIMENSIONS, EXTERNAL:
- Wing span over tip-tanks: 12·04 m (39 ft 6 in)
- Wing chord at root: 2·74 m (9 ft 0 in)
- Wing chord at tip: 1·55 m (5 ft 1 in)
- Wing aspect ratio: 5·74
- Length overall: 14·83 m (48 ft 8 in)
- Height overall: 3·73 m (12 ft 3 in)
- Tailplane span: 4·47 m (14 ft 8 in)
- Wheel track: 2·51 m (8 ft 3 in)
- Passenger door:
  - Standard:
    - Height: 1·57 m (5 ft 2 in)
    - Width: 0·61 m (2 ft 0 in)
  - Optional:
    - Height: 1·57 m (5 ft 2 in)
    - Width: 0·91 m (3 ft 0 in)
- Emergency exit:
  - Height: 0·71 m (2 ft 4 in)
  - Width: 0·48 m (1 ft 7 in)

DIMENSIONS, INTERNAL (A: Learjet 35A; B: Learjet 36A):
- Cabin: Length, incl flight deck:
  - A: 6·60 m (21 ft 8 in)
  - B: 5·77 m (18 ft 11 in)
- Max width: 1·50 m (4 ft 11 in)
- Max height: 1·31 m (4 ft 3½ in)
- Volume:
  - A: 7·59 m³ (268 cu ft)
  - B: 6·46 m³ (228 cu ft)
- Baggage compartment: 1·13 m³ (40 cu ft)

AREA:
- Wings, gross: 23·53 m² (253·3 sq ft)

WEIGHTS AND LOADINGS (A: Learjet 35A; B: Learjet 36A):
- Weight empty, equipped:
  - A: 4,132 kg (9,110 lb)
  - B: 4,152 kg (9,154 lb)
- Max payload:
  - A: 1,810 kg (3,990 lb)
  - B: 1,971 kg (4,346 lb)
- Max T-O weight:
  - A: 7,711 kg (17,000 lb)
  - B: 8,164 kg (18,000 lb)
- Max zero-fuel weight:
  - A, B: 6,123 kg (13,500 lb)
- Max landing weight:
  - A, B: 6,486 kg (14,300 lb)
- Max wing loading:
  - A: 327·6 kg/m² (67·1 lb/sq ft)
  - B: 347·1 kg/m² (71·1 lb/sq ft)
- Max power loading:
  - A: 247·1 kg/kN (2·43 lb/lb st)
  - B: 261·7 kg/kN (2·57 lb/lb st)

PERFORMANCE (at max T-O weight, except where indicated otherwise, A: Learjet 35A; B: Learjet 36A):
- Never-exceed speed:
  - A, B: Mach 0·83
- Max cruising speed:
  - A, B: 464 knots (859 km/h; 534 mph)
- Normal cruising speed:
  - A, B: 441 knots (817 km/h; 508 mph)
- Econ cruising speed:
  - A, B: 418 knots (774 km/h; 481 mph)
- Stalling speed, wheels and flaps down:
  - A, B: 92 knots (171 km/h; 106 mph) IAS
- Max rate of climb at S/L:
  - A: 1,554 m (5,100 ft)/min
  - B: 1,463 m (4,800 ft)/min
- Rate of climb at S/L, one engine out:
  - A: 450 m (1,475 ft)/min
  - B: 396 m (1,300 ft)/min
- Service ceiling:
  - A: 12,955 m (42,500 ft)
  - B: 12,650 m (41,500 ft)
- Service ceiling, one engine out:
  - A: 7,710 m (25,300 ft)
  - B: 7,165 m (23,500 ft)
- T-O to 15 m (50 ft):
  - A: 1,310 m (4,300 ft)
  - B: 1,510 m (4,950 ft)
- Landing from 15 m (50 ft), at typical landing weight of 5,080 kg (11,200 lb):
  - A, B: 872 m (2,860 ft)
- Range with 4 passengers, max fuel and 45 min reserves:
  - A: 2,410 nm (4,466 km; 2,775 miles)
  - B: 2,870 nm (5,318 km; 3,305 miles)

GENERAL DYNAMICS
GENERAL DYNAMICS CORPORATION

HEAD OFFICE:
Pierre Laclede Center, St Louis, Missouri 63105
Telephone: (314) 862-2440

CHAIRMAN AND CHIEF EXECUTIVE OFFICER:
David S. Lewis

EXECUTIVE VICE-PRESIDENTS:
Gordon E. MacDonald
Gene K. Beare
James M. Beggs

VICE-PRESIDENTS:
Leonard F. Buchanan (General Manager, Pomona Division)
M. C. Curtis (Deputy General Manager, Electric Boat Division)
Otto J. Glasser (International)
Max Golden (Contracts)
Algie A. Hendrix (Industrial Relations)
E. J. LeFevre (GD Field Offices)
Edward E. Lynn (General Counsel)
Frank Nugent (Chairman, Freeman Coal Mining Corporation and the United Electric Coal Companies)
Joseph D. Pierce (General Manager, Electric Boat Division)
P. Takis Veliotis (President and General Manager, Quincy Shipbuilding Division)
Wayne Wells (Treasurer)
Robert H. Widmer (Science and Engineering)

Convair Division:
PO Box 80847, San Diego, California 92138

VICE-PRESIDENTS:
Grant L. Hansen (General Manager)
James M. Adamson (Operations)
A. Kalitinsky (Research Engineering)
Kenneth E. Newton (Launch Vehicle Programmes)

Fort Worth Division:
PO Box 748, Fort Worth, Texas 76101

VICE-PRESIDENTS:
Richard E. Adams (General Manager)
William C. Dietz (F-16 Engineering)
Edward E. Hatchett (Finance)
Henry C. Jones (Contracts and Estimating)
Lyman C. Josephs (Deputy General Manager and F-16 Programme Director)
Julius Y. McClure (Quality Assurance)
Norman B. Robbins (111 Programmes)
Herbert F. Rogers (Marketing)
Gordon E. Sylvester (Operations)
Theodore S. Webb Jr (Research and Engineering)

In mid-June 1974 General Dynamics announced plans for the reorganisation of its Convair Aerospace Division, under which the Fort Worth operation became the Fort Worth Division, and the San Diego operation became the Convair Division. As a result of these changes the company now conducts its US aerospace activities at four divisions: Convair Division, with operations at San Diego, California; Fort Worth Division, with operations at Fort Worth, Texas; Pomona Division, at Pomona, California; and Electronics Division, with headquarters in San Diego and an operating unit at Orlando, Florida. Convair Division is responsible for the design, development and production of commercial aircraft and of systems for space exploration. Fort Worth Division is engaged in the design, development and production of military aircraft and electronics. Pomona Division is engaged in tactical missile and other ordnance programmes. Electronics Division is engaged in advanced radar development, and produces ocean data, location, communication and navigation systems, anti-submarine warfare sonobuoy receivers and ground support equipment.

Fort Worth is currently responsible for development and production of the F-111 series of combat aircraft and of the F-16 lightweight fighter, while Convair Division is responsible for production of a major portion of the fuselage for the McDonnell Douglas DC-10 commercial transport aircraft.

Convair Division also retains detailed tooling for high-usage spares for the Convair-Liner 240/340/440 series of piston-engined transports, and Convair 880 and 990 jet transports, and is manufacturing components for operators of these types.

GENERAL DYNAMICS F-111

Following a detailed evaluation of design proposals submitted by General Dynamics and Boeing, the US Department of Defense announced on 24 November 1962 that General Dynamics had been selected as prime contractor for development of the F-111 variable-geometry tactical fighter (known originally by the designation TFX), with Grumman Aircraft as an associate. An initial contract was placed for 23 development aircraft (18 F-111As for the USAF, five F-111Bs for the US Navy), of which the first were scheduled for delivery within 2½ years. Subsequently, further orders were placed, covering F-111D, E and F improved tactical fighters for the USAF, 24 F-111Cs for the Royal Australian Air Force, and the FB-111A strategic bomber version for the USAF.

A total of 562 F-111s of all types, including the 23 development models, were covered by these contracts.

The specification to which the F-111 was designed called for a maximum speed of about Mach 2·5, capability of supersonic speed at sea level, short take-off capability from rough airfields in forward areas and short landing capability. The F-111 had to be able to fly between any two airfields in the world in one day and to carry a full range of conventional and nuclear weapons including the latest air-to-surface tactical weapons.

Versions are as follows:

F-111A. USAF two-seat tactical fighter-bomber. Development models built with two P & W TF30-P-1 turbofan engines: production version has TF30-P-3 engines and Mk I electronics. First F-111A flew for the first time (with wings locked at sweepback of 26°) on 21 December 1964. Contracts covered the 18 development aircraft and 141 production models for the USAF Tactical Air Command. Production completed.

EF-111A. ECM jamming version. Under development by Grumman, which has converted two F-111As to this configuration for evaluation. Details under Grumman entry.

RF-111A. Reconnaissance conversion of No. 11 F-111A, tested in prototype form. No further development planned.

YF-111A. Two strike/reconnaissance fighters completed prior to cancellation of the British government's order for 50 aircraft, under the designation F-111K, were subsequently assigned to the USAF for use in its research, development, test and evaluation programme, with the designation YF-111A. Included in F-111A total of 141 production aircraft.

General Dynamics FB-111A of the USAF's Strategic Air Command, armed with four SRAM missiles

General Dynamics F-111E fighter-bomber of the USAF

F-111B. US Navy version, designed for carrier-based fleet defence duties. Powered initially by TF30-P-1 turbofan engines; production models were programmed to have more powerful TF30-P-12 engines. First F-111B, assembled by Grumman, flew for the first time on 18 May 1965. Original orders covered five development aircraft and 24 production models for the US Navy. The sixth aircraft, the first to be fitted with TF30-P-12 engines, flew on 29 June 1968. The seventh (and last) has been used as a testbed for the Phoenix missile. Continued development, production and support of the F-111B were halted by Congress in mid-1968.

F-111C. Strike aircraft. Outwardly similar to FB-111A, with Pratt & Whitney TF30-P-3 engines, Mk I electronics, cockpit ejection module and eight underwing attachments for stores. 24 built for RAAF.

F-111D. Similar to F-111A, but with Mk II electronics, offering improvements in navigation and in air-to-air weapon delivery. TF30-P-9 engines. Delivery of 96 completed in February 1973. Equips 27th Tactical Fighter Wing, Cannon AFB, New Mexico.

F-111E. Superseded F-111A from 160th aircraft. Modified air intakes improve engine performance above Mach 2·2. Total of 94 built; followed by F-111D. Most equip the 20th Tactical Fighter Wing, USAFE, at Upper Heyford, Oxon, England, the remainder serving with the 474th Tactical Fighter Wing, Nellis Air Force Base, Nevada.

F-111F. Fighter-bomber. Generally similar to F-111D, but with electronics that combine the best features of the F-111E and FB-111A systems, to provide effective tactical electronics at the lowest possible cost. TF30-P-100 engines, producing 25% more thrust than the basic TF30 and providing a significant improvement in T-O performance, single-engine rate of climb, payload capability, acceleration and max speed at low level without use of afterburning. 106 ordered for Tactical Air Command. The 366th Tactical Fighter Wing at Mountain Home AFB, Idaho, is equipped with F-111Fs.

It was announced in late 1970 that all F-111F aircraft would have a boron-epoxy doubler applied to the wing pivot fitting to increase fatigue life. This was to be retrofitted to all tactical F-111 aircraft already in service during each aircraft's inspect-and-repair-as-needed (IRAN) programme.

F-111K. See YF-111A.

FB-111A. Two-seat strategic bomber version for USAF Strategic Air Command with Mk IIB advanced electronics and TF30-P-7 engines. Requirement for 210 announced by US Secretary of Defense on 10 December 1965, to replace B-52C/F versions of the Stratofortress and B-58A Hustler. Initial contract for 64 signed in Spring of 1967. Subsequently, on 20 March 1969, the US Secretary of Defense stated that FB-111A production would total 76 aircraft. First of two prototypes converted from development F-111As flew on 30 July 1967, followed by first production FB-111A on 13 July 1968 (fitted temporarily with TF30-P-3 engines). Long-span wings. Strengthened landing gear. Increased braking capacity. Max load of six nuclear bombs, or six SRAM missiles (four under wings, two in weapons bay), or combinations of these weapons. Conventional weapon loadings of up to 14,288 kg (31,500 lb) of bombs can also be delivered.

First FB-111A was delivered to the 340th Bomb Group, a training unit of Strategic Air Command, at Carswell AFB, Texas, on 8 October 1969. FB-111A units (each two squadrons) are the 509th Bomb Wing at Pease AFB, New Hampshire, and the 380th Strategic Aerospace Wing at Plattsburgh AFB, New York. Production completed.

The following details apply to the F-111A except where otherwise indicated:

Type: Two-seat variable-geometry multi-purpose fighter.

Wings: Cantilever shoulder wing. Wing section of NACA 63 series, with conventional washout. Sweepback of outer portions variable in flight or on the ground from 16° to 72° 30′. Wing-actuating jacks by Jarry Hydraulics. Five-spar structure, with stressed and sculptured skin panels, each made in one piece between leading- and trailing-edge sections, from root to tip. Leading- and trailing-edge sections of honeycomb sandwich. Airbrake/lift dumpers above wing operate as spoilers for lateral control at low speeds. Full-span variable-camber leading-edge slats and full-span double-slotted trailing-edge flaps. General Electric flight control system.

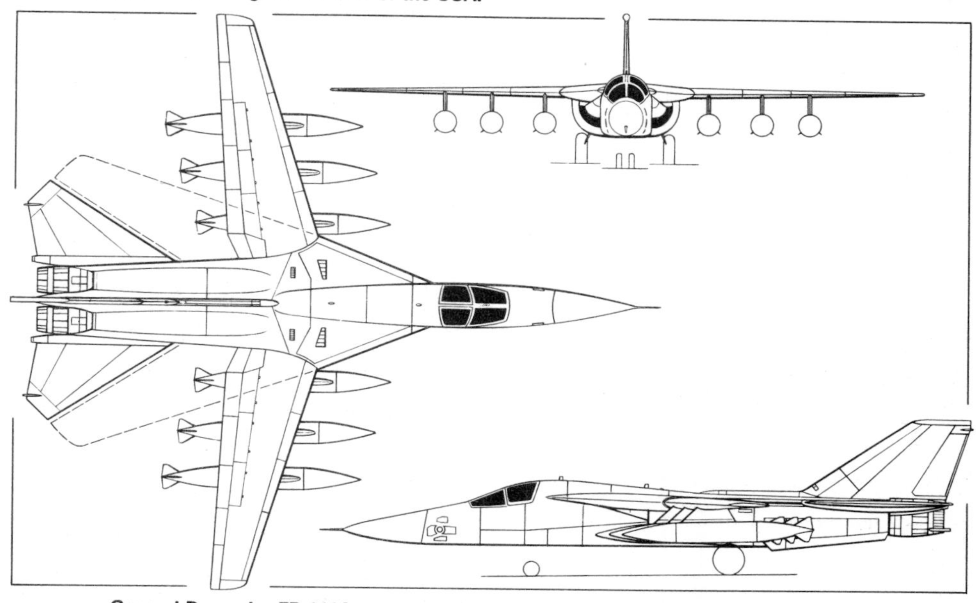
General Dynamics FB-111A two-seat variable-geometry strategic bomber *(Pilot Press)*

Fuselage: Semi-monocoque structure, mainly of aluminium alloy, with honeycomb sandwich skin. Some steel and titanium. Main structural member is a T-section keel, under the arms of which the engines are hung.

Tail Unit: Conventional cantilever sweptback surfaces, utilising honeycomb sandwich skin panels, except for tailplane tips and central area of fin on each side. All-moving horizontal surfaces operate both differentially and symmetrically to provide aileron and elevator functions. Two long narrow ventral stabilising fins.

Landing Gear: Hydraulically-retractable tricycle type. Single wheel on each main leg.Twin-wheel nose unit retracts forward. Main gear is a triangulated structure with hinged legs which are almost horizontal when the gear is extended. During retraction, the legs pivot downward, the wheels tilt to lie almost flat against them, and the whole gear rotates forward so that the wheels are stowed side by side in fuselage between engine air intake ducts. Low-pressure tyres on main wheels, size 47-18 in (42-13 in on F-111C and FB-111A). Disc brakes, with anti-skid system. Main landing gear door, in bottom of fuselage, hinges down to act as speed brake in flight.

Power Plant: Two Pratt & Whitney TF30-P-3 turbofan engines, each giving 82·3 kN (18,500 lb st) with afterburning. Fuel tanks in wings and fuselage. Pressure fuelling point in port side of fuselage, forward of engine air intake. Gravity fuel filler/in-flight refuelling receptacle in top of fuselage aft of cockpit. Hamilton Standard hydro-mechanical air intake system with movable shock-cone.

Accommodation: Crew of two side by side in air-conditioned and pressurised cabin. Portion of canopy over each seat is hinged on aircraft centreline and opens upward. Zero-speed, zero-altitude (including under-water) emergency escape module developed by McDonnell Douglas Corpn and utilising a 178 kN (40,000 lb st) Rocket Power Inc rocket motor. Emergency procedure calls for both crew members to remain in capsule cabin section, which is propelled away from aircraft by rocket motor and lowered to ground by parachute. Airbags cushion impact and form flotation gear in water. Entire capsule forms survival shelter.

Armament: Tactical fighter versions carry one M61 multi-barrel 20 mm gun or two 750 lb bombs in internal weapon bay. External stores are carried on four attachments under each wing. The two inboard pylons on each side pivot as the wings sweep back, to keep the stores parallel with the fuselage. The two outboard pylons on each wing are jettisonable and non-swivelling.

Dimensions:
Wing span:
F-111A, F-111D, F-111E, F-111F:
spread 19·20 m (63 ft 0 in)
fully swept 9·74 m (31 ft 11·4 in)
F-111B, F-111C, FB-111A:
spread 21·34 m (70 ft 0 in)
fully swept 10·34 m (33 ft 11 in)
Wing chord at root 2·11 m (6 ft 11 in)
Length overall:
F-111A, F-111C, F-111D, F-111E, F-111F, FB-111A 22·40 m (73 ft 6 in)
Height overall:
F-111A, F-111C, F-111D, F-111E, F-111F, FB-111A 5·22 m (17 ft 1·4 in)

Weights (F-111A):
Weight empty 20,943 kg (46,172 lb)
Max T-O weight 41,500 kg (91,500 lb)

Performance (F-111A):
Max speed at height Mach 2·2
Max speed at S/L Mach 1·2
Service ceiling over 15,500 m (51,000 ft)
T-O and landing run under 915 m (3,000 ft)
Range with max internal fuel over 2,750 nm (5,093 km; 3,165 miles)

GENERAL DYNAMICS F-16

USAF designations: F-16A and F-16B

Proposals from five competing companies (General Dynamics, Northrop, Boeing, LTV Aerospace and Lockheed) were submitted to the USAF on 28 February 1972 as candidates in the Lightweight Fighter (LWF) prototype programme.

On 13 April 1972, General Dynamics and Northrop were each awarded a contract to build two prototypes, the former pair being designated YF-16 and the latter pair YF-17. The prototypes were intended to determine the viability of a small, lightweight, low-cost air superiority fighter, and to aid evaluation of the operational potentional of such an aircraft as well as establishing its operational role.

The two General Dynamics YF-16 prototypes, built under a contract worth more than $37·9 million, were for evaluation in a twelve-month, 300-hour flight programme directed by the USAF Aeronautical Systems Division's Prototype Programs Office at Wright-Patterson AFB, Ohio, under the overall control of Col Lyle W. Cameron. Design priorities for this programme recognised cost as being of equal importance to schedule or performance. The USAF specified that the prototype aircraft must provide accurate information in respect of both cost to develop and cost to produce. Thus, each manufacturer had to consider how best to use advanced technology to provide very high performance within a price range considered acceptable to USAF planners. The concept chosen for the YF-16 blends advanced technology with a basically conservative configuration and a power plant offering high thrust/weight ratio.

The selection of a single-engined configuration meant that emphasis was placed on weight savings to meet the critical performance categories of high acceleration rates, high rate of climb and exceptional manoeuvrability. This dictated limitation of aircraft size, and the use of advanced concepts to obtain optimum lift.

More than 1,200 hours of wind tunnel testing of over 50 configurations led to the present design, with special emphasis on development of an optimum relationship between the wing leading-edges and the forebody strakes which provide vortex control. Similar in-depth study of potential requirements of a lightweight fighter resulted in the selection of manufacturing breaks, methods of attachment of external aerodynamic shapes and surfaces, structural provisions, and internal space, so that full advantage could be taken of any new features or concepts that might originate during progress of the prototype programmes. This ensured that changes could be made easily to a particular component, with minimum structural disruption to the rest of the airframe. The forward section of the engine air inlet, wings, tail surfaces and forebody strakes are examples of readily removable structures. This modular approach provides great flexibility, and could make it possible to flight test on the F-16 components such as supercritical wings, advanced composite wings, growth versions of the F100 engine, advanced armament, a more advanced high-*g* cockpit, and a variable-geometry engine air intake.

In other respects the structure is conventional, keeping material costs to a minimum; it consists of approx 78·3% aluminium alloy (of which 80% are sheet metal parts), 3·7% steel, 4·2% advanced composite materials, 2·2% titanium, and 10·6% other materials, including about 1% reinforced plastics. Despite the large-scale use of conventional materials, there has been no degradation of structural strength, the F-16 airframe being designed for a manoeuvre capability of 7·33*g* with full internal fuel, full ammunition, and two AIM-9 missiles.

USAF, NASA and company research all contributed to the technological advances built into the YF-16. They include vortex control, variable wing camber, a high-*g* cockpit, relaxed longitudinal static stability, fly-by-wire control system, a blended wing/body, and the use of advanced composites in the tail unit.

The first of the two YF-16 prototypes (72-01567) was rolled out at Fort Worth, Texas, on 13 December 1973, and was ferried in a USAF C-5 to Edwards AFB, California, on 8 January 1974. It made an unscheduled first flight on 20 January 1974, when test pilot Philip Oestricher elected to take off after the all-moving tailplane was damaged during high-speed taxi tests. The official first flight was made on 2 February 1974, and on 5 February a speed in excess of Mach 1 was recorded. A level speed of Mach 2 at 12,200 m (40,000 ft) was attained on 11 March 1974. The second YF-16 (72-01568) was ferried to Edwards AFB on 27 February 1974, where it flew for the first time on 9 May 1974.

During an eleven-month flight evaluation programme, from February to the end of December 1974, the YF-16s and YF-17s were flown competitively against each other and against other current USAF aircraft. In the course of this evaluation, the YF-16 prototypes flew at speeds in excess of Mach 2 and to heights of more than 18,300 m (60,000 ft); executed manoeuvres of up to 9*g*; made subsonic and supersonic firings of seven AIM-9 Sidewinder air-to-air missiles; fired 12,948 rds of 20 mm ammunition; dropped 10 Mk 84 2,000 lb bombs; made endurance flights of up to 4 hr 25 min with in-flight refuelling, and up to 2 hr 55 min without refuelling; flew a total of 330 missions, amounting to 417 hr in the air, of which 13 hrs 15 min were at supersonic speeds; and met or exceeded all design objectives.

On 11 September 1974 the Department of Defense announced that the winning design, now known as the Air Combat Fighter (ACF), would be declared in January 1975; and on 13 January the Secretary of the USAF announced that the F-16 had been selected and authorised for full-scale engineering development. Contracts were awarded to General Dynamics ($417·9 million) and Pratt & Whitney ($55·5 million), for fifteen F-16 engineering development aircraft and their F100 engines. A contract change on 9 April 1975 reduced the pre-production buy to

General Dynamics YF-16, with 18 underwing bombs and two Sidewinder air-to-air missiles on wing weapon stations

eight aircraft, comprising six single-seat F-16As and two two-seat F-16Bs, construction of which began in July 1975.

The first of these aircraft was scheduled to fly in the last quarter of 1976, and the last in 1978. They will be used to evaluate the potential of the F-16 under operational conditions, prior to full-scale production; the USAF has indicated its intention to procure an initial quantity of at least 650 such aircraft.

It was announced on 9 May 1975 that Marconi-Elliott Avionic Systems had been contracted to supply the HUD (head-up display) system for the pre-production full-scale development version of the F-16. This follows reports of outstanding performance of the HUD weapon-aiming computer system during flight evaluation of the second YF-16 prototype.

An important feature of this system, which has become known as a HUD sight, is its 'snapshoot' air-to-air gunsight display, believed to be the only combat-proven tracerline display in the world. It provides the pilot continuously with a simulated trace of the path which his bullets will take if the weapon is fired, superimposed on his view of the target, so enabling a burst of fire to be directed accurately and economically.

On 7 June 1975 a joint announcement by the four NATO countries of Belgium, Denmark, the Netherlands and Norway confirmed their selection of the F-16 to replace the F-104s in current service, the F-16 being selected in preference to offers of the Saab 37 Viggen and Dassault Mirage F1-E. The initial order is for 306 aircraft (Belgium 102, Denmark 48, the Netherlands 84 and Norway 72), with options for 42 more. Co-production arrangements with the consortium of the NATO countries provide for responsibility for production of 10% of the procurement value of the USAF's intended 650 aircraft, 40% of the procurement value of the aircraft required by their own air forces, and 15% of the procurement value of potential third nation sales. Final assembly lines will be established in Belgium and the Netherlands, and 56 potential subcontracts in the four European countries have been identified.

A joint proposal by LTV Aerospace (now Vought) and General Dynamics for derivatives of the F-16 (Models 1600 and 1601) to meet the US Navy's Air Combat Fighter (NACF) requirements, was submitted on 13 January 1975. However, the Navy chose instead the twin-engined F-18, a derivative of the YF-17 submitted by McDonnell Douglas and Northrop Corporation.

The eight full-scale development F-16A/Bs for the USAF will differ in a number of respects from the YF-16 prototypes. The structural strength of the YF-16s was deliberately increased by 25% to permit the fullest and most rapid possible exploration of high-*g* manoeuvres; F-16A and F-16B structural capability also allows full exploitation of the manoeuvring capability of the aircraft. The F-16 symmetric design load factor of 9·0 is applicable at gross weights corresponding to 100% internal fuel, full ammunition and two AIM-9 missiles (10,205 kg; 22,500 lb). Severe service life requirements have been applied to the structure. The service loads spectra contain a higher frequency of manoeuvre occurences in the intermediate and high load factor range (up to five times the current military specification requirements) than for any other fighter-type aircraft. In addition, newer, more stringent design concepts of durability and damage tolerance are used in the airframe design. The F-16A fuselage is 0·35 m (13·7 in) longer than that of the YF-16, but because of a shorter nose probe its overall length is 0·13 m (5·3 in) less. The tandem two-seat F-16B is the same length as the F-16A, but its internal fuel tankage is reduced by approximately 16% to make room for the second cockpit. The wing area of both full-scale development models (F-16A and F-16B) has been increased by 1·86 m² (20 sq ft) over that of the YF-16, and the horizontal tail area has been increased by 15%. Other modifications in the F-16 include the addition of a self-contained jet-fuel engine starter, and increased external stores-carrying capability on nine stores stations.

The following description applies to the YF-16, F-16A and F-16B, as indicated:

TYPE: Single-seat lightweight air combat fighter (F-16A) and two-seat fighter/trainer (F-16B).

WINGS: Cantilever mid-wing monoplane, of blended wing/body design and cropped-delta planform. The blended wing/body concept is achieved by flaring the wing/body intersection, thus not only providing lift from the body at high angles of attack but also giving less wetted area and increased internal fuel volume. In addition, thickening of the wing root gives a more rigid structure, with a weight saving of some 113 kg (250 lb). Basic wing is of NACA 64A-204 section, with 40° sweepback on leading-edges. Structure is mainly of aluminium alloy, with 12 spars, 5 ribs and single upper and lower skins, and is attached to fuselage by machined aluminium tension fittings. Vortex lift and control is provided by sharp, highly-swept strakes extending along the fuselage forebody. This permits significant reduction in wing area. Variable wing camber is achieved by the use of leading-edge manoeuvring flaps that are programmed automatically as a function of Mach number and angle of attack. The increased wing camber maintains effective lift coefficients at high angles of attack. These flaps are one-piece bonded aluminium honeycomb sandwich structures, actuated by an AiResearch drive system using rotary actuators. The trailing-edges carry large flaperons (flaps/ailerons), which are interchangeable left with right and are actuated by National Water Lift integrated servo-actuators. The maximum rate of flaperon movement is 80°/sec on the F-16.

FUSELAGE: Semi-monocoque all-metal structure of frames and longerons, built in three main modules: forward (to just aft of cockpit), centre and aft. Nose radome built by Brunswick Corporation. Highly-swept vortex control strakes along the fuselage forebody.

TAIL UNIT: Cantilever structure with sweptback surfaces, constructed largely of graphite-epoxy composite laminate skins with full-depth bonded aluminium honeycomb sandwich core. Steel leading-edge caps to tailplane. Glassfibre fin tip. Small glassfibre dorsal fin and root fairing. Interchangeable light alloy ventral fins. Interchangeable all-moving tailplane halves, with glassfibre tips. Split speed-brake inboard of rear portion of each horizontal tail surface to each side of nozzle, each deflecting 60° from the closed position. National Water Lift servo-actuators for rudder and tailplane.

LANDING GEAR: Menasco hydraulically-retractable type, nose unit retracting aft and main units forward into fuselage. Nosewheel is located aft of intake, to reduce the risk of foreign objects being drawn into the engine during ground operation, and rotates 90° during retraction to lie horizontally under engine air intake duct. Oleo-pneumatic struts on all units. Goodyear main wheels and brakes; B. F. Goodrich main-wheel tyres, size 25·5 × 8-14. Steerable nosewheel with B. F. Goodrich tyre, size 18 × 5·5-8. Eighty per cent of main unit components interchangeable. Brake-by-wire system on main gear, with Goodyear anti-skid units. Runway arrester hook under rear fuselage.

POWER PLANT: One Pratt & Whitney F100-PW-100(3) turbofan engine, rated at approx 111·2 kN (25,000 lb st) with afterburning, mounted within the rear fuselage.

Fixed-geometry intake, with boundary layer splitter plate, beneath fuselage. A fixed-geometry intake was chosen as it was calculated that it would be 181 kg (400 lb) lighter than a variable-geometry intake designed for optimum performance; but it can be changed to a variable-geometry intake later, without difficulty, if desirable to improve high-speed performance. The underfuselage intake position was chosen because here the airflow suffers least disturbance throughout the entire range of aircraft manoeuvres, and because it eliminates the problem of gun gas ingestion. Foreign object damage is avoided by placing the nose gear aft of the inlet lip. Standard fuel contained in wing and five fuselage cells which function as two tanks; internal fuel weight is 3,145 kg (6,934 lb) in F-16A, and approx 16% less in F-16B. In-flight refuelling receptacle in top of centre-fuselage, aft of cockpit. Auxiliary fuel can be carried in drop-tanks on underwing and underfuselage hardpoints.

ACCOMMODATION: Pilot only, in air-conditioned cockpit, on Stencel zero-zero ejection seat (McDonnell Douglas Escapac IH-8 in YF-16). Texstar transparent bubble canopy, made of polycarbonate, an advanced plastics material. The windscreen and forward canopy are an integral unit without a forward bow-frame, and are separated from the aft canopy by a simple support structure which serves also as the break-point where the forward section pivots upward and aft to give access to the cockpit. A redundant safety-lock feature prevents canopy loss. This new windscreen/canopy design provides 360° all-round vision, 195° fore and aft, 40° down over the side, and 15° down over the nose. While this canopy imposes a supersonic drag penalty, it is considered to be more than offset by the improved rearward view afforded to the pilot. To enable the pilot to sustain high-*g* forces, and for pilot comfort, the seat is inclined 30° aft and the heel-line is raised. In normal operation the canopy is pivoted upward and aft by electrical power; the pilot is also able to unlatch the canopy manually and open it with a backup handcrank. Emergency jettison is provided by explosive unlatching devices and two forward-mounted rockets. A limited-displacement, force-sensing control stick is provided on the right hand console, with a suitable armrest, to provide precise control inputs during combat manoeuvres. The F-16B has two cockpits arranged in tandem and equipped with all controls, displays, instruments, electronics and life-support systems required to perform both training and combat missions. The layout of the F-16B second station is essentially the same as that of the F-16A, and is fully systems-operational. A single-enclosure polycarbonate transparency, made in two pieces and spliced aft of the forward seat with a metal bow-frame and lateral support member, provides outstanding view from both cockpits.

SYSTEMS: Regenerative bootstrap air-cycle environmental control system by United Technologies' Hamilton Standard Division, using engine bleed air, for pressurisation and cooling. Two separate and independent hydraulic systems supply power for operation of the primary flight control surfaces and the utility functions. Electrical system powered by engine-driven integrated drive generator, rated at 40kVA in YF-16. Westinghouse 40kVA and Lear Siegler 5kVA generators and ground control units in F-16, with Sundstrand constant-speed drive. Four dedicated, sealed-cell batteries provide transient protection for the fly-by-wire flight control system. The use of relaxed static longitudinal stability in the YF-16 design resulted in the selection of a full fly-by-wire control system. The F-16 system, also full fly-by-wire, is an outgrowth of the YF-16 system, and the two systems are generally similar. Application of the control configured vehicle (CCV) principle of relaxed static stability produces a significant reduction in trim drag, especially at high load factors and supersonic speeds. The aircraft centre of gravity is allowed to move aft, reducing both the tail drag and the change in drag on the wing due to changes in lift required to balance the down-load on the tail. Relaxed static stability imposes a requirement for a highly-reliable, full-time-operating, stability augmentation system, including reliable electronic, electrical and hydraulic provisions. The signal paths in this quad-redundant system are used to control the aircraft, replacing the usual mechanical linkages. Direct electrical control is employed from pilot controls to surface actuators. An on-board Sundstrand/Solar jet fuel starter (not on the YF-16) is provided in the F-16 for engine self-start capability. Hamilton Standard turbine compressor, and Sundstrand accessory drive gearbox. Simmonds fuel measuring system.

ELECTRONICS AND EQUIPMENT (YF-16): The prototypes carried minimal electronics to restrict weight and costs, but ample space exists for installation of additional equipment in production aircraft, as detailed below. It was planned to utilise as much off-the-shelf equipment as possible in the prototypes. Thus, the horizontal tail and flaperon actuators, and electro-mechanical servos in the control system, are modified versions of units used in the F-111. The nose-mounted air data probe, feeding an air data converter and a central air data computer, is similar to that of the Lockheed SR-71. The stick-grip, embodying control force transducers, is a modified version of that used in the Vought A-7; the cockpit air-conditioning system also is similar to that used in the A-7. Other equipment fitted specifically in the YF-16s includes a General Electric SSR-1 nose-mounted pulse radar ranging system; Delco Carousel V (modified) inertial navigation system; and Marconi-Elliott digital head-up display and fire control sight system in the second YF-16.

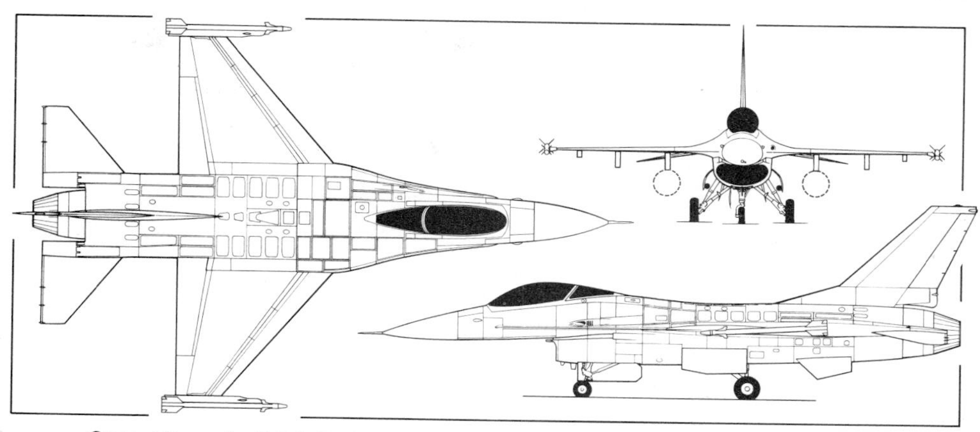
General Dynamics F-16A (Pratt & Whitney F100-PW-100(3) afterburning turbofan engine)

General Dynamics YF-16 single-seat air combat fighter, second prototype

ELECTRONICS AND EQUIPMENT (F-16): Westinghouse pulse-Doppler range and angle track radar, with planar array in nose. The radar has a lookdown range, in ground clutter, of 15-20 nm (28-37 km; 17-23 miles), and a lookup range of 20-25 nm (37-46 km; 23-29 miles). Forward electronics bay, immediately forward of cockpit, contains radar, air data equipment, inertial navigation system and flight control computer; rear electronics bay contains ILS, Tacan and IFF. A Dalmo Victor ALR-46 radar warning system, with Transco threat warning and beacon antennae, is installed. Communications equipment includes Magnavox ARC-164 UHF transceiver; Sylvania VHF; Andrea AN/AIC-18 intercom; provisions for a Magnavox KY-58 secure voice system and National Security Agency KIT-1A/TSEC cryptographic equipment; and Novatronics interference blanker. Sperry Flight Systems central air data computer. Singer-Kearfott modified SKN-2400 inertial navigation system; ILS; Collins ARN-118 Tacan; Teledyne Electronics APX-101 air-to-ground IFF transponder, with Hazeltine IFF control; Lear Siegler stick force sensors; Marconi-Elliott electronic head-up display set; a government-furnished horizontal situation indicator; Teledyne Avionics angle of attack transmitter; Gull Airborne angle of attack indicator; Clifton Precision attitude director indicator; Delco fire control computer; Photo-Sonics gun camera; Kaiser radar electro-optical display. Landing/taxying light on inside of each main-wheel door.

ARMAMENT: Armament was specified only for the second YF-16, which was fitted with a General Electric M61A-1 20 mm multi-barrel cannon in the port-side wing/body fairing, and was equipped with a 'snapshoot' gunsight (part of the head-up display system) and 500 rounds of ammunition. There is a mounting for an infra-red air-to-air missile at each wingtip, one underfuselage hardpoint and four underwing hardpoints for the carriage of additional stores. The M-61A1 gun installation, with General Electric ammunition handling system, is retained in the F-16, as are the two wingtip missile stations and the underfuselage station, but the number of underwing hardpoints is increased to six, making nine weapon stations in all. The underfuselage station is stressed for a load of up to 1,000 kg (2,200 lb), the two inboard underwing stations for 1,587 kg (3,500 lb) each, the two centre underwing stations for 1,134 kg (2,500 lb) each, all at 5·5*g*; the two outboard underwing stations and the two wingtip stations 113 kg (250 lb) each, all at 9·0*g*. Total possible external weapon load, with reduced internal fuel, is 6,894 kg (15,200 lb), and a load of more than 4,990 kg (11,000 lb) can be carried with full internal fuel. Typical stores loads can include two wingtip-mounted AIM-9J/L Sidewinders, with up to four more on the outer underwing stations; Sargent-Fletcher 1,400 litre (370 US gallon; 308 Imp gallon) drop-tanks on the inboard underwing stations; a 568 litre (150 US gallon; 125 Imp gallon) drop-tank or a 2,200 lb bomb on the underfuselage station; a Martin Marietta Pave Penny laser tracker pod along the starboard side of the nacelle; and single or cluster bombs, air-to-surface missiles, or flare pods, on the four inner underwing stations. Stores can be launched from Aircraft Hydro-Forming MAU-12C/A bomb ejector racks, Hughes LAU-88 launchers, or Orgen triple or multiple ejector racks. Westinghouse AN/ALQ-119 ECM (jammer) pods and pod control system have been listed among probable equipment, and can be carried on the centreline and two underwing stations. Other low-cost ECM systems are being studied by Sanders and ITT/Itek. Tracor ALE-40 internal pyrotechnic/chaff dispensers have also been specified. Weapon delivery capabilities include air-to-air combat with gun and Sidewinder missiles, and air-to-ground attack with gun, rockets, conventional bombs, special weapons, laser-guided and electro-optical weapons. Growth provisions are provided for radar-guided Sparrow air-to-air missiles.

DIMENSIONS, EXTERNAL:
- Wing span over missile launchers 9·45 m (31 ft 0 in)
- Wing span over missiles 10·01 m (32 ft 10 in)
- Wing aspect ratio 3·0
- Length overall, excl probe:
  - YF-16 14·175 m (46 ft 6 in)
  - F-16A/B 14·52 m (47 ft 7·7 in)
- Height overall:
  - YF-16 4·95 m (16 ft 3 in)
  - F-16A/B 5·01 m (16 ft 5·2 in)
- Tailplane span 5·495 m (18 ft 0·34 in)
- Wheel track 2·36 m (7 ft 9 in)
- Wheelbase 4·00 m (13 ft 1·52 in)

AREAS:
- Wings, gross:
  - YF-16 26·01 m² (280·0 sq ft)
  - F-16A/B 27·87 m² (300·0 sq ft)

WEIGHTS AND LOADINGS (F-16A):
- Operational weight empty approx 6,377 kg (14,060 lb)
- Max external load 6,894 kg (15,200 lb)
- Structural design gross weight (7·33*g*) with full internal fuel 10,205 kg (22,500 lb)
- Maximum symmetric design load factor with full internal fuel at 10,205 kg (22,500 lb) gross weight 9·0
- Max T-O weight:
  - YF-16, max weight at which flown 12,247 kg (27,000 lb)

F-16A with max external load 14,968 kg (33,000 lb)
Wing loading:
at 10,070 kg (22,200 lb) AUW 361 kg/m² (74 lb/sq ft)
at 14,968 kg (33,000 lb) AUW 537 kg/m² (110 lb/sq ft)
Thrust/weight ratio ('clean') 1·1 to 1

PERFORMANCE (YF-16, as assessed in NATO Steering Committee report, March 1975):
T-O weight, 'clean', with 2 Sidewinders 9,797 kg (21,600 lb)
External load with max internal fuel 5,216 kg (11,500 lb)
Thrust/weight ratio at 9,797 kg (21,600 lb) 1·1 to 1
Max level speed at 11,000 m (36,000 ft) with 2 Sidewinders Mach 1·95
Max rate of climb in 5g turn at low level at Mach 0·7, with Mk 82 bombs 12,802 m (42,000 ft)/min
Sustained turn rate at 6,100 m (20,000 ft), with max internal fuel and 2 Sidewinders 10·7°/sec
Sustained air turning radius at low level at Mach 0·7, with 6 Mk 82 bombs 1,372 m (4,500 ft)
T-O run with 1,814 kg (4,000 lb) external load 533 m (1,750 ft)
Landing run with 1,814 kg (4,000 lb) external load 808 m (2,650 ft)
Radius of action with 6 Mk 82 bombs 295 nm (547 km; 340 miles)

PERFORMANCE (F-16A, estimated):
Max level speed at 12,200 m (40,000 ft) above Mach 2·0
Service ceiling more than 15,240 m (50,000 ft)
Radius of action more than 500 nm (925 km; 575 miles)
Ferry range, with drop-tanks more than 2,000 nm (3,705 km; 2,303 miles)

GREAT LAKES

GREAT LAKES AIRCRAFT COMPANY

HEAD OFFICE:
PO Box 11132, Wichita, Kansas 67202
Telephone: (316) 265-0786

EUROPEAN OFFICE:
A. A. Williams, 89 Augsburg 21, Postfach 211227, Federal Republic of Germany

PRESIDENT:
Douglas L. Champlin

GENERAL MANAGER:
Brad Shelman

On 23 February 1972, Mr Douglas Champlin announced the purchase of Great Lakes Aircraft Company from Mr Harvey R. Swack.

Great Lakes Aircraft Company is building two versions of the well-known Sport Trainer at Wichita, Kansas, and in 1975 production was at a rate of two aircraft per month. Certificated components and assemblies are available to amateur constructors, but plans are no longer available.

GREAT LAKES SPORT TRAINER MODEL 2T-1A-1

The Great Lakes Sport Trainer was produced with Cirrus and Menasco engines by the original Great Lakes Company, founded on 2 January 1929.

Certification of the Model 2T-1A-1 with 104 kW (140 hp) Lycoming engine was obtained in May 1973, and delivery of production aircraft began in October 1973.

TYPE: Two-seat sporting biplane.

WINGS: Braced biplane, with N-type interplane struts, wire bracing and N-type centre-section support struts. Dual streamline-section landing and flying wires. Wing section M-12. No dihedral on upper wing, 2° on lower wings. Sweepback on upper wing 9° 13′. Composite structure, with Douglas fir spars, metal ribs and overall fabric covering. Ailerons on lower wings only. No flaps or tabs.

FUSELAGE: Welded chrome-molybdenum steel tube Warren girder structure, with fabric covering.

TAIL UNIT: Wire-braced welded chrome-molybdenum steel tube structure, fabric-covered. Tailplane incidence manually adjustable. No controllable tabs.

LANDING GEAR: Non-retractable type, with steerable Scott tailwheel. Divided main legs with spring-oleo shock-absorbers standard. Main wheels size 6·00-6 with hydraulic disc brakes. Parking brake. 7·00-6 tyres optional. Wheel fairings optional.

POWER PLANT: One 104 kW (140 hp) Lycoming O-320-E2A flat-four engine, driving a McCauley two-blade fixed-pitch propeller type 1C160/EGM 7654. Aluminium fuel tank in centre-section of upper wing, capacity 98·5 litres (26 US gallons). Refuelling point in upper wing surface. Propeller spinner and inverted fuel and oil systems optional.

ACCOMMODATION: Two seats in tandem in open cockpits. Dual controls standard. Compass, airspeed indicator, altimeter and engine speed indicator standard in rear cockpit, optional for front cockpit. Cockpit heating optional. Baggage compartment aft of rear cockpit, capacity 13·6 kg (30 lb), optional.

SYSTEMS: Engine-driven generator for electrical supply to navigation lights, two rotating beacons and rear cockpit instrument lights standard. Hydraulic system for brakes only.

ELECTRONICS AND EQUIPMENT: Emergency locator transmitter standard. Edo-Aire 551 transceiver, intercom, and Edo-Aire RT-553 with or without auto omni optional. Standard equipment includes map case, seat belts and shoulder harness, cockpit lights, navigation lights, two omni-flash beacons, and a choice of four paint trims. Optional equipment includes turn and bank indicator, rate of climb indicator, manifold pressure gauge, cylinder head temperature gauge and accelerometer for rear cockpit, glider tow hitch, Hobbs meter, and cockpit covers.

DIMENSIONS, EXTERNAL:
Wing span 8·13 m (26 ft 8 in)
Length overall 6·20 m (20 ft 4 in)
Height overall 2·24 m (7 ft 4 in)
Wheel track 1·78 m (5 ft 10 in)

AREAS:
Wings, gross 17·43 m² (187·6 sq ft)
Ailerons (total) 1·25 m² (13·5 sq ft)
Fin 0·55 m² (5·87 sq ft)
Rudder 0·63 m² (6·81 sq ft)
Tailplane 1·43 m² (15·44 sq ft)
Elevators 0·99 m² (10·68 sq ft)

WEIGHTS AND LOADINGS:
Weight empty 517 kg (1,140 lb)
Max T-O weight 793 kg (1,750 lb)
Max wing loading 45·5 kg/m² (9·32 lb/sq ft)
Max power loading 7·60 kg/kW (12·5 lb/hp)

PERFORMANCE (at max T-O weight):
Max level speed at S/L 104 knots (193 km/h; 120 mph)
Max cruising speed 96 knots (177 km/h; 110 mph)
Stalling speed 47 knots (87 km/h; 54 mph)
Max rate of climb at S/L 305 m (1,000 ft)/min
Service ceiling 3,780 m (12,400 ft)
T-O run 175 m (575 ft)
T-O to, and landing from, 15 m (50 ft) 366 m (1,200 ft)
Landing run 122 m (400 ft)
Range with max fuel 260 nm (482 km; 300 miles)

GREAT LAKES SPORT TRAINER MODEL 2T-1A-2

An improved version of the Sport Trainer, the Model 2T-1A-2 has a 134 kW (180 hp) engine, with constant-speed propeller and inverted fuel and oil systems as standard. Ailerons are fitted to both upper and lower wings and manifold pressure and fuel flow gauges are standard. Delivery of this version began in July 1974.

The description of the Model 2T-1A-1 applies generally also to the Model 2T-1A-2, except as follows:

WINGS: As for Model 2T-1A-1, except ailerons on both upper and lower wings.

FUSELAGE, TAIL UNIT AND LANDING GEAR: As for Model 2T-1A-1.

POWER PLANT: One 134 kW (180 hp) Lycoming IO-360-B1F6 flat-four engine, driving a Hartzell two-blade metal constant-speed propeller type HC-C2YK-4F/FC7666A-2 with spinner. Fully inverted fuel system standard, capacity as for Model 2T-1A-1, plus 5·3 litre (1·4 US gallon) inverted header tank. Fully inverted oil system standard, capacity as for Model 2T-1A-1.

ACCOMMODATION: As for Model 2T-1A-1, except optional baggage compartment capacity 18 kg (40 lb).

SYSTEMS, ELECTRONICS AND EQUIPMENT: As for Model 2T-1A-1.

DIMENSIONS, EXTERNAL: As for Model 2T-1A-1

AREAS: As for Model 2T-1A-1, except:
Ailerons (total) 2·70 m² (29·10 sq ft)

WEIGHTS AND LOADINGS:
Weight empty 558 kg (1,230 lb)
Max T-O weight 816 kg (1,800 lb)
Max wing loading 47·0 kg/m² (9·63 lb/sq ft)
Max power loading 6·09 kg/kW (10·0 lb/hp)

PERFORMANCE (at max T-O weight):
Max level speed at S/L 115 knots (212 km/h; 132 mph)
Max cruising speed 102 knots (190 km/h; 118 mph)
Stalling speed 50 knots (92 km/h; 57 mph)
Max rate of climb at S/L 427 m (1,400 ft)/min
Service ceiling 5,180 m (17,000 ft)
T-O run 252 m (825 ft)
T-O to 15 m (50 ft) 335 m (1,100 ft)
Landing from 15 m (50 ft) 366 m (1,200 ft)
Landing run 259 m (850 ft)
Range with max fuel, no reserves 260 nm (482 km; 300 miles)

Great Lakes Sport Trainer Model 2T-1A-2 (Lycoming IO-360-B1F6 engine)

GRUMMAN

GRUMMAN CORPORATION

HEAD OFFICE:
Bethpage, New York 11714
Telephone: (516) 575-0574
CHAIRMAN OF THE BOARD AND CHIEF EXECUTIVE OFFICER:
John C. Bierwirth
PRESIDENT AND CHIEF OPERATING OFFICER:
Joseph G. Gavin Jr
VICE-PRESIDENTS:
Edward Balinsky (Investment Management)
John F. Carr (Administration)
Patrick L. Cherry (Consultant)
Robert G. Freese (Treasurer)
Weyman B. Jones (Public Affairs)
Lawrence M. Pierce (General Counsel)
John B. Rettaliata
Peter E. Viemeister (Development)
SECRETARY: Robert W. Bradshaw
CONTROLLER: Nat P. Busi
DIRECTOR OF PUBLIC RELATIONS:
Edward V. Brookfield

GRUMMAN AEROSPACE CORPORATION

See below

GRUMMAN ALLIED INDUSTRIES INC

HEAD OFFICE AND WORKS:
600 Old Country Road, Garden City, New York 11530
Telephone: (516) 741-3500
CHAIRMAN OF THE BOARD AND CHIEF EXECUTIVE OFFICER:
Robert F. Loar
PRESIDENT AND CHIEF OPERATING OFFICER:
Robert W. Somerville

GRUMMAN AMERICAN AVIATION CORPORATION

See pages 296-299.

GRUMMAN DATA SYSTEMS CORPORATION

HEAD OFFICE AND WORKS:
Bethpage, New York 11714
Telephone: (516) 575-0574
CHAIRMAN OF THE BOARD:
John C. Bierwirth
PRESIDENT:
Robert A. Nafis

GRUMMAN ECOSYSTEMS CORPORATION

HEAD OFFICE AND WORKS:
Bethpage, New York 11714
Telephone: (516) 575-0574
CHAIRMAN OF THE BOARD:
John C. Bierwirth
PRESIDENT:
Clifford F. Jessberger

GRUMMAN INTERNATIONAL COMPANY

HEAD OFFICE:
Bethpage, New York 11714
Telephone: (516) 575-1101
CHAIRMAN:
Robert L. Townsend
PRESIDENT:
Peter B. Oram
SECRETARY: Pierre A. Frye

The Grumman Aircraft Engineering Corporation was incorporated on 6 December 1929. Important changes in the corporate structure of the company were announced in 1969, resulting in the formation of Grumman Corporation, a small holding company, with Grumman Aerospace Corporation, Grumman Allied Industries Inc and Grumman Data Systems Corporation. A new organisation known as Grumman Ecosystems Corporation was brought into operation in January 1971, and in the Autumn of 1972 a merger was planned with American Aviation Corporation of Cleveland, Ohio. This became effective on 2 January 1973 when Grumman American Aviation Corporation was announced as a new subsidiary of Grumman Corporation.

GRUMMAN AEROSPACE CORPORATION

HEAD OFFICE AND WORKS:
Bethpage, New York 11714
Telephone: (516) 575-0574
CHAIRMAN OF THE BOARD AND PRESIDENT:
George M. Skurla
EXECUTIVE VICE-PRESIDENT (Operations):
Ralph H. Tripp
SENIOR VICE-PRESIDENTS:
Edward Dalva (Long Island Production Operations)
Ira G. Hedrick (Presidential Assistant for all Corporate Technology)
Lawrence M. Mead Jr (Departmental Operations)
Ross S. Mickey (Corporate Programme Management)
John O'Brien (Administration and Resources)
Michael Pelehach (Business Development)
VICE-PRESIDENTS:
A. D. Alexandrovich (Product and Technology Development)
John M. Buxton (President, Grumman Houston Corporation)
Thomas A. Guarino (E-2C Programme)
Thomas J. Kane Jr (Customer Requirements)
Thomas J. Kelly (Space Programmes)
Robert C.Miller (Electronic Warfare Programmes)
Gordon H. Ochenrider (Washington Operations)
Carl A. Paladino (Treasurer)
G.Thomas Rozzi (Security and Personnel Services)
Philip S. Vassallo (Corporate Procurement Operations)
William M. Zarkowsky (Milledgeville Operations)
A. James Zusi (Stuart Operations)
SECRETARY AND GENERAL COUNSEL:
Raphael Mur

Current products of this subsidiary of Grumman Corporation include versions of the A-6 Intruder, EA-6B Prowler, E-2C Hawkeye and F-14 Tomcat for the US Navy.

It was announced in April 1974 that Grumman was collaborating with Rhein-Flugzeugbau GmbH of West Germany (which see), a subsidiary of VFW-Fokker GmbH, in the design and development of a Wankel-powered ducted-fan light aircraft named the Fanliner. All available details of this are given under RFB in the German aircraft section.

In early 1975 it was announced that Grumman had been awarded a contract to develop a tactical jamming version of the General Dynamics F-111, under the designation EF-111A. It will support the tactical strike forces by providing high-power jamming of acquisition, early-warning or GCI radars.

At the beginning of January 1976 Grumman Aerospace Corporation had a total of approximately 25,000 employees.

GRUMMAN HAWKEYE

US Navy designation: E-2

The E-2 Hawkeye was evolved as a carrier-borne early-warning aircraft, but is suitable also for land-based operations from unimproved fields. The prototype flew for the first time on 21 October 1960, since when the following versions have been built:

E-2A (formerly W2F-1). Initial production version, the first of which, equipped with full early-warning and command electronics system, flew on 19 April 1961. Delivery to the US Navy began officially on 19 January 1964, when the first Hawkeye was accepted at San Diego for training of air and ground crews of airborne early warning squadron VAW-11. This unit became operational on USS *Kitty Hawk* in 1965. Second Hawkeye unit was VAW-12. Total of 62 built; delivery of 59 production aircraft completed in Spring 1967. All those in service updated to E-2B by end of 1971.

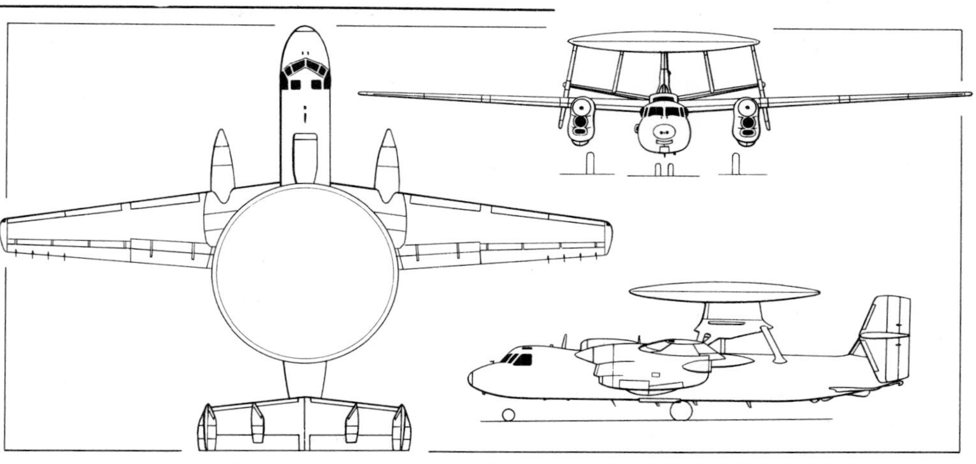

Grumman E-2C Hawkeye twin-turboprop airborne early-warning aircraft *(Pilot Press)*

E-2B. The prototype of this version flew for the first time on 20 February 1969. It differs from the E-2A by having a Litton Industries L-304 microelectronic general-purpose computer and several reliability improvements. A retrofit programme, completed in December 1971, updated all operational E-2As to E-2B standard. In service with RVAW-110, VAW-112, VAW-113, VAW-114, VAW-115, VAW-116 and VAW-117 in 1975.

E-2C. First of two E-2C prototypes flew on 20 January 1971. Production began in mid-1971 and the first flight of a production aircraft was made on 23 September 1972; 29 had been delivered by the end of 1975. Firm orders exist for a total of 47 aircraft, with procurement of 18 more planned by the end of 1984. Sales to other countries are anticipated, and an agreement for the supply of four aircraft to Israel was being finalised in mid-March 1976.

The E-2C has an advanced Grumman/General Electric-developed radar that is capable of detecting airborne targets in a land-clutter environment. Improvements for increased reliability and easier maintenance have been provided. First entered service, with airborne early-warning squadron VAW-123 at NAS Norfolk, Va, in November 1973, and went to sea on board the USS *Saratoga* in late 1974 and on the USS *John F. Kennedy* in mid-1975. In service with RVAW-120, VAW-121, VAW-122, VAW-123, VAW-124, VAW-125 and VAW-126 in 1975. Training version is **TE-2C.**

Teams of Hawkeyes are able to maintain patrols on naval task force defence perimeters in all weathers, and are capable of detecting and assessing any threat from approaching high-Mach-number enemy aircraft early enough to ensure successful interception. To make this possible highly sophisticated equipment is caried by the aircraft, including a Randtron Systems AN/APA-171 antenna system housed in a 7·32 m (24 ft) diameter saucer-shaped rotodome mounted above the rear fuselage of the aircraft. The rotodome revolves in flight at 6 rpm, and can be lowered 0·64 m (1 ft 10¼ in) to facilitate aircraft stowage on board ship. The Yagi type radar arrays within the rotodome are interfaced to the on-board electronic systems, providing radar sum and difference signals plus IFF.

Major detection capability stems from the General Electric AN/APS-120 radar and OL-93/AP radar detector processor (RDP) . The radar is able to spot distant airborne targets despite heavy sea or land echo 'clutter', as well as surface targets. It is linked to the tracking and intercept computer via the RDP, which carries out automatic detection and signals target reports which the computer needs for automatic tracking.

To provide the Combat Information Center (CIC) staff with the essential man/machine interface, the Hazeltine Corporation's AN/APA-172 control indicator group consists of three identical display stations, each with a 0·25 m (10 in) main and a 0·13 m (5 in) auxiliary display. The main display shows target track information, while the auxiliary provides alpha-numeric information with random-write capability. Station controls allow each of the three CIC operators to select independently specific information for their displays, so that each may have the same or a different perspective on any tactical situation. Other Hazeltine equipment includes an OL-76/AP IFF detector processor, providing automatic Mk X SIF processing capability in a single integrated system. Signals generated by the OL-76/AP enable the CIC operators to obtain instant range, azimuth and altitude positions of a friendly target. In order to identify that target as friend or foe, an RT-988/A IFF interrogator 'challenges' and identifies the aircraft, feeding its information direct to the OL-76/AP for processing.

Accurate navigation is critical for an aircraft which, after hours on patrol, needs to find without delay its mobile carrier base. Such a requirement is catered for by Litton Industries' AN/ASN-92 (LN-15C) carrier aircraft inertial navigation system (CAINS), an important feature being its capability of rapid alignment and orientation following take-off from a rolling and pitching carrier deck. Litton's Amecom division's AN/ALR-59 passive detection system provides early-warning capability. Able to capture short-duration signals in real time, its four-band simultaneous coverage ensures highly-accurate direction finding, even in an environment cluttered with enemy signals.

Linking all this advanced equipment is Litton Data Systems division's L-304 computer, which processes radar, Link 4 and Link 11 communications, navigation and passive detection data in real time. It comprises two L-304 processors, eight 8,192-word memory units (expandable to ten), power supplies, a recorder producer, power converter, system test module, a 4,096 word refresh memory for the displays, input/output buffers for each function, plus display, radar, navigation, communications and passive detection converter modules.

In addition to to the L-304 computer, the E-2C has also a Conrac Corporation CP-1085/AS air data computer (ADC). Combining solid-state pressure transducers with a special preprogrammed digital computer, it provides outputs of altitude, altitude hold, indicated airspeed, true

Grumman TE-2C Hawkeye airborne early-warning aircraft of US Navy Training Squadron 120 (RVAW-120), NAS North Island, San Diego

airspeed and Mach number in analogue and digital format to interface with the navigation, flight control and display subsystems.

An Advanced Radar Processing System (ARPS), designated AN/APS-125, has been developed for the E-2C and has demonstrated detection capability over land and water, even under conditions of intentional jamming. The first production aircraft to include ARPS will be No. 34, scheduled for delivery in late 1976, but earlier E-2Cs will be retrofitted to ARPS standard later.

The following details apply to the E-2C Hawkeye:

TYPE: Airborne early-warning aircraft.

WINGS: Cantilever high-wing monoplane of all-metal construction. Centre-section is a structural box consisting of three beams, ribs and machined skins. Hinged leading-edge is non-structural and provides access to flying and engine controls. The outer panels fold rearward about skewed-axis hinge fittings mounted on the rear beams, to stow parallel with the rear fuselage on each side. Folding is done through a double-acting hydraulic cylinder. Trailing-edges of outer panels and part of centre-section consist of long-span ailerons and hydraulically-actuated Fowler flaps. When flaps are lowered, ailerons are drooped automatically. All control surfaces of E-2C are power-operated and incorporate devices to produce artificial feel forces. Automatic flight control system (AFCS) can be assigned sole control of the system hydraulic actuators, or AFCS signals can be superimposed on the pilot's mechanical inputs for stability augmentation. Pneumatically-inflated rubber de-icing boots on leading-edges.

FUSELAGE: Conventional all-metal semi-monocoque structure.

TAIL UNIT: Cantilever structure, with four fins and three double-hinged rudders. Tailplane dihedral 11°. Portions of tail unit made of glassfibre to reduce radar reflection. Power control and artificial feel systems as for ailerons. Pneumatically-inflated rubber de-icing boots on all leading-edges.

LANDING GEAR: Hydraulically-retractable tricycle type. Pneumatic emergency extension. Steerable nosewheel unit retracts rearward. Main wheels retract forward, and rotate to lie flat in bottom of nacelles. Twin wheels on nose unit only. Oleo-pneumatic shock-absorbers. Main wheel tyres size 36 × 11 Type VII 24-ply, pressure 17·93 bars (260 lb/sq in) on ship, 14·48 bars (210 lb/sq in) ashore. Hydraulic brakes. Hydraulically-operated retractable tailskid. A-frame arrester hook under tail.

POWER PLANT: Two 3,661 kW (4,591 shp; 4,910 ehp) Allison T56-A-422/-425 turboprop engines, fitted originally with Aeroproducts A 6441 FN-248 four-blade fully-feathering reversible-pitch constant-speed propellers. All E-2 aircraft in service are being refitted with Hamilton Standard propellers with foam-filled blades which have a steel spar and glassfibre shell. Spinners and blades incorporate electrical anti-icers.

ACCOMMODATION: Crew of five on flight deck and in ATDS compartment in main cabin, consisting of pilot, co-pilot, combat information centre officer, air control officer and radar operator. Downward-hinged door, with built-in steps, on port side of centre-fuselage.

ELECTRONICS: AN/APA-171 rotodome and antenna, AN/APS-120 search radar, IFF interrogator type RT-988/A, OL-93/AP radar detector processor, OL-76/AP IFF detector processor. AN/APA-172 control indicator group, OL-77/ASQ computer programmer, L-304 airborne computer, ARC-158 UHF data link, ARQ-34 HF data link, ASM-440 in-flight performance monitor, ARC-51A UHF com, ARQ-34 HF com, AIC-14A intercom, AN/ASN-92 (LN-15C) CAINS carrier aircraft inertial navigation system, CP-1085/AS air data computer, APN-153 (V) Doppler, ASN-50 heading and attitude reference system, ARN-52 (V) Tacan, ARA-50 UHF ADF, ASW-25B ACLS and APN-171 (V) radar altimeter.

DIMENSIONS, EXTERNAL:
- Wing span: 24·56 m (80 ft 7 in)
- Length overall: 17·55 m (57 ft 7 in)
- Height overall: 5·59 m (18 ft 4 in)
- Diameter of rotodome: 7·32 m (24 ft 0 in)
- Propeller diameter: 4·11 m (13 ft 6 in)

AREA:
- Wings, gross: 65·03 m² (700 sq ft)

WEIGHTS:
- Weight empty: 17,090 kg (37,678 lb)
- Max fuel (internal): 5,624 kg (12,400 lb)
- Max T-O weight: 23,391 kg (51,569 lb)

PERFORMANCE (at max T-O weight):
- Max level speed: 325 knots (602 km/h; 374 mph)
- Cruising speed: 270 knots (500 km/h; 311 mph)
- Stalling speed (landing configuration): 74 knots (137·5 km/h; 85·5 mph)
- Service ceiling: 9,390 m (30,800 ft)
- T-O run: 576 m (1,890 ft)
- T-O to 15 m (50 ft): 768 m (2,520 ft)
- Ferry range: 1,394 nm (2,583 km; 1,605 miles)

GRUMMAN INTRUDER

US Navy designations: A-6, EA-6 and KA-6

The basic A-6A (originally A2F-1) Intruder was conceived as a carrier-borne low-level attack bomber equipped specifically to deliver nuclear or conventional weapons on targets completely obscured by weather or darkness. Performance is subsonic, but the Intruder possesses outstanding range and endurance and carries a heavier and more varied load of stores than any previous US naval attack aircraft.

Competition for the original A-6 contract was conducted from May to December 1957, among eight aircraft companies. Of the 11 designs submitted, Grumman's was adjudged the best on 31 December 1957. The A-6 was developed subsequently under the first 'cost plus incentive fee' contract placed by the US Navy. Seven variants of the basic design have been announced to date, as follows:

A-6A Intruder. Initial carrier-based attack bomber, described in 1972-73 *Jane's*. The first A-6A flew on 19 April 1960, and this version entered service officially on 1 February 1963, when the first aircraft was accepted for the US Navy's VA-42 squadron at NAS Oceana. A total of 482 were built, the last delivery taking place in December 1969. A-6As saw considerable service with the Navy and Marine Corps in Vietnam. Most other versions of the Intruder are modifications of A-6As.

Grumman has US Navy contracts covering the modernisation of A-6As by fitting A-6E advanced weapon systems. It is anticipated that the programme will continue until all remaining A-6As in the inventory have been modified to A-6E standard.

EA-6A. First flown in prototype form in 1963, this version retains partial strike capability, but is equipped primarily to support strike aircraft and ground forces by suppressing enemy electronic activity and obtaining tactical electronic intelligence within a combat area. Elements of the A-6A's bombing/navigation system are deleted and the EA-6A carries more than 30 different antennae to detect, locate, classify, record and jam enemy radiation. Externally-evident features include a radome at the top of the tail-fin, and attachment points under the wings and fuselage for ECM pods, fuel tanks and/or weapons. A total of 27 EA-6As were built for the US Marine Corps, including six A-6As modified into EA-6As.

EA-6B Prowler. Advanced electronics development of the EA-6A, described separately.

A-6B Intruder. Conversion of the A-6A to provide Standard ARM missile capability. Though primarily an electronics modification, it has three different configurations ranging from limited to full strike capability. A total of 19 A-6As were modified to A-6B configuration.

A-6C Intruder. Derived from the A-6A but differing externally by having an underfuselage turret housing forward-looking infra-red (FLIR) sensors and low-light-level television camera, providing additional night attack capability. This equipment is intended to permit detailed identification and acquisition of targets not discernible by the aircraft's radar. A total of 12 A-6As were modified to A-6C configuration.

KA-6D Intruder. An A-6A was modified into a prototype flight refuelling tanker, with hose and reel in the rear fuselage, and flew for the first time on 23 May 1966. The KA-6D production model is fitted with Tacan and can

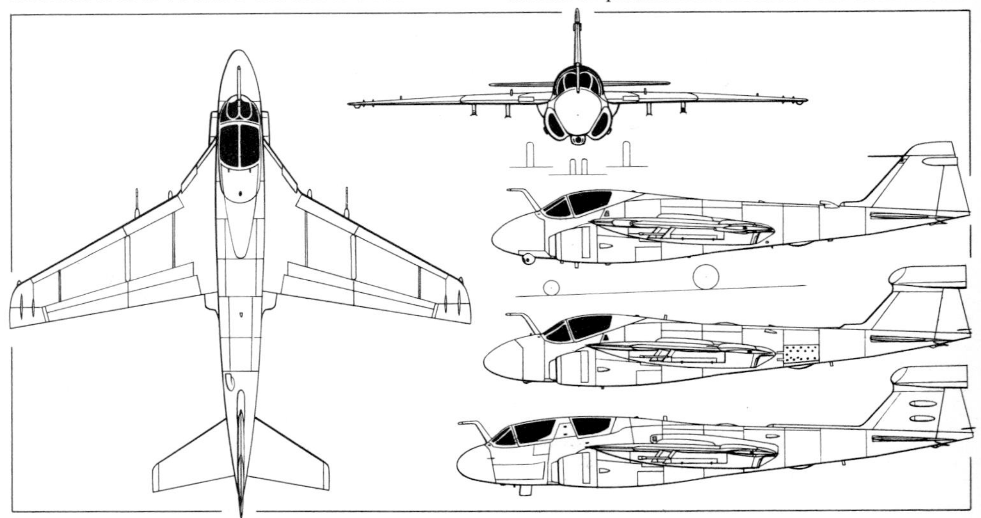

Grumman A-6E/TRAM, with additional side views of EA-6A (centre) and EA-6B (bottom) *(Pilot Press)*

transfer more than 9,500 kg (21,000 lb) of fuel immediately after take-off or 6,800 kg (15,000 lb) at a distance of 250 nm (463 km; 288 miles) from its carrier base. In addition, the KA-6D could act as a control aircraft for air-sea rescue operations or as a day bomber. A total of 62 A-6As were modified to KA-6D configuration.

A-6E Intruder. An advanced conversion of the A-6A with multi-mode radar and an IBM computer similar to that first tested in the EA-6B. First flight of an A-6E was made on 27 February 1970. First squadron deployment was made in September 1972, and the A-6E was approved officially for service use in December 1972. By the end of that month 24 A-6Es had been delivered to the US Navy, and a total procurement of 318 of these aircraft is planned. A total of 119 converted from A-6A, and 58 built as A-6E, delivered by 1 January 1976. Described separately.

GRUMMAN A-6E INTRUDER

Development of the A-6E began with the substitution of a single simultaneous multi-mode navigation and attack radar, developed by the Norden Division of United Aircraft Corporation, for the two earlier radar systems in the A-6A. Following the concepts of the EA-6B, the IBM Corporation and Fairchild Camera and Instrument Corporation have supplied a new attack and navigation computer system and an interfacing data converter. Conrac Corporation has designed an armament control unit and RCA has developed a video tape recorder for post-strike assessment of attacks.

The Norden Division's AN/APQ-148 multi-mode radar provides simultaneous ground mapping; identification, tracking, and rangefinding of fixed or moving targets; and terrain-clearance or terrain-following manoeuvres. It can also detect, locate and track radar beacons used by forward air controllers when providing close support for ground forces. The APQ-148 has mechanical scanning in azimuth and utilises a newly-developed electronics system for simultaneous vertical scanning. There are two cockpit displays, one for the pilot and one for the bombardier/navigator, and terrain data is also presented on a vertical display indicator ahead of the pilot. Built-in test equipment (BITE), that provides automatic fault location, is an integral part of the system, permitting repairs to be carried out at squadron level.

IBM's AN/ASQ-133 solid-state digital computer is coupled to the A-6E's radar, inertial and Doppler navigational equipment, communications and automatic flight control system. This computer receives its inputs from programmes written specifically for the A-6E and stored in the computer's memory circuits. As mission data is measured in flight by on-board aerodynamic and electronic sensors, the computer compares the data with the programmed information, computes any differences, and provides corrective data that can be used to alter the parameters of the mission. Pedestal control and digital display units form part of the complete computer subsystem. The digital display unit shows the basic parameters needed for operation of the system. The pedestal control, mounted at the navigator's position, has a slew stick which is used to place radar display cursors on the selected target for tracking purposes. Input keys mounted on top of the pedestal can be used to update or add to mission information to be factored into the flight path by the computer subsystem. The entire subsystem also features BITE circuitry to simplify maintenance at squadron level.

Fairchild Camera and Instrument Corporation's signal data converter for the A-6E accepts analogue input data from up to sixty sensors, and converts that information to a digital output that is fed into the computer of the navigation and attack system. A necessary interface between the analogue sensors and displays and the digital computer, it is in effect an interpretation system able to communicate in both analogue and digital language, and can translate from one to the other almost instantaneously. Updating (the processing of complete successive sets of data for comparison with previous information) is accomplished eight times a second. Like the other electronic equipment of the A-6E, the ASQ-133 converter has BITE circuitry.

Conrac Corporation's armament control unit (ACU) for the A-6E provides in a single unit all the inputs and outputs necessary to select and release the Intruder's weapons. It receives and sends appropriate signals to arm, control and release weapons individually or simultaneously in any of the standard attack patterns. The master arming switch has a 'practice' position that allows the ACU to be cycled up to the point of firing command. Safety circuits in the unit prevent accidental firing during the practice mode.

The multi-mode AN/AVA-1 display developed by Kaiser Aerospace and Electronics Corporation serves as a primary flight aid for navigation, approach, landing and weapons delivery. The basic vertical display indicator (VDI) is a 0·20 m (8 in) cathode-ray tube which shows a synthetic landscape, sky, and electronically-generated command flight path that move to simulate the motion of these features as they would be seen by the pilot through the windscreen of the aircraft. Symbols are superimposed to augment the basic attitude data, and for attack a second set of superimposed information provides a target symbol, steering symbol, and release and pull-up markers. A solid-state radar data scan converter can provide on the same display an apparent real-world perspective of terrain, ten shades of grey defining terrain elevation at ten different segmented contour intervals up to 8·7 nm (16 km; 10 miles) ahead of the aircraft. This makes it possible for the pilot to fly the Intruder in either a terrain-following or terrain-avoidance mode at low altitude. Flight path and attack symbols can be superimposed over the terrain elevation data on the VDI, enabling the pilot to make his attack while avoiding or following terrain in the target area. Kaiser has also developed a micromesh filter to prevent 'washout' of the data displayed on the VDI in sunlight conditions. Naval pilots currently use the VDI as a primary flight instrument, for precise steering in navigation, and for weapons cues, progress, and status information during an attack. For carrier landing the unit is used as a flight director and, linked to the APQ-148 radar, it presents steering information, allowing the pilot to select a descent angle for the final approach.

Grumman KA-6D Intruder tanker of US Navy Squadron VA-145, based on the USS *Ranger (Peter M. Bowers)*

Grumman A-6E/TRAM (target recognition attack multisensor) version of the A-6E Intruder

An **A-6E/TRAM** (target recognition attack multisensor) version of the A-6E flew for the first time on 22 March 1974 without sensors, and on 22 October 1974 with sensors. The initial flight test programme was carried out by converting this single aircraft, loaned by the Navy, but three other aircraft were converted and began flight testing in 1975. The conversion adds a turreted electro-optical sensor package, containing both infra-red and laser equipment, to a full-system Intruder, updates the inertial navigation system with CAINS, provides a new CNI system, automatic carrier landing capability and provisions for the Condor missile. The sensor package is integrated with the multi-mode radar, providing the capability of detecting, identifying and attacking a wide range of targets under adverse weather conditions, and with an improved degree of accuracy. It makes possible the viewing of terrain, in addition to conventional radar targets. A high-resolution image is presented on a FLIR (forward-looking infra-red) display by a newly-developed detecting-ranging set (DRS) designed and built by Hughes Aircraft Corporation, allowing the delivery of both conventional and laser-guided weapons. Also of importance is a capability to acquire and attack targets designated by a forward air controller on the ground. A production decision was to be taken by the US Navy in Summer 1976.

The following description applies to the standard A-6E:

TYPE: Two-seat carrier-based attack bomber.

WINGS: Cantilever mid-wing monoplane, with 25° sweepback at quarter-chord. All-metal structure. Hydraulically-operated almost-full-span leading-edge and trailing-edge flaps, with inset spoilers (flaperons) of same span as flaps forward of trailing-edge flaps. Trailing-edge of each wingtip, outboard of flap, splits to form speed-brakes which project above and below wing when extended. Two short fences above each wing. Outer panels fold upward and inward.

FUSELAGE: Conventional all-metal semi-monocoque structure. Bottom is recessed between engines to carry semi-exposed store.

TAIL UNIT: Cantilever all-metal structure. All-moving tailplane, without separate elevators. Electronic antenna in rear part of fin, immediately above rudder.

LANDING GEAR: Hydraulically-retractable tricycle type. Twin-wheel nose unit retracts rearward. Single-wheel main units retract forward and inward into air intake fairings. A-frame arrester hook under rear fuselage.

POWER PLANT: Two 41·4 kN (9,300 lb st) Pratt & Whitney J52-P-8A turbojet engines. Provision for up to four external fuel tanks under wings. Removable flight refuelling probe projects upward immediately forward of windscreen.

ACCOMMODATION: Crew of two on Martin-Baker Mk GRU7 ejection seats, which can be reclined to reduce fatigue during low-level operations. Bombardier/navigator slightly behind and below pilot to starboard. Hydraulically-operated rearward-sliding canopy.

SYSTEMS: AiResearch environmental control system for cockpit and electronics bay. Electrical system powered by two AiResearch constant-speed drive starters that combine engine starting and electrical power generation, each delivering 30kVA. An AiResearch ram-air turbine, mounted so that it can be projected into the airstream above the port wing-root, provides in-flight emergency electrical power for essential equipment. Dual hydraulic systems for operation of flight controls, leading-edge and trailing-edge flaps, wingtip speed-brakes, landing gear brakes and cockpit canopy. One electrically-driven hydraulic pump provides restricted flight capability by supplying the tailplane and rudder actuators only.

ARMAMENT: Five weapon attachment points, each with a 1,633 kg (3,600 lb) capacity. Typical weapon loads are thirty 500 lb bombs in clusters of six, or three 2,000 lb general purpose bombs plus two 1,135 litre (300 US gallon) drop-tanks.

DIMENSIONS, EXTERNAL:
- Wing span: 16·15 m (53 ft 0 in)
- Wing mean aerodynamic chord: 3·32 m (10 ft 10¾ in)
- Width folded: 7·72 m (25 ft 4 in)
- Length overall: 16·64 m (54 ft 7 in)
- Height overall: 4·93 m (16 ft 2 in)
- Height folded: 4·95 m (16 ft 3 in)
- Tailplane span: 6·21 m (20 ft 4½ in)
- Wheel track: 3·32 m (10 ft 10½ in)

AREAS:
- Wings, gross: 49·1 m² (528·9 sq ft)
- Flaperons (total): 3·81 m² (41·0 sq ft)
- Trailing-edge flaps (total): 9·66 m² (104·0 sq ft)
- Leading-edge slats (total): 4·63 m² (49·8 sq ft)
- Fin: 5·85 m² (62·93 sq ft)
- Rudder: 1·52 m² (16·32 sq ft)

WEIGHTS AND LOADING:
- Weight empty: 11,625 kg (25,630 lb)
- Max payload: 7,838 kg (17,280 lb)
- Max T-O weight:
  - catapult: 26,580 kg (58,600 lb)
  - field: 27,397 kg (60,400 lb)
- Max zero-fuel weight: 20,166 kg (44,460 lb)
- Max landing weight:
  - carrier: 16,329 kg (36,000 lb)
  - field: 20,411 kg (45,000 lb)
- Max wing loading: 557·6 kg/m² (114·2 lb/sq ft)

PERFORMANCE (no stores):
- Never-exceed speed: 689 knots (1,276 km/h; 793 mph)
- Max level speed at S/L: 558 knots (1,035 km/h; 643 mph)
- Cruising speed at optimum altitude: 413 knots (766 km/h; 476 mph)
- Stalling speed, flaps up: 121 knots (225 km/h; 140 mph)
- Stalling speed, flaps down: 84 knots (156 km/h; 97 mph)
- Max rate of climb at S/L: 2,804 m (9,200 ft)/min
- Rate of climb at S/L, one engine out: 945 m (3,100 ft)/min
- Service ceiling: 14,265 m (46,800 ft)
- Service ceiling, one engine out: 9,083 m (29,800 ft)
- T-O run: 579 m (1,900 ft)
- T-O run to 15 m (50 ft): 795 m (2,610 ft)
- Landing from 15 m (50 ft): 689 m (2,260 ft)
- Landing run: 500 m (1,640 ft)
- Combat range with max external fuel: 2,365 nm (4,382 km; 2,723 miles)
- Range with max payload, 5% reserves plus 20 min at S/L: 1,671 nm (3,096 km; 1,924 miles)

GRUMMAN EA-6B PROWLER

The EA-6B is an advanced electronics development of the EA-6A for which Grumman received a prototype design and development contract in the Autumn of 1966. Except for a 1·37 m (4 ft 6 in) longer nose section and large fin pod, the external configuration of this version is the same as that of the basic A-6A.

The longer nose section provides accommodation for a total crew of four, the two additional crewmen being necessary to operate the more advanced ECM equipment. This comprises high-powered electronic jammers and modern computer-directed receivers, providing the US Navy with its first aircraft designed and built specifically for tactical electronic warfare. The prototype EA-6B flew for the first time on 25 May 1968.

The US Administration Fiscal 1969 defence budget allocated a sum of $139 million for the initial purchase of eight EA-6Bs, and the total programme is expected to cover the supply of 77 aircraft (this including the four pre-production and one R and D aircraft), to equip 12 EA-6B squadrons. An EXCAP (Expanded Capability) version of the EA-6B, which doubles the capability of each aircraft to jam enemy electronics activity, has been developed, and current production aircraft are delivered with EXCAP as standard.

The description of the standard A-6E Intruder applies also to the EA-6B, except as follows:

TYPE: Four-seat carrier- or land-based advanced ECM aircraft.

WINGS: As for A-6E, but reinforced to cater for increased gross weight and 5·5*g* load factor.

FUSELAGE: As for A-6E, but reinforcement of underfuselage structure in areas of arrester hook and landing gear attachments, and lengthened by 1·37 m (4 ft 6 in).

TAIL UNIT: As for A-6E, except for provision of a large fin-tip pod to house ECM equipment.

LANDING GEAR: As for A-6E, except for reinforcement of attachments, A-frame arrester hook, and upgrading of structure to cater for increased gross weight.

POWER PLANT: As for A-6E.

ACCOMMODATION: Crew of four under two separate upward-opening canopies. The two additional crewmen operate the advanced ECM equipment.

SYSTEMS: Generally as for A-6E.

Grumman EA-6B Prowler of Squadron VAQ-133 *Wizards* **landing on the USS** *John F. Kennedy (Brian M. Service)*

ELECTRONICS: Newly-developed advanced electronic countermeasures (ECM) to enable the EA-6B to fulfil a tactical electronic warfare role.

DIMENSIONS, EXTERNAL:
- As for A-6E, except:
  - Length overall: 18·11 m (59 ft 5 in)

AREAS:
- As for A-6E

WEIGHTS AND LOADING:
- Weight empty: 15,096 kg (33,282 lb)
- T-O weight in stand-off jamming configuration: 25,255 kg (55,678 lb)
- T-O weight in ferry range configuration: 28,024 kg (61,782 lb)
- Max T-O weight, catapult or field: 29,483 kg (65,000 lb)
- Max zero-fuel weight: 18,260 kg (40,256 lb)
- Max landing weight, carrier or field: 20,638 kg (45,500 lb)
- Max wing loading: 600·5 kg/m² (123 lb/sq ft)

PERFORMANCE (no stores):
- Never-exceed speed: 718 knots (1,329 km/h; 826 mph)
- Max level speed at S/L: 570 knots (1,055 km/h; 656 mph)
- Cruising speed at optimum altitude: 425 knots (787 km/h; 489 mph)
- Stalling speed, flaps up: 127 knots (236 km/h; 147 mph)
- Stalling speed, flaps down: 84 knots (156 km/h; 97 mph)
- Max rate of climb at S/L: 3,810 m (12,500 ft)/min
- Rate of climb at S/L, one engine out: 1,128 m (3,700 ft)/min
- Service ceiling: 14,110 m (46,300 ft)
- Service ceiling, one engine out: 9,540 m (31,300 ft)
- T-O to 15 m (50 ft): 762 m (2,500 ft)
- Landing from 15 m (50 ft): 792 m (2,600 ft)
- Landing run: 549 m (1,800 ft)
- Combat range, with max external fuel: 2,182 nm (4,042 km; 2,512 miles)
- Range with max payload, 5% reserves plus 20 min at S/L: 1,024 nm (1,897 km; 1,179 miles)

GRUMMAN TOMCAT

US Navy designation: F-14

Grumman announced on 15 January 1969 that it had been selected as winner of the design competition for a new carrier-based fighter for the US Navy. Known as the VFX during the competitive phase of the programme, this aircraft was later designated officially F-14.

First flight of the F-14A Tomcat prototype took place on 21 December 1970, more than a month ahead of schedule. It was lost in a non-fatal accident, and flight testing was resumed on 24 May 1971 with the second aircraft. Seven more F-14As were flying before the end of 1971 and by early 1973 20 aircraft had logged almost 3,000 hours in more than 1,500 flights. Weapons system testing accounted for half of the total flight time. By the end of 1975, a total of 187 aircraft had logged more than 50,000 hours.

The F-14 is designed to fulfil three primary missions. The first of these, fighter sweep/escort, involves clearing contested airspace of enemy fighters and protecting the strike force, with support from early-warning aircraft, surface ships and communications networks to co-ordinate penetration and escape.

Second mission is to defend carrier task forces via Combat Air Patrol (CAP) and Deck Launched Intercept (DLI) operations. Third role is secondary attack of tactical targets on the ground, supported by electronic countermeasures and fighter escort.

Emphasis has been placed on producing a high-performance aircraft able to fulfil both 'dogfight' and air defence roles, and offering significant advantages compared with other current US and Soviet first-line combat aircraft. In terms of airframe design, the F-14 uses advanced constructional techniques and titanium for optimum strength/weight ratio. Structural strength and a high thrust/weight ratio enable it to combine a maximum speed in excess of Mach 2 with great agility in close-in air-to-air combat. Development time and risk were reduced by use of an already-existing electronics system, a landing gear evolved from that of the A-6 Intruder and proven high-performance engines in the initial version. Armament includes an M-61A1 20 mm multi-barrel gun,

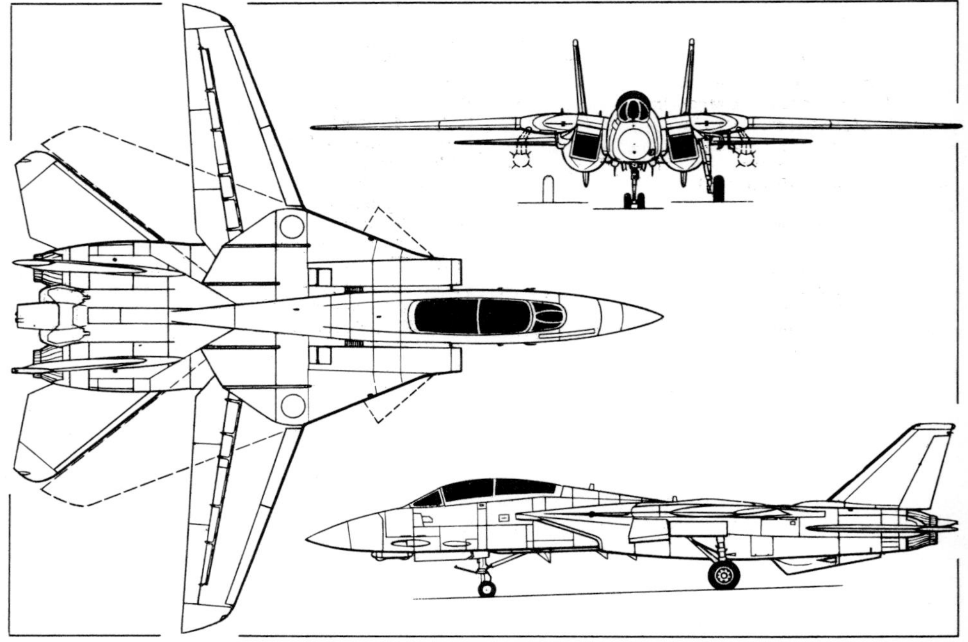

Grumman F-14A Tomcat carrier-based multi-mission fighter *(Pilot Press)*

Grumman F-14A Tomcat, one of the first for delivery to the Imperial Iranian Air Force

Sparrow and Sidewinder missiles, and the F-14/Phoenix AWG-9 missile system which has been demonstrated successfully against air-to-surface and surface-to-surface missiles, extreme high-altitude/high-speed interceptors, and in a simultaneous four-missile launch against five targets.

The Pratt & Whitney F401 (571R-1) afterburning turbofan engine is under development for the F-14B and is expected to produce 115·7-120·1 kN (26-27,000 lb st). It weighs some 91 kg (200 lb) less than the Pratt & Whitney TF30-P-412A engines which power current production models.

The configuration of the F-14 is unique, with variable-geometry wings, small foreplanes (glove vanes) which are extended as the wings sweep back to control centre-of-pressure shift, manoeuvring slats and flaps to create a lower effective wing loading, and twin outward-canted fins and rudders.

The variable-geometry wing in particular has contributed to exceptional aerodynamic performance. F-14s have made field take-offs in less than 305 m (1,000 ft); landings in under 457 m (1,500 ft) at normal 20° sweep and under 1,525 m (5,000 ft) at full 68° sweep; and have been manoeuvred at an angle of attack of ±90°, with no tendency towards uncontrolled flight.

Optimum sweep of the wing is controlled automatically by a Mach Sweep Programmer, which relates sweep to Mach number and altitude.

The engines are mounted in ducts under the fixed inner wings, with simple two-dimensional inlets and straight-line airflow for maximum efficiency over a wide range of altitudes and Mach numbers. The ducts have multiple-shock ramp systems for good pressure-recovery at high Mach numbers.

Three versions of the F-14 have been projected:

F-14A. Current production version, as described in detail.

F-14B. Airframe and electronics basically the same as those of the F-14A, but powered by Pratt & Whitney F401 turbofans. Intended to be capable of acceleration from Mach 0·8 to Mach 1·6 in 1·38 minutes. Flown for the first time on 12 September 1973.

F-14C. Development of F-14B, with new electronics and weapons. Fiscal restraints have retarded development of the electronics systems, and there are no current plans for F-14C production.

Under the initial contracts, Grumman was required to provide the US Navy with a mockup of the F-14A in May 1969, and to build 12 research and development aircraft. Subsequently, the US Navy ordered an initial series of 26 production F-14As, and is now expected to acquire a total of 390 Tomcats, including the 12 development aircraft. Carrier trials were started in June 1972, and initial deployment with the fleet began in October 1972. Replacement Training Squadron (RTS) VF-124 at Miramar NAS, San Diego, California, was responsible for working up ground and air crews for the new aircraft, and the first two operational squadrons, VF-1 and VF-2, were serving on board the USS *Enterprise* in the Western Pacific in September 1974. VF-14 and VF-32 formed in early 1975 and were deployed on board the USS *John F. Kennedy* in July 1975. Squadrons VF-142 and VF-143, formed at NAS Miramar in January 1975, started their initial fleet deployment on board the USS *America* in April 1976. Squadrons VF-24 and VF-211 were formed in February 1976. An aircraft was delivered to VF-101 at Oceana NAS, Norfolk, Virginia, in December 1975 to initiate an east coast RTS.

A total of 184 F-14As had been delivered by the end of 1975, and deliveries planned for 1976, 1977 and 1978 were 74, 72 and 56 respectively.

The Imperial Iranian Air Force has ordered a total of 80

Grumman F-14A Tomcat carrier-based multi-role fighter on the USS *Enterprise*

F-14As; first flight by one of these aircraft was made on 5 December 1975, and the first three aircraft were ferried to Khatami AFB, Isfahan, in January 1976. Deliveries to Iran were to continue throughout 1976 at the rate of two per month, rising to three per month from the beginning of 1977 until mid-1978.

Type: Two-seat carrier-based multi-role fighter.

Wings: Variable-geometry mid-wing monoplane, with 20° of leading-edge sweep in the fully-forward position and 68° when fully swept. Oversweep position of 75° for carrier stowage. Wing position is programmed automatically for optimum performance throughout the flight regime, but manual override is provided. A short movable wing outer panel, needing only a comparatively light pivot structure, results from the wide fuselage and fixed centre-section 'glove', with pivot points 2·72 m (8 ft 11 in) from the centreline of the airframe. The inboard wing sections, adjacent to the fuselage, arc upward slightly to minimise cross-sectional area and wave-drag, and consist basically of a one-piece electron beam-welded titanium assembly, 6·70 m (22 ft) in span, made from Ti-6A1-4V titanium alloy. Small canard surfaces, known as glove vanes, swing out from the leading-edge of the fixed portion of the wing as sweep of outer panels is increased. Spoilers on upper surfaces of wing. Stabilisation in pitch, provided by the canard surfaces, leaves the differential tailplane free to perform its primary control function. Trailing-edge control surfaces extend over almost entire span. Leading-edge slats.

Fuselage: The centre-fuselage section is a simple, fuel-carrying box structure; forward fuselage section comprises cockpit and nose. The aft section has a tapered aerofoil shape to minimise drag, with a fuel dump pipe projecting from the rear. Speed brakes located on the upper and lower surfaces, between the bases of the vertical tail fins.

Tail Unit: Twin vertical fins, mounted at the rear of each engine nacelle. Outward-canted ventral fin under each nacelle. The all-flying horizontal surfaces have skins of boron-epoxy composite material.

Landing Gear: Retractable tricycle type. Twin-wheel nose unit and single-wheel main units retract forward and upward. Arrester hook under rear fuselage, housed in small ventral fairing. Nose-tow catapult attachment on nose unit.

Engine Intakes: Straight two-dimensional external compression inlets. A double-hinged ramp extends down from the top of each intake, and these are programmed to provide the correct airflow to the engines automatically under all flight conditions. Each intake is canted slightly away from the fuselage, from which it is separated by some 0·25 m (10 in) to allow sufficient clearance for the turbulent fuselage boundary layer to pass between fuselage and intake without causing turbulence within the intake. Engine inlet ducts and aft nacelle structures are manufactured by Rohr Corporation. The inlet duct, constructed largely of aluminium honeycomb, is about 4·27 m (14 ft) long, while the aft nacelle structure, of bonded aluminium honeycomb and conventional aluminium, is about 4·88 m (16 ft) in length.

Power Plant: Two Pratt & Whitney TF30-P-412A turbofan engines of 93 kN (20,900 lb st) with afterburning, mounted in ducts which open to provide 180° access for ease of maintenance. AiResearch ATS200-50 air-turbine starter.

Accommodation: Pilot and naval flight officer seated in tandem on Martin-Baker GRU-7A rocket-assisted zero-zero ejection seats, under a one-piece bubble canopy, hinged at the rear and offering all-round view.

Armament: One General Electric M-61A1 Vulcan 20 mm gun mounted in the port side of forward fuselage. Four Sparrow air-to-air missiles mounted partially submerged in the underfuselage. Two wing pylons, one under each fixed wing section, can carry four Sidewinder missiles or two additional Sparrow or Phoenix missiles with two Sidewinders. For Phoenix and later missiles, Grumman has developed a concept in which removable pallets can be attached to the present Sparrow missile positions, the missiles then being attached to the pallets. Four Phoenix missiles can be accommodated. Various combinations of missiles and bombs to a max external weapon load of 6,577 kg (14,500 lb). A tactical reconnaissance pod has been developed by Grumman. This can accommodate a number of high- or low-altitude cameras and advanced electro-optical sensors. ECM equipment includes Goodyear AN/ALE-39 chaff and flare dispensers, with integral jammers.

Electronics: Hughes AN/AWG-9 weapons control system. Kaiser Aerospace AN/AVG-12 vertical and head-up display system.

Dimensions, external:

Wing span:	
unswept	19·45 m (64 ft 1½ in)
swept	11·64 m (38 ft 2¼ in)
overswept	9·97 m (32 ft 8½ in)
Wing aspect ratio	7·28
Length overall	18·86 m (61 ft 10½ in)
Height overall	4·88 m (16 ft 0 in)
Tailplane span	10·15 m (33 ft 3½ in)
Distance between fin tips	3·25 m (10 ft 8 in)
Wheel track	5·00 m (16 ft 5 in)
Wheelbase	7·02 m (23 ft 0½ in)

AREAS:

Wings, gross	52·49 m² (565·0 sq ft)
Horizontal tail surfaces (total)	13·01 m² (140·0 sq ft)
Vertical tail surfaces (total)	10·96 m² (118·0 sq ft)

WEIGHTS:

Weight empty	17,659 kg (38,930 lb)
Fuel (usable):	
internal	7,348 kg (16,200 lb)
external	1,724 kg (3,800 lb)
Normal T-O weight	26,553 kg (58,539 lb)
Max T-O weight	33,724 kg (74,348 lb)
Design landing weight	23,510 kg (51,830 lb)

PERFORMANCE:

Max design speed	Mach 2·40
Service ceiling	above 15,240 m (50,000 ft)
Min T-O distance at 26,533 kg (58,495 lb) AUW	366 m (1,200 ft)

Initial flight of F-111A with new weapons bay radome, one of two modified by Grumman to EF-111A prototype configuration

GRUMMAN (GENERAL DYNAMICS) EF-111A

The EF-111A conversion of the F-111A is intended to provide ECM jamming coverage for attack forces over a wide area. Basic equipment comprises AN/ALQ-99A jammers of the type fitted to the Grumman EA-6B Prowler. In addition, the EF-111A has a modified AN/ALQ-137 Self-Protection System and a modified AN/ALR-62 Terminal Threat Warning System. The ALQ-99A jammers are mounted in the weapons bay, with their antennae covered by a 4·9 m (16 ft) long canoe-shape radome. The fin-tip pod, similar in shape to that of the EA-6B Prowler, houses the receiver system and antennae. Total weight of the new equipment is about three tons.

The EF-111A will be able to locate enemy emitting radars and provide electronic cover for attacking forces.

Design study contracts were awarded to General Dynamics and Grumman by the USAF in 1974, and in January 1975 it was announced that Grumman had been awarded an $85·9 million contract to convert two existing F-111As to EF-111A prototype configuration. The first of these made its first flight in EF-111A configuration on 15 December 1975, and is shown in the accompanying illustration. It is planned to convert 40 F-111As to EF-111A configuration, to equip two USAF squadrons in the late 1970s.

GRUMMAN AMERICAN AVIATION CORPORATION

HEAD OFFICE AND WORKS:
PO Box 2206, Savannah, Georgia 31402
Telephone: (912) 964-1454
Telex: 54-6470

PRESIDENT:
Corwin H. Meyer

EXECUTIVE VICE-PRESIDENT:
Alan B. Lemlein

SENIOR VICE-PRESIDENTS:
Roy C. Garrison (Commercial Light Aircraft Marketing)
Charles G. Vogeley (Commercial Jet Aircraft Marketing)

VICE-PRESIDENTS:
Albert H. Glenn (General Manager, Savannah Operations)
Richard Kemper (Corporate Planning)
Frank Wisekel (Finance)

TREASURER:
Robert G. Freese

SECRETARY;
Fred D. Kidder

CUSTOMER SERVICE MANAGER:
Russell E. Belles

This subsidiary of Grumman Corporation builds the Trainer, Tr-2, Cheetah and Tiger, and is responsible for manufacture and marketing of the Grumman-designed Gulfstream II. It markets the Grumman-designed Super Ag-Cat, built by Schweizer Aircraft (which see).

Sales in 1975 included 19 Gulfstream IIs, 226 Ag-Cats, 113 two-seat Trainers and Tr-2s, 203 four-seat Travelers and Cheetahs, and 196 four-seat Tigers. Total sales of lightplanes passed the 2,500 mark.

GRUMMAN AMERICAN GULFSTREAM II

US Coast Guard designation: VC-11A

The decision to start production of this twin-turbofan executive transport was announced by Grumman on 17 May 1965. The first production Gulfstream II (no prototype was built) flew for the first time on 2 October 1966. FAA certification was gained on 19 October 1967, and the first production aircraft was delivered to National Distillers & Chemical Corporation on 6 December 1967.

Custom interiors and electronics, with the exception of the Sperry SP-50G automatic flight control system, which is standard, are installed by specialist agencies.

Deliveries totalled 163 by 1 June 1975, including a single Gulfstream II operated by the US Coast Guard under the designation **VC-11A**. Two other Gulfstream IIs have been converted as flying simulators for the Space Shuttle Orbiter vehicle.

From aircraft No. 166, delivered in July 1975, production aircraft incorporate an engine 'hush-kit', for which Grumman American received FAA certification on 2 May 1975; this modification can be retrofitted to earlier Gulfstream IIs if required.

Flight testing began in 1975 of a Gulfstream II with wingtip tanks, increasing the total fuel capacity to 12,247 kg (27,000 lb). This version has a max ramp weight of 29,937 kg (66,000 lb) and max T-O weight of 29,711 kg (65,500 lb). Increased fuel, amounting to 1,415 kg (3,120 lb), is carried in new wingtip tanks which serve as an extension of the main integral wing tanks. Wind tunnel tests have shown that the tip-tanks do not affect high-speed handling, that they offer a slight improvement in aircraft stability and lateral control, and that cruise performance penalty is approximately 3 per cent against a range improvement of 14 per cent with max fuel.

The description which follows applies to the standard version:

TYPE: Twin-turbofan executive transport aircraft.

WINGS: Cantilever low-wing monoplane of all-metal construction. Thickness/chord ratio 12% at wing station 50, 9·5% at wing station 145 and 8·5% at wing station 414. Dihedral 3°. Incidence 3° 30′ at wing station 50, 1° 30′ at wing station 145 and −0° 30′ at wing station 414. Sweepback 25° at quarter-chord. One-piece single-slotted Fowler-type trailing-edge flaps. Spoilers forward of flaps assist in lateral control and can be extended for use as airbrakes. All control surfaces actuated hydraulically.

FUSELAGE: Conventional all-metal semi-monocoque structure. Glassfibre nosecone hinged for access to radar, etc.

TAIL UNIT: Cantilever all-metal T-tail. All surfaces swept-back. Trim tab in rudder. Powered controls (see under 'Systems' paragraph).

LANDING GEAR: Retractable tricycle type, with twin wheels on each unit. Inward-retracting main units, with tyres size 34 × 8·25-32, pressure 10·34 bars (150 lb/sq in). Forward-retracting steerable nose unit. Nosewheel tyres size 21 × 7·25-22, pressure 6·55 bars (95 lb/sq in). Goodyear aircooled brakes with Goodyear fully-modulating anti-skid units.

POWER PLANT: Two Rolls-Royce Spey Mk 511-8 turbofan engines, each 50·7 kN (11,400 lb st), mounted in pod on each side of rear fuselage. Rohr target-type thrust reversers form aft portions of nacelles when in stowed position. All fuel in integral tanks in wings, capacity 10,568 kg (23,300 lb).

ACCOMMODATION: Crew of two or three. Certificated for 19 passengers in pressurised and air-conditioned cabin. Large baggage compartment at rear of cabin, capacity 907 kg (2,000 lb). Integral airstair door at front of cabin on port side.

SYSTEMS: Cabin pressurisation system, with max differential of 0·65 bars (9·45 lb/sq in). Two independent hydraulic systems, each 103·5 bars (1,500 lb/sq in). All flying controls hydraulically powered, with manual reversion. APU in tail compartment. Basic 28V DC electrical system, using two 300A generators and a 200A transformer-rectifier. Two 20kVA alternators provide AC power for secondary and auxiliary systems. Third (APU-driven) 20kVA alternator for on-ground power. Three 2·5kVA inverters, powered by the

Grumman American Gulfstream II executive transport (two Rolls-Royce Spey Mk 511-8 turbofan engines)

transformer-rectifiers, provide 400Hz fixed-frequency power. Two 24V batteries.

DIMENSIONS, EXTERNAL:
- Wing span 20·98 m (68 ft 10 in)
- Wing span over tip-tanks 21·87 m (71 ft 9 in)
- Length overall 24·36 m (79 ft 11 in)
- Length of fuselage 21·74 m (71 ft 4 in)
- Height overall 7·47 m (24 ft 6 in)
- Tailplane span 8·23 m (27 ft 0 in)
- Wheel track 4·16 m (13 ft 8 in)
- Wheelbase 10·16 m (33 ft 4 in)
- Passenger door:
  - Height 1·57 m (5 ft 2 in)
  - Width 0·91 m (3 ft 0 in)
- Baggage door:
  - Height 0·72 m (2 ft 4½ in)
  - Width 0·91 m (2 ft 11¾ in)
- Ventral door:
  - Width 0·46 m (1 ft 6 in)
  - Length 0·71 m (2 ft 4 in)

DIMENSIONS, INTERNAL:
- Cabin: Length 10·36 m (34 ft 0 in)
  - Width 2·24 m (7 ft 4 in)
  - Height 1·85 m (6 ft 1 in)
  - Volume 35·97 m³ (1,270·2 cu ft)
- Baggage compartment 4·44 m³ (156·85 cu ft)

AREA:
- Wings, gross 75·21 m² (809·6 sq ft)

WEIGHTS AND LOADINGS:
- Manufacturer's weight empty 13,934 kg (30,719 lb)
- Typical operating weight empty 16,737 kg (36,900 lb)
- Max T-O weight 29,711 kg (65,500 lb)
- Max ramp weight 29,937 kg (66,000 lb)
- Max landing weight 26,535 kg (58,500 lb)
- Max zero-fuel weight 19,050 kg (42,000 lb)
- Max wing loading 394·8 kg/m² (80·9 lb/sq ft)
- Max power loading 293 kg/kN (2·87 lb/lb st)

PERFORMANCE (at max T-O weight except where indicated):
- Max cruising speed at 7,620 m (25,000 ft) Mach 0·85 (505 knots; 936 km/h; 581 mph)
- Econ cruising speed at 13,105 m (43,000 ft) Mach 0·75 (430 knots; 796 km/h; 495 mph)
- Approach speed at max landing weight 140 knots (259 km/h; 161 mph)
- Max rate of climb at S/L 1,325 m (4,350 ft)/min
- Rate of climb at S/L, one engine out 457 m (1,500 ft)/min
- Service ceiling 13,100 m (43,000 ft)
- Service ceiling, one engine out 7,467 m (24,500 ft)
- FAA T-O field length 1,737 m (5,700 ft)
- FAA landing field length 972 m (3,190 ft)
- Range with max fuel, 30 min reserves 3,712 nm (6,880 km; 4,275 miles)

OPERATIONAL NOISE CHARACTERISTICS (FAR Pt 36):
- T-O noise level with thrust cutback 94·4 EPNdB
- Approach noise level 99·5 EPNdB
- Sideline noise level 107·5 EPNdB

GRUMMAN AMERICAN AA-1B TRAINER

Designed originally as a specialised trainer version of the American Aviation AA-1 American Yankee, the prototype AA-1A Trainer first flew on 25 March 1970; FAA certification in the Normal and Utility categories was granted on 14 January 1971. In 1973 this aircraft was redesignated as the AA-1B Trainer, and the 1976 version introduces as standard an improved landing light mounting, a canopy cable/pulley system, increased engine cooling and six different coloured interior trims. Also available are a camouflage paint scheme and 'invasion stripes'.

Three versions of the Trainer are available, differing in installed equipment, any item of which may be added as optional to the Standard Trainer.

Standard Trainer. As described in detail.

Basic Trainer. As Standard Trainer, plus sensitive altimeter, electric clock, dual controls, Narco Escort 110 nav/com radio with M-700 microphone, headset, and antenna, tinted windows, turn co-ordinator and rate of climb indicators.

Advanced Trainer. As Basic Trainer, plus vacuum system, de luxe interior, landing light, omni-flash beacon, outside air temperature gauge, heated pitot, and towbar.

TYPE: Two-seat trainer/utility monoplane.

WINGS: Cantilever low-wing monoplane. Wing section NACA 64_2415 (modified). Dihedral 5°. Incidence 1° 25′. No sweep. Alclad aluminium skin and ribs, attached to main spar by adhesive bonding. Tube-type circular-section main spar serves as integral fuel tank. Plain ailerons of bonded construction, with honeycomb ribs and Alclad aluminium skin. Electrically-actuated plain trailing-edge flaps of bonded construction, with honeycomb ribs and aluminium skin, and RAE Motors Corporation actuators. Ground-adjustable trim tab on each aileron.

FUSELAGE: Aluminium honeycomb cabin section and aluminium semi-monocoque rear fuselage structure, utilising adhesive bonding. The use of honeycomb eliminates false floors, resulting in greater usable cabin space relative to cross-sectional area.

TAIL UNIT: Cantilever adhesive-bonded aluminium structure. Movable surfaces built up of honeycomb ribs bonded to sheet aluminium. All three fixed surfaces interchangeable. Combined trim and anti-servo tab in starboard elevator. Ground-adjustable rudder trim tab.

Grumman American AA-1B Advanced Trainer (Lycoming O-235-C2C engine)

LANDING GEAR: Non-retractable tricycle type. Nose gear of 4340 tubular steel, with large free-swivelling fork. Main legs are cantilever leaf springs of glassfibre. Shock-mounted wheel fairings standard. US Royal main wheels and tyres Type III, size 17 × 6·00-6, pressure 1·31 bars (19 lb/sq in). US Royal nosewheel tyre Type III LP, size 5·00-5, pressure 1·52 bars (22 lb/sq in). Cleveland single-disc hydraulic brakes. Parking brake.

POWER PLANT: One 80·5 kW (108 hp) Lycoming O-235-C2C flat-four engine, driving a McCauley two-blade fixed-pitch propeller (type 1A105/SCM 7157 for cruise or 1A105/SCM 7154 for climb performance) with spinner. Two integral fuel tanks in wing spar, with total capacity of 91 litres (24 US gallons), of which 83 litres (22 US gallons) are usable. Refuelling points at wingtips. Oil capacity 5·7 litres (1·5 US gallons).

ACCOMMODATION: Two individual seats side by side in enclosed cabin, under large transparent sliding canopy. Aircraft certificated for open-canopy flight. Optional seat for child. Cabin heated and ventilated, with windscreeen defroster on pilot's side. Centre console, between seats, accommodates trim wheel and electric flap operating switch. Space for 45 kg (100 lb) baggage aft of seats.

SYSTEMS: Hydraulic system for brakes only. Electrical system includes 60A 14V engine-driven alternator and a 12V 25Ah battery to supply flap motor, lights, navigation/communication equipment and flight instrumentation.

ELECTRONICS AND EQUIPMENT: Standard equipment includes windscreen defroster, aileron and elevator control lock, cabin dome light, instrument lights, navigation lights, audible stall warning device, cabin heating system, air ventilators, cargo tie-down rings, seat belts, shoulder harness, baggage straps, chart holders, instrument panel glareshield, wheel hub covers, map holder and glove compartment. Optional items include sensitive altimeter, dual controls, blind-flying instrumentation, engine-driven vacuum pump and suction gauge, electric clock, outside air temperature gauge, hour recorder, cabin-mounted fire extinguisher, pitot heating system, landing light, omni-flash beacon, high intensity strobe lights, external power socket, canopy cover, alternative propeller for improved cruising performance, internal corrosion-proofing, towbar, canopy sun curtain, de luxe interior, intercom system, true airspeed indicator, tinted windows and dual windscreen defrosters. Optional electronics include Narco Escort 110 nav/com, with 110-channel VHF transceiver and 100-channel nav receiver with VOR/LOC and indicator; Narco Com 10A/Nav 10 360-channel VHF transceiver with 200-channel remote VOR/LOC indicator; Narco Com 11A/Nav 11 360-channel VHF transceiver with 200-channel nav receiver and indicator; Narco AT-50A transponder with 4096 code capability; Narco Com 11B 720-channel transceiver; Narco Com 111B 720-channel transceiver; Narco Nav 111B 200-channel remote nav indicator; Narco Com 11A/Nav 12 with UGR 3 glideslope and receiver; ADF 140 digital; marker beacon receiver; CP-125 audio switch panel; King KX-170B/KI-201C nav/com with 720-channel com and 200-channel nav with remote VOR/LOC indicator; KX-175/KI-201C nav/com with 720-channel transceiver and 200-channel nav with remote VOR/LOC indicator; KT-78 transponder with 4096 code capability; KR-85 ADF, including remote KI-225 indicator; KMA-20 audio control panel with marker beacon receiver; and emergency locator transmitter.

DIMENSIONS, EXTERNAL:
- Wing span 7·47 m (24 ft 6 in)
- Wing chord (constant) 1·25 m (4 ft 1¼ in)
- Wing aspect ratio 5·975
- Length overall 5·86 m (19 ft 3 in)
- Height overall 2·32 m (7 ft 7¼ in)
- Tailplane span 2·34 m (7 ft 8¼ in)
- Wheel track 2·45 m (8 ft 3 in)
- Wheelbase 1·33 m (4 ft 4½ in)
- Propeller diameter 1·80 m (5 ft 11 in)
- Propeller ground clearance 0·25 m (9¾ in)

DIMENSIONS, INTERNAL:
- Cabin: Length 1·37 m (4 ft 6 in)
  - Max width 1·04 m (3 ft 5 in)
  - Max height 1·15 m (3 ft 9¼ in)
  - Floor area 1·55 m² (16·7 sq ft)

AREAS:
- Wings, gross 9·38 m² (100·92 sq ft)
- Ailerons (total) 0·48 m² (5·20 sq ft)
- Trailing-edge flaps (total) 0·50 m² (5·44 sq ft)
- Fin 0·44 m² (4·76 sq ft)
- Rudder 0·34 m² (3·61 sq ft)
- Tailplane 0·88 m² (9·52 sq ft)
- Elevators, incl tab 0·67 m² (7·22 sq ft)

WEIGHTS AND LOADINGS:
- Weight empty 472 kg (1,041 lb)
- Max T-O and landing weight 707 kg (1,560 lb)
- Max zero-fuel weight 620 kg (1,368 lb)
- Max wing loading 75·7 kg/m² (15·5 lb/sq ft)
- Max power loading 8·78 kg/kW (14·4 lb/hp)

PERFORMANCE (at max T-O weight. A: standard propeller, B: with optional cruise propeller):
- Never-exceed speed 170 knots (313 km/h; 195 mph)
- Max level speed at S/L:
  - A 120 knots (222 km/h; 138 mph)
  - B 125 knots (232 km/h; 144 mph)
- Max cruising speed (75% power) at 915 m (3,000 ft):
  - A 108 knots (200 km/h; 124 mph)
- Econ cruising speed at 3,050 m (10,000 ft):
  - A 97 knots (180 km/h; 112 mph)
  - B 99 knots (183 km/h; 114 mph)
- Stalling speed, flaps up 54 knots (100 km/h; 62 mph)
- Stalling speed, flaps down 52·5 knots (97 km/h; 60 mph)
- Max rate of climb at S/L:
  - A 215 m (705 ft)/min
  - B 201 m (660 ft)/min
- Service ceiling:
  - A 3,885 m (12,750 ft)
  - B 3,520 m (11,550 ft)
- T-O run:
  - A 247 m (810 ft)
  - B 271 m (890 ft)
- T-O to 15 m (50 ft):
  - A 472 m (1,550 ft)
  - B 485 m (1,590 ft)
- Landing from 15 m (50 ft) 335 m (1,100 ft)
- Landing run 125 m (410 ft)
- Range with max fuel at max cruising speed, no reserves:
  - A 377 nm (700 km; 435 miles)
  - B 402 nm (745 km; 463 miles)
- Range with max fuel, no reserves, at 3,050 m (10,000 ft):
  - A 425 nm (788 km; 490 miles)
  - B 441 nm (817 km; 508 miles)

GRUMMAN AMERICAN Tr-2

Generally similar to the Grumman American Trainer, the Tr-2 is intended to satisfy a dual requirement: as an advanced trainer or as a sports aircraft with de luxe equipment.

It is generally similar to the Advanced Trainer version of the AA-1B, but has in addition the following equipment

Grumman American Trainer with 'invasion stripes'

Grumman American Tr-2 sporting aircraft/advanced trainer

as standard: carpeted floor to cabin and baggage area, de luxe vinyl/fabric interior, and polyurethane external trim in nine combinations; Narco Com 10A/Nav 10 radio in lieu of Escort 110, with M-700 microphone, headset, loudspeaker and antenna. The 1·45 m (57 in) pitch McCauley cruise propeller is standard on the Tr-2, the climb propeller as fitted to the AA-1B being available optionally. A three-tone exterior finish is also standard on this model.

WEIGHTS:
- Weight empty 469 kg (1,035 lb)
- Max T-O and landing weight 707 kg (1,560 lb)
- Max zero-fuel weight 620 kg (1,368 lb)

PERFORMANCE: As for AA-1B with optional cruise propeller

GRUMMAN AMERICAN AA-5A and CHEETAH

The AA-5A is an enlarged version of the AA-1B, with increased wing span, a more powerful engine, and an extended fuselage to provide accommodation for a pilot and three passengers. The first flight of the original AA-5 was made on 21 August 1970, and FAA certification was awarded on 12 November 1971. The 1976 version of the de luxe Traveler version has been renamed Cheetah.

Two versions of the AA-5A are available, as follows:

AA-5A. Standard version, to which the detailed description applies.

Cheetah. De luxe version; as standard model, plus the following additional equipment: sensitive altimeter, omni-flash beacon, dual controls, vacuum system, landing light, outside air temperature gauge, heated pitot, tinted windows, turn co-ordinator and rate of climb indicators, towbar, Narco Com 10A/Nav 10 nav/com transceiver with VOR/LOC indicator, M-700 microphone, headset, speaker and antenna.

The 1976 versions of the AA-5A and Cheetah introduce a number of improvements as standard including increased propeller ground clearance, larger tailplane and new interior trims and exterior finish. New options include a long-range fuel tank, sun visor, map light and alternate static source.

The general description of the AA-1B applies also to the AA-5A, except as follows:

TYPE: Four-seat cabin monoplane.

WINGS: Generally as for AA-1B, except that wing span and chord are increased, and flap electrical actuators are supplied by Commercial Aircraft Products.

FUSELAGE: As for AA-1B, except length increased.

TAIL UNIT: As for AA-1B, except general dimensions increased, and the addition of a dorsal fin, and spin fillets on inboard leading-edges of tailplane. Combined trim and anti-servo tab in port and starboard elevators.

LANDING GEAR: As for AA-1B.

POWER PLANT: One 112 kW (150 hp) Lycoming O-320-E2G flat-four engine, driving a McCauley fixed-pitch two-blade metal propeller with spinner. Two integral fuel tanks in wing spars, with a total capacity of 144 litres (38 US gallons), of which 140 litres (37 US gallons) are usable. Optionally, two integral fuel tanks in wings with total capacity of 199 litres (52·6 US gallons), of which 193 litres (51 US gallons) are usable. Refuelling point in upper surface of each wing. Oil capacity 7·5 litres (2 US gallons).

ACCOMMODATION: Pilot and three passengers in enclosed cabin on four separate seats, in pairs. Baggage area aft of rear seats, which may be folded forward when unoccupied to increase baggage space, providing a capacity of 1·18 m^3 (41·5 cu ft) for maximum load of 154 kg (340 lb). Max normal baggage load 54·4 kg (120 lb).

SYSTEMS: As for AA-1B.

ELECTRONICS AND EQUIPMENT: Standard equipment as for AA-1B, plus carpeted floor, electric clock and hat shelf. Options as for AA-1B, plus emergency locator transmitter, rear seat air ventilator, access steps, quick oil-drain valve, oil access door and alternate static source.

Cheetah de luxe version of the Grumman American AA-5A four-seat lightplane

DIMENSIONS, EXTERNAL:
- Wing span 9·60 m (31 ft 6 in)
- Wing chord (constant) 1·35 m (4 ft 5¼ in)
- Wing aspect ratio 7·10
- Length overall 6·71 m (22 ft 0 in)
- Height overall 2·39 m (7 ft 10 in)
- Tailplane span 3·86 m (12 ft 8 in)
- Wheel track 2·51 m (8 ft 3 in)
- Wheelbase 1·64 m (5 ft 4½ in)
- Propeller diameter 1·85 m (6 ft 1 in)
- Propeller ground clearance 0·24 m (9½ in)

DIMENSIONS, INTERNAL:
- Cabin: Length 1·98 m (6 ft 6 in)
  - Max width 1·04 m (3 ft 5 in)
  - Max height 1·23 m (4 ft 0¼ in)
  - Floor area 2·18 m^2 (23·5 sq ft)
- Baggage space 0·34 m^3 (12 cu ft)

AREAS:
- Wings, gross 13·02 m^2 (140·12 sq ft)
- Ailerons (total) 0·72 m^2 (7·74 sq ft)
- Trailing-edge flaps (total) 1·51 m^2 (16·26 sq ft)
- Fin 0·44 m^2 (4·76 sq ft)
- Rudder 0·34 m^2 (3·61 sq ft)
- Elevators, incl tabs 0·99 m^2 (10·68 sq ft)

WEIGHTS AND LOADINGS:
- Weight empty 572 kg (1,262 lb)
- Max T-O weight 998 kg (2,200 lb)
- Max zero-fuel weight 900 kg (1,984 lb)
- Max wing loading 76·6 kg/m^2 (15·7 lb/sq ft)
- Max power loading 8·91 kg/kW (14·7 lb/hp)

PERFORMANCE (at max T-O weight):
- Max level speed at S/L 136 knots (253 km/h; 157 mph)
- Max cruising speed, 75% power at 2,590 m (8,500 ft) 128 knots (237 km/h; 147 mph)
- Econ cruising speed, 65% power at 2,590 m (8,500 ft) 118 knots (219 km/h; 136 mph)
- Stalling speed, flaps up 54 knots (100 km/h; 62 mph)
- Stalling speed, flaps down 50·5 knots (93·5 km/h; 58 mph)
- Max rate of climb at S/L 201 m (660 ft)/min
- Service ceiling 3,855 m (12,650 ft)
- T-O run 268 m (880 ft)
- T-O to 15 m (50 ft) 488 m (1,600 ft)
- Landing from 15 m (50 ft) 335 m (1,100 ft)
- Landing run 116 m (380 ft)
- Range at max cruising speed with standard fuel, no reserves 554 nm (1,026 km; 638 miles)
- Range at max cruising speed with max optional fuel, no reserves 763 nm (1,414 km; 879 miles)
- Range at econ cruising speed with standard fuel, no reserves 585 nm (1,084 km; 674 miles)
- Range at econ cruising speed with max optional fuel, no reserves 807 nm (1,496 km; 930 miles)
- Optimum range at 3,050 m (10,000 ft) with standard fuel, no reserves 590 nm (1,094 km; 680 miles)
- Optimum range at 3,050 m (10,000 ft) with max optional fuel, no reserves 814 nm (1,509 km; 938 miles)

GRUMMAN AMERICAN AA-5B and TIGER

In late 1974 Grumman American announced the introduction of a new model to its range of single-engined light aircraft, and this was given the designation AA-5B. It differs from the AA-5A by having a more powerful engine and increased fuel capacity as standard.

Grumman American Tiger (Lycoming O-360-A4K engine)

Two versions of the AA-5B are available, as follows:

AA-5B. Standard version, to which the detailed description applies.

Tiger. De luxe version; as standard model, plus the additional equipment detailed for the Cheetah.

The description of the AA-5A and Cheetah applies also to the AA-5B and Tiger, except as follows:

TYPE, WINGS, FUSELAGE, TAIL UNIT AND LANDING GEAR: As for AA-5A.

POWER PLANT: One 134 kW (180 hp) Lycoming O-360-A4K flat-four engine, driving a McCauley two-blade metal fixed-pitch propeller with spinner. Two integral fuel tanks in wings with total capacity of 199 litres (52·6 US gallons), of which 193 litres (51 US gallons) are usable. Refuelling point in upper surface of each wing. Oil capacity 7·5 litres (2 US gallons).

ACCOMMODATION, SYSTEMS, ELECTRONICS AND EQUIPMENT AND DIMENSIONS: As for AA-5A

WEIGHTS AND LOADINGS:

Weight empty	623 kg (1,373 lb)
Max T-O weight	1,088 kg (2,400 lb)
Max wing loading	83·5 kg/m² (17·1 lb/sq ft)
Max power loading	8·12 kg/kW (13·3 lb/hp)

PERFORMANCE (at max T-O weight):

Max level speed at S/L	148 knots (274 km/h; 170 mph)
Max cruising speed, 75% power at 2,590 m (8,500 ft)	139 knots (257 km/h; 160 mph)
Econ cruising speed, 65% power at 2,590 m (8,500 ft)	129 knots (238 km/h; 148 mph)
Stalling speed, flaps up	57 knots (105 km/h; 65 mph)
Stalling speed, flaps down	53 knots (99 km/h; 61 mph)
Max rate of climb at S/L	259 m (850 ft)/min
Service ceiling	4,205 m (13,800 ft)
T-O run	264 m (865 ft)
T-O to 15 m (50 ft)	472 m (1,550 ft)
Landing from 15 m (50 ft)	341 m (1,120 ft)
Landing run	125 m (410 ft)
Range at max cruising speed with max fuel, no reserves	614 nm (1,139 km; 708 miles)
Range at econ cruising speed with max fuel, no reserves	655 nm (1,215 km; 755 miles)
Optimum range at 2,590 m (8,500 ft) with max fuel, no reserves	694 nm (1,287 km; 800 miles)

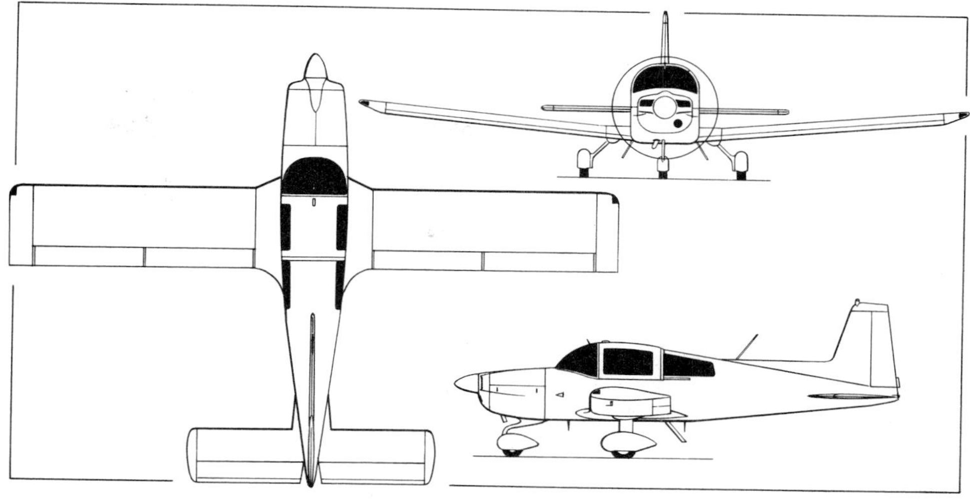

Grumman American Tiger four-seat light aircraft *(Pilot Press)*

GRUMMAN AMERICAN GA-7 COUGAR

Grumman American announced on 20 December 1974 the first flight of a new twin-engined light aircraft, which has the designation Model GA-7 and has been given the name Cougar. Representing the company's first entry into the lightweight twin-engine market, it is intended primarily for business use and for private pilots already possessing IFR experience in high-powered single-engined aircraft. It is considered also that the Cougar will be attractive to flying schools requiring an economical trainer for conversion to twin-engined flight.

The wings are of cantilever low-wing monoplane configuration, with cambered tips; the tail unit has swept vertical surfaces and a dorsal fin. The landing gear is retractable, with the nosewheel retracting forward and the main wheels inward into the undersurface of the wings. Power plant consists of two 119·3 kW (160 hp) Lycoming O-320 flat-four engines, each driving a two-blade metal propeller with spinner. There is seating for four persons, in pairs, in an enclosed cabin.

No details of dimensions, weights and performance had been released by Grumman American up to mid-1976; it was known only that the design cruising speed of the Cougar is approximately 165 knots (306 km/h; 190 mph), and that the aircraft would not be in production before 1977.

Prototype of the Grumman American Cougar, a new contender in the lightweight twin-engine category

HAMILTON
HAMILTON AVIATION

HEAD OFFICE:
PO Box 11746, Tucson, Arizona 85734
Telephone: (602) 294-3481

PRESIDENT:
Gordon B. Hamilton

CHIEF ENGINEER:
John D. Burnham Jr

Hamilton Aviation, which is engaged primarily in the procurement and overhaul of various types of military aircraft for foreign governments, is also building and marketing turboprop conversions of the Beech Model 18 (last described in the 1969-70 *Jane's*). Two versions are currently available, known as the Westwind III and Westwind II STD; these differ in power plant, internal layout and in the arrangement of passenger and cargo doors. A third version, designated Westwind IV, was under development in early 1976.

HAMILTON WESTWIND III

The Westwind III is a passenger/cargo aircraft, in which the passenger seats can be removed easily to make the whole cabin space available for cargo. Design began in 1961 and the prototype flew for the first time in 1963. FAA certification under CAR Part 3 was awarded in 1964.

The Westwind III can have agricultural or military applications.

TYPE: Utility passenger/cargo commuter airliner.

WINGS, FUSELAGE, TAIL UNIT: As for Beech Model 18.

LANDING GEAR: Electrically-retractable tailwheel type. Oleo-pneumatic shock-absorbers on all units. Main wheels retract aft. Main-wheel tyre pressure 4·14 bars (60 lb/sq in). Tailwheel tyre pressure 5·52 bars (80 lb/sq in). Goodyear single-disc multi-puck hydraulic brakes. Main wheels fully enclosed by wheel-well doors when retracted.

POWER PLANT: Two 432 kW (579 ehp) Pratt & Whitney Aircraft of Canada PT6A-20 turboprop engines standard. Optional power plants include two 533 kW (715 ehp) (derated to 470 kW; 630 ehp) PT6A-27s, two 533 kW (715 ehp) (derated to 470 kW; 630 ehp) PT6A-28s or two 455 kW (610 ehp) Lycoming LTP-101s. Hartzell three-blade metal fully-feathering constant-speed propellers. Standard fuel capacity 1,544 litres (408 US gallons), contained in outer wing, inner wing and centre-section tanks. Optional total fuel in larger-capacity tanks 2,801 litres (740 US gallons). Refuelling points on wing upper surface. Engine air intakes have an inertial separator system, including foreign object and hail bypass, and heating of the leading-edges by engine bleed air.

ACCOMMODATION: Pilot and co-pilot on flight deck, with cabin seating eight passengers. Door on port side, aft of wing, with built-in airstair, can be replaced by larger cargo door. Separate door to flight deck, on port side of fuselage, is optional. Emergency exit (push-out type) on starboard side of cabin. Passenger seating quickly removable for conversion to all-cargo role. Baggage or cargo space aft of cabin and in extended fuselage nose. Cabin is heated by bleed air, and can be cooled by a bleed air converter. Windscreen de-icing standard.

SYSTEMS: Cabin cooling by AiResearch engine bleed air converter. Cabin heater manufactured by Hamilton Aircraft. Pneumatic system, for flight instruments and wing and tail unit de-icing, supplied by engine bleed air. Electrical system powered by two 200A starter/generators and nickel-cadmium battery. Oxygen system optional.

ELECTRONICS AND EQUIPMENT: Radio com/nav and radar to customer's requirements. Blind-flying instrumentation standard.

ARMAMENT: Optional armament for military versions includes a cargo pod containing two General Electric Miniguns, and hardpoints on the wings for the carriage of bombs or rockets.

DIMENSIONS, EXTERNAL:

Wing span	14·02 m (46 ft 0 in)
Wing chord at root	4·19 m (13 ft 9 in)
Wing chord at tip	1·07 m (3 ft 6 in)
Wing aspect ratio	6·5
Length overall	10·85 m (35 ft 7¼ in)
Tailplane span	4·56 m (14 ft 11½ in)
Passenger door (port, aft): Height	1·22 m (4 ft 0 in)

Hamilton Westwind III conversion of the Beech Model 18 (two Pratt & Whitney Aircraft of Canada PT6A-27 derated turboprop engines)

Width 0·56 m (1 ft 10 in)
Height to sill 0·79 m (2 ft 7 in)
Cargo door (port, aft, optional):
Max height (forward edge) 1·52 m (4 ft 11¾ in)
Min height (rear edge) 1·19 m (3 ft 11 in)
Width 1·47 m (4 ft 9¾ in)
Height to sill 0·79 m (2 ft 7 in)
Emergency exit (stbd):
Height 0·64 m (2 ft 1 in)
Width 0·48 m (1 ft 7 in)
DIMENSIONS, INTERNAL:
Cabin (bare cargo configuration):
Length 4·57 m (15 ft 0 in)
Max width 1·32 m (4 ft 4 in)
Max height 1·55 m (5 ft 1 in)
Floor area 7·80 m^2 (84 sq ft)
Baggage/cargo hold (aft cabin) 0·85 m^3 (30 cu ft)
Baggage/cargo hold (fuselage nose) 1·53 m^3 (54 cu ft)
AREAS:
Wings, gross 30·32 m^2 (326·4 sq ft)
Ailerons (total) 2·47 m^2 (26·6 sq ft)
Trailing-edge flaps (total) 3·49 m^2 (37·6 sq ft)
Fins (total) 3·03 m^2 (32·6 sq ft)
Rudders (total) 3·21 m^2 (34·56 sq ft)
Tailplane 6·08 m^2 (65·4 sq ft)
Elevator 2·53 m^2 (27·22 sq ft)
WEIGHTS AND LOADINGS:
Weight empty 2,495 kg (5,500 lb)
Max payload 1,814 kg (4,000 lb)
Max T-O weight 5,094 kg (11,230 lb)
Max zero-fuel weight 4,854 kg (10,700 lb)
Max landing weight 4,763 kg (10,500 lb)
Max wing loading 167·9 kg/m^2 (34·4 lb/sq ft)
Max power loading 5·90 kg/kW (19·4 lb/ehp)
PERFORMANCE (at max T-O weight. A: PT6A-20 engines; B: PT6A-27):
Max level speed at 3,660 m (12,000 ft):
A 234 knots (435 km/h; 270 mph)
B 269 knots (499 km/h; 310 mph)
Max cruising speed at 3,660 m (12,000 ft):
A 217 knots (402 km/h; 250 mph)
B 252 knots (467 km/h; 290 mph)
Econ cruising speed at 3,050 m (10,000 ft):
A 204 knots (378 km/h; 235 mph)
B 234 knots (435 km/h; 270 mph)
Max rate of climb at S/L:
A 549 m (1,800 ft)/min
B 823 m (2,700 ft)/min
Rate of climb at S/L, one engine out:
A 183 m (600 ft)/min
B 335 m (1,100 ft)/min
Service ceiling:
A 7,315 m (24,000 ft)
B 8,535 m (28,000 ft)
Service ceiling, one engine out:
A 2,745 m (9,000 ft)
B 3,960 m (13,000 ft)
T-O run:
A 549 m (1,800 ft)
B 366 m (1,200 ft)
T-O to 15 m (50 ft):
A 1,005 m (3,300 ft)
B 731 m (2,400 ft)
Landing from 15 m (50 ft):
A, B 549 m (1,800 ft)
Landing run:
A, B 366 m (1,200 ft)
Range with max optional fuel:
A 3,240 nm (6,004 km; 3,731 miles)
Range with max payload:
A 810 nm (1,501 km; 933 miles)

HAMILTON WESTWIND II STD

The Westwind II STD is a 'stretched' version of the Beech 18, providing accommodation for a maximum of 17 passengers. Otherwise it is generally similar to the Westwind III except that a version with tricycle landing gear is available optionally. The Westwind II STD is intended primarily as a commuter airliner, but is convertible for freight carrying or military uses.

The description of the Westwind III applies also to the Westwind II STD, except as follows:

LANDING GEAR: Retractable tricycle type available optionally.

POWER PLANT: Two 626 kW (840 ehp) Pratt & Whitney Aircraft of Canada PT6A-34 turboprop engines, derated to 470 kW (630 ehp), are standard, each driving a Hartzell constant-speed fully-feathering and reversible-pitch propeller. Optional power plants include two 579 kW (776 ehp) AiResearch TPE 331-6-251 or two 820 kW (1,100 ehp) Lycoming T5307A turboprop engines, in each case derated to 470 kW (630 ehp).

ACCOMMODATION: Pilot and co-pilot or passenger on flight deck, with seating in main cabin for a maximum of 17 passengers. Two emergency exits on starboard side of fuselage.

DIMENSIONS, EXTERNAL:
As for Westwind III, except:
Length overall (standard) 13·72 m (45 ft 0 in)
Length overall (tricycle landing gear) 13·46 m (44 ft 2 in)
DIMENSIONS, INTERNAL:
As for Westwind III, except:
Cabin: Length 6·10 m (20 ft 0 in)
WEIGHTS AND LOADING (estimated):
Weight empty 2,712 kg (6,000 lb)
Max payload 2,041 kg (4,500 lb)
Max T-O weight 5,667 kg (12,495 lb)
Max zero-fuel weight 5,217 kg (11,500 lb)
Max landing weight 5,217 kg (11,500 lb)
Max wing loading 186·9 kg/m^2 (38·3 lb/sq ft)
PERFORMANCE (estimated at max T-O weight):
Max level speed at 4,265 m (14,000 ft) 278 knots (515 km/h; 320 mph)
Max cruising speed at 3,660 m (12,000 ft) 261 knots (483 km/h; 300 mph)
Econ cruising speed at 6,705 m (22,000 ft) 234 knots (435 km/h; 270 mph)
Stalling speed, flaps up 87 knots (161 km/h; 100 mph)
Stalling speed, flaps down 74 knots (137 km/h; 85 mph)
Max rate of climb at S/L 945 m (3,100 ft)/min
Rate of climb at S/L, one engine out 274 m (900 ft)/min
Service ceiling 9,755 m (32,000 ft)
Service ceiling, one engine out 5,485 m (18,000 ft)
T-O run 427 m (1,400 ft)
T-O to 15 m (50 ft) 1,036 m (3,400 ft)
Landing from 15 m (50 ft):
without propeller reversal 671 m (2,200 ft)
with propeller reversal 335 m (1,100 ft)
Range with max optional fuel 3,240 nm (6,004 km; 3,731 miles)
Range with max payload 810 nm (1,501 km; 933 miles)

HAMILTON WESTWIND IV

Latest conversion of the Beech Model 18 by Hamilton Aviation is the Westwind IV. It is generally similar to the Westwind III but has the fuselage lengthened by 0·76 m (2 ft 6 in) and will be available with six alternative power plants and a large cargo door. Construction of the prototype began in January 1976.

TYPE: Twin-engined utility aircraft.

WINGS: Standard Beech Model 18 wings with Hamilton spar modification and crack detector system. Pneumatic de-icing boots on wing leading-edges.

FUSELAGE: Beech Model 18 fuselage lengthened 0·76 m (2 ft 6 in) aft of the wing. Heavier gauge skins and strengthened structure to fit aircraft for cargo role.

TAIL UNIT: Standard Beech Model 18 structure. Pneumatic de-icing boots on fin and tailplane leading-edges.

LANDING GEAR: Electrically-retractable tailwheel type. Main units retract aft. Oleo-pneumatic shock-absorbers. Goodyear wheels and tyres with single wheel on each unit. Single-disc hydraulic brakes.

POWER PLANT: Optional turboprop power plants include 429 kW (575 shp) Lycoming LTP-101; 507 kW (680 shp), 559 kW (750 shp) and 761 kW (1,020 shp) Pratt & Whitney Aircraft of Canada PT6A-28, -34 and -45 respectively; and 526 kW (705 shp) and 626 kW (840 shp) Garrett-AiResearch TPE 331-101 and TPE 331-251 respectively. Standard fuel in wing tanks with capacity of 1,219 litres (322 US gallons). Optional auxiliary header tanks for PT6 engine installations with capacity of 416 litres (110 US gallons). Alternatively, optional integral outer panel tanks with capacity of 1,279 litres (338 US gallons). Max fuel capacity (all models) 2,498 litres (660 US gallons).

ACCOMMODATION: Large cargo door on port side of fuselage, aft of wing, able to accept standard 'D'-type cargo containers. Cargo hold in nose with volume of 1·98 m^3 (70 cu ft). Accommodation air-conditioned and heated.

SYSTEMS: Garrett-AiResearch air-conditioning. Cabin heated by engine bleed air. Hydraulic system for brakes only. Electrical system supplied from starter/generators and storage battery. Pneumatic system for wing and tail unit de-icing and for flight instruments. Oxygen system optional.

ELECTRONICS AND EQUIPMENT: To customer's requirements.

DIMENSIONS, EXTERNAL:
Cargo door (port, aft):
Max width 2·16 m (7 ft 1 in)
Max height 1·52 m (5 ft 0 in)
DIMENSIONS, INTERNAL:
Cabin volume 9·77 m^3 (345 cu ft)
Nose hold volume 1·98 m^3 (70 cu ft)
WEIGHTS:
Weight empty, equipped 2,767 kg (6,100 lb)
Max payload 2,222 kg (4,900 lb)
Max T-O and ramp weight 5,670 kg (12,500 lb)
Max zero-fuel weight 4,990 kg (11,000 lb)
Max landing weight 5,216 kg (11,500 lb)
PERFORMANCE (estimated):
Max cruising speed, depending on power plant 213-282 knots (394-523 km/h; 245-325 mph)
Max range with 1,635 litres (432 US gallons) fuel, 1,452 kg (3,202 lb) payload and 30 min reserves 1,302 nm (2,414 km; 1,500 miles)
Max range with 2,498 litres (660 US gallons) fuel, 785 kg (1,731 lb) payload and 30 min reserves 2,010 nm (3,724 km; 2,314 miles)

Prototype of the Hamilton Westwind IV utility aircraft

HELIO

HELIO AIRCRAFT COMPANY (a division of General Aircraft Corporation)

HEAD OFFICE:
Hanscom Field, Bedford, Massachusetts 01730
Telephone: (617) 274-9130
WORKS:
Pittsburg, Kansas
PRESIDENT:
R. B. Kimnach
DIRECTOR OF ENGINEERING:
R. L. Devine
CONTROLLER: J. R. Cray
SALES AND PUBLIC RELATIONS:
H. A. Wheeler Jr (Asst to the President)

The original Helio Aircraft Corporation was founded in 1948 by Dr Lynn L. Bollinger of the Harvard Graduate School of Business Administration and Professor Otto C. Koppen of the Massachusetts Institute of Technology, to develop a light aircraft in the STOL category. In 1969 this company became a division of General Aircraft Corporation and was renamed Helio Aircraft Company.

After considerable flight testing of a prototype, converted from a Piper Vagabond light aircraft, Helio designed the original Courier prototype, which first flew in 1953. The first production Courier was the four-seat Model H-391B, which was certificated in 1954. It was followed by the 4/5-seat Model H-395 in 1958 and the Model H-395A in 1959. These models were superseded by the 186·5 kW (250 hp) six-seat Model H-250 in 1964 and the 220 kW (295 hp) H-295 in 1965. Production of the Model H-250 has ended, but the H-295 Super Courier continues in production in an improved form. An alternative version with a non-retractable tricycle landing gear was introduced in 1974, and has since been designated Trigear Courier Model HT-295.

Also in production is the 8/10-seat turboprop H-550A Stallion, which incorporates the same aerodynamic features as the Super Courier and is specifically designed for safe operation from unprepared fields. In December 1971 the USAF ordered a number of the armed version of the H-550A Stallion, and these were delivered during 1972.

Helio refers to all its products as C/STOL aircraft, signifying 'controlled short take-off and landing'.

Over 550 Helio Courier and Stallion aircraft of all types have been sold and are in service throughout the world. More than 150 have been delivered to the USAF; others are used by various US and foreign government agencies.

HELIO SUPER COURIER MODEL H-295 and TRIGEAR COURIER MODEL HT-295

USAF designation: U-10

The original version of the Super Courier was flown for

the first time in 1958 and received FAA Type Approval on 17 November that year. Three were supplied to the USAF for evaluation, under the designation L-28A. Further substantial orders were received subsequently, some aircraft being assigned to Tactical Air Command for counter-insurgency duties.

The current commercial versions of the Courier are the Super Courier Model **H-295** with non-retractable tail-wheel landing gear and the Trigear Courier Model **HT-295** with non-retractable tricycle-type landing gear.

Design and construction of the prototype H-295 began in late 1964 and it flew for the first time on 24 February 1965, FAA certification being received in the following month. Certification for the Trigear Courier HT-295 was received in March 1974, with the first production deliveries beginning immediately afterwards.

USAF Super Couriers are of three types, as follows:

U-10A. Standard model with fuel capacity of 227 litres (60 US gallons).

U-10B. Long-range version with standard internal fuel capacity of 455 litres (120 US gallons). This version has been operated in South-east Asia, South America, and in other parts of the world, on a wide variety of military missions and has an endurance of more than 10 hours. Paratroop doors standard.

U-10D. Improved long-range version with max AUW increased to 1,633 kg (3,600 lb). Standard internal fuel capacity of 455 litres (120 US gallons). Accommodation for pilot and five passengers.

The following details refer to the standard commercial Super Courier and Trigear Courier, except as noted:

Type: Six-seat light STOL personal, corporate and utility monoplane.

Wings: Cantilever high-wing monoplane. NACA 23012 wing section. Dihedral 1°. Incidence 3°. All-metal single-spar structure. Frise ailerons have duralumin frames and fabric covering and are supplemented by Arc-type aluminium spoilers, located at 15·5% chord on upper surface of each wing and geared to ailerons for control at low speeds. Ground-adjustable tab on ailerons. Full-span automatic all-metal Handley Page leading-edge slats. Electrically-operated NACA slotted all-metal trailing-edge flaps over 74% of span. No anti-icing equipment.

Fuselage: All-metal structure. Cabin section has welded steel tube framework, covered with aluminium; rear section is an aluminium monocoque.

Tail Unit: Cantilever all-metal structure. All-moving one-piece horizontal surface is fitted with trim and anti-balance tabs. Electrically-operated elevator trim optional.

Landing Gear (H-295): Non-retractable tailwheel type. Cantilever main legs. Oleo-pneumatic shock-absorbers of Helio design and manufacture on all three units. Goodyear crosswind landing gear with main-wheel tyres size 6·50-8, pressure 1·93 bars (28 lb/sq in). Goodyear 254 mm (10 in) tailwheel tyre, pressure 2·75 bars (40 lb/sq in). Goodyear hydraulic disc brakes. Edo 582-3430 floats, Edo Flying Dolphin amphibious floats or AirGlas Model LW3600 glassfibre wheel-skis optional.

Landing Gear (HT-295): Non-retractable tricycle type. Cantilever spring steel main gear, with wheels and tyres size 8·00-6, pressure 2·41 bars (35 lb/sq in). Nosewheel carried on oleo-pneumatic shock-strut, with wheel and tyre size 6·00-6, pressure 2·90 bars (42 lb/sq in).

Power Plant: One 220 kW (295 hp) Lycoming GO-480-G1A6 flat-six geared engine, driving a Hartzell three-blade constant-speed propeller. Rajay turbocharger system optional. Two 113·7 litre (30 US gallon) bladder-type fuel tanks in wings. Two further 113·7 litre (30 US gallon) tanks may be fitted to give total fuel capacity of 455 litres (120 US gallons). Oil capacity 11·4 litres (3 US gallons).

Accommodation: Cabin seats six in three pairs. Front and centre pair of seats individually adjustable. Rear pair comprises double sling seat. FAA standard instrument panel. Special over-strength cabin and seats, stressed to 15*g* and all fitted with safety harness, are based on Flight Safety Foundation recommendations. Two large doors, by pilot's seat on port side and opposite centre row of seats on starboard side. Baggage compartment aft of rear seats. Second- and third-row seats are removable for carrying over 454 kg (1,000 lb) of freight.

Electronics and Equipment: Radio and blind-flying instrumentation to customer's requirements.

Dimensions, external:
Wing span 11·89 m (39 ft 0 in)
Wing chord (constant) 1·83 m (6 ft 0 in)
Wing aspect ratio 6·58
Length overall 9·45 m (31 ft 0 in)
Height overall:
H-295 2·69 m (8 ft 10 in)
HT-295 4·52 m (14 ft 10 in)
Tailplane span 4·72 m (15 ft 6 in)
Wheel track 2·74 m (9 ft 0 in)
Wheelbase 7·14 m (23 ft 5 in)
Propeller diameter 2·44 m (8 ft 0 in)
Cabin door (fwd, port):
Height 1·04 m (3 ft 5 in)

Helio Super Courier Model H-295 six-seat STOL monoplane (Lycoming GO-480-G1A6 engine)

Helio Trigear Courier Model HT-295 with non-retractable tricycle landing gear

Width 0·85 m (2 ft 9½ in)
Height to sill 0·91 m (3 ft 0 in)
Cabin door (stbd, rear):
Height 0·98 m (3 ft 2½ in)
Width 0·85 m (2 ft 9½ in)
Height to sill 0·67 m (2 ft 2½ in)

Dimensions, internal:
Cabin: Length 3·05 m (10 ft 0 in)
Max width 1·14 m (3 ft 9 in)
Max height 1·22 m (4 ft 0 in)
Floor area 2·79 m² (30 sq ft)
Volume 3·96 m³ (140 cu ft)
Baggage space 0·42 m³ (15 cu ft)

Areas:
Wings, gross 21·46 m² (231 sq ft)
Ailerons (total) 1·92 m² (20·7 sq ft)
Flaps (total) 3·54 m² (38·1 sq ft)
Leading-edge slats (total) 2·91 m² (31·3 sq ft)
Spoilers (total) 0·16 m² (1·68 sq ft)
Fin 1·41 m² (15·2 sq ft)
Rudder 0·99 m² (10·6 sq ft)
Tailplane 3·48 m² (37·5 sq ft)

Weights and Loadings:
Weight empty:
H-295 943 kg (2,080 lb)
HT-295 970 kg (2,140 lb)
Max T-O and landing weight 1,542 kg (3,400 lb)
Max wing loading 71·8 kg/m² (14·7 lb/sq ft)
Max power loading 7·01 kg/kW (11·5 lb/hp)

Performance (at max T-O weight):
Never-exceed speed 174 knots (322 km/h; 200 mph)
Max level speed at S/L 145 knots (269 km/h; 167 mph)
Max cruising speed (75% power) at 2,600 m (8,500 ft) 143 knots (265 km/h; 165 mph)
Econ cruising speed (60% power) 130 knots (241 km/h; 150 mph)
Min flying speed, power on 26 knots (48 km/h; 30 mph)
Max rate of climb at S/L 350 m (1,150 ft)/min
Service ceiling 6,250 m (20,500 ft)
T-O run 102 m (335 ft)
T-O to 15 m (50 ft) 186 m (610 ft)
Landing from 15 m (50 ft) 158 m (520 ft)
Landing run 82 m (270 ft)
Range with standard tanks 573 nm (1,062 km; 660 miles)
Range with optional tanks 1,198 nm (2,220 km; 1,380 miles)

HELIO STALLION MODEL H-550A

USAF designation: AU-24A

Design of the turboprop Stallion was started in July 1963 and construction of the prototype Model HST-550 began in November 1963. First flight took place on 5 June 1964 and FAA certification was received in August 1965.

Construction of the first production version, known as the Stallion Model H-550A, began in April 1966 and FAA certification of this was received in August 1969.

The Model H-550A has full-span automatic leading-edge slats, an augmented lateral control system, slotted flaps to enhance STOL performance, and crash-resistant cabin structure. It is designed to operate over a wide speed range, to allow flexibility in operation.

The AU-24A armed version of the Stallion, acquired by the USAF, has an increased max T-O weight of 2,857 kg (6,300 lb) and incorporates a number of modifications to meet specific military requirements. These include two hardpoint pylons under each wing and a fuselage centre-line hardpoint for the mounting of MA4A bomb racks. Each outboard wing station has a capacity of 158 kg (350 lb), each inboard station 227 kg (500 lb), and the fuselage station has a 245 kg (540 lb) capacity. Equipment includes an armament control panel, gunsight, a cabin mounting for a side-firing gun as large as the M-197 20 mm cannon and ammunition magazines, and complete military VHF, UHF, FM and HF electronics. Numerous combinations of rockets, bombs and flares can be carried on the five external stores stations, and the cantilever high-wing design provides a virtually unobstructed field of fire for cabin-mounted rapid-firing guns.

The AU-24A can be configured for a variety of missions, including armed reconnaissance, COIN operations, close air support, transportation, and other special missions including forward air control.

A squadron comprising fourteen AU-24As was supplied to Cambodia by the USAF.

The following details refer to the commercial version of the Model H-550A:

TYPE: Eight/ten-seat general-utility STOL turboprop aircraft.

WINGS: Cantilever high-wing monoplane. Wing section slatted NACA 23012 (constant). Dihedral 1°. Incidence 3°. No sweepback. All-aluminium single-spar structure. Each wing unbolts at side of fuselage. Dacron-covered Frise balanced metal ailerons. NACA high-lift slotted all-metal flaps, electrically actuated. Arc-type all-metal spoilers at front of wing upper surface, interconnected with ailerons. Ground-adjustable tab on starboard aileron. Fully-automatic Handley Page full-span leading-edge slats.

FUSELAGE: Aluminium semi-monocoque structure, with welded steel tube framework forward of pilot's position.

TAIL UNIT: Cantilever all-aluminium structure, with sweptback vertical surfaces. All-moving one-piece horizontal surface with combined trim and anti-balance tab and separate flap trim interconnect tab.

LANDING GEAR: Non-retractable tailwheel type. Rearwardly-inclined cantilever main legs. Oleo-pneumatic shock-absorbers, designed and manufactured by Helio, on all three units. Goodyear tyres. Main wheels size 7·50-10, tyre pressure 1·52 bars (22 lb/sq in). Steerable tailwheel with 5·00-5 Type II tyre, pressure 3·79 bars (55 lb/sq in). Goodyear disc brakes. Wheel-ski landing gear available.

POWER PLANT: One 507 kW (680 ehp) Pratt & Whitney Aircraft of Canada PT6A-27 turboprop engine, driving a Hartzell three-blade reversible-pitch propeller. Fuel tanks in wings, with total capacity of 455 litres (120 US gallons). Refuelling points above wing. Oil capacity 8·75 litres (2·3 US gallons).

ACCOMMODATION: Pilot and co-pilot or passenger side by side at front, on fully-adjustable seats. Eight passengers in three rows, or up to six passengers two-abreast in individual seats with reclining backs and headrests. All passenger seats can be removed for cargo carrying. Full-length rails in floor for cargo restraint. Jettisonable door on each side of cabin by pilot. Double door, without central pillar, on port side of main cabin. Forward section of this door is hinged, rear portion slides. When sliding section is in place, forward section can be used alone as forward-hinged door. When rear (sliding) section is moved aft, the entire double door opening, size 1·55 m × 1·09 m (61 in × 43 in), is available for cargo loading. In the air, the sliding section moves aft to provide a parachuting or cargo-drop doorway. Similar double door on starboard side optional. Doors are non-structural. Hatches in wall of rear cabin enable pieces of freight up to 3·65 m (12 ft) long to be carried. Hatch size 0·58 m × 1·09 m (23 in × 43 in) in floor. Seats and harness stressed for 15*g*. Walls lined with fireproofing and soundproofing. Heating and ventilation standard.

SYSTEMS: Hydraulic system for brakes only. No pneumatic system. 24V electrical system supplied by 150A (optionally 200A) generator.

ELECTRONICS AND EQUIPMENT: To customer's requirements.

DIMENSIONS, EXTERNAL:
- Wing span 12·50 m (41 ft 0 in)
- Wing span over tip-tanks 12·72 m (41 ft 9 in)
- Wing chord (constant) 1·83 m (6 ft 0 in)
- Wing aspect ratio (without tip-tanks) 6·93
- Length overall 12·07 m (39 ft 7 in)
- Height overall 2·81 m (9 ft 3 in)
- Tailplane span 5·49 m (18 ft 0 in)
- Wheel track 2·94 m (9 ft 8 in)
- Wheelbase 7·52 m (24 ft 8 in)
- Propeller diameter 2·44 m (8 ft 0 in)
- Pilot's compartment doors (each):
  - Height 1·35 m (4 ft 5 in)
  - Width 1·03 m (3 ft 4 in)
  - Height to sill (mean) 1·14 m (3 ft 9 in)
- Hinged portion of double door:
  - Height 1·22 m (4 ft 0 in)
  - Width 0·79 m (2 ft 7 in)
  - Height to sill 0·89 m (2 ft 11 in)
- Sliding portion of double door:
  - Height 1·12 m (3 ft 8 in)
  - Width 0·79 m (2 ft 7 in)
  - Height to sill 0·91 m (3 ft 0 in)

DIMENSIONS, INTERNAL:
- Cabin: Length 4·11 m (13 ft 6 in)
  - Max width 1·28 m (4 ft 2½ in)
  - Max height 1·56 m (5 ft 1¼ in)
  - Floor area 4·03 m² (43·4 sq ft)
  - Volume 5·14 m³ (181·4 cu ft)

AREAS:
- Wings, gross 22·48 m² (242 sq ft)
- Wings, incl tip-tanks 23·04 m² (248 sq ft)
- Ailerons (total) 1·92 m² (20·7 sq ft)
- Trailing-edge flaps (total) 3·75 m² (40·32 sq ft)
- Leading-edge slats (total) 3·56 m² (38·3 sq ft)
- Spoilers (total) 0·29 m² (3·1 sq ft)
- Fin 1·58 m² (17·0 sq ft)
- Rudder, incl tab 1·82 m² (19·62 sq ft)
- Tailplane, incl tabs 5·33 m² (57·43 sq ft)

WEIGHTS AND LOADINGS:
- Weight empty 1,297 kg (2,860 lb)
- Max payload (with 455 litres; 120 US gallons fuel and pilot) 585 kg (1,290 lb)
- Max T-O and landing weight 2,313 kg (5,100 lb)
- Max wing loading 103·0 kg/m² (21·1 lb/sq ft)
- Max power loading 4·56 kg/kW (7·5 lb/hp)

PERFORMANCE (at max T-O weight):
- Never-exceed speed 190 knots (351 km/h; 218 mph) CAS
- Max level speed at 3,050 m (10,000 ft) 188 knots (348 km/h; 216 mph)
- Max cruising speed at 3,050 m (10,000 ft) 179 knots (332 km/h; 206 mph)
- Econ cruising speed at 3,050 m (10,000 ft) 139 knots (257 km/h; 160 mph)
- Min fully-manoeuvrable descent speed, power on 37 knots (68 km/h; 42 mph)
- Max rate of climb at S/L 671 m (2,200 ft)/min
- Service ceiling 7,620 m (25,000 ft)
- T-O run 98 m (320 ft)
- T-O to 15 m (50 ft) 201 m (660 ft)
- Landing from 15 m (50 ft) 229 m (750 ft)
- Landing run 76 m (250 ft)
- Range with max fuel, allowances for warm-up, taxying, take-off and climb to 3,050 m (10,000 ft) 557 nm (1,031 km; 641 miles)
- Range with max payload, allowances as above 386 nm (716 km; 445 miles)

Helio Stallion Model H-550A eight/ten-seat general-utility STOL aircraft

Helio AU-24A armed version of the Stallion with side-mounted XM-197 gun system and pylon-mounted rocket launchers

HELITEC
HELITEC CORPORATION

HEAD OFFICE:
4930 East Falcon Drive, Falcon Field, Mesa, Arizona 85205
Telephone: (602) 832-0600
Telex: 668-446

MARKETING MANAGER:
Arthur P. Murphy

Aviation Specialties developed a turbine-powered conversion of the Sikorsky S-55 helicopter which was awarded FAA certification in the Transport Category on 19 January 1971. Since that time a new company has been formed to carry out the conversions, Helitec Corporation, which is responsible also for marketing these aircraft in the USA. Aviation Specialties International is now purely a marketing company, responsible for international sales and for export of the finished aircraft. Production of at least 100 conversions is anticipated, and in early 1976 examples were operational in Alaska, Canada, Europe, South America and the USA.

HELITEC (SIKORSKY) S-55T

The S-55T conversion entails removing the existing Wright R-1300 or Pratt & Whitney R-1340 piston engine and replacing it with a Garrett-AiResearch TSE 331-3U-303 turboshaft engine of 626 kW (840 shp) derated to 485 kW (650 shp). This is mounted at an angle of approximately 35°, with the exhaust outlet facing to starboard, so that in flight it partially unloads the tail rotor. The new engine is connected to the engine mount by three gearbox mount fittings, and this in turn mates to the original attachment fittings on the firewall of the helicopter structure. The output-flange of the turbine connects directly to the existing fluid-drive clutch unit. The throttle-collective interconnection system has been redesigned to eliminate the interconnection, so that the throttle can be connected to the underspeed governor shaft of the engine's fuel control.

Various mechanical and electronic components formerly located in the tailboom of the S-55 have been resited in the forward compartment. This improves accessibility for maintenance and allows a wider range of loads to be carried within the CG limitations. A weight saving of approximately 408 kg (900 lb) is made by the conversion, and the large cabin of this helicopter makes it ideal for deployment in ambulance and rescue roles.

WEIGHTS AND LOADINGS:
- Weight empty, equipped 2,132 kg (4,700 lb)
- Max T-O and landing weight 3,265 kg (7,200 lb)
- Max disc loading 16·25 kg/m² (3·33 lb/sq ft)
- Max power loading 6·73 kg/kW (11·1 lb/shp)

PERFORMANCE (at max T-O weight):
- Max level speed at S/L 99 knots (183 km/h; 114 mph)
- Econ cruising speed at S/L 85 knots (157 km/h; 98 mph)
- Max rate of climb at S/L 366 m (1,200 ft)/min
- Max vertical rate of climb at S/L 244 m (800 ft)/min
- Service ceiling 3,780 m (12,400 ft)
- Hovering ceiling in ground effect 3,050 m (10,000 ft)
- Hovering ceiling out of ground effect 2,040 m (6,700 ft)
- Range with 681 litres (180 US gallons) fuel, 20 min reserves 321 nm (595 km; 370 miles)

Helitec Corporation S-55T turbine-powered conversion of the Sikorsky S-55 helicopter

HILLER

HILLER AVIATION (Division of Heliparts Inc)

ADDRESS:
2075 West Scranton Avenue, Porterville, California 93257
Telephone: (209) 781-2261
Telex: 682454

EXECUTIVE VICE-PRESIDENT:
Edwin L. Trupe

Hiller Aviation, formed in March 1973, acquired from Fairchild Industries the design rights, production tooling and spares of the Hiller 12E piston-engined light helicopter. Initially, the company provided product support for UH-12 helicopters in service throughout the world, a total then estimated as being in excess of 2,200 aircraft. Service and repair facilities were added as a first move to expand the company's business. It was then decided to begin the manufacture of new aircraft from existing components, incorporating all modifications approved for the type since closure of the production line in the late 1960s.

The company plans to build completely new aircraft if there is sufficient demand for the type, and has been working in conjunction with Soloy Conversions, of The Dalles, Oregon, to develop a turbine-powered version of the UH-12. The aim has been to evolve an installation which could not only be applied to a new production version, but would be suitable also for retrofit to existing UH-12s.

All available details of the basic UH-12E follow:

Hiller UH-12E three-seat utility helicopter

HILLER AVIATION UH-12E

TYPE: Three-seat utility helicopter.

ROTOR SYSTEM: Two-blade main rotor mounted universally on driveshaft, with small servo rotor; the latter is connected directly to the pilot's cyclic control stick, through a universally-mounted transfer bearing and simple linkage. Movement of the control stick introduces positive or negative pitch changes to the servo rotor paddles. The resulting aerodynamic forces tilt the rotor head and produce cylic pitch changes to the rotor blades. Each main rotor blade has a steel spar and leading-edge, with light alloy trailing-edge skin, extrusion and channels. Blades are interchangeable individually and are bolted to forks which are retained at the rotor head by tension-torsion bars. Blades do not fold. Rotor brake optional. Two-blade tail rotor of light alloy construction, mounted on port side of tailboom.

ROTOR DRIVE: Mechanical drive through two-stage planetary main transmission. Bevel gear drive to auxiliaries. Tail rotor gearbox and fan gearbox. Main rotor/engine rpm ratio 1 : 8·64. Tail rotor/engine rpm ratio 1 : 1·45.

FUSELAGE: Light alloy fully-stressed semi-monocoque platform structure supporting the non-stressed cabin enclosure, engine mounting and landing gear. Tailboom of beaded light alloy sheet with no internal stiffeners.

TAIL UNIT: Horizontal stabiliser on starboard side of tailboom, with steel tube spar, light alloy ribs and skin. Incidence ground-adjustable.

LANDING GEAR: Wide-track steel tube skids carried on spring steel cross members. Optional 'zip-on' pontoons, attached above the skids, permit water or land operations.

POWER PLANT: One 253·5 kW (340 hp) Lycoming VO-540 flat-six engine, installed vertically and derated to 227·5 kW (305 hp). Single bladder fuel cell, capacity 174 litres (46 US gallons), mounted in lower portion of rear fuselage, beneath engine. Two optional 75·7 litre (20 US gallon) auxiliary fuel tanks, mounted in fuselage on each side of engine. Oil capacity 12·5 litres (3·3 US gallons).

ACCOMMODATION: Three persons side by side on bench seat. Dual controls optional. Forward-hinged door on each side. Baggage compartment immediately aft of engine. Heater-defroster optional.

ELECTRONICS AND EQUIPMENT: A range of optional electronics is available. Optional equipment includes 453·6 kg (1,000 lb) capacity rescue hook, hydraulically-driven rescue hoist, twin heavy-duty cargo racks, twin litters, agricultural spray equipment, loudspeaker/siren, engine hour meter, additional engine oil filter, extended landing gear legs, navigation lights, inertia reel shoulder harness, Mason cyclic control grip, cabin fire extinguisher and first aid kit.

DIMENSIONS, EXTERNAL:

Diameter of main rotor	10·80 m (35 ft 5 in)
Diameter of tail rotor	1·68 m (5 ft 6 in)
Distance between rotor centres	6·17 m (20 ft 3 in)
Length overall, rotors turning	12·41 m (40 ft 8½ in)
Length of fuselage	8·69 m (28 ft 6 in)
Height to top of rotor hub	3·08 m (10 ft 1¼ in)
Skid track	2·29 m (7 ft 6 in)
Cabin doors (each):	
Height	1·13 m (3 ft 8½ in)
Max width	0·76 m (2 ft 6 in)
Height to sill	0·58 m (1 ft 11 in)

DIMENSIONS, INTERNAL:

Cabin: Length	1·52 m (5 ft 0 in)
Max width	1·50 m (4 ft 11 in)
Max height	1·35 m (4 ft 5 in)
Floor area	1·16 m² (12·5 sq ft)

AREAS:

Main rotor blades (each)	1·51 m² (16·3 sq ft)
Tail rotor blades (each)	0·094 m² (1·01 sq ft)
Main rotor disc	91·97 m² (990 sq ft)
Tail rotor disc	2·57 m² (27·7 sq ft)

WEIGHTS AND LOADINGS:

Weight empty	798 kg (1,759 lb)
Max T-O weight	1,270 kg (2,800 lb)
Max disc loading	13·82 kg/m² (2·83 lb/sq ft)
Max power loading	5·01 kg/kW (9·2 lb/hp)

PERFORMANCE (at max T-O weight):

Max level speed at S/L	83·5 knots (154 km/h; 96 mph)
Max cruising speed at S/L	78·5 knots (145 km/h; 90 mph)
Max rate of climb at S/L	393 m (1,290 ft)/min
Vertical rate of climb at S/L	226 m (740 ft)/min
Service ceiling	4,940 m (16,200 ft)
Hovering ceiling in ground effect	3,290 m (10,800 ft)
Hovering ceiling out of ground effect	2,195 m (7,200 ft)
Range with standard fuel	187 nm (346 km; 215 miles)
Range with max auxiliary fuel	364 nm (676 km; 420 miles)

HUGHES

HUGHES HELICOPTERS (Division of Summa Corporation)

HEAD OFFICE AND WORKS:
Culver City, California 90230
Telephone: (213) 390-4451
Telex: 67-222
VICE-PRESIDENT AND GENERAL MANAGER:
Thomas R. Stuelpnagel
VICE-PRESIDENTS:
William E. Rankin (Finance)
C. D. Perry (Marketing)
DIRECTORS:
W. J. Blackburn (Manufacturing)
R. E. Brix (Ordnance Engineering)
J. N. Kerr (Military Helicopters)
L. P. Sonsini (Quality Assurance)
F. C. Strible (Commercial Helicopters)
PUBLIC RELATIONS AND ADVERTISING MANAGER:
Robert L. Parrish

Following reorganisation of Hughes Tool Company as the Summa Corporation, its former Aircraft Division is now known as Hughes Helicopters.

Products comprise light helicopters powered by reciprocating and turboshaft engines. Current research activities include work on composite rotor blades and tailbooms, metal insulation and IR suppression systems, as well as chain gun and lockless ordnance systems for air or ground applications.

Kawasaki Heavy Industries Ltd in Japan (which see) has assembled a number of Model 369HM helicopters, this being the designation of the uprated version of the Hughes OH-6 available to foreign military customers. Other licence manufacture is undertaken by RACA in the Argentine and BredaNardi in Italy.

In the Summer of 1973 Hughes was awarded a US Army contract for the development of two Advanced Attack Helicopter prototypes to compete against those being developed by Bell Helicopter.

HUGHES MODEL 300

US Army designation: TH-55A Osage

Design and development of the original Hughes Model 269 two-seat light helicopter began in September 1965 and the first of two prototypes was flown 13 months later.

The design was then re-engineered for production with the emphasis on simplicity and ease of maintenance. The resulting Model 269A offered an overall life of over 1,000 hours for all major components.

Five Model 269A pre-production helicopters were purchased by the US Army under the designation YHO-2HU, and completed a highly successful evaluation programme in the command and observation roles.

The Model 269A was then put into production and deliveries began in October 1961. The design has since undergone considerable development, leading to new versions, the latest commercial and military models being as follows:

Model 300. Three-seat version developed under the engineering designation 269B. Received FAA Type Approval 30 December 1963. In production at a rate of one a day by 1964. Hughes engineers have perfected a quiet tail rotor (QTR) for the Model 300, which reducees the sound level of the aircraft by 80%. At cruise rpm, the QTR-equipped version operates at a noise level comparable with that of a fixed-wing light aircraft. QTR has been standard factory-installed equipment on production Model 300s since June 1967, and retrofit kits are available to all Model 269A and 300 owners.

Model 300C. This three-seat version of the Hughes Model 300 was developed under the engineering designation 269C. It is described separately.

TH-55A. The Hughes 269A was selected by the US Army as a light helicopter primary trainer in mid-1964, under the designation TH-55A. A total of 792 were eventually delivered, production being completed by the end of March 1969.

It was reported in March 1973 that Hughes had been evaluating a TH-55A powered by a 138 kW (185 hp) Wankel RC 2-60 rotating-piston engine. Another has been fitted experimentally with an Allison 250-C18 turboshaft, similar to the standard engine of the US Army's OH-6As (see page 305).

TYPE: One-, two- or three-seat light helicopter.

ROTOR SYSTEM (all models): Fully-articulated metal three-blade main rotor. Blades are of bonded construction, with constant-section extruded aluminium spar, wraparound skin and a trailing-edge section. Blade section NACA 0015. Two-blade teetering tail rotor, each blade comprising a steel tube spar with glassfibre skin. Blades do not fold. No rotor brake.

ROTOR DRIVE: Vee-belt drive system eliminates need for conventional clutch. Metal-coated and hard-anodised sheaves. Spiral bevel angular drive-shaft. Tail rotor shaft-driven directly from belt-drive. Main rotor/engine rpm ratio 1 : 6.

FUSELAGE: Welded steel tube structure, with aluminium and Plexiglas cabin and one-piece aluminium tube tailboom.

TAIL UNIT: Horizontal and vertical fixed stabilisers made up of aluminium ribs and skin.

LANDING GEAR: Skids carried on Hughes oleo-pneumatic shock-absorbers. Two cast magnesium ground handling wheels with 0·25 m (10 in) balloon tyres, pressure 4·14-5·17 bars (60-75 lb/sq in). Model 300 is available on floats made of polyurethane coated nylon fabric, 4·70 m (15 ft 5 in) long and with total installed weight of 27·2 kg (60 lb).

POWER PLANT: One 134 kW (180 hp) Lycoming HIO-360-A1A (HIO-360-B1A in TH-55A) flat-four engine, mounted horizontally below seats. Aluminium fuel tank, capacity 103·5 litres (30 US gallons), mounted externally aft of cockpit. Provision for aluminium auxiliary fuel tank, capacity 72 litres (19 US gallons), mounted opposite standard tank. Oil capacity 7·5 litres (2 US gallons).

ACCOMMODATION: Two seats (TH-55A) or three seats (Model 300) side by side in Plexiglas-enclosed cabin. Door on each side. Dual controls optional. Baggage capacity 45 kg (100 lb). Exhaust muff or gasoline-heating and ventilation kits available.

ELECTRONICS AND EQUIPMENT (Model 300): Optional equipment includes King KY 90 radio, welded aluminium Stokes litter kit, cargo rack, external load sling of 272 kg (600 lb) capacity.

ELECTRONICS AND EQUIPMENT (TH-55A): Provision for ARC-524M VHF radio.

DIMENSIONS, EXTERNAL:
Diameter of main rotor 7·71 m (25 ft 3½ in)
Main rotor blade chord 0·173 m (6·83 in)
Diameter of tail rotor 1·17 m (3 ft 10 in)
Distance between rotor centres 4·29 m (14 ft 1 in)
Length overall 8·80 m (28 ft 10¾ in)
Length of fuselage 6·80 m (21 ft 11¾ in)
Height overall 2·50 m (8 ft 2¾ in)
Skid track 2·00 m (6 ft 6½ in)
Cabin doors (each):
Height 1·12 m (3 ft 8 in)
Width 0·81 m (2 ft 8 in)
Height to sill 0·89 m (2 ft 11 in)

DIMENSIONS, INTERNAL:
Cabin: Length 1·40 m (4 ft 7 in)
Max width 1·30 m (4 ft 3 in)
Max height 1·32 m (4 ft 4 in)
Floor area 1·21 m² (13·0 sq ft)

AREAS:
Main rotor blades (each) 0·66 m² (7·1 sq ft)
Tail rotor blades (each) 0·07 m² (0·77 sq ft)
Main rotor disc 46·73 m² (503 sq ft)
Tail rotor disc 0·81 m² (8·70 sq ft)
Fin 0·11 m² (1·22 sq ft)
Horizontal stabiliser 0·32 m² (3·44 sq ft)

WEIGHTS AND LOADINGS:
Weight empty:
300 434 kg (958 lb)
TH-55A 457 kg (1,008 lb)
Max certificated T-O and landing weight:
300, TH-55A 757 kg (1,670 lb)
Max recommended weight (restricted operation):
300, TH-55A 839 kg (1,850 lb)
Max disc loading (at certificated AUW):
300, TH-55A 16·1 kg/m² (3·3 lb/sq ft)
Max power loading (at certificated AUW):
300, TH-55A 5·65 kg/kW (9·3 lb/hp)

PERFORMANCE (at max certificated T-O weight):
Never-exceed speed:
300 75·5 knots (140 km/h; 87 mph)
TH-55A 75 knots (138 km/h; 86 mph)
Max level speed at S/L:
300 75·5 knots (140 km/h; 87 mph)
TH-55A 75 knots (138 km/h; 86 mph)
Max cruising speed:
300 69 knots (129 km/h; 80 mph)
TH-55A 65 knots (121 km/h; 75 mph)
Econ cruising speed:
300, TH-55A 57 knots (106 km/h; 66 mph)
Max water contact speed (on floats)
17 knots (32 km/h; 20 mph)
Max water taxying speed (on floats)
9 knots (16 km/h; 10 mph)
Max rate of climb at S/L:
300 347 m (1,140 ft)/min
TH-55A (mission weight) 347 m (1,140 ft)/min
Service ceiling:
300 3,960 m (13,000 ft)
TH-55A (mission weight) 3,625 m (11,900 ft)
Hovering ceiling in ground effect:
300 2,350 m (7,700 ft)
TH-55A 1,675 m (5,500 ft)
Hovering ceiling out of ground effect:
300 1,770 m (5,800 ft)
TH-55A 1,145 m (3,750 ft)
Range with max fuel, no reserves:
300 260 nm (480 km; 300 miles)
TH-55A 177 nm (328 km; 204 miles)
Endurance with max fuel:
300 3 hr 30 min
TH-55A 2 hr 35 min

HUGHES MODEL 300C

This is a developed version of the Model 300, with improvements to allow an increase in payload of 45 per cent. Construction of the prototype started in July 1968, and this made its first flight in August 1969, followed by the first production model in December 1969. FAA certification was received in May 1970. A total of 101 Model 300Cs were delivered during 1974.

The Model 300C is also manufactured in Italy by BredaNardi (which see).

Hughes TH-55A (modified) with Allison Model 250-C18 turboshaft engine, similar to the standard engine of the Army OH-6A *(Henry Artof)*

Special bubble window over ends of two stretchers in Hughes Model 500C ambulance *(Henry Artof)*

The introduction of a more powerful engine and an increase of main rotor diameter required a number of related structural changes, including use of a larger tail rotor and fin of greater area. The main rotor mast and tailboom were lengthened to accommodate the longer and heavier rotor blades.

Following upon the research that produced a modified version of the OH-6A known as 'The Quiet One', Hughes used similar techniques to develop and obtain full FAA certification of a quiet version of the Model 300, and this has the designation Model **300CQ**. In this new configuration, emission of audible sound is 75 per cent less than with earlier models, and it is possible for the necessary modifications to be retrofitted to existing 300Cs. Max T-O weight of the 300CQ for quiet operation is 873 kg (1,925 lb), with a useful load of 397 kg (875 lb), and there is little change in range and endurance by comparison with the standard Model 300C.

The description of the standard Model 300 applies also to the Model 300C, except in the following details:

ROTOR SYSTEM: As Model 300 except that limited folding is possible. Tracking tabs on main rotor blades at three-quarters radius.

ROTOR DRIVE: Combination Vee-belt/pulley and reduction gear drive system. Main rotor and tail rotor gearbox have spiral bevel right-angle drive. Main rotor/engine rpm ratio 1 : 6·8. Tail rotor/engine rpm ratio 0·97 : 1.

POWER PLANT: One 142 kW (190 hp) Lycoming HIO-360-D1A flat-four engine. Oil capacity 9·5 litres (2·5 US gallons).

ACCOMMODATION: Three persons seated side by side on sculptured and cushioned bench seat.

ELECTRONICS AND EQUIPMENT: Optional electronics include King KY-95 VHF radio and headsets. Optional equipment includes amphibious floats, litters, cargo rack, external load sling of 272 kg (600 lb) capacity, agricultural spray or dry powder dispersion kits, 72 litre (19 US gallon) auxiliary fuel tank, fire extinguisher, dual baggage case, night flying kit, external power socket, dual controls, all-weather cover, heavy-duty skid plates, exhaust muffler, main rotor tie-down kit, door lock, dual oil cooler, tinted glass for cabin windows, gasoline or exhaust mainfold cabin heating.

DIMENSIONS, EXTERNAL:
- Diameter of main rotor — 8·18 m (26 ft 10 in)
- Main rotor blade chord — 0·171 m (6¾ in)
- Diameter of tail rotor — 1·30 m (4 ft 3 in)
- Length overall, rotor blades fore and aft — 9·42 m (30 ft 11 in)
- Height over rotor hub — 2·67 m (8 ft 9 in)
- Width, rotor partially folded — 2·44 m (8 ft 0 in)
- Skid track — 1·91 m (6 ft 3 in)
- Passenger doors (each):
  - Height — 1·09 m (3 ft 7 in)
  - Width — 0·97 m (3 ft 2 in)
  - Height to sill — 0·91 m (3 ft 0 in)

AREAS:
- Main rotor blades (each) — 0·70 m² (7·55 sq ft)
- Tail rotor blades (each) — 0·08 m² (0·86 sq ft)
- Main rotor disc — 52·5 m² (565·5 sq ft)
- Tail rotor disc — 1·32 m² (14·2 sq ft)
- Fin — 0·26 m² (2·8 sq ft)
- Horizontal stabiliser — 0·32 m² (3·44 sq ft)

WEIGHTS AND LOADINGS (A: 300C; B: 300CQ):
- Weight empty — 476 kg (1,050 lb)
- Max T-O weight:
  - A — 930 kg (2,050 lb)
  - B — 921 kg (2,030 lb)
- Max disc loading:
  - A — 17·67 kg/m² (3·62 lb/sq ft)
  - B — 17·48 kg/m² (3·58 lb/sq ft)

PERFORMANCE (at max T-O weight, ISA. A: 300C; B: 300CQ normal operation; C: 300CQ quiet operation at AUW of 873 kg; 1,925 lb):
- Never-exceed speed:
  - A, B, C — 91 knots (169 km/h; 105 mph)
- Max cruising speed at S/L:
  - A, B — 78 knots (145 km/h; 90 mph)
- Max speed at S/L:
  - C — 61 knots (113 km/h; 70 mph)
- Max cruising speed at 1,220 m (4,000 ft):
  - A — 82·5 knots (153 km/h; 95 mph)
  - B — 78 knots (145 km/h; 90 mph)
- Max speed at 1,220 m (4,000 ft):
  - C — 62·5 knots (116 km/h; 72 mph)
- Max rate of climb at S/L:
  - A — 229 m (750 ft)/min
  - C — 113 m (370 ft)/min
- Service ceiling:
  - A — 3,110 m (10,200 ft)
  - B — 3,050 m (10,000 ft)
  - C — 2,440 m (8,000 ft)
- Hovering ceiling in ground effect:
  - A — 1,830 m (6,000 ft)
  - B — 1,220 m (4,000 ft)
- Hovering ceiling out of ground effect:
  - A — 823 m (2,700 ft)
- Range at 1,220 m (4,000 ft), 2 min warm-up, max fuel, no reserves — 200 nm (370 km; 230 miles)

Hughes Model 300C three-seat light helicopter (Lycoming HIO-360-D1A engine)

HUGHES OH-6

US Army designation: OH-6A (formerly HO-6) Cayuse

This aircraft was chosen for development following a US Army design competition for a light observation helicopter in 1961. Five prototypes were ordered for evaluation in competition with the Bell OH-4A and Hillèr OH-5A, and the first of these flew on 27 February 1963.

On 26 May 1965 it was announced that the OH-6A had been chosen, as a result of the evaluation, and an initial order for 714 was placed by the US Army; this was increased by subsequent orders to a total of 1,434, all of which were delivered by August 1970.

In March and April 1966, US Army and civilian pilots set up 23 international records in OH-6A helicopters. Among Class E1 (covering all classes of helicopters) records established was one for a distance of 1,922 nm (3,561·55 km; 2,213 miles) in a straight line (California to Florida) nonstop with one pilot.

On 8 April 1971 Hughes announced the existence of a modified OH-6A light observation helicopter known as 'The Quiet One'. Product of a research project funded by the Department of Defense Advanced Research Projects Agency, Hughes claims that it is the world's quietest helicopter.

Modifications include the installation of a five-blade main rotor, four-blade tail rotor and engine exhaust muffler; sound blanketing of the complete power plant assembly, including engine air intake; and reshaping of the tips of the main rotor blades. The modified aircraft can operate with engine and rotor speeds reduced to 67 per cent of normal in-flight levels and is able to offer a 272 kg (600 lb) increase in payload and 20 knot (37 km/h; 23 mph) increase in airspeed.

A similar aircraft, designated **OH-6C** and powered by a 298 kW (400 shp) Allison 250-C20 turboshaft engine, achieved a speed of 173 knots (322 km/h; 200 mph) during tests at Edwards AFB.

It is reported that an improved version of the OH-6C, with a four-blade tail rotor and max T-O weight of 1,315 kg (2,900 lb), has been offered to the US Army. Designated **OH-6D**, this has been proposed to meet the Army's requirements for an Advanced Scout Helicopter (ASH).

Full-scale production of the Hughes 500 commercial and 500M international military versions of the OH-6A (which are described separately) began in November 1968.

TYPE: Turbine-powered light observation helicopter.

ROTOR SYSTEM: Four-blade fully-articulated main rotor, with blades attached to laminated strap retention system by means of folding quick-disconnect pins. Each blade consists of an extruded aluminium spar hot-bonded to one-piece wraparound aluminium skin. Trim tab outboard on each blade. Main rotor blades can be folded. Two-blade tail rotor, each blade comprising a swaged steel tube spar and glassfibre skin covering. No rotor brake.

ROTOR DRIVE: Three sets of bevel gears, three drive-shafts and one overrunning clutch. Main rotor/engine rpm ratio 1 : 12·806. Tail rotor/engine rpm ratio 1 : 1·987.

FUSELAGE: Aluminium semi-monocoque structure of pod and boom type. Clamshell doors at rear of pod give access to engine and accessories.

TAIL UNIT: Fixed fin, horizontal stabiliser and ventral fin.

LANDING GEAR: Tubular skids carried on Hughes single-acting shock-absorbers.

POWER PLANT: One 236·5 kW (317 shp) Allison T63-A-5A turboshaft engine, derated to 188·3 kW (252·5 shp) for take-off and 160 kW (214·5 shp) max continuous rating. Two 50% self-sealing bladder fuel tanks under rear cabin floor, capacity 232 litres (61·5 US gallons). Refuelling point aft of cargo door on starboard side. Oil capacity 4·75 litres (1·25 US gallons).

ACCOMMODATION: Crew of two side by side in front of cabin. Two seats in rear cargo compartment can be folded to make room for four fully-equipped soldiers, seated on floor. Crew door and cargo compartment door on each side. Fourteen cargo tie-down points.

ELECTRONICS AND EQUIPMENT: Government-furnished electronics. Sylvania SLAE electronics package installed in 1969/70 production aircraft. ARC-114 VHF-FM and ARC-116 UHF radios, ARN-89 ADF, ASN-43 gyro compass, ID 1351 bearing-heading indicator and ARC-6533 intercoms are standard. ARC-115 may be substituted for ARC-116.

ARMAMENT: Provision for carrying packaged armament on port side of fuselage, comprising XM-27 7·62 mm machine-gun, with 2,000-4,000 rds/min capability, or XM-75 grenade launcher.

DIMENSIONS, EXTERNAL:
- Diameter of main rotor — 8·03 m (26 ft 4 in)
- Main rotor blade chord — 0·171 m (6¾ in)
- Diameter of tail rotor — 1·30 m (4 ft 3 in)
- Distance between rotor centres — 4·58 m (15 ft 0¼ in)
- Length overall, rotors fore and aft — 9·24 m (30 ft 3¾ in)
- Length of fuselage — 7·01 m (23 ft 0 in)
- Height to top of rotor hub — 2·48 m (8 ft 1½ in)
- Skid track — 2·06 m (6 ft 9 in)
- Cabin doors (fwd, each):
  - Height — 1·19 m (3 ft 11 in)
  - Width — 0·89 m (2 ft 11 in)
- Cargo compartment doors (each):
  - Height — 1·04 m (3 ft 5 in)
  - Width — 0·88 m (2 ft 10½ in)
  - Height to sill — 0·57 m (1 ft 10½ in)

DIMENSIONS, INTERNAL:
- Cabin: Length — 2·44 m (8 ft 0 in)
  - Max width — 1·37 m (4 ft 6 in)
  - Max height — 1·31 m (4 ft 3½ in)

AREAS:
- Main rotor blades (each) — 0·69 m² (7·41 sq ft)
- Tail rotor blades (each) — 0·079 m² (0·85 sq ft)
- Main rotor disc — 50·60 m² (544·63 sq ft)
- Tail rotor disc — 1·32 m² (14·19 sq ft)
- Fin — 0·52 m² (5·65 sq ft)
- Horizontal stabiliser — 0·72 m² (7·70 sq ft)

WEIGHTS AND LOADINGS:
- Weight empty, equipped — 557 kg (1,229 lb)
- Design gross weight — 1,090 kg (2,400 lb)
- Overload gross weight — 1,225 kg (2,700 lb)
- Design disc loading — 21·48 kg/m² (4·4 lb/sq ft)
- Design power loading — 5·79 kg/kW (9·5 lb/shp)

PERFORMANCE (at design gross weight):
- Never-exceed speed and max cruising speed at S/L — 130 knots (241 km/h; 150 mph)
- Cruising speed for max range at S/L — 116 knots (216 km/h; 134 mph)
- Max rate of climb at S/L (military power) — 560 m (1,840 ft)/min
- Max rate of climb at S/L (max continuous power) — 381 m (1,250 ft)/min
- Service ceiling — 4,815 m (15,800 ft)
- Hovering ceiling in ground effect — 3,595 m (11,800 ft)
- Hovering ceiling out of ground effect — 2,225 m (7,300 ft)
- Normal range at 1,525 m (5,000 ft) — 330 nm (611 km; 380 miles)

Ferry range (590 kg; 1,300 lb fuel)
1,354 nm (2,510 km; 1,560 miles)

HUGHES MODEL 500, 500C, 500D and 500M

These are the commercial and foreign military counterparts of the OH-6A military helicopter:

Model 500. Commercial helicopter, with accommodation for pilot and four passengers or equivalent freight. Optional accommodation for seven with litter kit in use or with four in passenger compartment.

Model 500C. As Model 500, except for installation of 298 kW (400 shp) engine for improved hot-day/altitude performance.

Model 500D. Commercial version of the OH-6C, described separately.

Model 500M. Uprated version of OH-6A, available to foreign military customers. First deliveries to Colombian Air Force in April 1968. Now in service also in Japan, Argentina, Denmark, Spain, Mexico and the Philippines. The Model 500Ms delivered to the Spanish Navy for ASW duties have an AN/ASQ-81 magnetic anomaly detector installed on the starboard side of the fuselage, and can carry two Mk 44 torpedoes beneath the fuselage. Control boxes for the MAD equipment are mounted on the instrument panel and centre pedestal, and special instrumentation includes a 6 in attitude indicator and radar altimeter.

Licence manufacture of the Model 500 Series is undertaken by RACA in the Argentine, Kawasaki in Japan and BredaNardi in Italy (which see).

Although similar in basic design and construction to the OH-6A, the Model 500s have been substantially uprated. The 236·5 kW (317 shp) Allison Model 250-C18A turbine engine (civil version of the T63-A-5A), installed in the 500 and 500M, is derated only to 207 kW (278 shp) for T-O and has a maximum continuous rating of 181 kW (243 shp). The Model 500C is powered by a 298 kW (400 shp) Allison Model 250-C20 turboshaft engine; this also is derated to 207 kW (278 shp) for T-O and has a maximum continuous rating of 181 kW (243 shp). The Models 500 and 500C have a fuel capacity of 242 litres (64 US gallons); the Model 500M has a capacity of 227 litres (60 US gallons).

Optional equipment available for the Model 500 includes shatterproof glass, heating system, radios and intercom, attitude and directional gyros, rate of climb indicator, inertia reels and shoulder harnesses for pilot and co-pilot, fire extinguisher, dual controls, cargo hook, hoist, auxiliary fuel system, heated pitot tube, extended landing gear, blade storage rack, litter kit, emergency inflatable floats, inflated utility floats, rotor brake, seating for four in passenger compartment, and first aid kit. Standard equipment includes engine hour meter, navigation lights, clock and ground handling wheels.

DIMENSIONS AND AREAS:
As for OH-6A

WEIGHTS:
Weight empty:
500 — 493 kg (1,088 lb)
500C — 501 kg (1,105 lb)
500M — 512 kg (1,130 lb)
Max normal T-O weight — 1,157 kg (2,550 lb)
Max overload T-O weight — 1,360 kg (3,000 lb)

PERFORMANCE (at max T-O weight):
Max level speed at 305 m (1,000 ft):
500, 500M — 132 knots (244 km/h; 152 mph)
Max cruising speed at S/L:
500C — 125 knots (232 km/h; 144 mph)
Max cruising speed at 1,220 m (4,000 ft):
500C — 126 knots (233 km/h; 145 mph)
Cruising speed for max range at S/L:
500, 500M — 117 knots (217 km/h; 135 mph)
500C — 124 knots (230 km/h; 143 mph)
Max rate of climb at S/L:
500, 500C, 500M — 518 m (1,700 ft)/min
Service ceiling:
500, 500M — 4,390 m (14,400 ft)
500C — 4,420 m (14,500 ft)
Hovering ceiling in ground effect:
500, 500M — 2,500 m (8,200 ft)
500C — 3,960 m (13,000 ft)
Hovering ceiling out of ground effect:
500, 500M — 1,615 m (5,300 ft)
500C — 2,040 m (6,700 ft)
Range at 1,220 m (4,000 ft):
500 — 327 nm (606 km; 377 miles)
500M — 318 nm (589 km; 366 miles)
Range at 1,220 m (4,000 ft), 2 min warm-up with max fuel, no reserves:
500C — 325 nm (603 km; 375 miles)

HUGHES MODEL 500D

Announced in February 1975, the Model 500D is similar in size and general appearance to the Hughes Model 500C. It differs in having a 313 kW (420 shp) Allison 250-C20B engine, plus the modifications detailed for the OH-6C and which have been developed and test-flown extensively on that aircraft. It introduces also a small T-tail which gives greater flight stability in both high and low speed regimes, as well as better handling characteristics in abnormal manoeuvres.

Hughes 500C equipped with fire suppression equipment developed by Chadwick Inc *(Howard Levy)*

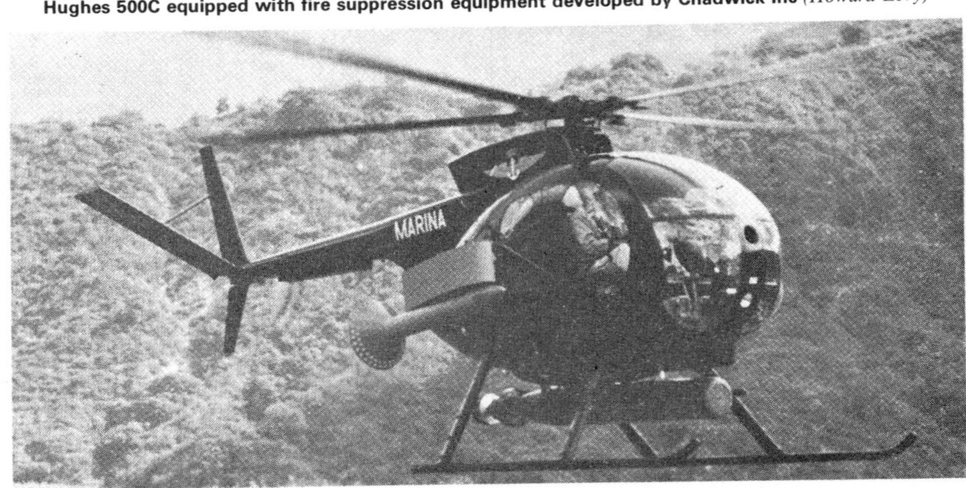

Anti-submarine version of the Hughes 500M in Spanish Navy insignia

Construction of the prototype and its first flight took place in August 1974, and the first flight of a production aircraft was scheduled during 1975. The flight test and development programme was aimed at achieving FAA Normal and Utility production certification by January 1976.

TYPE: Turbine-powered light commercial helicopter.

ROTOR SYSTEM: Generally the same as for Model 500C, except five-blade main rotor and teetering tail rotor with light alloy blades.

ROTOR DRIVE: Generally the same as for Model 500C, but entire drive train strengthened to give longer service life. Main rotor/engine rpm ratio 0·0794; tail rotor/engine rpm ratio 0·5111.

FUSELAGE: as for Model 500C.

TAIL UNIT: Fixed fin with T-tail; fixed ventral fin.

LANDING GEAR: Tubular skids carried on Hughes oleo-pneumatic shock-absorbers.

POWER PLANT: One 313 kW (420 shp) Allison 250-C20B turboshaft engine. Two interconnected bladder fuel tanks with combined usable capacity of 240 litres (63·4 US gallons). Refuelling point on starboard side of fuselage. Oil capacity 5·7 litres (1·5 US gallons).

ACCOMMODATION: Forward bench seat for pilot and two passengers, with two or four passengers seated in aft portion of cabin. Baggage compartment of 1·19 m³ (42 cu ft) capacity beneath seats. Two doors on each side of cabin.

SYSTEM: Electric power supplied by 150A engine-driven generator.

ELECTRONICS AND EQUIPMENT: Nav/com optional. Optional cargo sling has capacity of 907 kg (2,000 lb). Rescue hoist optional.

DIMENSIONS, EXTERNAL:
As for Model 500C, except:
Diameter of main rotor — 8·05 m (26 ft 5 in)
Distance between rotor centres — 4·62 m (15 ft 2 in)
Length overall, rotors fore and aft — 9·30 m (30 ft 6 in)
Height to top of rotor hub — 2·53 m (8 ft 3½ in)

Hughes Model 500D one/seven-seat commercial helicopter (Allison 250-C20B turboshaft engine)

Cabin doors (each):
Height 1·16 m (3 ft 9½ in)
Width 0·76 m (2 ft 6 in)
Height to sill 0·76 m (2 ft 6 in)

DIMENSIONS, INTERNAL:
As for Model 500C, except:
Max width 1·31 m (4 ft 3½ in)
Max height 1·52 m (5 ft 0 in)

AREAS:
Main rotor blades (each) 0·69 m² (7·43 sq ft)
Tail rotor blades (each) 0·09 m² (0·94 sq ft)
Main rotor disc 50·89 m² (547·81 sq ft)
Tail rotor disc 1·32 m² (14·19 sq ft)
Fin 0·56 m² (6·05 sq ft)
Tailplane 0·61 m² (6·52 sq ft)

WEIGHTS AND LOADINGS:
Weight empty 598 kg (1,320 lb)
Max T-O and landing weight 1,360 kg (3,000 lb)
Max disc loading 26·76 kg/m² (5·48 lb/sq ft)
Max power loading 4·35 kg/kW (7·14 lb/hp)

PERFORMANCE (at max T-O weight, ISA):
Never-exceed speed 152 knots (282 km/h; 175 mph)
Max cruising speed at S/L 139 knots (258 km/h; 160 mph)
Econ cruising speed at S/L 130 knots (241 km/h; 150 mph)
Econ cruising speed at 1,525 m (5,000 ft) 126 knots (233 km/h; 145 mph)
Max rate of climb at S/L 518 m (1,700 ft)/min
Service ceiling 4,570 m (15,000 ft)
Hovering ceiling in ground effect:
ISA 2,745 m (9,000 ft)
ISA +20°C 1,830 m (6,000 ft)
Hovering ceiling out of ground effect:
ISA 2,135 m (7,000 ft)
ISA +20°C 1,370 m (4,500 ft)
Range, 2 min warm-up, standard fuel, no reserves:
S/L 260 nm (482 km; 300 miles)
at 1,525 m (5,000 ft) 291 nm (539 km; 335 miles)

HUGHES MODEL 77

US Army designation: YAH-64

The YAH-64 is Hughes' entry in the US Army's Advanced Attack Helicopter (AAH) programme, in which it is in competition with the Bell YAH-63. The US Army announced on 22 June 1973 the award to Hughes of a $70·3 million contract to build two flight test prototype helicopters and a ground test vehicle, for competitive evaluation against the Bell contender. The disparity between the Hughes contract and that for Bell ($44·7 million) is due to the fact that Hughes had done less preliminary work than Bell; the Hughes contract covers, in addition, the development of the XM-230 chain gun to be installed in the YAH-64. The Defense Department has in any case emphasised that final unit costs and overall programme costs for the selected helicopter are more important than those for prototype development. The recurring flyaway cost per unit has to remain within a target figure, in 1972 dollars, of $1·6 million, based on a stated US Army requirement for 472 AAHs; this requirement has since been increased to 536.

The YAH-64 ground test vehicle began ground running in late June 1975, and by 19 September had completed the first 50 hours of its test programme. It was followed by the first flights of the first prototype (22248) at Palomar Airport, California, on 30 September and the second on 22 November 1975. By 1 January 1976 these two aircraft had completed more than 65 hours of flight test, of a total of 300 hours due to be flown before they were handed over to the US Army in May 1976 for competitive fly-off with the YAH-63. The ground test programme included static test, rotor flutter and vibration; and firing tests of the XM-230 gun, rockets, and TOW missiles were also completed. Hughes is teamed with Teledyne Ryan Aeronautical, which built the fuselage structure of the prototypes, and a 12-company major subcontracting team; systems installation was carried out by Hughes.

As noted under the Bell entry, announcement of the winning AAH design was expected by November 1976, and will be followed by Phase 2 of the programme involving fitting the winning prototypes with advanced electronics, electro-optical equipment and weapon fire control systems, for further evaluation; continued development of the airframe; and the manufacture of three more aircraft.

The following description applies to the YAH-64 prototypes:

TYPE: Prototype armed helicopter.

ROTOR SYSTEM: Four-blade fully-articulated main rotor and four-blade tail rotor, with blades manufactured by Tool Research and Engineering Corpn (Advance Structures Division). Main rotor blades are of high-camber aerofoil section and broad chord. Each blade has five aluminium spars, a laminated stainless steel skin and a fixed trailing-edge flap. Blades are attached to hub by a laminated strap retention system similar to that of the OH-6A, and are fitted with elastomeric lead/lag dampers and offset flapping hinges. Four-blade tail rotor comprises two sets of two blades, mounted on port side of pylon/fin support structure at optimum quiet setting of approx 60°/120° to each other. Rotor mast similar to that of OH-6A, with driveshaft turning within a hollow, fixed outer shaft. Entire system capable of flight in zero g conditions.

Hughes Advanced Attack Helicopter prototype (two General Electric T700-GE-700 turboshaft engines)

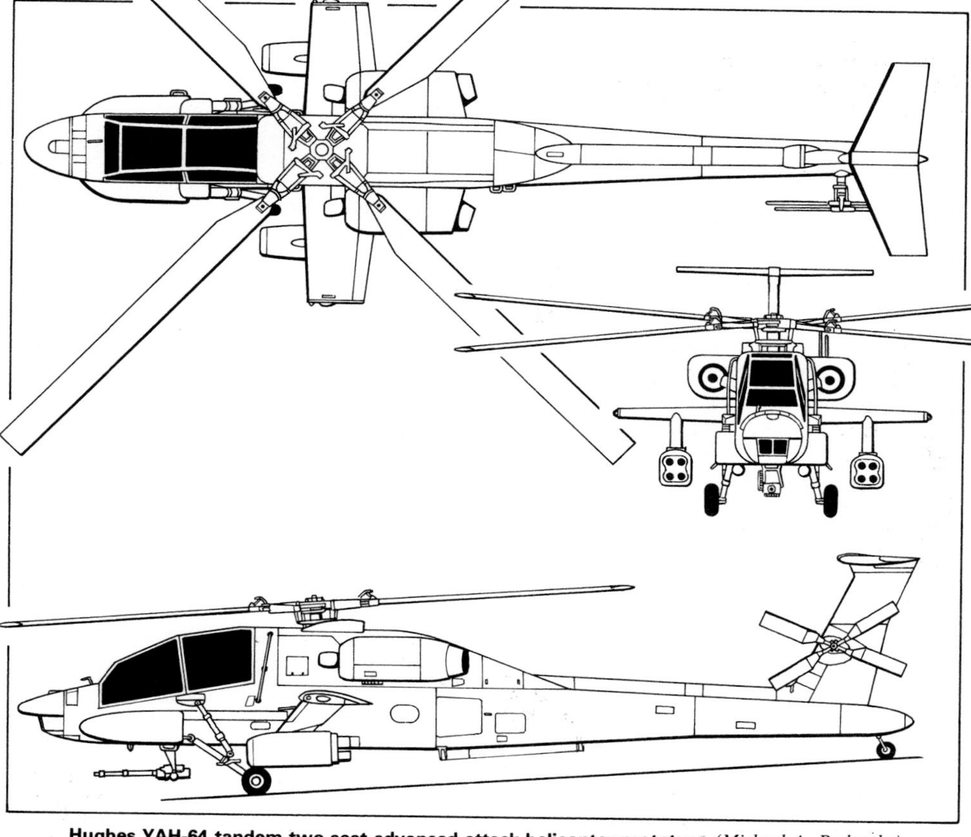

Hughes YAH-64 tandem two-seat advanced attack helicopter prototype *(Michael A. Badrocke)*

ROTOR DRIVE: Transmission to main rotor via Litton (Precision Gear Division) engine nose gearboxes, and to tail rotor via Western Gear intermediate and tail rotor gearboxes, with Bendix driveshafts and couplings. Redundant flight control system for both rotors. Selected dynamic components constructed of 70/49 aluminium and electro-slag remelt (ESR) steel; critical parts of transmission (eg, bearings) have ESR collars for protection against hits by 12·7 mm or 23 mm ammunition. Rotor/engine rpm ratios approx 66·7 for main rotor, approx 14·3 for tail rotor.

WINGS: Cantilever mid-wing monoplane, of low aspect ratio, aft of cockpit. Trailing-edge flaps deploy automatically as function of control attitude and airspeed (max deflection 20°), and can be deflected 45° upward to offload wings in an emergency autorotative landing. Wings are removable, and attach to sides of cabin for transport and storage. Two hardpoints beneath each wing for the carriage of mixed ordnance.

FUSELAGE: Conventional semi-monocoque aluminium structure, built by Teledyne Ryan Aeronautical. Designed to survive hits by 12·7 mm and 23 mm ammunition.

TAIL UNIT: Fixed fin and cantilever T tailplane.

LANDING GEAR: Menasco tailwheel type, with single wheel on each unit. Main legs fold rearward to reduce overall height for storage and transportation. Fully-castoring, self-centering tailwheel. Main-wheel tyres size 22 × 8; tailwheel tyre size 13 × 5. Hydraulic brakes.

POWER PLANT: Two 1,145 kW (1,536 shp) General Electric T700-GE-700 turboshaft engines, derated for normal operations to provide reserve power for combat emergencies. Engines mounted one on each side of fuselage, above wings. Two crashproof fuel cells, capacity 1,366 litres (361 US gallons).

ACCOMMODATION: Crew of two in tandem, co-pilot/gunner in front and pilot aft on 483 mm (19 in) elevated seat. Large, curved transparent cockpit enclosure for optimum field of view. Canopy and crew escape system by Hi-Shear Corporation. Lightweight boron armour shields in cockpit floor and sides. Cockpits separated by armour plating and an anti-23 mm inner plastics shield.

SYSTEMS AND EQUIPMENT: Large electronics bay adjacent to gunner's position, in lower fuselage. Bertea hydraulic control system. Bendix electrical power system, with two fully-redundant engine-driven generators and standby DC battery. Sperry Flight Systems automatic stabilisation equipment. Garrett infra-red suppression and integrated pressurised air systems. Solar APU.

ARMAMENT AND OPERATIONAL EQUIPMENT: Fixed armament consists of a Hughes-developed XM-230 30 mm chain gun, mounted in an underfuselage turret between the main-wheel legs, and having a normal rate of fire of 700 rds/min. Ammunition standard load 1,086 rds. Turret designed to collapse into fuselage floor in the event of a crash-landing. Four underwing hardpoints, on which can be carried up to sixteen Hughes BGM-71A TOW anti-tank missiles, housed in pylon-mounted streamlined pods; or up to seventy-six 2·75 in folding-fin rockets in their launchers; or a combination of TOW missiles and rockets. CPG stabilised sight in forward fuselage, ahead of cockpit, incorporates day and night (FLIR: forward-looking infra-red) sighting equipment, laser ranger and target designator, and TOW tracking equipment. Co-pilot/gunner has primary responsibility for firing all weapons, but pilot can override his controls to fire gun or launch rockets. Space and power provision made for pilot's night vision system (PNVS) in extreme tip of nose. Forward bay includes electronics for stabilised sight, missiles and fire control computer; design assistance in fire control computer provided by Teledyne Systems Inc.

DIMENSIONS, EXTERNAL:
- Diameter of main rotor: 14·63 m (48 ft 0 in)
- Diameter of tail rotor: 2·54 m (8 ft 4 in)
- Main rotor blade chord: 0·53 m (1 ft 9 in)
- Length of fuselage: 15·05 m (49 ft 4½ in)
- Wing span: 3·63 m (11 ft 10¾ in)
- Height overall: 3·69 m (12 ft 1¼ in)
- Wheel track: 2·03 m (6 ft 8 in)

AREA:
- Main rotor disc: 168·06 m² (1,809 sq ft)

WEIGHTS:
- Weight empty: 4,309 kg (9,500 lb)
- Primary mission gross weight: 5,987 kg (13,200 lb)
- Structural design gross weight: 6,328 kg (13,950 lb)
- Max T-O weight: 7,892 kg (17,400 lb)

PERFORMANCE (estimated, at 5,987 kg; 13,200 lb AUW, ISA except where indicated):
- Never-exceed speed: 204 knots (378 km/h; 235 mph)
- Max level speed: 166 knots (307 km/h; 191 mph)
- Max cruising speed: 156 knots (289 km/h; 180 mph)
- Max vertical rate of climb at S/L: 975 m (3,200 ft)/min
- Max vertical rate of climb at 1,220 m (4,000 ft) at 35°C: 424 m (1,390 ft)/min
- Service ceiling: 6,250 m (20,500 ft)
- Service ceiling, one engine out: 3,630 m (11,900 ft)
- Hovering ceiling in ground effect: 4,450 m (14,600 ft)
- Hovering ceiling out of ground effect: 3,600 m (11,800 ft)
- Max range, internal fuel: 312 nm (578 km; 359 miles)
- Ferry range, max internal and external fuel: 1,015 nm (1,880 km; 1,168 miles)
- Endurance at 1,220 m (4,000 ft) at 35°C: 1 hr 54 min
- Max endurance, internal fuel: 3 hr 12 min

PERFORMANCE (envelope explored to 2 December 1975):
- Total flight time: more than 40 hr
- Max T-O weight: 6,758 kg (14,900 lb)
- Forward speed: 130 knots (241 km/h; 150 mph)
- Sideways speed: 35 knots (65 km/h; 40 mph)
- Backward speed: 30 knots (55·5 km/h; 34·5 mph)
- Altitude: 3,660 m (12,000 ft)

INTERCEPTOR

INTERCEPTOR COMPANY

ADDRESS:
2950 Pearl Street, Boulder, Colorado 80301
Telephone: (303) 442-3877

PARTNERS:
Ted D. Malpass
Peter Paul Luce

The original Interceptor Corporation was formed on 18 November 1968 for the purpose of designing, manufacturing, distributing and servicing aircraft for the general aviation market. It had a manufacturing plant at Max Westheimer Field, Norman, Oklahoma.

First aircraft to be produced by that company was the Interceptor 400, an advanced turbine-engined development of what was known originally as the Meyers 200B and was produced subsequently by Aero Commander as its Model 200. Interceptor Corporation acquired all design drawings, production jigs and tools for this latter aircraft and redesigned the power plant installation and tail unit. Construction of the prototype started in January 1969 and the first flight was made on 27 June 1969. FAA certification was received on 20 August 1971.

The first Interceptor 400 was sold to a Mr. F. Lee Bailey who leased it back to the company for demonstration flights. Following a forced landing the company ran into financial difficulties, but it has since been re-formed as Interceptor Company with Mr Ted Malpass and Mr Peter Luce as partners. The Type Certificate for the Interceptor 400 was re-issued to Interceptor Company on 31 December 1974.

Interceptor 400 pressurised four-seat light aircraft (AiResearch TPE 331-1-101 turboprop engine)

INTERCEPTOR 400

The description refers to the current demonstrator. Production Interceptor 400s will have a more powerful 626 kW (840 shp) engine, flat rated at 298 kW (400 shp), and higher performance.

TYPE: Four-seat light cabin monoplane.

WINGS: Cantilever low-wing monoplane. Wing section NACA 23015 at root, NACA 23012 at station 62, NACA 4412 at tip. Dihedral 6°. Incidence 2° at root, 2° at station 62, −3° at tip. Sweepback at quarter-chord approx 1°. Root section outboard to station 62 of welded steel tube construction with aluminium skins. Outer wings are of conventional two-spar flush-riveted light alloy construction. Conventional all-metal ailerons with ground-adjustable trim tab. All-metal Fowler trailing-edge flaps. Wing leading-edge de-icing optional.

FUSELAGE: All-metal structure. Cabin section of welded steel tube, with overturn structure and light alloy covering. Light alloy semi-monocoque flush-riveted rear fuselage. Fuselage pressurised from firewall to fuselage station 110.

TAIL UNIT: Cantilever all-metal structure. Fixed-incidence tailplane. Rudder and elevators have controllable trim tabs. De-icing of leading-edges optional.

LANDING GEAR: Hydraulically-retractable tricycle type. Nose unit retracts aft, main units inward. Interceptor oleo-pneumatic shock-absorber struts; single wheel on each unit. Goodyear main wheels and tyres size 7·00-6, 6-ply rating, pressure 2·62 bars (38 lb/sq in). Goodyear nosewheel and tyre size 5·00-5, 6-ply rating, pressure 3·38 bars (49 lb/sq in). Goodyear caliper drum brakes.

POWER PLANT: One 496 kW (665 shp) Garrett-AiResearch TPE 331-1-101 turboprop engine, flat rated to 298 kW (400 shp) and driving a Hartzell three-blade metal constant-speed fully-feathering reversible-pitch propeller with spinner. Fuel contained in integral tank in each outer wing panel, with total capacity of 548 litres (145 US gallons). Oil capacity 14·8 litres (3·9 US gallons). Propeller de-icing optional.

ACCOMMODATION: Four seats in pairs in enclosed cabin. Door on starboard side, above wing, hinged at forward edge. Baggage compartment aft of cabin with external door on starboard side. Baggage capacity 90·5 kg (200 lb). Cabin air-conditioned and pressurised.

SYSTEMS: AiResearch air-cycle air-conditioning and pressurisation by engine bleed air; max differential 0·19 bars (2·8 lb/sq in). Pneumatic system for pressurisation control and instruments. Hydraulic system, pressure 69-89·5 bars (1,000-1,300 lb/sq in), for operation of flaps, landing gear and passenger step, which retracts simultaneously with landing gear. Electrical DC power supplied by 28V 100A starter/generator and two 22Ah nickel-cadmium batteries. Oxygen system optional.

ELECTRONICS AND EQUIPMENT: Nav/com transceivers, ADF, DME, transponder and autopilot are available to customer's requirements. Optional equipment includes Grimes 3-light strobe system, electric propeller de-icing, wing and tail unit de-icing and 4-outlet oxygen system. Full IFR instrumentation standard.

DIMENSIONS, EXTERNAL:
- Wing span: 9·29 m (30 ft 6 in)
- Wing chord at root: 2·29 m (7 ft 6 in)
- Wing chord at tip: 1·02 m (3 ft 4 in)
- Wing aspect ratio: 5·81
- Length overall: 8·35 m (27 ft 4¾ in)
- Height overall: 3·20 m (10 ft 6 in)
- Tailplane span: 4·47 m (14 ft 8 in)
- Wheel track: 2·72 m (8 ft 11 in)
- Wheelbase (approx): 2·07 m (6 ft 9½ in)
- Propeller diameter: 2·18 m (7 ft 2 in)
- Cabin door (approx):
  - Height: 1·02 m (3 ft 4 in)
  - Width: 0·91 m (3 ft 0 in)
- Baggage compartment door (approx):
  - Height: 0·51 m (1 ft 8 in)
  - Width: 0·58 m (1 ft 11 in)
  - Height to sill: 0·76 m (2 ft 6 in)

DIMENSIONS, INTERNAL:
- Cabin:
  - Length: 2·13 m (7 ft 0 in)
  - Max width: 1·07 m (3 ft 6 in)
  - Max height: 1·22 m (4 ft 0 in)
  - Floor area: 2·23 m² (24 sq ft)
  - Volume: 1·70 m³ (60 cu ft)
- Baggage hold: 0·28 m³ (10 cu ft)

AREAS:
- Wings, gross: 14·86 m² (160 sq ft)
- Ailerons (total): 1·00 m² (10·8 sq ft)
- Trailing-edge flaps (total): 2·08 m² (22·4 sq ft)
- Fin (approx): 0·83 m² (8·94 sq ft)
- Rudder, incl tab (approx): 0·66 m² (7·06 sq ft)
- Tailplane: 2·31 m² (24·88 sq ft)
- Elevators, incl tabs: 1·22 m² (13·10 sq ft)

WEIGHTS AND LOADINGS:
- Weight empty, equipped: 1,088 kg (2,400 lb)
- Max payload with full fuel and IFR equipment: 272 kg (600 lb)
- Max T-O weight: 1,828 kg (4,030 lb)
- Max wing loading: 122·94 kg/m² (25·18 lb/sq ft)
- Max power loading: 6·13 kg/kW (10·7 lb/shp)

PERFORMANCE (at max T-O weight):
- Never-exceed speed: 260 knots (481 km/h; 300 mph)
- Max level speed at 4,875 m (16,000 ft): 250 knots (463 km/h; 287 mph)
- Max cruising speed at 5,485 m (18,000 ft): 243 knots (450 km/h; 280 mph)
- Stalling speed, flaps and landing gear down: 59 knots (110 km/h; 68 mph)
- Max rate of climb at S/L: 457 m (1,500 ft)/min
- Service ceiling: 7,315 m (24,000 ft)
- T-O run: 262 m (860 ft)
- T-O to 15 m (50 ft): 430 m (1,410 ft)
- Landing from 15 m (50 ft): 366 m (1,200 ft)
- Landing run: 155 m (510 ft)
- Range at max cruising speed, with allowances for warm-up, taxi, T-O, climb and 45 min reserves: 868 nm (1,610 km; 1,000 miles)
- Max range, allowances as above: 998 nm (1,850 km; 1,150 miles)
- Range with max payload, allowances as above: 738 nm (1,368 km; 850 miles)

KAMAN

KAMAN AEROSPACE CORPORATION
(a subsidiary of Kaman Corporation)

HEAD OFFICE:
Old Windsor Road, Bloomfield, Connecticut 06002
Telephone: (203) 242-4461

PRESIDENT:
Charles H. Kaman

VICE-PRESIDENTS:
Weston B. Haskell Jr (Communications)
Eamon N. Kelly (Contracts)
James T. King Jr (Manufacturing)
Robert D. Moses (Marketing)
Donald W. Robinson (Engineering)
Fred L. Smith (Programme Management)

SECRETARY: J. S. Murtha

COMPTROLLER: Walter R. Kozlow

MANAGER, WASHINGTON OFFICE:
T. E. Glass Jr

MANAGER, PUBLIC RELATIONS AND INTERNATIONAL:
Bruce A. Goodale

The original Kaman Aircraft Corporation was founded in 1945 by Mr Charles H. Kaman, who continues as President and Chairman of the Board of Kaman Corporation and as President of Kaman Aerospace Corporation. Its initial programme was to develop and test a novel servo-flap control system for helicopter rotors, and the Kaman K-125 of 1947 was the first in a series of 'synchropter' designs with intermeshing contra-rotating rotors and the servo-flap control system. The later H-2 Seasprite naval helicopter utilises the servo-flap control system on a conventional single main rotor.

Under a succession of US Navy contracts which began in 1970, and which totalled approximately $50 million by the end of 1975, Kaman modified and converted earlier-model H-2 helicopters to SH-2D/F configuration for the Navy's LAMPS (Light Airborne Multi-Purpose System) programme, which teams these aircraft with frigates, destroyers and escorts for broader capabilities in anti-submarine warfare (ASW) and anti-ship missile defence (ASMD). By early 1976, all but three of the 105 available H-2s in the Navy's inventory had been converted to LAMPS configuration.

Kaman is anticipating further production of the SH-2, and this helicopter will also be made available internationally. New SH-2s would be qualified for a maximum gross weight of 6,033 kg (13,300 lb), compared to the 5,805 kg (12,800 lb) of the current SH-2F. Capacity of the external auxiliary fuel tanks would be almost doubled, at about 454 litres (120 US gallons) each, to offer increased endurance.

Kaman Aerospace operates two plants in Connecticut, located at Bloomfield and Moosup. Its capabilities include major production work in the fields of airframe structures, sheet metal, glassfibre, composites, reinforced honeycomb bonding and chemical milling. It is also a leading company in the field of helicopter dynamics, research and development, and in the application of new materials and technology.

Kamatics Corporation (formerly Kacarb Products Corporation), a subsidiary of Kaman Aerospace Corporation, was reorganised in January 1973 to reflect its developing potential in the bearings industry. Products include bearings for a variety of aerospace applications.

In research and development, Kaman Aerospace is working on several advanced concepts. One of the most promising programmes is the development of a Circulation Control Rotor (CCR) concept, under a Naval Air Systems Command contract awarded in February 1975. This project is concerned with a slotted rotor blade through which compressed air is blown to provide primary control, enhance aerodynamic performance and improve the lift characteristics of the retreating blade. Several other R & D programmes are under sponsorship of the US Army Air Mobility Research and Development Laboratory, Eustis Directorate, Fort Eustis, Virginia. They include a controllable twist rotor (CTR); dynamic antiresonant vibration isolator; composite rotor hub; repairable/expendable main rotor blade concept; new concepts in structural dynamic testing; maintainability of major helicopter components; elastic pitch beam tail rotor; and new design, fabrication and inspection techniques for helicopter structures. Kaman is also modifying CTR hardware to incorporate a multicyclic flap system at the request of NASA's Ames Research Center, Moffett Field, California.

For the Office of Naval Research, Kaman is under contract to design, build and demonstrate the feasibility of the Ship Tethered Aerial Platform (STAPL), a ship-towed drone autogyro with automatic flight controls and data recording equipment, which is visualised as an elevated sensor or instrument platform. A similar analysis of a tethered drone helicopter concept for battlefield surveillance has been carried out for Eustis Directorate.

In addition to its R and D programmes related directly to aerospace technology, Kaman is involved in new energy-related programmes. Among these are a wind generator system study for NASA's Lewis Research Center, Cleveland, Ohio; fabrication of a vertical axis wind generator (Darrieus rotor) for Sandia Laboratories, Albuquerque, New Mexico; and fabrication of several toroidal field coils to be used in fusion research by Princeton University Plasma Physics Laboratory.

Kaman Aerospace is engaged in several major airframe programmes under subcontract. In July 1972 Rockwell International selected the company to build rudders and the tailplane fairing assembly for the three original prototypes of the USAF's B-1 strategic bomber, plus test articles. This was followed in December by an additional contract for three aircraft sets of engine access doors. In December 1973, Fairchild Republic awarded Kaman a contract covering tooling for and construction of aft fuselage sections for the A-10A close-support aircraft.

Kaman is an associate contractor with Grumman on the US Navy's F-14 Tomcat fighter, and is constructing flaps, slats, spoilers and cove doors for the outer VG wing. Kaman has delivered wing control surfaces for the first 115 F-14s, and has been authorised to build an additional 115 aircraft sets on an accelerated delivery schedule. Other major airframe contracts have included production of components for the Lockheed C-5A and C-130; Grumman A-6, EA-6B and OV-1; McDonnell Douglas DC-8 and DC-9; Boeing B-52; and Boeing Vertol YUH-61A; thrust reversers for the General Electric TF39 turbofan engine; and space hardware under subcontracts to General Electric and Perkin Elmer. Under a $4·8 million contract announced in May 1975, Kaman designed and is producing 33 prototypes of a new rotor blade for the Bell AH-1Q HueyCobra in service with the US Army, funded by the Army's Aviation Systems Command, St Louis, Missouri. The first AH-1Q fitted with the new blades made its first flight on 26 July 1976.

In August 1975, Kaman was selected as a subcontractor for the slosh baffle of the external tank of the Space Shuttle, under a multi-year contract with Martin Marietta Corporation. Kaman also supplies design and fabrication services to Lockheed Missiles and Space Division for various items of Trident ground support equipment under a multi-year contract.

In May 1971 Kaman was awarded a contract by the USAF's Electronic Systems Division, Air Force Systems Command, for the design, fabrication and testing of the Airborne Weather Reconnaissance System (AWRS). Intended to improve collection, measurement, analysis, recording and communication of meteorological data, to increase the accuracy of predicting the force and direction of hurricanes, typhoons and lesser storms, AWRS was fitted in two Hercules aircraft delivered to the USAF Air Weather Service and the US Commerce Department's National Oceanic and Atmospheric Administration (NOAA) Research Flight Facility. AWRS may be installed in additional Air Weather Service C-130s for worldwide coverage of weather phenomena. Details can be found in the 1975-76 *Jane's*.

Kaman announced in October 1972 that, under a $648,000 contract from the US Naval Engineering Center, Philadelphia, an HH-2D Seasprite was to be modified for the installation, testing and evaluation of Beartrap and Harpoon rapid-capture and securing systems to make possible helicopter landings on small, non-aviation ships in high sea states. The same Seasprite is now being used for testing the Recovery Assist/Secure/Traverse (RAST) system selected by the US Navy for its LAMPS ships.

In recent years the parent Kaman Corporation has diversified its activities considerably, offering products and services in six principal markets: aerospace, science and technology, general aviation, music, bearings, and industrial supplies and services. These operations produced total corporate sales of approximately $158 million in 1975, of which $40 million resulted from aerospace activities.

Kaman Sciences Corporation, with headquarters at Colorado Springs, Colorado, is engaged in nuclear research, weapons studies, advanced aerodynamics, computer programming and time sharing, neutron generators, advanced materials, measuring devices and systems analysis. Kaman's general aviation subsidiaries comprise AirKaman Inc, a fixed-base operator at Bradley International Airport, Windsor Locks, Connecticut; AirKaman of Omaha, Nebraska, and AirKaman of Jacksonville, Florida, all of which provide sales and service of light and twin-engined business aircraft, repairs, fuel sales, charter service, flight training and airline services.

At the beginning of 1976 total corporate employment was approximately 3,000, of whom about 1,075 were engaged in aerospace activities.

KAMAN SEASPRITE

US Navy designations: UH-2 (formerly HU2K-1), HH-2 and SH-2

The prototype Seasprite flew for the first time on 2 July 1959, and many versions (described in previous editions of *Jane's*) were produced subsequently for the US Navy.

From 1967, all of the original UH-2A/B Seasprites were converted progressively to UH-2C twin-engined configuration, with two 932 kW (1,250 shp) General Electric T58-GE-8F turboshaft engines in place of the former single T58. They have since undergone further modification, under the US Navy's important LAMPS (Light Airborne Multi-Purpose System) programme, to provide helicopters for ASW (Anti-Submarine Warfare) and ASMD (Anti-Ship Missile Defence) operations.

The following versions remained available in early 1976:

NHH-2D. Test aircraft assigned to the 1976 Recovery Assist/Secure/Traverse (RAST) programme.

SH-2D. LAMPS version, for ASW, ASMD and a utility role. The first of 20 SH-2Ds, modified from HH-2D unarmed search and rescue helicopters, made its first flight on 16 March 1971, and by March 1972 Kaman completed the modification of all of these aircraft. This involved the installation of Canadian Marconi LN 66 high-power surface search radar in a glassfibre honeycomb dome under the chin; ASQ-81 MAD deployed by winch from a pylon on the starboard side of the fuselage; 15 AN/SSQ-47 active or AN/SSQ-41 passive sonobuoys launched by a small explosive charge from a removable rack on the port side; ALR-54 electronic support measure; eight Mk 25 marine flares/smoke markers; data link; tactical navigation system, and associated command/control units, recorders, displays and antennae. Auxiliary fuel tank mounts on each side of the fuselage were hardened for the added purpose of launching Mk 44 and Mk 46 ASW homing torpedoes.

As the LAMPS helicopters became operational, the Navy reorganised squadrons to provide detachments to fleet units and to train additional personnel to operate and maintain them. First was Helicopter Combat Support Squadron 4 (HC-4) at Naval Air Station Lakehurst, New Jersey, which became Helicopter Antisubmarine Squadron Light 30 (HSL-30).

Operational deployment of SH-2Ds began on 7 December 1971, with assignment of the first unit from HSL-30 to the guided missile cruiser USS *Belknap* (CG-26), the detachment reporting on board in Crete. The second detachment, from HSL-31, was assigned to the guided missile cruiser USS *Sterett* (CG-31) and deployed to the Pacific in January 1972. The USS *Joseph Hewes* (FF-1078) was the first Atlantic Fleet frigate of the FF-1052 'Knox' class to become operational with a LAMPS detachment, and the USS *Harold E. Holt* (FF-1074) was the first of this class in the Pacific. First FF-1040 'Garcia' class frigates operational with LAMPS were the USS *O'Callahan* (FF-1051) in the Pacific and USS *Edward*

Kaman SH-2F LAMPS ASW helicopter operating from the deck of DD-963 'Spruance' destroyer

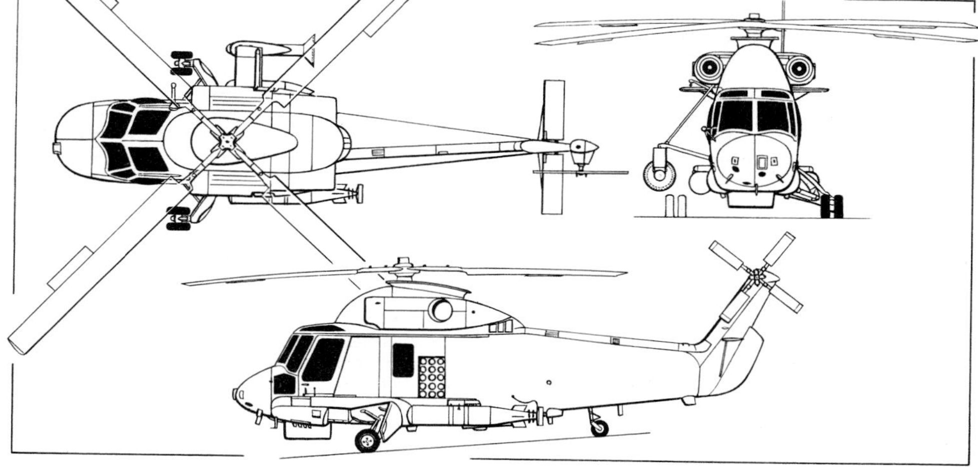

Kaman SH-2F Seasprite Light Airborne Multi-Purpose System (LAMPS) helicopter *(Pilot Press)*

McDonnell (FF-1043) in the Atlantic; the USS *Schofield* (FFG-3) was the first of its class to operate with LAMPS helicopters. By February 1976, 59 LAMPS detachments had been deployed (not simultaneously) in the Mediterranean and in the Pacific. The new DD-963 and FFG-7 classes are designed to operate with two LAMPS helicopters per ship.

YSH-2E. The two YSH-2Es continued to participate in the Mk III LAMPS development programme in 1975, as testbeds for equipment installation and integration. They were being converted to SH-2F LAMPS configuration at Kaman in early 1976.

SH-2F. Deliveries of this further-developed LAMPS version began in May 1973 and the first unit became operational with squadron HSL-33, deployed to the Pacific on board USS *Bagley* (FF-1069), on 11 September 1973. By December 1975 a total of 85 SH-2Fs had been delivered. The earlier SH-2Ds will be uprated to SH-2F configuration.

The SH-2F is fitted with Kaman's '101' rotor, which provides substantially increased performance in all flight regimes, while practically eliminating rotor vibrations at all speeds and weights, thus improving system reliability and maintainability. The new simplified rotor control system utilises titanium hub and retention assemblies, reduces the number of control elements by two-thirds, and offers a 3,000-hour life for the entire rotor system.

Other features of the SH-2F include increased-strength landing gear; a shortened wheelbase by relocation of the tailwheel; twin 1,007 kW (1,350 shp) General Electric T58-GE-8F turboshaft engines; an improved LN 66HP radar, improved tactical navigation and communications systems, and other modifications. In January-February 1973 Kaman flight-tested the prototype for flight qualification to a maximum gross weight of 6,033 kg (13,300 lb), which is 227 kg (500 lb) more than the current SH-2F. This may be utilised as increased payload, or in the form of additional fuel in larger auxiliary tanks to provide extended range and endurance in a new production version of the SH-2.

The following details apply to the SH-2D and SH-2F versions of the Seasprite:

TYPE: Naval anti-submarine warfare and anti-missile defence helicopter, with secondary capability for search and rescue, observation and utility missions.

ROTOR SYSTEM: Four-blade main and tail rotors. Blades of aluminium and glassfibre construction, with servo-flap controls. Blades folded manually. Main rotor rpm 287. Performance figures given for SH-2F aircraft are with the Kaman '101' rotor modification.

FUSELAGE: All-metal semi-monocoque structure, with flotation hull housing main fuel tanks. Nose fairing split on centreline, to fold rearward on each side to reduce stowage space required. Fixed horizontal stabiliser on tail rotor pylon.

LANDING GEAR: Tailwheel type, with forward-retracting dual main wheels and non-retractable tailwheel. Liquid spring shock-absorbers.

POWER PLANT: Two 1,007 kW (1,350 shp) General Electric T58-GE-8F turboshaft engines, mounted on each side of rotor pylon structure. Normal fuel capacity of 1,499 litres (396 US gallons), including external auxiliary tanks with a capacity of 454·6 litres (120 US gallons).

ACCOMMODATION: Crew of three, consisting of pilot, co-pilot and sensor operator. One passenger or litter patient with LAMPS equipment installed; four passengers or two litters with sonobuoy launcher removed. Provision for transportation of internal or external cargo.

DIMENSIONS, EXTERNAL:
- Diameter of main rotor: 13·41 m (44 ft 0 in)
- Main rotor blade chord: 0·55 m (21·6 in)
- Diameter of tail rotor: 2·49 m (8 ft 2 in)
- Tail rotor blade chord: 0·236 m (9·3 in)
- Length overall (rotors turning): 16·03 m (52 ft 7 in)
- Length overall, nose and blades folded: 11·68 m (38 ft 4 in)
- Height overall (rotors turning): 4·72 m (15 ft 6 in)
- Height to top of rotor head: 4·14 m (13 ft 7 in)
- Stabiliser span: 2·97 m (9 ft 9 in)
- Wheel track (outer wheels): 3·30 m (10 ft 10 in)
- Wheelbase:
  - SH-2D: 6·88 m (22 ft 7 in)
  - SH-2F: 5·11 m (16 ft 9 in)

WEIGHTS:
- Weight empty:
  - SH-2D: 3,153 kg (6,953 lb)
  - SH-2F: 3,193 kg (7,040 lb)
- Normal T-O weight:
  - SH-2D: 5,670 kg (12,500 lb)
- Overload T-O weight:
  - *SH-2D, SH-2F: 5,805 kg (12,800 lb)

**Although not yet certificated for a T-O gross weight of 6,033 kg (13,300 lb), all testing has been accomplished at that weight*

PERFORMANCE (at normal AUW, except where indicated):
- Max level speed at S/L:
  - SH-2D, SH-2F: 143 knots (265 km/h; 165 mph)
- Normal cruising speed: 130 knots (241 km/h; 150 mph)
- Max rate of climb at S/L:
  - SH-2D, SH-2F: 744 m (2,440 ft)/min
- Service ceiling:
  - SH-2D, SH-2F: 6,858 m (22,500 ft)
- Hovering ceiling in ground effect:
  - SH-2D, SH-2F: 5,670 m (18,600 ft)
- Hovering ceiling out of ground effect:
  - SH-2D, SH-2F: 4,695 m (15,400 ft)
- Normal range with max fuel:
  - SH-2D, SH-2F: 367 nm (679 km; 422 miles)

LAKE

LAKE AIRCRAFT DIVISION OF CONSOLIDATED AERONAUTICS INC

EXECUTIVE OFFICES:
PO Box 399, Tomball, Texas 77375

SALES OFFICES:
David Hooks Memorial Airport, Tomball, Texas 77375
Telephone: (713) 376-5421
Telex: 76-2054 Lake Air Hou

PRESIDENT:
John J. O'Toole

EXECUTIVE VICE-PRESIDENT:
M. L. Alson

VICE-PRESIDENT:
Laurin Darrell (Marketing)

SECRETARY: B. A. Sigsbee

TREASURER: Herbert P. Lindblad

In 1962 Consolidated Aeronautics merged with Lake Aircraft Corporation of Sandford, Maine, as a result of which it operates Lake Aircraft as a division.

It is continuing production of the LA-4 amphibian, which Lake Aircraft developed from the original Colonial C-2 Skimmer IV after purchasing manufacturing rights from Colonial Aircraft Corporation in October 1959.

LAKE LA-4-200 BUCCANEER

Lake LA-4-200 Buccaneer four-seat light amphibian (Lycoming IO-360-A1B flat-four engine)

Design of the original C-1 Skimmer was started in August 1946. Construction of the prototype began in January 1947 and it flew for the first time in May 1948. Versions of the Lake LA-4 developed from the improved C-2 Skimmer IV have included the LA-4, LA-4A, LA-4P, LA-4S and LA-4T, as described in previous editions of *Jane's*.

The LA-4-200 current production version, described here, received FAA certification in 1970.

A total of 723 LA-4s of all versions had been built by 1 January 1976.

TYPE: Single-engined four-seat amphibian.

WINGS: Cantilever shoulder-wing monoplane with tapered wing panels attached directly to sides of hull. Wing section NACA 4415 at root, NACA 4409 at tip. Dihedral 5° 30′. Incidence 3° 15′. Structure consists of duralumin leading- and trailing-edge torsion boxes separated by a single duralumin main spar. All-metal ailerons and hydraulically-operated slotted flaps over 80% of span. Ground-adjustable trim tabs on ailerons.

HULL: Single-step all-metal structure, with double-sealed boat hull. Alodined and zinc chromated inside and out against corrosion; hot-enamel painted.

TAIL UNIT: Cantilever all-metal structure. Outboard elevator section separate from inboard section and actuated hydraulically for trimming. Retractable water rudder in base of aerodynamic rudder.

LANDING GEAR: Hydraulically-retractable tricycle type. Consolidated oleo-pneumatic shock-absorbers on main gear, which retracts inward into wings. Long-stroke nosewheel oleo retracts forward. Goodyear or Cleveland main wheels and tyres, size 6·00-6, pressure 2·41 bars (35 lb/sq in). Goodyear nosewheel and tyre size 5·00-5, pressure 1·38 bars (20 lb/sq in). Goodyear or Cleveland disc brakes. Parking brake. Nosewheel is free to swivel 30° each side. Floats are aluminium alloy monocoque structures.

POWER PLANT: One 149 kW (200 hp) Lycoming IO-360-A1B flat-four engine, mounted on pylon above hull and driving a Hartzell constant-speed pusher propeller. US Rubber DL10 fuel tank in hull, capacity 151 litres (40 US gallons). Refuelling point above hull. Auxiliary fuel in stabilising floats, 28·4 litres (7·5 US gallons) each, optional. Total fuel capacity with optional tanks 208 litres (55 US gallons). Oil capacity 7·5 litres (2 US gallons).

ACCOMMODATION: Enclosed cabin seating pilot and three passengers. Dual controls. Entry through two forward-hinged windscreen sections. Baggage compartment, capacity 90·5 kg (200 lb), aft of cabin. Dual windscreen defroster system.

SYSTEMS: Vacuum system for flight instruments. Hydraulic system, pressure 86·2 bars (1,250 lb/sq in), for flaps, horizontal trim and landing gear actuation. Engine-driven 12V 60A alternator and 12V 35Ah battery. Stewart Warner 20,000 BTU heater optional.

ELECTRONICS AND EQUIPMENT: Standard equipment includes full blind-flying instrumentation, outside air temperature gauge, electric clock, stall warning indicator, tinted windscreen and windows, instrument lights, heated pitot, navigation lights, landing lights, rotating beacon, full flow oil filter, quick fuel drains and paddle. A cabin speaker, microphone, VHF antennae and broad-band com antenna are standard. Three basic electronics packages include (1): Narco 11A 360-channel com transceiver, Narco 11 200-channel nav receiver, AT 50A transponder with 4096 code capability and ELT; (2): King KX-175B 720-channel com transceiver, KI-201C 200-channel nav receiver with VOR/LOC, KT-76 transponder with 4096 code capability and ELT; (3): Collins VHF-251 720-channel micro line com transceiver, VIR-351 200-channel nav with VOR/LOC and IND-350 indicator, TDR-950 transponder with 4096 code capability and ELT.

DIMENSIONS, EXTERNAL:
- Wing span: 11·58 m (38 ft 0 in)
- Wing chord, mean: 1·35 m (4 ft 5·1 in)
- Wing aspect ratio: 8·67
- Length overall: 7·60 m (24 ft 11 in)
- Height overall: 2·84 m (9 ft 4 in)
- Tailplane span: 3·05 m (10 ft 0 in)
- Wheel track: 3·40 m (11 ft 2 in)
- Wheelbase: 2·69 m (8 ft 10 in)
- Propeller diameter: 1·88 m (6 ft 2 in)

DIMENSIONS, INTERNAL:
- Cabin: Length: 1·57 m (5 ft 2 in)
  - Max width: 1·05 m (3 ft 5½ in)
  - Max height: 1·32 m (3 ft 11½ in)
  - Floor area: approx 1·53 m² (16·5 sq ft)
  - Volume: approx 1·70 m³ (60·0 cu ft)
- Baggage hold: 0·24 m³ (8·5 cu ft)

AREAS:
- Wings, gross: 15·8 m² (170 sq ft)
- Ailerons (total): 1·16 m² (12·5 sq ft)
- Trailing-edge flaps (total): 2·28 m² (24·5 sq ft)
- Fin: 1·25 m² (13·5 sq ft)
- Rudder: 0·79 m² (8·5 sq ft)
- Tailplane: 1·45 m² (15·6 sq ft)
- Elevators: 0·78 m² (8·4 sq ft)

WEIGHTS AND LOADINGS:
- Weight empty, equipped: 705 kg (1,555 lb)
- Max T-O and landing weight: 1,220 kg (2,690 lb)
- Max wing loading: 74·2 kg/m² (15·2 lb/sq ft)
- Max power loading: 8·19 kg/kW (13·45 lb/hp)

PERFORMANCE (at max T-O weight):
- Max level speed at S/L: 126·5 knots (235 km/h; 146 mph)
- Max cruising speed, 75% power at 2,440 m (8,000 ft): 130 knots (241 km/h; 150 mph)
- Stalling speed: 39 knots (72·5 km/h; 45 mph)
- Max rate of climb at S/L: 366 m (1,200 ft)/min
- Service ceiling: 4,480 m (14,700 ft)
- T-O run on land: 183 m (600 ft)
- T-O run on water: 335 m (1,100 ft)
- Landing run on land: 145 m (475 ft)
- Alighting run on water: 183 m (600 ft)
- Range with max fuel, at normal cruising speed, with reserves: 564 nm (1,046 km; 650 miles)
- Max range with max fuel, with reserves: 716 nm (1,327 km; 825 miles)

LAS

LOCKHEED AIRCRAFT SERVICE COMPANY (Division of Lockheed Aircraft Corporation)

HEAD OFFICE AND WORKS:
Ontario International Airport, Ontario, California 91761
Telephone: (714) 988-2411
BASE:
Luke Air Force Base, Arizona
SPECIAL DEVICES DIVISION:
Ontario, California
MARINE SERVICES DIVISION:
Ontario, California
JETPLAN AVIATION SERVICES:
Ontario, California
PRESIDENT:
Charles T. Thum
VICE-PRESIDENTS:
M. H. Greene (Asst to President)
K. E. Neudoerffer (Operations)
C. M. Schnepp (International Marketing)
R. L. Vader (International Operations)
DIRECTOR OF PUBLIC RELATIONS:
John R. Dailey

Lockheed Aircraft Service Company is claimed to be the world's largest independent aircraft maintenance and modification company. It has designed and installed major modifications for such aircraft as the Boeing KC-135 and 707; Douglas DC-8; and Lockheed C-130, C-141, L-188 Electra, C-121, L-1649, L-1011 and P-3. It has also designed and installed interiors for various transport aircraft.

LAS has diversified into many other fields, including aircraft maintenance training devices, aircraft maintenance recording systems and airborne integrated data systems, marine anti-corrosion systems, aircraft ground support equipment and water pollution monitoring.

The company introduced the Model 209 digital flight recorder during 1970, and by 1 January 1976 had sold and delivered more than 600 of these units to equip wide-body jets. In 1974 LAS began delivery of its new Model 280 maintenance recorder, which records 50 hours of flight data on an easily accessible cassette. LAS has also developed JETPLAN, a computerised flight planning and worldwide weather service for airlines and corporate jet operations, and this became available on a worldwide basis in 1971.

During 1975 the company completed and delivered a fourth Electra cargo conversion to Ansett Airlines, and three L-188 Electras converted to a passenger/cargo configuration to the Argentine Naval Service. By the beginning of 1976, LAS had delivered 41 Electras in a cargo configuration; details of this conversion can be found in the 1970-71 *Jane's*.

Brief details of the company's C-130 conversion programme and available details of the LAS/McDonnell Douglas A-4S/TA-4S Skyhawk conversion follow:

LAS DC-130H taking off with four underwing RPV test units having a combined weight of 20,189 kg (44,510 lb). This is believed to be an unofficial world record for external load lifted by a turboprop-engined aircraft

LAS/McDonnell Douglas TA-4S Skyhawk trainer, with tandem cockpits

LOCKHEED C-130 CONVERSIONS

LAS specialises in complex aircraft modification, and recent work included conversion of Lockheed C-130E Hercules aircraft to DC-130E drone launch configuration for the US Air Force. The converted aircraft can each launch and control up to two RPVs for reconnaissance operations. The LAS modification programme included design, structural strengthening of the aircraft's wings, installation of drone pylons, extension of the nose to accommodate a chin radome housing a drone control antenna, and installation of a noise abating operations compartment, a two-man launch operator's console and a two-man tracking and control station within the compartment, and a stellar inertial Doppler navigation system, as well as related electronics equipment for command and control of the RPVs.

LAS received a USAF contract in December 1974 to modify an HC-130H Hercules to DC-130H configuration. This has four pylons, each capable of carrying a 4,535 kg (10,000 lb) new-generation RPV, and LAS has built also five control station consoles for the aircraft.

LAS (McDONNELL DOUGLAS) A-4S/TA-4S SKYHAWK CONVERSION

Expansion of the Singapore Air Defence Command's operational element began in mid-1972 when 40 ex-United States Navy McDonnell Douglas A-4B Skyhawks were ordered, these being taken from storage at Davis-Monthan Air Force Base in Arizona. The first eight were sent to the LAS works at Ontario for the embodiment of more than 100 modifications to convert them to A-4S standard. They were also refurbished, repaired as necessary, and received a full inspection of the entire airframe. First flight of an A-4S took place on 14 July 1973.

The remaining 32 aircraft were dismantled at Davis-Monthan AFB for shipment direct to Singapore, where they were refurbished and equipped to A-4S standard at the LAS facility on the island. The final aircraft modified in Singapore was delivered in November 1975.

Based on the A-4B (formerly A4D-2), the A-4S has improved electronics, weapon delivery capability and performance which make it comparable with present-generation aircraft. Primary changes include the addition of split wing spoilers above the flaps, a braking parachute canister beneath the aft fuselage, a longer nose to house advanced electronics equipment of British origin, so that the aircraft is compatible with Singapore ADC's Hawker Hunters, and replacement of the 20 mm guns in the wing roots by 30 mm Aden cannon. Newly installed equipment includes a Ferranti lightweight lead-computing gunsight, and solid-state electronics packages for the communications, radio and navigation systems. The cockpit is completely redesigned to accommodate the new instrumentation and control boxes; and the original 34·25 kN (7,700 lb st) Curtiss-Wright J65-W-16A turbojet engine is replaced by a 37·4 kN (8,400 lb st) J65-W-20.

The initial batch of eight aircraft, modified at Ontario, were used initially in a pilot training programme, carried out with the support of LAS at Lemoore NAS, California, since the company's contract called also for maintenance, pilot training and logistics support. The 40 aircraft now equip two fighter-bomber squadrons on the Singapore ADC airfield at Changi.

The Singapore Air Defence Command ordered also a two-seat training version of the A-4S, and both versions are now in service, as follows:

A-4S. Single-seat combat aircraft, as described above and below.

TA-4S. Two-seat trainer, of which three were produced. They differ from the A-4S by having two individual cockpits in tandem, a steerable nosewheel and new brake system with dual disc brakes; armour plate is removed to reduce weight, and the A-4S fuselage fuel bladder cell is replaced by 317·5 kg (700 lb) capacity integral tanks. First flight of a TA-4S took place in early 1975.

The description of the McDonnell Douglas A-4M in this edition applies also to the A-4S, except as follows:

FUSELAGE: As for A-4M, except fixed nose with detachable nose radome over communications and navigation equipment. Integral flak-resistant armour in cockpit area, including internal armour plate below, forward and aft of cockpit.

POWER PLANT: As for A-4M, except one 37·4 kN (8,400 lb st) Wright J65-W-20 turbojet engine.

ACCOMMODATION: Pilot on zero-zero lightweight ejection seat.

SYSTEMS: Dual hydraulic system with manual backup. Electrical system powered by a 9kVA generator, with wind-driven generator to provide emergency power.

ELECTRONICS: Plessey PTR-377 UHF/VHF radio transceiver, with UHF homing; Collins ARC-159 UHF radio transceiver; Plessey PTR-442 IFF; Collins DF-206 low frequency ADF; Arvin ARN-52 Tacan; Rodale APN-141 radar altimeter; Stewart-Warner APQ-145 air-to-ground mapping and ranging radar; Decca Type 72 Doppler and TANS digital navigation computer system; Lear Siegler AJB-7 AHRS; Ferranti ISIS D-101 lead-computing gunsight, weapons release programmer and weapons delivery computer.

EQUIPMENT: Ring-slot braking parachute, 4·88 m (16 ft) in diameter, contained in canister secured in aft fuselage below engine efflux duct. Arrester hook for SATS operation.

ARMAMENT: As for A-4M, except no provision for Bullpup air-to-surface missiles. Two 30 mm Mk 4 Aden guns in wing roots replace the 20 mm Mk 12 guns of the A-4M, each with 150 rounds of ammunition.

DIMENSIONS, EXTERNAL:
Wing span 8·38 m (27 ft 6 in)
Wing chord at root 4·72 m (15 ft 6 in)
Length overall (excl flight refuelling probe) 12·01 m (39 ft 5 in)
Height overall 4·57 m (15 ft 0 in)
Tailplane span 3·44 m (11 ft 3½ in)
Wheel track 2·38 m (7 ft 9½ in)

AREAS:
Wings, gross 24·16 m² (260 sq ft)
Vertical tail surfaces (total) 4·65 m² (50 sq ft)
Horizontal tail surfaces (total) 4·54 m² (48·85 sq ft)

WEIGHTS:
Weight empty 4,356 kg (9,603 lb)
Max T-O weight 10,206 kg (22,500 lb)
Max landing weight 7,257 kg (16,000 lb)

PERFORMANCE (at design T-O weight except where indicated):
Max level speed 573 knots (1,062 km/h; 660 mph)
T-O run at max T-O weight 1,187 m (3,895 ft)
Landing distance (at 6,577 kg; 14,500 lb AUW):
without braking parachute 1,052 m (3,450 ft)
with braking parachute 631 m (2,070 ft)
Max ferry range 1,680 nm (3,114 km; 1,935 miles)

LOCKHEED

LOCKHEED AIRCRAFT CORPORATION

HEAD OFFICE:
Burbank, California 91520
Telephone: (213) 847-6121

CHAIRMAN OF THE BOARD:
Robert W. Haack
VICE-CHAIRMAN:
Roy A. Anderson (Finance and Administration)
PRESIDENT AND CHIEF OPERATING OFFICER:
L. O. Kitchen

DIRECTORS:
Roy A. Anderson
Michael Berberian
Cyril Chappellet
D. M. Cochran
James E. Cross

Joseph P. Downer
Houston J. Flourney
C. S. Gross
Robert W. Haack
Willis M. Hawkins
Ellison L. Hazard
H. L. Hibbard (Honorary)
J. K. Horton
C. L. Johnson
L. O. Kitchen
J. Wilson Newman
Leslie N. Shaw
Fred M. Vinson Jr

CORPORATE SENIOR VICE-PRESIDENTS:
Robert B. Ormsby Jr (President, Lockheed-Georgia Company)
W. B. Rieke (President, Lockheed Missiles and Space Co Inc)
W. R. Wilson (Marketing)
Duane O. Wood (President, Lockheed-California Company)

CORPORATE VICE-PRESIDENTS:
F. A. Cleveland (Engineering)
Richard K. Cook (Washington Area)
James Everington (Industrial Relations)
L. E. Frisbee (General Manager, Engineering and Operations, Lockheed-California Co)
Robert A. Fuhrman (Executive Vice-President, Lockheed Missiles and Space Co Inc)
R. R. Heppe (Vice-President, Lockheed-California Co)
G. M. Kalember (International Marketing)
Dr H. Potter Kerfoot (General Manager, Research and Development Division of Lockheed Missiles and Space Co Inc)
J. Fred Lashley (Executive Vice-President, Lockheed-California Co)
A. H. Lorch (Executive Vice-President, Lockheed-Georgia Co)
Dr F. C. E. Oder (General Manager, Space Systems Division of Lockheed Missiles and Space Co Inc)
Thomas J. O'Hara (Contracts and Pricing)
W. D. Perreault Snr (Public Relations)
Dr R. Smelt (Chief Scientist)
William A. Stevenson (President, Lockheed Electronics Co Inc)
Dr D. A. Stuart (General Manager, Missile Systems Division)
R. W. Taylor (General Manager, Commercial Programs, Lockheed-California Co)
C. T. Thum (President, Lockheed Aircraft Service Co)
G. G. Whipple (President, Lockheed Shipbuilding and Construction Co)

SECRETARY:
James J. Ryan (Assistant General Counsel)

VICE-PRESIDENT AND GENERAL COUNSEL:
J. E. Cavanagh

CHIEF COUNSEL AND ASST SECRETARY:
John H. Martin

VICE-PRESIDENT AND CONTROLLER:
V. N. Marafino

VICE-PRESIDENT AND TREASURER:
Robert R. McKirahan

VICE-PRESIDENT AND ASST TREASURER:
L. T. Barrow (International Finance)

DIVISIONS:

LOCKHEED-CALIFORNIA COMPANY:
Burbank, California 91520
PRESIDENT: D. O. Wood

LOCKHEED-GEORGIA COMPANY:
Marietta, Georgia 30063
PRESIDENT: R. B. Ormsby Jr

LOCKHEED AIRCRAFT SERVICE COMPANY:
Ontario, California 91761
PRESIDENT: Charles T. Thum

SUBSIDIARIES:

AVIQUIPO INC:
Lyndhurst, New Jersey 07071
PRESIDENT: L. Craig Woodhouse

LOCKHEED AIR TERMINAL INC:
Burbank, California 91505
PRESIDENT: D. M. Simmons

LOCKHEED ELECTRONICS COMPANY INC:
Plainfield, New Jersey 07061
PRESIDENT: W. A. Stevenson

LOCKHEED MISSILES AND SPACE COMPANY INC:
Sunnyvale, California 94088
PRESIDENT: W. B. Rieke

LOCKHEED PETROLEUM SERVICES LTD:
New Westminster, British Columbia, Canada
PRESIDENT: Walter L. Weber

LOCKHEED SHIPBUILDING AND CONSTRUCTION COMPANY:
Seattle, Washington 98134
PRESIDENT: G. G. Whipple

MURDOCK MACHINE & ENGINEERING CO OF TEXAS:
Irving, Texas 75060
PRESIDENT: Paul R. Holmes

Built by the brothers Allan and Malcolm Lockheed, the first Lockheed aircraft, a tractor seaplane, first flew in 1913. Three years later the brothers established a company at Santa Barbara, California, to manufacture a twin-engined flying-boat, two seaplanes for the Navy and a small sport biplane that was a forerunner of the true streamlined aeroplane. Lockheed Aircraft Co, formed in 1926, moved to Burbank, California, in 1928 and was reorganised as Lockheed Aircraft Corporation in 1932.

On 30 November 1943, the Vega Aircraft Corporation, which had been formed in 1937 as an affiliate and in 1941 became a wholly-owned subsidiary of the Lockheed Aircraft Corporation, was absorbed and the name Vega abandoned.

Lockheed's aircraft and missile activities are now handled by three separate companies, which were evolved from the former California, Georgia and Missiles and Space Divisions in the Summer of 1961.

The current products of the Lockheed-California and Lockheed-Georgia Companies are described hereafter under the individual company headings.

Lockheed has diversified into many fields of industry since 1959. Following the acquisition of Stavid Engineering Inc, it combined this company and its own Electronics and Avionics Division into Lockheed Electronics Company Inc.

Lockheed Air Terminal Inc (LAT), a wholly-owned subsidiary, manages, operates and maintains the Hollywood-Burbank Airport, acquired in 1940, as well as providing fuelling and related services at 20 other locations in 11 states. Lockheed Aircraft Service Co (LAS) designs and manufactures products for the aerospace and marine industries, and Lockheed International Company provides international marketing services to all other Lockheed units; it co-ordinates efforts outside of the USA, and assists the development of new international marketing opportunities.

Since April 1959 Lockheed has also had an interest in shipbuilding and heavy construction, following its purchase of the Puget Sound Bridge and Dry Dock Company (now Lockheed Shipbuilding and Construction Co).

The total number of people employed by Lockheed Aircraft Corporation at the end of 1975 was 57,600.

LOCKHEED-CALIFORNIA COMPANY

Burbank, California 91520

Lockheed-California has responsibility for the land-based P-3 Orion, carrier-based S-3A Viking naval anti-submarine aircraft, the Carrier On-board Delivery US-3A, developed from the Viking, and the YF-12A/SR-71 military aircraft. Also in production is the L-1011 TriStar three-turbofan transport.

LOCKHEED F-104 STARFIGHTER

The Starfighter continues in production, under licence, by Aeritalia of Italy (which see).

LOCKHEED SR-71A

A Lockheed SR-71A, last described in the 1974-75 *Jane's,* has set six new world records, subject to ratification by the FAI.

In three flights, made by USAF crews on 27 July 1976, the following records were set:

Flown by Major Adolphus H. Bledsoe Jr, with Major John T. Fuller, a closed-circuit speed record and a class record of 1,812 knots (3,357 km/h; 2,086 mph) in a 1,000 km closed circuit.

Flown by Captain Eldon W. Joersz, with Major George T. Morgan Jr, a world absolute speed record and a class record of 1,901 knots (3,523 km/h; 2,189 mph) over a 15-25 km straight course.

Flown by Captain Robert C. Helt, with Major Larry A. Elliott, a sustained altitude record and class record of 26,213 m (86,000 ft) in horizontal flight.

LOCKHEED MODEL 85 ORION

US Navy designation: P-3
CAF designation: CP-140 Aurora

In April 1958 it was announced that Lockheed had been successful in winning with a developed version of the civil Electra four-turboprop airliner a US Navy competition for an 'off-the-shelf' ASW aircraft. The two original contracts provided for initial research, development and pre-production activities; further contracts provided for purchase by the Navy of a standard commercial Electra and its modification, development and testing as a tactical testbed for anti-submarine warfare systems.

An aerodynamic prototype, produced by modifying the airframe of the third civil Electra, flew for the first time on 19 August 1958. A second aircraft, designated YP-3A (formerly YP3V-1), with full electronics, flew on 25 November 1959.

Production versions are as follows:

P-3A. Initial production version for US Navy, with 3,356 kW (4,500 ehp) (with water-alcohol injection) Allison T56-A-10W turboprop engines. First P-3A flew for the first time on 15 April 1961. Deliveries to the US Navy began on 13 August 1962, to replace the P-2 Neptune. P-3As from the 110th aircraft are known as Deltic P-3As, as they are fitted with the Deltic system, including more sensitive ASW detection devices and improved tactical display equipment. Three Deltic P-3As were supplied to the Spanish Air Force. Production completed.

WP-3A. Weather reconnaissance version of P-3A. Four delivered to US Navy during 1970 to re-equip squadron previously flying WC-121Ns. WP-3A has meteorological equipment of the type used in the WC-121Ns and a radome of different configuration.

P-3B. Follow-on production version with 3,661 kW (4,910 ehp) Allison T56-A-14 turboprop engines, which do not need water-alcohol injection. USN contracts covered 286 P-3As and P-3Bs. In addition, five P-3Bs were delivered to the Royal New Zealand Air Force in 1966, ten to the Royal Australian Air Force during 1968 and five to Norway in the Spring of 1969. USN-P-3Bs were modified retrospectively to carry Bullpup missiles. Others became **EP-3Bs**. The US Navy and Lear Siegler were developing in early 1976 a modification kit for retrofitting to P-3B aircraft. This includes a 32K Rorn computer, ASN-84 inertial navigation system, Omega, ASA-66 displays and ASN-124 navigation controller, together with the necessary controls and equipment to integrate the new and existing systems. Kits were scheduled to be available for installation in 1977.

P-3C. Advanced version with the A-NEW system of sensors and control equipment, built around a Univac digital computer that integrates all ASW information and permits retrieval, display and transmission of tactical data in order to eliminate routine log-keeping functions. This increases crew effectiveness by allowing them sufficient time to consider all tactical data and devise the best action to resolve problems. First flight of this version was made on 18 September 1968 and the P-3C entered service in 1969. A total of 132 of this version had been delivered to the US Navy by the end of January 1976.

Under a programme designated P-3C Update, new electronics and software were developed to enhance the effectiveness of this aircraft. Equipment includes a magnetic drum that gives a sevenfold increase in computer memory capacity, a new versatile computer language, Omega navigation system, improved acoustic processing sensitivity, a tactical display for two of the sensor stations, and an improved magnetic tape transport. A prototype with this equipment was handed over to the US Navy on 29 April 1974, and the first production aircraft was delivered in January 1975. All subsequent production aircraft for the US Navy have this equipment.

Eight P-3Cs with Update modifications were ordered for the Royal Australian Air Force in May 1975, for delivery in 1977-78.

The US Navy and Lockheed were proceeding in early

Lockheed EP-3E Orion of US Navy Fleet Air Reconnaissance Squadron 2 (VQ-2)

Lockheed P-3F Orion long-range maritime patrol aircraft for the Imperial Iranian Air Force

1976 with a further electronics improvement programme for the P-3C. Known as Update II, this adds a forward-looking infra-red (FLIR) and a sonobuoy reference system (SRS). The Harpoon missile and control system are included in Update II, which will be incorporated into production aircraft from August 1977. Update III, under development in 1976, chiefly involves new ASW electronics, and is scheduled for introduction on aircraft due for delivery in 1979.

Five of the eight international records for turboprop aircraft, set up in a P-3C by Cdr Donald H. Lilienthal in early 1971, had not been beaten by mid-1976. They include a speed of 434·97 knots (806·10 km/h; 500·89 mph) over a 15/25 km course; and four time-to-climb records, to 3,000 m in 2 min 51·75 sec; 6,000 m in 5 min 46·36 sec; 9,000 m in 10 min 26·12 sec; and 12,000 m in 19 min 42·24 sec.

RP-3D. One P-3C was reconfigured during manufacture for a five-year mission to map the Earth's magnetic field, under Project Magnet, which is controlled by the US Naval Oceanographic Office. Crew of 17. Range increased to more than 5,000 nm (9,265 km; 5,755 miles) by installation of 4,545 litre (1,200 US gallon) fuel tank in weapons bay. ASW equipment removed, making way for dual inertial navigation systems, dual computers for recording and analysis of magnetic data and for providing co-ordinate readout of Loran A and C, a satellite tracker, an Omega navigation system, a gyro-stabilised sextant capable of giving six lines of position almost simultaneously, highly sensitive magnetometers, and equipment for time correlation by cesium beam time standard. Seven bunks provided for off-duty crew.

The RP-3D is operated by the US Navy's Oceanographic Development Squadron 8 (VXN-8) based at the Naval Air Test Center, Patuxent River, Maryland. On 4 November 1972, Cdr Philip R. Hite used it to set up an international closed-circuit distance record for turboprop aircraft, covering 5,451·97 nm (10,103·51 km; 6,278·03 miles).

WP-3D. Two aircraft equipped as airborne research centres, ordered by the US National Oceanic and Atmospheric Administration. Equipped to carry out atmospheric research and weather modification experiments. Scheduled to become operational during the Summer of 1976.

EP-3E. Ten P-3As and two EP-3Bs were converted to EP-3E configuration to replace Lockheed EC-121s in service with VQ-1 and VQ-2 squadrons. They can be identified by large canoe radars on the upper and lower surfaces of the fuselage and a ventral radome forward of the wing.

P-3F. Six aircraft, similar to the US Navy's P-3Cs, for the Imperial Iranian Air Force. To be used initially for long-range surface surveillance and subsequently also for ASW missions. Delivery of these aircraft was completed in January 1975.

CP-140 Aurora. Version for Canadian Armed Forces; 18 ordered in July 1976, for delivery from 1980. Will incorporate many of the P-3C Update improvements, most of the ASW electronics developed for the S-3A Viking, and other special sensors specified by Canadian government.

By the beginning of 1976 Lockheed-California had delivered 440 P-3s of all versions. The following data refer to the P-3C, but are generally applicable to other versions, except for the details noted:

TYPE: Four-turboprop ASW aircraft.

WINGS: Cantilever low-wing monoplane. Wing section NACA 0014 (modified) at root, NACA 0012 (modified) at tip. Dihedral 6°. Incidence 3° at root, 0° 30′ at tip. Fail-safe box beam structure of extruded integrally-stiffened aluminium alloy. Lockheed-Fowler trailing-edge flaps. Hydraulically-boosted aluminium ailerons. Anti-icing by engine bleed air ducted into leading-edges.

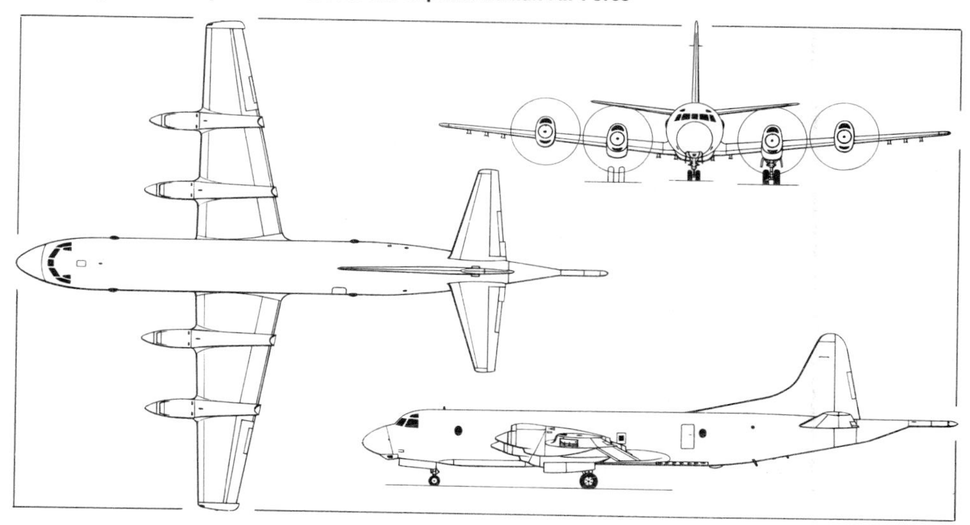

Lockheed P-3C Orion four-turboprop anti-submarine aircraft *(Pilot Press)*

FUSELAGE: Conventional aluminium alloy semi-monocoque fail-safe structure.

TAIL UNIT: Cantilever aluminium alloy structure with dihedral tailplane and dorsal fin. Fixed-incidence tailplane. Hydraulically-boosted rudder and elevators. Leading-edges of fin and tailplane have electric anti-icing system.

LANDING GEAR: Hydraulically-retractable tricycle type, with twin wheels on each unit. All units retract forward, main wheels into inner engine nacelles. Oleo-pneumatic shock-absorbers. Main wheels have size 40-14 type VII 26-ply tubeless tyres. Nosewheels have size 28-7·7 type VII tubeless tyres. Hydraulic brakes. No anti-skid units.

POWER PLANT: Four 3,661 kW (4,910 ehp) Allison T56-A-14 turboprop engines, each driving a Hamilton Standard 54H60 four-blade constant-speed propeller. Fuel in one tank in fuselage and four wing integral tanks, with total usable capacity of 34,826 litres (9,200 US gallons). Four overwing gravity fuelling points and central pressure refuelling point. Oil capacity (min usable) 111 litres (29·4 US gallons) in four tanks. Electrically de-iced propeller spinners.

ACCOMMODATION: Normal ten-man crew. Flight deck has wide-vision windows, and circular windows for observers are provided fore and aft in the main cabin, each bulged to give 180° view. Main cabin is fitted out as a five-man tactical compartment containing advanced electronic, magnetic and sonic detection equipment, an all-electric galley and large crew rest area.

SYSTEMS: Air-conditioning and pressurisation system supplied by two engine-driven compressors. Pressure differential 0·37 bars (5·4 lb/sq in). Hydraulic system, pressure 207 bars (3,000 lb/sq in), for flaps, control surface boosters, landing gear actuation, brakes and bomb bay doors. Pneumatic system, pressure 207/83 bars (3,000/1,200 lb/sq in), for ASW store launchers (P-3A/B only). Electrical system utilises three 60kVA generators for 120/208V 400Hz AC supply. 24V DC supply. Integral APU with 60kVA generator for ground air-conditioning and electrical supply and engine starting.

ELECTRONICS AND EQUIPMENT (P-3A/B): Communications and navigation equipment comprises two ARC-94 HF transceivers, ARC-84 VHF transmitter, two ARC-84 VHF receivers, two ARC-51A UHF transceivers, AIC-22 interphone, TT-264/AG teletypewriter, UNH-6 communications tape recorder, two CU-351 HF couplers, ASN-42 inertial navigation system, APN-153 Doppler navigation system, ASA-47 Doppler air mass computer, ASN-50 AHRS, APN-70 Loran, ARN-52 Tacan, DF-202 radio compass, ARN-32 marker beacon receiver, APN-141 radar altimeter, ARA-50 UHF direction finder, two ARN-87 VOR installations, two HSI, A/A24G-9 true airspeed computer, PB-20N autopilot, ID-888/U latitude/longitude indicator, and APQ-107 radar altitude warning system. ASW equipment includes ASA-16 tactical display, two APS-80 radar antennae for 360° coverage, APA-125A radar display, ASR-3 trail detector, 31 ARR-72 sonobuoy signal receivers, AQA-7 acoustic processer (DIFAR), ASQ-10A magnetic anomaly detector in plastics tail 'sting', modified ALD-2B ECM direction finder for detecting and locating electronic emissions from submarines, ULA-2 ECM signal analyser, APX-6 IFF identification, APX-7 IFF recognition, APA-89 IFF coder group, AQH-1(V) tactical tape recorder, ASA-50 ground speed and bearing computer, R-1047/A on-top position indicator, TD-441/A intervalometer, PT396/AS ground track plotter, ASA-13 tactical plot board, bearing-distance-heading indicator and KY 364 video decoder. Equipment for day or night photographic reconnaissance. Searchlight under starboard wing. RAAF P-3Bs have Marconi-Elliott AQS 901 acoustic signal processing and display system.

ELECTRONICS AND EQUIPMENT (P-3C): The ASQ-114 general-purpose digital computer is the heart of the P-3C system. Together with the AYA-8 data processing equipment and computer-controlled display systems, it permits rapid analysis and utilisation of electronic, magnetic and sonic data. Nav/com system comprises two ASN-84 inertial navigation systems, with latitude and longitude indicators; APN-187 Doppler; ARN-81 Loran A and C; ARN-84 Tacan; two ARN-87 VOR receivers; ARN-32 marker beacon receiver; ARN-83 LF-ADF; ARA-50 UHF direction finder; AJN-15 flight director indicator for tactical directions; HSI for long-range flight directions; glideslope indicator; on-top position indicator; two ARC-161 HF transceivers; two ARC-143 UHF transceivers; ARC-101 VHF receiver/transmitter; AGC-6 teletype and high-speed printer; HF and UHF secure communication units; ACQ-5 data link communication set and AIC-22 interphone set; APX-72 IFF transponder and APX-76 SIF interrogator. Electronic computer-controlled display equipment includes ASA-70 tactical display; ASA-66 pilot's display; ASA-70 radar display and two auxiliary readout (computer-stored data) displays. ASW equipment includes two ARR-72 sono receivers; two AQA-7 DIFAR sonobuoy indicator sets; hyperbolic fix unit; acoustic source signal generator; time code generator and AQH-4(V) sonar tape recorder; ASQ-81 magnetic anomaly detector; ASA-64 submarine anomaly detector; ASA-65 magnetic compensator; ALQ-78 electronic countermeasures set; APS-115 radar set (360° coverage); ASA-69 radar scan converter; KA-74 forward computer-assisted camera; KB-18A automatic strike assessment camera with horizon-to-horizon coverage; RO-308 bathythermograph recorder. Additional equipment includes APN-141(V) radar altimeter; two APQ-107 radar altimeter warning systems; A/A24G-9 true airspeed computer and ASW-31 automatic flight control system. P-3Cs delivered from 1975 have the electronics package updated by addition of an extra 393K memory drum and fourth logic unit, Omega navigation, new magnetic tape transport, and an ASA-66 tactical display for the sonar operators. To accommodate the new systems a new operational software computer programme will be written in CMS-2 language.

ARMAMENT: Bomb bay, 2·03 m wide, 0·88 m deep and 3·91 m long (80 in × 34·5 in × 154 in), forward of wing, can accommodate a 2,000 lb MK 25/39/55/56 mine, three 1,000 lb MK 36/52 mines, three MK 57 depth bombs, eight MK 54 depth bombs, eight MK 43/44/46 torpedoes or a combination of two MK 101 nuclear depth bombs and four MK 43/44/46 torpedoes. Ten underwing pylons for stores: two under centre-section each side can carry torpedoes or 2,000 lb mines; three under outer wing each side can carry respectively (inboard to outboard) a torpedo or 2,000 lb mine (or searchlight on starboard wing), a torpedo or 1,000 lb mine or rockets singly or in pods; a torpedo or 500 lb mine or rockets singly or in pods. Torpedoes can be carried underwing only for ferrying; mines can be carried and released. Search stores, such as sonobuoys and sound signals, are launched from inside cabin area in the P-3A/B. In the P-3C sonobuoys are loaded and launched externally and internally. Max total weapon load includes six 2,000 lb mines under wings and a 3,290 kg (7,252 lb) internal load made up of two MK 101 depth bombs, four MK 44 torpedoes, pyrotechnic pistol and 12 signals, 87 sonobuoys, 100 MK 50 underwater sound signals (P-3A/B), 18 MK 3A marine markers (P-3A/B), 42 MK 7 marine markers, two B.T. buoys, and two MK 5 parachute flares. Sonobuoys are ejected from P-3C aircraft with explosive cartridge actuating devices (CAD), eliminating the need for a pneumatic system.

DIMENSIONS, EXTERNAL:
- Wing span: 30·37 m (99 ft 8 in)
- Wing chord at root: 5·77 m (18 ft 11 in)
- Wing chord at tip: 2·31 m (7 ft 7 in)
- Wing aspect ratio: 7·5
- Length overall: 35·61 m (116 ft 10 in)
- Height overall: 10·29 m (33 ft 8½ in)

Artist's impression of the CP-140 version of the Lockheed Orion for the Canadian Armed Forces **(see Addenda)**

- Fuselage outside diameter: 3·45 m (11 ft 4 in)
- Tailplane span: 13·06 m (42 ft 10 in)
- Wheel track (c/l shock-absorbers): 9·50 m (31 ft 2 in)
- Wheelbase: 9·07 m (29 ft 9 in)
- Propeller diameter: 4·11 m (13 ft 6 in)
- Cabin door:
  - Height: 1·83 m (6 ft 0 in)
  - Width: 0·69 m (2 ft 3 in)

DIMENSIONS, INTERNAL:
- Cabin, excl flight deck and electrical load centre:
  - Length: 21·06 m (69 ft 1 in)
  - Max width: 3·30 m (10 ft 10 in)
  - Max height: 2·29 m (7 ft 6 in)
  - Floor area: 61·13 m² (658 sq ft)
  - Volume: 120·6 m³ (4,260 cu ft)

AREAS:
- Wings, gross: 120·77 m² (1,300 sq ft)
- Ailerons (total): 8·36 m² (90 sq ft)
- Trailing-edge flaps (total): 19·32 m² (208 sq ft)
- Fin, incl dorsal fin: 10·78 m² (116 sq ft)
- Rudder, incl tab: 5·57 m² (60 sq ft)
- Tailplane: 22·39 m² (241 sq ft)
- Elevators, incl tabs: 7·53 m² (81 sq ft)

WEIGHTS (P-3B/C):
- Weight empty: 27,890 kg (61,491 lb)
- Max expendable load: 9,071 kg (20,000 lb)
- Max normal T-O weight: 61,235 kg (135,000 lb)
- Max permissible weight: 64,410 kg (142,000 lb)
- Design landing weight: 51,709 kg (114,000 lb)

PERFORMANCE (P-3B/C, at max T-O weight, except where indicated otherwise):
- Max level speed at 4,570 m (15,000 ft) at AUW of 47,625 kg (105,000 lb): 411 knots (761 km/h; 473 mph)
- Econ cruising speed at 7,620 m (25,000 ft) at AUW of 49,895 kg (110,000 lb): 328 knots (608 km/h; 378 mph)
- Patrol speed at 457 m (1,500 ft) at AUW of 49,895 kg (110,000 lb): 206 knots (381 km/h; 237 mph)
- Stalling speed, flaps up: 133 knots (248 km/h; 154 mph)
- Stalling speed, flaps down: 112 knots (208 km/h; 129 mph)
- Max rate of climb at 457 m (1,500 ft): 594 m (1,950 ft)/min
- Service ceiling: 8,625 m (28,300 ft)
- Service ceiling, one engine out: 5,790 m (19,000 ft)
- T-O run: 1,290 m (4,240 ft)
- T-O to 15 m (50 ft): 1,673 m (5,490 ft)
- Landing from 15 m (50 ft) at design landing weight: 845 m (2,770 ft)
- Max mission radius (no time on station) at 61,235 kg (135,000 lb): 2,070 nm (3,835 km; 2,383 miles)
- Mission radius (3 hr on station at 457 m; 1,500 ft): 1,346 nm (2,494 km; 1,550 miles)

LOCKHEED S-3A VIKING

US Navy designation: S-3A

On 4 August 1969 Lockheed announced the receipt of a $461 million contract from the US Navy to develop a new anti-submarine aircraft under the designation S-3A. Development was carried out by Lockheed in partnership with Vought Systems Division of LTV and Univac Federal Systems Division of Sperry Rand. Vought designed and is building the wing, engine pods, tail unit and landing gear, and Univac is responsible for the digital computer, the heart of the weapon system, which provides high-speed processing of data essential for the S-3A's ASW role. Lockheed is building the fuselage, integrating the electronics, and is responsible for final assembly at Burbank, California.

The selection of Lockheed-California as contractor for this aircraft followed more than a year of intensive competition between North American Rockwell, McDonnell Douglas, Grumman, Convair Division of General Dynamics, and Lockheed-California in conjunction with LTV.

The Lockheed team was responsible for development, test and demonstration of the aircraft and its weapon systems. The first prototype was rolled out on schedule on 8 November 1971 at Burbank, California, and the first flight was made on 21 January 1972. An increased ceiling of $494 million on the contract, funded over a five-year period, provided for production of eight research and development aircraft in two lots.

On 4 May 1972 the US Navy announced an order for the first production lot of 13 S-3As, and orders for 35 and 45 more were received in April and October 1973 respectively. Other orders have followed.

The Viking was introduced into the Fleet officially on 20 February 1974, during ceremonies held at North Island NAS, near San Diego, California. Initial deliveries were made to Squadron VS-41, the S-3A training squadron based at North Island NAS. By 1 January 1976 Lockheed had delivered 101 of the 187 Vikings then on order for the US Navy.

At North Island, USN personnel from Squadron VS-21, VS-22 and VS-29 completed conversion to the S-3A by mid-1975, and VS-28, 31 and 32 personnel had also qualified by the end of that year. First operational deployment of the Viking, with Squadron VS-21, was made in July 1975, on board the USS *John F. Kennedy*.

Lockheed S-3A Viking of US Navy Squadron VS-21 landing on board the carrier *John F. Kennedy*

VS-21's deployment had been completed and the squadron had returned to its home base at North Island NAS by early 1976.

The S-3A has a crew of four, comprising a pilot, co-pilot, tactical co-ordinator (Tacco) and acoustic sensor operator (Senso). The pilot maintains command of the aircraft, while the Tacco formulates strategy and instructs the pilots on the necessary manoeuvres for a successful submarine attack. In addition to flying duties, the co-pilot is responsible for the non-acoustic sensors (such as radar and infra-red) and navigation: the Senso controls the acoustic sensors.

The development of quieter submarines has led to the design of sonobuoys of increased sensitivity, and advanced cathode ray tube displays are provided in the S-3A to maintain flexibility of operation with a limited crew. In particular, a cathode ray tube is utilised to monitor the acoustic sensors. The information formerly stowed in roll form from paper plotters is now stored in the Univac 1832A computer and is available for instant recall. Other functions of the computer include weapon trajectory calculations and pre-flight navigation. Magnetic anomaly detection (MAD) equipment of increased sensitivity makes it possible to detect submarines at greater depths than has been possible until the present time.

Shipboard maintenance is simplified by the provision of computerised fault-finding equipment, built-in test equipment (BITE), and versatile avionic shop test (VAST) compatibility. Complete deck-level servicing accessibility contributes to the attainment of a quick turnaround time.

A utility transport variant, for carrier on-board delivery, designated **US-3A**, has been ordered and is described separately. The performance characteristics of the S-3A will make possible other future design variants, including tanker, ASW command and control, and a variety of electronic countermeasures aircraft. To cater for future growth, the fuselage volume allows for 50 per cent expansion of electronics equipment. KS-3A tanker and ES-3 electronics patrol versions are under study by the US Navy.

The following description applies to the current production S-3A:

TYPE: Twin-turbofan carrier-borne anti-submarine aircraft.

WINGS: Cantilever shoulder-wing monoplane. Sweepback at quarter-chord 15°. No dihedral. Incidence 3° 15′ at root, −3° 50′ at tip. All-metal fail-safe structure. Wings fold upward and inward hydraulically, outboard of engine pylons, for carrier stowage. Single-slotted Fowler-type trailing-edge flaps, operated by hydraulic power with an integral electric motor for emergency operation. Electrically-operated leading-edge flaps, extending from engine pylons to wingtips, are fully extended after 15° of trailing-edge flap movement. Ailerons augmented by under- and over-wing spoilers for roll control. All primary flight control surfaces are actuated by irreversible servos powered by dual hydraulic systems. Loss of either hydraulic system results in loss of half the available hinge movement, but the remaining system can meet all control requirements. Automatic reversion to manual control in the event of failure of both hydraulic systems. In emergency operation the spoilers are inoperative. Wing anti-icing by engine bleed air, but portions of wing leading-edges are cyclically heated to reduce consumption of bleed air.

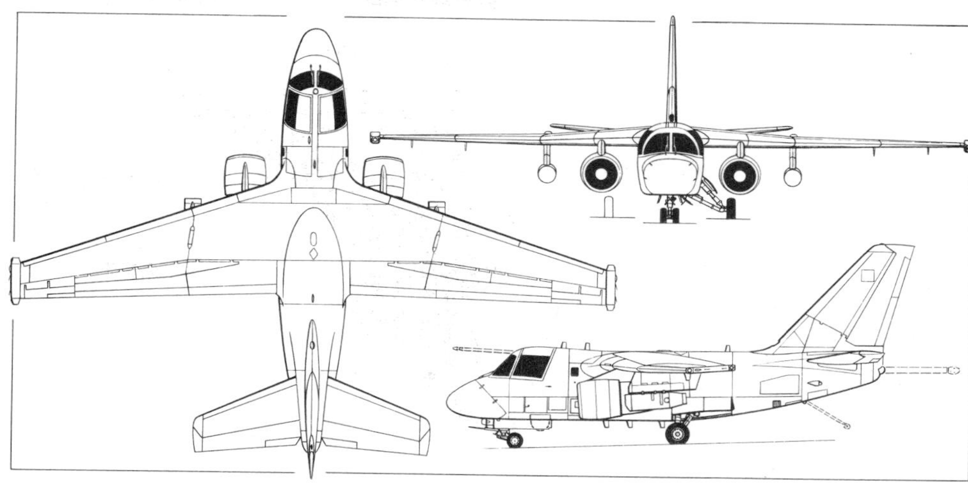

Lockheed S-3A Viking carrier-based anti-submarine aircraft *(Pilot Press)*

FUSELAGE: Semi-monocoque all-metal fail-safe structure, incorporating split weapons bays with clamshell doors. Two parallel beams form a keelson from nose gear to tail-hook, strengthening the fuselage and improving cabin structural integrity by distributing catapult and arrester loads throughout the airframe. Launch tubes for 60 sonobuoys in belly. No provision for in-flight reloading of these launch tubes. Frangible canopies in top of fuselage are so designed that the crew can eject through them in emergency. Electronics bays with external access doors in forward and aft fuselage. An illuminated in-flight refuelling probe, mounted within the fuselage on the top centreline, is operated by an electric drive and protected by a positive-seal door. It can be extended or retracted in emergency by a hand crank. MAD boom, extensible in flight, housed in fuselage tail.

TAIL UNIT: Cantilever all-metal structure with swept vertical and horizontal surfaces. Fin and rudder are folded downward by hydraulic servos for carrier stowage. During fin-folding sequence the pedal input to the rudder servo is disconnected to allow the pilot to steer the nosewheel by the rudder pedals. Variable-incidence tailplane, electrically controlled. Elevator and rudder controlled by hydraulic servos. Trim tabs in elevators and rudder. Anti-icing of tailplane leading-edges by engine bleed air.

LANDING GEAR: Hydraulically-retractable tricycle type. Main units, similar to those of the Vought F-8 Crusader, are fitted with single wheels and retract rearward into wheel wells immediately aft of the split weapons bays. Nose unit similar to that of the Vought A-7 Corsair II, with twin wheels and catapult towbar, retracts rearward into fuselage. Nosewheel steering by hydraulic power. Hydraulic brakes. Arrester hook.

POWER PLANT: Two General Electric TF34-GE-2 high bypass ratio turbofan engines, each rated at 41·25 kN (9,275 lb st), pylon-mounted beneath the wings. Fuel in integral wing tanks, entirely within the wing box beam, one on each side of the fuselage centreline and inboard of the wing fold-line. Usable fuel capacity approximately 7,192 litres (1,900 US gallons). Two 1,136 litre (300 US gallon) jettisonable fuel tanks can be carried on underwing pylons. Single-point pressure refuelling adapter located on starboard side of fuselage aft of main landing gear door. Internal tanks may also be gravity fuelled through overwing connections. Fuel jettison system. Anti-icing of engine inlet nozzles by engine bleed air.

ACCOMMODATION: Crew of four. Pilot and co-pilot side by side on flight deck with transparent canopy. Tacco and Senso accommodated in aft cabin, with individual polarised side windows. All crew on McDonnell Douglas Escapac 1-E zero-zero ejection seats. Each seat has a rigid seat survival kit (RSSK), which can be opened during descent for inflation of life raft. Electric windscreen wipers. Windscreen surfaces electrically heated; side canopy is demisted with conditioned air. Liquid rain-repellent system to augment action of windscreen wipers. Cabin pressurised and air-conditioned, and each crewman's anti-exposure suit is ventilated with conditioned air from this system.

SYSTEMS: Garrett-AiResearch environmental control system, with engine bleed air supply and air-cycle refrigeration unit. Pressurisation system operates on a differential of 0·41-0·55 bars (6-8 lb/sq in), maintaining a cabin altitude of 1,525 m (5,000 ft) to a height of 7,620 m (25,000 ft) and 3,505 m (11,500 ft) cabin altitude to 12,200 m (40,000 ft). Two engine-driven pumps supply hydraulic power for two completely independent 207 bar (3,000 lb/sq in) systems. Port system supplies landing gear, flaps, brakes, wing and tail fold, arrester hook and weapon bay doors. Its secondary function is to power one side of the primary flight control servos. Starboard system powers only the primary flight controls, energising one side of the dual servo actuators; port system energises the other. Electrical system includes two 75kVA generators supplying 115-120V AC at a frequency of 400Hz. Secondary DC power is obtained from two transformer-rectifiers that deliver 28V DC at 200A. Williams Research Corporation gas turbine APU has a 5kVA generator for emergency elec-

Lockheed S-3A Viking of US Navy Squadron VS-21, USS *John F. Kennedy*

trical power, providing 115-120V AC at 400Hz to the essential AC bus and 28V DC at 30A through the transformer-rectifiers. Emergency electrical power is adequate only for essential capabilities such as those required for night flight under instrument conditions.

Electronics: ASW data processing, control and display includes Univac 1832A general-purpose digital computer, acoustic data processer, sonobuoy receiver, command signal generator and analogue tape recorder. Non-acoustic sensors comprise AN/APS-116 high-resolution radar, OR-89/AA forward-looking infra-red (FLIR) scanner in retractable turret, AN/ASQ-81 MAD and compensation equipment, and ALR-47 passive ECM receiving and instantaneous frequency-measuring system housed in wingtip pods. Primary navigation system composed of ASN-92(V) CAINS inertial navigator, AN/APN-200 Doppler ground velocity system (DGVS), AYN-5 airspeed/altitude computing set (AACS), ASN-107 attitude heading reference system (AHRS), ARS-2 sonobuoy reference system (SRS), APN-201 radar altimeter and altitude warning system (RAAWS), ARN-83 LF/ADF and ARA-50 UHF/DF radio navigation aids, ARN-84(V) Tacan, and the aircraft's flight displays and interface system (FDIS). Communications equipment includes a 1kW ARC-153 HF transceiver for long-range communications, dual ARC-156 UHF transceivers, AN/ARA-63 receiver/decoder set for use with shipboard ILS, data terminal set (DTS), OK-173 integral intercom system (ICS) and APX-72 IFF/APX-76A SIF units with altitude reporting, and AN/ASW-25B automatic carrier landing system (ACLS) communication set. Search stores are designated as LOFAR (SSQ-41), R/O (SSQ-47), DIFAR (SSQ-53), CASS (SSQ-50), DICASS (SSQ-62) and BT (SSQ-36) sonobuoys.

Armament: Split weapons bays equipped with BRU-14/A bomb rack assemblies can deploy either four MK-36 destructors, four MK-46 torpedoes, four MK-82 bombs, two MK-57 or four MK-54 depth bombs, or four MK-53 mines. BRU-11/A bomb racks installed on the two wing pylons permit carriage of SUU-44/A flare launchers, MK-52, MK-55 or MK-56 mines, MK-20-2 cluster bombs, Aero 1D auxiliary fuel tanks, or two rockets pods of type LAU-68/A (7 FFAR 2·75 in), LAU-61/A (19 FFAR 2·75 in), LAU-69/A (19 FFAR 2·75 in), or LAU-10A/A (4 FFAR 5·0 in). Alternatively, installation of TER-7 triple ejector racks on the BRU-11/A bomb racks makes it possible to carry three rocket pods, flare launchers, MK-20 cluster bombs, MK-82 bombs, MK-36 destructors, or MK-76-5 or MK-106-4 practice bombs under each wing.

Dimensions, external:

Wing span	20·93 m (68 ft 8 in)
Wing span, wings folded	8·99 m (29 ft 6 in)
Length overall	16·26 m (53 ft 4 in)
Length overall, tail folded	15·06 m (49 ft 5 in)
Height overall	6·93 m (22 ft 9 in)
Height overall, tail folded	4·65 m (15 ft 3 in)
Tailplane span	8·23 m (27 ft 0 in)

Dimensions, internal:

Max height	2·29 m (7 ft 6 in)
Max width	2·18 m (7 ft 2 in)

Area:

Wings, gross	55·56 m² (598 sq ft)

Weights and Loading:

Weight empty	12,088 kg (26,650 lb)
Max design gross weight	23,831 kg (52,539 lb)
Normal ASW T-O weight	19,277 kg (42,500 lb)
Max landing weight	20,826 kg (45,914 lb)
Max carrier landing weight	17,098 kg (37,695 lb)
Max wing loading	429·2 kg/m² (87·9 lb/sq ft)

Performance (at normal ASW T-O weight, unless otherwise indicated):

Max level speed	450 knots (834 km/h; 518 mph)
Max cruising speed	370 knots (686 km/h; 426 mph)
Loiter speed	160 knots (296 km/h; 184 mph)
Approach speed	100 knots (185 km/h; 115 mph)
Stalling speed	84 knots (157 km/h; 97 mph)
Max rate of climb at S/L	over 1,280 m (4,200 ft)/min
Service ceiling	above 10,670 m (35,000 ft)
T-O run	671 m (2,200 ft)
Landing run at landing weight of 16,556 kg (36,500 lb)	488 m (1,600 ft)
Combat range	more than 2,000 nm (3,705 km; 2,303 miles)
Ferry range	more than 3,000 nm (5,558 km; 3,454 miles)

LOCKHEED US-3A VIKING

US Navy designation: US-3A

On 15 December 1975 Lockheed-California announced the receipt of a $3 million contract from the US Navy to begin development of a carrier on-board delivery (COD) version of the S-3A Viking, for the transport of passengers and/or cargo between shore bases and aircraft carriers.

Designated US-3A Viking, the new aircraft is a utility transport, with cargo and/or seats for up to six passengers occupying the cabin space which in the S-3A is allocated to the control stations and equipment of the tactical co-ordinator and acoustic sensor operator. Additional cargo space is provided by deletion of certain ASW equipment, accommodated in the S-3A in the upper fuselage below the starboard wing, and in six forward and aft lower-fuselage compartments. With no requirement for a weapon load, the split bomb bays also are used for cargo, the bomb bay doors being deleted. Special streamlined cargo pods, with a diameter of 1·07 m (3 ft 6 in) and overall length of 5·08 m (16 ft 8 in), have been designed for attachment to the underwing pylons, used on the S-3A for the carriage of weapons and other stores. Each pod has an internal volume of 2·55 m³ (90 cu ft) and can accommodate up to 454 kg (1,000 lb) of cargo.

These changes, by comparison with the S-3A, give the US-3A a total cargo volume of 12·74 m³ (450 cu ft), which can be utilised for the carriage of a net maximum payload of 2,608 kg (5,750 lb). For basic passenger/cargo or all-cargo missions, maximum cruise altitude is restricted to 10,670 m (35,000 ft) to maintain a cabin altitude of approximately 3,000 m (10,000 ft); a total of 1,700 kg (3,750 lb) can be carried in the all-cargo configuration, or six passengers (or five plus crew chief/loadmaster) and 1,275 kg (2,810 lb) of cargo. En-route loading equipment weighing 345 kg (760 lb) is standard.

The starboard cargo bay beneath the wing, which is environmentally controlled, could be used in an emergency for the carriage of one or two litter patients.

Construction of the prototype US-3A began in August 1975, and it flew for the first time on 2 July 1976. If the US-3A is shown to meet the Navy's requirements, production contracts for 30 are likely to follow. The first six are included in FY 1977 budget funding.

The description of the S-3A applies also to the US-3A, except for the following details:

Type: Cargo/passenger transport for carrier on-board delivery (COD).

Fuselage: Basically as for S-3A, with weapons bays deleted. Cargo bays with external access doors in forward, centre and aft fuselage. No MAD.

Landing Gear: Basically as for S-3A. Hand pump for emergency retraction of landing gear. Main-wheel tyres size 30 × 11·5-14·5, Type VIII 24-ply rating, pressure 22 bars (320 lb/sq in) for carrier landings, 16·9 bars (245 lb/sq in) for land operation. Nosewheel tyres size 22 × 6·75-10, Type VII 18-ply rating, pressure 22 bars (320 lb/sq in) for carrier landings, 8·27 bars (120 lb/sq in) for land operation. Hydraulic brakes. Arrester hook.

Accommodation: Crew of three, comprising pilot, co-pilot and crew chief/loadmaster. Pilot and co-pilot side by side on flight deck with transparent canopy. Crew chief/loadmaster and up to five passengers (or six without crew chief/loadmaster) on two three-abreast rows of seats in cabin aft of flight deck. Windows in cabin sides. Electric windscreen wipers. Windscreen surfaces electrically heated; side canopy is demisted with conditioned air. Liquid rain-repellent system to augment action of windscreen wipers. Cabin pressurised and air-conditioned.

Electronics: Communication systems comprise dual AN/ARC-156 UHF transceivers; AN/ARC-153 HF transceiver for long-range communications; AN/ARC-175(V) VHF; TSEC/KY-28 UHF secure voice; OK-248A(V)AI internal communication system and radio control, with cabin speaker. AN/ASN-92(V) CAINS inertial navigation system; AN/ASA-84 inertial navigation system interface; AN/ASN-107 attitude and heading reference system; AN/APN-200 Doppler ground velocity system; AN/APN-201 radar altimeter and altitude warning system; AN/ARN-84 Tacan; two AN/ARN-126 VOR/ILS marker beacons; AN/ARN-83 LF/ADF; AN/ARA-50 UHF/ADF; AN/AYN-5 airspeed/altitude computing set; AN/OD-59 navigation indicator group; AN/APX-72 IFF transponder; AN/ASW-33 automatic flight control system; AN/ASW-25 automatic carrier landing system; AN/APN-202 radar beacon; AN/ARA-63 carrier instrument landing system; and weather radar.

Dimensions, external:

Wing span	20·93 m (68 ft 8 in)
Wing span, wings folded	8·99 m (29 ft 6 in)
Wing chord at root	4·29 m (14 ft 1 in)
Wing chord at tip	1·07 m (3 ft 6 in)
Wing aspect ratio	7·73
Length overall	16·26 m (53 ft 4 in)
Length overall, tail folded	15·06 m (49 ft 5 in)
Height overall	6·93 m (22 ft 9 in)
Height overall, tail folded	4·65 m (15 ft 3 in)
Tailplane span	8·23 m (27 ft 0 in)
Wheel track	4·19 m (13 ft 9 in)
Wheelbase	5·72 m (18 ft 9 in)

Prototype Lockheed US-3A COD transport, produced by modifying an S-3A development aircraft

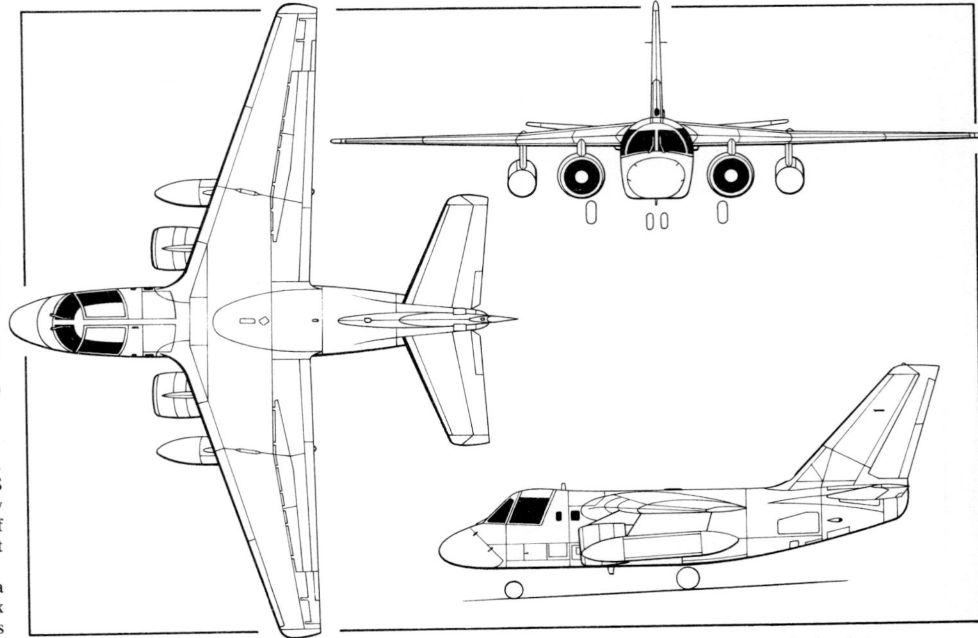

Lockheed US-3A Viking carrier on-board delivery transport *(Michael A. Badrocke)*

DIMENSIONS, INTERNAL:
- Passenger cabin:
  - Max height 2·29 m (7 ft 6 in)
  - Max width 2·18 m (7 ft 2 in)

AREAS:
- Wings, gross 55·56 m² (598 sq ft)
- Ailerons (total) 1·23 m² (13·3 sq ft)
- Fin 8·51 m² (91·6 sq ft)
- Rudder, incl tab 3·48 m² (37·4 sq ft)
- Elevators, incl tabs 4·32 m² (46·5 sq ft)

WEIGHTS AND LOADINGS:
- Weight empty 10,954 kg (24,150 lb)
- Max T-O weight 21,592 kg (47,602 lb)
- Max zero-fuel weight 13,290 kg (29,299 lb)
- Max carrier landing weight 16,676 kg (36,766 lb)
- Max wing loading 388·6 kg/m² (79·6 lb/sq ft)
- Max power loading 262 kg/kN (2·56 lb/lb st)

PERFORMANCE (at max T-O weight, unless otherwise indicated):
- Max level speed at 6,100 m (20,000 ft) 450 knots (834 km/h; 518 mph)
- Service ceiling 12,200 m (40,000 ft)
- Operational ceiling (with passengers) 10,670 m (35,000 ft)
- T-O to 15 m (50 ft) at T-O weight of 19,251 kg (42,441 lb) 807 m (2,650 ft)
- Range with max payload 2,000 nm (3,706 km; 2,303 miles)
- Max ferry range 3,230 nm (6,085 km; 3,719 miles)

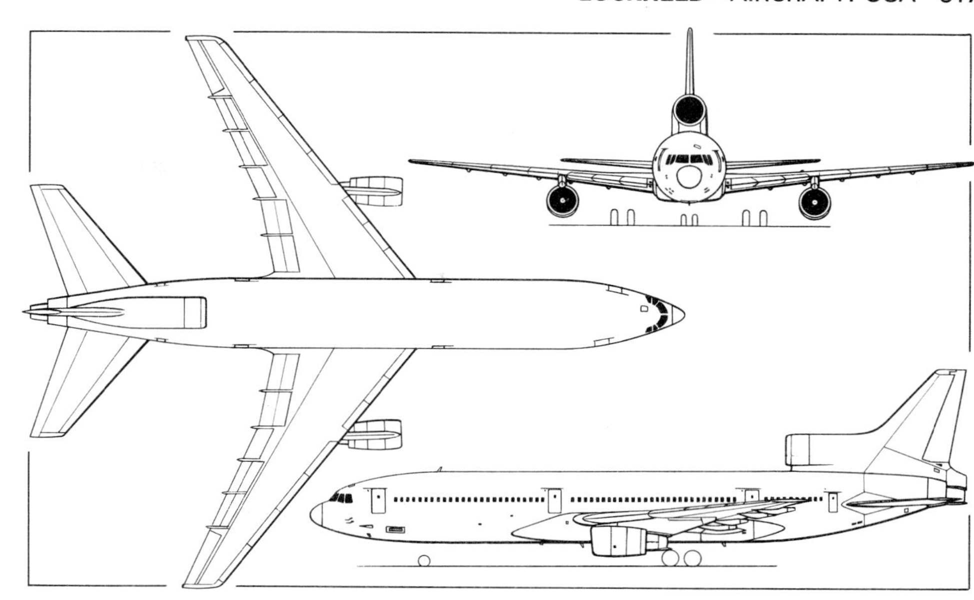

Lockheed L-1011-1 TriStar wide-bodied short/medium-range transport *(Pilot Press)*

LOCKHEED L-1011-1 (MODEL 385) TRISTAR

In January 1966, Lockheed-California began a study of future requirements in the short/medium-haul airliner market. The design which emerged, known as the L-1011 (Lockheed Model 385 TriStar), was influenced by the published requirements of American Airlines, which specified optimum payload/range performance over the Chicago-Los Angeles route, coupled with an ability to take off from comparatively short runways with full payload.

The original design centred around a twin-turbofan configuration. Discussions which followed with American domestic carriers led to the eventual selection of a three-engined configuration, and the Rolls-Royce RB.211 high bypass ratio turbofan was chosen as power plant.

In June 1968 the L-1011 TriStar moved to the production design stage. Construction of the first aircraft began in March 1969, and this was rolled out in September 1970. The first flight was made on 16 November 1970. On 22 December 1971 a class II provisional Type Certification was received, permitting delivery of aircraft to customers for route proving and demonstration purposes.

This original version of the TriStar is now known as the L-1011-1. It has been followed by several variants of the same basic airframe, and five models were available in early 1976, as follows:

L-1011-1. Basic TriStar, as described in detail. Initial delivery of the L-1011-1, to Eastern Air Lines for crew training, made on 6 April 1972, followed by a similar delivery to TWA. FAA certification was granted in the same month and the first passenger service with the TriStar was flown by Eastern on 15 April. Scheduled services began eleven days later.

L-1011-100. Longer-range version. Outward configuration identical with that of L-1011-1. Available with RB.211-22B engines (each 187 kN; 42,000 lb st) or RB.211-22F engines (each 193·5 kN; 43,500 lb st). Max T-O weight of 204,120 kg (450,000 lb) can be increased to 211,375 kg (466,000 lb) with additional 8,165 kg (18,000 lb) of fuel in new centre-section tanks. Nominal range with -22F engines, AUW of 204,120 kg (450,000 lb) and payload of 24,250 kg (55,000 lb) is 3,585 nm (6,645 km; 4,130 miles). Ordered by Air Canada, Cathay Pacific, Gulf Air and Saudi Arabian Airlines.

L-1011-200. Longer-range version, with improved take-off and climb performance, offering particular benefits to operators serving 'hot or high' areas. Outward configuration identical with that of L-1011-1. Powered by RB.211-524 engines (each 213·5 kN; 48,000 lb st). Optional max T-O weights of 204,120 kg (450,000 lb) or 211,375 kg (466,000 lb) according to whether new centre-section tankage is fitted. Nominal range at AUW of 211,375 kg (466,000 lb), with payload of 24,950 kg (55,000 lb) is 4,020 nm (7,450 km; 4,630 miles). Ordered by Saudi Arabian Airlines.

L-1011-250. Long-range version, with further increase in max T-O weight to 224,980 kg (496,000 lb) and max fuel capacity of 96,160 kg (212,000 lb), through added centre-section tankage. Outward configuration identical with that of L-1011-1. Wings, fuselage and fin front spar web reinforced to cater for higher design loads. New nosewheel unit and strengthened main landing gear axles. Larger tyres with increased ply rating on all units. Braking capacity increased. Powered by RB.211-524B engines (each 213·5 kN; 48,000 lb st). Galley can be below-deck, as on other versions, or dispersed on main deck, which doubles available space in forward cargo hold. Expanded forward hold accommodates 16 LD-3 half-width containers or 5 pallets, each measuring 2·23 m × 3·17 m (88 in × 125 in). For pallet loading, the forward cargo door is replaced by a 1·72 m × 2·64 m (68 in × 104 in) power-operated upward-opening door. Main-deck galleys reduce passenger accommodation from typical 273 to 253 in eight-abreast coach configuration, and from typical 302 to 284 in nine-abreast coach configuration, in each case with 10 per cent first class forward. Nominal range at max AUW with 24,950 kg (55,000 lb) payload is 4,303 nm (7,975 km; 4,955 miles).

L-1011-500. Extended-range version, with a max T-O weight of 224,980 kg (496,000 lb) and max fuel capacity of 96,160 kg (212,000 lb) through added centre-section tankage. Fuselage is shortened by 4·11 m (13 ft 6 in); all other external dimensions are the same as for other models. Three RB.211-524B engines (each 222·4 kN; 50,000 lb st). Galley located on main deck. Forward cargo hold accommodates 12 LD-3 containers or four pallets each measuring 2·24 m × 3·17 m (88 in × 125 in). Centre hold takes 7 LD-3 containers. In a mixed class configuration, with 24 first class passengers in six-abreast seating and 222 economy passengers in nine-abreast seating, the aircraft carries 246 passengers. Max accommodation for 300 passengers. Max range 5,297 nm (9,815 km; 6,100 miles). Ordered by British Airways.

By September 1976, Lockheed had delivered 136 L-1011 TriStars to Air Canada, All Nippon Airways, British Airways, Cathay Pacific Airways, Court Line, Delta Air Lines, Eastern Air Lines, Gulf Air, Lufttransport Unternehmen, Pacific Southwest Airlines, Saudi Arabian Airlines and Trans World Airlines.

Approximately 12,000 workers were employed on the L-1011 programme in early 1976, at which time orders and options for 213 aircraft of all models had been received, as follows:

	Orders	*Options*
Air Canada	10	9
All Nippon Airways	21	0
British Airways	15	6
Cathay Pacific Airways	2	0
Court Line Aviation	2	0
Delta Air Lines	21	9
Eastern Air Lines	37	13
Gulf Air	4	4
Haas/Turner Investment Group	2	0
Lufttransport Unternehmen	1	1
Pacific Southwest Airlines	5	0
Saudi Arabian Airlines	7	0
Trans World Airlines	33	11

The description which follows applies to the L-1011-1 TriStar in its initial operational form, except where indicated:

TYPE: Three-turbofan commercial transport.

WINGS: Cantilever low-wing monoplane. Special Lockheed aerofoil sections. Dihedral at trailing-edge: 7° 31′ on inner wings, 5° 30′ outboard. Sweepback at quarter-chord 35°. The wing consists of a centre-section, passing through the lower fuselage, and an outer wing panel on each side. It is of conventional fail-safe construction, with aluminium surfaces, ribs and spars, and integral fuel tanks. Hydraulically-powered aluminium ailerons of conventional two-spar box construction, with aluminium honeycomb trailing-edge, in inboard and outboard sections on each wing, operate in conjunction

Lockheed L-1011 TriStar medium-range transport in the insignia of Gulf Air

with flight spoilers. The low-speed ailerons extend from approximately 80% of semi-span to within 0·25 m (10 in) of the wingtips, the high-speed ailerons extend from approximately WBL 387 to WBL 480 on each wing. Double-slotted Fowler trailing-edge flaps, constructed of aluminium and aluminium honeycomb. Each flap segment consists of a honeycomb trailing-edge, a front spar, ribs, skin panels, carriages, and tracks mounted on the forward segment to provide for extension and rotation of the aft segment. A sheet metal vane surface, actuated by a linkage system during flap rotation, forms the forward section of the extended flap. Four aluminium leading-edge slats outboard of engine pylon on each wing. Each segment is mounted to two roller-supported tracks and extends in a circular motion down and forward for take-off and landing. Three leading-edge slats inboard of engine pylon on each wing, made of aluminium alloy honeycomb and sheet metal fairings. Six spoilers on the upper surface of each wing, two inboard and four outboard of the inboard aileron, constructed from bonded aluminium tapered honeycomb. No trim tabs. Flight controls fully powered. Each control surface system is controlled by a multiple redundant servo actuator system that is powered by four independent and separate hydraulic sources. Thermal de-icing of outboard wing leading-edge slats by engine bleed air.

Fuselage: Semi-monocoque structure of aluminium alloy. Constant cross-sectional diameter of 5·97 m (19 ft 7 in) for most of the length. Bonding utilised in skin joints, for attaching skin-doublers at joints and around openings to improve fatigue life. Skins and stringers supported by frames spaced at 0·51 m (20 in) intervals, with fail-safe straps midway between frames. These frames, with the exception of main frames and door-edge members, are 0·076 m (3 in) deep at the sides of the cabin increasing progressively to a depth of 0·15 m (6 in) at the top of the fuselage and below the floor. Fuselage length reduced on L-1011-500.

Tail Unit: Conventional cantilever structure, consisting of variable-incidence horizontal tailplane-elevator assembly and vertical fin and rudder. Primary loads of the fin are carried by a conventional box-beam structure, with ribs spaced at approx 0·51 m (20 in) centres. The rudder comprises forward and aft spars, glassfibre trailing-edges, hinge and actuator backup ribs, sheet metal formers, box surface panels and leading-edge fairings. Elevators are of similar construction. Truss members for the tailplane centre-section are built up from forged and extruded sections. Outboard of the centre-section, construction is similar to that of the fin box-beam, leading- and trailing-edges, except that the surface structure is integrally stiffened. The elevators are linked mechanically to the tailplane actuation gear, to modify its camber and improve its effectiveness. No trim tabs. Controls are fully powered, the hydraulic servo actuators receiving power from four independent hydraulic sources, under control of electronic flight control system. Control feel is provided, with the force gradient scheduled as a function of flight condition. No de-icing equipment.

Landing Gear: Hydraulically-retractable tricycle type, produced by Menasco Manufacturing. Twin-wheel units in tandem on each main gear; twin wheels on nose gear, which is steerable 65° on each side. Nosewheels retract forward into fuselage. Main wheels retract inward into fuselage wheel-wells. Oleo-pneumatic shock-absorbers on all units. B. F. Goodrich forged aluminium alloy wheels of split construction. Main wheels have tubeless tyres size 50 × 20-20, Type VIII, pressure 10·34-11·38 bars (150-165 lb/sq in) for short- to medium-range operational weights, 12·41 bars (180 lb/sq in) for max-range weight. Nosewheels have tubeless tyres size 36 × 11-16, Type VII, pressure 12·76 bars (185 lb/sq in). Hydraulically-operated brakes, controlled by the rudder pedals. Anti-skid units, with individual wheel skid and modulated control, installed in the normal and alternative braking systems.

Power Plant (L-1011-1): Three Rolls-Royce RB.211-22B turbofan engines, each rated at 187 kN (42,000 lb st). Two engines mounted in pods on pylons under the wings, the third mounted in the rear fuselage at the base of the fin. Engine bleed air is used to anti-ice the engine inlet lips. Two integral fuel tanks in each wing; inboard tank capacity 30,581 litres (8,079 US gallons), outboard tank capacity 14,489 litres (3,828 US gallons). Total fuel capacity 90,140 litres (23,814 US gallons). Pressure refuelling points in wing leading-edges. Oil capacity approx 34 litres (9 US gallons) per engine. A detachable pylon can be fitted between the starboard engine nacelle and fuselage to permit carriage of a replacement engine for another TriStar. Alternative power plants for -100, -200, -250 and -500 detailed under model listings. These four models each have provision for additional centre-section tankage, raising total fuel capacity to 100,317 litres (26,502 US gallons) in -100 and -200, and 119,774 litres (31,642 US gallons) in -250 and -500.

Accommodation: Crew of 13. First class and coach mixed accommodation for 256 passengers, with a maximum of 400 in all-economy configuration. Alternative intermediate seating capacities are provided by using eight seat-tracks which permit 6, 8, 9 or 10-abreast seating, with two full-length aisles. Underfloor galley. Seven lavatories are provided, two forward and five aft. Three Type A passenger doors of the upward-opening plug type on each side of the fuselage, one pair immediately aft of flight deck, one pair forward of wing, one pair aft of wing. Two Type I emergency exit doors, one each side of fuselage, at rear of cabin, replaced by two Type A doors for 10-abreast seating. Baggage and freight compartments beneath the floor able to accommodate 16 containers, totalling 71·58 m³ (2,528 cu ft), and 19·8 m³ (700 cu ft) bulk cargo (19 containers and 14·2 m³; 500 cu ft in -500).

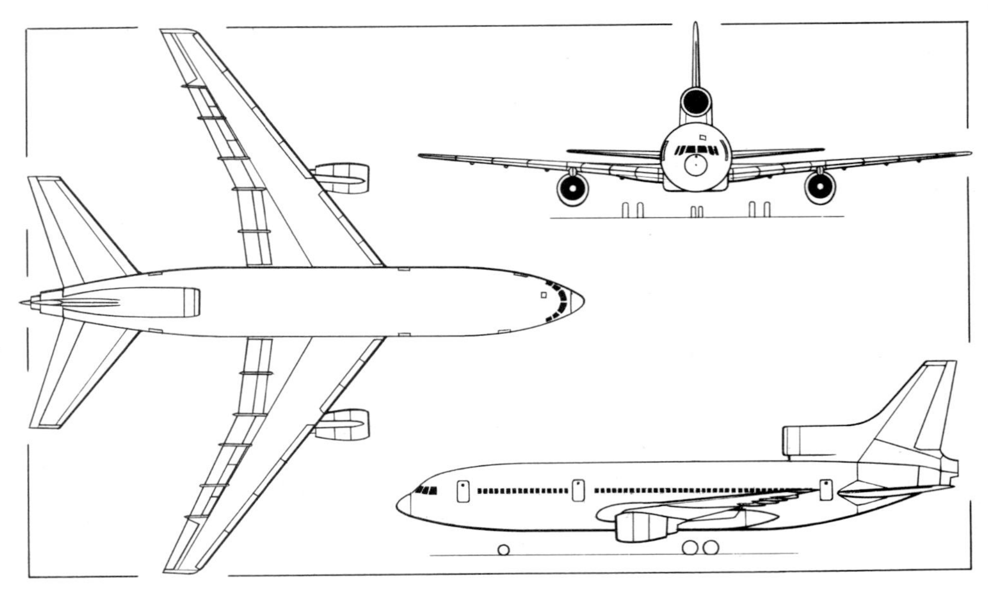

L-1011-500 short-fuselage version of the Lockheed TriStar *(Pilot Press)*

Systems: Air-conditioning and pressurisation system, using engine bleed air or APU air combined with air-cycle refrigeration. Pressurisation system maintains equivalent of 2,440 m (8,000 ft) conditions to 12,800 m (42,000 ft). Normal cabin pressure differential 0·582 bars (8·44 lb/sq in). Four independent 207 bars (3,000 lb/sq in) hydraulic systems provide power for primary flight control surfaces, normal brake power, landing gear retraction and nosewheel steering, etc. Electrical system includes four 120/208V 400Hz alternators, one on each engine and one driven by the APU, which is sited in the aft fusealge. APU provides ground and in-flight power, to an altitude of 9,145 m (30,000 ft), producing both shaft and pneumatic power for utilisation by the electrical, environmental control and hydraulic systems. Integral electric heaters are used to anti-ice windscreens, pitot masts and total temperature probes.

Electronics and Equipment: Standard equipment includes two ARINC 546 VHF communication transceivers, two ARINC 547 VHF navigation systems, two ARINC 568 interrogator units, an ARINC 564 weather radar system, two ARINC 572 air traffic control transponders, partial provision for a dual collision system, three vertical gyros, and full blind-flying instrumentation. Space is provided for installation of two ARINC 533A HF transceivers and a dual SATCOM system.

Dimensions, external:	
Wing span	47·34 m (155 ft 4 in)
Wing chord at root	10·46 m (34 ft 4 in)
Wing chord at tip	3·12 m (10 ft 3 in)
Wing aspect ratio	6·95
Length overall:	
-1, -100, -200, -250	54·17 m (177 ft 8½ in)
-500	50·05 m (164 ft 2½ in)
Height overall	16·87 m (55 ft 4 in)
Tailplane span	21·82 m (71 ft 7 in)
Wheel track	10·97 m (36 ft 0 in)
Wheelbase	21·34 m (70 ft 0 in)
Passenger doors (each):	
Height	1·93 m (6 ft 4 in)
Width	1·07 m (3 ft 6 in)
Height to sill	4·60 m (15 ft 1 in)
Emergency passenger doors (each):	
Height	1·52 m (5 ft 0 in)
Width	0·61 m (2 ft 0 in)
Height to sill	4·60 m (15 ft 1 in)
Baggage and freight compartment doors (forward and centre):	
Height	1·73 m (5 ft 8 in)
Width	1·78 m (5 ft 10 in)
Height to sill	2·72 m (8 ft 11 in)
Baggage and freight compartment doors (aft):	
Height	1·22 m (4 ft 0 in)
Width	1·12 m (3 ft 8 in)
Height to sill	2·92 m (9 ft 7 in)

Dimensions, internal:	
Cabin, excl flight deck and underfloor galley:	
Length	41·43 m (135 ft 11 in)
Max width	5·77 m (18 ft 11 in)
Max height	2·41 m (7 ft 11 in)
Floor area:	
-1, -100, -200, -250	215·5 m² (2,320 sq ft)
-500	192·6 m² (2,073 sq ft)
Volume	453 m³ (16,000 cu ft)
Baggage/cargo holds, bulk capacity:	
-1, -100, -200, -250	110·4 m³ (3,900 cu ft)
-500	118·9 m³ (4,200 cu ft)

Areas:	
Wings, gross	320·0 m² (3,456 sq ft)
Ailerons (total)	14·86 m² (160 sq ft)
Trailing-edge flaps (total)	49·80 m² (536 sq ft)
Leading-edge slats (total):	
inboard slats	11·52 m² (124 sq ft)
outboard slats	21·93 m² (236 sq ft)
Spoilers (total)	19·88 m² (214 sq ft)
Fin	51·10 m² (550 sq ft)
Rudder	11·89 m² (128 sq ft)
Tailplane	119·10 m² (1,282 sq ft)

Weights:	
Operating weight empty:	
-1	109,045 kg (240,400 lb)
-100	110,720 kg (244,100 lb)
-200	111,495 kg (245,800 lb)
-250	112,969 kg (249,054 lb)
-500	108,925 kg (240,139 lb)
Max payload:	
-1	38,373 kg (84,600 lb)
-100	34,427 kg (75,900 lb)
-200	33,020 kg (72,800 lb)
-250	40,345 kg (88,946 lb)
-500	44,390 kg (97,861 lb)
Max T-O weight:	
-1	195,045 kg (430,000 lb)
-100, -200	211,375 kg (466,000 lb)
-250, -500	224,980 kg (496,000 lb)
Max zero-fuel weight:	
-1	147,417 kg (325,000 lb)
-100, -200	145,150 kg (320,000 lb)
-250, -500	153,315 kg (338,000 lb)
Max landing weight:	
-1	162,385 kg (358,000 lb)
-100, -200, -250, -500	166,920 kg (368,000 lb)

Performance (-1, -100, -200 with 273 passengers, -250 with 284 passengers and -500 with 246 passengers):	
Cruising speed, all versions	Mach 0·84
T-O field length:	
-1	2,425 m (7,960 ft)
-100	3,245 m (10,640 ft)
-200	2,460 m (8,070 ft)
-250	2,760 m (9,060 ft)
-500	2,845 m (9,330 ft)
Landing field length:	
-1	1,735 m (5,690 ft)
-100, -200, -250	1,770 m (5,800 ft)
-500	1,955 m (6,420 ft)
Max range:	
-1	3,110 nm (5,760 km; 3,580 miles)
-100	3,820 nm (7,080 km; 4,400 miles)
-200	3,864 nm (7,160 km; 4,450 miles)
-250	4,533 nm (8,400 km; 5,220 miles)
-500	5,297 nm (9,815 km; 6,100 miles)

Operational Noise Characteristics (FAR Pt 36):	
T-O noise level	97 EPNdB
Approach noise level	103 EPNdB
Sideline noise level	95 EPNdB

LOCKHEED-GEORGIA COMPANY

86 South Cobb Drive, Marietta, Georgia 30063

Lockheed-Georgia's main building at Marietta covers 76 acres and is one of the world's largest aircraft production plants under a single roof. Aircraft in current production on its assembly lines are the C-130 Hercules turboprop transport, its commercial counterpart, the L 100, and the JetStar II turbofan light transport. Manufacture of the C-5 Galaxy heavy logistics transport, the largest aeroplane yet ordered into production anywhere in the world, has been completed.

Lockheed-Georgia had a total of approximately 9,600 employees at the beginning of 1976.

LOCKHEED MODEL 82 HERCULES

USAF designations: C-130, AC-130, DC-130, HC-130, JC-130, RC-130 and WC-130
US Navy designations: C-130, DC-130, EC-130 and LC-130
US Marine Corps designation: KC-130
US Coast Guard designations: EC-130 and HC-130
Canadian Armed Forces designation: CC-130
RAF designations: Hercules C.Mk 1 and W.Mk 2

The C-130 was designed to a specification issued by the USAF Tactical Air Command in 1951. Lockheed was awarded its first production contract for the C-130A in September 1952, and a total of 461 C-130As and C-130Bs was manufactured. Details of these basic versions and of many variants for special duties can be found in the 1967-68 and 1975-76 *Jane's*. Later military versions of the C-130 are as follows:

C-130E (Lockheed Model 382-44). Extended-range development of C-130B, with four 3,020 kW (4,050 ehp) T56-A-7 turboprop engines and two 5,145 litre (1,360 US gallon) underwing fuel tanks. Normal max T-O weight is 70,310 kg (155,000 lb). Take-off at overload gross weight of 79,380 kg (175,000 lb) increases the range and endurance capabilities, with certain operating restrictions at this higher weight. Deliveries of this version began in April 1962, and by February 1975 the planned production of a total of 503 C-130Es had been completed. Details of the basic C-130E can be found in the 1973-74 *Jane's*.

EC-130G. Redesignation of four C-130Gs acquired by US Navy. Equipped with VLF radio to relay emergency action messages to Fleet Ballistic Missile submarines anywhere in the world.

C-130H. Similar to earlier Hercules models except for more powerful engines: T56-A-15 turboprops rated at 3,661 kW (4,910 ehp) for take-off, but limited to 3,362 kW (4,508 ehp). By January 1976 a total of 408 C-130Hs, or variants, were on order or had been delivered, including 66 C-130Ks for the UK. A total of 24 countries have ordered versions of the C-130H. Deliveries to USAF began in April 1975.

HC-130H. Lockheed was awarded two initial contracts in September 1963 for this extended-range air search, rescue and recovery version to be utilised by the Aerospace Rescue and Recovery Service of the USAF for aerial recovery of personnel or equipment and other duties. The US Coast Guard subsequently ordered three. New folding nose-mounted recovery system makes possible repeated pickups from ground of persons or objects weighing up to 227 kg (500 lb) including the recoverable gear. Four 3,661 kW (4,910 ehp) (limited to 3,355 kW; 4,500 ehp) Allison T56-A-15 turboprop engines, each driving a Hamilton Standard 54H60-91 four-blade constant-speed propeller. Normal fuel tankage as for C-130H. Provision for installing two 6,184 litre (1,800 US gallon) tanks in cargo compartment. Normal crew of 10, consisting of pilot, co-pilot, navigator, 2 flight mechanics, radio operator, 2 loadmasters and 2 para-rescue technicians, with provision for additional pilot and navigator for long missions. Standard equipment includes four 6-man rafts, two litters, bunks, 16 personnel kits, recovery winches, 10 flare launchers. Total of 66 delivered, of which the first one flew on 8 December 1964. Four modified as **JHC-130H** with added equipment for aerial recovery of re-entering space capsules. One modified by LAS to **DC-130H**.

KC-130H. A tanker version of the C-130H, very similar to the KC-130R. Exported to four countries, including Saudi Arabia (4) and Spain (3).

C-130K (C.Mk 1). This is basically a C-130H, modified for use by the Royal Air Force. Much of the electronics and instrumentation is of UK manufacture. Sixty-six delivered for service with RAF Air Support (now Strike) Command. First of these flew on 19 October 1966. One modified by Marshall of Cambridge (Engineering) Ltd in the UK for use by the RAF Meteorological Research Flight, under the designation Hercules W. Mk 2.

C-130M. Basically as C-130E but with forward freight door, flight deck equipment and electronics of C-130B. For overseas delivery under MAP.

HC-130N. Search and rescue version for recovery of aircrew and retrieval of space capsules after re-entry, using advanced direction-finding equipment. Fifteen ordered for USAF in 1969.

HC-130P. Twenty HC-130Hs were modified into HC-130Ps with capability of refuelling helicopters in flight, and for mid-air retrieval of parachute-borne payloads. Modification involved the addition of refuelling drogue pods and associated plumbing. Typical helicopter refuelling mission involves taking off at an AUW of 70,310 kg (155,000 lb), with 33,385 kg (73,600 lb) of fuel on board, meeting up with the helicopters at a radius of499 nm (925 km; 575 miles), transferring 22,000 kg (48,500 lb) of fuel to the helicopters and returning 499 nm (925 km; 575 miles) to the point of origin.

Lockheed HC-130H search and rescue aircraft assigned to a US Coast Guard unit in Alaska

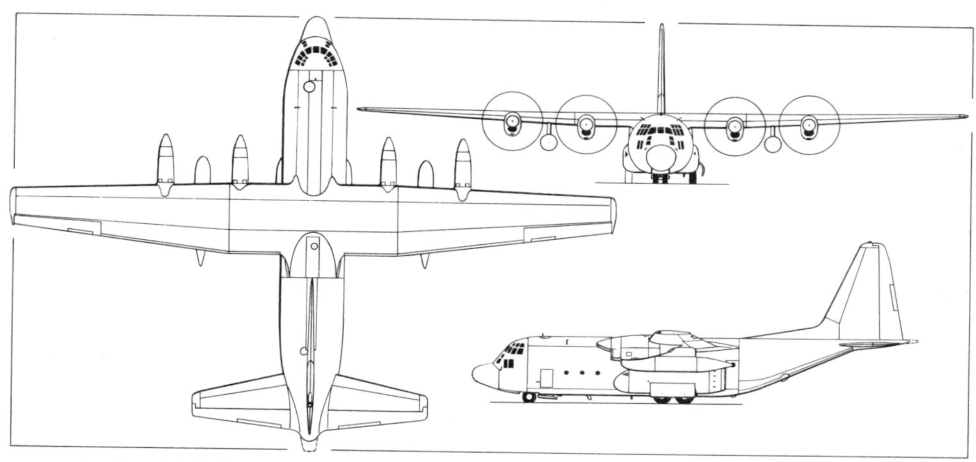

Lockheed C-130E Hercules four-turboprop medium/long-range combat transport *(Pilot Press)*

EC-130Q. Ten aircraft similar to EC-130G but with improved equipment and crew accommodation, for USN command communications duties.

KC-130R. Tanker version of the C-130H for US Marine Corps. Major changes from the earlier KC-130F include engines of 3,362 kW (4,508 ehp), increased T-O and landing weights, pylon-mounted fuel tanks to provide an additional 10,296 litres (2,720 US gallons) of fuel, plus a removable 13,627 litre (3,600 US gallon) fuel tank located in the cargo compartment. Four ordered in early 1974; four more later.

LC-130R. Basically a C-130H with wheel-ski gear for US Navy. Main skis each approximately 6·10 m (20 ft 0 in) long by 1·68 m (5 ft 6 in) wide. The nose ski is approximately 3·05 m (10 ft 0 in) long by 1·68 m (5 ft 6 in) wide. The total ski installation weighs approximately 2,540 kg (5,600 lb). The main skis have 8° nose-up and nose-down pitch and the nose skis have 15° nose-up and nose-down pitch, to enable them to follow uneven terrain. The load-bearing surfaces of the skis are coated with Teflon plastic to reduce friction and resist ice adhesion. Provision is made for fitting JATO units. Four converted, for service in the Antarctic. Two more aircraft in this configuration were ordered in late 1975 by the National Science Foundation, for use in the Antarctic.

RC-130S. Variant of RC-130A aerial survey and reconnaissance aircraft with additional special equipment, including searchlight pods on the main landing gear housings. One only; not operational.

Commercial versions of the Hercules are described separately.

The C-130 is able to deliver single loads of up to 11,340 kg (25,000 lb) by the ground proximity extraction method. This involves making a fly-past 1·2-1·5 m (4-5 ft) above the ground with the rear loading ramp open. The aircraft trails a hook which is attached by cable to the palletised cargo. The hook engages a steel cable on the ground and the cargo is extracted from the aircraft and brought to a stop on the ground in about 30 m (100 ft) by an energy absorption system manufactured by All American Engineering of Wilmington, Delaware. An alternative extraction technique involves deploying a 6·70 m (22 ft) ribbon parachute to drag the pallet from the cabin. Loads of up to 22,680 kg (50,000 lb) have been delivered by this method.

By January 1976 firm orders for all versions of the C-130 totalled 1,463, and the 1,400th Hercules was delivered in July 1976. Production rate continued during 1976 at six aircraft per month.

The following details refer specifically to the C-130H, except where indicated otherwise:

TYPE: Medium/long-range combat transport.

WINGS: Cantilever high-wing monoplane. Wing section NACA 64A318 at root, NACA 64A412 at tip. Dihedral 2° 30′. Incidence 3° at root, 0° at tip. Sweepback at quarter-chord 0°. All-metal two-spar stressed-skin structure, with integrally-stiffened tapered machined skin panels up to 14·63 m (48 ft 0 in) long. Conventional aluminium alloy ailerons have tandem-piston hydraulic boost, operated by either of two independent hydraulic systems. Lockheed-Fowler aluminium alloy trailing-edge flaps. Trim tabs in ailerons. Leading-edge anti-iced by hot air bled from engines.

FUSELAGE: Semi-monocoque structure of aluminium and magnesium alloys.

TAIL UNIT: Cantilever all-metal stressed-skin structure. Fixed-incidence tailplane. Trim tabs in elevators and rudder. Elevator tabs use AC electrical power as primary source and DC as emergency source. Control surfaces have tandem-piston hydraulic boost. Hot-air anti-icing of tailplane leading-edge, by engine bleed air.

LANDING GEAR: Hydraulically-retractable tricycle type. Each main unit has two wheels in tandem, retracting into fairings built on to the sides of the fuselage. Nose unit has twin wheels and is steerable through 60° each side of centre. Oleo shock-absorbers. Main-wheel tyres size 56 × 20-20, pressure 5·52 bars (80 lb/sq in). Nose-wheel tyres size 39 × 13-16, pressure 4·14 bars (60 lb/sq in). Goodyear aircooled hydraulic brakes with anti-skid units. Retractable combination wheel-skis available.

POWER PLANT: Four 3,362 kW (4,508 ehp) Allison T56-A-15 turboprop engines, each driving a Hamilton Standard type 54H60 four-blade constant-speed fully-feathering reversible-pitch propeller. Eight Aerojet-General 15KS-1000 JATO units (each 4·45 kN; 1,000 lb st for 15 sec) can be carried. Fuel in six integral tanks in wings, with total capacity of 26,344 litres (6,960 US

Lockheed EC-130Q command communications aircraft of the US Navy *(T. Matsuzaki)*

gallons) and two underwing pylon tanks, each with capacity of 5,146 litres (1,360 US gallons). Total fuel capacity 36,636 litres (9,680 US gallons). Single pressure refuelling point in starboard wheel well. Fillers for overwing gravity fuelling. Oil capacity 182 litres (48 US gallons).

Accommodation: Crew of four on flight deck, comprising pilot, co-pilot, navigator and systems manager. Provision for fifth man to supervise loading. Sleeping quarters for relief crew, and galley. Flight deck and main cabin pressurised and air-conditioned. Standard complements are as follows: troops (max) 92, paratroops (max) 64, litters 74 and 2 attendants. As a cargo carrier, loads can include heavy equipment such as a 12,080 kg (26,640 lb) type F.6 refuelling trailer or a 155 mm howitzer and its high-speed tractor. Up to six preloaded pallets of freight can be carried. Hydraulically-operated main loading door and ramp at rear of cabin. Paratroop door on each side aft of landing gear fairing.

Systems: Air-conditioning and pressurisation system max pressure differential 0·52 bars (7·5 lb/sq in). Two independent hydraulic systems, pressure 207 bars (3,000 lb/sq in). Electrical system supplied by four 40kVA AC generators, plus one 40kVA auxiliary generator driven by APU. Current production aircraft incorporate many systems and component design changes for increased reliability. There continue to be some differences between the installed components for US government and export versions.

Dimensions, external:

Wing span	40·41 m (132 ft 7 in)
Wing chord at root	4·88 m (16 ft 0 in)
Wing chord, mean	4·16 m (13 ft 8½ in)
Wing aspect ratio	10·09
Length overall:	
all except HC-130H	29·78 m (97 ft 9 in)
HC-130H, recovery system folded	30·10 m (98 ft 9 in)
HC-130H, recovery system spread	32·41 m (106 ft 4 in)
Height overall	11·66 m (38 ft 3 in)
Tailplane span	16·05 m (52 ft 8 in)
Wheel track	4·35 m (14 ft 3 in)
Wheelbase	9·77 m (32 ft 0¾ in)
Propeller diameter	4·11 m (13 ft 6 in)
Main cargo door (rear of cabin):	
Height	2·77 m (9 ft 1 in)
Width	3·05 m (10 ft 0 in)
Height to sill	1·03 m (3 ft 5 in)
Paratroop doors (each):	
Height	1·83 m (6 ft 0 in)
Width	0·91 m (3 ft 0 in)
Height to sill	1·03 m (3 ft 5 in)

Dimensions, internal:

Cabin, excl flight deck:	
Length without ramp	12·60 m (41 ft 5 in)
Length with ramp	15·73 m (51 ft 8½ in)
Max width	3·13 m (10 ft 3 in)
Max height	2·81 m (9 ft 2¾ in)
Floor area, excl ramp	39·5 m² (425 sq ft)
Volume, incl ramp	127·4 m³ (4,500 cu ft)

Areas:

Wings, gross	162·12 m² (1,745 sq ft)
Ailerons (total)	10·22 m² (110 sq ft)
Trailing-edge flaps (total)	31·77 m² (342 sq ft)
Fin	20·90 m² (225 sq ft)
Rudder, incl tab	6·97 m² (75 sq ft)
Tailplane	35·40 m² (381 sq ft)
Elevators, incl tabs	14·40 m² (155 sq ft)

Weights and Loadings:

Operating weight empty	34,169 kg (75,331 lb)
Max payload	19,872 kg (43,811 lb)
Max normal T-O weight	70,310 kg (155,000 lb)
Max overload T-O weight	79,380 kg (175,000 lb)
Max landing weight	58,970 kg (130,000 lb)
Max zero-fuel weight	54,040 kg (119,142 lb)
Max wing loading	434·5 kg/m² (89 lb/sq ft)
Max power loading	5·23 kg/kW (8·6 lb/ehp)

Performance (at max T-O weight, unless indicated otherwise):

Max cruising speed:	
C-130H	335 knots (621 km/h; 386 mph)
HC-130H	318 knots (589 km/h; 366 mph)
Econ cruising speed	300 knots (556 km/h; 345 mph)
Stalling speed	100 knots (185 km/h; 115 mph)
Max rate of climb at S/L:	
C-130H	579 m (1,900 ft)/min
HC-130H	555 m (1,820 ft)/min
Service ceiling at 58,970 kg (130,000 lb) AUW	10,060 m (33,000 ft)
Service ceiling, one engine out, at 58,970 kg (130,000 lb) AUW	8,075 m (26,500 ft)
Min ground turning radius	19·2 m (63 ft)
Runway LCN at 70,310 kg (155,000 lb) AUW:	
asphalt	37
concrete	42
T-O run	1,091 m (3,580 ft)
T-O to 15 m (50 ft)	1,573 m (5,160 ft)
Landing from 15 m (50 ft) at 45,360 kg (100,000 lb) AUW	741 m (2,430 ft)
Landing from 15 m (50 ft) at max landing weight	838 m (2,750 ft)
Landing run at max landing weight	533 m (1,750 ft)
Range with max payload, with 5% reserves and allowance for 30 min at S/L	2,160 nm (4,002 km; 2,487 miles)
Range with max fuel, incl external tanks, 9,070 kg (20,000 lb) payload and reserves of 5% initial fuel plus 30 min at S/L	4,460 nm (8,264 km; 5,135 miles)

LOCKHEED L 100 SERIES COMMERCIAL HERCULES

Details of earlier versions of the commercial Hercules have appeared in previous editions of *Jane's;* current models are as follows:

Model 382E (L 100-20). Certificated on 4 October 1968, this 'stretched' version of the Hercules has a 2·54 m (100 in) fuselage extension. A 1·52 m (60 in) fuselage plug is inserted aft of the forward crew door and a 1·02 m (40 in) plug aft of the paratroop doors. Allison 501-D22A engines. Operators have included Alaska International Air, Delta Air Lines, Saturn Airways and Southern Air Transport in the USA; Pacific Western Airlines in Canada; SATCO in Peru; Safair Freighters in the Republic of South Africa; the Kuwait Air Force (2); the Peruvian Air Force (5); Philippine Aerotransport; and the Philippine Air Force (4).

Model 382G (L 100-30). Generally similar to the Model 382E, but with the fuselage extended a further 2·03 m (80 in). Rear cargo windows, paratroop doors and provision for JATO eliminated. Saturn Airways was the first operator of this model, with services beginning in December 1970. Alaska International Air, Safair Freighters, the Republic of Gabon and Southern Air are among those who have operated the Model 382G.

L 100-50. Proposed 'stretched' version, with 10·66 m (35 ft) longer fuselage.

Details given for the C-130H apply also to the L 100-20 and L 100-30, except as follows:

Type: Medium/long-range transport.

Landing Gear: As for C-130H, except main-wheel tyre pressure 3·24-7·38 bars (47-107 lb/sq in) and nose-wheel tyre pressure 4·14 bars (60 lb/sq in).

Power Plant: Either four 3,020 kW (4,050 ehp) Allison 501-D22 or four 3,362 kW (4,508 ehp) Allison 501-D22A turboprop engines.

Dimensions, external:

Length overall:	
L 100-20	32·33 m (106 ft 1 in)
L 100-30	34·37 m (112 ft 9 in)
Wheelbase:	
L 100-20	11·30 m (37 ft 1 in)
L 100-30	12·32 m (40 ft 5 in)

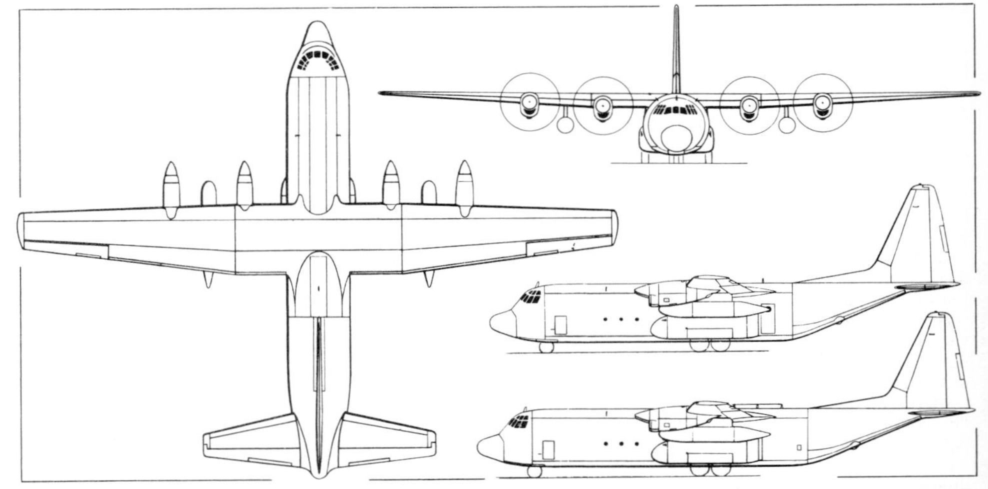

Lockheed L 100-20 version of the commercial Hercules, with additional side view (bottom) of L 100-30 *(Pilot Press)*

Lockheed L 100-30 Hercules commercial transport in service with Safair Freighters of South Africa *(Austin J. Brown)*

Crew door (integral steps):
- Height: 1·14 m (3 ft 9 in)
- Width: 0·76 m (2 ft 6 in)
- Height to sill: 1·04 m (3 ft 5 in)

DIMENSIONS, INTERNAL:
- Cabin, excl flight deck:
  - Length:
    - L 100-20: 15·04 m (49 ft 4 in)
    - L 100-30: 17·07 m (56 ft 0 in)
  - Max height: 2·74 m (9 ft 0 in)
- Floor area, excl ramp:
  - L 100-20: 46·36 m² (499 sq ft)
  - L 100-30: 52·30 m² (563 sq ft)
- Floor area, ramp: 9·57 m² (103 sq ft)
- Volume, incl ramp:
  - L 100-20: 150·28 m³ (5,307 cu ft)
  - L 100-30: 171·5 m³ (6,057 cu ft)

WEIGHTS AND LOADINGS:
- Operating weight empty:
  - L 100-20: 33,305 kg (73,428 lb)
  - L 100-30: 33,550 kg (73,964 lb)
- Max payload:
  - L 100-20: 21,125 kg (46,572 lb)
  - L 100-30: 23,150 kg (51,036 lb)
- Max ramp weight: 70,670 kg (155,800 lb)
- Max T-O weight: 70,308 kg (155,000 lb)
- Max landing weight:
  - L 100-20: 58,970 kg (130,000 lb)
  - L 100-30: 61,235 kg (135,000 lb)
- Max zero-fuel weight:
  - L 100-20: 54,430 kg (120,000 lb)
  - L 100-30: 56,700 kg (125,000 lb)
- Max wing loading: 433·5 kg/m² (88·8 lb/sq ft)
- Max power loading: 5·23 kg/kW (8·6 lb/ehp)

PERFORMANCE (at max T-O weight):
- Max cruising speed at 6,100 m (20,000 ft) at 54,430 kg (120,000 lb) AUW: 327 knots (607 km/h; 377 mph)
- Landing speed:
  - L 100-20: 126 knots (233 km/h; 145 mph)
  - L 100-30: 128 knots (237 km/h; 147 mph)
- Max rate of climb at S/L: 579 m (1,900 ft)/min
- Min ground turning radius:
  - L 100-20: 26·8 m (88 ft)
  - L 100-30: 27·5 m (90 ft)
- Runway LCN:
  - asphalt: 37
  - concrete: 42
- FAA T-O field length: 1,830 m (6,000 ft)
- FAA landing field length, at max landing weight:
  - L 100-20: 1,450 m (4,760 ft)
  - L 100-30: 1,472 m (4,830 ft)
- Range with max payload, 45 min reserves:
  - L 100-20: 2,220 nm (4,113 km; 2,556 miles)
  - L 100-30: 1,815 nm (3,363 km; 2,090 miles)
- Range with zero payload, 45 min reserves:
  - L 100-20: 4,100 nm (7,597 km; 4,721 miles)
  - L 100-30: 4,081 nm (7,562 km; 4,699 miles)

OPERATIONAL NOISE CHARACTERISTICS (FAR Pt 36):
- T-O noise level: 98·4 EPNdB
- Approach noise level: 99·1 EPNdB
- Sideline noise level: 93·9 EPNdB

LOCKHEED MODEL 1329-25 JETSTAR II

The Lockheed JetStar II, which was first announced in the Summer of 1973, has an airframe generally similar to that of the earlier JetStar (1973-74 *Jane's*), but with detail changes in configuration and equipment.

Design of the Model 1329-25 began in October 1972. The major change involved the installation of four AiResearch TFE 731-3 turbofan engines, flat rated at 16·5 kN (3,700 lb st) to 76°F (24·4°C), to replace the 14·7 kN (3,300 lb st) Pratt & Whitney JT12 turbojet engines of the JetStar. The new power plant offers significant improvement in both range and noise levels, as well as allowing an increase in maximum take-off weight.

Production of the JetStar II, at Lockheed-Georgia's Marietta plant, began in the Spring of 1975, and the first aircraft off the line flew on 18 August 1976. Delivery of the 18 aircraft ordered by that time began in September 1976. In addition, AiResearch offers a re-engining scheme to convert early JetStars to JetStar II standard. The first AiResearch Aviation Company 731 JetStar conversion flew for the first time on 10 July 1974, and the first production conversion on 18 March 1976.

TYPE: Four-turbofan light utility transport.

WINGS: Cantilever low-wing monoplane. Wing section NACA 63A112 at root, NACA 63A309 (modified) at tip. Dihedral 2°. Incidence 1° at root, −1° at tip. Sweepback at quarter-chord 30°. Conventional fail-safe stressed-skin structure of high-strength aluminium. Bending loads carried by integral skin/stringer extrusion and sheet ribs, shear loads by three beams. Plain aluminium alloy ailerons; trim tab located near the centre of the trailing-edge of the port aileron. An electrically-powered dual trim actuator is located within the aileron directly forward of the trim tab. Hydraulically-boosted aileron controls are powered by both normal and standby hydraulic systems, either of which is capable of operating the ailerons independently. Manual aileron control is possible in the event of complete hydraulic failure. Aileron booster actuators manufactured by National Water Lift Company. Double-slotted all-metal trailing-edge flaps. Hinged leading-edge flaps. No spoilers. Rubber-boot de-icers on leading-edge.

FUSELAGE: Semi-monocoque fail-safe structure of light alloy. The nose section, crew compartment and cabin are pressurised. The aft section, where most of the aircraft's system components are mounted, is unpressurised. Hydraulically-operated speed-brake on underside of fuselage aft of pressurised compartment.

TAIL UNIT: Cantilever light alloy structure with tailplane mounted part-way up fin. Variable incidence is achieved by the fin being pivoted, thus allowing an electro-mechanical dual actuator to move the entire tail unit to rotate the tailplane. No trim tabs in elevators. Mechanically-operated elevator control system is hydraulically-boosted, using a National Water Lift Company actuator, sited in the aft fuselage equipment area. The rudder is mechanically controlled, with servo tab assistance. Two pneumatic cylinders, biased by engine bleed air, automatically assist directional control in the event of a power loss from either engine. Details of tail unit de-icing system not yet finalised.

LANDING GEAR: Hydraulically-retractable tricycle type, with twin wheels on all units. Pneumatic emergency extension. Main units retract inward, nosewheels forward. Oleo-pneumatic shock-absorbers. Main wheels with tubeless tyres size 26 × 6·60, EHP Type VII, 14-ply rating with reinforced tread, pressure 15·17 bars (220 lb/sq in). Nosewheels with tubeless chine tyres size 18× 4·40, EHP Type VII, 12-ply rating with reinforced tread, pressure 15·17 bars (220 lb/sq in). Hytrol fully-modulated anti-skid units.

POWER PLANT: Four AiResearch TFE 731-3 turbofan engines, flat rated at 16·5 kN (3,700 lb st) to 76°F (24·4°C), mounted in lateral pairs on sides of rear fuselage. Thrust reversers fitted. Air intake anti-icing provided by engine bleed air. Fuel in four integral wing tanks and two non-removable external auxiliary tanks glove-mounted on the wings. Capacity of numbers 1 and 4 internal tanks each 1,420 litres (375 US gallons); numbers 2 and 3 internal tanks each 1,476 litres (390 US gallons), auxiliary tanks each 2,139 litres (565 US gallons). Total fuel capacity 10,070 litres (2,660 US gallons). Gravity refuelling point above each tank, or single-point pressure refuelling from starboard wing root. Oil capacity 24·2 litres (6·4 US gallons).

ACCOMMODATION: Normal accommodation for crew of two and ten passengers, with wardrobe, galley and toilet aft of cabin and baggage compartments fore and aft. Layout and furnishing can be varied to suit customer's requirements. Optional jump-seat available for crew compartment. Door at forward end of fuselage, on port side, opens by moving inward and sliding aft. The fourth window aft on each side of the cabin is a CAR Type IV emergency exit, of plug type and removed inward. Accommodation heated, ventilated, air-conditioned and pressurised. High-pressure oxygen system for passengers and crew standard. Integral electric heaters for windscreen anti-icing and demisting.

Lockheed JetStar II executive transport (four AiResearch TFE 731-3 turbofan engines) *(Pilot Press)*

First Lockheed Model 1329-25 JetStar II executive transport (four AiResearch TFE 731-3 turbofan engines)

SYSTEMS: Two independent hydraulic systems with engine-driven pumps, pressure 207 bars (3,000 lb/sq in), to operate landing gear, wheel brakes, nosewheel steering, flight control booster units, flaps, speed-brake and thrust reversers. Separate pneumatic systems installed for emergency extension of the landing gear. Air bottles can be manually discharged into the down ports of the landing gear actuators. Two pneumatic cylinders provided to assist directional control if engine power lost. Four 28V 300A engine-driven starter/generators power main DC buses. Two high-discharge 24V 34Ah nickel-cadmium batteries for engine starting and emergency power. Three 2,500VA static inverters provide AC power for electronics equipment, flight and engine instruments, and windscreen anti-icing, two being on-load and one on standby. A 250VA rotary inverter is used to power the engine instruments during engine starting. High-pressure oxygen system, 124 bars (1,800 lb/sq in) reduced to 4·83-6·21 bars (70-90 lb/sq in) at the cylinder, provides selective dilution demand or 100 per cent positive pressure demand for crew. A separate 100 per cent demand system with safety pressure and manual control for dilution is installed for passengers. An altitude control valve activates the passenger system when cabin altitude exceeds 4,267 m (14,000 ft), the masks being presented automatically. APU for ground air-conditioning and electrical power is optional.

DIMENSIONS, EXTERNAL:
Wing span 16·60 m (54 ft 5 in)
Wing chord at root 4·16 m (13 ft 7¾ in)
Wing chord at tip 1·55 m (5 ft 1 in)
Wing aspect ratio 5·27
Length overall 18·42 m (60 ft 5 in)
Length of fuselage 17·92 m (58 ft 9½ in)
Height overall 6·23 m (20 ft 5 in)
Tailplane span 7·55 m (24 ft 9 in)
Wheel track 3·75 m (12 ft 3½ in)
Wheelbase 6·28 m (20 ft 7 in)
Cabin door:
Height 1·50 m (4 ft 11 in)
Width 0·67 m (2 ft 2½ in)
Height to sill approx 1·37 m (4 ft 6 in)
Servicing door (underfuselage), diameter 0·61 m (2 ft 0 in)
Emergency exits, each:
Height 0·49 m (1 ft 7¼ in)
Width 0·66 m (2 ft 2½ in)

DIMENSIONS, INTERNAL:
Cabin, excl flight deck:
Length 8·59 m (28 ft 2½ in)
Max width 1·89 m (6 ft 2½ in)
Max height 1·85 m (6 ft 1 in)
Volume 24·07 m^3 (850 cu ft)
Baggage hold volume:
stbd forward 1·25 m^3 (43·1 cu ft)
port forward 0·70 m^3 (24·8 cu ft)
centre aft 1·05 m^3 (37·0 cu ft)

AREAS:
Wings, gross 50·40 m^2 (542·5 sq ft)
Ailerons (total) 2·27 m^2 (24·4 sq ft)
Trailing-edge flaps (extended, total) 5·82 m^2 (62·6 sq ft)
Leading-edge flaps (total) 3·16 m^2 (34·0 sq ft)
Speed-brake 0·85 m^2 (9·2 sq ft)
Fin 8·73 m^2 (94·0 sq ft)
Rudder, incl tab 1·51 m^2 (16·2 sq ft)
Tailplane 10·94 m^2 (117·8 sq ft)
Elevators 2·90 m^2 (31·2 sq ft)

WEIGHTS AND LOADINGS:
Basic operating weight 10,967 kg (24,178 lb)
Max payload 1,280 kg (2,822 lb)
Max ramp weight 19,958 kg (44,000 lb)
Max T-O weight 19,844 kg (43,750 lb)
Max landing weight 16,329 kg (36,000 lb)
Max zero-fuel weight 12,247 kg (27,000 lb)
Max wing loading 393·5 kg/m^2 (80·6 lb/sq ft)
Max power loading 300·7 kg/kN (2·96 lb/lb st)

PERFORMANCE (at max T-O weight except where indicated):
Never-exceed speed Mach 0·87
Max level and cruising speed at 9,145 m (30,000 ft) 475 knots (880 km/h; 547 mph)
Econ cruising speed at 10,670 m (35,000 ft) 441 knots (817 km/h; 508 mph)
Stalling speed, T-O flap setting 123 knots (229 km/h; 142 mph)
Max rate of climb at S/L 1,280 m (4,200 ft)/min
Rate of climb at S/L, one engine out 762 m (2,500 ft)/min
Service ceiling 11,580 m (38,000 ft)
Service ceiling, one engine out 9,145 m (30,000 ft)
T-O to 15 m (50 ft) 1,600 m (5,250 ft)
Landing from 15 m (50 ft) at max landing weight 1,189 m (3,900 ft)
Landing run at max landing weight 777 m (2,550 ft)
Range with max fuel, 30 min reserves 2,770 nm (5,132 km; 3,189 miles)
Range with max payload, 30 min reserves 2,600 nm (4,818 km; 2,994 miles)

LOCKHEED C-141 STARLIFTER

The USAF has awarded Lockheed-Georgia a $24·3 million contract for prototype conversion and testing of a 'stretched' version of the C-141 StarLifter logistics transport aircraft. Two additional fuselage plugs will increase the aircraft's length by 7·11 m (23 ft 4 in) and the internal volume to 180·3 m^3 (6,368 cu ft), enabling cargo capacity to be increased from 10 to 13 standard pallets.

The prototype, designated **YC-141B**, will provide several options, including an in-flight refuelling capability, upon which the USAF will base a decision whether to seek funds to modify its entire fleet of 274 C-141s. Roll-out is scheduled for February 1977.

WEIGHTS:
Max ramp weight 156,490 kg (345,000 lb)
Max T-O weight 155,580 kg (343,000 lb)
Max landing weight 152,860 kg (337,000 lb)

LOCKHEED C-5 GALAXY

USAF designation: C-5A

Design studies for a very large logistics transport for Military Airlift Command (then MATS) began in 1963, when the requirement was for a 272,200 kg (600,000 lb) aircraft known by the designation CX-4. Eventually, this and other requirements evolved into a specification known as CX-HLS (Cargo, Experimental—Heavy Logistics System).

Following an initial design competition in May 1964, contracts were awarded to Boeing, Douglas and Lockheed to develop their designs further. At this time, the requirement was for an aircraft with a gross weight of about 317,500 kg (700,000 lb), to which the definitive designation C-5A and the name Galaxy were allocated. Large contracts also went to Pratt & Whitney and General Electric, to finance the development of prototype power plants for the C-5A.

In August 1965, the General Electric GE1/6 turbofan was selected for continued development. In October, Lockheed was nominated as prime contractor for the airframe. Construction of the first C-5A was started in August 1966, and it flew for the first time on 30 June 1968; the first operational aircraft (the ninth C-5A built) was delivered to Military Airlift Command on 17 December 1969. Lockheed-Georgia and the USAF assigned the first eight aircraft to a flight test programme that extended into mid-1971. Contracts were placed covering the manufacture of 81 C-5As for the USAF and delivery of these was completed in May 1973. Full structural and specification details can be found in the 1975-76 *Jane's*.

POWER PLANT: Four General Electric TF39-GE-1 turbofan engines, each rated at 182·4 kN (41,000 lb st). Twelve integral fuel tanks in wings between front and rear spars, comprising four main tanks (each 13,721 litres; 3,625 US gallons), four auxiliary tanks (each 17,507 litres; 4,625 US gallons) and four extended-range tanks (each 15,142 litres; 4,000 US gallons). Total usable capacity 185,480 litres (49,000 US gallons). Two refuelling points each side, in forward part of main landing gear pods. Flight refuelling capability, via inlet in upper forward fuselage, over flight engineer's station (compatible with KC-135 tanker). Oil capacity 138 litres (36·4 US gallons).

ACCOMMODATION: Normal crew of five, consisting of pilot,

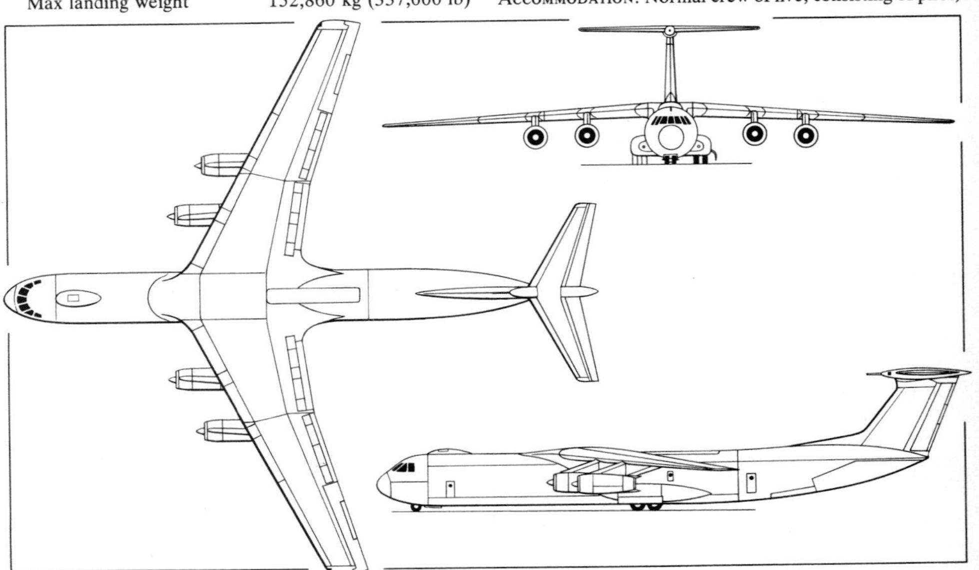

Lockheed YC-141B lengthened version of the StarLifter logistics transport *(Pilot Press)*

Lockheed C-5A Galaxy long-range military heavy transport (four General Electric TF39-GE-1 turbofan engines) *(Brian M. Service)*

co-pilot, flight engineer, navigator and loadmaster, with rest area for 15 people (relief crew, couriers, etc) at front of upper deck. Basic version has seats for 75 troops on rear part of upper deck, aft of wing box. Provision for carrying 270 troops on lower deck, but aircraft is employed primarily as a freighter. Typical freight loads include two M-60 tanks or sixteen ¾ ton lorries; or one M-60 and two Bell Iroquois helicopters, five M-113 personnel carriers, one M-59 2½ ton truck and an M-151 ¼ ton truck; or 10 Pershing missiles with tow and launch vehicles; or 36 standard 463L load pallets. Visor-type upward-hinged nose, and loading ramp, permit straight-in loading into front of hold, under flight deck. Rear straight-in loading via ramp which forms undersurface of rear fuselage. Side panels of rear fuselage, by ramp, hinge outward to improve access on ground but do not need to open for air-drop operations in view of width of ramp. Provision for Aerial Delivery System (ADS) kits for paratroops or cargo. Two passenger doors on port side, at rear end of upper and lower decks. Two crew doors on port side, at forward end of upper and lower decks.

DIMENSIONS, EXTERNAL:
- Wing span — 67·88 m (222 ft 8½ in)
- Wing chord at root — 13·85 m (45 ft 5·3 in)
- Wing chord at tip — 4·67 m (15 ft 4 in)
- Wing aspect ratio — 7·75
- Length overall — 75·54 m (247 ft 10 in)
- Length of fuselage — 70·29 m (230 ft 7¼ in)
- Height overall — 19·85 m (65 ft 1½ in)
- Tailplane span — 20·94 m (68 ft 8½ in)
- Wheel track (between outer wheels) — 11·42 m (37 ft 5½ in)
- Wheelbase (c/l main gear to c/l nose gear) — 22·23 m (72 ft 11 in)
- Crew door (lower deck):
  - Height — 1·80 m (5 ft 11 in)
  - Width — 1·02 m (3 ft 4 in)
  - Height to sill — 3·94 m (12 ft 11 in)
- Passenger door (lower deck):
  - Height — 1·83 m (6 ft 0 in)
  - Width — 0·91 m (3 ft 0 in)
  - Height to sill — 3·56 m (11 ft 8 in)
- Aft loading opening (ramp lowered):
  - Max height — 3·93 m (12 ft 10¾ in)
  - Max width — 5·79 m (19 ft 0 in)
- Aft straight-in loading:
  - Max height — 2·90 m (9 ft 6 in)
  - Max width — 5·79 m (19 ft 0 in)

DIMENSIONS, INTERNAL:
- Cabins, excl flight deck:
  - Length:
    - upper deck, forward — 11·99 m (39 ft 4 in)
    - upper deck, aft — 18·20 m (59 ft 8½ in)
    - lower deck, without ramp — 36·91 m (121 ft 1 in)
    - lower deck, with ramp — 44·07 m (144 ft 7 in)
  - Max width:
    - upper deck, forward — 4·20 m (13 ft 9½ in)
    - upper deck, aft — 3·96 m (13 ft 0 in)
    - lower deck — 5·79 m (19 ft 0 in)
  - Max height:
    - upper deck — 2·29 m (7 ft 6 in)
    - lower deck — 4·11 m (13 ft 6 in)
  - Floor area:
    - upper deck, forward — 50·17 m² (540 sq ft)
    - upper deck, aft — 72·10 m² (776·1 sq ft)
    - lower deck, without ramp — 213·76 m² (2,300·9 sq ft)
  - Height to floor (kneeled):
    - forward — 1·34 m (4 ft 4¾ in)
    - aft — 1·45 m (4 ft 9 in)
  - Volume:
    - upper deck, forward — 56·91 m³ (2,010 cu ft)
    - upper deck, aft — 170·46 m³ (6,020 cu ft)
    - lower deck — 985·29 m³ (34,795 cu ft)

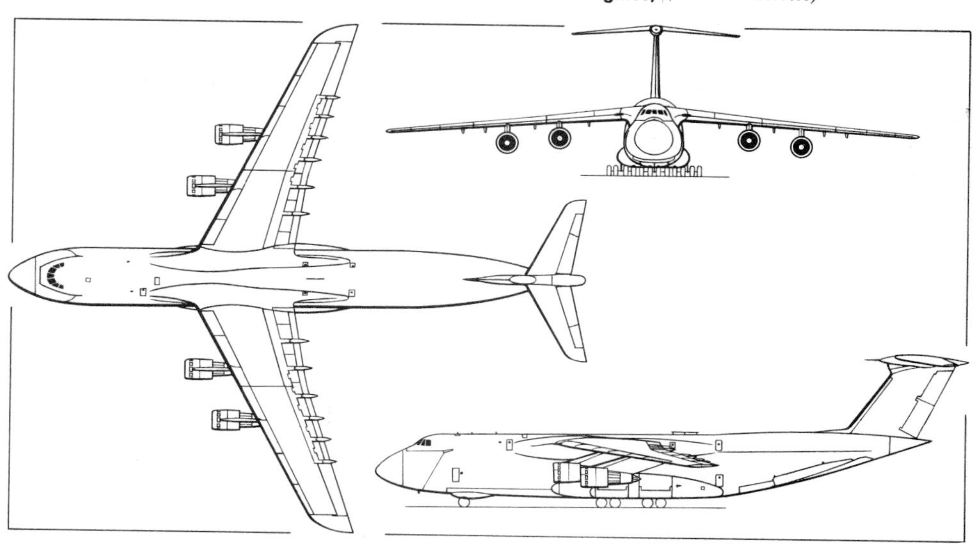

Lockheed C-5A Galaxy four-turbofan military heavy transport aircraft *(Pilot Press)*

AREAS:
- Wings, gross — 576·0 m² (6,200 sq ft)
- Ailerons (total) — 23·49 m² (252·8 sq ft)
- Trailing-edge flaps (total) — 92·13 m² (991·7 sq ft)
- Leading-edge slats (total) — 60·25 m² (648·5 sq ft)
- Spoilers (total) — 40·01 m² (430·7 sq ft)
- Fin — 89·29 m² (961·1 sq ft)
- Rudder — 21·06 m² (226·7 sq ft)
- Tailplane — 89·73 m² (965·8 sq ft)
- Elevators — 24·03 m² (258·7 sq ft)

WEIGHTS AND LOADINGS (for 2·25*g*):
- Basic operating weight — 153,285 kg (337,937 lb)
- Design payload — 100,228 kg (220,967 lb)
- Max T-O and ramp weight — 348,810 kg (769,000 lb)
- Max landing weight — 288,416 kg (635,850 lb)
- Max zero-fuel weight — 253,515 kg (558,904 lb)
- Max wing loading — 605·4 kg/m² (124·0 lb/sq ft)
- Max power loading — 478 kg/kN (4·69 lb/lb st)

PERFORMANCE (at max T-O weight, except where indicated):
- Never-exceed speed — Mach 0·875 (409·5 knots; 760 km/h; 472 mph CAS)
- Max level speed at 7,620 m (25,000 ft) — 496 knots (919 km/h; 571 mph)
- High-speed cruise at 7,620 m (25,000 ft) at normal rated thrust — 460-480 knots (853-890 km/h; 530-553 mph)
- Average cruising speed — 450 knots (834 km/h; 518 mph)
- Aerial delivery drop speed — 130-150 knots (241-278 km/h; 150-173 mph)
- Stalling speed, 40° flap at max landing weight — 104 knots (194 km/h; 120 mph) EAS
- Rate of climb at S/L, ISA, at max rated thrust — 549 m (1,800 ft)/min
- Service ceiling at AUW of 278,950 kg (615,000 lb) — 10,360 m (34,000 ft)
- Min ground turning radius — 22·86 m (75 ft 0 in)
- Runway LCN:
  - concrete — 40
  - asphalt — 64
- T-O run — 2,134 m (7,000 ft)
- T-O to 15 m (50 ft) — 2,560 m (8,400 ft)
- Landing from 15 m (50 ft) — 1,097 m (3,600 ft)
- Landing run — 680 m (2,230 ft)
- Range with 100,228 kg (220,967 lb) payload — 3,256 nm (6,033 km; 3,749 miles)
- Range with 51,074 kg (112,600 lb) payload — 5,670 nm (10,505 km; 6,529 miles)
- Ferry range — 6,940 nm (12,860 km; 7,991 miles)

LTV—*See 'Vought'*

MARTIN MARIETTA
MARTIN MARIETTA CORPORATION

Aerospace Headquarters:
International Club Building, 1800 K Street NW, Washington, DC 20006

Denver Division:
PO Box 179, Denver, Colorado 80201
Telephone: (303) 794-5211

Officers: See Missiles section

Martin Marietta has been engaged in lifting-body research and development since 1959, during which time more than two million man-hours have been devoted to engineering design studies, materials investigation and wind tunnel testing. Its current activities in this field are aimed towards the development of manoeuvring manned re-entry vehicles able to perform as spacecraft in orbit, fly in Earth's atmosphere like aircraft and land at conventional airports.

The small unmanned X-23A (described in the 1967-68 *Jane's*) first proved the aerodynamic characteristics of the design evolved by Martin Marietta. In three flights from orbital altitude and hypersonic speed, the X-23A's stability and manoeuvrability were demonstrated successfully through re-entry conditions down to a speed of Mach 2 and altitude of 30,480 m (100,000 ft). It was followed by a manned, rocket-powered research aircraft known as the X-24A.

An illustrated description of the X-24A appeared in the 1972-73 *Jane's*. The aircraft logged a total of 28 flights from Edwards AFB, California, being air-launched from under the wing of a B-52 'mother-plane'. Several of the flights were supersonic, and a maximum speed of Mach 1·62 and altitude of 21,765 m (71,407 ft) were attained in 1971.

The X-24A was then stripped down to its basic structure and rebuilt as the X-24B, with completely new external lines. In its new form, the aircraft made 36 successful flights, completing its research programme in November 1975. Flight to an altitude of 22,400 m (73,500 ft) and a speed of Mach 1·76 was recorded during 21 supersonic flights. Details of the X-24B can be found in the 1975-76 *Jane's*. Studies of a hypersonic version of the X-24 were being conducted in 1976.

MAULE
MAULE AIRCRAFT CORPORATION

Head Office and Works:
Spence Air Base, Moultrie, Georgia 31768
Telephone: (912) 985-2045
President: B. D. Maule
Vice-President: Mrs B. D. (June) Maule (Treasurer)
Engineering Manager: Lewis E. Blomeley
Sales Manager: Brenda Corbin

This company was formed to manufacture the Maule M-4 four-seat light aircraft in various versions. It transferred to new facilities in Moultrie, Georgia, in September 1968, and has since added the M-5 Lunar Rocket to its range of products.

The company has also designed auxiliary fuel transfer tanks for installation in the outboard wing bays of M-4 and M-5 aircraft. Providing a total usable additional fuel capacity of 87 litres (23 US gallons), these tanks offer owners of the M-4 or M-5 a minimum payload range of 650 nm (1,200 km; 750 miles). FAA approval of the modification was given on 31 October 1973.

Maule M-5-210C four-seat light aircraft mounted on Pee Kay Model 2300 floats

MAULE M-4 JETASEN AND ROCKET

Design of the M-4 was started in 1956. Construction of the prototype began in 1960 and it flew for the first time on 8 September 1960. Production began in early 1962 and four versions have been marketed, as the **Jetasen, Astro-Rocket, Rocket** and **Strata-Rocket.** Production of these aircraft ended in 1975; details can be found in the 1975-76 and earlier editions of *Jane's*.

MAULE M-5 LUNAR ROCKET

Developed from the M-4 Strata-Rocket, the M-5 series have a 30% increase in flap area and enlarged tail surfaces to improve field performance and rate of climb. Two prototypes were built originally. The one powered by a 164 kW (220 hp) Franklin 6A-350-C1 engine was known as the M-5-220C Lunar Rocket, and has since been discontinued as the Franklin engine is no longer available. The second, with a 156·5 kW (210 hp) Continental IO-360-D engine, is designated M-5-210C. The first flight of the M-5-220C prototype was made on 1 November 1971, followed by the first M-5-210C on 16 October 1973; FAA certification was awarded on 28 December 1973. In early 1976 a third version, designated M-5-235C, was completing its certification programme, so that the versions available currently are:

M-5-210C. Basic model with 156·5 kW (210 hp) Continental IO-360-D flat-six engine.

M-5-235C. New model for 1976 with 175 kW (235 hp) Lycoming O-540-J1A5D flat-six engine, driving a larger-diameter propeller.

Maule Patroler. Civil patrol version of either of the above aircraft, which can have any or all of the following modifications: Plexiglas-covered doors for improved view; port side rear observation window; 1·5-3·5 million candlepower belly-mounted manually-controlled searchlight; 100/200W siren or public address system; 28V electrical system; specialised radios to customer's requirements.

Maule Patroler, special-purpose version of the M-5, with underbelly searchlight as operated by Monroe, Louisiana, Police Department

The M-5-210C/M-5-235C is a STOL aircraft, the 'C' in its designation implying that it has double aft doors on the starboard side to facilitate the loading of cargo.

Type: Four-seat light aircraft.

Wings: Braced high-wing monoplane. Streamline-section Vee bracing strut each side. USA 35B (modified) wing section. Dihedral 1°. Incidence 30′. All-metal two-spar structure with metal covering and glassfibre tips. All-metal ailerons and two-position flaps. Ailerons linked with rudder tab, so that aircraft can be controlled in flight by using only the control wheel in the cockpit. Cambered wingtips standard.

Fuselage: Welded 4130 steel tube structure. Covered with glassfibre, except for metal doors and aluminium skin around cabin.

Tail Unit: Braced steel tube structure with glassfibre covering. Trim tabs in port elevator and rudder.

Landing Gear: Non-retractable tailwheel type. Maule oleo-pneumatic shock-absorbers on main units. Maule steerable tailwheel. Cleveland main wheels with Goodyear, General or Schenuit tyres size 17 × 6·00-6, pressure 1·79 bars (26 lb/sq in). Tailwheel tyre size 8 × 3·50-4, pressure 1·03-1·38 bars (15-20 lb/sq in). Cleveland hydraulic disc brakes. Parking brake. Oversize tyres, size 20 × 7·50-6 (pressure 1·24 bars; 18 lb/sq in), and fairings aft of main wheels optional. Provisions for fitting optional Edo Model 248B2440, Pee Kay Model 2300 or Aqua Model 2400 floats, or Federal skis Model C2200H.

Power Plant: One flat-six engine, as detailed in model listings, driving a McCauley two-blade metal constant-speed propeller. Two fuel tanks in wings with total usable capacity of 151 litres (40 US gallons). Optional auxiliary fuel tanks in outer wings, each with usable capacity of 43·5 litres (11·5 US gallons). Maximum usable fuel capacity 238 litres (63 US gallons). Refuelling points on wing upper surface. Oil capacity 8·3 litres (2·2 US gallons).

Accommodation: Pilot and three passengers on two front bucket seats and rear bench seat, or optional quickly-removed rear sling seat. One door on port side of fuselage, hinged at front edge and opening forward. Three doors on starboard side of fuselage, the forward and centre doors hinged at the front edge, the rear baggage door hinged at the rear edge. The centre and aft doors can be opened together to provide an opening 1·24 m (4 ft 1 in) wide to facilitate loading of bulky cargo. Accommodation heated and ventilated.

Systems: Hydraulic system for brakes only. Electrical system powered by 60A engine-driven alternator. 28V electrical system optional.

Electronics and Equipment: A wide range of Collins Micro Line, King, Genave and Narco communication and navigation equipment is available to customer's requirements. Blind-flying instrumentation, autopilot, wing levelling system, automatic glideslope and systems failure detector optional.

Dimensions, external:

Wing span	9·40 m (30 ft 10 in)
Wing chord, constant	1·60 m (5 ft 3 in)
Wing aspect ratio	5·71
Length overall	6·88 m (22 ft 7 in)
Height overall	1·89 m (6 ft 2½ in)
Tailplane span	3·28 m (10 ft 9 in)
Wheel track	1·83 m (6 ft 0 in)
Wheelbase	4·82 m (15 ft 10 in)
Propeller diameter:	
M-5-210C	1·88 m (6 ft 2 in)
M-5-235C	1·98 m (6 ft 6 in)
Cabin doors (fwd, each):	
Height	0·84 m (2 ft 9 in)
Width	0·76 m (2 ft 6 in)
Height to sill	0·94 m (3 ft 1 in)
Cabin door (centre, stbd):	
Height	0·75 m (2 ft 5½ in)
Width	0·69 m (2 ft 3 in)
Height to sill	0·76 m (2 ft 6 in)
Baggage door (aft, stbd):	
Height	0·58 m (1 ft 11 in)
Width	0·56 m (1 ft 10 in)
Height to sill	0·61 m (2 ft 0 in)

Areas:

Wings, gross	14·67 m² (157·9 sq ft)
Ailerons (total)	1·19 m² (12·8 sq ft)
Trailing-edge flaps (total)	1·75 m² (18·8 sq ft)
Fin	1·22 m² (13·14 sq ft)
Rudder, incl tab	0·54 m² (5·83 sq ft)
Tailplane	1·32 m² (14·2 sq ft)
Elevators, incl tab	1·58 m² (17·0 sq ft)

Weights and Loadings (A: M-5-210C; B: M-5-235C):

Basic operating weight:	
A	601 kg (1,325 lb)
Weight empty:	
B	624 kg (1,375 lb)

Max T-O and landing weight:
A, B 1,043 kg (2,300 lb)
Max wing loading:
A, B 71·3 kg/m² (14·6 lb/sq ft)
Max power loading:
A 6·66 kg/kW (10·95 lb/hp)
B 5·96 kg/kW (9·79 lb/hp)
PERFORMANCE (at max T-O weight. A: M-5-210C; B: M-5-235C):
Never-exceed speed:
A, B 156 knots (290 km/h; 180 mph) TAS
Max level speed at 2,440 m (8,000 ft):
A 143 knots (266 km/h; 165 mph) TAS
B 152 knots (282 km/h; 175 mph) TAS
Max cruising speed, 75% power at 2,440 m (8,000 ft):
A 139 knots (257 km/h; 160 mph) CAS
B 148 knots (274 km/h; 170 mph) CAS
Econ cruising speed, 65% power at 2,440 m (8,000 ft):
A 130 knots (241 km/h; 150 mph) CAS
B 139 knots (257 km/h; 160 mph) CAS
Stalling speed, flaps up:
A 44 knots (80·5 km/h; 50 mph)
Stalling speed, flaps down:
A 36 knots (66 km/h; 41 mph)
Max rate of climb at S/L:
A 380 m (1,250 ft)/min
B 411 m (1,350 ft)/min
Service ceiling:
A 5,485 m (18,000 ft)
T-O and landing run:
A 122 m (400 ft)
T-O to and landing from 15 m (50 ft):
A 183 m (600 ft)
Range with max standard fuel:
A 538 nm (997 km; 620 miles)
B 486 nm (901 km; 560 miles)
Range with max fuel, 30 min reserves:
A 781 nm (1,448 km; 900 miles)
B 760 nm (1,408 km; 875 miles)

MCDONNELL DOUGLAS
MCDONNELL DOUGLAS CORPORATION

HEAD OFFICE AND WORKS:
Box 516, St Louis, Missouri 63166
Telephone: (314) 232-0232
DIRECTORS:
John C. Brizendine
George H. Capps
Donald W. Douglas Jr
George S. Graff
Edwin S. Jones
Richard Lloyd Jones Jr
Robert C. Little
James S. McDonnell (Chairman)
James S. McDonnell III
John F. McDonnell
Sanford N. McDonnell (President)
Dolor P. Murray
William R. Orthwein Jr
CHAIRMAN:
James S. McDonnell
PRESIDENT AND CHIEF EXECUTIVE OFFICER:
Sanford N. McDonnell
CORPORATE EXECUTIVE VICE-PRESIDENT:
Dolor P. Murray
CORPORATE OFFICERS:
John R. Allen (Vice-President, Eastern Region)
David C. Arnold (Vice-President)
John C. Brizendine (Vice-President)
Ben G. Bromberg (Vice-President)
Jerry G. Brown (Vice-President, Treasurer)
Robert F. Cortinovis (Vice-President, Material)
Richard J. Davis (Vice-President, External Relations)
John E. Forry (Vice-President, Controller)
Charles M. Forsyth (Vice-President)
George S. Graff (Vice-President)
Gordon M. Graham (Vice-President, Far East)
Robert E. Hage (Vice-President)
Robert L. Harmon (Vice-President)
O. Lee Howser (Vice-President)
Robert L. Johnson (Vice-President)
Warren E. Kraemer (Vice-President, Europe)
Robert C. Krone (Vice-President, Personnel)
Robert C. Little (Vice-President, Marketing)
James S. McDonnell III (Vice-President, Marketing Administration)
John F. McDonnell (Vice-President, Finance and Development)
James T. McMillan (Vice-President)
Donald Malvern (Vice-President)
Gilbert D. Masters (Vice-President, Manufacturing)
John R. Moore (Vice-President)
T. Wynne Morriss (Assistant Secretary and Counsel)
William R. Orthwein Jr (Vice-President)
A. Joseph Quackenbush (Vice-President)
Albert J. Redway Jr (Vice-President, Washington, DC)
John T. Sant (Vice-President, General Counsel and Secretary)
Stanley J. Sheinbein (Asst Treasurer)
Harry I. Sieferman (Tax Officer)
Albert H. Smith Jr (Vice-President, Contracts and Pricing)
STAFF OFFICERS:
Russell G. Adamson (Vice-President, Personnel MDC West)
Donald P. Ames (Vice-President, McDonnell Douglas Research Laboratories)
Charles A. Gaskill (Vice-President, Properties and Facilities)
Alfred V. Guillou (Vice-President, Corporate Planning)
Arthur W. Hyland (Vice-President, Accounting)
Leo I. Mirowitz (Vice-President, Corporate Diversification)
William E. Schowengerdt (Vice-President, Corporate Auditor)
Howard C. Todt (Vice-President, Quality Assurance)
Lupton A. Wilkinson (Vice-President, Manufacturing Systems)
Michael Witunski (Vice-President)

McDonnell Douglas Research Laboratories
Box 516, St Louis, Missouri 63166
DIRECTOR: Dr Donald P. Ames

McDonnell Douglas Corporation was formed on 28 April 1967, by the merger of the former Douglas Aircraft Company Inc and McDonnell Company. It encompasses both of the original companies and their subsidiaries.

There are eight major operating components of McDonnell Douglas Corporation, as follows:

Douglas Aircraft Company
See pages 330-337 of this section
McDonnell Douglas Astronautics Company
See RPVs, Missiles and Spaceflight sections
McDonnell Aircraft Company
See below
McDonnell Douglas Automation Company
Box 516, St Louis, Missouri 63166
PRESIDENT:
William R. Orthwein Jr
McDonnell Douglas Electronics Company
St Charles, Missouri
PRESIDENT:
David C. Arnold
McDonnell Douglas—Tulsa
Tulsa, Oklahama
VICE-PRESIDENT AND GENERAL MANAGER:
O. Lee Howser
Actron
Monrovia, California
PRESIDENT:
John R. Moore
Nitron
Cupertino, California
PRESIDENT:
Louay E. Sharif
Subsidiaries:

Subsidiaries of McDonnell Douglas Corporation include Douglas Aircraft Company of Canada Ltd, Malton, Ontario; McDonnell Douglas (Japan) Ltd, Tokyo; McDonnell Douglas International Sales Corporation, St Louis, Missouri; MDC Realty Company, Irvine, California; and McDonnell Douglas Finance Corporation, Long Beach, California.

At 1 January 1976, McDonnell Douglas employed a total of 62,830 people, working in 48 communities in 21 states, the District of Columbia, Canada, England, Germany, Italy and Japan. Total office, engineering, laboratory and manufacturing floor area was 2,259,145 m² (24,317,271 sq ft).

MCDONNELL AIRCRAFT COMPANY (A Division of McDonnell Douglas Corporation)

HEADQUARTERS:
Box 516, St Louis, Missouri 63166
Telephone: (314) 232-0232
PRESIDENT:
George S. Graff
EXECUTIVE VICE-PRESIDENT:
Donald Malvern
VICE-PRESIDENTS:
Aksel R. Andersen (Avionics Engineering)
William J. Blatz (General Manager F-18)
Alvin L. Boyd (Fiscal Management)
Chester V. Braun (General Manager F-15)
Denver D. Clark (Marketing)
Robert H. Koenig (Controller)
Edward B. Kuhlmann (Quality Assurance)
Nate Molinarro (Personnel)
Herbert Perlmutter (Manufacturing)
Madison L. Ramey (Engineering)
William S. Ross (Flight and Laboratory Development)
John N. Schuler (Contracts and Pricing)
John F. Sutherland (Product Support)

Production at St Louis continues to be concentrated on versions of the F-4 Phantom II and the F-15 Eagle air superiority fighter.

MCDONNELL DOUGLAS PHANTOM II
US Navy and USAF designations: F-4 and RF-4

The Phantom II was developed initially as a twin-engined two-seat long-range all weather attack fighter for service with the US Navy. A letter of intent to order two prototypes was issued on 18 October 1954, at which time the aircraft was designated AH-1. The designation was changed to F4H-1 on 26 May 1955, with change of mission to missile fighter, and the prototype XF4H-1 flew for the first time on 27 May 1958. The first production Phantom II was delivered to US Navy Squadron VF-101 in December 1960. Trials in a ground attack role led to USAF orders, and the basic USN and USAF versions became the F-4B and F-4C respectively. Many other variants have appeared, as follows:

F-4A (formerly F4H-1F): Basic power plant comprised two General Electric J79-GE-2 turbojet engines, with afterburning. Total of 23 pre-production and 24 production aircraft built. After evaluation of this version, the USAF decided to order land-based versions of the F-4B under the designation F-4C.

F-4B (formerly F4H-1). All-weather fighter for US Navy and Marine Corps, powered by two General Electric J79-GE-8 turbojet engines. Total of 649 built. (See F-4G and F-4N.)

QF-4B. See under US Navy entry in RPVs and Targets section.

RF-4B (formerly F4H-1P). Multi-sensor reconnaissance version of F-4B for US Marine Corps. No dual controls or armament. Reconnaissance system as for RF-4C. J79-GE-8 engines. High-frequency single sideband radio. First flown on 12 March 1965. Overall length increased to 19·2 m (63 ft). Total of 46 built.

F-4C (formerly F-110A). Variant of F-4B for USAF, with J79-GE-15 turbojets, cartridge starting, wider-tread low-pressure tyres size 30 × 11·5, larger brakes, Litton type LN-12A/B (ASN-48) inertial navigation system, APQ-100 radar, APQ-100 PPI scope, LADD timer, Lear Siegler AJB-7 bombing system, GAM-83 controls, dual controls and boom flight refuelling instead of drogue (receptacle in top of fuselage, aft of cockpit). Folding wings and arrester gear retained. For close support and attack duties with Tactical Air Command, PACAF and USAFE, and with the Air National Guard (ANG) from January 1972. Sufficient F-4Cs were modified to equip two squadrons for a defence suppression role under the USAF's **Wild Weasel** programme. These aircraft carry ECM warning sensors, jamming pods, chaff dispensers and anti-radiation missiles. First F-4C flew on 27 May 1963; 36 supplied to Spanish Air Force. The last of 583 was delivered to TAC on 4 May 1966. Replaced in production by F-4D.

RF-4C (formerly RF-110A). Multi-sensor reconnaissance version of F-4C for USAF, with radar and photo-

McDonnell Douglas F-4E Phantom II, with leading-edge slats

graphic systems in modified nose which increases overall length by 0·84 m (2 ft 9 in). Three basic reconnaissance systems are: side-looking radar to record high-definition radar picture of terrain on each side of flight path on film; infra-red detector to locate enemy forces under cover or at night by detecting exhaust gases and other heat sources; forward and side-looking cameras, including panoramic models with moving-lens elements for horizon-to-horizon pictures. Systems are operated from rear seat. HF single sideband radio. YRF-4C flew on 9 August 1963; first production RF-4C on 18 May 1964. Taken into service with ANG in February 1972. Production ended December 1973. Total of 505 built.

F-4D. Development of F-4C for USAF, with J79-GE-15 turbojets, APQ-109 fire control radar, ASG-22 servoed sight, ASQ-91 weapon release computer, ASG-22 lead computing amplifier, ASG-22 lead computing gyro, 30kVA generators, and ASN-63 inertial navigation system. First F-4D flew on 8 December 1965. Two squadrons of F-4Ds (32 aircraft) delivered to the Imperial Iranian Air Force and 18 to the Republic of Korea. Production completed. Total of 843 built.

F-4E. Multi-role fighter for air superiority, close support and interdiction missions with USAF. Has internally-mounted M-61A1 20 mm multi-barrel gun, improved (AN/APQ-120) fire-control system and J79-GE-17 turbojets (each 79·6 kN; 17,900 lb st). Additional fuselage fuel cell. First production F-4E delivered to USAF on 3 October 1967. Supplied to the Israeli Air Force, Hellenic Air Force, Turkish Air Force, Republic of Korea Air Force and Imperial Iranian Air Force. All F-4Es being fitted retrospectively with leading-edge manoeuvring slats.

In early 1973 F-4Es began to be fitted with Northrop's target identification system electro-optical (TISEO). Essentially a vidicon TV camera with a zoom lens, it aids positive visual identification of airborne or ground targets at long range. The ASX-1 TISEO is mounted in a cylindrical housing on the leading-edge of the port wing of the F-4E.

F-4EJ. On 1 November 1968, the Japan Defence Agency selected the F-4E as the main fighter for the JASDF. Except for the first two, these aircraft were built in Japan under a licence agreement, with some components being supplied from St Louis. The first US-built F-4EJ flew on 14 January 1971. Equipment includes tail warning radar and launchers for Mitsubishi AAM-2 air-to-air missiles. Total of 158 built.

RF-4E. Multi-sensor reconnaissance version in service with the Federal Republic of Germany (which ordered 88), Iran, Israel and Japan. Generally similar to the RF-4C, it differs by having the J79-GE-17 turbojets of the F-4E and changed reconnaissance equipment.

F-4F. Two-seat fighter, with leading-edge slats to improve manoeuvrability and modified electronics. 175 ordered by Federal Germany for the Luftwaffe. First one rolled out on 24 May 1973; last was delivered in July 1976.

F-4G (Navy; no longer operational). Development of F-4B for US Navy, with AN/ASW-21 data link communications equipment, first flown on 20 March 1963. In service over Vietnam with Squadron VF-213 from USS *Kitty Hawk* in Spring of 1966. Only 12 were built and these are included in the total quoted for F-4B production.

F-4G (Wild Weasel). The USAF's Wild Weasel programme is concerned primarily with the suppression of hostile weapon radar guidance systems. The provision of airborne equipment able to fulfil such a role, and modification of the necessary aircraft to create an effective force for deployment against such targets, had first priority in tactical Air Force planning in the Spring of 1975. The requirement for such a weapon system had been appreciated by Tactical Air Command as early as 1968, and feasibility studies were initiated in September of that year, following which eight sets of equipment were acquired for development, qualification testing and flight testing in two F-4D aircraft. In the interests of force standardisation and airframe life, the F-4E Phantom has now been selected for modification to fulfil the Advanced Wild Weasel role. Technical studies of the F-4D and F-4E showed the latter aircraft to be easier to modify, resulting in a more satisfactory installation. This includes the addition of a torpedo-shape fairing to the top of the tail fin to carry APR-38 antennae, with other APR-38 antennae installed on the side of the fin and along the upper surface of the fuselage. Other modifications include changes to the LCOSS amplifier in the upper equipment bay, APR-38 CIS installation in the aft cockpit, APR-38 CIS installation in the forward cockpit, removal of the M-61A1 gun system to allow sufficient room for installation of APR-38 subsystems (receiver, HAWC, CIS), and the provision of suitable cockpit displays. The changes give the F-4G Wild Weasel the capability to detect, identify and locate hostile electromagnetic emitters, and to deploy against them suitable weapons for their suppression or destruction. Such aircraft would be able to operate independently in a hunter-killer role, but their main utilisation is likely to be as a component of a strike force where they would provide warning and suppression of hostile emitters, and have the capability of deploying their weapons against such targets.

The USAF sought funding in FY 1976 for the Advanced Wild Weasel concept, which would provide an expansion in the memory of the airborne processer and extended low-frequency emission coverage. The programme provided for the first F-4G operational kit installation in the Spring of 1976 and the second in the Autumn of that year, followed by 15 installations in 1977, 60 in 1978 and 39 in 1979, to provide a force of 116 aircraft.

(F-4H designation not used, to avoid confusion with original F4H.)

F-4J. Development of F-4B for US Navy and Marine Corps, primarily as interceptor but with full ground attack capability. J79-GE-10 turbojets. Use of 16½° drooping ailerons and slotted tail gives reduced approach speed in spite of increased landing weight. Westinghouse AN/AWG-10 pulse Doppler fire-control system. Lear Siegler AJB-7 bombing system; 30kVA generators. First F-4J demonstrated publicly on 27 May 1966. Production of 518 completed in December 1972.

F-4K. Development of F-4B for Royal Navy, with improvements evolved for F-4J plus other changes. Westinghouse AN/AWG-10 pulse Doppler fire-control radar system modified to allow the antenna to swing around with the radome. This 'foldable radome' reduces the length of the aircraft, making it compatible with the deck elevators on British aircraft carriers. Two Rolls-Royce Spey RB.168-25R Mk 201 turbofans (each rated at 55·6 kN; 12,500 lb st dry) with 70% afterburning. Air intake ducts 0·15 m (6 in) wider than on US models to cater for more powerful engines. Drooped ailerons. Tailplane has leading-edge fixed slot. Strengthened main landing gear. Nose landing gear strut extends to 1·02 m (40 in), compared to 0·51 m (20 in) on the F-4J, to permit optimum-incidence catapulting. Martin-Baker ejection seats. Weapons include Sparrow air-to-air missiles. Initial contracts for two YF-4Ks and two F-4Ks; ordered as Phantom FG. Mk 1. First flight 27 June 1966. First operational Phantom unit, 892 Squadron, commissioned at RNAS Yeovilton on 31 March 1969. Total of 52 built.

F-4M. For Royal Air Force. Generally similar to F-4K, but with larger brakes and low-pressure tyres of F-4C, and no tailplane leading-edge slot. Folding wings and arrester gear retained. Up to 50% of the components manufactured in the UK. First F-4M flew on 17 February 1967. Deliveries began on 23 August 1968. RAF designation is Phantom FGR. Mk 2. Total of 118 built. Some delivered with dual controls for use as conversion trainers.

F-4N. The US Navy is updating 178 F-4Bs under this designation. The first (150430) was delivered on 21 February 1973.

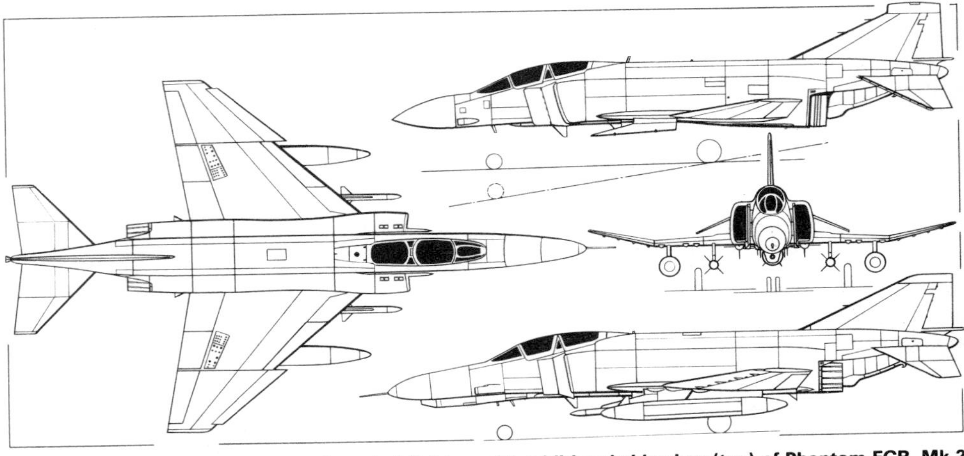

McDonnell Douglas F-4E Phantom II tactical fighter, with additional side view (top) of Phantom FGR. Mk 2 (F-4M) *(Pilot Press)*

McDonnell Douglas Phantom FGR. Mk 2 of No. 41 Squadron, Royal Air Force, armed with seven BL 755 cluster bombs, four Sparrow and four Sidewinder air-to-air missiles

F-4S. The US Navy plans to modify its F-4Js under this designation, with initial structural strengthening to increase operational life and, later, the addition of leading-edge slats.

F-4CCV. This Phantom, shown in an accompanying illustration, has been fitted with experimental canard foreplanes under a joint USAF/McDonnell Douglas research programme to evaluate CCV (Control Configured Vehicle) techniques for combat aircraft. A full-authority fly-by-wire flight control system is also installed. As on the Swedish Viggen, the foreplanes are so positioned, at the top front of the engine air intakes, that they generate a favourable vortex over the wings. The outboard flap sections have been modified to operate as 'flaperons' for manoeuvre-load and direct-lift control. To reduce stability for the CCV research, a slab of lead has been fitted under the rear fuselage to move the CG rearward. The canard foreplanes have an area of approximately 3·72 m² (40 sq ft) and 20° of movement. There is no current suggestion that they might be applied to production F-4s. The F-4CCV flew for the first time on 29 April 1974.

A total of 4,742 Phantoms had been delivered by 1 January 1976.

The Phantom II has set up many official records, including a speed of 783·92 knots (1,452·777 km/h; 902·72 mph: Mach 1·2) over a hazardous 3 km low-level course (maximum altitude 100 m; 328 ft), by Lt Hunt Hardisty and Lt E. De Esch in one of the F-4As on 28 August 1961. This exceeded the previous (subsonic) record, set up eight years earlier, by more than 130 knots (240 km/h; 149 mph); and had not been beaten by mid-1976.

The following details apply to the F-4B:

TYPE: Twin-engined two-seat all-weather fighter.

WINGS: Cantilever low-wing monoplane. Average thickness/chord ratio 5·1%. Sweepback 45° on leading-edges. Outer panels have extended chord and dihedral of 12°. Centre-section and centre wings form one-piece structure from wing fold to wing fold. Portion that passes through fuselage comprises a torsion-box between the front and main spars (at 15% and 40% chord) and is sealed to form two integral fuel tanks. Spars are machined from large forgings. Centre wings also have forged rear spar. Centreline rib, wing-fold ribs, two intermediate ribs forward of main spar and two aft of main spar are also made from forgings. Wing skins machined from aluminium panels 0·0635 m (2½ in) thick, with integral stiffening. Trailing-edge is a one-piece aluminium honeycomb structure. Flaps and ailerons of all-metal construction, with aluminium honeycomb trailing-edges. Inset ailerons limited to down movement only, the 'up' function being supplied by hydraulically-operated spoilers on upper surface of each wing. Ailerons and spoilers fully powered by two independent hydraulic systems. Hydraulically-operated trailing-edge flaps and leading-edge flap on outboard half of each inner wing panel are 'blown'. Hydraulically-operated airbrake under each wing aft of wheel well. Outer wing panels fold upward for stowage.

FUSELAGE: All-metal semi-monocoque structure, built in forward, centre and rear sections. Forward fuselage fabricated in port and starboard halves, so that most internal wiring and finishing can be done before assembly. Keel and rear sections make extensive use of steel and titanium. Double-wall construction under fuel tanks and for lower section of rear fuselage, with ram-air cooling.

TAIL UNIT: Cantilever all-metal structure, with 23° of anhedral on one-piece all-moving tailplane. Ribs and stringers of tailplane are of steel, skin of titanium and trailing-edge of steel honeycomb. Rudder interconnected with ailerons at low speeds.

LANDING GEAR: Hydraulically-retractable tricycle type, main wheels retracting inward into wings, nose unit rearward. Single wheel on each main unit, with tyres size 30 × 7·70; twin wheels on nose unit, which is steerable and self-centering and can be lengthened pneumatically to increase the aircraft's angle of attack for take-off. Brake-chute housed in fuselage tailcone.

POWER PLANT: Two General Electric J79-GE-8 turbojet engines (each 75·6 kN; 17,000 lb st with afterburning). Variable-area inlet ducts monitored by air data computer. Integral fuel tankage in wings, between front and main spars, and in six fuselage tanks, with total capacity of 7,569 litres (2,000 US gallons). Provision for one 2,270 litre (600 US gallon) external tank under fuselage and two 1,400 litre (370 US gallon) underwing tanks. Equipment for probe-and-drogue and 'buddy tank' flight refuelling, with retractable probe in starboard side of fuselage.

ACCOMMODATION: Crew of two in tandem on Martin-Baker Mk H7 ejection seats, under individual rearward-hinged canopies. Optional dual controls.

SYSTEMS: Three independent hydraulic systems, each of 207 bars (3,000 lb/sq in). Pneumatic system for canopy operation, nosewheel strut extension and ram-air turbine extension. Primary electrical source is AC generator. No battery.

ELECTRONICS: Eclipse-Pioneer dead-reckoning navigation computer, Collins AN/ASQ-19 communications-navigation-identification package, AiResearch A/A 24G central air data computer, Raytheon radar altimeter, General Electric ASA-32 autopilot, RCA data link, Lear attitude indicator and AJB-3 bombing system. Westinghouse APQ-72 automatic radar fire-control system in nose. ACF Electronics AAA-4 infra-red detector under nose.

ARMAMENT: Six Sparrow III, or four Sparrow III and four Sidewinder, air-to-air missiles on four semi-submerged mountings under fuselage and two underwing mountings. Provision for carrying alternative loads of up to about 7,250 kg (16,000 lb) of nuclear or conventional bombs and missiles on five attachments under wings and fuselage. Typical loads include eighteen 750 lb bombs, fifteen 680 lb mines, eleven 1,000 lb bombs, seven smoke bombs, eleven 150 US gallon napalm bombs, four Bullpup air-to-surface missiles or fifteen packs of air-to-surface rockets.

DIMENSIONS, EXTERNAL:
- Wing span: 11·70 m (38 ft 5 in)
- Width, wings folded: 8·39 m (27 ft 6½ in)
- Length overall: 17·76 m (58 ft 3 in)
- Height overall: 4·96 m (16 ft 3 in)
- Wheel track: 5·30 m (17 ft 10½ in)

AREA:
- Wings, gross: 49·2 m² (530 sq ft)

WEIGHTS:
- T-O weight ('clean'): 20,865 kg (46,000 lb)
- Max T-O weight: 24,765 kg (54,600 lb)

PERFORMANCE:
- Max level speed with external stores: over Mach 2
- Approach speed: 130 knots (240 km/h; 150 mph)
- T-O run (interceptor): 1,525 m (5,000 ft)
- Landing run (interceptor): 915 m (3,000 ft)
- Combat radius:
  - interceptor: over 781 nm (1,450 km; 900 miles)
  - ground attack: over 868 nm (1,600 km; 1,000 miles)
- Ferry range: 1,997 nm (3,700 km; 2,300 miles)

MCDONNELL DOUGLAS F-15 EAGLE

The USAF requested development funding for a new air superiority fighter in 1965, and in due course design proposals were sought from three airframe manufacturers: Fairchild Hiller Corporation, McDonnell Douglas Corporation, and North American Rockwell Corporation. On 23 December 1969 it was announced that McDonnell Douglas had been selected as prime airframe contractor. The contract called for the design and manufacture of 20 aircraft for development testing, these to comprise 18 single-seat **F-15As** and two **TF-15A** two-seat trainers, with production scheduled at a rate of one aircraft every other month.

McDonnell Douglas F-4CCV research aircraft with canard foreplanes

First flight of the F-15A was made on 27 July 1972, and the first flight of a two-seat TF-15A trainer on 7 July 1973.

A production go-ahead for the first 30 operational aircraft (FY 1973 funds) was announced on 1 March 1973. The FY 1974 Defense Procurement Bill authorised production of 62 aircraft, and the Defense Procurement Bills for FY 1975 and 1976/7T authorised further production of 72 and 135 aircraft respectively. The FY 1973-74 production contracts included 13 of the two-seat TF-15A version; one of these (the 21st Eagle built) was the first Eagle delivered to the USAF on 14 November 1974. Structural weight of the TF-15A is approx 363 kg (800 lb) more than that of the single-seater.

By early 1976, Eagles were in operational service with the 58th TFTW at Luke AFB, Arizona, and the 1st TFW at Langley AFB, Virginia. By 30 June 1976 a total of 77 Eagles were in service with Tactical Air Command squadrons, and more than 100 had been delivered. It is planned to procure 749 for the USAF by 1981, including the 20 R & D models. Others are to be sold to Israel.

Design of the F-15 remained stable from the outset, with relatively few changes. Two significant changes made as a result of flight experience involved improvements to the landing gear and provision of a larger airbrake. The wingtip shape was also modified, and the basic control system was refined to optimise the Eagle's air combat tracking capabilities. The results of combat exercises conducted by USAF and contractor pilots demonstrated a tracking capability far superior to that of other known fighter aircraft.

Designed specifically as an air superiority fighter, the F-15A Eagle has proved equally suitable for air-to-ground missions without degradation of its primary role. It is able to carry a variety of air-to-air and air-to-ground weapons.

A participant in the Farnborough air show in September 1974 was one of the TF-15 development aircraft. It had been ferried 2,660 nm (4,930 km; 3,063 miles) nonstop and unrefuelled from Loring AFB, Maine, where it took off at an AUW of 30,390 kg (67,000 lb), including 14,970 kg (33,000 lb) of fuel. Time of flight was 5·4 hours, at Mach 0·85 average speed, with 1,950 kg (4,300 lb) of fuel remaining when the TF-15 began its let-down over the UK, towards the RAF station at Bentwaters where it landed.

Key to this large increase in the normal ferry range of a 'clean' F-15 was the use of two low-drag fuel pallets known as Fast Packs (Fuel And Sensor Tactical Packs) developed specially for the F-15 by McDonnell Aircraft Company. Each Fast Pack contains approximately 3,228 litres (114 cu ft) of usable volume, which can accommodate 2,268 kg (5,000 lb) of JP-4 fuel. It attaches to the side of either the port or starboard engine air intake trunk (being made in handed pairs), is designed to the same load factors and airspeed limits as the basic aircraft, and can be removed in 15 minutes.

In addition to virtually eliminating the requirement for tanker support during global deployments, the use of Fast Packs would permit the carriage of a heavier bomb-load to distant targets. For reconnaissance missions, part of the available volume could be used for cameras and other sensors. By adding Wild Weasel equipment to the Fast Pack, the F-15 could be used for surface-to-air missile site suppression. A laser designator or low light level TV system could be installed in the Pack to enhance day/night strike effectiveness, or an infra-red search and track system for interceptor missions. All external stores stations remain available with the Packs in use. AIM-7F missiles attach to the corners of the Fast Packs.

First flight of the Fast Pack was made on 27 July 1974, only 139 days after the prototype programme go-ahead on 11 March. Testing included flight at speeds in excess of Mach 2 and at load factors greater than 5*g*, and showed that aircraft handling qualities are essentially unchanged with the Pack installed.

Under a programme dubbed 'Streak Eagle', the F-15 demonstrated its climb capability by capturing eight time-to-height records, between 16 January and 1 February

Demonstrating the air-to-surface capability of the McDonnell Douglas F-15A Eagle

1975. Three USAF pilots were involved: Majors Willard Macfarlane, David Peterson and Roger Smith. The new records were:

Altitude (metres)	*Previous record time to height*	*New record time to height*
3,000	34·5 sec	27·57 sec
6,000	48·8 sec	39·33 sec
9,000	1 min 1·7 sec	48·86 sec
12,000	1 min 17·1 sec	59·38 sec
15,000	1 min 54·5 sec	1 min 17·04 sec
20,000	2 min 49·8 sec	2 min 2·94 sec
25,000*	3 min 12·6 sec	2 min 41·02 sec
30,000*	4 min 3·9 sec	3 min 27·80 sec

*These two records have since been reclaimed by the E-266M version of the MiG-25 (which see).

The following description applies to the F-15A:

TYPE: Single-seat twin-turbofan air superiority fighter, with secondary attack role.

WINGS: Cantilever shoulder-wing monoplane. Leading-edge swept back at approximately 45°. Outboard aileron actuators by Ozone Metal Products.

FUSELAGE: All-metal semi-monocoque structure.

TAIL UNIT: Cantilever structure with twin fins and rudders. All-moving horizontal tail surfaces outboard of fins, with extended chord on outer leading-edges. Rudder servo actuators by Ronson Hydraulic Units Corporation. Actuators for horizontal surfaces by National Water Lift Company. Boost and pitch compensator for control stick by Moog Inc, Controls Division.

LANDING GEAR: Hydraulically-retractable tricycle type, with single wheel on each unit. Nose and main landing gear by Cleveland Pneumatic Tool Company. Wheels and brake assemblies by Goodyear Tire and Rubber Company. Main and nosewheel tyres by B. F. Goodrich Company. Wheel braking skid control system by Hydro-Aire Division of Crane Company. All units retract forward.

POWER PLANT: Two Pratt & Whitney F100-PW-100 turbofan engines of approximately 111·2 kN (25,000 lb st). Internal fuel load 5,260 kg (11,600 lb). Fuel tanks by Goodyear Aviation Products Division. Refuelling control valve by Ronson Hydraulic Units Corporation. Fuel tank valves and check valves by Parker Hannifin Corporation. Fuel gauge system by Simmonds Precision Products Inc. Fuel tank pressure regulators by Vap-Air Division of Vapor Corporation. Fast Pack conformal fuel pallets attached to side of engine air intakes, beneath wing, can be removed within 15 min. Each has usable volume of 3·23 m³ (114 cu ft) and can contain 2,268 kg (5,000 lb) of JP-4 fuel.

ENGINE INTAKES: Straight two-dimensional external compression inlets, on each side of the fuselage. Air inlet controllers by Hamilton Standard. Air inlet actuators by National Water Lift Company.

ACCOMMODATION: Pilot only, on ejection seat developed by McDonnell Douglas. Stretched acrylic canopy and windscreen. Windscreen anti-icing valve by Dynasciences Corporation.

SYSTEMS: Electric power generating system by Lear Siegler Power Equipment Division; transformer-rectifiers by Electro Development Corporation; 40/50kVA generator constant-speed drive units by Sundstrand Corporation, Aviation Division. Three independent hydraulic systems (each 207 bars; 3,000 lb/sq in) powered by Abex engine-driven pumps; modular hydraulic packages by Hydraulic Research and Manufacturing Company. The oxygen system includes a liquid oxygen indicator by Simmonds Precision Products Inc. Air-conditioning system by AiResearch Manufacturing Company. Automatic flight control system by General Electric, Aircraft Equipment Division. Auxiliary power unit for engine starting, and for the provision of electric or hydraulic power on the ground independently of the main engines, supplied by AiResearch Manufacturing Company.

ELECTRONICS: Lightweight APG-63 pulse-Doppler radar developed by Hughes Aircraft Company provides long-range detection and tracking of small high-speed targets operating at all altitudes down to treetop level, and feeds accurate tracking information to the airborne central computer to ensure effective launch of the aircraft's missiles or the firing of its internal gun. For close-in dogfights, the radar automatically acquires the target on a head-up display. International Business Machines, Electronic Systems Center, is subcontractor for the central computer, and McDonnell Douglas Electronics Company for the head-up display. This latter unit projects all essential flight information in the form of symbols on to a combining glass positioned above the instrument panel at pilot's eye level. The display presents the pilot with all the information required to intercept and destroy an enemy aircraft without need for him to remove his eyes from the target. The display also provides navigation and other steering control information under all flight conditions. A transponder for the IFF system, developed by Teledyne Electronics Company, informs ground stations and other suitably equipped aircraft that the F-15 is a friendly aircraft. It also supplies data on the F-15's range, azimuth, altitude and identification to air traffic controllers. The F-15 carries an AN/APX-76 interrogator receiver-transmitter, built by Hazeltine Corporation, to inform the pilot if an aircraft seen visually or on radar is friendly. A reply evaluator for the IFF system, which operates with the AN/APX-76, was developed by Litton Systems Inc, Van Nuys. A vertical situation display set, that uses a cathode-ray tube to present radar, electro-optical identification and attitude director indicator formats to the pilot, has been developed by Sperry Rand Corporation, Sperry Flight Systems Division. This permits inputs received from the aircraft's sensors and the central computer to be visible to the pilot under any light conditions. This company has also developed an air data computer for the F-15, as well as an attitude and heading reference set to provide information on the aircraft's pitch, roll and magnetic heading that is fed to cockpit displays. This latter unit also serves as a backup to the inertial navigation set developed by Litton Systems Inc. This provides the basic navigation data and is the aircraft's primary attitude reference, enabling the F-15 to navigate anywhere in the world. In addition to giving the aircraft's position at all times, the inertial navigation system provides pitch, roll, heading, acceleration and speed information.

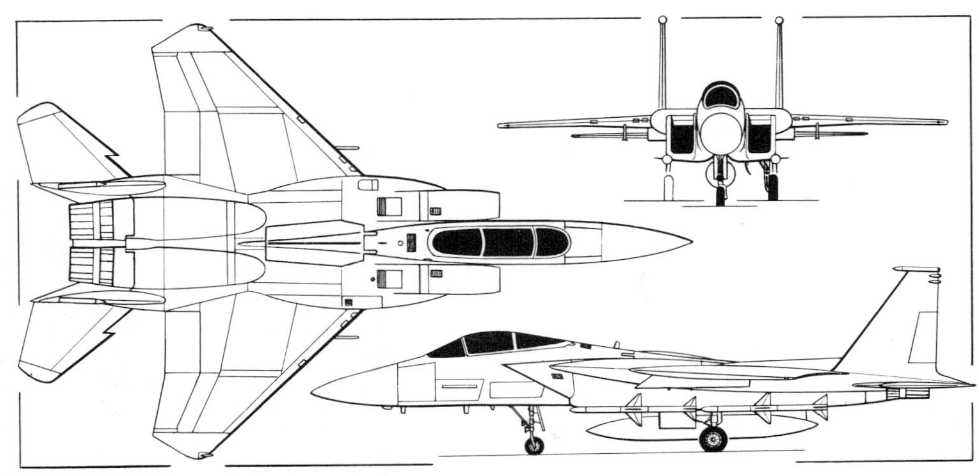

McDonnell Douglas F-15A Eagle twin-turbofan air superiority fighter *(Pilot Press)*

McDonnell Douglas TF-15A Eagle two-seat trainer, fitted with Fast Packs

Other specialised equipment for flight control, navigation and communications includes a micro-miniaturised Tacan system by Hoffman Electronics Corporation; a horizontal situation indicator to present aircraft navigation information on a symbolic pictorial display, by Collins Radio Company, which is also responsible for the ADF and ILS receivers, UHF transceiver and UHF auxiliary-fix receiver. The communications sets have cryptographic capability. Dorne and Margolin Aviation Products is responsible for the glideslope localiser antenna, and Teledyne Avionics Company for angle of attack sensors. A special nose radome has been developed by Brunswick Corporation, Technical Products Division. This is fabricated from syntactic foam material sandwiched between outer skins, and offers a weight saving of 35% by comparison with conventional radome structures, as well as providing heat resistance up to 500°F (260°C), undistorted passage for signals from the nose radar, and the strength of a primary structure.

An internal countermeasures set is supplied by Hallicrafters Company; radar warning systems by Loral Electronic Systems; and an electronic warfare warning set by Magnavox.

EQUIPMENT: Tachometer, fuel and oil indicators by Bendix Corporation, Flight and Engine Instrument Division. Feel trim actuators by Plessey Airborne Corporation.

ARMAMENT: Provision for carriage and launch of a variety of air-to-air weapons over short and medium ranges, including four AIM-9L Sidewinders, four AIM-7F Sparrows, and a 20 mm M-61A1 six-barrel gun with 940 rounds of ammunition. A lead-computing gunsight has been developed by the General Electric Co. To keep the pilot informed of the status of his weapons and provide for their management, an integrated stores monitoring and management system has been developed by Dynamic Controls Corporation. Five weapon stations allow for the carriage of up to 7,257 kg (16,000 lb) of ordnance or additional ECM equipment.

DIMENSIONS, EXTERNAL:
- Wing span: 13·05 m (42 ft 9¾ in)
- Length overall: 19·43 m (63 ft 9 in)
- Height overall: 5·63 m (18 ft 5½ in)
- Tailplane span: 8·61 m (28 ft 3 in)
- Wheel track: 2·75 m (9 ft 0¼ in)
- Wheelbase: 5·42 m (17 ft 9½ in)

AREA:
- Wings, gross: 56·5 m² (608 sq ft)

WEIGHTS:
- T-O weight (interceptor, full internal fuel and 4 Sparrows): 18,824 kg (41,500 lb)
- T-O weight (incl three 600 US gallon drop-tanks): 24,675 kg (54,400 lb)
- Max T-O weight: 25,401 kg (56,000 lb)

PERFORMANCE:
- Max level speed: more than Mach 2·5 (800 knots; 1,482 km/h; 921 mph CAS)
- Approach speed: 125 knots (232 km/h; 144 mph) CAS
- T-O run (interceptor): 274 m (900 ft)
- Landing run (interceptor), without braking parachute: 762 m (2,500 ft)
- Absolute ceiling: 30,500 m (100,000 ft)
- Ferry range:
  - without Fast Pack: more than 2,500 nm (4,631 km; 2,878 miles)
  - with Fast Pack: more than 3,000 nm (5,560 km; 3,450 miles)
- *g* limits: +9·0; −3·0

AV-8B ADVANCED HARRIER

In late 1973 and early 1974 the British and US governments received for approval various proposals for an advanced version of the Hawker Siddeley Harrier (see UK section). Subsequent to this came the announcement, on 15 May 1975, of a British order for 25 Sea Harrier FRS. Mk 1s for the Royal Navy.

Two months before the announcement of this order, the British Secretary of State for Defence, Mr Roy Mason, had stated that there was "not enough common ground on the Advanced Harrier for us to join in the programme with the US", and development studies for a US version have therefore been continued primarily by McDonnell Douglas to meet requirements of the US Navy and Marine Corps. These have broadly been concerned with two alternatives: the so-called 'AV-16A' based on the Rolls-Royce Pegasus 15 engine, and the lower-cost AV-8B utilising an existing or future growth version of the present Pegasus 11 power plant.

Essentially, the objective of the Advanced Harrier programme is to evolve a version which, without too much of a departure from the existing Harrier airframe, would virtually double the aircraft's weapons payload/combat radius. Major changes envisaged for the originally-projected AV-16A version included installation of the 109 kN (24,500 lb st) Rolls-Royce Pegasus 15 turbofan engine, use of a supercritical wing, increased fuel capacity, strengthened exhaust nozzles for thrust vectoring in forward flight (VIFF), and strengthened landing gear to cater for the increased gross weight.

The USMC has stated a requirement for 336 Advanced Harriers, and initially McDonnell Douglas and the USMC are modifying several existing AV-8As as prototype YAV-8Bs. These are to fly in late 1978 and early 1979, with USN preliminary evaluation beginning in mid-1979. Aim of this proposal is to achieve the improved performance capability required of the AV-16A by aerodynamic means, while retaining the existing F402-RR-402 (Pegasus 11) engine, thus saving the cost of developing the Pegasus 15 engine that was originally considered necessary for the advanced version.

Proposed aerodynamic changes include use of a supercritical wing; the addition of under-gun-pod strakes and a movable flap panel forward of the pods, to increase lift for vertical take-off; the use of larger wing trailing-edge flaps and drooped ailerons; and redesigned engine air intakes. The landing gear will be strengthened to cater for the higher operating weights and greater external stores loads made possible by these changes. Also under consideration is a programme to uprate the Pegasus 11 engine by approximately 4·5 kN (1,000 lb st).

Two production-standard AV-8Bs may next be built, to fly in early 1981, with Navy BIS (Bureau of Inspection and Survey) trials following in Spring/Summer 1982. By this time deliveries would have begun (in late 1981) of an initial batch of 10 production aircraft. The first of the main production batch would follow in Autumn 1982, the AV-8B becoming operational by the beginning of 1983. McDonnell Douglas would be prime contractor for the airframe, with Hawker Siddeley as subcontractor; prime engine contractor would be either Pratt & Whitney or Rolls-Royce, with the other as subcontractor.

WINGS: Cantilever shoulder-wing monoplane, of broadly similar planform to Harrier/AV-8A but of supercritical section, approx 20% greater in span and 14% greater in area. 10° less sweepback on leading-edges, and non-swept inboard trailing-edges. Composite construction, of aluminium alloy and titanium and making extensive use of graphite epoxy in the main multi-spar torsion box, ribs, skins, outrigger fairings and wingtips. Trailing-edge single-slotted flaps, of substantially greater chord than those of AV-8A, and drooping ailerons, also of graphite epoxy construction.

FUSELAGE: Generally similar to AV-8A, but with additional lift-augmenting surfaces. These comprise a fixed strake on each of the two underfuselage gun packs, and a retractable forward flap just aft of the nosewheel unit. During VTOL modes the 'box' formed by the ventral strakes and the lowered nose flap would serve to augment lift by trapping the cushion of air bounced off the ground by the engine exhaust. This additional lift would allow the AV-8B to take off vertically at a gross weight equal to its maximum hovering gross weight.

LANDING GEAR: Main landing gear strengthened to cater for higher operating weights. Outrigger wheels and fairings moved inboard, to approx mid-span beneath each wing between flaps and ailerons.

POWER PLANT: One Rolls-Royce Bristol Pegasus 11 vectored-thrust turbofan engine. This may be either the standard 95·64 kN (21,500 lb st) Mk 803 (F402-RR-402) as in the AV-8A, or more powerful proposed versions designated Pegasus 11D or Pegasus 11+. Engine air intakes redesigned, with elliptical lip shape (to reduce nozzle loss) and double instead of single row of suction relief doors. Increased fuel tankage available in wings, raising total internal fuel capacity (fuselage and wing tanks) from approx 2,268 kg (5,000 lb) in the AV-8A to 3,402 kg (7,500 lb) in the AV-8B. Each of the four inner underwing stations capable of carrying an auxiliary fuel tank.

ELECTRONICS AND EQUIPMENT: Improved attitude and heading reference system.

ARMAMENT AND OPERATIONAL EQUIPMENT: Twin underfuselage gun/ammunition packs, as in AV-8A, each mounting a US 20 mm cannon instead of a 30 mm Aden gun. Single 454 kg (1,000 lb) stores point on fuselage centreline, between gun packs. Three stores stations under each wing, the inner one capable of carrying a 907 kg (2,000 lb) store, the centre one 454 kg (1,000 lb), and the outer one an AIM-9D Sidewinder missile. Including fuel, stores, weapons and ammunition, and water injection for the engine, the maximum useful load for vertical take-off would be more than 3,175 kg (7,000 lb), and for short take-off nearly 4,080 kg (9,000 lb). Typical weapons may include Mk 82 Snakeye bombs, and laser or electro-optical guided weapons. Main weapon delivery by Angle Rate Bombing System (ARBS), comprising a dual-mode (TV and laser) target seeker linked to a Marconi-Elliott head-up display via an IBM digital computer. Passive ECM equipment.

DIMENSIONS, EXTERNAL:
Wing span:
AV-8A 7·70 m (25 ft 3 in)
AV-8B approx 9·23 m (30 ft 3½ in)
Length overall:
AV-8A 13·87 m (45 ft 6 in)
AV-8B 13·08 m (42 ft 10¾ in)
Height overall:
AV-8A 3·43 m (11 ft 3 in)
AV-8B 3·44 m (11 ft 3½ in)
Wing area (gross):
AV-8A 18·68 m² (201·1 sq ft)
AV-8B approx 21·37 m² (230 sq ft)

WEIGHTS:
Basic operating weight, empty:
AV-8A 5,533 kg (12,200 lb)
AV-8B 5,624 kg (12,400 lb)
Combat T-O weight:
AV-8B with 7 Mk 82 bombs 11,790 kg (25,994 lb)
Max T-O weight:
AV-8A over 11,340 kg (25,000 lb)
AV-8B 13,154-13,608 kg (29,000-30,000 lb)
Max landing weight:
AV-8B 8,799 kg (19,400 lb)

PERFORMANCE (AV-8B data estimated):
AV-8A operational radius with external loads shown:
vertical T-O, 1,360 kg (3,000 lb)
50 nm (92 km; 57 miles)
short T-O (185 m; 600 ft), 2,268 kg (5,000 lb)
125 nm (231 km; 144 miles)
short T-O (457 m; 1,500 ft), 3,630 kg (8,000 lb)
222 nm (411 km; 255 miles)
short T-O (305 m; 1,000 ft), 1,360 kg (3,000 lb)
360 nm (667 km; 414 miles)
AV-8B operational radius with external loads shown:
vertical T-O, 3,538 kg (7,800 lb)
100 nm (185 km; 115 miles)
short T-O (305 m; 1,000 ft), twelve Mk 82 Snakeye bombs, internal fuel, 1 hr loiter
more than 150 nm (278 km; 172 miles)
short T-O (305 m; 1,000 ft), seven Mk 82 Snakeye bombs, external fuel tanks, no loiter
more than 650 nm (1,204 km; 748 miles)

Mockup of the AV-8B Advanced Harrier of the US Marine Corps, showing the additional stores-carrying capability of this version

MCDONNELL DOUGLAS F-18

In the Spring of 1974 the US Department of Defense accepted a proposal from the US Navy to study a low-cost lightweight multi-mission fighter, then identified as the VFAX. In June 1974 the USN approached the US aircraft industry to submit critiques and comments on such an aircraft. Six companies responded, including McDonnell Aircraft Company; but in August of that year Congress terminated the VFAX concept, directing instead that the Navy should investigate versions of the General Dynamics YF-16 and Northrop YF-17 lightweight fighter prototypes then under evaluation for the USAF.

McDonnell Douglas made a study of the configuration of these two aircraft and concluded that Northrop's contender not only met most nearly the Navy's requirements, but would also prove the easiest to convert to a combat fighter suitable for operation from aircraft carriers.

As a result of this review, McDonnell Douglas teamed with Northrop to propose a derivative of the YF-17 to meet the Navy's requirement, with McDonnell Douglas as the prime contractor. Identified as the Navy Air Combat Fighter (NACF), this received the designation F-18 when selected for further development. The initial short-term contracts, announced on 2 May 1975, allocated $4·4 million to McDonnell Douglas/Northrop and $2·0 million to General Electric, for continued engineering studies and refinement of the projected airframe and power plant.

On 28 January 1976 it was announced that full-scale development had been initiated by the US Navy, with initial funding of $16 million. Total programme funding to mid-1977 was expected to be $133 million, including allocations for development of the General Electric F404-GE-400 turbojet engine. Total cost of the development programme is expected to be about $1·4 billion, including the production of 11 F-18s for the flight test programme.

The F-18 is scheduled to make its first flight in 1978 and to become operational in 1982.

The F-18 derives from development work carried out by Northrop during recent years to evolve an advanced tactical fighter, and stems from the P-530 Cobra concept of 1968-73, which formed the basis of the company's YF-17 prototype. The F-18 airframe differs from that of the latter aircraft by having increased wing area, a wider and longer fuselage to provide greater internal fuel capacity, an enlarged nose to accommodate the 0·71 m (28 in) radar dish to meet the Navy's search range requirement of over 30 nm (56 km; 35 miles), and strengthening of the airframe structure to cater for the increased loads caused by catapult launches and arrested landings. The foregoing modifications, plus electronics, will result in an increase of approximately 2,720 kg (6,000 lb) in take-off weight, allocated as 1,360 kg (3,000 lb) to structure, 1,088 kg (2,400 lb) for additional fuel and 272 kg (600 lb) for electronics.

A team of Northrop engineers is established at the St Louis headquarters of McDonnell Douglas, responsible for some 30 per cent of the development engineering. Northrop's share of the production will be about 40 per cent, with responsibility for developing and building the centre and aft fuselage. McDonnell Douglas will build the rest of the airframe and carry out final assembly.

It is estimated that approximately 800 production aircraft may be required, as the F-18 is intended to replace both USN and US Marine Corps F-4 Phantoms for the primary missions of fighter escort and interdiction. There would be a proportion of two-seat trainers. Additionally, an attack version of the F-18 is being developed to replace the Navy's A-7 Corsair II aircraft in the mid-1980s, under the designation **A-18**.

TYPE: Single-seat carrier-based air combat fighter.

WINGS: Cantilever mid-wing monoplane. Multi-spar

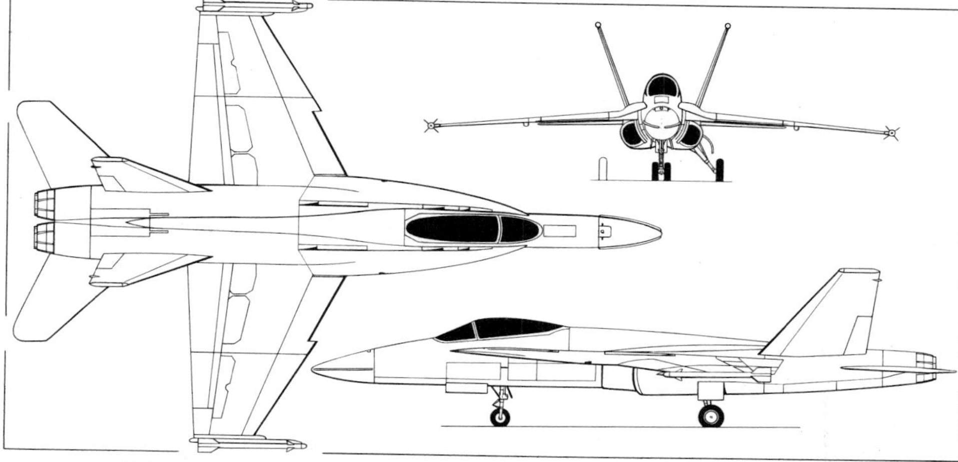

McDonnell Douglas F-18 twin-engined carrier-based air combat fighter *(Pilot Press)*

structure, primarily of light alloy. Boundary layer control achieved by wing root slots. Leading-edge manoeuvring flaps have a maximum extension angle of 35°. Trailing-edge flaps deploy to a maximum of 45°. Ailerons can be drooped to 45°, providing the advantages of full-span flaps for low approach speeds. Notched sections on outer wing leading-edge to enhance aileron effectiveness. Wings fold at the inboard end of each aileron.

FUSELAGE: Semi-monocoque basic structure. Airbrake in upper surface of fuselage between tail fins. Pressurised cockpit section of fail-safe construction.

TAIL UNIT: Cantilever structure with swept vertical and horizontal surfaces. Twin outward-canted fins and rudders, mounted forward of all-moving tailplane.

LANDING GEAR: Retractable tricycle type, manufactured by Cleveland for 11 development aircraft, with twin-wheel nose and single-wheel main units. Nose unit retracts forward, main wheels aft, turning 90° to stow horizontally inside the lower surface of the engine air ducts.

POWER PLANT: Two General Electric F404-GE-400 low bypass turbojet engines, each producing approx 71·2 kN (16,000 lb) thrust and developed from the YJ101 turbojets that power the YF-17. Internal fuel load 4,926 kg (10,860 lb); provision for up to three 1,135 litre (300 US gallon) external tanks, increasing total fuel capacity to more than 7,257 kg (16,000 lb).

ACCOMMODATION: Pilot only, on ejection seat in pressurised, heated and air-conditioned cockpit. Upward-opening canopy, hinged at rear.

SYSTEMS: Fly-by-wire flight control system, with mechanical backup. An APU will provide self-contained start and maintenance facilities.

ELECTRONICS: Will include an Automatic Carrier Landing System (ACLS) for all-weather carrier operations, and a Hughes multi-mode radar.

ARMAMENT: Nine weapon stations with a combined capacity in excess of 5,900 kg (13,000 lb) of mixed ordnance. These comprise two wingtip stations for AIM-9 Sidewinder air-to-air missiles; two outboard wing stations for an assortment of air-to-ground or air-to-air weapons; two inboard wing stations for external fuel tanks or air-to-ground weapons; two nacelle fuselage stations for Sparrows or sensor pods, and a centreline fuselage station for external fuel or weapons. In addition, an M-61 20 mm six-barrel gun, with 540 rounds of ammunition, is mounted in the nose.

DIMENSIONS, EXTERNAL:
- Wing span: 11·43 m (37 ft 6 in)
- Width, wings folded: 7·62 m (25 ft 0 in)
- Length overall: 16·94 m (55 ft 7 in)
- Height overall: 4·51 m (14 ft 9½ in)
- Tailplane span: 6·92 m (22 ft 8½ in)
- Wheel track: 3·11 m (10 ft 2½ in)
- Wheelbase: 5·25 m (17 ft 2½ in)

AREA:
- Wings, gross: 37·16 m² (400 sq ft)

WEIGHTS:
- Fighter mission T-O weight: 15,240 kg (33,600 lb)
- Max T-O weight: more than 19,960 kg (44,000 lb)

PERFORMANCE (estimated):
- Max level speed: more than Mach 1·8
- Max speed, intermediate power: Mach 1·0
- Approach speed: 130 knots (240 km/h; 150 mph)
- Combat ceiling: approx 15,240 m (50,000 ft)
- Combat radius (internal fuel): over 400 nm (740 km; 460 miles)
- Ferry range, unrefuelled: more than 2,000 nm (3,706 km; 2,303 miles)

DOUGLAS AIRCRAFT COMPANY (a Division of McDonnell Douglas Corporation)

HEADQUARTERS: 3855 Lakewood Boulevard, Long Beach, California 90846

Telephone: (213) 593-5511

CHAIRMAN: Sanford N. McDonnell

PRESIDENT:
John C. Brizendine

EXECUTIVE VICE-PRESIDENTS:
Charles M. Forsyth
Robert E. Hage (Marketing)

VICE-PRESIDENTS:
Ray E. Bates (Engineering Design and Development)
Harold Bayer (Product Support)
Howard W. Cleveland (Manufacturing)
Charles Conrad Jr (International Commercial Sales)
John E. Crosthwait (Government Marketing)
Edward Curtis (Contracts and Pricing)
Eugene F. Dubil (Engineering)
Joseph S. Dunning (Personnel and Administration)
Gilbert G. Fleming (Production)
Tom Gabbert (Fiscal Management)
William T. Gross (Programme Manager DC-10)
Robert C. P. Jackson (Planning)
John C. Londelius (Flight and Laboratory Development)
Marvin D. Marks (Programme Manager YC-15)
Jack W. Stillwell (Quality and Reliability Assurance)
Howell L. Walker (Commercial Sales, Americas)
William R. Worrell (Material)

CHIEF COUNSEL: John H. Carroll Jr

DIRECTOR, EXTERNAL RELATIONS: Raymond L. Towne

The Douglas Aircraft Company operates plants at Long Beach, Palmdale and Torrance, California.

MCDONNELL DOUGLAS DC-9

USAF designations: C-9A and VC-9C
US Navy designation: C-9B

Design study data on the DC-9, then known as the Douglas Model 2086, were released in 1962. Preliminary design work began during that year. Fabrication was started on 26 July 1963 and assembly of the first airframe began on 6 March 1964. It flew for the first time on 25 February 1965 and five DC-9s were flying by the end of June 1965. These aircraft were of the basic version now known as the DC-9 Series 10. The full range of DC-9 variants so far announced is as follows:

Series 10 Model 11. Initial version, powered by two 54·5 kN (12,250 lb st) Pratt & Whitney JT8D-5 turbofan engines. Max accommodation for 80 passengers at 86 cm (34 in) seat pitch, with normal facilities, or 90 passengers with reduced facilities. This version received FAA Type Approval on 23 November 1965 and entered scheduled service with Delta Air Lines on 8 December 1965. Production completed.

Series 10 Model 15. Generally similar to Srs 10 Model 11 but with 62·3 kN (14,000 lb st) JT8D-1 turbofan engines, increased fuel capacity and increased all-up weight. Production completed.

Series 20. For operation in hot climate/high-altitude conditions, combining long-span wings of Series 30 with short fuselage of Series 10. Up to 90 passengers. Two 64·5 kN (14,500 lb st) JT8D-9 turbofans. The Series 20 flew for the first time on 18 September 1968, and was certificated on 11 December 1968. The first Series 20 was delivered to SAS on the same day and entered commercial service on 23 January 1969. Production completed.

Series 30. Developed version, initially with 62·3 kN (14,000 lb st) JT8D-7s, increased wing span, longer fuselage accommodating up to 105 passengers (normal) or 115 (with reduced facilities), and new high-lift devices including full-span leading-edge slats and double-slotted flaps. First Srs 30 flew for first time on 1 August 1966. First delivery, to Eastern Air Lines, was made on 27 January 1967 and scheduled services began on 1 February 1967.

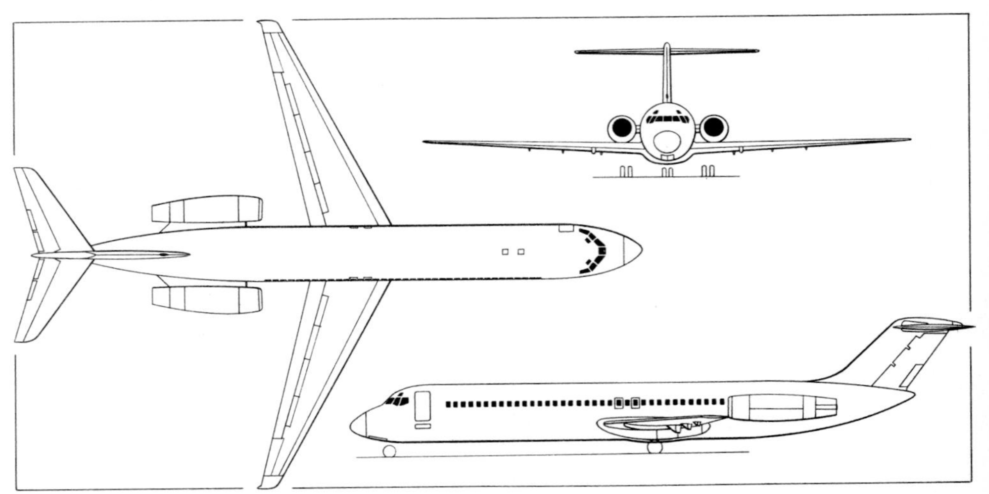

McDonnell Douglas DC-9 Srs 50, the latest 'stretched' version of this twin-turbofan transport *(Pilot Press)*

Engine options available include JT8D-9 of 64·5 kN (14,500 lb st); JT8D-11 of 66·7 kN (15,000 lb st); JT8D-15 of 69 kN (15,500 lb st); and JT8D-17 of 71·2 kN (16,000 lb st). All engines have sound-treated nacelles that comply with FAA FAR Pt 36 noise regulations.

Series 40. As Series 30, but with 64·5 kN (14,500 lb st) JT8D-9, 69 kN (15,500 lb st) JT8D-15 or 71·2 kN (16,000 lb st) JT8D-17 turbofans, increased fuel capacity, longer fuselage accommodating up to 125 passengers, and greater AUW. First flight was made on 28 November 1967 and FAA certification was received on 27 February 1968. The first Series 40 was delivered to SAS on 29 February 1968 and entered commercial service with that airline on 12 March 1968.

Series 50. 'Stretched' short/medium-range development of the Series 30, announced on 5 July 1973. High-density seating for up to 139 passengers made possible by a 4·34 m (14 ft 3 in) fuselage extension. A 'new look' interior features enclosed overhead racks for carry-on baggage, sculptured wall panels, acoustically-treated ceiling panels and indirect lighting. Available with either Pratt & Whitney JT8D-15 or -17 turbofan engines, rated at 69 kN (15,500 lb st) and 71·2 kN (16,000 lb st) respectively, and embodying sound-absorption materials as developed for the engines and nacelles of the DC-10, the Series 50 meets FAR Pt 36 noise requirements. The engines are smokeless and have thrust reversers rotated 17° from the vertical to reduce the possibility of exhaust gas ingestion. The landing gear is fitted with an improved anti-skid braking system. First flight of a Series 50 was made on 17 December 1974. First deliveries were made to Swissair, with whom it entered service in August 1975.

Current versions are offered in passenger, cargo (**DC-9F**), convertible (**DC-9CF**) or passenger-cargo (**DC-9RC**) configurations. The cargo and convertible models have a main cabin cargo door measuring 3·45 m (11 ft 4 in) wide and 2·06 m (6 ft 9 in) high. An executive transport version is also offered, with increased fuel, enabling up to 15 persons to be carried nonstop over 2,865 nm (5,300 km; 3,300 mile) transcontinental or transocean stages. First delivery of an all-cargo model, a DC-9 Srs 30F, was made to Alitalia on 13 May 1968. This model has 122·1 m³ (4,313 cu ft) of cargo space in main cabin, plus the underfloor hold, enabling it to carry eight full cargo pallets and two half-pallets with total weight of nearly 18,144 kg (40,000 lb).

There are also three military versions of the DC-9, as follows:

C-9A Nightingale. Aeromedical airlift transport, of which eight were ordered in 1967 for operation by the 375th Aeromedical Wing of the USAF Military Airlift Command. Essentially an 'off-the-shelf' DC-9 Srs 30 commercial transport, but with JT8D-9 engines, the C-9A is able to carry 30 to 40 litter patients, more than 40 ambulatory patients or a combination of the two, together with two nurses and three aeromedical technicians. The interior includes a special-care compartment, with separate atmospheric and ventilation controls. Galleys and toilets are provided fore and aft. There are three entrances, two with hydraulically-operated stairways. The third has a forward door 2·06 m (6 ft 9 in) high and 3·45 m (11 ft 4 in) wide, with a hydraulically-operated ramp, to facilitate loading of litters. First C-9A was rolled out on 17 June 1968 and was delivered to the US Air Force at Scott Air Force Base on 10 August 1968. Orders for the C-9A totalled 21, which were all delivered by February 1973.

C-9B Skytrain II. Fleet logistic support transport, of which five were ordered by the USN under a $25·3 million contract announced on 19 April 1972, increased subsequently to fourteen. Described separately.

VC-9C. DC-9-30 type aircraft with special configuration, ordered in December 1973 by the USAF for service in the Special Air Missions Wing based at Andrews AFB, Maryland, near Washington, DC. Three delivered in 1975.

Announced orders and leases for the commercial DC-9 up to mid-1976 were as follows:

Series 10	137
Series 20	10
Series 30	579
Series 40	57
Series 50	74
Leased	4

A total of 824 DC-9s (including 21 C-9As, 14 C-9Bs and 3 VC-9Cs) had been delivered by 1 July 1976.

In June 1966, the FAA certificated three Category 2 all-weather landing systems for the DC-9, comprising the Collins FD-108 flight director system, Sperry AD-200 flight director system, and coupled approach utilising the Sperry SP-50A autopilot.

The following structural details apply to the DC-9 Series 10:

TYPE: Twin-turbofan short/medium-range airliner.

WINGS: Cantilever low-wing monoplane. Mean thickness/chord ratio 11·6%. Sweepback 24° at quarter-chord. All-metal construction, with three spars inboard,

McDonnell Douglas DC-9 Series 50 in the insignia of Hawaiian Air

two spars outboard and spanwise stringers riveted to skin. Glassfibre trailing-edges on wings, ailerons and flaps. Hydraulically-controlled ailerons, each in two sections, outer sections used at low speed only. Wing-mounted speed brakes. Hydraulically-actuated double-slotted flaps over 67% of semi-span. (Leading-edge slats on Srs 20/30/40/50.) Single boundary-layer fence (vortillon) under each wing. Detachable wingtips. Thermal anti-icing of leading-edges.

FUSELAGE: Conventional all-metal semi-monocoque structure.

TAIL UNIT: Cantilever all-metal structure with hydraulically-actuated variable-incidence T-tailplane. Manually-controlled elevators with servo tabs. Hydraulically-controlled rudder with manual override. Glassfibre trailing-edges on control surfaces.

LANDING GEAR: Retractable tricycle type of Menasco manufacture, with steerable nosewheel. Hydraulic retraction, nose unit forward, main units inward. Twin Goodyear wheels on each unit. Main-wheel tyres size 40 × 14. Nosewheel tyres size 26 × 6·60. Goodyear brakes. Hydro-Aire Hytrol Mk II anti-skid units.

POWER PLANT: Two Pratt & Whitney JT8D turbofan engines (details given under individual model listings), mounted on sides of rear fuselage. Engines fitted with 40% target-type thrust reversers for ground operation only. Standard fuel capacity 10,546 litres (2,786 US gallons) in Srs 10 Model 11, 14,000 litres (3,700 US gallons) in Srs 10 Model 15, 13,925 litres (3,679 US gallons) in Srs 20, 30 and 40, 19,073 litres (5,038 US gallons) in Srs 50.

ACCOMMODATION (Srs 10): Crew of two on flight deck, plus cabin attendants. Accommodation in main cabin for 56-68 first class passengers four-abreast, or up to 90 tourist class five-abreast. Mixed class versions include one with 16 first class and 40 tourist seats. Fully pressurised and air-conditioned. Toilets at rear of cabin. Provision for galley. Passenger door at front of cabin on port side, with electrically-operated built-in airstairs. Optional ventral stairway. Servicing and emergency exit door opposite on starboard side. Underfloor freight and baggage holds, with forward door on starboard side, rear door on port side.

DIMENSIONS, EXTERNAL:
Wing span:
Srs 10 — 27·25 m (89 ft 5 in)
Srs 20, 30, 40, 50 — 28·47 m (93 ft 5 in)
Wing aspect ratio:
Srs 10 — 8·55
Srs 20, 30, 40, 50 — 8·71
Length overall:
Srs 10, 20 — 31·82 m (104 ft 4¾ in)
Srs 30 — 36·37 m (119 ft 3½ in)
Srs 40 — 38·28 m (125 ft 7¼ in)
Srs 50 — 40·72 m (133 ft 7¼ in)
Height overall:
Srs 10, 20, 30 — 8·38 m (27 ft 6 in)
Srs 40, 50 — 8·53 m (28 ft 0 in)
Tailplane span:
Srs 10, 50 — 11·23 m (36 ft 10¼ in)
Wheel track:
Srs 10, 20, 30, 40, 50 — 5·03 m (16 ft 6 in)
Wheelbase:
Srs 10, 20 — 13·32 m (43 ft 8½ in)
Srs 30 — 16·22 m (53 ft 2½ in)
Srs 40 — 17·10 m (56 ft 1¼ in)
Srs 50 — 18·56 m (60 ft 11 in)
Passenger door (port, fwd):
Height — 1·83 m (6 ft 0 in)
Width — 0·85 m (2 ft 9½ in)
Height to sill — 2·13 m (7 ft 2 in)
Servicing door (stbd, fwd):
Height — 1·22 m (4 ft 0 in)
Width — 0·69 m (2 ft 3 in)
Height to sill — 2·18 m (7 ft 2 in)
Freight and baggage hold doors:
Height — 1·27 m (4 ft 2 in)
Width:
fwd — 1·35 m (4 ft 5 in)
rear — 0·91 m (3 ft 0 in)
Height to sill — 1·07 m (3 ft 6 in)

DIMENSIONS, INTERNAL:
Cabin (Srs 10): Length — 16·99 m (55 ft 9 in)
Max width — 3·07 m (10 ft 1 in)
Floor width — 2·87 m (9 ft 5 in)
Max height — 2·06 m (6 ft 9 in)
Floor area — 47·4 m^2 (510 sq ft)
Volume — 97·7 m^3 (3,450 cu ft)
Carry-on baggage compartment:
Srs 10, 20 — 1·42 m^3 (50 cu ft)
Freight hold (underfloor):
Srs 10, 20 — 17·0 m^3 (600 cu ft)
Srs 30 — 25·3 m^3 (895 cu ft)
Srs 40 — 28·9 m^3 (1,019 cu ft)
Srs 50 — 29·3 m^3 (1,034 cu ft)

AREAS:
Wings, gross:
Srs 10 — 86·77 m^2 (934·3 sq ft)
Srs 20, 30, 40, 50 — 92·97 m^2 (1,000·7 sq ft)
Ailerons (total):
Srs 50 — 3·53 m^2 (38·0 sq ft)
Trailing-edge flaps (total):
Srs 50 — 19·58 m^2 (210·8 sq ft)
Leading-edge slats (total):
Srs 50 — 11·22 m^2 (120·8 sq ft)
Spoilers (total):
Srs 50 — 3·22 m^2 (34·7 sq ft)
Fin:
Srs 50 — 14·96 m^2 (161·0 sq ft)
Rudder:
Srs 50 — 6·07 m^2 (65·3 sq ft)
Tailplane:
Srs 10, 50 — 25·60 m^2 (275·6 sq ft)
Elevators, incl tabs:
Srs 50 — 9·83 m^2 (105·8 sq ft)

WEIGHTS AND LOADINGS:
Manufacturer's empty weight:
Srs 10 Model 11 — 20,550 kg (45,300 lb)
Srs 10 Model 15 — 22,235 kg (49,020 lb)
Srs 20 — 23,985 kg (52,880 lb)
Srs 30 — 25,940 kg (57,190 lb)
Srs 40 — 26,612 kg (58,670 lb)
Srs 50 — 28,068 kg (61,880 lb)
Max space-limited payload:
Srs 10 Model 11 — 8,188 kg (18,050 lb)
Srs 10 Model 15 — 9,698 kg (21,381 lb)
Srs 20 — 9,925 kg (21,885 lb)
Srs 30 — 14,118 kg (31,125 lb)
Srs 40 — 15,610 kg (34,415 lb)
Max weight-limited payload:
Srs 10 Model 15 — 9,325 kg (20,560 lb)
Srs 20 — 10,565 kg (23,295 lb)
Srs 30 — 12,743 kg (28,094 lb)
Srs 40 — 14,363 kg (31,665 lb)
Srs 50 — 15,617 kg (34,430 lb)
Max T-O weight:
Srs 10 Model 11 — 35,245 kg (77,700 lb)
Srs 10 Model 15 — 41,140 kg (90,700 lb)
Srs 20 — 44,450 kg (98,000 lb)
Srs 30, 40, 50 — 54,884 kg (121,000 lb)
Max ramp weight:
Srs 10 Model 11 — 35,605 kg (78,500 lb)
Srs 10 Model 15 — 41,500 kg (91,500 lb)
Srs 20, 30, 40, 50 — 55,338 kg (122,000 lb)
Max zero-fuel weight:
Srs 10 Model 11 — 30,120 kg (66,400 lb)
Srs 10 Model 15 — 32,386 kg (71,400 lb)
Srs 20 — 35,380 kg (78,000 lb)
Srs 30, 40, 50 — 44,678 kg (98,500 lb)
Max landing weight:
Srs 10 Model 11 — 33,565 kg (74,000 lb)
Srs 10 Model 15 — 37,060 kg (81,700 lb)
Srs 20 — 42,365 kg (93,400 lb)
Srs 30, 40, 50 — 49,895 kg (110,000 lb)
Max wing loading:
Srs 10 Model 11 — 406·2 kg/m^2 (83·2 lb/sq ft)
Max power loading:
Srs 10 Model 15 — 330·2 kg/kN (3·24 lb/lb st)

PERFORMANCE (at max T-O weight, except where indicated):
Never-exceed speed:
Srs 50 — 537 knots (994 km/h; 618 mph)
Max level speed:
Srs 50 — 500 knots (927 km/h; 576 mph)
Max cruising speed at 7,620 m (25,000 ft):
Srs 10 Model 11 and 15 and Srs 30: — 490 knots (907 km/h; 564 mph)
Srs 20 — 494 knots (915 km/h; 569 mph)
Srs 40, 50 — 485 knots (898 km/h; 558 mph)
Average long-range cruising speed at 9,145-10,675 m (30,000-35,000 ft) 443 knots (821 km/h; 510 mph)
Max rate of climb at S/L:
Srs 10 Model 11 — 1,097 m (3,600 ft)/min
Srs 20 — 1,035 m (3,400 ft)/min
Srs 30 — 885 m (2,900 ft)/min
Srs 40 — 865 m (2,850 ft)/min
Srs 50 — 792 m (2,600 ft)/min
FAA T-O field length:
Srs 10 Model 11 — 1,387 m (4,550 ft)
Srs 20 — 1,555 m (5,100 ft)
Srs 30 — 1,685 m (5,530 ft)
Srs 40 — 2,088 m (6,850 ft)
FAR T-O distance to 10·7 m (35 ft):
Srs 10 Model 15 — 1,970 m (6,470 ft)
Srs 20 — 1,495 m (4,900 ft)
Srs 30, 40 — 2,255 m (7,400 ft)
Srs 50 — 2,445 m (8,020 ft)
FAA landing field length:
Srs 10 — 1,535 m (5,030 ft)
Srs 20 — 1,355 m (4,450 ft)
Srs 30 — 1,425 m (4,680 ft)
Srs 40 — 1,440 m (4,720 ft)
Srs 50 — 1,485 m (4,880 ft)
FAR landing distance from 15 m (50 ft):
Srs 50 — 1,440 m (4,720 ft)
Range at Mach 0·8, with reserves for 200 nm (370 km; 230 mile) flight to alternate and 60 min hold at 3,050 m (10,000 ft):
Srs 10 Model 11 at 7,620 m (25,000 ft) with 50 passengers and baggage — 864 nm (1,601 km; 995 miles)
Srs 20 at 7,620 m (25,000 ft) with 50 passengers and baggage — 1,140 nm (2,111 km; 1,312 miles)
Srs 30 at 9,150 m (30,000 ft) with 64 passengers and baggage — 1,160 nm (2,148 km; 1,335 miles)
Srs 40 at 7,620 m (25,000 ft) with 70 passengers and baggage — 930 nm (1,723 km; 1,071 miles)
Range at long-range cruising speed at 9,150 m (30,000 ft), reserves for 200 nm (370 km; 230 mile) flight to alternate and 45 min continued cruise at 9,150 m (30,000 ft):
Srs 10 with 63 passengers and baggage — 1,590 nm (2,946 km; 1,831 miles)
Srs 20 with 63 passengers and baggage — 1,605 nm (2,974 km; 1,848 miles)
Srs 30 with 80 passengers and baggage — 1,670 nm (3,095 km; 1,923 miles)
Srs 40 with 87 passengers and baggage — 1,555 nm (2,880 km; 1,790 miles)
Srs 50 with 97 passengers and baggage — 1,795 nm (3,326 km; 2,067 miles)
Ferry range, reserves as above:
Srs 10 — 1,910 nm (3,539 km; 2,199 miles)
Srs 20 — 1,865 nm (3,455 km; 2,147 miles)
Srs 30 — 1,980 nm (3,669 km; 2,280 miles)
Srs 40 — 1,850 nm (3,428 km; 2,130 miles)
Srs 50 — 2,185 nm (4,049 km; 2,516 miles)

McDONNELL DOUGLAS C-9B SKYTRAIN II

The US Navy's C-9B Skytrain II is a special convertible passenger/cargo version of the DC-9 Series 30 commercial transport named after the long-enduring Navy R4D Skytrain, a DC-3 variant of which 624 were procured by that service.

The contract for five (increased subsequently to eight) C-9Bs was signed by Naval Air Systems Command on 24 April 1972, and the first of these aircraft made its initial flight on 7 February 1973, two months ahead of schedule. The first two aircraft were delivered on 8 May 1973, to Fleet Tactical Support Squadrons 1 (VR-1) at NAS Norfolk, Virginia, and 30 (VR-30) at NAS Alameda, California. All eight were delivered during 1973. A further six C-9Bs were ordered in late 1974, and delivery of these was

McDonnell Douglas C-9B Skytrain II (two Pratt & Whitney JT8D-9 turbofan engines)

completed by mid-1976.

A compromise between the DC-9 Series 30 and 40, the C-9B has the overall dimensions of the former, 64·5 kN (14,500 lb st) Pratt & Whitney JT8D-9 turbofan engines, and the optional 3·45 m (11 ft 4 in) by 2·06 m (6 ft 9 in) cargo door at the port forward end of the cabin. This allows loading of standard military pallets measuring 2·24 m (7 ft 4 in) by 2·74 m (9 ft 0 in), and in an all-cargo configuration eight of these can be accommodated, representing a total cargo load of 14,716 kg (32,444 lb). When loading, each pallet is first elevated to door sill height, and then rolled forward on to a ball transfer system before being positioned finally by means of roller tracks.

Normal flight crew consists of pilot, co-pilot, crew chief and two cabin attendants, and standard accommodation is for 90 passengers on five-abreast seating at 97 cm (38 in) pitch, or up to 107 passengers at 86 cm (34 in) pitch. In a typical passenger/cargo configuration, three pallets are carried in the forward area, with 45 passengers in the rear section. A galley and toilet are located at each end of the cabin. In all-cargo or mixed passenger/cargo configuration, a cargo barrier net can be erected at the forward end of the cabin; in the latter configuration a smoke barrier curtain is placed between the cargo section and the passengers.

Normal passenger access is by means of forward port and aft ventral doors, each with hydraulically-operated airstairs to make the C-9B independent of ground facilities. The ventral door allows passengers to board while cargo is being loaded in the forward area. Two Type III emergency exits, each 0·91 m (3 ft 0 in) by 0·51 m (1 ft 8 in), are positioned on each side of the fuselage to permit overwing escape, and four 25-man life rafts are carried in stowage racks. To complete the C-9B's independence of ground facilities, an auxiliary power unit provides both electrical and hydraulic services when the aircraft is on the ground. An environmental control system maintains a sea level cabin altitude to a height of 5,640 m (18,500 ft) and a 2,440 m (8,000 ft) cabin altitude to 10,670 m (35,000 ft).

A maximum fuel capacity of 22,443 litres (5,929 US gallons) provides a ferry range of 2,953 nm (5,472 km; 3,400 miles), the standard wing fuel tanks being supplemented by a 4,732 litre (1,250 US gallon) tank in the forward underfloor freight hold, and a 3,785 litre (1,000 US gallon) tank in the aft hold.

Advanced nav/com equipment is installed, including Omega and inertial navigation systems, and FAA certification has been received for both manual and automatic approaches under Category II weather conditions.

Dimensions, external:
As for DC-9 Series 30

Dimensions, internal:
Cabin: Length 20·73 m (68 ft 0 in)
Width 3·05 m (10 ft 0 in)
Volume (cargo) 118·9 m³ (4,200 cu ft)
Baggage holds (underfloor):
forward 8·44 m³ (298 cu ft)
aft 3·82 m³ (135 cu ft)

Weights:
Operating weight, empty:
passenger configuration 29,612 kg (65,283 lb)
cargo configuration 27,082 kg (59,706 lb)
Max ramp weight 50,350 kg (111,000 lb)
Max T-O weight 49,900 kg (110,000 lb)
Max landing weight 44,906 kg (99,000 lb)

Performance (at max T-O weight unless otherwise specified):
Max cruising speed 500 knots (927 km/h; 576 mph)
Long-range cruising speed 438 knots (811 km/h; 504 mph)
Military critical field length 2,259 m (7,410 ft)
Landing distance, at max landing weight 786 m (2,580 ft)
Range, long-range cruising speed at 9,145 m (30,000 ft) with 4,535 kg (10,000 lb) payload 2,538 nm (4,704 km; 2,923 miles)

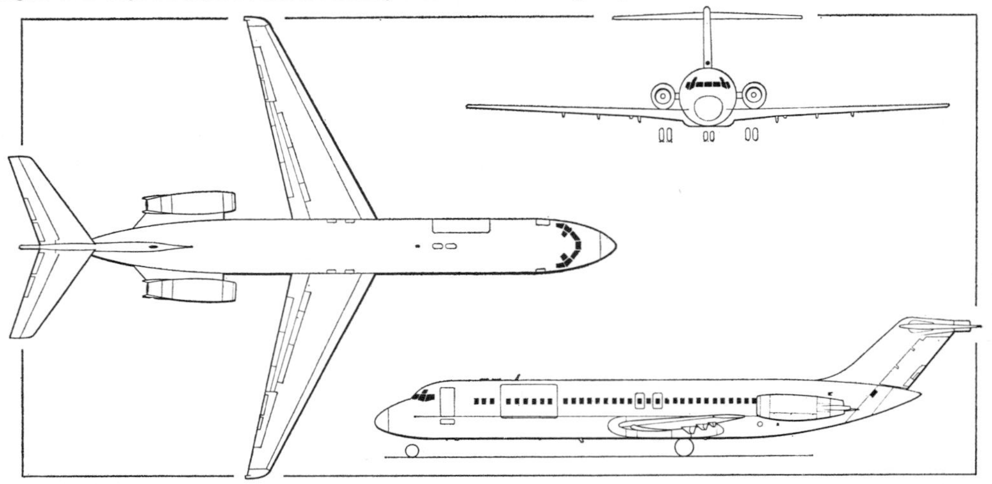

McDonnell Douglas C-9B Skytrain II convertible passenger/cargo transport, based on the DC-9 Srs 30 *(Michael A. Badrocke)*

MCDONNELL DOUGLAS DC-10

In April 1966, American Airlines circulated to seven airframe manufacturers a statement of the company's requirements, based on traffic forecasts. It appreciated that increasing airport congestion would be alleviated by the introduction of commercial transport aircraft of greater passenger-carrying capacity, but considered it essential that such aircraft should not be restricted to operation from those airports with very long runways. At that time, it visualised a twin-turbofan aircraft with dimensions and performance tailored specifically for operation from smaller airports. During the evolutionary period, the major change was a decision to use three instead of two turbofan engines.

The aircraft which McDonnell Douglas evolved to meet this specification was designated DC-10, an all-purpose commercial transport able to operate economically over ranges from 260 nm to 5,300 nm (480 to 9,815 km; 300 to 6,100 miles), according to Series, and able to carry 270 mixed class passengers, or a maximum of 380 passengers in an all-economy configuration.

It is being produced in three versions, as follows:

Series 10. Initial version, powered by three General Electric CF6-6D or -6D1 turbofan engines, each rated at 178 kN (40,000 lb st) or 182·4 kN (41,000 lb st) respectively. Intended for service on domestic routes of 260-3,125 nm (480-5,795 km; 300-3,600 miles). First ordered by American Airlines on 19 February 1968.

Manufacture of the first aircraft started on 6 January 1969, and assembly at the McDonnell Douglas plant at Long Beach, California, began in the Summer of 1969. The first DC-10 Srs 10 made its first flight on 29 August 1970.

On 29 July 1971 simultaneous deliveries of the first two McDonnell Douglas DC-10 Srs 10 commercial transport aircraft were made to American Airlines and United Air Lines at Long Beach, California. The FAA awarded the company a type certificate for, the Srs 10 on the same day.

First scheduled passenger flight of the DC-10 Srs 10 was made on 5 August 1971, when American Airlines began a daily DC-10 service between Los Angeles and Chicago.

Series 30. Extended-range version for inter-continental operations, powered by three General Electric CF6-50A or -50C turbofan engines, each rated at 218 kN (49,000 lb) and 227 kN (51,000 lb st) respectively with exhaust nozzles. Increased fuel capacity. Wing span increased by 3·05 m (10 ft 0 in). Landing gear supplemented by additional dual-wheel unit, mounted on the fuselage centreline between the four-wheel bogie main units. Version with Rolls-Royce RB.211-524 engines is available.

First flight of the Series 30 was made on 21 June 1972. FAA certification was granted on 21 November 1972, simultaneously with the first deliveries of production aircraft to KLM and Swissair.

Series 40. As Series 30, but power plant is three Pratt & Whitney JT9D-20 turbofan engines, each rated at 220 kN

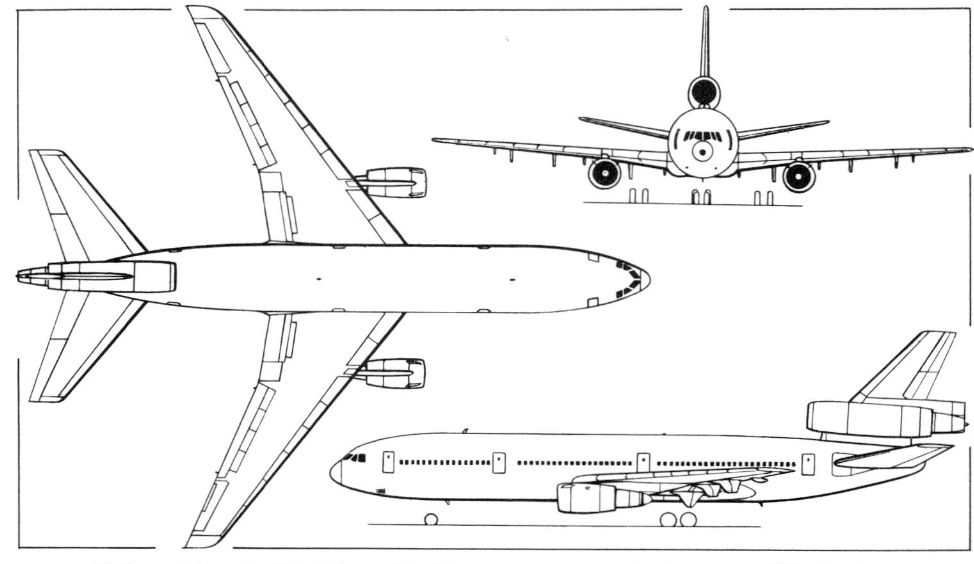

McDonnell Douglas DC-10 Series 30 high-capacity three-engined transport *(Pilot Press)*

McDonnell Douglas DC-10 Series 40 in the insignia of Japan Air Lines

(49,400 lb st) with water injection and exhaust nozzle installed. Japan Air Lines ordered six Srs 40s for delivery in 1976; these are equipped with JT9D-59A engines rated at 236 kN (53,000 lb st). First flight of a DC-10-40 with JT9D-59A engines was made on 25 July 1975.

First flight of the Series 40 (known originally as Series 20) was made on 28 February 1972. FAA certification was received on 20 October 1972, and the first delivery of a production aircraft, to Northwest Orient Airlines, was made on 10 November 1972.

There are also convertible cargo versions designated Model 10CF and 30CF, of which the latter is described separately. A tanker/cargo version has been proposed to the USAF.

By 1 July 1976 orders and options totalled 268 aircraft (239 orders, 29 options) for the following airlines; of these, 223 had been delivered.

Aeromexico	Series 30
Air Afrique	Series 30
Air New Zealand	Series 30
Air Siam	Series 30
Air Zaïre	Series 30
Alitalia	Series 30
American Airlines	Series 10
British Caledonian Airways	Series 30
Continental Air Lines	Series 10/10CF
Finnair	Series 30
Garuda Indonesian Airways	Series 30
Iberia	Series 30
Japan Air Lines	Series 40
KLM	Series 30
Korean Air Lines	Series 30
Laker Airways	Series 10
Lufthansa	Series 30
Malaysian Airline System	Series 30
Martinair Holland	Series 30CF
National Airlines	Series 10
National Airlines	Series 30
Northwest Orient Airlines	Series 40
Overseas National Airways	Series 30CF
Pakistan International Airlines	Series 30
Philippine Air Lines	Series 30
Sabena	Series 30CF
Scandinavian Airlines System	Series 30
Swissair	Series 30
Thai Airways International	Series 30
Trans International Airlines	Series 30CF
Turkish Airlines	Series 10
Union de Transports Aériens	Series 30
United Air Lines	Series 10
VARIG	Series 30
VIASA	Series 30
Western Airlines	Series 10

The DC-10 manufacturing plan calls for subassemblies and components to be brought together at Long Beach for final assembly. Certain major subassemblies are produced at other divisions of McDonnell Douglas, and Convair Division of General Dynamics Corporation at San Diego, California, is subcontractor for the fuselage, being responsible for five sections totalling 39·01 m (128 ft).

The 200th DC-10 was delivered, to National Airlines, on 20 June 1975; on 30 June 1976 the DC-10 carried its 100 millionth passenger.

TYPE: Three-turbofan commercial transport.

WINGS: Cantilever low-wing monoplane of all-metal fail-safe construction. Several different wing sections of Douglas design are used between wing root and tip. Thickness/chord ratio varies from slightly more than 12·2% at root to less than 8·4% at tip. Dihedral 5° 14·4′ inboard, 3° 1·8′ outboard. Incidence ranges from positive at wing root to negative at tip. Sweepback at quarter-chord 35°. All-metal inboard and outboard ailerons, the former used conventionally, the latter only when the leading-edge slats are extended. Double-slotted all-metal trailing-edge flaps mounted on external hinges, with an inboard and outboard flap panel on each wing. On Series 30 and 40 aircraft the inboard ailerons droop symmetrically with the flaps to a maximum of 13° 12′: their differential operation as ailerons is superimposed on top of their symmetrical deployment as flaps. Five all-metal spoiler panels on each wing, at the rear edge of the fixed wing structure, forward of the flaps. All spoilers operate in unison as lateral control, speed brake, direct lift control and ground spoilers. Full-span two-position all-metal leading-edge slats. Ailerons are powered by hydraulic actuators manufactured by Bertea Corporation, spoilers by hydraulic actuators manufactured by Parker-Hannifin Corporation. Each aileron is powered by either of two hydraulic systems; each spoiler is powered by a single system. All leading-edge slat segments outboard of the engines are anti-iced with engine bleed air.

FUSELAGE: Aluminium semi-monocoque fail-safe structure of circular cross-section. Except for auxiliary areas the entire fuselage is pressurised.

TAIL UNIT: Cantilever all-metal structure. Variable-incidence tailplane, actuated by Vickers hydraulic motors. Longitudinal and directional controls are fully powered and comprise inboard and outboard elevators, each segment powered by a Bertea tandem actuator; upper and lower rudder each powered by a Bertea actuator. Rudder standby power supplied by two transfer motor pumps manufactured by Abex Corporation.

LANDING GEAR (Srs 10): Hydraulically-retractable tricycle type, with gravity free-fall for emergency extension. Nosewheel unit retracts forward, main units inward into fuselage. Twin-wheel steerable nose unit. Main gear comprises two four-wheel bogies. Oleo-pneumatic shock-absorbers on all units. Goodyear nosewheels and tyres size 37 × 14-14, pressure 11·03 bars (160 lb/sq in). Goodyear main wheels and tyres size 50 × 20-20, pressure 11·72 bars (170 lb/sq in). Goodyear disc brakes and anti-skid system, with individual wheel control.

LANDING GEAR (Srs 30 and 40): These versions have an additional dual-wheel main unit mounted on the fuselage centreline between the four-wheel bogie units, and this retracts forward. Goodyear nosewheels and tyres size 40 × 15·5-16, pressure 12·41 bars (180 lb/sq in). Four-wheel bogie main units and centreline unit have Goodyear wheels and tyres size 52 × 20·5-23. The former have a pressure of 11·38 bars (165 lb/sq in), the latter 9·65 bars (140 lb/sq in). Otherwise as Srs 10.

POWER PLANT: Three turbofan engines (details under Series descriptions), two of which are mounted on underwing pylons, the third above the rear fuselage at the base of the fin. All engines are fitted with both fan and turbine reversers for ground operation. Engine air inlets have load-carrying acoustically-treated panels for noise attenuation, and each engine fan case and fan exhaust is similarly treated. Three integral wing fuel tanks with a total capacity of approximately 82,518 litres (21,800 US gallons). Four standard pressure refuelling adapters, two in each wing outboard of the engine pylons. Series 30 and 40 aircraft have four integral wing fuel tanks and an auxiliary tank in the wing centre-section with a connected structural compartment fitted with a bladder cell, giving increased total capacity of approximately 135,510 litres (35,800 US gallons). Oil capacity, Series 10 and 30: 34·1 litres (9 usable US gallons); Series 40: 56·8 litres (15 usable US gallons).

ACCOMMODATION: Crew of five (pilot, first officer, flight engineer, two observers) plus cabin attendants. Standard seating for 255 or 270 in mixed class versions, with a maximum of 380 passengers in an economy class arrangement. Two aisles run the length of the cabin, which is separated into sections by cloakroom dividers. In the first class section, with three pairs of reclining seats abreast, the aisles are 0·78 m (2 ft 7 in) wide. In the coach class section, four pairs of seats, with a table between the centre pairs, also have two aisles, these being 0·51 m (1 ft 8 in) wide. One pair of seats is exchanged for a three-seat unit in the nine-abreast high-density layout. Up to nine lavatories located throughout the passenger cabin. Cloakrooms of standard and elevating type distributed throughout the cabin. Cabin windows, 0·28 × 0·41 m (11 in × 16 in), are spaced at 0·51 m (20 in) centres. Overhead stowage modules, fully enclosed and providing stowage for passengers' personal effects, are located on the sidewalls and extend the full length of the cabin. Eight passenger doors, four on each side, open by sliding inward and upward into the above-ceiling area. Containerised or bulk cargo compartments located immediately forward and aft of the wing, with outward-opening doors on the starboard side. A bulk cargo compartment is located in the lower aft section of the fuselage, with its door on the port side. Entire accommodation is fully air-conditioned, with five separate control zones for the standard below-floor galley configuration. Series 30 and 40 aircraft have an optional main cabin galley to replace the lower galley, and in this configuration there are four separate control zones for the air-conditioning. The lower-deck galley is provided with eight high-temperature ovens, refrigerators, storage space for linen, china and other accessories. Serving carts are taken to cabin level by two electric elevators, to a buffet service centre, from where stewardesses serve passengers. To permit quick turnround at terminals, without interference to passenger movement in the main cabin, the kitchen is provisioned through the cargo doors at ground level.

SYSTEMS: Three parallel continuously-operating and completely separate hydraulic systems supply the fully-powered flight controls and wheel brakes. Two of these systems supply power for nosewheel steering. Normally, one of the systems supplies power for landing gear actuation; two reversible motor pumps deliver power from the other two systems for standby operation of landing gear. Each hydraulic system is powered by two identical engine-driven pumps, capable of delivering a total of 265 litres (70 US gallons)/min at 207 bars (3,000 lb/sq in) at take-off. An AiResearch TSCP-700-4 APU provides ground electrical and pneumatic power, including main engine starting, and auxiliary electric power in flight.

ELECTRONICS AND EQUIPMENT: A dual fail-operative landing system is installed to meet Category IIIA weather minima. Digital air data computer meeting ARINC 576 requirements on Srs 10. Triple inertial navigation system with dual area navigation on Srs 30. Triple inertial navigation system meeting ARINC 561 requirements on Srs 40.

DIMENSIONS, EXTERNAL:

Wing span:	
Series 10	47·34 m (155 ft 4 in)
Series 30, 40	50·41 m (165 ft 4·4 in)
Wing chord at root	10·71 m (35 ft 1·8 in)

McDonnell Douglas DC-10 Series 30 wide-bodied passenger transport in the insignia of KLM

Wing chord at tip:
Series 10 — 3·21 m (10 ft 6½ in)
Series 30, 40 — 2·73 m (8 ft 11½ in)
Wing aspect ratio:
Series 10 — 6·8
Series 30, 40 — 7·5
Length overall:
Series 10 — 55·30 m (181 ft 5 in)
Series 30, 40 — 55·50 m (182 ft 1 in)
Length of fuselage — 51·97 m (170 ft 6 in)
Height overall — 17·70 m (58 ft 1 in)
Tailplane span — 21·69 m (71 ft 2 in)
Wheel track — 10·67 m (35 ft 0 in)
Wheelbase:
Series 10, 40 — 22·07 m (72 ft 5 in)
Series 30 — 22·05 m (72 ft 4 in)

DIMENSIONS, INTERNAL:
Cabin: Length, from aft bulkhead of flight deck to aft cabin bulkhead — approx 41·45 m (136 ft 0 in)
Max width — 5·72 m (18 ft 9 in)
Height (basic) — 2·41 m (7 ft 11 in)
Series 10, 30, 40 in lower-galley configuration:
Forward baggage and/or freight hold (forward of wing):
Containerised volume — 27·2 m³ (960 cu ft)
Bulk volume — 38·9 m³ (1,375 cu ft)
Centre baggage and/or freight hold (aft of wing):
Containerised volume — 36·2 m³ (1,280 cu ft)
Bulk volume — 44·9 m³ (1,585 cu ft)
Aft hold:
Bulk volume — 22·8 m³ (805 cu ft)
Series 30, 40 in upper-galley configuration:
Forward baggage and/or freight hold (forward of wing):
Containerised volume — 72·5 m³ (2,560 cu ft)
Bulk volume — 86·2 m³ (3,045 cu ft)
Centre baggage and/or freight hold (aft of wing):
Containerised volume — 45·3 m³ (1,600 cu ft)
Bulk volume — 54·8 m³ (1,935 cu ft)
Aft hold:
Bulk volume — 14·4 m³ (510 cu ft)

AREAS:
Wings, gross:
Series 10 — 358·7 m² (3,861 sq ft)
Series 30, 40 — 367·7 m² (3,958 sq ft)
Ailerons, inboard (total) — 7·68 m² (82·7 sq ft)
Ailerons, outboard (total) — 9·76 m² (105·1 sq ft)
Trailing-edge flaps (total) — 62·1 m² (668·2 sq ft)
Leading-edge slats (total):
Series 10 — 42·05 m² (452·6 sq ft)
Series 30, 40 — 43·84 m² (471·9 sq ft)
Spoilers (total) — 12·73 m² (137·0 sq ft)
Fin — 45·92 m² (494·29 sq ft)
Rudders (total) — 10·29 m² (110·71 sq ft)
Tailplane — 96·6 m² (1,040·2 sq ft)
Elevators (total) — 27·7 m² (298·1 sq ft)

WEIGHTS AND LOADINGS:
Basic operating weight:
Series 10 — 107,274 kg (236,500 lb)
Series 30 — 118,590 kg (261,450 lb)
Series 40 — 120,678 kg (266,050 lb)
Max payload:
Series 10 — 44,678 kg (98,500 lb)
Series 30 — 48,330 kg (106,550 lb)
Series 40 — 46,243 kg (101,950 lb)
Max T-O weight:
Series 10 — 199,580 kg (440,000 lb)
Series 30 — 256,280 kg (565,000 lb)
Series 40 (-20 engines) — 251,744 kg (555,000 lb)
Series 40 (-59A engines) — 259,450 kg (572,000 lb)
Max ramp weight:
Series 10 — 200,940 kg (443,000 lb)
Series 30 — 257,640 kg (568,000 lb)
Series 40 (-20 engines) — 253,105 kg (558,000 lb)
Series 40 (-59A engines) — 260,815 kg (575,000 lb)
Max zero-fuel weight:
Series 10 — 151,953 kg (335,000 lb)
Series 30, 40 — 166,922 kg (368,000 lb)
Max landing weight:
Series 10 — 164,880 kg (363,500 lb)
Series 30, 40 — 182,798 kg (403,000 lb)
Max wing loading:
Series 10 — 605·4 kg/m² (124 lb/sq ft)
Series 30 — 756·3 kg/m² (154·9 lb/sq ft)
Series 40 (-20 engines) — 743·1 kg/m² (152·2 lb/sq ft)
Series 40 (-59A engines) — 765·6 kg/m² (156·8 lb/sq ft)

PERFORMANCE (at max T-O weight unless specified):
Never-exceed speed — Mach 0·95
Max level speed at 7,620 m (25,000 ft) — Mach 0·88 (530 knots; 982 km/h; 610 mph)
Max cruising speed at 9,145 m (30,000 ft):
Series 10 (-6D engines) — 499 knots (925 km/h; 575 mph)
Series 10 (-6D1 engines) — 501 knots (928 km/h; 577 mph)
Series 30 — 490 knots (908 km/h; 564 mph)
Series 40 (-20 engines) — 489 knots (906 km/h; 563 mph)
Series 40 (-59A engines) — 498 knots (922 km/h; 573 mph)
T-O speed (V_2):
Series 10 (-6D engines) — 168 knots (311 km/h; 193 mph)
Series 10 (-6D1 engines) — 163 knots (302 km/h; 188 mph)
Series 30 (-50A engines) at 251,744 kg (555,000 lb) AUW — 181 knots (335 km/h; 208 mph)
Series 30 (-50C engines) at 256,280 kg (565,000 lb) AUW — 179 knots (332 km/h; 206 mph)
Series 40 (-20 engines) — 183·5 knots (340 km/h; 211 mph)
Series 40 (-59A engines) — 180·6 knots (334 km/h; 208 mph)
Landing speed at max landing weight:
Series 10 — 136 knots (252 km/h; 157 mph)
Series 30 — 145 knots (269 km/h; 167 mph)
Series 40 (-20 engines) — 146·5 knots (271 km/h; 168·5 mph)
Series 40 (-59A engines) — 147 knots (272 km/h; 169 mph)
Max rate of climb at S/L:
Series 10 (-6D engines) — 817 m (2,680 ft)/min
Series 10 (-6D1 engines) — 838 m (2,750 ft)/min
Series 30 — 884 m (2,900 ft)/min
Series 40 (-20 engines) — 829 m (2,720 ft)/min
Series 40 (-59A engines) — 762 m (2,500 ft)/min
Service ceiling:
Series 10, -6D engines, at 192,775 kg (425,000 lb) AUW — 10,605 m (34,800 ft)
Series 10, -6D1 engines, at 192,775 kg (425,000 lb) AUW — 10,730 m (35,200 ft)
Series 30, at 249,475 kg (550,000 lb) AUW — 10,180 m (33,400 ft)
Series 40 (-20 engines) at 242,670 kg (535,000 lb) AUW — 9,660 m (31,700 ft)
Series 40 (-59A engines) — 9,965 m (32,700 ft)
En-route climb altitude, one engine out:
Series 10 at 195,045 kg (430,000 lb) AUW — 4,145 m (13,600 ft)
Series 30 at 251,744 kg (555,000 lb) AUW — 4,360 m (14,300 ft)
Series 40 (-20 engines) at 247,205 kg (545,000 lb) AUW — 3,565 m (11,700 ft)
Series 40 (-59A engines) at 254,010 kg (560,000 lb) AUW — 5,135 m (16,850 ft)
FAR T-O distance to 10·7 m (35 ft):
Series 10 (-6D engines) — 2,987 m (9,800 ft)
Series 10 (-6D1 engines) — 2,743 m (9,000 ft)
Series 30 (-50A engines) — 3,840 m (12,600 ft)
Series 30 (-50C engines) — 3,383 m (11,100 ft)
Series 40 (-20 engines) — 3,734 m (12,250 ft)
Series 40 (-59A engines) — 3,261 m (10,700 ft)
FAR landing distance from 15 m (50 ft) at max landing weight:
Series 10 — 1,777 m (5,830 ft)
Series 30 — 1,817 m (5,960 ft)
Series 40 — 1,835 m (6,020 ft)
Range with max fuel:
Series 10 — 5,150 nm (9,543 km; 5,930 miles)
Series 30 — 6,250 nm (11,580 km; 7,197 miles)
Series 40 (-20 engines) — 5,850 nm (10,840 km; 6,736 miles)
Series 40 (-59A engines) — 6,100 nm (11,305 km; 7,024 miles)
Range with max payload:
Series 10 — 2,350 nm (4,355 km; 2,706 miles)
Series 30 — 4,000 nm (7,413 km; 4,606 miles)
Series 40 (-20 engines) — 3,500 nm (6,485 km; 4,030 miles)
Series 40 (-59A engines) — 4,050 nm (7,505 km; 4,663 miles)

MCDONNELL DOUGLAS DC-10 SERIES 30CF

The Series 30CF is a convertible freighter version of the McDonnell Douglas DC-10 transport. Generally similar to the basic DC-10 Series 30 and 40, it is designed for easy conversion to either passenger or cargo configuration. Its payload can consist of 380 passengers and baggage or 64,860 kg (143,000 lb) of cargo over full intercontinental range; or up to 70,626 kg (155,700 lb) of cargo on domestic transcontinental routes.

The Series 30CF is powered by three General Electric CF-50A turbofans. It flew for the first time on 28 February 1973 and initial deliveries were made to Overseas National Airways and Trans International Airlines on 17 April 1973.

In the passenger configuration, interior layout is generally similar to that of the DC-10, but the Series 30CF was designed to permit overnight conversion to an all-cargo configuration. This entails removal of seats, overhead baggage racks, forward food service centre, cloakrooms and carpeting from the main cabin, and installation of freight loading tracks and rollers, a cargo tie-down system and restraint nets. Coffee service fixtures and lavatories in the aft cabin may also be removed but are retained normally for regular cargo flights.

The cargo loading system for the Series 30CF is based on that in use in the DC-8 Super Sixty Series freighters. A two-channel network of roller conveyors, adjustable guide rails and pallet restraint fittings is installed in the seat tracks in the cabin floor, by use of simple stud and locking pin devices. A 2·59 m × 3·56 m (8 ft 6 in × 11 ft 8 in) cargo door in the side of the fuselage swings upward and allows easy loading of bulky freight.

A total of 30 standard 2·24 m × 2·74 m (7 ft 4 in × 9 ft) cargo pallets, or 22 larger pallets measuring 2·24 m × 3·18 m (7 ft 4 in × 10 ft 5 in) or 2·44 m × 3·05 m (8 ft × 10 ft), can be accommodated in the main cabin. The Series 30CF also has 86·08 m³ (3,040 cu ft) of cargo space in its two lower baggage compartments for bulk freight or 16 half-size or 8 full-size pallets. The entire cargo loading and restraint system can be stowed in this lower hold, and is thus available for conversion of the aircraft to cargo configuration at any airport.

A DC-10 Series 30CF delivered to Sabena Belgian World Airlines in 1973 is certificated for carrying combination loads of freight and passengers in the main cabin. Other DC-10 series aircraft are also offered in convertible versions. Continental Air Lines is operating the DC-10 Series 10CF.

MCDONNELL DOUGLAS YC-15

The McDonnell Douglas contender for the USAF's AMST prototype fly-off programme, the requirement for which is described under The Boeing Company's entry

McDonnell Douglas YC-15 prototype seen during cargo-loading demonstration

(which see), has the designation YC-15. Two prototypes were ordered under an $85·9 million contract. The first of these (01875) was rolled out on 5 August 1975 and made its first flight on 26 August, three months ahead of schedule. The second prototype joined the flight development programme on 5 December 1975.

The McDonnell Douglas YC-15 differs considerably from the Boeing YC-14: powered by four turbofan engines, it represents a different aerodynamic approach to the STOL requirement.

It was announced on 18 August 1976 that the two YC-15 AMST prototypes had been returned to the company's facilities at Long Beach, California, at the termination of the first phase of their flight test and evaluation programme at Edwards AFB. During this phase of the programme the two prototypes had completed a combined total of 226 flights involving 473·2 flight hours.

The effectiveness of the high-lift blown flap system has been shown to limit short-take-off and landing distances to well within the specification requirement of 610 m (2,000 ft). The joint USAF/McDonnell Douglas test team has completed landing gear tests on soft fields, in-flight drops of military loads and of anthropomorphic dummies, night flights, airborne tactical manoeuvres, formation flights and evaluation of performance, stability and control. Cargo drops have included parachuting of pallet loads weighing up to 9,072 kg (20,000 lb) from an altitude of 610 m (2,000 ft), and low-altitude parachute extraction of 4,536 kg (10,000 lb) loads.

While at Long Beach the first prototype is to be fitted with an alternative wing of increased span (40·41 m; 132 ft 7 in), and one of its JT8D-17 engines is being removed and replaced by a General Electric/SNECMA CFM56 turbofan. The second prototype will have one of its engines replaced by a refanned Pratt & Whitney JT8D-209. When these changes have been completed, the test programme is scheduled for resumption during the first quarter of 1977.

McDonnell Douglas believes that there will also be a commercial requirement for an aircraft in this category, and expects that the YC-15 design will be commercially acceptable without significant changes. At the termination of the fly-off programme, one of the YC-15 prototypes is expected to be made available to the company for development and evaluation in a commercial role.

Type: Advanced military STOL transport.

Wings: Cantilever high-wing monoplane. All-metal structure. Sweepback 5° 54′. Lateral control provided by a combination of aileron and triple inboard spoilers on each wing. For STOL landings the spoilers are used also as direct-lift controls, speed brakes and ground lift spoilers. Wide-span two-section titanium trailing-edge flaps. The engines, mounted on pylons extending forward from the wing leading-edge, are positioned so that the exhaust nozzles are close to the undersurface of the leading-edge. This provides a high-velocity airflow which can be used to blow externally the two-segment flaps.

Fuselage: Conventional semi-monocoque all-metal structure, the protoype utilising the flight deck enclosure of the DC-10.

Tail Unit: Cantilever all-metal structure, with T-tail and swept vertical surfaces.

Landing Gear: Retractable tricycle type. Twin wheels on nose unit. Each main unit comprises a four-wheel bogie, made up of twin-wheel units in tandem. Long-stroke main units to allow for high sink rates.

McDonnell Douglas YC-15 prototype advanced military STOL transport

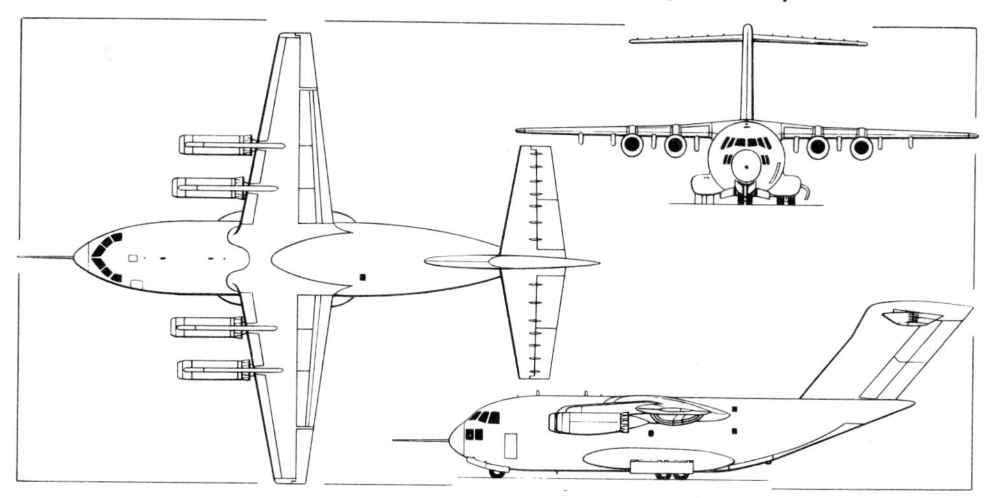

McDonnell Douglas YC-15 (four Pratt & Whitney JT8D-17 turbofan engines) *(Pilot Press)*

Power Plant: Four Pratt & Whitney JT8D-17 turbofan engines, each of 71·2 kN (16,000 lb st). The engines are fitted with nozzles that mix cool ambient air with the hot core exhaust gases, reducing the outflow temperature to a level that does not require the use of special materials for the wing structure. Total fuel capacity 30,396 litres (8,030 US gallons). Provision for in-flight refuelling on second prototype.

Accommodation: Flight deck layout allows for operation by a crew of two. Main cabin for about 150 fully-equipped troops. Passenger doors on each side of fuselage. Cargo loading ramp in undersurface of rear fuselage. Second prototype has soundproofed cabin.

System: Fully-powered control system, boosted by a stability and control augmentation system.

Dimensions, external:

Wing span	33·63 m (110 ft 4 in)
Length overall	37·87 m (124 ft 3 in)
Height overall	13·21 m (43 ft 4 in)
Fuselage width	5·49 m (18 ft 0 in)
Tailplane span	17·07 m (56 ft 0 in)
Wheel track	6·05 m (19 ft 10 in)
Wheelbase	12·17 m (39 ft 11 in)

DIMENSIONS, INTERNAL:
Cargo compartment:
Length 42·98 m (47 ft 0 in)
Max width 3·56 m (11 ft 8 in)
Max height 3·45 m (11 ft 4 in)
Volume, excl ramp 175·96 m³ (6,214 cu ft)

AREA:
Wings, gross 161·66 m² (1,740 sq ft)

WEIGHTS (estimated):
Max T-O weight and design gross weight 98,284 kg (216,680 lb)
Max weight-limited payload 28,122 kg (62,000 lb)
Design landing weight 68,040 kg (150,000 lb)

PERFORMANCE:
Max level speed 434 knots (805 km/h; 500 mph)
Approach speed 85 knots (157·5 km/h; 98 mph)
T-O field length with payload of 12,247 kg (27,000 lb) 610 m (2,000 ft)
Landing field length at design landing weight 610 m (2,000 ft)
Design operational radius, with 12,247 kg (27,000 lb) payload and 610 m (2,000 ft) mid-point field length, or 28,122 kg (62,000 lb) payload and runway of conventional length 400 nm (742 km; 461 miles)
Design ferry range 2,600 nm (4,818 km; 2,994 miles)

MCDONNELL DOUGLAS SKYHAWK

US Navy designation: A-4

Designed originally to provide the US Navy and Marine Corps with a simple low-cost lightweight attack and ground suppport aircraft, the Skyhawk was based on experience gained during the Korean War. Since the initial requirement called for operation by the US Navy, special design consideration was given to providing low-speed control and stability during take-off and landing, added strength for catapult launch and arrested landings, and dimensions that would permit it to negotiate standard aircraft carrier lifts without the complexity of folding wings.

Construction of the XA-4A (originally XA4D-1) prototype Skyhawk began in September 1953 and the first flight of this aircraft, powered by a Wright J65-W-2 engine (32 kN; 7,200 lb st), took place on 22 June 1954.

Early Skyhawk versions included the A-4A, -4B, -4C and -4E, of which 1,845 examples were built. These were described in the 1973-74 *Jane's*; more recent versions are as follows:

TA-4E. Original designation of prototypes of TA-4F.

A-4F. Attack bomber with J52-P-8A turbojet (41·4 kN; 9,300 lb st), new lift-spoilers on wings to shorten landing run by up to 305 m (1,000 ft), nosewheel steering, low-pressure tyres, zero-zero ejection seat, additional bullet- and flak-resistant materials to protect pilot, updated electronics contained in fairing 'hump' aft of cockpit. Prototype flew for the first time on 31 August 1966. Deliveries to US Navy began on 20 June 1967, and were completed in 1968. 146 built.

TA-4F. Tandem two-seat dual-control trainer version of A-4F for US Navy. Fuselage extended 0·71 m (2 ft 4 in), fuselage fuel tankage reduced to 379 litres (100 US gallons), Pratt & Whitney J52-P-6 or -8A engine optional, Douglas Escapac rocket ejection seats. Provision to carry full range of weapons available for A-4F. Reduced electronics.First prototype flew on 30 June 1965. Deliveries to the US Navy began in May 1966.

A-4G. Similar to A-4F for Royal Australian Navy. Equipped to carry Sidewinder air-to-air missiles. First of eight delivered on 26 July 1967.

TA-4G. Similar to TA-4F for Royal Australian Navy. First of two delivered on 26 July 1967.

A-4H. Designation of version supplied to Israel. Delivery of an initial batch of 48 in 1967-68, followed by 60 more by early 1972.

TA-4H. Tandem two-seat trainer version of the A-4H for Israel. Ten delivered.

TA-4J. Tandem two-seat trainer, basically a simplified version of the TA-4F. Ordered for US Naval Air Advanced Training Command, under $26,834,000 contract, followed by further contract in mid-1971. Deletion of the following equipment, although provisions retained: radar, dead reckoning navigation system, low-altitude bombing system, air-to-ground missile systems, weapons delivery computer and automatic release, intervalometer, gun pod, standard stores pylons, in-flight refuelling system and spray tank provisions. Addition and relocation of certain instruments. J52-P-6 engine standard. Provision for J52-P-8A engine and combat electronics. Prototype flew in May 1969 and the first four were delivered to the US Navy on 6 June 1969. In production.

A-4K. Similar to A-4F, for Royal New Zealand Air Force. Different radio, and braking parachute. First of ten handed over to the RNZAF on 16 January 1970.

TA-4K. Similar to TA-4F, for Royal New Zealand Air Force. The first of four was handed over to the RNZAF on 16 January 1970.

TA-4KU. Designation of six aircraft, similar to the TA-4F, for Kuwait Air Force.

A-4L. Modification of A-4C with uprated engine, bombing computing system and electronics relocated in fairing 'hump' aft of cockpit as on A-4F. Delivery to US Navy Reserve carrier air wing began in December 1969.

A-4M Skyhawk II. Similar to A-4F, but with J52-P-408 turbojet (50 kN; 11,200 lb st) and braking parachute standard, making possible combat operation from 1,220 m (4,000 ft) fields and claimed to increase combat effectiveness by 30%. Larger windscreen and canopy; windscreen bullet-resistant. Increased ammunition capacity for 20 mm cannon. More powerful generator, provision of wind-driven backup generator and self-contained engine starter. First of two prototypes flew for the first time on 10 April 1970. About 50 ordered initially for US Marine Corps, the first of which was delivered on 3 November 1970. Further order was placed subsequently, and the FY 1976 budget includes $70 million for the procurement of a final 24 aircraft. Funds have been allocated also for the installation of improved electronic warfare equipment in service aircraft, and for the continued development of an Angle Rate Bombing System (ARBS) for future installation in A-4Ms (see A-4Y). In production for USMC and Kuwait (36).

A-4N Skyhawk II. Light attack version ordered by US Navy for export. Basically similar to A-4M. First flown on 8 June 1972. In production.

A-4P. Revised A-4B for Argentine Air Force (50).

A-4Q. Revised A-4B for Argentine Navy (16).

A-4S. Designation of 40 Skyhawks for service with Singapore Air Defence Command. Conversion from ex-USN A-4Bs began in 1973, carried out by Lockheed Aircraft Service Company, under which heading all available details are given.

TA-4S. Three two-seat A-4B conversions for Singapore, by Lockheed Aircraft Service Company (which see).

A-4Y. USMC A-4M with updated HUD, redesigned cockpit and Hughes Angle Rate Bombing System (ARBS). All A-4Ms to be modified. Final procurement probably will be new-build A-4Ys.

Current US Navy planning calls for continued production of the Skyhawk into 1977, and logistic support for its continued usage into the 1980s.

The following structural description refers specifically to the A-4M:

TYPE: Single-seat attack bomber.

WINGS: Cantilever low-wing monoplane. Sweepback 33° at quarter-chord. All-metal three-spar structure. Spars machined from solid plate in one piece tip-to-tip. One-piece wing skins. Hydraulically-powered all-metal ailerons, with servo trim tab in port aileron. All-metal split flaps. Automatic leading-edge slats with fences. Hydraulically-actuated lift spoilers above flaps.

FUSELAGE: All-metal semi-monocoque structure in two sections. Rear section removable for engine servicing. Outward-hinged hydraulically-actuated airbrake on each side of rear fuselage. Detachable nose over communications and navigation equipment. Integral flak-resistant armour in cockpit area, with internal armour plate below and forward of cockpit.

TAIL UNIT: Cantilever all-metal structure. Electrically-actuated variable-incidence tailplane. Hydraulically-powered elevators. Powered rudder with unique central skin and external stiffeners.

LANDING GEAR: Hydraulically-retractable tricycle type, with single wheel on each unit. All units retract forward. Free-fall emergency extension. Main legs pre-shorten for retraction and wheels turn through 90° to stow horizontally in wings. Menasco shock-absorbers. Hydraulic nosewheel steering. Ribbon-type braking parachute of 4·88 m (16 ft) diameter contained in canister secured in rear fuselage below engine exhaust. Arrester hook for carrier operation.

POWER PLANT: One 50 kN (11,200 lb st) Pratt & Whitney J52-P-408 turbojet engine. Fuel in integral wing tanks and self-sealing fuselage tank aft of cockpit, total capacity 3,028 litres (800 US gallons). One 568, 1,136 or 1,514 litre (150, 300 or 400 US gallon) auxiliary tank can be carried on the underfuselage bomb-rack, and one 150 or 300 US gallon auxiliary tank on each of the inboard underwing racks. Maximum fuel capacity, internal plus auxiliary tanks, 6,814 litres (1,800 US gallons). Large flight refuelling probe on starboard side of nose. Douglas-developed self-contained flight refuelling unit can be carried on the underfuselage standard bomb shackles. Provisions for JATO.

ACCOMMODATION: Pilot on Douglas Escapac 1-G3 zero-zero lightweight ejection seat. Enlarged cockpit enclosure to improve pilot's view, with rectangular bullet-resistant windscreen.

SYSTEMS: Dual hydraulic system. Oxygen system. Electrical system powered by 20kVA generator, with wind-driven generator to provide emergency power.

McDonnell Douglas TA-4J tandem two-seat training version of the Skyhawk

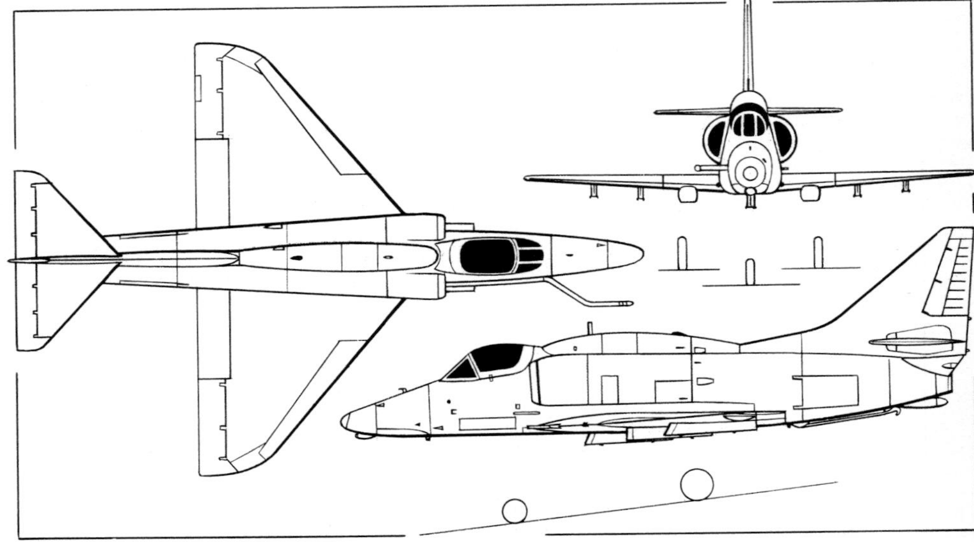
McDonnell Douglas A-4M Skyhawk II single-seat light attack aircraft *(Pilot Press)*

ELECTRONICS: Include Bendix Automatic Flight Control, ARC-159 UHF radio transceiver, ARA-50 UHF direction finder, APX-72 IFF, Marconi-Elliott AVQ-24 head-up display system, Douglas angle of attack indicator, electronic countermeasures, ASN-41 nav computer, APN-153(V) radar nav, ARC-114 VHF/FM radio transceiver, ARR-69 auxiliary radio receiver, ARN-84 Tacan and APN-194 radar altimeter.

ARMAMENT: Provision for several hundred variations of military load, carried externally on one underfuselage rack, capacity 1,588 kg (3,500 lb); two inboard underwing racks, capacity of each 1,020 kg (2,250 lb); and two outboard underwing racks, capacity of each 450 kg (1,000 lb). Weapons that can be deployed include nuclear or HE bombs, air-to-surface and air-to-air rockets, Sidewinder infra-red missiles, Bullpup air-to-surface missiles, ground attack gun pods, torpedoes, countermeasures equipment, etc. Two 20 mm Mk 12 cannon in wing roots standard, each with 200 rounds of ammunition. DEFA 30 mm cannon available as optional on international versions, with 150 rounds of ammunition per gun.

DIMENSIONS, EXTERNAL:
- Wing span: 8·38 m (27 ft 6 in)
- Wing chord at root: 4·72 m (15 ft 6 in)
- Length overall (excl flight refuelling probe):
  - A-4M: 12·29 m (40 ft 4 in)
  - TA-4F: 12·98 m (42 ft 7¼ in)
- Height overall:
  - A-4M: 4·57 m (15 ft 0 in)
  - TA-4F: 4·66 m (15 ft 3 in)
- Tailplane span: 3·45 m (11 ft 4 in)
- Wheel track: 2·37 m (7 ft 9½ in)

AREAS:
- Wings, gross: 24·16 m² (260 sq ft)
- Vertical tail surfaces (total): 4·65 m² (50 sq ft)
- Horizontal tail surfaces (total): 4·54 m² (48·85 sq ft)

McDonnell Douglas A-4M Skyhawk II attack aircraft of the US Marine Corps

WEIGHTS:
- Weight empty:
  - A-4F: 4,581 kg (10,100 lb)
  - TA-4F: 4,853 kg (10,700 lb)
  - A-4M: 4,899 kg (10,800 lb)
- Normal T-O weight:
  - A-4F, M, TA-4F: 11,113 kg (24,500 lb)
  - *A-4F from land base: 12,437 kg (27,420 lb)

**export version only: overload condition not authorised by US Navy*

PERFORMANCE (at combat weight):
- Max level speed:
  - TA-4F: 568 knots (1,052 km/h; 654 mph)
- Max level speed (with 1,814 kg; 4,000 lb bomb load):
  - A-4F: 548 knots (1,015 km/h; 631 mph)
  - A-4M: 561 knots (1,040 km/h; 646 mph)
- Max rate of climb (ISA at S/L):
  - A-4F: 2,440 m (8,000 ft)/min
  - A-4M: 3,140 m (10,300 ft)/min
- Rate of climb (ISA at 7,620 m; 25,000 ft):
  - A-4F: 1,097 m (3,600 ft)/min
  - A-4M: 1,463 m (4,800 ft)/min
- T-O run (at 10,433 kg; 23,000 lb T-O weight):
  - A-4F: 1,030 m (3,380 ft)
  - A-4M: 832 m (2,730 ft)
- Max ferry range, A-4M at 11,113 kg (24,500 lb) T-O weight with max fuel, standard reserves: 1,740 nm (3,225 km; 2,000 miles)

McKINNON

McKINNON ENTERPRISES INC

HEAD OFFICE AND WORKS: 12960 Southeast Ten Eyck Road, Sandy, Oregon 97055

Telephone: (503) 668-4154

PRESIDENT: A. G. McKinnon

OWNER AND MANAGER: A. G. McKinnon

McKinnon Enterprises (formerly McKinnon-Hickman Company) entered the aircraft conversion field in 1953 when it undertook the conversion of the Grumman Widgeon twin-engined light amphibian into an executive aircraft. The success of this conversion, which is still being manufactured, led to the development and manufacture of a larger four-engined amphibian, known as the McKinnon G-21, which is a much improved and more luxurious conversion of the Grumman Goose.

Details of the four-engined McKinnon G-21 Goose conversion can be found in the 1966-67 *Jane's*. It has since been superseded by the turboprop-powered G-21C, D and G, as described here. McKinnon is also offering a minimum conversion scheme by which the standard Goose can be re-engined with turboprops and fitted with any other parts of the G-21C/D conversion specified by the customer.

McKinnon has received official approval for a modification scheme to fit retractable wingtip floats to standard Goose amphibians. This offers increased cruising speeds, reduced landing speed, better stability on both land and water, and greatly improved accessibility for loading and unloading on water.

Also offered by McKinnon is an officially-approved conversion kit by which the max T-O weight of the standard Goose can be increased by 545 kg (1,200 lb) to 4,173 kg (9,200 lb).

McKINNON G-21C and G-21D TURBO-GOOSE

In these current versions of the G-21, the two normal 335·5 kW (450 hp) Pratt & Whitney R-985 radial engines of the Goose amphibian are replaced by two 507 kW (680 shp) Pratt & Whitney Aircraft of Canada PT6A-27 turboprop engines, moved further inboard and driving constant-speed and reversible-pitch propellers. The internal fuel capacity is increased. Other modifications include the fitting of retractable wingtip floats, extended radar nose, dorsal fin, one-piece windscreen, larger cabin windows, a new instrument panel, oxygen system, and a 24V electrical system. Landing gear and wingtip float retraction and flaps are all electrically operated.

The fully-modified G-21C and D will operate with ease from 610 m (2,000 ft) fields or small lakes, at even the highest altitudes.

The G-21C is fitted out to accommodate from nine to twelve people, including pilot. It received FAA Type Approval in February 1967.

TYPE: Twin-turboprop light amphibian.

WINGS: Cantilever high-wing monoplane. Wing section NACA 23000. Dihedral 2° 30′. All-metal structure with metal covering. Fabric-covered metal ailerons.

FUSELAGE: All-metal semi-monocoque flying-boat hull with two steps.

McKinnon G-21 Turboprop Goose of the Bureau of Land Management (two Pratt & Whitney Aircraft of Canada PT6A engines) *(Norman E. Taylor)*

TAIL UNIT: Braced all-metal structure.

LANDING GEAR: Retractable tailwheel type. All wheels retract electrically into hull, with manual extension. Bendix oleo-pneumatic shock-absorbers. Goodyear wheels and double-disc brakes. Retractable wingtip stabilising floats.

POWER PLANT: Two 507 kW (680 shp) Pratt & Whitney Aircraft of Canada PT6A-27 turboprop engines, driving three-blade constant-speed reversible-pitch and fully-feathering propellers. Fuel tanks in wings, total capacity 2,218 litres (586 US gallons).

ACCOMMODATION: Pilot and up to 11 passengers in standard version. Bow-loading entrance and baggage space in nose. Main cabin, forward of the standard rear door, seats seven people, with four more in a cabin aft of the door. One baggage compartment, capacity 136 kg (300 lb).

DIMENSIONS, EXTERNAL:
- Wing span: 15·49 m (50 ft 10 in)
- Wing chord at root: 3·05 m (10 ft 0 in)
- Wing chord at tip: 1·52 m (5 ft 0 in)
- Wing aspect ratio: 6·101
- Length overall: 12·07 m (39 ft 7 in)
- Width of hull: 1·52 m (5 ft 0 in)
- Tailplane span: 6·02 m (19 ft 9 in)
- Wheel track: 2·29 m (7 ft 6 in)
- Wheelbase: 5·23 m (17 ft 2 in)

AREAS:
- Wings, gross: 34·44 m² (377·64 sq ft)
- Ailerons (total): 2·75 m² (29·64 sq ft)
- Fin: 1·97 m² (21·20 sq ft)
- Rudder: 2·49 m² (26·80 sq ft)
- Tailplane: 3·67 m² (39·48 sq ft)
- Elevators: 3·99 m² (42·92 sq ft)
- Elevator tab: 0·195 m² (2·10 sq ft)

WEIGHTS:
- Weight empty, equipped: 3,009 kg (6,635 lb)
- Max T-O weight: 5,670 kg (12,500 lb)
- Max landing weight, on land: 5,445 kg (12,000 lb)
- Max landing weight, on water: 5,670 kg (12,500 lb)

PERFORMANCE (at max T-O weight):
- Max level speed: 191 knots (355 km/h; 220 mph)
- Range with max fuel: approx 1,390 nm (2,575 km; 1,600 miles)

McKINNON G-21G TURBO-GOOSE

This version of the Turbo-Goose is an 8/12-seat conversion of the standard Grumman G-21A. The power plant comprises two 507 kW (680 shp) Pratt & Whitney Aircraft of Canada PT6A-27 turboprop engines, driving Hartzell three-blade metal constant-speed fully-feathering reversible-pitch propellers.

Modifications to the airframe include a 0·38 m (15 in) nose extension to accommodate radar, metallising treatment of the wings and provision of a wraparound windscreen, retractable wingtip floats, rotating beacon on top of the fin, a small dorsal fin, hull vents, and auxiliary wing tanks that increase total fuel capacity to 2,218 litres (586 US gallons). Optional improvements include provision of picture windows for the cabin, a centre main fuel tank of increased capacity, dual landing lights in wing leading-edges, electrically-operated retraction of landing gear and enlargement of the cabin by removing the bulkhead at station 26.

McKinnon has received FAA approval for this conversion. All available specification details follow:

WEIGHTS:
- Weight empty, equipped (approx): 3,039 kg (6,700 lb)
- Max T-O weight: 5,670 kg (12,500 lb)

Performance (at max T-O weight):
Max operating speed
211 knots (391 km/h; 243 mph)
Service ceiling 6,100 m (20,000 ft)
Service ceiling, one engine out 3,660 m (12,000 ft)
Range with 2,218 litres (586 US gallons) fuel
1,390 nm (2,575 km; 1,600 miles)

McKINNON TURBOPROP GOOSE CONVERSION

For owners of Goose amphibians who do not require a full conversion of their aircraft to G-21C, D or G standard, McKinnon offers a simple conversion which involves only replacement of the original R-985 piston engines with two 507 kW (680 shp) Pratt & Whitney Aircraft of Canada PT6A-27 turboprop engines in the original location, driving three-blade constant-speed reversible-pitch propellers.

Any of the other modifications incorporated on the G-21C/D can be made during this conversion. Speed and take-off performance are comparable with those of the G-21C/D. Range is also comparable after fitment of the optional auxiliary tanks.

Weights (minimum conversion):
Weight empty, equipped 3,009 kg (6,635 lb)
Max T-O weight 5,670 kg (12,500 lb)
Max landing weight, on land or water
5,445 kg (12,000 lb)

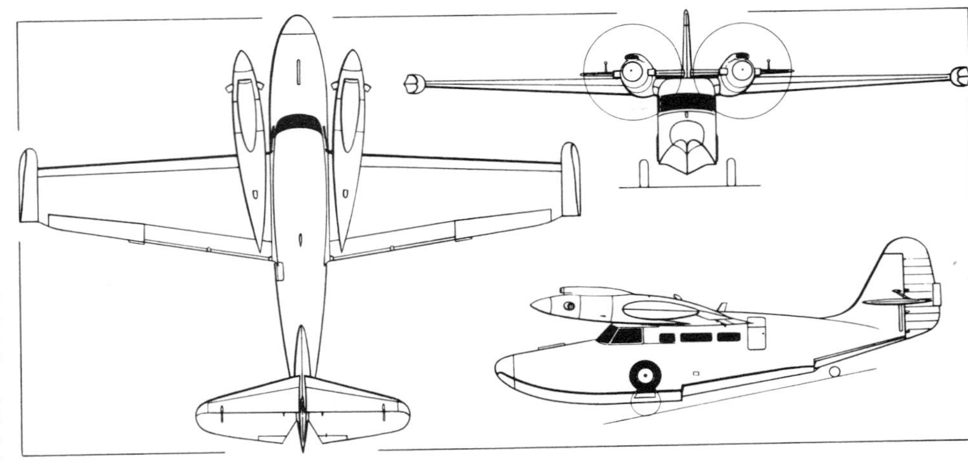

McKinnon G-21G Turbo-Goose, an 8/12-seat conversion of the Grumman G-21A *(Pilot Press)*

McKINNON SUPER WIDGEON

The Super Widgeon is an executive conversion of the Grumman Widgeon light amphibian, with the two original 149 kW (200 hp) Ranger six-cylinder in-line inverted engines replaced by two 201 kW (270 hp) Lycoming GO-480-B1D flat-six engines driving Hartzell three-blade fully-feathering propellers. Modifications to the hull and landing gear permit an increase in loaded weight. Extra tanks are provided in the outer wings to increase the fuel capacity from 408 to 582 litres (108 to 154 US gallons). Other new features include picture windows, a modern IFR instrument panel, improved soundproofing and the provision of an emergency escape hatch. Approval to install retractable floats was obtained in 1960.

The cabin is arranged to accommodate a pilot, co-pilot and three or four passengers.

Well over 70 Widgeons have been converted to Super Widgeon standard by McKinnon, and several retractable float installations have been completed.

Dimensions, external:
Wing span 12·19 m (40 ft 0 in)
Length overall 9·47 m (31 ft 1 in)

McKinnon Super Widgeon with retractable wingtip floats

Height overall 3·48 m (11 ft 5 in)
Weight:
Max T-O weight 2,500 kg (5,500 lb)
Performance (at max T-O weight):
Max level speed at S/L
165 knots (306 km/h; 190 mph)
Cruising speed at 3,050 m (10,000 ft) (62½% power)
156 knots (290 km/h; 180 mph)
Cruising speed at S/L (70% power)
152 knots (282 km/h; 175 mph)
Landing speed 54 knots (100 km/h; 62 mph)
Max rate of climb at S/L 534 m (1,750 ft)/min
Service ceiling 5,490 m (18,000 ft)
Service ceiling, one engine out 1,525 m (5,000 ft)
T-O run on land 183 m (600 ft)
T-O from smooth water 10 sec
Range with max fuel, 30 min reserves
868 nm (1,600 km; 1,000 miles)

MITSUBISHI
MITSUBISHI AIRCRAFT INTERNATIONAL INC

Works: PO Box 3848, San Angelo, Texas 76901
Telephone: (915) 944-1511
President: Wataru Ikeda

This wholly-owned subsidiary of the well-known Japanese Mitsubishi Heavy Industries Ltd was established at San Angelo in 1967, to assemble and equip the MU-2 twin-turboprop utility STOL transport designed by the parent company.

Wings, fuselage and tail unit components are manufactured in Japan and shipped to the USA. At San Angelo they go back on to an assembly production line, where American-built components which include engines, electronics, tyres, brakes and interiors are installed. A description of current versions of the MU-2 can be found in the Japanese aircraft section of this edition.

Mitsubishi Aircraft International in San Angelo has worldwide marketing responsibility for the MU-2. Aircraft of this type produced at San Angelo are in service in Africa, Canada, Europe, Mexico, the Middle East, South America and in the USA.

MOHAWK
MOHAWK AIR SERVICES

Address:
c/o Allegheny Airlines Inc, Washington National Airport, Hangar 12, Washington, DC 20001
President: Walter J. Short

Mohawk Air Services, a subsidiary of Allegheny Airlines, was formed to organise the updating of the fleet of Nord 262s serving with Allegheny's commuter operators. Mohawk, in turn, contracted with Frakes Aviation Inc to be responsible for modification of the aircraft and completion of the certification programme. The resulting aircraft has the designation Mohawk 298, reflecting its certification in accordance with FAR Part 298, which is applicable to aircraft with accommodation for up to 30 passengers, or with a payload not exceeding 3,400 kg (7,500 lb).

The prototype conversion flew for the first time on 7 January 1975, and following certification it was planned to convert 15 Nord 262s to Mohawk 298 standard by the Summer of 1976.

MOHAWK 298

Modifications carried out by Frakes Aviation to the basic Nord 262, last described in the 1969-70 *Jane's,* include the fitting of new wingtips to improve low-speed manoeuvrability, take-off performance and rate of climb; installation of a Hamilton Standard air-conditioning system; a completely new Collins solid-state electronics system which includes communications and navigation transceivers, autopilot and ground proximity warning; a new electrical system; a new unheated windscreen; and a new cabin layout to accommodate 28 passengers, with provision of a rear toilet and lockers for hand baggage. Improved standards of performance result from the installation of two 875·5 kW (1,174 ehp) Pratt & Whitney Aircraft of Canada PT6A-45 free-turbine turboprop engines, each driving a Hartzell five-blade metal low-speed fully-feathering propeller. The modifications result in a weight saving of 227 kg (500 lb), by comparison with the Nord 262.

Mohawk Air Services Mohawk 298, a 28-passenger conversion of the Nord 262 *(Howard Levy)*

Weights:
Weight empty 6,778 kg (14,943 lb)
Max T-O weight 10,600 kg (23,370 lb)
Performance (at max T-O weight):
Max cruising speed at 3,050 m (10,000 ft)
214 knots (397 km/h; 246 mph)
Operational ceiling 6,000 m (19,640 ft)
T-O run 995 m (3,260 ft)
Range with 26 passengers and baggage, FAA standard reserves 515 nm (954 km; 593 miles)

MOONEY

MOONEY AIRCRAFT CORPORATION

HEAD OFFICE AND WORKS:
PO Box 72, Kerrville, Texas 78028
Telephone: (512) 257-4043
PRESIDENT: J. C. Vaverck
VICE-PRESIDENTS:
Don K. Cox (Marketing)
L. P. Lopresti (Engineering)
T. J. Smith (Operations)
D. R. White (Finance)

The original Mooney Aircraft Inc was formed in June 1948, in Wichita, Kansas, from where the single-seat Model M-18 Mooney Mite was produced until 1952. The company transferred to Kerrville, Texas, in 1953 and completed a merger with Alon Inc of McPherson, Kansas, in October 1967. Subsequently, in late 1969, Butler Aviation International Inc and American Electronic Laboratories Inc entered into an agreement whereby the former company acquired 100 per cent stock ownership of Mooney Aircraft, the company name being changed to Aerostar Aircraft Corporation on 1 July 1970. Production of Aerostar aircraft was suspended in early 1972, at which time a number of companies were discussing the purchase of Aerostar Aircraft Corporation. On 4 October 1973 came news that the Republic Steel Corporation of Cleveland, Ohio, had assumed control of the company, once again named Mooney Aircraft.

Details of the company's products follow:

MOONEY RANGER

The prototype Ranger (known initially as Mark 21) flew on 23 September 1961 and the first production model on 7 November 1961. FAA Type Approval was received on 7 November 1961.

The Ranger has a Positive Control (PC) system which co-ordinates yaw/roll stability. Developed in association with Brittain Industries, PC employs a sensor in the form of a tilted-axis rate gyro. Any deviation in roll or yaw causes the gyro to emit a mechanical signal to a master vacuum control valve. This valve is connected to the standard vacuum system of the engine and by lines to servo cans. One vacuum servo is located at each aileron for roll control and two in the tailcone to activate the rudder for yaw control. Interconnected control linkages, combined with the sensitivity of the rate gyro, bring pressure on both ailerons and rudder to produce co-ordinated corrective action and restore the aircraft to straight and level flight. PC can be overridden by normal control pressures and can be disengaged by depressing a button on the control yoke.

TYPE: Four-seat cabin monoplane.

WINGS: Cantilever low-wing monoplane. Wing section NACA 63_2-215 at root, NACA 64_1-412 at tip. Dihedral 5° 30′. Incidence 2° 30′ at root, 1° at tip. Sweepforward 2° 29′. Light alloy structure with flush-riveted stretch-formed wraparound skins. Full-span main spar; rear spar terminates at mid-span of flaps. Sealed-gap differentially-operated light alloy ailerons. Electrically-operated single-slotted light alloy flaps over 70% of trailing-edge. No tabs.

FUSELAGE: Composite all-metal structure. Cabin section is of welded 4130 chrome-molybdenum steel tube with sheet light alloy covering. Rear section is of semi-monocoque construction, with sheet light alloy bulkheads and skin and extruded alloy stringers.

TAIL UNIT: Cantilever light alloy structure, with variable-incidence tailplane. All surfaces covered with wraparound metal skin.

LANDING GEAR: Electrically-retractable levered-suspension tricycle type. Nosewheel retracts rearward, main units inward into wings. Rubber disc shock-absorbers on main units. Delco hydraulic shock-absorber on nose unit. Cleveland main wheels, size 6·00-6, and steerable nosewheel, size 5·00-5. Tyre pressure (all units) 2·07 bars (30 lb/sq in). Cleveland hydraulic single-disc brakes on main wheels. Parking brakes.

POWER PLANT: One 134 kW (180 hp) Lycoming O-360-A1D flat-four engine, driving a Hartzell HC-C2YK-1/7662-2 two-blade metal constant-speed propeller. Two integral fuel tanks, with total capacity of 197 litres (52 US gallons), in wing roots. Flush refuelling point above each tank. Oil capacity 7·5 litres (2 US gallons).

ACCOMMODATION: Cabin accommodates four in two pairs of individual seats, front pair with dual controls. Rear pair of seats reclinable. Standard rudder pedals optionally removable to allow more leg room for passenger. Overhead ventilation system. Cabin heating and cooling system, with adjustable outlets and illuminated control. One-piece wraparound windscreen. Tinted Plexiglas windows. Starboard front and both rear seats removable for freight stowage. Single door on starboard side. Compartment for 54 kg (120 lb) baggage behind cabin, with access from cabin or through door on starboard side. Windscreen defrosting system standard.

SYSTEMS: Hydraulic system for brakes only. Electrical system includes 60A alternator, 12V 35Ah battery, voltage regulator and warning light, together with protective circuit breakers.

ELECTRONICS: An extensive range of optional equipment is available to customer's requirements, manufactured by King and Narco.

EQUIPMENT: Standard equipment includes many de luxe features as well as basic instruments, sensitive altimeter, streamlined spinner, alternate static source, full flow oil filter, quick oil drain, annunciator lights, navigation lights, landing/taxi light, wingtip strobe lights, cabin light, map light, heated pitot, sun visors, glareshield, zinc chromate anti-corrosion treatment and dual controls. Optional equipment includes rotating beacon, all-leather interior trim, auxiliary power socket, curtains, dual brakes, exhaust gas temperature gauge, set of headrests, emergency locator transmitter, inertia-reel shoulder harness, cabin fire extinguisher, hour meter and altitude encoder.

DIMENSIONS, EXTERNAL:
Wing span 10·67 m (35 ft 0 in)
Wing chord, mean 1·45 m (4 ft 9¼ in)
Wing aspect ratio 7·338
Length overall 7·34 m (24 ft 1 in)
Height overall 2·59 m (8 ft 6 in)
Tailplane span 3·55 m (11 ft 8 in)
Wheel track 2·76 m (9 ft 0¾ in)
Wheelbase 1·68 m (5 ft 6½ in)
Propeller diameter 1·88 m (6 ft 2 in)
Cabin door:
Height 0·95 m (3 ft 1¼ in)
Width 0·78 m (2 ft 6½ in)
Height to sill 0·34 m (1 ft 1½ in)
Baggage compartment door:
Height 0·61 m (2 ft 0 in)
Width 0·48 m (1 ft 7 in)

DIMENSIONS, INTERNAL:
Cabin: Length 2·64 m (8 ft 8 in)
Max width 1·04 m (3 ft 4½ in)
Max height 1·13 m (3 ft 8½ in)
Baggage compartment 0·38 m^3 (13·5 cu ft)

AREAS:
Wings, gross 15·51 m^2 (167·00 sq ft)
Ailerons (total) 1·03 m^2 (11·05 sq ft)
Trailing-edge flaps (total) 1·62 m^2 (17·48 sq ft)
Fin 0·73 m^2 (7·88 sq ft)
Rudder 0·46 m^2 (5·01 sq ft)
Tailplane 2·00 m^2 (21·50 sq ft)
Elevators 1·11 m^2 (12·02 sq ft)

WEIGHTS AND LOADINGS:
Weight empty 691 kg (1,525 lb)
Max T-O and landing weight 1,168 kg (2,575 lb)
Max wing loading 75·2 kg/m^2 (15·4 lb/sq ft)
Max power loading 8·72 kg/kW (14·3 lb/hp)

PERFORMANCE:
Max level speed at S/L 153 knots (283 km/h; 176 mph)
Stalling speed (flaps and wheels down, power off) 50 knots (92 km/h; 57 mph) IAS
Rate of climb at S/L 262 m (860 ft)/min
Service ceiling 5,743 m (19,500 ft)
T-O run, zero wind, ISA 248 m (815 ft)
Landing run, zero wind, ISA 181 m (595 ft)
Range, with allowance for taxi, climb and 45 min reserves 713 nm (1,322 km; 822 miles)

Mooney Chaparral four-seat light aircraft (Lycoming IO-360-A1A engine)

MOONEY CHAPARRAL

The Mooney Chaparral is an updated version of the Super-21, which flew for the first time in July 1963. It is generally similar to the current Mooney Ranger, but is fitted with a more powerful engine. It has a Positive Control system as described for the Ranger.

The description of the Ranger applies also to the Chaparral except in the following details:

POWER PLANT: One 149 kW (200 hp) Lycoming IO-360-A1A flat-four engine, driving a Hartzell two-blade metal constant-speed propeller. Fuel injection, tuned induction manifold, exhaust gas temperature gauge and ram-air boost. Other details as for Ranger.

WEIGHTS AND LOADINGS:
As for Ranger, except:
Weight empty 725 kg (1,600 lb)
Max power loading 7·84 kg/kW (12·9 lb/hp)

PERFORMANCE:
Max level speed at S/L 165 knots (306 km/h; 190 mph)
Stalling speed (flaps and wheels down, power off) 50 knots (92 km/h; 57 mph) IAS
Max rate of climb at S/L 343 m (1,125 ft)/min
Service ceiling 6,460 m (21,200 ft)
T-O run, zero wind, ISA 232 m (760 ft)
Landing run, zero wind, ISA 181 m (595 ft)
Range, allowance for taxi, climb and 45 min reserves 693 nm (1,284 km; 798 miles)

MOONEY EXECUTIVE

This member of the Mooney family of aircraft is basically similar to the Ranger. It differs in having a 149 kW (200 hp) Lycoming fuel-injection engine, a longer fuselage, providing more leg room and baggage stowage, and additional cabin windows. Differences compared with the Ranger are as follows:

POWER PLANT: One 149 kW (200 hp) Lycoming IO-360-A1A flat-four engine, driving a Hartzell two-blade metal constant-speed propeller. Fuel injection, tuned induction manifold, exhaust gas temperature gauge and ram-air power boost. Total fuel capacity 242 litres (64 US gallons).

DIMENSIONS, EXTERNAL:
As for Ranger, except:
Length overall 7·59 m (24 ft 11 in)

DIMENSIONS, INTERNAL:
As for Ranger, except:
Cabin: Length 2·90 m (9 ft 6 in)

Mooney Executive four-seat cabin monoplane (Lycoming IO-360-A1A engine)

WEIGHTS AND LOADING:
- Weight empty 743 kg (1,640 lb)
- Max baggage 54 kg (120 lb)
- Max T-O weight 1,243 kg (2,740 lb)
- Max wing loading 80·1 kg/m² (16·4 lb/sq ft)

PERFORMANCE:
- Max level speed 161 knots (298 km/h; 185 mph)
- Stalling speed (flaps and wheels down, power off) 54 knots (100 km/h; 62 mph) IAS
- Max rate of climb at S/L 322 m (1,055 ft)/min
- Service ceiling 5,730 m (18,800 ft)
- T-O run, zero wind, ISA 268 m (879 ft)
- Landing run, zero wind, ISA 239 m (785 ft)
- Range, with allowance for taxi, climb and 45 min reserves 848 nm (1,572 km; 977 miles)

MOONEY 201

The Mooney 201, a faster development of the Executive, was introduced in 1976. The improved performance is achieved by aerodynamic refinements, which include a more compact engine cowling, a redesigned windscreen of greater area and more streamlined contour, and double main landing gear doors. The interior has also been improved by the use of thicker carpeting and fabrics, and new optional interior trims are available. Power plant is the 149 kW (200 hp) Lycoming IO-360-A1B6D.

DIMENSIONS, EXTERNAL:
- As for Ranger, except:
- Length overall 7·52 m (24 ft 8 in)

WEIGHTS AND LOADING:
- As for Executive

PERFORMANCE (at max T-O weight):
- Max cruising speed (75% power) 169 knots (314 km/h; 195 mph)
- Range at max cruising speed, no reserves 1,032 nm (1,913 km; 1,189 miles)

NAL
NATIONAL AIRCRAFT LEASING LTD

HEAD OFFICE:
Tiger International Building, 1888 Century Park East, Los Angeles, California 90067
Telephone: (213) 552-6311
Telex: 696198

National Aircraft Leasing, a member of the Tiger Leasing Group, has purchased from American Airlines the remainder of the BAC One-Eleven Series 401AK aircraft operated formerly by that airline. (A description of the basic aircraft can be found in the UK Aircraft section.) A total of 18 aircraft are involved; these have been, or are being, modified into luxurious corporate aircraft, and are being offered to major corporations under direct sale or flexible lease terms.

National Aircraft Leasing One-Eleven luxury corporate aircraft conversion

NAL ONE-ELEVEN

As part of the purchase agreement, American Airlines is responsible for giving each aircraft a major overhaul before delivery to NAL. This includes complete overhaul of the landing gear, with replacement of wheels, brakes, tyres and anti-skid system; of the pneumatic system, with replacement of cold air units, ducting and valves; of the APU and APU generator; and of the powered flying controls and flap system; as well as bench check and overhaul or replacement of all radios, electronics and autopilots, and inspection and reconditioning of fuel tanks as necessary. Additionally, X-ray inspection and weighing is carried out. Both the hot and cold sections of the Rolls-Royce Spey turbofan engines receive a complete overhaul, during which all recommended modifications are embodied. Following completion of this work, the aircraft are flight tested before being ferried to the modification centre at San Antonio, Texas.

At this point in the programme the One-Elevens are virtually 'new' aircraft, with all time or cycle controlled components replaced or overhauled to provide 7,000 hours TBO, and with approximately 30,000 hours of airframe life expectancy.

The Dee Howard Company, under contract to NAL, is responsible for converting these aircraft to a luxury corporate configuration, after which they are designated NAL One-Elevens. The work includes installation of a NAL-designed long-range fuel system in the underfloor freight holds; wiring for a dual inertial navigation system and dual HF radios; and a de luxe interior in colours and fabrics of the customer's choice. This comprises 11 individual fully reclining and adjustable 360° swivelling chairs; one four-seat and two three-seat couches, the latter converting to provide sleeping accommodation; two lavatories and a separate dressing room; an executive desk complete with telephone, calculator and provision for an electric typewriter; a centrally located bar complete with lighting and cabinet storage; a fully-equipped all-electric galley, with lighting and ventilation isolated from the main cabin; cabin overhead lighting, table lamps and reading lights; four-place conference table with provision for slide or film projector; video tape system; and an 8-track stereo system. The aircraft's exterior paintwork is finished in colours of the customer's choice.

The long-range fuel installation, consisting of seven tanks, provides an additional 4,618 litres (1,220 US gallons) capacity, to give a total usable fuel supply of 14,965 kg (32,994 lb) and so making possible a nonstop range of approximately 3,259 nm (6,040 km; 3,753 miles). Following installation of these tanks, 3·88 m³ (137 cu ft) of underfloor baggage space remains. An optional long-range fuel system, consisting of ten tanks, provides an additional 6,056 litres (1,600 US gallons) capacity, to give a total usable fuel supply of 16,126 kg (35,553 lb), making possible a maximum nonstop range of approximately 3,431 nm (6,357 km; 3,950 miles).

NAL's complete One-Eleven programme includes crew training, airframe and engine support and a full maintenance programme.

On 11 September 1975, NAL obtained FAA Supplemental Type Certification for en-route and terminal area navigation (RNAV) based on the installation of dual Delco Carousel IVa Inertial Navigation Systems (INS). DME update of the INS allows the NAL One-Eleven to use direct RNAV routes, thus saving time and fuel.

WEIGHTS:
- Manufacturer's empty weight 23,050 kg (50,822 lb)
- Operating weight empty 23,505 kg (51,822 lb)
- Max T-O weight 40,142 kg (88,500 lb)
- Max ramp weight 40,370 kg (89,000 lb)
- Max zero-fuel weight 31,070 kg (68,500 lb)
- Max landing weight 35,380 kg (78,000 lb)

PERFORMANCE (at max T-O weight):
- Never-exceed speed 345 knots (639 km/h; 397 mph) EAS
- Max level speed and max cruising speed Mach 0·78
- Econ cruising speed at 11,885 m (39,000 ft) Mach 0·72
- Service ceiling 12,200 m (40,000 ft)
- Range with max standard fuel, 10 passengers, 45 min reserves 3,259 nm (6,040 km; 3,753 miles)
- Range with max optional fuel, 8 passengers, 45 min reserves 3,431 nm (6,357 km; 3,950 miles)

NASA
NATIONAL AERONAUTICS AND SPACE ADMINISTRATION

ADDRESS:
Washington, DC 20546

NASA has several research programmes of general aviation interest, including an augmentor wing jet STOL research aircraft, digital fly-by-wire techniques, research with the YF-12 which is capable of sustained cruise flight at Mach 3, and creation of an airborne infra-red observatory.

NASA also contracted with the Robertson Aircraft Corporation to design, develop and test a new concept in general aviation aircraft, employing a wing with advanced aerofoil section developed by NASA. Details of this are given in the entry for Robertson Aircraft Corporation.

NASA / DITC AUGMENTOR WING JET STOL RESEARCH AIRCRAFT

In a co-operative venture between NASA and the Canadian government, as represented by the Department of Industry, Trade and Commerce (DITC), a de Havilland Canada C-8A Buffalo transport aircraft has been modified extensively to serve as an augmentor wing research aircraft.

De Havilland Canada began theoretical studies of the augmentor wing principle in early 1960, and a co-operative NASA/Canadian government research programme was started in 1965. By early 1970 studies and tests had advanced to the extent that a proof-of-concept aircraft was warranted to test the principle in flight.

The US and Canadian governments entered into an international agreement whereby NASA and the DITC would modify a C-8A Buffalo to flight test the concept. This aircraft was selected because it enabled the primary research objective to be achieved at a reasonable cost and within an acceptable time span. In particular, the high wing and T-tail of the C-8A made it very suitable for modification into a powered-lift jet STOL transport, and its wing planform was basically similar to that of the large-scale wind-tunnel test model.

The DITC contracted with de Havilland Aircraft of Canada Ltd and its subcontractor, Rolls-Royce of Canada Ltd, to provide the propulsion system and modify the engine nacelles. NASA contracted with The Boeing Company to modify the aircraft, install the propulsion system and perform the initial flight tests.

The major modifications and additions to the aircraft included a reduction of wing span to 24·0 m (78 ft 9 in); replacement of all the original wing structure aft of the rear spar, by installation of an augmentor flap system, including augmentor chokes, installation of drooped ailerons with boundary layer control (BLC), and repositioning and redesign of spoilers; installation of fixed full-span leading-edge slats; installation of two Rolls-Royce Spey Mk 801SF turbofan engines (40·03 kN; 9,000 lb st); installation of an air-distribution duct system to supply fan air to the augmentor flaps and for aileron and fuselage BLC; installation of lateral and directional stability augmentation systems, increased-capacity hydraulic systems and extensive flight test instrumentation; and fixing the landing gear in the extended position, with the normal two

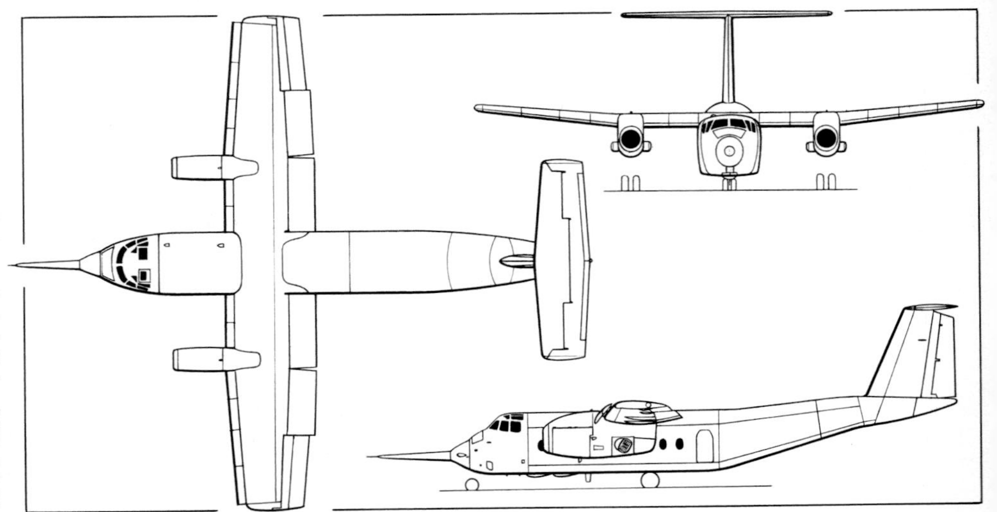

NASA/DITC XC-8A augmentor wing jet STOL research aircraft *(Pilot Press)*

main wheels on each unit replaced by two Boeing 727 nosewheels size 32 × 10-15, with tyres at 6·21 bars (90 lb/sq in) pressure and fitted with Goodrich brakes.

The augmentor flaps, with a constant chord of 1·07 m (3 ft 6 in), are made in four equal spanwise sections, two on each wing, and have a maximum deflection of 75°. They are designed to be efficient ejectors of the fan air and consist of upper and lower segments, each of which is slotted. When extended these flaps deflect the primary jet flow downward and mix it with induced flow coming over the upper wing surface. At the same time, air from above the upper flap is pulled through the slotted surface of the upper flap, and air from below the lower surface of the lower flap is pulled up through that surface's slot, increasing the airflow between the two flap segments. The net effect of this is to combine four different airflows into one jet stream between the two flap segments, increasing both lift and thrust and providing suction-type BLC to prevent or delay airflow separation from the upper flap surface.

Three surfaces are used on each wing to produce the rolling moments required for lateral control; a drooped BLC aileron, a spoiler in front of the drooped aileron, and an augmentor choke in the trailing-edge flaps. The ailerons are drooped mechanically as a function of the flap deflection, with a full droop of 30° attained at a flap deflection of 60°. The aileron deflection is ±17° from the droop position. A large volume of blowing BLC is used on the aileron to give maximum effectiveness for both aileron and spoiler. The augmentor chokes, which are used to restrict the fan air outflow area of the augmentor flaps, are designed to control the lift of the flap system. Although there are augmentor chokes in each section of the flaps, only the choke in the outboard section of each wing is used for lateral control. All four chokes are activated on the ground after landing, for lift dump.

The Rolls-Royce Spey engines were modified extensively by Rolls-Royce of Canada, the main changes including a new bypass duct that collects the fan air and directs it to two 0·33 m (13 in) diameter offtake ducts on top of the engine, and installation of a vectorable nozzle assembly (as used on the Pegasus engine of the Hawker Siddeley Harrier) in place of the normal tailpipe. The nozzles, one on each side of each engine nacelle, provide vectored thrust from 18° 30' to 116° below the aircraft's datum line. Because the engines are installed low in the nacelles there is insufficient room for the main landing gear to retract; consequently the landing gear is locked down.

The air distribution system directs the fan air from the engines to the flaps, ailerons and fuselage blowing nozzles. One of the two offtake ducts on top of each engine directs 36% of the mass flow to feed the inner flap section aft of that engine. The other duct carries 64% of the mass flow, 7·1% of this being used for fuselage BLC, the remainder being ducted to the outer section of the augmentor flaps (44%) and aileron BLC (12·9%) in the opposite wing. This layout was adopted so that in the event of an engine failure on approach, the aileron in the opposite wing would lose its BLC, compensating for the rolling moment due to the loss of thrust from the vectored nozzles, deflected to 90° on approach.

The research programme is not directed primarily toward lessening the noise problem of STOL aircraft, except in terms of the operational aspect of steep approach and landing. A separate programme to determine methods of reducing the noise of augmentor wing flaps has been conducted for NASA by The Boeing Company. Results suggest that with proper design the noise level of the augmentor wing flap could be reduced to a figure of 95 EPNdB for a large commercial STOL aircraft.

The modified Buffalo, which is designated XC-8A, was rolled out from the Boeing factory at Seattle, Washington, on 5 February, and the first flight was made on 1 May 1972.

DIMENSIONS, EXTERNAL:
- Wing span: 24·00 m (78 ft 9 in)
- Wing chord at root: 3·83 m (12 ft 7 in)
- Wing chord at tip: 2·36 m (7 ft 9 in)
- Wing aspect ratio: 7·2
- Length overall (incl 4·88 m; 16 ft probe): 28·44 m (93 ft 4 in)
- Height overall: 8·75 m (28 ft 8½ in)
- Tailplane span: 9·75 m (32 ft 0 in)
- Wheel track: 9·29 m (30 ft 6 in)
- Wheelbase: 8·48 m (27 ft 10 in)

AREAS:
- Wings, gross: 80·36 m² (865 sq ft)
- Trailing-edge flaps (projected, incl ailerons aft of wing line): 17·38 m² (187·10 sq ft)
- Ailerons (total aft of hinge line, incl tab): 4·30 m² (46·30 sq ft)
- Spoilers: 2·48 m² (26·7 sq ft)
- Fin: 8·55 m² (92·0 sq ft)
- Rudder: 5·57 m² (60·0 sq ft)
- Tailplane: 14·08 m² (151·5 sq ft)
- Elevators: 7·57 m² (81·5 sq ft)

WEIGHTS AND LOADINGS:
- Weight empty: 14,515 kg (32,000 lb)
- Max T-O weight: 20,412 kg (45,000 lb)
- Max landing weight: 19,504 kg (43,000 lb)
- Max wing loading: 254 kg/m² (52 lb/sq ft)
- Max power loading: 255 kg/kN (2·5 lb/lb st)

NASA/DITC DHC XC-8A Buffalo modified by Boeing for augmentor wing jet STOL research

PERFORMANCE (designed):
- Never-exceed speed: 180 knots (333 km/h; 207 mph)
- Max cruising speed: 160 knots (296 km/h; 184 mph)
- Max manoeuvring speed: 140 knots (259 km/h; 161 mph)
- Max flaps-down speed, 50° flap: 100 knots (185 km/h; 115 mph)
- Max flaps-down speed, 75° flap: 95 knots (176 km/h; 109 mph)
- Stalling speed: 41 knots (76 km/h; 47·5 mph)
- Max rate of climb at S/L: 1,065 m (3,500 ft)/min
- Take-off to 10·7 m (35 ft), S/L, ISA at max T-O weight, with 18° hot-thrust deflection:
  - 60° flap: 295 m (965 ft)
  - 30° flap: 320 m (1,050 ft)
- Max ferry range at max cruising speed at 3,050 m (10,000 ft), including climb, descent, taxying and 862 kg (1,900 lb) fuel reserves: 300 nm (555 km; 345 miles)

NASA DIGITAL FLY-BY-WIRE SYSTEM

Under its Digital Fly-By-Wire (DFBW) programme, NASA modified extensively a Vought F-8 Crusader jet fighter for research into this important field of flight control. It is believed that a number of advantages will accrue if, as a result of a detailed test and evaluation programme, it is proved conclusively that the system is both robust and operationally viable. These may include smoother air travel, a reduction of the pilot's workload, improvements in aircraft payload and/or flight performance and, in the case of military aircraft, provision of a flight control system that is less vulnerable to battle damage.

In the research aircraft the mechanical flight controls, consisting of the usual collection of pushrods, bellcranks and control cables, have been removed completely. They have been replaced by an electronic system in which movements of the pilot's controls initiate signals that are fed via wire circuits to an on-board digital computer. Simultaneously, an inertial measuring unit senses the motion of the aircraft in flight and the resulting aerodynamic forces, and these are also fed to the computer. The inputs from these two sources provide the data required for the computer to evaluate the most appropriate control surface positions, which it signals by wires to electro-mechanical actuators which respond by setting their related controls in the optimum position.

The F-8 research aircraft used in NASA's DFBW programme has a secondary flight control system, consisting of three separate fly-by-wire analogue channels, which serves as a backup system. In this respect it differs fundamentally from earlier fly-by-wire research aircraft, which have retained the mechanical flight controls to serve as a backup in the event of failure of the new system.

The airborne computer and inertial measuring unit are similar to those developed for the flight control system of the Apollo Lunar Module, already proved to be reliable under the most demanding conditions. Their use together for the control of a conventional aircraft in Earth's atmosphere will ensure fast and accurate positioning of the aircraft's control surfaces, which means that aircraft vibration induced by turbulent air will be reduced to a minimum.

More importantly for the future, it is believed that this faster and more accurate response, which will set control surfaces at their optimum position more effectively than a human pilot, may make it possible to reduce the size of control surfaces or even relocate them. This could reduce the basic weight and drag of new-generation aircraft and result in increased payload and/or flight performance.

NASA YF-12 PROGRAMME

NASA is operating two Lockheed YF-12 aircraft in the second phase of a joint USAF/NASA research programme. Basic purpose is to obtain information from sustained cruising flight at a speed of Mach 3 (approximately 1,735 knots; 3,220 km/h; 2,000 mph) at altitudes in the 22,860 m (75,000 ft) range to assist in the development and operation of future supersonic aircraft and the Space Shuttle. Major areas of interest include structural and performance research, stability, control, aerodynamics, the physiological and biomedical aspects of sustained high-speed cruise flight, as well as the physics of the upper atmosphere. Prior to flight testing, a series of ground tests was carried out at the High Temperature Loads Calibration Laboratory. In these tests quartz lamps were used to heat an aircraft to the temperatures achieved in flight, so

Vought F-8 Crusader modified under NASA's Digital Fly-By-Wire programme

Underbelly module on this YF-12 permits correlation of flight test and wind tunnel test data concerning flight at Mach 3

that aerodynamic loads on the aircraft's structure could be recorded.

The effects of kinetic heating in flight are being studied closely, and structural deformations or bending, structural dynamics and gust response are being examined. Changing airflow within the propulsion system and the resulting effects on high-performance aircraft form the major aspect of performance research, and since certain basic aerodynamics can only be examined realistically in flight, boundary layer flow, boundary layer noise, heat transfer and skin friction are all being measured. The aerodynamic results recorded in flight tests will be correlated with wind tunnel tests of the module that is shown on the underside of the aircraft in the accompanying illustration. In the area of stability and control, particular emphasis is being placed on the altitude excursions, or oscillations, at high-speed cruise conditions.

Apart from the aircraft itself, flight crews are being studied to gain a better understanding of physiological stress; while temperature, pressure and other physical characteristics of the upper atmosphere are being carefully evaluated. It is believed that a better appreciation of these factors will have great impact on the performance and operation of future aircraft.

SPACE SHUTTLE ORBITER FERRY AIRCRAFT

In June 1974 NASA acquired from American Airlines a Boeing 747-123 to be modified for use as a Space Shuttle orbiter ferry aircraft. It is being modified for this role by The Boeing Company, and it was anticipated that the conversion would be completed in late 1976. Pylon structures on the upper surface of the fuselage of the 747 will carry the orbiter, and initial tests will involve take-offs, level flight and landings with the orbiter being carried. The next stage will involve air launches of the orbiter, which will land at Edwards AFB, California. Ferry T-O weight complete with orbiter and added fittings is estimated at 351,530 kg (775,000 lb).

Artist's impression showing how the Space Shuttle orbiter will be ferried 'piggyback' above a Boeing 747 carrier aircraft that is being modified for the task by Boeing

To facilitate the mating and/or demating of the orbiter and its ferry, three special rigs are to be built; these will be erected at the Rockwell International factory at Palmdale, California; NASA's Flight Research Center at Edwards AFB; and Kennedy Space Center, Florida.

For further details see under Boeing entry.

NAVION

NAVION RANGEMASTER AIRCRAFT CORPORATION

HEAD OFFICE:
PO Box 311, Wharton, Texas 77488
Telephone: (713) 532-4444

CHAIRMAN OF THE BOARD AND CHIEF EXECUTIVE OFFICER:
John C. Dalton Jr

PRESIDENT:
Cedric Kotowicz

In late 1972 Mr Kotowicz purchased the assets of the bankrupt Navion Aircraft Corporation, including jigs, machine tools and a large parts inventory, and these were moved to a new facility at Wharton, Texas. Navion Rangemaster Aircraft started by providing support for the approximate total of 1,800 Navion aircraft which remain in operation throughout the world. In November 1973 it was announced that the Navion would be put back into production, with primary subassembly at facilities in Wharton, and wing and fuselage mating and installation of engine, electronics and interiors at Wharton Airport.

The first production Navion Rangemaster Model G made its first flight on 16 November 1974, but due to economic problems only one other Rangemaster was produced before control of the company passed to Consolidated Holding Inc in late 1975. Since then five more aircraft have been built and, following an exclusive distributor arrangement with Two Jacks Inc of Olive Branch, Mississippi, it was planned to produce one aircraft per week from late 1976.

Artist's impression of the new Navion Rangemaster Model H (Continental IO-520 engine)

NAVION RANGEMASTER MODEL H

The history of this aircraft stretches back to the mid-1940s when the first Navion was designed and produced by North American Aviation.

By comparison with former Navions, current aircraft have detail improvements. Improved field of view stems from elimination of the windscreen centrepost, and a 'cleaner' aircraft results from a change to flush riveting. Further improvements in the 1976 version, which has the designation Model H, include a quickly-removable instrument panel to simplify maintenance, new interior and exterior finish, and provision of a King Silver Crown electronics package, complete with autopilot and flight director system.

TYPE: Five-seat light cabin monoplane.

WINGS: Cantilever low-wing monoplane. Wing section NACA 4415R at root, NACA 6410R at tip. Dihedral 7° 30′. Incidence 1° at root, −2° at tip. Conventional structure of light alloy. Frise-type ailerons and single-slotted trailing-edge flaps of light alloy construction. Ground-adjustable trim tab on starboard aileron.

FUSELAGE: Light alloy semi-monocoque structure.

TAIL UNIT: Cantilever structure of light alloy. Fixed-incidence tailplane. Trim tab in each elevator. Ground-adjustable trim tab in rudder.

LANDING GEAR: Hydraulically-retractable tricycle type with single wheel on each unit. Oleo-pneumatic shock-absorbers. Disc brakes.

POWER PLANT: One 212·5 kW (285 hp) Continental IO-520 flat-six engine, driving a McCauley two-blade metal constant-speed propeller. Fuel in 151 litre (40 US gallon) centre main tank and two wingtip tanks each of 129 litres (34 US gallons). Total capacity 409 litres (108 US gallons). Three-blade propeller optional.

ACCOMMODATION: Pilot and four passengers in two pairs of individual seats and single rear seat. Dual controls, headrests and heater-defroster standard. Forward-hinged door on port side. Baggage space aft of rear seat.

ELECTRONICS: Standard King Silver Crown package includes dual KX-175B nav/com; glideslope; KR-85 ADF; DME; KT-76 transponder with encoding altimeter; King KFC-200 autopilot and flight director system.

DIMENSIONS, EXTERNAL:
- Wing span: 10·59 m (34 ft 9 in)
- Wing chord at root: 2·19 m (7 ft 2½ in)
- Wing chord at tip: 1·15 m (3 ft 9½ in)
- Wing aspect ratio: 6·04
- Length overall: 8·38 m (27 ft 6 in)
- Height overall: 2·54 m (8 ft 4 in)
- Tailplane span: 4·01 m (13 ft 2 in)

AREA:
- Wings, gross: 17·13 m² (184·34 sq ft)

WEIGHTS AND LOADINGS:
- Weight empty: 882 kg (1,945 lb)
- Max T-O weight: 1,504 kg (3,315 lb)
- Max wing loading: 87·4 kg/m² (17·9 lb/sq ft)
- Max power loading: 7·08 kg/kW (11·6 lb/hp)

PERFORMANCE (at max T-O weight):
- Max level speed: 166 knots (307 km/h; 191 mph)
- Max cruising speed, 75% power: 161 knots (298 km/h; 185 mph)
- Stalling speed, flaps down: 48 knots (88·5 km/h; 55 mph)
- Max rate of climb at S/L: 396 m (1,300 ft)/min
- Service ceiling: 6,555 m (21,500 ft)
- T-O and landing run: 130 m (425 ft)
- T-O to 15 m (50 ft): 221 m (725 ft)
- Landing from 15 m (50 ft): 229 m (750 ft)
- Range with max fuel: 1,397 nm (2,589 km; 1,609 miles)

NORTHROP

NORTHROP CORPORATION

CORPORATE OFFICE:
1800 Century Park East, Century City, Los Angeles, California 90067
Telephone: (213) 553-6262

CHAIRMAN OF THE BOARD AND CHIEF EXECUTIVE OFFICER:
Thomas V. Jones

PRESIDENT AND CHIEF OPERATING OFFICER:
Dr Thomas O. Paine

VICE-CHAIRMAN OF THE BOARD:
Richard W. Millar

SENIOR VICE-PRESIDENTS:
Donald A. Hicks (Technical)
Frank W. Lynch (Operations)
James D. Willson (Finance and Treasurer)

GROUP VICE-PRESIDENTS:
R. F. Miller (Communications and Electronics)
F. Stevens (Construction)

VICE-PRESIDENTS:
J. R. Alison (Customer Relations)
J. H. Bruce (Industrial Relations)
D. A. Burchinal (Europe, Middle East, Near East and Africa)

J. B. Campbell (Controller)
L. Daly (Public Affairs)
W. B. Dennis (Forward Planning)
D. N. Ferguson (General Manager, Electronics Division)
W. E. Gasich (General Manager, Aircraft Division)
C. R. Gates (Northrop International)
J. V. Holcombe (Corporate Domestic Field Offices)
H. J. Jablonski (Iran)
R. P. Jackson (Programme Management)
K. Kresa (Manager, Northrop Research and Technology Center)
D. L. Lewis (Business Analysis and Management Services)
F. J. Manzella (President, Page Communications Engineers Inc)
T. A. McDougall (International Business Operations)
D. C. McPherson (Assistant Treasurer)
R. W. Page (President, George A. Fuller Co)
J. M. Ricketts (General Manager, Electro-Mechanical Division)
H. E. Riggins Jr (Material and Facilities)
J. E. Ware (President, Northrop Architectural Systems)
P. O. Wierk (Manager, Northrop Data Processing)
W. E. Woolwine (General Manager, Ventura Division)

SECRETARY: D. H. Olson

ACTING GENERAL COUNSEL: R. B. Watts Jr

ASST TO PRESIDENT—ANALYSIS:
C. H. Bernstein

ASST TO PRESIDENT—AERONAUTICAL SYSTEMS:
J. C. Jones

ASST TO PRESIDENT—COMMUNITY RELATIONS:
W. H. Habblett

EXECUTIVE ASST TO PRESIDENT: W. H. Gurnee

DIRECTOR OF PUBLIC AFFAIRS—EUROPE:
J. K. Corfield

DIRECTOR OF CORPORATE COMMUNICATIONS:
W. A. Schoneberger

This company was formed in 1939 by John K. Northrop and others to undertake the design and manufacture of military aircraft. During the second World War it built 1,131 aircraft of its own design and was engaged in extensive subcontract work. It also devoted considerable attention to the design and construction of aircraft of the 'Flying Wing' type.

Although continuing its activities in the design, development and production of aircraft, missiles and target drone systems, Northrop has broadened its scope of operation to include electronics, space technology, communications, construction, support services and commercial products. To reflect this changing character of its business, the company changed its name from Northrop Aircraft Inc to Northrop Corporation in 1959.

Divisions of Northrop Corporation now include Aircraft Division, specialising in aircraft, missiles, aeronautical systems and weapon systems management; Ventura Division, engaged primarily in the design, development and manufacture of remotely piloted vehicles, aircraft target drones, mobile underwater targets and target range support services; Electronics Division and Electro-Mechanical Division, which handle Northrop activities in the design, development and manufacture of electronic, electro-mechanical and optical-mechanical products, components and electronic countermeasures equipment; and George A. Fuller Company, a leading construction-management and building company.

Northrop's Electronics Division, under subcontract to The Boeing Company, has responsibility for the integration, production and testing of special navigation and guidance equipment for the Boeing E-3A AWACS. This division announced on 26 March 1976 that the first production navigation system, comprising a Northrop AN/ARN-120 Omega radio navigation set, Delco AN/ASN-119 inertial platform and Ryan AN/APN-213 Doppler velocity sensor, had been delivered on schedule.

In 1959 Northrop expanded into the field of advanced systems for long-range radio communications with the purchase of Page Communications Engineers Inc as a wholly-owned subsidiary.

In 1961 Northrop combined the operations of Acme Metal Molding Co and Arcadia Metal Products in a single organisation, Northrop Architectural Systems (a wholly-owned subsidiary).

Northrop Pacific Inc was formed from Northrop Architectural Systems in 1969 as a new subsidiary of the company to continue the manufacture of floor panels for commercial aircraft.

To further expand its research and development work, Northrop has divided its Research and Technology Center into three organisations: the Corporate Laboratories, Laser System Laboratories and Laser Technology Laboratories. Current programmes include research and development in such fields as information sciences, electronic devices and materials, nuclear radiation effects, high-energy laser development and laser systems applications.

In March 1971 Northrop formed the Thai Communications Co in Bangkok, Thailand, to repair, maintain and manufacture two-way radio equipment as well as to design and install communications systems for Thailand and, eventually, for adjoining countries.

In June 1971 Northrop acquired American Standard's Wilcox Division of Kansas City and all outstanding stock of its international sales affiliate. Wilcox business is primarily in commercial aviation, communication and navigation areas. In another major acquisition Northrop acquired the George A. Fuller Company of New York, a construction company which is now a Division of Northrop Corporation.

In 1972 Northrop acquired Page Aircraft Maintenance Inc, Lawton, Oklahoma. Since renamed Northrop Worldwide Aircraft Services Inc, it is engaged in aircraft maintenance and support services. Northrop acquired also the entire assets of Berkeley Scientific Laboratories Inc, Hayward, California, a major supplier of medical and business data systems.

The number of employees of Northrop Corporation totalled about 24,000 in early 1976.

NORTHROP CORPORATION AIRCRAFT DIVISION

ADDRESS:
3901 West Broadway, Hawthorne, California 90250
Telephone: (213) 970-2000

CORPORATE VICE-PRESIDENT AND GENERAL MANAGER:
Welko E. Gasich

CORPORATE VICE-PRESIDENT:
Roy P. Jackson (F-18 Programme)

SENIOR VICE-PRESIDENT:
R. M. McNamara (Business Operations)

VICE-PRESIDENTS:
Charles W. Benson (Materiel and Procurement)
Carl H. Burris (Administration)
Ben F. Collins Jr (Saudi Arabia Operations)
Grif B. Doyle (Asst General Manager)
Manuel G. Gonzalez (Marketing)
H. B. Gunther (Commercial Programmes)
Milton Kuska (F-5 Programmes)
R. D. Lovell (Manufacturing Support)
Rex H. Madeira (Support Operations)
Jack Mannion (Public Affairs)
John L. McCoy (Product Support)
B. T. Moser (Assembly Operations)
G. S. Shackelford (Contracts and Pricing)
Stephen R. Smith (Iran Operations)
R. S. Taylor (Finance)
Donald D. Warner (Technical)

Current production at Northrop's Aircraft Division is centred on the F-5E Tiger II International Fighter Aircraft, the F-5F two-seat fighter/trainer, and major Boeing 747 subcontract work which includes the main fuselage section and the extra large side-loading cargo door. During 1975, deliveries totalled 220 F-5Es and 17 F-5Bs. The two-seat F-5B was superseded in production in 1976 by the F-5F, a version of the F-5E.

Following selection of the General Dynamics F-16 for production, after a fly-off evaluation against the Northrop YF-17, under the USAF's Air Combat Fighter prototype programme, the US Navy made an evaluation of the YF-17. As a result, the Secretary of the Navy announced that McDonnell Douglas Corporation (which see) and Northrop Corporation had been selected to develop for the Navy a new air combat fighter under the designation F-18. This is, in effect, an improved and somewhat larger version of the F-17. A land-based version of the F-18 is also available.

In addition to its main factory at Hawthorne, the Aircraft Division has facilities at El Segundo, Long Beach, Palmdale and Edwards Air Force Base, California.

NORTHROP F-5

USAF designations: F-5 and RF-5
CAF designations: CF-5A/D
R Netherlands AF designations: NF-5A/B
R Norwegian AF designations: F-5G/RF-5G
Spanish AF designations: C-9/CE-9

Design of this light tactical fighter started in 1955 and construction of the prototype of the single-seat version (then designated N-156C) began in 1958. It flew for the first time on 30 July 1959, exceeding Mach 1 on its maiden flight. Two more prototypes were built, followed by several production versions, as follows:

F-5A. Basic single-seat fighter. Two General Electric J85-GE-13 afterburning turbojets. First production F-5A flew in October 1963. Norwegian version has ATO and arrester hook for short-field operation.

F-5B. Generally similar to F-5A, but with two seats in tandem for dual fighter/trainer duties. First F-5B flew on 24 February 1964. Production was to be terminated during 1976.

CF-5A/D. These are the designations of the versions of the F-5A/B that were produced for the Canadian Armed Forces, the first of them entering service in 1968. Several improvements were incorporated in the CF series, including higher-thrust engines (J85-CAN-15), and flight refuelling capability. 115 built by Canadair, under licence. Described under Canadair entry in 1972-73 *Jane's*.

NF-5A/B. Versions of the F-5 produced for the Royal Netherlands Air Force with a Doppler navigation system, 1,040 litre (275 US gallon) fuel tanks and manoeuvring flaps. Manufacture and assembly of the 105 aircraft ordered were integrated with CF-5 production by Canadair Ltd, as described in the 1972-73 *Jane's*, and the first of them entered service in 1969.

RF-5A. Reconnaissance version of the F-5; initial deliveries were made in mid-1968. Its four KS-92 cameras, each with a 100 ft film magazine, can provide forward oblique, trimetrogon and split vertical coverage, including horizon-to-horizon with overlap. Associated equipment includes four light sensors, defogging and cooling systems, a pitot static nose boom and a computer/'J' box, all housed in a nose compartment with forward-hinged clamshell top cover.

SF-5A/B (C-9/CE-9). Spanish versions of the F-5, as described under the CASA entry in the 1972-73 *Jane's*. Total of 70 built.

F-5E. Advanced version of F-5A. Described separately.

F-5F. Two-seat tactical fighter/trainer version of the F-5E, scheduled to replace the F-5B on the production line. Described separately under F-5E (Tiger II) entry.

F-5G. Royal Norwegian Air Force designation for its 78 F-5As.

RF-5G. Royal Norwegian Air Force designation for its 16 RF-5As.

On 4 November 1974, Northrop announced delivery of the 2,500th aircraft in its F-5/T-38 series. The total exceeded 2,800 by early 1976, including more than 300 built under licence in other countries. Eighteen versions of the F-5 are flown by the air forces of 22 countries, including six NATO nations. In addition to those built by Northrop Aircraft Division facilities in California, F-5s have been produced under licensing agreements in Canada and Spain. The F-5 was first ordered into production by the US government, through the USAF, in October 1962, to meet the defence requirements of allied and friendly nations. Initial deliveries, beginning April 1964, were made to Williams AFB, Chandler, Arizona, where the USAF Tactical Air Command has since trained pilots and maintenance personnel of countries receiving F-5s. The first allied air force to receive F-5s was the Imperial Iranian Air Force, which put into service its initial squadron of 13 aircraft on 1 February 1965. The Republic of China, Greece, Republic of Korea, the Philippines and Turkey received F-5s in 1965. Ethiopia, Morocco, Norway and Thailand first received F-5s in 1966, the Republic of Vietnam in 1967 and Libya in 1968.

TYPE: Light tactical fighter and reconnaissance aircraft.

WINGS: Cantilever low-wing monoplane. Wing section NACA 65A004·8 (modified). No dihedral or incidence. Sweepback at quarter-chord 24°. Multi-spar light alloy structure with heavy plate machined skins. Hydraulically-powered sealed-gap ailerons at approximately mid-span with light alloy single-slotted flaps inboard. Continuous-hinge leading-edge flaps of full-depth honeycomb construction. No trim tabs. No de-icing system.

FUSELAGE: Semi-monocoque basic structure of light alloy, with steel, magnesium and titanium used in certain areas. 'Waisted' area rule lines. Two hydraulically-actuated airbrakes on underside of fuselage forward of wheel wells.

TAIL UNIT: Cantilever all-metal structure, with

Northrop RF-5A single-seat reconnaissance fighter, showing the reconnaissance nose unit

hydraulically-powered rudder and one-piece all-moving tailplane. Single spars with full depth light alloy honeycomb secondary structure. No trim tabs. Longitudinal and directional stability augmentors installed in series with control system.

LANDING GEAR: Hydraulically-retractable tricycle type with steerable nosewheel. Emergency gravity extension. Main units retract inward into fuselage, nosewheel forward. Oleo-pneumatic shock-absorbers. Main wheels fitted with tubeless tyres size 22 × 8·5, pressure 5·86-14·48 bars (85-210 lb/sq in). Nosewheel fitted with tubeless tyre size 18 × 6·5, pressure 4·14-12·41 bars (60-180 lb/sq in). Multiple-disc hydraulic brakes.

POWER PLANT: Two General Electric J85-GE-13 turbojets (each with max rating of 18·15 kN; 4,080 lb st with afterburning). Two internal fuel tanks composed of integral cells with total usable capacity of 2,207 litres (583 US gallons). Provision for one 568 litre (150 US gallon) jettisonable tank on fuselage centreline pylon, two 568 litre (150 US gallon) jettisonable tanks on underwing pylons and two 189 litre (50 US gallon) wingtip tanks. Total fuel, with external tanks, 4,289 litres (1,133 US gallons). Single pressure refuelling point on lower fuselage. Oil capacity 4·5 litres (4·7 US quarts) each engine.

ACCOMMODATION (F-5A): Pilot only, on rocket-powered ejection seat in pressurised and air-conditioned cockpit. (F-5B): Pupil and instructor in tandem on rocket-powered ejection seats in pressurised and air-conditioned cockpits separated by windscreen. Separate manually-operated rearward-hinged jettisonable canopies. Instructor's seat at rear raised 0·25 m (10 in) higher than that of pupil to give improved forward view.

SYSTEMS: Electrical system includes two 8kVA engine-driven generators, providing 115V 400Hz AC power, and 24V battery.

ELECTRONICS AND EQUIPMENT: Standard equipment includes AN/ARC-34C UHF radio, PP-2024 SWIA-Missile AVX, AN/AIC-18 interphone, J-4 compass, Norsight optical sight, AN/APX-46 IFF, and AN/ARN-65 Tacan. Space provision for AN/ARW-77 Bullpup AUX. Blind-flying instrumentation not standard.

ARMAMENT: Basic interception weapons are two Sidewinder missiles on wingtip launchers and two 20 mm guns in the fuselage nose. Five pylons, one under the fuselage and two under each wing, permit the carriage of a wide variety of other operational warloads. A bomb of more than 910 kg (2,000 lb) or high-rate-of-fire gun pack can be suspended from the centre pylon. Underwing loads can include four air-to-air missiles, Bullpup air-to-surface missiles, bombs, up to 20 air-to-surface rockets, gun packs or external fuel tanks. The reconnaissance nose does not eliminate the 20 mm nose gun capability.

DIMENSIONS, EXTERNAL:
- Wing span: 7·70 m (25 ft 3 in)
- Wing span over tip-tanks: 7·87 m (25 ft 10 in)
- Wing chord at root: 3·43 m (11 ft 3 in)
- Wing chord at tip: 0·69 m (2 ft 3 in)
- Length overall:
  - F-5A: 14·38 m (47 ft 2 in)
  - F-5B: 14·12 m (46 ft 4 in)
- Height overall:
  - F-5A: 4·01 m (13 ft 2 in)
  - F-5B: 3·99 m (13 ft 1 in)
- Tailplane span: 4·28 m (14 ft 1 in)
- Wheel track: 3·35 m (11 ft 0 in)
- Wheelbase:
  - F-5A: 4·67 m (15 ft 4 in)
  - F-5B: 5·94 m (19 ft 6 in)

AREAS:
- Wings, gross: 15·79 m² (170 sq ft)
- Ailerons (total): 0·86 m² (9·24 sq ft)
- Trailing-edge flaps (total): 1·77 m² (19·0 sq ft)
- Leading-edge flaps (total): 1·14 m² (12·3 sq ft)
- Fin: 3·85 m² (41·42 sq ft)
- Rudder: 0·57 m² (6·1 sq ft)
- Tailplane: 5·48 m² (59·0 sq ft)

WEIGHTS AND LOADINGS:
- Weight empty, equipped:
  - F-5A: 3,667 kg (8,085 lb)
  - F-5B: 3,792 kg (8,361 lb)
- Max military load: 2,812 kg (6,200 lb)
- Max T-O weight:
  - F-5A: 9,379 kg (20,677 lb)
  - F-5B: 9,298 kg (20,500 lb)
- Max design landing weight: 9,006 kg (19,857 lb)
- Max zero-fuel weight:
  - F-5A: 6,446 kg (14,212 lb)
  - F-5B: 6,237 kg (13,752 lb)
- Max wing loading:
  - F-5A: 590·8 kg/m² (121 lb/sq ft)
  - F-5B: 576 kg/m² (118 lb/sq ft)

PERFORMANCE (F-5A at AUW of 5,193 kg; 11,450 lb: F-5B at AUW of 4,916 kg; 10,840 lb, unless indicated otherwise):
- Never-exceed speed: 710 knots (1,315 km/h; 818 mph) IAS
- Max level speed at 11,000 m (36,000 ft):
  - F-5A: Mach 1·4
  - F-5B: Mach 1·34
- Max cruising speed without afterburning, at 11,000 m (36,000 ft): Mach 0·97
- Econ cruising speed: Mach 0·87
- Stalling speed, 50% fuel, flaps extended:
  - F-5A: 128 knots (237 km/h; 147 mph)
  - F-5B: 120 knots (223 km/h; 138 mph)
- Max rate of climb at S/L:
  - F-5A: 8,750 m (28,700 ft)/min
  - F-5B: 9,265 m (30,400 ft)/min
- Service ceiling:
  - F-5A: 15,390 m (50,500 ft)
  - F-5B: 15,850 m (52,000 ft)
- Service ceiling, one engine out:
  - F-5A, F-5B: over 10,365 m (34,000 ft)
- T-O run (with two Sidewinder missiles):
  - F-5A at AUW of 6,203 kg (13,677 lb): 808 m (2,650 ft)
  - F-5B at AUW of 5,924 kg (13,061 lb): 671 m (2,200 ft)
- T-O to 15 m (50 ft) (with two Sidewinder missiles):
  - F-5A at AUW of 6,203 kg (13,677 lb): 1,113 m (3,650 ft)
  - F-5B at AUW of 5,924 kg (13,061 lb): 960 m (3,150 ft)
- Landing from 15 m (50 ft), with brake-chute:
  - F-5A at AUW of 4,504 kg (9,931 lb): 1,189 m (3,900 ft)
  - F-5B at AUW of 4,363 kg (9,619 lb): 1,158 m (3,800 ft)
- Landing run, with brake-chute:
  - F-5A at AUW of 4,504 kg (9,931 lb): 701 m (2,300 ft)
  - F-5B at AUW of 4,363 kg (9,619 lb): 671 m (2,200 ft)
- Range with max fuel, with reserve fuel for 20 min max endurance at S/L:
  - F-5A, tanks retained: 1,205 nm (2,232 km; 1,387 miles)
  - F-5B, tanks retained: 1,210 nm (2,241 km; 1,393 miles)
  - F-5A, tanks dropped: 1,400 nm (2,594 km; 1,612 miles)
  - F-5B, tanks dropped: 1,405 nm (2,602 km; 1,617 miles)
- Combat radius with max payload, allowances as above and five minutes combat at S/L:
  - F-5A: 170 nm (314 km; 195 miles)
  - F-5B: 175 nm (323 km; 201 miles)
- Combat radius with max fuel, two 530 bombs, allowances as above and five minutes combat at S/L:
  - F-5A: 485 nm (898 km; 558 miles)
  - F-5B: 495 nm (917 km; 570 miles)
- Operational hi-lo-lo-hi reconnaissance radius with max fuel, 50 nm (93 km; 58 mile) S/L dash to and from target and allowances as for combat radius with max fuel:
  - RF-5A: 560 nm (1,036 km; 644 miles)

Northrop SF-5B two-seat fighter/trainer, assembled in Spain by CASA and operated under the designation CE-9 by the Spanish Air Force *(Howard Levy)*

NORTHROP TIGER II

USAF designations: F-5E and F-5F

The F-5E was selected in November 1970 by the US government as the winner of a competition to determine the International Fighter Aircraft (IFA) which was to succeed Northrop's F-5A aircraft.

Before initiation of the IFA competition, Northrop had proposed a follow-on version of the F-5, and had flown a prototype of this new type at the end of March 1969. It consisted of a two-seat F-5B powered by two General Electric YJ85-GE-21 engines rated at 22·24 kN (5,000 lb st) each, representing an increase of 23 per cent over the rated thrust of the J85-GE-13 engines that power the F-5A/B series.

More than 70 flights were made with this aircraft, and Northrop was able to explore the flight envelope, including operation at altitudes up to 15,240 m (50,000 ft), a maximum speed of Mach 1·6, and aerial combat manoeuvres.

When making the announcement that the Northrop design had been selected as the winner of the IFA competition, the USAF stated that the aircraft would be built under a fixed-price-plus-incentive contract with an initial value of $21 million. This programme was considered initially to involve production of 325 aircraft; but subsequent sales have required revision of this figure, and production is expected to exceed 1,500 aircraft.

The F-5E design places particular emphasis on manoeuvrability rather than high speed, notably by the incorporation of manoeuvring flaps, based on the design of a similar system for the Netherlands Air Force's NF-5A/Bs. Full-span leading-edge flaps work in conjunction with conventional trailing-edge flaps, and are operated by a control on the pilot's throttle quadrant.

Wing loading on the F-5E is maintained at approximately the same value as on the F-5A, as the result of an increase in wing area to 17·30 m² (186 sq ft). This is due principally to the widened fuselage, which also increases wing span. The tapered wing leading-edge extension, between the inboard leading-edge and fuselage, was modified to enhance airflow over the wing at high angles of attack.

The F-5E incorporates other features developed for the Canadian, Dutch and Norwegian F-5s. These include two-position nosewheel gear, which increases wing angle of attack on the ground by 3° 22′ and which, in conjunction with the more powerful engines, has improved F-5E take-off performance some 30% by comparison with earlier F-5s. ATO provision and arrester gear permit operation from short runways. It is qualified to carry two 1,040 litre (275 US gallon) underwing fuel tanks, in addition to

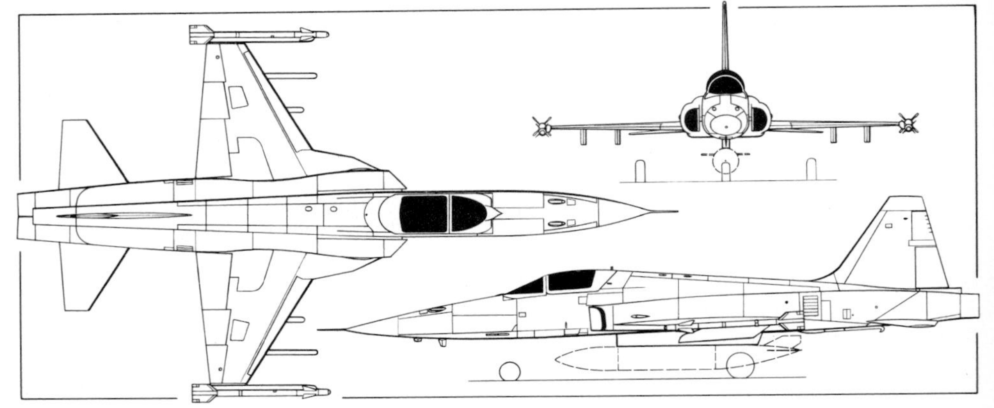

Northrop F-5E Tiger II single-seat twin-jet tactical fighter aircraft *(Pilot Press)*

Northrop F-5E Tiger II 'aggressor' aircraft in service with the USAF

Northrop F-5F tandem two-seat trainer version of the F-5E Tiger II, which retains combat capability

the centreline 1,040 litre (275 US gallon) tank, and up to nine 500 lb MK-82 bombs, following the addition of a Multiple Ejector Rack (MER) on the centreline stores station.

The first F-5E was rolled out on 23 June 1972, and made its first flight on 11 August 1972. USAF Tactical Air Command, with assistance from Air Training Command, has been assigned responsibility for training pilots and technicians of user countries. First deliveries of the F-5E, to the USAF's 425th Tactical Fighter Squadron, were made in the Spring of 1973. Twenty aircraft had been supplied for the USAF training programme by the end of September 1973, and deliveries to foreign countries began in late 1973. By the middle of August 1976 orders for the F-5E/F totalled 947 aircraft, following an announcement that the government of Switzerland had signed contracts with the US Department of Defense and Northrop Corporation for the supply of 72 F-5E and F-5F aircraft. Production was then at a rate of 18 aircraft per month, and more than 500 had been delivered. Customers include the USAF (112 F-5Es), US Navy (10 F-5Es and 3 F-5Fs), Brazil, Chile, Republic of China, Iran, Jordan, Kenya, South Korea, Malaysia, Saudi Arabia, Switzerland and Thailand.

The US Navy Fighter Weapons School and USAF Aggressor Squadron, the top air combat training schools for US fighter pilots, both fly F-5Es as 'enemy' aircraft against other first-line operational US tactical fighters.

Two versions of the Tiger II are produced:

F-5E. Standard production version, to which the detailed description applies. In production also, under licence, by AIDC in Taiwan (which see). Can be fitted with an R-843A/ARN-58 localiser receiver and a reconnaissance nose containing four KS-121A 70 mm framing cameras and related equipment. Intended for low/medium-altitude photo-reconnaissance, the nose is similar to that described for the RF-5A.

The F-5Es for the Royal Saudi Air Force (RSAF) have a Litton LN-33 inertial navigation system, capable of accuracy exceeding 1·5 nm (2·7 km; 1·7 miles) CEP per flight hour, which provides attitude reference, range and bearing to ten pre-set destinations, as well as true ground track steering. The system is self-aligning in 10 min in the gyro compass mode, and can be aligned in 3 min to a stored heading. In-flight refuelling capability.

F-5F. Tandem two-seat version of F-5E, with fuselage lengthened by 1·04 m (3 ft 5 in). Fire control system retained, enabling aircraft to be used for both training and combat duties, but one M-39 gun deleted. Development approved by USAF in early 1974. First flight was made on 25 September 1974. Two F-5Fs completed flight test and qualification in early 1976. More than 90 ordered; deliveries began in the Summer of 1976. Max T-O weight 11,442 kg (25,225 lb).

The following details refer specifically to the F-5E, but are generally applicable to both versions except for details noted under model listings:

TYPE: Single-seat light tactical fighter.

WINGS: Cantilever low-wing monoplane. Wing section NACA 65A004·8 (modified). No dihedral. No incidence. Sweepback at quarter-chord 24°. Multi-spar light alloy structure with heavy plate machined skins. Hydraulically-powered sealed-gap ailerons at approximately mid-span. Electrically-operated light alloy single-slotted trailing-edge flaps inboard of ailerons. Electrically-operated leading-edge manoeuvring flaps. No de-icing system.

FUSELAGE: Light alloy semi-monocoque basic structure, with steel, magnesium and titanium used in certain areas. Two hydraulically-actuated airbrakes of magnesium alloy construction, mounted on underside of fuselage forward of main-wheel wells. Electronics bay and cockpit pressurised; fail-safe structure in pressurised sections.

TAIL UNIT: Cantilever all-metal structure, with hydraulically-powered rudder and one-piece all-moving tailplane. Tailplane incidence varied by hydraulic actuators. No trim tabs. Dual hydraulic actuators of Northrop design for control of rudder and tailplane.

LANDING GEAR: Hydraulically-retractable tricycle type, main units retracting inward into fuselage, nosewheel forward. Oleo-pneumatic struts of Northrop design on all units. Two-position extending nose unit increases static angle of attack by 3° 22′ to reduce T-O distance, and is shortened automatically during the retraction cycle. Gravity-operated emergency extension. Main wheels and tyres size 24 × 8·00-13, pressure 14·48 bars (210 lb/sq in). Steerable nose unit with wheel and tyre size 18 × 6·50-8, pressure 8·27 bars (120 lb/sq in). All-metal multiple-disc brakes of Northrop design.

POWER PLANT: Two General Electric J85-GE-21 turbojet engines, each rated at 22·24 kN (5,000 lb st) with afterburning. Two independent fuel systems, one for each engine. Fuel for starboard engine supplied from two rubber-impregnated bladder-type nylon fabric cells, comprising a centre-fuselage cell of 802 litre (212 US gallon) capacity, and an aft-fuselage cell of 640 litre (169 US gallon) capacity. Port engine supplied from a forward fuselage cell of 1,120 litre (296 US gallon) capacity. Total fuel capacity 2,562 litres (677 US gallons). No fuel is carried in the wings. Fuel crossfeed system allows fuel from either or both cell systems to be fed to either or both engines. A 1,040 litre (275 US gallon) jettisonable fuel tank can be carried on the fuselage centreline pylon. Auxiliary fuel tanks of 568 or 1,040 litres (150 or 275 US gallons) can be carried on the inboard underwing pylons. Single refuelling point on lower fuselage for fuselage fuel cell and external tank installation. Provision for in-flight refuelling by means of a detachable probe. Oil capacity 3·8 litres (1 US gallon) per engine.

ENGINE INTAKES: Intakes are supplemented with auxiliary air inlet doors for use during T-O and low-speed flight, to improve compressor face pressure recovery and to decrease distortion. Each door consists of a set of six pivot-mounted louvres in removable panels on each side of the fuselage. The doors are actuated by the pilot at T-O, and controlled automatically in flight by Mach sensor switches, and are maintained in the open position at airspeeds below Mach 0·35-0·4.

ACCOMMODATION: Pilot only in pressurised, heated and air-conditioned cockpit, on rocket-powered ejection seat. Upward-opening canopy, hinged at rear.

SYSTEMS: Cockpit and electronics bay pressurised, heated and air-conditioned by engine bleed air, maximum pressure differential 0·34 bars (5 lb/sq in). Hydraulic power supplied by two independent systems at a pressure of 207 bars (3,000 lb/sq in). Flight control system provides power solely for operation of primary flight control surfaces. Utility system provides hydraulic backup power for the primary flight control surfaces and operating power for the landing gear, landing gear doors, airbrakes, wheel brakes, nosewheel steering, gun bay purge doors, gun gas deflectors and stability augmentation system. Electrical power supplied by two 13/15kVA 115/200V three-phase 320-480Hz non-paralleled engine-driven alternators. Each alternator has the capacity to accept full aircraft power load via an automatic transfer function. 250VA 115V 400Hz single-phase solid-state static inverter provides secondary AC source for engine starting. Two 33A 26-32V transformer-rectifiers and a 24V 11Ah nickel-cadmium battery provide DC power. Liquid oxygen system with capacity of 5 litres.

ELECTRONICS AND EQUIPMENT: AN/ARC-164 UHF command radio, 3,500-channel with 50kHz spacing. Lightweight microminiature X-band radar for air-to-air search and range tracking; target information, at a range of up to 20 nm (37 km; 23 miles), is displayed on a 0·13 m (5 in) DVST in cockpit. AN/ARA-50 UHF ADF; AN/AIC-18 intercom system; AN/APX-72 IFF/SIF system; AN/ARN-118 Tacan; SST-181 X-band radar transponder (Skyspot); attitude and heading reference system; angle of attack system; and central air data computer. Full blind-flying instrumentation. Optional electronics include LN-33 inertial navigation system; AN/ARN-108 instrument landing system; CPU-80 flight director computer; VHF; VOR/ILS with DME; LF ADF; CRT with scan converter for radar or electro-optical weapon (AGM-65 Maverick); radar warning receiver; and improved radar, with angle track. Optional equipment includes assisted take-off rockets; photo-reconnaissance nose; in-flight refuelling system; pylon jettison conversion kits; improved-performance ejection seats; anti-skid brakes; and chaff/flare countermeasures package.

ARMAMENT: Two AIM-9 Sidewinder missiles on wingtip launchers. Two M-39A2 20 mm cannon mounted in fuselage nose, with 280 rounds per gun. Up to 3,175 kg (7,000 lb) of mixed ordnance can be carried on four underwing and one underfuselage stations, including M129 leaflet bombs; MK-82 GP and Snakeye 500 lb bombs; MK-36 destructors; MK-84 2,000 lb bombs; BLU-1, -27 or -32 U or F napalm; LAU-68 (7) 2·75 in rockets; LAU-3 (19) 2·75 in rockets; CBU-24, -49, -52 or -58 cluster bomb units; SUU-20 bomb and rocket packs; SUU-25 flare dispensers; TDU-10 tow targets (Dart); and RMU-10 reel (Dart). Lead-computing optical gunsight uses inputs from airborne radar for air-to-air missiles and cannon, and provides a roll-stabilised manually-depressible reticle aiming reference for air-to-ground delivery. A 'snapshoot' capability is included for attack on violently manoeuvring and fleeting targets. The gunsight incorporates also a detachable 16 mm reticle camera with 15 m (50 ft) film magazine. Optional ordnance capability includes the AGM-65 Maverick; centreline multiple ejector rack; laser guided bombs; and (for F-5F only) a laser designator.

DIMENSIONS, EXTERNAL:

Wing span	8·13 m (26 ft 8 in)
Span over missiles	8·53 m (27 ft 11⅞ in)
Wing chord at root	3·57 m (11 ft 8⅝ in)
Wing chord at tip	0·68 m (2 ft 2⅞ in)
Wing aspect ratio	3·82
Length overall (incl nose-probe):	
F-5E	14·68 m (48 ft 2 in)
F-5F	15·72 m (51 ft 7 in)
Height overall:	
F-5E	4·06 m (13 ft 4 in)
F-5F	4·01 m (13 ft 1¾ in)
Tailplane span	4·31 m (14 ft 1½ in)
Wheel track	3·80 m (12 ft 5½ in)
Wheelbase	5·17 m (16 ft 11½ in)

AREAS:

Wings, gross	17·3 m^2 (186 sq ft)
Ailerons (total)	0·86 m^2 (9·24 sq ft)
Trailing-edge flaps (total)	1·95 m^2 (21·0 sq ft)
Leading-edge flaps (total)	1·14 m^2 (12·3 sq ft)
Fin	3·85 m^2 (41·42 sq ft)
Rudder	0·57 m^2 (6·10 sq ft)
Tailplane	5·48 m^2 (59·0 sq ft)

WEIGHTS AND LOADINGS:

Weight empty	4,346 kg (9,583 lb)
Max T-O weight	11,192 kg (24,675 lb)
Max zero-fuel weight	7,953 kg (17,534 lb)
Max wing loading	649·4 kg/m^2 (133 lb/sq ft)
Max power loading	251·6 kg/kN (2·5 lb/lb st)

PERFORMANCE (F-5E, at AUW of 6,010 kg; 13,250 lb, unless stated otherwise):

Never-exceed speed	710 knots (1,314 km/h; 817 mph) IAS
Max level speed at 11,000 m (36,000 ft):	
F-5E	Mach 1·63
F-5F	Mach 1·55
Max cruising speed at 11,000 m (36,000 ft), without afterburning	Mach 0·98
Econ cruising speed	Mach 0·80

Stalling speed, 50% fuel, flaps and wheels down
124 knots (230 km/h; 143 mph)
Max rate of climb at S/L:
F-5E 10,515 m (34,500 ft)/min
F-5F 10,030 m (32,905 ft)/min
Service ceiling:
F-5E 15,790 m (51,800 ft)
F-5F 15,484 m (50,800 ft)
Service ceiling, one engine out
over 12,495 m (41,000 ft)
Min ground turning radius 11·13 m (36 ft 6 in)
T-O run with two Sidewinder missiles, at 7,141 kg (15,745 lb) AUW 610 m (2,000 ft)
T-O to 15 m (50 ft), loaded as above 884 m (2,900 ft)
Landing from 15 m (50 ft) without brake-chute, at 5,187 kg (11,436 lb) AUW 1,128 m (3,700 ft)
Landing run with brake-chute, weight as above
762 m (2,500 ft)
Range with max fuel, with reserve fuel for 20 min max endurance at S/L:
tanks retained 1,385 nm (2,567 km; 1,595 miles)
tanks dropped 1,590 nm (2,946 km; 1,831 miles)
Combat radius with two Sidewinder missiles and max fuel, allowances as above and five minutes combat with max afterburning power at 4,570 m (15,000 ft)
585 nm (1,083 km; 673 miles)
Combat radius with 2,358 kg (5,200 lb) ordnance load and two Sidewinder missiles, max fuel, allowances as above and five minutes combat at military power at S/L 120 nm (222 km; 138 miles)
Combat radius with max fuel, two Sidewinders and two 530 lb bombs, allowances as above and five minutes combat at military power at S/L
470 nm (870 km; 541 miles)
Operational hi-lo-hi reconnaissance radius with max fuel, 50 nm (92·5 km; 57·5 miles) S/L dash to and from target, allowances as above, no combat
550 nm (1,018 km; 633 miles)

NORTHROP YF-17

Northrop's YF-17 is a twin-engined fighter prototype developed to demonstrate advanced technology applicable to air combat. Distinguishing features include mid-wing configuration, a moderately-swept wing, highly-swept leading-edge root extensions, underwing intakes and twin vertical fins.

The basic wing planform, combined with the highly swept leading-edge root extensions, is identified as a hybrid wing. The vortex flow generated by the extensions significantly increases lift, reduces drag, and improves handling characteristics. Leading-edge and trailing-edge flaps are used to vary the wing camber for maximum manoeuvrability.

The horizontal tail surface is located lower than the wing to provide increasing longitudinal stability at high angles of attack approaching maximum lift, and to preclude buffet from the wing wake under high-*g* flight conditions. The vertical tail surfaces are sized and located to provide positive directional stability beyond the maximum trimmed angles of attack across the speed range. The forward location was chosen to eliminate reduction of horizontal tail surface effectiveness due to the outward cant of the vertical surfaces, and to provide low supersonic drag through favourable influence on the area distribution of the aircraft.

Location of the engine intakes beneath the wing minimises flow angularity, placing the intakes in a position to take advantage of the compression effects of the wing leading-edge root extensions during supersonic flight. The key feature of airframe/intake integration is a longitudinal slot through each wing root, which allows a portion of the fuselage boundary layer air to flow over the upper surface of the wings. Thus, a narrow fuselage boundary layer gutter can be used, which results in a low-drag installation.

Outstanding visibility for the pilot is achieved by the canopy shape and location, with full aft vision at eye level and above.

Extensive use is made of graphite composite materials in the aircraft's structure, to the extent that approximately 10 per cent, by weight, is composed of such material.

Northrop YF-17 lightweight fighter (two General Electric YJ101 turbojet engines)

Design of Northrop's lightweight fighter began in May 1966, and construction of two prototypes started in early 1973, following the award of a $39 million contract by the USAF. The first of them was rolled out on 4 April 1974 and flew for the first time on 9 June 1974; the second prototype made its first flight on 21 August 1974. Within a period of six months these two aircraft had accumulated a total in excess of 200 flights.

Evaluation in competition with the General Dynamics YF-16 prototypes resulted in USAF selection of the latter aircraft for production as an operational air combat fighter. Subsequent evaluation by the US Navy led to the teaming of Northrop with McDonnell Douglas Corporation to develop a new air combat fighter for the Navy. Evolved from the YF-17 the new aircraft, designated F-18, will be some 3,628 kg (8,000 lb) heavier to satisfy the USN's operational requirements. A description of the F-18 can be found under the McDonnell Douglas heading.

The following details refer to the YF-17 prototypes, one of which was used by NASA in 1976 for an eight-week flight programme to study manoeuvrability in the transonic speed range and to collect data that will lead to improved wind tunnel predictions for future fighter aircraft:

Type: Single-seat lightweight fighter prototype.

Wings: Cantilever mid-wing monoplane. Anhedral 5°. Sweepback at quarter-chord 20°. Multi-spar structure, primarily of light alloy. Boundary layer control achieved by wing root slots. Sealed-gap ailerons. Plain trailing-edge flaps, of graphite composite material, inboard of ailerons. Aileron actuators manufactured by Parker-Hannifin. Leading-edge manoeuvring flaps.

Fuselage: Semi-monocoque basic structure, primarily of light alloy, but with some graphite composite material used in structure of fuselage and airbrakes. Airbrake above fuselage between tail fins.

Tail Unit: Cantilever structure, primarily of light alloy, with swept vertical and horizontal surfaces. Twin outward-canted fins and rudders set forward of all-moving tailplane. Tailplane actuators manufactured by Parker-Hannifin.

Landing Gear: Retractable tricycle type, main units retracting aft, nose unit forward, with single wheel on each unit. Goodyear wheels, tyres and anti-skid brakes.

Power Plant: Two General Electric YJ101-GE-100 two-spool continuous-bleed turbojet engines with afterburning, each rated in the 62·25 kN (14,000 lb) thrust class. Provision for two 2,273 litre (600 US gallon) underwing fuel tanks.

Accommodation: Pilot only, on Stencel Aero IIIC ejection seat, in pressurised, heated and air-conditioned cockpit. Seat tilted back 18° from vertical. Upward-opening canopy, hinged at rear.

Systems: Environmental control system by Hamilton Standard. APU by Westinghouse Electric. Control stability augmentation system by Sperry Rand.

Electronics: Radar by Rockwell International, air data computer by Bendix, inertial navigation system by Litton Industries and transponder by Teledyne Electronics.

Armament: One General Electric M-61 multi-barrel 20 mm cannon mounted in fuselage nose. One heat-seeking infra-red Sidewinder missile mounted on each wingtip. Gunsight head-up display by JLM International. Snapshoot capability.

Dimensions, external:
Wing span 10·67 m (35 ft 0 in)
Wing aspect ratio 3·5
Length overall 17·07 m (56 ft 0 in)
Height overall 4·42 m (14 ft 6 in)
Tailplane span 6·77 m (22 ft 2½ in)
Wheel track 2·08 m (6 ft 10 in)
Wheelbase 5·25 m (17 ft 2¾ in)
Area:
Wings, gross 32·5 m² (350 sq ft)
Weight:
T-O weight, 'clean' 10,430 kg (23,000 lb)
Performance:
Max level speed above Mach 2
Max height attained during test programme
15,240 m (50,000 ft)
T-O to 15 m (50 ft) less than 305 m (1,000 ft)
Landing from 15 m (50 ft), without braking parachute
less than 610 m (2,000 ft)
Radius of action
more than 500 nm (927 km; 576 miles)
Ferry range
more than 2,600 nm (4,816 km; 2,993 miles)

PIASECKI

PIASECKI AIRCRAFT CORPORATION

Head Office and Works:
Island Road, International Airport, Philadelphia, Pennsylvania 19153
Telephone: (215) 365-2222
Directors:
Virgil Kauffman
Donald N. Meyers
Arthur J. Kania
J. Mecallef
F. K. Weyerhaeuser
F. N. Piasecki
President:
Frank N. Piasecki
Vice-President:
Donald N. Meyers (Engineering)
Secretary-Treasurer: Arthur J. Kania
Industrial Engineering: K. R. Meenen

The Piasecki Aircraft Corporation was formed in 1955 by Mr Frank Piasecki, who was formerly Chairman of the Board and President of the Piasecki Helicopter Corporation (now the Boeing Vertol Company).

PIASECKI HELI-STAT 97-212B

Recent work carried out by the company, under contract to the US Navy, is concerned with the investigation of a hybrid VTOL vehicle, named a Heli-Stat. This concept links the envelope of a lighter-than-air craft with current-technology helicopters, the aerostat providing static lift to support approximately the full empty weight of the entire assembly. The helicopters furnish the lift to support the payload, as well as providing propulsion and control, with adequate control forces to enable the Heli-Stat to hover with precision, a characteristic with which conventional airships cannot comply.

The project is now beyond the investigation stage, and Piasecki plans to have a prototype flying in 1977. The rigid structure supporting the envelope, and that carrying the landing gear and conventional helicopter/lifting units, will be of light alloy. The four twin-wheel units of the landing gear will be non-castoring, steerable, and will have hydraulic shock-absorption.

Power plant will consist, essentially, of four Bell UH-1N

helicopters, each powered by a Pratt & Whitney Aircraft of Canada PT6T Turbo Twin Pac, rated at 1,398 kW (1,875 shp). Total combined fuel capacity will be approximately 7,873 litres (2,080 US gallons).

Cargo space is provided in the central keel, which can be seen clearly in the accompanying illustration. Its dimensions are 4·88 m × 4·88 m × 34·75 m (16 ft × 16 ft × 114 ft), providing a volume of approximately 826·4 m³ (29,184 cu ft).

DIMENSIONS, EXTERNAL:
- Width over rotor blades: 45·42 m (149 ft)
- Length overall: 89·92 m (295 ft)
- Height overall: 37·19 m (122 ft)
- Wheel track: 24·38 m (80 ft)
- Wheelbase: 34·75 m (114 ft)

WEIGHTS:
- Weight empty: 23,517 kg (51,846 lb)
- Max payload: 32,023 kg (70,600 lb)
- Max T-O weight: 57,018 kg (125,704 lb)

PERFORMANCE (at max T-O weight):
- Max cruising speed: 60 knots (111 km/h; 69 mph)
- Max rate of climb at S/L: 305 m (1,000 ft)/min
- Service ceiling: 1,525 m (5,000 ft)
- Range with max fuel: 1,738 nm (3,218 km; 2,000 miles)
- Range with max payload: 51 nm (95 km; 59 miles)

Model of the proposed Piasecki Heli-Stat, which combines an aerostat with conventional helicopters

PIPER
PIPER AIRCRAFT CORPORATION

HEAD OFFICE AND WORKS:
Lock Haven, Pennsylvania 17745
Telephone: (717) 748-6711
Telex: 841425

OTHER WORKS:
Vero Beach, Florida 32960
Lakeland, Florida 33801
Piper, Pennsylvania 16845
Renovo, Pennsylvania 17764

BOARD OF DIRECTORS:
Lawrence R. Barnett
William G. Gunn
David F. Linowes
John J. Martin
Dudley C. Phillips
William T. Piper Jr
James J. Rochlis
Herbert J. Siegel
David W. Wallace

CHAIRMAN OF THE BOARD:
David W. Wallace

PRESIDENT AND CHIEF EXECUTIVE OFFICER:
J. Lynn Helms

SENIOR VICE-PRESIDENT:
Thomas W. Gillespie Jr (Marketing and Sales)

VICE-PRESIDENTS:
Marion J. Dees Jr (Engineering)
Findley A. Estlick (General Manager, Lock Haven plant)
Walter C. Jamouneau (Product Assurance and Safety)
Vincent J. Montuoro (General Manager, Vero Beach plant)
Dudley C. Phillips (General Counsel)
Richard B. Stockton (Finance and Administration)

CONTROLLER: Jack J. Cattoni
TREASURER: John R. Leeson
SECRETARY: John J. Martin
MANAGER, PUBLIC RELATIONS:
J. R. Skinner

In April 1976, Piper produced its 100,000th aircraft, a twin-turboprop Cheyenne.

Piper makes annual changes to all Cherokee models, but incorporates improvements in other types as they become available. Since 1964, when the PA-28-140 Cherokee 140 superseded the PA-22-108 Colt, the entire range of Piper products has been low-wing except for the PA-18 Super Cub. All types in current production are described in detail in the following pages.

Vero Beach is responsible for the experimental development of Piper aircraft and also houses one of the company's Plastics Divisions. Lock Haven also has R & D facilities for aircraft built at Lock Haven.

Piper operates also three other plants. The first two are at Piper, Pennsylvania, where sheet metal parts are formed, and Renovo, Pennsylvania, which makes plastics components; the third is at Lakeland, Florida, where PA-31 aircraft are in production.

Optional equipment on several current models includes a choice of four automatic flight systems. Simplest of these is the Piper-developed AutoFlite II, a solid-state system which holds the wings level and has turn-command capability for up to standard rate turns. This flight system is integrated with the pictorial turn rate indicator, which serves as the sensing element. If the Nav Tracker II is added to the AutoFlite II, it allows automatic tracking to and from Omni stations and is equipped with a two-position sensitivity switch.

Latest version of the established two-control autopilot is the Piper AutoControl III, which features positive heading lock and course selector, and is available with automatic VOR/ILS radio coupling.

A full three-control system, the Piper AltiMatic III-C, provides course preselection and positive heading lock; altitude preselection and hold, with automatic pitch trim. It is also available with automatic VOR/ILS radio coupling.

The fourth of the range of flight systems is the AltiMatic V/FD-1, an autopilot and flight director system, replacing the model V which was offered previously as an option for Aztecs, Navajos and Pressurised Navajos only. It consists of three basic panel components: a horizontal situation indicator, a steering horizon and a compact flight programmer console. The AltiMatic V-1 autopilot unit is available without the flight director unit.

The horizontal situation indicator combines directional gyro, VOR/LOC and glideslope information in a single instrument. The electrically-driven directional gyro unit is magnetically slaved. A fast-slaving function is automatic and conventional flag alarms are provided.

The steering horizon replaces the standard gyro horizon, providing normal attitude information in pictorial form. Computed information, required to direct the aircraft on its proper course, appears in the form of a command disc at each wingtip of a miniature aircraft in the gyro horizon indicator. To follow a programmed flight sequence the discs are held in alignment with the wingtips of the miniature aircraft, either by autopilot or by manual control.

The flight programmer console, which has five push-buttons, three annunciator lights and a pitch attitude control disc, can be programmed for autopilot action and/or flight director display. Once a sequence, such as an ILS approach, is initiated, logic circuits automatically execute sequential mode switching and dynamic response changes as required. Altitude is set by the pilot on the pitch attitude disc; when this is attained the disc is locked on the altitude, holding plus or minus 6 m (20 ft).

On 24 November 1971 Piper announced signature of an agreement with Chincul SA for the manufacture of a broad range of Piper products in Argentina. Chincul, a wholly-owned subsidiary of Piper's Argentine distributor, La Macarena SRL, is progressing through a series of manufacturing phases of increasing complexity, to allow the gradual assimilation of current aircraft manufacturing technology.

On 19 August 1974 Piper announced the signature of an agreement with Empresa Brasileira de Aeronautica SA (EMBRAER) for the development, production and marketing of Piper aircraft in Brazil. Further details can be found under the EMBRAER entry.

PIPER PA-18 SUPER CUB 150

There are two versions of the Super Cub 150, as follows:

Standard Super Cub 150. As described in detail.

De luxe Super Cub 150. As Standard model, but with addition of electric starter, generator, battery, navigation lights, sensitive altimeter, tie-down rings, control locks, parking brake and propeller spinner.

The original PA-18 with 67 kW (90 hp) Continental C90-12F engine received FAA Type Approval on 18 November 1949. The PA-18-150, PA-18A-150 agricultural aircraft and PA-18S and PA-18AS seaplanes were all approved on 1 October 1954.

The current international height record in Class C1b (aircraft with T-O weight of 500-1,000 kg) is held by Miss C. Bayley of the USA, who climbed to a height of 9,206 m (30,203 ft) in a Super Cub with 93·2 kW (125 hp) Lycoming engine, on 4 January 1951.

By mid-February 1976 more than 40,000 examples of the PA-18 Cub and its predecessors had been delivered.

TYPE: Two-seat light cabin monoplane.

WINGS: Braced high-wing monoplane, with steel tube Vee bracing struts each side. Wing section USA 35B. Thickness/chord ratio 12%. Dihedral 1°. No incidence at mean aerodynamic chord. Total washout of 3° 18'. Aluminium spars and ribs, aluminium sheet leading-edge and aileron false spar, wingtip bow of ash, with fabric covering overall and fire-resistant Duraclad plastic finish. Plain aluminium ailerons and flaps with fabric covering. No trim tabs.

FUSELAGE: Rectangular welded steel tube structure covered with fabric. Fire-resistant Duraclad plastic finish.

TAIL UNIT: Wire-braced structure of welded steel tubes and channels, covered with fabric. Fire-resistant Duraclad plastic finish. Tailplane incidence variable for trimming. Balanced rudder and elevators. No trim tabs.

LANDING GEAR: Non-retractable tailwheel type. Two side Vees and half axles hinged to cabane below fuselage. Rubber cord shock-absorption. Goodrich main-wheel tyres, size 8·00-4 four-ply, pressure 1·24 bars (18 lb/sq in). Steerable leaf-spring tailwheel by Maule (standard) or Scott (optional 8 in). Goodrich D-2-113 dual expanding brakes. Tandem-wheel landing gear, special

Piper PA-18 Super Cub 150 two-seat light cabin monoplane (Lycoming O-320 engine)

915 mm (36 in) low-pressure tyres, Federal skis or wheel-skis, or Edo 2000 standard or amphibious floats may be fitted.

Power Plant: One 112 kW (150 hp) Lycoming O-320 flat-four engine, driving a Sensenich two-blade metal fixed-pitch propeller. Steel tube engine mounting is hinged at firewall, allowing it to be swung to port for access to rear of engine. One 68 litre (18 US gallon) metal fuel tank in each wing. Total fuel capacity 136 litres (36 US gallons). Refuelling points on top of wing.

Accommodation: Enclosed cabin seating two in tandem with dual controls. Adjustable front seat. Rear seat quickly removable for cargo carrying. Inertia-reel shoulder harness standard for front and rear seats. Heater and adjustable cool-air vent. Downward-hinged door on starboard side, and upward-hinged window above, can be opened in flight. Sliding windows on port side. Baggage compartments aft of rear seat, capacity 22 kg (50 lb).

Electronics and Equipment: Equipment may be installed for spraying, dusting, fertilising, etc. Stall warning device standard. Optional extras include Narco Com 10A/Nav 10, Narco Com 11A, Narco Nav 11, automatic locator beacon, blind-flying instruments, vacuum system, outside air temperature gauge, dome light, radio speaker, electric fuel gauge, metal fuselage bottom panels, cabin fire extinguisher, 8-day clock, landing light, metallising and stainless steel control cables.

Dimensions, external:
Wing span 10·73 m (35 ft 2½ in)
Wing chord (constant) 1·60 m (5 ft 3 in)
Wing aspect ratio 7
Length overall:
landplane 6·88 m (22 ft 7 in)
seaplane 7·28 m (23 ft 11 in)
Height overall:
landplane 2·02 m (6 ft 8½ in)
seaplane 3·14 m (10 ft 3½ in)
Tailplane span 3·20 m (10 ft 6 in)
Wheel track 1·84 m (6 ft 0½ in)

Dimension, internal:
Baggage compartment 0·51 m³ (18 cu ft)

Areas:
Wings, gross 16·58 m² (178·5 sq ft)
Ailerons (total) 1·75 m² (18·80 sq ft)
Trailing-edge flaps (total) 1·07 m² (11·50 sq ft)
Fin 0·43 m² (4·66 sq ft)
Rudder 0·63 m² (6·76 sq ft)
Tailplane 1·40 m² (15·10 sq ft)
Elevators 1·09 m² (11·70 sq ft)

Weights and Loadings (N: Normal category; R: Restricted, agricultural, category):
Weight empty:
N landplane, R 422 kg (930 lb)
N seaplane 540 kg (1,190 lb)
Max T-O and landing weight:
N landplane 794 kg (1,750 lb)
N seaplane 798 kg (1,760 lb)
R 939 kg (2,070 lb)
Max wing loading:
N landplane 48·8 kg/m² (10·0 lb/sq ft)
N seaplane 48·8 kg/m² (10·0 lb/sq ft)
R 56·64 kg/m² (11·6 lb/sq ft)
Max power loading:
N landplane 7·09 kg/kW (11·6 lb/hp)
N seaplane 7·13 kg/kW (11·7 lb/hp)
R 8·38 kg/kW (13·8 lb/hp)

Performance (at max T-O weight: N: Normal category; R: Restricted, agricultural, category):
Never-exceed speed:
N, R 132 knots (246 km/h; 153 mph)
Max level speed at S/L:
N landplane 113 knots (208 km/h; 130 mph)
N seaplane 100 knots (185 km/h; 115 mph)
R 91 knots (169 km/h; 105 mph)
Max cruising speed (75% power):
N landplane 100 knots (185 km/h; 115 mph)
N seaplane 89 knots (166 km/h; 103 mph)
R 78 knots (145 km/h; 90 mph)
Econ cruising speed:
N landplane 91 knots (169 km/h; 105 mph)
R 78 knots (145 km/h; 90 mph)
Stalling speed, flaps down:
N landplane 38 knots (69 km/h; 43 mph)
N seaplane 37 knots (67 km/h; 42 mph)
R 39 knots (73 km/h; 45 mph)
Max rate of climb at S/L:
N landplane 293 m (960 ft)/min
N seaplane 253 m (830 ft)/min
R 232 m (760 ft)/min
Service ceiling:
N landplane 5,795 m (19,000 ft)
N seaplane 5,335 m (17,500 ft)
R 5,180 m (17,000 ft)
Absolute ceiling:
N landplane 6,492 m (21,300 ft)
N seaplane 5,943 m (19,500 ft)
T-O run:
N landplane 61 m (200 ft)
N seaplane 214 m (700 ft)
R 92 m (300 ft)
T-O to 15 m (50 ft):
N landplane 153 m (500 ft)
N seaplane 300 m (990 ft)
R 290 m (950 ft)
Landing from 15 m (50 ft):
N landplane 221 m (725 ft)
N seaplane 223 m (730 ft)
R 267 m (875 ft)
Landing run:
N landplane 107 m (350 ft)
N seaplane 131 m (430 ft)
R 125 m (410 ft)
Range with max fuel and max payload:
N landplane 399 nm (735 km; 460 miles)
N seaplane 375 nm (663 km; 412 miles)
R 312 nm (580 km; 360 miles)

PIPER PA-23-250 AZTEC F

The 1976 version of the Aztec, designated Model F, has a number of improvements as standard. These include a flap and tailplane interconnection, which automatically trims the aircraft to neutralise pitch control pressures when the flaps are lowered; reduced control forces to improve feel and handling; new brake assembly; a redesigned fuel system; new easy-to-scan instrument panel; new front seats with increased leg room; and new interior styling. It is available in several configurations, as follows:

Custom Aztec. Basic model, as described in detail.

Sportsman Aztec. As Custom model, with addition of Piper external power socket and de luxe Palm Beach interior; adding total of 3·08 kg (6·8 lb) to basic empty weight.

Professional Aztec. As Sportsman model, with addition of electrical propeller de-icing and pneumatic de-icing boots on wings and tail, but without de luxe interior; adding a total of 22·14 kg (48·8 lb) to basic empty weight.

Turbo Aztec. Turbocharged version, described separately.

Each of the above versions can be fitted with one of six electronics packages, as follows:

Electronic Group N 1-23. Two Narco Com 11A 360-channel VHF transceivers; Narco Nav 11 200-channel VOR/LOC receiver with converter indicator; Narco Nav 12 200-channel VOR/LOC receiver with glideslope deviation and indicator; UGR-3 glideslope receiver; ADF-140; AT-50A transponder; MBT-24R marker beacon receiver; CP-125 audio panel; Piper AutoControl IIIB; Piper electric trim; Piper VOR/LOC coupler to autopilot; Piper anti-static antennae and wicks, headset, noise-cancelling microphone, radio selector panel; adding total of 22·95 kg (50·6 lb) to basic empty weight. Dual ADF and HF optional.

Electronic Group N 2-23. As Group N 1-23, with deletion of Narco Com 11As, Nav 11 and Nav 12 and Piper AutoControl IIIB. Replaced by dual Narco Com 11B 720-channel VHF transceivers; Nav 14 VOR receiver; DME-190; Piper AltiMatic IIIC autopilot and glideslope coupler to autopilot; adding 35·11 kg (77·4 lb) to basic empty weight. Dual ADF and HF optional.

Electronic Group NT 3-23. As Group N 2-23, except for deletion of dual Com 11B, Nav 11 and Nav 14, and their replacement by dual Com 111B 720-channel VHF transceivers and Nav 111 and 114; adding 34·65 kg (76·4 lb) to basic empty weight. Dual ADF and HF optional.

Electronic Group KS 1-23. Dual King KX 170B nav/com transceivers with 720-channel com and 200-channel nav; KI 214 VOR/LOC/glideslope indicator with VOR/LOC converter and 40-channel glideslope receiver; KI 201C VOR/LOC converter indicator; KR 86 ADF; KT 76 transponder; KMA 20 audio panel; marker beacon receiver with indicator lights; KN 60C DME; Piper AutoControl IIIB; Piper VOR/LOC coupler to autopilot; Piper 100T microphone; Piper anti-static kit; and Piper electric trim; adding 24·81 kg (54·7 lb) to basic empty weight. Dual ADF and HF optional.

Electronic Group KTS 2-23. Dual King KX 175B nav/com transceivers with 720-channel com and 200-channel nav; dual KNI 520 VOR/LOC/glideslope indicators with VOR/LOC/RNAV/glideslope deviation, to-from indication and warning flags; dual KN 77 VOR/LOC converters; KN 73 40-channel glideslope receiver; KR 85 ADF with KI-225 indicator; KT 76 transponder; KMA 20 audio panel, marker beacon receiver and indicator; KN 65 DME; Piper AltiMatic IIIC with electric trim; Piper VOR/LOC coupler to autopilot; Piper glideslope coupler to autopilot; Piper 100T noise-cancelling microphone; and Piper anti-static kit; adding 43·1 kg (95 lb) to basic empty weight. Dual ADF and HF optional.

Electronic Group CTM-1-23. Dual Collins VHF-251 720-channel com transceivers, with Piper noise-cancelling microphone, headset and broad band com antenna; dual Collins VIR-351 nav receivers with VOR/LOC converter and standard VOR antenna; HSI-NSD-360 VOR/LOC/GS indicator No. 1; Collins GLS-350 glideslope receiver; Collins IND-350 VOR/LOC indicator No. 2; Collins AMR-350 audio selector panel, audio amplifier, marker beacon lights and marker beacon receiver; Collins ADF-650 and IND-650 indicator; Collins TDR-950 transponder with 4096 code capability; King KN-65A DME with KA-43 tuning adaptor and KI-266 digital display; Piper AltiMatic IIIC with VOR/LOC/GS coupling and longitudinal electric trim; Piper electronics master switch; Piper static discharge wicks; and dual Collins PWC-150 power converters to supply 14V DC for dual com/nav systems.

Most items covered under model and electronic group listings are available individually and, in addition, a number of items of nav/com equipment are available as options, including the AVQ-47 weather-avoidance radar.

The Aztec received FAA Type Approval as a five-seat aircraft on 18 September 1959, and with six seats on 15 December 1961.

The prototype of a floatplane version of the Aztec was produced as a joint project by Melridge Aviation of Vancouver, Washington, and Jobmaster Company Inc of Seattle. Fitted with Edo 4930 floats, this aircraft can take off from calm water in 20 seconds at max T-O weight of 2,360 kg (5,200 lb). Useful load is 816 kg (1,800 lb), permitting a six-passenger load with 455 litres (120 US gallons) of fuel. To simplify docking and loading from either side, a door was designed for installation on the port side, by the pilot's seat, and is part of the conversion kit offered by Melridge Aviation to permit conversion in the field.

Twenty Aztecs were supplied to the US Navy as 'off-the-shelf' utility transports, under the designation U-11A (formerly UO-1). Several South and Central American governments and armed services have also acquired Aztecs, notably the Argentine Army, which took delivery of six in 1964. The French and Spanish Air Forces have each acquired two.

Approximately 4,000 Aztecs have been produced.

Type: Six-seat twin-engined executive transport.

Wings: Cantilever low-wing monoplane. Wing section USA 35-B (modified). Thickness/chord ratio 14%. Dihedral 5°. Incidence 0° at root, −1° 12′ at mean chord. All-metal stressed-skin structure, with heavy stepped-down main spar, front and rear auxiliary spars, ribs, stringers and detachable wingtips. Plain all-metal ailerons and hydraulically-actuated flaps. Optional Goodrich de-icing system.

Fuselage: Basic aluminium semi-monocoque structure with welded steel tube truss around cabin.

Tail Unit: Cantilever all-metal structure with swept fin and all-moving horizontal surfaces. Trim tab in rudder. Geared anti-servo tab in horizontal surfaces. Optional Goodrich de-icing system.

Landing Gear: Retractable tricycle type. Hydraulic retraction, with CO_2 emergency extension system. Nosewheel retracts rearward, main wheels forward. Wheel doors enclose landing gear fully when retracted. Electrol oleo shock-absorber struts. Cleveland main wheels, size 6·00-6, with size 7·00-6 8-ply type III tyres.

Piper PA-23-250 Aztec F six-seat cabin monoplane (two Lycoming IO-540-C4B5 engines)

Cleveland steerable nosewheel, size 6·00-6, with 6·00-6 4-ply type III tyre. Hydraulic disc brakes. Parking brake.

POWER PLANT: Two 186·5 kW (250 hp) Lycoming IO-540-C4B5 flat-six engines, each driving a Hartzell HC-E2YK-2RB constant-speed fully-feathering two-blade metal propeller. Two rubber fuel cells in each wing with NACA type anti-icing non-siphoning vents. Total standard fuel capacity 544 litres (144 US gallons); 530 litres (140 US gallons) usable. Refuelling points above wings. Two internal wingtip fuel tanks optional, with total capacity of 151 litres (40 US gallons). Propeller synchroniser and electrical de-icing system optional.

ACCOMMODATION: Six persons on two pairs of adjustable individual seats and rear bench seat. Dual controls standard. Individual seat lights and controllable overhead ventilation. Southwind 35,000 BTU heater with four adjustable cool/warm air outlets and two windscreen defrosters. Heated windscreen optional. Double windows. Passenger step. Door at front of cabin on starboard side. Emergency exit at rear on port side. Centre and rear seats removable to provide space for stretcher, survey camera or up to 725 kg (1,600 lb) freight. Rear cabin bulkhead removable for stretcher and cargo loading via rear baggage door. Baggage compartments at rear of cabin and in nose, with tie-down fittings, each with capacity of 68 kg (150 lb). Baggage doors on starboard side; rear one enlarged on current aircraft, for stretcher loading. Armrests, cabin dome light, individual reading lights, coat hooks, complete soundproofing and two sun visors. Seat headrests optional. Choice of five interior trims.

SYSTEMS: Hydraulic system, pressure 79 bars (1,150 lb/sq in), for landing gear and flaps. Two 70A 28V alternators. 28V 17Ah battery.

ELECTRONICS AND EQUIPMENT: Standard equipment includes full blind-flying instrumentation with 3 in pictorial rate of turn indicator, artificial horizon and directional gyro (flight instruments arranged in 'T' configuration), sensitive altimeter, Piper TruSpeed Indicator, alternate static source, clock, gyro air filter, outside air temperature gauge, dual vacuum gauges, stall warning indicator, dual recording tachometers, oil pressure, oil temperature, cylinder head temperature and fuel quantity gauges, ammeter (with test switch), dual fuel flow and manifold pressure gauges, flap position indicator, navigation lights, glare-ban instrument lights, landing light, taxi light, white wingtip anti-collision lights, two map lights, two door-ajar indicator lights, baggage compartment courtesy lights, heated pitot tube, two quick oil drains, NACA-type anti-icing non-siphoning fuel tank vents, towbar, tie-down rings, jack pads, nosewheel safety mirror, cabin and baggage door locks, cabin curtains and a wide choice of three-tone exterior trims. Optional items listed under descriptions of individual models and under electronic groups, plus altimeter and toe-brakes for co-pilot, dual tachometer and hour meter, fire extinguisher, blind-flying instrumentation for co-pilot, oxygen system with 3·23 m³ (114 cu ft) bottle and six outlets, Palm Beach interior trim, Piper mixture control indicator, Piper automatic locator, and propeller ice protection shields.

DIMENSIONS, EXTERNAL:
- Wing span: 11·34 m (37 ft 2½ in)
- Wing chord (constant): 1·70 m (5 ft 7 in)
- Wing aspect ratio: 6·8
- Length overall: 9·52 m (31 ft 2¾ in)
- Height overall: 3·15 m (10 ft 4 in)
- Tailplane span: 3·81 m (12 ft 6 in)
- Wheel track: 3·45 m (11 ft 4 in)
- Wheelbase: 2·29 m (7 ft 6 in)
- Propeller diameter: 1·96 m (6 ft 5 in)
- Cabin door: Height: 0·97 m (3 ft 2 in)
  - Width: 0·84 m (2 ft 9 in)
- Baggage compartment door (front):
  - Height: 0·51 m (1 ft 8 in)
  - Width: 0·76 m (2 ft 6 in)
- Baggage compartment door (rear):
  - Height: 0·76 m (2 ft 6 in)
  - Width: 0·79 m (2 ft 7 in)

DIMENSIONS, INTERNAL:
- Baggage compartments:
  - front: 0·60 m³ (21·3 cu ft)
  - rear: 0·72 m³ (25·4 cu ft)
- Max cargo space, incl baggage compartments: 3·45 m³ (122 cu ft)

AREAS:
- Wings, gross: 19·28 m² (207·56 sq ft)
- Ailerons (total): 0·77 m² (8·38 sq ft)
- Trailing-edge flaps (total): 1·54 m² (16·60 sq ft)
- Fin: 1·37 m² (14·80 sq ft)
- Rudder: 0·96 m² (10·30 sq ft)
- Horizontal surfaces (total): 3·70 m² (39·80 sq ft)

WEIGHTS AND LOADINGS:
- Weight empty (Custom): 1,442 kg (3,180 lb)
- Max T-O and landing weight: 2,360 kg (5,200 lb)
- Max wing loading: 122·3 kg/m² (25·05 lb/sq ft)
- Max power loading: 6·33 kg/kW (10·4 lb/hp)

PERFORMANCE (at max T-O weight):
- Never-exceed speed: 240 knots (446 km/h; 277 mph)
- Max level speed: 187 knots (346 km/h; 215 mph)
- Normal cruising speed at 1,220 m (4,000 ft): 182 knots (338 km/h; 210 mph)
- Intermediate cruising speed at 1,830 m (6,000 ft): 175 knots (324 km/h; 201 mph)
- Econ cruising speed at 1,950 m (6,400 ft): 167 knots (309 km/h; 192 mph)
- Long-range cruising speed at 3,110 m (10,200 ft): 150 knots (278 km/h; 172 mph)
- Stalling speed, flaps down: 59 knots (109 km/h; 68 mph)
- Max rate of climb at S/L: 426 m (1,400 ft)/min
- Rate of climb at S/L, one engine out: 72 m (235 ft)/min
- Absolute ceiling: 5,775 m (18,950 ft)
- Absolute ceiling, one engine out: 1,905 m (6,250 ft)
- T-O run: 288 m (945 ft)
- T-O to and landing from 15 m (50 ft): 517 m (1,695 ft)
- Landing run: 400 m (1,310 ft)
- Accelerate-stop distance: 605 m (1,985 ft)
- Range with max fuel, 45 min reserves:
  - Normal cruising speed: 880 nm (1,630 km; 1,012 miles)
  - Intermediate cruising speed: 1,060 nm (1,963 km; 1,220 miles)
  - Econ cruising speed: 1,110 nm (2,055 km; 1,277 miles)
  - Long-range cruising speed: 1,320 nm (2,445 km; 1,519 miles)

PIPER PA-23-250 TURBO AZTEC F

The Turbo Aztec F is identical in every way with the Aztec F, except that it has 186·5 kW (250 hp) Lycoming TIO-540-C1A engines, fitted with the AiResearch turbocharging system. These specially modified engines allow a turbo cruise setting at 2,400 rpm, providing a constant manifold pressure from sea level to 6,700 m (22,000 ft), and result in considerably improved performance.

Standard equipment includes a density controller to prevent inadvertent overboost of the engines at full throttle, and a differential pressure controller to provide constant manifold pressure during cruising flight. An oxygen system with 3·23 m³ (114 cu ft) bottle and six outlets is optional.

During the Summer of 1972 the Spanish Air Force took delivery of six Turbo Aztecs which, together with two Aztecs acquired earlier, are used largely for instrument flight training.

WEIGHTS AND LOADINGS:
- As for Aztec F, except:
- Weight empty (standard): 1,464 kg (3,229 lb)

PERFORMANCE (at max T-O weight):
- As for Aztec F, except:
- Max level speed at 5,639 m (18,500 ft): 220 knots (407 km/h; 253 mph)
- Turbo cruising speed at 3,050 m (10,000 ft): 192 knots (355 km/h; 221 mph)
- Turbo cruising speed at 6,705 m (22,000 ft): 210 knots (389 km/h; 241 mph)
- Intermediate cruising speed at 3,050 m (10,000 ft): 182 knots (337 km/h; 209 mph)
- Intermediate cruising speed at 7,315 m (24,000 ft): 203 knots (376 km/h; 233 mph)
- Econ cruising speed at 3,050 m (10,000 ft): 171 knots (316 km/h; 196 mph)
- Econ cruising speed at 7,315 m (24,000 ft): 193 knots (357 km/h; 222 mph)
- Long-range cruising speed at 3,050 m (10,000 ft): 156 knots (289 km/h; 179 mph)
- Long-range cruising speed at 7,315 m (24,000 ft): 165 knots (305 km/h; 190 mph)
- Max rate of climb at S/L: 448 m (1,470 ft)/min
- Rate of climb at S/L, one engine out: 68 m (225 ft)/min
- Absolute ceiling: over 7,315 m (24,000 ft)
- Absolute ceiling, one engine out: 5,180 m (17,000 ft)
- Range with max fuel, 45 min reserves:
  - Turbo cruising speed at 3,050 m (10,000 ft): 885 nm (1,639 km; 1,018 miles)
  - Turbo cruising speed at 6,705 m (22,000 ft): 945 nm (1,750 km; 1,087 miles)
  - Intermediate cruising speed at 3,050 m (10,000 ft): 945 nm (1,750 km; 1,087 miles)
  - Intermediate cruising speed at 7,315 m (24,000 ft): 1,020 nm (1,889 km; 1,173 miles)
  - Econ cruising speed at 3,050 m (10,000 ft): 1,005 nm (1,861 km; 1,156 miles)
  - Econ cruising speed at 7,315 m (24,000 ft): 1,075 nm (1,990 km; 1,237 miles)
  - Long-range cruising speed at 3,050 m (10,000 ft): 1,110 nm (2,055 km; 1,277 miles)
  - Long-range cruising speed at 6,100 m (20,000 ft): 1,145 nm (2,120 km; 1,317 miles)

PIPER PA-25 PAWNEE D

The PA-25 Pawnee was developed by Piper's Vero Beach Development Center as a specialised agricultural aircraft for dispersal of chemical dusts and sprays. Special attention was paid to pilot safety, bearing in mind the recommendations of the Crash Injury Research Unit of Cornell Medical College. Thus, the pilot is placed high to ensure a good view, including rearward, during low flying. Extra-strong seat belt and shoulder harness are fitted, and a rounded sheet metal cushion is provided above the instrument panel to prevent the pilot's head from striking the instruments in a severe crash.

The fuselage is designed to fail progressively from the front to reduce the deceleration of the cockpit, and in ordinary low-speed crashes of the kind usually associated with crop-spraying and crop-dusting the pilot's compartment should remain substantially undamaged. The top longerons in the cockpit bay are given a slight outward bulge, so that they would fail outwards in a severe head-on crash. All heavy objects or loads are forward of the cockpit and there is a 0·25 m (10 in) space between the metal floor and the bottom of the fuselage to provide additional safety in a relatively flat crash.

The initial production version of the Pawnee had a 112 kW (150 hp) Lycoming engine, but production is now concentrated on the 175 or 194 kW (235 hp or 260 hp) Pawnee D, to which the detailed description applies.

The D was introduced in 1973, with a number of improved features, including wing fuel tanks replacing the rubberised fuselage fuel cell of the Pawnee C, and introducing an engine bay fire extinguisher as standard equipment. The entire top of the fuselage from the cockpit to the fin can be removed in 60 seconds to provide easy access for inspection and cleaning. A high-capacity cockpit ventilation system is installed. Ventilating air taken in through an intake in the top of the canopy is used also to pressurise lightly the rear fuselage to keep out dust and chemicals. The engine installation is modified to permit efficient operation under the most severe hot weather conditions. The landing gear is fitted with oleo-pneumatic shock-absorbers. A pilot's adjustable seat is installed and there is a safety exit on each side of the cockpit.

The first five production Pawnees were delivered in August 1959, with subsequent aircraft leaving the assembly line at the rate of one a day, later increased to two a day.

TYPE: Single-seat agricultural monoplane.

WINGS: Braced low-wing monoplane, based on wings of Super Cub. Streamlined Vee bracing struts on each side of fuselage, with additional short support struts. Wing

Piper PA-25 Pawnee D agricultural aircraft

section USA 35B (modified). Thickness/chord ratio 12%. Dihedral 7°. Incidence 1° 18′ at mean aerodynamic chord. Wings, ailerons and flaps are all of fabric-covered aluminium construction, with fire-resistant Duraclad plastic finish. No trim tabs.

FUSELAGE: Basically rectangular-section welded steel tube structure, with fabric covering and Duraclad plastic finish, except for removable metal underskin and removable metal top of rear fuselage. Glassfibre engine cowling.

TAIL UNIT: Wire-braced steel tube structure with fabric covering and Duraclad plastic finish. Fixed-incidence tailplane. Balanced rudder and elevators. No trim tabs. Cable from top of cockpit to top of rudder to deflect wires and cables.

LANDING GEAR: Non-retractable tailwheel type. Oleo-pneumatic shock-absorbers. Main gear has two side Vees and half-axles hinged to centreline of underside of fuselage. Cleveland 40-61 main wheels, with 8·00-6 4-ply tyres, pressure 1·72 bars (25 lb/sq in). Cleveland type 30-41 toe-actuated hydraulic brakes. Parking brake. Wire-cutters on leading-edge of each side Vee. Scott 200 mm (8 in) steerable tailwheel, tyre pressure 3·45 bars (50 lb/sq in).

POWER PLANT: One 175 kW (235 hp) (derated) Lycoming O-540-B2B5 flat-six engine, driving a McCauley Type 1A200/FA84 two-blade metal fixed-pitch propeller. Optionally, one 194 kW (260 hp) Lycoming O-540-E engine, with two-blade fixed-pitch propeller or optional constant-speed propeller. Fuel tank in each wing, with combined capacity of 145·7 litres (38·5 US gallons), of which 136 litres (36 US gallons) are usable. Oil capacity 11·4 litres (3 US gallons).

ACCOMMODATION: Pilot on adjustable seat in specially-strengthened enclosed cockpit, with steel tube overturn structure. Heavy-duty safety belt and shoulder harness with inertia reel. Wire-cutter mounted on centre of windscreen. Combined window and door on each side, hinged at bottom. Window assemblies jettisonable for emergency exit. Cabin is heated and ventilated. Adjustable cool air vents. Air-conditioning unit optional. A jump-seat can be fitted in the hopper to transport a mechanic or loader between operations. Utility compartment under seat.

SYSTEMS: Electrical system includes a 60A alternator, 14V 35Ah battery and a battery charging diode. Hydraulic system for brakes only.

ELECTRONICS AND EQUIPMENT: Standard equipment includes a non-corrosive hopper/tank, installed forward of cockpit and approximately on CG, volume of which is 0·59 m³ (21 cu ft) or 568 litres (150 US gallons), with capacity for 544 kg (1,200 lb) of dust; quick-change boom brackets, quick-drain gascolator, low-quantity fuel warning light, full-flow oil filter, quick-drain oil sump, quick-dump valve to jettison hopper contents in emergency, quick-release hinge pins in side windows, tie-down rings, top-deck loading door, sensitive altimeter, engine bay fire extinguisher and provision for automatic locator beacon. Spray system uses similar 1 in Simplex centrifugal pump to that on PA-18A, with spraybars. The venturi distributor used for dry chemicals gives a total effective swath width of up to 18·3 m (60 ft). Changeover from dust to spray, and vice versa, takes less than five minutes. Optional side loading nozzle for liquid chemicals. Optional equipment includes Narco Com 10A/Nav 10 transceiver, multi-directional inertia-reel shoulder harness (exchange), automatic locator beacon, control lock, hand fire extinguisher, landing lights, navigation lights, rotating beacon, electric turn and bank indicator and metallisation.

DIMENSIONS, EXTERNAL:
- Wing span: 11·02 m (36 ft 2 in)
- Wing chord (constant): 1·60 m (5 ft 3 in)
- Wing aspect ratio: 7·15
- Length overall: 7·53 m (24 ft 8½ in)
- Height overall: 2·21 m (7 ft 3 in)
- Tailplane span: 2·90 m (9 ft 6 in)
- Wheel track: 2·13 m (7 ft 0 in)
- Wheelbase: 5·52 m (18 ft 1¼ in)
- Propeller diameter: 2·13 m (7 ft 0 in)

AREAS:
- Wings, gross: 17·0 m² (183 sq ft)
- Ailerons (total): 1·78 m² (19·2 sq ft)
- Trailing-edge flaps (total): 0·78 m² (8·4 sq ft)
- Fin: 0·35 m² (3·8 sq ft)
- Rudder: 0·64 m² (6·9 sq ft)
- Tailplane: 1·21 m² (13·0 sq ft)
- Elevators: 1·27 m² (13·7 sq ft)

WEIGHTS AND LOADINGS (A: 175 kW; 235 hp engine; B: 194 kW; 260 hp engine, fixed-pitch propeller; C: 194 kW; 260 hp engine, constant-speed propeller):

Weight empty:
- A no dispersal equipment: 644 kg (1,420 lb)
- B no dispersal equipment: 668 kg (1,472 lb)
- C no dispersal equipment: 675 kg (1,488 lb)
- A duster: 671 kg (1,479 lb)
- B duster: 694 kg (1,531 lb)
- C duster: 702 kg (1,547 lb)
- A sprayer: 675 kg (1,488 lb)
- B sprayer: 698 kg (1,540 lb)
- C sprayer: 706 kg (1,556 lb)

- Max T-O and landing weight: 1,315 kg (2,900 lb)
- Max wing loading: 77·15 kg/m² (15·8 lb/sq ft)
- Max power loading: 7·51 kg/kW (12·3 lb/hp)

PERFORMANCE (at max T-O weight, except where indicated):
- Never-exceed speed: 135 knots (251 km/h; 156 mph)

Max level speed at S/L:
- A no dispersal equipment: 108 knots (200 km/h; 124 mph)
- B, C no dispersal equipment: 111 knots (206 km/h; 128 mph)
- A duster: 96 knots (177 km/h; 110 mph)
- B, C duster: 98 knots (182 km/h; 113 mph)
- A sprayer: 102 knots (188 km/h; 117 mph)
- B, C sprayer: 104 knots (193 km/h; 120 mph)

Max cruising speed (75% power):
- A no dispersal equipment: 99 knots (183 km/h; 114 mph)
- B no dispersal equipment: 100 knots (185 km/h; 115 mph)
- C no dispersal equipment: 101 knots (187 km/h; 116 mph)
- A, B, C duster: 87 knots (161 km/h; 100 mph)
- A sprayer: 91 knots (169 km/h; 105 mph)
- B, C sprayer: 92 knots (171 km/h; 106 mph)

- Stalling speed, flaps down: 53 knots (98 km/h; 61 mph)
- Stalling speed at normal landing weight of 771 kg (1,700 lb): 40 knots (74 km/h; 46 mph)

Max rate of climb at S/L:
- A no dispersal equipment: 213 m (700 ft)/min
- B no dispersal equipment: 230 m (755 ft)/min
- C no dispersal equipment: 236 m (775 ft)/min
- A duster: 152 m (500 ft)/min
- B duster: 169 m (555 ft)/min
- C duster: 175 m (575 ft)/min
- A sprayer: 192 m (630 ft)/min
- B sprayer: 209 m (685 ft)/min
- C sprayer: 215 m (705 ft)/min

T-O run:
- A no dispersal equipment: 239 m (785 ft)
- B no dispersal equipment: 223 m (730 ft)
- C no dispersal equipment: 201 m (660 ft)
- A duster: 291 m (956 ft)
- B duster: 271 m (890 ft)
- C duster: 253 m (830 ft)
- A sprayer: 244 m (800 ft)
- B sprayer: 226 m (740 ft)
- C sprayer: 207 m (680 ft)

T-O to 15 m (50 ft):
- A no dispersal equipment: 411 m (1,350 ft)
- B no dispersal equipment: 381 m (1,250 ft)
- C no dispersal equipment: 366 m (1,200 ft)
- A duster: 428 m (1,470 ft)
- B duster: 433 m (1,420 ft)
- C duster: 418 m (1,370 ft)
- A sprayer: 418 m (1,370 ft)
- B sprayer: 387 m (1,270 ft)
- C sprayer: 372 m (1,220 ft)

- Max landing run: 259 m (850 ft)

Range (75% power) with max fuel:
- A no dispersal equipment: 251 nm (467 km; 290 miles)
- B no dispersal equipment: 247 nm (459 km; 285 miles)
- C no dispersal equipment: 251 nm (467 km; 290 miles)
- A duster: 221 nm (410 km; 255 miles)
- B, C duster: 199 nm (370 km; 230 miles)
- A sprayer: 234 nm (434 km; 270 miles)
- B, C sprayer: 230 nm (426 km; 265 miles)

PIPER PA-28 CHEROKEE SERIES

The Cherokee is a low-cost all-metal low-wing monoplane which is available in 2/4, 4- and 6/7-seat versions. All except the Cherokee Lance and Arrow II have a non-retractable tricycle landing gear. The Cherokee Lance and Arrow II gear is retractable.

Only 1,200 parts go into the manufacture of a Cherokee, compared with over 1,600 in the four-seat high-wing Tri-Pacer, which preceded it. The first production Cherokee flew on 10 February 1961.

Models currently available are the Cherokee Cruiser (2/4-seat); Warrior, Archer II, Pathfinder and Arrow II (4-seat); Cherokee SIX and Lance (6/7-seat). During 1975 Piper delivered a total of 1,714 Cherokees of all versions.

Descriptions of all current models follow:

PIPER CHEROKEE CRUISER

Piper introduced in 1972 a de luxe version of the Cherokee 140 which was named Cherokee Cruiser 2 plus 2 and had as standard two easily-removable rear family seats and other refinements. In 1973 this became the standard production model, replacing the Cherokee 140, which was described in the 1972-73 *Jane's*; in 1974 it was renamed Cherokee Cruiser. The two rear family seats, with seat-belts, two additional cabin fresh air vents, de luxe baggage compartment and hatshelf, now form an optional Family Group, adding 12·25 kg (27 lb) to the basic empty weight.

The 1976 version of the Cruiser introduces as standard a new cabin door with improved latching and construction which reduces cabin noise significantly, improved seats and new exterior finish.

Two groups of optional equipment are available for the Cruiser as basic packages:

Custom. Comprises Piper TruSpeed indicator; instrument panel white backlighting and overhead red spotlight, cabin dome, navigation, landing/taxi, and radio dimming lights; wheel speed fairings; rotating beacon; sensitive altimeter; assist strap; aircraft step; adding 8·75 kg (19·3 lb) to basic empty weight.

Executive. As Custom package, plus vacuum system with engine-driven pump and advanced instrument panel comprising 3 in pictorial gyro horizon, 3 in directional gyro, Piper pictorial turn rate indicator, rate of climb indicator, outside air temperature gauge and eight-day clock; adding 16 kg (35·3 lb) to basic empty weight.

The Cherokee Cruiser can be fitted with one of six electronics packages as follows:

Electronic Group N 1-28/32. Narco Com 11A 360-channel VHF transceiver; Nav 11 200-channel VOR/LOC receiver with converter indicator; AT-50A transponder; M-700A noise-cancelling microphone and headset; adding 6·8 kg (15 lb) to basic empty weight.

Electronic Group N 2-28/32. Dual Narco Com 11A 360-channel VHF transceivers; dual Nav 11 200-channel VOR/LOC receivers with converter indicators: CP-125 audio selector panel; MBT-12R marker beacon receiver; ADF-140; AT-50A transponder; Piper AutoControl IIIB autopilot with VOR/LOC coupling and electric trim; M-700A noise-cancelling microphone, headset and two broad band com antennae; adding 22·2 kg (49 lb) to basic empty weight.

Electronic Group NT 3-28/32. As Group N 2-28/32, except dual Com 11As replaced by dual Com 111B 360-channel VHF transceivers; dual Nav 11s replaced by one Nav 111 200-channel VOR/LOC receiver with converter indicator, including 40 localiser channels, and one Nav 112 200-channel VOR/LOC receiver with converter indicator, including UGR-2A glideslope receiver; adding 25 kg (55 lb) to basic empty weight.

Electronic Group KS 1-28/32. King KX 170B 720-channel com transceiver, with 200-channel nav receiver; KI 201C VOR/LOC indicator; KT-78 transponder; Piper headset and 66C microphone; adding 8·6 kg (19 lb) to basic empty weight.

Electronic Group KS 2-28/32. Dual KX 170B 720-channel com transceivers, with 200-channel nav receivers; dual KI 201C VOR/LOC converter indicators; KMA 20 audio panel with marker beacon receiver and indicator lights; KR 68 ADF with BFO; KT 78 transponder; Piper AutoControl IIIB autopilot; Piper VOR/LOC coupler;

Piper Cherokee Cruiser, basic version of the Cherokee series (Lycoming O-320 engine)

Piper electric trim; Piper static wicks; dual Piper broad band com antennae; and Piper 66C microphone and headset; adding 26·3 kg (58 lb) to basic empty weight.

Electronic Group KTS 3-28/32. Dual KX 175B nav/com transceivers, with 720-channel com transceivers and 200-channel nav receivers; KNI 520 VOR/LOC/glideslope indicator with to-from indication and appropriate warning flags; KNI 520 VOR/LOC indicator; dual KN 77 VOR/LOC converters; KN 73 glideslope receiver; KR 85 ADF; KT 76 transponder; Piper AutoControl IIIB autopilot; Piper VOR/LOC coupler; Piper electric trim; dual Piper broad band com antennae; Piper 100T noise-cancelling microphone and headset; adding 31 kg (68 lb) to basic empty weight.

The structural description of the Cherokee Warrior (which see) applies also to the Cherokee Cruiser, except for the following details:

TYPE: Two/four-seat sporting and training light aircraft.

POWER PLANT: One 112 kW (150 hp) Lycoming O-320 flat-four engine, driving a two-blade fixed-pitch metal propeller. Two fuel tanks in wing leading-edges, with total capacity of 189 litres (50 US gallons), of which 181·5 litres (48 US gallons) are usable. Standard fuel capacity of 136 litres (36 US gallons), of which 128·5 litres (34 US gallons) are usable, the remaining tankage for 53 litres (14 US gallons) being regarded as a reserve.

ACCOMMODATION: Two reclinable individual front seats side by side in enclosed cabin, with dual controls. Adjustable seats, which raise, lower and tilt, are available optionally. Two optional full-size family seats in rear of cabin, which are easily removable to accommodate up to 90 kg (200 lb) of freight or baggage. Inertia-reel shoulder harness is standard for the two front seats, optional for rear seats. Door on starboard side. Cabin heated and ventilated.

EQUIPMENT: The following items of optional equipment are available: Piper Aire air-conditioning system, alternate static source, super soundproofing, Piper automatic locator, ventilation fan for overhead vent system, fire extinguisher, 35Ah battery, external power socket, heated pitot tube, adjustable front seats, headrests, solar control windows, red tail and white wingtip strobe lights, zinc chromate application and stainless steel control cables.

DIMENSIONS AND AREAS:

As for Cherokee Archer, except:

Length overall	7·10 m (23 ft 3½ in)
Propeller diameter	1·88 m (6 ft 2 in)

WEIGHTS AND LOADINGS:

Weight empty	578 kg (1,275 lb)
Max T-O weight	975 kg (2,150 lb)
Max wing loading	65·5 kg/m² (13·4 lb/sq ft)
Max power loading	8·71 kg/kW (14·3 lb/hp)

PERFORMANCE:

*Max level speed at S/L	123 knots (229 km/h; 142 mph)
*Max cruising speed (75% power) at 2,135 m (7,000 ft)	117 knots (217 km/h; 135 mph)
Econ cruising speed (60% power) at 1,220 m (4,000 ft)	100 knots (185 km/h; 115 mph)
Stalling speed, flaps down	48 knots (89 km/h; 55 mph)
Max rate of climb at S/L	192 m (631 ft)/min
Service ceiling	3,340 m (10,950 ft)
Absolute ceiling	3,960 m (13,000 ft)
T-O run	244 m (800 ft)
Landing run	163 m (535 ft)
Range, 75% power at optimum altitude, standard fuel, no reserves	443 nm (820 km; 510 miles)
Range, 75% power at optimum altitude, max fuel, no reserves	625 nm (1,158 km; 720 miles)

**With optional speed fairings*

PIPER PA-28-151 CHEROKEE WARRIOR

Piper Aircraft announced on 26 October 1973 the introduction of a new 112 kW (150 hp) four-seat model in the Cherokee series. Named Cherokee Warrior, it combined the fuselage of the original Cherokee Archer with a completely new wing.

Two groups of optional equipment are available for the Warrior as basic packages:

Custom. Comprises Piper TruSpeed indicator; instrument panel white backlighting and overhead red spotlight, cabin dome, navigation, landing/taxi, and radio dimming lights; wheel speed fairings; rotating beacon; sensitive altimeter; assist strap; aircraft step; engine primer system; and cabin speaker; adding 9·5 kg (20·9 lb) to basic empty weight.

Executive. As Custom package, plus vacuum system with engine-driven pump; and advanced instrument panel comprising 3 in pictorial gyro horizon, 3 in directional gyro, Piper pictorial turn rate indicator, rate of climb indicator, outside air temperature gauge and eight-day clock; adding 16·75 kg (36·9 lb) to basic empty weight.

In addition, the six optional electronics groups as detailed for the Cherokee Cruiser are available also for this member of the Cherokee family.

Design of the Cherokee Warrior began in June 1972, and an important feature of this version is the increased-span tapered wing. As a result of its introduction the Warrior, which has essentially the same 112 kW (150 hp) engine as the Cruiser, is certificated at a maximum T-O weight 79 kg (175 lb) greater.

First flight of a prototype was made on 17 October 1972, and FAA certification was granted on 9 August 1973.

The 1976 version of the Cherokee Warrior introduces the same standard improvements as described for the Cruiser.

TYPE: Four-seat cabin monoplane.

WINGS: Cantilever low-wing monoplane. Wing section NACA 65_2-415 on inboard panels; outboard leading-edge incorporates modification No. 5 of NACA TN 2228. Dihedral 7°. Incidence 2° at root, −1° at tip. Sweepback at quarter-chord 5°. Light alloy single-spar structure with glassfibre wingtips. Plain ailerons of light alloy construction. Trailing-edge flaps constructed of light alloy with ribbed skins.

FUSELAGE: Light alloy semi-monocoque structure with glassfibre nose cowl and tailcone.

TAIL UNIT: Cantilever structure of light alloy, except for glassfibre tips on fin and tailplane. Fin and rudder have ribbed light alloy skins. One-piece all-moving tailplane, with combined anti-servo and trim tab. Rudder trimmable, but no trim tab in rudder.

LANDING GEAR: Non-retractable tricycle type. Steerable nosewheel. Piper oleo-pneumatic shock-absorbers; single wheel on each unit. Cleveland wheels with tyres size 17·50 × 6·30-6 on main units, pressure 1·65 bars (24 lb/sq in). Cleveland nosewheel and tyre size 14·20 × 4·95-5, pressure 1·65 bars (24 lb/sq in). Cleveland disc brakes. Parking brake. Wheel fairings optional.

POWER PLANT: One 112 kW (150 hp) Lycoming O-320-E3D flat-four engine, driving a Sensenich two-blade metal fixed-pitch propeller type 74DM6-O-58. Fuel in two wing tanks, with total capacity of 189 litres (50 US gallons). Refuelling point on upper surface of each wing. Oil capacity 7·5 litres (2 US gallons).

ACCOMMODATION: Four persons in pairs in enclosed cabin. Individual adjustable front seats, bench type rear seat. Dual controls standard. Large door on starboard side. Baggage compartment at rear of cabin, with volume of 0·68 m³ (24 cu ft) and capacity of 90 kg (200 lb). External access door on starboard side. Heating, ventilation and windscreen defrosting standard.

SYSTEMS: Hydraulic system for brakes only. Electrical system powered by 60A engine-driven alternator. 12V 25Ah battery standard, 12V 35Ah battery optional. Vacuum system for blind-flying instrumentation optional.

ELECTRONICS AND EQUIPMENT: King and Narco electronics in optional groups detailed in entry on Cherokee Cruiser, or an extensive range of King, Narco, Bendix and Piper electronics. Optional equipment as detailed for Cherokee Cruiser.

DIMENSIONS, EXTERNAL:

Wing span	10·67 m (35 ft 0 in)
Wing chord at root	1·60 m (5 ft 3 in)
Wing chord at tip	1·07 m (3 ft 6¼ in)
Wing aspect ratio	7·24
Length overall	7·25 m (23 ft 9½ in)
Height overall	2·22 m (7 ft 3½ in)
Tailplane span	3·92 m (12 ft 10½ in)
Wheel track	3·05 m (10 ft 0 in)
Wheelbase	2·03 m (6 ft 8 in)
Propeller diameter	1·88 m (6 ft 2 in)
Propeller ground clearance	0·21 m (8¼ in)
Cabin door: Height	0·89 m (2 ft 11 in)
Width	0·99 m (3 ft 3 in)
Baggage door: Height	0·48 m (1 ft 7 in)
Max width	0·66 m (2 ft 2 in)
Height to sill	0·71 m (2 ft 4 in)

DIMENSIONS, INTERNAL:

Cabin: Length	2·74 m (9 ft 0 in)
Max width	1·07 m (3 ft 6 in)
Max height	1·22 m (4 ft 0 in)
Floor area	2·28 m² (24·5 sq ft)
Volume	2·61 m³ (92 cu ft)

AREAS:

Wings, gross	15·8 m² (170 sq ft)
Ailerons (total)	1·23 m² (13·2 sq ft)
Trailing-edge flaps (total)	1·36 m² (14·6 sq ft)
Fin	0·69 m² (7·4 sq ft)
Rudder	0·38 m² (4·1 sq ft)
Tailplane, incl tab	2·46 m² (26·5 sq ft)

Piper PA-28-151 Cherokee Warrior four-seat cabin monoplane

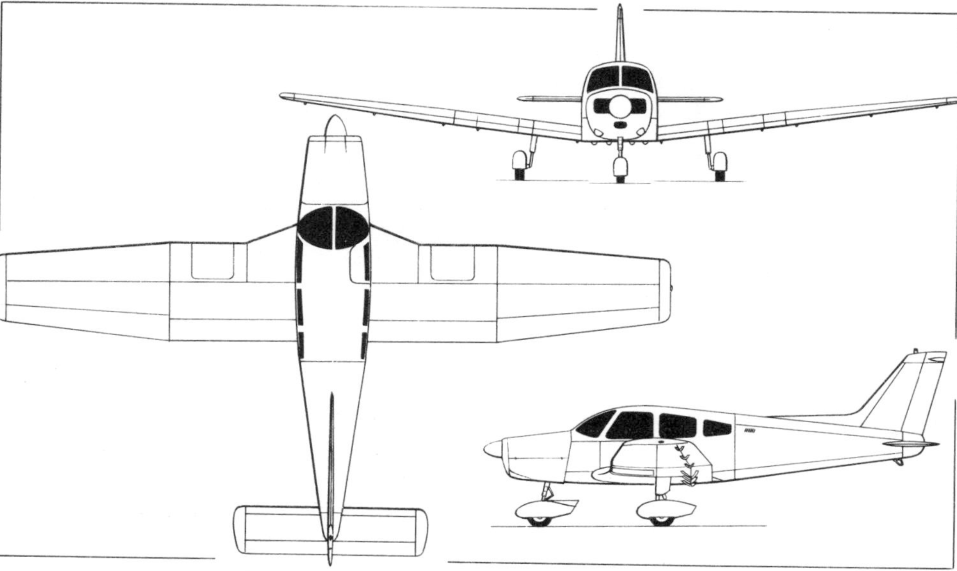

Piper Cherokee Warrior (Lycoming O-320-E3D engine) *(Pilot Press)*

WEIGHTS AND LOADINGS:
- Weight empty, standard — 604 kg (1,331 lb)
- Max T-O and landing weight — 1,054 kg (2,325 lb)
- Max wing loading — 66·74 kg/m² (13·67 lb/sq ft)
- Max power loading — 9·41 kg/kW (15·5 lb/hp)

PERFORMANCE (at max T-O weight):
- Never-exceed speed 152 knots (283 km/h; 176 mph)
- Max level and cruising speed at S/L — 117 knots (217 km/h; 135 mph)
- Econ cruising speed at optimum altitude — 115 knots (214 km/h; 133 mph)
- Stalling speed, flaps up — 56 knots (104 km/h; 64·5 mph)
- Stalling speed, flaps down — 50·5 knots (94 km/h; 58 mph)
- Max rate of climb at S/L — 198 m (649 ft)/min
- Service ceiling — 3,870 m (12,700 ft)
- Absolute ceiling — 4,560 m (14,960 ft)
- T-O run — 325 m (1,065 ft)
- T-O to 15 m (50 ft) — 536 m (1,760 ft)
- Landing from 15 m (50 ft) — 340 m (1,115 ft)
- Landing run — 181 m (595 ft)
- Range with max fuel at 75% power — 625 nm (1,158 km; 720 miles)

PIPER PA-28-181 CHEROKEE ARCHER II

On 9 October 1972 Piper introduced the Cherokee Challenger as successor to the Cherokee 180. In 1974 this was superseded by the Cherokee Archer, with the same basic airframe and power plant, but introducing many new equipment and electronics options. For 1976 this aircraft has been redesignated PA-28-181 Cherokee Archer II, and has the tapered wings of the Cherokee Warrior.

The 1976 Cherokee Archer II has the improvements listed for the 1976 Cherokee Cruiser. It can be fitted with the same optional Custom and Executive equipment packages, as well as the six optional electronics packages.

TYPE: Four-seat cabin monoplane.

WINGS: Cantilever low-wing monoplane. Wing section NACA 65_2-415 on inboard panels; outboard leading-edge has modification No. 5 of NACA TN 2228. Dihedral 7°. Incidence 2° at root, −1° at tip. Sweepback at quarter-chord 5°. Light alloy single-spar structure with glassfibre wingtips. Plain ailerons of light alloy construction. Trailing-edge flaps constructed of light alloy with ribbed skins.

FUSELAGE: Aluminium alloy semi-monocoque structure. Glassfibre engine cowling.

TAIL UNIT: Cantilever structure of aluminium alloy, except for glassfibre tips on fin and tailplane. Fin and rudder have corrugated metal skin. One-piece all-moving horizontal surface with combined anti-servo and trim tab. Trim tab in rudder.

LANDING GEAR: Non-retractable tricycle type. Steerable nosewheel. Piper oleo-pneumatic shock-absorbers. Cleveland wheels and Schenuit tyres, size 6·00-6, 4-ply rating, on all three wheels. Cleveland disc brakes. Parking brake.

POWER PLANT: One 134 kW (180 hp) Lycoming O-360-A3A flat-four engine, driving a Sensenich two-blade fixed-pitch propeller with spinner. Fuel in two tanks in wing leading-edges, with total capacity of 189 litres (50 US gallons), of which 181·5 litres (48 US gallons) are usable.

ACCOMMODATION: Four persons in pairs in enclosed cabin. Individual adjustable front seats, with dual controls; individual rear seats. Large door on starboard side. Heater and ventilation. Windscreen defrosting. Baggage compartment aft of cabin, with volume of 0·68 m³ (24 cu ft) and capacity of 90 kg (200 lb); door on starboard side. Rear seats removable to provide 1·25 m³ (44 cu ft) cargo space.

SYSTEMS: Optional Piper Aire air-conditioning system. Electrical system includes 60A alternator and 12V 25Ah battery. Hydraulic system for brakes only. Vacuum system optional.

ELECTRONICS AND EQUIPMENT: Details of electronics under group listings for the Cherokee Cruiser. Standard equipment includes external tie-down points, wing jack points and Piper automatic locator. Optional equipment available includes fire extinguisher, 35Ah battery, external power socket, inertia-reel shoulder harness for rear seats, adjustable front seats, alternate static source, overhead vent system, ventilation fan for air vent system, super soundproofing, outside air temperature gauge, heated pitot, Piper mixture control indicator, headrests, solar control windows, white wingtip strobe lights, zinc chromate application, and stainless steel control cables.

DIMENSIONS, EXTERNAL:
- Wing span — 10·67 m (35 ft 0 in)
- Wing chord at root — 1·60 m (5 ft 3 in)
- Wing chord at tip — 1·07 m (3 ft 6¼ in)
- Length overall — 7·25 m (23 ft 9½ in)
- Height overall — 2·23 m (7 ft 3½ in)
- Tailplane span — 3·92 m (12 ft 10½ in)
- Wheel track — 3·05 m (10 ft 0 in)
- Wheelbase — 2·00 m (6 ft 7 in)
- Propeller diameter — 1·93 m (6 ft 4 in)

AREAS:
- Wings, gross — 15·79 m² (170 sq ft)
- Ailerons (total) — 1·23 m² (13·20 sq ft)
- Trailing-edge flaps (total) — 1·36 m² (14·60 sq ft)
- Fin — 0·70 m² (7·50 sq ft)
- Rudder — 0·38 m² (4·10 sq ft)
- Tailplane — 2·46 m² (26·50 sq ft)

Piper PA-28-181 Cherokee Archer II, which now has the tapered wings introduced on the Cherokee Warrior

WEIGHTS AND LOADINGS:
- Weight empty, equipped (standard) 642 kg (1,416 lb)
- Max T-O and landing weight — 1,156 kg (2,550 lb)
- Max wing loading — 73·2 kg/m² (15·0 lb/sq ft)
- Max power loading — 8·63 kg/kW (14·17 lb/hp)

PERFORMANCE (at max T-O weight):
- Never-exceed speed — 148 knots (275 km/h; 171 mph) CAS
- Max level speed at S/L — 128 knots (237 km/h; 147 mph) CAS
- Max cruising speed, 75% power at 2,745 m (9,000 ft) — 124 knots (230 km/h; 143 mph) TAS
- Econ cruising speed, 65% power at 3,810 m (12,500 ft) — 115 knots (214 km/h; 133 mph) TAS
- Stalling speed, flaps up — 58·5 knots (108 km/h; 67 mph) CAS
- Stalling speed, flaps down — 52·5 knots (97 km/h; 60 mph) CAS
- Max rate of climb at S/L — 226 m (740 ft)/min
- Service ceiling — 4,175 m (13,700 ft)
- T-O run, S/L, ISA, 25° flap — 265 m (870 ft)
- T-O to 15 m (50 ft), S/L, ISA, 25° flap — 488 m (1,600 ft)
- Landing from 15 m (50 ft), S/L, ISA, 40° flap, max braking on paved dry runway — 424 m (1,390 ft)
- Landing run, conditions as above — 282 m (925 ft)
- Range with 181·5 litres (48 US gallons) usable fuel, 65% power at 3,810 m (12,500 ft), no reserves — 703 nm (1,303 km; 810 miles)
- Range, conditions as above, with max payload — 612 nm (1,134 km; 705 miles)
- Range with 181·5 litres (48 US gallons) usable fuel, 55% power at 3,810 m (12,500 ft) and 45 min reserves — 595 nm (1,102 km; 685 miles)

PIPER PA-28R-200 CHEROKEE ARROW II

The Cherokee Arrow II is generally similar to the Cherokee Archer II, but has a retractable landing gear, more powerful engine, and the untapered wing of the 1975 PA-28-180 Archer.

The tricycle landing gear is retracted hydraulically, with an electrically-operated pump supplying the hydraulic pressure. In addition to the usual 'gear up' warning horn and red light, the Cherokee Arrow II has an automatic extension system which drops the landing gear automatically if power is reduced and airspeed drops below 91 knots (169 km/h; 105 mph). The sensing system consists of a small probe mounted on the port side of the fuselage. Being located in the propeller slipstream, it can differentiate between a climb with power on and an approach to land with power reduced. A free-fall emergency extension system is also fitted. An 'anti-retraction' system guards against premature retraction of the landing gear below an airspeed of 74 knots (137 km/h; 85 mph) at take-off, or accidental retraction on the ground. There is also a manual override lever by which the pilot can hold the landing gear retracted as airspeed falls below 91 knots (169 km/h; 105 mph).

The 1976 Cherokee Arrow II has the improvements listed for the current Cherokee Cruiser, and the optional equipment and electronics packages listed for the Cruiser are available for the Arrow II.

The description of the Cherokee Archer applies also to the Cherokee Arrow II, except for the following details:

WINGS: Cantilever low-wing monoplane. Wing section NACA 65_2-415. Dihedral 7°. Incidence 2°. Single-spar wings, plain ailerons and slotted flaps of light alloy construction. Glassfibre wingtips. Ailerons and four-position flaps have corrugated skin. Ground-adjustable tab in port aileron.

LANDING GEAR: Retractable tricycle type, with single wheel on each unit. Hydraulic retraction, main units inward into wings, nose unit rearward. All units fitted with oleo-pneumatic shock-absorbers. Main wheels and tyres size 6·00-6, 4-ply rating. Nosewheel and tyre size 5·00-5, 4-ply rating. High-capacity dual hydraulic brakes and parking brake.

POWER PLANT: One 149 kW (200 hp) Lycoming IO-360-C1C flat-four engine, driving a two-blade constant-speed propeller with spinner. Fuel system as for Cherokee Archer II.

DIMENSIONS, EXTERNAL:
- Wing span — 9·82 m (32 ft 2½ in)
- Length overall — 7·50 m (24 ft 7¼ in)
- Height overall — 2·44 m (8 ft 0 in)
- Wheel track — 3·20 m (10 ft 6 in)
- Wheelbase — 2·40 m (7 ft 10½ in)
- Propeller diameter — 1·88 m (6 ft 2 in)

AREA:
- Wings, gross — 15·79 m² (170 sq ft)

Piper PA-28R-200 Cherokee Arrow II four-seat cabin monoplane with retractable landing gear

WEIGHTS AND LOADINGS:
Weight empty 694 kg (1,531 lb)
Max T-O weight 1,202 kg (2,650 lb)
Max wing loading 76·17 kg/m² (15·6 lb/sq ft)
Max power loading 6·01 kg/hp (13·25 lb/hp)
PERFORMANCE (at max T-O weight):
Max level speed 152 knots (282 km/h; 175 mph)
Max cruising speed (75% power) at optimum altitude 143 knots (266 km/h; 165 mph)
Stalling speed, wheels and flaps down 56 knots (103 km/h; 64 mph)
Max rate of climb at S/L 274 m (900 ft)/min
Service ceiling 4,575 m (15,000 ft)
Absolute ceiling 5,181 m (17,000 ft)
T-O run 312 m (1,025 ft)
Landing run 238 m (780 ft)
Range with max fuel, 75% power at optimum altitude 642 nm (1,191 km; 740 miles)
Range with max fuel, 55% power at optimum altitude 738 nm (1,368 km; 850 miles)

PIPER PA-28-235 CHEROKEE PATHFINDER

Simultaneously with announcement of the Cherokee Archer in October 1973, Piper gave details of the Cherokee Pathfinder, which superseded the earlier Cherokee 235.

The Pathfinder has a wing similar to that of the Arrow II. By comparison with the Cherokee 235, it has a 'stretch' of 0·13 m (5 in) in the fuselage length which, in addition to providing 50 per cent more leg room for rear-seat passengers, makes possible a wider cabin door, wider forward side windows, and improved access to the rear seats. The interior decor of the cabin has been revised, and there is a choice of four different colour schemes. Rear seats of new design are individually reclinable, and are quickly removable.

The power plant is a 175 kW (235 hp) Lycoming O-540-B4B5 flat-six engine, driving a Hartzell HC-C2YK-1/8468A-4 constant-speed propeller. Normal fuel capacity of 189 litres (50 US gallons) is supplemented by two tanks in the wingtips, containing a total of 129 litres (34 US gallons) of fuel. Total fuel capacity 318 litres (84 US gallons), of which 310 litres (82 US gallons) are usable.

The Cherokee Pathfinder has a glassfibre engine cowling made in two pieces (top and bottom) which can be removed easily to expose the entire engine for servicing. A landing light is incorporated in the nose directly under the propeller spinner and the ram-air scoop for the carburettor is offset to accommodate the exhaust system.

In most other airframe details, the Archer II and Pathfinder are similar, but propeller governor pad, high capacity electric fuel pump, four fuel contents gauges, central drain in cabin, double-pane side windows, super soundproofing, centre-mounted handbrake, main tyres of 6-ply rating, wheel fairings and Palm Beach exterior trim are standard. The Pathfinder is available with the same optional electronic packages and has the improvements listed for the 1976 Cherokee Cruiser.

DIMENSIONS, EXTERNAL:
Wing span 9·75 m (32 ft 0 in)
Wing chord (constant) 1·60 m (5 ft 3 in)
Length overall 7·35 m (24 ft 1¼ in)
Height overall 2·38 m (7 ft 9¾ in)
Tailplane span 3·05 m (10 ft 0 in)
Wheel track 3·05 m (10 ft 0 in)
Wheelbase 2·04 m (6 ft 8½ in)
WEIGHTS AND LOADINGS:
Weight empty, equipped 710 kg (1,565 lb)
Max T-O weight 1,360 kg (3,000 lb)
Max wing loading 85·9 kg/m² (17·6 lb/sq ft)
Max power loading 7·77 kg/kW (12·8 lb/hp)
PERFORMANCE (at max T-O weight):
Max level speed at S/L 140 knots (259 km/h; 161 mph)
Max cruising speed, 75% power at optimum altitude 133 knots (246 km/h; 153 mph)
Stalling speed, full flap 57 knots (105 km/h; 65 mph)
Max rate of climb at S/L 244 m (800 ft)/min
Service ceiling 4,130 m (13,550 ft)
Absolute ceiling 4,725 m (15,500 ft)
T-O run, 25° flap 259 m (850 ft)
T-O to 15 m (50 ft), 25° flap 430 m (1,410 ft)
Landing from 15 m (50 ft) 530 m (1,740 ft)
Landing run 317 m (1,040 ft)
Range, 75% power at optimum altitude 794 nm (1,472 km; 915 miles)
Range, 55% power at optimum altitude 964 nm (1,786 km; 1,110 miles)

PIPER PA-32 CHEROKEE SIX

The prototype of the PA-32 Cherokee SIX (N9999W) was flown for the first time on 6 December 1963, followed by the first production model (N9998W) on 17 September 1964. FAA Type Approval was received on 4 March 1965 (SIX 260) and 27 May 1966 (SIX 300).

The original version was a six-seater, but the model introduced in October 1966 (FAA Type Approval 15 November 1966) offered an optional seventh seat. The 1969 Cherokee SIX B introduced increased cabin space, achieved by moving the instrument panel forward. Additional shoulder, hip and leg room was provided by moving the seats one inch away from the fuselage walls. The new features incorporated in the 1976 Cherokee Cruiser apply also to the Cherokee SIX. In addition, the SIX 300 has a new ram-air induction system with an airscoop on the lower port side of the cowling, and has three exhaust stacks replacing the original single-stack exhaust.

Piper Cherokee Pathfinder four-seat light aircraft (Lycoming O-540-B4B5 engine)

The PA-32 is available with two alternative power plants, as follows:

Cherokee SIX 260. Basic version with 194 kW (260 hp) Lycoming O-540-E flat-six engine, driving a two-blade constant-speed propeller. Carburettor heat control and engine primer standard.

Cherokee SIX 300. More powerful version with 224 kW (300 hp) Lycoming IO-540-K flat-six engine, driving a two-blade constant-speed propeller. Two oil coolers and alternate air source are standard. Piper Aire air-conditioning optional on this version.

Optional Custom and Executive equipment packages, as listed for the Warrior, are available for both versions of the Cherokee SIX. They include also map lights, four individual reading lights and a forward baggage compartment light.

Overall dimensions are increased by comparison with the two/four- and four-seat versions of the Cherokee, but the basic structural description of the Cherokee Archer II and Pathfinder applies generally to the Cherokee SIX also. It is available with similar equipment and with the same optional electronics packages. The optional range of autopilots is extended by availability of the Piper Alti-Matic IIIC, a full three-control system.

POWER PLANT: One 194 kW or 224 kW (260 hp or 300 hp) Lycoming flat-six engine, driving a two-blade propeller. Two fuel tanks in inner wings, total capacity 189 litres (50 US gallons). Two auxiliary tanks in glassfibre wing-tips, with total capacity of 129 litres (34 US gallons). Total standard fuel capacity, with auxiliary tanks, 318 litres (84 US gallons), of which 315 litres (83·3 US gallons) are usable. Refuelling point above each tank. Oil capacity 11·5 litres (3 US gallons).

ACCOMMODATION: Enclosed cabin, seating six people in pairs. Optional seventh seat between two centre seats. Dual controls standard. Two forward-hinged doors, one on starboard side at front and other on port side at rear. Space for 45 kg (100 lb) baggage at rear of cabin, and another 45 kg (100 lb) forward, between engine and instrument panel. A set of four pieces of matched luggage to fit the nose baggage compartment is available as an option. Passenger seats easily removable to provide up to 3·11 m³ (110 cu ft) of cargo space inside cabin, or room for stretcher and one or two attendants. Large upward-hinged utility door adjacent to rear door provides loading entrance nearly 1·5 m (5 ft) wide. Ten silent fresh air outlets, cabin heater with eight warm air outlets including two defrosters, and two cabin air exhaust vents are standard.

DIMENSIONS, EXTERNAL:
Wing span 9·98 m (32 ft 8¾ in)
Length overall 8·45 m (27 ft 8¾ in)
Height overall 2·50 m (8 ft 2½ in)
Tailplane span 3·92 m (12 ft 10½ in)
Wheel track 3·22 m (10 ft 7 in)
Wheelbase 2·39 m (7 ft 10 in)
Propeller diameter:
SIX 260 2·08 m (6 ft 10 in)
SIX 300 2·03 m (6 ft 8 in)
Cabin door (rear, port):
Height 0·86 m (2 ft 10 in)
Width 0·94 m (3 ft 1 in)
DIMENSIONS, INTERNAL:
Cabin:
Length, panel to rear wall 3·02 m (9 ft 11 in)
Max width 1·24 m (4 ft 1 in)
Max height 1·23 m (4 ft 0½ in)
Baggage compartment volume:
Forward 0·23 m³ (8 cu ft)
Aft 0·62 m³ (22 cu ft)
AREA:
Wings, gross 16·21 m² (174·5 sq ft)
WEIGHTS AND LOADINGS (A: SIX 260; B: SIX 300):
Weight empty, equipped:
A 801 kg (1,766 lb)
B 826 kg (1,822 lb)
Max T-O weight:
A, B 1,542 kg (3,400 lb)
Max wing loading:
A, B 95 kg/m² (19·5 lb/sq ft)
Max power loading:
A 7·95 kg/kW (13·1 lb/hp)
B 6·88 kg/kW (11·3 lb/hp)
PERFORMANCE (at max T-O weight; A: SIX 260; B: SIX 300):
Max level speed:
A 142 knots (264 km/h; 164 mph)
B 151 knots (279 km/h; 174 mph)
Max cruising speed, 75% power at optimum altitude:
A 133 knots (246 km/h; 153 mph)
B 146 knots (270 km/h; 168 mph)
Stalling speed, flaps down:
A, B 55 knots (102 km/h; 63 mph)
Max rate of climb at S/L:
A 236 m (775 ft)/min
B 320 m (1,050 ft)/min
Service ceiling:
A 3,900 m (12,800 ft)

Piper PA-32-300 Cherokee SIX 300 six/seven-seat light cabin monoplane

B 4,950 m (16,250 ft)
Absolute ceiling:
A 4,495 m (14,750 ft)
B 5,485 m (18,000 ft)
T-O run, 10° flap:
A 404 m (1,325 ft)
B 320 m (1,050 ft)
T-O to 15 m (50 ft), 10° flap:
A 579 m (1,900 ft)
B 457 m (1,500 ft)
Landing from 15 m (50 ft):
A, B 305 m (1,000 ft)
Landing run:
A, B 192 m (630 ft)
Range (75% power), with auxiliary fuel:
A 773 nm (1,432 km; 890 miles)
B 738 nm (1,368 km; 850 miles)
Range (55% power), with auxiliary fuel:
A 898 nm (1,665 km; 1,035 miles)
B 894 nm (1,657 km; 1,030 miles)

PIPER PA-32R-300 CHEROKEE LANCE

On 30 August 1974 Piper flew the prototype of a new 6/7-seat single-engined aircraft, which is included in the Cherokee range with the designation PA-32R-300 (the R indicating retractable landing gear) and the name Lance. It combines the fuselage and heavy-duty landing gear of the Seneca II with other components of the Cherokee SIX 300. FAA certification in the CAR 3-8 Normal category was granted on 25 February 1975, and the first production aircraft made its first flight on 17 July 1975. Between that date and the end of May 1976 Piper had completed 127 of the 186 Cherokee Lances then on order.

Type: Six/seven-seat cabin monoplane.
Wings: As for Cherokee SIX 300.
Fuselage: As for Seneca II.
Tail Unit: As for Cherokee SIX 300.
Landing Gear: Hydraulically-retractable tricycle type. Steerable nosewheel. Emergency free-fall extension system. Piper oleo-pneumatic shock-absorbers on each unit. Cleveland main wheels, with tyres size 17·8 × 6·25-8·4, pressure 2·62 bars (38 lb/sq in). Cleveland nosewheel, with tyre size 14·4 × 5·0-6·6, pressure 2·41 bars (35 lb/sq in). Cleveland aircooled wheel brakes.
Power Plant: One 224 kW (300 hp) Lycoming IO-540-K1A5D flat-six engine, driving a Hartzell two-blade metal constant-speed propeller. Two interconnected metal fuel tanks in the leading-edge of each wing with a total capacity of 371 litres (98 US gallons) of which 356 litres (94 US gallons) are usable. Refuelling points in upper surface of outboard wing panels. Oil capacity 11·5 litres (3 US gallons).
Accommodation, Systems, Electronics and Equipment: As for Cherokee SIX 300.
Dimensions, external:
Wing span 10·00 m (32 ft 9⅞ in)
Wing chord (constant) 1·60 m (5 ft 3 in)
Wing aspect ratio 6·17
Length overall 8·42 m (27 ft 7½ in)
Height overall 2·74 m (9 ft 0 in)
Tailplane span 3·92 m (12 ft 10½ in)
Wheel track 3·38 m (11 ft 0⅞ in)
Wheelbase 2·42 m (7 ft 11⅜ in)
Propeller diameter 2·03 m (6 ft 8 in)
Propeller ground clearance 0·18 m (7¼ in)
Areas: As for Cherokee SIX 300
Weights and Loadings:
Weight empty 895 kg (1,973 lb)
Max T-O and landing weight 1,633 kg (3,600 lb)
Max wing loading 100·6 kg/m² (20·6 lb/sq ft)
Max power loading 7·29 kg/kW (12·0 lb/hp)
Performance (at max T-O weight):
Never-exceed speed 188 knots (349 km/h; 217 mph) CAS
Max level speed at S/L 165 knots (306 km/h; 190 mph) CAS
Max cruising speed, 75% power at 2,315 m (7,600 ft) 158 knots (293 km/h; 182 mph) TAS
Econ cruising speed, 65% power at 3,500 m (11,500 ft) 148 knots (273 km/h; 170 mph) TAS
Stalling speed, flaps up 67 knots (124 km/h; 77 mph) CAS
Stalling speed, flaps down 61 knots (113 km/h; 70 mph) CAS
Max rate of climb at S/L 305 m (1,000 ft)/min
Service ceiling 4,450 m (14,600 ft)
T-O run, S/L, ISA, 25° flap 295 m (970 ft)
T-O to 15 m (50 ft), conditions as above 494 m (1,620 ft)
Landing from 15 m (50 ft), S/L, ISA, 40° flap, max braking, paved dry runway 510 m (1,670 ft)
Landing run, conditions as above 265 m (870 ft)
Range with 356 litres (94 US gallons) usable fuel, 65% power at 3,500 m (11,500 ft), no reserves 920 nm (1,706 km; 1,060 miles)
Range, conditions as above with max payload 495 nm (917 km; 570 miles)

Piper PA-32R-300 Cherokee Lance, which has the fuselage and landing gear of the Seneca II

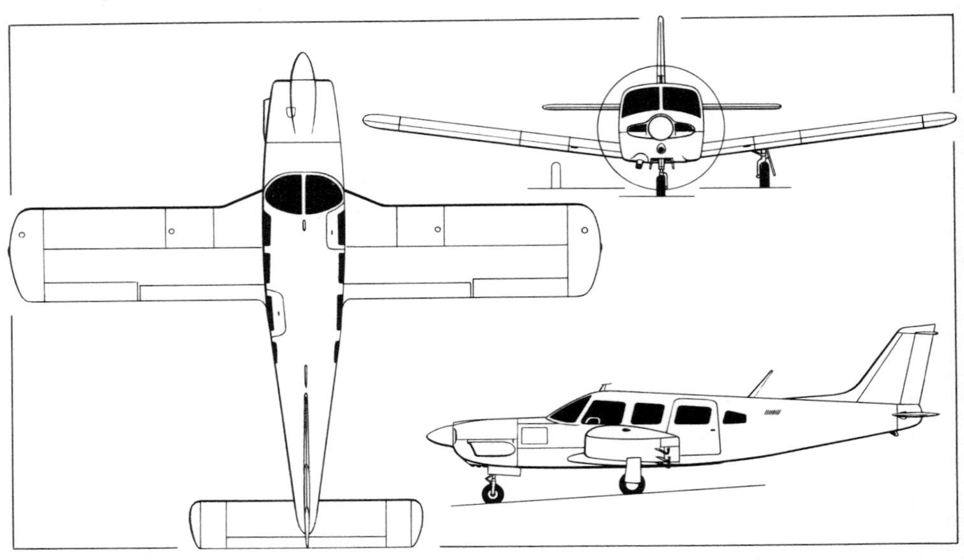
Piper PA-32R-300 Cherokee Lance six/seven-seat cabin monoplane *(Pilot Press)*

PIPER PA-31-310 TURBO NAVAJO C

On 30 September 1964 Piper flew the first of what it described as a new series of larger executive aircraft for corporate and commuter airline service. Named Navajo, it was then available with normally-aspirated or turbo-charged engines, the latter being known as the Turbo Navajo. Subsequently, two additional versions were introduced, the Pressurised Navajo and Navajo Chieftain, and these are described separately.

The following description applies to the Turbo Navajo C, which is available in the following versions:

Standard Turbo Navajo C. Six individual seats, with headrests and armrests, in pairs with centre aisle. Seventh and eighth seats optional.

Commuter Turbo Navajo C. Eight individual seats in pairs, with headrests and armrests. Standard equipment includes pilot cabin divider with two pilot manual racks, two cabin magazine racks, lighted 'No smoking' and 'Fasten seat belts' signs and aft cabin divider with luggage shelf.

Executive Turbo Navajo C. Six individual seats, in pairs, with headrests and armrests. Standard equipment as for Commuter version, plus 3·23 m³ (114 cu ft) oxygen system with eight individual outlets, rearward-facing third and fourth seats with folding tables, refreshment unit and toilet. Seventh and eighth seats may be installed in place of refreshment unit and toilet.

These versions of the Turbo Navajo C are available with a choice of electronics groups and operational groups as follows:

KTS 1-31. Dual King KX-175B nav/coms with broad band com antennae and standard VOR antenna; FD/HSI VOR/LOC/GS indicator; KNI-520 VOR/LOC indicator; dual KN-77 VOR/LOC converters; KN-73 glideslope receiver; KR-85 ADF; KI-225 ADF indicator; KT-76 transponder; KMA-20 audio selector panel including marker beacon receiver and lights; KN-65 DME; KWX-40 weather radar; AltiMatic V/FD-1 automatic flight system with couplers and HSI; Piper passenger address system, 100T noise-cancelling microphone, headset, static discharge wicks, insulated ADF sense antenna and electronics master switch; adding 65·6 kg (144·6 lb) to basic empty weight.

KTS 2-31. As KTS 1-31, except dual KX-175Bs replaced by KX-175BEs; duplicated KR-85 ADF; KI-225 indicator replaced by RCA-AVI-202 RMI; duplicated insulated ADF sense antennae and addition of Sunair ASB-130 10-channel HF radio with SCU-13 remote tuner and CU-110 antenna coupler; adding 75 kg (165 lb) to basic empty weight.

KTG 3-31. Gables master control panel with remote tuning and audio selector; dual KTR-905 transceivers with broad band com antennae; KNR-630 analog VOR/LOC/GS receiver/converter with RMI output, 200-channel nav, 40-channel GS and VOR antenna; KNR-632 analog VOR/LOC receiver/converter with RMI output, 200- channel nav; FD/HSI VOR/LOC/GS indicator; KNI-520 VOR/LOC converter; KMR-675 marker beacon receiver; KDF-805/KNI-580 ADF and indicator; KXP-755 transponder; KA-35A marker beacon lights; KAA-445 audio amplifier; KDM-705/KDI-570 DME and indicator; KWX-40 weather radar; AltiMatic V/FD-1 automatic flight system with couplers and HSI; Piper 100T noise-cancelling microphone, headset, static discharge wicks, insulated ADF sense antenna and electronics master switch; adding 79·7 kg (175·6 lb) to basic empty weight.

KTG 4-31. As KTG 3-31, except dual KDF-805 ADF with Collins RMI 332C-10 ADF indicator; KXP-755 transponder deleted; addition of Sunair ASB-130 10-channel HF transceiver with MCU-33 remote tuner and CU-110 antenna coupler; and duplicated insulated ADF sense antennae.

CT 1-31. Gables single unit remote control panel with audio selector; dual Collins VHF-20 720-channel com transceivers and broad band antennae; VIR-30MGM VHF nav receiver with 200-channel VOR/LOC and 40-channel GS, with GS and marker beacon receivers and VOR antenna; VIR-30M VHF nav receiver with VOR antenna; FD/HSI VOR/LOC indicator; Collins 331H-3G VOR/LOC indicator; KDF-805/KNI-580 ADF and indicator; TDR-90 transponder; KA-35A marker beacon lights; Collins DME-40; Collins 346-B3 audio amplifier; KWX-40 weather radar; AltiMatic V/FD-1 automatic flight system; Piper PA system, 100T noise-cancelling microphone, headset, static discharge wicks, insulated ADF sense antenna and electronics master switch; adding 79 kg (174·2 lb) to basic empty weight.

CT/D 2-31. As CT 1-31, except Gables single unit remote control panel with audio selector deleted and replaced by Collins NCS-31 RNAV system with digital tuning and keyboard. Piper noise-cancelling microphone, headset, insulated ADF sense antenna deleted, hard wired thumb wheel backup controls added; adding 80·9 kg (178·3 lb) to basic empty weight.

CTM-1-31. Dual Collins VHF-251 720-channel com transceivers, with Piper noise-cancelling microphone, headset and broad band com antenna; dual Collins VIR-

351 200-channel nav receivers with VOR/LOC converter and a blade antenna; HSI-NSD-360 VOR/LOC/GS indicator No. 1; Collins GLS-350 glideslope receiver; Collins IND-350 VOR/LOC indicator No. 2; Collins AMR-350 audio selector panel, audio amplifier, marker beacon lights and marker beacon receiver; Collins ADF-650 and IND-650 indicator; Collins TDR-950 transponder with 4096 code capability; King KN-65A DME with KA-43 tuning adaptor and KI-266 digital display; King KWX-40 radar; Piper AltiMatic IIIC with VOR/LOC/GS coupling and longitudinal electric trim; Piper electronics master switch; Piper static discharge wicks; and dual Collins PWC-150 power converters to supply 14V DC for dual com/nav systems.

Co-pilot Flight Instrument Group. Includes Piper TruSpeed indicator; Piper pictorial turn rate indicator; sensitive altimeter; 3 in attitude gyro and directional gyro; 8-day clock; rate of climb indicator; heated pitot tube; and separate static system; adding 5·4 kg (12 lb) to basic empty weight. Electrical gyros or vacuum gyros optional.

De-icing Group. Pneumatic de-icing boot installation for wing and tail unit leading-edges; electrical propeller de-icing; ice detection light; and electrical windscreen de-icing and windscreen wiper port side; adding 33·3 kg (73·5 lb) to basic empty weight.

Other combinations of the above equipment are available optionally, together with an extensive range of radio and radar equipment.

TYPE: Six/nine-seat corporate and commuter airline transport.

WINGS: Cantilever low-wing monoplane. Wing section NACA 63_2415 at root, NACA 63_1212 at tip. 1° aerodynamic twist. 2° 30′ geometric twist. All-metal structure, with heavy stepped-down main spar, front and rear spars, lateral stringers, ribs and stressed skin. Wings spliced on centreline with heavy steel plates. Flush riveted forward of main spar. Wing-root leading-edge extended forward between nacelle and fuselage. Glassfibre wingtips. Balanced ailerons are interconnected with rudder. Trim tab in starboard aileron. Electrically-operated flaps. Pneumatic de-icing boots optional.

FUSELAGE: Conventional all-metal semi-monocoque structure.

TAIL UNIT: Cantilever all-metal structure, with sweptback vertical surfaces. Variable-incidence tailplane. Trim tabs in rudder and starboard elevator. Optional pneumatic de-icing boots.

LANDING GEAR: Hydraulically-actuated retractable tricycle type, with single wheel on each unit. Manual hydraulic emergency extension. Main wheels and tyres size 6·50-10, eight-ply rating, pressure 4·14 bars (60 lb/sq in). Steerable nosewheel and tyre size 6·00-6, six-ply rating, pressure 2·90 bars (42 lb/sq in). Toe-controlled hydraulic disc brakes. Main-wheel doors close when gear is fully extended.

POWER PLANT: Two 231 kW (310 hp) Lycoming TIO-540-A flat-six turbocharged engines. Hartzell three-blade fully-feathering metal propellers. Propeller de-icing optional. Four rubber fuel cells in wings; inboard cells each contain 208 litres (55 US gallons), outboard cells 151·5 litres (40 US gallons) each. Total fuel capacity 719 litres (190 US gallons), of which 708 litres (187 US gallons) are usable. Fuel cells equipped with NACA-type anti-icing non-siphoning fuel vents. Two-piece glassfibre engine nacelles.

ACCOMMODATION: Six to nine seats, as described under notes on individual models. Dual controls standard. Thermostatically-controlled Janitrol 35,000 BTU combustion heater, windscreen defrosters and fresh air system standard. Double-glazed windows. Electrical de-icing and windscreen wiper for port side of windscreen optional. 'Dutch' door at rear of cabin on port side. Top half hinges upward; lower half hinges down and has built-in steps. Baggage compartments in nose, capacity 68 kg (150 lb), and in rear of cabin, capacity 91 kg (200 lb). Cargo door and cockpit door available as optional items.

SYSTEMS: Hydraulic system utilises two engine-driven pumps. 24V electrical system supplied by two engine-driven 28V 70A alternators and 24V 17Ah battery; 25Ah battery optional. External power socket standard. Oxygen system optional.

ELECTRONICS AND EQUIPMENT: Optional electronics are described under the standard groupings. Blind-flying instrumentation standard, with optional dual installation for co-pilot. Optional equipment includes cabin ground ventilation fan, Piper Aire air-conditioning system, cabin fire extinguisher, cold-weather heater for rear cabin, propeller synchroniser, nacelle baggage compartments, aft cabin divider with curtain and shelf, propeller ice protection shields, beverage dispensers, folding tables, forward cabin divider with curtain and magazine racks, Piper automatic locator, ice inspection light, toilet, utility door, tinted windows, toe-brakes for co-pilot, and pilot's windscreen wiper, Standard electrical equipment includes navigation, landing, taxying, cockpit, cabin dome and passenger reading lights, two rotating beacons, stall warning light, courtesy lights, cabin and cockpit speakers and heated pitot tube.

DIMENSIONS, EXTERNAL:
- Wing span: 12·40 m (40 ft 8 in)
- Length overall: 9·94 m (32 ft 7½ in)
- Height overall: 3·96 m (13 ft 0 in)
- Tailplane span: 5·52 m (18 ft 1½ in)
- Wheel track: 4·19 m (13 ft 9 in)
- Wheelbase: 2·64 m (8 ft 8 in)
- Propeller diameter: 2·03 m (6 ft 8 in)

DIMENSIONS, INTERNAL:
- Cabin: Length: 4·88 m (16 ft 0 in)
- Height: 1·31 m (4 ft 3½ in)
- Baggage compartments:
  - Nose: 0·40 m^3 (14 cu ft)
  - Aft: 0·62 m^3 (22 cu ft)

AREA:
- Wings, gross: 21·3 m^2 (229 sq ft)

WEIGHTS AND LOADINGS:
- Weight empty (standard): 1,782 kg (3,930 lb)
- Max T-O and landing weight: 2,948 kg (6,500 lb)
- Max zero-fuel weight: 2,812 kg (6,200 lb)
- Max wing loading: 138·7 kg/m^2 (28·4 lb/sq ft)
- Max power loading: 6·38 kg/kW (10·5 lb/hp)

PERFORMANCE (at max T-O weight):
- Max level speed at 4,570 m (15,000 ft): 227 knots (420 km/h; 261 mph)
- Max cruising speed (75% power) at 6,705 m (22,000 ft): 215 knots (399 km/h; 248 mph)
- Max cruising speed (75% power) at 3,660 m (12,000 ft): 195 knots (362 km/h; 225 mph)
- Econ cruising speed at 6,100 m (20,000 ft): 207 knots (383 km/h; 238 mph)
- Econ cruising speed at 3,660 m (12,000 ft): 192 knots (356 km/h; 221 mph)
- Stalling speed, flaps down: 63·5 knots (118 km/h; 73 mph)
- Max rate of climb at S/L: 440 m (1,445 ft)/min
- Rate of climb at S/L, one engine out: 75 m (245 ft)/min
- Service ceiling: 8,015 m (26,300 ft)
- Service ceiling, one engine out: 4,635 m (15,200 ft)
- Normal T-O run: 314 m (1,030 ft)
- Short-field T-O run: 262 m (860 ft)
- Normal T-O to 15 m (50 ft): 668 m (2,190 ft)
- Short-field T-O to 15 m (50 ft): 518 m (1,700 ft)
- Normal landing from 15 m (50 ft) at max landing weight: 713 m (2,340 ft)
- Short-field landing from 15 m (50 ft): 552 m (1,810 ft)
- Normal landing run: 584 m (1,915 ft)
- Short-field landing run: 376 m (1,235 ft)
- Accelerate/stop distance: 636 m (2,085 ft)
- Range with max fuel, 45 min reserves:
  - at max cruising speed at 6,705 m (22,000 ft): 855 nm (1,585 km; 985 miles)
  - at max cruising speed at 3,660 m (12,000 ft): 812 nm (1,504 km; 935 miles)
  - at econ cruising speed at 6,100 m (20,000 ft): 1,011 nm (1,875 km; 1,165 miles)
  - at econ cruising speed at 3,660 m (12,000 ft): 990 nm (1,834 km; 1,140 miles)

PIPER PA-31-325 TURBO NAVAJO C/R

Identical to the PA-31-310 Turbo Navajo C except for having counter-rotating engines as introduced on the Navajo Chieftain in 1972. The Turbo Navajo C/R is available in the same optional versions, and with the same electronics and operational package options, as the Turbo Navajo C. The description of this latter aircraft applies also to the Turbo Navajo C/R, except as follows:

POWER PLANT: One 242·5 kW (325 hp) Lycoming LTIO-540-F2BD and one 242·5 kW (325 hp) Lycoming TIO-540-F2BD flat-six turbocharged counter-rotating engines.

WEIGHTS AND LOADINGS:
- As for Turbo Navajo C, except:
- Weight empty: 1,814 kg (4,000 lb)

PERFORMANCE (at max T-O weight):
- Max level speed at 4,570 m (15,000 ft): 230 knots (426 km/h; 265 mph)
- Max cruising speed at 6,705 m (22,000 ft): 220 knots (407 km/h; 253 mph)
- Max cruising speed at 3,660 m (12,000 ft): 201 knots (373 km/h; 232 mph)
- Econ cruising speed at 6,100 m (20,000 ft): 212 knots (393 km/h; 244 mph)
- Econ cruising speed at 3,660 m (12,000 ft): 197 knots (365 km/h; 227 mph)
- Stalling speed, flaps down: 63·5 knots (118 km/h; 73 mph)
- Max rate of climb at S/L: 457 m (1,500 ft)/min
- Rate of climb at S/L, one engine out: 78 m (255 ft)/min
- Absolute ceiling: above 7,315 m (24,000 ft)
- Absolute ceiling, one engine out: 4,875 m (16,000 ft)
- Service ceiling: above 7,315 m (24,000 ft)
- Service ceiling, one engine out: 4,665 m (15,300 ft)
- Normal T-O run: 302 m (990 ft)
- Short-field T-O run: 256 m (840 ft)
- Normal T-O to 15 m (50 ft): 634 m (2,080 ft)
- Short-field T-O to 15 m (50 ft): 497 m (1,630 ft)
- Normal landing from 15 m (50 ft): 713 m (2,340 ft)
- Short-field landing from 15 m (50 ft): 552 m (1,810 ft)
- Normal landing run: 584 m (1,915 ft)
- Short-field landing run: 376 m (1,235 ft)
- Accelerate/stop distance: 620 m (2,035 ft)
- Range with max fuel, 45 min reserves:
  - at max cruising speed at 6,705 m (22,000 ft): 790 nm (1,464 km; 910 miles)
  - at max cruising speed at 3,660 m (12,000 ft): 742 nm (1,376 km; 855 miles)
  - at econ cruising speed at 6,100 m (20,000 ft): 955 nm (1,770 km; 1,100 miles)
  - at econ cruising speed at 3,660 m (12,000 ft): 916 nm (1,698 km; 1,055 miles)

PIPER PA-31P PRESSURISED NAVAJO

Piper announced on 6 March 1970 that a new pressurised version of the Navajo was to be marketed by the company. Generally similar to the PA-31-300 Navajo and Turbo Navajo already in service, the new version began as a company project in January 1966, and the first prototype was flown in March 1968. From that time to the roll-out of the first production aircraft in March 1970, more than 4,000 hours of flight and ground-testing of the new model were completed, including 850 hours at altitudes up to 8,840 m (29,000 ft), the aircraft's certificated maximum operating altitude.

Two optional interior groups are available:

Standard. Six individual seats in pairs, with headrests and armrests. Pilot/co-pilot seats four-way adjustable with shoulder harness inertia reels; third and fourth cabin seats aft-facing and all cabin seats have seat belts; window curtains and wall to wall carpet. Rear cabin divider with clothes bar and baggage security net. Forward cabin divider curtain. 'No smoking/Fasten seat belt' sign. Oxygen outlets and masks at each seat position. Options available include pneumatic door extender; forward cabin combination unit; storage cabinets; folding tables; aft cabin combination unit, which includes side-facing seventh seat/toilet; seventh and eighth seats; tinted cabin windows; cabin fire extinguisher; stereo system; and all-leather seat covering.

Executive. Six individual seats, comprising two crew seats and four reclining chairs in the arrangement detailed above. Other standard equipment as above, plus forward cabin combination unit which includes cabin dividers and curtain, electrically-heated Thermos unit, cup dispenser, storage for ice, beverages and manuals; two folding tables; pneumatic door extender and aft cabin combination unit which includes side-facing seventh seat/toilet, cabin divider with mirror, privacy curtain, refreshment centre,

Piper PA-31P Pressurised Navajo six/eight-seat corporate or commuter light transport

AC power outlet for electric razor. Options include cabin fire extinguisher, storage cabinets, eighth seat, tinted cabin windows, stereo system and all-leather seat covering.

The above versions of the Pressurised Navajo are available with a choice of electronics groups and operational groups as follows:

KTG 1-31P. Dual King KTR-905 720-channel transceivers with KFS-590B manual frequency selectors and broad band antennae; KNR-665 digital VOR/LOC/GS receiver/converter with RMI and built-in RNAV computer with KCU-565 RNAV memory/control display and keyboard selector; KNR-630 analogue VOR/LOC/GS receiver/converter with RMI, 200-channel nav and 40-channel GS; KFS-560B manual tuner; KPI-553 pictorial nav indicator with DME readout and KDA-335 display adapter; KNI-520 VOR/LOC/GS indicator; KMR-675 marker beacon receiver with KA-35A indicator lights; KDF-805 ADF with KFS-580B manual selector and Collins 332C-10 indicator; KXP-755 transponder with KFS-570B selector; KDM-705A DME; KAA-445 audio amplifier; KA-37 audio selector panel; RCA AVQ-47 weather radar; Piper AltiMatic V/FD-1 automatic flight system, electronics master switch, PA system, 100T noise-cancelling microphone, headset, static discharge wicks and insulated ADF sense antenna; adding 86·7 kg (191·1 lb) to basic empty weight.

KTG 2-31P. As KTG 1-31P with deletion of dual KFS-590B, single KFS-560B, KFS-570B and KFS-580B selectors, and KA-37 audio selector panel. Replaced by Gables single unit remote control, audio selector and KDI-570 mechanical DME display/readout. KNR-665/KCU-565 deleted, KNR-632 analogue VOR/LOC receiver/converter with 200-channel nav added. KPI-553/KDA-335 deleted, replaced by FD/HSI VOR/LOC/GS indicator; adding 84·4 kg (186 lb) to basic empty weight.

KTS 3-31P. Dual King KXB-175 transceivers with 720-channel com, 200-channel nav, broad band antennae and VOR antenna; FD/HSI VOR/LOC/GS indicator; KNI-520 VOR/LOC indicator; dual KN-77 VOR/LOC converters; KN-73 GS receiver; KR-85 ADF; KI-225 ADF indicator; KT-76 transponder; KMA-20 audio selector panel and amplifier; KN-65 DME; RCA AVQ-47 weather radar; AltiMatic V/FD-1 automatic flight system with couplers and HSI; Piper PA system, 100T noise-cancelling microphone, headset, static discharge wicks, insulated ADF sense antenna and electronics master switch; adding 68·1 kg (150·2 lb) to basic empty weight.

CT 1-31P. Gables single unit remote control panel and audio selector; dual Collins VHF-20 com transceivers with broad band antennae; VIR-30A6M nav receiver with 200-channel VOR/LOC, 40-channel GS and VOR antenna; VIR-30A nav receiver; FD/HSI VOR/LOC indicator; 331H-3G VOR/LOC indicator; KDF-805 ADF; Collins 332C-10 ADF indicator; TDR-90 transponder; KA-35A marker beacon lights; DME-40 DME; 332C-10 RMI with dual VOR and ADF displays; 346B-3 audio amplifier; RCA AVQ-47 weather radar; Piper PA system, AltiMatic V/FD-1 automatic flight system; static discharge wicks, insulated ADF sense antenna and electronics master switch; adding 83 kg (183 lb) to basic empty weight.

CT/D 2-31P. As CT 1-31P with deletion of Gables remote control and addition of Collins NCS-31 digital tuning and keyboard RNAV system; KA-37 audio panel; Piper 100T noise-cancelling microphone, headset and hard-wired thumb wheel backup controls; adding 84·5 kg (186·4 lb) to basic empty weight.

***Full-time Pressurisation Group**. Comprises two 28V 100A gear-driven alternators, 24,000 BTU freon vapour-cycle air-conditioner with automatic two-position inlet scoop; cabin altitude selector with variable rate control; 1·36 m³ (48 cu ft) oxygen system with outlets and masks; forward cabin divider with adjustable vertical blind, 'Fasten seat belts' and 'No smoking' signs; and rear cabin divider with baggage retaining net and clothes hanger support bar with hangers; adding 67 kg (149 lb) to basic empty weight.

***Pilot/Operations Utility Group**. Comprises pilot's heated windscreen (provisions for co-pilot); two 94·5 litre (25 US gallon) auxiliary fuel tanks; two sump and two fuel cell drains; four electric fuel transfer pumps with integral filters; pilot's electric windscreen wiper; adding 19·5 kg (43 lb) to basic empty weight.

Co-pilot Flight Instruments Group. Comprises Tru-speed, pictorial turn rate and rate of climb indicators; altimeter; electric attitude gyro, electric directional gyro; clock; heated pitot; static system; brakes and windscreen wiper; adding 9·1 kg (20 lb) to basic empty weight.

De-icing Group. Comprises pneumatic de-icing boot installation for wing and tail unit leading-edges; electric propeller de-icing; and ice detection light; adding 24 kg (53 lb) to basic empty weight.

Other combinations of the foregoing equipment are available optionally, together with an extensive range of radio and radar equipment.

**Because of customer preference, these optional groups are included in all Pressurised Navajos*

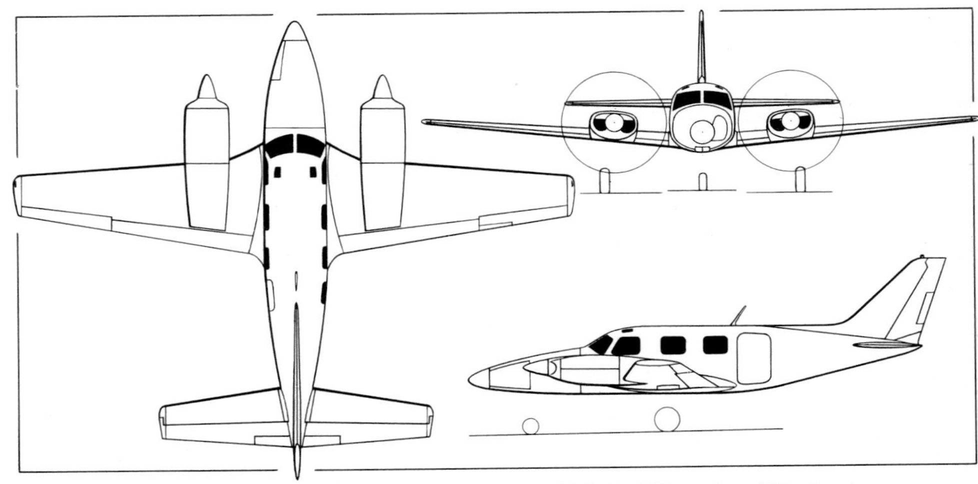

Piper PA-31P Pressurised Navajo (two Lycoming TIGO-541-E1A engines *(Pilot Press)*

The Navajo's pressurisation system, with a maximum cabin differential of 0·38 bars (5·5 lb/sq in), provides a sea level cabin atmosphere up to 3,770 m (12,375 ft), and can maintain a cabin altitude of 2,440 m (8,000 ft) to a height of 7,620 m (25,000 ft) and of 3,050 m (10,000 ft) to the aircraft's certificated maximum operating altitude of 8,840 m (29,000 ft). Pressurisation is obtained from four different air sources. The primary source is bleed air from the engine turbochargers which enters the system through sonic nozzles. This is supplemented by relief flow from two engine-driven dry pneumatic pumps which have a main function of providing air for the aircraft's pneumatic system. Over-pressurisation is prevented automatically by an isobaric control valve, with backup safety valve. A three-position cabin air control lever directs the flow of air either to the cabin or to atmosphere. When the control is in the pressurised position air is directed into the cabin; Outside Air mode provides unpressurised air for normal ventilation; Recirculated Air mode, used only on the ground, allows the cabin air to be recirculated through the air-conditioning evaporator/heater units.

The environmental control system provides automatically thermostatic control of cabin temperature, air purification, circulation/recirculation and de-humidification. The normal flow of air for heating, cooling, ventilation and defrosting is taken primarily through the main pressure line and the recirculating air duct. A ram-air duct in the nose section provides an alternative source of air for unpressurised flight. Airflow is heated by a 35,000 BTU in-line gasoline combustion heater or cooled by a similarly integrated 24,000 BTU freon type air conditioner, depending on demand of cabin thermostat. The air-conditioning compressor is driven by the starboard engine. An evaporator, mounted in the nose section, acts as a heat exchanger which cools and automatically dehumidifies the air before it enters the cabin. Baseboard outlets on both sides of the flight deck and main cabin provide warm air, pressurised or unpressurised, and individual eyeball outlets provide air-conditioned cool air. When the aircraft is not pressurised, these latter outlets circulate fresh air from outside throughout the cabin.

One example of the Pressurised Navajo was acquired by the Spanish Air Force during 1972.

The description of the Turbo Navajo C applies also to the PA-31P, except in the following details:

FUSELAGE: Conventional all-metal semi-monocoque structure, with fail-safe structure in the pressurised section. Swing-open nosecone of reinforced glassfibre.

LANDING GEAR: Steerable nosewheel, with tyre size 6·00-6, eight-ply rating. Goodyear toe-controlled heavy-duty hydraulic brakes, six-puck system with multiple discs.

POWER PLANT: Two 317 kW (425 hp) Lycoming TIGO-541-E1A flat-six turbocharged and geared engines, each driving a Hartzell three-blade metal constant-speed fully-feathering propeller. Electric propeller synchronisation optional. Four rubber fuel cells in wings; inboard cells each contain 212 litres (56 US gallons), outboard cells each 151·5 litres (40 US gallons). Total fuel capacity 727 litres (192 US gallons) of which 704 litres (186 US gallons) are usable. Two optional 94·5 litre (25 US gallon) transfer cells can be installed in the engine nacelles to provide a maximum fuel capacity of 916 litres (242 US gallons), of which 893 litres (236 US gallons) are usable. Inboard fuel tanks have NACA-type non-icing non-siphoning fuel tank vents, outboard tanks have heated vents.

ACCOMMODATION: Six to eight seats, as described under notes on individual models. Inertia-reel shoulder harness for pilot and co-pilot. Dual controls standard. One-piece passenger door with integral airstair; when closed the door is secured by seven locking pins. Each seat position has an individual fresh air outlet, reading lamp, headrest, full-length armrests and storage pocket in the seatback. Walk-in baggage compartment aft of cabin can accommodate 90·7 kg (200 lb) of baggage. Nose compartment can hold 90·7 kg (200 lb) of baggage. Birdproof windscreen and windscreen defrosting standard.

SYSTEMS: Hydraulic system supplied by two engine-driven pumps, pressure 124 bars (1,800 lb/sq in). Electrical system supplied by two 28V 50A engine-driven alternators, 24V 25Ah battery. Fuel system has positive fuel flow at all altitudes and temperatures supplied by two engine-driven fuel pumps, two auxiliary electric fuel pumps and four submerged electric fuel pumps, pressure 3·79 bars (55 lb/sq in).

ELECTRONICS AND EQUIPMENT: Optional electronics described under standard groupings above, but a wide range of alternative optional electronics is also available. Standard equipment includes flight control lock, baggage tie-down straps, courtesy lights for cabin entrance door and nose baggage compartment, jack pads, tie-down rings, towbar, corrosion proofing, and nose-gear safety mirror. Optional equipment includes beverage dispensers, pneumatic door extender, tinted windows, co-pilot's toe-brakes, refreshment centre, folding tables, electrically-heated window for co-pilot, windscreen wiper for co-pilot, propeller synchroniser, true airspeed indicator, ice protection shields, lightning-resistant fuel filler caps, Piper automatic locator, cabin utility tie-down rings, hand fire extinguisher, strobe lights, torque-meter and stereo tape system.

DIMENSIONS, EXTERNAL:
- As for Turbo Navajo, except:
- Length overall — 10·52 m (34 ft 6 in)
- Height overall — 3·99 m (13 ft 1 in)
- Tailplane span — 6·05 m (19 ft 10 in)
- Propeller diameter — 2·36 m (7 ft 9 in)

DIMENSIONS, INTERNAL:
- Cabin, incl flight deck and rear baggage compartment:
  - Length — 4·90 m (16 ft 1 in)
  - Max height — 1·32 m (4 ft 4 in)
  - Max width — 1·30 m (4 ft 3 in)
- Baggage compartments:
  - Aft — 0·62 m³ (22 cu ft)
  - Nose — 0·57 m³ (20 cu ft)

WEIGHTS AND LOADINGS:
- Weight empty — 2,222 kg (4,900 lb)
- Max T-O weight — 3,538 kg (7,800 lb)
- Max wing loading — 166·4 kg/m² (34·1 lb/sq ft)
- Max power loading — 5·58 kg/kW (9·18 lb/hp)

PERFORMANCE (at max T-O weight):
- Max level speed at 5,475 m (18,000 ft) — 243 knots (451 km/h; 280 mph)
- Max cruising speed, 75% power at 7,315 m (24,000 ft) — 231 knots (428 km/h; 266 mph)
- Intermediate cruising speed, 65% power at 7,315 m (24,000 ft) — 212 knots (393 km/h; 244 mph)
- Econ cruising speed, 55% power at 7,315 m (24,000 ft) — 193 knots (357 km/h; 222 mph)
- Long-range cruising speed, 45% power at 7,315 m (24,000 ft) — 165 knots (306 km/h; 190 mph)
- Stalling speed, flaps up — 80 knots (148 km/h; 92 mph)
- Stalling speed, flaps down — 72 knots (134 km/h; 83 mph)
- Single-engine min control speed — 83 knots (153 km/h; 95 mph)
- Max rate of climb at S/L — 530 m (1,740 ft)/min
- Rate of climb at S/L, one engine out — 73 m (240 ft)/min
- Max operationally approved ceiling — 8,840 m (29,000 ft)
- Ceiling, one engine out — 4,675 m (15,300 ft)
- T-O run — 439 m (1,440 ft)
- T-O to 15 m (50 ft) — 671 m (2,200 ft)
- Landing from 15 m (50 ft) — 823 m (2,700 ft)
- Landing run — 418 m (1,370 ft)
- Accelerate/stop distance — 863 m (2,830 ft)

Max range at 7,315 m (24,000 ft) with max fuel, no reserves, at speeds shown:
- Max cruising speed 951 nm (1,760 km; 1,095 miles)
- Intermediate cruising speed 1,090 nm (2,020 km; 1,255 miles)
- Econ cruising speed 1,237 nm (2,290 km;1,425 miles)
- Long-range cruising speed 1,302 nm (2,414 km; 1,500 miles)

Max range at 7,315 m (24,000 ft) with max fuel, 45 min reserves, at speeds shown:
- Max cruising speed 868 nm (1,609 km; 1,000 miles)
- Intermediate cruising speed 981 nm (1,818 km; 1,130 miles)
- Econ cruising speed 1,116 nm (2,065 km; 1,285 miles)
- Long-range cruising speed 1,172 nm (2,170 km; 1,350 miles)

PIPER PA-31-350 NAVAJO CHIEFTAIN

First announced on 11 September 1972, the Navajo Chieftain is a lengthened version of the Turbo Navajo, with the fuselage extended by 0·61 m (2 ft 0 in) and the provision of 261 kW (350 hp) counter-rotating turbocharged engines. The Chieftain does not replace the existing Turbo Navajo, but represents an extension of the Navajo series.

The main cabin floor is designed to carry heavy concentrated loads of up to 976 kg/m² (200 lb/sq ft) and, in addition to the 6·14 m³ (217 cu ft) of cargo space in the main cabin, 91 kg (200 lb) of cargo or baggage can be carried in the forward nose compartment, and 68 kg (150 lb) in the rear of each engine nacelle.

Three optional interior groups of equipment are available, depending upon the proposed use of the aircraft:

Standard Interior Group. Comprises six forward-facing seats.

Commuter Interior Group. Comprises 10 forward-facing seats and 'No smoking/Fasten seat belt' sign; adding 93 kg (205 lb) to basic empty weight.

Executive Interior Group. Comprises six executive-type seats, the third and fourth aft-facing, with headrests and armrests; forward cabin divider with curtain and manual and map stowage; two folding tables, one each side of cabin; 3·23 m³ (114 cu ft) oxygen system with 8 outlets and masks; 'No smoking/Fasten seat belt' sign; adding 112·5 kg (248 lb) to basic empty weight.

Operational groups, comprising Co-pilot Instrument Group and De-icing Group, as detailed for the Turbo Navajo C. Electronics groups as detailed for the Turbo Navajo C are available for the Chieftain.

The description of the Turbo Navajo C applies also to the Navajo Chieftain, except as follows:

FUSELAGE: As for Turbo Navajo C, except length increased by 0·61 m (2 ft 0 in).

POWER PLANT: Two 261 kW (350 hp) Lycoming TIO-540-J2BD flat-six turbocharged engines, each driving a three-blade fully-feathering metal propeller. Four rubber fuel cells in wings; inboard cells each contain 212 litres (56 US gallons), outboard cells each 151·5 litres (40 US gallons). Total fuel capacity 727 litres (192 US gallons), of which 689 litres (182 US gallons) are usable.

ACCOMMODATION: Pilot and co-pilot on individually adjustable and reclining seats. Dual controls standard. Interior seating and equipment as detailed in optional interior groups. Cabin heated by thermostatically-controlled Janitrol 50,000 BTU combustion heater. Piper Aire 18,000 BTU air-conditioning system optional. Baggage/cargo compartments in nose, capacity 91 kg (200 lb), and in the rear of each engine nacelle, each 68 kg (150 lb).

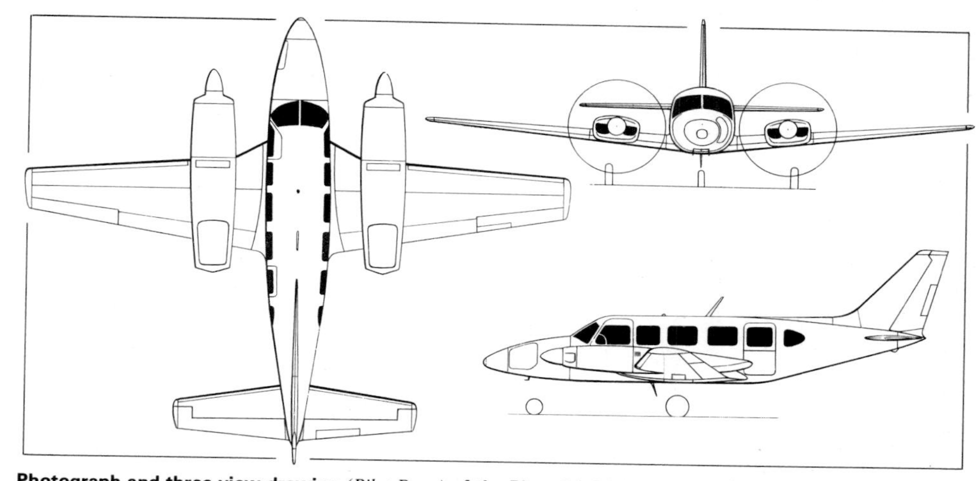

Photograph and three-view drawing *(Pilot Press)* **of the Piper PA-31-350 Navajo Chieftain six/ten-seat executive/commuter/cargo aircraft**

ELECTRONICS AND EQUIPMENT: A wide range of optional electronics is available, as well as full dual instrumentation, a flight director system integrated with alternative autopilots, weather radar, pneumatic wing de-icing, and electric propeller and windscreen anti-icing. Also available optionally for all three interior groups is a cargo kit which includes cargo barrier, sidewall protection, tie-down rings and net, eight seat-track tie-down rings, four plug-in tie-down rings, four tie-down straps, four track cargo rollers and two strap tie-down pouches.

DIMENSIONS, EXTERNAL:
- Wing span 12·40 m (40 ft 8 in)
- Length overall 10·55 m (34 ft 7½ in)
- Height overall 3·96 m (13 ft 0 in)
- Tailplane span 5·52 m (18 ft 1½ in)
- Wheel track 4·19 m (13 ft 9 in)
- Wheelbase 3·25 m (10 ft 8 in)
- Propeller diameter 2·03 m (6 ft 8 in)

DIMENSIONS, INTERNAL:
- Cabin: Length 5·49 m (18 ft 0 in)
  - Height 1·31 m (4 ft 3½ in)
- Baggage/cargo compartments:
  - Nose 0·40 m³ (14 cu ft)
  - Engine nacelles (each) 0·37 m³ (13·25 cu ft)

AREA:
- Wings, gross 21·3 m² (229 sq ft)

WEIGHTS AND LOADINGS:
- Weight empty (standard) 1,866 kg (4,114 lb)
- Max T-O and landing weight 3,175 kg (7,000 lb)
- Max wing loading 149·4 kg/m² (30·6 lb/sq ft)
- Max power loading 6·08 kg/kW (10·0 lb/hp)

PERFORMANCE (at max T-O weight):
- Max level speed at 4,575 m (15,000 ft) 234 knots (435 km/h; 270 mph)
- Max cruising speed at 7,315 m (24,000 ft) 226 knots (418 km/h; 260 mph)
- Econ cruising speed at 6,100 m (20,000 ft) 214 knots (396 km/h; 246 mph)
- Single-engine min control speed 78 knots (145 km/h; 90 mph)
- Stalling speed, flaps up 80 knots (148 km/h; 92 mph)
- Stalling speed, flaps down 74 knots (137 km/h; 85 mph)
- Max rate of climb at S/L 424 m (1,390 ft)/min
- Rate of climb at S/L, one engine out 70 m (230 ft)/min
- Service ceiling 8,290 m (27,200 ft)
- Service ceiling, one engine out 4,175 m (13,700 ft)
- Absolute ceiling 8,625 m (28,300 ft)
- Absolute ceiling, one engine out 4,725 m (15,500 ft)
- Normal T-O run 415 m (1,360 ft)
- Short-field T-O run 320 m (1,050 ft)
- Normal T-O to 15 m (50 ft) 759 m (2,490 ft)
- Short-field T-O to 15 m (50 ft) 543 m (1,780 ft)
- Normal landing from 15 m (50 ft) 831 m (2,725 ft)
- Short-field landing from 15 m (50 ft) 655 m (2,150 ft)
- Normal landing run 480 m (1,575 ft)
- Short-field landing run 360 m (1,180 ft)
- Accelerate/stop distance 695 m (2,280 ft)
- Range at max cruising speed, with max fuel and allowances for taxi, T-O, climb, cruise at best power mixture setting at 7,315 m (24,000 ft), descent and 45 min reserves 920 nm (1,706 km; 1,060 miles)
- Range at econ cruising speed, allowances and altitude as above, 45 min reserves 1,103 nm (2,044 km; 1,270 miles)

PIPER PA-31T CHEYENNE

The PA-31T, with an airframe similar to the Pressurised Navajo, introduced turboprop power to the Piper line for the first time.

Design of the PA-31T began at the end of 1965, with construction of the prototype beginning in May 1967. First flight of the prototype was made on 20 August 1969, with FAA certification being granted on 3 May 1972. The first production aircraft flew for the first time on 22 October 1973.

It is available with the same Standard and Executive interior options detailed for the Pressurised Navajo, and with the following operational group options:

De-icing Group. Pneumatic de-icing boots for wing and tail unit leading-edges, and wing ice inspection light; adding 17·9 kg (39·4 lb) to basic empty weight.

Co-pilot Flight Group. Airspeed and rate of climb indicator, altimeter, electric turn rate indicator, attitude and directional gyro, clock, heated pitot, static system with alternate source, toe-brakes and windscreen wiper; adding 8·9 kg (19·7 lb) to basic empty weight.

Five optional factory-installed electronics packages are available:

Group KTG/D 1-31T. King KCU-565 keyboard and display; dual KTR-905 720-channel com transceivers with KCU-591 control and display units and broad band antennae; KNR-665 digital VOR/LOC/GS receiver/converter with RNAV computer, RMI, 200-channel nav and 40-channel GS; KNR-615 standby digital VOR/LOC/GS

PA-31T Cheyenne *Heritage of '76,* **the 100,000th Piper production aircraft**

receiver/converter, without RNAV computer; KPI-553 pictorial navigation indicator with KDA-335 display data adapter; KDM-705A DME; KCU-561 control/display unit; KPI-552 pictorial navigation indicator; KMR-675 marker beacon receiver with dual KA-35A marker beacon light displays; KDF-805 ADF with Collins 332C-10 RMI ADF indicator; KXP-775 transponder; KCU-578 ADF and transponder control unit; KAA-445 audio amplifier with KA-37 selector panel; Bendix RDR-1200 Weather-vision; Bendix M-4D integrated AP/FD system with 4 in displays; Piper electronics master switch, PA system, static discharge wicks, ramp hailer, external microphone and phone jack, 100T noise-cancelling microphone, headset and insulated ADF sense antenna; adding 131 kg (288·7 lb) to basic empty weight.

Group KTG 2-31T. As KTG/D 1-31T except for deletion of KNR-615, KPI-552, KCU-561 and one unit KA-35A, and addition of KNR-630 analogue VOR/LOC/GS receiver/converter with RMI, 200-channel nav and 40-channel GS; KFS-560B manual tuner; KNI-520 VOR/LOC/GS indicator; KFS-580B manual ADF tuner and KFS-570B manual transponder tuner; adding 126·9 kg (279·8 lb) to basic empty weight.

Group KTG 3-31T. As KTG 2-31T, except for deletion of KFS-560B, KFS-570B and KFS-580B tuners, KNR-665/KCU-565 and KPI-553/KDA-335; and addition of second KNR-630; FD/HSI VOR/LOC indicator; KDI-570 digital DME readout; and Gables master control panel; adding 123·6 kg (272·4 lb) to basic empty weight.

Group CT 1-31T. Gables master control panel; dual Collins VHF-20 com transceivers with broad band antennae; dual VIR-30 nav receivers with 200-channel VOR/LOC and 40-channel GS; FD/HSI and Collins 331H-3G VOR/LOC indicators; KDF-805 ADF; Collins 332C-10 ADF indicator; TDR-90 transponder; KA-35A marker beacon lights; DME-40 DME; Collins 332C-10 RMI and dual VOR and ADF displays; Collins 346B-3 audio amplifier; Bendix RDR-1200 Weather-vision; Bendix M-4D integrated AP/FD system with 4 in displays; Piper electronics master switch, PA system, ramp hailer, external mike and phone jack, static discharge wicks, insulated ADF sense antenna, 100T noise-cancelling microphone and headset; adding 126 kg (277·8 lb) to basic empty weight.

Group CT/D 2-31T. As Group CT 1-31T, except for deletion of Gables master control panel, Collins 332C-10 ADF indicator and Piper PA system; and addition of Collins NCS-31 navigation control system with digital keyboard control; and hard-wired thumb wheel backup controls; adding 125·7 kg (277·2 lb) to basic empty weight.

TYPE: Six/eight-seat cabin monoplane.

WINGS: Cantilever low-wing monoplane. Wing section NACA 63_2-415 at root, NACA 63A212 at tip. Dihedral 5°. Incidence 1° 30′ at root, −1° at tip. Sweepback 0° at 30% chord. Three-spar structure of 2024ST light alloy. Balanced ailerons and single-slotted trailing-edge flaps of 2024ST light alloy. Trim tab in starboard aileron. Pneumatic de-icing boots on wing leading-edges optional.

FUSELAGE: Semi-monocoque structure of 2024ST light alloy, with fail-safe structure in the pressurised areas.

TAIL UNIT: Cantilever structure of 2024ST light alloy with sweptback vertical surfaces. Fixed-incidence tailplane. Trim tabs in elevators and rudder. Pneumatic de-icing of tailplane leading-edges optional.

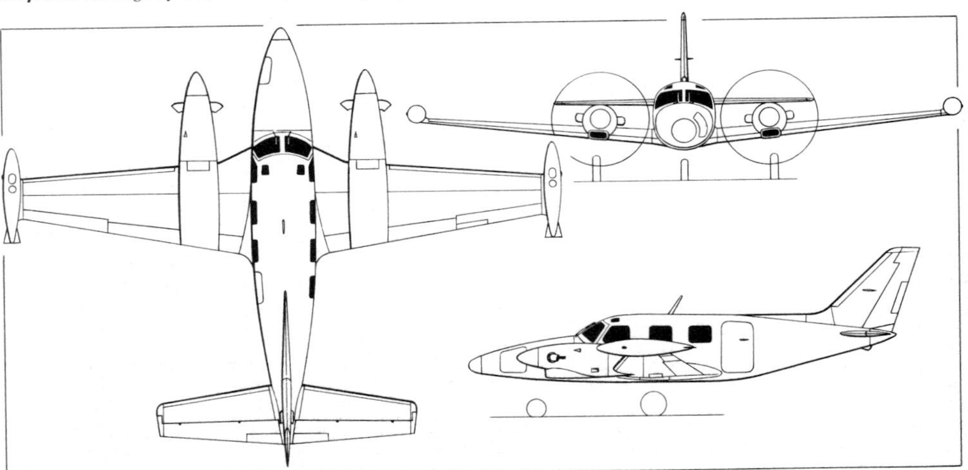

Piper PA-31T Cheyenne six/eight-seat light transport aircraft *(Pilot Press)*

LANDING GEAR: Hydraulically-retractable tricycle type with single wheel on each unit, main units retracting inward and nosewheel aft. Piper oleo-pneumatic shock-absorbers. Main wheels and tyres size 6·50 × 10, 10-ply rating. Nosewheel with Type VII tyre size 18 × 4·4, 6-ply rating. Goodyear disc-type hydraulic brakes. Parking brake.

POWER PLANT: Two 462 kW (620 ehp) Pratt & Whitney Aircraft of Canada PT6A-28 turboprop engines, each driving a Hartzell three-blade metal constant-speed reversible and fully-feathering propeller type HC-BTN-3B. Each wing has three interconnected fuel cells and a tip-tank, with combined total capacity of 1,476 litres (390 US gallons), of which 1,446 litres (382 US gallons) are usable. Refuelling points in engine nacelles and on upper surface of each tip-tank. Oil capacity 24·6 litres (6·5 US gallons).

ACCOMMODATION: Pilot and co-pilot on two individual adjustable seats. Dual controls standard. Pilot's storm window. Heated windscreen and windscreen wiper for pilot, optional for co-pilot. Cabin seating for four to six passengers on individual seats. Door with built-in airstair on port side, which has seven locking pins and inflatable pressurisation seal. Dual-pane windows. Emergency exit window on starboard side. Cabin heated and air-conditioned. Forward and aft cabin dividers. A wide range of options for cabin includes folding tables, beverage dispensers, pneumatic door extender, storage cabinets and tinted windows. Baggage compartments in nose and rear of cabin, each with 91 kg (200 lb) capacity. External access door to nose compartment.

SYSTEMS: Air-conditioning and pressurisation, with pressure differential of 0·38 bars (5·5 lb/sq in). Freon-type air-conditioner of 23,000 BTU capacity. Janitrol combustion heater of 35,000 BTU capacity with automatic windscreen defroster. Hydraulic system supplied by dual engine-driven pumps for landing gear retraction and brakes. Pneumatic system provided by engine bleed air. Electrical system supplied by two 28V 200A generators and 24V 43Ah nickel-cadmium battery. Oxygen system of 1·37 m^3 (48·3 cu ft) capacity. De-icing system comprises electric anti-icing boots for air intakes, heated pitot and electric propeller de-icing. Fire detection system with six sensors.

ELECTRONICS: In addition to the five optional factory-installed packages, an extensive range of optional electronics is available, including radar, communications, area navigation, autopilot and flight director systems by Bendix, Collins, King, Piper and Sperry.

EQUIPMENT: Installed standard equipment is extensive, and optional items include toe-brakes for co-pilot, heated windscreen and windscreen wiper for co-pilot, wing and tail pneumatic de-icing boots, engine fire-extinguisher system, fuselage ice protection plates, ice inspection lights, propeller synchronisers, automatic locator beacon and co-pilot's flight instrument group.

DIMENSIONS, EXTERNAL:	
Wing span over tip-tanks	13·01 m (42 ft 8¼ in)
Length overall	10·57 m (34 ft 8 in)
Height overall	3·89 m (12 ft 9 in)
Tailplane span	6·05 m (19 ft 10 in)
Wheel track	4·19 m (13 ft 9 in)
Wheelbase	2·64 m (8 ft 8 in)
Propeller diameter	2·36 m (7 ft 9 in)
Propeller ground clearance	0·27 m (10½ in)
Passenger door (port, aft):	
Height	1·17 m (3 ft 10 in)
Width	0·71 m (2 ft 4 in)
Height to sill	0·94 m (3 ft 1 in)
Baggage door (fwd):	
Height	0·53 m (1 ft 9 in)
Width	0·66 m (2 ft 2 in)
Height to sill	1·10 m (3 ft 7½ in)
Emergency exit (stbd, fwd):	
Height	0·64 m (2 ft 1 in)
Width	0·48 m (1 ft 7 in)
DIMENSIONS, INTERNAL:	
Cabin (incl flight deck):	
Length	4·90 m (16 ft 1 in)
Max width	1·30 m (4 ft 3 in)
Max height	1·32 m (4 ft 4 in)
Floor area	4·37 m^2 (47 sq ft)
Volume	6·29 m^3 (222 cu ft)

Forward baggage compartment 0·57 m³ (20 cu ft)
Aft baggage compartment 0·62 m³ (22 cu ft)

AREAS:
Wings, gross 21·3 m² (229 sq ft)
Ailerons (total) 1·21 m² (13 sq ft)
Trailing-edge flaps (total) 3·12 m² (33·6 sq ft)
Fin 1·48 m² (15·9 sq ft)
Rudder, incl tab 0·98 m² (10·6 sq ft)
Tailplane 6·55 m² (70·5 sq ft)
Elevators, incl tab 2·63 m² (28·3 sq ft)

WEIGHTS AND LOADINGS:
Weight empty, standard, equipped 2,209 kg (4,870 lb)
Max T-O and landing weight 4,082 kg (9,000 lb)
Max ramp weight 4,105 kg (9,050 lb)
Max zero-fuel weight 3,265 kg (7,200 lb)
Max wing loading 191·9 kg/m² (39·3 lb/sq ft)
Max power loading 4·42 kg/kW (7·26 lb/ehp)

PERFORMANCE (at 3,447 kg; 7,600 lb AUW):
Max level and cruising speed at 3,355 m (11,000 ft) 283 knots (525 km/h; 326 mph)
Econ cruising speed at 7,620 m (25,000 ft) 212 knots (393 km/h; 244 mph)
Stalling speed, flaps down 77 knots (142 km/h; 88 mph)
Max rate of climb at S/L 853 m (2,800 ft)/min
Rate of climb at S/L, one engine out 201 m (660 ft)/min
Max approved operating altitude 8,840 m (29,000 ft)
Service ceiling, one engine out 4,450 m (14,600 ft)
Min ground turning radius 9·63 m (31 ft 7 in)
T-O to 15 m (50 ft) 604 m (1,980 ft)
Landing from 15 m (50 ft) 567 m (1,860 ft)
Accelerate/stop distance 957 m (3,140 ft)
Range with max fuel at econ cruising power, allowances for taxi, T-O, climb, descent and 45 min reserves 1,478 nm (2,739 km; 1,702 miles)
Range at max cruising speed, allowances as above 1,408 nm (2,608 km; 1,621 miles)

PIPER PA-34 SENECA II

On 23 September 1971, Piper announced a new twin-engined light aircraft which had the company designation PA-34 and, following Piper tradition, had the Indian name Seneca. Built at Piper's Vero Beach, Florida, factory, the 1975 version of this aircraft was redesignated Seneca II.

The Seneca II has a counter-rotating (C/R) engine and propeller installation. The retractable landing gear is operated by an electro-hydraulic system and includes an emergency extension system which allows the wheels to free-fall into the down and locked position. A dual-vane stall warning system provides warning by horn well in advance of the stall in either 'clean' or gear/flaps-down configuration.

The 1976 Seneca II introduces as standard a dual range electric fuel boost pump for cold weather priming and emergency backup. In the event of engine-driven fuel pump failure the electric boost pump can provide fuel to maintain 75% power. Also new is a 45,000 BTU combustion heater; auxiliary fuel pump lights are added to the annunciator panel. A new option is the availability of 57 litre (15 US gallon) auxiliary tanks to give a max optional usable fuel capacity of 466 litres (123 US gallons).

The extensive range of electronics available for the Seneca II includes many options, plus six groups as follows:

Group N 1-34. Narco Com 11A 360-channel VHF transceiver; Nav 11 200-channel VOR/LOC receiver with VOR/LOC converter indicator; ADF-140 ADF; AT-50A transponder; Piper AutoControl IIIB autopilot; Piper VOR/LOC coupler; Piper electric trim; M-700A noise-cancelling microphone and MBT-12R marker beacon receiver and lights; adding 19·5 kg (43 lb) to basic empty weight.

Group N 2-34. As above, plus second Com 11A transceiver; Nav 12 200-channel VOR/LOC receiver, converter indicator and UGR-3 40-channel glideslope receiver; VOR/LOC converter; CP-125 audio panel; and dual Piper broad band com antennae; adding 26·8 kg (59 lb) to basic empty weight.

Group NT 3-34. Dual Narco Com 111B 360-channel VHF transceivers; Nav 111 200-channel VOR/LOC receiver with 40-channel localiser and VOR/LOC converter indicator; Nav 112 200-channel VOR/LOC receiver with converter indicator, and UGR-2A glideslope receiver; MBT-12R marker beacon receiver and lights; CP-125 audio panel; ADF-140 ADF; AT-50A transponder; Piper AltiMatic IIIC autopilot; Piper VOR/LOC coupler; Piper electric trim; Piper dual broad band com antennae, static wicks and M-700A noise-cancelling microphone; adding 34 kg (75 lb) to basic empty weight.

Group KS 1-34. King KX-170B 720-channel nav/com transceiver with 200-channel nav; KI-201C VOR/LOC converter indicator; KR-86 ADF; KT-76 transponder; KR-21 marker beacon receiver and lights; Piper AutoControl IIIB autopilot; Piper VOR/LOC coupler; Piper electric trim; and Piper 66C microphone; adding 20 kg (44 lb) to basic empty weight.

Group KS 2-34. As Group KS 1-34, plus second KX-170B; KI-214 VOR/LOC/glideslope indicator with VOR/LOC converter and 40-channel glideslope receiver; KI-201C VOR/LOC indicator; KMA-20 audio panel and amplifier; and Piper dual broad band antennae; adding 30 kg (66 lb) to basic empty weight.

Group KTS 3-34. Dual King KX-175B nav/com transceivers with 720-channel com and 200-channel nav; dual KNI-520 VOR/LOC/glideslope indicators; dual KN-77 VOR/LOC converters; KN-73 40-channel glideslope receiver; KMA-20 audio panel with marker beacon receiver and indicator lights; KR-85 ADF with KI-225 indicator; KT-76 transponder; Piper AltiMatic IIIC autopilot with VOR/LOC coupler; Piper electric trim; Piper dual broad band antennae and static discharge wicks; and Piper 100T noise-cancelling microphone; adding 40 kg (88 lb) to basic empty weight.

Two operational groups are also available:

Executive. Comprising Piper TruSpeed indicator, four individual reading lights, anti-collision lights, heated pitot, towbar, dual vacuum system and advanced instrument panel with 3 in pictorial gyro horizon, 3 in directional gyro, Piper turn rate indicator, rate of climb indicator, outside air temperature gauge and eight-day clock; adding 13·2 kg (29·1 lb) to basic empty weight.

Sportsman. As Executive group, plus inertia-reel safety belts for centre and rear seats, six headrests, solar control windows and external power socket; adding 19 kg (41·9 lb) to basic empty weight.

A total of 397 PA-34 Senecas were delivered during 1975.

TYPE: Six/seven-seat twin-engined light aircraft.

WINGS: Cantilever low-wing monoplane. Single-spar wings, Frise ailerons, and wide-span slotted flaps, of light alloy construction. Glassfibre wingtips. Flaps manually operated.

FUSELAGE: Light alloy semi-monocoque structure.

TAIL UNIT: Cantilever structure of light alloy. One-piece all-moving horizontal surface with combined anti-balance and trim tab. Anti-servo tab in rudder.

LANDING GEAR: Hydraulically-retractable tricycle type. Steerable nosewheel. Emergency free-fall extension system. Main wheels with tyre size 6·00-6, 8-ply rating; nosewheel and tyre size 6·00-6, 6-ply rating. High-capacity disc brakes. Parking brake. Heavy-duty tyres and brakes optional, which reduce landing run by 25% and landing distances by 12½%.

POWER PLANT: Two 149 kW (200 hp) Continental TSIO-360-E flat-four turbocharged counter-rotating engines, driving Hartzell two-blade metal constant-speed fully-feathering propellers. Fuel in two tanks in wings, with a total capacity of 371 litres (98 US gallons) of which 352 litres (93 US gallons) are usable. Optional 57 litre (15 US gallon) auxiliary tank in each wing to provide a max capacity of 485 litres (128 US gallons) of which 466 litres (123 US gallons) are usable. Glassfibre engine cowlings.

ACCOMMODATION: Enclosed cabin, seating six people in pairs on individual seats with 0·25 m (10 in) centre aisle. Optional seventh seat between two centre seats. Dual controls standard. Pilot's storm window. Two forward-hinged doors, one on starboard side at front, the other on port side at rear. Large optional door adjacent to rear cabin door provides an extra-wide opening for loading bulky items. Passenger seats removable easily to provide different seating/baggage/cargo combinations. Space for 45 kg (100 lb) baggage at rear of cabin, and for 45 kg (100 lb) in nose compartment with external access door on port side. Cabin heated and ventilated.

SYSTEMS: Electro-hydraulic system for landing gear retraction. Electrical system powered by dual 12V 65A alternators. 12V 35Ah battery. Oxygen system with six outlets optional. Dual engine-driven vacuum pumps for flight instruments optional. Janitrol 45,000 BTU combustion heater.

ELECTRONICS AND EQUIPMENT: Factory-installed electronics packages as group listing in introductory copy. Standard equipment includes sun visors, soundproofing, tie-down rings, aircraft step, nose gear safety mirror and dual stall warning sensors. Optional items include individual reading lights, zinc chromate finish, stainless steel cables, cabin fire extinguisher, headrests, inertia-reel safety belts for centre and rear seats, seventh seat, solar control windows, external power socket, a complete de-icing group, anti-collision lights, ventilation fan, super soundproofing, and automatic locator beacon.

DIMENSIONS, EXTERNAL:
Wing span 11·85 m (38 ft 10¾ in)
Length overall 8·73 m (28 ft 7½ in)
Height overall 3·02 m (9 ft 10¾ in)
Wheel track 3·38 m (11 ft 1¼ in)
Wheelbase 2·13 m (7 ft 0 in)
Propeller diameter 1·93 m (6 ft 4 in)

AREA:
Wings, gross 19·39 m² (208·7 sq ft)

WEIGHTS AND LOADINGS:
Weight empty 1,264 kg (2,788 lb)
Max T-O weight 2,073 kg (4,570 lb)
Max landing weight 1,969 kg (4,342 lb)
Max wing loading 107·4 kg/m² (22 lb/sq ft)
Max power loading 6·96 kg/kW (11·4 lb/hp)

PERFORMANCE (at max T-O weight, except where indicated):
Max level speed at 4,265 m (14,000 ft) 198 knots (367 km/h; 228 mph)
Max cruising speed, 75% power at 6,100 m (20,000 ft) 189 knots (351 km/h; 218 mph)
Normal cruising speed, 65% power at 6,705 m (22,000 ft) 180 knots (333 km/h; 207 mph)
Econ cruising speed, 55% power at 7,315 m (24,000 ft) 162 knots (301 km/h; 187 mph)
Stalling speed, wheels and flaps down 60 knots (111 km/h; 69 mph)
Max rate of climb at S/L 408 m (1,340 ft)/min
Rate of climb at S/L, one engine out 69 m (225 ft)/min
Max approved operating altitude 7,620 m (25,000 ft)
Service ceiling, one engine out 4,085 m (13,400 ft)
Absolute ceiling, one engine out 4,510 m (14,800 ft)
T-O run, flaps up 335 m (1,100 ft)
T-O run, 25° flap 274 m (900 ft)
T-O to 15 m (50 ft), flaps up 445 m (1,460 ft)
T-O to 15 m (50 ft), 25° flap 378 m (1,240 ft)
Landing from 15 m (50 ft) at max landing weight 637 m (2,090 ft)
Landing run at max landing weight 421 m (1,380 ft)
Range at best power mixture, 75% power at 4,880 m (16,000 ft), no reserves:
Standard fuel 644 nm (1,194 km; 742 miles)
Max optional fuel 885 nm (1,641 km; 1,020 miles)
Range, conditions as above, with 45 min reserves:
Standard fuel 543 nm (1,007 km; 626 miles)
Max optional fuel 781 nm (1,448 km; 900 miles)
Range at best power mixture, 55% power at 4,880 m (16,000 ft), no reserves:
Standard fuel 720 nm (1,335 km; 830 miles)
Max optional fuel 990 nm (1,834 km; 1,140 miles)
Range, conditions as above, with 45 min reserves:
Standard fuel 608 nm (1,128 km; 701 miles)
Max optional fuel 877 nm (1,625 km; 1,010 miles)

Piper PA-34 Seneca II, powered by Continental TSIO-360-E turbocharged counter-rotating engines

PIPER PA-36 PAWNEE BRAVE

On 9 October 1972 Piper Aircraft Corporation released details of a new agricultural aircraft named the Pawnee Brave, which has a more powerful engine than the PA-25 Pawnee D, is larger, and has increased capacity for either liquid or dry chemicals.

More than 4,250 PA-25 Pawnees were built by Piper; experience gained in their construction, progressive refinement, and operation led to design of the Brave. Primary consideration was to provide an aircraft able to offer high standards of safety and comfort for the pilot.

The basic configuration seats the pilot well aft. The long nose is designed to collapse progressively in an emergency. The fuselage is a welded truss structure of chrome-molybdenum steel, which is graded in strength to provide excellent energy absorption and progressive collapse. A sturdy overturn pylon is an integral part of the fuselage structure. The wing is of conventional cantilever construction, with laminated spars to provide structural redundancy. The wing leading-edges each comprise two glassfibre sections, reinforced by a foam insert beam running spanwise. Normal impacts are absorbed by the leading-edge, more serious contacts by ribs designed to collapse with minimal impact transference to the basic wing structure.

The pilot is located in an isolated cockpit capsule which keeps him well clear of main structural members. The floor, for example, is 0·30 m (1 ft 0 in) above the lower longerons, and a cockpit width of 0·97 m (3 ft 2 in) allows for substantial deformation of the fuselage structure without hazard to the pilot. The seat is attached to the overturn pylon, and is articulated to allow the pilot's position to change with fuselage deformation. The cockpit capsule is sealed to prevent the ingress of toxic chemicals; and all protrusions, knobs and levers which might cause injury have been eliminated. The instrument panel is equipped with a large energy-absorbing crash roll.

Ventilation of the cockpit capsule is provided by an airscoop in the top of the canopy, which filters the incoming air before discharge through two adjustable diffusers. A heating system is standard, and the inflow of ventilating and/or heated air has the effect of pressurising the cockpit, further discouraging any inflow of toxic fumes or chemicals.

Power plant consists of a 212·5 kW (285 hp) Teledyne Continental Tiara engine which, having a 2 : 1 reduction gear, permits the use of a large-diameter propeller. Turning at only 1,700 rpm at normal cruising speed, this ensures that the Brave is quiet in operation.

Several fire suppression provisions have been introduced which are unique for an agricultural aircraft. The fuel tanks, located in the wing roots, are filled with reticulated polyurethane foam to serve both as a fire suppressant and as a constant baffle to reduce fuel surge. Fire-resistant fuel pipes are wire-reinforced at potential rupture points.

To meet varying requirements, two hopper sizes are available. The larger hopper has a maximum dry chemicals capacity of 862 kg (1,900 lb), and is compatible with applicators designed to spread chemicals at rates of up to 181 kg (400 lb) per acre.

Spray equipment for the PA-36 has a capability of up to 863 litres (228 US gallons) per minute, which is the equivalent of 64 litres (17 US gallons) per acre at 117 knots (217 km/h; 135 mph) and with a 15·25 m (50 ft) swath width. The spray equipment consists of a quickly-removable pylon-mounted wind-driven spraypump, and spraybooms located just aft of the wing trailing-edges. This location reduces drag and allows the pilot to make visual checks of their operation.

All parts of the Brave's airframe are treated to prevent corrosion damage, with extensive use of polyurethane coating, selection of stainless steel for cables and other moving components in vulnerable areas, and internal oiling of lower truss sections. The design eliminates dust traps and inaccessible areas, and fuselage covering is spaced away from the frame to permit thorough hosing down. To facilitate washing, inspection and maintenance, the plastics side panels and entire belly covering are attached by quick-release fasteners.

TYPE: Single-seat agricultural aircraft.

WINGS: Cantilever low-wing monoplane. Conventional two-spar metal structure. Light alloy laminated spars with two-bolt main spar attachment to fuselage structure. Light alloy covering, except for detachable leading-edges of glassfibre, reinforced by foam inserts, and glassfibre wingtips. Conventional ailerons and trailing-edge flaps. Landing lights in wing leading-edges.

FUSELAGE: Welded chrome-molybdenum steel tube structure. Removable metal underskin and removable side panels of plastics material. Glassfibre engine cowling.

TAIL UNIT: Cantilever all-metal structure. Tab on rudder and in each elevator. Cable from top of cockpit structure to tip of fin to deflect cables.

LANDING GEAR: Non-retractable tailwheel type. Interchangeable cantilever spring steel main-gear struts, with wire-cutters on leading-edges. Main wheels and tyres size 8·50-10. Steerable tailwheel with tyre of 0·25 m (10 in) diameter. Parking brake.

POWER PLANT: One 212·5 kW (285 hp) Teledyne Continental Tiara 6-285 flat-six engine, driving a Hartzell two-blade metal constant-speed propeller. Three-blade constant-speed propeller optional. One fuel tank in each wing root, capacity 170·3 litres (45 US gallons). Total fuel capacity 340·6 litres (90 US gallons), of which 322 litres (85 US gallons) are usable. Refuelling point on upper surface of each wing. Fuel tanks filled with reticulated polyurethane safety foam (Safom).

ACCOMMODATION: Pilot only, on adjustable seat in an isolated cockpit capsule, with steel tube overturn structure. Seat, equipped with double shoulder harness and inertia reel, is attached to overturn structure. Wire-cutter mounted in centre of windscreen. Combined window and door on each side, hinged at bottom. Cockpit capsule is heated and ventilated.

SYSTEMS: Electrical system supplied by 28V 60A alternator, with 24V 17Ah battery. Hydraulic system for brakes only.

EQUIPMENT: Standard equipment includes a non-corrosive hopper/tank of translucent glassfibre-reinforced plastics, installed forward of cockpit and approximately on CG, of 0·85 m³ (30 cu ft) capacity, containing 852 litres (225 US gallons). Optional hopper/tank of 1·08 m³ (38 cu ft) capacity, containing 1,041 litres (275 US gallons). The latter has a maximum capacity for dry chemicals of 862 kg (1,900 lb). Venturi-type dry material spreaders of either stainless steel or aluminium available, including a basic design capable of application rates of 2·3 to 91 kg (5 to 200 lb) per acre. Spray system comprises an easily-removable wind-driven spraypump and 38 mm (1½ in) diameter spraybooms equipped with 60 nozzles. Other optional equipment includes sensitive altimeter; 8-day clock; turn co-ordinator; landing and taxi lights; navigation, instrument panel and anti-collision lights; cockpit fire extinguisher; and heater.

DIMENSIONS, EXTERNAL:
- Wing span: 11·89 m (39 ft 0 in)
- Length overall: 8·34 m (27 ft 4¼ in)
- Propeller diameter: 2·41 m (7 ft 11 in)
- Propeller ground clearance: 0·25 m (10 in)

WEIGHTS AND LOADINGS:
- Weight empty:
  - Normal category: 930 kg (2,050 lb)
  - Restricted category, 2-blade propeller: 991 kg (2,185 lb)
  - Restricted category, 3-blade propeller: 999 kg (2,203 lb)
- Max T-O weight:
  - Normal category: 1,769 kg (3,900 lb)
  - Restricted category: 1,996 kg (4,400 lb)
- Wing loading:
  - Normal category: 84·4 kg/m² (17·3 lb/sq ft)
  - Restricted category: 93·2 kg/m² (19·1 lb/sq ft)

PERFORMANCE (A: Normal category at 1,769 kg; 3,900 lb max T-O weight; B: Restricted category at 1,996 kg; 4,400 lb with two-blade propeller; C: Restricted category at 1,996 kg; 4,400 lb with three-blade propeller):
- Max level speed at S/L:
  - A: 131 knots (243 km/h; 151 mph)
  - B: 108 knots (200 km/h; 124 mph)
  - C: 105 knots (195 km/h; 121 mph)
- Max cruising speed, 75% power at optimum altitude:
  - A: 128 knots (236 km/h; 147 mph)
- Max cruising speed, 75% power at 610 m (2,000 ft):
  - B: 88 knots (163 km/h; 101 mph)
  - C: 83·5 knots (154 km/h; 96 mph)
- Stalling speed, flaps down, power off:
  - A: 54 knots (100 km/h; 62 mph)
  - B, C: 58·5 knots (108 km/h; 67 mph)
- Max rate of climb at S/L:
  - A: 241 m (790 ft)/min
  - B, C: 108 m (355 ft)/min
- Service ceiling:
  - A: 3,960 m (13,000 ft)
  - B, C: 1,800 m (5,900 ft)
- Absolute ceiling:
  - A: 4,570 m (15,000 ft)
  - B, C: 2,470 m (8,100 ft)
- T-O run, 15° flap:
  - A: 267 m (875 ft)
  - B: 557 m (1,829 ft)
  - C: 449 m (1,473 ft)
- T-O to 15 m (50 ft), 15° flap:
  - B: 792 m (2,600 ft)
  - C: 686 m (2,250 ft)
- Landing from 15 m (50 ft):
  - A: 503 m (1,650 ft)
  - B, C: 448 m (1,470 ft)
- Landing run:
  - A: 213 m (700 ft)
  - B: 218 m (715 ft)
- Range, 75% power at 610 m (2,000 ft), max fuel, no reserves:
  - B: 428 nm (793 km; 493 miles)
  - C: 404 nm (750 km; 466 miles)

Piper PA-36 Pawnee Brave agricultural aircraft (Teledyne Continental Tiara 6-285 engine)

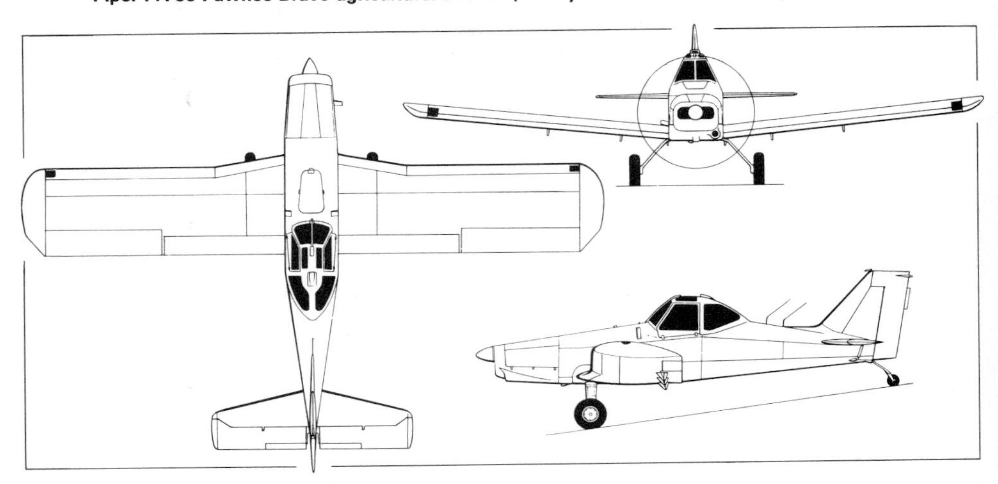

Piper PA-36 Pawnee Brave, latest version of the Pawnee agricultural aircraft *(Pilot Press)*

PITTS

PITTS AVIATION ENTERPRISES INC

ADDRESS:
PO Box 548, Homestead, Florida 33030
Telephone: (305) 247-5423

PRESIDENT: C. H. Pitts

One of the best-known US designers of high-performance sporting aircraft, Mr Curtis Pitts is responsible for the single-seat and two-seat Pitts Special biplanes. Detailed construction drawings for the single-seat S-1 version are available to amateur constructors (see Homebuilt section). The two-seat S-2 is available only as a factory-built aircraft, being produced for Pitts Aviation by Aerotek Inc of Afton, Wyoming. This latter company is also constructing for Pitts examples of the single-seat

S-1S, supplied to pilots who do not wish to build their own aircraft.

Kelly Aeroplane Ltd in the UK has world sales rights in the Pitts Specials. After-sales maintenance is assigned to Personal Plane Services Ltd of Wycombe Air Park, Booker, Marlow, Buckinghamshire.

PITTS S-1 SPECIAL

The original single-seat Pitts Special was designed in 1943-44. Construction of a prototype began in 1944 and it flew in September of that year. One of the most successful early models was *Little Stinker,* powered by a 67 kW (90 hp) Continental engine, and built by Mr Pitts in 1947 for Miss Betty Skelton, then an internationally-known aerobatic display pilot. The *Black Beauty* biplane, built by Pitts for Miss Caro Bailey, was of similar design, but powered by a 93 kW (125 hp) Lycoming O-290-D engine.

Since then even more powerful engines have been installed in Pitts Specials built by both the designer and other people, and the version of the single-seat Special for which drawings are supplied by Mr Pitts is designed to take a Lycoming engine of up to 134 kW (180 hp), as noted in the Homebuilts section.

Current versions of the S-1 are as follows:

S-1D. Intended for homebuilders only, with plans available.

S-1S. Production aircraft, FAA type certificated and built at rate of 25 aircraft per year. It is available also in kit form, parts, materials and components being produced under an FAA Approved Production Certificate.

By early 1976, about 300 S-1s and 31 S-1Ss were under construction or flying.

There is also an S-1C version, but Pitts can no longer supply plans for its construction.

Details of some of the major successes achieved by US pilots of Pitts Specials in national and international aerobatic competitions, since 1966, were given in the 1972-73 *Jane's.* After that edition went to press, a US team, all flying Pitts Specials, recorded a major success in the 7th World Aerobatic Championships, held at Salon-en-Provence, France, on 18-31 July 1972. Charles Hillard became men's champion, with Gene Soucy third in another Pitts and the remaining Americans in sixth and ninth positions. Mary Gaffaney won the women's contest, and the US team carried off the team prize, with the Soviet Yak-18s second.

The details which follow apply to the S-1S factory-built aircraft with 134 kW (180 hp) engine, but engines of 74·5-134 kW (100-180 hp) can be fitted to the S-1 Special.

TYPE: Single-seat sporting biplane.

WINGS: Braced biplane type, with single faired interplane strut each side and N-type cabane struts. Dual streamline flying and landing wires. Wing section M6. Thickness/chord ratio 12%. Dihedral 0° on upper wing, 3° on lower wings. Incidence 1° 30′ on upper wing, 0° on lower wings. Sweepback at quarter-chord 6° 40′ on upper wing only. Wooden structure, with fabric covering. Frise-type ailerons on both upper and lower wings, of similar construction to wings. No flaps or tabs.

FUSELAGE: Welded steel tube structure, covered with fabric.

TAIL UNIT: Wire-braced steel tube structure, fabric-covered. Fixed-incidence tailplane. Trim tab in each elevator.

LANDING GEAR: Non-retractable tailwheel type. Rubber-cord shock-absorption. Cleveland main wheels with 6-ply tyres, size 5·00-5, pressure 2·07 bars (30 lb/sq in). Cleveland hydraulic disc brakes. Steerable tailwheel. Glassfibre fairings on main wheels.

POWER PLANT: One 134 kW (180 hp) Lycoming IO-360-B4A flat-four engine, driving a Sensenich type 76EM8-O-56/62 two-blade metal fixed-pitch propeller. Fuel tank aft of firewall, capacity 75 litres (20 US gallons). Refuelling point on upper surface of fuselage, immediately forward of windscreen. Oil capacity 7·5 litres (2 US gallons). Inverted fuel and oil systems standard.

ACCOMMODATION: Single seat in open cockpit.

DIMENSIONS, EXTERNAL:

Wing span, upper	5·28 m (17 ft 4 in)
Wing chord (constant, both)	0·91 m (3 ft 0 in)
Wing aspect ratio	5·77
Length overall	4·71 m (15 ft 5½ in)
Height overall	1·92 m (6 ft 3½ in)
Tailplane span	1·98 m (6 ft 6 in)
Propeller diameter	1·93 m (6 ft 4 in)

AREA:

Wings, gross	9·15 m² (98·5 sq ft)

WEIGHTS AND LOADING:

Weight empty	326 kg (720 lb)
Max T-O weight	521 kg (1,150 lb)
Max power loading	3·89 kg/kW (6·38 lb/hp)

PERFORMANCE (at max T-O weight):

Never-exceed speed	176 knots (326 km/h; 203 mph)
Max level speed at S/L	153 knots (283 km/h; 176 mph)
Max cruising speed at S/L	122 knots (227 km/h; 141 mph)
Stalling speed	54 knots (100 km/h; 62 mph)
Max rate of climb at S/L	792 m (2,600 ft)/min
Service ceiling	6,795 m (22,300 ft)
T-O to 15 m (50 ft)	331 m (1,085 ft)
Range with max fuel, no reserves	273 nm (507 km; 315 miles)

Pitts S-1S Special single-seat sporting biplane

PITTS S-2A SPECIAL

First flown in 1967, the S-2A is a two-seat version of the Pitts Special. It is similar to the single-seat S-1 in basic configuration and construction, but is slightly larger in overall dimensions, with no attempt at commonality of components. The increased size and power, coupled with aerodynamic changes, give the two-seater improved aerobatic and landing characteristics, and make it extremely stable in rough air conditions. Control responses are better than on the S-1. The ailerons are aerodynamically balanced for higher rate of roll at low speeds, and full vertical rolls can be made with ease. The different wing sections used on the S-2A provide inverted performance equal to conventional flight, and facilitate outside loops.

The S-2A is FAA type certificated in the Normal and Aerobatic categories. It is a production aeroplane, not intended for the homebuilder, and plans are not available.

Five S-2As, each fitted with a 149 kW (200 hp) Lycoming engine, were supplied in early 1973 to the British aerobatic display team financed by the Rothman Tobacco Company. During displays, the front cockpit of each aircraft is covered by a removable panel. Five similar aircraft were supplied to the Carling Black Label aerobatic team operating from Toronto, Canada.

S-2As have been exported to Australia, Brazil, Sweden and Venezuela, and a total of 119 of this model had been built by early 1976. In mid-1976, the Peruvian Air Force announced the order of six Pitts Specials, for 'an unspecified training role'.

TYPE: Two-seat aerobatic biplane.

WINGS: Braced biplane type, with single faired interplane strut each side and N-type cabane. Wing section NACA 6400 series on upper wing, 00 series on bottom wings. Two-spar wooden (spruce) structure with fabric covering. Aerodynamically-balanced ailerons on both upper and lower wings. No flaps or tabs.

FUSELAGE: Welded steel tube structure with wooden stringers, covered with Dacron fabric except for metal top-decking.

TAIL UNIT: Wire-braced welded steel tube structure. Fixed surfaces metal-covered, control surfaces fabric-covered. Trim tab in each elevator.

LANDING GEAR: Non-retractable tailwheel type. Rubber-cord shock-absorption. Steerable tailwheel. Fairings on main wheels.

POWER PLANT: One 149 kW (200 hp) Lycoming IO-360-A1A flat-four engine, driving a Hartzell type HC-C2YK-4/C7666A-2 two-blade metal constant-speed propeller with spinner. Fuel tank in fuselage, immediately aft of firewall, capacity 90·5 litres (24 US gallons). Refuelling point on fuselage upper surface forward of front windscreen. Oil capacity 9 litres (2·4 US gallons). Inverted fuel and oil systems standard.

ACCOMMODATION: Two seats in tandem cockpits with dual controls. Rear cockpit can be enclosed by a transparent canopy, if required. Space for 9 kg (20 lb) baggage aft of rear cockpit when flown in non-aerobatic category.

SYSTEM: Electrical system powered by 12V 40A alternator and non-spill 12V battery.

DIMENSIONS, EXTERNAL:

Wing span, upper	6·10 m (20 ft 0 in)
Wing span, lower	5·79 m (19 ft 0 in)
Wing chord (constant, both)	1·02 m (3 ft 4 in)
Length overall	5·41 m (17 ft 9 in)
Height overall	1·94 m (6 ft 4½ in)

AREA:

Wings, gross	11·6 m² (125 sq ft)

WEIGHTS AND LOADINGS (A: Aerobatic; B: Normal category):

Weight empty:	
A, B	453 kg (1,000 lb)
Max T-O weight:	
A	680 kg (1,500 lb)
B	714 kg (1,575 lb)
Max wing loading:	
A	58·6 kg/m² (12·0 lb/sq ft)
B	61·5 kg/m² (12·6 lb/sq ft)
Max power loading:	
A	4·56 kg/kW (7·5 lb/hp)
B	5·33 kg/kW (7·87 lb/hp)

PERFORMANCE (at max T-O weight. A: Aerobatic; B: Normal category):

Never-exceed speed:	
A, B	176 knots (326 km/h; 203 mph)

Pitts S-2A Special, a factory-built two-seater with enclosed rear cockpit

Max level speed at S/L:
A, B 136 knots (253 km/h; 157 mph)
Max cruising speed at S/L:
A, B 132 knots (245 km/h; 152 mph)
Stalling speed:
A 51 knots (94 km/h; 58 mph)
B 51·5 knots (95 km/h; 59 mph)
Max rate of climb at S/L:
A 579 m (1,900 ft)/min
B 549 m (1,800 ft)/min
Service ceiling:
A 6,125 m (20,100 ft)
B 4,875 m (16,000 ft)
T-O to 15 m (50 ft):
A 351 m (1,150 ft)
Range with max fuel:
A 297 nm (552 km; 343 miles)

RAISBECK
THE RAISBECK GROUP

CORPORATE OFFICES:
84 N.E. Loop 410, Suite 234E, San Antonio, Texas 78216
Telephone: (512) 349-1771
NORTHWEST ENGINEERING DIVISION:
7777 Perimeter Road, Seattle, Washington 98108
Telephone: (206) 762-5156
MANUFACTURING DIVISION:
8833 Shirley Avenue, Northridge, California 91324
Telephone: (213) 885-9200
PRESIDENT AND BOARD CHAIRMAN:
James D. Raisbeck
EXECUTIVE VICE-PRESIDENT AND CORPORATE TREASURER:
Dwight D. Christy, Jr
VICE-PRESIDENTS:
R. Kim Frinell (Engineering)
A. G. Willauer (Manufacturing)

The Raisbeck Group was formed in 1974 by many of the team members who had developed high-lift systems for Boeing, McDonnell Douglas and Robertson Aircraft Corporation.

Its aim is to improve the overall productivity of business and commercial jet aircraft by the infusion of current and advanced technology into existing airframes. Heavy emphasis is placed on aerodynamic refinements to reduce cruise drag at normal operating speeds, and to reduce required runway field length through an associated reduction in certificated stalling speeds in take-off and landing configurations.

Personnel employed in the Group's corporate offices, in San Antonio, Texas, are engaged primarily on feasibility studies involving the application of advanced technology to current types of business and commercial jet aircraft. The Group's Northwest Engineering Division, located in Seattle, Washington, is responsible for the detailed development and certification of all the Group's systems, as well as the maintenance of previously certificated systems, and including the provision of flight manual supplements. The Manufacturing Division, located in Northridge, California, has been expanded to allow virtually full in-house fabrication of prototype and production aircraft components.

The first programme by The Raisbeck Group involved the Gates Learjet family of aircraft. Work on a 'Mark II System' began in late 1973. Subsequent flight test and development, initially in conjunction with Gates Learjet Corporation and completed later through a joint effort with The Dee Howard Company of San Antonio, Texas, has brought to the market an advanced technology high-lift system for Learjet aircraft. This makes the following contribution to the aircraft's performance:

(i) Reduction in T-O and approach speeds of 16-23 knots (30-43 km/h; 18·5-26·5 mph), depending on configuration and the specific model.

(ii) Reduction in the aircraft's cruise drag of measurably significant proportions.

(iii) Certification of the turbojet versions of the Learjet Models 23, 24 and 25 to FAR Part 36 acoustic requirements without the use of sound suppressors or any change to the engine nacelle configuration.

(iv) Elimination of the stick-pusher as a performance-limiting item.

(v) Aircraft has inherent aerodynamic stall-warning characteristics.

Certification of the Mark II System was received by the Howard/Raisbeck joint venture in October 1975, and production of retrofit systems for the Learjet family of aircraft was launched immediately. By May 1976 more than 60 production Mark II Systems had been retrofitted to existing Learjet aircraft, and production was continuing at a rate in excess of ten systems per month.

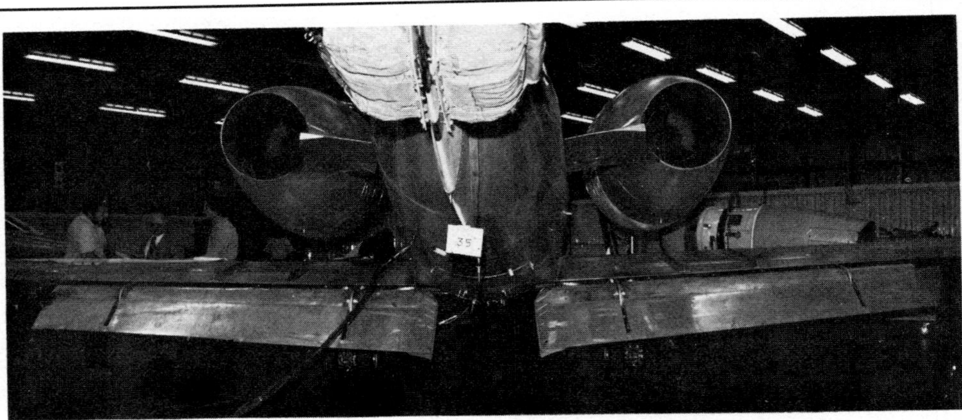

Learjet Model 36 fitted with Raisbeck wide-chord Fowler trailing-edge flaps

	LEARJET 24B/24D		LEARJET 25/25B/25C	
	Basic Aircraft knots-km/h-mph	Mk II System knots-km/h-mph	Basic Aircraft knots-km/h-mph	Mk II System knots-km/h-mph
T-O speed at max T-O weight, 10° flap	*	123-229-142	*	130-241-150
T-O speed at max T-O weight, 20° flap	131-243-151	115-212-132	144-267-166	121-224-139
Minimum control speed (VMC)	108-200-124	80-148-92	102-188-117	Below stall speed
Normal cruising speed	441-816-507	458-848-527	441-816-507	458-848-527
Long range cruising speed	405-750-466	430-797-495	408-756-470	434-805-500
Approach speed at max landing weight, flaps down	133-246-153	115-212-132	139-257-160	122-225-140
Approach speed at 120% of empty weight, flaps down	114-211-131	95-175-109	118-219-136	101-187-116
Stalling speed at max T-O weight, flaps up	126-233-145	106-196-122	137-254-158	112-208-129
Stalling speed at max T-O weight, 10° flap	*	101-187-116	*	106-196-122
Stalling speed at max T-O weight, 20° flap	109-203-126	94-174-108	118-219-136	99-183-114
Stalling speed at max landing weight, flaps down	100-185-115	85-158-98	107-198-123	91-169-105

NOTE: Figures quoted are indicated airspeeds (IAS) and are stated by Howard/Raisbeck to be typical of the improvements under all operating conditions. *Indicates new capability in this configuration.

Feasibility studies for the next development and certification programme to be undertaken by The Raisbeck Group were being conducted during the Summer of 1976, involving aircraft such as the Rockwell Sabreliner, Grumman Gulfstream II, Douglas DC-8 Series 30 and DC-9 Series 10, and Boeing 707/720 series. Initiation of a formal programme on one of these aircraft was expected to begin in the Autumn of 1976.

HOWARD / RAISBECK MARK II SYSTEM

Features of the Mark II System include the installation of full-span supercritical wing leading-edges over the first 5% of wing chord, while retaining the standard Learjet anti-icing system; addition of strakes inboard of the wingtip tanks to enhance aileron response during slow-speed and engine-out operation; removal of vortex generators; provision of new aileron gap seals; extension of the upper surface of the basic wing to cover the existing trailing-edge flap, combined with new flap tracks, to allow a significant increase in aft flap travel prior to downward deflection; addition of stall turbulators on wing leading-edges; provision of a flap-actuated pitch trim system which puts an 'up' pre-load on the elevators automatically as the flaps are deflected; installation of improved angle of attack indicator and new flap position sensor/indicator systems; and resetting of the stick shaker and pusher to conform with enhanced aerodynamic stall characteristics.

A summary of Howard/Raisbeck Mark II System performance improvements for Gates Learjet aircraft is given in the accompanying table.

RILEY
RILEY TURBOSTREAM CORPORATION

HEAD OFFICE:
PO Box 5247, Waco Madison Cooper Airport, Waco, Texas 76708
Telephone: (817) 752-9781
Telex: 73-0900
PRESIDENT: Jack M. Riley

Mr J. Riley was responsible for the Riley 55 Twin-Navion conversion scheme in 1952 and the Riley Rocket conversion of the Cessna 310 in 1962. His company is marketing currently a Cessna 310/320 conversion known as the Riley Super 310, a Cessna 340 conversion which is designated Riley Super 340, and a conversion of the Cessna 414 designated Riley Super 414-8.

Riley Dove, Heron and Cessna 310/320 conversions known respectively as the Riley Turbo-Exec 400, Turbo Skyliner and Turbostream have been discontinued.

FAA approved export conversion kits are available for the three aircraft in the current production programme, and assistance is available if required to gain approval by the civil aviation authority in the country to which the kit has been exported.

RILEY SUPER 310

The Super 310 represents a similar conversion to that of the Super 340, the existing power plants in the Cessna 310/320 being replaced by 231 kW (310 hp) Continental TSIO-520-Js. The conversion is FAA approved for the Cessna 310I to 310R, 320B to 320F and T310P to T310R. Installation of the complete kit requires 12 working days.

An aerial mapping version is available with either single or dual Wild RC-8 or RC-9 camera provisions.

A description of the current Cessna Model 310 appears under the Cessna entry in this edition. It applies also to the Riley Super 310, except as follows:

POWER PLANT: As described for Riley Super 340.

WEIGHTS AND LOADING: As for the appropriate Cessna 310/320 model except as follows:
Weight empty increased by 10 kg (22 lb)
Max payload reduced by 10 kg (22 lb)
Max power loading 5·11 kg/kW (8·4 lb/hp)

PERFORMANCE (at max T-O weight):
Max level speed at 7,315 m (24,000 ft)
270 knots (500 km/h; 311 mph)
Max cruising speed at 7,315 m (24,000 ft)
255 knots (473 km/h; 294 mph)

Max cruising speed at 3,660 m (12,000 ft)
225 knots (417 km/h; 259 mph)
Econ cruising speed at 3,660 m (12,000 ft)
200 knots (370 km/h; 230 mph)
Max rate of climb at S/L 671 m (2,200 ft)/min
Cruise rate of climb at max cruising power
457 m (1,500 ft)/min
Rate of climb at S/L, one engine out
152 m (500 ft)/min
Service ceiling 10,980 m (36,000 ft)
Service ceiling, one engine out 7,315 m (24,000 ft)
T-O run 335 m (1,100 ft)
T-O to 15 m (50 ft) 472 m (1,550 ft)
Landing from 15 m (50 ft) 546 m (1,790 ft)
Range with 757 litres (200 US gallons) fuel, max cruising speed at 7,315 m (24,000 ft), no reserves
1,476 nm (2,735 km; 1,700 miles)
Range with 757 litres (200 US gallons) fuel, econ cruising speed at 7,315 m (24,000 ft), no reserves
1,815 nm (3,363 km; 2,090 miles)

RILEY SUPER 340

The Riley Super 340 differs from the standard Cessna 340 primarily through replacement of the latter's TSIO-520-K engines by TSIO-520-Js. The most important change associated with this new engine installation is the addition of an intercooler which allows a higher power output and also improves specific fuel consumption, critical altitude, engine life and reliability.

The prototype Super 340 received FAA certification in June 1974 and 100 aircraft had been modified by January 1976. The standard Cessna 340A for 1976 has been updated in a manner similar to that of the Riley Super 340. Consequently, only Cessna 340 aircraft constructed from 1972 until the last of the 1975 models were completed are suitable for this conversion, which is completed by Riley engineers in five working days.

A description of the current Cessna Model 340A appears under that company's entry in this edition. It applies also to the Riley Super 340, except as follows:

Power Plant: Two 231 kW (310 hp) Continental TSIO-520-J flat-six turbocharged and intercooled engines, each driving a McCauley three-blade propeller. Engine installation upgraded from Cessna 340 to Cessna 414 configuration. Fuel system as for Cessna 340. Optional aft nacelle fuel tanks, each of 75·7 litres (20 US gallons), provide max optional fuel capacity of 908 litres (240 US gallons) when used in conjunction with the Cessna optional 757 litre (200 US gallon) system.

Weights and Loading:
Weight empty 1,754 kg (3,868 lb)
Max payload 904 kg (1,993 lb)
Max T-O weight 2,710 kg (5,975 lb)
Max power loading 5·87 kg/kW (9·64 lb/hp)

Performance (at max T-O weight):
Max level speed at 7,315 m (24,000 ft)
252 knots (467 km/h; 290 mph)
Max cruising speed at 7,315 m (24,000 ft)
234 knots (435 km/h; 270 mph)
Max cruising speed at 5,485 m (18,000 ft)
226 knots (418 km/h; 260 mph)
Max cruising speed at 4,265 m (14,000 ft)
217 knots (402 km/h; 250 mph)
Econ cruising speed at 7,315 m (24,000 ft)
208 knots (386 km/h; 240 mph)
Max rate of climb at S/L 549 m (1,800 ft)/min
Cruise rate of climb at max cruising power
305 m (1,000 ft)/min
Rate of climb at S/L, one engine out
107 m (350 ft)/min
Service ceiling 9,755 m (32,000 ft)
Service ceiling, one engine out 4,875 m (16,000 ft)
T-O run 457 m (1,500 ft)
T-O to 15 m (50 ft) 594 m (1,950 ft)
Landing from 15 m (50 ft) 561 m (1,840 ft)
Landing run 233 m (765 ft)
Range with 757 litres (200 US gallons) fuel, max cruising speed at 7,315 m (24,000 ft), no reserves
1,302 nm (2,415 km; 1,500 miles)

Riley Super 340 conversion of the Cessna Model 340

Range with 908 litres (240 US gallons) fuel, econ cruising speed at 7,315 m (24,000 ft), no reserves
2,170 nm (4,025 km; 2,500 miles)
Range with pilot, co-pilot, four passengers, 91 kg (200 lb) baggage, fuelled to max T-O weight, econ cruising speed at 7,315 m (24,000 ft), no reserves
868 nm (1,610 km; 1,000 miles)

RILEY SUPER 414-8

Latest product of the Riley Corporation is a conversion of the Cessna Model 414 in which the original 231 kW (310 hp) flat-six engines are replaced by two 298 kW (400 hp) Lycoming flat-eight turbocharged engines. The prototype of the Super 414-8 was scheduled to make its first flight in August 1976, with FAA certification and production following in December 1976.

A description of the Cessna Model 414 appears under the Cessna entry in this edition. It applies also to the Riley Super 414-8, except as follows:

Power Plant: Two 298 kW (400 hp) Lycoming IO-720-B1BD flat-eight turbocharged and intercooled engines, driving Hartzell four-blade metal constant-speed propellers. Standard fuel capacity of 757 litres (200 US gallons), with optional capacity of 908 litres (240 US gallons). A fuel dump system able to jettison 283 kg (625 lb) is standard.

Weights and Loadings:
Weight empty increased by 170 kg (375 lb)
Max payload increased by 113 kg (250 lb)
Max T-O weight increased by 283 kg (625 lb)
Max wing loading 173·8 kg/m² (35·6 lb/sq ft)
Max power loading 5·31 kg/kW (8·72 lb/hp)

Performance (estimated, at max T-O weight):
Max level speed at 7,315 m (24,000 ft)
270 knots (500 km/h; 311 mph)
Max cruising speed at 7,315 m (24,000 ft)
248 knots (460 km/h; 286 mph)
Max cruising speed at 3,660 m (12,000 ft)
208 knots (386 km/h; 240 mph)
Econ cruising speed at 7,315 m (24,000 ft)
222 knots (412 km/h; 256 mph)
Max rate of climb at S/L 716 m (2,350 ft)/min
Cruise rate of climb at max cruising power
411 m (1,350 ft)/min
Rate of climb at S/L, one engine out
160 m (525 ft)/min
Service ceiling 10,980 m (36,000 ft)
Service ceiling, one engine out 7,315 m (24,000 ft)
Range with 908 litres (240 US gallons) fuel, max cruising speed at 7,315 m (24,000 ft), 45 min reserves
1,172 nm (2,172 km; 1,350 miles)
Range with 908 litres (240 US gallons) fuel, econ cruising speed at 7,315 m (24,000 ft), 45 min reserves
1,476 nm (2,735 km; 1,700 miles)

RILEY JETSTREAM

Riley was converting in 1975 a number of Mk 1 (Handley Page built) Jetsreams to take 701 kW (940 shp) Turboméca Astazou XVI turboprop engines, a conversion which permits operation from 'hot and high' airfields at 5,670 kg (12,500 lb) gross weight. Three such conversions have been delivered to Sierra Pacific Airlines.

Riley conversion of the Handley Page Mk 1 Jetstream, with Turboméca Astazou XVI turboprop engines

ROBERTSON

ROBERTSON AIRCRAFT CORPORATION

Headquarters:
839 West Perimeter Road, Renton Municipal Airport, Renton, Washington 98055
Telephone: (206) 228-5000
(800) 426-7692 (Wats Toll-Free)

President: Ronald L. Lien
Executive Vice-President and General Manager: Leland R. Lynch
Senior Vice-President (Engineering): John T. Calhoun
Vice-President (Marketing) and Chief Pilot: Henry I. McKay
Sales Manager: David E. Parvin
Operations Manager: Earl Severns
Production Manager: Ted Cederblom

Robertson Aircraft Corporation was formed by the late Mr James L. Robertson, who had long been a pioneer in the development of STOL aircraft, having been responsible for the Skycraft Skyshark, Wren 460 and STOL modifications to the IMCO CallAir A-9 and B-1. It has designed, built and certificated a series of R/S and R/STOL advanced technology safety and performance systems for standard single- and multi-engined Cessna and Piper aircraft.

Work is also in progress on the design and integration of individually-proven STOL concepts for application to other general aviation light aircraft.

Research contracts with Kansas University and NASA were completed in 1974, covering the design of an Advanced Technology Light Twin (ATLIT). A standard Piper Seneca fuselage was used as the basic structure and provided with an experimental wing that includes a new wing section and 30% chord full-span Fowler flaps. There are no ailerons and roll control is achieved by the use of upper-surface spoilers. Construction of the aircraft was completed by Piper Aircraft Corporation and the first flight was recorded in late 1974.

Since the ATLIT is purely an experimental aircraft, designed for the investigation of advanced aerodynamic concepts, it is not included among the list of Robertson conversions. Some of the technology developed for this programme has led to an R/STOL system for the Piper Seneca which is described. The ATLIT contract with Robertson included also the design and construction of propeller blades incorporating a supercritical aerofoil section.

ROBERTSON / CESSNA and ROBERTSON / PIPER SAFETY and STOL CONVERSIONS

Continuous product improvement has been made on Robertson's line of R/STOL systems for Cessna and Piper single-engined and smaller twin-engined aircraft. The Robertson conversion, first applied to a Cessna 182, comprises full-span wing leading-edge and trailing-edge high-lift systems which greatly reduce the take-off and landing distances normally required by such aircraft.

The existing ailerons are used as an integral part of the full-span trailing-edge flap system. When the conventional inboard flaps are lowered for take-off or landing, the ailerons droop with them, virtually doubling the wing lift at low speeds. The ailerons retain their differential operation for roll control when drooped.

In addition, the wing is fitted with a full-span distributed-camber leading-edge to provide an optimum spanwise lift distribution for maximum cruise efficiency. The cambered leading-edge also reduces the aerofoil leading-edge pressure peak at high angles of attack, to impart maximum resistance to stall and to provide highly-responsive manoeuvrability at low airspeeds. Most Cessna single-engined aircraft built since 1973 include this Robertson leading-edge as a standard production feature.

The full-span flap system, in combination with Robertson's conical-cambered wingtips, dorsal and ventral fins, belly-mounted vortex generators, and flap/elevator automatic trim system, are combined in various models to offer increased performance.

To improve controllability at low speeds, stall fences are provided between flaps and ailerons, and to complete the STOL modification the aileron gap is sealed with a strip of aluminium sheet or rubberised canvas. These modifications permit safe STOL landings and take-offs by even novice pilots, and cruising speed and range are increased by 2-4%.

Maximum gross weight increases have accompanied the certification of all Robertson conversions of twin-engined aircraft such as the Cessna Super Skymaster and Piper Twin Comanche. This is due primarily to their increased climb performance and slower take-off and landing speeds.

One of Robertson's most ambitious developments has been the R/STOL system for the entire Cessna 400 series of twin-engined business aircraft. Complete redesign of the wings of these aircraft from the rear spar aft has allowed installation of 100% Fowler flaps, together with flap-actuated drooping ailerons. This system, in combination with Robertson's automatic pitch trim system and double-hinged rudder, has led to FAA-certificated decreases in take-off and landing field lengths of approximately 40%.

During 1974 many new developments were completed at the Robertson plant at Renton, including certification of the Robertson-equipped Piper Seneca. This conversion stemmed from the ATLIT technology, and comprises wing upper-surface spoilers for roll control, full-span slotted flaps, cambered wingtips and an anti-servo rudder tab. These modifications allow shorter take-off and landing distances; minimum control speed is reduced by 16%, the best rate of climb speed is lowered by 19%, single-engine service ceiling is raised by 213 m (700 ft) and roll response is increased greatly at all speeds and configurations.

The Robertson integrated high-lift and safety systems have been designed for easy field maintenance. They are designed to be applicable to almost the entire range of Cessna and Piper aircraft, as detailed in the accompanying tables.

Full details of the basic Cessna and Piper airframes are given under the appropriate company headings in this edition (Piper Twin Comanche in the 1973-74 edition), and apply also to the Robertson versions, except for the added R/STOL systems as described. Weights and performance details of the entire range of R/STOL conversions are given in the accompanying tables. The conversions can be fitted as a retrospective modification to any of the models listed, irrespective of year.

The complete line of Robertson STOL systems is available throughout the world from twenty-three installation centres and more than sixty dealers.

Right, top to bottom: Robertson STOL-equipped Cessna 180, showing the cambered wingtips, stall fences and drooped ailerons; Robertson STOL installation on a Piper Cherokee SIX 300, showing the fuselage flap and cuffed leading-edge; Robertson-equipped Cessna 402B, showing Fowler flaps and drooped ailerons; and upper-surface spoilers, full-span slotted flaps and cambered wingtips on a Robertson-modified Seneca

Robertson STOL conversion of the Piper Aztec; the drooped ailerons and cambered wingtips are seen clearly in this view

R/STOL VERSIONS OF CESSNA MODELS

	Weight empty equipped kg (lb)	Weight gross kg (lb)	Max level speed knots (km/h; mph)	Max cruising speed knots (km/h; mph)	Stalling speed, wheels and flaps down knots (km/h; mph)	Max rate of climb at S/L m (ft)/min	Single-engine rate of climb at S/L m (ft)/min	Service ceiling m (ft)	T-O run m (ft)	T-O to 15 m (50 ft) m (ft)	Landing from 15 m (50 ft) m (ft)	Landing run m (ft)	Max range** nm (km; miles)
Model 150 and Commuter	449 (990)	725 (1,600)	110 (204; 127)	105 (195; 121)	20 (48·3; 30)	213 (700)		3,932 (12,900)	A 129 (422) B 161 (527)	A 248 (815) B 273 (895)	A 193 (632) B 230 (755)	A 90 (295) B 106 (348)	790 (1,464; 910)
Model 172 and Skyhawk†	572 (1,263)	1,043 (2,300)	126 (233; 145)	118 (219; 136)	28 (51·5; 32)	206 (675)		4,150 (13,600)	A 140 (460) B 175 (575)	A 274 (900) B 302 (990)	A 223 (730) B 267 (875)	A 92 (302) B 109 (356)	738 (1,367; 850)
Model 172 and Skyhawk floatplane*	646 (1,425)	1,007 (2,220)	97 (180; 112)	94 (174; 108)	28 (51·5; 32)	191 (625)		3,764 (12,350)	A 256 (840) B 320 (1,050)	A 405 (1,330) B 451 (1,480)	A 267 (875) B 296 (970)	A 145 (475) B 171 (560)	477 (885; 550)
Model 180 Skywagon†	707 (1,560)	1,270 (2,800)	152 (282; 175)	144 (267; 166)	32·2 (60; 37)	364 (1,195)		6,217 (20,400)	A 110 (360) B 137 (450)	A 216 (710) B 239 (785)	A 207 (680) B 254 (835)	A 88 (290) B 104 (342)	1,098 (2,035; 1,265)
Model 180 Skywagon floatplane	850 (1,875)	1,338 (2,950)	142 (264; 164)	133 (246; 153)	32·2 (60; 37)	332 (1,090)		5,425 (17,800)	A 245 (805) B 307 (1,006)	A 369 (1,210) B 402 (1,320)	A 254 (832) B 308 (1,010)	A 145 (475) B 171 (560)	1,063 (1,971; 1,225)
Model 182 and Skylane†	725 (1,599)	1,338 (2,950)	150 (278; 173)	143 (266; 165)	33 (62; 38)	288 (945)		5,608 (18,400)	A 131 (430) B 164 (537)	A 248 (815) B 270 (885)	A 237 (777) B 280 (920)	A 99 (325) B 117 (384)	1,050 (1,947; 1,210)
Model 185 Skywagon†	721 (1,590)	1,519 (3,350)	159 (295; 183)	149 (277; 172)	34 (63; 39)	320 (1,050)		5,440 (17,850)	A 114 (375) B 143 (469)	A 233 (763) B 265 (870)	A 230 (755) B 271 (890)	A 95 (310) B 111 (365)	972 (1,802; 1,120)
Model 185 Skywagon floatplane	866 (1,910)	1,505 (3,320)	149 (277; 172)	140 (259; 161)	34 (63; 39)	311 (1,020)		5,212 (17,100)	A 182 (596) B 227 (745)	A 332 (1,090) B 364 (1,195)	A 265 (870) B 326 (1,070)	A 146 (480) B 173 (566)	903 (1,673; 1,040)
Model 188 AGwagon 230	836 (1,844)	1,723 (3,800)	124 (230; 143)	116 (214; 133)	38·5 (71; 44)	245 (805)		4,328 (14,200)	A 207 (680) B 259 (850)	A 338 (1,110) B 431 (1,420)	A 186 (610) B 256 (840)	A 94 (308) B 111 (363)	303 (563; 350)
Model A188 AGwagon 300†	843 (1,859)	1,814 (4,000)	135 (251; 156)	127 (235; 146)	40 (74; 46)	302 (990)		4,907 (16,100)	A 183 (600) B 229 (750)	A 293 (960) B 381 (1,250)	A 186 (610) B 256 (840)	A 94 (308) B 111 (363)	390 (724; 450)

A, B, *, **, † see notes under table on page 366.

CESSNA MODELS—*continued*

	Weight empty equipped kg (lb)	Weight gross kg (lb)	Max level speed knots (km/h; mph)	Max cruising speed knots (km/h; mph)	Stalling speed, wheels and flaps down knots (km/h; mph)	Max rate of climb at S/L m (ft)/min	Single-engine rate of climb at S/L m (ft)/min	Service ceiling m (ft)	T-O run m (ft)	T-O to 15 m (50 ft) m (ft)	Landing from 15 m (50 ft) m (ft)	Landing run m (ft)	Max range** nm (km; miles)	Min control speed knots (km/h; mph)
Model 206 Stationair†	785 (1,732)	1,633 (3,600)	155 (288; 179)	148 (274; 170)	36 (66; 41)	296 (970)		4,695 (15,400)	A 147 (482) B 184 (603)	A 302 (990) B 352 (1,155)	A 226 (740) B 270 (885)	A 92 (301) B 108 (355)	916 (1,697; 1,055)	
Model 206 Stationair floatplane	943 (2,080)	1,587 (3,500)	144 (267; 166)	136 (253; 157)	36 (66; 41)	276 (905)		4,450 (14,600)	A 248 (815) B 311 (1,019)	A 454 (1,490) B 486 (1,595)	A 280 (917) B 343 (1,125)	A 148 (485) B 174 (572)	833 (1,544; 960)	
Model T206 Turbo Stationair	831 (1,832)	1,633 (3,600)	178 (330; 205)	163 (303; 188)	36 (66; 41)	322 (1,055)		8,260 (27,100)	A 148 (485) B 185 (606)	A 303 (995) B 358 (1,175)	A 226 (740) B 270 (885)	A 92 (301) B 108 (355)	963 (1,786; 1,110)	
Model T206 Turbo Stationair floatplane	979 (2,160)	1,633 (3,600)	164 (304; 189)	148 (274; 170)	36 (66; 41)	311 (1,020)		7,650 (25,100)	A 241 (790) B 301 (987)	A 442 (1,450) B 475 (1,560)	A 283 (928) B 344 (1,130)	A 151 (495) B 178 (584)	911 (1,690; 1,050)	
Model 207 Skywagon	862 (1,902)	1,723 (3,800)	150 (278; 173)	142 (264; 164)	38·5 (71; 44)	262 (860)		4,206 (13,800)	A 155 (510) B 194 (637)	A 332 (1,090) B 390 (1,280)	A 244 (800) B 298 (975)	A 97 (318) B 114 (375)	829 (1,536; 955)	
Model T207 Turbo Skywagon	908 (2,002)	1,723 (3,800)	168 (312; 194)	156 (290; 180)	38·5 (71; 44)	277 (910)		7,650 (25,100)	A 155 (510) B 194 (637)	A 332 (1,090) B 390 (1,280)	A 244 (800) B 298 (975)	A 97 (318) B 114 (375)	812 (1,504; 935)	
Model 210 Centurion II	953 (2,102)	1,723 (3,800)	178 (330; 205)	167 (309; 192)	38·5 (71; 44)	274 (900)		4,907 (16,100)	A 155 (510) B 194 (637)	A 326 (1,070) B 376 (1,232)	A 239 (783) B 293 (960)	A 93 (305) B 110 (360)	1,137 (2,108; 1,310)	
Model T210 Turbo Centurion II	998 (2,202)	1,723 (3,800)	204 (378; 235)	192 (356; 221)	38·5 (71; 44)	293 (960)		9,022 (29,600)	A 160 (525) B 200 (656)	A 328 (1,075) B 401 (1,318)	A 239 (783) B 293 (960)	A 93 (305) B 110 (360)	1,133 (2,100; 1,305)	
Model 337 Super Skymaster	1,196 (2,638)	2,100 (4,630)	177 (328; 204)	168 (312; 194)	39 (72·5; 45)	369 (1,210)	99 (325)	6,126 (20,100)	A 130 (428) B 163 (535)	A 265 (870) B 322 (1,055)	A 273 (895) B 323 (1,060)	A 105 (343) B 123 (405)	1,207 (2,236; 1,390)	
Model T337 Turbo Super Skymaster	1,289 (2,843)	2,131 (4,700)	204 (378; 235)	200 (370; 230)	40 (74; 46)	353 (1,160)	93 (305)	9,266 (30,400)	A 136 (445) B 169 (556)	A 280 (920) B 332 (1,088)	A 273 (895) B 323 (1,060)	A 107 (352) B 126 (415)	1,406 (2,607; 1,620)	
Model T337 Pressurised Super Skymaster	1,315 (2,900)	2,132 (4,700)	217 (402; 250)	208 (385; 239)	40 (74; 46)	353 (1,160)	126 (415)	6,095 (20,000)	A 126 (413) B 157 (516)	A 280 (920) B 332 (1,088)	A 273 (895) B 323 (1,060)	A 107 (352) B 126 (415)	1,307 (2,422; 1,505)	
Model 401	1,673 (3,690)	2,858 (6,300)	226 (420; 261)	208 (386; 240)	65·5 (121; 75)	491 (1,610)	69 (225)	7,980 (26,180)	A 240 (786) B 300 (983)	A 378 (1,240) B 472 (1,550)	A 354 (1,160) B 442 (1,450)	A 155 (510) B 183 (600)	1,263 (2,340; 1,454)	72 (134; 83)
Model 402	1,673 (3,690)	2,858 (6,300)	226 (420; 261)	208 (386; 240)	65·5 (121; 75)	491 (1,610)	69 (225)	7,980 (26,180)	A 240 (786) B 300 (983)	A 378 (1,240) B 472 (1,550)	A 354 (1,160) B 442 (1,450)	A 155 (510) B 183 (600)	1,263 (2,340; 1,454)	72 (134; 83)
Model 411	1,764 (3,890)	2,948 (6,500)	233 (431; 268)	212 (393; 244)	61·5 (114; 71)	579 (1,900)	98 (320)	7,925 (26,000)	A 278 (912) B 347 (1,140)	A 372 (1,220) B 465 (1,525)	A 340 (1,115) B 425 (1,395)	A 183 (600) B 276 (905)	1,303 (2,414; 1,500)	76 (142; 88)
Model 414	1,764 (3,890)	2,880 (6,350)	236 (438; 272)	217 (402; 250)	68·5 (126·5; 78·4)	482 (1,580)	73 (240)	9,175 (30,100)	A 271 (888) B 338 (1,110)	A 397 (1,304) B 497 (1,630)	A 353 (1,160) B 442 (1,450)	A 145 (476) B 171 (560)	1,402 (2,599; 1,615)	74 (137; 85)
Model 421A	1,932 (4,260)	3,102 (6,840)	240 (444; 276)	224 (414; 258)	69 (128; 79)	512 (1,680)	88 (290)	8,230 (27,000)	A 307 (1,008) B 384 (1,260)	A 443 (1,452) B 553 (1,815)	A 419 (1,375) B 524 (1,720)	A 208 (683) B 245 (804)	1,488 (2,756; 1,713)	83 (153; 95)
Model 421B	2,011 (4,435)	3,379 (7,450)	245 (454; 282)	230 (426; 265)	71 (132; 81·8)	564 (1,850)	93 (305)	9,450 (31,000)	A 313 (1,028) B 365 (1,196)	A 428 (1,403) B 535 (1,754)	A 427 (1,400) B 534 (1,752)	A 134 (440) B 158 (517)	1,490 (2,762; 1,716)	78 (145; 90)

A: Robertson STOL operation. B: Robertson Normal operation. *Available also with engines of increased horsepower. **With optional long-range tanks fitted, if available.
†Leading-edge already installed by Cessna on current models.

R/STOL VERSIONS OF PIPER MODELS

	Weight empty equipped kg (lb)	Weight gross kg (lb)	Max level speed knots (km/h; mph)	Max cruising speed knots (km/h; mph)	Stalling speed, wheels and flaps down knots (km/h; mph)	Max rate of climb at S/L m (ft)/min	Single-engine rate of climb at S/L m (ft)/min	Service ceiling m (ft)	T-O run m (ft)	T-O to 15 m (50 ft) m (ft)	Landing from 15 m (50 ft) m (ft)	Landing run m (ft)	Max range** nm (km; miles)	Min control speed knots (km/h; mph)
PA-28-140 Cherokee	558 (1,232)	975 (2,150)	126 (223; 145)	120 (222; 138)	29 (53·2; 33)	206 (675)		4,480 (14,700)	A 171 (560) B 189 (620)	A 354 (1,160) B 404 (1,325)	A 192 (630) B 221 (725)	A 94 (310) B 110 (360)	825 (1,529; 950)	
PA-28-160 Cherokee	576 (1,270)	997 (2,200)	128 (237; 147)	121 (224; 139)	31 (56·5; 35)	216 (710)		4,970 (16,300)	A 152 (500) B 177 (580)	A 341 (1,120) B 390 (1,280)	A 204 (670) B 226 (740)	A 104 (340) B 114 (375)	838 (1,553; 965)	
PA-28-180 Cherokee	638 (1,406)	1,111 (2,450)	129 (238; 148)	122 (227; 141)	36 (66; 41)	221 (725)		4,313 (14,510)	A 165 (540) B 186 (610)	A 351 (1,150) B 399 (1,310)	A 238 (780) B 262 (860)	A 131 (430) B 146 (480)	596 (1,104; 686)	
PA-28R-180 Cherokee Arrow II	611 (1,349)	1,134 (2,500)	149 (275; 171)	142 (262; 163)	35 (64·5; 40)	270 (885)		4,695 (15,400)	A 171 (560) B 195 (640)	A 347 (1,140) B 396 (1,300)	A 259 (850) B 299 (980)	A 145 (475) B 168 (550)	911 (1,690; 1,050)	
PA-28R-200 Cherokee Arrow II	693 (1,528)	1,202 (2,650)	152 (282; 175)	143 (266; 165)	37 (69; 43)	274 (900)		4,572 (15,000)	A 175 (575) B 198 (650)	A 344 (1,130) B 393 (1,290)	A 283 (930) B 319 (1,045)	A 168 (550) B 187 (615)	782 (1,448; 900)	
PA-28-235 Cherokee 235	712 (1,570)	1,361 (3,000)	140 (259; 161)	132 (245; 152)	39 (72; 45)	244 (800)		3,658 (12,000)	A 168 (550) B 191 (625)	A 265 (870) B 302 (990)	A 274 (900) B 312 (1,025)	A 149 (490) B 171 (560)	926 (1,716; 1,066)	
PA-32-260 Cherokee SIX	783 (1,726)	1,542 (3,400)	144 (267; 166)	137 (254; 158)	38 (71; 44)	259 (850)		4,420 (14,500)	A 180 (590) B 238 (780)	A 317 (1,040) B 341 (1,120)	A 247 (810) B 267 (875)	A 155 (510) B 171 (560)	964 (1,786; 1,110)	
PA-32-300 Cherokee SIX	825 (1,819)	1,542 (3,400)	151 (280; 174)	146 (270; 168)	38 (71; 44)	320 (1,050)		4,953 (16,250)	A 171 (560) B 226 (740)	A 299 (980) B 320 (1,050)	A 247 (810) B 267 (875)	A 158 (520) B 168 (550)	921 (1,706; 1,060)	
PA-24-180 Comanche	694 (1,530)	1,157 (2,550)	149 (277; 172)	143 (266; 165)	35 (64; 40)	293 (960)		5,852 (19,200)	A 183 (600) B 302 (990)	A 324 (1,065) B 475 (1,560)	A 262 (860) B 326 (1,070)	A 128 (420) B 149 (490)	868 (1,609; 1,000)	
PA-24-250 Comanche	776 (1,710)	1,315 (2,900)	168 (311; 193)	161 (298; 185)	35·5 (66; 41)	427 (1,400)		6,309 (20,700)	A 187 (615) B 290 (950)	A 296 (970) B 389 (1,275)	A 256 (840) B 347 (1,140)	A 140 (460) B 213 (700)	1,537 (2,848; 1,770)	
PA-24-260 Comanche	812 (1,792)	1,451 (3,200)	174 (322; 200)	164 (304; 189)	36 (66; 41)	410 (1,345)		6,355 (20,850)	A 200 (655) B 302 (990)	A 338 (1,110) B 401 (1,315)	A 268 (880) B 360 (1,180)	A 152 (500) B 226 (740)	1,137 (2,108; 1,310)	
PA-24-260 Turbo Comanche	821 (1,810)	1,451 (3,200)	213 (394; 245)	201 (372; 231)	36 (66; 41)	410 (1,345)		7,620 (25,000)	A 200 (655) B 302 (990)	A 338 (1,110) B 401 (1,315)	A 268 (880) B 360 (1,180)	A 152 (500) B 226 (740)	1,306 (2,422; 1,505)	
PA-24-400 Comanche	966 (2,130)	1,633 (3,600)	196 (364; 226)	189 (351; 218)	39 (72; 45)	506 (1,660)		6,157 (20,200)	A 126 (415) B 168 (550)	A 233 (765) B 271 (890)	A 303 (995) B 379 (1,245)	A 184 (605) B 216 (710)	1,568 (2,905; 1,805)	
PA-30 Twin Comanche*	1,022 (2,253)	1,724 (3,800)	158 (293; 182)	153 (283; 176)	45 (84; 52)	427 (1,400)	79 (260)	6,096 (20,000)	B 206 (675)	B 341 (1,120)	B 355 (1,165)	B 186 (610)	1,481 (2,744; 1,705)	69 (129; 80)
PA-30 Turbo Twin Comanche*	1,088 (2,399)	1,724 (3,800)	213 (394; 245)	197 (365; 227)	45 (84; 52)	427 (1,400)	69 (225)	7,620 (25,000)	B 206 (675)	B 341 (1,120)	B 355 (1,165)	B 186 (610)	1,528 (2,832; 1,760)	69 (129; 80)
PA-39 Twin Comanche C/R*	1,022 (2,253)	1,724 (3,800)	181 (336; 209)	174 (322; 200)	45 (84; 52)	445 (1,460)	79 (260)	6,096 (20,000)	B 206 (675)	B 320 (1,050)	B 355 (1,165)	B 186 (610)	1,468 (2,720; 1,690)	65 (121; 75)
PA-39 Turbo Twin Comanche C/R*	1,088 (2,399)	1,724 (3,800)	212 (393; 244)	197 (365; 227)	45 (84; 52)	427 (1,400)	69 (225)	7,620 (25,000)	B 158 (520)	B 323 (1,060)	B 355 (1,165)	B 189 (620)	1,515 (2,808; 1,745)	65 (121; 75)
PA-23-235 Aztec	1,241 (2,735)	2,177 (4,800)	182 (338; 210)	175 (325; 202)	40 (74; 46)	465 (1,525)	62 (205)	5,517 (18,100)	B 210 (690)	B 331 (1,085)	B 381 (1,250)	B 195 (640)	1,090 (2,020; 1,255)	54 (98; 62)
PA-E23-250 Aztec	1,339 (2,953)	2,266 (4,995)	197 (365; 227)	191 (354; 220)	41 (76; 47)	509 (1,670)	99 (325)	6,614 (21,700)	B 190 (625)	B 315 (1,035)	B 395 (1,295)	B 203 (665)	1,112 (2,060; 1,280)	56 (103; 64)
PA-23-250 Aztec	1,326 (2,925)	2,359 (5,200)	188 (348; 216)	179 (332; 206)	45·5 (84; 52)	491 (1,610)	85 (280)	6,035 (19,800)	B 195 (640)	B 323 (1,060)	B 395 (1,295)	B 203 (665)	916 (1,697; 1,055)	56 (105; 65)
PA-23-250 Turbo Aztec	1,397 (3,080)	2,359 (5,200)	222 (412; 256)	182 (388; 210)	43 (79; 49)	372 (1,220)	64 (210)	9,144 (30,000)	B 195 (640)	B 323 (1,060)	B 395 (1,295)	B 203 (665)	1,050 (1,947; 1,210)	56 (105; 65)
PA-34 Seneca	1,160 (2,557)	1,905 (4,200)	170 (315; 196)	162 (300; 187)	58 (106; 66)	414 (1,360)	57 (190)	5,730 (18,800)	B 195 (640)	B 320 (1,050)	B 381 (1,250)	B 196 (645)	743 (1,378; 856)	75 (139; 87)

A: Robertson STOL operation. B: Robertson Normal operation. *Available also with engines of increased horsepower.
**With optional long-range tanks fitted, if available.

ROBINSON
ROBINSON HELICOPTER COMPANY INC

HEAD OFFICE AND WORKS:
2903 Earhart Apron, Torrance Airport, Torrance, California 90505
Telephone: (213) 539-0508
CHAIRMAN OF THE BOARD:
C. K. LeFiell
PRESIDENT:
Franklin D. Robinson
VICE-PRESIDENT: C. K. LeFiell
SECRETARY: Karen L. Walling
TREASURER: R. C. Ragland

Robinson Helicopter Company was formed specifically to design and manufacture a lightweight helicopter which could be competitive in price with current two/four-seat fixed-wing light aircraft. The design of this aircraft, the Robinson R22, began in July 1973 with emphasis on efficiency, low noise emission and minimum maintenance. The first prototype flew for the first time on 28 August 1975. During the ensuing four months approximately 35 hours of flight testing were completed, and in January 1976 the R22 began its FAA certification programme. It was anticipated that this would be completed by late 1976 and that production would begin in mid-1977.

Prototype of the Robinson Model R22 two-seat lightweight helicopter (Lycoming O-235-C2A engine)

ROBINSON MODEL R22

TYPE: Two-seat lightweight helicopter.

ROTOR SYSTEM: Two-blade semi-rigid main rotor, with a tri-hinged underslung rotor hub to reduce blade flexing, rotor vibration and control force feedback. Blades are of bonded all-metal construction, with a stainless steel leading-edge, light alloy skin and light alloy honeycomb core. The two-blade tail rotor, mounted on the port side, is of light alloy bonded construction.

ROTOR DRIVE: Belt drive to automatic centrifugal clutch and sprag-type overrunning clutch. The main and tail gearboxes each utilise spiral bevel gears. Maintenance-free flexible couplings of proprietary manufacture are used in both the main and tail rotor drive systems.

FUSELAGE: Welded steel tube primary structure for cabin, rotor pylon and engine mounting, with full monocoque tailcone. Cabin skins of light alloy and glassfibre.

TAIL UNIT: Cruciform light alloy structure with horizontal stabiliser and vertical fin. Small spring skid beneath lower half of fin to give protection in a tail-down landing.

LANDING GEAR: Welded steel tube skid landing gear, with energy-absorbing cross tubes.

POWER PLANT: One 85·75 kW (115 hp) Lycoming O-235-C2A flat-four engine, mounted in the lower aft section of the main fuselage, and partially exposed to improve cooling and simplify maintenance. Light alloy fuel tank in upper rear section of the fuselage on port side, capacity 64·3 litres (17 US gallons). Oil capacity 5·7 litres (1·5 US gallons).

ACCOMMODATION: Two seats side by side in enclosed cabin. Cyclic control stick mounted between seats, with dual grips on yoke so that aircraft can be flown from either seat. Conventional dual collective and throttle controls mounted at the port side of each seat. Cyclic control pivots to either side to simplify entry and exit. Curved two-panel windscreen. Door, with window, on each side. Baggage space beneath each seat. Accommodation heated and ventilated.

ELECTRONICS: One 360-channel com transceiver. Space provisions for optional VOR nav receiver.

DIMENSIONS, EXTERNAL:

Diameter of main rotor	7·67 m (25 ft 2 in)
Diameter of tail rotor	1·07 m (3 ft 6 in)
Length overall (rotors turning)	8·76 m (28 ft 9 in)
Length of fuselage	6·30 m (20 ft 8 in)
Height overall	2·67 m (8 ft 9 in)
Skid track	1·96 m (6 ft 5 in)

DIMENSION, INTERNAL:

Cabin: Max width	1·12 m (3 ft 8 in)

AREAS:

Main rotor blades (each)	0·70 m² (7·55 sq ft)
Tail rotor blades (each)	0·05 m² (0·57 sq ft)
Main rotor disc	46·21 m² (497·4 sq ft)
Tail rotor disc	0·89 m² (9·63 sq ft)
Fin	0·21 m² (2·28 sq ft)
Stabiliser	0·14 m² (1·53 sq ft)

WEIGHTS AND LOADINGS:

Weight empty	327 kg (720 lb)
Max T-O weight	558 kg (1,230 lb)
Max disc loading	12·07 kg/m² (2·47 lb/sq ft)
Max power loading	6·51 kg/kW (10·7 lb/hp)

PERFORMANCE (preliminary figures at max T-O weight):

Cruising speed	86·8 knots (161 km/h; 100 mph)
Max rate of climb at S/L	455 m (1,500 ft)/min
Service ceiling	4,265 m (14,000 ft)
Hovering ceiling	1,585 m (5,200 ft)
Range, no reserves	approx 217 nm (402 km; 250 miles)

ROCKWELL INTERNATIONAL
ROCKWELL INTERNATIONAL CORPORATION

GENERAL OFFICES:
2230 East Imperial Highway, El Segundo, California 90245
600 Grant Street, Pittsburgh, Pennsylvania 15219
CHAIRMAN OF THE BOARD:
Willard F. Rockwell Jr
PRESIDENT AND CHIEF EXECUTIVE OFFICER:
Robert Anderson
SENIOR VICE-PRESIDENT:
John J. Henry (Corporate Staffs)
CORPORATE VICE-PRESIDENTS:
Robert De Palma (Treasurer)
J. A. Earley (Corporate Development)
Charles Fazio (Operations)
Kenneth B. Gay (Purchasing)
George W. Jeffs (President, North American Space Operations)
Crosby M. Kelly (Communications)
A. B. Kight (International)
Donald S. MacLeod (Investor Relations)
C. James Meechan (Research and Engineering)
Dale D. Myers (President, North American Aircraft Operations)
Carl J. Oles (Personnel)
W. B. Panny (President, Automotive Operations)
Louis Putze (President, Utility and Industrial Operations)
Robert M. Rice (Finance and Chief Financial Officer)
John J. Roscia (General Counsel)
C. E. Ryker (Controller)
W. F. Swanson Jr (Secretary)
Charles J. Urban (Consumer Operations)
Donn L. Williams (President, Electronics Operations)

North American Aviation Inc, incorporated in Delaware in 1928 and a manufacturer of aircraft of various kinds from 1934, and Rockwell-Standard Corporation of Pittsburgh, Pennsylvania, a manufacturer of automotive components and builder of the Aero Commander line of civilian aircraft, merged on 22 September 1967 to form North American Rockwell Corporation.

During 1971 the Corporation was reorganised into four principal parts: the North American Aerospace Group (formerly the Aerospace and Systems Office); the Industrial Products Group (formerly the Commercial Products Group); the Automotive Group; and the Electronics Group. There is, in addition, a further component known as the Utility and Consumer Products Group. The constitution of the Corporation was changed on 16 February 1973, when North American Rockwell and Rockwell Manufacturing Company merged to become Rockwell International Corporation.

A further change in the organisation was announced on 15 February 1974. The former North American Aerospace Group was replaced by two groups, now known as North American Aircraft Operations and North American Space Operations. North American Aircraft Operations comprises the Atomics International Division, the B-1 Division, Los Angeles Aircraft Division, Columbus Aircraft Division, Sabreliner Division, General Aviation Division and Tulsa Division. North American Space Operations comprises Rocketdyne Division and Space Division.

Divisions and products of the North American Aircraft and Industrial Products Groups are detailed hereafter:

NORTH AMERICAN AIRCRAFT OPERATIONS

EXECUTIVE OFFICES:
2230 East Imperial Highway, El Segundo, California 90245
Telephone: (213) 647-5000
PRESIDENT:
Dale D. Myers

Atomics International Division

8900 De Soto Avenue, Canoga Park, California 91304
Telephone: (213) 341-1000
PRESIDENT: S. F. Iacobellis

B-1 Division

5701 West Imperial Highway, Los Angeles, California 90009
Telephone: (213) 670-9151
PRESIDENT: Bastian J. Hello

Columbus Aircraft Division

4300 East Fifth Avenue, Columbus, Ohio 43216
Telephone: (614) 239-3344
PRESIDENT: J. P. Fosness

Tulsa Division

2000 North Memorial Drive, Tulsa, Oklahoma 74151
Telephone: (918) 835-31111
PRESIDENT: W. J. Cecka Jr

NORTH AMERICAN SPACE OPERATIONS

EXECUTIVE OFFICES:
2230 East Imperial Highway, El Segundo, California 90245
PRESIDENT: George W. Jeffs

Rocketdyne Division

6633 Canoga Avenue, Canoga Park, California 91304
Telephone: (213) 884-4000
PRESIDENT: Norman J. Ryker

Space Division

12214 Lakewood Boulevard, Downey, California 90241
Telephone: (213) 922-2111
PRESIDENT: George B. Merrick

Current aircraft products of North American Aircraft Operations are as follows:

ROCKWELL INTERNATIONAL BUCKEYE

US Navy designation: T-2

After a design competition among several leading US manufacturers, what was then North American's Columbus Division was awarded a contract in 1956 to develop and build a jet training aircraft for the US Navy. The first T2J-1 flew on 31 January 1958. Five versions of the aircraft have since been produced, as follows:

T-2A (formerly T2J-1). Initial version, with single 15·12 kN (3,400 lb st) Westinghouse J34-WE-48 turbojet engine. Initial orders were for 26 production T-2As. Follow-up contracts were awarded in 1958 and 1959, and 217 had been built when production ended in January

1961. The T-2A was used by US Naval Air Training Command, Pensacola, Florida, but was phased out of service in early 1973, having been replaced completely by T-2Bs or -2Cs.

T-2B (formerly T2J-2). To evaluate the potential of the Buckeye airframe, two T-2As were each re-engined with two Pratt & Whitney J60-P-6 turbojets (each 13·34 kN; 3,000 lb st), under US Navy contract, with the designation T-2B. First one flew on 30 August 1962. A US Navy production contract for 10 new T-2Bs was announced in March 1964, and further contracts brought the total ordered to 97. The first production T-2B flew on 21 May 1965; deliveries to Naval Air Training Command were completed in February 1969.

T-2C. Generally similar to T-2B, but powered by two General Electric J85-GE-4 turbojet engines, each rated at 13·12 kN (2,950 lb st). The T-2C entered production in late 1968, following extensive evaluation of J85-GE-4 engines in a T-2B which was redesignated T-2C No. 1. T-2C production began as an amendment of an existing contract. First production T-2C flew on 10 December 1968. Total of 231 T-2Cs ordered by Naval Air Training Command, deliveries of which were completed by the end of 1975.

T-2D. Generally similar to T-2C; differs only in electronics equipment and by deletion of carrier landing capability. A total of 12 were supplied to the Venezuelan Air Force and are used as advanced jet trainers for student pilots; delivery was completed during 1973. An additional 12 T-2D trainers with the attack kit of a T-2E were ordered for delivery during 1976-77.

T-2E. Generally similar to the T-2C, except for new electronics equipment and the provision of an accessory kit which permits utilisation in an attack role. A total of 40 are on order for Greece under a contract managed by US Naval Air Systems Command. The accessory kit provides six wing store stations with a combined capacity of 1,587 kg (3,500 lb), and protection against small arms fire for the fuel tanks. The aircraft will be used by the Hellenic Air Force Training Command as advanced and tactical jet trainers for student pilots in their final stages of training. Delivery was scheduled to be made during 1976.

The following details apply to the standard T-2C:

TYPE: Two-seat general-purpose jet trainer.

WINGS: Cantilever mid-wing monoplane. Wing section NACA 64A212 (modified). Thickness/chord ratio 12%. All-metal two-spar structure. Interchangeable all-metal ailerons, with hydraulic boost. Large all-metal trailing-edge flaps.

FUSELAGE: All-metal semi-monocoque structure in three main sections: forward fuselage containing equipment bay and cockpit; centre fuselage housing power plant, fuel and wing carry-through structure; and rear fuselage, carrying the arrester hook and a hydraulically-actuated airbrake on each side of the fuselage.

TAIL UNIT: Cantilever all-metal structure. Each half of tailplane and elevators interchangeable. Elevators boosted hydraulically. Rudder manually controlled. Trim tabs in elevators and rudder.

LANDING GEAR: Retractable tricycle type. Oleo-pneumatic shock-absorbers. Hydraulic retraction. Main units retract inward into wings. Nosewheel retracts forward into fuselage. Main wheels size 24 × 5·50. Nose-wheel size 20 × 4·40. Main-wheel tyre pressure 10·34 bars (150 lb/sq in), nosewheel tyre pressure 5·17 bars (75 lb/sq in). Goodyear aircooled single-disc hydraulic brakes. Retractable sting-type, universal joint mounted, arrester hook.

POWER PLANT: Two 13·12 kN (2,950 lb st) General Electric J85-GE-4 turbojet engines, with jet outlets under rear fuselage. Fuel in main tanks over engines with capacity of 1,465 litres (387 US gallons), two wingtip tanks each of 386 litres (102 US gallons) capacity, and two tanks in the inboard sections of the wings. Total fuel capacity 2,616 litres (691 US gallons).

ACCOMMODATION: Pupil and instructor in tandem in enclosed cabin, on rocket-powered LS-1 ejection seats, under clamshell canopy. Instructor is raised 0·25 m (10 in) above level of pupil.

ARMAMENT: Optional packaged installations of guns, target-towing gear, 100 lb practice bombs, M-5 or MK76 practice bomb clusters, Aero 4B practice bomb containers, 2·25 in rocket launchers or seven 2·75 in rockets in Aero 6A-1 rocket containers, can be carried on two store stations, one beneath each wing, with a combined capacity of 290 kg (640 lb).

DIMENSIONS, EXTERNAL:

Wing span over tip-tanks	11·62 m (38 ft 1½ in)
Length overall	11·67 m (38 ft 3½ in)
Height overall	4·51 m (14 ft 9½ in)
Tailplane span	5·46 m (17 ft 11 in)
Wheel track	5·61 m (18 ft 4¾ in)

AREAS:

Wings, gross	23·69 m² (255 sq ft)
Trailing-edge flaps (total)	4·23 m² (45·56 sq ft)
Fin	2·54 m² (27·29 sq ft)
Rudder, incl tab	0·84 m² (9·01 sq ft)
Tailplane	3·95 m² (42·55 sq ft)
Elevators, incl tabs	1·95 m² (21·00 sq ft)

WEIGHTS:

Weight empty	3,680 kg (8,115 lb)
Max T-O weight	5,977 kg (13,179 lb)

PERFORMANCE (at max T-O weight):

Max level speed at 7,620 m (25,000 ft)	453 knots (840 km/h; 522 mph)
Stalling speed	86·6 knots (161 km/h; 100 mph)
Max rate of climb at S/L	1,890 m (6,200 ft)/min
Service ceiling	12,315 m (40,400 ft)
Max range	909 nm (1,685 km; 1,047 miles)

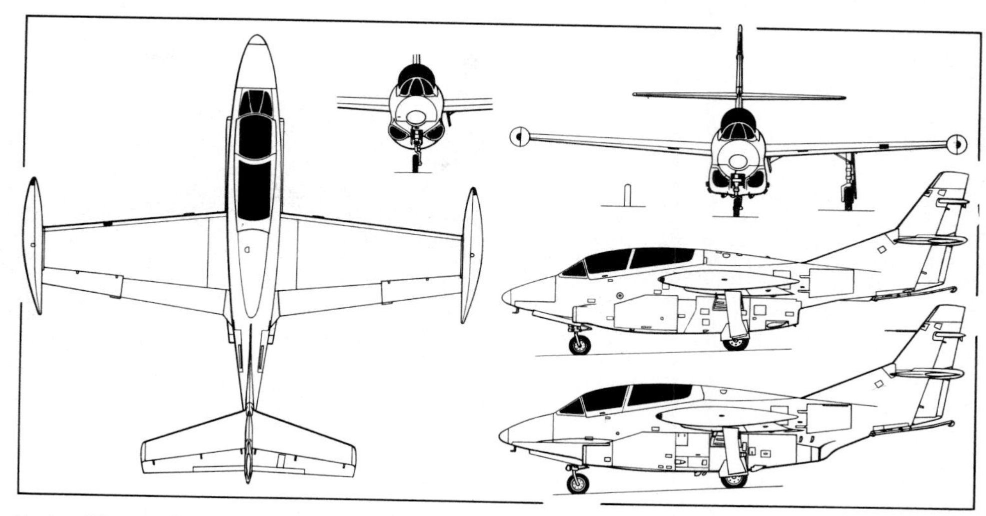
Rockwell International T-2C Buckeye, with additional side view (centre) and scrap front view of T-2A *(Pilot Press)*

Rockwell International T-2D Buckeye in service with the Venezuelan Air Force

ROCKWELL INTERNATIONAL BRONCO

US military designation: OV-10

This aircraft was North American's entry for the US Navy's design competition for a light armed reconnaissance aeroplane (LARA) specifically suited for counter-insurgency missions. Nine US airframe manufacturers entered for the competition and the NA-300 was declared the winning design in August 1964. Seven prototypes were then built by the company's Columbus Division, under the designation YOV-10A Bronco. The first of these flew on 16 July 1965, followed by the second in December 1965.

A number of modifications were made as a result of flight experience with the prototypes. In particular, the wing span was increased by 3·05 m (10 ft 0 in), the T76 turboprop engines were uprated from 492 kW (660 shp) to 534 kW (716 shp), and the engine nacelles were moved outboard approximately 0·15 m (6 in) to reduce noise in the cockpit.

A prototype with lengthened span flew for the first time on 15 August 1966. The seventh prototype had Pratt & Whitney (Canada) T74 (PT6A) turboprops for comparative testing.

The following versions have been built:

OV-10A. Initial production version ordered in October 1966 and first flown on 6 August 1967. US Marine Corps had 114 in service in September 1969, of which 18 were on loan to the USN; used for light armed reconnaissance, helicopter escort and forward air control duties. At the same date the USAF had 157 OV-10As for use in the forward air control role, as well as for limited quick-response ground support pending the arrival of tactical fighters.

Production of the OV-10A for the US services ended in April 1969; but 15 aircraft were modified by LTV Electrosystems Inc, under the USAF Pave Nail programme, to permit their use in a night forward air control and strike designation role in 1971.

Equipment installed by LTV included a stabilised night periscopic sight, a combination laser rangefinder and target illuminator, a Loran receiver and a Lear Siegler Loran co-ordinate converter. This combination of equipment generates an offset vector to enable an accompanying strike aircraft to attack the target or, alternatively, illuminate the target, enabling a laser-seeking missile to home on to it. These specially configured aircraft reverted to the OV-10A configuration in 1974 by removal of the LTV-installed equipment.

Under the designation **YOV-10A** a single OV-10A was equipped with rotating cylinder flaps for evaluation in a STOL flight test programme by NASA.

OV-10B. Generally similar to the OV-10A; six supplied to the Federal German government for target towing duties.

Rockwell International OV-10A Bronco of US Navy Squadron VAL-4 on a combat sortie

OV-10B(Z). Structurally similar to the OV-10B, except that a General Electric J85-GE-4 turbojet engine of 13·12 kN (2,950 lb st) is mounted above the wing, on a pylon attached to existing hoisting points, to increase performance for target towing duties. First flown on 21 September 1970. Delivery of 18 OV-10B(Z) aircraft to the Federal German government was completed in November 1970. The jet pods were fitted by RFB, following the prototype installation by Rockwell.

OV-10C. Version of the OV-10A for the Royal Thai Air Force. Deliveries of 32 completed in September 1973.

YOV-10D/OV-10D. Two YOV-10Ds were OV-10As modified under a 1970 contract from the US Navy to provide a new concept in night operational capability for the US Marine Corps. Distinguishing features of the YOV-10D Night Observation/Gunship System (NOGS) are a 20 mm gun turret mounted beneath the aft fuselage and a forward-looking infra-red (FLIR) sensor installed beneath the extended nose. A laser target designator is incorporated within the FLIR sensor turret. Two wing pylons are installed at the Sidewinder missile wing stations which are cabable of carrying a variety of rocket pods, flare pods and free-fall stores. In 1974, Rockwell received a US Navy contract to establish and test a production OV-10D configuration.

It is anticipated that 18 to 24 OV-10A aircraft will be converted to the Night Observation System (NOS) role. In addition to retention of the basic weapon system capability, the OV-10D NOS will have an uprated 775·5 kW (1,040 shp) power plant and will be able to carry 378·5 litre (100 US gallon) drop-tanks on the wing pylons when extended radius/loiter time is required.

OV-10E. Version of the OV-10A for the Fuerzas Aéreas Venezolana. Sixteen ordered through the US Department of Defense foreign military sales programme. The first of these was delivered in March 1973.

OV-10F. Version of the OV-10A for the government of Indonesia. Sixteen aircraft ordered through the US Department of Defense military sales programme. Delivery of the first of these aircraft was scheduled to be made in mid-1976.

The following details apply to the standard OV-10A, except where stated:

TYPE: Two-seat multi-purpose counter-insurgency aircraft.

WINGS: Cantilever shoulder-wing monoplane. Constant-chord wing without dihedral or sweep. Conventional aluminium alloy two-spar structure. Manually-operated ailerons, supplemented by four small manually-operated spoilers forward of outer flap on each wing, for lateral control at all speeds. Hydraulically operated double-slotted flaps in two sections on each wing, separated by tailbooms.

FUSELAGE: Short pod-type fuselage of conventional aluminium semi-monocoque construction, suspended from wing. Glassfibre nosecone.

TAIL UNIT: Cantilever all-metal structure carried on twin booms of semi-monocoque construction. Tailplane mounted near tips of fins. Manually-operated rudders and elevator.

LANDING GEAR: Retractable tricycle type, with single wheel on each unit, developed by Cleveland Pneumatic Tool Co. Hydraulic actuation, nosewheel retracting forward, main units rearward into tailbooms. Two-stage oleo-pneumatic shock-absorbers. Forged aluminium wheels. Main wheels with tyres size 29 × 11-10, pressure 4·48 bars (65 lb/sq in). Nosewheel tyre size 7·50-10, pressure 5·52 bars (80 lb/sq in). Cleveland hydraulic disc brakes.

POWER PLANT: Two 533 kW (715 ehp) AiResearch T76-G-416/417 turboprops, with Hamilton Standard three-blade propellers. Inter-spar fuel tank in centre portion of wing, capacity 976 litres (258 US gallons). Refuelling points above tank. Provision for carrying 568 or 871 litre (150 or 230 US gallon) jettisonable ferry tank on underfuselage pylon.

ACCOMMODATION: Crew of two in tandem, on ejection seats, under canopy with two large upward-opening transparent door panels on each side. Dual controls standard. Cargo compartment aft of rear seat, with rear-loading door at end of fuselage pod. Rear seat removable to provide increased space for up to 1,452 kg (3,200 lb) of freight, or for carriage of five paratroops, or two stretcher patients and attendant.

ELECTRONICS AND EQUIPMENT: UHF, VHF, HF radios and Tacan are standard, provision of a radar altimeter optional. Gunsight above pilot's instrument panel, the latter having space provisions beneath the gunsight for a TV-type display with applications for night sensor reconnaissance, weapon delivery and navigation, such as in the YOV-10D configuration.

ARMAMENT: Two weapon attachment points, each with capacity of 272 kg (600 lb), under short sponson extending from bottom of fuselage on each side, under wings. Fifth attachment point, capacity 544 kg (1,200 lb) under centre fuselage. Two 7·62 mm M-60C machine-guns carried in each sponson. Provision for carrying one Sidewinder missile under each wing and, by use of a wing pylon kit, various stores including rocket pods, flare pods and free-fall ordnance. Max weapon load 1,633 kg (3,600 lb).

Rockwell International OV-10B(Z) Bronco, with auxiliary turbojet *(Ing Hans Redemann)*

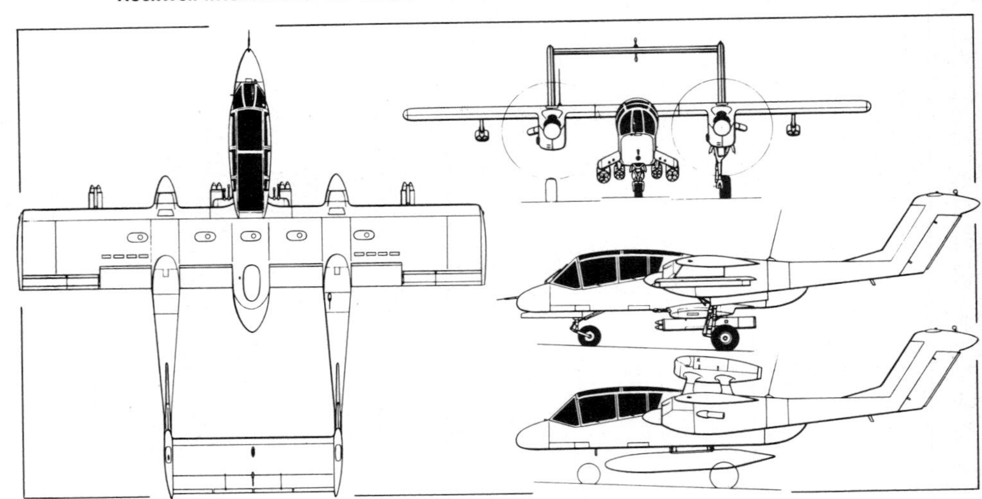

Rockwell International OV-10A Bronco, with additional side view (bottom) of OV-10B(Z) *(Pilot Press)*

Rockwell International YOV-10D Night Observation/Gunship System (NOGS)

DIMENSIONS, EXTERNAL:
- Wing span: 12·19 m (40 ft 0 in)
- Length overall: 12·67 m (41 ft 7 in)
- Height overall: 4·62 m (15 ft 2 in)
- Tailplane span: 4·45 m (14 ft 7 in)
- Wheel track: 4·52 m (14 ft 10 in)
- Rear loading door: Height: 0·99 m (3 ft 3 in)
  - Width: 0·76 m (2 ft 6 in)

DIMENSIONS, INTERNAL:
- Cargo compartment: 2·12 m³ (75 cu ft)
- Cargo compartment, rear seat removed: 3·14 m³ (111 cu ft)

AREA:
- Wings, gross: 27·03 m² (291 sq ft)

WEIGHTS:
- Weight empty: 3,161 kg (6,969 lb)
- Normal T-O weight: 4,494 kg (9,908 lb)
- Overload T-O weight: 6,563 kg (14,466 lb)

PERFORMANCE (A: OV-10A/C/E/F; B: OV-10B; C: OV-10B(Z); D: OV-10D, with internal 20 mm ammunition only):

Max level speed at S/L, without weapons:
- A: 244 knots (452 km/h; 281 mph)
- D: 240 knots (444 km/h; 276 mph)

Max level speed at 3,050 m (10,000 ft) at AUW of 4,536 kg (10,000 lb):
- B: 241 knots (447 km/h; 278 mph)
- C: 341 knots (632 km/h; 393 mph)

Max rate of climb at S/L at AUW of 4,494 kg (9,908 lb):
- A: 808 m (2,650 ft)/min

Max rate of climb at S/L at AUW of 5,443 kg (12,000 lb):
- B: 701 m (2,300 ft)/min
- C: 2,073 m (6,800 ft)/min

Max rate of climb at S/L at AUW of 5,644 kg (12,443 lb):
- D: 812 m (2,665 ft)/min

T-O run:
- A, at normal AUW: 226 m (740 ft)
- B, at 5,443 kg (12,000 lb) AUW: 344 m (1,130 ft)
- C, at 5,443 kg (12,000 lb) AUW: 168 m (550 ft)
- D, at 6,025 kg (13,284 lb) AUW: 338 m (1,110 ft)

T-O to 15 m (50 ft):
- A, at normal AUW — 341 m (1,120 ft)
- A, at overload AUW — 853 m (2,800 ft)

Landing from 15 m (50 ft):
- A, at normal AUW — 372 m (1,220 ft)

Landing run:
- A, at normal AUW — 226 m (740 ft)
- A, at overload AUW — 381 m (1,250 ft)
- D, at landing AUW — 244 m (800 ft)

Combat radius with max weapon load, no loiter:
- A — 198 nm (367 km; 228 miles)
- D — 265 nm (491 km; 305 miles)

Ferry range with auxiliary fuel:
- A — 1,240 nm (2,298 km; 1,428 miles)

ROCKWELL INTERNATIONAL B-1

The B-1 is the outcome of a succession of defence studies, begun in 1962 and leading to the AMSA (Advanced Manned Strategic Aircraft) requirement of 1965, for a low-altitude penetration bomber to replace the Boeing B-52s of USAF Strategic Air Command by 1980. It is the third and most flexible component of the US Triad defence system, which comprises also land-based and submarine-launched ballistic missiles.

To meet the B-1 requirement, the Department of Defense issued RFPs (Requests For Proposals) to the US aerospace industry on 3 November 1969, and from three airframe and two engine finalists it awarded research, development, test and evaluation contracts on 5 June 1970 to North American Rockwell's Los Angeles Division (now the B-1 Division of Rockwell International's North American Aircraft Operations) for the airframe and to the General Electric Company for the F101 turbofan engine. The original cost-plus-incentive contracts were for five flying prototypes, two structural test airframes and 40 engines; in January 1971, in which month the essential design of the B-1 was frozen, these quantities were reduced to three flight test aircraft, one ground test aircraft and 27 engines. Procurement of a fourth flight test aircraft, as a pre-production prototype, was approved under the FY 1976 budget; this is due to fly in early 1979.

Completion of a full-scale engineering mockup of the B-1 was announced on 4 November 1971. This had been used during the period 18-31 October 1971 by the USAF's mockup review team including key personnel from the B-1 Systems Program Office, Strategic Air Command, AF Systems Command, Air Training Command and the General Electric Company. During the review period the team was able to resolve no fewer than 257 of a total of 297 Requests For Alteration (RFAs).

The B-1 prototypes are assembled in USAF facilities known as Plant 42 at Palmdale, California. Assembly of the first aircraft began in late 1972; this aircraft (71-40158) was rolled out on 26 October 1974, and made its first flight, at Palmdale, on 23 December 1974. This occasion was also the first flight of the YF101 engine. The third B-1, which is being used as testbed for the electronics systems, made its first flight on 1 April 1976, and was followed by the first flight of the second B-1 on 14 June 1976. The second aircraft had been used for proof loads testing; these tests occupied approximately eight months during 1975, after which No. 2 was converted to a flight test aircraft. By mid-July 1976 the three aircraft involved in the test programme had accumulated a total of just over 200 flight hours. The maximum speed attained to that date was Mach 2·1 at an altitude of 15,240 m (50,000 ft) and a total of more than 6½ hr had been flown at supersonic speed. A production decision is scheduled to be taken in February 1977, which would permit the delivery of production aircraft to begin in 1978. The USAF hopes to order 244 B-1s, including prototypes, to replace B-52s now in service.

Designed to operate at treetop heights at near-sonic speed, and at supersonic speeds at high altitude, the B-1 has a radar signature about 5% as large as that of the B-52, carries nearly twice the payload, and requires shorter runways. The design incorporates the blended wing/body configuration developed previously for Rockwell's submission in the F-15 fighter competition. It also features a unique structural mode control system (SMCS) to minimise the effects of turbulence on crew and airframe likely to be encountered during high-speed, low-level penetration flights. The SMCS is a sensor-controlled automatic system which utilises movable foreplane vanes, in conjunction with the bottom rudder section, to produce aerodynamic forces to compensate for or suppress motion in the forward fuselage, so ensuring that crew efficiency is not impaired in a turbulent environment.

The crew escape capsule embodied in the prototypes has been deleted from later aircraft as an economy measure, together with the originally-planned variable-geometry engine inlets.

The principal materials used in the construction of the B-1 are aluminium alloys and titanium (41·3% and 21% by weight respectively of the aircraft's structure), the latter being used primarily in the wing carry-through structure, engine nacelles and aft fuselage. The balance is made up of steel (6·6%); composites (0·3%); and glassfibre, polyimide quartz and other non-metallic materials (30·8%). The entire structure is hardened to withstand nuclear blast. In terms of cash value, between 35 and 40% of the manufacturing work is subcontracted, and to date the B-1 programme has involved more than 5,000 manufacturers in 47 States. Major structural assemblies built in this manner are the tailplane, fin, leading-edge wing slats, trailing-edge flaps and landing gear.

TYPE: Strategic heavy bomber.

WINGS: Cantilever low-wing fail-safe blended wing/body structure, with variable geometry on outer panels. The wing carry-through structure, which is sealed as an integral fuel tank, is mainly (about 80%) of diffusion-bonded 6AL-4V titanium, and is built by Rockwell International's B-1 Division at Los Angeles. The wing pivot mechanism is also of diffusion-bonded titanium, with a pin made from a single 6AL-4V forging on each side, in spherical steel bearings, above and below which are integrally stiffened double cover plates of machined titanium. Wing sweep is actuated by screwjacks, driven by four hydraulic motors; it can be powered by any two of the aircraft's four hydraulic systems, asymmetric movement being prevented by a torque shaft between the two screwjacks. Sweep actuators are covered by a leading-edge 'knuckle' fairing which prevents a gap from opening when the outer panels are swept back. Aft of the wing pivot on each side are a hinged panel and two fixed fairings which blend the wing trailing-edges and engine nacelles. Each of the outer wing panels, which have 15° of leading-edge sweep when fully forward and 67° 30′ when fully swept, is a conventional two-spar aluminium alloy torsion-box structure, with machined spars, ribs, and one-piece integrally stiffened top and bottom skin panels. Wingtips, wing/body fairings, and some outer-wing skin panels, are of glassfibre-reinforced plastics. The outer panels are built at Palmdale, except for the control surfaces, which are manufactured by North American Aircraft Operations' Tulsa Division. Full-span seven-segment leading-edge slats on each outer panel can be drooped 20° for take-off and landing. Six-segment single-slotted trailing-edge flaps on each outer panel, with maximum downward deflection of 40°. There are no ailerons; instead, lateral control is provided by four-segment spoilers on each outer wing, forward of the outer four flap segments, with a maximum upward deflection of 70°. Outer sections are locked in at speeds in excess of Mach 1. All control surfaces are actuated electro-hydraulically by rods, cables, pulleys and bellcrank levers, except for the two outboard spoilers on each side. These are actuated by an electrical fly-by-wire system.

FUSELAGE: Conventional area-ruled fail-safe stressed-skin structure of closely-spaced frames and longerons, built mainly of 2024 and 7075 aluminium alloys. Built in six main sections: forward, forward intermediate, and (on first three aircraft only) crew escape capsule, all forward of the wing carry-through structure; and aft intermediate, aft fuselage and tailcone, to the rear of this structure. Titanium used for engine bays and firewalls, tail support structure, aft fuselage skins and other high-load or high-heat areas. Dorsal spine of steel/boron-filled titanium sandwich construction. Nose radome of polyimide quartz; dielectric panels of glassfibre-reinforced plastics. Forward section (excluding nosecone), crew escape capsule and engine nacelles are built by the B-1 Division; remainder of fuselage at Palmdale; and nosecone by Brunswick Corporation. Small sweptback movable foreplane, with 30° anhedral, on each side of nose, actuated by structural mode control system (SMCS) accelerometers in the fuselage which 'feel' up-and-down and side-to-side motion of the forward fuselage in turbulent conditions and compensate for or suppress it by relaying electrical signals which move the foreplanes and the bottom segment of the rudder. Forward section of fuselage houses the nosewheel unit, radar, most other electronics bays, and part of the forward fuel tank. Forward intermediate section contains two weapons bays, fuel tanks, and has space for side-looking airborne radar (SLAR) and/or other electronics equipment. Aft intermediate section incorporates the third weapons bay, main landing gear units and fuel tanks; and provides the main support structure for the engines. Aft section is occupied almost entirely by fuel tanks, and also provides support for the tail unit. Tailcone houses the rear electronics bay.

TAIL UNIT: Cantilever fail-safe structure, with sweepback on all surfaces; built under subcontract by Martin Marietta. Fin is a conventional titanium and aluminium alloy torsion-box structure, attached to the aft fuselage by a double shear attachment, bolts on the tailplane spindle, a vertical shear pin in the tailplane spindle fitting, and a shear-bolt joint on the front beam of the box. Aluminium alloy rudder is in three sections, of which the lowest one (below the tailplane) forms part of the SMCS and is actuated by the motion sensors in the fuselage, acting in conjunction with the foreplane vanes to provide control in yaw. All rudder sections have 25° travel to left and right. Variable-incidence tailplane is operated differentially (±20°) for roll control, symmetrically (between 10° up and 25° down) for pitch control, the two halves moving independently on bearings on the steel spindle. Tailplane is of aluminium alloy, with leading- and trailing-edges, and tips, of glassfibre-reinforced plastics. Rudder and tailplane actuated hydraulically, with electrical fly-by-wire backup system for use in the event of a mechanical system failure.

LANDING GEAR: Electrically-controlled hydraulically-

Rockwell International B-1 strategic bomber prototype at take-off

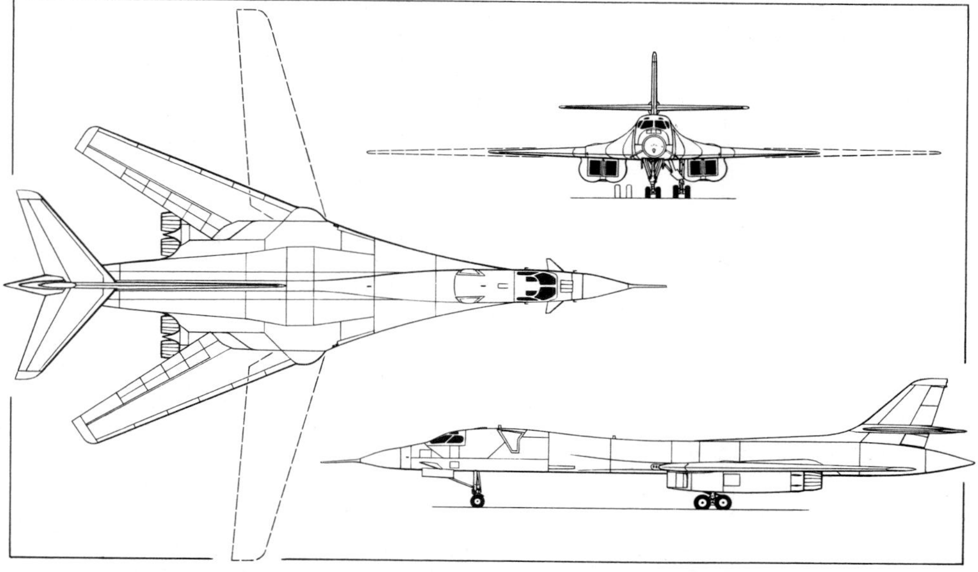

Rockwell International B-1 supersonic variable-geometry strategic bomber *(Pilot Press)*

Rockwell International B-1 strategic bomber (four General Electric YF101-GE-100 turbofan engines) with its wings in the swept position

retractable tricycle type, with single-stage oleo-pneumatic shock-absorbers on all units. Nose unit retracts forward, main units inward and rearward. Menasco steerable nose unit, with twin Goodyear wheels and tyres, size 35 in × 11·5 in. Twin landing and taxying lights on nosewheel leg. Cleveland main units, each with twin tandem pairs of Goodyear wheels and tyres, size 44·5 in × 16 in, and five-rotor structural carbon brakes.

Power Plant: Four General Electric YF101-GE-100 turbofan engines, each rated at approx 75·62 kN (17,000 lb st) dry and 133·4 kN (30,000 lb st) with afterburning, mounted in pairs beneath the fixed centre-section of the wing, close to the CG, to provide optimum stability in low-altitude turbulence conditions. Wedge-shaped external-shock inlets, with lips which open outward to increase air mass flow for take-off. Inward- and outward-opening dump doors in outer walls of each nacelle. Boundary layer bleed, to reduce drag, is provided via a two-position louvre beneath each engine. Eight integral fuel tanks (two in forward intermediate fuselage, one in wing carry-through structure, one in each outer wing panel, two in aft intermediate fuselage and one in rear fuselage). Computerised fuel transfer system, with manual backup, to maintain CG trim as fuel is depleted. Fuel is used from the mid-fuselage tanks first, followed by the wing tanks and finally by the fore and aft tanks. Cross-feed provision to supply all four engines from either of the two main fuselage tanks. An additional 9,980 kg (22,000 lb) of fuel can be carried in a removable auxiliary tank in one or other of the two forward weapons bays. Receptacle in upper nose section, forward of windscreen, for in-flight refuelling; aircraft is compatible with KC-135A tanker.

Accommodation: Crew access is via a downward-opening door and retractable ladder under the fuselage, aft of the nosewheel unit. First crewman to board presses the APU-powered engine start switch on the nosewheel leg, so that by the time all members are seated the engines are at start-up revolutions ready for take-off. Four-man operational crew, on Aircraft Mechanics Inc seats, comprises pilot, co-pilot, offensive systems operator and defensive systems operator. Fold-down seats are provided for fifth and sixth crew members (eg, flight instructor and systems instructor), if required; there are bunks on the cabin floor for crew rest on extended flights; and a toilet is provided. On first three aircraft only, all crew members are accommodated in a 4½ ton AV-1 self-contained compartment that is designed to function also as an escape capsule. This is a heated, air-conditioned and pressurised compartment (2,440 m; 8,000 ft environment) in which the crew can work without being restricted by personal parachutes. In the event of an emergency, requiring the crew to evacuate the aircraft, separation of the capsule is initiated by an OEA Inc severance system (incorporating a crew restraint assembly); a G & H Technology Inc explosive-cord system severs all electrical connections to the compartment. A Teledyne McCormick Selph energy transfer system maintains essential services within the escape capsule, which is then fired clear of the airframe by two Rocketdyne solid-fuel gimballed rocket motors. If separation is initiated while the aircraft is inverted, the escape capsule can be righted and stabilised by a Northrop Electronics manoeuvring rocket control system, using the two Rocketdyne gimballed rocket motors. At a suitable altitude, and actuated by a mortar-deployed drogue 'chute, three Pioneer Parachute Apollo-type ring-sail main 'chutes (stowed in the roof) are deployed to lower the capsule to the surface. A Goodyear Aerospace system, using inflatable rubber bags, will then serve as an impact reducer on land. A separate Goodyear flotation system can be activated if the capsule falls into water. Aerodynamic control of the capsule during descent can be effected by the deployment of twin side-fins at the rear, a spoiler at the front, and two side spoilers. The windscreen is birdproof, the flight deck side windows serve as emergency exits; and all flight deck windows have an electrically conductive coating to restrict any contribution to the radar signature from inside the aircraft.

In October 1974 it was announced that, for budgetary reasons, the escape capsule crew compartment would not be incorporated in the fourth or any subsequent B-1 aircraft built. Instead, a more conventional system would be employed, with crew members each having an individual high-performance ejection seat. A subcontract for these advanced concept ejection seats (ACES) was awarded to McDonnell Douglas Corporation in early 1976; use of these instead of the escape capsule compartment will result in a weight saving of approx 2,270 kg (5,000 lb) per aircraft.

Systems: All systems and subsystems are either fail-operative or fail-safe, to ensure that no single system failure prevents accomplishment of the primary mission, and that no second failure in the same system prevents a safe return to base.

Hamilton Standard air-conditioning and pressurisation systems. Four independent hydraulic systems, each 276 bars (4,000 lb/sq in), for actuation of wing sweep, control surfaces, landing gear and weapons bay doors. Nos. 1 and 2 systems are powered by the two port engines, and Nos. 3 and 4 by the starboard pair. No pneumatic system. Sundstrand main electrical system has three 115kVA integrated engine-driven constant-speed generators, supplying 230/400V three-phase AC power at 400Hz through four main buses. Provision for fourth generator in prototypes. Four accessory drive gearboxes, one for each engine, mounted in pairs between the two engines in each nacelle. Westinghouse generator and controls subsystem. A Harris Intertype (Radiation Division) self-testing electronic multiplexing system (EMUX) is interfaced with the computer-based Central Integrated Test System (CITS) in the prototypes, and replaces much of the signal/control wire and relay-logic found in conventional aircraft systems. It is designed to control electrical power distribution to subsystem and electronics equipment, landing gear, engine instruments, fuel system instruments, air inlet control system, weapons system operation, lights and heaters. Two 298 kW (400 hp) Garrett-AiResearch APUs, one in each nacelle between the two main engines, are started by hydraulic power from an accumulator; they provide self-start capability for operation from advance airfields (engine start switch mounted on the nosewheel oleo), and power a 15kVA emergency generator to drive the essential bus. Quadruplex automatic flight control system (AFCS) controls flight path, roll attitude, altitude, airspeed, autothrottle, Mach holds and terrain following. Flight director panel has heading hold, navigation and automatic approach modes. Central air data computer; gyro stabilisation system; stability control augmentation system; and structural mode control subsystem (SMCS). Vap-Air engine bleed air control system. Stewart-Warner (Southwind) fuel heat sink subsystem. Liquid nitrogen bottles for maintaining fuel tank pressure and for inflation of escape capsule flotation bags. Two fire extinguishers for each pair of engines.

Electronics and Equipment: Standard GFE (government furnished equipment) includes Collins ARA-50 UHF/ADF; Collins ARC-109 UHF com; Florida Communications PRC-90 UHF rescue beacon; Avco ARC-123 HF com; Motorola APX-78 X-band tracking transponder; Stewart-Warner APX-64 IFF; Hoffman ARN-84 Tacan; Collins ARN-108 ILS; Hughes AIC-27XA-3 intercom; Sandia DCK-175/A-37A(V) code enabling switch and coded switch system control.

Boeing Aerospace acts as overall electronics systems interface contractor, and is also subcontractor for the nav/attack system and other offensive electronics. These include Raytheon AN/APQ-140 nose radar; General Electric APQ-144 (modified) forward-looking radar; Texas Instruments APQ-146 (modified) computerised terrain-following radar; Singer-Kearfott APN-185 Doppler radar; Hughes forward looking infra-red; two Singer-Kearfott SKC-2070 general-purpose computers (one for navigation, one for weapon aiming); IBM AP-1 mass memory unit; Litton LN-15S redundant twin inertial platforms; and Honeywell APN-194 radar altimeter.

The AIL Division of Cutler-Hammer has responsibility for defensive electronics in the B-1. Most of these have still to be specified in detail, but items under consideration include active and passive electronic countermeasures (ECM); electronic jamming or other counter-countermeasures (ECCM); radio frequency surveillance equipment; homing and warning systems; and other countermeasures such as expendable types (ie, chaff) or infra-red.

Armament: Three identical internal weapons bays in fuselage, two forward and one aft of the wing carry-through structure. Each bay is 4·57 m (15 ft 0 in) long and has hydraulically-actuated three-position doors. Each bay can accommodate up to eight 1,016 kg (2,240 lb) Boeing AGM-69A SRAMs (Short Range Attack Missiles) on a rotary launcher, or up to 11,340 kg (25,000 lb) of nuclear or conventional weapons, although the rear bay is more likely to be utilised for penetration aids. In addition, there are four underfuselage hardpoints, each capable of carrying two additional SRAMs or 4,535 kg (10,000 lb) of other ordnance. Max possible weapons load 52,160 kg (115,000 lb). Other stores which the B-1 is capable of carrying include the proposed BDM (Bomber Defense Missile), ALCM (Air Launched Cruise Missile), decoy missiles or RPVs.

Dimensions, external:

Wing span:	
fully spread	41·67 m (136 ft 8½ in)
fully swept	23·84 m (78 ft 2½ in)
Length overall:	
incl nose probe	45·78 m (150 ft 2½ in)
excl nose probe	43·68 m (143 ft 3½ in)
Height overall	10·24 m (33 ft 7¼ in)
Tailplane span	13·67 m (44 ft 10 in)

Wheel track (c/l of shock-absorbers) 4·42 m (14 ft 6 in)
Wheelbase 17·53 m (57 ft 6 in)
AREA:
Wings, gross approx 181·2 m² (1,950 sq ft)
WEIGHTS AND LOADING:
Design max T-O weight 176,810 kg (389,800 lb)
Design max ramp weight 179,168 kg (395,000 lb)
Max landing weight approx 158,757 kg (350,000 lb)
Max wing loading approx 976 kg/m² (200 lb/sq ft)
PERFORMANCE (estimated, with VG inlets):
Max level speed at 15,240 m (50,000 ft)
approx Mach 2·0 (1,145 knots; 2,125 km/h; 1,320 mph)
Max level speed at 152 m (500 ft)
approx 650 knots (1,205 km/h; 750 mph)
Cruising speed at 15,240 m (50,000 ft)
Mach 0·85 (562 knots; 1,042 km/h; 648 mph)
Max range without refuelling
5,300 nm (9,815 km; 6,100 miles)

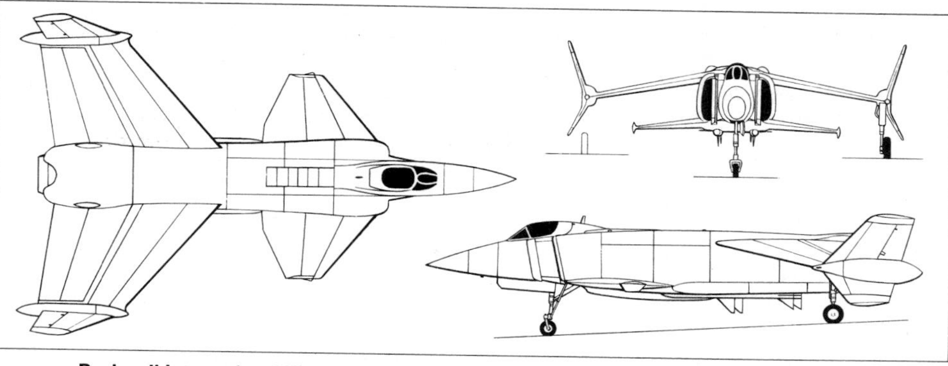

Rockwell International XFV-12A single-seat V/STOL fighter/attack prototype *(Pilot Press)*

ROCKWELL INTERNATIONAL XFV-12A

The US Navy initiated the XFV-12A V/STOL Fighter/Attack Technology Prototype programme to develop the capability of V/STOL operation from comparatively small carrier decks that would have neither catapult nor arrester gear.

The two prototype aircraft being built are roughly the size of a McDonnell Douglas A-4 Skyhawk. They employ an augmentor-wing concept with forward canard and aft semi-delta wings, and will be powered by a single, special version of the Pratt & Whitney F401-PW-400 advanced-technology turbofan engine.

The augmentor system has a diverter valve to block off the turbofan nozzle and divert the exhaust gases through ducts to nozzles in the wings and canards for V/STOL operations. A full-span ejector-flap system on each wing and canard allows ambient air to be drawn in over the flaps and ejected downward, mixed with the primary exhaust flow in a 7·5 : 1 ratio to provide the required jet-lift.

Cost considerations have limited the amount of test hardware associated with the development programme. To evaluate thrust augmentor components, a complete flight wing and canard with diffuser flaps were mounted on a rotary test rig that can be operated at speeds of up to 150 knots (278 km/h; 173 mph). An F401 engine with thrust diverter was incorporated in the rig in January 1974. This allows engine exhaust air to be ducted along the rig and exhausted through the augmentor components for static lift measurements, or while the rig is rotated at high speed.

Another economy measure is the use of a 'free air wind tunnel', one of the prototype aircraft, complete with engine, in flying attitude being mounted on a flat railway truck. Travelling at speeds of up to 70 knots (130 km/h; 81 mph), this enables the aircraft's controls to be put through a full transition to evaluate control effectiveness and harmony.

A full-size mockup was built, embodying existing airframe assemblies from other aircraft that had been selected to limit development costs. These were assembled in their correct physical relationship, allowing full and careful study of the integration of the structures, systems and power plant, before construction of the flying prototypes was started.

First flight of an XFV-12A prototype was scheduled for late 1976.

TYPE: Single-seat all-weather V/STOL fighter/attack prototype.

WINGS: Cantilever shoulder-wing monoplane. Wing section NACA 64 series (modified). Thickness/chord ratio 0·076 at root, 0·045 at tip. Anhedral 10°. Incidence 1° 30′. Sweepback at quarter-chord 35°. Light alloy structure of semi-delta configuration, forward portion of wing structure embodying an F-4 wing box. Titanium honeycomb is used for construction of the ejector flaps. Hydraulically-powered controls with irreversible actuators of Rockwell design and manufacture. Full-span trailing-edge flaps provide a lifting force for manoeuvrability in high-speed flight. Vertical endplate surfaces are mounted at each wingtip, comprising a fixed fin below the wing, outward-canted at 35°, and a fixed fin and rudder above the wing, outward-canted at 19°. Wing augmentor (ejector) flaps extend almost full span. They provide control of the vertical lift propulsion, acting as thrust vectors and so giving attitude and height control in hover and low-speed flight. The aft ejector flaps (together with those in the canard surfaces) serve as conventional flight controls in cruising flight. The fore and aft ejector flaps can be used together as speed brakes.

CANARD SURFACES: Cantilever low-wing monoplane. Anhedral 5°. Full-span trailing-edge flaps provide a lifting force for manoeuvrability in high-speed flight. Full-span augmentor (ejector) flaps function in combination with those on wings.

FUSELAGE: Forward fuselage, to aft of cockpit, is that of an A-4. Broad-section fuselage aft of cockpit, to house engine intake ducts and augmentor system ducting, is of light alloy semi-monocoque construction. Engine mounted in aft fuselage, which incorporates titanium material in its structure.

LANDING GEAR: Hydraulically-retractable tricycle type. Main units retract rearward into wingtip fairings, nose-wheel unit forward. Oleo-pneumatic shock-absorption. Hydraulic nosewheel steering. All units as for McDonnell Douglas A-4. Main wheels with tyres size 24 × 5·5-14, pressure 20·7 bars (300 lb/sq in). Nosewheel with tyre size 18 × 5·7-8, pressure 14·82 bars (215 lb/sq in). Goodyear dual disc brakes.

POWER PLANT: One modified Pratt & Whitney F401-PW-400 afterburning turbofan engine in the 133·4 kN (30,000 lb) thrust class. Engine inlet ducts are modified from the F-4. Auxiliary inlet in fuselage upper surface, aft of cockpit, augments air mass flow when aircraft is operating in vertical mode. A special electro-hydraulically actuated diverter valve, designed by Pratt & Whitney, will be installed in the tailpipe of the engine. When open, in the horizontal flight mode, it will allow free passage of engine exhaust gases for conventional propulsion. When closed, for vertical flight, the exhaust gases will be diverted to the ducts that feed the wing and canard augmentor nozzles. Fuel contained in two fuselage bladder tanks, capacity 1,590 litres (420 US gallons), and integral wing tanks, capacity 1,173 litres (310 US gallons). Total fuel capacity 2,763 litres (730 US gallons). Single-point refuelling. Oil capacity 11·4 litres (3 US gallons).

ACCOMMODATION: Pilot only, on McDonnell Douglas Escapac zero-zero ejection seat. Cockpit pressurised and air-conditioned.

SYSTEMS: AiResearch air cycle air-conditioning and pressurisation system, maintaining sea level cockpit altitude to 2,440 m (8,000 ft). Two independent and simultaneously operating hydraulic systems, at a pressure of 207 bars (3,000 lb/sq in), to operate flight controls, landing gear, ejector flaps and inlet ramps. Primary power source of the electrical system is a 30kVA integrated drive generator, the system providing 115/200V 400Hz AC power and 28V DC power. Emergency oxygen system with capacity of 5 litres (0·18 cu ft) of liquid oxygen, with converter. Anti-icing by engine bleed air.

ELECTRONICS AND EQUIPMENT: Collins AV/ARC-159 UHF radio. Radar system under study. Bendix RN-242A VOR; King KN-65 DME. Blind-flying instrumentation standard.

ARMAMENT: Ability to carry air-to-air and air-to-ground weapons. Space for internal gun in lower fuselage. Associated equipment is under study.

DIMENSIONS, EXTERNAL:
Wing span 8·69 m (28 ft 6¼ in)
Wing chord at root 4·98 m (16 ft 4¼ in)
Wing chord at tip 2·25 m (7 ft 4½ in)
Wing aspect ratio 2·09
Length overall 13·39 m (43 ft 11 in)
Height overall 3·15 m (10 ft 4 in)
Canard surface span 3·69 m (12 ft 1¼ in)
Wheel track 7·34 m (24 ft 1 in)
Wheelbase 7·62 m (25 ft 0 in)
AREAS:
Wings, gross 27·2 m² (293 sq ft)
Elevons (total) 1·91 m² (20·57 sq ft)
Fins (total) 5·08 m² (54·64 sq ft)
Rudders (total) 1·23 m² (13·20 sq ft)
Canard surface, gross 7·72 m² (83·05 sq ft)
Elevators (total) 2·75 m² (29·62 sq ft)
WEIGHTS (estimated):
Basic operating weight 6,259 kg (13,800 lb)
Max vertical T-O weight 8,845 kg (19,500 lb)
Max short-field T-O weight 11,000 kg (24,250 lb)
PERFORMANCE (estimated, at max T-O weight):
Max level speed in excess of Mach 2
T-O run at 11,000 kg (24,250 lb) 91 m (300 ft)

SABRELINER DIVISION

EXECUTIVE OFFICES:
827 Lapham Street, El Segundo, California 90245
PRESIDENT: J. J. Edwards Jr
EXECUTIVES:
P. Alexander (Manager, Missouri Completion Center)
P. Wickham (Chief Engineer)
MARKETING AND SUPPORT HEADQUARTERS:
6161 Aviation Drive, St Louis, Missouri 63134
Telephone: (314) 731-2260
Telex: 44-7227
EXECUTIVES:
E. J. Brandreth Jr (Vice-President, Marketing)
R. L. Chatley (Manager, Marketing Services)
D. Denison (Manager, Aircraft Sales)
J. Hamilton (Manager, International Sales)
J. Medeiros (Manager, Sales Administration)
P. Picciano (Manager, Customer Support)
T. Reilly (Manager, Eastern Regional Sales)
W. Smiley (Manager, Western Regional Sales)
F. Smith (Director, US Government Sales)
R. Alexander (Public Relations/Advertising Administrator)

The Sabreliner Division of Rockwell International continues to produce the twin-turbojet Sabreliner 60 and twin-turbofan Sabreliner 75A business transport aircraft.

In 1976 two new projects were under consideration. Designated provisionally as Sabre X and Sabre Y, they are intended to provide turbofan engine retrofits to basic T-39 and Sabreliner 40, 40A or 60 aircraft. The Sabre X proposal will provide for installation of three Pratt & Whitney Aircraft of Canada JT15D-4 two-shaft turbofan engines, each of 11·12 kN (2,500 lb st), while the Sabre Y proposal provides for installation of two AiResearch TFE 731-3 two-shaft geared turbofan engines each of 16·46 kN (3,700 lb st). The former is the more complicated engineering development, and it is considered unlikely that a production aircraft could be available before 1980. The two-engine Sabre Y proposal could be in production in 1978.

ROCKWELL SABRELINER 60

USAF and US Navy designation: T-39

To meet the USAF's UTX requirements for a combat readiness trainer and utility aircraft, Rockwell built as a private venture the prototype of a small, sweptwing twin-jet monoplane named Sabreliner. Design work began on 30 March 1956 and the prototype, powered by two General Electric J85 turbojet engines, flew for the first time on 16 September 1958.

In January 1959, the USAF ordered the first of 143 **T-39A** pilot proficiency/administrative support aircraft.

Other military versions have included six **T-39B** radar trainers for the USAF, 42 **T-39D** radar interception officer trainers for the US Navy, 7 **CT-39E** rapid response airlift aircraft for the USN, and 12 **CT-39Gs** for fleet tactical support. Production of the Sabreliner 40 has ended; details may be found in the 1974-75 *Jane's*. Two commercial versions remain in production:

Sabreliner 60. Generally similar to Sabreliner 40, but fuselage lengthened by 0·97 m (3 ft 2 in). Accommodation for crew of two and up to ten passengers. Five windows on each side of passenger cabin.

Sabreliner 75A. Described separately.

The following details refer to the current Sabreliner 60 production version, built at Rockwell's Los Angeles plant:

TYPE: Twin-engined jet business transport.

WINGS: Cantilever low-wing monoplane. Sweepback 28° 33′. All-metal two-spar milled-skin structure. Electrically-operated trim tab in each aileron. Electrically-operated trailing-edge flaps. Aerodynamically-operated leading-edge slats in five sections on each wing. Optional full-span pneumatically-operated de-icing boots.

FUSELAGE: All-metal semi-monocoque structure. Large hydraulically-operated airbrake under centre-fuselage.

TAIL UNIT: Cantilever all-metal structure, with flush antennae forming tip of fin and inset in dorsal fin. Moderate sweepback on all surfaces. Direct mechanical flight controls with electrically-operated variable-incidence tailplane. Electrically-operated trim tab in rudder. Optional full-span pneumatically-operated leading-edge de-icing boots.

LANDING GEAR: Retractable tricycle type. Twin-wheel

nose unit retracts forward. Single wheel on each main unit, retracting inward into fuselage. Main-wheel tyres size 26 × 6·60-14, pressure 12·41 bars (180 lb/sq in). Nosewheel tyres size 18 × 4·40-10, pressure 6·90 bars (100 lb/sq in). Hydraulic brakes with anti-skid units. Optional kit for operation from gravel runways.

POWER PLANT: Two Pratt & Whitney JT12A-8 turbojet engines (each 14·68 kN; 3,300 lb st) in pods on sides of rear fuselage. Integral fuel tanks in wings, with total capacity of 3,418 litres (903 US gallons). Fuselage tank, capacity 606 litres (160 US gallons). Total fuel capacity 4,024 litres (1,063 US gallons). Single-point refuelling.

ACCOMMODATION: Crew of two and up to ten passengers in pressurised air-conditioned cabin. Downward-hinged plug-type door, with built-in steps, forward of wing on port side. Emergency exits on both sides of cabin. Baggage space at front of cabin opposite door, with adjacent coat rack specified in many interior configurations. Beverage galley at forward end of cabin, two folding tables, door between cabin and flight deck, cold/hot air outlets and reading lights at each seat, two magazine racks. Toilet at rear of cabin. With seats removed there is room for 1,135 kg (2,500 lb) of freight.

SYSTEMS: AiResearch air-conditioning. Hydraulic system with audible failure warning system. Electrical system powered by engine-driven generators with 34Ah battery. Automatic oxygen system.

ELECTRONICS AND EQUIPMENT: Standard electronics include dual Collins VHF-20A com transceivers, dual Collins VIR-30A nav receivers with VOR/LOC/GS, Collins AP-105 autopilot, Collins FD-108Z pilot's flight director, Collins FD-108Y co-pilot's flight director, Collins NCS-31 navigation control system, dual Collins MC-103 compass systems, dual Collins 332C-10 radio magnetic indicators, Collins TDR-90 ATC transponder, Collins DME-40 DME, Collins DF-206 ADF, dual Collins 346B-3 audio system control centres, and Bendix RDR-1200C weather radar with digital data display. Standard equipment includes white instrument lights, encoding altimeter and altitude alerting, pilot's and co-pilot's clocks, synchroscope and cockpit floor and sidewall heating pads.

DIMENSIONS, EXTERNAL:
- Wing span: 13·61 m (44 ft 8 in)
- Length overall: 14·73 m (48 ft 4 in)
- Height overall: 4·88 m (16 ft 0 in)
- Tailplane span: 5·35 m (17 ft 6½ in)
- Wheel track: 2·20 m (7 ft 2½ in)
- Wheelbase: 4·85 m (15 ft 10¾ in)
- Cabin door:
  - Height: 1·19 m (3 ft 11 in)
  - Width: 0·71 m (2 ft 4 in)

DIMENSIONS, INTERNAL:
- Cabin (excl flight deck):
  - Length: 5·79 m (19 ft 0 in)
  - Max width: 1·59 m (5 ft 2½ in)
  - Max height: 1·71 m (5 ft 7½ in)
  - Volume: 13·59 m³ (480 cu ft)

AREAS:
- Wings, gross: 31·78 m² (342·05 sq ft)
- Ailerons (total): 1·52 m² (16·42 sq ft)
- Flaps (total): 3·74 m² (40·26 sq ft)
- Slats (total): 3·38 m² (36·34 sq ft)
- Fin: 3·86 m² (41·58 sq ft)
- Rudder: 0·83 m² (8·95 sq ft)
- Tailplane: 7·15 m² (77·0 sq ft)
- Elevators: 1·53 m² (16·52 sq ft)

WEIGHTS AND LOADINGS:
- Weight empty, equipped: 5,103 kg (11,250 lb)
- Max payload, incl crew: 1,156 kg (2,550 lb)
- T-O weight with four passengers, baggage and max fuel: 8,877 kg (19,572 lb)
- Max T-O weight: 9,150 kg (20,172 lb)
- Max ramp weight: 9,221 kg (20,372 lb)
- Max zero-fuel weight: 6,259 kg (13,800 lb)
- Max landing weight: 7,938 kg (17,500 lb)
- Landing weight with four passengers, baggage and 1 hr reserve fuel: 6,094 kg (13,435 lb)
- Max wing loading: 285·5 kg/m² (58·47 lb/sq ft)
- Max power loading: 311·65 kg/kN (3·03 lb/lb st)

PERFORMANCE (at max T-O weight, except where indicated):
- Max diving speed: Mach 0·85
- Max level speed at 6,550 m (21,500 ft): Mach 0·8 (489 knots; 906 km/h; 563 mph)
- Max cruising speed: Mach 0·8
- Econ cruising speed at 11,900-13,715 m (39,000-45,000 ft): Mach 0·75
- Stalling speed, landing configuration at 6,094 kg (13,435 lb) AUW: 83·5 knots (156 km/h; 96·5 mph)
- Max rate of climb at S/L: 1,433 m (4,700 ft)/min
- Max certificated operating altitude: 13,715 m (45,000 ft)
- Service ceiling, one engine out at AUW of 7,257 kg (16,000 lb): 7,925 m (26,000 ft)
- Min ground turning radius: 8·69 m (28 ft 6 in)
- T-O balanced field length: 1,539 m (5,050 ft)
- Landing distance at landing weight with four passengers, baggage and 1 hr reserve fuel: 693 m (2,275 ft)
- Max range with four passengers, baggage, max fuel and 45 min reserves: 1,748 nm (3,239 km; 2,013 miles)

Rockwell Sabreliner 60 business transport (two Pratt & Whitney JT12A-8 turbojet engines)

Rockwell T-39D of US Navy Training Squadron 86, at NAS Miramar, California *(Robert L. Lawson)*

ROCKWELL SABRELINER 75A

The Sabreliner 75A differs from the earlier Sabre 75 (which was described in the 1974-75 *Jane's*) by having General Electric turbofan engines; increased tailplane span; a new landing gear anti-skid system; improved galley, seating and toilet; and a new air-conditioning system.

In May 1973 Rockwell received from the FAA a $33·9 million contract for the lease/purchase of 11 Model 75A Sabreliners, with options on four more. All 15 aircraft were delivered by March 1976. Equipped with an extensive range of solid-state electronics, and measurement devices which include an advanced inertial area navigation computer, they are used for flight testing the accuracy of en route and terminal navigation aids.

A version designated Sea Sabre 75B, with lengthened fuselage, ALF 502R turbofan engines and other changes, has been proposed to meet a US Coast Guard requirement.

The following description refers to the standard commercial Sabreliner 75A:

TYPE: Twin-engined jet business transport.

WINGS: Cantilever low-wing monoplane. Wing section NACA 64₁A212 (modified) at wing station 62·90, NACA 64₁A012 (modified) at wing station 254·94. Dihedral 3° 9′. Incidence 0° at root, 2° 54′ at construction tip. Sweepback at quarter-chord 28° 33′. Two-spar milled-skin light alloy structure. Conventional ailerons of light alloy construction with electrically-operated trim tab in port aileron. Aerodynamically-operated leading-edge slats of light alloy construction. Electrically-operated slotted trailing-edge flaps. Optional full-span pneumatically-operated de-icing boots.

FUSELAGE: Light alloy semi-monocoque structure. Large hydraulically-operated airbrake under centre-fuselage.

TAIL UNIT: Cantilever light alloy structure. Electrically-operated variable-incidence tailplane. Electrically-operated trim tab in rudder. Elevator has electrically-operated trim tab, mechanically interconnected. Optional pneumatic de-icing boots.

LANDING GEAR: Hydraulically-retractable tricycle type with twin wheels on each unit. Nose unit retracts forward into fuselage nose, main units inward into undersurface of wings. Loud oleo-pneumatic shock-absorbers. Twin main wheels with Goodrich 10-ply tyres size 22 × 5·75-12, pressure 12·41 bars (180 lb/sq in). Steerable nose unit with twin wheels and Goodrich Type VII tyres size 18 × 4·40-10, pressure 6·90 bars (100 lb/sq in). Goodyear multiple-disc brakes. Fully-modulating anti-skid units. Optional kit for operation from gravel runways.

POWER PLANT: Two General Electric CF700-2D-2 turbofan engines, each 20·02 kN (4,500 lb st), mounted in pod on each side of rear fuselage. Cascade-type vertically-orientated thrust reversers. Integral fuel

Rockwell Sabreliner 75A eight/twelve-seat business transport (two General Electric CF700-2D-2 turbofan engines)

tanks in wings, with capacity of 3,418 litres (903 US gallons). Bladder-type fuel tank in aft fuselage with capacity of 753 litres (199 US gallons). Total fuel capacity of 4,171 litres (1,102 US gallons). Single pressure refuelling point in lower surface of starboard inboard wing leading-edge. Alternative gravity refuelling points at each wingtip and on top of aft fuselage tank. Oil capacity 3·8 litres (1 US gallon).

ACCOMMODATION: Crew of two and 8-10 passengers in pressurised air-conditioned cabin, with a variety of optional seating layouts. Improved seating, galley and toilet. Dual controls standard. Downward-hinged door, with built-in steps, forward of wing on port side. Emergency exit on each side of cabin, over wing. Baggage compartment at forward end of cabin, opposite door. Electrically-operated windscreen wipers.

SYSTEMS: Cabin pressurisation and heating by bleed air from both engines. Emergency pressurisation provided by starboard engine bleed air. Air-conditioning system incorporates a three-wheel bootstrap refrigeration unit, with separate ducting and temperature controls for cabin and flight deck. Hydraulic system powered by a single electrically-driven hydraulic pump, pressure 207 bars (3,000 lb/sq in). Auxiliary hydraulic accumulator for use in event of pump failure. Electrical system of 28V DC and 110V 400Hz constant-frequency AC. Primary DC power supplied by engine-driven starter/generators, with 34Ah batteries interconnected in the system. Emergency cabin and exit lighting by standby battery. Oxygen system supplied from 2·10 m³ (74 cu ft) cylinder with quick-donning masks for crew and dropout masks for passengers. Pneumatic system optional for optional wing and tail unit de-icing boots. Solar APU for ground heating and cooling and electrical power generation.

ELECTRONICS AND EQUIPMENT: Standard cabin equipment includes folding tables, cold/hot air outlets and reading lights at each seat position, fluorescent cabin lights, two magazine racks, and doors to isolate flight deck and toilet from cabin. Equipped to Cat II IFR requirements with Collins electronics comprising FD-109Y and FD-109Z flight directors; AP-105 autopilot; NCS-31 nav/com control/computer system; dual VHF-20A com transceivers; dual VIR-30A VHF nav receivers with VOR/LOC/GS; DF-206 ADF; dual 346B-3 audio systems, public address and intercom; DME-40 DME; dual TDR-90 ATC transponders; Bendix RDR-1200C weather avoidance radar; dual MC-103 compass systems; dual 332C-10 radio magnetic indicators; 54W-1C comparator warning monitor; ALT-50 radio altimeter; IDC-16007 encoding altimeter and altitude alerting system; plus dual Teledyne SLZ-9123 instant vertical speed indicators; and dual Mach airspeed indicators.

DIMENSIONS, EXTERNAL:
- Wing span: 13·61 m (44 ft 8 in)
- Wing aspect ratio: 5·77

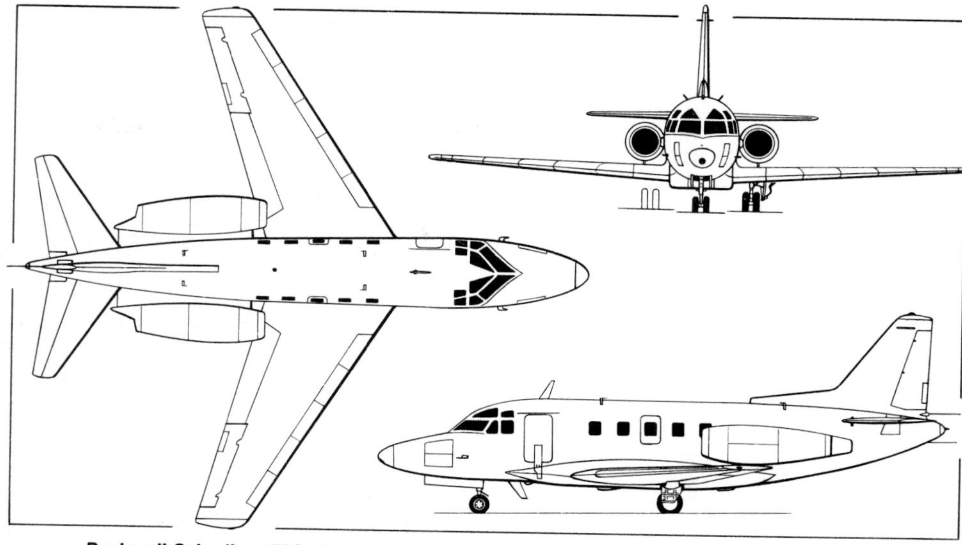

Rockwell Sabreliner 75A eight/twelve-seat twin-turbofan business transport *(Pilot Press)*

- Length overall: 14·38 m (47 ft 2 in)
- Height overall: 5·26 m (17 ft 3 in)
- Tailplane span: 5·91 m (19 ft 4½ in)
- Wheel track: 2·54 m (8 ft 4 in)
- Wheelbase: 4·85 m (15 ft 11 in)
- Cabin door:
  - Height: 1·19 m (3 ft 11 in)
  - Width: 0·71 m (2 ft 4 in)
  - Height to sill: 0·30 m (1 ft 0 in)
- Emergency exits (each):
  - Height: 0·74 m (2 ft 5 in)
  - Width: 0·51 m (1 ft 8 in)

DIMENSIONS, INTERNAL:
- Cabin (excl flight deck):
  - Length: 5·89 m (19 ft 4 in)
  - Max width: 1·62 m (5 ft 3¾ in)
  - Max height: 1·88 m (6 ft 2 in)
  - Volume: 15·57 m³ (550 cu ft)

AREAS:
- Wings, gross: 31·8 m² (342·05 sq ft)
- Ailerons (total): 1·53 m² (16·42 sq ft)
- Trailing-edge flaps (total): 3·74 m² (40·26 sq ft)
- Leading-edge slats (total): 3·38 m² (36·34 sq ft)
- Airbrake: 0·70 m² (7·54 sq ft)
- Fin: 4·85 m² (52·24 sq ft)
- Rudder, incl tab: 0·91 m² (9·75 sq ft)
- Tailplane: 8·37 m² (90·08 sq ft)
- Elevators, incl tab: 1·81 m² (19·43 sq ft)

WEIGHTS AND LOADINGS:
- Weight empty: 5,896 kg (13,000 lb)
- T-O weight with four passengers, baggage and max fuel: 9,879 kg (21,780 lb)
- Max T-O and ramp weight: 10,432 kg (23,000 lb)
- Max zero-fuel weight: 7,085 kg (15,620 lb)
- Max landing weight: 9,979 kg (22,000 lb)
- Landing weight with four passengers, baggage and 1 hr reserve fuel: 7,008 kg (15,450 lb)
- Max wing loading: 328·3 kg/m² (67·25 lb/sq ft)
- Max power loading: 260·54 kg/kN (2·66 lb/lb st)

PERFORMANCE (at max T-O weight, except where indicated):
- Max diving speed: Mach 0·85
- Max level and cruising speed: Mach 0·80
- Econ cruising speed: Mach 0·74
- Stalling speed, landing configuration, at 7,008 kg (15,450 lb) AUW: 86 knots (160 km/h; 99 mph)
- Max rate of climb at S/L: 1,372 m (4,500 ft)/min
- Max certificated operating altitude: 13,715 m (45,000 ft)
- Min ground turning radius: 8·69 m (28 ft 6 in)
- T-O balanced field length: 1,326 m (4,350 ft)
- Landing distance at 7,008 kg (15,450 lb) landing weight: 770 m (2,525 ft)
- Max range, with four passengers, baggage, max fuel and 45 min reserves: 1,712 nm (3,173 km; 1,972 miles)

GENERAL AVIATION DIVISION

EXECUTIVE OFFICES:
5001 North Rockwell Avenue, Bethany, Oklahoma 73008
Telephone: (405) 789-5000

PRESIDENT:
Cornell J. Slivinsky

EXECUTIVES:
A. F. Balaban (Manager, Public Relations)
D. E. Bradford (Director, International Sales)
R. L. Chatley (Manager, Communications)
D. L. Closs (Manager, Single Engine Sales)
J. Cobb (Acting Director, Industrial Relations)
John Keller (Director, Engineering)
R. A. Nielson (Director, Multi Engine Sales)
C. J. Slivinsky (Acting Director, Marketing)
Robert Ward (Manager, Marketing Support)

Albany, Georgia Facility
One Rockwell Avenue, Albany, Georgia 31702
PLANT MANAGER: Sherman Griffiths

Rockwell Aviation Services
Love Field, Dallas, Texas 75235

Rockwell International's General Aviation Division is a part of North American Aircraft Operations, and incorporates the Corporation's manufacturing, marketing and servicing of general aviation aircraft. It manufactures and markets Rockwell Commander turboprop and piston-engined multi-engine aircraft and single-engine personal, business and agricultural aircraft.

ROCKWELL COMMANDER 112

Following extensive consumer research in the lightplane market, the company initiated design of the Model 112 in December 1969. Construction of the first of five prototypes began in February 1970 and the first flight was made on 17 December 1970. Deliveries to customers of the Model 112 began in 1972.

In late 1973 manufacture of subassemblies and components was moved to Bethany, Oklahoma, and the first aircraft assembled from this production source was delivered in January 1974.

Known as the Model 112A, it introduced a number of improvements, including metal cabin doors, revised ventilation system and an increase of 45 kg (100 lb) in maximum gross weight. Subsequently, in October 1974, all assembly operations were transferred to Bethany.

Two additional versions were introduced for 1976. The three versions currently available are as follows:

Commander 112A. Basic version with one 149 kW (200 hp) Lycoming IO-360-C1D6 flat-four engine, driving a Hartzell type HC-E2YR-1BF/F7666A two-blade metal constant-speed propeller with spinner.

Commander 112TC. Generally similar to the Model 112A, except for the installation of a 156·6 kW (210 hp) Lycoming TIO-360-C1A6D turbocharged flat-four engine, driving a Hartzell type HC-E2YR-1BF/F8467-7R two-blade metal constant-speed propeller, wing of increased area and landing gear changes noted for Commander 114.

Commander 114. Version with more powerful engine and some equipment changes. Described separately.

The description which follows applies to both the Commander 112A and 112TC:

TYPE: Four-seat lightweight cabin monoplane.

WINGS: Cantilever low-wing monoplane. Wing section NACA 63.415 (modified). Dihedral 7°. Incidence at root 2°. Sweepforward at quarter-chord 2° 30′. Conventional light alloy structure. Ailerons of light alloy construction, using a channel spar and one-piece beaded skin. Electrically-operated light alloy single-slotted trailing-edge flaps, extending from wing station 25 to 121.20 with chord of 0·33 m (1 ft 1 in). Ground-adjustable trim tab on port aileron.

FUSELAGE: Conventional semi-monocoque light alloy structure.

TAIL UNIT: Cantilever light alloy structure with swept vertical surfaces. Dorsal fin faired into fuselage. Fixed-incidence tailplane. Adjustable trim tab in each elevator.

LANDING GEAR: Hydraulically-retractable tricycle type. Main wheels retract inward, nosewheel aft. Trailing-beam type main units, with oleo-pneumatic shock-absorbers. Commander 112A has Cleveland main wheel assemblies type 40-113A with tyres size 6·00-6, pressure 2·0 bars (29 lb/sq in); Cleveland nosewheel and tyre size 5·00-5, pressure 2·14 bars (31 lb/sq in); and Cleveland type 30-79 hydraulic brakes. Commander 112TC main wheels and brakes as for Commander 114.

POWER PLANT: One flat-four engine as detailed in Model listings. Two integral fuel tanks in the wing leading-edges, capacity of each 94·5 litres (25 US gallons) standard, 132·5 litres (35 US gallons) optional. Total fuel capacity 189 litres (50 US gallons) standard, 265 litres (70 US gallons) optional. Refuelling point in upper surface of each wing. Oil capacity 7·5 litres (2 US gallons).

ACCOMMODATION: Pilot and three passengers seated in pairs in enclosed cabin. Individual forward seats, bench seat for two aft. Passenger door on each side of cabin, over wing, hinged at forward edge. External baggage door on port side of fuselage aft of wing. Provisions for heating and ventilation.

SYSTEMS: Hydraulic system powered by a single electrically-driven hydraulic pump. Electrical system supplied by 70A 12V engine-driven alternator and 12V 35Ah battery.

ELECTRONICS: A wide range of Collins, King and Narco 360-channel com transceivers and 200-channel nav receivers are available to customers' requirements. Additional options for Commander 112TC include ADF, DME, transponder, encoding altimeter and autopilot.

DIMENSIONS, EXTERNAL (A: Commander 112A; B: Commander 112TC):
- Wing span:
  - A: 9·98 m (32 ft 9 in)
  - B: 10·86 m (35 ft 7·4 in)
- Wing chord at centreline:
  - A: 1·78 m (5 ft 10¼ in)
  - B: 2·39 m (7 ft 10·2 in)
- Wing chord at tip:
  - A: 0·89 m (2 ft 11 in)
  - B: 0·813 m (2 ft 8·1 in)
- Wing chord, mean aerodynamic:
  - A: 1·40 m (4 ft 7 in)
  - B: 1·52 m (4 ft 11·8 in)

Wing aspect ratio:
A 7
B 7·9
Length overall:
A, B 7·57 m (24 ft 10 in)
Height overall:
A, B 2·57 m (8 ft 5 in)
Tailplane span:
A, B 4·11 m (13 ft 6 in)
Wheel track:
A 3·25 m (10 ft 8 in)
B 3·33 m (10 ft 11 in)
Wheelbase:
A, B 2·11 m (6 ft 11 in)
Propeller diameter:
A 1·93 m (6 ft 4 in)
B 1·96 m (6 ft 5 in)
Propeller ground clearance:
A 0·22 m (8½ in)
B 0·19 m (7·4 in)
Passenger doors (each):
Height:
A, B 0·91 m (3 ft 0 in)
Width:
A, B 0·86 m (2 ft 10 in)
Baggage door:
Height:
A, B 0·46 m (1 ft 6 in)
Width:
A, B 0·64 m (2 ft 1 in)

DIMENSIONS, INTERNAL:
Cabin:
Length:
A, B 2·77 m (9 ft 1 in)
Max width:
A, B 1·17 m (3 ft 10 in)
Max height:
A, B 1·22 m (4 ft 0 in)
Baggage compartment:
A, B 0·62 m³ (22 cu ft)

AREAS:
Wings, gross:
A 14·12 m² (152 sq ft)
B 15·22 m² (163·8 sq ft)
Ailerons (total):
A, B 1·02 m² (11 sq ft)
Trailing-edge flaps (total):
A, B 1·67 m² (18 sq ft)
Fin:
A, B 1·58 m² (17 sq ft)

WEIGHTS AND LOADINGS:
Weight empty:
A 781 kg (1,723 lb)
B 794 kg (1,750 lb)
Max T-O weight, Normal category:
A 1,202 kg (2,650 lb)
Max T-O and landing weight, Utility category:
A 1,128 kg (2,488 lb)
Max T-O weight:
B 1,293 kg (2,850 lb)
Max landing weight, Normal category:
A 1,157 kg (2,550 lb)
Max zero-fuel weight, Utility category:
A 1,043 kg (2,300 lb)
Max zero-fuel weight:
B 1,188 kg (2,620 lb)
Max wing loading:
A, B 85·0 kg/m² (17·4 lb/sq ft)
Max power loading:
A 8·07 kg/kW (13·25 lb/hp)
B 8·26 kg/kW (13·6 lb/hp)

PERFORMANCE (at max T-O weight):
Never-exceed speed:
A, B 180 knots (333 km/h; 207 mph)
Max level speed:
A 148 knots (275 km/h; 171 mph)
B 166 knots (307 km/h; 191 mph)
Max cruising speed, 75% power at 2,285 m (7,500 ft):
A 140 knots (259 km/h; 161 mph)
Max cruising speed, 75% power at 6,100 m (20,000 ft):
B 148 knots (273 km/h; 170 mph)
Stalling speed, flaps up:
A 61 knots (113 km/h; 70·5 mph)
B 56 knots (104 km/h; 64·5 mph)
Stalling speed, flaps down:
A 54 knots (100 km/h; 62·5 mph)
B 50 knots (93·5 km/h; 58 mph)
Max rate of climb at S/L:
A 311 m (1,020 ft)/min
B 312 m (1,023 ft)/min
Service ceiling:
A 4,235 m (13,900 ft)
Max operating altitude:
B 6,100 m (20,000 ft)
Absolute ceiling:
B 7,620 m (25,000 ft)
T-O run:
A, B 363 m (1,190 ft)
T-O to 15 m (50 ft):
A 483 m (1,585 ft)
B 543 m (1,780 ft)

Rockwell's range of single-engine Commanders: Commander 112TC (rear), Commander 112A (centre) and Commander 114

Landing from 15 m (50 ft):
A 399 m (1,310 ft)
B 372 m (1,221 ft)
Landing run:
A 207 m (680 ft)
B 220 m (723 ft)
Range, with max optional fuel, 45 min reserves:
A 846 nm (1,569 km; 975 miles)
B 882 nm (1,633 km; 1,015 miles)

ROCKWELL COMMANDER 114

The Commander 114, introduced in 1976, is basically similar to the Commander 112A, but has as standard a 194 kW (260 hp) Lycoming IO-540 flat-six engine and certain equipment changes.

The description of the Commander 112A applies also to the Commander 114, except as follows:

LANDING GEAR: As for Commander 112A, except Cleveland main-wheel assemblies type 40-75H with tyres size 7·00-6, pressure 2·62 bars (38 lb/sq in). Nosewheel tyre as for Commander 112A, but pressure 3·45 bars (50 lb/sq in). Cleveland type 30-52H hydraulic brakes.

POWER PLANT: One 194 kW (260 hp) Lycoming IO-540-T4A5D flat-four engine, driving a Hartzell type HC-C2YR-1BF/F8467-7R two blade metal constant-speed propeller with spinner. Two integral fuel tanks in the wing leading-edges, capacity of each 132·5 litres (35 US gallons). Total fuel capacity 265 litres (70 US gallons). Refuelling points in upper surface of each wing.

ELECTRONICS: A wide range of Collins, King and Narco 360-channel com transceivers and 200-channel nav receivers, ADF, DME, transponder, encoding altimeter and autopilots available to customers' requirements.

DIMENSIONS, EXTERNAL:
As for Commander 112A, except:
Wheel track 3·33 m (10 ft 11 in)
Propeller diameter 1·96 m (6 ft 5 in)
Propeller ground clearance 0·19 m (7·4 in)

DIMENSIONS, INTERNAL:
As for Commander 112A

WEIGHTS AND LOADINGS:
Weight empty 812 kg (1,790 lb)
Max T-O weight 1,424 kg (3,140 lb)
Max zero-fuel weight 1,239 kg (2,732 lb)
Max wing loading 101·1 kg/m² (20·7 lb/sq ft)
Max power loading 7·34 kg/kW (12·1 lb/hp)

PERFORMANCE (at max T-O weight):
Never-exceed speed 186 knots (344 km/h; 214 mph)
Max level speed 161 knots (298 km/h; 185 mph)
Max cruising speed, 75% power at 2,135 m (7,000 ft) 156 knots (290 km/h; 180 mph)
Stalling speed, flaps up 59 knots (110 km/h; 68 mph)
Stalling speed, flaps down 55 knots (102 km/h; 63·5 mph)
Max rate of climb at S/L 321 m (1,054 ft)/min
Service ceiling 5,120 m (16,800 ft)
T-O run 335 m (1,100 ft)
T-O to 15 m (50 ft) 503 m (1,650 ft)
Landing from 15 m (50 ft) 402 m (1,318 ft)
Landing run 238 m (780 ft)
Range with max fuel, 45 min reserves 713 nm (1,321 km; 821 miles)

ROCKWELL SHRIKE COMMANDER 500S

The Rockwell Shrike Commander is a twin-engined aircraft designed for the businessman-pilot and certificated for Utility category operation.

TYPE: Twin-engined light transport.

WINGS: Cantilever high-wing monoplane. Wing section NACA 23012 modified. Dihedral 4°. Incidence 3° at root, −3° 30′ at tip. All-metal two-spar flush-riveted structure. Frise statically-balanced all-metal ailerons. Hydraulically-operated all-metal slotted flaps. Ground-adjustable tab in starboard aileron. Pneumatic de-icing boots optional.

FUSELAGE: All-metal semi-monocoque structure with flush-riveted skin.

TAIL UNIT: Cantilever all-metal structure with 10° dihedral on tailplane. Trim tabs in each elevator and rudder. Pneumatic de-icing boots optional.

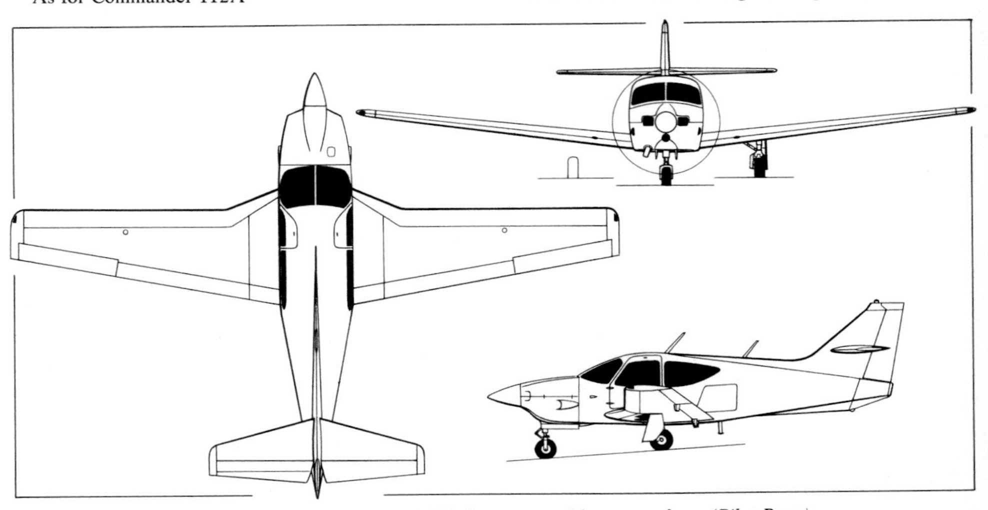

Rockwell Commander 112A four-seat cabin monoplane *(Pilot Press)*

LANDING GEAR: Retractable tricycle type, with single wheel on each unit. All wheels retract rearward hydraulically, main wheels turning through 90° to stow horizontally in nacelles. Oleo-pneumatic shock-absorbers. Hydraulically-steerable nosewheel. Goodyear main wheels with tyres size 25·65 × 8·70-10, pressure 3·79 bars (55 lb/sq in). Nosewheel tyre size 17·5 × 6·30-6, pressure 2·90 bars (42 lb/sq in). Goodyear aircooled hydraulic disc brakes.

POWER PLANT: Two 216 kW (290 hp) Lycoming IO-540-E1B5 flat-six engines, each driving a Hartzell HC-C3YR-2/C8468-6R three-blade constant-speed fully-feathering metal propeller. Bag-type fuel tanks in wings, capacity 590 litres (156 US gallons). Overwing refuelling. Oil capacity 22·7 litres (6 US gallons). Electrically-heated fuel vents and propeller anti-icing boots optional.

ACCOMMODATION: Four individual seats: two in front with dual controls and two at rear. Curtains divide pilot's compartment from cabin. Swivel-mounted fresh air vents above each seat, window curtains, emergency exit, announcement signs, adjustable heating and fresh air ventilation ports at cabin floor level, double-glazed windows in cabin and a hatbox shelf in aft cabin bulkhead are standard. Optional seating layouts for up to eight persons, some with rear bench seat for two or three. Optional refreshment cabinet for hot and cold drinks. Forward-opening passenger door under wing on port side. Forward-opening door by pilot's seat at front of cabin on port side. All equipment can be removed to permit cabin to be used for freight carrying. Compartment for 227 kg (500 lb) baggage aft of cabin, with outside door. Windscreen wiper and alcohol de-icing system for port side optional.

SYSTEMS: Hydraulic system, pressure 86 bars (1,250 lb/sq in), for landing gear, flaps, brakes and nosewheel steering. Electrical system includes two 70A alternators and two 35Ah batteries. 100A alternators optional.

ELECTRONICS AND EQUIPMENT: Standard equipment includes flight and engine instrumentation, clock, Janitrol 35,000 BTU cabin heater, rotating beacon, landing lights, reading lights, position lights, vacuum warning lights, instrument lighting system, air filter for vacuum instruments, dual vacuum pumps, electrically-adjustable cowl flap, external power socket, stall warning indicator and alternative static source. Optional equipment includes more advanced instruments, and extensive range of electronics which are available in package form or as individual items, underfuselage rotating beacon, pilot and co-pilot vertically-adjustable seats, inertia reels and shoulder harnesses for all seats, storage drawers under aft couch, cabin and/or cockpit fire extinguishers, low-fuel warning light, glass-holders, seat headrests, lavatory chair for starboard side of aft cabin, complete with curtain, 1·37 m³ (48·3 cu ft) or 2·74 m³ (96·6 cu ft) oxygen system, propeller synchronising equipment, dual relief tubes, sidewall-mounted stereo console, extra seat tracks, polished spinners, cabin table, storm window for co-pilot, and wing ice lights.

DIMENSIONS, EXTERNAL:

Wing span	14·95 m (49 ft 0½ in)
Wing chord at root	2·54 m (8 ft 4 in)
Wing chord at tip	0·65 m (2 ft 1½ in)
Wing aspect ratio	9·45
Length overall	11·22 m (36 ft 9¾ in)
Height overall	4·42 m (14 ft 6 in)
Tailplane span	5·10 m (16 ft 9 in)
Wheel track	3·95 m (12 ft 11 in)
Wheelbase	4·26 m (13 ft 11¾ in)
Propeller diameter	2·03 m (6 ft 8 in)
Crew door (fwd):	
Height	1·17 m (3 ft 10 in)
Width	0·58 m (1 ft 11 in)
Passenger door (aft):	
Height	1·14 m (3 ft 9 in)
Width	0·71 m (2 ft 4 in)
Baggage door:	
Height	0·60 m (1 ft 11½ in)
Width	0·50 m (1 ft 7½ in)

DIMENSIONS, INTERNAL:

Cabin: Length	3·23 m (10 ft 7 in)
Max width	1·32 m (4 ft 4 in)
Max height	1·35 m (4 ft 5 in)
Volume	5·01 m³ (177 cu ft)
Baggage hold	0·93 m³ (33 cu ft)

AREAS:

Wings, gross	23·69 m² (255 sq ft)
Ailerons (total)	1·90 m² (20·52 sq ft)
Trailing-edge flaps (total)	1·97 m² (21·20 sq ft)
Fin	2·23 m² (24·00 sq ft)
Rudder, incl tab	1·43 m² (15·40 sq ft)
Tailplane	3·07 m² (33·06 sq ft)
Elevators, incl tabs	1·91 m² (20·54 sq ft)

WEIGHTS AND LOADINGS:

Weight empty, equipped	2,102 kg (4,635 lb)
Max T-O and landing weight	3,062 kg (6,750 lb)
Max wing loading	129·2 kg/m² (26·47 lb/sq ft)
Max power loading	7·09 kg/kW (11·64 lb/hp)

Rockwell Shrike Commander 500S executive transport (two Lycoming IO-540-E1B5 engines)

PERFORMANCE (at max T-O weight):

Max cruising speed at S/L
187 knots (346 km/h; 215 mph) TAS

Max cruising speed (75% power) at 2,745 m (9,000 ft) 176 knots (326 km/h; 203 mph) TAS

Stalling speed, flaps and landing gear up
68 knots (126 km/h; 78 mph) CAS

Stalling speed, flaps and landing gear down
59 knots (109 km/h; 68 mph) CAS

Min single-engine control speed
65·5 knots (121 km/h; 75 mph)

Max rate of climb at S/L	408 m (1,340 ft)/min
Rate of climb at S/L, one engine out	81 m (266 ft)/min
Service ceiling	5,913 m (19,400 ft)
Service ceiling, one engine out	1,981 m (6,500 ft)
Min ground turning radius	11·63 m (38 ft 2 in)
T-O to 15 m (50 ft)	584 m (1,915 ft)
Landing from 15 m (50 ft)	681 m (2,235 ft)

Range with standard fuel at 2,745 m (9,000 ft) at 178 knots (330 km/h; 205 mph) TAS, 45 min reserves
693 nm (1,282 km; 797 miles)

Range, conditions as above, no reserves
824 nm (1,525 km; 948 miles)

Absolute range, standard fuel, at 4,570 m (15,000 ft) at 45% power and TAS of 148 knots (273 km/h; 170 mph), no reserves 936 nm (1,735 km; 1,078 miles)

ROCKWELL COMMANDER 685

Announced in April 1972, the Rockwell Commander 685 is a pressurised seven/nine-seat business transport evolved, like the Turbo Commander 690A, from the Turbo Commander 690 but powered by two 325 kW (435 hp) Continental GTSIO-520-F flat-six engines, each driving a Hartzell three-blade metal constant-speed and fully-feathering propeller. Standard fuel capacity is 969 litres (256 US gallons), with provision for optional auxiliary tanks raising total capacity to 1,218 litres (322 US gallons).

DIMENSIONS, EXTERNAL: As for Turbo Commander 690A except:

Wing span	14·19 m (46 ft 6·64 in)
Length overall	13·10 m (42 ft 11¾ in)
Propeller diameter	2·24 m (7 ft 4 in)
Propeller ground clearance	0·61 m (2 ft 0 in)

WEIGHTS AND LOADINGS:

Weight empty, standard	2,731 kg (6,021 lb)
Weight empty, with optional fuel tanks	2,742 kg (6,046 lb)
Max T-O and landing weight	4,082 kg (9,000 lb)
Max ramp weight	4,105 kg (9,050 lb)
Max wing loading	165·2 kg/m² (33·83 lb/sq ft)
Max power loading	6·29 kg/kW (10·34 lb/hp)

PERFORMANCE (at max T-O weight, unless detailed otherwise):

Max level speed at 6,100 m (20,000 ft)
242 knots (449 km/h; 279 mph) TAS

Max cruising speed at 7,315 m (24,000 ft)
222 knots (412 km/h; 256 mph) TAS

Econ cruising speed at 4,265 m (14,000 ft)
152 knots (281 km/h; 175 mph) TAS

Stalling speed, wheels and flaps up
81 knots (150·5 km/h; 93·5 mph) CAS

Stalling speed, wheels and flaps down
75 knots (139 km/h; 86·5 mph) CAS

Max rate of climb at S/L	454 m (1,490 ft)/min
Rate of climb at S/L, one engine out	75 m (247 ft)/min
Service ceiling	8,380 m (27,500 ft)
Service ceiling, one engine out	3,780 m (12,400 ft)
FAA operational ceiling	7,620 m (25,000 ft)
Min ground turning radius	12·47 m (40 ft 11 in)
Normal T-O run	594 m (1,949 ft)
Short-field T-O run	447 m (1,467 ft)
Normal T-O to 15 m (50 ft)	826 m (2,711 ft)
Short-field T-O to 15 m (50 ft)	570 m (1,869 ft)
Short-field landing from 15 m (50 ft)	592 m (1,943 ft)
Normal landing from 15 m (50 ft)	705 m (2,312 ft)
Short-field landing run	303 m (993 ft)
Normal landing run	362 m (1,188 ft)

Range with max standard fuel, allowances for start, taxi, climb to cruise altitude, descent and 45 min reserves, with 677 kg (1,493 lb) payload:
at 7,315 m (24,000 ft), max recommended cruising power 848 nm (1,570 km; 976 miles)
at 5,485 m (18,000 ft), econ cruising power
1,147 nm (2,124 km; 1,320 miles)

Range with fuel, allowances and payload as above, no reserves:
at 7,315 m (24,000 ft), max recommended cruising power 943 nm (1,746 km; 1,085 miles)
at 5,485 m (18,000 ft), econ cruising power
1,276 nm (2,364 km; 1,469 miles)

Range with max standard and optional fuel, allowances as above, with 486 kg (1,072 lb) payload and 45 min reserves:
at 7,315 m (24,000 ft), max recommended cruising power 1,115 nm (2,066 km; 1,284 miles)
at 6,100 m (20,000 ft), econ cruising power
1,534 nm (2,842 km; 1,766 miles)

Range with fuel, allowances and payload as above, no reserves:
at 7,315 m (24,000 ft), max recommended cruising power 1,211 nm (2,243 km; 1,394 miles)
at 6,100 m (20,000 ft), econ cruising power
1,668 nm (3,090 km; 1,920 miles)

Rockwell Commander 685 seven/nine-seat pressurised business transport

ROCKWELL TURBO COMMANDER 690A

The Rockwell Turbo Commander 690A is a pressurised transport aircraft powered by two AiResearch turboprop engines. It is an advanced development of the Turbo Commander 690, and improvements include an increase in cabin pressure differential from 0·29-0·36 bars (4·2 to 5·2 lb/sq in), an increase in FAA operational ceiling to 9,450 m (31,000 ft), and higher permitted speeds with landing gear and flaps down. Electrically heated windscreens, better flight deck lighting, new interior decor, ground cooling increase to 26,000 BTU, and systems for flight in known icing conditions are standard equipment.

The prototype of the Turbo Commander 690A was flown for the first time in June 1972, with FAA certification being awarded on 25 April 1973. Although certification was granted under CAR Part 3, many portions of the systems and structure exceed the requirements of CAR Part 4, SR422B and FAR 25. Nearly 150 Turbo Commander 690s and 690As had been built by the Autumn of 1976.

The description of the Shrike Commander applies also to the Turbo Commander 690A except in the following details:

Wings: Span reduced. Incidence at tip −1°. Electrically-operated trim tab in starboard aileron.

Tail Unit: Tailplane increased in span and area. Vertical surfaces increased in height and area.

Landing Gear: Wheelbase increased. Main-wheel tyre pressure 4·83 bars (70 lb/sq in).

Power Plant: Two 522 kW (700 ehp) AiResearch TPE 331-5-251K turboprop engines, each driving a Hartzell HC-B3TN-5FL/LT 10282H+4 three-blade constant-speed fully-feathering and reversible-pitch propeller. Fuel contained in bag-type wing tanks with a total usable capacity of 1,453 litres (384 US gallons). Refuelling points on upper surface of each wing. Oil capacity 11·5 litres (3 US gallons).

Accommodation: Standard seating for pilot and six passengers, on two adjustable seats on flight deck, two forward-facing single seats and a three-place forward-facing bench seat in main cabin. A variety of optional seating layouts offer accommodation for up to 11 persons. Cabin pressurised, heated and air-conditioned. Forward-hinged outward-opening cabin door on port side, with retractable cabin step. Emergency exit on starboard side of fuselage. Baggage compartment of 1·22 m³ (43 cu ft) with 272 kg (600 lb) capacity aft of rear pressure bulkhead with external access door on port side of fuselage.

Systems: Cabin pressurisation, heating and air-conditioning by engine bleed air; max pressure differential 0·36 bars (5·2 lb/sq in). Hydraulic system supplied by two pumps at 86 bars (1,250 lb/sq in), with hydraulic reservoir for emergency use. Electrical system powered by two 300A starter/generators. Two 44Ah nickel-cadmium batteries. Scott constant-flow emergency oxygen system of 0·62 m³ (22 cu ft) capacity, with individual outlets.

Electronics and Equipment: Full blind-flying instrumentation and a selection of IFR electronics packages by King, Collins, Sperry and Bendix are available as options. Automatic propeller synchronisation, fuel flow system and gauges with digital readout, heated fuel vents, stall warning and engine fire detection system standard.

Dimensions, external:

Wing span	14·22 m (46 ft 8 in)
Wing chord at root	2·64 m (8 ft 7¾ in)
Wing chord at tip	0·84 m (2 ft 9 in)
Wing aspect ratio	8·19
Length overall	13·52 m (44 ft 4¼ in)
Height overall	4·56 m (14 ft 11½ in)
Tailplane span	6·03 m (19 ft 9¼ in)
Wheel track	4·70 m (15 ft 5 in)
Wheelbase	5·38 m (17 ft 7¾ in)
Propeller diameter	2·69 m (8 ft 10 in)
Propeller ground clearance	0·36 m (1 ft 2¼ in)
Cabin door:	
Height	1·19 m (3 ft 11 in)
Width	0·67 m (2 ft 2½ in)
Height to sill	0·47 m (1 ft 6½ in)
Baggage door:	
Height	0·80 m (2 ft 7½ in)
Width	0·51 m (1 ft 8 in)
Height to sill	0·52 m (1 ft 8½ in)
Emergency exit:	
Height	0·48 m (1 ft 7 in)
Width	0·68 m (2 ft 2¾ in)

Areas:

Wings, gross	24·7 m² (266·0 sq ft)
Ailerons (total)	1·83 m² (19·74 sq ft)
Trailing-edge flaps (total)	1·65 m² (17·74 sq ft)
Fin	2·25 m² (24·23 sq ft)
Rudder (incl tab)	1·92 m² (20·66 sq ft)
Tailplane	3·51 m² (37·80 sq ft)
Elevators (incl tab)	1·91 m² (20·57 sq ft)

Weights and Loadings:

Weight empty, standard	2,778 kg (6,126 lb)
Max T-O weight	4,649 kg (10,250 lb)
Max ramp weight	4,672 kg (10,300 lb)
Max zero-fuel weight	3,969 kg (8,750 lb)
Max landing weight	4,354 kg (9,600 lb)
Max wing loading	188·1 kg/m² (38·53 lb/sq ft)
Max power loading	4·54 kg/kW (7·32 lb/ehp)

Performance (at max T-O weight, unless specified otherwise):

Max level speed at 3,660 m (12,000 ft)	285 knots (528 km/h; 328 mph) TAS
Max cruising speed at 5,335 m (17,500 ft)	280 knots (518 km/h; 322 mph) TAS
Econ cruising speed at 9,450 m (31,000 ft)	251 knots (465 km/h; 289 mph) TAS
Stalling speed, flaps and wheels up	82 knots (152 km/h; 94·5 mph) CAS
Stalling speed, flaps and wheels down	77 knots (143·5 km/h; 89 mph) CAS
Max rate of climb at S/L	868 m (2,849 ft)/min
Rate of climb at S/L, one engine out	272 m (893 ft)/min
Service ceiling	10,060 m (33,000 ft)
Service ceiling, one engine out	6,005 m (19,700 ft)
FAA operational ceiling	9,450 m (31,000 ft)
Min ground turning radius	12·47 m (40 ft 11 in)
Normal T-O run	437 m (1,434 ft)
Short-field T-O run	360 m (1,180 ft)
Normal T-O to 15 m (50 ft)	675 m (2,216 ft)
Short-field T-O to 15 m (50 ft)	508 m (1,666 ft)
Landing from 15 m (50 ft) with propeller reversal, at max landing weight	490 m (1,606 ft)
Landing from 15 m (50 ft) without propeller reversal, at max landing weight	635 m (2,084 ft)
Landing run with propeller reversal, at max landing weight	275 m (902 ft)
Landing run without propeller reversal, at max landing weight	422 m (1,385 ft)

Range with max fuel, allowances for start, taxi, T-O and climb to 9,450 m (31,000 ft), 45 min reserves, with 726 kg (1,601 lb) payload:

Max cruising power	1,460 nm (2,705 km; 1,681 miles)
Econ cruising power	1,471 nm (2,725 km; 1,693 miles)

Range with max payload, allowances as above, with 1,190 kg (2,624 lb) payload:

Max cruising power	740 nm (1,370 km; 852 miles)
Econ cruising power	741 nm (1,372 km; 853 miles)

Rockwell Turbo Commander 690A business aircraft (two AiResearch TPE 331-5-251K turboprop engines)

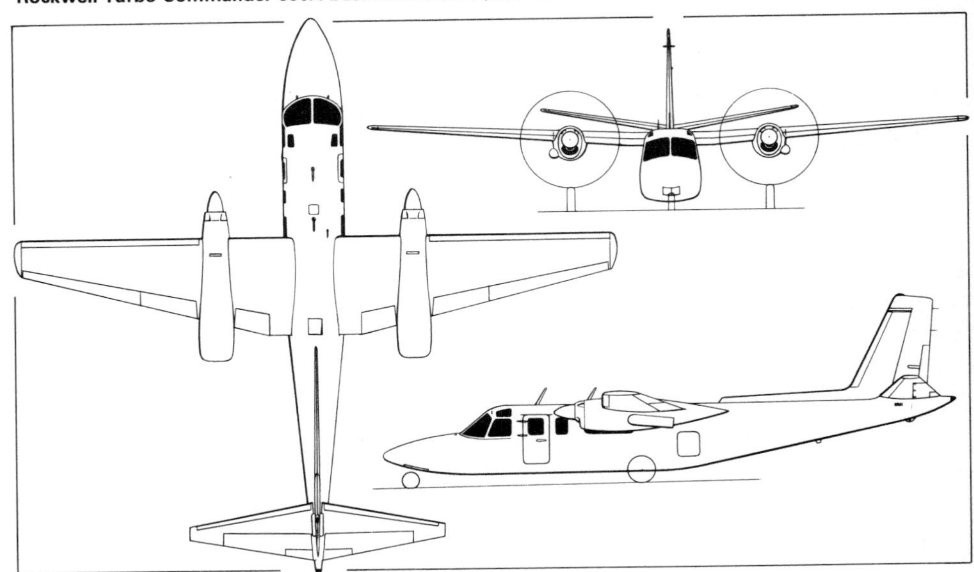

Rockwell Turbo Commander 690A seven/eleven-seat twin-turboprop transport *(Pilot Press)*

ROCKWELL COMMANDER 700

This twin-engined six/eight-seat light transport aircraft is being developed in collaboration with Fuji in Japan. A

First Rockwell-assembled Commander 700 (two Lycoming TIO-540-R2AD engines)

description appears under that company's entry on page 121 of this edition.

ROCKWELL THRUSH COMMANDER-600 and -800

The Thrush Commander is the largest specially-designed agricultural aircraft in production in the USA at the present time.

Two versions are available:

Thrush Commander-600. Basic version with 447·5 kW (600 hp) Pratt & Whitney R-1340 Wasp nine-cylinder aircooled radial engine, driving a Hamilton Standard type 12D40 two-blade metal constant-speed propeller.

Thrush Commander-800. As Thrush Commander-600, but with 596·5 kW (800 hp) Wright R-1300-1B Cyclone nine-cylinder aircooled radial engine, driving a Hamilton Standard type 3D40 or 23D40 three-blade metal constant-speed propeller. Additional standard equipment includes engine hour meter, electrically-operated auxiliary fuel pump, larger-diameter tyres, 24V 105A electrical system and external gust locks.

Both versions have a 1·50 m³ (53 cu ft) hopper able to contain up to 1,514 litres (400 US gallons) of liquid or 1,487 kg (3,280 lb) of dry chemicals. They have also corrosion proofing of activated Copon and are certificated to both CAR 3 Normal category and CAM 8 Restricted category requirements.

Improvements introduced on the 1976 models of the Thrush Commander include port and starboard boarding steps, quick release door hinges, lightweight battery, and corrosion proofing on the interior surface of all removable skins.

TYPE: Single-seat agricultural aircraft.

WINGS: Cantilever low-wing monoplane. Dihedral 3° 30′. Two-spar structure of light alloy throughout, except for main spar caps of heat-treated SAE 4000 Series steel. Leading-edge formed by heavy main spar and the nose-skin. Light alloy plain ailerons. Electrically-operated flaps. Wing roots sealed against chemical entry.

FUSELAGE: Welded chrome-molybdenum steel tube structure covered with quickly-removable light alloy panels. Underfuselage skin of stainless steel.

TAIL UNIT: Wire-braced welded chrome-molybdenum steel tube structure, fabric-covered. Streamline-section heavy-duty stainless steel wire bracing and heavy-duty stainless steel attachment fittings. Light alloy controllable trim tab in each elevator. Deflector cable from cockpit to fin-tip.

LANDING GEAR: Non-retractable tailwheel type. Main units have rubber-in-compression shock-absorption and 25·65 × 8·50-10 wheels with 10-ply tyres standard on Commander-600. Main-wheel tyres size 29 × 11·00-10 optional for Commander-600, standard for 800. Hydraulically-operated disc brakes. Parking brakes. Wire cutters on main gear. Steerable, locking tailwheel, size 12·5 × 4·5 in.

POWER PLANT: One 447·5 kW (600 hp) or 596·5 kW (800 hp) nine-cylinder aircooled radial engine as detailed in model listings. One 200·5 litre (53 US gallon) integral tank in each wing, giving total fuel capacity of 401 litres (106 US gallons). Oil capacity 43 litres (11·4 US gallons).

ACCOMMODATION: Single adjustable seat in 40*g* 'safety pod' sealed cockpit enclosure, with steel tube overturn structure. Two overhead windows for improved view in turns. Downward-hinged door on each side. Tempered safety-glass windscreen. Inertia-reel safety harness standard. Baggage compartment. Openable windscreen optional. Hopper forward of cockpit with capacity of 1·50 m³ (53 cu ft) or 1,514 litres (400 US gallons). Hopper has a 0·33 m² (3·56 sq ft) lid, openable by a single handle.

SYSTEMS: Electrical system of Commander-600 is powered by a 24V 50A engine-driven alternator, that of the Commander-800 by a 24V 105A engine-driven generator. 70A alternator optional for Commander-600. Lightweight 24V 35Ah battery. Freon air-conditioning system optional.

ELECTRONICS AND EQUIPMENT: Standard equipment includes Universal spray system with external 50 mm (2 in) stainless steel plumbing, 50 mm (2 in) Root Model 67 pump with wooden fan, Transland gate, 50 mm (2 in) valve, quick-disconnect pump mount and strainer. Streamlined spraybooms with outlets for 68 nozzles, 36 nozzles installed. Micro-adjust valve control (spray) and calibrator (dry). A 63 mm (2·5 in) side-loading system is installed on the port side. Navigation lights, instrument lights and two rotating beacons. Optional equipment includes Transland high-volume spreader with micro-adjust calibrator, agitator installation, extra-high-density spray configuration with 70 nozzles installed; Agrinautics electrically-operated 3-way valve, emergency shut-off valve, pump in lieu of Root pump and strainer in lieu of Transland strainer; Agevenco 6520 pump in lieu of Root pump. Six-unit AU3000 Micronair installation in lieu of standard booms and nozzles. Transland S-2 Boommaster with Q-D flange in lieu of standard gate and Root pump, and S-2R Boommaster replacement units; night working lights including wingtip turn lights, cockpit fire extinguisher, windscreen wiper and washer and water bomber configuration. Optional electronics include basic installation kit, Bendix T-12C ADF or T-12D digital ADF; King KX-170B or KX-170BE or KX-175B or KX-175BE nav/com, KI-201C converter indicator; Narco Com-11A or Com-11B com transceiver and Nav-11 nav receiver.

DIMENSIONS, EXTERNAL:

Wing span	13·51 m (44 ft 4 in)
Length overall	8·89 m (29 ft 2 in)
Height overall	2·79 m (9 ft 2 in)
Tailplane span	4·86 m (15 ft 11½ in)
Wheel track	2·72 m (8 ft 11 in)

AREA:

Wings, gross	30·34 m² (326·6 sq ft)

WEIGHTS AND LOADINGS (A: Thrush Commander-600; B: Thrush Commander-800):

Weight empty, equipped:	
A	1,678 kg (3,700 lb)
B	1,860 kg (4,100 lb)
Max T-O weight (CAR.3):	
A, B	2,721 kg (6,000 lb)
Max T-O weight (CAM.8):	
A	3,130 kg (6,900 lb)
B	3,538 kg (7,800 lb)
Max wing loading:	
A	103·0 kg/m² (21·1 lb/sq ft)
B	115·2 kg/m² (23·6 lb/sq ft)
Max power loading:	
A	6·99 kg/kW (11·5 lb/hp)
B	5·93 kg/kW (9·75 lb/hp)

PERFORMANCE (with spray equipment installed and at CAR.3 max T-O weight, unless indicated otherwise. A: Thrush Commander-600; B: Thrush Commander-800):

Max level speed:	
A	122 knots (225 km/h; 140 mph)
B	135 knots (249 km/h; 155 mph)
Max cruising speed, 70% power:	
A	108 knots (200 km/h; 124 mph)
B	119 knots (220 km/h; 137 mph)
Working speed, 70% power:	
A	91-100 knots (169-185 km/h; 105-115 mph)
B	100-109 knots (185-201 km/h; 115-125 mph)
Stalling speed, flaps up:	
A, B	61 knots (113 km/h; 70 mph)
Stalling speed, flaps down:	
A, B	57·5 knots (107 km/h; 66 mph)
Stalling speed at normal landing weight, flaps up:	
A	50 knots (92 km/h; 57 mph)
B	51·5 knots (95 km/h; 59 mph)
Stalling speed at normal landing weight, flaps down:	
A	48 knots (89 km/h; 55 mph)
B	50 knots (92 km/h; 57 mph)
Max rate of climb at S/L:	
A	274 m (900 ft)/min
B	335 m (1,100 ft)/min
Service ceiling:	
A	4,575 m (15,000 ft)
B	7,620 m (25,000 ft)
T-O run:	
A	236 m (775 ft)
B	305 m (1,000 ft)
Landing run:	
A, B	152 m (500 ft)
Ferry range with max fuel at 70% power:	
A	350 nm (648 km; 403 miles)
B	286 nm (531 km; 330 miles)

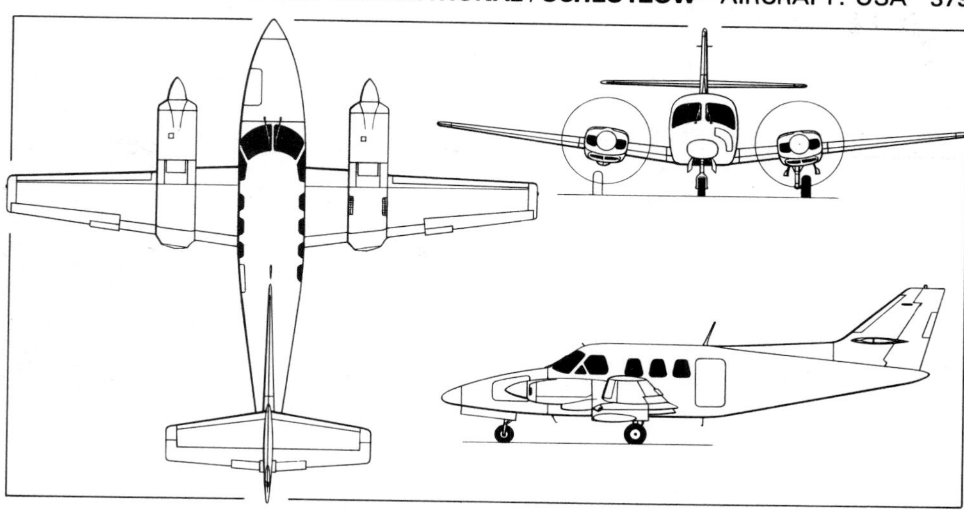

Rockwell Commander 700 six/eight-seat pressurised transport *(Pilot Press)*

Rockwell Thrush Commander-800 agricultural aircraft (Wright Cyclone R-1300-1B engine)

SCHEUTZOW

SCHEUTZOW HELICOPTER CORPORATION

POSTAL ADDRESS:
PO Box 27, Columbia Station, Ohio 44028
WORKS:
27100 Royalton Road, Columbia Station, Ohio 44028, and Sweetwater, Texas
Telephone: (216) 236-8470

PRESIDENT: Webb Scheutzow
VICE-PRESIDENT: Elmer J. Scheutzow

This company was formed by Mr Webb Scheutzow to develop and build a light helicopter incorporating a new type of rotor head in which the blades are carried on rubber bushings.

Initial development of the new head was carried out successfully on a small testbed helicopter, which was described and illustrated in the 1964-65 *Jane's.* It is now being used on the production-type Scheutzow Bee.

Two helicopters are being used in the FAA Type Certification programme. The 100-hour durability test was completed at full throttle for the entire duration of the test, and rotor system qualification tests have been completed. Initial flight trials were completed during 1971 and towards the end of 1973 the Bee was almost at the final stage of FAA Type Certification. Non-availability of capital brought operations to a halt throughout 1974, but the certification programme was resumed in February 1975. No further news has been received since that time.

SCHEUTZOW BEE

Design of this small lightweight side-by-side two-seat helicopter began in early 1964 and construction of three prototypes was started in March 1965. The first of these flew for the first time in 1966 and had completed 14 hours' flying by 26 January 1967. By that date, forward speeds of up to 78 knots (145 km/h; 90 mph) had been attained. Autorotative tests revealed good power-off handling qualities and the aircraft has proved very stable in hovering flight.

The Scheutzow Bee has a welded steel tube structure with a completely enclosed metal cabin. The landing gear is of the skid type, with cross-tubes on which the skids are carried.

Power plant is a 134 kW (180 hp) Lycoming IVO-360-A1A flat-four engine, which drives the two-blade flapping-type main rotor (with integral control gyro-bar) through a centrifugal clutch and multiple V-belt drive system. Use of this drive and isolation of the engine from the cabin area have resulted in an extremely low cabin noise level, permitting normal operation without an intercom system. The conventional two-blade tail rotor is driven through a shaft and bevel gear.

The main rotor incorporates Mr Scheutzow's patented 'Flexhub'. Elastomeric bearings with an offset flapping hinge are claimed to provide exceptional control power and precise response in all manoeuvres. This is claimed to reduce vibration as well as cost and lubrication requirements. Standard fuel capacity is 83 litres (22 US gallons). A 56 litre (15 US gallon) auxiliary fuel tank is available as an optional accessory, extending range to an estimated 248 nm (458 km; 285 miles).

Scheutzow Bee two-seat light helicopter (Lycoming IVO-360-A1A engine)

There is a large baggage compartment aft of the seats, on the centre of gravity, with a door on each side. Optional items include dual controls, radio, lights, heater, internal litter, ground handling wheels and agricultural spray equipment.

DIMENSIONS, EXTERNAL:
- Diameter of main rotor: 8·23 m (27 ft 0 in)
- Main rotor blade chord: 0·20 m (8 in)
- Diameter of tail rotor: 1·22 m (4 ft 0 in)
- Length overall (rotors fore and aft): 9·50 m (31 ft 2 in)
- Length of fuselage: 7·34 m (24 ft 1 in)
- Width overall, rotor fore and aft: 2·13 m (7 ft 0 in)
- Height overall: 2·59 m (8 ft 6 in)

WEIGHTS:
- Weight empty: 514 kg (1,135 lb)
- Max T-O weight: 764 kg (1,685 lb)

PERFORMANCE (at max T-O weight):
- Max level speed at S/L: 81 knots (150 km/h; 93·5 mph)
- Max cruising speed: 69 knots (128 km/h; 80 mph)
- Max rate of climb at S/L: 366 m (1,200 ft)/min
- Service ceiling: 3,960 m (13,000 ft)
- Hovering ceiling in ground effect: 2,225 m (7,300 ft)
- Range with standard fuel at max cruising speed: 152 nm (280 km; 175 miles)

SCHWEIZER
SCHWEIZER AIRCRAFT CORPORATION

HEAD OFFICE AND WORKS:
Box 147, Elmira, New York 14902
Telephone: (607) 739-3821

PRESIDENT AND CHIEF ENGINEER:
Ernest Schweizer

VICE-PRESIDENT AND GENERAL MANAGER:
Paul A. Schweizer

VICE-PRESIDENT IN CHARGE OF MANUFACTURING:
William Schweizer

SALES MANAGER:
W. E. Doherty Jr

ASSISTANT TREASURER:
Joseph Kroczynski

SECRETARY:
Kenneth Tifft

Schweizer Aircraft Corporation is the leading American designer and manufacturer of sailplanes (which see in Sailplanes section). In early 1972 the company acquired the design and production rights of the Teal amphibian from Mr David B. Thurston, and the production tools and jigs were transferred to Elmira, New York. Mr Thurston joined Schweizer Aircraft as Engineering Manager.

Schweizer also manufactures the Super Ag-Cat agricultural biplane for Grumman American at the rate of 265 aircraft per year.

SCHWEIZER MODEL TSC-1A2 TEAL II*

The Teal amphibian was developed to provide a simple, economical and easily-handled two-seat aircraft for cross-country and sporting flying, seaplane training and business use. The entire structure and covering are of aluminium alloy, except for the bow deck and cabin top skins which are of glassfibre.

The prototype flew for the first time in June 1968. FAA certification was awarded on 28 August 1969. Two versions have since been built, of which the TSC-1A1 was described in the 1974-75 *Jane's*.

The current production version is the TSC-1A2 Teal II, an improved model which received certification in June 1973 and is approved for day or night IFR operations in non-icing conditions. The following description applies to this version:

TYPE: Two-seat cabin monoplane amphibian.

WINGS: Cantilever shoulder-wing monoplane. Wing section NACA 4415. Dihedral 4°. Incidence 4°. All-metal D-spar structure. Single-slotted trailing-edge flaps of light alloy construction, extending outboard from hull to ailerons.

HULL: All-metal semi-monocoque structure, with glassfibre foredeck and cabin top skins.

TAIL UNIT: Cantilever all-metal T-tail. Trim tabs on elevator and rudder.

LANDING GEAR: Retractable tailwheel type; manually actuated. Spring steel main struts. Tailwheel integral with water rudder. Tailwheel retractable independently of main gear to provide water rudder control when needed. Main wheels size 6·00-6. Tailwheel size 8·00-3. Single-disc brakes on main wheels.

**All rights in the Teal were purchased by Teal Aircraft Corporation in mid-1976 (see Addenda).*

POWER PLANT: One 112 kW (150 hp) Lycoming O-320-A3B flat-four engine, driving a Hartzell two-blade constant-speed propeller. Two integral fuel tanks in wing leading-edges, each containing 87 litres (23 US gallons), of which 75·7 litres (20 US gallons) are usable. Optional all-metal fuel tank in hull, aft of main bulkhead, capacity 94·6 litres (25 US gallons). Total optional fuel capacity 268·6 litres (71 US gallons), of which 246 litres (65 US gallons) are usable. Oil capacity 7·5 litres (2 US gallons).

ACCOMMODATION: Enclosed cabin seating two persons side by side. Baggage compartment behind seats, capacity 104 kg (230 lb). Seat backs fold down for access to baggage compartment and for stand-up fishing from cabin. Door on each side. May be flown with window open. Ventilation system standard. An auxiliary side-facing third seat installation has been approved and is available as an option.

SYSTEMS: Electrical system has 12V 60A alternator and 12V 37Ah battery. Janitrol heating and defrosting system optional.

ELECTRONICS: A wide choice of navigation and communications equipment is available to customer's requirements.

EQUIPMENT: Standard equipment includes heated pitot, stall warning indicator, soundproofing, map pockets, tinted glass overhead panels, corrosion proofing, anchor, mooring line and paddle. Optional items include dual controls, strobe light, blind-flying instrumentation, 8-day clock, outside air temperature gauge and navigation, instrument, landing and cabin lights.

DIMENSIONS, EXTERNAL:
- Wing span: 9·73 m (31 ft 11 in)
- Wing chord, constant: 1·52 m (5 ft 0 in)
- Wing aspect ratio: 6·5
- Length overall: 7·19 m (23 ft 7 in)
- Height overall on land: 2·87 m (9 ft 5 in)
- Tailplane span: 2·44 m (8 ft 0 in)
- Wheel track: 2·51 m (8 ft 3 in)
- Propeller diameter: 1·83 m (6 ft 0 in)

AREAS:
- Wings, gross: 14·59 m² (157 sq ft)
- Ailerons (total): 0·93 m² (10 sq ft)
- Trailing-edge flaps (total): 1·94 m² (20·9 sq ft)
- Fin: 0·99 m² (10·7 sq ft)
- Rudder: 0·63 m² (6·8 sq ft)
- Tailplane: 1·43 m² (15·4 sq ft)
- Elevator: 1·17 m² (12·6 sq ft)

WEIGHTS AND LOADINGS:
- Weight empty: 651 kg (1,435 lb)
- Max T-O weight, land or water: 998 kg (2,200 lb)
- Max wing loading: 68·3 kg/m² (14·0 lb/sq ft)
- Max power loading: 8·91 kg/kW (14·7 lb/hp)

PERFORMANCE (at max T-O weight):
- Never-exceed speed: 117 knots (217 km/h; 135 mph) IAS
- Max level speed at 915 m (3,000 ft): 104 knots (193 km/h; 120 mph)
- Max cruising speed at 1,525 m (5,000 ft): 101 knots (187 km/h; 116 mph) IAS
- Econ cruising speed at 1,525 m (5,000 ft): 96 knots (177 km/h; 110 mph) IAS
- Stalling speed, flaps down: 45·5 knots (84 km/h; 52 mph)
- Max rate of climb at S/L: 198 m (650 ft)/min
- Service ceiling: 3,660 m (12,000 ft)
- T-O run:
  - land: 213 m (700 ft)
  - water: 290 m (950 ft)
- T-O to 15 m (50 ft):
  - land: 351 m (1,150 ft)
  - water: 427 m (1,400 ft)
- Landing from 15 m (50 ft):
  - land: 259 m (850 ft)
  - water: 198 m (650 ft)
- Landing run:
  - land: 137 m (450 ft)
  - water: 107 m (350 ft)

Schweizer Model TSC-1A2 Teal II amphibian (Lycoming O-320-A3B engine)

Range at econ cruising speed with max standard fuel and 45 min reserves 410 nm (759 km; 472 miles)
Range at econ cruising speed, with max standard plus optional fuel and 45 min reserves
650 nm (1,203 km; 748 miles)

SCHWEIZER (GRUMMAN AMERICAN) SUPER AG-CAT

The prototype of the original Ag-Cat agricultural biplane flew for the first time on 22 May 1957. Series production was entrusted to Schweizer, under subcontract from Grumman. First deliveries were made in 1959, and a total of 1,642 Ag-Cats had been built by 1 March 1976, when the type was in service in 34 countries.

The Ag-Cat was certificated in the Restricted (agricultural) category on 20 January 1959, with a 106 kW (220 hp) Continental engine, and received additional approval in this category for patrolling and surveying on 9 April 1962. Other engines for which FAA Type Approval was received are the 179 kW (240 hp) Gulf Coast W-670-240, 183 kW (245 hp) Jacobs L-4M or L-4MB, 205-224 kW (275-300 hp) Jacobs R-755, 335·5 kW (450 hp) Pratt & Whitney R-985 and 447·5 kW (600 hp) Pratt & Whitney R-1340.

A new version of the Super Ag-Cat was introduced for 1976. This is essentially similar to the Ag-Cat A with 335·5 kW (450 hp) engine, but has the wing span increased by 1·93 m (6 ft 4 in) to permit use of increased-span sprayblooms for greater swath width. Three versions are therefore available:

Super Ag-Cat A/450. Basic version, powered by a 335·5 kW (450 hp) Pratt & Whitney R-985 nine-cylinder radial aircooled engine, driving a Hamilton Standard two-blade metal constant-speed propeller with type 2D30 hub and AG-100-2 blades.

Super Ag-Cat A/600. As basic version, except for installation of 447·5 kW (600 hp) Pratt & Whitney R-1340 nine-cylinder radial aircooled engine, driving a Hamilton Standard two-blade metal constant-speed propeller with type 12D40 hub and AG-100-2 blades.

Super Ag-Cat B/450. As basic version, but with wings of increased span, longer fuselage and taller fin and rudder. 13 built by 1 March 1976.

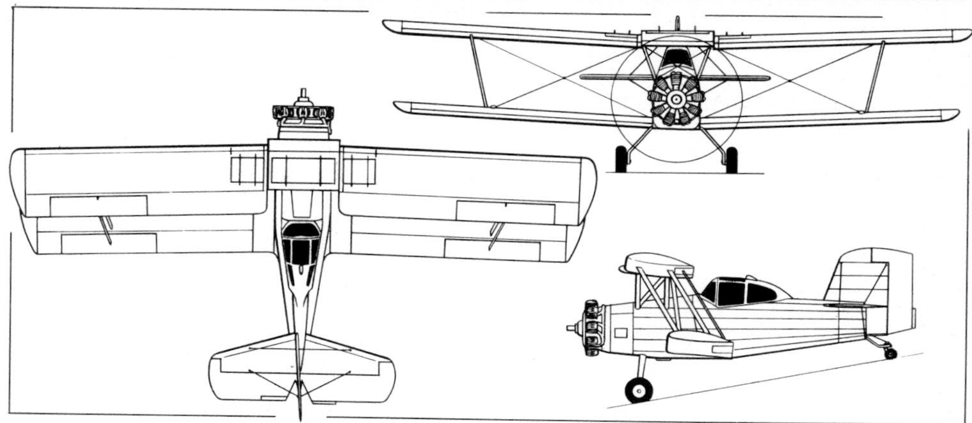

Photograph and three-view drawing *(Pilot Press)* **of the Schweizer (Grumman American) Super Ag-Cat**

TYPE: Single-seat agricultural biplane.

WINGS: Single-bay staggered biplane. NACA 4412 (modified) wing section. Dihedral 3°. Incidence 6°. Aluminium alloy (6061-T6) two-spar structure with 6061-T6 skins on entire top surface, around leading-edge and back to front spar on undersurface. Remainder of undersurface fabric-covered. Each D leading-edge is made of five separate sections to facilitate replacement if damaged. Glassfibre wingtips. N-type interplane struts. Ailerons of light alloy construction on all four wings. Ground-adjustable tab in lower port aileron. No flaps.

FUSELAGE: Welded 4130 chrome-molybdenum steel tube structure, covered with duralumin sheet. Removable side panels.

TAIL UNIT: Welded 4130 chrome-molybdenum steel tube structure, covered with fabric and wire-braced. Cable deflector wire from tip of fin to top of cockpit canopy. Controllable trim tab in port elevator. Ground-adjustable tabs on rudder and starboard elevator.

LANDING GEAR: Non-retractable tailwheel type. Cantilever spring steel legs. Cleveland wheels with tyres size 8·50-10 6-ply. Steerable tailwheel with tyre of 10 in diameter. Heavy duty disc brakes. Parking brake.

POWER PLANT: One Pratt & Whitney nine-cylinder aircooled radial engine with Hamilton Standard constant-speed propeller, as detailed in model listings. Fuel tank in upper centre-section with standard usable capacity of 174 litres (46 US gallons). Optional tanks, installed in wings on one or both sides of centre-section, are available for total usable capacities of 241 or 302 litres (64 or 80 US gallons). Oil capacity 33 litres (8·7 US gallons).

ACCOMMODATION: Single seat beneath enclosed cockpit canopy. Reinforced fairing aft of cockpit for turnover protection. Baggage compartment. Cockpit ventilation by ram air. Safety-padded instrument panel. Forward of cockpit, over CG, is a 1·13 m³ (40 cu ft) glassfibre hopper for agricultural chemicals (dry or liquid) with distributor beneath fuselage. Low-volume, ULV or high-volume spray system, with leading- or trailing-edge booms.

SYSTEMS: Spray or dust distribution systems available. Hydraulic system for brakes only. Emergency dump system for hopper load; can be used also for water-bomber operations. Optional electrical system with 12 or 24V alternator, external power socket, navigation lights and/or strobe lights and electric engine starter.

EQUIPMENT: Standard equipment includes refuelling steps and assist handles, tie-down rings, control column lock, instrument glareshield, seat belt and shoulder harness, stall warning light, tinted windshield and urethane paint in high-visibility yellow.

DIMENSIONS, EXTERNAL (A: Ag-Cat A/450; B: Ag-Cat A/600; C: Ag-Cat B/450):
Wing span:
A, B 10·95 m (35 ft 11 in)
C 12·88 m (42 ft 3 in)
Wing chord (constant):
A, B, C 1·47 m (4 ft 10 in)
Wing aspect ratio, upper wing:
A, B 7·81
Biplane, estimated mean:
A, B 5·29
Length overall:
A, B 7·42 m (24 ft 4 in)
C 7·90 m (25 ft 11 in)
Height overall:
A, B 3·30 m (10 ft 10 in)
C 3·35 m (11 ft 0 in)
Tailplane span:
A, B, C 3·96 m (13 ft 0 in)
Wheel track:
A, B, C 2·44 m (8 ft 0 in)
Wheelbase:
A, B 5·64 m (18 ft 6 in)

AREAS:
Wings, gross:
A, B 30·47 m² (328 sq ft)
C 36·42 m² (392 sq ft)
Ailerons (total):
A, B 2·93 m² (31·5 sq ft)
Fin:
A, B 0·84 m² (9·0 sq ft)
Rudder:
A, B 1·12 m² (12·0 sq ft)
Tailplane:
A, B, C 2·12 m² (22·8 sq ft)
Elevators:
A, B, C 2·06 m² (22·2 sq ft)

WEIGHTS AND LOADINGS:
Weight empty, equipped, spray version:
A 1,301 kg (2,870 lb)
B 1,426 kg (3,145 lb)
C 1,372 kg (3,025 lb)
Max certificated T-O weight:
A, B, C 2,041 kg (4,500 lb)
Max T-O weight (CAM.8):
A, B, C 2,755 kg (6,075 lb)
Max wing loading at certificated max T-O weight:
A, B 66·89 kg/m² (13·7 lb/sq ft)
C 56·15 kg/m² (11·5 lb/sq ft)
Power loading at certificated max T-O weight:
A, C 8·21 kg/kW (10·0 lb/hp)
B 6·16 kg/kW (7·5 lb/hp)

PERFORMANCE (at weights indicated):
Normal working speed:
A, C 91 knots (169 km/h; 105 mph) IAS
B 95·5 knots (177 km/h; 110 mph) IAS
Stalling speed, power off, at 2,041 kg (4,500 lb) AUW:
A 58·5 knots (108 km/h; 67 mph) IAS
B 59 knots (110 km/h; 68 mph) IAS
C 52·5 knots (97 km/h; 60 mph) IAS
Max rate of climb at S/L, at 2,041 kg (4,500 lb) AUW:
A 302 m (990 ft)/min
B 488 m (1,600 ft)/min
C 323 m (1,060 ft)/min
Max rate of climb at S/L, at 2,755 kg (6,075 lb) AUW:
A 140 m (460 ft)/min
B 244 m (800 ft)/min
C 186 m (610 ft)/min
T-O run, paved surface, at 2,041 kg (4,500 lb) AUW:
A 194 m (635 ft)
B 154 m (505 ft)
C 142 m (465 ft)
T-O run, paved surface, at 2,755 kg (6,075 lb) AUW:
A 396 m (1,300 ft)
B 335 m (1,100 ft)
C 308 m (1,010 ft)
T-O run, unprepared strip, at 2,041 kg (4,500 lb) AUW:
A 349 m (1,145 ft)
B 262 m (860 ft)
C 242 m (795 ft)
T-O run, unprepared strip, at 2,755 kg (6,075 lb) AUW:
A 736 m (2,415 ft)
B 634 m (2,080 ft)
C 555 m (1,820 ft)

SERVO-AIRE

SERVO-AIRE ENGINEERING

ADDRESS: Salinas, California

Servo-Aire is one of a number of companies involved in programmes to re-engine standard types of agricultural aircraft which are now powered by Pratt & Whitney radial engines. The R-1340 radial, in particular, has been out of production since the mid-fifties, and spare parts for this engine are becoming increasingly difficult to obtain.

THRUSH COMMANDER / PT6

On behalf of Fred Ayres of Albany, Georgia, Servo-

Aire has converted a Rockwell Thrush Commander to turbine power by installation of a 559 kW (750 shp) Pratt & Whitney Aircraft of Canada PT6A-34 turboprop engine, driving a three-blade metal constant-speed and reversible-pitch propeller. To compensate for the small size and light weight of the turboprop, it is mounted well forward of the firewall, in a slender cowling. Advantages claimed for the conversion include greatly improved take-off and climb performance; improved short landing capability; a 454 kg (1,000 lb) increase in payload due to reduced power plant weight; ability to operate on aviation turbine fuel, avgas or diesel fuel; a TBO of more than 3,000 hours; quieter operation; and the ability to stop the propeller without shutting down the engine, because of the free-turbine configuration.

First flight of the re-engined prototype was made on 9 September 1975. Two conversion kits have been purchased by a Swedish agricultural aviation operator. Address of Fred Ayres is PO Box 3090, Albany, Georgia 31706.

AG-CAT / LEONIDES and THRUSH COMMANDER / LEONIDES

Two further conversions for which Servo-Aire was responsible involved the installation of 417 kW (560 hp) Alvis Leonides radial engines in a Grumman Ag-Cat and a Rockwell Thrush Commander. In each case, the engine drives a Dowty Rotol R289 three-blade propeller.

Despite the slightly lower power of the Leonides, by comparison with the R-1340 that was fitted originally to each aircraft, its reduced drag and increased propeller efficiency has improved overall performance.

The Leonides engine is itself out of production, but Scottish Aviation of Prestwick, Ayrshire, Scotland (which see), has acquired and is overhauling a number of engines of this type, removed from retired airframes. Scottish Aviation also has an option to acquire manufacturing rights and tooling for the Leonides should the engine's sales potential justify putting it back into production.

Rockwell Thrush Commander re-engined with a PT6A-34 turboprop by Servo-Aire Engineering

SIKORSKY

SIKORSKY AIRCRAFT, DIVISION OF UNITED TECHNOLOGIES CORPORATION

HEAD OFFICE AND WORKS:
Stratford, Connecticut 06602
Telephone: (203) 378-6361

OTHER WORKS:
South Avenue, Bridgeport, Connecticut; and Sikorsky Memorial Airport, Stratford, Connecticut

PRESIDENT:
Gerald J. Tobias

SENIOR VICE-PRESIDENTS:
John R. Graham (Finance and Administration)
Robert J. Torok (Military Programmes)

VICE-PRESIDENTS:
Richard C. Abington (Planning)
Robert F. Daniell (Commercial)
Harry T. Jensen (Technology)
Derek J. Jonson (International Marketing)
Phil Locke (Contracts and Counsel)
William H. Parry (Operations)
William F. Paul (Engineering)
Allen K. Poole (Product Support—Government Programmes)

DIVISION CONTROLLER: William Flaherty
FACTORY MANAGER: James W. Dunn
DIVISIONAL AUDITOR: H. W. Engstrom
PUBLIC RELATIONS MANAGER: Frank J. Delear

On 5 March 1973 Sikorsky Aircraft celebrated the 50th anniversary of its foundation, and a sculpture of its founder, the late Igor I. Sikorsky, who died on 26 October 1972, was unveiled at the company's main works at Stratford.

Mr Sikorsky incorporated the organisation which is now the Sikorsky Aircraft division of United Technologies Corporation on 5 March 1923 in New York State, to build an all-metal aircraft designated S-29-A. Previously, he had followed a successful aviation career in Russia, leaving for the USA after the Soviet Revolution. In the late 1920s and 1930s he designed and built amphibians and flying-boats that pioneered intercontinental and transoceanic commercial flight. He designed and flew his VS-300 helicopter for the first time in 1939: when fully developed in 1942 it represented the world's first fully practical single-rotor helicopter, leading to establishment of the helicopter industry.

Sikorsky's main plant at Stratford, which has 120,775 m² (1,300,000 sq ft) of working space, produces the S-61 twin-turbine amphibious transport helicopter and its military counterparts, the rear-loading S-61R, the S-64 Skycrane and the S-65 multi-purpose helicopter. The company's original 55,740 m² (600,000 sq ft) plant at Bridgeport is utilised for detail fabrication, and for overhaul and repair.

Production of the S-58 and S-62 ended some years ago, but a twin-turbine re-engined S-58, designated S-58T, is in production.

Under development is the S-70 (YUH-60A) utility transport helicopter, intended to meet the US Army's UTTAS requirement.

Sikorsky's new ABC (advancing blade concept) rotor system has operated successfully in NASA's wind tunnel at Ames Research Center. Tested at simulated speeds of up to 304 knots (563 km/h; 350 mph) the system, which consists of two contra-rotating three-blade rotors, demonstrated substantial improvements in speed and lift capabilities in comparison with conventional rotor systems.

On 7 February 1972 Sikorsky announced that it was designing and building a research helicopter designated S-69 to flight test the ABC rotor system in a programme funded under a $9·9 million contract. The contract was awarded to Sikorsky by the Eustis Directorate, US Army Air Mobility Research and Development Laboratory, Fort Eustis, Virginia.

Sikorsky announced in October 1973 that the company had been selected by NASA to design and build two high-speed multi-purpose research helicopters. Known as S-72 Rotor Systems Research Aircraft (RSRA), these aircraft will be used in a joint NASA/Army research programme aimed at developing and testing a variety of existing and future helicopter rotor systems.

Sikorsky licensees include Westland of Great Britain, Agusta of Italy, Aérospatiale of France, Mitsubishi of Japan, and Pratt & Whitney Aircraft of Canada Ltd.

SIKORSKY S-58T

In January 1970 Sikorsky announced plans to produce and market kits for conversion of the piston-engined S-58 into a twin-turbine helicopter. The turbine version, designated S-58T, provides increased safety and reliability, greater speed and lifting power, and improved performance at high altitude on hot days. It also has lower operating costs than the piston-engined models.

Design of the S-58T began in January 1970 and construction of the prototype started in May, with the first flight following on 19 August 1970. Construction of production aircraft and delivery of conversion kits began in January 1971 and FAA certification was awarded in April. Certification for IFR operation was received in June 1973.

Sikorsky will deliver FAA-certificated retrofit kits to S-58 operators for their own installation, or will install the kits in customers' aircraft at the Sikorsky plant. The company has also obtained used S-58s and offers them with the turbine engines installed.

Two versions are available:

S-58T. Original conversion with Pratt & Whitney Aircraft of Canada PT6T-3 Twin Pac turbine power plant. Available since mid-1974 with PT6T-6 Twin Pac, with max rating of 1,398 kW (1,875 shp) at sea level and max continuous rating of 1,249 kW (1,675 shp).

S-58T Mark II. Improved version with airframe improvements and PT6T-6 engine. Improvements include strengthened tail gearbox housing; addition of 8 windows in cabin; increased cabin space; pilot and co-pilot bubble windows; outside step for pilot and co-pilot above landing gear V support; addition of access door to No. 1 engine fuel control; addition of pylon access steps; 16 airline-type passenger seats, sidewalls and improved cabin lighting; improved fuel boost pumps and fuel cell covers; removal of main gearbox oil cooler blower; deletion of tail gearbox fairing and relocation of rotating beacon; EAPS option and air-conditioning option. Any of the foregoing configuration improvements can be retrofitted into existing or newly-converted S-58T aircraft.

TYPE: General-purpose helicopter.

ROTOR SYSTEM: Four-blade all-metal main and tail rotors, both with servo control. Fully-articulated main rotor blades, each made up of a hollow extruded aluminium spar and trailing-edge pockets of aluminium. Main rotor blade abrasion strips optional. Each tail rotor blade has an aluminium spar, sheet aluminium skin and honeycomb trailing-edge. Blades of each rotor interchangeable. Main rotor blades fold. Main and tail rotor brakes.

ROTOR DRIVE: Direct gear drive from forward angle-change gearbox. Steel tube drive-shafts with rubber couplings. Main gearbox below main rotor, intermediate gearbox at base of tail pylon and tail gearbox behind tail rotor. Main rotor/engine rpm ratio 1 : 0·0376. Tail rotor/engine rpm ratio 1 : 0·2257.

FUSELAGE: Semi-monocoque structure, primarily of magnesium and aluminium alloys, with some titanium and stainless steel.

TAIL SURFACE: Ground-adjustable stabiliser made of magnesium skin over magnesium and aluminium structure.

LANDING GEAR: Non-retractable three-wheel undercarriage, with tailwheel towards rear of fuselage. Sikorsky oleo-pneumatic shock-absorber struts. Tailwheel is fully-castoring and self-centering, with an anti-swivelling lock. Goodyear main wheels with tyres size 32·2 × 11-12, pressure 3·10 bars (45 lb/sq in). Tailwheel tyre size 18·3 × 6-6, pressure 3·10 bars (45 lb/sq in). Toe-operated Goodyear disc brakes. Provision for amphibious gear, pontoons, 'doughnut' or pop-out flotation bags.

POWER PLANT: Initially, Pratt & Whitney Aircraft of Canada PT6T-3 Twin Pac turboshaft engine developing 1,342 kW (1,800 shp) for take-off and having a maximum continuous rating of 1,193 kW (1,600 shp).

Sikorsky S-58T (PT6T-6 Twin Pac turboshaft engine) converted for British Airways Helicopters

PT6T-6 Twin Pac from June 1974, developing 1,398 kW (1,875 shp) for take-off and with a maximum continuous rating of 1,249 kW (1,675 shp). Fuel contained in twelve cells, with total internal capacity of 1,109 litres (293 US gallons). Provision for 568 litre (150 US gallon) external metal tank. Oil capacity 17 litres (4·5 US gallons).

ACCOMMODATION: Pilot's compartment above main cabin seats two side by side with dual controls. Cabin seats 10-16 passengers, with door on starboard side. Entire accommodation heated and ventilated.

SYSTEMS: Heating and ventilation systems. Primary and secondary hydraulic systems. Electrical power supplied by two 200A starter/generators.

ELECTRONICS AND EQUIPMENT: Navigation and communications equipment available to customer's requirements. Litters, rescue hoist, cargo sling, cargo handling equipment and air-conditioning optional.

DIMENSIONS, EXTERNAL:

Diameter of main rotor	17·07 m (56 ft 0 in)
Main rotor blade chord	0·42 m (1 ft 4⅜ in)
Diameter of tail rotor	2·90 m (9 ft 6 in)
Tail rotor blade chord	0·18 m (7¼ in)
Distance between rotor centres	10·08 m (33 ft 1 in)
Length overall	20·06 m (65 ft 10 in)
Length of fuselage	14·40 m (47 ft 3 in)
Width of fuselage	1·73 m (5 ft 8 in)
Height to top of rotor hub	4·36 m (14 ft 3½ in)
Height overall	4·85 m (15 ft 11 in)
Wheel track	3·66 m (12 ft 0 in)
Wheelbase	8·75 m (28 ft 3 in)
Cabin door:	
Height	1·22 m (4 ft 0 in)
Width	1·32 m (4 ft 4 in)

DIMENSIONS, INTERNAL:

Cabin:	
Length: S-58T	3·91 m (12 ft 10 in)
S-58T Mk II	4·67 m (15 ft 4 in)
Max width	1·52 m (5 ft 0 in)
Max height	1·75 m (5 ft 9 in)

AREAS:

Main rotor blades (each)	3·25 m² (35·00 sq ft)
Tail rotor blades (each)	0·26 m² (2·75 sq ft)
Main rotor disc	228·54 m² (2,460 sq ft)
Tail rotor disc	6·59 m² (70·9 sq ft)

WEIGHTS AND LOADINGS:

Weight empty:	
S-58T	3,437 kg (7,577 lb)
S-58T Mk II	3,789 kg (8,354 lb)
Max T-O and landing weight	5,896 kg (13,000 lb)
Max disc loading	25·8 kg/m² (5·29 lb/sq ft)
Max power loading	4·72 kg/kW (7·76 lb/shp)

PERFORMANCE (at max T-O weight. A: PT6T-3; B: PT6T-6):

Max level speed at S/L:	
A, B	120 knots (222 km/h; 138 mph)
Cruising speed:	
A, B	110 knots (204 km/h; 127 mph)
Hovering ceiling out of ground effect:	
A	1,433 m (4,700 ft)
B	1,980 m (6,500 ft)
Single-engine absolute ceiling:	
A	640 m (2,100 ft)
B	1,280 m (4,200 ft)
Min ground turning radius:	
A, B	12·50 m (41 ft 0 in)
Runway LCN:	
A, B	approx 3·7
Range with 1,071 litres (283 US gallons) usable fuel, including 20 min reserves at cruising speed:	
A	260 nm (481 km; 299 miles)
B	242 nm (447 km; 278 miles)

SIKORSKY S-61A, S-61B and S-61F

US military designations: RH-3 and SH-3 Sea King, HH-3A, VH-3A, CH-3B

CAF designation: CH-124

The first version of the S-61 ordered into production was the SH-3A (formerly HSS-2) Sea King amphibious anti-submarine helicopter. The original US Navy contract for this aircraft was received on 23 September 1957, the prototype flew for the first time on 11 March 1959 and first deliveries to the Fleet were made in September 1961.

Sikorsky's S-61 series of twin-turbine helicopters now includes the following military and commercial variants:

SH-3A (formerly HSS-2) Sea King. Initial anti-submarine version for the US Navy, powered by 932 kW (1,250 shp) General Electric T58-GE-8B turboshaft engines. A total of 255 were produced by Sikorsky. Also standard equipment in the Japan Maritime Self-Defence Force (details under entry for Mitsubishi, which also converted two SH-3As to S-61A standard for use during Antarctic expeditions).

CH-124. Designation of 41 aircraft, similar to SH-3A, ordered for the Canadian Armed Forces. First of these was delivered in May 1963: fifth and subsequent aircraft were assembled by United Aircraft of Canada Ltd. Originally designated CHSS-2.

S-61A. Amphibious transport, generally similar to the US Navy's SH-3A. Accommodates 26 troops, 15 litters, cargo, or 12 passengers in VIP configuration. Rolls-Royce Gnome H.1200 turboshafts available as alternative to standard General Electric T58 engines. Nine delivered to Royal Danish Air Force for long-range air-sea rescue duties, with additional fuel tankage. The Decca Navigator Co in the UK received a £100,000 contract in 1973 to supply Mk 19 Decca Navigator airborne receivers, Danac computers and pictorial displays for the Royal Danish Air Force S-61A fleet. One S-61A delivered to Construction Helicopters.

S-61A-4 Nuri. Six aircraft for the Royal Malaysian Air Force, each with 31 seats, rescue hoists and auxiliary fuel tanks as standard equipment. Delivery began in August 1971, to supplement 10 delivered in 1967-68 and used for troop transport, cargo carrying and rescue.

HH-3A. A further modified version of the SH-3A, tested by the US Navy as a search and rescue helicopter, with armament, armour and a high-speed rescue hoist. HH-3A conversion kits were subsequently supplied to the Navy's overhaul and repair base at Quonset Point, Rhode Island, where 12 conversions were carried out. The first modified aircraft was delivered to the US Navy's HC-7 squadron at Subic Point in the Philippines. Changes in HH-3As modified at Quonset Point include the installation of two electrically-powered Minigun turrets behind the sponsons, T58-GE-8F turbine engines, a high-speed refuelling and fuel dumping system, a high-speed rescue hoist, modified electronics package, external auxiliary fuel tanks and complete armour installation. The SH-3A's sonar well is covered and a reinforced cabin floor substituted.

RH-3A. Nine SH-3As converted for mine countermeasures duty with the US Navy. Each has two cargo doors, one on each side, instead of one; bubble windows on each side aft of the cargo doors; rearview mirrors for pilot and co-pilot; and a pivoting tow-tube and hook assembly attached to the fuselage above the tailwheel. The RH-3A is designed to carry, stream, tow and retrieve a variety of mine countermeasures (MCM) gear. Deliveries were made in 1965, on the basis of three for testing and three each to USN helicopter squadrons on the east and west coasts of the USA, four of these aircraft being assigned for service on board two MCM ships.

VH-3A (formerly HSS-2Z). Ten specially-equipped aircraft used by the Executive Flight Detachment which provides a VIP transport and emergency evacuation service for the US President and other key personnel. Five operated by the US Army and five by the US Marine Corps.

CH-3B. Six 'borrowed' HSS-2s (S-61As) operated by the USAF for missile site support and drone recovery duties.

SH-3D Sea King. Standard anti-submarine helicopter of the US Navy, powered by 1,044 kW (1,400 shp) General Electric T58-GE-10 turboshaft engines, and with 530 litres (140 US gallons) more fuel than SH-3A. First SH-3D, delivered in June 1966, was one of 10 for the Spanish Navy, which later ordered 12 more. Four were delivered to the Brazilian Navy and 72 to the US Navy. Versions with Rolls-Royce Gnome turboshaft engines and British anti-submarine equipment are manufactured by Westland Helicopters Ltd (which see). SH-3Ds are also manufactured under licence by Agusta in Italy.

S-61D-4. Four aircraft for the Argentine Navy, similar to the SH-3D.

VH-3D. Eleven ordered to replace VH-3As of Executive Flight Detachment, becoming operational in the primary VIP mission on 19 January 1976. A VH-3D was used for transport by President Ford for the first time on 31 January 1976.

SH-3G. US Navy conversion of 105 SH-3As into utility helicopters, by removing anti-submarine warfare equipment. Three SH-3Gs, assigned to HC-1's Detachment 1, entered service on board the USS *Ranger* in the West Pacific; more than 20 others delivered to HC-2, at NAS Lakehurst, New Jersey. Six of these were equipped with Minigun pods for use on search and rescue missions in combat conditions.

SH-3H. Multi-purpose version of the SH-3G. Initial 1971 contract from the US Navy called for conversion of 11 aircraft, to increase fleet helicopter capability against submarines and low-flying enemy missiles.

New anti-submarine warfare equipment includes lightweight sonar, active and passive sonobuoys, magnetic anomaly detection equipment and radar. Electronic surveillance measurement (ESM) equipment and LN 66HP radar enable the SH-3H to make an important contribution to the missile defence of the fleet. General Electric T58-GE-10 engines are fitted.

Initially, existing SH-3Gs were flown to Stratford from the Navy's facility at North Island, California, for conversion to SH-3H configuration, but the programme entails also the conversion of SH-3As and SH-3Ds to equip 10 active squadrons and four Reserve squadrons with eight SH-3Hs each. A total of 35 aircraft had been converted to SH-3H configuration by 1 March 1976. An FY 1975 contract called for 10 more conversions for delivery during 1976. The ESM equipment and LN 66HP radar are deleted from FY 1975 and subsequent aircraft. An attitude heading reference system and centre fuel cell have been added, and FY 1977 aircraft are to include a TAC NAV system.

S-61L. Non-amphibious civil transport with longer fuselage than S-61A/B. Described separately.

S-61N. Amphibious counterpart of S-61L, with which it is described.

Sikorsky SH-3H multi-purpose helicopter for ASW and expansion of fleet missile defence

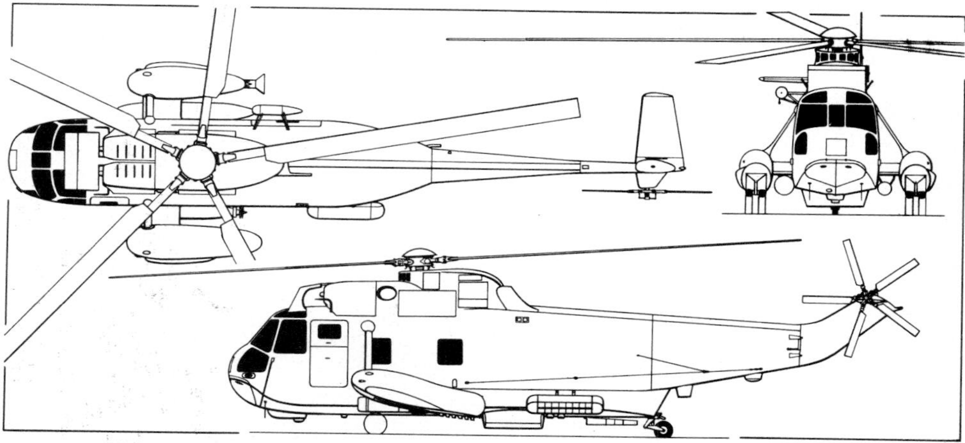

Sikorsky SH-3H twin-engined multi-purpose amphibious helicopter *(Pilot Press)*

S-61R. Development of S-61B for transport duties with USAF, under the designations **CH-3C** and **E**. Rear loading ramp, new landing gear and other changes. Described separately.

The following details apply to the SH-3D Sea King, but are generally applicable to the other versions except for accommodation and equipment:

TYPE: Twin-engined amphibious all-weather anti-submarine helicopter.

ROTOR SYSTEM: Five-blade main and tail rotors. All-metal fully-articulated oil-lubricated main rotor. Flanged cuffs on blades bolted to matching flanges on all-steel rotor head. Main rotor blades are interchangeable and are provided with an automatic powered folding system. Rotor brake standard. All-metal tail rotor.

ROTOR DRIVE: Both engines drive through freewheel units and rotor brake to main gearbox. Steel drive-shafts. Tail rotor shaft-driven through intermediate and tail gearboxes. Accessories driven by power take-off on tail rotor shaft. Additional freewheel units between accessories and port engine, and between accessories and tail rotor shaft. Main rotor/engine rpm ratio 1 : 93·43. Tail rotor/engine rpm ratio 1 : 16·7.

FUSELAGE: Boat hull of all-metal semi-monocoque construction. Single step. Tail section folds to reduce stowage requirements.

TAIL SURFACE: Fixed stabiliser on starboard side of tail section.

LANDING GEAR: Amphibious. Land gear consists of two twin-wheel main units, which are retracted rearward hydraulically into stabilising floats, and non-retractable tailwheel. Oleo-pneumatic shock-absorbers. Goodyear main wheels and tubeless tyres size 6·50-10 type III, pressure 4·83 bars (70 lb/sq in). Goodyear tailwheel and tyre size 6·00-6. Goodyear hydraulic disc brakes. Boat hull and pop-out flotation bags in stabilising floats permit emergency operation from water.

POWER PLANT: Two 1,044 kW (1,400 shp) General Electric T58-GE-10 turboshaft engines. Three bladder-type fuel tanks in hull; forward tank capacity 1,314 litres (347 US gallons), centre tank capacity 530 litres (140 US gallons), rear tank capacity 1,336 litres (353 US gallons). Total fuel capacity 3,180 litres (840 US gallons). Refuelling point on port side of fuselage. Oil capacity 26·5 litres (7 US gallons).

ACCOMMODATION: Pilot and co-pilot on flight deck, two sonar operators in main cabin. Dual controls. Crew door at rear of flight deck on port side. Large loading door at rear of cabin on starboard side.

SYSTEMS: Primary and auxiliary hydraulic systems, pressure 103·5 bars (1,500 lb/sq in), for flying controls. Utility hydraulic system, pressure 207 bars (3,000 lb/sq in), for landing gear, winches and blade folding. Pneumatic system, pressure 207 bars (3,000 lb/sq in), for blow-down emergency landing gear extension. Electrical system includes one 300A DC generator, two 20kVA 115A AC generators and 24V 22Ah battery. APU optional.

ELECTRONICS AND EQUIPMENT: Bendix AQS-13 sonar with 180° search beam width. Hamilton Standard auto-stabilisation equipment. Automatic transition into hover. Sonar coupler holds altitude automatically in conjunction with Teledyne APN-130 Doppler radar and radar altimeter. Provision for 272 kg (600 lb) capacity rescue hoist and 3,630 kg (8,000 lb) capacity automatic touchdown-release low-response cargo sling for external loads.

ARMAMENT: Provision for 381 kg (840 lb) of weapons, including homing torpedoes.

DIMENSIONS, EXTERNAL:

Diameter of main rotor	18·90 m (62 ft 0 in)
Main rotor blade chord	0·46 m (1 ft 6¼ in)
Diameter of tail rotor	3·23 m (10 ft 7 in)
Distance between rotor centres	11·10 m (36 ft 5 in)
Length overall	22·15 m (72 ft 8 in)
Length of fuselage	16·69 m (54 ft 9 in)
Length, tail pylon folded	14·40 m (47 ft 3 in)
Width, rotors folded	4·98 m (16 ft 4 in)
Height to top of rotor hub	4·72 m (15 ft 6 in)
Height overall	5·13 m (16 ft 10 in)
Wheel track	3·96 m (13 ft 0 in)
Wheelbase	7·18 m (23 ft 6½ in)
Crew door (fwd, port):	
Height	1·68 m (5 ft 6 in)
Width	0·91 m (3 ft 0 in)
Height to sill	1·14 m (3 ft 9 in)
Main cabin door (stbd):	
Height	1·52 m (5 ft 0 in)
Width	1·73 m (5 ft 8 in)
Height to sill	1·14 m (3 ft 9 in)

DIMENSIONS, INTERNAL (S-61A):

Cabin: Length	7·60 m (24 ft 11 in)
Max width	1·98 m (6 ft 6 in)
Max height	1·92 m (6 ft 3½ in)
Floor area	15·1 m² (162 sq ft)
Volume	28·9 m³ (1,020 cu ft)

AREAS:

Main rotor blades (each)	4·14 m² (44·54 sq ft)
Tail rotor blades (each)	0·22 m² (2·38 sq ft)
Main rotor disc	280·5 m² (3,019 sq ft)
Tail rotor disc	8·20 m² (88·30 sq ft)
Stabiliser	1·86 m² (20·00 sq ft)

WEIGHTS:

Weight empty:	
S-61A	4,428 kg (9,763 lb)
S-61B	5,382 kg (11,865 lb)
Normal T-O weight:	
S-61A	9,300 kg (20,500 lb)
SH-3A (ASW)	8,185 kg (18,044 lb)
SH-3D (ASW)	8,449 kg (18,626 lb)
Max T-O weight:	
S-61A	9,750 kg (21,500 lb)
S-61B	9,300 kg (20,500 lb)
SH-3H	9,525 kg (21,000 lb)

PERFORMANCE (at 9,300 kg; 20,500 lb AUW):

Max level speed	144 knots (267 km/h; 166 mph)
Cruising speed for max range	118 knots (219 km/h; 136 mph)
Max rate of climb at S/L	670 m (2,200 ft)/min
Service ceiling	4,480 m (14,700 ft)
Hovering ceiling in ground effect	3,200 m (10,500 ft)
Hovering ceiling out of ground effect	2,500 m (8,200 ft)
Range with max fuel, 10% reserves	542 nm (1,005 km; 625 miles)

Sikorsky VH-3D of the US Executive Flight Detachment which provides VIP transport *(Howard Levy)*

SIKORSKY S-61L and S-61N

Although basically similar to the S-61A and B, the S-61L and N commercial transports incorporate a number of changes, including a longer fuselage. Other details are as follows:

S-61L. Non-amphibious configuration. Modified landing gear, rotor head and stabiliser. Accommodation for up to 30 passengers. First flight of the prototype S-61L was made on 6 December 1960, and it received FAA Type Approval on 2 November 1961.

S-61N. Similar to S-61L, but with sealed hull for amphibious operation and stabilising floats as on SH-3. Accommodation for 26-28 passengers. First flight of the first S-61N was made on 7 August 1962.

Payloader. Stripped-down version of S-61N, weighing nearly 907 kg (2,000 lb) less than standard version but capable of lifting a payload of more than 4,990 kg (11,000 lb). Intended for logging, general construction, powerline installation and similar operations. Sponsons replaced by fixed main-wheel landing gear; sealed-off rear airstair door.

On 6 October 1964 the S-61L and S-61N became the first transport helicopters to receive FAA approval for instrument flight operations.

Both models are now offered in **Mark II** versions, with General Electric CT58-140-1 or -2 turboshaft engines (earlier aircraft have 1,007 kW; 1,350 shp engines), enabling them to carry 22 passengers on an 86°F (30°C) day, compared with the former 10. There are six individual cargo bins to speed baggage handling. Other changes include improved vibration damping.

A total of 88 aircraft had been delivered by 1 January 1976. Bristow Helicopters, with a fleet of 19 S-61s, is the world's largest private operator of the type.

TYPE: Twin-turbine all-weather helicopter airliners.

ROTOR SYSTEM AND ROTOR DRIVE: As for SH-3A/D, S-61A and S-61B, except blades do not fold.

FUSELAGE: All-metal semi-monocoque structure of boat-hull form.

TAIL SURFACE: Stabiliser on starboard side of tail section.

LANDING GEAR (S-61L): Non-amphibious non-retractable tailwheel type with twin wheels on main units. Oleo-pneumatic shock-absorbers. Goodyear main wheels and tubeless tyres, size 22·1 × 6·50-10 Type III, pressure 6·55 bars (95 lb/sq in). Goodyear tailwheel with tyre size 18·3 × 6·00-10, pressure 5·17 bars (75 lb/sq in). Goodyear hydraulic disc brakes.

LANDING GEAR (S-61N): Amphibious hydraulically-retractable type. Twin wheels on main units, which retract rearward into stabilising floats. Non-retractable tailwheel. Each float provides 1,506 kg (3,320 lb) buoyancy and, with the sealed hull, permits operation from water. Shock-absorbers, wheels, tyres and brakes as for S-61L.

POWER PLANT: Two 1,118 kW (1,500 shp) General Electric CT58-140-1, -2 turboshaft engines. Two bladder-type fuel tanks in hull; forward tank capacity 796 litres (210 US gallons), rear tank capacity 757 litres (200 US gallons). Total fuel capacity 1,553 litres (410 US gallons). Additional 924 litre (244 US gallon) tank optionally available for S-61N. Refuelling point on port side of fuselage. Oil capacity 26·5 litres (7 US gallons).

ACCOMMODATION: Crew of three: pilot, co-pilot and flight attendant. Main cabin accommodates up to 30 passengers (22 at 86°F; 30°C). Standard arrangement has eight single seats and one double seat on port side of cabin, seven double seats on starboard side and one double seat at rear. Rear seat may be replaced by a toilet. Galley may be installed in forward baggage compartment area on starboard side. Forward half of cabin may be provided with folding seats and tie-down rings for convertible passenger/freight operations. Two doors on starboard side of cabin: main cabin door of airstair type. Baggage space above and below floor at front, on starboard side of cabin, and below floor in area of airstair door (aft, starboard side).

SYSTEMS: As for SH-3D, S-61A and S-61B.

ELECTRONICS AND EQUIPMENT: Radio and radar to customer's specification. Blind-flying instrumentation standard.

DIMENSIONS, EXTERNAL:

Diameter of main rotor	18·90 m (62 ft 0 in)
Diameter of tail rotor	3·23 m (10 ft 7 in)
Distance between rotor centres:	
S-61L	11·10 m (36 ft 5 in)
S-61N	11·17 m (36 ft 8 in)
Length overall (rotors fore and aft):	
S-61L	22·21 m (72 ft 10½ in)
S-61N	22·20 m (72 ft 10 in)
Width, over landing gear:	
S-61L	4·47 m (14 ft 8 in)
S-61N	6·02 m (19 ft 9 in)
Height to top of rotor hub pitot head:	
S-61L	5·18 m (17 ft 0 in)
S-61N	5·32 m (17 ft 5½ in)
Height overall:	
S-61L	5·18 m (17 ft 0 in)
S-61N	5·63 m (18 ft 5½ in)
Wheel track:	
S-61L	3·96 m (13 ft 0 in)
S-61N	4·27 m (14 ft 0 in)
Wheelbase:	
S-61L	7·15 m (23 ft 5½ in)
S-61N	7·30 m (23 ft 11½ in)
Cabin door (airstair):	
Height	1·68 m (5 ft 6 in)
Width	0·81 m (2 ft 8 in)
Height to sill	1·14 m (3 ft 9 in)
Cargo door:	
Height	1·68 m (5 ft 6 in)
Width	1·27 m (4 ft 2 in)
Height to sill	1·14 m (3 ft 9 in)

DIMENSIONS, INTERNAL:

Cabin: Length	9·73 m (31 ft 11 in)
Max width	1·98 m (6 ft 6 in)
Max height	1·92 m (6 ft 3½ in)
Floor area	approx 20·16 m² (217 sq ft)
Volume	approx 36·95 m³ (1,305 cu ft)
Freight hold (above floor)	approx 3·54 m³ (125 cu ft)
Freight hold (underfloor)	approx 0·71 m³ (25 cu ft)

AREAS:

Main rotor blades (each)	3·75 m² (40·4 sq ft)
Tail rotor blades (each)	0·22 m² (2·38 sq ft)
Main rotor disc	280·5 m² (3,019 sq ft)
Tail rotor disc	8·20 m² (88·3 sq ft)

Sikorsky S-61N of British Airways Helicopters, used to support North Sea oil rig operations *(Dr Alan Beaumont)*

Stabiliser 2·51 m² (27·0 sq ft)

WEIGHTS AND LOADINGS:

Weight empty:

- S-61L 5,308 kg (11,704 lb)
- S-61N 5,674 kg (12,510 lb)

Max T-O weight (S-61N):

- FAA 8,620 kg (19,000 lb)
- CAA 9,300 kg (20,500 lb)
- FAA with external load 9,980 kg (22,000 lb)

Max disc loading 35·59 kg/m² (7·29 lb/sq ft)

Max power loading 4·46 kg/kW (7·33 lb/shp)

PERFORMANCE (at max T-O weight):

Max level speed at S/L 127 knots (235 km/h; 146 mph)

Average cruising speed 120 knots (222 km/h; 138 mph)

Max rate of climb at S/L 395 m (1,300 ft)/min

Service ceiling 3,810 m (12,500 ft)

Hovering ceiling in ground effect:

- S-61L 2,743 m (9,000 ft)
- S-61N 2,652 m (8,700 ft)

Hovering ceiling out of ground effect:

- S-61L 1,189 m (3,900 ft)
- S-61N 1,158 m (3,800 ft)

Min ground turning radius:

- S-61L 13·70 m (44 ft 11½ in)
- S-61N 13·93 m (45 ft 8½ in)

Runway LCN at max T-O weight approx 4·4

Range with max fuel, 30 min reserves:

- S-61L 230 nm (426 km; 265 miles)
- S-61N 430 nm (796 km; 495 miles)

SIKORSKY S-61R

US military designations: CH-3 and HH-3 Jolly Green Giant

Although based on the SH-3A, this amphibious transport helicopter introduced many important design changes. They include provision of a hydraulically-operated rear ramp for straight-in loading of wheeled vehicles, a 907 kg (2,000 lb) capacity winch for internal cargo handling, retractable tricycle-type landing gear, pressurised rotor blades for quick and easy inspection, gas-turbine auxiliary power supply for independent field operations, self-lubricating main and tail rotors, and built-in equipment for the removal and replacement of all major components in remote areas.

The first S-61R flew on 17 June 1963, followed by the first CH-3C a few weeks later. FAA Type Approval was received on 30 December 1963, and the first delivery of an operational CH-3C was made on the same day, for drone recovery duties at Tyndall AFB, Florida. Subsequent deliveries made to USAF Aerospace Defense Command, Air Training Command, Tactical Air Command, Strategic Air Command and Aerospace Rescue and Recovery Service.

There have been four versions, as follows:

CH-3C. Two 969·5 kW (1,300 shp) T58-GE-1 turboshaft engines. After a total of 41 had been built for the USAF, production was switched to the CH-3E. All aircraft delivered as CH-3Cs were modified to CH-3E standard.

CH-3E. Designation applicable since February 1966, following introduction of uprated engines (1,118 kW; 1,500 shp T58-GE-5s). A total of 42 were built as new aircraft to this standard. Of the 83 new and uprated aircraft, 50 were adapted as HH-3Es (which see).

HH-3E. For USAF Aerospace Rescue and Recovery Service. Additional equipment comprises armour, self-sealing fuel tanks, retractable flight refuelling probe, defensive armament and rescue hoist. Two 1,118 kW (1,500 shp) T58-GE-5 turboshafts. A total of 50 HH-3Es were converted from CH-3Es, and are known as **Jolly Green Giants**.

On 31 May-1 June 1967, two HH-3Es made the first non-stop transatlantic flights by helicopters, en route to the Paris Air Show. Nine aerial refuellings were made by each aircraft. The 3,708 nm (6,870 km; 4,270 miles) from New York to Paris were flown in 30 hr 46 min.

HH-3F. Similar to HH-3E, for US Coast Guard, which has given them the name **Pelican**. Advanced electronic equipment for search and rescue duties. No armour plate, armament or self-sealing tanks. First order announced in August 1965. Deliveries began in 1968 and a total of 40 were built.

The following details apply to the CH-3E:

TYPE: Twin-engined amphibious transport helicopter.

ROTOR SYSTEM: Five-blade fully-articulated main rotor of all-metal construction. Flanged cuffs on blades bolted to matching flanges on rotor head. Control by rotating and stationary swashplates. Blades do not fold. Rotor brake standard. Conventional tail rotor with five aluminium blades.

ROTOR DRIVE: Twin turbines drive through freewheeling units and rotor brake to main gearbox. Steel drive-shafts. Tail rotor shaft-driven through intermediate gearbox and tail gearbox. Main rotor/engine rpm ratio 1 : 93·43. Tail rotor/engine rpm ratio 1 : 16·7.

FUSELAGE: All-metal semi-monocoque structure of pod and boom type. Cabin of basic square section.

TAIL SURFACE: Horizontal stabiliser on starboard side of tail rotor pylon.

LANDING GEAR: Hydraulically-retractable tricycle type, with twin wheels on each unit. Main wheels retract forward into sponsons, each of which provides 2,176 kg (4,797 lb) of buoyancy and, with boat hull, permits amphibious operation. Oleo-pneumatic shock-absorbers. All wheels and tyres tubeless Type III rib, size 22·1 × 6·50-10, manufactured by Goodyear. Tyre pressure 6·55 bars (95 lb/sq in). Goodyear hydraulic disc brakes.

POWER PLANT: Two 1,118 kW (1,500 shp) General Electric T58-GE-5 turboshaft engines, mounted side by side above cabin, immediately forward of main transmission. Fuel in two bladder-type tanks beneath cabin floor; forward tank capacity 1,204 litres (318 US gallons), rear tank capacity 1,226 litres (324 US gallons). Total fuel capacity 2,430 litres (642 US gallons). Refuelling point on port side of fuselage. Total oil capacity 26·5 litres (7 US gallons).

ACCOMMODATION: Crew of two side by side on flight deck, with dual controls. Provision for flight engineer or attendant. Normal accommodation for 25 fully-equipped troops. Alternative arrangements for 30 troops, 15 stretchers or 2,270 kg (5,000 lb) of cargo. Jettisonable sliding door on starboard side at front of cabin. Internal door between cabin and flight deck. Hydraulically-operated rear loading ramp for vehicles, in two hinged sections, giving opening with minimum width of 1·73 m (5 ft 8 in) and headroom of up to 2·21 m (7 ft 3 in).

SYSTEMS: Primary and auxiliary hydraulic systems, pressure 103·5 bars (1,500 lb/sq in), for flying control servos. Utility hydraulic system, pressure 207 bars (3,000 lb/sq in), for landing gear, rear ramp and winches. Pneumatic system, pressure 207 bars (3,000 lb/sq in), for emergency blow-down landing gear extension. Electrical system includes 24V 22Ah battery, two 20kVA 115V AC generators and one 300A DC generator. APU standard.

DIMENSIONS, EXTERNAL:

- Diameter of main rotor 18·90 m (62 ft 0 in)
- Main rotor blade chord 0·46 m (1 ft 6¼ in)
- Diameter of tail rotor 3·15 m (10 ft 4 in)
- Distance between rotor centres 11·22 m (36 ft 10 in)
- Length overall 22·25 m (73 ft 0 in)
- Length of fuselage 17·45 m (57 ft 3 in)
- Width, over landing gear 4·82 m (15 ft 10 in)
- Height to top of rotor hub 4·90 m (16 ft 1 in)
- Height overall 5·51 m (18 ft 1 in)
- Wheel track 4·06 m (13 ft 4 in)
- Wheelbase 5·21 m (17 ft 1 in)
- Cabin door (fwd, stbd):
  - Height 1·65 m (5 ft 4¾ in)
  - Width 1·22 m (4 ft 0 in)
  - Height to sill 1·27 m (4 ft 2 in)
- Rear ramp:
  - Length 4·29 m (14 ft 1 in)
  - Width 1·85 m (6 ft 1 in)

DIMENSIONS, INTERNAL:

- Cabin (excl flight deck):
  - Length 7·89 m (25 ft 10½ in)
  - Max width 1·98 m (6 ft 6 in)
  - Max height 1·91 m (6 ft 3 in)

Sikorsky HH-3F Pelican search and rescue helicopter of the US Coast Guard

Floor area approx 15·61 m² (168 sq ft)
Volume approx 29·73 m³ (1,050 cu ft)

AREAS:
Main rotor blades (each) 3·71 m² (39·9 sq ft)
Tail rotor blades (each) 0·22 m² (2·35 sq ft)
Main rotor disc 280·5 m² (3,019 sq ft)
Tail rotor disc 7·80 m² (83·9 sq ft)
Stabiliser 2·51 m² (27·0 sq ft)

WEIGHTS:
Weight empty 6,010 kg (13,255 lb)
Normal T-O weight 9,635 kg (21,247 lb)
Max T-O weight 10,000 kg (22,050 lb)

PERFORMANCE (at normal T-O weight):
Max level speed at S/L 141 knots (261 km/h; 162 mph)
Cruising speed for max range 125 knots (232 km/h; 144 mph)
Max rate of climb at S/L 400 m (1,310 ft)/min
Service ceiling 3,385 m (11,100 ft)
Hovering ceiling in ground effect 1,250 m (4,100 ft)
Min ground turning radius 11·29 m (37 ft 0½ in)
Runway LCN at max T-O weight approx 4·75
Range with max fuel, 10% reserves 404 nm (748 km; 465 miles)

SIKORSKY S-64 SKYCRANE

US military designation: CH-54 Tarhe

The S-64 flying crane was designed initially for military transport duties. Equipped with interchangeable pods, it is suitable for use as a troop transport, and for minesweeping, cargo and missile transport, anti-submarine or field hospital operations. Equipment includes a removable 9,072 kg (20,000 lb) hoist, a sling attachment and a load stabiliser to prevent undue sway in cargo winch operations. Attachment points are provided on the fuselage and landing gear to facilitate securing of bulky loads.

Versions of the S-64 are as follows:

S-64A. Under this designation the first of three prototypes flew for the first time on 9 May 1962 and was used by the US Army at Fort Benning, Georgia, for testing and demonstration. The second and third prototypes were evaluated by the German armed forces.

CH-54A. Six ordered by US Army in 1963 to investigate the heavy lift concept, with emphasis on increasing mobility in the battlefield. Delivery of five CH-54As (originally YCH-54As) to the US Army took place in late 1964 and early 1965. A sixth CH-54A remained at Stratford, with a company-owned S-64, for a programme leading toward a restricted FAA certification, which was awarded on 30 July 1965. Further US Army orders followed.

The CH-54As were assigned to the US Army's 478th Aviation Company, and performed outstanding service in support of the Army's First Cavalry Division, Airmobile, in Vietnam. On 29 April 1965, a CH-54A of this unit lifted 90 persons, including 87 combat-equipped troops in a detachable van. This is believed to be the largest number of people ever carried by a helicopter at one time. Other Skycranes in Vietnam transported bulldozers and road graders weighing up to 7,937 kg (17,500 lb), 9,072 kg (20,000 lb) armoured vehicles and a large variety of heavy hardware. They retrieved more than 380 damaged aircraft, involving savings estimated at $210 million.

Sikorsky Aircraft developed an all-purpose van, known as the Universal Military Pod, for carriage by the US Army's CH-54As, and received an order, worth $2·9 million, to supply 22 to the Army. The pods were delivered complete with communications, ventilation and lighting systems, and with wheels to simplify ground handling. They superseded earlier pods which were not approved for the carriage of personnel. The first pod was accepted by the US Army on 28 June 1968, following approval for personnel transport.

Internal dimensions of the pod are length 8·36 m (27 ft 5 in), width 2·69 m (8 ft 10 in) and height 1·98 m (6 ft 6 in). Doors are provided on each side of the forward area of the pod, and a double-panelled ramp is located aft. With a max loaded weight of 9,072 kg (20,000 lb), each pod accommodates 45 combat-equipped troops, or 24 litters, and in the field may be adapted for a variety of uses, such as surgical unit, field command post and communications post.

On 18 April 1969, two commercial Skycranes were delivered to Rowan Drilling Company Inc of Houston, Texas, for operation in support of oil exploration and drilling operations in Alaska.

S-64E. FAA certification of the improved S-64E for civil use was announced in 1969, for the transportation of external cargo weighing up to 9,072 kg (20,000 lb).

In January 1972 Erickson Air-Crane Company of Marysville, California, purchased the first S-64E, for logging and other heavy-lift tasks. On 1 November 1972 this company ordered three additional S-64Es, the first of which was delivered in the following month, to extend its operations on a worldwide basis, offering heavy-lift capability to the logging, petroleum, power line, shipping and general construction industries. Three S-64Es are operated by Evergreen Helicopters Inc of McMinnville, Oregon. One was delivered to Tri-Eagle Company in 1975, for timber harvesting in California. Production in 1975 included two more aircraft for Evergreen and one for Tri-Eagle.

CH-54B. On 4 November 1968 Sikorsky announced that it had received a US Army contract to increase the payload capacity of the CH-54 from 10 to 12½ short tons. The contract called for a number of design improvements to the engine, gearbox, rotor head and structure; altitude performance and hot weather operating capability were also to be improved. Two of the improved flying cranes, designated CH-54B, were accepted by the US Army during 1969.

The original JFTD12-4A engines were replaced by two Pratt & Whitney JFTD12-5As, each rated at 3,579 kW (4,800 shp), and a gearbox capable of receiving 5,891 kW (7,900 hp) from the two engines was introduced. Single-engine performance was increased, since the new gearbox receives 3,579 kW (4,800 hp) from one engine, compared with 3,020 kW (4,050 hp) on the CH-54A.

A new rotor system was also introduced, utilising a high-lift rotor blade with a chord some 0·064 m (2·5 in) greater than that of the blades used formerly.

Other changes included the provision of twin wheels on the main landing gear, an improved automatic flight control system and some general structural strengthening throughout the aircraft. Gross weight was increased from 19,050 kg (42,000 lb) to 21,318 kg (47,000 lb).

In October 1970, two US Army CH-54Bs lifted an 18,488 kg (40,760 lb) load during a series of tests being conducted to evaluate the technical feasibility and cost of a twin-lift system for potential application to military requirements for greater helicopter external load capacity. Later in the same month, a single US Army CH-54B lifted an 18,497 kg (40,780 lb) load during tests being conducted to evaluate maximum hover lift capability.

Nine international helicopter records in Class E1 are held by the CH-54B. On 26 October 1971, piloted by B. P. Blackwell, a payload of 1,000 kg was lifted to a height of 9,499 m (31,165 ft). On 29 October CWO E. E. Price flew to 9,595 m (31,480 ft) with 2,000 kg. On 27 October the same pilot had reached 7,778 m (25,518 ft) with a 5,000 kg payload. CWO J. K. Church flew to 5,246 m (17,211 ft) with 10,000 kg on 29 October, and on 12 April 1972 CWO D. L. Spivey reached 3,307 m (10,850 ft) with a 15,000 kg payload. CWO Church set an earlier record on 4 November 1971, by maintaining a height of 11,010 m (36,122 ft) in horizontal flight. Major J. C. Henderson set up two time-to-height climb records in a CH-54B on 12 April 1972, reaching 3,000 m in 1 min 22·2 sec and 6,000 m in 2 min 58·9 sec. Earlier, on 4 November 1971, CWO D. W. Hunt had climbed to 9,000 m in 5 min 57·7 sec.

S-64F. Designation of a commercial version of the military CH-54B. The improvements distinguishing this upgraded commercial helicopter were introduced into production CH-54s in late 1969, and FAA certification tests were completed in late 1970. Not yet in production for commercial use.

TYPE: Twin-turbine heavy flying crane helicopter.

ROTOR SYSTEM: Six-blade fully-articulated main rotor with aluminium blades and aluminium and steel head. Four-blade tail rotor with titanium head and aluminium blades. Rotor brake standard.

ROTOR DRIVE: Steel tube drive-shafts. Main gearbox below main rotor, intermediate gearbox at base of tail pylon, tail gearbox at top of pylon. Main gearbox rated at 4,922 kW (6,600 shp) on CH-54A and S-64E, 5,891 kW (7,900 shp) on S-64F.

FUSELAGE: Pod and boom type of aluminium and steel semi-monocoque construction.

LANDING GEAR: Non-retractable tricycle type, with single wheel on each unit of CH-54A/S-64E, twin wheels on main units of S-64F. CH-54A/S-64E main-wheel tyres size 38·45 × 12·50-16, pressure 6·55 bars (95 lb/sq in). S-64F main-wheel tyres size 25·65 × 8·50-10, pressure 6·90 bars (100 lb/sq in). Nosewheels and tyres of all versions size 25·65 × 8·50-10, pressure 6·90 bars (100 lb/sq in).

POWER PLANT (CH-54A/S-64E): Two Pratt & Whitney JFTD12-4A (military T73-P-1) turboshaft engines, each rated at 3,356 kW (4,500 shp) for take-off and with max continous rating of 2,983 kW (4,000 shp). Two fuel tanks in fuselage, forward and aft of transmission, each with capacity of 1,664 litres (440 US gallons). Total standard fuel capacity 3,328 litres (880 US gallons). Provision for auxiliary fuel tank of 1,664 litres (440 US gallons) capacity, raising total fuel capacity to 4,992 litres (1,320 US gallons).

POWER PLANT (CH-54B/S-64F): Two Pratt & Whitney JFTD12-5A turboshaft engines, each rated at 3,579 kW (4,800 shp) for take-off and with max continuous rating of 3,303·5 kW (4,430 shp). Fuel tanks as for CH-54A/S-64E.

ACCOMMODATION: Pilot and co-pilot side by side at front of cabin. Aft-facing seat for third pilot at rear of cabin, with flying controls. The occupant of this third seat is able to take over control of the aircraft during loading and unloading. Two additional jump seats available in cabin. Payload in interchangeable pods.

DIMENSIONS, EXTERNAL:
Diameter of main rotor 21·95 m (72 ft 0 in)
Diameter of tail rotor 4·88 m (16 ft 0 in)
Distance between rotor centres 13·56 m (44 ft 6 in)
Length overall 26·97 m (88 ft 6 in)
Length of fuselage 21·41 m (70 ft 3 in)

Sikorsky S-64 Skycrane heavy-lift helicopter, operated in Alaska by Rowan Air Cranes *(Norman E. Taylor)*

Width, rotors folded 6·65 m (21 ft 10 in)
Height to top of rotor hub 5·67 m (18 ft 7 in)
Height overall 7·75 m (25 ft 5 in)
Ground clearance under fuselage boom 2·84 m (9 ft 4 in)
Wheel track 6·02 m (19 ft 9 in)
Wheelbase 7·44 m (24 ft 5 in)

AREAS:
Main rotor disc 378·1 m² (4,070 sq ft)
Tail rotor disc 18·67 m² (201 sq ft)

WEIGHTS (CH-54A/S-64E):
Weight empty 8,724 kg (19,234 lb)
Max T-O weight 19,050 kg (42,000 lb)

PERFORMANCE (CH-54A/S-64E at normal T-O weight of 17,237 kg; 38,000 lb):
Max level speed at S/L 109 knots (203 km/h; 126 mph)
Max cruising speed 91 knots (169 km/h; 105 mph)
Max rate of climb at S/L 405 m (1,330 ft)/min
Service ceiling 2,475 m (9,000 ft)
Hovering ceiling in ground effect 3,230 m (10,600 ft)
Hovering ceiling out of ground effect 2,100 m (6,900 ft)
Min ground turning radius:
CH-54A, S-64E, S-64F 16·4 m (54 ft 0 in)
Runway LCN:
CH-54A, S-64E at max T-O weight of 19,050 kg (42,000 lb) 7·1
S-64F at max T-O weight of 21,318 kg (47,000 lb) 7·7
Range with max fuel, 10% reserves 200 nm (370 km; 230 miles)

SIKORSKY S-65A

US Navy designation: CH-53A Sea Stallion
USAF designations: HH-53B/C
US Marine Corps designations: CH-53A/D

On 27 August 1962, it was announced that Sikorsky had been selected by the US Navy to produce a heavy assault transport helicopter for use by the Marine Corps. First flight was made on 14 October 1964, and deliveries began in mid-1966. Versions are as follows:

CH-53A. This initial version uses many components based on those of the S-64A Skycrane, but is powered by two General Electric T64 turboshaft engines and has a watertight hull. A full-size rear opening, with built-in ramp, permits easy loading and unloading, with the aid of a special hydraulically-operated internal cargo loading system and floor rollers.

Typical cargo loads include two Jeeps, or two Hawk missiles with cable reels and control console, or a 105 mm howitzer and carriage. An external cargo system permits in-flight pickup and release without ground assistance.

The CH-53A is able to operate under all weather and climatic conditions. Its main rotor blades and tail pylon fold hydraulically for stowage on board ship.

On 17 February 1968, a CH-53A, with General Electric T64-6 (modified) engines, flew at a gross weight of 23,541 kg (51,900 lb) carrying 12,927 kg (28,500 lb) of payload and fuel, establishing new unofficial payload and gross weight records for a production helicopter built outside the Soviet Union.

On 26 April 1968, a Marine Corps CH-53A made the first automatic terrain clearance flight in helicopter history and subsequently concluded flight tests of an Integrated Helicopter Avionics System (IHAS). Prime contractor for the IHAS programme was Teledyne Systems Company. Norden Division of United Aircraft Corporation provided the terrain-clearance radar and vertical structure display.

On 23 October 1968, a Marine Corps CH-53A performed a series of loops and rolls, as part of a joint Naval Air Systems Command and Sikorsky flight test programme, aimed at investigating the CH-53A's rotor system dynamics and manoeuvrability characteristics. Details of these trials have appeared in previous editions of *Jane's*.

Six S-65As were ordered in 1975 by the Imperial Iranian Navy.

RH-53A. Fifteen CH-53As borrowed by US Navy for mine countermeasures duties, with T64-GE-413 engines.

HH-53B. Eight ordered by USAF in September 1966 for Aerospace Rescue and Recovery Service. The first of these flew on 15 March 1967 and deliveries began in June 1967. Production completed.

The HH-53B is generally similar to the CH-53A, but is powered by 2,297 kW (3,080 shp) T64-GE-3 turboshaft engines. It has the same general equipment as the HH-3E, including a retractable flight refuelling probe, jettisonable auxiliary fuel tanks, rescue hoist, all-weather electronics and armament.

HH-53C. Improved version of the HH-53B, with 2,927 kW (3,925 shp) T64-GE-7 engines, auxiliary jettisonable fuel tanks each of 1,703 litres (450 US gallons) capacity on new cantilever mounts, flight refuelling probe, and rescue hoist with 76 m (250 ft) of cable. External cargo hook of 9,070 kg (20,000 lb) capacity. First HH-53C was delivered to the USAF on 30 August 1968. A total of 72 HH-53B/Cs were built. Production completed.

HH-53 Pave Low III. The prototype Pave Low III rescue helicopter is a Sikorsky HH-53 which was modified by the Aeronautical Systems Division of the USAF's Air Force Systems Command to permit operation at night and in adverse weather over all kinds of terrain. It made its first flight from Wright Field to Patterson Field, Ohio, in July 1975. After preliminary testing at Patterson Field through that month, it was fitted with the remainder of its search and navigational equipment, comprising infra-red scanner and radar systems linked by an airborne computer, before resuming its tests in fully operational form. Designed entirely within ASD, the Pave Low III helicopter represents the most extensive project undertaken by the Division in recent years. If testing in jungle areas of Panama, in the second half of 1976, is successful, up to 30 helicopters may be modified to similar standard.

CH-53D. Improved CH-53A for US Marine Corps, the first of which was delivered on 3 March 1969. Two T64-GE-413 engines, each with a maximum rating of 2,927 kW (3,925 shp). A total of 55 troops can be carried in a high-density arrangement. An integral cargo handling system makes it possible for one man to load or unload one short ton of palletised cargo a minute. Main rotor and tail pylon fold automatically for carrier stowage.

Last CH-53D (the 265th CH-53 built) was delivered on 31 January 1972. All but the first 34 CH-53s were provided with hardpoints for supporting towing equipment and transferring tow loads to the airframe, so that the US Marines could utilise the aircraft as airborne minesweepers, giving an assault commander the capability of clearing enemy mines from harbours and off beaches without having to wait for surface minesweepers. Tow kits installed in the 15 CH-53Ds operated by US Navy Squadron HM-12 included automatic flight control system interconnections to provide automatic cable yaw angle retention and aircraft attitude and heading hold; rearview mirrors for pilot and co-pilot; tow cable tension and yaw angle indicator; automatic emergency cable release; towboom and hook system with 6,803 kg (15,000 lb) load capacity when cable was locked to internal towboom; dam to prevent cabin flooding in emergency water landing with lower ramp open; dual hydraulically-powered cable winches; racks and cradles for stowage of minesweeping gear; auxiliary fuel tanks in cabin to increase endurance.

RH-53D. Specially-equipped minesweeping version for the US Navy, described separately.

YCH-53E. Three-engined development of the CH-53D. Described separately.

VH-53F. Proposed VIP transport version for Presidential Flight, with T64-GE-414 engines. Not built.

CH-53G. Version of the CH-53 for the German armed forces, with T64-GE-7 engines. A total of 112 were produced, the first of two built by Sikorsky being delivered on 31 March 1969. The next 20 were assembled in Germany from American-built components. The remainder embody some 50% components of German manufacture. Prime contractor in Germany was VFW-Fokker, whose first CH-53G flew for the first time on 11 October 1971. Deliveries to the German Army were completed during 1975.

S-65-Oe. Two ordered in 1969 by Austrian Air Force and delivered in 1970. Used for rescue duties in the Alps, they have the same rescue hoist as the HH-53B/C, fittings for auxiliary fuel tanks and accommodation for 38 passengers.

S-65C. Commercial inter-city helicopter proposal based on the military CH-53. Announced in January 1975, it would differ from the CH-53 by having a lengthened nose section, larger sponsons carrying more than 3,175 kg (7,000 lb) of fuel, airliner-type cabin windows, an airstair or other passenger door, and a cargo or baggage loading door. Power plant, rotors and transmission system would be generally similar to those of the CH-53. For normal passenger services a total of 44 people and a crew of two would be accommodated; for servicing offshore oil or gas platforms, the S-65C would carry 30 passengers and their baggage, or 2,722 kg (6,000 lb) of cargo. Cruising at 160 knots (296 km/h; 184 mph), the respective ranges for these two versions would be 200 nm (482 km; 300 miles) and 620 nm (1,150 km; 714 miles).

As a preliminary to the S-65C programme, Sikorsky converted an existing CH-53, under contract to NASA, into a prototype/demonstration aircraft with a 16-seat airline interior. In the Spring of 1976 this aircraft was undergoing an evaluation programme conducted by NASA at Langley Field, Virginia.

The following details refer to the CH-53A:

TYPE: Twin-turbine heavy assault transport helicopter.

ROTOR SYSTEM AND DRIVE: Generally similar to those of S-64A Skycrane, but main rotor head is of titanium and steel, and has folding blades.

FUSELAGE: Conventional semi-monocoque structure of aluminium, steel and titanium. Folding tail pylon.

TAIL SURFACE: Large horizontal stabiliser on starboard side of tail rotor pylon.

LANDING GEAR: Retractable tricycle type, with twin wheels on each unit. Main units retract into the rear of sponsons on each side of fuselage. Fully-castoring nose unit. Main wheels and nosewheels have tyres size 25·65 × 8·50-10, pressure 6·55 bars (95 lb/sq in).

POWER PLANT: Normally two 2,125 kW (2,850 shp) General Electric T64-GE-6 turboshaft engines, mounted in pod on each side of main rotor pylon. The CH-53A can also utilise, without airframe modification, the T64-GE-1 engine of 2,297 kW (3,080 shp) or the later T64-GE-16 (mod) engine of 2,561·5 kW (3,435 shp). Two self-sealing bladder fuel tanks, each with capacity of 1,192 litres (315 US gallons), housed in forward part of sponsons. Total fuel capacity 2,384 litres (630 US gallons).

ACCOMMODATION: Crew of three. Main cabin accommodates 37 combat-equipped troops on inward-facing seats. Provision for carrying 24 stretchers and four attendants. Roller-skid track combination in floor for handling heavy freight. Door on starboard side of cabin at front. Rear loading ramp.

DIMENSIONS, EXTERNAL:
Diameter of main rotor 22·02 m (72 ft 3 in)
Diameter of tail rotor 4·88 m (16 ft 0 in)
Length overall, rotors turning 26·90 m (88 ft 3 in)
Length of fuselage, excl refuelling probe 20·47 m (67 ft 2 in)
Width overall, rotors folded 4·72 m (15 ft 6 in)
Width of fuselage 2·69 m (8 ft 10 in)

Sikorsky CH-53D operated in a heavy transport role by the Israeli Defence Force *(Stephen P. Peltz)*

Height to top of rotor hub	5·22 m (17 ft 1½ in)
Height overall	7·60 m (24 ft 11 in)
Wheel track	3·96 m (13 ft 0 in)
Wheelbase	8·23 m (27 ft 0 in)
DIMENSIONS, INTERNAL:	
Cabin: Length	9·14 m (30 ft 0 in)
Max width	2·29 m (7 ft 6 in)
Max height	1·98 m (6 ft 6 in)
AREAS:	
Main rotor disc	378·1 m² (4,070 sq ft)
Tail rotor disc	18·67 m² (201 sq ft)
WEIGHTS:	
Weight empty:	
CH-53A	10,180 kg (22,444 lb)
HH-53B	10,490 kg (23,125 lb)
HH-53C	10,690 kg (23,569 lb)
CH-53D	10,653 kg (23,485 lb)
Normal T-O weight:	
CH-53A	15,875 kg (35,000 lb)
Mission T-O weight:	
HH-53B	16,964 kg (37,400 lb)
HH-53C	17,344 kg (38,238 lb)
CH-53D	16,510 kg (36,400 lb)
Max T-O weight:	
HH-53B/C, CH-53D	19,050 kg (42,000 lb)
PERFORMANCE:	
Max level speed at S/L:	
HH-53B	162 knots (299 km/h; 186 mph)
HH-53C, CH-53D	170 knots (315 km/h; 196 mph)
Cruising speed:	
HH-53B/C, CH-53D	150 knots (278 km/h; 173 mph)
Max rate of climb at S/L:	
HH-53B	440 m (1,440 ft)/min
HH-53C	631 m (2,070 ft)/min
CH-53D	664 m (2,180 ft)/min
Service ceiling:	
HH-53B	5,610 m (18,400 ft)
HH-53C	6,220 m (20,400 ft)
CH-53D	6,400 m (21,000 ft)
Hovering ceiling in ground effect:	
HH-53B	2,470 m (8,100 ft)
HH-53C	3,565 m (11,700 ft)
CH-53D	4,080 m (13,400 ft)
Hovering ceiling out of ground effect:	
HH-53B	490 m (1,600 ft)
HH-53C	1,310 m (4,300 ft)
CH-53D	1,980 m (6,500 ft)
Min ground turning radius	13·46 m (44 ft 2 in)
Runway LCN at max T-O weight	7·1

Range:

HH-53B/C, with 4,502 kg (9,926 lb) fuel (two 1,703 litre; 450 US gallon auxiliary tanks), including 10% reserves and 2 min warm-up
468 nm (869 km; 540 miles)

CH-53D, with 1,849 kg (4,076 lb) fuel, 10% reserves at cruising speed and 2 min warm-up
223 nm (413 km; 257 miles)

SIKORSKY CH-53E

The Sikorsky S-65A was chosen in 1973 for development with a three-engined power plant to provide the US Navy and Marine Corps with a heavy-duty multi-purpose helicopter. Other changes to increase performance included installation of a new seven-blade main rotor of increased diameter, with blades of titanium construction, and an uprated transmission of 8,628 kW (11,570 shp) capacity to cater for future development.

Development was initiated by the award of a $1·7 million US Navy cost-plus-fixed-fee contract; in May 1973 Sikorsky announced that construction of two prototypes was to go ahead, with the objective of a first flight in April 1974. Bettering this by a month, the first of these two helicopters, with the designation YCH-53E, made a successful half-hour flight on 1 March 1974, during which low-altitude hovering and limited manoeuvres were carried out. It was lost subsequently in an accident on the ground, but the programme was resumed on 24 January 1975 with the second YCH-53E. This aircraft has flown at an AUW of 31,751 kg (70,000 lb), the highest gross weight achieved by any helicopter outside the USSR. It has been used for preliminary evaluation and testing under Phase I of the development programme. Phase II, which was contingent upon successful completion of Phase I, called for the construction of a static test vehicle and two pre-production prototypes, embodying changes or modifications evolving from Phase I. Construction of these pre-production prototypes began in late 1974; the first of them flew on 8 December 1975 and both have been involved in flight testing since early 1976. The static test example was also completed by the end of 1975. A production decision was expected to follow the completion of Phase II, in late 1976.

The US Navy plans to use the CH-53E for vertical on-board delivery operations, to support mobile construction battalions, and for the removal of battle-damaged aircraft from carrier decks. In amphibious operations, it would be able to airlift 93 per cent of a US Marine division's combat items, and would be able to retrieve 98 per cent of the Marine Corps' tactical aircraft without disassembly. Features of the helicopter include extended-range fuel tanks, flight refuelling capability, an on-board all-weather navigation system, and an advanced automatic flight control system.

CH-53 converted by Sikorsky, under NASA contract, as a 16-seat demonstration prototype of the projected S-65C commercial inter-city transport

A significant milestone in the development programme of the CH-53E began in May 1976, when the structural demonstration tests began at the US Naval Air Test Center, Patuxent River, Maryland. The tests involved one of the pre-production prototypes being flown at its design limits to verify performance. Completion of the programme, later in the Summer, was expected to signal the release of production funds to procure long-lead items, preparation of production drawings and fabrication of production tooling.

Largest and most powerful helicopter in the west, the CH-53E is also the largest helicopter capable of full operation from the Navy's existing and planned ships, requiring only 10 per cent more deck space than the twin-turbine H-53. It offers double the lift of the latter aircraft with an increase of only 50 per cent in engine power.

It is anticipated that CH-53Es will begin to join the US fleet in late 1978.

TYPE: Triple-turbine heavy-duty multi-purpose helicopter.

ROTOR SYSTEM AND TRANSMISSION: Seven-blade main rotor with blades of titanium construction. Titanium and steel main rotor head. Four-blade tail rotor mounted on pylon canted 20° to port. Rotor transmission rated at 10,067 kW (13,500 shp) for ten seconds, 8,628 kW (11,570 shp) for 30 minutes.

FUSELAGE: Conventional semi-monocoque structure of light alloy, steel and titanium.

TAIL SURFACE: Initial fixed tailplane on undersurface of fuselage, superseded successively by single high-mounted stabiliser on starboard side and lightweight gull-wing type.

LANDING GEAR: Retractable tricycle type, with twin

Sikorsky YCH-53E heavy-lift helicopter prototype (three General Electric T64-GE-415 turboshaft engines)

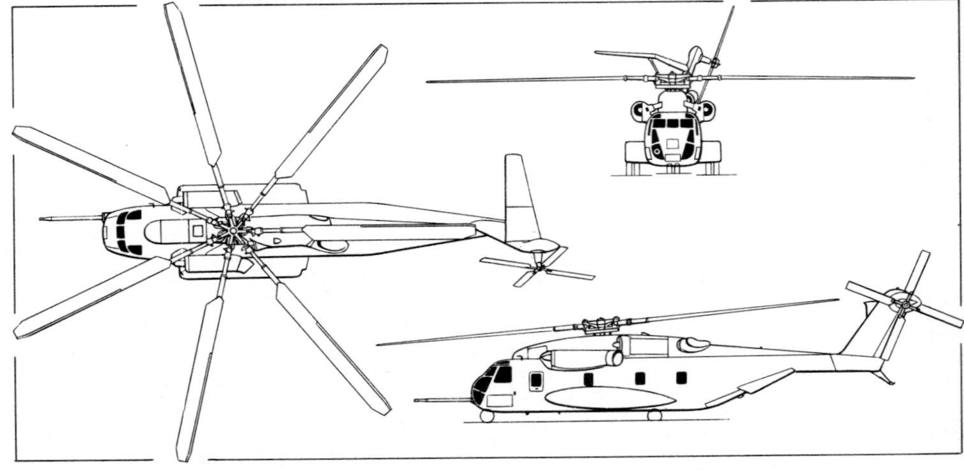

Sikorsky CH-53E heavy-duty multi-purpose helicopter *(Pilot Press)*

Sikorsky YCH-53E prototype heavy-lift helicopter, under development for the US Navy and Marine Corps

wheels on each unit. Main units retract into rear of sponsons on each side of fuselage.

POWER PLANT: Three General Electric T64-GE-415 turboshaft engines, each with a max rating of 3,266 kW (4,380 shp) for 10 min, intermediate rating of 3,065 kW (4,110 shp) for 30 min and max continuous power rating of 2,737 kW (3,670 shp).

ACCOMMODATION: Crew of three. Main cabin will accommodate up to 55 troops in a high-density seating arrangement.

DIMENSIONS, EXTERNAL:

Main rotor diameter	24·08 m (79 ft 0 in)
Tail rotor diameter	6·10 m (20 ft 0 in)
Length overall	30·20 m (99 ft 1 in)
Length of fuselage	22·48 m (73 ft 9 in)
Width of fuselage	2·69 m (8 ft 10 in)
Height overall	8·46 m (27 ft 9 in)
Wheel track	3·96 m (13 ft 0 in)
Wheelbase	8·31 m (27 ft 3 in)

WEIGHTS:

Weight empty	14,536 kg (32,048 lb)
Internal payload (100 nm; 185 km; 115 miles radius)	13,607 kg (30,000 lb)
External payload (50 nm; 92·5 km; 57·5 miles radius)	14,605 kg (32,200 lb)
Max T-O weight	31,638 kg (69,750 lb)

PERFORMANCE (ISA, at T-O weight of 25,400 kg; 56,000 lb):

Max level speed at S/L	170 knots (315 km/h; 196 mph)
Cruising speed at S/L	150 knots (278 km/h; 173 mph)
Max rate of climb at S/L	725 m (2,380 ft)/min
Hovering ceiling in ground effect, at max power	3,265 m (10,720 ft)
Hovering ceiling out of ground effect, at max power	2,280 m (7,480 ft)
Service ceiling, at max continuous power	3,780 m (12,400 ft)
Range, at optimum cruise condition for best range, 20 min reserves	266 nm (492 km; 306 miles)

SIKORSKY S-65 (MCM)

US Navy designation: RH-53D

On 27 October 1970 the US Navy announced plans to establish helicopter mine countermeasures (MCM) squadrons. The first unit, Helicopter Mine Countermeasures Squadron 12 (HM-12), borrowed 15 CH-53As from the US Marine Corps, pending production of specially equipped helicopters. Details of the tow kits installed in these aircraft are given under the CH-53D entry.

Congress gave approval subsequently for the development of a new and more powerful version of the CH-53 for service with the Navy's mine countermeasures squadrons, and in February 1972 Sikorsky announced that the US Navy had awarded the company an advanced procurement authorisation for 30 helicopters under the designation RH-53D. Production began in October 1972, at a rate of two per month, under a programme extending to December 1973. The first RH-53D flew on 27 October 1972 and first deliveries were made to HM-12 in September 1973. The delivery of six aircraft for the Iranian Navy was to be completed during 1976.

The RH-53D is designed to tow existing and future equipment evolved to sweep mechanical, acoustic and magnetic mines. That for mechanical and acoustic mines can be carried on board the aircraft, and deployed and retrieved in flight. Magnetic sweep equipment, too large to be carried internally by the helicopter, is first streamed behind a surface vessel and then transferred to the aircraft's tow hook. It can also be carried on the external cargo hook, and is lifted from ship to sea or shore to sea by this means. Basic design gross weight is increased to 19,050 kg (42,000 lb), mission gross weight is 18,656 kg (41,130 lb), and the alternate design gross weight is increased to 22,680 kg (50,000 lb). Space, weight and power provisions have been made for installation of a projected advanced navigation system and an approach and hover coupler.

The description of the CH-53A applies also to the RH-53D, except as follows:

ROTOR SYSTEM AND DRIVE: Generally similar to those of the CH-53A, but transmission uprated to 6,458 kW (8,660 shp).

FUSELAGE: As for CH-53A, but heavier-gauge skins are used aft of the transmission area and heavier-gauge stringers around the landing gear.

LANDING GEAR: Stronger landing gear and brakes to cater for the increased gross weights.

POWER PLANT: Two General Electric T64-GE-413A turboshafts with a combined rating of 5,637 kW (7,560 shp) in early models; but it is planned to modify these engines to 3,266 kW (4,380 shp) T64-GE-415 standard by retrofit kits. Standard fuel tankage supplemented by two 1,892 litre (500 US gallon) external tanks. These are standard USAF 2,460 litre (650 US gallon) auxiliary fuel tanks, modified to reduced capacity for better roll control. Flight refuelling capability provided by an HH-53 nose-mounted refuelling probe. In addition, the RH-53D is equipped for ship-to-helicopter refuelling while airborne, with a sensing filter in the helicopter to cut off fuel flow if liquid or solid impurities are detected in the incoming fuel.

EQUIPMENT: Automatic flight control system interconnections to give automatic tow cable yaw angle retention, and aircraft attitude and heading hold. Indicator for tow cable tension and yaw angle, with automatic cable release if limits are exceeded. Nose-mounted adjustable rearview mirrors for pilot and co-pilot. Dual hydraulically-powered winches for streaming tow. Tow system comprising a separate winch and hook, rated at 3,175 kg (7,000 lb) capacity. Towboom rated at 9,072 kg (20,000 lb) capacity with hook locked in retention jaw. External cargo hook capacity rated at 11,340 kg (25,000 lb). Stowage racks and cradles for Mk 103 mechanical and Mk 104 acoustic mine countermeasures gear. Can tow Mk 105 magnetic and Mk 106 magnetic/acoustic ship-based gear. Anti-exposure suit ventilation system for crew working in cabin. Variable-speed rescue hoist rated at 272 kg (600 lb) capacity.

ARMAMENT: Provision for two 0·50 in machine-guns to detonate surfaced mines.

WEIGHTS:

Normal T-O weight	19,050 kg (42,000 lb)
Mission T-O weight	18,656 kg (41,130 lb)
Max T-O weight	22,680 kg (50,000 lb)

PERFORMANCE:

Min ground turning radius	13·46 m (44 ft 2 in)
Runway LCN at max T-O weight	8·0
Endurance	over 4 hr

SIKORSKY S-69

US Army designation: XH-59A

On 7 February 1972 Sikorsky announced that the company was designing and building a research aircraft, designated S-69, to flight test the Advancing Blade Concept (ABC) rotor system, under a contract awarded by the Eustis Directorate, US Army Air Mobility Research and Development Laboratory, Fort Eustis, Virginia. Subsequently, the value of the contract was increased to cover detail design changes and the construction of two demonstrator aircraft under the Army designation XH-59A.

The ABC rotor system, consisting of two co-axial counter-rotating rigid rotors, takes advantage of the aerodynamic lift potential of the advancing blades. At high speed, the retreating blades are unloaded, the majority of the load being carried on the advancing sides of both rotors, and the usual penalties of retreating blade stall are eliminated. This removes the need for a wing to supplement the rotor to provide speed and agility, even at high density altitude. Another advantage of the concept is the

Sikorsky RH-53D minesweeping and multi-mission helicopter of the US Navy

elimination of a conventional anti-torque tail rotor and its drive system.

The purpose of this programme is to evaluate the performance of the ABC system in flight, following successful full-scale wind tunnel tests of a 12·2 m (40 ft) diameter rotor at NASA's Ames Research Center. The first aircraft (21941) made its first flight on 26 July 1973, as a pure helicopter, but was damaged in a flight accident at Sikorsky's Stratford works in the following month. Detailed investigation led to a number of design changes and the installation of a modified control system. The test programme was resumed on 21 July 1975, with the second aircraft (21942) also flying as a pure helicopter.

Eventually, it will have two additional Pratt & Whitney J60 turbojet engines for auxiliary forward thrust in a high-speed compound helicopter configuration. Speed objective for the pure helicopter configuration is 170 knots (315 km/h; 196 mph) with a 2·5*g* load factor. With turbojet auxiliary propulsion, design speed is 300 knots (555 km/h; 345 mph) at a 2*g* load factor.

Task II flight testing of the XH-59A began in November 1975, and the two-year programme is concerned with continued expansion of the flight envelope. Since the start of the programme the aircraft has been flown to a height of 1,220 m (4,000 ft), has demonstrated sideways and rearward flight at 20 knots (37 km/h; 23 mph), as well as full autorotation at 60, 80 and 100 knots (111, 148 and 185 km/h; 69, 92 and 115 mph). Hard-over tests of the stability augmentation system have been carried out at speeds ranging from 40 to 150 knots (74 to 278 km/h; 46 to 173 mph), and single engine cuts have been made at 80 to 150 knots (148 to 278 km/h; 92 to 173 mph). By 1 March 1976 the demonstrator had attained a speed of 160 knots (296 km/h; 184 mph) in level flight at sea level and a load factor of 2·2*g*. A speed of 193 knots (357 km/h; 222 mph) was achieved during a shallow descent in June 1976.

TYPE: Two-seat research helicopter.

ROTOR SYSTEM: Two contra-rotating three-blade main rotors mounted co-axially. No tail rotor.

FUSELAGE: All-metal semi-monocoque structure of circular cross-section.

TAIL UNIT: Cantilever all-metal structure of conventional fixed-wing aircraft type. Twin endplate fins and rudders.

LANDING GEAR: Retractable tricycle type, with twin wheels on nose unit and single wheels on main units. Nosewheels retract aft into fuselage nose, main wheels inward into fuselage.

POWER PLANT: One Pratt & Whitney Aircraft of Canada PT6T-3 Turbo Twin Pac mounted within the fuselage. Provision for later installation of two Pratt & Whitney J60 turbojet engines in pod on each side of the fuselage.

ACCOMMODATION: Crew of two on flight deck, with door on each side. Fuselage access door on port side of fuselage, aft of flight deck.

DIMENSIONS, EXTERNAL:

Diameter of rotors (each)	10·97 m (36 ft 0 in)
Length overall, rotors fore and aft	12·62 m (41 ft 5 in)
Length of fuselage	12·42 m (40 ft 9 in)
Height over fins	3·94 m (12 ft 11 in)
Tailplane span	4·72 m (15 ft 6 in)
Wheel track	2·44 m (8 ft 0 in)

Sikorsky S-69 (XH-59A) prototype for evaluation of the ABC rotor system

SIKORSKY S-70

US Army designation: YUH-60A

At the end of August 1972, the US Army selected Sikorsky and Boeing Vertol (which see) as competitors to build three prototypes each, plus one ground test vehicle, of their submissions for the Utility Tactical Transport Aircraft System (UTTAS) requirement. Sikorsky's $61 million contract has an option for the purchase of six prototypes and called for flight trials to begin in November 1974. However, the first YUH-60A made its first flight on 17 October 1974, six weeks ahead of schedule. The second prototype flew for the first time on 21 January 1975, followed by the third on 28 February 1975. Fly-off evaluation against Boeing Vertol's YUH-61A prototypes began in early 1976. Two of the aircraft, in an operational configuration, were being tested under tactical conditions by US Army pilots at Fort Rucker, Alabama, and Fort Campbell, Kentucky. The third prototype, which is fully instrumented, began its tests at Edwards AFB, California, with the US Army Aviation Engineering Flight Activity carrying out the evaluation as part of the government competitive tests.

The YUH-60A has a single main rotor, canted tail rotor, twin turbine engines, and is of compact design, so that one helicopter could be carried over long range in a C-130 transport, or as many as six in a C-5A.

A variant of the UTTAS aircraft, having 83% commonality with the YUH-60A, is entered as Sikorsky's contender for the US Navy's Mk III LAMPS requirement. The principal airframe changes include provision for automatic folding of the rotor blades, wheelbase reduced by moving the tailwheel forward, and the installation of magnetic anomaly detection (MAD) gear and an undernose surface search radar. Mission equipment would include sonobuoys and two Mk 46 torpedoes. Selection of a LAMPS helicopter for the Mk III requirements was scheduled to be made in late 1976, at about the same time as that of the successful UTTAS design.

Sikorsky is also building the prototype of a commercial helicopter with the designation S-78 (described separately) which is based on the UTTAS design and incorporates many of the same advanced design features. The fourth UTTAS prototype, built with Sikorsky funds and first flown on 23 May 1975, is intended to obtain FAA certification in support of this programme.

The following description applies to the YUH-60A:

TYPE: Twin-turbine combat assault squad transport.

ROTOR SYSTEM: Four-blade main rotor. Ballistically-tolerant main rotor blades with titanium spars, glassfibre skins and aft-swept tips. Blades pressurised and equipped with gauges providing fail-safe confirmation of blade structural integrity. Elastomeric rotor hub bearings require no lubrication, reduce hub maintenance by 60%. Bifilar self-tuning vibration absorbers on rotor head. Canted tail rotor (to port) increases vertical lift and allows greater CG travel. 'Cross beam' four-blade tail rotor of composite materials, eliminating all rotor head bearings.

ROTOR DRIVE: Conventional transmission system with both turbines driving through freewheeling units to main gearbox. This is of modular construction to simplify maintenance. Intermediate and tail rotor gearboxes are grease-packed, reducing both vulnerability and maintenance.

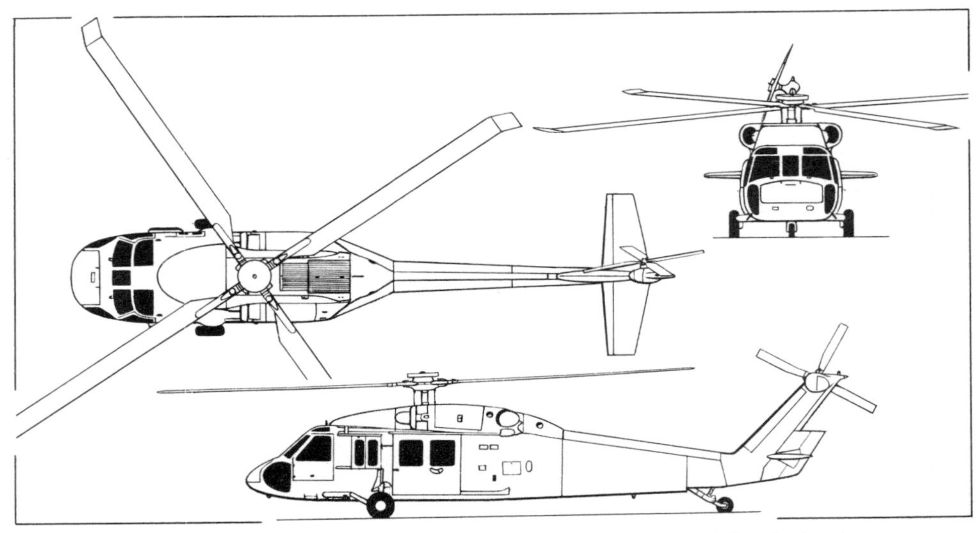
Sikorsky YUH-60A Utility Tactical Transport Aircraft System (UTTAS) *(Pilot Press)*

Sikorsky YUH-60A prototype twin-turboshaft combat assault transport helicopter

Fuselage: Conventional semi-monocoque light alloy structure.

Tail Unit: Pylon structure with port-canted tail rotor mounted on the starboard side. Large fixed tailplane, which serves also to keep troops clear of tail rotor.

Landing Gear: Non-retractable tailwheel type with single wheel on each unit. Energy-absorbing main gear with a tailwheel which gives protection for the tail rotor in taxying over rough terrain or during a high-flare landing.

Power Plant: Two 1,145 kW (1,536 shp) General Electric T700-GE-700 advanced technology turboshaft engines.

Accommodation: Pilot and co-pilot on armour protected seats. Main cabin area open to cockpit to provide good communication with flight crew and forward view for squad commander. Accommodation for 11 troops and crew of three. Eight troop seats can be removed and replaced by four litters for medivac missions, or to make room for internal cargo. External cargo hook, having a 3,175 kg (7,000 lb) lift capability. Large aft-sliding door on each side of fuselage for rapid entry and exit.

Armament: Provision for side-firing machine-gun in forward area of cabin.

Dimensions, external:

Main rotor diameter	16·36 m (53 ft 8 in)
Tail rotor diameter	3·35 m (11 ft 0 in)
Length overall (rotors turning)	19·76 m (64 ft 10 in)
Height overall	5·00 m (16 ft 5 in)
Fuselage length	15·26 m (50 ft 0¾ in)
Fuselage width	2·36 m (7 ft 9 in)
Wheel track	2·75 m (9 ft 0½ in)
Wheelbase	8·89 m (29 ft 2 in)

Weights:

Max T-O weight	9,707 kg (21,400 lb)
Mission T-O weight	7,597 kg (16,750 lb)

Performance:

Max level speed at S/L, at mission T-O weight
172 knots (318 km/h; 198 mph)

Max level speed at 9,072 kg (20,000 lb) AUW
147 knots (272 km/h; 169 mph)

Service ceiling 5,945 m (19,500 ft)

Sikorsky YUH-60A utility transport helicopter prototype demonstrating bank angle in excess of 90°

SIKORSKY S-72 (RSRA)

Sikorsky announced in October 1973 that, following a design competition in which Bell Helicopter Company also took part, it had been selected by NASA/US Army as prime contractor for a high-speed multi-purpose research helicopter which has since received the company designation S-72. A $25 million contract for the construction of two prototypes, one set of removable wings and a pair of podded turbofan engines was awarded to Sikorsky by NASA/US Army in January 1974. Known officially as Rotor Systems Research Aircraft (RSRA), these prototypes will be used by NASA and the US Army to develop and test a wide variety of rotor systems and integrated propulsion systems. They will provide test facilities that cannot be met in existing aircraft or wind tunnels, and will serve as a standardised base for comparing the various rotor systems. Present NASA and Army plans call for the evaluation of composite bearingless, variable-geometry, gimballed, articulated, hingeless, circulation control, reverse velocity and jet flap rotors.

Rollout of the first S-72 took place on 7 June 1976, with first flight planned for September 1976. The second aircraft, a compound helicopter with two auxiliary turbofan engines and a full-span wing, was expected to make its first flight in December. Sikorsky will test both aircraft for approximately 100 hours before turning them over to NASA and the Army. The aircraft will have a potential service life of 12 years, and will be able to fly as pure helicopters, compound helicopters or fixed-wing aircraft, as required.

The fuselage of the S-72 is entirely new, designed to meet the unique requirements of the RSRA. It includes tail surfaces like those of a fixed-wing aircraft. The vertical surfaces are swept; conventional rudder and elevators are fitted, and there is a large ventral fin which carries the tailwheel. The five-blade anti-torque tail rotor is mounted on the port side of the fin.

Initially each S-72 will be equipped with a Sikorsky S-61 rotor system and two 1,118·5 kW (1,500 shp) General Electric T58-GE-5 turboshaft engines.

In addition to flying with a variety of rotor systems, the S-72s will be able to operate without any rotor at all, using full-length cantilever low-wing monoplane wings and two 41·26 kN (9,275 lb st) General Electric TF34-GE-2 turbofan cruise engines in Lockheed S-3A Viking pods. The wings will be fitted with conventional ailerons and flaps, and will have adjustable incidence, over the range of −9° to +15°.

The wings and auxiliary engines will permit the S-72 to test rotor systems that might be too small to support the aircraft, and will provide an extra margin of safety for the crew, comprising two pilots, side by side, and a flight engineer. In the event of trouble with a rotor system, the crew will be able to jettison the main blades by means of explosives and return to base by flying the S-72 as a conventional aircraft.

The S-72 will also be equipped with a crew escape system that first severs the rotor blades and then extracts the three crewmen by igniting rockets on the backs of their Stanley Aviation ejection seats. This is an independent system that does not rely upon the aircraft for power. In order to test the escape system, Sikorsky has designed a special sled that will simulate the in-flight blade separation and crew extraction during rocket-propelled runs along a 10·4 nm (19 km; 12 mile) test track at Holloman AFB, New Mexico. Equipped with the same dynamic components as the RSRA, the 10 m (33 ft) sled, weighing 10,886 kg (24,000 lb), will be used in tests at speeds ranging from a hover to 230 knots (426 km/h; 265 mph).

Other features of the S-72 include accurate measuring systems to measure forces and moments on the rotor, wing, tail rotor and auxiliary engines; a fly-by-wire control system that operates through a mechanical backup system; an active vibration isolation system; and an electronic flight control system with an on-board digital computer to control the aircraft during research missions and carry out automatic, pre-programmed manoeuvres.

Dimensions, external:

Diameter of main rotor	18·90 m (62 ft 0 in)
Diameter of tail rotor	3·25 m (10 ft 8 in)
Wing span	13·74 m (45 ft 1 in)

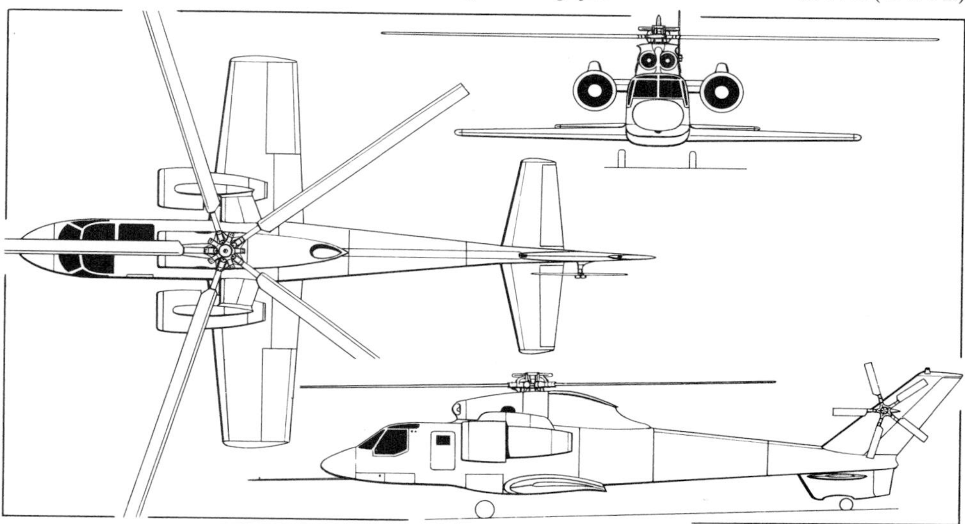

Sikorsky S-72 in fully-developed form, with fixed wings and turbofan cruise engines *(Pilot Press)*

Sikorsky S-72 prototype NASA/US Army Rotor Systems Research Aircraft

AREAS:
- Main rotor disc: 280·5 m² (3,019 sq ft)
- Tail rotor disc: 8·29 m² (89·2 sq ft)
- Wings, gross: 34·37 m² (370 sq ft)

WEIGHTS (estimated):
- Weight empty, helicopter configuration: 6,572 kg (14,490 lb)
- Weight empty, compound configuration: 9,535 kg (21,022 lb)
- Max T-O weight, helicopter configuration: 8,346 kg (18,400 lb)
- Max T-O weight, compound configuration: 11,884 kg (26,200 lb)

PERFORMANCE (estimated):
- Max level speed: 300 knots (555 km/h; 345 mph)

SIKORSKY S-76

Sikorsky Aircraft announced on 19 January 1975 the company's decision to build a new 12-passenger twin-turbine commercial helicopter, as the first stage of a programme intended to give the company a bigger share of the civil aircraft market.

The go-ahead for prototype construction followed a period of market research during which firm contracts were signed with numerous commercial operators, both in the USA and abroad. By 1 April 1976 a total of 80 aircraft had been ordered by 21 operators. Fabrication of four prototypes began in May 1976, with a first flight scheduled for May 1977 and initial deliveries of fully certificated IFR aircraft beginning in July 1978. Designated Sikorsky S-76, the design will conform with FAR Part 29 Category A IFR, or appropriate military specifications, the deciding factor being which of these two specifications has the more stringent requirement. By designing and building to this latter standard, Sikorsky plans not only to produce a rugged and reliable civil helicopter, but one which could be taken 'off-the-shelf' to satisfy a military role without any major modifications to airframe structure or dynamic system.

The S-76 will benefit from the design, research and development work carried out on the dynamic system of Sikorsky's YUH-60A UTTAS. The main rotor, for example, is a scaled-down version of that developed for the YUH-60A. The power plant is to consist of two Allison 250-C30 turboshaft engines, a growth version of the current production Model 250-C20 which powers a number of important helicopters.

TYPE: Twin-turbine general-purpose all-weather helicopter.

ROTOR SYSTEM: Four-blade main rotor. Each blade consists of a titanium spar, titanium leading-edge cover, nickel leading-edge abrasion strip, and glassfibre/nylon honeycomb trailing-edge. Blades have swept tips. Blades pressurised and equipped with gauges to provide fail-safe indication of blade structural integrity. Elastomeric rotor hub bearings which need no lubrication. Bifilar vibration absorbers on rotor head. Cross-beam four-blade tail rotor of composite materials. Rotor brake optional.

ROTOR DRIVE: Conventional transmission system, with both turbines driving through freewheeling units to main gearbox. Intermediate and tail rotor gearboxes are grease-packed to reduce maintenance.

FUSELAGE: Conventional semi-monocoque light alloy structure.

TAIL UNIT: Pylon structure with tail rotor on port side. Large fixed tailplane, which serves also to protect passengers or ground crew from contact with tail rotor.

LANDING GEAR: Hydraulically-retractable tricycle type, with single wheel on each unit. Nosewheel retracts aft, main units inward into rear fuselage; all three units are enclosed by wheel doors when retracted. Main-wheel tyres size 14·5 × 5·5-6, nosewheel tyre size 13 × 5·0-4. Hydraulic brakes; hydraulic main-wheel parking brake.

POWER PLANT: Two 522 kW (700 shp) Allison 250-C30 turboshaft engines, mounted above the cabin aft of the main rotor shaft. Standard fuel system has a capacity of 1,030 litres (272 US gallons). Extended range fuel tanks and pressure refuelling optional.

ACCOMMODATION: Pilot and co-pilot plus a maximum of 12 passengers. In this configuration passengers are seated on three four-abreast rows of seats, floor-mounted at a pitch of 79 cm (31 in). A number of executive layouts are available, including a four-passenger 'Office-in-the-Sky' configuration. Executive versions will have luxurious interior trim, full carpeting, special soundproofing, radio-telephone, and co-ordinated furniture. Dual controls optional. Two large doors on each side of fuselage, hinged at their forward edges. Baggage hold aft of cabin, with external access door on each side of the fuselage; capacity 1·19 m³ (42 cu ft). Cabin heated and ventilated. Windscreen demisting and dual windscreen wipers. Windscreen anti-icing optional. Optional external cargo hook with capacity of 2,268 kg (5,000 lb).

SYSTEMS: Hydraulic system at pressure of 207 bars (3,000 lb/sq in) supplied by two engine-driven pumps. Electrical system comprises two Lucas 6/7·5kVA engine-driven generators, two 200A transformer-rectifiers, one 115V 200VA 400Hz static inverter and 24V 22Ah nickel-cadmium battery. Engine fire detection and extinguishing system.

Full-scale mockup of Sikorsky S-76 (two Allison 250-C30 turboshaft engines)

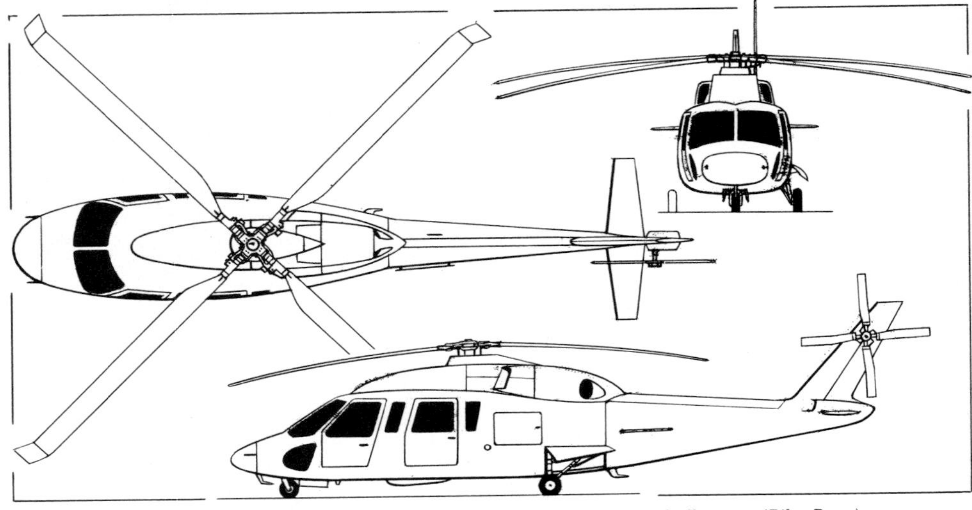

Sikorsky S-76 eight/twelve-passenger commercial transport helicopter *(Pilot Press)*

ELECTRONICS AND EQUIPMENT: Standard equipment includes provision for dual controls; cabin fire extinguishers; cockpit, cabin, instrument, navigation and anti-collision lights; landing light; external power socket; first aid kit; and utility soundproofing. VHF com transceiver and intercom system standard. Optional equipment includes air-conditioning, cargo hook, rescue hoist, emergency flotation gear, engine air particle separators, full IFR, litter installation and stability augmentation system. A wide range of optional electronics available, according to configuration, including VHF nav receivers, transponder, compass system, weather radar, flight director system, radar altimeter, ADF, DME, VLF nav system and ELT and sonic transmitters.

DIMENSIONS, EXTERNAL:
- Diameter of main rotor: 13·41 m (44 ft 0 in)
- Main rotor blade chord: 0·39 m (1 ft 3½ in)
- Diameter of tail rotor: 2·44 m (8 ft 0 in)
- Tail rotor blade chord: 0·16 m (6½ in)
- Length overall, rotors turning: 16·00 m (52 ft 6 in)
- Height overall: 4·41 m (14 ft 5¾ in)
- Tailplane span: 3·15 m (10 ft 4 in)
- Length of fuselage: 13·44 m (44 ft 1 in)
- Width of fuselage: 2·13 m (7 ft 0 in)
- Wheel track: 2·44 m (8 ft 0 in)
- Wheelbase: 5·00 m (16 ft 5 in)

DIMENSIONS, INTERNAL:
- Cabin: Length: 2·49 m (8 ft 2 in)
  - Max width: 1·93 m (6 ft 4 in)
  - Max height: 1·37 m (4 ft 6 in)
  - Floor area: 5·30 m² (57 sq ft)
  - Volume: 5·61 m³ (198 cu ft)
- Baggage compartment volume: 1·19 m³ (42 cu ft)

AREAS:
- Main rotor disc: 116·77 m² (1,257 sq ft)
- Tail rotor disc: 4·67 m² (50·27 sq ft)
- Tailplane: 2·00 m² (21·5 sq ft)

WEIGHTS AND LOADING:
- Weight empty, standard equipment: 2,241 kg (4,942 lb)
- Max T-O weight: 4,399 kg (9,700 lb)
- Max disc loading: 37·7 kg/m² (7·72 lb/sq ft)

PERFORMANCE (A: at gross weight of 4,399 kg; 9,700 lb. B: at gross weight of 3,810 kg; 8,400 lb):
- Max cruising speed:
  - A: 145 knots (269 km/h; 167 mph)
  - B: 155 knots (286 km/h; 178 mph)
- Cruising speed for max range:
  - A: 125 knots (232 km/h; 144 mph)
- Hovering ceiling in ground effect:
  - A: 1,554 m (5,100 ft)
  - B: 2,743 m (9,000 ft)
- Service ceiling, one engine out:
  - A: 1,465 m (4,800 ft)
  - B: 2,530 m (8,300 ft)
- Range with 12 passengers, standard fuel, 30 min reserves: 400 nm (742 km; 461 miles)
- Range with 8 passengers, auxiliary fuel and offshore equipment: 600 nm (1,112 km; 691 miles)
- Max ferry range, VFR equipment, 30 min reserves: 750 nm (1,390 km; 864 miles)

SIKORSKY S-78

Sikorsky has projected two commercial variants of the YUH-60A UTTAS helicopter. Originally designated S-70C-20 and S-70C-29, these were later re-designated as follows:

S-78-20. Generally similar to the YUH-60A, with accommodation for 20 passengers. Cabin headroom restricted to 1·37 m (4 ft 6 in), as the YUH-60A is designed for shipment by air within the USAF's C-130 Hercules transports. External lift capability of 5,443 kg (12,000 lb).

S-78-29. Developed version with larger fuselage; length, width and height increased, to provide improved accommodation for 29 passengers.

TYPE: Twin-turbine commercial transport helicopter.

ROTOR SYSTEM: Four-blade main rotor of advanced design, with titanium spars, Nomex honeycomb filling and glassfibre trailing-edge skins. Main rotor hub has spherical elastomeric bearings. Four-blade tail rotor is canted to port.

ROTOR DRIVE: Newly-developed lighter-weight main gearbox. Tail rotor and intermediate gearboxes are grease-packed and sealed-for-life.

FUSELAGE: Semi-monocoque structure of light alloy.

TAIL UNIT: Large fixed and cambered fin, serving also as mounting for tail rotor. Fixed horizontal tailplane. All tail surfaces swept.

LANDING GEAR (S-78-20): Non-retractable tailwheel type with single wheel on each unit.

LANDING GEAR (S-78-29): Retractable tricycle type with main wheels retracting into sponson on each side of the fuselage.

POWER PLANT: Two 1,059 kW (1,420 shp) General Electric T700 turboshaft engines. Fuel capacity 1,752 litres (463 US gallons).

ACCOMMODATION (S-78-20): Pilot and co-pilot, with seating for 20 passengers in cabin.

ACCOMMODATION (S-78-29): Pilot and co-pilot, with seating for 29 passengers in cabin.

DIMENSIONS, EXTERNAL (A: S-78-20; B: S-78-29):

Main rotor diameter:	
A, B	16·15 m (53 ft 0 in)
Tail rotor diameter:	
A, B	3·35 m (11 ft 0 in)
Length of fuselage:	
A	15·53 m (50 ft 11½ in)
B	16·71 m (54 ft 10 in)
Width of fuselage:	
A	2·36 m (7 ft 9 in)
B	2·62 m (8 ft 7 in)
Height overall:	
A	5·17 m (16 ft 11½ in)
B	6·30 m (20 ft 8 in)
Wheel track:	
A	2·79 m (9 ft 2 in)
B	3·18 m (10 ft 5 in)
Wheelbase:	
A	8·81 m (28 ft 11 in)
B	5·84 m (19 ft 2 in)

DIMENSIONS, INTERNAL (A: S-78-20; B: S-78-29):

Cabin: Length:	
A	4·95 m (16 ft 3 in)
B	6·55 m (21 ft 6 in)
Max height:	
A	1·37 m (4 ft 6 in)
B	1·52 m (5 ft 0 in)

WEIGHTS (estimated. A: S-78-20; B: S-78-29):

Weight empty, equipped:	
A	4,139 kg (9,126 lb)
B	4,704 kg (10,372 lb)
Max T-O weight:	
A	7,947 kg (17,520 lb)
B	9,070 kg (19,997 lb)

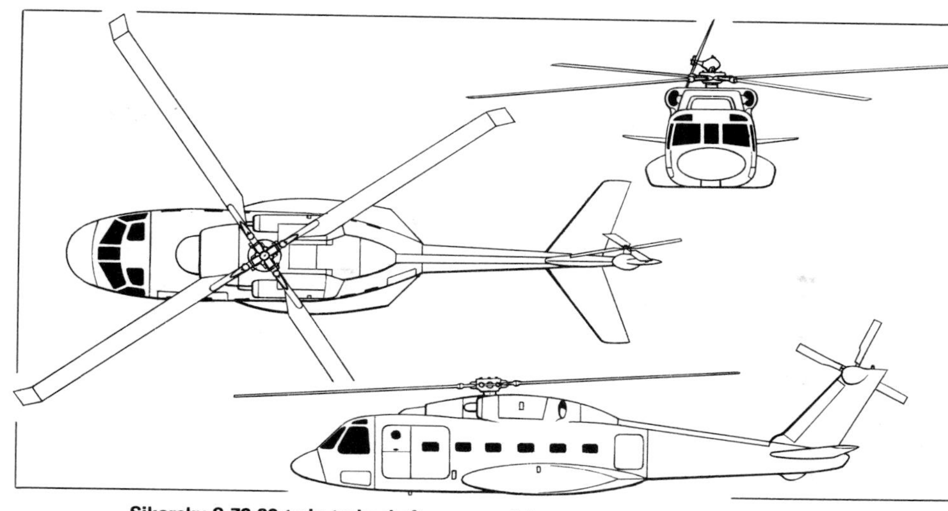

Sikorsky S-78-29 twin-turboshaft commercial transport helicopter *(Pilot Press)*

PERFORMANCE (estimated. A: S-78-20; B: S-78-29):

Max level speed at S/L:	
A, B	168 knots (311 km/h; 193 mph)
Cruising speed:	
A, B	160 knots (296 km/h; 184 mph)
Single-engine service ceiling:	
A	1,615 m (5,300 ft)
B	745 m (2,450 ft)
Hovering ceiling in ground effect:	
A	3,050 m (10,000 ft)
B	1,830 m (6,000 ft)
Hovering ceiling out of ground effect:	
A	1,770 m (5,800 ft)
B	395 m (1,300 ft)
Range with max fuel:	
A, B	399 nm (740 km; 460 miles)

SPITFIRE—*See Addenda*

SWEARINGEN

SWEARINGEN AVIATION CORPORATION (a subsidiary of Fairchild Industries)

ADDRESS:
PO Box 32486, San Antonio, Texas 78284
Telephone: (512) 824-9421
Telex: 767-315

CHAIRMAN: E. J. Swearingen
PRESIDENT: G. Stathis
VICE-PRESIDENTS:
R. N. Robinson (Marketing)
J. O'Connell (Finance)
R. E. McKelvey (Engineering)

In a joint announcement, made on 2 November 1971, Fairchild Industries and Swearingen Aircraft gave details of an agreement under which a new subsidiary, to be known as Swearingen Aviation Corporation and of which 90% of the stock was owned by Fairchild Industries, would acquire the assets of Swearingen Aircraft. Mr Edward J. Swearingen, founder and Chief Executive Officer of Swearingen Aircraft, became Chairman of the new company.

Since 1966 Swearingen has been engaged in the manufacture of the Merlin series of twin-turboprop pressurised executive transport aircraft. Early models were the Merlin IIA and generally similar Merlin IIB, details of which can be found in the 1969-70 and 1972-73 *Jane's* respectively. Current production versions are the Merlin IIIA and IVA and Metro II. A total of 236 Merlins and Metros of all models had been delivered throughout the world by January 1976.

SWEARINGEN MERLIN IIIA

The Merlin IIIA is an eight/eleven-seat all-weather pressurised executive transport which differs from the earlier Merlin III by having additional windows, major system and flight deck improvements and a wide selection of new interiors. The original Merlin III received FAA certification on 27 July 1970, and was described in the 1974-75 *Jane's*.

A total of 57 Merlin IIIs and IIIAs had been delivered by 1 January 1976. Customers include the Belgian Air Force, which took delivery of six in 1976.

TYPE: Eight/eleven-seat twin-turboprop executive transport.

WINGS: As for Metro II.

FUSELAGE: Cylindrical all-metal fail-safe structure, flush-riveted throughout. Glassfibre honeycomb nose-cap will accommodate a 0·45 m (18 in) weather radar antenna.

TAIL UNIT: Cantilever all-metal structure with sweptback vertical and horizontal surfaces. Dorsal fin, with tailplane mounted approximately one-third up from base of fin. Small ventral fin. Pneumatic de-icing boots on tailplane leading-edges.

LANDING GEAR: As for Metro II.

POWER PLANT: Two 626·5 kW (840 shp) AiResearch TPE 331-3U-303G turboprop engines, each driving a Hartzell three-blade fully-feathering and reversible-pitch metal propeller. Integral fuel tank in each wing, each with a usable capacity of 1,226 litres (324 US gallons): total usable fuel capacity 2,452 litres (648 US gallons). Refuelling points in each outer wing panel. Engine inlet de-icing by bleed air.

ACCOMMODATION: Crew of two on flight deck, with dual controls. Bulkhead with sliding door divides flight deck from cabin. Standard accommodation for six to nine passengers with seats disposed on each side of a central aisle. Passenger door at rear of cabin on port side, with integral airstair. Emergency exit on starboard side of cabin.

SYSTEMS: As for Metro II except that the automatic water-methanol system is not installed.

DIMENSIONS, EXTERNAL:

Wing span	14·10 m (46 ft 3 in)
Wing chord at root	2·62 m (8 ft 7 in)
Wing chord at tip	1·04 m (3 ft 5 in)
Wing aspect ratio	7·71
Length overall	12·85 m (42 ft 1·9 in)
Height overall	5·12 m (16 ft 9½ in)
Tailplane span	4·61 m (15 ft 1½ in)
Wheel track	4·57 m (15 ft 0 in)
Wheelbase	3·23 m (10 ft 7 in)
Propeller diameter	2·49 m (8 ft 2 in)
Propeller ground clearance	0·305 m (1 ft 0 in)
Passenger door:	
Height	1·35 m (4 ft 5 in)
Width	0·64 m (2 ft 1 in)

DIMENSIONS, INTERNAL:

Length, incl flight deck and rear utility section	6·93 m (22 ft 9 in)
Cabin:	
Length between front and rear bulkheads	3·23 m (10 ft 7 in)
Width	1·57 m (5 ft 2 in)
Height	1·45 m (4 ft 9 in)
Baggage volume	2·97 m³ (105 cu ft)

AREA:

Wings, gross	25·78 m² (277·50 sq ft)

WEIGHTS AND LOADINGS:

Weight empty, equipped	3,356 kg (7,400 lb)
Max T-O weight	5,670 kg (12,500 lb)
Max ramp weight	5,697 kg (12,560 lb)
Max zero-fuel weight	4,535 kg (10,000 lb)
Max landing weight	5,217 kg (11,500 lb)
Max wing loading	219·7 kg/m² (45·0 lb/sq ft)
Max power loading	4·53 kg/kW (7·44 lb/shp)

PERFORMANCE (at max T-O weight except where indicated):

Max cruising speed at 4,875 m (16,000 ft)	282 knots (523 km/h; 325 mph) TAS
Econ cruising speed at 8,535 m (28,000 ft)	250 knots (463 km/h; 288 mph) TAS
Max speed, flaps and wheels down	153 knots (283 km/h; 176 mph) CAS
Stalling speed, flaps and wheels up	97 knots (179 km/h; 111 mph) IAS
Stalling speed, flaps and wheels down	84 knots (155 km/h; 96 mph) IAS
Max rate of climb at S/L	771 m (2,530 ft)/min
Rate of climb at S/L, one engine out	189 m (620 ft)/min

Swearingen Merlin IIIA eight/eleven-seat executive transport (two AiResearch TPE 331-3U-303G turboprop engines)

Service ceiling at 5,443 kg (12,000 lb) 8,810 m (28,900 ft)
Service ceiling at 5,443 kg (12,000 lb), one engine out 4,575 m (15,000 ft)
Max operating altitude 9,450 m (31,000 ft)
T-O run, short field 655 m (2,150 ft)
FAA T-O field length 930 m (3,050 ft)
Landing from 15 m (50 ft) with propeller reversal 479 m (1,570 ft)
Range with max fuel at max cruising speed 1,709 nm (3,167 km; 1,968 miles)
Ferry range with max fuel at econ cruising speed, 45 min reserves 2,483 nm (4,602 km; 2,860 miles)

SWEARINGEN MERLIN IVA

The Merlin IVA is a corporate version of the Metro II commuter airliner, to which it is generally similar. It differs principally in its internal configuration, which provides accommodation for 12 to 15 passengers, with a private toilet and a baggage volume of 4·05 m³ (143 cu ft). Initial deliveries of the earlier Merlin IV (1974-75 *Jane's*) were made in 1970, following FAA certification on 23 September that year. A total of 38 Merlin IV/IVAs had been delivered by 1 January 1976.

The description of the Metro applies also to the Merlin IVA, except for some variation in the systems. In particular, usable fuel capacity is 2,096 litres (554 US gallons).

DIMENSIONS, EXTERNAL:
As for Metro II

DIMENSIONS, INTERNAL:
Cabin, excl flight deck:
Length 7·75 m (25 ft 5 in)
Baggage compartment:
Length 2·34 m (7 ft 8 in)
Volume 4·05 m³ (143 cu ft)

AREAS:
As for Metro II

WEIGHTS AND LOADINGS:
Weight empty, equipped 3,719 kg (8,200 lb)
Max T-O, landing and zero-fuel weight 5,670 kg (12,500 lb)
Max ramp weight 5,697 kg (12,560 lb)
Max wing loading 219·7 kg/m² (45·0 lb/sq ft)
Max power loading 4·53 kg/kW (7·44 lb/shp)

PERFORMANCE (at max T-O weight):
As for Metro II, except:
Max cruising speed at 4,875 m (16,000 ft) 269 knots (499 km/h; 310 mph)
Econ cruising speed at 8,535 m (28,000 ft) 240 knots (444 km/h; 276 mph)
Range with max fuel at max cruising speed, 45 min reserves 1,367 nm (2,534 km; 1,575 miles)
Ferry range with max fuel at econ cruising speed, 45 min reserves 1,819 nm (3,371 km; 2,095 miles)

Swearingen Merlin IVA, a corporate version of the Metro II commuter airliner

SWEARINGEN MODEL SA-226TC METRO II

The Metro II is a 19/20-passenger all-weather pressurised, air-conditioned airliner, certificated under Special Federal Air Regulation 23. The standard aircraft has an easily convertible passenger/cargo interior with special kits available for conversion to an executive interior, air ambulance with up to 10 litters, hospital equipped for on-board emergency operations, and single or dual camera photographic configuration. An all-cargo configuration is available, with 286 kg (630 lb) additional payload and without windows and passenger amenities.

The Metro II differs from the earlier Metro (1974-75 *Jane's*) by the introduction of larger windows and major systems and flight deck improvements.

A total of 20 Metros/Metro IIs had been delivered by 1 January 1976. Orders include ten for the Chilean Police Department and one for the Royal Oman Police Air Wing. It was announced in March 1976 that European Air Transport of Belgium had acquired one Metro II and held options on two more.

TYPE: Twin-turboprop 19/20-passenger commuter airliner.

WINGS: Cantilever low-wing monoplane. Wing section NACA 65_2A215 at root, NACA 64_2A415 at tip. Dihedral 5°. Incidence 1° at root, −1° at tip. Sweepback at quarter-chord 0·9°. All-metal two-spar semi-monocoque fail-safe structure of aluminium alloy. Hydraulically-operated double-slotted trailing-edge flaps. Manually-controlled trim tab in port aileron. Goodrich pneumatic de-icing boots on wing leading-edges, with automatic bleed air cycling system.

FUSELAGE: All-metal cylindrical semi-monocoque fail-safe structure of aluminium alloy. Glassfibre honeycomb nose cap can accommodate a 0·45 m (18 in) weather radar antenna.

TAIL UNIT: Cantilever all-metal structure with sweptback vertical surfaces and dorsal fin. Electrically-adjustable variable-incidence tailplane. Manually-controlled rudder trim. Goodrich pneumatic de-icing boots on tailplane leading-edges, with automatic bleed air cycling system.

LANDING GEAR: Retractable tricycle type with twin wheels on each unit. Hydraulic retraction, with dual actuators on each unit. All wheels retract forward, main gear into engine nacelles, nosewheels into fuselage. Ozone Aircraft Systems oleo-pneumatic shock-absorber struts. Nosewheel steerable. Free-fall emergency extension system, with backup of hand-operated hydraulic pump. B.F. Goodrich main wheels and tyres, size 18 × 5·50, type VII, pressure 6·90 bars (100 lb/sq in). Jay-Em nosewheels and Goodyear tyres, size 16 × 4·40, type VII, pressure 5·86 bars (85 lb/sq in). Low-pressure tyres optional. Goodrich self-adjusting hydraulically-operated disc brakes. Heavy duty brakes and anti-skid system optional.

POWER PLANT: Two 701 kW (940 shp) AiResearch TPE 331-3UW-303G turboprop engines with automatic emergency water-methanol system, each driving a Hartzell three-blade fully-feathering and reversible propeller with automatic synchronisation. Integral fuel tank in each wing, each with a usable capacity of 1,226 litres (324 US gallons). Total usable fuel capacity 2,452 litres (648 US gallons). Refuelling point on each outer wing panel. Oil capacity 15·1 litres (4 US gallons). Engine inlet de-icing by bleed air. Oil cooler inlet anti-icing by hot oil. Electrical propeller de-icing. Automatic fuel heating system to prevent filter icing. Flush-mounted fuel vents. Single-point rapid drain provisions. Optional JATO and continuous water injection.

ACCOMMODATION: Crew of two on flight deck, separated from passenger/cargo area by arm-level curtain. Dual controls standard. Bulkhead between cabin and flight deck optional. Standard accommodation for 19-20 passengers seated two abreast, on each side of centre aisle. Double pane tinted windows. 'No Smoking' and 'Fasten seat belt' signs. Stowing fold-up seats for rapid conversion to cargo or mixed passenger/cargo configuration. Movable bulkhead between passenger and cargo sections. Snap-in carpeting. Self-stowing aisle filler. Tie-down fittings for cargo at 0·76 m (30 in) spacing. Integral-step passenger door on port side of fuselage, immediately aft of flight deck. Large cargo loading door on port side of fuselage at rear of cabin, hinged at top. Three window emergency exits, one on the port, two on the starboard side. Forward baggage compartment in nose, capacity 1·27 m³ (45 cu ft). Pressurised rear cargo compartment in passenger version, capacity 3·85 m³ (136 cu ft). Cabin air-conditioned and pressurised. Electrical windscreen de-icing. Windscreen wipers.

SYSTEMS: AiResearch automatic cabin pressure control system maintains a differential of 0·48 bars (7·0 lb/sq in). Engine bleed air heating, dual air-cycle cooling system, with automatic temperature control. Air blower

Swearingen Metro 19/20-passenger commuter airliner (two AiResearch TPE-331-3UW-303G turboprop engines)

system for on-ground ventilation. Independent hydraulic system for brakes. Dual engine-driven hydraulic pumps provide 138 bars (2,000 lb/sq in) to operate flaps and landing gear. Electrical system supplied by two 300A 28V DC starter/generators. Fail-safe system with overload and over-voltage protection. Redundant circuits for essential systems. Two 650VA static inverters supply 115V and 26V AC. Two 25Ah nickel-cadmium batteries for main services. One small nickel-cadmium battery for utility lights only. Engine fire detection system standard. Engine fire extinguisher systems optional. Oxygen system of 1·39 m³ (49 cu ft) capacity with flush outlets at each seat. Stall avoidance system comprising angle indicator, visual and aural warning and stick pusher.

ELECTRONICS AND EQUIPMENT: Standard equipment includes individual reading lights and air vents for each passenger, electrically-heated pitot heads, external power socket, automatic engine start cycle, static wicks, internally-operated control locks, pilot and co-pilot foot warmers, navigation lights, retractable landing lights, taxi light, ice inspection light, rotating beacon, instrument and map lights, baggage compartment light, cargo compartment lights and entrance lights, ice-free instrument static sources. Provisions for installation of remotely-mounted or panel-mounted electronics, customer-furnished antennae, weather radar and autopilot. Two flight deck and four cabin speakers standard.

DIMENSIONS, EXTERNAL:

Wing span	14·10 m (46 ft 3 in)
Wing chord at root	2·62 m (8 ft 7 in)
Wing chord at tip	1·04 m (3 ft 5 in)
Wing aspect ratio	7·71
Length overall	18·10 m (59 ft 4¾ in)
Height overall	5·12 m (16 ft 9¾ in)
Tailplane span	4·61 m (15 ft 1½ in)
Wheel track	4·57 m (15 ft 0 in)
Wheelbase	5·83 m (19 ft 1½ in)
Propeller diameter	2·59 m (8 ft 6 in)
Propeller ground clearance	0·254 m (10 in)
Passenger door (fwd):	
Height	1·35 m (4 ft 5 in)
Width	0·64 m (2 ft 1 in)
Cargo door (aft):	
Height	1·30 m (4 ft 3¼ in)
Width	1·35 m (4 ft 5 in)
Height to sill	1·30 m (4 ft 3¼ in)

DIMENSIONS, INTERNAL:

Cabin, excl flight deck and aft cargo compartment:	
Length	7·75 m (25 ft 5 in)
Max width	1·57 m (5 ft 2 in)
Max height (aisle)	1·45 m (4 ft 9 in)
Floor area	13·01 m² (140 sq ft)
Volume	13·88 m³ (490 cu ft)
Aft cargo compartment (pressurised):	
Length	2·34 m (7 ft 8 in)
Max width	1·57 m (5 ft 2 in)
Max height	1·32 m (4 ft 4 in)
Volume	3·85 m³ (136 cu ft)
Nose cargo compartment (unpressurised):	
Length	1·75 m (5 ft 9 in)
Volume	1·27 m³ (45 cu ft)

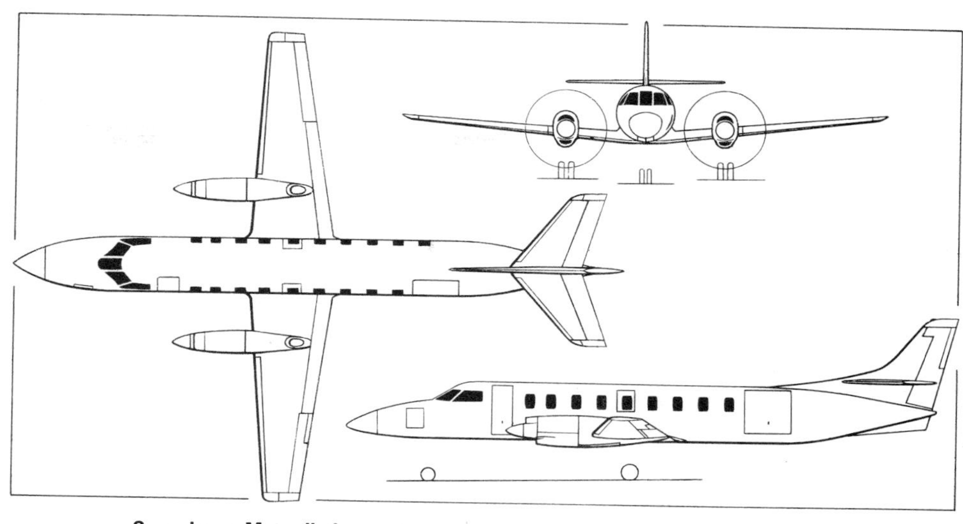

Swearingen Metro II nineteen/twenty-passenger commuter airliner *(Pilot Press)*

AREAS:

Wings, gross	25·78 m² (277·50 sq ft)
Ailerons (total)	1·31 m² (14·12 sq ft)
Trailing-edge flaps (total)	3·78 m² (40·66 sq ft)
Fin	5·20 m² (56·00 sq ft)
Rudder, incl tab	1·80 m² (19·38 sq ft)
Tailplane	7·06 m² (75·97 sq ft)
Elevators	1·98 m² (21·27 sq ft)

WEIGHTS AND LOADINGS:

Weight empty	3,379 kg (7,450 lb)
Max T-O and landing weight	5,670 kg (12,500 lb)
Max ramp weight	5,697 kg (12,560 lb)
Max wing loading	219·7 kg/m² (45·0 lb/sq ft)
Max power loading	4·04 kg/kW (6·65 lb/shp)

PERFORMANCE (at max T-O weight except where indicated):

Max cruising speed at 3,050 m (10,000 ft), at 5,445 kg (12,000 lb) AUW	255 knots (473 km/h; 294 mph)
Long range cruising speed at 6,100 m (20,000 ft), at 5,445 kg (12,000 lb) AUW	242 knots (449 km/h; 279 mph)
Max speed, flaps and wheels down	153 knots (283 km/h; 176 mph)
Stalling speed, flaps and wheels up	97 knots (180 km/h; 112 mph)
Stalling speed, flaps and wheels down	86 knots (160 km/h; 99 mph)
Max rate of climb at S/L	732 m (2,400 ft)/min
Rate of climb at S/L, one engine out	198 m (650 ft)/min
Service ceiling at 5,445 kg (12,000 lb)	8,230 m (27,000 ft)
Service ceiling at 5,445 kg (12,000 lb), one engine out	4,480 m (14,700 ft)
T-O to 15 m (50 ft)	799 m (2,620 ft)
Landing from 15 m (50 ft)	1,082 m (3,550 ft)
Range at max cruising speed, with 19 passengers and 45 min reserves	187 nm (346 km; 215 miles)
Range at max cruising speed, with 15 passengers and 45 min reserves	595 nm (1,102 km; 685 miles)
Max ferry range (allowance for taxi, T-O, climb, cruise, descent and 45 min reserves)	2,132 nm (3,952 km; 2,456 miles)

TAYLORCRAFT

TAYLORCRAFT AVIATION CORPORATION

ADDRESS:
14600 Commerce Avenue NE, Alliance, Ohio 44601
Telephone: (216) 823-6675

PRESIDENT: Charles Feris

Taylorcraft Aviation Corporation, which was re-formed on 1 April 1968 primarily to provide product support, has in production a two-seat trainer/sporting aircraft which has the designation Model F-19 Sportsman 100. Design originated in August 1967, based on the well-known Taylorcraft Model B of second World War origin; but it was not until 1973 that construction of pre-production and production aircraft started. The company had received orders for 75 aircraft by early 1976, of which 58 had been completed.

Canadian-registered Taylorcraft Model F-19 Sportsman 100 *(Neil A. Macdougall)*

TAYLORCRAFT MODEL F-19 SPORTSMAN 100

TYPE: Two-seat trainer/sporting aircraft.

WINGS: Braced high-wing monoplane with Vee bracing struts each side. Wing section NACA 23012. Dihedral 1°. Composite structure with spruce spars, stamped metal ribs and fabric covering. Plain wide-span ailerons of similar construction. No flaps. No trim tabs.

FUSELAGE: Welded structure of 4130 steel tube with fabric covering.

TAIL UNIT: Wire-braced welded steel tube structure with fabric covering. Trim tab in port elevator.

LANDING GEAR: Non-retractable tailwheel type. Two side Vees and half-axles. Main wheels fitted with 6·00-6 4-ply tyres. Cleveland heavy-duty caliper brakes. Parking brake. Float and ski landing gear to be made available optionally.

POWER PLANT: One 74·5 kW (100 hp) Continental O-200-A flat-four engine, driving a McCauley Type 1A105SCM two-blade metal fixed-pitch propeller with spinner. One fuel tank in each wing, with combined capacity of 45·4 litres (12 US gallons), and one fuel tank in fuselage, immediately aft of firewall, with capacity of 45·4 litres (12 US gallons). Total fuel capacity 90·8 litres (24 US gallons). Oil capacity 2·84 litres (0·75 US gallons).

ACCOMMODATION: Two seats side by side in enclosed cabin. Dual controls standard. Seat belts, carpeted floor. Metal door each side with sliding windows. Baggage compartment aft of seats, capacity 32·7 kg (72 lb). Accommodation heated and ventilated.

SYSTEM: Electrical system powered by engine-driven generator.

ELECTRONICS AND EQUIPMENT: Optional electronics include Genave Alpha 200B com transceiver, Narco Escort 110 com transceiver, EBC-102A emergency locator transmitter, microphone, headsets, speakers and com antenna. Standard equipment includes navigation lights, wiring for landing and anti-collision lights, cargo tie-down straps and glove compartment. Optional equipment includes cockpit, landing and anti-collision lights, streamlined fairings for main landing gear legs, streamlined wheel fairings, full blind-flying instrumentation with vacuum system, sensitive altimeter, 8-day clock, tiedown rings and razorback or Ceconite covering.

DIMENSIONS, EXTERNAL:

Wing span	10·97 m (36 ft 0 in)
Wing chord (constant)	1·60 m (5 ft 3 in)
Length overall	6·74 m (22 ft 1¼ in)
Height overall	1·98 m (6 ft 6 in)
Tailplane span	3·05 m (10 ft 0 in)
Wheel track	1·83 m (6 ft 0 in)

AREAS:

Wings, gross	17·07 m² (183·71 sq ft)
Ailerons (total)	1·86 m² (20·0 sq ft)
Fin	0·34 m² (3·7 sq ft)
Rudder	0·59 m² (6·3 sq ft)
Tailplane	1·21 m² (13·0 sq ft)
Elevators, incl tab	0·99 m² (10·66 sq ft)

WEIGHTS AND LOADINGS:

Weight empty	408 kg (900 lb)
Max T-O weight	680 kg (1,500 lb)
Max wing loading	39·9 kg/m² (8·17 lb/sq ft)
Max power loading	9·12 kg/kW (15 lb/hp)

PERFORMANCE (at max T-O weight):

Max level speed	110 knots (204 km/h; 127 mph)
Cruising speed	100 knots (185 km/h; 115 mph)
Stalling speed	37·5 knots (69·5 km/h; 43 mph)
Max rate of climb at S/L	236 m (775 ft)/min
Service ceiling	5,485 m (18,000 ft)
T-O run	91·5 m (300 ft)
Range	347 nm (643 km; 400 miles)

TED SMITH

TED SMITH AEROSTAR CORPORATION

Address:
2560 Skyway Drive, PO Box 2009, Santa Maria, California 93454
Telephone: (805) 922-8411
Chairman of the Board: Ted R. Smith
President: Ron W. Smith
Vice-Presidents:
Niels W. Andersen (Engineering)
Paul Brouillet (Controller)
Lloyd Cox (Operations)
Boyd W. Lydick (Marketing)

In 1967 limited production began of the first two of a projected series of small business aircraft, known as Aerostars, designed by Ted Smith Aircraft Company, a subsidiary of the American Cement Corporation. The entire assets of the former company were acquired by Butler Aviation International on 16 February 1970. Production of Aerostars was subsequently suspended; but during 1972 Ted R. Smith and Associates Inc was formed and re-acquired from Butler Aviation all existing airframe components. This enabled the new company to resume production of the original Models 600 and 601 Aerostars, as well as a pressurised Model 601P for which Ted R. Smith and Associates gained certification in 1972.

The company also designed a slightly larger and more powerful version of the Model 600, designated Aerostar 700, the prototype of which flew for the first time on 22 November 1972. At that time the company had plans to develop this aircraft, as well as a 'stretched' nine-seat version of the Model 700. This programme has now been placed in abeyance, and the company is concentrating its efforts instead on the development of a new turbofan-powered six/seven-seat pressurised light transport designated Model 4000 Turbofan Aerostar.

TED SMITH AEROSTAR 600/601/601P

Design work on this series began in November 1964 and the first Model 600/601 prototype made its first flight in October 1967. FAA Type Approval of the Model 600 was awarded in March 1968, and of the Model 601 in November 1968.

It is claimed that the Aerostar airframe contains only 50% as many components as are used in other designs of comparable size. Construction involves extensive use of monocoque assemblies, in which unstiffened sections of heavy-gauge skin carry loads. The horizontal and vertical fixed tail surfaces, together with their related control surfaces, are interchangeable.

On 22 January 1975, over southern and central California, an Aerostar 601 piloted by Hal Fishman set up a new Class C1d 1,000 km closed-circuit speed record for piston-engined landplanes of 264 knots (489 km/h; 304 mph). On the same day the same aircraft, flown by Capt Barry Schiff, also captured the 500 km closed-circuit record with a speed of 265 knots (491 km/h; 305 mph). Both records had previously been held since 1951 by a Soviet Yak-11.

Current production versions of the Aerostar comprise:

Model 600. Powered by two 216 kW (290 hp) Lycoming IO-540-K1F5 flat-six engines.

Model 601. As Model 600, but powered by two 216 kW (290 hp) Lycoming TIO-540-S1A5 flat-six turbocharged engines.

Model 601P. As Model 601, but with increased wing span, higher flow-rate turbochargers to supply bleed air for cabin pressurisation, and certificated for a maximum T-O and landing weight of 2,721 kg (6,000 lb)

Type: Twin-engined light transport aircraft.

Wings: Cantilever mid-wing monoplane. Wing section NACA 64_1A212. Dihedral 2°. Incidence 1°. No sweepback. All-metal structure using heavy gauge skins attached to three spars, several bulkheads and stringers. Entire wing assembly, excluding attachments for ailerons and flaps, contains fewer than 50 detail parts. Ailerons and flaps each comprise a spar, ribs, nose skin and one-piece wraparound light alloy skin aft of spar. No trim tabs. Model 610P has increased wing span.

Fuselage: All-metal fail-safe monocoque structure. Skin composed of large segments of light alloy sheet over stringers and frames. Entire fuselage contains fewer than 100 parts, including skin panels. All fuselage assemblies designed basically for pressurisation.

Tail Unit: Cantilever all-metal structure, with swept vertical and horizontal surfaces. Both fixed and control surfaces are interchangeable. Electrically-operated trim tab in rudder and each elevator.

Landing Gear: Hydraulically-retractable tricycle type. Main units retract inward, nosewheel forward. Steerable nosewheel. Hydraulically-operated dual caliper brakes. Parking brake.

Power Plant: Two Lycoming engines, as detailed in model listings, each driving a Hartzell three-blade metal constant-speed and fully-feathering propeller with spinner. Fuel in integral wing tanks and fuselage tank with total capacity of 669 litres (177 US gallons), of which 660·5 litres (174·5 US gallons) are usable. Oil capacity 22·7 litres (6 US gallons).

Ted Smith Aerostar 601 light transport (two turbocharged Lycoming TIO-540 engines)

Accommodation: Cabin seats six people on track-mounted individual reclining seats, in pairs. Dual controls standard. Door on port side by pilot's seat; top half hinges upward, bottom half downward. Emergency escape windows at rear of cabin. Tinted windscreen and dual pane tinted cabin windows. Large utility shelf in aft cabin. Baggage compartment, capacity 109 kg (240 lb), aft of cabin, with external access. Individual air vents and reading lights for each seat. Cabin air-conditioned in 600/601, and also pressurised in 601P. Control locks. Windscreen defrosting.

Systems: Air-conditioning system includes a Janitrol 35,000 BTU heater. Pressurisation system for Model 601P supplied by engine bleed air; pressure differential 0·29 bars (4·25 lb/sq in). Hydraulic system for landing gear actuation, wheel brakes and flaps, powered by an engine-driven pump, pressure 69 bars (1,000 lb/sq in). Dual pneumatic systems for instrument gyros and de-icing boots.Electrical system powered by two 28V 70A engine-driven alternators with failure warning lights. Two 12V 24Ah batteries. Models 601 and 601P each have an oxygen system, with individual outlets at each seat, as standard; capacity 601 3·26 m^3 (115 cu ft), 601P 0·31 m^3 (11 cu ft).

Electronics and Equipment: Standard electronics by King for Models 600 and 601 include dual KX-175B nav/coms, dual KNI-520 VOR/ILS, dual KN-77 VOR/LOC converters, KN-73 glideslope receiver, KR-85 ADF, KI-225 ADF indicator, KN-266 digital DME, KN-65 DME remote unit, KT-76 transponder, KMA-20 control console including audio amplifier and marker beacon receiver, and associated antennae, filters, microphone and cabin speaker. Standard electronics for Model 601P include dual Collins VHF-251 com transceivers and antennae, dual VIR-351 nav receivers with VOR/LOC converters and antenna, GLS-350 glideslope with antenna, IND-351 course deviation indicator with glideslope indicator, IND-350 course deviation indicator, ADF-650 ADF system with indicator and antenna, TDR-950 transponder, AMR-350 marker/audio panel, dual PWC-150 power converters, radio master switch, Narco DME-195, Narco AR-500 altitude encoder, dual cabin speakers, microphones and associated equipment. A wide range of optional electronics is available to customer's requirements. Standard equipment includes map lights, red and white instrument panel lights, baggage compartment light, dual taxi and landing lights, navigation lights, strobe lights, full blind-flying instrumentation, outside air temperature gauge, flight hour recorder, emergency locator transmitter, heated pitot, jack pads and tow bar. 601P has in addition as standard an alternate static source, eight-day clock and a sensitive altimeter.

Dimensions, external:
Wing span:
600, 601 — 10·41 m (34 ft 2 in)
601P — 11·18 m (36 ft 8 in)
Wing chord at root — 2·18 m (7 ft 2 in)
Wing chord at tip — 0·87 m (2 ft 10⅜ in)
Wing aspect ratio — 6·83
Length overall — 10·61 m (34 ft 9¾ in)
Height overall — 3·70 m (12 ft 1½ in)
Tailplane span — 4·37 m (14 ft 4 in)
Wheel track — 3·11 m (10 ft 2½ in)
Propeller diameter — 1·98 m (6 ft 6 in)
Passenger door:
Height — 1·14 m (3 ft 9 in)
Width — 0·71 m (2 ft 4 in)
Baggage compartment door:
Height — 0·61 m (2 ft 0 in)
Width — 0·56 m (1 ft 10 in)

Dimensions, internal:
Cabin: Length — 3·81 m (12 ft 6 in)
Width — 1·17 m (3 ft 10 in)
Height — 1·22 m (4 ft 0 in)
Baggage space — 0·85 m^3 (30 cu ft)

Areas:
Wings, gross:
600, 601 — 15·79 m^2 (170 sq ft)
601P — 16·54 m^2 (178 sq ft)
Tailplane — 4·20 m^2 (45·2 sq ft)

Weights and Loadings:
Weight empty:
600 — 1,655 kg (3,650 lb)
601 — 1,740 kg (3,837 lb)
601P — 1,814 kg (4,000 lb)
Max T-O and landing weight:
600 — 2,495 kg (5,500 lb)
601 — 2,585 kg (5,700 lb)
601P — 2,721 kg (6,000 lb)
Max wing loading:
600 — 157·7 kg/m^2 (32·3 lb/sq ft)
601 — 163·6 kg/m^2 (35·5 lb/sq ft)
601P — 164·5 kg/m^2 (33·7 lb/sq ft)
Max power loading:
600 — 5·77 kg/kW (9·5 lb/hp)
601 — 5·98 kg/kW (9·8 lb/hp)
601P — 6·30 kg/kW (10·3 lb/hp)

Performance (at max T-O weight):
Max level speed:
600 at S/L — 226 knots (418 km/h; 260 mph)
601 at 7,620 m (25,000 ft) — 271 knots (502 km/h; 312 mph)
601P at 7,620 m (25,000 ft) — 257 knots (476 km/h; 296 mph)
Cruising speed:
600, 70% power at 3,050 m (10,000 ft) — 217 knots (402 km/h; 250 mph)
601, 601P, 70% power at 6,100 m (20,000 ft) — 235 knots (436 km/h; 271 mph)
600, 65% power at 3,050 m (10,000 ft) — 209 knots (388 km/h; 241 mph)
601, 65% power at 6,100 m (20,000 ft) — 228 knots (422 km/h; 262 mph)
601P, 65% power at 6,100 m (20,000 ft) — 227 knots (420 km/h; 261 mph)
Stalling speed, wheels and flaps down:
600 — 67 knots (124 km/h; 77 mph)
601 — 69 knots (128 km/h; 79 mph)
601P — 71·5 knots (132 km/h; 82 mph)
Max rate of climb at S/L:
601, 601P — 549 m (1,800 ft)/min
601 — 518 m (1,700 ft)/min
Service ceiling:
600 — 6,430 m (21,100 ft)
601 — 9,175 m (30,100 ft)
601P — 8,535 m (28,000 ft)
T-O to 15 m (50 ft):
600 — 427 m (1,400 ft)
601 — 488 m (1,600 ft)
601P — 512 m (1,680 ft)
Landing from 15 m (50 ft):
600 — 589 m (1,932 ft)
601 — 624 m (2,047 ft)
601P — 664 m (2,180 ft)
Range with max fuel, 65% power with 45 min reserves:
600 at 2,745 m (9,000 ft) — 1,222 nm (2,266 km; 1,408 miles)
601 at 6,100 m (20,000 ft) — 1,233 nm (2,285 km; 1,420 miles)
601P at 6,100 m (20,000 ft) — 1,263 nm (2,341 km; 1,455 miles)

TEXAS AIRPLANE

TEXAS AIRPLANE MANUFACTURING COMPANY INC

ADDRESS:
PO Box 230, Addison Airport, Dallas, Texas 75001
Telephone: (214) 661-8355
PRESIDENT: W. S. Thomas

TEXAS AIRPLANE CJ600

Texas Airplane Manufacturing Company Inc acquired the assets of the former Carstedt Inc of Long Beach, California, and under the designation CJ600 is continuing to market the turboprop conversion of the Hawker Siddeley (D.H.104) Dove light transport designed by Carstedt Inc, which was known as the JetLiner 600. This was last described in the 1972-73 *Jane's*.

Since that time the aircraft has been certificated for a maximum T-O weight of 4,762 kg (10,500 lb); the major changes involved in the current conversion can be summarised briefly as follows:

WINGS: Each wing is completely rebuilt and sealed outboard of the engine nacelles to provide an integral tank with a capacity of 454 litres (120 US gallons). Total usable fuel capacity 851 litres (225 US gallons). Wing spars reinforced. Controllable aileron trim tabs installed. Optional TKS liquid or Goodrich pneumatic de-icing system can be fitted to wing leading-edges.

FUSELAGE: Fuselage lengthened by 0·94 m (3 ft 1 in) forward and 1·27 m (4 ft 2 in) aft of main spar. Canopy over cockpit lengthened, streamlined and lowered.

TAIL UNIT: Optional TKS liquid or Goodrich pneumatic de-icing system can be fitted to leading-edges of fin and tailplane.

LANDING GEAR: Oleo-pneumatic shock-absorbers fitted, and landing gear components strengthened to cater for increased gross weight.

POWER PLANT: Two 526 kW (705 ehp) Garrett-AiResearch TPE 331-101E turboprop engines, derated to 429 kW (575 shp), are installed in Volpar cowlings. Each drives a Hartzell three-blade metal constant-speed fully-feathering and reversible propeller. Electric propeller anti-icing. Bleed air anti-icing system for engine air intakes.

ACCOMMODATION: Pilot, co-pilot, and high-density seating for 18 passengers in commuter version. Quickly-removable seats allow rapid conversion to passenger/cargo configuration. Optionally a heavy-duty floor and a 1·78 m (5 ft 10 in) wide and 1·35 m (4 ft 5 in) high cargo door on the port side can be installed for all-cargo operation. Air-conditioning and cabin heating by engine bleed air. Ground ventilation blower. Electric windscreen wipers.

SYSTEMS: Electrical system has dual 200A starter/generators, dual 24Ah batteries and dual solid-state inverters. Pneumatic system supplied by dual electrically-driven air compressors. Engine fire detection and extinguishing system.

ELECTRONICS AND EQUIPMENT: Standard electronics package consists of dual King KX-175B transceivers with 720 com and 200 nav channels, KI-214 VOR/LOC/glideslope indicator and KI-201C VOR/LOC indicator; KR-85 ADF; KA-40 remote marker; KMA-20 audio system; and KT-76 transponder with Aero Mech encoder. Standard equipment includes electrically-heated pitot heads, provisions for dual retractable landing lights, and ground power socket.

Texas Airplane CJ600 'stretched' conversion of the de Havilland Dove

DIMENSIONS, EXTERNAL:

Wing span	17·37 m (57 ft 0 in)
Length overall	14·17 m (46 ft 6 in)
Height overall	4·06 m (13 ft 4 in)
Wheel track	4·17 m (13 ft 8 in)

DIMENSIONS, INTERNAL:

Cabin: Length (excl flight deck)	9·02 m (29 ft 7 in)
Max width	1·37 m (4 ft 6 in)
Max height	1·60 m (5 ft 3 in)

WEIGHTS:

Weight empty	approx 2,721 kg (6,000 lb)
Max T-O weight	4,762 kg (10,500 lb)
Max zero-fuel weight	4,604 kg (10,150 lb)

PERFORMANCE (at max T-O weight):

Max cruising speed at 3,050 m (10,000 ft), ISA	250 knots (463 km/h; 288 mph)
Nominal cruising speed	230 knots (426 km/h; 265 mph)
Max rate of climb at S/L	853 m (2,800 ft)/min
Rate of climb at S/L, one engine out	296 m (970 ft)/min
T-O to 15 m (50 ft)	678 m (2,225 ft)

UNIVAIR

UNIVAIR AIRCRAFT CORPORATION

HEAD OFFICE AND WORKS:
Route 3, PO Box 59, Aurora, Colorado 80011
Telephone: (303) 364-7661
CHAIRMAN OF THE BOARD:
Veda Dyer Williams
PRESIDENT:
Stephen E. Dyer
VICE-PRESIDENTS:
F. A. King (Sales)
Robert White
Robert M. Williams (Production)
SECRETARY: Janice M. Dyer

Univair Aircraft Corporation, founded in 1946, is a company specialising in the manufacture of propellers and components for light aircraft. It holds the type certificates, drawings, production tooling and rights to manufacture the Globe and Temco Swift Models GC-1A and GC-1B; the Ercoupe, Alon, Forney and Mooney M-10 Cadet series; and the Stinson 108. Details of these aircraft have appeared in earlier editions of *Jane's,* under their original manufacturers. All structural parts for each type are available now from Univair, together with wing lift struts for fabric-wing Cessna 120 and 140 aircraft.

Propellers produced by Univair under type certificate include Flottorp wooden fixed-pitch, Flottorp variable-pitch (of Beech-Roby design), Flottorp F1 fixed-pitch and F12 constant-speed metal, Flottorp F200 and Aeromatic 220 automatic propellers.

The company is also building to order an updated version of the Stinson 108, which has the designation Univair Stinson 108-5. A prototype and 17 production aircraft had been completed by 1 March 1976. All available details follow:

UNIVAIR STINSON 108-5

Design of Univair's version of the Stinson 108 began in November 1962, and the first flight of the prototype was made on 21 April 1964; FAA certification was granted almost three months later, on 14 July. Apart from building examples to order, Univair markets a kit to convert the earlier Stinson Model 108-3 to Univair Stinson standard.

TYPE: Four-seat cabin monoplane.

WINGS: Braced high-wing monoplane, with light alloy Vee bracing struts each side. Wing section NACA 4412. Dihedral 2° 30′. Incidence 1° 31′ at root, 0° 7′ at tip. No sweepback. Constant-chord two-spar structure of light alloy with fabric covering. Plain ailerons of light alloy and fabric covering, with a fixed slot forward of each aileron. Plain trailing-edge flaps of light alloy stressed-skin construction. Ground-adjustable tab on each aileron.

Univair Stinson 108-5 four-seat cabin monoplane (Franklin 6A-335-B1 engine)

FUSELAGE: Welded steel tube structure, with fabric covering.

TAIL UNIT: Cantilever all-metal structure of light alloy. Fixed-incidence tailplane with horn-balanced elevators. Servo tab in port elevator. Horn-balanced rudder with servo tab.

LANDING GEAR: Non-retractable tailwheel type. Cantilever main units with spring/hydraulic shock-absorbers. Cleveland main wheels with tyres size 7·00-6, pressure 1·38 bars (20 lb/sq in). Maule P8 steerable tailwheel, with tyre size 2·80 × 2·50-4, pressure 1·72 bars (25 lb/sq in). Cleveland hydraulic shoe or disc brakes.

POWER PLANT: One 134 kW (180 hp) Franklin 6A-335-B1 flat-six engine, driving a McCauley two-blade metal constant-speed propeller type 2A31C21/84S with spinner. One fuel tank in each wing; total capacity 189·3 litres (50 US gallons). Refuelling point on inboard upper surface of each wing. Oil capacity 8·5 litres (2·25 US gallons).

ACCOMMODATION: Standard seating for four in two pairs, with rear seats easily removable to accommodate cargo or extra baggage. Door, with window, on each side of cabin. Accommodation heated and ventilated. Space for 45 kg (100 lb) baggage, with external access door.

SYSTEMS: Electrical system powered by 12V DC engine-driven generator, with storage battery. Scott Mark II oxygen system.

ELECTRONICS AND EQUIPMENT: Narco Mark 12 VHF nav/com with VOR. Blind-flying instrumentation standard.

DIMENSIONS, EXTERNAL:

Wing span	10·34 m (33 ft 11 in)
Wing chord, constant	1·45 m (4 ft 9 in)
Wing aspect ratio	6·8
Length overall	7·67 m (25 ft 2 in)
Height overall	2·29 m (7 ft 6 in)
Tailplane span	3·40 m (11 ft 2 in)
Wheel track	2·16 m (7 ft 1 in)
Wheelbase	5·66 m (18 ft 7 in)
Propeller diameter	1·93 m (6 ft 4 in)
Propeller ground clearance	0·23 m (9·14 in)
Passenger doors (each):	
Height	1·22 m (4 ft 0 in)
Width	0·76 m (2 ft 6 in)
Height to sill	0·81 m (2 ft 8 in)
Baggage door:	
Height	0·61 m (2 ft 0 in)
Width	0·56 m (1 ft 10 in)
Height to sill	0·99 m (3 ft 3 in)

DIMENSIONS, INTERNAL:

Cabin: Length	1·75 m (5 ft 9 in)
Max width	0·97 m (3 ft 2 in)
Max height	1·42 m (4 ft 8 in)

AREAS:

Wings, gross	14·4 m² (155 sq ft)
Ailerons (total)	1·675 m² (18·02 sq ft)
Trailing-edge flaps (total)	1·14 m² (12·22 sq ft)
Fin	1·33 m² (14·28 sq ft)
Rudder, incl tab	0·63 m² (6·78 sq ft)
Tailplane	1·36 m² (14·66 sq ft)
Elevators, incl tab	1·60 m² (17·24 sq ft)

WEIGHTS AND LOADINGS:
- Weight empty, equipped: 590 kg (1,300 lb)
- Max T-O and landing weight: 1,088 kg (2,400 lb)
- Max wing loading: 75·7 kg/m² (15·5 lb/sq ft)
- Max power loading: 8·12 kg/kW (13·3 lb/hp)

PERFORMANCE (at max T-O weight):
- Never-exceed speed: 147 knots (273 km/h; 170 mph)
- Max level speed at S/L: 132 knots (245 km/h; 152 mph)
- Max cruising speed, 75% power at 1,525 m (5,000 ft): 117 knots (217 km/h; 135 mph)
- Econ cruising speed at 2,285 m (7,500 ft): 115 knots (212 km/h; 132 mph)
- Stalling speed, flaps up: 56·5 knots (105 km/h; 65 mph)
- Stalling speed, flaps down: 53 knots (98·5 km/h; 61 mph)
- Max rate of climb at S/L: 305 m (1,000 ft)/min
- Service ceiling: 5,790 m (19,000 ft)
- T-O run: 122 m (400 ft)
- T-O to 15 m (50 ft): 213 m (700 ft)
- Landing from 15 m (50 ft): 198 m (650 ft)
- Landing run: 91·5 m (300 ft)
- Range with max payload, 45 min reserves: 412 nm (764 km; 475 miles)

UTC
UNITED TECHNOLOGIES CORPORATION

HEAD OFFICE:
United Technologies Building, Hartford, Connecticut 06101

Telephone: (203) 728-7000

DIRECTORS:
Hubert Faure
T. Mitchell Ford
Harry J. Gray
Edward L. Hennessy Jr
David C. Hewitt
Paul W. O'Malley
Walter F. Probst
Arthur E. Smith
Olcott D. Smith
Richard S. Smith
William I. Spencer
Robert L. Sproull
Alfred W. Van Sinderen
Ralph A. Weller
Roger C. Wilkins

EXECUTIVES:
CHAIRMAN, PRESIDENT AND CHIEF EXECUTIVE OFFICER:
Harry J. Gray

SENIOR VICE-PRESIDENTS:
Joseph H. Allen (Communications and Marketing)
William E. Diefenderfer
Edward L. Hennessy Jr (Finance and Administration)
Robert F. Stewart (Strategic Planning)

GROUP VICE-PRESIDENTS:
Paul W. O'Malley
Robert F. Stewart (Flight Systems and Equipment)
Bruce N. Torell (Propulsion)

VICE-PRESIDENTS:
Robert A. Aspinwall
Rolf D. Bibow (International)
Robert L. Cole
James Ferguson
Wesley A. Kuhrt (Technology)
Edward W. Large (Corporation Counsel)
Clark MacGregor
N. B. Morse (Industrial Relations)
Francis L. Murphy (Public Relations and Advertising)
Kenneth L. Otto (Personnel Resources)
Dale W. Van Winkle

CONTROLLER: Charles B. Preston
TREASURER: Joseph A. Biernat
SECRETARY AND DEPUTY CORPORATION COUNSEL:
Martin R. Lewis Jr

DIVISIONS AND SUBSIDIARIES:
Pratt & Whitney Aircraft Group (see Engines section, Canada and USA)
Otis
Essex Group
Sikorsky Aircraft (see this section)
Hamilton Standard
Norden
Chemical Systems Division (see Engines section)
Power Systems Division
Turbo Power and Marine Systems
Pratt & Whitney Aircraft of West Virginia
Pratt & Whitney Aircraft of Canada Ltd (see Engines section)
United Technologies International
United Technologies Research Center

VOLPAR
VOLPAR INC

HEAD OFFICE AND WORKS:
7929 Hayvenhurst Avenue, Van Nuys, California 91406

Telephone: (213) 787-4393 and 873-5599

PRESIDENT: Frank V. Nixon Jr
GENERAL MANAGER: Albert B. Seed
CHIEF ENGINEER: John Cadrobbi
CHIEF INSPECTOR: F. F. Taylor
PURCHASING AGENT: Carl Jones
PUBLIC RELATIONS MANAGER: R. M. Byrne

Volpar Inc was formed in 1960 to market in kit form a tricycle landing gear modification which is suitable for all models of the Beechcraft Model 18 light twin-engined transport.

The Volpar kit, which has full FAA certification, was designed by Thorp Engineering to require a minimum of modification to the basic airframe. The machined parts are manufactured by Paragon Precision Products, a producer of rocket engine components, tools and dies. The various assemblies formerly fabricated by Volitan Aviation Inc have been manufactured by Volpar following a merger between the two companies.

As a follow-up to the above modification, Volpar produces kits to convert the Model 18 to turboprop power, using AiResearch TPE 331 engines. The converted aircraft is known as the Turbo 18. After wide acceptance of this latter aircraft, Volpar introduced the Turboliner, and subsequently the Turboliner II, which is approved under SFAR 23 for commuter airline operation. Both of these aircraft are 'stretched' versions of the Turbo 18.

Using the nacelles that were developed for the Turbo 18 and Turboliner, Volpar then produced an engine installation which it markets under the name of Packaged Power. This is available with any of the AiResearch TPE 331 series of turboprop engines, with either over-engine or under-engine air intake as required by the particular installation. These units have been fitted to such aircraft as the Beechcraft Model 18, de Havilland Dove, Grumman Goose and de Havilland Beaver. It is understood that Packaged Power units are being considered for installation in several aircraft currently in the development stage. Volpar developed a turboprop version of the de Havilland Beaver, utilising the Packaged Power concept, and known as the Volpar Model 4000. Details of this appeared in the 1974-75 *Jane's*.

In February 1975 Volpar moved to larger facilities at Van Nuys Airport, to allow for increased production of the Turboliner II as well as for expansion of modification programmes involving larger aircraft.

In early 1976 Volpar was constructing the prototype of a commuter aircraft named the Volpar Centennial.

VOLPAR (BEECHCRAFT) MODEL 18

The Volpar modification converts the Beechcraft Model 18 to a tricycle landing gear configuration, offering substantially slower approach speeds, greatly improved braking and easier ground handling. Cruising speed is improved, as all three wheels are completely retracted. Futhermore, the aircraft can be kept in hangars with a lower roof clearance, since the overall height is reduced to 2·79 m (9 ft 2 in).

The Volpar kit, which has passed all FAA static tests for a maximum landing weight of 4,433 kg (9,772 lb), utilises basic components of the existing main landing gear. The new nose gear is connected to the existing retraction system, where the tailwheel connection was removed. The complete modification can be made without removing the wings or stripping any of the wing skin. All cockpit controls and emergency procedures are unchanged, including the instrument panel wheel position indicator. Existing airstair doors can be retained with only minor modification.

Volpar 'Packaged Power' units, embodying AiResearch TPE 331-2U-203 engines, installed in a Grumman Goose

Basically, the modification moves the main landing gear 1·22 m (4 ft 0 in) aft of the original position, attaching it to a welded tube truss that increases the torsional strength of the centre wing structure by 60% in landing configuration. The nose assembly is completely new and includes a streamlined nose fairing which adds 0·67 m (2 ft 2½ in) to the fuselage length. Space inside the fairing can be used for additional equipment, including a weather radar dish of up to 0·305 m (12 in) diameter.

All three wheels are of aluminium and can be fitted with either Goodrich or Goodyear tubed or tubeless tyres, size 8·50-10, ten-ply rating. Main-wheel tyre pressure 4·48 bars (65 lb/sq in), nosewheel tyre pressure 3·10 bars (45 lb/sq in). Shock-absorption is provided by hydraulic oleo struts of Volpar manufacture. Goodrich multiple disc brakes. All three wheels retract forward in less than eight seconds. On the ground the cabin floor is only 1·07 m (3 ft 6 in) off the ground at the door. Wheelbase is 2·62 m (8 ft 7 in). The aircraft will turn on a 1·22 m (4 ft) radius of the inside wheel and a centering device is incorporated on the shimmy damper for take-off and landing.

The current Mk IV Volpar conversion incorporates Goodrich nine-piston full-circle brakes with twice the braking energy and three times the service life of the two-piston type fitted formerly. The new brakes fit on the original gear and are obtainable from either Volpar or Goodrich.

A total of more than 400 sets of Volpar tri-gear have been delivered.

VOLPAR (BEECHCRAFT) TURBO 18

The Turbo 18 is a Beechcraft Model 18 fitted with the Volpar Mk IV tricycle landing gear described above and re-engined with two 526 kW (705 ehp) AiResearch TPE 331-1-101B turboprop engines, flat rated to 451 kW (605 ehp). The wing planform is changed, by extending forward the entire leading-edge inboard of each engine nacelle and carrying the new leading-edge line past the nacelle, so increasing the chord and sweepback to a point some distance outboard of the nacelle. The rectangular wingtip panels of the standard Super 18 are replaced by smaller tips which decrease the wing span and maintain the normal leading-edge sweep to the tip.

Installation of TPE 331 engines and Hartzell Model HC-B3TN-5 three-blade reversible-pitch propellers reduces the empty weight, permitting an increase in fuel or payload. Internal fuel capacity is increased by 379 litres (100 US gallons) by installing new integral tanks in the leading-edge immediately outboard of each engine nacelle. These become the main tanks, each delivering fuel directly to the adjacent engine. They increase the maximum fuel capacity to 2,385 litres (630 US gallons), with a normal capacity of 1,159 litres (306 US gallons).

Air-conditioning and heating installations are available, using engine bleed air. A large cargo door, 1·57 m (5 ft 2 in) wide, with a max height of 1·09 m (3 ft 7 in), can be provided, incorporating the existing airstair door.

The detailed description of the Turboliner (which follows), applies also to the Turbo 18, except that the Turbo 18 does not have the 'stretched' fuselage.

FAA Supplemental Type Approval of the Turbo 18 was received on 17 February 1966. Two were in service with the US Public Health Service at the end of that month and

Volpar Turbo 18 conversion of the Beechcraft Model 18, operated by the US Public Health Service for air sampling

conversion kits are in full production. Customers include Air Asia of Taiwan, which has been supplied with 15 kits.

DIMENSIONS, EXTERNAL:
- Wing span: 14·02 m (46 ft 0 in)
- Length overall: 11·40 m (37 ft 5 in)
- Height overall: 2·92 m (9 ft 7 in)
- Wheelbase: 2·62 m (8 ft 7 in)

DIMENSIONS, INTERNAL:
- Cabin, excl flight deck:
  - Length: 3·87 m (12 ft 8½ in)
  - Max width: 1·32 m (4 ft 4 in)
  - Max height: 1·68 m (5 ft 6 in)
  - Volume: 7·36 m³ (260 cu ft)

WEIGHTS AND LOADINGS:
- Weight empty, basic: 2,495 kg (5,500 lb)
- Max payload: 2,171 kg (4,786 lb)
- Max T-O weight: 4,666 kg (10,286 lb)
- Max zero-fuel weight: 4,082 kg (9,000 lb)
- Max landing weight: 4,433 kg (9,772 lb)
- Max wing loading: 134·3 kg/m² (27·51 lb/sq ft)
- Max power loading: 5·17 kg/kW (8·50 lb/ehp)

PERFORMANCE (at max T-O weight):
- Max cruising speed at 3,050 m (10,000 ft): 243 knots (451 km/h; 280 mph)
- Econ cruising speed at 3,050 m (10,000 ft): 222 knots (412 km/h; 256 mph)
- Stalling speed, wheels and flaps up, power off: 80 knots (148 km/h; 92 mph)
- Stalling speed, wheels and flaps down, power off: 77 knots (142 km/h; 88 mph)
- Max rate of climb at S/L: 521 m (1,710 ft)/min
- Service ceiling: 7,925 m (26,000 ft)
- Service ceiling, one engine out: 4,265 m (14,000 ft)
- T-O run: 507 m (1,665 ft)
- T-O to 15 m (50 ft): 725 m (2,380 ft)
- Landing from 15 m (50 ft): 642 m (2,107 ft)
- Landing run with reverse thrust: 265 m (870 ft)
- Range with max fuel at 222 knots (412 km/h; 256 mph), 45 min reserves: 1,884 nm (3,492 km; 2,170 miles)
- Range with max payload, 45 min reserves: 400 nm (741 km; 461 miles)

VOLPAR (BEECHCRAFT) TURBOLINER

This is a 'stretched' 15-passenger version of the Volpar (Beechcraft) Turbo 18, intended for the third-level airline market. Design was started in August 1966 and construction of the prototype began in December 1966. The prototype flew for the first time on 12 April 1967 and FAA certification was granted on 29 March 1968, the Turboliner being approved for operation at a gross weight of 5,216 kg (11,500 lb).

By the end of February 1975 a total of 26 Turboliners had been delivered and were in service with small airlines throughout the world. In March 1970 a Turboliner (N353V), on a delivery flight from Los Angeles to Singapore, set six official international speed records. It carried on board during the flight all necessary spares for one year's normal operation, together with a 1,515 litre (400 US gallon) ferry tank in the fuselage, and was in operation with a commuter airline two days after arrival in Singapore.

TYPE: Twin-turboprop light transport aircraft.

WINGS: Cantilever low-wing monoplane. Wing section NACA 63-015 at station 28·0, NACA 23014 at station 144·5, NACA 23012 at station 260·4. Dihedral 6°. Incidence 5° 20′ at root, 1° at tip. Sweepback 16° 21′ on inner wings, 8° 23′ on outer panels. Steel truss centre-section spar; remainder of structure aluminium semi-monocoque. Plain differential ailerons and plain trailing-edge flaps of conventional aluminium construction. Trim tab in port aileron. Optional Goodrich pneumatic de-icing boots on leading-edges.

FUSELAGE: Conventional aluminium semi-monocoque structure.

TAIL UNIT: Cantilever aluminium semi-monocoque structure with twin endplate fins and rudders. Fixed-incidence tailplane. Trim tabs in rudder and elevators. Optional Goodrich pneumatic de-icing boots on leading-edges.

LANDING GEAR: Volpar electrically-retractable tricycle type. All units retract forward, main wheels into engine nacelles. Volpar hydraulic shock-absorbers. All three wheels size 8·50-10 with Goodrich or Goodyear tubeless or tube-type tyres. Main-wheel tyre pressure 5·52 bars (80 lb/sq in); nosewheel tyre pressure 3·10 bars (45 lb/sq in). Goodrich multiple-disc brakes.

POWER PLANT: Two 526 kW (705 ehp) AiResearch TPE 331-1-101B turboprop engines, each driving a Hartzell HC-B3TN-5 three-blade reversible-pitch propeller with T10176H blades. Four to eight fuel tanks in wings, including new integral main tanks in wing leading-edges outboard of nacelles. Normal fuel capacity 1,159 litres (306 US gallons); max capacity 2,385 litres (630 US gallons). Refuelling points in upper surface of wings. Total oil capacity 11·4 litres (3 US gallons).

ACCOMMODATION: Crew of two and up to 15 passengers. Downward-hinged airstair door on port side at rear of cabin. Optional double-door for freight loading. Seats removable to enable aircraft to be used for freight-carrying. Heating and air-conditioning optional. Baggage space aft of cabin and in each wing.

SYSTEMS: Hydraulic system for brakes only. Electrical supply from two 200A starter/generators and two 24V batteries, for landing gear and flap operation, propeller anti-icing, landing lights, radio and lighting.

ELECTRONICS AND EQUIPMENT: Blind-flying instrumentation, radio and radar to customer's specification.

DIMENSIONS, EXTERNAL:
- Wing span: 14·02 m (46 ft 0 in)
- Wing chord at root: 4·15 m (13 ft 7·36 in)
- Wing chord at tip: 1·14 m (3 ft 8·94 in)
- Wing aspect ratio: 5·67
- Length overall: 13·47 m (44 ft 2½ in)
- Height overall: 2·92 m (9 ft 7 in)
- Tailplane span: 4·57 m (15 ft 0 in)
- Wheel track: 3·94 m (12 ft 11 in)
- Wheelbase: 3·84 m (12 ft 7 in)
- Propeller diameter: 2·46 to 2·57 m (8 ft 0⅜ in to 8 ft 5⅜ in)
- Passenger door:
  - Height: 1·22 m (4 ft 0 in)
  - Width: 0·69 m (2 ft 3 in)
  - Height to sill: 1·07 m (3 ft 6 in)

DIMENSIONS, INTERNAL:
- Cabin, excl flight deck:
  - Length: 5·94 m (19 ft 6 in)
  - Max width: 1·32 m (4 ft 4 in)
  - Max height: 1·68 m (5 ft 6 in)
  - Floor area: 7·43 m² (80 sq ft)
  - Volume: 11·16 m³ (394 cu ft)
- Freight hold (aft of cabin) volume: 0·65 m³ (23 cu ft)
- Freight holds (wings) volume (total): 0·91 m³ (32 cu ft)

AREAS:
- Wings, gross: 34·75 m² (374 sq ft)
- Ailerons (total): 2·47 m² (26·6 sq ft)
- Trailing-edge flaps (total): 2·62 m² (28·2 sq ft)
- Fins (total): 1·51 m² (16·3 sq ft)
- Rudders (total): 16·05 m² (17·28 sq ft)

Volpar Turboliner II, a passenger conversion of the Beechcraft Model 18, in operation with Winship Air Service, Anchorage, Alaska

Tailplane 35·49 m² (38·2 sq ft)
Elevators, incl tab 25·28 m² (27·22 sq ft)

WEIGHTS AND LOADINGS:
Weight empty:
Cargo version 2,676 kg (5,900 lb)
Airliner 2,993 kg (6,600 lb)
Max T-O weight 5,216 kg (11,500 lb)
Max zero-fuel weight 4,762 kg (10,500 lb)
Max landing weight 4,989 kg (11,000 lb)
Max wing loading 150·1 kg/m² (30·75 lb/sq ft)
Max power loading 4·96 kg/kW (8·15 lb/ehp)

PERFORMANCE (at max T-O weight):
Max level and cruising speed at 3,050 m (10,000 ft) 243 knots (451 km/h; 280 mph)
Econ cruising speed at 3,050 m (10,000 ft) 222 knots (412 km/h; 256 mph)
Stalling speed, wheels and flaps up, power off 84 knots (154·5 km/h; 96 mph)
Stalling speed, wheels and flaps down, power off 80 knots (148·5 km/h; 92 mph)
Max rate of climb at S/L 463 m (1,520 ft)/min
Service ceiling 7,315 m (24,000 ft)
Service ceiling, one engine out 3,960 m (13,000 ft)
T-O run 570 m (1,870 ft)
T-O to 15 m (50 ft) 989 m (3,245 ft)
Landing from 15 m (50 ft) 762 m (2,500 ft)
Landing run 317 m (1,040 ft)
Range with max fuel, 45 min reserves 1,802 nm (3,340 km; 2,076 miles)
Range with max payload, 45 min reserves 300 nm (556 km; 346 miles)

VOLPAR (BEECHCRAFT) TURBOLINER II

The Turboliner II is basically a Turboliner that has been modified to meet the new requirements of SFAR 23. The prototype was completed in February 1970 and received certification in July 1970. Dimensions and performance are the same as those given for the Turboliner.

Recent conversions incorporate a number of improvements, including battery temperature indicators, a fail-safe Hydro-Aire Hytrol anti-skid braking system, installation of a 38,000 BTU Janitrol heater in the nose for ground heating of cockpit and engine nacelles, and modification to the standard Volpar side-opening cargo door. This now incorporates an inward-opening door 0·66 m (2 ft 2 in) in width and with a minimum height of 1·16 m (3 ft 9½ in), which may be opened in flight to permit the air drop of firefighting personnel or cargo. During 1974 Volpar, in conjunction with Sierracin Manufacturing Company, developed electrically heated windscreens for the Turboliner II. Complete installation kits are available also for other Volpar conversions.

VOLPAR CENTENNIAL

Artist's impression of the Volpar Centennial 19-passenger transport

In May 1975 Volpar began the design of a lengthened and much improved version of the Turboliner, known as the Centennial. This has a new fail-safe wing, a swept fin and rudder in place of the familiar Beech twin endplate fins and rudders, a fuselage lengthened by 2·22 m (7 ft 3½ in) to provide four additional seats, and Pratt & Whitney Aircraft of Canada PT6A turboprop engines.

Volpar had received orders for two Centennials by early 1976. Construction of these aircraft and the prototype began almost simultaneously in August 1975. Roll-out of the prototype was scheduled for December 1976.

TYPE: Twin-turboprop light transport aircraft.

WINGS: Cantilever low-wing monoplane. Wing section (constant) NACA 64_2A415. Constant chord from root to outboard of engine nacelles at wing station 108. Dihedral 7°. Incidence 0° at root, −4° at tip. Sweepback at quarter-chord 6° 14′ on outboard wing panels. Two-spar light alloy structure of fail-safe construction. Plain cable-actuated ailerons of similar construction. Single-slotted trailing-edge flaps of conventional light alloy construction. Trim tab in port aileron. Optional leading-edge de-icing boots.

FUSELAGE: Conventional semi-monocoque structure of light alloy.

TAIL UNIT: Cantilever structure of light alloy with swept vertical surfaces. Fixed-incidence tailplane. Trim tabs in rudder and elevators. Optional leading-edge de-icing boots.

LANDING GEAR: Volpar electrically-retractable tricycle type. All units retract forward, main wheels into engine nacelles. Volpar hydraulic shock-absorbers and nose unit shimmy-damper. All three wheels size 8·50-10 with Goodrich or Goodyear tyres. Main-wheel tyre pressure 5·52 bars (80 lb/sq in), nosewheel tyre pressure 3·10 bars (45 lb/sq in). Goodrich multiple-disc brakes. Hydro-Aire Hytrol anti-skid units available optionally.

POWER PLANT: Two 634 kW (850 shp) Pratt & Whitney Aircraft of Canada PT6A-34 turboprop engines, flat rated to 559 kW (750 shp), each driving a Hartzell three-blade metal constant-speed fully-feathering and reversible-pitch propeller. Integral fuel tanks in outboard wing panels, with combined capacity of 2,650 litres (700 US gallons). Refuelling points in outboard ends of wing upper surface. Oil capacity 8·7 litres (2·3 US gallons). Propeller de-icing standard.

ACCOMMODATION: Pilot and co-pilot on flight deck. Up to 19 passengers in individual seats, with central aisle; 9 seats on the port side and 10 on the starboard side. Downward-hinged airstair door on port side at rear of cabin. Optional double door for cargo loading. Seats removable to permit use in varying passenger/cargo or all-cargo configurations. Door to flight deck on port side, forward of wing. Three emergency exits; one overwing on each side, the third on starboard side opposite airstair door. Baggage space aft of cabin, and in fuselage belly forward of the wing with external access door. Heating and air-conditioning optional. Windscreen de-icing standard.

SYSTEMS: Hydraulic system for brakes only. Electrical system powered by two 200A 28V DC starter/generators, with two 25Ah storage batteries. Dual static inverters for AC power.

ELECTRONICS AND EQUIPMENT: Communication transceivers, navigation receivers, radar, autopilot and blind-flying instrumentation to customer's specification.

DIMENSIONS, EXTERNAL:
Wing span 15·24 m (50 ft 0 in)
Wing chord at root, constant to station 108 3·05 m (10 ft 0 in)
Wing chord at tip 1·522 m (5 ft 0 in)
Wing aspect ratio 5·95
Length overall 15·82 m (51 ft 10¾ in)
Height overall 5·03 m (16 ft 6 in)
Tailplane span 5·49 m (18 ft 0 in)
Wheel track 4·37 m (14 ft 4 in)
Wheelbase 5·31 m (17 ft 5 in)
Propeller diameter 2·44 m (8 ft 0 in)
Propeller ground clearance 0·36 m (1 ft 2 in)
Passenger door (port, aft):
Height 1·22 m (4 ft 0 in)
Width 0·61 m (2 ft 0 in)
Height to sill 1·52 m (5 ft 0 in)
Crew door (port, fwd):
Height 0·84 m (2 ft 9 in)
Width 0·56 m (1 ft 10 in)
Height to sill 1·85 m (6 ft 1 in)
Cargo door (optional, port, aft):
Height 1·22 m (4 ft 0 in)
Width 1·57 m (5 ft 2 in)
Height to sill 1·42 m (4 ft 8 in)
Ventral baggage door:
Height 0·305 m (1 ft 0 in)
Width 1·30 m (4 ft 3 in)
Height to sill 1·14 m (3 ft 9 in)
Emergency exits (one port, two stbd):
Height 0·58 m (1 ft 11 in)
Width 0·71 m (2 ft 4 in)

DIMENSIONS, INTERNAL:
Cabin, excl flight deck:
Length 7·92 m (26 ft 0 in)
Max width 1·30 m (4 ft 3¼ in)
Max height:
fwd cabin 1·45 m (4 ft 9 in)
aft cabin 1·68 m (5 ft 6 in)
Floor area 10·39 m² (111·8 sq ft)
Volume 14·53 m³ (513 cu ft)
Baggage compartment, aft 0·68 m³ (24 cu ft)
Baggage compartment, belly 0·50 m³ (17·5 cu ft)

AREAS:
Wings, gross 39·0 m² (420 sq ft)
Trailing-edge flaps (total) 4·67 m² (50·25 sq ft)
Rudder, incl tab 1·24 m² (13·4 sq ft)
Tailplane 2·60 m² (28·0 sq ft)
Elevators 2·53 m² (27·23 sq ft)

WEIGHTS AND LOADINGS (estimated):
Weight empty 3,184 kg (7,020 lb)
Max T-O weight 5,670 kg (12,500 lb)
Max zero-fuel weight 5,216 kg (11,500 lb)
Max landing weight 5,443 kg (12,000 lb)
Max wing loading 145·3 kg/m² (29·76 lb/sq ft)
Max power loading 5·07 kg/kW (8·33 lb/shp)

PERFORMANCE (estimated, at max T-O weight except where indicated):
Max cruising speed at 3,050 m (10,000 ft) 225 knots (417 km/h; 259 mph)
Stalling speed, wheels and flaps up 80 knots (148 km/h; 92 mph)
Stalling speed, wheels and flaps down 68 knots (126 km/h; 78 mph)
Max rate of climb at S/L 500 m (1,640 ft)/min
Rate of climb at S/L, one engine out 75 m (245 ft)/min
T-O to 15 m (50 ft) 930 m (3,050 ft)
Landing from 15 m (50 ft) at max landing weight 975 m (3,200 ft)
Range with max fuel 1,563 nm (2,896 km; 1,800 miles)
Range with max payload 434 nm (805 km; 500 miles)

VOUGHT

VOUGHT CORPORATION (a subsidiary of THE LTV CORPORATION)

HEAD OFFICE:
PO Box 5907, Dallas, Texas 75222
Telephone: (214) 266-4171

PRESIDENT AND CHIEF EXECUTIVE OFFICER:
Sol Love (to retire end 1976)

SENIOR VICE-PRESIDENTS:
R. S. Buzard
D. G. Gilmore
J. J. Welch Jr

VICE-PRESIDENTS:
J. B. Allyn (Washington Operations)
J. B. Andrasko (Personnel)
E. M. Reyno (International Operations)
E. R. Spiegel Jr (Controller)

CORPORATE DIRECTOR OF PUBLIC RELATIONS AND ADVERTISING:
Beal Box

DIVISIONS OF VOUGHT CORPORATION:

Systems Division

DIVISION HEADQUARTERS:
PO Box 5907, Dallas, Texas 75222

PRESIDENT: Sol Love (to retire end 1976)

SENIOR VICE-PRESIDENTS:
R. S. Buzard
D. G. Gilmore
J. J. Welch Jr

VICE-PRESIDENTS:
J. B. Andrasko (Administration)
D. P. Appleby (Materials)
J. W. Casey (Logistics)
E. F. Cvetko (Operations and Manufacturing)
P. W. Hare (Advanced Systems)
F. W. Randall (Advanced Planning)
E. R. Spiegel Jr (Financial and Controller)
G. T. Upton (Engineering)

Service Center

DIVISIONAL HEADQUARTERS:
PO Box 5907, Dallas, Texas 75222

DIRECTOR: W. J. Bosworth

Hampton Technical Center

HEADQUARTERS:
3221 N. Armistead Avenue, Hampton, Virginia 23666

MANAGER: Jack McLain

Michigan Division

Division Headquarters:
38111 Van Dyke, Sterling Heights, Michigan 48077

Vice-President and General Manager: B. M. Smith

Advanced Technology Center Inc

Headquarters:
PO Box 6144, Dallas, Texas 75222

Chairman of the Board and President: F. W. Fenter

The former Chance Vought Aircraft Inc, founded in 1917 and a leading producer of aircraft for the US Navy throughout its history, became the Chance Vought Corporation on 31 December 1960. On 31 August 1961, Chance Vought Corporation merged with Ling-Temco Electronics Inc, to form a combined company known as Ling-Temco-Vought Inc.

What had been Chance Vought Corporation was renamed the Aerospace Division of LTV. In February 1965, the Vought Aeronautics Division, Astronautics Division, Range Systems Division, Michigan Division and Kentron Hawaii Ltd were all grouped to form LTV Aerospace Corporation as a new subsidiary of LTV. In 1972, LTV Aerospace Corporation became a wholly-owned subsidiary of the LTV Corporation (formerly Ling-Temco-Vought Inc), responsible for all aerospace activities of the Corporation.

In September 1972 a reorganisation of the aerospace divisions and subsidiaries of LTV Aerospace Corporation entered its first phase with a merger of Vought Aeronautics Company and Vought Missiles and Space Company into a new Vought Systems Division. The reorganisation continued into early 1974, resulting in a merger of the Ground Transportation Division with Vought Systems Division, the sale of Vought Helicopter Incorporated to Aérospatiale of France, and the creation of Vought Service Center, to handle the company's support service programmes.

With effect from 1 January 1976 LTV Aerospace Corporation changed its name to Vought Corporation. The divisions and subsidiaries of Vought Corporation consist of Systems Division, Michigan Division, Hampton Technical Center, Service Center, and Advanced Technology Center Inc. Employment totalled 11,400 in mid-1976.

As a division of the Vought Corporation, Systems Division has responsibility for all aircraft and spacecraft work, aerospace support and training equipment; production of wings, engine nacelles, pylons, aft fuselage, tail unit and landing gear for the S-3A Viking, and responsibility for assurance of carrier suitability (in conjunction with Lockheed-California, which see); construction of Boeing 747 tail assemblies, McDonnell Douglas DC-10 tailplanes and elevators, Lockheed C-130 and P-3 control surfaces. Additionally, the division has in production the A-7D fighter and A-7E attack aircraft for the USAF and USN respectively, and the A-7H for the Hellenic Air Force. Other current products of Systems Division include the Scout launch vehicle for NASA; components for manned and unmanned space vehicles; advanced missile, guidance, control and environmental systems; Airtrans automatic transit systems; and advanced thermal protection systems.

In December 1974, Vought Systems' Low Volume RamJet (LVRJ) propulsion system test vehicle made a successful first flight, covering a distance of more than 30 nm (56 km; 35 miles) and attaining a speed in excess of 1,259 knots (2,334 km/h; 1,450 mph). Under development for the US Navy, the LVRJ is regarded as a highly promising propulsion system for a new generation of higher-performance tactical missiles. Air-launched from a US Navy A-7 Corsair II, the flight test vehicle is 4·57 m (15 ft) long and 0·38 m (1 ft 3 in) in diameter. It can be scaled up or down for air-to-air, surface-to-air or surface-to-surface applications.

The Michigan Division of Vought Corporation is prime contractor for the US Army's Lance battlefield missile system.

VOUGHT CORSAIR II

US military designation: A-7

On 11 February 1964 the US Navy named the former LTV Aerospace Corporation winner of a design competition for a single-seat carrier-based light attack aircraft. The requirement was for a subsonic aircraft able to carry a greater load of non-nuclear weapons than the A-4E Skyhawk. To keep costs to a minimum and speed delivery it had been stipulated by the Navy that the new aircraft should be based on an existing design; the LTV design study was based, therefore, on the F-8 Crusader. An initial contract to develop and build three aircraft, under the designation A-7A, was awarded on 19 March 1964; first flight was made on 27 September 1965.

Since that time several versions of the A-7 have been evolved as Corsair IIs, for the US Navy, the USAF and the Hellenic Air Force, as follows:

A-7A. Initial attack version for the US Navy, powered by a non-afterburning Pratt & Whitney TF30-P-6 turbofan engine, rated at 50·5 kN (11,350 lb st). The first four were delivered to US Naval Air Test Center on 13-15 September 1966. Deliveries to user squadrons began on 14 October 1966. The A-7A went into combat for the first time with Squadron VA-147 on 3 December 1967, off the USS *Ranger* in the Gulf of Tonkin. Delivery of 199 A-7As to the US Navy was completed in the Spring of 1968.

A-7B. Developed version for the US Navy with non-afterburning TF30-P-8 engine, rated at 54·3 kN (12,200 lb st). First production aircraft flew on 6 February 1968 and A-7B entered combat in Vietnam on 4 March 1969. Last of 196 A-7Bs was delivered to the US Navy on 7 May 1969.

A-7C. Designation applied in late 1971 to the first 67 A-7Es (which see) to eliminate confusion with subsequent Allison-powered A-7Es.

TA-7C. Sixty-five A-7Bs and A-7Cs are being converted into tandem two-seat trainers, with operational capability, under this designation, with deliveries to begin in early 1977. Gun and weapon pylons retained. Configuration similar to YA-7E.

A-7D. Tactical fighter version for the USAF, with a continuous-solution navigation and weapon delivery system, including the capability of all-weather radar bomb delivery. First two were powered by TF30-P-8 engine. Subsequent aircraft have a non-afterburning Allison TF41-A-1 (Spey) turbofan engine. First flight of an A-7D was made on 5 April 1968, and first flight with the TF41 engine on 26 September 1968. First A-7D was accepted by the USAF on 23 December 1968. First unit equipped with the A-7D was the 54th Tactical Fighter Wing at Luke AFB, Arizona. The A-7D entered combat in Southeast Asia in October 1972 with the 354th Tactical Fighter Wing, deployed from Myrtle Beach AFB, South Carolina. Delivery of A-7Ds to Air National Guard units in Colorado, New Mexico, Ohio, Pennsylvania, Puerto Rico and South Carolina began in 1973. Production continues.

In early 1976, an A-7D was being test-flown with a port wing manufactured by Vought Corporation almost entirely from composite materials. The basic material consists of reinforcing graphite and boron fibres supported in an epoxy resin matrix. Eight similar wings are to be tested on A-7Ds operated by ANG units.

A-7E. Developed version for the US Navy equipped as a light attack/close air support/interdiction aircraft. First 67 aircraft (since redesignated A-7C, as indicated) powered by TF30-P-8 non-afterburning turbofan engine; 68th and subsequent aircraft by Allison TF41-A-2 (Spey) non-afterburning turbofan engine rated at 66·7 kN (15,000 lb st). Max internal fuel 5,678 litres (1,500 US gallons), max external fuel 4,542 litres (1,200 US gallons). Max speed at S/L 600 knots (1,112 km/h; 691 mph) and ferry range up to 2,800 nm (5,188 km; 3,224 miles). First flight of an A-7E was made on 25 November 1968 and deliveries began on 14 July 1969. Airframe and equipment are virtually identical to those of the A-7D, except that the gas-turbine self-starter is replaced by an air-turbine starter. Weight empty is 8,800 kg (19,403 lb). The primary navigation/weapon delivery systems for the A-7E are identical to those of the A-7D except for a different ASCU, type C-8185/AWE. The automatic controls differ only by inclusion of an AN/ASN-54 approach power compensator. Radio communications sets include the following: AN/ARC-51A UHF radio; AN/ARR-69 auxiliary UHF receiver; AN/ARN-51 Tacan; AN/APX-72 IFF transponder; AN/ADN-154 radar beacon; AN/ARA-50 ADF; AN/AIC-25 audio system and AN/ASW-25 data link. ECM equipment consists of APR-25 and -27 internal homing and warning systems and ALR-100 active ECM, also external pod-pounted systems compatible with the aircraft internal systems. The A-7E entered combat service in Southeast Asia with Attack Squadrons 146 and 147 in May 1970, operating from the aircraft carrier USS *America*. Production continues at a rate of about two a month.

In early 1976, two A-7Es were being tested with a Vought Systems Division TRAM (target recognition and attack multi-sensor system) pod under the starboard wing, for improved night capability. Equipment includes a Texas Instruments FLIR sensor, and a Marconi-Elliott raster-HUD cockpit display.

YA-7E. Prototype two-seat version, produced as a company funded project by modification of a US Navy A-7E. Described separately.

KA-7F. Proposed carrier-based tanker to replace the KA-3B Skywarrior. Not built.

A-7G. Proposed version for the Swiss Air Force, based on the A-7D with uprated TF41-A-3 engine and equipment changes. Not put into production.

A-7H. Land-based version of A-7E, retaining the folding wings. Total of 60 ordered for the Hellenic Air Force, for delivery between August 1975 and mid-1977. First A-7H flew for first time on 6 May 1975.

Deliveries of all versions totalled 1,318 by 1 January 1976.

The following description, which applies in particular to the A-7D, is generally applicable to other versions of the A-7 except as detailed under the individual model listings:

Type: Subsonic single-seat tactical fighter.

Wings: Cantilever high-wing monoplane. Wing section NACA 65A007. Anhedral 5°. Incidence −1°. Wing

Vought A-7E Corsair II close air support/interdiction aircraft of the US Navy

Vought A-7H Corsair II combat aircraft of the Hellenic Air Force

sweepback at quarter-chord 35°. Outer wing sections fold upward to allow best utilisation of revetments at combat airfields. All-metal multi-spar structure with integrally-stiffened aluminium alloy upper and lower skins. Plain sealed inset aluminium ailerons, outboard of wing fold, are actuated by fully-triplicated hydraulic system. Leading-edge flaps. Large single-slotted trailing-edge flaps. Spoiler above each wing forward of flaps.

FUSELAGE: All-metal semi-monocoque structure. Large door-type ventral speed-brake under centre fuselage.

TAIL UNIT: Large vertical fin and rudder, swept back 44·28° at quarter-chord. One-piece horizontal all-moving tailplane, swept back 45° at quarter-chord and set at dihedral angle of 5° 25′. Tailplane and rudder are operated by triplicated hydraulic systems.

LANDING GEAR: Hydraulically-retractable tricycle type, with single wheel on each main unit and twin-wheel nose unit. Main wheels retract forward into fuselage, nosewheels aft. Main wheels and tyres size 28 × 9-12; nosewheels and tyres size 22 × 5·50. Sting-type arrester hook under rear fuselage for emergency landings or aborted take-offs. Anti-skid brake system.

POWER PLANT: One Allison TF41-A-1 (Rolls-Royce Spey 168-62) non-afterburning turbofan engine, rated at 63·4 kN (14,250 lb st). Engine has self-start capability through the medium of battery-powered electric motor that spins an air-breathing gas-turbine starter. The starter unit includes a turbine-driven compressor that compresses air for combustion, a free turbine for accelerating the engine to a self-sustaining speed, and an integral control system. The engine has self-contained ignition for start/airstart, automatic relight and selective ignition. Integral fuel tanks in wings and additional fuselage tanks. Maximum internal fuel 5,394 litres (1,425 US gallons). Maximum external fuel 4,542 litres (1,200 US gallons). All fuel tanks filled with polyurethane fire-suppressing foam. Some fuselage tanks and fuel lines self-sealing. Alternate fuel feed system. Flight refuelling capability of first 26 A-7Ds provided by a probe and drogue system; 27th and subsequent aircraft have boom receptacle above fuselage on port side in line with wing leading-edge. Boron carbide (HFC) engine armour.

ACCOMMODATION: Pilot on McDonnell Douglas Escapac rocket-powered ejection system, complete with USAF life support system, that provides a fully-inflated parachute three seconds after sequence initiation; positive seat/man separation and stabilisation of the ejected seat and pilot. Boron carbide (HFC) cockpit armour.

SYSTEMS: Triple-redundant hydraulic system for flight controls; double-redundant system for flaps, brakes and landing gear retraction. Electrical system includes storage batteries for engine starting and maintenance of ground-alert radio communications without need for the engine to be running. Liquid oxygen system. An air-conditioning unit using engine bleed air provides pressurisation and cooling for the cockpit and cooling for certain electronics components. Automatic flight control system provides control-stick steering, altitude hold, heading hold, heading pre-select and attitude hold. Ram-air turbine provides hydraulic pressure and electrical power down to airspeeds below those used in normal landing approaches.

ELECTRONICS AND EQUIPMENT: The primary navigation/weapon-delivery system comprises AN/APQ-126 forward-looking radar, which is utilised for air-to-ground ranging for weapons delivery, ground mapping for navigation or adverse-weather bombing, manual terrain following and terrain avoidance, circular polarisation to enhance adverse-weather penetration; beacon mode for use in conjunction with AN/APN-134 rendezvous beacons in other aircraft; direct-view storage tube (DVST) for Walleye or radar presentation; AN/ASN-91 digital computer; AN/APN-141 radar altimeter; AN/AVQ-7 head-up display (HUD) by Marconi-Elliott (UK); CP-953/A air data computer; AN/ASN-90 inertial measuring set; CV-2622(ASN-99 projected map; AN/APN-190 Doppler radar; C-8230/AWE armament station and control unit (ASCU); and provisions for Loran. Automatic controls comprise nose-gear steering and AN/ASW-30 automatic flight control system. Radio communication and navigation aids include FM-622 VHF radio; AN/ARC-51 BX UHF radio; AN/ARR-69 auxiliary UHF receiver; AN/ARN-52 Tacan; AN/APX-72 IFF transponder; AN/ARN-58A ILS; AN/APN-154 radar beacon; CPU-80A flight director computer; AN/ARA-50 ADF and AN/AIC-26 audio system, ECM equipment consists of internal homing and warning systems and external pod-mounted systems compatible with the aircraft's internal systems.

ARMAMENT: A wide range of stores, to a total weight of more than 6,805 kg (15,000 lb), can be carried on six underwing pylons and two fuselage weapon stations. Two outboard pylons on each wing can each accommodate a load of 1,587 kg (3,500 lb). Inboard pylon on each wing can carry 1,134 kg (2,500 lb). Two fuselage weapon stations, one on each side, can each carry load of 227 kg (500 lb). Weapons carried include air-to-air and air-to-ground missiles; general-purpose bombs; rockets; gun pods; and auxiliary fuel tanks. In addition, an M-61A1 Vulcan 20 mm cannon is mounted in the port side of the fuselage. This has 1,000-round ammunition storage and selected firing rates of 4,000 or 6,000 rds/min. Strike camera in lower rear fuselage for damage assessment.

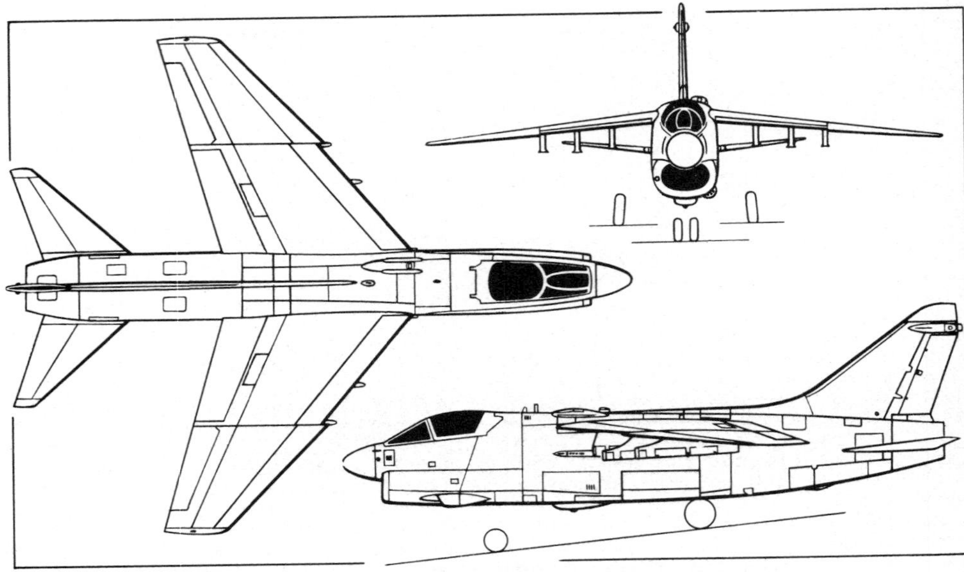

Vought A-7D tactical fighter version of the Corsair II for the USAF *(Pilot Press)*

Vought YA-7E, a prototype two-seat trainer version of the US Navy's A-7E Corsair II

DIMENSIONS, EXTERNAL:
- Wing span: 11·80 m (38 ft 9 in)
- Width, wings folded: 7·24 m (23 ft 9 in)
- Wing chord at root: 4·72 m (15 ft 6 in)
- Wing chord at tip: 1·18 m (3 ft 10¼ in)
- Wing aspect ratio: 4
- Length overall: 14·06 m (46 ft 1½ in)
- Height overall: 4·90 m (16 ft 0¾ in)
- Tailplane span: 5·52 m (18 ft 1½ in)
- Wheel track: 2·90 m (9 ft 6 in)

AREAS:
- Wings, gross: 34·83 m² (375 sq ft)
- Ailerons (total): 1·85 m² (19·94 sq ft)
- Trailing-edge flaps (total): 4·04 m² (43·48 sq ft)
- Leading-edge flaps (total): 4·53 m² (48·74 sq ft)
- Spoiler: 0·43 m² (4·60 sq ft)
- Deflector: 0·32 m² (3·44 sq ft)
- Fin: 10·33 m² (111·20 sq ft)
- Rudder: 1·40 m² (15·04 sq ft)
- Horizontal tail surfaces: 5·24 m² (56·39 sq ft)
- Speed-brake: 2·32 m² (25·00 sq ft)

WEIGHTS:
- Weight empty: 8,988 kg (19,815 lb)
- Max T-O weight: 19,050 kg (42,000 lb)

PERFORMANCE:
- Max level speed at S/L: 606 knots (1,123 km/h; 698 mph)
- Max level speed at 1,525 m (5,000 ft):
  - with 12 Mk 82 bombs: 562 knots (1,040 km/h; 646 mph)
  - after dropping bombs: 575 knots (1,065 km/h; 661 mph)
- Sustained manoeuvring performance at 1,525 m (5,000 ft), at AUW of 13,047 kg (28,765 lb) with 6 pylons and 2 Sidewinder missiles
  - 1,770 m (5,800 ft) turning radius at 4g and 500 knots (925 km/h; 575 mph)
- T-O run at max T-O weight: 1,525 m (5,000 ft)
- Ferry range:
  - max internal fuel: 1,981 nm (3,671 km; 2,281 miles)
  - max internal and external fuel: 2,494 nm (4,621 km; 2,871 miles)

VOUGHT CORSAIR II₂

US military designation: YA-7E

Vought's Systems Division, under a company funded project and by modification of an aircraft furnished by the US Navy, designed and built the prototype of a two-seat version of the A-7 Corsair II. Envisaged as an advanced trainer, or as a new operational configuration for tactical duties such as electronic countermeasures, it was known originally as the Vought Project V-159, and later as the YAH-7H, but now has the designation YA-7E to avoid confusion with the single-seat A-7H ordered by the Hellenic Air Force.

The YA-7E has the same basic structure and equipment as the A-7, with a 0·41 m (1 ft 4 in) longer nose section to make room for the second cockpit. A 0·46 m (1 ft 6 in) section is inserted in the rear fuselage, in line with the trailing-edge of the wing, and the overwing fairing is modified to maintain the fuselage profile. The aft fuselage is also modified to cant upward 1° 19′, so making possible approach and landing attitudes identical to those of other A-7s, despite the longer fuselage.

The new cockpit section has two McDonnell Douglas Escapac ejection seats in tandem, under an electrically-actuated sideways-opening (to starboard) cockpit canopy. It is intended that ejection would be made through the canopy; so the seat headrests carry strikers to shatter the canopy and provide a clear egress. Each cockpit has full flying controls, communications and navigation equipment. The rear cockpit, occupied by an instructor or check pilot, has a repeater-type head-up display, and the seat is raised slightly above the level of the front seat to enhance forward view. A brake parachute has been installed to permit short-field operations.

The YA-7E, which was the first production A-7E fitted with the Allison TF41-A-2 (Spey) turbofan engine, was modified at the Systems Division's works in Dallas. It was not fitted with an in-flight refuelling probe or gun, but is able to accept such equipment if required.

First flight of the YA-7E was made on 29 August 1972. Despite an increase in empty weight, there is little difference in performance between the YA-7E and the A-7E. The 81 TA-7C Corsair II trainers undergoing conversion from A-7B/C are similar in configuration to the YA-7E.

DIMENSIONS, EXTERNAL:
- Wing span: 11·80 m (38 ft 9 in)
- Length overall: 14·68 m (48 ft 2 in)
- Height overall: 5·00 m (16 ft 5 in)

WEIGHTS:
- Weight empty: 8,938 kg (19,705 lb)
- Normal T-O weight (no military stores): 13,725 kg (30,259 lb)

PERFORMANCE (at T-O weight of 15,450 kg; 34,062 lb):
- Max level speed at S/L: 537 knots (994 km/h; 618 mph)
- Catapult T-O speed: 142 knots (264 km/h; 164 mph)
- T-O run: 1,097 m (3,600 ft)
- Landing run:
  - without brake parachute: 960 m (3,150 ft)
  - with brake parachute: 670 m (2,200 ft)
- Radius of action: 405 nm (750 km; 466 miles)
- Ferry range (no external stores): 1,970 nm (3,650 km; 2,268 miles)

WILLIAMS
WILLIAMS RESEARCH CORPORATION

ADDRESS:
2280 West Maple Road, Walled Lake, Michigan 48088
Telephone: (313) 624-5200

On 26 January 1970, Bell Aerospace announced that it had granted to the Williams Research Corporation a licence to manufacture, use and sell certain small lift device systems in the USA and Canada. They included the Jet Flying Belt which Bell had developed for the US Army under a $3 million contract, and which was described in the 1972-73 *Jane's*.

Since that time Williams has been working on a new and more advanced version, and announced on 14 February 1974 that it had tested successfully a two-man turbine-powered flying platform. Known as **WASP** (Williams Aerial Systems Platform), its flight tests were conducted with the vehicle attached to a safety tether line, under a US Navy contract to demonstrate its suitability to meet a US Marine Corps STAMP (Small Tactical Aerial Mobility Platform) requirement.

Powered by a Williams WR19-9 miniature turbofan engine, developing 3·11 kN (700 lb) thrust, the WASP is essentially a platform to which the WR19 engine has been mounted vertically. This means that the pilot has merely to mount the platform, start the engine and fly off, using simple hand controls so designed that the vehicle can be controlled in flight using one hand.

It is anticipated that the WASP will prove capable of carrying two men at speeds of up to 52 knots (97 km/h; 60 mph) for a duration of approximately 30 minutes. The vehicle will be able to accelerate rapidly, move in any direction, hover and rotate on its own axis. The company believes that WASP will be suitable for a number of military and civil applications, including law enforcement, firefighting, rescue and medical aid. Empty weight of the vehicle is 123 kg (370 lb).

The company stated in early 1976 that no contract was then in existence for further development, but that the next stage would be to demonstrate free flight.

Williams Research prototype WASP flying platform (one Williams WR19-9 turbofan engine)

THE UNION OF SOVIET SOCIALIST REPUBLICS

ANTONOV

GENERAL DESIGNER IN CHARGE OF BUREAU:
Oleg Konstantinovich Antonov

After establishing his reputation with a series of successful glider and sailplane designs, Oleg K. Antonov became one of Russia's leading designers of transport aircraft, particularly those types intended for short-field operation.

Details of the current products of his design bureau, which is situated in Kiev, are given hereafter.

ANTONOV An-2

NATO reporting name: *Colt*

Following manufacture of the An-2M specialised agricultural version of this large single-engined biplane, in the mid-sixties, production of the An-2 came to an end in the Soviet Union. Details of the various versions that were built can be found in the 1971-72 *Jane's*.

Several versions of the An-2 continued in production under licence in Poland (see WSK-PZL-Mielec entry).

ANTONOV An-3

It was reported in the Spring of 1972 that the Antonov design bureau was engaged on design studies for a turboprop development of the An-2 biplane (see WSK-PZL-Mielec in Polish section). Further details were given by Mr Oleg Antonov during the 1975 Paris Air Show.

Designated An-3, the new aircraft is intended specifically for agricultural duties and will compete directly with the Polish turbofan-engined WSK-Mielec M-15 as the next-generation agricultural aircraft for use throughout the countries of eastern Europe and the Soviet Union. The prototype was produced by converting an An-2 to have a 716 kW (960 shp) Glushenkov TVD-10A turboprop engine, driving a slow-turning large-diameter propeller optimised for an aircraft operating speed of 75-97 knots (140-180 km/h; 87-112 mph). The prototype An-3 was scheduled to fly for the first time in 1976.

ANTONOV An-12

NATO reporting name: *Cub*

The basic An-12 is a military and civil freight-carrying version of the now-retired An-10 passenger transport, with redesigned rear fuselage and tail unit. A loading ramp for freight and vehicles, in the underside of the upswept rear fuselage, can be lowered in flight for air-drop operations. The built-in freight-handling gantry has a capacity of 2,300 kg (5,070 lb). The cargo floor is designed for loadings of up to 1,500 kg/m² (307 lb/sq ft).

The military transport version (known to NATO as *Cub-A*) is a standard paratroop and freight transport in the Soviet Air Force. Sixteen were supplied to the Indian Air Force. Others are operated by the air forces of Algeria, Bangladesh, Egypt, Indonesia, Iraq, Poland and Syria. Civil An-12s serve with Aeroflot, Polish Air Lines (LOT), Bulair and Cubana. Altogether, more than 900 are reported to have been built for military and civil use. They equip nearly half of the Soviet military air transport fleet, and have the capability of carrying two army divisions, totalling 14,000 men and equipment, over a radius of 651 nm (1,207 km; 750 miles).

Cub-A has a tail gunner's position. In the An-12 which Ghana Airways operated for a time, this was fitted out as a toilet. In the refined commercial version, first demonstrated at the 1965 Paris Air Show, the turret is removed and replaced by a streamlined fairing.

The special *Cub-C* version of the An-12, used by the Soviet Air Force and Navy for ECM duties, also has a 'solid' tail, housing electronic equipment, instead of a gun position. It retains the glazed nose and undernose radar of other versions, and has additional electronic pods faired into the forward fuselage and ventral surfaces.

Equipment for all-weather operation is standard on all versions. Current Soviet Air Force An-12s have a larger undernose radome than that originally fitted.

One of the An-12s operated by Aeroflot's Polar aviation service was used to test skis of an entirely new design. Of unusually wide and deep section, these have a shallow curved Vee lower surface, like a flattened version of the planing bottom of a seaplane float. The skis are equipped with braking devices and warming equipment and are claimed to permit landings at prepared fields as well as on

ECM version of the Antonov An-12 *Cub-C)* **in Egyptian Air Force insignia**

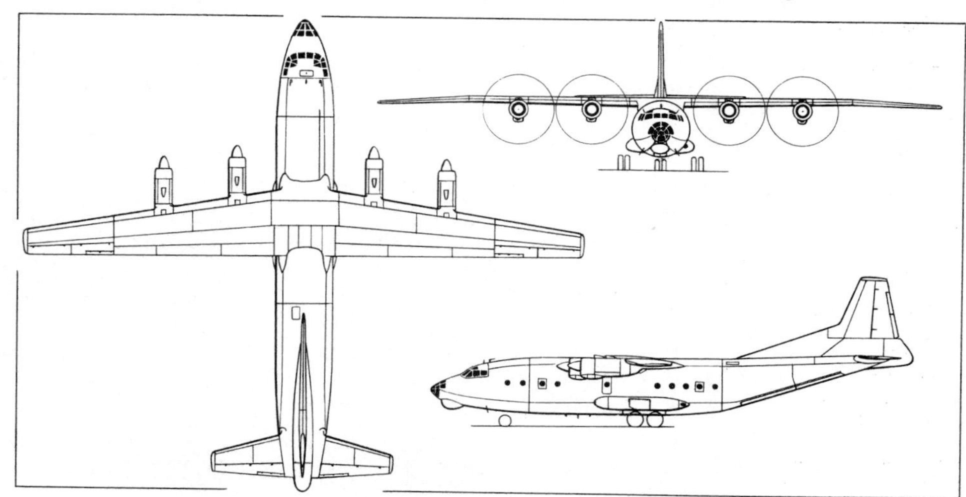

Antonov An-12 four-turboprop general-purpose commercial transport aircraft *(Pilot Press)*

Antonov An-12 rear-loading transport (four Ivchenko AI-20K turboprop engines) of Aeroflot *(M. D. West)*

virgin snow. Each main ski is supported by a primary oleo strut, with scissor-arm system, and fore and aft secondary oleos to absorb pitching (and possibly rolling) moments. They were intended as standard equipment on aircraft used in the Arctic and Antarctic and for Winter services.

TYPE: Four-engined cargo transport.

WINGS: Cantilever high-wing monoplane. All-metal two-spar structure in five panels, comprising centre-section, intermediate wings and tip sections. TsAGI wing sections: S-5-18 at centreline, S-3-16 at intermediate station, S-3-14 at tip. Anhedral 4° on tip sections. Sweepback 6° 30′ at quarter-chord. Manually-operated aerodynamically-balanced ailerons. Double-slotted Fowler flaps in two portions each side, hydraulically-actuated. Electro-thermal de-icing.

FUSELAGE: Stressed-skin semi-monocoque structure of circular section.

TAIL UNIT: Cantilever all-metal structure. Electrically-operated trim tabs. All controls are manually operated and aerodynamically balanced. Electro-thermal de-icing of fin and tailplane.

LANDING GEAR: Retractable tricycle type. Hydraulic actuation. Shock-absorbers use nitrogen instead of air and have stroke of 340 mm (13·4 in). Four-wheel bogie on each side retracts into blister on side of fuselage. Hydraulically-steerable dual nosewheels. Main-wheel tyres size 1,050 × 300 mm; pressure 5·52-6·55 bars (80-95 lb/sq in). Hydraulic disc brakes.

POWER PLANT: Four 2,983 kW (4,000 ehp) Ivchenko AI-20K turboprops, driving AV-68 four-blade reversible-pitch propellers. All fuel in 22 bag-type tanks in wings, total normal capacity 13,900 litres (3,058 Imp gallons). Max capacity (military) 18,100 litres (3,981 Imp gallons).

ACCOMMODATION: Pilot and co-pilot side by side on flight deck. Engineer's station on starboard side, behind co-pilot. Radio operator in well behind pilot, facing outboard. Navigator in glazed nose compartment. Rear gunner in tail turret of military version. Crew door on port side forward of wing. Access to freight hold via ramp-door at rear, under upswept rear fuselage. Ramp-door is divided into two longitudinal halves, which can be hinged upward inside cabin to provide access for direct loading of freight from trucks. Under-surface of fuselage aft of ramp is formed by door which hinges upward into fuselage to facilitate loading and unloading. Commercial version can carry 14 passengers in compartment aft of flight deck. Military version can carry 100 paratroops, all of whom can be despatched in under one minute, with ramp-doors folded upward.

SYSTEMS: Entire accommodation air-conditioned and pressurised to differential of 0·49 bars (7·1 lb/sq in). Hydraulic system operates landing gear retraction, nosewheel steering, flaps, brakes and rear loading ramp and door.

ARMAMENT (military version): Two 23 mm NR-23 guns in tail turret.

DIMENSIONS, EXTERNAL:

Wing span	38·00 m (124 ft 8 in)
Wing chord (mean)	3·452 m (11 ft 4 in)
Wing aspect ratio	11·85
Length overall	33·10 m (108 ft 7¼ in)
Height overall	10·53 m (34 ft 6½ in)
Tailplane span	12·20 m (40 ft 0¼ in)
Wheel track	5·42 m (17 ft 9½ in)
Wheelbase	10·82 m (35 ft 6 in)
Propeller diameter	4·50 m (14 ft 9 in)
Rear loading hatch:	
Length	7·70 m (25 ft 3 in)
Width	2·95 m (9 ft 8 in)

DIMENSIONS, INTERNAL:

Cargo hold:	
Length	13·50 m (44 ft 3½ in)
Max width	3·50 m (11 ft 5¾ in)
Max height	2·60 m (8 ft 6¼ in)
Volume	97·2 m³ (3,432·6 cu ft)

AREAS:

Wings, gross	121·70 m² (1,310 sq ft)
Ailerons (total)	7·84 m² (84·39 sq ft)
Trailing-edge flaps (total)	27·00 m² (290·63 sq ft)
Vertical tail surfaces	21·53 m² (231·75 sq ft)
Rudder	6·53 m² (70·29 sq ft)
Horizontal tail surfaces	26·95 m² (290·09 sq ft)
Elevators	7·11 m² (76·53 sq ft)
Elevator trim tabs (total)	0·78 m² (8·40 sq ft)

WEIGHTS AND LOADINGS (military model):

Weight empty	28,000 kg (61,730 lb)
Max payload	20,000 kg (44,090 lb)
Normal T-O weight	55,100 kg (121,475 lb)
Max T-O weight	61,000 kg (134,480 lb)
Normal wing loading	461 kg/m² (94·4 lb/sq ft)
Normal power loading	4·62 kg/kW (7·5 lb/ehp)

WEIGHTS (late civil model):

Max payload	20,000 kg (44,090 lb)
Normal T-O weight	54,000 kg (119,050 lb)
Max T-O weight	61,000 kg (134,480 lb)

PERFORMANCE (military model):

Max level speed	419 knots (777 km/h; 482 mph)
Max cruising speed	361 knots (670 km/h; 416 mph)
Min flying speed	88 knots (163 km/h; 101 mph)
Landing speed	108 knots (200 km/h; 124 mph)
Rate of climb at S/L	600 m (1,970 ft)/min
Service ceiling	10,200 m (33,500 ft)
T-O run	700 m (2,300 ft)
Landing run	500 m (1,640 ft)
Range with max payload	1,942 nm (3,600 km; 2,236 miles)
Range with max fuel	3,075 nm (5,700 km; 3,540 miles)

PERFORMANCE (late civil model, at normal T-O weight):

Max cruising speed	324 knots (600 km/h; 373 mph)
Normal cruising speed at 7,500 m (25,000 ft)	297 knots (550 km/h; 342 mph)
Max rate of climb at S/L	600 m (1,970 ft)/min
Service ceiling	10,200 m (33,500 ft)
T-O run	850 m (2,790 ft)
Landing run	860 m (2,820 ft)
Range with 10,000 kg (22,050 lb) cargo, 1 hour reserves	1,832 nm (3,400 km; 2,110 miles)

ANTONOV An-14 PCHELKA (LITTLE BEE)

NATO reporting name: *Clod*

The An-14 Pchelka is a twin-engined light general-purpose aircraft, the first prototype of which made its first flight on 15 March 1958.

The two prototypes were each powered by two 194 kW (260 hp) Ivchenko AI-14R radial engines and accommodated six passengers and 150 kg (330 lb) of baggage, or 600 kg (1,320 lb) of freight. They flew originally with a straight tailplane and with V-shape leading-edges on the twin tail-fins. This type of tail unit was superseded by a dihedral tailplane and rectangular fins of increased area on production aircraft, which also have 224 kW (300 hp) engines, increased wing span and accommodation for a pilot and six to eight passengers, plus baggage.

Production began in 1965 at the Progress Plant at Arsenyev in the far east of the Soviet Union, for both Aeroflot and the Soviet armed forces; about 300 had been delivered by the Spring of 1975.

The military An-14 was first seen at the Domodedovo air display in July 1967 and does not appear to differ externally from the civilian passenger version. It serves also with the air forces of Bulgaria, the German Democratic Republic and Guinea, and with Balkan Bulgarian Airlines.

An executive version is available with de luxe accommodation for five passengers and their baggage, with tables between facing seats. All seats are quickly removable to provide an unobstructed cabin for cargo carrying. An ambulance version, which is in production, can accommodate six stretchers, in tiers of three on each side of the cabin, with an attendant. Dual controls are available for pilot training and a variety of equipment can be fitted for geological survey, aerial photography and agricultural duties.

Photographs were issued showing aircraft No. CCCP-L1053 (with original fins and probably a prototype) equipped for agricultural duties. It had spraybars attached to the aileron-flap brackets under each wing, from the wingtip to a point immediately aft of the wing/bracing strut junction. From there the spraybars ran down the bracing strut and along the stub-wings to meet under the fuselage. The chemical tank was housed in the main cabin and the standard rear loading doors were replaced by a larger removable fairing panel. Entry to the flight deck was via a forward-opening door on the starboard side of the nose, hinged on the centreline of the aircraft.

Great emphasis has been placed on simplicity of servicing and handling, and the An-14 is said to be suitable for

Antonov An-14 Pchelka (two Ivchenko AI-14RF engines) of the Bulgarian Air Force

operation by "pilots of average skill". It will maintain height on one engine at its maximum T-O weight. Radio, navigation equipment, instrumentation, landing light and de-icing equipment make possible operation at night and in bad weather.

The following data apply to the standard production An-14:

TYPE: Twin-engined light general-purpose aircraft.

WINGS: Braced high-wing monoplane with single streamline-section bracing strut each side. Dihedral 2°. Conventional all-metal two-spar structure. Full-span leading-edge slats, in three sections on each wing. Section between fuselage and engine nacelle on each wing is extended pneumatically when the flaps are lowered 35-40°. The remaining four sections are actuated automatically by the airflow over the wing. Entire trailing-edges hinged; each comprising a double-slotted flap, with the slat of the flap extending to the wingtip and built into the single-slotted aileron as a leading-edge structure. Flaps operated pneumatically. Trim tab in port aileron. Small stub-wing carries each main landing gear unit and provides lower attachment for bracing strut.

FUSELAGE: Conventional all-metal semi-monocoque pod and boom structure.

TAIL UNIT: Cantilever all-metal structure, with 9° dihedral on tailplane. Twin fins and rudders, mounted at right-angles to the tips of the tailplane, so that they toe inwards at the top. Trim tab in port rudder and in each elevator. Leading-edges of fins and tailplane embody both warm air and electrical anti-icing systems.

LANDING GEAR: Non-retractable tricycle type, with single wheel on each unit. Main units carried on short stub-wings. Wide-tread tyres, size 700 × 250, on all three units. Tyre pressure: main units 3·45 bars (50 lb/sq in); nosewheel 2·93 bars (42·5 lb/sq in). Nosewheel is fully-castoring and steerable to 70° each way, and is self-centering for take-off. Pneumatic brakes on main wheels. Skis can be fitted for operation from snow, or floats for operation from water.

POWER PLANT: Two 224 kW (300 hp) Ivchenko AI-14RF nine-cylinder radial aircooled engines, each driving a V-530 two-blade or three-blade variable-pitch propeller. Each engine draws its fuel normally from an inboard metal tank and outer flexible tank in the wing on which it is mounted, with a cross-feed to enable either engine to be supplied from both tank groups in the event of an engine failure. Total fuel capacity 383 litres (84 Imp gallons).

ACCOMMODATION: Pilot and one passenger side by side on flight deck. Main cabin normally seats six persons in pairs in individual forward-facing armchair seats, each by a large window and with central aisle. Provision for seven seats in main cabin in high-density version. Cabin soundproofed and provided with heating and ventilation systems. Door from cabin to flight deck. Passengers enter cabin through clamshell rear doors which form underside of upswept rear fuselage. Chemical tank capacity of agricultural version is 1,000 litres (220 Imp gallons). Optional dual controls.

SYSTEMS: Pneumatic system, pressure 49 bars (710 lb/sq in), for wheel brakes, flap and slat actuation, engine starting, cabin heating and anti-icing. Main 27V DC electrical system supplied by two 3kW engine-driven generators. Two converters to supply 115V 400Hz AC power.

ELECTRONICS AND EQUIPMENT: Standard equipment includes an artificial horizon, gyro-compass, magnetic compass, altimeter, airspeed indicator, clock, rate-of-climb indicator, turn indicator, outside air temperature gauge and ice-warning device. Com/nav equipment can include duplicated transceivers, ADF, VOR, marker beacon and radio altimeter.

DIMENSIONS, EXTERNAL:

Wing span	21·99 m (72 ft 2 in)
Wing chord, mean	1·89 m (6 ft 2½ in)
Wing aspect ratio	12·15
Length overall	11·44 m (37 ft 6½ in)
Height overall	4·63 m (15 ft 2½ in)
Tailplane span	5·00 m (16 ft 4¾ in)
Wheel track	3·60 m (11 ft 9¾ in)
Wheelbase	3·71 m (12 ft 2 in)
Propeller diameter	2·90 m (9 ft 6 in)
Cabin door:	
Length	1·90 m (6 ft 3 in)
Width	0·85 m (2 ft 9½ in)

DIMENSIONS, INTERNAL:

Cabin, excl flight deck:	
Length	3·10 m (10 ft 2 in)
Width	1·53 m (5 ft 0 in)
Height	1·60 m (5 ft 3 in)

AREA:

Wings, gross	39·72 m² (427·5 sq ft)

WEIGHTS AND LOADINGS:

Weight empty	2,000 kg (4,409 lb)
Max payload (normal)	720 kg (1,590 lb)
Max T-O weight	3,600 kg (7,935 lb)
Max wing loading	90·8 kg/m² (18·6 lb/sq ft)
Max power loading	8·04 kg/kW (13·2 lb/hp)

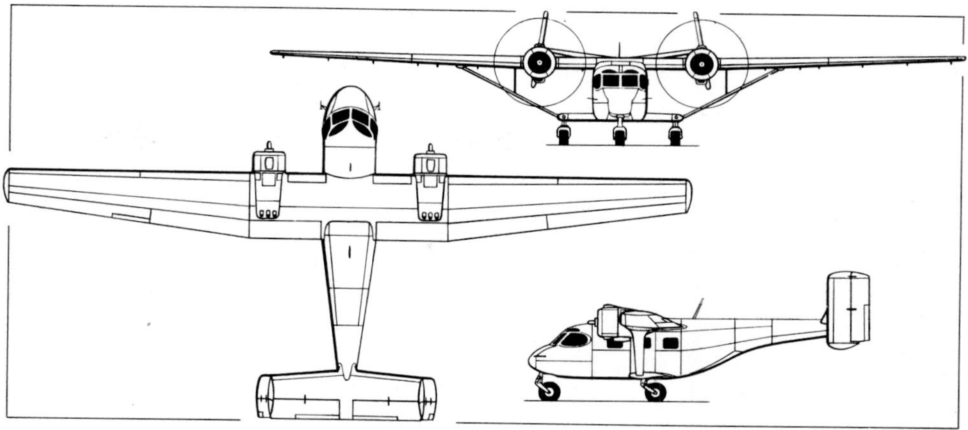

Antonov An-14 Pchelka twin-engined light general-purpose aircraft *(Pilot Press)*

PERFORMANCE (at max T-O weight):

Max level speed at 1,000 m (3,280 ft)	120 knots (222 km/h; 138 mph)
Normal cruising speed at 2,000 m (6,560 ft)	97 knots (180 km/h; 112 mph)
Operating speed, agricultural duties	76 knots (140 km/h; 87 mph)
Landing speed	46 knots (85 km/h; 53 mph)
Max rate of climb at S/L	306 m (1,000 ft)/min
Service ceiling	5,200 m (17,060 ft)
T-O run, on concrete	100 m (328 ft)
T-O to 15 m (50 ft)	200 m (656 ft)
Landing from 15 m (50 ft)	300 m (985 ft)
Landing run	70 m (230 ft)
Range:	
with max payload	350 nm (650 km; 404 miles)
with 550 kg (1,212 lb) payload	385 nm (715 km; 444 miles)
with max fuel	431 nm (800 km; 497 miles)

ANTONOV An-22 ANTHEUS

NATO reporting name: *Cock*

Nothing was known of this very large transport aircraft until the first prototype arrived in Paris on 16 June 1965, during the Salon. It had flown for the first time on 27 February 1965 and the walls of its main cabin were lined with test equipment, recorders, etc. Four more prototypes and the first production An-22 were flying by mid-1967. Two of the prototypes were then being operated by Aeroflot on experimental freight services. Three An-22s, in military insignia, took part in the air display at Domodedovo on 9 July 1967, landing batteries of *Frog-3* and *Ganef* missiles on tracked launchers. Production aircraft have been delivered to both the Soviet Air Force and Aeroflot, which uses the An-22 mainly in underdeveloped areas of the northern USSR, Siberia and the Far East. Deliveries are believed to have been completed during 1974.

In general configuration, the An-22 is similar to its much smaller predecessor, the An-12, with the outer wing anhedral that is a characteristic of many Antonov designs. It is intended primarily for long-distance transportation of heavy bulk cargoes and civil and military equipment, accompanied by operating personnel.

The later production version shown in the accompanying illustrations differs in a number of details from early An-22s. The main navigation radar, housed originally under the starboard landing gear fairing, is mounted at the nose, which carries two radars, in a thimble-type fairing above the modified nose windows and in a large under-fuselage radome. A fairing extends forward from the lower radome and might be fitted with shutters, but its purpose is unknown. One aircraft of this type took part in military manoeuvres at Dvina in NW Byelorussia in early 1970. Four (serial numbers 09302/3/4/6) were used to airlift relief supplies to Peru after the severe earthquake of July 1970; 09303 was lost over the Atlantic, south of Greenland, on 18 July.

On 26 October 1967, an An-22 set up fourteen payload-to-height records, piloted by I. Davydov and with a crew of seven. It reached a height of 7,848 m (25,748 ft) with a payload of 100,000 kg of metal blocks, qualifying also for records with 35,000, 40,000, 45,000, 50,000, 55,000, 60,000, 65,000, 70,000, 75,000, 85,000, 90,000 and 95,000 kg. Max payload lifted to a height of 2,000 m was 100,444·6 kg (221,443 lb). Take-off run with this load was stated to be just over one kilometre. The flight lasted 78 minutes.

A further series of ten records, for speed with payload, was set up in February 1972 by an An-22 captained by Marina Popovich, wife of the Soviet cosmonaut Pavel Popovich. The aircraft averaged 320·161 knots (593·318 km/h; 368·671 mph) around a 2,000 km closed circuit with a 50,000 kg payload, qualifying also for records with 30,000, 35,000, 40,000 and 45,000 kg, on 19 February. Two days later, it averaged 328·326 knots (608·449 km/h; 378·073 mph) around 1,000 km with the same payload.

The An-22 also holds three records for speed with payload over a 5,000 km closed circuit. On 21 October 1974, piloted by S. Dedoukh, a payload of 30,000 kg was carried at a speed of 322·300 knots (597·283 km/h; 371·134 mph). On 24 October 1974, piloted by Y. Romanov, a payload of 35,000 kg was carried at a speed of 317·636 knots (588·639 km/h; 365·763 mph). On 17 April 1975, piloted by G. Pakilev, a payload of 40,000 kg was carried at a speed of 315·156 knots (584·042 km/h; 362·907 mph).

The following details refer to the production version as illustrated:

TYPE: Long-range heavy turboprop transport.

WINGS: Cantilever high-wing monoplane. Marked anhedral on outer panels. All-metal structure, appearing to have three main spars which attach to three strong fuselage ring-frames. Double-slotted trailing-edge flaps. Tab in each aileron.

FUSELAGE: All-metal semi-monocoque structure, with upswept rear fuselage containing loading-ramp/door for direct loading. Retractable jacks support rear fuselage

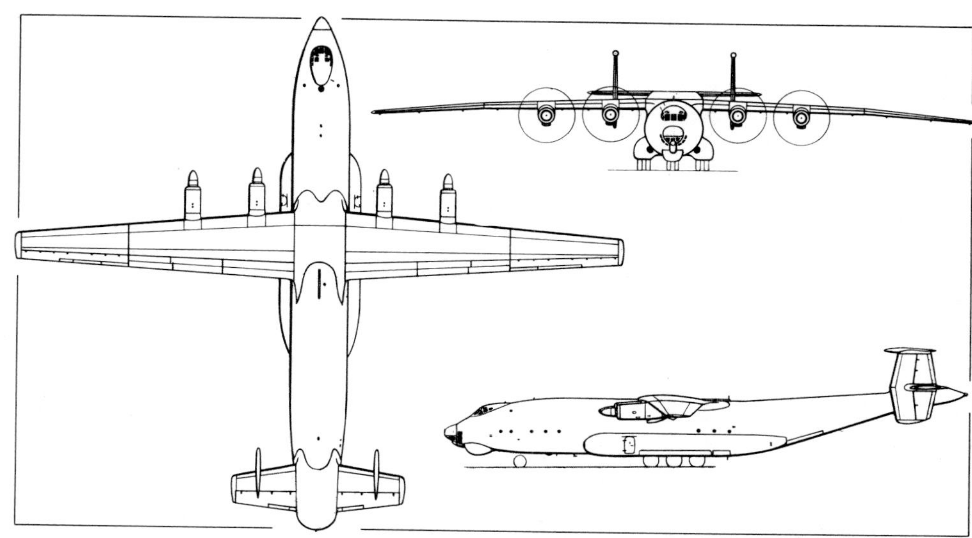

Antonov An-22 Antheus long-range heavy transport aircraft *(Pilot Press)*

Antonov An-22 Antheus long-range heavy transport aircraft (four Kuznetsov NK-12MA turboprop engines) *(D. J. Holford)*

at point where rear loading ramp is hinged.

TAIL UNIT: Cantilever all-metal structure. Twin fins and rudders (each in two sections, above and below tailplane) mounted outboard of mid-span. Tabs in each elevator and in each of the four rudder sections.

LANDING GEAR: Retractable tricycle type, designed to permit off-runway operation. Steerable twin-wheel nose unit. Each main gear consists of three twin-wheel levered-suspension units in tandem, each unit mounted at the bottom of one of the fuselage ring frames that also picks up a wing spar. Main units retract upward into fairings built on to sides of fuselage. Tyre pressure adjustable in flight or on ground to suit airfield surface.

POWER PLANT: Four 11,186 kW (15,000 shp) Kuznetsov NK-12MA turboprop engines, each driving a pair of four-blade contra-rotating propellers.

ACCOMMODATION: Crew of five or six. Navigator's station in nose. Cabin for 28-29 passengers aft of flight deck, separated from main cabin by bulkhead containing two doors. Uninterrupted main cabin, with reinforced titanium floor, tiedown fittings and rear loading ramp. When ramp lowers, a large door which forms the underside of the rear fuselage retracts upward inside fuselage to permit easy loading of tall vehicles. Rails in roof of cabin for four travelling gantries continue rearward on underside of this door. Two winches, used in conjunction with the gantries, each have a capacity of 2,500 kg (5,500 lb). Door in each landing gear fairing, forward of wheels, for crew and passengers.

ELECTRONICS AND EQUIPMENT: Pressurisation equipment and APU in forward part of starboard landing gear fairing. Two radars, in nose 'thimble' and undernose fairings.

DIMENSIONS, EXTERNAL:

Wing span	64·40 m (211 ft 4 in)
Length overall (prototype)	57·80 m (189 ft 7 in)
Height overall	12·53 m (41 ft 1½ in)
Propeller diameter	6·20 m (20 ft 4 in)

DIMENSIONS, INTERNAL:

Main cabin: Length	33·0 m (108 ft 3 in)
Max width	4·4 m (14 ft 5 in)
Max height	4·4 m (14 ft 5 in)

AREA:

Wings, gross	345 m² (3,713 sq ft)

WEIGHTS:

Weight empty, equipped	114,000 kg (251,325 lb)
Max payload	80,000 kg (176,350 lb)
Max fuel	43,000 kg (94,800 lb)
Max T-O weight	250,000 kg (551,160 lb)

PERFORMANCE:

Max level speed	399 knots (740 km/h; 460 mph)
T-O run	1,300 m (4,260 ft)
Landing run	800 m (2,620 ft)
Range with max fuel and 45,000 kg (99,200 lb) payload	5,905 nm (10,950 km; 6,800 miles)
Range with max payload	2,692 nm (5,000 km; 3,100 miles)

ANTONOV An-24

NATO reporting name: *Coke*

Development of this twin-turboprop transport was started in 1958, to replace piston-engined types on Aeroflot's internal feederline routes. The An-24 was intended originally to carry 32-40 passengers, but when the prototype flew in April 1960 it had been developed into a 44-seater. It was followed by a second prototype and five pre-production An-24s. Flight trials were stated to be complete in September 1962 and the An-24 entered service on Aeroflot's routes from Moscow to Voronezh and Saratov in September 1963. More than 50 million passengers and 500,000 tonnes of cargo had been carried by Aeroflot An-24Vs by 1971.

The An-24 is designed to operate from airfields of limited size, with paved or natural runways, and can be fitted with rocket-assisted take-off units to permit operation with a full load of cargo at ambient temperatures above 30°C. Two were taken to the Antarctic in late 1969, to replace piston-engined Il-14s used previously for flights between Antarctic stations.

Export orders have been received from the following airlines:

Air Guinée	3
Air Mali	2
Balkan Bulgarian Airlines	8
CAAC (China)	2
Cubana	8
Interflug (E Germany)	8
Iraqi Airways	1
Lebanese Air Transport	1*
Lina Congo	2
LOT (Poland)	15
Misrair (EgyptAir)	10
Mongolian Airlines	4
Pan African Air Services (Tanzania)	2
Tarom (Romania)	9

**Sold to Misrair and included in latter's total of ten*

The An-24 has also been supplied for military service, usually in small numbers, with the air forces of the USSR, Bangladesh, the Republic of Congo (Brazzaville), Czechoslovakia, Egypt, East Germany, Hungary, Iraq, North Korea, Mongolia, Poland, Romania, the Somali Republic, North Vietnam and South Yemen.

On the prototype, the engine nacelles extended only a little past the wing trailing-edges: production An-24s have lengthened nacelles with conical rear fairings. A ventral tail-fin was also added on production models, which have been followed by the An-24T and An-26 specialised freight-carrying versions of the same basic design. The An-26 is described separately, as is the An-30, a camera-carrying derivative for air survey work.

The current production versions of the An-24 are designed for a service life of 30,000 hours and 15,000 landings, and are available in a variety of forms, as follows:

An-24V Srs II. Standard version, seating up to 50 passengers. Superseded Srs I (with 2,550 ehp AI-24 engines) in 1968, and described in detail below. Basically as Srs I, powered by two Ivchenko AI-24A turboprop engines, with water injection. Can have crew of up to five (two pilots, navigator, radio operator and, on jump seat, an engineer or cargo handler) on flight deck. TG-16 self-contained starter/generator in rear of starboard engine nacelle. Mixed passenger/freight, convertible cargo/passenger, all-freight and executive versions available.

An-24P. Firefighting *(Pozharny)* version, which underwent evaluation in the USSR in 1971. Special provisions for enabling firefighters to be parachuted from a height of 800-1,200 m (2,625-3,940 ft) to deal with forest fires.

An-24RV. Generally similar to Srs II version of An-24V, but with an 8·83 kN (1,985 lb st) Type RU 19-300 auxiliary turbojet engine in starboard nacelle instead of starter/generator. This turbojet is used for engine starting, to improve take-off performance and to improve performance in the air. It permits take-off with a full payload from airfields up to 3,000 m (9,840 ft) above S/L and at temperatures up to ISA + 30°C. It also ensures considerably improved stability and handling characteristics after a failure of one of the turboprop engines in flight. Max T-O weight is increased by 800 kg (1,760 lb) at S/L ISA and by 2,000 kg (4,410 lb) at S/L ISA + 30°C by use of the auxiliary turbojet. An An-24RV was demonstrated at the 1967 Paris Air Show.

An-24T. Generally similar to An-24V Srs II but equip-

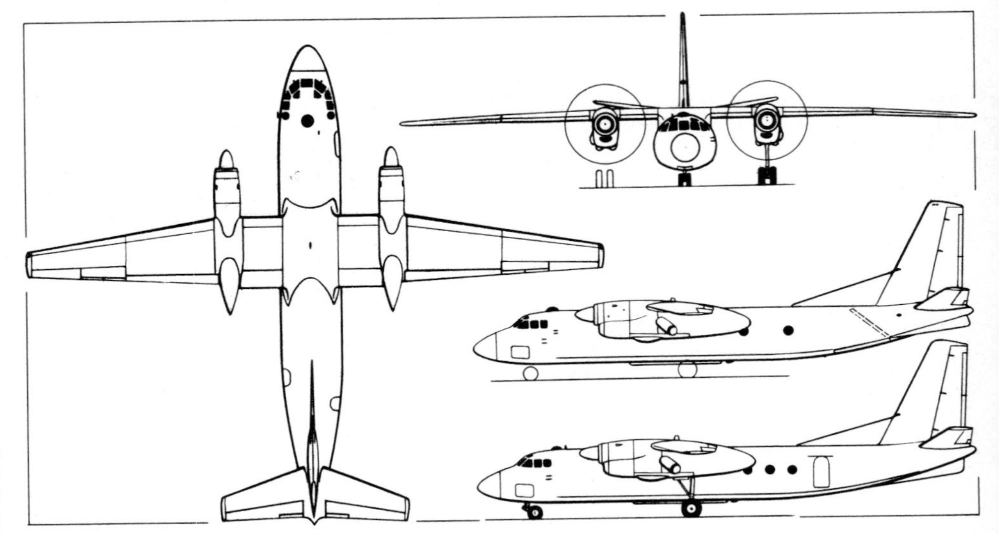

Antonov An-24V transport, with additional side view of the An-24T (centre) *(Pilot Press)*

Antonov An-24T light transport (two Ivchenko AI-24A turboprop engines) of Tarom *(John Wegg)*

ped as specialised freighter. Crew of five, consisting of pilot, co-pilot, navigator, radio operator and flight engineer. Normal passenger door at rear of cabin is deleted and replaced by a belly freight door at the rear of the cabin. This hinges upward and to the rear, providing a hatchway for cargo loading. An electrically-powered winch, capacity 1,500 kg (3,300 lb), is used to hoist crates through the hatch and runs on a rail in the cabin ceiling to position the payload inside the cabin. Electrically or manually powered conveyor, capacity 4,500 kg (9,920 lb), flush with cabin floor. Fewer windows. Folding seats along walls of cabin. Emergency exit hatches in side and in floor at front of cabin. Rear cargo door permits air-dropping of payload or parachutists. Provision for stretcher-carrying in air ambulance role. Single ventral fin replaced by twin ventral fins, forming Vee, aft of cargo door. An An-24T was displayed at the 1967 Paris Air Show, and this version serves with several of the airlines listed earlier.

An-24RT. Generally similar to An-24T but with Type RU auxiliary turbojet in starboard nacelle, as on An-24RV.

The following description refers to the basic An-24V Srs II, unless otherwise noted, but is generally applicable to all versions except for the detailed differences noted above:

TYPE: Twin-turboprop short-range transport.

WINGS: Cantilever high-wing monoplane, with 2° anhedral on outer panels. Incidence 3°. Sweepback at quarter-chord on outer panels 6° 50′. All-metal two-spar structure, built in five sections: centre-section, two inner wings and two outer wings. Wing skin is attached by electrical spot-welding. Mass-balanced servo-compensated ailerons, with large trim tabs of glassfibre construction. Hydraulically-operated Fowler flaps along entire wing trailing-edges inboard of unpowered ailerons; single-slotted flaps on centre-section, double-slotted outboard of nacelles. Servo and trim tabs in each aileron. Thermal de-icing system.

FUSELAGE: All-metal semi-monocoque structure in front, centre and rear portions, of bonded/welded construction.

TAIL UNIT: Cantilever all-metal structure, with ventral fin (two ventral fins on An-24T/RT versions). 9° dihedral on tailplane. All controls manually operated. Trim tabs in elevators. Trim tab and spring tab in rudder. All leading-edges incorporate thermal de-icing system.

LANDING GEAR: Retractable tricycle type with twin wheels on all units. Hydraulic retraction. Emergency extension by gravity. All units retract forward. Main wheels size 900 × 300-370, tyre pressure 3·45-4·90 bars (50-71 lb/sq in). Nosewheels size 700 × 250, tyre pressure 2·45-3·45 bars (35·5-50 lb/sq in). Tyre pressures variable to cater for different types of runway. Disc brakes on main wheels. Steerable and castoring nosewheel unit.

POWER PLANT (all versions): Two 1,902 kW (2,550 ehp) Ivchenko AI-24A turboprop engines (with provision for water injection; weight of water 68 kg; 150 lb), each driving an AV-72 four-blade constant-speed fully-feathering propeller. Electrical de-icing system for propeller blades and hubs; hot air system for engine air intakes. Fuel in integral tanks immediately outboard of nacelles, and four bag-type tanks in centre-section, total capacity 5,550 litres (1,220 Imp gallons). Provision for four additional tanks in centre-section. Pressure refuelling socket in starboard engine nacelle. Gravity fuelling point above each tank. Carbon dioxide inert gas system to create fireproof condition inside fuel tanks. Oil capacity 53 litres (11·5 Imp gallons). One 8·83 kN (1,985 lb st) Type RU 19-300 auxiliary turbojet in starboard nacelle of An-24RV and An-24RT. Provision for fitting rocket-assisted take-off units on cargo versions.

ACCOMMODATION (An-24V/RV): Crew of three (pilot, co-pilot/radio operator/navigator and one stewardess). Provision for carrying navigator, radio operator and engineer. Normal accommodation for 44-52 passengers in air-conditioned and pressurised cabin. Standard layout has baggage and freight compartments on each side aft of flight deck; then the main cabin with 52 forward-facing reclining seats, in pairs at a pitch of 72 cm (28·3 in), on each side of centre aisle (optionally two small sofas for babies at rear, instead of two of the seats); buffet and stewardess's seat, and toilet, opposite door to rear of cabin; and wardrobes at rear. Passenger door on port side, aft of cabin, is of airstair type. Door on starboard side for freight hold (front). All doors open inward. The 46-seat version has a removable partition aft of the fifth row of seats, instead of one row of seats. The mixed passenger/cargo version is laid out normally for 36 passengers, with 14 m³ (495 cu ft) forward hold for baggage, freight and mail, and rear wardrobe and baggage hold (2·8 m³; 99 cu ft). A typical de luxe or executive layout retains the forward and aft baggage and freight holds of the airliner version but has the main cabin divided into three compartments. The forward compartment contains four pairs of seats, in aft-facing and forward-facing pairs with tables between, and a buffet. Next comes a similar cabin without the buffet, followed by a sleeping compartment, with a sofa, two seats and table. At the rear is the standard toilet compartment opposite the airstair door, and a large wardrobe space.

ACCOMMODATION (An-24T/RT): Provision for crew of up to five, with optional cargo handler. Door at front of cabin on starboard side. Upward-opening cargo door in belly at rear of cabin. Max overall dimensions of cargo packages that can be handled are 1·1 × 1·5 × 2·6 m (43·3 × 59 × 102 in) or 1·3 × 1·5 × 2·1 m (51·2 × 59 × 82·7 in). Toilet (port side) and emergency exit door in belly, immediately aft of flight deck. Folding seats, in two-, three- and four-place units, for 30 paratroops or 38 equipped soldiers along walls of main cabin. Ambulance configuration is equipped to carry 24 stretcher cases and one medical attendant. Cargo loading system includes rails in floor, electric winch, overhead gantry, tie-down fittings, nets and harness. Electrical de-icing system for windscreens.

SYSTEMS: Air-conditioning system uses hot air tapped from the 10th compressor stage of each engine, with a heat exchanger and turbocooler in each nacelle. Cabin pressure differential 0·29 bars (4·27 lb/sq in). Main and emergency hydraulic systems, pressure 151·7 bars (2,200 lb/sq in), for landing gear retraction, nosewheel steering, flaps, brakes, windscreen wipers, propeller feathering and, on An-24T, operation of cargo and emergency escape doors. Hand-pump to operate doors only and build up pressure in main system. Electrical system includes two 27V DC starter/generators, two alternators to provide 115V 400Hz AC supply and two inverters for 36V 400Hz three-phase AC supply. An-24T has permanent oxygen system for pilot, installed equipment for other crew members and three portable bottles for personnel in cargo hold.

ELECTRONICS AND EQUIPMENT (An-24T/RT): Standard radio equipment includes two R-802V VHF transceivers, R-836 HF transmitter and US-8 receiver, SPU-7 intercom, two ARK-11 ADF, RV-2 radio altimeter, SP-50 ILS with KRP-F glidepath receiver, GRP-2 glideslope receiver and MRP-56 marker receiver, and RPSN-2AN weather, obstruction and navigation radar. Flight and navigational equipment includes an AP-28L1 autopilot, TsGV-4 master vertical gyro, GPK-52AP directional gyro, GIK-1 gyro compass, two ZK-2 course setting devices, two AGD-1 artificial horizons, AK-59P astro-compass, NI-50BM-K ground position indicator and other standard blind-flying instruments, plus three clocks. Optional OPB-1R sight for pinpoint dropping of cargo and determination of navigational data.

DIMENSIONS, EXTERNAL:

Wing span	29·20 m (95 ft 9½ in)
Wing aspect ratio	11·7
Length overall	23·53 m (77 ft 2½ in)
Height overall	8·32 m (27 ft 3½ in)
Width of fuselage	2·90 m (9 ft 6 in)
Depth of fuselage	2·50 m (8 ft 2½ in)
Tailplane span	9·08 m (29 ft 9½ in)
Wheel track (c/l shock-struts)	7·90 m (25 ft 11 in)
Wheelbase	7·89 m (25 ft 10½ in)
Propeller diameter	3·90 m (12 ft 9½ in)
Propeller ground clearance	1·145 m (3 ft 9 in)
Passenger door (port, aft, except on An-24T):	
Height	1·40 m (4 ft 7 in)
Width	0·75 m (2 ft 5½ in)
Height to sill	1·40 m (4 ft 7 in)
Freight compartment door (stbd, fwd):	
Height	1·10 m (3 ft 7¼ in)
Width	1·20 m (3 ft 11¼ in)
Height to sill	1·30 m (4 ft 3 in)
Baggage compartment door (stbd, aft, except on An-24T):	
Height	1·41 m (4 ft 7½ in)
Width	0·75 m (2 ft 5½ in)
Cargo door (belly, rear, An-24T only):	
Length	2·85 m (9 ft 4 in)
Width:	
max	1·40 m (4 ft 7 in)
min	1·10 m (3 ft 7¼ in)
Height above ground	1·25-1·62 m (4 ft 1 in to 5 ft 4 in)
Emergency exit (An-24T, side):	
Height	0·60 m (1 ft 11½ in)
Width	0·50 m (1 ft 7½ in)
Emergency exit (An-24T, underfuselage):	
Length	1·155 m (3 ft 9½ in)
Width	0·70 m (2 ft 3½ in)

DIMENSIONS, INTERNAL:

Main passenger cabin (52-seater):	
Length	9·69 m (31 ft 9½ in)
Max width	2·76 m (9 ft 1 in)
Max height	1·91 m (6 ft 3 in)
Floor area	39·95 m² (430 sq ft)
Cargo hold (An-24T):	
Length	15·68 m (51 ft 5½ in)
Width	2·17 m (7 ft 1½ in)
Height	1·765 m (5 ft 9½ in)
Volume	50 m³ (1,765 cu ft)

AREAS:

Wings, gross	74·98 m² (807·1 sq ft)
Horizontal tail surfaces (total)	17·23 m² (185·5 sq ft)
Vertical tail surfaces (total, excl dorsal fin)	13·38 m² (144·0 sq ft)

WEIGHTS AND LOADINGS:

Weight empty:	
An-24V	13,300 kg (29,320 lb)
An-24T	14,060 kg (30,997 lb)

Basic operating weight:
An-24T 14,698 kg (32,404 lb)
Fuel weight:
An-24T with max payload 1,800 kg (3,968 lb)
An-24T for max range 4,760 kg (10,494 lb)
Max payload (ISA, S/L):
An-24V, An-24RV 5,500 kg (12,125 lb)
An-24T 4,612 kg (10,168 lb)
An-24RT 5,700 kg (12,566 lb)
Max ramp weight:
An-24T 21,110 kg (46,540 lb)
Max T-O and landing weight:
An-24V, An-24T, ISA, S/L 21,000 kg (46,300 lb)
An-24V, An-24T, S/L, ISA + 30°C 19,800 kg (43,650 lb)
An-24RV, An-24RT, S/L, ISA or ISA + 30°C 21,800 kg (48,060 lb)
Max wing loading:
An-24V 276 kg/m² (56·53 lb/sq ft)

Performance (at max T-O weight):
Normal cruising speed at 6,000 m (19,700 ft) 243 knots (450 km/h; 280 mph)
Max range cruising speed at 7,000 m (23,000 ft) 243 knots (450 km/h; 280 mph)
T-O speed:
An-24T 97-100 knots (180-185 km/h; 112-115 mph)
Landing speed:
An-24V 89 knots (165 km/h; 103 mph) CAS
An-24T 87-95 knots (160-175 km/h; 100-109 mph)
Max rate of climb at S/L:
An-24V 114 m (375 ft)/min
An-24RV 204 m (670 ft)/min
Rate of climb at S/L, one engine out:
An-24V, ISA 84 m (275 ft)/min
An-24V, ISA + 30°C, with water injection 84 m (275 ft)/min
An-24RV, ISA 174 m (570 ft)/min
An-24RV, ISA + 30° 90 m (295 ft)/min
Service ceiling:
An-24V, An-24T 8,400 m (27,560 ft)
An-24RV, An-24RT 9,000 m (29,525 ft)
Service ceiling, one engine out:
An-24T 2,750 m (9,020 ft)
T-O run:
An-24V 600 m (1,970 ft)
An-24T 640 m (2,100 ft)
Balanced T-O runway:
An-24T, ISA 1,720 m (5,645 ft)
An-24T, ISA + 15°C 1,750 m (5,745 ft)
Landing run at AUW of 20,000 kg (44,100 lb):
An-24T 880 m (1,903 ft)
Landing from 15 m (50 ft) at AUW of 20,000 kg (44,100 lb):
An-24T 1,590 m (5,217 ft)
Range with max payload, with reserves:
An-24V, An-24RV 296 nm (550 km; 341 miles)
An-24T, An-24RT 344 nm (640 km; 397 miles)
Range with max fuel:
An-24V, 45 min fuel reserves 1,293 nm (2,400 km; 1,490 miles)
An-24T, with 1,612 kg (3,554 lb) payload, no reserves 1,618 nm (3,000 km; 1,864 miles)

ANTONOV An-26

NATO reporting name: *Curl*

First displayed in public at the 1969 Paris Air Show, the An-26 was known initially as the 'An-24T with an enlarged freight door'. It is, in fact, generally similar to the An-24RT, but has more powerful AI-24T turboprop engines and a completely redesigned rear fuselage of the 'beaver-tail' type.

Although intended primarily for cargo-carrying, with air-drop capability, the An-26 can be adapted easily for passenger-carrying, ambulance or paratroop transport duties.

The Air Wing of the Bangladesh Defence Force has a small number of An-26s, as well as one An-24. Other An-26s serve with the Polish Air Force.

The basic structural description of the An-24 applies also to the An-26, except for the following details:

Type: Twin-turboprop short-haul transport.

Wings: Made in three sections: centre-section and two outer panels which contain integral fuel tanks.

Fuselage: Skin on lower portion of fuselage is made of 'bimetal' (duralumin-titanium) sheet for protection during operations from unpaved airfields.

Landing Gear: Shock-absorbers are of oleo-nitrogen type. Main wheels are fitted with hydraulic disc brakes and anti-skid units. Nosewheels can be steered hydraulically through 45° each side while taxying and are controllable through ±10° during take-off and landing. Main-wheel tyres size 1,050 × 400, pressure 3·93 bars (57 lb/sq in). Nosewheel tyres size 700 × 250, pressure 4·41 bars (64 lb/sq in).

Power Plant: Two 2,103 kW (2,820 ehp) Ivchenko AI-24T turboprop engines, each driving a four-blade constant-speed fully-feathering propeller. One 8·83 kN (1,985 lb st) RU 19-300 auxiliary turbojet in starboard nacelle for use, as required, at take-off, during climb and in level flight, and for self-contained starting of main engines. Fuel load 5,500 kg (12,125 lb).

Accommodation: Basic crew of five (pilot, co-pilot, radio operator, flight engineer and navigator), with station at rear of cabin on starboard side for loading supervisor or load despatcher. Toilet on port side aft of flight deck; small galley and oxygen bottle stowage on starboard side. Emergency escape hatch in floor immediately aft of flight deck. Large downward-hinged rear ramp-door, which can also slide forward under fuselage for direct loading on to cabin floor or for air-dropping of freight. Electrically-powered mobile winch, capacity 1,500 kg (3,300 lb), hoists crates through rear entrance and runs on a rail in the cabin ceiling to position payload in cabin. Electrically- or manually-operated conveyor, capacity 4,500 kg (9,920 lb), built-in flush with cabin floor, facilitates loading and air-dropping of freight. Can accommodate a variety of motor vehicles, including GAZ-69 and UAZ-469 military vehicles, or cargo items up to 1·50 m (59 in) high by 2·10 m (82·6 in) wide. Height of rear edge of cargo door surround above the cabin floor is 1·50 m (4 ft 11 in). Cabin is pressurised and air-conditioned, and is fitted with a row of tip-up seats along each wall to accommodate up to 40 paratroops. Conversion to troop transport role, or to an ambulance for 24 stretcher patients and a medical attendant, takes 20 to 30 minutes in the field.

Systems: Basically as for An-24. Electrical system includes two 27V DC starter/generators on engines, a standby generator on the auxiliary turbojet, and three storage batteries for emergency use. Two engine-driven alternators provide 115V 400Hz single-phase AC supply, with standby inverter. Basic source of 36V 400Hz three-phase AC supply is two inverters, with standby transformer.

Electronics and Equipment: Standard com/nav equipment comprises two VHF transceivers, HF, intercom, two ADF, radio altimeter, glidepath receiver, glideslope receiver, marker receiver, weather/navigation radar, directional gyro and flight recorder. Optional equipment includes a flight director system, astrocompass and autopilot. Standard operational equipment includes parachute static line attachments and retraction devices, tie-downs, jack to support ramp sill, flight deck curtains, sun visors and windscreen wipers. Optional items include a navigator's observation blister on port side of flight deck, OPB-1R sight for pinpoint dropping of freight, medical equipment and liquid heating system.

Dimensions, external:
As for An-24, except:
Length overall 23·80 m (78 ft 1 in)
Height overall 8·575 m (28 ft 1½ in)
Tailplane span 9·973 m (32 ft 8¾ in)
Wheelbase 7·651 m (25 ft 1¼ in)
Propeller ground clearance 1·227 m (4 ft 0¼ in)
Crew door (stbd, front):
Height 1·40 m (4 ft 7 in)
Width 0·60 m (1 ft 11¾ in)
Height to sill 1·47 m (4 ft 9¾ in)
Loading hatch (rear):
Length 3·40 m (11 ft 1¾ in)
Width at front 2·40 m (7 ft 10½ in)
Width at rear 2·00 m (6 ft 6¾ in)
Height to sill 1·47 m (4 ft 9¾ in)
Height to top edge of hatchway 3·014 m (9 ft 10¾ in)
Emergency exit (floor at front):
Length 1·02 m (3 ft 4¼ in)
Width 0·70 m (2 ft 3½ in)
Emergency exit (top):
Diameter 0·65 m (2 ft 1½ in)
Emergency exits (each side of hold, two):
Height 0·60 m (1 ft 11¾ in)
Width 0·50 m (1 ft 7½ in)

Dimensions, internal:
Cargo hold:
Length of floor 11·50 m (37 ft 8¾ in)
Width of floor 2·40 m (7 ft 10½ in)
Max height 1·91 m (6 ft 3 in)

Areas:
Wings, gross 74·98 m² (807·1 sq ft)
Horizontal tail surfaces (total) 19·83 m² (213·45 sq ft)

Antonov An-26, a development of the An-24T freight transport with enlarged rear-loading ramp-door

Vertical tail surfaces (total, incl dorsal fin) 15·85 m² (170·61 sq ft)

WEIGHTS:
- Weight empty 15,020 kg (33,113 lb)
- Normal payload 4,500 kg (9,920 lb)
- Max payload 5,500 kg (12,125 lb)
- Normal T-O and landing weight 23,000 kg (50,706 lb)
- Max T-O and landing weight 24,000 kg (52,911 lb)

PERFORMANCE (at normal T-O weight):
- Cruising speed at 6,000 m (19,675 ft) 229-234 knots (425-435 km/h; 264-270 mph)
- T-O speed 108 knots (200 km/h; 124 mph) CAS
- Landing speed 102 knots (190 km/h; 118 mph) CAS
- Max rate of climb at S/L 480 m (1,575 ft)/min
- Service ceiling 8,100 m (26,575 ft)
- T-O run, on concrete 780 m (2,559 ft)
- T-O to 15 m (50 ft) 1,240 m (4,068 ft)
- Landing from 15 m (50 ft) 1,740 m (5,709 ft)
- Landing run, on concrete 730 m (2,395 ft)
- Min ground turning radius 22·3 m (73 ft 2 in)
- Range, with allowance for taxying and 580 kg (1,278 lb) reserve fuel:
  - with 4,500 kg (9,920 lb) payload 485 nm (900 km; 559 miles)
  - with 2,126 kg (4,687 lb) payload 1,214 nm (2,250 km; 1,398 miles)

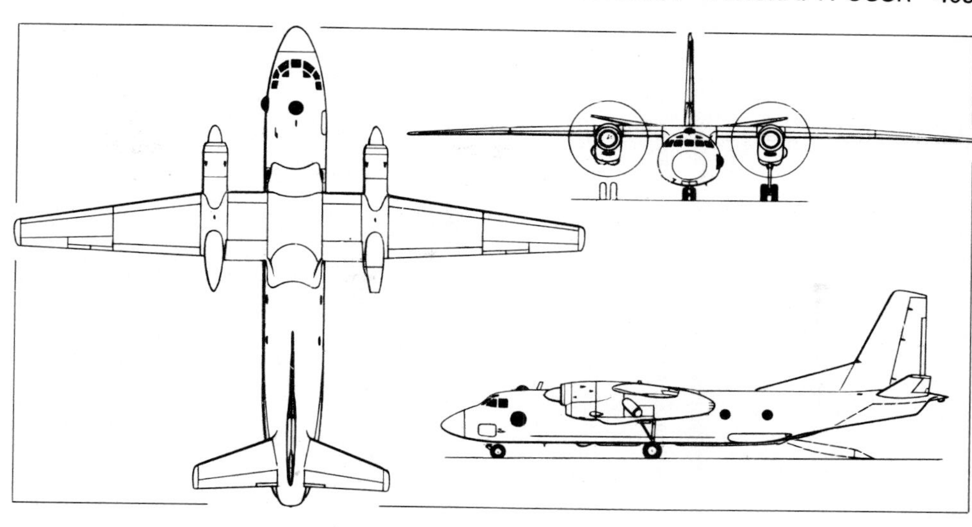

Antonov An-26 twin-turboprop short-haul transport *(Pilot Press)*

ANTONOV An-28

Although Oleg Antonov first referred to planned production of an enlarged turboprop version of the piston-engined An-14 light general-purpose aircraft in the early 'sixties, there was no proof that such an aircraft had been built until the Spring of 1972. Photographs of the prototype (CCCP-1968) were then published in the Polish press. It had flown for the first time in September 1969, powered by two 604 kW (810 shp) Isotov TVD-850 turboprop engines, and was described in this form in the 1974-75 and previous editions of *Jane's*.

Initially, the new aircraft was designated An-14M. Its official flight testing was completed in 1972, and during 1973 it was allocated the production designation An-28; the accompanying illustration shows the development aircraft CCCP-19723 with TVD-850 engines and bearing this designation. Flight trials suggested that field performance and climb, in particular, could be improved by fitting more powerful engines. Thus, in April 1975, the same development aircraft (re-registered CCCP-19753) flew for the first time with 716 kW (960 shp) Glushenkov TVD-10A turboprop engines, which are specified also for production An-28s. Flight tests were described as nearing completion in December 1975.

In general configuration the An-28 differs from the piston-engined An-14 mainly in having a much-enlarged fuselage to carry up to 15 passengers or equivalent alternative payloads. The first prototype had a retractable landing gear, with small fairings on the sides of the fuselage into which the main units retracted. Subsequently it was decided that retraction was unnecessary for flights over short distances at low speeds, and later prototypes have fixed gear. The shape of the vertical tail surfaces has also changed during flight testing.

The Antonov design bureau evolved the An-28 for service on Aeroflot's shortest routes, particularly those operated by An-2 biplanes into places which are relatively inaccessible to other types of fixed-wing aircraft. The turbporop engines make possible full-payload operation under high-temperature conditions and in mountainous regions; and the An-28 is described as being suitable for carrying passengers, cargo and mail, for scientific expeditions, geological surveying, forest fire patrol, firefighting, air ambulance or rescue operations, and parachute training. In agricultural form it can carry an 800 kg (1,760 lb) chemical payload for dusting and spraying operations.

TYPE: Twin-turboprop light general-purpose aircraft.

WINGS: Braced high-wing monoplane, with single streamline-section bracing strut each side. Entire trailing-edges hinged, ailerons being designed to droop with the large flaps. On first prototype a short spar-beam extended from each side of the lower fuselage, carrying the main landing gear units, providing lower attachments for the wing bracing struts and supporting the fairings into which the main wheels retracted. De-icing system for wing leading-edges.

FUSELAGE: Conventional all-metal semi-monocoque structure, longer, wider and deeper than that of the piston-engined An-14. Underside of rear fuselage upswept and made up of clamshell doors.

TAIL UNIT: Cantilever all-metal structure. Twin fins and rudders mounted vertically on a tailplane that lacks the dihedral of that on the An-14. Latest version has fixed leading-edge slot over full span of tailplane leading-edge. De-icing system for leading-edges.

LANDING GEAR: Non-retractable (except on first prototype) tricycle type, with single wheel on each unit. Wide-tread balloon tyres of same size on all units. Steerable nosewheel. Brakes on main wheels. Provision for skis or floats.

POWER PLANT: Two 716 kW (960 shp) Glushenkov TVD-10A turboprop engines.

ACCOMMODATION: Crew of one or two on flight deck. Cabin of passenger version contains 15 seats in five rows, with double units on starboard side of aisle. Seats fold back against walls when aircraft is operated as a freighter or in mixed passenger/cargo role. Provision for baggage and toilet compartments and wardrobe space. Clamshell rear doors, under upswept fuselage, for use by passengers and for cargo loading. Winch of 250 kg (550 lb) capacity for handling cargo. Six/seven-passenger executive version has four folding tables, which can be joined together in pairs to give working tops measuring 160 × 55 cm (63 in × 21·5 in). Ambulance version accommodates six stretcher patients, a medical attendant and medical equipment.

ELECTRONICS AND EQUIPMENT: Flight and navigation equipment, and de-icing system, for all-weather operation. Landing light in nose.

DIMENSIONS, EXTERNAL:
- Wing span 21·99 m (72 ft 2 in)
- Length overall 12·98 m (42 ft 7 in)
- Height overall 4·60 m (15 ft 1 in)

DIMENSIONS, INTERNAL:
- Cabin: Length 5·26 m (17 ft 3 in)
  - Max width 1·66 m (5 ft 5 in)
  - Max height 1·70 m (5 ft 7 in)

WEIGHTS:
- Max payload 1,500 kg (3,306 lb)
- Max T-O weight 5,700 kg (12,566 lb)

PERFORMANCE (at max T-O weight):
- Max cruising speed 188 knots (350 km/h; 217 mph)
- Max rate of climb at S/L 720 m (2,360 ft)/min
- Rate of climb at S/L, one engine out 240 m (785 ft)/min
- T-O to 15 m (50 ft) 270 m (885 ft)
- Landing from 15 m (50 ft) 270 m (885 ft)
- Range with max payload 540 nm (1,000 km; 620 miles)

ANTONOV An-30

NATO reporting name: *Clank*

Described as the first specialised aerial survey aeroplane produced in the Soviet Union, the An-30 is evolved from the An-24 twin-turboprop transport, to which it is generally similar. The major modifications are made to the nose, which is extensively glazed to give the navigator a wide field of vision, and to the flight deck, which is raised to improve the pilots' view and increase the size of the navigator's compartment. There are fewer windows in the main cabin, which contains a darkroom and film storage cupboard, as well as survey cameras and a control desk.

Photography can be automatic or semi-automatic if required, but two photographer/surveyors are normally carried, in addition to a flight crew of five.

For the primary task of air photography for map-

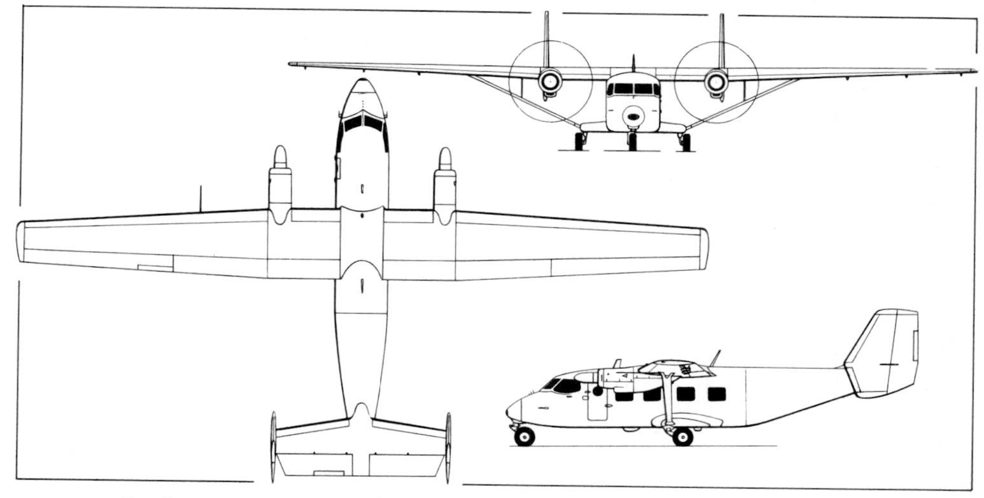

Development version of Antonov An-28 with TVD-850 turboprop engines *(Pilot Press)*

Antonov An-28 light general-purpose transport (TVD-850 turboprop engines) *(Tass)*

Antonov An-30 aerial survey development of the An-24 twin-turboprop transport aircraft, with glazed nose and other modifications *(M. D. West)*

making, the An-30 is equipped with four large survey cameras. These are mounted in the cabin above apertures which are each covered by a door. A crew photographer uncovers the apertures, as required, by remote control from his desk in the aircraft. A fifth window is provided for an exposure meter.

Details of the An-30 published in the Far East suggest that one of the survey cameras can be stabilised, in gimbal mountings, to ensure precise photographic coverage of the desired area in turbulent conditions.

The pre-programmed flight path of the aircraft over the area to be photographed is fed into an on-board computer which controls the speed, altitude, and direction of flight throughout the mission. If required, the cameras can be replaced by other kinds of survey equipment, such as those used for mineral prospecting or for microwave radiometer survey, which measures the heat emission of land and ocean to obtain data on ocean surface characteristics, sea and lake ice, snow cover, flooding, seasonal vegetation changes, and soil types.

The power plant comprises two 2,103 kW (2,820 ehp) Ivchenko AI-24VT turboprop engines, supplemented by a 7·8 kN (1,765 lb st) RU 19A-300 auxiliary turbojet in the rear of the starboard engine nacelle. The latter is used for engine starting, and for take-off, climb and cruise power in the event of failure of the primary power plant.

Conversion into a transport aircraft is provided for, with cover plates to place over the camera apertures.

Dimensions, external:

Wing span	29·20 m (95 ft 9½ in)
Length overall	24·26 m (79 ft 7 in)
Height overall	8·32 m (27 ft 3½ in)
Tailplane span	9·973 m (32 ft 8¾ in)

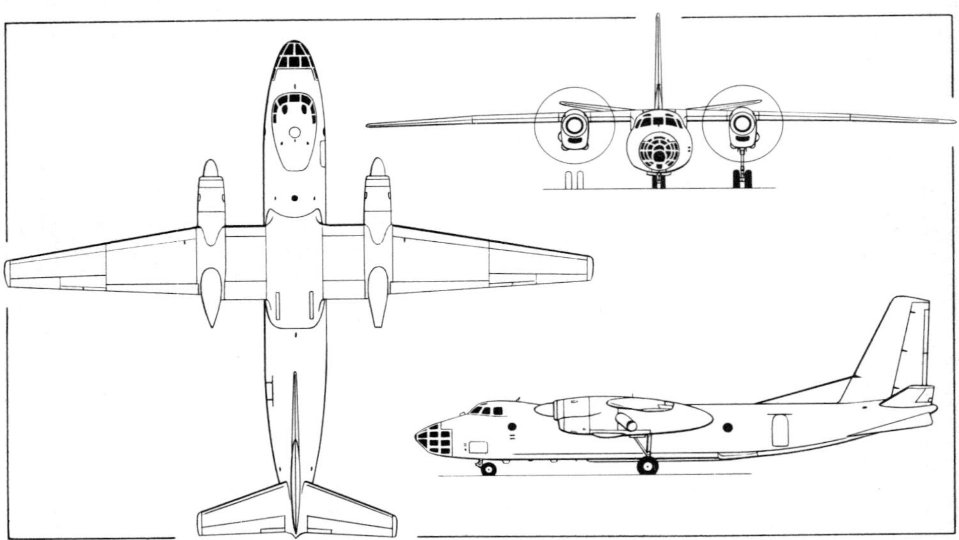

Antonov An-30 (two AI-24VT turboprops and RU 19A-300 auxiliary turbojet) *(Pilot Press)*

Weight:

Max T-O weight 23,000 kg (50,705 lb)

Performance:

Max cruising speed 232 knots (430 km/h; 267 mph)

BERIEV

General Designer in Charge of Bureau:

Georgi Mikhailovich Beriev

G. M. Beriev, a graduate of the Leningrad Polytechnic Institute, took up seaplane design in 1928 and has since become the best-known Soviet designer of water-based aircraft. He was appointed chief designer of the seaplane group at the TsKB (Central Design Bureau of Aviatrust) in 1930, and his first complete design, the twin-engined MBR-2 flying-boat, was flown for the first time two years later, entering production in 1934. Other pre-war designs included the KOR-1 twin-float shipboard reconnaissance seaplane and the KOR-2 flying-boat, later redesignated Be-2 and Be-4 respectively.

In 1945 the Beriev bureau at Taganrog became the centre for all Soviet seaplane development, and the piston-engined Be-6 (first flown in 1947) was a standard military flying-boat during the 1950s. It was described in the 1959-60 *Jane's.* Only limited production was undertaken of the Be-8, also flown in 1947; and Beriev's next major flying-boat was the sweptwing twin-jet Be-10, based on the Be-R-1 prototype of 1949. This entered service in about 1960, and was described in the 1966-67 *Jane's.*

The latest maritime aircraft of Beriev design is the M-12 (Be-12) twin-turboprop amphibian, which is in standard service with Soviet Naval Air Force units as a successor to the Be-6. The more recent Be-30 twin-turboprop light transport (see 1972-73 *Jane's*) did not enter series production. It has, however, been developed as the Be-32, which set up two time-to-height records, as noted briefly in this entry.

BERIEV M-12 (Be-12) TCHAIKA (SEAGULL)

NATO reporting name: *Mail*

This twin-turboprop medium-range maritime reconnaissance amphibian was displayed for the first time in the 1961 Aviation Day flypast at Tushino Airport, Moscow. Subsequently, during the period 23-27 October 1964, it established six officially-recognised international height records in Class C.3 Group II. Data submitted in respect of these records revealed that the designation of the aircraft was M-12 and the power plant two 4,000 shp Ivchenko AI-20D turboprop engines. The aircraft was also, clearly, able to lift a payload of around 10 tons under record conditions.

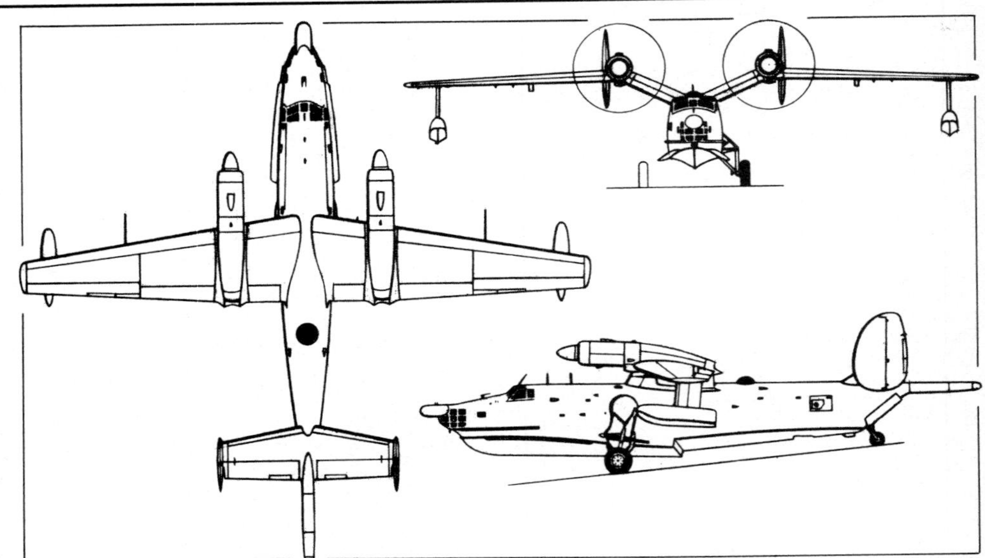

Beriev M-12 (Be-12) Tchaika twin-turboprop maritime reconnaissance amphibian *(Pilot Press)*

The records set by the M-12 in 1964 were altitude of 12,185 m (39,977 ft) without payload, altitude of 11,366 m (37,290 ft) with payloads of 1,000 kg and 2,000 kg, altitude of 10,685 m (35,055 ft) with 5,000 kg payload, altitude of 9,352 m (30,682 ft) with 10,000 kg payload, and maximum payload of 10,100 kg (22,266 lb) lifted to a height of 2,000 m (6,560 ft). In each case, the crew consisted of M. Mikhailov, I. Kouprianov and L. Kuznetsov.

On 24 April 1968, A. Souchko set a Class C.3 speed

Beriev M-12 (Be-12) Tchaika maritime patrol amphibian flying-boat of the Soviet Naval Air Force *(Tass)*

record of 298·013 knots (552·279 km/h; 343·169 mph) over a 500 km closed circuit in an M-12. On 9 October 1968, the same pilot set a speed record of 293·919 knots (544·693 km/h; 338·456 mph) over a 1,000 km circuit and a closed-circuit distance record in this class. The latter record was beaten by Vladimir Svyatochnur and crew in an M-12, on 25 October 1973, when they covered a closed-circuit distance of 1,382·968 nm (2,562·897 km; 1,592·510 miles).

Three speed-with-payload records over a 1,000 km closed circuit were set in 1970. A. Suchov attained 283·841 knots (526·011 km/h; 326·848 mph) with 1,000 kg on 21 April; A. Smirnov averaged 286·267 knots (530·504 km/h; 329·640 mph) with 2,000 kg on 8 July; and A. Zakharov averaged 284·163 knots (526·606 km/h; 327·218 mph) with 5,000 kg on 9 July.

Subsequent record attempts, improving on earlier performances by the M-12, have ensured that this aircraft retains all 20 records listed in Class C.3 Group II. On 31 October 1972, P. Yakouchine and crew of three achieved 300·450 knots (556·789 km/h; 345·973 mph) over a 2,000 km closed circuit, raising the records for no payload and 1,000 kg payload, and setting a new record with 2,000 kg. The current record of 263·720 knots (488·722 km/h; 303·678 mph) with a 5,000 kg payload was set by A. Souchko and crew on 30 October 1973. On 5 November 1974, V. Averchine set a time to height record to 6,000 m in 12 min 24·6 sec; on 14 November 1974, A. Zakharov climbed to 3,000 m in 5 min 6·2 sec. On 28 April 1975, V. Efimov set a record of 8,289 m (27,200 ft) for altitude maintained in horizontal flight. Next day he climbed to 9,000 m in an M-12 in 27 min 3·4 sec.

The M-12 also holds all 14 current records in Class C.2 Group II, for turboprop flying-boats. On 25 April 1968, E. Nikitine set a 500 km closed-circuit speed record of 305·064 knots (565·347 km/h; 351·290 mph) in an M-12, followed on 12 October 1968 by a speed record of 297·793 knots (551·871 km/h; 342·916 mph) over a 1,000 km closed circuit. On 21 April 1970, A. Zakharov set a speed record of 289·272 knots (536·074 km/h; 333·101 mph) over a 1,000 km closed circuit with a 1,000 kg payload. P. Yakushin averaged 288·848 knots (535·288 km/h; 332·613 mph) over the same distance with 2,000 kg on 8 July 1970; and on the following day E. Nikitine averaged 285·454 knots (528·998 km/h; 328·704 mph) over 1,000 km with a 5,000 kg payload.

On 30 October 1972, A. Zakharov and crew of three averaged 300·015 knots (555·983 km/h; 345·472 mph) over a 2,000 km circuit, claiming also the speed record over this distance with a 1,000 kg payload. Over this same distance, V. Averchine and crew averaged 295·999 knots (548·542 km/h; 340·848 mph) with a 2,000 kg payload on 28 October 1973, followed by E. Nikitine and crew who averaged 258·727 knots (479·470 km/h; 297·929 mph) with 5,000 kg on the next day. The closed-circuit distance record was raised to 1,393·071 nm (2,581·62 km; 1,604·144 miles) on 20 November 1973, by G. Efimov and crew.

On 5 November 1974, V. Belov set a time to height record to 3,000 m in 5 min 9·8 sec; on 14 November 1974, E. Nikitine climbed to 6,000 m in 11 min 57·4 sec. The current record of 8,223 m (26,975 ft) for sustained altitude in horizontal flight was set by V. Averchine in an M-12 on 28 April 1975. The same pilot set a time-to-height record next day, climbing to 9,000 m in 22 min 9·8 sec.

Layout and construction of the M-12 are conventional. The single-step hull has a high length-to-beam ratio and is fitted with two long strakes, one above the other, on each side of the nose to prevent spray from enveloping the propellers at take-off. There is a glazed observation and navigation station in the nose, with a long radar 'thimble' built into it, and an astrodome type of observation position above the rear fuselage. The nose radome on current aircraft is wider and somewhat flatter in section than that on early M-12s. A long MAD (magnetic anomaly detection) 'sting' extends from the tail, and there appears to be an APU exhaust on the port side of the rear fuselage.

Fuelling a Beriev M-12 (Be-12) Tchaika (two Ivchenko AI-20D turboprop engines) of the Red Banner Northern Fleet *(Tass)*

The sharply-cranked high-set wing, with non-retractable wingtip floats, is reminiscent of that of the Be-6, and is intended to raise the AI-20 turboprop engines well clear of the water. The cowlings of the turboprops open downward in two halves, so that they may be used as servicing platforms. The tail unit, with twin fins and rudders at the tips of a 'dihedral' tailplane, is also similar to that of the Be-6.

The tailwheel landing gear consists of single-wheel main units, which retract upward through 180° to lie flush within the sides of the hull, and a rearward-retracting tailwheel.

In addition to an internal bomb bay aft of the step, there is provision for one large and one small external stores pylon under each outer wing panel.

When three M-12s took part in the 1967 air display at Domodedovo, the commentator said that the unit to which they belonged was "one of those serving where the country's military air force began", implying that the aircraft were then in operational service. M-12s have since been identified in standard service at Soviet Northern and Black Sea Fleet air bases and were reported to be operational for a period from bases in Egypt.

DIMENSIONS, EXTERNAL (approx):
- Wing span: 29·70 m (97 ft 6 in)
- Length overall: 30·20 m (99 ft 0 in)
- Height overall: 7·00 m (22 ft 11½ in)
- Propeller diameter: 4·85 m (16 ft 0 in)

WEIGHT (estimated):
- Max T-O weight: 29,500 kg (65,035 lb)

PERFORMANCE (estimated):
- Max level speed: 329 knots (610 km/h; 379 mph)
- Normal operating speed: 172 knots (320 km/h; 199 mph)
- Max range: 2,158 nm (4,000 km; 2,485 miles)

BERIEV Be-32

The official Soviet news agency, Novosti, has reported that two of the time-to-height records for turboprop aircraft held by Cdr Donald H. Lilienthal, USN, in a Lockheed P-3C Orion, have been beaten by a Soviet test pilot named Yevgeni Lakhmostov in a Be-32. This is described as a twin-turboprop transport evolved from the Be-30, with a payload of 18 passengers or 1,900 kg (4,189 lb) of

Photo-copy of the only picture of the Beriev Be-32 yet published

freight. Its records, set from Podmoskovnoe aerodrome, are claimed as a climb to 3,000 m in 2 min 24·8 sec, and to 6,000 m in 5 min 18 sec.

The accompanying poor-quality photograph of a Be-32 is the only one that had been published by the Summer of 1976. It shows that the aircraft is virtually identical in configuration to the Be-30. The caption states that the aircraft is designed for operation on wheel, float or ski landing gear.

ILYUSHIN

GENERAL DESIGNER IN CHARGE OF BUREAU:
Sergei Vladimirovich Ilyushin

Sergei Ilyushin was awarded the Order of Lenin and a third Hammer and Sickle gold medal on 29 March 1974, at the age of 80, in recognition of his service in the development of aviation technology and the Soviet aircraft industry.

Aircraft designed by Ilyushin and currently in service include the veteran Il-28 twin-jet bomber and Il-12 and Il-14 piston-engined light transports, of which details have been given in earlier editions of *Jane's*. The four-turboprop Il-18 transport has been in scheduled service with Aeroflot and other airlines and air forces for many years. More recent types include an anti-submarine variant of the same design, designated Il-38; a four-jet rear-engined airliner known as the Il-62; and the Il-76 turbofan-engined heavy freighter.

Under development is a large wide-bodied transport designated Il-86.

ILYUSHIN Il-18

NATO reporting name: *Coot*

The Il-18 prototype, named Moskva (Moscow), flew for the first time in July 1957 and was followed by two pre-production models. Production began while these were completing their flight trials, enabling the Il-18 to enter service with Aeroflot on 20 April 1959. In its first ten years of operation by Aeroflot, it carried 60 million passengers and was being utilised on 800 domestic services by the Spring of 1969.

The initial production version was equipped to carry 84 passengers, and could be powered by either Kuznetsov NK-4 or Ivchenko AI-20 turboprops. All aircraft from the 21st built have had AI-20 engines. Production is believed to have exceeded 700 aircraft, of which more than 100 were exported for use by commercial airlines. Known civilian operators were listed in the 1975-76 *Jane's*. Military operators of Il-18s include the air forces of Afghanistan, Algeria, Bulgaria, China, Czechoslovakia, Poland, the Soviet Union, Syria and Yugoslavia, mostly in comparatively small numbers. An anti-submarine derivative, the Il-38 (NATO reporting name *May*), is also in service and is described separately.

Testing of all-weather landing systems on the Il-18 began in 1963 and current versions of the aircraft can be fitted with the Polosa automatic landing system, which meets ICAO Cat III specifications.

An Il-18 was used for flight evaluation over a two-year period of the EI POS de-icing system, announced in April 1972. This operates on the principle of converting electrical impulses into mechanical impulses powerful enough to remove ice of any thickness from the skin of an airliner in flight. It is said to require hundreds of times less energy than a hot-air or electrical-heating de-icing system, to weight only 35 kg (77 lb) fully installed, and to be effective in temperatures from zero to 50°C below zero.

Versions of the Il-18 commercial transport are as follows:

Il-18V. Standard version for Aeroflot, with four 2,983 kW (4,000 ehp) AI-20K turboprops and fuel capacity of 23,700 litres (5,213 Imp gallons). Accommodation for 110 mixed tourist/economy class passengers, or 90 in all-tourist configuration.

Il-18E. Developed version with 3,169 kW (4,250 ehp) AI-20M engines. Same fuel capacity as Il-18V. Accommodation can be increased to 122 mixed class or 110 tourist class in Summer, by deleting coat storage space essential in Winter time.

Il-18D. Generally similar to Il-18E, but with additional centre-section fuel tankage, increasing total capacity to 30,000 litres (6,600 Imp gallons). Increased all-up weight. Offered also with 65 seats, equivalent to first class seating standards, and in executive transport versions.

By the Spring of 1960, the Il-18 had established 12 officially-recognised international records, piloted in each case by Vladimar Kokkinaki. The nine closed-circuit speed-with-payload records were beaten subsequently by the Tu-114, but the Il-18 retains records for climb to 13,154 m (43,156 ft) with a 10,000 kg payload, on 15 November 1958; climb to 12,471 m (40,915 ft) with a 15,000 kg payload on 14 November 1958; and climb to 12,118 m (39,757 ft) with a 20,000 kg payload on 25 November 1959.

On 6 May 1968, an Il-18 piloted by B. Konstantinov set up a still-unbeaten speed record of 380·962 knots (706 km/h; 438·7 mph) over a 100 km closed circuit. Miss L. Ulanova and an all-woman crew set up an international straight-line distance record of 4,134·427 nm (7,661·949 km; 4,760·89 miles) on 14-15 October 1967, an altitude record of 13,513 m (44,334 ft) on 20 October 1967, a record for sustained altitude of 12,900 m (42,323 ft) in horizontal flight on 13 June 1969, a closed-circuit distance record of 4,329·333 nm (8,023·153 km; 4,985·35 miles) on 18-19 June 1969, and a speed record of 378·304 knots (701·068 km/h; 435·623 mph) over a 5,000 km circuit on 12 June 1969.

TYPE: Four-engined passenger transport.

WINGS: Cantilever low-wing monoplane. Mean thickness/chord ratio 14%. All-metal structure. Three spars in centre-section, two in outer wings. All-metal ailerons are mass-balanced and aerodynamically-compensated, and fitted with spring tabs. Manually-operated flying controls. Electrically-actuated double-slotted flaps. Electro-thermal de-icing.

FUSELAGE: Circular-section all-metal monocoque structure. The structure is of the fail-safe type, and appears to employ rip stop doublers around window cutouts, door frames and the more-heavily loaded skin panels.

TAIL UNIT: Cantilever all-metal structure. Trim tabs on rudder and elevators. Additional spring tab on rudder. Manually-operated flying controls. Electro-thermal de-icing.

LANDING GEAR: Retractable tricycle type. Hydraulic actuation. Four-wheel bogie main units, with 930 mm × 305 mm tyres and hydraulic brakes. Steerable (45° each way) twin nosewheel unit, with 700 mm × 250 mm tyres. Tyre pressures: main 7·86 bars (114 lb/sq in), nose 5·86 bars (85 lb/sq in). Hydraulic brakes and nosewheel steering. Pneumatic emergency braking system, using nitrogen gas.

POWER PLANT: Four Ivchenko AI-20 turboprops (details under model listings), driving AV-68I four-blade reversible-pitch propellers. Ten flexible bag-type fuel tanks in inboard panel of each wing and integral tank in outboard panel, with a total capacity of 23,700 litres (5,213 Imp gallons). The Il-18D has additional bag tanks in centre-section, giving a total capacity of 30,000 litres (6,600 Imp gallons). Pressure fuelling through four international standard connections in inner nacelles. Provision for overwing fuelling. Oil capacity 58·5 litres (12·85 Imp gallons) per engine.

ACCOMMODATION: Crew of five, comprising two pilots, navigator, wireless operator and flight engineer. Flight deck is separated from remainder of fuselage by a pressure bulkhead to reduce the hazards following a sudden decompression of either. Standard 110-seat high-density version has a forward cabin containing 24 seats six-abreast; then, successively, an entrance lobby with two toilets on the starboard side, two large wardrobes in line with the propellers, the main cabin containing 71 seats in six-abreast rows, a galley/pantry opposite the rear door, a rear cabin containing 15 seats five-abreast, and a rear toilet compartment. Deletion of the wardrobes enables two more rows of seats to be installed in the main cabin in Summer, increasing max capacity to 122 seats. In 90-seat configuration, all seating is five-abreast, with 20 passengers in the front cabin, 55 in centre cabin and 15 in rear cabin. Again, two more rows of seats can replace the wardrobes in Summer, increasing the capacity to 100 seats. The 65-seat layout of the Il-18D has 14 seats (5-5-4) in front cabin, 43 seats (4-5-5-5-5-5-5-5-4) in centre cabin and 8 seats (4-4) in rear cabin. Pressurised cargo holds under floor forward and aft of the wing, and a further,unpressurised, hold aft of the rear pressure bulkhead.

SYSTEMS: Cabin pressurised to max differential of 0·49 bars (7·1 lb/sq in). Electrical system includes eight 12kW DC generators and 28·5V single-phase AC inverters. Hydraulic system, pressure 207 bars (3,000 lb/sq in), for landing gear retraction, nosewheel steering, brakes and flaps.

ELECTRONICS AND EQUIPMENT: Equipment includes dual controls and blind-flying panels, weather radar and ILS indicators, automatic navigation equipment, two automatic radio compasses, radio altimeter.

DIMENSIONS, EXTERNAL:

Wing span	37·4 m (122 ft 8½ in)
Wing chord at root	5·61 m (18 ft 5 in)
Wing chord at tip	1·87 m (6 ft 2 in)
Wing aspect ratio	10
Length overall	35·9 m (117 ft 9 in)
Height overall	10·17 m (33 ft 4 in)
Tailplane span	11·8 m (38 ft 8½ in)
Wheel track	9·0 m (29 ft 6 in)
Wheelbase	12·78 m (41 ft 10 in)
Propeller diameter	4·50 m (14 ft 9 in)
Passenger doors (each):	
Height	1·40 m (4 ft 7 in)
Width	0·76 m (2 ft 6 in)
Height to sill	2·90 m (9 ft 6 in)

Ilyushin Il-18 medium-range transport (four Ivchenko AI-20 turboprop engines) in service with Interflug of East Germany *(John Wegg)*

Freight hold doors (underfloor, each):
Height 0·90 m (2 ft 11 in)
Width 1·20 m (3 ft 11 in)
DIMENSIONS, INTERNAL:
Flight deck:
Volume 9·36 m³ (330 cu ft)
Cabin, excl flight deck:
Length approx 24·0 m (79 ft 0 in)
Max width 3·23 m (10 ft 7 in)
Max height 2·00 m (6 ft 6 in)
Volume 238 m³ (8,405 cu ft)
Baggage and freight holds (underfloor and aft of cabin:
total) 29·3 m³ (1,035 cu ft)
AREAS:
Wings, gross 140 m² (1,507 sq ft)
Ailerons (total) 9·11 m² (98·05 sq ft)
Trailing-edge flaps (total) 27·15 m² (292·2 sq ft)
Vertical tail surfaces (total) 17·93 m² (193·0 sq ft)
Rudder 6·83 m² (73·52 sq ft)
Horizontal tail surfaces (total)
27·79 m² (299·13 sq ft)
Elevators (total) 11·80 m² (127·0 sq ft)
WEIGHTS AND LOADINGS:
Weight empty, equipped (90-seater):
Il-18E 34,630 kg (76,350 lb)
Il-18D 35,000 kg (77,160 lb)
Max payload 13,500 kg (29,750 lb)
Max T-O weight:
Il-18V, E 61,200 kg (134,925 lb)
Il-18D 64,000 kg (141,100 lb)
Max wing loading (Il-18D) 457 kg/m² (93·6 lb/sq ft)
Max power loading (Il-18D) 5·05 kg/kW (8·30 lb/ehp)
PERFORMANCE (at max T-O weight):
Max cruising speed:
Il-18V 351 knots (650 km/h; 404 mph)
Il-18E, D 364 knots (675 km/h; 419 mph)
Econ cruising speed:
Il-18V 324 knots (600 km/h; 373 mph)
Il-18E, D 337 knots (625 km/h; 388 mph)
Operating height:
Il-18D 8,000-10,000 m (26,250-32,800 ft)
T-O run:
Il-18E 1,100 m (3,610 ft)
Il-18D 1,300 m (4,265 ft)
Landing run:
Il-18E, D 850 m (2,790 ft)
Range with max fuel, 1-hour reserves:
Il-18E 2,805 nm (5,200 km; 3,230 miles)
Il-18D 3,508 nm (6,500 km; 4,040 miles)
Range with max payload, 1-hour reserves:
Il-18E 1,728 nm (3,200 km; 1,990 miles)
Il-18D 1,997 nm (3,700 km; 2,300 miles)

ILYUSHIN Il-38

NATO reporting name: *May*

This anti-submarine/maritime patrol development of the Il-18 airliner represents a conversion similar to that by which the US Navy's P-3 Orion was evolved from the Lockheed Electra transport. It has a lengthened fuselage fitted with an undernose radome similar in shape to that of the Ka-25 ASW helicopter but housing a different radar, an MAD tail 'sting', other specialised electronic equipment and a weapon-carrying capability.

The main cabin of the Il-38 has few windows. The complete wing assembly is much further forward than on the Il-18, to cater for the effect of internal equipment and stores on the CG position.

The Il-38 is a standard shore-based maritime patrol aircraft of the Soviet Naval Air Force, operating widely over the Atlantic and Mediterranean. In the latter area, some aircraft carried Egyptian Air Force insignia for a period, but are believed to have been manned by Soviet aircrew, operating from North African bases such as Matru, near Cairo.

In 1975, the Indian Navy ordered an initial batch of three Il-38s.

DIMENSIONS, EXTERNAL:
Wing span 37·4 m (122 ft 8½ in)
Length overall 39·6 m (129 ft 10 in)
Height overall 10·17 m (33 ft 4 in)
PERFORMANCE (estimated):
Max cruising speed at 8,230 m (27,000 ft)
347 knots (645 km/h; 400 mph)
Max range 3,900 nm (7,250 km; 4,500 miles)

ILYUSHIN Il-62

NATO reporting name: *Classic*

Announced on 24 September 1962, when the first prototype (CCCP-06156) was inspected by Premier Krushchev, the standard Il-62 is a long-range airliner, with four Kuznetsov turbofan engines mounted in horizontal pairs on each side of the rear fuselage. It accommodates up to 186 passengers and was designed to fly on ranges equivalent to Moscow-New York (about 4,155 nm; 7,700 km; 4,800 miles) with more than 150 passengers and reserve fuel.

The Kuznetsov engines were not ready in time for the first flight of the first prototype, which took place in January 1963, with four 73·55 kN (16,535 lb st) Lyulka AL-7 engines installed. This aircraft was followed by a second prototype and three pre-production aircraft. Series production then started at Kazan, and Aeroflot introduced the Il-62 on to its Moscow-Montreal service on 15 September 1967, as a replacement for the Tu-114.

The Il-62 inaugurated Aeroflot's Moscow-New York service in July 1968, and has been used subsequently on many other routes, including Moscow-Paris and Moscow-Tokyo.

A leased Il-62 was introduced on the Prague-London service of CSA Czechoslovakian Airlines on 11 May 1968. CSA announced subsequently that it had placed the first export order for the Il-62, for three aircraft (increased later to seven). Delivery of these began in October 1969. Six began to enter service with the East German airline Interflug in the Spring of 1970, each equipped to carry 150 passengers; and EgyptAir operated seven for a period. The Polish airline LOT placed initial orders for three for delivery in 1972, and had five in 1976. China ordered five, under a 1970 trade agreement, for operation by CAAC. Tarom of Romania has three and Cubana one.

The Il-62's automatic flight control system is capable of taking over from a height of 200 m (650 ft) after take-off to a similar height during the landing approach. It can maintain a predetermined speed during climb and descent, and a selected cruising height, and can follow automatically a programmed track under command of the navigation computer.

The Il-62 is designed for an airframe service life of 25,000-30,000 flying hours, including 7,000-8,000 take-offs and landings.

A high-density version designated Il-62M, able to accommodate 198 passengers, was flown for the first time in 1971. Details of this aircraft are given separately, after the following description of the standard Il-62:

TYPE: Four-turbofan long-range airliner.

WINGS: Cantilever low-wing monoplane. Sweepback 35° at quarter-chord. Extended-chord 'dog-tooth' leading-edge on outer two-thirds of each wing. All-metal three-spar structure. Each wing fitted with three-section manually-operated ailerons, electrically-actuated double-slotted flaps and two hydraulically-

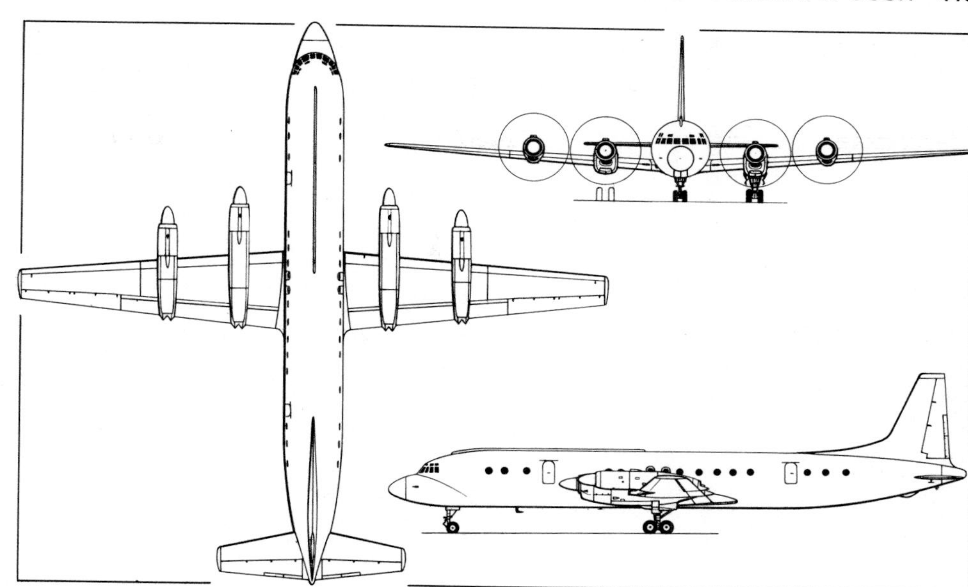

Ilyushin Il-18 four-turboprop medium-range airliner *(Pilot Press)*

Ilyushin Il-38 anti-submarine/maritime patrol aircraft (four Ivchenko AI-20 turboprop engines)

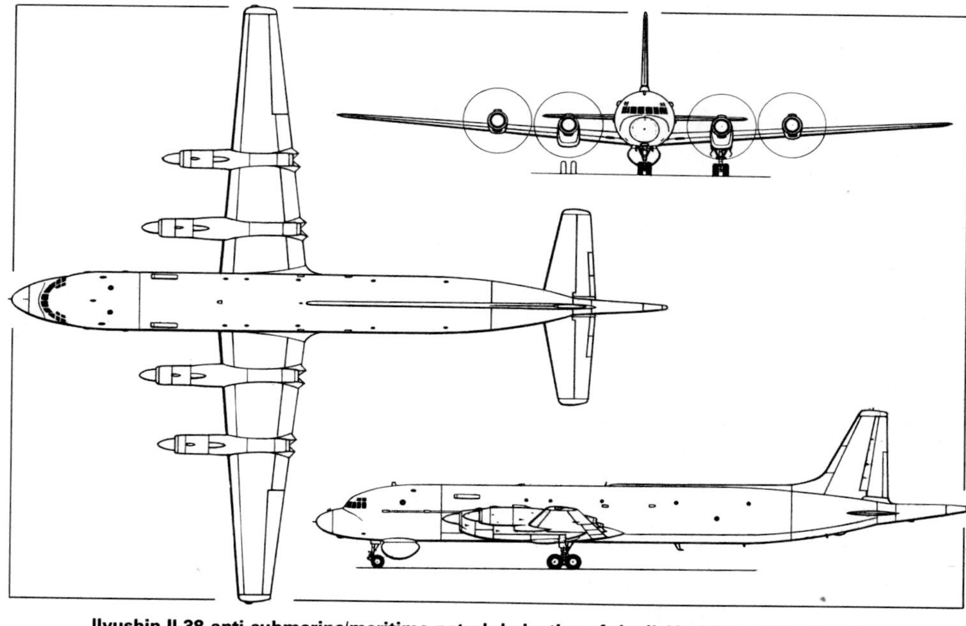

Ilyushin Il-38 anti-submarine/maritime patrol derivative of the Il-18 airliner *(Pilot Press)*

Ilyushin Il-62 long-range transport aircraft (four Kuznetsov NK-8-4 turbofan engines) in service with Aeroflot

operated spoiler sections forward of flaps. Trim tab and spring-loaded servo tab in each centre aileron, spring-loaded servo tab in each inner aileron. Hot-air anti-icing of leading-edges.

FUSELAGE: Conventional all-metal semi-monocoque structure. Frames are duralumin stampings and pressings. Integrally pressed skin panels at highly-stressed areas. Floors are sandwich panels with foam plastics filler. Nosecone hinges upward for access to radar.

TAIL UNIT: Cantilever all-metal structure, with electrically-actuated variable-incidence tailplane mounted at tip of fin. All surfaces sweptback. Manually-operated rudder, fitted with yaw damper, trim tab and spring servo tab. Manually-operated elevators have two automatic trim tabs and two manual trim tabs. Hot-air leading-edge anti-icing system.

LANDING GEAR: Hydraulically-retractable tricycle type. Forward-retracting twin-wheel steerable nose unit. Emergency extension by gravity. Oleo-nitrogen shock-absorber on each unit. Each main unit carries a four-wheel bogie and retracts inward into wing-roots. Main-wheel tyres size 1450 × 450, pressure 9·31 bars (135 lb/sq in). Nosewheel tyres size 930 × 305, pressure 7·86 bars (114 lb/sq in). Hydraulic disc brake and inertia-type electric anti-skid unit on each main wheel, supplemented by large tail parachute. Parking brakes. Hydraulic twin-wheel strut is extended downward to support rear fuselage during loading and unloading.

POWER PLANT: Four Kuznetsov NK-8-4 turbofan engines, each rated at 103 kN (23,150 lb st), mounted in horizontal pairs on each side of rear fuselage. Thrust reverser on each outboard engine. Hot-air anti-icing system for engine intakes. Automatically-controlled fuel system, with seven integral tanks extending through entire wing from tip to tip. Each engine has its own independent fuel system, with cross-feed. Total fuel capacity 100,000 litres (21,998 Imp gallons). Four standard international underwing pressure refuelling sockets. Eight gravity refuelling sockets. Total oil capacity 204 litres (45 Imp gallons).

ACCOMMODATION: Crew of five (two pilots, navigator, radio operator and flight engineer) on flight deck. Provision for two supernumerary pilot/navigators. Basic two-cabin layout, and galley, toilet and wardrobe facilities, are unchanged in the three main versions, only the width and pitch of the seats being varied. In the 186-passenger version, there are 72 seats in the forward cabin and 114 in the rear cabin, all six-abreast and all at a seat pitch of 86 cm (34 in). In the 168-seat configuration, increased pitch reduces capacity to 66 in the forward cabin and 102 in the rear cabin. The 114-passenger version has 45 seats in the forward cabin and 69 in the rear cabin, all five-abreast, except for four-abreast rear row by door. A first class/de luxe version for 85 passengers is available, with 45 seats in forward cabin and 40 four-abreast sleeperette chairs with footrests in rear cabin. Passenger doors forward of front cabin and between cabins on port side. Total of five toilets, opposite forward door, between cabins (starboard) and aft of rear cabin (both sides). Electrically-powered galley/pantry amidships and wardrobes in each version. Two pressurised baggage and freight compartments under cabin floor, forward and aft of wing. Unpressurised baggage/cargo compartment at extreme rear of fuselage. All compartments have tie-down fittings and rails in floor, and removable nets to restrain cargo.

SYSTEMS: Air-conditioning and pressurisation system maintains sea level conditions up to 7,000 m (23,000 ft) and gives equivalent of 2,100 m (6,900 ft) at 13,000 m (42,600 ft). Pressure differential 0·62 bars (9·0 lb/sq in). Hydraulic system, pressure 207 bars (3,000 lb/sq in), for landing gear retraction, nosewheel steering, brakes, spoilers and windscreen wipers. Three-phase 200/115V AC electrical supply from four 40kVA engine-driven generators (optional 27V DC system with eight 18kW engine-driven generators). Four transformer-rectifiers and four batteries for DC supply. Electrical windscreen de-icing. Type TA-6 APU in tailcone.

ELECTRONICS AND EQUIPMENT: Standard equipment includes two-channel autopilot, navigation computer, air data system, HF and UHF radio, VOR/ILS, RMI, Doppler, radio altimeter and weather radar. Polyot automatic flight control system optional.

DIMENSIONS, EXTERNAL:
- Wing span: 43·20 m (141 ft 9 in)
- Length overall: 53·12 m (174 ft 3½ in)
- Length of fuselage: 49·00 m (160 ft 9 in)
- Height overall: 12·35 m (40 ft 6¼ in)
- Tailplane span: 12·23 m (40 ft 1½ in)
- Fuselage width: 4·10 m (13 ft 5½ in)
- Fuselage height: 3·75 m (12 ft 3½ in)
- Wheel track: 6·80 m (22 ft 3½ in)
- Wheelbase: 24·49 m (80 ft 4½ in)
- Passenger doors (each):
  - Height: 1·83 m (6 ft 0 in)
  - Width: 0·86 m (2 ft 9¾ in)
  - Height to sill: 3·55 m (11 ft 8 in)
- Emergency exit (galley service) door:
  - Height: 1·37 m (4 ft 6 in)
  - Width: 0·61 m (2 ft 0 in)
- Front cargo hold door:
  - Height: 1·31 m (4 ft 3½ in)
  - Width: 1·26 m (4 ft 1½ in)
  - Height to sill: 1·92 m (6 ft 3½ in)
- Second cargo hold door:
  - Height: 1·00 m (3 ft 3½ in)
  - Width: 1·26 m (4 ft 1½ in)
  - Height to sill: 1·92 m (6 ft 3½ in)
- Third cargo hold door:
  - Height: 0·70 m (2 ft 3½ in)
  - Width: 0·70 m (2 ft 3½ in)
  - Height to sill: 2·32 m (7 ft 7¼ in)
- Rear cargo hold door:
  - Height: 1·15 m (3 ft 9 in)
  - Width: 1·07 m (3 ft 6 in)
  - Height to sill: 3·62 m (11 ft 10½ in)

DIMENSIONS, INTERNAL:
- Cabin:
  - Max height: 2·12 m (6 ft 11½ in)
  - Max width: 3·49 m (11 ft 5¼ in)
  - Volume: 163 m³ (5,756 cu ft)
- Total volume of pressure cell: 396 m³ (13,985 cu ft)
- Cargo hold volume:
  - Underfloor (two, total): 39·1 m³ (1,380 cu ft)
  - Rear fuselage: 5·8 m³ (205 cu ft)

AREAS:
- Wings, gross: 279·6 m² (3,010 sq ft)
- Ailerons (total): 16·25 m² (174·9 sq ft)
- Spoilers (total): 9·54 m² (102·7 sq ft)
- Flaps (total): 43·48 m² (468·0 sq ft)
- Horizontal tail surfaces (total): 40·00 m² (430·5 sq ft)
- Vertical tail surfaces (total): 35·60 m² (383·2 sq ft)

WEIGHTS AND LOADING:
- Weight empty: 66,400 kg (146,390 lb)
- Operating weight, empty: 69,400 kg (153,000 lb)
- Max payload: 23,000 kg (50,700 lb)
- Max fuel: 83,325 kg (183,700 lb)
- Max ramp weight: 167,000 kg (368,000 lb)
- Max T-O weight: 162,000 kg (357,000 lb)
- Max landing weight: 105,000 kg (232,000 lb)
- Max zero-fuel weight: 93,500 kg (206,000 lb)
- Max wing loading: 572 kg/m² (117·2 lb/sq ft)

PERFORMANCE (at max T-O weight):
- Normal cruising speed: 442-486 knots (820-900 km/h; 510-560 mph)
- Normal cruising height: 10,000-12,000 m (33,000-39,400 ft)
- Landing speed: 119-129 knots (220-240 km/h; 137-149 mph)
- Max rate of climb at S/L: 1,080 m (3,540 ft)/min
- FAR T-O field length:
  - ISA at S/L: 3,250 m (10,660 ft)
  - ISA+20°C at S/L: 3,915 m (12,840 ft)
- FAR landing field length:
  - ISA at S/L: 2,800 m (9,185 ft)
  - ISA+20°C at S/L: 2,950 m (9,680 ft)
- Range with max payload, 66,700 kg (147,050 lb) fuel, 1 hour fuel reserves: 3,612 nm (6,700 km; 4,160 miles)
- Range with 80,000 kg (176,370 lb) fuel and 10,000 kg (22,050 lb) payload, 1 hour fuel reserves: 4,963 nm (9,200 km; 5,715 miles)

ILYUSHIN Il-62M

First displayed publicly at the 1971 Paris Air Show, the Il-62M is a high-density, developed version of the Il-62 able to seat up to 198 passengers, with no dimensional changes to the airframe. It is fitted with more powerful turbofans, of a different type, with clamshell thrust reversers on the outboard engine of each pair, offering a lower approach speed and improved airflow over the rear of the nacelles. An additional fuel tank is installed in the tail-fin, contributing (with the improved specific fuel consumption of the engines) to the longer range of this version.

Revised layout of the flight deck equipment, and new navigation and radio communications equipment, are features of the Il-62M. Control wheels of new design allow the pilots a better field of view, and the aircraft's automatic flight control system permits automatic landings in ICAO Category II conditions, with extension to Category III conditions envisaged later. The wing spoilers of this version can be utilised differentially to enhance roll control.

Additional emergency and rescue equipment is installed on the Il-62M. The electrical, hydraulic and radio equipment in the rear fuselage has been repositioned. Together with the elimination of a wardrobe and transfer further aft of one central toilet and two rear toilets, this has permitted the installation of extra seats in the passenger cabin and optional provision of a compartment for buffet serving trolleys.

Unlike the Il-62, this version has a containerised baggage and freight system, with mechanised loading and unloading.

The Il-62M exhibited in Paris in 1971 and 1973 was the prototype (CCCP-86673). Production models entered service on Aeroflot's Moscow-Havana route in 1974 and have taken over progressively all of the airline's very-long-distance services.

Ilyushin Il-62M, the high-density version of this long-range airliner (four Soloviev D-30KU turbofan engines) *(Tass)*

The basic structural description of the Il-62 applies also to the Il-62M. The main innovations are as follows:

POWER PLANT: Four Soloviev D-30KU turbofan engines, each rated at 112·8 kN (25,350 lb st), mounted in horizontal pairs on each side of rear fuselage. Clamshell-type thrust reverser on each outboard engine. Remainder of power plant installation basically as for Il-62, but additional fuel tank in tail-fin with capacity of 5,000 litres (1,100 Imp gallons).

ACCOMMODATION: Alternative configurations for up to 198 economy class, 186 tourist class or 161 mixed class passengers. In the economy class version there are two toilets opposite the forward door, on the starboard side, aft of the flight deck. The forward cabin contains 72 seats, all six-abreast in threes with centre aisle. Galley/pantry amidships as on Il-62. Rear cabin contains 126 seats, six-abreast in threes with centre aisle. Three toilets and wardrobe to rear of this cabin. Doors as on Il-62. Forward underfloor baggage and freight hold accommodates nine containers, each weighing approximately 45 kg (100 lb) empty and with a capacity of 600 kg (1,322 lb) and 1·6 m³ (56·5 cu ft). Rear hold accommodates five similar containers. Two compartments for non-containerised cargo.

SYSTEMS AND EQUIPMENT: See introductory notes.

DIMENSIONS AND AREAS:
Same as for Il-62

WEIGHTS:
Max payload 23,000 kg (50,700 lb)
Max T-O weight 165,000 kg (363,760 lb)

PERFORMANCE (at max T-O weight):
Normal cruising speed
458-486 knots (850-900 km/h; 528-560 mph)
Normal cruising height
10,000-12,000 m (33,000-39,400 ft)
Balanced T-O distance (ISA, S/L) 3,000 m (9,845 ft)
Landing run (ISA, S/L) 2,800 m (9,185 ft)
Range with max payload, with reserves
4,315 nm (8,000 km; 4,970 miles)
Range with 10,000 kg (22,045 lb) payload, with reserves 5,555 nm (10,300 km; 6,400 miles)

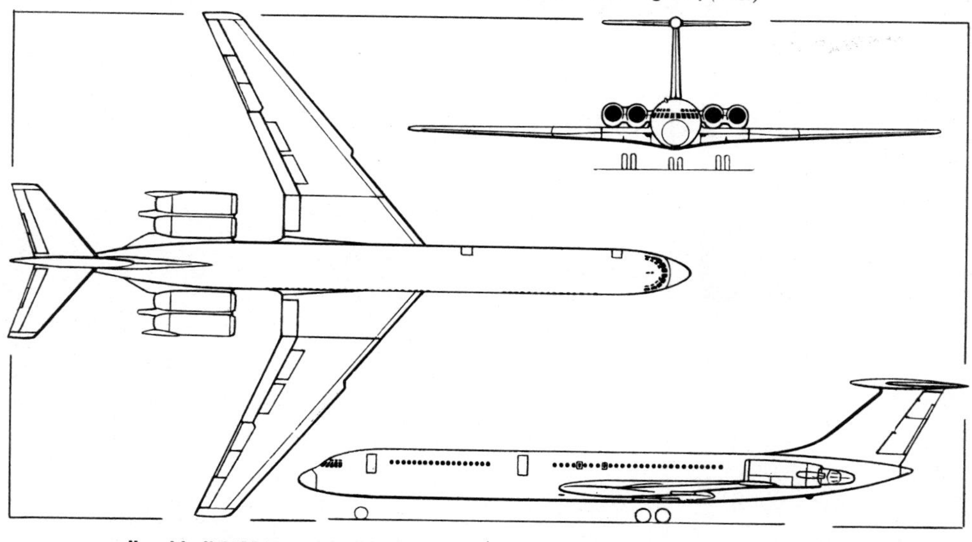

Ilyushin Il-62M high-density long-range four-turbofan transport *(Pilot Press)*

ILYUSHIN Il-76

NATO reporting name: *Candid*

Flown for the first time on 25 March 1971, the Il-76 prototype (CCCP-86712) made its public debut at the 29th Salon de l'Aéronautique et de l'Espace in Paris in May 1971.

It is a high-performance pressurised heavy transport of conventional layout, powered by four turbofan engines of similar basic type to those installed in the Il-62M. The clamshell thrust reversers, fitted to all four engines, are of different configuration, stowing above and below the nozzle when not in use, instead of to each side.

Nominal task of the Il-76 is to transport 40 tonnes of freight for a distance of 2,700 nm (5,000 km; 3,100 miles) in less than six hours. It can take off from short unprepared airstrips and an official statement in 1971 said that it would be used first during the period of the then-current five-year plan (1971-75) in Siberia, the north of the Soviet Union and the Far East, where operation of other types of transport is difficult. However, it was clear from the start that the Il-76 had considerable potential as a military transport, and the aircraft is known to be in first-line squadron service with the Soviet Air Force. The military Il-76 has a rear gun turret at the tail.

There is evidence that a version of the Il-76 has been evaluated as a flight refuelling tanker for the *Backfire* supersonic strategic bombers of the Soviet Air Force and Naval Air Fleet, as a potential successor to modified M-4 *(Bison)* aircraft in current service.

Aircraft seen and photographed during 1973-76 embody a number of modifications compared with the prototype. Most important is that the hinge-line of each rear clamshell door is higher on the fuselage of at least one aircraft, giving a larger door and permitting taller and wider loads to pass between the doors when they are open. Other new features include a modified rear fin fillet and strengthening of the upper fuselage.

In July 1975, the Il-76 set a total of 24 officially recognised records for speed and altitude with payload. Piloted by Yakov I. Vernikov, on 4 July, it raised to 70,121 kg (154,590 lb) the record for greatest payload carried to a height of 2,000 m. The same flight recorded an altitude of 11,875 m (38,960 ft) with payloads of 60,000 kg, 65,000 kg and 70,000 kg. Also on 4 July, Alexander Turumine averaged 462·283 knots (856·697 km/h; 532·327 mph) around a 2,000 km circuit in an Il-76 carrying a payload of 55,000 kg, qualifying for additional records with 35,000 kg, 40,000 kg, 45,000 kg and 50,000 kg. On 7 July, Turumine averaged 462·801 knots (857·657 km/h; 532·923 mph) around 1,000 km, claiming nine records for payloads from 30,000 kg to 70,000 kg. The same pilot averaged 440·305 knots (815·968 km/h; 507·019 mph) around a 5,000 km circuit on 10 July, claiming records with 15,000 kg, 20,000 kg, 25,000 kg, 30,000 kg, 35,000 kg and 40,000 kg payloads.

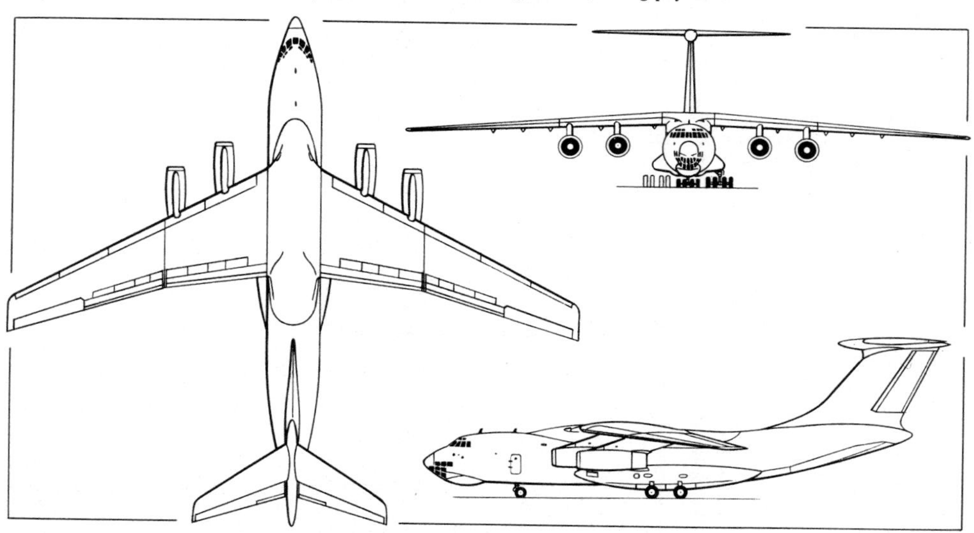

Ilyushin Il-76 four-turbofan heavy freight-carrying transport *(Pilot Press)*

Ilyushin Il-76 freight transport (four Soloviev D-30KP turbofan engines) photographed during proving flights to oil and natural gas producing areas in the Tyumen region of Siberia *(Tass)*

The dished fairing on the fuselage tail-cone of this Il-76 suggests that it may have been fitted at some time with a tail gun turret or flight refuelling equipment, as a testbed for the military versions *(Tass)*

Type: Four-turbofan medium/long-range freight transport.

Wings: Cantilever monoplane, mounted above fuselage to leave interior unobstructed, and with marked anhedral from roots. Sweepback 25° at quarter-chord. All-metal structure. Two-section double-slotted flaps over full span from wing root to inboard edge of aileron each side. Spoilers forward of inboard flaps. Leading-edge slots over almost entire span. Tabs in each aileron.

Fuselage: All-metal semi-monocoque structure of basically circular section. Underside of upswept rear fuselage made up of two outward-hinged clamshell doors, upward-hinged panel between these doors, and downward-hinged loading ramp.

Tail Unit: Cantilever all-metal structure, with tailplane mounted at tip of fin. All surfaces sweptback. Tabs in rudder and each elevator.

Landing Gear: Retractable tricycle type, designed for operation from prepared and unprepared runways. Nose unit made up of two pairs of wheels, side by side with central oleo. Each main-wheel bogie made up of four pairs of wheels in two rows. Low-pressure tyres size 1,300 × 480 on main wheels, 1,100 × 330 on nosewheels. Main units retract inward into two large ventral fairings under fuselage, with an additional large fairing on each side of lower fuselage over actuating gear. During retraction main-wheel axles rotate around leg, so that wheels stow with axles parallel to fuselage axis (ie: wheels remain vertical but at 90° to direction of flight).

Power Plant: Four Soloviev D-30KP turbofan engines, each rated at 117·7 kN (26,455 lb st), in individual underwing pods. Each pod is carried on a large forwardly-inclined pylon and is fitted with a clamshell thrust reverser. No fuel is carried in the wings.

Accommodation: Conventional side-by-side seating for pilot and co-pilot on spacious flight deck. Station for navigator below flight deck in glazed nose. Forward-hinged door on each side of fuselage forward of wing. Cabin loaded via rear ramp. Entire accommodation is pressurised, and advanced mechanical handling systems are provided for containerised and other freight.

Electronics and Equipment: Full equipment for all-weather operation by day and night, including a computer for automatic flight control and automatic landing approach. Large ground-mapping radar in undernose radome. APU in port side landing gear fairing.

Dimensions, external:
- Wing span: 50·50 m (165 ft 8 in)
- Length overall: 46·59 m (152 ft 10½ in)
- Height overall: 14·76 m (48 ft 5 in)

Area:
- Wings, gross: 300·0 m² (3,229·2 sq ft)

Weights:
- Max payload: 40,000 kg (88,185 lb)
- Max T-O weight: 157,000 kg (346,125 lb)

Performance:
- Normal cruising speed: 458 knots (850 km/h; 528 mph)
- Normal cruising height: 13,000 m (42,650 ft)
- T-O run (unpaved runway): 850 m (2,790 ft)
- Landing run (unpaved runway): 450 m (1,476 ft)
- Nominal range with max payload: 2,700 nm (5,000 km; 3,100 miles)

ILYUSHIN Il-86

First indication that this aircraft was under development was given at the 1971 Paris Air Show. Mr Genrikh Novozhilov, successor to the semi-retired Sergei Ilyushin as chief of the Ilyushin design bureau, told visitors that a new wide-bodied transport known as the Il-86 was then in the early project design stage.

No final decision on the configuration, or number of engines, had been taken at that time; but in the Spring of 1972 a model of one projected configuration was displayed publicly in Moscow. This design was similar in layout to the Il-62, with four rear-mounted turbofan engines and a T-tail, but was intended to be much larger, with a two-deck fuselage. It was described and illustrated in the 1972-73 *Jane's*.

Simultaneously with the display of this original model, it became known that the Il-86 had been chosen for development, after a competition in which it was matched against proposals from the Antonov and Tupolev design teams. If it proves successful, it is expected to follow the 'stretched' Tu-154 interim airbus in service with Aeroflot in the late 'seventies.

By the end of 1972, it became clear that the design of the Il-86 had evolved along different lines to those suggested by the model displayed six months earlier. In particular the engines had been repositioned into four underwing pods, permitting the tailplane to be lowered on to the rear fuselage, as shown in the accompanying drawing.

The fuselage is circular in cross-section, with the dividing floor positioned just below the widest point. The upper deck, on which all seats are located, is divided into three separate passenger cabins by wardrobes, galleys and cabin staff accommodation, with toilets at front and rear of the aircraft. Up to 350 passengers could be carried in basic nine-abreast seating throughout, with two aisles. A suggested mixed class alternative provides for 28 passengers six-abreast in the front cabin and 206 passengers eight-abreast in the other two cabins. Films will be shown in flight, and there will be a choice of 12 tape-recorded programmes to listen to.

Passengers are intended to enter the aircraft via three airstair-type doors which hinge down from the port side of the lower deck. Two of these doors are forward of the wing; the other is aft of the wing. There are four further doors at upper-deck level on each side, presumably for emergency use.

Coats and hand baggage are intended to be stowed on the lower deck before passengers climb one of the three fixed staircases to the main deck. The cargo holds are designed to accommodate baggage and freight in 16 standard LD3 containers. Access is via upward-hinged doors forward of the starboard wing-root leading-edge and at

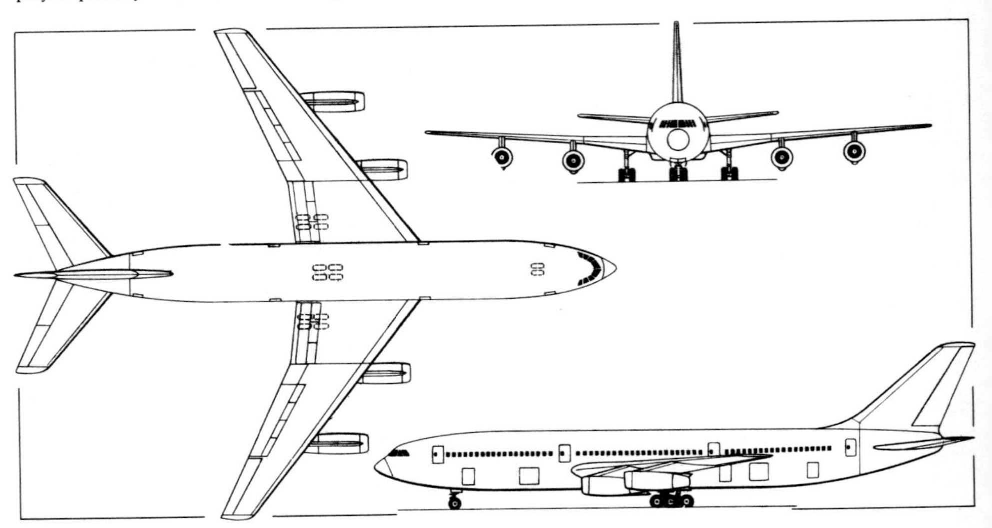

Ilyushin Il-86 four-turbofan wide-bodied passenger transport *(Pilot Press; provisional)*

the side of the rear hold, and containers can be loaded and unloaded by means of a self-propelled truck with built-in roller conveyor.

There appear to be high-lift devices on both the leading- and trailing-edges of the wings to improve field performance. Wing sweep is 35° at quarter-chord. The landing gear comprises a rearward-retracting twin-wheel nose unit and three four-wheel bogie main units. Two of the latter retract inward into the enlarged wing-root fairings; the third unit is mounted centrally under the fuselage, slightly forward of the others.

It was expected originally that the power plants of the Il-86 would be Soloviev D-30KP high by-pass ratio turbofans, each rated at 117·7 kN (26,455 lb st). More recently, it has been suggested that choice of engine will follow evaluation of competing turbofan designs from the Soloviev and Lotarev engine teams. Fuel capacity will be 70,000-80,000 litres (15,400-17,600 Imp gallons).

Standard flight crew is to comprise two pilots and a flight engineer, with provision for a navigator if required.

The prototype Il-86 was nearing completion in the Spring of 1976.

DIMENSIONS, EXTERNAL:

Wing span	48·33 m (158 ft 6½ in)
Length overall	58·50 m (191 ft 11 in)
Diameter of fuselage	6·08 m (19 ft 11½ in)
Height overall	15·70 m (51 ft 6 in)
Tailplane span	19·00 m (62 ft 4 in)

DIMENSIONS, INTERNAL:

Main cabins: Height	2·61 m (8 ft 7 in)
Max width	approx 5·70 m (18 ft 8½ in)

AREA:

Wings, gross	320 m² (3,444 sq ft)

WEIGHTS:

Max payload	40,000 kg (88,185 lb)
Max T-O weight	188,000 kg (414,470 lb)

PERFORMANCE (estimated):

Normal cruising speed at 9,000-10,000 m (30,000-33,000 ft)
485-512 knots (900-950 km/h; 560-590 mph)

Landing speed
130-135 knots (240-250 km/h; 149-155 mph)

Range with max payload
1,268 nm (2,350 km; 1,460 miles)

Range with max fuel
2,480 nm (4,600 km; 2,858 miles)

KAMOV

Nikolai I. Kamov, who died on 24 November 1973, aged 71, had been a leading designer of rotating-wing aircraft since the late 1920s and, with N. K. Skrzhinskii, was responsible for the first successful Soviet rotorcraft, the KaSkr-I, in 1929. He became well known internationally when he designed a series of one-man lightweight helicopters of the 'flying motorcycle' type in the late 1940s.

The Ka-15 and Ka-18 helicopters, developed by Kamov and his design team, under chief engineer Vladimir Barshevskii, were both put into large-scale production and service. Details of them can be found in the 1962-63 and 1963-64 editions of *Jane's* respectively.

Later Kamov types are the Ka-25 turbine-powered anti-submarine helicopter; a flying-crane version of the same design, designated Ka-25K; and a twin-engined general-purpose helicopter designated Ka-26. All available details of these types follow:

KAMOV Ka-25

NATO reporting name: *Hormone*

The prototype of this military helicopter was first shown in public in the Soviet Aviation Day flypast over Tushino Airport, Moscow, in July 1961. It was allocated the NATO code name *Harp,* but this was changed to *Hormone* for the production versions, which have replaced the Mi-4 in service with the Soviet Navy, ashore and at sea. Nine are operated on coastal anti-submarine duties by the Syrian Air Force.

Basically, the Ka-25 follows the formula established by earlier Kamov designs such as the Ka-15 and Ka-18, with two three-blade co-axial contra-rotating rotors, a pod-and-boom fuselage, multi-fin tail unit, and four-wheel landing gear. It is powered by two small turboshaft engines mounted side by side above the cabin, and this has left the cabin space clear for personnel, operational equipment, fuel and payload.

In its anti-submarine version, the Ka-25 operates from ships of the Soviet Navy, including cruisers of the *Kresta* and *Kara* classes, and the helicopter carrier/cruisers *Moskva* and *Leningrad,* each of which accommodates about 20 aircraft. It has a search radar installation in a radome under the nose. This exists in two forms: as illustrated, with a diameter of 1·25 m (4 ft 1 in), and in a considerably larger form. Other equipment includes a towed magnetic anomaly detector, dipping sonar housed in a compartment at the rear of the cabin, and an electro-optical sensor. Each landing wheel can be surrounded by an inflatable pontoon surmounted by inflation bottles to provide flotation in the event of an emergency alighting on the water. The rear legs are pivoted to retract upward about their wishbone supports, so that the wheels can be moved to a position where they offer least interference to signals from the nose radar.

The two so-called 'air-to-surface missiles' carried on outriggers on each side of the cabin of the prototype during its Tushino appearance were dummies, and there is no evidence that externally-mounted weapons are carried. The production Ka-25 has an internal weapons bay for stores, including ASW torpedoes and nuclear depth charges.

As well as serving with the Red Banner Black Sea Fleet as an anti-submarine aircraft, based on the *Moskva* and *Leningrad,* and on the new carrier/cruiser *Kiev,* the Ka-25 fulfils a variety of other military roles. Only two versions may be identified at present by NATO reporting names:

Hormone-A. Basic anti-submarine version, as described above and below.

Hormone-B. Special electronics variant. No details available for publication.

The Ka-25K, described separately, is a commercial counterpart of the Ka-25. It can be assumed that the two types are similar in details such as basic structure, overall dimensions, power plant, weights and performance, except that the military version is a little shorter, with an estimated overall length of 9·75 m (32 ft 0 in).

TYPE: Twin-turbine anti-submarine and general-purpose helicopter.

AIRFRAME AND POWER PLANT: Basically as for Ka-25K, except for pivoted main landing gear. Provision for carrying an external fuel tank on each side of main cabin.

ACCOMMODATION: Pilot and co-pilot side by side on flight deck, with rearward-sliding door on each side. Entry to main cabin is via a rearward-sliding door to rear of main landing gear on port side. Cabin is large enough to contain 12 folding seats for passengers in transport version.

ELECTRONICS AND EQUIPMENT: Equipment available for all versions includes autopilot, navigational system, radio compass, radio communications installations, and lighting system for all-weather operation by day or night. Dipping sonar housed in compartment at rear of main cabin, immediately forward of tailboom, and search radar under nose of anti-submarine version, which carries also a towed magnetic anomaly detector. Some aircraft have a blister fairing over equipment mounted at the base of the centre tail-fin; others have a cylindrical housing, with a transparent top, above the central point of the tailboom (see illustration), with a shallow blister fairing to the rear of this. Doors under the fuselage

The version of the Kamov Ka-25 anti-submarine helicopter with a blister fairing at the base of the central tail-fin

Kamov Ka-25 anti-submarine helicopter (two Glushenkov turboshaft engines) *(US Navy)*

enclose a weapons bay for ASW torpedoes, nuclear depth charges and other stores.

KAMOV Ka-25K

NATO reporting name: *Hormone*

This flying-crane helicopter was shown publicly for the first time at the 1967 Paris Air Show. Instead of the undernose radome of the anti-submarine version of the Ka-25, it has a removable gondola giving an exceptional field of view for the occupant.

One of the pilots occupies this gondola during loading, unloading and positioning of externally-slung cargoes, while the helicopter is hovering. His seat faces rearward, giving him an unobstructed view of the operation, and he is able to control the aircraft by means of a set of dual flying controls fitted in the gondola. This distribution of duty, with one pilot controlling the aircraft during loading and unloading operations and the other pilot controlling it in cruising flight, is claimed to increase the precision and safety of payload handling and to offer a considerable reduction in the overall time required to do a particular job.

The Ka-25K is claimed to combine high payload-to-AUW ratio with good manoeuvrability and minimum dimensions. The rotors, transmission and engines, with their auxiliaries, form a single self-contained assembly, which can be removed in one hour.

TYPE: Twin-turbine flying-crane helicopter.

ROTOR SYSTEM: Two three-blade co-axial contra-rotating rotors. Automatic blade-folding.

FUSELAGE: Conventional all-metal semi-monocoque structure of pod and boom type. Detachable gondola under nose.

TAIL UNIT: Cantilever all-metal structure, with central fin, ventral fin and twin endplate fins and rudders which are toed inward.

LANDING GEAR: Non-retractable four-wheel type. Oleo-pneumatic shock-absorbers. Nosewheels are smaller than main wheels and are of castoring type. Each wheel can be enclosed in an inflatable pontoon surmounted by inflation bottles.

POWER PLANT: Two 671 kW (900 shp) Glushenkov GTD-3 turboshaft engines, mounted side by side above cabin, forward of rotor driveshaft.

ACCOMMODATION: Crew of two side by side on flight deck. Rearward-facing pilot's seat with dual flying controls in undernose gondola for use during loading and unloading. Main cabin, normally used for freight carrying, contains 12 folding seats for passengers. Rearward-sliding door on each side of flight deck. Large rearward-sliding door at rear of main cabin on port side. Hatchway in cabin floor, with two downward-opening doors, through which sling cable passes from winch on CG.

ELECTRONICS AND EQUIPMENT: Optional equipment includes autopilot, navigational system, radio compass, radio communications installation, and lighting system for all-weather operation by day or night.

DIMENSIONS, EXTERNAL:
Diameter of rotors (each) 15·74 m (51 ft 8 in)
Length overall 9·83 m (32 ft 3 in)
Height to top of rotor head 5·37 m (17 ft 7½ in)
Width over tail-fins 3·76 m (12 ft 4 in)
Wheel track:
front 1·41 m (4 ft 7½ in)
rear 3·52 m (11 ft 6½ in)
Cabin door: Height 1·10 m (3 ft 7¼ in)
Width 1·20 m (3 ft 11¼ in)

WEIGHTS:
Weight empty 4,400 kg (9,700 lb)
Max payload 2,000 kg (4,400 lb)
Max T-O weight 7,300 kg (16,100 lb)

PERFORMANCE:
Max level speed 119 knots (220 km/h; 137 mph)
Normal cruising speed 104 knots (193 km/h; 120 mph)
Service ceiling 3,500 m (11,500 ft)
Range with standard fuel, with reserves 217 nm (400 km; 250 miles)
Range with max fuel, with reserves 351 nm (650 km; 405 miles)

KAMOV Ka-26

NATO reporting name: *Hoodlum*

First details of this twin-engined light helicopter were announced in January 1964, and the prototype flew for the first time in the following year. Kamov described it as an ideal helicopter for agriculture, possessing all the virtues of the Ka-15 (which was used in about a dozen countries) but able to lift three times as much chemical payload, and the Ka-26 entered large-scale service as an agricultural aircraft in the Soviet Union in 1970, being used primarily over orchards and vineyards. It is also used widely on Aeroflot's air ambulance services and is suitable for many other applications, including cargo and passenger transport, forest firefighting, mineral prospecting, pipeline construction and laying transmission lines.

The usual Kamov contra-rotating co-axial three-blade rotor system is retained, with hydraulic dampers fitted to each rotor head and the rotor shafts inclined forward at 6° to the vertical. The blades, made of glass-textolyte (plastics) materials, weigh only 25 kg (55 lb) each and are completely interchangeable. They, and the cabin windscreen, are equipped with an anti-icing system, activated automatically by a radioisotope ice warning device and utilising an alcohol glycerine mixture.

A powered control system is standard. The jacks are actuated by a single hydraulic system, with manual override in case of system failure.

The fully-enclosed cabin, with a door on each side, is fitted out normally for operation by a single pilot, but a second seat and dual controls are optional. The cabin is warmed and demisted by air from a combustion heater, which also heats the passenger compartment when fitted.

The tailplane, with twin fins and rudders toed inward at 15°, is carried on two plastics tailbooms. Short high-mounted stub-wings carry the two podded 242·5 kW (325 hp) M-14V-26 aircooled radial piston engines, designed by I. M. Vedeneev, and the main units of the non-retractable four-wheel landing gear. Each engine is cooled by a fan in the front of its nacelle, which absorbs about 18·6 kW (25 hp) from the engine output. Dust filters are fitted in the air delivery ducts, to protect the engines, each of which is connected to the rotor transmission by a shaft and two flexible couplings. Both rotors can be driven by either engine if the other fails; disengagement of the failed engine is automatic, and an autorotative landing can be made if both engines fail.

All four landing gear units embody oleo-pneumatic shock-absorbers. The forward wheels are of the castoring type and are not fitted with brakes. The rear wheels are

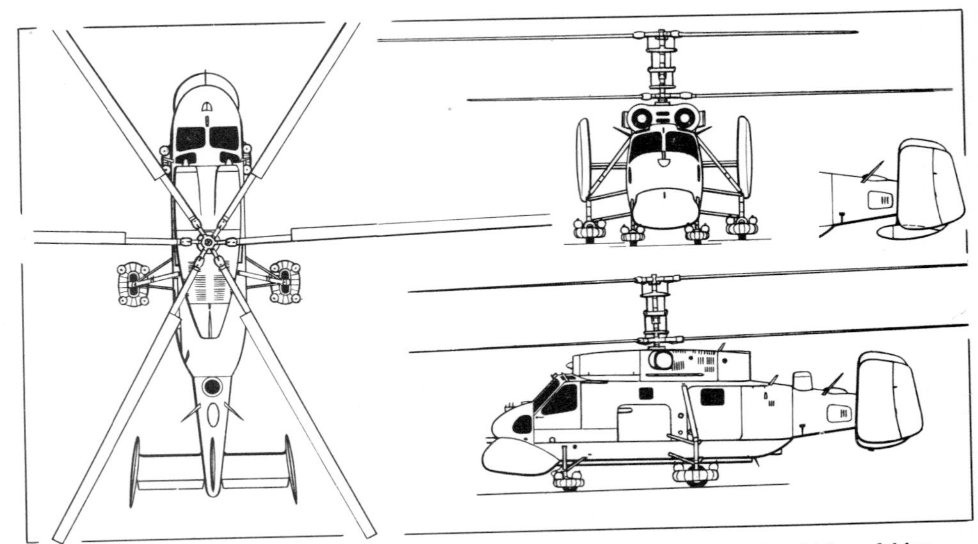

Anti-submarine version of the Kamov Ka-25 helicopter. Scrap view shows optional blister fairing at base of central tail-fin *(Pilot Press)*

Kamov Ka-25K flying-crane helicopter (two Glushenkov GTD-3 turboshaft engines) *(S. P. Peltz)*

Kamov Ka-26 in service with Interflug of East Germany

fitted with pneumatically-operated brakes. Tyre size is 595 × 185 on the main wheels, 300 × 125 on the forward wheels.

The space aft of the cabin, between the main landing gear units and under the rotor transmission, is able to accommodate a variety of interchangeable payloads. For agricultural work the chemical hopper (capacity 900 kg; 1,985 lb) and dust-spreader or spraybars are fitted in this position, on the aircraft's centre of gravity. This equipment is quickly removable and can be replaced by a cargo/passenger pod accommodating six persons, with provision for a seventh passenger beside the pilot. Alternatively, the Ka-26 can be operated with either an open platform for hauling freight or a hook for slinging bulky loads at the end of a cable or in a cargo net.

A version for geophysical survey has an electromagnetic pulse generator in the cabin and is encircled by a huge 'hoop' antenna. It carries on the port side of the fuselage a mounting for the receiver 'bird' which is towed at the end of a cable, beneath the helicopter, when in use. The receiver is lowered by an electric winch and the cable is cut by automatic shears if its traction should exceed the authorised limit.

An aerial survey model is available with an AFA-31-MA camera mounted in the cabin. This aircraft can photograph 5 km² (2 sq miles) per hour at a scale of 1 : 10,000.

As an air ambulance, the Ka-26 can carry two stretcher patients, two seated casualties and a medical attendant. A winch, with a capacity of up to 150 kg (330 lb), enables it to be used for rescue duties.

When operating as an agricultural sprayer, the Ka-26 discharges its chemical payload at 1·5-12 litres/sec (0·33-2·65 Imp gallons/sec). The rate of discharge in a dusting role is 1·5-12 kg/sec (3·3-26·5 lb/sec). Up to 120 hectares (296 acres) can be sprayed during each flying hour at the rate of 50 kg/ha (44·5 lb/acre). As a duster, 140 ha (346 acres) can be treated at the same discharge rate. 50 ha (123 acres) can be topdressed with chemical fertilisers each flying hour, at a rate of 100 kg/ha (89 lb/acre).

To protect the pilot against toxic chemicals in the agricultural role, the cabin is lightly pressurised by a blower and air filter system which ensures that the cabin air is always clean. The flying and navigation equipment are adequate for all-weather operation, by day and night. VHF and HF radio are fitted, together with a radio compass and radio altimeter.

Because of its small size and manoeuvrability, the Ka-26 can be operated from platforms on small ships such as whalers and icebreakers, and a Soviet fishing boat operating in the North Atlantic in early 1970 carried a Ka-26 for fish-spotting duties. This aircraft was equipped with inflated pontoons to permit alighting on the water.

In mid-1969, a Ka-26 was tested in Siberia and the north-west USSR in a forest protection version able to deliver six firemen and their equipment speedily to the site of a forest fire. In the Spring of 1972, Ka-26s joined Mi-1, Mi-2 and Mi-4 helicopters in operations to clear ice from Soviet rivers, by landing demolition teams on thick ice-floes and destroying thinner ice-fields from the air.

Ka-26s are in civilian service in Bulgaria, East Germany, West Germany, Hungary, Romania and Sweden, as well as in the USSR. Military operators include the air forces of Hungary and Sri Lanka.

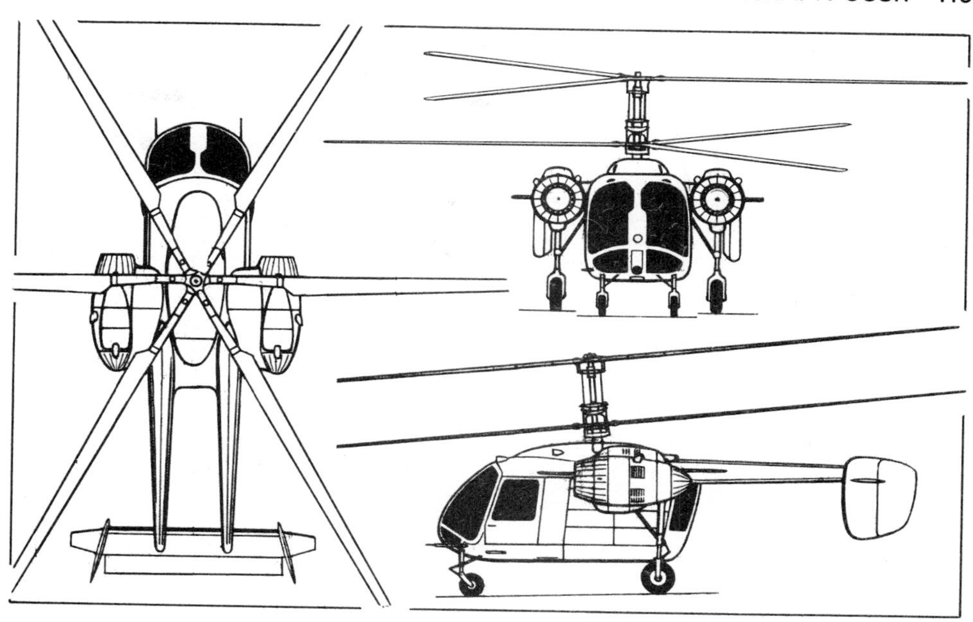

Kamov Ka-26 twin-engined light general-purpose helicopter *(Pilot Press)*

Agricultural spraying version of the Kamov Ka-26 *(Tass)*

DIMENSIONS, EXTERNAL:
- Diameter of rotors (each): 13·00 m (42 ft 8 in)
- Vertical separation between rotors: 1·17 m (3 ft 10 in)
- Length of fuselage: 7·75 m (25 ft 5 in)
- Height overall: 4·05 m (13 ft 3½ in)
- Width over engine pods: 3·64 m (11 ft 11½ in)
- Width over agricultural spraybars: 11·20 m (36 ft 9 in)
- Tailplane span: 4·60 m (15 ft 1 in)
- Wheel track:
  - Main wheels: 2·42 m (7 ft 11½ in)
  - Nosewheels: 0·90 m (2 ft 11½ in)
- Wheelbase: 3·48 m (11 ft 5 in)
- Passenger pod door:
  - Height: 1·40 m (4 ft 7 in)
  - Width: 1·25 m (4 ft 1¼ in)

DIMENSIONS, INTERNAL:
- Passenger pod:
  - Length, floor level: 1·83 m (6 ft 0 in)
  - Width, floor level: 1·25 m (4 ft 1¼ in)
  - Headroom: 1·40 m (4 ft 7 in)

WEIGHTS:
- Operating weight, empty:
  - Stripped: 1,950 kg (4,300 lb)
  - Cargo/platform: 2,085 kg (4,597 lb)
  - Cargo/hook: 2,050 kg (4,519 lb)
  - Passenger: 2,100 kg (4,630 lb)
  - Agricultural: 2,216 kg (4,885 lb)
- Fuel weight:
  - Transport: 360 kg (794 lb)
  - Other versions: 100 kg (220 lb)
- Payload:
  - Transport: 900 kg (1,985 lb)
  - Agricultural duster: 1,065 kg (2,348 lb)
  - Agricultural sprayer: 900 kg (1,985 lb)
  - With cargo platform: 1,065 kg (2,348 lb)
  - Flying crane: 1,100 kg (2,425 lb)
- Normal T-O weight:
  - Transport: 3,076 kg (6,780 lb)
  - Agricultural: 2,980 kg (6,570 lb)
- Max T-O weight:
  - all versions: 3,250 kg (7,165 lb)

PERFORMANCE (at max T-O weight):
- Max level speed: 91 knots (170 km/h; 105 mph)
- Max cruising speed: 81 knots (150 km/h; 93 mph)
- Econ cruising speed: 49-59 knots (90-110 km/h; 56-68 mph)
- Agricultural operating speed range: 16-62 knots (30-115 km/h; 19-71 mph)
- Service ceiling: 3,000 m (9,840 ft)
- Service ceiling, one engine out: 500 m (1,640 ft)
- Hovering ceiling in ground effect at AUW of 3,000 kg (6,615 lb): 1,300 m (4,265 ft)
- Hovering ceiling out of ground effect at AUW of 3,000 kg (6,615 lb): 800 m (2,625 ft)
- Range with 7 passengers, 30 min fuel reserves: 215 nm (400 km; 248 miles)
- Max range with auxiliary tanks: 647 nm (1,200 km; 745 miles)
- Endurance at econ cruising speed: 3 hr 42 min

MiG

Colonel-General Artem I. Mikoyan, who died on 9 December 1970 at the age of 65, was head of the design bureau responsible for the MiG series of fighter aircraft. With Mikhail I. Gurevich, a mathematician, he collaborated in the design of the first of the really-modern Soviet jet-fighters, the MiG-15, which began to appear in squadron service in numbers in 1949.

The MiG-17, a progressive development of the MiG-15, appeared in Soviet squadrons in 1953 or 1954, and was followed into service by the supersonic MiG-19, which appeared in 1955 and has been manufactured also in large numbers in China (which see).

All available details of Mikoyan designs currently in production or under development follow:

MIKOYAN MiG-21

NATO reporting names: *Fishbed* **and** *Mongol*

The Soviet design bureau that was led by the late Colonel-General Artem I. Mikoyan developed the MiG-21 air superiority fighter on the basis of experience of jet-to-jet combat between MiG-15s and US aircraft during the war in Korea. The emphasis was placed on good transonic and supersonic handling, high rate of climb, small size and light weight, using a turbojet engine of medium power, in contrast with the heavier and much more powerful Sukhoi Su-7 and Su-9 fighters that were developed simultaneously. The first versions of the MiG-21 were, therefore, day fighters of limited range, with comparatively light armament and limited avionics. Subsequent development of the type has been aimed primarily at improvements in range, weapons and all-weather capability.

The E-5 prototype of the MiG-21 flew for the first time in 1955, and made its public debut during the flypast in the Soviet Aviation Day display at Tushino Airport, Moscow, on 24 June 1956. The initial production version (NATO *Fishbed-A*) was built in only limited numbers, with a Tumansky R-11 turbojet engine rated at 38·25 kN (8,600 lb st) dry and 50 kN (11,240 lb st) with afterburning, and with an armament of two 30 mm NR-30 cannon. Meanwhile, the Soviet Union had been developing a small infra-red homing air-to-air missile, designated K-13 (NATO *Atoll*) and generally similar to the US AIM-9B Sidewinder 1A. Underwing pylons for two K-13s were fitted on the MiG-21F, the suffix 'F' standing for *Forsirovanny* (boosted) and indicating that this model also had a slightly more powerful turbojet. To save weight and provide room for electronics associated with the missiles, the port NR-30 cannon was removed and its blast-tube fairing on the lower fuselage was blanked off. Further details of this and subsequent versions of the MiG-21 are as follows:

MiG-21F *(Fishbed-C)*. First major production version, built also in Czechoslovakia. Short-range clear-weather fighter, with radar ranging equipment and a Tumansky R-11 turbojet rated at 42·25 kN (9,500 lb st) dry and 56·4 kN (12,676 lb st) with afterburning (designation of engine given in Soviet press statements as TDR Mk R37F). Two underwing pylons for UV-16-57 pods, each containing sixteen 57 mm rockets, or K-13 air-to-air missiles, and one NR-30 cannon in starboard side of fuselage (one each side on early aircraft and on the ten supplied to India). Internal fuel capacity of 2,340 litres (515 Imp gallons), plus underfuselage pylon for external fuel tank of 490 litres (108 Imp gallons) capacity. Small nose air intake of approximately 69 cm (27 in) diameter, with small movable three-shock centrebody housing the radar ranging equipment. Undernose pitot boom, which folds upward on the ground to reduce risk of ground personnel walking into it. Transparent blister cockpit canopy which hinges upward about base of integral flat bulletproof windscreen. Transparent rearview panel (not on aircraft built in Czechoslovakia) aft of canopy at front of shallow dorsal spine fairing. Large blade antenna at rear of this panel, with small secondary antenna midway along spine. Fowler-type flap between fuselage and aileron on each trailing-edge, with fairing plate under wing at outer extremity. Small forward-hinged airbrake under fuselage, forward of ventral fin; two further forward-hinged airbrakes, on each side of underfuselage in line with wing-root leading-edges, integral with part of cannon fairings. Brake-parachute housed inside small door on port underside of rear fuselage, with cable attachment under rear part of ventral fin. Semi-encapsulated escape system, in which canopy is ejected with seat, forming shield to protect pilot from slipstream, until the seat has been slowed by its drogue chute. Leading-edge of fin extended forward on all but early aircraft, to increase chord.

MiG-21PF *(Fishbed-D)*. Basic model of a new series of operational versions with forward fuselage of less-tapered form. Intake enlarged to diameter of approximately 91 cm (36 in) and housing larger centrebody for R1L search/track radar (NATO *Spin Scan*) to enhance all-weather capability (designation suffix letter 'P', standing for *Perekhvatchik*, is applied to aircraft adapted for all-weather interception from an earlier designed role). Remainder of airframe generally similar to that of MiG-21F, but pitot boom repositioned above air intake; cannon armament and fairings deleted, permitting simplified design for forward airbrakes; larger main wheels and tyres, requiring enlarged blister fairing on each side of fuselage, over wing, to accommodate wheel in retracted position; dorsal spine fairing widened and deepened aft of canopy, to reduce drag and house additional fuel tankage, and rear-view panel deleted; primary blade antenna repositioned to mid-spine and secondary antenna deleted. Uprated R-11 turbojet, giving 58·4 kN (13,120 lb st) with afterburning. Internal fuel capacity increased to 2,850 litres (627 Imp gallons) in seven fuselage tanks. Late production aircraft have attachments for a rocket-assisted take-off unit (RATOG) aft of each main landing gear bay, and provision for a flap-blowing system known as *Sduva Pogranichnovo Sloya* (SPS), which reduces the normal landing speed by some 22 knots (40 km/h; 25 mph). Flaps are larger than original Fowler type, do not move aft, and lack outboard fairing plates. Prototype shown at Tushino in 1961 had dummy metal centrebody. Production aircraft in service with many air forces.

Fishbed-E. Basically similar to *Fishbed-C* with broad-chord vertical tail surfaces. Parachute-brake repositioned into acorn fairing, made up of clamshell doors, at base of rudder, above jet nozzle. Provision for GP-9 underbelly pack, housing GSh-23 twin-barrel 23 mm gun, in place of centreline pylon, with associated predictor sight and electrical ranging system.

MiG-21FL. Export version of late-model MiG-21PF series, with broad-chord vertical tail surfaces and parachute-brake housing at base of rudder but no provision for SPS or RATOG. About 200 were initially assembled and later built under licence in India by Hindustan Aeronautics Ltd (which see), with the IAF designation

A two-seat training version of the MiG-21, in Czechoslovakian insignia *(Letectvi + Kosmonautika)*

Mikoyan MiG-21 multi-role fighter; version known to NATO as *Fishbed-L (Tass)*

Type 77. R-11-300 turbojet rated at 38·25 kN (8,598 lb st) dry and 60·8 kN (13,668 lb) with afterburning. Suffix letter 'L' *(Lokator)* indicates the installation of Type R2L search/track radar. Can be fitted with GP-9 underbelly gun pack.

MiG-21PFS or MiG-21PF(SPS). Similar to *Fishbed-D*, but with SPS as standard production installation.

MiG-21PFM *(Fishbed-F)*. Successor to interim MiG-21PFS, embodying all the improvements introduced progressively on the PF and PFS, the suffix letter 'M' indicating an exportable version of an existing design. Leading-edge of fin extended forward a further 45 cm (18 in). Small dorsal fin fillet eliminated. Additional refinements, including sideways-hinged (to starboard) canopy and conventional windscreen quarter-lights; simple ejection seat instead of semi-encapsulated type; and large dielectric portion at tip of tail-fin. R2L radar reported to have lock-on range of under 7 nm (13 km; 8 miles) and to be ineffective at heights below about 915 m (3,000 ft) because of ground 'clutter'. Max permissible speed at low altitude is reported to be 593 knots (1,100 km/h; 683 mph). Built also in Czechoslovakia.

Analogue. Based on a standard MiG-21PF airframe, this aircraft was fitted with a scaled-down replica of the original 'ogee' delta wing of the Tu-144 supersonic transport, for aerodynamic flight testing and development before the Tu-144 prototype was completed. It had no horizontal tail surfaces. Following its several dozen research flights, modifications were made to the full-size wing. One only.

Fishbed-G. Experimental STOL version of MiG-21PFM, with a pair of vertically-mounted lift-jet engines in lengthened centre-fuselage. Demonstrated in the air display at Domodedovo in July 1967, and described and illustrated in the 1970-71 *Jane's*. Prototype only.

MiG-21PFMA *(Fishbed-J)*. Multi-role version. Basically similar to MiG-21PFM but with deeper dorsal fairing containing fuel tankage above fuselage, giving straight line from top of canopy to fin. Pitot tube remains above air intake but is offset to starboard. Provision for GP-9 underbelly gun pack as alternative to centreline fuel tank. Four underwing pylons, instead of former two, for a variety of ground attack weapons and stores, as alternative or supplementary to two or four air-to-air missiles. Latter can include radar-homing *'Advanced Atoll'* as well as infra-red K-13A *Atoll*. Able to carry two underwing tanks in addition to standard underbelly tank, offsetting reduced internal fuel capacity of 2,600 litres (572 Imp gallons). Zero-speed zero-altitude ejection seat. Small boat-shape fairing with angle-of-attack indicator on port side of nose. Later production aircraft can have the GSh-23 gun installed inside the fuselage, with a shallow underbelly fairing for the twin barrels and splayed cartridge-ejection chutes to clear each side of centreline store.

MiG-21M. Generally similar to MiG-21PFMA with internal GSh-23 gun pack. Has superseded MiG-21FL on Hindustan Aeronautics production line in India, with IAF designation Type 88. First Indian-built MiG-21M was handed over officially to IAF on 14 February 1973.

MiG-21R *(Fishbed-H)*. Tactical reconnaissance version, basically similar to MiG-21PFMA. Equipment includes an external pod for forward-facing or oblique cameras, infra-red sensors or ECM devices, and fuel, on fuselage centreline pylon. Suppressed antenna at mid-fuselage and optional ECM equipment in wingtip fairings.

MiG-21MF *(Fishbed-J)*. Generally similar to MiG-21PFMA but re-engined with a Tumansky R-13-300 turbojet, lighter in weight and with higher performance ratings. Small rearview mirror above cockpit canopy. Debris deflector beneath each suction relief door forward of wing root. Entered service with Soviet Air Force in 1970.

MiG-21RF *(Fishbed-H)*. Tactical reconnaissance version of MiG-21MF. Equipment as for MiG-21R.

MiG-21SMT *(Fishbed-K)*. Similar to MiG-21MF, except for having deep dorsal spine extended rearward as far as parachute-brake housing, to provide maximum possible fuel tankage and optimum aerodynamic form. Able to carry ECM equipment in small removable wingtip pods. Deliveries to Warsaw Pact air forces reported to have begun in 1971. Like the MiG-21PFMA and MiG-21MF, this version can carry K-13A *Atoll* infra-red missiles and/or radar-homing *'Advanced Atolls'*.

Fishbed-L. Generally similar to MiG-21MF, but with short-range navigation and landing system similar to Tacan.

MiG-21U *(Mongol)*. Two-seat training versions. Initial version, sometimes referred to as *Mongol-A*, is generally similar to the MiG-21F but has two cockpits in tandem with sideways-hinged (to starboard) double canopy, larger main wheels and tyres of MiG-21PF, one-piece forward airbrake, and pitot boom repositioned above intake. Cannon armament is deleted. Later models, sometimes called *Mongol-B*, have the broader-chord vertical tail surfaces and under-rudder brake-parachute housing of the later operational variants, with a deeper dorsal spine and no dorsal fin fillet.

MiG-21US *(Mongol)*. Similar to later MiG-21U but with provision for SPS flap-blowing, and retractable periscope for instructor in rear seat.

MiG-21UM *(Mongol)*. Two-seat trainer counterpart of MiG-21MF with R-13 turbojet and four underwing stores pylons.

Alternative designations, allocated by the Soviet authorities to MiG-21s used to set up FAI-recognised international records, are as follows:

E-33. This designation has been applied to training versions of the MiG-21 *(Mongol)* used to establish women's records. Those confirmed by the FAI include an altitude of 24,336 m (79,842 ft) set up by Natalya Prokhanova on 22 May 1965, and a sustained altitude of 19,020 m (62,402 ft) in horizontal flight established by Lydia Zaitseva on 23 June 1965.

E-66. Aircraft of basic MiG-21F series, used by Col Georgi Mossolov to set up a world absolute speed record (since beaten) of 1,288·6 knots (2,388 km/h; 1,484 mph) over a 15/25 km course on 31 October 1959. Engine described as a 58·35 kN (13,120 lb st) Type TDR Mk R37F.

E-66A. Variant of E-66 used by Mossolov to raise the world height record to 34,714 m (113,892 ft) on 28 April 1961, from Podmoskovnœ aerodrome. Powered additionally by a 29·4 kN (6,615 lb st) GRD Mk U2 rocket engine in underbelly pack, exhausting between twin ventral fins. Other changes compared with then-standard operational model included a widened dorsal spine and repositioned blade antenna, as standardised for the MiG-21PF, and a blister fairing above the nose.

E-66B. Used by Svetlana Savitskaya to set four women's time-to-height records on 15 November 1974, from Podmoskovnœ aerodrome. Described as having one 68·7 kN (15,432 lb st) PDM engine (presumably afterburning turbojet) and two 22·6 kN (5,070 lb st) TTPDs (possibly assisted take-off rockets). Times recorded were 41·2 sec to 3,000 m, 1 min 0·1 sec to 6,000 m, 1 min 21 sec to 9,000 m, and 1 min 59·3 sec to 12,000 m.

E-76. Designation allocated to apparently-standard MiG-21PFs used by Soviet women pilots to establish international records. Those confirmed by the FAI are for a speed of 1,112·7 knots (2,062 km/h; 1,281·27 mph) over a 500 km closed circuit by Marina Solovyova on 16 September 1966; a speed of 485·78 knots (900·267 km/h; 559·40 mph) over a 2,000 km closed circuit by Yevgenia Martova on 11 October 1966; a speed of 1,148·7 knots (2,128·7 km/h; 1,322·7 mph) over a 100 km closed circuit by Miss Martova on 18 February 1967; and a speed of 700·5 knots (1,298·16 km/h; 806·64 mph) over a 1,000 km closed circuit by Lydia Zaitseva on 28 March 1967.

There is reason to believe that the similar designations E-74, E-77 and E-88 apply to versions of the export MiG-21F, MiG-21FL and MiG-21M respectively.

MiG-21s have been supplied to the Afghan, Algerian, Bangladesh, Bulgarian, Chinese, Cuban, Czech, Egyptian, Finnish, East German, Hungarian, Indian, Indonesian, Iraqi, North Korean, Polish, Romanian, South Yemen, Syrian, North Vietnamese, Yemen Arab Republic and Yugoslav air forces.

In mid-1976 the Iraqi Air Force had about 90 MiG-21s and the Syrian Air Force about 200, while the Egyptian Air Force was believed to have 210 MiG-21s of the latest models. The Chinese industry is assisting with spares and overhaul services for the MiG-21s, MiG-17s and MiG-19s operated by Egypt. Ferranti and Smiths Industries of the UK are collaborating in a scheme to provide the Egyptian MiG-21s with a digital inertial navigation, weapon aiming and head-up display system (Hudwac).

A version of the MiG-21 has been built in China under the Chinese designation F-8, and about 16 of these were supplied to Tanzania.

The following details refer to the MiG-21MF *(Fishbed-J)*:

Type: Single-seat multi-role fighter.

Wings: Cantilever mid-wing monoplane of clipped-delta planform, with slight anhedral from roots. No leading-edge camber. Sweepback approximately 53°. Small pointed fairing on each side of fuselage forward of wing-root leading-edge. Small boundary-layer fence above each wing near tip. All-metal construction. Inset ailerons, actuated hydraulically. Large 'blown' plain trailing-edge flaps, actuated hydraulically.

Fuselage: Circular-section all-metal semi-monocoque structure. Ram-air intake in nose, with three-position movable centrebody. Large dorsal spine fairing along top of fuselage from canopy to fin. Forward-hinged door-type airbrake on each side of underfuselage below wing leading-edge. A further forward-hinged airbrake under fuselage forward of ventral fin. All airbrakes actuated hydraulically. Blister fairings above and below wing on each side to accommodate main wheels when retracted.

Tail Unit: Cantilever all-metal structure, with all surfaces sharply swept. Conventional fin and hydraulically-powered rudder. Hydraulically-actuated one-piece all-moving horizontal surface, with two gearing ratios for use at varying combinations of altitude and airspeed. Tailplane trim switch on control column. No trim tabs. Single large ventral fin.

Landing Gear: Hydraulically-retractable tricycle type, with single wheel on each unit; all units housed in fuselage when retracted. Forward-retracting non-steerable nosewheel unit; inward-retracting main wheels which turn to stow vertically inside fuselage. Tyres on main wheels inflated to approximately 7·93 bars (115 lb/sq in), ruling out normal operation from grass runways. Pneumatic braking on all three wheels, supplied from compressed-air bottles. Steering by differential main-wheel braking. Wheel doors remain open when legs are extended. Brake parachute housed inside acorn fairing at base of rudder.

Power Plant: One Tumansky R-13-300 turbojet engine, rated at 50 kN (11,240 lb st) dry and 64·73 kN (14,550 lb st) with afterburning. Fuel tanks in fuselage, with total capacity of 2,600 litres (572 Imp gallons), of which approx 1,800 litres (396 Imp gallons) are usable within CG limits at low speed. Provision for carrying one finned external fuel tank, capacity 490 litres (108 Imp gallons), on underfuselage pylon and two similar drop-tanks on outboard underwing pylons. Two jettisonable solid-propellant JATO rockets can be fitted under rear fuselage, aft of wheel doors.

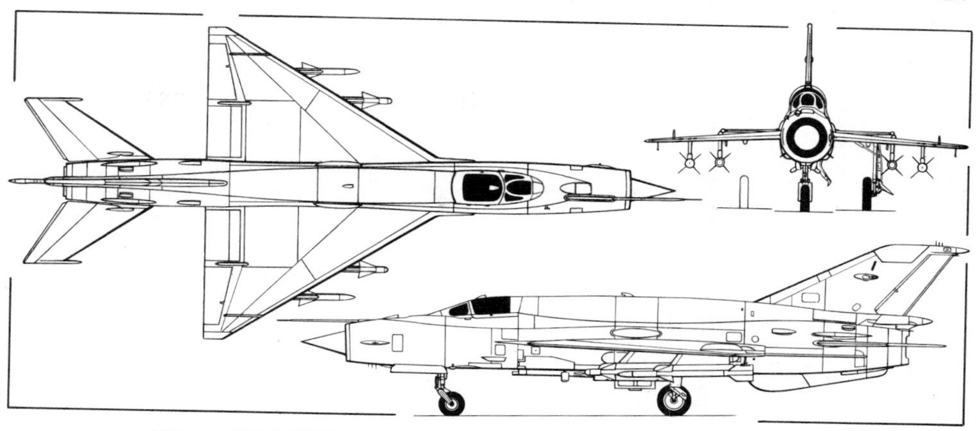

Mikoyan MiG-21SMT *(Fishbed-K)* **single-seat multi-role fighter** *(Pilot Press)*

Accommodation: Pilot only, on ejection seat with spring-loaded arm at top which ensures that seat cannot be operated unless hood is closed. Canopy is sideways-hinged, to starboard, and is surmounted by a small rearview mirror. Flat bullet-proof windscreen. Cabin air-conditioned. Armour plating forward and aft of cockpit.

Systems: Duplicated hydraulic system, supplied by engine-driven pump, with backup by battery-powered electric pump, and emergency electric tailplane trim and manual operation of flying controls. Autostabilisation in pitch and roll only.

Electronics and Equipment: Search and track radar in intake centrebody, with search range of approx 10·4 nm (19·3 km; 12 miles). Other standard electronics include VOR/ADF and warning radar with an indicator marked in 45° sectors in front of and behind the aircraft. Gyro gunsight is reported to topple at 2·75*g*. Automatic ranging can be fed into gunsight. Full blind-flying instrumentation, with attitude and heading indicators driven by remote central gyro platform.

Armament: One twin-barrel 23 mm GSh-23 gun, with 200 rounds, in belly pack. Four underwing pylons for weapons or drop-tanks. Typical loads for interceptor role include two K-13A (*Atoll*) air-to-air missiles on inner pylons and two radar-homing *'Advanced Atolls'* or two UV-16-57 rocket packs (each sixteen 57 mm rockets) on outer pylons; four K-13As/*'Advanced Atolls'*; or two drop-tanks and two K-13As or *'Advanced Atolls'*. Typical loads for ground attack role are four UV-16-57 rocket packs; two 500 kg and two 250 kg bombs; or four S-24 240 mm air-to-surface missiles.

Dimensions, external:

Wing span	7·15 m (23 ft 5½ in)
Length, incl pitot boom	15·76 m (51 ft 8½ in)
Length, excl pitot boom and intake centre-body	13·46 m (44 ft 2 in)
Height overall	4·50 m (14 ft 9 in)
Wheel track	2·69 m (8 ft 10 in)

Area:

Wings, gross	23 m² (247 sq ft)

Weights:

T-O weight:

with four K-13A missiles	8,200 kg (18,078 lb)
with two K-13A missiles and two 490 litre drop-tanks	8,950 kg (19,730 lb)
with two K-13As and three drop-tanks	9,400 kg (20,725 lb)

Performance:

Max level speed above 11,000 m (36,000 ft)	Mach 2·1 (1,203 knots; 2,230 km/h; 1,385 mph)
Max level speed at low altitude	Mach 1·06 (701 knots; 1,300 km/h; 807 mph)
Landing speed	146 knots (270 km/h; 168 mph)
Service ceiling	18,000 m (59,050 ft)
T-O run at normal AUW	800 m (2,625 ft)
Landing run	550 m (1,805 ft)
Range, internal fuel only	593 nm (1,100 km; 683 miles)
Ferry range, with three external tanks	971 nm (1,800 km; 1,118 miles)

MIKOYAN MiG-23

NATO reporting names: *Flogger-A, B, C* and *E*

The production configuration of this variable-geometry tactical fighter represents an almost total redesign by comparison with the prototype, which was first displayed in public during the 1967 Aviation Day flypast at Domodedovo Airport, Moscow. Initial deliveries to the Soviet Air Force are believed to have been made in 1971; but problems encountered subsequently prevented the type from becoming fully operational until early 1972. Two Soviet fighter regiments, with a total of about 75 aircraft, have been based in East Germany since 1973/74, and deliveries of all versions to the Soviet Air Force exceeded 500 by 1976. Export versions, with a lower equipment standard, are operated by the Egyptian, Iraqi, Libyan and Syrian Air Forces.

There appear to be at least six versions of which details can be published:

MiG-23 *(Flogger-A)*. Original version, of which prototype was shown at Domodedovo on 9 July 1967. On that occasion, during a display by test pilot Alexander Fedotov,

MiG-23S *(Flogger-B)* **single-seat variable-geometry air combat fighter of the Soviet Air Force. Note the large splitter plates forward of variable-geometry intakes on this version, on the two-seat MiG-23U and on all export models of this family of fighters**

Flogger-E, **the export version of** *Flogger-B*, **with smaller radome and other changes. This one is operated by the Libyan Air Force**

the wings were moved from fully-forward to fully-swept position in about four seconds. The commentator credited the aircraft with supersonic speed at ground level and Mach 2 at medium and high altitudes. Illustrated in 1973-74 *Jane's*.

MiG-23S *(Flogger-B)*. Single-seat air combat fighter in service with Soviet Air Force. Design changes compared with prototype include movement further rearward of all tail surfaces except ventral fin, giving much increased gap between wing and tailplane; a considerable increase in the size of the dorsal fin; and the introduction of fixed inboard wing leading-edges. Detailed description applies to this version.

MiG-23U *(Flogger-C)*. Tandem two-seat version suitable for both operational training and combat use. Individual canopy over each seat. Rear seat slightly higher than forward seat, with retractable periscopic sight for occupant. Dorsal fairing of increased depth aft of rear canopy. Otherwise identical to MiG-23S. In service.

Flogger-D. Much-modified single-seat ground-attack/interdictor version of same basic design. Described separately.

MiG-23S *(Flogger-E)*. Export version of *Flogger-B*. Generally similar to Soviet Air Force version, but equipped to a lower standard. Smaller radar in shorter nose radome.

Flogger-F. Export counterpart of *Flogger-D*. Basically similar to Soviet Air Force version, but equipped to a lower standard. Described with *Flogger-D*.

The following description refers specifically to the single-seat MiG-23S as supplied to the Soviet Air Force, but is generally applicable also to the two-seat MiG-23U:

Type: Single-seat variable-geometry tactical fighter.

Wings: Cantilever shoulder wing. Sweepback of main panels variable in flight or on the ground by manual control, from approximately 19° to approximately 72°. Fixed triangular inboard panels, with leading-edges swept at approximately 72°. Full-span trailing-edge single-slotted flaps, each in three sections, permitting independent actuation of outboard sections when wings are fully swept. Top-surface spoilers/lift dumpers forward of flaps, for differential operation in conjunction with horizontal tail surfaces, and for collective operation for improved runway adherence and braking after touchdown. Extended-chord leading-edge flap on outboard two-thirds of each main (variable-geometry) panel.

Fuselage: Conventional semi-monocoque structure of basic circular section; flattened on each side of cockpit, forward of lateral air intake trunks which blend into circular shape of rear fuselage. Large flat boundary layer splitter plate (similar to that of US F-4 Phantom II) forms inboard face of each intake. Two small rectangular 'blow-in' air intakes in each trunk, under inboard wing leading-edge. Perforations under rear fuselage, aft of main wheel bays, are pressure-relief vents. Door-type airbrake mounted on each side of rear fuselage.

Tail Unit: All-moving horizontal surfaces, swept back at approximately 57° on leading-edge, operate both differentially and symmetrically to provide aileron and elevator function respectively. Conventional fin, swept back at approximately 65° on leading-edge, with large dorsal fin and inset rudder. No tabs. Large ventral fin in two portions. Lower portion is hinged to fold to starboard when landing gear is extended, to increase ground clearance.

Landing Gear: Retractable tricycle type, with single wheel on each main unit and steerable twin-wheel nose unit. Main units retract inward into rear of air intake trunks. Main fairings to enclose these units are attached to legs. Small inboard fairing for each wheel bay hinged to fuselage belly. Nose unit, fitted with small mudguard, retracts rearward. Brake parachute housed in cylindrical fairing at base of rudder.

The two-seat MiG-23U, identical to the MiG-23S except for second cockpit *(Flug Revue)*

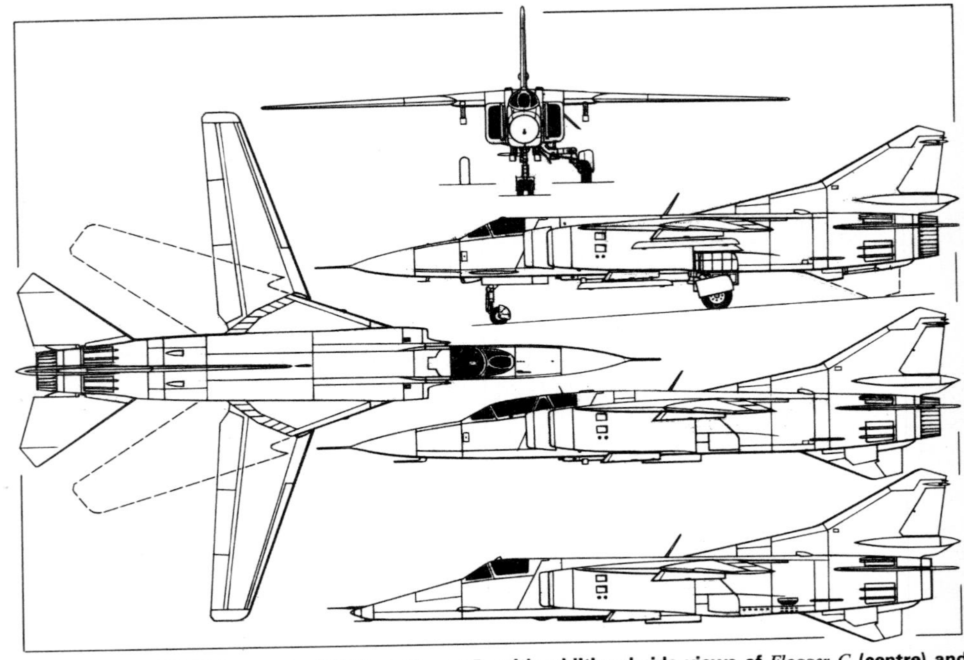

Version of the MiG-23 known to NATO as *Flogger-B*, **with additional side views of** *Flogger-C* **(centre) and** *Flogger-D* **(bottom)** *(Pilot Press)*

Power Plant: One large afterburning turbojet engine of unknown type. Thrust has been estimated at 63·7 kN (14,330 lb st) dry and 91·2 kN (20,500 lb st) with afterburning. Variable-geometry air intakes and variable nozzle. Provision for carrying external fuel tank on underfuselage centreline pylon.

Accommodation: Single seat in air-conditioned cockpit, under small rearward-hinged canopy.

Electronics and Equipment: Radar dish (NATO *High Lark*) behind dielectric nosecone. Small cylindrical fairings forward of starboard underwing pylon and above rudder are believed to contain ECM equipment. Undernose laser rangefinder and Doppler equipment standard on Soviet Air Force version. Dr Robert C. Seamans, then US Secretary of the Air Force, stated his belief in early 1973 that the radar and missile systems are comparable with those of the USAF's F-4 Phantom II. Retractable landing light under nose, aft of radome.

Armament: One 23 mm GSh-23 twin-barrel gun in fuselage belly pack, with large flash eliminator around nozzles. One pylon under centre-fuselage, one under each engine air intake duct, and one under each fixed inboard wing panel, for air-to-air missiles (NATO *Apex* and *Aphid*) or other external stores.

Dimensions (estimated):
- Wing span:
  - fully spread: 14·25 m (46 ft 9 in)
  - fully swept: 8·17 m (26 ft 9½ in)
- Length overall: 16·80 m (55 ft 1½ in)

Weight (estimated):
- T-O weight: 12,700-15,000 kg (28,000-33,050 lb)

Performance (estimated):
- Max level speed at height with external stores: Mach 2·3
- Max level speed at S/L: Mach 1·1
- Service ceiling: 18,000 m (59,000 ft)
- Combat radius: 520 nm (960 km; 600 miles)

MIKOYAN V-G INTERDICTOR

NATO reporting names: *Flogger-D* and *F*

Although the single-seat ground attack aircraft known to NATO as *Flogger-D* has many airframe features in common with the MiG-23, it differs in important respects

Flogger-D **landing, with wings extended and ventral fin folded** *(Flug Revue)*

and is believed to have a different official Soviet MiG designation. Use of fixed air intakes on the version in service with the Soviet Air Force is consistent with the primary requirement of high subsonic speed at low altitude. An uprated turbojet engine (107·9 kN; 24,250 lb st with afterburning) is believed to be fitted.

The forward portion of the fuselage is completely redesigned by comparison with the interceptor versions of the MiG-23. Instead of an ogival radome, *Flogger-D* has a sharply-tapered nose in side elevation, with a small sloping window covering a laser rangefinder and marked target seeker, and additional armour on the flat sides of the cockpit. A six-barrel 23 mm Gatling-type underbelly gun replaces the GSh-23 of the interceptor, and there are five pylons for external stores, including tactical nuclear weapons. There is provision for carrying an external fuel tank for ferry flights under each outer wing, which must be kept in a fully-forward position when the tank is in place. Equipment includes an ECM antenna above the port glove pylon.

The aircraft of this basic type shown in one accompanying illustration (NATO *Flogger-F*) appears to retain features of the MiG-23 interceptor, such as variable-geometry intakes and GSh-23 twin-barrel gun, although having the nose-shape and larger, low-pressure tyres of *Flogger-D*. This suggests that it could be a prototype/pre-production model or an export version with lower standards of equipment and performance. The following amended and additional specification data are estimated for the *Flogger-D* operated by the Soviet Air Force:

Power Plant: Total internal fuel capacity 5,380 litres (1,183 Imp gallons), with 4,040 litres (889 Imp gallons) in the fuselage, 890 litres (195 Imp gallons) in the wings and 450 litres (99 Imp gallons) in the tail fin. Provision for three external tanks, under fuselage and each outer wing.

Dimensions, external:
- Wing aspect ratio (spread) 7·45
- Tailplane span 5·75 m (18 ft 10¼ in)

Areas:
- Wings, gross (spread) 27·26 m² (293·4 sq ft)
- Horizontal tail surfaces 6·88 m² (74·06 sq ft)

Weights:
- Max weapon load 1,900 kg (4,200 lb)
- Max T-O weight 17,750 kg (39,130 lb)

Performance:
- T-O to 15 m (50 ft) at AUW of 15,700 kg (34,600 lb) 800 m (2,625 ft)
- Max ferry range with three external tanks 1,350 nm (2,500 km; 1,550 miles)

MIKOYAN MiG-25 (E-266)

NATO reporting name: *Foxbat*

First news of the existence of this aircraft came in a Soviet claim, in April 1965, that a twin-engined aircraft designated E-266 had set a 1,000 km closed-circuit speed record of 1,251·9 knots (2,320 km/h; 1,441·5 mph), carrying a 2,000 kg payload. The attempt was made at a height of 21,000-22,000 m (69,000-72,200 ft) by Alexander Fedotov, who had earlier set up a 100 km record in the E-166 research aircraft (described in 1967-68 *Jane's*).

The same pilot set a payload-to-height record of 29,997 m (98,349 ft) with a 2,000 kg payload in the E-266 on 5 October 1967, after a rocket-assisted take-off. This qualified also for the record with a 1,000 kg payload. Photographs of the E-266 issued officially in the Soviet Union identified it subsequently as the twin-finned Mikoyan single-seat fighter of which four examples had taken part in the Domodedovo display in July 1967 and which is now known to be designated MiG-25 in the Soviet Air Force.

Its performance in level flight was demonstrated further on 5 October 1967, when M. Komarov set up a speed record of 1,608·83 knots (2,981·5 km/h; 1,852·61 mph) over a 500 km closed circuit. On 27 October, P. Ostapenko raised the 1,000 km closed-circuit record to 1,576·00 knots (2,920·67 km/h; 1,814·81 mph) in an E-266, carrying a 2,000 kg payload and qualifying also for records with 1,000 kg payload and no payload.

On 8 April 1973, Fedotov achieved a speed of 1,405·741 knots (2,605·1 km/h; 1,618·734 mph) over a 100 km closed circuit, beating his own earlier record in the E-166. Next, on 25 July 1973, Fedotov set a new World absolute height record by climbing to 36,240 m (118,898 ft) in an E-266. On the same day, he climbed to 35,230 m (115,584 ft) with a 2,000 kg payload, qualifying also for the record with 1,000 kg.

Three time-to-height records were established by the E-266 on 4 June 1973, when Boris Orlov climbed to 20,000 m in 2 min 49·8 sec, and P. Ostapenko climbed to 25,000 m in 3 min 12·6 sec and 30,000 m in 4 min 3·86 sec. All three records were beaten by the McDonnell Douglas F-15 *Streak Eagle* in January-February 1975; but two of them were recaptured by an E-266M (presumably with an uprated power plant) on 17 May 1975. Subject to confirmation, Fedotov climbed to 25,000 m in 2 min 34 sec and Ostapenko reached 30,000 m in 3 min 9 sec. Fedotov also set a new record by climbing to 35,000 m in 4 min 11 sec.

There are known to be at least three versions of the MiG-25, identified by the following NATO reporting names:

Foxbat-A. Basic interceptor, with large radar (NATO *Fox Fire*) in nose and armed with four air-to-air missiles (NATO *Acrid*) on underwing attachments. Slightly reduced wing leading-edge sweep towards tips. Wingtip balance-weight fairings appear to house missile guidance equipment in their nose.

Mikoyan variable-geometry interdictor *(Flogger-D)* **of the Soviet Air Force. Note the fixed-geometry intakes, laser rangefinder under nose, and dielectric panels** *(Tass)*

Variable-geometry intakes and other features suggest that this is either an export *Flogger-F* **or pre-production** *Flogger-D*

MiG-25 *(Foxbat-A)* **interceptors, each armed with four air-to-air missiles (NATO** *Acrid)*

Foxbat-B. Basic reconnaissance version, with five camera windows and various flush dielectric panels aft of very small dielectric nosecap. SLAR (side-looking airborne radar) on some, if not all, aircraft. Slightly reduced span. Wing leading-edge sweep constant from root to tip.

Foxbat-C **(MiG-25U).** Trainer, of which first photographs were published towards the end of 1975. Generally similar to combat versions, but with new nose, containing separate cockpit with individual canopy, forward of standard cockpit and at a lower level. No radar or reconnaissance sensors in nose. The aircraft designated **E-133** in which Svetlana Savitskaya set a women's world speed record of 1,448·942 knots (2,683·44 km/h; 1,667·412 mph) on 2 June 1975 is believed to have been a MiG-25U.

The low aspect ratio square-tipped wings of the MiG-25 are mounted high on the fuselage. They have anhedral over the full span, with a slight reduction in the angle of anhedral on the outer wing panels. Leading-edge sweep of approx 42° is constant from root to tip on *Foxbat-B,* but is slightly reduced near the wingtips of *Foxbat-A.* Triangular endplates have been seen fitted to the rear of the wingtip balance-weight fairings on some aircraft, presumably to improve stability. The fairings themselves are missing on aircraft shown in some photographs, but are standard on operational MiG-25s.

The twin tail fins were almost certainly adopted as being preferable to the single large and tall fin that would otherwise have been essential with such a wide-bodied supersonic design. The fins incline outward, as do the large ventral fins.

The basic fuselage is quite slim, but is blended into the two huge rectangular air intake trunks, which have wedge inlets. The inner walls of the intakes are curved at the top and do not run parallel with the outer walls; a hinged panel forms the lower lip of each intake, enabling the intake area to be varied. The ejection seat in *Foxbat-A* is identical to that in current MiG-21s.

The landing gear is a retractable tricycle type, with the main wheels retracting forward into the air intake trunks. A bulge in the undersurface of each trunk was made necessary by the single large-diameter wheel on each main leg. Twin wheels are fitted to the nose unit.

The power plant of the MiG-25 consists of a pair of large afterburning turbojet engines, designated R-266 (each rated at 107·9 kN; 24,250 lb st and attributed to the Tumansky design bureau), mounted side by side in the rear fuselage. To each side of the jet nozzles are low-set all-moving horizontal tail surfaces of characteristic MiG shape. Under the nozzles is a one-piece narrow-chord airbrake, which follows the curvature of the tailpipes. Some photographs suggest that there might also be a similar airbrake above the nozzles.

Total internal fuel capacity of *Foxbat-A* is approximately 14,000 kg (30,865 lb).

The only visible external weapon attachments are four underwing hardpoints, for air-to-air guided weapons and other stores. ECM pods can be carried on the two inboard pylons. The missiles shown on a *Foxbat-A* in an accompanying photograph have the NATO reporting name of *Acrid.* The aircraft flown to Japan by a defecting pilot in 1976 (see below) was said to be equipped to carry the missiles known as *Apex.*

In early 1973 Dr Robert C. Seamans, then US Secretary of the Air Force, described the MiG-25 as "probably the best interceptor in production in the world today", and added "This Mach 3 aircraft performs both interceptor and reconnaissance missions, can operate at 24,400 m (80,000 ft), and has a highly capable avionics and missile system". In his FY 1975 US Defense Department Report, then-Secretary Schlesinger commented: "Should the Soviet Union develop and deploy an AWACS-*Foxbat* 'look-down, shoot-down' air defence system, we would have to counter it with new penetration devices and techniques such as the cruise missile, bomber defence missiles and improved ECM".

The interceptor version had not been seen outside the Soviet Union until Lt Viktor Belenko defected in one from the Soviet air base of Sikharovka, 200 km (120 miles) from Vladivostok, to Hakodate airport, Japan, on 6 September 1976. Statements attributed to this pilot suggest that more than 400 MiG-25s had been built by that time, and that his particular aircraft left the production line less than three years earlier. Japanese and US military technicians who examined the aircraft reported that the airframe is constructed mainly of steel, with titanium only in places subjected to extreme heating such as the wing leading-edges. The inevitable weight penalty restricts the amount of equipment that can be carried. Belenko was reported to have said, incorrectly, that no ejection seat was fitted, to save weight. He added that the aircraft took a considerable time to accelerate to high speeds, which were then difficult to maintain.

Examination of the aircraft is said to have shown that the fuselage weighs about 13,600 kg (30,000 lb) with the wings, tail surfaces and afterburners removed; the fire control system is bulky and lacking in advanced technology, with vacuum tubes rather than solid-state circuitry throughout the electronics; electrically-controlled variable ramps are fitted in the engine air intakes; the turbojets employ water methanol injection; the number of cockpit instruments was described as 50% of those in

The reconnaissance version of the MiG-25, known to NATO as *Foxbat-B.* **Note the camera-carrying nose and dielectric panels** *(Flug Revue)*

Landing shot of *Foxbat-B (Flug Revue)*

Close-up of tandem cockpits in nose of MiG-25U *(Foxbat-C) (Flug Revue)*

F-4EJ Phantoms of the JASDF, with a smaller and less versatile weapon sight; and the Machmeter has a 'red-line' limit at Mach 2·8, which almost certainly represents a never-exceed speed when carrying missiles and pylons rather than the maximum speed of which the 'clean' aircraft is capable.

Four MiG-25 reconnaissance aircraft *(Foxbat-B)* were operational with Soviet Air Force units in Egypt as early as the Spring of 1971, having been airlifted to that country in An-22 transports. Between the Autumn of 1971 and the Spring of 1972, MiG-25s from Cairo West airfield were despatched in pairs on at least four reported occasions to carry out high-speed reconnaissance missions off the Israeli coastline or down the full length of the Israeli-occupied Sinai Peninsula. Phantom interceptors sent up by the Israeli defence forces failed to make contact with the MiGs, which remained in Egypt until September 1975. Similar overflights of Iran have been made regularly, without hindrance; other MiG-25 reconnaissance aircraft fly from bases in East Germany, Poland and Syria.

DIMENSIONS (estimated):

Wing span:	
Foxbat-A	14·00 m (46 ft 0 in)
Foxbat-B	13·40 m (44 ft 0 in)
Length overall	22·3 m (73 ft 2 in)
Height overall	5·6 m (18 ft 4¼ in)

AREA:

Wings, gross	56 m² (603 sq ft)

WEIGHTS (*Foxbat-A,* estimated):

Basic operating weight at least	20,000 kg (44,100 lb)
Max T-O weight	35,000 kg (77,150 lb)

PERFORMANCE (estimated):

Max level speed at height:	
Foxbat-B, 'clean'	Mach 3·2
Never-exceed combat speed:	
Foxbat-A, with missiles	Mach 2·8
Service ceiling	24,400 m (80,000 ft)
Time to 11,000 m (36,000 ft) with afterburning	2 min 30 sec
Normal combat radius	610 nm (1,130 km; 700 miles)
Max combat radius, econ power	700 nm (1,300 km; 805 miles)

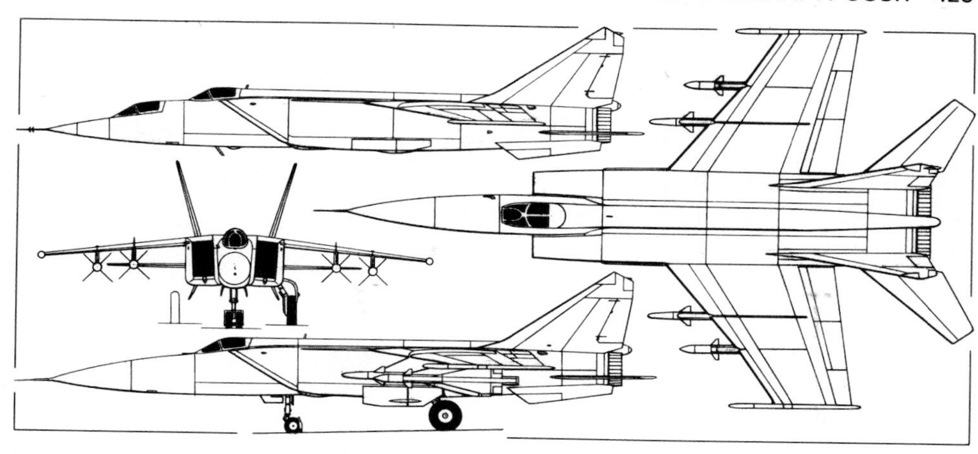

Mikoyan MiG-25 single-seat fighter (NATO *Foxbat-A),* with additional side view (top) of two-seat MiG-25U *(Pilot Press)*

MIL

GENERAL DESIGNER IN CHARGE OF BUREAU:
Marat N. Tishchenko

M. L. Mil was connected with Soviet gyroplane and helicopter development from at least 1930. His achievements were recognised by the award of the Order of Lenin on his 60th birthday in November 1969. He died on 31 January 1970.

His original Mi-l, which was designed in 1949, first flown in 1950 and introduced into squadron service in 1951, was the first helicopter to enter series production in the Soviet Union. It was followed by the larger Mi-4 in a number of variants, and these types are still in service, in civil and military forms.

The Mi-l was also produced in Poland, under the designation SM-1.

Subsequent products of the bureau that was headed by Mikhail Mil include the Mi-6, a very large passenger and freight helicopter, the Mi-10 (V-10) and Mi-10K flying crane versions of the Mi-6, the smaller turbine-powered Mi-2 (V-2) and Mi-8 (V-8) passenger helicopters, the Mi-12 (V-12), which is the largest helicopter currently flying anywhere in the world, and the Mi-24 military assault helicopter. Aviaexport has sold helicopters of Mil design in 38 countries.

MIL Mi-2 (V-2)

Built exclusively in Poland and described under Polish aircraft industry entry for WSK-PZL-Swidnik.

MIL Mi-6

NATO reporting name: *Hook*

First announced in the Autumn of 1957, the Mi-6 was then the largest helicopter flying anywhere in the world. From it were evolved the Mi-10 and Mi-10K flying crane helicopters, and its dynamic components are used in duplicated form on the V-12 (Mi-12).

Layout of the Mi-6 is conventional. Clamshell rear loading doors and folding ramps facilitate the loading of bulky freight and vehicles. Freight can also be carried externally, suspended from a hook on the CG. When the aircraft is operated in this flying crane role, the small wings which normally offload the rotor in flight can be removed, permitting an increase in payload.

The stub-wings are deleted also from the fire-fighting version. First demonstrated at the 1967 Paris Air Show, this carries several tons of water in tanks inside its cabin and can either spray this slowly from nozzles or dump it through the hoist cutout in its belly.

In setting up 14 FAI-recognised records in Class E-1, the Mi-6 has lifted payloads of up to 20,117 kg (44,350 lb). Records still standing in mid-1976 included a 100 km closed-circuit speed record of 183·54 knots (340·15 km/h; 211·36 mph), set up by Boris Galitsky on 26 August 1964.

On 15 September 1962 the same pilot, and crew, in an Mi-6 flew at 162·08 knots (300·377 km/h; 186·64 mph) over a 1,000 km circuit, setting the current records for speed with payload of 1,000 kg and payload of 2,000 kg.

On 11 September 1962 Vasily Kolochenko and crew of four averaged 153·44 knots (284·354 km/h; 176·69 mph) over a 1,000 km closed circuit, setting a record for speed with a payload of 5,000 kg.

Five Mi-6s are reported to have been built for development testing, followed by an initial pre-series of 30 and subsequent manufacture of some 500 for military and civil use. Rate of production was reported to be eight per month in 1970. Six were supplied to the Indonesian Air Force; many others have been delivered to the Bulgarian, Egyptian, Iraqi, Syrian and North Vietnamese air forces and to the government of Peru.

TYPE: Heavy transport helicopter.

ROTOR SYSTEM: Five-blade main rotor and four-blade tail rotor. Main rotor blades each have a tapered steel tube spar, to which are bonded built-up metal aerofoil sections. Blades have coincident flapping and drag hinges and fixed tabs. Main rotor shaft inclined forward at 5° to vertical. Control via large welded swashplate. Hydraulically-actuated powered controls. All rotor blades incorporate electro-thermal de-icing system.

FUSELAGE: Conventional all-metal riveted semi-monocoque structure of pod and boom type.

WINGS: Two small cantilever shoulder wings, mounted above main landing gear struts, off-load rotor by providing some 20% of total lift in cruising flight. Removed when aircraft is operated as flying crane.

TAIL UNIT: Tail rotor support acts as vertical stabiliser. Variable-incidence horizontal stabiliser, near end of tailboom, for trim purposes.

LANDING GEAR: Non-retractable tricycle type, with steerable twin-wheel nose unit and single wheel on each main unit. Twin-chamber oleo-pneumatic (high-presssure and low-pressure) main landing gear shock-struts. High-pressure chambers interconnected through overflow system incorporating spring damper, to damp out oscillations at full landing gear loading and so eliminate ground resonance. Main wheels size 1,325 × 480 mm. Nosewheels size 720-310. Brakes on main wheels. Small tail-bumper under end of tailboom.

POWER PLANT: Two 4,101 kW (5,500 shp) Soloviev D-25V (TV-2BM) turboshaft engines, mounted side by side above cabin, forward of main rotor shaft. Eleven internal fuel tanks, with total capacity of 6,315 kg (13,922 lb), and two external tanks, on each side of cabin, with total capacity of 3,490 kg (7,695 lb). Provision for two additional ferry tanks inside cabin, with total capacity of 3,490 kg (7,695 lb). Automatic fuel-flow control system with manual override. Side panels of engine cowlings are opened and closed hydraulically and are used as platforms for inspection and maintenance of engines and rotor head.

ACCOMMODATION: Crew of five, consisting of two pilots, navigator, flight engineer and radio operator. Four jettisonable doors on flight deck. Equipped normally for cargo operation, with tip-up seats along side walls. When these seats are supplemented by additional seats installed in centre of cabin, 65 passengers can be carried, with cargo or baggage in the aisles. As an air ambulance, 41 stretcher cases and two medical attendants on tip-up seats can be carried. One of attendant's stations is provided with intercom to flight deck, and provision is made for portable oxygen installations for the patients. Cabin floor is stressed for loadings of 2,000 kg/m² (410 lb/sq ft), with provision for cargo tie-down rings. Rear clamshell doors and ramps are operated hydraulically. Standard equipment includes an electric winch of 800 kg (1,765 lb) capacity and pulley block system. External cargo sling system for bulky loads. Central hatch in cargo floor. Two passengers doors, fore and aft of main landing gear on port side.

ELECTRONICS AND EQUIPMENT: Standard equipment includes VHF and HF communications radio, intercom, radio altimeter, radio compass, autopilot, marker beacon, directional gyro and full all-weather instrumentation.

SYSTEMS: Main, standby and auxiliary hydraulic systems, each with separate pump mounted on main gearbox. Operating pressure 118-152 bars (1,705-2,205 lb/sq in). Main 27V DC electrical system, supplied by two

Mil Mi-6 heavy general-purpose helicopter (two Soloviev D-25V turboshaft engines) *(Martin Fricke)*

12kW starter/generators, with batteries for 30 min emergency supply. De-icing system and some radio equipment supplied by three-phase 360V 400Hz AC system, utilising two 90kVA generators. Trolley-mounted APU, consisting of 74·5 kW (100 hp) AI-8 gas turbine and 24kW generator, carried on board.

ARMAMENT: A few Mi-6s are fitted with a gun of unknown calibre in the fuselage nose.

DIMENSIONS, EXTERNAL:
- Diameter of main rotor: 35·00 m (114 ft 10 in)
- Diameter of tail rotor: 6·30 m (20 ft 8 in)
- Distance between rotor centres: 21·09 m (69 ft 2½ in)
- Length overall, rotors turning: 41·74 m (136 ft 11½ in)
- Length of fuselage: 33·18 m (108 ft 10½ in)
- Height overall: 9·86 m (32 ft 4 in)
- Wing span: 15·30 m (50 ft 2½ in)
- Span of horizontal stabiliser: 5·04 m (16 ft 6½ in)
- Wheel track: 7·50 m (24 ft 7¼ in)
- Wheelbase: 9·10 m (29 ft 10½ in)
- Rear loading doors:
  - Height: 2·70 m (8 ft 10¼ in)
  - Width: 2·65 m (8 ft 8¼ in)
- Passenger doors:
  - Height: front 1·71 m (5 ft 7¼ in)
  - rear 1·62 m (5 ft 3¾ in)
  - Width: 0·81 m (2 ft 7¾ in)
  - Sill height: front 1·40 m (4 ft 7¼ in)
  - rear 1·30 m (4 ft 3¼ in)
- Central hatch in floor: 1·44 m (4 ft 9 in) × 1·93 m (6 ft 4 in)

DIMENSIONS, INTERNAL:
- Cabin:
  - Length: 12·00 m (39 ft 4½ in)
  - Max width: 2·65 m (8 ft 8¼ in)
  - Max height:
    - at front: 2·01 m (6 ft 7 in)
    - at rear: 2·50 m (8 ft 2½ in)
  - Cabin volume: 80 m³ (2,825 cu ft)

WEIGHTS:
- Weight empty: 27,240 kg (60,055 lb)
- Max internal payload: 12,000 kg (26,450 lb)
- Max slung cargo: 9,000 kg (19,840 lb)
- Max T-O weight with slung cargo at altitudes under 1,000 m (3,280 ft): 37,500 kg (82,675 lb)
- Normal T-O weight: 40,500 kg (89,285 lb)
- Max T-O weight for VTO: 42,500 kg (93,700 lb)

PERFORMANCE (at max T-O weight):
- Max level speed: 162 knots (300 km/h; 186 mph)
- Max cruising speed: 135 knots (250 km/h; 155 mph)
- Service ceiling: 4,500 m (14,750 ft)
- Range with 6,000 kg (13,228 lb) payload: 350 nm (650 km; 404 miles)
- Range with external tanks and 4,300 kg (9,480 lb) payload: 566 nm (1,050 km; 652 miles)
- Max ferry range (tanks in cabin): 781 nm (1,450 km; 900 miles)

MIL Mi-8 (V-8)

NATO reporting name: *Hip*

This turbine-powered transport helicopter was shown in public for the first time during the 1961 Soviet Aviation Day display. Its overall dimensions are similar to those of the Mi-4, which it superseded; but the power plant is mounted above the cabin, leaving a clear unobstructed interior, and the Mi-8 (often referred to in the Soviet Union as the V-8) is able to carry a greatly increased payload.

More than 1,000 Mi-8s had been built by mid-1974, mainly for military use, and about 300 of these had been exported.

The Mi-8 serves with the Soviet armed forces. It appeared in military insignia for the first time at the Domodedovo air show in July 1967 and can be equipped to carry external stores to support assault landings by airborne troops. Military Mi-8s have also been supplied to the Bangladesh, Bulgarian, Czechoslovakian, Egyptian, Ethiopian, Finnish, East German, Hungarian, Indian, Iraqi, Libyan, Pakistani, Polish, South Yemen, Sudanese and Syrian armed forces.

Mil Mi-8 military helicopter. This differs from the commercial version in having circular cabin windows and twin weapon-carriers on outriggers *(Tass)*

The commercial Mi-8, with larger, square windows in place of the circular cabin windows of the military version, is in service with Aeroflot for transport and air ambulance duties, and is operated by this airline in support of Soviet activities in the Antarctic. Standard Mi-8s are used there for ice patrol and reconnaissance, for rescue operations, and for carrying supplies and equipment to Vostok Station, near the South Pole.

Three international women's helicopter records for distance and speed in a 2,000 km closed circuit were credited to the Mi-8 in mid-1976.

The original prototype had a single 2,013 kW (2,700 shp) Soloviev turboshaft engine. The second prototype, which flew for the first time on 17 September 1962, introduced the now-standard Isotov twin-turbine power plant.

Early Mi-8s had a four-blade main rotor, but this was superseded by a five-blade rotor in 1964. In an emergency, the blades and intermediate and tail gearboxes are interchangeable with those of the piston-engined Mi-4, although this prevents use of the de-icing system.

The controls of the Mi-8 are hydraulically-powered in the cyclic and collective pitch channels. It is claimed that the autopilot, with barometric height lock, can control all flight modes, including transition.

There are three civil versions, as follows:

Mi-8. Passenger version, with standard seating for 28 persons in main cabin.

Mi-8T. General utility version, equipped normally to

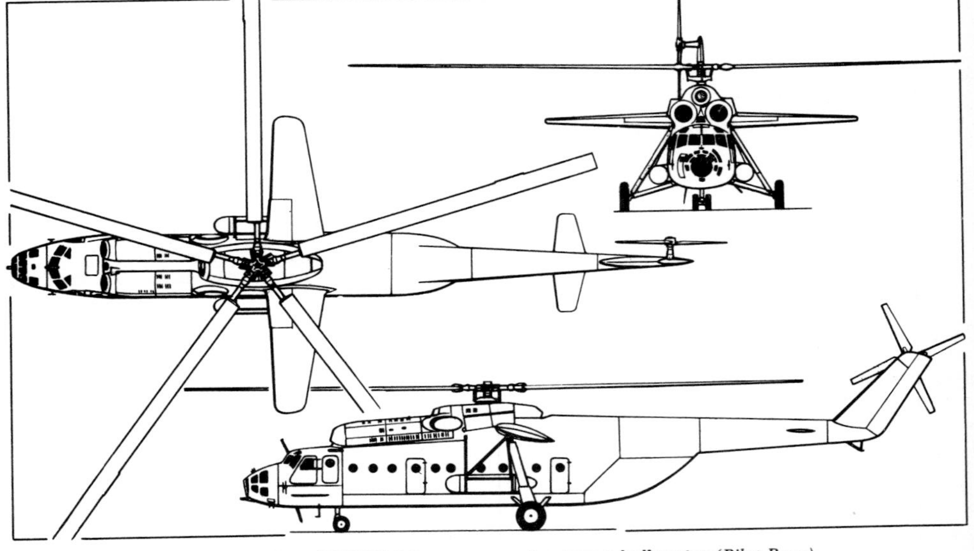

Military version of Mil Mi-6 heavy general-purpose helicopter *(Pilot Press)*

Troops boarding an Mi-6 in Soviet military service. This aircraft has a nose gun *(Tass)*

Mil Mi-8 (V-8) passenger helicopter (two Isotov TV2-117A turboshaft engines) *(Brian M. Service)*

carry internal or external freight, but able to accommodate 24 passenger seats.

Mi-8 Salon. de luxe version. Main cabin is furnished for eleven passengers, with an eight-place couch facing inward on the port side, and two chairs and a swivelling seat on the starboard side. There is a table on each side. An air-to-ground radio telephone and removable ventilation fans are standard equipment. Forward of the main cabin is a compartment for a hostess, with buffet and crew wardrobe. Aft of the main cabin are a toilet (port) and passenger wardrobe (starboard), to each side of the entrance. The Mi-8 Salon has a max T-O weight of 10,400 kg (22,928 lb) and range of 205 nm (380 km; 236 miles) with 30 min fuel reserve. In other respects it is similar to the standard Mi-8.

A float-equipped version was reported to be under test in early 1974, under the designation V-14.

Type: Twin-engined transport helicopter.

Rotor System: Five-blade main rotor and three-blade tail rotor. Transmission comprises a type VR-8 main gearbox giving main rotor shaft/engine rpm ratio of 0·016 : 1, intermediate and tail gearboxes, main rotor brake and drives off the main gearbox for the tail rotor, fan, AC generator, hydraulic pumps and tachometer generators. Main rotor shaft inclined forward at 4° 30′ to vertical. All-metal main rotor blades of basic NACA 230 section; solidity 0·0777. Each main blade is made up of an extruded light alloy spar carrying the blade root fitting, 21 trailing-edge pockets and the blade tip. Pockets are honeycomb-filled. Main rotor blades are fitted with balance tabs, and are interchangeable. Their drag and flapping hinges are a few inches apart, and they are carried on a machined spider. Controls hydraulically-powered. All-metal tail rotor blades, each made up of a spar and honeycomb-filled trailing-edge. Automatically-controlled electro-thermal de-icing system on all blades.

Fuselage: Conventional all-metal semi-monocoque structure of pod and boom type.

Tail Unit: Tail rotor support acts as small vertical stabiliser. Horizontal stabiliser near end of tailboom.

Landing Gear: Non-retractable tricycle type, with steerable twin-wheel nose unit and single wheel on each main unit. All units embody oleo-pneumatic (gas) shock-absorbers. Main-wheel tyres size 865 × 280; nosewheel tyres size 595 × 185. Pneumatic brakes on main wheels. Pneumatic system can also recharge tyres in the field, using air stored in main landing gear struts. Optional main-wheel fairings.

Power Plant: Two 1,118·5 kW (1,500 shp) Isotov TV2-117A turboshaft engines. Main rotor speed governed automatically, with manual override. Single flexible internal fuel tank, capacity 445 litres (98 Imp gallons), and two external tanks, on each side of cabin, with capacity of 745 litres (164 Imp gallons) in the port tank and 680 litres (149·5 Imp gallons) in the starboard tank. Total standard fuel capacity 1,870 litres (411·5 Imp gallons). Provision for carrying one or two additional ferry tanks in cabin, raising max total capacity to 3,700 litres (814 Imp gallons). Fairing over starboard external tank houses optional cabin air-conditioning equipment at front. Engine cowling side panels form maintenance platforms when open, with access via hatch on flight deck. Engine air intake de-icing standard. Total oil capacity 60 kg (132 lb).

Accommodation: Two pilots side by side on flight deck, with provision for a flight engineer's station. Windscreen de-icing standard. Basic passenger version is furnished with 28 four-abreast track-mounted tip-up seats at a pitch of 72-75 cm (28·3-29·5 in), with a centre aisle 32 cm (12·6 in) wide, a wardrobe and baggage compartment; or 32 seats without wardrobe. Seats and bulkheads of basic version are quickly removable for cargo-carrying. Mi-8T has cargo tie-down rings in floor, a winch of 200 kg (440 lb) capacity and pulley block system to facilitate the loading of heavy freight, an external cargo sling system, and 24 tip-up seats along the side walls of the cabin. All versions can be converted for air ambulance duties, with accommodation for 12 stretchers and a tip-up seat for a medical attendant. The large windows on each side of the flight deck slide rearward. The sliding, jettisonable main passenger door is at the front of the cabin on the port side. An electrically-operated rescue hoist can be installed at this doorway. The rear of the cabin is made up of large clamshell freight-loading doors, with a downward-hinged passenger airstair door inset centrally at the rear. Hook-on ramps are used for vehicle loading.

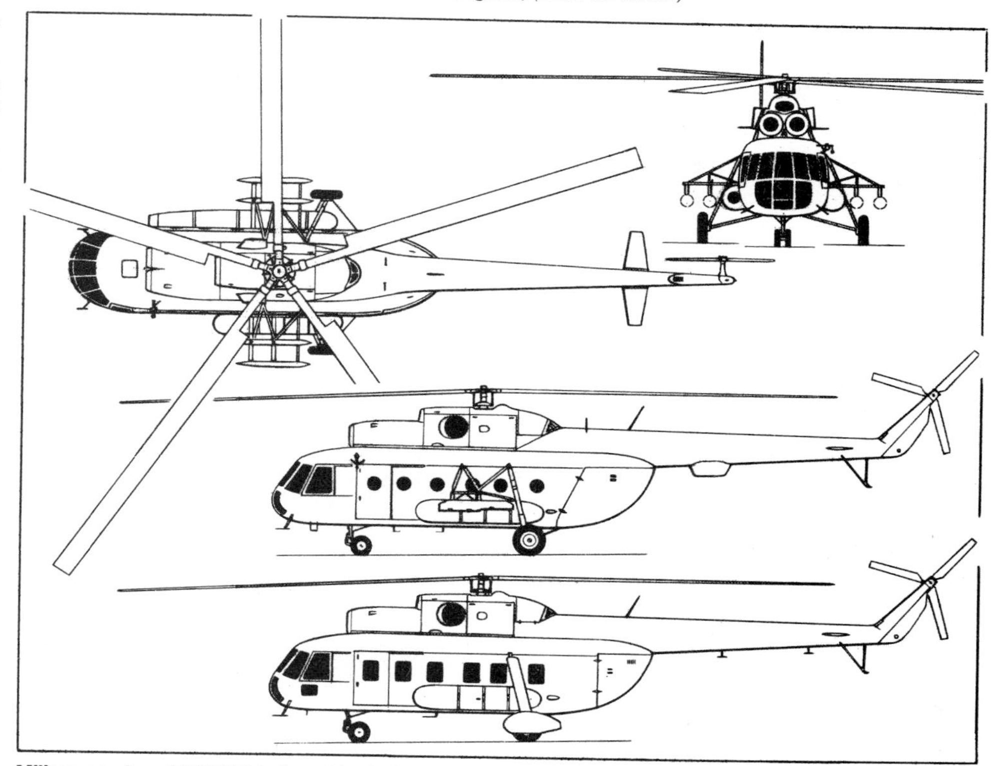

Military version of Mil Mi-8 twin-turbine helicopter, with additional side view (bottom) of commercial version *(Pilot Press)*

Systems: Standard heating system can be replaced by full air-conditioning system. Two independent hydraulic systems, each with own pump; operating pressure 44-64 bars (640-925 lb/sq in). DC electrical supply from two 27V 18kW starter/generators and six 28Ah storage batteries. AC supply for de-icing system and some radio equipment supplied by 208/115/36/7·5V 400Hz generator, with 36V three-phase standby system. Provision for oxygen system for crew and, in ambulance version, for patients. Freon fire-extinguishing system in power plant bays and service fuel tank compartments, actuated automatically or manually. Two portable fire extinguishers for use in cabin.

Electronics and Equipment: Standard equipment includes a type R-842 HF transceiver with frequency range of 2 to 8 Mc/s and range of up to 540 nm (1,000 km; 620 miles), type R-860 VHF transceiver operating on 118 to 135·9 Mc/s over ranges of up to 54 nm (100 km; 62 miles), intercom, radio telephone, type ARK-9 automatic radio compass, type RV-3 radio altimeter with 'dangerous height' warning, and four-axis autopilot to give yaw, roll and pitch stabilisation under any flight conditions, stabilisation of altitude in level flight or hover, and stabilisation of pre-set flying speed, navigation equipment and instrumentation for all-weather flying by day and night, including two gyro horizons, two airspeed indicators, two main rotor speed indicators, turn indicator, two altimeters, two rate of climb indicators, magnetic compass, radio altimeter, radio compass and astro-compass for Polar flying.

Armament: Military versions can be equipped with a twin rack for external stores, including pods each containing sixteen 57 mm rockets, on an outrigger structure on each side of the main cabin.

Dimensions, external:	
Diameter of main rotor	21·29 m (69 ft 10¼ in)
Diameter of tail rotor	3·90 m (12 ft 9½ in)
Distance between rotor centres	12·65 m (41 ft 6 in)
Length overall, rotors turning	25·24 m (82 ft 9¾ in)
Length of fuselage	18·31 m (60 ft 0¾ in)
Height overall	5·65 m (18 ft 6½ in)
Wheel track	4·50 m (14 ft 9 in)
Wheelbase	4·26 m (13 ft 11¾ in)
Fwd passenger door:	
Height	1·41 m (4 ft 7¼ in)
Width	0·82 m (2 ft 8¼ in)
Rear passenger door:	
Height	1·70 m (5 ft 7 in)
Width	0·84 m (2 ft 9 in)
Rear cargo door:	
Height	1·82 m (5 ft 11½ in)
Width	2·34 m (7 ft 8¼ in)
Dimensions, internal:	
Passenger cabin:	
Length	6·30 m (20 ft 7¾ in)
Width	2·34 m (7 ft 8¼ in)

Height 1·82 m (5 ft 11¾ in)
Cargo hold (freighter):
Length at floor 5·34 m (17 ft 6¼ in)
Width 2·34 m (7 ft 8¼ in)
Height 1·82 m (5 ft 11¾ in)
Volume approx 23 m³ (812 cu ft)
AREA:
Main rotor disc 355 m² (3,828 sq ft)
WEIGHTS:
Weight empty:
Passenger version 7,261 kg (16,007 lb)
Cargo version 6,816 kg (15,026 lb)
Max payload:
internal 4,000 kg (8,820 lb)
external 3,000 kg (6,614 lb)
Normal T-O weight 11,100 kg (24,470 lb)
T-O weight with 28 passengers, each with 15 kg (33 lb) of baggage 11,570 kg (25,508 lb)
T-O weight with 2,500 kg (5,510 lb) of slung cargo 11,428 kg (25,195 lb)
Max T-O weight for VTO 12,000 kg (26,455 lb)
PERFORMANCE:
Max level speed at 1,000 m (3,280 ft):
Normal AUW 140 knots (260 km/h; 161 mph)
Max level speed at S/L:
Normal AUW 135 knots (250 km/h; 155 mph)
Max AUW 119 knots (220 km/h; 137 mph)
With 2,500 kg (5,510 lb) of slung cargo 97 knots (180 km/h; 112 mph)
Max cruising speed:
Normal AUW 122 knots (225 km/h; 140 mph)
Max AUW 97 knots (180 km/h; 112 mph)
Service ceiling 4,500 m (14,760 ft)
Hovering ceiling in ground effect at normal AUW 1,900 m (6,233 ft)
Hovering ceiling out of ground effect at normal AUW 800 m (2,625 ft)
Ranges:
Cargo version at 1,000 m (3,280 ft), with standard fuel, 5% reserves:
Normal AUW 259 nm (480 km; 298 miles)
Max AUW 248 nm (460 km; 285 miles)
Passenger version at 1,000 m (3,280 ft), with 20 min fuel reserves 229 nm (425 km; 264 miles)
Ferry range of cargo version, with auxiliary fuel, 5% reserves 647 nm (1,200 km; 745 miles)

MIL Mi-10 (V-10)

NATO reporting name: *Harke*

This flying crane development of the Mi-6 was demonstrated at the 1961 Soviet Aviation Day display at Tushino, having flown for the first time in the previous year. Above the line of the cabin windows the two helicopters are almost identical, but the depth of the fuselage is reduced considerably on the Mi-10, and the tailboom is deepened so that the flattened undersurface runs unbroken to the tail. The Mi-10 also lacks the fixed wings of the Mi-6.

Items which are interchangeable between the Mi-6 and Mi-10 include the power plant, transmission system and reduction gearboxes, swashplate assembly, main and tail rotors, control system and most items of equipment.

The tall long-stroke quadricycle landing gear, with wheel track exceeding 6·0 m (19 ft 8 in) and clearance under the fuselage of 3·75 m (12 ft 3½ in) with the aircraft fully loaded, enables the Mi-10 to taxi over a load it is to carry and to accommodate loads as bulky as a prefabricated building.

Use can be made of interchangeable wheeled cargo platforms which are held in place by hydraulic grips controllable from either the cockpit or a remote panel. Using these grips without a platform, cargoes up to 20 m (65 ft 7 in) long, 10 m (32 ft 9½ in) wide and 3·1 m (10 ft 2 in) high can be lifted and secured in 1½ to 2 minutes. The cabin can accommodate additional freight or passengers.

A closed-circuit TV system, with cameras scanning forward from under the rear fuselage and downward through the sling hatch, is used to observe the payload and main landing gear, touchdown being by this reference. The TV system replaces the retractable undernose 'dustbin' fitted originally.

The power of the Soloviev turboshaft engines remains constant up to 3,000 m (9,850 ft) and to an ambient air temperature of 40°C at sea level. The aircraft will maintain level flight on one engine. Full navigation equipment and an autopilot permit all-weather operation, by day and night.

The following details refer to the standard Mi-10, which has been operated by both Aeroflot and the Soviet armed forces and is also available for export.

Tasks performed by the Mi-10 have included the transport of complete wing assemblies for Tu-144 supersonic airliners from the factory at Voronezh, where they are built, to the assembly works near Moscow.

TYPE: Heavy flying-crane helicopter.

ROTOR SYSTEM: Same as for Mi-6, except that main rotor shaft is inclined forward at an angle of only 45′.

FUSELAGE: Conventional all-metal riveted semi-monocoque structure.

TAIL UNIT: Same as for Mi-6.

LANDING GEAR: Non-retractable quadricycle type, with twin wheels on each unit. All units fitted with oleo-pneumatic shock-absorbers. Telescopic main legs. Main wheels size 1,230 × 260 mm, each with brake. Levered-suspension castoring nose units. Nosewheels size 950 × 250. All landing gear struts are faired. The port nose gear fairing incorporates steps to the crew entry door. Despite the height of the gear, the Mi-10 can make stable landing and take-off runs at speeds up to 54 knots (100 km/h; 62 mph).

POWER PLANT: Two 4,101 kW (5,500 shp) Soloviev D-25V turboshaft engines, mounted side by side above cabin, forward of main rotor drive-shaft. Single fuel tank in fuselage and two external tanks, on sides of cabin, with total capacity of 6,340 kg (13,975 lb). Provision for carrying two auxiliary tanks in cabin, to give total fuel capacity of 8,260 kg (18,210 lb). Engine cowling side panels (opened and closed hydraulically) can be used as maintenance platforms when open.

ACCOMMODATION: Two pilots and flight engineer accommodated on flight deck, which has bulged side windows to provide an improved downward view. Flight deck is heated and ventilated and has provision for oxygen equipment. Crew door is immediately aft of flight deck on port side. Main cabin can be used for freight and/or passengers, 28 tip-up seats being installed along the side walls. Freight is loaded into this cabin through a door on the starboard side, aft of the rear landing gear struts, with the aid of a boom and 200 kg (440 lb) capacity electric winch. In addition to the cargo platform described earlier, the Mi-10 has external sling gear as standard equipment. This can be used in conjunction with a winch controlled from a portable control panel inside the cabin. The winch can be also be used to raise loads of up to 500 kg (1,100 lb) while the aircraft is hovering on rescue and other duties, via a hatch in the cabin floor.

ELECTRONICS, EQUIPMENT AND SYSTEMS: Generally as for Mi-6, including APU.

DIMENSIONS, EXTERNAL:
Diameter of main rotor 35·00 m (114 ft 10 in)
Diameter of tail rotor 6·30 m (20 ft 8 in)
Distance between rotor centres 21·24 m (69 ft 8 in)
Length overall, rotors turning 41·89 m (137 ft 5½ in)
Length of fuselage 32·86 m (107 ft 9¾ in)
Ground clearance under fuselage 3·75 m (12 ft 3½ in)
Height overall 9·80 m (32 ft 2 in)
Wheel track (c/l shock-struts):
nosewheels 6·01 m (19 ft 8¾ in)
main wheels 6·92 m (22 ft 8½ in)
Wheelbase 8·29 m (27 ft 2½ in)
Cargo platform:
Length 8·53 m (28 ft 0 in)
Width 3·54 m (11 ft 7¼ in)
Crew door:
Height 1·35 m (4 ft 5¼ in)
Width 0·78 m (2 ft 6¾ in)
Height to sill 3·91 m (12 ft 10¼ in)
Freight loading door:
Height 1·56 m (5 ft 1½ in)
Width 1·26 m (4 ft 1½ in)
Height to sill 3·92 m (12 ft 10½ in)
Cabin floor hatch:
Diameter 1·00 m (3 ft 3½ in)

Mil Mi-10 flying crane helicopter, with load platform, in Soviet military service *(photocopied from Repules)*

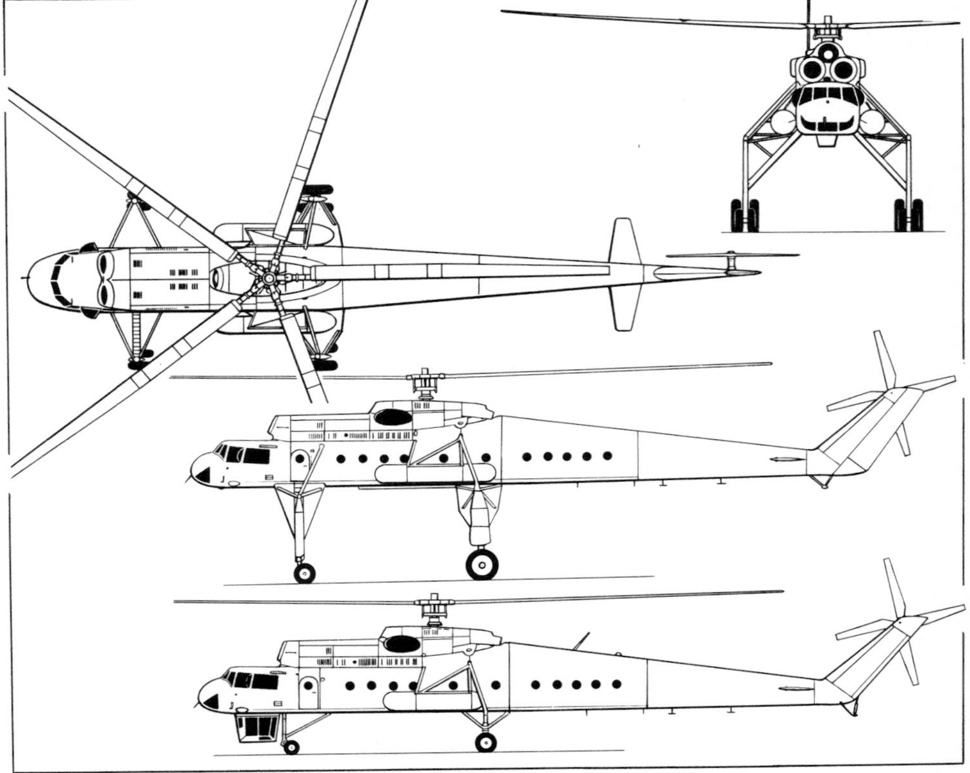

Mil Mi-10 flying crane derivative of the Mi-6, with additional side view (bottom) of Mi-10K *(Pilot Press)*

DIMENSIONS, INTERNAL:
Cabin: Length 14·04 m (46 ft 0¾ in)
Width 2·50 m (8 ft 2½ in)
Height 1·68 m (5 ft 6 in)
Volume approx 60 m³ (2,120 cu ft)
WEIGHTS:
Weight empty 27,300 kg (60,185 lb)
Max payload on platform, incl platform 15,000 kg (33,070 lb)
Max slung payload 8,000 kg (17,635 lb)
T-O weight with slung cargo 38,000 kg (83,775 lb)
Max T-O weight 43,700 kg (96,340 lb)
PERFORMANCE:
Max level speed at max T-O weight 108 knots (200 km/h; 124 mph)
Cruising speed at max T-O weight 97 knots (180 km/h; 112 mph)
Service ceiling (limited) 3,000 m (9,850 ft)
Range with platform payload of 12,000 kg (26,455 lb) 135 nm (250 km; 155 miles)

MIL Mi-10K

First displayed publicly in Moscow on 26 March 1966, the Mi-10K is a development of the Mi-10 with a number of important design changes, most apparent of which are a reduction in the height of the landing gear and a more slender tail rotor support structure.

It can be operated by a crew of only two pilots. This is made possible by the provision of an additional cockpit gondola under the front fuselage, with full flying controls and a rearward-facing seat. By occupying this seat, one of the pilots can control the aircraft in hovering flight and, at the same time, have an unrestricted view of cargo loading, unloading and hoisting, which are also under his control.

In the Mi-10K, the maximum slung payload is 11,000 kg (24,250 lb) and is expected to be increased further to 14,000 kg (30,865 lb) by using Soloviev D-25VF turboshaft engines, uprated to 6,500 shp each, in due course. Fuel capacity of the Mi-10K, in standard internal and external tanks, is 9,000 litres (1,980 Imp gallons). The rotor turns at 120 rpm.

DIMENSIONS, EXTERNAL:
Generally as for Mi-10, except:
Height overall 7·80 m (25 ft 7 in)
Wheel track 5·00 m (16 ft 4¾ in)
Wheelbase 8·74 m (28 ft 8 in)
Door sill heights:
Crew door 1·81 m (5 ft 11 in)
Freight door 1·82 m (5 ft 11½ in)
WEIGHTS:
Weight empty 24,680 kg (54,410 lb)
Max payload, slung cargo 11,000 kg (24,250 lb)
Max fuel load with ferry tanks in cabin 8,670 kg (19,114 lb)
Max T-O weight with slung cargo 38,000 kg (83,776 lb)
PERFORMANCE:
Cruising speed, empty 135 knots (250 km/h; 155 mph)
Max cruising speed with slung load 109 knots (202 km/h; 125 mph)
Service ceiling 3,000 m (9,850 ft)
Ferry range with auxiliary fuel 428 nm (795 km; 494 miles)

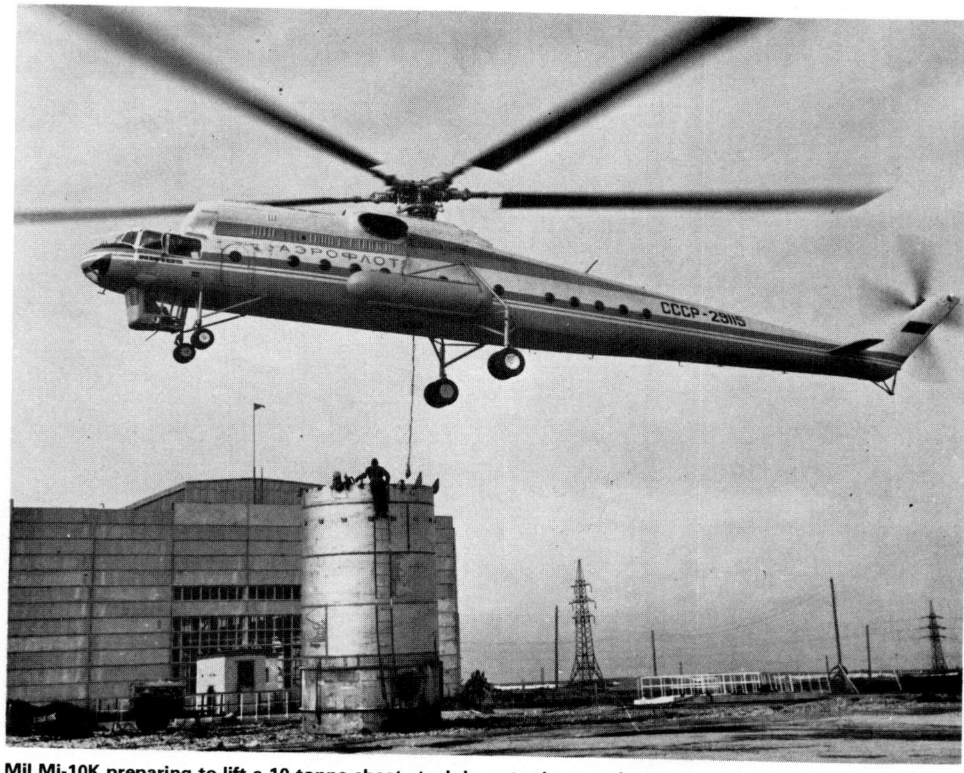

Mil Mi-10K preparing to lift a 10-tonne sheet steel drum to the top of a tower at the Sinarski pipe works in Kamensk-Uralski *(Tass)*

MIL V-12 (Mi-12)

NATO reporting name: *Homer*

First confirmation of the existence of this aircraft was given in a statement in March 1969 that it had set a number of payload-to-height records which exceeded by some 20 per cent the records established previously by the Mi-6 and Mi-10K.

Flying from the airfield at Podmoskovnœ on 22 February 1969, the V-12 climbed at a rate of more than 180 m (600 ft)/min to an altitude of 2,951 m (9,682 ft) carrying a payload of 31,030 kg (68,410 lb). This represented new records for maximum load lifted to a height of 2,000 m, and for height attained with payloads of 20,000, 25,000 and 30,000 kg. The pilot was Vasily Kolochenko who, on 6 August 1969, far exceeded his own record for payload raised to 2,000 m by lifting 40,204·5 kg (88,636 lb) to a height of 2,255 m (7,398 ft) in the V-12, which carried a full crew of six. This flight also qualified for new payload-to-height records with 35,000 kg and 40,000 kg.

Work on the V-12 had started in 1965, the basic requirement being for a VTOL aircraft that could accommodate missiles and other payloads compatible with those carried by the An-22 fixed-wing transport. The original specification called for a tandem-rotor configuration, using existing dynamic components. Instead, the Mil design bureau obtained approval for a side-by-side rotor layout, claimed to offer better stability, reliability and fatigue life. Thus, the V-12 utilises two power plant/rotor packages similar to those of the Mi-6/Mi-10 series, mounted at the tips of its fixed wings.

The D-25VF engines are uprated by comparison with the D-25Vs fitted to the earlier helicopters, by the addition of a zero stage on the compressor and by acceptance of higher operating temperatures.

The prototype V-12 is reported to have crashed in 1969, largely as a result of engine failure, without fatalities. Two prototypes were flying in mid-1971, at which time production was expected to begin before the end of the year.

There has been no recent news of progress with the V-12, but any production aircraft are likely to embody a number of modifications. In particular the fixed wings will probably have increased camber in place of the present trailing-edge flaps, which have been fixed (or possibly ground-adjustable) since their original function to improve autorotation performance was proved unnecessary.

In addition to its military applications, the V-12 is intended for operation by Aeroflot, notably for supporting oil and natural gas production and for hauling geophysical survey equipment, vehicles and heavy freight in remote regions of the Soviet Union. It is claimed to be easy to fly by average pilots with experience of handling other types of helicopter and to have an extremely low vibration level, particularly on the flight deck. No special ground equipment is needed for servicing.

TYPE: Heavy general-purpose helicopter.

ROTOR SYSTEM: Two five-blade opposite-rotating rotors, mounted side by side at the tips of fixed wings. Port rotor turns in clockwise direction, starboard rotor anti-clockwise, viewed from below. All-metal blades, similar to those of Mi-6/Mi-10, with trailing-edge tabs. Rotors are cross-shafted to ensure synchronisation and to maintain rotation following the failure of engines on either side. Rotor rpm 112.

WINGS: High-mounted strut-braced wings, with considerable dihedral and inverse taper to give increasing chord from root to tip. All-metal construction. Long-span two-section fixed or ground-adjustable (originally three-position) trailing-edge flaps on each wing.

FUSELAGE: Conventional all-metal semi-monocoque structure, with clamshell rear loading doors and ramp. Two side-by-side 'bumpers' under ramp.

TAIL UNIT: Cantilever all-metal structure comprising central main fin and rudder, small dorsal fin, tailplane, elevator and endplate auxiliary fins. Tailplane has considerable dihedral. Auxiliary fins are toed inward at leading-edges. Tabs in rudder and elevators.

LANDING GEAR: Non-retractable tricycle type, with twin wheels on each unit. Steerable nosewheels. Main-wheel tyres size 1,750 × 730; nosewheel tyres size 1,200 × 450.

POWER PLANT: Four 4,847 kW (6,500 shp) Soloviev D-25VF turboshaft engines, mounted in side-by-side pairs under tips of fixed wings. Each pair is coupled to drive one rotor, with cross-shafting. Lower part of cowling under each pair of engines can be lowered about 1·8 m (6 ft) by hand-crank to form working platform for up to three men and to provide access for servicing of power plant and rotor head. Cowling side panels hinge downward for same purpose. Cylindrical external fuel tank mounted on each side of main cabin.

ACCOMMODATION: Main flight deck in nose has side-by-side seats for pilot (port) and co-pilot in front. Flight engineer's station behind pilot; electrician seated

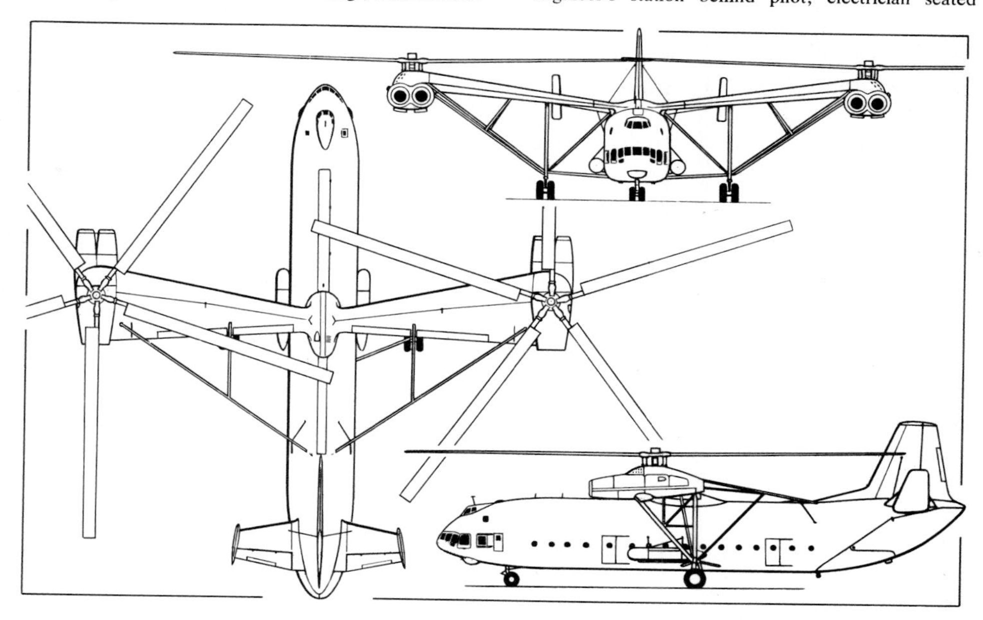

Mil V-12 four-turboshaft heavy-duty freight-carrying helicopter *(Pilot Press)*

Mil V-12 (Mi-12) heavy general-purpose helicopter (four Soloviev D-25VF turboshaft engines) *(Brian M. Service)*

behind co-pilot. Upper cockpit seats navigator and radio operator in tandem. Windscreen panels forward of pilot, co-pilot and navigator fitted with wipers. Rubber-bladed fans for cooling crew accommodation. Unobstructed main cargo hold has rails in roof for electrically-operated platform-mounted travelling crane with four loading points each capable of lifting 2,500 kg (5,500 lb) and max capacity of 10,000 kg (22,000 lb) for a single item. About 50 upward-folding seats along side walls for work crews or troops accompanying freight loads. Primary access to hold between rear clamshell doors which hinge outward and upward, via downward-hinged ramp. Rearward-sliding door forward of fuel tank on port side. Emergency exit door on each side at rear of hold. Downward-hinged emergency exits on starboard side of main flight deck and upper cockpit.

Systems and Equipment: Electrical system has 480kW capacity. Ground mapping radar in undernose blister fairing. Fail-safe powered control system and automatic stabilisation system standard, but aircraft can be landed manually. Ivchenko AI-8V APU for independent engine starting.

Dimensions, external:
Diameter of main rotors (each) 35·00 m (114 ft 10 in)
Span over rotor tips 67·00 m (219 ft 10 in)
Length of fuselage 37·00 m (121 ft 4½ in)
Height overall 12·50 m (41 ft 0 in)

Dimensions, internal:
Freight compartment:
Length 28·15 m (92 ft 4 in)
Max width 4·40 m (14 ft 5 in)
Max height 4·40 m (14 ft 5 in)

Weights:
Normal payload:
VTOL 25,000 kg (55,000 lb)
STOL 30,000 kg (66,000 lb)
Normal T-O weight 97,000 kg (213,850 lb)
Max T-O weight 105,000 kg (231,500 lb)

Performance:
Max level speed 140 knots (260 km/h; 161 mph)
Max cruising speed 130 knots (240 km/h; 150 mph)
Service ceiling 3,500 m (11,500 ft)
Range with 35,400 kg (78,000 lb) payload 270 nm (500 km; 310 miles)

MIL V-14

A float-equipped version of the Mi-8 was reported to be under test in the Soviet Union in early 1974, with the designation V-14. No details are available.

MIL Mi-24

NATO reporting name: *Hind*

This assault helicopter was known to exist for some two years before photographs became available to the technical press in early 1974. Two versions were shown in these photographs and are identified by the following NATO code names:

Hind-A. The auxiliary wings of this version have considerable anhedral and each carry three weapon stations. The two inboard stations on each side are used normally as attachments for large rocket pods, each containing thirty-two 57 mm rockets. The wingtip stations take the form of deep rectangular pylons, each carrying two missile rails for air-to-surface adaptations of a standard Soviet anti-tank weapon. A 12·7 mm machine-gun is flexibly mounted beneath a flat panel of bulletproof glass in the nose.

Hind-B. Generally similar to *Hind-A* except that the auxiliary wings have no anhedral or dihedral, and carry only the two inboard weapon stations on each side. Paradoxically, *Hind-B* preceded *Hind-A* in development and is not a major production variant.

The general appearance of the Mi-24 is shown in the accompanying illustrations. It is of conventional all-metal pod and boom design, with the comparatively low profile associated with gunship helicopters. In addition to the crew, on side-by-side seats, it is estimated that eight assault troops can be accommodated in the main cabin. Access to the flight deck is via a large rearward-sliding blistered transparent panel which forms the aft flight deck window on the port side. At the front of the passenger cabin on each side is a large door, divided horizontally into two sections which are hinged to open upward and downward respectively.

The tapered auxiliary wings are set at an incidence of

Two views of the Mi-24 military assault helicopter known by the NATO reporting name *Hind-A (Flug Revue)*

about 20°. There is a variable-incidence horizontal stabiliser at the base of the sweptback fin, which is offset a few degrees to port and serves also as a pylon to carry the tail anti-torque rotor. The tricycle landing gear is retractable, and comprises a twin-wheel nose unit and single-wheel main units. The latter retract rearward and inward into the aft end of the fuselage pod, turning through 90° to stow almost vertically, discwise to the longitudinal axis of the fuselage, under prominent blister fairings. A tubular tripod skid assembly protects the tail rotor in a tail-down take-off or landing.

It was suggested initially that the Mi-24 utlises the power plant and rotor system of the Mi-8; but only the three-blade tail rotor appears to be common to the two designs. Using its assumed diameter to scale other dimensions of the Mi-24, it becomes clear that both the turboshaft engines and the five-blade main rotor are smaller in size than their counterparts on the Mi-8. The engines are mounted conventionally, side by side above the cabin, with their output shafts driving rearward to the main rotor shaft through a combining gearbox.

The Mi-24 has been operational with two units of approximate squadron strength based in East Germany since the early months of 1974.

DIMENSIONS, EXTERNAL:

Diameter of main rotor	17·00 m (55 ft 9 in)
Diameter of tail rotor	3·90 m (12 ft 9½ in)
Length overall	17·00 m (55 ft 9 in)
Height overall	4·25 m (14 ft 0 in)

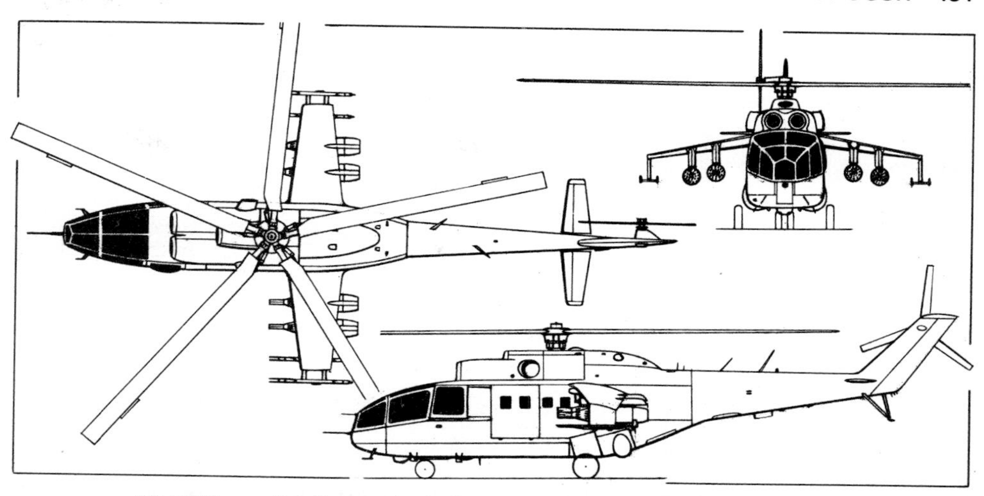

Mil Mi-24 assault helicopter, in the form known to NATO as *Hind-A (Pilot Press)*

MIL A-10

In the Summer of 1975, a total of seven helicopter records in class E1 were set by a Soviet women's crew in a helicopter designated A-10. All that is known about the A-10 is that it was designed by the Mil bureau and is powered by two 1,118·5 kW (1,500 shp) Isotov TV2-117A turboshaft engines, as fitted to the Mi-8. The performance capability represented by the records suggests that the A-10 might be related to the Mi-24.

Pilot on all the record flights was Galina Rastorgoueva, a test pilot and engineer. She was accompanied by Ludmila Polyanskaya, who is employed as navigator on Il-18 airliners of Aeroflot. On 18 July 1975, they averaged 180·480 knots (334·464 km/h; 207·826 mph) around a 100 km circuit, setting a new women's speed record. On 1 August, a speed of 178·624 knots (331·023 km/h; 205·688 mph) around 500 km set both general and women's records, as did an average speed of 179·500 knots (332·646 km/h; 206·697 mph) over 1,000 km on 13 August. Two women's time-to-height helicopter records followed, with a time of 2 min 33·5 sec to 3,000 m on 8 August, and 7 min 43 sec to 6,000 m on 26 August.

A further record by the same crew in an A-10 was reported by the Novosti press agency in October 1975, but had not received FAI recognition by mid-1976. Representing the first claim for a women's helicopter speed record over a 15/25 km course, it was quoted as 184·196 knots (341·35 km/h; 212·105 mph).

MYASISHCHEV

GENERAL DESIGNER IN CHARGE OF BUREAU:
Vladimir M. Myasishchev

Although Myasishchev's work has been little publicised, he is believed to have been responsible for the development of several important types. They include the four-jet M-4 bomber (known in the West by the reporting name of *Bison*) which remains in service for maritime reconnaissance and as a flight refuelling tanker.

The aircraft referred to by the Soviet authorities as the 201-M when it set up a number of officially-recognised records, in 1959, was identified subsequently as a variant of the M-4 and was described briefly and illustrated in the 1972-73 *Jane's*.

Myasishchev was reported to have designed a long-range four-jet heavy bomber to replace the M-4. This aircraft is believed to have been the delta-wing type which was allocated the NATO code name of *Bounder* and was described in the 1964-65 *Jane's*. Changing requirements limited *Bounder* to a research role.

MYASISHCHEV M-4

NATO reporting name: *Bison*

Three major production versions of this four-jet aircraft have been identified by NATO code names, as follows:

Bison-A. The Soviet Union's first operational four-jet strategic bomber, displayed initially over Moscow in May 1954. Comparable with early versions of Boeing B-52 Stratofortress. Powered by four 85·3 kN (19,180 lb st) Mikulin AM-3D turbojets, buried in wing-roots. Range reported to be 6,075 nm (11,250 km; 7,000 miles) at 450 knots (835 km/h; 520 mph) with 4,500 kg (10,000 lb) of nuclear or conventional free-fall bombs. Defensive armament of ten 23 mm cannon in twin-gun turrets in tail, above fuselage fore and aft of wing and under fuselage fore and aft of bomb bays, believed necessary because of aircraft's operational ceiling of only 13,700 m (45,000 ft). In early 1975, the Chairman of the US Joint Chiefs of Staff commented that all 85 M-4s then serving in the Soviet bomber force were available for conversion into flight refuelling tankers for the *Backfire* supersonic bomber force, but that a tanker version of the Il-76 transport would probably be preferred. About 50 *Bison-As* have, in fact, been adapted as tankers, carrying a hose-reel unit in the bomb bay.

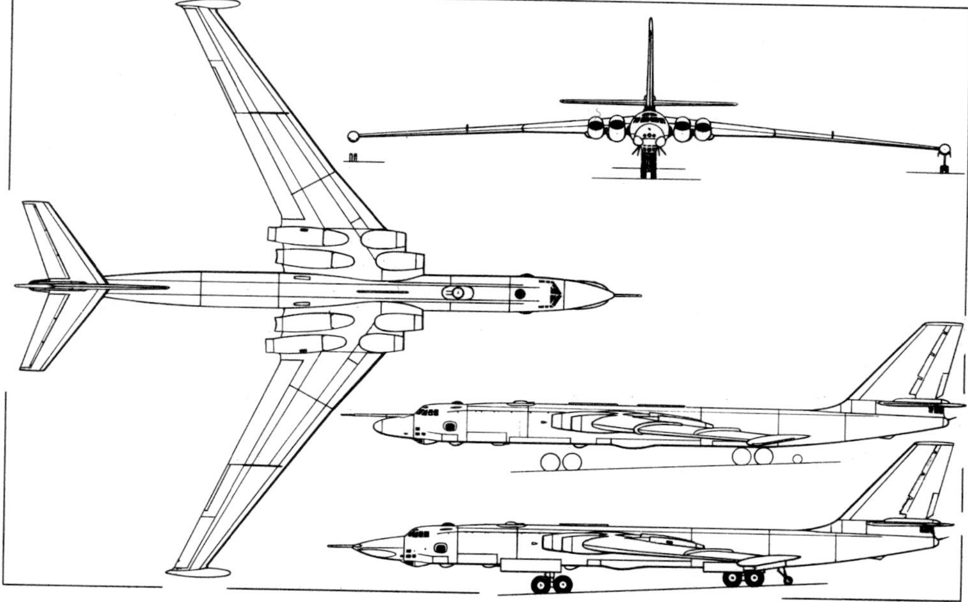

Myasishchev M-4, in the form known to NATO as *Bison-C*, **with additional side view (centre) of** *Bison-B* *(Pilot Press)*

Bison-B. Maritime reconnaissance version identified in service in 1964. 'Solid' nose radome in place of hemispherical glazed nose of *Bison-A*, with large superimposed flight refuelling probe. Numerous underfuselage blister fairings for specialised electronic equipment. Forward portion of centre bomb bay doors bulged. Aft gun turrets above and below fuselage deleted, reducing armament to six 23 mm cannon.

Bison-C. Generally similar configuration to *Bison-B* but with large search radar faired neatly into new and longer nose, aft of centrally-mounted flight refuelling probe. Prone bombing/observation station, with optically-flat glass panels, below and to rear of radar; further small windows and a domed observation (and probably gunnery aiming) window on each side; underfuselage blister fairings, bulged bomb bay and armament; all as *Bison-B*. An example of this version with the experimental aircraft designation 201-M was used to set up a number of official records in 1959 and was exhibited statically in the Soviet

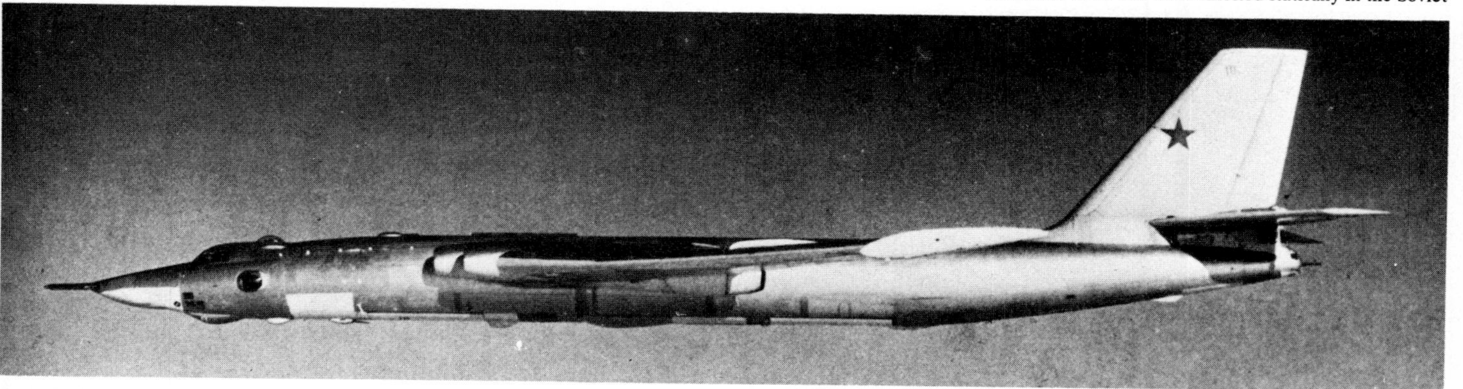

This photograph of the *Bison-C* **version of the M-4 shows clearly its modified nose** *(Royal Air Force)*

Aviation Day display at Domodedovo Airport, Moscow, in 1967. Powered by four 127·5 kN (28,660 lb st) Type D-15 turbojet engines, this testbed aircraft established seven payload-to-height records, including a weight of 55,220 kg (121,480 lb) lifted to 2,000 m (6,560 ft) and height of 15,317 m (50,253 ft) with a 10,000 kg payload.

DIMENSIONS, EXTERNAL (*Bison-A*):

Wing span	50·48 m (165 ft 7½ in)
Length overall	47·20 m (154 ft 10 in)
Tailplane span	15·00 m (49 ft 2½ in)

WEIGHT (*Bison-A*):

Max T-O weight	158,750 kg (350,000 lb)

PERFORMANCE (*Bison-B*, estimated):

Max level speed at 11,000 m (36,000 ft)
485 knots (900 km/h; 560 mph)

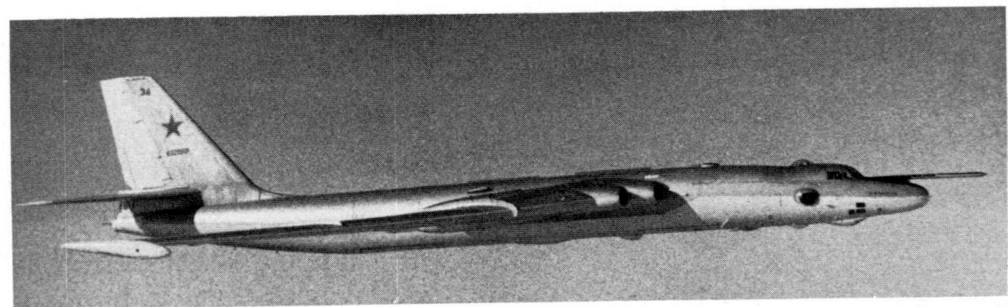

Myasishchev M-4 maritime reconnaissance aircraft in the form known to NATO as *Bison-B (Royal Air Force)*

SUKHOI

Pavel Osipovich Sukhoi, who headed this design bureau until his death in September 1975, helped to design the ANT-25 and had a share in the construction of the 'Rodina' before the second World War; his Su-2 attack aeroplane was used in the war. He was also responsible for one of the jet aircraft in the 1947 Soviet Aviation Day display.

Nearly a decade later, on 24 June 1956, there appeared over Tushino new swept-wing and delta-wing fighters from Sukhoi's design team. Both aircraft subsequently entered squadron service with the Soviet Air Force, as the Su-7 and Su-9, and have been followed by other Sukhoi designs.

SUKHOI Su-7B

NATO reporting names: *Fitter-A* **and** *Moujik*

The Su-7B single-seat ground attack fighter (NATO reporting name *Fitter-A*) was first seen in prototype form during the 1956 Soviet Aviation Day Display and appeared in formations of up to 21 aircraft at the 1961 Tushino display. It subsequently became the standard tactical fighter-bomber of the Soviet Air Force, with which about 500 continue in service. Others have been supplied to Cuba, Czechoslovakia, Egypt, East Germany, Hungary, India, Iraq, Poland, Syria and North Vietnam.

The fuselage and tail unit of the Su-7B are almost identical with those of the delta-wing Su-11. Its wings are swept back at an angle of approximately 60° and each is fitted with two boundary-layer fences, at approximately mid-span and immediately inboard of the tip. The wing-root chord is extended, giving a straight trailing-edge on the inboard section of each wing. Very large area-increasing flaps are fitted, extending over the entire trailing-edge of each wing from the root to the inboard end of the aileron.

The Su-7B carries a variety of external stores, including rocket packs and bombs (usually two 750 kg and two 500 kg) under its wings. It can carry a pair of external fuel tanks under its centre fuselage, but these reduce the max external weapon load to 1,000 kg (2,200 lb). A 30 mm NR-30 cannon, with 70 rounds of ammunition, is installed in each wing-root leading-edge.

The power plant of the standard Su-7B is a Lyulka AL-7F-1 turbojet engine (referred to by the designation TRD31 in the Soviet press), rated at 68·65 kN (15,432 lb st) dry or 98·1 kN (22,046 lb st) with afterburning. Total internal fuel capacity is 3,175 kg (7,000 lb). Capacity of the twin external tanks totals 952 kg (2,100 lb).

Two JATO solid-propellant rockets can be fitted under the rear fuselage of late production Su-7Bs to shorten the aircraft's take-off run.

Early production models had the pitot boom mounted centrally above the air intake, but it is offset to starboard on current versions. Another change was made in the brake-chute installation. Early aircraft had a single ribbon-type parachute, attached under the rear fuselage. Later Su-7Bs have a large fairing, housing twin brake-chutes, at the base of their rudder. The size of the blast panels on the sides of the front fuselage by the wing-root guns was also increased, implying that the cannon now fitted have a higher muzzle velocity or rate of fire.

Among other changes that led to use of the revised designation **Su-7BM** (for *Modifikatsirovanny:* modified) was the introduction of a low-pressure nosewheel tyre, requiring blistered doors to enclose it when retracted.

A variant of the Su-7 seen first at Domodedovo in 1967 is the two-seat **Su-7U**, with the second cockpit in tandem, aft of the standard cockpit and with a slightly raised canopy. A prominent dorsal 'spine' extends from the rear of the aft canopy to the base of the tail fin. The two-seater is a standard operational trainer and has the NATO reporting name *Moujik*.

Sukhoi Su-7B close support fighter of the Soviet Air Force preparing for a night take-off *(Tass)*

Su-7U (NATO *Moujik***), in service with the Egyptian Air Force** *(Flight International)*

Sukhoi Su-7BM single-seat close support fighter, with additional side view (bottom) of two-seat Su-7U *(Pilot Press)*

DIMENSIONS, EXTERNAL:

Wing span	8·93 m (29 ft 3½ in)
Length overall, incl probe	17·37 m (57 ft 0 in)
Height overall	4·57 m (15 ft 0 in)

WEIGHTS:

Weight empty	8,620 kg (19,000 lb)
Normal T-O weight	12,000 kg (26,450 lb)
Max T-O weight	13,500 kg (29,750 lb)

PERFORMANCE:

Max level speed at 11,000 m (36,000 ft):
'clean' Mach 1·6 (917 knots; 1,700 km/h; 1,055 mph)
with external stores
Mach 1·2 (685 knots; 1,270 km/h; 788 mph)

Max level speed at S/L without afterburning
approx 460 knots (850 km/h; 530 mph)

Max rate of climb at S/L
approx 9,120 m (29,900 ft)/min

Service ceiling	15,150 m (49,700 ft)

Combat radius
172-260 nm (320-480 km; 200-300 miles)

Max range	780 nm (1,450 km; 900 miles)

SUKHOI Su-9 and Su-11

NATO reporting names: *Fishpot* **and** *Maiden*

First seen at Tushino during the 1956 Aviation Day Display, the prototype of these single-seat all-weather fighters (allocated the NATO reporting name *Fishpot-A*) had a small conical radome above its engine air intake. This was replaced by a centrebody air intake on the production version, which entered standard service in the Soviet Air Force in two forms as follows:

Su-9 *(Fishpot-B)*. Initial version, operational since

The only photograph yet made available of the tandem two-seat training version of the Su-9 (NATO *Maiden*)

1959 and still in service in considerable numbers. Powered by 88·25 kN (19,840 lb st) Lyulka AL-7F afterburning turbojet. Small-diameter air intake and centrebody. Examples included in the Tushino display of 1961 carried four of the Soviet Air Force's then-standard radar-homing air-to-air missiles (NATO reporting name *Alkali*) on underwing attachments, plus two underfuselage fuel tanks side by side. No fixed armament.

Su-11 *(Fishpot-C)*. First seen publicly at the Domodedovo Aviation Day display in 1967, the Su-11 is a much-improved development of the Su-9, with a Lyulka AL-7F-1 turbojet (98·1 kN; 22,046 lb st with afterburning) and a standard armament of two underwing missiles (NATO *Anab*), one with radar homing head and one with infra-red homing head. It also has a lengthened nose of less-tapered form than that of the Su-9, with an enlarged centrebody, and two slim duct fairings along the top of the centre-fuselage, as on the Su-7B. The fuselage and tail unit of the two types are, in fact, almost identical.

There is also a tandem two-seat training version (NATO reporting name: *Maiden)*, with a cockpit layout similar to that of the two-seat Su-7 *(Moujik)*.

Although the Su-9 and Su-11 are generally similar in layout to their Mikoyan contemporary, the MiG-21, they are larger and heavier aircraft, with a much more powerful afterburning turbojet. They are less limited in all-weather capability than early versions of the MiG-21. The Sukhoi and Mikoyan 'tailed deltas' were, therefore, regarded as complementary rather than competitive when ordered into production in the late 'fifties. In 1976, the Su-9 and Su-11 continued to form 25 per cent of the Soviet defence interceptor force.

The Su-9 and Su-11 can be distinguished from the MiG-21 by their cleaner airframe, and the absence of both a ventral stabilising fin and fairings on the fuselage forward of the wing-root leading-edges. Their pitot boom is mounted above the nose air intake. The cockpit canopy of the standard single-seat versions is rearward-sliding, whereas that of the MiG-21 is hinged to open either forward about the base of the windscreen or sideways.

The tricycle landing gear of the Su-9 and Su-11 has a wide track, with a single wheel on each unit. The main units retract inward into the wings, the nosewheel forward. Control surfaces appear to be conventional, with a one-piece all-moving tailplane, carrying the balance-weight projection at each tip that is found on many Soviet combat aircraft. There are four petal-type airbrakes, in pairs on each side of the rear fuselage.

DIMENSIONS, EXTERNAL (Su-11, estimated):
- Wing span 8·43 m (27 ft 8 in)
- Length overall, incl probe 17·0 m (56 ft 0 in)

WEIGHT (Su-11, estimated):
- Max T-O weight 13,600 kg (30,000 lb)

PERFORMANCE (Su-11, estimated):
- Max level speed at 11,000 m (36,000 ft) Mach 1·8 (1,033 knots; 1,915 km/h; 1,190 mph)
- Service ceiling 17,000 m (55,700 ft)

SUKHOI Su-15

NATO reporting name: *Flagon*

Ten examples of this single-seat twin-jet delta-wing fighter participated in the flying display at Domodedovo in July 1967. First to appear was a single black-painted machine, piloted by Vladimir Ilyushin, son of the famous designer and known to be a test pilot for Sukhoi. When a formation of nine similar aircraft appeared later, the identity of the design bureau responsible for them was confirmed by the obvious 'family likeness' to the Su-9 and Su-11 in the shape of the wings and tail unit.

It seems possible that this aircraft was developed to meet a Soviet Air Force requirement for a Mach 2·5 interceptor to replace the Su-11. It is in service with the Soviet Air Force in several forms:

Flagon-A. Described in detail. Has simple delta wings, identical in form to those of the Su-11.

Flagon-B. This STOL version appeared at Domodedovo in 1967, with three lift-jet engines mounted vertically in the centre-fuselage and wings of compound

Sukhoi Su-9 all-weather fighter, with four underwing mountings for *Alkali* missiles *(Tass)*

Sukhoi Su-11 single-seat fighter, armed with two of the missiles known to NATO as *Anab (Novosti)*

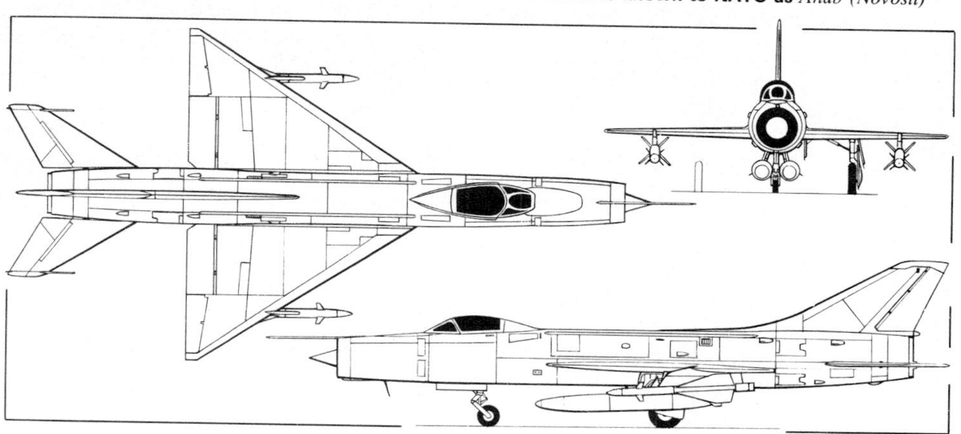

Sukhoi Su-11 single-seat all-weather interceptor *(Pilot Press)*

Sukhoi Su-15 *(Flagon-A)* **single-seat tactical fighter (two turbojet engines with afterburners)**

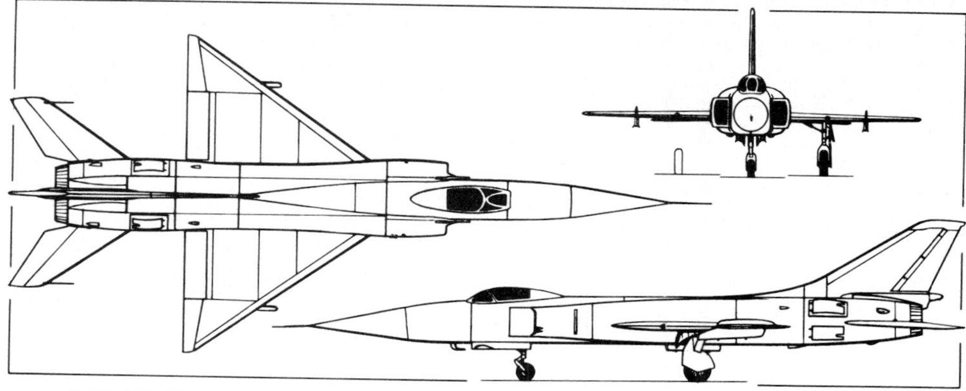

Sukhoi Su-15 single-seat twin-jet all-weather fighter, known to NATO as *Flagon-A (Pilot Press)*

sweep similar to, but different in detail from, those of the *Flagon D/E* combat aircraft. This version is unlikely to be more than an R and D prototype; it was described briefly in the 1970-71 *Jane's*.

Flagon-C. Two-seat training version, with probable combat capability.

Flagon-D. Generally similar to *Flagon-A*, but with wings of compound sweep, produced by reducing the sweepback at the tips without increasing the span.

Flagon-E. Wings similar to those of *Flagon-D*. New and more powerful propulsion system, increasing speed and range. Uprated electronics. Operational since second half of 1973.

A total of about 650 Su-15s were believed to be in service in 1976, all based in the Soviet Union.

TYPE: Single-seat twin-jet all-weather interceptor.

WINGS: Cantilever mid-wing monoplane, basically similar to those of Su-11. Sweepback approx 53°. No dihedral. All-metal structure. Single boundary-layer fence above each wing at approx 70% span. Large area-increasing flap extends from inboard end of aileron to fuselage on each side.

FUSELAGE: Cockpit section is basically circular with large ogival dielectric nosecone. Centre fuselage is faired into rectangular-section air intake ducts. Two door-type airbrakes on each side of rear fuselage, forward of tailplane.

TAIL UNIT: Cantilever all-metal structure, with sweepback on all surfaces. All-moving tailplane, with anhedral, mounted slightly below mid position. Conventional rudder. No trim tabs.

LANDING GEAR: Tricycle type, with single wheel on each unit. Main wheels retract inward into wings and intake ducts; nosewheel retracts forward.

POWER PLANT: Two afterburning turbojets, with variable-area nozzles, mounted side by side in rear fuselage. Ram-type air intakes, with splitter plates; blow-in auxiliary inlets midway between main intake and wing leading-edge in each duct.

ACCOMMODATION: Single seat in enclosed cockpit, with blister canopy.

ARMAMENT: Single pylon for external store under each wing, in line with boundary-layer fence. Normal armament comprises one radar-homing and one infra-red homing air-to-air missile (NATO *Anab*). Side-by-side pylons under centre-fuselage for further weapons or external fuel tanks.

DIMENSIONS, EXTERNAL (estimated):
- Wing span: 9·15 m (30 ft 0 in)
- Length overall: 20·5 m (68 ft 0 in)

WEIGHT (estimated):
- Max T-O weight: 16,000 kg (35,275 lb)

PERFORMANCE (estimated):
- Max level speed above 11,000 m (36,000 ft):
  - with external stores: Mach 2·3
  - 'clean': Mach 2·5
- Combat radius: 390 nm (725 km; 450 miles)

SUKHOI Su-17

NATO reporting name: *Fitter-C*

First of two variable-geometry fighter aircraft demonstrated at Domodedovo in July 1967 was an adaptation of the Su-7, which was allocated the NATO reporting name *Fitter-B*. It was externally identical with the standard operational Su-7 except for the movable outer wing panels and associated fences, outboard of the main landing gear.

This variable-geometry Su-7 was thought at first to be no more than an economically-produced aerodynamic testbed aircraft, built to gain experience with the technique. However, this simple adaptation of a standard fighter offers much improved take-off and landing performance, and it was announced in the USA in 1972 that at least one or two squadrons of 'improved Fitter-Bs' had been identified as operational with the Soviet Air Force. They were allocated the NATO reporting name *Fitter-C*, and the type is now deployed in large numbers, under the Soviet designation Su-17.

The movable part of each wing is about 4·0 m (13 ft) long and is fitted with a full-span leading-edge slat. Its entire trailing-edge is also hinged, forming slotted ailerons and flaps. The large main fence on each side is square-cut at the front and incorporates attachments for external stores. There are two shorter and shallower fences inboard of the main fence on each side, on the sweptback portion of the centre-section trailing-edge which aligns with the trailing-edge of the outer panel when it is fully swept. The standard flap is retained on the inner portion of the centre-section on each side.

The Su-17 has an uprated Lyulka AL-21F-3 turbojet engine (76·5 kN; 17,200 lb st dry: 111·2 kN; 25,000 lb st with afterburning), with a better sfc than the Su-7's AL-7F-1. Variations in rear fuselage contours suggest that the Su-20 'export' version (which see) may retain the AL-7F-1.

Including the wing fence stores attachments, eight weapon pylons can be fitted under the wings and fuselage, for up to 5,000 kg (11,023 lb) of bombs, rocket pods and guided missiles such as the air-to-surface AS-7 (NATO *Kerry*); the 30 mm wing-root guns of the Su-7 are retained. There is a prominent dorsal spine fairing on the fuselage, in place of the Su-7's twin ducts. This, presumably, houses the control runs, some electronics and fuel. Equipment is thought to include SRD-5M (NATO *High Fix*) centre-body radar, ASP-5ND fire control system, Sirena 3 radar homing and warning system, IFF, SOD-57M ATC/SIF, RSIU-5/R-831 VHF/UHF and R5B-70 HF.

DIMENSIONS, EXTERNAL (estimated):
- Wing span:
  - spread (28° sweep): 14·00 m (45 ft 11¼ in)
  - swept (62° sweep): 10·60 m (34 ft 9½ in)
- Length overall, incl probe: 18·75 m (61 ft 6¼ in)
- Fuselage length: 15·40 m (15 ft 6¼ in)
- Height overall: 4·75 m (15 ft 7 in)

AREAS (estimated):
- Wings, gross:
  - spread: 40·1 m² (431·6 sq ft)
  - swept: 37·2 m² (400·4 sq ft)

WEIGHTS (estimated):
- Weight empty: 10,000 kg (22,046 lb)
- Max internal fuel: 3,700 kg (8,157 lb)
- T-O weight, 'clean': 14,000 kg (30,865 lb)
- Max T-O weight: 19,000 kg (41,887 lb)

PERFORMANCE (estimated for 'clean' aircraft, 60% internal fuel, except where indicated):
- Max level speed at height: Mach 2·17
- Max level speed at S/L: Mach 1·05
- Touchdown speed: 143 knots (265 km/h; 165 mph)
- Max rate of climb at S/L: 13,800 m (45,275 ft)/min
- Service ceiling: 18,000 m (59,050 ft)
- T-O run at AUW of 17,000 kg (37,478 lb): 620 m (2,035 ft)

The Sukhoi Su-15 fighter in its experimental STOL version (NATO *Flagon-B*) *(Tass)*

Sukhoi Su-17 single-seat variable-geometry fighter with wings extended

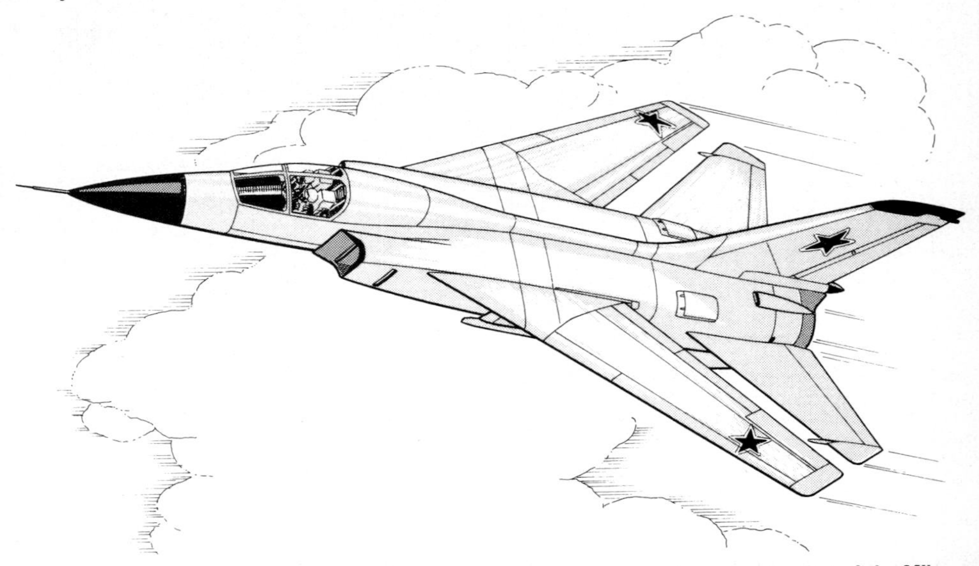

Artist's impression of Sukhoi Su-19. The nose of the basic version may be more like that of the Mikoyan *Flogger-D* **(see page 422)** *(Michael A. Badrocke)*

T-O to 15 m (50 ft) at AUW of 17,000 kg (37,478 lb) 835 m (2,740 ft)
Landing run 600 m (1,970 ft)
Combat radius with 2,000 kg (4,409 lb) external stores:
hi-lo-hi 340 nm (630 km; 391 miles)
lo-lo-lo 195 nm (360 km; 224 miles)

SUKHOI Su-19

NATO reporting name: *Fencer*

This variable-geometry attack aircraft was identified as a major new operational type by Admiral Thomas H. Moorer, Chairman of the US Joint Chiefs of Staff, in early 1974, when he described it as "the first modern Soviet fighter to be developed specifically as a fighter-bomber for the ground attack mission". Designated Su-19 in the Soviet Union, it is in the same class as the USAF's F-111.

No photographs or officially-released details of the Su-19 had been made available by mid-1975. However, it is believed that the accompanying three-view drawing reflects the aircraft's major characteristics.

Seating for the crew of two is shown to be side by side in a slim and clean fuselage typical of Sukhoi designs. The wings are pivoted much further inboard than on the Su-17/20 or Tupolev *Backfire*. Each is shown with flying control surfaces and high-lift devices similar in principle to those of the F-111, comprising full-span leading-edge and trailing-edge flaps, with airbrake/lift dumpers forward of the latter operating also as spoilers for lateral control at low speeds. This implies that the all-moving horizontal tail surfaces operate both differentially and symmetrically to provide aileron and elevator functions.

Wing leading-edge sweep appears to be approximately 23° in the fully-spread position, and 70° fully swept, the latter angle being similar to that on the centre-section glove. The wings are shown without dihedral or anhedral.

Except for the two-seat cockpit, the overall lines of the fuselage, air intake trunks and vertical tail surfaces have much in common with those of the Su-15. It is further suggested that the Su-19 may be powered by two Lyulka AL-21F turbojets of the kind fitted to the single-engined Su-17.

The Su-19 was entering squadron service in early 1975, with some examples based in East Germany, presumably for operational evaluation. Armament includes more than 4,535 kg (10,000 lb) of guided and unguided air-to-surface weapons on six pylons under the fuselage and wing-root gloves, in addition to a GSh-23 twin-barrel 23 mm gun.

DIMENSIONS, EXTERNAL (estimated):
Wing span:
spread 17·15 m (56 ft 3 in)
swept 9·53 m (31 ft 3 in)
Length overall 21·29 m (69 ft 10 in)
WEIGHT (estimated):
Max T-O weight 30,850 kg (68,000 lb)
PERFORMANCE (estimated):
Combat radius, lo-lo-lo over 174 nm (322 km; 200 miles)

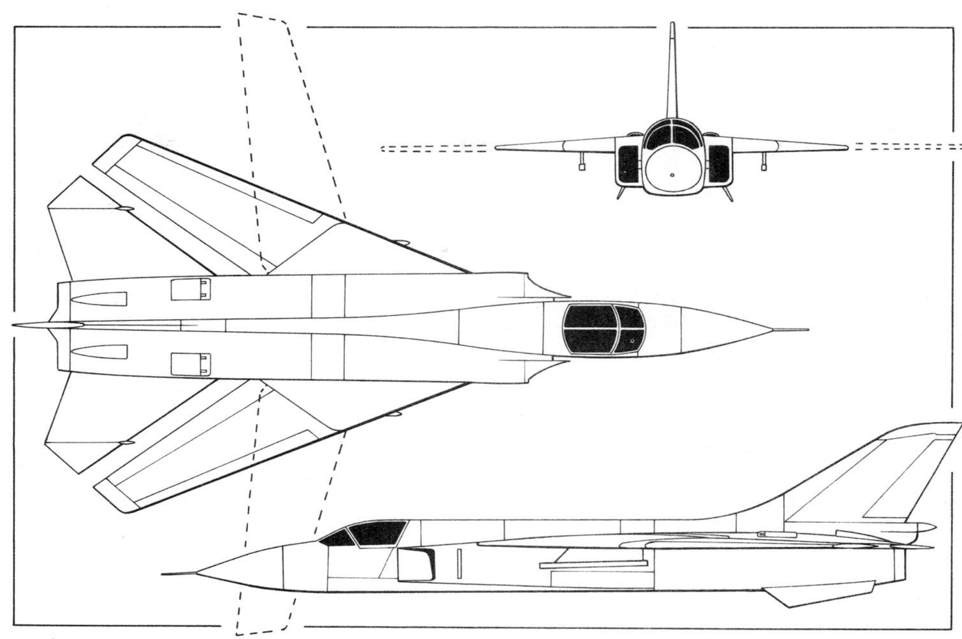
Sukhoi Su-19 two-seat twin-engined variable-geometry fighter-bomber *(Roy J. Grainge, provisional)*

Sukhoi Su-20 variable-geometry ground attack fighter in service with the Polish Air Force

SUKHOI Su-20

NATO reporting name: *Fitter-C*

The operational Su-20 is an 'export' version of the Su-17, to which it is generally similar except, perhaps, for having a lower-rated engine and different equipment. Operators include the Polish Air Force, whose aircraft fly normally with two very large jettisonable fuel tanks on the wing fence attachments, instead of the twin centreline tanks usually carried by the Su-7 and Su-9/11.

It was reported in the second half of 1976 that the Peruvian Air Force is to have 36 Su-20s (referred to in some news items as Su-22s).

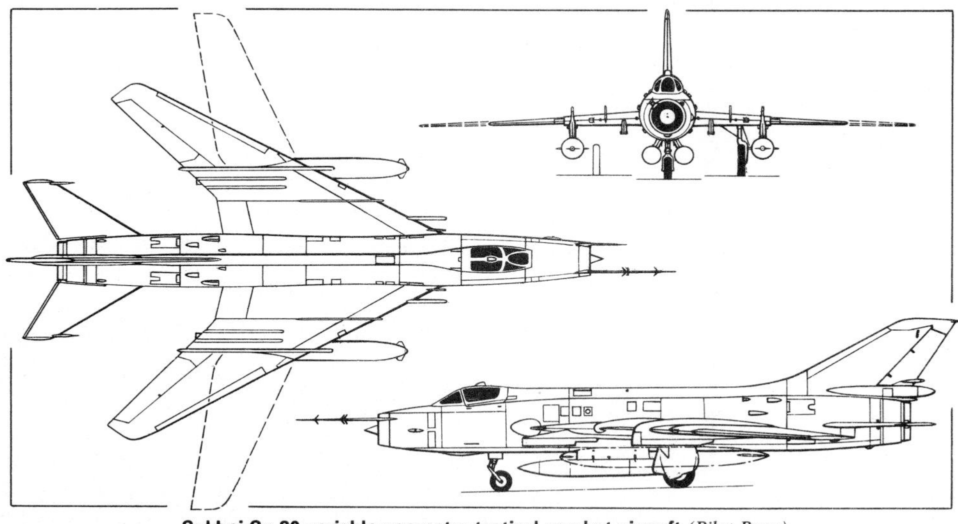
Sukhoi Su-20 variable-geometry tactical combat aircraft *(Pilot Press)*

TUPOLEV

CHIEF DESIGNER (Tu-144):
Dr Alexei A. Tupolev
CHIEF DESIGNER (Tu-154):
Dmitry Markov
DEPUTY CHIEF OF BUREAU:
Andrei Kandolov

Andrei Tupolev, born in 1888, was a leading figure in the Central Aero-Hydrodynamic Institute (TsAGI) in Moscow from the time when it was founded, in 1929, until his death on 23 December 1972. He was for long the Soviet Union's outstanding designer, and the recent products of his design team range from turbofan civil transports to the first Soviet supersonic bomber to enter service and the first supersonic transport aircraft. Also in production in the Soviet Union are small amphibious aerosleighs of Tupolev design, powered by aircraft piston engines and capable of travelling over both water and snow.

Current chief designers of the Tupolev bureau include Dr Alexei A. Tupolev (son of Andrei Tupolev), who is responsible for the Tu-144 supersonic transport; and Dmitry Markov, who is responsible for the Tu-154 airliner.

TUPOLEV Tu-16

NATO reporting name: *Badger*

This Tupolev bomber, from which the Tu-104 airliner was derived, made its first public appearance in some numbers in 1954. In July 1955 a formation of 54 flew over Moscow on Aviation Day, and the Tu-16 has since been standard equipment in the Soviet Air Force and Naval Air Force. About 2,000 are believed to have been built, of which 500 still fly with medium-range squadrons of the Soviet strategic bomber force, and 400 with the Soviet Naval Air Force.

In addition, about 60 Tu-16s were built in China, where production began in 1968.

Seven versions of the Tu-16 are identified by unclassified NATO reporting names, as described below. All except *Badger-B* (see 1975-76 *Jane's*) remain in first-line service.

Badger-A. First Soviet long-range strategic jet bomber. Crew of seven. Glazed nose, with small undernose radome fairing. Defensive armament of seven 23 mm cannon. Nine supplied to Iraq.

Badger-C. Missile-carrier first seen at 1961 Soviet Aviation Day display. Large stand-off bomb (NATO reporting name *Kipper*), similar in configuration to USAF Hound Dog, carried under fuselage and stated to be for anti-shipping use. Radar in wide nose radome, displacing normal glazing and nose gun.

Badger-D. Maritime/electronic reconnaissance version. Nose similar to that of *Badger-C*, with slightly enlarged undernose radome fairing, and three more blister fairings in tandem under centre-fuselage.

Badger-E. Similar to *Badger-A* but with cameras in bomb bay.

Badger-F. Basically similar to *Badger-E* but with electronic intelligence pod on a pylon under each wing.

Badger-G. Similar to *Badger-A* but with large underwing pylons for rocket-powered missiles (NATO reporting name *Kelt*). Up to 275 believed to have been supplied to Soviet Naval Air Force for anti-shipping strike duties; and about 25 were in first-line service with the Egyptian Air Force in mid-1976. Aircraft of this type launched about 25 *Kelts* against Israeli targets during the October 1973 war. The Israelis claimed that only five penetrated the defences to hit two radar sites and a supply dump in Sinai.

Maritime reconnaissance versions of *Badger* make regular flights over units of the US Navy and other NATO

Badger-G **version of the Tupolev Tu-16 twin-jet bomber, in Egyptian Air Force insignia, with two** *Kelt* **missiles on underwing launchers**

naval forces at sea in the Atlantic, Pacific and elsewhere, and have been photographed while doing so. The aircraft often operate in pairs, with one *Badger-F* accompanied by a different version. They also make electronic intelligence (elint) sorties around the coastlines of NATO and other non-Communist countries.

TYPE: Twin-jet medium bomber and maritime reconnaissance/attack aircraft.

WINGS: Cantilever high mid-wing monoplane with slight anhedral and with 37° of sweep. Thickness/chord ratio 12½%.

FUSELAGE: All-metal semi-monocoque structure of circular cross-section.

TAIL UNIT: Cantilever all-metal structure, with sweepback on all surfaces. Trim tabs in rudder and each elevator.

LANDING GEAR: Retractable tricycle type. Twin-wheel nose unit retracts rearward. Main four-wheel bogies retract into housings projecting beyond the wing trailing-edge.

POWER PLANT: Two Mikulin AM-3M turbojet engines, each rated at about 93·2 kN (20,950 lb st) at sea level. Fuel in wing and fuselage tanks, with total capacity of approx 45,450 litres (10,000 Imp gallons). Provision for underwing auxiliary fuel tanks and for flight refuelling. Tu-16 tankers trail hose from starboard wingtip; receiving equipment is in port wingtip extension.

ACCOMMODATION: Crew of about seven, with two pilots side by side on flight deck. Navigator in glazed nose of *Badger-A, E* and *F.* Manned tail position plus lateral observation blisters in rear fuselage under tailplane.

ARMAMENT: Forward dorsal and rear ventral barbettes each containing two 23 mm cannon. Two further cannon in tail position controlled by an automatic gun-ranging radar set. Seventh, fixed, cannon on starboard side of nose of versions without nose radome. Bomb load of up to 9,000 kg (19,800 lb) delivered from weapons bay about 6·5 m (21 ft) long in standard bomber. Naval versions can carry air-to-surface winged stand-off missiles.

ELECTRONICS AND EQUIPMENT: Radio and radar aids probably include HF and VHF R/T equipment, as well as IFF and a radio compass and radio altimeter. Other equipment differs according to role.

DIMENSIONS, EXTERNAL:

Wing span	33·5 m (110 ft 0 in)
Length overall	36·5 m (120 ft 0 in)
Height overall	10·8 m (35 ft 6 in)

AREA:

Wings, gross	approx 169 m² (1,820 sq ft)

WEIGHT:

Normal T-O weight	approx 68,000 kg (150,000 lb)

PERFORMANCE (estimated, at max T-O weight):

Max level speed at 10,700 m (35,000 ft)
510 knots (945 km/h; 587 mph)

Service ceiling 13,000 m (42,650 ft)

Range with max bomb load
2,605 nm (4,800 km; 3,000 miles)

Range at 417 knots (770 km/h; 480 mph) with 3,000 kg (6,600 lb) of bombs
3,450 nm (6,400 km; 3,975 miles)

TUPOLEV Tu-95

NATO reporting name: *Bear*

This huge Tupolev bomber flew for the first time in the late Summer of 1954, was first seen at Tushino in July 1955, and subsequently became standard equipment in the Soviet Air Force. It is often referred to as the Tu-20, but its correct Soviet designation is understood to be Tu-95.

As well as maintaining its important strategic attack role, as the Soviet counterpart of the USAF's B-52 Stratofortress, the Tu-95 is in major service with the Soviet Naval Air Force for maritime reconnaissance and to provide targeting data to the launch control and guidance stations responsible for both air-to-surface and surface-to-surface anti-shipping missiles.

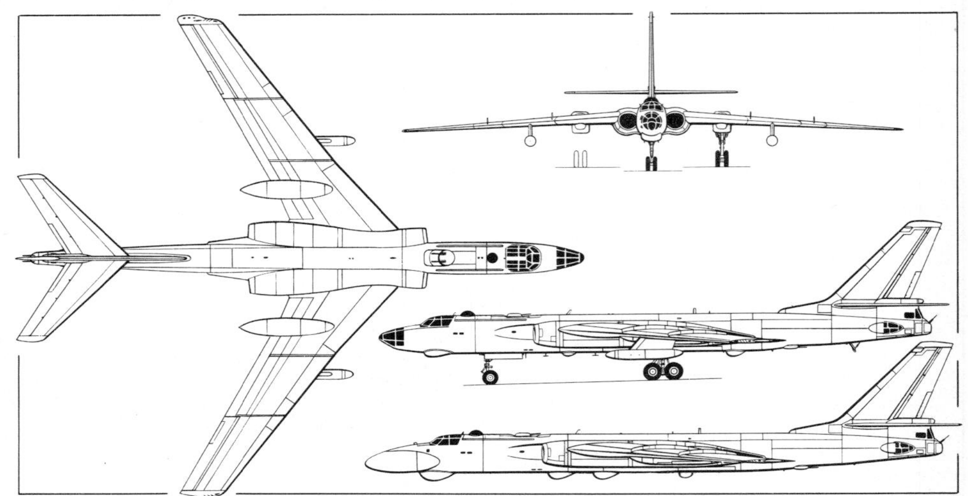

Tupolev Tu-16, in the form known to NATO as *Badger-F,* **with additional side view (bottom) of** *Badger-D (Pilot Press)*

Tupolev Tu-16 reconnaissance bomber *(Badger-F),* **being investigated by a Phantom II interceptor of the US Navy**

Tupolev Tu-95 maritime reconnaissance bomber in the form known to NATO as *Bear-C (US Navy)*

Camera ports under the bomb bay are a recognition feature of the *Bear-E* **version of the Tu-95** *(Royal Air Force)*

Six versions have been identified by NATO reporting names, and all remain operational:

Bear-A. Basic strategic bomber, with chin radar, and defensive armament comprising three pairs of 23 mm cannon in remotely-controlled dorsal and ventral barbettes and manned tail gun turret. Two glazed blisters on rear fuselage, under tailplane, are used for sighting by the gunner controlling all these weapons. The dorsal and ventral barbettes can also be controlled from a station aft of the flight deck. A braking parachute may be used to reduce landing run. Total of about 110 *Bear-As* and *Bear-Bs* remain operational with the Soviet bomber force.

Bear-B. First seen in 1961 Aviation Day flypast, with additional radar equipment in wide undernose radome, replacing the original glazing, and carrying a large air-to-surface missile (NATO reporting name *Kangaroo*) with estimated range of 350 nm (650 km; 400 miles). Now used mainly for maritime patrol, with flight refuelling nose-probe and, sometimes, a streamlined blister fairing on the starboard side of the rear fuselage. Defensive armament retained.

Bear-C. First identified when it appeared in vicinity of NATO naval forces during Exercise Teamwork in September 1964. Generally similar to *Bear-B* but with streamlined blister fairing on *both* sides of rear fuselage. Refuelling probe standard.

Bear-D. This version was first photographed extensively when several examples (together with Tu-16s) made low passes over the US Coast Guard icebreakers *Edisto* and *Eastwind* off Severnaya Zemlya, in the Soviet Arctic, in August 1967. These aircraft differed in detail, but each had a glazed nose, an undernose radar scanner, a large underbelly radome for X-band radar, a blister fairing on each side of the rear fuselage like *Bear-C,* a nose refuelling probe, and a variety of other blisters and antennae, including a streamlined fairing on each tailplane tip. The rearward-facing radar above the tail turret is much larger than on previous versions of the Tu-95. It is now known that *Bear-D* has an extremely important function in support of operations involving surface-to-surface and air-to-surface missiles. It provides data on the location and nature of potential targets to missile launch crews on board ships and aircraft which are themselves too distant from the target to ensure precise missile aiming and guidance. About 50 serve with Soviet Naval air fleet.

Bear-E. Version basically similar in configuration to *Bear-A* but with a refuelling probe above its glazed nose and the rear fuselage blister fairings of *Bear-C*. Six bomb bay windows, in pairs in line with the wing flaps, indicate the presence of reconnaissance cameras, sometimes with a seventh window to the rear on the starboard side.

Bear-F. First identified in 1973, this version has enlarged and lengthened fairings aft of its inboard engine nacelles, for purely aerodynamic reasons. The undernose radar of *Bear-D* is missing on some aircraft; others have a radome in this position, but of considerably modified form. On both models the main underfuselage X-band radar housing is considerably further forward than on *Bear-D* and smaller in size; the forward portion of the fuselage is longer; there are no large blister fairings under and on the sides of the rear fuselage; and the nosewheel doors are bulged prominently, suggesting the use of larger or low-pressure tyres. *Bear-F* has two stores bays in its rear fuselage, one of them replacing the usual rear ventral gun turret and leaving the tail turret as the sole defensive gun position.

Examples of all versions of the Tu-95 have made reconnaissance flights over units of the US Fleet at sea and have been photographed by US naval fighters whilst doing so. They are also encountered frequently over the North Sea by the RAF and Royal Navy.

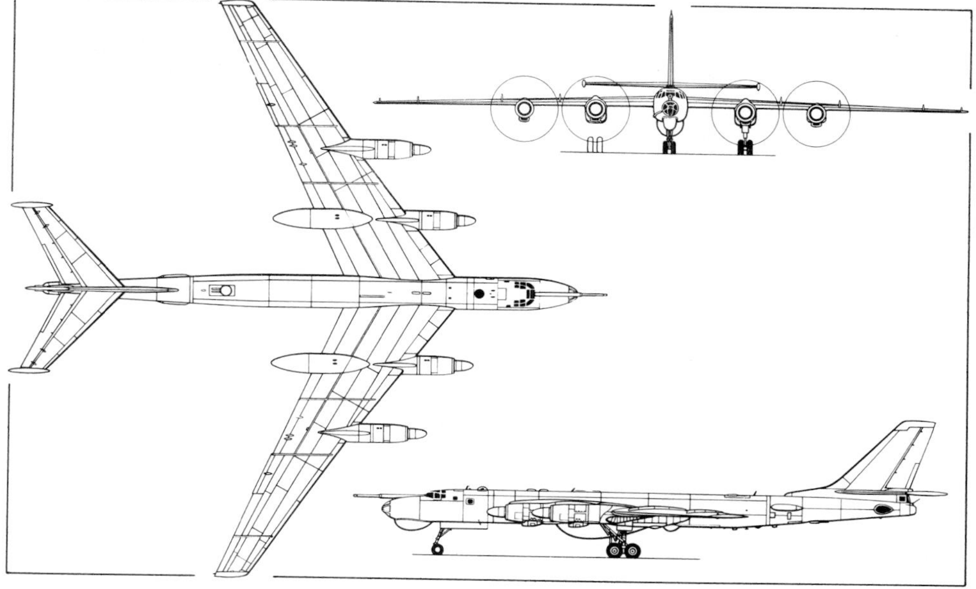

The version of the Tupolev Tu-95 known to NATO as *Bear-D (Pilot Press)*

The version of *Bear-F* **with an undernose radome**

The brief performance details quoted were given by the US Secretary of Defense, Mr Robert McNamara, in 1963.

Type: Four-turboprop long-range bomber and maritime reconnaissance aircraft.

Wings: Cantilever mid-wing monoplane. Sweepback 37° at quarter-chord on inner panels, 35° at quarter-chord on outer panels. All-metal three-spar structure. All-metal hydraulically-powered ailerons and Fowler flaps. Trim tabs in ailerons. Spoilers in top surface of wing forward of inboard end of ailerons. Three boundary layer fences on top surface of each wing. Thermal anti-icing system in leading-edges.

Fuselage: All-metal semi-monocoque structure of circular section, containing three pressurised compartments. Those forward and aft of the weapons bay are linked by a crawlway tunnel. The tail gunner's compartment is not accessible from the other compartments.

Tail Unit: Cantilever all-metal structure, with sweepback

The Tu-126 airborne warning and control system (AWACS) aircraft, known to NATO as *Moss*

on all surfaces. Adjustable tailplane incidence. Hydraulically-powered rudder and elevators. Trim tabs in rudder and each elevator.

Landing Gear: Retractable tricycle type. Main units consist of four-wheel bogies, with tyres approx 1·50 m (5 ft) diameter and hydraulic internal expanding brakes. Twin wheels on nose unit. All units retract rearward, main units into nacelles built on to wing trailing-edge. Retractable tail bumper consisting of two small wheels.

Power Plant: Four Kuznetsov NK-12MV turboprop engines, each originally with max rating of approx 8,948 kW (12,000 ehp) but now uprated to 11,033 kW (14,795 ehp) and driving eight-blade contra-rotating reversible-pitch Type AV-60N propellers. Fuel in wing tanks, with normal capacity of 72,980 litres (16,540 Imp gallons).

Accommodation and Armament: See notes applicable to individual versions and under 'Fuselage'.

Operational Equipment (*Bear-D*): Large X-band radar in blister fairing under centre fuselage, for reconnaissance and to provide data on potential targets for anti-shipping aircraft or surface vessels. In latter mode, PPI presentation is data-linked to missile launch station. Four-PRF range J-band circular and sector scan bombing and navigation radar (NATO *Short Horn*). I-band tail warning radar (NATO *Bee Hind*) in housing at base of rudder.

Dimensions, external (approx):

Wing span	48·50 m (159 ft 0 in)
Length overall	47·50 m (155 ft 10 in)
Height overall	12·12 m (39 ft 9 in)

Weight (estimated):

Max T-O weight	154,220 kg (340,000 lb)

Performance (*Bear-A*):

Over-target speed at 12,500 m (41,000 ft)
435 knots (805 km/h; 500 mph)

Max range with 11,340 kg (25,000 lb) bomb load
6,775 nm (12,550 km; 7,800 miles)

TUPOLEV Tu-126

NATO reporting name: *Moss*

An officially-released Soviet documentary film, shown in the West in 1968, included sequences depicting a military version of the Tu-114 four-turboprop transport (see 1972-73 *Jane's*), carrying above its fuselage a rotating 'saucer' type early warning radar with a diameter of about 11 m (36 ft). This was a logical development, as the Tu-114 had a fuselage of larger diameter than the military Tu-95, and could accommodate more easily the extensive electronic equipment and large crew required by a long-endurance early-warning and fighter control aircraft. The new aircraft, designated Tu-126 in the Soviet Union, also has wings similar to those of the Tu-114, with extended-chord trailing-edge flaps, rather than the 'straight' trailing-edge of the Tu-95. Its power plant comprises four 11,033 kW (14,795 ehp) Kuznetsov NK-12MV turboprop engines.

The general appearance of the Tu-126, which has the NATO reporting name *Moss,* is shown in the accompanying illustrations. It can be seen to have a flight refuelling nose-probe, ventral tail-fin and numerous additional antennae and blisters for electronic equipment.

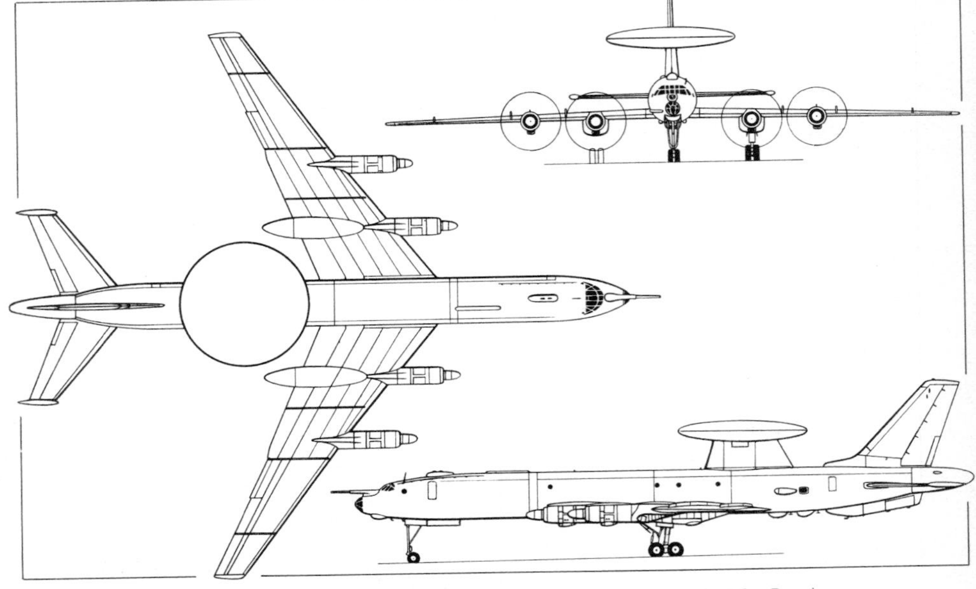

Tupolev Tu-126 (four Kuznetsov NK-12MV turboprops) *(Pilot Press)*

In the AWACS (airborne warning and control system) role, the Tu-126 is intended to work in conjunction with advanced interceptors. After locating incoming low-level strike aircraft, the Tu-126 would ideally direct towards them fighters armed with 'snap-down' air-to-air missiles able to be fired from a cruising height of 6,100 m (20,000 ft) or higher. It has a further, obvious application in assisting strike aircraft to elude enemy interceptors picked up by its radar.

At least ten or twelve Tu-126s were operational with the Soviet air defence forces in 1976. They are said, by US defence experts, to have only limited effectiveness over water and to be ineffective over land at the present stage of development.

Dimensions, external:

Wing span	51·10 m (167 ft 8 in)
Wing aspect ratio	10·4
Length	57·30 m (188 ft 0 in)
Wheel track	13·70 m (44 ft 11½ in)
Propeller diameter	5·60 m (18 ft 4½ in)

Area:

Wings, gross	311·1 m² (3,349 sq ft)

TUPOLEV Tu-22

NATO reporting name: *Blinder*

First shown publicly in the 1961 Aviation Day flypast over Moscow, this twin-turbojet bomber and maritime patrol aircraft was the first operational Soviet supersonic bomber, with an estimated maximum speed of Mach 1·4 at height.

Its wings have some 45° of sweepback on the outer panels, 50° on the inner panels and an acute sweep at the extreme root. They are low-set on an area-ruled fuselage, which has a nose radome and accommodates a crew of three in tandem. The pilot has an upward-ejection seat; the other crew members have downward-ejection seats. There is a row of windows in the bottom of the fuselage aft of the nose radome.

The slab tailplane is also low-set, and the large turbojet engines (reported to be rated at 115·7 kN; 26,000 lb st each, with afterburning) are mounted in pods above the rear fuselage, on each side of the vertical fin. The lip of each pod is in the form of a ring which can be extended forward by jacks for take-off. Air entering the ram intake is then supplemented by air injected through the annular slot between the ring and the main body of the pod.

The original nozzles had a short fluted final section aft of a short fixed section, with an annular space between this and the outer fairing; they have been superseded by new nozzles, with a longer-chord convergent-divergent nozzle inside the outer fairing. These are believed to offer increased thrust and range.

The wide-track four-wheel bogie main landing gear units retract into fairings built on to the wing trailing edges. As well as embodying oleo-pneumatic shock absorbers, the legs are designed to swing rearward for

Latest photograph of a Tupolev Tu-22 issued by *Tass* shows the outline of a *Kitchen* missile on the folding bomb-bay doors, and damage assessment cameras in the rear of the landing gear fairings

additional cushioning during taxying and landing on rough runways. The twin-wheel nose unit retracts rearward.

Of the ten Tu-22s shown in 1961, only one carried visible weapons, in the form of an air-to-surface missile (NATO reporting name *Kitchen*), some 11 m (36 ft) long, semi-submerged in the underside of its fuselage. This aircraft had also a wider nose radome, and a tail radome above a radar-directed turret mounting a single gun.

A total of 22 Tu-22s took part in the 1967 display at Domodedovo. One was escorted by six MiG-21PFs, permitting a more accurate calculation of its overall dimensions than had previously been possible. Most carried *Kitchen* missiles; all had a partially-retractable nose refuelling probe and the wide radome seen on the single missile-armed aircraft in 1961.

There are at least four versions of the Tu-22, as follows:

Blinder-A. Basic reconnaissance bomber, with fuselage weapon bay for free-fall bombs. *Blinder-A* entered only limited service, its max range of 1,215 nm (2,250 km; 1,400 miles) being inadequate for the originally intended strategic role.

Blinder-B. Generally similar to *Blinder-A* but equipped to carry air-to-surface stand-off missile (NATO reporting name *Kitchen)* with estimated range of 400 nm (740 km; 460 miles) recessed in weapon bay. Larger radar in nose. Partially-retractable flight refuelling probe on nose. Total of about 200 *Blinder-As* and *Blinder-Bs* believed to have been delivered, including about 12 serving with the Libyan Air Force.

Blinder-C. Maritime reconnaissance version, with battery of six cameras in weapon bay and camera windows in weapon bay doors. Modifications to nosecone, dielectric panels, etc, suggest possible electronic intelligence role or equipment for electronic countermeasures (ECM) duties. About 60 delivered, for operation primarily over sea approaches to the Soviet Union, from bases in the Southern Ukraine and Estonia.

Blinder-D. This is a tandem-cockpit training version, in which the rear pilot sits in a raised position, with a stepped-up canopy.

A missile-armed long-range interceptor version of *Blinder* has been reported in service, as a possible replacement for the Tu-28P.

Dimensions, external (estimated):

Wing span	27·70 m (90 ft 10½ in)
Length overall	40·53 m (132 ft 11½ in)
Height overall	5·18 m (17 ft 0 in)

Weight (estimated):

Max T-O weight	83,900 kg (185,000 lb)

Performance (estimated):

Max level speed at 12,200 m (40,000 ft)
Mach 1·4 (800 knots; 1,480 km/h; 920 mph)

Service ceiling	18,300 m (60,000 ft)
Max range	1,215 nm (2,250 km; 1,400 miles)

The tandem-cockpit training version of the Tu-22, known to NATO as *Blinder-D*

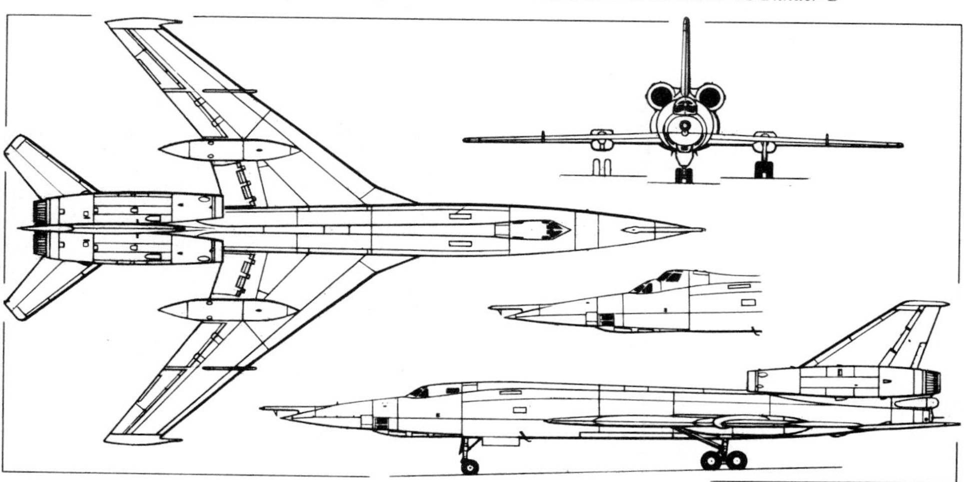

Tupolev Tu-22 twin-jet supersonic bomber *(Blinder-A)* **with additional view of nose of** *Blinder-D* **training version** *(Pilot Press)*

TUPOLEV V-G BOMBER

NATO reporting name: *Backfire*

Official US sources first acknowledged the existence of a Soviet variable-geometry ('swing-wing') medium bomber in the Autumn of 1969. Such an aircraft was not unexpected, as the Tu-22 (NATO *Blinder*) was clearly incapable of fulfilling a long-range strategic bombing role in the 'seventies.

A prototype of the new bomber is said to have been observed in July 1970, on the ground near the Tupolev works at Kazan in Central Asia. Subsequent official statements confirmed the aircraft as a twin-engined design by the Tupolev bureau. At least two prototypes were built initially; up to twelve pre-production models followed, for development testing, weapons trials and evaluation, by the beginning of 1973. The NATO reporting name allocated to the aircraft is *Backfire*.

When drawing up the basic parameters for the bomber, the Tupolev bureau is believed to have aimed at a design over-target speed in the range of Mach 2·25 to Mach 2·5, with a maximum unrefuelled range of 4,775-5,200 nm (8,850-9,650 km; 5,500-6,000 miles) at high altitude, and a low-level penetration capability at supersonic speed.

There is reason to believe that *Backfire* fell short of such an unrefuelled range in its original form, but that, following some redesign, the latest version has a non-refuelled maximum combat radius of about 3,100 nm (5,745 km; 3,570 miles).

Admiral Thomas H. Moorer, former Chairman of the US Joint Chiefs of Staff, has said that *Backfire* "weighs two and one-half times as much as an FB-111 and is about four-fifths as large as the B-1".

Unwillingness to depart initially from the Tupolev practice of retracting the main landing gear bogies into fairings on the wing trailing-edges limited the variable geometry to the outer wings, as on the Sukhoi Su-17 and Su-20. There is evidence to believe that the large size of these fairings, with the wheels stowed beneath the wing, caused excessive drag. Redesign almost eliminated the fairings from subsequent aircraft, requiring a revised main landing gear, retracting inside the wing. This accounts for the two versions of the bomber:

Backfire-A. Initial version, with large landing gear fairing pods on the wing trailing-edges.

Backfire-B. Developed version, with landing gear fairing pods eliminated except for shallow fairings under the wings, no longer protruding beyond the trailing-edge. Increased wing span.

The leading-edge of each outer wing may be fitted with a slat, as on the Su-20. There is no dihedral on the upper surface of the wing, but the section is so thin that considerable flexing of the outer panels takes place in flight.

In contrast with the fin-side engine installation on the Tu-22, the engines of *Backfire* are housed inside large square-section trunks, built on to each side of the fuselage. There is no reason to expect external area-rule 'waisting' of these trunks, but the intakes are fitted with splitter plates and must embody complex internal variable geometry.

It is not yet possible to identify positively the type of engine fitted to *Backfire*, but US sources have suggested the use of two Kuznetsov turbofans similar to those installed in Tupolev's Tu-144 supersonic transport. This would be logical, as each engine is rated at 196·1 kN (44,090 lb st) with reheat in the Tu-144. Uprated for military use, such engines would give an increase of at least 70 per cent over the installed power in the Tu-22.

A flight refuelling nose-probe is fitted and, after one observed refuelling, a *Backfire* prototype is said to have remained airborne for a further 10 hours.

Backfire can be expected to carry the full range of Soviet free-fall weapons and an air-to-surface stand-off missile at least as advanced and formidable as the *Kitchen* carried semi-submerged in the belly of the Tu-22. A *Kitchen* could be carried under the fixed glove portion of each wing of *Backfire*. Alternatively, one of the new AS-6 missiles could be carried on each of these mountings. Little is known yet about the AS-6; but it is said to be a rocket missile with a range of 135 nm (250 km; 155 miles) at low altitude or 378-434 nm (700-805 km; 435-500 miles) at high altitude. US reports have suggested that the Soviet Union is also developing small nuclear weapons like the American SRAM (short-range attack missile) and decoy missiles to assist penetration of advanced defence systems.

Loaded weight of *Backfire* has been reported as 123,350 kg (272,000 lb).

Two units of approximate squadron size are believed to have been formed by early 1975, one equipped with pre-series *Backfire-As* and the other with initial operational *Backfire-Bs*. Soviet Naval Aviation also began to operate *Backfires* in a maritime reconnaissance role during 1975. By early 1976, more than 50 operational *Backfire-Bs* had been delivered to the Soviet Air Force and Navy, with production continuing. In his FY 1977 US Defense Department Report, Secretary Rumsfeld commented: "Even without aerial refuelling or staging from bases in the Arctic, *Backfire* bombers could cover virtually all of the US on one-way missions, with recovery in third countries. Using Arctic staging and refuelling, they could achieve a similar target coverage and still return to their staging bases in the Soviet Union". The RAF's Chief of Staff said in December 1975: "Russian fast, wide-ranging and high-performance aircraft like *Backfire*, armed with stand-off missiles, may soon become an even greater danger to allied shipping than the relatively slow-moving Russian submarines".

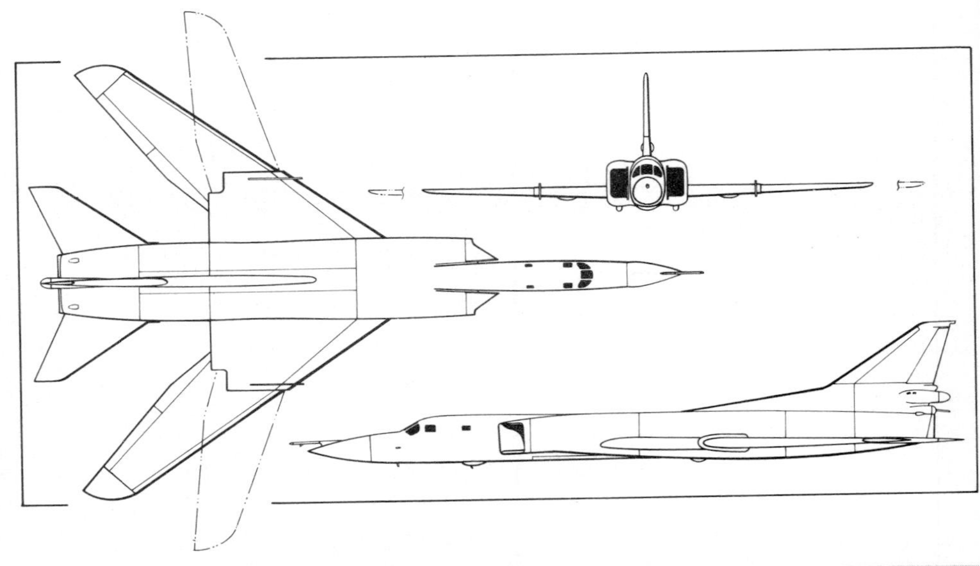

Tupolev variable-geometry supersonic bomber known to NATO as *Backfire-B (Roy J. Grainge)*

TUPOLEV Tu-28P

NATO reporting name: *Fiddler*

This supersonic twin-jet interceptor was seen for the first time at Tushino in July 1961, with a large delta-wing air-to-air missile (NATO *Ash*) mounted under each wing. It is thought to have the service designation Tu-28P. Its NATO reporting name is *Fiddler*.

The Tu-28P has a large ogival nose radome and carries a crew of two in tandem. The shoulder intakes for its two afterburning turbojet engines have half-cone shock-bodies, and the jet-pipes are side by side in the bulged tail. Each engine is estimated to have a max rating of about 120·1 kN (27,000 lb st).

The sharply-swept wings are mid-set, with slight anhedral, and have considerably increased chord on the inboard panels, which have both increased sweep and a straight trailing-edge. The wide-track main landing gear units, comprising four-wheel bogies, retract into large fairings built on to the wing trailing-edges.

The tail unit is also sharply swept, and the two aircraft seen in 1961 were each fitted with two ventral fins. These were missing on the three Tu-28Ps which flew past at Domodedovo in July 1967, as was the large bulged fairing fitted under the fuselage in 1961.

The armament has been doubled since 1961, each aircraft now being equipped to carry two *Ash* missiles under each wing, one usually of the radar homing type and the other of the infra-red homing type. This was confirmed as the standard armament of current first-line service aircraft in a film released in 1969, showing units of the Soviet armed forces taking part in defence exercises.

It has been suggested that the Tu-28P is being replaced in service by an interceptor version of the Tu-22.

Tu-28P all-weather interceptor with *Ash* **infra-red missile on inboard pylon under port wing**

DIMENSIONS, EXTERNAL (estimated):
- Wing span: 20·00 m (65 ft 0 in)
- Length overall: 26·00 m (85 ft 0 in)

WEIGHT (estimated):
- Max T-O weight: 45,000 kg (100,000 lb)

PERFORMANCE (estimated):
- Max speed at 11,000 m (36,000 ft): Mach 1·75 (1,000 knots; 1,850 km/h; 1,150 mph)
- Service ceiling: 20,000 m (65,620 ft)
- Range with max fuel: 2,692 nm (4,989 km; 3,100 miles)

TUPOLEV Tu-134

NATO reporting name: *Crusty*

Known originally as the Tu-124A, this aircraft is a rear-engined twin-turbofan development of the Tu-124 (described in earlier editions of *Jane's*). It had completed more than 100 test flights when first details and photographs were released in mid-September 1964. The prototype was followed by five pre-production aircraft and the Tu-134 then went into series production in the factory at Kharkov where the Tu-104 had been manufactured. It entered international service on Aeroflot's Moscow-Stockholm route in September 1967, after a period on internal services, and was joined by the 'stretched' Tu-134A in the Autumn of 1970.

The Tu-134 was developed by Tupolev's design team, under the direct leadership of chief designer Leonid Selyakov. Deputy designer Alexander Arkhangelsky has said that the aircraft can be operated from earth runways. He added that it is equipped for fully-automatic landing and has navigation aids that enable the pilot to land in fog with horizontal visibility down to 50 m (165 ft).

The Tu-134 is designed for a service life of 30,000 flying hours. Airframe overhaul life is 5,000 hours.

Two versions are in service, as follows:

Tu-134. Initial version, with Soloviev D-30 turbofans, accommodating 64-72 passengers. Export orders included eleven for Interflug (East Germany), six for Balkan Bulgarian Airlines, five for LOT (Poland), six for Malev (Hungary), three for Aviogenex (Yugoslavia) and one for Iraqi Airways.

Tu-134A. Fuselage lengthened by 2·10 m (6 ft 10½ in) to accommodate 76-80 passengers and increase baggage space by 2·5 m³ (89 cu ft). Wider seats. Wings strengthened locally. Main landing gear units strengthened and fitted with Il-18 wheels and brakes. Thrust reversers on Soloviev D-30-2 engines. New radio and navigation equipment to international standards. APU for self-contained engine starting, electrical power supply and air-conditioning on the ground. Export orders included eleven for CSA (Czechoslovakia), three for Malev, seven for Balkan Bulgarian Airlines, three for LOT, and four for Aviogenex. In some cases these replace Tu-134s operated earlier.

The third aircraft delivered to Aviogenex differed from all Tu-134s seen previously in having the usual glazed nose and undernose radome replaced by a more conventional conical nose radome. This is optional on both the Tu-134 and the Tu-134A.

The following details apply to both versions:

Type: Twin-turbofan short/medium-range transport aircraft.

Wings: Cantilever low-wing monoplane. Sweepback at quarter-chord 35°. Anhedral 1° 30′. Conventional all-metal two-spar structure. Two-section aileron on each wing, operated manually through geared tabs, and fitted also with trim tabs. Electro-mechanically-actuated all-metal double-slotted flaps. Hydraulically-actuated spoilers. Hot-air de-icing system.

Fuselage: Conventional all-metal semi-monocoque structure of circular section. Electro-mechanically-actuated airbrake under fuselage, to steepen angle of approach.

Tail Unit: Cantilever all-metal structure, with variable-incidence tailplane mounted at top of fin. Elevators operated manually through geared tabs. Rudder control is hydraulically powered, with yaw damper. Trim tabs in elevators. Fin leading-edge de-iced by hot air; tailplane leading-edge de-iced electrically.

Landing Gear: Retractable tricycle type. All units retract rearward. Main units consist of four-wheel bogies retracting into fairings built on to wing trailing-edge. Oleo-pneumatic shock-absorbers, supplemented by ability of legs to swing rearward to cushion taxying and landing on rough runways. Main wheels size 930 × 305, tyre pressure 5·86 bars (85 lb/sq in). Steerable twin nosewheels size 660 × 200, tyre pressure 6·38-6·90 bars (92·5-100 lb/sq in). Disc brakes and anti-skid units standard. Brake-chute stowed in fuselage tailcone of Tu-134 (APU exhaust in this position on Tu-134A).

Power Plant: Two Soloviev D-30 turbofan engines, each rated at 66·7 kN (14,990 lb st), in pod on each side of rear fuselage, available with thrust reversers, constant-speed drives and AC generators. Three fuel tanks in each wing, total capacity 16,500 litres (3,630 Imp gallons) when gravity fuelled, 16,000 litres (3,520 Imp gallons) when pressure fuelled. Single-point refuelling socket in starboard wing-root leading-edge, Gravity fuelling point above each tank. Hot-air de-icing system for nacelle intakes. Fire-warning and freon extinguishing system.

Accommodation (Tu-134): Flight crew of three, consisting of two pilots and a navigator, plus two stewardesses. Mixed class version accommodates 64 passengers in four-abreast seats, with 0·45 m (17·5 in) centre aisle. 16 first class passengers in front cabin have seats at 930 mm (36·6 in) pitch, with tables between first two rows; 20 tourist class in centre cabin and 28 tourist class in rear cabin. Economy class version accommodates 72 passengers in four-abreast seating, with 44 seats (at 720 mm; 28·35 in pitch) and two tables in forward cabin, and 28 seats (at 750 mm; 29·5 in pitch) in aft cabin. In each version there is a galley on the starboard side and baggage compartment and galley on the port side immediately aft of the flight deck, two toilets at the rear and a large baggage and freight compartment aft, in line with the engines. Max loading on floor of freight compartment 400 kg/m² (82 lb/sq ft). The passenger door is on the port side, forward of the front cabin. There are two cargo doors, on the starboard side by the baggage compartments, and an emergency exit on each side over the wing. Crew cabin and canopy observation panel de-iced by electric heater and hot air.

Accommodation (Tu-134A): Generally similar to Tu-134 except for lengthened cabins. All versions have 28 seats in four-abreast rows in rear cabin. Front cabin seats 44, 48 or 52 passengers, four-abreast, with tables between front two rows. Seat pitch 750 mm (29·5 in) in all versions. Three wardrobes forward of main cabin on 72-seat version, one on other versions. Reduced forward baggage space on 80-seater.

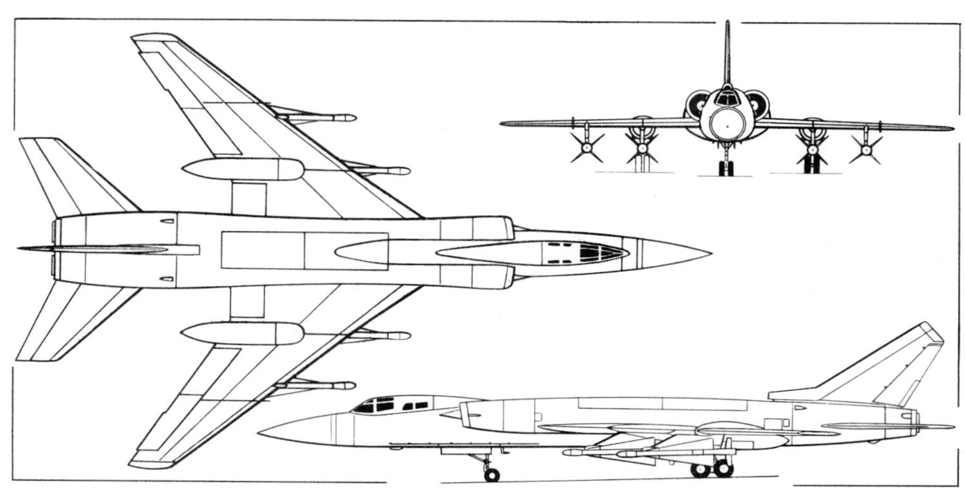

Tupolev Tu-28P supersonic twin-jet all-weather interceptor *(Pilot Press)*

Systems: Air-conditioning system, pressure differential 0·56 bars (8·10 lb/sq in), fed with bleed air from engine compressors. Hydraulic system operating pressure 207 bars (3,000 lb/sq in). Electrical system includes 27V DC supply from four 12kW starter/generators and two batteries, single-phase 115V 400Hz AC supply from two inverters and three-phase 36V 400Hz AC supply. APU available since 1969-70. Oxygen available continuously for pilot, from 92 litre bottle, with 1 hr supply for other crew members and portable supply for emergency use by passengers.

Electronics and Equipment (Tu-134): Provision for full range of radio and radar communications and navigation equipment, including R-807 HF communications radio, Lotos VHF radio, SPU-7 intercom, RO3-1 weather/navigation radar, SOM-64 transponder, AGD-1 remote-reading artificial horizon, AUASP-3 angle-of-attack and *g* load control unit, KS-8 direction finder, NAS-1A6 navigation system (including Trassa-A Doppler), BSU-3P automatic flight control and landing system (including AP-6EM-3P autopilot) for automatic control in flight and automatic or semi-automatic landing approaches down to 40-60 m (130-200 ft), ARK-11 radio compass, RV-UM low-altitude (0-600 m; 0-2,000 ft) radio altimeter and 'Course

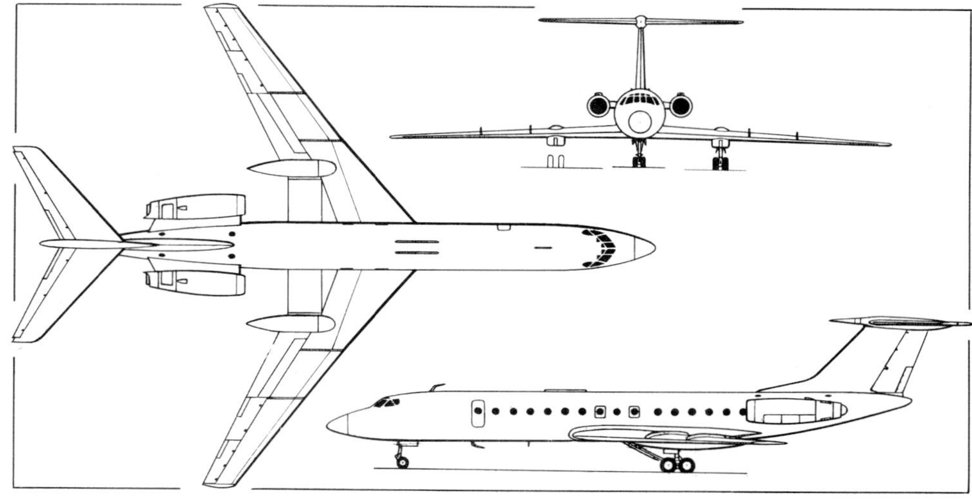

Tupolev Tu-134A twin-turbofan short/medium-range transport *(Pilot Press)*

Tupolev Tu-134A medium-range transport (two Soloviev D-30 turbofan engines) of CSA Czechoslovakian Airlines

MP-1' VOR/ILS/SP-50 navigation and landing system.

Electronics and Equipment (Tu-134A): Typical installation includes two ARK-15 radio compasses, Mikron HF communications radio, two UHF transceivers, RV-5 radio altimeter, two 'Course MP-2' VOR/ILS, two SO-70 transponders, ROZ-1 weather radar and DISS-013 Doppler.

Dimensions, external:
Wing span 29·01 m (95 ft 2 in)
Wing chord at root 8·66 m (28 ft 5 in)
Wing chord at tip 1·92 m (6 ft 3½ in)
Wing aspect ratio 7·3
Length overall:
Tu-134 34·95 m (114 ft 8 in)
Tu-134A 37·05 m (121 ft 6½ in)
Fuselage max diameter 2·90 m (9 ft 6 in)
Height overall:
Tu-134 9·02 m (29 ft 7 in)
Tu-134A 9·14 m (30 ft 0 in)
Tailplane span 11·80 m (38 ft 8½ in)
Wheel track 9·45 m (31 ft 0 in)
Wheelbase:
Tu-134 13·93 m (45 ft 8½ in)
Tu-134A 16·40 m (53 ft 9½ in)
Passenger door:
Height 1·30 m (4 ft 3 in)
Width 0·70 m (2 ft 3½ in)
Height to sill 2·60 m (8 ft 6½ in)
Baggage compartment doors:
Height 0·90 m (2 ft 11½ in)
Width: fwd 1·10 m (3 ft 7¼ in)
aft 1·20 m (3 ft 11¼ in)
Height to sill 2·40 m (7 ft 10½ in)

Dimensions, internal:
Cabin (portion containing seats only):
Length:
Tu-134 13·85 m (45 ft 5½ in)
Width 2·71 m (8 ft 10½ in)
Height 1·96 m (6 ft 5 in)
Floor area:
Tu-134 31·85 m^2 (343 sq ft)
Volume:
Tu-134 58·7 m^3 (2,073 cu ft)
Tu-134A 68·0 m^3 (2,400 cu ft)
Max usable floor area, excl flight deck:
Tu-134 47·00 m^2 (506 sq ft)
Max usable volume, excl flight deck:
Tu-134 86·10 m^3 (3,040 cu ft)
Baggage compartment, Tu-134 (fwd):
Height (mean) 1·875 m (6 ft 1¾ in)
Length (mean) 1·875 m (6 ft 1¾ in)
Width (mean) 1·28 m (4 ft 2½ in)
Floor area 2·4 m^2 (25·8 sq ft)
Volume 3·50 m^3 (123 cu ft)
Baggage compartment, Tu-134A (fwd):
Volume 4·0-6·0 m^3 (141-212 cu ft)
Baggage compartment, Tu-134/134A (aft):
Height (mean) 1·75 m (5 ft 9 in)
Length (mean) 2·80 m (9 ft 2 in)
Width (mean) 1·75 m (5 ft 9 in)
Floor area 4·5 m^2 (48·4 sq ft)
Volume 8·50 m^3 (300 cu ft)

Areas:
Wings, gross 127·3 m^2 (1,370·3 sq ft)
Ailerons (total) 9·68 m^2 (104·2 sq ft)
Trailing-edge flaps (total) 22·50 m^2 (242·2 sq ft)
Spoilers (total) 4·48 m^2 (48·2 sq ft)
Vertical tail surfaces (total) 20·03 m^2 (215·6 sq ft)
Rudder 5·76 m^2 (62·0 sq ft)
Horizontal tail surfaces (total) 30·68 m^2 (330·2 sq ft)
Elevators 6·42 m^2 (69·1 sq ft)

Weights:
Operating weight, empty:
Tu-134 27,500 kg (60,627 lb)
Tu-134A 29,000 kg (63,950 lb)
Max fuel:
Tu-134 13,000 kg (28,660 lb)
Tu-134A 14,400 kg (31,800 lb)
Max payload:
Tu-134 7,700 kg (16,975 lb)
Tu-134A 8,200 kg (18,000 lb)
Max ramp weight:
Tu-134 45,200 kg (99,650 lb)
Tu-134A 47,200 kg (104,000 lb)
Max T-O weight:
Tu-134 45,000 kg (99,200 lb)
Tu-134A 47,000 kg (103,600 lb)
Max landing weight:
Tu-134, standard 40,000 kg (88,185 lb)
Tu-134, emergency 44,000 kg (97,000 lb)
Tu-134A, standard 43,000 kg (94,800 lb)
Max zero-fuel weight:
Tu-134 35,200 kg (77,603 lb)

Performance (Tu-134, at T-O weight of 44,000 kg; 97,000 lb unless otherwise stated):
Max cruising speed:
at 11,000 m (36,000 ft) 469 knots (870 km/h; 540 mph)
at 8,500 m (28,000 ft) 485 knots (900 km/h; 559 mph)
Long-range cruising speed at 11,000 m (36,000 ft) 405 knots (750 km/h; 466 mph)
T-O safety speed, one engine out 141 knots (260 km/h; 162 mph)
Approach speed 133 knots (247 km/h; 153 mph)
Landing speed 116-122 knots (215-225 km/h; 134-140 mph)
Stalling speed, wheels and flaps down 103 knots (190 km/h; 118 mph)
Max rate of climb at S/L 888 m (2,913 ft)/min
Rate of climb at S/L, one engine out 180 m (590 ft)/min
Service ceiling at AUW of 42,000 kg (92,600 lb) 12,000 m (39,370 ft)
Service ceiling, one engine out, at AUW of 43,000 kg (94,800 lb) 5,600 m (18,375 ft)
T-O run 1,000 m (3,280 ft)
Balanced field length for T-O, FAR standard:
at S/L, ISA, max T-O weight 2,180 m (7,152 ft)
Balanced field length for landing, FAR standard:
at S/L, ISA, max landing weight 2,050 m (6,726 ft)
Landing from 15 m (50 ft) at AUW of 37,000 kg (81,570 lb) 1,200 m (3,937 ft)
Landing run at AUW of 37,000 kg (81,570 lb) 800-865 m (2,625-2,838 ft)
Min ground turning radius 35 m (115 ft)
Range, against 27 knots (50 km/h; 31 mph) headwind, 60 min fuel reserve:
with 7,000 kg (15,430 lb) payload at 459 knots (850 km/h; 528 mph) at 11,000 m (36,000 ft) 1,293 nm (2,400 km; 1,490 miles)
with 5,190 kg (11,442 lb) payload at above speed 1,656 nm (3,070 km; 1,907 miles)
with 3,000 kg (6,600 lb) payload at long-range cruising speed 1,888 nm (3,500 km; 2,175 miles)

Performance (Tu-134A):
Max cruising speed at AUW of 42,000 kg (92,600 lb) at 10,000 m (32,800 ft) 477 knots (885 km/h; 550 mph)
Normal cruising speed 405-458 knots (750-850 km/h; 466-528 mph)
Service ceiling at max T-O weight 11,900 m (39,000 ft)
Landing run at max landing weight 780 m (2,560 ft)
Range at max AUW, cruising at 458 knots (850 km/h; 528 mph) at 11,000 m (36,000 ft), with 1 hour fuel reserve:
with max payload 938 nm (1,740 km; 1,081 miles)
with max fuel 1,495 nm (2,770 km; 1,720 miles)

TUPOLEV Tu-144

NATO reporting name: *Charger*

Since this supersonic transport aircraft was first shown in model form at the 1965 Paris Salon de l'Aéronautique, it has undergone considerable development. Its general configuration has become more like that of the Anglo-French Concorde, with a fully-cambered delta wing and two separate underwing ducts for the engines. However, it has larger overall dimensions than the Concorde and is intended to carry a slightly larger number of passengers initially. It also embodies in its production form retractable 'moustache' foreplanes, which were introduced to enhance its take-off and landing characteristics.

Three airframes were laid down initially, of which two were regarded as prototypes, plus a structure test version. In addition, an otherwise-standard MiG-21 was fitted with a scaled-down replica of the Tu-144's original ogival wing, in place of its normal delta wing and horizontal tail surfaces. This aircraft made several dozen research flights, leading to modifications in the design of the full-size wing.

The first of the two prototypes of the Tu-144 (CCCP-68001) was assembled and ground-tested at the Zhukovsky Plant, near Moscow, and flew for the first time on 31 December 1968, this being the first flight by a supersonic airliner anywhere in the world. Its landing gear remained extended throughout the 38-minute flight, as it did during the 50-minute second test flight on 8 January 1969. The crew comprised Eduard Elyan, pilot, Mikhail Kozlov, co-pilot, and two engineers. The pilots occupied upward-ejection seats, side by side on the flight deck. Two further escape hatches in the top of the fuselage further aft indicated the positions of the crew ejection seats.

On 5 June 1969 the Tu-144 exceeded Mach 1 for the first time, at a height of 11,000 m (36,000 ft), half-an-hour after take-off. Only a slight tremble was said to be discernible as it passed through the transonic region. On 26 May 1970 this prototype became the first commercial transport to exceed Mach 2, by flying at 1,160 knots (2,150 km/h; 1,335 mph) at a height of 16,300 m (53,475 ft) for several minutes. The pilot was again Eduard Elyan. Highest speed reported subsequently was Mach 2·4, probably with the aircraft in its production form.

At the first public showing of the Tu-144, at Sheremetyevo Airport, Moscow, on 21 May 1970, the Soviet Deputy Minister for the Aviation Industry, Alexander Kobzarev, said that series production had already started at Voronezh. By May 1972 the prototype had logged a total of about 200 flying hours in nearly 150 flights, of which more than 100 hours were at supersonic speed. The second and third aircraft had each completed only a few flights at that time.

There were no pre-production Tu-144s, and the aircraft (CCCP-77102) exhibited at the 1973 Paris Salon de l'Aéronautique was No. 2 of the initial series production models, representing almost a total redesign by comparison with the prototype described and illustrated in the 1972-73 *Jane's*. It was lost during its flight demonstration at Paris; but the subsequent enquiry did not make necessary any significant modifications to the Tu-144 and the aircraft exhibited at the 1975 Paris Salon (CCCP-77144) was said by its designer, Alexei Tupolev, to differ from CCCP-77102 only in having some updated equipment. It was described as the eighth aircraft of the type, including prototypes but excluding two ground test airframes. Total flying time accumulated by the eight aircraft was not announced, but they were stated to have completed 1,000 flights by the end of May 1975. A further twelve Tu-144s had been ordered at that time.

Initial route proving flights began in the first half of 1974, primarily between Moscow and Vladivostock, via Tyumen. Four production Tu-144s were reported to be available for these operations, and were used to deliver urgent freight and mail on some occasions. Regular supersonic flights by Aeroflot began on 26 December 1975, between Moscow's Domodedovo Airport and Alma-Ata, capital of Kazakhstan. Carrying a payload of freight and mail, Tu-144 No. CCCP-77106 took 1 hr 59 min for the journey of more than 1,619 nm (3,000 km; 1,864 miles), flying for most of the time at 1,187 knots (2,200 km/h; 1,367 mph) at a height of 16,000-18,000 m (52,500-59,000 ft). Take-off from Moscow was at 8.30 am local time. It returned to Domodedovo at 3 pm on the same day. Passenger services are expected to begin in the first half of 1977.

Construction of the Tu-144 is mainly of VAD-23 light alloy, with extensive use of integrally-stiffened panels, produced by both chemical milling and machining from solid metal. Stainless steel and titanium are used for the leading-edges, elevons, rudder and undersurface of the

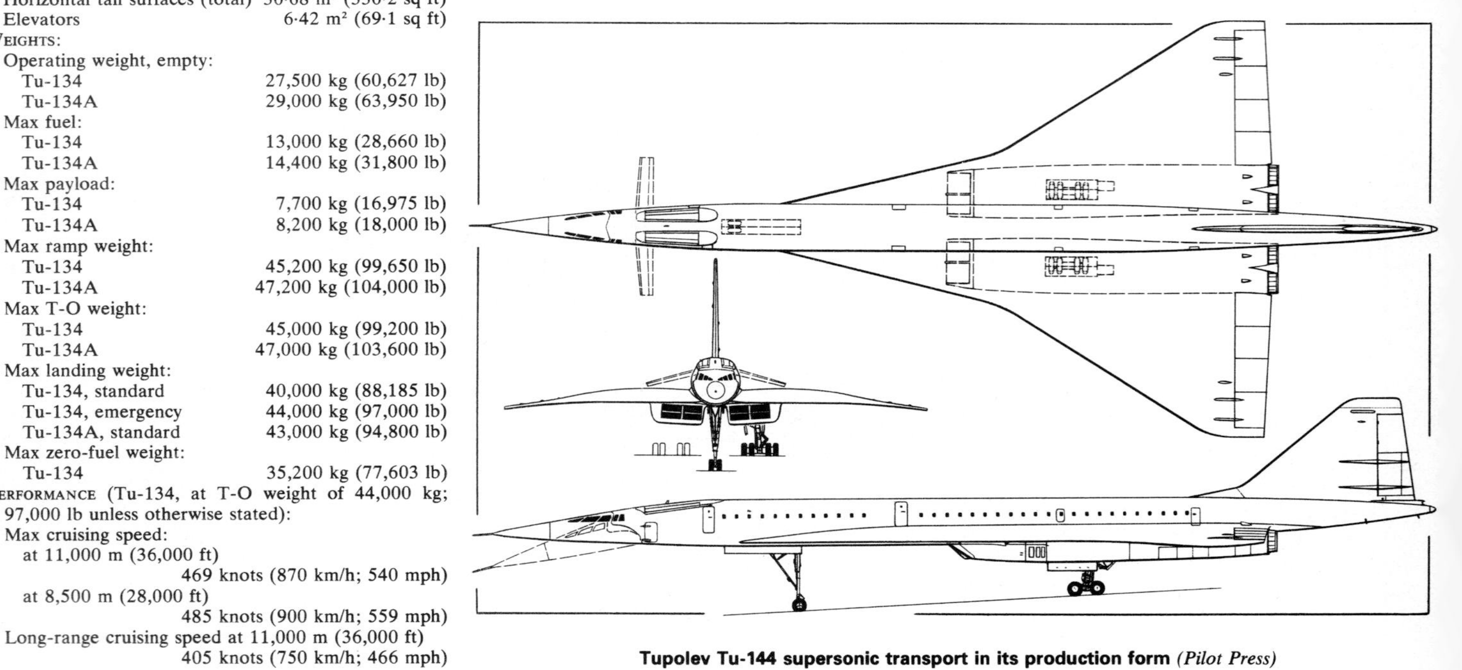

Tupolev Tu-144 supersonic transport in its production form *(Pilot Press)*

Tupolev Tu-144 supersonic transport, in production form. This aircraft was displayed at the 1975 Paris Air Show *(Brian M. Service)*

rear fuselage, and the aircraft is stated to embody 10,000 parts made of plastics.

The wings have a double-delta planform, with a sweepback in the order of 76° on the inboard portions and 57° on the main panels. The prototype had marked conical camber on the highly-swept inboard leading-edges, but flat trailing-edges. The production aircraft has wings increased in span by nearly 1·15 m (4 ft) and cambered over the full area, with a downward-curving trailing-edge like that of the Concorde. The structure is multi-spar, with large honeycomb panels. The powered control surfaces consist of four separate elevons on each wing and a two-section rudder, each operated by two separate actuators.

The fuselage (nearly 5·7 m; 19 ft longer on production aircraft) blends with the low-set wings, giving a flat undersurface which contributes to fuselage lift and directional stability. The number of cabin windows is increased from 25 each side on the prototype to 34 on production aircraft. There are doors forward of the passenger cabins and in the centre on the port side; the number of emergency exits has been increased from four to six.

The 'moustache' foreplanes are pivoted from points near the top of the fuselage, immediately aft of the flight deck. Each is fitted with a double-slotted trailing-edge flap and a fixed leading-edge double-slat. the foreplanes retract rearward, protruding only a little externally but restricting to a narrow passage the space between flight deck and cabin. When extended they have anhedral but no sweep.

It was reported that flight development during 1973-74 had indicated the value of extending these foreplanes during transonic acceleration from just below Mach 1 to Mach 1·3; this was said to reduce aerodynamic drag, and, hence, fuel consumption. However, Alexei Tupolev stated in Paris in 1975 that the foreplanes are used only during take-off and landing.

Following relocation of the engines (see below) all three units of the landing gear have been redesigned. The twin-wheel steerable nose unit now retracts forward into the fuselage. Each main eight-wheel bogie (two rows of four, compared with three rows of four on prototype) now retracts forward and up into one of the engine ducts, between the divided air-intake trunks. This requires the bogie first to pivot sideways through 90° about the base of the leg, before retraction. Nosewheel tyres are size 950 × 300. The main wheels are fitted with size 950 × 400 tyres and quadruple steel disc brakes. All wheel-bays are thermally insulated, and the nosewheel tyres are blown with cooling air after retraction, throughout cruising flight.

The first flight of the Tu-144 prototype was also the first time that the Kuznetsov NK-144 turbofan engine had been tested in the air. At that time the engine max ratings were 127·5 kN (28,660 lb st) without afterburning and 171·6 kN (38,580 lb) with full afterburning; and the four turbofans were mounted side by side in the rear of a single large underbelly duct with bifurcated twin intake trunks. On production aircraft the rating with full afterburning has been increased to 196·1 kN (44,090 lb st), and the engines are paired in two separate ducts, further outboard. As before, each intake trunk contains a central vertical wall, giving an individual flow of air to each engine. The intakes have fully-automatic movable ramps, with manual reversion, and with airflow dump doors midway from the inlet to the engines. Afterburning is normally maintained at 30% to 40% of its maximum additional thrust throughout cruising flight. No thrust reversers are installed, but a twin brake-parachute is fitted solely for use on short runways.

Total fuel capacity has been increased from 70,000 kg (154,325 lb) on the prototype to about 95,000 kg (209,440 lb) on production aircraft, with a transfer tank in the fuselage tailcone to counterbalance CG movement in flight.

Retractable foreplanes enhance the take-off and landing performance of the production Tu-144 *(Air Portraits)*

A flight crew of three is normally carried, consisting of two pilots and a flight engineer. The pilots have fully-adjustable armchair seats. During cruising flight, their windscreen is faired in by a retractable visor which has birdproof side windows and a 'solid' top. The entire nose can be drooped for improved visibility during take-off and landing.

The basic interior layout is for a total of 140 passengers in three cabins. The front cabin contains 11 seats for first class passengers, basically three-abreast, with tables between the front two rows. It is divided by a movable partition from the forward tourist class cabin, which contains six rows of five-abreast seats, with the three-seat units on the port side of the centre aisle. The rear tourist class cabin contains 15 rows of five-abreast seating at the front and six rows of four-abreast seating at the rear. Seat pitch is normally 1,020 mm (40 in) for first class and 870 mm (34·25 in) for tourist class; but alternative layouts are available. The Tu-144 shown at Paris in 1973 contained fewer than 100 seats, but the aircraft displayed in 1975 featured the standard accommodation for 140 passengers.

Forward of the passenger accommodation there are toilet (starboard) and cloakroom compartments (port), with a bench seat for two cabin staff by the forward door. A second cloakroom, toilet and buffet kitchen are located between the two tourist class cabins, with two further toilets at the rear. Aft of these, in line with the engines, is a large compartment for containerised baggage and freight, which are loaded and unloaded semi-automatically through a large door on the starboard side of the hold, at the rear. There are no underfloor holds.

Little information is yet available on aircraft systems. The prototype had three independent hydraulic systems and two separate systems for pressurisation and air-conditioning. Preparation for flight, ground air-conditioning and engine starting can be performed independently of airport services. Advanced automatic flight control and navigation systems are standard, with the intention of progressing eventually to full automatic landing under all weather conditions. Six landing and taxi lights are mounted on the nosewheel leg.

DIMENSIONS, EXTERNAL:
- Wing span: 28·80 m (94 ft 6 in)
- Length overall: 65·70 m (215 ft 6½ in)
- Height, wheels up: 12·85 m (42 ft 2 in)
- Wheel track: 6·05 m (19 ft 10¼ in)
- Wheelbase: 19·60 m (64 ft 3½ in)

DIMENSIONS, INTERNAL:
- Cabin: Headroom: 1·93 m (6 ft 4 in)
- Baggage/cargo hold capacity: 20 m³ (706 cu ft)

AREA:
- Wings, gross: 438 m² (4,714·5 sq ft)

WEIGHTS:
- Operating weight, empty: 85,000 kg (187,400 lb)
- Max fuel: 95,000 kg (209,440 lb)
- Max payload (space limited): 14,000 kg (30,865 lb)
- Max payload (structure limited): 15,000 kg (33,070 lb)
- Max ramp weight: 185,000 kg (407,850 lb)
- Max T-O weight: 180,000 kg (396,830 lb)
- Max zero-fuel weight: 100,000 kg (220,460 lb)
- Max landing weight: 110,000-120,000 kg (242,500-264,550 lb)

PERFORMANCE (nominal):
- Max cruising speed: Mach 2·35 (1,350 knots; 2,500 km/h; 1,550 mph)
- Normal cruising speed: Mach 2·2 (1,240 knots; 2,300 km/h; 1,430 mph)

Landing speed 151 knots (280 km/h; 174 mph)
Cruising height 16,000-18,000 m (52,500-59,000 ft)
Balanced field length at max T-O weight (approx):
ISA, S/L 3,000 m (9,845)
ISA+15°C, S/L 3,200 m (10,500 ft)
Landing run 2,600 m (8,530 ft)
Max range with 140 passengers, at an average speed of Mach 1·9 (1,080 knots; 2,000 km/h; 1,243 mph) 3,500 nm (6,500 km; 4,030 miles)

TUPOLEV Tu-154

NATO reporting name: *Careless*

The three-engined Tu-154, announced in the Spring of 1966, was intended to replace the Tu-104, Il-18 and An-10 on medium/long stage lengths of up to 3,240 nm (6,000 km; 3,725 miles). It is able to operate from airfields with a class B surface, including packed earth and gravel. Normal flight can be maintained after shutdown of any one engine. Single-engine flight is possible at a lower altitude.

The first of six prototype and pre-production models flew for the first time on 4 October 1968. The seventh Tu-154 was delivered to Aeroflot for initial route proving and crew training in early 1971. Mail and cargo flights began in May. Initial passenger-carrying services were flown for a few days in the early Summer of 1971 between Moscow and Tbilisi. Regular services began on 9 February 1972, over the 700 nm (1,300 km; 800 mile) route between Moscow and Mineralnye Vody, in the North Caucasus. International services began with a proving flight between Moscow and Prague on 1 August 1972.

It is believed that more than 100 Tu-154s had entered service with Aeroflot by the Spring of 1974. Others had been ordered by, or delivered to, Balkan Bulgarian Airlines (4), Malev (3), EgyptAir (8), Aviogenex (2) and CSA Czechoslovakian Airlines (7). The EgyptAir Tu-154s were returned to the Soviet Union after one had been lost in a training accident.

An article in the April 1975 issue of the Soviet magazine *Grazhdanskaya Aviatsiya* stated that Tu-154s had logged 120,000 flying hours and 65,000 landings in airline service, and had carried some 7 million passengers. It also gave details of an improved version known as the Tu-154A, which is described separately.

TYPE: Three-engined medium/long-range transport aircraft.

WINGS: Cantilever low-wing monoplane. Sweepback 35° at quarter-chord. Conventional all-metal three-spar fail-safe structure; centre spar extending to just outboard of inner edge of aileron on each wing. Five-section slat on outer 80% of each wing leading-edge. Triple-slotted flaps. Four-section spoilers on each wing. Outboard sections supplement ailerons for roll control. Section inboard of landing gear housing serves as airbrake and lift-dumper; two middle sections can be used as airbrakes in flight. All control surfaces hydraulically actuated and of honeycomb construction. Hot-air de-icing of wing leading-edge. Slats are electrically heated.

FUSELAGE: Conventional all-metal semi-monocoque fail-safe structure of circular section.

TAIL UNIT: Cantilever all-metal structure, with variable-incidence tailplane mounted at tip of fin. Rudder and elevator of honeycomb construction. Sweepback of 40° at quarter-chord on horizontal surfaces, 45° on leading-edge of vertical surfaces. Control surfaces hydraulically actuated by irreversible servo-controls. Leading-edges of fin and tailplane and engine air intake de-iced by hot air.

LANDING GEAR: Retractable tricycle type. Hydraulic actuation. Main units retract rearward into fairings on wing trailing-edge. Each consists of a bogie made up of three pairs of wheels, size 930 × 305, in tandem; tyre pressure 7·86 bars (114 lb/sq in). Steerable anti-shimmy twin-wheel nose unit has wheels size 800 × 225 and retracts forward. Disc brakes and anti-skid units on main wheels.

POWER PLANT: Three Kuznetsov NK-8-2 turbofan engines, each rated at 93·2 kN (20,950 lb st), one on each side of rear fuselage and one inside extreme rear of fuselage. Two lateral engines fitted with upper and lower thrust-reversal grilles. Integral fuel tanks in wings; standard capacity 41,140 litres (9,050 Imp gallons). Max fuel capacity 46,825 litres (10,300 Imp gallons). Single-point refuelling standard.

ACCOMMODATION: Flight crew of two pilots and flight engineer; provision for navigator aft of pilot and folding seats for additional pilots or instructors. There are basic passenger versions for a total of 167, 158, 152, 146 and 128 passengers. Each has a toilet at the front (starboard), removable galley amidships and three toilets aft. Coat storage, folding seat and inflatable evacuation chute in each entrance lobby. Standard economy class version has 54 seats in six-abreast rows, with two tables between front rows, in forward cabin; and 104 seats in six-abreast rows (rear two rows four-abreast) in rear cabin at seat pitch of 750 mm (29·5 in). The 167-seat high-density version differs in having one further row of six seats in the forward cabin and reduced galley facilities. The tourist class versions carry 146 passengers at a seat pitch of 810 mm (31·9 in) or 152 at a pitch of 870 mm (34·25 in) with reduced galley facilities. The 128-seat version has only 24 first class seats, four-abreast at a pitch of 1,020 mm (40 in), in the forward cabin. There is also an all-cargo version. Passenger doors are forward of front cabin and between cabins on the port side, with emergency and service doors opposite. All four doors open outwards. Four emergency exits, two over wing on each side. Two pressurised baggage holds under main cabin floor, with two inward-opening doors. Normal provision for mechanised loading and unloading of baggage and freight in containers. Smaller unpressurised hold under rear cabin for carrying spare parts or special cargo such as radioactive isotopes.

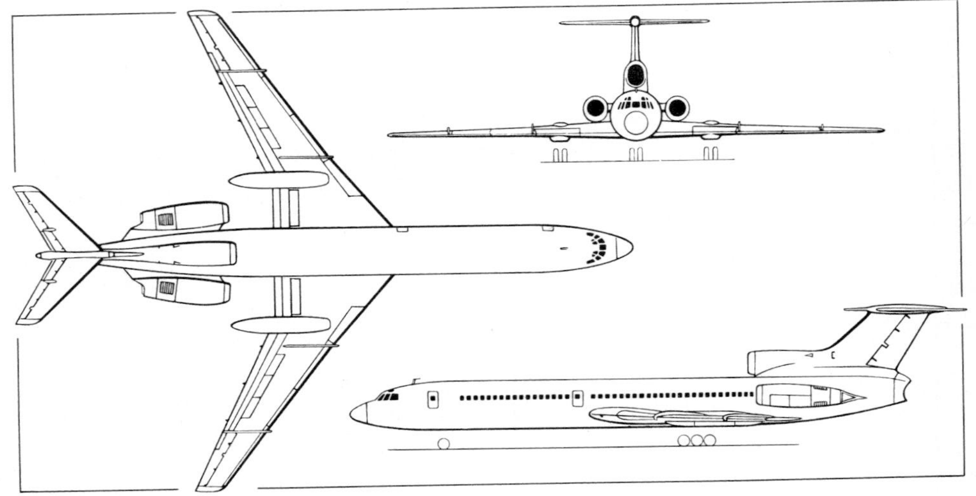

Tupolev Tu-154 medium/long-range three-turbofan transport aircraft *(Pilot Press)*

ELECTRONICS AND EQUIPMENT: Automatic flight control system standard, including automatic navigation on pre-programmed route under control of navigational computer with en-route checks by ground radio beacons (including VOR, VOR/DME) or radar, and automatic approach by ILS to ICAO category II standards (development to category III standard in hand). Moving-map ground position indicator, HF and VHF radio, and radar standard. Safety equipment includes four inflatable life-rafts, each for 26 persons.

SYSTEMS: Air-conditioning system pressure differential 0·62 bars (9·0 lb/sq in). Three independent hydraulic systems; working pressure 207 bars (3,000 lb/sq in). No. 1 system, powered by two pumps driven by centre engine and port engine, operates landing gear, brakes and all control surfaces. No. 2 system, powered by a pump driven by centre engine, actuates nosewheel steering, the second flying controls circuit and landing gear emergency extension. No. 3 system, powered by pump on starboard engine, actuates the third flying controls circuit and second landing gear emergency extension circuit. Three-phase 200/115V AC electrical system, supplied by three 40kVA alternators. 28V DC system. APU standard, driving 40kVA alternator and 12kW starter/generator.

DIMENSIONS, EXTERNAL:
Wing span 37·55 m (123 ft 2½ in)
Length overall 47·90 m (157 ft 1¾ in)
Height overall 11·40 m (37 ft 4¾ in)
Diameter of fuselage 3·80 m (12 ft 5½ in)
Tailplane span 13·40 m (43 ft 11½ in)
Wheel track 11·50 m (37 ft 9 in)
Wheelbase 18·92 m (62 ft 1 in)
Passenger doors (each):
Height 1·73 m (5 ft 7 in)
Width 0·80 m (2 ft 7½ in)
Height to sill 3·10 m (10 ft 2 in)
Servicing door:
Height 1·28 m (4 ft 2½ in)
Width 0·61 m (2 ft 0 in)
Emergency door:
Height 1·28 m (4 ft 2½ in)
Width 0·64 m (2 ft 1¼ in)
Emergency exits (each):
Height 0·90 m (2 ft 11½ in)
Width 0·48 m (1 ft 7 in)
Main baggage hold doors (each):
Height 1·20 m (3 ft 11¼ in)
Width 1·35 m (4 ft 5 in)
Height to sill 1·80 m (5 ft 11 in)
Rear (unpressurised) hold:
Height 0·90 m (2 ft 11½ in)
Width 1·10 m (3 ft 7¼ in)
Height to sill 2·20 m (7 ft 2½ in)

DIMENSIONS, INTERNAL:
Cabin: Width 3·58 m (11 ft 9 in)
Height 2·02 m (6 ft 7½ in)
Volume 163·2 m³ (5,763 cu ft)
Main baggage holds:
Front 21·5 m³ (759 cu ft)
Rear 16·5 m³ (582 cu ft)
Rear underfloor hold 5·0 m³ (176 cu ft)

AREAS:
Wings, gross 201·45 m² (2,169 sq ft)
Horizontal tail surfaces 40·55 m² (436·48 sq ft)
Vertical tail surfaces 31·72 m² (341·43 sq ft)

WEIGHTS:
Operating weight empty 43,500 kg (95,900 lb)
Normal payload 16,000 kg (35,275 lb)
Max payload 20,000 kg (44,090 lb)
Max fuel 33,150 kg (73,085 lb)
Max ramp weight 90,300 kg (199,077 lb)
Normal T-O weight 84,000 kg (185,188 lb)
Max T-O weight 90,000 kg (198,416 lb)
Normal landing weight 68,000 kg (149,915 lb)
Max landing weight 80,000 kg (176,370 lb)
Max zero-fuel weight 63,500 kg (139,994 lb)

PERFORMANCE (at max T-O weight, except where indicated):
Max level speed:
above 11,000 m (36,000 ft) Mach 0·90
at low altitudes 283 knots (525 km/h; 326 mph) IAS
Max cruising speed at 9,500 m (31,150 ft) 526 knots (975 km/h; 605 mph)
Best-cost cruising speed at 11,000-12,000 m (36,000-39,350 ft) Mach 0·85 (486 knots; 900 km/h; 560 mph)
Long-range cruising speed at 11,000-12,000 m (36,000-39,350 ft) Mach 0·80 (459 knots; 850 km/h; 528 mph)
Approach speed 127 knots (235 km/h; 146 mph)
Min ground turning radius 24·60 m (80 ft 8½ in)
T-O run at normal T-O weight, ISA 1,140 m (3,740 ft)
Balanced runway length at max T-O weight, FAR standard:
ISA, S/L 2,100 m (6,890 ft)
ISA+20°C, S/L 2,420 m (7,940 ft)
Landing field length, at max landing weight, FAR standard:
ISA, S/L 2,060 m (6,758 ft)
ISA+20°C, S/L 2,217 m (7,273 ft)
Range at 11,000 m (36,000 ft) with standard fuel, reserves for 1 hour and 6% of total fuel:
at 486 knots (900 km/h; 560 mph), with T-O weight of 84,000 kg and max payload (158 passengers, baggage and 5 tonnes of cargo and mail) 1,360 nm (2,520 km; 1,565 miles)
as above, T-O weight of 90,000 kg 1,867 nm (3,460 km; 2,150 miles)
at 459 knots (850 km/h; 528 mph), with T-O weight of 84,000 kg and max payload as above 1,510 nm (2,800 km; 1,740 miles)
as above, T-O weight of 90,000 kg 2,050 nm (3,800 km; 2,360 miles)
max range with 13,650 kg (30,100 lb) payload 2,850 nm (5,280 km; 3,280 miles)
Range at 11,000 m (36,000 ft) with optional centre-wing tanks, reserves as above:
with 9,000 kg (19,840 lb) payload (95 passengers) 3,453 nm (6,400 km; 3,977 miles)
with 6,700 kg (14,770 lb) payload (70 passengers) 3,723 nm (6,900 km; 4,287 miles)

TUPOLEV Tu-154A and Tu-154B

NATO reporting name: *Careless*

A developed version of the Tu-154, with the designation **Tu-154A,** was reported in early 1973, with the first flight scheduled for later that year. An article in the April 1975 issue of the Soviet magazine *Grazhdanskaya Aviatsiya* recorded that this aircraft entered service with Aeroflot in April 1974 and that production Tu-154As were expected to be put into scheduled operation during 1975. By 1976, they were also operating on the scheduled services of Balkan Bulgarian Airlines (3).

The Tu-154A is dimensionally unchanged by comparison with the original model, and is intended to carry a

Tupolev Tu-154A medium/long-range transport aircraft (three Kuznetsov NK-8-2U turbofan engines) in service with Balkan Bulgarian Airlines *(S. G. Richards)*

normal payload of 152 passengers in Summer and 144 in Winter. Alternative configurations provide seats for 168 passengers on high-density routes, or 12 first class and 128 tourist class. Changes have centred mainly on the power plant, equipment and systems, to permit an increased gross weight, improve performance and reliability, and reduce servicing requirements.

It was expected that the NK-8-2 turbofans of the Tu-154 would be replaced by Soloviev D-30KUs in the Tu-154A; but this major change of power plant seems to have been deferred, as the Tu-154A has uprated NK-8-2Us of unspecified thrust.

Increased max take-off and landing weights allow extra fuel to be carried, raising the maximum capacity to 39,750 kg (87,630 lb). An additional tank, capacity 6,600 kg (14,550 lb), is mounted between the front and centre spars in the centre-section. It is intended primarily as a ballast tank for ferrying, and the fuel it contains can be pumped into the main system only on the ground. When the aircraft carries less than a full payload, this tank can be filled and its contents can be transferred to the main tanks at a destination airport, so reducing purchases of fuel outside the operator's home country. Other fuel system improvements have been made to the anti-icing fluid additive system; the centre-section tanks can be purged with CO_2 in the event of a forced landing with the wheels retracted.

The controls for the flaps, leading-edge slats and tailplane are interconnected, so that when the flaps are operated the tailplane is trimmed 3° down. An override switch caters for CG conditions which require a movement of more than 3°.

Additional emergency exits in the rear fuselage meet international requirements. The floor of the baggage holds has been strengthened to prevent damage by sharp-edged packages and baggage; and a smoke warning system has been introduced in the holds.

The electrical system has been modified by comparison with the Tu-154 and employs three alternators, on separate supply circuits, to provide 200/115V AC power. Two circuits supply all electrical services; the third supplies the electrical anti-icing system for the leading-edge slats. If one alternator fails, the remaining primary alternator can provide for all essential services, supplemented by the alternator on the APU. The duplicated DC electrical system embodies three rectifiers, of which one is for emergency use in the event of a failure of either of the others.

An ABSU automatic approach and landing system is fitted. This met ICAO Category I requirements initially, but was to be uprated to Category II on production aircraft before the end of 1975. Other equipment changes include the provision of duplicated radio compass, radio altimeter and DME; and the introduction of two-speed windscreen wipers and a system to indicate angle-of-bank limitations. A new MSRP-64 flight recorder covers some 80 parameters, and a Mars-B voice recorder with open microphone is now standard.

Servicing requirements and costs have been reduced considerably on the Tu-154A, for which the servicing cycle is 300/900/1,800 hours.

Also in service with Aeroflot and Malev (3) is a further version, designated **Tu-154B**. This offers a further increase in max take-off and zero-fuel weights. Six seats have been removed from the rear part of the cabin, and two more emergency exits have been installed in front of the engines on the starboard side. A typical seating arrangement, employed by Malev, has 8 first class passengers at the front, 36 economy class in the forward cabin and 98 in the rear. Improvements have been made to the electronics, notably to simplify take-off and landing procedures. A different radar is fitted, and there is an additional fuel tank.

Weights (Tu-154A):
- Normal payload 16,000 kg (35,275 lb)
- Max payload 18,000 kg (39,680 lb)
- Max T-O weight 94,000 kg (207,235 lb)
- Max landing weight (normal) 78,000 kg (171,960 lb)
- Max landing weight (emergency) 92,000-94,000 kg (202,825-207,235 lb)

Weights (Tu-154B):
- Basic operating weight 50,775 kg (111,940 lb)
- Max T-O weight 96,000 kg (211,650 lb)
- Max zero-fuel weight 71,000 kg (156,525 lb)

Performance (Tu-154A, at max T-O weight, except where indicated):
- Max level speed 310 knots (575 km/h; 357 mph) IAS, except with less than 7,150 kg (15,763 lb) fuel at heights above 7,000 m (23,000 ft)
- Normal cruising speed Mach 0·85 (486 knots; 900 km/h; 560 mph)
- Range with payload of 16,000 kg (35,275 lb) 1,725-1,780 nm (3,200-3,300 km; 1,985-2,050 miles)

YAKOVLEV

General Designer in Charge of Bureau:
Alexander Sergeivich Yakovlev

Yakovlev is one of the most versatile Russian designers and products of his design bureau have ranged from transonic long-range fighters to the Yak-24 tandem-rotor helicopter, an operational VTOL carrier-based fighter and a variety of training and light general-purpose aircraft. Types in current production and service, or under development, are described hereafter.

YAKOVLEV Yak-18T

Details of this extensively-redesigned cabin version of the Yak-18 were given for the first time at the 1967 Paris Air Show, where an unregistered example was displayed statically.

The prototype flew for the first time in the Summer of that year, powered, like the Yak-18A and -18PM, with a 224 kW (300 hp) Ivchenko AI-14RF nine-cylinder radial engine, driving a two-blade variable-pitch propeller. Its braced tail unit and retractable tricycle landing gear, with inward-retracting main wheels (size 500-150 tyres) and rearward-retracting nosewheel (size 400-150 tyre), were similar in configuration to those of the Yak-18PM (described in the 1972-73 *Jane's*).

The wing span was increased by extending the constant-chord centre-section; the fuselage was entirely new, being built as an all-metal semi-monocoque of square section. Four persons could be carried in pairs in the enclosed cabin, which had a large forward-hinged door on each side. Dual controls, heating and ventilation were standard. The rear bench seat was removable to enable the Yak-18T to be used for cargo-carrying. As an ambulance, it was designed to accommodate the pilot, one stretcher patient and an attendant.

Standard equipment included ILS, VHF radio, radio compass, radio altimeter and intercom.

An initial evaluation programme of 450 test flights was completed by two prototypes during 1968-69. Together with experience gained during several months of operation at the Sasovo flying school, this suggested that a number of improvements would be worthwhile. Most important of these was the installation of a 268·5 kW (360 hp) M-14P radial engine of the type evolved by Vedeneev from the AI-14 and chosen also for the Kamov Ka-26 helicopter. In addition, improvements were made to the cabin layout and ventilation.

An aircraft embodying the modifications was approved by the Research Institute of the Soviet Ministry of Civil Aviation. Four others logged more than 600 flying hours over a five-month period of testing on every kind of airfield, under a wide variety of weather conditions. The Yak-18T was pronounced superior to earlier versions of the Yak-18 for basic training, and far more economical to operate than the An-2, which was also widely used.

Full production was ordered, and by 1974 it was possible to train the complete intake of 100 pupil pilots at Sasovo on the new aircraft. Seventeen were then available for use at the school, of which two were being flown at double the normal rate of utilisation under a Research Institute programme.

Yakovlev Yak-18T basic trainer (Vedeneev M-14P engine) *(B. M. Service)*

As the standard basic trainer at Aeroflot flying schools, the Yak-18T is used for circuits, instrument training and navigation training, and as a flying classroom for an instructor and three pupils. Only one pupil accompanies the instructor on aerobatic flights.

Next to enter service, as a successor to the Yak-12, will be the ambulance version, with light communications and forest fire patrol versions under consideration. A float-plane version is under development and the Yak-18T will also operate eventually on skis.

DIMENSIONS, EXTERNAL:
- Wing span: 11·16 m (36 ft 7¼ in)
- Length overall: 8·35 m (27 ft 4¾ in)

AREA:
- Wings, gross: 18·75 m² (201·8 sq ft)

WEIGHTS AND LOADINGS (A, with instructor and one pupil; B, with instructor and three pupils):
- Max payload:
  - A: 306 kg (675 lb)
  - B: 436 kg (960 lb)
- Max T-O weight:
  - A: 1,500 kg (3,307 lb)
  - B: 1,650 kg (3,637 lb)
- Max wing loading:
  - A: 80 kg/m² (16·4 lb/sq ft)
  - B: 88 kg/m² (18·0 lb/sq ft)
- Max power loading:
  - A: 5·59 kg/kW (11·0 lb/hp)
  - B: 6·15 kg/kW (12·1 lb/hp)

PERFORMANCE (at max T-O weight: A, with instructor and one pupil; B, with instructor and three pupils):
- Max level speed:
  - A, B: 162 knots (300 km/h; 186 mph)
- Service ceiling:
  - A, B: 5,500 m (18,000 ft)
- T-O run:
  - A: 330 m (1,085 ft)
  - B: 400 m (1,315 ft)
- Landing run:
  - A: 400 m (1,315 ft)
  - B: 500 m (1,640 ft)
- Range with max fuel, with reserves:
  - A: 350 nm (650 km; 403 miles)
  - B: 485 nm (900 km; 560 miles)

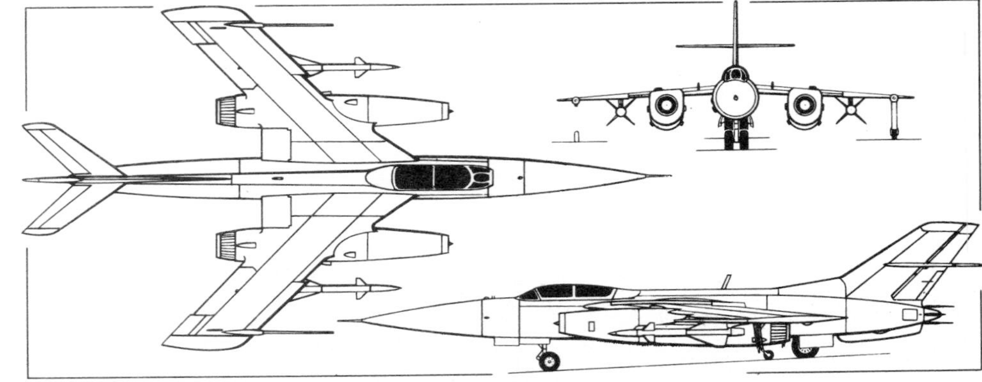

The current long-nose version of the Yakovlev Yak-28P two-seat all-weather fighter *(Firebar) (Pilot Press)*

Version of the Yak-28 (NATO *Brewer-E*) **equipped for active ECM operations**

YAKOVLEV Yak-28

NATO reporting names: *Brewer, Firebar* **and** *Maestro*

First seen in considerable numbers in the 1961 Soviet Aviation Day flypast were three successors to the Yak-25/27 series (see 1971-72 *Jane's*), described by the commentator as supersonic multi-purpose aircraft and identified subsequently by the designation Yak-28. These aircraft are shoulder-wing monoplanes, whereas all versions of the Yak-25, 26 and 27 were mid-wing. The Yak-28 series were, in fact, produced as entirely new designs, following only the general configuration of the earlier types.

The landing gear comprises two twin-wheel units in tandem, with the forward unit under the pilot's cockpit and the rear unit moved further aft than on the Yak-25/27, to a point immediately in front of the ventral fin. Wingtip balancer wheels are retained. The entire wing-root leading-edge has been extended forward and the height of the fin and rudder increased. Tailplane sweep is also increased.

Several versions of the basic design have been reported, with the following NATO reporting names:

Brewer-A to C **(Yak-28).** Two-seat tactical attack versions. Single cockpit for pilot, with blister canopy, and glazed nose for navigator/bomb aimer. Corresponding to Yak-26 (*Mangrove*) and produced to replace the Il-28 in the Soviet Air Force. Most examples have blister radome under fuselage just forward of wings. On some aircraft, long engine nacelles extend forward as far as the front of this radome. Others have shorter nacelles. Guns semi-submerged in each side of the fuselage on some aircraft; on starboard side only on others. Internal bomb bay between the underfuselage radome and the rear main landing gear unit.

Brewer-D. Reconnaissance version, with cameras in bomb bay.

Brewer-E. First Soviet operational ECM escort aircraft, deployed in 1970 and shown in accompanying illustration. Underfuselage radome deleted. Active ECM pack built into bomb bay, from which it projects in form of a semi-cylindrical pack. Attachment under each outer wing, outboard of external fuel tank, for a rocket pod.

Firebar. Tandem two-seat all-weather fighter derivative of Yak-28, corresponding to Yak-27. Nose radome. Internal weapons bay deleted. *Anab* air-to-air missile under each wing instead of guns. Identified as **Yak-28P** (Perekhvatchik; interceptor) at 1967 Domodedovo display, the suffix 'P' indicating that the design had been *adapted* for the fighter role. Example shown in static park had a much longer dielectric nosecone than the standard operational *Firebars* in the flying display and had two missile pylons under each wing, one for an *Atoll* and one for an *Anab*. This suggested that it was a weapons development aircraft. However, the lengthened nosecone has since been fitted retrospectively on Yak-28Ps in squadron service, as shown in an accompanying illustration. This does not indicate any increase in radar capability or aircraft performance.

Maestro **(Yak-28U).** Trainer version of *Firebar.* Normal cockpit layout replaced by two individual single-seat cockpits in tandem, each with its own canopy. Front canopy sideways-hinged to starboard; rear canopy rearward-sliding. Large conical nose-probe.

In early 1974, Admiral Thomas H. Moorer, Chairman of the US Joint Chiefs of Staff, reported to Congress that the Yak-25 had been phased out of service during the previous year. The Yakovlev high-altitude reconnaissance aircraft known to NATO as *Mandrake* (see 1973-74 *Jane's*) is also understood to have been retired from first-line service, although some appear to have been converted into pilotless targets (see RPVs and Targets section). The Yak-28P *(Firebar),* Tupolev Tu-28P, Sukhoi Su-15 and MiG-25, now account for 50 per cent of the Soviet defence interceptor force of more than 2,500 aircraft.

By contrast, the Yak-28 *Brewer* series is changing gradually from first-line attack to support roles, with the emphasis on ECM, reconnaissance and operational training.

The following details refer specifically to the Yak-28P, but are generally applicable to the other versions of the Yak-28:

Yakovlev Yak-28U *(Maestro)* **trainer** *(Tass)*

Yakovlev Yak-28P *(Firebar)* **two-seat all-weather interceptor with current lengthened nosecone**

TYPE: Two-seat transonic all-weather interceptor.

WINGS: Cantilever shoulder-wing monoplane of basically constant chord. Extended leading-edge on outer wings and also between fuselage and each engine nacelle. Outer extensions are drooped. Slotted flap, with unswept trailing-edge, between fuselage and each engine nacelle. Basic wing sweepback 45°. Anhedral from root. Single fence on upper surface of each wing, between fuselage and engine nacelle. Large trailing-edge flap and short aileron, with tab, outboard of nacelle on each wing. Balancer-wheel fairings, inset from wing-tips, are extended forward as lead-filled wing balance weights.

FUSELAGE: All-metal semi-monocoque structure of basically circular section. Finely-tapered dielectric nosecone over radar scanner.

TAIL UNIT: Cantilever all-metal structure. Variable-incidence tailplane mounted midway up fin. All surfaces sweptback. Trim tab in rudder. Dorsal fin fairs into spine along top of fuselage. Shallow ventral stabilising fin.

LANDING GEAR: Two twin-wheel main units in tandem, retracting into fuselage. Front unit retracts forward, rear unit rearward. Small balancer wheel near each wingtip, retracting rearward under wing; fairing integral with leg.

POWER PLANT: Two afterburning turbojet engines, believed to be of same basic type as Tumansky R-11 fitted to MiG-21, with rating of 58·35 kN (13,120 lb st). Each fitted with centrebody shock-cone. A pointed slipper-type external fuel tank can be carried under the leading-edge of each wing, outboard of the engine nacelle.

ACCOMMODATION: Crew of two in tandem on ejection seats in pressurised cabin under long transparent blister canopy.

ARMAMENT: Pylon under each outer wing for *Anab* air-to-air missile, with alternative infra-red or semi-active radar homing heads.

OPERATIONAL EQUIPMENT: Reported to include tail-warning radar.

DIMENSIONS, EXTERNAL (estimated):
Wing span 12·95 m (42 ft 6 in)
Length overall:
Yak-28 21·65 m (71 ft 0½ in)
Height overall 3·95 m (12 ft 11½ in)

WEIGHT (estimated):
Max T-O weight:
Yak-28P 15,875 kg (35,000 lb)

PERFORMANCE (Yak-28P, estimated):
Max level speed at 10,670 m (35,000 ft)
Mach 1·1 (636 knots ; 1,180 km/h; 733 mph)
Cruising speed 496 knots (920 km/h; 571 mph)
Service ceiling 16,750 m (55,000 ft)
Max combat radius 500 nm (925 km; 575 miles)
Max range 1,040-1,390 nm (1,930-2,575 km; 1,200-1,600 miles)

YAKOVLEV *FREEHAND*

The Yakovlev experimental V/STOL aircraft demonstrated at the 1967 Soviet Aviation Day display was allocated the NATO reporting name *Freehand* and was generally believed to have the Soviet designation Yak-36. This designation is now questionable, as the US Department of Defense has suggested that Yak-36 is the official Soviet military designation of the new VTOL combat aircraft (NATO *Forger*) deployed on the Soviet Navy's carrier/cruiser *Kiev*. Until the situation is clarified, it is advisable to refer to the 1967 experimental aircraft only by its NATO reporting name of *Freehand*.

Details of *Freehand* can be found in previous editions of *Jane's*. It is believed to have been used for initial experiments in the shipboard operation of jet-powered V/STOL aircraft, from a specially-installed platform on the helicopter cruiser *Moskva*. Production is thought to have totalled about six or seven aircraft.

YAKOVLEV Yak-36 (?)

NATO reporting name: *Forger*

This is the VTOL combat aircraft deployed by the Soviet Navy on the *Kiev*, first of its new class of 40,000 ton carrier/cruisers to put to sea. It has been referred to as the Yak-36 by the US Department of Defense; its NATO reporting name is *Forger*.

Two versions have been observed on the *Kiev*, as follows:

Forger-A. Basic single-seat combat aircraft. About ten or twelve appear to be operational on the *Kiev*, in addition to Kamov Ka-25 anti-submarine helicopters. Primary operational roles are assumed to be attack and reconnaissance.

Forger-B. Two-seat training version, of which one example was seen on the *Kiev*. A second cockpit is located forward of the normal cockpit, with the blister canopy at a lower level, as on the training version of the MiG-25. To compensate for the longer nose, a 'plug' is inserted in the fuselage aft of the wing, lengthening the constant-section portion without requiring modification of the tapering rear fuselage assembly. In other respects this version appears to be identical to *Forger-A*, but has no ranging radar or weapon pylons.

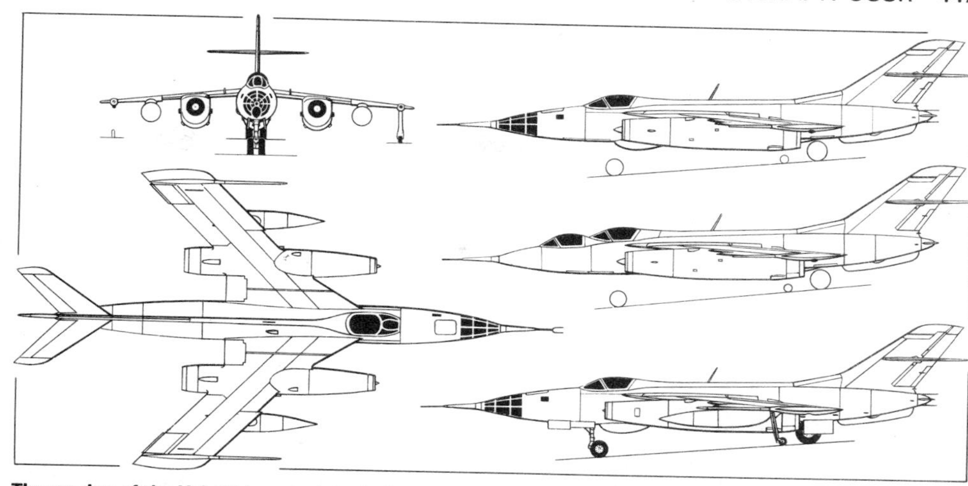

The version of the Yak-28 two-seat tactical attack aircraft known to NATO as *Brewer-C*, **with additional side views of** *Brewer-B* **(top) and the Yak-28U** *(Maestro)* **tandem-cockpit trainer (centre)** *(Pilot Press)*

The likelihood that an aircraft of this type was under development in the Soviet Union was first confirmed in 1974 by Admiral Thomas H. Moorer, then Chairman of the US Joint Chiefs of Staff. In his annual report, he said of the *Kiev:* "This ship is over 900 ft in length and should displace 30-40,000 tons. The deck configuration and the lack of catapults or arresting gear indicate that this ship apparently is designed to operate V/STOL aircraft and helicopters. It should be capable of carrying 25 V/STOL aircraft or 36 helicopters. It is believed, however, that a mixture of new V/STOL tactical aircraft and *Hormone* (Ka-25) helicopters is the most likely complement".

The 1975-76 *Jane's* contained the remark that "A strike/reconnaissance V/STOL aircraft is thought to have been evolved from the Yak-36 *(Freehand)* by the Yakovlev bureau, utilising a mixture of vectored thrust and direct jet-lift". This belief was confirmed when the *Kiev* entered the Mediterranean in July 1976 and subsequently operated its complement of *Forgers* extensively during passage through that sea and the Atlantic en route to Murmansk. These aircraft are assumed to be operated by a development squadron.

The general appearance of the single-seat *Forger-A* is shown in several of the accompanying illustrations. Its basic configuration is conventional, except that VTOL capability has permitted the mid-set wings to be made relatively small in area. They fold upward at approximately mid-span, for stowage on board ship. No leading-edge devices are fitted, but the entire trailing-edge of each wing is made up of an aileron on the outboard (folding) panel and a large Fowler-type flap on the inboard panel. Sweepback is approximately 45° on the leading-edge, and there is considerable anhedral from the wing roots.

All tail surfaces are swept, with conventional rudder and elevators. 'Puffer-jet' stability control orifices are apparent above and below the tailcone and at each wingtip, but there is no indication of a nose jet.

Each leg of the trailing-link tricycle landing gear carries a single wheel. The nose gear retracts rearward, the main units forward into the fuselage. A small bumper is fitted under the upward-curving rear fuselage.

Precise details of the engine installation are unknown. Primary propulsion appears to be by a single large turbojet, exhausting through a single pair of vectoring side-nozzles aft of the wing. No afterburner is fitted. The large lateral air intake ducts do not appear to embody splitter plates.

Two lift-jets are installed in tandem in the fuselage immediately aft of the cockpit, under a rearward-hinged louvred door of the kind fitted to the Mikoyan and Sukhoi STOL prototypes demonstrated in 1967. The position of the corresponding underfuselage doors implies that the lift-jets are mounted at an angle, in such a way that their thrust is exerted both upward and slightly forward. As the main vectored-thrust nozzles also turn up to 10° forward of vertical during take-off and landing, the total of four

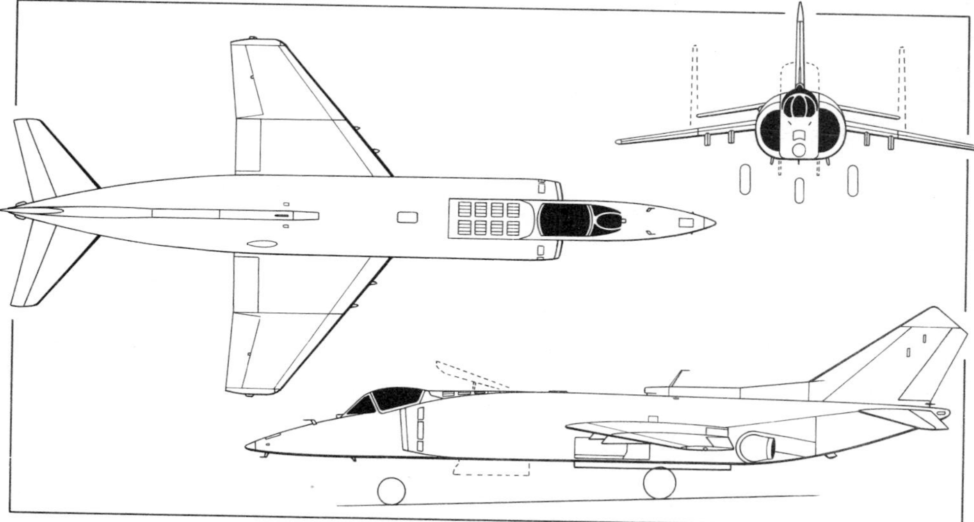

Yakovlev single-seat VTOL carrier-based combat aircraft (NATO *Forger-A*) *(Michael A. Badrocke)*

Yak-36 *(Forger-A)* **approaching to land on the carrier/cruiser** *Kiev (US Navy)*

exhaust effluxes can be envisaged as forming a V under the fuselage. There appears to be a small intake for cooling air at the front of the dorsal fin fairing.

Observers of deck flying by *Forger-As* from the *Kiev* report that the aircraft appeared to be extremely stable during take-off and landing. Take-offs were made vertically, with a smooth conversion about 5 to 6 m (15-20 ft) above the deck, followed by a fairly shallow climb-out as forward speed increased. Landings were so precise that some form of control from the ship during take-off and approach has been suggested, perhaps in association with laser devices lining each side of the rear deck. The purpose of the aircraft's small dielectric nose cap is as yet conjectural.

At no time was a STOL take-off observed, as practised by the Hawker Siddeley Harrier/AV-8A combat aircraft of the Royal Air Force and US Marine Corps to increase their load-carrying capability. It is suggested that anything but direct vertical take-off might be difficult for the pilot of *Forger-A*, as take-off with forward speed over the deck would impose formidable stability and safety problems. The Soviet aircraft must also lack the Harrier's ability to increase its combat manoeuvrability by the use of thrust vectoring in forward flight (VIFF).

Initial estimates put the thrust of *Forger-A's* primary power plant at around 75 kN (17,000 lb), and the thrust of each lift-jet at 25 kN (5,600 lb). This would appear adequate to permit a considerable weight of fuel and weapons to be carried. Gun pods and rocket packs have been photographed on four pylons under the aircraft's inner wing panels on deck, but no stores have yet been seen on these stations in flight. Performance of *Forger-A* is estimated to include a maximum level speed of Mach 1·3 at altitude.

DIMENSIONS, EXTERNAL (estimated):
- Wing span: 7·00 m (23 ft 0 in)
- Length overall:
  - *Forger-A*: 15·00 m (49 ft 3 in)
  - *Forger-B*: 17·66 m (58 ft 0 in)

WEIGHT (estimated):
- Max T-O weight:
  - *Forger-A*: 10,000 kg (22,050 lb)

Forger-A **single-seat version of the Yak-36 taking off** *(Royal Navy)*

First photograph of the two seat training version of the Yak-36 *(Forger-B)*

YAKOLEV Yak-40

NATO reporting name: *Codling*

This short-haul jet transport was designed to replace the Li-2 (Soviet-built DC-3) and to operate from Class 5 (grass) airfields. Although comparatively small, it is powered by three turbofan engines, mounted at the tail. The prototype flew for the first time on 21 October 1966 and was followed quickly by four more prototypes. Production was initiated in 1967 and the Yak-40 made its first passenger flight in Aeroflot service on 30 September 1968. By the Spring of 1973, more than eight million passengers had been carried, and Aeroflot Yak-40s had flown 108,000,000 nm (200,000,000 km; 124,270,000 miles). They had become the most widely used aircraft on Soviet domestic routes up to 810 nm (1,500 km; 930 miles) long by 1975, operating over several thousand short routes, some of them in mountain areas.

By the Summer of 1976, more than 800 Yak-40s had been built. Most are in service with Aeroflot, some as air ambulances carrying patients to medical centres and to Black Sea convalescent centres. Production continues in a factory at Saratov, about 500 km (300 miles) south-east of Moscow. Two were delivered to Aertirrena of Italy, who are also distributors of the Yak-40. Others have been sold in Afghanistan, Bulgaria, Czechoslovakia, France, West Germany, Poland and Yugoslavia. Military operators include the Soviet and Yugoslav air forces.

There has been no recent news of the high-density 40-seat version of the Yak-40, shown at the 1971 Paris Air Show, or of versions with uprated AI-25T engines, to which A. S. Yakovlev referred at that time. All current production Yak-40s are structurally similar, with AI-25 engines, and differ only in their standard of accommodation, for 27 or 32 airline passengers in a single class, 16 or 20 passengers in two-class layouts, or up to 11 passengers in an executive layout. All have clamshell thrust reversers aft of the centre engine. The pointed fairing forward of the fin/tailplane intersection on earlier aircraft is absent on current versions.

At the 1974 Hanover Air Show, a member of the Yakovlev bureau revealed that the prototype of a freighter version of the Yak-40 was under test at Saratov. This has a cargo door, size approximately 1·50 m × 1·60 m (5 ft × 5 ft 2½ in), in the port side of the fuselage. Novosti Press Agency reported in May 1975 that this freighter version would enter service with Aeroflot later in the year.

The Yak-40 is designed for a service life of 30,000 hours. It can take off and climb on any two engines and maintain height in cruising flight with two engines inoperative.

When the Yak-40 first entered service, it was the standard Soviet technique to cruise with the centre engine throttled back to idling thrust. It is now more usual to set all three engines at cruise thrust.

TYPE: Three-turbofan short-haul transport.

WINGS: Cantilever low-wing monoplane. Thickness/chord ratio 15% at root, 10% at tip. No sweepback at quarter-chord. All-duralumin structure consists of a main spar, fore and aft auxiliary spars, ribs and stringers, covered with skin of varying thickness for which chemical milling is utilised. Wing made in two sections, joined at aircraft centreline. Manually-operated ailerons, each in two sections. Hydraulically-operated plain flaps, each in three sections linked together by perforated plates, reportedly to cater for wing flexing in severe turbulence. Electrically-actuated trim tabs in ailerons. Automatic or manually-controlled hot-air de-icing system.

FUSELAGE: Semi-monocoque duralumin structure of frames, longerons and stringers. Floor of foam plastics with veneer covering. Skin panels spot-welded and bonded in place, then flush-riveted at ends.

TAIL UNIT: Cantilever structure of duralumin, with electrically-controlled hydraulically-actuated variable-incidence tailplane mounted at tip of fin. Manually-operated control surfaces. Electrically-actuated trim tab in rudder. Automatic or manually-controlled hot-air de-icing system.

LANDING GEAR: Hydraulically-actuated retractable type, with single wheel on each main unit. Emergency extension by gravity. Main wheels retract inward and are unfaired in flight. Long-stroke oleo-nitrogen shock-absorbers. Main-wheel tyres size 1,120 × 450, pressure 3·45-3·93 bars (50-57 lb/sq in). Hydraulically-steerable (55° each side) forward-retracting nosewheel, with tyre size 720 × 310, pressure 3·93-4·41 bars (57-64 lb/sq in). Hydraulic disc brakes.

POWER PLANT: Three Ivchenko AI-25 turbofan engines, each rated at 14·7 kN (3,300 lb st). Fin and boundary layer splitter beneath and forward of intake for centre engine. Clamshell thrust reverser fitted to airframe aft of this engine. Hot-air anti-icing system for all three engine air intakes. Fire warning and extinguishing systems standard. Fuel in integral tanks between front auxiliary spar and main spar in each wing, from outboard of the fuselage to the inner end of the aileron, total capacity 3,910 litres (860 Imp gallons). Type AI-9 turbine APU mounted in rear of top engine intake fairing for engine starting. Provision for starting from ground compressed air supply.

ACCOMMODATION: Two pilots side by side on flight deck, on adjustable seats, with dual controls. Central jump seat at rear for third person. Automatically-actuated electrical windscreen de-icing system. Main cabin normally laid out for 27 passengers in three-abreast rows, with two-chair units on starboard side of aisle. Seat pitch 755 mm (29·7 in). Individual ventilator by each seat. Rack for hand baggage on starboard side of cabin ceiling. Cloakroom (port), buffet, baggage compartment and toilet (starboard) aft of main cabin. Seat for stewardess against rear face of partition separating cabin from rear compartments, on port side. Normal access via hydraulically-actuated ventral airstair door at rear. Service door on port side of cabin, at front. For high-density services, twin-seat units can be installed on each side of aisle, giving a total of 32 seats in only eight rows. Seat pitch is unchanged, enabling the rear cabin partition to be moved forward and so giving a larger baggage compartment. The 16-seat mixed class version has two passenger cabins, separated by a partition. The forward cabin has two swivelling chairs, on each side of a table, on the port side; and an inward-facing four-place settee on the starboard side, with small cupboards fore and aft. Alternatively, two more swivelling seats and a table can replace the settee. The rear cabin contains 12 seats in standard three-abreast rows. A lobby between the flight deck and forward cabin provides access for the crew without passing through the main cabins, and contains a

Yakovlev Yak-40 short-range transport (three Ivchenko AI-25 turbofan engines) in Aeroflot markings *(Gordon S. Williams)*

cloakroom (starboard) and seat for the stewardess. Compartments aft of the rear cabin are as in the standard 27-seat version. The 20-seat version differs from the 16-seater in having 16 seats four-abreast in the rear cabin. The executive version has a toilet and other facilities in a large compartment aft of the flight deck; a centre lounge furnished with a four-place settee, three armchairs, a writing desk and a sideboard; and rear cabin containing six seats, three-abreast. There is a further toilet aft, and the galley is equipped to special standards. A bar with two adjustable tables is built into the wall between the flight deck and lounge, on the port side. A freight-carrying version is also available.

Electronics and Equipment: Standard equipment includes full blind-flying instrumentation, two Landysh-5 VHF radio communications installations, an ARK-10 automatic radio compass, KURS-MP-2 VOR/ILS system, RV-5 radio altimeter, Type SO 70 transponder, Grosa-40 weather radar, PRIVOD-ANE-1 flight director system, Kremenj 40E autopilot, AGD 1 artificial horizon and GMK-1GE gyro-compass, permitting automatic approach to ICAO Category II standards. A Collins avionics package is available optionally to Western operators.

Systems: Cabin fully pressurised and air-conditioned, with air bleed from final compressor stages of AI-25 engines; max pressure differential 0·294 bars (4·26 lb/sq in). Two independent hydraulic systems, pressure 147 bars (2,135 lb/sq in), supplied by Type NP-72M pumps on centre and port engines, with electrical standby pump and emergency handpump. Type AMG 10 hydraulic fluid. Electrical supply from three Type VG-7.500-1a DC generators and two Type 20 KNVN 25 batteries, each generator being driven by one of the turbofan engines. Two Type PO-1500 inverters provide 115V single-phase AC supply; Type PT-500 and PT-1000 inverters supply a 36V three-phase AC system. The Type AI-9 APU can also be used for cabin air-conditioning when the aircraft is on the ground. Installed and portable oxygen systems.

Dimensions, external:
- Wing span: 25·0 m (82 ft 0¼ in)
- Wing aspect ratio: 9
- Length overall: 20·36 m (66 ft 9½ in)
- Length of fuselage: 17·00 m (55 ft 9 in)
- Diameter of fuselage: 2·40 m (7 ft 10½ in)
- Height overall: 6·50 m (21 ft 4 in)
- Tailplane span: 7·50 m (24 ft 7¼ in)
- Wheel track: 4·52 m (14 ft 10 in)
- Wheelbase: 7·47 m (24 ft 6 in)
- Rear cabin door:
  - Height: 1·74 m (5 ft 8½ in)
  - Width: 0·94 m (3 ft 1 in)
- Service door:
  - Height: 1·20 m (3 ft 11¼ in)
  - Width: 0·55 m (1 ft 9½ in)

Dimensions, internal:
- Cabin: Length: 7·07 m (23 ft 2½ in)
  - Max width: 2·15 m (7 ft 0¾ in)
  - Max height: 1·85 m (6 ft 0¾ in)

Area:
- Wings, gross: 70·00 m² (753·5 sq ft)

Weights and Loadings (A: 27 seats, B: 32 seats, C: 16 seats, D: executive version):
- Weight empty:
  - A: 9,010-9,400 kg (19,865-20,725 lb)
  - B: 9,400 kg (20,725 lb)
  - D: 9,560-9,850 kg (21,075-21,715 lb)
- Max payload:
  - A: 2,300 kg (5,070 lb)
  - B: 2,720 kg (6,000 lb)
  - C: 1,360 kg (3,000 lb)
  - D: 990 kg (2,180 lb)
- Max fuel weight:
  - A: 2,125-4,000 kg (4,685-8,820 lb)
  - B: 4,000 kg (8,820 lb)
  - D: 3,000-4,000 kg (6,615-8,820 lb)
- Normal T-O weight:
  - A: 12,360-15,500 kg (27,250-34,170 lb)
  - B: 15,500 kg (34,170 lb)
  - D: 12,360-15,000 kg (27,250-33,070 lb)
- Max T-O weight:
  - A, B: 16,000 kg (35,275 lb)
  - C: 15,310 kg (33,750 lb)
  - D: 15,400 kg (33,950 lb)
- Max wing loading:
  - A, B: 230 kg/m² (47·1 lb/sq ft)

Performance (corresponding to weights given above):
- Max level speed at S/L: Mach 0·7 (324 knots; 600 km/h; 373 mph) IAS
- Max cruising speed at 7,000 m (23,000 ft): 297 knots (550 km/h; 342 mph)
- Max rate of climb at S/L: 480 m (1,575 ft)/min
- T-O speed:
  - A, B: 86 knots (160 km/h; 100 mph)
  - D: 81-84 knots (150-156 km/h; 93-97 mph)
- Normal T-O run:
  - A, B: 700 m (2,297 ft)
  - C: 650 m (2,133 ft)
  - D: 660 m (2,165 ft)
- Normal landing run:
  - A, C, D: 320 m (1,050 ft)
  - B: 360 m (1,182 ft)
- Range with max payload at 254 knots (470 km/h; 292 mph) at 8,000 m (31,500 ft), with reserves:
  - A, C, D: 971 nm (1,800 km; 1,118 miles)
  - B: 782 nm (1,450 km; 900 miles)
- Range with max fuel at 254 knots (470 km/h; 292 mph) at 8,000 m (31,500 ft), with reserves:
  - All versions: 971 nm (1,800 km; 1,118 miles)
- Max range at 254 knots (470 km/h; 292 mph) at 8,000 m (31,500 ft), no reserves:
  - All versions: 1,080 nm (2,000 km; 1,240 miles)

Yak-40 operated by the Polish aviation industry union PZL *(M. J. Axe)*

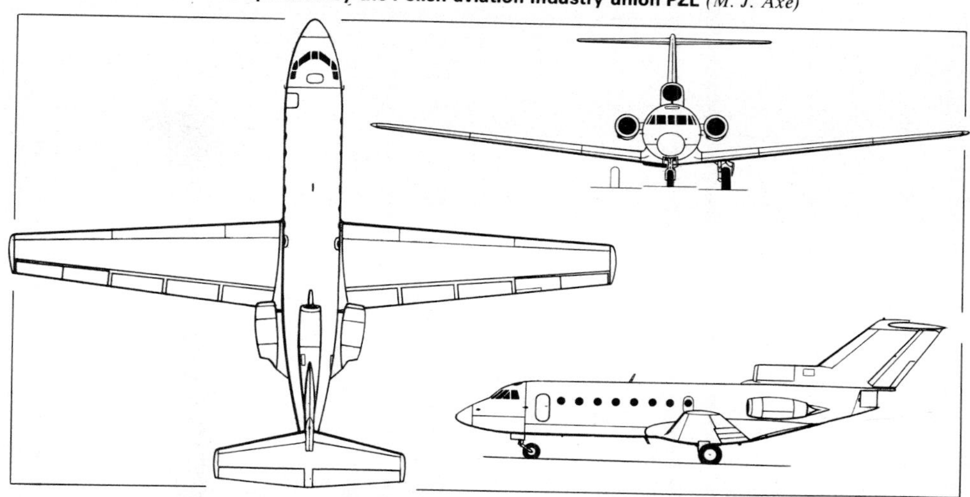

Yakovlev Yak-40 three-turbofan short-range transport *((Pilot Press)*

YAKOVLEV Yak-42

On the basis of experience with the Yak-40, the Yakovlev design bureau is developing for Aeroflot this larger civil airliner with a similar three-engined layout. A full-scale mockup existed in the bureau's prototype hangar in Moscow in mid-1973, but the design of the Yak-42 had not been finalised in detail by that time. First photographs of a model were released officially through *Tass* in 1974.

According to Alexander Yakovlev, the basic design objectives were simple construction, reliability in operation, economy and the ability to operate in remote areas with widely differing climatic conditions. Up to 2,000 aircraft in this category are needed, for use particularly on feederline services extending north and south from the main east-west trans-Siberian trunk routes.

The Yak-42 can accommodate 120 passengers in six-abreast seats at a pitch of 800 mm (31·5 in). Replacement of the front ten rows by four-abreast first class seats offers an alternative mixed class version for 100 passengers. Access to the cabin is by airstair doors under the rear fuselage and at the front of the cabin on the port side, making the aircraft independent of airport ground equipment. Immediately inside each lobby are carry-on baggage and coat compartments for use by the passengers.

Two freight holds are provided for cargo-carrying, with a chain-drive handling system built into the aircraft's floor. The forward hold can accommodate six containers, each

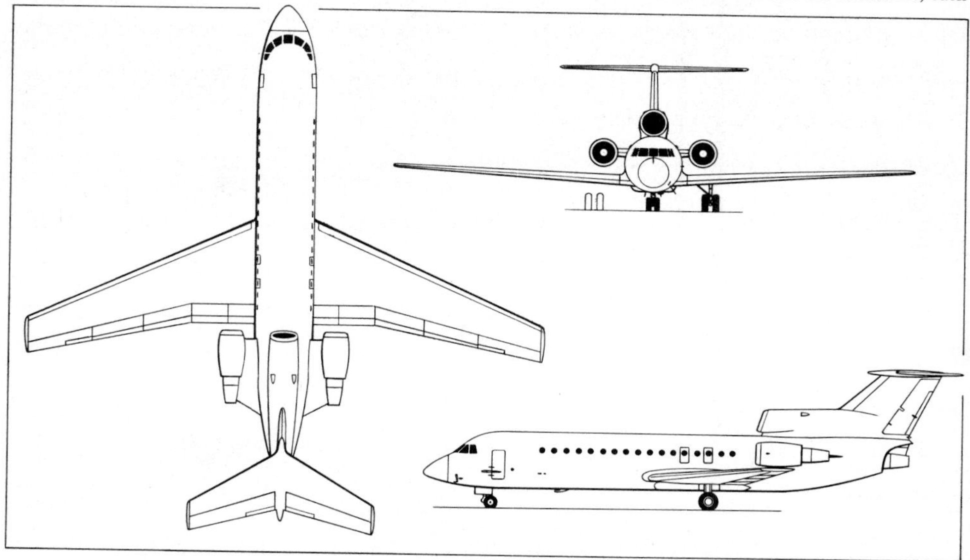

Yakovlev Yak-42 three-turbofan short-range passenger transport *(Pilot Press)*

with a capacity of 2·2 m³ (77·7 cu ft); the rear hold takes three similar containers.

A flight crew of two is normal, with provision for a high degree of automation, including an area navigation system. Control surfaces are actuated hydraulically, and high-lift devices include wing leading-edge slats. To cater for rough-field operations, a heavy-duty tricycle landing gear is fitted, with twin wheels on each unit.

The Yak-42 is powered by three D-36 high by-pass ratio (5·34 : 1) turbofan engines, designed under the leadership of Vladimir Lotarev at the Zaporozhye engine works. Take-off rating of each engine is 63·2 kN (14,200 lb st), and the Yak-42 is intended to use all three engines at cruise power during flight. Special care has been taken during design to ensure that the D-36 will conform with national and international limits on smoke and noise; and the Yak-42 is intended to operate in temperatures ranging from −50°C to +50°C.

Three prototypes of the Yak-42 were ordered initially. The first of these (CCCP-1974) flew for the first time on 7 March 1975 and has a wing sweepback of 11°. The second prototype has 25° of wing sweep for simultaneous evaluation. The configuration offering the better combination of high-speed cruise, economic and low-speed handling characteristics will be adopted for production Yak-42s, of which deliveries to Aeroflot are scheduled to begin in late 1978 or early 1979. The following data should be regarded as provisional:

DIMENSIONS, EXTERNAL:
Wing span 35·00 m (114 ft 10 in)
Wing sweepback 25°
Length overall 35·00 m (114 ft 10 in)

DIMENSION, INTERNAL:
Cabin: Max width 3·80 m (12 ft 6 in)

WEIGHTS:
Max payload 14,000 kg (30,850 lb)
Max T-O weight 52,000 kg (114,640 lb)

PERFORMANCE (estimated):
Max cruising speed 470 knots (870 km/h; 540 mph)
Econ cruising speed at 8,000 m (26,250 ft) 442 knots (820 km/h; 510 mph)
T-O run 800 m (2,625 ft)
Range with max payload 970 nm (1,800 km; 1,118 miles)
Range with max fuel 1,725 nm (3,200 km; 1,985 miles)

Yakovlev Yak-42 short-range transport (three Lotarev D-36 turbofan engines) *(Tass)*

YAKOVLEV Yak-50

A Novosti Press Agency bulletin, dated 30 June 1975, stated that tests of a new Yakovlev sporting aircraft, designated Yak-50, had been carried out near Arsenyev in the Soviet far east. Mr Nikolai Sazykin, director of the Progress Engineering Works in which all Yakovlev sporting aircraft are assembled, was quoted as saying that the Yak-50 was intended to participate in the 1976 world aerobatic championships.

No descriptive details, specification or illustrations of the new aircraft were released. Test pilot Anatoly Sergeyev stated only that it was more advanced than the familiar Yak-18 training and aerobatic monoplane, with a more powerful engine, better manoeuvrability, a speed of over 215 knots (400 km/h; 248 mph) in a dive, and the ability to perform all aerobatics with its landing gear retracted or extended.

When six Yak-50s participated in the 1976 world aerobatic championships at Kiev, their evolution from the Yak-18 was apparent, but with significant changes. Basic configuration is little different from that of the single-seat Yak-18PS, with tailwheel-type landing gear. This was deliberate, to keep the handling characteristics of the two types as similar as possible. However, overall dimensions are reduced; control surface hinge-lines have been moved to keep control forces light; and overall structural strength has been increased by switching entirely to metal covering. In particular, the fuselage is now semi-monocoque instead of steel tube with fabric covering to the rear of the cockpit.

The wings dispense with the Yak-18's centre-section and have no dihedral, but retain an asymmetric section. To ensure a high power:weight ratio in such a large aircraft, the power plant is a 268 kW (360 hp) Vedeneev (Ivchenko) M-14P aircooled radial piston engine, instead of the 224 kW (300 hp) Ivchenko AI-14RF of the Yak-18PS. Observers at the world championships reported that the Yak-50s performed the all-important Aresti manoeuvres with smooth precision, their primary shortcoming being excessive directional stability.

At Kiev, Yak-50s flown by V. Letsko and I. Egorov finished first and second in the men's competition. Others came fifth, seventh and ninth, to win the team prize. First five places in the women's championship were taken by Yak-50s.

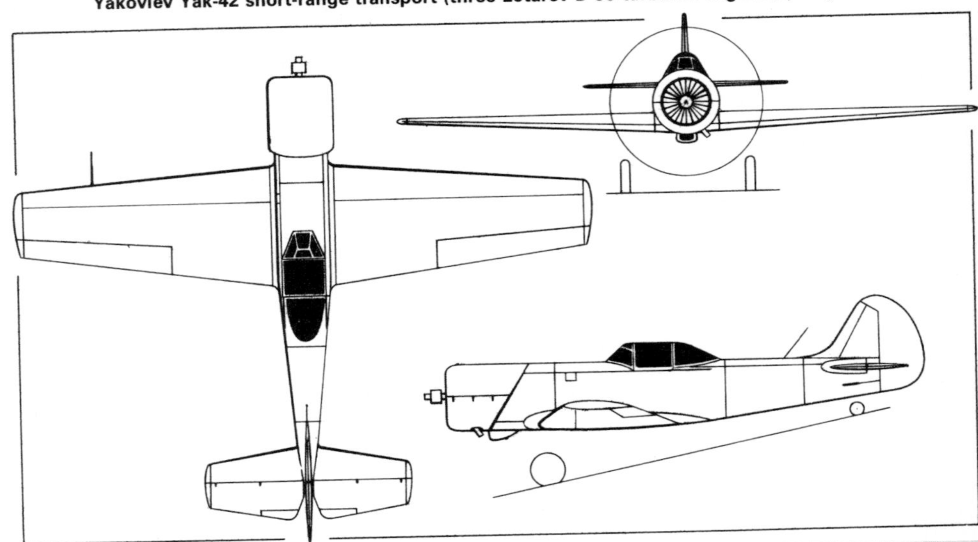

Photograph *(Tony Smith via Flight International)* **and three-view drawing** *(Pilot Press)* **of the Yakovlev Yak-50, winner of the 1976 world aerobatic championships**

DIMENSIONS, EXTERNAL:
Wing span 9·50 m (31 ft 2 in)
Length overall 7·46 m (24 ft 5¾ in)

AREA:
Wings, gross 15·0 m² (161·5 sq ft)

WEIGHTS:
Weight empty, equipped 740 kg (1,632 lb)
Max T-O weight 875 kg (1,930 lb)

PERFORMANCE:
Max permissible diving speed 253 knots (470 km/h; 292 mph)
Max level speed 173 knots (320 km/h; 199 mph)
Stalling speed 54 knots (100 km/h; 63 mph)

YUGOSLAVIA

SOKO

"SOKO" METALOPRERADIVACKA INDUSTRIJA OOUR FABRIKA VAZDUHOPLOVA BEZ OGRANICENE ODGOVORNOSTI

ADDRESS:
Mostar
Telephone: 22-121/125, 22-139, 22-156/157, 22-183
Telex: 46-180

GENERAL MANAGER: Dipl-Ing Ivan Sert
ASSISTANT MANAGING DIRECTOR:
Dipl-Ing Tomislav Miric
DIRECTOR, AIRCRAFT DIVISION:
Dipl-Ing Sulejman Gosto
MARKETING DIRECTOR: Fuad Bijedic

Founded in 1951, this company is manufacturing a two-seat jet basic trainer named the Galeb, a single-seat light attack version of the same design, named the J-1 Jastreb, and a two-seat conversion trainer version of the Jastreb.

Soko is participating, with Romania, in developing the Orao strike aircraft described under the 'VTI/CIAR' heading in the International section.

Soko is also building under licence the Aérospatiale/Westland Gazelle helicopter, on behalf of the Yugoslav government.

SOKO G2-A GALEB (SEAGULL)

Design of the Galeb was started in 1957. Construction of two prototypes began in 1959 and the first of these flew for the first time in May 1961. Development was carried out in collaboration with the Yugoslav Aeronautical Research Establishments and construction is in accordance with current military airworthiness requirements. Production began in 1963 for the Yugoslav Air Force and has continued to fulfil repeat Yugoslav and export orders. First overseas operator was the Zambian Air Force, in early 1971.

There are two current versions of the Galeb:

G2-A. Standard version for Yugoslav Air Force. Progressive design improvements include availability of optional cockpit air-conditioning system.

G-2A-E. Export version, with updated equipment. First flown in late 1974. Series production began in 1975, reportedly to fulfil an order from the Libyan Arab Air Force.

TYPE: Two-seat armed jet basic trainer, designed for load factors of +8*g* and −4*g*.

WINGS: Cantilever low-wing monoplane. Wing section NACA 64A213·5 at root, NACA 64A212·0 at tip. Dihedral 1° 30′. No incidence. Sweepback at quarter-chord 4° 19′. Conventional light alloy two-spar stressed-skin structure, consisting of a centre-section, integral with the fuselage, and two outer panels which can be removed easily. Manually-operated internally-sealed light alloy ailerons. Trim tab on port aileron. Hydraulically-actuated Fowler flaps. No de-icing system.

FUSELAGE: Light alloy semi-monocoque structure in two

portions, joined together by four bolts at frame aft of wing trailing-edge. Rear portion removable for engine servicing. Two hydraulically-actuated door-type airbrakes under centre-fuselage.

TAIL UNIT: Cantilever light alloy stressed-skin structure. Fixed-incidence tailplane. Rudder and elevators statically and dynamically balanced and manually operated. Manually-operated trim tab in each elevator. VHF radio aerial forms tip of fin.

LANDING GEAR: Hydraulically-retractable tricycle type, with single wheel on each unit. Nosewheel retracts forward, main units inward into wings. Oleo-pneumatic shock-absorbers manufactured by Prva Petoletka of Trstenik. Dunlop main wheels and tyres size 23 × 7·25-10, pressure 4·41 bars (64 lb/sq in). Dunlop nosewheel and tyre size 6·50-5·5 TC, pressure 3·43 bars (49·8 lb/sq in). Prva Petoletka hydraulic differential disc brakes, toe-operated from both cockpits.

POWER PLANT: One Rolls-Royce Bristol Viper 11 Mk 22-6 turbojet engine, rated at 11·12 kN (2,500 lb st). Two flexible fuel tanks aft of cockpits, with total capacity of 780 kg (1,720 lb). Two jettisonable wingtip tanks, each with capacity of 170 kg (375 lb). Refuelling point on upper part of fuselage aft of cockpits. Fuel system designed to permit up to 15 seconds of inverted flight. Oil capacity 6·25 litres (1·4 Imp gallons).

ACCOMMODATION: Crew of two in tandem on HSA (Folland) Type 1-B fully-automatic lightweight ejection seats. Separate sideways-hinged (to starboard) jettisonable canopy over each cockpit. Cockpit air-conditioning to special order only.

SYSTEMS: Hydraulic system, pressure 58·5-69 bars (850-1,000 lb/sq in), for landing gear, airbrakes and flaps. Separate system for wheel brakes. Pneumatic system for armament cocking. Electrical system includes 6kW 24V generator, 24V battery, and inverter to provide 115V 400Hz AC supply for instruments. G2-A has low-pressure oxygen system, capacity 1,450 litres. G-2A-E has high-pressure oxygen system.

ELECTRONICS AND EQUIPMENT (G2-A): Blind-flying instrumentation, Marconi radio compass (licence-built by Rudi Cajavec), intercom and STR-9Z1 VHF radio transceiver standard. Standard electrical equipment includes navigation lights, 250W landing lamp in nose, and 50W taxying lamp on nose landing gear. Camera, with focal length of 178 mm (7 in) and 125-exposure magazine, can be fitted in fuselage, under rear cockpit floor. Flares can be carried on the underwing bomb racks for night photography. Target towing hook under centre-fuselage.

ELECTRONICS AND EQUIPMENT (G-2A-E): Full IFR instrumentation. Electronique Aérospatiale (EAS) Type TVU-740 VHF/UHF com radio transceiver, Marconi AD 370B radio compass, EAS RNA-720 VOR/LOC and ILS, Iskra 75R4 marker beacon receiver, and intercom. Otherwise as G2-A.

ARMAMENT: All production aircraft have two 0·50 in machine-guns in nose (with 80 rds/gun); and underwing pylons for two 50 kg or 100 kg bombs and four 57 mm rockets or two 127 mm rockets; or clusters of small bombs and expendable bomblet containers of up to 150 kg (330 lb) weight (300 kg; 660 lb total).

DIMENSIONS, EXTERNAL:

Wing span	10·47 m (34 ft 4½ in)
Wing span over tip-tanks	11·62 m (38 ft 1½ in)
Wing chord at root	2·36 m (7 ft 9 in)
Wing chord at tip	1·40 m (4 ft 7 in)
Wing aspect ratio	5·55
Length overall	10·34 m (33 ft 11 in)
Height overall	3·28 m (10 ft 9 in)
Tailplane span	4·27 m (14 ft 0 in)
Wheel track	3·89 m (12 ft 9 in)
Wheelbase	3·59 m (11 ft 9½ in)

AREAS:

Wings, gross	19·43 m² (209·14 sq ft)
Ailerons (total)	2·36 m² (25·40 sq ft)
Trailing-edge flaps (total)	2·02 m² (21·75 sq ft)
Airbrake	0·34 m² (3·66 sq ft)
Fin	1·34 m² (14·42 sq ft)
Rudder, incl tab	0·56 m² (6·03 sq ft)
Tailplane	3·66 m² (39·40 sq ft)
Elevators, incl tabs	0·83 m² (8·93 sq ft)

WEIGHTS:

Weight empty, equipped	2,620 kg (5,775 lb)
Max T-O weight:	
Fully-aerobatic trainer ('clean')	3,374 kg (7,438 lb)
Basic trainer (no tip-tanks)	3,488 kg (7,690 lb)
Navigational trainer (with tip-tanks)	3,828 kg (8,439 lb)
Weapons trainer	3,988 kg (8,792 lb)
Strike version	4,300 kg (9,480 lb)

PERFORMANCE (at normal T-O weight):

Max level speed at S/L	408 knots (756 km/h; 470 mph)
Max level speed at 6,200 m (20,350 ft)	438 knots (812 km/h; 505 mph)
Max cruising speed at 6,000 m (19,680 ft)	394 knots (730 km/h; 453 mph)
Stalling speed:	
flaps and airbrakes down	85 knots (158 km/h; 98 mph)
flaps and airbrakes up	97 knots (180 km/h; 112 mph)
Max rate of climb at S/L	1,370 m (4,500 ft)/min
Time to 3,000 m (9,840 ft)	2·4 min
Time to 6,000 m (19,680 ft)	5·5 min
Time to 9,000 m (29,520 ft)	10·2 min
Service ceiling	12,000 m (39,375 ft)
T-O run on grass	490 m (1,610 ft)
T-O to 15 m (50 ft)	640 m (2,100 ft)
Landing from 15 m (50 ft)	710 m (2,330 ft)
Landing run on grass	400 m (1,310 ft)
Max range at 9,000 m (29,520 ft), with tip-tanks full	669 nm (1,240 km; 770 miles)
Max endurance at 7,000 m (23,000 ft)	2 hr 30 min

Soko G2-A Galeb two-seat basic training aircraft (Rolls-Royce Bristol Viper 11 turbojet engine)

SOKO J-1/RJ-1 JASTREB (HAWK)

The basic J-1 Jastreb is a single-seat light attack version of the G2-A Galeb, developed and produced for service with the Yugoslav Air Force. An export version is available, and the first overseas operator was the Zambian Air Force, which received four Jastrebs in early 1971.

In the J-1 Jastreb, the front cockpit of the G2-A Galeb trainer, with sideways-hinged (to starboard) canopy, is retained, a metal fairing replacing the rear canopy. The engine is the more powerful Rolls-Royce Bristol Viper 531. Other changes include the installation of improved day and night reconnaissance equipment, electronic navigation and communications equipment, and self-contained engine starting. In other respects the airframe and power plant remain essentially unchanged except for some local strengthening and the provision of strongpoints for heavier underwing stores.

Currently in production and service are two attack versions and a tactical reconnaissance version of the Jastreb, as follows:

J-1. Standard attack version for Yugoslav Air Force.

J-1-E. Export attack version with updated equipment. A number of J-1-Es have been ordered by foreign operators.

RJ-1. Tactical reconnaissance version for Yugoslav Air Force.

An export reconnaissance version with updated equipment was scheduled to fly for the first time in Summer 1976. In addition, there is a two-seat operational conversion and pilot proficiency training version, designated TJ-1. This is described separately.

The details given for the G2-A Galeb apply equally to the J-1, J-1-E and RJ-1 Jastreb, with the following exceptions:

TYPE: Single-seat light attack and tactical reconnaissance aircraft.

POWER PLANT: One Rolls-Royce Bristol Viper 531 turbojet engine, rated at 13·32 kN (3,000 lb st). Capacity of each wingtip tank 220 kg (485 lb). Provision for attaching two 4·44 kN (1,000 lb st) JATO rockets under fuselage for use at take-off or in flight.

ACCOMMODATION: Pilot only, on HSA (Folland) Type 1-B fully-automatic lightweight ejection seat. Cockpit air-conditioning to special order only.

SYSTEMS: Electrical system includes 6kW 24V generator and second battery, permitting independent engine starting without ground electrical supply. Two oxygen bottles supply high-pressure oxygen system of nominal 1,900 litres (67 cu ft) capacity, at a pressure of 138 bars (2,000 lb/sq in).

ELECTRONICS AND EQUIPMENT (J-1 and RJ-1): Full IFR instrumentation. Standard Telephones & Cables STR-9Z1 VHF com transceiver and Marconi AD 370B radio compass. The fuselage camera of the RJ-1 is supplemented by two further cameras in nose of tip-tanks, which are also available for the J-1 and J-1-E attack versions. An aerial target can be towed from a hook under the centre-fuselage. Brake parachute housed in fairing above jet nozzle.

ELECTRONICS (J-1-E): Nav/com equipment same as for G-2A-E Galeb export version.

ARMAMENT (J-1 and J-1-E): Three 0·50 in Colt-Browning machine-guns in nose (with 135 rds/gun). Total of eight underwing weapon attachments. Two inboard attachments can carry two bombs of up to 250 kg each, two clusters of small bombs, two 200 litre napalm tanks, two pods each with twelve or sixteen 57 mm or four 128 mm rockets, two multiple carriers each with three 50 kg bombs, two bomblet containers, or two 45 kg photo flares. Other attachments can each carry a 127 mm rocket. Semi-automatic gyro gunsight and camera gun standard.

ARMAMENT (RJ-1): Four underwing attachments, intended basically for carrying flash bombs for night photography, can be used also for carrying high-explosive or other types of bombs. The inboard pylons can each carry a single bomb of up to 250 kg, the outboard pylons up to 150 kg. No rocket armament. Otherwise same as for J-1 and J-1-E.

DIMENSIONS, EXTERNAL:

As for Galeb, except:

Wing span over tip-tanks	11·68 m (38 ft 4 in)
Length overall	10·88 m (35 ft 8½ in)
Height overall	3·64 m (11 ft 11½ in)
Wheelbase	3·61 m (11 ft 10 in)

AREAS:

As for Galeb

WEIGHTS:

Weight empty, equipped	2,820 kg (6,217 lb)
Max ramp weight	5,287 kg (11,655 lb)
Max T-O weight	5,100 kg (11,243 lb)

Soko J-1 Jastreb single-seat light attack aircraft, developed from the Galeb

Max landing weight 3,750 kg (8,267 lb)
PERFORMANCE (T-O and landing runs on concrete):
Max level speed at 6,000 m (19,680 ft) at AUW of 3,968 kg (8,748 lb) 442 knots (820 km/h; 510 mph)
Max cruising speed at 5,000 m (16,400 ft), at AUW of 3,968 kg (8,748 lb) 399 knots (740 km/h; 460 mph)
Stalling speed, wheels down:
flaps and airbrakes down 82 knots (152 km/h; 95 mph)
flaps and airbrakes up 94 knots (174 km/h; 108 mph)
Max rate of climb at S/L, at AUW of 3,968 kg (8,748 lb) 1,260 m (4,135 ft)/min
Service ceiling at AUW of 3,968 kg (8,748 lb) 12,000 m (39,375 ft)
T-O run at AUW of 3,968 kg (8,748 lb) 700 m (2,300 ft)
T-O run, rocket-assisted, at max T-O weight 404 m (1,325 ft)
T-O to 15 m (50 ft) at AUW of 3,968 kg (8,748 lb) 960 m (3,150 ft)
T-O to 15 m (50 ft), rocket-assisted, at max T-O weight 593 m (1,945 ft)
Landing from 15 m (50 ft) 1,100 m (3,610 ft)
Landing run 600 m (1,970 ft)
Max range at 9,000 m (29,520 ft), with tip-tanks full 820 nm (1,520 km; 945 miles)

SOKO TJ-1 JASTREB TRAINER

This two-seat operational conversion and pilot proficiency training version of the Jastreb is designed for maximum commonality with the J-1, retaining the full operational capability of the ground attack version. The prototype TJ-1 flew for the first time in mid-1974. Deliveries of production aircraft began in January 1975, to fulfil Yugoslav and export orders.

The details given for the J-1 Jastreb apply equally to the TJ-1 Jastreb Trainer, with the following exceptions:

TYPE: Two-seat operational conversion trainer.

ACCOMMODATION: Crew of two in tandem on HSA (Folland) Type 1-B ejection seats. Separate sideways-hinged (to starboard) jettisonable canopy over each cockpit.

ELECTRONICS AND EQUIPMENT: Same as for J-1 Jastreb, plus intercom and Iskra 75R4 marker beacon receiver. Only two cameras, in tip-tank nosecones.

Cockpit section of the TJ-1 Jastreb Trainer

WEIGHTS:
Weight empty, equipped 2,980 kg (6,570 lb)
Typical training mission T-O weight 4,350 kg (9,590 lb)
Max landing weight 3,950 kg (8,708 lb)

UTVA

FABRIKA AVIONA

HEAD OFFICE AND WORKS:
Utva Zlatokrila 9, Pancevo
Telephone: (013) 44-755
Telex: 131-16
GENERAL MANAGER: Marko Saranović
MANAGER OF AIRCRAFT DIVISION: Zdravko Rapaić
ASSISTANT MANAGER: Dipl Ing Dragoslav Dimić

Following completion of series production of the UTVA-65 Super Privrednik-350 agricultural aircraft and UTVA-66 utility aircraft (both described fully in 1975-76 *Jane's*), UTVA Fabrika Aviona is now concentrating its effort on development of the new UTVA-75 two-seat light aircraft.

UTVA-75

The UTVA-75 is a side-by-side two-seat training, glider towing and utility lightplane, of conventional low-wing monoplane configuration with non-retractable tricycle landing gear. It has been projected, designed and built in partnership by UTVA-Pancevo, Prva Petoletka-Trstenik, Vazduhoplovnotehnicki Institut and Institut Masinskog Fakulteta of Belgrade. Design was started in 1974. Construction of two prototypes was undertaken in 1975, and the first of these flew for the first time on 20 May 1976. It is intended to certificate the UTVA-75 to FAR Pt 23 Aerobatic category standard, at load factors of $+6g$ to $-3g$.

TYPE: Two-seat light aircraft.

WINGS: Cantilever low-wing monoplane, with short-span centre-section and two constant-chord outer panels. Wing section NACA 65_2415. Dihedral 0° on centre-section, 6° on outer panels. Conventional all-metal structure. Ailerons and flaps along entire trailing-edge of outer panels, except for tips. No tabs.

FUSELAGE: Conventional all-metal semi-monocoque structure.

TAIL UNIT: Cantilever all-metal structure, with sweptback vertical surfaces. Fluted skin on fin and rudder. Elevator and rudder horn-balanced. Trim tab in elevator.

LANDING GEAR: Non-retractable tricycle type, with single wheel on each unit, and small tail bumper. PPT oleo-pneumatic shock-absorbers. Goodyear tyres, size 6·00-6 on main wheels, 5·00-5 on nosewheel.

POWER PLANT: One 134 kW (180 hp) Lycoming IO-360-B1F flat-four engine, driving a Hartzell HC-C2YK-1BF/F7666A two-blade metal variable-pitch propeller. Integral fuel tanks in wings, total capacity 150 litres (33 Imp gallons). Provision for carrying two 100 litre (22 Imp gallon) drop-tanks under wings, raising max total capacity to 350 litres (77 Imp gallons).

ACCOMMODATION: Two seats side-by-side in enclosed cabin, with large upward-opening canopy door over each seat, hinged on centreline. Cabin heated and ventilated.

SYSTEMS: Dual hydraulic system. 12V electrical system, with 35Ah battery, navigation lights, rotating beacon and landing lights as standard equipment.

ELECTRONICS AND EQUIPMENT: King KY-195 BE radio optional. Standard equipment includes radio compass, VOR and ILS.

ARMAMENT AND MILITARY EQUIPMENT: Provision for various weapon loads on underwing stores attachments.

DIMENSIONS, EXTERNAL:
Wing span 9·69 m (31 ft 9½ in)
Wing chord (constant) 1·55 m (5 ft 1 in)
Length overall 7·11 m (23 ft 4 in)
Height overall 3·15 m (10 ft 4 in)
Tailplane span 3·60 m (11 ft 9¾ in)
Wheel track 2·58 m (8 ft 5½ in)

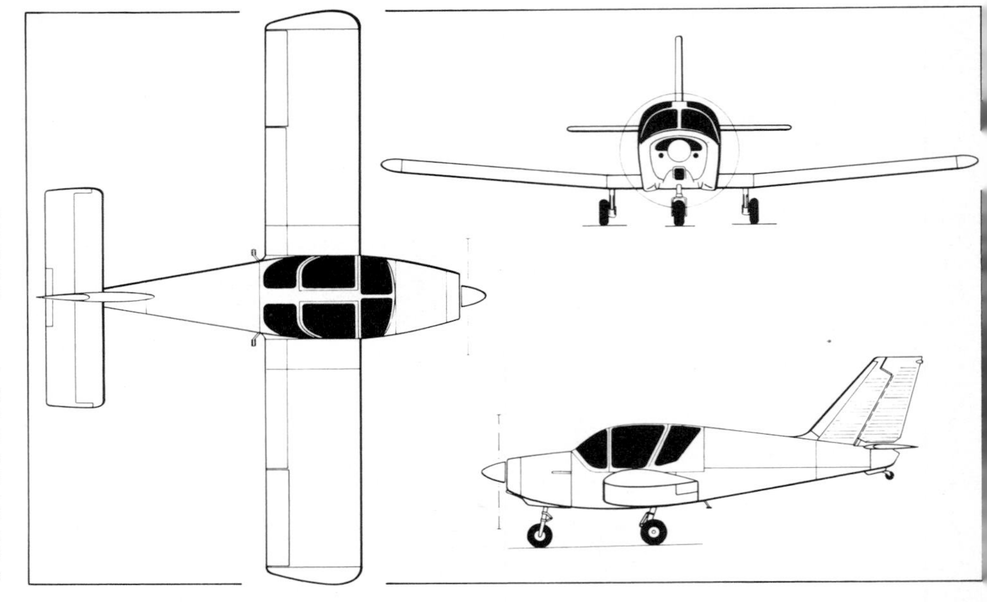

Photograph and three-view drawing *(Roy J. Grainge)* **of the UTVA-75 light aircraft**

Propeller diameter 1·93 m (6 ft 4 in)
Propeller ground clearance 0·295 m (11¾ in)
AREA:
Wings, gross 14·63 m² (157·5 sq ft)
WEIGHTS AND LOADINGS:
Weight empty, equipped 620 kg (1,367 lb)
Max T-O weight 900 kg (1,984 lb)
T-O weight, glider towing version 730 kg (1,609 lb)
PERFORMANCE (estimated at max T-O weight):
Max level speed 129 knots (240 km/h; 149 mph)
Max cruising speed (75% power) 114 knots (212 km/h; 132 mph)
Cruising speed (60% power) 103 knots (191 km/h; 119 mph)
Stalling speed, engine idling:
flaps up 55 knots (101 km/h; 63 mph)
20° flaps 47 knots (87 km/h; 54 mph)
40° flaps 44·5 knots (82 km/h; 51 mph)
Max rate of climb at S/L 348 m (1,140 ft)/min
Service ceiling 5,400 m (17,725 ft)
T-O run:
on concrete 138 m (455 ft)
on grass 160 m (525 ft)
T-O to 15 m (50 ft):
on concrete 237 m (780 ft)
on grass 257 m (845 ft)
Landing from 15 m (50 ft):
on concrete 240 m (790 ft)
on grass 317 m (1,040 ft)
Landing run:
on concrete 109 m (360 ft)
on grass 186 m (610 ft)
Range with max standard fuel, at 90·5 knots (167 km/h; 104 mph), no reserves 502 nm (931 km; 578 miles)
Range with drop-tanks, at 90·5 knots (167 km/h; 104 mph), no reserves 1,078 nm (2,000 km; 1,242 miles)

HOMEBUILT AIRCRAFT
(including Man-powered and Racing Aircraft)

ARGENTINE REPUBLIC

AVEX
ASOCIACION ARGENTINA DE CONSTRUCTORES DE AVIONES EXPERIMENTALES

ADDRESS:
Accasusso 1640, Olivos-FCNGBM, Buenos Aires
Telephone: 797-1629
PRESIDENT: Yves Arrambide
SECRETARY: Norberto Marino

AVEX is an Argentine light aircraft association for amateur constructors, similar in concept to the Experimental Aircraft Association in the USA. It was formed in 1968 and its members include many people well known among the Argentine aircraft industry, including specialists in most aspects of materials and construction, including the use of glassfibre and plastics.

Recent AVEX activities have been described in the 1971-72 and subsequent editions of *Jane's*. Of 42 current aircraft projects by AVEX members, two had flown and six others were nearing completion in 1975: two Gorrions, the Yakstas racer, and single examples of the Evans VP-1, Jodel D.9 and Mignet H-14. Details of the more important Argentinian designs follow:

ARRAMBIDE / MARINO ARMAR I GORRION (SPARROW)

The Gorrion single-seat ultra-light aircraft was designed in collaboration by Mr Yves Arrambide and Sr Norberto Marino in 1971. Trial flights were made with a rubber-propelled one-fortieth scale model; construction of a full-scale prototype began on 30 April 1972, and this was scheduled to make its first flight in 1975. A second Gorrion, shown in the accompanying photograph, was also expected to make its first flight in 1975. Built by Sr E. Puglisi of Rosario, Santa Fé, it differs from the prototype in having rounded wingtips and top-decking, and a 30 kW (40 hp) Continental A40 engine.

All available details of the prototype Gorrion follow:

TYPE: Single-seat ultra-light aircraft.

WINGS: Parasol-wing monoplane. Centre-section braced by N strut on each side of upper fuselage, and outer panels by Vee struts from bottom of fuselage. Wing section NACA 4412 (constant). Dihedral 3° from roots. Incidence 3°. No sweepback. Two-spar wooden structure, of constant chord except for cutout in centre of trailing-edge. Leading-edges plywood-covered; remainder fabric-covered. Frise-type fabric-covered wooden ailerons. No tabs.

FUSELAGE: Conventional wooden box structure. Aluminium cowling panels; remainder plywood-covered except for fabric-covered top-decking.

TAIL UNIT: Cantilever wooden structure; plywood-covered fin and one-piece tailplane, fabric-covered rudder and elevators. No tabs. Rudder control by cables.

LANDING GEAR: Non-retractable tailwheel type. Glassfibre legs provide all necessary shock-absorption. Main units have scooter wheels and brakes.

POWER PLANT: One 29 kW (39 hp) Citröen 3 CV motor-car engine, with reduction gear, driving a two-blade fixed-pitch wooden propeller. Single fuel tank in fuselage, capacity 35 litres (7·5 Imp gallons). Refuelling point on top of fuselage aft of firewall. Oil capacity 3 litres (0·66 Imp gallons).

ACCOMMODATION: Single seat in open cockpit. Windscreen fitted. Headrest faired into top of fuselage.

DIMENSIONS, EXTERNAL:
Wing span (excl tip fairings) 7·00 m (22 ft 11¾ in)
Wing chord (constant) 1·25 m (4 ft 1¼ in)
Wing area, gross 8·50 m² (91·5 sq ft)
Wing aspect ratio 5·6
Length overall 4·72 m (15 ft 5¾ in)
Height overall (tail up) 1·96 m (6 ft 5¼ in)
Tailplane span 2·20 m (7 ft 2¾ in)
Wheel track 1·40 m (4 ft 7 in)
Propeller diameter 1·60 m (5 ft 3 in)

DIMENSIONS, INTERNAL:
Cabin:
Width (constant) 0·56 m (1 ft 10 in)

WEIGHTS AND LOADINGS:
Weight empty 166 kg (366 lb)
Max T-O weight 276 kg (608 lb)
Max wing loading 32·0 kg/m² (6·6 lb/sq ft)
Max power loading 9·52 kg/kW (15·59 lb/hp)

PERFORMANCE (estimated, at max T-O weight):
Max level speed at 1,500 m (5,000 ft) 75·5 knots (140 km/h; 87 mph)
Max cruising speed at 1,500 m (5,000 ft) 65 knots (120 km/h; 75 mph)
Stalling speed 30·5 knots (56 km/h; 35 mph)
Service ceiling 3,000 m (9,845 ft)
Range with max fuel 194 nm (360 km; 223 miles)

GHINASSI HELICOPTERS

Sr Sesto Ghinassi is a specialist in, and racer of, motor-cycles. He built a small single-seat helicopter, using unapproved materials and a 22·4 kW (30 hp) engine developed by himself.

This aircraft (described and illustrated in the 1973-74 *Jane's*) was later scrapped, but Sr Ghinassi currently has a new helicopter under construction, for which he is using aircraft quality materials.

ROTOR SYSTEM AND DRIVE: Variable-pitch main and tail rotors, the former driven by chain drive from engine. Symmetrical-section blades, with 7% thickness/chord ratio, of wooden construction with aluminium skin. Max rpm of main rotor 400.

FUSELAGE: Welded steel tube structure.

POWER PLANT: One 22·4 kW (30 hp) 470 cc two-cylinder four-stroke turbine-cooled engine of own design; max rpm 6,300. Fuel tank capacity 20 litres (4·4 Imp gallons).

ACCOMMODATION: Single seat.

DIMENSIONS, EXTERNAL:
Main rotor diameter 6·00 m (19 ft 8¼ in)
Tail rotor diameter 0·60 m (1 ft 11¾ in)
Main rotor blade chord 0·22 m (8¾ in)
Fuselage length 4·00 m (13 ft 1½ in)
Fuselage width 1·00 m (3 ft 3¼ in)
Height overall 1·55 m (5 ft 1 in)

PERFORMANCE (estimated):
Range 162 nm (300 km; 186 miles)

YAKSTAS RACER

Begun by Prof Adolfo Yakstas as a much-modified development of the Baserga H.B.1 (see 1970-71 *Jane's*), this has evolved into virtually a new design having only the two-cylinder Praga engine in common with its predecessor.

Work was restarted in the Spring of 1974, and by early 1975 the basic structure was almost complete and ready for covering. There has been no subsequent news of the aircraft; but in the Spring of 1976 Prof Yakstas was reported to be engaged on completing assembly of the AL-2 Tijerete single-seat twin-boom pusher-engined light aircraft of which manufacture was started by the former Al-Aire company (see 1974-75 *Jane's*).

The following details apply to the Yakstas Racer:

TYPE: Single-seat racing monoplane.

FUSELAGE: Aluminium-skinned steel tube forward section. Rear section, aft of main landing gear legs, is of wooden construction with fabric-covered upper and plywood-covered lower surfaces.

TAIL UNIT: Fabric-covered steel tube structure, without sweepback.

LANDING GEAR: Non-retractable tailwheel type. Spring steel main gear legs.

POWER PLANT: One 33·5 kW (45 hp) Praga B-2 two-cylinder engine, driving a two-blade fixed-pitch propeller with spinner.

ACCOMMODATION: Single seat under two-piece moulded Plexiglas canopy.

DIMENSIONS, EXTERNAL:
Wing span 7·60 m (24 ft 11¼ in)
Wing chord 1·20 m (3 ft 11¼ in)
Wing area, gross 9·12 m² (98·2 sq ft)
Length overall 5·10 m (16 ft 8¾ in)
Tailplane span 2·50 m (8 ft 2½ in)

AUSTRALIA

CORBY
JOHN C. CORBY

ADDRESS:
86 Eton Street, Sutherland, NSW 2232

Mr Corby, a consultant aero engineer, has designed and is marketing plans for a single-seat ultra-light homebuilt aircraft known as the Starlet.

By March 1976 nine Starlets had been completed and a further 40 were known to be under construction in Australia, Tasmania and New Zealand, including a metal-construction Starlet.

CORBY CJ-1 STARLET

The first Starlet (VH-ULV) was built by a group of about ten members of the Latrobe Valley division of the Australian Ultra Light Aircraft Association. Mr Erle Jones (Secretary and former President of the ULAA, Latrobe Valley Aero Club) was responsible for test flying the completed aircraft, with Mr John Brown acting as instructor in building techniques. Details of this aircraft were given in the 1974-75 *Jane's*.

The following description applies to the standard Corby Starlet, except where indicated:

TYPE: Single-seat ultra-light homebuilt aircraft.

WINGS: Cantilever low-wing monoplane of wooden construction. Wing section NACA 43012A. Dihedral 6°. Incidence 2° 30′ at root, −1° at tip. Laminated main spar of solid spruce, subspars of spruce, built-up girder-type ribs and D-shaped nose section. Plywood covering from leading-edge to main spar, remainder fabric covered. Provision for dismantling into two equal halves. Ailerons, of spruce with birch plywood covering, deflect 15° up and down.

FUSELAGE: Plywood-covered spruce structure.

TAIL UNIT: Cantilever type, of similar construction to wings. Fixed-incidence tailplane. Plywood-covered fixed surfaces; fabric-covered rudder and elevators. Elevators deflect 30° up, 20° down; rudder deflects 25° to left and right.

LANDING GEAR: Non-retractable two-wheel type standard. Separate spring steel leaf-type shock-absorbing main legs, attached directly to fuselage via a solid spruce/ash beam which also serves as the wing leading-edge attachment member. Wheels, tyres and brakes of customer's choice, subject to main wheels of 89 mm (3½ in) minimum diameter with 4·00-4 tyres and Olympic go-kart hubs. Sturmey Archer cycle drum/shoe brakes may be used. Leaf-spring tailskid, or tailwheel at customer's option. Wheel fairings optional.

POWER PLANT: Any suitable engine of up to 56 kW (75 hp) and 72 kg (160 lb) weight, driving a two-blade propeller. Fuel tank, capacity 36-45 litres (8-10 Imp gallons), aft of engine firewall. Oil capacity 2·25 kg (5 lb)

Accommodation: Single seat. Sliding canopy optional. Baggage locker behind seat.

DIMENSIONS, EXTERNAL:
Wing span 5·64 m (18 ft 6 in)
Wing chord at root 1·32 m (4 ft 4 in)
Wing area, gross 6·36 m² (68·50 sq ft)
Length overall 4·50 m (14 ft 9 in)
Fuselage: Max width 0·55 m (1 ft 9¾ in)
Height overall 1·47 m (4 ft 10 in)
Tailplane span 1·98 m (6 ft 6 in)
Wheel track 1·37 m (4 ft 6 in)
Propeller diameter:
standard 1·37 m (4 ft 6 in)
prototype 1·32 m (4 ft 4 in)
Propeller ground clearance 25·5 cm (10 in)

WEIGHTS:
Weight empty 183-190 kg (405-420 lb)
Max T-O weight (semi-aerobatic) 295 kg (650 lb)

PERFORMANCE (prototype, with 36·5 kW (49 hp) engine, at 295 kg; 650 lb AUW):
Never-exceed speed 138 knots (255 km/h; 159 mph) IAS
Max level speed 117 knots (217 km/h; 135 mph) TAS
Max cruising speed 107 knots (198 km/h; 123 mph) TAS
Stalling speed, power off: 42 knots (79 km/h; 49 mph) TAS
30 knots (57 km/h; 35 mph) IAS
Typical rate of climb at S/L 213-259 m (700-850 ft)/min
Service ceiling 4,420 m (14,500 ft)
T-O to, and landing from, 15 m (50 ft) 305-335 m (1,000-1,100 ft)
g limits ±4·5

LOBET-DE-ROUVRAY
LOBET-DE-ROUVRAY AVIATION PTY LTD

ADDRESS:
Suite 7/506 Miller Street, 2062 Cammeray, New South Wales
Telephone: 922 2599 and 922 2960
MANAGER:
James Lobet
TECHNICAL ASSISTANT:
George Jacquemin

The original Ganagobie was designed and built by the brothers William and James Lobet at Lille, France, and made its first flight in 1953, powered by an old Clerget engine. After modification and redesignation as Ganagobie 02 it flew for a further 30 hours before being grounded by engine failure in 1954, and later became Ganagobie 2 when fitted with a two-stroke target drone engine. Further modifications were then incorporated.

The second aircraft, Ganagobie 3, was constructed in Alberta, Canada, by Mr La Rue Smith; built of birch plywood, it was somewhat heavier than the first aircraft and was powered by a 53·7 kW (72 hp) McCulloch engine. A later Ganagobie 3 was fitted with a 30 kW (40 hp) Continental engine, and this version is also suitable for converted Volkswagen engines of 1,500 cc and above. The 'ultra-light' version, known as the Ganagobie 4, is suitable for 35·8 kW (48 hp) Nelson and other small two-stroke engines. Very light okoumé mahogany, and other weight-saving features, may be used in its construction. Latest version is the Ganagobie 05.

GANAGOBIE 05

The Ganagobie 05 is a small, high-wing single-seat aircraft, designed primarily for amateur construction. A prototype was under construction in the Spring of 1976.

Construction of the basic homebuilt aircraft is all-wooden; but production versions are under consideration, either in kit form with a fabric-covered steel tube fuselage and wooden wings and tail, or in factory-built form with all-metal fabric-covered wing and tail control surfaces. The following description applies to the all-wood homebuilt version:

TYPE: Single-seat homebuilt light aircraft.

WINGS: Braced high-wing monoplane. Wing section NACA 23012. Constant-chord wings, with main and auxiliary spars and semi-circular tips. Centre-section integral with top of fuselage. Main panels have dihedral and can be detached for storage and transit, being carried in frames attached to the landing gear and cabane fittings on each side of the fuselage. A special frame fits over the rear of the fuselage to provide added support. Wooden spars and ribs, with non-structural plywood or aluminium leading-edge and fabric covering. Cable-operated plain ailerons. No flaps. Wings braced to fuselage by streamline-section steel tube Vee strut assembly on each side.

FUSELAGE: Basically wooden structure, of diamond-shaped cross-section, consisting of spruce longerons, 12 bulkheads and formers, and plywood covering. Steel tube engine mounting frame and wing root cabane structure.

TAIL UNIT: Plywood-covered wooden fin and strut-braced tailplane; fabric-covered wooden elevators and horn-balanced rudder. Struts detachable to permit horizontal surfaces to fold upwards for storage and transit. Rudder and elevators cable-operated. Trim tab on port elevator.

LANDING GEAR: Non-retractable main wheels and tailskid or tailwheel. Main wheels are mounted on tripod struts, the main legs of which have rubber-in-compression shock-absorption, and are of 203 mm (8 in) diameter with 8·00-4 tyres. Normally, spoon-type tailskid mounted at end of flat spring beneath rear fuselage. Front-wheel brakes are necessary if a tailwheel is fitted.

POWER PLANT: Aircraft is designed for a modified VW engine of at least 26 kW (35 hp), with direct-drive two-blade propeller; a typical engine is the Limbach SL 1700D. Alternatively, geared-down VW engines with either Vee-belt or gear reduction drive may be installed, provided that aircraft does not exceed its weight and CG range limitations. Fuel in two wing tanks between main and auxiliary spars; almost all of fuel load is usable.

ACCOMMODATION: Single seat in fully-enclosed cabin, with upward-opening door on each side. Ventilation devices in each transparent door panel. Seat belt attached to bulkhead.

DIMENSIONS, EXTERNAL:

Wing span	7·40 m (24 ft 3¼ in)
Wing chord (constant)	1·20 m (3 ft 11¼ in)
Wing area, gross	8·57 m² (92·25 sq ft)
Wing aspect ratio	6·25
Length overall	4·92 m (16 ft 1¾ in)
Height overall	1·83 m (6 ft 0 in)
Tailplane span	2·38 m (7 ft 9¾ in)
Wheel track	1·50 m (4 ft 11 in)
Propeller diameter (direct drive)	1·50 m (4 ft 11 in)

WEIGHTS AND LOADING:

Weight empty	285 kg (630 lb)
Max T-O weight	362 kg (800 lb)
Max wing loading	43·4 kg/m² (8·88 lb/sq ft)

PERFORMANCE (at max T-O weight, estimated, with SL 1700D engine):

Max level speed	98 knots (182 km/h; 113 mph)
Max cruising speed (75% power)	87 knots (161 km/h; 100 mph)
Stalling speed	41 knots (76 km/h; 47 mph)
Service ceiling	3,050 m (10,000 ft)

MILLICER

HENRY K. MILLICER

ADDRESS:

12 Murdoch Street, Camberwell, Victoria 3124

Mr Millicer, who in 1953 designed the original Airtourer light aircraft (see under Aerospace heading in New Zealand section), has recently designed an ultra-light aeroplane known as the Airmite.

MILLICER AIRMITE

Mr Millicer, who was for 10 years Chief Aerodynamicist of the Australian Government Aircraft Factories, is principal lecturer in aeronautics at the Royal Melbourne Institute of Technology. Assisted by students at the RMIT, he is building a prototype of the Airmite, an all-metal single-seat ultra-light aeroplane built mainly of aluminium alloy.

It is designed for marketing in kit form, and to be capable of assembly in a normal-sized garage or workshop by homebuilders of average abilities, using standard commercially-available pop-riveting tools. First flight was provisionally scheduled for 1976. Powered by an engine in the class of the Rolls-Royce O-240 or Lycoming O-320, the Airmite is a cantilever low-wing monoplane with a retractable tailwheel-type landing gear and enclosed cockpit. The wings are swept forward at the roots, and the trailing-edges of the outer portions are tapered slightly. The tailplane is mounted midway up the fin.

DIMENSIONS, EXTERNAL:

Wing span	5·79 m (19 ft 0 in)
Length overall	5·79 m (19 ft 0 in)

WEIGHT:

Max T-O weight	476 kg (1,050 lb)

PERFORMANCE (approx):

Max level speed	200 knots (370 km/h; 230 mph)
Stalling speed	42 knots (78 km/h; 48·5 mph)
Range	570 nm (1,055 km; 656 miles)

YAGER

KARL YAGER

ADDRESS:

43 Victoria Street, Lewisham, New South Wales 2049
Telephone: 56-2775

Mr Yager, whose first Libellula two-seat light aircraft was described briefly in the 1971-72 *Jane's,* advised subsequently that this was a design study only. He built a prototype of his second design; but this was scrapped after shortcomings had been revealed during initial taxying runs and short hop-flights. Mr Yager then began the development of a third design, the KY-03 Libellula, which was expected to fly for the first time in 1977 and was described in the 1975-76 *Jane's.* Due to circumstances beyond his control, development of this four-seat tandem-engined light aircraft has had to be suspended for about two years. At present Mr Yager is working on a new design which, if all goes well, should be sufficiently advanced to be included in the 1977-78 *Jane's.*

Armar I Gorrion 002 single-seat ultra-light aircraft

Corby Starlet single-seat homebuilt aircraft

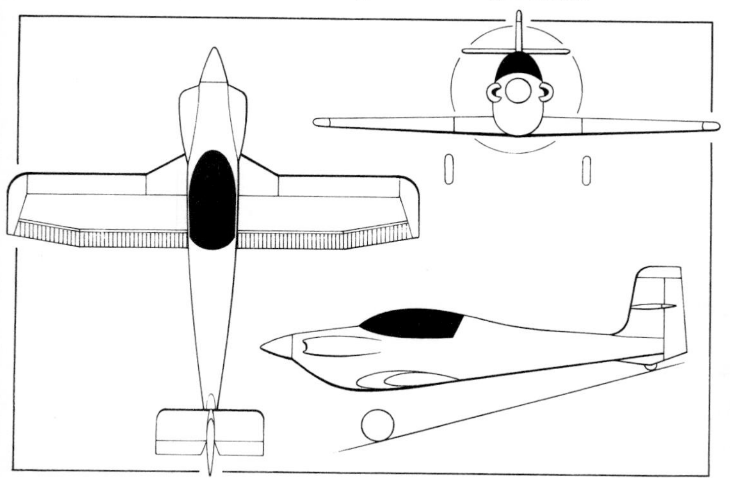
Millicer Airmite all-metal single-seat light aircraft *(Michael A. Badrocke)*

Ganagobie 03 (CF-REZ), built in Canada

BELGIUM

CW
CW HELICOPTER RESEARCH

ADDRESS:
Kloosterhof 6B, 8200 Brugge 2
Telephone: 317138
DIRECTOR:
Willy Clybouw

CW 205

The CW 205 is a high-speed three-seat all-metal light helicopter, designed both for factory production and for marketing in kit form for amateur constructors. Design began in 1972, and construction of a prototype started in December 1973.

TYPE: Three-seat light helicopter.

ROTOR SYSTEM: Two counter-rotating and intermeshing four-blade rigid main rotors, mounted side by side on pylons above the fuselage and canted outward from the vertical. Blades on each rotor are mounted in pairs, with included angle of only 10° instead of conventional 90°, enabling the rear blade of each pair to act as a kind of 'Fowler flap' for the front one. Blades are of NACA 23015 section, constant chord, and have 5° twist. Construction is of 2024T3 aluminium alloy. The intermeshing configuration is fully controllable cyclically, and the blades of each pair are actuated collectively. Each rotor has a coning angle of 170° and is canted forward at 6°. Each rotor drive-shaft is inclined at a 12° angle. A rotor brake is fitted.

ROTOR DRIVE: Direct drive, via a centrifugal clutch and a dual rotor gearbox in the fuselage roof. Rotor rpm 400.

FUSELAGE: Streamlined semi-monocoque, of all-metal construction.

TAIL UNIT: Cantilever structure of 2024T3 aluminium alloy. Variable-incidence tailplane, with elevators. Twin sweptback endplate fins; no rudders.

LANDING GEAR: Retractable tailwheel type, with spring steel shock-absorbers. Manual or hydraulic retraction, main wheels inwards into wells in fuselage beneath cabin floor, tailwheel forward.

POWER PLANT: One 183 kW (245 hp) Lycoming O-435 flat-six engine, mounted at 39° from the horizontal. Fuel capacity 180 litres (39·6 Imp gallons).

ACCOMMODATION: One-piece moulded Plexiglas canopy, which slides forward to give access to three side-by-side seats in cabin. Dual controls at outer seats.

DIMENSIONS, EXTERNAL:
Rotor diameter (each) 7·24 m (23 ft 9 in)
Rotor blade chord (constant, each) 0·15 m (6 in)
Rotor disc area (total) 82·00 m² (882·6 sq ft)
Rotor blade area (total, 8 blades) 3·60 m² (38·75 sq ft)
Distance between rotor hubs 2·13 m (7 ft 0 in)
Span over rotor tips 9·38 m (30 ft 9¼ in)
Length overall 6·50 m (21 ft 4 in)
Length of fuselage 6·30 m (20 ft 8 in)
Height overall 2·20 m (7 ft 2¾ in)
Tailplane span (c/l of vertical fins) 2·23 m (7 ft 3¾ in)
Fuselage: Max width 1·79 m (5 ft 10½ in)
Wheel track 1·65 m (5 ft 5 in)

WEIGHTS AND LOADINGS:
Weight empty 740 kg (1,631 lb)
Max T-O weight 1,080 kg (2,381 lb)
Max blade loading 300 kg/m² (61·4 lb/sq ft)
Max power loading 5·90 kg/kW (9·92 lb/hp)

PERFORMANCE (estimated, at max T-O weight):
Max cruising speed 129 knots (240 km/h; 149 mph)
Max rate of climb at S/L 300 m (984 ft)/min
Max range 324 nm (600 km; 373 miles)

MASSCHELEIN / VERSTRAETE
PAUL and STEPHAAN MASSCHELEIN and ERIC VERSTRAETE

ADDRESS:
Bellestraat 71, B-8960 Reningelst

MM Masschelein and Verstraete have designed and flown a series of five man-powered aircraft at Wevelgem Airport in Belgium and Calais-Dunkirk Airport in France. First flights, during which a height of 6 m (20 ft) was reached during a 50 m (164 ft) flight, were made in May 1974 with M Verstraete as the pilot. On 27 July 1974, a flight of 250 m (820 ft) was achieved. The fifth aircraft was completed in the Spring of 1976.

No other details had been received at the time of closing for press.

BULGARIA

Reports have appeared in the East European press during the past year of a number of light aircraft said to be of Bulgarian origin. All known details are given below; the photographs are, unfortunately, not of a good enough standard for reproduction.

BONIEV HELICOPTER

This small helicopter, designed by Boryslav Boniev, has a simple open-framework skeletal metal fuselage of pod and boom outline, two-blade main and tail rotors, a non-retractable tricycle landing gear, and a single open seat, behind a curved windscreen, for the pilot. The tail rotor is mounted on the port side, and there is a tailskid beneath the tailboom at the rear. The report, in *Skrzydlata Polska,* claimed that it is the first Bulgarian helicopter.

MILANOV XL-13T

A small and rather indistinct photograph, appearing in *Letectvi & Kosmonautika* in mid-1975, showed this to be a small, single-seat light aircraft with a strut-braced high wing, T-tail, non-retractable tricycle landing gear, and one-piece flush-fitting cockpit canopy. An uncowled Trabant motor car engine, with a two-blade tractor propeller, was mounted on a pylon above the wing centre-section. The general appearance of the XL-13T, which is attributed to Angel Milanov of Sofia, suggests that it may have been built using sailplane components.

MILANOV / VASILEVA TANGRA

Master of Sport Angel Milanov is also credited, in a 1976 report in *Skrzydlata Polska,* with having collaborated with Ing Ludmila Vasileva in the design and construction of a tandem two-seat light amphibian known as the Tangra. This aircraft also is powered by a Trabant motor car engine, which in this case drives a two-blade tractor propeller mounted on the fin leading-edge. The Tangra has a high-wing configuration, apparently unbraced, a T-tail, a tailwheel landing gear, and a large, framed canopy giving all-round view from the tandem cockpits. The following data are given:

DIMENSIONS, EXTERNAL:
Wing span 8·40 m (27 ft 6¾ in)
Length 5·60 m (18 ft 4½ in)
Height 1·42 m (4 ft 8 in)

PERFORMANCE:
Max level speed 86 knots (160 km/h; 99 mph)
Stalling speed 38 knots (70 km/h; 43·5 mph)
Max range 432 nm (800 km; 497 miles)

SLAVIEJ

The Slaviej is a cantilever low-wing monoplane, powered by a flat-four engine driving a two-blade propeller. Of extremely clean appearance, it has a single cockpit, a T-tail, and a non-retractable tailwheel landing gear with cantilever spring steel main legs. It is said to have a maximum speed of 135 knots (250 km/h; 155 mph) and a ceiling of 4,000 m (13,125 ft). The report, in *Skrzydlata Polska,* refers to it as a 'sporting' aircraft (ie presumably aerobatic), and adds that it is adaptable also for agricultural duties.

CANADA

BOUGIE
YVAN C. BOUGIE

BOUGIE HAUSCAT

Yvan C. Bougie, a maintenance foreman from Nitro-Valleyfield, Quebec, designed and built the Hauscat at a cost of $2,500. Construction took two and a half years, and the aircraft made its first flight in June 1972. Built to resemble the Grumman Bearcat fighter of the mid-forties, the Hauscat carries the markings of the 80th Reconnaissance Squadron of the French Armée de l'Air which served in Vietnam.

Plans for the Hauscat are available to amateur constructors.

TYPE: Single-seat homebuilt aircraft.

WINGS: Cantilever low-wing monoplane. Laminar-flow section, with 1° 15′ washout; 0·24 m (9⅝ in) thickness at root and 0·16 m (6¼ in) at tip. No flaps. Wooden structure, Dacron covered.

FUSELAGE: Wooden structure of circular section, Dacron covered.

TAIL UNIT: Conventional cantilever unit, with tailplane, elevators, fin and horn-balanced rudder of wooden construction, Dacron covered. Trim tabs on elevators.

LANDING GEAR: Non-retractable tailwheel type. Faired main legs and wheels. Spring-loaded shock-absorbers. Main wheels have 5·00-5 tyres and disc brakes.

POWER PLANT: One 1,800 cc Bougie-converted Volkswagen motor car engine, developing 56 kW (75 hp) and driving a Bougie 54-40 constant-speed propeller.

ACCOMMODATION: Single seat under rearward-sliding bubble canopy.

DIMENSIONS, EXTERNAL:
Wing span 7·85 m (25 ft 9 in)
Wing chord: at root 1·52 m (5 ft 0 in)
at tip 0·91 m (3 ft 0 in)
Length overall 5·64 m (18 ft 6 in)
Height overall 1·63 m (5 ft 4 in)

WEIGHTS:
Weight empty 255 kg (562 lb)
Max T-O weight 445 kg (980 lb)

PERFORMANCE:
Max speed 113 knots (209 km/h; 130 mph)
Max cruising speed 104 knots (193 km/h; 120 mph)
Landing speed 35 knots (64 km/h; 40 mph)
Max rate of climb at S/L 366 m (1,200 ft)/min
Service ceiling 3,660 m (12,000 ft)
T-O run 76 m (250 ft)
Range 304 nm (563 km; 350 miles)

CASI
CANADIAN AERONAUTICS AND SPACE INSTITUTE

ADDRESS:
Commonwealth Building, 77 Metcalfe Street, Ottawa 4, Ontario

CASI MAN-POWERED AIRCRAFT

Construction was started by the CASI in 1975 of a two-seat man-powered aircraft. It is a shoulder-wing monoplane with sweptback vertical tail surfaces, a pod-and-boom fuselage, a tandem-wheel drive and landing gear arrangement, and two pusher propellers mounted on fairings on the inboard wing trailing-edges. The wings have a long centre-section, of constant chord, and dihedral on the outer panels, which are fitted with ailerons.

DIMENSIONS:
Wing span 27·43 m (90 ft 0 in)
Wing area, gross 41·62 m² (448·0 sq ft)

WEIGHTS:
Weight empty 95 kg (209 lb)
T-O weight 237 kg (522 lb)

K & S
KAYE & STAN McLEOD

K & S Aircraft Supply acquired from Mr Rim Kaminskas all rights in the latter's Papoose Jungster I and Jungster II. Sets of plans of both designs were made available to amateur constructors from K & S, together with plans for the SA 102 Point 5 Cavalier. Descriptions of all three types can be found in the 1975-76 *Jane's.*

Mr McLeod informed *Jane's* that he expected to terminate this plans marketing service by the end of 1976.

REPLICA PLANS
REPLICA PLANS

ADDRESS: 953 Kirkmond Crescent, Richmond, BC

The SE-5A replica was designed to be an easy-to-build and inexpensive 85% scale representation of the famous First World War fighter, although exact reproduction was waived in favour of making the aircraft simple to construct, using modern and more readily available materials.

Design of the aircraft began in 1969, in which year construction of the first prototype also started. The SE-5A prototypes were designed for Continental engines ranging from 48·5 to 74·5 kW (65-100 hp), but larger engines can be installed to the individual homebuilder's preference. The first prototype flew for the first time in 1970 and certification has been granted by the FAA in the Experimental (homebuilt) category.

Plans are available to amateur builders, and by April 1976 some 209 sets had been sold.

REPLICA PLANS SE-5A REPLICA

TYPE: Single-seat sporting biplane.

WINGS: Braced biplane wings of Clark CYH section. Dihedral 3°. Incidence 3°. Ailerons on lower wings only. Centre-section of upper wing houses a small tank which can be used as an auxiliary fuel tank or smoke tank, or can be left out at building stage. The centre-section is carried on four spruce cabane struts and is braced with stainless steel cables and turnbuckles. Spruce interplane struts, with 4130 steel end fittings. Stainless steel flying, landing and incidence wires. Wing ribs of mahogany plywood, with cap strips; spruce spars. From the front spar forward, the leading-edge is covered with glassfibre or aluminium. Wings are fabric-covered.

FUSELAGE: Ply-skinned box structure, with fabric-covered turtledeck and aluminium-covered forward top decking. Dummy Vickers machine-gun in housing on the port side of fuselage, and gunsights on decking forward of windscreen.

TAIL UNIT: Tail surfaces built on spruce spars, with structure similar to that of wings except for drag bracing. Push-rod operated elevators. Cable-operated rudder, with a horn for tailwheel steering.

LANDING GEAR: Non-retractable tailwheel type. Bungee cord shock-absorption. Converted motorcycle wheels on main units, size 3·25-16. Size 6·00-2 tailwheel. Mechanical brakes.

POWER PLANT: Various engines can be installed. Performance figures quoted below relate to aircraft with 63·5 kW (85 hp) Continental C85 flat-four engine, driving a two-blade fixed-pitch wooden propeller. Fuel capacity 72 litres (19 US gallons). Oil capacity 3·8 litres (1 US gallon).

ACCOMMODATION: Single seat in open cockpit.

DIMENSIONS, EXTERNAL:

Wing span	6·96 m (22 ft 10 in)
Wing chord, constant	1·27 m (4 ft 2 in)
Wing area, gross	13·01 m² (140 sq ft)
Height overall	2·18 m (7 ft 2 in)
Wheel track	1·52 m (5 ft 0 in)
Propeller diameter	1·83 m (6 ft 0 in)

WEIGHTS:

Weight empty	358 kg (790 lb)
Max T-O weight	499 kg (1,100 lb)

PERFORMANCE:

Max level speed at S/L	78 knots (145 km/h; 90 mph)
Max cruising speed	74 knots (137 km/h; 85 mph)
Stalling speed	30·5 knots (57 km/h; 35 mph)
Max rate of climb at S/L	152 m (500 ft)/min

WESTERN
WESTERN AIRCRAFT SUPPLIES

ADDRESS:
623 Markerville Road NE, Calgary T2E SX1, Alberta
Telephone: (403) 276 3087

DIRECTOR:
Jean J. Peters

Western Aircraft Supplies markets materials for amateur aircraft constructors, and is also selling plans for construction of the RL-3 Monsoon, a two-seat light aircraft originally designed in India.

Western is also developing another aircraft, of wooden construction, of which no details are yet available.

RL-3 MONSOON

The RL-3 Monsoon, designed by Mr Renato Levi of Afco (Private) Ltd, originated in India; the prototype was described on page 105 of the 1960-61 *Jane's*, and this description was repeated under the Western heading in the 1972-73 edition. At least one more example of the RL-3 was completed in India.

Mr Jean J. Peters of Western Aircraft Supplies acquired from Mr Levi in 1969 rights to market plans of the aircraft, and at least 12 sets of these plans and five kits of materials have been sold. Five Monsoons are known to be under construction in Canada and the USA, including one by Mr Peters himself.

The plans for the Monsoon are accepted by the EAA, and the Canadian Ministry of Transport had approved all components built up to March 1973 by Mr Peters and another Calgary constructor. It is not considered appropriate to publish a detailed description of the currently-marketed Monsoon until the Western-built prototype (which will have a Lycoming O-235 engine) is completed.

TYPE: Two-seat homebuilt light aircraft.

WINGS: Cantilever low-wing monoplane. Constant-chord single-spar wooden structure, using Canadian spruce.

TAIL UNIT: Cantilever wooden structure.

LANDING GEAR: Non-retractable tailwheel type.

POWER PLANT: One flat-four engine, driving a two-blade propeller.

ACCOMMODATION: Two seats side by side in fully-enclosed cabin.

ZENAIR
ZENAIR LTD

ADDRESS:
236 Richmond Street, Richmond Hill L4C 3Y8, Ontario
Telephone: (416) 884-9044

PRESIDENT:
Christophe Heintz

HEINTZ ZÉNITH

M Heintz, a professional aeronautical engineer, participated in the design of several of the aircraft produced by Avions Pierre Robin. While in France, he also designed and built the prototype of a two-seat light aircraft named the Zénith, intended for amateur construction.

Work on the Zénith began in October 1968; the prototype, registered F-WPZY (later C-FEYC), flew for the first time on 22 March 1970 and was granted French CNRA (homebuilt experimental aircraft) certification. In October 1970 the original wing of NACA 64A315 (modified) section was replaced by one offering improved low-speed characteristics.

In June 1971, the prototype won a handicap race at Iverdon, Switzerland, at an average speed of 124 knots (230 km/h; 143 mph) from a standing start.

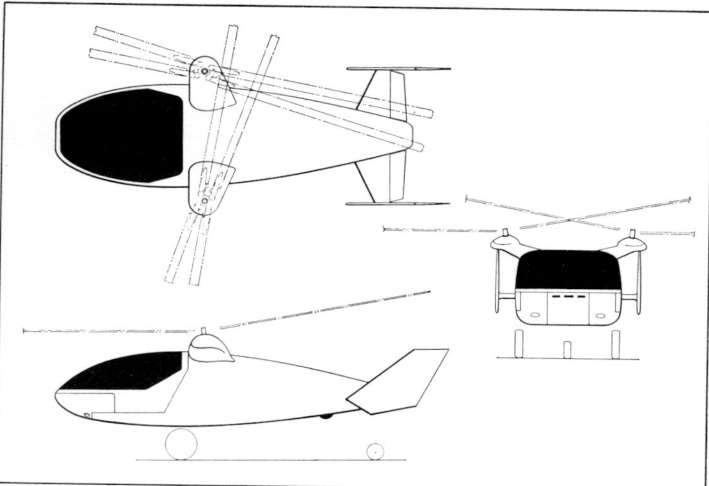

CW 205 three-seat light helicopter *(Roy J. Grainge)*

Fourth in the series of Masschelein/Verstraete man-powered aircraft, being flown by M. Eric Verstraete *(Bureau de Commerce, Calais)*

Bougie Hanscat, painted in the markings of a French reconnaissance squadron *(Howard Levy)*

Replica Plans SE-5A, an 85% scale representation of a First World War fighter *(Howard Levy)*

First Zénith built in North America

Prototype of the Mono Zénith, single-seat version of the Heintz Zénith homebuilt light aircraft

In 1974 the Zénith was granted the National Association of Sport Aircraft Designers (NASAD, USA) 'seal of quality' No. 108.

Sets of plans and a constructional manual for the Zénith are available to amateur builders, as follows:

France: French manual and metric measurements from D. Triques, 23 Ave Edouard Belin, Fontaine d'Ouche, F21 Dijon.

Germany: German manual and metric measurements from K. Arens, Rollstrasse 26, D-3392 Clausthal-Zellerfeld 1.

USA and Canada: English manual and drawings to US standards, with English measurements, from Zenair, which offers materials, parts and complete kits for the Zénith. Zenair also designs and manufactures wooden propellers for engines of up to 149 kW (200 hp).

By 1976 more than 300 sets of plans had been sold in all parts of the world, and several aircraft were flying in Europe. The first Zénith built in North America flew for the first time in October 1975. Constructed by Mr Orgel Dever, it has a 108 kW (145 hp) Continental O-300 flat-four engine, and has modified wings, carrying external wingtip fuel tanks.

The single-seat Mono Zénith and three-seat Tri-Zénith are variations of the basic design and are described separately.

The following description applies to the standard Zénith, of which Zenair was producing three complete kits each month in early 1976:

TYPE: Two-seat all-metal homebuilt light aircraft, with ultimate stress factor of 9*g*.

WINGS: Cantilever low-wing monoplane. Constant-chord wings, of NACA 64A515 (modified) section. Dihedral 6° from roots. Single-spar aluminium alloy structure, with blind riveted aluminium alloy skin. Hoerner wing-tips. Aluminium alloy piano-hinged ailerons and electrically-actuated plain flaps on trailing-edge.

FUSELAGE: Conventional aluminium alloy stressed-skin structure, of basically rectangular section with rounded top-decking.

TAIL UNIT: Rectangular one-piece all-moving tailplane, with combined trim and anti-servo tabs. Plans show rudder only, with slight sweepback. Conventional fin and rudder can be fitted if desired. Tailplane and rudder are both single-spar structures with ribs and skin of aluminium alloy.

LANDING GEAR: Non-retractable tricycle type, with rubber-block shock-absorbers. Manual locking of nosewheel. All three wheels and tyres size 380 × 150 mm. Hydraulically-actuated drum brakes on main units. Streamlined glassfibre fairings over all three wheels and legs.

POWER PLANT: One 74·5 kW (100 hp) Rolls-Royce Continental O-200-A flat-four engine, driving a McCauley ECM-72-50 two-blade fixed-pitch metal propeller. Design suitable for engines from 63·5 kW (85 hp) to 119 kW (160 hp). Fuel tank in fuselage, aft of passenger seat, capacity 90 litres (20 Imp gallons). Optional fuel tanks in wing leading-edges, total capacity 72·5 litres (16 Imp gallons).

ACCOMMODATION: Side-by-side seating for pilot and one passenger under sideways-opening (to starboard) Plexiglas canopy. Dual controls, with single control column located centrally between seats. Space for 35 kg (77 lb) of baggage aft of seats. Cabin heated and ventilated.

SYSTEMS: 12V battery and generator provide power for engine starting, fuel pump and flap actuation. VHF radio.

DIMENSIONS, EXTERNAL:
- Wing span — 7·00 m (22 ft 11¾ in)
- Wing chord (constant) — 1·40 m (4 ft 7 in)
- Wing area, gross — 9·80 m² (105·9 sq ft)
- Wing aspect ratio — 5
- Length overall — 6·30 m (20 ft 8 in)
- Height overall — 1·85 m (6 ft 0¾ in)
- Tailplane span — 2·30 m (7 ft 6½ in)
- Wheel track — 2·25 m (7 ft 4½ in)
- Wheelbase — 1·42 m (4 ft 8 in)
- Min ground turning radius — 4·00 m (13 ft 1¾ in)
- Propeller diameter — 1·83 m (6 ft 0 in)
- Propeller ground clearance — 0·25 m (9¾ in)

DIMENSION, INTERNAL:
- Cabin: Max width — 1·01 m (3 ft 3¾ in)

WEIGHTS AND LOADINGS:
- Weight empty, equipped — 400 kg (881 lb)
- Normal T-O and landing weight — 650 kg (1,433 lb)
- Max T-O weight — 680 kg (1,499 lb)
- Max wing loading — 65 kg/m² (13·3 lb/sq ft)
- Max power loading — 8·72 kg/kW (14·33 lb/hp)

PERFORMANCE (at max T-O weight):
- Max level speed at S/L — 126 knots (233 km/h; 145 mph)
- Cruising speed (75% power) at S/L — 110 knots (205 km/h; 127 mph)
- Cruising speed (75% power) at 2,750 m (9,000 ft) — 116 knots (215 km/h; 134 mph)
- Stalling speed, flaps down — 46 knots (85 km/h; 53 mph)
- Max rate of climb at S/L — 240 m (787 ft)/min
- Service ceiling — 4,600 m (15,100 ft)
- Range with max fuel, no reserves (75% power) — 432 nm (800 km; 497 miles)

HEINTZ MONO ZÉNITH

The single-seat Mono Zénith is of generally similar all-metal construction to the two-seat Zénith, but is slightly smaller overall and is designed to be powered by engines in the 37·25 to 74·5 kW (50 to 100 hp) range. The prototype (C-GNYM), which has a Volkswagen engine, made its first flight on 8 May 1975. Like the two-seat Zénith, it is stressed to ± 9*g* ultimate at normal max T-O weight.

Construction drawings and manual, materials, parts and complete kits to build the Mono Zénith are available from Zenair.

TYPE: Single-seat homebuilt light aircraft.

WINGS: Cantilever low-wing monoplane. Wing section GA(PC) 1. Thickness/chord ratio 15%. Dihedral 6°. Incidence 7° 30′. Single-spar aluminium alloy structure, with aluminium alloy skin, blind riveted. Aluminium alloy piano-hinged ailerons.

FUSELAGE: Conventional aluminium alloy stressed-skin structure, of basically rectangular section, with rounded top-decking.

TAIL UNIT: Rectangular one-piece all-moving tailplane, with automatic and controllable trim tab. Single-spar structures, with ribs and skins of aluminium.

LANDING GEAR: Non-retractable tricycle type, with rubber-block shock-absorbers. All three wheels and tyres size 5·00-5. Hydraulically-actuated Gerdes disc brakes.

POWER PLANT: One 41 kW (55 hp) 1,700 cc converted Volkswagen motor car engine in prototype, driving a Zenair wooden propeller. Fuel tank in fuselage, capacity 54·5 litres (12 Imp gallons).

ACCOMMODATION: Single seat under Plexiglas canopy. Baggage compartment aft of seat, capacity 11·3 kg (25 lb).

DIMENSIONS, EXTERNAL:
- Wing span — 6·71 m (22 ft 0 in)
- Wing chord, constant — 1·27 m (4 ft 2 in)
- Wing area, gross — 8·50 m² (91·5 sq ft)
- Wing aspect ratio — 5·27
- Length overall — 5·94 m (19 ft 6 in)
- Tailplane span — 2·26 m (7 ft 5 in)
- Wheel track — 2·13 m (7 ft 0 in)
- Wheelbase — 1·27 m (4 ft 2 in)
- Propeller diameter — 1·47 m (4 ft 10 in)

WEIGHTS:
- Weight empty — 263 kg (580 lb)
- Max T-O weight:
  - VW engine — 413 kg (910 lb)
  - 74·5 kW; 100 hp engine — 444 kg (980 lb)

PERFORMANCE (A: 1,700 cc Volkswagen, B (estimated): 48·5 kW; 65 hp, C (estimated): 74·5 kW; 100 hp engine):
- Max level speed:
  - A — 103 knots (190 km/h; 118 mph)
  - B — 109 knots (200 km/h; 125 mph)
  - C — 130 knots (240 km/h; 150 mph)
- Cruising speed (75% power):
  - A — 91 knots (170 km/h; 105 mph)
  - B — 96 knots (177 km/h; 110 mph)
  - C — 118 knots (218 km/h; 135 mph)
- Stalling speed:
  - A, B, C — 41 knots (76 km/h; 47 mph)
- Max rate of climb at S/L:
  - A — 200 m (610 ft)/min
  - B — 230 m (700 ft)/min
  - C — 490 m (1,500 ft)/min
- Service ceiling: A — 2,745 m (9,000 ft)
- T-O run: A — 183 m (600 ft)
- T-O to 15 m (50 ft): A — 335 m (1,100 ft)
- Landing run: A — 152 m (500 ft)
- Range with 55 litres (12 Imp gallons) fuel:
  - A, B — 350 nm (645 km; 400 miles)
  - C — 312 nm (580 km; 360 miles)
- Endurance with 55 litres (12 Imp gallons) fuel:
  - A, B — 4 hr
  - C — 2 hr 30 min

HEINTZ TRI-ZÉNITH

The three-seat Tri-Zénith, prototypes of which are under construction, is somewhat larger than the two-seat Zénith. It has a greater wing span, longer fuselage and enlarged cabin, with a rear seat for one adult, two children or baggage, up to a weight of 95 kg (209 lb). The tailplane is larger, and a fin and rudder vertical assembly is standard. Recommended power is in the 93-119 kW (125-160 hp) range, though engines of between 86 and 134 kW (115 and 180 hp) may be installed. Limiting load factors, at max T-O weight, are +3·8*g* and −1·9*g*. Fuel is carried in two 60 litre (13 Imp gallon) tanks, one in each wing leading-edge. The wing trailing-edge is fitted with electrically-actuated slotted flaps and aerodynamically-balanced ailerons. The wings are not detachable.

The first prototype of the Tri-Zénith is expected to make its first flight in early 1977.

DIMENSIONS, EXTERNAL:
- Wing span — 8·10 m (26 ft 6¾ in)
- Wing chord (constant) — 1·48 m (4 ft 10¼ in)
- Wing area, gross — 12·00 m² (129·2 sq ft)
- Wing aspect ratio — 5·48
- Length overall — 6·80 m (22 ft 3¾ in)
- Tailplane span — 2·60 m (8 ft 6¼ in)
- Wheel track — 2·25 m (7 ft 4½ in)
- Wheelbase — 1·42 m (4 ft 8 in)

WEIGHTS (estimated):
- Weight empty — 470-500 kg (1,036-1,102 lb)
- Max T-O weight — 810-840 kg (1,785-1,851 lb)

PERFORMANCE (estimated. A: 93 kW; 125 hp at 810 kg; 1,785 lb AUW; B: 119 kW; 160 hp at 840 kg; 1,851 lb AUW):
- Max level speed:
  - A — 132 knots (245 km/h; 152 mph)
  - B — 140 knots (260 km/h; 162 mph)
  - B — 46 knots (85 km/h; 53 mph)
- Max rate of climb at S/L:
  - A — 252 m (827 ft)/min
  - B — 372 m (1,220 ft)/min
- Range at max cruising speed:
  - A — 474 nm (880 km; 546 miles)
  - B — 442 nm (820 km; 509 miles)
- Endurance at max cruising speed:
  - A — 4 hr 0 min
  - B — 3 hr 30 min
- Max cruising speed (75% power):
  - A — 119 knots (220 km/h; 137 mph)
  - B — 127 knots (235 km/h; 146 mph)
- Stalling speed, flaps down:
  - A — 45 knots (83 km/h; 52 mph)

DENMARK

SEREMET

W. VINCENT SEREMET

ADDRESS: Godsparken 50, 2670 Greve Strand
Telephone: (01) 90 29 49

Mr W. Vincent Seremet, a Danish engineer and amateur constructor, has designed and built a number of small rotating-wing aircraft, the first trials being carried out in 1962 with two aircraft designated W.S.1 and W.S.2. More recent designs include the W.S.3, W.S.4/4A, W.S.5, W.S.6 and W.S.7, described in the 1970-71 and subsequent editions of *Jane's*.

The W.S.4 has been modified and is now known as the W.S.8. Work is also continuing on the W.S.6 and new W.S.9 autogyros. Mr Seremet's latest design is the W.S.10 parawing aircraft, which has been fitted with an engine. No other details are available.

SEREMET W.S.8

Modification of the W.S.4 led to its redesignation as the W.S.8. It was seen first as a strap-on helicopter, with the pilot's left hand controlling the tail rotor. Since then it has been modified to embody a pilot's seat, wide-track tripod landing gear legs, and foot control of the tail rotor. By March 1976, trials had been carried out with the aircraft suspended from wires, and work was continuing.

TYPE: Single-engined ultra-light helicopter.

ROTOR SYSTEM: Two-blade main rotor. Blade section NACA 0015. Two-blade tail rotor, controlled by foot pedals.

Seremet W.S.10 powered parawing undergoing early towed testing

POWER PLANT: One 26 kW (35 hp) Kiekhaefer engine.

ACCOMMODATION: Open seat for pilot only.

DIMENSIONS:

Diameter of main rotor	4·50 m (14 ft 9 in)
Diameter of tail rotor	0·80 m (2 ft 7½ in)
Length overall	3·05 m (10 ft 0 in)
Length folded	1·55 m (5 ft 1 in)
Height overall	1·80 m (5 ft 11 in)
Width overall, rotor fore and aft	1·40 m (4 ft 7 in)

WEIGHTS:

Weight empty	52 kg (115 lb)
Max T-O weight	150 kg (330 lb)

Seremet W.S.8 helicopter in latest, modified, form

FINLAND

PIK

POLYTEKNIKKOJEN ILMAILUKERHO

(The Flying Club of the Helsinki University of Technology)

ADDRESS:
Dipoli, 02150 Otaniemi

PROJECT ENGINEER:
Kai Mellén

Mr K. Mellén, an engineer with Finnair, has designed and is building, under the auspices of the Polyteknikkojen Ilmailukerho, a Formula V racing aircraft known as the Super Sytky. Design work started at the end of 1973, and construction of the prototype began on 15 March 1975.

PIK-21 SUPER SYTKY

TYPE: Single-seat racing monoplane.

WINGS: Cantilever constant-chord wooden wings, with no anhedral or dihedral. Wing section NACA 64212. Full-span narrow-chord ailerons. No tabs.

FUSELAGE: Wooden structure.

TAIL UNIT: Conventional cantilever wooden construction. Constant-chord tailplane, fitted with elevators. Fin integral with rear fuselage. No tabs.

LANDING GEAR: Non-retractable tailwheel type. Steel leaf spring main legs, carrying Azuza wheels, with tyres size 5·00-5. Wheel fairings fitted. Azuza mechanical brakes.

POWER PLANT: One 1,600 cc modified Volkswagen motor car engine, close-cowled and driving a two-blade propeller with large spinner.

ACCOMMODATION: Enclosed cabin seating pilot only. One-piece fully-transparent windscreen/canopy.

DIMENSIONS, EXTERNAL:

Wing span	5·19 m (17 ft 0½ in)
Wing area, gross	7·1 m² (76·4 sq ft)

WEIGHTS (estimated):

Weight empty	200 kg (441 lb)
Max T-O weight	320 kg (705 lb)

TERVAMÄKI

JUKKA TERVAMÄKI

ADDRESS:
Aidasmäentie 16-20E, 00650 Helsinki 65

Mr Tervamäki, who is currently Technical Manager of Wihuri-Yhtymä Oy, Lentohuolto, Finland's largest private aviation company, first became interested in autogyros in 1956. In 1959 he worked briefly for the Bensen Aircraft Corporation in the USA. He obtained a Diploma in Aeronautical Engineering at the Helsinki Institute of Technology in 1963, and later served in the helicopter section of the Finnish Air Force. He was for two years project manager and chief designer of the PIK-19 Muhinu glider-towing aircraft. More recently, he has modified a Schleicher ASK 14 powered sailplane to make it capable of taxying on large airports; and has designed the engine installation for the JT-6 prototype powered sailplane, which is now in production as the PIK-20E (see Sailplanes section).

Early autogyros designed by Mr Tervamäki were completed in 1958 (JT-1) and 1965 (JT-2). More recent designs are the Tervamäki-Eerola ATE-3 and Tervamäki JT-5.

TERVAMÄKI-EEROLA ATE-3

Design of this single-seat light autogyro was begun in May 1966 by Mr Tervamäki, assisted by Mr Eerola, a former helicopter mechanic with the Finnish Air Force. Construction of a prototype began in the following September, and this aircraft (OH-XYV) flew for the first time on 11 May 1968.

The ATE-3 was intended for amateur construction, and several examples were under construction in Finland in 1976. However, no further plans for the ATE-3 have been sold since the JT-5 plans were completed.

A full description of the ATE-3 was published in the 1974-75 *Jane's,* followed by a somewhat condensed version in 1975-76.

TERVAMÄKI JT-5

The JT-5 is a development of the ATE-3 design, the major visible differences being the use of a triple tail assembly, to improve static and dynamic stability; a fully-enclosed cockpit; improved, low-drag fuselage contours; and the extensive use of plastics materials in the basic structure and main components. Other features include an upward-directed exhaust, to reduce engine noise, and a simplified carburettor installation and heating system of Tervamäki design.

The prototype JT-5 (OH-XYS) was flown for the first time on 7 January 1973. It was later sold, together with all production rights, tools and moulds, to Sr Vittorio Magni of Italy (see 1975-76 *Jane's*), by whom the production of rotor blades and other parts began in 1974. In addition, Sr Magni has rights for the sale of drawings of the JT-5 to amateur constructors in French, Italian and Spanish-speaking countries. For those in other countries, drawings (with metric dimensions and English annotations) are available directly from Mr Tervamäki. At least two JT-5s were under construction by homebuilders in Finland in early 1976; about 30 sets of plans had been sold by Mr Tervamäki at that time.

With minor modifications, glassfibre rotor blades of the type fitted to the JT-5 can be installed on other autogyros and rotor head designs, and several sets are already flying on aircraft built in Scandinavia.

TYPE: Single-seat light autogyro.

ROTOR SYSTEM: Two-blade semi-rigid rotor of glassfibre-reinforced epoxy resin, with polyurethane plastics foam core. Blades, of constant chord and NACA 8-H-12 section, are each attached to hub by two bolts. A lead bar in each blade leading-edge forms the chordwise balance weight. Rotor mast of streamlined SAE 4130 steel tubing. Rotor head is of a compact offset-gimbal type with centrifugal teeter stops and rotor brake installed. There are two spiral springs for trim adjustment, which is effected via the control stick twist-grip handle. Normal rotor rpm is 400, maximum 600. Designed for the JT-5, but not yet fitted, is a modified Cierva-type inclined drag hinge which would allow the blades to move to zero pitch when pre-rotation torque is applied, permitting an increase of 100 rpm in pre-spin speed and, consequently, a shorter take-off.

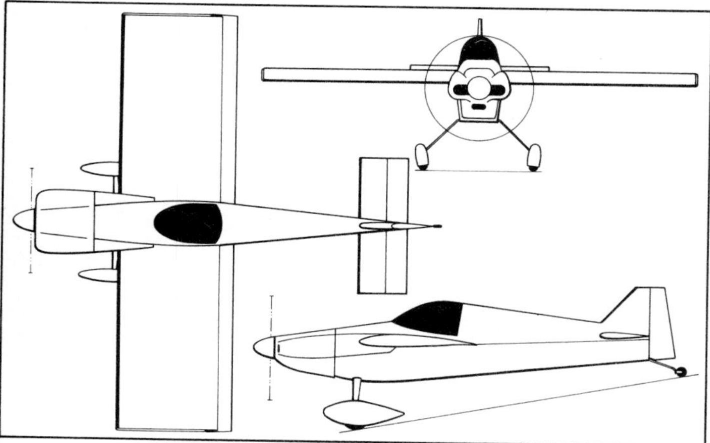

PIK-21 Super Sytky single-seat racing aircraft *(Roy J. Grainge)*

Tervamäki JT-5 single-seat autogyro, with canopy open

ROTOR DRIVE: Rotor spin-up by Vee-belt, clutch, 90° gearbox, sliding universal shaft and inertia-operated Bendix drive. Overall reduction ratio 8. Rotor spin-up of 300 rpm can be achieved.

FUSELAGE: Basic structure of welded 4130 steel tubing with a glassfibre/HFB honeycomb sandwich cockpit. All internal cockpit structures of glassfibre-reinforced epoxy resin. One-piece aluminium engine cowling.

TAIL UNIT: Central main fin and rudder, of glassfibre-reinforced epoxy resin, with rigid PVC-foam ribs and Courtauld carbon-fibre stiffeners. Horizontal tail and auxiliary endplate fins of glassfibre sandwich construction with honeycomb core. Tail assembly attached to fuselage by a single streamlined steel tube. Small tail-wheel beneath base of fin.

LANDING GEAR: Non-retractable tricycle type. Main gear legs consist of 4 × 4 cm (1·6 × 1·6 in) glassfibre-reinforced epoxy resin springs, encased in streamlined fairings of the same material. Cables inside these fairings to main gear drum brakes. Main-wheel tyres size 300 × 100. Compression rubber shock-absorption in nose gear. Nosewheel tyre size 260 × 80. Nosewheel steerable by rudder pedals.

POWER PLANT: One 56 kW (75 hp) 1·7 litre Volkswagen engine, converted for autogyro use by Limbach Motorenbau. No oil cooler, generator or electric starter. Two-blade pusher propeller, of glassfibre-reinforced epoxy. Glassfibre fuel tank, integrally built into fuselage aft of pilot's seat, capacity 50 litres (11 Imp gallons).

ACCOMMODATION: Single seat under sideways-opening Plexiglas canopy. Instrument panel cover and pilot's seat back (the latter also forming a firewall to the engine compartment) open together with the canopy.

EQUIPMENT: Standard equipment and controls are as listed in the 1974-75 *Jane's*. A 6-channel radio is installed.

DIMENSIONS, EXTERNAL:

Rotor diameter	7·00 m (22 ft 11½ in)
Rotor blade chord (constant, each)	0·18 m (7·1 in)
Length of fuselage	3·50 m (11 ft 5¾ in)
Height overall	2·00 m (6 ft 6¾ in)
Wheel track	1·70 m (5 ft 7 in)
Propeller diameter	1·20 m (3 ft 11¼ in)

AREAS:

Rotor blades (each)	0·63 m² (6·78 sq ft)
Rotor disc	38·50 m² (414·4 sq ft)

WEIGHTS AND LOADINGS:

Weight empty, equipped	167 kg (368 lb)
Max T-O weight	290 kg (639 lb)
Max disc loading	7·5 kg/m² (1·54 lb/sq ft)
Max power loading	5·18 kg/kW (8·52 lb/hp)

PERFORMANCE (at max T-O weight):

Never-exceed speed	97 knots (180 km/h; 111·5 mph)
Max level speed at S/L	92 knots (170 km/h; 106 mph)
Max cruising speed at S/L	81 knots (150 km/h; 93 mph)
Econ cruising speed	70 knots (130 km/h; 81 mph)
Min level speed	19 knots (35 km/h; 22 mph)
Max rate of climb at S/L	180 m (590 ft)/min
Service ceiling	4,000 m (13,125 ft)
T-O run	70 m (230 ft)
T-O to 15 m (50 ft)	120 m (394 ft)
Landing from 15 m (50 ft)	50 m (165 ft)
Landing run	5 m (16 ft)
Range with max fuel, no reserves	189 nm (350 km; 217 miles)

FRANCE

AM

ÉCOLE NATIONALE SUPÉRIEURE D'ARTS ET MÉTIERS

ADDRESS:
Cluny

Interest in the possibility of constructing a two-seat training aircraft at the École Nationale Supérieure d'Arts et Métiers at Cluny dates from 1965. Work on a prototype, designated AM-69, began in 1969, on the basis of incomplete plans of a tandem two-seat light aircraft known as the Gaucher RG-662 and design studies and experiments conducted at the school. Some 3,000 hours of new design and testing by a group of twelve students, and 4,000 hours of construction by ten other students, went into this all-wooden prototype (F-PTXB), which flew for the first time on 6 May 1973.

It was certificated by the CNRA in July 1973, and was placed subsequently at the disposal of members of the Aero Club of Issoudun. The designers are, meanwhile, working on an all-metal version, with larger vertical tail surfaces, a moulded canopy, revised engine cowling and other changes. Plans of this version are available to amateur constructors through the RSA, rue Sauffroy, 75-Paris.

AM-69 GEORGES PAYRE

TYPE: Two-seat homebuilt light aircraft.

WINGS: Cantilever low-wing monoplane. Modified Mureau 234 wing section, flat over 65% of the chord of the upper surface and over 70% of the chord of the lower surface. Thickness/chord ratio 13·2%. Dihedral from roots. All-wood structure, covered with birch plywood. Flaps and rod-actuated ailerons over full span of each wing trailing-edge.

FUSELAGE: Conventional wood structure of basic rectangular section.

TAIL UNIT: Conventional all-wood structure with swept-back vertical surfaces. Horn-balanced control surfaces; elevators rod-actuated and rudder cable-actuated. Large tab on port elevator.

LANDING GEAR: Non-retractable tailwheel type. Main units modified from Robin (Jodel) DR 220 components and mounted on wing front spar. Tailwheel modified from that of a Stampe SV.4.

POWER PLANT: One 67 kW (90 hp) Continental C90-14F flat-four engine, driving a Ratier two-blade ground-adjustable propeller. One fuel tank in each wing; total capacity 84 litres (18·5 Imp gallons).

ACCOMMODATION: Two persons in tandem under continuous transparent canopy, with separate side-hinged (to starboard) section over each seat. When flown solo, pilot occupies front seat.

DIMENSIONS, EXTERNAL:

Wing span	8·94 m (29 ft 4 in)
Wing area, gross	11·20 m² (120·5 sq ft)
Length overall	7·90 m (25 ft 11 in)

WEIGHTS:

Weight empty	524 kg (1,155 lb)
Max T-O weight	728 kg (1,605 lb)

PERFORMANCE:

Never-exceed speed	135 knots (250 km/h; 155 mph)
Max level speed	105 knots (195 km/h; 121 mph)
Normal cruising speed	86 knots (160 km/h; 99 mph)
Stalling speed, flaps up	46 knots (85 km/h; 53 mph)
Stalling speed, flaps down	40·5 knots (75 km/h; 47 mph)
Max rate of climb at S/L	180 m (590 ft)/min
Service ceiling	3,500 m (11,500 ft)
T-O to 15 m (50 ft)	500 m (1,640 ft)
Landing from 15 m (50 ft)	380 m (1,250 ft)
Range with max fuel	323 nm (600 km; 373 miles)

BESNEUX

ALAIN BESNEUX

ADDRESS:
28 rue de la Sergenterie, 61140 Tesse-la-Madeleine

M Besneux has built, to the design of Jean Pottier, a small single-seat light aircraft designated P.70B. In its original form, with enclosed cockpit and 30 kW (40 hp) Volkswagen engine, this was exhibited at the 1973 Paris Air Show under the auspices of the RSA. It had not flown at that time.

The description that follows applies to the P.70B in its current form, with open cockpit and up-rated engine:

BESNEUX P.70B

M Besneux began constructing the P.70B in February 1972, and it flew for the first time on 19 July 1974.

The details below apply to the prototype in its current form. Twelve modified P.70Ss, under construction in early 1975, will have new wings of NACA 4415 section, with a span of 5·90 m (19 ft 4¼ in) and fitted with flaps; a tricycle landing gear; and max T-O weight of 290 kg (639 lb) (see under 'Pottier').

TYPE: Single-seat all-metal homebuilt light aircraft.

WINGS: Cantilever mid-wing monoplane. Wing section NACA 23012. No dihedral. Incidence 3° 30′. All-metal stressed-skin pop-riveted structure of AU4G light alloy. Full-span ailerons. No flaps or tabs.

FUSELAGE: Conventional all-metal semi-monocoque structure, with pop-riveted AU4G light alloy skin.

TAIL UNIT: Cantilever all-metal structure, similar in construction to wings. One-piece all-moving vertical surface, with ground-adjustable tab. Tailplane incidence (fixed) 1°.

LANDING GEAR: Non-retractable tailwheel type. Cantilever main legs of AU4G, 100 mm wide by 18 mm thick (3·9 in × 0·71 in). Main-wheel tyre diameter 310 mm (12·2 in), pressure 1·77 bars (25·6 lb/sq in). No brakes. Steerable solid tailwheel.

POWER PLANT: One 1,500 cc Volkswagen motor car engine, modified by M Besneux to give 33·6 kW (45 hp) at 3,600 rpm, driving a two-blade fixed-pitch wooden propeller of 0·90 m (2 ft 11½ in) pitch. Fuel tank in fuselage, aft of firewall, capacity 34 litres (7·5 Imp gallons). Oil capacity 2·5 litres (0·55 Imp gallons).

ACCOMMODATION: Single seat in open cockpit.

DIMENSIONS, EXTERNAL:

Wing span	5·58 m (18 ft 3¾ in)
Wing chord, constant	1·26 m (4 ft 1½ in)
Wing area, gross	6·80 m² (73·2 sq ft)
Length overall	4·90 m (16 ft 1 in)
Tailplane span	1·80 m (5 ft 11 in)
Wheel track	1·20 m (3 ft 11¼ in)
Propeller diameter	1·30 m (4 ft 3¼ in)
Propeller ground clearance	0·25 m (9¾ in)

WEIGHTS:

Weight empty	171 kg (377 lb)
Max T-O weight	272 kg (599 lb)

PERFORMANCE (at max T-O weight):

Never-exceed speed	118 knots (220 km/h; 136 mph)
Max level speed at 305 m (1,000 ft)	108 knots (200 km/h; 124 mph)
Max cruising speed at 305 m (1,000 ft)	97 knots (180 km/h; 112 mph)
Econ cruising speed at 305 m (1,000 ft)	92 knots (170 km/h; 106 mph)
Stalling speed	38·5 knots (70 km/h; 44 mph)
Service ceiling	3,500 m (11,475 ft)
T-O to 15 m (50 ft)	400 m (1,310 ft)
Landing from 15 m (50 ft)	200 m (655 ft)
Endurance with max fuel, 30 min reserve	2 hr 15 min

AM-69 Georges Payre two-seat light aircraft *(Aviation Magazine International)*

Besneux P.70B single-seat amateur-built aircraft in its latest, modified form

BRANDT
MICHEL BRANDT

ADDRESS: Résidence Biancotto, 21000-Darois

Mr Brandt, a former aeronautical student at the Swiss Federal Institute of Technology, has been Chief Engineer of Avions Pierre Robin since 1972. He has also designed a single-seat aerobatic monoplane known as the Kochab, intended initially as a competition aircraft for his own use.

BRANDT KOCHAB

TYPE: Single-seat light aircraft for competitive aerobatics.

WINGS: Cantilever mid-wing monoplane of straight-tapered planform. Symmetrical wing section of 16% thickness/chord ratio. No dihedral. Semi-monocoque wood structure. Conventional slotted ailerons.

FUSELAGE: Fabric-covered steel tube structure.

TAIL UNIT: Wire-braced steel tube structure, fabric-covered. Horn-balanced rudder and elevators.

LANDING GEAR: Non-retractable tailwheel type. Sandow shock-absorption. Fairings over main legs and wheels.

POWER PLANT: One modified 134 kW (180 hp) Teledyne Continental Tiara 4-180 flat-four engine, driving a three-blade Hoffmann constant-speed propeller or two-blade fixed-pitch propeller.

ACCOMMODATION: Pilot only, beneath transparent 'bubble' canopy.

DIMENSIONS, EXTERNAL:
- Wing span: 6·60 m (21 ft 8 in)
- Wing area, gross: 7·2 m² (77·5 sq ft)
- Wing aspect ratio: 6
- Length overall: 4·80 m (15 ft 9 in)

WEIGHT AND LOADINGS:
- Max T-O weight: 450 kg (992 lb)
- Max wing loading: 62·5 kg/m² (12·8 lb/sq ft)
- Max power loading: 3·36 kg/kW (5·5 lb/hp)

PERFORMANCE (estimated, at max T-O weight):
- Never-exceed speed: 205 knots (380 km/h; 236 mph)
- Max level speed: 156 knots (290 km/h; 180 mph)
- Max rate of climb at S/L: 840 m (2,755 ft)/min
- *g* limits: +9; −6

CHASLE
YVES CHASLE

ADDRESS:
Le Goya, rue de Traynes, 65-Tarbes

M Chasle, a stress engineer with Aérospatiale, designed and built a light aircraft named the YC-12 Tourbillon. Its dimensions were governed by the maximum size that could be accommodated in his garage workshop. First flight was made on 9 October 1965. As a result of the flight tests leading to its restricted C of A, the height of the vertical tail surfaces was later increased slightly. Plans are available to amateur constructors.

More recently, M Chasle designed a tandem two-seat light aircraft known as the YC-20, of which details can be found in the 1972-73 *Jane's*. He also designed the LMC-1 side-by-side two-seater, of which a prototype has been built by members of the Aero Club Léon Morane.

CHASLE YC-12 TOURBILLON (WHIRLWIND)

The Tourbillon can be built in a variety of forms, differing mainly in the type of engine fitted, as follows:

YC-121. With 48·5 kW (65 hp) Continental A65 engine. Generally similar to prototype (see 1970-71 *Jane's*) except for detail changes.

YC-122. Similar to YC-121, but with 71 kW (95 hp) Continental C90 or 74·5 kW (100 hp) Rolls-Royce Continental O-200-A engine.

YC-123. Similar to YC-121, but with 78·3 kW (105 hp) Potez 4E-20b engine.

Construction of YC-12s has been undertaken in several countries, including France, Canada, the USA, New Zealand and the UK. Marketing of the YC-12 in North America is by E. Littner, 546 83rd Avenue, Laval-Chomedey, Quebec, Canada.

TYPE: Single-seat amateur-built light aircraft.

WINGS: Cantilever low-wing monoplane. Wing section NACA Srs 7. Dihedral 6°. Incidence 3° 30′. All-wood structure, with main box spar of spruce and okoumé, spruce plank rear spar, girder-type ribs and okoumé plywood covering. All-wood ailerons and three-position slotted flaps.

FUSELAGE: Conventional plywood-covered wood structure, built around four spruce longerons, four main frames, five secondary frames and stringers.

TAIL UNIT: Cantilever all-wood structure, with swept vertical surfaces. Fixed tailplane and conventional elevators.

LANDING GEAR: Non-retractable tailwheel type. Steerable tailwheel linked with rudder. Main units have ERAM oleo-pneumatic suspension. Vespa wheels, size 400-100, mounted on L-shape legs. Independent mechanical brakes. Tailwheel carried on leaf spring. Provision for changing to a tricycle configuration, by switching main legs port and starboard, with lower arm of L facing rearward, and mounting nose unit on firewall.

POWER PLANT: One flat-four engine (see introductory copy) driving an EVRA two-blade propeller. Fuel tank, capacity 60·5 litres (13·3 Imp gallons), aft of firewall. Oil capacity 3·75 litres (0·83 Imp gallons).

ACCOMMODATION: Single seat under large transparent rearward-sliding canopy. Baggage space aft of seat.

ELECTRONICS AND EQUIPMENT: Optional items include Radiomaster radio, generator, starter, and night-flying equipment.

DIMENSIONS, EXTERNAL:
- Wing span: 6·70 m (22 ft 0 in)
- Wing chord at root: 1·40 m (4 ft 7¼ in)
- Wing chord at tip: 0·79 m (2 ft 7¼ in)
- Wing area, gross: 7·50 m² (80·7 sq ft)
- Wing aspect ratio: 6·0
- Length overall:
  - YC-121: 5·95 m (19 ft 6 in)
  - YC-122, YC-123: 5·85 m (19 ft 2¼ in)
- Height overall: 2·40 m (7 ft 10½ in)
- Tailplane span: 2·00 m (6 ft 6¾ in)
- Wheel track: 1·60 m (5 ft 3 in)
- Wheelbase: 3·65 m (11 ft 11¾ in)
- Propeller diameter: 1·75 m (5 ft 9 in)

WEIGHTS AND LOADINGS:
- Weight empty:
  - YC-121: 285 kg (628 lb)
  - YC-122, YC-123: 313 kg (690 lb)
- Max T-O weight, without radio:
  - YC-121: 432 kg (952 lb)
  - YC-122, YC-123: 460 kg (1,015 lb)
- Max wing loading:
  - YC-121: 57·5 kg/m² (11·77 lb/sq ft)
  - YC-122, YC-123: 61·3 kg/m² (12·55 lb/sq ft)
- Max power loading:
  - YC-121: 8·91 kg/kW (14·64 lb/hp)
  - YC-122: 6·48 kg/kW (10·69 lb/hp)
  - YC-123: 5·87 kg/kW (9·63 lb/hp)

PERFORMANCE (estimated):
- Max level speed at S/L:
  - YC-121: 127 knots (235 km/h; 146 mph)
  - YC-122: 146 knots (270 km/h; 168 mph)
  - YC-123: 151 knots (280 km/h; 174 mph)
- Max cruising speed (70% power):
  - YC-121: 110 knots (205 km/h; 127 mph)
  - YC-122: 129 knots (240 km/h; 149 mph)
  - YC-123: 135 knots (250 km/h; 155 mph)
- Stalling speed:
  - YC-121: 41 knots (75 km/h; 47 mph)
  - YC-122, YC-123: 44 knots (80 km/h; 50 mph)
- Max rate of climb at S/L:
  - YC-121: 276 m (905 ft)/min
  - YC-122: 420 m (1,380 ft)/min
  - YC-123: 480 m (1,575 ft)/min
- T-O run:
  - YC-121: 260 m (855 ft)
  - YC-122: 200 m (660 ft)
  - YC-123: 180 m (593 ft)
- Max range:
  - YC-121: 434 nm (800 km; 500 miles)
  - YC-122, YC-123: 377 nm (700 km; 435 miles)

CHASLE LMC-1 SPRINTAIR

This side-by-side two-seat all-metal light aircraft was built by some twenty members of the Léon Morane Club, hence its LMC designation. The high quality of the workmanship is explained by the fact that most members of the club are employees of Socata, the light aircraft subsidiary of Aérospatiale, at Tarbes.

Design of the Sprintair was started by M Chasle in February 1973. Construction of the prototype (F-PXKD) began in the Autumn of the same year, and it flew for the first time on 18 June 1975. It has been employed since that time as a training aircraft at the Léon Morane Club. No plans to market it in kit form, or to produce it commercially, have yet been announced.

TYPE: Two-seat all-metal light aircraft.

WINGS: Cantilever low-wing monoplane. Wing section NACA 63A218. Dihedral 4°. Incidence 3° (constant). All-metal torsion-box structure, with a main spar at 30% chord, light rear spar and nine pressed ribs in each wing, covered with AU4G-1/A5 light alloy skin, attached with countersunk rivets. Entire trailing-edge of each wing, except for tip, occupied by all-metal aileron and manually-operated three-position flap. Flaps and ailerons are identical except that they are bottom-hinged and top-hinged respectively. Ailerons actuated by control rods. No tabs.

FUSELAGE: Conventional all-metal semi-monocoque structure.

TAIL UNIT: Cantilever all-metal structure, with swept (30°) vertical surfaces. Tailplane incidence adjustable on the ground. Rudder actuated by cables, elevator by control rods. Controllable trim tab on port side of elevator. Ground-adjustable tab on rudder.

LANDING GEAR: Non-retractable tricycle type. Cantilever main legs each comprise an AU4G-1 light alloy leaf-spring. Steerable nosewheel leg embodies a telescopic compression spring shock-absorber and shimmy damper. Disc brakes on main wheels. Small spring tailskid.

POWER PLANT: One 74·5 kW (100 hp) Rolls-Royce Continental O-200-A flat-four engine, driving a two-blade fixed-pitch McCauley metal propeller with spinner. Two metal fuel tanks of the kind fitted to the Socata Rallye 100T, in the wing roots between the spars, with total capacity of 105 litres (23 Imp gallons).

ACCOMMODATION: Two seats side by side in enclosed cabin, with dual controls. Forward-hinged and jettisonable one-piece canopy, with rear-view transparent panel aft of seats. Provision for electronics, including radio-navigation aids.

DIMENSIONS, EXTERNAL:
- Wing span: 8·60 m (28 ft 2½ in)
- Wing area, gross: 10·00 m² (107·6 sq ft)
- Wing chord (constant): 1·20 m (3 ft 11¼ in)
- Wing aspect ratio: 7·5
- Length overall: 6·45 m (21 ft 2 in)
- Tailplane span: 3·00 m (9 ft 10 in)
- Propeller diameter: 1·80 m (5 ft 10¾ in)

DIMENSION, INTERNAL:
- Cabin: Width at seats: 1·20 m (3 ft 11¼ in)

WEIGHTS:
- Weight empty: 450 kg (992 lb)
- Max T-O weight: 700 kg (1,543 lb)

PERFORMANCE:
- Max level speed at S/L: 116 knots (215 km/h; 133 mph)
- Max cruising speed: 102 knots (190 km/h; 118 mph)
- Stalling speed, flaps down: 43·5 knots (80 km/h; 50 mph)
- T-O run: 230 m (755 ft)
- T-O to 15 m (50 ft): 380 m (1,250 ft)
- *g* limits at AUW of 680 kg (1,500 lb): +9; −4·5

Prototype Chasle YC-12 Tourbillon single-seat light aircraft (Continental A65 engine)

Chasle LMC-1 Sprintair two-seat light aircraft

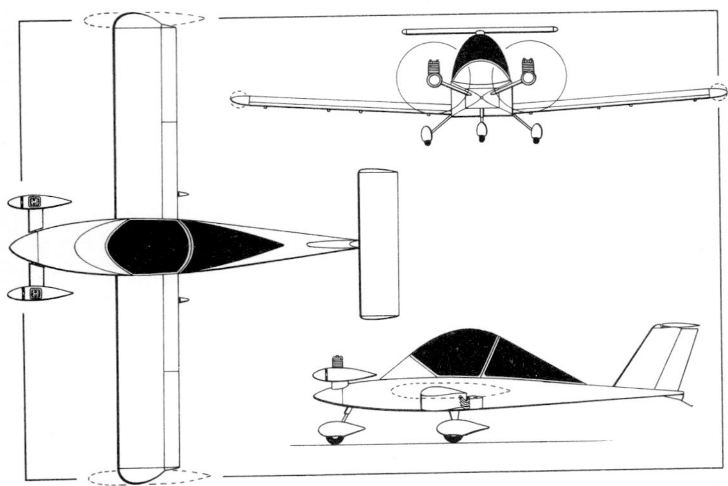

Colomban MC 10 Cricri, production version with optional wing-tip tanks indicated *(Michael A. Badrocke)*

Prototype Colomban MC 10 Cricri single-seat ultra-light amateur-built aircraft

COLOMBAN
MICHEL COLOMBAN

ADDRESS:

37*bis* rue Lakanal, 92500-Rueil-Malmaison

Telephone: 967-88-76

Formerly with the Morane and Potez companies, and now an aerodynamicist with Aérospatiale, M Colomban has designed and built a very small and unique twin-engined lightplane named the Cricri. Its construction required some 1,200 hours of work and cost only 5,000 francs (1971-72 prices), including the engines.

Plans of the Cricri are available to amateur constructors in both French and English language.

COLOMBAN MC 10 CRICRI (CRICKET)

Initial design studies for an aeroplane of only 15 kW (20 hp), for economical operation, were completed by M Colomban in 1958. His circumstances at that time did not permit its construction, and it was not until September 1970 that manufacture of the Cricri began. In the intervening years, the design was refined to take advantage of new developments in technology and aerodynamics.

The prototype (F-WTXJ) is powered by two Rowena 6507J single-cylinder two-stroke engines of 137cc, each giving 6·7 kW (9 hp) and weighing 6·5 kg (14·3 lb). It is claimed to be the smallest twin-engined aeroplane currently flying, and the only one able to lift a useful load equivalent to 170 per cent of its own empty weight. Special constructional features permit assembly or disassembly in only five minutes. Its light weight and small size make it particularly easy to transport on a trailer towed by car and to store in a garage or shed.

The Cricri was flown for the first time on 19 July 1973 by Robert Buisson, a 68-year-old pilot who had already logged 12,000 flying hours. A number of design refinements were made later that year, after which testing was resumed. Within fifteen days the Cricri had logged a total of 13 trouble-free flying hours, including rolls, reversements, 'split S' manoeuvres and inverted flight, made possible by its Tillotson diaphragm carburettor. Flight tests at up to 119 knots (220 km/h; 137 mph) confirmed that no special piloting skills are needed to fly this aircraft.

In particular, the Cricri handles like a single-engined design. This results from the fact that the two small engines are mounted close together, and from the carefully-conceived shape of the cockpit canopy which deflects the propeller slipstream over the tail surfaces in such a way that an engine failure produces no dangerous handling problems. If one engine is throttled back fiercely, with hands and feet off the controls, the Cricri is said to do no more than begin a gentle turn.

The version of the Cricri for which plans are available, and to which the following detailed description applies, differs in certain respects from the prototype. The type of construction is unchanged, but modifications have been introduced to reduce the time required to build the aircraft and to assemble it for flight. The power plant has also been changed.

Ten Cricris were under construction in early 1976, in advance of the availability of definitive plans.

TYPE: Twin-engined single-seat ultra-light aircraft, stressed to +10*g* and −5*g*.

WINGS: Cantilever low-wing monoplane of constant chord. Laminar-flow aerofoil derived from a Wortmann section. Thickness/chord ratio 21·7%. Dihedral 4° from roots. Incidence 1° at root, −30′ at tip. No sweep. Single-spar box structure. Spar comprises a web riveted to AU4G angle-section booms. Inboard end of spar in each wing is of 'forked-tongue' form, like that of many sailplanes, to permit rapid assembly and disassembly of wings (2 minutes). Closely-spaced Klégécel ribs are bonded fore and aft of the spar. Skin consists of a single sheet of AU4G, bonded to structure under pressure after its leading-edge has been formed. No rear spar. Wing box is closed at each end by a riveted metal rib. Entire trailing-edge is occupied by two-section external flaps of the kind fitted to many wartime Junkers aircraft, operating collectively as high-lift devices (movement −5° to +30°) and differentially as ailerons (+5° to −8°). Flaps are spar-less, consisting of a metal monocoque structure, with four metal ribs per section (at each tip and each pivot point), filled with Klégécel over the entire span and over 20% of the chord. Flaps are each actuated via a ball-joint at the root. No controls pass through the wing box, which contains only an AU4G tube as provision for any future installation of fuel tanks in wingtips.

FUSELAGE: Simple metal box structure of AU4G sheet made in two parts to reduce space required for manufacture. Rear portion is of inverted triangular section. Rectangular-section forward portion is riveted together with four angle sections at the corners. Structure is stiffened by Klégécel stringers, bonded in place. AU4G frames riveted in position in line with the attachments for the wings, landing gear, tail unit and engine mountings.

TAIL UNIT: Cantilever T type, with sweptback vertical surfaces and all-moving constant-chord horizontal surface. Construction similar to that of wings. No tabs. Tailplane actuated by control rods, rudder by cables. Tailplane provided with artificial loading by bungee cord.

LANDING GEAR: Non-retractable tricycle type. Nosewheel fitted with bungee shock-absorption and linked to rudder bar for steering. Each main wheel carried on cantilever leg of glassfibre/epoxy laminations. Main-wheel tyres size 210-70, pressure 0·98 bars (14·2 lb/sq in). Nosewheel tyre size 200-50, pressure 0·98 bars (14·2 lb/sq in). Colomban disc brakes. Provision for fairing on all three wheels.

POWER PLANT: Two Solo W424 single-cylinder two-stroke engines of 210 cc, each giving 8·95 kW (12 hp) at 5,300 rpm and weighing 9 kg (20 lb). Dual ignition. Tillotson diaphragm carburettors to permit inverted flight. Each engine drives a Hoffmann MCH-2 two-blade propeller, made of laminated wood and sheathed in plastics. Plastics fuel tank in fuselage, capacity 20 litres (4·4 Imp gallons). Provision for tank on each wingtip, total capacity 24 litres (5·25 Imp gallons).

ACCOMMODATION: Single seat under large transparent canopy, hinged to open sideways, to starboard. Ventilation through port in side of fuselage. No heating.

SYSTEM: Electrical system supplied by two 19W 6V and two 5W 6V batteries.

DIMENSIONS, EXTERNAL:

Wing span, with or without tip-tanks	4·90 m (16 ft 0¾ in)
Wing chord, incl flap (constant)	0·63 m (2 ft 0¾ in)
Wing chord, excl flap (constant)	0·48 m (1 ft 6¾ in)
Wing area, gross	3·10 m² (33·4 sq ft)
Wing aspect ratio	7·75
Length overall	3·91 m (12 ft 10 in)
Height overall	1·20 m (3 ft 11¼ in)
Tailplane span	1·55 m (5 ft 1 in)
Wheel track	1·10 m (3 ft 7¼ in)
Wheelbase	1·15 m (3 ft 9¼ in)
Propeller diameter	0·75 m (2 ft 5½ in)
Distance between propeller centres	0·95 m (3 ft 1½ in)

DIMENSIONS, INTERNAL:

Cabin: Length	1·30 m (4 ft 3¼ in)
Max width	0·55 m (1 ft 9½ in)
Max height	0·82 m (2 ft 8¼ in)

WEIGHTS AND LOADINGS:

Weight empty	75 kg (165 lb)
Max T-O and landing weight	180 kg (397 lb)
Max wing loading	58·1 kg/m² (11·89 lb/sq ft)
Max power loading	10·06 kg/kW (16·54 lb/hp)

PERFORMANCE (estimated at AUW of 160 kg; 352 lb):

Never-exceed speed	158 knots (293 km/h; 182 mph)
Max level speed	118 knots (220 km/h; 136 mph)
Max cruising speed (75% power)	105 knots (195 km/h; 121 mph)
Stalling speed, flaps down	42 knots (77 km/h; 48 mph)
Stalling speed, flaps up	50 knots (93 km/h; 58 mph)
Max rate of climb at S/L	336 m (1,100 ft)/min
Rate of climb at S/L, one engine out	80 m (262 ft)/min
Service ceiling	4,600 m (15,090 ft)
T-O run	170 m (558 ft)
T-O to 15 m (50 ft)	420 m (1,380 ft)
Landing from 15 m (50 ft)	350 m (1,150 ft)
Landing run	150 m (495 ft)
Range with max fuel	430 nm (800 km; 496 miles)

CROSES
EMILIEN and ALAIN CROSES

ADDRESS:

Route de Davayé, 71-Charnay les Macon

The 1960-61 *Jane's* contained details of the Croses EC-1-02 side-by-side two-seat lightplane of the Mignet tandem-wing type. M Emilien Croses subsequently built and flew an improved version of this aircraft known as the EC-6 Criquet, a three-seat lightplane/air ambulance development designated B-EC 7, with more powerful engine, and the EAC-3 Pouplume, an ultra-light single-seat aeroplane of the same general configuration. The B-EC 7 was described and illustrated in the 1970-71 *Jane's*.

CROSES EAC-3 POUPLUME

As in the familiar Mignet designs, the Pouplume single-seat tandem-wing biplane has a fixed rear wing and a pivoted forward wing which dispenses with the need for ailerons and elevators. A conventional rudder is fitted, with a large tailwheel built into its lower edge.

Construction is conventional, with spruce wing structure and a square-section spruce fuselage covered with okoumé ply. The main landing gear consists of Vespa scooter wheels carried on a wooden cross-member.

The power unit in the prototype (EAC-3-01) is a 7·8 kW (10·5 hp) Moto 232 cc two-stroke motorcycle engine, with chain reduction drive to the propeller shaft. The reduction ratio is 3·5 : 1, giving a propeller speed of 1,300 rpm. Fuel capacity is 10 litres (2·2 Imp gallons).

The EAC-3-01 Pouplume took 600 hours to build and flew for the first time in June 1961. This machine was followed, in 1967, by a second prototype (EAC-3-02), with an 20 cm (8 in) longer fuselage. M Croses is offering sets of plans to other constructors, and the Pouplume shown in the accompanying illustration was built in France by an amateur constructor.

Alternative engines that may be fitted in the Pouplume include the various Volkswagen conversions.

DIMENSIONS, EXTERNAL (EAC-3-01):

Span of forward wing	7·8 m (25 ft 7 in)
Span of rear wing	7·0 m (23 ft 0 in)
Wing area, gross	16·0 m² (172 sq ft)
Length overall	3·0 m (9 ft 10 in)
Height overall	1·8 m (5 ft 11 in)

WEIGHTS:

Weight empty	110-140 kg (243-310 lb)
Max T-O weight	220-260 kg (485-573 lb)

PERFORMANCE (A: 7·8 kW; 10·5 hp engine. B: 13·4 kW; 18 hp engine):

Max level speed:
- A: 38 knots (70 km/h; 43·5 mph)
- B: 65 knots (120 km/h; 75 mph)

Econ cruising speed:
- A: 27 knots (50 km/h; 31 mph)
- B: 38 knots (70 km/h; 43·5 mph)

T-O speed:
- A: 13·5 knots (25 km/h; 15·5 mph)

Landing speed:
- A: 9·7 knots (18 km/h; 11 mph)

T-O run:
- A: 60 m (200 ft)
- B: 40 m (131 ft)

Landing run:
- A: 24 m (80 ft)

Fuel consumption:
- A: 4·5 litres (1 Imp gallon)/hr

CROSES EC-6 CRIQUET (LOCUST)

This design by Emilien Croses is a development of his earlier EC-1-02 prototype and is a side-by-side two-seater based on the familiar Mignet tandem-wing formula. Construction was started in March 1964 and the EC-6-01 flew for the first time on 6 July 1965.

In 1975, M Croses was completing the prototype of an all-plastics version of the Criquet, in the works of M Millet, who is also President of the Aero Club of l'Aude. The fuselage, engine cowling, fin, wing spars, upper wing support struts and main landing gear legs are made from polyester, resin and glassfibre. The wing ribs and certain other components are of Klégécel. Other parts are made from a balsa/glassfibre composite material.

Plans of the original wooden version of the Criquet, of which details follow, are available to amateur constructors. It is hoped to make available later both plans and kits of the plastics version, with some 40% of the components finished and ready for assembly.

TYPE: Two-seat tandem-wing light aircraft.

WINGS: Forward wing built in one piece and pivoted on two streamlined supports, giving variable incidence between −2° and +12°. Fixed rear (lower) wing. Wing section NACA 23012 (modified). Both wings have two-spar wooden structure, with plywood leading-edge, overall fabric covering and some components of glassfibre. No ailerons.

FUSELAGE: Spruce structure, covered with plywood. Glassfibre engine cowling.

TAIL UNIT: Plywood-covered spruce fin and rudder. No tailplane or elevators.

LANDING GEAR: Non-retractable tailwheel type. Main wheels, size 420-150, carried on single cantilever arch structure made from ash wood on a forme and covered with glassfibre. Tailwheel, size 420-150, semi-enclosed in bottom of rudder.

POWER PLANT: One 67 kW (90 hp) Continental flat-four engine, driving a modified SIPA two-blade propeller. Fuel capacity originally 60 litres (13 Imp gallons); planned to be increased to 90 litres (20 Imp gallons).

ACCOMMODATION: Two seats side by side in enclosed cabin. Door on starboard side.

DIMENSIONS, EXTERNAL:
- Span of forward wing: 7·80 m (25 ft 7 in)
- Span of rear wing: 7·00 m (22 ft 11½ in)
- Wing chord (constant, each): 1·20 m (3 ft 11¼ in)
- Wing area, gross: 16·0 m² (172 sq ft)
- Length overall: 4·65 m (15 ft 3 in)

WEIGHTS:
- Weight empty: 290 kg (639 lb)
- Max T-O weight: 550 kg (1,213 lb)

PERFORMANCE (officially certificated, at max T-O weight):
- Max level speed at S/L: 115 knots (213 km/h; 132 mph)
- Max cruising speed: 92 knots (170 km/h; 106 mph)
- Econ cruising speed: 86 knots (160 km/h; 99 mph)
- Min flying speed: 22 knots (40 km/h; 25 mph)
- Will not stall
- T-O time (max): 6 sec
- Climb to 2,000 m (6,560 ft): 6 min 14 sec

DURUBLE

ROLAND DURUBLE

ADDRESS: 40 rue de Paradis, Les Essarts, 76530-Grand-Couronne

Telephone: 92-20-63

M Roland Durable, with MM Guy Chanut and Legrand, of Rouen, designed and built a two-seat all-metal light aircraft named the RD-02 Edelweiss, which flew for the first time on 7 July 1962. Full details of this aircraft can be found in the 1972-73 *Jane's*.

Plans of an enlarged and improved version, known as the RD-03 Edelweiss, are available to other constructors.

DURUBLE RD-03 EDELWEISS

The RD-03 Edelweiss is designed to AIR 2052 (CAR 3) standards, and is projected in three versions, as follows:

RD-03A. With 74·5 kW (100 hp) Continental O-200 flat-four engine and fuel capacity of 100 litres (22 Imp gallons) in two wing tanks. Can be fitted with 67 kW (90 hp) engine or 100·5 kW (135 hp) Lycoming O-320 engine. Side-by-side seats for pilot and one passenger (total weight 172 kg; 380 lb) in cabin.

RD-03B. With 100·5 kW (135 hp) Lycoming O-320 or Franklin Sport 4B engine and same fuel capacity as RD-03A. Seating as RD-03A for Utility category operation, or in '2 + 2' arrangement for pilot and three passengers (154 kg; 340 lb on front seats, 110 kg; 240 lb on rear seats) in Normal category.

RD-03C. With 112 kW (150 hp) Lycoming engine and additional wing tanks, increasing total fuel capacity to 150 litres (33 Imp gallons). In Utility two-seat form (as RD-03A) or with seating for a pilot and three passengers (total weight 308 kg; 680 lb) in Normal category.

Plans of the RD-03 have been available since the Autumn of 1970. By early 1976, a total of 30 sets of plans had been sold. Four Edelweiss are known to be under construction in France, four in Belgium, one in Luxembourg, two in Canada and the USA, and one in New Zealand. The first of these is not expected to fly until 1978.

TYPE: Two/four-seat light aircraft.

WINGS: Cantilever low-wing monoplane. Wing section NACA 23000 series. Thickness/chord ratio 18% at root, 12% at tip. Dihedral 6° 5′ from roots. Incidence 3° at root, 0° at tip. No sweepback. All-metal two-spar duralumin structure, with metal slotted trailing-edge flaps and slotted ailerons. No trim tabs.

FUSELAGE: Conventional semi-monocoque duralumin structure.

TAIL UNIT: Cantilever all-metal structure, with sweptback vertical surfaces. Fixed-incidence tailplane. Trim tab in each elevator, one actuated by flap linkage and the other manually.

LANDING GEAR: Retractable tricycle type. Hydraulic retraction, nosewheel rearward, main units inward into wings. Durable hydro-air shock-absorbers on all three units. Main-wheel tyres size 355 × 150, nosewheel tyre size 330 × 130. Pressure (all tyres) 1·24 bars (18 lb/sq in). Hydraulic disc brakes.

POWER PLANT: One 67, 74·5, 100·5 or 112 kW (90, 100, 135 or 150 hp) flat-four engine (details under individual model listings). Refuelling point above wing.

ACCOMMODATION: Side-by-side seats for two, three or four persons (details under individual model listings) in fully-enclosed cabin.

SYSTEMS: Hydraulic system, pressure 69 bars (1,000 lb/sq in), for flap and landing gear actuation.

ELECTRONICS AND EQUIPMENT: Radio optional. Blind-flying instrumentation not fitted.

DIMENSIONS, EXTERNAL:
- Wing span: 8·75 m (28 ft 8½ in)
- Wing chord at root: 1·70 m (5 ft 7 in)
- Wing chord at tip: 0·86 m (2 ft 10 in)
- Wing area, gross: 11·04 m² (118·5 sq ft)
- Wing aspect ratio: 6·95
- Length overall (RD-03A): 6·27 m (20 ft 7 in)
- Height overall: 2·35 m (7 ft 8½ in)
- Tailplane span: 3·05 m (10 ft 0 in)

DIMENSIONS, INTERNAL:
- Cabin: Max length: 2·44 m (8 ft 0 in)
- Max width: 1·10 m (3 ft 7¼ in)

WEIGHTS AND LOADINGS (estimated. A: RD-03A; B: RD-03B; C: RD-03C):

Weight empty, equipped:
- A: 406·5 kg (896 lb)
- B: 416 kg (917 lb)
- C: 421·5 kg (929 lb)

Max T-O and landing weight:
- A (Utility): 699 kg (1,541 lb)
- B (Utility): 715 kg (1,576 lb)
- B (Normal): 796 kg (1,754 lb)
- C (Utility): 740 kg (1,631 lb)
- C (Normal): 866 kg (1,909 lb)

Max wing loading:
- A (Utility): 63·0 kg/m² (12·90 lb/sq ft)
- B (Utility): 64·0 kg/m² (13·11 lb/sq ft)
- B (Normal): 71·5 kg/m² (14·64 lb/sq ft)
- C (Utility): 66·4 kg/m² (13·60 lb/sq ft)
- C (Normal): 78·0 kg/m² (16·00 lb/sq ft)

Max power loading:
- A (Utility): 9·37 kg/kW (15·41 lb/hp)
- B (Utility): 7·11 kg/kW (11·68 lb/hp)
- B (Normal): 7·92 kg/kW (13·01 lb/hp)
- C (Utility): 6·61 kg/kW (10·85 lb/hp)
- C (Normal): 7·73 kg/kW (12·79 lb/hp)

PERFORMANCE (estimated, at max T-O weight):

Never-exceed speed:
- A (Utility), B (Utility), C (Utility): 170 knots (316 km/h; 196 mph)
- B (Normal), C (Normal): 182 knots (339 km/h; 210 mph)

Max level speed at S/L:
- A (Utility): 139 knots (257 km/h; 160 mph)
- B (Utility): 143 knots (265 km/h; 165 mph)
- B (Normal): 141 knots (262 km/h; 163 mph)
- C (Utility): 149 knots (277 km/h; 172 mph)
- C (Normal): 146 knots (270 km/h; 168 mph)

Max cruising speed at S/L:
- A (Utility): 126·5 knots (234 km/h; 145·5 mph)
- B (Utility): 129 knots (240 km/h; 149 mph)
- B (Normal): 128 knots (238 km/h; 148 mph)
- C (Utility): 136 knots (252 km/h; 157 mph)
- C (Normal): 133 knots (246 km/h; 153 mph)

Econ cruising speed at S/L:
- A (Utility), B (Utility): 121 knots (224 km/h; 139 mph)
- B (Normal): 120 knots (222 km/h; 138 mph)
- C (Utility): 128 knots (237 km/h; 147·5 mph)
- C (Normal): 125 knots (233 km/h; 144·5 mph)

Stalling speed, flaps up:
- A (Utility), B (Utility): 50 knots (92 km/h; 57·5 mph)
- B (Normal), C (Utility): 56 knots (104 km/h; 65 mph)
- C (Normal): 59 knots (109 km/h; 68 mph)

Stalling speed, flaps down:
- A (Utility), B (Utility): 41 knots (76·5 km/h; 47·5 mph)

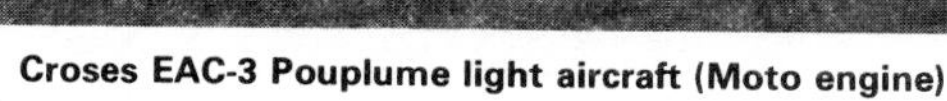
Croses EAC-3 Pouplume light aircraft (Moto engine)

Croses EC-6 Criquet two-seat light aircraft *(Geoffrey P. Jones)*

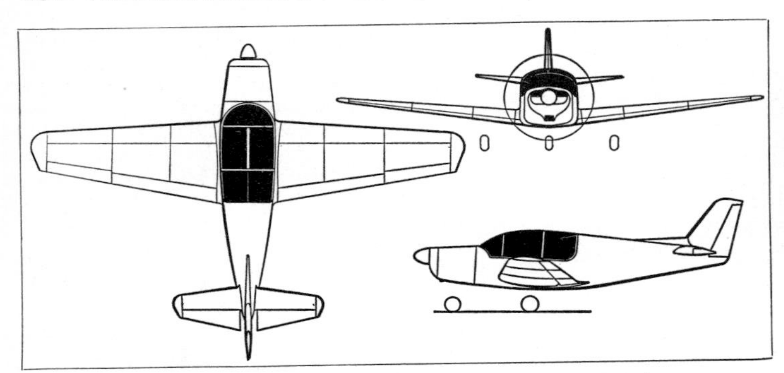

Durable RD-03C Edelweiss two-four-seat light aircraft

Duruble RD-02 Edelweiss, from which the RD-03 was evolved

B (Normal), C (Utility)
47 knots (86·5 km/h; 54 mph)
C (Normal) 48 knots (89 km/h; 55·5 mph)

Max rate of climb at S/L:
- A 198 m (650 ft)/min
- B 213 m (700 ft)/min
- C 244 m (800 ft)/min

Service ceiling:
- A, B 4,570 m (15,000 ft)
- C 5,030 m (16,500 ft)

T-O run:
- A 250 m (820 ft)
- B 274 m (900 ft)
- C 305 m (1,000 ft)

T-O to 15 m (50 ft):
- A, B, C 457 m (1,500 ft)

Landing from 15 m (50 ft):
- A 186 m (610 ft)
- B 244 m (800 ft)
- C 287 m (940 ft)

Landing run:
- A 305 m (1,000 ft)
- B 335 m (1,100 ft)
- C 427 m (1,400 ft)

Range with max fuel, 30 min reserve:
- A 607 nm (1,125 km; 700 miles)
- B 521 nm (965 km; 600 miles)
- C 547 nm (1,010 km; 630 miles)

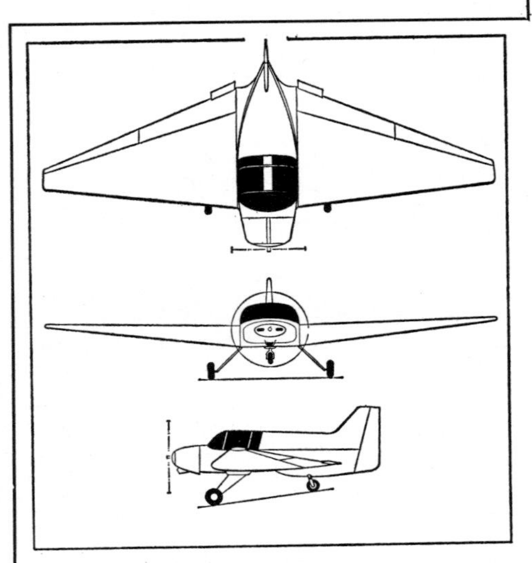

Fauvel AV.44 *(Roy J. Grainge)*

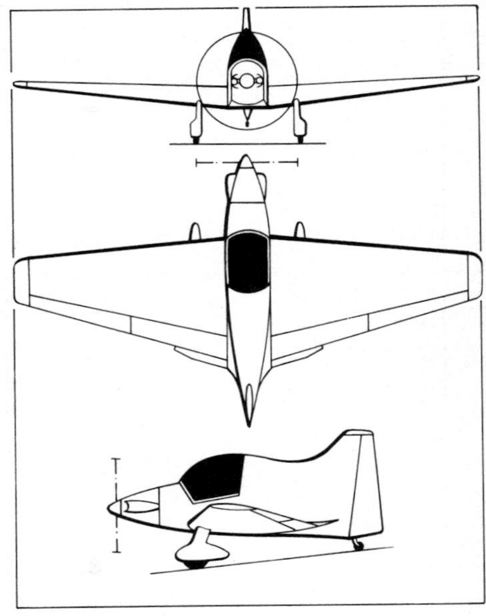

Fauvel AV.50(61) Lutin *(Tony Mitchell)*

FAUVEL
"SURVOL"-CHARLES FAUVEL

HEAD OFFICE:
72 Boulevard Carnot, 06400-Cannes AM
Telephone: (93) 39-83-32

In addition to the sailplanes and self-launching sailplanes described in the relevant section of this edition, Charles Fauvel has designed several powered lightplanes. Plans of these are available to amateur constructors.

M Fauvel is also perfecting a unique form of wooden integral fuel tank for potential embodiment in all of his powered designs.

FAUVEL AV.44

The general appearance of this all-wood side-by-side two-seat, or three-seat, tailless monoplane can be seen in the accompanying illustration. It is a direct development of M Fauvel's AV.10 aircraft which flew for the first time in 1935, with a 56 kW (75 hp) Pobjoy R engine, and subsequently set up altitude records in single-seat and two-seat categories, in competition with aircraft of considerably greater power.

The AV.44 can be powered by a variety of engines in the 67-97 kW (90-130 hp) range, and is classed as a STOL (ADAC) type. Alternatively, it can be powered by two engines of 41-52 kW (55-70 hp), mounted in the nose and driving two co-axial propellers. This would involve no penalties in terms of drag or dissymmetry; nor would it require any additional 'twin-engine' training for the pilot. Examples of the AV.44 are expected to be built by several amateurs, but none had flown by early 1976. All available details follow:

DIMENSIONS, EXTERNAL:
- Wing span 10·70 m (35 ft 1¼ in)
- Wing area, gross 19·8 m² (213·1 sq ft)
- Wing aspect ratio 5·8
- Length overall 5·00 m (16 ft 4¾ in)

WEIGHTS:
- Weight empty 340 kg (749 lb)
- Fuel and oil 90 kg (198 lb)
- Normal T-O weight, two-seat 604 kg (1,331 lb)
- Max T-O weight, three-seat 681 kg (1,501 lb)

PERFORMANCE (74·5 kW; 100 hp Continental engine):
- Max level speed at S/L 113 knots (210 km/h; 130 mph)
- Max cruising speed at S/L 102 knots (190 km/h; 118 mph)
- Max rate of climb at S/L 294 m (965 ft)/min
- Endurance with max fuel at econ cruising speed 5½ hours

FAUVEL AV.50 (61) LUTIN (ELF)

The AV.50 (61) is a single-seat all-wood light aircraft of tailless configuration. Its wing section can be either Fauvel F2 or a special Wortmann FX-66-H-159 laminar aerofoil. Suitable power plants include a modified Volkswagen motor car engine developing 30 kW (40 hp) and the 41 kW (55 hp) Hirth O 28 two-cylinder aircooled two-stroke engine. A tandem-wheel, tailwheel or tricycle landing gear can be fitted.

Predecessor of the AV.50 (61) was the AV.60 Leprechaun, of which an example built and flown in the USA was illustrated in the 1975-76 *Jane's*. Only details of the AV.50 (61) available in mid-1976 were the following:

DIMENSIONS, EXTERNAL:
- Wing span 7·50 m (24 ft 7¼ in)
- Wing area, gross 10·80 m² (116·25 sq ft)
- Wing aspect ratio 5·2
- Length overall:
  - Hirth engine 4·40 m (14 ft 5¼ in)
  - VW engine 4·10 m (13 ft 5½ in)

WEIGHTS:
- Weight empty, equipped:
  - Hirth engine 160 kg (352 lb)
  - VW engine 190 kg (419 lb)
- Fuel 45 kg (99 lb)
- Max T-O weight:
  - Hirth engine 299 kg (659 lb)
  - VW engine 329 kg (725 lb)

PERFORMANCE (estimated, at max T-O weight):
- Max level speed (Hirth engine):
  - tandem landing gear 116 knots (215 km/h; 134 mph)
  - tailwheel landing gear 110 knots (205 km/h; 127 mph)
  - tricycle landing gear 102 knots (190 km/h; 118 mph)
- Max level speed (VW engine):
  - tandem landing gear 102 knots (190 km/h; 118 mph)
  - tailwheel landing gear 98 knots (182 km/h; 113 mph)
  - tricycle landing gear 92 knots (170 km/h; 106 mph)
- Max rate of climb at S/L:
  - Hirth engine 330 m (1,085 ft)/min
  - VW engine 216 m (710 ft)/min
- T-O run:
  - Hirth engine 70 m (230 ft)
  - VW engine 100 m (330 ft)
- Endurance with max fuel:
  - Hirth or VW engine 5 hours

GATARD
AVIONS A. GATARD

ADDRESS:
52 route de Jonzac, 17130-Montendre
Telephone: 183

M Albert Gatard has developed a control system for aeroplanes which involves the use of a variable-incidence lifting tailplane of large area, and has built a series of aircraft, including the Alouette, Poussin and Pigeon, incorporating his ideas. The Alouette (described in the 1959-60 *Jane's*) was purely experimental, and the Pigeon is at an early stage of development, but plans of the Poussin are available to amateur constructors.

Instead of altering the wing angle of attack to increase lift on these aircraft, the pilot lowers full-span slotted aileron/flaps and adjusts the tailplane to maintain pitching equilibrium. In consequence, the aircraft climb with the fuselage datum at no more than 4° to the horizontal, which preserves a good forward view and low body drag.

GATARD STATOPLAN AG 02 POUSSIN (CHICK)

M Gatard built two prototypes of the Poussin and the detailed description applies to the second of these, which introduced a number of design improvements, in its original form. Flight tests revealed excellent aerobatic qualities and the power plant is modified to permit up to 20 seconds of inverted flying.

This prototype was extensively flight-tested at the Centre d'Essais en Vol at Istres, and the performance figures quoted are those which were obtained during the tests. As a result of recommendations by the CEV, a 27 kW (36 hp) Rectimo (modified Volkswagen VW 1200) engine is suggested as the most suitable power plant for use by amateur constructors of the Poussin. The second prototype has been re-engined with a 1,200 cc Volkswagen, by M Mathevet of Mollard-Chateauneuf (Loirs), on behalf of M Gatard. Installation of this engine was expected to improve the CG position and make possible a max speed of approx 92 knots (170 km/h; 106 mph), a max cruising speed of approx 83 knots (155 km/h; 96 mph) and a rate of climb at S/L of 210 m (690 ft)/min.

Several Poussins are being built by amateur constructors, and three were nearing completion in 1975. One of them is the work of Mr Gomès of Lubumbashi in the Zaïre Republic.

TYPE: Single-seat ultra-light monoplane.

WINGS: Cantilever low-wing monoplane. NACA 23012 wing section. Dihedral 4°. Incidence 3° 30′ at root, 2° at tip. Plywood-covered single-spar all-wood structure. Full-span slotted aileron/flaps, each in two sections which are moved together but at different angles (inboard sections up to 35°, outboard up to 20°) to give the effect of increased aerodynamic twist of the complete wing/aileron/flap assemblies. Aileron/flaps are linked with the variable-incidence tailplane.

FUSELAGE: Plywood-covered wood structure. Perforated airbrake, under fuselage, operates automatically when the main aileron/flaps are lowered at large angles, as during landing.

TAIL UNIT: Braced all-wood structure, with variable-incidence all-moving tailplane of NACA 2309 section. Endplates fitted to tailplane to increase vertical fin area and effective tailplane span. No elevators. Rudder trim tab actuated by lateral movement of control column, permitting full control by means of the control column alone in normal flight.

LANDING GEAR: Non-retractable tailwheel type. Cantilever levered-suspension main units with rubber-band shock-absorption. Modified Dunlop brakes. Steerable tailwheel.

POWER PLANT: One 18 kW (24 hp) modified Volkswagen flat-four engine, driving a Gatard two-blade fixed-pitch wooden propeller. Provision for fitting any alternative engine of up to 30 kW (40 hp) weighing between 50 and 60 kg (110-132 lb). Fuel tank aft of firewall, capacity 30 litres (6·6 Imp gallons). Oil capacity 2 litres (0·45 Imp gallons).

ACCOMMODATION: Single seat under large rearward-sliding transparent canopy. Baggage space aft of seat. Two map pockets.

DIMENSIONS, EXTERNAL:

Wing span	6·40 m (21 ft 0 in)
Wing chord (constant)	1·00 m (3 ft 3¼ in)
Wing area, gross	6·15 m² (66·2 sq ft)
Length overall	4·53 m (14 ft 10½ in)
Height overall	1·50m (4 ft 11 in)
Wheel track	1·50 m (4 ft 11 in)
Wheelbase	3·20 m (10 ft 6 in)

WEIGHTS:

Weight empty	170 kg (375 lb)
Max T-O weight	280 kg (617 lb)

PERFORMANCE (at max T-O weight):

Never-exceed speed	116 knots (216 km/h; 134 mph)
Max cruising speed	77 knots (144 km/h; 89 mph)
Max speed for aerobatics	69 knots (130 km/h; 80 mph)
Stalling speed	35 knots (65 km/h; 40·5 mph)
Max rate of climb at S/L	132 m (435 ft)/min
T-O run	190 m (625 ft)
T-O to 15 m (50 ft)	435 m (1,425 ft)
Landing from 15 m (50 ft)	320 m (1,050 ft)
Landing run	200 m (655 ft)

GATARD STATOPLAN AG 04 PIGEON

The AG 04 Pigeon is a three/four-seat high-wing monoplane, powered by a 67 kW (90 hp) Continental engine. It utilises the same type of control system as its predecessors, except that the Statoplan aileron/flaps are in three sections on each wing. As on the Poussin, they move together but through different angles (successively 45°, 30° and 20°) to give the effect of increased aerodynamic twist of the complete wing/aileron/flap assemblies.

The wings are braced by a single streamline-section strut each side and can be folded by two people in seven or eight minutes, to permit the aircraft to be towed along roads behind a motor car. Overall dimensions under tow are: length 6·00 m (19 ft 8¼ in), width 2·10 m (6 ft 10¾ in) and height 3·0 m (9 ft 10 in).

A non-retractable tailwheel-type landing gear is fitted, with long-stroke rubber-ring shock-absorbers on all three units. Both manual and toe-operated brakes are specified, the former serving also as a parking brake.

The two front seats are adjustable fore and aft and tilt forward to facilitate access to the rear of the cabin. All controls are so mounted that the main cabin area is completely clear, with a suspended control column. Full control is possible in normal flight by means of this control column alone, as on the Poussin. Dual controls are provided for a second pilot, but can be disconnected easily. Entry is via a two-section door on each side; the lower section opens forward, the top section upward under the wing. The top section can be opened to a mid-position for additional ventilation while taxying. An optional door on the starboard side permits loading of a stretcher 1·80 m long by 0·45 m wide (5 ft 11 in by 1 ft 6 in) when the aircraft is used in an ambulance role. The rear seat on this side can be folded down to make room for the stretcher.

Fuel is carried in two completely independent tanks, with capacities of 70 litres (15·4 Imp gallons) and 14 litres (3·08 Imp gallons) respectively. Should the fuel pump fail, a special cock enables the engine to be supplied by gravity feed.

The prototype Pigeon was flying in mid-1976. It was expected to offer automatic stability in flight, and to take off and land at about 38 knots (70 km/h; 44 mph).

WEIGHTS:

Weight empty	approx 360 kg (793 lb)
Normal T-O weight	620 kg (1,366 lb)
Max T-O weight	700 kg (1,543 lb)

GATARD STATOPLAN AG 05 MÉSANGE (TOMTIT)

The AG 05 Mésange is essentially an enlarged development of the Poussin with side-by-side seating for two persons. It is intended primarily as a training or aerobatic aircraft, but will have provision either for seating a third occupant or for installing a supplementary fuel tank aft of the two front seats, to make the aircraft suitable for touring. The control system will be similar to that of the Poussin, but the Mésange will have larger, broader-chord wings without the rounded tips of its predecessor, and will have leading-edge fuel tanks.

Completion of the prototype will follow successful development of the AG 04 Pigeon.

TYPE: Two/three-seat light aircraft.

WINGS: Cantilever low-wing monoplane. NACA 23012 wing section. Dihedral 4°. Incidence 3° at root, −1° 30′ at tip. Plywood-covered all-wood structure. Full-span slotted aileron/flaps of similar type to those of Poussin, inboard sections movable between 35° and −20°, outboard sections between 20° and −12°. Aileron/flaps are linked with the variable-incidence tailplane.

FUSELAGE: Plywood-covered steel tube structure. Airbrake beneath centre-section, length 1·50 m (4 ft 11 in), operates automatically in similar manner to that on Poussin.

TAIL UNIT: Plywood-covered all-wood structure. Variable-incidence all-moving tailplane.

LANDING GEAR: Non-retractable tailwheel type, with rubber-band shock-absorbers on main units. Steerable tailwheel. Main-wheel brakes and parking brake.

POWER PLANT: Installations envisaged at present are either a 1,600 cc modified Volkswagen horizontally-opposed aircooled engine or a 67/78·3 kW (90/105 hp) Continental flat-four engine. Fuel in two tanks in wing leading-edge, with total capacity of 70 litres (15 Imp gallons). Provision for installing 100 litre (21·5 Imp gallon) auxiliary fuel tank aft of two front seats.

ACCOMMODATION: Normal seating for pilot and one passenger on side-by-side seats in trainer version. For club or private use a third seat may be installed aft of the two front seats when no auxiliary fuselage fuel tank is fitted.

DIMENSIONS, EXTERNAL:

Wing span	8·60 m (28 ft 2½ in)
Wing chord (constant)	1·35 m (4 ft 5¼ in)
Wing area, gross	11·30 m² (121·6 sq ft)
Length overall	6·00 m (19 ft 8¼ in)
Wheel track	2·00 m (6 ft 6¾ in)

WEIGHTS:

Weight empty	360 kg (793 lb)
Max T-O weight	600 kg (1,322 lb)

PERFORMANCE (estimated, at max T-O weight):

Max cruising speed	95 knots (175 km/h; 109 mph)
Landing speed	39 knots (72 km/h; 44·7 mph)
Max rate of climb at S/L	300-360 m (985-1,180 ft)/min
Range (3-seat version)	approx 807 nm (1,500 km; 930 miles)
Max endurance:	
with standard fuel	3 hr 30 min
with auxiliary fuel	8 hr 30 min

GEISER

JEAN MARC GEISER

ADDRESS

95 Boulevard Saint-Michel, 75005 Paris

Telephone: 326 50 99

GEISER MOTO DELTA G.10

The Moto Delta G.10, as the accompanying photograph shows, is a simple, ultra-light powered Rogallo-type light aircraft. It has been produced as a low-cost, economical-to-run machine for amateur construction, and is road-transportable.

The fuselage and landing gear legs are built entirely of laminated glassfibre and Klégécel foam plastics. The tail unit consists of a fin and rudder only. A 13·5 kW (18 hp) German ECE flat-twin two-stroke engine is mounted behind the pilot's seat and drives a pusher propeller.

The delta-shaped Rogallo wing is made of Dacron, mounted on a tubular duralumin frame. An overhead control arm regulates the movement of the wing; a separate stick control actuates rudder movement.

The following details apply to the G.10 as illustrated; an alternative wing, of greater span and improved aerodynamic properties, is being developed.

DIMENSIONS, EXTERNAL:

Wing span	7·00 m (22 ft 11¾ in)
Wing area, gross	18·00 m² (193·8 sq ft)
Length overall	5·00 m (16 ft 4¾ in)

WEIGHTS:

Weight empty	45 kg (99 lb)
Max T-O weight	135 kg (297 lb)

PERFORMANCE:

Max level speed	49 knots (90 km/h; 56 mph)
Cruising speed	35 knots (65 km/h; 40·5 mph)
Stalling speed	19 knots (35 km/h; 22 mph)
T-O run	20 m (65 ft)
Landing run	0-10 m (0-33 ft)

HUREL

MAURICE HUREL

HUREL AVIETTE

Thought to be the largest single-seat man-powered aircraft yet built, the Aviette is a parasol-wing monoplane with a two-blade tractor propeller, chain- or belt-driven from a pair of small wheels mounted side by side beneath (and semi-recessed in) the front of the fuselage. It was designed by M Maurice Hurel and built by M Jacques Martinache, and made its first man-powered flight in 1974 after a series of towed flights.

The Aviette's extremely long-span wing, of Wortmann FX-61-184 section, is supported above the fuselage by a pylon, and braced by an inverted-Vee kingpost above and by streamlined rigging wires above and below. The tapered outer panels are fitted with conventional ailerons. In addition, to prevent aerodynamic twisting of the wings, large auxiliary surfaces, performing the function of aileron balance tabs, are carried on outriggers aft of the wing trailing-edge and inboard of the ailerons, each with a mass balance weight projecting forward of the wing. The fuselage is of hollow box construction, with balsa longerons. All movable surfaces are hinged on one side only by the Mylar material with which the aircraft is covered.

DIMENSIONS:

Wing span	40·25 m (132 ft 0¾ in)
Wing aspect ratio	30
Wing area, gross	54·00 m² (581·25 sq ft)
Propeller diameter	3·20 m (10 ft 6 in)

WEIGHTS AND LOADING:

Weight empty	66 kg (145·5 lb)
T-O weight	134 kg (295 lb)
Wing loading	2·49 kg/m² (0·51 lb/sq ft)

JODEL

AVIONS JODEL SA

HEAD OFFICE:

36 Route de Seurre, 21-Beaune

DESIGN OFFICE:

21-Darois

PRESIDENT-DIRECTOR GENERAL:

J. Delemontez

The Société des Avions Jodel was formed in March 1946, by MM Jean Delemontez and Edouard Joly, with the former acting as business and technical manager and the latter as test pilot.

Its first activities were concerned with the repair of gliders and light aircraft of the Service d'Aviation Légère et Sportive, on behalf of the State. Simultaneously, the company designed and built the D.9 Bébé Jodel single-seat light monoplane, which made its first flight in January 1948. This aeroplane, which is certificated with various power plants, is intended for amateur construction and can be built in as little as 500 man-hours.

As the result of official tests with the D.9, the French authorities placed an order for the development and construction of two prototypes of a two-seat model, the D.11 fitted with a 33·6 kW (45 hp) Salmson, and the D.111 with a 56 kW (75 hp) Minié engine. Subsequent developments of the D.11 are the D.112 and D.117, which have a 48·5 kW and 67 kW (65 hp and 90 hp) Continental engine respectively.

These designs also have been built in large numbers, both commercially and by amateurs.

Avions Jodel now devotes its activities mainly to designing advanced developments of its established types and to acting as a consultant to those building and developing its designs.

JODEL D.9 and D.92 BÉBÉ

The type designation of the Bébé varies according to the type of engine fitted. The original version, with 18·6 kW (25 hp) Poinsard engine, was designated D.9; the D.92 has a modified Volkswagen engine.

Second Gatard Statoplan AG 02 Poussin in its latest form *(F. E. v. Bruggen)*

Jodel D.9 single-seat homebuilt aircraft (Volkswagen engine) *(Peter J. Bish)*

Geiser Moto Delta G.10 homebuilt powered Rogallo

Gatard Statoplan AG 04 Pigeon, with engine cowling removed *(Geoffrey P. Jones)*

The following details refer to all standard versions of the Bébé:

Type: Single-seat light monoplane.

Wings: Cantilever low-wing monoplane. Single-spar one-piece wing with wide-span centre-section of constant chord and thickness and two tapering outer portions set at a coarse dihedral angle (14°). Spar and ribs of spruce and plywood, with fabric covering. Ailerons similar in construction.

Fuselage: Rectangular spruce and plywood structure.

Tail Unit: Cantilever structure of spruce and plywood, with plywood covering on tailplane and fabric-covered rudder and elevators. No fin.

Landing Gear: Non-retractable cantilever main legs with rubber-in-compression springing. Leaf-spring tailskid or tailwheel. Cable brakes.

Power Plant: One 18·6 kW (25 hp) Poinsard (D.9) or modified Volkswagen (D.92) flat-four engine; but other engines of 18·5 to 48·5 kW (25 to 65 hp) may be fitted, including the 27 kW (36 hp) Aeronca JAP and Continental A40. Fuel tank in fuselage, capacity 25 litres (5·5 Imp gallons).

Accommodation: Single seat in open cockpit.

Dimensions, external:
Wing span 7·00 m (22 ft 11 in)
Wing chord (centre-section, constant) 1·40 m (4 ft 7 in)
Wing area, gross 9·0 m² (96·8 sq ft)
Wing aspect ratio 5·45
Length overall 5·45 m (17 ft 10½ in)

Weights:
Weight empty 190 kg (420 lb)
Max T-O weight 320 kg (705 lb)

Performance (30 kW; 40 hp engine, at max T-O weight):
Max level speed at S/L 87 knots (160 km/h; 100 mph)
Cruising speed 74 knots (137 km/h; 85 mph)
Stalling speed 35 knots (65 km/h; 40 mph)
Max rate of climb at S/L 180 m (590 ft)/min
T-O run 110 m (360 ft)
Landing run 100 m (330 ft)
Range with max fuel 217 nm (400 km; 250 miles)

JODEL D.11 and D.119

The D.11, with 33·5 kW (45 hp) Salmson engine, was the basic model in the series of Jodel two-seaters for amateur and commercial production.

The version for amateur construction with 67 kW (90 hp) Continental engine is designated D.119.

A typical D.11 was built over an eight-year period by Wayne Nelson, an aeronautical engineer of Bountiful, Utah, at a cost of $2,000. The wing is of wood, covered with Dacron, the fuselage and tail unit of wood covered with glassfibre. Changes from the standard design include the fitting of a fixed tail-fin forward of the rudder, and of cantilever spring main landing gear legs. This D.11 spans 8·23 m (27 ft 0 in), has an empty weight of 340 kg (750 lb) and loaded weight of 562 kg (1,240 lb), and is powered by a 48·5 kW (65 hp) Continental A65-8 flat-four engine. Performance is as follows:

Performance:
Max level speed at S/L 93 knots (173 km/h; 108 mph)
Cruising speed 86 knots (161 km/h; 100 mph)
Landing speed 35 knots (64·5 km/h; 40 mph)
Max rate of climb at S/L 152 m (500 ft)/min
Service ceiling 4,875 m (16,000 ft)
T-O run 152 m (500 ft)
Landing run 244 m (800 ft)
Range with max fuel 260 nm (482 km; 300 miles)

JODEL D.112 CLUB

The D.112 is a two-seat dual-control version of the D.9. Except for increased overall dimensions, a wider fuselage and enclosed side-by-side cockpit, the D.112 conforms in layout and structure to the D.9, but is fitted normally with a 48·5 kW (65 hp) Continental flat-four engine. Fuel capacity is 60 litres (13 Imp gallons).

Dimensions, external:
Wing span 8·2 m (26 ft 10 in)
Wing area, gross 12·72 m² (137 sq ft)
Length overall 6·36 m (20 ft 10 in)
Dihedral on outer wings 19°

Weights:
Weight empty 270 kg (600 lb)
Max T-O weight 520 kg (1,145 lb)

Performance (at max T-O weight):
Max level speed at S/L 102 knots (190 km/h; 118 mph)
Max cruising speed 92 knots (170 km/h; 105·5 mph)
Econ cruising speed 81 knots (150 km/h; 93 mph)
Stalling speed 38 knots (70 km/h; 43 mph)
Max rate of climb at S/L 193 m (632 ft)/min
T-O run 137 m (450 ft)
Landing run 120 m (395 ft)
Range with max fuel 323 nm (600 km; 373 miles)

JURCA
MARCEL JURCA

Address:
2, rue des Champs Philippe, 92-La Garenne-Colombes (Seine)

Telephone: 242.9633 and 551.6306

M Marcel Jurca, an ex-military pilot and hydraulics engineer, has designed a series of high-performance light aircraft of which plans are available to amateur constructors.

A prototype of his first design, the M.J.1, was built but did not fly. To gain experience, M Jurca next built a two-seat Jodel light aircraft, with the help of members of the Aero Club of Courbevoie, and this flew for the first time in 1954.

The same team then built a prototype of M Jurca's second design, the M.J.2 Tempête single-seat light aircraft, incorporating many Jodel components. It proved so successful that sets of plans were offered to amateur constructors and many more Tempêtes are now flying or under assembly throughout the world.

M Jurca developed from the Tempête the two-seat M.J.5 Sirocco and the M.J.51 Sperocco, and has produced a further series of designs by scaling down the basic airframes of second World War fighters to two-thirds or three-quarters of the original size.

A point of interest is that Jurca designs are considered to be suitable for the entire range of basic and advanced flying training duties. The M.J.5 Sirocco is the two-seat basic trainer, the M.J.2 Tempête the single-seat basic trainer, the M.J.51 Sperocco the two-seat advanced trainer and the M.J.7S Solo the single-seat advanced trainer.

For the North American market, Jurca plans are available from Jurca Plans Office, 581 Helen Street, Mt Morris, Michigan 48458, USA. Representative for Australia and New Zealand is Mr Steve Rankin, RD 9, Whangarei, New Zealand.

JURCA M.J.2 and M.J.20 TEMPÊTE

The prototype Tempête was flown for the first time, by its designer, on 27 June 1956. It obtained its certificate of airworthiness very quickly, and a total of at least 28 Tempêtes are now flying, with 20 more under construction, in France, Denmark, Luxembourg, Portugal, the UK, the United States and Canada, all amateur built.

The type of engine fitted to a particular aircraft is indicated by a suffix letter in its designation. Suffix letters are A for the 48·5 kW (65 hp) Continental A65, B for the 56 kW (75 hp) Continental A75, C for the 63·5 kW (85 hp) Continental C85, D for the 67 kW (90 hp) Continental C90-14F, E for the 74·5 kW (100 hp) Continental O-200-A, F for the 78·5 kW (105 hp) Potez 4 E-20, G for the 86 kW (115 hp) Potez 4 E-30, and H for the 93 kW (125 hp) Lycoming.

The standard version is the M.J.2A with A65 engine. The M.J.2D, with 67 kW (90 hp) C90-14F, cruises at 105 knots (195 km/h; 121 mph) and climbs to 1,000 m (3,280 ft) in 3 minutes. It can also perform aerobatics without loss of height. The Tempête built in Portugal is an M.J.2D with 67 kW (90 hp) Continental; that under construction in Denmark is designated **M.J.20**, and has a 134 kW (180 hp) engine and a strengthened airframe.

The Tempête is basically a single-seat aircraft, but the 112 and 134 kW (150 and 180 hp) versions have provision for carrying on cross-country flights a second person weighing not more than 70 kg (154 lb). They are intended to have an aerobatic capability adequate to compete with the American Pitts Specials in international competitions.

The following details apply generally to all basic single-seat M.J.2 models:

Type: Single-seat light monoplane.

Wings: Cantilever low-wing monoplane. NACA 23012 wing section. Incidence varies according to engine power. The 48·5 kW (65 hp) version has an incidence of 4° at root, 2° at tip; the 134 kW (180 hp) version has no incidence. No dihedral. All-wood one-piece single-spar structure with fabric covering. Fabric-covered wooden ailerons.

Jodell D.11 *(Neil A. Macdougall)*

Registration CS-AXB indicates that this Jurca M.J.2 Tempête was Portugal's second modern homebuilt aircraft. The first was registered CS-AXA, with 'X' signifying 'experimental' *(Howard Levy)*

FUSELAGE: All-wood structure of basic rectangular section, plywood-covered.
TAIL UNIT: Cantilever all-wood structure. Tailplane and fin plywood-covered, elevators and rudder fabric-covered. Trim tab on starboard elevator.
LANDING GEAR: Non-retractable tailwheel type. Jodel D.112 cantilever legs with rubber-in-compression springing. Jodel D.112 wheels and Dunlop 420 × 150 tyres. Jodel D.112 tailskid or tailwheel.
POWER PLANT: One 48·5 kW (65 hp) Continental A65 flat-four engine, driving a Ratier two-blade wooden propeller with ground-adjustable pitch. Provision for fitting 56, 63·5, 67, or 74·5 kW (75, 85, 90 or 100 hp) Continental, 78·5 or 86 kW (105 or 115 hp) Potez or 93 kW (125 hp) Lycoming engine. Jodel engine mounting and cowling. Jodel fuel tank, capacity 60 litres (13·2 Imp gallons), aft of firewall in fuselage.
ACCOMMODATION: Single seat under long rearward-sliding transparent canopy.
DIMENSIONS, EXTERNAL:

Wing span	6·00 m (19 ft 8 in)
Wing chord (basic)	1·40 m (4 ft 7 in)
Wing area, gross	7·98 m² (85·90 sq ft)
Wing aspect ratio	4·5
Length overall	5·855 m (19 ft 2½ in)
Height overall	2·40 m (7 ft 10 in)
Tailplane span	2·50 m (8 ft 2 in)
Wheel track	2·30 m (7 ft 6½ in)

WEIGHTS:

Weight empty	290 kg (639 lb)
Max T-O weight	430 kg (950 lb)

PERFORMANCE (48·5 kW; 65 hp engine):

Max level speed	104 knots (193 km/h; 120 mph)
Cruising speed	89 knots (165 km/h; 102 mph)
Landing speed	43 knots (80 km/h; 50 mph)
Max rate of climb at S/L	170 m (555 ft)/min
Service ceiling	3,500 m (11,500 ft)
T-O run	250 m (820 ft)
Endurance	3 hr 20 min

JURCA M.J.5 SIROCCO

The M.J.5 Sirocco is a tandem two-seat monoplane, developed from the M.J.2 Tempête as a potential club training and touring aircraft. It is fully aerobatic when flown as a two-seater.

The longer-span wings have an extended leading-edge inboard of the fence on each side and a completely new tip shape. A sweptback fin and rudder are standard.

The prototype M.J.5 flew for the first time on 3 August 1962, powered by a 78·5 kW (105 hp) Potez 4 E-20 engine. It was fitted originally with a non-retractable landing gear, but retractable landing gear and a 119·5 kW (160 hp) Lycoming O-320 engine were fitted in 1966, followed by a 134 kW (180 hp) Lycoming engine later. Its fuel capacity is 116 litres (25·5 Imp gallons).

By mid-February 1967, five more Siroccos were flying, one of them factory-built at Nancy. This aircraft, powered by a 74·5 kW (100 hp) Continental engine, concluded tests at Istres in January 1969. The French government then concluded an agreement with Constructions Aéronautiques Lorraines, François et Cie of Nancy, which built an airframe for static tests, in March 1971. These were required in view of the fact that the Sirocco is regarded as a basic trainer suitable for amateur construction; and it was awarded subsequently a certificate of airworthiness in the Utility category. Supplementary tests were conducted at the CEV with another Sirocco, powered by a 100·5 kW (135 hp) Lycoming O-320 engine.

A full C of A, covering Aerobatic requirements and unlimited spinning, is applicable only when a power plant of 86 kW (115 hp) minimum rating is installed.

The version of the Sirocco for amateur construction is generally similar to the factory-built version, with optional retractable landing gear.

At least 40 Siroccos are reported to be flying or under construction by amateurs in France, Canada, Germany, Switzerland, England and the USA, with various engines.

The type of engine fitted to a particular aircraft is indicated by a suffix letter in its designation. Suffix letters are A for the 67 kW (90 hp) Continental C90-8 or -14F, B for the 74·5 kW (100 hp) Continental O-200-A, C for the 78·5 kW (105 hp) Potez 4 E-20, D for the Potez 4 E-30, E for the 78·5 kW (105 hp) Hirth, F for the 93 kW (125 hp) Lycoming, G for the 100·5 kW (135 hp) Regnier, H for the 119·5 kW (160 hp) Lycoming, K for the 134 kW (180 hp) Lycoming and L for the 164 kW (220 hp) Franklin. Addition of the numeral 1 indicates a non-retractable landing gear and the numeral 2 indicates a retractable landing gear. Thus, the designation of the original prototype in its current form is M.J.5K2. The example built at Nancy for certification has a 74·5 kW (100 hp) Continental engine and so is designated M.J.5B1.

Two examples of the M.J.5L Sirocco, with 164 kW (220 hp) Franklin engine, were under construction in 1975, one in the USA and one in France. These are intended for use in international aerobatic championships.

A Sirocco with 86 kW (115 hp) Lycoming O-235-C2B engine and 1·85 m (6 ft 0¾ in) diameter propeller has been completed by Luftsportgruppe Liebherr-Aero-Technik (LAT) in Germany. This has a modified rudder of reduced height and greater chord, and a jettisonable, sideways-hinged cockpit canopy, and is intended for certification for aerobatic flying. The details which follow apply to this aircraft, but are generally typical of all versions.

Two developed versions, known as the M.J.50 Windy and M.J.51 Sperocco, are described separately.

DIMENSIONS, EXTERNAL:

Wing span	7·00 m (23 ft 0 in)
Wing area, gross	10·00 m² (107·64 sq ft)
Wing aspect ratio	4·9
Length overall	6·15 m (20 ft 2 in)
Height overall, tail up:	
standard model	2·80 m (9 ft 2¼ in)
LAT version	2·60 m (8 ft 6¼ in)
Tailplane span	3·24 m (10 ft 7½ in)
Wheel track	2·80 m (9 ft 2¼ in)

WEIGHTS AND LOADINGS:

Weight empty	430 kg (947 lb)
Max T-O weight	680 kg (1,499 lb)
Max wing loading	68·0 kg/m² (13·9 lb/sq ft)
Max power loading	7·93 kg/kW (13·03 lb/hp)

PERFORMANCE (at max T-O weight):

Max level speed	127 knots (235 km/h; 146 mph)
Cruising speed	116 knots (215 km/h; 134 mph)
Stalling speed	44 knots (80 km/h; 50 mph)
Climb to 1,000 m (3,280 ft)	4 min
Service ceiling	5,000 m (16,400 ft)
T-O run	250 m (820 ft)
Landing run	200 m (655 ft)
Endurance	4 hr 20 min

JURCA M.J.5 SIROCCO (SPORT WING)

A special version of the Sirocco, with 86 kW (115 hp) engine and increased wing span, has been evolved for the New Zealand and Australian market. Known as a 'Sport' wing, the wing of this aircraft embodies one additional rib and inter-rib bay each side. The modification is available in the English-language set of Sirocco plans.

JURCA M.J.50 WINDY

The M.J.50 Windy is generally similar to the M.J.5 Sirocco, but is of all-metal construction. It embodies the retractable landing gear that is available for the M.J.5, and the M.J.7 flying control system.

A prototype is under construction by Mr Wesolowsky of Luçon. Plans will be available to amateur constructors.

The wing section is NACA 23018 on the upper surface and NACA 23012 on the lower surface at the root, and NACA 23012 at the tip. The structure is stressed for engines in the 112-149 kW (150-200 hp) range.

DIMENSIONS, EXTERNAL:

Wing span	7·30 m (23 ft 11½ in)
Wing chord at root	1·814 m (5 ft 11½ in)
Wing chord at tip	1·40 m (4 ft 7 in)
Length overall, tail up	6·673 m (21 ft 10½ in)
Tailplane span	3·26 m (10 ft 8¼ in)
Wheel track	2·58 m (8 ft 5½ in)

JURCA M.J.51 SPEROCCO

Using knowledge gained from flight experience with the M.J.5 and the Canadian prototype M.J.7, M Jurca evolved, with the assistance of M J. Lecarme, a design incorporating features of each aircraft. It is known as the M.J.51 Sperocco, the name being a contraction of 'Special Sirocco', and is intended for high-performance aerobatic and competition flying. Like other Jurca designs, the M.J.51 is suitable for amateur construction.

The wings, of Habib 64000 748 laminar-flow profile, are essentially those of the M.J.7 Gnatsum. They are without dihedral, and the angle of incidence varies according to the rating of the engine that is installed, as in the Tempête. The fuselage is of completely new design, with a basically triangular cross-section, but is of similar construction to the M.J.5. The tail unit consists of M.J.7 horizontal surfaces with a shorter and wider-chord fin and rudder. Landing gear is of the M.J.5 type and is fully retractable.

Any horizontally-opposed engine of 112-179 kW (150-240 hp) may be installed. Fuel is contained in two wing tanks, each of 55 litres (12 Imp gallons) capacity, and one fuselage tank of 45 or 100 litres (10 or 22 Imp gallons) capacity.

The M.J.51 seats two persons in tandem under a one-piece sliding canopy, the rear seat being 10 cm (3·9 in) higher than the front seat.

The first M.J.51, powered by a 134 kW (180 hp) Lycoming AIO-360 engine, is under construction by M Serge Brilliant at Melun, and was expected to fly in 1976.

DIMENSIONS, EXTERNAL:

Wing span	7·623 m (25 ft 0 in)
Wing area, gross	11·00 m² (118 sq ft)
Length overall	7·24 m (23 ft 9 in)

WEIGHT:

Max T-O weight	730 kg (1,653 lb)

PERFORMANCE (estimated, with 112 kW; 150 hp Lycoming engine):

Max level speed	149 knots (275 km/h; 171 mph)
Max cruising speed (75% power)	135 knots (250 km/h; 155 mph)
Stalling speed	49 knots (90 km/h; 56 mph)
Time to 1,000 m (3,280 ft)	1 min 30 sec

JURCA M.J.7 and M.J.77 GNATSUM

The Gnatsum is a scale replica, for amateur construction, of the North American P-51 Mustang single-seat fighter of the second World War. Its name 'Gnatsum' is 'Mustang' reversed.

Initially, M Jurca designed the wings, fuselage, tail surfaces and manually-retractable landing gear. The engine installation was deliberately not designed, to permit constructors to utilise any of the suitable Lycoming, Continental, Ranger or other power plants that are available.

During construction of the M.J.7 prototype in Canada (see below), a number of modifications and improvements were made to the basic design. These were embodied in the drawings, which are available from M Jurca in two forms, as follows:

M.J.7. To two-thirds scale. Prototype (CF-XZI, now N51HR) built in the works of Falconar Aircraft Ltd on the Industrial Airport, Edmonton, Alberta, Canada, and first flown on 31 July 1969. Granted DoT type approval by early 1970. Described under the SAL entry in the Canadian section of the 1973-74 *Jane's*. Further examples under construction by Mr J. P. Dloyer of Torrance, California, and three others.

M.J.77. To three-quarters scale. Prototype under construction by Mr Gilbert C. McAdams of Victorville, California, and others by Mr Bob Aughton in Michigan, M Glorieux of Pau and M Piazola of Egletons, France.

Unlike previous small-scale replicas of this aircraft, the Gnatsum is scaled down precisely. Use of an in-line engine, such as the 119·5 kW (160 hp) Walter Minor 6-III or 149 kW (200 hp) Ranger, permits the fuselage cowling lines to follow closely those of the original. Alternative installation of a 149 kW (200 hp) Lycoming horizontally-opposed aircooled engine requires fairing blisters over the cylinders.

M Jurca's plans provide for alternative plywood-covered semi-monocoque fuselage construction or a square wooden box structure covered with two plastics shells supplied by Rattray of the USA.

JURCA M.J.7S SOLO

Intended as a single-seat advanced trainer, the M.J.7S Solo is basically similiar to the M.J.7 Gnatsum but does not retain the underbelly scoop which the latter inherited from the original P-51 Mustang design. A prototype was under construction by M Duhamel of Strasbourg in 1975, with a 134 kW (180 hp) Lycoming AIO-360 flat-four engine.

The wing section of the M.J.7S is quoted as Habib 64-000 748-MJ7-104.

DIMENSIONS, EXTERNAL:

Wing span	7·523 m (24 ft 8½ in)
Wing area, gross	10·8 m² (116·2 sq ft)

Jurca M.J.8 1-Nine-O prototype built by Mr Ronald Kitchen of Carson City, Nevada

Jurca M.J.5H2 Sirocco, with Lycoming IO-320-B engine and retractable landing gear, built by Mr J. D. Schmidt of Wagon Mound, New Mexico *(Howard Levy)*

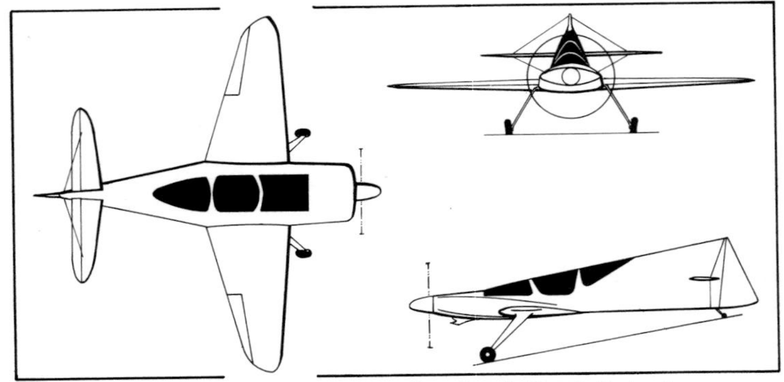

Jurca M.J.14 single-seat racing aircraft *(Roy J. Grainge)*

Jurca M.J.7 Gnatsum built by Major Foster, RCAF

Lederlin 380-L two-seat light aircraft (Continental C90-14F engine)

Lefebvre MP.205 Busard in its latest form, with 67 kW; 90 hp Continental engine *(Geoffrey P. Jones)*

Length overall, tail up	6·664 m (21 ft 10 in)
Tailplane span	3·00 m (9 ft 10 in)

JURCA M.J.8 1-NINE-O

The M.J.8 is a single-seat sporting aircraft which has been designed by M Jurca by scaling down to three-quarters of the original dimensions the airframe of the Focke-Wulf Fw 190 fighter. Its general appearance is shown in the accompanying illustration.

This prototype, built by Mr Ronald Kitchen of Carson City, Nevada, flew for the first time on 30 March 1975. A second example is under construction by Mr J. Kiska of Norwalk, Connecticut.

The M.J.8 prototype has a 216 kW (290 hp) Lycoming IO-540 engine, but the design is suitable for the alternative use of any horizontally-opposed or radial engine in the 74·5-149 kW (100-200 hp) range. The landing gear is retractable.

DIMENSIONS, EXTERNAL:

Wing span	7·87 m (25 ft 10 in)
Wing chord at root	1·70 m (5 ft 7 in)
Wing chord at tip	0·90 m (2 ft 11½ in)
Wing area, gross	10·2 m² (109·8 sq ft)
Length overall	6·63 m (21 ft 9 in)
Tailplane span	2·84 m (9 ft 4 in)

WEIGHTS (119·5 kW; 160 hp engine):

Weight empty	400 kg (880 lb)
Max T-O weight	626 kg (1,380 lb)

PERFORMANCE (estimated, with 119·5 kW; 160 hp engine):

Max level speed at S/L	139 knots (257 km/h; 160 mph)
Max cruising speed	124 knots (230 km/h; 143 mph)
Stalling speed	49 knots (90 km/h; 56 mph)
Max rate of climb at S/L	503 m (1,650 ft)/min

JURCA M.J.10 SPIT

The M.J.10 is a single-seat, three-quarter scale representation of the Supermarine Spitfire which can also be modified as a two-seater. It is suitable for any horizontally-opposed or in-line engine of 89·5-164 kW (120-220 hp), although some slight variations from the Spitfire's contours are necessary in the former case. Construction is entirely of wood, except for the glassfibre engine cowling and fabric covering on the control surfaces. The single-spar wing is similar in construction to that of the Sirocco. The manually-operated retractable landing gear is fitted with helicoidal spring shock-absorbers.

The basic plans adopted the Spitfire Mk IX as the standard M.J.10 version, but alternative detail plans are available for representing both Merlin- and Griffon-engined models, including the Mks VC and XIV, and for clipped, standard or extended-span wings.

A prototype is under construction by Mr Pendlebury of the Chesterfield Air Touring Group at West Bridgford, Nottingham, England, and another by Mr Ed Storo of New York, USA.

DIMENSIONS, EXTERNAL:

Wing span:	
standard	8·40 m (27 ft 6¾ in)
clipped	7·46 m (24 ft 5½ in)
Wing area, gross	12·60 m² (135·6 sq ft)
Length overall	7·125 m (23 ft 4½ in)

WEIGHTS (119·5 kW; 160 hp engine):

Weight empty	658 kg (1,450 lb)
Max T-O weight	907 kg (2,000 lb)

PERFORMANCE (estimated, with 119·5 kW; 160 hp engine):

Max level speed at S/L	139 knots (257 km/h; 160 mph)
Cruising speed	124 knots (230 km/h; 143 mph)
Stalling speed	49 knots (90 km/h; 56 mph)
Max rate of climb at S/L	503 m (1,650 ft)/min
T-O run	200 m (660 ft)

JURCA M.J.12 PEE-40

The M.J.12 is a three-quarter scale representation of the Curtiss P-40 single-seat fighter of the second World War. It spans 8·524 m (27 ft 11½ in) and has an overall length (tail up) of 7·62 m (25 ft 0 in)

Two M.J.12s are under construction in the USA.

JURCA M.J.14 RACER

Designed in 1971, the M.J.14 will be a small single-seat racing aircraft of unorthodox configuration, with a semi-reclining seat for the pilot. Construction of a prototype has begun at Strasbourg, in France. Its general appearance is shown in the accompanying three-view drawing.

The standard tapered wings can be replaced by constant-chord wings of the same span if the aircraft is intended for Class III racing. Tailplane incidence is adjustable on the ground. The following data apply to the aircraft as illustrated, with a 67 kW (90 hp) Continental C90-8F flat-four engine. Fuel capacity is 70 litres (15·5 Imp gallons).

DIMENSIONS, EXTERNAL:

Wing span	6·00 m (19 ft 8½ in)
Wing area, gross	6·1 m² (65·66 sq ft)
Length overall	5·68 m (18 ft 7½ in)
Tailplane span	2·97 m (9 ft 9 in)

WEIGHTS:

Weight empty	250 kg (550 lb)
Max T-O weight	420 kg (925 lb)

LEDERLIN

FRANÇOIS LEDERLIN

ADDRESS:
2 rue Charles Peguy, 38-Grenoble

M Lederlin, an architect, designed and built a two-seat light aeroplane based on the familiar Mignet 'Pou-du-Ciel' formula. Although derived from the Mignet HM-380 and designated 380-L, it retains little of the original except for the wing section. First flight was made on 14 September 1965, a restricted C of A being granted in the following month.

Plans of the 380-L, annotated in English and with both English and metric measurements, are available to amateur constructors, and several examples are being completed.

LEDERLIN 380-L

TYPE: Two-seat amateur-built light aircraft.

WINGS: Tandem-wing biplane. Wing section 3·40-13. Dihedral 3° 30′ on outer sections only (both wings). Incidence variable from 0° to 12° (forward wing). Incidence of rear wing 6°. No sweepback. Each wing is made in two parts, bolted together at the centreline. Construction is conventional, with wooden box-spar and trellis ribs, plywood leading-edge and overall fabric covering. The variable-incidence front wing is pivoted on the cabane structure by ball-joints and on the bracing struts (one each side) by cardan-joints. No ailerons or flaps. Long-span tab on trailing-edge of rear wing, controllable in flight.

FUSELAGE: Welded steel tube structure, covered with light alloy to front of cabin and with fabric on rear fuselage, over light spruce formers.

TAIL UNIT: Fin and rudder only. Spruce and ply structure, covered with fabric. Ground-adjustable tab in rudder.

LANDING GEAR: Non-retractable tailwheel type. Cantilever main legs consist of conical spring steel rods, inclined rearward. Fournier main wheels and tyres, size 380 × 150, with mechanical brakes. Large tailwheel, carried on telescopic leg with spring shock-absorber, can be steered by the rudder controls through a linkage engaged by the pilot.

Power Plant: One 67 kW (90 hp) Continental C90-14F flat-four engine, driving a McCauley two-blade metal fixed-pitch propeller. Single fuel tank, capacity 85 litres (18·75 Imp gallons). Oil capacity 4·5 litres (1 Imp gallon).

Accommodation: Two seats side by side in enclosed cabin. Forward-hinged door on each side. Controls comprise a rudder bar for directional control and a stick, suspended from the roof of the cabin and free laterally, to control the incidence of the forward wing. A further lever, suspended from the roof, controls the tab on the rear wing. Baggage space aft of seats.

Dimensions, external:
- Wing span:
  - forward: 7·92 m (26 ft 0 in)
  - rear: 6·00 m (19 ft 8¼ in)
- Wing chord (constant, each): 1·30 m (4 ft 3¼ in)
- Wing area, gross:
  - forward: 9·92 m² (106·8 sq ft)
  - rear: 7·43 m² (80·0 sq ft)
- Length overall: 4·77 m (15 ft 7¾ in)
- Height overall: 2·08 m (6 ft 10 in)
- Wheel track: 2·05 m (6 ft 8¾ in)
- Wheelbase: 3·10 m (10 ft 2 in)
- Propeller diameter: 1·83 m (6 ft 0 in)
- Doors (each): Height: 0·90 m (2 ft 11½ in)
  - Width: 0·75 m (2 ft 5½ in)
  - Height to sill: 0·50 m (1 ft 7½ in)

Dimensions, internal:
- Cabin:
  - Max width: 1·07 m (3 ft 6 in)
  - Max height: 1·03 m (3 ft 4 in)
  - Baggage space: 0·20 m³ (7 cu ft)

Weights and Loadings:
- Weight empty: 360 kg (794 lb)
- Max T-O weight: 600 kg (1,323 lb)
- Max wing loading: 34 kg/m² (6·96 lb/sq ft)
- Max power loading: 8·96 kg/kW (14·8 lb/hp)

Performance (at max T-O weight):
- Never-exceed speed: 126 knots (233 km/h; 145 mph)
- Max level speed at 305 m (1,000 ft): 109 knots (201 km/h; 125 mph)
- Max cruising speed: 97 knots (180 km/h; 112 mph)
- Econ cruising speed at 610 m (2,000 ft): 87 knots (161 km/h; 100 mph)
- Stalling speed, power off: 26 knots (49 km/h; 30 mph)
- Max rate of climb at S/L: 275 m (900 ft)/min
- Service ceiling: over 3,660 m (12,000 ft)
- T-O run: 122 m (400 ft)
- Landing run: 153 m (500 ft)
- Range at econ cruising speed: 477 nm (885 km; 550 miles)

LEFEBVRE
ROBERT LEFEBVRE

Address:
CES A. Camus, rue Adeline, 76100-Rouen

M Lefebvre has built and flown a small single-seat racing aircraft named the Busard, assisted by pupils of the A. Camus technical school at Rouen. Basis of the design was the MP.204 prototype racer with 56 kW (75 hp) Minié engine, designed by Max Plan and first flown on 5 June 1952. By comparison with the MP.204, the Busard has been lightened, and simplified for construction by amateurs.

LEFEBVRE MP.205 BUSARD

The description below applies to the prototype Busard (F-PTXT) built and flown by M. Lefebvre. This aircraft was powered originally with a 48·5 kW (65 hp) Continental engine and was illustrated in this form in the 1974-75 *Jane's*. After 20 flying hours it was re-engined with a 67 kW (90 hp) Continental and underwent several refinements, including the fitting of main-wheel fairings.

At least fifteen sets of plans have been sold to amateur constructors. One Busard is being built with a 48·5 kW (65 hp) Walter engine, two with versions of the Volkswagen, and the remainder with Continentals.

Type: Single-seat amateur-built racing aircraft.

Wings: Cantilever low-wing monoplane. Wing section NACA 23012. Constant incidence of 1°. Slight dihedral. Conventional single-spar wood structure, covered entirely with plywood. Fabric-covered wooden slotted ailerons, operated by control rods. Fabric-covered wooden slotted three-position trailing-edge flaps.

Fuselage: Conventional wooden structure, covered with plywood. Domed plywood decking. Plastics engine cowling, built in top and bottom sections.

Tail Unit: Cantilever wood structure. Fixed surfaces plywood-covered; control surfaces fabric-covered. Neither rudder nor elevators are aerodynamically balanced. Flettner tab in starboard elevator. Rudder is cable-operated.

Landing Gear: Non-retractable tailwheel type. Cessna-type aluminium leaf-spring cantilever main legs, with plastics wheel fairings. Steerable tailwheel. Brakes on main wheels.

Power Plant: One 67 kW (90 hp) Continental flat-four engine, driving a two-blade fixed-pitch propeller. Light alloy fuel tank aft of firewall in fuselage, capacity 40 litres (8·8 Imp gallons).

Accommodation: Single seat in enclosed cabin, with max width of 0·58 m (1 ft 10¾ in). Sideways-opening canopy, hinged on starboard side. Baggage space aft of seat.

Dimensions, external:
- Wing span: 6·00 m (19 ft 8¼ in)
- Wing chord at root: 1·50 m (4 ft 11 in)
- Wing chord at tip: 0·75 m (2 ft 5½ in)
- Wing area, gross: 6·00 m² (64·6 sq ft)
- Wing aspect ratio: 6·00
- Length overall:
  - Continental: 5·35 m (17 ft 6¾ in)
  - Volkswagen: 5·20 m (17 ft 0¾ in)
- Height overall: 1·50 m (4 ft 11 in)

Weights (A, 48·5 kW; 65 hp Continental; B, 67 kW; 90 hp Continental):
- Weight empty:
  - A: 233 kg (514 lb)
  - B: 239 kg (527 lb)
- Max T-O weight:
  - A: 339 kg (747 lb)
  - B: 345 kg (760 lb)

Performance (A, 48·5 kW; 65 hp Continental; B, 67 kW; 90 hp Continental; C, 1,600 cc Volkswagen):
- Max level speed at S/L:
  - A: 127 knots (235 km/h; 146 mph)
  - B: 156 knots (290 km/h; 180 mph)
  - C: 113 knots (210 km/h; 130 mph)
- Landing speed:
  - A: 43 knots (80 km/h; 50 mph)
- Range with max fuel:
  - A: 242 nm (450 km; 279 miles)

PIEL
AVIONS CLAUDE PIEL

Address:
104 Côte de Beulle, 78580-Maule
Telephone: 478.82.49

M Claude Piel has designed several light aircraft, including the Emeraude, Diamant and Beryl, of which sets of plans are available to amateur constructors.

In addition, M Piel has granted licence rights for their manufacture by several commercial concerns. Four French companies, listed in the 1968-69 *Jane's*, built versions of the Emeraude under licence, as did Binder Aviatik KG (in association with Schempp-Hirth KG) in Germany, Durban Aircraft Corporation in South Africa, Aeronasa in Spain and Fairtravel in the UK. Over 200 factory-built Emeraudes were completed by these manufacturers, in addition to those built by amateur constructors.

The authorised distributor for plans of all Piel designs available to amateur constructors is:

E. Littner, CP 272, Saint Laurent, Montreal, Quebec H4L 4V6, Canada.

In addition, servicing and constructional facilities for Emeraude variants are available at the works of M Choisel at Abbeville.

PIEL EMERAUDE and SUPER EMERAUDE

There have been several factory-built versions of the Emeraude and Super Emeraude, but the aircraft are no longer being produced in this form. The designs continue to be available for amateur construction, and the following amateur-built versions have flown:

C.P.301. With 67 kW (90 hp) Continental engine.

C.P.302. With 67 kW (90 hp) Salmson engine.

C.P.303. With 63·4 kW (85 hp) Salmson engine.

C.P.304. With 63·4 kW (85 hp) Continental C85-12F engine and wing flaps.

C.P.305. With 86 kW (115 hp) Lycoming engine.

C.P.308. With 56 kW (75 hp) Continental engine.

C.P.320. With Super Emeraude wings and 74·5 kW (100 hp) Continental engine. **C.P.320A** has sweptback fin.

C.P.321. As C.P.320, with 78·5 kW (105 hp) Potez engine.

C.P.323A. With 112 kW (150 hp) Lycoming engine and sweptback fin. **C.P.323AB** has tricycle landing gear.

The Emeraude is one of the types approved by the Popular Flying Association for amateur construction in the United Kingdom.

The following details refer to the basic C.P.301 Emeraude and C.P.320 Super Emeraude, but are generally applicable to all versions:

Type: Two-seat light monoplane.

Wings: Cantilever low-wing monoplane. NACA 23012 wing section. Dihedral 5° 40′. Incidence 4° 10′. Inner half of each wing is rectangular in plan, outer half elliptical. All-wood single-spar structure with fabric covering overall. Slotted ailerons and flaps.

Fuselage: Conventional wood structure, covered with fabric.

Tail Unit: Cantilever wood structure. Fin integral with fuselage. Single-piece all-wood tailplane. Elevators and rudder fabric-covered. Trim tab in starboard elevator.

Landing Gear: Non-retractable tailwheel type. Cantilever main legs have rubber-in-compression springing. Hydraulic brakes.

Power Plant: (C.P.301): One 67 kW (90 hp) Continental C90-12F flat-four engine. Two-blade fixed-pitch wooden propeller. Fuel tank in fuselage, behind fireproof bulkhead, capacity 80 litres (17·6 Imp gallons). Provision for auxiliary tank, capacity 40 litres (8·8 Imp gallons).

Power Plant (C.P.320): One 74·5 kW (100 hp) Continental O-200 flat-four engine, driving a two-blade fixed-pitch wooden propeller. Fuel as for C.P.301.

Accommodation: Enclosed cockpit seating two side by side with dual controls. Sides of canopy hinge forward for access and exit. Heating and ventilation.

Dimensions, external:
- Wing span: 8·04 m (26 ft 4½ in)
- Wing chord at root: 1·50 m (4 ft 11 in)
- Wing chord at tip: 0·55 m (1 ft 9½ in)
- Wing area, gross: 10·85 m² (116·7 sq ft)
- Wing aspect ratio: 5·95
- Length overall:
  - C.P.301: 6·30 m (20 ft 8 in)
  - C.P.320: 6·45 m (21 ft 2 in)
- Height overall:
  - C.P.301: 1·85 m (6 ft 0¾ in)
  - C.P.320: 1·90 m (6 ft 2¾ in)
- Wheel track: 2·05 m (6 ft 8¾ in)
- Propeller diameter:
  - C.P.301: 1·80 m (5 ft 11 in)
  - C.P.320: 1·78 m (5 ft 10 in)

Weights and Loadings:
- Weight empty:
  - C.P.301: 380 kg (838 lb)
  - C.P.320: 410 kg (903 lb)
- Max T-O weight:
  - C.P.301: 650 kg (1,433 lb)
  - C.P.320: 700 kg (1,543 lb)
- Max wing loading:
  - C.P.301: 60·0 kg/m² (12·3 lb/sq ft)
  - C.P.320: 64·5 kg/m² (13·2 lb/sq ft)
- Max power loading:
  - C.P.301: 9·70 kg/kW (15·87 lb/hp)
  - C.P.320: 9·40 kg/kW (15·43 lb/hp)

Performance (at max T-O weight):
- Never-exceed speed:
  - C.P.301: 118·5 knots (220 km/h; 136·5 mph)
  - C.P.320: 149 knots (277 km/h; 172 mph)
- Max level speed:
  - C.P.301: 110 knots (205 km/h; 127 mph)
  - C.P.320: 124 knots (230 km/h; 143 mph)
- Max cruising speed (75% power) at 1,200 m (3,940 ft):
  - C.P.301: 108 knots (200 km/h; 124 mph)
  - C.P.320: 119 knots (220 km/h; 137 mph)
- Econ cruising speed (65% power) at 1,200 m (3,940 ft):
  - C.P.301: 101 knots (187 km/h; 116 mph)
  - C.P.320: 110 knots (205 km/h; 127 mph)
- Approach speed, flaps down:
  - C.P.301, C.P.320: 65 knots (120 km/h; 75 mph)
- Stalling speed, flaps up:
  - C.P.301: 51 knots (92 km/h; 58 mph)
  - C.P.320: 53 knots (97 km/h; 61 mph)
- Stalling speed, flaps down:
  - C.P.301: 46 knots (85 km/h; 53 mph)
  - C.P.320: 49 knots (90 km/h; 56 mph)
- Max rate of climb at S/L:
  - C.P.301: 168 m (551 ft)/min
  - C.P.320: 240 m (787 ft)/min
- Service ceiling:
  - C.P.301: 4,000 m (13,125 ft)
  - C.P.320: 4,300 m (14,100 ft)
- T-O run:
  - C.P.301: 250 m (820 ft)
  - C.P.320: 230 m (755 ft)
- T-O to 15 m (50 ft):
  - C.P.301: 440 m (1,443 ft)
  - C.P.320: 400 m (1,312 ft)
- Landing from 15 m (50 ft):
  - C.P.301: 475 m (1,558 ft)
  - C.P.320: 490 m (1,608 ft)
- Landing run:
  - C.P.301: 250 m (820 ft)
  - C.P.320: 260 m (853 ft)

The Piel Emeraude C.P.301 *(Neil A. Macdougall)*

Piel Super Diamant (150 hp Lycoming engine) built by Mr Eric R. Glew of Agincourt Ontario, Canada *(Neil A. Macdougall)*

Piel C.P.80 Zef with all-plastics fuselage, built by M Pierre Calvel *(Geoffrey P. Jones)*

Piel C.P.750 Beryl (Lycoming O-320 engine) built by Mr Norman Taylor of Ontario, Oregon *(Howard Levy)*

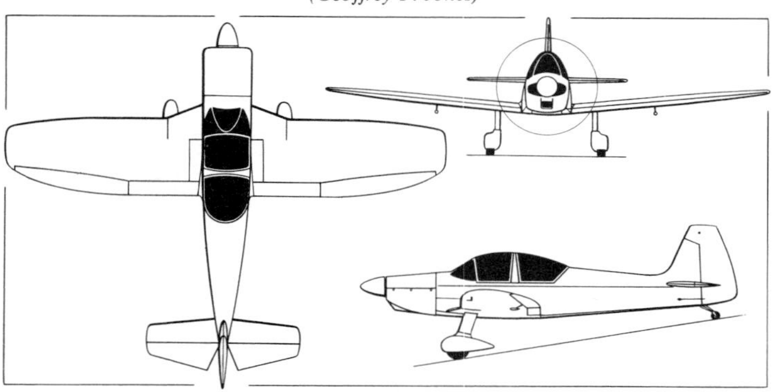

Piel C.P.750 Beryl two-seat light aircraft (Lycoming O-320-E2A engine) *(Pilot Press)*

Piel C.P.90 Pinocchio single-seat light sporting aircraft *(Pilot Press)*

Range at econ cruising speed:
C.P.301, C.P.320 538 nm (1,000 km; 620 miles)

PIEL C.P. 1320

This aircraft combines the general characteristics of the Super Emeraude with the Diamant's three-seat cabin and fuel tanks in the wings. It can be fitted with engines of up to 149 kW (200 hp) and the prototype, built by an amateur constructor, will have a 119·5 kW (160 hp) Lycoming, driving a two-blade wooden propeller. The airframe will be of all-wood construction, with a non-retractable tailwheel-type landing gear. Slotted flaps will be standard. Fuel capacity 160 litres (35 Imp gallons).

Normal load factors will be +5*g* and −2·5*g*. For aerobatics in two-seat form at a T-O weight of 720 kg (1,585 lb), the permissible load factors will be +6*g* and −3*g*.

Dimensions, external:
- Wing span 8·04 m (26 ft 4½ in)
- Wing area, gross 10·85 m² (116·7 sq ft)
- Wing aspect ratio 5·95
- Wing dihedral 5° 40′
- Length overall 6·60 m (21 ft 8 in)
- Height overall 1·90 m (6 ft 2¾ in)
- Wheel track 2·05 m (6 ft 8¾ in)
- Propeller diameter 1·80 m (5 ft 11 in)

Weights and Loadings:
- Weight empty 500 kg (1,102 lb)
- Max T-O weight 840 kg (1,852 lb)
- Max wing loading 77·5 kg/m² (15·87 lb/sq ft)
- Max power loading 7·04 kg/kW (11·6 lb/hp)

Performance (estimated, at max T-O weight):
- Never-exceed speed 183 knots (340 km/h; 211 mph)
- Max level speed at S/L 145 knots (270 km/h; 167 mph)
- Max cruising speed (75% power) at 1,200 m (3,940 ft) 135 knots (250 km/h; 155 mph)
- Cruising speed (65% power) at 1,200 m (3,940 ft) 127 knots (235 km/h; 146 mph)
- Approach speed, flaps down 70 knots (130 km/h; 81 mph)
- Stalling speed, flaps up 54 knots (100 km/h; 62 mph)
- Stalling speed, flaps down 51·5 knots (95 km/h; 59 mph)
- Max rate of climb at S/L 600 m (1,968 ft)/min
- Service ceiling 5,000 m (16,400 ft)
- T-O run 200 m (657 ft)
- T-O to 15 m (50 ft) 420 m (1,378 ft)
- Landing from 15 m (50 ft) 600 m (1,968 ft)
- Landing run 300 m (984 ft)
- Range with max fuel at 65% power 593 nm (1,100 km; 683 miles)

PIEL DIAMANT and SUPER DIAMANT

The Diamant is essentially a three/four-seat version of the Emeraude. It is fully certificated for commercial production and available also in plan form for construction by amateurs.

The C.P.60, C.P.601, and C.P.602 versions, with engines in the 67-86 kW (90-115 hp) range, are no longer built. Current versions are as follows:

C.P.604 Super Diamant. Prototype (F-PMEC) flown in Summer of 1964, with a 108 kW (145 hp) Continental engine. Current version has swept vertical tail surfaces.

C.P.605 Super Diamant. Much-modified four-seat ('2+2') version, with 112 kW (150 hp) Lycoming O-320-E2A engine. Fully certificated for commercial production, as well as for amateur construction. Details in 1973-74 *Jane's.*

C.P.605B Super Diamant. Version of C.P.605 with retractable tricycle landing gear.

Type: Three/four-seat light monoplane.

Wings: Cantilever low-wing monoplane. Wing section NACA 23012. Dihedral 5° 40′. Incidence 4° 10′. All-wood single-spar structure, made in one piece, with fabric covering. Slotted ailerons and slotted flaps of wood construction, with fabric covering.

Fuselage: Wood structure, covered with fabric.

Tail Unit: Cantilever wood structure, with sweptback fin and rudder. Fixed surfaces plywood-covered. Control surfaces fabric-covered. Ground-adjustable tab on each elevator.

Landing Gear (C.P.604): Non-retractable tailwheel type. Main wheels size 420 × 150. Hydraulic brakes. Wheel spats. Steerable tailwheel, size 155 × 50.

Landing Gear (C.P.605B): Retractable tricycle type. Main wheels retract inward. All three wheels and tyres size 400 × 100.

Power Plant: One flat-four engine, driving an EVRA two-blade fixed-pitch wooden propeller. Fuel tank in fuselage, capacity 85 litres (18·7 Imp gallons). Provision for additional tankage to give total capacity of 160 litres (35 Imp gallons). Oil capacity 4 litres (0·9 Imp gallons).

Accommodation: Four seats ('2+2') in enclosed cabin under large rearward-sliding transparent canopy.

Dimensions, external (C.P.605B):
- Wing span 9·20 m (30 ft 2¼ in)
- Wing chord at root 1·50 m (4 ft 11 in)
- Wing area, gross 13·30 m² (143·2 sq ft
- Wing aspect ratio 6·
- Length overall 7·00 m (22 ft 11¾ in
- Height overall 2·00 m (6 ft 6¾ in
- Wheel track 3·00 m (9 ft 10 in
- Propeller diameter 1·80 m (5 ft 11 in

Weights and Loadings (C.P.605B):
- Weight empty 520 kg (1,146 lb
- Max T-O weight 850 kg (1,873 lb
- Max wing loading 64·00 kg/m² (13·1 lb/sq ft
- Max power loading 7·59 kg/kW (12·35 lb/hp

Performance (C.P.605B, at max T-O weight):
- Never-exceed speed 151 knots (280 km/h; 174 mph
- Max level speed 141 knots (260 km/h; 162 mph
- Max cruising speed (75% power) at 1,200 m (3,94 ft) 132 knots (245 km/h; 152 mph
- Econ cruising speed (65% power) at 1,200 m (3,94 ft) 124 knots (230 km/h; 143 mph
- Approach speed, flaps down 68 knots (125 km/h; 78 mph
- Stalling speed, flaps up 49 knots (90 km/h; 56 mph
- Stalling speed, flaps down 45 knots (82 km/h; 51 mph
- Max rate of climb at S/L 330 m (1,082 ft)/mi
- Service ceiling 5,000 m (16,400 ft
- T-O run 160 m (525 ft
- T-O to 15 m (50 ft) 380 m (1,247 ft
- Landing from 15 m (50 ft) 600 m (1,969 ft
- Landing run 270 m (886 ft
- Range at econ cruising speed 620 nm (1,150 km; 714 miles

PIEL C.P.70 and C.P.750 BERYL

The prototype of the **C.P.70 Beryl** tandem two-sea light aircraft was displayed publicly for the first time i August 1965. It retains the wing of the C.P.30 Emeraud virtually unchanged, combining this wing with a modifie fuselage and non-retractable tricycle landing gear.

The fuselage of the C P.70 is a fabric-covered woode structure, of slimmer section than that of the Emeraude Each main landing gear unit is articulated, with the whee aft of the oleo-pneumatic shock-absorber. The steerabl nosewheel is carried on a conventional fork.

Intended for aerobatic flying, the **C.P.750 Beryl** is als similar in general appearance to the Emeraude but has longer, steel tube fuselage seating two persons in tandem slightly reduced span, a non-retractable tailwheel-typ landing gear and other changes.

The C.P.750 has so far been built principally by amate

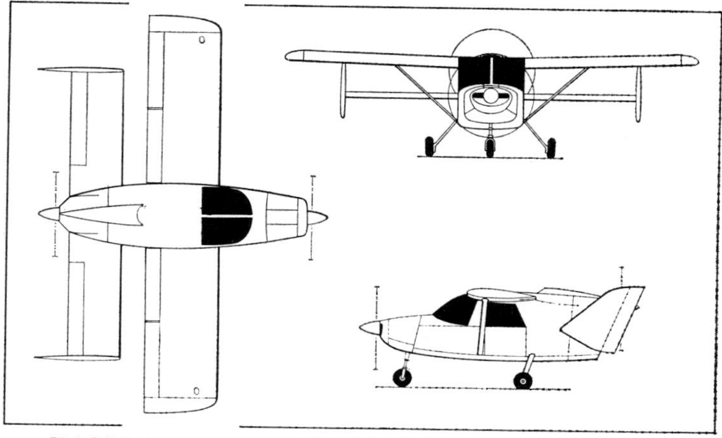

Piel C.P.500 tandem-wing twin-engined light aircraft *(Roy J. Grainge)*

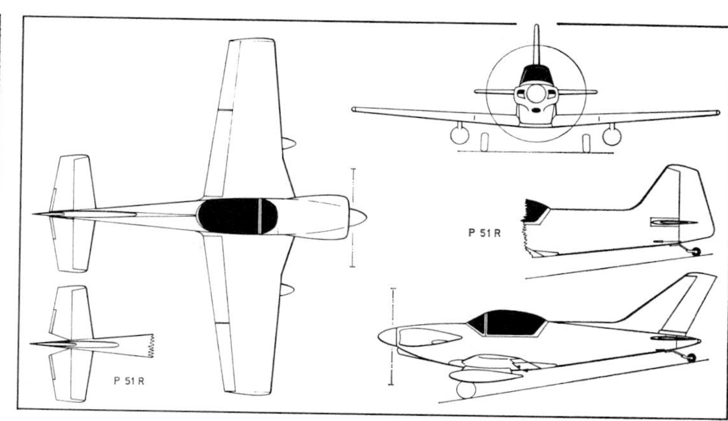

Pottier P.50R Bouvreuil single-seat racing aircraft. Scrap views show tail of P.51R

Berger FB2 single-seat sporting biplane

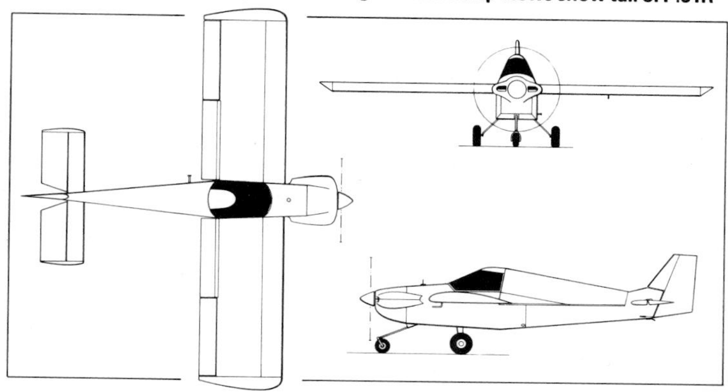

Pottier P.70S single-seat sporting aircraft *(Roy J. Grainge)*

constructors in Canada, but may also be built in France through the facilities offered by M Choisel at Abbeville.

The example (N7NT) shown in an accompanying illustration was built by Mr Norman Taylor of Ontario, Oregon, over a period of 4 years and 10 months, at a cost of $8,000. Constructed according to plans, with a 119·5 kW (160 hp) Lycoming O-320 engine, it has an empty weight of 605 kg (1,332 lb), max take-off weight of 905 kg (1,996 lb), cruising speed of 139 knots (257 km/h; 160 mph), landing speed of 52 knots (97 km/h; 60 mph), rate of climb of 457 m (1,500 ft)/min at sea level, take-off and landing run of 145 m (475 ft), and range of 390 nm (725 km; 450 miles) with max fuel.

Type: Two-seat aerobatic monoplane.

Wings: Cantilever low-wing monoplane. Wing section NACA 23012. Dihedral 5° 40′. Incidence 4° 10′. All-wood single-spar structure, made in one piece, with fabric covering. Slotted ailerons and slotted flaps of wood construction with fabric covering.

Fuselage: Fabric-covered structure of wood (C.P.70) or welded steel tube (C.P.750).

Tail Unit: Cantilever wood structure. Fixed surfaces plywood-covered, control surfaces fabric-covered. Ground-adjustable tab on each elevator.

Landing Gear (C.P.70): Non-retractable tricycle type.

Landing Gear (C.P.750): Non-retractable tailwheel type. Main wheels size 420 × 150, pressure 1·65 bars (24 lb/sq in). Hydraulic brakes. Wheel fairings. Steerable tailwheel.

Power Plant (C.P.70): One 48·5 kW (65 hp) Continental C65-8F flat-four engine, driving a two-blade wooden propeller. Fuel tank in fuselage, capacity 70 litres (15·4 Imp gallons).

Power Plant (C.P.750): One 112 kW (150 hp) Lycoming O-320-E2A flat-four engine, driving an EVRA two-blade fixed-pitch wooden propeller. Fuel tank in fuselage, capacity 70 litres (15·4 Imp gallons), with provision for two auxiliary tanks in wings to give total capacity of 140 litres (30·75 Imp gallons). Oil capacity 5 litres (1·0 Imp gallon).

Accommodation: Two seats in tandem under rearward-sliding transparent canopy. Rear seat of C.P.70 is wide enough to accommodate one adult and a child, or two children.

Dimensions, external:
Wing span:
C.P.70 8·25 m (27 ft 0¼ in)
C.P.750 8·04 m (26 ft 4½ in)
Wing chord at root 1·50 m (4 ft 11 in)
Wing area, gross:
C.P.70 10·85 m² (116·8 sq ft)
C.P.750 11·00 m² (118 sq ft)
Wing aspect ratio:
C.P.70 5·95
C.P.750 5·85
Length overall:
C.P.70 6·45 m (21 ft 2 in)
C.P.750 6·90 m (22 ft 7¾ in)
Height overall:
C.P.70 1·60 m (5 ft 3 in)
C.P.750 2·10 m (6 ft 10¾ in)
Wheel track:
C.P.70 2·00 m (6 ft 6¾ in)
C.P.750 2·40 m (7 ft 10½ in)
Propeller diameter 1·80 m (5 ft 11 in)

Weights and Loadings:
Weight empty:
C.P.70 320 kg (705 lb)
C.P.750 480 kg (1,058 lb)
Max T-O weight:
C.P.70 540 kg (1,190 lb)
C.P.750 760 kg (1,675 lb)
Max wing loading:
C.P.70 50·0 kg/m² (10·2 lb/sq ft)
C.P.750 69·0 kg/m² (14·1 lb/sq ft)
Max power loading:
C.P.70 11·13 kg/kW (18·3 lb/hp)
C.P.750 6·79 kg/kW (11·0 lb/hp)

Performance (at max T-O weight):
Never-exceed speed:
C.P.70 118·5 knots (220 km/h; 136·5 mph)
C.P.750 183 knots (340 km/h; 211 mph)
Max level speed:
C.P.70 95 knots (175 km/h; 109 mph)
C.P.750 151 knots (280 km/h; 174 mph)
Max cruising speed (75% power) at 1,200 m (3,940 ft):
C.P.70 84 knots (156 km/h; 97 mph)
C.P.750 143 knots (265 km/h; 165 mph)
Econ cruising speed (65% power) at 1,200 m (3,940 ft):
C.P.70 78 knots (145 km/h; 90 mph)
C.P.750 135 knots (250 km/h; 155 mph)
Approach speed, flaps down:
C.P.70 54 knots (100 km/h; 62·5 mph)
C.P.750 70 knots (130 km/h; 81 mph)
Stalling speed, flaps up:
C.P.70 41 knots (75 km/h; 47 mph)
C.P.750 54 knots (100 km/h; 62·5 mph)
Stalling speed, flaps down:
C.P.70 39 knots (70 km/h; 44 mph)
C.P.750 52 knots (95 km/h; 59 mph)
Max rate of climb at S/L:
C.P.70 120 m (394 ft)/min
C.P.750 390 m (1,280 ft)/min
Service ceiling:
C.P.70 3,000 m (9,850 ft)
C.P.750 5,200 m (17,060 ft)
T-O run:
C.P.70 280 m (919 ft)
C.P.750 190 m (623 ft)
T-O to 15 m (50 ft):
C.P.70 420 m (1,378 ft)
C.P.750 350 m (1,148 ft)
Landing from 15 m (50 ft):
C.P.70 280 m (919 ft)
C.P.750 520 m (1,706 ft)
Landing run:
C.P.70 140 m (459 ft)
C.P.750 280 m (919 ft)
Range at econ cruising speed:
C.P.70 323 nm (600 km; 372 miles)
C.P.750 593 nm (1,100 km; 683 miles)

PIEL C.P.80/ZEF

The C.P.80 was designed as a single-seat racing aircraft for amateur construction. The basic version is made of wood, as described; but M Calvel of l'Hospitalet du Larzac adapted the design to enable his C.P.80 Zef to be constructed of laminated plastics. This was the first C.P.80 to fly, followed in July 1974 by the C.P.80 Racer No. 01 built by M Claude Piel.

About 20 wooden C.P.80s are under construction by amateurs. The general appearance of the aircraft is shown in the accompanying illustration.

M Piel's prototype has confirmed the accuracy of the estimated weight and performance figures quoted. It has attained a maximum level speed of 162 knots (300 km/h; 186 mph) without main wheel fairings, suggesting an eventual maximum level speed of 172 knots (320 km/h; 199 mph) when these are fitted and development is completed.

Type: Single-seat amateur-built racing aircraft.

Wings: Cantilever low-wing monoplane. Wing section NACA 23012. Dihedral 3°. Incidence 2° (constant). No sweep at quarter-chord. Conventional single-spar wood structure, plywood-covered and with polyester plastics tips. Ailerons mass-balanced and cable-actuated. No flaps or tabs.

Fuselage: Conventional plywood-covered wood structure of basic rectangular section, with four longerons, nine frames and domed rear decking. Polyester plastics engine cowling. Steel tube engine mounting attached to fireproof bulkhead.

Tail Unit: Cantilever plywood-covered all-wood structure, with vertical surfaces swept back at 50° on leading-edge. All-moving constant-chord horizontal surfaces, with centrally-positioned anti-balance and trim tab, and with mass-balance arm projecting forward inside fuselage. Horn-balanced rudder. Control surfaces cable-operated.

Landing Gear: Non-retractable tailwheel type. Main wheels carried on cantilever spring legs of treated AU4SG alloy. Steerable tailwheel carried on steel spring. Hydraulic brakes on main wheels.

Power Plant: One 67 kW (90 hp) Continental C90-8F flat-four engine, driving through a short extension shaft a two-blade fixed-pitch wooden propeller. Provision for other engines, including 48·5 kW (65 hp) Continental. Fuel tank of AG-3 alloy, capacity 40 litres (8·8 Imp gallons), aft of firewall, with refuelling point in top-decking.

Accommodation: Pilot only, in enclosed cockpit, under sideways-hinged transparent canopy.

Dimensions, external:
Wing span 6·00 m (19 ft 8¼ in)
Wing chord at aircraft centreline 1·35 m (4 ft 5¼ in)
Wing chord at tip 0·90 m (2 ft 11½ in)
Wing area, gross 6·20 m² (66·7 sq ft)
Wing aspect ratio 5·8
Length overall 5·30 m (17 ft 4¾ in)
Height overall 1·70 m (5 ft 7 in)

Tailplane span 1·58 m (5 ft 2¼ in)
Wheel track 1·60 m (5 ft 3 in)
Wheelbase 3·50 m (11 ft 5¾ in)
Propeller diameter 1·52 m (5 ft 0 in)

WEIGHTS AND LOADINGS (67 kW; 90 hp engine):
Weight empty 260 kg (573 lb)
Max T-O weight 380 kg (837 lb)
Max wing loading 61·2 kg/m² (12·5 lb/sq ft)
Max power loading 5·67 kg/kW (9·3 lb/hp)

PERFORMANCE (estimated, with 67 kW; 90 hp engine, at max T-O weight):
Never-exceed speed 205 knots (380 km/h; 236 mph)
Max level speed 167 knots (310 km/h; 193 mph)
Max cruising speed (75% power) at 1,200 m (3,940 ft) 151 knots (280 km/h; 174 mph)
Econ cruising speed (65% power) at 1,200 m (3,940 ft) 129·5 knots (240 km/h; 149 mph)
Approach speed 70 knots (130 km/h; 81 mph)
Stalling speed 51·5 knots (95 km/h; 59 mph)
Max rate of climb at S/L 720 m (2,360 ft)/min
Service ceiling 6,000 m (19,685 ft)
T-O run 200 m (656 ft)
T-O to 15 m (50 ft) 400 m (1,312 ft)
Landing from 15 m (50 ft) 360 m (1,181 ft)
Landing run 200 m (656 ft)
Range at econ cruising speed 243 nm (450 km; 280 miles)
g limits +8g; −6g

PIEL C.P.90 PINOCCHIO

The C.P.90 Pinocchio is essentially a slightly smaller, single-seat development of the basic Emeraude, intended for aerobatic and general sporting flying.

WINGS: Cantilever low-wing monoplane, of similar general planform and construction to Emeraude. Dihedral 5° 40′. Incidence 3°. Ailerons only, no flaps.

FUSELAGE: Fabric-covered wooden structure of basically rectangular cross-section with domed decking.

TAIL UNIT: Cantilever fabric-covered wooden structure, similar to that of Emeraude.

LANDING GEAR: Non-retractable tailwheel type. Streamlined leg and wheel fairings on main units.

POWER PLANT: One 74·5 kW (100 hp) Continental O-200 flat-four engine, driving a two-blade wooden propeller. Fuel capacity 60 litres (13·2 Imp gallons).

ACCOMMODATION: Single seat under fully-transparent canopy.

DIMENSIONS, EXTERNAL:
Wing span 7·20 m (23 ft 7½ in)
Wing area, gross 9·65 m² (103·9 sq ft)
Wing aspect ratio 5·4
Length overall 6·00 m (19 ft 8¼ in)
Height overall 1·80 m (5 ft 11 in)
Wheel track 1·60 m (5 ft 3 in)
Propeller diameter 1·80 m (5 ft 11 in)

WEIGHTS AND LOADINGS:
Weight empty 335 kg (738 lb)
Max T-O weight 460 kg (1,014 lb)
Max wing loading 47·7 kg/m² (9·8 lb/sq ft)
Max power loading 6·17 kg/kW (10·14 lb/hp)

PERFORMANCE (estimated, at max T-O weight):
Never-exceed speed 171 knots (320 km/h; 198 mph)
Max level speed 141 knots (260 km/h; 162 mph)
Max cruising speed (75% power) at 1,200 m (3,940 ft) 132 knots (245 km/h; 152 mph)
Econ cruising speed (65% power) at 1,200 m (3,940 ft) 124 knots (230 km/h; 143 mph)
Approach speed 59 knots (110 km/h; 68 mph)
Stalling speed 41 knots (75 km/h; 47 mph)
Max rate of climb at S/L 480 m (1,575 ft)/min
Service ceiling 6,000 m (19,685 ft)
T-O run 180 m (590 ft)
T-O to 15 m (50 ft) 400 m (1,312 ft)
Landing from 15 m (50 ft) 300 m (984 ft)
Landing run 160 m (525 ft)
Range at econ cruising speed 296 nm (550 km; 341 miles)

PIEL C.P.500

As can be seen in the accompanying illustration, the C.P.500 will be a 'push and pull' twin-engined aircraft of staggered tandem-wing configuration. Although this gives it some similarity to the Mignet formula, the wings will be fixed, and the pilot's controls conventional.

The strut-braced forward wing will have four-section slotted trailing-edge flaps over 75% of the span and 25% of the chord, actuated electrically through 35°. The rear wing will carry two elevons, actuated by control rods from 40° up to 35° down; these will function differentially for roll control and collectively for pitch control. Endplate fins and rudders on the rear wing will provide yaw control. The relative position of the wings is expected to permit steep 'parachute' descents of the kind possible with Mignet designs.

Wing section will be NACA 23015. The front wing will have a dihedral of 1° 30′ and incidence of 2° 30′ constant; the rear wing will have a constant incidence of 4° 30′ but no dihedral.

The prototype C.P.500 was intended to be built of wood. However, when construction is started, probably in 1976, this aircraft may now be made of metal, as specified for any future series production of the type. The engine cowlings, wingtips and fairings will be of laminated plastics. Two 112/119·5 kW (150/160 hp) Lycoming O-320 flat-four engines are specified, with the rear engine driving its propeller through an extension shaft, 15 cm (5·9 in) long. Fuel tanks in the tips of the forward wing will have a combined capacity of 300 litres (66 Imp gallons).

A non-retractable tricycle landing gear will be standard, with Wittman-type cantilever steel spring main legs and a steerable nosewheel. Each main wheel will be fitted with a hydraulic brake.

Basic accommodation will be provided for two persons side by side in front, with optional dual controls, and three passengers on a rear bench seat. Aft of the rear seat will be space for a sixth person or a considerable quantity of baggage.

DIMENSIONS, EXTERNAL:
Wing span:
front 8·80 m (28 ft 10½ in)
rear 6·43 m (21 ft 1¼ in)
Wing chord (constant):
front 1·50 m (4 ft 11 in)
rear 1·10 m (3 ft 7¼ in)
Wing area, gross:
front 13·20 m² (142·1 sq ft)
rear 7·10 m² (76·42 sq ft)
Wing aspect ratio:
front 5·85
rear 5·88
Wing stagger 0·46 m (1 ft 6 in)
Length overall 6·10 m (20 ft 0 in)
Height overall 2·25 m (7 ft 4½ in)
Fuselage depth (max) 1·56 m (5 ft 1½ in)
Fuselage width (max) 1·40 m (4 ft 7 in)
Wheel track 2·56 m (8 ft 4¾ in)
Wheelbase 2·60 m (8 ft 6¼ in)

WEIGHTS:
Weight empty 866 kg (1,909 lb)
Max T-O weight 1,500 kg (3,307 lb)

PERFORMANCE (estimated):
Max level speed 162 knots (300 km/h; 186 mph)
Max level speed, one engine out 129 knots (240 km/h; 149 mph)
Max cruising speed (75% power) 143 knots (265 km/h; 165 mph)
Max cruising speed, one engine out (75% power) 108 knots (200 km/h; 124 mph)
Stalling speed, flaps down 49 knots (90 km/h; 56 mph)
Max rate of climb at S/L 540 m (1,770 ft)/min
Max rate of climb at S/L, one engine out 180 m (590 ft)/min
Service ceiling 6,800 m (22,300 ft)
Service ceiling, one engine out 3,000 m (9,850 ft)
Range with max fuel 647 nm (1,200 km; 745 miles)

POTTIER

JEAN POTTIER

ADDRESS:
4 rue Emilio Castelar, 75012-Paris
Telephone: 343 63-16

In addition to the light aircraft and sailplanes that he designed jointly with M Robert Jacquet, during his period as technical director at Société CARMAM, M Pottier is responsible for the purely amateur projects described below.

POTTIER P.50 BOUVREUIL (BULLFINCH)

Designed by M Jean Pottier, the Bouvreuil is a single-seat racing monoplane, intended for construction by amateurs. Construction is entirely of wood, except for the plastics engine cowling and main-wheel fairings.

The Bouvreuil can be fitted with a variety of engines in the 48·5-86 kW (65-115 hp) category. It has also been designed from the start to have either a non-retractable (P.50) or retractable (P.50R) landing gear. Design load factors are ± 10.

Six Bouvreuils are thought to be under construction, numbers 01 and 02 in France, 03 in the Netherlands, 04 in Switzerland, and 05 and 06 in Germany. Four of these aircraft will have a 67 kW (90 hp) Continental C90 engine; 02 will have a 48·5 kW (65 hp) Continental and 06 a 61 kW (82 hp) Porsche. Numbers 03, 04 and 06 will be P.50Rs, with retractable landing gear, offering a 10% improvement in performance.

Numbers 01 and 04 were expected to be completed first.

TYPE: Single-seat racing monoplane.

WINGS: Cantilever low-wing monoplane. Wing section NACA 23015 at root, NACA 23012 at tip. Dihedral from roots. All-wood structure, with full-span ailerons and flaps. No tabs.

FUSELAGE: Conventional wood semi-monocoque structure, with plastics engine cowling.

TAIL UNIT: Cantilever all-wood structure, with swept vertical surfaces. Trim tab in each elevator.

LANDING GEAR: Alternative retractable or non-retractable tailwheel type. Wheel fairings standard on non-retractable main wheels. Steerable tailwheel. Independent main-wheel brakes.

POWER PLANT: Standard power plant is a 67 kW (90 hp) Continental C90 flat-four engine, driving a two-blade fixed-pitch propeller with spinner. Other engines of 48·5 to 86 kW (65-115 hp) are optional. Fuel capacity 60 litres (13 Imp gallons) for racing, 100 litres (22 Imp gallons) for touring. Provision for carrying one removable auxiliary fuel tank under each wing.

ACCOMMODATION: Single seat in enclosed cabin, under large rearward-sliding transparent canopy.

DIMENSIONS, EXTERNAL:
Wing span 6·20 m (20 ft 4 in)
Wing area, gross 7·50 m² (80·7 sq ft)
Wing aspect ratio 5·10
Length overall 5·65 m (18 ft 6½ in)

WEIGHTS AND LOADING (67 kW; 90 hp engine):
Weight empty 270 kg (595 lb)
Max T-O weight 400 kg (882 lb)
Max wing loading 53·5 kg/m² (10·95 lb/sq ft)

PERFORMANCE (estimated, with 67 kW; 90 hp engine and non-retractable landing gear):
Max level speed 167 knots (310 km/h; 192 mph)
Max cruising speed (75% power) 151 knots (280 km/h; 174 mph)
Min speed 43 knots (80 km/h; 50 mph)

POTTIER P.51R

The P.51R, designed in 1973, is generally similar to the version of the P.50 with retractable landing gear and optional underwing auxiliary tanks, but has unswept vertical tail surfaces. The wings, forward fuselage structure and flying controls of the two types are identical, and the same variety of power plants may be fitted.

DIMENSIONS, AREA AND WEIGHTS:
As for P.50

PERFORMANCE (estimated):
Max level speed 167 knots (310 km/h; 192 mph)
Max cruising speed (75% power) 151 knots (280 km/h; 174 mph)
Landing speed 43 knots (80 km/h; 50 mph)

POTTIER P.70S

This small sporting aircraft is derived from the P.70B, designed by M Pottier and built by M Alain Besneux. Design was started in January 1974, and twelve P.70S are known to be under construction by amateurs, of which two or three were expected to fly during 1976.

Major changes by comparison with the P.70B include an increased wing span and installation of tricycle landing gear.

TYPE: Single-seat amateur-built sporting aircraft.

WINGS: Cantilever mid-wing monoplane. Wing section NACA 4415. No dihedral. Incidence 2°. No sweep. Constant-chord all-metal structure of 2024 alloy, with I-beam main spar and channel-section rear spar. Entire trailing-edge of each wing formed by aileron hinged to upper surface and plain flap hinged to bottom surface. No tabs.

FUSELAGE: All-metal structure of 2024 alloy, built up on five frames.

TAIL UNIT: Cantilever all-metal structure, with sweptback vertical surfaces. Minimal fixed fin. No tabs.

LANDING GEAR: Non-retractable tricycle type. Cantilever main legs.

POWER PLANT: One 30/37·3 kW (40/50 hp) Volkswagen converted motor car engine, driving a two-blade fixed-pitch propeller. Single fuel tank in fuselage, aft of firewall, capacity 40 litres (8·75 Imp gallons).

ACCOMMODATION: Pilot only, in enclosed cockpit.

DIMENSIONS, EXTERNAL:
Wing span 5·90 m (19 ft 4¼ in)
Wing chord (constant) 1·25 m (4 ft 1¼ in)
Wing area, gross 7·2 m² (77·5 sq ft)
Wing aspect ratio 4·8
Length overall 5·00 m (16 ft 4¾ in)
Height overall 1·60 m (5 ft 3 in)
Tailplane span 2·10 m (6 ft 10¾ in)
Wheel track 1·20 m (3 ft 11¼ in)
Propeller diameter 1·30 m (4 ft 3¼ in)
Propeller ground clearance 0·20 m (8 in)

WEIGHTS AND LOADING:
Weight empty, equipped 180 kg (397 lb)
Max T-O and landing weight 290 kg (639 lb)
Max wing loading 40 kg/m² (8·19 lb/sq ft)

PERFORMANCE (30 kW; 40 hp engine, at max T-O weight):
Never-exceed speed at S/L 129 knots (240 km/h; 149 mph)
Max level speed at S/L 97 knots (180 km/h; 112 mph)
Max cruising speed at S/L 89 knots (165 km/h; 103 mph)
Econ cruising speed at S/L 65 knots (120 km/h; 75 mph)
Stalling speed, flaps down 38 knots (70 km/h; 44 mph)
Max rate of climb at S/L 300 m (985 ft)/min
Service ceiling 4,500 m (14,775 ft)
T-O to 15 m (50 ft) 320 m (1,050 ft)
Range with max fuel 269 nm (500 km; 310 miles)

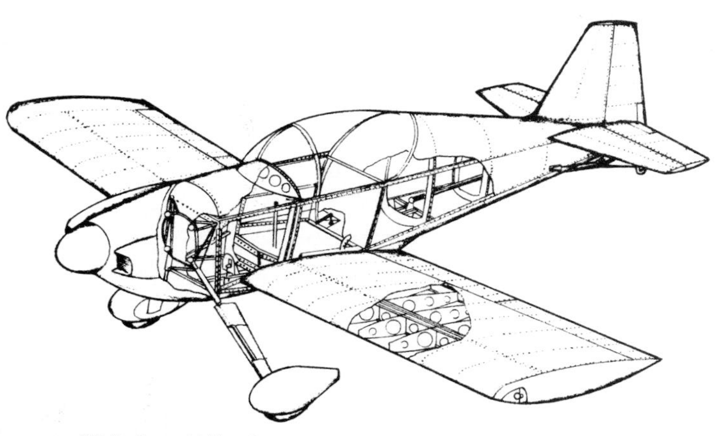

Wallerkowski Hornisse aerobatic monoplane *(Flug Revue)*

Bagalini Leonardino powered Rogallo (Citroen AM6 engine)

GERMANY
(FEDERAL REPUBLIC)

BERGER
FRANZ BERGER

ADDRESS:
Töging/Inn, Weichselstrasse 20
Telephone: 08631/99151

Herr Franz Berger has designed and built the prototype of a lightweight sporting biplane, designated FB2. Design and construction began simultaneously in July 1972, and the first flight was made on 31 October 1974.

BERGER FB2

TYPE: Single-seat homebuilt sporting aircraft.

WINGS: Braced biplane structure. Single-bay, with N-type interplane and N-type centre-section struts each side. Dual landing and flying wires. Wing section NACA 4412. Dihedral 0° on upper wing, 2° on lower wing. Conventional structure, with two wooden spars, wooden leading-edge, built-up ribs and fabric covering. Ailerons of light alloy construction on lower wings only.

FUSELAGE: Welded steel tube structure with fabric covering.

TAIL UNIT: Wire-braced welded steel tube structure with fabric covering. Ground adjustable tabs on rudder and elevators.

LANDING GEAR: Non-retractable tailwheel type. Main wheels carried on two side Vees and half-axles. Shock-absorption by rubber in compression. Main wheels with tyres size 8·00-4, pressure 0·79 bars (11·5 lb/sq in). Hydraulic brakes.

POWER PLANT: One 93 kW (125 hp) Lycoming O-235-F2B flat-four engine, driving a two-blade wooden fixed-pitch propeller with spinner.

ACCOMMODATION: Single seat in open cockpit.

DIMENSIONS, EXTERNAL:

Wing span, upper	5·16 m (16 ft 11¼ in)
Wing span, lower	4·80 m (15 ft 9 in)
Wing chord, constant (both)	0·92 m (3 ft 0¼ in)
Wing area, gross (both)	9·5 m^2 (102·3 sq ft)
Length overall	4·63 m (15 ft 2¼ in)
Height overall	2·35 m (7 ft 8½ in)
Tailplane span	1·94 m (6 ft 4¼ in)
Propeller diameter	1·80 m (5 ft 11 in)

WEIGHTS AND LOADINGS:

Weight empty	251 kg (553 lb)
Max T-O weight	500 kg (1,102 lb)
Max wing loading	53·47 kg/m^2 (10·95 lb/sq ft)
Max power loading	5·38 kg/kW (8·82 lb/hp)

PERFORMANCE (at max T-O weight):

Never-exceed speed	167 knots (310 km/h; 192 mph)
Max level speed	129 knots (240 km/h; 149 mph)
Cruising speed	81 knots (150 km/h; 93 mph)

WALLERKOWSKI
HEINZ WALLERKOWSKI

ADDRESS:
D-8019 Assling, Hochreit 16
Telephone: (08092) 96 16

Heinz Wallerkowski is an airline pilot, and flies a BAC One-Eleven for a charter company. He began the design of the Hornisse in 1974; construction of the prototype was initiated in the following year. First flight was scheduled for the Autumn of 1976 and the aircraft will be certificated by the LBA in the Restricted Special Category. Mr Wallerkowski is a member of the Oskar-Ursinus-Vereinigung, the German Chapter of the EAA.

WALLERKOWSKI HORNISSE (HORNET)

TYPE: Single-seat homebuilt monoplane.

WINGS: Cantilever low-wing monoplane. NACA 23012 wing section. No sweepback. All-metal construction, flush-riveted. All-metal Frise-type ailerons and all-metal flaps along entire trailing-edges.

FUSELAGE: All-metal semi-monocoque structure, flush-riveted.

TAIL UNIT: Conventional cantilever assembly of all-metal construction. Two-spar tailplane, with elevators. Trim tabs in rudder and elevators.

LANDING GEAR: Non-retractable tailwheel type. Cantilever steel-tube main legs, with streamline fairings. Wheel fairings. Steerable tailwheel.

POWER PLANT: One 112 kW (150 hp) Lycoming O-320 flat-four engine. Fuel capacity 85 kg (187 lb).

ACCOMMODATION: Single seat under transparent canopy.

DIMENSIONS, EXTERNAL:

Wing span	6·14 m (20 ft 1¾ in)
Wing area, gross	8·4 m^2 (90·4 sq ft)
Wing aspect ratio	4·45
Length overall	5·80 m (19 ft 1 in)
Height overall	1·60 m (5 ft 3 in)

DIMENSION, INTERNAL:

Cockpit: Max width	0·64 m (2 ft 1¼ in)

WEIGHTS AND LOADINGS:

Weight empty	340 kg (750 lb)
Max payload	90 kg (198 lb)
Max T-O weight: normal	500 kg (1,102 lb)
aerobatic	460 kg (1,014 lb)
Max wing loading	59·5 kg/m^2 (12·19 lb/sq ft)
Max power loading	4·46 kg/kW (7·35 lb/hp)

PERFORMANCE (estimated, at max T-O weight):

Never-exceed speed	207 knots (385 km/h; 239 mph)
Max level speed	178 knots (330 km/h; 205 mph)
Max cruising speed at S/L	162 knots (300 km/h; 186 mph)
Stalling speed	46 knots (85 km/h; 53 mph)
Max rate of climb at S/L	480 m (1,575 ft)/min
Service ceiling	6,000 m (19,700 ft)
T-O run	150 m (490 ft)
T-O to 15 m (50 ft)	300 m (985 ft)
Landing from 15 m (50 ft)	300 m (985 ft)
Landing run	200 m (655 ft)
Range with max fuel	485 nm (900 km; 559 miles)

ITALY

BAGALINI
WALTER BAGALINI

Among various types of aircraft built and flown by members of the Club Aviazione Popolare (the Italian Chapter of the EAA) is a powered Rogallo-type aircraft known as Leonardino (Little Leonardo). This was designed and constructed by Walter Bagalini, after experiments using radio-controlled models. The work took three months, at a cost equivalent to $500, including purchase of the engine. Walter Bagalini then learned to fly and tested the aircraft simultaneously.

BAGALINI LEONARDINO

TYPE: Single-seat homebuilt light aircraft.

WINGS: Rogallo wing made of Dacron with a specific weight of 0·13 kg/m^2 (0·0274 lb/sq ft). Normal Rogallo leading-edge spars and keel supplemented by lateral cross-member which keeps the wing rigidly extended in flight. Bracing wires from fuselage longeron to apex and each end of cross-member. All structure is of round and square section aluminium tubing.

FUSELAGE: Basic T-structure comprising longeron and vertical mast supporting wing and power plant, made of square section aluminium tubing, with some home-made metal bonding.

TAIL UNIT: Conventional cantilever tailplane and elevator, positioned aft of fin and rudder.

LANDING GEAR: Each main wheel is carried on a steel-tube side Vee and a spring steel shock-strut attached at its upper end to the wing support mast. Small steerable tailwheel. Nosewheel at front end of fuselage longeron.

POWER PLANT: One 15·4 kW (20·7 hp) Citroen AM6 engine, mounted at shoulder height on rear face of wing support mast, and driving a pusher propeller.

ACCOMMODATION: Single seat in open position. Conventional control stick and rudder bar.

DIMENSIONS, EXTERNAL:

Wing span	6·70 m (21 ft 11¾ in)
Wing area, gross	15·70 m^2 (168·92 sq ft)
Length overall	5·55 m (18 ft 2¾ in)
Height overall	2·60 m (8 ft 6¼ in)

WEIGHTS AND LOADING:

Weight empty	100 kg (220 lb)
Max T-O weight	165 kg (364 lb)
Max wing loading	10·50 kg/m^2 (2·15 lb/sq ft)

PERFORMANCE:

Max cruising speed	33 knots (61 km/h; 38 mph)
T-O speed	23·5 knots (44 km/h; 27 mph)
Max rate of climb at S/L	71·5 m (235 ft)/min
T-O run	120 m (394 ft)
Endurance	30 min
Best glide ratio	4

Silvestri Aquilottero powered Rogallo (Praga engine)

Fukuoka High School (Bensen) Lark III autogyro *(Howard Levy)*

SILVESTRI
OLINDO SILVESTRI

ADDRESS:
58010 Talamone (Gr), Fonteblanda

After ten years experience of being towed under a Rogallo-type kite, over water, by a motor boat, Mr Silvestri decided to make himself independent of towing, by building a powered Rogallo light aircraft. In its original form this aircraft, known as the Aquilottero, had a DAF 44 converted motor car engine and was designed for water operation, with a large central single-step float and a small stabilising float on a long multi-strut assembly under each wingtip. Subsequently, Mr Silvestri fitted a different power plant and wheeled landing gear, and had completed a total of 70 flying hours, without accident, by the Spring of 1976.

SILVESTRI AQUILOTTERO (EAGLET)

TYPE: Single-seat homebuilt light aircraft.

WINGS: Rogallo-type fabric wing, supported by normal leading-edge spars and keel and rigid cross-member.

FUSELAGE: Basic T-structure comprising fuselage longeron and twin side-by-side vertical masts supporting wing and power plant.

TAIL UNIT: Strut-braced tailplane and elevators mounted at top of fin and rudder assembly.

LANDING GEAR: Two main wheels carried on cross-axle and side Vees, one arm of each Vee extending upward to wing mast to act as shock-absorber. Small nosewheel at front end of fuselage longeron. Tailskid.

POWER PLANT: One 30 kW (40 hp) Praga engine, mounted on aft face of wing support masts and driving a two-blade pusher propeller. Streamlined fuel tank mounted between masts, above engine.

ACCOMMODATION: Pilot only, on open seat.

DIMENSIONS, EXTERNAL:
Wing span 7·50 m (24 ft 7¼ in)
Length overall 5·50 m (18 ft 0½ in)

WEIGHT:
Weight empty 120 kg (265 lb)

PERFORMANCE:
Max level speed 43 knots (80 km/h; 50 mph)
Min flying speed 8 knots (15 km/h; 9·5 mph)
Will not stall
T-O run 7-10 m (23-33 ft)
Max L/D ratio 6

JAPAN

FUKUOKA
FUKUOKA HIGH SCHOOL

ADDRESS:
Near Fukuoka

FUKUOKA LARK III

The Lark III autogyro has been constructed by a group of students from the Fukuoka High School. It is basically a modified Bensen Gyro-Copter built from plans supplied from the United States, but is unique in being powered by a 60 kW (80 hp) 1,400 cc Subaru watercooled flat-four motor car engine, the first time a Japanese engine has been used in an autogyro.

The Lark III first flew in 1974.

JEAA
JAPAN EXPERIMENTAL AIRCRAFT ASSOCIATION (Chapter 306 of EAA International)

ADDRESS:
2-27 Uehara, Shibuya-ku, Tokyo 151

PRESIDENT:
Asahi Miyahara

Telephone: (03) 467-8522

Various fixed- and rotating-wing aircraft have been designed and/or built and flown by members of the JEAA, and several of these have appeared in recent editions of *Jane's*. Others are examples of US homebuilt types such as the Baby Great Lakes, Flaglor Scooter and Hovey Whing Ding II, as described in the US section.

ABE MIZET II (MIDGET)

The Mizet II single-seat homebuilt light aircraft was built between 1968 and 1973 by a team of students at Kushiro Technical High School, Hokkaido, led by Mr Keiichi Abe, its designer. It made its first flight in December 1973 and is described briefly below. A Mizet III is under construction.

WINGS: High-wing monoplane, with Vee bracing struts. Göttingen 387 wing section, of 15% thickness/chord ratio and constant chord. Dihedral 2° 30′. Incidence 3°. Wooden two-spar structure with fabric covering. Ailerons of similar construction. No flaps.

FUSELAGE: Cockpit and wing support structure of square-section aluminium alloy. Rear fuselage of plywood-covered wooden construction.

TAIL UNIT: Fabric-covered wooden structure comprising braced tailplane, elevators, rudder and ventral fin.

LANDING GEAR: Non-retractable tricycle type, with steerable nosewheel. Main wheels and tyres size 300 × 200 mm.

POWER PLANT: One 26 kW (35 hp) 785 cc Toyota 2U-1 motor car engine, driving a two-blade fixed-pitch walnut pusher propeller. Fuel capacity 7 litres (1·5 Imp gallons); oil capacity 3 litres (0·66 Imp gallons).

ACCOMMODATION: Single open seat.

DIMENSIONS, EXTERNAL:
Wing span 7·50 m (24 ft 7¼ in)
Wing chord (constant) 1·20 m (3 ft 11¼ in)
Wing area, gross 9·00 m^2 (96·9 sq ft)
Length overall 5·50 m (18 ft 0½ in)
Height over tail 2·20 m (7 ft 2¾ in)
Tailplane span 2·40 m (7 ft 10½ in)
Wheel track 1·30 m (4 ft 3¼ in)
Propeller diameter 1·10 m (3 ft 7⅜ in)

WEIGHTS:
Weight empty 180 kg (397 lb)
Max T-O weight 265 kg (584 lb)

PERFORMANCE (at max T-O weight):
Max level speed 42·5 knots (79 km/h; 49 mph)
Stalling speed 35·5 knots (65·5 km/h; 41 mph)
Max rate of climb at S/L 72 m (236 ft)/min
Service ceiling 1,500 m (4,920 ft)
T-O run 170 m (560 ft)
Landing run 140 m (460 ft)
Max range 37·8 nm (70 km; 43·5 miles)

TERUO KAGO TK-1

Mr T. Kago began construction of the TK-1 in April 1973, assisted by Mr K. Kubota. It was 70% complete by early 1975 and by early 1976 the wings had also been constructed but not covered or assembled to the fuselage.

TYPE: Single-seat homebuilt aircraft.

WINGS: Cantilever low-wing monoplane. RAF 48 wing section, with 14% thickness/chord ratio. Dihedral 7°. Incidence 5°. Wooden structure, with plywood and fabric covering. No flaps.

FUSELAGE: Wooden box structure, with plywood and fabric covering.

TAIL UNIT: Fabric-covered wooden cantilever structure.

LANDING GEAR: Non-retractable tailwheel type, with steerable tailwheel of 200 mm diameter. Balloon tyres on main wheels, diameter 600 mm. No shock-absorbers.

POWER PLANT: One 33·6 kW (45 hp) 1,500 cc modified Volkswagen engine, driving a two-blade propeller. Fuel capacity 20 litres (4·4 Imp gallons); oil capacity 4 litres (7 Imp pints).

ACCOMMODATION: Single seat under one-piece bubble canopy.

DIMENSIONS, EXTERNAL:
Wing span 9·90 m (32 ft 5¾ in)
Wing chord (constant) 1·51 m (4 ft 11½ in)
Wing area, gross 14·85 m^2 (159·8 sq ft)
Length overall 6·36 m (20 ft 10½ in)
Height overall 2·70 m (8 ft 10¼ in)
Wheel track 1·50 m (4 ft 11 in)
Propeller diameter 1·40 m (4 ft 7 in)

WEIGHTS AND LOADINGS:
Weight empty 280 kg (617 lb)
Max T-O weight 365 kg (804 lb)
Max wing loading 24·6 kg/m^2 (5·04 lb/sq ft)
Max power loading 10·86 kg/kW (17·86 lb/hp)

PERFORMANCE (estimated):
Max level speed 86 knots (160 km/h; 99 mph)
Max cruising speed 54 knots (100 km/h; 62 mph)
Stalling speed 35·5 knots (65 km/h; 40·5 mph)
Max rate of climb at S/L 180 m (590 ft)/min
T-O run 70 m (230 ft)
Landing run 120 m (395 ft)

HAMAO SIOKARA TOMBO

This small single-seat biplane, designed and built by Mr T. Hamao, is powered by a 112 kW (150 hp) Lycoming O-320 engine. Test flying was expected to begin in the Summer of 1975, but no news of this has been received.

The Tombo is of wooden construction, fabric covered, and its general layout can be seen in the accompanying photograph. The wing section is NACA 4218.

DIMENSIONS, EXTERNAL:
Wing span (upper) 6·40 m (21 ft 0 in)
Wing span (lower) 5·60 m (18 ft 4½ in)
Wing chord, mean 0·90 m (2 ft 11½ in)
Wing area, gross 10 m^2 (107·6 sq ft)
Length overall 5·70 m (18 ft 8¼ in)
Height overall 2·40 m (7 ft 10½ in)
Tailplane span 2·40 m (7 ft 10½ in)
Propeller diameter 1·88 m (6 ft 2 in)

WEIGHTS AND LOADINGS:
Weight empty 380 kg (838 lb)
Max T-O weight 500 kg (1,102 lb)
Max wing loading 50 kg/m^2 (10·24 lb/sq ft)
Max power loading 4·46 kg/kW (7·35 lb/hp)

PERFORMANCE:
No figures available

Abe Mizet II single-seat ultra-light aircraft (Toyota motor car engine)

Swano Mr Smoothie, a modified Bowers Fly Baby

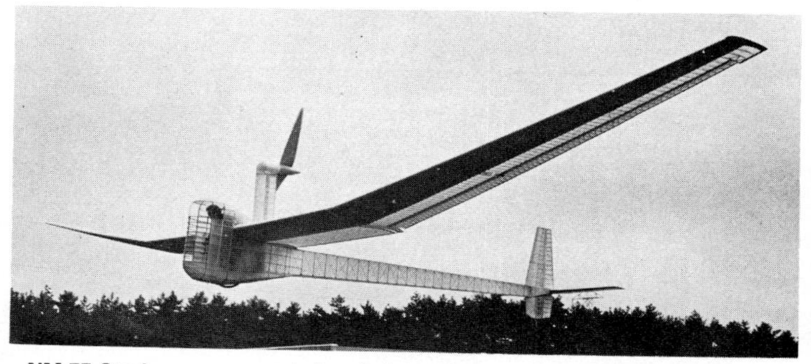

NM-75 Stork man-powered aircraft, latest in a series of designs by Nihon University *(Hiroshi Seo)*

Tombo single-seat biplane designed by Mr T. Hamao (Lycoming O-320 engine)

NIHON UNIVERSITY
COLLEGE OF SCIENCE AND ENGINEERING (DEPARTMENT OF MECHANICAL ENGINEERING), NIHON UNIVERSITY

ADDRESS:
8 Kanda-Surugadai, 1-chome, Chiyoda-ku, Tokyo 101
Telephone: Tokyo (03) 293-3251

CHIEF PROFESSOR: Dr Hidemasa Kimura

Under the leadership of Dr Kimura, students of Nihon University designed and built several aircraft, including the Okamura N-52 and N-58 Cygnet lightplanes and a STOL lightplane designated N-62, in collaboration with the Itoh company, which built the prototype. The N-62 was put into production by Itoh, as the Eaglet, and was described in the 1968-69 *Jane's*.

This design team has also built a series of successful man-powered aircraft named Linnet and Egret, all of which have been described fully in previous editions of *Jane's*. A powered sailplane, the N-70 Cygnus, is described in the appropriate section of the 1974-75 edition.

The latest aircraft of Nihon University design to fly is another man-powered aircraft, the NM-75 Stork.

NIHON UNIVERSITY NM-75 STORK

Design of the Stork began in April 1975 and construction was started in the following September. The aircraft made its first flight on 12 March 1976, and within the next 12 days had flown 11 times. The best of these flights, limited by the 600 m (1,968 ft) length of the campus runway, covered 446 m (1,463 ft) in 57 sec on 14 March 1976.

TYPE: Single-seat man-powered aircraft.

WINGS: Cantilever shoulder-wing monoplane. Wing section Wortmann FX-61-184 at root, FX-63-137 at tip. Constant-chord, no-dihedral centre-section. Tapered outer panels have 7° dihedral, 8° washout, and 1° 53′ of forward sweep at quarter-chord. Three-piece wooden structure. Two-spar construction, with spruce flanges and balsa stringers and ribs, covered with styrene paper and Japanese tissue. Rib spacing 100 mm (3·94 in). Conventional ailerons. No flaps or spoilers.

FUSELAGE: Front portion, including cabin and propeller pylon, is a welded chrome-molybdenum steel tube truss structure, covered with styrene paper and Japanese tissue. Aft portion, of square section, is a spruce and balsa truss and is similarly covered.

TAIL UNIT: Cantilever wooden structure, of similar construction to wings. All-moving horizontal surface of NACA 0009 section; single fin and rudder of NACA 0015 section.

LANDING GEAR: Non-retractable 686 mm (27 in) diameter bicycle wheel, driven by pedals on ground. No shock-absorbers. Bicycle-type brakes fitted after first flight.

POWER SYSTEM: Man-power on bicycle pedals, transmitted by chain drive to a two-blade all-balsa solid pusher propeller mounted aft of a pylon behind the cabin. Foot pedal rpm 72; propeller rpm 210.

ACCOMMODATION: Pilot only, in enclosed cabin, with a specially-designed control bar with a twist-grip at each end (right-hand grip for tailplane control; complete bar pivoted on universal joint for aileron and rudder control).

DIMENSIONS, EXTERNAL:

Wing span	21·00 m (68 ft 11 in)
Wing chord, centre-section (constant)	1·30 m (4 ft 3¼ in)
Wing chord at tip	0·55 m (1 ft 9¾ in)
Wing aspect ratio	20·3
Length overall	8·85 m (29 ft 0½ in)
Height overall	2·40 m (7 ft 10½ in)
Tailplane span	3·44 m (11 ft 3½ in)
Propeller diameter	2·50 m (8 ft 2½ in)

AREAS:

Wings, gross	21·70 m² (233·6 sq ft)
Ailerons (total)	2·52 m² (27·13 sq ft)
Fin	0·81 m² (8·72 sq ft)
Rudder	0·35 m² (3·77 sq ft)
Tailplane	1·71 m² (18·4 sq ft)

WEIGHTS AND LOADING:

Weight empty	36 kg (79·4 lb)
T-O weight	94 kg (207 lb)
Wing loading	4·33 kg/m² (0·89 lb/sq ft)

PERFORMANCE:

Speed	16·7 knots (31 km/h; 19·2 mph)

SWANO
SHIRO SWANO

SWANO MR SMOOTHIE

Built by Mr Shiro Swano of Yokohama, this single-seat wire-braced low-wing monoplane is a modified Bowers Fly Baby (see US section) and is constructed of wood.

POWER PLANT: One 30 kW (40 hp) 1,200 cc modified Volkswagen engine.

DIMENSIONS, EXTERNAL:

Wing span	7·60 m (24 ft 11½ in)
Wing chord (constant)	1·20 m (3 ft 11¼ in)
Wing area, gross	9·20 m² (99·0 sq ft)
Length overall	5·40 m (17 ft 8¾ in)
Height overall	2·15 m (7 ft 0¾ in)

WEIGHTS AND LOADINGS:

Weight empty	247 kg (544 lb)
Max T-O weight	345 kg (760 lb)
Max wing loading	37·5 kg/m² (7·68 lb/sq ft)
Max power loading	11·5 kg/kW (19·0 lb/hp)

PERFORMANCE:

Max level speed at S/L	54 knots (100 km/h; 62 mph)
Max cruising speed	46 knots (85 km/h; 53 mph)
Stalling speed	30·5 knots (56 km/h; 35 mph)
T-O run	80 m (260 ft)

POLAND

JANOWSKI
JAROSLAW JANOWSKI

ADDRESS:
ul. Nowomiejska 2M29, 91-061 Lodz

Mr Janowski, assisted by Mr Witold Kalita, designed and built a light single-seat amateur-built aircraft, the J-1 Don Kichot, of which brief details follow. Its power plant, the 17 kW (23 hp) Saturn two-cylinder engine, was also designed by Mr Janowski, and built by Mr S. Polawski. The aircraft has also been referred to by the Polish name Przasniczka.

A smaller development of the J-1, known as the J-2 Polonez, is also described.

JANOWSKI J-1 DON KICHOT (DON QUIXOTE)

Design and construction of the J-1 were started in 1967, and it flew for the first time on 30 July 1970.

Two additional J-1s are reportedly being built by amateur constructors; one by Mr Michal Offierski, a pre-war Polish holder of international powered sailplane records, and one by Mr T. Wood in the UK.

A full description appeared in the 1970-71 *Jane's*.

LANDING GEAR: Non-retractable. Cantilever spring steel main legs, each with single wheel and size 350 × 100 tyre, pressure 1·47 bars (21·3 lb/sq in). Rubber-sprung self-castoring tailwheel, with 120 × 30 solid tyre.

POWER PLANT: One 17 kW (23 hp) Janowski Saturn 500B two-stroke two-cylinder engine, driving a two-blade fixed-pitch wooden pusher propeller designed by Mr Janowski. Fuel capacity 18 litres (4 Imp gallons).

DIMENSIONS, EXTERNAL:

Wing span	7·60 m (24 ft 11¼ in)
Wing chord (constant)	1·00 m (3 ft 3¼ in)
Wing area, gross	7·50 m² (80·7 sq ft)
Wing aspect ratio	7·7
Length overall	4·88 m (16 ft 0 in)
Height overall	1·40 m (4 ft 7 in)
Tailplane span	2·05 m (6 ft 8¾ in)
Wheel track	1·15 m (3 ft 9¼ in)
Propeller diameter	1·06 m (3 ft 5¾ in)

WEIGHTS AND LOADINGS:

Weight empty	163 kg (359 lb)
Max T-O weight	270 kg (595 lb)
Max wing loading	36·0 kg/m² (7·37 lb/sq ft)
Max power loading	15·74 kg/kW (25·87 lb/hp)

PERFORMANCE (at max T-O weight):

Never-exceed speed	96·5 knots (180 km/h; 111 mph)
Max level speed at S/L	73 knots (135 km/h; 84 mph)
Max cruising speed	65 knots (120 km/h; 75 mph)
Stalling speed	31·5 knots (58 km/h; 36 mph)
Max rate of climb at S/L	120 m (390 ft)/min
Service ceiling	2,500 m (8,200 ft)
T-O run	200 m (656 ft)

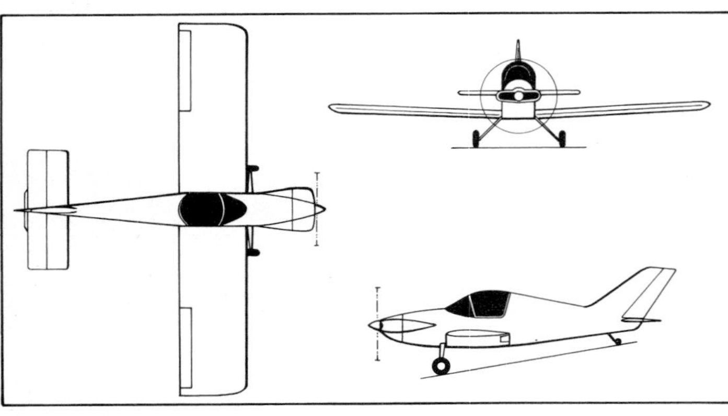

Olszewski-Obarewicz Aerosport single-seat monoplane *(Roy J. Grainge)*

Model of the Janowski J-2 Polonez light aircraft

Landing run	150 m (492 ft)
Range	134 nm (250 km; 155 miles)
g limits	+4·0; −1·5

JANOWSKI J-2 POLONEZ

Mr Janowski designed in 1971 an amateur-built single-seat aircraft known as the J-2 Polonez. This is smaller than the J-1, but is of generally similar configuration except for its mid-mounted wings and a T tail. A prototype was nearing completion in mid-1976.

TYPE: Single-seat ultra-light aircraft.

WINGS: Cantilever mid-wing monoplane. Wing section NACA 23015. Constant-chord wings. All-wood single-spar structure. Leading-edge plywood-covered, rest of wing fabric-covered. Fabric-covered ailerons. No tabs.

FUSELAGE: Pod and boom type. Enclosed cabin faired into front fuselage. Wooden single-boom structure supporting tail unit.

TAIL UNIT: Cantilever wooden structure, with T tailplane. Sweptback vertical surfaces and constant-chord non-swept horizontal surfaces. No tabs.

LANDING GEAR: Non-retractable single main wheel and tailskid.

POWER PLANT: One 22·5 kW (30 hp) Janowski Saturn 500 two-stroke two-cylinder engine, mounted at top of fuselage aft of cabin and driving a two-blade fixed-pitch wooden pusher propeller. The aircraft may be fitted with any other suitable engine of 18·5-30 kW (25-40 hp).

DIMENSIONS, EXTERNAL:

Wing span	6·40 m (21 ft 0 in)
Wing chord (constant)	1·00 m (3 ft 3¼ in)
Wing area, gross	6·35 m² (68·4 sq ft)
Wing aspect ratio	6·4
Length overall	4·80 m (15 ft 9 in)
Height overall	1·30 m (4 ft 3¼ in)

WEIGHTS:

Weight empty	110 kg (242 lb)
Normal T-O weight	205 kg (452 lb)

OLSZEWSKI-OBAREWICZ
KAZIMIERZ OLSZEWSKI AND J. OBAREWICZ

ADDRESS:
Lodz-Stoki 22, ul. Podgorze 10/6

OLSZEWSKI-OBAREWICZ AEROSPORT

Mr K. Olszewski and Mr J. Obarewicz designed a light single-seat amateur-built aircraft known as the Aerosport, of which construction began in 1971. General appearance of the Aerosport can be seen in the accompanying three-view drawing. Its first flight was made in the Summer of 1975.

TYPE: Single-seat light aircraft.

WINGS: Cantilever low-wing monoplane. Wing section NACA 23012. Constant-chord wings. All-wood single-spar structure. Leading-edge plywood-covered, rest of wing fabric-covered. Fabric-covered ailerons. No flaps or tabs.

FUSELAGE: Wooden structure, plywood-covered. Glassfibre engine cowling.

TAIL UNIT: Cantilever all-wood structure. Plywood-covered sweptback fin and unswept tailplane; fabric-covered rudder and elevator. Tab on elevator.

LANDING GEAR: Non-retractable tailwheel type, with spring-steel cantilever main legs and 300 mm main-wheel tyres.

POWER PLANT: One 22·5 kW (30 hp) modified 1200 cc Volkswagen engine, driving a two-blade fixed-pitch wooden propeller.

ACCOMMODATION: Single seat in enclosed cockpit.

DIMENSIONS, EXTERNAL:

Wing span	6·50 m (21 ft 4 in)
Wing chord (constant)	1·15 m (3 ft 9¼ in)
Wing area, gross	7·5 m² (80·7 sq ft)
Wing aspect ratio	5·7
Length overall	5·25 m (17 ft 2¾ in)
Height overall	1·90 m (6 ft 2¾ in)
Tailplane span	2·10 m (6 ft 10¾ in)
Wheel track	1·60 m (5 ft 3 in)
Propeller diameter	1·35 m (4 ft 5¼ in)

WEIGHTS:

Weight empty	175 kg (385 lb)
Normal T-O weight	270 kg (595 lb)

PERFORMANCE (estimated):

Max level speed at S/L	75·5 knots (140 km/h; 87 mph)
Cruising speed	65 knots (120 km/h; 74·5 mph)
Stalling speed	49 knots (90 km/h; 56 mph)

POLNIAK
LEON POLNIAK

ADDRESS:
ul Czerwinskiego 5a/1, 40-123 Katowice

POLNIAK LP DEDAL-2

A brief description of Mr Polniak's Dedal-1 man-powered aircraft appeared in the 1972-73 *Jane's*.

The later Dedal-2 had originally a tractor propeller, as shown in the accompanying three-view drawing. Photographs appearing in the Polish press in the Spring of 1976, however, show it with a pusher propeller and other modifications. In this form, its empty weight is said to be 27 kg (59·5 lb).

The following description applies to the Dedal-2 as originally designed:

TYPE: Single-seat man-powered aircraft.

WINGS: Wire-braced mid-wing monoplane. Eiffel 400 wing section, thickness/chord ratio 13·1%. Six-piece single-spar spruce/balsa structure, with slight forward sweep. Incidence and outer-wing dihedral can be varied, the latter raising wingtips up to 1·00 m (3 ft 3¼ in) above centre-section datum line. No flaps or ailerons: lateral control by pilot's body movement.

FUSELAGE: Spruce (cabin) and balsa structure.

TAIL UNIT: One-piece balsa tailplane, aft of fuselage, incidence of which is adjustable on ground. Oval balsa rudder aft of tailplane. No fin.

LANDING GEAR: Non-retractable balsa monowheel with size 12·5-2·25 pneumatic tyre, and small tailwheel.

POWER PLANT: Muscle-power (leg and arm), transmitted by belt drive from monowheel to a two-blade fixed-pitch propeller.

ACCOMMODATION: Pilot only, in enclosed cockpit.

DIMENSIONS, EXTERNAL:

Wing span	23·00 m (75 ft 5½ in)
Wing chord (constant)	1·90 m (6 ft 2¾ in)
Wing aspect ratio	approx 12
Length overall	9·20 m (30 ft 2¼ in)
Length of fuselage	6·85 m (22 ft 5¾ in)
Height overall	2·10 m (6 ft 10¾ in)
Tailplane span	5·30 m (17 ft 4¾ in)
Propeller diameter	2·60 m (8 ft 6¼ in)

DIMENSIONS, INTERNAL:

Cabin: Length	0·83 m (2 ft 8¾ in)
Width	0·62 m (2 ft 0¼ in)
Height	1·75 m (5 ft 9 in)

AREAS:

Wings, gross	40·00 m² (430·6 sq ft)
Rudder	2·25 m² (24·22 sq ft)
Tailplane	5·30 m² (57·00 sq ft)

WEIGHTS AND LOADING:

Weight empty	27 kg (60 lb)
T-O weight	85 kg (187 lb)
Wing + tailplane loading	approx 2·0 kg/m² (0·41 lb/sq ft)

PERFORMANCE:

Speed (estimated)	11 knots (20·5 km/h; 12·75 mph)

SOUTH AFRICA

CRUTCHLEY
S. CRUTCHLEY

ADDRESS:
21 Charles Boniface Road, Pietermaritzburg 3201
Telephone: 61049

CRUTCHLEY SPECIAL

The Crutchley Special was designed by Mr S. Crutchley as an easy-to-build all-metal aircraft that could be powered by a modified Volkswagen engine. Its angular shape reflected the fact that all work had to be carried out with hand tools. Design of the Special was started in June 1970 and construction began two months later. The first flight took place on 28 November 1975, and by June 1976 the Crutchley Special had completed 66 flying hours. Although no passenger had flown in the aircraft at that time (a stipulation of the permit to fly), the equivalent weight in sandbag ballast had been carried. The aircraft is designed to 6*g* limit load factor at maximum weight; but this was mainly to provide additional safety as it is intended to gain certification at 3·8*g* in the Normal category.

TYPE: Two-seat homebuilt monoplane.

WINGS: Cantilever low-wing monoplane. NACA 4415 wing section. No dihedral. Incidence 3°. Main spar fabricated as I-beam with flanges of B51S-TF extruded light alloy angle sections and web of 2024-T3 Alclad. Ribs and skins of Alclad. Rod-actuated plain ailerons, with B51S-TF spars and Alclad ribs and skins. Plain flaps of similar construction, with mechanical actuation.

FUSELAGE: Rectangular-section four-longeron box. B51S-TF extruded angle-sections, with formers and skins of 2024-T3 Alclad.

TAIL UNIT: Cantilever assembly, of similar construction to ailerons. Rudder cable-operated. Elevator actuated by pushrods. Fixed tab on elevator.

LANDING GEAR: Non-retractable tricycle type. Rubber-disc shock-absorbers in telescopic units of steel tube construction. German Continental go-kart steerable nosewheel, with tyre size 10 × 3·00-5, pressure 1·72 bars (25 lb/sq in). Swedish Varnamo industrial main wheels, each with tyre size 12 × 4·00-4, pressure 1·72 bars (25 lb/sq in). Aircooled Yamaha motorcycle drum brakes on main wheels and special disc brake for parking on nosewheel.

POWER PLANT: One 2,100 cc Volkswagen/Revmaster 2100S engine, developing 55 kW (74 hp) at 3,500 rpm and driving a Hegy wooden propeller of 0·97 m (3 ft 2 in) pitch. Fuel tank between instrument panel and firewall with capacity of 50 litres (11 Imp gallons). Refuelling point forward of windscreen on starboard

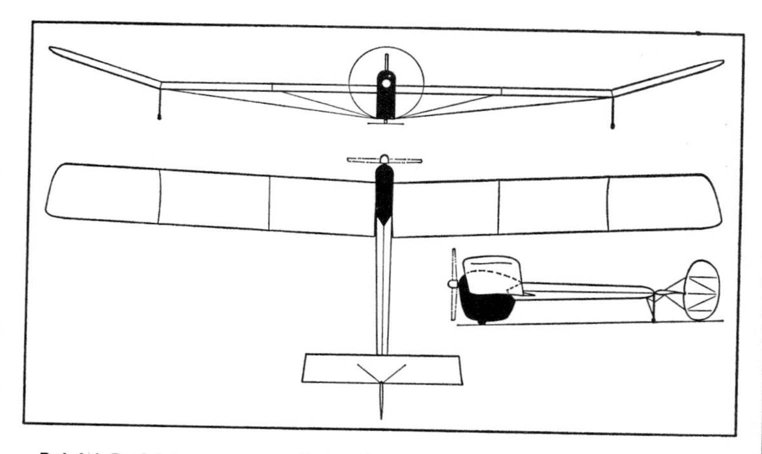

Polniak Dedal-2 man-powered aircraft as originally designed *(Roy J. Grainge)*

Janowski J-1 Don Kichot, described on page 475 *(Z. Szulc)*

York Swift, first of two Globe Swifts modified by Mr T. R. York *(R. Watts)*

Crutchley Special being flown by its designer/builder

Andreasson BA-4B single-seat fully-aerobatic homebuilt biplane

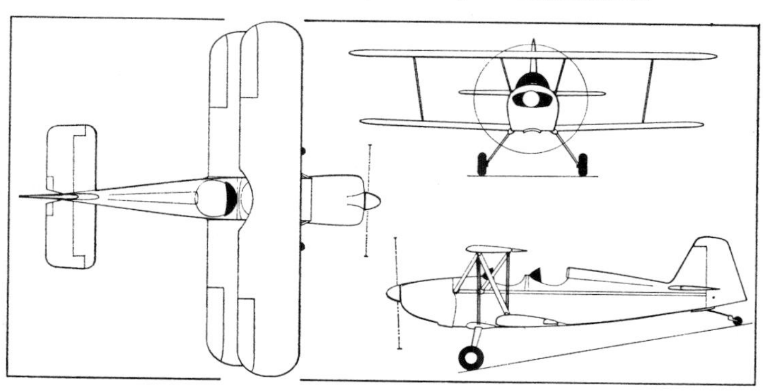

Andreasson BA-11 two-seat biplane *(Roy J. Grainge)*

side. Oil capacity 2·5 litres (0·55 Imp gallons). Engine fitted with Posa low-pressure injector carburettor.

ACCOMMODATION: Two seats side by side under rearward-hinged Perspex canopy. Ventilated by adjustable ram-air scoop.

SYSTEM: 12V 5Ah Yuasa motorcycle battery for radio and emergency electric fuel pump.

ELECTRONICS AND EQUIPMENT: Genave Alpha 10 transceiver.

DIMENSIONS, EXTERNAL:

Wing span	6·00 m (19 ft 8 in)
Wing chord (constant)	1·32 m (4 ft 4 in)
Wing area, gross	7·90 m² (85 sq ft)
Wing aspect ratio	4·5
Length overall	5·64 m (18 ft 6 in)
Height overall	1·83 m (6 ft 0 in)
Tailplane span	2·03 m (6 ft 8 in)
Wheel track	1·70 m (5 ft 7 in)
Wheelbase	1·12 m (3 ft 8 in)
Propeller diameter	1·52 m (5 ft 0 in)

DIMENSION, INTERNAL:

Cabin: Max width	1·02 m (3 ft 4 in)

WEIGHTS AND LOADINGS:

Weight empty	264 kg (581 lb)
Max T-O weight	450 kg (992 lb)
Max wing loading	57·1 kg/m² (11·7 lb/sq ft)
Max power loading	8·18 kg/kW (13·4 lb/hp)

PERFORMANCE (at max T-O weight):

Never-exceed speed	113 knots (209 km/h; 130 mph)
Max level speed at 915 m (3,000 ft)	92 knots (171 km/h; 106 mph)
Max cruising speed at 915 m (3,000 ft)	78 knots (145 km/h; 90 mph)
Econ cruising speed at 915 m (3,000 ft)	73 knots (135 km/h; 84 mph)
Stalling speed, flaps down	50·5 knots (94 km/h; 58 mph)
Max rate of climb at S/L	183 m (600 ft)/min
Service ceiling	3,660 m (12,000 ft)
T-O run	244 m (800 ft)
Landing run	274 m (900 ft)
Range with max fuel and max payload	269 nm (500 km; 310 miles)

YORK

T. R. YORK

ADDRESS:
13 Swawel Road, Birch Acres, Kempton Park 1620, Transvaal

Telephone: 976 1915

Mr York has extensively modified a Globe GC-1B Swift light aircraft (ZS-FZH), and this is now known in South Africa as the York Swift. It is shown in the accompanying illustration. Another Swift is undergoing conversion (ZS-JJF) and was scheduled to fly in its new form during 1976.

YORK (GLOBE) SWIFT

The original GC-1 Swift side-by-side two-seat cabin monoplane was developed in 1941 by the Globe Aircraft Corporation of Fort Worth, Texas, USA, and was certificated in the Spring of 1942. Of welded steel tube and wooden construction, it was powered by a 60 kW (80 hp) Continental A80 flat-four engine and had retractable main wheels and a fixed tailwheel. The all-metal GC-1B of 1945 was powered by a 93 kW (125 hp) Continental C125 engine.

The principal modifications made by Mr York to his first GC-1B involved the installation of a 108 kW (145 hp) Continental O-300-A flat-six engine, with accompanying modifications to the cowling and to the fuel, cooling and exhaust systems; modification of the cockpit instrumentation, layout and canopy; and the fitting of a non-retractable tricycle landing gear. All these modifications on ZS-FZH (described in 1975-76 *Jane's*) are being processed for certification in the USA, and both this aircraft and ZS-JJF are certifiable within the Aerobatic categories up to 776 kg (1,710 lb) AUW and up to 998 kg (2,200 lb) AUW in the Utility category.

A description of the basic GC-1B can be found under the Temco heading in the 1951-52 *Jane's*; the following details apply to the second aircraft modified by Mr York as the York Swift R GC-1B (ZS-JJF):

TYPE: Side-by-side two-seat light aircraft.

WINGS: Cantilever low-wing monoplane. NACA 23015 wing section at root, NACA 23009 at tip. Dihedral 6° from roots. Incidence 2°. Sweepback at quarter-chord 4°. Manual actuation of trailing-edge flaps. Drooped wingtips.

LANDING GEAR: Retractable nosewheel. Forward retraction of all wheels. Rubber disc shock-absorption. Small tailskid under base of fin. Pedal steering of nosewheel from both seats. Piper Cherokee-type handbrake system with vertical brake cylinders.

POWER PLANT: One 108 kW (145 hp) Rolls-Royce Continental O-300-D flat-six engine, driving a McCauley two-blade fixed-pitch propeller with spinner. Six fuel tanks in wings, total capacity approx 273 litres (60 Imp gallons). Electric fuel pump. Cessna 172 exhaust system. Downdraught cooling system. Oil capacity 7·5 litres (2 US gallons).

ACCOMMODATION: Side-by-side bucket seats for two persons in fully-enclosed heated and ventilated cabin. Original fully-transparent framed canopy replaced by a sliding canopy. Baggage compartment aft of seats.

SYSTEMS AND ELECTRONICS: Hydro-electric landing gear control with manual override. Instrumentation includes electric turn co-ordinator and vacuum gauge, installed in shock-mounted instrument panel. Cherokee-type yoke fitted to control column. Fully-modernised electrical system, with all-new equipment, including alternator. Battery box aft of baggage compartment. King KY-95 VHF radio and Bendix T-12D ADF.

DIMENSIONS, EXTERNAL:
- Wing span (approx) 9·75 m (32 ft 0 in)
- Wing chord at root 1·91 m (6 ft 3 in)
- Wing chord at tip 0·84 m (2 ft 9 in)
- Wing aspect ratio 6·9
- Length overall 6·40 m (21 ft 0 in)
- Height overall (approx) 1·73 m (5 ft 8 in)
- Tailplane span (approx) 3·66 m (12 ft 0 in)
- Wheel track (approx) 3·35 m (11 ft 0 in)
- Propeller diameter 1·93 m (6 ft 4 in)
- Propeller ground clearance 0·203 m (8 in)

WEIGHTS AND LOADINGS:
- Weight empty 544 kg (1,200 lb)
- Max T-O weight 998 kg (2,200 lb)
- Max wing loading 73·2 kg/m² (15 lb/sq ft)
- Max power loading 8·40 kg/kW (13·8 lb/hp)

PERFORMANCE (estimated, at max T-O weight):
- Max level speed at 3,050 m (10,000 ft) 156 knots (290 km/h; 180 mph)
- Max cruising speed at 3,050 m (10,000 ft) 122 knots (225 km/h; 140 mph)
- Stalling speed 45 knots (84 km/h; 52 mph)
- Max rate of climb at S/L 229 m (750 ft)/min
- Service ceiling 4,875 m (16,000 ft)
- Range at 75% power 868 nm (1,609 km; 1,000 miles)

SWEDEN

ANDREASSON
BJÖRN ANDREASSON

ADDRESS:
c/o Saab-Scania, Box 463, S-201 24, Malmö 1

Mr Andreasson has designed eleven different types of light aircraft. Of these, the BA-7 was built in series by AB Malmö Flygindustri as the MFI-9B Trainer/Militrainer and by MBB in Germany as the BO 208 C Junior (see 1970-71 *Jane's*).

An earlier design, the BA-4 biplane, was modernised by Mr Andreasson for members of the Swedish branch of the Experimental Aircraft Association, and a prototype was built by apprentices of the MFI apprentice school as part of their training programme. To distinguish it from the original BA-4, it is designated BA-4B.

Mr Andreasson's latest design is the BA-11.

ANDREASSON BA-4B

The prototype BA-4B, built by MFI apprentices, was of all-metal construction. The design provides for alternative all-wooden wings.

World manufacturing rights in the BA-4B are held by Mr P. J. C. Phillips of Down House, Cocking, Midhurst, Sussex, and the aircraft is being built in small numbers in the UK by Crosby Aviation Ltd (which see). Plans for homebuilders continue to be available from Mr Andreasson.

TYPE: Single-seat fully-aerobatic light biplane.

WINGS: Braced biplane type, with a single streamline-section interplane strut each side. A streamline-section bracing strut runs from the bottom fuselage longeron on each side to the top of the interplane strut, and an N-type cabane structure supports the centre-section. Incidence, upper wing 3°, lower wings 4°. Stagger 20°. Dihedral, upper wing 2°, lower wings 4°. Alternative all-metal structure or all-wood structure, with solid spars, covered with heavy plywood skin. Pop-riveted ailerons, of simplified sheet metal construction, on lower wings only. No flaps. Provision for fitting detachable plastics wingtips.

FUSELAGE: Sheet metal structure, with external stringers, making extensive use of pop-riveting. Turtledeck either sheet metal or reinforced plastics.

TAIL UNIT: Cantilever structure of pop-riveted sheet metal construction.

LANDING GEAR: Non-retractable tailwheel type. Cantilever spring steel main legs. Main wheels size 5·00-4 or 5·00-5. Hydraulic brakes. Steerable tailwheel carried on leaf spring.

POWER PLANT: Prototype has 74·5 kW (100 hp) Rolls-Royce Continental O-200-A flat-four engine. Provision for other engines, including Volkswagen conversions. Standard fuel tank, capacity 50 litres (11 Imp gallons), forward of cockpit. Provision for carrying external 'bullet' tank of 50 litres (11 Imp gallons) capacity under fuselage.

ACCOMMODATION: Single seat in open cockpit.

ELECTRONICS AND EQUIPMENT: Provision for battery, VHF radio and IFR instrumentation.

DIMENSIONS, EXTERNAL:
- Wing span:
  - upper 5·34 m (17 ft 7 in)
  - lower 5·14 m (16 ft 11 in)
- Wing chord (upper and lower, constant) 0·80 m (2 ft 7½ in)
- Wing area, gross 8·3 m² (90 sq ft)
- Wing aspect ratio (upper and lower) 6
- Length overall 4·60 m (15 ft 0 in)
- Tailplane span 2·00 m (6 ft 6 in)

WEIGHT:
- Max T-O weight 375 kg (827 lb)

PERFORMANCE (prototype, at max T-O weight):
- Max level speed 122 knots (225 km/h; 140 mph)
- Max cruising speed 104 knots (193 km/h; 120 mph)
- Min flying speed 35 knots (64 km/h; 40 mph)
- Max rate of climb at S/L 610 m (2,000 ft)/min
- T-O and landing run less than 100 m (330 ft)
- Range with standard fuel 152 nm (280 km; 175 miles)

ANDREASSON BA-11

The BA-11 is an all-metal biplane, intended for single-seat aerobatic, two-seat training or competition flying. It is designed generally to FAR Pt 23 Appendix A category A (aerobatic) requirements, but will have enhanced limiting load factors. These will be +9 to −6*g* as a single-seater, and in excess of +6 to −3*g* as a two-seater.

WINGS: Biplane type, braced with dual sets of streamlined tie-rods. Ailerons, of simplified pop-riveted sheet metal construction, on both upper and lower wings. Positive stagger.

FUSELAGE: Metal structure, with one-piece moulded glassfibre turtledeck.

TAIL UNIT: All-metal structure. Control surfaces of similar construction to ailerons.

LANDING GEAR: Non-retractable tailwheel type. Main legs consist of two steel leaf springs attached to bottom of fuselage. Size 5·50-5 main wheels, with hydraulic disc brakes. Tailwheel also uses leaf spring and is steerable.

POWER PLANT: Designed for one 149 kW (200 hp) Lycoming fuel-injection engine, driving a 1·88 m (6 ft 2 in) diameter Hartzell constant-speed propeller. Main fuel tank in upper front fuselage, capacity approx 60 litres (13 Imp gallons). Auxiliary fuel tank, capacity approx 50 litres (11 Imp gallons), in upper wing centre-section.

ACCOMMODATION: Two seats in tandem, each designed to accommodate a back-type parachute. Basic instrumentation only in forward cockpit. Instrument panel of rear cockpit is large enough to accommodate a limited IFR panel in addition to the normal engine instruments. Electrical equipment, including starter, alternator and battery, can be fitted.

DIMENSIONS, WEIGHTS AND PERFORMANCE:
No details received for publication

EKSTRÖM
STAFFAN W. EKSTRÖM

ADDRESS:
Tivedsvägen 1, S-181 64 Lidingö
Telephone: (08) 7663448

EKSTRÖM HUMLAN 2

Mr Ekström began the design of this single-seat autogyro in June 1971. Construction began in April 1972, and it flew for the first time in June 1973. Three more Humlans are under construction in Sweden, and the first of these was expected to fly towards the end of 1976.

ROTOR SYSTEM: Single two-blade semi-rigid rotor, attached to hub by a single bolt. Ztan Zee rotor blades. Rotor brake added 1975.

ROTOR DRIVE: Flexible shaft for rotor spin-up only. Via gearbox and two Vee-belts.

FUSELAGE: Cruciform chassis of 6061 T6 square-section aluminium tube, on which is mounted a pod-type nacelle.

TAIL UNIT: Conventional single fin and rudder, and fixed tailplane with dihedral, built of 0·4 mm and 0·8 mm aluminium sheet.

LANDING GEAR: Non-retractable tricycle type, with additional small wheel beneath tail. Rubber shock-absorption on tailwheel only. Go-kart wheels on main and nose units, tyre pressure 0·88 bars (12·8 lb/sq in). Nosewheel is steerable, self-centering, and is fitted with cycle-type brake.

POWER PLANT: One 67 kW (90 hp) McCulloch AF 100-X3 four-cylinder engine, driving a two-blade fixed-pitch pusher propeller. Fuel tank, capacity 48 litres (10·5 Imp gallons), behind pilot's seat. Fuel is a petrol/oil mixture, with 5% oil.

ACCOMMODATION: Single seat in open cockpit. One-piece curved windscreen. Shoulder harness fitted.

DIMENSIONS, EXTERNAL:
- Rotor diameter 6·80 m (22 ft 3¾ in)
- Length overall 3·42 m (11 ft 2¾ in)
- Height overall 1·98 m (6 ft 6 in)
- Width over wheels 1·65 m (5 ft 5 in)
- Propeller diameter 1·20 m (3 ft 11¼ in)

WEIGHTS AND LOADINGS:
- Weight empty 146 kg (321 lb)
- Normal max T-O weight 260 kg (573 lb)
- Max T-O weight 295 kg (650 lb)
- Normal max disc loading 7·16 kg/m² (1·47 lb/sq ft)
- Normal max power loading 4·40 kg/kW (7·22 lb/hp)

PERFORMANCE (at 260 kg; 573 lb AUW):
- Never-exceed speed 97 knots (180 km/h; 111·5 mph)
- Max cruising speed 81 knots (150 km/h; 93 mph)
- Econ cruising speed 64·5 knots (120 km/h; 74·5 mph)
- Max rate of climb at S/L 300 m (984 ft)/min
- T-O run 60 m (197 ft)
- T-O to 15 m (50 ft) 100 m (328 ft)
- Landing from 15 m (50 ft), zero wind 30 m (98 ft)
- Landing run, zero wind 5 m (16 ft)
- Max range 140 nm (260 km; 161 miles)

SWITZERLAND

BERGER
BERGER-HELIKOPTER

ADDRESS:
CH-6573 Magadino TI
Telephone: 092-64 21 71
DIRECTOR: Hans Berger

Details have now been received of the Berger BX-110 homebuilt helicopter, a brief reference to which appeared in the 1974-75 *Jane's*.

BERGER BX-110

The BX-110 (HB-YAK) is a two-seat homebuilt light helicopter, powered by a Wankel rotating-piston engine. It flew for the first time on 3 June 1974, and has been awarded a permit for experimental flying by the Swiss Board of Aviation.

TYPE: Two-seat light helicopter.

ROTOR SYSTEM: Single three-blade semi-rigid main rotor and two-blade tail rotor. Main rotor blades are of NACA 8 H 12 section, and are foldable. Max pitch of main rotor 12°. Main and tail rotors of alloy construction. No rotor brake.

ROTOR DRIVE: By toothed belt, from specially designed gearbox. Main rotor/engine rpm ratio 0·09 : 1; tail rotor/engine rpm ratio 0·6 : 1.

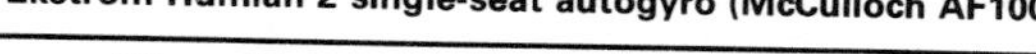

Ekström Humlan 2 single-seat autogyro (McCulloch AF100-X3 engine)

Brügger MB-2 Colibri 2 single-seat light aircraft (VW engine)

FUSELAGE: All-metal frame, of square and circular tubular alloy construction. Large 'goldfish bowl' cabin, with framed transparencies.

TAIL UNIT: Half-tailplane on starboard side of tailboom, forward of tail rotor, with approx 45° dihedral.

LANDING GEAR: Original skid-type replaced later by non-retractable tricycle wheeled gear, with main-wheel brakes.

POWER PLANT: One 97 kW (130 hp) Wankel rotating-piston engine. Fuel in two saddle tanks aft of cabin, total capacity 108 litres (23·75 Imp gallons). Refuelling point on top of each tank. Oil capacities: engine 4 litres (0·9 Imp gallons); gearbox 4 litres (0·9 Imp gallons); tail rotor 0·2 litres (0·04 Imp gallons).

ACCOMMODATION: Side-by-side seats for pilot and one passenger in 'goldfish bowl' cabin.

ELECTRICAL SYSTEM: 12V battery.

DIMENSIONS, EXTERNAL:
- Diameter of main rotor: 7·40 m (24 ft 3¼ in)
- Diameter of tail rotor: 1·20 m (3 ft 11¼ in)
- Distance between rotor centres: 4·35 m (14 ft 3¼ in)
- Length overall: 6·40 m (21 ft 0 in)
- Height to top of rotor hub: 2·52 m (8 ft 3¼ in)
- Wheel track: 1·85 m (6 ft 0¾ in)

DIMENSIONS, INTERNAL:
- Cabin: Length: 1·30 m (4 ft 3¼ in)
- Max width: 1·25 m (4 ft 1¼ in)
- Max height: 1·28 m (4 ft 2½ in)

AREAS:
- Main rotor disc: 43·00 m² (462·85 sq ft)
- Tail rotor disc: 1·13 m² (12·16 sq ft)

WEIGHTS:
- Weight empty: 460 kg (1,014 lb)
- Max T-O weight: 720 kg (1,587 lb)

PERFORMANCE (at max T-O weight):
- Never-exceed speed: 86 knots (160 km/h; 99 mph)
- Max cruising speed: 81 knots (150 km/h; 93 mph)
- Max rate of climb at S/L: 240 m (787 ft)/min

BRÜGGER
MAX BRÜGGER

ADDRESS:
1751 Villarsel-le-Gibloux, CH Fribourg
Telephone: (037) 31 16 20

Brief details of the Brügger Colibri 1 single-seat ultra-light aircraft, which flew for the first time on 30 October 1965, were given in the 1967-68 and 1971-72 *Jane's*.

A more recent design is the Colibri 2, of which a number have been built by amateur constructors.

BRÜGGER MB-2 COLIBRI 2

Mr Brügger began design of the Colibri 2 in January 1966. Construction was started a year later, and the first of two prototypes flew for the first time on 1 May 1970. By early 1975, about 65 Colibri 2s were under construction by amateur builders in Europe and at least three others were flying.

TYPE: Single-seat homebuilt light aircraft.

WINGS: Cantilever low-wing monoplane. Wing section NACA 23012. Dihedral from roots. Two-spar constant-chord wings. Wings and ailerons built of spruce with fabric covering. No flaps or tabs.

FUSELAGE: Plywood-covered wooden structure.

TAIL UNIT: Cantilever all-wood structure. Rudder only: no fin. All-moving horizontal surfaces, with Flettner-type elevators.

LANDING GEAR: Non-retractable tailwheel type, with coil spring shock-absorption on main units. Main wheels size 400 × 100, with streamline fairings. Tailwheel mounted on leaf spring. Mechanically-operated disc brakes.

POWER PLANT: One 30 kW (40 hp) 1,600 cc Volkswagen engine (Brügger modification), driving a Brügger two-blade fixed-pitch wooden propeller with plastics-coated blades. Fuel in single fuselage tank, capacity 33 litres (7·25 Imp gallons). Oil capacity 2·5 litres (0·55 Imp gallons).

ACCOMMODATION: Single seat under one-piece moulded transparent canopy, with quarter-lights to rear.

DIMENSIONS, EXTERNAL:
- Wing span: 6·00 m (19 ft 8¼ in)
- Wing chord (constant): 1·40 m (4 ft 7 in)
- Wing area, gross: 8·20 m² (88·25 sq ft)
- Length overall: 4·80 m (15 ft 9 in)
- Height overall: 1·60 m (5 ft 3 in)
- Tailplane span: 2·00 m (6 ft 6¾ in)
- Wheel track: 1·80 m (5 ft 11 in)
- Propeller diameter: 1·38 m (4 ft 6⅓ in)

WEIGHTS:
- Weight empty: 215 kg (474 lb)
- Max T-O and landing weight: 330 kg (727 lb)

PERFORMANCE (at max T-O weight):
- Max speed at 4,000 m (13,125 ft): 97 knots (180 km/h; 111 mph)
- Econ cruising speed (70% power) at 4,000 m (13,125 ft): 86 knots (160 km/h; 99 mph)
- Stalling speed: 32·5 knots (60 km/h; 37·5 mph)
- Max rate of climb at S/L: 180 m (590 ft)/min
- Service ceiling: 4,500 m (14,760 ft)
- T-O and landing run: 200 m (656 ft)
- Range with max fuel: 270 nm (500 km; 310 miles)

UNITED KINGDOM

CLUTTON-TABENOR
ERIC CLUTTON

ADDRESS:
92 Newlands Street, Shelton, Stoke-on-Trent, Staffordshire ST4 2RF

Mr E. Clutton and Mr E. Sherry designed and built a single-seat light aircraft known as FRED. Details of the original version were given in the 1967-68 *Jane's*. Since then, it has undergone considerable development, including several changes of power plant, and during 1974 the aircraft was further modified by the design and installation of folding wings. Plans of the current FRED Series 2, in folding-wing form, are available to other constructors.

Another aeroplane, the Easy Too, has been designed by Mr Clutton and Mr Tabenor to make full use of a geared Volkswagen engine that was flight-tested on FRED Series 2 and is described briefly in the Aero-Engines section.

CLUTTON-TABENOR FRED SERIES 2

FRED (Flying Runabout Experimental Design) was designed as a powered aircraft that could be flown by any reasonably experienced glider pilot without further training. Other aims were that it should be able to operate from small, rough fields and be roadable.

First flight was made on 3 November 1963, with a 20 kW (27 hp) 500 cc Triumph 5T motorcycle engine; this was replaced later by a Scott A2S engine, and later still (1966) by a converted American-built Lawrance radial engine from an APU.

Another change made after the first flights was replacement of the original bungee-in-tension landing gear shock-absorbers by steel springs.

During 1968 the Lawrance engine was replaced by a 49 kW (66 hp) Volkswagen 1,500 cc engine, modified by the provision of a toothed belt to drive the propeller at half engine speed. Early in 1970 it was running with a reduction ratio of 2 : 1, driving a 1·73 m (5 ft 8 in) American-style propeller. Ignition was by two Lucas SR4 magnetos, chain-driven from the clutch end of the crankshaft. Since then, FRED has again been re-engined, with a Franklin, as described below.

FRED is described by its builders as being virtually unstallable with power on. Stall warning is a pronounced Dutch roll. It can be rigged in 20 minutes and has been road-towed behind a motorcycle combination.

To meet numerous requests, sets of plans for FRED have been available to amateur constructors since February 1970, and at least two examples built from plans were close to completion in mid-1976.

TYPE: Single-seat amateur-built light aircraft.

WINGS: Wire-braced parasol monoplane. Wing section Göttingen 535. Thickness/chord ratio 17·2%. No dihedral or incidence. 1° washout on tips. Spruce and plywood structure, with torsion-box leading-edge, auxiliary rear spar and drag spar, fabric-covered. Non-differential ailerons. No flaps or trim tabs.

FUSELAGE: Spruce longerons. Plywood covered to rear of cockpit, except for aluminium top decking. Fabric covering on rear fuselage, except for plywood top decking, front portion of which is removable for access to baggage locker.

TAIL UNIT: Cantilever structure of spruce and plywood. No fixed fin. Tailplane incidence adjustable on ground. Pushrod-operated elevators. No tabs.

LANDING GEAR: Non-retractable main wheels and tailskid. Main units sprung with motorcycle rear suspension springs. Industrial truck wheels. Tyre pressure 1·79 bars (26 lb/sq in). No brakes.

POWER PLANT: One 37·25 kW (50 hp) Franklin 4AC-150 engine in prototype, giving performance roughly equivalent to a geared 1,600 cc Volkswagen modified motor car engine, driving a two-blade fixed-pitch propeller. Single fuel tank in centre-section, capacity 34 litres (7·5 Imp gallons). Provision for second centre-section tank. Oil capacity 3·5 litres (0·75 Imp gallons).

ACCOMMODATION: Single seat in open cockpit.

DIMENSIONS, EXTERNAL:
- Wing span: 6·86 m (22 ft 6 in)
- Wing chord (constant): 1·52 m (5 ft 0 in)
- Wing area, gross: 10·22 m² (110 sq ft)
- Wing aspect ratio: 4·4
- Length overall: 5·18 m (17 ft 0 in)
- Height overall: 1·83 m (6 ft 0 in)
- Tailplane span: 2·74 m (9 ft 0 in)
- Wheel track: 1·22 m (4 ft 0 in)
- Wheelbase: 3·20 m (10 ft 6 in)
- Propeller diameter: 1·83 m (6 ft 0 in)

WEIGHTS:
- Weight empty: 242 kg (533 lb)
- Max T-O weight: 350 kg (773 lb)

PERFORMANCE (at max T-O weight):
- Max cruising speed: 65 knots (120 km/h; 75 mph)
- Econ cruising speed: 55 knots (101 km/h; 63 mph)
- Approach speed: 45 knots (84 km/h; 52 mph)
- Stalling speed: approx 35 knots (63 km/h; 40 mph)
- Range with max fuel: 173 nm (320 km; 200 miles)

CLUTTON-TABENOR E.C.2 EASY TOO

The Easy Too design was started in 1969 to utilise the geared Volkswagen power plant developed by Mr Clutton and Mr Tabenor. It is a single-seat folding-wing aircraft, plywood-covered with a polyester resin finish, and is stressed for aerobatics. A prototype is under construction, but its completion has been delayed by other work.

In mid-1976, the Easy Too was undergoing some redesign, to enable it to qualify as an entrant for a PFA competition, on the lines of the Light Aeroplane Competitions organised at Lympne in the 1920s.

Berger BX-110 two-seat helicopter, described on page 478

FRED series 2 with 1,500 cc Volkswagen engine *(Ron Moulton)*

The general appearance of the Easy Too can be seen in the accompanying three-view drawing. The prototype is to have the new geared 1,500 cc Volkswagen engine, but the aircraft is equally suited to a direct-drive VW engine. The outer wing panels can be folded back by one person, by withdrawing pins and replacing them by an irreversible screwjack arrangement which locks them in position. The ailerons and flaps are coupled automatically. The folding hinge and support are entirely separate from the flying fittings, and the aeroplane can be towed on the road behind a motor car. The wingtips, of glassfibre, are of similar type and size to those fitted to the Druine Turbulent and the Taylor Monoplane. A one-piece sliding cockpit canopy is fitted. Wing span of Easy Too is 7·11 m (23 ft 4 in) and the estimated empty weight is 220 kg (485 lb). It is intended to make plans of the aircraft available after the successful conclusion of flight testing.

COATES

J. R. COATES

ADDRESS:

The Spinney, Breachwood Green, Hitchin, Hertfordshire SG4 8PL

Mr Coates designed and built a two-seat light aircraft known as the S.A.II Swalesong. Drawings for amateur construction are not available, but a simplified version, the S.A.III, is under development and will be suitable for homebuilding.

The prototype S.A.III is being built in all-wood form. Following flight development, some parts may be changed to metal or structural foam, the latter being preferred as it enables weight reductions to be achieved and allows engines like the Volkswagen to be considered as adequate power sources.

COATES S.A.II SWALESONG

The Swalesong was designed for sporting and touring purposes, with an emphasis on a good short-field performance, to enable the aircraft to operate from unprepared surfaces. It is stressed to +3½g and −2½g. The prototype (G-AYDV) made its first flight on 2 September 1973 and a Special Category C of A was issued for this aircraft on 4 December 1974.

A full description appeared in the 1975-76 *Jane's*.

TYPE: Two-seat light aircraft.

POWER PLANT: One 67 kW (90 hp) Continental C90 flat-four engine, driving a two-blade wooden fixed-pitch propeller. Fuel in main tank in nose, capacity 64 litres (14 Imp gallons), with 45 litre (10 Imp gallon) reserve tank aft of seats.

DIMENSIONS, EXTERNAL:

Wing span	8·05 m (26 ft 5 in)
Wing chord at root	1·45 m (4 ft 9 in)
Wing chord at tip	1·22 m (4 ft 0 in)
Wing area, gross	11·15 m² (120 sq ft)
Length overall	5·79 m (19 ft 0 in)
Height overall	2·21 m (7 ft 3 in)
Wheel track	1·98 m (6 ft 6 in)
Wheelbase	1·22 m (4 ft 0 in)

WEIGHTS:

Weight empty	331 kg (730 lb)
Max T-O weight	547 kg (1,207 lb)

PERFORMANCE (at max T-O weight):

Max level speed at 305 m (1,000 ft)	113 knots (209 km/h; 130 mph)
Max cruising speed at 305 m (1,000 ft)	100 knots (185 km/h; 115 mph)
Econ cruising speed at 305 m (1,000 ft)	82·5 knots (153 km/h; 95 mph)
Stalling speed, flaps down	43 knots (79 km/h; 49 mph)
Max rate of climb at S/L	244 m (800 ft)/min
T-O run	137 m (450 ft)
Landing from 15 m (50 ft)	274 m (900 ft)
Landing run	92 m (300 ft)
Range with max fuel	390 nm (724 km; 450 miles)

COATES S.A.III SWALESONG

This aircraft (G-BAAH) is a development of the S.A.II, and the best features of that aircraft have been maintained. Changes include a fuselage of more rounded form, a simplified wing of constant chord, and a vertical tail unit of reduced height. All control surfaces are mass and aerodynamically balanced.

First flight was scheduled to take place in 1976.

TYPE: Two-seat light aircraft.

WINGS: Cantilever low-wing monoplane. Wing section NACA 63415. Dihedral 4°. Incidence 1° 30′. All-wood (spruce) structure with plywood covering, built in three pieces. All-wood slotted ailerons. Slotted all-wood flaps.

FUSELAGE: Semi-monocoque spruce structure with plywood covering.

TAIL UNIT: Cantilever structure, with sweptback vertical surfaces. Tailplane incidence adjustable on ground. One-piece fabric-covered wooden elevator, with tab.

LANDING GEAR: Non-retractable tricycle type. Cantilever light alloy main legs. Steerable nosewheel, with motor car shock-absorber and glassfibre fairing. Size 5·00-5 tyres on main wheels; 130-1300 tyre on nosewheel. Disc brakes.

POWER PLANT: One 63·4-80·5 kW (85-108 hp) Continental or Lycoming flat-four engine, driving a fixed-pitch wooden propeller. Fuel capacity 72·7 litres (16 Imp gallons).

ACCOMMODATION: Two seats side by side in enclosed cockpit, with sliding one-piece hood. Baggage space aft of seats. Cockpit heated and ventilated.

DIMENSIONS, EXTERNAL:

Wing span	7·62 m (25 ft 0 in)
Wing chord (constant)	1·37 m (4 ft 6 in)
Wing area, gross	10·41 m² (112 sq ft)
Wing aspect ratio	5·6
Length overall	5·64 m (18 ft 6 in)
Width, wings folded	1·73 m (5 ft 8 in)
Height overall	2·08 m (6 ft 10 in)
Tailplane span	2·44 m (8 ft 0 in)
Wheel track	1·68 m (5 ft 6 in)
Wheelbase	1·30 m (4 ft 3 in)
Propeller diameter	1·63 m (5 ft 4 in)

DIMENSIONS, INTERNAL:

Cockpit: Max width	1·07 m (3 ft 6 in)

WEIGHTS AND LOADINGS:

Weight empty	317 kg (700 lb)
Max T-O and landing weight	544 kg (1,200 lb)
Max wing loading	48·8 kg/m² (10 lb/sq ft)
Max power loading	7·30 kg/kW (12 lb/hp)

PERFORMANCE (estimated):

Never-exceed speed at 305 m (1,000 ft)	165 knots (305 km/h; 190 mph)
Max level speed at 305 m (1,000 ft)	130 knots (241 km/h; 150 mph)
Max cruising speed at 305 m (1,000 ft)	113 knots (209 km/h; 130 mph)
Econ cruising speed at 305 m (1,000 ft)	95·5 knots (177 km/h; 110 mph)
Stalling speed, flaps down	44 knots (81 km/h; 50 mph)
Max rate of climb at S/L	260 m (850 ft)/min

CRANWELL

RAF CRANWELL MAN-POWERED AIRCRAFT GROUP

ADDRESS:

Electrical Engineering Wing, RAF College, Cranwell, Sleaford, Lincolnshire NG34 8HB

Telephone: Cranwell (040 06) 201, extn T231

OFFICER IN CHARGE OF PROJECT:

Sqn Ldr J. Potter, MA, MSc, MRAeS, RAF

The efforts of the Man-Powered Aircraft Group at RAF Cranwell in 1975 were concentrated upon displaying the Group's aircraft during a Royal visit and during an Open University engineering programme. Consequently, there is little to report of flight progress with the Jupiter. However, experience with the Mercury aircraft has confirmed the superiority of the Jupiter concept in terms of controllability and practicability, and the many new designs in the UK, USA and Japan built to a similar concept demonstrate a widespread confidence in this design philosophy.

CRANWELL (HALTON) JUPITER

Under the direction of Sqn Ldr (then Flt Lt) J. Potter, staff and apprentices at No. 1 School of Technical Training, RAF Halton, built a single-seat man-powered aircraft named Jupiter. The original aircraft was designed in 1960 by Mr C.H. Roper of Woodford, Essex; but in 1969, when nearly completed, it was seriously damaged by fire. The remains were acquired in September 1970 by the Halton group, who rebuilt it completely.

The Jupiter made its first flight, of approx 183 m (200 yd), at RAF Benson on 9 February 1972, piloted by Flt Lt Potter. Subsequent flights included several in 1972 of more than half a mile, as noted in earlier editions of *Jane's*. On 29 June 1972 the Jupiter, flown by Flt Lt Potter, set new British national records for distance and duration by a man-powered aircraft. (These records have not yet been ratified by the FAI, whose rules relating to man-powered aircraft are still under review.) The flight lasted for 1 min 47·4 sec and covered a distance of 1,071 m (1,171 yd). The previous record distance, of 908 m (993 yd), was set up in the HMPAC Puffin in 1962.

In 1972 an electrical attitude indicator was fitted, which assisted precise pitch control and helped the pilot to fly steadily at the minimum power cruise conditions.

A full description of the Jupiter appeared in the 1972-73 and 1973-74 *Jane's*. Work on the programme was suspended between August 1972 and November 1973, but 19 flights were made during 1974.

At Cranwell in 1974 Sqn Ldr Potter formed a new man-powered aircraft group, and shortly afterwards acquired the Weybridge man-powered aircraft. This aircraft has now been renamed Mercury.

CRANWELL (WEYBRIDGE) MERCURY

Design of the Mercury single-seat man-powered aircraft was started in July 1967 by the Weybridge Man Powered Aircraft Group, with financial assistance from the Royal Aeronautical Society and in premises provided by BAC. Construction began in June 1968. The aircraft was completed at the end of 1970 and flew for the first time at Weybridge on 18 September 1971, piloted by Mr Christopher Lovell of the Surrey Gliding Club; it covered about 46 m (50 yd) at a height of approximately 0·9 m (3 ft). Following the first flight, modifications were made to the rudder and transmission system to improve control and reliability.

In April 1974 the aircraft was acquired on loan by the RAF Cranwell Man-Powered Aircraft Group, which renamed it Mercury. After a 900 man-hour extensive renovation and reskinning programme, two flights were made in 1974. These indicated a need for modifications to the propeller and to pilot position, and for a reassessment of the original cruise power requirement.

The cruise power requirement has now been estimated at 0·7 to 0·75 hp; although transmission efficiency is good (93%, static), the wings generate higher drag than had been calculated from the Wortmann characteristics. This can be attributed to the poor profile of the leading-edge section, inherent in the original design, and clearly demon-

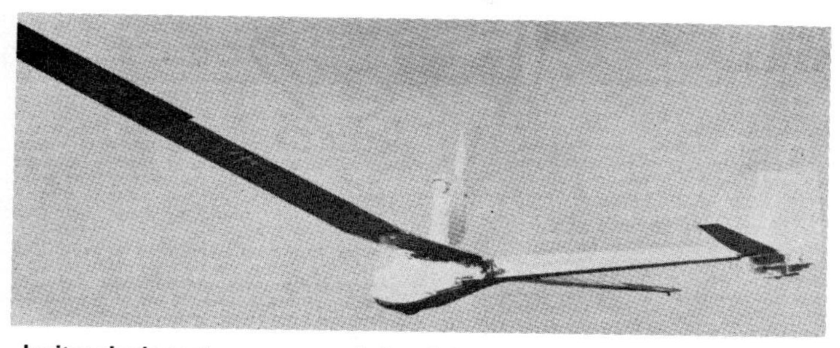

Jupiter single-seat man-powered aircraft built by apprentices at RAF Halton

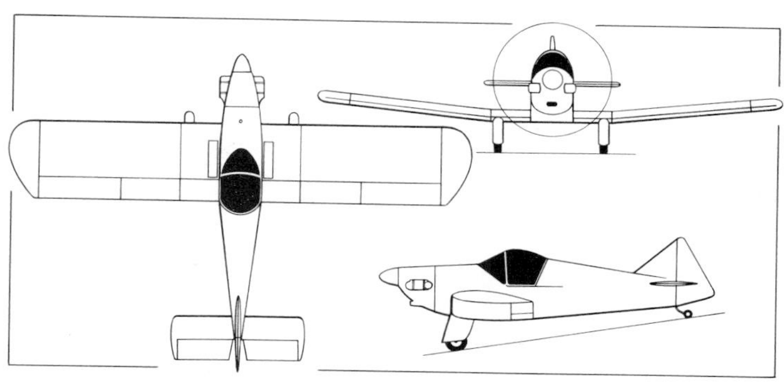

E.C.2 Easy Too, designed by Eric Clutton *(Pilot Press)*

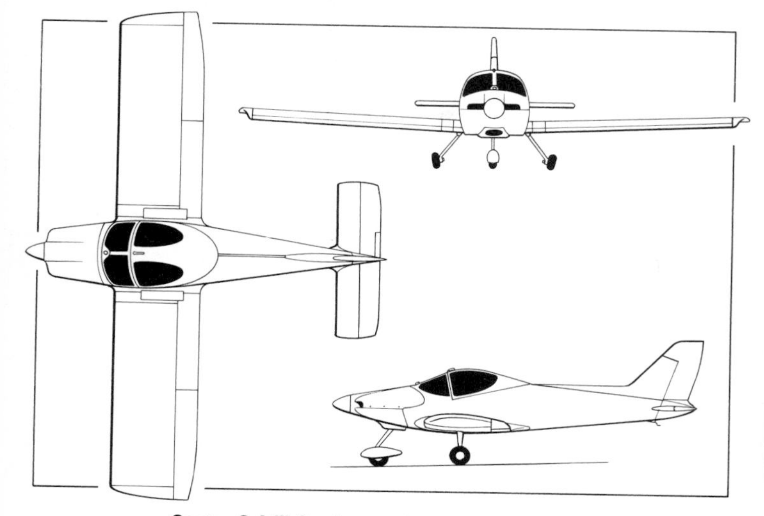

Coates S.A.III Swalesong *(Michael A. Badrocke)*

Coates S.A.II Swalesong two-seat light aircraft *(Air Portraits)*

strates the importance of a structural technique which reproduces aerofoil co-ordinates accurately. Handling difficulties are also aggravated by the aft CG position (approx 45% of mean chord) built into the original aircraft. With regard to the efficacy of the differential wing control (in lieu of ailerons), theoretical studies (Potter 1973) predicted the dominance of the secondary yaw effects over the intended primary roll effects. Such predictions were confirmed in practice, when highly unconventional control techniques were found necessary in order to maintain level flight.

Although these stability and control problems may be surmounted, the conclusions of Project Mercury favour a concentration of effort into a Jupiter style of design. Although the long-span configuration offers a potential reduction in power requirements, this advantage is offset by problems of storage and ground handling. In 1976 it seemed likely that Mercury would be dismantled to make way for a smaller successor.

A full description of the aircraft in its 1974 configuration can be found in the 1975-76 *Jane's*.

TYPE: Single-seat man-powered aircraft.

WINGS: Cantilever low-wing monoplane. Wortmann wing sections: FX-68-M-180 at root, FX-68-M-160 at mid-span, FX-68-M-140 at tip. Dihedral 11° 12′ at roots. Incidence 3° 18′. Differential wing incidence change instead of ailerons.

FUSELAGE: Basic framework of L63 aluminium alloy tubing, with balsa frames and stringers and Melinex covering.

TAIL UNIT: Plywood and spruce box-spar structure, with balsa ribs and leading- and trailing-edges and Melinex covering.

LANDING GEAR: Non-retractable tandem arrangement, using standard bicycle wheels (size 18 in × 1⅜ in) and brakes. No shock-absorbers. Front wheel is chain-driven from the pedal axle.

POWER PLANT: Man power on bicycle pedals, driving a two-blade laminated balsa pusher propeller aft of the tail unit.

ACCOMMODATION: Pilot only, in enclosed cockpit.

DIMENSIONS, EXTERNAL:
- Wing span: 36·68 m (120 ft 4 in)
- Wing aspect ratio: 30
- Length overall: 6·40 m (21 ft 0 in)
- Height overall: 3·51 m (11 ft 6 in)
- Propeller diameter: 2·13 m (7 ft 0 in)

AREA:
- Wings, gross: 45·06 m² (485·0 sq ft)

WEIGHTS AND LOADINGS:
- Weight empty: 81 kg (178 lb)
- Max T-O weight: 154 kg (340 lb)
- Max wing loading: 3·42 kg/m² (0·70 lb/sq ft)
- Max power loading: 193 kg/hp (425 lb/hp)

PERFORMANCE (approx):
- Cruising speed: 15·3 knots (28·3 km/h; 17·6 mph)

CROSBY
CROSBY AVIATION LTD

ADDRESS:
Archery House, Leycester Road, Knutsford, Cheshire
Telephone: 0565-4254

DIRECTORS:
John Crosby
P. J. C. Phillips

CROSBY (ANDREASSON) BA-4B

The Andreasson BA-4B single-seat biplane, described in the Swedish section, is being produced in the UK by Crosby Aviation. Mr P. J. C. Phillips holds the world rights for commercially-manufactured examples of this aircraft, and has vested these rights in Crosby Aviation, of which he is a director. Crosby also markets plans and kits for amateur constructors wishing to build their own aircraft; plans (only) can be obtained from the BA-4B's Swedish designer, Mr Björn Andreasson.

Three versions are available from Crosby Aviation:

BA-4B. Basic model with 74·5 kW (100 hp) Rolls-Royce Continental O-200-A flat-four engine. Standard fuel capacity 56·3 litres (12·4 Imp gallons). Provision for carrying an external 'bullet' tank of 50 litres (11 Imp gallons) capacity.

Super BA-4B. Structurally identical to BA-4B, but powered by a 97 kW (130 hp) Rolls-Royce Continental O-240-A flat-four engine.

Super BA-4B Srs 2. Identical to Super BA-4B except for having modified fuel and oil systems for inverted flight.

One example of each of the three versions had been completed by 1975, when two more aircraft were under construction. The Super BA-4B is able to tow a modern two-seat sailplane to 610 m (2,000 ft) in 4½ minutes.

The description of the BA-4B given in the Swedish section applies also to the aircraft produced in the UK, except in the following details:

ELECTRONICS AND EQUIPMENT: Aircraft can be equipped with electric starter for engine, engine-driven alternator, battery, 360-channel VHF com radio, cabin heater, disc brakes, fully-castoring tailwheel, corrosion proofing and *g*-meter.

DIMENSIONS, EXTERNAL (A: BA-4B; B: Super BA-4B):
- Wing span: 5·64 m (18 ft 6 in)
- Wing area, gross: 8·36 m² (90 sq ft)
- Length overall:
  - A: 4·67 m (15 ft 4 in)
  - B: 4·72 m (15 ft 6 in)

WEIGHTS AND LOADINGS (A: BA-4B; B: Super BA-4B):
- Weight empty, basic:
  - A: 295 kg (650 lb)
  - B: 304 kg (670 lb)
- T-O weight, aerobatic: 451 kg (996 lb)
- Max T-O weight: 460 kg (1,014 lb)
- Wing loading, aerobatic: 54·0 kg/m² (11·06 lb/sq ft)
- Max wing loading: 54·7 kg/m² (11·2 lb/sq ft)
- Max power loading:
  - A: 6·17 kg/kW (10·14 lb/hp)
  - B: 4·74 kg/kW (7·8 lb/hp)

PERFORMANCE (at max T-O weight. A: BA-4B; B: Super BA-4B):
- Never-exceed speed:
  - A, B: 161 knots (299 km/h; 186 mph)
- Max level speed at S/L:
  - A: 130 knots (241 km/h; 150 mph)
  - B: 139 knots (257 km/h; 160 mph)
- Max cruising speed, 75% power at 2,135 m (7,000 ft):
  - A: 117 knots (217 km/h; 135 mph)
  - B: 126 knots (233 km/h; 145 mph)
- Stalling speed, power off:
  - A,B: 51 knots (94 km/h; 58 mph)
- Stalling speed, power on:
  - A, B: 39 knots (73 km/h; 45 mph)
- Max rate of climb at S/L:
  - A: 366 m (1,200 ft)/min
  - B: 549 m (1,800 ft)/min
- Range, standard fuel, 75% power:
  - A: 282 nm (523 km; 325 miles)
  - B: 273 nm (507 km; 315 miles)
- Range, max optional fuel, 75% power:
  - A: 529 nm (981 km; 610 miles)
  - B: 521 nm (965 km; 600 miles)

HPA
HERTFORDSHIRE PEDAL AERONAUTS

ADDRESS:
48 Orchard Drive, Park Street, St Albans, Hertfordshire AL2 2QG
Telephone: Park Street 72486

OFFICERS:
M. S. Pressnell, BSc, CEng, MRAeS (Chairman)
P. R. Sladden, BSc (Vice-Chairman)
R. E. Harris, BSc (Treasurer)
P. L. Jones, BSc, CEng, MRAeS (Hon Secretary)

The Hertfordshire Pedal Aeronauts group was formed in September 1965, mainly from engineers of Handley Page Ltd, to design and build a man-powered aircraft to compete for the Kremer prizes. With the aid of a grant from the Royal Aeronautical Society, construction of the group's first aircraft was completed in mid-1972. This aircraft (named Toucan, the pun being deliberate) made its first flight at Radlett on 23 December 1972. Its longest flight, of 640 m (2,100 ft), was made on 3 July 1973.

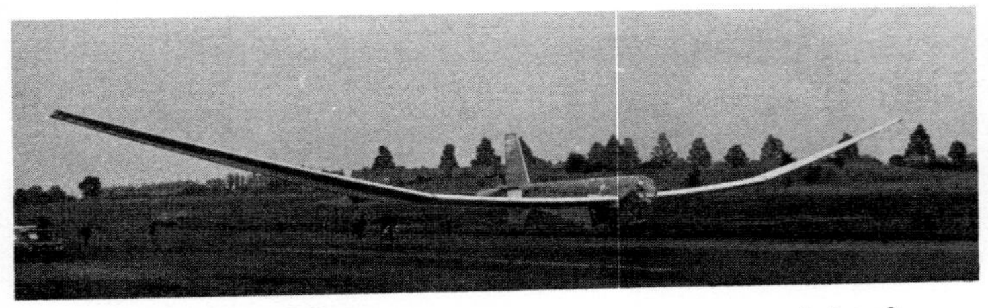

HPA Toucan Mk 2 extended-span two-seat man powered aircraft

Cranwell (formerly Weybridge) Mercury single-seat man-powered aircraft *(Royal Air Force)*

This original version was the largest man-powered aircraft then to have flown, and also the first two-man-powered aircraft to fly. A description can be found in the 1974-75 *Jane's*.

HPA TOUCAN Mk 2

This is a modified version of the Toucan Mk 1 (1974-75 *Jane's*), with a 4·88 m (16 ft 0 in) greater wing span. Early flight testing began in October and November 1975, the best flight being one of 261 m (855 ft). The aircraft was then damaged when an outrigger wheel struck the ground heavily during a landing, causing a torsional drag failure of the inboard section of the starboard wing. Repairs were in hand in the Spring of 1976, and it was planned to resume flight testing in the latter part of that year.

TYPE: Two-seat man-powered aircraft.

WINGS: Cantilever mid-wing monoplane. Wing section NACA 63_3618. Dihedral 6° on outer panels. Incidence 8°. Spars have spruce booms and plywood webs. Balsa ribs. Leading-edges and other forward areas of wing are covered with expanded polystyrene. Melinex plastics film covering overall. Slot-lip ailerons, of similar construction to wings. No flaps. Yaw control by extreme movement of ailerons.

FUSELAGE: Braced structure of spruce and balsa, with Melinex plastics film covering. Crew supported on 30SWG L72 light alloy box framework.

TAIL UNIT: Cantilever structure of Melinex-covered balsa. One-piece non-reversible trimming tailplane, cable-operated through a spiral cam, for pitch control. Fin extends above and below fuselage. No rudder.

LANDING GEAR: Main wheel (with Raleigh RSW wheel and 16 × 2 in diameter balloon tyre, pressure 4·14 bars; 60 lb/sq in) and nosewheel on fuselage centreline, in tandem; outrigger wheels under wings, at ends of flat centre-section. Wheels do not retract, but are recessed into fuselage. Twin lightweight bicycle-type brakes.

POWER PLANT: Two-man crew provide, by pedalling, power via an alloy chain to a road wheel and a shaft to a two-blade fixed-pitch pusher propeller mounted aft of the tail unit. Propeller rpm 180 at cruise.

ACCOMMODATION: Pilot (on rear seat) and second crewman, in tandem, under transparent lift-off canopy. Aperture in front fuselage fairing for breathing and cooling air.

INSTRUMENTS: Ultra-low-speed (0-26 knots; 0-48 km/h; 0-30 mph) ASI and yaw meter.

DIMENSIONS, EXTERNAL:
- Wing span: 42·37 m (139 ft 0 in)
- Wing chord at root: 1·83 m (6 ft 0 in)
- Wing chord at tip: 0·73 m (2 ft 4¾ in)
- Wing aspect ratio: 27·76
- Length overall: 8·74 m (28 ft 8 in)
- Height overall: 4·11 m (13 ft 6 in)
- Tailplane span: 5·49 m (18 ft 0 in)
- Wheelbase: 1·22 m (4 ft 0 in)
- Propeller diameter: 3·05 m (10 ft 0 in)

AREAS:
- Wings, gross: 64·7 m² (696·0 sq ft)
- Fin (total): 4·46 m² (48·0 sq ft)
- Tailplane: 5·02 m² (54·0 sq ft)

WEIGHTS AND LOADING:
- Weight empty: 109 kg (240 lb)
- Max T-O weight: 249 kg (550 lb)
- Max wing loading: 3·86 kg/m² (0·79 lb/sq ft)

PERFORMANCE (estimated):
- Never-exceed speed: 29·5 knots (54·5 km/h; 34 mph)
- Min cruising speed at 0-1·5 m (0-5 ft): 17·5 knots (32 km/h; 20 mph)
- Measured T-O run on level ground: 74·4 m (244 ft)
- Landing run: approx 46 m (150 ft)

ISAACS
JOHN O. ISAACS

ADDRESS:
42 Landguard Road, Southampton, Hampshire SO1 5DP
Telephone: Southampton 25853

Mr Isaacs designed and built a single-seat light aircraft, the airframe of which is basically a $^7/_{10}$th scale wooden version of that of the Hawker Fury fighter of the 1930s. Constructional drawings are available to amateur builders.

He has also designed and built an all-wood scaled-down version of the Supermarine Spitfire single-seat fighter of the second World War.

ISAACS FURY II

Design of the Isaacs Fury was started in January 1961 and construction of the aircraft began in April 1961. It flew for the first time on 30 August 1963, powered by a 48·5 kW (65 hp) Walter Mikron engine (see 1965-66 *Jane's*).

In 1966-67 Mr Isaacs modified the Fury prototype to Mk II standard, by re-stressing the airframe and installing a 93 kW (125 hp) Lycoming O-290 engine, and flew the aircraft in this form in the Summer of 1967. It was acquired subsequently by Mr W. Raper of Wrotham, Kent, who made further refinements, including the addition of blister fairings over the engine cylinders. It is now owned by Mr D. Toms, and is based at Land's End airfield.

The aircraft illustrated was built by Mr A. V. Francis of Dunstable, Bedfordshire, and was the fourth Isaacs Fury II to fly, on 12 October 1975. Examples built and flown in Canada and New Zealand were illustrated in the 1973-74 and 1975-76 editions of *Jane's* respectively.

Several other Furies are under construction in the UK; three are known to be under construction in New Zealand; one in Jersey; and others in the USA and Canada.

TYPE: Single-seat ultra-light biplane, stressed to 9*g* for aerobatics.

WINGS: Staggered biplane, with N type interplane strut each side and two N strut assemblies supporting centre-section of top wing above fuselage. Conventional wire bracing. Wing section RAF 28. Thickness/chord ratio 9·75%. Dihedral 1° on top wing, 3° 30′ on bottom wings. Incidence 3° 20′ on top wing, 3° 50′ on bottom wings. Spruce 'plank' spars and Warren girder ribs, with fabric covering. Fabric-covered spruce ailerons on top wing only. No flaps.

FUSELAGE: Spruce structure, covered with birch plywood.

TAIL UNIT: Strut-braced spruce structure of 'plank' spars and girder ribs, fabric-covered. Ground-adjustable tab in port elevator.

LANDING GEAR: Non-retractable type, with tailskid. Cross-axle tied to Vees with rubber-cord shock-absorption. Main wheels consist of WM.2 35·5 cm (14 in) rims spoked to home-made hubs. Dunlop tyre, size 3·25-14, pressure approx 2·28 bars (33 lb/sq in). Brakes optional.

POWER PLANT (prototype): One 93 kW (125 hp) Lycoming O-290 flat-four engine. Two-blade fixed-pitch propeller. Fuel tank in fuselage, aft of fireproof bulkhead, capacity 45·5 litres (10 Imp gallons) or 54·5 litres (12 Imp gallons).

ACCOMMODATION: Single seat in open cockpit. Small door above top longeron on port side opens downward. Space for light baggage aft of seat. Radio optional.

DIMENSIONS, EXTERNAL:
- Wing span:
  - upper: 6·40 m (21 ft 0 in)
  - lower: 5·54 m (18 ft 2 in)
- Wing chord (both, constant): 1·07 m (3 ft 6 in)
- Wing area (total): 11·50 m² (123·8 sq ft)
- Wing aspect ratio (upper): 6
- Length overall: 5·87 m (19 ft 3 in)
- Height over tail (flying attitude): 2·16 m (7 ft 1 in)
- Tailplane span: 2·13 m (7 ft 0 in)
- Wheel track: 1·27 m (4 ft 2 in)

WEIGHTS AND LOADINGS (93 kW; 125 hp Lycoming):
- Weight empty: 322 kg (710 lb)
- Max permissible T-O weight: 450 kg (1,000 lb)
- Max wing loading: 39·3 kg/m² (8·05 lb/sq ft)
- Max power loading: 4·83 kg/kW (8·00 lb/hp)

PERFORMANCE (with uncowled 93kW; 125 hp engine):
- Max level speed: 100 knots (185 km/h; 115 mph)
- Stalling speed: 33 knots (61 km/h; 38 mph)
- Max rate of climb at S/L: 488 m (1,600 ft)/min

ISAACS SPITFIRE

Construction of this prototype $^6/_{10}$-scale Spitfire (G-BBJI) began in the Summer of 1969, and it flew for the first time on 5 May 1975. The airframe is stressed to meet the aerobatic requirements of +9*g* and −4·5*g* (factored) as laid down in BCAR.

TYPE: Single-seat homebuilt sporting aircraft.

WINGS: Cantilever low-wing monoplane of semi-elliptical planform. Wing section NACA 2200 series. Thickness/chord ratio 13·2% at root, 6% at tip. Dihedral 6°. Incidence 2° at root, −30′ at tip. Two-spar wing built in one piece, mainly of spruce, with birch plywood covering, except for ailerons which are fabric-covered.

FUSELAGE: Spruce structure, covered with birch plywood.

TAIL UNIT: Cantilever structure of plywood-covered spruce.

LANDING GEAR: Non-retractable tailwheel type on prototype. Cantilever main legs. Dunlop 5·00-5 tyres, wheels and hydraulic disc brakes.

POWER PLANT: One 74·5 kW (100 hp) Continental O-200 flat-four engine, or alternative engine in same category. Two-blade ground-adjustable Ratier metal propeller. Fuel tank in fuselage, aft of fireproof bulkhead, capacity 45·5 litres (10 Imp gallons).

ACCOMMODATION: Single seat under blister-type transparent canopy. Space for light baggage aft of seat.

DIMENSIONS, EXTERNAL:
- Wing span: 6·75 m (22 ft 1½ in)
- Wing chord at root: 1·52 m (5 ft 0 in)
- Wing area, gross: 8·08 m² (87 sq ft)
- Length overall: 5·88 m (19 ft 3 in)
- Height overall: 1·73 m (5 ft 8 in)
- Tailplane span: 1·92 m (6 ft 3½ in)
- Wheel track: 1·80 m (5 ft 11 in)

WEIGHTS AND LOADINGS:
- Weight empty: 366 kg (805 lb)
- Max T-O weight: 499 kg (1,100 lb)
- Max wing loading: 62 kg/m² (12·64 lb/sq ft)
- Max power loading: 6·70 kg/kW (11 lb/hp)

PERFORMANCE (at max T-O weight):
- Max level speed: 130 knots (240 km/h; 150 mph)
- Cruising speed: 116 knots (215 km/h; 134 mph)
- Stalling speed, clean: 45-47 knots (84-87 km/h; 52-54 mph)
- Stalling speed, with optional fuselage airbrake extended: 41 knots (76 km/h; 47 mph)
- Max rate of climb at S/L: 336 m (1,100 ft)

KNOWLES
GP CAPT A. S. KNOWLES

ADDRESS:
17 East Meads, Guildford, Surrey GU2 5SW
Telephone: Guildford (0483) 4242

As reported in previous editions of *Jane's*, under the heading of the now-defunct Phoenix Aircraft Ltd, Gp Capt Knowles was responsible for developing a variant of the Phoenix Luton Minor with side-by-side seating. This aircraft is named Duet.

KNOWLES DUET

The Duet, known originally as the Minor III, is a version of the Luton L.A.4a Minor (see description under PFA heading in this section) with side-by-side seating for two persons, and was developed originally by Gp Capt A. S. Knowles in association with the former Phoenix Aircraft Ltd. Design began in 1968 and the prototype Duet (registration G-AYTT) flew for the first time on 22 June 1973. A special category C of A was granted on 24 June 1974. Although only the prototype Duet had flown by mid-1976, at least one other is under construction.

TYPE: Two-seat light aircraft.

WINGS: Strut-braced parasol monoplane. Fabric-covered wooden wings, braced with steel tube struts, as described for L.A.4a.

FUSELAGE: Wood- and fabric-covered wooden structure, as L.A.4a except for widened cockpit.

TAIL UNIT: Cantilever fabric-covered wooden structure, as L.A.4a. Plain tab on elevator.

LANDING GEAR: Non-retractable tailwheel type, as L.A.4a, with rubber-block shock-absorption. Main-wheel tyres size 6 in × 6 in, pressure 0·83 bars (12 lb/sq in). Goodyear caliper brakes. Solid tailwheel.

POWER PLANT: One 71 kW (95 hp) Continental C90 flat-four engine, driving a two-blade fixed-pitch propeller. Single fuel tank in upper forward fuselage, capacity 63·6 litres (14 Imp gallons). Refuelling point above tank. Oil capacity 4·5 litres (1 Imp gallon).

ACCOMMODATION: Side-by-side seats for two persons, with dual controls, in open cockpit under wing centre-section. Baggage locker aft of seats.

ELECTRONICS AND EQUIPMENT: Venturi-driven full blind-flying panel. Avionic Systems (Heathrow) Ltd 360-channel VHF radio.

DIMENSIONS, EXTERNAL:

Wing span	8·18 m (26 ft 10 in)
Wing chord (constant)	1·60 m (5 ft 3 in)
Wing aspect ratio	5·1
Wing area, gross	13·01 m² (140·0 sq ft)
Length overall	6·55 m (21 ft 6 in)
Height overall	2·01 m (6 ft 7 in)
Tailplane span	2·44 m (8 ft 0 in)
Wheel track	1·68 m (5 ft 6 in)
Wheelbase	4·50 m (14 ft 9 in)
Propeller diameter	1·75 m (5 ft 9 in)

DIMENSIONS, INTERNAL:

Cockpit: Max width	1·07 m (3 ft 6 in)
Baggage compartment volume	0·06 m³ (2·0 cu ft)

WEIGHTS AND LOADINGS:

Max payload	170 kg (375 lb)
Max T-O and landing weight	544 kg (1,200 lb)
Max wing loading	41·5 kg/m² (8·5 lb/sq ft)
Max power loading	7·66 kg/kW (12·63 lb/hp)

PERFORMANCE (at max T-O weight):

Never-exceed speed	130 knots (240 km/h; 149 mph)
Max cruising speed at S/L	85 knots (157 km/h; 98 mph)
Econ cruising speed at S/L	70 knots (130 km/h; 81 mph)
Stalling speed	35 knots (66 km/h; 41 mph)
Max rate of climb at S/L	134 m (440 ft)/min
Service ceiling	2,440 m (8,000 ft)
T-O run	128 m (420 ft)
Landing run	110 m (360 ft)
Max range with 9·1 litres (2 Imp gallons) reserve fuel	185 nm (342 km; 213 miles)

LEISURE SPORT

LEISURE SPORT LTD (Member Company of the Ready Mixed Concrete Ltd Group)

ADDRESS: Eastley End House, Coldharbour Lane, Thorpe, Egham, Surrey TW20 8TD

Telephone: Chertsey (09328) 64142

LEISURE SPORT S.5 REPLICA

This company has completed a full-size reproduction of the Supermarine S.5 Schneider Trophy seaplane of 1927. Construction took place at Thruxton, Wiltshire, over a 2½ year period, and the aircraft (registered G-BDFF) began preliminary water taxying tests at Water Park, Thorpe, Surrey, on 13 June 1975. The first flight was made on 28 August 1975 and lasted 14 minutes. General impressions were that the S.5 replica was more pleasant to handle than expected; apart from a very light elevator and stiff ailerons, the controls were effective and well harmonised from the start.

The aircraft is not a true replica, differing from the original S.5 principally in its power plant, which is a 156·5 kW (210 hp) Rolls-Royce Continental engine, driving a two-blade variable-pitch propeller. Construction of the reproduction S.5 is of wood; other detail differences from the original include a different wing section and a slightly wider fuselage.

Leisure Sport planned to have its second replica, a Fokker Dr I triplane, airborne in August 1976, with a Sopwith Camel to follow. A Hawker Nimrod biplane and a Supermarine S.6B may be built at a later date.

MAIN DIMENSIONS: as for original aircraft

WEIGHTS:

Weight empty	about 499 kg (1,100 lb)
Max T-O weight	681 kg (1,500 lb)

PERFORMANCE:

Max level speed	approx 150 knots (278 km/h; 173 mph)
Max cruising speed (75% power)	approx 120 knots (222 km/h; 138 mph)
T-O speed	approx 60 knots (111 km/h; 69 mph)
Stalling speed	55 knots (102 km/h; 63·5 mph)
Max rate of climb at S/L	396 m (1,300 ft)/min
Endurance (average)	2¼ hr

LIVESEY

DAVID M. LIVESEY

ADDRESS:

'Rawhiti', 12 Kenwood Drive, Burwood Park, Walton on Thames, Surrey KT12 5AU

LIVESEY D.L.5

The D.L.5 has been designed for amateur construction by Mr David Livesey, who had begun construction of a prototype by the Spring of 1974. Design was started in 1971, to a 'minimum aeroplane' concept that could be easily accomplished by unskilled constructors, using easily obtainable materials and a minimum of basic tools.

Progress during the past two years, although slow, has included the design of an enclosed cockpit which will be offered optionally.

TYPE: Single-seat ultra-light homebuilt aircraft.

WINGS: Cantilever low-wing monoplane. Wing section NACA 632-615. Dihedral 7° from roots. Incidence 3°. Constant-chord wings. No sweep. Sitka spruce primary structure, covered mainly in marine plywood, with rigid foam filling. Trailing-edge plain flaps (optional) and ailerons of glassfibre, with foam filling. No tabs: trimming is by variable tension on control column.

FUSELAGE: Basic wooden keel structure of 279 mm (11 in) square constant section, bearing nacelle-type cockpit and tail surfaces. Spruce primary structure, with plywood skin.

TAIL UNIT: Cantilever surfaces, fin and tailplane each having a central ply panel, to which are glued half-ribs with foam core infills. Conventional rudder and one-piece elevator, without tabs.

LANDING GEAR: Non-retractable tailwheel type. Main gear on single curved cantilever unit of laminated ash, moulded round a circular former and reinforced with glassfibre. Main wheels and tyres size 5·00-5, with go-kart hubs and brakes. Steerable tailwheel.

POWER PLANT: One 1,300-1,600 cc modified Volkswagen engine, driving a two-blade fixed-pitch wooden propeller. Single 36 litre (8 Imp gallon) fuel tank forward of instrument panel.

ACCOMMODATION: Single seat in open cockpit with large one-piece curved windscreen. Cockpit heated by air supply from engine. Enclosed cockpit optional.

DIMENSIONS, EXTERNAL:

Wing span	7·62 m (25 ft 0 in)
Wing chord (constant)	1·22 m (4 ft 0 in)
Wing area, gross	9·29 m² (100·0 sq ft)
Wing aspect ratio	6
Length overall	5·18 m (17 ft 0 in)
Width, wings removed	1·60 m (5 ft 3 in)
Height overall	1·88 m (6 ft 2 in)
Tailplane span	2·44 m (8 ft 0 in)
Wheel track	1·68 m (5 ft 6 in)
Wheelbase	3·73 m (12 ft 3 in)

WEIGHTS AND LOADINGS:

Weight empty	204 kg (450 lb)
Max T-O weight	295 kg (650 lb)
Max wing loading	31·7 kg/m² (6·5 lb/sq ft)
Max power loading	9·74 kg/kW (16 lb/hp)

PERFORMANCE (estimated, at max T-O weight):

Never-exceed speed	104 knots (193 km/h; 120 mph)
Max cruising speed at S/L	85 knots (158 km/h; 98 mph)
Econ cruising speed at S/L	65 knots (121 km/h; 75 mph)
Stalling speed, without flaps	35 knots (64·5 km/h; 40 mph)
Max rate of climb at S/L	168 m (550 ft)/min

MACDONALD

D. J. MACDONALD

ADDRESS:

23 South Drive, Stoney Stanton, Leicestershire LE9 6JP

MACDONALD MAC 6 MERCURY

Mr Macdonald, who is the Chairman of the British Association of Inventors (BRAIN), has designed a man-powered gyroplane known as the Mercury. His partner in the venture is Mr Brett Mason, who has been responsible for the aircraft's construction.

It was hoped to make the first flight during the late summer or early Autumn of 1976. The aircraft is not expected to achieve vertical flight, an initial ground speed of some 4·6 m (15 ft)/sec being considered necessary for take-off, which should be accomplished within 92 m (300 ft).

As can be seen from the accompanying drawing, the basic configuration of the Mercury is that of a co-axial counter-rotating rotor system, with flexible delta-shaped lifting surfaces mounted at the blade tips. The dynamic thrust unit has already been tested, and has indicated that ample propulsion power should be available, probably permitting a substantial reduction in the area of the deltas.

TYPE: Man-powered gyroplane.

ROTOR SYSTEM: Contra-rotating, co-axial twin two-blade rotors. Blades are of aerofoil section balsa, with a triangular fabric auxiliary lifting surface at each blade tip.

ROTOR DRIVE: Chain and sprocket system, horizontal and then vertical within a tubular rotor shaft, drives rotors via a bevel gear assembly. Optimum rotation 60 rpm; tip speed approx 39 knots (72·5 km/h; 45 mph).

LANDING GEAR: Non-retractable, with transverse axle and fore-and-aft castoring main wheels, both of which are clear of the ground when the aircraft is in a horizontal flying attitude.

POWER: Pilot's weight is used rather than muscle power: pilot stands on pedals, which are non-reciprocal, and 'marks time'. It has been established that this system makes the most effective use of pilot's energy, and that a fit pilot could operate the aircraft for more than 20 min without fatigue or distress. The initial torque is the result of the pilot's weight operating through a 381 mm (15 in) lever. Established lift at 17·5 knots (32 km/h; 20 mph) is 27·2 kg (60 lb), and drag 2·7 kg (6 lb). Cross-wind lift exceeds 45·4 kg (100 lb), and there is effective drag. Forward motion is achieved progressively as the rotors accelerate, and at 60 rpm has proved to be in excess of 2·44 m (8 ft)/sec; it appears that this characteristic is due largely to precession.

ACCOMMODATION: Tubular frame and seat, allowing pilot movement for in-flight control.

DIMENSIONS, EXTERNAL:

Diameter of rotors (each)	6·10 m (20 ft 0 in)
Length of deltas (each)	2·74 m (9 ft 0 in)
Max width of deltas (each)	1·83 m (6 ft 0 in)
Area of deltas (four, total)	10·03 m² (108·0 sq ft)

WEIGHT:

Weight empty	not more than 36·3 kg (80 lb)

NIPPER

NIPPER KITS AND COMPONENTS LTD

HEAD OFFICE:

1 Ridgeway Drive, Bromley, Kent BR1 5DG

Telephone: 01-857 7821

CHAIRMAN: D. P. L. Antill

Complete worldwide rights for the Nipper aircraft were purchased from Belgium in 1966, and the aircraft was marketed in both factory-built form and in the form of several stages of kits for amateur construction.

The former Nipper Aircraft Ltd went into receivership in May 1971. Prior to this Mr D. P. L. Antill, formerly Managing Director of Nipper Aircraft Ltd, acquired all rights in the Nipper aircraft, and on 20 October 1971 formed a new company, Nipper Kits and Components Ltd, to supply spares for existing aircraft and to encourage and support amateur construction of the Nipper. Plans and an advisory service for amateur constructors continue to be available.

NIPPER Mk III and IIIA

The Mk III Nipper is powered by a 1·5 litre Rollason Ardem engine; when fitted with the 1·6 litre Ardem engine it is known as the Mk IIIA. The Nipper may also be fitted with wingtip fuel tanks which almost double the standard fuel capacity. With these tanks fitted, but empty, the aircraft remains aerobatic. Flutter tests have been completed satisfactorily at speeds up to 156 knots (290 km/h; 180 mph).

TYPE: Single-seat ultra-light monoplane.

WINGS: Cantilever mid-wing monoplane. Modified NACA 43012A wing section. Dihedral 5° 30'. Incidence 2°. All-wood one-piece single-spar structure, with plywood-covered leading-edge and overall fabric covering. Wooden ailerons with fabric covering. No flaps. Portion of port wing-root trailing-edge is made of light alloy and hinged, with built-in foot-rest, so that it can be folded down to assist access to cockpit. Wing is quickly removable, to permit aircraft to be towed behind a motor car.

FUSELAGE: Welded steel tube structure. Underfuselage fairing of glassfibre. Rear fuselage fabric-covered.

TAIL UNIT: Braced tailplane and elevators of wood construction. No fin. Rudder of steel tube construction with fabric covering.

LANDING GEAR: Non-retractable tricycle type. Nieman transverse rubber-ring shock-absorbers. Steerable nosewheel. Continental tyres, size 4·00-4, pressure 1·79 bars (26 lb/sq in). Disc brakes.

POWER PLANT: Standard power plant is one 33·5 kW (45 hp) Rollason Ardem X flat-four engine, driving a two-blade fixed-pitch wooden propeller with glassfibre spinner. More powerful versions of Ardem engine can be fitted optionally. Fuel tank between engine and cockpit, capacity 34 litres (7·5 Imp gallons). Provision for two 16·5 litre (3·6 Imp gallon) wingtip fuel tanks. Oil capacity 3·5 litres (0·77 Imp gallons).

Artist's impression of the Macdonald Mercury man-powered gyroplane *(Michael A. Badrocke)*

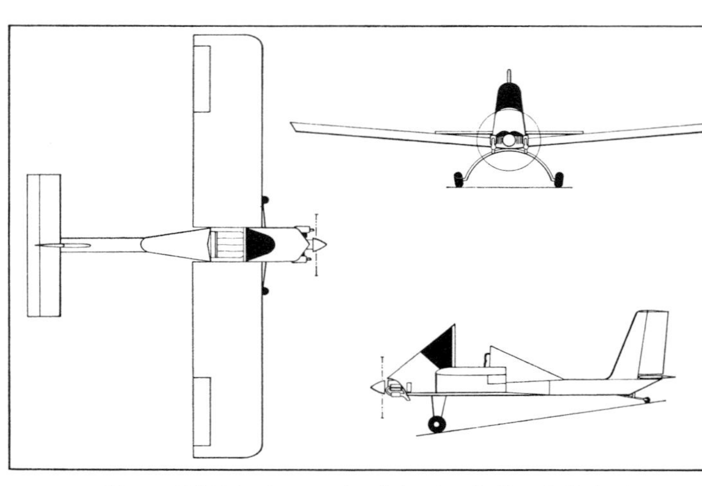

Livesey D.L.5 single-seat ultra-light aircraft *(Roy J. Grainge)*

Prototype Isaacs Spitfire single-seat light sporting aircraft

Isaacs Fury II (Rolls Royce Continental C90-12F engine) built by Mr A. V. Francis

Leisure Sport S5 Replica *(Austin J. Brown)*

Knowles Duet, adapted from Luton Minor *(P. R. March)*

ACCOMMODATION: Single seat under blown Perspex canopy which hinges sideways to starboard. Small baggage space aft of seat.

ELECTRONICS: Pye Bantam, Bayside BEI 990P and various other radio installations available.

DIMENSIONS, EXTERNAL:

Wing span (without tip-tanks)	6·00 m (19 ft 8 in)
Wing span (with tip-tanks)	6·25 m (20 ft 6 in)
Wing chord at c/l	1·40 m (4 ft 7¼ in)
Wing chord at tip	1·10 m (3 ft 7¼ in)
Wing area, gross	7·50 m² (80·70 sq ft)
Wing aspect ratio	4·8
Length overall	4·56 m (15 ft 0 in)
Height overall	1·91 m (6 ft 3 in)
Tailplane span	2·14 m (7 ft 0 in)
Wheel track	1·40 m (4 ft 7 in)
Wheelbase	1·13 m (3 ft 8 in)

WEIGHTS AND LOADINGS:

Weight empty	210 kg (465 lb)
Max T-O weight:	
Aerobatic	310 kg (685 lb)
Normal	340 kg (750 lb)
Max wing loading	45·4 kg/m² (9·3 lb/sq ft)
Max power loading	10·12 kg/kW (16·6 lb/hp)

PERFORMANCE (at max T-O weight):

Never-exceed speed	126 knots (235 km/h; 146 mph)
Max level speed at S/L:	
without tip-tanks	93 knots (173 km/h; 107 mph)
with tip-tanks	83 knots (155 km/h; 96 mph)
Max cruising speed (75% power) at S/L:	
without tip-tanks	81 knots (150 km/h; 93 mph
Econ cruising speed at S/L	78 knots (145 km/h; 90 mph
Stalling speed, power off	33 knots (61 km/h; 38 mph
Max rate of climb at S/L	198 m (650 ft)/mi
Service ceiling	3,660 m (12,000 ft
T-O run	85 m (280 ft
T-O to 15 m (50 ft)	338 m (1,110 ft
Landing from 15 m (50 ft)	457 m (1,500 ft
Landing run	110 m (360 ft
Range with max internal fuel, 30 min reserves	173 nm (320 km; 200 miles
Range with tip-tanks	390 nm (720 km; 450 miles

ORD-HUME

ARTHUR W. J. G. ORD-HUME

ADDRESS:
14 Elmwood Road, Chiswick, London W4
Telephone: 01-994-3292

Mr Ord-Hume was a co-founder and (until 1962) a director of the former Phoenix Aircraft Ltd. During this time he was responsible for redesigning the pre-war Luton Minor and Luton Major aircraft. As an amateur aircraft constructor, he was one of the first in the UK to construct his own aircraft after the second World War and has since built or restored 11 aeroplanes.

ORD-HUME GY-201 MINICAB

The original two-seat GY-20 Minicab was designed in France by M Yves Gardan and flew for the first time in February 1949. A small number of production aircraft were built by Constructions Aéronautiques du Béarn (see 1956-57 *Jane's*). Mr Ord-Hume subsequently acquired original drawings for both the Minicab and its precursor, the Babyclub, and redesigned and redrafted the plans to make the aircraft suitable for amateur construction in the UK, United States and Australia. The resulting variant was first introduced in 1963, and amateur construction of the Ord-Hume GY-201 Minicab, for which Mr Ord-Hume holds the sole and exclusive rights, is now approved in the USA, Canada, Australia, New Zealand, France, Germany and the UK. At least nine examples built to the English plans have flown in the UK and others are under construction. A large number are flying in the US, and also in Australia, where it was one of the first postwar-designed aircraft to gain DCA approval for amateur construction.

Owing to non-availability of Sitka spruce in Australia and New Zealand, aircraft built in that part of the world from local materials are unable to conform to the normal max T-O weight of 560 kg (1,235 lb). For these countrie only, the design has been re-stressed for a max T-O weigh of 635 kg (1,400 lb), for which weight a 74·5 kW (100 hp Continental engine is mandatory.

The standard Ord-Hume GY-201 Minicab is suitabl for engines from 48·5-89·5 kW (65 to 120 hp), althoug for practical operation at optimum weight 67 kW (90 hp is strongly advised as a minimum.

TYPE: Two-seat light monoplane.

POWER PLANT: One 48·5 kW (65 hp) Continental flat-fou engine; or one 67 kW (90 hp) (recommende minimum), 74·5 kW (100 hp) (Australian/New Zealan minimum), or 89·5 kW (120 hp) Continental engine Standard fuel capacity 50 litres (11 Imp gallons) i fuselage tank aft of engine firewall. Provision for 23 litr (5 Imp gallon) internal slipper tank in baggage com partment.

ACCOMMODATION: Side-by-side seating for two person

under forward-hinged canopy. Space for 11 kg (25 lb) of baggage aft of seats. Dual controls optional.

DIMENSIONS, EXTERNAL:

Wing span	7·62 m (25 ft 0 in)
Wing area, gross	10·0 m² (107·6 sq ft)
Length overall	5·44 m (17 ft 10 in)
Height overall	1·65 m (5 ft 5 in)

WEIGHTS AND LOADINGS:

Weight empty	270 kg (595 lb)
Max T-O weight (except Australia and NZ)	560 kg (1,235 lb)
Max T-O weight (Australia and NZ only)	635 kg (1,400 lb)
Max wing loading	48·5 kg/m² (9·84 lb/sq ft)
Max power loading (74·5 kW; 100 hp engine)	8·52 kg/kW (14·0 lb/hp)

*PERFORMANCE (at max T-O weight):

Max level speed at S/L, fixed-pitch propeller	108 knots (200 km/h; 124 mph)
Max level speed at S/L, 74·5 kW; 100 hp engine, fixed-pitch propeller	112 knots (208 km/h; 129 mph)
Cruising speed at S/L, fixed-pitch propeller	97 knots (180 km/h; 112 mph)
Stalling speed, flaps up	41 knots (76 km/h; 47 mph)
Max rate of climb at S/L	207 m (680 ft)/min
Max rate of climb at S/L, 74·5 kW (100 hp) engine	366 m (1,200 ft)/min
Service ceiling	4,000 m (13,100 ft)
Range with normal fuel capacity	404 nm (750 km; 466 miles)

**Minimum performance: a number of examples have demonstrated considerably more favourable speeds, and much depends on the engine/propeller combination used*

ORD-HUME O-H4B MINOR

The pre-war Luton Minor light aircraft was redesigned by Mr Ord-Hume for post-war use and was for many years marketed by the former Phoenix company, of which he was a director and co-founder. It is described under the 'PFA' heading in this edition.

A development of the Ord-Hume version of the L.A.4a Minor, known as the O-H4B Minor, was announced at the end of August 1975, and production drawings for this were first offered in 1976. Six were under construction in the UK, one in South Africa and eight in the United States by the Spring of 1976.

The O-H4B Minor is a foldable-wing aircraft, and can be towed on the road behind a motor car, without the need for a trailer. A unique aileron control system eliminates the need to break control runs when the wings are folded back alongside the fuselage, and the two halves of the tailplane can be folded upward alongside the fin by disconnecting the bracing struts, thus reducing the overall width for towing to only 1·37 m (4 ft 6 in).

The recommended power plant is a 1,700 cc Volkswagen engine (Peacock conversion); the 1,600 cc conversion is also suitable, affecting only the take-off and climb characteristics. A full set of plan drawings and a construction booklet are available.

TYPE: Single-seat parasol-wing light monoplane.

WINGS: Parasol-wing monoplane. RAF 48 (modified) wing section. Thickness/chord ratio 14·8%. Dihedral 3°. Incidence 3°. Wing structure comprises two built-up spruce and plywood spars, truss-type ribs and D-shaped leading-edge torsion-box. Two wing panels attached to rigid centre-section and braced by tubular lift struts. Wings foldable about rear spar swivel hinge. Aft portion of centre-section detachable as a triangular-section executive briefcase. Main wings fabric-covered. Wing structure is fail-safe. Plain semi-differential ailerons. No flaps or tabs.

FUSELAGE: All-wooden structure with fabric covering on top decking. Four spruce longerons; plywood-covered (non-structural).

TAIL UNIT: Tailplane built in two halves; plywood covered. Strut-braced to fin-post; by disconnecting the struts, both halves of the tailplane can be folded upwards flat against the fin and rudder. Elevators attached to fuselage-mounted operating horn by single pin, removal of which allows them to be folded. Variable tailplane incidence, by changing simple fittings. Bowden cable-operated trim on starboard elevator. No tabs.

LANDING GEAR: Non-retractable tailwheel type. Rubber block shock-absorption. Any type of wheel can be fitted. Cable-operated shoe brakes or hydraulic disc brakes. Floats or skis can be fitted.

POWER PLANT: One modified Volkswagen engine of 1,300-1,800 cc. One fuselage tank for 32 litres (7 Imp gallons) of fuel. Reserve tank can be fitted in centre-section of wings, capacity 32 litres (7 Imp gallons).

ACCOMMODATION: Pilot only, in open cockpit. Door on port side of fuselage.

DIMENSIONS, EXTERNAL:

Wing span	7·62 m (25 ft 0 in)
Wing chord (constant)	1·60 m (5 ft 3 in)
Wing area, gross	11·61 m² (125 sq ft)
Wing aspect ratio	5
Length overall	6·17 m (20 ft 3 in)
Height overall	2·20 m (7 ft 3 in)
Tailplane span	2·44 m (8 ft 0 in)
Wheel track	1·22 m (4 ft 0 in)
Wheelbase	1·37 m (4 ft 6 in)

WEIGHTS AND LOADINGS:

Weight empty	177 kg (390 lb)
Max T-O weight	340 kg (750 lb)
Max wing loading	30 kg/m² (6·15 lb/sq ft)
Max power loading	8·45 kg/kW (13·9 lb/hp)

PERFORMANCE (at 340 kg; 750 lb max T-O weight, 1,700 cc engine):

Never-exceed speed	95 knots (177 km/h; 110 mph)
Max level speed at S/L	70 knots (128 km/h; 80 mph)
Max cruising speed	65 knots (120 km/h; 75 mph)
Econ cruising speed	56 knots (104 km/h; 65 mph)
Stalling speed	23 knots (42 km/h; 26 mph)
Max rate of climb at S/L	250 m (820 ft)/min
Service ceiling	3,350 m (11,000 ft)
T-O run	50 m (165 ft)
T-O to 15 m (50 ft)	155 m (510 ft)
Landing run	61 m (200 ft)
Range with standard fuel	260 nm (483 km; 300 miles)
Range with centre-section long-range tank	538 nm (1,000 km; 620 miles)

PFA
POPULAR FLYING ASSOCIATION

ADDRESS:

Terminal Building, Shoreham Airport, Shoreham-by-Sea, Sussex BN4 5FF

Telephone: Shoreham-by-Sea 61616

Following the collapse of Phoenix Aircraft Ltd, the PFA assumed responsibility for marketing plans of the Luton L.A.4a Minor. Other types of which plans are obtainable from the Association include the Isaacs Fury, Evans VP-1 and Pazmany PL-4, described under their designers' names in this section of *Jane's*, and the Currie Wot and Druine Turbulent, described below with the L.A.4a Minor.

CURRIE WOT

This aircraft was designed originally by Mr J. R. Currie in 1937. Two examples were built at Lympne in that year, but were destroyed in a wartime bombing raid. Mr V. H. Bellamy took over the design after the war, at the Hampshire Aeroplane Club, and the first Wot built by members of this club, (G-APNT) flew for the first time on 11 September 1958. The second example built at the club (G-APWT) was powered by a 45 kW (60 hp) Walter Mikron flat-four engine. Both aircraft were described in earlier editions of *Jane's*.

Further Wots have since been completed, including Dr J. H. B. Urmston's G-ARZW with a 48·5 kW (65 hp) Walter Mikron III engine.

Dr Urmston purchased all rights in the design from Mr Bellamy, and the following details refer to his Wot, built to standard plans, as obtainable from the PFA. Data on the versions with Aeronca-JAP and Mikron engines can be found in the 1961-62 *Jane's*.

TYPE: Single-seat fully-aerobatic light biplane.

WINGS: Braced biplane type, with two parallel interplane struts each side and N-type centre section support struts. Wing section Clark Y. Dihedral (both wings) 3°. No incidence. Conventional spruce and plywood structure, with fabric covering. Fabric-covered ailerons on lower wings only. No flaps.

FUSELAGE: All-wood structure. Plywood-box construction, with overall fabric covering.

TAIL UNIT: Cantilever structure of spruce and plywood, with fabric covering. Fixed-incidence tailplane. Adjustable tab on rudder. Trim tab in port elevator.

LANDING GEAR: Non-retractable two-wheel type. Rubber-cord shock-absorbtion. Main wheels fitted with Dunlop tyres, size 400 × 8, pressure 1·24 bars (18 lb/sq in). No brakes.

POWER PLANT: One 48·5 kW (65 hp) Walter Mikron III flat-four engine, driving a two-blade fixed-pitch wooden propeller. Fuel tank aft of firewall, capacity 54·5 litres (12 Imp gallons). Oil capacity 7 litres (1·5 Imp gallons).

ACCOMMODATION: Single seat in open cockpit.

DIMENSIONS, EXTERNAL:

Wing span (both)	6·73 m (22 ft 1 in)
Wing chord (both, constant)	1·07 m (3 ft 6 in)
Wing area, gross	13·0 m² (140 sq ft)
Wing aspect ratio	6·3
Length overall	5·58 m (18 ft 3½in)
Height overall	2·06 m (6 ft 9 in)
Wheel track	1·38 m (4 ft 6½ in)

WEIGHTS:

Weight empty	250 kg (550 lb)
Max T-O weight	408 kg (900 lb)

PERFORMANCE (at max T-O weight):

Never-exceed speed	112 knots (209 km/h; 130 mph)
Max level speed at 610 m (2,000 ft)	83 knots (153 km/h; 95 mph)
Max cruising speed at 610 m (2,000 ft)	78 knots (145 km/h; 90 mph)
Econ cruising speed at 610 m (2,000 ft))	69 knots (129 km/h; 80 mph)
Stalling speed	35 knots (65 km/h; 40 mph)
Max rate of climb at S/L	183 m (600 ft)/min
Range with max fuel	208 nm (385 km; 240 miles)

LUTON L.A.4a MINOR

The first Luton Minor flew in 1936 and proved entirely suitable for construction and operation by amateur builders and pilots. Examples were built pre-war in England and other parts of the world.

In 1960, the design was modernised and restressed completely to the latest British Airworthiness Requirements, allowing for a power increase to 41 kW (55 hp) and a maximum flying weight of 340 kg (750 lb).

Minors are under construction in many parts of the world, and a considerable number of amateur-built examples have been completed and flown successfully since mid-1962. At least one of them, built in Australia by R. A. Pearman and H. Nash, has obtained a full Certificate of Airworthiness.

Under consideration is the possibility of introducing a few small modifications to the design. At the time of closing for press *Jane's* had not been advised of the nature of these changes, if any; the following description therefore applies to the aircraft in the form previously marketed by Phoenix.

TYPE: Single-seat light monoplane.

WINGS: Strut-braced parasol monoplane. Wing section RAF 48. No dihedral. Wooden two-spar structure in two halves, attached to the fuselage by tubular centre-section pylons and braced by parallel lift struts of streamline-section steel tubing. Wings removable for ground transport and storage. Leading-edge and tips plywood-covered, remainder fabric-covered. Plain ailerons of wood construction, fabric-covered, hinged directly from rear spar. No flaps.

FUSELAGE: Rectangular all-wood structure. Sides and bottom plywood-covered. Curved decking aft of cockpit fabric-covered.

TAIL UNIT: Cantilever all-wood structure, fabric-covered. Fixed fin. Aerodynamically-balanced rudder.

LANDING GEAR: Non-retractable tailwheel type with divided main legs of tubular steel construction. Rubber disc shock-absorbers. Brakes and wheel fairings optional. Fully-castoring tailwheel.

POWER PLANT: One aircooled engine in the 27·5-41 kW (37-55 hp) range, driving a two-blade fixed-pitch wooden propeller. Fuel tank forward of cockpit, capacity 29·5 litres (6·5 Imp gallons). Provision for additional tanks in wings.

ACCOMMODATION: Single seat in open cockpit. Coupé top optional. Luggage space aft of seat.

DIMENSIONS, EXTERNAL:

Wing span	7·62 m (25 ft 0 in)
Wing chord (constant)	1·60 m (5 ft 3 in)
Wing area, gross	11·6 m² (125 sq ft)
Wing aspect ratio	5
Length overall	6·32 m (20 ft 9 in)
Height overall	2·29 m (7 ft 6 in)

WEIGHTS:

Weight empty	177 kg (390 lb)
Max T-O weight	340 kg (750 lb)

PERFORMANCE (27·5 kW; 37 hp Aeronca-JAP J.99 engine, at normal T-O weight):

Max level speed at 450 m (1,500 ft)	60 knots (111 km/h; 69 mph)
Normal cruising speed	55 knots (102 km/h; 63 mph)
Stalling speed	25 knots (45 km/h; 28 mph)
Max rate of climb at S/L	76 m (250 ft)/min
T-O run	92 m (300 ft)
Landing run	36·5 m (120 ft)
Range with standard fuel	155 nm (290 km; 180 miles)
Range with auxiliary tanks	340 nm (645 km; 400 miles)

DRUINE D31 TURBULENT

The following data apply to the D31 Turbulent as factory-built in the UK for many years by Rollason Aircraft and Engines Ltd. Rollason's version was generally similar to the standard Druine design. Main differences were that it had wheels of slightly greater size and a tailskid instead of a tailwheel, although a tailwheel was available optionally.

The fittings of optional wheel spats and a sliding canopy increases speed by about 7 knots (13 km/h; 8 mph).

TYPE: Single-seat ultra-light monoplane.

WINGS: Cantilever low-wing monoplane. Wing section NACA 23012. Dihedral 4°. Incidence 3° 40'. All-wood two-spar structure, covered with fabric. Built-in leading-edge slot on outer 45% of half-span. Wooden slotted ailerons with fabric covering. No flaps or tabs.

FUSELAGE: Conventional rectangular four-longeron wood structure with domed decking. Plywood-covered.

TAIL UNIT: Cantilever wooden structure. Fixed surfaces plywood-covered, movable surfaces fabric-covered. No tabs.

Ord-Hume GY 201 Minicab built at Niton, Isle of Wight, by Mr Barry Evans

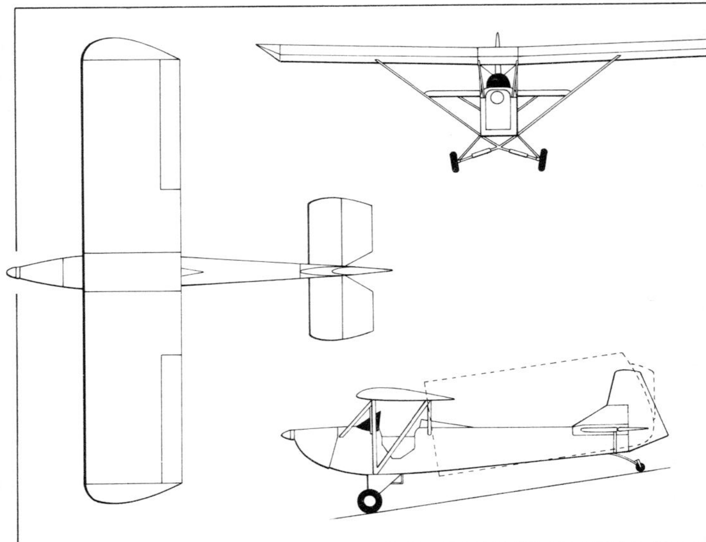

Ord-Hume O-H4B Minor, evolved from Luton Minor *(Michael A. Badrocke)*

Nipper Mk III single-seat ultra-light aircraft

Currie Wot fully-aerobatic biplane *(Air Portraits)*

Landing Gear: Non-retractable tailwheel type (optional tailskid). Compression-spring shock-absorbers. Dunlop or Goodyear main wheels and tyres, size 14 × 3, pressure 1·95 bars (28 lb/sq in). Vespa mechanical brakes. Wheel spats, taxying and parking brakes optional. Skis or floats may be fitted as alternative to wheels.

Power Plant: One 33·5 kW (45 hp) Rollason Ardem 4CO2 Mk IV or 41 kW (55 hp) Ardem Mk V flat-four engine, driving a two-blade fixed-pitch wooden propeller. Fuel tank in fuselage forward of cockpit, capacity 39 litres (8·5 Imp gallons). Oil capacity 2·25 litres (0·5 Imp gallons).

Accommodation: Pilot only, in open cockpit (sliding canopy optional). Baggage locker aft of seat, capacity 11·5 kg (25 lb).

Electronics: Provision for lightweight radio.

Dimensions, external:

Wing span	6·58 m (21 ft 7 in)
Wing chord (constant)	1·90 m (3 ft 11 in)
Wing area, gross	7·20 m² (77·5 sq ft)
Wing aspect ratio	5·4
Length overall	5·33 m (17 ft 6 in)
Height overall	1·52 m (5 ft 0 in)
Tailplane span	1·98 m (6 ft 6 in)
Wheel track	1·73 m (5 ft 8 in)
Wheel/tailskid base	3·81 m (12 ft 6 in)

Weights and Loadings:

Weight empty	179 kg (395 lb)
Max T-O weight	281 kg (620 lb)
Max wing loading	39·1 kg/m² (8·0 lb/sq ft)
Max power loading	8·39 kg/kW (13·8 lb/hp

Performance (with 33·5 kW; 45 hp engine, at max T-(weight):

Never-exceed speed	108 knots (202 km/h; 125 mph
Max level speed	95 knots (176 km/h; 109 mph
Max cruising speed	87 knots (161 km/h; 100 mph
Econ cruising speed	76 knots (141 km/h; 87 mph
Stalling speed	39 knots (71 km/h; 44 mph
Max rate of climb at S/L	137 m (450 ft)/mi
Service ceiling	2,740 m (9,000 ft
T-O run from grass	95 m (310 ft
T-O to 15 m (50 ft) from grass	125 m (410 ft
Landing from 15 m (50 ft) on grass	98 m (320 ft
Landing run on grass	52 m (170 ft
Range with max fuel, normal allowances	217 nm (400 km; 250 miles

PRACTAVIA

PRACTAVIA LTD

Address:
Wycombe Air Park, Booker, near Marlow, Buckinghamshire
Telephone: (01) 836-9036
Managing Director:
C. B. Healey

This company was formed to market plans and kits of a two-seat all-metal aerobatic aircraft known as the Sprite, the design of which was initiated by *Pilot* magazine.

PRACTAVIA SPRITE

Initial design work on the Sprite was started by the staff of *Pilot* magazine in early 1968, after consultation with many experienced light aircraft constructors, and a design and development panel was set up to foster the project. Detailed design began in November 1968 and construction of the first Sprite was started in 1969. Mr Brian Healey, former editor of *Pilot* magazine, is project executive, and Mr Lloyd Jenkinson and Mr Peter Sharman, lecturers at Loughborough University, are the designers.

First flight of a Sprite, built by Mr Peter Burril of Snape, Yorkshire, took place on 16 June 1976. Two other Sprites are being constructed by British Airways apprentices, under their instructor, Mr Bert Page, at London Airport, and about seven more privately-built examples were approaching completion in mid-1976. Plans are available for amateur construction, and more than 100 sets have been sold. Kits are also available.

Type: Two-seat all-metal aerobatic aircraft, suitable for amateur construction.

Wings: Cantilever low-wing monoplane. Wing section NACA 64315. Dihedral 6° on outer panels only. No incidence or sweepback. All-metal structure of aluminium alloy, built in three equal-length sections. Single main spar with light rear spar forming central torsion box. Skins and ribs of L72 alloy, extrusions of L65 alloy and spar caps of L73 alloy. Single-slotted flaps and plain ailerons of L72 alloy extend over full span. No trim tabs. Outer wing panels detachable for transit.

Fuselage: All-metal semi-monocoque structure, with no double curvature. Longerons of L65 aluminium alloy, skins and frames of L72 alloy. Sides and top curved to avoid drumming.

Tail Unit: Cantilever all-metal structure with swept vertical surfaces, constructed of L72 alloy. Fixed-incidence tailplane. Trim tab in centre of elevator trailing-edge, of one-third span; outer one-third on each side comprises anti-balance tab.

Landing Gear: Non-retractable tricycle type standard, although design of wing structure will allow for retractable gear as a future development. Shock-absorption by rubber in compression. Wheels and tyres size 5·00-5. Hydraulic disc brakes.

Power Plant: One 97 kW (130 hp) Rolls-Royce Continental O-240-A flat-four engine, driving a two-blade fixed-pitch propeller. Other suitable power plants include an 86 or 112 kW (115 or 150 hp) Lycoming or a 74·5 kW (100 hp) Rolls-Royce Continental engine. Fuel contained in one fuselage tank, aft of firewall, capacity 54·5 litres (12 Imp gallons). Wingtip fuel tanks optional, capacity 54·5 litres (12 Imp gallons) each. Maximum total capacity 163·5 litres (36 Imp gallons). Oil capacity 4·5 litres (1 Imp gallon)

Accommodation: Two seats, side by side, in enclosed cockpit, with rearward-sliding transparent canopy. Space for baggage behind seats.

Systems: Hydraulic system for brakes only. 12V electrica system.

Electronics and Equipment: Radio, blind-flyin instrumentation and special equipment to individua builder's requirements.

Dimensions, external:

Wing span	7·32 m (24 ft 0 in
Wing span over tip-tanks	8·23 m (27 ft 0 in
Wing chord (constant)	1·22 m (4 ft 0 in
Wing area, gross	8·92 m² (96 sq ft
Wing aspect ratio	
Length overall	6·10 m (20 ft 0 in
Width, outer panels removed	2·44 m (8 ft 0 in
Height overall	2·51 m (8 ft 3 in
Tailplane span	2·44 m (8 ft 0 in
Wheel track	2·29 m (7 ft 6 in
Wheelbase	1·40 m (4 ft 7 in

Dimension, internal:

Cabin: Max width	1·17 m (3 ft 10 in

Weights and Loadings (74·5 kW; 100 hp engine):

Weight empty	385 kg (850 lb
Max T-O weight	635 kg (1,400 lb
Max wing loading	71·2 kg/m² (14·6 lb/sq ft
Max power loading	8·52 kg/kW (14·00 lb/hp

Performance (estimated, at max T-O weight with 93 kW 125 hp engine):

Never-exceed speed	212 knots (394 km/h; 245 mph
Max cruising speed	120 knots (222 km/h; 138 mph
Economical cruising speed	111 knots (206 km/h; 128 mph
Stalling speed, flaps down	48 knots (89 km/h; 55 mph

PRESTWICK

PRESTWICK MAN POWERED AIRCRAFT GROUP

ADDRESS:

Project Department, Scottish Aviation Ltd, Prestwick International Airport, Ayrshire KA9 2RW

Telephone: Prestwick (0292) 79888

OFFICERS:

R. J. Hardy (Project Leader)
A. Furner (Structure)
R. Churcher (Construction)

PRESTWICK DRAGONFLY MPA Mk 1

The Dragonfly is the latest of three designs specifically aimed at the Kremer prize course and started at Southampton University under the supervision of Prof G. M. Lilley. The current design was completed at RAF Cranwell in 1974, and makes use of practical features and experience gained there while operating the Jupiter and Mercury man-powered aircraft (which see).

The Dragonfly is unique in that it is designed to fly outside ground effect, being optimised for flight at a height of 18·3 m (60 ft). This should give a small but realistic rate of climb, and it has been estimated that a 63·5 kg (140 lb) pilot could reach 60 ft in 5 minutes from the start of take-off roll.

The aircraft is basically very simple, and has no unconventional features. It is also fairly small compared to some other designs, thus minimising control problems and promising a good performance through low structural weight and an efficient structure.

Construction began in February 1975, and the aircraft was completed 22 weeks later. Taxying trials were started early in August, and the following two months were spent in modifying the aircraft. Bad weather and poor visibility prevented any further trials from October 1975 onwards, and no attempts to fly will be made before the Spring of 1977.

TYPE: Single-seat man-powered aircraft.

WINGS: Wire-braced high-wing monoplane. Wortmann FX-63-137 wing section. Dihedral 6° static, 10° in flight. Incidence 3°. No sweepback at quarter-chord. Single main box spar, with spruce flanges and ply webs. Ribs of balsa/expanded polystyrene foam/balsa sandwich. Leading-edge of sheet balsa. Melinex covering. Differentially-operating ailerons, of Melinex-covered balsa.

FUSELAGE: Aluminium alloy longerons, cross-braced with balsa and covered with Melinex. Triangular section aft of cockpit. Some racing bicycle components incorporated.

TAIL UNIT: Cantilever structure, with I-section balsa spar, of similar construction to wings. All-moving horizontal surfaces, with elevator fixed at upward angle of 15°.

LANDING GEAR: Single main wheel (of 406 mm; 16 in diameter) and tailwheel, plus outriggers beneath outer wings. All units non-retractable. Bicycle-type brakes.

POWER PLANT: Man-power on bicycle pedals, providing an estimated 0·5 hp (max) to drive a two-blade fixed-pitch tractor propeller mounted on a pylon above and behind the cockpit.

ACCOMMODATION: Pilot only, in enclosed and ventilated cockpit. Door in starboard side of fuselage. Instrumentation includes attitude indicator for pitch trim.

DIMENSIONS, EXTERNAL:

Wing span	24·38 m (80 ft 0 in)
Wing chord at root	1·04 m (3 ft 4·8 in)
Wing aspect ratio	30
Length overall	7·35 m (24 ft 1½ in)
Height overall:	
propeller vertical	4·18 m (13 ft 8·7 in)
propeller horizontal	2·92 m (9 ft 7 in)
Tailplane span	3·45 m (11 ft 3·8 in)
Wheelbase	5·41 m (17 ft 9 in)
Propeller diameter	2·74 m (9 ft 0 in)

DIMENSION, INTERNAL:

Cockpit: Max width	0·61 m (2 ft 0 in)

AREAS:

Wings, gross	19·83 m² (213·5 sq ft)
Ailerons (total)	2·74 m² (29·50 sq ft)
Fin	1·01 m² (10·85 sq ft)
Rudder	1·01 m² (10·85 sq ft)
Tailplane	1·84 m² (19·77 sq ft)

WEIGHTS AND LOADINGS:

Weight empty	43 kg (95 lb)
Normal T-O weight	105 kg (232 lb)
Max T-O weight	114 kg (252 lb)
Max wing loading	5·76 kg/m² (1·18 lb/sq ft)
Max power loading	229 kg/hp (504 lb/hp)

PERFORMANCE (estimated, at normal T-O weight):

Never-exceed speed	27 knots (50 km/h; 31 mph)
Max level speed	20 knots (37 km/h; 23 mph)
Cruising speed	16·5 knots (31 km/h; 19 mph)
Stalling speed	13·5 knots (26 km/h; 16 mph)
Max rate of climb at S/L (out of ground effect)	4·1 m (13·5 ft)/min
Absolute ceiling	27·5 m (90 ft)
T-O run	122 m (400 ft)
T-O to 15 m (50 ft)	1,890 m (6,200 ft)
Landing from 15 m (50 ft)	671 m (2,200 ft)
Landing run	61 m (200 ft)

PROCTER

PROCTER AIRCRAFT ASSOCIATES LTD

HEAD OFFICE:

Greenball, Crawley Ridge, Camberley, Surrey GU15 2AJ

Telephone: Camberley 25566

DIRECTORS:

Roy G. Procter, CEng, MRAeS
Roger H. White-Smith
Mrs Barbara Alexander

SECRETARY:

Mrs Ann Procter

This company changed its name from Mitchell-Procter Aircraft Ltd in November 1968. The latter company comprised a group of enthusiasts who designed and built the prototype Kittiwake I, plans and parts for which are available from Procter Aircraft Associates.

Aircraft of this type are under construction in the UK, USA and Canada.

Procter Aircraft Associates has designed a larger aircraft on the same lines, known as the Petrel.

MITCHELL-PROCTER KITTIWAKE I

The Kittiwake I was designed by Dr C. G. B. Mitchell to make full use of modern materials and constructional techniques while retaining a simplicity of design making it possible for the aircraft to be built without special tooling. The wings attach directly to the sides of the fuselage, so that construction and storage can take place in a normal-sized garage.

Design of the Kittiwake I was started in February 1965. Construction of the prototype (G-ATXN) began in June 1965 and it flew for the first time on 23 May 1967.

Dr Mitchell no longer retains the design rights in the Kittiwake I. Plans of the aircraft, and parts or kits for its construction, are available from Procter Aircraft Associates.

Kittiwake Is are under construction in the UK, USA and Canada. One, built by the Royal Navy Air Engineering School at Gosport, Hampshire (illustrated) has the serial number XW784 and was flown for the first time at RNAS Lee-on-Solent on 21 October 1971. It is powered by the standard O-200-A engine.

The details that follow apply to the prototype in its original form, with a similar power plant. It was fitted subsequently with a Lycoming O-290 engine, in a glassfibre cowling, and a new nosewheel leg with rubber-in-compression shock-absorption.

TYPE: Single-seat glider-towing and sporting light aircraft.

WINGS: Cantilever low-wing monoplane. Wing section NACA 3415. Dihedral 5°. Incidence 2° 30′. No washout. All-metal (L72 and L64 aluminium alloys) structure, with single main spar at 30% chord and light false spar at 66%. Multiple ribs. No spanwise stiffeners. Wings attach at fuselage sides; centre-section is integral with fuselage. All-metal (L72) NACA single-slotted flaps.

FUSELAGE: All-metal (L72) structure. Four-longeron box with flat sides and bottom, and single-curvature top decking. Integral wing centre-section forms seat and landing gear attachment structure.

TAIL UNIT: Cantilever all-metal (L72) structure. Fixed-incidence tailplane. Manually-operated tab on elevator. Tab on rudder.

LANDING GEAR: Non-retractable tricycle type. Cantilever spring steel main legs. Nose unit has rubber torsion-bush shock-absorption. All three units fitted with Goodyear wheels and tyres, size 5·00-5. Tyre pressure 1·72 bars (25 lb/sq in). Goodyear hydraulically-operated disc brakes on main wheels.

POWER PLANT: One 74·5 kW (100 hp) Rolls-Royce Continental O-200-A flat-four engine, driving a McCauley 69CM52 two-blade fixed-pitch metal propeller for general use or a McCauley 76CM36 two-blade fixed-pitch metal propeller for glider-towing. Two integral leading-edge fuel tanks, total capacity 100 litres (22 Imp gallons). Space for 54·5 litre (12 Imp gallon) tank forward of instrument panel. Oil capacity 7 litres (1·5 Imp gallons).

ACCOMMODATION: Single seat under rearward-sliding canopy.

ELECTRONICS AND EQUIPMENT: Prototype has full blind-flying instrumentation and electrical system, but no radio.

DIMENSIONS, EXTERNAL:

Wing span	7·32 m (24 ft 0 in)
Wing chord (constant)	1·38 m (4 ft 6½ in)
Wing area, gross	9·75 m² (105 sq ft)
Wing aspect ratio	5·28
Length overall	5·97 m (19 ft 7 in)
Height overall	2·29 m (7 ft 6 in)
Tailplane span	2·44 m (8 ft 0 in)
Wheel track	1·75 m (5 ft 9 in)
Wheelbase	1·52 m (5 ft 0 in)
Propeller diameter:	
69CM52	1·75 m (5 ft 9 in)
76CM36	1·93 m (6 ft 4 in)

DIMENSIONS, INTERNAL:

Cabin: Length	1·52 m (5 ft 0 in)
Max width	0·64 m (2 ft 1 in)
Max height, seat to canopy	1·04 m (3 ft 5 in)

WEIGHTS AND LOADINGS (prototype, incl some test equipment):

Weight empty, equipped	413 kg (910 lb)
Max T-O and landing weight	612 kg (1,350 lb)
Max aerobatic weight	567 kg (1,250 lb)
Max wing loading	63·0 kg/m² (12·9 lb/sq ft)
Max power loading	8·20 kg/kW (13·5 lb/hp)

PERFORMANCE (at AUW of 567 kg; 1,250 lb):

Max level speed	114 knots (211 km/h; 131 mph)
Max cruising speed (75% power)	106 knots (196 km/h; 122 mph)
Max rate of climb at S/L:	
69CM52 propeller	259 m (850 ft)/min
76CM36 propeller	320 m (1,050 ft)/min
Range at 100 knots (185 km/h; 115 mph)	425 nm (790 km; 490 miles)
Range at 80 knots (148 km/h; 92 mph)	468 nm (870 km; 540 miles)

PROCTER PETREL

This two-seat light aircraft is based upon the Kittiwake I single-seat lightplane, with many components in common, but has increased wing area and has been optimised for glider towing. It has a number of improvements and simplifications to the detail mechanical design, compared with Kittiwake I. Materials used throughout are L72 clad dural and S510 mild steel.

Completion of the prototype Petrel has been subcontracted, on an opportunity basis, to Mr Brian Swales of Thirsk, Yorkshire.

Construction of two other Petrels was undertaken by apprentice organisations, and the aircraft illustrated was expected to make its first flight during 1976.

It is intended to offer the Petrel for amateur construction.

TYPE: Two-seat light aircraft.

WINGS: Cantilever low-wing monoplane. Wing section NACA 3415. Dihedral 5° on outer panels. No sweepback or washout. All-metal constant-chord structure, built in three sections: centre-section, integral with fuselage, to which outer panels are each attached with three bolts. Single main spar at 30% chord and lightweight auxiliary spar at 66% chord. Multiple ribs, with no spanwise stiffeners. All-metal NACA slotted flaps and ailerons. Flaps are operated manually by pushrod and torque tube; ailerons are mass-balanced and operated by cables.

FUSELAGE: All-metal structure. Four-longeron basic structure, with flat sides and bottom and single-curvature top-decking. Integral wing centre-section forms seat and main landing gear attachment structure.

TAIL UNIT: Cantilever all-metal structure. Fixed-incidence tailplane. Manually-operated tab in starboard elevator. Tab on rudder. Control surfaces mass-balanced and operated by cables.

LANDING GEAR: Non-retractable tricycle type. Nose unit is an oleo-pneumatic strut with Goodyear 5·00-6 wheel, and is steerable from the rudder pedals. Main gear is of cantilever spring type, with Goodyear 6·00-6 wheels and hydraulic disc brakes. Tyre pressure (all) 1·72 bars (25 lb/sq in).

POWER PLANT: One 97 kW (130 hp) Rolls-Royce Continental O-240 flat-four engine, driving a McCauley fixed-pitch two-blade metal propeller. Fuel capacity 73 litres (16 Imp gallons).

ACCOMMODATION: Two persons side by side, on seats with individually adjustable backs. Baggage space aft of seats.

EQUIPMENT: Starter, generator and basic instrumentation. Radio, navigation and other equipment to builder's requirements.

DIMENSIONS, EXTERNAL:

Wing span	9·14 m (30 ft 0 in)
Wing chord (constant)	1·38 m (4 ft 6½ in)
Wing area, gross	12·5 m² (135·0 sq ft)
Wing aspect ratio	6·6
Length overall	6·30 m (20 ft 8 in)
Height overall	2·33 m (7 ft 8 in)
Tailplane span	2·79 m (9 ft 2 in)
Wheel track	2·24 m (7 ft 4 in)
Wheelbase	1·52 m (5 ft 0 in)

WEIGHTS:

Weight empty	515·5 kg (1,137 lb)
Max T-O weight	762 kg (1,680 lb)

PERFORMANCE (at max T-O weight: estimated, based on measured performance of Kittiwake I):

Max level speed	113 knots (209 km/h; 130 mph)
Cruising speed	104 knots (193 km/h; 120 mph)
Max rate of climb at S/L	305 m (1,000 ft)/min

Taylor J.T.1 monoplane *(R. Kunert)*

Prestwick Dragonfly MPA Mk 1 single-seat man-powered aircraft

Luton L.A.4a Minor, described under PFA entry *(P.F.A.)*

Practavia Sprite built by Mr Peter Burril of Snape, Yorkshire

Druine D.31 Turbulent, described under PFA entry *(Peter R. March)*

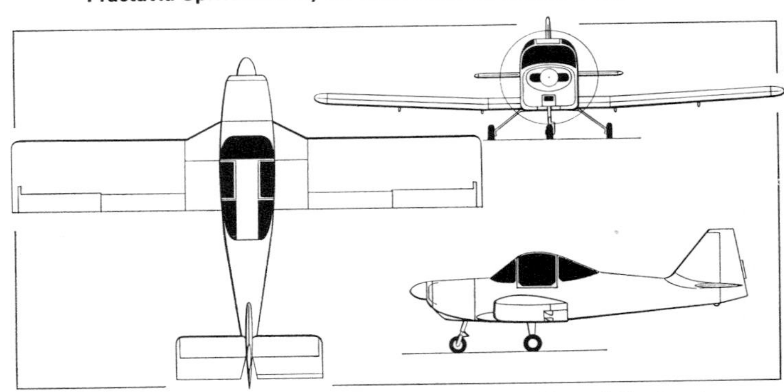
Procter Petrel two-seat light aircraft in its revised form *(Pilot Press)*

TAYLOR

Mrs JOHN F. TAYLOR

Address: 25, Chesterfield Crescent, Leigh-on-Sea, Essex SS9 5PD
Telephone: Southend (0702) 521063

The late Mr John Taylor, AMIED, an amateur constructor, designed and built the prototype of a single-seat ultra-light sporting monoplane, designated J.T.1. His object was to produce the airframe for not more than £100. Construction took about 14 months, and it flew for the first time on 4 July 1959.

A second design by Mr Taylor, the J.T.2 Titch, was awarded second prize in the Midget Racer Design Competition organised by Mr Norman Jones of the Rollason company in 1964. A prototype was built and flown successfully, but crashed on 16 May 1967, killing its designer.

Mrs J. F. Taylor is continuing to market plans of both these aircraft to amateur constructors.

TAYLOR J.T.1 MONOPLANE

Plans of the Taylor monoplane have been sold to amateur constructors in the United Kingdom and nearly 20 other countries in all parts of the world. Forty-eight J.T.1s are known to be flying, including 11 in the UK, 9 in Canada, 19 in the USA, 4 in Australia and 5 in New Zealand.

Aircraft currently flying or under construction are fitted with a variety of engines, including the 30 kW (40 hp) Aeronca E 113, 48·5 kW (65 hp) Continental A65, 48·5 kW (65 hp) Lycoming, 53·7 kW (72 hp) two-stroke McCulloch and the modified Volkswagen series. Aircraft with the 48·5 kW (65 hp) engines have a 10 cm (4 in) longer nose and 25 cm (10 in) longer rear fuselage to maintain the correct CG position.

Type: Single-seat fully-aerobatic ultra-light monoplane.

Wings: Cantilever low-wing monoplane. Wing section RAF 48. Constant chord. Dihedral on outer panels 4°. Incidence 3°. Wooden two-spar structure, comprising centre-section and outer panels. Plywood and fabric covering. Differential ailerons. Split trailing-edge flaps. No tabs.

Fuselage: Conventional plywood-covered wood structure of four main longerons and curved formers. Centre-section integral with fuselage.

Tail Unit: Cantilever fin and fixed-incidence tailplane are plywood-covered wood structures. Elevators and rudder are fabric-covered wood structures.

Landing Gear: Non-retractable two-wheel type. Cantilever main legs, with coil spring shock-absorption. Size 5 × 4 (305 mm; 12 in) main wheels, tyre pressure 1·65 bars (24 lb/sq in). Leaf spring tailskid with steerable skid-pad. One example, built by Mr R. Ladd in the US (see 1972-73 *Jane's*), has manually-actuated inward-retracting main gear.

Power Plant (prototype): One 28·3 kW (38 hp) JAP two-cylinder engine, driving a modified Flottorp two-blade wooden fixed-pitch propeller. Fuel tank aft of firewall, capacity 27 litres (6 Imp gallons).

Accommodation: Single seat under transparent Perspex canopy. Aerobatic harness. Small locker aft of seat.

Dimensions, external:

Wing span	6·40 m (21 ft 0 in)
Wing area, gross	7·06 m² (76 sq ft)
Wing aspect ratio	6
Length overall	4·57 m (15 ft 0 in)
Height over tail	1·47 m (4 ft 10 in)
Tailplane span	1·98 m (6 ft 6 in)
Wheel track	1·52 m (5 ft 0 in)

Weights and Loadings:

Weight empty	186 kg (410 lb)
Max T-O weight	276 kg (610 lb)
Max wing loading	39 kg/m² (8 lb/sq ft)
Max power loading	9·75 kg/kW (16 lb/hp)

Performance:

Never-exceed speed	113 knots (209 km/h; 130 mph)
Max level speed at S/L	91 knots (169 km/h; 105 mph)
Econ cruising speed	78 knots (145 km/h; 90 mph)
Never-exceed speed with flaps down	56 knots (105 km/h; 65 mph)
Stalling speed, flaps up	40 knots (75 km/h; 46 mph)
Stalling speed, flaps down	33 knots (62 km/h; 38 mph)
Max rate of climb at S/L	305 m (1,000 ft)/min
Range	200 nm (370 km; 230 miles)
g limits	+9; −9

TAYLOR J.T.2 TITCH

Construction of the prototype Titch was started in February 1965 and it flew for the first time on 22 January 1967.

Fourteen Titches are known to be flying, including eight in the USA, two in New Zealand, one in France, one in Rhodesia and two in England; plans have been supplied to amateur constructors in Brazil, Iceland. Italy, Mexico and Spain.

The following description refers to the prototype:

Type: Single-seat light monoplane.

Wings: Cantilever low-wing monoplane. Taylor-modified NACA 23012 wing section. Dihedral 5° on top surface. Incidence 3°. Spruce structure with main box-spar and 'plank' auxiliary spar. Plywood and fabric covering. Plain manually-operated ply-covered flaps over half-span and fabric-covered differential ailerons.

Fuselage: All-wood structure, with four main longerons, four secondary longerons and double-curvature ply covering. Aluminium cockpit side panels.

Tail Unit: All-wood structure, with fixed-incidence tailplane. Fixed surfaces plywood-covered, control surfaces fabric-covered.

Landing Gear: Non-retractable tailwheel type. Steerable tailwheel. Chrome-vanadium compression coil-spring shock-absorbers. Wheels of own manufacture with 4-ply tyres size 5·00-4 and drum brakes.

Power Plant: One 63·5 kW (85 hp) Continental C85 flat-four engine, driving a Hegy wooden two-blade scimitar propeller. Glassfibre fuel tank between firewall and instrument panel, capacity 45·5 litres (10 Imp gallons).

Accommodation: Single seat, with aerobatic harness, under bubble canopy hinged along starboard side.

Dimensions, external:

Wing span	5·72 m (18 ft 9 in)
Wing chord at root	1·37 m (4 ft 6 in)
Wing chord at tip	0·91 m (3 ft 0 in)
Wing area, gross	6·32 m² (68 sq ft)
Wing aspect ratio	5·14
Length overall	4·91 m (16 ft 1½ in)
Height overall	1·42 m (4 ft 8 in)
Tailplane span	1·98 m (6 ft 6 in)
Wheel track	1·52 m (5 ft 0 in)
Propeller diameter	1·52 m (5 ft 0 in)

Weights (A: 63·5 kW; 85 hp Continental engine; B: Volkswagen engine):

Weight empty:	
A	227 kg (500 lb)
B	185 kg (410 lb)
Max T-O weight:	
A	338 kg (745 lb)
B	290 kg (640 lb)

Kittiwake 1 built by Royal Navy apprentices *(Roy G. Procter)*

Taylor J.T.2 Titch *(P.F.A.)*

PERFORMANCE (prototype, at max T-O weight except where indicated):
Never-exceed speed 195 knots (362 km/h; 225 mph)
Max level speed 174 knots (322 km/h; 200 mph)
Normal cruising speed
135 knots (250 km/h; 155 mph)
Econ cruising speed 95·5 knots (177 km/h; 110 mph)
Best approach speed 65 knots (121 km/h; 75 mph)
Stalling speed, flaps up:
A 50·5 knots (93·5 km/h; 58 mph)
B 46 knots (86 km/h; 53 mph)
Stalling speed, flaps down:
A 43·5 knots (80·5 km/h; 50 mph)
B 40 knots (74 km/h; 46 mph)
T-O speed 54 knots (100 km/h; 62 mph)
Touchdown speed 48 knots (89 km/h; 55 mph)
Max rate of climb at S/L 335 m (1,100 ft)/min

WHITTAKER

MICHAEL WHITTAKER

ADDRESS:
Bodmin Airfield, Cardinham, Cornwall
Telephone: Cardinham 255

WHITTAKER MW2 EXCALIBUR

Sponsored by Mr C. M. Robertson of Charles Robertson (Developments) Ltd, Mr Whittaker has completed the prototype (G-BDDX) of his MW2 design for a single-seat ultra-light aircraft. Design began in 1972, and construction started in September 1974; first flight of the aircraft was due to take place in the Summer of 1976. The prototype has been stressed to BCAR (K) requirements, and it is intended to obtain PFA certification initially; CAA certification will be sought before any series production is undertaken.

TYPE: Single-seat ultra-light aircraft.

WINGS: Cantilever low-wing monoplane. Wing section NACA 63A615. Dihedral 4°. Incidence 0°. Aluminium alloy and glassfibre stressed-skin structure. Aluminium alloy differential ailerons, with fluted skins. No flaps.

FUSELAGE: Pod-type central nacelle, consisting of a moulded glassfibre shell carried on an aluminium alloy box beam. Twin aluminium alloy beams carry tail unit.

TAIL UNIT: Fixed-incidence tailplane, and elevator of equal size, contained within twin endplate fins and rudders at ends of tailbooms. Aluminium alloy structure, with glassfibre fairings. Fixed tab at centre of elevator. Tailplane, elevator and rudders have fluted skins.

LANDING GEAR: Non-retractable tricycle type, with rubber-disc shock-absorption. Go-kart wheels, size 13 × 3½ in on main units, 9 × 3 in on nose unit; and go-kart brakes. Glassfibre wheel fairings.

POWER PLANT: One 41 kW (55 hp) 1,600 cc Volkswagen watercooled engine, driving a Dowty Rotol four-blade fixed-pitch glassfibre pusher propeller shrouded by an annular duct. Fuel (petrol/oil mixture) in two integral wing tanks outboard of tailbooms, total capacity 36·4 litres (8 Imp gallons). Refuelling point on top of each tank.

ACCOMMODATION: Single seat under one-piece flush-mounted sideways-opening (to starboard) moulded canopy, with 0·06 m³ (2 cu ft) of baggage space to rear of seat. Cockpit ventilated.

SYSTEMS: Vacuum pump on engine. 12V alternator.

DIMENSIONS, EXTERNAL:
Wing span 7·32 m (24 ft 0 in)
Wing chord (constant) 1·22 m (4 ft 0 in)
Wing area, gross 8·83 m² (95·0 sq ft)
Wing aspect ratio 6
Length overall 5·49 m (18 ft 0 in)
Height overall 1·60 m (5 ft 3 in)
Tailplane span 2·44 m (8 ft 0 in)
Wheel track 2·29 m (7 ft 6 in)
Wheelbase 1·98 m (6 ft 6 in)
Propeller diameter 0·86 m (2 ft 10 in)

DIMENSIONS, INTERNAL:
Cabin: Length 1·22 m (4 ft 0 in)
Max width 0·61 m (2 ft 0 in)
Max height 1·02 m (3 ft 4 in)

WEIGHTS AND LOADINGS:
Weight empty 258·5 kg (570 lb)
Max T-O weight 385·5 kg (850 lb)
Max wing loading 42·94 kg/m² (8·8 lb/sq ft)
Max power loading 9·40 kg/kW (15·45 lb/hp)

PERFORMANCE (estimated, at max T-O weight):
Never-exceed speed 173 knots (321 km/h; 200 mph)
Max level speed 96 knots (177 km/h; 110 mph)
Max cruising speed 83 knots (153 km/h; 95 mph)
Econ cruising speed 78 knots (145 km/h; 90 mph)
Stalling speed 39 knots (73 km/h; 45 mph)
Max rate of climb at S/L 152 m (500 ft)/min
Service ceiling 4,265 m (14,000 ft)
T-O run 152 m (500 ft)
Landing from 15 m (50 ft) 254 m (833 ft)
Range with max fuel 173 nm (320 km; 200 miles)

UNITED STATES OF AMERICA

ACE

ACE AIRCRAFT MANUFACTURING CO

ADDRESS:
106 Arthur Road, Asheville, North Carolina 28806
Telephone: (704) 252-4325
OWNER: Thurman G. Baird
SALES AND PUBLIC RELATIONS MANAGER:
V. P. Baird

This company is a successor to the original Corben Aircraft Company, which was established in 1923 and began manufacturing the Baby Ace single-seat ultra-light monoplane in kit form in 1931.

The Corben assets were acquired in 1953 by Mr Paul Poberezny, President of the Experimental Aircraft Association. With Mr S. J. Dzik, a former Waco engineer, he completely redesigned the Baby Ace, with the intention of offering it in the form of plans and kits of parts for amateur construction. All rights in the new version, known as the Model C, were sold to Mr Cliff DuCharme of West Bend, Wisconsin, to dispel any suggestions of the Experimental Aircraft Association being concerned with a profit-making venture. Again the Baby Ace was redesigned, as the Model D, and special tools were built to produce Baby Ace components in quantity. At the same time, the side-by-side two-seat Junior Ace was redesigned as the Junior Ace Model E.

In 1961, the company was acquired by Mr Edwin T. Jacob of McFarland, Wisconsin, from whom all rights were purchased by the present owner in 1965. Plans, kits and parts of the Baby Ace and Junior Ace are available to amateur builders from Ace Aircraft, together with plans of the Flaglor Scooter, described in this section of *Jane's*.

Ace Aircraft also has full rights in the American Flea Ship and Heath Parasol light aircraft, of which plans are available; descriptions of these two aircraft appeared in the 1970-71 *Jane's*.

BABY ACE MODEL D

The prototype of the redesigned Baby Ace Model D flew for the first time on 15 November 1956. Large numbers have since been built by amateurs, some of whom have introduced authorised refinements to the basic design. At least one Baby Ace is flying with a float landing gear. That shown in the accompanying illustration was built by EAA Chapter 60 of Janesville, Wisconsin. The floats were constructed of glassfibre, using a wood pattern and glassfibre mould, and have spruce struts wrapped in glassfibre, with cast aluminium end fittings. Twin water rudders are fitted for surface handling, and performance is claimed to be comparable with that of a Piper Cub seaplane with 48·5 kW (65 hp) engine.

TYPE: Single-seat ultra-light monoplane.

WINGS: Braced parasol monoplane. Wing section Clark Y (modified). Dihedral 1°. Incidence 1°. Fabric-covered two-spar wood structure. Fabric-covered wood ailerons. No flaps.

FUSELAGE: Welded steel tube structure, fabric-covered.

TAIL UNIT: Wire-braced steel tube structure, fabric-covered.

LANDING GEAR: Non-retractable tailwheel type. Combination special tubing and spring shock-absorption. Goodrich 8·00-4 main wheels. Scott hydraulic brakes. Wheel spats optional. Steerable tailwheel. Edo 1140 floats on seaplane version.

POWER PLANT: One Continental A65, A85, C65 or C85 flat-four engine of 48·5-63·5 kW (65-85 hp), driving a two-blade wood fixed-pitch propeller. Fuel in tank aft of firewall with capacity of 63·6 litres (16·8 US gallons). Oil capacity 3·8 litres (1 US gallon).

ACCOMMODATION: Single seat in open cockpit. Wide door on starboard side. Space for 4·5 kg (10 lb) baggage.

DIMENSIONS, EXTERNAL:
Wing span 8·05 m (26 ft 5 in)
Wing chord (constant) 1·37 m (4 ft 6 in)
Wing area, gross 10·43 m² (112·3 sq ft)
Wing aspect ratio 5·95
Length overall 5·40 m (17 ft 8¾ in)
Height overall 2·02 m (6 ft 7¾ in)
Tailplane span 2·13 m (7 ft 0 in)
Wheel track 1·83 m (6 ft 0 in)
Wheelbase 3·96 m (13 ft 0 in)

WEIGHTS:
Weight empty, equipped 261 kg (575 lb)
Max T-O weight:
48·5 kW (65 hp) 431 kg (950 lb)
63·5 kW (85 hp) landplane or seaplane
522 kg (1,150 lb)

PERFORMANCE (48·5 kW; 65 hp engine, at max T-O weight):
Max level speed at S/L 96 knots (177 km/h; 110 mph)
Max cruising speed
87-91 knots (160-169 km/h; 100-105 mph)
Stalling speed 30 knots (54·7 km/h; 34 mph)
Max rate of climb at S/L 365 m (1,200 ft)/min
Service ceiling 4,875 m (16,000 ft)
T-O run 60 m (200 ft)
Landing run 76 m (250 ft)
Range with max fuel 303·5 nm (560 km; 350 miles)

JUNIOR ACE MODEL E

The Junior Ace Model E differs from the Baby Ace Model D in being a side-by-side two-seater. It is powered usually by a 63·5 kW (85 hp) Continental C85 flat-four engine, and the details refer to an aircraft with this power plant that was built by Mr Louis C. Seno of Melrose Park, Illinois.

First flown on 2 August 1966, it has a cockpit 7·5 cm (3 in) wider and 10 cm (4 in) deeper than that of the standard Model E, a full electrical system and increased fuel capacity of 85 litres (22·5 US gallons).

DIMENSIONS, EXTERNAL:
Wing span 7·92 m (26 ft 0 in)
Wing chord (constant) 1·37 m (4 ft 6 in)
Wing aspect ratio 5·95
Length overall 5·50 m (18 ft 0 in)
Height overall 2·00 m (6 ft 7 in)

WEIGHTS:
Weight empty 367 kg (809 lb)
Max T-O weight 606 kg (1,335 lb)

PERFORMANCE:
Max level speed at S/L 113 knots (209 km/h; 130 mph)

Whittaker MW2 Excalibur, with ducted propeller

Baby Ace Model D (Continental A65 engine) *(Ray Bohrear)*

Cruising speed	91 knots (169 km/h; 105 mph)	Service ceiling	3,050 m (10,000 ft)	Landing run	183 m (600 ft)
Landing speed	57 knots (105 km/h; 65 mph)	T-O run	122 m (400 ft)	Range with max fuel	303·5 nm (560 km; 350 miles)

AEROCAR

AEROCAR INC

Head Office and Works:
Box 1171, Longview, Washington 98632
Telephone: (206) 423-8260

President and General Manager:
Moulton B. Taylor

Aerocar Inc began developing in February 1948 a flying automobile designed by Mr M. B. Taylor. The prototype Aerocar, with a Lycoming O-290 engine, was completed in October 1949. It was followed by a pre-production Aerocar Model I, with Lycoming O-320 engine, and this was used for tests which led to FAA airworthiness certification of the Aerocar on 13 December 1956.

Four additional Model I Aerocars were completed subsequently for demonstration tours of the United States and for sale to customers. One of these is fitted with the more powerful Lycoming O-360 engine and has also been certificated by the FAA.

The accumulated road travel on the six Model I Aerocars is well over 200,000 miles and they have logged a total of more than 5,000 flying hours. No further production of this model is planned, but development of the Aerocar has continued, and many changes have been made to the hand-built Model Is, enhancing both the flight performance and the road operation. A prototype of a refined version known as the Model III Aerocar was also built, and details of this were given in the 1970-71 *Jane's*. It was intended to be followed by a four-seat Aerocar IV, which would be based on an automobile in the class of the Ford Pinto and would have a normal cruising speed in flight of 150 knots (278 km/h; 172 mph). Current and projected US government safety and environmental requirements for automobiles suggest that an Aerocar able to conform with them would be so heavy and expensive that it could be neither practical nor economical. As a result, any future development will depend on whether the US government decides to exempt vehicles of this kind from the regulations which govern the operation of automobiles. In the meantime, Aerocar Inc has continued to devote most of its activity to other projects, notably the single-seat Mini-Imp, which embodies features of the Aerocar and is suitable for amateur construction.

In 1966 Aerocar built the prototype of a light flying-boat for a private customer. It was followed by two further aircraft, named Coot, on similar lines but provided with tricycle landing gear for amphibious operation. Many sets of plans for this aircraft, now known as the Sooper-Coot, have been sold to amateur constructors.

AEROCAR SOOPER-COOT MODEL A

Known originally as the Coot Model A, but since renamed Sooper-Coot Model A, the prototype of this aircraft flew for the first time in February 1971. It logged approximately 100 hours powered by an 89·5 kW (120 hp) Franklin 225 engine, driving a Sensenich fixed-pitch metal propeller. This was subsequently replaced by a 134 kW (180 hp) Franklin 335 engine, driving a Hartzell constant-speed metal propeller, and the aircraft has been flown extensively since 1972 with this more powerful engine.

The 'float-wing' configuration of the Sooper-Coot permits rough-water operation and, since the close proximity of the wings to the water forms a 'pressure wedge', unusually low take-off and landing speeds are possible without recourse to flaps or other lift-enhancing devices.

The structure is basically of wood, but the tailboom and tail unit can be of steel tube and fabric, wood monocoque or all-metal construction. The rearward-folding wings, of NACA 4415 section, can be folded by one person. The fabric-covered ailerons are of metal construction and statically balanced. There are no tip floats. Construction of the hull, which has only seven bulkheads, is straightforward, without the complication of wheel-well doors. Tailplane and elevators also fold and the elevators have trim tabs. All control surfaces are statically balanced. The tricycle-type landing gear is manually retractable into the wings, but an alternative powered retraction system is shown on the plans.

Recommended power plant for the Sooper Coot is a Franklin flat-six engine of either 134 or 164 kW (180 or 220 hp); until these engines again become available in quantity, from Poland, many builders of the Sooper-Coot are fitting a 157 kW (210 hp) Continental IO-360. Maximum fuel capacity is 189 litres (50 US gallons).

Certain component parts (including the glassfibre engine cowls, glassfibre hull shell, foredeck, instrument panel, tail fairings, engine cooling-fan blades and spring steel main landing gear legs), and plans, are available to amateur constructors. The company also maintains a list of recommended suppliers of welded assemblies and machined components. More than 400 Sooper-Coots were under construction in the Spring of 1976, with 14 already flying.

Dimensions, external:
- Wing span: 10·97 m (36 ft 0 in)
- Wing chord (constant): 1·52 m (5 ft 0 in)
- Wing area, gross: 16·72 m² (180 sq ft)
- Length overall: 6·10 to 6·71 m (20 ft 0 in to 22 ft 0 in)
- Height overall: 2·44 m (8 ft 0 in)
- Width folded: 2·44 m (8 ft 0 in)
- Tailplane span: 3·04 m (10 ft 0 in)

Weights:
- Weight empty: 499 kg (1,100 lb)
- Max T-O weight: 884 kg (1,950 lb)

Performance (134 kW; 180 hp engine, at max T-O weight):
- Never-exceed speed: 120 knots (223 km/h; 139 mph)
- Max cruising speed: 113 knots (209 km/h; 130 mph)
- Econ cruising speed at 50% power: 95·5 knots (177 km/h; 110 mph)
- Max rate of climb at S/L: 381 m (1,250 ft)/min
- T-O run (land): 61 m (200 ft)
- T-O (water): 6-8 sec

AEROCAR IMP

Design of the Imp (an acronym derived from Independently Made Plane) began in January 1972. Mr Taylor's aim was to evolve an aircraft suitable for the homebuilder, that could be constructed easily and quickly.

Many design features of the Aerocar are embodied in the Imp, including folding wings and torsion-bar type wheel suspension to allow easy and fast towing, and a pusher propeller aft of the tail unit. All-metal construction, using pop-rivets, bolt-together assemblies and limited welding is specified to reduce building time, which Mr Taylor estimates would average approximately 750 hours.

Construction of the prototype began in August 1972, but completion was delayed initially by the need for further investigation into the torsion and vibration characteristics of the long propeller shaft, as well as by a decision to design new-technology wings employing the results of recent NASA research. Subsequent non-availability of the Franklin Sport 4R engine which was intended to power the aircraft led to a shift of priority to the Mini-Imp programme (which see).

The future of the Imp was uncertain in the Spring of 1976. If the prototype is completed, it is likely that it will be fitted with wings of GA(PC)-1 aerofoil section (as on the Mini-Imp) in place of the wings of NACA 4415 section that are currently fitted. It is also hoped that the Imp might later be converted to 2/4-seat configuration, with a 119 kW (160 hp) turbocharged version of the Franklin engine.

Type: Two-seat homebuilt light aircraft.

Wings: Cantilever high-wing monoplane. Wing section currently NACA 4415. No dihedral. Incidence 4°. No sweepback. All-metal structure of constant chord. Wings fold aft, alongside fuselage, for towing or storage. All-metal ailerons, each with trim tab. No flaps.

Fuselage: All-metal structure with glassfibre shell.

Tail Unit: Inverted 'Vee' type, of all-metal construction, with trim tab on each rudder/elevator.

Landing Gear: Electrically-retractable tricycle type, main units retracting inward. Torsion-bar suspension. Special highway-type wheels and 6-ply tyres suitable for road towing. Tyre pressure 2·76 bars (40 lb/sq in). Wheel fenders for road-towing. Rosenhan wheel brakes. Nosewheel retracts separately from main wheels for road towing.

Power Plant: One 89·5 kW (120 hp) Franklin Sport 4R flat-four engine, driving a two-blade controllable-pitch Beech Roby pusher propeller. One fuel tank of 53 litres (14 US gallons) capacity in each wing root. Total fuel capacity 106 litres (28 US gallons). Refuelling points in upper surface of wings, inboard of wing fold. Oil capacity 7·5 litres (2 US gallons).

Accommodation: Two seats, side by side, under transparent cockpit canopy which opens upward and forward. Space for baggage aft of seats.

System: Electrical system powered by 12V 60A engine-driven alternator.

Electronics: Narco Escort 110 radio and Type 50A transponder.

Dimensions, external:
- Wing span: 8·84 m (29 ft 0 in)
- Wing chord, constant: 1·22 m (4 ft 0 in)
- Wing area, gross: 10·41 m² (112 sq ft)
- Wing aspect ratio: 7
- Length overall: 6·10 m (20 ft 0 in)
- Width, wings folded: 2·34 m (7 ft 8 in)
- Height overall (propeller blades horizontal): 1·68 m (5 ft 6 in)
- Tailplane span: 2·44 m (8 ft 0 in)
- Wheel track: 1·78 m (5 ft 10 in)
- Wheelbase: 2·13 m (7 ft 0 in)
- Propeller diameter: 1·83 m (6 ft 0 in)

Dimension, internal:
- Cockpit:
  - Max width: 1·12 m (3 ft 8 in)

Weights and Loadings (estimated):
- Weight empty: 431 kg (950 lb)
- Max T-O and landing weight: 703 kg (1,550 lb)
- Max wing loading: 67·4 kg/m² (13·8 lb/sq ft)
- Max power loading: 7·85 kg/kW (12·9 lb/hp)

Performance (estimated):
- Max cruising speed: more than 130 knots (241 km/h; 150 mph)
- Stalling speed: 44 knots (81 km/h; 50 mph)
- Max rate of climb at S/L: 244 m (800 ft)/min

AEROCAR MINI-IMP

The Mini-Imp is a single-seat version of the Aerocar Imp, to which it is generally similar. The basic structure, which is stressed to $\pm 9g$, is all-metal and is assembled with bolts and pop-rivets, thereby eliminating the need for welding skill. It is claimed that only a bench drill, sander and bandsaw are needed for home construction of the Mini-Imp, in addition to the usual hand tools and a pop-riveter. Average time taken for construction is 700 working hours, at a cost of $3,000, excluding engine, instruments and electronics.

The prototype is powered by a 1,900 cc Limbach engine, but the aircraft is suitable for power plants of 44·5 to 74·5 kW (60 to 100 hp), and it is intended at a later stage of development to experiment with a 2·54 kN (570 lb st) Williams turbofan installation.

The plans, when they become fully available, are expected to give details of extended wing panels which will increase overall span by 1·83 m (6 ft 0 in). Their object is to simplify the owner/builder's task during early solo flights, by permitting lower take-off and landing speeds. When experience has been gained, the outer panels can be removed easily.

By early 1976 the Mini-Imp prototype had completed about 40 flying hours. During the winter of 1975/76 Aerocar built a folding wing for this aircraft, which folds in a similar way to that of the Imp. This permits the builder to construct the aircraft in small areas. Aerocar is now releasing drawings for construction of this wing, and several people are already building wings in anticipation of getting their Mini-Imp flying by late 1976.

While all drawings for the Mini-Imp were not available by mid-1976, Aerocar was releasing drawings as quickly as they could be checked. To speed up this process, two more prototypes were under construction, one to be powered by a 61 kW (82 hp) Kawasaki four-cylinder motor-

Aerocar Mini-Imp single-seat homebuilt aircraft (1,900 cc Limbach engine)

Aerocar Sooper-Coot Model A lightweight homebuilt amphibian

cycle engine, with reduction gearing, and the other by a 74·5 kW (100 hp) Continental O-200 engine.

Aerocar intends eventually to supply plans, difficult-to-make parts and glassfibre components for the Mini-Imp.

The description of the Imp applies also to the Mini-Imp, except as detailed:

TYPE: Single-seat homebuilt light sporting aircraft.

WINGS: As for Imp, except wing section GA(PC)-1. Wing pivots 90° to a fore and aft position for towing by motor car. Outboard wing panels, to increase overall wing span by 1·83 m (6 ft 0 in), are optional.

FUSELAGE, TAIL UNIT AND LANDING GEAR: As for Imp.

POWER PLANT: Suitable for installation of engines from 44·5-74·5 kW (60 to 100 hp). Prototype has 51 kW (68 hp) Limbach 1,900 cc flat-four engine, driving a Hoffmann two-blade pusher propeller via an extended drive shaft incorporating a Flexidyne dry fluid coupling. Fuel tank in wing centre-section, capacity 45·4 litres (12 US gallons). Refuelling point on wing upper surface.

ACCOMMODATION: Single semi-reclining moulded glassfibre bucket seat beneath sideways-opening (to port) transparent canopy. Seat folds forward to provide access to space for 23 kg (50 lb) of baggage.

DIMENSIONS, EXTERNAL:
- Wing span, standard — 5·94 m (19 ft 6 in)
- Wing span, with optional extended wing panels — 7·77 m (25 ft 6 in)
- Wing chord, constant — 0·91 m (3 ft 0 in)
- Length overall — 4·88 m (16 ft 0 in)
- Height over fuselage — 1·24 m (4 ft 1 in)
- Height over propeller — 1·89 m (6 ft 2½ in)
- Tailplane span — 1·98 m (6 ft 6 in)
- Propeller diameter — 1·45 m (4 ft 9 in)

WEIGHTS:
- Weight empty (extended wing) — 236 kg (520 lb)
- Max T-O weight — 362 kg (800 lb)

PERFORMANCE (Limbach engine):
- Never-exceed speed — 260 knots (482 km/h; 300 mph)
- Max cruising speed at S/L — 130 knots (241 km/h; 150 mph)
- Stalling speed — 39 knots (73 km/h; 45 mph)
- Max rate of climb at S/L — 366 m (1,200 ft)/min
- T-O run — 244 m (800 ft)
- Range — over 434 nm (804 km; 500 miles)

AERO DESIGN
AERO DESIGN ASSOCIATES

ADDRESS:
Hangar One, Building 147, Opa-Locka Airport, Opa-Locka, Florida 33054
Telephone: (305) 685-8128
DIRECTOR: David Garber

AERO DESIGN DG-1

Nearing completion in mid-1976 was the first custom-built Unlimited Class racing aircraft to be constructed in the United States since the 1930s. Designated DG-1, it is intended to compete directly with the highly modified P-51 Mustangs, F8F Bearcats and Hawker Sea Furies, and also to be capable of breaking the existing world speed record for propeller-driven aircraft, which stands at 418·980 knots (776·447 km/h; 482·462 mph).

The DG-1 is of unusual design, with a cigar-shaped fuselage embodying a flush cockpit canopy. It is powered by two engines—one at each end of the fuselage—with respective tractor and pusher propellers. The tailplane has marked anhedral and a ventral tail fin of low profile is fitted. Stress limits are ±12*g*.

The director of Aero Design Associates, Mr David Garber, a pilot with Pan American World Airways, has several years' experience of racing in the Sport Biplane Class in modified Pitts Specials.

Construction of the DG-1 began in late 1974.

TYPE: Single-seat racing monoplane.

WINGS: Cantilever monoplane. NACA 065 wing section. Thickness/chord ratio 12%. Glassfibre-reinforced spar; foam-filled wing interior and Dacron covering.

FUSELAGE: Steel tube structure, covered with 2·4 mm (3/32 in) plywood, Dacron cloth and epoxy plastics.

TAIL UNIT: Vertical surface beneath fuselage. Tailplane is modified BD-5 unit with 10° anhedral.

LANDING GEAR: Fully retractable tricycle type, with single 0·25 m (10 in) diameter wheel on each unit. All three wheels retract rearward hydraulically into fuselage. Nosewheel lock coupled to control stick.

POWER PLANT: Two Mazda rotary-combustion motor car engines, modified to produce an estimated 254 kW (340 hp) each at more than 8,000 rpm. Engines have fuel injection, and custom-built Air Research turbosuperchargers. Each drives a Hartzell three-blade Aeromatic propeller. Usable fuel capacity 208 litres (55 US gallons) plus externally-carried 113·5 litre (30 US gallon) optional tank.

SYSTEMS: 14V DC electrical system, with AC for electric gyros. Full blind-flying, high-altitude capability.

ACCOMMODATION: Single semi-reclining seat under two-section Plexiglas canopy.

DIMENSIONS, EXTERNAL:
- Wing span — 6·10 m (20 ft 0 in)
- Wing chord at root — 1·07 m (3 ft 6 in)
- Wing chord at tip — 0·53 m (1 ft 9 in)
- Wing area, gross — 4·88 m² (52·5 sq ft)
- Wing aspect ratio — 7·6
- Length overall — 6·10 m (20 ft 0 in)
- Height overall — 1·68 m (5 ft 6 in)
- Tailplane span — 2·13 m (7 ft 0 in)
- Wheel track — 1·22 m (4 ft 0 in)
- Wheelbase — 2·24 m (7 ft 4 in)
- Propeller diameter — 1·73 m (5 ft 8 in)

WEIGHTS:
- Weight empty — 612 kg (1,350 lb)
- Max T-O weight, racing — 794 kg (1,750 lb)
- Max T-O weight — 862 kg (1,900 lb)

PERFORMANCE (estimated):
- Never-exceed speed — Mach 0·75
- Max level speed at S/L with racing wing — 391 knots (724 km/h; 450 mph)
- Max level speed with special speed record wing — over 434 knots (804 km/h; 500 mph)
- Normal cruising speed at 5,180 m (17,000 ft) — 304 knots (563 km/h; 350 mph)
- Stalling speed — 69·5 knots (129 km/h; 80 mph)
- Service ceiling — 7,315 m (24,000 ft)
- Service ceiling, one engine out — 5,180 m (17,000 ft)
- Range with max fuel, 15 min reserve, at 60% power — 600 nm (1,112 km; 691 miles)
- Range with internal fuel, 15 min reserve, at 60% power — 400 nm (740 km; 460 miles)
- Range, 100% power, no reserves — 290 nm (537 km; 334 miles)

AERONEERING
AERONEERING INC

ADDRESS:
PO Box 8, Claxton, Georgia 30417
Telephone: (912) 739-1930
PRESIDENT: Merle B. Miller

Mr Miller designed and built a lightweight sporting aircraft known as the Red Bare-un, of which plans were made available to amateur constructors. Design began in June 1970; construction was started in the following month and the prototype Red Bare-un flew for the first time on 3 June 1971. Total cost of construction was approximately $600.

Details of the Red Bare-un can be found in the 1974-75 *Jane's*. In early 1976 it was undergoing modification to reduce drag and height. Major modifications include lowering of the thrust line, repositioning the wings to a mid-wing configuration (retaining the cable bracing), and replacing the original power plant with a more powerful 1,300 cc Volkswagen engine. Completion of the modifications and flight testing were scheduled for Spring 1976.

A second aircraft has been designed by Mr Miller, and details of this follow:

MILLER LIL' RASCAL

This aircraft has been designed primarily for construction by amateurs and by schools. The prototype, which was scheduled for completion in June 1976, was built by a class of 15 students, under the direction of Mr B. G. Tippins of Claxton High School, specially formed for this project under the sponsorship of Chapter 330 of the Experimental Aircraft Association (Savannah, Georgia). Work on the prototype began in November 1975.

TYPE: Two-seat sporting aircraft.

WINGS: Strut-braced biplane wings of Clark Y section. Sweepback 9° at quarter-chord on upper wing only. Conventional structure of spruce and plywood, covered with Stits Poly-Fiber Dacron. Wire-braced internally. No landing or flying wires. Wooden ailerons. No flaps.

FUSELAGE: Conventional 4130 steel tube structure, with wooden stringers, covered with Stits Poly-Fiber Dacron.

TAIL UNIT: Structure of 4130 steel tube, covered with Stits Poly-Fiber Dacron. Elevators fitted. No tabs.

LANDING GEAR: Non-retractable tailwheel type. Steel

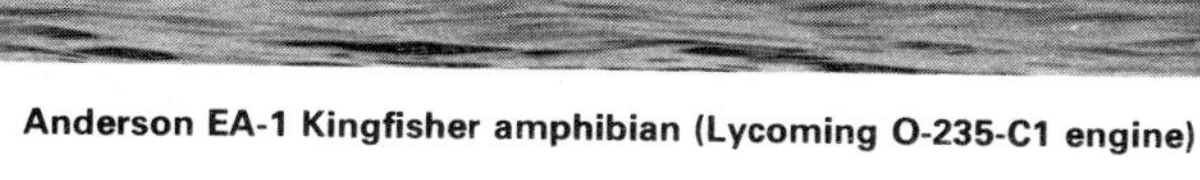

Anderson EA-1 Kingfisher amphibian (Lycoming O-235-C1 engine)

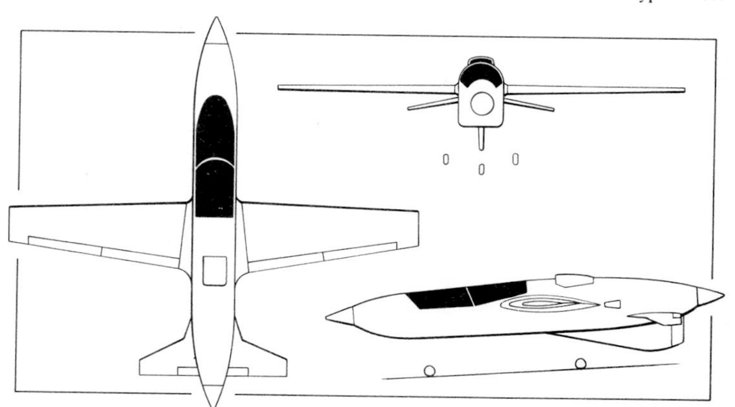

Aero Design DG-1 Unlimited Class racing aircraft *(Michael A. Badrocke)*

spring main legs, with wheels size 5·00-5. Disc brakes.

POWER PLANT: One 63·5 kW (85 hp) Continental C85-8 flat-four engine, driving a two-blade fixed-pitch propeller. Main fuel tank forward of cockpit; auxiliary tanks in upper-wing roots. Oil capacity 4·75 litres (1·25 US gallons).

ACCOMMODATION: Two seats side by side in open cockpit.

DIMENSIONS, EXTERNAL:

Wing span (upper wing)	6·30 m (20 ft 8 in)
Wing chord (constant)	1·07 m (3 ft 6 in)
Wing area, gross	11·15 m² (120 sq ft)
Propeller diameter	1·75 m (5 ft 9 in)
Propeller ground clearance	0·203 m (8 in)

DIMENSION, INTERNAL:

Cockpit:

Max width	0·85 m (2 ft 9⅜ in)

WEIGHTS AND LOADINGS:

Weight empty (approx)	281 kg (620 lb)
Max T-O weight	499 kg (1,100 lb)
Max wing loading	44·7 kg/m² (9·16 lb/sq ft)
Max power loading	7·87 kg/kW (12·95 lb/hp)

PERFORMANCE (estimated):

Max cruising speed	78 knots (145 km/h; 90 mph)

AEROSPORT
AEROSPORT INC

ADDRESS:

Box 278, Holly Springs, NC 27540

Telephone: (919) 552-6375

PRESIDENT: E. B. Trent

The late Mr H. L. Woods, former chief engineer of Bensen Aircraft Corporation, designed and built a number of aircraft and air cushion vehicles, including flex-wing and rotating-wing aircraft, and a variety of gliders.

In 1970 Mr Woods formed a company known as Aerosport Inc to design, manufacture and maintain aircraft, and to continue the sale of plans and kits. Details of designs marketed by this company follow:

AEROSPORT WOODY PUSHER

The prototype Woody Pusher was designed originally with a fuselage of wooden construction, plywood-covered, with fabric covering overall, and was powered by a 48·5 kW (65 hp) Lycoming engine. Mr Woods subsequently redesigned the fuselage and landing gear and increased the engine power.

Several hundred sets of construction plans for the Woody Pusher have been sold, and at least 27 aircraft have been completed and flown.

TYPE: Two-seat amateur-built light aircraft.

WINGS: Braced parasol monoplane, with Vee streamline-section main bracing struts each side and multi-strut centre-section cabane structure. Wing section NACA 4412. Two-spar wood structure, with metal leading-edge and fabric covering overall. Conventional ailerons and trailing-edge flaps.

FUSELAGE: Welded steel tube structure with fabric covering.

TAIL UNIT: Wire and strut-braced type. Ground-adjustable tab on rudder.

LANDING GEAR: Non-retractable tailwheel type. Cantilever spring steel main gear. Champion wheels. Wheel fairings on main gear optional.

POWER PLANT: One 56 kW (75 hp) Continental flat-four engine, driving a two-blade wooden fixed-pitch pusher propeller type LYL 36-68 SEN. Provision for other engines in 48·5-63·5 kW (65-85 hp) range. Fuel tank above wing, forward of engine, capacity 45 litres (12 US gallons).

ACCOMMODATION: Two seats in tandem in open cockpit.

DIMENSIONS, EXTERNAL:

Wing span	8·84 m (29 ft 0 in)
Wing chord, constant	1·37 m (4 ft 6 in)
Wing area, gross	12·07 m² (130 sq ft)
Length overall	6·22 m (20 ft 5 in)
Height overall	2·13 m (7 ft 0 in)
Tailplane span	2·62 m (8 ft 7⅛ in)

WEIGHTS:

Weight empty	285 kg (630 lb)
Max T-O weight	522 kg (1,150 lb)

PERFORMANCE (at max T-O weight):

Max level speed at S/L	85 knots (158 km/h; 98 mph)
Cruising speed	76 knots (140 km/h; 87 mph)
Stalling speed	39 knots (72 km/h; 45 mph)
Max rate of climb at S/L	183 m (600 ft)/min
T-O to 15 m (50 ft)	457 m (1,500 ft)
Landing from 15 m (50 ft)	305 m (1,000 ft)
Endurance with max fuel	2 hr 30 min

AEROSPORT RAIL Mk II

In January 1970 Mr Woods set out to design an aircraft that would be as simple as possible for construction by amateur builders, requiring no specialised knowledge of constructional techniques or the need for a comprehensive selection of tools. Special emphasis was also placed on safety, ease of handling and maintenance, and economy in operation. The appearance of the resulting design led to its being named Rail.

Construction of the prototype began in May 1970 and the first flight was made on 4 November of that year. FAA certification in the Experimental category was awarded on 24 June 1971.

The prototype Rail was powered originally by two 24·5 kW (33 hp) modified snowmobile engines. Later Mr Woods modified the design to utilise a single Volkswagen modified motor car engine with a capacity of 1,600 to 2,100 cc, and this is now standard.

Plans, a construction manual and kits of materials and components are available to amateur constructors, and 175 sets of plans had been sold by early 1976. At least six Rails are under construction and two are flying.

TYPE: Single-seat homebuilt lightweight aircraft. Aerobatics and spins are prohibited.

WINGS: Cantilever low-wing monoplane. Wing section NACA 23015. Dihedral 4°. Incidence 3°. No sweepback. All-metal two-spar structure of 2024-T3 and 6061-T6 light alloy. Plain ailerons of light alloy construction with piano hinge at upper surface. No flaps. No trim tabs. Endplates at wingtips.

FUSELAGE: Consists of a 5 cm by 12·5 cm (2 in by 5 in) extruded fuselage boom of 6061-T6 light alloy. Boom folds sideways aft of wing.

TAIL UNIT: Cantilever T-tail of light alloy construction. Fixed-incidence tailplane. Elevator trimmed by spring-loading of control column. Trim tab on rudder.

LANDING GEAR: Non-retractable tricycle type. Cantilever spring main gear struts of light alloy. Nosewheel shock-absorption by rudder in compression. Main wheels and tyres size 5·30 × 4·50-6; nosewheel and tyre size 2·80 × 2·50-4. Tyre pressure 1·38 bars (20 lb/sq in). Barrel-type wheel brakes. Nosewheel steerable.

POWER PLANT: One Volkswagen modified motor car engine, pylon-mounted from fuselage boom, driving a two-blade pusher propeller. Fuel tank mounted on pylon structure, forward of engine, capacity 38 litres (10 US gallons). Filler cap on top of tank.

ACCOMMODATION: Open seat for pilot.

SYSTEM: Electrical power supplied by engine-driven alternator.

DIMENSIONS, EXTERNAL:

Wing span	7·10 m (23 ft 3½ in)
Wing chord, constant	1·07 m (3 ft 6 in)
Wing area, gross	7·57 m² (81·5 sq ft)
Wing aspect ratio	6·66
Length overall	4·80 m (15 ft 9 in)
Length, fuselage folded	2·44 m (8 ft 0 in)
Height overall	1·83 m (6 ft 0 in)
Tailplane span	1·83 m (6 ft 0 in)
Wheel track	1·52 m (5 ft 0 in)
Wheelbase	1·63 m (5 ft 4 in)
Propeller diameter	1·37 m (4 ft 6 in)

WEIGHTS AND LOADING:

Weight empty	203 kg (446 lb)
Max T-O and landing weight	331 kg (730 lb)
Max wing loading	43·5 kg/m² (8·9 lb/sq ft)

PERFORMANCE (at max T-O weight):

Never-exceed speed	108 knots (201 km/h; 125 mph) IAS
Max level speed at S/L	78 knots (145 km/h; 90 mph)
Max cruising speed	69 knots (129 km/h; 80 mph)
Stalling speed, power off	39 knots (72·5 km/h; 45 mph) IAS
Max rate of climb at S/L	335 m (1,100 ft)/min
Service ceiling (calculated)	3,660 m (12,000 ft)
Range	191 nm (354 km; 220 miles)

AEROSPORT QUAIL

Though very different in appearance to the Rail, the Quail has a similar wing and the same type of landing gear. Its design also began in January 1970, and construction of the prototype was started in July 1971. It flew for the first time in December 1971.

The prototype had an all-moving tailplane, but Mr Woods designed a fixed-incidence tailplane with elevators before plans and construction kits were made available to amateur constructors. About 250 sets of plans had been sold by early 1976. Twenty-two Quails are under construction and six are flying.

TYPE: Single-seat lightweight homebuilt cabin monoplane.

WINGS: Cantilever high-wing monoplane, otherwise as described for Rail except for slightly increased span and trailing-edge flaps. Endplates optional.

FUSELAGE: Semi-monocoque all-metal structure of 2024-T3 light alloy.

TAIL UNIT: Cantilever all-metal structure of light alloy, with swept vertical surfaces. Fixed-incidence tailplane. Trim tab on rudder and elevator of prototype; optional for later aircraft.

LANDING GEAR: As for Rail. Optional wheel fairings.

POWER PLANT: One 1,600 cc modified Volkswagen motor car engine, driving an Aymar 54-36 two-blade fixed-pitch wooden propeller. Provision for installation of Volkswagen engines from 1,500 cc to 1,800 cc capacity. Two fuel tanks in wings, and 7·5 litre (2 US gallon) meter tank, with total capacity of 38 litres (10 US gallons). Refuelling point in top of each wing.

ACCOMMODATION: Single seat in enclosed cabin, which is heated. Door on starboard side is hinged at top and opens upwards. Stowage for 9 kg (20 lb) baggage.

EQUIPMENT: Electric starter for engine.

DIMENSIONS, EXTERNAL:

Wing span	7·32 m (24 ft 0 in)
Wing chord, constant	1·07 m (3 ft 6 in)
Wing area, gross	7·8 m² (84 sq ft)
Wing aspect ratio	6·87
Length overall	4·85 m (15 ft 11 in)
Height overall	1·69 m (5 ft 6½ in)
Tailplane span	1·83 m (6 ft 0 in)
Wheel track	1·52 m (5 ft 0 in)
Wheelbase	1·22 m (4 ft 0 in)
Propeller diameter	1·37 m (4 ft 6 in)

WEIGHTS AND LOADINGS:

Weight empty	242 kg (534 lb)
Normal T-O weight	345 kg (762 lb)
Max T-O weight	359 kg (792 lb)
Normal wing loading	44·4 kg/m² (9·1 lb/sq ft)
Normal power loading	9·26 kg /kW (15·2 lb/hp)

PERFORMANCE (at max T-O weight):

Max level speed at S/L	113 knots (209 km/h; 130 mph)
Max cruising speed	100 knots (185 km/h; 115 mph)
Econ cruising speed	95·5 knots (177 km/h; 110 mph)
Max manoeuvring speed	78 knots (145 km/h; 90 mph)
Stalling speed, flaps down	42 knots (78 km/h; 48 mph)
Max rate of climb at S/L	259 m (850 ft)/min

Aerosport Rail prototype, re-engined with Volkswagen modified motor car engine and re-designated Mk II

Prototype Aerosport Scamp aerobatic biplane (1,834 cc Volkswagen modified motor car engine)

Aerosport Quail single-seat lightweight cabin monoplane

Aerosport Woody Pusher built in Canada (65 hp Continental engine) *(Neil A. Macdougall)*

Service ceiling (estimated) 3,660 m (12,000 ft)
T-O run 91 m (300 ft)
Landing run 122 m (400 ft)
Range with max fuel, no reserve
200 nm (370 km; 230 miles)

AEROSPORT SCAMP

The prototype of the single-seat all-metal Scamp flew for the first time on 21 August 1973. It was intended primarily for operation from grass strips, and tricycle landing gear was chosen as being more rational for a generation of amateur pilots who had received their initial flight training on aircraft equipped with landing gear of this configuration. Stressed to +6*g* and −3*g*, the Scamp can be used for limited aerobatics; and emphasis has been placed on simple construction techniques to make it an easy project for the homebuilder.

Plans and kits of parts, except for the Volkswagen engine, are available to amateur constructors. More than 500 sets of plans had been sold by early 1976. Four Scamps are flying and a further 18 are under construction from Aerosport kits.

TYPE: Single-seat homebuilt light aircraft.

WINGS: Braced biplane structure, with Vee-type interplane strut each side. Flying and landing wires of streamline section. Single 5 cm by 12·5 cm (2 in by 5 in) extruded section of 6061-T6 light alloy forms a pylon to support the centre-section of the upper wing. All-metal two-spar structures of light alloy. Plain ailerons of light alloy construction, with piano hinge at upper surface, on upper wing only. No flaps. No trim tabs.

FUSELAGE: Semi-monocoque all-metal structure of light alloy.

TAIL UNIT: Braced T-tail of light alloy construction. Single bracing strut on each side. Fixed-incidence tailplane. Ground-adjustable trim tab on rudder.

LANDING GEAR: Non-retractable tricycle type. Cantilever spring main-gear struts of light alloy. Wheel fairing on each unit.

POWER PLANT: Prototype has one 44·5 kW (60 hp) 1,834 cc Volkswagen modified motor car engine. Design suitable for Volkswagen engines of 1,600 cc to 2,100 cc. Aymar 54-38 two-blade fixed-pitch wooden propeller. Fuel tank in fuselage nose, aft of firewall, capacity 30·5 litres (8 US gallons). Refuelling point on fuselage upper surface, forward of windscreen.

ACCOMMODATION: Single seat in open cockpit.

DIMENSIONS, EXTERNAL:
Wing span 5·33 m (17 ft 6 in)
Wing area, gross 9·52 m² (102·5 sq ft)
Length overall 4·27 m (14 ft 0 in)
Height overall 1·69 m (5 ft 6½ in)

WEIGHTS AND LOADINGS (prototype):
Weight empty 236 kg (520 lb)
Max T-O weight 348 kg (768 lb)
Max wing loading 36·6 kg/m² (7·49 lb/sq ft)
Max power loading 7·79 kg/kW (12·8 lb/hp)

PERFORMANCE:
Never-exceed speed 108 knots (201 km/h; 125 mph)
Max level speed 82 knots (153 km/h; 95 mph)
Cruising speed 74 knots (137 km/h; 85 mph)
Max manoeuvring speed
72 knots (134 km/h; 83 mph)
Stalling speed 39 knots (73 km/h; 45 mph)
Service ceiling (estimated) 3,660 m (12,000 ft)
T-O run 91 m (300 ft)
Landing run 122 m (400 ft)
Range at cruising speed 130 nm (241 km; 150 miles)

ANDERSON
ANDERSON AIRCRAFT CORPORATION

ADDRESS:
PO Box 422, Raymond, Maine 04071

Mr Earl Anderson, a Boeing 747 captain flying for Pan American World Airways, designed and built an original light amphibian which he named EA-1 Kingfisher. The project occupied a period of nine years from start of design to completion, at a cost of around $5,500, and the first flight was made on 24 April 1969.

After a time, Mr Anderson replaced the original 74·5 kW (100 hp) Continental O-200 engine by an 86 kW (115 hp) Lycoming O-235-C1, driving a Sensenich M76AM-4-44 propeller. With this power plant the Kingfisher has an empty weight of 495 kg (1,092 lb), a max T-O weight of 725 kg (1,600 lb) and improved performance. By January 1975 the prototype had accumulated a total of more than 600 flying hours.

Plans available to amateur constructors cover the increase in weight. The Kingfisher was designed originally to accept alternative power plants up to a maximum of 104·4 kW (140 hp), but on the basis of experience with the 86 kW (115 hp) Lycoming engine, Mr Anderson is discouraging homebuilders from installing more powerful engines than this.

Mr Anderson formed Anderson Aircraft Corporation to market plans of the Kingfisher. By January 1976 well over 200 sets of plans had been sold, and more than 100 Kingfishers were under construction in the USA, Canada, Mexico, Sweden, Germany and Panama. At least ten homebuilt Kingfishers are known to be flying.

The following details apply to Mr Anderson's Kingfisher in its original configuration:

ANDERSON EA-1 KINGFISHER

TYPE: Two-seat light amphibian.

WINGS: Braced high-wing monoplane with streamline-section Vee bracing struts each side (standard Piper Cub wing). Stabilising floats mounted beneath wings, adjacent to wingtips, are constructed of ⅜ in square mahogany stringers, covered with ¹/₁₆ in mahogany plywood coated with glassfibre. Each float weighs 4·1 kg (9 lb).

FUSELAGE: Conventional flying-boat hull of wooden construction with spruce frames and longerons, covered with ¹/₁₆ in and ¼ in mahogany plywood coated with glassfibre.

TAIL UNIT: Conventional strut-braced tail unit.

LANDING GEAR: Retractable tailwheel type. Each main unit is retracted forward, manually and individually, with spring-loaded assist mechanism.

POWER PLANT: One 74·5 kW (100 hp) Continental O-200 flat-four engine, driving a fixed-pitch two-blade tractor propeller. Single fuel tank in hull, immediately forward of windscreen, capacity 76 litres (20 US gallons).

ACCOMMODATION: Two seats, side by side, in enclosed cabin. Piper Tri-Pacer windscreen.

DIMENSIONS, EXTERNAL:
Wing span 11·00 m (36 ft 1 in)
Length overall 7·16 m (23 ft 6 in)
Height overall 2·44 m (8 ft 0 in)
Wheel track 1·52 m (5 ft 0 in)

WEIGHTS:
Weight empty 468 kg (1,032 lb)
Max T-O weight 680 kg (1,500 lb)

PERFORMANCE (at max T-O weight):
Cruising speed 74 knots (136 km/h; 85 mph)
Max rate of climb at S/L 150-180 m (500-600 ft)/min
Service ceiling 3,050 m (10,000 ft)

BACKSTROM
AL BACKSTROM

ADDRESS:
Fort Worth, Texas

BACKSTROM WPB-1 PLANK

In the 1950s Al Backstrom designed, constructed and flew an aircraft which he named the Plank sailplane, with help from Jack Powell and Phil Easley. In 1969 he began studies for a powered version of the Plank, and construction was initiated in 1972, most of the work being undertaken by Van White, a Director of the EAA, at his electrical workshop. It was intended originally to fit a Sachs Wankel or OMC snowmobile Wankel engine, but as neither was available a fan-cooled single-ignition Kiekhaefer Aeromarine 440 had to be installed, requiring shock mountings and a belt drive system.

Based at Lubblock, Texas, the aircraft has made a number of flights along a runway, as a result of which it has undergone considerable refinement. First flight of the Plank with a new tricycle landing gear, replacing the original, too fragile tandem-wheel and outriggers arrangement, was scheduled for the latter part of 1976.

A two-seat version of the Plank and a more refined single-seat version are planned for the future.

TYPE: Single-seat tailless monoplane.

WINGS: Shoulder-mounted unswept constant-chord wings with cambered tips and tip-mounted fins and rudders. Ribs are spaced at 0·15 m (6 in) intervals. Moulded plywood leading-edge. Fabric covering glued to wooden structure. Control by elevons.

FUSELAGE: Tubular structure with fabric covering. Flush engine air-ducts forward of wing.

LANDING GEAR: Non-retractable tricycle type.

POWER PLANT: One single-ignition Kiekhaefer Aeromarine 440 engine, driving a two-blade pusher propeller.

ACCOMMODATION: Single seat under transparent canopy.

DIMENSIONS, EXTERNAL:
Wing span 6·60 m (21 ft 8 in)
Wing chord, constant 1·37 m (4 ft 6 in)
Wing area 9·06 m² (97·5 sq ft)
Length 3·38 m (11 ft 1 in)

WEIGHT (with tandem landing gear):
Weight empty, with radio and battery 177 kg (390 lb)

PERFORMANCE:
Max level speed 95 knots (175 km/h; 109 mph) IAS
T-O speed 43·5-48 knots (81-89 km/h; 50-55 mph)

BAKENG
BAKENG AIRCRAFT

ADDRESS:
19025 92nd W, Edmonds, Washington 98020

Mr Gerald Bakeng designed and built a two-seat high-performance parasol-wing monoplane known as the Duce, which received the EAA's 'Outstanding New Design' and 'Design Improvement' awards in 1971. It was followed in 1972 by a tandem two-seat biplane known as the Double Duce.

Plans of both aircraft are available to other builders.

BAKENG DUCE

Design and construction of the original Duce began in October 1969; it was completed six months later at a cost of approximately $1,500. The first flight was made on 2 April 1970. Plans are available to amateur constructors, and more than 200 sets have been sold.

TYPE: Two-seat homebuilt light sporting aircraft.

WINGS: Braced parasol-wing monoplane, with Vee bracing struts each side. Streamline centre-section struts. Wing section Clark Y modified. Dihedral 1°. Incidence 3°. No sweepback. Composite structure of spruce and 4130 steel tube, fabric-covered. Frise-type ailerons of wooden construction, fabric-covered. Trailing-edge flaps of wooden construction, fabric-covered. Ground-adjustable tabs on ailerons.

FUSELAGE: Welded 4130 steel tube Warren truss structure, with fabric covering.

TAIL UNIT: Conventional wire-braced structure of welded 4130 steel tube, fabric-covered. Incidence of tailplane ground-adjustable. Ground-adjustable trim tab on rudder.

LANDING GEAR: Non-retractable tailwheel type. Cantilever spring steel main units. Main-wheel tyres size 6·00-6, pressure 0·34 bars (5 lb/sq in). Goodyear puck-type wheel brakes. Glassfibre wheel fairings on main units. Design may be adapted for float or ski landing gear.

POWER PLANT: One 93 kW (125 hp) Lycoming O-290-G (GPU) flat-four engine in prototype, driving a Sensenich two-blade fixed-pitch metal propeller with spin-

Backstrom WPB-1 Plank tailless monoplane *(Howard Levy)*

Baker Special *Aquarius* single-seat Formula One racing aircraft

Bakeng Duce two-seat sporting monoplane *(Peter M. Bowers)*

Bakeng Double Duce two-seat sporting biplane (Continental R-670 radial engine) *(Peter M. Bowers)*

J-3M utility version of Barnett gyroplane

Barnett J-4B single-seat autogyro. A fixed fin is now fitted forward of the rudder

ner. Design suitable for other engines in 56-93 kW (75-125 hp) range. One fuel tank in wing centre-section, capacity 42 litres (11 US gallons), and one tank in forward fuselage, immediately aft of firewall, capacity 64 litres (17 US gallons). Total fuel capacity 106 litres (28 US gallons). Refuelling points on upper surface of wing centre-section and front fuselage. Oil capacity 7·5 litres (2 US gallons).

Accommodation: Two persons in tandem in open cockpits. Fold-down doors on starboard side.

Dimensions, external:

Wing span	9·25 m (30 ft 4 in)
Wing area, gross	12·8 m² (138 sq ft)
Wing chord, constant	1·37 m (4 ft 6 in)
Length overall	6·32 m (20 ft 9 in)
Height overall	2·44 m (8 ft 0 in)
Tailplane span	2·44 m (8 ft 0 in)
Wheel track	2·13 m (7 ft 0 in)
Wheelbase	4·88 m (16 ft 0 in)
Propeller diameter	1·88 m (6 ft 2 in)

Dimension, internal:

Cockpit: Max width	0·64 m (2 ft 1¼ in)

Weights and Loadings:

Weight empty	407 kg (898 lb)
Max T-O and landing weight	680 kg (1,500 lb)
Max wing loading	53·1 kg/m² (10·87 lb/sq ft)
Max power loading	7·31 kg/kW (12 lb/hp)

Performance (at max T-O weight):

Never-exceed speed	130 knots (241 km/h; 150 mph)
Max level speed at S/L	126 knots (233 km/h; 145 mph)
Max cruising speed at S/L	122 knots (225 km/h; 140 mph)
Econ cruising speed at S/L	91 knots (169 km/h; 105 mph)
Stalling speed, flaps down	31·5 knots (58 km/h; 36 mph)
Max rate of climb at S/L	610 m (2,000 ft)/min
Service ceiling	5,180 m (17,000 ft)
T-O run	46 m (150 ft)
Landing run	46 m (150 ft)
Range with max fuel	260 nm (482 km; 300 miles)

BAKENG DOUBLE DUCE

The Double Duce is similar to the Duce in all respects except for its biplane configuration. It has a fabric-covered steel tube fuselage and wooden wings, braced with N-type interplane struts and with ailerons on all four wings. The power plant can be almost any horizontally-opposed or radial engine in the 93-164 kW (125-220 hp) range.

No additional details had been received from Mr Bakeng by mid-1976.

BAKER

MARION E. BAKER

Address:

912 Salem Drive, Huron, Ohio 44839

Mr Marion Baker, a veteran racing pilot, is remembered as designer of the small delta-winged Delta Kitten, described in the 1962-63 *Jane's*. Since that time he has built three Formula One racing aircraft. First was the Cassutt *Ole Yaller,* built in partnership with Mr Robert Grieger in 1965. This was followed by the original-design *Boo Ray,* which has been flown very successfully since 1967. His latest racer is the Baker Special *Aquarius,* of which construction was started in November 1969 and completed in August 1970. It was first raced a few weeks later, at Reno. Best performance to date was a third place at Miami, Florida in 1973, with a recorded speed of 184 knots (341 km/h; 212 mph).

BAKER SPECIAL, *Aquarius*

Type: Single-seat Formula One racing aircraft.

Wings: Cantilever mid-wing monoplane. Wing section NACA 66-209. No dihedral. Incidence 1°. All-metal structure, with laminated spar, ribs and skins of 2024 T-3 aluminium alloy. Plain ailerons of similar construction. No trailing-edge flaps.

Fuselage: Welded steel tube structure of 4130 chrome molybdenum, with 2024 T-3 aluminium alloy skins.

Tail Unit: Cantilever structure, with spars, ribs and skins of 2024 T-3 light alloy.

Landing Gear: Non-retractable tailwheel type. Main legs are cantilever light alloy leaf springs. Main-wheel tyres size 5·00-5. Goodyear hydraulically operated disc brakes. Close-fitting light alloy fairings over main landing gear struts and main wheels.

Power Plant: One Continental O-200 flat-four engine, driving a two-blade metal fixed-pitch propeller with spinner. Engine rated at 74·5 kW (100 hp) at 2,600 rpm, and at 97 kW (130 hp) at 3,900 rpm for racing. Fuel tank in fuselage with capacity of 56 litres (14·8 US gallons).

Accommodation: Single seat under Plexiglas canopy. Nose-over protection bars built into pilot's headrest.

Systems: Hydraulic system for brakes only. No electrical system installed.

Dimensions, external:

Wing span	5·84 m (19 ft 2 in)
Wing area, gross	6·13 m² (66·0 sq ft)
Wing aspect ratio	5
Length overall	5·08 m (16 ft 8 in)

Height overall 1·47 m (4 ft 10 in)
WEIGHTS AND LOADINGS:
Weight empty 245 kg (540 lb)
Normal T-O weight 335 kg (740 lb)
Max T-O weight 403 kg (890 lb)
Max wing loading 65·8 kg/m² (13·48 lb/sq ft)
Max power loading 5·41 kg/kW (8·9 lb/hp)
PERFORMANCE (at normal T-O weight):
Max level speed 226 knots (418 km/h; 260 mph)
Cruising speed 156 knots (290 km/h; 180 mph)
Landing speed 56 knots (105 km/h; 65 mph)
Max rate of climb at S/L 610 m (2,000 ft)/min
T-O run 91 m (300 ft)
Range 304 nm (563 km; 350 miles)

BARNETT
BARNETT ROTORCRAFT COMPANY

ADDRESS:
4307 Olivehurst Avenue, Olivehurst, California 95961
Telephone: (916) 742-7416
PROPRIETOR AND DESIGNER: K. J. Barnett

This company has designed and built prototypes of two generally-similar ultra-light gyroplanes, designated J-3M and J-4B. The basic difference between the two lies in the power plant and the degree of skill needed to construct them. Plans, materials and kits of parts are available to amateur constructors.

Barnett Rotorcraft is reported to be developing a 71 kW (95 hp) Barnett-Corvair engine for installation on the basic J-4B airframe. It is also developing a two-seat autogyro, of which no details are available.

BARNETT J-3M and J-4B

The essential differences between these two versions of the Barnett gyroplane are as follows:

J-3M. Utility model with 48·5 kW (65 hp) Continental A65 engine and flat-sided fabric-covered cabin enclosure.

J-4B. Higher-performance version, with 63·5 kW (85 hp) Continental C85 engine and more-streamlined glassfibre nacelle, resulting in some changes to overall dimensions. Prototype flew for the first time on 15 July 1968.

The following details apply generally to both models:

TYPE: Single-seat light autogyro.

ROTOR SYSTEM: Single two-blade autorotating main rotor. Offset gimbal control head. Rotor blades of bonded aluminium with extruded spar. Pre-spin device, to spin up the rotor to about 250 rpm, to be made available as a bolt-on optional extra. Normal rotor rpm: J-3M 400; J-4B 425.

FUSELAGE: All-metal structure of welded 4130 steel tube. Bolt-together aluminium airframe under development for J-4B. Nacelle of J-3M has streamlined glassfibre nose, and fabric covering elsewhere. J-4B has all-glassfibre streamlined nacelle.

TAIL UNIT: Fin, rudder and fixed one-piece horizontal surfaces, carried on tubular steel extension from main fuselage structure. All surfaces of welded 4130 steel tube, with fabric covering.

LANDING GEAR: Non-retractable tricycle type, with 100 mm (4 in) solid rubber tailwheel or bumper. J-3M has no shock-absorbers; J-4B has compound spring-lever shock-absorption. Steerable nosewheel linked to rudder pedals. Main-wheel tyres size 14·5 × 5·00-6, pressure 0·69 bars (10 lb/sq in). Nosewheel tyre size 4·10 × 3·50-4. Disc brakes. Light alloy or plastics floats to be made available as optional items. Parking brake.

POWER PLANT: J-3M has one 48·5 kW (65 hp) Continental A65 flat-four engine, driving a Barnett two-blade resin-laminated birch fixed-pitch propeller. J-4B has one 63·5 kW (85 hp) Continental C85 flat-four engine, driving a Barnett 5756F two-blade wooden fixed-pitch propeller of 1·42 m (56 in) pitch. Other Continental engines of 48·5 to 74·5 kW (65-100 hp) may be fitted to J-4B. Single aluminium fuel tank under pilot's seat of J-3M, in rotor mast of J-4B. J-3M tank has capacity of 26·5 litres (7 US gallons); J-4B tank has a capacity of 56 litres (15 US gallons).

ACCOMMODATION: Single seat in open or fully enclosed cockpit. Moulded Plexiglas windscreen and optional cockpit canopy. Door can be fitted to fully-enclosed version. Baggage space under and behind seat of J-4B, which has much wider cockpit than J-3M.

ELECTRONICS: Prototype J-4B has Hughes 338 23-channel radio.

DIMENSIONS, EXTERNAL (A: J-3M; B: J-4B):
Diameter of rotor: A, B 7·32 m (24 ft 0 in)
Rotor blade chord (each): A, B 0·19 m (7½ in)
Rotor disc area, A, B 38·6 m² (415·5 sq ft)
Length overall:
A 3·45 m (11 ft 4 in)
B 3·71 m (12 ft 2 in)
Height overall:
A 2·18 m (7 ft 2 in)
B 2·34 m (7 ft 8 in)
Wheel track: A, B 2·00 m (6 ft 6¾ in)
Wheelbase: A, B 1·96 m (6 ft 5 in)
Propeller diameter: B 1·45 m (4 ft 9 in)
WEIGHTS (A: J-3M; B: J-4B):
Weight empty:
A 181 kg (400 lb)
B 200 kg (441 lb)
Max T-O weight:
A 294 kg (650 lb)
B 340 kg (750 lb)
PERFORMANCE (A: J-3M; B: J-4B):
Max level speed:
A 74 knots (137 km/h; 85 mph)
B 100 knots (185 km/h; 115 mph)
Econ cruising speed:
A 61 knots (113 km/h; 70 mph)
B 78 knots (145 km/h; 90 mph)
Max rate of climb at S/L:
A 152 m (500 ft)/min
B 213 m (700 ft)/min
Service ceiling:
A 1,830 m (6,000 ft)
B 4,265 m (14,000 ft)
T-O run: A, B 61 m (200 ft)
Landing run: A, B 0-6 m (0-20 ft)
Range:
A 104 nm (193 km; 120 miles)
B 217 nm (402 km; 250 miles)

BARNEY OLDFIELD
BARNEY OLDFIELD AIRCRAFT CO

ADDRESS:
PO Box 5974, Cleveland, Ohio 44101
Telephone: (216) 449-6300
PRESIDENT: Harvey R. Swack

On 23 February 1972 Mr Harvey Swack announced formation of this company, following sale of the Great Lakes Aircraft Company to Windward Aviation Inc of Enid, Oklahoma. This latter company is continuing to operate under the name of Great Lakes Aircraft Co (which see).

Barney Oldfield Aircraft Co markets plans and material kits for the 'Baby' Lakes, a scaled down version of the Great Lakes Sport Trainer, the prototype of which was designed and built by Mr Andrew Oldfield, who died during 1970. Over 600 sets of drawings had been sold by early 1976, and many 'Baby' Lakes are under construction. About 40 were flying by 1976.

The company is developing alternative power plants for the 'Baby' Lakes, and in early 1976 was flying a standard airframe fitted experimentally with an 80·5 kW (108 hp) Lycoming engine. With this power plant the aircraft has a rate of climb of 1,065 m (3,500 ft)/min and a cruising speed of 130 knots (241 km/h; 150 mph). Empty weight is increased to only 218 kg (480 lb).

Also under development are inverted oil and fuel systems for the 'Baby' Lakes, so that it can be used as a low-cost aerobatic aircraft. It is stressed for ±9g at the recommended 385 kg (850 lb) gross weight.

OLDFIELD 'BABY' LAKES

TYPE: Single-seat amateur-built sporting biplane.

WINGS: Braced biplane, with N-type interplane struts, double landing and flying wires and N-type centre-section support struts. Wing section modified M6, tapering to USA 27 46 cm (18 in) from tips. Incidence 2° 30′ on top wing, 1° 30′ on bottom wing. Wood structure of spruce spars and Warren truss ribs, with overall fabric covering. Ailerons on lower wings only. No flaps.

FUSELAGE: Welded steel tube structure, fabric-covered.

TAIL UNIT: Wire-braced welded steel tube structure, fabric-covered.

LANDING GEAR: Non-retractable tailwheel type. Oleo main legs with size 5·00-4 wheels. Steerable tailwheel.

POWER PLANT: One 59·5 kW (80 hp) Continental A80 flat-four engine, driving a two-blade fixed-pitch propeller. Provision for alternative engines of between 37·25 and 74·5 kW (50 and 100 hp), and several aircraft now under construction will have 1,500 and 1,600 cc Volkswagen engines. Fuel tank in front fuselage, capacity 45 litres (12 US gallons).

ACCOMMODATION: Single seat, normally in open cockpit. Cockpit canopy optional.

DIMENSIONS, EXTERNAL:
Wing span: top 5·08 m (16 ft 8 in)
Wing chord (both wings, constant)
0·91 m (3 ft 0 in)
Wing area, gross 7·99 m² (86 sq ft)
Length overall 4·19 m (13 ft 9 in)
Height overall 1·37 m (4 ft 6 in)
WEIGHTS (A80 engine):
Weight empty 215 kg (475 lb)
Max T-O weight 385 kg (850 lb)
PERFORMANCE (A80 engine, at max T-O weight):
Max level speed at S/L
117 knots (217 km/h; 135 mph)
Cruising speed at S/L
102 knots (190 km/h; 118 mph)
Stalling speed 43·5 knots (81 km/h; 50 mph)
Max rate of climb at S/L 610 m (2,000 ft)/min
Service ceiling 5,200 m (17,000 ft)
T-O run 91 m (300 ft)
Landing run (no brakes) 122 m (400 ft)
Max range 217 nm (400 km; 250 miles)

BEAUJON
HERBERT BEAUJON

ADDRESS:
1119 12th Avenue NE, Ardmore, Oklahoma 73401
Telephone: (405) 223-0587

Mr Beaujon has designed and built two prototypes of an ultra lightweight sporting monoplane named the Flybike which he claims to be the world's simplest and cheapest aircraft. Construction is such that first-time homebuilders should experience little difficulty, and flying controls are simple in the extreme. Forward and backward movement of a handlebar control arm operates the elevator, and side to side movement of the same handlebar controls rudder/spoilers mounted on the upper surface of each wingtip. As an option, instead of the rudder/spoilers, drag ailerons can be mounted at the lower surface of each wingtip.

The aircraft is supported on the ground by a single go-kart wheel beneath the open pilot's seat and a glassfibre tailskid at the rear. The necessity to keep one foot on the ground, to keep it balanced until forward speed is such that it can be balanced like a bicycle, is largely responsible for the name Flybike.

Design of the aircraft began in March 1973 and construction of the first prototype started in December 1974, with the first flight only three months later. Plans are available to amateur constructors, and 190 sets had been sold by 16 March 1975.

BEAUJON FLYBIKE

TYPE: Ultra lightweight homebuilt sporting aircraft.

WINGS: Braced monoplane. Wing section Clark Y modified. Dihedral 2°. Incidence 4°. No sweepback. Composite structure, with two Sitka spruce spars sandwiched between solid styrofoam aerofoil sections. Dynel or cotton fabric covering applied with epoxy resin as adhesive and coated with epoxy. Steel tube (4130 chrome molybdenum) cabane structure, integral with fuselage, projects above upper surface of wing and serves as anchor point for one landing wire each side. Two flying wires each side, attached one to each spar at approximately mid-span, are secured to the fuselage, adjacent to the base of the pilot's seat. Rudder/spoilers, constructed of a plywood-styrofoam-plywood sandwich, are mounted on the upper surface of each wing adjacent to the wingtips. Optional alternative drag ailerons, of similar construction, are affixed to the undersurface of each wing at its tip. Glassfibre skid, with aluminium tube tip, mounted near each wingtip.

FUSELAGE: Simple welded steel tube structure of 4130 chrome molybdenum. At the lower end is mounted the single landing wheel, and this structure serves also as mounting for the pilot's seat, the wings, power plant and tailboom.

TAIL UNIT: Strut-braced composite structure, mounted on tailboom which consists of 0·102 m (4 in) outside diameter 6061-T6 light alloy tube. For reinforcement, a spruce spar, to which dowels have been affixed by epoxy resin, is inserted within the tube, the dowels being located at bolt securing points. Tailplane formed from spruce box frame, filled with solid styrofoam and covered with epoxy resin. Elevator and twin fins carved from styrofoam block and covered with epoxy resin. Tailskid formed from glassfibre rod with tubular aluminium tip. Ground-adjustable trim tabs on port fin and elevator.

LANDING GEAR: Non-retractable monowheel, mounted at base of fuselage structure. No shock-absorption. Go-kart wheel with tyre size 10 × 3·00-4. No brakes.

POWER PLANT: One 6 kW (8 hp) Briggs and Stratton single-cylinder four-stroke aircooled engine, driving a two-blade wooden fixed-pitch pusher propeller. Fuel tank attached to side of engine, capacity 3·8 litres (1 US gallon). Refuelling point on top of tank. Oil capacity 1·30 litres (2¾ US pints).

ACCOMMODATION: Single seat in open position forward of fuselage vertical frame. Aircraft controlled by handlebar unit hinged from cabane structure, controlling

Bede BD-4, showing optional wheel fairings completely enclosing main wheels

Bede BD-5B, the basic homebuilt version of the Micro

Artist's impression of Beaujon Flybike

Oldfield 'Baby' Lakes built by Russ Sholle of Ohio

elevator and rudder/spoilers, with twist-grip for throttling of power plant.

DIMENSIONS, EXTERNAL:
- Wing span: 8·73 m (28 ft 7¼ in)
- Wing chord (constant): 0·99 m (3 ft 3 in)
- Wing area, gross: 8·55 m² (92 sq ft)
- Wing aspect ratio: 8·8
- Length overall: 4·29 m (14 ft 1 in)
- Height overall: 2·03 m (6 ft 8 in)
- Tailplane span: 1·83 m (6 ft 0 in)
- Propeller diameter: 0·97 m (3 ft 2 in)

WEIGHTS AND LOADINGS:
- Weight empty: 66 kg (145 lb)
- Max T-O weight: 170 kg (375 lb)
- Max wing loading: 19·5 kg/m² (4 lb/sq ft)
- Max power loading: 28·3 kg/kW (46·9 lb/hp)

PERFORMANCE (at max T-O weight):
- Never-exceed speed: 78 knots (145 km/h; 90 mph)
- Max level speed at S/L: 43 knots (80 km/h; 50 mph)
- Max cruising speed at S/L: 42 knots (77 km/h; 48 mph)
- Econ cruising speed at S/L: 36 knots (68 km/h; 42 mph)
- Stalling speed: 23 knots (42 km/h; 26 mph)
- Max rate of climb at S/L: 61 m (200 ft)/min
- Service ceiling: 1,525 m (5,000 ft)
- T-O run: 76 m (250 ft)
- Landing run: 30 m (100 ft)
- Range with max fuel, no allowances: 87 nm (161 km; 100 miles)

BEDE
BEDE AIRCRAFT INC

HEAD OFFICE:
Newton Municipal Airport, PO Box 706, Newton, Kansas 67114

Telephone: (316) 283-8870

PRESIDENT: James R. Bede

Mr James R. Bede formed Bede Aircraft Inc as a successor to the former Bede Aviation Corporation, to undertake production of plans and kits of parts for construction of his BD-4 two/four-seat light aircraft. Since then he has devoted further efforts to the development of sporting aircraft for the homebuilder, and the range of designs now available includes a lightweight monoplane known as the Bede BD-5 Micro, a jet-powered version designated BD-5J, and a single-seat version of the BD-4 which has the designation BD-6. New projects under development include the BD-7, a two/four-seat piston-engined version of the BD-5, and a new single-seat aircraft which has the designation BD-8.

Factory-built versions of both the piston-engined BD-5 and jet-powered BD-5J are available, as mentioned in the Aircraft section of this edition.

BEDE BD-4

The Bede BD-4 can be built as a two-seater with an 80·5 kW (108 hp) engine or as a four-seater with an engine of 112-149 kW (150-200 hp). By 1976 well over 2,000 sets of plans for the BD-4 had been sold; 736 were under construction, and 90 completed aircraft were known to have flown. At least five have non-standard tailwheel-type landing gear.

TYPE: Two/four-seat sporting and utility monoplane for amateur construction.

WINGS: Cantilever high-wing monoplane. Wing section NACA 64-415 (modified). No dihedral. Incidence 3°. Each wing is built up of 12 glassfibre 'panel-ribs' which are slid over a 0·165 m (6½ in) diameter extruded aluminium tubular spar, to which they are secured by epoxy resin and large-diameter tube clamps. Each 3·05 m (10 ft) wing section then slides for 0·30 m (1 ft) over a smaller-diameter 2024-T3 centre-section tube. Identical and interchangeable trailing-edge flaps and ailerons are attached to the trailing-edges of the ribs. Design allows for either removable or folding wings.

FUSELAGE: All-metal structure of bolt-together design. Formed 2024-T3 aluminium angle-sections, made from 0·063 in sheet stock in 2 in × 2 in, 1½ in × 1½ in and 1 in × 1 in sizes, are used for the basic structure of the fuselage. Simple metal gussets are used to form joints, bolted together with AN3 bolts and AN509 flush screws and lock nuts. Skin of either aluminium sheet or glassfibre panels can be pop-riveted or bonded to the primary structure. Cabin door on each side, under wing.

TAIL UNIT: Cantilever all-metal structure with swept vertical surfaces. All-moving tailplane, consisting of a single 6·35 cm (2½ in) diameter tubular spar, with six metal ribs and aluminium skin. Fin and rudder are constructed on U-section metal spars, with aluminium skin. Rudder pivots on a one-piece piano hinge. All control surfaces statically mass-balanced.

LANDING GEAR: Non-retractable tricycle type. Each main leg is formed from a single piece of 2024-T3 aluminium plate, which rotates on a pivot point inside the fuselage. Shock loads are absorbed by rubber in compression. Main wheels size 6·00-6 with 0·38 m (15 in) diameter tubed tyres. Non-steerable fully-castoring nosewheel carried on a 3·81 cm (1½ in) diameter 4130 steel tube which pivots on the firewall and has similar shock-absorption to the main legs. Nosewheel of either 0·20 m or 0·25 m (8 in or 10 in) diameter. Hydraulic brakes. Optional wheel fairings, with wheel covers that retract for take-off and landing, and which are lowered to enclose the wheels completely in flight. These fairings add 14·7 knots (27 km/h; 17 mph) to the aircraft's max cruising speed.

POWER PLANT: One 80·5 kW (108 hp) Lycoming O-235-C1 flat-four engine, driving a McCauley 1B-90/CM two-blade fixed-pitch propeller for two-seat configuration; or one 112 kW (150 hp) Lycoming O-320 engine, driving a Sensenich type 74DM6-0-60 fixed-pitch propeller; or one 134 kW (180 hp) Lycoming O-360 engine, driving a Hartzell 7666-A2 constant-speed propeller for four-seat configuration. Alternative engines of up to 149 kW (200 hp) and McCauley or Hartzell constant-speed propellers, are optional. Propeller spinner standard. Simple three-piece cowling with glassfibre nose section. Engine mounting of swing-out type for easy maintenance. Fuel contained between wing panel ribs, with standard capacity of 97 litres (25·8 US gallons) in each wing. Total standard fuel capacity 194 litres (51·6 US gallons). Max fuel capacity 322 litres (85 US gallons). Refuelling points above each wing.

ACCOMMODATION: Two or four seats, in pairs, in enclosed cabin.

SYSTEMS: Hydraulic system for brakes only. Electrical system includes engine-driven generator, 12V battery and navigation lights.

DIMENSIONS, EXTERNAL:
- Wing span: 7·80 m (25 ft 7 in)
- Wing chord (constant): 1·22 m (4 ft 0 in)
- Wing area, gross: 9·51 m² (102·33 sq ft)
- Wing aspect ratio: 6·4
- Length overall: 6·52 m (21 ft 4½ in)
- Width, wings folded: 2·20 m (7 ft 2½ in)
- Height overall: 2·20 m (7 ft 2¾ in)
- Tailplane span: 2·22 m (7 ft 3½ in)
- Wheel track: 2·13 m (7 ft 0 in)
- Propeller diameter:
  - 80·5 kW (108 hp): 1·75 m (5 ft 9 in)
  - 112 kW (150 hp): 1·88 m (6 ft 2 in)
  - 134 kW (180 hp): 1·93 m (6 ft 4 in)

DIMENSION, INTERNAL:
- Cabin: Max width: 1·07 m (3 ft 6 in)

WEIGHTS AND LOADINGS: (A: 80·5 kW; 108 hp, B: 112 kW; 150 hp, C: 134 kW; 180 hp, D: 149 kW; 200 hp):
- Weight empty:
  - A: 435 kg (960 lb)
  - B: 458 kg (1,010 lb)
  - C: 489 kg (1,080 lb)
  - D: 510 kg (1,125 lb)
- Max T-O weight:
  - A: 725 kg (1,600 lb)
  - B, C, D: 907 kg (2,000 lb)
- Max wing loading:
  - A: 74·0 kg/m² (15·15 lb/sq ft)
  - B, C, D: 93·0 kg/m² (19·06 lb/sq ft)
- Max power loading:
  - A: 9·0 kg/kW (14·81 lb/hp)
  - B: 8·10 kg/kW (13·33 lb/hp)
  - C: 6·77 kg/kW (11·11 lb/hp)
  - D: 6·09 kg/kW (10·00 lb/hp)

PERFORMANCE (at max T-O weight; A: 80·5 kW; 108 hp, B: 112 kW; 150 hp, C: 134 kW; 180 hp, D: 149 kW; 200 hp):
- Max level speed at S/L:
  - A: 135 knots (251 km/h; 156 mph)
  - B: 149 knots (277 km/h; 172 mph)
  - C: 159 knots (295 km/h; 183 mph)
  - D: 176 knots (327 km/h; 203 mph)
- Cruising speed 75% power):
  - A: 126 knots (233 km/h; 145 mph)
  - B: 143 knots (266 km/h; 165 mph)
  - C: 151 knots (280 km/h; 174 mph)

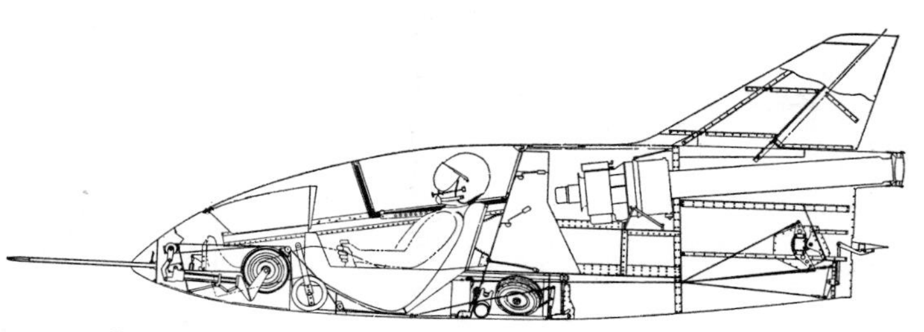

Sectional drawing of the BD-5J homebuilt light jet aircraft

Prototype of the Bede BD-6 single-seat lightweight monoplane

Bede BD-5J, the jet-powered derivative of the BD-5D *(J. M. G. Gradidge)*

Bede BD-7 two/four-seat homebuilt sporting monoplane *(Howard Levy)*

D 167 knots (309 km/h; 192 mph)
Cruising speed (65% power):
A 118 knots (219 km/h; 136 mph)
B 135 knots (249 km/h; 155 mph)
C 147 knots (272 km/h; 169 mph)
D 156 knots (290 km/h; 180 mph)
Stalling speed, flaps up:
A 50·5 knots (93·5 km/h; 58 mph)
B 55 knots (101·5 km/h; 63 mph)
C 57 knots (105 km/h; 65 mph)
D 59 knots (108 km/h; 67 mph)
Stalling speed, flaps down:
A 47 knots (87 km/h; 54 mph)
B 50·5 knots (93·5 km/h; 58 mph)
C 53 knots (98·5 km/h; 61 mph)
D 55 knots (102 km/h; 63 mph)
Max rate of climb at S/L:
A 274 m (900 ft)/min
B 381 m (1,250 ft)/min
C 427 m (1,400 ft)/min
D 518 m (1,700 ft)/min
T-O run:
A, B 198 m (650 ft)
C, D 183 m (600 ft)
T-O to 15 m (50 ft):
A 366 m (1,200 ft)
B 335 m (1,100 ft)
C 305 m (1,000 ft)
D 229 m (750 ft)
Landing run:
A 152 m (500 ft)
B, C, D 183 m (600 ft)
Max range, 45 min reserve:
A, B, C, D 781 nm (1,448 km; 900 miles)

BEDE BD-5 MICRO

Design of this unusual single-seat sporting monoplane began in February 1967, and construction of the prototype was started in December 1970. First flight of the BD-5 was made on 12 September 1971, and a succession of developed versions has since appeared, as follows:

BD-5A. Designation of original prototype which had a wing span of 4·37 m (14 ft 4 in), 30 kW (40 hp) Kiekhaefer Aeromarine two-cylinder two-stroke aircooled engine and 'butterfly' tail unit.

BD-5B. Original designation of alternative version of the BD-5A with increased wing span of 6·55 m (21 ft 6 in). As a result of early flight tests the 'butterfly' tail unit was replaced by a conventional fin and rudder, with an all-moving horizontal 'stabilator' on the lower portion of the rear fuselage. In early 1974 Bede Aircraft decided to discontinue the short-span wing and 30 kW (40 hp) power plant. From that time the designation BD-5B referred to the version with 52 kW (70 hp) engine, NACA 64_1-212/218 wing sections and 106 litres (28 US gallons) of fuel, as described in detail in the 1975-76 *Jane's*. Superseded by BD-5G.

BD-5D. Factory-built version of the BD-5G. Described in main Aircraft section.

BD-5G. Improved version of BD-5B, with GA(W)-2 wing section. Fully aerobatic; stressed for ±6*g*. Increased fuel capacity. Reduced wing span.

BD-5J. Jet-powered version. Plans, materials and some components available to amateur constructors. Described separately.

BD-5JP. Factory-built version of BD-5J. Described in main Aircraft section of this edition.

BD-5S. Sailplane version of the BD-5, reference to which can be found in the Sailplanes section.

Bede Aircraft reported that by the beginning of 1975 it had received more than 5,000 orders for plans and kits of the BD-5 Micro.

The details which follow apply to the BD-5G:

TYPE: Single-seat lightweight homebuilt monoplane.

WINGS: Cantilever low-wing monoplane. Wing section NASA GA(W)-2. Thickness/chord ratio 13% (average). Dihedral 5°. Incidence 1°. Sweepback 0°. Light alloy structure; moulded wingtips. Plain aerodynamically-balanced ailerons and trailing-edge flaps of light alloy construction. No trim tabs.

FUSELAGE: Light alloy. Aft end of fuselage terminates as a deep knife-edge section to provide directional stability.

TAIL UNIT: Conventional cantilever light alloy structure, comprising fin, rudder and all-moving tailplane. Combination trim tab and anti-servo tab in tailplane.

LANDING GEAR: Manually-retractable tricycle type. Nosewheel retracts aft, main wheels inward. Shock-absorption of main gear by glassfibre cantilever struts. Nosewheel leg embodies oleo shock-absorber. Mainwheel and nosewheel tyres size 2·80-4. Disc brakes.

POWER PLANT: One 52 kW (70 hp) Xenoah three-cylinder two-stroke aircooled engine, mounted in aft fuselage and driving a two-blade fixed-pitch wooden pusher propeller. Fuel in two inboard wing tanks, total capacity 132·5 litres (35 US gallons). Refuelling point in rear fuselage. Special air intakes.

ACCOMMODATION: Single seat under upward-opening transparent canopy. Cockpit heated and ventilated.

SYSTEM: Electrical power supplied by 12V DC battery.

DIMENSIONS, EXTERNAL:
Wing span 5·18 m (17 ft 0 in)
Wing area, gross 3·53 m² (38·0 sq ft)
Wing aspect ratio 7·6
Length overall, excl nose probe 4·05 m (13 ft 3½ in)
Height overall 1·28 m (4 ft 2½ in)

DIMENSIONS, INTERNAL:
Cabin:
Length 1·63 m (5 ft 4 in)
Width 0·60 m (1 ft 11½ in)
Height 0·91 m (3 ft 0 in)

WEIGHTS AND LOADING:
Weight empty 186 kg (410 lb)
Max T-O weight 385 kg (850 lb)
Max wing loading 109·4 kg/m² (22·4 lb/sq ft)

PERFORMANCE (at max T-O weight):
Max level speed at S/L
200 knots (370 km/h; 230 mph)
Max cruising speed at 2,285 m (7,500 ft)
182 knots (338 km/h; 210 mph)
Max manoeuvring speed
139 knots (257 km/h; 160 mph) IAS
Stalling speed, clean 67 knots (124 km/h; 77 mph)
Stalling speed, flaps and landing gear down
60 knots (111 km/h; 69 mph)
Max rate of climb at S/L 533 m (1,750 ft)/min
Service ceiling 5,730 m (18,800 ft)
Range (75% power)
981 nm (1,818 km; 1,130 miles)

BEDE BD-5J

This designation identifies a jet-powered version of the BD-5 which was introduced in the Summer of 1973. Similar to the BD-5G, it has the 5·18 m (17 ft) wing, increased 'wet wing' fuel capacity, engine intakes on each side of the fuselage, over the wing, and power plant that consists of a US-built Microturbo TRS 18 lightweight axial-flow turbojet of 0·90 kN (202 lb st).

The US military services have evaluated a BD-5J, their interest ranging from RPV applications to a cheap-to-buy-and-operate vehicle suitable for tactical training and the maintenance of pilot efficiency. In addition, a BD-5J demonstration team has performed aerobatics at many US air shows and other functions.

Plans, materials and certain finished components and assemblies are available to amateur constructors. There is also a factory-built version, designated BD-5JP, of which details can be found in the main Aircraft section of this edition.

The description of the BD-5G applies also to the BD-5J, except in the following respects:

FUSELAGE: As for BD-5G, except for provision of engine air intakes in each side of fuselage, above the wing.

POWER PLANT: One 0·90 kN (202 lb st) Microturbo TRS 18 turbojet engine, mounted in rear fuselage with tail-pipe extending below and aft of rudder. Integral fuel tank in each wing, with total capacity of 151·4 litres (40 US gallons), plus 56·8 litre (15 US gallon) fuselage tank. Total fuel capacity 208·2 litres (55 US gallons).

SYSTEMS: Automatic electrically-operated fuel control system. Electrical system supplied by 28V engine-driven generator.

ELECTRONICS AND EQUIPMENT: Optional electronics include nav/com transceiver with 360-channel com and 200-channel nav; VOR/LOC heading indicator with glideslope and marker beacon; transponder; and DME. Optional equipment includes instrument, navigation and landing lights.

DIMENSIONS, EXTERNAL: As BD-5G, except:
Length overall 3·78 m (12 ft 4¾ in)
Height overall 1·83 m (6 ft 0 in)

DIMENSIONS, INTERNAL:
Cabin: Length 1·63 m (5 ft 4 in)
Max width 0·60 m (1 ft 11½ in)
Max height 0·91 m (3 ft 0 in)

WEIGHTS AND LOADINGS:
Weight empty 204 kg (450 lb)
Max T-O weight 435 kg (960 lb)
Normal landing weight 304 kg (670 lb)
Max wing loading 124·3 kg/m² (25·4 lb/sq ft)
Max power loading 483 kg/kN (4·8 lb/lb st)

PERFORMANCE (at max T-O weight except where indicated):
Max level speed at S/L
240 knots (444 km/h; 276 mph)
Max level speed at 3,050 m (10,000 ft)
232 knots (429 km/h; 267 mph)
Stalling speed, flaps down
66 knots (123 km/h; 76 mph)
Stalling speed at normal landing weight, flaps down
56 knots (103 km/h; 64 mph)
Max rate of climb at S/L 472 m (1,550 ft)/min
Service ceiling 7,620 m (25,000 ft)
T-O run 549 m (1,800 ft)

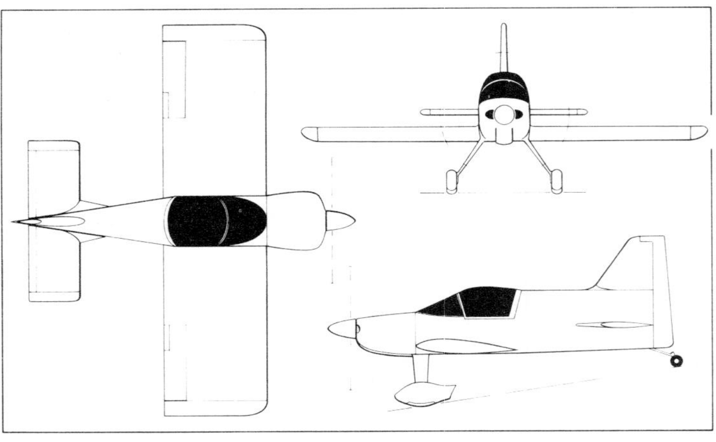

Bede BD-8 single-seat sporting aircraft *(Roy J. Grainge)*

Beets G/B Special lightweight sporting aircraft (Volkswagen engine) *(Howard Levy)*

Bensen Model B-8M Gyro-Copter *(Peter J. Bish)*

Bensen Model B-16S (two Kiekhaefer KAM 525 snowmobile engines)

Landing run 305 m (1,000 ft)
Range at 3,050 m (10,000 ft), 20 min reserve 508 nm (941 km; 585 miles)

BEDE BD-6

Simultaneously with announcement of the BD-5J came news of the BD-6, which is essentially a single-seat development of the BD-4, with reduced overall dimensions and power. Pressure of work on the BD-5 programme has delayed further development of the BD-6. However, it is planned that following certification of the BD-5D and -5JP work will begin again on the BD-6, which is to be evaluated with both the 52 kW (70 hp) Xenoah engine and a 48·5 kW (65 hp) Continental.

Type: Single-seat lightweight homebuilt sporting aircraft.

Wings: Similar to those of BD-4, except dimensions reduced.

Fuselage and Tail Unit: Similar to those of BD-4.

Landing Gear: Non-retractable tricycle type. Cantilever main-gear struts of glassfibre construction. Non-steerable fully-castoring nosewheel. Hydraulic brakes.

Power Plant: Prototype has one 41 kW (55 hp) Hirth two-cylinder two-stroke engine, driving a two-blade metal fixed-pitch propeller with spinner. Fuel contained between wing panel ribs, with total capacity of 79·4 litres (21 US gallons).

Accommodation: Single seat in enclosed cabin. Baggage space aft of seat.

Dimensions, external:
- Wing span 6·55 m (21 ft 6 in)
- Wing area, gross 5·16 m^2 (55·5 sq ft)
- Wing aspect ratio 8·32
- Length overall 5·11 m (16 ft 9 in)
- Height overall 1·98 m (6 ft 6 in)

Dimension, internal:
- Cabin: Max width 0·61 m (2 ft 0 in)

Weights and Loadings:
- Weight empty 170 kg (375 lb)
- Max T-O weight 295 kg (650 lb)
- Max wing loading 57·1 kg/m^2 (11·7 lb/sq ft)
- Max power loading 7·20 kg/kW (11·8 lb/hp)

Performance (prototype with 41 kW; 55 hp engine):
- Max level speed at S/L more than 122 knots (225 km/h; 140 mph)
- Cruising speed at 2,285 m (7,500 ft) more than 122 knots (225 km/h; 140 mph)
- Stalling speed, flaps down 43·5 knots (80·5 km/h; 50 mph)
- Max rate of climb at S/L 274 m (900 ft)/min
- Service ceiling 4,265 m (14,000 ft)
- T-O run 183 m (600 ft)
- Landing run 122 m (400 ft)
- Range at 75% power with 30 min reserve more than 390 nm (724 km; 450 miles)

BEDE BD-7

This aircraft is essentially a two/four-seat version of the BD-5, designed to utilise one or two conventional piston engines in the 74·5 to 149 kW (100 to 200 hp) range. Design originated in January 1974 and construction of a prototype began in May 1975. First flight of this aircraft was scheduled for the second half of 1976.

Plans, raw materials and prefabricated components are available to amateur constructors.

Type: Two/four-seat homebuilt sporting monoplane.

Wings: Cantilever low-wing monoplane. Wing section NACA 64_2-415 (modified) at root, 64_3-418 (modified) at tip. Dihedral 5°. Incidence 3°. No sweepback. Two different designs offered for inboard section of wings; as for BD-4, with glassfibre 'panel ribs' and extruded aluminium tubular spar; or all-metal spar, ribs and skin. Outboard panels have tubular aluminium spar with light alloy ribs and skins. Plain ailerons of light alloy construction operated by torque tubes, the latter serving as pivots for the light alloy trailing-edge flaps.

Fuselage: Light alloy structure as for BD-5, with aft end of fuselage terminating as a deep knife-edge section to provide directional stability.

Tail Unit: Cantilever light alloy structure, with swept vertical surfaces and all-moving tailplane, similar to that of BD-4. Combination trim tab and anti-servo tab in tailplane.

Landing Gear: Manually-retractable tricycle type, similar to that of BD-5. Nosewheel retracts aft, main wheels inward. Shock-absorption of main gear by cantilever glassfibre struts. Nosewheel carried on hydraulic strut.

Power Plant: Airframe designed to accept engines of 74·5 to 149 kW (100 to 200 hp), mounted in aft fuselage and driving a two-blade fixed-pitch pusher propeller. The design, at present, specifies two 52 kW (70 hp) Xenoah three-cylinder engines, connected to one drive-shaft. Each engine can work independently in an emergency. Variable-pitch propeller optional. Fuel in wing tanks with total capacity of 189 to 303 litres (50 to 80 US gallons); refuelling point in upper surface of each wing. Oil capacity 7·6 litres (2 US gallons).

Accommodation: Two or four seats, in pairs, in enclosed cabin. Clamshell canopy, hinged on centreline of aircraft, and opening upward on each side. Baggage space aft of rear seats. Accommodation heated and ventilated.

Systems: Hydraulic system for brakes only. Electrical system powered by engine-driven generator and 12V storage battery. Oxygen system optional.

Electronics and Equipment: Navigation lights standard. IFR capability optional.

Dimensions, external:
- Wing span 7·32 m (24 ft 0 in)
- Wing chord at root 1·22 m (4 ft 0 in)
- Wing chord at tip 1·01 m (3 ft 3¾ in)
- Wing area, gross 8·69 m^2 (93·5 sq ft)
- Wing aspect ratio 6·15
- Length overall 6·25 m (20 ft 6 in)
- Height overall 2·29 m (7 ft 6 in)
- Tailplane span 2·80 m (9 ft 2¼ in)
- Wheel track 1·93 m (6 ft 4 in)
- Wheelbase 2·47 m (8 ft 1¼ in)
- Propeller diameter 1·52 m (5 ft 0 in)

Dimensions, internal:
- Cabin: Max width 1·12 m (3 ft 8 in)
- Baggage capacity 0·20 m^3 (7 cu ft)

Weights and Loadings (calculated for aircraft with 134 kW; 180 hp engine):
- Basic operating weight empty 435 kg (960 lb)
- Max T-O and landing weight 907 kg (2,000 lb)
- Max wing loading 104·0 kg/m^2 (21·3 lb/sq ft)
- Max power loading 6·77 kg/kW (11·1 lb/hp)

Performance (estimated for aircraft with 134 kW; 180 hp engine):
- Never-exceed speed 217 knots (402 km/h; 250 mph)
- Max level speed at S/L 191 knots (354 km/h; 220 mph)
- Max cruising speed at 2,285 m (7,500 ft) 182 knots (338 km/h; 210 mph)
- Econ cruising speed at 2,285 m (7,500 ft) 165 knots (306 km/h; 190 mph)
- Stalling speed, flaps down 57 knots (105 km/h; 65 mph)
- Max rate of climb at S/L 457 m (1,500 ft)/min
- Service ceiling 7,620 m (25,000 ft)
- T-O to 15 m (50 ft) 366 m (1,200 ft)
- Landing from 15 m (50 ft) 213 m (700 ft)
- Landing run 152 m (500 ft)
- Range with max fuel 868 nm (1,609 km; 1,000 miles)

BEDE BD-8

Designed as an easy-to-build single-seat aerobatic light aircraft with tailwheel landing gear, this low-wing monoplane has constructional features of earlier Bede designs. Its development originated in November 1973, and construction of two prototypes began in January 1974; little work had been completed by early 1976, due to the demands of the BD-5 programme.

Type: Single-seat homebuilt sporting aircraft.

Wings: Cantilever low-wing monoplane. Wing section NACA 63_2-015. Incidence 3°. Wing construction as for BD-4, with tubular aluminium spar and glassfibre 'panel ribs'. Plain ailerons of light alloy construction, each with trim tab. No flaps.

Fuselage: Light alloy structure of bolt-together design, as described for the BD-4.

Tail Unit: Cantilever light alloy structure. All-moving tailplane, with combination trim tab and anti-servo tab.

Landing Gear: Non-retractable tailwheel type. Main wheels carried on cantilever glassfibre struts. Hydraulic brakes.

Power Plant: One Lycoming engine of 134 or 149 kW (180 or 200 hp), driving a two-blade constant-speed tractor propeller. Fuel tanks in wings, with combined capacity of 189 litres (50 US gallons). Refuelling points in upper surface of wings. Oil capacity of 7·6 litres (2 US gallons).

ACCOMMODATION: Single seat beneath upward-opening transparent cockpit canopy. Baggage space aft of seat. Accommodation heated and ventilated.

SYSTEMS: Hydraulic system for brakes only. Electrical system powered by engine-driven generator.

DIMENSIONS, EXTERNAL:

Wing span	5·89 m (19 ft 4 in)
Wing chord, constant	1·52 m (5 ft 0 in)
Wing area, gross	8·98 m² (96·67 sq ft)
Wing aspect ratio	3·87
Length overall	5·13 m (16 ft 10 in)
Height overall	2·06 m (6 ft 9 in)
Tailplane span	2·44 m (8 ft 0 in)
Wheel track	2·13 m (7 ft 0 in)
Wheelbase	3·73 m (12 ft 3 in)
Propeller diameter	1·88 m (6 ft 2 in)
Propeller ground clearance	0·24 m (9½ in)

DIMENSIONS, INTERNAL:

Cockpit: Length	1·37 m (4 ft 6 in)
Max width	0·74 m (2 ft 5 in)
Max height	1·03 m (3 ft 4¾ in)

WEIGHTS (estimated):

Weight empty	372 kg (820 lb)
Max T-O weight, Aerobatic category	494 kg (1,090 lb)
Max T-O weight, Utility category	589 kg (1,300 lb)

PERFORMANCE (estimated, Utility category with 149 kW; 200 hp engine):

Never-exceed speed	238 knots (442 km/h; 275 mph)
Max level speed at S/L	182 knots (338 km/h; 210 mph)
Max cruising speed at 2,285 m (7,500 ft)	178 knots (330 km/h; 205 mph)
Stalling speed	42 knots (78 km/h; 48 mph)
Max rate of climb at S/L	671 m (2,200 ft)/min
Service ceiling	5,485 m (18,000 ft)
T-O to 15 m (50 ft)	229 m (750 ft)
Landing from 15 m (50 ft)	152 m (500 ft)
Landing run	122 m (400 ft)
Range with max fuel, 45 min reserve	607 nm (1,126 km; 700 miles)

BEETS

GLENN BEETS

Mr G. Beets of Riverside, California, designed a light-weight sporting aircraft which he named the G/B Special. Construction of the prototype occupied two years and cost approximately $2,500, the first flight being made on 25 July 1973. Plans of the G/B Special, kits of components and materials are available from Stolp Starduster Corporation (which see).

BEETS G/B SPECIAL

TYPE: Single-seat homebuilt sporting aircraft.

WINGS: Braced parasol monoplane. Wing section Curtis 72. Vee bracing struts each side with auxiliary struts. N-type cabane struts. Conventional wood structure with spruce spars and truss ribs, Dacron covered. Plain ailerons of similar construction. No trim tabs. Cut-out in wing trailing-edge.

FUSELAGE: Welded structure of 4130 steel tube with Dacron covering.

TAIL UNIT: Cantilever structure of wood, with foam filling and covering of 1/16 in mahogany plywood.

LANDING GEAR: Non-retractable tailwheel type, with main wheels carried on braced tubular steel struts. Glassfibre wheel fairings on main wheels.

POWER PLANT: One 52 kW (70 hp) Volkswagen 1,641 cc modified motorcar engine, driving a two-blade fixed-pitch propeller through a 2½ : 1 reduction drive. Design will accept engines of 37·25-74·5 kW (50-100 hp).

ACCOMMODATION: Single seat in open cockpit.

DIMENSIONS, EXTERNAL:

Wing span	7·62 m (25 ft 0 in)
Wing chord	1·27 m (4 ft 2 in)
Length overall	4·98 m (16 ft 4 in)
Height overall	1·83 m (6 ft 0 in)

WEIGHTS:

Weight empty	274 kg (603 lb)
Max T-O weight	420 kg (925 lb)

PERFORMANCE (at max T-O weight with cruise performance propeller):

Max level speed at S/L	135 knots (251 km/h; 156 mph)
Cruising speed	104 knots (193 km/h; 120 mph)
Landing speed	30·5 knots (56 km/h; 35 mph)
Max rate of climb at S/L	610 m (2,000 ft)/min
T-O run	61 m (200 ft)
Landing run	92 m (300 ft)
Range	520 nm (965 km; 600 miles)

BENSEN

BENSEN AIRCRAFT CORPORATION

HEAD OFFICE AND WORKS:
Box 31047, Raleigh-Durham Airport, Raleigh, North Carolina 27612

Telephone: (919) 787-4224/0945

PRESIDENT: Igor B. Bensen

The Bensen Aircraft Corporation was formed by Dr Igor B. Bensen, formerly Chief of Research of the Kaman company, to develop a series of lightweight helicopters and rotary-wing gliders suitable for production in kit form for amateur construction as well as in ready-to-fly condition. More than 2,000 were flying by 1976.

Production is centred on the B-8M/V and Super Bug Gyro-Copter powered autogyros, the twin-engined B-16S Gyro-Copter, the B-8MA Agricopter agricultural spraying version and various land and waterborne versions of the B-8 rotor-kite. Research into new models and concepts continues, and Bensen expected to introduce its Model B-8H, described as a Hovering Gyro-Copter, in 1976.

BENSEN MODEL B-8 GYRO-GLIDER

USAF designation: X-25

The Gyro-Glider is a simple unpowered rotor-kite which can be towed behind even a small motorcar and has achieved free gliding with the towline released. It is available as either a completed aircraft or kit of parts for amateur construction. Alternatively, would-be constructors can purchase a set of plans, with building and flying instructions. No pilot's licence is required to fly it in the United States and many hundreds of kits and plans have been sold. Application has been made for an Approved Type Certificate.

The original Model B-7 Gyro-Glider was described in the 1958-59 *Jane's*. It was followed by the Model B-8, which is offered as either a single-seater or two-seater, the latter version being suitable for use as a pilot trainer.

The Model B-8 consists basically of an inverted square-section tubular aluminium T-frame structure, of which the forward arm supports the lightweight seat, towing arm, rudder bar and landing gear nosewheel. The rear arm supports a large stabilising fin and rudder, made normally of plywood, but optionally of metal. The main landing gear wheels are carried on a tubular axle near the junction of the T-frame. The free-turning two-blade rotor is universally-mounted at the top of the T-frame and is normally operated directly by a hanging-stick control. A floor-type control column is available as optional equipment. Pedal controls for the rudder are standard.

The Gyro-Glider rotor is made normally of laminated plywood, with a steel spar. Factory-built all-metal rotor blades are available as optional items.

The two-seat trainer version of the Gyro-Glider is fitted with castoring crosswind landing gear and has an extra-wide wheel track. It will maintain level flight down to 16·5 knots (30·5 km/h; 19 mph).

Under contract to the USAF, Bensen delivered a single-seat and a two-seat Gyro-Glider to the USAF Flight Dynamics Laboratory at Wright-Patterson AFB, Dayton, Ohio, where, designated X-25, they were used to explore the feasibility of using folding rotors on a Discretionary Descent Vehicle (DDV).

A new concept in rescue devices, the DDV would have a set of rotor blades folded into an ejection system, in addition to the normal parachute, enabling a pilot to attain any pre-selected site within gliding range.

DIMENSIONS, EXTERNAL:

Diameter of rotor	6·10 m (20 ft 0 in)
Length of fuselage	3·45 m (11 ft 4 in)
Height overall	1·90 m (6 ft 3 in)

BENSEN MODEL B-8W HYDRO-GLIDER

The basic structure of this floatplane rotor-kite is similar to that of the B-8 Gyro-Glider and conversion from one to the other is simple. Main change is that the nosewheel landing gear is replaced by two floats. The original round-type floats have been superseded by flat-bottomed pontoons of polyurethane foam covered by glassfibre, which give better planing, with less spray.

The Hydro-Glider is towed by a motorboat.

BENSEN MODEL B-8M, B-8V and SUPER BUG GYRO-COPTERS and B-8MW HYDRO-COPTER

First flown on 6 December 1955, the Gyro-Copter is a powered autogyro conversion of the Gyro-Glider, designed for home construction from kits or plans. When fitted with floats it is known as a Hydro-Copter.

The current **B-8M** version of the Gyro-Copter has a more powerful engine than the original B-7M and can be equipped with an optional mechanical rotor drive. By engaging this drive, the rotor can be accelerated to flying speed while the aircraft is stationary. Then, by transferring the power to the pusher propeller, it is possible to take off in only 15 m (50 ft), with the rotor autorotating normally. Alternatively, a 1 hp Ohlsson & Rice Compact III two-stroke engine can be attached to the rotor for pre-rotation, automatically disengaging itself at take-off rpm.

Other non-standard items available optionally include a 67 kW (90 hp) engine instead of the normal 53·5 kW (72 hp) engine, a larger-diameter rotor, an offset gimbal rotor head, a floor-type control column instead of the normal overhead type of column, dual ignition, nosewheel arrester and Bensen-manufactured pontoons of polyurethane foam covered with glassfibre. All-metal rotor blades and tail surfaces are available as alternatives to the standard wooden components.

The prototype Model B-8M Gyro-Copter flew for the first time on 8 July 1957 and the first production model on 9 October 1957.

The B-8M is roadable, requiring no removal of, or changes in, its equipment for transition from air to ground travel. The rotor is merely stopped in a fore-and-aft position by a lock. Gyro-Copters have been driven on highways and have negotiated heavy city traffic with ease in a number of public demonstrations in the USA.

The **B-8V**, which flew for the first time in the Autumn of 1967, is basically a standard B-8M, but is powered by a 1,600 cc Volkswagen engine. In unmodified form the VW1600 yields just adequate flight performance at 272 kg (600 lb) gross weight. Since Bensen engineers considered that most Gyro-Copters would not have a gross weight as high as 600 lb, the VW engine justified inclusion as an alternative power plant to the standard McCulloch engine.

Kits and parts for the B-8V, excluding engine and mounting, are available, as are plans and an instruction manual for converting the B-8M to a B-8V, or for mounting a VW engine on a standard B-8 airframe.

In May 1971, Bensen announced introduction of the **Super Bug**, an advanced version of the standard Model B-8M. This features a twin-engine installation to spin up the rotor prior to take-off. Bensen claims this as an intermediate step towards full VTOL capability, as this more powerful pre-rotation enables the Super Bug to take off and clear a 15 m (50 ft) obstacle within 137 m (450 ft) of starting its T-O run in zero wind conditions at max T-O weight. Other new standard equipment of the Super Bug includes rotor brake, parking brake on main wheels, single control of rudder and nosewheel steering, soft suspension of the auxiliary tailwheel and an increase of 45·5 kg (100 lb) in max T-O weight.

The following description applies to the Models B-8M and B-8V:

TYPE: Single-seat light autogyro.

ROTOR SYSTEM: B-8M has single two-blade rotor of laminated plywood construction, with steel spar (optional all-metal rotor). Blade section Bensen G2. Teetering hub, with no lag hinges or collective pitch control. A similar rotor, of all-metal construction, is provided for the B-8V. A larger-diameter rotor is not available as an alternative for this latter model. No anti-torque rotor. Rotor speed 400 rpm.

ROTOR DRIVE (optional): An auxiliary 0·75 kW (1 hp) Ohlsson & Rice engine is available to spin up the rotor.

FUSELAGE: Square-section tubular 6061-T6 aluminium structure.

TAIL SURFACES: Vertical fin and rudder of ¼ in plywood. Optional all-metal tail surfaces.

LANDING GEAR: Non-retractable tricycle type, with auxiliary tailwheel. No shock-absorbers. Steerable nosewheel. General Tire wheels, size 30·5-10·15 cm (12-4 in). Tyre pressure 0·69 bars (10 lb/sq in). Brake on nosewheel.

POWER PLANT: One 53·5 kW (72 hp) McCulloch Model 4318AX flat-four two-stroke engine (or, optionally, a 67 kW; 90 hp McCulloch 4318GX engine of similar weight and dimensions), driving a two-blade wooden fixed-pitch Aero Prop Model BA 48-A2 pusher propeller with leading-edges covered with stainless steel. Alternatively, one 47·5 kW (64 hp) Volkswagen 1,600 cc flat-four four-stroke engine, driving a Troyer Model 50-24-65 two-blade wooden fixed-pitch pusher propeller. Fuel tank under pilot's seat, capacity 22·75 litres (6·0 US gallons). Can be fitted with auxiliary tank for ferrying.

ACCOMMODATION: Open seat. Overhead azimuth stick and rudder pedal controls. Optional floor-type control column. Safety belt.

DIMENSIONS, EXTERNAL:

Diameter of rotor:	
standard	6·10 m (20 ft 0 in)
optional (on B-8M and B-8MW)	6·70 m (22 ft 0 in)
Rotor blade chord	0·18 m (7 in)
Length of fuselage	3·45 m (11 ft 4 in)
Height overall	1·90 m (6 ft 3 in)
Wheel track	1·52 m (5 ft 0 in)
Propeller diameter:	
53·5 kW (72 hp) McCulloch	1·22 m (4 ft 0 in)
47·5 kW (64 hp) Volkswagen	1·27 m (4 ft 2 in)

AREAS (standard rotor):

Rotor blades (each)	0·54 m² (5·83 sq ft)
Rotor disc	29·17 m² (314 sq ft)

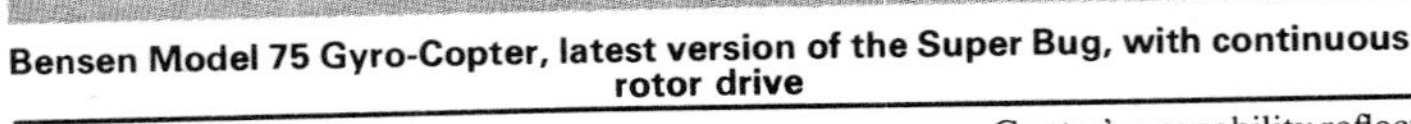
Bensen Model 75 Gyro-Copter, latest version of the Super Bug, with continuous rotor drive

Production version of Birdman TL-1 with V-tail

Weights (standard rotor):

Weight empty:	
B-8M	112 kg (247 lb)
B-8V	158 kg (348 lb)
Max T-O weight:	
B-8M	227 kg (500 lb)
B-8V	272 kg (600 lb)

Performance (at max T-O weight, with standard rotor):

Max level speed at S/L:	
B-8M	74 knots (137 km/h; 85 mph)
B-8V	52 knots (96·5 km/h; 60 mph)
Max cruising speed at S/L:	
B-8M	52 knots (96·5 km/h; 60 mph)
B-8V	43 knots (80·5 km/h; 50 mph)
Econ cruising speed:	
B-8M, B-8V	39 knots (72·5 km/h; 45 mph)
Min speed in level flight:	
B-8M	13 knots (24 km/h; 15 mph)
B-8V	17·5 knots (32 km/h; 20 mph)
T-O speed at S/L:	
B-8M	17·5 knots (32 km/h; 20 mph)
B-8V	22 knots (40 km/h; 25 mph)
Landing speed:	
B-8M	6 knots (11·5 km/h; 7 mph)
B-8V	9 knots (16 km/h; 10 mph)
Max rate of climb at S/L:	
B-8M	305 m (1,000 ft)/min
B-8V	198 m (650 ft)/min
Service ceiling:	
B-8M	3,800 m (12,500 ft)
B-8V	2,440 m (8,000 ft)
T-O run, unpowered rotor, zero wind:	
B-8M	92 m (300 ft)
B-8V	122 m (400 ft)
T-O run, powered rotor, zero wind:	
B-8M	15 m (50 ft)
Landing run in 9 knot (16 km/h; 10 mph) wind:	
B-8M, B-8V	0 ft
Landing run in zero wind:	
B-8M	6 m (20 ft)
B-8V	7·5 m (25 ft)
Normal range:	
B-8M	86 nm (160 km; 100 miles)
B-8V	130 nm (241 km; 150 miles)
Ferry range:	
B-8M	260 nm (482 km; 300 miles)
B-8V	345 nm (643 km; 400 miles)
Endurance:	
B-8M	1·5 hr
B-8V	2·25 hr

BENSEN MODEL B-8H

Scheduled to be demonstrated in public for the first time in 1976, the B-8H is described as a 'Hovering Gyro-Copter', a capability reflected by the 'H' in its designation.

Development dates from 1973. On 29 June that year, Dr Igor Bensen piloted a Gyro-Glider with two two-blade co-axial rotors during its first flight at Raleigh-Durham Airport. Nineteen further flights followed, during which the aircraft executed a full range of prescribed flight manoeuvres. In other tests, a pair of co-axial rotors, powered by separate engines, were shown to be capable of hovering flight. The B-8H adds to the co-axial powered rotors a conventional Gyro-Copter pusher engine installation. No other details were available in mid-1976.

BENSEN MODEL B-8MA AGRICOPTER

Bensen announced in 1970 the introduction of an agricultural version of the B-8M Gyro-Copter. Known as the B-8MA Agricopter, this is basically the same as the B-8M but has a 19 litre (5 US gallon) tank for chemicals mounted beneath the engine. Introduction of ultra-low-volume insecticides has made it possible for this small-capacity system to offer effective cover of an area of 160 to 400 acres during a single 40-minute flight. When first introduced, the Agricopter had only a single spray nozzle mounted at the tip of a tail 'stinger'. This has been supplemented by conventional spraybooms, allowing the aircraft to apply agricultural chemicals at the low altitudes normal for a spray-control role.

First flight of the prototype was made in September 1969.

Subsequently, the Agricopter received FAA certification in an Experimental (Market Survey) category, to permit its demonstration in the US. Bensen is hoping to gain certification in four utility fields of the FAA Restricted category, namely aerial survey, aerial application, forest patrol and pipeline patrol. At the present time operators of such aircraft in the USA must hold an agricultural aircraft operator's certificate.

The description of the B-8M applies also to the Model B-8MA except in the following details:

Power Plant: One 67 kW (90 hp) McCulloch 4318GX flat-four two-stroke engine, driving an Aerial Prop Model BA 48-A8-90 two-blade wooden fixed-pitch propeller.

Dimensions, external:

Diameter of rotor	6·60 m (21 ft 8 in)
Length of fuselage	4·04 m (13 ft 3 in)
Propeller diameter	1·22 m (4 ft 0 in)

Weight:

Max T-O and landing weight	272 kg (600 lb)

Performance (at AUW of 227 kg; 500 lb):
As for the Model B-8M with standard rotor

BENSEN MODEL B-16S GYRO-COPTER

This advanced version of the Bensen Model B-8M Gyro-Copter is described as the world's first flying snowmobile when equipped with skis. Basically a B-8M, it is converted to the new configuration by installation of twin Kiekhaefer snowmobile engines, and ski landing gear, a larger-diameter rotor, snow baffle on the tail, new plastics fuel tank and pipelines, gimbal head and double-tube mast.

The B-16S can be used on wheels, skis or floats, and conversion from wheel to ski landing gear can be accomplished in ten minutes. It is claimed to be particularly economical in operation, as the Kiekhaefer engines run on low-octane automobile fuel.

The description of the B-8M applies also to the B-16S, except as follows:

Rotor System: As B-8M, except rotor diameter increased. Rotor speed 380 rpm.

Landing Gear: Ski landing gear; optionally wheel gear as described for the B-8M, or floats.

Power Plant: Two 35·75 kW (48 hp) Kiekhaefer KAM 525 snowmobile engines, mounted side by side aft of the rotor mast, each driving a two-blade wooden fixed-pitch pusher propeller. Plastics fuel tank beneath pilot's seat, capacity 22·5 litres (6 US gallons). Provision for fitting second tank of same capacity for extended range.

Dimensions, external:

Rotor diameter	6·92 m (22 ft 8½ in)
Length overall	3·54 m (11 ft 7¼ in)
Height overall	1·98 m (6 ft 6 in)
Width, rotor blades fore and aft	1·77 m (5 ft 9½ in)

Weights:

Weight empty	204 kg (450 lb)
Max T-O weight	317 kg (700 lb)

Performance (at max T-O weight):

Max level speed at S/L	69 knots (129 km/h; 80 mph)
Cruising speed	43·5 knots (80·5 km/h; 50 mph)
Landing speed	6 knots (11·5 km/h; 7 mph)
T-O run on skis	183 m (600 ft)
Landing run	under 30 m (100 ft)
Range, standard fuel	52 nm (96 km; 60 miles)

BENSEN MODEL 75 GYRO-COPTER

Bensen Aircraft Corporation has released brief details of a new version of the Gyro-Copter. The basic airframe is similar to that of the Super Bug, except for a change in the rotor control system, and has a similar twin-engine installation. Instead of using the auxiliary engine only to spin up the rotor prior to take-off, the auxiliary engine of the Model 75 drives the rotor continuously through an infinite-ratio automatic transmission. Advantages claimed for the gyroplane with a driven rotor include quick starts with full power from a standstill, with a T-O run of less than 61 m (200 ft); absence of rotor slowdown during zero *g* manoeuvres; increased lifting capacity; and a shallower glide angle when the main engine is throttled back. Rotor drive torque is claimed to be light and offset easily by the conventional rudder.

BIRDMAN
BIRDMAN AIRCRAFT INC

Address:
480 Midway (Airport), Daytona Beach, Florida 32014
Telephone: (904) 252-4053
Engineering: J. G. Ladesig
Production: W. P. Prettyman
Comptroller: C. D. Owens
Sales: C. B. Chapman

Birdman Aircraft Inc has developed an ultra lightweight sporting aircraft which, despite a wing span of 10·36 m (34 ft 0 in), has an empty weight of only 45 kg (100 lb). The company's aim was to design a strong, lightweight and inexpensive aircraft that could be assembled easily by a novice builder. In achieving this aim, it introduced some unusual ideas and construction materials. The covering, for example, consists of a bi-axially oriented synthetic film known as aircraft Monokote, and this offers the key to the combination of light weight and strength.

Design of what proved to be the world's lightest powered aircraft, known as the Birdman TL-1, began in April 1969. The first flight was recorded almost six years later, on 25 January 1975, after considerable structural research and testing. The company is marketing the aircraft in kit form, with all materials, components, engine and accessories provided, together with plans for its construction and a builder's manual. First orders were taken in June 1975 and by February 1976 over 200 kits were under construction.

BIRDMAN TL-1

The prototype TL-1 (N111ET) had full-span trailing-edge flaps. These are deleted on the production version, which also has a V-tail instead of the T-tail fitted to the prototype.

Type: Ultra lightweight sporting aircraft for home assembly.

Wings: Cantilever monoplane, built in three sections, with detachable outer panels. Wing section USA 35B modified. Thickness/chord ratio 14%. Dihedral on outer wing panels 5°. Incidence 3°. No sweepback. Composite structure, with single built-up spruce spar, wooden ribs, leading-edge D cell of reinforced expanded synthetic foam and aircraft Monokote covering. No ailerons. Sequentially-operated spoilers, immediately aft of main spar, for both yaw and roll control; the more the stick is moved, the more panels come into operation.

Fuselage: Three-part structure: forward portion of riveted semi-monocoque light alloy, centre section of spruce and plywood, and an aft section comprising a tapered plywood monocoque.

Tail Unit: Braced V-tail. Two-spar structure with laminated plywood ribs and aircraft Monokote covering. No rudders.

Bowers Fly Baby 1-B biplane (Continental C85 engine) *(Chuck Billingsley)*

Bowers Namu II fitted with streamline leg and wheel fairings *(Howard Levy)*

Bowers Fly Baby 1-A built by Jan Eriksson of Uppsala, Sweden, and first flown on 7 May 1975

Brokaw-Jones BJ-520 homebuilt cabin monoplane (Continental TSIO-520-B engine)

LANDING GEAR: Non-retractable main units only, sited at point of balance. Pilot's legs and feet substitute as retractable nose gear. Castor at aft end of fuselage to support aircraft on ground. Shock-absorption by rubber bungee. Main-wheel tyres size 12 × 1·25, pressure 3·10 bars (45 lb/sq in). No brakes.

POWER PLANT: One 11·2 kW (15 hp) Tally Aircraft M.C. 101 DT single-cylinder aircooled two-stroke engine, pylon-mounted from centre-section structure aft of pilot, and driving a Tally Aircraft 28-9 two-blade wooden fixed-pitch pusher propeller. Fuel contained in two aluminium tanks with combined capacity of 15 litres (4 US gallons). Lubricating oil mixed in fuel at ratio of 1 : 16.

ACCOMMODATION: Exposed seat with harness, adjustable to allow pilot to fly the aircraft in an upright or semi-reclining position.

DIMENSIONS, EXTERNAL:
- Wing span: 10·36 m (34 ft 0 in)
- Wing chord (constant): 1·30 m (4 ft 3 in)
- Wing area, gross: 13·42 m² (144·5 sq ft)
- Wing aspect ratio: 8
- Length overall: 6·10 m (20 ft 0 in)
- Height overall: 2·18 m (7 ft 2 in)
- Tailplane span: 2·59 m (8 ft 6 in)
- Wheel track: 1·83 m (6 ft 0 in)
- Propeller diameter: 0·71 m (2 ft 4 in)

WEIGHTS AND LOADINGS:
- Weight empty: 45 kg (100 lb)
- Max T-O and landing weight: 159 kg (350 lb)
- Normal T-O and landing weight: 131 kg (288 lb)
- Max wing loading: 11·81 kg/m² (2·42 lb/sq ft)
- Max power loading: 14·2 kg/kW (23·3 lb/hp)

PERFORMANCE (at max T-O weight):
- Never-exceed speed: 69 knots (129 km/h; 80 mph)
- Max level speed at S/L: 52 knots (97 km/h; 60 mph)
- Max cruising speed at 1,980 m (6,500 ft): 47 knots (87 km/h; 54 mph)
- Econ cruising speed at 1,220 m (4,000 ft): 30 knots (56 km/h; 35 mph)
- Stalling speed: 16·5 knots (31 km/h; 19 mph)
- Max rate of climb at S/L: 107 m (350 ft)/min
- Service ceiling: 3,810 m (12,500 ft)
- T-O run: 23 m (75 ft)
- T-O to 15 m (50 ft): 61 m (200 ft)
- Landing from 15 m (50 ft): 30 m (100 ft)
- Landing run: 7·62 m (25 ft)
- Range with max fuel: 174 nm (322 km; 200 miles)
- Range with max payload: 65 nm (121 km; 75 miles)

BOWERS
PETER M. BOWERS

ADDRESS:
10458 16th Avenue South, Seattle, Washington 98168

Mr Peter Bowers, an aeronautical engineer with Boeing in Seattle, is a principal source of detailed information on vintage aircraft in the United States, and has provided much of the data for a number of replicas of first World War aircraft now under construction or flying.

Among several aircraft built by Mr Bowers is a full-scale replica of the Wright Model EX of 1911, the first aeroplane to cross the American continent. This machine was tested as a towed sailplane in the Autumn of 1961 and during 1967/68 was converted into a replica of the Wright Model 'A', the US Army's first aeroplane. Powered by an 18·6 kW (25 hp) Ford Model 'T' engine, it was intended to take part in celebrations to mark the anniversary of the 1908/09 flights of the original machine.

In addition to this work on replicas, Mr Bowers has designed and built a single-seat light aircraft known as the Fly Baby, which has flown in both monoplane and biplane forms, and a side-by-side two-seat light monoplane named Namu II. Details of both types follow:

BOWERS FLY BABY 1-A

The prototype Fly Baby monoplane was produced to compete in an Experimental Aircraft Association design contest, organised to encourage the development of a simple, low-cost, easy-to-fly aeroplane that could be built by inexperienced amateurs for recreational flying. It was built in 720 working hours, at a cost of $1,050, and flew for the first time on 27 July 1960. As only one other aircraft was completed by the specified closing date, the contest was postponed for two years.

Following a crash in April 1962, when a pilot borrowed the Fly Baby and became lost in mountain country in bad weather, an entirely new fuselage was built. This is 15 cm (6 in) longer than the original and features minor structural improvements. In addition, the original Continental A65 engine, converted to give 56 kW (75 hp), was replaced by a C75 engine converted to give 63·5 kW (85 hp), and the capacity of the fuel tank was increased from 45·5 litres (12 US gallons) to 60·5 litres (16 US gallons).

When the EAA contest was finally held in the Summer of 1962, Fly Baby was placed first and won a prize of $2,500. Home construction plans of the aircraft are available and 3,725 sets had been sold by May 1976. Construction of well over 670 Fly Babies is known to have been undertaken, of which over 165 had been completed and flown by May 1976, including some based on detailed drawings and instructions published in *Sport Aviation*, journal of the EAA.

The Fly Baby monoplane has been tested as a twin-float seaplane, in which configuration it has a max AUW of 454 kg (1,000 lb) and cruising speed of 84 knots (156 km/h; 97 mph).

During 1968, Mr Bowers designed and built biplane wings for the Fly Baby, which are interchangeable with the monoplane wings. The monoplane version is now known as the Fly Baby 1-A, and the biplane as the Fly Baby 1-B (described separately).

Mr Bowers received repeated requests to provide plans for a two-seat version of the Fly Baby, but declined initially on the grounds that an extensive redesign would be necessary to maintain structural integrity. Despite this a number of homebuilders completed two-seat versions, usually by confining their modifications to increased fuselage width, for side-by-side seats and a single, central control column.

Mr Bowers amended his plans during 1973 to allow for construction of a two-seat version of the Fly Baby. The changes include a 0·97 m (3 ft 2 in) wide fuselage, a 1·52 m (5 ft 0 in) span wing centre-section to support a shock-absorbing landing gear similar to that of the Ryan ST/PT-22 of 1934-42, the use of heavier flying wires with swaged fork ends, a raised aft turtledeck to offset the drag of the larger cockpit and a recommendation to use an engine of 63·5 to 80 kW (85 to 107 hp). The outer wing panels are unchanged, giving the two-seat Fly Baby a span of 9·45 m (31 ft 0 in) and wing area of 12·4 m² (133·5 sq ft).

The following description applies to the original single-seat Fly Baby 1-A:

TYPE: Single-seat light monoplane.

WINGS: Wire-braced low-wing monoplane. Double ½ in 1 × 19 stainless steel bracing wires. Wing section NACA 4412. Wooden two-spar structure, covered with Dacron fabric and finished with two coats of nitrate dope and one coat of automotive enamel. Wings rotate about a special fitting to fold back alongside the fuselage for towing.

FUSELAGE: Conventional plywood-covered wood structure of rectangular section. Decking behind cockpit, including pilot's headrest, is removable and can be replaced with higher transparent section matched with a sliding transparent cockpit canopy for enclosed cockpit operation.

TAIL UNIT: Wire-braced wood structure, fabric-covered.

LANDING GEAR: Non-retractable tailwheel type. Main landing gear struts of laminated wood, braced by crossed steel wires. Steel tube straight-across axle faired with streamline-section steel tube. Ends of axles project beyond wheel hubs to serve as anchor points for wing bracing wires. Shock-absorption by low-pressure 8·00-4 tyres, carried on Piper Cub wheels, with hydraulic brakes.

POWER PLANT: One 63·5 kW (85 hp) Continental C75 flat-four engine, driving a two-blade fixed-pitch propeller. Fuel tank from Piper J-3 Cub, capacity 60·5 litres (16 US gallons).

ACCOMMODATION: Single seat in open or enclosed cockpit. Baggage in underfuselage 'tank' which can be removed and carried like a suitcase.

DIMENSIONS, EXTERNAL:
- Wing span: 8·53 m (28 ft 0 in)
- Wing chord (constant): 1·37 m (4 ft 6 in)

Length overall 5·64 m (18 ft 6 in)
Height, wings folded 1·98 m (6 ft 6 in)

WEIGHTS:
Weight empty 274 kg (605 lb)
Max T-O weight 419 kg (924 lb)

PERFORMANCE (at max T-O weight):
Max level speed at S/L over 104 knots (193 km/h; 120 mph)
Cruising speed 91-96 knots (169-177 km/h; 105-110 mph)
Landing speed 39 knots (72·5 km/h; 45 mph)
Max rate of climb at S/L 335 m (1,100 ft)/min
T-O run 76 m (250 ft)
Landing run 76 m (250 ft)
Range with max fuel 277 nm (515 km; 320 miles)

BOWERS FLY BABY 1-B

During 1968 Mr Bowers designed and built a set of interchangeable biplane wings for the original prototype Fly Baby and with these fitted it flew for the first time on 27 March 1969.

The new wings have the same aerofoil section and incidence as those of the monoplane version, but the rib webs are made of 1/16 in instead of 1/8 in plywood and the wingtip bows are formed from 1/2 in aluminium tube instead of laminated wood strips. This lightweight construction limits weight increase to only 21 kg (46 lb) for an increase of 2·79 m² (30 sq ft) in wing area. Span is reduced by 1·83 m (6 ft) and chord by 0·30 m (1 ft). Ailerons are fitted to the lower wings only.

To facilitate entry to the cockpit the upper wing has been located well forward, and in order to bring the new centre of lift in line with the original CG, both planes have been given 11° of sweepback. Changeover from monoplane to biplane configuration can be accomplished by two people in approximately one hour. (See also Cronk Fly Baby.)

The description of the Fly Baby 1-A applies also to the 1-B, except in the following details:

TYPE: Single-seat light biplane.

WINGS: Forward-stagger single-bay biplane with N-type interplane and centre-section struts. Landing and flying bracing wires. Sweepback 11°. Wooden structure with Dacron covering. Rib webs constructed of 1/16 in plywood, wingtip bows formed of 1/2 in aluminium tube. Ailerons on lower wings only.

POWER PLANT: One 63·5 kW (85 hp) Continental C85 flat-four engine.

DIMENSIONS, EXTERNAL:
Wing span 6·71 m (22 ft 0 in)
Wing area, gross 13·94 m² (150 sq ft)
Wing chord, both wings (constant) 1·07 m (3 ft 6 in)
Height overall 2·08 m (6 ft 10 in)

WEIGHTS AND LOADING:
Weight empty 295 kg (651 lb)
Max T-O weight 440 kg (972 lb)
Max wing loading 31·74 kg/m² (6·5 lb/sq ft)

PERFORMANCE (at max T-O weight):
Cruising speed 75·5 knots (140 km/h; 87 mph)
Max rate of climb at S/L 267 m (875 ft)/min

BOWERS NAMU II

Because of the bulky appearance of this side-by-side two-seater, by comparison with the standard Fly Baby, Mr Bowers decided to name it Namu II, after Seattle's famous captive whale. His own prototype (N75PA), shown in the accompanying illustration, flew for the first time on 2 July 1975. Another Namu II, with a different type of canopy and without wheel fairings, has been completed by Richard Lowe and Tom Godbey of Seattle, and is powered by a 112 kW (150 hp) Lycoming O-320 engine.

The following details refer to Mr Bowers' prototype:

TYPE: Two-seat amateur-built sporting aircraft.

WINGS: Cantilever low-wing monoplane. Wing section NACA 4415 at root, NACA 4412 at tip. Wing centre-section has anhedral, outer wing has dihedral. Composite three-piece monospar structure. The spar has laminated spruce flanges and 1/8 in mahogany marine plywood webs. The leading-edge torsion box is a sandwich of two sheets of 2024-T3 aluminium with glassfibre cloth between, epoxy-cemented together and formed over 3/8 in mahogany marine plywood nose ribs. The ribs aft of the spar are cut from 1/8 in plywood and have 1/4 in by 1/2 in slotted spruce cap strips. The structure is Dacron-covered. Slotted ailerons. No flaps. No trim tabs.

FUSELAGE: Wooden structure, plywood-covered except for Dacron on rear fuselage top decking.

TAIL UNIT: Dacron-covered wooden structure. Controllable trim tab in starboard elevator.

LANDING GEAR: Non-retractable tailwheel type. Main wheels are mounted in symmetrical bent-steel forks welded directly to the cantilever main legs. Main wheels size 6·00-6. Steerable tailwheel. Disc-type hydraulic brakes. Glassfibre fairings on main wheels.

POWER PLANT: One 93 kW (125 hp) Lycoming O-290-G (GPU) flat-four engine. One fuselage fuel tank, mounted immediately aft of the firewall, capacity 121 litres (32 US gallons).

ACCOMMODATION: Two seats side by side under transparent canopy, with upward sliding panels over seats. Dual controls. Stowage for 45 kg (100 lb) baggage under and aft of the seats.

DIMENSIONS, EXTERNAL:
Wing span 10·06 m (33 ft 0 in)
Wing area, gross 13·9 m² (150 sq ft)
Length overall 6·55 m (21 ft 6 in)

WEIGHTS AND LOADINGS:
Weight empty 544 kg (1,200 lb)
Max T-O weight 839 kg (1,850 lb)
Max wing loading 60·20 kg/m² (12·33 lb/sq ft)
Max power loading 9·02 kg/kW (14·8 lb/hp)

PERFORMANCE (at max T-O weight):
Max level speed 122 knots (225 km/h; 140 mph)
Cruising speed 109 knots (203 km/h; 126 mph)
Landing speed 43·5 knots (80·5 km/h; 50 mph)
Max rate of climb at S/L 290 m (950 ft)/min
Service ceiling 4,570 m (15,000 ft)
Range with max fuel 434 nm (804 km; 500 miles)

BROKAW

BROKAW AVIATION INC

ADDRESS:
2625 Johnson Point, Leesburg, Florida 32748
Telephone: (904) 787-1324/2329

PRESIDENT:
Bergon F. Brokaw, MD, FACFP

BROKAW-JONES BJ-520

Dr B. F. Brokaw, a former US Navy pilot, and Dr E. Jones, who has a PhD in aeronautical engineering, combined their talents to design and build a two-seat low-wing monoplane which was claimed to be the world's fastest homebuilt. Dr Brokaw was concerned primarily with overall design and construction, Dr Jones with stress analysis and structural design.

Basic intention was to evolve a high-speed all-weather two-seat homebuilt suitable for cross-country flying. Aerobatic potential was of secondary consideration, but the BJ-520 is stressed to ±6*g* for aerobatics and 9*g* ultimate.

Design began in September 1966 and construction of the prototype started five months later. First flight of the BJ-520 was made on 18 November 1972, and during the Summer of 1973 work was carried out to clean up the airframe to take better advantage of the design potential. This included the provision of wheel-well doors, and reduction of the drag of the engine cooling system, achieved by the use of baffles, ducts and direct ram-air cooling.

Further major redesign and modification were under way in early 1976, and the BJ-520 is expected to resume flying in 1977. The latest refinements include changes to the wing span and section, and power plant, as described below. (For performance and other data relating to the original version, powered by a 212·5 kW (285 hp) Continental TSIO-520-B turbocharged engine, see 1975-76 *Jane's.*)

Dr Brokaw has formed Brokaw Aviation Inc to market information and plans to amateur constructors. If there is sufficient demand, it is intended also to supply certain components such as the canopy, fillets, cowling and, possibly, bulkheads and ribs, to simplify the task of the homebuilder.

TYPE: Two-seat homebuilt sporting aircraft.

WINGS: Cantilever low-wing monoplane. Laminar-flow wing section: NACA 64A412 at root, NACA 64A410 at tip. Dihedral 3°. Incidence 1° at root, −2° at tip. Conventional structure of 2024-T3 light alloy. Ailerons and trailing-edge flaps of light alloy.

FUSELAGE: Semi-monocoque structure of light alloy.

TAIL UNIT: Cantilever light alloy structure. Fixed-incidence tailplane. Electrically-operated trim tabs in elevators and rudder.

LANDING GEAR: Hydraulically-retractable tricycle type. Main and nose units and associated hydraulic system are from a Navion aircraft. Single wheel on each unit, with 6·00 × 15 low-profile tyre, pressure 4·48 bars (65 lb/sq in). Goodyear hydraulic brakes on main wheels.

POWER PLANT: One 283 kW (380 hp) Avco Lycoming TIO-541-A flat-six engine, driving a McCauley three-blade metal constant-speed propeller with spinner. Four integral wing fuel tanks, with usable capacity of 240·7 litres (63·6 US gallons). Refuelling points at wingtips. Oil capacity 11·3 litres (3 US gallons).

ACCOMMODATION: Two seats in tandem beneath transparent individual canopies. Port half of each canopy hinged at centreline to open upwards. Baggage space, which accommodates 18 kg (40 lb), aft of rear seat. Cabin heated and ventilated.

SYSTEMS: Hydraulic system at 103·4 bars (1,500 lb/sq in) for landing gear retraction and brakes. 24V electrical system. Oxygen system for pilot and passenger, with capacity in excess of 4 hours.

ELECTRONICS AND EQUIPMENT: Full IFR instrumentation, including DGO 10, DME 70, dual 360-channel transceivers and dual VORs, ADF, transponder, marker beacon and ILS. Landing gear warning lights, navigation lights and external power socket.

DIMENSIONS, EXTERNAL:
Wing span 7·21 m (23 ft 8 in)
Wing chord at root 1·52 m (5 ft 0 in)
Wing area, gross 8·95 m² (96·3 sq ft)
Length overall 6·94 m (22 ft 9 in)
Height overall 2·69 m (8 ft 10 in)
Tailplane span 2·44 m (8 ft 0 in)

PERFORMANCE (estimated):
Max cruising speed (70% power) at 7,300 m (24,000 ft) over 278 knots (515 km/h; 320 mph)
Rate of climb at S/L over 915 m (3,000 ft)/min

BUSHBY

BUSHBY AIRCRAFT INC

ADDRESS:
Route 1, PO Box 13B, Minooka, Illinois 60447
Telephone: (815) 462-2346

Mr Robert W. Bushby, a research engineer with Sinclair Oil Co, began by building a Midget Mustang single-seat sporting monoplane, using drawings, jigs and certain components produced by the aircraft's designer, the late David Long. He has since produced the aircraft in kit form and also offers sets of plans of the Midget Mustang and a two-seat derivative known as the Mustang II to amateur constructors.

BUSHBY/LONG MM-1 MIDGET MUSTANG

The prototype of the Midget Mustang was completed in 1948 by David Long, then chief engineer of the Piper company. He flew it in the National Air Races that year, and in 1949 was placed fourth in the Continental Trophy Race at Miami.

Two basic versions have been developed by Robert Bushby, as follows:

MM-1-85. Powered by 63·5 kW (85 hp) Continental C85-8FJ engine. Flew for the first time on 9 September 1959.

MM-1-125. Powered by 101 kW (135 hp) Lycoming O-290-D2 engine. Otherwise similar to MM-1-85. Flew for first time in July 1963. New propeller introduced during 1973 has improved max speed and cruising speed.

Ninety Midget Mustangs had been completed by the Spring of 1976, with 800 more under construction throughout the world. Several of those now flying have a 112 kW (150 hp) Lycoming O-320 engine, providing a cruising speed of 234 knots (435 km/h; 270 mph) at 2,440 m (8,000 ft). Some have been fitted with retractable main landing gear.

The following details apply to the two basic versions:

TYPE: Single-seat fully-aerobatic sporting monoplane.

WINGS: Cantilever low-wing monoplane. Wing section NACA 64A212 at root, NACA 64A210 at tip. Dihedral 5°. Incidence 1° 30′. Two-spar flush-riveted stressed-skin aluminium structure. Aluminium statically-balanced ailerons and plain trailing-edge flaps.

FUSELAGE: Aluminium flush-riveted stressed-skin monocoque structure.

TAIL UNIT: Cantilever all-metal structure. Controllable trim tab in port elevator.

LANDING GEAR: Non-retractable tailwheel type. Cantilever spring steel main legs. Steerable tailwheel. Goodyear wheels and tyres, size 5·00-5, pressure 1·24 bars (18 lb/sq in). Goodyear hydraulic disc brakes.

POWER PLANT (MM-1-85): One 63·5 kW (85 hp) Continental C85-8FJ or -12 flat-four engine, driving a McCauley two-blade metal fixed-pitch propeller. Fuel tank aft of firewall, capacity 57 litres (15 US gallons). Optional integral wing fuel tanks, each with capacity of 57 litres (15 US gallons). Optional wingtip tanks, each with capacity of 13 litres (3·5 US gallons). Oil capacity 3·75 litres (1 US gallon).

POWER PLANT (MM-1-125): One 101 kW (135 hp) Lycoming O-290-D2 flat-four engine, driving a Sensenich two-blade metal fixed-pitch propeller. Fuel tank aft of firewall, capacity 57 litres (15 US gallons). No provision for wingtip tanks. Oil capacity 5·75 litres (1·5 US gallons).

ACCOMMODATION: Single seat in enclosed cabin. Canopy hinged on starboard side. Space for 5·5 kg (12 lb) of baggage aft of seat. Room for back parachute.

ELECTRONICS AND EQUIPMENT: Radio optional. No provision for blind-flying instrumentation. Electrical system available on MM-1-85 only.

DIMENSIONS, EXTERNAL:
Wing span 5·64 m (18 ft 6 in)
Span over tip-tanks (MM-1-85) 5·99 m (19 ft 8 in)

Bushby/Long MM-1 Midget Mustang *(Jean Seele)*

Bushby M-II Mustang II with retractable landing gear, built by Mr Oby Tolman *(Henry Artof)*

Wing chord at root 1·53 m (5 ft 0 in)
Wing chord at tip 0·76 m (2 ft 6 in)
Wing area, gross 6·32 m² (68 sq ft)
Wing aspect ratio 4
Length overall 5·00 m (16 ft 5 in)
Height overall 1·37 m (4 ft 6 in)
Tailplane span 1·98 m (6 ft 6 in)
Wheel track 1·55 m (5 ft 1 in)

DIMENSIONS, INTERNAL:
Cabin:
Max width 0·56 m (1 ft 10 in)
Baggage space 0·057 m³ (2 cu ft)

WEIGHTS AND LOADINGS:
Weight empty:
MM-1-85 261 kg (575 lb)
MM-1-125 268 kg (590 lb)
Max T-O and landing weight:
MM-1-85 397 kg (875 lb)
MM-1-125 408 kg (900 lb)
Max wing loading:
MM-1-85 63·00 kg/m² (12·9 lb/sq ft)
MM-1-125 64·45 kg/m² (13·2 lb/sq ft)
Max power loading:
MM-1-85 6·25 kg/kW (10·3 lb/hp)
MM-1-125 4·04 kg/kW (6·67 lb/hp)

PERFORMANCE (at max T-O weight):
Never-exceed speed:
243 knots (450 km/h; 280 mph)
Max level speed at S/L:
MM-1-85 165 knots (306 km/h; 190 mph)
MM-1-125 195 knots (362 km/h; 225 mph)
Max cruising speed at 2,440 m (8,000 ft):
MM-1-85 171 knots (317 km/h; 197 mph)
MM-1-125 187 knots (346 km/h; 215 mph)
Econ cruising speed:
MM-1-85 129 knots (238 km/h; 148 mph)
MM-1-125 143 knots (265 km/h; 165 mph)
Stalling speed, flaps down:
MM-1-85 50 knots (92 km/h; 57 mph)
MM-1-125 53 knots (97 km/h; 60 mph)
Max rate of climb at S/L:
MM-1-85 533 m (1,750 ft)/min
MM-1-125 670 m (2,200 ft)/min
Service ceiling:
MM-1-85 over 4,875 m (16,000 ft)
MM-1-125 5,790 m (19,000 ft)
T-O run:
MM-1-85 137 m (450 ft)
MM-1-125 122 m (400 ft)
T-O to 15 m (50 ft):
MM-1-85 274 m (900 ft)
MM-1-125 213 m (700 ft)
Landing from 15 m (50 ft) 365 m (1,200 ft)
Landing run 152 m (500 ft)
Range with max fuel:
MM-1-85 347 nm (640 km; 400 miles)
MM-1-125 325 nm (603 km; 375 miles)
Range with max fuel and tip-tanks:
MM-1-85 651 nm (1,200 km; 750 miles)

BUSHBY M-II MUSTANG II

Design of this side-by-side two-seat derivative of the Midget Mustang was started in 1963. Construction of a prototype began in 1965 and it flew for the first time on 9 July 1966. During 1968 Mr Bushby designed an alternative non-retractable tricycle landing gear for the Mustang II, and amateur constructors have the option of either configuration. About 700 Mustang IIs were being built by amateurs in the Spring of 1976, at which time 40 were flying.

The Mustang II illustrated was built by Mr Oby Tolman of Reseda, California. It differs from the standard Bushby plans by having mechanically-actuated retractable main landing gear, a Thorp T-18 cockpit canopy and 106 litre (28 US gallon) fuel tank. First flight was made on 8 April 1975, and Mr Tolman's Mustang II has a max level speed of 166 knots (307 km/h; 191 mph) at 2,440 m (8,000 ft); cruising speed of 156 knots (290 km/h; 180 mph) at the same altitude; stalling speed of 59 knots (110 km/h; 68 mph); max rate of climb at S/L of 457 m (1,500 ft)/min; and range of 434 nm (804 km; 500 miles) with 30 min reserve. Power plant is a 112 kW (150 hp) Lycoming O-320-E2D engine, driving a two-blade metal constant-speed propeller.

The details below apply to the de luxe model, and the empty weight quoted includes IFR instrumentation and nav/com equipment. The M-II can also be operated as an aerobatic aircraft in what Bushby Aircraft calls the 'Sport' configuration. This is identical to the de luxe model except that the electrical system, radio and additional IFR instrumentation are deleted. The 'Sport' model has an empty weight of 340 kg (750 lb) and T-O weight of 567 kg (1,250 lb).

TYPE: Two-seat amateur-built light aircraft.

WINGS: Cantilever low-wing monoplane. Outer wings similar to those of Midget Mustang, attached to new constant-chord centre-section of short span. Wing section NACA 64A212 at root, NACA 64A210 at tip. Dihedral 5° on outer wings only. Incidence 1° 30'. Two-spar flush-riveted stressed-skin aluminium structure. Aluminium statically-balanced ailerons and plain trailing-edge flaps. No trim tabs.

FUSELAGE: Aluminium flush-riveted stressed-skin monocoque structure.

TAIL UNIT: Cantilever all-metal structure. Fixed-incidence tailplane. Controllable trim tab in starboard elevator.

LANDING GEAR: Standard version has non-retractable tailwheel type. Cantilever spring steel main legs. Goodyear 5·00-5 main wheels and tyres, pressure 1·38 bars (20 lb/sq in). Goodyear hydraulic disc brakes. Steerable tailwheel. Alternatively, non-retractable tricycle type. Cantilever spring steel main legs. Cleveland or Goodyear wheels and tyres size 5·00-5. Non-steerable nosewheel, mounted on oleo-pneumatic shock strut and free to swivel up to 16° either side. Goodyear wheel and tyre size 5·00-5. Goodyear or Cleveland hydraulic disc brakes. Wheel fairings optional on either type of landing gear.

POWER PLANT: Normally one 119 kW (160 hp) Lycoming O-320 flat-four engine, driving a two-blade fixed-pitch metal propeller. Provision for other engines including a 93 kW (125 hp) Lycoming O-290 engine, driving a two-blade fixed-pitch metal propeller. Fuel tank aft of firewall, capacity 94·6 litres (25 US gallons). Optional integral wing fuel tanks, each with a capacity of 45 litres (12 US gallons). Refuelling point on starboard side of fuselage aft of firewall. Provision for wingtip tanks. Oil capacity 7·5 litres (2 US gallons).

ACCOMMODATION: Two seats side by side, under large rearward-sliding transparent canopy. Dual controls. Baggage space aft of seats, capacity 34 kg (75 lb).

SYSTEMS: 12V electrical system, supplied by Delco-Remy 15A generator and Exide 33A battery.

ELECTRONICS AND EQUIPMENT: Provision for full IFR instrumentation and dual nav/com system.

DIMENSIONS, EXTERNAL:
Wing span 7·37 m (24 ft 2 in)
Wing chord at root 1·47 m (4 ft 10 in)
Wing chord at tip 0·79 m (2 ft 7 in)
Wing area, gross 9·02 m² (97·12 sq ft)
Wing aspect ratio 5·5
Length overall 5·94 m (19 ft 6 in)
Height overall 1·60 m (5 ft 3 in)
Tailplane span 2·29 m (7 ft 6 in)
Wheel track 2·08 m (6 ft 10 in)
Propeller diameter:
93 kW (125 hp) 1·73 m (5 ft 8 in)
119 kW (160 hp) 1·82 m (6 ft 0 in)

DIMENSIONS, INTERNAL:
Cabin:
Max width 1·02 m (3 ft 4 in)
Baggage space 0·16 m³ (5·5 cu ft)

WEIGHTS:
Weight empty, equipped (N: nosewheel, T: tailwheel landing gear):
N 93 kW (125 hp) engine 413 kg (911 lb)
T 93 kW (125 hp) engine 408 kg (900 lb)
N 119 kW (160 hp) engine 425 kg (938 lb)
T 119 kW (160 hp) engine 420 kg (927 lb)
*Max T-O and landing weight 680 kg (1,500 lb)
**Except for countries that restrict max wing loading to 73·2 kg/m² (15 lb/sq ft), where T-O weight of 658 kg (1,450 lb) applies*

PERFORMANCE (with tailwheel, at max T-O weight):
Never-exceed speed:
93 kW (125 hp) 173 knots (322 km/h; 200 mph)
119 kW (160 hp) 211 knots (391 km/h; 243 mph)
Max level speed at S/L:
93 kW (125 hp) 156 knots (290 km/h; 180 mph)
119 kW (160 hp) 177 knots (328 km/h; 204 mph)
Max cruising speed at 2,285 m (7,500 ft):
93 kW (125 hp) 152 knots (282 km/h; 175 mph)
119 kW (160 hp) 175 knots (323 km/h; 201 mph)
Stalling speed, flaps down:
93 kW (125 hp) 47 knots (87 km/h; 54 mph)
119 kW (160 hp) 51 knots (94 km/h; 58 mph)
Stalling speed, flaps up:
93 kW (125 hp) 51 knots (94 km/h; 58 mph)
119 kW (160 hp) 53 knots (96 km/h; 60 mph)
Max rate of climb at S/L:
93 kW (125 hp) 305 m (1,000 ft)/min
119 kW (160 hp) 425 m (1,400 ft)/min
Service ceiling:
93 kW (125 hp) 4,875 m (16,000 ft)
119 kW (160 hp) 5,485 m (18,000 ft)
T-O run 198 m (650 ft)
T-O to 15 m (50 ft):
93 kW (125 hp) 320 m (1,050 ft)
119 kW (160 hp) 305 m (1,000 ft)
Landing from 15 m (50 ft):
93 kW (125 hp) 290 m (950 ft)
119 kW (160 hp) 305 m (1,000 ft)
Landing run:
93 kW (125 hp) 215 m (700 ft)
119 kW (160 hp) 228 m (750 ft)
Range with standard fuel (75% power):
93 kW (125 hp) 416 nm (770 km; 480 miles)
119 kW (160 hp) 373 nm (692 km; 430 miles)
Range with optional wingtip tanks:
119 kW (160 hp) 542 nm (1,005 km; 625 miles)

CASSUTT
THOMAS K. CASSUTT

ADDRESS:
11718 Persuasion Drive, San Antonio, Texas 78216

While employed as an airline pilot, Capt Tom Cassutt designed and built in 1954 a small single-seat racing monoplane known as the Cassutt Special I (No. 111), in which he won the 1958 National Air Racing Championships. In 1959, he completed a smaller aircraft on the same lines, known as the Cassutt Special II (No. 11).

Plans of both aircraft, and of a sporting version of No. 111 with a larger cockpit, are available to amateur constructors. As a result, many Cassutt Specials are flying and under construction, and brief details of a number of these are given in the accompanying table.

Following his retirement from airline flying, Tom Cassutt plans to develop a number of different aircraft to meet Formula I specifications. He envisages an aircraft which, powered by an overspeeded Continental O-200 engine developing approximately 97 kW (130 hp), should have a maximum level speed of around 221 knots (410 km/h; 255 mph). No specific details of this, or other projects, are yet available.

CASSUTT SPECIAL I

The original Cassutt Special I (No. 111), named *Jersey Skeeter,* is described in detail on the next page.

Airmark Ltd in the UK built several slightly modified examples of the Cassutt Special I; these are known as Airmark/Cassutt 111Ms in Britain. First Cassutt Special to be completed in Australia was built during 1973 by Peter Furlong, an aircraft maintenance engineer at the Latrobe Valley Airfield near Traralgon, Victoria. Powered by a 74·5 kW (100 hp) Rolls-Royce Continental engine, this aircraft flew for the first time in early 1974.

TYPE: Single-seat racing monoplane.

WINGS: Cantilever mid-wing monoplane. Wing section Cassutt 1107. No incidence or dihedral. All-wood two-spar structure with spruce ribs, solid spars and plywood skin, fabric-covered. Ailerons are of welded steel tube construction, fabric-covered. No flaps.

FUSELAGE: Steel tube structure, fabric-covered.

TAIL UNIT: Cantilever steel tube structure, with fabric covering.

LANDING GEAR: Non-retractable tailwheel type. Wittman cantilever spring steel main legs. Main-wheel tyres size 5·00-5. Wheel fairings standard.

POWER PLANT: One Continental C85-8F flat-four engine, rated normally at 63·5 kW (85 hp) but capable of developing 83·5-85·75 kW (112-115 hp) in racing trim. Sensenich two-blade fixed-pitch propeller. Fuel capacity 57 litres (15 US gallons). Oil capacity 3·8 litres (1 US gallon).

ACCOMMODATION: Single seat in enclosed cockpit.

DIMENSIONS, EXTERNAL:

Wing span	4·54 m (14 ft 11 in)
Wing chord (mean)	1·37 m (4 ft 6 in)
Wing area, gross	6·13 m² (66·0 sq ft)
Wing aspect ratio	3·37
Length overall	4·88 m (16 ft 0 in)
Height overall	1·30 m (4 ft 3 in)
Tailplane span	1·19 m (3 ft 11 in)
Wheel track	1·37 m (4 ft 6 in)

WEIGHTS:

Weight empty	234 kg (516 lb)
Max T-O weight	331 kg (730 lb)

PERFORMANCE (at max T-O weight):

Max level speed	200 knots (370 km/h; 230 mph)
Max cruising speed	165 knots (306 km/h; 190 mph)
Stalling speed	61 knots (113 km/h; 70 mph)
Max rate of climb at S/L	610 m (2,000 ft)/min
Endurance with max fuel	3 hours

CASSUTT SPECIAL II

TYPE: Single-seat racing monoplane.

WINGS: Cantilever mid-wing monoplane. Wing section Cassutt 13106. No incidence or dihedral. All-wood structure. No flaps.

FUSELAGE: Steel tube structure, fabric-covered.

TAIL UNIT: Cantilever steel tube structure, fabric-covered. The prototype (No. 11) has small centre fin and auxiliary fins on the tailplane tips.

LANDING GEAR: Non-retractable tailwheel type. Wittman cantilever spring steel main legs. Main-wheel tyre size 5·00-5. Wheel fairings standard.

POWER PLANT AND ACCOMMODATION: As for Cassutt Special I.

DIMENSIONS, EXTERNAL:

Wing span	4·16 m (13 ft 8 in)
Wing chord (constant)	1·47 m (4 ft 10 in)
Wing area, gross	6·13 m² (66 sq ft)
Wing aspect ratio	2·83
Length overall	4·88 m (16 ft 0 in)
Height overall	1·16 m (3 ft 10 in)
Tailplane span	1·14 m (3 ft 9 in)
Wheel track	0·97 m (3 ft 2 in)

WEIGHTS:

Weight empty	196 kg (433 lb)
Max T-O weight	363 kg (800 lb)

PERFORMANCE (at max T-O weight):

Max level speed at S/L	204 knots (378 km/h; 235 mph)
Max cruising speed	174 knots (322 km/h; 200 mph)
Stalling speed	54 knots (100 km/h; 62 mph)
Max rate of climb at S/L	915 m (3,000 ft)/min
Endurance with max fuel	3 hours

Builder (B) *Owner* (O)	*Name of Aircraft*	*Racing Number*	*Wing Span*	*Length Overall*	*Height Overall*	*Normal Loaded Weight*	*Max Speed*
(B, O) Marion Baker & Associates	Boo-Ray	81	4·57 m (15 ft 0 in)	5·08 m (16 ft 8 in)	1·22 m (4 ft 0 in)	340 kg (750 lb)	182 knots (337 km/h; 210 mph)
(B, O) Ken Burmeister	Firefly	4	4·17 m (13 ft 8 in)	5·21 m (17 ft 1 in)	1·30 m (4 ft 3 in)	335 kg (738 lb)	N.A.
(B) R. Grieger (O) E. E. Stover	Ole Yaller	58	4·57 m (15 ft 0 in)	5·03 m (16 ft 6 in)	1·50 m (4 ft 11 in)	378 kg (834 lb)	N.A.
(B) D. Hoffman R. Philbrick (O) Harold Lund	Chabasco	76	4·17 m (13 ft 8 in)	5·04 m (16 ft 6½ in)	N.A.	351 kg (774 lb)	182 knots (337 km/h; 210 mph)
(B, O) Ray O. Morris	Miss A-Go-Go	12	4·55 m (14 ft 11 in)	4·88 m (16 ft 0 in)	1·37 m (4 ft 6 in)	291 kg (642 lb)	195 knots (362 km/h; 225 mph)
(B, O) Fred Wofford	Gold Dust	7	4·37 m (14 ft 4 in)	5·26 m (17 ft 3 in)	1·31 m (4 ft 3½ in)	357 kg (787 lb)	169 knots (314 km/h; 195 mph)
(B, O) James H. Wilson	Snoopy	—	4·42 m (14 ft 6 in)	5·11 m (16 ft 9 in)	1·27 m (4 ft 2 in)	363 kg (800 lb)	208 knots (386 km/h; 240 mph)

CB

CB ENTERPRISES

ADDRESS:

2022 N Acoma, Hobbs, New Mexico 88240

This company markets plans of a sporting monoplane named El Gringo. The external contours of the aircraft are formed with overlays of styrofoam. To facilitate cutting and contouring of this material, CB Enterprises also offers for sale to amateur constructors an electric hot wire foam cutting unit.

CB ENTERPRISES EL GRINGO

TYPE: Single-seat homebuilt sporting monoplane.

WINGS: Cantilever low-wing monoplane. Metal structure, with overlay of contoured styrofoam coated with Dynel-epoxy.

FUSELAGE: Steel tube structure; overlay as for wings.

TAIL UNIT: Conventional cantilever type, of similar construction to wings.

LANDING GEAR: Non-retractable tailwheel type. Cantilever main units, with streamline fairings over wheels.

POWER PLANT: One Volkswagen flat-four modified motor car engine, driving a two-blade fixed-pitch propeller.

ACCOMMODATION: Single seat, under one-piece moulded canopy.

DIMENSIONS AND WEIGHTS:

No details available

PERFORMANCE:

Max level speed	130 knots (241 km/h; 150 mph) IAS
Stalling speed	under 35 knots (65 km/h; 40 mph)

CHRIS TENA

CHRIS TENA AIRCRAFT ASSOCIATION

ADDRESS:

PO Box 1, Hillsboro, Oregon 97123

Telephone: (503) 985-7612

Chris Tena Aircraft has designed a lightweight all-metal single-seat sporting aircraft which is known as the Mini Coupe. Design originated in June 1968, and construction of the first prototype began in July 1970. This aircraft made its first flight in September 1971 and FAA certification in the Experimental category was awarded on 2 June 1972. Kits of components and materials, less engine, are available to amateur constructors.

CHRIS TENA MINI COUPE

Glassfibre wingtips have been added to the Mini Coupe during the past year, to increase the wing area by 0·47 m² (5·1 sq ft). This improves the glide ratio, makes the aircraft more stable during banks, and has reduced stalling speed. At least 156 sets of plans of the Mini Coupe had been sold by early 1976, when 26 aircraft were known to be flying.

TYPE: Single-seat lightweight sporting aircraft.

WINGS: Cantilever low-wing monoplane. Wing section modified Clark Y. Thickness/chord ratio 7%. Incidence 0°. Conventional metal stressed-skin structure of constant chord, with glassfibre wingtips. Plain all-metal ailerons. No flaps. No trim tabs.

FUSELAGE: All-metal semi-monocoque structure.

TAIL UNIT: Cantilever all-metal structure with twin endplate fins and rudders of constant chord. Fixed-incidence tailplane. Manually controlled trim tab in centre of elevator.

LANDING GEAR: Non-retractable tricycle type. Shock-absorption on all units depends on use of oversize tyres. Main wheels with tyres size 14 × 6·00-6, pressure 0·55 bars (8 lb/sq in). Nosewheel tyre size 12 × 6·00-5, pressure 0·34 bars (5 lb/sq in). Asuze drum and band brakes.

POWER PLANT: One 48·5 kW (65 hp) modified Volkswagen 1,600 cc motor car engine, driving a Reese-Shores two-blade fixed-pitch wooden propeller. (Two aircraft fitted with 48·5 kW; 65 hp Continental.) One metal fuel tank in fuselage, immediately aft of firewall, capacity 49 litres (13 US gallons). Refuelling point on fuselage upper surface, forward of windscreen.

ACCOMMODATION: Single seat for pilot in open cockpit. Transparent cockpit canopy optional. Baggage compartment aft of headrest, volume 0·085 m³ (3 cu ft).

SYSTEM: Provision for electrical system, for engine starter and radio.

DIMENSIONS, EXTERNAL:

Wing span	7·32 m (24 ft 0 in)
Wing chord, constant	1·07 m (3 ft 6 in)
Wing area, gross	7·76 m² (83·5 sq ft)
Length overall	4·98 m (16 ft 4 in)
Height overall	1·80 m (5 ft 11 in)
Tailplane span	1·83 m (6 ft 0 in)
Wheel track	1·83 m (6 ft 0 in)
Wheelbase	0·97 m (3 ft 2 in)
Propeller diameter	1·37 m (4 ft 6 in)
Propeller ground clearance	0·25 m (10 in)

WEIGHTS AND LOADINGS:

Weight empty	225 kg (497 lb)
Max T-O weight	385 kg (850 lb)
Max wing loading	50·7 kg/m² (10·39 lb/sq ft)
Max power loading	7·94 kg/kW (12·6 lb/hp)

PERFORMANCE (at max T-O weight with 1,600 cc engine):

Never-exceed speed	125 knots (231 km/h; 145 mph)
Max level speed at 610 m (2,000 ft)	91 knots (169 km/h; 105 mph)
Max cruising speed at 610 m (2,000 ft)	78 knots (145 km/h; 90 mph)
Stalling speed:	
power off	42 knots (78 km/h; 48 mph)
power on	38 knots (70 km/h; 43 mph)
Max rate of climb at S/L	229 m (750 ft)/min
Service ceiling	3,810 m (12,500 ft)
T-O run	122 m (400 ft)
T-O to 15 m (50 ft)	274 m (900 ft)
Landing from 15 m (50 ft)	274 m (900 ft)
Landing run	152 m (500 ft)
Range with max fuel, 20 min reserve	260 nm (482 km; 300 miles)

CONDOR

CONDOR AERO INC

ADDRESS:

PO Box 762, Vero Beach, Florida 32960

PRESIDENT:

Landis G. Ketner

CONDOR SHOESTRING

The original Shoestring was designed and built by the Mercury Air Group, consisting of Mr Vincent Ast, Mr Carl Ast and Mr Rodney Kreimendahl, in 1949. As a racing monoplane (No. 16) it first competed at Miami in 1950 where, piloted by Bob Downey, it achieved third place at a speed of 157·474 knots (291·844 km/h; 181·334 mph). At San Jose, California, in the same year it finished first in the hands of Vincent Ast, at 152 knots (282 km/h; 175 mph). By 1974, when a new wing was fitted to Shoestring, the aircraft had achieved fourteen first places in air races, three seconds and four thirds.

A full description of the original Shoestring appeared in the 1959-60 *Jane's*.

In 1965 Shoestring was purchased by John Anderson and taken to Miami, where a set of drawings of the aircraft was made. The first new racer to be built from these drawings was *Yellow Jacket*, constructed by Jim Strode, which flew for the first time on 15 July 1970. Although this aircraft was very similar to the original Shoestring, the fuselage structure was made 5 cm (2 in) wider and 10 cm (4 in) longer to provide a more comfortable cockpit. The power plant installation, canopy and fairings were also of Jim Strode's design.

Since then other Shoestrings have been built in modified form, and plans are available to amateur constructors.

TYPE: Single-seat Formula One racing monoplane.

WINGS: Cantilever shoulder-mounted monoplane of wood construction, built as one unit and fabric covered. Front spar built of three pieces of spruce bonded together. Rear spar can be cut from one piece of spruce or built up from several pieces. Plywood ribs with cap strips bonded

Cassutt Special with modified tail unit *(Peter M. Bowers)*

Chris Tena Mini Coupe with extended glassfibre wingtips

to top and bottom. Skins bonded to spars and ribs. Ailerons built as part of wing and cut out after the top skin has been bonded in place.

FUSELAGE: Steel tube structure, with plywood formers and fabric covering from instrument panel aft. Thin-gauge aluminium skin forward of panel and on each side of cockpit above wing. Roll-over structure of heavy gauge steel.

TAIL UNIT: Conventional cantilever spruce structure, with fabric-covered plywood skin.

LANDING GEAR: Non-retractable tailwheel type. Main wheels carried on one-piece formed aluminium member bolted directly to bottom of fuselage. Sheet aluminium fairings over legs; aluminium wheel fairings. Solid rubber tailwheel. Brakes fitted.

POWER PLANT: Originally one 63·5 kW (85 hp) Continental flat-four engine. Alternative 74·5 kW (100 hp) engine to meet current PRPA regulations. Recommended engines are Continental C85, C90 and O-200 flat-fours. Fuel tank aft of firewall, capacity 38 litres (10 US gallons). Provision for auxiliary tank behind seat.

ACCOMMODATION: Single seat under blown Plexiglas canopy.

DIMENSIONS, EXTERNAL:

Wing span	5·79 m (19 ft 0 in)
Wing area, gross	6·13 m² (66 sq ft)
Length overall	5·38 m (17 ft 7¾ in)
Height overall	1·42 m (4 ft 8 in)

WEIGHT:

Weight empty	238 kg (525 lb)

PERFORMANCE:

Max level speed	over 208 knots (386 km/h; 240 mph)
Max cruising speed	156 knots (290 km/h; 180 mph)
Stalling speed	57 knots (105 km/h; 65 mph)
Max rate of climb at S/L	915 m (3,000 ft)/min
T-O run	245 m (800 ft)

CRONK

DAVID CRONK

ADDRESS: Schoolcraft, Michigan

CRONK (BOWERS) FLY BABY 1-B

David Cronk, owner of a Dairly Elsie Drive-In in Michigan, was first to complete a Bowers Fly Baby 1-B biplane from plans. In order to give the aircraft the appearance of a first World War Fokker D.VII, some departures from the plans have been made, including the fitting of modified wingtips, a new engine cowling, spoked wheels, and dummy machine-guns mounted on the engine decking. Construction of the aircraft took nearly five years and cost $2,500.

Details given under the basic Bowers Fly Baby 1-B entry apply to this aircraft, except for the following:

WINGS: Dacron fabric is finished with 12 coats of nitrate dope, 12 coats of motorcar body primer and 3 coats of acrylic enamel.

TAIL UNIT: Basically a Taylorcraft welded steel-tube structure with fabric covering.

POWER PLANT: One 93 kW (125 hp) Lycoming O-290-G flat-four engine.

DIMENSIONS, EXTERNAL:

As for standard Fly Baby 1-B, except:

Length	5·99 m (19 ft 8 in)
Height overall	2·13 m (7 ft 0 in)

WEIGHTS:

Weight empty	375 kg (827 lb)
Max T-O weight	544 kg (1,200 lb)

PERFORMANCE:

Max speed	96 knots (177 km/h; 110 mph)
Cruising speed	78 knots (145 km/h; 90 mph)
Landing speed	37 knots (68 km/h; 42 mph)
Max rate of climb at S/L	549 m (1,800 ft)/min

CVJETKOVIC

ANTON CVJETKOVIC

ADDRESS:
624 Fowler Avenue, PO Box 323, Newbury Park, California 91320
Telephone: (805) 498 3092

When living in Yugoslavia, Mr Anton Cvjetkovic designed a single-seat light aeroplane designated CA-51 and powered by a modified Volkswagen engine. A prototype was built by members of Zagreb Aeroclub in 1951, and was followed by five more aircraft of the same type.

After moving to the United States, Mr Cvjetkovic began work, in May 1960, on the design of an improved light aircraft which he designated CA-61. Construction of a prototype was started in February 1961 and it flew for the first time in August 1962. Plans of both single-seat and two-seat all-wood versions are available to amateur constructors, together with plans of a two-seat aircraft designated CA-65, of which the prototype was completed in 1965. Plans of an all-metal version of the CA-65, designated CA-65A, are also available.

By early 1976 a total of more than 300 sets of plans of these aircraft had been sold, and completed aircraft were flying in Canada, South Africa and the United States.

CVJETKOVIC CA-61/-61R MINI ACE

The CA-61 can be built as a single-seat or side-by-side two-seat light aircraft, with any Continental engine of between 48·5 and 63·5 kW (65 and 85 hp). Alternatively, the single-seater can be fitted with a modified Volkswagen engine. Construction takes less than 1,000 hours.

The design was modified during 1973 to allow for installation of retractable landing gear; when constructed in this form, the aircraft is designated **CA-61R**.

The following details refer specifically to the single-seat CA-61 prototype:

TYPE: Single-seat light aircraft.

WINGS: Cantilever low-wing monoplane. Wing section NACA 4415. Dihedral 3°. No incidence. Structure consists of two spruce spars, each built in one piece, built-up spruce girder-type ribs and plywood-covered leading-edge torsion box, with fabric covering overall. Fabric-covered spruce ailerons. No flaps.

FUSELAGE: Conventional wooden structure of basic square section, plywood-covered.

TAIL UNIT: Cantilever wooden structure, covered with plywood. Fixed-incidence tailplane. Trim tab in elevator.

LANDING GEAR: Non-retractable tailwheel type. Cantilever main legs, with helical spring shock-absorption. Goodyear main wheels and tyres, size 5·00-5 Type III, and Model L5 brakes. Steerable tailwheel.

POWER PLANT: One 48·5 kW (65 hp) Continental A65 flat-four engine, driving a Flottorp 63-55 two-blade fixed-pitch propeller. Fuel in two steel tanks in fuselage, with capacities of 45 litres (12 US gallons) and 19 litres (5 US gallons) respectively. Total fuel capacity 64 litres (17 US gallons). Oil capacity 4·5 litres (1·25 US gallons).

ACCOMMODATION: Single seat in enclosed cockpit.

ELECTRONICS AND EQUIPMENT: Prototype fitted with Nova Star radio and Omni.

DIMENSIONS, EXTERNAL:

Wing span	8·38 m (27 ft 6 in)
Wing chord (constant)	1·40 m (4 ft 7 in)
Wing area, gross	11·75 m² (126·5 sq ft)
Wing aspect ratio	6·0
Length overall	5·77 m (18 ft 11 in)
Height overall (in flying position)	2·08 m (6 ft 10 in)
Wheel track:	
Single-seat	2·49 m (8 ft 2 in)
Two-seat	2·62 m (8 ft 7 in)

WEIGHTS AND LOADINGS:

Weight empty:	
Single-seat	275 kg (606 lb)
Two-seat	363 kg (800 lb)
Max T-O weight:	
Single-seat	430 kg (950 lb)
Two-seat	590 kg (1,300 lb)
Max wing loading:	
Single-seat	36·6 kg/m² (7·5 lb/sq ft)
Two-seat	50·0 kg/m² (10·25 lb/sq ft)

PERFORMANCE:

Max level speed at S/L	104 knots (193 km/h; 120 mph)
Normal cruising speed	87 knots (161 km/h; 100 mph)
Min flying speed:	
Single-seat	37 knots (67·5 km/h; 42 mph)
Two-seat	44 knots (80·5 km/h; 50 mph)
Range with max fuel:	
Single-seat	369 nm (685 km; 425 miles)
Two-seat	321 nm (595 km; 370 miles)

CVJETKOVIC CA-65

Design of this side-by-side two-seat light aircraft was started in September 1963. Construction of the prototype began in March 1964 and it flew for the first time in July 1965. Plans are available, and the accompanying illustration shows a CA-65 with O-290-G engine built by Mr John Hickle.

The CA-65 closely resembles the CA-61 in general appearance, but has a more powerful engine and retractable landing gear. A folding-wing version was introduced during 1967.

TYPE: Two-seat light aircraft.

WINGS: Cantilever low-wing monoplane. Modified NACA 4415 wing section. Dihedral 0° on centre-section, 3° on outer wings. Structure consists of two spruce spars, each built in one piece, and built-up spruce girder-type ribs, completely plywood-covered. Fabric-covered spruce ailerons. On the folding-wing version, the outer wings fold upward from their junction with the centre-section.

FUSELAGE: Conventional wooden structure of basically square section, plywood-covered. Manually-operated landing flap under fuselage.

TAIL UNIT: Cantilever wooden structure. Fixed surfaces covered with plywood. Elevator and rudder fabric-covered. Fixed-incidence tailplane.

LANDING GEAR: Mechanically-retractable tailwheel type. Main wheels retract inward. Goodyear main wheels and tyres, size 5·00-5 Type III. Goodyear type L5 brakes. Steerable tailwheel.

POWER PLANT: One 93 kW (125 hp) Lycoming O-290-G flat-four engine, driving a Sensenich 66-68 two-blade fixed-pitch propeller. Two aluminium fuel tanks in fuselage, each with capacity of 53 litres (14 US gallons). Total fuel capacity 106 litres (28 US gallons).

ACCOMMODATION: Two seats side by side in enclosed cockpit, with dual controls; although hydraulic brakes can be operated only by the pilot. Forward-opening canopy.

RADIO: Bayside BEI-990 radio fitted in prototype.

DIMENSIONS, EXTERNAL:

Wing span	7·62 m (25 ft 0 in)
Wing area, gross	10·03 m² (108 sq ft)
Width, wings folded	2·74 m (9 ft 0 in)
Length overall	5·79 m (19 ft 0 in)

Condor Shoestring Yellow Jacket, built from current plans

David Cronk's adaptation of a Bowers Fly Baby 1-B *(Howard Levy)*

The single-seat Cvjetkovic CA-61 (Continental A65 engine)

Cvjetkovic CA-65 two-seat light aircraft (Lycoming O-290-G engine) *(Peter M. Bowers)*

Height overall (in flying position) 2·24 m (7 ft 4 in)
Height, wings folded 3·05 m (10 ft 0 in)
Wheel track 2·11 m (6 ft 11 in)
Propeller diameter 1·73 m (5 ft 8 in)

Weights and Loadings:
Weight empty 408 kg (900 lb)
Max T-O weight 680 kg (1,500 lb)
Max wing loading 67·9 kg/m^2 (13·9 lb/sq ft)
Max power loading 7·3 kg/kW (12·0 lb/hp)

Performance (at max T-O weight):
Max level speed 139 knots (257 km/h; 160 mph)
Normal cruising speed 117 knots (217 km/h; 135 mph)
Stalling speed 48 knots (89 km/h; 55 mph)
Max rate of climb at S/L 305 m (1,000 ft)/min
Service ceiling 4,575 m (15,000 ft)
T-O run 137 m (450 ft)
Landing run 183 m (600 ft)
Range with max fuel 434 nm (804 km; 500 miles)

CVJETKOVIC CA-65A

This aircraft is essentially similar to the all-wood CA-65; it differs by having swept vertical tail surfaces and is of all-metal construction. It is designed for +9*g* and −6*g* ultimate loading.

The general description of the CA-65 applies also to the CA-65A, except in the following details:

Wings: The wing structure consists of a single main spar and an auxiliary wing spar, with aluminium sheet ribs and skin, riveted throughout. The main wing spar cap is made of extruded and bent-up sheet aluminium angles, tapered towards the tip to produce a wing of uniform bending strength. Ribs are formed from 0·025 in aluminium sheet. Wing skin is of 2024T-3 aluminium alloy sheet.

Fuselage: All-metal structure with four aluminium angle longerons and built-up frames. Fuselage skin is of 0·025-0·032 in 2024T-3 aluminium alloy sheet. To simplify formation of the curvature on the upper fuselage, the skins are broken up into small sections of flat panels.

Tail Unit: Cantilever all-metal structure, with swept vertical surfaces. Construction similar to that of the wings.

Power Plant: The structure is designed to accommodate a Lycoming engine of 80·5-112 kW (108-150 hp).

Dimensions, external: As for Model CA-65 except:
Wing span 7·75 m (25 ft 5 in)
Wing area, gross 10·16 m^2 (109·4 sq ft)
Length overall 5·99 m (19 ft 8 in)
Height overall 2·29 m (7 ft 6 in)

Loading:
Max wing loading 66·9 kg/m^2 (13·7 lb/sq ft)

Performance (112 kW; 150 hp engine):
Max level speed 151 knots (280 km/h; 174 mph)
Normal cruising speed 130 knots (241 km/h; 150 mph)
Stalling speed 48 knots (89 km/h; 55 mph)
Max rate of climb at S/L 466 m (1,530 ft)/min
Service ceiling 4,570 m (15,000 ft)
T-O run 99 m (325 ft)
Landing run 183 m (600 ft)
Range with max fuel 460 nm (853 km; 530 miles)

D'APUZZO

NICHOLAS E. D'APUZZO

Address:
1029 Blue Rock Lane, Blue Bell, Pennsylvania 19422
Telephone: (215) 646-4792

Mr D'Apuzzo, who was formerly employed by the Naval Air Development Center, Warminster, Pennsylvania, as a project manager on specialised projects, retired from the Navy Department during 1973. He retains an association with the Navy on a consultant basis.

He has designed several sporting aircraft for amateur construction, among the best known of which are the Denight Special midget racer, described in the 1962-63 *Jane's,* and the PJ-260 single-seat aerobatic biplane described under the Parsons-Jocelyn heading in the 1974-75 *Jane's.*

His other designs include the Senior Aero Sport, which is a two-seat version of the PJ-260, and the smaller single-seat Junior Aero Sport.

In early 1976 a total of 28 PJ-260s and Senior Aero Sports were known to have been completed by amateur constructors in the United States, with a further 95 under construction.

D'APUZZO D-260 SENIOR AERO SPORT

There are five basic versions of the Senior Aero Sport, as follows:

D-260(1). With Lycoming O-435 series engine.

D-260(2). With Continental O-470/E-185 series engine. Construction of prototype started by Mr C. L. McHolland of Sheridan, Wyoming, in May 1962 and the first flight made on 17 July 1965, with a 168 kW (225 hp) E-185 (modified) engine, driving an Aeromatic F-200H-O-85 propeller. Fuel in one 60·5 litre (16 US gallon) tank in fuselage and one 79·5 litre (21 US gallon) streamlined external tank under fuselage.

D-260(3). With Lycoming GO-435 series engine. Construction of prototype started by Mr G. A. Shallbetter of Minneapolis in April 1961 and first flight made on 17 July 1965, with a 194 kW (260 hp) GO-435-C2 engine, driving a Hartzell controllable-pitch propeller of 2·29 m (7 ft 6 in) diameter. Four fuel tanks in fuselage, with total capacity of 136 litres (36 US gallons). Second D-260(3), completed by Mr Alfred Fessenden of Lafayette, New York, is fitted with a Hartzell constant-speed propeller.

D-260(4). With Ranger 6-440-C six-cylinder inverted aircooled engine.

D-260(5). With 224 kW (300 hp) Lycoming R-680-E3 engine. Prototype built by Mr Henry Neys of Lake Stevens, Washington.

Type: Two-seat sporting biplane.

Wings: Conventional braced biplane type. Wing section NACA M12 (modified). Dihedral 30′ on lower wings only. Incidence 2°. Sweepback 9° 15′ on upper wings. Two wood spars, metal ribs, fabric covering. Metal Frise ailerons on all four wings, with fabric covering. No flaps.

Fuselage: Steel tube structure, with aluminium alloy access panels forward of cockpit and fabric covering aft.

Tail Unit: Wire-braced fabric-covered steel tube structure. Trim tab in starboard elevator.

Landing Gear: Non-retractable tailwheel type. Cantilever spring steel main units. Goodyear 6·00-6 main wheels and tyres, pressure 1·38 bars (20 lb/sq in). Goodyear disc brakes.

Power Plant: One flat-six engine; details given under individual model listings. Fuel in 125 litre (33 US gallon) main tank and 41·5 litre (11 US gallon) aerobatic tank, both in fuselage. Total fuel capacity 166·5 litres (44 US gallons). Oil capacity 11 litres (3 US gallons).

Accommodation: Two seats in tandem in open cockpits. Baggage space behind headrest.

System: 12V electrical system, with optional starter and navigation lights.

Electronics: Prototypes fitted with two-way radio and Omni.

Dimensions, external:
Wing span 8·23 m (27 ft 0 in)
Wing chord (constant, both) 1·17 m (3 ft 10 in)
Wing area, gross 17·2 m^2 (185 sq ft)
Wing aspect ratio 4·22
Length overall 6·40 m (21 ft 0 in)
Height overall 2·32 m (7 ft 7½ in)
Tailplane span 3·10 m (10 ft 2 in)
Wheel track 2·57 m (8 ft 5 in)
Wheelbase 4·80 m (15 ft 9 in)

Weights and Loadings:
Normal T-O weight:
D-260 (2) 930 kg (2,050 lb)
D-260 (3) 975 kg (2,150 lb)
Max T-O and landing weight:
All versions 975 kg (2,150 lb)
Max wing loading:
All versions 56·1 kg/m^2 (11·5 lb/sq ft)
Max power loading:
D-260 (2) 5·80 kg/kW (9·56 lb/hp)
D-260 (3) 5·02 kg/kW (8·27 lb/hp)

Performance (D-260 (3) at max T-O weight. D-260 (2) comparable):
Never-exceed speed 165 knots (305 km/h; 190 mph)
Max level speed at 2,135 m (7,000 ft) 135 knots (250 km/h; 155 mph)
Max cruising speed at 2,135 m (7,000 ft) 122 knots (225 km/h; 140 mph)
Econ cruising speed at 2,135 m (7,000 ft) 113 knots (209 km/h; 130 mph)
Stalling speed 48 knots (89 km/h; 55 mph)
Max rate of climb at S/L 610 m (2,000 ft)/min
Service ceiling 6,100 m (20,000 ft)
T-O run 122 m (400 ft)
T-O to 15 m (50 ft) 213 m (700 ft)
Landing from 15 m (50 ft) 275 m (900 ft)
Landing run 183 m (600 ft)
Range with max fuel and max payload 434 nm (805 km; 500 miles)

D'Apuzzo D-260(5) Senior Aero Sport built by Mr Henry Neys

Partially completed airframe of D'Apuzzo D-200 Junior Aero Sport

Davis DA-2A built by 72 year old Brice Rohrer of Montague, California

Davis DA-5A single-seat lightweight sporting aircraft (Continental A65 engine)

D'APUZZO D-200 JUNIOR AERO SPORT

The Junior Aero Sport is a smaller single-seat version of the PJ-260. Its design was started in September 1963 and construction of two prototypes began in September 1964. The first of these was almost completed in early 1973 when the building in which it was being constructed was destroyed by fire. The D-200 was not damaged severely, but due to the delay in obtaining new workshop premises completion of the prototype did not get under way until early 1975, and the first flight was not expected to take place before the Autumn of 1976.

TYPE: Single-seat sporting biplane.

WINGS: Conventional braced biplane type. Wing section NACA M-12 (mod). Dihedral 0° on top wing, 0° 30′ on bottom wing. Incidence 2° on both wings. No sweepback. Spruce spars, wooden ribs and light alloy nose skin, with fabric covering. Fabric-covered aluminium alloy ailerons. No flaps.

FUSELAGE: Welded steel tube structure, with aluminium alloy panels forward of cockpit and fabric covering aft.

TAIL UNIT: Wire-braced welded steel tube structure with fabric covering. Fixed-incidence tailplane. Trim tab in each elevator.

LANDING GEAR: Non-retractable tailwheel type. Cantilever spring steel main legs. Goodrich-Hayes main wheels and tyres, size 5·00-4, pressure 1·38 bars (20 lb/sq in). Goodrich-Hayes brakes.

POWER PLANT: One 134 kW (180 hp) Lycoming O-360 flat-four engine, driving a Hartzell two-blade constant-speed metal propeller. Fuel tank in fuselage, capacity 75 litres (20 US gallons). Oil capacity 7·5 litres (2 US gallons).

ACCOMMODATION: Single seat in open cockpit.

DIMENSIONS, EXTERNAL:

Wing span (both)	6·60 m (21 ft 8 in)
Wing chord (constant)	1·02 m (3 ft 4 in)
Wing area, gross	13·0 m² (140 sq ft)
Wing aspect ratio	4·2
Length overall	5·56 m (18 ft 3 in)
Height overall	1·93 m (6 ft 4 in)
Tailplane span	2·18 m (7 ft 2 in)
Wheel track	1·52 m (5 ft 0 in)
Wheelbase	4·11 m (13 ft 6 in)

WEIGHTS:

Weight empty	381 kg (840 lb)
Max T-O and landing weight	578 kg (1,275 lb)

PERFORMANCE (estimated, at max T-O weight):

Never-exceed speed	191 knots (354 km/h; 220 mph)
Max level speed at 2,135 m (7,000 ft)	139 knots (257 km/h; 160 mph)
Max cruising speed at 2,135 m (7,000 ft)	122 knots (225 km/h; 140 mph)
Stalling speed	48 knots (89 km/h; 55 mph)
Max rate of climb at S/L	762 m (2,500 ft)/min
Service ceiling	6,100 m (20,000 ft)
T-O run	122 m (400 ft)
T-O to 15 m (50 ft)	198 m (650 ft)
Landing from 15 m (50 ft)	260 m (850 ft)
Landing run	168 m (550 ft)
Range with max fuel	260 nm (480 km; 300 miles)

DAVIS
LEEON D. DAVIS

ADDRESS:

PO Box 207, 405 North St Paul, Stanton, Texas 79782

Telephone: (915) 756-2100

Mr Davis designed and built his first light aircraft, the DA-1A five-seat high-wing monoplane, fourteen years ago. Details can be found in the 1960-61 *Jane's*.

He completed subsequently a prototype of a two-seat low-wing monoplane designated DA-2A, of which plans are available to other builders.

Lack of time prevented completion of the DA-3, a four-seat development of the DA-2A of which details appeared in the 1973-74 *Jane's*, but Mr Davis and his son built a new single-seat sporting aircraft during 1974 and this has the designation DA-5A. It was constructed in only 67 working days.

DAVIS DA-2A

This side-by-side two-seat light aircraft was flown for the first time on 21 May 1966, after 18 months of spare-time work and an expenditure of $1,600. At the Experimental Aircraft Association's annual Fly-in a few weeks later, it gained the awards for both the most outstanding design and the most popular aircraft. Plans are available to other amateur constructors.

By early 1976 the prototype had flown a total of 860 hours, and the twenty DA-2As that were then flying had accumulated approximately 4,000 flying hours. It is believed that about 100 other DA-2As are under construction.

An extensively modified Davis DA-2A is described under the Gonserkevis heading in this section.

The DA-2A is of simple all-metal construction and has an all-moving Vee-tail (included angle 100°) like Mr Davis's earlier DA-1A. The wings are of constant chord, without flaps. Dihedral is 5°. The non-retractable tricycle landing gear has cantilever spring steel main legs and a steerable nosewheel. Power plant is a 48·5 kW (65 hp) Continental A65-8 flat-four engine; but the DA-2A is stressed for engines of up to 74·5 kW (100 hp). Total fuel capacity is 75 litres (20 US gallons) and oil capacity 3·75 litres (1 US gallon).

There is baggage space aft of the side-by-side seats or, alternatively, a child's seat may be located in this position.

DIMENSIONS, EXTERNAL:

Wing span	5·86 m (19 ft 2¾ in)
Wing chord (constant)	1·31 m (4 ft 3½ in)
Wing area, gross	7·66 m² (82·5 sq ft)
Wing aspect ratio	4·48
Length overall	5·44 m (17 ft 10¼ in)
Height overall	1·65 m (5 ft 5 in)

DIMENSIONS, INTERNAL:

Cabin: Length	1·49 m (4 ft 6¾ in)
Max width	1·04 m (3 ft 5 in)
Max height	1·14 m (3 ft 8¾ in)

WEIGHTS AND LOADINGS:

Weight empty	277 kg (610 lb)
Max T-O weight	510 kg (1,125 lb)
Max wing loading	70·7 kg/m² (14·5 lb/sq ft)
Max power loading	10·52 kg/kW (17·3 lb/hp)

PERFORMANCE (at max T-O weight):

Max level speed at S/L	104 knots (193 km/h; 120 mph)
Cruising speed	100 knots (185 km/h; 115 mph)
Landing speed	54 knots (100 km/h; 62 mph)
Range with max fuel	390 nm (725 km; 450 miles)

DAVIS DA-5A

Design of this aircraft began in October 1972, but it was not until 4 May 1974 that Mr Davis and his son were able to begin construction of the prototype. It was intended to power the aircraft originally with a two-cylinder Franklin Sport 2 engine, but non-availability of this power plant brought a design change mid-stream to utilise a 48·5 kW (65 hp) Continental A65. Despite the extra work involved, the first flight of the DA-5A, as Mr Davis has designated this aircraft, was made on 22 July 1974.

Plans are available to amateur constructors.

TYPE: Single-seat homebuilt sporting aircraft.

WINGS: Cantilever low-wing monoplane. Wing section Clark Y. Thickness/chord ratio 12%. Dihedral 5°. Incidence 0°. Light alloy structure with single spar, ribs of 2024T-3 alloy, and stressed skins. Plain ailerons of light alloy construction. No flaps. No trim tabs.

FUSELAGE: Light alloy stressed-skin structure with four frames and stainless steel firewall. Steel tube overturn structure aft of pilot's seat.

TAIL UNIT: All-moving Vee tail (included angle 100°) with steel tube spar and light alloy ribs and skins. Anti-servo tabs in trailing-edges.

LANDING GEAR: Non-retractable tricycle type. Main wheel legs of light alloy streamline section mounted rigidly to wing spar. Shock-absorption by tyres and rubber grommets. Main wheels have tyres size 14 × 5-4, pressure 1·03 bars (15 lb/sq in). Steerable nosewheel with tyre size 10 × 3-4, pressure 1·03 bars (15 lb/sq in). Rosenhan drum brakes.

POWER PLANT: One 48·5 kW (65 hp) Continental A65 flat-four engine, driving a Hegy type 60-70 two-blade fixed-pitch wooden propeller with spinner. Fuel tank in fuselage, immediately aft of firewall, capacity 64·3 litres (17 US gallons). Refuelling point on fuselage upper surface forward of windscreen. Oil capacity 3·75 litres (1 US gallon).

ACCOMMODATION: Single seat beneath canopy hinged on port side.

DIMENSIONS, EXTERNAL:

Wing span	4·76 m (15 ft 7¼ in)
Wing chord, constant	1·12 m (3 ft 8 in)
Wing area, gross	5·31 m² (57·20 sq ft)
Wing aspect ratio	4·26
Length overall	4·80 m (15 ft 9 in)
Height overall	1·35 m (4 ft 5¼ in)
Span over Vee tail	1·60 m (5 ft 3 in)
Wheel track	1·55 m (5 ft 1 in)
Wheelbase	1·12 m (3 ft 8 in)
Propeller diameter	1·52 m (5 ft 0 in)
Propeller ground clearance	0·203 m (8 in)

DIMENSION, INTERNAL:

Cockpit:

Max width	0·53 m (1 ft 9 in)

WEIGHTS AND LOADINGS:

Weight empty	208 kg (460 lb)
Max T-O weight	351 kg (775 lb)
Max wing loading	65·9 kg/m² (13·5 lb/sq ft)
Max power loading	7·24 kg/kW (11·9 lb/hp)

PERFORMANCE (at max T-O weight):

Never-exceed speed	147 knots (273 km/h; 170 mph)
Max level speed at S/L	139 knots (257 km/h; 160 mph)
Max cruising speed at S/L	122 knots (225 km/h; 140 mph)
Econ cruising speed at S/L	104 knots (193 km/h; 120 mph)
Stalling speed	52·5 knots (97 km/h; 60 mph)
Max rate of climb at S/L	244 m (800 ft)/min
Service ceiling	4,420 m (14,500 ft)
T-O run	183 m (600 ft)
T-O to 15 m (50 ft)	259 m (850 ft)
Landing from 15 m (50 ft)	335 m (1,100 ft)
Landing run	183 m (600 ft)
Range with max fuel	390 nm (724 km; 450 miles)

DSK

DSK AIRMOTIVE INC

ADDRESS:

PO Box 629, Fort Walton Beach, Florida 32548

Telephone: (904) 243-4160

PRESIDENT: Sarah R. Killingsworth

Mr Richard Killingsworth, then a USAF Chief Master Sergeant, designed and built a single-seat sporting aircraft which he designated DSK-1 Hawk. It flew for the first time on 26 May 1973, and interest in the design was such that he formed DSK Airmotive Inc to market plans and partial kits to amateur constructors. The business is being carried on by Mrs Killingsworth following her husband's death on 12 April 1975, when the prototype Hawk crashed as the result of an engine failure at take-off.

The fuselage structure of the prototype was basically a surplus Air Force/Navy 200 US gallon drop-tank; but Mr Killingsworth's plans provide for alternative bulkhead/stressed-skin construction for builders unable to obtain a suitable drop-tank. In this form the aircraft is designated DSK-2 Golden Hawk and is described separately.

By early 1976, about 45 Hawks were known to be under construction.

DSK AIRMOTIVE DSK-1 HAWK

TYPE: Single-seat homebuilt sporting aircraft. Structure stressed to 9*g* ultimate.

WINGS: Cantilever low-wing monoplane. Clark Y section. Thickness/chord ratio 15%. Slight dihedral on outer panels. No incidence. Conventional two-spar structure of 2024T-3 light alloy. Trailing-edge flaps of light alloy construction extend from inner edge of ailerons and continue beneath fuselage. Plain ailerons of light alloy construction. Optional plans for aileron droop and drooped, raked wingtips. No tabs.

FUSELAGE: All-metal structure utilising a surplus military drop-tank as its basis, with 2024T-3 alloy sheet and 6061T-6 structural angle reinforcement.

TAIL UNIT: Cantilever all-metal structure with swept vertical surfaces. Incidence of tailplane ground-adjustable. Trim tab on elevator.

LANDING GEAR: Non-retractable tricycle type. All units have a tubular steel strut with compression springs for shock-absorption. Nosewheel and main-wheel tyres size 14 × 5·00-4, pressure 2·07 bars (30 lb/sq in). Mechanical disc brakes. Parking brake. Wheel fairings optional.

POWER PLANT: Prototype has one 48·5 kW (65 hp) Lycoming O-145-B2 flat-four engine, driving a McCauley two-blade metal fixed-pitch propeller refurbished by Aero Prop. Provision for alternative engines of up to 93 kW (125 hp). Fuel contained in forward fuselage tank, capacity 34 litres (9 US gallons). Aft fuselage tank of 45·4 litres (12 US gallons) capacity optional. Refuelling point for front tank forward of windscreen, for rear tank in turtledeck aft of cockpit. Oil capacity 4·7 litres (1·25 US gallons).

ACCOMMODATION: Single seat in enclosed cockpit. Canopy slides aft for access. Baggage space aft of pilot's seat. Cockpit heated and ventilated.

SYSTEM: Simple air-driven alternator under design by DSK Airmotive.

ELECTRONICS: Battery-powered Narco VHT-3 nav/com and Omni.

DIMENSIONS, EXTERNAL:

Wing span	6·21 m (20 ft 4½ in)
Wing chord, constant	0·97 m (3 ft 2 in)
Wing area, gross	5·95 m² (64 sq ft)
Wing aspect ratio	6·4
Length overall	4·57 m (15 ft 0 in)
Height overall	1·83 m (6 ft 0 in)
Tailplane span	1·88 m (6 ft 2 in)
Wheel track	1·63 m (5 ft 4 in)
Wheelbase	1·14 m (3 ft 9 in)
Propeller diameter	1·60 m (5 ft 3 in)
Propeller ground clearance	0·20 m (8 in)

WEIGHTS AND LOADINGS:

Weight empty	238 kg (525 lb)
Max T-O and landing weight	405 kg (893 lb)
Max wing loading	67·7 kg/m² (13·87 lb/sq ft)
Max power loading	8·35 kg/kW (13·7 lb/hp)

PERFORMANCE (at max T-O weight):

Max level speed at 1,525 m (5,000 ft)	127 knots (235 km/h; 146 mph)
Max cruising speed at 1,525 m (5,000 ft)	114 knots (211 km/h; 131 mph)
Econ cruising speed at 1,525 m (5,000 ft)	96 knots (177 km/h; 110 mph)
Stalling speed, flaps up	52·5 knots (97 km/h; 60 mph)
Stalling speed, flaps down	44 knots (81 km/h; 50 mph)
Max rate of climb at S/L	457 m (1,500 ft)/min
Service ceiling	3,660 m (12,000 ft)
T-O run	183 m (600 ft)
T-O to 15 m (50 ft)	244 m (800 ft)
Landing from 15 m (50 ft)	229 m (750 ft)
Landing run	168 m (550 ft)
Range with max fuel, 45 min reserve	477 nm (885 km; 550 miles)

DSK AIRMOTIVE DSK-2 GOLDEN HAWK

Generally similar to the DSK-1 Hawk, the DSK-2 was designed to eliminate the use of a military drop-tank as the basic fuselage structure and is 0·46 m (1 ft 6 in) longer overall. Design began in January 1974 and construction of the prototype was started four months later. First flight was expected to take place in 1976.

TYPE: Single-seat homebuilt sporting aircraft.

WINGS: As for DSK-1, with 6061T-6 light alloy structural angle reinforcement of spars. Ailerons droop collectively with flaps.

FUSELAGE: Light alloy semi-monocoque structure with five bulkheads, 6061T-6 structural angle longerons, and 2024T-3 alloy bulkheads and stressed skin.

TAIL UNIT: Cantilever structure of light alloy, with swept tail surfaces. Box-beam/rib/stressed skin structure of 2024T-3 alloy. Incidence of tailplane ground-adjustable. Manually-adjustable trim tab on elevator.

LANDING GEAR: Non-retractable tricycle type. Main legs of steel tube, with springs in compression. Corded neoprene for absorption of side loads. Self-centering steerable nosewheel. Main wheels and tyres size 14 × 5·00-5, pressure 2·07 bars (30 lb/sq in). Nosewheel and tyre size 13 × 5·00-5, 14 × 5·00-5 optional, pressure 2·07 bars (30 lb/sq in). Cleveland hydraulic brakes. Fairings on main legs.

POWER PLANT: As for DSK-1, except forward fuselage and optional aft fuselage fuel tanks of 38-53 litres (10-14 US gallons) capacity.

ACCOMMODATION: As for DSK-1, except seatback and control column removable to provide room for pilot to sleep, between flights, during sporting trips. Aircraft can be flown with canopy open or closed.

SYSTEMS: Hydraulic system for brakes only. Power for electronics provided by rechargeable battery. DSK air-driven alternator optional.

ELECTRONICS: Prototype has, currently, Narco VHT-2 nav/com. It is intended to install a 90-channel nav/com.

DIMENSIONS, EXTERNAL:

As for DSK-1, except:

Wing area, gross	5·99 m² (64·5 sq ft)
Wing aspect ratio	6·43
Length overall	5·03 m (16 ft 6 in)
Height overall	1·91 m (6 ft 3 in)
Wheel track	1·64 m (5 ft 4½ in)

WEIGHTS AND LOADINGS (estimated):

Weight empty	249 kg (550 lb)
Max T-O weight	414 kg (914 lb)
Max wing loading	69·18 kg/m² (14·17 lb/sq ft)
Max power loading	8·54 kg/kW (14·06 lb/hp)

PERFORMANCE (estimated, at max T-O weight):

Max level speed at 1,525 m (5,000 ft)	135 knots (251 km/h; 156 mph)
Max cruising speed at 1,525 m (5,000 ft)	122 knots (227 km/h; 141 mph)
Econ cruising speed	104 knots (193 km/h; 120 mph)
Stalling speed, flaps up	52·5 knots (97 km/h; 60 mph)
Stalling speed, flaps down	43·5 knots (81 km/h; 50 mph)
Max rate of climb at S/L	457 m (1,500 ft)/min
Service ceiling	3,660 m (12,000 ft)
T-O run	183 m (600 ft)
T-O to 15 m (50 ft)	320 m (1,050 ft)
Landing from 15 m (50 ft)	320 m (1,050 ft)
Landing run	168 m (550 ft)
Range with 53 litres (14 US gallons) fuel at econ cruising speed	416 nm (772 km; 480 miles)

DYKE

DYKE AIRCRAFT

ADDRESS:

2840 Old Yellow Springs Road, Fairborn, Ohio 45324

Telephone: (513) 878-9832

Mr John W. Dyke was the designer of a then-unique delta-wing aircraft, designated JD-1 Delta, which he built with Jennie Dyke. This was described in the 1964-65 *Jane's*. Since that time Mr Dyke has developed the design and the first flight of the JD-2 Delta was made in July 1966.

DYKE AIRCRAFT JD-2 DELTA

Plans of the JD-2 are available to amateur constructors, and Mr Dyke has formed Dyke Aircraft to market them. A total of 288 JD-2s were reported to be under construction by early 1976; examples are under construction or flying in Australia, Brazil, Canada, France, Germany, Japan, New Zealand, South Africa, the United Kingdom and the United States.

As a further help to the homebuilder, hardware and tubing kits are available.

TYPE: Delta-winged homebuilt sporting aircraft.

WINGS: Delta wing of modified NACA 63012 and 66015 wing section. No dihedral. No incidence. Sweepback of wing centre-section 61°. Sweepback of outer wing panels 31°. Welded steel tube structure, with stainless steel cap-strips to which the laminated glassfibre skins are secured by Dupont explosive rivets. Aluminium skin optional; attached by aluminium pop rivets. Trailing-edge elevons extending from centreline to approximately two-thirds span. Elevons have a basic structure of metal but are fabric covered. Trim tab at inboard edge of each elevon. Wing outer panels fold upward and lay above the fuselage so that the JD-2 can be towed on the landing gear. Folding and unfolding can be carried out by one person.

FUSELAGE: Welded steel tube structure with stainless steel capstrips and glassfibre skins, except for fuselage under-surface which has 4130 steel tube capstrips and fabric covering.

TAIL UNIT: Welded steel tube structure with swept fin and rudder, both fabric covered.

LANDING GEAR: Manually retractable tricycle type, all units retracting aft. Shock-absorption by torsion bars of 6150 heat-treated steel. Main wheels have tyres size 14 × 5-6, pressure 3·10 bars (45 lb/sq in). Nosewheel tyre size 12 × 5-4, pressure 2·07 bars (30 lb/sq in). Firestone hydraulic brakes.

POWER PLANT: Prototype has one 134 kW (180 hp) Lycoming O-360 flat-four engine, driving a McCauley two-blade fixed-pitch propeller with spinner. Constant-speed propeller and 149 kW (200 hp) engine installation optional. Glassfibre fuel tank in fuselage, immediately aft of cabin, capacity 178 litres (47 US gallons). Refuelling point on port side of aft fuselage upper surface. Oil capacity 7·57 litres (2 US gallons).

ACCOMMODATION: Standard accommodation for pilot, on single forward seat, and three passengers on aft bench seat. Access by upward-opening canopy, hinged on starboard side. Accommodation is ventilated. Space for limited baggage in starboard wing centre-section.

SYSTEMS: Hydraulic system for brakes only. Vacuum system for instruments. Electrical supply from 12V DC generator; solid-state inverter provides 110V AC at 400Hz.

DSK Airmotive DSK-1 Hawk single-seat sporting aircraft (Lycoming O-145-B2 engine)

Dyke JD-2 Delta built by John and Jennie Dyke *(Howard Levy)*

EAA Biplane Model P built by Mr Charles E. Jensen *(Howard Levy)*

EAA Super Acro-Sport aerobatic biplane

ELECTRONICS AND EQUIPMENT: Genave Series 200A 200-channel nav and com transceiver. Blind-flying instrumentation standard.

DIMENSIONS, EXTERNAL:
- Wing span 6·71 m (22 ft 0 in)
- Wing chord at root (centre-section) 4·27 m (14 ft 0 in)
- Wing chord at root (outer panels) 2·90 m (9 ft 6 in)
- Wing chord at tip 0·51 m (1 ft 8 in)
- Wing area, gross 16·07 m² (173 sq ft)
- Wing aspect ratio 2·7
- Length overall 5·79 m (19 ft 0 in)
- Width, wings folded 2·24 m (7 ft 4 in)
- Height overall 1·83 m (6 ft 0 in)
- Wheel track 1·83 m (6 ft 0 in)
- Wheelbase 2·13 m (7 ft 0 in)
- Propeller diameter 1·88 m (6 ft 2 in)
- Propeller ground clearance 0·15 m (6 in)

DIMENSIONS, INTERNAL:
- Cabin: Length 1·88 m (6 ft 2 in)
  - Max width 1·22 m (4 ft 0 in)
  - Max height 1·02 m (3 ft 4 in)
  - Floor area 1·67 m² (18 sq ft)
- Baggage space 0·17 m³ (6 cu ft)

WEIGHTS AND LOADINGS (A: 134 kW; 180 hp, B: 149 kW; 200 hp engine):
- Weight empty, equipped:
  - A 435 kg (960 lb)
- Max T-O weight:
  - A 816 kg (1,800 lb)
  - B 862 kg (1,900 lb)
- Max landing weight:
  - A, B 794 kg (1,750 lb)
- Max wing loading:
  - A 46·9 kg/m² (9·6 lb/sq ft)
- Max power loading:
  - A 6·09 kg/kW (10 lb/hp)

PERFORMANCE (at max T-O weight with 134 kW; 180 hp engine):
- Never-exceed speed 191 knots (354 km/h; 220 mph)
- Max level speed at 2,285 m (7,500 ft) 165 knots (306 km/h; 190 mph)
- Max cruising speed at 2,285 m (7,500 ft), fixed-pitch propeller 156 knots (290 km/h; 180 mph)
- Max cruising speed, constant-speed propeller 161 knots (298 km/h; 185 mph)
- Econ cruising speed at 2,285 m (7,500 ft) 135 knots (249 km/h; 155 mph)
- Max rate of climb at S/L 670 m (2,200 ft)/min
- Service ceiling 4,420 m (14,500 ft)
- T-O run 213 m (700 ft)
- T-O to 15 m (50 ft) 488 m (1,600 ft)
- Landing from 15 m (50 ft) 549 m (1,800 ft)
- Landing run 244-305 m (800-1,000 ft)
- Range with max fuel and pilot only 625 nm (1,158 km; 720 miles)
- Range with max payload 390 nm (724 km; 450 miles)

EAA
EXPERIMENTAL AIRCRAFT ASSOCIATION INC

ADDRESS:
PO Box 229, Hales Corners, Wisconsin 53130
Telephone: (414) 425-4860
PRESIDENT: Paul H. Poberezny
VICE-PRESIDENT: Ray Scholler
SECRETARY: S. H. Schmid
TREASURER: Arthur Kilps

As a service to its members, the EAA decided in 1955 to develop a modern single-seat sporting biplane suitable for home construction by amateurs. The design drawings were prepared by Mr J. D. Stewart and Mr T. Seely of the Allison Division, General Motors Corporation, with the assistance of Mr Paul H. Poberezny, President of the EAA, and this aircraft became known as the EAA Biplane. It was followed in 1972 by the Acro-Sport, an aerobatic aircraft designed by Mr Poberezny. These two designs were supplemented in 1973 by a more advanced aerobatic aircraft, the Super Acro-Sport.

Latest design to emanate from the EAA President is the lightweight Pober P-9 Pixie, which flew for the first time in July 1974.

EAA BIPLANE

The prototype of the EAA Biplane was built between 1957 and May 1960 as a classroom project by students of St Rita's High School, Chicago, under the supervision of Mr Robert Blacker. It flew for the first time on 10 June 1960, powered by a 48·5 kW (65 hp) Continental A65 engine.

After this prototype had been taken over by the EAA, a number of modifications were made, including installation of a 63·5 kW (85 hp) engine. Mr Poberezny subsequently introduced further changes, including lighter wing spars and fittings, which reduced the empty weight of the aircraft, together with increases in the area of the tail surfaces and many other improvements. This modified version is known as the EAA Biplane Model P.

Over 7,000 sets of plans of the EAA Biplane have been sold and many examples are flying and under construction. That illustrated was built by Mr Charles E. Jensen of Alliance, Nebraska. It differs from plans by having square-tipped wings and tail, and larger cockpit. Power plant is a 93 kW (125 hp) Lycoming O-290-D engine.

The following details apply to the standard EAA Biplane Model P:

TYPE: Single-seat sporting biplane.

WINGS: Braced biplane, with N-shape streamline-section cabane struts and interplane struts. Dihedral 0° on upper wing, 2° on lower wings. Incidence 2° on upper wing, 2° on lower wings. All-wood two-spar structure, with aluminium leading-edge and overall fabric covering. Ailerons of similar construction to wings, on lower wings only. No flaps.

FUSELAGE: Welded steel tube structure, fabric-covered.

TAIL UNIT: Wire-braced welded steel tube structure, fabric-covered.

LANDING GEAR: Non-retractable tailwheel type, modified from standard Piper J-3 components. Rubber cord shock-absorption. Brakes on main wheels. Wheel fairings optional. Piper J-3 steerable tailwheel.

POWER PLANT: Prototype Model P has a 63·5 kW (85 hp) Continental C85-8 flat-four engine, driving a two-blade metal fixed-pitch propeller. Provision for engines of up to 112 kW (150 hp); but most Model Ps have a 93 kW (125 hp) Lycoming. Piper J-3 fuel tank aft of firewall in fuselage, capacity 68 litres (18 US gallons).

ACCOMMODATION: Single seat in open cockpit.

DIMENSIONS, EXTERNAL:
- Wing span 6·10 m (20 ft 0 in)
- Wing chord, both (constant) 0·91 m (3 ft 0 in)
- Wing area, gross 10·03 m² (108 sq ft)
- Length overall 5·18 m (17 ft 0 in)
- Height overall 1·83 m (6 ft 0 in)

WEIGHTS (63·5 kW; 85 hp engine):
- Weight empty 322 kg (710 lb)
- Max T-O weight 522 kg (1,150 lb)

PERFORMANCE (63·5 kW; 85 hp engine, at max T-O weight):
- Max level speed at S/L 109 knots (201 km/h; 125 mph)
- Econ cruising speed 96 knots (177 km/h; 110 mph)
- Stalling speed 44 knots (80 km/h; 50 mph)
- Max rate of climb at S/L 305 m (1,000 ft)/min
- Service ceiling 3,500 m (11,500 ft)
- T-O run 152 m (500 ft)
- Landing run 245 m (800 ft)
- Range with max fuel 304 nm (560 km; 350 miles)

EAA ACRO-SPORT

The Acro-Sport was designed by Mr Paul Poberezny, President of the EAA, specifically for construction by school students, as a pupils' project. The EAA considers that such a project enables those who participate to discover their capabilities and potential in the manual crafts necessary to build an aircraft, possibly helping the individual to choose a career more wisely.

First flight of the prototype Acro-Sport (N1AC) was made on 11 January 1972, only 352 days after its design was started, although it represented a completely new design, unrelated to the EAA Biplane.

The prototype has demonstrated good flight characteristics, and represents a versatile aircraft for sport or aerobatic use. The provision of ailerons on both wings gives positive aileron response and a high rate of roll.

Plans and construction manuals are available to homebuilders, and 700 sets had been sold by January 1976.

The following details apply to the prototype:

TYPE: Single-seat homebuilt aerobatic biplane.

WINGS: Braced single-bay biplane, with single streamline-section interplane strut each side. N-type centre-section struts. Double streamline-section flying and landing wires. Wing section M-6. Dihedral: upper 0°, lower 2°. Incidence (both) 1° 30′. Conventional two-spar structure, with spruce spars and ribs, single wire drag and anti-drag truss, fabric-covered. Glassfibre wingtips. Ailerons, on all four wings, of wood construction with fabric covering. No flaps. Cutout in trailing-edge of upper wing.

FUSELAGE: Composite structure of welded steel tube, with wooden stringers, fabric-covered. Glassfibre nose cowl and light alloy engine cowlings.

TAIL UNIT: Wire-braced welded steel tube structure with fabric covering. Controllable trim tab on port side of elevator, servo tab on starboard side.

LANDING GEAR: Non-retractable tailwheel type, modified from Piper J-3 components. Two side Vees and half axles. Rubber bungee shock-absorption. Main-wheel tyres size 5·00-5, pressure 1·86-1·93 bars (27-28 lb/sq in). Cleveland or Goodyear hydraulic brakes. Glassfibre wheel fairings. Steerable tailwheel. Parking brake.

POWER PLANT: Prototype has a 134 kW (180 hp) Lycoming engine, driving a Sensenich two-blade metal fixed-pitch propeller with spinner. Basic power plant is a 74·5 kW (100 hp) Continental O-200 flat-four engine. Single fuel tank immediately aft of firewall, capacity 104 litres (20 US gallons). Refuelling point on upper surface of fuselage. Small smoke oil tank, capacity 19 litres (5 US gallons), forward of instrument panel, could also be used for fuel.

ACCOMMODATION: Single seat in open cockpit, which is large enough to accommodate a pilot 1·95 m (6 ft 5 in) tall and weighing 115 kg (250 lb). Baggage space behind headrest, capacity 16 kg (35 lb).

DIMENSIONS, EXTERNAL:

Wing span, upper	5·97 m (19 ft 7 in)
Wing span, lower	5·82 m (19 ft 1 in)
Wing chord, constant, both	0·91 m (3 ft 0 in)
Wing area, gross	10·73 m² (115·5 sq ft)
Wing aspect ratio, upper	6·6
Length overall	5·33 m (17 ft 6 in)
Height overall	1·83 m (6 ft 0 in)
Tailplane span	2·16 m (7 ft 1 in)
Wheel track	1·78 m (5 ft 10 in)
Propeller diameter	1·93 m (6 ft 4 in)

WEIGHTS AND LOADINGS:

Weight empty, equipped	332 kg (733 lb)
Max T-O and landing weight	612 kg (1,350 lb)
Max wing loading	57·0 kg/m² (11·68 lb/sq ft)
Max power loading	4·57 kg/kW (7·5 lb/hp)

PERFORMANCE (at max T-O weight):

Never-exceed speed	156 knots (289 km/h; 180 mph)
Max level speed	132 knots (245 km/h; 152 mph)
Max cruising speed	113 knots (209 km/h; 130 mph)
Econ cruising speed	91 knots (169 km/h; 105 mph)
Stalling speed	43·5 knots (80·5 km/h; 50 mph)
Max rate of climb at S/L	1,067 m (3,500 ft)/min
T-O run	46 m (150 ft)
T-O to 15 m (50 ft)	107 m (350 ft)
Landing from 15 m (50 ft)	267 m (875 ft)
Landing run	244 m (800 ft)
Range with max fuel	260 nm (482 km; 300 miles)

EAA SUPER ACRO-SPORT

Design of a developed version of the Acro-Sport was started in January 1971. Construction began in the following year and the first flight of this prototype was made on 28 March 1973. Known as the Super Acro-Sport, it is intended for unlimited International Class aerobatic competition at a world championship level. Generally similar in external appearance to the Acro-Sport, it has a more powerful engine, nearly symmetrical aerofoil sections, and an improved rate of roll and inverted flight capability.

The differences by comparison with the standard Acro-Sport are covered in a supplement to the basic plans. The description of the Acro-Sport applies also to the Super Acro-Sport, except as follows:

TYPE: Single-seat homebuilt advanced aerobatic biplane.

WINGS: As Acro-Sport, except wing section NACA 23012.

TAIL UNIT: As Acro-Sport, except controllable trim tab on starboard side of elevator, servo tab on port side.

LANDING GEAR: As Acro-Sport, except Cleveland hydraulic brakes. Steerable tailwheel has a solid tyre.

POWER PLANT: Prototype has a 149 kW (200 hp) Lycoming IO-360-A2A flat-four engine, driving a Sensenich two-blade metal fixed-pitch propeller type 76EM8-0-60 with spinner. Fuel system and capacity as for Acro-Sport. Oil capacity 7·5 litres (2 US gallons).

DIMENSIONS, EXTERNAL: As for Acro-Sport, except:

Length overall	5·30 m (17 ft 4½ in)

WEIGHTS AND LOADINGS:

Weight empty	401 kg (884 lb)
Max T-O weight	612 kg (1,350 lb)
Max wing loading	57·0 kg/m² (11·68 lb/sq ft)
Max power loading	4·11 kg/kW (6·75 lb/hp)

PERFORMANCE (at max T-O weight):

Never-exceed speed	156 knots (289 km/h; 180 mph)
Max level speed at S/L	135 knots (251 km/h; 156 mph)
Max cruising speed	117 knots (217 km/h; 135 mph)
Stalling speed	43·5 knots (80·5 km/h; 50 mph)
Max rate of climb at S/L	1,128 m (3,700 ft)/min
Service ceiling	4,570 m (15,000 ft)
T-O run	38 m (125 ft)
T-O to 15 m (50 ft)	91 m (300 ft)
Landing from 15 m (50 ft)	274 m (900 ft)
Landing run	244 m (800 ft)
Range with max fuel	260 nm (482 km; 300 miles)

EAA POBER P-9 PIXIE

Design and construction of this new lightweight sporting aircraft began simultaneously in January 1974, and the first flight was made in July 1974. Plans will be available to amateur constructors.

TYPE: Single-seat lightweight sporting aircraft.

WINGS: Braced parasol monoplane. Two streamline-section struts and wire bracing each side. Wing section Clark Y. Dihedral 2°. Incidence 2°. Conventional wooden structure utilising spruce spars, mahogany plywood and fabric covering. Full-span Frise-type ailerons. No flaps. No trim tabs.

FUSELAGE: Welded structure of 4130 chrome molybdenum steel tubing with wood formers and fabric covering.

TAIL UNIT: Wire-braced structure of welded 4130 chrome molybdenum steel tubing, fabric covered. Fixed-incidence tailplane. No trim tabs.

LANDING GEAR: Non-retractable tailwheel type. Side Vees with half-axles carry main wheels. Shock-absorption by rubber bungee. Main wheels with tyres size 5·00-5, pressure 2·07 bars (30 lb/sq in). Tailwheel tyre of solid type. Cleveland brakes. Glassfibre fairings over main wheels.

POWER PLANT: One 44·5 kW (60 hp) Limbach SL 1700 EA flat-four engine, driving a Sensenich two-blade fixed-pitch propeller. Fuel tank in wing centre-section, capacity 46·6 litres (12·3 US gallons). Refuelling point on wing upper surface. Oil capacity 2·84 litres (0·75 US gallons).

ACCOMMODATION: Single seat in open cockpit. Door on starboard side.

DIMENSIONS, EXTERNAL:

Wing span	9·09 m (29 ft 10 in)
Wing chord, constant	1·37 m (4 ft 6 in)
Wing area, gross	12·47 m² (134·25 sq ft)
Length overall	5·26 m (17 ft 3 in)
Height overall	1·88 m (6 ft 2 in)
Wheel track	1·60 m (5 ft 3 in)
Propeller diameter	1·35 m (4 ft 5 in)

WEIGHTS AND LOADINGS:

Weight empty	246 kg (543 lb)
Max T-O weight	408 kg (900 lb)
Max wing loading	32·7 kg/m² (6·7 lb/sq ft)
Max power loading	9·13 kg/kW (15 lb/hp)

PERFORMANCE (at max T-O weight):

Never-exceed speed	113 knots (209 km/h; 130 mph)
Max level speed at S/L	89 knots (166 km/h; 103 mph)
Max cruising speed	74 knots (137 km/h; 85 mph)
Stalling speed	26 knots (49 km/h; 30 mph)
Max rate of climb at S/L	152 m (500 ft)/min
T-O and landing run	91 m (300 ft)
T-O to 15 m (50 ft)	305 m (1,000 ft)
Landing from 15 m (50 ft)	152 m (500 ft)
Range with max fuel	251 nm (466 km; 290 miles)

EAA NESMITH COUGAR

A prototype side-by-side two-seat light monoplane named the Cougar, designed by Mr Robert Nesmith, flew for the first time in March 1957. Sets of plans are available to amateur constructors (originally from Mr Nesmith and since the beginning of 1977 from the EAA), and about 250 Cougars were flying or under construction in early 1976.

The data given below apply to the standard Cougar, built according to the plans produced originally by Mr Nesmith. Some aircraft now flying incorporate detail modifications. One has a T-tail and others have been completed with folding wings.

TYPE: Two-seat sporting monoplane.

WINGS: Braced high-wing monoplane, with single bracing strut each side. Wing section NACA 4309 (modified). Dihedral 1° 30′. No incidence. All-wood two-spar structure, plywood-covered except for fabric-covered trailing-edge. Ailerons have steel tube structure, fabric-covered. No flaps.

FUSELAGE: Steel tube structure, fabric-covered.

TAIL UNIT: Cantilever steel tube structure, fabric-covered, with comparatively small rudder.

LANDING GEAR: Non-retractable tailwheel type. Cantilever spring steel main legs. Goodyear main wheels and tyres, size 5·00-5. Brakes. Steerable tailwheel can be unlocked for 360° castoring on ground.

POWER PLANT: One 86 kW (115 hp) Lycoming O-235 flat-four engine, driving a McCauley two-blade fixed-pitch propeller. Suitable for installation of engines from 63·5 to 112 kW (85 to 150 hp). Fuel tank in fuselage, capacity 94·6 litres (25 US gallons). Oil capacity 5·7 litres (6 US quarts).

ACCOMMODATION: Two seats side by side in enclosed cabin, with dual controls. Space for 41 kg (90 lb) baggage. Provision for radio.

DIMENSIONS, EXTERNAL:

Wing span	6·25 m (20 ft 6 in)
Wing chord (constant)	1·22 m (4 ft 0 in)
Wing area, gross	7·66 m² (82·5 sq ft)
Wing aspect ratio	5·16
Length overall	5·76 m (18 ft 11 in)
Height overall	1·68 m (5 ft 6 in)
Tailplane span	1·93 m (6 ft 4 in)
Wheel track	1·57 m (5 ft 2 in)

DIMENSION, INTERNAL:

Cabin: Width	0·96 m (3 ft 2 in)

WEIGHTS:

Weight empty	283 kg (624 lb)
Max T-O weight	567 kg (1,250 lb)

PERFORMANCE (at max T-O weight):

Max level speed at S/L	169 knots (314 km/h; 195 mph)
Max cruising speed at 2,130 m (7,000 ft)	144 knots (267 km/h; 166 mph)
Econ cruising speed	135 knots (249 km/h; 155 mph)
Stalling speed	46 knots (85 km/h; 53 mph)
Max rate of climb at S/L	395 m (1,300 ft)/min
Service ceiling	3,950 m (13,000 ft)
T-O run	137 m (450 ft)
T-O to 15 m (50 ft)	335 m (1,100 ft)
Landing from 15 m (50 ft)	305 m (1,000 ft)
Landing run	107 m (350 ft)
Range with max fuel	651 nm (1,207 km; 750 miles)
Range with max payload	607 nm (1,125 km; 700 miles)

EVANS

EVANS AIRCRAFT

ADDRESS:

PO Box 744, La Jolla, California 92037

Mr W. S. Evans, while a design engineer with the Convair Division of General Dynamics Corporation, set out to design for the novice homebuilder an all-wood aircraft that would be easy to build and safe to fly. He was prepared to sacrifice both appearance and performance to achieve this aim. Two years of spare-time design and a year of construction produced a strut-braced low-wing monoplane with an all-moving tail unit, powered by a 30 kW (40 hp) Volkswagen engine, which Mr Evans called the VP (originally Volksplane).

The prototype was subsequently re-engined with a 39·5 kW (53 hp) Volkswagen, giving a rate of climb of 183 m (600 ft)/min at sea level.

Mr Evans developed subsequently a two-seat version of the VP; this is generally similar to the single-seat model which is now designated VP-1. The two-seat VP-2 is powered by a 44·5/48·5 kW (60/65 hp) Volkswagen engine, but in other respects has only minor constructional variations from the VP-1.

Plans of both models are available to amateur constructors and the VP-1 is an approved type by the PFA.

EVANS VP-1

TYPE: Single-seat homebuilt aircraft.

WINGS: Strut-braced low-wing monoplane. Two streamline-section bracing struts on each side. Wing section NACA 4412. Square tips. Dihedral 5°. Conventional wood structure with two rectangular spar beams, internal wooden compression struts and diagonal wire bracing, dispensing with the need for a complicated box spar. Fabric covering. Ailerons of wooden construction, fabric covered. No trim tabs. No flaps.

FUSELAGE: Rectangular-section all-wood stressed-skin structure, consisting essentially of three bulkheads, four longerons and plywood skin. Stressed-skin design eliminates the need for any diagonal bracing. Glassfibre fairing aft of pilot's seat.

TAIL UNIT: No fixed fin. The rudder is constructed of plywood ribs clamped to a 5·08 cm (2 in) aluminium tube which is mounted vertically through the rear fusel-

Nesmith Cougar two-seat sporting monoplane *(Howard Levy)*

EAA Pober P-9 Pixie (Limbach SL 1700 EA engine)

Evans VP-1 photographed at Pomona Lake Airport, Kansas *(Jean Seele)*

Evans VP-2 two-seat homebuilt monoplane *(Jean Seele)*

age and pivots in two nylon bushes. Leading- and trailing-edges are of wood and the whole unit is fabric-covered. The fabric-covered all-moving tailplane is a wooden cantilever structure, comprising ply ribs blocked and glued to a simple constant-section box spar. Both rudder and tailplane have anti-servo tabs.

LANDING GEAR: Non-retractable main wheels and tailskid. Main wheels carried on a bent section of heavy-gauge 24ST-3 aluminium bar, wire-braced by diagonal cables. Shock-absorption by low-pressure tyres. Main wheels and tyres size 6·00-6. Tyre pressure 0·83 bars (12 lb/sq in). Hydraulic brakes operated by single hand lever.

POWER PLANT: One 30 kW, 39·5 kW or 44·5 kW (40 hp, 53 hp or 60 hp) modified Volkswagen motor car engine, driving a Hegy two-blade propeller, with pitch of 0·61 m (24 in) for 30 kW (40 hp) engine, 0·76 m (30 in) for 39·5 kW (53 hp) and 0·91 m (36 in) for 44·7 kW (60 hp). Glassfibre fuel tank aft of firewall and integral with the forward fuselage cowlings, capacity 30 litres (8 US gallons). Filling point on top of fuselage, forward of windscreen.

ACCOMMODATION: Single seat in open cockpit. No baggage stowage.

DIMENSIONS, EXTERNAL:

Wing span	7·32 m (24 ft 0 in)
Wing chord (constant)	1·27 m (4 ft 2 in)
Wing area, gross	9·29 m² (100 sq ft)
Length overall	5·49 m (18 ft 0 in)
Height overall	1·56 m (5 ft 1½ in)
Tailplane span	2·13 m (7 ft 0 in)
Wheel track	1·50 m (4 ft 11 in)
Propeller diameter	1·37 m (4 ft 6 in)

WEIGHTS:

Weight empty	200 kg (440 lb)
Max T-O weight	340 kg (750 lb)

PERFORMANCE (with 30 kW; 40 hp engine, at T-O weight of 295 kg; 650 lb):

Never-exceed speed	104 knots (193 km/h; 120 mph)
Cruising speed	65 knots (121 km/h; 75 mph)
Stalling speed	35 knots (65 km/h; 40 mph)
Max rate of climb at S/L	122 m (400 ft)/min
T-O run (average breeze)	137 m (450 ft)
Landing run (average breeze)	61 m (200 ft)

EVANS VP-2

TYPE: Two-seat homebuilt aircraft.

WINGS: Generally similar to VP-1, except for NACA 4415 wing section and increased wing span and chord.

FUSELAGE: Similar to VP-1, but width increased by 0·305 m (1 ft 0 in).

TAIL UNIT: Similar to VP-1. No fin. Increased rudder area; all-moving tailplane of increased span and chord.

LANDING GEAR: Similar to VP-1. Wheel track increased by 0·23 m (9 in).

POWER PLANT: One 44·5 kW (60 hp) 1,834 cc or 48·5 kW (65 hp) 2,100 cc modified Volkswagen motor car engine, driving a two-blade propeller. Glassfibre fuel tank aft of firewall, capacity 53 litres (14 US gallons).

ACCOMMODATION: Two seats side by side in open cockpit.

DIMENSIONS, EXTERNAL:

Wing span	8·23 m (27 ft 0 in)
Wing chord (constant)	1·47 m (4 ft 10 in)
Wing area, gross	12·08 m² (130 sq ft)
Length overall	5·87 m (19 ft 3 in)
Tailplane span	2·44 m (8 ft 0 in)
Wheel track	1·73 m (5 ft 8 in)
Propeller diameter	1·52 m (5 ft 0 in)

WEIGHTS:

Weight empty	290 kg (640 lb)
Max T-O weight	471 kg (1,040 lb)

PERFORMANCE (44·5 kW; 60 hp engine, at max T-O weight):

Never-exceed speed	104 knots (193 km/h; 120 mph)
Max level speed	87 knots (161 km/h; 100 mph)
Max cruising speed	65 knots (121 km/h; 75 mph)
Stalling speed	35 knots (64·5 km/h; 40 mph)
Max rate of climb at S/L (pilot only)	213 m (700 ft)/min
Max rate of climb at S/L (pilot and passenger)	122 m (400 ft)/min

FIKE
WILLIAM J. FIKE

ADDRESS:
PO Box 683, Anchorage, Alaska 99510
Telephone: (907) 272-7069

Mr W. J. Fike, whose 16,000 hours logged as a pilot include thousands of hours of 'bush flying' in Alaska, has designed and built five light aircraft since 1929. His Model 'B' of 1935 was a tiny parasol-wing single-seat monoplane, powered by a 26 kW (35 hp) Long Harlequin engine and with an empty weight of only 136 kg (300 lb). In the following year he produced the Model 'C' of similar configuration. A later design was the single- or two-seat Model 'D' high-wing cabin monoplane, of which full details were given in the 1961-62 *Jane's*.

This was followed by the Model 'E', completed in 1970, a description of which follows. In early 1976 Mr Fike had almost completed the construction of a new lightweight cabin monoplane which he has designated Model 'F'. This needed only covering and a few minor details before flight testing could begin. The 'F' is a braced sweptwing monoplane which may be powered by engines of 30-74·5 kW (40-100 hp). A 48·5 kW (65 hp) Continental A65-8 has been installed in the prototype.

Plans of the Model 'D' and Model 'E' are available to amateur constructors.

FIKE MODEL 'E'

In 1953 Mr Fike began design of an aircraft to evaluate the flight characteristics of a low aspect ratio (3·0) wing of only 9% thickness/chord ratio when applied to a low-power monoplane of high-wing configuration. The wing, of wooden geodetic construction, is so designed that various wingtips may be installed for evaluation. A standard Piper J-3 tail unit is utilised, but this is modified by keeping the tailplane span within the 2·44 m (8 ft) limit allowed by US highway regulations for towed vehicles.

A secondary objective of the Model 'E' project was to develop a low-cost easy-to-build two-seat lightplane. The wing can be removed within ten minutes to enable the aircraft to be towed by a motor vehicle or for storage in an ordinary garage.

Mr Fike's former heavy commitments as an airline pilot were responsible for the extended construction time, over a period of seven years; but construction was completed in early 1970 and the first flight was made on 22 March 1970.

During 1971 Mr Fike re-engined the Model 'E' with a 63·5 kW (85 hp) Continental C85-8 engine, and it was first flown with this power plant on 6 June 1971. It had accumulated a total of 195 flight hours by early 1976.

During 1974 Mr Fike added 0·36 m (1 ft 2 in) wingtip extensions and, at a later stage of flight testing, flat endplates were added to the wingtips. The extensions to the wing have been shown to enhance performance but the effect of the endplates has proved negligible.

TYPE: High-wing sporting monoplane.

WINGS: Cantilever high-wing monoplane. Wing section NACA 4409. No dihedral. Incidence 1° 15'. No sweepback. All-wood geodetic structure, fabric-covered. Conventional wooden ailerons. No flaps. No trim tabs.

FUSELAGE: Welded steel tube structure; fabric-covered.

TAIL UNIT: Cantilever welded steel tube structure with fabric covering. Adjustable-incidence tailplane. No trim tabs.

LANDING GEAR: Standard Piper J-3 gear of non-retractable tailwheel type, with rubber-cord shock-absorption. Main wheels and tyres size 8·00-4, pressure 1·03-1·38 bars (15-20 lb/sq in). Max speed can be increased by 7 knots (13 km/h; 8 mph) by fitting 5·00-4 wheels, tyres and tubes. Goodrich toe-operated hydraulic brakes. Federal A-1500 skis available optionally.

POWER PLANT: One 63·5 kW (85 hp) Continental C85-8 flat-four engine, driving a Sensenich Type 76AK two-blade fixed-pitch metal propeller. One (Cessna 140 type) fuel tank in each wing, capacity 47·5 litres (12·5 US gallons). Total fuel capacity 95 litres (25 US gallons). Refuelling point on top of each wing. Oil capacity 3·8 litres (1 US gallon).

ACCOMMODATION: Pilot (with provision for passenger seated in tandem) in enclosed cabin. Door on each side of fuselage, hinged at forward edge. Cabin heated and ventilated. Baggage compartment aft of cabin. Overhead control assembly permits sleeping in the cabin after removing or collapsing seat.

EQUIPMENT: Radair 10 10-channel VHF communications transceiver.

DIMENSIONS, EXTERNAL:

Wing span	6·82 m (22 ft 4½ in)
Wing chord (constant, except for tips)	2·03 m (6 ft 8 in)
Wing chord at tip	1·73 m (5 ft 8 in)
Wing area, gross	13·29 m² (143·10 sq ft)
Wing aspect ratio	3·0

Length overall	5·84 m (19 ft 2 in)
Height overall	1·73 m (5 ft 8 in)
Tailplane span	2·39 m (7 ft 10 in)
Wheel track	1·83 m (6 ft 0 in)
Wheelbase	4·32 m (14 ft 2 in)
Propeller diameter	1·88 m (6 ft 2 in)

DIMENSIONS, INTERNAL:

Cabin:	
Max width	0·61 m (2 ft 0 in)
Volume	0·99 m³ (35 cu ft)
Baggage compartment	0·34 m³ (12 cu ft)

WEIGHTS AND LOADINGS:

Weight empty	313 kg (690 lb)
Normal T-O weight	499 kg (1,100 lb)
Max wing loading	40·5 kg/m² (8·3 lb/sq ft)
Max power loading	7·86 kg/kW (12·94 lb/hp)

PERFORMANCE (with 63·5 kW; 85 hp engine and 5·00-4 wheels, at normal T-O weight):

Max level speed at S/L	104 knots (193 km/h; 120 mph)
Cruising speed	82·5 knots (153 km/h; 95 mph)
Landing speed	30·5 knots (56 km/h; 35 mph)
Max rate of climb at S/L	305 m (1,000 ft)/min
Service ceiling	over 3,050 m (10,000 ft)
T-O run	76-92 m (250-300 ft)
Landing run	under 92 m (300 ft)
Range with max fuel, 15 min reserve	390 nm (724 km; 450 miles)

FLAGLOR

FLAGLOR AIRCRAFT

ADDRESS:

1550A Sanders Road, Northbrook, Illinois 60062

The latest of a series of light aircraft designed and built in prototype form by Mr K. Flaglor is an ultra-light sporting monoplane named the Scooter. Design work began in July 1965, and construction was started in November of the same year.

The prototype Scooter was powered originally by a 13·5 kW (18 hp) Cushman golf-kart engine, and it was with this power plant that the first flight was made in June 1967. Performance was marginal and, as a result, Mr Flaglor replaced the Cushman with a nominal 27 kW (18·5-21 kW output) 36 hp (25-28 hp output) Volkswagen engine. Current power plant is a 1,500 cc Volkswagen engine developing 30 kW (40 hp). When flown to the 1967 EAA meet at Rockford, Illinois, the prototype Scooter won the 'Outstanding Ultra-light' and 'Outstanding Volkswagen-Powered Airplane' awards. Plans are available to amateur constructors from Ace Aircraft Manufacturing Company (which see), and many Scooters are now flying.

FLAGLOR SCOOTER

The following description applies to the prototype Scooter in its current form:

TYPE: Ultra-light sporting monoplane.

WINGS: High-wing monoplane, braced by wires attached to fuselage and to kingpost mounted above centre-section. Wing section NACA 23012. Dihedral 2°. Incidence 3°. Two-spar all-wood structure with wood drag and anti-drag bracing. Aluminium leading-edge and plywood covering. Conventional wooden ailerons. No flaps. No trim tabs.

FUSELAGE: Wooden structure, plywood-covered in the forward cockpit area, fabric-covered aft. Fuselage of triangular section aft of the wing. Wing centre-section and engine mounting constructed of 4130 steel tube.

TAIL UNIT: All-wood construction with strut bracing. No fixed fin. No trim tabs.

LANDING GEAR: Non-retractable tailwheel type. Fixed spring steel main units. Steerable tailwheel. Main wheels of go-kart type, size 4·10 × 3·50-5. Tyre pressure 1·38 bars (20 lb/sq in). Vespa or Sears motor scooter brakes.

POWER PLANT: One 30 kW (40 hp) Volkswagen 1,500 cc flat-four engine, driving a two-blade Troyer 54-28 propeller. Single fuel tank in fuselage nose, capacity 19 litres (5 US gallons). Refuelling point on top of fuselage forward of windscreen. Oil capacity 2·37 litres (2·5 US quarts).

ACCOMMODATION: Single seat in cockpit protected by deep windscreen.

DIMENSIONS, EXTERNAL:

Wing span	8·64 m (28 ft 0 in)
Wing chord (constant)	1·27 m (4 ft 2 in)
Wing area, gross	10·68 m² (115 sq ft)
Wing aspect ratio	6·7
Length overall	4·78 m (15 ft 8 in)
Height overall	2·13 m (7 ft 0 in)
Tailplane span	2·18 m (7 ft 2 in)
Wheel track	1·37 m (4 ft 6 in)

WEIGHTS AND LOADINGS:

Weight empty	177 kg (390 lb)
Max T-O and landing weight	295 kg (650 lb)
Max wing loading	27·8 kg/m² (5·7 lb/sq ft)
Max power loading	9·83 kg/kW (16·25 lb/hp)

PERFORMANCE:

Never-exceed speed	82 knots (153 km/h; 95 mph)
Max level speed	78 knots (145 km/h; 90 mph)
Max cruising speed	69 knots (129 km/h; 80 mph)
Econ cruising speed	56 knots (105 km/h; 65 mph)
Stalling speed	30 knots (55 km/h; 34 mph)
Max rate of climb at S/L	183 m (600 ft)/min
T-O run	76 m (250 ft)
Landing run	76 m (250 ft)
Range with max fuel	152 nm (282 km; 175 miles)

FLIGHT DYNAMICS

FLIGHT DYNAMICS INC

ADDRESS:

PO Box 5070, State University Station, Raleigh, North Carolina 27607

Telephone: (919) 834-6806

PRESIDENT:

Thomas H. Purcell Jr

Flight Dynamics Inc has designed, built and developed an unusual two-seat lightweight amphibian of which plans are available to amateur constructors. It can be built in Stage I configuration with a flexible wing and without power plant, and be towed into the air by a water-ski tow boat. In Stage II configuration a power plant is added, enabling the aircraft to be flown from land or water. The final Stage III configuration substitutes a conventional fixed wing in place of the flexible wing.

Flight Dynamics claims that the simplified method of construction enables the homebuilder to get the aircraft into the air in a minimum of time, and that it can be completed in more sophisticated form as time and finances allow.

The unpowered version, which is known as the Seasprite, is described in the Sailplanes section.

Design began in January 1966, with construction of the prototype starting in March 1967, and the first flight was made in October 1970. The details which follow apply to the aircraft in Stage III (Flightsail VII) configuration, with fixed wing and power plant:

FLIGHT DYNAMICS FLIGHTSAIL VII

TYPE: Two-seat lightweight homebuilt amphibian.

WINGS: Braced high-wing monoplane, with Vee bracing struts and auxiliary struts on each side. Wing section NACA 23012. Dihedral 2°. Incidence 6°. No sweepback. Built-up light alloy main spar, bolted together; aerofoil of solid sculptured styrofoam; the whole covered in glassfibre. Plain trailing-edge flaps and ailerons of styrofoam construction, glassfibre-covered. No trim tabs.

FUSELAGE: Light alloy structure bolted together, stiffened with styrofoam and covered with glassfibre. Specially designed floats, one each side of fuselage undersurface, are integral with fuselage structure. Fuselage terminates in twin booms, wire-braced.

TAIL UNIT: Twin fins and rudders mounted on fuselage booms. Narrow-chord fixed horizontal surface braces tailbooms to each other. Large balanced elevator.

LANDING GEAR: Mechanically-retractable tricycle type. Shock-absorption by coil springs. Modified commercial wheels. Main-wheel tyre pressure 3·45 bars (50 lb/sq in). Hearst-Aerheart wheel brakes.

POWER PLANT: One 67 kW (90 hp) Continental C90 flat-four engine, carried on two pylons above the wing, and driving a two-blade fixed-pitch pusher propeller. Provision for engines of 63·5-93 kW (85-125 hp). Present fuel tank has capacity of 75·5 litres (20 US gallons), but optimum size not yet determined. Oil capacity 4·7 litres (1·25 US gallons).

ACCOMMODATION: Two seats side by side in enclosed cockpit. Transparent canopy hinged at top, opening upward. Provision for cockpit heating system.

SYSTEMS: Hydraulic system for brakes only. Electrical system powered by engine-driven generator.

ELECTRONICS: Com transceiver only.

DIMENSIONS, EXTERNAL:

Wing span	11·89 m (39 ft 0 in)
Wing chord (constant)	1·52 m (5 ft 0 in)
Wing area, gross	18·1 m² (195 sq ft)
Wing aspect ratio	7·8
Length overall	7·62 m (25 ft 0 in)
Height overall	2·68 m (8 ft 9½ in)
Elevator span	4·88 m (16 ft 0 in)
Wheel track	2·29 m (7 ft 6 in)
Wheelbase	2·62 m (8 ft 7 in)
Propeller diameter	1·88 m (6 ft 2 in)

DIMENSIONS, INTERNAL:

Cockpit: Length	2·16 m (7 ft 1 in)
Max width	1·78 m (5 ft 10 in)

WEIGHTS AND LOADINGS:

Weight empty	544 kg (1,200 lb)
Max T-O weight	771 kg (1,700 lb)
Max wing loading	42·5 kg/m² (8·7 lb/sq ft)
Max power loading	11·51 kg/kW (18·8 lb/hp)

PERFORMANCE (at max T-O weight):

Never-exceed speed	112 knots (209 km/h; 130 mph)
Max level speed at S/L	82 knots (153 km/h; 95 mph)
Max cruising speed	74 knots (137 km/h; 85 mph)
Stalling speed, flaps down	35 knots (65 km/h; 40 mph)
Max rate of climb at S/L	152 m (500 ft)/min
Service ceiling	3,660 m (12,000 ft)
T-O time on water	20 sec
T-O to 15 m (50 ft)	335 m (1,100 ft)
Landing from 15 m (50 ft)	305 m (1,000 ft)
Landing run	122 m (400 ft)
Range with max fuel	260 nm (482 km; 300 miles)
Range with max payload	173 nm (321 km; 200 miles)

GARRISON

PETER GARRISON

ADDRESS:

4019 Cumberland Avenue, Los Angeles, California 90027

Telephone: (213) 665-0934

Mr Garrison was a contributor to the initial design of the Practavia Sprite (see UK section) when employed in London by the *Pilot and Light Aeroplane* magazine. After his return to the USA he modified and developed this basic design, and first flight of the resulting aircraft, which he has named Melmoth, was made on 6 September 1973.

Comparison of the designs reveals to a small degree their common parentage; but whereas the Sprite is a simply-constructed lightplane for amateur builders, Mr Garrison's Melmoth could be described more accurately as a lightplane research prototype. It includes such aerodynamic features as double-slotted Fowler trailing-edge flaps and adjustable-incidence ailerons. Special systems and equipment include automatic fuel balancing, remote compass, Century I autopilot, full IFR panel and Collins Micro Line electronics.

In August 1975 Mr Garrison flew the Melmoth nonstop from Gander, Newfoundland, to Shannon, Ireland, in 10 hours 55 minutes. He is contemplating a fuselage stretch which would make the aircraft a full four-seater, and possible installation of a turbocharged power plant.

GARRISON OM-1 MELMOTH

TYPE: Two/three-seat homebuilt lightplane.

WINGS: Cantilever low-wing monoplane. Wing section NACA 65A-316. Dihedral 8° on outer panels only. Incidence 1°. No sweepback. Constant-chord wing of single-spar light alloy construction. Plain all-metal ailerons with piano-type hinge at lower surface. Aileron incidence variable in flight from +10° to −10°. Double-slotted Fowler-type trailing-edge flaps of light alloy construction.

FUSELAGE: Conventional semi-monocoque structure of light alloy. Glassfibre nose cowl and fairings.

TAIL UNIT: Cantilever light alloy structure. T-tail configuration with all-moving tailplane of single-spar construction. Anti-servo tab in tailplane. Manually-adjustable trim tab on rudder.

LANDING GEAR: Hydraulically-retractable tricycle type. Main units retract inward into fully-enclosed wheel wells in wing centre-section. Nosewheel retracts aft. Oleo-pneumatic shock-absorber on each unit. All three tyres size 5·00-5. Cleveland hydraulic disc brakes.

POWER PLANT: One 156·5 kW (210 hp) Continental IO-360-A flat-six engine, driving a Hartzell two-blade metal constant-speed propeller, with 19·7 cm (7¾ in) hub extension. One integral fuel tank in each outer wing panel, with capacity of 155 litres (41 US gallons). One glassfibre wingtip tank on each wing, capacity 132 litres (35 US gallons). Total fuel capacity 575 litres (152 US gallons). Refuelling points in upper surface of each wing and on top of tip-tanks. Oil capacity 9·4 litres (2½ US gallons).

ACCOMMODATION: Two seats side by side beneath transparent canopy. Each side window is hinged at top centre, opening outward and upward to provide access. Space for 90·7 kg (200 lb) baggage aft of seats, or for third passenger on removable jump-seat. Cabin is heated and ventilated.

SYSTEMS: Electrical system powered by a 28V 25A engine-driven alternator. External power socket.

Garrison OM-1 Melmoth two/three-seat lightplane (Continental IO-360-A engine)

Flaglor Scooter single-seat ultra-light monoplane *(F. van Bruggen)*

Flight Dynamics Flightsail VII amphibian (Continental C90 engine)

Fike Model 'E' with wingtip extensions and endplates

Electrically-driven hydraulic pump provides pressure of 34·5 bars (500 lb/sq in) for landing gear and flaps. Constant-flow oxygen system, capacity 1·08 m³ (38 cu ft) at pressure of 138 bars (2,000 lb/sq in).

Electronics and Equipment: Collins communication radios, Mitchell Century I autopilot with couplers, full blind-flying instrumentation, navigation lights, strobe lights, heated pitot and angle of attack transmitter, and electronically-timed fuel switching for fuel balancing.

Dimensions, external:

Wing span	7·04 m (23 ft 1 in)
Wing chord, constant	1·22 m (4 ft 0 in)
Wing area, gross	8·55 m² (92 sq ft)
Wing aspect ratio	5·78
Length overall	6·50 m (21 ft 4 in)
Height overall	2·57 m (8 ft 5 in)
Tailplane span	2·69 m (8 ft 10 in)
Wheel track	3·25 m (10 ft 8 in)
Wheelbase	1·83 m (6 ft 0 in)
Propeller diameter	1·93 m (6 ft 4 in)
Propeller ground clearance	0·23 m (9 in)

Dimensions, internal:

Cabin: Length	2·44 m (8 ft 0 in)
Max width	1·12 m (3 ft 8 in)
Baggage hold, volume	0·85 m³ (30 cu ft)

Weights and Loadings:

Weight empty, equipped	682 kg (1,500 lb)
Max T-O and landing weight	1,270 kg (2,800 lb)
Max wing loading	148·6 kg/m² (30·43 lb/sq ft)
Max power loading	8·12 kg/kW (13·33 lb/hp)

Performance: (at T-O weight of 952 kg; 2,100 lb)

Never-exceed speed	205 knots (379 km/h; 236 mph)
Max level speed at S/L	183 knots (340 km/h; 211 mph)
Max cruising speed at 1,980 m (6,500 ft)	176 knots (327 km/h; 203 mph)
Econ cruising speed at 3,050 m (10,000 ft)	148 knots (274 km/h; 170 mph)
Stalling speed, flaps down	65 knots (121 km/h; 75 mph)
Max rate of climb at S/L	549 m (1,800 ft)/min
Service ceiling	6,400 m (21,000 ft)
T-O run	366 m (1,200 ft)
T-O to 15 m (50 ft)	549 m (1,800 ft)
Landing from 15 m (50 ft)	457 m (1,500 ft)
Landing run	183 m (600 ft)
Range with max fuel, no reserve	2,952 nm (5,470 km; 3,400 miles)

GEORGIAS

GEORGIAS SPECIAL

The Georgias Special is a conventional braced parasol monoplane, which was photographed for *Jane's* at the Municipal Airport, Atchison, Kansas in June 1975. All available details follow:

Type: Single-seat homebuilt aircraft.

Wings: Braced parasol monoplane. Six cabane struts; two parallel bracing struts and jury struts under each wing; and wire bracing. Constant chord. Wings and cable-operated ailerons fabric-covered. No flaps. Trailing-edge cutout over cockpit. No tabs.

Tail Unit: Wire-braced conventional structure. Control surfaces cable-actuated. No tabs.

Landing Gear: Non-retractable tailwheel type. Spoked main wheels carried on side Vees and half-axles. Steerable tailwheel.

Power Plant: One flat-four engine, driving a two-blade wooden fixed-pitch propeller, with small spinner.

Accommodation: Single seat in open cockpit.

GREAT LAKES

GREAT LAKES AIRCRAFT COMPANY

Head Office:
PO Box 1113w, Wichita, Kansas 67202

In addition to marketing factory-built **Great Lakes Sport Trainers**, this company offers certificated components and assemblies, but not plans, to amateur constructors. The aircraft is described fully in the main US Aircraft section of this edition.

HANSEN

WILLIAM HANSEN

Address:
Minneapolis, Minnesota

HANSEN 'SUMP'N ELSE'

This racing aircraft was flown for the first time on 6 July 1971, as *Thunderchicken* (N35WH), by its designer and builder William Hansen. It first competed at Cleveland, Ohio, on 15 July 1971; its pilot, Ken Haas, qualified at 166 knots (307 km/h; 191 mph) over a 2·2 nm (4 km; 2·5 mile) course. Due to a variety of mechanical problems, it failed to complete any of its subsequent heats. In September 1971, Tom Cooney flew it in the Reno National Championship Air Races, qualifying at 180 knots (333 km/h; 207 mph), and was placed fifth in the Consolation Race at 176 knots (327 km/h; 203 mph).

The aircraft was purchased by the current owner, Tom Summers of Friendswood, Texas, in late 1974, and was first raced by him at Mojave, California. Its best performances to date have been a 182 knot (336 km/h; 209 mph) qualifying speed, and 178 knots (330 km/h; 205 mph) in the Consolation Race, at Reno in 1975.

Type: Single-seat Formula One racing monoplane.

Wings: Cantilever low-wing monoplane, based on plans for *Shoestring* Formula One racer (which see). No dihedral. Wooden ribs and spars, covered with 2·4 mm (3/32 in) plywood, with fabric covering overall. No flaps.

Fuselage: Chrome-molybdenum 4130 steel tube structure. Moulded glassfibre cowling and fuselage sides to rear of cockpit. Cotton fabric covering aft.

Tail Unit: Steel tube T-tail, fabric covered.

Landing Gear: Non-retractable tailwheel type, with fairings over wheels and aluminium spring leaf main gear legs. Size 5·00-5 main wheels, with Cleveland brakes.

Power Plant: One Continental O-200 flat-four engine, driving a two-blade fixed-pitch propeller of 1·57 m (62 in) pitch. Fuel capacity 41·6 litres (11 US gallons).

Accommodation: Single seat under Plexiglas canopy. Nose-over structure built into headrest.

Dimensions, external:

Wing span	5·69 m (18 ft 8 in)
Wing area, gross	6·13 m² (66 sq ft)
Wing aspect ratio	5·5
Length overall	5·56 m (18 ft 3 in)
Height overall	1·45 m (4 ft 9 in)
Propeller diameter	1·47 m (4 ft 10 in)

Weights:

Weight empty	283 kg (625 lb)
Normal T-O weight	381 kg (840 lb)
Max T-O weight	397 kg (875 lb)

Performance:

Max level speed	over 204 knots (378 km/h; 235 mph)
Landing speed	65 knots (121 km/h; 75 mph)

HARMON

HARMON ENGINEERING COMPANY

Address:
Route 1, PO Box 186, Howe, Texas 75059

Telephone: (214) 532-6551

President: James B. Harmon

Harmon Engineering Company designed and built the prototype of a single-seat sporting aircraft named Der Donnerschlag (The Thunderclap). Design began in December 1972, and construction of the prototype started six months later. The first flight of this aircraft was recorded in June 1974.

Georgias Special single-seat monoplane *(Jean Seele)*

Hansen Sump'n Else Formula One racing aircraft

Prototype of the Harmon Der Donnerschlag

Harmon Mister America sporting aircraft

Four months later, the aircraft suffered an engine failure while being flown by Mr James Harmon. He hit the ground at 52 knots (96 km/h; 60 mph) at a 45° impact angle, but was able to walk away from the accident. The aircraft was so little damaged that the decision was made to rebuild it as the prototype of a variant of Der Donnerschlag named Mister America. Design of the variant was started in November 1974, simultaneously with the rebuilding, and first flight of Mister America took place on 31 October 1975.

Drawings of both Der Donnerschlag and Mister America are available.

HARMON DER DONNERSCHLAG

Type: Single-seat lightweight sporting aircraft.

Wings: Wire-braced shoulder-wing monoplane. Original wing section, with thickness/chord ratio of 14·5%. Dihedral 1°-2°. Incidence 0°. Wooden structure with fabric covering. Two simple beam spars; built-up ribs. Plain ailerons of similar construction. No flaps. No trim tabs. Dual flying and landing wires each side, landing wires attached to cabane structure at forward end of fuselage.

Fuselage: Welded structure of 4130 chrome-molybdenum steel tube with fabric covering.

Tail Unit: Wire-braced welded structure of 4130 chrome-molybdenum steel tube with fabric covering. Tailplane incidence ground-adjustable. Elevators fitted.

Landing Gear: Non-retractable tailwheel type. Main units consist of two Vees, attached to fuselage structure, wire-braced and with a through axle. Shock-absorption by rubber bungee. Motor cycle wire wheels with 3·0-13 tyres.

Power Plant: One 56 kW (75 hp) 1,800 cc Volkswagen modified motor car engine, driving a two-blade wooden fixed-pitch propeller. Fuel tank in fuselage with maximum capacity of 37·8 litres (10 US gallons).

Accommodation: Single seat in open cockpit.

Dimensions, external:
Wing span 5·94 m (19 ft 6 in)
Wing chord (constant) 1·32 m (4 ft 4 in)
Wing area, gross 7·25 m² (78 sq ft)
Wing aspect ratio 5
Length overall 4·42 m (14 ft 6 in)
Height overall 1·83 m (6 ft 0 in)
Tailplane span 1·98 m (6 ft 6 in)
Wheel track 1·37 m (4 ft 6 in)
Propeller diameter 1·27 m (4 ft 2 in)
Propeller ground clearance 0·41 m (1 ft 4 in)

Weights and Loadings:
Weight empty 159 kg (350 lb)
Max T-O weight 272 kg (600 lb)
Max wing loading 37·6 kg/m² (7·7 lb/sq ft)
Max power loading 4·86 kg/kW (8·0 lb/hp)

Performance (at max T-O weight):
Never-exceed speed 139 knots (257 km/h; 160 mph)
Max level speed 104 knots (193 km/h; 120 mph)
Max cruising speed at 305 m (1,000 ft) 96 knots (177 km/h; 110 mph)
Stalling speed 48 knots (89 km/h; 55 mph)
Max rate of climb at S/L 244 m (800 ft)/min
Service ceiling above 3,050 m (10,000 ft)
T-O run 46 m (150 ft)
T-O to 15 m (50 ft) 91 m (300 ft)
Landing from 15 m (50 ft) 152 m (500 ft)
Landing run 76 m (250 ft)
Range with max fuel 434 nm (804 km; 500 miles)
Range with max payload 260 nm (482 km; 300 miles)

HARMON 1-2 MISTER AMERICA

Type: Single-seat lightweight sporting aircraft.

Wings: As for Der Donnerschlag. Provision for ground-adjustable tabs in ailerons, if required. Cabane forms windscreen support.

Fuselage: Welded structure of 4130 chrome-molybdenum steel tube with fabric covering. Wooden stringer turtledeck, with fabric covering. Aluminium and glassfibre engine cowling.

Tail Unit: As for Der Donnerschlag. Provision for ground-adjustable trim tabs, if required.

Landing Gear: Basically as for Der Donnerschlag. Cleveland 5·00-5 main wheels and tyres, pressure 1·03 bars (15 lb/sq in). Aviation Products 10 cm (4 in) steerable tailwheel. Cleveland disc brakes with air cooling. Faire main legs and wheels.

Power Plant: One 44·7-48·5 kW (60-65 hp) 1,650 c Volkswagen modified motor car engine, driving a two blade fixed-pitch 54-36 propeller. Fuel tank aft o engine firewall, capacity 34 litres (9 US gallons). Refuel ling point forward of windscreen.

Accommodation: Single seat in open cockpit.

Dimensions, external:
Wing span 5·99 m (19 ft 8 in
Wing chord (constant) 1·22 m (4 ft 0 in
Wing area, gross 7·06 m² (76·0 sq ft
Wing aspect ratio 4·91
Length overall 4·62 m (15 ft 2 in
Height overall 1·52 m (5 ft 0 in
Tailplane span 1·78 m (5 ft 10 in
Wheel track 1·52 m (5 ft 0 in
Wheelbase 3·51 m (11 ft 6 in
Propeller diameter 1·37 m (4 ft 6 in
Propeller ground clearance 0·305 m (1 ft 0 in

Dimensions, internal:
Cockpit:
Max width 0·52 m (1 ft 8½ in

Weights and Loading:
Weight empty 195 kg (430 lb
Max T-O weight 295 kg (650 lb
Max wing loading 41·74 kg/m² (8·55 lb/sq ft

Performance :
Never-exceed speed 143 knots (265 km/h; 165 mph
Max level speed at S/L 109 knots (201 km/h; 125 mph
Max cruising speed at S/L 95·5 knots (177 km/h; 110 mph
Econ cruising speed at S/L 87 knots (161 km/h; 100 mph
Stalling speed 42 knots (77·5 km/h; 48 mph
Max rate of climb at S/L 244 m (800 ft)/mi
Service ceiling 3,660 m (12,000 ft
T-O run 61 m (200 ft
T-O to 15 m (50 ft) 122 m (400 ft
Landing from 15 m (50 ft) 152 m (500 ft
Landing run 92 m (300 ft
Range with max fuel 347 nm (643 km; 400 miles

HARRIS

J. WARREN HARRIS

Address:
283 East-500 North, Vernal, Utah 84078
Telephone: (801) 789-2481

HARRIS GEODETIC LW 108

Design of this aircraft began in 1971. The only example to be produced was completed on 5 March 1975 and is certificated in the Experimental category. By April 1976 the LW 108 had logged over 119 flying hours; no alterations to the airframe have been made since construction.

The most unusual feature of the aircraft is its geodetically-constructed fuselage, which is claimed to be light in weight and economical to build, while offering high strength and a low-drag external shape.

Type: Two-seat homebuilt aircraft.

Wings: Low-wing monoplane. Wing section NACA 4412. Dihedral on outer panels only. Two-spar structure of Sitka spruce, with plywood covering. Box spar at front; channel section rear spar. Hinged ailerons.

Fuselage: Geodetic structure of circular section, using Sitka spruce strips. Laminated spruce bulkhead ring-frames, faced with ply at stress points.

Tail Unit: Spruce and plywood construction. One-piece all-moving horizontal surfaces, with servo tab.

Landing Gear: Non-retractable tailwheel type. Rubber disc shock-absorbers. Main-wheel tyres size 5·00-5 pressure 0·97 bars (14 lb/sq in). Band-type brakes.

Power Plant: One 59·5 kW (80 hp) Continental A80- flat-four engine, driving a fixed-pitch McCaule CM68-50 two-blade propeller. One fuel tank in fuse age, capacity 60·5 litres (16 US gallons). Oil capacit 3·8 litres (1 US gallon).

Accommodation: Two seats side by side in enclosed cock pit. Upward-hinged canopy door on starboard side.

Dimensions, external:
Wing span 8·56 m (28 ft 1 in
Wing chord at root 1·52 m (5 ft 0 in

Wing chord at tip 0·91 m (3 ft 0 in)
Wing area, gross 10·03 m² (108 sq ft)
Wing aspect ratio 6·22
Length overall 5·89 m (19 ft 4 in)
Tailplane span 2·44 m (8 ft 0 in)
Propeller diameter 1·73 m (5 ft 8 in)

VEIGHTS AND LOADINGS:
Weight empty 265 kg (585 lb)
Max T-O weight 453 kg (1,000 lb)
Max wing loading 45·2 kg/m² (9·25 lb/sq ft)
Max power loading 7·61 kg/kW (12·5 lb/hp)

PERFORMANCE (at max T-O weight):
Never-exceed speed
156 knots (290 km/h; 180 mph) IAS
Max level speed at 1,525 m (5,000 ft)
130 knots (241 km/h; 150 mph) IAS
Max cruising speed at 1,525 m (5,000 ft)
113 knots (209 km/h; 130 mph)
Econ cruising speed at 1,525 m (5,000 ft)
97 knots (180 km/h; 112 mph)
Stalling speed 35 knots (65 km/h; 40 mph)
Max rate of climb at S/L 457 m (1,500 ft)/min
T-O to 15 m (50 ft) approx 153 m (500 ft)
Range with max fuel, no reserve
347 nm (643 km; 400 miles)

IATZ

IOHN D. HATZ

ADDRESS:
Merrill Airways, Municipal Airport, Merrill, Wisconsin 54452

Mr Hatz designed and built the prototype of a two-seat ghtweight biplane, designated CB-1, of which plans are vailable to amateur constructors. Design and construcion started in September 1959, and FAA certification in he Experimental category was awarded on 18 April 1968. irst flight of the CB-1 was made on the following day, owered by a 63·5 kW (85 hp) Continental C85-12 ngine, but this has since been replaced by a 112 kW (150 p) Lycoming O-320. Many more CB-1s are under contruction.

HATZ CB-1 BIPLANE

YPE: Two-seat lightweight homebuilt biplane.

VINGS: Braced single-bay biplane, with N-type interplane struts each side. N-type centre-section struts and streamline-section flying and landing wires. Wing section Clark Y. Dihedral 2°, on lower wings only. Incidence (both wings) 2°. Wooden two-spar structure with fabric covering. Cutout in trailing-edge of upper wing. Plain unbalanced ailerons of wood construction, with fabric covering, on both wings. No flaps. No trim tabs.

FUSELAGE: Welded steel tube structure, with fabric covering.

TAIL UNIT: Wire-braced welded steel tube structure, with fabric covering. Tailplane incidence adjustable by screwjack.

LANDING GEAR: Non-retractable tailwheel type. Two side Vees and half-axles hinged to fuselage structure. No shock-absorption. Main wheels from Piper J-3 Cub, with Goodyear tyres size 8·00-4, pressure 1·03 bars (15 lb/sq in). Wheel brakes from J-3 Cub. Glassfibre fairings on main wheels.

POWER PLANT: One 112 kW (150 hp) Lycoming O-320 flat-four engine, driving a Sensenich Type M74DM two-blade metal fixed-pitch propeller. Two fuel tanks in centre-section of upper wing, each with capacity of 60·5 litres (16 US gallons). Total fuel capacity 121 litres (32 US gallons). Refuelling points on upper surface of upper wing centre-section. Oil capacity 7·5 litres (2 US gallons).

ACCOMMODATION: Two seats in tandem in open cockpits.

SYSTEMS: Electrical system, with 12V 12A DC engine-driven generator and 12V battery, for engine starting, navigation and instrument lights.

DIMENSIONS, EXTERNAL:
Wing span (both) 7·92 m (26 ft 0 in)
Wing chord (constant, both) 1·37 m (4 ft 6 in)
Wing area, gross 17·65 m² (190 sq ft)
Wing aspect ratio 5·77
Length overall 5·64 m (18 ft 6 in)
Height overall 2·39 m (7 ft 10 in)
Tailplane span 2·74 m (9 ft 0 in)
Wheel track 1·83 m (6 ft 0 in)
Propeller diameter 1·88 m (6 ft 2 in)

WEIGHTS AND LOADINGS:
Weight empty 438 kg (966 lb)
Max T-O weight 726 kg (1,600 lb)
Max wing loading 41·1 kg/m² (8·42 lb/sq ft)
Max power loading 6·48 kg/kW (10·7 lb/hp)

PERFORMANCE (at max T-O weight):
Never-exceed speed 130 knots (241 km/h; 150 mph)
Max cruising speed 87 knots (161 km/h; 100 mph)
Stalling speed 39 knots (72·5 km/h; 45 mph)
Max rate of climb at S/L 366 m (1,200 ft)/min
T-O run approx 122 m (400 ft)
Range with max fuel, 30 min reserve
234 nm (434 km; 270 miles)

IELIGYRO

IELIGYRO CORPORATION

ADDRESS:
PO Box 2242, Scottsdale, Arizona 85252
PRESIDENT: Bruno Nagler
MANAGEMENT/ENGINEERING CONSULTANT: E. K. Liberatore, 567 Fairway Road, Ridgewood, New Jersey 07450

The Heligyro Corporation has been formed to market n air-jet rotor helicopter, named the Phoenix, that is uitable for amateur construction. The Phoenix has volved from several years of experimenting with three rototype helicopters, the first of which was the single-seat Ioncho 100 'cold-jet' tip-driven helicopter described in he 1974-75 *Jane's*. Following the two-seat Honcho 200, he company, then known as Nagler Helicopters Inc, built he Model 202 which was basically a pre-production verion of the Phoenix. Details of the Model 202 can be found n the 1975-76 *Jane's*.

More than 200 test flights were made by the three elicopters, which accumulated between them nearly 100 ying hours. Like these earlier research helicopters, the hoenix uses tip-mounted 'cold-jets' to drive the rotor; his method of propulsion eliminates the need for a tail otor and for a complicated drive mechanism for the main otor. Because of this, only a simple and comparatively lightweight structure is required to carry the rotor shaft.

HELIGYRO PHOENIX

TYPE: Two-seat lightweight helicopter.

ROTOR: Two-blade rotor, driven by tip-mounted 'cold-jets' supplied with air from a bleed air compressor. Rotor system designed to be operable over wide rpm range without significant effect on propulsive thrust or lift capability. Rotor blades, which are hollow light alloy extrusions, are attached to the rotor hub by laminated stainless steel tension-torsion straps. Simple pylon structure immediately aft of cabin bulkhead, incorporating shock mounts. No rotor brake. No tail rotor. Rotor rpm range 200-380. Normal rotor speed 290 rpm.

FUSELAGE: Rectangular-section forward fuselage, comprising basic steel tube structure and glassfibre shell. Rotor pylon consists of an 'A' frame.

TAIL UNIT: Large tail fin. Movable rudder, beneath fin, mounted centrally in efflux from turbine compressor. Narrow-chord horizontal stabiliser mounted at base of fin trailing-edge.

POWER PLANT: One 164 kW (220 hp) Victor water-cooled flat-six engine. Single-stage centrifugal air compressor driven through clutch, coupling and simple gearbox. Single fuel tank in fuselage, capacity 91 litres (24 US gallons). Radiator capacity 11 litres (3 US gallons).

ACCOMMODATION: Two seats side by side in enclosed cabin. Conventional helicopter controls.

DIMENSIONS, EXTERNAL:
Rotor diameter 10·97 m (36 ft 0 in)
Rotor blade chord 25·4 cm (10 in)
Length of fuselage 3·69 m (12 ft 1 in)
Width overall, less rotor 1·66 m (5 ft 5¼ in)
Height overall 2·13 m (7 ft 0 in)

WEIGHTS:
Weight empty 333 kg (735 lb)
Normal T-O weight 553 kg (1,220 lb)
Design limit load factor 3·0*g*

PERFORMANCE:
Max cruising speed at S/L
83 knots (153 km/h; 95 mph)
Normal cruising speed at S/L
74 knots (137 km/h; 85 mph)
Max rate of climb at S/L 314 m (1,030 ft)/min
Vertical rate of climb at S/L 244 m (800 ft)/min
Hovering ceiling out of ground effect
1,525 m (5,000 ft)
Hovering ceiling in ground effect 3,353 m (11,000 ft)
Service ceiling 4,575 m (15,000 ft)
Range at 60 knots (113 km/h; 70 mph)
130 nm (241 km; 150 miles)

IOLLMANN

IOLLMANN AIRCRAFT

ADDRESS:
7917 Festival Court, Cupertino, California 95014
elephone: (408) 255-2194

The HA-2M Sportster two-seat gyroplane has been eveloped by Mr Martin Hollmann, a senior design ngineer in the aerospace industry. It is claimed to be the rst aircraft of its type designed for the homebuilder who as access to a minimum of power tools. About 90% of the tructure is bolted and riveted together, and a minimum of achined parts are used.

Two average-size people can fly in the Sportster, which s suitable for pilot training and for flying on short cross-ountry journeys of up to 78 nm (145 km; 90 miles). It has een designed for towing behind a car, with the rotor towed in a box attached to the car's roof. From the towed ondition, the Sportster can be ready for its pre-flight valk-round inspection in ten minutes.

Plans, materials and many components are available to mateur constructors, and a network of distributors is eing organised.

HOLLMANN HA-2M SPORTSTER

Design of this gyroplane began in June 1969, and contruction of the first prototype was started in December 972. The first flight of the prototype was made in October 974, with FAA certification in the Experimental cateory. The first passenger was Dr Tom Butler, Vice-President Engineering and Research of AMF Incorporated, who flew in October 1975; test flying was completed by January 1976. Three other Sportsters were being constructed in mid-1976, the first of them by Dr Butler and Walter 'Skip' Tyler, who planned to have the aircraft ready before the 1976 EAA Fly-in. It is hoped to set new official performance records for this type of aircraft with the Sportster.

TYPE: Two-seat homebuilt gyroplane.

ROTOR SYSTEM: Two-blade rotor of NACA 8-H-12 section. Solidity ratio 0·034. Pre-cone angle 2°. Blade pitch +2½°. Metal blades, each made up of a 2024-T8511 leading-edge extrusion, formed ribs and 2024-T3 Alclad skin.

FUSELAGE: Square-tube 6061-T6 aluminium structure, with Alclad skin. Glassfibre fairings. Fail-safe structure, with two mast tubes. Large rear window for 360° visibility.

TAIL UNIT: Twin fins and rudders carried on short tail-booms. Aluminium structure, with Alclad skins pop-riveted in place.

LANDING GEAR: Non-retractable tricycle type, with single wheel on each unit. Two small tailwheels. Main-wheel tyres size 18 × 6, pressure 1·8 bars (26 lb/sq in). Nose-wheel tyre of 0·25 m (10 in) diameter, pressure 1·8 bars (26 lb/sq in). Mechanical drum brake on nosewheel only.

POWER PLANT: One 97 kW (130 hp) Franklin Sport 4B flat-four engine, driving a Banks-Maxwell 66 CA 36 pusher propeller. One fuel tank of 45·4 litres (12 US gallons) capacity, with refuelling point inside cockpit. Oil capacity 5·2 litres (1·35 US gallons).

ACCOMMODATION: Two seats side by side in cabin with open sides.

SYSTEMS: Standard aircraft instruments.

DIMENSIONS, EXTERNAL:
Rotor diameter 8·53 m (28 ft 0 in)
Blade chord, constant 0·23 m (9 in)
Length overall 3·96 m (13 ft 0 in)
Max width 2·06 m (6 ft 9 in)
Height to top of rotor hub 2·34 m (7 ft 8 in)
Wheel track 1·88 m (6 ft 2 in)
Wheelbase 1·42 m (4 ft 8 in)
Propeller diameter 1·68 m (5 ft 6 in)

DIMENSIONS, INTERNAL:
Max width 0·914 m (3 ft 0 in)
Max height 1·45 m (4 ft 9 in)

WEIGHTS AND LOADINGS:
Weight empty, equipped 281 kg (620 lb)
Max T-O and landing weight 476 kg (1,050 lb)
Max disc loading 8·33 kg/m² (1·71 lb/sq ft)
Max power loading 4·91 kg/kW (8·0 lb/hp)

PERFORMANCE:
Never-exceed speed at S/L
78 knots (145 km/h; 90 mph)
Max cruising speed at S/L
65 knots (121 km/h; 75 mph)
Economic cruising speed at S/L
52 knots (97 km/h; 60 mph)
Max rate of climb at S/L 152 m (500 ft)/min
Service ceiling 2,135 m (7,000 ft)
T-O run 107 m (350 ft)
Range with max fuel 78 nm (145 km; 90 miles)
Range with max payload 61 nm (112 km; 70 miles)

Harris Geodetic LW 108 two-seat homebuilt aircraft

Hatz CB-1 biplane (Lycoming O-320 engine) *(Howard Levy)*

Hollmann HA-2M Sportster two-seat gyroplane

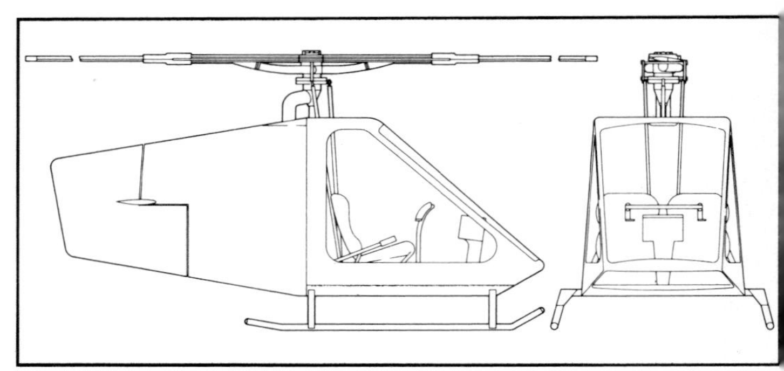

Heligyro Phoenix two-seat lightweight helicopter

HOVEY
ROBERT W. HOVEY

Address:
Aircraft Specialties, PO Box 1074, Saugus, California 91350

Telephone: (805) 252-4054

Mr Hovey has designed and built an ultra-lightweight biplane of which plans are available to amateur constructors. His objective was to produce an aircraft which would require minimal construction time and have STOL performance, and which could be quickly disassembled for transportation. To achieve these ends the design has some unusual features, such as wing warping for roll control, use of an aluminium tube tailboom which has high-strength light alloy sheet and urethane foam stiffening, and the use of styrofoam core sandwiched in craft paper for horizontal and vertical tail surfaces.

Design began in October 1970, construction starting in the following month. First flight was made in February 1971, at which time the aircraft, known as Whing Ding II, received FAA certification in the Experimental category.

The original prototype was sold in Japan, where considerable interest has been aroused among homebuilders. A second prototype was completed subsequently. About 4,800 sets of plans had been sold and many Whing Dings were under construction by early 1976.

Mr Hovey is currently working on an enclosed single-seat mid-wing monoplane, which he has designated WD-V, powered by a 31·3 kW (42 hp) Snowmobile engine. Construction of the prototype was well advanced in early 1976.

HOVEY WHING DING II (WD-II)

Type: Single-seat ultra-lightweight homebuilt biplane.

Wings: Braced single-bay biplane with parallel streamline-section interplane struts. Aircraft's fuselage, into which wing spars are socketed, gives location of inboard ends of wings. Landing and flying wires, the rear flying wires being used to control warping of upper wing. Wing section Hovey-10. Thickness/chord ratio 10%. Dihedral, both 1°. Incidence, both 4°. No sweepback. Wooden two-spar structure with ribs formed of 9·5 mm (3/8 in) light alloy tube. Wingtip bows of 9·5 mm (3/8 in) light alloy tube. Leading-edge faired in with rigid urethane foam. Wing structure fabric covered, tension of which retains the ribs in position. A plasticised fabric dope is used to ensure adequate flexibility for wing warping. No ailerons. No flaps. No trim tabs.

Fuselage: A closed box structure of 3·2 mm (1/8 in) mahogany plywood glued to 12·7 mm (1/2 in) square pine stringers, which is filled with urethane foam to stiffen and stabilise the plywood skin. This narrow fuselage provides attachment points for the seat, rudder bar and controls, and sockets for the wing spars. A reinforced extension at the top of the fuselage carries the engine. Aluminium tube tailboom is reinforced by high-strength alloy sheet at the forward end, this being wrapped around the tube and bonded with epoxy resin. The entire tube is filled with free-foam urethane.

Tail Unit: Strut-braced structure with all-moving tailplane. Tailplane consists of one piece of Foam Core sheet, which has a styrofoam core sandwiched between high-strength craft paper. Leading- and trailing-edges are pressed together and taped to form a streamline shape. Tailplane attached to tailboom by piano hinge. Stressed areas reinforced by 3·2 mm (1/8 in) plywood sheet. Fin and rudder of similar construction, utilising Foam Core reinforced with 1/8 in plywood sheet. Rudder attached to fin by cloth hinges. No trim tabs.

Landing Gear: Non-retractable tailwheel type. Main wheels carried on spring-type strut of laminated fir covered with a layer of polyester glassfibre. Go-kart type main wheels with 28 cm (11 in) diameter tyres. Tyre pressure 1·38 bars (20 lb/sq in). Tailwheel has solid rubber tyre. Alternative steel tube landing gear available.

Power Plant: One 10·44 kW (14 hp) McCulloch 101A single-cylinder two-stroke aircooled go-kart engine, driving a two-blade hand-carved laminated birch or beechwood fixed-pitch pusher propeller. Fuel tank integral with engine, capacity 1·9 litres (0·5 US gallons).

Accommodation: Pilot only on open seat.

Equipment: Basic instrumentation only, comprising airspeed and engine speed indicators and cylinder head temperature gauge.

Dimensions, external:	
Wing span (both)	5·18 m (17 ft 0 in)
Wing chord (both, constant)	0·91 m (3 ft 0 in)
Wing area, gross	9·10 m² (98 sq ft)
Wing aspect ratio	5·66
Length overall	4·27 m (14 ft 0 in)
Height overall	1·68 m (5 ft 6 in)
Tailplane span	1·93 m (6 ft 4 in)
Wheel track	1·22 m (4 ft 0 in)
Wheelbase	2·97 m (9 ft 9 in)
Propeller diameter	1·22 m (4 ft 0 in)
Weights and Loadings:	
Weight empty, incl fuel	55·5 kg (123 lb)
Max T-O weight	140 kg (310 lb)
Max wing loading	15·7 kg/m² (3·22 lb/sq ft)
Max power loading	13·41 kg/kW (22·2 lb/hp)
Performance (at max T-O weight):	
Never-exceed speed	52 knots (96·5 km/h; 60 mph)
Max level speed at S/L	43·5 knots (80·5 km/h; 50 mph)
Econ cruising speed at S/L	35 knots (64·5 km/h; 40 mph)
Stalling speed	23 knots (42 km/h; 26 mph)
Service ceiling	1,220 m (4,000 ft)
T-O run	76 m (250 ft)
T-O to 15 m (50 ft)	107 m (350 ft)
Landing from 15 m (50 ft)	76 m (250 ft)
Landing run	46 m (150 ft)
Range	17 nm (32 km; 20 miles)

JAVELIN
JAVELIN AIRCRAFT COMPANY INC

Address:
9175 East Douglas, Wichita, Kansas 67207

Telephone: (316) 682-0111

President: David D. Blanton

Javelin Aircraft Company was founded on 1 March 1953 to manufacture a low-cost automatic pilot for small aircraft; this was followed by equipment manufacture, and aircraft development work. Following restoration of a Curtiss Robin between 1957 and 1961, which won the National Championship award for the best restored antique aircraft in 1961, Javelin sought an Arrow Sport biplane for similar treatment. Unable to find a suitable aircraft, the company began design and development of a biplane on 1 January 1964.

The resulting aircraft, designated Wichawk, has structural geometry similar to a Stearman biplane, as well as some of its aerodynamic features, and is stressed for +12 and −6*g*. It flew for the first time on 24 May 1971 and has received FAA certification in the homebuilt category. Javelin Aircraft does not build the Wichawk, but plans, wing ribs and fuel tanks are available to amateur constructors. More than 100 Wichawks are known to be under construction.

JAVELIN WICHAWK

Type: Two/three-seat homebuilt sporting biplane.

Wings: Braced single-bay biplane, with N-shape streamline-section cabane and interplane struts. Streamline-section landing and flying wires. Wing section NACA 23015. 2° dihedral on lower wings only. Incidence 0°. No sweepback. Composite structure, with two wooden spars and 2024T3 light alloy ribs, fabric-covered. Simple sealed ailerons on lower wings only. No flaps. Geared trim tab.

Fuselage: Welded structure of 4130 chrome-molybdenum steel tube with light alloy tubular stringers, fabric-covered.

Tail Unit: Wire-braced welded structure of 4130 chrome-molybdenum steel tube with fabric covering. Fixed-incidence tailplane. Trim tab in starboard elevator.

Landing Gear: Non-retractable tailwheel type. Main wheels carried on side Vees hinged to lower fuselage longerons. Shock-absorption by automotive-type shock-struts, similar to those of Piper PA-20, and rubber shock cord. Main wheels and tyres size 6·00-6. Cleveland toe brakes. Steerable tailwheel. Fittings available for float or ski landing gear.

Power Plant: Prototype has one 134 kW (180 hp) Lycoming O-360 flat-four engine, driving a Sensenich two-blade fixed-pitch propeller type 76EM8-0-56. McCauley propellers optional. Provision for alternative engines up to 156 kW (210 hp), and Javelin can provide installation drawings for various horizontally-opposed or radial engines. Fuel tank of 94·5 litre (25 US gallon) capacity in upper wing centre-section, and one of 56·5 litre (14 US gallon) capacity in fuselage aft of firewall. Refuelling points above tanks. Oil capacity of prototype 7·5 litres (2 US gallons).

Accommodation: Two seats, side by side, in open cockpit. Provision for tandem two-seat or three-seat configurations. Drawings available for rearward-sliding transparent cockpit canopy. Dual controls standard. Baggage compartment aft of seats, capacity 54 kg (120 lb). Baggage locker, in turtleback, capacity 9 kg (20 lb).

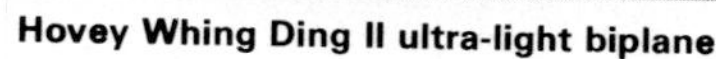

Hovey Whing Ding II ultra-light biplane

Javelin Wichawk two-seat biplane (Lycoming O-360 engine)

SYSTEMS: Electrical system powered by 12V 50A engine-driven generator. Hydraulic system for brakes only.

DIMENSIONS, EXTERNAL:

Wing span (upper)	7·32 m (24 ft 0 in)
Wing chord (constant, both)	1·27 m (4 ft 2 in)
Wing area, gross	17·2 m² (185 sq ft)
Wing aspect ratio	5·76
Length overall	5·87 m (19 ft 3 in)
Height overall	2·18 m (7 ft 2 in)
Tailplane span	2·44 m (8 ft 0 in)
Wheel track	1·87 m (6 ft 1½ in)
Propeller diameter	1·93 m (6 ft 4 in)

DIMENSIONS, INTERNAL:

Cockpit: Max width	0·93 m (3 ft 0½ in)
Baggage compartment	0·34 m³ (12 cu ft)

WEIGHTS AND LOADINGS (prototype with 134 kW; 180 hp engine):

Weight empty	580 kg (1,280 lb)
Max T-O weight	907 kg (2,000 lb)
Max wing loading	52·7 kg/m² (10·8 lb/sq ft)
Max power loading	6·77 kg/kW (11·1 lb/hp)

PERFORMANCE (prototype with 134 kW; 180 hp engine):

Never-exceed speed	156 knots (289 km/h; 180 mph)
Max level speed at S/L	121·5 knots (225 km/h; 140 mph)
Max cruising speed	110 knots (204 km/h; 127 mph)
Landing speed	39 knots (72·5 km/h; 45 mph)
Max rate of climb at S/L	518 m (1,700 ft)/min
T-O run	46 m (150 ft)

JEFFAIR

JEFFAIR CORPORATION

ADDRESS:
PO Box 975, Renton, Washington 98055
Telephone: (206) 863-7992
PRESIDENT: Geoffrey L. Siers

This company was formed by Mr Geoffrey Siers, a former RAF fighter pilot and design engineer with BAC, who emigrated to the USA in 1964. In 1966 he began designing a high-performance all-wooden two-seat light aircraft, of which construction was started in June 1969. Now known as the Barracuda, this aircraft (N19GS) flew for the first time on 29 June 1975, and is certificated in the FAA's Experimental category. Plans are available to amateur constructors, and several more Barracudas were being built by early 1976.

JEFFAIR BARRACUDA

TYPE: Two-seat all-wooden sporting monoplane.

WINGS: Cantilever low-wing monoplane, of 'inverted gull' configuration, made in three pieces: two constant-chord outer panels and a centre-section which is integral with the fuselage. Wing section NACA 64_2415. Slight anhedral on centre-section; dihedral of 5° on outer panels. No incidence. Basic structure of spruce, with mahogany and birch plywood covering. Box-section main spar with spruce booms and ply webs; truss-type ribs of spruce. Frise-type ailerons of wood and glassfibre. Electrically-actuated wooden plain flap on entire centre-section trailing-edge, extending under fuselage.

FUSELAGE: Conventional spruce structure, covered with birch plywood stressed skin.

TAIL UNIT: Cantilever spruce structure, plywood-covered. Glassfibre tips. Electrically-actuated trim tab in port elevator. Rudder has bungee trim.

LANDING GEAR: Retractable tricycle type. Electro-hydraulic retraction, main wheels inward, nosewheel rearward; all wheels fully enclosed by doors when retracted. Emergency extension by gravity. Legs made from 4130 steel tubing, with steel coil spring shock-absorption. Cleveland wheels, size 5·00-5, with 0·36 m (14 in) diameter tyres. Cleveland aircooled brakes.

POWER PLANT: One 164 kW (220 hp) Lycoming GO-435-2 flat-six engine, driving a Hartzell three-blade controllable-pitch propeller. Two glassfibre fuel tanks in centre-section, total capacity 166·5 litres (44 US gallons). Oil capacity 11·5 litres (3 US gallons).

ACCOMMODATION: Two armchair seats, with thigh support, side by side under upward-hinged individual 'gull-wing' canopy doors. Dual controls. Baggage space behind seats. Cabin heated and ventilated.

SYSTEMS: Electric pump supplies hydraulic system. Electrical supply via 12V DC system.

ELECTRONICS: Edo-Air 360-channel nav/com.

DIMENSIONS, EXTERNAL:

Wing span	7·54 m (24 ft 9 in)
Wing area, gross	11·15 m² (120 sq ft)
Length overall	6·55 m (21 ft 6 in)
Wheel track	2·54 m (8 ft 4 in)
Propeller diameter	2·18 m (7 ft 2 in)

DIMENSIONS, INTERNAL:

Cabin: Max width	1·07 m (3 ft 6 in)
Baggage hold capacity	18 kg (40 lb)

WEIGHTS:

Weight empty	678 kg (1,495 lb)
Max T-O weight	998 kg (2,200 lb)

PERFORMANCE:

Never-exceed speed	260 knots (482 km/h; 300 mph)
Max level speed at 2,130 m (7,000 ft)	189 knots (351 km/h; 218 mph)
Max cruising speed at 2,130 m (7,000 ft)	174 knots (322 km/h; 200 mph)
Stalling speed, flaps down	54 knots (100 km/h; 62 mph)
Max rate of climb at S/L	670 m (2,200 ft)/min

KAHN

DAVID KAHN

ADDRESS:
18760 Frankfort Street, Northridge, California 91324
Telephone: (213) 887-0880

Mr Kahn and Mr Earl Allen are co-owners of a high-performance single-seat light aircraft named Piranha, which was conceived originally as a counter-insurgency combat aircraft, carrying bombs and rockets, for use by the USAF in Vietnam.

The idea of a low-cost military aircraft of this type was put forward in 1962 by the late Milt Blair and Dick Ennis. American Electric became interested in the project; the total cost of subsequent research and development, and of producing a single prototype for presentation to the USAF for evaluation, was more than $250,000. Flight testing at Eglin AFB, Florida, was abandoned following Blair's death in another aircraft. The Piranha was then stored in a hangar for seven years, before being acquired and refurbished by David Kahn and Earl Allen. The original camouflage finish was changed for more sporting trim, and an electrical system was added.

PIRANHA

TYPE: Single-seat high-performance sporting monoplane.

WINGS: Cantilever low-wing monoplane, of the kind fitted to the LeVier Cosmic Wind racing aircraft. Laminar-flow section. Conventional all-aluminium flush-riveted structure and covering. Long-span ailerons with corrugated aluminium skin. No flaps.

FUSELAGE: Conventional flush-riveted aluminium semi-monocoque structure, of basic circular cross-section. Hand-formed aluminium cowling and fairings.

TAIL UNIT: Conventional cantilever structure of aluminium, with tailplane dihedral. Tabs on elevator and rudder.

LANDING GEAR: Non-retractable tailwheel type. Faired cantilever main legs and wheels. Tailwheel steerable with rudder. Brakes on main wheels.

POWER PLANT: One 134 kW (180 hp) Lycoming flat-four engine, with fuel injection, driving a special two-blade constant-speed propeller. Fuel tank in fuselage, capacity 76 litres (20 US gallons); integral tankage in entire leading-edge of each wing. Total fuel capacity 197 litres (52 US gallons).

ACCOMMODATION: Single seat under moulded bubble canopy. All controls fully balanced.

SYSTEM: Electrical system includes alternator, battery and voltage regulator.

ELECTRONICS: Narco Escort 110 radio.

DIMENSIONS, EXTERNAL:

Wing span	6·10 m (20 ft 0 in)
Length overall	5·49 m (18 ft 0 in)

WEIGHTS:

Weight empty	401 kg (884 lb)
Max T-O weight	907 kg (2,000 lb)
Max landing weight	726 kg (1,600 lb)

PERFORMANCE:

Never-exceed speed	295 knots (547 km/h; 340 mph) IAS
Max cruising speed	over 208 knots (386 km/h; 240 mph)
Stalling speed, full tanks	78 knots (145 km/h; 90 mph) IAS
Max rate of climb at S/L	915 m (3,000 ft)/min
Range with max fuel	1,390 nm (2,575 km; 1,600 miles)

KELEHER

JAMES J. KELEHER

ADDRESS:
4321 Ogden Drive, Fremont, California 94538

In the early 1960s Mr J. Keleher designed and built a mid-wing sporting monoplane which he called the Lark. The design was revised in 1963, and the current model, for which plans are available to amateur constructors, is designated Lark-1B.

The accompanying illustration shows the Lark-1B built by Mr Parker Warren of Pompano Beach, Florida. Constructed over a five-year period at a cost of approximately $4,000, Mr Warren's Lark differs from standard Keleher plans by having a modified tail planform. Powered by a 74·5 kW (100 hp) Continental O-200 engine, it has an empty weight of 347 kg (765 lb) and max T-O weight of 476 kg (1,050 lb). Max level speed is 130 knots (241 km/h; 150 mph), cruising speed 109 knots (201 km/h; 125 mph), and landing speed 61 knots (113 km/h; 70 mph). With a sea level rate of climb of 305 m (1,000 ft)/min, service ceiling is over 3,870 m (12,700 ft).

KELEHER LARK-1B

The following description applies to the Lark-1B with a 48·5 kW (65 hp) Continental engine. Examples are flying with 56 kW (75 hp) A75-8 and 74·5 kW (100 hp) O-200 Continental engines.

Jeffair Barracuda two-seat high-performance monoplane

Keleher Lark-1B built by Mr Parker Warren *(Howard Levy)*

Piranha as a COIN combat aircraft with underwing ordnance

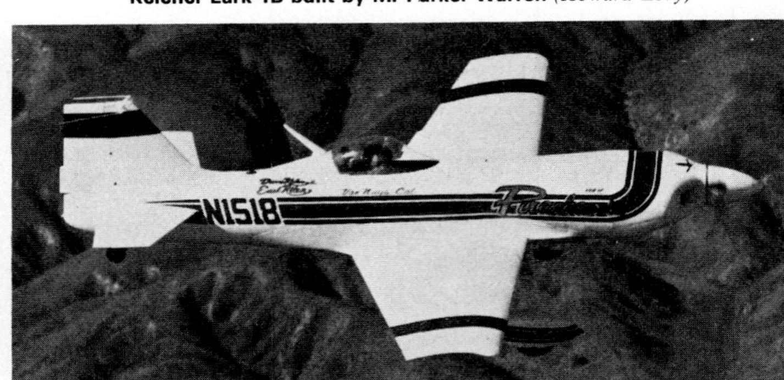

Piranha in its current form, as a sporting aircraft

Kraft Super Fli aerobatic monoplane *(Howard Levy)*

Lane Flycycle, a replica of the French DePischoff lightplane of 1921 *(Howard Levy)*

TYPE: Single-seat amateur-built sporting monoplane.

WINGS: Braced mid-wing monoplane with streamline-section Vee bracing struts each side. Wing section NACA 2R₁12. No dihedral. Incidence 4° at root, 2° 30′ at tip. All-wood structure of Sitka spruce, with built-up I beam front spar and ribs, fabric-covered. Stressed to 6g plus. Fabric-covered wooden ailerons; no trim tabs or flaps.

FUSELAGE: Welded steel tube structure, fabric-covered, stressed to 6g plus.

TAIL UNIT: Wire-braced welded steel tube structure with sheet steel ribs and fabric covering. Adjustable-incidence tailplane. Swept fin. No trim tabs.

LANDING GEAR: Non-retractable tailwheel type. Divided main landing gear with shock-absorption by rubber cord in fuselage. Cleveland main wheels and tyres size 5·00-5, pressure 1·38-1·72 bars (20-25 lb/sq in). Cleveland disc brakes. Wheel fairings optional.

POWER PLANT: Provision for alternative flat-four engines of 48·5-74·5 kW (65-100 hp), driving a two-blade metal fixed-pitch propeller. One galvanised steel fuel tank in the fuselage, aft of the firewall, capacity 56 litres (15 US gallons). Refuelling point on top of cowl, forward of windscreen. Oil capacity 5·7 litres (1·5 US gallons).

ACCOMMODATION: Single seat in enclosed cockpit under sliding canopy. Lowered turtledeck and bubble canopy optional. Stowage for 9 kg (20 lb) baggage aft of seat.

DIMENSIONS, EXTERNAL:

Wing span	7·01 m (23 ft 0 in)
Wing chord, constant	1·22 m (4 ft 0 in)
Wing area, gross	7·48 m² (80·5 sq ft)
Wing aspect ratio	5·75
Length overall	5·18 m (17 ft 0 in)
Height overall	1·65 m (5 ft 5 in)
Tailplane span	2·03 m (6 ft 8 in)
Wheel track	1·57 m (5 ft 2 in)
Propeller diameter:	
A65-8	1·70 m (5 ft 7 in
A75-8	1·66 m (5 ft 5½ in

WEIGHTS AND LOADINGS (48·5 kW: 65 hp engine):

Weight empty	249 kg (550 lb
Max T-O and landing weight	387 kg (855 lb
Max wing loading	51·8 kg/m² (10·61 lb/sq ft
Max power loading	7·98 kg/kW (13·15 lb/hp

PERFORMANCE (with 48·5 kW; 65 hp engine, at max T-O weight):

Never-exceed speed	160 knots (297·5 km/h; 185 mph
Max level speed	115 knots (212 km/h; 132 mph
Max cruising speed	103 knots (192 km/h; 119 mph
Stalling speed	48 knots (89 km/h; 55 mph
Max rate of climb at S/L	274 m (900 ft)/min
Service ceiling	5,950 m (19,500 ft
T-O run	183 m (600 ft
Range with max payload, with reserve	303 nm (563 km; 350 miles

KRAFT

PHIL KRAFT

ADDRESS:
Kraft Systems, Vista, California

KRAFT K-1 SUPER FLI

Designed and built by Phil Kraft, with help from Paul White, the Super Fli was produced to model aircraft standards in terms of wing design, areas and moments, as Phil Kraft is a world champion model aircraft builder. After 1½ years of work, the Super Fli was first flown in December 1974 and had accumulated 95 flying hours by the time it was taken to the EAA Fly-in at Oshkosh in 1975 to evaluate the potential of producing plans and kits for amateur construction. Aerobatic pilots at Oshkosh who flew the aircraft found it most satisfactory.

TYPE: Single-seat aerobatic monoplane.

WINGS: Cantilever low-wing monoplane. Spruce spars, plywood ribs and plywood covering. Horn-balanced ailerons.

FUSELAGE: Oval section fuselage of steel tube construction, aluminium covered.

TAIL UNIT: Wire-braced steel tube structure, fabric covered.

LANDING GEAR: Non-retractable tailwheel type. Faired main legs and faired wheels. Steerable tailwheel.

POWER PLANT: One 149 kW (200 hp) Lycoming IO-360-A1D flat-four engine.

ACCOMMODATION: Single seat under transparent canopy hinged on starboard side.

DIMENSION, EXTERNAL:

Wing span	7·32 m (24 ft 0 in)

WEIGHTS:

Weight empty	445 kg (980 lb)
Max T-O weight	635 kg (1,400 lb)

PERFORMANCE:

Max level speed	174 knots (322 km/h; 200 mph)
Cruising speed	143 knots (265 km/h; 165 mph)
Landing speed	35 knots (65 km/h; 40 mph)
Max rate of climb at S/L	914 m (3,000 ft)/min
Service ceiling	3,660 m (12,000 ft)
Range with max fuel	260 nm (483 km; 300 miles)

LANE

DICK LANE

ADDRESS:
Fulton, New York

LANE FLYCYCLE

Dick Lane, an industrial arts teacher, and students from the Auburn High School have constructed this aircraft, which is described as a replica of a French DePischoff lightplane of 1921. Construction took four and a ha[lf] school terms at a cost of $350. First flight was expected [to] take place in mid-1976. If the aircraft is successful, pla[ns] will be offered for sale through the Barney Oldfield Ai[r] craft Company (which see).

TYPE: Single-seat homebuilt biplane.

WINGS: Braced biplane wings of wooden construction, fabric covered. Upper wings mounted at shoulder level and lower at wheel-axle level. Ailerons fitted to upper wings. Slight dihedral on lower wing only. I-type interplane struts.

FUSELAGE: Wire-braced uncovered Vee-section structure of steel tube, tapering towards tail.

TAIL UNIT: Wire-braced tailplane and elevators mounted above rear fuselage. Triangular fin and narrow-chord rudder. Wooden structure, fabric covered.

LANDING GEAR: Non-retractable tailwheel type. Two spoked main wheels positioned so as to allow half of each to be inset into lower wing leading-edge.

POWER PLANT: One Lane-modified Volkswagen motor-car engine, driving a two-blade propeller. Engine enclosed in a cowling which forms a fairing forward of pilot.

ACCOMMODATION: Single seat in a fully-exposed position. Small windscreen attached to rear of engine cowling.

DIMENSIONS, EXTERNAL:

Wing span	4·11 m (13 ft 6 in)
Wing chord, constant	0·76 m (2 ft 6 in)
Length overall	3·81 m (12 ft 6 in)
Height overall	1·07 m (3 ft 6 in)

WEIGHTS:

Weight empty	145 kg (320 lb)
Max T-O weight	250 kg (550 lb)

PERFORMANCE (estimated):

Max level speed	65 knots (121 km/h; 75 mph)
Cruising speed	61 knots (113 km/h; 70 mph)
Landing speed	35 knots (64 km/h; 40 mph)

LARKIN

LARKIN AIRCRAFT CORPORATION

ADDRESS:
PO Box 66899, Scotts Valley, California 95066
Telephone: (408) 475-1234

PRESIDENT: Keith Larkin

Larkin Aircraft Corporation designed and built the prototype of a two-seat light aircraft known as the Model KC-3 Skylark. Design began in 1961 and prototype construction started in 1970, with the first flight on 8 June 1972. Emphasis was placed on producing a robust structure suitable for homebuilders, the primary skill necessary to build a Skylark being that of sheet metal riveting.

A feature of the aircraft's basic structure is use of a square tubular aluminium keel, which carries the loads from the landing gear and the fuselage weight. At its aft end is a vertical mast of similar section, with the engine mounted at the top. Forward of the mast, and attached to it and the keel, is a triangular support structure to carry the wings.

Plans and material kits are available to amateur constructors, who may also order from the company any part or parts which they consider beyond their ability to fabricate. By March 1976 one further Skylark had been constructed, in addition to the prototype.

LARKIN KC-3 SKYLARK

TYPE: Two-seat homebuilt light aircraft.

WINGS: Braced shoulder-wing monoplane, with single bracing strut each side. Wing section NACA 4412. Light alloy two-spar structure, with metal ribs and skins. Cambered wingtips of glassfibre. Ailerons of aluminium construction. No flaps or tabs.

FUSELAGE: Light alloy square-section tubular keel, to which glassfibre lower and upper fairings are attached. Optional Vee-shaped stepped lower hull of glassfibre for amphibious operations.

TAIL UNIT: Twin-boom structure of light alloy, attached to wings, with twin fins and rudders. Large trim tab in centre of elevator trailing-edge.

LANDING GEAR: Non-retractable tricycle type. Shock-absorption by low-pressure tyres. Nosewheel steerable. Goodyear main wheels with tyres size 16 × 8·00-6, pressure 1·38 bars (20 lb/sq in). Nosewheel tyre size 14 × 6·00-5, pressure 1·38 bars (20 lb/sq in). Goodyear disc brakes.

POWER PLANT: Prototype has one 74·5 kW (100 hp) 1,600 cc Volkswagen modified motor car engine, driving a Fahlin two-blade fixed-pitch wooden pusher propeller via Larkin seven-belt V-drive reduction gearing. Fuel tank aft of pilot's seat, capacity 64·35 litres (17 US gallons). Refuelling point on tank's upper surface. Oil capacity 1·9 litres (0·5 US gallons).

ACCOMMODATION: Two seats, side by side, in enclosed cockpit. Dual controls standard. Port side of transparent canopy is hinged on centreline to provide access. Cockpit heating optional.

SYSTEMS: Electrical system standard, comprising engine-driven generator and storage battery for engine starting.

EQUIPMENT AND ELECTRONICS: Navigation lights, radio and additional electronics optional.

DIMENSIONS, EXTERNAL:

Wing span	8·08 m (26 ft 6 in)
Wing chord (constant)	1·12 m (3 ft 8 in)
Wing area, gross	10·59 m² (114 sq ft)
Wing aspect ratio	7·0
Length overall	5·94 m (19 ft 6 in)
Height overall	1·89 m (6 ft 2½ in)
Tailplane span	2·44 m (8 ft 0 in)
Wheel track	1·57 m (5 ft 2 in)
Wheelbase	1·55 m (5 ft 1 in)
Propeller diameter	1·83 m (6 ft 0 in)

DIMENSIONS, INTERNAL:

Cabin: Max width	1·04 m (3 ft 5 in)
Baggage capacity	45 kg (100 lb)

WEIGHTS AND LOADINGS:

Weight empty	358 kg (790 lb)
Max T-O weight	565 kg (1,246 lb)
Max wing loading	53·2 kg/m² (10·9 lb/sq ft)
Max power loading	7·58 kg/kW (12·4 lb/hp)

PERFORMANCE:

Never-exceed speed	156 knots (289 km/h; 180 mph)
Max level speed	100 knots (185 km/h; 115 mph)
Max cruising speed	91 knots (169 km/h; 105 mph)
Econ cruising speed	87 knots (161 km/h; 100 mph)
Stalling speed	37 knots (68 km/h; 42 mph)
Max rate of climb at S/L	168 m (550 ft)/min
Service ceiling (estimated)	3,660 m (12,000 ft)
T-O run	183 m (600 ft)
T-O to 15 m (50 ft)	225 m (740 ft)
Landing from 15 m (50 ft)	168 m (550 ft)
Landing run	122 m (400 ft)
Range with max fuel and max payload	456 nm (845 km; 525 miles)

McCARLEY

CHARLES E. McCARLEY

ADDRESS:
Hueytown, Alabama

McCARLEY MINI-MAC

Designed by Charles E. McCarley, the prototype Mini-Mac took 1,600 working hours to build, at a cost of $1,800, and first flew in July 1970. It is capable of limited aerobatics, and recent additions to the aircraft include a radio, electrical system and cockpit canopy.

A modified version of the Mini-Mac (N75GH), built by George M. Harrison, was completed on 4 July 1974 and had accumulated 122 flying hours by Summer 1975. As well as having a shortened wing span, N75GH has 10° drooping ailerons.

Plans for the Mini-Mac are available to amateur constructors.

The following description generally applies to both aircraft mentioned above, except where indicated:

TYPE: Single-seat limited aerobatic monoplane.

WINGS: Cantilever low-wing monoplane. Clark Y wing section. All-metal construction. 10° drooping ailerons fitted to N75GH.

FUSELAGE: Conventional all-metal structure.

TAIL UNIT: Conventional cantilever all-metal structure. Tab in elevators.

LANDING GEAR: Non-retractable tricycle type. Wheels enclosed in streamline fairings. Small tail bumper skid.

POWER PLANT: One 1,834 cc (N152CM prototype) or 1,600 cc (N75GH) modified Volkswagen motor-car engine, driving a two-blade propeller. Conventional fuel and oil systems, not modified for inverted flight.

ACCOMMODATION: Single seat in open or enclosed cockpit.

DIMENSIONS, EXTERNAL (A: prototype; B: N75GH):

Wing span:	
A	6·25 m (20 ft 6 in)
B	5·79 m (19 ft 0 in)
Wing chord:	
A	1·02 m (3 ft 4 in)
B	1·04 m (3 ft 5 in)
Length overall:	
A	4·47 m (14 ft 8 in)
B	4·39 m (14 ft 5 in)
Height overall:	
A and B	1·73 m (5 ft 8 in)

WEIGHTS:

Weight empty:	
B	208 kg (458 lb)
Max T-O weight:	
B	318 kg (700 lb)

PERFORMANCE:

Max level speed:	
A and B	139 knots (257 km/h; 160 mph)
Cruising speed:	
A	109 knots (201 km/h; 125 mph)
B	104 knots (193 km/h; 120 mph)
Landing speed:	
A	44 knots (81 km/h; 50 mph)
B	48 knots (89 km/h; 55 mph)
Max rate of climb at S/L:	
A	305 m (1,000 ft)/min
B	213 m (700 ft)/min
Service ceiling:	
A and B	2,440 m (8,000 ft)
T-O run:	
A	152 m (500 ft)
B	183 m (600 ft)
Landing run:	
A and B	152 m (500 ft)
Range with max fuel:	
A	434 nm (804 km; 500 miles)
B	260 nm (483 km; 300 miles)

MacDONALD

MacDONALD AIRCRAFT COMPANY

ADDRESS:
1282 Fowler Creek Road, Sonoma, California 95476
Telephone: (707) 996-7897

PROPRIETOR: Robert A. MacDonald

Mr MacDonald, an aircraft engineer, has designed and built a single-seat lightweight sporting aircraft of which plans are available to amateur constructors. His aim was to evolve a design that would be simple to build, easy to fly and economical in operation. It is of all-metal construction, and fabrication is simplified by extensive use of pop rivets.

Design and construction of the aircraft, which is designated MacDonald S-20, began simultaneously in March 1969. First flight was made on 9 March 1972. Since then the S-20 has been flown from Sonoma, California, to the EAA Fly-in at Oshkosh, Wisconsin, a distance of 1,606 nm (2,977 km; 1,850 miles).

The designation S-20 applies to the prototype, which is described. Aircraft built to Mr MacDonald's plans have the designation **S-21**.

Sales of plans totalled around 100 sets by early 1976, and the first aircraft to be built from the plans was expected to make its first flight during that year. Other aircraft were in an advanced stage of construction.

MacDONALD S-20

TYPE: Single-seat homebuilt lightweight sporting aircraft.

WINGS: Cantilever low-wing monoplane. Wing section NACA 747A315. Small centre-section without dihedral. Dihedral on outer wing panels 5°. Incidence 1°. No sweepback. Constant-chord light alloy structure, with main and auxiliary spars and skins of 2024-T3 aluminium. Modified Frise-type ailerons of 2024-T3 light alloy. No flaps. No trim tabs.

FUSELAGE: Forward fuselage of welded steel tube truss construction. Aft fuselage has light alloy bulkheads and longerons. Skin is of 2024-T3 light alloy.

TAIL UNIT: Cantilever light alloy structure of 2024-T3 aluminium, with two spars in both fin and tailplane. Fixed-incidence tailplane. Combination trim and anti-servo tab on starboard elevator. Ground-adjustable tab on rudder (integral with port-side skin on S-21).

LANDING GEAR: Non-retractable tailwheel type. Shock-absorption of main-gear struts by neoprene discs in compression. Go-kart wheels with tyres size 4·00-5. Mechanical caliper wheel brakes.

POWER PLANT: One 1,500 cc Volkswagen modified motor car engine, developing 39·5 kW (53 hp) and driving a Hegy two-blade wooden fixed-pitch propeller. Aluminium fuel tank in fuselage, immediately aft of firewall, capacity 37 litres (9·8 US gallons). Refuelling point on fuselage upper surface, forward of windscreen. Oil capacity 1·27 litres (0·34 US gallons).

ACCOMMODATION: Single seat for pilot in open cockpit. Space for baggage aft of seat, nominally 9 kg (20 lb) capacity, but prototype has storage battery and radio weighing 5 kg (11 lb) stowed in this area.

ELECTRONICS: Battery-powered Genave Alpha 100 com transceiver in prototype.

DIMENSIONS, EXTERNAL:

Wing span	7·62 m (25 ft 0 in)
Wing chord, constant	1·14 m (3 ft 9 in)
Wing area, gross	8·73 m² (94 sq ft)
Wing aspect ratio	6·65
Length overall	5·64 m (18 ft 6 in)
Height overall	1·60 m (5 ft 3 in)
Tailplane span	2·13 m (7 ft 0 in)
Wheel track	1·57 m (5 ft 2 in)
Wheelbase	4·70 m (15 ft 5 in)

Prototype Larkin KC-3 Skylark (1,600 cc Volkswagen engine)

Merkel Mark II aerobatic biplane (Franklin 6A-350-C1 engine)

Modified Mini-Mac N75GH in foreground, with the prototype N152GM in background *(Howard Levy)*

MacDonald S-20 lightweight single-seat sporting aircraft

Propeller diameter	1·35 m (4 ft 5 in)
Propeller ground clearance	0·31 m (1 ft 0 in)
WEIGHTS AND LOADINGS:	
Weight empty	206 kg (456 lb)
Design T-O weight	326 kg (720 lb)
Design wing loading	37·4 kg/m² (7·66 lb/sq ft)
Design power loading	8·25 kg/kW (13·5 lb/hp)
PERFORMANCE (at T-O weight of 313 kg; 690 lb):	
Never-exceed speed	140 knots (260 km/h; 162 mph)
Max level speed, from S/L to 610 m (2,000 ft)	96 knots (177 km/h; 110 mph)
Max cruising speed, from S/L to 610 m (2,000 ft)	78 knots (145 km/h; 90 mph)
Econ cruising speed, from S/L to 610 m (2,000 ft)	69·5 knots (129 km/h; 80 mph)
Stalling speed	33 knots (61·5 km/h; 38 mph)
Max rate of climb at S/L	259 m (850 ft)/min
T-O run	91 m (300 ft)
Landing run	approx 91 m (300 ft)
Range with max fuel, no reserve	218 nm (404 km; 251 miles)

MERKEL
MERKEL AIRPLANE COMPANY

ADDRESS:
3920 North Charles Street, Wichita, Kansas 67204
Telephone: (316) 838-6767

Mr Edwin W. Merkel, a professional aeronautical engineer, has designed, built and developed over a ten-year period an aerobatic biplane which he has named the Merkel Mark II.

The aircraft evolved from a Master's Thesis on lightplanes and has a background of engineering development that includes wind tunnel testing of a one-fifth scale model. It is designed to FAA Part 23 aerobatic requirements, and its flight characteristics were developed to meet the level 1 of military specification MIL-F-8785B. It is stressed to +6 and −3*g* limit and +9 and −4·5*g* ultimate.

Mr Merkel began the design of the Mark II in 1963, and construction began three years later, in February 1966. First flight was made on 11 April 1973 and FAA certification in the Experimental category was awarded. It is Mr Merkel's intention to gain Type Certification under FAR 23 Aerobatic category and begin production of the aircraft if adequate financing can be arranged. If this aim is realised it is planned to produce a single-seat version for competitive aerobatics, and a two-seat version to serve as an aerobatic trainer.

MERKEL MARK II

TYPE: Two-seat aerobatic biplane.

WINGS: Braced single-bay biplane with I-type interplane and N-type centre-section struts. Streamline flying and landing wires. Wing section NACA 23012. Dihedral 3° 30′ on lower wings only. Incidence (both wings) 1° 30′. Sweepback at quarter-chord 12° on upper wing only. Wing panels are single-spar two-cell torsional structures with conventional biplane wire-braced truss. Wing skins of 2024-T3 light alloy. Plain ailerons on both wings of similar construction. No flaps.

FUSELAGE: Forward fuselage is a welded structure of 4130-N steel tube to rear side of cockpit. Semi-monocoque light alloy structure aft of cockpit with 2024-T3 light alloy skin. Fuselage-mounted airbrake planned for single-seat competition version.

TAIL UNIT: Cantilever light alloy structure with 2024-T3 skins on all surfaces. All-moving tailplane with trimmable anti-servo tab. Ground-adjustable trim tab in rudder.

LANDING GEAR: Non-retractable tailwheel type. Main wheels carried on cantilever leaf spring struts. Main wheels have US Rubber tyres type 5935-056-AD3, size 17 × 6-6, pressure 1·65 bars (24 lb/sq in). Maule tailwheel type SFS 1-4 with 150 mm (6 in) diameter solid rubber tyre. Cleveland hydraulic disc brakes.

POWER PLANT: One 164 kW (220 hp) Franklin 6A-350-C1 flat-six engine, driving a Hartzell type HC-C2YF-4/C8459-4 two-blade metal constant-speed propeller. Light alloy fuel tank immediately aft of firewall, capacity 68 litres (18 US gallons). Refuelling point on upper surface of forward fuselage. An optional belly tank is planned for extended cross-country flights. Oil capacity 8·3 litres (2·2 US gallons).

ACCOMMODATION: Two seats in tandem in open cockpits. Canopy optional. Cockpits heated.

SYSTEM: Electrical power supplied by Prestolite F363 12V engine-driven alternator.

DIMENSIONS, EXTERNAL:	
Wing span (upper)	7·77 m (25 ft 6 in)
Wing span (lower)	7·47 m (24 ft 6 in)
Wing chord (constant, both)	0·91 m (3 ft 0 in)
Wing area, gross	13·56 m² (146 sq ft)
Wing aspect ratio (biplane cell)	5·79
Length overall	6·95 m (22 ft 9½ in)
Height overall	2·47 m (8 ft 1¼ in)
Tailplane span	2·90 m (9 ft 6 in)
Wheel track	1·98 m (6 ft 6 in)
Propeller diameter	2·03 m (6 ft 8 in)
WEIGHTS AND LOADINGS (single seat occupied in accordance with FAR 23 Aerobatic category):	
Weight empty	544 kg (1,200 lb)
Max T-O and landing weight	698 kg (1,540 lb)
Max wing loading	51·5 kg/m² (10·55 lb/sq ft)
Max power loading	4·26 kg/kW (7·0 lb/hp)
PERFORMANCE (at max Aerobatic T-O weight, ISA):	
Never-exceed speed	178 knots (331 km/h; 206 mph) IAS
Max level speed at S/L	142 knots (262 km/h; 163 mph) IAS
Max cruising speed, 75% power at 2,135 m (7,000 ft)	139 knots (257 km/h; 160 mph) TAS
Econ cruising speed, 55% power above 2,135 m (7,000 ft)	109 knots (201 km/h; 125 mph) IAS
Stalling speed	48 knots (89 km/h; 55 mph) IAS
Max rate of climb at S/L	762 m (2,500 ft)/min
T-O run	91 m (300 ft)
T-O to 15 m (50 ft)	124 m (406 ft)
Range at econ cruising speed with max fuel, no reserve	221 nm (410 km; 255 miles)

MILLER
J. W. MILLER AVIATION INC

ADDRESS:
Horseshoe Bay Airport, Route 3, PO Box 757, Marble Falls, Texas 78654

To evaluate the potential of a somewhat radical four/six-seat light aircraft, Mr Jim Miller of J. W. Miller Aviation Inc decided to build for research purposes a half-scale single-seat version. When flight tested it was soon apparent that the aircraft was easy to fly, and its performance and eye-catching appearance resulted in the company receiving a large number of enquiries as to the availability of plans and/or kits. The demand was such that Jim Miller has decided to market kits of components suitable for assembly by advanced amateur constructors.

Known as the Miller JM-2, the aircraft in its original form had a composite structure of metal and glassfibre. Unusual features of the configuration included a needle nose, monoplane mid-wing mounted at the aft end of the fuselage, and a rear fuselage-mounted engine which drove a shrouded pusher propeller. There was no conventional tail unit. Instead, two opposite vertical sections of the propeller shroud served as rudders. The extension of an upper arc of the shroud carried in an elevator, above which was mounted a swept fin.

Ideas under consideration for inclusion in production kits include leading-edge slats and trailing-edge flaps to reduce landing speed, larger-capacity fuel tank, retractable landing gear, and an adjustable seat. A tandem two-seat version is reported to be at the design stage.

The prototype, *Texas Gem,* has since been modified with a T-tail for Formula One racing. Although retaining the same power plant, the propeller is no longer shrouded and other modifications include a less pointed nose and smaller wheel fairings.

The description below applies to the JM-2 in its original form:

MILLER JM-2

TYPE: Single-seat lightweight sporting aircraft.

WINGS: Cantilever mid-wing monoplane mounted at the aft end of the fuselage. Wing centre-section is an integral part of the fuselage structure, extending to just outboard of the propeller shroud. Wing section NACA 6400 modified. Wing outer panels have two light alloy I-section spars and a root and tip rib of the same material. Wing skins are of honeycomb construction, with glassfibre and epoxy resin reinforcement. Ailerons of similar construction.

FUSELAGE: Welded steel tube structure, of which the wing centre-section is an integral part. Structure is covered by four moulded glassfibre panels, which have bonded joints at the vertical and horizontal centrelines.

TAIL UNIT: Fixed fin extending from rear fuselage and terminating above shroud-mounted elevator. Twin rudders are formed by movable vertical sections of the propeller shroud. Elevator, conforming to circular section of the shroud, carried by an extension of the upper arc of the shroud.

LANDING GEAR: Non-retractable tricycle type. Main units are of light alloy with glassfibre wheel fairings. Shock-absorber fitted in the steerable nosewheel unit is from a Beech Bonanza. Main-wheel tyres size 5·00-5. Nosewheel tyre size 2·50-4. Brakes on main wheels.

POWER PLANT: One 74·5 kW (100 hp) Continental O-200-B flat-four engine, driving a four-blade wooden fixed-pitch shrouded pusher propeller with spinner. Fuel tank aft of pilot's seat, capacity 47·3 litres (12·5 US gallons).

ACCOMMODATION: Pilot only, beneath transparent canopy.

SYSTEM: Electrical system powered by engine-driven generator and storage battery.

DIMENSIONS, EXTERNAL:

Wing span	4·57 m (15 ft 0 in)
Wing chord at root	1·83 m (6 ft 0 in)
Wing chord at tip	0·91 m (3 ft 0 in)
Wing area, gross	6·13 m² (66 sq ft)
Length overall	5·79 m (19 ft 0 in)
Height overall	1·73 m (5 ft 8 in)
Propeller shroud diameter	1·02 m (3 ft 4 in)
Propeller shroud chord	0·56 m (1 ft 10 in)
Propeller diameter	0·99 m (3 ft 3 in)

WEIGHTS AND LOADINGS:

Weight empty	286 kg (630 lb)
Max T-O weight	498 kg (1,100 lb)
Max wing loading	81·5 kg/m² (16·7 lb/sq ft)
Max power loading	6·68 kg /kW (11·0 lb/hp)

PERFORMANCE (at max T-O weight):

Max level speed	204 knots (378 km/h; 235 mph)
Cruising speed	165 knots (306 km/h; 190 mph)
Stalling speed, power off	65 knots (119 km/h; 74 mph) IAS
Stalling speed, power on	61 knots (113 km/h; 70 mph) IAS
Max rate of climb at S/L	488 m (1,600 ft)/min
T-O run	457 m (1,500 ft)
Landing run	more than 610 m (2,000 ft)
Range (75% power, no reserve)	260 nm (482 km; 300 miles)

MILLER

W. TERRY MILLER

ADDRESS:

Box 570, RR1, Furlong, Pennsylvania 18925

Mr Miller is the designer of the WM-2 sport aircraft, described below, and the Tern and Tern II sailplanes (which see). All of these aircraft were designed for amateur construction.

MILLER WM-2

The WM-2 is a low-powered, high-performance aircraft, conceived for the exploration of wave soaring conditions, thermal soaring with the engine stopped, and high-altitude, economical powered sport flying. The prototype (N24832) was built by Mr William Miller between 1969 and 1972, and made its first flight in August 1972. Flight testing was undertaken during 1973-74, and plans have been prepared for amateur construction of this aircraft.

TYPE: Single-seat sport aircraft.

WINGS: Cantilever low-wing monoplane. Modified NACA laminar-flow series wing sections. Thickness/chord ratio 15%. Dihedral 4°. Incidence 1°. Sweepback 0° 53′ at quarter-chord. Conventional structure of spruce spars, with birch plywood, glassfibre and fabric covering. Wooden ailerons. No flaps or tabs. Metal spoiler in each upper surface.

FUSELAGE: Conventional spruce structure, with birch plywood and glassfibre covering.

TAIL UNIT: Plywood- and fabric-covered spruce cantilever structure.

LANDING GEAR: Manually-retractable monowheel (wheel and tyre size 6·00-6) and tailskid. Hydraulic brake.

POWER PLANT: One 48·5 kW (65 hp) Continental flat-four engine, driving a two-blade fixed-pitch metal propeller. Fuel tank, capacity 37·8 litres (10 US gallons; 8·3 Imp gallons), aft of firewall.

ACCOMMODATION: Single seat under one-piece sideways-opening bubble canopy.

DIMENSIONS, EXTERNAL:

Wing span	12·19 m (40 ft 0 in)
Wing chord at root	1·37 m (4 ft 6 in)
Wing chord at tip	0·76 m (2 ft 6 in)
Length overall	6·10 m (20 ft 0 in)
Height over tail	1·60 m (5 ft 3 in)
Wing area, gross	13·38 m² (144·0 sq ft)
Wing aspect ratio	11·11
Tailplane span	2·44 m (8 ft 0 in)
Propeller diameter	1·88 m (6 ft 2 in)

WEIGHTS AND LOADING:

Weight empty, equipped	351 kg (775 lb)
Max T-O weight	476 kg (1,050 lb)
Max wing loading	35·62 kg/m² (7·3 lb/sq ft)

PERFORMANCE (at max T-O weight):

Never-exceed speed	130 knots (241 km/h; 150 mph)
Max level speed at S/L	118 knots (219 km/h; 136 mph)
Normal cruising speed at 3,050 m (10,000 ft), 50% power	109 knots (203 km/h; 126 mph)
Stalling speed	39·5 knots (72·5 km/h; 45 mph)
Max rate of climb at S/L	271 m (890 ft)/min
Service ceiling (computed)	7,315 m (24,000 ft)
Range at normal cruising speed at 3,050 m (10,000 ft) with 30 min reserves	291 nm (540 km; 336 miles)
Best glide ratio at 54 knots (100 km/h; 62 mph), power off	15

MINI-HAWK

MINI-HAWK INTERNATIONAL INC

ADDRESS:

1930 Stewart Street, Santa Monica, California 90404

Telephone: (213) 828-4078

DIRECTOR OF MARKETING:

E. Y. Treffinger

Mini-Hawk International was formed to market plans and kits for construction of an all-metal single-seat monoplane known as the Mini-Hawk TH.E.01 Tiger-Hawk. Three officers of the corporation, designer William B. Taylor, engineer Thomas E. Maloney and pilot E. Y. Treffinger, combined their efforts to design and construct the prototype. First flight of the prototype was made during 1974 and the test programme was scheduled for completion in 1976.

The Tiger-Hawk features simplified construction and has detachable wings for towing or storage. Removal of the wings takes only ten minutes.

Mini-Hawk offers amateur constructors a complete set of plans and a construction manual or a complete kit package with or without an engine.

MINI-HAWK TH.E.01 TIGER-HAWK

TYPE: Single-seat lightweight homebuilt aircraft.

WINGS: Cantilever low-wing monoplane of all-metal construction. Constant-chord wing. Dihedral 4° on outer panels. Full-span ailerons of all-metal construction, operated by push/pull rods. All-metal trailing-edge flaps.

FUSELAGE: Built-up structure of light alloy.

TAIL UNIT: Cantilever all-metal structure. Fixed-incidence tailplane. Elevator of light alloy construction. No trim tabs.

LANDING GEAR: Non-retractable tricycle type with steerable nosewheel. Hurst/Airheart hydraulic disc brakes. Single wheel with speed fairing on each unit.

POWER PLANT: One 53·5 kW (72 hp) Revmaster Model 1831D modified Volkswagen motor car engine, with dual ignition, driving an Eng. Duplicating 54/42 two-blade fixed-pitch propeller with spinner. Fuel tank in fuselage with capacity of 45·4 litres (12 US gallons). Suitable for Volkswagen engines of up to 67 kW (90 hp). Oil capacity 2·35 litres (5 US pints).

ACCOMMODATION: Single seat under transparent cockpit canopy.

SYSTEM: Hydraulic system for brakes only.

EQUIPMENT: Genave radio optional.

DIMENSIONS, EXTERNAL:

Wing span	5·49 m (18 ft 0 in)
Wing span for travelling	1·83 m (6 ft 0 in)
Wing chord, constant	0·99 m (3 ft 3 in)
Wing area, gross	5·30 m² (57 sq ft)
Length overall	4·04 m (13 ft 3 in)
Width overall, wings removed	1·83 m (6 ft 0 in)
Height overall	2·08 m (6 ft 10 in)
Tailplane span	1·83 m (6 ft 0 in)
Wheel track	1·78 m (5 ft 10 in)
Wheelbase	1·93 m (6 ft 4 in)
Propeller diameter	1·37 m (4 ft 6 in)

WEIGHTS AND LOADINGS:

Weight empty	238 kg (525 lb)
Max T-O weight	362 kg (800 lb)
Max wing loading	44·9 kg/m² (9·2 lb/sq ft)
Max power loading	6·74 kg/kW (10·6 lb/hp)

PERFORMANCE (estimated, at max T-O weight):

Never-exceed speed	173 knots (321 km/h; 200 mph)
Max level speed	152 knots (282 km/h; 175 mph)
Max cruising speed	139 knots (257 km/h; 160 mph)
Stalling speed, flaps up	54 knots (100 km/h; 62 mph)
Stalling speed, flaps down	44 knots (81 km/h; 50 mph)
Max rate of climb at S/L	275-305 m (900-1,000 ft)/min
Service ceiling	3,050 m (10,000 ft)
Absolute ceiling	3,660 m (12,000 ft)
T-O and landing run	122 m (400 ft)
Max range	608 nm (1,126 km; 700 miles)

MIT

MIT MANPOWERED AIRCRAFT ASSOCIATION INC

ADDRESS: Massachusetts Institute of Technology, 77 Massachusetts Avenue, Cambridge, Massachusetts 02139

Telephone: (617) 253-6159

The Department of Aeronautics and Astronautics of the Massachusetts Institute of Technology has designed, built and started testing a second man-powered aircraft, following the destruction of BURD I (Biplane Ultralightweight Research Device) during its flight test programme. Designated BURD II, its construction began in June 1976 and taxi tests started three months later.

MIT BURD II

TYPE: Ultra-lightweight man-powered aircraft.

WINGS: Braced biplane structure with 15° negative stagger. Wing section Wortmann FX-61-164. Thickness/chord ratio 16%. No dihedral. Incidence 5°. No sweepback. Composite structure with carbon fibre spars, balsa wood ribs, polystyrene and polypropylene skins. Spoilers at two-thirds span on lower wing for roll control, 80% height spoilers on wingtip endplates for yaw control and 20% split flaps on lower outboard wing panels. Controls retained in neutral setting by bungee balance. Fixed vertical surface mounted centrally above upper wing.

FUSELAGE: Primary structure of light alloy tube, with secondary structure of bamboo and balsa with polypropylene skins.

TAIL UNIT: All-moving cantilever canard horizontal surface of similar construction to wing. Aerofoil section Wortmann FX-61-124. Surface retained in neutral position by bungee balance cable.

LANDING GEAR: Two bicycle wheels mounted in tandem on aircraft's centreline. Aft wheel is spring-mounted for shock-absorption. Small castoring balancer wheel mounted at each wingtip.

POWER PLANT: Two-man crew provide, by pedalling, power via a chain drive to a two-blade fixed-pitch pusher propeller.

ACCOMMODATION: Pilot on forward bicycle-type seat, crewman in tandem on aft seat. Both members of crew on open seats between tubular fuselage members.

DIMENSIONS, EXTERNAL:

Wing span (both)	18·90 m (62 ft 0 in)
Wing chord at root (both)	1·67 m (5 ft 6 in)
Wing aspect ratio	11·8
Length overall	8·08 m (26 ft 6 in)
Height overall	4·06 m (13 ft 4 in)
Canard surface span	6·71 m (22 ft 0 in)
Wheelbase	2·44 m (8 ft 0 in)
Propeller diameter	3·05 m (10 ft 0 in)
Propeller ground clearance	0·61 m (2 ft 0 in)

AREAS:

Wings, gross	59·46 m² (640 sq ft)
Ailerons (total)	3·34 m² (36 sq ft)
Spoilers (total)	0·37 m² (4 sq ft)
Fin	1·67 m² (18 sq ft)
Canard surface	6·13 m² (66 sq ft)

WEIGHTS AND LOADING:

Weight empty	57·2 kg (126 lb)
Max T-O weight	200 kg (440 lb)
Max wing loading	3·37 kg/m² (0·69 lb/sq ft)

Prototype WM-2 sport aircraft built by Mr William Miller

Mini-Hawk TH.E.01 Tiger-Hawk, with current, revised canopy

Two views of the MIT Manpowered Aircraft Association's BURD II two-seat man-powered aircraft

Miller JM-2 modified for Formula One racing *(Henry Artof)*

Monnett Sonerai homebuilt Formula V racer in towing configuration

PERFORMANCE (estimated at max T-O weight at S/L):
- Max level speed 16·5 knots (30·5 km/h; 19 mph)
- Never-exceed speed 18 knots (33 km/h; 20·5 mph)
- Max cruising speed 15 knots (27·5 km/h; 17 mph)
- Stalling speed 12 knots (22 km/h; 13·5 mph)

MONNETT
MONNETT EXPERIMENTAL AIRCRAFT INC

ADDRESS:
955 Grace, Elgin, Illinois 60120

Mr John T. Monnett formed this company to market plans and certain components of an original-design Formula V racer. Known originally as the Monnett II Sonerai, this received the Best in Class Formula V Racer award at the EAA Fly-in at Oshkosh in 1971, as well as an award for its outstanding contribution to low-cost flying. Since that time Mr Monnett has designed a two-seat version of the Sonerai, with the result that the original single-seat model is now known simply as Sonerai, the two-seat model as Sonerai II.

MONNETT SONERAI

Mr Monnett began design of the Sonerai in September 1970, construction starting two months later. First flight was made in July 1971, with FAA certification in the Experimental category. Plans and certain components are available to amateur constructors, including glassfibre engine cowlings, clear or tinted Plexiglas cockpit canopy, main landing gear struts, formed aluminium ribs, tapered rod tail spring, fuel tanks, spar kits, instruments, injector carburettor and wheels and brakes.

Approximately 280 sets of plans are known to have been sold, and more than 100 Sonerais are under construction or flying.

TYPE: Single-seat Formula V homebuilt racing aircraft.

WINGS: Cantilever mid-wing monoplane. Wing section NACA 64212. No dihedral, incidence or sweepback. Conventional light alloy structure. Full-span light alloy ailerons. No flaps or tabs. Wings fold on each side of the fuselage to allow the aircraft to be towed tail-first.

FUSELAGE: Welded chrome-molybdenum steel tube structure with fabric covering. Glassfibre engine cowling.

TAIL UNIT: Cantilever structure of welded chrome-molybdenum steel tube with fabric covering. Tailplane incidence ground-adjustable. No trim tabs.

LANDING GEAR: Non-retractable tailwheel type. Cantilever spring main gear of light alloy. Main wheels and tyres size 5·00-5. Caliper type wheel-brakes. Glassfibre fairings on main wheels.

POWER PLANT: One 44·7 kW (60 hp) Volkswagen 1,600 cc modified motor car engine, driving a Hegy two-blade propeller with spinner. Fuel tank in fuselage, immediately aft of firewall, capacity 41·5 litres (11 US gallons). Refuelling point on fuselage upper surface forward of canopy. Oil capacity 2·82 litres (0·75 US gallons).

ACCOMMODATION: Single seat under jettisonable Plexiglas bubble canopy, hinged at the starboard side.

ELECTRONICS: Battery-powered 10-channel com transceiver.

DIMENSIONS, EXTERNAL:
- Wing span 5·08 m (16 ft 8 in)
- Wing chord, constant 1·37 m (4 ft 6 in)
- Wing area, gross 6·97 m² (75 sq ft)
- Length overall 5·08 m (16 ft 8 in)
- Height overall 1·52 m (5 ft 0 in)
- Tailplane span 1·98 m (6 ft 6 in)
- Wheel track 1·22 m (4 ft 0 in)
- Propeller diameter 1·27 m (4 ft 2 in)

WEIGHTS:
- Weight empty 199 kg (440 lb)
- Max T-O weight 317 kg (700 lb)

PERFORMANCE (at max T-O weight):
- Max level speed at S/L over 139 knots (257 km/h; 160 mph)
- Max cruising speed 130 knots (241 km/h; 150 mph)
- Econ cruising speed 109 knots (201 km/h; 125 mph)
- Stalling speed 40 knots (74 km/h; 46 mph)
- Landing run 183 m (600 ft)
- Range over 260 nm (482 km; 300 miles)

MONNETT SONERAI II

The success of the Sonerai encouraged Mr Monnett to begin the design and construction of a two-seat version in December 1972. Generally similar to the Sonerai, it differs by being slightly larger and by having a more powerful Volkswagen engine. It is stressed to $\pm 4{\cdot}4g$ in the Utility category and to $\pm 6g$ in a single-seat Aerobatic category. The prototype made its first flight in July 1973, and orders have been received for at least 225 sets of plans. Many components, complete kits for fuselage and wings, and materials, are also available to amateur constructors.

The description of the Sonerai applies also to Sonerai II, except as follows:

Type: Two-seat homebuilt high-performance sporting aircraft.

Wings: As for Sonerai, except span increased.

Fuselage: As for Sonerai, except length increased.

Tail Unit: As for Sonerai, except tailplane has fixed incidence and reduced span.

Power Plant: One 1,700 cc Volkswagen modified motor-car engine, developing 48·5-52·2 kW (65-70 hp), driving a two-blade wooden fixed-pitch propeller. Fuel capacity 37·8 litres (10 US gallons); oil as for Sonerai.

Accommodation: Two seats in tandem beneath transparent bubble canopy, hinged on starboard side.

Electronics: Prototype has battery-powered 10-channel com transceiver.

Dimensions, external: As Sonerai, except:

Wing span	5·69 m (18 ft 8 in)
Wing area, gross	7·80 m² (84 sq ft)
Length overall	5·74 m (18 ft 10 in)
Tailplane span	1·83 m (6 ft 0 in)
Propeller diameter	1·32-1·37 m (4 ft 4 in to 4 ft 6 in)

Dimension, internal:

Cockpit: Max width	0·61 m (2 ft 0 in)

Weights and Loading:

Weight empty	230 kg (506 lb)
Max T-O weight	419 kg (925 lb)
Max wing loading	53·7 kg/m² (11·0 lb/sq ft)

Performance (at max T-O weight):

Max level speed at S/L	approx 143 knots (266 km/h; 165 mph)
Max cruising speed at S/L	122 knots (225 km/h; 140 mph)
Econ cruising speed at S/L	113 knots (209 km/h; 130 mph)
Stalling speed	39 knots (73 km/h; 45 mph)
Max rate of climb at S/L	229 m (750 ft)/min
T-O run, ISA	274 m (900 ft)
Landing run	152 m (500 ft)
Range with max fuel and max payload, 30 min reserve	364 nm (676 km; 420 miles)

OSADCHY

PAUL OSADCHY

Address:
Irvington, New Jersey

OSADCHY WING CHARMER MODEL 1

Completed in late 1974, the Wing Charmer 1 is a single-seat man-powered aircraft of tractor monoplane configuration. Compared with other man-powered aircraft, it is a relatively heavy aircraft, having a fuselage and wing spar of welded chrome-molybdenum steel tube and wing ribs of spruce, although other components are made of Styrofoam. The entire structure is covered with polyurethane sheet. Maximum depth of the wing is 450 mm (17·7 in) at the root and 280 mm (11 in) at the tip; the outer panels have 5° dihedral and are fitted with conventional ailerons. The tail unit comprises a single angular fin and rudder and an all-moving tailplane. A single wheel, from a tricycle, is situated forward of the wing, from which a system similar to that of a child's pedal car provides drive to the two-blade propeller. There is a small tailwheel beneath the rear of the tailboom. Estimated power requirement is 1 hp for 10 seconds for take-off, and 0·33 hp for sustained flight.

Dimensions, external:

Wing span	17·98 m (59 ft 0 in)
Wing chord at root	1·52 m (5 ft 0 in)
Length overall	7·92 m (26 ft 0 in)
*Propeller diameter	2·39 m (7 ft 10 in)

**A 3·05 m (10 ft 0 in) diameter propeller may have been fitted since original completion*

Weights:

Weight empty	127 kg (280 lb)
T-O weight	195 kg (430 lb)

OSPREY

OSPREY AIRCRAFT

Address:
3741 El Ricon Way, Sacramento, California 95825
Telephone: (917) 483-3004

Osprey Aircraft was formed originally to market to amateur constructors plans of the Osprey I aircraft designed and built by Mr George Pereira. This was an unusual project for the homebuilder, being a flying-boat, intended for operation on and from enclosed waters rather than the open sea. The plans drawn up by Mr Pereira included drawings of a special trailer for carriage of the aircraft, which allowed the pilot to launch and recover the Osprey unassisted.

On 27 July 1971, following study of seven different aircraft, the US Navy purchased the prototype Osprey from Mr Pereira. Under the designation X-28A Air Skimmer, it was evaluated by the Naval Air Development Center, at the request of the Director of Navy Laboratories, to study the potential of a small single-seat seaplane for civil police patrol duties in Southeast Asia. Details of this aircraft can be found in the 1974-75 *Jane's*.

Mr Pereira has since completed the prototype of a two-seat amphibian version designated Osprey II.

PEREIRA GP3 OSPREY II

Design and construction of the Osprey II, a two-seat amphibian development of the Osprey I, began in January 1972.

Mr Pereira evolved an unusual form of hull construction for this aircraft. When the all-wood fuselage structure had been completed and controls installed, the undersurface was given a deep coating of polyurethane foam. This was then sculptured to the requisite hull form before being covered with several protective layers of glassfibre cloth bonded with resin. The resulting structure is light, but extremely strong, with good shock resisting characteristics.

First flight of the Osprey II from water was made in April 1973, the amphibian becoming airborne in less than 244 m (800 ft), with no tendency to porpoise at any speed. In later tests from land it was found that with the landing gear retracted and at a speed of about 104 knots (193 km/h; 120 mph), there was slight buffet aft of the cabin and the noise level was unacceptably high. Modifications carried out in early 1974 included lengthening of the cabin by 0·18 m (7 in), and installation of a Lycoming O-320 engine in place of the original Franklin Sport, in a new cowling. Testing was resumed and completed satisfactorily during 1974, since when the shape of the tail fin has been changed. Sets of plans are now available to amateur constructors.

Type: Two-seat lightweight homebuilt amphibian.

Wings: Cantilever mid-wing monoplane, of constant chord. Wing section NACA 23012. Dihedral 4° 30′. Incidence 5°. All-wood structure, with single box spar and auxiliary rear spar for aileron attachment. Forward of the main spar the wing is plywood-covered to form a rigid 'D' section. Aft of the spar the wing is fabric-covered. Conventional ailerons, 100% mass-balanced, will be fitted with a ground-adjustable tab if this proves desirable. No trailing-edge flaps. Wingtip stabilising floats of polyurethane foam covered with glassfibre.

Hull: All-wood structure of longerons and frames, covered with 3/32 in marine plywood. Hull undersurface contours formed from polyurethane foam, protected by several layers of glassfibre cloth bonded with resin.

Tail Unit: Cantilever all-wood structure, with swept vertical surfaces; tailplane mounted high on fin, which is integral with hull. Incidence of tailplane ground-adjustable. Controllable trim tab in starboard elevator. Water rudder, contained within the base of the aerodynamic rudder, is spring-loaded in the down position and retracted by cable.

Landing Gear: Retractable tricycle type, with single wheel on each unit. Main units retract inward into the wing roots, the wheel wells being covered by doors in the retracted position. Nosewheel retracts forward into the nosecone and is also enclosed by a door. Manual retraction system. Shock-absorption by coil springs. Cleveland main wheels and tyres size 5·00-5. Nosewheel, of industrial type with roller bearings, has a tyre of 10 in diameter. Cleveland hydraulic disc brakes.

Power Plant: One 112 kW (150 hp) Lycoming O-320 flat-four engine, mounted on a steel tube pylon structure which is bolted to the wing truss. Hendrickson 66 × 52 three-blade fixed-pitch wooden pusher propeller. One glassfibre fuel tank mounted beneath the main spar at the wing centre-section, usable capacity 98·4 litres (26 US gallons). Refuelling point on starboard side of hull, just aft of cabin.

Accommodation: Two seats side by side beneath transparent canopy, which is hinged at rear and swings upward. Dual controls standard; but toe-operated wheel brakes on starboard side only. Baggage compartment aft of seats, capacity 41 kg (90 lb).

Systems: Hydraulic system for brakes only. Electrical system powered by engine-driven generator.

Dimensions, external:

Wing span	7·92 m (26 ft 0 in)
Wing chord, constant	1·52 m (5 ft 0 in)
Wings area, gross	12·08 m² (130 sq ft)
Wing area, gross	12·08 m² (130 sq ft)
Length overall	6·25 m (20 ft 6 in)
Height overall (wheels down)	1·83 m (6 ft 0 in)
Tailplane span	2·44 m (8 ft 0 in)
Wheel track	2·59 m (8 ft 6 in)
Wheelbase	2·13 m (7 ft 0 in)
Propeller diameter	1·68 m (5 ft 6 in)

Weights and Loadings:

Weight empty	440 kg (970 lb)
Max T-O weight	707 kg (1,560 lb)
Max wing loading	58·6 kg/m² (12 lb/sq ft)
Max power loading	6·31 kg/kW (10·4 lb/hp)

Performance (at max T-O weight unless specified otherwise):

Never-exceed speed	130 knots (241 km/h; 150 mph)
Max cruising speed at 75% power	113 knots (209 km/h; 130 mph)
Econ cruising speed at 55% power	94 knots (175 km/h; 109 mph)
Stalling speed	53 knots (97 km/h; 60 mph)
Max rate of climb at S/L, with pilot only	365 m (1,200 ft)/min
Rate of climb at S/L	305 m (1,000 ft)/min
T-O run, land	122 m (400 ft)
T-O run, water	159 m (520 ft)

OWL

GEORGE A. OWL Jr

Address:
17700 S. Western Avenue, Apartment 195, Gardena, California 90248
Telephone: (213) 323-3385

Mr George Owl, a member of the preliminary design staff of Rockwell International, is also the designer of two Formula One racing aircraft known as the Owl Racers OR-70 and OR-71. The former was designed to order for Bernadine and Jim Stevenson, both racing pilots, and was built by them. The OR-71 was produced in co-operation with Mr Vince DeLuca, proprietor of Vin-Del Aircraft, who has built a prototype and is making plans available to amateur constructors.

Prototypes of a version of the OR-71 with a new wing were under construction in 1975. This wing, of all-wood construction and utilising a special laminar flow aerofoil section, is calculated to raise the maximum speed of the OR-71 by 5 knots (10 km/h; 6 mph). The designation of these aircraft will be OR-71B.

Stevenson, Bernadine and Jim

Address:
6850 Vineland Avenue, Apartment 7A, N. Hollywood, California 91605
Telephone: 766-5074

Bernadine and Jim Stevenson began construction of their Owl Racer, designated OR-70 *Fang,* in October 1970. Preliminary construction was completed in September 1971, and the first flight was made on the eighth of that month. FAA certification in the Experimental category was awarded on 11 September and ten days later *Fang* was flown to third place in the Reno National Races. Construction was finalised during 1972, since which time the aircraft has been placed in several national air races. Normal load factor of the OR-70 is ±6·67*g*; ultimate load factor is ±10·0*g*.

STEVENSON/OWL OR-70 FANG

Type: Single-seat Formula One racing aircraft.

Wings: Cantilever mid-wing monoplane. Owl laminar flow aerofoil section, tapered in chord. Thickness/chord ratio 13·7%. No dihedral. No incidence. All-wood structure. Laminated spruce one-piece main spar, mahogany plywood ribs and skins. Ailerons of spruce and plywood construction, mass-balanced at tip. No flaps. No trim tabs.

Fuselage: Welded steel tube structure, with light alloy fairings and Dacron covering. Glassfibre nose cowl incorporates an annular cooling inlet. Engine cowl has controllable air exit flap.

Tail Unit: Cantilever wooden structure, with spruce spars and mahogany plywood ribs and skins. Fixed-incidence tailplane. No trim tabs.

Landing Gear: Non-retractable tailwheel type. Cantilever spring steel main legs. Main wheels and tyres size 5·00-5, pressure 2·41 bars (35 lb/sq in). Cleveland hydraulic disc brakes. Light alloy fairings on main wheels. Tailwheel is a ball bearing castor.

Power Plant: One 74·5 kW (100 hp) Continental O-200 flat-four engine, driving an Anderson, McCauley or Sensenich two-blade fixed-pitch propeller. Fuel tank in fuselage, immediately aft of firewall, capacity 24·6 litres (6·5 US gallons). Auxiliary integral fuel tank in each wing leading-edge, with combined capacity of 22·7 litres

Monnett Sonerai II two-seat sporting aircraft *(Howard Levy)*

Osadchy Wing Charmer Model 1 man-powered aircraft *(Howard Levy)*

Pereira GP3 Osprey II two-seat amphibian, with newly-raised tail fin

Stevenson/Owl OR-70 Fang Formula One racing aircraft

Vin-Del/Owl OR-71 *Lil Quickie* **Formula One racing aircraft**

Parker Teenie, amateur-built in Switzerland *(Air Portraits)*

(6 US gallons). Total fuel capacity 47·3 litres (12·5 US gallons). Refuelling point on upper surface of fuselage, forward of windscreen. Oil capacity 5·7 litres (1·5 US gallons).

ACCOMMODATION: Single seat beneath transparent bubble canopy. Access through removable hatch aft of the bubble canopy. False cockpit floor folds up and forward to form seatback. Pilot enters, slides forward and then erects seatback. Cockpit ventilated.

ELECTRONICS: Battery-powered Radair 360 360-channel com transceiver.

DIMENSIONS, EXTERNAL:
- Wing span: 6·10 m (20 ft 0 in)
- Wing chord at root: 1·27 m (4 ft 2 in)
- Wing chord at tip: 0·76 m (2 ft 6 in)
- Wing area, gross: 6·13 m² (66 sq ft)
- Wing aspect ratio: 6
- Length overall: 5·94 m (19 ft 6 in)
- Height overall: 1·40 m (4 ft 7 in)
- Wheel track: 1·30 m (4 ft 3 in)
- Wheelbase: 3·63 m (11 ft 11 in)
- Propeller diameter: 1·42 m-1·47 m (4 ft 8 in-4 ft 10 in)

WEIGHTS AND LOADINGS:
- Weight empty: 263 kg (580 lb)
- Max T-O weight: 381 kg (840 lb)
- Max wing loading: 62·15 kg/m² (12·73 lb/sq ft)
- Max power loading: 5·1 kg/kW (8·4 lb/hp)

PERFORMANCE:
- Never-exceed speed: 239 knots (443 km/h; 275 mph)
- Max level speed at S/L: 217 knots (402 km/h; 250 mph)
- Max cruising speed at 1,525 m (5,000 ft): 182 knots (338 km/h; 210 mph)
- Econ cruising speed at 1,525 m (5,000 ft): 143 knots (266 km/h; 165 mph)
- Stalling speed: No stall or buffet at 34·7 knots (64·3 km/h; 40 mph) IAS
- Max rate of climb at S/L: 914 m (3,000 ft)/min
- Min T-O run: 91 m (300 ft)
- Normal T-O run: 305 m (1,000 ft)
- Landing from 15 m (50 ft): 914 m (3,000 ft)
- Landing run: 762 m (2,500 ft)
- Range with standard fuel: 174 nm (322 km; 200 miles)
- Range with standard plus auxiliary fuel: 347 nm (644 km; 400 miles)

Vin-Del Aircraft

ADDRESS:
29718 Knollview Drive, Miraleste, California 90732

Mr Vince DeLuca, proprietor of Vin-Del Aircraft, has built a prototype of a Formula One racer of Mr Owl's design which he has designated OR-71 *Lil Quickie.* Construction began on 2 December 1971 and the first flight was made on 6 June 1972. Plans (entirely different from those of OR-70) are available to amateur constructors.

VIN-DEL/OWL OR-71 LIL QUICKIE

The OR-71 is generally similar to the OR-70, the description of which applies also to the OR-71 except as detailed. Limit load factor of the OR-71 is ±7·33*g*, ultimate load factor ±11·0*g*.

WINGS: As for OR-70, except dihedral 0° 51′. Non-linear thickness distribution. Thickness/chord ratio is 13·7% at root, 10% on outer 60% of wing. Plywood ribs with spruce caps. Plans show optional high-lift leading-edge of larger radius and increased camber.

LANDING GEAR: As for OR-70, except prototype has cantilever light alloy main legs. Spring steel legs optional. Main-wheel fairings of glassfibre.

POWER PLANT: As for OR-70, except propeller diameter increased. Fuselage fuel tank has a capacity of 34 litres (9 US gallons).

ACCOMMODATION: As for OR-70, except for a conventional canopy, hinged on starboard side, opening upwards and to starboard. No folding seatback or access panel aft of canopy.

DIMENSIONS, EXTERNAL:
As for OR-70, except:
- Length overall: 4·98 m (16 ft 4 in)
- Propeller diameter (max): 1·52 m (5 ft 0 in)

WEIGHTS:
- Weight empty: 251 kg (553 lb)
- Max T-O weight: 386 kg (850 lb)

PERFORMANCE (at max T-O weight):
- Never-exceed speed: 260 knots (482 km/h; 300 mph) IAS
- Max level speed: more than 221 knots (410 km/h; 255 mph)
- Stalling speed: 60 knots (111 km/h; 69 mph)

PARKER

C. Y. PARKER

ADDRESS:
PO Box 181, Dragoon, Arizona 85609
Telephone: (602) 586-3836

PARKER TEENIE TWO

Mr Cal Parker has completed and flown the prototype of an improved version of the small lightweight all-metal homebuilt aircraft which he designated Jeanie's Teenie. With completion of the new prototype, the original model became known as Teenie One.

Mr Parker's original aim was to build an aircraft specifically to utilise the Volkswagen motor car engine and, at the same time, to evolve an all-metal design that would present few constructional problems even to homebuilders with virtually no metal-working experience. This has been achieved, and no special tools or jigs are needed beyond a tool to close and form the cadmium-plated steel pop rivets that are used for practically all assembly. One gauge of aluminium sheet and one size of light alloy angle section is used for almost all of the structure, except for chromoly steel tube and sheet which are used for construction of the landing gear and control actuation tubes respectively. For simplicity and economy, push/pull tubes are used for all flying controls.

Teenie One conformed to these ideas, but the Teenie Two, which first flew in 1969, was considerably refined to produce a much cleaner aeroplane. The prototype cost approximately $650 to build, over a period of six months. Its structure is stressed for full aerobatics, but the fuel and oil systems are not suitable for inverted flight.

Teenie Two has been tested with various propellers, but a computer-designed propeller is available in the parts kit which gives optimum performance for take-off, climb and cruise.

Plans, complete kits of parts, and details of modifications for the Volkswagen engine are available to amateur constructors. Some Teenie Twos built from plans had completed 750 flying hours by early 1976.

In June 1975 an aircooled Wankel engine was installed experimentally in the Teenie Two prototype. Although taxying took place, the modified aircraft was never flown and work on this variant has ended. A two-seat version of the Teenie, named **Double Teenie,** is under construction.

The following details apply to the standard Teenie Two:

TYPE: Single-seat homebuilt light aircraft.

WINGS: Cantilever low-wing monoplane. Wing section NACA 4415. All-metal two-spar structure, with detachable outer wing panels. Light alloy ribs and skin. Plain ailerons of metal construction. No flaps.

FUSELAGE: All-metal semi-monocoque structure with longerons of light alloy angle, three built-up bulkheads and light alloy skin.

TAIL UNIT: Cantilever all-metal structure with swept vertical surfaces. Small dorsal fin eliminates the need for a fourth bulkhead by carrying loads from fin leading-edge to centre bulkhead. Conventional rudder and elevators of metal construction.

LANDING GEAR: Non-retractable tricycle type. Shock-absorption provided by springs in compression and rubber hose. All three wheels same size, with tyres size 10·5 × 4·00-4, pressure 1·72 bars (25 lb/sq in). Mechanically-actuated wheel brakes.

POWER PLANT: One 31·5 kW (42 hp) 1,600 cc or 30 kW (40 hp) 1,500 cc Volkswagen modified motor car engine (conversion parts sold by Parker), driving a two-blade fixed-pitch wooden propeller. Single fuselage fuel tank, immediately aft of firewall, capacity 34 litres (9 US gallons). Refuelling point on top of fuselage, forward of windscreen. Oil capacity 2·5 litres (0·66 US gallons).

ACCOMMODATION: Single seat in open cockpit. Drawings of optional canopy available.

DIMENSIONS, EXTERNAL:

Wing span	5·49 m (18 ft 0 in)
Wing chord, constant	1·02 m (3 ft 4 in)
Width, wings detached	1·83 m (6 ft 0 in)
Length overall	3·91 m (12 ft 10 in)

WEIGHTS:

Weight empty	140 kg (310 lb)
Max T-O weight	267 kg (590 lb)

PERFORMANCE (at max T-O weight, 1,600 cc engine):

Max level speed	104 knots (193 km/h; 120 mph)
Max cruising speed (75% power)	95·5 knots (177 km/h; 110 mph)
Landing speed	43·5 knots (80·5 km/h; 50 mph)
Max rate of climb at S/L	244 m (800 ft)/min
Service ceiling	4,575 m (15,000 ft)
Min ground turning radius	9·14 m (30 ft 0 in)
Range	347 nm (643 km; 400 miles)

PAYNE

VERNON W. PAYNE

ADDRESS:
Route No. 4, PO Box 319M, Escondido, California 92025
Telephone: (714) 746-4465

Mr Vernon Payne is the designer of the Knight Twister, a light sporting biplane of which plans and kits are available for amateur construction. It exists in four main versions.

KNIGHT TWISTER KT-85

The original prototype of the Knight Twister KT-85 single-seat sporting biplane flew in 1933. Considerable refinement of the design since that time has improved both the appearance and the performance of later models, which have been built in substantial numbers by amateur constructors in the United States and elsewhere.

Standard power plant is a Continental flat-four engine of 63·5-67 kW (85-90 hp), but alternative engines have been fitted by some constructors. Most powerful Knight Twister flown to date is one owned by Mr Charles Williams of Mount Prospect, Illinois, which has a 134 kW (180 hp) Lycoming O-360 engine installed.

During 1971 Mr Payne modified the design of the tailplane, which now has increased span and reduced chord, the area remaining unchanged.

The following details refer to the standard Knight Twister built from Mr Payne's plans:

TYPE: Single-seat light biplane.

WINGS: Braced biplane type. Wing section NACA M-6. No dihedral. Incidence 1° 30′. All-wood two-spar structure, plywood-covered and with fabric covering overall. Ailerons on lower wings only, of fabric-covered wood construction. No flaps.

FUSELAGE: Steel tube truss structure with wood stringers and fabric covering.

TAIL UNIT: Cantilever type. Vertical surfaces have fabric-covered steel tube structure. Horizontal surfaces have plywood-covered wood structure, with fabric covering overall.

LANDING GEAR: Non-retractable tailwheel type. Cantilever main units. Rubber cord or hydraulic shock-absorption. Wheels size 6·00-6 with Goodyear tyres, pressure 0·345-0·69 bars (5-10 lb/sq in). Goodyear disc brakes.

POWER PLANT: One 67 kW (90 hp) Continental C90 flat-four engine, driving a two-blade wood or metal fixed-pitch propeller. Alternatively any other Continental or Lycoming flat-four engine of 63·5-108 kW (85-145 hp). Fuel tank aft of engine firewall, capacity 68 litres (18 US gallons). Oil capacity 3·7-5·7 litres (1-1·5 US gallons).

ACCOMMODATION: Single seat, normally in open cockpit. Baggage compartment capacity 9 kg (20 lb). Radio optional.

DIMENSIONS, EXTERNAL:

Wing span:	
upper	4·57 m (15 ft 0 in)
lower	3·96 m (13 ft 0 in)
Wing chord (mean, both)	0·65 m (2 ft 1·6 in)
Wing area, gross	5·57 m² (60 sq ft)
Wing aspect ratio:	
upper	6·87
lower	6·13
Length overall	4·27 m (14 ft 0 in)
Height overall	1·60 m (5 ft 3 in)
Tailplane span	2·13 m (7 ft 0 in)
Wheel track	1·52 m (5 ft 0 in)
Wheelbase	5·23 m (17 ft 2 in)

DIMENSION, INTERNAL:

Cockpit: Width	0·53 m (1 ft 9 in)

WEIGHTS AND LOADINGS (67 kW; 90 hp engine):

Weight empty	243 kg (535 lb)
Max T-O weight	435 kg (960 lb)
Max wing loading	78 kg/m² (16 lb/sq ft)
Max power loading	6·49 kg/kW (10·7 lb/hp)

PERFORMANCE (67 kW; 90 hp engine, at max T-O weight):

Max level speed at S/L	139 knots (257 km/h; 160 mph)
Max cruising speed	122 knots (225 km/h; 140 mph)
Econ cruising speed	109 knots (201 km/h; 125 mph)
Stalling speed	53 knots (97 km/h; 60 mph)
Max rate of climb at S/L	275 m (900 ft)/min
T-O run	125 m (410 ft)
T-O to 15 m (50 ft)	305 m (1,000 ft)
Landing from 15 m (50 ft)	366 m (1,200 ft)
Landing run	205 m (670 ft)
Range with max fuel	338 nm (625 km; 390 miles)

KNIGHT TWISTER IMPERIAL

At the request of Don Fairbanks, owner of a flying training school at Lunken Airport, Cincinnati, Ohio, Mr Payne modified the design of the original Knight Twister to enable Mr Fairbanks to compete in US National Air Races in the Sport Biplane class. The resulting aircraft won the Silver Biplane Race at Reno, Nevada, in 1971. Variations from the standard Knight Twister include a change in wing section, increased wing and tailplane span and increased fuel tankage. First flight was made on 19 June 1970.

Dubbed the Knight Twister Imperial by Mr Payne, this aircraft is generally similar to the Knight Twister KT-85, except as follows:

WINGS: Wing section NACA 21. Span and chord of both upper and lower wings increased. Flying and landing wires deleted. Incidence 0°.

POWER PLANT: One 101 kW (135 hp) Lycoming O-290-D2 flat-four engine, driving a two-blade fixed-pitch propeller with spinner. Fuel contained in an upper tank of 85 litres (22·5 US gallons) capacity and a lower tank of 47 litres (12·5 US gallons) capacity, aft of firewall. Total fuel capacity 132 litres (35 US gallons).

DIMENSIONS, EXTERNAL:

Wing span:	
upper	5·33 m (17 ft 6 in)
lower	4·72 m (15 ft 6 in)
Wing area, gross	6·99 m² (75·25 sq ft)

WEIGHTS AND LOADINGS:

Normal T-O weight	408 kg (900 lb)
Max T-O weight	520 kg (1,125 lb)
Max wing loading	72·7 kg/m² (14·9 lb/sq ft)
Max power loading	5·14 kg/kW (8·3 lb/hp)

PERFORMANCE (at normal T-O weight):

Max level speed	156 knots (290 km/h; 180 mph)
Max cruising speed	148 knots (274 km/h; 170 mph)
Stalling speed	45·5 knots (84 km/h; 52 mph)
Max rate of climb at S/L	366 m (1,200 ft)/min
T-O run	130 m (425 ft)
Landing run	244 m (800 ft)
Range at max cruising speed	590 nm (1,094 km; 680 miles)

SUNDAY KNIGHT TWISTER SKT-125

This developed version of the Knight Twister has a 93 kW (125 hp) Lycoming engine. Increased wing area makes it easier to fly and its name is meant to imply that it is for the 'Sunday flyer'.

A fully aerobatic Sunday Knight Twister, with a 134 kW (180 hp) engine, has been built by Mr J. F. Carter of Drewry, Alabama.

DIMENSIONS, EXTERNAL:

Wing span	5·94 m (19 ft 6 in)
Wing area, gross	7·71 m² (83 sq ft)
Length overall	4·72 m (15 ft 6 in)
Height overall	1·68 m (5 ft 6 in)

WEIGHTS:

Weight empty	318 kg (700 lb)
Max T-O weight	461 kg (1,016 lb)

PERFORMANCE (at max T-O weight):

Max level speed at S/L	146 knots (270 km/h; 168 mph)
Max cruising speed	144 knots (267 km/h; 166 mph)
Econ cruising speed	126 knots (233 km/h; 145 mph)
Stalling speed	43·5 knots (80·5 km/h; 50 mph)
Max rate of climb at S/L	366 m (1,200 ft)/min
T-O run	113 m (370 ft)
T-O to 15 m (50 ft)	232 m (760 ft)
Landing run	265 m (870 ft)
Range at max cruising speed with max fuel	307 nm (570 km; 354 miles)

KNIGHT TWISTER JUNIOR KT-75

The Knight Twister Junior has the same fuselage, tail unit and landing gear as the KT-85, but its tapered wings have a larger area. The prototype flew in 1947. Details are as for the KT-85, except for the following:

WINGS: Incidence 2°.

POWER PLANT: One Continental or Lycoming flat-four engine of 56-93 kW (75-125 hp). Fuel capacity with 56 kW (75 hp) engine 45 litres (12 US gallons).

DIMENSIONS, EXTERNAL:

Wing span:	
upper	5·33 m (17 ft 6 in)
lower	4·11 m (13 ft 6 in)
Wing area, gross	6·76 m² (72·8 sq ft)
Wing aspect ratio:	
upper	7·78
lower	7·10

WEIGHTS AND LOADINGS (56 kW; 75 hp engine):

Weight empty	227 kg (500 lb)
Max T-O weight	404 kg (890 lb)
Max wing loading	58·6 kg/m² (12 lb/sq ft)
Max power loading	7·21 kg/kW (11·87 lb/hp)

PERFORMANCE (56 kW; 75 hp engine, at max T-O weight):

Max level speed at S/L	117 knots (217 km/h; 135 mph)
Max cruising speed	109 knots (201 km/h; 125 mph)
Econ cruising speed	97 knots (180 km/h; 112 mph)
Stalling speed	42 knots (77·5 km/h; 48 mph)
Max rate of climb at S/L	275 m (900 ft)/min
T-O run	114 m (375 ft)
T-O to 15 m (50 ft)	277 m (910 ft)
Landing from 15 m (50 ft)	311 m (1,020 ft)
Landing run	190 m (625 ft)
Range with max fuel	247 nm (460 km; 285 miles)

Knight Twister Imperial, in current form with revised landing gear *(Howard Levy)*

Pazmany PL-1 built by Mr Dieter Bochmann *(Niel A. Macdougall)*

Pazmany PL-2 built by Mr Kenneth Arnold *(Howard Levy)*

Pazmany PL-4A prototype (1,600 cc modified Volkswagen motor car engine)

PAZMANY

PAZMANY AIRCRAFT CORPORATION

ADDRESS:

Box 80051, San Diego, California 92138

Telephone: (714) 276-0424

This company was formed by Mr Ladislao Pazmany, designer of a two-seat light aircraft known as the PL-1 Laminar. A prototype, constructed by Mr John Green and Mr Keith Fowler, was flown for the first time on 23 March 1962, the test pilots being Cdr Paul Hayek, USN, and Lieut Richard Gordon, who is best known as one of the Gemini/Apollo astronauts.

Some 5,000 design hours and 4,000 hours of construction went into the prototype PL-1, which had logged more than 1,500 flying hours by January 1976.

Pazmany Aircraft Corporation is no longer marketing plans of the PL-1: instead, plans and instructions for building the improved PL-2 are available to amateur constructors and many aircraft of this type are being built. A total of 276 sets of plans had been sold by early 1975.

Mr Pazmany also designed and built the prototype of another lightweight, low-cost monoplane, designated PL-4A.

PAZMANY PL-1 LAMINAR

A total of 375 sets of plans and instructions for building the PL-1 have been sold, and PL-1s are being built in the USA, Canada, Australia, Norway and other countries.

In early 1968 the Aeronautical Research Laboratory of the Chinese Nationalist Air Force, at Taichung, Taiwan, acquired a set of PL-1 drawings. Under the supervision of General K. F. Ku and Colonel C. Y. Lee, personnel of the ARL built a PL-1 in a record time of 100 days. It was flown for the first time on 26 October 1968 and on 30 October was presented to Generalissimo Chiang Kai-Shek. Extensive flight testing resulted in the decision to utilise the PL-1 as a basic trainer for CAF cadets, and 58 additional aircraft, designated PL-1B, were constructed between 1970 and 1974, powered by the 112 kW (150 hp) Lycoming O-320 engine.

The details below apply to the prototype PL-1, which was stressed to 9*g* (ultimate) for aerobatics and to permit the fitting of more powerful engines.

TYPE: Two-seat light aircraft.

WINGS: Cantilever low-wing monoplane. Wing section NACA $63_{2}615$. Dihedral 3°. Incidence −1° 20′. All-metal single-spar structure in one piece, with leading-edge torsion box. Plain piano-hinged ailerons and flaps of all-metal construction. No trim tabs.

FUSELAGE: Conventional all-metal semi-monocoque structure, with flat or single-curvature skins.

TAIL UNIT: Cantilever all-metal structure. One-piece horizontal surface, with anti-servo tab which serves also as a trim tab.

LANDING GEAR: Non-retractable tricycle type, with all three oleo-pneumatic shock-absorbers interchangeable. Goodyear wheels and tyres, size 5·00-5. Tyre pressure 2·14 bars (31 lb/sq in). Goodyear brakes. Steerable nosewheel.

POWER PLANT: One 71 kW (95 hp) Continental C90-12F flat-four engine, driving a McCauley Model IA100/MCM 6663 two-blade metal fixed-pitch propeller. Fuel in two glassfibre wingtip tanks, each of 47 litres (12·5 US gallons) capacity. Total fuel capacity 94 litres (25 US gallons). Oil capacity 4·5 litres (5 US quarts).

ACCOMMODATION: Two seats side by side under rearward-sliding transparent canopy. Dual controls. Space for 18 kg (40 lb) baggage aft of seats. Heater and airscoops for ventilation. VHF radio.

DIMENSIONS, EXTERNAL:

Wing span	8·53 m (28 ft 0 in)
Wing chord (constant)	1·27 m (4 ft 2 in)
Wing area, gross	10·78 m² (116 sq ft)
Wing aspect ratio	6·7
Length overall	5·77 m (18 ft 11 in)
Height overall	2·64 m (8 ft 8 in)
Tailplane span	2·44 m (8 ft 0 in)
Wheel track	2·50 m (8 ft 2½ in)
Wheelbase	1·30 m (4 ft 3 in)

DIMENSIONS, INTERNAL:

Cabin: Length	1·27 m (4 ft 2 in)
Width	1·02 m (3 ft 4 in)
Height	1·02 m (3 ft 4 in)

WEIGHTS AND LOADINGS:

Weight empty, equipped	363 kg (800 lb)
Max T-O weight	602 kg (1,326 lb)
Max wing loading	55·7 kg/m² (11·4 lb/sq ft)
Max power loading	8·48 kg/kW (14 lb/hp)

PERFORMANCE (at max T-O weight):

Never-exceed speed	178 knots (330 km/h; 205 mph)
Max level speed at S/L	104 knots (193 km/h; 120 mph)
Max cruising speed at S/L	100 knots (185 km/h; 115 mph)
Econ cruising speed at S/L	91 knots (169 km/h; 105 mph)
Stalling speed, flaps down	44 knots (82 km/h; 51 mph)
Max rate of climb at S/L	305 m (1,000 ft)/min
Service ceiling	5,500 m (18,000 ft)
T-O run	168 m (550 ft)
T-O to 15 m (50 ft)	239 m (784 ft)
Landing from 15 m (50 ft)	335 m (1,100 ft)
Landing run	54 m (175 ft)
Range with max fuel	521 nm (965 km; 600 miles)

PAZMANY PL-1B

This is the version of the PL-1 built in Taiwan. It differs from the basic PL-1 mainly in having a 112 kW (150 hp) Lycoming O-320 engine.

PAZMANY PL-2

Shortly after flight trials of the PL-1 began, Mr Pazmany initiated a complete redesign of the aircraft. The developed design, known as the PL-2, is almost identical with the PL-1 in external configuration. Cockpit width is increased by 5 cm (2 in) and wing dihedral is increased from 3° to 5°. The internal structure is extensively changed, to simplify construction and reduce weight. Suitable Lycoming power plants are the 80·5 kW (108 hp) O-235-C1, 93 kW (125 hp) O-290-G (ground power unit), 101 kW (135 hp) O-290-D2B or 112 kW (150 hp) O-320-A.

Static tests of every major assembly up to ultimate loads had been made by early 1967. The first PL-2 to be completed was built by Mr H. Pio of Ramona, California, and this aircraft made its first flight on 4 April 1969, piloted by Mr Pio. It has an O-290-G engine.

A single example of the PL-2 was built by the Vietnam Air Force, each VNAF base contributing towards its construction. This flew for the first time on 1 July 1971, and it was reported that production of at least ten more PL-2s for use at the VNAF Air Training Center was being considered. The Royal Thai Air Force was also known to have two PL-2s under construction; but no further news of this project had been received by early 1976. The Republic of Korea Air Force built one PL-2 prototype for flight testing, and later completed three more for evaluation as trainers. In Japan, the Miyauchi Manufacturing Co Ltd, Tokyo, completed a prototype of the PL-2 in the Autumn of 1971, this being exhibited at the Nagoya International Air Show during October-November 1971.

The Indonesian Air Force began construction of an example of the PL-2 at the Lipnur factory in late 1973, for evaluation as a military trainer. This was completed during October 1974, and its first flight was made on 9 November. The designation LT-200 has now been allocated to this version, of which 6 pre-production examples were due to be built in 1976. There are plans to build at least 30 LT-200s, although it is possible that as many as 60 might be built.

DIMENSIONS, EXTERNAL:

As for PL-1, except:

Length overall	5·90 m (19 ft 3½ in)
Height overall	2·44 m (8 ft 0 in)
Wheel track	2·60 m (8 ft 5½ in)

WEIGHTS:

Weight empty:	
80·5 kW (108 hp)	396 kg (875 lb)
93, 101 kW (125, 135 hp)	408 kg (900 lb)
112 kW (150 hp)	409 kg (902 lb)
Max T-O weight:	
80·5 kW (108 hp)	642 kg (1,416 lb)
93, 101 kW (125, 135 hp)	655 kg (1,445 lb)
112 kW (150 hp)	656 kg (1,447 lb)

PERFORMANCE (at max T-O weight):

Max level speed at S/L:	
80·5 kW (108 hp)	120 knots (222 km/h; 138 mph)
93 kW (125 hp)	125 knots (232 km/h; 144 mph)
101 kW (135 hp)	128 knots (238 km/h; 148 mph)
112 kW (150 hp)	133 knots (246 km/h; 153 mph)
Econ cruising speed:	
80·5 kW (108 hp)	103 knots (192 km/h; 119 mph)
93 kW (125 hp)	111 knots (206 km/h; 128 mph)
101 kW (135 hp)	113 knots (209 km/h; 130 mph)
112 kW (150 hp)	118 knots (219 km/h; 136 mph)
Stalling speed (flaps down):	
80·5 kW (108 hp)	45·5 knots (84 km/h; 52 mph)
93, 101, 112 kW (125, 135, 150 hp)	47 knots (87 km/h; 54 mph)
Max rate of climb at S/L:	
80·5 kW (108 hp)	390 m (1,280 ft)/min
93 kW (125 hp)	457 m (1,500 ft)/min
101 kW (135 hp)	488 m (1,600 ft)/min
112 kW (150 hp)	518 m (1,700 ft)/min
Range at econ cruising speed:	
80·5 kW (108 hp)	427 nm (790 km; 492 miles)

93 kW (125 hp)	422 nm (780 km; 486 miles)
101 kW (135 hp)	428 nm (792 km; 493 miles)
112 kW (150 hp)	330 nm (610 km; 381 miles)

PAZMANY PL-4A

The prototype of a lightweight single-seat low-wing monoplane designated PL-4A flew for the first time on 12 July 1972. It was designed specifically for easy low-cost construction by amateur builders, to provide a safe aircraft that would be economical in operation. The prototype had completed 312 hours of flight by January 1976. Sets of plans, kits of prefabricated components, glassfibre wing-tips and fuel tank and transparent cockpit canopy are available to amateur constructors.

By February 1976, approximately 500 sets of plans had been sold, and the PL-4A had received approval in Australia for construction by amateurs.

In November 1973, Lt Col Roy Windover, Director of the Air Cadets Programme, Canadian Ministry of Defence, made a flight evaluation of the PL-4A prototype. As a result of this, it is planned to provide 200 of these aircraft for the Air Cadets.

Two pre-production aircraft are being built to evaluate three different power plant installations, comprising a Volkswagen engine with 2¼ : 1 Vee-belt reduction; a Limbach SL 1700 E Volkswagen with direct drive to the propeller; and a Continental A65 aircooled engine. The first of these aircraft was expected to be completed by mid-1976.

Components for the pre-production aircraft, as well as for the production version, are being made by inmates of a civil prison, who are responsible also for construction of the first two aircraft. The remainder will be assembled by Air Cadets, who will use them for cross-country, aerobatic and IFR flying.

TYPE: Single-seat lightweight homebuilt sporting aircraft.

WINGS: Cantilever low-wing monoplane. Wing section NACA 63_3418. Dihedral 5°. Incidence 3°. No sweep-back. All-metal structure, with main spar, 'Z' section rear beam, sheet metal ribs and skins. Wings fold along-side fuselage for towing or storage. Plain piano-hinged ailerons of all-metal construction. Glassfibre wingtips. No flaps. No trim tabs.

FUSELAGE: All-metal structure, with bulkheads built up from bent sheet metal channels and standard extruded angles for longerons, and with sheet metal skins.

TAIL UNIT: All-metal cantilever T-tail. All-moving tail-plane with large anti-servo tab which serves also as a trim tab.

LANDING GEAR: Non-retractable tailwheel type. Spring steel cantilever main legs. Single go-kart type wheel on each main unit, with 4·10 × 3·50-6 four-ply tyre, pressure 4·48 bars (65 lb/sq in). Steerable and castoring tailwheel with solid tyre size 5 × 1·5-1·5. Go-kart type hydraulic disc brakes by Hurst-Airheart.

POWER PLANT: One 1,600 cc modified Volkswagen motor car engine with Becar V-belt reduction of 2¼ : 1, developing approximately 37·5 kW (50 hp) and driving a two-blade fixed-pitch wooden propeller of Pazmany design, manufactured by Ted Hendricksen. Glassfibre fuel tank immediately aft of firewall, usable capacity 45 litres (12 US gallons). Refuelling point on upper fuselage forward of windscreen. Oil capacity 2·8 litres (0·75 US gallons).

ACCOMMODATION: Single seat under transparent Plexiglas canopy, hinged on starboard side. Compartment aft of seat for 9 kg (20 lb) baggage. Cabin heated and ventilated.

SYSTEMS: Hydraulic system for brakes only. Electrical system powered by 12V 25Ah battery situated in baggage compartment.

DIMENSIONS, EXTERNAL:

Wing span	8·13 m (26 ft 8 in)
Wing chord (constant)	1·02 m (3 ft 4 in)
Wing area, gross	8·27 m² (89·0 sq ft)
Wing aspect ratio	8·0
Length overall	5·04 m (16 ft 6½ in)
Width, wings folded	2·44 m (8 ft 0 in)
Height overall	1·73 m (5 ft 8 in)
Tailplane span	2·29 m (7 ft 6 in)
Wheel track	2·06 m (6 ft 9 in)
Wheelbase	3·56 m (11 ft 8 in)
Propeller diameter	1·73 m (5 ft 8 in)
Propeller ground clearance	0·25 m (10 in)

WEIGHTS AND LOADINGS:

Weight empty	262 kg (578 lb)
Max T-O and landing weight	385 kg (850 lb)
Max wing loading	46·4 kg/m² (9·5 lb/sq ft)
Max power loading	10·27 kg/kW (17·0 lb/hp)

PERFORMANCE (at max T-O weight):

Never-exceed speed	161 knots (299 km/h; 186 mph)
Max level speed at S/L	109 knots (201 km/h; 125 mph)
Max cruising speed at S/L	85 knots (158 km/h; 98 mph)
Econ cruising speed at S/L	78 knots (145 km/h; 90 mph)
Stalling speed, power on	40 knots (74 km/h; 46 mph)
Stalling speed, power off	42 knots (77·5 km/h; 48 mph)
Max rate of climb at S/L	198 m (650 ft)/min
Service ceiling	3,960 m (13,000 ft)
Min ground turning radius	3·05 m (10 ft 0 in)
T-O run	148 m (486 ft)
Landing run	133 m (436 ft)
Range with max fuel, no allowances	295 nm (545 km; 340 miles)

PDQ
PDQ AIRCRAFT PRODUCTS

ADDRESS:
28975 Alpine Lane, Elkhart, Indiana 46514
Telephone: (219) 264-2906

Mr Wayne Ison formed this company to market plans of the PDQ-2 lightweight sporting aircraft which he had designed and built. The PDQ-2 is intended to provide a cheap, robust, easily and quickly built aircraft which is easy for an average pilot to fly.

Design of the prototype had begun in September 1972, and construction was started on 5 January 1973. Excluding the 27 kW (36 hp) Rockwell (Venture) JLO-LB-600-2 engine that was originally fitted, the cost of construction was only $350, and the first flight was made on 30 May 1973.

During 1975 it became clear that an alternative power plant was needed, owing to the increasing scarcity of the JLO engine. One amateur constructor has fitted a BMW motorcycle engine; but Mr Ison decided to test the prototype PDQ-2 with a converted Volkswagen motor car engine, a type of power plant which is readily available and inexpensive. The increased engine weight required a number of structural modifications, notably to the engine mounting, wing spars and landing gear; but flight testing proved satisfactory, and details of the new engine installation have been sent to all past and current purchasers of plans, in 30 countries.

PDQ AIRCRAFT PRODUCTS PDQ-2

The following details refer to the standard PDQ-2, as shown on Mr Ison's plans. At least two aircraft are being fitted with wings of the new NASA GAW-1 section.

TYPE: Lightweight homebuilt sporting aircraft.

WINGS: Wire-braced monoplane. Wing section NACA 63_2A615. Dihedral 5°. Incidence 3°. Composite structure with wooden spars, polyurethane foam ribs, sheet foam skins, covered with Dynel fabric and impregnated with epoxy resin. Wing-root fences and cambered tips now standard. Full-span plain ailerons. No flaps or trim tabs.

FUSELAGE: Basic structure consists of 0·05 m (2 in) square tubes of 6061-T6 light alloy. A lower forward tube carries the landing gear; attached to it is a vertical king-post, to which the wing is attached and wire-braced. An aft horizontal tube attached to the kingpost carries the T-tail.

TAIL UNIT: Strut-braced T-tail, with swept vertical surfaces. Fixed-incidence tailplane. No trim tabs.

LANDING GEAR: Non-retractable tricycle type. Shock-absorption by spring steel leaf mounting struts. Main-wheel diameter 0·28 m (11 in); tailwheel diameter 0·15 m (6 in). No brakes.

POWER PLANT: Prototype flew originally with Rockwell (Venture) JLO-LB-600-2 engine, mounted on top of the kingpost and driving a two-blade fixed-pitch wooden pusher propeller. This engine remains suitable, but prototype now has a 1,385 cc converted Volkswagen motor car engine, developing 30 kW (40 hp). Plastics fuel tank mounted alongside pilot's seat on port side, capacity 22·7 litres (6 US gallons). Refuelling point on tank upper surface.

ACCOMMODATION: Single open seat mounted immediately forward of kingpost.

DIMENSIONS, EXTERNAL:

Wing span, with cambered wingtips	6·71 m (22 ft 0 in)
Wing chord, constant	1·07 m (3 ft 6 in)
Length overall	4·42 m (14 ft 6 in)
Height overall	1·27 m (4 ft 2 in)
Tailplane span	1·83 m (6 ft 0 in)
Wheel track	1·27 m (4 ft 2 in)

WEIGHTS AND LOADING (Volkswagen engine):

Weight empty	164 kg (360 lb)
Max T-O weight	272 kg (600 lb)
Max power loading	9·07 kg/kW (15·0 lb/hp)

PERFORMANCE (at max T-O weight, with JLO engine):

Never-exceed speed	86·5 knots (161 km/h; 100 mph)
Max level speed at S/L	69 knots (129 km/h; 80 mph)
Max cruising speed at S/L	61 knots (113 km/h; 70 mph)
Stalling speed	35 knots (65 km/h; 40 mph)
Max rate of climb at S/L	152 m (500 ft)/min
Service ceiling	3,050 m (10,000 ft)
T-O run	91 m (300 ft)
Landing run	122 m (400 ft)

PITTS
PITTS AVIATION ENTERPRISES INC

ADDRESS:
PO Box 548, Homestead, Florida 33030
Telephone: (305) 247-5423
PRESIDENT: C. H. Pitts

In addition to marketing factory-built examples of his single-seat and two-seat Pitts Special biplanes, Mr Curtis Pitts supplies plans of the single-seater to amateur constructors.

PITTS S-1D SPECIAL

The version of the single-seat S-1 for which plans are currently available is designated S-1D. Details are generally similar to those given for the factory-built S-1S in the main US Aircraft section of this edition.

There is also an S-1C version, but Mr Pitts can no longer supply plans for its construction. By comparison with earlier models, this has flat-bottomed wings, and ailerons on the lower wing only, and is suitable for the installation of engines of between 74·5 and 134 kW (100 to 180 hp).

POWELL
JOHN C. POWELL

ADDRESS:
4 Donald Drive, Middletown, Rhode Island 02840

John Powell, formerly a Commander in the US Navy, designed and built a two-seat parasol-wing monoplane, of which plans are available to amateur constructors. Known as the P-70 Acey Deucy, its design was started in 1966 and construction began during 1967. FAA certification in the Experimental homebuilt category was awarded on 19 June 1970 and the first flight was recorded on the following day. By the beginning of 1976 this prototype had accumulated 535 hours' flying time.

More than 80 sets of plans have been sold, and the first aircraft built from plans was flying by the Autumn of 1973. It is believed that about 25 more Acey Deucys are either flying or under construction.

POWELL P-70 ACEY DEUCY

TYPE: Two-seat homebuilt monoplane.

WINGS: Braced parasol-wing monoplane with steel tube Vee bracing struts on each side, auxiliary bracing struts and N-type centre-section struts. Wing section NACA 4412. Dihedral 1° on outer panels. Incidence 2°. No sweepback. Composite structure of steel tube and wood, fabric covered. Frise-type ailerons of wooden construction, fabric covered. No flaps. No trim tabs.

FUSELAGE: Welded 4130 steel tube structure with wooden stringers, fabric covered.

TAIL UNIT: Wire-braced welded steel tube structure with 'U' channel ribs. Tailplane incidence adjustable by screwjack at leading-edge. No trim tabs.

LANDING GEAR: Non-retractable tailwheel type. Two side Vees and half axles hinged to fuselage structure. Shock-absorption by springs in compression. Goodyear main wheels and tyres size 8·00-4, pressure 0·83 bars (12 lb/sq in). Motor scooter type caliper brakes.

POWER PLANT: Suitable for installation of engines from 48·5-67 kW (65 to 90 hp). Prototype has one 48·5 kW (65 hp) Continental A65 flat-four engine, driving a McCauley two-blade metal fixed-pitch propeller. One fuel tank in fuselage, immediately aft of firewall, capacity 53 litres (14 US gallons). Refuelling point on top of fuselage, forward of front cockpit. Oil capacity 3·75 litres (1 US gallon).

ACCOMMODATION: Two persons in tandem in open cockpits. Small door by front cockpit on starboard side.

DIMENSIONS, EXTERNAL:

Wing span	9·91 m (32 ft 6 in)
Wing chord (constant)	1·52 m (5 ft 0 in)
Wing area, gross	14·4 m² (155 sq ft)
Wing aspect ratio	6·5
Length overall	6·32 m (20 ft 9 in)
Height overall	2·06 m (6 ft 9 in)
Tailplane span	2·59 m (8 ft 6 in)
Wheel track	1·83 m (6 ft 0 in)

PDQ-2 lightweight homebuilt sporting aircraft in its original form, with JLO engine and uncambered wingtips *(Howard Levy)*

British-built single-seat Pitts Special *(Air Portraits)*

Powell P-70 Acey Deucy two-seat homebuilt lightweight aircraft

Rand Robinson KR-1 single-seat homebuilt aircraft *(Howard Levy)*

RAND ROBINSON
RAND ROBINSON ENGINEERING INC

ADDRESS:
6171 Cornell Drive, Huntington Beach, California 92647

During 1974 Mr Rand formed Rand Robinson Engineering Inc to market plans for the KR-1 single-seat lightweight sporting aircraft, and of a slightly larger two-seat version, designated KR-2.

RAND ROBINSON KR-1

Mr Kenneth Rand, a flight test engineer with the Douglas Aircraft Company, designed and built the prototype of a single-seat lightweight sporting aircraft known as the Rand KR-1. The design originated in 1969; construction of the prototype was started in 1970 and the first flight was made in February 1972. Plans are available to amateur constructors; 5,770 sets had been sold by the beginning of 1976 and about 200 KR-1s were known to be flying.

The performance figures that are quoted relate generally to the re-engined prototype fitted with a 43·25 kW (58 hp) 1,700 cc Volkswagen engine. Performance figures with the original 27 kW (36 hp) 1,200 cc VW engine can be found in the 1975-76 *Jane's*.

TYPE: Single-seat homebuilt lightweight sporting aircraft.

WINGS: Cantilever low-wing monoplane. Wing section RAF 48. Thickness/chord ratio 18%. Dihedral 5°. Incidence 5° at root, 2° at tip. No sweepback. Composite two-spar structure. Front spar of spruce; rear spar built of spruce and plywood. Most ribs formed from Styrofoam plastics, spaces between ribs being filled with Styrofoam slab. Structure covered with Dynel reinforced epoxy. Outer wing panels removable for storage. Ailerons constructed of Styrofoam, with Dynel reinforced epoxy covering, over full span of outer panels. No flaps.

FUSELAGE: Composite structure, lower half of spruce longerons with plywood skin, upper surface of carved Styrofoam covered with Dynel epoxy. Firewall is a plywood, asbestos and aluminium lamination.

TAIL UNIT: Cantilever structure with spruce spars, the remainder of the structure being carved Styrofoam, Dynel epoxy covered. Fixed-incidence tailplane. Trim tabs in rudder and elevator.

LANDING GEAR: Tailwheel type. Main units retract aft manually into wing centre-section. Shock-absorption by flat spring crossbar to which main units are attached. Main-wheel tyres size 10½ × 4·00-5, pressure 1·38 bars (20 lb/sq in). Steerable tailwheel with solid tyre of 7·6 cm (3 in) diameter. Manual drum brakes.

POWER PLANT: One Volkswagen modified motor-car engine, driving a two-blade fixed-pitch propeller with spinner. Prototype had initially a 27 kW (36 hp) 1,200 cc VW and Hegy propeller, but has been re-engined with a 43·25 kW (58 hp) 1,700 cc VW. Fuel tankage with larger engine comprises one tank immediately aft of firewall, capacity 38 litres (10 US gallons), and one 76 litre (20 US gallon) tank in each wing, giving total capacity of 190 litres (50 US gallons). Refuelling point on fuselage upper surface, forward of windscreen. Oil capacity 2·8 litres (0·75 US gallon).

ACCOMMODATION: Pilot only, beneath transparent cockpit canopy, built integrally with centre-fuselage decking, which is hinged on starboard side and opens upward and to starboard. Baggage space aft of seat.

SYSTEM: Electrical power supplied by 4·5A 12V Honda motorcycle engine-driven alternator and 12V 7Ah storage battery.

ELECTRONICS: Genave Alpha 200B 200-channel nav/com transceiver.

DIMENSIONS, EXTERNAL:

Wing span	5·23 m (17 ft 2 in)
Wing chord at root	1·22 m (4 ft 0 in)
Wing chord at tip	0·91 m (3 ft 0 in)
Wing area, gross	5·95 m² (64 sq ft)
Wing aspect ratio	4·5
Length overall	3·81 m (12 ft 6 in)
Width, wings removed	1·52 m (5 ft 0 in)
Height overall	1·07 m (3 ft 6 in)
Tailplane span	1·52 m (5 ft 0 in)
Wheel track	1·27 m (4 ft 2 in)
Propeller diameter	1·35 m (4 ft 5 in)
Propeller ground clearance	0·15 m (6 in)

DIMENSIONS, INTERNAL:

Cockpit:	
Length	1·22 m (4 ft 0 in)
Max width	0·51 m (1 ft 8 in)
Max height	0·76 m (2 ft 6 in)
Baggage hold	0·11 m³ (4 cu ft)

WEIGHTS AND LOADINGS (A: 27 kW; 36 hp. B: 43·25 kW; 58 hp):

Weight empty, equipped:	
A	154 kg (340 lb)
B	172 kg (380 lb)
Max T-O and landing weight:	
A	272 kg (600 lb)
B	408 kg (900 lb)
Max wing loading:	
A	45·7 kg/m² (9·4 lb/sq ft)
B	68·6 kg/m² (14·1 lb/sq ft)
Max power loading:	
A	10·1 kg/kW (16·7 lb/hp)
B	9·43 kg/kW (15·5 lb/hp)

PERFORMANCE (A: 27 kW; 36 hp. B: 43·25 kW; 58 hp, at max T-O weight):

Never-exceed speed:	
A	140 knots (259 km/h; 161 mph)
B	187 knots (346 km/h; 215 mph)
Max level speed at S/L:	
A	130 knots (241 km/h; 150 mph)
B	174 knots (322 km/h; 200 mph)
Max cruising speed at 1,525 m (5,000 ft):	
A	130 knots (241 km/h; 150 mph)
B	156 knots (290 km/h; 180 mph)
Econ cruising speed at 1,830 m (6,000 ft):	
A	100 knots (185 km/h; 115 mph)
B	130 knots (241 km/h; 150 mph)
Stalling speed:	
A	39 knots (73 km/h; 45 mph)
Max rate of climb at S/L:	
A	182 m (600 ft)/min
B	274 m (900 ft)/min
Service ceiling:	
A	3,660 m (12,000 ft)
B	4,570 m (15,000 ft)
T-O run:	
A	122 m (400 ft)
T-O to 15 m (50 ft):	
A	244 m (800 ft)
Landing from 15 m (50 ft):	
A	305 m (1,000 ft)
Landing run:	
A	152 m (500 ft)
Range with max fuel:	
B	2,600 nm (4,825 km; 3,000 miles)

RAND ROBINSON KR-2

The KR-2 is a slightly larger two-seat version of the KR-1, to which it is generally similar in construction. Design began in 1973 and the prototype flew for the first time in July 1974. Construction occupied approximately 800 man hours, at a cost of about $2,000. Plans and kits of parts are available to amateur constructors.

By March 1976, a total of 2,030 sets of plans and 1,000 kits had been sold; about 100 KR-2s were flying at that time.

TYPE: Two-seat homebuilt lightweight sporting aircraft.

WINGS: As for KR-1, except span increased by 1·07 m (3 ft 6 in).

FUSELAGE: As for KR-1, except dimensions increased.

TAIL UNIT AND LANDING GEAR: As for KR-1, except wheel track increased.

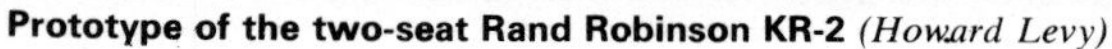

Prototype of the two-seat Rand Robinson KR-2 *(Howard Levy)*

Breezy Model RLU-1 built in France *(Austin J. Brown)*

POWER PLANT: Airframe designed to accept Volkswagen modified motor car engines of 1,600 to 2,200 cc. Prototype has a Rajay turbocharged 1,834 cc Volkswagen engine, driving a Warnke two-blade adjustable-pitch propeller with spinner. Fuel tank immediately aft of firewall, capacity 38 litres (10 US gallons). One fuel tank in each wing, capacity 53 litres (14 US gallons). Total fuel capacity 144 litres (38 US gallons). Refuelling point on fuselage upper surface, forward of windscreen.

ACCOMMODATION: Two persons, side by side, beneath transparent cockpit canopy.

SYSTEM AND ELECTRONICS: Provision for wide range of electronics and blind flying instruments.

DIMENSIONS, EXTERNAL:

Wing span	6·30 m (20 ft 8 in)
Wing chord at root	1·22 m (4 ft 0 in)
Wing chord at tip	0·91 m (3 ft 0 in)
Wing area, gross	7·43 m² (80 sq ft)
Wing aspect ratio	5·5
Length overall	4·42 m (14 ft 6 in)
Height overall	1·07 m (3 ft 6 in)
Tailplane span	1·52 m (5 ft 0 in)
Wheel track	1·52 m (5 ft 0 in)
Propeller diameter	1·32 m (4 ft 4 in)
Propeller ground clearance	0·15 m (6 in)

DIMENSIONS, INTERNAL:

Cockpit:	
Length	1·22 m (4 ft 0 in)
Max width	0·91 m (3 ft 0 in)
Max height	0·76 m (2 ft 6 in)
Baggage hold	0·11 m³ (4 cu ft)

WEIGHTS AND LOADINGS (prototype, A: without turbocharger, B: with turbocharger):

Weight empty, equipped:	
A	200 kg (440 lb)
B	218 kg (480 lb)
Max T-O and landing weight:	
A	363 kg (800 lb)
B	408 kg (900 lb)
Max wing loading:	
A	48·8 kg/m² (10 lb/sq ft)
B	54·9 kg/m² (11·25 lb/sq ft)
Max power loading:	
A	7·56 kg/kW (12·3 lb/hp)

PERFORMANCE (prototype, at max T-O weight. A: without turbocharger, B: with turbocharger):

Never-exceed speed:	
A	186 knots (346 km/h; 215 mph)
Max level speed at S/L:	
A	156 knots (290 km/h; 180 mph)
Max cruising speed:	
A at 1,525 m (5,000 ft)	156 knots (290 km/h; 180 mph)
B at 7,620 m (25,000 ft)	217 knots (402 km/h; 250 mph)
Econ cruising speed at 3,660 m (12,000 ft):	
A	148 knots (274 km/h; 170 mph)
Stalling speed:	
A	39 knots (73 km/h; 45 mph)
Max rate of climb at S/L:	
A	244 m (800 ft)/min
Service ceiling:	
A	4,875 m (16,000 ft)
T-O run:	
A	122 m (400 ft)
T-O to 15 m (50 ft):	
A	244 m (800 ft)
Landing from 15 m (50 ft):	
A	305 m (1,000 ft)
Landing run:	
A	152 m (500 ft)
Range with max fuel:	
B	1,735 nm (3,215 km; 2,000 miles)

RLU

CHARLES ROLOFF, ROBERT LIPOSKY and CARL UNGER

ADDRESS:
c/o Charles B. Roloff, 8025 West 90th Street, Hickory Hills, Illinois 60457
Telephone: (312) 471-4480

Three professional pilots designed and built a unique light aircraft known as the Breezy Model RLU-1, the designation being made up of the initial letters of the surnames of the designers.

Well over 500 sets of plans have been sold, and many Breezys are flying, including examples built in Australia, Canada and South Africa.

BREEZY MODEL RLU-1

Described as being of vintage configuration with all modern facilities, such as full radio, instruments and hydraulic brakes, the prototype Breezy is an open three-seat light aircraft powered by a 67 kW (90 hp) Continental engine.

Construction took six months, at a cost of $3,400, including radio. First flight was made on 7 August 1964.

First Breezy to be built from the published plans was that constructed by Airpark Aero of Santa Rosa, California, for Mr Jack Gardiner of Pandora, Ohio. It differed from the prototype only by having two bucket seats, one of these replacing the usual two-place bench seat behind the pilot.

The extensively modified Breezy shown in the accompanying illustration was built by Mr R. Fabian. It utilises the wing, tail unit, wheels and fairings, wheel brakes and seats of a Cessna Model 172, and is powered by a 108 kW (145 hp) Continental O-300-D engine.

The following description applies to the prototype Breezy:

TYPE: Three-seat homebuilt parasol-wing monoplane.

WINGS: Strut-braced parasol-wing monoplane. Standard Piper PA-12 wing, with Vee streamline-section bracing struts each side.

FUSELAGE: Triangular-section welded chrome-molybdenum steel tube structure, without any covering.

TAIL UNIT: Welded chrome-molybdenum steel tube braced structure; all surfaces fabric-covered.

LANDING GEAR: Non-retractable tricycle type. Main wheels and tyres size 6·00-6, 4-ply; nosewheel and tyre size 5·00-5. Cleveland hydraulic brakes.

POWER PLANT: One 67 kW (90 hp) Continental C90-8F-P flat-four engine, driving a Flottorp 72A50 two-blade pusher propeller. Single fuel tank, capacity 68 litres (18 US gallons), in wing centre-section. Oil capacity 4·5 litres (1·25 US gallons).

ACCOMMODATION: Seats for three in tandem. Pilot on single seat forward, two passengers on bench seat aft.

DIMENSIONS, EXTERNAL:

Wing span	10·06 m (33 ft 0 in)
Wing area, gross	15·3 m² (165 sq ft)
Length overall	6·86 m (22 ft 6 in)
Height overall	2·59 m (8 ft 6 in)
Wheel track	1·83 m (6 ft 0 in)
Wheelbase	3·05 m (10 ft 0 in)

WEIGHTS AND LOADINGS:

Weight empty	317 kg (700 lb)
Max T-O weight	544 kg (1,200 lb)
Max wing loading	35·5 kg/m² (7·27 lb/sq ft)
Max power loading	8·12 kg/kW (13·3 lb/hp)

PERFORMANCE:

Never-exceed speed	91 knots (168·5 km/h; 105 mph)
Cruising speed, 70% power	65 knots (121 km/h; 75 mph)
Stalling speed	26 knots (49 km/h; 30 mph)
Service ceiling	4,572 m (15,000 ft)
T-O run (grass)	137 m (450 ft)
T-O to 15 m (50 ft)	335 m (1,100 ft)
Landing from 15 m (50 ft)	335 m (1,100 ft)
Landing run (grass)	91 m (300 ft)
Range with max fuel	217 nm (402 km; 250 miles)

ROTORWAY

ROTORWAY INC

ADDRESS:
14805 S. Interstate 10, Tempe, Arizona 85281
Telephone: (602) 963-6652

Mr B. J. Schramm formed the Schramm Aircraft Company to market, in both ready-to-fly and prefabricated component form, a single-seat helicopter of his own design, named the Javelin. Details of this aircraft, which flew for the first time in August 1965, can be found in the 1967-68 *Jane's*.

Subsequently, a new company named RotorWay Inc was formed to market to amateur constructors plans and kits of components to build Mr Schramm's Scorpion helicopter, described as a production version of the Javelin. The company has now ended production of components for this helicopter, of which details can be found in the 1972-73 *Jane's*. It has been superseded by the two-seat Scorpion Too, which now uses a new power plant in place of its original 104·5 kW (140 hp) Vulcan V-4 four-cylinder watercooled marine engine.

RotorWay offers comprehensive plans, technical advice from its engineers, a preflight training school, a complete kit to build the Scorpion Too, or a series of small progressive kits, allowing the constructor to proceed as finance allows. The company will also supply plans and rotor blades to those builders wishing to provide their own materials and power plant.

ROTORWAY SCORPION TOO

TYPE: Two-seat light helicopter.

ROTOR SYSTEM: Two-blade semi-rigid main rotor, incorporating Schramm Tractable Control rotor system. Blade section NACA 0015. Blades, which do not fold, are attached to aluminium rotor hub by retention straps. Two-blade aluminium teetering tail rotor. Swashplate for cyclic pitch control. Cable through rotor shaft to blades for collective pitch control.

ROTOR DRIVE: Drive from engine to vertical shaft via eight Vee-belts. Drive from vertical shaft to main rotor shaft via three chain sprockets. Tail rotor driven by Vee-belt from first stage of reduction pulleys.

FUSELAGE: Basic steel tube structure of simplified form. Removable glassfibre body fairing.

TAIL UNIT: Braced steel tube tailboom only, to carry tail rotor.

LANDING GEAR: Tubular skid type.

POWER PLANT: One RotorWay horizontally-opposed four-stroke liquid-cooled engine, designed and produced for the Scorpion Too by RotorWay Inc and mounted aft of cabin area. Standard fuel capacity 37·5 litres (10 US gallons) in tank mounted above drive chain, aft of main rotor shaft. Optional increased capacity fuel tanks available.

ACCOMMODATION: Two individual bucket seats, side by side, in enclosed cabin.

DIMENSIONS, EXTERNAL:

Diameter of main rotor	7·32 m (24 ft 0 in)
Diameter of tail rotor	1·10 m (3 ft 7¼ in)
Length, nose to tail rotor axis	6·18 m (20 ft 3½ in)
Height to top of main rotor	2·22 m (7 ft 3½ in)
Width of cabin	1·22 m (4 ft 0 in)
Landing skid track	1·64 m (5 ft 4¾ in)

WEIGHTS:

Weight empty	340 kg (750 lb)
Max T-O weight	544 kg (1,200 lb)

PERFORMANCE (at max T-O weight, except where indicated):

Cruising speed	74-78 knots (136-145 km/h; 85-90 mph)
Max rate of climb at S/L	244 m (800 ft)/min
Hovering ceiling in ground effect	1,675 m (5,500 ft)
Service ceiling	3,050 m (10,000 ft)
Range, standard fuel	104 nm (193 km; 120 miles)
Range, auxiliary fuel	156 nm (290 km; 180 miles)

RUTAN
RUTAN AIRCRAFT FACTORY

ADDRESS:
PO Box 656, Mojave Airport, Mojave, California 93501
Telephone: (805) 824-2645
PRESIDENT: Elbert L. Rutan

RUTAN VARIVIGGEN

The prototype of a new light aircraft, known as the VariViggen (N27VV), was rolled out on 27 February 1972. Mr Rutan had begun its design in 1963 and its configuration was developed via a low-cost automobile-mounted test system. This involved construction of a one-fifth scale model which was mounted on a specially-built test rig attached to the roof of a motor car. Ailerons, rudders and canard elevators were operated by remote control from within the car, and transducers in the test rig allowed measurement of airspeed, angle of attack, lift, drag, sideslip, side force, roll moment and elevator/aileron/rudder positions. An extra data channel provided for measurement of stick forces and structural load.

A one-fifth scale radio-controlled model was used to confirm the design's spin-proof characteristics. Construction of the prototype began during 1968, and the first flight was made in May 1972. By early 1976 the VariViggen had accumulated a total flying time of nearly 600 hours, and flight testing had confirmed the spin-free characteristics demonstrated by the free-flying scale model. The prototype had no conventional stall in its original form, and could climb, cruise, glide, turn and land with continuous full aft stick, giving a stable speed of 45 knots (83·5 km/h; 52 mph) throughout. Rate of climb at this speed is 152 m (500 ft)/min. The full-span ailerons provide a high-rate of roll, and a 360° roll can be accomplished at a speed of only 80 knots (148 km/h; 92 mph) without loss of height. Manoeuvrability is such that the aircraft's turn radius is less than 61 m (200 ft) at speeds of 60-110 knots (111-204 km/h; 69-127 mph). Nosewheel rotation on take-off occurs at 50 knots (93 km/h; 58 mph), at which speed the aircraft will enter an immediate stable climb.

The full-span ailerons are described as 'reflexerons', since they serve both as ailerons and as an adjustable 'reflex' control for the main wing. Differential aileron motion is related mechanically to the stick, but the collective 'reflex' is electrically controlled. This is achieved by electrically controlled aileron droop, which causes the resulting nose-down trim to be countered by a nose-up deflection of the elevators on the forward canard surface. Thus both ailerons and elevators serve also as flaps. The ailerons are also adjusted at cruising speed to minimise trim drag and optimise the lift/drag ratio. Trim is achieved by an electrically-controlled bungee device on the mechanical elevator system.

In 1975 Mr Rutan began experimenting with a new SP (special performance) wing outer panel, constructed from urethane foam and unidirectional glassfibre, and with a Wortmann FX-60-126 section. With an increase of wing span to 7·23 m (23 ft 8½ in), and of area to 11·61 m² (125 sq ft), it was anticipated that this would provide a 25% increase in the max rate of climb and give a cruising speed 4·5-6 knots (8-11 km/h; 5-7 mph) greater than with the standard wing. Tests showed that the new wing increased rate of climb, but that the former 'no stall' characteristic had been sacrificed. With the SP wing, the VariViggen will 'roll off' if uncoordinated at full aft stick. It is, however, easier to build and stronger than the aluminium outer panel, and is covered on sets of plans currently available to amateur constructors.

A further refinement on the SP wing is the addition of NASA-developed 'winglets', designed by Dr Whitcomb, to enhance directional stability at no cost in performance.

The SP wing panels attach to the standard inboard wing panels, without modification. Each SP outer wing houses a 28·4 litre (7·5 US gallon) fuel tank, giving an additional 56·8 litres (15 US gallons) of auxiliary fuel.

About 500 sets of VariViggen plans have been sold, and it is believed that approximately 245 aircraft are being built.

The following details apply to the VariViggen with the original type of outer wing panels:

TYPE: Two-seat (or 2+2) homebuilt light aircraft.

WINGS: Cantilever low-wing monoplane of cropped delta configuration. Rutan wing section. Thickness/chord ratio 7% at root, 9% at tip. Dihedral 3° on outer wing panels. Incidence 0°. Sweepback at quarter-chord 27°. Composite structure with spruce spars, plywood ribs and skin, Ceconite-covered, except for outboard aft wing panels which are of flush-riveted metal construction. Inward-canted fin and rudder each side at approximately one-third span. Full-span ailerons, extending between fins and wingtips, constructed as a shell of ·016 in aluminium with foam filling. Cutout in inboard trailing-edges to accommodate pusher propeller.

CANARD SURFACES: Cantilever structure mounted high on the nose, forward of the windscreen. Aerofoil section NACA 4414 (modified). Slotted flap-type elevator in trailing-edge.

FUSELAGE: Basically square-section fuselage of wooden construction, Ceconite-covered. Canard surfaces mounted high on nose. Landing light in nosecone, which is hinged at top and opens upwards for access to equipment. Engine mounted in aft end of fuselage.

LANDING GEAR: Electrically-retractable tricycle type. Nosewheel retracts forward, main wheels inward into wings. Nosewheel mounted on oleo-pneumatic shock-strut. Shock-absorption of main units by rubber discs in compression. Goodyear main wheels, with tyres size 14 × 5·00-5, pressure 2·07 bars (30 lb/sq in). Scott nosewheel with 228 mm (9 in) diameter tyre, pressure 2·07 bars (30 lb/sq in). Goodyear caliper brakes.

POWER PLANT: One 112 kW (150 hp) Lycoming O-320-A2A flat-four engine, mounted in rear fuselage and driving a Hegy two-blade wooden fixed-pitch propeller. One fuel tank in fuselage, capacity 87 litres (23 US gallons) and one external fuel tank, mounted on aircraft centreline under fuselage, capacity 45 litres (12 US gallons). Total capacity 132 litres (35 US gallons). Refuelling point on fuselage upper surface. Oil capacity 7·5 litres (2 US gallons).

ACCOMMODATION: Two seats in tandem in individual cockpits, beneath transparent canopies which are hinged on the starboard side. Space for two children, each weighing not more than 22·5 kg (50 lb), or 45 kg (100 lb) of baggage aft of rear seat.

SYSTEMS: Dual 12V electrical systems. Storage battery. Hydraulic system for brakes only.

ELECTRONICS AND EQUIPMENT: ARC 360-channel VHF com transceiver, ARC 200-channel VHF nav receiver. Edo Air transponder. Angle of attack indicator.

DIMENSIONS, EXTERNAL:

Wing span	5·79 m (19 ft 0 in)
Wing chord at root	2·26 m (7 ft 5 in)
Wing chord at tip	0·89 m (2 ft 11 in)
Wing area, gross	11·06 m² (119 sq ft)
Wing aspect ratio	3
Length overall	5·79 m (19 ft 0 in)
Canard surface span	2·44 m (8 ft 0 in)
Wheel track	2·20 m (7 ft 2½ in)
Wheelbase	2·44 m (8 ft 0 in)
Propeller diameter	1·78 m (5 ft 10 in)
Propeller ground clearance	0·41 m (1 ft 4 in)

DIMENSIONS, INTERNAL:

Cabin: Length	2·54 m (8 ft 4 in)
Max width	0·64 m (2 ft 1 in)

WEIGHTS AND LOADINGS:

Weight empty, equipped	431 kg (950 lb)
Max T-O and landing weight	771 kg (1,700 lb)
Max wing loading	69·8 kg/m² (14·3 lb/sq ft)
Max power loading	6·88 kg/kW (11·3 lb/hp)

PERFORMANCE (with original type of wings, at max T-O weight, except as indicated):

Never-exceed speed	156 knots (289 km/h; 180 mph)
Max level speed at S/L	142 knots (262 km/h; 163 mph)
Max cruising speed at 2,135 m (7,000 ft)	130 knots (241 km/h; 150 mph)
Econ cruising speed at 2,135 m (7,000 ft)	109 knots (201 km/h; 125 mph)
Max rate of climb at S/L	366 m (1,200 ft)/min
Service ceiling	4,265 m (14,000 ft)
T-O run	244 m (800 ft)
T-O to 15 m (50 ft)	290 m (950 ft)
Landing from 15 m (50 ft) at max landing weight	183 m (600 ft)
Landing run at max landing weight	146 m (480 ft)
Range with max fuel, 30 min reserve	347 nm (643 km; 400 miles)

RUTAN VARIEZE

Mr Rutan has designed and built two prototypes of a high-performance two-seat aircraft of canard configuration, named the VariEze, of which plans are available. The name stems from its simplicity of construction, the entire structure being a composite of high-strength, primarily unidirectional glassfibre with rigid urethane foam as core material.

The configuration of the VariEze is based on that of the VariViggen, with canard foreplanes; but the cropped-delta wings of that aircraft are replaced by more conventional swept wings of high aspect ratio. At the tip of each wing is a vertical fin, known as a 'winglet'. Developed by Dr Richard Whitcomb at NASA's Langley Research Center, each 'winglet' consists of large above-wing and small below-wing surfaces. That beneath the wing extends aft from the leading-edge to 33% of the tip chord, is cambered inward and inclined outward at 30° from vertical. The above-wing surface, extending aft from 33% of the tip chord, is cambered outward and also inclined outward at 15° from vertical. This 'winglet' system has been shown to 'unwind' the wingtip vortex to a maximum, limiting induced drag and resulting in a fuel saving of 6%; the inclination of the upper and lower surfaces also offsets 40% of the parasite drag of the vertical fins. These vertical surfaces include rudders, which are moved outward individually by single cables and centralised by return springs. The rudder control cables include in their run slotted bellcranks, so that extended movement of either rudder pedal causes the bellcrank to actuate its respective wheel-brake master cylinder. Roll and pitch control are combined in a single control stick, operating elevons mounted on the trailing-edge of the canard foreplanes.

The tricycle landing gear has another unusual feature: the main units are fixed but the nosegear retracts both on the ground and in the air. Retracting the nose unit on the ground, termed kneeled parking, not only facilitates access to the cockpits but raises the propeller at a convenient height for hand-swinging and, at the same time, dispenses with a need for wheel chocks. The gear is designed to extend and retract with both pilot and passenger in the aircraft.

Designed in late 1974, the first VariEze (N7EZ) was built over a ten-week period in the Spring of 1975 and made its first flight on 21 May, powered by a 47 kW (63 hp; 1,834 cc) Volkswagen engine. By early 1976 it had logged approximately 220 flying hours. Optimum economy cruise performance was a primary design aim, so that the prototype could be used to attack existing world distance records in the under 500 kg gross weight class C1a. On 4 August 1975 the aircraft set a new closed-circuit distance record in this class, by covering 1,415·119 nm (2,620·80 km; 1,628·49 miles).

A second prototype (N4EZ), embodying some modifications and powered by a Continental O-200 engine, was built during the Winter of 1975-76. It represents the prototype of the version for which plans are available, and is described in detail below.

All raw materials and certain component parts of the VariEze are also available to homebuilders, including the Plexiglas canopy, moulded glassfibre nosewheel and main landing gear struts, glassfibre cowling, moulded urethane foam combined centre bulkhead/seat, a three-piece wing spar/centre-section spar system permitting removal of each wing after withdrawal of a single retaining pin, and the engine mounting.

TYPE: Two-seat homebuilt sporting aircraft.

WINGS: Cantilever mid-wing monoplane with swept surfaces. Single-spar structure of unidirectional glassfibre with rigid urethane foam core. Vertical above-wing 'winglet' surface at each wingtip includes rudder.

CANARD FOREPLANE: Cantilever structure of unidirectional glassfibre with rigid urethane foam core. Trailing-edge elevons of similar construction provide pitch and roll control.

FUSELAGE: Composite structure comprising large sheets of rigid urethane foam, with wood strips as corner fillers, and internal and external covering of unidirectional glassfibre. A standard-size light alloy extrusion is used for engine, landing gear, and control stick mounts, as well as for canopy latches and other parts.

LANDING GEAR: Tricycle landing gear, with fixed main units and mechanically-retractable nosewheel which is carried on a glassfibre strut moulded to conform to the outside contour of the fuselage, so eliminating need for a fairing door. Nose gear retracted by handcrank or optionally, a small electric motor. Main wheels carried on one-piece moulded glassfibre strut. Fairings on main wheels. Hydraulically-operated brakes.

POWER PLANT: One 74·5 kW (100 hp) Continental O-200-B flat-four engine, mounted in the aft fuselage and driving a two-blade fixed-pitch pusher propeller with spinner. Provision for Volkswagen engines of 1,700 cc to 2,100 cc capacity, or Continental C65, C75, C85 or C90 engines. Fuel tanks, of glassfibre/foam/glassfibre sandwich construction, form strakes on each side of the fuselage that fair the wing roots; total capacity 75 litres (20 US gallons). Both tanks drain to a lower central sump tank mounted above the engine carburettor. There are no fuel control cocks. Refuelling points on each side of fuselage, on the upper surface of tanks.

ACCOMMODATION: Pilot and passenger on semi-reclining seats in individual cockpits. Side-stick controls. One-piece bubble canopy of moulded Plexiglas covers both cockpits and is hinged on starboard side. Roll-over structure. Space for 14 kg (30 lb) baggage in two specially designed suitcases which fit in rear seat area.

SYSTEM: Hydraulic system for wheel brakes only.

DIMENSIONS, EXTERNAL:

Wing span	6·81 m (22 ft 4 in)
Wing chord at root	0·71 m (2 ft 4 in)
Wing chord at tip	0·305 m (1 ft 0 in)
Foreplane span	4·02 m (13 ft 2½ in)
Foreplane chord, constant	0·32 m (1 ft 0½ in)
Wing area, gross	4·98 m² (53·6 sq ft)
Length overall	3·78 m (12 ft 5 in)

WEIGHTS AND LOADINGS (Continental O-200 engine):

Weight empty	222 kg (490 lb)
Max T-O weight	444 kg (980 lb)
Max wing loading	71·3 kg/m² (14·6 lb/sq ft)
Max power loading	5·96 kg/kW (9·8 lb/hp)

PERFORMANCE (Continental O-200 engine):

Max cruising speed	181 knots (335 km/h; 208 mph)
Econ cruising speed	126 knots (233 km/h; 145 mph)
Stalling speed	52·5 knots (97 km/h; 60 mph)
Max rate of climb at S/L	549 m (1,800 ft)
T-O run	230 m (750 ft)
Landing run	244 m (800 ft)
Range at 75% power	607 nm (1,126 km; 700 miles)
Range at econ cruising speed	955 nm (1,770 km; 1,100 miles)

RotorWay Scorpion Too two-seat lightweight helicopter

Shober Willie II two-seat homebuilt light aircraft (Lycoming O-360-A3A engine)

Rutan VariViggen two-seat canard delta

Rutan VariEze two-seat high-performance tail-first monoplane

Two views of the Sawyer Skyjacker II two-seat concept demonstrator *(Howard Levy)*

SALVAY-STARK

SKYHOPPER AIRPLANES INC

This company was formed by Mr M. E. Salvay, who is currently director of structural design for Rockwell International's B-1 bomber programme, and Mr George Stark, to market plans of a light aeroplane named the Skyhopper which they designed and built in 1944-45 and later developed for amateur construction. Some 500 sets were sold, and about 75 Skyhoppers are under construction or have been completed.

Plans are no longer available. For full details of the Skyhopper see the 1975-76 *Jane's*.

SAWYER

RALPH SAWYER

Address: Lancaster, California

SAWYER SKYJACKER II

Mr Ralph Sawyer, a technician in the aircraft industry for more than 35 years and designer/builder of radio-controlled scale models, has built an aeroplane named Skyjacker II, based on a unique design concept for which he holds a US patent. It has what is virtually a lifting-body configuration, evolved after prolonged study of twin-boomed arrangements for model aircraft. Mr Sawyer found that it was possible to overcome flutter and other problems associated with twin-boom designs by extending the wing chord aft and so eliminating the need for booms.

The Skyjacker II has a short fuselage nacelle, conventional centre-section and two short-span, very wide-chord outer panels. These replace tailbooms; but the gap between their trailing-edges is spanned by a conventional tailplane and elevator. An aileron is hinged to each wing trailing-edge, in line with the elevator. Wing endplates replace the usual tail fins, and a yaw panel is inset near the leading-edge of each endplate to act as a rudder. The endplates and control surfaces all have thick trailing-edges.

The design concept was flight tested with scale models, after which Mr Sawyer attempted to interest several major US aerospace companies in the configuration. Meeting with no success, he constructed the all-metal Skyjacker II over a period of 32 months, as a two-seat demonstrator, at a cost of $17,000. A 93 kW (125 hp) Lycoming O-290-G engine was installed initially, driving a high-pitch propeller in pusher configuration. Mounted at the rear of the fuselage nacelle, engine and propeller were located in the rectangular space forward of the tailplane. Low-level runway hops showed that the aircraft was underpowered; so the O-290-G was replaced by a 149 kW (200 hp) Lycoming IO-360-A1B6D engine, driving a Hartzell constant-speed propeller of lower pitch and larger diameter. The first hop with the new power plant. was made in March 1975, and the first true flight on 3 July 1975. Problems with engine cooling were encountered, and the Skyjacker II was being modified in the Spring of 1976 by the addition of airscoops under the centre-section. Further modifications were being made to accommodate a three-blade propeller, which will be positioned 152 mm (6 in) aft of the location of the original propeller and should remedy performance deficiencies.

Type: Two-seat demonstrator for new design concept.

Wings: Cantilever mid-wing monoplane of large chord. Steel tube centre-section, with 0·635 mm (0·025 in) aluminium skin. All-aluminium outer panels, with internal structure of built-up ribs and girder-like reinforcing, and heavy skin. Endplates at tips, with yaw panels near leading-edge. Aileron on trailing-edge of each outer panel. No tabs or flaps.

Fuselage: Conventional aluminium structure of frames and skin panels, of rectangular section. Two engine airscoops at rear, to each side of canopy fairing.

Tail Unit: Fixed horizontal surface and elevator between trailing-edges of outer wing panels.

Landing Gear: Non-retractable tricycle type, with single wheel on each unit. Each main unit comprises an inclined spring steel leg and vertical shock-absorber. Steerable nosewheel.

POWER PLANT: See introductory notes.

ACCOMMODATION: Two seats in tandem under large transparent canopy. Dual controls, but instruments only at front.

DIMENSIONS, EXTERNAL:

Wing span	5·49 m (18 ft 0 in)
Wing chord, constant	5·33 m (17 ft 6 in)
Length overall	5·33 m (17 ft 6 in)
Height overall	1·88 m (6 ft 2 in)

WEIGHTS:

Weight empty	730 kg (1,610 lb)
Max T-O weight	1,041 kg (2,295 lb)

PERFORMANCE (one pilot and 50% fuel):

T-O speed	48 knots (89 km/h; 55 mph)
Landing speed	39 knots (73 km/h; 45 mph)

SHOBER

SHOBER AIRCRAFT ENTERPRISES

ADDRESS:

PO Box 111, Gaithersburg, Maryland 20760

Mr William C. Shober has designed, built and flown the prototype of a two-seat sporting biplane which he has named Willie II. Being stressed for a loading of ±9*g*, the aircraft is suitable for limited aerobatics in standard homebuilt configuration. The plans which are available to amateur constructors give details of the necessary fuel system conversions to make Willie II capable of inverted flight.

The designer estimates that 2,500 to 3,000 hours of work are involved in building this aircraft and that, using all new materials, the cost is approximately $1,800 plus the price of the engine selected to power it.

SHOBER WILLIE II

TYPE: Two-seat sporting or aerobatic biplane.

WINGS: Braced single-bay biplane, with single streamline-section interplane strut each side and N-type centre-section struts. Dual streamline-section flying and landing wires. Constant-chord wings of M-6 aerofoil section. Dihedral: upper wing 0°, lower wing 3°. Upper wing is sweptback and has a cutout in the trailing-edge of the centre-section. Conventional two-spar structures, with solid spruce spars, and ribs built up from 5/16 in square strip and plywood gussets. Each wing has compression tubes and drag wires, the entire structure being fabric-covered. Wide-span ailerons on lower wings only, of wooden construction with fabric covering. Ailerons controlled by push-rods via bell-crank. No flaps. No trim tabs.

FUSELAGE: Welded structure of 4130 steel tube, with fabric covering. Light alloy engine cowling.

TAIL UNIT: Wire-braced structure of welded 4130 steel tube with spruce tips. Fixed-incidence tailplane. All surfaces fabric-covered. No trim tabs.

LANDING GEAR: Non-retractable tailwheel type. Two side Vees and half-axles hinged to bottom of fuselage. Rubber cord shock-absorption. Cleveland main wheels with tyres size 6·00-6. Cleveland brakes.

POWER PLANT: Prototype has 134 kW (180 hp) Lycoming O-360-A3A flat-four engine, driving a two-blade fixed-pitch propeller. Installation designed to take Lycoming engines of 112 kW to 149 kW (150 hp to 200 hp). Fuel tank in fuselage, immediately aft of firewall, capacity 94·5 litres (25 US gallons). Refuelling point on fuselage upper surface, forward of windscreen.

ACCOMMODATION: Two persons in tandem in open cockpits. Space for 9 kg (20 lb) baggage.

DIMENSIONS, EXTERNAL:

Wing span, upper	6·10 m (20 ft 0 in)
Wing span, lower	5·79 m (19 ft 0 in)
Wing chord, constant	1·02 m (3 ft 4 in)
Wing area, gross	13·75 m² (148 sq ft)
Length overall	5·79 m (19 ft 0 in)

WEIGHTS AND LOADINGS (prototype):

Weight empty	388 kg (856 lb)
Max T-O weight	612 kg (1,350 lb)
Max wing loading	44·4 kg/m² (9·1 lb/sq ft)
Power loading	4·57 kg/kW (7·5 lb/hp)

PERFORMANCE (prototype, at max T-O weight):

Cruising speed, 75% power at 1,525 m (5,000 ft)	130 knots (241 km/h; 150 mph)
Stalling speed	52 knots (96·5 km/h; 60 mph)
Max rate of climb at S/L	915 m (3,000 ft)/min
Service ceiling	4,570 m (15,000 ft)
T-O run	137 m (450 ft)
Landing run	213 m (700 ft)
Range with max fuel	325 nm (603 km; 375 miles)

SIEGRIST

RUDOLF SIEGRIST

ADDRESS:

6451 Myrtle Hill, Valley City, Ohio 44280

Mr R. Siegrist designed and built the prototype of a four-seat cabin monoplane, intending to make plans available to amateur constructors. The design originated in 1965. Construction of the prototype began in 1966, and the first flight was made in June 1971. FAA certification in the Experimental category was awarded in August of the same year.

While en route from California in July 1973, at which time the RS1 had completed approximately 180 flight hours, two-thirds of one propeller blade sheared off in flight. A successful forced landing caused no further damage, but the violent vibration before the engine was shut down had sheared many engine mounting tubes.

A new mount has been designed, differing from the original, to provide space between engine and firewall to mount a governor for a constant-speed propeller that will be 0·10 m (4 in) greater in diameter than the original.

Plans for the RS1 Ilse were delayed by this incident, but were 80% complete in 1975. The RS1 was expected to be flying again in 1976.

SIEGRIST RS1 ILSE

TYPE: Four-seat homebuilt cabin monoplane.

WINGS: Braced high-wing monoplane, with single streamline-section bracing strut each side. Wing section NACA 64₂215. Dihedral 0°. Incidence 1° 30′. Forward sweep 4°. Conventional two-spar structure, with spruce spars and ribs and mahogany plywood skin. The entire surface is covered with glassfibre and epoxy resin. Ailerons of all-metal construction, with light alloy ribs riveted to a steel torque tube, with light alloy skins. Trailing-edge flaps of similar construction to ailerons. No trim tabs.

FUSELAGE: Welded structure of 4130 steel tube with wood stringers. Cockpit area has light alloy skin, the remainder being covered with glassfibre.

TAIL UNIT: Welded 4130 steel tube structure, covered with glassfibre. Swept vertical surfaces. All-moving tailplane with anti-servo tab extending almost full span.

LANDING GEAR: Non-retractable tailwheel type. Main wheels carried on cantilever spring steel legs of modified Wittman type. Goodyear main wheels and tyres size 6·00-6, pressure 1·59 bars (23 lb/sq in). Goodyear hydraulic disc brakes.

POWER PLANT: One 134 kW (180 hp) Lycoming O-360 flat-four engine, driving a two-blade propeller. Four interconnected fuel cells in each wing root, total capacity 189 litres (50 US gallons). Refuelling points on upper surface of wing. Oil capacity 7·5 litres (2 US gallons).

ACCOMMODATION: Four seats in enclosed cabin. Two individual buckets forward, bench seat for two aft. Door in each side of fuselage, hinged at forward edge. Cabin heated and ventilated.

SYSTEMS: Hydraulic system for brakes only. Electric power supplied by 50A engine-driven generator.

ELECTRONICS: Narco Escort 110 com transceiver.

DIMENSIONS, EXTERNAL:

Wing span	8·53 m (28 ft 0 in)
Wing chord, constant	1·45 m (4 ft 9 in)
Wing area, gross	12·36 m² (133 sq ft)
Wing aspect ratio	6
Length overall	6·55 m (21 ft 6 in)
Height overall	1·91 m (6 ft 3 in)
Tailplane span	2·49 m (8 ft 2 in)
Wheel track	1·80 m (5 ft 11 in)
Wheelbase	5·03 m (16 ft 6 in)
Propeller diameter	1·93 m (6 ft 4 in)

WEIGHTS AND LOADINGS:

Weight empty	532 kg (1,173 lb)
Max T-O weight	943 kg (2,080 lb)
Max wing loading	76·36 kg/m² (15·64 lb/sq ft)
Max power loading	7·04 kg/kW (11·56 lb/hp)

PERFORMANCE (at max T-O weight):

Never-exceed speed	165 knots (306 km/h; 190 mph)
Max level speed at S/L	148 knots (274 km/h; 170 mph)
Cruising speed at 60% power	130 knots (241 km/h; 150 mph)
Landing speed	63 knots (117 km/h; 73 mph)
Stalling speed, power off	58 knots (108 km/h; 67 mph)
T-O run	approx 396 m (1,300 ft)
Landing run	approx 457 m (1,500 ft)

SINDLINGER

FRED G. SINDLINGER

ADDRESS:

5923 9th Street NW, Puyallup, Washington 98371

Mr Sindlinger began the design of a 5/8-scale replica of the second World War Hawker Hurricane IIC fighter in April 1969. Construction of the prototype was started three months later and the first flight was made in January 1972. By February 1973, Mr Sindlinger's Hurricane had accumulated a total of approximately 160 flying hours, and a full stress analysis of the aircraft had been completed. Plans and certain component parts were made available to amateur constructors from September 1973; by January 1976 about thirty sets had been sold and approximately 12 aircraft were under construction, including examples being built in Africa, Australia and Germany.

SINDLINGER HH-1 HAWKER HURRICANE

TYPE: Homebuilt single-seat sporting aircraft.

WINGS: Cantilever low-wing monoplane. Wing section NACA 2418 in centre-section, with progressive change to NACA 2412 at tip. Dihedral 3° 30′ on outer panels only. Incidence 0° 36′. Sweepback at quarter-chord 3°. Two-spar structure of wood. Front spar is of I-beam construction in the centre-section and of built-up box section in the outer panels. Rear spar is an I-beam in the centre-section, and of U-channel form in the outer panels. Built-up truss ribs and 2·4 mm (3/32 in) plywood skin, fabric-covered overall. Frise-type ailerons of wood construction, fabric-covered and statically balanced. Split trailing-edge flaps of wood.

FUSELAGE: All-wood monocoque structure of 3·2 mm (1/8 in) plywood from firewall to aft of cockpit. Rear fuselage is a built-up box truss frame, with formers and stringers. Entire structure fabric-covered.

TAIL UNIT: All-wood cantilever structure. Two-spar tailplane and fin have plywood skins, covered with fabric overall. Rudder and elevators are of wood construction with fabric covering. Elevator trimmed by internal spring tension. Ground-adjustable trim tab on rudder.

LANDING GEAR: Manually-retractable tailwheel type. Main wheels retract inward. Shock-absorption of main and tail units by coil spring inside steel tubes. Tailwheel steerable. Goodyear main wheels and tyres size 5·00-5. Goodyear hydraulic brakes.

POWER PLANT: One 112 kW (150 hp) Lycoming O-320 flat-four engine, driving a Hartzell two-blade metal constant-speed propeller with spinner. Three fuel tanks: one in fuselage aft of firewall with capacity of 54 litres (14 US gallons), and one in each wing root, with capacity of 30 litres (8 US gallons) each. Total fuel capacity 114 litres (30 US gallons). Refuelling point in fuselage upper surface, forward of windscreen. Oil capacity 7·5 litres (2 US gallons). Glassfibre engine cowlings.

ACCOMMODATION: Pilot only, beneath rearward-sliding transparent canopy. Cockpit heated and ventilated. Space for 18 kg (40 lb) baggage aft of pilot's seat.

SYSTEMS: Hydraulic system for brakes only. Electrical system powered by 12V DC engine-driven generator. 12V 35Ah battery.

ELECTRONICS AND EQUIPMENT: 90-channel VHF com transceiver, VOR Omni nav receiver. Partial IFR instrumentation. Wooden imitation cannon in wing leading-edges.

DIMENSIONS, EXTERNAL:

Wing span	7·62 m (25 ft 0 in)
Wing chord on centre-section (constant)	1·52 m (5 ft 0 in)
Wing chord at tip	0·91 m (3 ft 0 in)
Wing area, gross	9·38 m² (101 sq ft)
Wing aspect ratio	6·2
Length overall	5·99 m (19 ft 8 in)
Height overall	1·78 m (5 ft 10 in)
Tailplane span	2·24 m (7 ft 4 in)
Wheel track	1·83 m (6 ft 0 in)
Propeller diameter	1·93 m (6 ft 4 in)
Propeller ground clearance	0·28 m (11 in)

DIMENSION, INTERNAL:

Cockpit: Max width	0·66 m (2 ft 2 in)

WEIGHTS AND LOADINGS:

Weight empty	456 kg (1,005 lb)
Max T-O weight	624 kg (1,375 lb)
Max wing loading	66·4 kg/m² (13·6 lb/sq ft)
Max power loading	5·58 kg/kW (9·2 lb/hp)

PERFORMANCE (at max T-O weight):

Never-exceed speed	208 knots (386 km/h; 240 mph)
Max level speed at S/L	174 knots (322 km/h; 200 mph)
Max cruising speed, 65% power at 1,830 m (6,000 ft)	143 knots (265 km/h; 165 mph)
Econ cruising speed, 60% power at 2,745 m (9,000 ft)	135 knots (249 km/h; 155 mph)
Stalling speed, flaps up	58 knots (108 km/h; 67 mph)

Siegrist RS1 Ilse four-seat cabin monoplane (Lycoming O-360 engine)

Smith DSA-1 Miniplane *(Gordon S. Williams)*

Sindlinger ⅝-scale replica of a Hawker Hurricane IIC

Smyth Sidewinder *(Howard Levy)*

Stalling speed, flaps down
54 knots (100 km/h; 62 mph)
Max rate of climb at S/L 564 m (1,850 ft)/min
T-O run 107 m (350 ft)
Landing run 168 m (550 ft)
Range, 55% power at 2,285 m (7,500 ft), 30 min reserve 477 nm (885 km; 550 miles)
Max range, no reserve fuel
542 nm (1,005 km; 625 miles)

SMITH

MRS FRANK W. (DOROTHY) SMITH

ADDRESS:
3502 Sunny Hills Drive, Norco, California 91760

The late Frank W. Smith built and flew in October 1956 the prototype of a single-seat fully-aerobatic sporting biplane which he designated the DSA-1 (Darn Small Aeroplane) Miniplane. Plans of this aircraft continue to be marketed by Mrs Smith, and about 350 sets have been sold to constructors in several countries. Two Miniplanes are known to be flying in France, one in West Germany and others in Sweden and England.

Mrs Smith's son designed a two-seat version of the DSA-1, which was provisionally designated Miniplane + 1; all available details of this yet-uncompleted aircraft appeared in the 1974-75 *Jane's*.

SMITH DSA-1 MINIPLANE

The following details refer to the standard Miniplane, built according to Frank Smith's original plans:

TYPE: Single-seat sporting biplane.

WINGS: Braced biplane with N-type interplane struts each side and two N-type strut assemblies supporting centre of top wing above fuselage. NACA 4412 wing section. Dihedral 2° on lower wings only. Incidence 0° on top wing, 2° on lower wings. All-wood structure, fabric-covered. Fabric-covered wooden ailerons on lower wings only. No flaps.

FUSELAGE: Welded steel tube structure, fabric-covered.

TAIL UNIT: Wire-braced welded steel tube structure, fabric-covered. Adjustable-incidence tailplane.

LANDING GEAR: Non-retractable tailwheel type. Tripod streamlined-tube main legs. Compression-spring shock-absorbers optional (now fitted on prototype). Goodyear main wheels and tyres, size 7·00-4, pressure 1·38 bars (20 lb/sq in). Goodyear shoe-type brakes. Scott tailwheel.

POWER PLANT: Designed to take any engine in 48·5-93 kW (65-125 hp) category. Prototype has 80·5 kW (108 hp) Lycoming O-235-C flat-four engine, driving a Sensenich two-blade metal fixed-pitch propeller. Most aircraft have a 48·5 kW (65 hp) Continental A65, 56 kW (75 hp) Continental A75 or 93 kW (125 hp) Lycoming flat-four engine. Fuel in tank in fuselage, capacity 64·5 litres (17 US gallons). Oil capacity 5·7 litres (1·5 US gallons).

ACCOMMODATION: Single seat in open cockpit. Space for 27 kg (60 lb) baggage.

DIMENSIONS, EXTERNAL:
Wing span (upper) 5·18 m (17 ft 0 in)
Wing span (lower) 4·80 m (15 ft 9 in)
Wing chord, constant (both) 0·91 m (3 ft 0 in)
Wing area, gross 9·29 m² (100 sq ft)
Length overall 4·65 m (15 ft 3 in)
Height overall 1·52 m (5 ft 0 in)
Wheel track 1·52 m (5 ft 0 in)
Propeller diameter 1·80 m (5 ft 11 in)

WEIGHTS AND LOADING (prototype):
Weight empty, equipped 279 kg (616 lb)
Max T-O weight 454 kg (1,000 lb)
Max wing loading 48·8 kg/m² (10 lb/sq ft)

PERFORMANCE (prototype, at max T-O weight):
Max level speed at S/L
117 knots (217 km/h; 135 mph)
Max cruising speed 102 knots (190 km/h; 118 mph)
Econ cruising speed 96 knots (177 km/h; 110 mph)
Stalling speed 48 knots (88·5 km/h; 55 mph)
Max rate of climb at S/L 380 m (1,250 ft)/min
Service ceiling 3,960 m (13,000 ft)
T-O run 107 m (350 ft)
Landing run 152 m (500 ft)
Endurance with max fuel 2 hr 30 min

SMYTH

JERRY SMYTH

ADDRESS:
Box 308, Huntington, Indiana 46750

In February 1958 Mr Smyth began the design of a sporting monoplane, setting out to evolve an aircraft that would be reasonably easy to construct, easy to fly, stressed to 9*g* for limited aerobatics, of good appearance and offering economic operation. Construction of the prototype began in January 1967, and occupied two years before completion, at a cost of around $2,500. First flight of what Mr Smyth named the Model 'S' Sidewinder was made on 21 February 1969, and this aircraft received the 'Outstanding Design' award at the EAA Fly-in at Rockford, Illinois, in 1969.

Construction was simplified by utilising a number of standard and readily-obtainable items of equipment. For example, the bubble canopy is that of a Thorp T-18, and Wittman landing gear is used. Plans are available to amateur constructors, and Mr Smyth can also supply a glassfibre nosewheel fairing and two-piece engine cowling to those constructors who do not wish to mould their own.

An illustration in the 1975-76 *Jane's* showed a Sidewinder with retractable landing gear, built by Mr Donald Adams of Columbia City, Indiana. Powered, like the prototype, by a 93 kW (125 hp) Lycoming O-290-G engine, the retractable gear allows a max cruising speed of 156 knots (290 km/h; 180 mph), an increase of 17 knots (32 km/h; 20 mph) by comparison with Mr Smyth's prototype. Construction of Mr Adams' Sidewinder occupied 20 months at a cost of approximately $4,300, with first flight on 11 June 1973.

SMYTH MODEL 'S' SIDEWINDER

The following description applies to Mr Smyth's prototype:

TYPE: Two-seat homebuilt sporting monoplane.

WINGS: Cantilever low-wing monoplane. Wing section NACA 64-612 at root, NACA 64-210 at tip. Dihedral 4°. Incidence 1° 30′. No sweepback. All-metal structure comprising a centre-section and two outer wing panels. Built-up main spar of ·040 in 2024-T3 aluminium 'U'-sections, to which flat aluminium cap strips are riveted; secondary spar is of formed sections. Eleven equally-spaced ribs in each wing panel are made of ·025 in 6061-T4 aluminium. The wing skin, of ·025 in 2024-T3 aluminium, is in three sections: leading-edge, lower and upper skin, and is flush-riveted. Wings filled with epoxy. Simple sealed-gap ailerons of aluminium construction, attached to secondary spar by piano-type hinge. No trim tabs. No flaps.

FUSELAGE: Welded steel tube structure with aluminium formers and skin. Electrically-operated speed brake may be fitted on lower fuselage.

TAIL UNIT: Cantilever all-metal structure with swept vertical surfaces. All-moving horizontal surface with electrically-operated anti-servo tab.

LANDING GEAR: Non-retractable nosewheel type. Wittman cantilever spring steel main gear. Main wheels and tyres size 5·00-5, pressure 1·72 bars (25 lb/sq in). Nose unit carries a 25·4 cm (10 in) diameter tailwheel and smooth tyre, free-castoring and non-steerable, pressure 1·72 bars (25 lb/sq in). Cleveland hydraulic brakes. Glassfibre fairings on all wheels.

POWER PLANT: Provision for installation of engines from 67-134 kW (90-180 hp). Prototype has a 93 kW (125 hp) Lycoming O-290-G flat-four engine, driving a two-blade fixed-pitch aluminium propeller with spinner. Fuel tank in fuselage, forward of instrument panel, capacity 66·2 litres (17·5 US gallons). Refuelling point on top of fuselage, forward of windscreen. Provision for wingtip tanks. Oil capacity 7·5 litres (2 US gallons).

ACCOMMODATION: Pilot and passenger, seated side by side under rearward-sliding bubble canopy. Compartment for 40·8 kg (90 lb) of baggage aft of seats. Cabin heated and ventilated.

SYSTEMS: Hydraulic system for brakes and, optionally, for operation of aerodynamic speed brake. Engine-driven generator provides 35A 12V DC for instruments, lights, electrically-operated stabilator tab and optional electrically-driven hydraulic pump to operate aerodynamic speed brake.

ELECTRONICS: Simple 10-channel VHF communications transceiver.

DIMENSIONS, EXTERNAL:

Wing span	7·57 m (24 ft 10 in)
Wing chord at root	1·52 m (5 ft 0 in)
Wing chord at tip	0·91 m (3 ft 0 in)
Wing area, gross	8·92 m² (96 sq ft)
Wing aspect ratio	6·85
Length overall	5·89 m (19 ft 4 in)
Height overall	1·66 m (5 ft 5½ in)
Tailplane span	2·33 m (7 ft 7¾ in)
Wheel track	1·70 m (5 ft 7 in)
Wheelbase	1·30 m (4 ft 3 in)
Propeller diameter	1·70 m (5 ft 7 in)

DIMENSIONS, INTERNAL:

Cabin: Max width	0·97 m (3 ft 2 in)
Baggage compartment	0·25 m³ (9 cu ft)

WEIGHTS AND LOADINGS:

Weight empty	393 kg (867 lb)
Max T-O and landing weight	657 kg (1,450 lb)
Max wing loading	77 kg/m² (15·8 lb/sq ft)
Max power loading	7·05 kg/kW (11·6 lb/hp)

PERFORMANCE (at max T-O weight):

Never-exceed speed	173 knots (321 km/h; 200 mph)
Max level speed at 610 m (2,000 ft)	161 knots (298 km/h; 185 mph)
Max cruising speed, 75% power at 610 m (2,000 ft)	139 knots (257 km/h; 160 mph)
Stalling speed	48 knots (89 km/h; 55 mph)
Max rate of climb at S/L, 32°F (0°C)	366 m (1,200 ft)/min
Max rate of climb at S/L, 75°F (24°C)	274 m (900 ft)/min
Service ceiling	4,570 m (15,000 ft)
T-O run	244 m (800 ft)
T-O to and landing from 15 m (50 ft)	610 m (2,000 ft)
Landing run	457 m (1,500 ft)
Range with max fuel, no reserve	369 nm (684 km; 425 miles)

SORRELL
SORRELL AVIATION

ADDRESS:
Route 1, Box 660, Tenino, Washington 98589
Telephone: (206) 264-2866

Sorrell Aviation designed and built a two-seat aerobatic biplane named the SNS-6 Hiperbipe, with the intention of providing a true HIgh PERformance BIPlane that would be suitable for construction by amateurs. Flight testing confirmed that the aircraft had an oustanding aerobatic performance and, when demonstrated and displayed at the EAA 1973 Fly-in at Oshkosh, it received the Outstanding New Design of 1973 award. Plans for the fully-developed SNS-7 'production' version are not sold separately; but a basic kit package, containing construction drawings for the aircraft and certain completed components, is available.

SORRELL SNS-7 HIPERBIPE

Design of the Hiperbipe began in 1964; construction of the SNS-6 first prototype started in June 1971. This aircraft made its initial flight in March 1973; a second prototype, modified to the SNS-7 standard offered to homebuilders, followed in March 1975.

TYPE: Two-seat homebuilt aerobatic biplane.

WINGS: Braced single-bay biplane, of modified 0012 wing sections. Dihedral 1° 30′ on lower wings only. Incidence 1° 30′ on upper wing, 2° 30′ on lower wings. Sweepback 4° 30′ on lower wings only. Wide-chord welded 4130 steel I-type interplane struts and dual streamline-section landing and flying wires. Conventional structure of spruce spars, web ribs and stressed plywood skin, fabric covered overall. Centre-section of upper wing is skinned with transparent plastics to allow improved visibility for aerobatics. Cambered wingtips. Four full-span 'flaperons' of aluminium alloy sheet pop-riveted to aluminium torque tubes. No tabs.

FUSELAGE: Conventional structure of welded 4130 chrome-molybdenum steel tube, fabric-covered. Glassfibre engine cowling.

TAIL UNIT: Wire-braced structure of welded 4130 chrome-molybdenum steel tube, fabric-covered. Large dorsal fin. Tailplane and inset elevator extend aft of rudder trailing-edge. No tabs.

LANDING GEAR: Non-retractable tailwheel type. Wittman-type tapered spring steel rod main gear, with single wheel, glassfibre wheel fairing, and leg fairing on each unit. Cleveland 6·00-6 main wheels, with low-profile tyres. Maule 0·15 m (6 in) steerable tailwheel. Cleveland hydraulic spot disc brakes.

POWER PLANT: One 134 kW (180 hp) Lycoming IO-360-B1E flat-four engine, driving a Hartzell HC-C2YK-4AF two-blade metal constant-speed propeller with spinner. Fuel tank in fuselage, capacity 147·5 litres (39 US gallons). Christen 801 inverted oil system, capacity 7·5 litres (2 US gallons).

ACCOMMODATION: Two seats side by side in enclosed cabin. Dual controls standard. Forward-hinged door on each side. Baggage capacity 36 kg (80 lb). Cabin heated and ventilated.

SYSTEM: Full electrical system, with lights.

ELECTRONICS AND EQUIPMENT: Narco Comm 11A 360-channel radio.

DIMENSIONS, EXTERNAL:

Wing span	6·96 m (22 ft 10 in)
Wing chord, constant	1·016 m (3 ft 3·9 in)
Wing area (projected)	13·9 m² (150 sq ft)
Length overall	6·35 m (20 ft 10 in)
Height overall	1·80 m (5 ft 10¾ in)
Tailplane span	2·87 m (9 ft 5 in)
Wheel track	2·16 m (7 ft 1 in)
Propeller diameter	1·93 m (6 ft 4 in)

DIMENSIONS, INTERNAL:

Cabin: Length	1·40 m (4 ft 7 in)
Max width	1·07 m (3 ft 6 in)

WEIGHTS AND LOADINGS:

Weight empty	561 kg (1,236 lb)
Max T-O weight: Aerobatic	766 kg (1,690 lb)
Normal	867 kg (1,911 lb)
Wing loading (Aerobatic)	55·5 kg/m² (11·36 lb/sq ft)
Power loading (Aerobatic)	6·47 kg/kW (9·38 lb/hp)

PERFORMANCE (at max T-O weight):

Never-exceed speed	195 knots (362 km/h; 225 mph)
Max level speed at S/L	149 knots (277 km/h; 172 mph)
Max cruising speed at S/L	139 knots (257 km/h; 160 mph)
Econ cruising speed	130 knots (241 km/h; 150 mph)
Stalling speed, flaperons down	43 knots (79 km/h; 49 mph)
Stalling speed, flaperons up	51 knots (94 km/h; 58 mph)
Max rate of climb at S/L	457 m (1,500 ft)/min
Service ceiling, estimated	6,100 m (20,000 ft)
T-O run	122 m (400 ft)
Landing run	181 m (595 ft)
Range	436 nm (807 km; 502 miles)

SPENCER
P. H. SPENCER

ADDRESS:
8725 Oland Avenue, Sun Valley, California 91352
Telephone: (213) 767-7042

Mr P. H. Spencer, who made his first solo flight in a powered aircraft on 15 May 1914, has been associated with the design of several single-engined amphibians, dating back to 1930, when Amphibians Inc of Garden City, Long Island, NY, put the Privateer amphibian into production. This was followed by the Spencer-Larsen, Spencer Air Car S-12, Republic Seabee RC 1, RC 2 and RC 3.

All of the above designs, as well as the Trident TR-1 amphibian prototype (see entry in main Canadian section), are variations of Mr Spencer's basic Air Car configuration, on which he was granted a patent on 3 January 1950. This was originally a two-seat amphibian powered by an 82 kW (110 hp) engine, and was developed into a four-seat version, known as the S-12-C. Mr Spencer then completed the design of the more advanced S-12-D, of which plans are available to homebuilders as well as certain glassfibre mouldings and metal assemblies. Since that time development has continued, the installation of a Teledyne Continental Tiara 6-285-B engine resulting in a change of designation to S-12-E for the prototype. This had accumulated a total of 750 hours flying time by early 1976, 370 of them with the Tiara engine. The wing sweepback has also been increased, from 3° to 5°.

By early 1976 at least 25 Air Cars were known to be under construction, with a variety of power plants ranging from 149 to 212·5 kW (200 hp to 285 hp), and 110 sets of plans had been sold.

The first S-12-E Air Car to be completed from Mr Spencer's plans made its first flight on 1 August 1974. Built by Mr Peter Breinig of Sausalito, California, it had logged 200 flying hours by early 1976. Four more Air Cars flew for the first time in 1975, and another eight or nine were expected to follow in 1976.

SPENCER AMPHIBIAN AIR CAR MODEL S-12-E

TYPE: Four-seat homebuilt amphibian.

WINGS: Braced high-wing monoplane with single streamline-section bracing strut on each side. Specially-designed STOL wing section. Thickness/chord ratio 15%. Dihedral 1°. Incidence 2°. Sweepback at quarter-chord 5°. Conventional two-spar structure of wood, steel and glassfibre. Frise-type ailerons of wooden construction. Electrically-operated trailing-edge flaps. Glassfibre stabilising float mounted on strut beneath each wing at approximately two-thirds span.

HULL: Conventional single-stepped hull with wood frames, longerons and skin. Welded steel tube structure to provide wing and engine mounting and attachment points for landing gear.

TAIL UNIT: Cantilever structure, comprising conventional fin and rudder, and all-moving tailplane set approximately midway up fin. Combined anti-servo and trim tab in tailplane. Retractable water rudder in base of aerodynamic rudder.

LANDING GEAR: Manually-retractable tricycle type. Main wheels retract aft to take up a near-vertical position on each side of the cabin. Nosewheel retracts forward through almost 180° and is partially housed in the nose of the hull, above the waterline, to form a nose fender. Cantilever spring steel main gear. Main wheels and tyres size 7·00-6. Nosewheel and tyre size 6·00-6. Cleveland hydraulic disc brakes.

POWER PLANT: One 212·5 kW (285 hp) Continental Tiara 6-285-B flat-six engine, driving a Hartzell three-blade metal constant-speed reversible pusher propeller. Fuel tanks in fuselage and wing stabilising floats. Total fuel capacity 355 litres (94 US gallons). Oil capacity 8·5 litres (2·25 US gallons).

ACCOMMODATION: Four seats in pairs in enclosed cabin. Backs of front seats fold forward to improve access. Rear seats fold back against bulkhead to provide cargo or baggage space. Baggage space in rear fuselage, aft of rear cabin bulkhead. Door on each side of fuselage, hinged at forward edge. Bow access door on starboard side, hinged on centreline and opening upward. Dual controls standard. Accommodation heated and ventilated.

SYSTEMS: Hydraulic system for brakes only. Electrical system supplied by 24V 50A engine-driven alternator.

ELECTRONICS AND EQUIPMENT: Complete IFR instrumentation. Bendix 360 nav/com transceiver.

DIMENSIONS, EXTERNAL:

Wing span	11·38 m (37 ft 4 in)
Wing chord, constant	1·52 m (5 ft 0 in)
Wing area, gross	17·1 m² (184 sq ft)
Wing aspect ratio	7·4
Length overall	8·05 m (26 ft 5 in)
Height overall	2·90 m (9 ft 6 in)
Tailplane span	3·66 m (12 ft 0 in)
Wheel track	2·54 m (8 ft 4 in)
Wheelbase	3·10 m (10 ft 2 in)
Propeller diameter	2·13 m (7 ft 0 in)
Cabin doors (port and starboard, each): Height	0·97 m (3 ft 2 in)
Width	1·02 m (3 ft 4 in)
Height to sill	0·86 m (2 ft 10 in)
Cabin door (bow, starboard): Length	0·89 m (2 ft 11 in)
Width	0·51 m (1 ft 8 in)

DIMENSIONS, INTERNAL:

Cabin: Length	2·59 m (8 ft 6 in)
Max width	1·14 m (3 ft 9 in)

WEIGHTS AND LOADINGS:

Weight empty	993 kg (2,190 lb)
Max T-O weight	1,451 kg (3,200 lb)
Max wing loading	85·0 kg/m² (17·4 lb/sq ft)
Max power loading	6·83 kg/kW (11·2 lb/hp)

PERFORMANCE (at max T-O weight):

Max level speed at S/L	128 knots (237 km/h; 147 mph)
Max cruising speed at 1,675 m (5,500 ft)	122 knots (225 km/h; 140 mph)
Econ cruising speed, 65% power at 2,315 m (7,600 ft)	117 knots (217 km/h; 135 mph)
Stalling speed, flaps up	46 knots (86 km/h; 53 mph)
Stalling speed, 35° flap	37·5 knots (70 km/h; 43 mph)
Max rate of climb at S/L	305 m (1,000 ft)/min
T-O time from calm water at S/L	16 sec
Range, 65% power at 2,375 m (7,800 ft), 20 min reserve	695 nm (1,285 km; 800 miles)

SPEZIO
WILLIAM EDWARDS

ADDRESS:
25 Madison Avenue, Northampton, Massachusetts 01060

Mr and Mrs Spezio designed and built a two-seat light aircraft named the Tuholer, all rights of which were acquired by Mr Edwards in August 1973. He is continuing

Sorrell Aviation Hiperbipe aerobatic biplane (Lycoming IO-360-B1E engine)

Spencer Amphibian Air Car Model S-12-E *(Don Dwiggens)*

Spezio Tuholer two-seat sporting aircraft

Spratt Model 107 two-seat movable-wing flying-boat

to market plans of the Tuholer to amateur constructors.

SPEZIO DAL-1 TUHOLER

Named Tuholer because of its two open cockpits, the prototype flew for the first time on 2 May 1961.

Folding wings enable the Tuholer to be kept in a normal home garage and it is towed behind a car on its own landing gear. It can be made ready for flight by two people in about 10 minutes or by one person in 20 minutes.

The following description applies to the Tuholer built by Mr Edwards to the current plans:

TYPE: Two-seat homebuilt sporting aircraft.

WINGS: Strut-braced low-wing monoplane, with streamline-section Vee bracing struts each side. Jury struts brace centre of these struts. Clark Y wing section. Dihedral 3°. Incidence 1°. Washout at wingtip 1°. Two-spar spruce structure, with plywood leading-edge and overall fabric covering. Conventional wooden ailerons. Drawings for Frise type ailerons are available. No flaps. Wings fold back along sides of fuselage for stowage.

FUSELAGE: Steel tube structure with wood or light alloy stringers and fabric covering.

TAIL UNIT: Braced steel tube structure, fabric covered. Tailplane incidence adjustable by screwjack.

LANDING GEAR: Non-retractable tailwheel type. Coil spring shock-absorption. Main units fitted with Cleveland wheels and tyres, size 6·00-6. Cleveland brakes. Tyre pressure 2·76 bars (40 lb/sq in). Steerable tailwheel. Wheel fairings optional.

POWER PLANT: One 112 kW (150 hp) Lycoming O-320 flat-four engine. Sensenich two-blade metal fixed-pitch propeller. Glassfibre fuel tank aft of firewall, capacity 90·5 litres (24 US gallons). Oil capacity 7·5 litres (2 US gallons).

ACCOMMODATION: Two persons in tandem in open cockpits. Small baggage compartment aft of rear seat.

EQUIPMENT: Nova-Tech TR-102 radio.

DIMENSIONS, EXTERNAL:

Wing span	7·55 m (24 ft 9 in)
Wing chord, constant	1·52 m (5 ft 0 in)
Wing area, gross	11·21 m² (120·7 sq ft)
Wing aspect ratio	5
Length overall	5·56 m (18 ft 3 in)
Height overall	1·57 m (5 ft 2 in)
Tailplane span	2·26 m (7 ft 5 in)
Wheel track	1·57 m (5 ft 2 in)

WEIGHTS:

Weight empty	408 kg (900 lb)
Max T-O weight	680 kg (1,500 lb)

PERFORMANCE (at max T-O weight):

Max level speed at S/L	130 knots (241 km/h; 150 mph)
Cruising speed	109-117 knots (201-217 km/h; 125-135 mph)
Stalling speed	48 knots (89 km/h; 55 mph)
Max rate of climb at S/L	732 m (2,400 ft)/min
T-O and landing run	61 m (200 ft)
Endurance with max fuel	3 hr

SPRATT

SPRATT AND COMPANY INC

ADDRESS:

PO Box 351, Media, Pennsylvania 19063

Mr George G. Spratt, formerly a design engineer with The Boeing Company and Consolidated Vultee (now Convair), has completed more than 30 years' work on developing a two-piece movable-wing control system, which he claims provides improved safety factors compared with the conventional aileron, elevator and rudder control system.

While he was with Consolidated Vultee, Mr Spratt designed a roadable aircraft which featured an earlier version of his wing control system, but this did not enter production. Since that time Mr Spratt has concentrated on perfecting his idea as a private venture.

To flight test his movable-wing control system, Mr Spratt built a lightweight experimental flying-boat (N910Z) constructed almost entirely of moulded plastics; details of this can be found in the 1974-75 *Jane's*.

SPRATT MODEL 107

Following construction and flight testing of his first experimental flying-boat, Mr Spratt designed and constructed a more advanced prototype known as the Model 107 (N2236) for public demonstration. Dimensions and weights are essentially the same as those of the test vehicle, but construction has been simplified to facilitate fabrication. The Mercury 800 modified outboard marine engine of the Model 107 is of slightly increased capacity, and produces greater horsepower with better fuel economy as a result of improved combustion chamber and inlet port design.

Mr Spratt claims that the Model 107 will neither stall nor spin, and displays 75% less reaction to turbulence than a conventional design.

Plans of this aircraft are available to amateur constructors. By early 1976 nearly 60 sets had been sold, and the first amateur-built Model 107 flew in the Autumn of 1975. Several more were expected to fly by the Summer of 1976.

TYPE: Two-seat lightweight homebuilt flying -boat.

WINGS: Pivoted controllable parasol wings, with inverted Vee bracing struts each side. Wing section NACA 23112. No dihedral. No sweepback. Reinforced plastics structure. No ailerons, flaps or trim tabs. Flying controls so arranged that the wings are allowed to move freely and collectively in incidence, while their incidence is controlled differentially by a steering wheel. The wings' angle of attack can be adjusted by a separate control.

HULL: Structure of polyurethane foam with reinforced plastics skin. Small water rudder interconnected with the control wheel.

TAIL UNIT: Butterfly-type tail unit, with no movable surfaces, constructed of reinforced plastics.

POWER PLANT: One 59·7 kW (80 hp) Mercury 800 modified outboard two-stroke marine engine, driving a two-blade plastics pusher propeller through an extended drive shaft, which locates the propeller between the butterfly tail surfaces. The pitch of the propeller, which is of Mr Spratt's design, is adjustable on the ground. Outboard engine type of fuel tank.

ACCOMMODATION: Two persons side by side, in open cockpit.

DIMENSIONS, EXTERNAL:

Wing span	7·32 m (24 ft 0 in)
Wing chord (constant)	1·22 m (4 ft 0 in)
Wing area, gross	8·92 m² (96 sq ft)
Wing aspect ratio	6
Length overall	5·18 m (17 ft 0 in)
Height overall	1·52 m (5 ft 0 in)
Propeller diameter	1·52 m (5 ft 0 in)

WEIGHTS AND LOADINGS:

Weight empty	226 kg (500 lb)
Max T-O weight	453 kg (1,000 lb)
Max wing loading	50·8 kg/m² (10·4 lb/sq ft)
Max power loading	7·59 kg/kW (12·5 lb/hp)

SPRATT MODEL 105

Mr Spratt has designed a landplane version of the movable-wing flying-boats already described, and a prototype of this aircraft (N49888) has been built by Mr Robert Quaintance of Coatsville, Pennsylvania, as the Spratt Controlwing or Model 105.

Since it first flew, several refinements have been made to the aircraft, aimed at improving its efficiency by reduction of fuselage drag.

Few details of the Model 105 are known, but all available information follows:

TYPE: Two-seat lightweight experimental landplane.

WINGS: As described for Model 107, except wings can be folded alongside fuselage for towing or storage.

FUSELAGE: Primarily a composite structure of polyurethane foam and glassfibre.

TAIL UNIT: Fixed vertical surface only, of polyurethane foam and glassfibre. No movable surfaces.

LANDING GEAR: Non-retractable tricycle type. Steerable nosewheel is of conventional aircraft type. Main wheels are of the automotive type to allow extensive road towing, and are now fitted with fairings. Nosewheel designed to attach to a trailer hitch for road towing.

POWER PLANT: One Mercury outboard motor boat engine of 983 cc capacity mounted in a mid-fuselage position. Two-blade fixed-pitch pusher propeller driven via an extended shaft.

ACCOMMODATION: Two seats side by side in semi-enclosed cockpit.

DIMENSIONS, EXTERNAL:

Wing span	6·71 m (22 ft 0 in)
Length overall	3·81 m (12 ft 6 in)

WEIGHT:

Weight empty	204 kg (450 lb)

Spratt Model 105 landplane built and owned by Mr Robert Quaintance

Steen Skybolt built by Mr Dick Blair of Vincetown, New Jersey *(Howard Levy)*

STEEN
STEEN AERO LAB

Address:
15623 De Gaulle Cir, Brighton, Colorado 80601
Telephone: (303) 659-7182

Mr Lamar Steen, an aerospace teacher in a Denver, Colorado, high school, designed a two-seat fully-aerobatic biplane named Skybolt which was built as a class project in the school. Simplicity of construction was a primary aim of the design, begun in June 1968, and it is stressed to +12 and −10*g*. Construction began on 19 August 1969, costing approximately $5,000, and the first flight was made in October 1970. The Skybolt received an EAA award for Best School Project. Plans are available to amateur constructors together with fuselage and wing kits, and well over 1,000 sets of plans have been sold.

The adjacent illustration shows the Skybolt built by Mr Dick Blair of Vincetown, New Jersey. Powered by a 149 kW (200 hp) Lycoming IO-360 engine, and equipped with an inverted fuel system, its construction cost $7,900 and the first flight was recorded in July 1973. When Mr Blair exhibited his aircraft at the EAA 1973 Fly-in at Oshkosh, it received the Best Skybolt award.

STEEN SKYBOLT

The following description applies to the prototype with 134 kW (180 hp) Lycoming engine, built under Mr Steen's supervision:

Type: Two-seat aerobatic homebuilt biplane.

Wings: Braced biplane with single interplane strut each side. N-type centre-section struts. Streamline-section landing and flying wires. Wing sections: upper wing NACA 63_2A015, lower wing NACA 0012. Incidence (both) 1° 30′. Sweepback 6° on upper wing only. Wooden two-spar structures with spruce spars, built-up ribs and fabric covering. Fabric-covered Frise-type ailerons on upper and lower wings. Cutout in centre-section trailing-edge of upper wing.

Fuselage: Welded structure of 4130 chrome-molybdenum steel tube, with fabric covering.

Tail Unit: Wire-braced welded structure of 4130 chrome-molybdenum steel tube, with fabric covering. Adjustable trim tab in port elevator.

Landing Gear: Non-retractable main wheels and tailwheel. Two side Vees and half axles hinged to fuselage structure. Shock-absorption by rubber bungee. Cleveland wheels with tyres size 6·00-6, pressure 1·72 bars (25 lb/sq in). Cleveland hydraulic disc brakes. Glassfibre fairings for main wheels.

Power Plant: One 134 kW (180 hp) Lycoming HO-360-B1B flat-four engine, driving a McCauley two-blade fixed-pitch propeller with spinner. Provision for alternative engines of 93-194 kW (125-260 hp). Fuselage fuel tank, immediately aft of firewall, capacity 113·4 litres (30 US gallons). Optional tank of 37·8 litres (10 US gallons) capacity can be installed in centre-section of upper wing. Total optional fuel capacity 151·2 litres (40 US gallons). Refuelling points on fuselage upper surface, forward of windscreen, and on top surface of upper wing. Oil capacity 7·5 litres (2 US gallons).

Accommodation: Two seats in open cockpits. Space for 13·6 kg (30 lb) baggage aft of rear seat.

System: Hydraulic system for brakes only.

Electronics: Battery-powered Alpha 200 nav/com transceiver.

Dimensions, external:	
Wing span, upper	7·32 m (24 ft 0 in)
Wing span, lower	7·01 m (23 ft 0 in)
Wing chord (constant, both)	1·07 m (3 ft 6 in)
Wing area, gross	14·2 m² (152·7 sq ft)
Length overall	5·79 m (19 ft 0 in)
Height overall	2·13 m (7 ft 0 in)
Propeller diameter	1·88 m (6 ft 2 in)
Propeller ground clearance	0·31 m (1 ft 0 in)
Weights and Loadings:	
Weight empty	490 kg (1,080 lb)
Max T-O weight	762 kg (1,680 lb)
Max wing loading	53·7 kg/m² (11 lb/sq ft)
Max power loading	5·69 kg/kW (9·3 lb/hp)
Performance (at max T-O weight):	
Max level speed	126 knots (233 km/h; 145 mph)
Cruising speed	113 knots (209 km/h; 130 mph)
Landing speed	43 knots (80·5 km/h; 50 mph)
Max rate of climb at S/L	762 m (2,500 ft)/min
Service ceiling	5,500 m (18,000 ft)
T-O run	122 m (400 ft)
Range with max fuel	390 nm (720 km; 450 miles)

STEPHENS
STEPHENS AIRCRAFT

Address for Plans:
Gerry Zimmerman, 8563 West Sixty-Eighth Place, Arvada, Colorado 80004

Mr C. L. Stephens designed a single-seat aerobatic monoplane specifically for homebuilders who wish to own an aircraft for competitive aerobatics. The prototype, designated Model A, was designed for Margaret Ritchie, US National Women's Aerobatic Champion in 1966, and the second aircraft, the Model B, for Dean S. Engelhardt of Garden Grove, California.

Stressed to +12*g* and −11*g*, it was the first US aircraft known to be designed around the Aresti Aerocriptografic System for competitive aerobatics. All control surfaces are fully static-balanced and the entire aircraft comes very close to being aerodynamically symmetrical. Design of the **Model A** started in July 1966 and construction of the prototype began a month later. First flight of this version was made on 27 July 1967, and of the **Model B**, with wings and ailerons of increased area and reduced fuel tankage, on 9 July 1969. Plans of the Stephens Akro are available to amateur constructors.

STEPHENS AKRO

The following description applies to the prototype with 134 kW (180 hp) Lycoming engine:

Type: Single-seat homebuilt monoplane.

Wings: Cantilever mid-wing monoplane. Wing section NACA 23012. No dihedral, incidence or sweepback. All-wood two-spar structure. One-piece wing, with solid spar passing through fuselage and positioned by means of removable top longeron sections. Rear spar in two pieces. No internal wires or compression struts. Wing covered with mahogany skin. Plain ailerons have a 4130 steel spar, and spruce ribs and trailing-edge, and are fabric-covered. Ground-adjustable trim tabs on ailerons, which are statically balanced. No flaps.

Fuselage: Welded 4130 steel tube structure, mostly of 0·75 in outside diameter tubing, with Ceconite covering.

Tail Unit: Wire-braced welded 4130 steel tube structure with swept surfaces, fabric-covered. Tailplane has variable incidence. Ground-adjustable trim tab on rudder; controllable trim tab in elevator. All control surfaces statically balanced.

Landing Gear: Non-retractable tailwheel type. Cantilever spring steel main gear. Goodyear main wheels and tyres size 5·00-5, pressure 1·93 bars (28 lb/sq in). Cleveland disc brakes. Maule steerable tailwheel. Glassfibre fairings on main wheels.

Power Plant: One 134 kW (180 hp) Lycoming AIO-360-A1A flat-four engine, driving a Sensenich Type 7660 two-blade fixed-pitch metal propeller. Model A has fuel system for prolonged inverted flight, Model B has both fuel and oil system so modified. Model B can also have optional constant-speed propeller. Fuel tank in fuselage, forward of instrument panel. Model A has fuel capacity of 121 litres (32 US gallons), Model B has capacity of 102 litres (27 US gallons). Refuelling point on top of fuselage, forward of windscreen. Oil capacity 7·6 litres (2 US gallons).

Accommodation: Single seat for pilot under rearward-sliding bubble canopy. Large window in underfuselage, forward of control column. Model B has, in addition, a quarter window in each side of the fuselage, beneath the wings. Forced-air ventilation.

System: Hydraulic system for brakes only.

Electronics: Battery-operated Bayside transceiver.

Dimensions, external:	
Wing span	7·47 m (24 ft 6 in)
Wing chord at root	1·60 m (5 ft 3 in)
Wing chord at tip:	
Model A	0·76 m (2 ft 6 in)
Model B	0·91 m (3 ft 0 in)
Wing area, gross:	
Model A	8·73 m² (94 sq ft)
Model B	9·29 m² (100 sq ft)
Length overall	5·82 m (19 ft 1 in)
Height overall	1·73 m (5 ft 8 in)
Tailplane span	2·44 m (8 ft 0 in)
Wheel track	1·37 m (4 ft 6 in)
Propeller diameter	1·93 m (6 ft 4 in)
Weights and Loadings:	
Weight empty:	
Model A	385 kg (850 lb)
Model B	431 kg (950 lb)
Max T-O weight:	
Model A	544 kg (1,200 lb)
Model B	589 kg (1,300 lb)
Max wing loading	63·5 kg/m² (13 lb/sq ft)
Max power loading	4·40 kg/kW (7 lb/hp)
Performance (at 544 kg; 1,200 lb T-O weight):	
Never-exceed speed	191 knots (354 km/h; 220 mph)
Max level speed at 610 m (2,000 ft)	148 knots (274 km/h; 170 mph)
Max cruising speed at 610 m (2,000 ft)	139 knots (257 km/h; 160 mph)
Econ cruising speed at 610 m (2,000 ft)	109 knots (201 km/h; 125 mph)
Stalling speed	48 knots (89 km/h; 55 mph)
Max rate of climb at S/L	1,220 m (4,000 ft)/min
Service ceiling	6,705 m (22,000 ft)
T-O run	61 m (200 ft)
T-O to 15 m (50 ft)	122 m (400 ft)
Landing from 15 m (50 ft)	457 m (1,500 ft)
Landing run	183 m (600 ft)
Range with max fuel	303 nm (563 km; 350 miles)

STEWART
STEWART AIRCRAFT CORPORATION

Address:
11420 Route 165, Salem, Ohio 44460
Telephone: (216) 332-0865

Mr Donald Stewart formed this company to market plans of a simple single-seat light aircraft named the Headwind, of which he designed and built a prototype. During 1969 he designed a new wing for the Headwind and this is an integral part of the plans available to homebuilders. A two-seat version was still under construction in March 1976. Mr Stewart expects to power this aircraft with either a 44·5 kW (60 hp) Franklin or 1,600 cc Volkswagen modified motor car engine with Maximizer. Work on the single-seat Headwind was followed by design and construction of the JD$_2$FF Foo Fighter, which is described separately.

STEWART JD$_1$ HW 1·7 HEADWIND

Built in only five months, the prototype Headwind flew for the first time on 28 March 1962. Plans are available to amateur constructors, who can build the aircraft with either an open or enclosed cockpit. The wings can be removed or fitted in about 20 minutes by two people.

During 1970, the fuselage, tail unit and landing gear of the Headwind were redesigned, resulting in the new designation JD$_1$ HW 1·7.

Fuselage changes included a reduction of 6·4 cm (2½

in) in the depth of the structure, while increasing the headroom by 2·5 cm (1 in). The cowling forward of the instrument panel was lowered by 7·6 cm (3 in), the aft seat support was moved back 5·08 cm (2 in) and the landing gear bracing truss was resited, resulting in increased cockpit volume.

The landing gear was increased in height by 15·2 cm (6 in) and an optional rubber-in-compression shock-strut was designed. All of these changes are shown on current plans and were made available to existing plan holders.

The power plant of the prototype Headwind has a belt-driven propeller reduction drive designed by Mr Stewart. Given the name Maximizer, this unit was put into production in the Spring of 1972 and is available to amateur constructors for use with Volkswagen power plants.

Several thousand sets of plans for the Headwind have been sold. About 150 aircraft are believed to be under construction, with approximately 15 already flying.

The following details apply to Mr Stewart's basic design with the new wing:

Type: Single-seat homebuilt light aircraft.

Wings: Strut-braced high-wing monoplane, with streamline-section Vee bracing strut each side. Wing section NACA 4412. Dihedral 2°. Incidence 2°. Two spruce spars, steel tube compression members, drag and anti-drag wires, plywood ribs, fabric covering. Frise-type ailerons of similar construction to wings. No flaps.

Fuselage: Welded steel tube structure, fabric-covered.

Tail Unit: Braced steel tube structure, fabric-covered. Ground-adjustable tailplane incidence. Fixed tabs on rudder and starboard elevator.

Landing Gear: Non-retractable tailwheel type. Shock-absorption by low-pressure tyres. Hayes main wheels with Goodyear tyres size 8·00-4. Tyre pressure 0·83 bars (12 lb/sq in). Alternatively, rubber-in-compression shock-struts, with Cleveland or Goodyear wheels and tyres size 6·00-6, tyre pressure 1·24 bars (18 lb/sq in). No brakes. Steerable tailwheel.

Power Plant: One 27 kW (36 hp) modified Volkswagen 1,192 cc motor car engine, driving a special Stewart/Kirk two-blade fixed-pitch propeller via a Stewart belt-driven reduction unit, ratio 1·6 : 1. Engines weighing up to 84 kg (185 lb) can be utilised. Fuel tank aft of firewall, capacity 19 litres (5 US gallons). Oil capacity 2·4 litres (5 US pints).

Accommodation: Single seat in open or enclosed cockpit, with door on starboard side. Provision for up to 4·5 kg (10 lb) baggage in net, directly aft of cockpit.

Dimensions, external:

Wing span	8·61 m (28 ft 3 in)
Wing chord, constant	1·22 m (4 ft 0 in)
Wing area, gross	10·3 m^2 (110·95 sq ft)
Wing aspect ratio	7
Length overall	5·41 m (17 ft 9 in)
Height overall	1·68 m (5 ft 6 in)
Tailplane span	2·13 m (7 ft 0 in)
Wheel track	1·52 m (5 ft 0 in)
Wheelbase	4·11 m (13 ft 6 in)
Propeller diameter	1·57 m (5 ft 2 in)

Weights and Loading:

Weight empty	198 kg (437 lb)
Max T-O and landing weight	317 kg (700 lb)
Max wing loading	30·8 kg/m^2 (6·3 lb/sq ft)

Performance (at max T-O weight):

Never-exceed speed	95·5 knots (177 km/h; 110 mph)
Max level speed at S/L	69·5 knots (129 km/h; 80 mph)
Cruising speed	65 knots (121 km/h; 75 mph)
Stalling speed	32-33 knots (58-61 km/h; 36-38 mph)
Max rate of climb at S/L	122 m (400 ft)/min
Absolute ceiling	3,355 m (11,000 ft)
T-O run	91 m (300 ft)
T-O to 15 m (50 ft)	365 m (1,200 ft)
Landing from 15 m (50 ft)	490 m (1,600 ft)
Landing run	137 m (450 ft)
Endurance with max fuel, no reserve	2½ hours

STEWART JD$_2$FF FOO FIGHTER

Design of Mr Stewart's Foo Fighter began in October 1967 and the first prototype (N2123) made its first flight in June 1971 powered by a six-cylinder Ford Falcon motor-car engine developing 89·5 kW (120 hp) at 3,800 rpm. This engine was replaced subsequently by a 93 kW (125 hp) Franklin Sport 4.

A second prototype of the Foo Fighter (M2124), also with a 93 kW (125 hp) Franklin engine, was sold for exhibition flights at Lafayette Escadrille '76 in Pennsylvania. During 1972, Mr Stewart designed new wings for the Foo Fighter, of increased span and chord. Since that time he has modified the design to allow for installation of Lycoming flat-four engines of up to 112 kW (150 hp), and has made refinements to the fuselage, tail unit and landing gear. Sets of plans of the Foo Fighter are available to amateur constructors.

The following description is applicable to the aircraft in fully-updated form:

Type: Single-seat lightweight sporting biplane.

Wings: Braced single-bay biplane. Wing section NACA 4412. Dihedral 0° upper wing, 1° lower wing. Incidence 2° upper wing, 0° lower wing. Conventional structure with two wooden spars and light alloy ribs, fabric-covered. N-type interplane struts each side; two N-type struts, joined at their upper ends, support the centre of the upper wing above fuselage. The lower wing extends below the fuselage, being attached to a cabane, and is faired over with light gauge light alloy sheet. Streamline-section landing and flying wires. Cutout in trailing-edges of both wings. Frise-type ailerons of similar construction to wings. No flaps. No trim tabs.

Fuselage: Welded steel tube structure with wood formers and stringers, fabric-covered.

Tail Unit: Wire-braced welded steel tube structure with fabric covering. Tailplane incidence ground-adjustable. Fixed tab in starboard elevator.

Landing Gear: Non-retractable tailwheel type, with steerable tailwheel or tailskid. Two side Vees with half-axles attached to fuselage structure. Shock-absorption by rubber cords in tension. Stewart wheels with tyres size 3·00-16, pressure 2·76 bars (40 lb/sq in). Stewart caliper brakes.

Power Plant: One 93 kW (125 hp) Franklin Sport 4 flat-four engine, driving a two-blade wooden fixed-pitch propeller. Provision for Lycoming flat-four engine of up to 112 kW (150 hp). Fuel contained in glassfibre tank mounted in fuselage immediately aft of firewall, capacity 72 litres (19 US gallons). Refuelling point on fuselage upper surface. Oil capacity 3·75 litres (1 US gallon).

Accommodation: Single seat in open cockpit. Space for 4·5 kg (10 lb) baggage aft of seat.

Dimensions, external:

Wing span (both)	6·30 m (20 ft 8 in)
Wing chord, constant (both)	1·02 m (3 ft 4 in)
Wing area, gross	13·0 m^2 (140 sq ft)
Wing aspect ratio	6·075
Length overall	5·72 m (18 ft 9 in)
Height overall	2·13 m (7 ft 0 in)
Tailplane span	1·93 m (6 ft 4 in)
Wheel track	1·75 m (5 ft 9 in)
Wheelbase	3·76 m (12 ft 4 in)
Propeller diameter	1·83 m (6 ft 0 in)

Weights and Loading:

Weight empty	328 kg (725 lb)
Max T-O weight	499 kg (1,100 lb)
Max power loading	5·37 kg/kW (8·8 lb/hp)

Performance (at max T-O weight):

Never-exceed speed	126 knots (233 km/h; 145 mph)
Max cruising speed	100 knots (185 km/h; 115 mph)
Stalling speed	42 knots (77·5 km/h; 48 mph)
Max rate of climb at S/L	366 m (1,200 ft)/min
T-O run	137 m (450 ft)
Landing run	168 m (550 ft)

STOLP

STOLP STARDUSTER CORPORATION

Address:

4301 Twining, Riverside, California 92509

Telephone: (714) 686-7943

President: Jim Osborne

General Manager: Eric Shilling

Secretary-Treasurer: Hanako Osborne

Mr Louis A. Stolp and Mr George M. Adams designed and built a light single-seat sporting biplane known as the Starduster, which flew for the first time in November 1957; and founded Stolp Starduster Corporation to market to amateur constructors plans, components and basic materials for the Starduster and subsequent designs. On 1 May 1972 this company was acquired by Jim and Hanako Osborne, who continue to trade under the original name.

Plans of the SA-100 Starduster are no longer available; but the company continues to market plans, kits and materials for the two-seat Starduster Too, single-seat Starlet, aerobatic V-Star, Acroduster 1 and Acroduster Too. In addition, it is marketing plans for a replica of the Fokker D.VII, the G/B Special designed by Glenn Beets and the Knight Twister designed by Vernon Payne.

STOLP SA-300 STARDUSTER TOO

The SA-300 Starduster Too is an enlarged two-seat version of the original SA-100 Starduster, and is suitable for engines of 93-194 kW (125-260 hp). The prototype has a 134 kW (180 hp) Lycoming O-360-A1A engine.

Type: Two-seat sporting biplane.

Wings: Biplane wings of unequal span with a single interplane strut each side. Multiple centre-section bracing struts. Streamline section landing and flying wires. Wing section M-6 modified. Dihedral 1° 30′ on lower wings only. Incidence 1° on lower wings only. Sweepback on leading-edge of upper wing 6°. All-wood structure with spruce spars and ribs of 6·5 mm (¼ in) plywood, fabric-covered. Ailerons of wooden construction, fabric-covered, on both upper and lower wings. No trailing-edge flaps.

Fuselage: Welded 4130 steel tube structure with fabric covering. Glassfibre turtleback.

Tail Unit: Welded 4130 steel tube structure with fabric covering. Wire-braced fixed-incidence tailplane.

Landing Gear: Non-retractable tailwheel type. Rubber cord shock-absorption. Wheel fairings on main units. Hydraulic brakes.

Power Plant (prototype): One 134 kW (180 hp) Lycoming O-360-A1A flat-four engine, driving a two-blade fixed-pitch propeller with spinner. Fuel tank in fuselage, immediately aft of firewall.

Accommodation: Two seats in tandem open cockpits.

The specification and performance details which follow apply to the radial-engined Starduster Too with a 123 kW (165 hp) Warner Super Scarab engine, built by Mr Jack Mills of Zionsville, Indiana, which was illustrated in the 1973-74 *Jane's:*

Dimensions, external:

Wing span, upper	7·32 m (24 ft 0 in)
Wing chord (constant, both)	1·22 m (4 ft 0 in)
Length overall	6·10 m (20 ft 0 in)
Height overall	2·29 m (7 ft 6 in)

Weights:

Weight empty	501 kg (1,105 lb)
Max T-O weight	748 kg (1,650 lb)

Performance (at max T-O weight):

Max level speed	156 knots (290 km/h; 180 mph)
Cruising speed	104 knots (193 km/h; 120 mph)
Landing speed	52 knots (97 km/h; 60 mph)
Max rate of climb at S/L	792 m (2,600 ft)/min
Absolute ceiling	3,050 m (10,000 ft)
T-O run	152 m (500 ft)
Landing run	305 m (1,000 ft)
Range	260 nm (482 km; 300 miles)

STOLP SA-500 STARLET

The SA-500 Starlet is a single-seat swept parasol-wing monoplane. The wing is of wooden construction with spruce spars, plywood web and capstrip ribs, with Dacron covering. It has a Clark YH section; sweepback is 9° and incidence 3° 30′. The fuselage is of welded 4130 steel tube with Dacron covering, and the tail unit is a braced structure of the same materials. The non-retractable tailwheel-type landing gear has cantilever main legs with wheel fairings. Power plant in the prototype consists of a 1,500 cc Volkswagen flat-four engine, driving a fixed-pitch two-blade propeller with spinner. Other engines of 63·5-93 kW (85-125 hp) may be fitted, the 80·5 kW (108 hp) Lycoming being recommended.

Construction of the prototype occupied three months and cost $1,500. First flight was made on 1 June 1969.

An illustration in the 1975-76 *Jane's* showed the Starlet built by Mr J. D. Hiller of Montgomery, Ohio, which differs from the standard plans in having a two-piece cantilever spring main landing gear. Powered by a 74·5 kW (100 hp) Continental O-200-A engine, Mr Hiller's Starlet has a max level speed of 116 knots (216 km/h; 134 mph). Max cruising speed is 103 knots (192 km/h; 119 mph); landing speed 48 knots (89 km/h; 55 mph); max rate of climb at S/L 404 m (1,325 ft)/min; service ceiling 3,350 m (11,000 ft); T-O run 107 m (350 ft); landing run 274 m (900 ft); range, with 11·4 litres (3 US gallons) reserve, 304 nm (563 km; 350 miles). Empty weight is 347 kg (766 lb) and max T-O weight 499 kg (1,100 lb).

The following details refer to the prototype:

Dimensions, external:

Wing span	7·62 m (25 ft 0 in)
Wing chord	0·91 m (3 ft 0 in)
Wing area, gross	7·71 m^2 (83 sq ft)
Length overall	5·18 m (17 ft 0 in)
Height overall	2·03 m (6 ft 8 in)

Weight:

Max T-O weight	340 kg (750 lb)

Performance (at max T-O weight):

Cruising speed	78 knots (145 km/h; 90 mph)
Landing speed	48-52 knots (89-97 km/h; 55-60 mph)

STOLP SA-700 ACRODUSTER 1

Introduced in 1973, the SA-700 is a single-seat fully-aerobatic biplane. Ailerons on both wings produce a roll rate in excess of 240° a second, and an interesting design feature is that the four ailerons are raised slightly when the control column is pulled back. This helps maintain aileron control when the aircraft is stalled in a normal attitude. Conversely, the ailerons are drooped slightly when the control column is pushed forward, which helps to maintain aileron control in an inverted stall. Plans and kits of components are available to amateur constructors.

Type: Single-seat homebuilt aerobatic biplane.

Wings: Braced single-bay biplane. Single I-type interplane strut each side. N-type centre-section struts. Streamline-section flying and landing wires. Aerofoil section Osborne A-1. Upper wing swept back 6°. Conventional two-spar structure. Spruce spars, plywood ribs and fabric covering. Ailerons on both wings. Upper wing built as two separate panels, joined by bolts at the centre. Stressed to ±9*g* ultimate.

Fuselage: All-metal semi-monocoque structure of light alloy.

Tail Unit: Cantilever structure of light alloy.

Landing Gear: Non-retractable tailwheel type. Main

Stephens Akro Model B flown by Mr Dean S. Engelhardt *(Peter M. Bowers)*

Stewart Headwind built by Mr Richard F. Geide Jr of Wichita, Kansas (Volkswagen modified motor car engine)

Stewart JD₂FF Foo Fighter (Franklin Sport 4 engine) *(Howard Levy)*

Stolp SA-300 Starduster Too built by Mr Francis Lundo (112 kW; 150 hp Lycoming engine) *(Peter M. Bowers)*

Stolp SA-500 Starlet built in England by Mr S. S. Miles *(Peter J. Bish)*

Stolp SA-700 Acroduster 1 with 134 kW (180 hp) engine *(Howard Levy)*

Stolp SA-750 Acroduster Too aerobatic (Lycoming IO-360 engine)

Stolp SA-900 V-Star single-seat biplane

wheels carried on sprung cantilever legs of 2024-0 T-4 light alloy. Fairings for main wheels.

Power Plant: Prototype had originally a 149 kW (200 hp) Lycoming flat-four engine, driving a two-blade fixed-pitch propeller with spinner, but was re-engined subsequently with a 134 kW (180 hp) engine. Design is suitable for engines of 93-149 kW (125-200 hp). Fuel tank in fuselage, aft of firewall, capacity 94·5 litres (25 US gallons). Refuelling point on upper fuselage forward of windscreen.

Accommodation: Single seat in open cockpit. Space for 23 kg (50 lb) of baggage in turtledeck compartment.

Dimensions, external:

Wing span, upper	5·79 m (19 ft 0 in)
Wing area, gross	9·75 m² (105 sq ft)
Length overall	4·80 m (15 ft 9 in)
Height overall	1·91 m (6 ft 3 in)

Weights and Loadings (prototype with 149 kW; 200 hp engine):

Weight empty	335 kg (740 lb)
Aerobatic T-O weight	476 kg (1,050 lb)
Max T-O weight	539 kg (1,190 lb)
Wing loading for aerobatics	48·8 kg/m² (10 lb/sq ft)
Max wing loading	55·3 kg/m² (11·33 lb/sq ft)
Power loading for aerobatics	3·19 kg/kW (5·25 lb/hp)
Max power loading	3·62 kg/kW (5·95 lb/hp)

Performance (prototype with 149 kW; 200 hp engine, at AUW of 476 kg; 1,050 lb):

Max level speed	156 knots (290 km/h; 180 mph)
Cruising speed	143 knots (266 km/h; 165 mph)
Stalling speed	61 knots (113 km/h; 70 mph)
Max rate of climb at S/L	more than 914 m (3,000 ft)/min
Endurance at cruising speed, with reserve	2 hr

STOLP SA-750 ACRODUSTER TOO

The SA-750 is basically a two-seat aerobatic biplane generally similar to the Starduster Too.

Few details of this aircraft have been made available. Stressed to $\pm 9g$, it has symmetrical wings, the upper wing being swept back 6°, and is powered by a 149 kW (200 hp) Lycoming IO-360 engine driving a two-blade constant-speed propeller. The front cockpit is open and has a small windscreen. A bubble canopy for the rear cockpit is faired neatly to the turtleback.

Dimensions, external:

Wing span, upper	6·53 m (21 ft 5 in)
Wing area, gross	12·1 m² (130 sq ft)
Length overall	5·64 m (18 ft 6 in)
Height overall	2·08 m (6 ft 10 in)

Performance (at max T-O weight):

Cruising speed	139 knots (257 km/h; 160 mph)
Stalling speed	48 knots (88·5 km/h; 55 mph)
Max rate of climb at S/L	701 m (2,300 ft)/min

STOLP SA-900 V-STAR

To meet the demand for low-cost, low-horsepower aircraft with aerobatic capability, Stolp has introduced the SA-900 V-Star, which is essentially a biplane version of the single-seat SA-500 Starlet.

It is stressed to $\pm 9g$. The wings, of Clark YH section, have N centre-section and I interplane struts. Incidence of the upper wing is 2° 30′ and that of the lower wings 2°. The upper wing is swept back 6°.

The prototype has a 48·5 kW (65 hp) Continental flat-four engine, driving a two-blade fixed-pitch propeller, but engines of 44·5-93 kW (60-125 hp) may be installed.

DIMENSIONS, EXTERNAL:
Wing span, upper 7·01 m (23 ft 0 in)
Wing area, gross 13·1 m² (141 sq ft)
Length overall 5·23 m (17 ft 2 in)
Height overall 2·26 m (7 ft 5 in)
PERFORMANCE (prototype, at max T-O weight):
Cruising speed 65 knots (121 km/h; 75 mph)
Stalling speed 30·5 knots (56·5 km/h; 35 mph)
Max rate of climb at S/L 183 m (600 ft)/min

TEDDE

Turner Educational Development Enterprises

ADDRESS:
PO Box 425, Stratford, Connecticut 06497
Telephone: (203) 377-3254
DIRECTOR: Steven C. Wieczorek

The 1966-67 *Jane's* contained details of a single-seat sporting aircraft designated T-40, which was designed and built by Mr E. L. Turner and flew for the first time on 3 April 1961. This aircraft was modified by Mr Turner and his son into a prototype of the two-seat T-40A and has since formed the basis of a succession of developed versions of the same general design, all of which are described in this edition.

Plans are now distributed by Turner Educational Development Enterprises (TEDDE), which is also responsible for continued design and development of the series of aircraft, with Mr Turner available as a consultant. A total of 380 sets of plans had been sold by February 1976.

In 1975 TEDDE, under its director, Steven C. Wieczorek, moved to its present location, with access to wind tunnel and computer facilities.

The latest version of the basic T-40 design is the T-40C, utilizing the TEDDE/2 supercritical wing section, an advanced derivative of the NASA GAW-2 supercritical aerofoil; spoilers without ailerons for roll control; and aerodynamically-operated leading-edge slats. The T-40C will undergo flight testing and development as a testbed for a new series of single-engined sporting aircraft, including a four-seater, a two-seater, a single-seater, and a single-seat aerobatic competition aircraft.

All Turner aircraft have folding wings, for reduced hangar space requirements and for transport by trailer. Approval for construction by amateur builders is being sought in Australia, England and West Germany.

TURNER T-40A

The prototype T-40A was produced by conversion of the original T-40. Modification took about four months and the aircraft flew for the first time in this form on 29 July 1966.

The T-40A is small enough to fit in a single-car garage and is transported on a small trailer. It has built-in skids in the fuselage, to protect the pilot in a minor crash landing, and an overturn structure.

TYPE: Two-seat sporting aircraft.
WINGS: Cantilever low-wing monoplane. Wing section NACA 65-215. Dihedral 4°. Incidence 1° 30'. All-wood (fir) two-spar structure with mahogany plywood covering. Hoerner low-drag tips. Plain ailerons. Large plain flaps. Wings fold rearward for stowage.
FUSELAGE: All-wood (fir) structure, covered with mahogany plywood. Glassfibre engine cowling.
TAIL UNIT: Cantilever all-wood (fir) structure with mahogany plywood covering. Horizontal surface of all-flying type with anti-servo tabs. Glassfibre dorsal fin.
LANDING GEAR: Non-retractable tailwheel type. Cantilever spring steel main units attached to front spar. Cleveland main wheels and tyres, size 5·00-5, pressure 3·10 bars (45 lb/sq in). Cleveland brakes.
POWER PLANT: One Continental flat-four engine of 63·5-74·5 kW (85 to 100 hp), driving a McCauley two-blade fixed-pitch propeller, type 65/57. Fuel tank in front fuselage, capacity 75 litres (20 US gallons). Oil capacity 3·75 litres (1 US gallon).
ACCOMMODATION: Pilot and passenger side by side. Each half of transparent canopy is hinged on centreline of aircraft to form a door, folding in two as it opens upward. Space for 11·5 kg (25 lb) baggage aft of seats.
EQUIPMENT: Prototype had Narco VHT-3 radio.
DIMENSIONS, EXTERNAL:
Wing span 7·67 m (25 ft 2 in)
Wing chord, constant 1·08 m (3 ft 6½ in)
Wing area, gross 8·35 m² (89·9 sq ft)
Wing aspect ratio 7·2
Length overall 6·02 m (19 ft 9 in)
Width, wings folded 2·39 m (7 ft 10 in)
Height overall 1·83 m (6 ft 0 in)
Tailplane span 1·96 m (6 ft 5 in)
Wheel track 2·24 m (7 ft 4 in)
DIMENSION, INTERNAL:
Cabin:
Max width 1·02 m (3 ft 4 in)
WEIGHTS:
Weight empty 376 kg (828 lb)
Max T-O and landing weight 640 kg (1,410 lb)
PERFORMANCE (63·5 kW; 85 hp engine, at max T-O weight):
Never-exceed speed 191 knots (354 km/h; 220 mph)
Max level speed at S/L
130 knots (241 km/h; 150 mph)
Max cruising speed at S/L
113 knots (209 km/h; 130 mph)
Econ cruising speed at S/L
104 knots (193 km/h; 120 mph)
Stalling speed, flaps up
51·5 knots (95 km/h; 59 mph)
Stalling speed, flaps down
47 knots (87 km/h; 54 mph)
Max rate of climb at S/L 229 m (750 ft)/min
Service ceiling 3,660 m (12,000 ft)
T-O run 380 m (1,250 ft)
T-O to 15 m (50 ft) 730 m (2,400 ft)
Landing from 15 m (50 ft) 520 m (1,700 ft)
Landing run 305 m (1,000 ft)
Range with max payload, 20 min reserve
412 nm (756 km; 475 miles)

TURNER SUPER T-40A

The Super T-40A differs from the standard T-40A by having a larger wing, more powerful engine, swept tail, bubble canopy and other improvements. The prototype made its first flight in early 1972.

WINGS: As for T-40A, except span and chord increased.
TAIL UNIT: As for T-40A, except swept vertical surfaces, and vertical and horizontal surfaces of increased area.
LANDING GEAR: Non-retractable tailwheel type standard, with optional non-retractable or retractable tricycle type.
POWER PLANT: One 93 kW (125 hp) flat-four engine standard; provision for engines of up to 112 kW (150 hp).
ACCOMMODATION: As for T-40A, except for having a bubble canopy.
DIMENSIONS, EXTERNAL:
Wing span 8·13 m (26 ft 8 in)
Wing chord, constant 1·17 m (3 ft 10 in)
Wing area, gross 9·5 m² (102·5 sq ft)
Wing aspect ratio 6·98
Length overall 6·12 m (20 ft 1 in)
WEIGHTS AND LOADINGS:
Weight empty 445 kg (980 lb)
Max T-O weight 703 kg (1,550 lb)
Max wing loading 71·3 kg /m² (14·6 lb/sq ft)
Max power loading 7·56 kg/kW (12 lb/hp)
PERFORMANCE (at max T-O weight):
Max level speed at S/L
152 knots (282 km/h; 175 mph)
Max cruising speed 135 knots (249 km/h; 155 mph)
Stalling speed, flaps down
43·5 knots (81 km/h; 50 mph)
Max rate of climb at S/L 425 m (1,400 ft)/min

TURNER T-40B

This aircraft is basically similar to the T-40A but has a tricycle landing gear and other refinements. Conversion of the prototype started in October 1966 and the first flight was made on 2 March 1969 with a 63·5 kW (85 hp) engine installed. Flight tests showed that high-altitude performance was below expectation and a 93 kW (125 hp) Lycoming O-320-E1C was then installed.

During 1972 Mr Turner modified his T-40B by fitting a bubble canopy as shown in the illustration of Dr Mandley's Super T-40A, and by installing a 112 kW (150 hp) flat-four Lycoming engine.

The description of the T-40A applies also to the T-40B, except in the following details:

WINGS: Wing section NACA 64-212. Fixed leading-edge droop. Hydraulically-operated double-slotted flaps.
LANDING GEAR: Non-retractable tricycle type. Nosewheel size 4·10-6.
DIMENSIONS, EXTERNAL:
As for T-40A except:
Wing span 6·91 m (22 ft 8 in)
Wing chord at root 1·32 m (4 ft 4 in)
Wing area, gross 9·85 m² (106 sq ft)
Wing aspect ratio 5·13
Length overall 6·23 m (20 ft 5 in)
Wheel track 2·18 m (7 ft 2 in)
Wheelbase 1·83 m (6 ft 0 in)
WEIGHTS AND LOADINGS:
Weight empty 481 kg (1,060 lb)
Max T-O weight 725 kg (1,600 lb)
Max wing loading 81·5 kg/m² (16·7 lb/sq ft)
Max power loading 9·73 kg /kW (12·8 lb/hp)
PERFORMANCE (at max T-O weight):
Max level speed at S/L
152 knots (282 km/h; 175 mph)
Max cruising speed 135 knots (249 km/h; 155 mph)
Cruising speed at 62% power
122 knots (225 km/h; 140 mph)
Stalling speed at 10° flap extension
56·5 knots (105 km/h; 65 mph)
Stalling speed at 50° flap extension
49·5 knots (92 km/h; 57 mph)
Max rate of climb at S/L 488 m (1,600 ft)/min
Service ceiling 5,485 m (18,000 ft)
T-O run 305 m (1,000 ft)
T-O to 15 m (50 ft) 460 m (1,500 ft)
Landing run 213 m (700 ft)

TURNER T-40C

The T-40C is a derivative of the T-40B, utilising the T-40B fuselage and incorporating simplified model aeroplane type construction. The wing has a highly modified and computer-developed version of the NASA GAW-2 supercritical section, known as TEDDE/2 supercritical. It incorporates a quick-release wing folding mechanism, with built-in electrical, flight control and seatbelt/shoulder harness interlock system. This is a fail-safe system, which prevents starting of the engine or actuation of the flight control system unless the wings are unfolded and locked, as well as requiring the seatbelt/shoulder harness to be fastened.

A retractable tandem-type landing gear is used for better load distribution during landing.

TYPE: Two-seat sporting and aerobatic aircraft.
WINGS: Cantilever low-wing monoplane. Wing section TEDDE/2 supercritical, an advanced derivative of the NASA GAW-2 supercritical aerofoil. All-wood (fir) two-spar structure, with mahogany plywood covering. Hoerner low-drag tips. Aerodynamically-operated leading-edge slats. Hydraulically-operated leading-edge Krueger flaps on second prototype. Electrically-actuated, single-slotted, Fowler flaps. Spoilers without supplemental ailerons, in six sections, are utilised for roll control, and as ground spoilers. Wings fold rearward, incorporating a quick-release folding mechanism with built-in electrical, flight control and seatbelt/shoulder harness interlock.
FUSELAGE: All-wood (fir) structure, covered with mahogany plywood. Glassfibre engine cowling. Central section embodies the wing centre-section structure, landing gear, engine mounting and cockpits. Rear fuselage carries the tail unit.
TAIL UNIT: Cantilever all-wood (fir) T-tail structure with mahogany plywood covering. Horizontal surface of all-flying type with anti-servo tabs, which serves also as a trim tab. Glassfibre dorsal fin.
LANDING GEAR: Electrically-retractable tandem type, with retractable balancer wheels at mid-span. Cantilever main units attached to front spar. Cleveland main wheels and tyres, size 5·00-5, pressure 3·10 bars (45 lb/sq in). Cleveland brakes.
POWER PLANT: One 112 kW (150 hp) Lycoming flat-four engine, driving a McCauley two-blade fixed-pitch propeller, type 65/57. Fuel tanks in front fuselage, capacity 95 litres (25 US gallons), and in centre-section, capacity 60 litres (16 US gallons). Oil capacity 3·75 litres (1 US gallon).
ACCOMMODATION: Pilot and passenger side by side under rearward-sliding transparent canopy. Space for 22·5 kg (50 lb) baggage aft of seats.
EQUIPMENT: Prototype has Narco VHT-3 radio.
DIMENSIONS, EXTERNAL:
Wing span 8·53 m (28 ft 0 in)
Wing chord 1·08 m (3 ft 6½ in)
Wing area, gross 9·48 m² (102 sq ft)
Wing aspect ratio 9·2
Length overall 6·12 m (20 ft 1 in)
Width, wings folded 2·39 m (7 ft 10 in)
Height overall 1·83 m (6 ft 0 in)
Tailplane span 1·96 m (6 ft 5 in)
DIMENSIONS, INTERNAL:
Cabin: Length 1·78 m (5 ft 10 in)
Max width 1·02 m (3 ft 4 in)
WEIGHTS:
Weight empty 376 kg (828 lb)
Max T-O and landing weight 748 kg (1,650 lb)
PERFORMANCE (112 kW; 150 hp engine, at max T-O weight):
Never-exceed speed 225 knots (418 km/h; 260 mph)
Max level speed at S/L
165 knots (306 km/h; 190 mph)
Max cruising speed at S/L
152 knots (282 km/h; 175 mph)
Econ cruising speed at S/L
122 knots (225 km/h; 140 mph)
Stalling speed, flaps up 51 knots (94 km/h; 58 mph)
Stalling speed, flaps down
41 knots (76 km/h; 47 mph)
Max rate of climb at S/L 457 m (1,500 ft)/min
Estimated service ceiling 6,400 m (21,000 ft)
T-O run 152 m (500 ft)
T-O to 15 m (50 ft) 305 m (1,000 ft)
Landing run 91 m (300 ft)
Range, max payload, 20 min reserve
521 nm (965 km; 600 miles)

Turner Super T-40A built by Dr Jim Mandley

Turner T-40B prototype, a conversion of the T-40A with tricycle landing gear *(Don Dwiggins)*

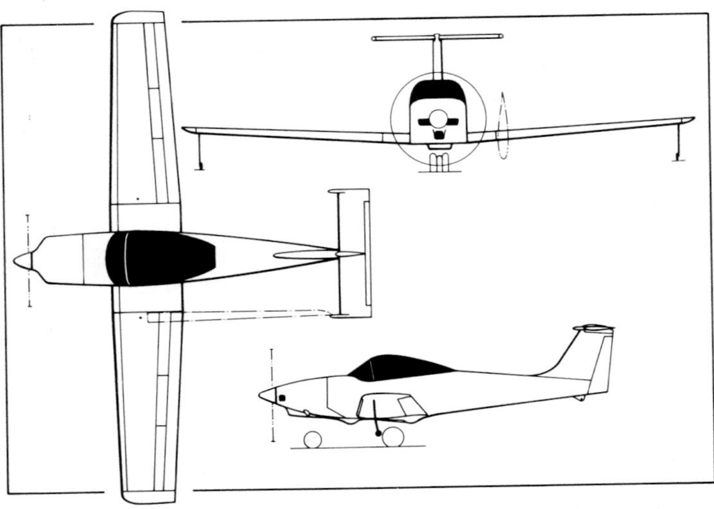

Turner T-40C two-seat sporting and aerobatic aircraft *(Roy J. Grainge)*

Folding-wing version of the Thorp T-18 built by Mr Ken Knowles of Palos Verdes, California *(Howard Levy)*

THORP
THORP ENGINEERING COMPANY

Address:
PO Box 516, Sun Valley, California 91352

This company was founded by Mr John W. Thorp, who is well known as a designer of light aircraft. It markets plans of the T-18 Tiger two-seat all-metal sporting aircraft, described below. Several hundred sets of drawings have been sold and many T-18s are flying.

THORP T-18 TIGER

First T-18 to be completed was N9675Z with a 134 kW (180 hp) Lycoming O-360 engine. Built by Mr W. Warwick, it flew for the first time on 12 May 1964 and was illustrated in the 1964-65 *Jane's*.

The Thorp T-18 shown in the accompanying illustration was built by Mr Ken Knowles of Palos Verdes, California. It differs from standard plans by having folding wings designed by Lon Sunderkind, giving a 2·44 m (8 ft 0 in) width when folded. The wing centre-section is shortened by comparison with Thorp plans, but the wing outer panels are lengthened to give an overall wing span of 7·67 m (25 ft 2 in). Powered by a 100·5 kW (135 hp) Lycoming O-290 engine, empty weight is 390 kg (859 lb); max T-O weight 680 kg (1,500 lb); max level speed 156 knots (290 km/h; 180 mph); cruising speed 139 knots (257 km/h; 160 mph); landing speed 55 knots (101 km/h; 63 mph); max rate of climb at S/L 305 m (1,000 ft)/min; T-O run 305 m (1,000 ft); landing run 457 m (1,500 ft); and range 556 nm (1,030 km; 640 miles).

The following details apply to the standard Thorp T-18:

Type: Two-seat high-performance sporting aircraft.

Wings: Cantilever low-wing monoplane, with 8° dihedral on outer panels only. All-metal two-spar structure. Normally no flaps, but a flap installation is under design.

Fuselage: All-metal structure, without double curvature.

Tail Unit: Cantilever all-metal structure.

Landing Gear: Non-retractable tailwheel type. Cantilever main legs. Steerable tailwheel. Main-wheel tyres size 5·00-5.

Power Plant: One Lycoming or Continental flat-four engine in 80·5-149 kW (108-200 hp) category, driving a two-blade fixed-pitch propeller. Fuel tank aft of firewall, capacity 110 litres (29 US gallons).

Accommodation: Two seats side by side in open cockpit, with dual controls. Space for 36 kg (80 lb) baggage. Canopy optional.

Dimensions, external:
Wing span 6·35 m (20 ft 10 in)
Wing chord, constant 1·27 m (4 ft 2 in)
Wing area, gross 8·0 m² (86 sq ft)
Length overall 5·54 m (18 ft 2 in)
Height overall 1·47 m (4 ft 10 in)
Tailplane span 2·10 m (6 ft 11 in)
Propeller diameter 1·60 m (5 ft 3 in)

Weights (134 kW; 180 hp Lycoming):
Weight empty 408 kg (900 lb)
Max T-O weight 683 kg (1,506 lb)

Performance (134 kW; 180 hp Lycoming):
Max level speed at S/L 174 knots (321 km/h; 200 mph)
Max cruising speed 152 knots (282 km/h; 175 mph)
Stalling speed 57 knots (105 km/h; 65 mph)
Max rate of climb at S/L 610 m (2,000 ft)/min
Service ceiling 6,100 m (20,000 ft)
T-O run 91 m (300 ft)
Landing run 275 m (900 ft)
Range with max fuel 434 nm (805 km; 500 miles)

VAN'S
VAN'S AIRCRAFT

Address:
22730 SW Francis, Beaverton, Oregon 97005
Telephone: (503) 649-5378

Mr Richard VanGrunsven designed and built a single-seat all-metal sporting aircraft known as Van's RV-3. It was built over a 2½-year period, from 1968, at a cost of approximately $2,000, and won its designer the Best Aerodynamic Detailing award at the 1972 EAA Fly-in. In addition to trailing-edge flaps, it has drooping ailerons to improve low-speed control.

After the RV-3's first flight, and subsequent EAA award, Mr VanGrunsven formed Van's Aircraft to market plans to amateur constructors. By March 1976, 340 sets of plans had been sold, with 100 aircraft reported to be under construction and eight RV-3s flying, including the prototype.

VAN'S RV-3

Type: Single-seat homebuilt sporting monoplane.

Wings: Cantilever low-wing monoplane. Wing section NACA 23012. Dihedral 3° 30′. Incidence 1°. Conventional 2024-T3 light alloy structure of constant chord, with I-beam main spar, light rear spar, pressed ribs and moulded glassfibre tips. All-metal bottom-hinged plain trailing-edge flaps. All-metal Frise-type ailerons, which can be drooped to augment flaps. No tabs.

Fuselage: All-metal semi-monocoque structure of 2024-T3 light alloy. Glassfibre engine cowling.

Tail Unit: Cantilever structure of light alloy, with glassfibre tips. Trim tab in port elevator.

Landing Gear: Non-retractable tailwheel type. Cantilever tapered steel-spring main gear struts, with streamline fairings. Cleveland main wheels with tyres size 14 × 5·00-5, pressure 1·38 bars (20 lb/sq in). Steerable tailwheel with 0·15 m (6 in) diameter tyre. Cleveland brakes. Glassfibre streamlined fairings on main wheels.

Power Plant: One 93 kW (125 hp) Lycoming O-290-G (GPU) flat-four engine, driving a Sensenich two-blade fixed-pitch propeller with spinner. Fuel capacity 91 litres (24 US gallons).

Accommodation: Pilot only, beneath rearward-sliding Plexiglas bubble canopy. Baggage space aft of seat, capacity 0·23 m³ (8 cu ft).

Dimensions, external:
Wing span 6·07 m (19 ft 11 in)
Wing chord, constant 1·37 m (4 ft 6 in)
Wing area, gross 8·36 m² (90 sq ft)
Wing aspect ratio 4·43
Length overall 5·79 m (19 ft 0 in)
Height overall 1·55 m (5 ft 1 in)
Tailplane span 2·13 m (7 ft 0 in)
Wheel track 1·73 m (5 ft 8 in)
Wheelbase 4·29 m (14 ft 1 in)
Propeller diameter 1·73 m (5 ft 8 in)

Dimension, internal:
Cabin: Width 0·64 m (2 ft 1 in)

Weights and Loadings:
Weight empty 315 kg (695 lb)
Max T-O weight 476 kg (1,050 lb)
Max wing loading 56·9 kg/m² (11·66 lb/sq ft)
Max power loading 5·12 kg/kW (8·4 lb/hp)

Performance (at max T-O weight):
Never-exceed speed 191 knots (354 km/h; 220 mph)
Max level speed at S/L 169 knots (314 km/h; 195 mph)
Max cruising speed at 2,440 m (8,000 ft) 161 knots (298 km/h; 185 mph)
Econ cruising speed at 3,050 m (10,000 ft) 139 knots (257 km/h; 160 mph)
Stalling speed, flaps up 45·5 knots (84 km/h; 52 mph)
Stalling speed, flaps down 42 knots (78 km/h; 48 mph)
Max rate of climb at S/L 579 m (1,900 ft)/min
Service ceiling 6,400 m (21,000 ft)
T-O run 61 m (200 ft)

Landing run	91·5 m (300 ft)
Range, no reserve	520 nm (965 km; 600 miles)

VAN'S RV-5 SWINGER

The latest lightplane from Van's Aircraft, the RV-5 Swinger, is so named because it has longitudinally pivoting wings to facilitate transport and storage. It is expected that plans and kits will be made available to amateur constructors.

TYPE: Single-seat sporting monoplane.

WINGS: Shoulder-mounted wings of constant chord. NACA 4414 wing section. Full-span 'flaperons'. All-metal construction. No tabs.

FUSELAGE: All-metal structure, with rounded undersurface and flat top-decking. Large fairing aft of cockpit canopy.

TAIL UNIT: Conventional all-metal cantilever structure. No tabs.

LANDING GEAR: Non-retractable tailwheel type. Wittman main legs. Main wheels have 0·36 m (14 in) diameter tyres.

POWER PLANT: One 917 cc Carr-converted Volkswagen motor car engine, giving 24 kW (32 hp) and driving a two-blade propeller. Fuel capacity 40 litres (10·5 US gallons).

ACCOMMODATION: Single seat under transparent cockpit canopy.

DIMENSIONS, EXTERNAL:

Wing span	6·10 m (20 ft 0 in)
Length overall	5·00 m (16 ft 5 in)

WEIGHTS AND LOADINGS:

Weight empty	148 kg (325 lb)
Max T-O weight	272 kg (600 lb)
Max wing loading	39·06 kg/m² (8 lb/sq ft)
Max power loading	11·33 kg/kW (18·7 lb/hp)

PERFORMANCE (estimated):

Max level speed	80 knots (148 km/h; 92 mph)
Cruising speed at 1,525 m (5,000 ft)	75 knots (138 km/h; 86 mph)
Max rate of climb at S/L	180 m (590 ft)/min
T-O run	180 m (590 ft)

VOLMER

VOLMER AIRCRAFT

ADDRESS:
Box 5222, Glendale, California 91201
Telephone: (213) 247-8718

Mr Volmer Jensen, well known as a designer of sailplanes and gliders, also designed and built a two-seat light amphibian named the Sportsman (formerly Chubasco), which flew for the first time on 22 December 1958 and has since logged over 1,600 flying hours. He is attempting to find a manufacturer who will produce and market the aircraft commercially.

Meanwhile, plans of the Sportsman are available to amateur constructors. Nearly 750 sets had been sold by the Spring of 1976 and approximately 100 Sportsman amphibians are flying. Some have tractor propellers, but this modification is not recommended by Mr Jensen.

VOLMER VJ-22 SPORTSMAN

The following details refer to Mr Jensen's prototype:

TYPE: Two-seat light amphibian.

WINGS: Braced high-wing monoplane. Dihedral 1°. Incidence 3°. Wings are standard Aeronca Chief or Champion assemblies with wooden spars, metal ribs and fabric covering, and carry stabilising floats under the tips. Streamline Vee bracing struts each side.

FUSELAGE: Conventional flying-boat hull of wooden construction, covered with mahogany plywood and coated with glassfibre.

TAIL UNIT: Strut-braced steel tube structure, fabric-covered.

LANDING GEAR: Retractable tailwheel type. Rubber-cord shock-absorption. Manual retraction. Cleveland wheels and mechanical brakes. Tyre pressure 1·38 bars (20 lb/sq in). Castoring retractable tailwheel with integral water rudder.

POWER PLANT: 63·5 kW (85 hp) Continental C85, 67 kW (90 hp) or 74·5 kW (100 hp) Continental O-200-B flat-four engine, driving a Sensenich two-blade fixed-pitch pusher propeller. Fuel in a single tank, capacity 76 litres (20 US gallons). Oil capacity 4·5 litres (4½ US quarts).

ACCOMMODATION: Two seats side by side in enclosed cabin with dual controls.

DIMENSIONS, EXTERNAL:

Wing span	11·12 m (36 ft 6 in)
Wing chord	1·52 m (5 ft 0 in)
Wing area, gross	16·3 m² (175 sq ft)
Wing aspect ratio	7·2
Length overall	7·32 m (24 ft 0 in)
Height overall	2·44 m (8 ft 0 in)

WEIGHTS (63·5 kW; 85 hp):

Weight empty	454 kg (1,000 lb)
Max T-O weight	680 kg (1,500 lb)

PERFORMANCE (63·5 kW; 85 hp, at max T-O weight):

Max level speed at S/L	83 knots (153 km/h; 95 mph)
Max cruising speed	74 knots (137 km/h; 85 mph)
Stalling speed	39 knots (72 km/h; 45 mph)
Max rate of climb at S/L	183 m (600 ft)/min
Service ceiling	3,960 m (13,000 ft)
Range with max fuel, no reserve	260 nm (480 km; 300 miles)

WAG-AERO

WAG-AERO INC

ADDRESS:
Box 181, 1216 North Road, Lyons, Wisconsin 53148
Telephone: (414) 763-9588
PRESIDENT: Richard H. Wagner

WAG-AERO CUBy

This company is making available to homebuilders plans and kits of parts to enable them to construct a modern version of the famous Piper Cub. Known as the CUBy, this two-seat sporting aircraft follows the original design, but benefits by utilising up-to-date constructional techniques. The wing has a wooden main spar and ribs, light alloy leading-edge and fabric covering. The fuselage and tail unit are of welded 4130 chrome molybdenum steel tube with fabric covering. The CUBy can be powered by any flat-four Continental or Lycoming engine of between 48·5 and 93 kW (65 and 125 hp).

Also available is the CUBy Acro Trainer which differs from the standard version by having shortened wings, modified lift struts, improved wing fittings and rib spacing, and a new leading-edge.

Design of the CUBy began in 1974 and construction of a prototype started in December of that year. First flight took place on 12 March 1975. By February 1976 plans for 154 CUBys had been ordered.

The following details apply to the standard CUBy:

DIMENSIONS, EXTERNAL:

Wing span	10·73 m (35 ft 2½ in)
Wing chord, constant	1·60 m (5 ft 3 in)
Length overall	6·82 m (22 ft 4½ in)
Height overall	2·03 m (6 ft 8 in)

WEIGHTS:

Weight empty	314 kg (692 lb)
Max T-O weight	608 kg (1,340 lb)

PERFORMANCE (at max T-O weight):

Max level speed at S/L	89 knots (164 km/h; 102 mph)
Cruising speed	82 knots (151 km/h; 94 mph)
Stalling speed	34 knots (63 km/h; 39 mph)
Max rate of climb at S/L	149 m (490 ft)/min
Service ceiling	3,415 m (11,200 ft)
T-O run	114 m (375 ft)
Range at cruising speed with standard fuel (45 litres; 12 US gallons)	191 nm (354 km; 220 miles)
Range with auxiliary fuel (98 litres; 26 US gallons)	395 nm (732 km; 455 miles)

WAR

WAR AIRCRAFT REPLICAS

ADDRESS:
348 South Eighth Street, Santa Paula, California 93060
PRESIDENT: Kenneth L. Thoms

War Aircraft Replicas is a new company formed to market plans and kits from which amateur constructors can build ½-scale replicas of a series of second World War aircraft. The term '½-scale' is not strictly accurate, but refers to the general overall dimensions of the aircraft. For example, to provide adequate accommodation for the pilot, the cockpit is considerably larger than ½-scale, and the area of the horizontal and vertical tail surfaces has been increased beyond scale to ensure adequate stability.

The basic concept involves the use of a common-design wooden fuselage box and spar structure. The desired contours to duplicate a particular aircraft are obtained by using carved polyurethane foam, covered with high-strength laminating fabric and epoxy resin to form a light-weight and rigid structure that is stressed to ±6*g*, allowing for aerobatic manoeuvres. By changing fuselage contours, using different engine cowlings and wingtips, and by shape changes to tail unit surfaces, it was considered that a number of different aircraft could be copied with reasonable similarity to the full-scale combat types.

The Focke-Wulf 190 was chosen as the first prototype to be completed, its design starting in July 1973 and construction in February 1974. The first flight of this aircraft was made on 21 August 1974 and it had logged a total of 200 flying hours by 20 February 1976. Sixty sets of plans had been sold by that date. A prototype replica of the Vought F4U Corsair was almost complete, with replicas of the Hawker Sea Fury and Republic P-47 Thunderbolt under construction.

The description which follows applies specifically to the Focke-Wulf 190 replica, but will be applicable generally to the range of aircraft for which the company is producing plans and kits initially.

WAR AIRCRAFT REPLICAS FOCKE-WULF 190

TYPE: Homebuilt ½-scale aircraft replica.

WINGS: Cantilever low-wing monoplane, built in three sections: nominal 2·44 m (8 ft) centre-section, integral with fuselage box, and two nominally 1·83 m (6 ft) outer panels. Wing section NACA 23015 at root, 23012 at tip. Dihedral 5°. Incidence 2°. Washout 2°. Primary structure of wood, with a laminated hollow plywood-covered front spar and solid laminated rear spar. Plywood ribs are used at the root, both faces of the centre-section joints and at the tip sections, with intermediate ribs of polyurethane foam. Aerofoil contours built up with carved polyurethane foam, bonded in place. High-strength laminating fabric and epoxy resin used for covering and for internal strengthening. Frise-type ailerons with wooden front spar bonded to a shaped form of urethane foam with fabric/epoxy covering. No flaps. Ground-adjustable tab on each aileron.

FUSELAGE: Of similar general construction to wings, with a standard four-longeron box built from ¾ in fir stringers, ¾ in by ½ in diagonals and cross pieces, 1/16 in birch plywood covering and a metal-faced ⅛ in plywood firewall. Fuselage contoured by carved polyurethane foam with fabric/epoxy covering.

TAIL UNIT: Cantilever wooden structure, utilising the same construction technique as for the wings. Fixed tailplane with elevators. Ground-adjustable trim tab on rudder and each elevator.

LANDING GEAR: Electrically-retractable tailwheel type, with manual emergency retraction system. Main wheels retract inward into wings. Fixed tailwheel. Oleo-pneumatic shock-struts on main units. Main wheels and tyres size 3·50 × 4·10-6. Cleveland hydraulic disc brakes.

POWER PLANT: One 74·5 kW (100 hp) Continental O-200 flat-four engine, driving a three-blade fixed-pitch wooden propeller with spinner. Fuel tank in fuselage, immediately aft of firewall, with capacity of 45·5 litres (12 US gallons). Refuelling point on upper surface of fuselage, forward of windscreen.

ACCOMMODATION: Single seat beneath rearward-sliding cockpit canopy. Accommodation heated and ventilated.

SYSTEMS: Hydraulic system for brakes only. Electrical system powered by 12V engine-driven alternator.

DIMENSIONS, EXTERNAL:

Wing span	6·10 m (20 ft 0 in)
Wing chord at root	1·37 m (4 ft 6 in)
Wing chord at tip	0·94 m (3 ft 1 in)
Wing area, gross	6·50 m² (70 sq ft)
Wing aspect ratio	5·7
Length overall	5·05 m (16 ft 7 in)
Height overall	2·13 m (7 ft 0 in)
Tailplane span	2·29 m (7 ft 6 in)
Wheel track	2·03 m (6 ft 8 in)
Wheelbase	3·25 m (10 ft 8 in)
Propeller diameter	1·52 m (5 ft 0 in)
Propeller ground clearance	0·38 m (1 ft 3 in)

WEIGHTS AND LOADINGS:

Weight empty	286 kg (630 lb)
Max T-O weight	408 kg (900 lb)
Max wing loading	62·7 kg/m² (12·85 lb/sq ft)
Max power loading	5·48 kg/kW (9 lb/hp)

PERFORMANCE (at max T-O weight):

Max level speed at 1,065 m (3,500 ft)	160 knots (298 km/h; 185 mph)
Max cruising speed at 1,065 m (3,500 ft)	126 knots (233 km/h; 145 mph)
Econ cruising speed at 1,065 m (3,500 ft)	108 knots (201 km/h; 125 mph)
Stalling speed	48 knots (89 km/h; 55 mph)
Max rate of climb at S/L	305 m (1,000 ft)/min
Service ceiling	3,810 m (12,500 ft)
T-O run	305 m (1,000 ft)
Landing from 15 m (50 ft)	550 m (1,800 ft)
Landing run	365 m (1,200 ft)
Range with max fuel	347 nm (643 km; 400 miles)

Van's RV-3 single-seat homebuilt sporting monoplane (Lycoming O-290-G engine)

Van's RV-5 Swinger *(Howard Levy)*

Volmer VJ 22 Sportsman (Lycoming O-290-G engine) built by Mr Stan Woods

Wag-Aero CUBy, an updated version of the Piper Cub for construction by homebuilders

Half-scale replica of Focke-Wulf 190 fighter built to WAR plans *(Howard Levy)*

Almost-completed WAR half-scale replica of the F4U Corsair (52 kW; 70 hp geared Volkswagen engine) *(Howard Levy)*

WENDT

WENDT AIRCRAFT ENGINEERING

ADDRESS:

9900 Alto Drive, La Mesa, California 92041

Wendt Aircraft Engineering designed and built the prototype of a two-seat sporting monoplane which is known as the WH-1 Traveler. The design originated on 4 September 1969, and construction of the prototype began on 26 November of the same year. The first flight was made on 15 March 1972, and FAA certification in the Experimental category was awarded on 30 May 1972. Plans of the Traveler are available to amateur constructors, and 43 sets had been sold by February 1976, at which time it was reported that eight aircraft were under construction.

WENDT WH-1 TRAVELER

TYPE: Two-seat homebuilt sporting aircraft.

WINGS: Cantilever low-wing monoplane. Wing section NACA 64_3A-418. Dihedral 5° 30′. Incidence 2°. No sweepback. Constant-chord two-spar structure. Spruce spars, marine plywood ribs, pine leading- and trailing-edges and 3/32 in mahogany plywood skin from leading-edge to 37% chord. Aft of main spar, wing is Dacron-covered. Plain ailerons, hinged at upper surface, made of spruce with plywood ribs, and Dacron-covered. No flaps. Bungee trim on control column. Glassfibre wing-tips.

FUSELAGE: Conventional structure of spruce frames and longerons, plywood formers and tension ties, with steel tube overturn structure in the cockpit section. Fuselage undersurface and sides covered with 1/8 in mahogany plywood. Upper surface Dacron-covered. Glassfibre nose cowl.

TAIL UNIT: Cantilever wooden structure with swept vertical surfaces and all-moving tailplane. Each surface has a spruce spar, spruce and plywood ribs, and a 1/16 in mahogany plywood torsion box. All surfaces Dacron-covered. Static balance weights near tips of tailplane leading-edge. Tailplane has a half-span trim and anti-balance tab. Tailplane tips of glassfibre.

LANDING GEAR: Non-retractable tricycle type. Cantilever spring steel main gear. Steerable nosewheel has coil spring shock-absorption. Cleveland 5·00-5 wheels with Armstrong tyres, pressure 2·07 bars (30 lb/sq in). Cleveland caliper-type brakes. Glassfibre wheel fairings.

POWER PLANT: Prototype has one 56 kW (75 hp) Continental A75 flat-four engine, driving a McCauley Type 1C90 two-blade metal fixed-pitch propeller with glassfibre spinner. Design is suitable for installation of engines from 48·5-74·5 kW (65 to 100 hp). One aerofoil-shaped glassfibre fuel tank at each wingtip, capacity 41·5 litres (11 US gallons). Total fuel capacity 83 litres (22 US gallons). Refuelling points on upper surface of each wingtip. Oil capacity 3·8 litres (1·0 US gallon).

ACCOMMODATION: Pilot and passenger in tandem, beneath canopy which has large transparent panels at each side. Canopy hinged on port side. Dual controls standard. Stowage for 23 kg (50 lb) baggage aft of rear seat.

SYSTEM: Electrical system powered by 30A engine-driven alternator. 12V 25Ah storage battery in glassfibre battery box in aft fuselage.

ELECTRONICS: Prototype has a Narco Escort 110 com transceiver.

DIMENSIONS, EXTERNAL:

Wing span	9·14 m (30 ft 0 in)
Wing chord, constant	1·20 m (3 ft 11¼ in)
Wing area, gross	10·96 m^2 (118 sq ft)
Wing aspect ratio	7·63
Length overall	5·94 m (19 ft 6 in)
Height overall	2·08 m (6 ft 10 in)
Tailplane span	2·44 m (8 ft 0 in)
Wheel track	1·93 m (6 ft 4 in)
Wheelbase	1·45 m (4 ft 9 in)
Propeller diameter	1·80 m (5 ft 11 in)

Propeller ground clearance 0·25 m (10 in)
DIMENSION, INTERNAL:
Max width 0·71 m (2 ft 4 in)
WEIGHTS AND LOADINGS:
Weight empty, equipped 408 kg (900 lb)
Max T-O and landing weight 635 kg (1,400 lb)
Max wing loading 57·9 kg/m² (11·86 lb/sq ft)
Max power loading 11·34 kg/kW (18·67 lb/hp)
PERFORMANCE (at max T-O weight):
Never-exceed speed 142 knots (264 km/h; 164 mph)
Max level speed at 1,220 m (4,000 ft) 114 knots (211 km/h; 131 mph)
Max cruising speed at 1,220 m (4,000 ft) 107 knots (198 km/h; 123 mph)
Econ cruising speed at 1,220 m (4,000 ft) 100 knots (185 km/h; 115 mph)
Stalling speed 50 knots (92 km/h; 57 mph)
Max rate of climb at S/L (no passenger) 229 m (750 ft)/min
Max rate of climb at S/L (with passenger) 152 m (500 ft)/min
Service ceiling 3,960 m (13,000 ft)
T-O run 244 m (800 ft)
Landing run 213 m (700 ft)
Range with max fuel, no reserve 503 nm (933 km; 580 miles)
Range with max payload, no reserve 416 nm (772 km; 480 miles)

WHITE

E. MARSHALL WHITE

ADDRESS:
Meadowlark Airport, 5141 Warner Avenue, Huntington Beach, California 92649
Telephone: (714) 846-2409

WHITE WW-1 DER JÄGER D.IX

Mr Marshall White, a staff engineer of TRW Systems at Redondo Beach, California, designed an unusual homebuilt aircraft named Der Jäger D.IX, which is reminiscent of several German designs, mainly of first World War vintage. The wings are patterned on those of an Albatros D.Va, with the landing gear fairings of the Focke-Wulf Stösser and tail unit of the Fokker D.VII.

Design and construction of the prototype started simultaneously at the beginning of 1969, as Mr White's fifth homebuilt, and first flight of the prototype was made on 7 September 1969.

Plans and kits of materials, as well as some of the more difficult-to-construct parts in finished form, are available to amateur constructors, and at least 75 Der Jäger D.IXs are under construction. The first completed aircraft to be seen at an EAA Fly-in at Oshkosh, in 1974, was N1007, built by Mr Ray D. Fulwiler of Algoma, Wisconsin, with a 112 kW (150 hp) Lycoming engine.

The following details apply to the prototype in its original form. It has since been re-engined with a 112 kW (150 hp) Lycoming, but no details have been received of performance with this more powerful engine.

TYPE: Single-seat homebuilt sporting biplane.

WINGS: Forward-stagger single-bay biplane with N-type interplane and centre-section struts. Single streamlined lift strut from each side of lower fuselage to attachment point of forward interplane strut on upper wing. No flying or landing wires. Aerofoil section M-6. Incidence 3° upper wing, 2° lower wings. Spruce spars and plywood ribs, fabric covered. Internal steel tube bracing. Ailerons in both top and bottom wings. Scalloped trailing-edge to both wings.

FUSELAGE: Welded 4130 steel tube structure, fabric covered. Aluminium engine cowling.

TAIL UNIT: Wire-braced welded 4130 steel tube structure, with sheet metal ribs, fabric covered. Balanced rudder and elevator. Ground-adjustable trim tabs in elevator.

LANDING GEAR: Non-retractable tailwheel type. Main legs each consist of an 'A' frame, welded into the fuselage, with tension springs in the centre-fuselage to cushion landing shock. Main wheels and tyres size 5·00-5. Glassfibre wheel fairings.

POWER PLANT: One 86 kW (115 hp) Lycoming O-235-C1 flat-four engine, driving a McCauley two-blade propeller. Structure suitable for alternative power plants from 1,600 cc Volkswagen up to 112 kW (150 hp). Fuel contained in two tanks, one in upper wing centre-section, capacity 53 litres (14 US gallons), one in fuselage, capacity 38 litres (10 US gallons); total 91 litres (24 US gallons).

ACCOMMODATION: Single seat in open cockpit, with headrest faired into wood or glassfibre fuselage turtleback.

EQUIPMENT: Two dummy machine-guns mounted on top of fuselage, forward of cockpit. Dummy bomb, carried between legs of main landing gear, can be adapted as oil tank for smoke discharge system.

DIMENSIONS, EXTERNAL:
Wing span, upper 6·10 m (20 ft 0 in)
Wing span, lower 4·88 m (16 ft 0 in)
Wing chord, upper at root 1·07 m (3 ft 6 in)
Wing chord, upper at tip 1·22 m (4 ft 0 in)
Wing chord, lower (constant) 0·91 m (3 ft 0 in)
Wing area, gross 10·68 m² (115 sq ft)
Length overall 5·18 m (17 ft 0 in)
Tailplane span 2·44 m (8 ft 0 in)
Wheel track 1·52 m (5 ft 0 in)
Propeller diameter 1·68 m (5 ft 6 in)
WEIGHTS:
Weight empty 242 kg (534 lb)
Max T-O weight 403 kg (888 lb)
PERFORMANCE (at max T-O weight):
Never-exceed speed 152 knots (282 km/h; 175 mph)
Max level speed at 610 m (2,000 ft) 126 knots (233 km/h; 145 mph)
Max cruising speed at 610 m (2,000 ft) 116 knots (214 km/h; 133 mph)
Stalling speed 47 knots (87 km/h; 54 mph)
Max rate of climb at S/L 732 m (2,400 ft)/min
T-O run 46 m (150 ft)

WILLIAMS

WILLIAMS AIRCRAFT DESIGN COMPANY

ADDRESS:
11301 Yolanda Avenue, Northridge, California 91324
Telephone: (213) 360-6322

Employed for more than eleven years as a designer in the advanced concepts group at Lockheed, California, Mr Arthur L. Williams is also well-known as a designer of racing aircraft. Details of his latest designs follow:

WILLIAMS W-17 *Stinger*

This Formula One racing aircraft designed by Art Williams is owned by Mr John P. Jones of Granada Hills, California, and others. Flown by Mr Jones, it recorded a best single-lap speed of 203·44 knots (377·02 km/h; 234·27 mph) and gained second place in the National Air Races at Reno, Nevada, in 1973. It was badly damaged in a practice-flying accident in early 1975, but is expected to race again after extensive reconstruction.

TYPE: Single-seat Formula One racing aircraft.

WINGS: Cantilever mid-wing monoplane of elliptical planform. Wing section NACA 64-008 at root, 64-010 at tip, with leading-edge camber. Incidence 1°. Dihedral 0°. All-wood structure with laminated spruce spars, plywood ribs and mahogany plywood skins. Wide-span ailerons of similar construction. No flaps.

FUSELAGE: Light alloy monocoque structure, with formers and stringers covered by skins of 2024-T43 alloy.

TAIL UNIT: Cantilever wooden structure of spruce spars and ribs, covered by mahogany plywood.

LANDING GEAR: Non-retractable tailwheel type. Main legs are leaf springs of 7075-T63 light alloy. Rosenham main wheels and brakes. Main-wheel tyres size 5·00-5. Glassfibre fairings over main wheels.

POWER PLANT: One Continental O-200 flat-four engine, rated at 74·5 kW (100 hp) at 2,750 rpm and approx 97 kW (130 hp) at 4,000 rpm for racing. Sensenich two-blade fixed-pitch metal propeller. 19 litre (5 US gallon) fuel tank aft of firewall and 30 litre (8 US gallon) tank aft of pilot's shoulders. Refuelling point for forward tank on fuselage upper surface, forward of canopy, and aft tank inside canopy behind headrest.

ACCOMMODATION: Single seat beneath transparent bubble canopy. Nose-over protection bars incorporated in headrest.

SYSTEM: Hydraulic system for brakes only. No electrical system.

DIMENSIONS, EXTERNAL:
Wing span 5·79 m (19 ft 0 in)
Wing area, gross 6·14 m² (66·10 sq ft)
Wing aspect ratio 5·5
Length overall 4·84 m (15 ft 10½ in)
Height overall 1·42 m (4 ft 8 in)
Wheel track 1·22 m (4 ft 0 in)
Propeller diameter 1·47 m (4 ft 10 in)
WEIGHTS AND LOADINGS:
Weight empty 265 kg (585 lb)
Normal T-O weight 356 kg (785 lb)
Max T-O weight 379 kg (835 lb)
Normal wing loading 58·00 kg/m² (11·88 lb/sq ft)
Max wing loading 61·67 kg/m² (12·63 lb/sq ft)
Rated power loading 4·78 kg/kW (7·85 lb/hp)
Racing power loading 3·67 kg/kW (6·04 lb/hp)
PERFORMANCE (at normal T-O weight):
Max level speed 226 knots (418 km/h; 260 mph)
Landing speed 57 knots (105 km/h; 65 mph)

WILLIAMS W-18 *Falcon*

The *Falcon* is a Formula One racing aircraft of very similar appearance to the W-17 *Stinger*. It was built by, and is owned by, Daniel LaLee of Lompoc, California, who is in his mid-70s. First flight was made on 27 September 1975, and *Falcon* was first raced in the California National Air Races at Mojave on 19-20 June 1976. Flown on this occasion by Bob Downey, it came third in the finals, but took fourth place overall because a pylon was not rounded correctly.

All details of *Falcon* are the same as for *Stinger*, except for the following:

TAIL UNIT: Slightly reshaped vertical surfaces. Horizontal surfaces moved forward by 0·10 m (4 in), avoiding need for cutouts at elevator roots to provide clearance for rudder.

LANDING GEAR: More streamlined wheel fairings. Cleveland main wheels and brakes.

POWER PLANT: Forward fuel tank holds 19 litres (5 US gallons). Rear tank holds 30·2 litres (8 US gallons).

DIMENSIONS, EXTERNAL:
Wing span 5·97 m (19 ft 7 in)
Wing area, gross 6·13 m² (66·0 sq ft)
Wing aspect ratio 5·8
Length overall 4·81 m (15 ft 9½ in)
Height overall 1·49 m (4 ft 10½ in)
Wheel track 1·22 m (4 ft 0 in)
WEIGHTS AND LOADINGS:
Weight empty 259 kg (570 lb)
Racing weight 349 kg (770 lb)
Design max T-O weight 376 kg (830 lb)
Normal wing loading 56·93 kg/m² (11·66 lb/sq ft)
Max wing loading 61·32 kg/m² (12·56 lb/sq ft)
Racing power loading 3·60 kg/kW (5·96 lb/hp)
Max power loading 5·05 kg/kW (7·70 lb/hp)
PERFORMANCE:
Max level speed 225 knots (418 km/h; 260 mph)
Landing speed 56·5 knots (105 km/h; 65 mph)

WILLIAMS-CANGIE WC-1 *Sundancer*

WC-1 *Sundancer* is a Sport Biplane class racing aircraft designed by Art Williams in conjunction with Carl Cangie, and built by the late Ralph Thenhaus and Jack Swan. It was evolved around the basic fuselage and tail unit structure of the Bushby/Long Midget Mustang, the components for their construction having been acquired from Robert Bushby. The first flight was recorded in August 1973. At Reno, in the following month, the aircraft was raced for the first time, flown by Dr Sidney White. He set a national class record for time trials of 168·69 knots (312·62 km/h; 194·25 mph) and then won the championship race at a record speed of 169·30 knots (313·74 km/h; 194·95 mph). Flown by White, *Sundancer* won all five races in which it was entered during 1974, and in early 1975 held the national lap record at 172·41 knots (319·50 km/h; 198·53 mph), and race record at 172·09 knots (318·92 km/h; 198·17 mph). Latest modifications include changes to the engine cowling and wheel axles moved aft by 0·11 m (4½ in).

There are no plans to sell drawings of this aircraft.

TYPE: Single-seat racing biplane.

WINGS: Cantilever biplane of unequal span. NACA 64-012 wing section with leading-edge camber from root to tip. Single streamline-section strut on each side to comply with racing rules. 12% thickness/chord ratio. Dihedral 1° 30′ on upper and lower wings. Conventional light alloy structure, with plain ailerons on upper wing only. No flaps.

FUSELAGE: Light alloy flush-riveted stressed-skin monocoque structure, modified from the Midget Mustang plans to cater for biplane configuration.

TAIL UNIT: Cantilever light alloy structure. Ground adjustable trim tab in port elevator.

LANDING GEAR: Non-retractable tailwheel type. Main legs are leaf springs of 7075-T6 light alloy, covered by glassfibre fairings. Hydraulic brakes. Light alloy fairings over main wheels.

POWER PLANT: One Lycoming O-290-D2 flat-four engine, rated at 100·5 kW (135 hp) at 2,800 rpm and 123 kW (165 hp) at 3,500 rpm for racing. Two-blade metal fixed-pitch propeller with spinner. Fuel tanks aft of firewall and aft of pilot's seat, total capacity 60·6 litres (16 US gallons). Refuelling point for front tank on fuselage upper surface forward of canopy, for rear tank in top of turtledeck.

ACCOMMODATION: Single seat beneath transparent bubble canopy, which is partially open and hinges to side for access. Nose-over protection bars incorporated in headrest.

SYSTEM: Hydraulic system for brakes only. No electrical system.

DIMENSIONS, EXTERNAL:
Wing span, upper 6·02 m (19 ft 9 in)
Wing span, lower 3·89 m (12 ft 9 in)
Wing chord, upper 0·97 m (3 ft 2 in)
Wing chord, lower 0·66 m (2 ft 2 in)
Wing area, gross 8·25 m² (88·8 sq ft)
Wing aspect ratio, upper 6·33
Wing aspect ratio, lower 5·97
Length overall 4·90 m (16 ft 1 in)
Height overall 1·85 m (6 ft 1 in)

Wendt WH-1 Traveler two-seat homebuilt aircraft

White Der Jäger D.IX homebuilt biplane (Lycoming O-235-C1 engine)

Williams W-17 *Stinger* Formula One racing aircraft *(Art Williams)*

Williams-Cangie WC-1 *Sundancer* Sport Biplane class racing aircraft *(Barry Bronson)*

Weights and Loadings:
- Weight empty: 379 kg (835 lb)
- Normal T-O weight: 476 kg (1,050 lb)
- Max T-O weight: 506 kg (1,115 lb)
- Normal wing loading: 57·71 kg/m² (11·82 lb/sq ft)
- Max wing loading: 61·27 kg/m² (12·55 lb/sq ft)
- Rated power loading: 4·74 kg/kW (7·75 lb/hp)
- Racing power loading: 4·11 kg/kW (6·75 lb/hp)

Performance (at normal T-O weight):
- Max level speed at 2,440 m (8,000 ft): 204 knots (378 km/h; 235 mph)
- Stalling speed: 57 knots (105 km/h; 65 mph)

WITTMAN

S. J. WITTMAN

Address:
Box 276, Oshkosh, Wisconsin 54901

Famous as a racing pilot since 1926, Steve Wittman has designed and built a large number of different racing and touring aeroplanes at Winnebago County Airport, of which he became manager in 1931.

Most popular current Wittman design is the W-8 Tailwind side-by-side two-seat light aeroplane. The prototype was built in 1952-53 and proved so successful that sets of plans and prefabricated components were made available to amateur builders. By the Spring of 1972, there were more than 150 Model W-8 Tailwinds flying, including a number built in foreign countries, and more than 100 were known to be under construction. In January 1968 Mr Wittman's plans were approved by the Australian Department of Civil Aviation.

In 1966, a more powerful six-cylinder Continental engine was installed in a Tailwind redesigned to take the added weight and power. This version is designated W-9.

A modified version of the Tailwind is being marketed in the UK by AJEP (which see).

WITTMAN TAILWIND MODEL W-8

Some Tailwinds have been built with tricycle landing gear, retractable main wheels and other design changes. The following data refer to the standard W-8 Tailwind built to Mr Wittman's plans:

Type: Two-seat cabin monoplane.

Wings: Braced high-wing monoplane. Wing section is a combination of NACA 4309 (upper surface) and NACA 0006 (lower surface). Thickness/chord ratio 11·5%. No dihedral. Incidence 1°. Wooden structure with plywood and fabric covering. Single bracing strut each side. Ailerons and flaps of steel and stainless steel construction.

Fuselage: Steel tube structure, fabric-covered.

Tail Unit: Cantilever structure of steel and stainless steel. Ground-adjustable trim tabs in control surfaces.

Landing Gear: Non-retractable tailwheel type. Spring steel cantilever main legs. Goodyear 15 × 5 main wheels and tyres, pressure 2·21 bars (32 lb/sq in). Goodyear brakes.

Power Plant: Normally one 67 kW (90 hp) Continental C90-12F flat-four engine, driving a Sensenich or Flottorp two-blade wood fixed-pitch propeller. Alternative engines are the 63·5 kW (85 hp) Continental C85, 74·5 kW (100 hp) Continental O-200, 86 kW (115 hp) Lycoming O-235 or 104·5 kW (140 hp) Lycoming O-290. One fuel tank of 94·5 litres (25 US gallons) capacity in fuselage. Oil capacity 1·85-2·8 litres (4-6 US quarts).

Accommodation: Two seats side by side in enclosed cabin, with door on each side. Space for 27 kg (60 lb) baggage.

Dimensions, external:
- Wing span: 6·86 m (22 ft 6 in)
- Wing chord, constant: 1·22 m (4 ft 0 in)
- Wing area, gross: 8·36 m² (90 sq ft)
- Wing aspect ratio: 5·5
- Length overall: 5·87 m (19 ft 3 in)
- Height overall: 1·73 m (5 ft 8 in)
- Tailplane span: 2·03 m (6 ft 8 in)
- Wheel track: 1·65 m (5 ft 5 in)
- Propeller diameter: 1·63 m (5 ft 4 in)

Weights and Loadings (74·5 kW; 100 hp Continental engine):
- Weight empty: 318 kg (700 lb)
- Max T-O weight: 590 kg (1,300 lb)
- Max wing loading: 70·3 kg/m² (14·4 lb/sq ft)
- Max power loading: 7·92 kg/kW (13·0 lb/hp)

Performance (74·5 kW; 100 hp Continental engine at max T-O weight):
- Never-exceed speed: 160 knots (297 km/h; 185 mph)
- Max level speed at S/L: 143 knots (265 km/h; 165 mph)
- Max cruising speed: 139 knots (257 km/h; 160 mph)
- Econ cruising speed: 113 knots (209 km/h; 130 mph)
- Stalling speed, flaps down: 48 knots (89 km/h; 55 mph)
- Max rate of climb at S/L: 275 m (900 ft)/min
- Service ceiling: 4,876 m (16,000 ft)
- T-O run: 245 m (800 ft)
- T-O to 15 m (50 ft): 405 m (1,325 ft)
- Landing from 15 m (50 ft): 350 m (1,150 ft)
- Landing run: 183 m (600 ft)
- Range with max payload at 3,050 m (10,000 ft), no reserve:
  - at 139 knots (257 km/h; 160 mph): 521 nm (965 km; 600 miles)
  - at 122 knots (225 km/h; 140 mph): 607 nm (1,125 km; 700 miles)

WITTMAN TAILWIND MODEL W-9

This aircraft was first flown in 1958, as the W-9L Tailwind with tricycle landing gear (see 1965-66 *Jane's*). In 1965 it was fitted with a new wing exactly the same as that used on the W-8 Tailwind. It has more recently been re-engined with a 108 kW (145 hp) Continental O-300 flat-six engine. The structure has been strengthened as necessary to take the extra power and weight, and the landing gear is now of the tailwheel type.

Dimensions, external:
- Same as W-8, except:
- Length overall: 6·10 m (20 ft 0 in)

Weights:
- Weight empty: 363 kg (800 lb)
- Max T-O weight: 644 kg (1,420 lb)

Performance (at max T-O weight):
- Max level speed at S/L: 172 knots (319 km/h; 198 mph)
- Max cruising speed: 156 knots (290 km/h; 180 mph)
- Landing speed: 48 knots (89 km/h; 55 mph)
- Max rate of climb at S/L: 425 m (1,400 ft)/min
- Service ceiling: 5,180 m (17,000 ft)
- T-O run: 183 m (600 ft)
- Range with max fuel: 564 nm (1,045 km; 650 miles)

WOOD

STANLEY L. WOOD

Address:
PO Box 894, Riverview, Florida 33569

Mr Stanley Wood designed and built a Formula V racing aircraft, designated Wood SL-1, while serving in the USAF. Construction occupied a period of 18 months, at a cost of approximately $2,200, before the first flight was recorded on 8 June 1973.

WOOD SL-1

Type: Single-seat homebuilt Formula V racing aircraft.

Wings: Wire-braced mid-wing monoplane. Modified NACA 1308 wing section with 1% camber, 8·25% thickness/chord ratio. No dihedral. Conventional wooden structure with plywood and fabric skin. Ailerons of steel tubing, fabric covered and counterbalanced at tip.

Fuselage: Welded steel tube structure with fabric covering.

Tail Unit: Cantilever welded steel tube structure with fabric covering.

LANDING GEAR: Non-retractable tailwheel type. Cantilever main-gear struts.

POWER PLANT: One 48·5 kW (65 hp) 1,600 cc Volkswagen motor car engine with Wittman conversion (0·305 m; 12 in propeller shaft extension), driving a two-blade homebuilt 1·22 m × 1·07 m (48 in × 42 in) propeller with spinner.

ACCOMMODATION: Single seat in enclosed cockpit.

DIMENSIONS, EXTERNAL:
- Wing span: 5·54 m (18 ft 2 in)
- Wing chord, constant: 1·27 m (4 ft 2 in)
- Wing area, gross: 7·02 m² (75·6 sq ft)
- Wing aspect ratio: 4·3
- Length overall: 4·67 m (15 ft 4 in)
- Height overall: 1·27 m (4 ft 2 in)
- Tailplane span: 1·37 m (4 ft 6 in)

WEIGHTS AND LOADINGS:
- Weight empty: 189 kg (416 lb)
- Max T-O weight: 295 kg (650 lb)
- Max wing loading: 42·0 kg/m² (8·6 lb/sq ft)
- Max power loading: 6·08 kg/kW (10 lb/hp)

PERFORMANCE:
- Never-exceed speed: 173 knots (322 km/h; 200 mph)
- Max level speed: 156 knots (290 km/h; 180 mph)
- Cruising speed: 139 knots (257 km/h; 160 mph)
- Landing speed: 48 knots (89 km/h; 55 mph)
- Max rate of climb at S/L: 366 m (1,200 ft)/min
- Service ceiling: 3,660 m (12,000 ft)
- T-O run: 183 m (600 ft)
- Landing run: 213 m (700 ft)
- Range: 399 nm (740 km; 460 miles)

ZINNO
JOSEPH A. ZINNO & ASSOCIATES

TECHNICAL DESIGN STUDIOS: 44 Woodhaven Boulevard, North Providence, Rhode Island 02911

Telephone: (401) 231-3135

Mr Joseph Zinno has designed, built and accomplished the first flight of a man-powered aircraft which has the designation Olympian ZB-1 and the model number 74-001. Design originated in January 1972; construction began fourteen months later. FAA Experimental certification was granted on 19 January 1976, and the first flight was made on 21 April 1976.

ZINNO OLYMPIAN ZB-1

TYPE: Single-seat man-powered aircraft.

WINGS: Cantilever mid-wing monoplane of high aspect ratio. Wing section Wortmann FX MS 72-150B inboard, Wortmann FX 63-137 outboard, with thickness/chord ratio of 15% and 12% respectively. Dihedral 3°. No incidence. No sweepback. Single box spar with spruce flanges and birch ply webs. Polyetha D nose section, balsa ribs and aluminium trailing-edge, with clear plastic covering. Wingtip ailerons of similar construction.

FUSELAGE: Enclosed pod for pilot, with light alloy tailboom. Primary structure of light alloy tube with some steel, balsa fairings and clear plastic covering.

TAIL UNIT: Cantilever structure of similar materials to those used for the wings. All-moving tailplane mounted at trailing-edge of fin.

LANDING GEAR: Non-retractable monowheel and steerable nosewheel. Main wheel is a conventional bicycle wheel with a 27 × 1·25 tyre, pressure 5·86 bars (85 lb/sq in). Nosewheel has solid tyre size 7·5 × 1·25. No brakes. Retractable outriggers with small solid wheels mounted in lower surface of inboard wing, approximately 1·22 m (4 ft 0 in) outboard from fuselage on each side. Small wheel in base of fin, to simplify ground handling. Small skid at each wingtip, inboard of ailerons.

POWER PLANT: Power is derived from the pilot, working a reciprocating pedal drive to turn a triple ratchet hub which, via a single chain, turns the propeller shaft. This drives a two-blade wooden variable-pitch pusher propeller, with spinner, of original design.

ACCOMMODATION: Single seat in enclosed cabin. Access via forward section of cabin, which is easily removable. Accommodation ventilated by two airscoops in the canopy.

DIMENSIONS, EXTERNAL:
- Wing span: 23·93 m (78 ft 6 in)
- Wing chord at root: 1·60 m (5 ft 3 in)
- Wing chord at tip: 0·61 m (2 ft 0 in)
- Wing aspect ratio: 19·5
- Length overall: 6·55 m (21 ft 6 in)
- Height overall: 2·44 m (8 ft 0 in)
- Tailplane span: 4·42 m (14 ft 6 in)
- Wheelbase: 1·47 m (4 ft 10 in)
- Propeller diameter: 2·64 m (8 ft 8 in)

DIMENSIONS, INTERNAL:
- Cockpit length: 2·44 m (8 ft 0 in)
- Max width: 0·61 m (2 ft 0 in)
- Max height: 1·68 m (5 ft 6 in)

AREAS:
- Wings, gross: 28·98 m² (312 sq ft)
- Ailerons (total): 1·86 m² (20 sq ft)
- Fin: 0·79 m² (8·5 sq ft)
- Rudder: 1·21 m² (13 sq ft)
- Tailplane: 3·90 m² (42 sq ft)

WEIGHTS AND LOADING:
- Weight empty: 67·13 kg (148 lb)
- Optimum design T-O weight: 131·54 kg (290 lb)
- Optimum wing loading: 4·54 kg/m² (0·93 lb/sq ft)

PERFORMANCE:
- Never-exceed speed at S/L: 33·9 knots (62·8 km/h; 39 mph)
- Max level speed and max cruising speed at S/L: 19·1 knots (35·4 km/h; 22 mph)
- Normal cruising speed at S/L: 16·5 knots (30·6 km/h; 19 mph)
- Stalling speed: 13 knots (24·1 km/h; 15 mph)
- Max rate of climb at S/L: 18·3 m (60 ft)/min
- Ceiling: approx 9·1 m (30 ft)
- T-O run (no wind): 183 m (600 ft)
- Endurance: 15 m

THE UNION OF SOVIET SOCIALIST REPUBLICS

ARTIOMOV
MIKHAIL ARTIOMOV

ARTIOMOV OMEGA

Mr Mikhail Artiomov designed and built a lightplane known as the AT-1 Mriy during the early 1970s, with the help of Mr Viktor Timofeyev. Following this project, both designers decided to produce their own aircraft, Mr Artiomov's work culminating in the Omega.

The first flight of this homebuilt amphibian was made in January 1975 from a frozen lake in the Soviet Union. It has been designed to operate from ice during Winter and from water during the warmer months.

Most unusual aspect of the design is use of a channel wing which, according to the designer, provides increased lift compared with an ordinary wing, due to the induced flow of air over the greatest practicable wing area. The aircraft's nosewheel can be replaced with a skid for take-off from ice, a technique already tested successfully during early flights. The Omega is reported to be stable in flight, without sacrificing manoeuvrability.

TYPE: Single-seat light homebuilt monoplane.

WINGS: Channel-wing monoplane. Thick-section wings which taper towards tips. Slight anhedral. Bracing struts between horizontal wing surfaces and annular surfaces.

FUSELAGE: Boat-like amphibious forward section, housing the cockpit, appears to be of similar construction to wings. Rear fuselage is an uncovered welded tube structure.

TAIL UNIT: Conventional horizontal and vertical tail surfaces, the rudder contributing approximately two-thirds of the total area of the vertical surfaces.

LANDING GEAR: Non-retractable tricycle type, with steerable nosewheel linked to rudder. Spring shock-absorption. Nosewheel can be replaced by skid for operation from ice.

POWER PLANT: One IZ-56 converted motorcycle engine, developing 23·5 kW (32 hp) and driving a glassfibre-coated wooden two-blade fixed-pitch pusher propeller inside the wing channel. Two fuel tanks, with total capacity of 20 litres (4·4 Imp gallons), in wing centre-section.

ACCOMMODATION: Single seat in open cockpit.

DIMENSIONS, EXTERNAL:
- Wing span: 8·22 m (26 ft 11½ in)
- Length overall: 5·40 m (17 ft 8½ in)

WEIGHT:
- Max T-O weight: 270 kg (595 lb)

PERFORMANCE:
- Max level speed: 54 knots (100 km/h; 62 mph)
- Landing speed: 35 knots (65 km/h; 40·5 mph)

DMITRIEV
V. DMITRIEV

ADDRESS: Frunze, Kirgizia

DMITRIEV MIKROSAMOLET (MINIPLANE)

This simple single-seat lightplane was built by Mr V. Dmitriev and colleagues at the town of Frunze, in Kirgizia.

TYPE: Light homebuilt monoplane.

WINGS: Cantilever low-wing monoplane of tapering chord and with rounded tips. Large-chord ailerons.

FUSELAGE: Square-section fuselage, with large turtledeck fairing aft of cockpit canopy.

TAIL UNIT: Cantilever structure, with conventional elevators and rudder.

POWER PLANT: One unspecified engine, developing 30 kW (40 hp) and driving a two-blade propeller.

ACCOMMODATION: Single seat under transparent cockpit canopy.

WEIGHTS:
- Weight empty: 250 kg (551 lb)

PERFORMANCE:
- Design max level speed: 146 knots (270 km/h; 168 mph)

KAI
KHARKOV AVIATION INSTITUTE

A series of light aircraft has been designed and built by students of this institute. Some details of the KAI-17 (or KhAI-17) and KAI-18 were given in the 1962-63 *Jane's.* A three-view drawing of the KAI-19 single-seat ultra-light monoplane appeared in the 1964-65 *Jane's.* A photograph and brief details of the KAI-24 two-seat light autogyro can be found in the 1969-70 edition. These aircraft were followed by a single-seat pusher-engined light aircraft designated KAI-20, and the single-seat KAI-22A and two-seat KAI-27 ultra-light helicopters. The KAI-27 was described and illustrated in the 1970-71 *Jane's,* and the others in the 1972-73 edition.

No details concerning the flight trials of any of these aircraft have ever been received.

The latest product of the Kharkov students is a powered Rogallo, of which brief details follow:

KAI-21 GIBKOLET (FLEXIFLIGHT)

This simple aircraft utilises a flexible wing of the Rogallo type, and is powered by a 28·3 kW (38 hp) MT-8 converted motorcycle engine. Control in flight is achieved by movement of the rudder and by tilting the wing.

DIMENSION:
- Wing area: 20 m² (215 sq ft)

WEIGHT:
- Max T-O weight: 250 kg (551 lb)

PERFORMANCE:
- Max level speed: 54 knots (100 km/h; 62 mph)
- T-O and landing speed: 22 knots (40 km/h; 25 mph)
- Service ceiling: 2,000 m (6,560 ft)
- T-O run: 50-100 m (164-328 ft)
- Range with max fuel: 162 nm (300 km; 186 miles)

LENIN KOMSOMOL INSTITUTE, RIGA
LENIN KOMSOMOL INSTITUTE OF CIVIL AVIATION ENGINEERS

ADDRESS: Riga, Latvia

LENIN KOMSOMOL INSTITUTE RKIIGA-74

Design and construction of this small flying-boat was begun in the Summer of 1972 by members of the Lenin Komsomol Institute at Riga, led by technical science candidate F. Muchamiedow, senior lecturer W. Cejtlin, and students W. Jagniuk, J. Przybylsky and A. Szwejgert of the mechanical science faculty. The intention was to evolve an aircraft that would be easy and inexpensive to construct, but might yet have adequate performance to break existing Soviet distance records for light piston engined aircraft of under 1,000 kg loaded weight.

As a first step it was decided to build the Rkiiga-74 as a flying-boat, although it was hoped to make it amphibious later. To keep costs to a minimum, extensive use was made of existing production components. Thus, the hull is basically a Progress motorboat, with the frames strengthened in the areas of wing and tail attachment points. The windscreen was moved back 0·15 m (6 in) and given

Wittman W-B Tailwind photographed at Patty Field, El Dorado *(Jean Seele)*

Williams W-18 Falcon *(Bill Mallet)*

Zinner Olympian ZB-1 single-seat man-powered aircraft

Artimov Omega single-seat channel-wing monoplane

Lenin Komsomol Institute RK11GA-74 two-seat flying-boat

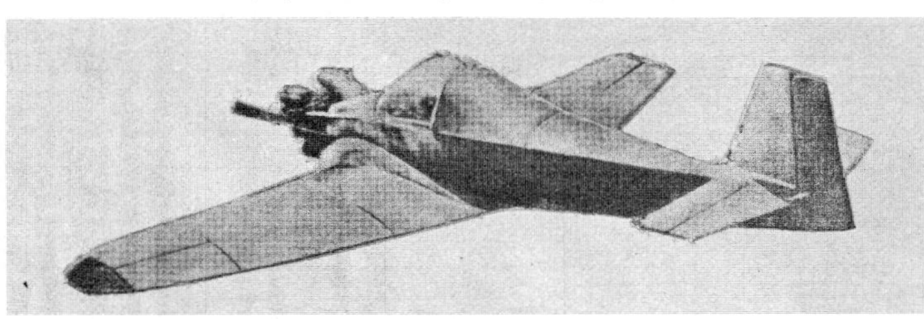

Dmitriev Mikrosamolet ultra-lift monoplane

Timofeyev T-1 Mustang single-seat monoplane

increased slope to reduce aerodynamic drag; for the same purpose a number of fairings were added, made of foam plastics covered with glassfibre. An aircraft-type instrument panel was installed, together with dual stick and rudder pedal controls; behind the two side by side seats, a fuel tank and electrical storage battery were installed. The wings and tail unit are components of a Primoriets primary training glider, and the power plant is a standard aircraft piston engine of Czech manufacture.

Following completion of the aircraft, it was taken to Lake Baltezerz, Riga, where an area was marked out for initial flight tests. Taxi trials and hops, begun on 6 September 1974, led to the addition of a step on the bottom of the hull. The first circuit, at a height of 150 m (500 ft) was accomplished by V. Abramov and W. Cejtlin on 17 September 1974. Altogether, fifteen flights were logged, totalling about two hours. They showed that the Rkiiga-74 is stable on the water and manoeuvrable in flight, with good performance both on the water and in the air, but that its water handling qualities would be improved by the addition of water rudders. At the same time, it was decided to improve the shape of the step and the engine cowling, and to install a retractable wheeled landing gear.

TYPE: Two-seat homebuilt flying-boat.

WINGS: Parasol monoplane, consisting essentially of the wing of a Primoriets glider. Wing section NACA 43012. Single-spar wooden structure, covered with plywood and with fabric covering overall. Wing supported above fuselage by welded tubular metal struts. Long-span ailerons. Detachable wingtip floats of foam plastics, covered with glassfibre and carried on metal struts.

HULL: Basically a Progress motorboat, modified as described in introductory notes.

TAIL UNIT: Conventional structure from a Primoriets glider.

POWER PLANT: One 104·5 kW (140 hp) M 332 four-cylinder inverted in-line aircooled engine, driving a two-blade propeller, with large spinner. Engine mounted under wing centre-section. Fuel tank aft of seats, capacity 90 litres (19·8 Imp gallons).

ACCOMMODATION: Two seats side by side in open cockpit. Dual controls. Basic aircraft instruments, including ASI, altimeter, rate of climb indicator, turn and ban indicator, compass and engine instruments.

DIMENSIONS, EXTERNAL:

Wing span	13·42 m (44 ft 0 i
Wing area, gross	20·2 m² (217·4 sq f
Length overall	8·10 m (26 ft 7 i
Height overall	2·34 m (7 ft 8¼ i

WEIGHTS AND LOADINGS:

Weight empty	550 kg (1,212 lt
Max T-O weight	750 kg (1,653 lt
Max wing loading	37·62 kg/m² (7·71 lb/sq
Max power loading	7·18 kg/kW (11·81 lb/hp

PERFORMANCE:

Max level speed	75 knots (140 km/h; 87 mpl
T-O speed	43·5 knots (80 km/h; 50 mpl
Landing speed	38-43·5 knots (70-80 km/h; 44-50 mpl
Max rate of climb at S/L	360 m (1,180 ft)/mi
Service ceiling	4,000 m (13,125 f
T-O run	250 m (820
Landing run	80 m (260 f
Normal range	269 nm (500 km; 310 mile

TIMOFEYEV

VIKTOR TIMOFEYEV

ADDRESS: Dneprodzerzhinsk, Ukraine

Since building the AT-1 Mriy lightplane in partnership with Mr Mikhail Artiomov, Mr Timofeyev has built and flown a single-seat high-wing monoplane resembling the Polish Janowski designs.

TIMOFEYEV T-1 MUSTANG

In general appearance and construction, this single-seat light aircraft is almost identical with the Janowski J-1 Don Kichot. External differences include the use of two bracing struts for each wing, instead of the single struts on the Polish aircraft, and the lack of fairings over the multi-tube engine support structure at the time of the Mustang's initial flight in 1975.

Power plant of the Mustang is a converted IZ-56 two-cylinder inverted motorcycle engine, developing 23·5 kW (32 hp) and driving a two-blade fixed-pitch wooden propeller.

DIMENSIONS, EXTERNAL:

Wing span	8·00 m (26 ft 3 in)
Length overall	5·60 m (18 ft 4½ i

WEIGHTS:

Weight empty	150 kg (331
Max T-O weight	270 kg (595

PERFORMANCE:

Max level speed	56 knots (105 km/h; 65 mp
T-O speed	35 knots (65 km/h; 41 mp
T-O run	100 m (328
Landing run	50-60 m (165-200

SAILPLANES
(including Hang Gliders)

SAILPLANES

ARGENTINE REPUBLIC

AS
AERO SALADILLO

ADDRESS:
Casilla Correo 24, Saladillo, Provincia de Buenos Aires
Telephone: 41-5260 (Comodoro A. R. Mantel, in Buenos Aires)

MANAGERS:
Comodoro Ildefonso Durana
Luis De Pizzol

This company was formed in 1973 with the object of producing aircraft built of glassfibre-reinforced plastics. Its first product, the Lenticular single-seat 15 m sailplane, is in production; development of the Biguá two-seat sailplane, described in the 1975-76 *Jane's*, has been discontinued.

AS LENTICULAR 15 S

The prototype of this single-seat sailplane, designed by Ing Téodoro Altinger, flew for the first time on 5 August 1971 and took part successfully in the national championships in 1972.

Series production began in 1974; orders have been placed for 20 Lenticulars.

TYPE: Single-seat Standard Class sailplane.

WINGS: Cantilever shoulder-wing monoplane. Wortmann FX-63-168 wing section. Dihedral 3°. Constant chord throughout span. Glassfibre/balsa sandwich construction. Glassfibre-reinforced plastics airbrakes on upper surfaces.

FUSELAGE: Glassfibre-reinforced plastics monocoque structure.

TAIL UNIT: Cantilever T-tail of glassfibre-reinforced plastics construction. Sweptback fin and rudder. One-piece all-moving tailplane, with anti-tab.

LANDING GEAR: Retractable monowheel (diameter 404 mm; 15·9 in), with internal brake, and tailskid.

ACCOMMODATION: Single seat under fully-transparent moulded Plexiglas canopy.

DIMENSIONS, EXTERNAL:
Wing span 15·00 m (49 ft 2½ in)
Wing aspect ratio 18·75
Length overall 6·50 m (21 ft 4 in)
Height over tail 1·40 m (4 ft 7 in)

DIMENSIONS, INTERNAL:
Cabin: Max height 0·90 m (2 ft 11½ in)
Max width 0·62 m (2 ft 0¼ in)

AREAS:
Wings, gross 12·00 m² (129·2 sq ft)
Airbrakes (total) 0·80 m² (8·61 sq ft)
Fin 0·60 m² (6·46 sq ft)
Rudder 0·60 m² (6·46 sq ft)
Tailplane, incl anti-tab 1·40 m² (15·07 sq ft)

WEIGHTS AND LOADING:
Weight empty 260 kg (573 lb)
Max T-O weight 410 kg (904 lb)
Max wing loading 34·2 kg/m² (7·00 lb/sq ft)

PERFORMANCE (at max T-O weight):
Best glide ratio at 46·5 knots (86 km/h; 53 mph) 32
Min sinking speed at 42·5 knots (78 km/h; 49 mph) 0·70 m (2·30 ft)/sec
Stalling speed 38 knots (70 km/h; 43·5 mph)
Max speed (smooth air) 132 knots (244 km/h; 152 mph)
Max speed (rough air) 94 knots (174 km/h; 108 mph)
Max aero-tow speed 86 knots (160 km/h; 99 mph)
Max winch-launching speed 65 knots (120 km/h; 74·5 mph)
g limits:
normal +5·3; −4·(
ultimate +7·9

AVEX
ASOCIACION ARGENTINA DE CONSTRUCTORES DE AVIONES EXPERIMENTALES

ADDRESS:
Acasusso 1640, Olivos-FCNGBM, Buenos Aires
Telephone: 797-1629

OFFICERS: See Homebuilt Aircraft section

LANZALONE AULANZ

The Aulanz is a single-seat powered sailplane which, with its 22·4 kW (30 hp) engine and propeller, has been designed and is being built by Sr Augusto Lanzalone of Rosario, Santa Fé. Sr Lanzalone has also evolved, for use in its construction, his own alloy of aluminium (with copper, nickel, magnesium, silicon and chrome), which he has named Alcusing; and also his own furnace (for heat-treating the fuselage) and rivet gun. By the Spring of 1973 the fuselage and tail unit were completed; construction of the remainder is proceeding slowly.

WINGS: Cantilever low-wing monoplane, of all-metal (Alcusing) construction. Wortmann FX-61-147 wing section. Incidence 4°.

FUSELAGE: All-metal (Alcusing) semi-monocoque structure.

TAIL UNIT: Cantilever all-metal structure, of similar construction to wings.

LANDING GEAR: Retractable monowheel and tailskid. Rubber shock-absorbers.

POWER PLANT: One 22·4 kW (30 hp) 700 cc Lanzalone two-cylinder two-stroke inverted in-line engine, driving a two-blade variable-pitch metal (Alcusing) propeller. Fuel capacity 20 litres (4·4 Imp gallons).

ACCOMMODATION: Single seat under one-piece transparent canopy.

DIMENSIONS, EXTERNAL:
Wing span 12·40 m (40 ft 8¼ in)
Wing chord at root 1·35 m (4 ft 5¼ in)
Wing chord at tip 0·70 m (2 ft 3½ in)
Length overall 5·40 m (17 ft 8½ in)
Height over tail 1·87 m (6 ft 1¾ in)
Tailplane span 2·55 m (8 ft 4½ in)
Propeller diameter 1·30 m (4 ft 3¼ in)

AREAS:
Wings, gross 11·70 m² (125·9 sq ft)
Ailerons (total) 1·00 m² (10·76 sq ft)
Trailing-edge flaps (total) 1·20 m² (12·92 sq ft)
Fin 0·36 m² (3·88 sq ft)
Rudder 0·38 m² (4·09 sq ft)
Tailplane 0·84 m² (9·04 sq ft)
Elevators 0·49 m² (5·27 sq ft)

WEIGHT:
Max T-O weight 280 kg (617 lb)

AUSTRALIA

SCHNEIDER
EDMUND SCHNEIDER PTY LTD

HEAD OFFICE AND WORKS:
Two Wells Road (Aerodrome), Gawler, South Australia 5118
Telephone: (085) 22-2978

CHIEF DESIGNER:
Harry Schneider

Edmund Schneider Pty Ltd, late of Grunau in Germany, was one of the pioneer sailplane manufacturing companies. It transferred its operations to Australia as a private venture, at the invitation of the Gliding Federation of Australia. Its first project in the Commonwealth was the Kangaroo two-seat sailplane, which flew during 1953, followed by the Grunau Baby 4, Club trainer, Kookaburra, Arrow, Boomerang and Series 2 Boomerang, described in previous editions of *Jane's*.

The company's latest product of its own design was the ES 60B Super Arrow, production of which ended in late 1975. Descriptions of the ES 60B can be found in *Jane's* for 1973-74 (Series 1 version) and 1975-76 (Series 2).

Schneider is currently developing a new side-by-side two-seat glassfibre sailplane, which will have a retractable landing gear and a glide ratio of 36. No further details are yet available for publication.

Schneider is the Australian and New Zealand agent for Glasflügel sailplanes (see German section), assembles semi-completed glassfibre sailplanes imported from Glasflügel, and also represents Start + Flug (which see) in Australia and New Zealand. Since July 1975 it has been the Australian agent for Schempp-Hirth of Germany, and its associated company, Sailplane Distributors Pty Ltd, of the same address, similarly represents Slingsby (see UK section). Other activities include repair and maintenance of sailplanes and the supply of sailplane materials and spares.

AUSTRIA

BRDITSCHKA
H. W. BRDITSCHKA OHG

ADDRESS:
A-4053 Haid, Dr Schärfstrasse 42, Postfach 12
Telephone: 07229/2093
Telex: 21909

DIRECTOR:
Heinz W. Brditschka

This company is producing a single-seat powered sailplane known as the HB-3B and an increased-span, two-seat development of it known as the HB-21.

BRDITSCHKA HB-3B

This single-seat powered sailplane utilises the basic wing design of the Krähe sailplane designed in Germany by Ing Fritz Raab. Design started in 1968.

Three prototypes were built, these making their first flights on 23 June 1971 and 5 June and 28 July 1972. By early 1976, nine production aircraft had been completed.

The Brditschka works was also responsible, in 1973, for the conversion of an HB-3 airframe to electric power as the MB-E1, described under the Militky heading in the German section of the 1975-76 *Jane's*.

The following description applies to the standard production HB-3B:

TYPE: Single-seat powered sailplane.

WINGS: Cantilever high-wing monoplane. Wing section Göttingen Gö 758 at root, Clark Y at tip. Thickness/chord ratio 13·8% at root, 11·7% at tip. Dihedral 2° on outer panels. Incidence 2°. Conventional all-wood structure, including ailerons and upper-surface spoilers. No flaps or tabs.

FUSELAGE: Tubular steel framework with glassfibre covering. Triangular cutout for propeller rotation aft of wing trailing-edge.

TAIL UNIT: Conventional all-wood structure. Fixed-incidence tailplane mounted on top of fuselage. Trim tab on starboard elevator.

LANDING GEAR: Non-retractable tricycle type. Glassfibre legs provide all necessary shock-absorption. Main wheels and tyres size 300 × 100, pressure 2·45 bars (35·5 lb/sq in); nosewheel and tyre size 200 × 50, pressure 2·94 bars (42·5 lb/sq in). Tost mechanical brakes on main wheels.

POWER PLANT: One 30·6 kW (41 hp) Rotax 642 flat-twin engine, driving a Hoffmann HO 11-150 B 100 LD two-blade fixed-pitch pusher propeller. Single fuselage fuel tank and aluminium tank in wing centre-section; total capacity 37 litres (8·15 Imp gallons). Fuel is 75/25% petrol/oil mixture.

ACCOMMODATION: Single seat in fully-enclosed cabin. Cockpit canopy opens sideways to starboard. Small baggage space behind seat. Ram-air intake in nose for cabin ventilation.

EQUIPMENT: 12V 15Ah battery standard. Radio and other equipment at customer's option.

DIMENSIONS, EXTERNAL:
Wing span 12·00 m (39 ft 4½ in)
Wing chord at root 1·40 m (4 ft 7 in)
Wing chord at tip 0·73 m (2 ft 4¾ in)
Wing aspect ratio 10·1
Length overall 7·00 m (22 ft 11¾ in)
Height overall 2·95 m (9 ft 8 in)
Tailplane span 2·55 m (8 ft 4½ in)
Wheel track 1·66 m (5 ft 5½ in)
Wheelbase 1·87 m (6 ft 1¾ in)
Propeller diameter 1·50 m (4 ft 11 in)

DIMENSION, INTERNAL:
Cabin: Max width 0·65 m (2 ft 1½ in)

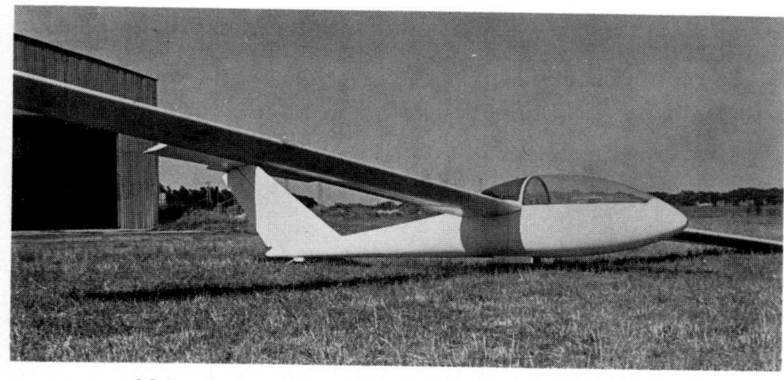

AS Lenticular 15 S single-seat Standard Class sailplane

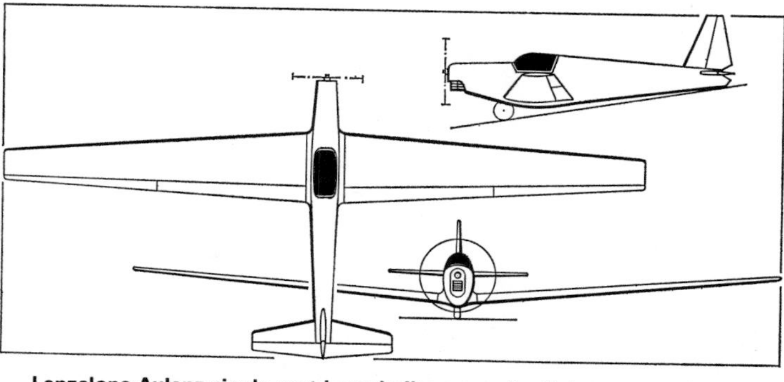

Lanzalone Aulanz single-seat homebuilt powered sailplane *(Tony Mitchell)*

New two-seat glassfibre sailplane being developed by Schneider *(The Age,* Melbourne)

Brditschka HB-3 single-seat powered sailplane

Second prototype of the Brditschka HB-21 two-seat powered sailplane

EEUFMG CB-2 Minuano single-seat high-performance sailplane

AREAS:
- Wings, gross — 14·22 m² (153·1 sq ft)
- Ailerons (total) — 1·50 m² (16·15 sq ft)
- Spoilers (total) — 0·18 m² (1·94 sq ft)
- Fin — 0·39 m² (4·20 sq ft)
- Rudder — 0·858 m² (9·24 sq ft)
- Tailplane — 1·754 m² (18·88 sq ft)
- Elevators, incl tab — 0·824 m² (8·87 sq ft)

WEIGHTS AND LOADINGS:
- Weight empty, equipped — 260 kg (573 lb)
- Max T-O and landing weight — 380 kg (837 lb)
- Max wing loading — 26·72 kg/m² (4·57 lb/sq ft)
- Max power loading — 12·42 kg/kW (19·95 lb/hp)

PERFORMANCE (at max T-O weight, powered):
- Max cruising speed — 86 knots (160 km/h; 99 mph)
- Econ cruising speed — 81 knots (150 km/h; 93 mph)
- Stalling speed — 32·5 knots (60 km/h; 37·5 mph)
- Max rate of climb at S/L — 180 m (590 ft)/min
- Service ceiling — 5,000 m (16,400 ft)
- T-O run — 100 m (330 ft)
- T-O to 15 m (50 ft) — 230 m (755 ft)
- Landing from 15 m (50 ft) — 100 m (330 ft)
- Landing run — 50 m (165 ft)
- Range with max fuel — 296 nm (550 km; 341 miles)

PERFORMANCE (at max T-O weight, power off):
- Best glide ratio at 43·5 knots (80 km/h; 50 mph) — 20
- Min sinking speed at 39·5 knots (73 km/h; 45·5 mph) — 1·15 m (3·77 ft)/sec
- Max speed (smooth air) — 94 knots (175 km/h; 108 mph)
- Max speed (rough air) — 69 knots (129 km/h; 80 mph)
- *g* limits — +5·3; −2·7

BRDITSCHKA HB-21

The HB-21, which first flew in 1973, is essentially an enlarged version of the HB-3, with increased-span wings, tandem seating for two persons, and a choice of either a Rotax (HB-21R) or Volkswagen (HB-21L) engine. Four had been built by early 1976, and two more were on order.

TYPE: Two-seat powered sailplane.

WINGS: Cantilever high-wing monoplane. Wortmann wing sections: FX-61-184 at root, FX-60-126 at tip. Dihedral 2°. Incidence 3°. Conventional wooden structure, with fabric covering. All-wood ailerons and upper-surface spoilers. No flaps or tabs.

FUSELAGE: Tubular steel framework with glassfibre covering, as HB-3, but of increased length.

TAIL UNIT: Similar to HB-3, but slightly larger.

LANDING GEAR: Non-retractable tricycle type, similar to HB-3. Main wheels and tyres size 330 × 130, pressure 2·45 bars (35·5 lb/sq in); nosewheel and tyre size 260 × 85, pressure 2·94 bars (42·5 lb/sq in). Tost mechanical brakes on main wheels.

POWER PLANT: One 48·5 kW (65 hp) VW-Westermayer 1600G flat-four (HB-21L) or 30·6 kW (41 hp) Rotax 642 flat-twin engine (HB-21R), driving a Hoffmann HO 14-175 B 117 LD two-blade fixed-pitch pusher propeller. Aluminium fuel tank in wing, capacity 54 litres (11·9 Imp gallons). Fuel is a 75/25% petrol/oil mixture.

ACCOMMODATION: Two seats in tandem in fully-enclosed cabin. Cockpit canopy opens sideways to starboard. Ram-air intake in nose for cabin ventilation. HB-21L has cabin heating.

EQUIPMENT: 12V 15Ah battery standard. Radio and other equipment at customer's option.

DIMENSIONS, EXTERNAL:
- Wing span — 16·24 m (53 ft 3½ in)
- Wing chord at root — 1·50 m (4 ft 11 in)
- Wing chord at tip — 0·60 m (1 ft 1¾ in)
- Wing aspect ratio — 13·89
- Length overall — 7·90 m (25 ft 11 in)
- Height overall — 2·60 m (8 ft 6¼ in)
- Tailplane span — 3·15 m (10 ft 4 in)
- Propeller diameter — 1·75 m (5 ft 8¾ in)

AREAS:
- Wings, gross — 18·98 m² (204·3 sq ft)
- Ailerons (total) — 1·50 m² (16·15 sq ft)
- Spoilers (total) — 0·34 m² (3·66 sq ft)
- Fin — 0·31 m² (3·34 sq ft)
- Rudder — 1·19 m² (12·81 sq ft)
- Tailplane — 2·15 m² (23·14 sq ft)
- Elevators, incl tab — 1·01 m² (10·87 sq ft)

WEIGHTS AND LOADINGS:
- Weight empty, equipped:
  - 21L — 481 kg (1,060 lb)
  - 21R — 430 kg (948 lb)
- Max T-O weight:
  - 21L — 700 kg (1,543 lb)
  - 21R — 650 kg (1,433 lb)
- Max wing loading:
  - 21L — 36·88 kg/m² (7·55 lb/sq ft)
  - 21R — 34·25 kg/m² (7·01 lb/sq ft)

PERFORMANCE (at max T-O weight, powered):
- Max cruising speed:
  - 21L — 86 knots (160 km/h; 99 mph)
  - 21R — 81 knots (150 km/h; 93 mph)
- Econ cruising speed:
  - 21L — 70·5 knots (130 km/h; 81 mph)
  - 21R — 65 knots (120 km/h; 74·5 mph)
- Stalling speed:
  - 21L — 39·5 knots (73 km/h; 45·5 mph)
  - 21R — 36·5 knots (67 km/h; 42 mph)
- Max rate of climb at S/L:
  - 21L — 198 m (650 ft)/min
  - 21R — 132 m (435 ft)/min
- Service ceiling:
  - 21L — 6,300 m (20,675 ft)
  - 21R — 4,000 m (13,125 ft)
- T-O run:
  - 21L — 100 m (330 ft)
  - 21R — 170 m (560 ft)
- T-O to 15 m (50 ft):
  - 21L — 200 m (655 ft)
  - 21R — 330 m (1,085 ft)
- Landing from 15 m (50 ft):
  - 21L, 21R — 150 m (490 ft)
- Landing run:
  - 21L, 21R — 80 m (262 ft)

Range with max fuel:
21L 431 nm (800 km; 497 miles)
21R 351 nm (650 km; 404 miles)
PERFORMANCE (at max T-O weight, power off):
Best glide ratio:
21L at 56·5 knots (105 km/h; 65 mph) 24-26
21R at 51 knots (95 km/h; 59 mph) 24-26
Min sinking speed:
21L at 45 knots (84 km/h; 52 mph) 1·0 m (3·28 ft)/sec
21R at 42 knots (78 km/h; 48·5 mph) 0·9 m (2·95 ft)/sec
Max speed (smooth air):
21L, 21R 113 knots (210 km/h; 130 mph)
Max speed (rough air):
21L, 21R 88·5 knots (165 km/h; 102 mph)
g limits:
21L, 21R +5·3; −2·7

BRAZIL

EEUFMG (CEA)

ESCOLA DE ENGENHARIA DA UNIVERSIDADE FEDERAL DE MINAS GERAIS (Centro de Estudos Aeronáuticos)

ADDRESS:
Rua Espirito Santo 35, 30.000 Belo Horizonte, Minas Gerais State
Telephone: 222-4011
HEAD OF CEA:
Prof Cláudio Pinto de Barros

CB-2 MINUANO

The Air Research Centre of the Engineering School at Minas Gerais Federal University began the design of this single-seat sailplane in 1961. Its designation, CB-2, indicates that it is the second design by Professor Cláudio Barros; his first, the CB-1 Gairota, was completed when he was an engineering student. Construction of the CB-2 prototype (PP-ZPZ) started in 1971, and it flew for the first time on 20 December 1975. In early 1976 a further four Minuanos were on order, the second aircraft being scheduled for completion by the end of the year. A description of the Minuano (named after a strong, cold wind in southern Brazil) follows:

TYPE: Single-seat high-performance sailplane.

WINGS: Cantilever high-wing monoplane. Wortmann wing sections: FX-61-163 at root, FX-60-126 at tip. No dihedral or anhedral. Incidence 3°. Single main spar of 2024 aluminium alloy, with plywood/glassfibre honeycomb sandwich covering. Plain flaps and ailerons on trailing-edge, of similar construction except for wooden spars. No spoilers; flaps deflect up to 90° to act as airbrakes.

FUSELAGE: All-wood semi-monocoque structure.

TAIL UNIT: Cantilever structure, with sweptback fin and rudder and non-swept horizontal surfaces. All surfaces covered in plywood, stiffened with foam plastics. All-moving tailplane, with trim tab in each half.

LANDING GEAR: Manually-retractable unsprung monowheel, size 210 × 120 mm, tyre pressure 2·45 bars (35·5 lb/sq in). External rubber friction brake on first prototype; internal shoe brake on second aircraft. Non-retractable tailskid, sprung with tennis balls.

ACCOMMODATION AND EQUIPMENT: Single semi-reclining seat under one-piece detachable blown Plexiglas canopy. Headrest for pilot. Rudder pedals adjustable in flight. Normal instrumentation includes radio and variometer with audio output unit. Provision for oxygen, water ballast and releasable drag parachute.

DIMENSIONS, EXTERNAL:
Wing span 15·00 m (49 ft 2½ in)
Wing chord at root 0·78 m (2 ft 6¾ in)
Wing chord at tip 0·39 m (1 ft 3¼ in)
Wing aspect ratio 22
Length overall 7·00 m (22 ft 11¾ in)
Height over tail 1·43 m (4 ft 8¼ in)
Tailplane span 2·20 m (7 ft 2¾ in)
AREAS:
Wings, gross 10·20 m² (109·8 sq ft)
Ailerons (total) 0·56 m² (6·03 sq ft)
Trailing-edge flaps (total) 0·83 m² (8·93 sq ft)
Fin 0·49 m² (5·27 sq ft)
Rudder 0·46 m² (4·95 sq ft)
Tailplane, incl tabs 0·83 m² (8·93 sq ft)
WEIGHTS AND LOADING:
Weight empty, equipped 214 kg (472 lb)
Max T-O weight 304 kg (670 lb)
Max wing loading 30 kg/m² (6·14 lb/sq ft)
PERFORMANCE (at max T-O weight):
Best glide ratio at 48·5 knots (90 km/h; 56 mph) 38
Min sinking speed at 39 knots (72 km/h; 45 mph) 0·55 m (1·80 ft)/sec
Stalling speed 35 knots (65 km/h; 40·5 mph)
Max speed (smooth air) 140 knots (260 km/h; 161 mph)
Max speed (rough air) and max aero-tow speed 86 knots (160 km/h; 99 mph)
g limits (at safety factor of 1·72) +6·07; −4·07

IPE

INDUSTRIA PARANAENSE DE ESTRUTURAS

ADDRESS:
Caixa Postal 2621, 80.000 Curitiba, Paraná State
SECRETARY:
Eng. J. C. Boscardin

This company is developing, for eventual series production, the KW 1 b 2 Quero Quero II single-seat training sailplane, of which a description follows. The programme for the KW 2 Biguá (see 1975-76 *Jane's*) has apparently been abandoned, but IPE is now working on a new two-seat training sailplane. It is also negotiating with Neiva (see Aircraft section) the possibility of licence production of an updated version of the latter company's P.56C Paulistinha aircraft for use as a glider tug.

IPE KW 1 b 2 QUERO QUERO II

The original KW 1 was designed by Ing Kuno Wiedmaier and built by the Aeroclube de N. Hamburgo, Rio Grande do Sul; it first flew in 1970, and its subsequent development into the present KW 1 b 2 has been described in previous editions of *Jane's*.

The KW 1 b 2 made its first flight on 1 October 1972. It has a Scheibe Spatz wing section, upper- and lower-surface spoilers, and is built of wood and plywood (Brazilian pine). There is a trim tab on the port elevator. Type approval by the Centro Tecnico Aeroespacial, in the semi-aerobatic category with an extension for cloud flying, was expected in the early part of 1976. In anticipation of this, the Brazilian Ministry of Aeronautics has approved the construction of a pre-production batch of four aircraft, and more extensive production is likely.

TYPE: Single-seat training glider.

LANDING GEAR: Non-retractable monowheel and tailwheel.

DIMENSIONS, EXTERNAL:
Wing span 15·00 m (49 ft 2½ in)
Wing aspect ratio 18
Length overall 6·47 m (21 ft 2¾ in)
Height over tail 1·34 m (4 ft 4¾ in)
Fuselage: Max depth 1·00 m (3 ft 3¼ in)
Tailplane span 2·40 m (7 ft 10½ in
AREA:
Wings, gross 11·70 m² (125·9 sq ft)
WEIGHTS AND LOADING:
Weight empty 170 kg (374 lb)
Max T-O weight 270 kg (595 lb)
Max wing loading 21·3 kg/m² (4·36 lb/sq ft)
PERFORMANCE:
Best glide ratio at 39 knots (73 km/h; 45 mph) 28
Min sinking speed at 33·5 knots (62 km/h; 38·5 mph) 0·64 m (2·10 ft)/sec
Max speed 81 knots (150 km/h; 93 mph)
g limit +8

IPE TWO-SEAT TRAINING SAILPLANE

The general appearance of this new IPE sailplane can be seen from the accompanying three-view drawing. It is hoped to complete the prototype by 1 April 1977. All known details follow:

DIMENSIONS, EXTERNAL:
Wing span 16·60 m (54 ft 5½ in)
Length overall 7·90 m (25 ft 11 in)
Height over tail 1·50 m (4 ft 11 in)
Tailplane span 3·00 m (9 ft 10 in)
WEIGHT:
Weight empty 250 kg (551 lb)
PERFORMANCE (estimated):
Best glide ratio at 43·5 knots (80 km/h; 50 mph) 30
Max speed (smooth air) 108 knots (200 km/h; 124 mph)

CZECHOSLOVAKIA

OMNIPOL

OMNIPOL FOREIGN TRADE CORPORATION

ADDRESS:
Washingtonova 11, Prague 1
Telephone: 2126
Telex: 121489, 121808 and 121077
GENERAL MANAGER:
Tomás Marecek, GE
SALES MANAGER:
Maroslav Vesely
PUBLICITY MANAGER:
Jiri Matula

This concern handles the export sales of the products of the Czechoslovak aircraft industry, including the L-13 Blaník sailplane.

LET

LET NÁRODNÍ PODNIK (Let National Corporation)

ADDRESS:
Uherské Hradiste-Kunovice
OFFICERS:
See Aircraft section

In addition to its work on powered aircraft, Let National Corporation manufactures the L-13 Blaník sailplane, of which full details follow:

LET L-13 BLANÍK

This tandem two-seat all-metal sailplane is designed for training in all categories from elementary to 'blind' flying and for high-performance flight. It is fully aerobatic when flown solo and capable of basic aerobatic manoeuvres when carrying a crew of two.

Design of the Blaník was started in January 1955, and construction of the first of two prototypes began in August of the same year. First flight was made in March 1956.

By the beginning of 1976, about 2,100 Blaníks had been sold, of which more than 1,800 had been exported to customers in more than 40 countries, including more than 800 to the USSR, several hundred to the USA, and more than 100 each to Australia, Canada and the UK. Production was continuing in 1976.

A small quantity was built of a powered version, the L-13J (see 1973-74 *Jane's*), but this version is not in series production.

No fewer than 13 international records have been set up by sailplanes of this type, in addition to numerous national records. In 1969 an FAI Gold Medal was awarded to the Chilean pilot Alejo Williamson for a 5 hr 51 min flight in a Blaník across the Andes, from Santiago de Chile to Mendoza in the Argentine Republic.

TYPE: Two-seat training sailplane.

WINGS: Cantilever shoulder-wing monoplane, with 5° forward sweep at quarter-chord. Wing section NACA 63_2A615 at root, NACA 63_2A612 at tip. Dihedral 3°. Incidence 4° at root, 1° at tip. All-metal two-spar structure. Main spar forms torsion box with leading-edge. Each wing secured by three fuselage attachments. Wingtip 'salmons'. Ailerons and slotted area-increasing flaps are fabric-covered metal structures. Rectangular light alloy airbrakes in the upper and lower surfaces of each wing.

FUSELAGE: All-metal semi-monocoque structure of oval cross-section, with riveted skin.

TAIL UNIT: Cantilever all-metal structure. Elevator and rudder fabric-covered. Controllable trim tab in elevator. Horizontal surfaces fold upward parallel to rudder for transport.

LANDING GEAR: Mechanically-retractable monowheel, type HP-4741-Z, located in lower part of fuselage on centreline. Wheel manufactured by Rudy Ríjen of Gottwaldov; tyre size 350 × 135 mm (13·8 × 5·3 in) by Moravan of Otrokovice, pressure 2·45 bars (35·5 lb/sq in). Oleo-pneumatic shock-absorber and mechanically-actuated brake.

ACCOMMODATION: Two seats in tandem in part-

IPE KW 1 b 2 Quero Quero II single-seat all-wood training sailplane

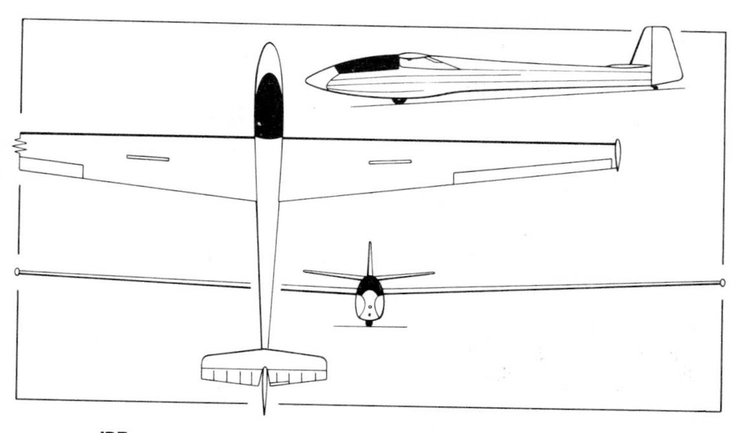

IPE two-seat training sailplane, currently under development *(Michael A. Badrocke)*

LET L-13 Blanik tandem two-seat all-metal sailplane

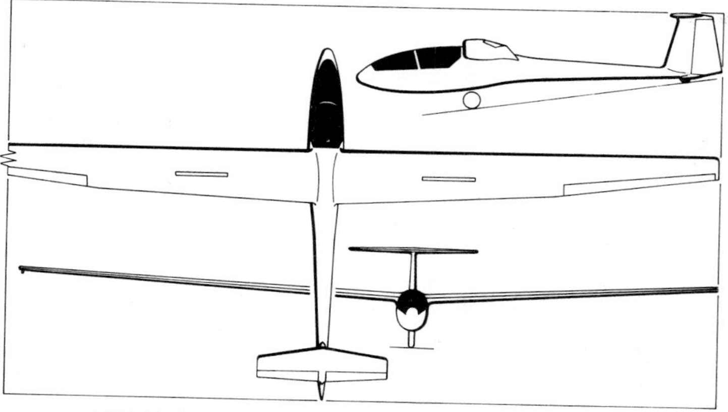

VSO 10 single-seat high-performance sailplane *(A. J. Smith)*

upholstered cabin, with heat-insulated walls. Sideways-opening transparent canopy, hinged on the starboard side, is jettisonable in flight.

EQUIPMENT: Standard equipment includes basic flight instruments on both front and rear instrument panels, towline and cockpit cover. Optional equipment includes electric gyros, second rate of climb indicator for rear instrument panel, rear compartment blinds for instrument flying instruction, navigation lights and 12V 10Ah battery, water ballast system to increase wing loading for solo flight, skis for operation on snow and a complete set of protective covers.

DIMENSIONS, EXTERNAL:
- Wing span: 16·20 m (53 ft 2 in)
- Wing chord at root: 1·65 m (5 ft 5 in)
- Wing chord at tip: 0·70 m (2 ft 3½ in)
- Wing aspect ratio: 13·7
- Length overall: 8·40 m (27 ft 6½ in)
- Height over tail: 2·09 m (6 ft 10 in)
- Tailplane span: 3·45 m (11 ft 3¾ in)

AREAS:
- Wings, gross: 19·15 m² (206·13 sq ft)
- Ailerons (total): 2·31 m² (24·87 sq ft)
- Flaps (total): 3·95 m² (42·52 sq ft)
- Spoilers (total): 0·65 m² (7·00 sq ft)
- Fin: 0·70 m² (7·58 sq ft)
- Rudder: 0·90 m² (9·73 sq ft)
- Tailplane: 1·56 m² (16·79 sq ft)
- Elevators, incl tab: 1·11 m² (11·95 sq ft)

WEIGHTS AND LOADINGS:
- Weight empty, standard equipment ±2%: 307 kg (677 lb)
- Max T-O weight: 500 kg (1,102 lb)
- Normal wing loading: 24·5 kg/m² (5·02 lb/sq ft)
- Max wing loading: 26·1 kg/m² (5·35 lb/sq ft)

PERFORMANCE (at max T-O weight):
- Best glide ratio, +5% at 48 knots (88 km/h; 55 mph) IAS: 28
- Min sinking speed at 44 knots (80 km/h; 50 mph) IAS: 0·85 m (2·79 ft)/sec
- Stalling speed: 31 knots (55 km/h; 35 mph) IAS
- Max speed (smooth air): 136 knots (253 km/h; 157 mph) IAS
- Max speed (rough air): 78 knots (145 km/h; 90 mph) IAS
- Max aero-tow speed: 76 knots (140 km/h; 87 mph) IAS
- Max winch-launching speed: 65 knots (120 km/h; 75 mph) IAS
- *g* limits: +5; −2·5

VSO

VÝVOJOVÁ SKUPINA ORLICAN

ADDRESS:
c/o Orlican Národní Podnik, 565 37 Chocen
Telephone: Chocen 70 and 80
Telex: 0 196 210

CHIEF DESIGNER:
Dipl Ing Jan Janovec

This design group was formed by several members of the former VSB (see 1973-74 *Jane's*) and some of the design staff of the Orlican National Works. Its first design is the single-seat VSO 10.

VSO 10

Design of the VSO 10 began in March 1972. Construction of three prototypes (one for structural test and two for flight test) began in 1975, and was due to be completed by the end of 1976.

TYPE: Single-seat high-performance sailplane.

WINGS: Cantilever shoulder-wing monoplane. Wortmann wing sections: FX-61-163 at root, FX-60-126 at tip. Dihedral 3°. Incidence 3° 30′. Sweepforward 1° 16′ at quarter-chord. All-wood single-spar structure, with sandwich skin. All-wood slotted ailerons. No flaps. All-metal double-shaft DFS-type airbrakes on upper surfaces at 46% chord and 38% of each half-span.

FUSELAGE: Glassfibre monocoque front and centre portions, the latter reinforced by a steel tube frame. Rear portion is an aluminium alloy sheet monocoque.

TAIL UNIT: Cantilever T-tail, of metal construction with fabric-covered elevators and rudder. Fixed-incidence tailplane. No tabs; elevator control includes torsion trim bar.

LANDING GEAR: Mechanically-retractable rubber-sprung monowheel and semi-recessed unsprung tailwheel. Monowheel tyre size 350 × 135-127 mm, pressure approx 2·45 bars (35·5 lb/sq in). Tailwheel diameter 160 mm. Moravan Otrokovice mechanically-operated drum brake.

ACCOMMODATION: Single moulded glassfibre seat under detachable transparent moulded canopy. Radio fitted.

DIMENSIONS, EXTERNAL:
- Wing span: 15·00 m (49 ft 2½ in)
- Wing chord at root: 1·75 m (5 ft 9 in)
- Wing chord at tip: 0·43 m (1 ft 5 in)
- Wing mean aerodynamic chord: 0·824 m (2 ft 8½ in)
- Wing aspect ratio: 18·75
- Length overall: 7·00 m (22 ft 11¾ in)
- Height over tail: 1·34 m (4 ft 4¾ in)
- Tailplane span: 2·48 m (8 ft 1¾ in)

AREAS:
- Wings, gross: 12·00 m² (129·2 sq ft)
- Ailerons (total): 0·875 m² (9·42 sq ft)
- Fin: 0·506 m² (5·45 sq ft)
- Rudder: 0·506 m² (5·45 sq ft)
- Tailplane: 0·923 m² (9·94 sq ft)
- Elevators: 0·384 m² (4·13 sq ft)

WEIGHTS AND LOADING:
- Weight empty, equipped: 234·4 kg (516·75 lb)
- Max T-O weight: 380 kg (837 lb)
- Max wing loading: 31·67 kg/m² (6·49 lb/sq ft)

PERFORMANCE (estimated, at 350 kg; 771 lb AUW):
- Best glide ratio at 51 knots (94 km/h; 58·5 mph): 36·2
- Min sinking speed at 39 knots (72 km/h; 45 mph): 0·63 m (2·07 ft)/sec
- Stalling speed: 37 knots (68 km/h; 42·5 mph)
- Max speed (smooth air): 140 knots (260 km/h; 161 mph)
- Max speed (rough air): 88 knots (163 km/h; 101 mph)
- Max aero-tow speed: 86 knots (160 km/h; 99 mph)
- Max winch-launching speed: 64·5 knots (120 km/h; 74·5 mph)
- *g* limits: +5·3; −2·65

DENMARK

PROJEKT 8

PROJEKT 8 I/S

ADDRESS:
Fynsvej 56, DK-4000 Roskilde

CHIEF DESIGNER:
Helge Petersen

DOLPHIN

The Projekt 8 I/S company, formed by Helge Petersen and 10 other glider pilots, is building a two-seat powered sailplane known as the Dolphin. Construction, which is taking place at three separate sites near Copenhagen, is well advanced, and the first flight was expected in mid-1976.

TYPE: Tandem two-seat powered sailplane.

WINGS: Cantilever mid-wing monoplane, built in three sections. Wing section Wortmann FX-67-K-170 on centre-section, outer panels varying through Wortmann FX-67-K-150 to NACA 64-212 at tip. Dihedral 4° on outer panels. Incidence 3°. Centre-section is of aluminium, outer panels of wood and glassfibre, including the top-hinged trailing-edge flaps and ailerons. Aluminium airbrakes on top surface of centre-section.

FUSELAGE: Forward portion is of welded steel tube covered by a light glassfibre shell. Rear portion of wood, reinforced by glassfibre.

TAIL UNIT: Cantilever wooden T-tail. Fixed-incidence tailplane. Central trim tab in elevator, inset tab at base of rudder.

LANDING GEAR: Semi-retractable monowheel, with mechanical retraction, nosewheel, and tailwheel. Rubber-block shock-absorption on main wheel, which is fitted with hand and parking brakes. Tost main (5 in × 350 mm) and nosewheel (50 × 200 mm). Steerable solid-tyre tailwheel, diameter 110 mm. Retractable wingtip balancer wheels.

POWER PLANT: One 40·25 kW (54 hp) VW 1600 engine, driving a two-blade propeller via Power Grip toothed-belt transmission with 4,000/2,000 rpm gearing. Propeller mounted on pylon which retracts rearward into top of fuselage when not in use.

ACCOMMODATION: Two seats in tandem under one-piece sideways-hinged Perspex canopy.

DIMENSIONS, EXTERNAL:

Wing span	18·72 m (61 ft 5 in)
Wing chord at root	1·35 m (4 ft 5¼ in)
Wing chord at tip	0·46 m (1 ft 6 in)
Wing mean aerodynamic chord	1·12 m (3 ft 8 in)
Wing aspect ratio	16·8
Length overall	8·50 m (27 ft 10¾ in)
Height over tail	1·30 m (4 ft 3¼ in)
Tailplane span	3·50 m (11 ft 5¾ in)
Propeller diameter	1·78 m (5 ft 10 in)

AREAS:

Wings, gross	20·80 m² (223·9 sq ft)
Ailerons (total)	1·80 m² (19·38 sq ft)
Trailing-edge flaps (total)	1·60 m² (17·22 sq ft)
Airbrakes (total)	0·40 m² (4·31 sq ft)
Fin	0·80 m² (8·61 sq ft)
Rudder, incl tab	0·70 m² (7·53 sq ft)
Tailplane	1·60 m² (17·22 sq ft)
Elevator, incl tab	0·80 m² (8·61 sq ft)

WEIGHTS AND LOADING:

Weight empty, equipped	480 kg (1,058 lb)
Max T-O weight	750 kg (1,653 lb)
Max wing loading	36·0 kg/m² (7·37 lb/sq ft)

PERFORMANCE (estimated, at max T-O weight, power off except where stated otherwise):

Best glide ratio at 54 knots (100 km/h; 62 mph)	32
Min sinking speed at 43·5 knots (80 km/h; 50 mph)	0·70 m (2·30 ft)/sec
Stalling speed	38 knots (70 km/h; 43·5 mph)
Max speed (rough and smooth air)	141·5 knots (263 km/h; 163 mph)
Max aero-tow speed	67·5 knots (125 km/h; 77·5 mph)
Max winch-launching speed	59 knots (110 km/h; 68 mph)
T-O run, powered	160 m (525 ft)
T-O to 15 m (50 ft), powered	420 m (1,380 ft)
g limits	+6; −4

FINLAND

PIK
TEKNILLINEN KORKEAKOULU (Helsinki University of Technology)

ADDRESS:
Konelaboratorio/Kevyttrakennetekniikka, Puumiehenkuja 5A, SF-02150 Espoo 15

EXECUTIVES:
Jukka Ahola (Chairman and Sales Manager)
Jukka Raunio (Secretary and Public Relations)
Pekka Tammi

The long series of sailplanes designed and built by PIK in recent years has included the PIK-3a, b and c (20 built); PIK-5a, b and c (35 built); PIK-7; PIK-12 (four built); PIK-16a, b and c (56 built); and PIK-17a and b (two built). Two additional PIK-16c Vasamas were built by Finnish gliding clubs.

Latest PIK sailplanes are those of the 15 metre Standard Class PIK-20 series.

PIK-20

Design of the PIK-20, by Tammi, Korhonen and Hiedanpää, began on 1 May 1971 in the aircraft research laboratory at Helsinki University of Technology. The first prototype flew for the first time on 10 October 1973, and in January 1974 was flown by Raimo Nurminen in the World Championships at Waikerie, Australia. A second prototype was delivered to the USA.

Finnish certification of the PIK-20 was granted on 20 June 1974. Further tests, to measure the aircraft's speed polar more precisely, were undertaken by PIK in 1975, although these were not required by the licensing authorities. FAA certification was granted on 20 June 1975, and on 15 January 1976 was extended to cover also the current-production PIK-20B.

By February 1976 a total of 150 PIK-20s had been ordered, of which 100 had been delivered. Production is undertaken by Eiriavion Oy (formerly known as Molino Oy). The original PIK-20, described in the 1975-76 *Jane's*, has now been replaced by the PIK-20B with higher gross weight, increased water ballast, and interconnected flaps and ailerons for improved performance. Versions of the PIK-20 currently in production or under development are now as follows:

PIK-20B. Current production version, to which the detailed description applies. Took first, second, third and fifth places at the 1976 World Championships at Räyskälä, Finland.

PIK-20C. As PIK-20B, but with carbon fibre spars as standard, glassfibre-reinforced plastics spars available optionally. Prototype due to fly during 1976.

PIK-20D. Fitted with airbrakes of Schempp-Hirth type. Reduced flap deflection (since flaps no longer function as airbrakes). Interconnection between flaps and ailerons can be disconnected to permit flight in both restricted and unrestricted Standard Class conditions. Prototype flew for first time on 19 April 1976.

PIK-20E. Powered version, in final design stage in early 1976 and due to fly before end of year. Carbon fibre main spar. Prototype, designated JT-6, has engine installation designed by Jukka Tervamäki, who has also modified the basic PIK-20 design to incorporate a steerable tailwheel and retractable outrigger wheels. This modification, to provide a taxying capability, will be optional on production PIK-20Es. Other features of the PIK-20E include approx 2° of wing sweepback, repositioning of the main wheel approx 0·25 m (10 in) further aft, and battery installation in the nose.

The following description applies primarily to the PIK-20B, except where indicated otherwise:

TYPE: Single-seat Standard Class sailplane.

WINGS: Cantilever shoulder-wing monoplane. Wortmann wing sections: FX-67-K-170 at root, FX-67-K-150 at tip. Dihedral 3°. Incidence 1°. No sweepback. Glassfibre/epoxy/PVC foam sandwich structure. PIK-20B also available with spars of carbon fibre reinforced epoxy. Plain ailerons, and interconnected trailing-edge flaps/airbrakes, of similar construction. Provision for 140 litres (30·8 Imp gallons) of water ballast.

FUSELAGE: Glassfibre/epoxy monocoque structure.

TAIL UNIT: Cantilever T-tail, of similar construction to wings. Fixed-incidence tailplane, with one-piece elevator.

LANDING GEAR: Manually-retractable Tost monowheel with size 5·00-5 Dunlop/Continental tyre. Tost drum brake. Continental 200 × 50 mm tailwheel.

POWER PLANT (PIK-20E): One 26 kW (35 hp) 440 cc two-stroke piston engine, retracting into fuselage aft of cockpit when not in use. Electric starter.

ACCOMMODATION: Single semi-reclining seat under one-piece sideways-opening transparent moulded canopy. Adjustable rudder pedals. Standard cockpit instrumentation plus special vario-integrator unit, and final approach computer and tracking unit.

DIMENSIONS, EXTERNAL:

Wing span	15·00 m (49 ft 2½ in)
Wing chord at root	0·90 m (2 ft 11½ in)
Wing chord (mean)	0·65 m (2 ft 1½ in)
Wing chord at tip	0·36 m (1 ft 2¼ in)
Wing aspect ratio	22·5
Length overall	6·45 m (21 ft 2 in)
Height over tail	1·36 m (4 ft 5½ in)
Tailplane span	2·00 m (6 ft 6¾ in)

AREAS:

Wings, gross	10·00 m² (107·6 sq ft)
Ailerons (total)	0·566 m² (6·09 sq ft)
Flaps/airbrakes (total)	1·12 m² (12·06 sq ft)
Fin	0·71 m² (7·64 sq ft)
Rudder	0·31 m² (3·34 sq ft)
Tailplane	0·80 m² (8·61 sq ft)
Elevator	0·20 m² (2·15 sq ft)

WEIGHTS AND LOADING (PIK-20B):

Weight empty, equipped:	
with standard spar	250 kg (551 lb)
with carbon fibre spar	230 kg (507 lb)
Max water ballast	140 kg (309 lb)
Max T-O weight (with water ballast)	450 kg (992 lb)
Max wing loading	45·0 kg/m² (9·22 lb/sq ft)

WEIGHTS AND LOADING (PIK-20C, D and E):
As PIK-20B, except:

Weight empty, equipped:	
C	210 kg (463 lb)
D	220 kg (485 lb)
E	290 kg (639 lb)

PERFORMANCE (PIK-20B, measured at 450 kg; 992 lb AUW except where indicated):

Best glide ratio at 58 knots (108 km/h; 67 mph)	42
Min sinking speed at 46·5 knots (86 km/h; 53·5 mph)	0·63 m (2·07 ft)/sec
Stalling speed (90° flap):	
at 320 kg (705 lb) AUW	32·5 knots (60 km/h; 37·5 mph)
at 400 kg (881 lb) AUW	36·5 knots (67 km/h; 42 mph)
Max speed (smooth air)	141·5 knots (262 km/h; 163 mph)
Max speed (rough air)	129·5 knots (240 km/h; 149 mph)
Max aero-tow speed	100 knots (185 km/h; 115 mph)
g limits	+7·1; −5·1

PERFORMANCE (PIK-20C, D and E, estimated):
As PIK-20B, except:

Min sinking speed at 46 knots (85 km/h; 53 mph):	
C	0·65 m (2·13 ft)/sec
D, E	0·63 m (2·07 ft)/sec
Stalling speed:	
C at 300 kg (661 lb) AUW	35·5 knots (65 km/h; 40·5 mph)
D at 300 kg (661 lb) AUW	32·5 knots (60 km/h; 37·5 mph)
E at 370 kg (815 lb) AUW	35·5 knots (65 km/h; 40·5 mph)
Max winch-launching speed:	
C, D	67·5 knots (125 km/h; 77·5 mph)

PERFORMANCE (PIK-20E, estimated, powered):

Cruising speed	81 knots (150 km/h; 93 mph)
Max rate of climb at S/L	150 m (490 ft)/min

FRANCE

CARMAM
SOCIÉTÉ ANONYME CARMAM

ADDRESS:
BP 201, Aérodrome, 03001 Moulins
Telephone: (70) 44.36.18

DIRECTOR:
Robert Jacquet

This company extended its work on sailplane repair and maintenance to include the construction of new sailplanes. It began by building under licence the M-100 S, designed in Italy by Alberto and Piero Morelli, and followed this with production of the two-seat M-200 by the same designers. Production of both types (described in the 1973-74 *Jane's*) has ended, as has the manufacture of sailplane components under subcontract for Glasflügel of Germany.

CARMAM is currently responsible for manufacture of the J.P.15.36 single-seat sailplane designed by M Robert Jacquet and M Jean Pottier.

The J.15-38 single-seat sailplane project, described under the Jacquet heading in the 1975-76 *Jane's*, has been placed in abeyance.

CARMAM J.P.15.36 AIGLON (EAGLET)

This Standard Class sailplane was designed as a private venture by M Robert Jacquet and M Jean Pottier.

Design of the J.P.15.36 began in September 1971, and prototype construction started at the end of 1972. The prototype (F-WCAP) made its first flight on 14 June 1974.

In the Spring of 1975, static testing was under way, and test flying was carried out in 1975. Series production, in the CARMAM works, was under way in early 1976. A version able to carry 80 litres (17·5 Imp gallons) of water ballast is to be available optionally.

TYPE: Single-seat Standard Class sailplane.

WINGS: Cantilever mid-wing monoplane. Wortmann wing sections: FX-67-K-170 at root, FX-60-126 at tip. Dihedral 3°. Incidence 7° at root, 4° at tip. Sweepback 0° at 30% chord. Structure consists of a single load-

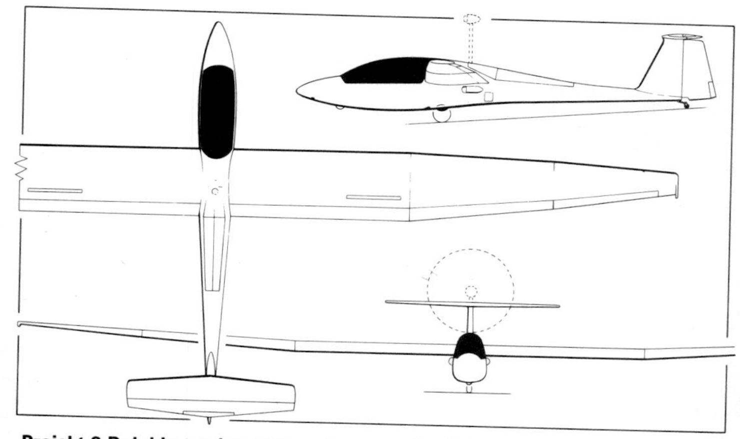

Projekt 8 Dolphin tandem two-seat powered sailplane *(Michael A. Badrocke)*

PIK-20B single-seat glassfibre Standard Class sailplane

CARMAM J.P.15.36 Aiglon Standard Class sailplane

CERVA CE 75 Silène side-by-side two-seat training sailplane

bearing glassfibre spar, 100 mm (3·94 in) deep, with four-point attachment to fuselage. Glassfibre/Rohacell/epoxy sandwich skin, 10 mm (0·4 in) thick. Steel-tipped wingtip 'salmons'. Plastics plain ailerons, deflecting 25° up and 15° down, can be operated differentially or in unison. Schempp-Hirth airbrakes on upper and lower surfaces.

FUSELAGE: Semi-monocoque glassfibre structure, moulded in two halves and joined at centreline. Single bulkhead, combining functions of cockpit backrest and shock-absorbing structure for main landing wheel. Special strengthening of the union between this bulkhead and the wing main spar.

TAIL UNIT: Cantilever type, with slight sweepback on vertical surfaces. Single-spar all-moving plastics tailplane and plastics rudder. Fin built integrally with fuselage.

LANDING GEAR: Non-retractable monowheel, size 330 × 130, and tail bumper. Cable-operated brake.

ACCOMMODATION: Single semi-reclining (25°) seat under one-piece removable transparent canopy.

DIMENSIONS, EXTERNAL:
- Wing span: 15·00 m (49 ft 2½ in)
- Wing chord at root: 1·00 m (3 ft 3¼ in)
- Wing chord at tip: 0·40 m (1 ft 3¾ in)
- Wing mean aerodynamic chord: 0·735 m (2 ft 5 in)
- Wing aspect ratio: 20·4
- Length overall: 6·18 m (20 ft 3¼ in)
- Height over tail: 1·40 m (4 ft 7 in)
- Tailplane span: 2·50 m (8 ft 2½ in)

AREAS:
- Wings, gross: 11·00 m² (118·4 sq ft)
- Ailerons (total): 0·956 m² (10·29 sq ft)
- Airbrakes (total): 1·10 m² (11·84 sq ft)
- Fin: 0·55 m² (5·92 sq ft)
- Rudder: 0·56 m² (6·03 sq ft)
- Tailplane: 1·05 m² (11·30 sq ft)

WEIGHTS AND LOADING:
- Weight empty, equipped: 180 kg (396 lb)
- Max T-O weight: 300 kg (661 lb)
- Max wing loading: 27 kg/m² (5·53 lb/sq ft)

PERFORMANCE:
- Best glide ratio at 43·5 knots (80 km/h; 50 mph): 36
- Min sinking speed at 39 knots (72 km/h; 45 mph): 0·62 m (2·03 ft)/sec
- Stalling speed: 33·5 knots (62 km/h; 39 mph)
- Max speed (smooth air): 135 knots (250 km/h; 155 mph)
- Max speed (rough air): 108 knots (200 km/h; 124 mph)
- Max aero-tow and winch-launching speed: 81 knots (150 km/h; 93 mph)
- *g* limit: +5·3

CERVA
CONSORTIUM EUROPÉEN DE RÉALISATION ET DE VENTES D'AVIONS (GROUPEMENT D'INTÉRÊTS ÉCONOMIQUES)

ADDRESS:
13 rue Saint-Honoré, BP 187, 78002 Versailles
Telephone: 950.63.95

As detailed in the Aircraft section, this company was formed, and is owned in equal proportions, by Siren SA and Wassmer-Aviation SA (which see).

It is currently responsible for the Siren-designed CE 75 Silène (formerly known as the Sagittaire).

CERVA CE 75 SILÈNE

The Silène is a side-by-side two-seat training glider, the design of which (by Siren SA) began on 1 January 1972. Construction by CERVA of a prototype started on 1 February 1973, and this aircraft (F-CCFF) made its first flight at Argenton on 2 July 1974.

Three pre-production aircraft were to be built, and deliveries of production aircraft were scheduled to begin in September 1975. An eventual output of four Silènes per month is planned. The wings are built by Wassmer-Aviation SA.

Siren SA is responsible for sales of the CE 75.

TYPE: Two-seat training sailplane.

WINGS: Cantilever mid-wing monoplane. Bertin E55 166 wing section, with thickness/chord ratio of 16·6%. Dihedral 2°. Incidence 3°. Sweepforward 2° at quarter-chord. Composite glassfibre/PMC foam sandwich construction. Two-section ailerons, of similar construction. No flaps. Schempp-Hirth airbrakes above and below each wing.

FUSELAGE: Semi-monocoque glassfibre/PMC foam sandwich structure.

TAIL UNIT: Cantilever structure, of similar construction to wings. Fixed-incidence tailplane. Trim tab in each elevator.

LANDING GEAR: Manually-retractable or non-retractable monowheel, size 330 × 130 mm, tyre pressure 1·96 bars (28·5 lb/sq in). Rubber-ring shock-absorption. Siren hydraulic brake. Tail bumper.

ACCOMMODATION: Seats for two persons side by side under two-piece transparent moulded canopy. Right hand (instructor's) seat is staggered 0·30 m (11¾ in) to rear. Rudder pedals in both positions are adjustable in flight. Oxygen equipment and/or radio, and second variometer, optional.

DIMENSIONS, EXTERNAL:
- Wing span: 18·00 m (59 ft 0½ in)
- Wing chord at root: 1·27 m (4 ft 2 in)
- Wing chord at tip: 0·50 m (1 ft 7¾ in)
- Wing aspect ratio: 18
- Length overall: 7·95 m (26 ft 1 in)
- Height over tail: 1·50 m (4 ft 11 in)
- Tailplane span: 3·20 m (10 ft 6 in)

DIMENSION, INTERNAL:
- Cockpit: Max width: 0·97 m (3 ft 2¼ in)

AREAS:
- Wings, gross: 18·00 m² (193·8 sq ft)
- Fin: 0·80 m² (8·61 sq ft)
- Rudder: 0·90 m² (9·69 sq ft)
- Tailplane: 0·90 m² (9·69 sq ft)
- Elevators, incl tabs: 0·80 m² (8·61 sq ft)

WEIGHTS AND LOADINGS:
- Weight empty, equipped: 320 kg (705 lb)
- Max T-O weight: 540 kg (1,190 lb)
- Wing loading:
  - single-seat: 25·0 kg/m² (5·12 lb/sq ft)
  - two-seat, with radio: 29·0 kg/m² (5·94 lb/sq ft)
  - two-seat, with radio and oxygen (max): 31·0 kg/m² (6·35 lb/sq ft)

PERFORMANCE (estimated, at relevant T-O weight):
- Best glide ratio, non-retractable monowheel:
  - single-seat at 45 knots (84 km/h; 52 mph): 36
  - two-seat, with radio, at 49 knots (91·5 km/h; 57 mph): 36
- Best glide ratio, retractable monowheel:
  - both versions: 38
- Min sinking speed:
  - single-seat at 45 knots (84 km/h; 52 mph): 0·59 m (1·94 ft)/sec
  - two-seat, with radio, at 49 knots (91·5 km/h (57 mph): 0·64 m (2·10 ft)/sec
  - two-seat, with radio and oxygen, at 49 knots (91·5 km/h; 57 mph): 0·67 m (2·20 ft)/sec
- Stalling speed:
  - single-seat: 34·5 knots (63 km/h; 39·5 mph)
  - two-seat, with radio: 37 knots (68 km/h; 42·5 mph)
  - two-seat, with radio and oxygen: 38·5 knots (70·5 km/h; 44 mph)
- Max speed (smooth air): 132 knots (245 km/h; 152 mph)
- Max aero-tow speed: 88·5 knots (164 km/h; 102 mph)
- *g* limits: +5·3 (normal); +8 (ultimate)

FAUVEL
'SURVOL'-CHARLES FAUVEL

HEAD OFFICE:
72 Boulevard Carnot, 06400-Cannes AM
Telephone: 39.83.32 and 39.55.21

Charles Fauvel has been developing and producing for many years a series of tail-less sailplanes. The original AV.36 Monobloc single-seater first flew in 1951 and more than 100 were sold to customers in 14 countries before this design was superseded by the improved AV.361.

Several powered versions of M Fauvel's sailplane designs have been produced. The prototype AV.45 has flight tested a lightweight turbojet power plant.

In early February 1971 M Fauvel decided to end commercial production of his sailplanes, but plans of the AV.361, AV.45 and AV.222 are available for their construction by gliding clubs or homebuilders. In addition, glassfibre component moulds for the AV.45, and cockpit transparency moulds for the AV.222, can be supplied if required. Formers or moulds for certain components of the AV.44 and AV.50 powered aircraft (see Homebuilt Aircraft section) are also available.

Details of recent Fauvel light aircraft designs can be found in the Homebuilt Aircraft section.

FAUVEL AV.361

It is known that a total of well over 100 AV.36 and AV.361 sailplanes are flying in 17 countries. Plans are available in French and English, and construction by amateurs continues, especially in the USA and Spain.

Details of this single-seat general-purpose sailplane may be found in the 1970-71 *Jane's*. The standard F2 section wing can, at builder's option, be replaced in the AV.361 by one with a Wortmann FX-66-H-159 laminar-flow section, which increases the best glide ratio to 30 at a speed of 46 knots (85 km/h; 53 mph).

The following particulars apply to an AV.361 completed by the Escuela de Aeromodelismo at Alicante, Spain:

DIMENSIONS, EXTERNAL:
- Wing span: 12·78 m (41 ft 11¼ in)
- Length overall: 3·24 m (10 ft 7½ in)
- Wing area, gross: 14·60 m² (157·15 sq ft)

WEIGHTS:
- Weight empty: 125 kg (275·5 lb)
- Max T-O weight: 258 kg (568 lb)

PERFORMANCE:
- Best glide ratio at 44 knots (82 km/h; 51 mph): 26
- Min sinking speed: 0·74 m (2·43 ft)/sec
- Max speed (smooth air): 118·5 knots (220 km/h; 136·5 mph)

FAUVEL AV.45

The AV.45 is a single-seat tail-less self-launching sailplane which first flew on 4 May 1960 with a 26 kW (35 hp) Nelson engine.

A second, slightly modified prototype, with 16·5 kW (22 hp) SOLO engine, was built by Société Aéronautique Normande (SAN) and is typical of several examples built by amateurs in France and other countries, including Japan, Martinique and the USA. The standard engine recommended for the AV.45 is the 30-41 kW (40-55 hp) modified Hirth O-280R, one of which was to undergo testing in the AV.45-02 in early 1976. Another French AV.45 was at that time being fitted with a 27 kW (36 hp) JLO-Rockwell engine. Fuel is contained in integral wooden tanks in the wing leading-edges.

As with the AV.361, the AV.45 may also be fitted with a Wortmann laminar-flow wing, with which the best glide ratio is increased to 30 at a speed of 47·5 knots (88 km/h; 55 mph).

FAUVEL AV.48

Brief details of the AV.48 as originally designed appeared in the 1970-71 *Jane's*, but construction of the prototype was suspended when the factory in which it was to have been built was destroyed by fire.

Advantage was taken of this delay to make further improvements in the design, of which brief details were given in the 1974-75 *Jane's*.

The AV.48 had not flown up to the Spring of 1976, but work on it continues.

FAUVEL AV.222

This side-by-side two-seat self-launching sailplane was developed from the AV.22 tail-less sailplane, of which details can be found in the 1960-61 *Jane's*. It flew for the first time on 8 April 1965. Brief details of the AV.221 prototype were given in the 1970-71 *Jane's*.

Plans are available to amateur constructors of a lighter and simplified version of the AV.221, designated AV.222, of which examples are being built in France, Germany, Italy and the USA.

The wings of the AV.222 are built in three sections, and a conventional twin-main-wheel landing gear has been fitted successfully to both the AV.221 prototype and the AV.222. This consisted of cantilever self-sprung laminated glassfibre legs, and Duruble 330 × 130 wheels with hydraulic brakes and streamlined wheel fairings, in place of the underfuselage monowheel and its fairing. For amateur construction, however, the monowheel landing gear is recommended as being lighter, cheaper and easier to install. Suitable power plants are considered to be the 30 kW (40 hp) Rectimo or 45 kW (60 hp) Limbach Volkswagen conversions.

DIMENSIONS, EXTERNAL:
- Wing span: 16·40 m (53 ft 9¾ in)
- Wing aspect ratio: 12
- Length overall: 5·22 m (17 ft 1½ in)

AREA:
- Wings, gross: 23·00 m² (247·6 sq ft)

WEIGHTS:
- Weight empty, equipped: 325 kg (716 lb)
- Max T-O weight: 550 kg (1,212 lb)

PERFORMANCE:
- Best glide ratio (propeller feathered) at 46 knots (85 km/h; 53 mph): 26
- Min sinking speed (2-seat) at 40 knots (74 km/h; 46 mph): 0·87 m (2·85 ft)/sec
- Min sinking speed (single-seat) at 38 knots (70 km/h; 43·5 mph): 0·78 m (2·56 ft)/sec
- Rate of climb, powered: 180 m (591 ft)/min
- T-O run, powered: 110 m (361 ft)
- T-O to 15 m (50 ft), powered: 230 m (755 ft)

FOURNIER
AVIONS FOURNIER

ADDRESS:
Aérodrome de Nitray, 37270-Montlouis
Telephone: (47) 50.68.30

OFFICERS:
See Aircraft section

Avions Fournier is building a two-seat powered sailplane designated RF-9. All known details follow:

FOURNIER RF-9

A prototype RF-9 is under construction, and was expected to make its first flight during 1976. It is intended for training, and has side-by-side seating for two persons. Power plant is a 44·7 kW (60 hp) Sportavia Limbach engine, and fuel capacity 30 litres (6·6 Imp gallons).

DIMENSIONS, EXTERNAL:
- Wing span: 17·00 m (55 ft 9½ in)
- Wing span, folded: 10·00 m (32 ft 9¾ in)
- Wing aspect ratio: 16

AREA:
- Wings, gross: 18·00 m² (193·75 sq ft)

WEIGHTS:
- Weight empty: 460 kg (1,014 lb)
- Max T-O weight: 700 kg (1,543 lb)

PERFORMANCE (estimated):
- Best glide ratio (propeller feathered):
  - 0° flap: 28
  - 12° flap: 31
- Min sinking speed: 0·78 m (2·56 ft)/sec
- Max level speed: 100 knots (185 km/h; 115 mph)
- Service ceiling: 6,000 m (19,675 ft)

GEP
GROUPE D'ÉTUDES GEORGES PAYRE

ADDRESS:
2 rue Abel, 75012 Paris
Telephone: 522 30 31

Dr Pierre Vaysse, head of the sailplane amateur construction department of the FFVV (Fédération Française de Vol à Voile), built and flew two single-seat sailplanes which he designated Trucavaysse TCV-01 and TCV-02. These flew for the first time on 7 August 1964 and 6 April 1969 respectively; brief details of both can be found in the 1974-75 *Jane's*.

Objective of this series of sailplanes was to evolve an aircraft that could be sold in kit form for amateur constructors or for club construction. Encouragement was given by both the FFVV and the RSA, and plans are available of the developed version, known as the TCV-03.

GEP TCV-03 TRUCAVAYSSE

Design of the TCV-03 takes into account work done by the GEP in modifying the original Breguet 905 design to make the aircraft suitable for amateur construction. Modifications include re-covered wings with improved control system and reinforced trailing-edges; a new, more slender fuselage outline; and deletion of the landing skid.

Design began in October 1968, and prototype construction started in February 1969. This aircraft (F-CRRH), built by the Aéro Club de Norois, flew for the first time on 14 July 1973 and by February 1974 eight others were under construction by amateur builders. Authorisation for about 15 had then been granted by the CNRA.

TYPE: Single-seat sailplane for amateur construction.

WINGS: Cantilever shoulder-wing monoplane. Wing section NACA 63-420 at root, NACA 63-513 at tip. Dihedral 3° from roots. Incidence 4°. No sweepback. Conventional single-spar structure, forming torsion box with plywood/Klégécel sandwich leading-edge. DFS metal airbrakes on upper and lower surfaces. Slotted wooden ailerons. No flaps.

FUSELAGE: Conventional plywood-covered wooden structure.

TAIL UNIT: Cantilever wooden structure with single fin and rudder and one-piece all-moving tailplane with anti-tabs.

LANDING GEAR: Non-retractable monowheel and tailskid. Rubber shock-absorbers. Tyre size 330 × 130 mm.

ACCOMMODATION: Single adjustable seat under one-piece transparent canopy.

DIMENSIONS, EXTERNAL:
- Wing span: 15·00 m (49 ft 2½ in)
- Wing chord at root: 1·10 m (3 ft 7¼ in)
- Wing chord at tip: 0·40 m (1 ft 3¾ in)
- Wing mean aerodynamic chord: 0·82 m (2 ft 8¼ in)
- Wing aspect ratio: 20
- Length overall: 6·70 m (21 ft 11¾ in)
- Height over tail: 1·80 m (5 ft 11 in)
- Fuselage: Max depth: 0·90 m (2 ft 11½ in)
- Tailplane span: 2·85 m (9 ft 4¼ in)

AREAS:
- Wings, gross: 11·25 m² (121·1 sq ft)
- Ailerons (total): 1·26 m² (13·56 sq ft)
- Fin: 0·60 m² (6·46 sq ft)
- Rudder: 0·50 m² (5·38 sq ft)
- Tailplane, incl tabs: 1·50 m² (16·15 sq ft)

WEIGHTS AND LOADING:
- Weight empty, equipped: 192 kg (423 lb)
- Max T-O weight: 302 kg (665 lb)
- Max wing loading: 26·9 kg/m² (5·5 lb/sq ft)

PERFORMANCE (at max T-O weight):
- Best glide ratio at 43·5 knots (80 km/h; 50 mph): 28
- Min sinking speed at 32·5 knots (60 km/h; 37·5 mph): 0·80 m (2·62 ft)/sec
- Stalling speed: 27·5 knots (50 km/h; 31·5 mph)
- Max speed (smooth air): 113 knots (210 km/h; 130 mph)
- Max speed (rough air): 81 knots (150 km/h; 93 mph)
- Max aero-tow and max winch-launching speed: 59 knots (110 km/h; 68 mph)
- *g* limit: +5

LORAVIA
LORRAINE AVIATION

ADDRESS:
Aérodrome de Thionville, 57110-Yutz, Moselle
Telephone: (87) 88.56.87

EXECUTIVES:
M. Schmitt
R. Kieger

This company took over from SLCA (1974-75 *Jane's*) continued production of the Scheibe SF-27 under licence in France, where it is known as the Topaze. German production of the SF-27 ended several years ago.

LA-11 TOPAZE

The Topaze is the Scheibe SF-27 Zugvogel V built under licence in France. It received a French certificate of airworthiness on 25 April 1972.

SLCA built nine examples of the SF-27 under the French designation SLCA-10 (now known as the LA-10). The LA-11 has the monowheel lowered 80 mm (3·15 in). It flew for the first time on 15 October 1973; six were completed by SLCA in 1973, and 12 in 1974. In early 1976, Loravia had completed 30 of 33 then on order.

The following description applies to the LA-11:

TYPE: Single-seat Standard Class sailplane.

WINGS: Cantilever shoulder-wing monoplane. Wing section Wortmann FX-61-184 at root, FX-60-126 at tip. Dihedral 3°. Wooden structure, with laminated beechwood box spar at about 43% chord. Plywood ribs. Leading-edge torsion box. Outboard half of wing plywood-covered; inboard half covered with plywood to 6 cm behind spar; remainder part-fabric and part-plywood covered. Wooden ailerons, plywood-covered. Schempp-Hirth glassfibre/metal airbrakes.

FUSELAGE: Welded steel tube structure. Nose section back to wing trailing-edge covered with moulded glassfibre shell. Rear section fabric-covered over wooden stringers. Moulded glassfibre fairing over wing/fuselage junction.

TAIL UNIT: Cantilever wooden structure, with all-moving horizontal surfaces. Tailplane covered with plywood and fabric. Fin plywood-covered; rudder fabric-covered. Anti-balance tab in port half of tailplane.

LANDING GEAR: Non-retractable and unsprung monowheel ahead of CG, tyre size 4·00-4. Wheel brake. No skid. Tailwheel diameter 200 mm (7·9 in).

ACCOMMODATION: Single inclined seat under moulded Plexiglas canopy. Rudder pedals adjustable. Baggage compartment behind seat.

DIMENSIONS, EXTERNAL:
- Wing span: 15·00 m (49 ft 2½ in)

Fauvel AV.222 side-by-side two-seat powered sailplane

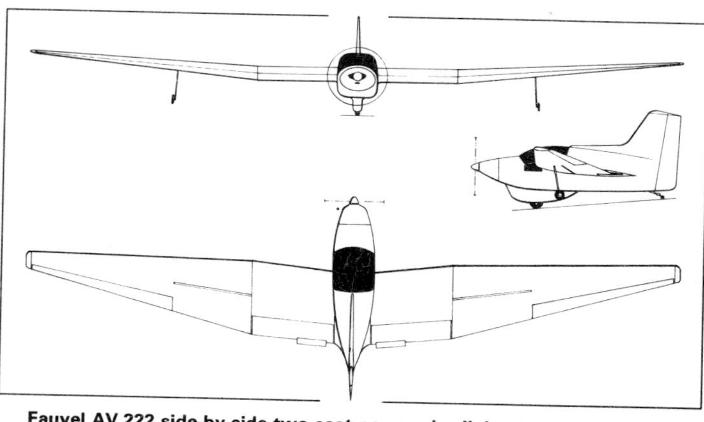

Fauvel AV.222 side-by-side two-seat powered sailplane *(Roy J. Grainge)*

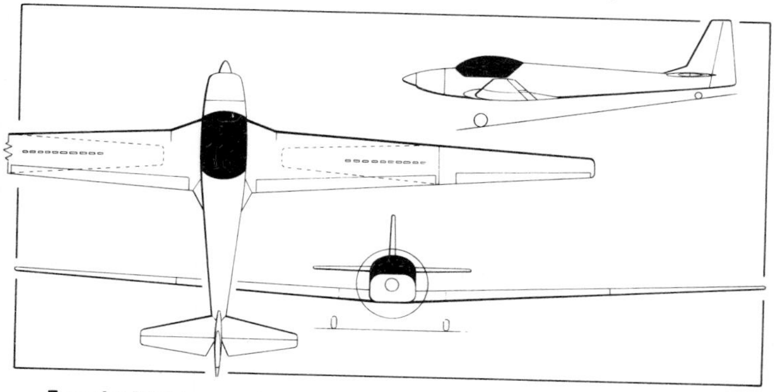

Fournier RF-9 two-seat powered training sailplane *(Michael A. Badrocke)*

GEP TCV-03 Trucavaysse single-seat sailplane for amateur constructors

LCA-11 production version of the Topaze single-seat Standard Class sailplane

Menin BM 1 prototype single-seat 15 metre sailplane

Wing chord at root	1·09 m (3 ft 7 in)
Wing chord at tip	0·44 m (1 ft 5½ in)
Wing aspect ratio	18·6
Length overall	7·09 m (23 ft 3¼ in)
Max width of fuselage	0·57 m (1 ft 10½ in)
Tailplane span	2·52 m (8 ft 3¼ in)
AREAS:	
Wings, gross	12·07 m² (129·9 sq ft)
Ailerons (total)	1·0 m² (10·76 sq ft)
Airbrakes (total)	0·38 m² (4·09 sq ft)
Fin	0·54 m² (5·81 sq ft)
Rudder	0·62 m² (6·67 sq ft)
Tailplane	1·55 m² (16·68 sq ft)
WEIGHTS AND LOADING:	
Weight empty, equipped	215 kg (474 lb)
Max T-O weight	345 kg (760 lb)
Max wing loading	28·7 kg/m² (5·88 lb/sq ft)
PERFORMANCE (at max T-O weight):	
Best glide ratio at 48 knots (88 km/h; 55 mph)	33·6
Min sinking speed at 40 knots (74 km/h; 46 mph)	0·64 m (2·10 ft)/sec
Stalling speed	31·5 knots (58 km/h; 36 mph)
Max speed (smooth air) and max aero-tow speed	108 knots (200 km/h; 124 mph)
Max speed (rough air)	91 knots (170 km/h; 105 mph)
Max winch-launching speed	59·5 knots (110 km/h; 68·5 mph)
g limit	+5·3

MENIN

JEAN-PAUL MENIN

ADDRESS:
16 rue Fouquet, 93700-Drancy
Telephone: 284.13.53

BM 1

This single-seat sailplane was designed and built by two amateur constructor members of the Réseau du Sport de l'Air. Original plans for its production by the Société des Ateliers du Pingouin had to be abandoned. However, in the Spring of 1975 M Menin was negotiating with other organisations which might undertake series production, and hoped also to make the BM 1 available for home-builders in countries where amateur construction is encouraged. No further news of progress had been received at the time of closing for press.

The prototype BM 1 (F-CRRI), which flew for the first time on 9 April 1974, had accumulated more than 100 hours of flying by March 1975, including much hill flying.

TYPE: Single-seat sailplane.

WINGS: Cantilever shoulder-wing monoplane. Wortmann wing section, of 19% thickness/chord ratio. Single-spar box structure, with Klégécel/epoxy sandwich skin, with two-point (steel bolt) attachment to fuselage. Sealed-gap ailerons, actuated by pushrods. DFS-type air-brakes. Space for 25 litres (5·5 Imp gallons) of water ballast in each leading-edge. Balancer wheel under each wingtip.

FUSELAGE: Forward section is a steel tube structure, with metal-reinforced laminated plastics covering. Aft section is of duralumin monocoque construction.

TAIL UNIT: Cantilever T-tail. Elevator actuated by push-rods, rudder by cables. Automatic elevator linkage in prototype and proposed production version; a simplified system has been designed for amateur construction.

LANDING GEAR: Retractable monowheel. Bumper fairing beneath tail unit.

ACCOMMODATION: Single seat under fully-transparent two-piece moulded canopy, rear portion of which is removable.

EQUIPMENT: Towing hook on prototype is attached to the monowheel, but will be relocated further forward on production version.

DIMENSIONS, EXTERNAL:	
Wing span	15·00 m (49 ft 2½ in)
Wing aspect ratio	22·5
Length overall	6·70 m (21 ft 11¾ in)
Height over tail	1·50 m (4 ft 11 in)
AREA:	
Wings, gross	10·00 m² (107·6 sq ft)
WEIGHTS AND LOADINGS:	
Weight empty	200 kg (441 lb)
Max T-O weight (incl ballast)	360 kg (793 lb)
Wing loading (prototype)	28-31 kg/m² (5·73-6·35 lb/sq ft)
Wing loading (proposed production aircraft)	27-35 kg/m² (5·53-7·17 lb/sq ft)
PERFORMANCE (at wing loading of 28 kg/m²; 5·73 lb/sq ft):	
Best glide ratio at 47·5 knots (88 km/h; 55 mph)	37
Min sinking speed at 38 knots (70 km/h; 43·5 mph)	0·60 m (1·97 ft)/sec
Sinking speed at 81 knots (150 km/h; 93 mph)	2·00 m (6·56 ft)/sec

POTTIER

JEAN POTTIER

ADDRESS:
4 rue Emilio Castelar, 75012-Paris
Telephone: 343 63 16

POTTIER P.A. 15-34 KIT-CLUB

This is essentially the same sailplane as the CARMAM J.P.15.36 Aiglon (which see), with some constructional simplification to make it suitable for amateur builders.

Prototype construction, by a group of students under M Jean Magne, a gliding instructor at the Centre National de Vol à Voile at St Auban sur Durance, began in November 1975; first flight was scheduled for Summer 1976. Subject to satisfactory progress, plans are expected to become available to homebuilders during the latter part of 1977. The prototype, after completion of tests, is to go into service with the SFA (Service de la Formation Aéronautique). See overleaf for general arrangement drawing.

The description of the Aiglon applies also to the P.A.15-34, except in the following respects:

TYPE: Single-seat sailplane for amateur construction.

WINGS: As Aiglon.

FUSELAGE: Of same contours and dimensions as Aiglon, but of mainly wooden construction (spruce longerons and plywood skin) except for glassfibre nosecone.

TAIL UNIT: As Aiglon, but of plywood-covered spruce construction. Rudder and rear part of tailplane fabric-covered.

LANDING GEAR: Non-retractable unsprung monowheel, size 300 × 100. Cable-operated drum brake for production aircraft.

DIMENSIONS, EXTERNAL: As Aiglon except:
- Wing aspect ratio 20·6
- Length overall 6·25 m (20 ft 6 in)
- Tailplane span 3·00 m (9 ft 10 in)

AREAS: As Aiglon except:
- Tailplane 1·50 m² (16·15 sq ft)

WEIGHTS AND LOADINGS:
- Weight empty, equipped 190 kg (419 lb)
- Max T-O weight:
  - without water ballast 300 kg (661 lb)
  - with water ballast 380 kg (838 lb)
- Max wing loading:
  - without water ballast 27·0 kg/m² (5·53 lb/sq ft)
  - with water ballast 34·5 kg/m² (7·07 lb/sq ft)

PERFORMANCE (estimated):
- Best glide ratio at 41·5 knots (77 km/h; 48 mph) 34
- Min sinking speed at 38 knots (70 km/h; 43·5 mph) 0·63 m (2·07 ft)/sec
- Stalling speed 32·5 knots (60 km/h; 37·5 mph)
- Max speed (rough or smooth air) and max aero-tow speed 135 knots (250 km/h; 155 mph)
- g limits +8·2; −4

SIREN
SIREN SA

WORKS AND OFFICES:
Route des Chambons, BP 42, 36200 Argenton-sur-Creuse
Telephone: (54) 04.14.47
Telex: 76534 Chamco-Châteauroux Siren 200-1

DIRECTOR:
Philippe Moniot

Well known as a manufacturer of aircraft components and equipment, Siren has in recent years built the C.30S Edelweiss single-seat Standard Class sailplane and an Open Class development designated Edelweiss IV, both of which were described in the 1971-72 *Jane's*.

A later design, the **CE 75 Silène,** is described under the CERVA heading in this section.

In 1976 Siren was building the prototype of a new single-seat sailplane known as the D 77 Iris, a description of which follows:

SIREN D 77 IRIS

Design of the Iris began in 1973. A prototype is under construction, and was expected to fly in 1976.

TYPE: Single-seat training sailplane.

WINGS: Cantilever mid-wing monoplane. Bertin E55 166 wing section, with thickness/chord ratio of 16·6%. Dihedral 3° at centre-chord line. Incidence 3°. PMC foam sandwich construction. Airbrake on each wing.

TAIL UNIT: Cantilever T-tail, of similar construction to wings. Fixed-incidence tailplane. Trim tab in starboard side of elevator.

LANDING GEAR: Non-retractable monowheel, without shock-absorption, and tailskid. Wheel size 330 × 130 mm, tyre pressure 1·37 bars (20 lb/sq in). Siren hydraulic brake.

ACCOMMODATION: Single adjustable semi-reclining seat under one-piece canopy.

DIMENSIONS, EXTERNAL:
- Wing span 13·50 m (44 ft 3½ in)
- Wing chord at root 0·92 m (3 ft 0¼ in)
- Wing aspect ratio 16
- Length overall 6·37 m (20 ft 10¾ in)
- Fuselage: Max width 0·60 m (1 ft 11¾ in)
- Height over tail 0·90 m (2 ft 11½ in)
- Tailplane span 2·40 m (7 ft 10½ in)

AREAS:
- Wings, gross 11·40 m² (122·7 sq ft)
- Fin 1·00 m² (10·76 sq ft)
- Rudder 1·00 m² (10·76 sq ft)
- Tailplane 0·65 m² (7·00 sq ft)
- Elevator, incl tab 0·35 m² (3·77 sq ft)

WEIGHTS AND LOADING:
- Weight empty, equipped 200 kg (441 lb)
- Max T-O weight 300 kg (661 lb)
- Max wing loading 26·2 kg/m² (5·37 lb/sq ft)

PERFORMANCE (estimated):
- Best glide ratio at 51 knots (95 km/h; 59 mph) 33
- Min sinking speed 0·65 m (2·13 ft)/sec
- Stalling speed 32·5 knots (60 km/h; 37·5 mph)
- Max speed (smooth air) 126 knots (234 km/h; 145 mph)

WASSMER
SOCIÉTÉ NOUVELLE WASSMER AVIATION

ADDRESS:
BP 7, 63501-Aérodrome d'Issoire
Telephone: (73) 89.19.15 and 89.01.54
Telex: 990.185 F

OFFICERS:
See Aircraft section

In addition to manufacturing light aeroplanes, as described in the Aircraft section, Wassmer has long been famous as a builder of gliders and sailplanes. These included the successful Javelot and Super Javelot single-seat sailplanes and the AV.36 tail-less single-seat sailplane designed by M Charles Fauvel, which are no longer in production; and, more recently, the WA 26 Squale and WA 30 Bijave. Details of these types can be found in previous editions of *Jane's*.

In current production is the WA 28 Espadon.

WASSMER WA 28 ESPADON (SWORDFISH)

The Espadon is a developed version of the WA 26 Squale, from which it differs primarily in having wings of all-plastics construction. The wings are geometrically identical to those of the WA 26, but are of glassfibre/PMC sandwich construction, like those of the CERVA CE 75 Silène (which see).

Design began in November 1972, and the WA 28 flew for the first time in May 1974. The first production example (F-CCBC) was flown in November 1974, and by the Spring of 1975 a total of 20 had been ordered, of which 10 had been completed. Wassmer expected to receive orders for about 30 from French customers in 1976. German certification is under way.

TYPE: Single-seat high-performance sailplane.

WINGS: Cantilever shoulder-wing monoplane. Wing section FX-61-163 at root, FX-60-126 at tip. Dihedral 3°. Glassfibre/PMC sandwich construction. Schempp-Hirth airbrakes.

FUSELAGE: Reinforced polyester plastics structure of oval section.

TAIL UNIT: Conventional wooden structure with fabric covering. All-moving horizontal surfaces, with spring-loaded anti-tab.

LANDING GEAR: Non-retractable (standard) or retractable (optional) monowheel mounted forward of CG. Wassmer size 330 × 130 mm wheel; tyre pressure 1·96 bars (28·5 lb/sq in). Satmo hydraulic brake.

ACCOMMODATION: Single semi-reclining adjustable seat under a long flush Plexiglas canopy which opens sideways to port.

EQUIPMENT: Jolliet ER 5 VHF radio and Eros oxygen equipment optional.

DIMENSIONS, EXTERNAL:
- Wing span 15·00 m (49 ft 2½ in)
- Wing mean aerodynamic chord 0·9365 m (3 ft 0¾ in)
- Wing aspect ratio 17·82
- Length overall 7·65 m (25 ft 1¼ in)
- Height over tail 1·66 m (5 ft 5½ in)
- Tailplane span 3·00 m (9 ft 10¾ in)

AREAS:
- Wings, gross 12·626 m² (135·9 sq ft)
- Rudder 1·10 m² (11·84 sq ft)
- Tailplane 1·50 m² (16·15 sq ft)

WEIGHTS:
- Weight empty 245 kg (540 lb)
- Max T-O weight 378 kg (833 lb)

PERFORMANCE:
- Best glide ratio at 48·5 knots (90 km/h; 56 mph) 38
- Stalling speed at max T-O weight 37 knots (68 km/h; 42·5 mph)
- Max speed (smooth air) 135 knots (250 km/h; 155 mph)
- Max aero-tow and winch-launching speed 84·5 knots (157 km/h; 97·5 mph)
- g limits:
  - at 135 knots (250 km/h; 155 mph) +4·0; −1·5
  - at 77 knots (143 km/h; 89 mph) −2·65
  - at 85 knots (157 km/h; 98 mph) +5·3

GERMANY
(FEDERAL REPUBLIC)

AKAFLIEG BERLIN
AKADEMISCHE FLIEGERGRUPPE BERLIN EV

ADDRESS:
Technische Universität Berlin, 1 Berlin 12, Strasse des 17 Juni 135
Telephone: 030 3141

AKAFLIEG BERLIN B-12

This high-performance two-seat sailplane utilises the wings of a Schempp-Hirth Janus (which see), combined with a new fuselage and tail unit, also of glassfibre-reinforced plastics construction. General appearance of the B-12 is shown in the accompanying three-view drawing.

DIMENSIONS, EXTERNAL:
- Wing span 18·20 m (59 ft 8½ in)
- Wing aspect ratio 19·97

AREA:
- Wings, gross 16·58 m² (178·5 sq ft)

WEIGHTS AND LOADINGS:
- Weight empty 400 kg (882 lb)
- Max T-O weight:
  - single-seat 496 kg (1,093 lb)
  - two-seat 582 kg (1,283 lb)
- Max wing loading:
  - single-seat 29·9 kg/m² (6·12 lb/sq ft)
  - two-seat 35·1 kg/m² (7·19 lb/sq ft)

PERFORMANCE (estimated):
- Best glide ratio at 39·5 knots (73 km/h; 45·5 mph) 41
- Min sinking speed 0·60 m (1·97 ft)/sec
- Max speed (smooth air) 135 knots (250 km/h; 155 mph)

AKAFLIEG BRAUNSCHWEIG
AKADEMISCHE FLIEGERGRUPPE BRAUNSCHWEIG EV

ADDRESS:
3300 Braunschweig, Flughafen Akafliegheim
Telephone: 0531 3952149

DIRECTOR:
Martin Hansen

The students of Brunswick University have built a series of high-performance sailplanes. Details of the SB-10 Schirokko were given in the 1973-74 *Jane's*.

AKAFLIEG BRAUNSCHWEIG SB-5E DANZIG

The prototype SB-5 flew for the first time on 3 June 1959. Licence production was undertaken by Fa Eichelsdörfer, 86 Bamberg, Hafenstrasse 6, and 15 more had been built by the spring of 1966. The SB-5c version, first flown in 1965, incorporated a number of design changes, and was described in the 1975-76 *Jane's*.

The earlier version continues in production, though in small numbers. The current version, which entered production in 1974, is the SB-5E, with an extended wing span of 16 m to comply with current Club Class regulations. Existing SB-5b or SB-5c Danzigs can be converted to SB-5E standard.

The following description applies to the SB-5E, of which nine had been completed by February 1976:

TYPE: Single-seat Club Class sailplane.

WINGS: Cantilever shoulder-wing monoplane. Wing section NACA 63_3-618. Dihedral 1° 30′. Incidence 0° 30′. Conventional single-spar wooden structure with plywood covering. Wooden ailerons. Schempp-Hirth airbrakes at 50% chord.

FUSELAGE: Plywood monocoque of circular section.

TAIL UNIT: Cantilever V tail, of wooden construction.

LANDING GEAR: Non-retractable unsprung monowheel, size 4·00-4, with friction brake. Tyre pressure 1·96 bars (28 lb/sq in). Tailskid.

ACCOMMODATION: Single seat under one-piece moulded Plexiglas canopy.

DIMENSIONS, EXTERNAL:
- Wing span 16·00 m (52 ft 6 in)
- Wing chord at root 1·00 m (3 ft 3¼ in)
- Wing chord at tip 0·42 m (1 ft 4½ in)
- Wing aspect ratio 18·4
- Length overall 6·65 m (21 ft 10 in)
- Tailplane span (horizontal projection) 2·80 m (9 ft 2¼ in)
- Max width of fuselage 0·60 m (1 ft 11½ in)

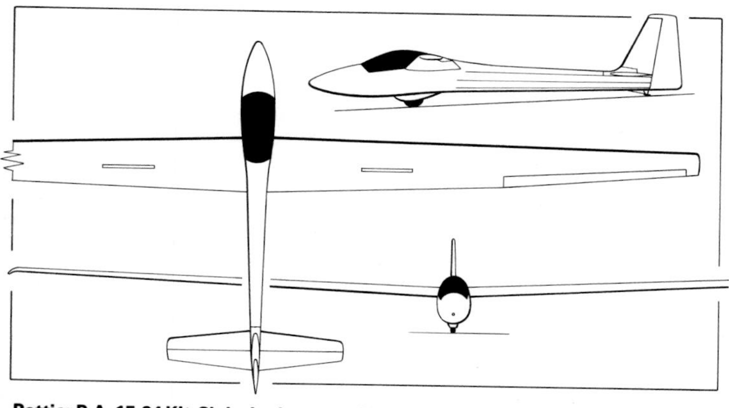

Pottier P.A. 15-34 Kit-Club single-seat sailplane for amateur construction *(Michael A. Badrocke)*

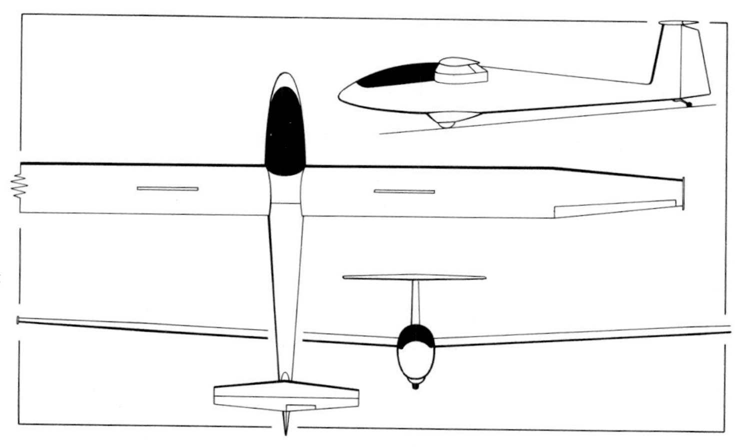

Siren D 77 Iris single-seat training sailplane

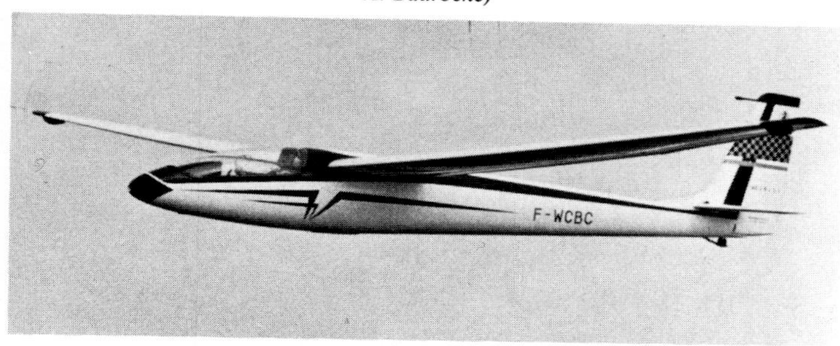

Wassmer WA 28 Espadon single-seat sailplane

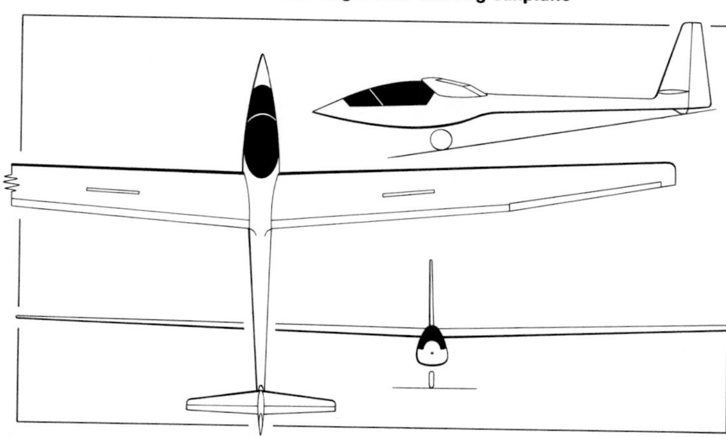

Akaflieg Berlin B-12 two-seat high-performance sailplane *(Michael A. Badrocke)*

Akaflieg Braunschweig SB-5E Danzig single-seat club class sailplane

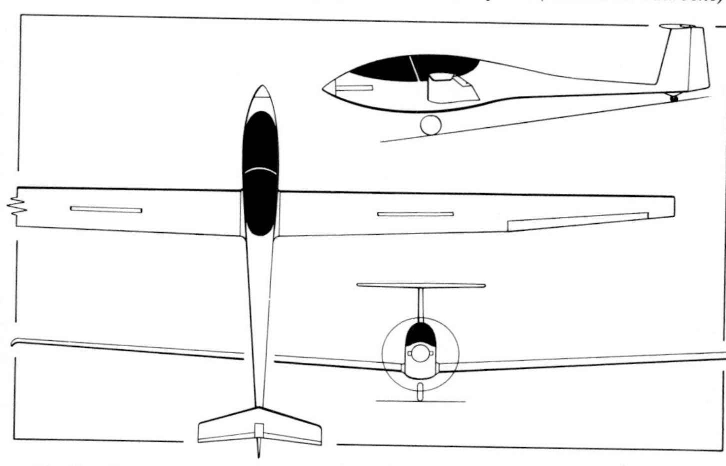

Akaflieg Darmstadt D-39 single-seat powered sailplane *(Michael A. Badrocke)*

AREAS:
- Wings, gross: 13·92 m² (149·8 sq ft)
- Airbrakes: 0·39 m² (4·20 sq ft)
- Fin (projected): 0·7 m² (7·53 sq ft)
- Rudder (projected): 0·5 m² (5·38 sq ft)
- Tailplane (projected): 1·0 m² (10·76 sq ft)
- Elevators (projected): 0·7 m² (7·53 sq ft)

WEIGHTS AND LOADING:
- Weight empty, equipped: 235 kg (518 lb)
- Max T-O weight: 325 kg (716 lb)
- Max wing loading: 24·0 kg/m² (4·92 lb/sq ft)

PERFORMANCE (at max T-O weight):
- Best glide ratio at 40·5 knots (75 km/h; 46·5 mph) 34
- Min sinking speed at 35 knots (64 km/h; 40 mph): 0·61 m (2·00 ft)/sec
- Stalling speed: 31·5 knots (58 km/h; 36 mph)
- Max speed (smooth air): 108 knots (200 km/h; 124 mph)
- Max speed (rough air): 76 knots (140 km/h; 87 mph)
- Max aero-tow speed: 59 knots (110 km/h; 68 mph)
- Max winch-launching speed: 49 knots (90 km/h; 56 mph)
- *g* limit: +5·7

AKAFLIEG BRAUNSCHWEIG SB-11

Construction of this single-seat high-performance sailplane was under way in the Spring of 1976. The main objective of the SB-11 is to achieve a practical solution to the problems presented by sailplane variable wing geometry.

Construction is of carbon-fibre-reinforced epoxy resin, and the SB-11 will be of T-tail configuration with an all-moving elevator, retractable monowheel landing gear, and a drag parachute. The wings will have a slotless Fowler-flap-type profile, and will be fitted with conventional flaps, ailerons and Schempp-Hirth airbrakes.

DIMENSIONS, EXTERNAL:
- Wing span: 15·00 m (49 ft 2½ in)
- Length of fuselage: 7·405 m (24 ft 3¼ in)

AKAFLIEG DARMSTADT
AKADEMISCHE FLIEGERGRUPPE DARMSTADT EV

ADDRESS:
6100 Darmstadt, Technische Hochschule, Hochschulestrasse

Telephone: 06151 162790

DIRECTOR:
H. Hertling

The Fliegergruppe of Darmstadt University has been designing, building and flying sailplanes since 1921. Its postwar products have included the D-34a single-seat high-performance sailplane, which flew for the first time in 1955; the D-34b, D-34c and D-34d, described in the 1962-63 *Jane's*; the D-36 Circe (1965-66 *Jane's*); the D-37b Artemis powered sailplane (1971-72 and 1972-73 *Jane's*); and the D-38 (1975-76 *Jane's*). The D-38 is now in production, in slightly modified form, as the Glaser-Dirks DG-100 (which see).

AKAFLIEG DARMSTADT D-39

The D-39 is a powered sailplane, which utilises the basic D-38 airframe, except for a low-mounted wing and a modified nose section in which are mounted a pair of piston engines driving a single folding propeller.

A prototype is under construction, and was due to fly in 1976.

TYPE: Single-seat high-performance powered sailplane.

WINGS: Cantilever low-wing monoplane. Wortmann wing sections: FX-61-184 at centreline, FX-60-126 at tip. Dihedral 4° from roots. Incidence 0°. Wing shell is a glassfibre/balsa sandwich, the ailerons being of glassfibre/Klégécel foam sandwich. Schempp-Hirth duralumin airbrake on each upper surface. No flaps.

FUSELAGE: Glassfibre/balsa sandwich monocoque structure.

TAIL UNIT: Cantilever T-tail of glassfibre/balsa sandwich construction. All-moving one-piece swept tailplane, with half-span Flettner tab.

LANDING GEAR: Manually-retractable sprung monowheel, with Dunlop 5·00-5 tyre and Tost drum brake, forward of CG. Small tailwheel.

POWER PLANT: Two 35 kW (47 hp) Fichtel & Sachs modified KM 914V piston engines, driving through a reduction gear a two-blade foldable wood and glassfibre propeller.

ACCOMMODATION AND EQUIPMENT: Single semi-reclining seat under flush transparent canopy, the rear section of which is removable. Standard instrumentation. Radio optional.

DIMENSIONS, EXTERNAL:
Wing span 15·00 m (49 ft 2½ in)
Wing chord at root 0·94 m (3 ft 1 in)
Wing chord (mean) 0·753 m (2 ft 5¾ in)
Wing chord at tip 0·376 m (1 ft 2¾ in)
Wing aspect ratio 20·5
Length overall 7·15 m (23 ft 5½ in)
Height over fuselage 1·02 m (3 ft 4¼ in)
Tailplane span 2·30 m (7 ft 6½ in)
Propeller diameter 1·35 m (4 ft 5¼ in)

AREAS:
Wings, gross 11·00 m² (118·4 sq ft)
Ailerons (total) 0·856 m² (9·21 sq ft)
Fin 0·92 m² (9·90 sq ft)
Rudder 0·368 m² (3·96 sq ft)
Tailplane, incl tab 1·00 m² (10·76 sq ft)

WEIGHTS AND LOADINGS:
Weight empty, equipped 290 kg (639 lb)
Max T-O weight 400 kg (882 lb)
Min wing loading 32·0 kg/m² (6·55 lb/sq ft)
Max wing loading 36·4 kg/m² (7·46 lb/sq ft)

PERFORMANCE (estimated, at max T-O weight, powered):
Max level speed 97 knots (180 km/h; 112 mph)
Max rate of climb at S/L 270 m (885 ft)/min
Range 269 nm (500 km; 310 miles)

PERFORMANCE (estimated, at max T-O weight, unpowered):
Best glide ratio at 56·5 knots (105 km/h; 65 mph) 36
Min sinking speed at 45 knots (84 km/h; 52 mph) 0·70 m (2·3 ft)/sec
Max speed (rough and smooth air) 135 knots (250 km/h; 155 mph)
Stalling speed 39 knots (72 km/h; 45 mph)
Max aero-tow speed 89 knots (165 km/h; 103 mph)
Max winch-launching speed 59 knots (110 km/h; 68 mph)
g limits +5·3; −3·6

AKAFLIEG MÜNCHEN
AKADEMISCHE FLIEGERGRUPPE MÜNCHEN EV

ADDRESS:
8 München 2, Arcisstrasse 21, Postfach 20 24 20
Telephone: 28 61 11

DIRECTOR:
Wolf D. Hoffmann

Students at Munich University have designed, built and flown a number of sailplanes, including the tandem two-seat Mü 23 Saurier of 20 m wing span, which made its first flight in 1959.

More recent designs have included the Mü 26 (1973-74 *Jane's*) and Mü 27.

AKAFLIEG MÜNCHEN Mü 27

The Mü 27 is a two-seat high-performance sailplane, the general appearance of which is shown in the accompanying three-view drawing. A prototype was under construction in 1976.

WINGS: Cantilever shoulder-wing monoplane. Wortmann FX-67-VC-170/136 section, with Fowler-type flaps which increase the wing area by 36% when fully extended. Mixed glassfibre/foam sandwich construction, with aluminium alloy spar and metal webs. Ailerons linked to flaps. Airbrake at 50% chord on upper surfaces.

FUSELAGE: All-glassfibre semi-monocoque structure.

TAIL UNIT: Cantilever T-tail of glassfibre/foam sandwich construction.

LANDING GEAR: Retractable monowheel and fixed tailwheel.

ACCOMMODATION: Two seats in tandem under two-piece moulded canopy.

DIMENSIONS, EXTERNAL:
Wing span 22·00 m (72 ft 2¼ in)
Wing aspect ratio:
flaps in 27·5
flaps out 20·2
Length overall 10·30 m (33 ft 9½ in)
Height over tail 1·80 m (5 ft 11 in)

AREAS:
Wings, gross:
flaps in 17·60 m² (189·4 sq ft)
flaps out 23·90 m² (257·3 sq ft)

WEIGHTS AND LOADINGS:
Weight empty 480 kg (1,058 lb)
Max T-O weight 700 kg (1,543 lb)
Max wing loading:
flaps in 40·0 kg/m² (8·2 lb/sq ft)
flaps out 29·3 kg/m² (6·0 lb/sq ft)

PERFORMANCE (estimated, at max T-O weight):
Best glide ratio:
at 54·5 knots (101 km/h; 63 mph), flaps in 47
at 47·5 knots (88 km/h; 54·5 mph), flaps out 39
Min sinking speed:
at 47 knots (87 km/h; 54 mph), flaps in 0·57 m (1·87 ft)/sec
at 32·5 knots (60 km/h; 37·5 mph), flaps out 0·56 m (1·84 ft)/sec
Max speed (smooth air) 151 knots (280 km/h; 174 mph)

AKAFLIEG STUTTGART
AKADEMISCHE FLIEGERGRUPPE STUTTGART EV

ADDRESS:
7000 Stuttgart 80, Pfaffenwaldring 35
Telephone: (0711) 784 2442 and 784 2443

TREASURER:
Gert Altpeter

The Akaflieg Stuttgart has been engaged continuously since the mid-1920s in aerodynamic research and experimental aircraft construction, mainly in the field of sailplane development, and it was in the wind tunnels at Stuttgart that the well-known Eppler and Wortmann wing sections were first developed. In more recent years Akaflieg Stuttgart has been in the forefront in developing the use of glassfibre in sailplane construction, and flew its first glassfibre sailplane, the FS-24 Phönix, on 25 November 1957. This aircraft, of which eight were built, was described under the Bölkow entry in the 1962-63 *Jane's*.

More recent sailplane designs have included the single-seat FS-25 Cuervo and the FS-26 Moseppl powered sailplane, of which descriptions and illustrations appeared in the 1973-74 *Jane's*.

Akaflieg Stuttgart's latest design is the FS-29, which has telescopic outer wing panels to increase or decrease the span.

AKAFLIEG STUTTGART FS-29

The FS-29 is a single-seat experimental variable-geometry sailplane, the design of which began in January 1972. Construction started on 15 November that year, and the prototype (D-2929) made its first flight on 15 June 1975.

So far as the design permits, the FS-29 has been built of already-available components. This is especially true of the fuselage, which utilises the retractable monowheel, tailskid, canopy, frame, control column and entire tail unit from a Schempp-Hirth Nimbus.

TYPE: Single-seat variable-geometry sailplane.

WINGS: Cantilever shoulder-wing monoplane, consisting of an inner, fixed wing and telescopic outer sections which slide in and out over it to vary the span and area. The inner wing is of Wortmann FX-73-170 section and is a box-spar structure, with skin and webs of glassfibre and Conticell foam sandwich. From this a stub spar protrudes into the outer wing panels, to carry the bending loads and to provide a mounting for the guide rails for the bearings on which the outer panels move in and out. Extension/retraction is actuated mechanically by pushrods, operated by a lever on the port side of the cockpit. The outer wing sections vary from FX-73-170 on the constant-chord inboard 'sleeve' to FX-73-K-170/22 on the tapered outer panels, so maintaining a 17% thickness/chord ratio over the entire wing span. The outer panels are monocoque structures of glassfibre/foam/carbon fibre sandwich. Dihedral 2° 30′. Plain ailerons of similar construction to outer wings. Schempp-Hirth airbrakes, on upper surface of inner wing, are operable only when the telescopic outer panels are fully extended. No flaps, spoilers or tabs.

FUSELAGE: Pod and boom type, the pod being a glassfibre monocoque shell and the boom a duralumin tube.

TAIL UNIT: Cantilever T tail, of glassfibre/foam sandwich monocoque construction, with all-moving tailplane.

LANDING GEAR: Mechanically-retractable unsprung monowheel, with Continental Type 3 tyre, size 5·00-5 (6 ply rating). Mechanical drum brake. Tailskid.

ACCOMMODATION: Single seat under one-piece flush-mounted transparent canopy which opens sideways to starboard.

DIMENSIONS, EXTERNAL:
Wing span:
min 13·30 m (43 ft 7¾ in)
max 19·00 m (62 ft 4 in)
Wing chord at root 0·752 m (2 ft 5¾ in)
Wing chord at tip 0·30 m (11¾ in)
Wing aspect ratio:
min 20·47
max 28·54
Length overall 7·16 m (23 ft 6 in)
Height over tail 1·27 m (4 ft 2 in)
Tailplane span 2·40 m (7 ft 10½ in)

AREAS:
Wings, gross:
min 8·56 m² (92·14 sq ft)
max 12·65 m² (136·2 sq ft)
Ailerons (total) 0·752 m² (8·09 sq ft)
Fin 0·854 m² (9·19 sq ft)
Rudder 0·256 m² (2·76 sq ft)
Tailplane 1·03 m² (11·1 sq ft)

WEIGHTS AND LOADINGS (without water ballast):
Weight empty 357 kg (787 lb)
Max T-O weight 450 kg (992 lb)
Max wing loading:
13·30 m span 52·6 kg/m² (10·77 lb/sq ft)
19·00 m span 35·6 kg/m² (7·29 lb/sq ft)

PERFORMANCE (at max T-O weight without water ballast):
Best glide ratio at 54 knots (100 km/h; 62 mph) 44
Min sinking speed at 40 knots (74 km/h; 46 mph) 0·54 m (1·77 ft)/sec
Stalling speed 39 knots (72 km/h; 45 mph)
Max speed (rough or smooth air) 135 knots (250 km/h; 155 mph)
Max aero-tow speed 97 knots (180 km/h; 112 mph)
Max winch-launching speed 65 knots (120 km/h; 74·5 mph)
g limit at 117 knots (218 km/h; 135 mph) +5·3

BLESSING
GERHARD BLESSING

ADDRESS:
205 Hamburg 80, Ochsenwerder Landstrasse 33

Herr Blessing, whose Gleiter Max two-seat powered sailplane was described and illustrated in the 1970-71 *Jane's*, has since designed a single/two-seat powered sailplane known as the Rebell, intended for amateur construction.

BLESSING REBELL

The Rebell is a single-seat powered sailplane, with provision for a second occupant. Its structure is of steel tube with wood covering. The wings have Wortmann FX-66-S-196 root and FX-66-17A 11 tip sections and can, at builder's option, be made in two, three or four parts. No component of the aircraft is more than 3·5 m (11 ft 5¾ in) long, to facilitate assembly in confined spaces. The fuel tank and several other components are standard items obtainable from the motor car industry.

Original power plant was a 40·3 kW (54 hp) Hirth M28 two-cylinder engine, driving a Hoffmann two-blade feathering pusher propeller, with which the Rebell made its first flight on 3 June 1973. Following the closure of the Hirth engine company, a modified Volkswagen was substituted, and a first flight with this engine was made in the Summer of 1975.

The landing gear consists of a semi-recessed monowheel, below the cockpit, and a tailwheel. Just inboard of each outer wing panel, which can be folded upward, there is an outrigger balancer wheel.

The prototype (D-KEBO) is shown in the accompanying illustration. The following data refer to this aircraft with the original Hirth engine:

DIMENSIONS, EXTERNAL:
Wing span 15·00 m (49 ft 2½ in)
Length overall 7·20 m (23 ft 7½ in)
Wing area, gross 17·00 m² (183·0 sq ft)

WEIGHTS:
Weight empty 420 kg (926 lb)
Max T-O weight 620 kg (1,366 lb)

PERFORMANCE (at max T-O weight, powered):
Max level speed 108 knots (200 km/h; 124 mph)
Cruising speed 81 knots (150 km/h; 93 mph)
Max rate of climb 180 m (591 ft)/min
Range 323 nm (600 km; 372 miles)

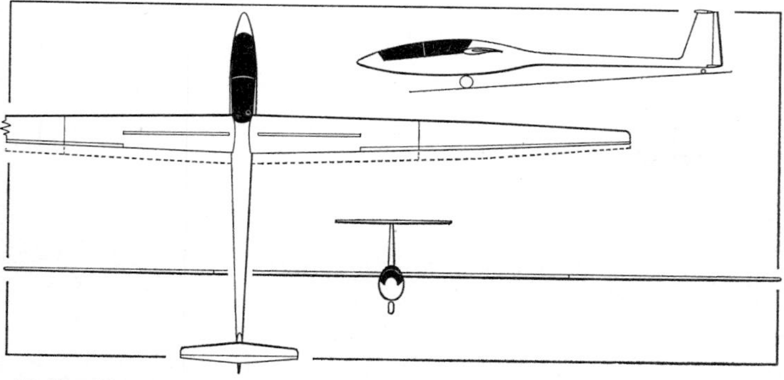

Akaflieg München Mü 27 two-seat high-performance sailplane *(Michael A. Badrocke)*

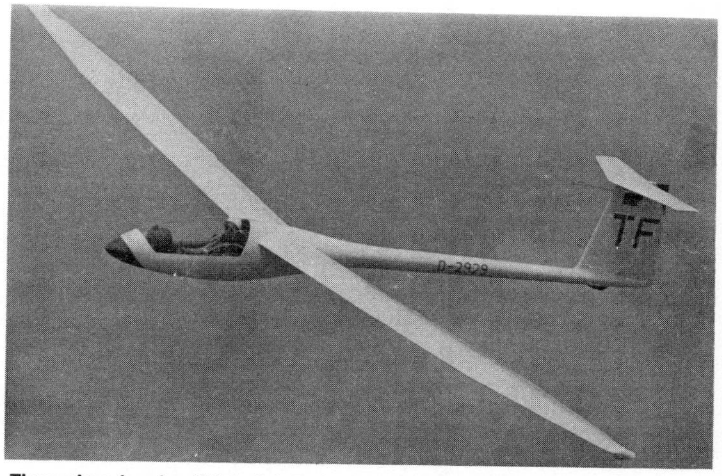

Three-view drawing *(Michael A. Badrocke)* **and photograph of the Akaflieg Stuttgart FS-29 variable-geometry sailplane**

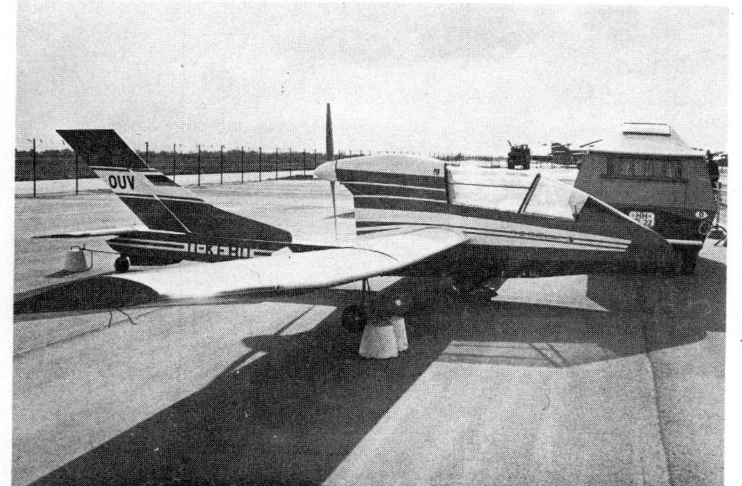

Prototype Blessing Rebell one/two-seat sailplane, powered by a Hirth engine *(Brian M. Service)*

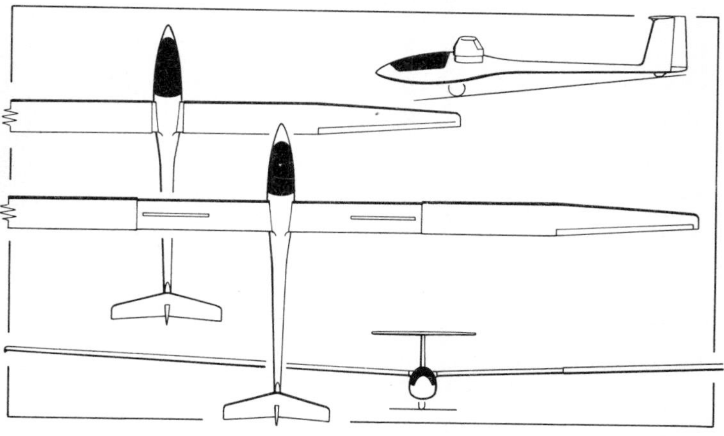

FLUWAG BREMEN

FLUGWISSENSCHAFTLICHE ARBEITS-GEMEINSCHAFT (FLUWAG) BREMEN

ADDRESS:
28 Bremen Oberneuland, Rockwinkeler Landstrasse 33

FLUWAG BREMEN FL-III

This single-seat Standard Class sailplane, announced at the end of 1973, is of all-plastics construction.

All known details, and a three-view drawing, were published in the 1975-76 *Jane's.*

FVA

FLUGWISSENSCHAFTLICHE VEREINIGUNG AACHEN AM INSTITUT FÜR LUFT- UND RAUMFAHRT (Aeronautical Research Association of the Rhine-Westfälia Technical College, Aachen)

ADDRESS:
5100 Aachen, Templergraben 55 (c/o U. Solies)

OFFICERS:
Prof Dr Ing A. W. Quick
Prof Dr Ing H. F. Thomac
Stud Ing R. Hartel

FOUNDER: Prof Dr Theodore von Kármán

The FVA designed and built a number of sailplanes between the first and second World Wars, beginning with the FVA-1 of 1920 and ending with the FVA-13 of 1940. It resumed these activities in 1951, and recent projects have included the construction of a Schleicher Ka 8 single-seat training sailplane. FVA also designed and built the FVA-18 described in the Aircraft section of the 1972-73 *Jane's.*

Its latest product is the FVA-20 Standard Class sailplane.

FVA-20

The FVA-20 is a single-seat Standard Class sailplane, designed to explore experimental methods of glassfibre construction. A prototype was scheduled to make its first flight in 1975. No subsequent production is intended; but after initial flight testing the prototype will be modified to have a slimmer fuselage and different wing/fuselage attachment system. See three-view drawing on next page.

The following description applies to the aircraft in its initial form:

TYPE: Single-seat Standard Class sailplane.

WINGS: Cantilever shoulder-wing monoplane, with taper on outer panels. Wortmann wing sections: FX-61-168 at root, FX-60-126 at tip. Schempp-Hirth airbrakes above and below each wing. Primary structure consists of a sandwich of 8 mm Conticell hard-foam PVC, coated with resin-bonded glassfibre, with a box spar. There are no ribs.

FUSELAGE: Semi-monocoque half-sandwich structure, consisting of a balsa-covered wooden frame with a resin-bonded outer skin of laminated glassfibre.

TAIL UNIT: Cantilever T-tail. Fin and rudder of half-sandwich construction, similar to fuselage. Tailplane and elevators of similar construction to wings.

LANDING GEAR: Retractable monowheel and tail bumper.

ACCOMMODATION: Single seat under moulded transparent canopy.

DIMENSIONS, EXTERNAL:
Wing span 15·00 m (49 ft 2½ in)
Wing aspect ratio 17·6
Length overall 7·00 m (22 ft 11¾ in)
Fuselage: Max width 0·65 m (2 ft 1½ in)
Max depth 0·90 m (2 ft 11½ in)

AREA:
Wings, gross 12·80 m² (137·8 sq ft)

WEIGHT AND LOADING:
Max T-O weight 320 kg (705 lb)
Max wing loading 25·0 kg/m² (5·1 lb/sq ft)

PERFORMANCE (estimated):
Best glide ratio at 48·5 knots (90 km/h; 56 mph) 35·3
Min sinking speed at 37 knots (68 km/h; 42·5 mph) 0·60 m (1·97 ft)/sec

GHH

GLASFLÜGEL, HOLIGHAUS & HILLENBRAND GmbH & Co KG

ADDRESS: 7311 Schlattstall Krs, Nürtingen
Telephone: 07026/855

This company was formed following the death on 5 October 1975 of Ing Eugen Hänle, former Director of the Glasflügel company. Glasflügel was responsible for the Standard Libelle, developed from the Hütter H 30 GFK and H 301 of which details appeared in earlier editions of *Jane's.* It completed 100 examples of the H 301, and a total of 601 examples of the Standard Libelle; a description of the latter type can be found in the 1974-75 *Jane's.* Glasflügel also delivered 18 examples of a developed version of Björn Stender's BS 1, utilising the same type of all-glassfibre construction as in the Libelle. Another type of which production has ended is the Glasflügel 604, also described in the 1974-75 *Jane's*; 10 were built.

Current production includes the Glasflügel 205 Club Libelle, the single-seat 17 m Kestrel and the Glasflügel 206 Mosquito. Slingsby in the UK (which see) is producing a 19 m version of the Kestrel.

Glasflügel Italiana (which see) produces sailplane components in Italy for the Kestrel designed by the parent company.

GLASFLÜGEL 205 CLUB LIBELLE

The Club Libelle is a direct derivative of the Standard Libelle, described in the 1974-75 edition of *Jane's,* and differs principally in having shoulder-mounted wings and a T tailplane. It also incorporates certain features of the Kestrel. Design began in the Autumn of 1972, and construction of a prototype started in the following February. This aircraft made its first flight on 14 September 1973; up to early 1975 approx 50 production examples had been built.

TYPE: Single-seat Standard Class sailplane.

WINGS: Cantilever shoulder-wing monoplane. Wortmann wing section with 18% thickness/chord ratio. Incidence 4° at root. Glassfibre and foam sandwich skin with unidirectional glassfibre spar caps, produced by HH method. Glassfibre and foam sandwich shear web. Fixed-hinge, partially mass-balanced ailerons, with all-glassfibre skin. Airbrakes over approx 60% of trailing-edge.

FUSELAGE: Glassfibre monocoque structure.

TAIL UNIT: Cantilever T tail. Tailplane of glassfibre and foam sandwich construction. Fin, rudder and elevator of glassfibre monocoque construction. Elevator fitted with spring trim.

LANDING GEAR: Non-retractable monowheel and tail-wheel. Gas spring and rubber spring shock-absorption. Main-wheel tyre size 300 × 100 mm, tailwheel tyre size 210 × 65 mm. Tyre pressure (both) 1·97 bars (28·5 lb/sq in). Internally-expanding brake.

ACCOMMODATION: Single seat under one-piece canopy.

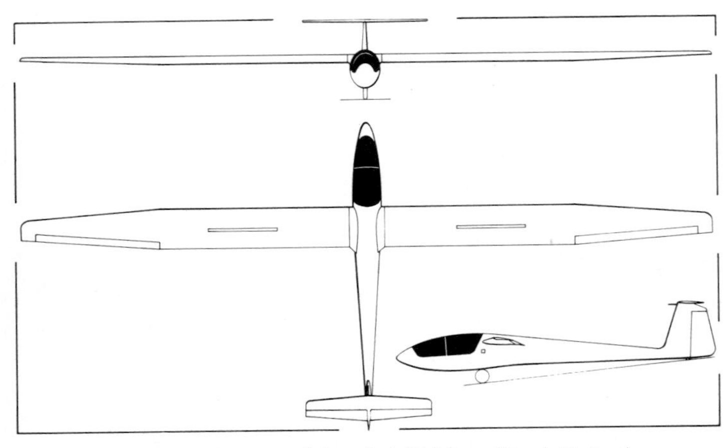

FVA-20 single-seat sailplane in initial form *(Roy J. Grainge)*

Glasflügel 205 Club Libelle single-seat Standard Class sailplane

Glasflügel Kestrel 17 metre single-seat high-performance sailplane

Glasflügel 206 Hornet prototype 15 metre competition sailplane

Standard equipment includes two towing hooks, seat cushion with inflatable knee supports, instrument panel, ASI and built-in VHF antenna. Provision for radio and oxygen, adjustable seat and rudder pedals.

DIMENSIONS, EXTERNAL:
- Wing span: 15·00 m (49 ft 2½ in)
- Wing chord at root: 0·90 m (2 ft 11½ in)
- Wing chord at tip: 0·36 m (1 ft 2¼ in)
- Wing aspect ratio: 23
- Length overall: 6·40 m (21 ft 0 in)
- Height over tail: 1·40 m (4 ft 7 in)
- Tailplane span: 2·50 m (8 ft 2½ in)

AREAS:
- Wings, gross: 9·80 m² (105·5 sq ft)
- Ailerons (total): 0·574 m² (6·18 sq ft)
- Airbrakes (total): 1·42 m² (15·28 sq ft)
- Fin: 0·89 m² (9·58 sq ft)
- Rudder: 0·27 m² (2·91 sq ft)
- Tailplane: 1·15 m² (12·38 sq ft)
- Elevator: 0·29 m² (3·12 sq ft)

WEIGHTS AND LOADING:
- Weight empty, equipped: approx 200 kg (441 lb)
- Max T-O weight: 330 kg (727 lb)
- Max wing loading: 33·67 kg/m² (6·9 lb/sq ft)

PERFORMANCE (at max T-O weight):
- Best glide ratio at 48·5 knots (90 km/h; 56 mph): 35
- Min sinking speed at 36·5 knots (67 km/h; 42 mph): 0·56 m (1·84 ft)/sec
- Stalling speed without airbrakes: 35 knots (64 km/h; 40 mph)
- Max speed (rough or smooth air): 108 knots (200 km/h; 124 mph)
- Max aero-tow speed: 81 knots (150 km/h; 93 mph)
- Max winch-launching speed: 70 knots (130 km/h; 80·5 mph)
- *g* limits: +5·7; −3·7 (safety factor 1·5)

GLASFLÜGEL KESTREL 17

The Kestrel is currently in production in 17 m form by Glasflügel and in a 19 m version by Slingsby in the UK (which see) as an advanced Open Class sailplane. It was known originally as the 17 metre Libelle. In addition to proven Libelle features, the Kestrel has a larger cockpit canopy, a new fuselage and wing profile, and a T tail.

Deliveries of production Kestrels began in 1969, and by January 1975 a total of 120 had been delivered. Kestrels (17 m version) were placed 17th and 26th in the 1974 World Gliding Championships at Waikerie, Australia. They also gained leading places in the Open Class of 1973 national gliding championships held in Australia (2nd and 3rd), Austria (first 4 places), Belgium (2nd), Sweden (2nd), South Africa (2nd), UK (2nd to 9th places shared by 17 m and 19 m Kestrels), Italy (3rd, 4th and 5th); and won the US Smirnoff Trophy.

Three world gliding records are held currently by the 17 metre version of the Kestrel. These are the men's single-seat 100 km closed-circuit speed record of 89·22 knots (165·348 km/h; 102·74 mph), set by K. B. Briegleb of the USA on 18 July 1974; the equivalent record for women, of 68·685 knots (127·204 km/h; 79·041 mph), set on 19 August 1975 by Adele Orsi of Italy; and the women's single-seat 300 km closed-circuit speed record of 61·76 knots (114·45 km/h; 71·12 mph), set by Miss Susan Martin of Australia on 11 February 1972.

TYPE: Single-seat high-performance Open Class sailplane.

WINGS: Two-piece glassfibre and balsa and/or foam sandwich skins, with unidirectional glassfibre spar caps produced by the HH method. Glassfibre and balsa sandwich shear web. Ailerons linked differentially with high-lift camber-changing flaps, all partially mass-balanced. Flush-fitting dive brakes.

FUSELAGE: Glassfibre (not sandwich) monocoque structure.

TAIL UNIT: Cantilever T tail of similar construction to wings. All control surfaces mass-balanced. Releasable drag parachute, diameter 0·91 m (3 ft), housed in rudder.

LANDING GEAR: Retractable monowheel, with internally-expanding brake. Interchangeable tailwheel or skid.

DIMENSIONS, EXTERNAL:
- Wing span: 17·00 m (55 ft 9¼ in)
- Wing aspect ratio: 25
- Length overall: 6·72 m (22 ft 0½ in)
- Height over tail: 1·52 m (5 ft 0 in)
- Tailplane span: 2·85 m (9 ft 4¼ in)

AREA:
- Wings, gross: 11·6 m² (124·8 sq ft)

WEIGHTS:
- Weight empty: 260 kg (574 lb)
- Max T-O weight (with water ballast): 400 kg (882 lb)

PERFORMANCE:
- Best glide ratio at 52·5 knots (97 km/h; 60·5 mph): 43
- Min sinking speed at 40 knots (74 km/h; 46 mph): 0·55 m (1·8 ft)/sec
- Stalling speed: 33·5 knots (62 km/h; 39 mph)
- Max speed (rough or smooth air): 135 knots (250 km/h; 155 mph)

GLASFLÜGEL 206 HORNET/MOSQUITO

The Hornet is a derivative of the Club Libelle, from which it differs principally in having a retractable monowheel, provision for water ballast, and an enlarged flush-fitting canopy. The prototype (D-9432) flew for the first time on 21 December 1974. Hornets flown by Australian and Argentinian pilots were placed 21st and 24th in the 1976 World Championships at Räyskälä, Finland.

Current production aircraft are named **Mosquito**.

TYPE: Single-seat Standard Class sailplane.

WINGS: Similar to Club Libelle, except for glassfibre/balsa sandwich shear web and rotating trailing-edge airbrakes.

FUSELAGE: Similar to Club Libelle except for redesigned nose section.

TAIL UNIT: Similar to Club Libelle, but with glassfibre monocoque fin and elevator, glassfibre/foam sandwich tailplane and rudder.

LANDING GEAR: Mechanically-retractable unsprung monowheel, size 300 × 100 mm, tyre pressure 3·45 bars (50 lb/sq in), with internally-expanding brake. Non-retractable tailwheel, size 210 × 65 mm, tyre pressure 1·97 bars (28·5 lb/sq in).

ACCOMMODATION AND EQUIPMENT: Single seat under one-piece fully-transparent moulded canopy. Standard equipment includes two towing hooks, seat cushion with inflatable knee supports, instrument panel, airspeed indicator, and built-in VHF antenna. Provision for radio, oxygen, adjustable seat and rudder pedals, and water ballast installation.

DIMENSIONS, EXTERNAL:
- As Club Libelle

AREA (Hornet):
- Wings, gross: 9·86 m² (106·1 sq ft)

WEIGHTS AND LOADINGS (Hornet):
- Weight empty: 227 kg (500 lb)
- Max T-O weight:
  - without water ballast: 345 kg (760 lb)
  - with water ballast: 420 kg (926 lb)
- Max wing loading:
  - without water ballast: 35·7 kg/m² (7·31 lb/sq ft)
  - with water ballast: 42·9 kg/m² (8·79 lb/sq ft)

WEIGHTS AND LOADING (Mosquito):
- Weight empty: 250 kg (551 lb)
- Max water ballast: 120 kg (264 lb)
- Max T-O weight with water ballast: 450 kg (992 lb)
- Max wing loading with water ballast: 46 kg/m² (9·42 lb/sq ft)

PERFORMANCE (Hornet, at max T-O weight except where indicated):
- Best glide ratio at 41 knots (75 km/h; 47 mph): 38
- Min sinking speed at 345 kg (760 lb) AUW, at 41 knots (75 km/h; 47 mph): 0·60 m (1·97 ft)/sec
- Stalling speed at 345 kg (760 lb) AUW:
  - airbrakes extended: 36·5 knots (67 km/h; 42 mph)
  - airbrakes retracted: 39 knots (72 km/h; 45 mph)

Max speed (rough and smooth air)
135 knots (250 km/h; 155 mph)
Max aero-tow speed 81 knots (150 km/h; 93 mph)
Max winch-launching speed
65 knots (120 km/h; 74·5 mph)
g limits (safety factor 1·5) +5·3; −3·25
PERFORMANCE (Mosquito. A: at 320 kg; 705 lb AUW; B: at 450 kg; 992 lb AUW):
Best glide ratio:
A at 42·5 knots (79 km/h; 49 mph) 42
B at 51 knots (94 km/h; 58·5 mph) 42
Min sinking speed:
A at 42·5 knots (79 km/h; 49 mph)
0·58 m (1·90 ft)/sec
B at 51 knots (94 km/h; 58·5 mph)
0·69 m (2·26 ft)/sec
Stalling speed:
A 36 knots (66 km/h; 41 mph)
B 42·5 knots (78 km/h; 48·5 mph)
Max speed (smooth air):
A, B 146 knots (270 km/h; 168 mph)

GLASER-DIRKS

GLASER-DIRKS FLUGZEUGBAU GmbH

ADDRESS:
752 Bruchsal-Untergrombach, Postfach 47, Im Schollengarten 19-20
Telephone: 07257 (1071)
DIRECTORS:
Gerhard Glaser
Dipl-Ing Wilhelm Dirks

This company is producing the DG-100 Standard Class sailplane, which is a modified and lighter-weight development of the Akaflieg Darmstadt D-38 described in the 1973-74 *Jane's*. A new prototype, designated DG-200, was due to fly during 1976.

GLASER-DIRKS DG-100

Design of the DG-100, by Dipl-Ing Wilhelm Dirks, began in August 1973. Construction of a prototype started in January 1974, and this aircraft (D-7100) flew for the first time on 10 May 1974. By the beginning of 1976, a total of 40 DG-100s had been delivered.

TYPE: Single-seat Standard Class sailplane.

WINGS: Cantilever shoulder-wing monoplane. Wortmann wing sections: FX-61-184 at centreline, FX-60-126 at tip. Dihedral 3° from roots. Incidence −1°. No sweepback. Glassfibre roving main spar. Glassfibre/Conticell/foam sandwich construction, including ailerons. Schempp-Hirth duralumin airbrakes on upper surfaces. Water ballast tank in each wing, combined capacity 100 kg (220 lb). Ballast can be jettisoned during flight.

FUSELAGE: All-glassfibre semi-monocoque structure.

TAIL UNIT: Cantilever T-tail. All-moving glassfibre/Conticell/foam sandwich tailplane with full-span Flettner tab. All-glassfibre fin and rudder.

LANDING GEAR: Manually-retractable monowheel, size 5·00-5, tyre pressure 2·0 bars (29 lb/sq in). Tost drum brake. Tailwheel size 200 × 50.

ACCOMMODATION: Single semi-reclining seat under flush transparent canopy, the rear section of which is removable. Standard instrumentation. Radio optional.

DIMENSIONS, EXTERNAL:
Wing span 15·00 m (49 ft 2½ in)
Wing chord at root 0·94 m (3 ft 1 in)
Wing chord at tip 0·376 m (1 ft 2¾ in)
Wing mean aerodynamic chord 0·753 m (2 ft 5¾ in)
Wing aspect ratio 20·5
Length overall 7·00 m (22 ft 11¾ in)
Height over tail 1·40 m (4 ft 7 in)
Tailplane span 2·30 m (7 ft 6½ in)
AREAS:
Wings, gross 11·00 m² (118·4 sq ft)
Ailerons (total) 0·856 m² (9·21 sq ft)
Fin 0·92 m² (9·90 sq ft)
Rudder 0·46 m² (4·95 sq ft)
Tailplane, incl tab 1·00 m² (10·76 sq ft)
WEIGHTS AND LOADINGS:
Weight empty, equipped 230 kg (507 lb)
Max T-O weight:
without water ballast 385 kg (849 lb)
with water ballast 418 kg (921 lb)
Max wing loading:
without water ballast 35·0 kg/m² (7·17 lb/sq ft)
with water ballast 38·0 kg/m² (7·78 lb/sq ft)
PERFORMANCE (at max T-O weight):
Best glide ratio at 51 knots (95 km/h; 59 mph) 38·7
Min sinking speed at 40·5 knots (75 km/h; 47 mph)
0·60 m (1·97 ft)/sec
Stalling speed 35·5 knots (66 km/h; 41 mph)
Max speed (rough or smooth air)
135 knots (250 km/h; 155 mph)
Max aero-tow speed 89 knots (165 km/h; 102·5 mph)
Max winch-launching speed
70 knots (130 km/h; 80·5 mph)
g limit +6·1

GLASER-DIRKS DG-200

The DG-200, which was due to fly for the first time in 1976, is an improved version of the DG-100 fitted with wing flaps, and was developed for Open Class international competition in 1976.

The description of the DG-100 applies generally also to the DG-200, with the following principal exceptions:

TYPE: Single-seat Open Class sailplane.

WINGS: Aerofoil section details not released. Incidence 0°. All-metal glassfibre flaps and ailerons. Water ballast increased to 110 kg (242 lb).

TAIL UNIT: Glassfibre/foam sandwich tailplane, with all-glassfibre elevator.

DIMENSIONS, EXTERNAL:
Wing chord at root 0·855 m (2 ft 9¾ in)
Wing chord at tip 0·342 m (1 ft 1½ in)
Wing aspect ratio 22·5
Tailplane span 2·24 m (7 ft 4¼ in)
AREAS:
Wings, gross 10·00 m² (107·64 sq ft)
Ailerons (total) 0·564 m² (6·07 sq ft)
Trailing-edge flaps (total) 1·12 m² (12·06 sq ft)
Fin 0·88 m² (9·47 sq ft)
Rudder 0·44 m² (4·74 sq ft)
Tailplane 0·946 m² (10·18 sq ft)
Elevator 0·255 m² (2·74 sq ft)
WEIGHTS AND LOADINGS:
Weight empty 325 kg (716 lb)
Max T-O weight:
without water ballast 360 kg (793 lb)
with water ballast 420 kg (926 lb)
Max wing loading:
without water ballast 30·0 kg/m² (6·14 lb/sq ft)
with water ballast 42·0 kg/m² (8·60 lb/sq ft)
PERFORMANCE (estimated, at 420 kg; 926 lb max T-O weight except where indicated):
Best glide ratio at 59·5 knots (110 km/h; 68·5 mph) 42
Min sinking speed at 39 knots (72 km/h; 45 mph)
0·55 m (1·80 ft)/sec
Stalling speed at 300 kg (661 lb) AUW
34 knots (62 km/h; 39 mph)
Max speed (rough and smooth air)
135 knots (250 km/h; 155 mph)
Max aero-tow speed 100 knots (185 km/h; 115 mph)
Max winch-launching speed
70 knots (130 km/h; 80·5 mph)

GROB

BURKHART GROB FLUGZEUGBAU

ADDRESS:
8948 Mindelheim, Industriestrasse
Telephone: 08261/8111 and 08268/411
Telex: 53 96 14

This company has, since 1972, been building the Schempp-Hirth Standard Cirrus under licence. It is now manufacturing a new sailplane of its own design, the G-102 Astir CS, and is developing the G-103 Twin Astir and G-104 Speed Astir.

GROB G-102 ASTIR CS

Prototype construction of the Astir CS (Club Standard) single-seat Standard Class sailplane began in March 1974, and this aircraft (D-6102) made its first flight on 19 December 1974. German certification by the LBA was awarded in August 1975.

By February 1976, a total of 350 had been ordered, of which 120 had been delivered. It was planned to complete delivery of the remaining 230 by October 1976.

Under development in mid-1976 was a powered version, the **Astir CSM,** which has an 18·6 kW (25 hp) Fichtel & Sachs Wankel KM 24 rotating-piston engine, mounted aft of the cockpit on a retractable pylon and driving a 1·20 m (3 ft 11¼ in) Hoffmann two-blade tractor propeller. A 30 litre (6·6 Imp gallon) fuel tank is installed. This installation, which adds 65 kg (143 lb) to the aircraft empty weight, enables the Astir CSM to cruise at 70 knots (130 km/h; 81 mph) for up to 280 nm (520 km; 323 miles).

The following description applies to the standard Astir CS sailplane:

TYPE: Single-seat Standard Class sailplane.

WINGS: Cantilever mid-wing monoplane. Eppler E 603 wing section, with thickness/chord ratio of 18·9%. Glassfibre roving main spar, and glassfibre/epoxy resin sandwich skin. Glassfibre ailerons. Schempp-Hirth aluminium airbrakes in upper surfaces. Provision for 100 kg (220 lb) of water ballast.

FUSELAGE: Glassfibre semi-monocoque structure. Hook fitted for aero-tow or winch launching.

TAIL UNIT: Cantilever T tail, of glassfibre/epoxy resin sandwich construction, with rudder and full-span elevator.

LANDING GEAR: Retractable Tost monowheel, size 5·00-5, tyre pressure 2·5 bars (36·3 lb/sq in), with internally-expanding drum brake. Rubber-sprung tailwheel.

ACCOMMODATION: Single seat under one-piece jettisonable moulded canopy which opens sideways to starboard.

DIMENSIONS, EXTERNAL:
Wing span 15·00 m (49 ft 2½ in)
Wing aspect ratio 18·2
Length overall 6·70 m (21 ft 11¾ in)
Height over tail 1·40 m (4 ft 7 in)
Tailplane span 2·80 m (9 ft 2¼ in)
AREAS:
Wings, gross 12·40 m² (133·5 sq ft)
Ailerons (total) 0·837 m² (9·01 sq ft)
Horizontal tail surface (total, incl tab)
1·47 m² (15·82 sq ft)
WEIGHTS AND LOADINGS:
Weight empty, equipped 250 kg (551 lb)
Max T-O weight:
without water ballast 370 kg (816 lb)
with water ballast 450 kg (992 lb)
Max wing loading:
without water ballast 25·0 kg/m² (5·12 lb/sq ft)
with water ballast 36·5 kg/m² (7·48 lb/sq ft)
PERFORMANCE (at 450 kg; 992 lb max T-O weight except where indicated):
Best glide ratio at 51·5 knots (95 km/h; 59 mph) and AUW of 370 kg (816 lb) 37·3
Best glide ratio at 56·5 knots (105 km/h; 65 mph) and AUW of 450 kg (992 lb) 38
Min sinking speed at 40·5 knots (75 km/h; 46·5 mph) and AUW of 370 kg (816 lb) 0·60 m (1·97 ft)/sec
Min sinking speed at 46 knots (85 km/h; 53 mph) and AUW of 450 kg (992 lb) 0·70 m (2·30 ft)/sec
Stalling speed 32·5 knots (60 km/h; 37·5 mph)
Max speed (rough and smooth air)
135 knots (250 km/h; 155 mph)
Max aero-tow speed 91 knots (170 km/h; 105 mph)
Max winch-launching speed
65 knots (120 km/h; 74·5 mph)

GROB G-103 TWIN ASTIR

The Twin Astir is essentially a tandem two-seat development of the Astir CS, from which it differs principally in having an enlarged cockpit and sweptforward wings of greater span. A prototype was expected to fly in the Spring of 1976.

TYPE: Two-seat training and competition sailplane.

WINGS: Cantilever mid-wing monoplane, of similar construction to G-102 but with forward sweep. Eppler E 603 wing section. Provision for 90 kg (198 lb) water ballast.

FUSELAGE: Glassfibre semi-monocoque structure.

TAIL UNIT: Cantilever T tail, with slightly-sweptback fin and statically balanced rudder.

LANDING GEAR: Similar to that of G-102.

ACCOMMODATION: Two seats in tandem under two-piece moulded canopy. Rear (instructor's) seat slightly elevated. Dual controls standard.

DIMENSIONS, EXTERNAL:
Wing span 17·50 m (57 ft 5 in)
Wing aspect ratio 17·1
Length overall 8·10 m (26 ft 8¾ in)
Height over tail 1·60 m (5 ft 3 in)
Tailplane span 1·60 m (5 ft 3 in)
AREA:
Wings, gross 17·90 m² (192·7 sq ft)
WEIGHTS AND LOADINGS:
Weight empty, equipped 330 kg (727 lb)
Max T-O weight:
single-seat 445 kg (981 lb)
two-seat, without water ballast 560 kg (1,234 lb)
two-seat, with water ballast 650 kg (1,433 lb)
Max wing loading:
single-seat 24·9 kg/m² (5·10 lb/sq ft)
two-seat, without water ballast
31·3 kg/m² (6·41 lb/sq ft)
two-seat, with water ballast
36·3 kg/m² (7·43 lb/sq ft)
PERFORMANCE (estimated, at relevant max T-O weight. A: single-seat; B: two-seat without water ballast; C: two-seat with water ballast):
Best glide ratio:
A at 51 knots (95 km/h; 59 mph) 38
B at 59·5 knots (110 km/h; 68·5 mph) 38·5
C at 59·5 knots (110 km/h; 68·5 mph) 39
Min sinking speed:
A at 40·5 knots (75 km/h; 46·5 mph)
0·62 m (2·03 ft)/sec
B at 43·5 knots (80 km/h; 50 mph)
0·68 m (2·23 ft)/sec
C at 49 knots (90 km/h; 56 mph)
0·73 m (2·40 ft)/sec
Stalling speed:
A 34 knots (62 km/h; 39 mph)
B 37·5 knots (68·5 km/h; 43 mph)
C 40·5 knots (74·2 km/h; 46·5 mph)
Max speed (rough and smooth air):
A 119 knots (220 km/h; 137 mph)
Max aero-tow speed:
A 91 knots (170 km/h; 105 mph)

Glaser-Dirks DG-100, production sailplane developed from the Akaflieg Darmstadt D-38

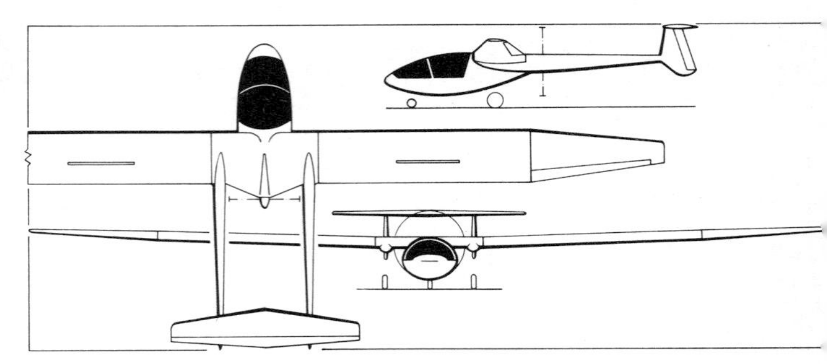

Kortenbach & Rauh Kora 1 two-seat powered sailplane *(Tony Mitchell)*

Grob Astir CS single-seat Standard Class sailplane

Second prototype of the Kora 1 powered sailplane, with non-retractable mair landing gear and streamlined wheel fairings *(Deutscher Aerokurier)*

Max winch-launching speed:
A 65 knots (120 km/h; 74·5 mph)

GROB G-104 SPEED ASTIR

The G-104 is a single-seat Racing Class version of the Astir CS, to which it is generally similar except in having a new wing of Wortmann section, with trailing-edge flaps, and water ballast capacity increased to 120 kg (264 lb).

Dimensions, external: As G-102 Astir CS except:
Wing aspect ratio 19
Length overall 6·47 m (21 ft 2¾ in)

Area:
Wings, gross 11·90 m² (128·1 sq ft)

Weights and Loadings:
Weight empty, equipped 270 kg (595 lb)
Max T-O weight:
without water ballast 380 kg (838 lb)
with water ballast 500 kg (1,102 lb)
Max wing loading:
without water ballast 30·0 kg/m² (6·14 lb/sq ft)
with water ballast 40·3 kg/m² (8·25 lb/sq ft)

Performance (estimated, at 500 kg; 1,102 lb max T-(weight):
Best glide ratio at 55 knots (102 km/h; 63·5 mph) 4:
Min sinking speed at 40 knots (74 km/h; 46 mph) 0·59 m (1·94 ft)/se
Max speed (rough and smooth air) 134 knots (250 km/h; 155 mph
Max aero-tow speed 91 knots (170 km/h; 105 mph
Max winch-launching speed 65 knots (120 km/h; 74·5 mph

KORTENBACH & RAUH
KORTENBACH & RAUH

Address:
5650 Solingen 15, Postfach 150121, Weyerstrasse 277
Telephone: (02122) 312031
Telex: 8514848 kora d

KORTENBACH & RAUH KORA 1

The Kora 1 is a two-seat twin-boom powered sailplane, the first prototype of which flew for the first time on 13 September 1973. It was designed by Herren Schultes, Seidel and Putz. The second prototype, which made its first flight on 9 April 1976, differs in having a non-retractable main landing gear.

About a dozen Koras have been ordered.

Type: Two-seat powered training sailplane.

Wings: Cantilever high-wing monoplane, of all-wood construction. Wortmann wing sections: FX-66-S-196 at root, FX-66-S-161 at tip. Schempp-Hirth airbrakes on upper surfaces.

Landing Gear: Forward-retracting nosewheel. Main units retract rearward into tailbooms on first prototype; on second prototype they are non-retractable, are cantilevered from the fuselage nacelle, and have streamlined wheel fairings.

Power Plant: One 48·5 kW (65 hp) Sportavia Limbach SL 1700 EA engine (SL 1700 ECI on second prototype), driving a Hoffmann two-blade variable-pitch feathering pusher propeller.

Accommodation: Side-by-side seats for two persons under fully-transparent canopy, which opens sideways to starboard. Space for parachutes.

Dimensions, external:
Wing span 18·00 m (59 ft 0¾ in)
Wing aspect ratio 16·65
Length overall:
first prototype 7·00 m (22 ft 11½ in)
second prototype 7·40 m (24 ft 3¼ in)
Height overall 1·85 m (6 ft 0¾ in)
Fuselage: Max width 1·20 m (3 ft 11¼ in)
Wheel track 2·00 m (6 ft 6¾ in)
Propeller diameter 1·60 m (5 ft 3 in)

Areas:
Wings, gross 19·44 m² (209·25 sq ft)
Vertical tail surfaces (total) 1·00 m² (10·76 sq ft)
Horizontal tail surfaces (total) 2·60 m² (27·99 sq ft)

Weights and Loadings:
Weight empty:
first prototype 600 kg (1,323 lb)
second prototype 510 kg (1,124 lb
Max T-O weight:
first prototype 713 kg (1,572 lb
second prototype 750 kg (1,653 lb
Max wing loading:
first prototype 36·68 kg/m² (7·51 lb/sq ft
second prototype 38·58 kg/m² (7·90 lb/sq ft
Max power loading:
first prototype 14·70 kg/kW (24·18 lb/hp
second prototype 15·46 kg/kW (25·43 lb/hp

Performance (first prototype, powered):
Max level speed at S/L 110 knots (205 km/h; 127 mph
Cruising speed (65% power) at S/L 94·5 knots (175 km/h; 109 mph
Stalling speed 35·5 knots (65 km/h; 40·5 mph
Max rate of climb at S/L 180 m (590 ft)/mi

Performance (unpowered):
Best glide ratio at 54 knots (100 km/h; 62 mph):
first prototype 31·
second prototype 3
Min sinking speed at 51 knots (95 km/h; 59 mph):
first prototype 0·76 m (2·49 ft)/se
second prototype 0·85 m (2·79 ft)/se

KUFFNER
WERNER KUFFNER

Address:
D-54 Koblenz 32, Aachener Strasse
Telephone: (0261) 24823

KUFFNER WK-1

The WK-1 is an all-metal powered sailplane, the design of which was started by Herr Kuffner at the end of 1973. It has a wing span of 18·80 m (61 ft 8¼ in), and accommodates two persons in a staggered side-by-side seating arrangement. The intended power plant is a 50 kW (67 hp) 900 cc BMW motor cycle engine, which will drive three-blade foldable pusher propeller mounted in th fuselage aft of the wings.

Owing to lack of time, funds and facilities, constructio of the airframe had not begun up to the spring of 1976 testing of the propeller drive mechanism was planned t take place in the Summer of 1976.

LCF
LUFTSPORTCLUB DER ZEPPELINSTADT FRIEDRICHSHAFEN

Address:
799 Friedrichshafen/Bodensee, Postfach 644

LCF II

The LCF II is a single-seat club sailplane suitable for training, competition and aerobatic flying. It was designed in 1971 and was built in approx 4,000 hr by five engineer members of the Luftsportclub Friedrichshafen, making its first flight on 22 March 1975. It won first prize at the 1975 meeting of the OUV (German EAA), and is reported to be going into production.

Type: Single-seat Club Class sailplane.

Wings: Cantilever shoulder-wing monoplane. Wortmann S-01, S-01/2 and FX-60-126 wing sections. Thickness/chord ratio 17%. Single-spar wooden structure, with Conticell foam ribs and plywood covering. Plain ailerons. Schempp-Hirth airbrakes on upper surfaces.

Fuselage: Oval-section steel tube structure, covered with glassfibre-reinforced plastics at front and fabric at rear.

Tail Unit: Cantilever structure of plywood, filled with Conticell foam.

Landing Gear: Semi-recessed non-retractable monowheel and tailwheel.

Accommodation: Single seat under one-piece flush-fitting transparent canopy.

Dimensions, external:
Wing span 13·00 m (42 ft 7¾ in)
Wing aspect ratio 16·9
Length overall 6·35 m (20 ft 10 in)
Height over tail 0·90 m (2 ft 11½ in)

Area:
Wings, gross 10·00 m² (107·6 sq f

Weights and Loading:
Weight empty 190 kg (419 lb
Max T-O weight 300 kg (661 lb
Max wing loading 30·0 kg/m² (6·14 lb/sq ft

Performance:
Best glide ratio at 46 knots (85 km/h; 53 mph) 3
Min sinking speed at 37 knots (68 km/h; 42·5 mph) 0·70 m (2·30 ft)/se
Max speed (rough or smooth air) 135 knots (250 km/h; 155 mph
Max aero-tow speed 91 knots (170 km/h; 105 mph
Max winch-launching speed 64·5 knots (120 km/h; 74·5 mph
Stalling speed 34 knots (62 km/h; 39 mph

Scheibe Bergfalke-IV fitted with a Lloyd piston engine by the Detmold Aero Club Flying Training School

Luftsportclub Friedichshafen LCF II single-seat Club Class sailplane

LVD
SCHÜLERFLUGGEMEINSCHAFT DER LUFTSPORTVEREIN DETMOLD EV

ADDRESS:
c/o Heinrich Hennigs, 4930 Detmold 19, Talstrasse 7
Telephone: 05231/58267

LVD BF IV-BIMO (BERGFALKE-IV CONVERSION)

As illustrated in the accompanying photograph, the Flying Training School of the Detmold Aero Club has converted a Scheibe Bergfalke-IV into a motor glider.

The engine is a Lloyd LS-400, and drives a pair of small two-blade pusher propellers which rotate within cutouts in each wing near the trailing-edge. A 12V battery is fitted for engine starting.

The standard Bergfalke-IV is described under the Scheibe heading in this section. The following data apply to the BF IV-BIMO:

WEIGHTS:
Weight empty, equipped 420 kg (926 lb)
Max T-O weight 600 kg (1,322 lb)

PERFORMANCE (at max T-O weight, unpowered):
Best glide ratio 30
Min sinking speed 0·80 m (2·62 ft)/sec
Stalling speed 38 knots (70 km/h; 43·5 mph)
Max speed (smooth air) 97·5 knots (180 km/h; 112 mph)
Max speed (rough air) 81 knots (150 km/h; 93 mph)
Max aero-tow speed 70 knots (130 km/h; 81 mph)
Max winch-launching speed 59·5 knots (110 km/h; 68·5 mph)

MILITKY
FRED MILITKY

ADDRESS:
7312 Kirchheim/Teck, Paradiesstrasse 27
Telephone: (07021) 45251

Fred Militky, an engineer with the Graupner model-building company in Kirchheim/Teck, applied his experience with electrically-powered radio-controlled models to the evolution of an electrically-powered propulsion system for a full-size manned aeroplane. The MB-E1, a prototype aircraft incorporating this system, was described and illustrated in the 1975-76 *Jane's*.

ROLLADEN-SCHNEIDER
ROLLADEN-SCHNEIDER OHG (Abteilung Segelflugzeugbau)

ADDRESS:
6073 Egelsbach, Mühlstrasse 10 (Schliessfach 1130)
Telephone: Langen (06103) 4126

OFFICERS:
Walter Schneider
Dipl-Ing Wolf Lemke

ROLLADEN-SCHNEIDER LS1-f

This sailplane, first flown in the 1972 World Championships at Vrsac, Yugoslavia, is an improved version of the LS1-c described in the 1974-75 *Jane's*. The LS1-f took 8th, 10th and 14th places at the 1976 World Championships at Räyskälä, Finland.

Compared with the LS1-c, the fuselage is of refined aerodynamic form, the rudder has been redesigned (although retaining the same surface area), a fixed tailplane with elevator replaces the all-moving tailplane of the LS1-c, and a one-piece hinged cockpit canopy is fitted. Other improvements include rubber shock-absorption for the monowheel; modifications to the tow release, cockpit interior and instrumentation; and location 25 mm (1 in) further forward of the rear point of the CG range. There is provision for 90 kg (198 lb) of water ballast.

A total of 170 LS1-fs had been built by January 1976.

TYPE: Single-seat high-performance Standard Class sailplane.

WINGS: Cantilever mid-wing monoplane. Wortmann wing section. Thickness/chord ratio 19%. Dihedral 4°. Incidence 3°. Sweepback 1° at quarter-chord. Glassfibre/foam sandwich structure. Airbrakes on upper surfaces.

FUSELAGE: Semi-monocoque structure, of glassfibre/foam sandwich construction.

TAIL UNIT: Cantilever T tailplane, with elevator. Construction similar to that of wings.

LANDING GEAR: Retractable Tost monowheel, size 300 × 100, tyre pressure 2·94 bars (43 lb/sq in), with rubber shock-absorption and drum brake. Tailskid.

ACCOMMODATION: Single semi-reclining seat under one-piece hinged transparent canopy.

DIMENSIONS, EXTERNAL:
Wing span 15·00 m (49 ft 2½ in)
Wing aspect ratio 23·1
Wing chord (mean) 0·65 m (2 ft 1½ in)
Length overall 6·70 m (21 ft 11¾ in)
Height over tail 1·20 m (3 ft 11¼ in)
Tailplane span 2·20 m (7 ft 2¾ in)

AREAS:
Wings, gross 9·75 m² (104·9 sq ft)
Ailerons (total) 0·70 m² (7·53 sq ft)
Airbrakes (total) 0·35 m² (3·77 sq ft)
Rudder 0·34 m² (3·66 sq ft)
Tailplane 0·84 m² (9·04 sq ft)

WEIGHTS AND LOADING:
Weight empty 230 kg (507 lb)
Max T-O weight 390 kg (859 lb)
Max wing loading 40·0 kg/m² (8·2 lb/sq ft)

PERFORMANCE:
Best glide ratio at 48·5 knots (90 km/h; 56 mph) 38
Min sinking speed at 38 knots (70 km/h; 43·5 mph) 0·65 m (2·13 ft)/sec
Stalling speed 35·5 knots (65 km/h; 40·5 mph)
Max speed (rough or smooth air) 135 knots (250 km/h; 155 mph)
Max aero-tow speed 91·5 knots (170 km/h; 105·5 mph)
Max winch-launching speed 70·5 knots (130 km/h; 81 mph)
g limit 6·1

ROLLADEN-SCHNEIDER LS3

The LS3 is a new Open Class sailplane, developed from the LS1-f. Design and construction began in 1975, and the prototype flew for the first time on 4 February 1976.

TYPE: Single-seat Open Class sailplane.

WINGS: Cantilever mid-wing monoplane, with Wortmann wing section. Dihedral 4°. Incidence 3°. Glassfibre/foam sandwich structure. Ailerons and flaps, of similar construction, over entire trailing-edge. Airbrakes on upper surfaces.

FUSELAGE: As LS1-f, with detail improvements.

TAIL UNIT: As LS1-f.

LANDING GEAR: As LS1-f.

ACCOMMODATION: As LS1-f.

DIMENSIONS, EXTERNAL:
Wing span 15·00 m (49 ft 2½ in)
Wing chord at root 0·94 m (3 ft 1 in)
Wing chord at tip 0·42 m (1 ft 4½ in)
Wing aspect ratio 22
Length overall 6·80 m (22 ft 3¾ in)
Height over tail 1·20 m (3 ft 11¼ in)
Tailplane span 2·20 m (7 ft 2¾ in)

AREAS:
Wings, gross 10·20 m² (109·8 sq ft)
Ailerons (total) 0·98 m² (10·55 sq ft)
Fin 0·881 m² (9·48 sq ft)
Rudder 0·329 m² (3·54 sq ft)
Tailplane 0·979 m² (10·54 sq ft)
Elevator 0·247 m² (2·66 sq ft)

WEIGHTS AND LOADING:
Weight empty, equipped 250 kg (551 lb)
Max T-O weight (incl approx 100 kg; 220 lb water ballast) 410 kg (904 lb)
Max wing loading 40·0 kg/m² (8·2 lb/sq ft)

PERFORMANCE (at max T-O weight):
Best glide ratio at 59·5 knots (110 km/h; 68·5 mph) 40
Min sinking speed at 38 knots (70 km/h; 43·5 mph) 0·60 m (1·97 ft)/sec
Stalling speed 35·5 knots (65 km/h; 40·5 mph)
Max speed (rough or smooth air) 135 knots (250 km/h; 155 mph)
Max aero-tow speed 91·5 knots (170 km/h; 105·5 mph)
Max winch-launching speed 70·5 knots (130 km/h; 81 mph)

SCHEIBE
SCHEIBE FLUGZEUGBAU GmbH

HEAD OFFICE AND WORKS:
D-8060 Dachau, August-Pfaltz-Strasse 23, Postfach 1829, near Munich
Telephone: Dachau 4047, 5794 and 6813

MANAGERS:
Dipl-Ing Egon Scheibe
Ing Christian Gad

Scheibe Flugzeugbau GmbH was founded at the end of 1951 by Dipl-Ing Scheibe, who had previously built a prototype two-seat general-purpose glider known as the Mü-13E Bergfalke in Austria. This aircraft flew for the first time on 5 August 1951 and was the first type produced in quantity by the newly-formed company.

Subsequently, Scheibe has built many new types of sailplane, and since developing the SF-24 Motorspatz in 1957 has also become the major producer of powered sailplanes in Germany. Currently in production are the SF-25C and C-S Falke, SF-25E Super-Falke, SF-28 Tandem-Falke and SF-32 powered sailplanes, in addition to the Bergfalke-IV and SF-30 Club-Spatz sailplanes. A motorised version of the Bergfalke-IV is under development.

Scheibe has built a total of nearly 2,000 aircraft of various types, in addition to many kits for home construction by amateurs. Gliders of Scheibe design are being built under licence by gliding clubs as well as by foreign companies, including Loravia in France and Slingsby in the UK, which see.

SCHEIBE BERGFALKE-IV

The Bergfalke-IV is a developed version of the Bergfalke-III, with a new wing which provides improved performance. Construction of the prototype began in early 1969 and first flight was accomplished a few months later. Fifty-eight had been built by the beginning of 1976 and production is continuing.

LS1-f single-seat Standard Class sailplane, developed from the LS1-c

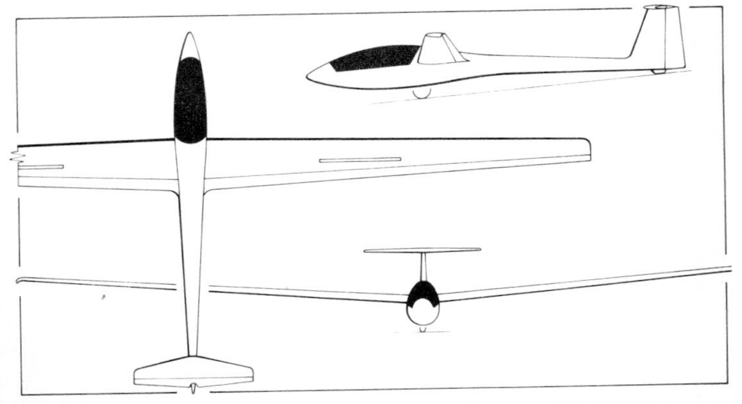
Rolladen-Schneider L53 single-seat Open Class sailplane *(Michael A. Badrocke)*

Scheibe Bergfalke-IV two-seat training and competition sailplane

Prototype Scheibe Bergfalke-IV M with Hirth engine

TYPE: Two-seat training and competition sailplane.
WINGS: Cantilever mid-wing monoplane. Wing section Wortmann SO 2 at root, SO 2/1 at tip. Thickness/chord ratio 19·4% at root, 15·8% at tip. Dihedral 3°. All-wood structure. Single laminated beechwood box spar. Plywood skin, fabric-covered. Wooden ailerons. Schempp-Hirth wooden airbrakes.
FUSELAGE: Welded steel tube structure. Nose section covered with a moulded glassfibre shell, remainder fabric-covered.
TAIL UNIT: Cantilever wooden structure. Tailplane mounted on top of fuselage, forward of fin. Flettner trim tab on starboard elevator.
LANDING GEAR: Non-retractable monowheel and tail-wheel.
ACCOMMODATION: Two laminated glassfibre seats in tandem beneath a blown Plexiglas canopy.
DIMENSIONS, EXTERNAL:
Wing span 17·20 m (56 ft 5¼ in)
Wing chord at root 1·38 m (4 ft 6¼ in)
Wing chord at tip 0·54 m (1 ft 9¼ in)
Wing aspect ratio 17·4
Length overall 8·00 m (26 ft 3 in)
Height over tail 1·50 m (4 ft 11 in)
AREA:
Wings, gross 17·00 m² (183 sq ft)
WEIGHTS AND LOADING:
Weight empty, equipped 300 kg (661 lb)
Max T-O weight 500 kg (1,102 lb)
Normal wing loading 28·0 kg/m² (5·7 lb/sq ft)
PERFORMANCE:
Best glide ratio at 46 knots (85 km/h; 53 mph) 34
Min sinking speed at 41 knots (75 km/h; 47 mph) 0·68 m (2·23 ft)/sec
Stalling speed 36 knots (65 km/h; 41 mph)
Max speed (smooth air) 108 knots (200 km/h; 124 mph)
Max speed (rough air) 92 knots (170 km/h; 106 mph)
Max aero-tow speed 76 knots (140 km/h; 87 mph)
Max winch-launching speed 59 knots (110 km/h; 68 mph)
g limit +8 (ultimate)

SCHEIBE BERGFALKE-IV M

In early 1976, Scheibe was test-flying a powered version of the Bergfalke-IV, fitted with a 38·8 kW (52 hp) Hirth O-28 flat-twin engine mounted on a retractable pylon aft of the cabin. The engine can be raised or lowered electrically in less than 20 sec.
DIMENSIONS, EXTERNAL:
Wing span 17·25 m (56 ft 7 in)
Wing aspect ratio 16·98
Length overall 7·90 m (25 ft 11 in)
Height over tail 1·50 m (4 ft 11 in)
AREA:
Wings, gross 17·52 m² (188·6 sq ft)
WEIGHTS AND LOADINGS:
Weight empty, equipped 400 kg (882 lb)
Max T-O weight 600 kg (1,322 lb)
Max wing loading 34·20 kg/m² (7·00 lb/sq ft)
Max power loading 15·46 kg/kW (25·4 lb/hp)
PERFORMANCE (at max T-O weight, powered):
Max level speed at S/L 86·5 knots (160 km/h; 99·5 mph)
Cruising speed at S/L 75·5 knots (140 km/h; 87 mph)
Max rate of climb at S/L 90 m (295 ft)/min
Service ceiling 2,000 m (6,550 ft)
Range 134 nm (250 km; 155 miles)
PERFORMANCE (at max T-O weight, unpowered):
Best glide ratio at 49 knots (91 km/h; 56·5 mph) 31
Min sinking speed at 43·5 knots (80 km/h; 50 mph) 0·80 m (2·62 ft)/sec
Stalling speed 38 knots (70 km/h; 43·5 mph)

SCHEIBE SF-25C and C-S FALKE '76 (FALCON)

The SF-25C is an improved version of the SF-25B Falke, to which it is structurally similar. The primary difference is in the use of a more powerful engine, giving an enhanced performance.

By January 1976 a total of 150 SF-25C Falkes had been built by Scheibe; a further 50 were built under licence by Sportavia in Germany. Type certification was granted in September 1972.

With a Hoffmann feathering propeller, adjustable engine cowl flap and slightly modified fuselage, the aircraft is known as the **SF-25C-S** and has a best glide ratio of 25 at 43·5 knots (80 km/h; 50 mph). Fifteen of this version had been built by January 1976.

Scheibe completed in March 1975 trials to reduce the overall noise level of the SF-25C by fitting an additional exhaust outlet and a slower-turning propeller. With this installation the nominal noise level is reduced to less than 60 dB.

Current models, which are known as **Falke '76**, introduce a number of design improvements. These include a domed canopy, giving a view to the rear; enlarged fin, smaller rudder and increased rudder sweep; a closable radiator valve, to improve engine cooling and reduce drag; new pre-heated carburettor, to improve engine starting and prevent icing; twin-pipe engine exhaust; redesigned instrument panel; increased baggage space, by relocating the fuselage fuel tank; optional 55 litre fuel tank; front fuselage coating of laminated glassfibre; and an optional twin-wheel main landing gear with streamlined wheel fairings.

In addition, Scheibe is developing for the Falke '76 a clipped wing of 9·50 m (31 ft 2 in) span, for easier hangar storage.

TYPE: Two-seat powered sailplane, particularly suitable for basic and advanced training.
WINGS: Two-piece cantilever forward-swept low wing of wooden construction, with airbrakes. Aerodynamically balanced ailerons. Design developed from the Motor-falke wing.
FUSELAGE: Fabric-covered welded steel tube structure. Forward fuselage coated with laminated glassfibre. Optional tow hitch for winch launching.
TAIL UNIT: Conventional wooden construction.
LANDING GEAR: Non-retractable unsprung monowheel with brake and aerodynamic fairing; steerable tail-wheel; spring outrigger stabilising wheel under each wing. Rubber-sprung monowheel optional. Alternative twin-wheel main gear available optionally, with stream-lined wheel fairings.
POWER PLANT: One 44·7 kW (60 hp) Limbach SL 1700 EA modified Volkswagen engine, driving a two-blade propeller. Starting on the ground and in the air is by means of an electric starter. Fuel in single fuselage tank, capacity 45 litres (9·9 Imp gallons) standard, 55 litres (12·1 Imp gallons) optional.
ACCOMMODATION: Two seats side by side in enclosed cabin. Dual controls standard.
DIMENSIONS, EXTERNAL:
Wing span 15·25 m (50 ft 0¼ in)
Wing aspect ratio 13·8
Length overall (tail up) 7·55 m (24 ft 9¼ in)
AREA:
Wings, gross 18·20 m² (195·9 sq ft)
WEIGHTS AND LOADINGS:
Weight empty approx 375 kg (826 lb)
Max T-O weight 610 kg (1,345 lb)
Max wing loading 33·5 kg/m² (6·86 lb/sq ft)
Max power loading 13·65 kg/kW (22·42 lb/hp)
PERFORMANCE (at max T-O weight, powered):
Max level speed 97 knots (180 km/h; 112 mph)
Cruising speed 86·5 knots (160 km/h; 99·5 mph)
Econ cruising speed 70·5 knots (130 km/h; 81 mph)
Stalling speed 35·5 knots (65 km/h; 40·5 mph)
Max rate of climb at S/L 138 m (453 ft)/min
T-O run approx 180 m (590 ft)
Econ endurance 4-5 hr
Range:
45 litres fuel 323 nm (600 km; 372 miles)
55 litres fuel 404 nm (750 km; 466 miles)
PERFORMANCE (at max T-O weight, unpowered):
Best glide ratio approx 23
Min sinking speed approx 1·0 m (3·3 ft)/sec

SCHEIBE SF-25E SUPER-FALKE

Developed from the SF-25C-S, the Super-Falke has a number of improvements and an increased wing span. Optionally, the outer wing panels can be made foldable to facilitate transportation and storage.

A more powerful (48·5 kW; 65 hp) Limbach SL 1700 engine is fitted, with a 12V battery and alternator for electrical engine starting. Other improvements include a rubber-sprung monowheel; a cabin heater is fitted as standard. Production aircraft have a tailwheel, and upper-surface Schempp-Hirth airbrakes.

The Super-Falke was flown for the first time in the Summer of 1974; 16 had been delivered by January 1976.

The description of the SF-25C/C-S applies also to the SF-25E, with the following principal exceptions:
DIMENSIONS, EXTERNAL:
Wing span 18·00 m (59 ft 0¾ in)
Wing aspect ratio 17·8

Scheibe SF-25C/C-S Falke '76 showing bulged canopy and modified vertical tail surfaces

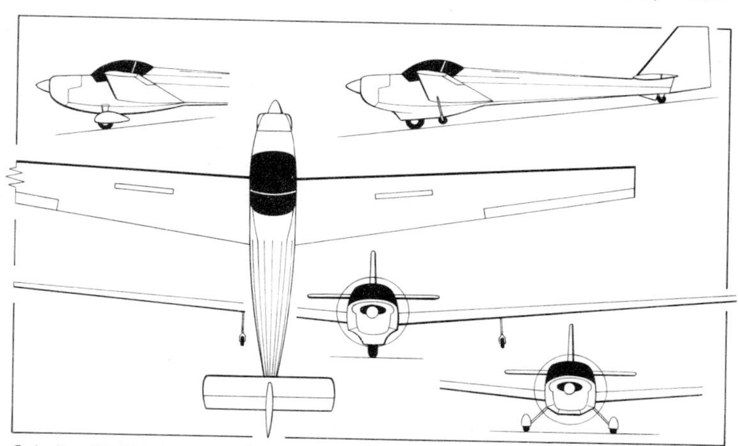

Scheibe SF-25C/C-S Falke'76 showing alternative monowheel and twin-wheel main landing gear *(Michael A. Badrocke)*

Scheibe SF-25E Super-Falke motor glider, with increased wing span

British-registered Scheibe SF-28 Tandem-Falke powered sailplane

Scheibe SF-30 Club-Spalz single-seat club sailplane

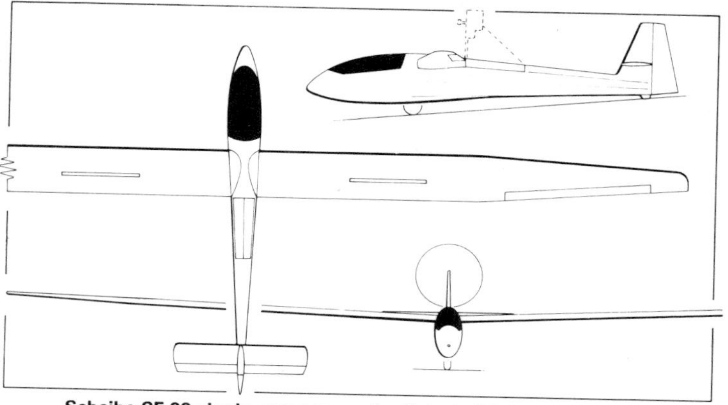

Scheibe SF-32 single-seat powered sailplane *(Michael A. Badrocke)*

Width, wings folded 10·00 m (32 ft 9¾ in)
Length overall 7·615 m (24 ft 11¾ in)

Weights and Loadings:
Weight empty approx 410 kg (904 lb)
Max T-O weight 630 kg (1,389 lb)
Max wing loading 35·0 kg/m² (7·17 lb/sq ft)
Max power loading 12·99 kg/kW (21·37 lb/hp)

Performance (at max T-O weight, powered):
Max cruising speed 86·5 knots (160 km/h; 99·5 mph)
Stalling speed 37 knots (68 km/h; 42·5 mph)
T-O run 150-200 m (490-655 ft)
Econ endurance approx 4 hr

Performance (at max T-O weight, unpowered):
Best glide ratio at 46 knots (85 km/h; 53 mph) 29
Min sinking speed at 41 knots (75 km/h; 47 mph) 0·85 m (2·79 ft)/sec

SCHEIBE SF-28 TANDEM-FALKE

The Tandem-Falke, as its name implies, is a further development of the Bergfalke and Falke series of sailplanes in which the two seats are arranged in tandem. Design began in 1970, and the prototype (D-KAFJ) flew for the first time in May 1971, powered by a 33·5 kW (45 hp) Stamo MS 1500 engine. Details apply to the production version, of which 80 had been built by January 1976.

Type: Tandem two-seat powered sailplane.

Wings: Cantilever low-wing monoplane. Wing section Gö 533. Single-spar wooden wings, with trailing-edge taper on outer panels. Wooden ailerons. No flaps. Spoiler on upper surface of each wing.

Fuselage: Fabric-covered steel tube structure.

Tail Unit: Conventional cantilever wooden structure. Trim tab in elevator.

Landing Gear: Non-retractable monowheel, with internal brake, and steerable tailwheel. Nylon leg with outrigger stabilising wheel under each wing. Main-wheel tyre size 8·00-4.

Power Plant: One 48·5 kW (65 hp) Limbach SL 1700 EAI engine, driving a Hoffmann two-blade feathering (optionally, fixed-pitch) propeller. Fuel capacity 34 litres (7·4 Imp gallons).

Accommodation: Two seats in tandem under one-piece blown Perspex canopy. Can be flown solo from front seat, with space for 90 kg (198 lb) of baggage on rear seat. Standard basic instrumentation, 12V electric starter and alternator.

Dimensions, external:
Wing span 16·30 m (53 ft 5¼ in)
Wing aspect ratio 14·5
Length overall 8·15 m (26 ft 9 in)
Height over tail 1·55 m (5 ft 1 in)

Area:
Wings, gross 18·35 m² (197·5 sq ft)

Weights and Loadings:
Weight empty, equipped 400 kg (881 lb)
Max T-O weight 590 kg (1,300 lb)
Max wing loading 32·18 kg/m² (6·59 lb/sq ft)
Max power loading 12·16 kg/kW (21·67 lb/hp)

Performance (at max T-O weight, powered):
Max level speed at S/L 97 knots (180 km/h; 112 mph)
Max cruising speed 86 knots (160 km/h; 99 mph)
Circling speed 37·5-43·5 knots (69-80·5 km/h; 43-50 mph)
Stalling speed 33·5 knots (62 km/h; 39 mph)
Max rate of climb at S/L 126 m (415 ft)/min
Service ceiling 5,000 m (16,400 ft)
T-O run 180 m (590 ft)
T-O to 15 m (50 ft) 300 m (985 ft)
Landing from 15 m (50 ft) 200 m (655 ft)
Landing run 100 m (328 ft)
Range 269 nm (500 km; 310 miles)
Endurance 4 hr

Performance (at max T-O weight, unpowered):
Best glide ratio at 46 knots (85 km/h; 53 mph) 26-27
Min sinking speed at 38 knots (70 km/h; 43·5 mph) 0·90 m (2·95 ft)/sec
Max speed (rough or smooth air) 102·5 knots (190 km/h; 118 mph)

SCHEIBE SF-30 CLUB-SPATZ

The SF-30 Club-Spatz is a single-seat Club Class sailplane, developed from the SF-27A. Design was started in 1973, and the prototype made its first flight on 20 May 1974. Six SF-30s had been built by the Spring of 1976.

Type: Single-seat Club Class sailplane.

Wings: Cantilever shoulder-wing monoplane. Wortmann wing sections, with thickness/chord ratio of 18% at root, varying through 16% to 12% at tip. Single wooden spar, plywood covering and glassfibre leading-edge. All-wood ailerons, with spring trim. Schempp-Hirth airbrakes on upper surfaces.

Fuselage: Welded steel tube structure. Nose and rear fuselage skin of glassfibre.

Tail Unit: Cantilever all-wood structure. Fixed-incidence tailplane. Elevators have spring trim.

Landing Gear: Non-retractable unsprung monowheel, size 4·00-4, with Tost brake, and 200 × 50 mm tailwheel.

ACCOMMODATION: Single glassfibre seat under hinged one-piece Plexiglas canopy.

DIMENSIONS, EXTERNAL:

Wing span	15·00 m (49 ft 2½ in)
Wing aspect ratio	24
Length overall	6·10 m (20 ft 0 in)

AREA:

Wings, gross	9·30 m² (100·1 sq ft)

WEIGHTS AND LOADING:

Weight empty	185 kg (408 lb)
Max T-O weight	295 kg (650 lb)
Normal wing loading	30 kg/m² (6·15 lb/sq ft)

PERFORMANCE:

Best glide ratio at 49 knots (90 km/h; 56 mph) 36
Min sinking speed at 40·5 knots (75 km/h; 46·5 mph) 0·59 m (1·94 ft)/sec
Stalling speed 35 knots (65 km/h; 40·5 mph)
Max speed (smooth air) 113 knots (210 km/h; 130 mph)
Max aero-tow and winch-launching speed 75·5 knots (140 km/h; 87 mph)
g limits +8·0; −4·0

SCHEIBE SF-32

In early 1976 a prototype of this new single-seat powered sailplane was under construction. Based on experience gained with the SF-27M (1975-76 *Jane's*), of which approx 30 were built, it has a retractable power plant and utilises the basic wing of the Neukom Elfe 17 (see Swiss section). The fuselage, apart from a few modifications, is that of the SF-27M.

TYPE: Single-seat powered sailplane.

WINGS: Cantilever shoulder-wing monoplane. Wing section Wortmann FX-61-163 at root, FX-60-126 at tip. Built in two parts, with aluminium alloy main spar and a 6 mm (0·24 in) outer shell of glassfibre and plywood/foam sandwich. Schempp-Hirth airbrakes on upper surfaces. Plain ailerons.

FUSELAGE: Welded steel tube structure. Nose section back to wing trailing-edge covered with moulded glassfibre shell. Rear section fabric-covered. Moulded glassfibre fairing over wing/fuselage junction.

TAIL UNIT: Cantilever wooden structure. Fixed surfaces plywood-covered, control surfaces fabric-covered.

LANDING GEAR: Retractable monowheel, size 380 × 150 mm, and non-retractable steerable tailwheel.

POWER PLANT: One 30 kW (40 hp) Rotax 642 flat-twin two-stroke engine, driving a two-blade pusher propeller, mounted on a retractable pylon aft of the wing trailing-edge. Electrical raising and lowering of pylon; electrical engine starting, with dual ignition. Fuel tank in fuselage, capacity 20 litres (4·4 Imp gallons).

ACCOMMODATION AND EQUIPMENT: Single seat under one-piece flush-fitting transparent canopy. Battery for engine starting and extension/retraction of engine pylon.

DIMENSIONS, EXTERNAL:

Wing span	17·00 m (55 ft 9¼ in)
Wing aspect ratio	21·73
Length overall	7·00 m (23 ft 0 in)
Height over tail	1·25 m (4 ft 1¼ in)

AREA:

Wings, gross	13·30 m² (143·2 sq ft)

WEIGHTS AND LOADINGS:

Weight empty	340 kg (749 lb)
Max T-O weight	450 kg (992 lb)
Max wing loading	33·8 kg/m² (6·92 lb/sq ft)
Max power loading	15·00 kg/kW (24·80 lb/hp)

PERFORMANCE (estimated, at max T-O weight, powered):

Max level speed at S/L 86·5 knots (160 km/h; 99·5 mph)
Cruising speed at S/L 75·7 knots (140 km/h; 87 mph)
Stalling speed 37 knots (68 km/h; 42·5 mph)
Max rate of climb at S/L 120 m (395 ft)/sec
Max cruising height 3,000 m (9,850 ft)
T-O run 200 m (655 ft)
Range 161 nm (300 km; 186 miles)
Endurance 2 hr

PERFORMANCE (estimated, at max T-O weight, unpowered):

Best glide ratio at 48·5 knots (90 km/h; 56 mph) 37
Min sinking speed at 43·5 knots (80 km/h; 50 mph) 0·65 m (2·13 ft)/sec

---

SCHEMPP-HIRTH
SCHEMPP-HIRTH KG

HEAD OFFICE:
7312 Kirchheim-Teck, Krebenstrasse 25, Postfach 143
Telephone: (07021) 2441 and 6097
Telex: 7267817 hate

DIRECTOR:
Dipl-Ing Klaus Holighaus

Schempp-Hirth specialises in the production of high-performance Open Class and Standard Class sailplanes. In 1970 Dipl-Ing Klaus Holighaus became a 50% shareholder of Schempp-Hirth KG, later becoming its director following the retirement of Herr Martin Schempp.

Production by Schempp-Hirth of the original Cirrus ended in late 1971, after 120 had been built; but manufacture has continued since early 1972 by VTC in Yugoslavia, under which heading the description of the Cirrus can now be found.

Descriptions of the company's current products follow. The company was producing 12 sailplanes per month in early 1976.

SCHEMPP-HIRTH STANDARD CIRRUS

Designed by Dipl-Ing Klaus Holighaus, the Standard Class version of the Schempp-Hirth Cirrus entered production during the Summer of 1969, following the first flight of the prototype in March 1969.

The Standard Cirrus was winner of the Standard Class at the International Soaring Competition at Hahnweide in 1969, and winner of the South African National Standard Class in 1970.

Standard Cirrus sailplanes took second and fourth places in the Standard Class of the 1974 World Championships at Waikerie, Australia; and nine of the first 25 places in the 1976 World Championships at Räyskälä, Finland. A glider of this type, piloted by Leonard R. McMaster (USA) set an out and return goal flight distance record of 701·387 nm (1,298·969 km; 807·142 miles), from Piper Memorial Airport, Lock Haven, to Mendota, Virginia and back, on 17 March 1976.

By 1 February 1976 a total of 630 Standard Cirrus had been built, including 200 under licence by Grob-Flugzeugbau of Mindelheim, Bavaria (which see), and production was continuing.

TYPE: Single-seat high-performance Standard Class sailplane.

WINGS: Cantilever mid-wing monoplane. Wortmann section. Thickness/chord ratio 19·6% at root, 17% at tip. Dihedral 3°. Incidence 3°. Sweepback 1°18′ at leading-edge. Wings and ailerons are glassfibre/foam sandwich structures. Schempp-Hirth glassfibre airbrakes on wing upper surface.

FUSELAGE: Glassfibre shell, 1·5 mm thick, stiffened with bonded-in foam rings.

TAIL UNIT: T-tail of glassfibre/foam sandwich construction. All-moving tailplane.

LANDING GEAR: Manually-retractable monowheel standard. Non-retractable faired monowheel optional. Tost wheel with drum brake and Continental 4·00-4 tyre, pressure 3·45 bars (50 lb/sq in).

ACCOMMODATION: Single semi-reclining seat under long flush Plexiglas canopy, hinged at starboard side. Adjustable rudder pedals.

DIMENSIONS, EXTERNAL:

Wing span	15·00 m (49 ft 2½ in)
Wing chord at root	0·93 m (3 ft 0½ in)
Wing chord at tip	0·36 m (1 ft 2¼ in)
Wing aspect ratio	22·5
Length overall	6·35 m (20 ft 9¾ in)
Height over tail	1·32 m (4 ft 4¾ in)
Tailplane span	2·40 m (7 ft 10½ in)

AREA:

Wings, gross	10·00 m² (107·6 sq ft)

WEIGHTS AND LOADINGS:

Weight empty	215 kg (474 lb)
Max T-O weight:	
with water ballast	390 kg (860 lb)
without water ballast	330 kg (728 lb)
Max wing loading:	
with water ballast	39 kg/m² (8·0 lb/sq ft)
without water ballast	33 kg/m² (6·8 lb/sq ft)

PERFORMANCE:

Best glide ratio at 48·5 knots (90 km/h; 56 mph) 38·5
Min sinking speed at 38·5 knots (71 km/h; 44 mph) 0·57 m (1·87 ft)/sec
Stalling speed 33·5 knots (62 km/h; 39 mph)
Max speed (rough or smooth air) 119 knots (220 km/h; 137 mph)
Max aero-tow speed 81 knots (150 km/h; 93 mph)
Max winch-launching speed 65 knots (120 km/h; 75 mph)
g limit 10

SCHEMPP-HIRTH NIMBUS II

Generally similar to the HS-3 Nimbus (1972-73 *Jane's*), from which it was developed, this single-seat high-performance sailplane differs principally by having reduced span and a wing built in four pieces to limit weight and dimensions for rigging, storage and trailer transport.

Design of the Nimbus II was initiated by Dipl-Ing Klaus Holighaus in January 1970 and construction of the prototype began in April of the same year. The first flight was made in April 1971; a Nimbus II took first place in the Open Class at the 1972 and 1974 World Championships at Vrsac, Yugoslavia, and Waikerie, Australia; and in the 1976 World Championships at Räyskälä, Finland, Nimbus IIs took no fewer than 14 of the first 25 places. On 5 January 1975, a Nimbus II flown by Georg Eckle of West Germany set a speed record of 66·293 knots (122·775 km/h; 76·289 mph) over a 750 km course. A similar record over 500 km was set in a Nimbus II on 31 January 1975 by Malcolm Jinks of Australia, at 75·772 knots (140·33 km/h; 87·197 mph).

By 1 February 1976 a total of 110 Nimbus IIs had been built, with production continuing.

TYPE: Single-seat high-performance sailplane.

WINGS: Cantilever mid-wing monoplane. Wortmann wing section. Thickness/chord ratio 17% at root, 15% at tip. Dihedral 2°. Incidence 1°. Sweepback at leading-edge 1° on inner panels, 2° on outer panels. Wings are of glassfibre/foam sandwich construction and built in four sections, with tongue-and-fork assembly. Water ballast valves connect automatically. Tip sections incorporate a quick-connect aileron fitting and locking pin. Ailerons and interconnected trailing-edge flaps are glassfibre shells. Schempp-Hirth airbrake of glassfibre construction on each upper surface. Provision for up to 160 kg (353 lb) of water ballast.

FUSELAGE: Central tubular steel framework. Glassfibre shell 1·5 to 2·0 mm thick, stiffened with bonded-in foam bulkheads.

TAIL UNIT: Cantilever structure of glassfibre/foam sandwich. All-moving T-tailplane.

LANDING GEAR: Manually-retractable monowheel type. Tost wheel has Continental tyre size 5·00-5, pressure 3·45 bars (50 lb/sq in). Tost drum brake. Shock-absorption of monowheel is provided by annular rubber springs. Tailskid designed to fail in the event of a ground loop, to relieve stress on the rear fuselage. Ribbon drogue 'chute in bottom of rudder for use in steep approaches or emergencies.

ACCOMMODATION: Single semi-reclining seat under long flush hinged canopy. Rudder pedals adjustable.

DIMENSIONS, EXTERNAL:

Wing span	20·30 m (66 ft 7¼ in)
Wing chord at root	0·96 m (3 ft 1¾ in)
Wing chord at tip	0·35 m (1 ft 1¾ in)
Wing aspect ratio	28·62
Length overall	7·33 m (24 ft 0½ in)
Height over tail	1·45 m (4 ft 9 in)
Tailplane span	2·40 m (7 ft 10½ in)

AREA:

Wings, gross	14·40 m² (155 sq ft)

WEIGHTS AND LOADINGS:

Weight empty, equipped	350 kg (771 lb)
Max T-O weight:	
with water ballast	580 kg (1,278 lb)
without water ballast	470 kg (1,036 lb)
Max wing loading:	
with water ballast	40·0 kg/m² (8·19 lb/sq ft)
without water ballast	32·6 kg/m² (6·68 lb/sq ft)

PERFORMANCE (at 403 kg; 888 lb AUW):

Best glide ratio at 56·5 knots (105 km/h; 65 mph) 49
Min sinking speed at 48·5 knots (90 km/h; 56 mph) 0·53 m (1·74 ft)/sec
Stalling speed 38 knots (70 km/h; 43·5 mph)
Max speed (rough or smooth air) 146 knots (270 km/h; 168 mph)
Max aero-tow speed 86 knots (160 km/h; 99 mph)
Max winch-launching speed 65 knots (120 km/h; 75 mph)
g limit +10·5

SCHEMPP-HIRTH NIMBUS II M

Design of a powered version of the Nimbus II, designated Nimbus II M and fitted with a retractable 37·3 kW (50 hp) Hirth O-28 engine, was started in early 1973. Construction of a prototype began towards the end of that year, and this aircraft made a successful first flight in June 1974.

Production of the Nimbus II M was due to begin in 1975.

WEIGHTS AND LOADINGS:

Weight empty	440 kg (970 lb)
Max T-O weight	580 kg (1,278 lb)
Max wing loading	40·30 kg/m² (8·25 lb/sq ft)
Max power loading	11·6 kg/hp (25·57 lb/hp)

PERFORMANCE (at max T-O weight, powered):

Max level speed at S/L	81 knots (150 km/h; 93 mph)
Cruising speed at S/L	70·5 knots (130 km/h; 81 mph)
Stalling speed	38 knots (70 km/h; 43·5 mph)
Max rate of climb at S/L	120 m (395 ft)/min
Service ceiling	5,000 m (16,400 ft)
T-O run	400 m (1,310 ft)
Landing run	150 m (490 ft)
Range	269 nm (500 km; 310 miles)

PERFORMANCE (at max T-O weight, unpowered):

Best glide ratio	48
Min sinking speed	0·54 m (1·77 ft)/sec

Above: Schempp-Hirth Janus two-seat all-glassfibre training sailplane

Left: Schempp-Hirth Standard Cirrus single-seat high-performance sailplane

Below: Schempp-Hirth Nimbus II high-performance Open Class sailplane

SCHEMPP-HIRTH JANUS

The Janus is a two-seat sailplane of all-glassfibre construction. Original design work, begun by Dipl-Ing Holighaus in 1969, was continued from early 1972 onward, and the prototype made its first flight in the Spring of 1974.

On 15 August 1974, flown by Dipl-Ing Klaus Holighaus and U. Plarre, the Janus set up in Switzerland a new world two-seater speed record over a 100 km closed circuit of 77·121 knots (142·919 km/h; 88·806 mph). Subsequently, it set a new German national record of 55·5 knots (103 km/h; 64 mph) over a 300 km circuit.

Production began in January 1975 with the second, improved aircraft.

TYPE: Two-seat high-performance training sailplane.

WINGS: Cantilever mid-wing monoplane, Wortmann wing sections: FX-67-K-170 at root, FX-67-K-15 at tip. Dihedral 4°. Incidence 2° 36′. Sweepforward 2° on leading-edge. Glassfibre/foam sandwich construction, with glassfibre monocoque ailerons, trailing-edge flaps and Schempp-Hirth upper-surface airbrakes.

FUSELAGE: Glassfibre monocoque structure, 1·5 mm to 2 mm thick, with bonded-in foam bulkheads.

TAIL UNIT: Cantilever structure of glassfibre/foam sandwich. All-moving one-piece T-tailplane.

LANDING GEAR: Non-retractable monowheel and nosewheel. Continental tyres: size 380 × 150 × 150 mm, pressure 2·69 bars (39 lb/sq in) on main wheel; size 260 × 85 × 123 mm, pressure 0·79 bars (11·5 lb/sq in) on nosewheel. Tost drum brake on main wheel. Bumper under rear fuselage. Tail drag parachute.

ACCOMMODATION: Two seats in tandem under hinged one-piece flush transparent canopy. Rudder pedals at front seat are adjustable.

DIMENSIONS, EXTERNAL:

Wing span	18·20 m (59 ft 8½ in)
Wing chord at root	1·18 m (3 ft 10½ in)
Wing chord at tip	0·48 m (1 ft 6¾ in)
Wing mean aerodynamic chord	0·912 m (3 ft 0 in)
Wing aspect ratio	19·97
Length overall	8·62 m (28 ft 3¼ in)
Height over tail	1·45 m (4 ft 9 in)
Tailplane span	2·70 m (8 ft 10¼ in)

AREAS:

Wings, gross	16·58 m² (178·5 sq ft)
Ailerons (total)	1·03 m² (11·09 sq ft)
Trailing-edge flaps (total)	1·82 m² (19·59 sq ft)
Fin	0·75 m² (8·07 sq ft)
Rudder	0·55 m² (5·92 sq ft)
Tailplane	1·24 m² (13·35 sq ft)

WEIGHTS AND LOADING:

Weight empty, equipped	380 kg (838 lb)
Max T-O weight	620 kg (1,366 lb)
Max wing loading	37·4 kg/m² (7·66 lb/sq ft)

PERFORMANCE (at max T-O weight except where indicated):

*Best glide ratio at 59·5 knots (110 km/h; 68·5 mph) 39·5

*Min sinking speed at 48·5 knots (90 km/h; 56 mph) 0·70 m (2·30 ft)/sec

*Stalling speed 38 knots (70 km/h; 43·5 mph)

Max speed (rough or smooth air) 118 knots (220 km/h; 136 mph)

Max aero-tow speed 91 knots (170 km/h; 105 mph)

Max winch-launching speed 64·5 knots (120 km/h; 74·5 mph)

g limit +9

**at wing loading of 37·0 kg/m² (7·58 lb/sq ft)*

SCHLEICHER
ALEXANDER SCHLEICHER SEGELFLUG-ZEUGBAU

HEAD OFFICE AND WORKS:
D-6416 Poppenhausen/Wasserkuppe
Telephone: (06658) 225

This company is one of the oldest manufacturers of sailplanes in the world. Its founder, Alexander Schleicher, was himself winner of the contest for training sailplanes at the 1927 meeting at the famous Wasserkuppe gliding centre. In the same year he built at Poppenhausen a small factory for manufacturing gliders and sailplanes, two of his best-known pre-war products being the Rhönbussard and Rhönadler, designed by Hans Jacobs.

During the second World War, the factory was engaged on the repair of Baby IIb sailplanes. For a time afterwards it became a furniture factory; but it began producing sailplanes once more in 1951.

Descriptions of the sailplanes and powered sailplanes in current production follow. In addition, the K 8 C, ASK 13, ASK 16 and ASK 18 are available in kit form for amateur constructors. Schleicher also manufactures and markets spare parts, constructional materials and dust- and weather-proof covers for sailplanes.

SCHLEICHER K 8

Designed by Rudolf Kaiser, the K 8 B (see 1974-75 *Jane's*) was developed from the Ka 6, but featured simplified construction throughout. As a result, it was suitable for amateur construction.

The prototype flew in November 1957 and well over 1,100 K 8s have since been built.

The version current in 1976 is the improved K 8 C, to which the following description applies:

TYPE: Single-seat training and sporting sailplane.

WINGS: Cantilever high-wing monoplane. Wing section Gö 533 from root to aileron, Gö 532 from aileron to tip. Dihedral 3°. Sweepforward 1° 18′ at quarter-chord. Single-spar structure with plywood D-type leading-edge. Glassfibre-reinforced plastics wingtips. Rear portion plywood and fabric-covered. Plywood-covered top-hinged wooden ailerons, actuated by push/pull rods. Schempp-Hirth metal airbrakes.

FUSELAGE: Welded steel tube structure with spruce formers, fabric-covered. Nosecone made of glassfibre.

TAIL UNIT: Cantilever type, with single-spar plywood-covered fin and low-set tailplane. Rudder and elevators are plywood torsion tube structures, fabric-covered at rear. Flettner tab in port elevator. Actuation of elevators by push/pull rods, of rudder by cables from adjustable rudder pedals.

LANDING GEAR: Non-retractable unsprung Continental monowheel, size 5·00-5, with internally-expanding drum brake. Tyre presssure 1·93-2·48 bars (28-36 lb/sq in). Foam tailskid.

ACCOMMODATION: Single glassfibre bucket seat under sideways-hinged blown Plexiglas canopy. Baggage space aft of seat.

EQUIPMENT: To customer's specification, including optional mounting of radio antenna internally in fin. Aero-tow release in nose. Kombi release at CG.

DIMENSIONS, EXTERNAL:

Wing span	15·00 m (49 ft 2½ in)
Wing chord at root	1·30 m (4 ft 3 in)
Wing aspect ratio	15·9
Length overall	7·05 m (23 ft 1½ in)
Height over tail	1·57 m (5 ft 1¾ in)
Tailplane span	2·80 m (9 ft 2¼ in)

AREAS:

Wings, gross	14·15 m² (152·3 sq ft)
Ailerons (total)	1·00 m² (10·76 sq ft)
Airbrakes (total)	0·34 m² (3·66 sq ft)
Fin	0·62 m² (6·67 sq ft)
Rudder	0·75 m² (8·07 sq ft)
Tailplane	0·96 m² (10·33 sq ft)
Elevators	0·94 m² (10·12 sq ft)

WEIGHTS AND LOADING:

Weight empty, equipped	190 kg (418 lb)
Max T-O weight	310 kg (683 lb)
Wing loading at 275 kg (606 lb) AUW	19·5 kg/m² (4·0 lb/sq ft)

PERFORMANCE (at max T-O weight):

Best glide ratio at 39·5 knots (73 km/h; 45·5 mph) 27

Min sinking speed at 32·5 knots (60 km/h; 37·5 mph) 0·65 m (2·13 ft)/sec

Stalling speed 29·5 knots (54 km/h; 34 mph)

Max speed (smooth air) 108 knots (200 km/h; 124 mph)

Max speed (rough air) and max aero-tow speed 70 knots (130 km/h; 81 mph)

Max winch-launching speed 54 knots (100 km/h; 62 mph)

SCHLEICHER ASK 13

This tandem-seat sailplane was developed from the K 7, which is in worldwide use by gliding clubs.

The prototype first flew in July 1966 and by January 1976 a total of 550 ASK 13s had been built. Production was continuing in 1976.

Compared with the K 7, the ASK 13 introduced many improvements, including a large full-blown canopy for all-round view, higher performance, improved comfort and a sprung landing wheel for softer touchdowns.

TYPE: Two-seat training and high-performance sailplane.

WINGS: Cantilever mid-wing monoplane. Wing section developed from Göttingen 535 and 549. Thick-

Schleicher K 8 single-seat training and sporting sailplane *(Lorna Minton)*

Schleicher ASK 13 tandem two-seat training and high-performance sailplane

Schleicher ASW 15B single-seat glassfibre sailplane

Schleicher ASK 16 two-seat powered sailplane Limbach SL 1700 EBI engine

Schleicher ASW 17 Super Orchidee single-seat Open Class sailplane

Schleicher ASK 18 single-seat sailplane, available for amateur construction

ness/chord ratio 16% at root, 12% at tip. Sweepforward at quarter-chord 6°. Dihedral 5°. Incidence 0°. Single-spar wooden structure, with plywood D-type leading-edge torsion box and fabric covering. Wooden ailerons, with plywood covering, actuated by push/pull rods. Schempp-Hirth metal airbrakes above and below each wing.

FUSELAGE: Welded steel tube structure with spruce formers and fabric main covering. Nose made of glassfibre. Turtledeck aft of canopy is plywood shell.

TAIL UNIT: Cantilever wooden structure. Single-spar fixed surfaces plywood-covered. Rear portion of rudder and elevators fabric-covered. Flettner tab in starboard elevator. Actuation of elevators by push/pull rods, of rudder by cables from adjustable rudder pedals.

LANDING GEAR: Non-retractable sprung monowheel, size 5·00-5 (350 × 125) with Tost drum brake, mounted aft of CG. Skid in front of wheel; steel tailskid.

ACCOMMODATION: Two seats in tandem under one-piece blown Mecaplex canopy, hinged to starboard. Glassfibre seat panels. Adjustable rudder pedals.

EQUIPMENT: Aero-tow release in nose. Kombi release at CG. Normal instrumentation; provision for radio and oxygen.

DIMENSIONS, EXTERNAL:
- Wing span: 16·00 m (52 ft 6 in)
- Wing chord (mean): 1·09 m (3 ft 7 in)
- Wing aspect ratio: 14·6
- Length overall: 8·18 m (26 ft 9½ in)
- Height over tail: 1·60 m (5 ft 3 in)
- Tailplane span: 3·00 m (9 ft 10 in)

AREAS:
- Wings, gross: 17·50 m² (188 sq ft)
- Ailerons (total): 1·48 m² (15·93 sq ft)
- Airbrakes (total): 0·43 m² (4·62 sq ft)
- Fin: 0·60 m² (6·45 sq ft)
- Rudder: 0·83 m² (8·93 sq ft)
- Tailplane: 2·25 m² (24·22 sq ft)
- Elevators: 1·05 m² (11·30 sq ft)

WEIGHTS AND LOADINGS:
- Weight empty: 290 kg (640 lb)
- Max T-O weight: 480 kg (1,060 lb)
- Wing loading:
  - single-seater: 21·7 kg/m² (4·45 lb/sq ft)
  - two-seater: 26·8 kg/m² (5·5 lb/sq ft)

PERFORMANCE (single-seater at 380 kg; 837 lb AUW, two-seater at 470 kg; 1,036 lb AUW):
- Best glide ratio
  - single-seater at 43·5 knots (80 km/h; 50 mph): 28
  - two-seater at 48·6 knots (90 km/h; 56 mph): 28
- Min sinking speed:
  - single-seater at 35 knots (64 km/h; 40 mph): 0·70 m (2·30 ft)/sec
  - two-seater at 38 knots (70 km/h; 43·5 mph): 0·80 m (2·62 ft)/sec
- Stalling speed:
  - single-seater: 30·5 knots (56 km/h; 35 mph)
  - two-seater: 33 knots (61 km/h; 38 mph)
- Max speed (smooth air): 108 knots (200 km/h; 124 mph)
- Max speed (rough air) and max aero-tow speed: 76 knots (140 km/h; 87 mph)
- Max winch-launching speed: 54 knots (100 km/h; 62 mph)
- *g* limit: 4*g* at safety factor of 2

SCHLEICHER ASW 15 B

Designed by Gerhard Waibel to meet Standard Class requirements, and built by Schleicher, the ASW 15 was first flown in April 1968. A French type certificate was awarded on 24 November 1971.

A total of 447 ASW 15s (all versions) had been built by January 1976. The original ASW 15 (184 built) was described in the 1972-73 *Jane's*.

The following description applies to the ASW 15 B current production version, which was introduced from the 185th aircraft. This has a strengthened main spar, an enlarged rudder, a bigger wheel and drum brake, an 80 litre (17·6 Imp gallon) water ballast installation and other improvements.

TYPE: Single-seat Standard Class sailplane.

WINGS: Cantilever shoulder-wing monoplane. Wing section Wortmann FX-61-163 at root/mean and FX-60-126 at tip. Dihedral 2°. Incidence 0°. No sweepback. Structure: glassfibre roving spar, glassfibre/foam sandwich torsion box. Ailerons of glassfibre/foam sandwich. Schempp-Hirth metal airbrakes above and below each wing in separate specially sealed compartments with spring-loaded cover plates. Reinforced structure, with provision for water ballast tank, capacity approx 45 kg (100 lb), in each leading-edge.

FUSELAGE: Glassfibre/honeycomb sandwich construction. Reinforced keel.

TAIL UNIT: All-moving horizontal surfaces of similar construction to wings. Fin construction same as fuselage, and rudder same as ailerons. Rudder control cables run within plastics conduits.

LANDING GEAR: Retractable landing gear with central monowheel, operated manually through push/pull rods. Dunlop/Continental size 5·00-5 (350 × 125) wheel and tyre, with Tost and own-production internal shoe brake. Tyre pressure 2·48 bars (36 lb/sq in).

ACCOMMODATION: Single semi-reclining seat, with adjustable backrest and headrest and integral parachute pan. Space for 11 kg (24 lb) of baggage. Adjustable ventilation. One-piece transparent canopy.

EQUIPMENT: Standard instrumentation. Provision for oxygen and radio transceiver. VHF antenna in fin.

DIMENSIONS, EXTERNAL:
- Wing span: 15·00 m (49 ft 2½ in)
- Wing chord at root: 0·92 m (3 ft 0 in)
- Wing chord at tip: 0·40 m (1 ft 3¾ in)
- Wing aspect ratio: 20·45
- Length overall: 6·48 m (21 ft 3 in)
- Height over tail: 1·56 m (5 ft 1½ in)
- Tailplane span: 2·62 m (8 ft 7¼ in)

DIMENSIONS, INTERNAL:
- Cockpit: Seating height: 0·80 m (2 ft 7½ in)
  - Width: 0·58 m (1 ft 10¾ in)
- Baggage volume: 0·04 m³ (1·5 cu ft)

AREAS:
- Wings, gross: 11·00 m² (118·40 sq ft)
- Ailerons (total): 0·85 m² (9·14 sq ft)
- Airbrakes (total): 0·40 m² (4·30 sq ft)
- Fin: 0·60 m² (6·46 sq ft)
- Tailplane: 1·15 m² (12·38 sq ft)

WEIGHTS AND LOADINGS:
- Weight empty, equipped: 230 kg (507 lb)
- Max T-O weight with 90 kg; 198 lb water ballast: 408 kg (899 lb)
- Max wing loading with 90 kg; 198 lb water ballast: 37·1 kg/m² (7·6 lb/sq ft)

PERFORMANCE (at wing loading of 28 kg/m^2; 5·73 lb/sq ft):
Best glide ratio at 48·5 knots (90 km/h; 56 mph) 38
Min sinking speed at 39·5 knots (73 km/h; 45·5 mph) 0·59 m (1·94 ft)/sec
Stalling speed 34·5 knots (63 km/h; 39·5 mph)
Max speed (rough or smooth air) 119 knots (220 km/h; 136·7 mph)
Max aero-tow speed 91 knots (170 km/h; 105 mph)
Max winch-launching speed 65 knots (120 km/h; 74·5 mph)
g limits:
normal +5·3; −2·65
ultimate (with water ballast) +8·4; −5·4
ultimate (without water ballast) +9·2; −6·2

ASW 15 M

This designation was given to an ASW 15 B fitted in 1976 with a 22 kW (30 hp) Wankel KM 27 300 cc rotating-piston engine by Ing Josef Vonderau of Fichtel & Sachs System-Technik, Schweinfurt. A 20 litre (4·4 Imp gallon) fuel tank is also installed, the empty equipped weight of this version being 288 kg (635 lb).

SCHLEICHER ASK 16

Design of the ASK 16 was started in 1969, construction of a prototype began in the following year, and this aircraft flew for the first time on 2 February 1971. The first production aircraft flew in 1972, and 38 had been built by January 1976.

TYPE: Two-seat powered sailplane.

WINGS: Cantilever low-wing monoplane. Wing section NACA 63618 at root, Joukowsky 12% at tip, with Wortmann modifications. Dihedral 5°. Incidence 2° 48′ at root. Sweepforward 1° at quarter-chord. Single-spar fabric-covered wooden structure, with plywood D-type leading-edge torsion box and glassfibre tips. Plain wooden ailerons and upper-surface spoilers, actuated by push/pull rods. No flaps or tabs. Outer panels detachable optionally.

FUSELAGE: Primary load-bearing structure of welded steel tube, with glassfibre, plywood and fabric covering.

TAIL UNIT: Cantilever structure. Plywood tailplane and fin, fabric-covered rudder and elevators. Combined trim and anti-balance tab in port elevator. Elevator actuated by push/pull rods, rudder by cables.

LANDING GEAR: Inward-retracting main wheels and non-retractable tailwheel. Rubber shock-absorbers. Wheels and tyres size 5·00-5 on main units, 210 × 65 on tail unit. Tyre pressure (main units) 2·45 bars (35·5 lb/sq in). Tost drum brakes.

POWER PLANT: One 53·7 kW (72 hp) Limbach SL 1700 EBI (modified Volkswagen) engine, driving a Hoffmann HO-V62 two-blade variable-pitch feathering propeller. 18Ah battery for electric starting.

ACCOMMODATION: Side-by-side seats for two persons under one-piece sideways-opening (to starboard) Plexiglas canopy. Seats designed for use of cushion- or back-type parachutes. Space for two suitcases aft of seats.

DIMENSIONS, EXTERNAL:
Wing span 16·00 m (52 ft 5¾ in)
Wing chord at root 1·80 m (5 ft 11 in)
Wing chord (mean) 1·19 m (3 ft 10¾ in)
Wing chord at tip 0·56 m (1 ft 10 in)
Wing aspect ratio 13·5
Width, outer wing panels removed 10·00 m (32 ft 9¾ in)
Length overall 7·32 m (24 ft 0¼ in)
Height overall 2·08 m (6 ft 9¾ in)
Tailplane span 3·20 m (10 ft 6 in)
Propeller diameter 1·60 m (5 ft 3 in)
AREAS:
Wings, gross 19·00 m^2 (204·5 sq ft)
Ailerons (total) 1·17 m^2 (12·59 sq ft)
Spoilers (total) 0·45 m^2 (4·84 sq ft)
Fin 0·65 m^2 (7·00 sq ft)
Rudder 0·705 m^2 (7·59 sq ft)
Tailplane 1·28 m^2 (13·78 sq ft)
Elevators, incl tab 1·09 m^2 (11·73 sq ft)
WEIGHTS AND LOADINGS:
Weight empty, equipped 460 kg (1,014 lb)
Max T-O weight 700 kg (1,543 lb)
Max wing loading 37 kg/m^2 (7·55 lb/sq ft)
Max power loading 13·04 kg/kW (25·8 lb/hp)
PERFORMANCE (at max T-O weight, powered):
Max level speed at S/L 97 knots (180 km/h; 112 mph)
Cruising speed 86 knots (160 km/h; 99 mph)
Stalling speed:
two-seat 37·5 knots (69 km/h; 43 mph)
single-seat 33·5 knots (62 km/h; 39 mph)
Max rate of climb at S/L 150 m (492 ft)/min
Service ceiling 5,000 m (16,400 ft)
T-O run 230 m (755 ft)
T-O to 15 m (50 ft) 335 m (1,099 ft)
Landing from 15 m (50 ft) 180 m (591 ft)
Landing run 120 m (394 ft)
Range 269 nm (500 km; 310 miles)
g limits +5·3; −2·65
PERFORMANCE (at max T-O weight, unpowered):
Best glide ratio 25
Min sinking speed at 40 knots (74 km/h; 46 mph) 1·00 m (3·3 ft)/sec
Max speed (smooth air) 108 knots (200 km/h; 124 mph)

SCHLEICHER ASW 17 SUPER ORCHIDEE

The ASW 17 is a single-seat Open Class sailplane, the prototype of which (D-1110) flew for the first time on 17 July 1971. ASW 17s took second and fifth places in the Open Class of the 1972 World Championships at Vrsac, Yugoslavia; third place in the 1974 World Championships at Waikerie, Australia; and six of the first 25 places (including first place) at the 1976 World Championships at Räyskälä, Finland. Forty-seven ASW 17s had been built by January 1976.

Four world records have been set in ASW 17s. On 16 April 1974, Werner Grosse of Germany set a straight-line distance to goal record of 665·118 nm (1,231·8 km; 765·405 miles). The same pilot set a distance record of 562·696 nm (1,042·114 km; 647·54 miles) over a triangular course, in Australia, on 6 February 1976; and a speed record of 47·60 knots (88·16 km/h; 54·78 mph) over a 1,000 km triangular course on 6 June 1975. Karl Striedieck (USA) set an out and return goal flight distance record of 701·387 nm (1,298·969 km; 807·142 miles) on 17 March 1976.

TYPE: Single-seat Open Class sailplane.

WINGS: Cantilever shoulder-wing monoplane. Wing section Wortmann FX-62-K-131 (modified). Structure: glassfibre roving spar, glassfibre/balsa sandwich torsion box. Full-span integrated camber-changing flaps/ailerons, of glassfibre/foam sandwich construction with mass balance. Schempp-Hirth aluminium airbrake above and below each wing, in separate specially sealed compartments with spring-loaded cover plates. Provision for 100 kg (220 lb) of water ballast.

FUSELAGE: Glassfibre/honeycomb sandwich construction.

TAIL UNIT: Cantilever type. Glassfibre/honeycomb sandwich fin, glassfibre/balsa sandwich tailplane. Elevators and rudder of glassfibre/foam sandwich construction, with mass balance.

LANDING GEAR: Manually-retractable rubber-sprung monowheel, size 5·00-5, and tail bumper. Internal shoe brake.

ACCOMMODATION AND EQUIPMENT: Single seat under one-piece detachable transparent canopy. Two towing hooks under fuselage. VHF antenna in fin.

DIMENSIONS, EXTERNAL:
Wing span 20·00 m (65 ft 7½ in)
Wing aspect ratio 27
Length overall 7·55 m (24 ft 9¼ in)
Height over tail 1·86 m (6 ft 1¼ in)
Tailplane span 2·90 m (9 ft 6 in)
DIMENSIONS, INTERNAL:
Cockpit: Max height above seat 0·80 m (2 ft 7½ in)
Max width 0·62 m (2 ft 0¼ in)
AREA:
Wings, gross 14·84 m^2 (159·7 sq ft)
WEIGHTS AND LOADINGS:
Weight empty, equipped 405 kg (893 lb)
Max T-O weight:
without water ballast 495 kg (1,091 lb)
with water ballast 570 kg (1,256 lb)
Max wing loading:
without water ballast 30 kg/m^2 (6·14 lb/sq ft)
with water ballast 35 kg/m^2 (7·17 lb/sq ft)
PERFORMANCE (at max T-O weight except where indicated):
Best glide ratio at 59·5 knots (110 km/h; 68·5 mph) 48·5
Min sinking speed at 40·5 knots (75 km/h; 46·5 mph) 0·50 m (1·64 ft)/sec
Max speed (rough or smooth air) 129 knots (240 km/h; 149 mph)
Stalling speed 37 knots (68 km/h; 42·5 mph)
Max aero-tow speed 91 knots (170 km/h; 105 mph)
Max winch-launching speed 65 knots (120 km/h; 74·5 mph)
g limit +8·56

SCHLEICHER ASK 18

The ASK 18, first flown in October 1974, is a Club Class single-seat sailplane, based on the design of the K 6 and K 8 B/C but with a 16 metre wing span.

A total of 21 had been built by January 1976.

TYPE: Single-seat Club Class sailplane.

WINGS: Cantilever high-wing monoplane. Wing section NACA 63618 at root, Joukowsky 12% at tip, with Wortmann modifications. Single-spar structure, with plywood D-type leading-edge torsion box and fabric-covered rear portion. Single-spar plywood-covered ailerons. Schempp-Hirth airbrakes above and below wings. Ailerons actuated by push/pull rods.

FUSELAGE: Welded steel tube structure with spruce formers, fabric-covered. Nose of glassfibre.

TAIL UNIT: Cantilever type, with single-spar plywood-covered tailplane and fin, fabric-covered elevators and rudder. Flettner tab in elevator. Elevators actuated by push/pull rods, rudder by cables from adjustable rudder pedals.

LANDING GEAR: Non-retractable monowheel, size 5·00-5, with internal drum brake. Foam tailskid.

ACCOMMODATION: Single glassfibre bucket seat under sideways-hinged blown Plexiglas canopy.

EQUIPMENT: Radio antenna in fin. Kombi release at CG.

DIMENSIONS, EXTERNAL:
Wing span 16·00 m (52 ft 5¾ in)
Wing aspect ratio 19·7
Length overall 7·00 m (22 ft 11¾ in)
Height over tail 1·68 m (5 ft 6 in)
AREA:
Wings, gross 12·99 m^2 (139·8 sq ft)
WEIGHTS AND LOADING:
Weight empty 215 kg (474 lb)
Max T-O weight 335 kg (738 lb)
Wing loading at 205 kg (452 lb) AUW 23·00 kg/m^2 (4·71 lb/sq ft)
PERFORMANCE (at max T-O weight):
Best glide ratio at 40·5 knots (75 km/h; 46·5 mph) 34
Min sinking speed at 35 knots (65 km/h; 40·5 mph) 0·60 m (1·97 ft)/sec
Stalling speed 32·5 knots (60 km/h; 37·5 mph)
Max speed (rough air) 108 knots (200 km/h; 124 mph)
Max manoeuvring speed 78 knots (145 km/h; 90 mph)
Max aero-tow speed 75·5 knots (140 km/h; 87 mph)
Max winch-launching speed 64·5 knots (120 km/h; 74·5 mph)

SCHLEICHER ASW 19

This Standard Class sailplane, which flew for the first time on 23 November 1975, is essentially an improved version of the ASW 15 B, from which it differs in having a T tail unit.

An ASW 19, flown by a New Zealand pilot, took 15th place in the 1976 World Championships at Räyskälä, Finland.

TYPE: Single-seat Standard Class sailplane.

WINGS: Cantilever mid-wing monoplane, generally similar to ASW 15 B.

FUSELAGE: Generally similar to ASW 15 B.

TAIL UNIT: Cantilever T tail. Tailplane section Wortmann FX-71-L.

LANDING GEAR: Retractable monowheel and tailskid.

ACCOMMODATION: Single seat under one-piece flush-fitting transparent canopy which is hinged at front and opens upward. Reinforced cockpit floor and canopy frame. Backrest and rudder pedals adjustable.

DIMENSIONS, EXTERNAL:
Wing span 15·00 m (49 ft 2½ in)
Wing aspect ratio 20·4
Length overall 6·82 m (22 ft 4½ in)
Height over tail 1·45 m (4 ft 9 in)
Tailplane span 2·50 m (8 ft 2½ in)
AREA:
Wings, gross 11·00 m^2 (118·40 sq ft)
WEIGHTS AND LOADINGS:
Weight empty 240 kg (529 lb)
Max water ballast 100 kg (220 lb)
Max T-O weight:
without water ballast 345 kg (760 lb)
with water ballast 410 kg (904 lb)
Max wing loading:
without water ballast 29·0 kg/m^2 (5·94 lb/sq ft)
with water ballast 37·2 kg/m^2 (7·62 lb/sq ft)
PERFORMANCE (at wing loading of 31 kg/m^2; 6·35 lb/sq ft):
Best glide ratio at 48·5 knots (90 km/h; 56 mph) 38
Min sinking speed at 40·5 knots (75 km/h; 46·5 mph) 0·65 m (2·13 ft)/sec
Stalling speed 36 knots (66 km/h; 41 mph)
Max speed (rough or smooth air) 129 knots (240 km/h; 149 mph)
Max aero-tow speed 94·5 knots (175 km/h; 109 mph)
Max winch-launching speed 64·5 knots (120 km/h; 74·5 mph)

SCHLEICHER ASW 20

The ASW 20 was designed by Gerhard Waibel, and bears a general resemblance to this designer's earlier ASW 12 and ASW 17. It is intended to take advantage of the March 1975 CIVV regulations for Open Class 15 metre sailplanes, and is fitted with trailing-edge flaps. Unlike the ASW 17, the ASW 20 will have an additional high-drag flap range incorporating a special mechanism to eliminate pitch and airspeed changes when changing flap position between 30° and 70°. A new device, of which details are not yet released, has been successfully tested and will automatically co-ordinate and optimise flap position to the prevailing airspeed. This eliminates the need for conventional underwing airbrakes, but large upper-surface spoilers are fitted.

The general appearance of the ASW 20 is shown in the accompanying three-view drawing.

DIMENSIONS, EXTERNAL:
Wing span 15·00 m (49 ft 2½ in)
Wing aspect ratio 21·43
Length overall 6·82 m (22 ft 4½ in)
Height over tail 1·45 m (4 ft 9 in)
AREA:
Wings, gross 10·50 m^2 (113·0 sq ft)
WEIGHTS AND LOADINGS:
Weight empty 240 kg (529 lb)
Max water ballast 120 kg (265 lb)
Max T-O weight:
without water ballast 355 kg (783 lb)
with water ballast 420 kg (926 lb)

Schleicher ASW 19 single-seat Standard Class sailplane

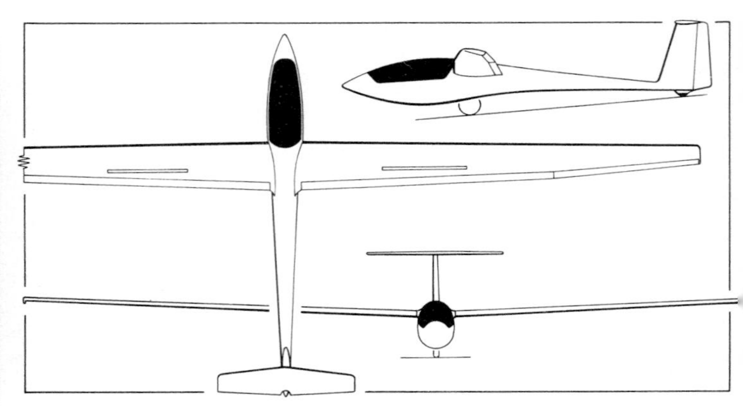
Schleicher ASW 20 Open Class sailplane *(Michael A. Badrocke)*

Siebert Sie 3 single-seat wooden-construction Standard Class sailplane

Sportavia RF5 tandem two-seat powered sailplane

Max wing loading:
- without water ballast — 30 kg/m² (6·14 lb/sq ft)
- with water ballast — 40 kg/m² (8·19 lb/sq ft)

PERFORMANCE (estimated, at wing loading of 33 kg/m²; 6·76 lb/sq ft):
- Best glide ratio at 48·5 knots (90 km/h; 56 mph) — 43
- Min sinking speed at 39·5 knots (73 km/h; 45·5 mph) — 0·60 m (1·97 ft)/sec
- Stalling speed — 37·5 knots (69 km/h; 43 mph)
- Max speed (smooth air) — 134 knots (250 km/h; 155 mph)
- Max speed (rough air) — 97 knots (180 km/h; 112 mph)
- Max aero-tow speed — 94·5 knots (175 km/h; 109 mph)
- Max winch-launching speed — 67 knots (125 km/h; 77·5 mph)

SIEBERT
PAUL SIEBERT SPORT- UND SEGELFLUG-ZEUGBAU

ADDRESS:
44 Münster-Mariendorf, Mariendorfer Strasse 38
Telephone: (0251) 32168

SIEBERT Sie 3

Paul Siebert has designed a single-seat wooden Standard Class sailplane designated Sie 3.

A total of 27 Sie 3s had been sold by January 1975, to customers in Belgium, Denmark, Germany, the Netherlands and Portugal.

Permission has been sought from the LBA to offer the Sie 3 for sale in a form suitable for amateur construction.

WINGS: Cantilever high-wing monoplane. Constant chord over approx two-thirds of span, with leading- and trailing-edge taper on outer panels. Incidence 1° 30′ at root, 0° from outer panels to tip. Schempp-Hirth aluminium airbrakes.

FUSELAGE: Oval-section monocoque structure.

TAIL UNIT: Cantilever structure, with sweptback fin and rudder and low-set non-swept all-moving tailplane.

LANDING GEAR: Monowheel, with brake, and tailskid.

ACCOMMODATION: Single seat. One-piece transparent canopy.

DIMENSIONS, EXTERNAL:
- Wing span — 15·00 m (49 ft 2½ in)
- Wing chord (centre-section, constant) — 0·90 m (2 ft 11½ in)
- Length overall — 6·70 m (22 ft 0 in)
- Height over tail — 1·20 m (3 ft 11 in)
- Tailplane span — 2·82 m (9 ft 3 in)

AREA:
- Wings, gross — 11·84 m² (127·44 sq ft)

WEIGHTS AND LOADING:
- Weight empty — 212 kg (467 lb)
- Max T-O weight — 340 kg (750 lb)
- Max wing loading — 28·7 kg/m² (5·88 lb/sq ft)

PERFORMANCE:
- Best glide ratio at 48·5 knots (90 km/h; 56 mph) — 34·3
- Min sinking speed at 42 knots (78 km/h; 48·5 mph) — 0·68 m (2·23 ft)/sec
- Max speed (smooth air) — 108 knots (200 km/h; 124 mph)
- Max speed (rough air) — 86 knots (160 km/h; 99 mph)

SPORTAVIA
SPORTAVIA-PÜTZER GmbH u Co KG

HEAD OFFICE AND WORKS:
D-5377 Dahlem-Schmidtheim, Flugplatz Dahlemer Binz
Telephone: (02447) 277/8
Telex: 08 33 602 spkg
SALES MANAGER: Alfons Pützer
PUBLIC RELATIONS MANAGER: Manfred Küppers

This company was formed in 1966 by Comte Antoine d'Assche, director of the French company Alpavia SA, and Herr Alfons Pützer, to take over from Alpavia manufacture of the Avion-Planeur series of powered sailplanes designed by M René Fournier.

In 1969, RFB (see Aircraft section), a subsidiary of VFW-Fokker GmbH, acquired a percentage holding in Sportavia.

Current Sportavia products are certificated as powered sailplanes with the exception of the RF6, described in the Aircraft section of this edition. Those in production in 1975 were the RF5 and RF5B.

In addition to its own designs, Sportavia has recently built under licence the SF-25B and SF-25C Falke powered sailplanes (see under the Scheibe heading in this section), completing 80 SF-25Bs and 50 SF-25Cs.

SPORTAVIA AVION-PLANEUR RF5

This tandem two-seat version of the Avion-Planeur differs from the earlier single-seaters mainly in having wings of increased span, with folding outer sections to facilitate hangarage, and a more powerful engine.

Construction of the prototype RF5 (D-KOLT) started in the Summer of 1967, and this aircraft flew for the first time in January 1968. Production began in late 1968. The RF5 was certificated by the LBA in the powered sailplane category in March 1969. A total of 125 RF5s had been delivered by the Spring of 1975.

TYPE: Two-seat powered sailplane.

WINGS: Cantilever low-wing monoplane. Wing section NACA 23015 at root, NACA 23012 at tip. Dihedral 3° 15′ at main spar centreline. Incidence 4° at root, 0° at tip. No sweepback. All-wood single-spar structure, with plywood and fabric covering. Fabric-covered wooden ailerons. Three-section metal-skinned spoilers on each wing at 50% chord, extended from slot in upper surface inboard of ailerons. Outer wing panels fold inward to facilitate hangarage. No flaps or tabs.

FUSELAGE: All-wood oval-section structure of bulkheads and stringers, plywood and fabric covered.

TAIL UNIT: Cantilever all-wood structure, plywood and fabric covered. Fixed-incidence tailplane. Flettner trim tab in port elevator. Entire unit detachable for transportation.

LANDING GEAR: Single Tost main wheel, with twin oleo-pneumatic shock-absorbers, manually retracted forward, with spring assistance, into front fuselage. Dunlop tyre, size 6·00-6, pressure 1·96 bars (28·4 lb/sq in), on main wheel, which has manually-operated brake. Single Rhombus 160 × 80 tailwheel, with Doetsch oleo-pneumatic shock-absorber, is steerable in conjunction with rudder movement. Outriggers beneath each wing, just inboard of fold line.

POWER PLANT: One 50·7 kW (68 hp) (max continuous rating 47 kW; 63 hp) Sportavia Limbach SL 1700 E Comet four-cylinder four-stroke engine, driving a Hoffmann HO-11-145-B80L two-blade fixed-pitch metal propeller. Fuel in two wing-root leading-edge metal tanks, total capacity 63 litres (13·8 Imp gallons). Refuelling point on top of port wing. Oil capacity 2·5 litres (0·55 Imp gallons).

ACCOMMODATION: Adjustable seats in tandem for pilot and one passenger under one-piece sideways-hinged Plexiglas canopy. Dual controls standard. The pupil sits in the forward seat during dual instruction, which is the pilot's seat when the aircraft is flown solo. Space for 10 kg (22 lb) of baggage aft of rear seat. Cabin heated and ventilated. Adjustable rudder pedals and canopy emergency release standard.

SYSTEMS: Electrical system includes alternator and 12V 25Ah battery.

ELECTRONICS AND EQUIPMENT: Optional equipment includes VHF radio nav/com equipment with intercom, radio compass, artificial horizon, VOR, ADF, oxygen, navigation and landing lights and rotating beacon.

DIMENSIONS, EXTERNAL:
- Wing span — 13·74 m (45 ft 1 in)
- Wing chord at root — 1·59 m (5 ft 2¾ in)
- Wing chord at tip — 0·60 m (1 ft 11½ in)
- Wing aspect ratio — 12·25
- Length overall — 7·80 m (25 ft 7¼ in)
- Width, wings folded — 8·74 m (28 ft 8 in)
- Height overall (tail down) — 1·96 m (6 ft 5 in)
- Tailplane span — 3·72 m (12 ft 2½ in)
- Distance between outriggers — 8·70 m (28 ft 6½ in)
- Propeller diameter — 1·47 m (4 ft 9¾ in)

AREAS:
- Wings, gross — 15·12 m² (162·8 sq ft)
- Ailerons (total) — 1·50 m² (16·15 sq ft)

Sportavia RF5B Sperber, developed from the RF5, with cut-down rear fuselage

Start + Flug H 101 Salto, based on the Glasflügel Standard Libelle

Start + Flug Hippie ultra-light glider

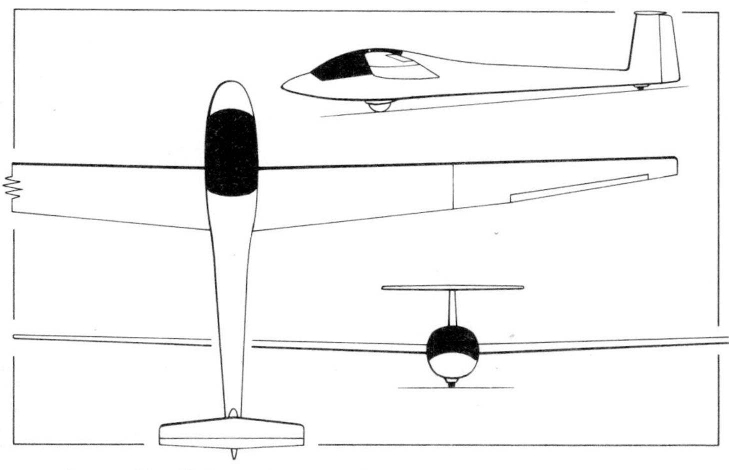

Start + Flug H 121 two-seat training sailplane *(Michael A. Badrocke)*

Spoilers (total) 0·75 m² (8·07 sq ft)
Fin 0·51 m² (5·49 sq ft)
Rudder 0·79 m² (8·50 sq ft)
Tailplane 1·62 m² (17·44 sq ft)
Elevators, incl tab 0·98 m² (10·55 sq ft)

WEIGHTS AND LOADINGS:
Weight empty, equipped 420 kg (926 lb)
Max T-O weight:
Aerobatic 605 kg (1,333 lb)
Utility 650 kg (1,433 lb)
Max wing loading 42·8 kg/m² (8·77 lb/sq ft)
Max power loading 12·82 kg/kW (21·05 lb/hp)

PERFORMANCE (at max T-O weight, powered):
Never-exceed speed 135 knots (250 km/h; 155 mph)
Max level speed at S/L 108 knots (200 km/h; 124 mph)
Max cruising speed at S/L 97 knots (180 km/h; 112 mph)
Econ cruising speed 65 knots (120 km/h; 75 mph)
Stalling speed 41 knots (75 km/h; 47 mph)
Max rate of climb at S/L 180 m (590 ft)/min
Service ceiling 5,000 m (16,400 ft)
T-O run 200 m (655 ft)
T-O to 15 m (50 ft) 480 m (1,575 ft)
Landing from 15 m (50 ft) 250 m (820 ft)
Landing run 180 m (590 ft)
Range with max fuel 410 nm (760 km; 472 miles)
Endurance with max fuel 4 hr

PERFORMANCE (at max T-O weight, unpowered):
Best glide ratio 22
Min sinking speed 1·40 m (4·59 ft)/sec

SPORTAVIA RF5B SPERBER

The RF5B Sperber is an improved, high-performance development of the RF5, differing chiefly in having increased wing span and area and reduced fuselage area. Construction began in early 1971, and the first flight was made in May of that year. The Sperber was certificated in the powered sailplane category by the LBA in March 1972, and by the Spring of 1975 a total of 70 RF5Bs had been delivered.

Modifications introduced in 1973 included an improved cabin heating system; engine muffler to decrease exterior and cabin noise levels; adjustable ventilation system; optional disc brakes; and a wider range of equipment options which include artificial horizon, electric compass and flight data computer.

TYPE: Two-seat powered sailplane.

WINGS: Similar to RF5, but with 3° 30′ dihedral at main spar centreline and increased span and area.

FUSELAGE: All-wood structure of pine bulkheads and stringers with birch plywood covering. Compared with RF5, the rear fuselage is cut down to reduce side area and improve rearward view from cockpit.

TAIL UNIT: Similar to RF5.

LANDING GEAR: As RF5.

POWER PLANT: As RF5, but all fuel is contained in a single fuselage tank of 39 litres (8·6 Imp gallons) capacity. Fully-feathering and HO-V62R variable-pitch propellers available optionally.

ACCOMMODATION: As in RF5, but cockpit canopy is of the bulged type giving an all-round field of view.

SYSTEMS: Electrical system includes alternator and 12V 20Ah battery.

DIMENSIONS, EXTERNAL:
Wing span 17·02 m (55 ft 10 in)
Wing chord at root 1·68 m (5 ft 6 in)
Wing chord at tip 0·54 m (1 ft 9¼ in)
Wing aspect ratio 15·25
Length overall 7·70 m (25 ft 3¼ in)
Width, wings folded 11·22 m (36 ft 9¾ in)
Height overall 1·96 m (6 ft 5 in)
Tailplane span 3·72 m (12 ft 2½ in)
Wheelbase 4·86 m (15 ft 11¼ in)
Propeller diameter 1·47 m (4 ft 9¾ in)
Propeller ground clearance 0·62 m (2 ft 0¼ in)

AREAS: As RF5, except:
Wings, gross 19·00 m² (204·5 sq ft)

WEIGHTS AND LOADINGS:
Weight empty, equipped 460 kg (1,014 lb)
Max T-O weight 680 kg (1,499 lb)
Max wing loading 35·8 kg/m² (7·33 lb/sq ft)
Max power loading 13·41 kg/kW (22·05 lb/hp)

PERFORMANCE (at max T-O weight, powered, variable-pitch propeller):
Never-exceed speed 121 knots (225 km/h; 139 mph)
Max level speed at S/L 102 knots (190 km/h; 118 mph)
Max cruising speed at S/L 97 knots (180 km/h; 112 mph)
Econ cruising speed 65 knots (120 km/h; 75 mph)
Stalling speed 37 knots (68 km/h; 42·5 mph)
Max rate of climb at S/L 180 m (590 ft)/min
Service ceiling 5,500 m (18,050 ft)
T-O run 187 m (615 ft)
T-O to 15 m (50 ft) 497 m (1,630 ft)
Landing from 15 m (50 ft) 303 m (995 ft)
Landing run 204 m (669 ft)
Range with max fuel 226 nm (420 km; 261 miles)

PERFORMANCE (at max T-O weight, unpowered):
Best glide ratio at 53 knots (98 km/h; 61 mph) 26
Min sinking speed at 40·5 knots (75 km/h; 46·5 mph) 0·89 m (2·92 ft)/sec

START + FLUG
START + FLUG GmbH

ADDRESS:
7968 Saulgau, Am Flugplatz, Postfach 126
Telephone: 0 75 81/71 65

This company is producing, as the H 101 Salto, a version of the Glasflügel Standard Libelle developed by Frau Ursula Hänle, the widow of Ing Eugen Hänle of Glasflügel.

Since 1974, it has also marketed an ultra-light glider of its own design, known as the Hippie.

START + FLUG H 101 SALTO

Design of the H 101 is based upon that of the Glasflügel Standard Libelle (see 1974-75 *Jane's*), from which it differs in having a V-type tail unit. It flew for the first time in 1971, and 50 had been delivered by the Spring of 1975. Initial LBA certification was granted on 28 April 1972; the Salto has been certificated by the LBA and FAA for Normal and Aerobatic category operation.

The description of the Glasflügel Standard Libelle applies generally to the H 101 Salto, except in the following respects:

WINGS: Ailerons and four flush-fitting airbrakes on trailing-edge.

TAIL UNIT: Cantilever V-tail (included angle 99°).

DIMENSIONS, EXTERNAL:

*Wing span	13·60 m (44 ft 7½ in)
Wing aspect ratio	21·6
Length overall	5·95 m (19 ft 6¼ in)
Height over tail	0·88 m (2 ft 10¾ in)
Tailplane span	2·14 m (7 ft 0¼ in)

**15·00 m (49 ft 2½ in) span wing available optionally*

AREA:

Wings, gross	8·58 m^2 (92·35 sq ft)

WEIGHTS AND LOADINGS:

Weight empty	180 kg (396 lb)
Max T-O weight:	
Aerobatic	280 kg (617 lb)
Normal	310 kg (683 lb)
Max wing loading:	
Aerobatic	32·6 kg/m^2 (6·68 lb/sq ft)
Normal	36·1 kg/m^2 (7·40 lb/sq ft)

PERFORMANCE (at Aerobatic max T-O weight):

Best glide ratio at 48·5 knots (90 km/h; 56 mph)	35
Min sinking speed at 40·5 knots (75 km/h; 46·5 mph)	0·60 m (1·97 ft)/sec
Never-exceed speed	161 knots (300 km/h; 186 mph)
Max level speed	135 knots (250 km/h; 155 mph)
Max aero-tow speed	81 knots (150 km/h; 93 mph)
Max winch-launching speed	70 knots (130 km/h; 81 mph)
Stalling speed	35·5 knots (65 km/h; 40·5 mph)
g limits (Aerobatic)	+7·0; −4·9

START + FLUG HIPPIE

The Hippie is a simple ultra-light single-seat glider weighing only 48 kg (106 lb) without its pilot. It was flown for the first time on 15 August 1974, and examples are now flying in five countries outside Germany.

Take-off can be by foot-launch, auto-tow, winch, or any other suitable form of assisted launch. After a foot-launch, the pilot draws up his legs into the hoop-like glassfibre seat fairing at the front of the aircraft. This fairing also serves as the landing skid.

The dihedral wings have a Wortmann aerofoil section. The airframe basic structure is of glassfibre, with carbon-fibre reinforcement; wings are plastics-covered, the tail surfaces fabric-covered. The airframe can be dismantled and packed into a carrying case 5·00 m (16 ft 4¾ in) long by 1·10 m (3 ft 7¼ in) by 0·60 m (1 ft 11¾ in).

A second version is now available, having honeycomb-construction wings, aluminium tube tail structure, enlarged tailplane and rudder, quick-connect fittings, and enlarged seating area for the pilot.

DIMENSIONS, EXTERNAL:

Wing span	10·00 m (32 ft 9¾ in)
Wing aspect ratio	11·1
Length overall	5·00 m (16 ft 4¾ in)
Height over tail	1·40 m (4 ft 7 in)
Tailplane span	2·40 m (7 ft 10½ in)

AREA:

Wings, gross	9·00 m^2 (96·9 sq ft)

WEIGHTS AND LOADING:

Weight empty	48 kg (106 lb)
Max T-O weight	133 kg (293 lb)
Max wing loading	14·8 kg/m^2 (3·03 lb/sq ft)

PERFORMANCE:

Glide ratio	approx 12
Max permitted speed	32 knots (60 km/h; 37 mph)
Stalling speed	16·5 knots (30 km/h; 19 mph)

START + FLUG H 121 SCHULMEISTER (SCHOOLMASTER)

A prototype of this two-seat training sailplane was under construction in early 1976, and was expected to fly in the following Summer. Construction is largely of glassfibre-reinforced plastics, and an Eppler 603 wing section is used. The side-by-side seats are staggered, that on the right being some 250 mm (10 in) further back than the left-hand seat.

The general appearance of the H 121 can be seen in the accompanying three-view drawing. No other details had been received at the time of closing for press.

DIMENSIONS, EXTERNAL:

Wing span	17·00 m (55 ft 9¼ in)
Length overall	7·66 m (25 ft 1½ in)
Tailplane span	2·75 m (9 ft 0¼ in)

WEIGHT AND LOADING:

Max T-O weight	500 kg (1,102 lb)
Max wing loading	31·7 kg/m^2 (6·49 lb/sq ft)

PERFORMANCE (estimated):

Best glide ratio	36
Min sinking speed	0·65 m (2·13 ft)/sec

STRAUBER
DIPL-ING MANFRED STRAUBER

ADDRESS:
6140 Bensheim-Auerbach, Wilhelm-Leuschner-Strasse 11
Telephone: 06251/73867

ASSOCIATES:
Alois Fries
Hartmut Frommhold
Horst Gaber

These four men have designed and built a high-performance Standard Class sailplane known as the Mistral.

MISTRAL

Design of the Mistral was initiated in January 1970. Construction began in May 1972, and the aircraft (D-4998) flew for the first time in July 1975. Series production is not intended.

TYPE: Single-seat Standard Class sailplane.

WINGS: Cantilever shoulder-wing monoplane. Wortmann wing sections: FX-66-S-196 at root, FX-66-S-161 at tip. Dihedral 0° 24′. Incidence 1°. Sweepback 0° 24′ at quarter-chord. Structure: main wings of glassfibre/balsa sandwich, ailerons of glassfibre/Rohacell sandwich. Schempp-Hirth aluminium airbrakes on upper surfaces.

FUSELAGE: Pod and boom monocoque, of glassfibre/8 mm balsa sandwich.

TAIL UNIT: Cantilever T tail of glassfibre/Rohacell sandwich construction. All-moving tailplane.

LANDING GEAR: Manually-retractable monowheel, size 4·00-4, tyre pressure 2·45 bars (35·5 lb/sq in), with Tost brake. Tailskid.

ACCOMMODATION: Single seat under one-piece Megaplex flush-fitting framed canopy.

DIMENSIONS, EXTERNAL:

Wing span	15·00 m (49 ft 2½ in)
Wing chord at root	0·78 m (2 ft 6¾ in)
Wing chord (mean)	0·62 m (2 ft 0½ in)
Wing chord at tip	0·344 m (1 ft 1½ in)
Wing aspect ratio	23·9
Length overall	6·67 m (21 ft 10½ in)
Fuselage: Max width	0·62 m (2 ft 0½ in)
Max depth	0·80 m (2 ft 7½ in)
Height over tail	1·33 m (4 ft 4½ in)
Tailplane span	2·35 m (7 ft 8½ in)

AREAS:

Wings, gross	9·40 m^2 (101·2 sq ft)
Ailerons (total)	0·63 m^2 (6·78 sq ft)
Vertical tail surfaces (total)	0·75 m^2 (8·07 sq ft)
Tailplane	0·786 m^2 (8·46 sq ft)

WEIGHTS AND LOADING:

Weight empty	213 kg (469 lb)
Max T-O weight	310 kg (683 lb)
Max wing loading	33 kg/m^2 (6·76 lb/sq ft)

PERFORMANCE (at max T-O weight):

Best glide ratio at 51 knots (95 km/h; 59 mph)	39
Min sinking speed at 46 knots (85 km/h; 53 mph)	0·60 m (1·97 ft)/sec
Stalling speed	37·5 knots (69 km/h; 43 mph)
Max speed (rough and smooth air)	119 knots (220 km/h; 137 mph)
Max aero-tow speed	75·5 knots (140 km/h; 87 mph)
Max winch-launching speed	59·5 knots (110 km/h; 68·5 mph)
g limits	+5·9; −3·9

INDIA

CIVIL AVIATION DEPARTMENT
TECHNICAL CENTRE, CIVIL AVIATION DEPARTMENT

HEAD OFFICE:
Civil Aviation Department, R. K. Puram, New Delhi 22

WORKS:
Technical Centre, opposite Safdarjung Airport, New Delhi 110003
Telephone: 611504

OFFICERS: See Aircraft section

The Technical Centre is the research and development establishment of the Indian Civil Aviation Department. It is equipped with all facilities necessary for the development of design, airworthiness and operational standards, operational research, development testing and standardisation of indigenous aircraft materials, type certification of prototype aircraft and equipment, and the scientific investigation of accidents.

Since 1950 the Technical Centre has undertaken the design and development of gliders, under the leadership of S. Ramamritham, utilising predominantly indigenous materials. The first of these gliders, of the open-cockpit primary type, was flown in November 1950. Since then the Technical Centre has built gliders of eight types for service at civil gliding centres in India, as listed in the 1972-73 *Jane's*. Of these eight types, five—the Ashvini (1964-65 *Jane's*; five built), Rohini (1974-75 *Jane's*; 107 built), Bharani (1965-66 *Jane's*; one built), Kartik, HS-I and HS-II Mrigasheer—are original designs.

The Technical Centre does not undertake quantity production of gliders. Complete sets of drawings of the designs developed at the Centre are supplied to interested organisations with permission to manufacture them in series.

KS-II KARTIK

The original KS-I Kartik, designed by Mr S. Ramamritham, flew for the first time on 18 March 1963. A description appeared in the 1972-73 *Jane's*.

The third Kartik, designated KS-II, first flew on 4 May 1965 and was type certificated in 1965. It introduced a conventional tapered wing, a reduction in the height of the cockpit, a slight increase in fuselage length, and modifications to the shape of the aileron leading-edge.

During the first Indian national gliding rally, held in 1967, the Kartik set up a national speed record over a 200 km triangular course.

Six more KS-IIs were test flown between February 1967 and November 1975. From the third onwards, further improvements to the original design include a reduction in fuselage height, improved forward view and seating, and larger airbrakes. The ninth Kartik, which began flight testing in November 1975, has trailing-edge fixed-hinge slotted flaps instead of airbrakes.

The following details apply to the ninth Kartik:

TYPE: Single-seat high-performance sailplane.

WINGS: Cantilever high-wing monoplane. Wing section NACA 64_3-618. Dihedral 1° 30′. Incidence 0°. Wooden structure, with one main spar, one rear spar and a diagonal spar at the root. Plywood-covered torsion box back to rear spar. Trailing-edge fabric-covered. Fabric-covered slotted wooden ailerons and slotted flaps.

FUSELAGE: Semi-monocoque wooden structure with plywood covering and glassfibre nosecap.

TAIL UNIT: Cantilever wooden structure. Fin plywood-covered. Remainder fabric-covered except for plywood leading-edges. Plywood trim tab in starboard elevator.

LANDING GEAR: Non-retractable Palmer/Dunlop unsprung monowheel and tyre, size 4·00-3·5, pressure 2·07 bars (30 lb/sq in). Drum-type brake, operated by a separate lever mounted on the control column. Rubber-sprung nose-skid with replaceable steel shoe. Tailskid sprung with tennis balls.

ACCOMMODATION: Single seat under upward-opening rearward-hinged Perspex canopy. Oxygen equipment optional.

DIMENSIONS, EXTERNAL:

Wing span	15·00 m (49 ft 2½ in)
Wing chord at root	1·00 m (3 ft 3¼ in)
Wing chord at tip	0·64 m (2 ft 1¼ in)
Wing aspect ratio	16·6
Length overall	7·37 m (24 ft 2 in)
Height over tail	2·26 m (7 ft 5 in)
Tailplane span	2·90 m (9 ft 6 in)

AREAS:

Wings, gross	13·54 m^2 (145·7 sq ft)
Ailerons (total)	1·32 m^2 (14·21 sq ft)
Trailing-edge flaps (total)	0·99 m^2 (10·65 sq ft)
Fin	0·47 m^2 (5·02 sq ft)
Rudder	0·89 m^2 (9·60 sq ft)
Tailplane	1·22 m^2 (13·20 sq ft)

Mistral single-seat Standard Class sailplane built by Dipl-Ing Strauber and three associates

KS-II Kartik single-seat high-performance sailplane

First prototype of the HS-II Mrigasheer single-seat Standard Class sailplane

Caproni Vizzola Calif A-21S two-seat high-performance sailplane

Elevators, incl tab 1·05 m² (11·30 sq ft)
WEIGHTS AND LOADING:
Weight empty, equipped 220 kg (485 lb)
Max T-O weight 320 kg (705 lb)
Max wing loading 23·63 kg/m² (4·86 lb/sq ft)
PERFORMANCE (at max T-O weight):
Best glide ratio at 41 knots (75 km/h; 47 mph) 31
Min sinking speed at 35 knots (65 km/h; 40 mph) 0·60 m (1·97 ft)/sec
Stalling speed 23 knots (42 km/h; 26 mph)
Max speed (smooth air) 108 knots (200 km/h; 124 mph)
Max speed (rough air) 76 knots (140 km/h; 87 mph)
Max aero-tow speed 62 knots (115 km/h; 71·5 mph)
Max winch-launching speed 54 knots (100 km/h; 62 mph)

HS-II MRIGASHEER

This single-seat Standard Class sailplane is the first to be designed and developed at the Technical Centre by the team of designers and engineers led by Mr K. B. Ganesan, Director of Research and Development. The original version, the HS-I, made its first flight in November 1970 and was described and illustrated in the 1973-74 *Jane's*.

The further-developed HS-II flew for the first time in April 1973, and in the following month was placed second in the first Indian national gliding championships at Kanpur.

Based on the aerodynamic design of the first HS-II, a second prototype is under construction. This has trailing-edge slotted flaps instead of airbrakes, and glassfibre-reinforced plastics horizontal tail surfaces of NACA 63₁212 inverted aerofoil section instead of the NACA 0009 section of the first HS-II.

The following description applies to this second aircraft:

TYPE: Single-seat high-performance Standard Class sailplane.

WINGS: Cantilever high-wing monoplane. Wortmann wing sections: FX-61-184 at root, FX-61-163 from 16·8% to 50% of each half-span, and FX-60-126 at tip. Dihedral 1° 30′. Incidence 3°. Wooden structure, comprising main spar, rear spar, and diagonal spar at the root. Plywood-covered torsion-box nose cell and plywood-covered rear cell. Plain wooden ailerons and slotted trailing-edge flaps. Wooden airbrake above and below each wing.

FUSELAGE: Semi-monocoque wooden structure with plywood covering.

TAIL UNIT: Cantilever structure. Fin plywood-covered rudder fabric-covered. Horizontal surfaces of glassfibre-reinforced plastics, with trim tab of same construction in starboard elevator.

LANDING GEAR: Retractable unsprung monowheel, tyre size 4·00-3·5, pressure 2·07 bars (30 lb/sq in). Drum-type brake. Rubber-sprung nose-skid with replaceable steel shoe. Rubber-sprung tailskid.

ACCOMMODATION: Single seat under forward-opening hinged jettisonable Perspex canopy. Oxygen equipment optional.

DIMENSIONS, EXTERNAL:
Wing span 15·00 m (49 ft 2½ in)
Wing chord at root 1·17 m (3 ft 10 in)
Wing chord at tip 0·40 m (1 ft 3¾ in)
Wing aspect ratio 19·85
Length overall 7·59 m (24 ft 10¾ in)
Height overall (tail up) 2·50 m (8 ft 2½ in)
Tailplane span 2·60 m (8 ft 6¼ in)
AREAS:
Wings, gross 11·24 m² (121·0 sq ft)
Ailerons (total) 1·04 m² (11·19 sq ft)
Trailing-edge flaps (total) 1·27 m² (13·70 sq ft)
Fin 0·46 m² (4·95 sq ft)
Rudder 0·89 m² (9·58 sq ft)
Tailplane 1·055 m² (11·36 sq ft)
Elevators, incl tab 0·955 m² (10·28 sq ft)
WEIGHTS AND LOADING:
Weight empty, equipped 237 kg (522 lb)
Max T-O weight 335 kg (738 lb)
Max wing loading 29·55 kg/m² (6·05 lb/sq ft)
PERFORMANCE (at max T-O weight):
Best glide ratio 32
Min sinking speed 0·58 m (1·90 ft)/sec
Max speed (smooth air) 115 knots (213 km/h; 132 mph)
Max speed (rough air) 80 knots (148 km/h; 92 mph)
Max aero-tow speed 62 knots (115 km/h; 71·5 mph)
Max winch-launching speed 53·5 knots (99 km/h; 61·5 mph)

ITALY

CAPRONI VIZZOLA

CAPRONI VIZZOLA COSTRUZIONI AERONAUTICHE SpA

HEAD OFFICE: 20122 Milano, Via Durini 24
Telephone: (02) 700826 and 781975
WORKS:
21010 Vizzola Ticino (Varese)
Telephone: (0331) 230826 and 230847
Telex: Caproni 32035
PRESIDENT: Dott Giovanni Caproni di Taliedo
VICE-PRESIDENT:
Rag Achille Caproni di Taliedo

The Caproni company, formed in 1910, is the oldest Italian aircraft manufacturer. Its works at Vizzola Ticino have approx 30,000 m² (322,917 sq ft) of covered space, and are adjacent to Malpensa Airport. They are equipped to manufacture complete structural subassemblies for helicopters and medium-sized fixed-wing aircraft. Caproni Vizzola also produces ground support equipment for General Electric T64/CT64, J79, J85 and CF6 turbojet and turbofan engines.

In 1969 Caproni Vizzola began producing a series of Calif sailplanes designed by Carlo Ferrarin and Livio Sonzio. Details of the earlier A-10 (one built), A-12 (two built), A-14 (one built), A-15 (one built), A-20 and A-21 have been given in the 1972-73 and subsequent *Jane's*.

CAPRONI VIZZOLA CALIF A-21S

The original A-21 was a two-seat version of the A-14, from which it differed principally in having a wider (to accommodate two side-by-side seats) and slightly longer fuselage.

The prototype A-21 made its first flight on 23 November 1970, and a total of 17 A-21s had been ordered by mid-February 1973. The original A-21 (1975-76 *Jane's*) is no longer being built, its place having been taken by the A-21S which has a conventional T tailplane with elevator, chemically milled flaps, ailerons and elevator, and modified cockpit heating and ventilation systems. The A-21S has been certificated by the RAI.

It was in an A-21S that, on 17 August 1974, Adele Orsi and Patrizia Golin set a new Class D2 speed record for women over a 100 km closed circuit of 54·909 knots (101·758 km/h; 63·229 mph). On the following day, also in an A-21S, Adele Orsi and Franca Bellingeri raised the corresponding record over a 300 km closed circuit to 52·742 knots (97·741 km/h; 60·733 mph).

TYPE: Two-seat high-performance sailplane.

WINGS: Cantilever mid-wing monoplane. Wing section Wortmann FX-67-K-170 at root, FX-60-126 at tip. Dihedral 0° at root, 1° 30′ on outer panels. Incidence 0°. Sweepback 4° on outer leading-edge only. Three-piece all-metal structure, with main spar and two auxiliary spars. Main spar forms torsion box with leading-edge. Glassfibre wingtips. Top-hinged partially-balanced differentially-operated plain ailerons of all-metal chemically milled construction. Automatic connection of controls when wings are assembled. Lower-hinged aerodynamically-balanced trailing-edge flaps/spoilers of all-metal chemically milled construction, manually operated by a single control, are utilised as camber-changing surfaces in the −8° to +12° range, and as airbrakes when lowered to a 90° position.

Fuselage: Low-drag tadpole-shaped fuselage. Monocoque forward section of glassfibre and foam plastics construction with load-carrying light alloy structure. Narrow-diameter all-metal stressed-skin tailboom.

Tail Unit: Cantilever all-metal non-swept T tailplane, with elevator, and sweptback vertical surfaces. Fin is a single-spar stressed-skin structure, elevator is chemically milled. Spring-adjusted tailplane trimming. Automatic control connections during assembly.

Landing Gear: Mechanically-retractable twin wheels, with rubber-in-compression shock-absorption; mechanical up and down lock. Non-retractable steerable tailwheel for ground handling. Main wheels and tyres size 3·50-5, pressure 5·07 bars (73·5 lb/sq in). Mechanically-operated Tost wheel brake.

Accommodation: Two seats, side by side, in enclosed cabin. Cockpit heating and ventilation standard.

Dimensions, external:
- Wing span: 20·38 m (66 ft 10¼ in)
- Wing chord at root: 0·90 m (2 ft 11½ in)
- Wing chord (mean): 0·794 m (2 ft 7¼ in)
- Wing chord at tip: 0·321 m (1 ft 0½ in)
- Wing aspect ratio: 25·65
- Length overall: 7·838 m (25 ft 8½ in)
- Height over tail: 1·61 m (5 ft 3½ in)
- Tailplane span: 2·89 m (9 ft 5¾ in)

Areas:
- Wings, gross: 16·19 m² (174·3 sq ft)
- Ailerons (total): 1·04 m² (11·19 sq ft)
- Trailing-edge flaps (total): 2·07 m² (22·28 sq ft)
- Spoilers (total): 0·91 m² (9·795 sq ft)
- Fin: 0·79 m² (8·50 sq ft)
- Rudder: 0·59 m² (6·35 sq ft)
- Tailplane: 1·36 m² (14·64 sq ft)
- Elevator: 0·34 m² (3·66 sq ft)

Weights and Loading:
- Weight empty, equipped: 436 kg (961 lb)
- Max T-O weight: 644 kg (1,419 lb)
- Max wing loading: 39·8 kg/m² (8·15 lb/sq ft)

Performance:
- Best glide ratio at 56·5 knots (105 km/h; 65 mph): 43
- Min sinking speed at 46 knots (85 km/h; 53 mph): 0·60 m (1·97 ft)/sec
- Stalling speed, flaps up: 38 knots (70 km/h; 43·5 mph)
- Stalling speed, flaps down: 34 knots (63 km/h; 39·5 mph)
- Max speed (rough or smooth air): 137·5 knots (255 km/h; 158 mph)
- Max aero-tow speed: 75·5 knots (140 km/h; 87 mph)
- *g* limit: +5·0 (ultimate)

CAPRONI VIZZOLA CALIF A-21J

This jet-powered version of the A-21 two-seat sailplane flew for the first time at the end of January 1972.

Only the prototype was built; a description and illustration of this can be found in the 1975-76 *Jane's*.

CVT
CENTRO DI VOLO A VELA DEL POLITECNICO DI TORINO

Address:
Corso Duca Degli Abruzzi 24, 10129 Turin

Office and Laboratory:
Corso Luigi Einaudi 44, 10129 Turin

Telephone: 011-511250

Director:
Piero Morelli

Gliders built at the CVT have included the CVT-2 Veltro (1960-61 *Jane's*); the M-100 S, which was manufactured under licence in Italy by Avionautica Rio and in France by CARMAM as the Mésange; the M-200, also manufactured by CARMAM; and the M-300, described in the 1972-73 *Jane's*.

Current activity at the CVT in 1976 was concerned with a research programme to develop a light alloy extruded wing structure. Pioneering work on this original concept, started in 1968, led to the extruded wing spar, aileron, airbrake and tailplane structure of the M-300 prototypes. A complete extruded wing structure was under study, and static tests of span portions were carried out. A research programme was also started into the feasibility of a powered aircraft with a lightweight engine, low fuel consumption and low noise level.

GLASFLÜGEL
GLASFLÜGEL ITALIANA SrL

Address:
24030 Valbrembo (Bergamo), Via Locatelli 1

Telephone: (035) 612617

President:
Dott Ing Sergio Aldo Capoferri

Vice-President:
Dott Ing Mario Moltrasio

Technical Director: Giampaolo Ghidotti

This company was established at Valbrembo Airport, where it is accommodated in a factory which occupies an area of 1,300 m² (13,993 sq ft) and which is insulated and heated to maintain a controlled temperature of 20°C, essential for work on glassfibre structures.

Glasflügel Italiana holds RAI licences for the construction of glassfibre sailplanes and for repair, maintenance and modification of sailplanes and training aircraft.

Glasflügel Italiana has already assembled 25 single-seat 17 metre Kestrel sailplanes, details of which may be found under the GHH entry in the German section of this edition. The company is manufacturing complete fuselage assemblies for the Kestrel and had built 120, with 10 more being manufactured and assembled, by January 1976.

It also assists the parent company in the production of the H 205 Club Libelle, Hornet and Mosquito.

Other activities include construction of glassfibre trailers for road transportation of sailplanes, the repair and maintenance of sailplanes of glassfibre, wood and metal construction, and installation and modification work on sailplanes of all types.

JAPAN

NIPPI
NIHON HIKOKI KABUSHIKI KAISHA (Japan Aircraft Manufacturing Co Ltd)

Head Office and Sugita Plant:
No. 3175 Showa-machi, Kanazawa-ku, Yokohama 236

Telephone: Yokohama (045) 771-1251

Telex: (3822) 267

Officers: See Aircraft section

NIPPI NP-100A

Nippi began the design of this side-by-side two-seat powered sailplane in late 1973, and the prototype (NP-100) made its first flight on 25 December 1975. A second prototype was under construction in early 1976, and JCAB certification was anticipated later in the same year.

Principal point of interest in the NP-100A is the power plant, which is of the ducted-fan type and is fully buried within the fuselage aft of the main landing gear.

The following description applies to the first prototype:

Type: Two-seat powered sailplane.

Wings: Cantilever shoulder-wing monoplane. Wing section Wortmann FX-67-K-170 (modified) from root to tip. No dihedral or sweepback. Incidence 3° at root, 0° at tip. All-metal structure, with single spar at 40% chord. Two-section metal-skinned flaps on each trailing-edge. Although linked mechanically, the inner flaps have a greater range of movement and serve also as airbrakes. With full flap selected, the inner flaps have up to 80° deflection, the outer ones 48°; both flaps are also designed for upward deflection (10° inner, 6° outer). All-metal plain ailerons. No spoilers.

Fuselage: All-metal semi-monocoque structure.

Tail Unit: Cantilever all-metal structure, with conventional elevators and rudder. Entire tailplane can be detached by removing a single pin.

Landing Gear: Forward-retracting twin-wheel main gear, operated mechanically with spring assistance. Steerable, non-retractable tailwheel, linked to rudder movement.

Power Plant (prototype): One 44·7 kW (60 hp at 6,000 rpm) Kawasaki HZI 748 cc modified motor-cycle engine, installed in centre of fuselage and driving a four-blade wooden ducted fan. Triple 'Venetian blind' type air intake doors, on each side of fuselage, are interconnected with engine starting circuit to prevent engine operating when doors are closed. Fuel in single tank in fuselage, capacity 40 litres (8·8 Imp gallons). Production aircraft will have a Xenoah engine of similar power.

Accommodation: Seats for two persons, side by side under rearward-sliding framed canopy. Rudder pedals adjustable in flight.

Dimensions, external:
- Wing span: 18·00 m (59 ft 0⅔ in)
- Wing chord at root: 1·45 m (4 ft 9 in)
- Wing chord at tip: 0·55 m (1 ft 9⅔ in)
- Wing aspect ratio: 18
- Length overall, tail up: 8·00 m (26 ft 3 in)
- Fuselage:
  - Max depth: 1·10 m (3 ft 7¼ in)
  - Max width: 1·10 m (3 ft 7¼ in)
- Height over tail: 2·23 m (7 ft 3¾ in)
- Tailplane span: 3·40 m (11 ft 1¾ in)
- Wheel track: 0·70 m (2 ft 3½ in)
- Wheelbase: 5·54 m (18 ft 2 in)
- Fan diameter: 0·60 m (1 ft 11½ in)

Areas:
- Wings, gross: 18·00 m² (193·75 sq ft)
- Ailerons (total): 0·852 m² (9·17 sq ft)
- Trailing-edge flaps (total): 2·40 m² (25·83 sq ft)
- Fin: 1·76 m² (18·94 sq ft)
- Rudder: 0·73 m² (7·86 sq ft
- Tailplane: 2·465 m² (26·53 sq ft)
- Elevators (total): 0·73 m² (7·86 sq ft)

Weights and Loadings:
- Weight empty: 420 kg (925 lb
- Max T-O weight: 600 kg (1,322 lb
- Max wing loading: 33·3 kg/m² (6·82 lb/sq ft
- Max power loading: 13·42 kg/kW (22·03 lb/hp

Performance (at max T-O weight, powered):
- Max level speed at S/L: 86 knots (160 km/h; 99 mph
- Max cruising speed at S/L: 65 knots (120 km/h; 74 mph
- Econ cruising speed at S/L: 48·5 knots (90 km/h; 56 mph
- Stalling speed, flaps down: 35·5 knots (65 km/h; 40·5 mph
- Max rate of climb at S/L: 120 m (394 ft)/mi
- T-O run: 365 m (1,200 ft
- T-O to 15 m (50 ft): 600 m (1,968 ft
- Landing from 15 m (50 ft): 400 m (1,312 ft
- Range with max fuel at 59 knots (110 km/h; 68 mph): 108 nm (200 km; 124 miles
- Endurance at 48·5 knots (90 km/h; 56 mph): 2 h

Performance (at max T-O weight, unpowered):
- Best glide ratio at 48·5 knots (90 km/h; 56 mph): 3
- Min sinking speed at 45 knots (83 km/h; 51·5 mph): 0·80 m (2·62 ft)/se

TAINAN
TAINAN KOGYO CO

Address:
5139-3 1-chome, Komathubara, Zama-shi, Kanagawa-ken 228

Telephone: 0462-54-2332/3/4

President:
H. Shinozaki

Manager, Light Aircraft Department:
Terutoshi Tanigutchi

This company took over the manufacture of sailplanes from LADCO (see 1971-72 *Jane's*), and is continuing production, under licence, of the former company's Mita III two-seat sailplane.

Its latest design is the TN-1, of which a description also follows.

TAINAN MITA III

By 1 January 1976, a total of 37 Mita IIIs had been built.

Type: Two-seat training and sporting sailplane.

Wings: Cantilever shoulder-wing monoplane. Wing section NACA 63_3-618. Three-piece wing, with constant-chord centre-section and tapered outer panels. All-wood box monospar construction, plywood-covered. Ailerons fabric-covered. Schempp-Hirth airbrakes on upper surfaces.

Fuselage: Steel tube frame with wooden stringers a fabric covering. Nose and front section of glassfib Towing hook under nose.

Tail Unit: Cantilever type, of wooden construction; ru der and elevators fabric-covered.

Landing Gear: Non-retractable monowheel with bra and rubber springing. Tailskid.

Accommodation: Two seats in tandem under two-pie blown canopy which opens sideways to starboard.

Dimensions, external:
- Wing span: 16·00 m (52 ft 5
- Wing aspect ratio: 16·

First prototype Nippi NP-100A, with Kawasaki HZI engine

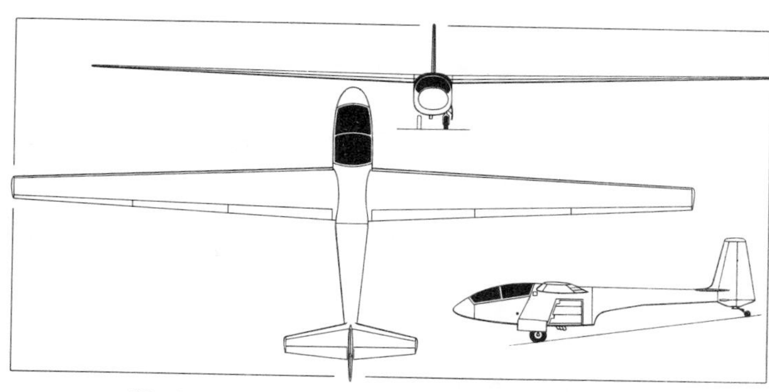

Nippi NP-100A two-seat powered sailplane *(Pilot Press)*

Tainan Mita III two-seat training and sporting sailplane

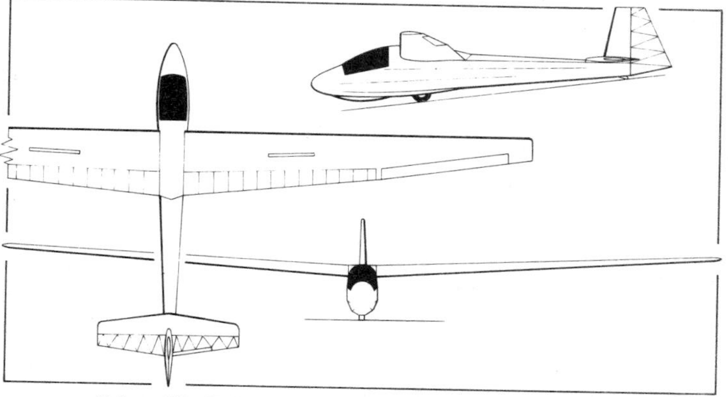

Tainan TN-1(F5) single-seat sailplane *(Michael A. Badrocke)*

Length overall — 7·96 m (26 ft 1½ in)
Height over tail — 1·28 m (4 ft 2½ in)

AREAS:
Wings, gross — 15·87 m² (170·82 sq ft)
Fin — 0·56 m² (6·03 sq ft)
Rudder — 0·77 m² (8·29 sq ft)
Tailplane — 1·44 m² (15·50 sq ft)
Elevators — 1·08 m² (11·63 sq ft)

WEIGHTS AND LOADING:
Weight empty — 300 kg (661 lb)
Max T-O weight — 450 kg (992 lb)
Max wing loading — 28·4 kg/m² (5·8 lb/sq ft)

PERFORMANCE:
Best glide ratio at 44 knots (82 km/h; 51 mph) — 30
Min sinking speed at 41 knots (75 km/h; 47 mph) — 0·72 m (2·36 ft)/sec
Stalling speed — 34 knots (62·3 km/h; 39 mph)
Max speed — 102 knots (190 km/h; 118 mph)
Max aero-tow speed — 70 knots (130 km/h; 81 mph)
Max winch-launching speed — 59 knots (110 km/h; 68 mph)

TAINAN TN-1

The TN-1 (manufacturer's designation F5) was designed by Yukio Tanaka. Construction of a prototype began in August 1974, and it was hoped to complete this during 1976.

TYPE: Single-seat sailplane.

WINGS: Cantilever high-wing monoplane. Wing section Göttingen Gö 533 at root, Gö 532 at tip. Thickness/chord ratio 17% at root, 14·4% at tip. Dihedral 3°. Incidence 0°. Sweepforward 2° 4′ at quarter-chord. Single-spar structure of spruce and plywood, with fabric covering. Top-hinged plywood-covered wooden ailerons. Schempp-Hirth aluminium airbrakes on upper surface.

FUSELAGE: Steel tube frame with wooden stringers and fabric covering. Forward section of glassfibre construction.

TAIL UNIT: Cantilever wooden structure, with fabric-covered elevators and rudder. Trim tab in starboard elevator.

LANDING GEAR: Non-retractable unsprung monowheel, at CG, with tyre size 318 × 140 mm, pressure 2·94 bars (42·7 lb/sq in). Tainan band brake. Tailskid.

ACCOMMODATION: Single seat under one-piece blown Plexiglas canopy.

EQUIPMENT: Altimeter, airspeed indicator, rate of climb indicator, winch and aero-tow release standard.

DIMENSIONS, EXTERNAL:
Wing span — 15·50 m (50 ft 10½ in)
Wing chord at root — 1·461 m (4 ft 9½ in)
Wing chord at tip — 0·517 m (1 ft 8½ in)
Wing aspect ratio — 15·03
Length overall — 7·64 m (25 ft 0¾ in)
Height over tail — 2·203 m (7 ft 2¾ in)
Tailplane span — 3·00 m (9 ft 10 in)

AREAS:
Wings, gross — 15·98 m² (172·0 sq ft)
Ailerons (total) — 1·21 m² (13·02 sq ft)
Airbrakes (total) — 0·48 m² (5·17 sq ft)
Fin — 0·54 m² (5·81 sq ft)
Rudder — 0·81 m² (8·72 sq ft)
Tailplane — 1·22 m² (13·13 sq ft)
Elevators, incl tab — 1·18 m² (12·70 sq ft)

WEIGHTS AND LOADING:
Weight empty, equipped — 250 kg (551 lb)
Max T-O weight — 360 kg (793 lb)
Max wing loading — 22·5 kg/m² (4·61 lb/sq ft)

PERFORMANCE (estimated, at max T-O weight):
Best glide ratio — 25·8
Min sinking speed — 0·74 m (2·43 ft)/sec
Stalling speed — 30 knots (54·9 km/h; 34·2 mph)
Max speed (smooth air) — 108 knots (200 km/h; 124 mph)
Max speed (rough air) — 79 knots (146 km/h; 91 mph)
Max aero-tow speed — 70 knots (130 km/h; 81 mph)
Max winch-launching speed — 59·5 knots (110 km/h; 68·5 mph)
g limits — +5·0; −2·5

TAKATORI-SANKEN

TAKATORI-SANKEN CO LTD (Aircraft Division)

ADDRESS:
6-32 Funakoshi-machi, Yokosuka, Kanagawa

Telephone: (0468) 61-7226

This company took over the sailplane development and construction work of the Yokohama Gliding Club (see 1973-74 *Jane's*).

No news of the company's activities has been received since early 1974. Descriptions of its recent products can be found in the 1974-75 and 1975-76 *Jane's*.

PHILIPPINES
(REPUBLIC OF)

PATS/NSDB

PHILIPPINE AERONAUTICS TRAINING SCHOOL/NATIONAL SCIENCE DEVELOPMENT BOARD

As mentioned briefly in the Aircraft section, these two bodies are developing jointly a two-seat training glider which has the designation **XG-001**. The aircraft is said to be constructed largely of native timber, but no other details were known at the time of closing for press.

POLAND

SZD

PRZEDSIEBIORSTWO DOSWIADCZALNO-PRODUKCYJNE SZYBOWNICTWA (Experimental and Production Concern for Gliders PZL-BIELSKO)

HEAD OFFICE AND WORKS:
43-300 Bielsko-Biala 1, ul. Cieszynska 325
Telephone: 250-21 to 250-26
Telex: 031-259 SZD PL

DIRECTOR:
Mgr Ing Wladyslaw Nowakowski

SALES REPRESENTATIVE:
Pezetel, 02-344 Warszawa, ul. Czestochowska 4a

The Instytut Szybownictwa (Gliding Institute), formed officially in January 1946 at Bielsko-Biala, was renamed two years later the Szybowcowy Zaklad Doswiadczalny —SZD (Experimental Glider Establishment). In July

1969 the name was changed to Zaklad Doswiadczalny Rozwoju i Budowy Szybowcow (Experimental Establishment for Development of Gliders); and in January 1972 to Zaklady Szybowcowe PZL-Bielsko (Glider Works—Bielsko) and Osrodek Badawczo-Rozwojowy Szybonictwa (Research and Development Centre for Gliders). The change to the present title took place in July 1975, but the well-known designation initials SZD are retained. This organisation is responsible for the design and development of all Polish gliders and sailplanes. Production plants are situated at Bielsko-Biala, Wroclaw and Jezów.

Between 1947 and 1975 the Polish aircraft industry produced some 3,250 gliders of about 88 different types, and SZD sailplanes have been exported all over the world in substantial numbers.

SZD-9bis BOCIAN 1E (STORK)

The original prototype of the Bocian flew for the first time on 11 March 1952. By the end of 1975, a total of 555 SZD-9 Bocian sailplanes had been built, in several versions, for customers in Argentina, Australia, Austria, Belgium, Bulgaria, China, Denmark, Egypt, Finland, France, Germany (Democratic Republic), Germany (Federal Republic), Greece, Hungary, India, Indonesia, Iraq, Italy, Japan, Korea, Norway, New Zealand, Poland, Portugal, Romania, Syria, Switzerland, Tunisia, Turkey, the UK, USSR, Venezuela and Zambia. The latest version is the Bocian 1E, which first flew on 6 December 1966. On 5 November 1966 an earlier version of the Bocian, the 1D, established an international gain of height record for multi-seat sailplanes of 11,680 m (38,320 ft). The corresponding record for ladies, of 8,430 m (27,650 ft), was set up in another Bocian on 17 October 1967.

The controls, instrument panel and other details were designed to make the aircraft suitable for sporting flight as well as for school and training duties. Cloud-flying, spinning and basic aerobatics are permitted.

The following description applies to the Bocian 1E, production of which was continuing in 1976:

TYPE: Tandem two-seat general-purpose sailplane.

WINGS: Cantilever mid-wing monoplane. Wing section NACA 43018A at root, NACA 43012A at tip. Dihedral 4°. Incidence 2° 30′. Sweepforward 1° 30′ at quarter-chord. Two-spar wooden structure, with plywood D-section leading-edge and fabric covering. Slotted ailerons. No flaps. SZD airbrakes inboard of ailerons.

FUSELAGE: Plywood-covered wooden structure of oval section.

TAIL UNIT: Cantilever wooden structure. Trim tab in port elevator.

LANDING GEAR: Non-retractable monowheel and front skid. Shock-absorber fitted. Wheel size 350 × 135, with brake.

ACCOMMODATION: Two seats in tandem under long transparent canopy.

DIMENSIONS, EXTERNAL:
- Wing span: 17·80 m (58 ft 4¾ in)
- Wing chord at root: 1·75 m (5 ft 8¾ in)
- Wing chord at tip: 0·50 m (1 ft 7·7 in)
- Wing aspect ratio: 16·2
- Length overall: 8·21 m (26 ft 11¼ in)
- Height over tail, excl wheel: 1·20 m (4 ft 0¼ in)
- Tailplane span: 3·10 m (10 ft 2 in)

AREAS:
- Wings, gross: 20·00 m^2 (215·3 sq ft)
- Ailerons (total): 2·74 m^2 (29·50 sq ft)
- Airbrakes (total): 0·63 m^2 (6·82 sq ft)
- Fin: 0·68 m^2 (7·32 sq ft)
- Rudder: 0·82 m^2 (8·83 sq ft)
- Tailplane: 1·00 m^2 (10·76 sq ft)
- Elevators, incl tab: 1·50 m^2 (16·15 sq ft)

WEIGHTS AND LOADING:
- Weight empty, equipped: 342 kg (754 lb)
- Max T-O weight: 540 kg (1,191 lb)
- Max wing loading: 27·0 kg/m^2 (5·58 lb/sq ft)

PERFORMANCE (at max T-O weight):
- Best glide ratio at 43·4 knots (80 km/h; 50 mph): 26
- Min sinking speed at 38·3 knots (71 km/h; 44 mph): 0·82 m (2·69 ft)/sec
- Stalling speed: 33 knots (60 km/h; 37·5 mph)
- Max speed (smooth air): 108 knots (200 km/h; 124 mph)
- Max speed (rough air): 81 knots (150 km/h; 93 mph)
- Max aero-tow speed: 76 knots (140 km/h; 87 mph)
- Max winch-launching speed: 62 knots (115 km/h; 71 mph)
- g limits: +6; −3

SZD-30A PIRAT

Designed by Ing Jerzy Smielkiewicz, this single-seat Standard Class sailplane flew for the first time on 19 May 1966. It is suitable for the full range of duties from training to competition flying and is cleared for cloud flying, spinning and basic aerobatics.

Production of the Pirat started in 1967 and 660 had been built by the end of 1975, including three which are in service with the Escuela de Aviacion Militar in Argentina. Production of the Pirat is undertaken also by the WSK-Swidnik works, which had completed 279 by the beginning of 1976. Pirats have been exported to 20 countries, including Argentina, Austria, Finland, France, Germany (Democratic Republic), Holland, Hungary, Italy, North Korea, New Zealand, Norway, Spain, Sweden, Switzerland, the UK, USA and USSR.

The current production version is designated SZD-30A.

TYPE: Single-seat Standard Class sailplane.

WINGS: Cantilever high-wing monoplane. Wing section Wortmann FX-61-168 at root, Wortmann FX-60-1261 at tip. Dihedral 2° 30′ on outer panels only. No sweep at quarter-chord. Wooden wing, built in three parts. Rectangular centre-section is a plywood-covered multi-spar structure. Tapered outer panels are of single-spar torsion-box construction. Mass-balanced ailerons. Double-plate airbrakes.

FUSELAGE: Plywood monocoque structure, with glassfibre nose and cockpit floor.

TAIL UNIT: Cantilever wooden T-tail. Tab on trailing-edge of elevator.

LANDING GEAR: Front skid with shock-absorber is easily removable. Non-retractable monowheel, size 350 × 135, with band brake.

ACCOMMODATION: Single seat under jettisonable sideways-hinged blown Perspex canopy. Two baggage compartments. Map pockets on each side of cockpit. Provision for radio and oxygen equipment.

DIMENSIONS, EXTERNAL:
- Wing span: 15·00 m (49 ft 2½ in)
- Wing chord at root: 1·03 m (3 ft 4½ in)
- Wing chord at tip: 0·60 m (1 ft 11½ in)
- Wing aspect ratio: 16·3
- Length overall: 6·86 m (22 ft 6 in)
- Height over tail: 1·67 m (5 ft 5¾ in)
- Tailplane span: 3·10 m (10 ft 2 in)

AREAS:
- Wings, gross: 13·80 m^2 (148·5 sq ft)
- Ailerons (total): 1·09 m^2 (11·72 sq ft)
- Airbrakes (total): 0·73 m^2 (7·86 sq ft)
- Fin: 0·48 m^2 (5·35 sq ft)
- Rudder: 0·77 m^2 (8·10 sq ft)
- Tailplane: 1·08 m^2 (11·68 sq ft)
- Elevator, incl tab: 0·78 m^2 (7·69 sq ft)

WEIGHTS AND LOADING:
- Weight empty, equipped: 261 kg (575 lb)
- Max T-O weight: 370 kg (816 lb)
- Max wing loading: 26·8 kg/m^2 (5·49 lb/sq ft)

PERFORMANCE (at AUW of 340 kg; 750 lb):
- Best glide ratio at 44·5 knots (82 km/h; 51 mph): 33
- Min sinking speed at 40·5 knots (75 km/h; 46·5 mph): 0·66 m (2·16 ft)/sec
- Stalling speed: 33 knots (59 km/h; 37 mph)
- Max speed (smooth air): 135 knots (250 km/h; 155 mph)
- Max speed (rough air): 76-89 knots (140-165 km/h; 87-103 mph)
- Max aero-tow speed: 71 knots (132 km/h; 82 mph)
- Max winch-launching speed: 74 knots (137 km/h; 85 mph)
- g limits: +6; −3

SZD-36A COBRA 15

The SZD-36 Cobra 15 is a single-seat Standard Class high-performance sailplane designed by Ing Wladyslaw Okarmus. Design was started on 15 October 1968, and construction of a prototype began on 10 November 1968. This aircraft flew for the first time on 30 December 1969. Polish competitors flying the Cobra 15 gained second and third places in the World Championships at Marfa in 1970.

A total of 209 Cobra 15s had been built by the beginning of 1976, for customers in Austria, Bulgaria, Czechoslovakia, Denmark, Finland, France, the German Democratic Republic, Holland, Hungary, Italy, Poland, Sweden, the UK, USA and USSR. Under a contract signed in 1972, Poland also supplied 20 Cobra 15s per year to the USSR, for Soviet aeroclubs, until 1975; this contract was subsequently extended into 1976. Some of these carry oxygen equipment and are used for wave soaring. Current production aircraft, which are designated SZD-36A, have the towing hook displaced approx 1 m (3 ft 3¼ in) forward of the CG, but retain the provision to locate this at the CG if required. The empty weight is 20 kg (44 lb) lighter than that of the original SZD-36.

Two prototypes were built of the SZD-39 Cobra 17, a 17 metre span Open Class version of the SZD-36 with water ballast, and the first flight by this version was made on 17 March 1970. A description of the SZD-39 appeared in the 1972-73 *Jane's*.

The following description applies to the standard 15 metre span SZD-36A:

TYPE: Single-seat high-performance Standard Class sailplane.

WINGS: Cantilever shoulder-wing monoplane. Wortmann wing sections: FX-61-168 at root, FX-60-1261 at tip. Dihedral 2°. Incidence 2°. No sweepback at quarter-chord. Single-spar wooden structure with heavy moulded plywood stressed skin covered with glassfibre. Plain ailerons, hinged at their upper surface and of plywood/polystyrene foam sandwich construction, are mass-balanced and actuated by pushrods. SZD type metal/glassfibre double-plate airbrake above and below each wing. No tabs.

FUSELAGE: All-wood semi-monocoque structure of oval section, covered with plywood at rear and glassfibre at front. Aero-tow hook in lower fuselage, forward of monowheel.

TAIL UNIT: Cantilever all-wood structure with swept vertical surfaces. T-tail with all-moving mass-balanced tailplane. Geared trim tab on tailplane trailing-edge.

LANDING GEAR: Mechanically-retractable monowheel which lies horizontally in bottom of fuselage when retracted. Stomil wheel and tyre size 300 × 125 mm, with SZD brake. Tyre pressure 2·48 bars (36 lb/sq in). Tailskid.

ACCOMMODATION: Single seat in enclosed cabin under vacuum-formed forward-sliding canopy which can be jettisoned in emergency. Baggage compartment aft of pilot's seat, size 0·27 × 0·72 m (10½ in × 2 ft 4½ in).

EQUIPMENT: Instrumentation includes airspeed indicator, altimeter, total energy variometer, rate of climb indicator, turn indicator, artificial horizon, variometer, compass and RS-3A radio. SAT-5 oxygen system with 4 litre (0·14 cu ft) cylinder in baggage compartment.

DIMENSIONS, EXTERNAL:
- Wing span: 15·00 m (49 ft 2½ in)
- Wing chord at root: 1·15 m (3 ft 9¼ in)
- Wing chord at tip: 0·38 m (1 ft 3 in)
- Wing chord (mean): 0·84 m (2 ft 9 in)
- Wing aspect ratio: 19·4
- Length overall: 7·05 m (23 ft 1½ in)
- Height over tail: 1·59 m (5 ft 2¾ in)
- Tailplane span: 2·40 m (7 ft 10½ in)

AREAS:
- Wings, gross: 11·60 m^2 (125 sq ft)
- Ailerons (total): 0·63 m^2 (6·78 sq ft)
- Airbrakes (total): 0·76 m^2 (8·18 sq ft)
- Fin: 0·64 m^2 (6·89 sq ft)
- Rudder: 0·46 m^2 (4·95 sq ft)
- Tailplane, incl tab: 1·40 m^2 (15·07 sq ft)

WEIGHTS AND LOADING:
- Weight empty, equipped: 237 kg (522 lb)
- Max T-O weight: 385 kg (848 lb)
- Max wing loading: 33·2 kg/m^2 (6·80 lb/sq ft)

PERFORMANCE (at max T-O weight):
- Best glide ratio at 52 knots (97 km/h; 60 mph): 38
- Min sinking speed at 39·5 knots (73 km/h; 45·5 mph): 0·68 m (2·23 ft)/sec
- Stalling speed: 36·5 knots (67 km/h; 42 mph)
- Max speed (smooth air): 135 knots (250 km/h; 155 mph)
- Max speed (rough air): 92 knots (170 km/h; 106 mph)
- Max aero-tow speed: 81 knots (150 km/h; 93 mph)
- Max winch-launching speed: 70 knots (130 km/h; 80·5 mph)
- g limits: +6; −3

SZD-38A JANTAR-1 (AMBER)

This high-performance single-seat Open Class sailplane was developed by Dipl Ing Adam Kurbiel from the prototype SZD-37 Jantar 19 (see 1973-74 *Jane's*) which, in the 1972 World Championships at Vrsac, Yugoslavia, gained 3rd place and was awarded the OSTIV prize for the best sailplane of up to 19 m span. It is the first Polish series-built sailplane of all-plastics construction. A Standard Class version, the SZD-41A, is described separately.

The SZD-38 flew for the first time on 7 August 1973. The first two examples were flown by the Polish team in the 1974 World Championships at Waikerie, Australia, in which they gained 15th and 18th places. The production version of the Jantar-1 is designated SZD-38A.

In 1975 the Polish pilot Adela Dankowska set three new women's international records while flying a Jantar-1: a distance record of 415·44 nm (769·4 km; 478·08 miles) over a triangular course; a speed record of 39·755 knots (73·627 km/h; 45·75 mph) over the same course; and a goal-and-return distance record of 362·7 nm (672·2 km; 417·7 miles).

A total of 57 Jantar-1s had been built by the beginning of 1976, for customers in Australia, Bulgaria, France, Germany (Democratic Republic), Hungary, Poland, Switzerland, the UK and the USA.

TYPE: Single-seat high-performance Open Class sailplane.

WINGS: Cantilever shoulder-wing monoplane. Wortmann wing sections: FX-67-K-170 at root, FX-67-K-150 at tip. Dihedral 1° 30′. No sweepback. Single-spar ribless structure, with foam-filled glassfibre/epoxy resin sandwich skin. Multi-hinged ailerons and hingeless trailing-edge flaps, hinged on the upper surface and actuated by pushrods. Provision in wings for 100 litres (22 Imp gallons) of water ballast. DFS glassfibre airbrake above and below each wing. No tabs.

FUSELAGE: All-glassfibre/epoxy resin shell structure; centre portion has a steel tube frame coupling together the wings, fuselage and landing gear.

TAIL UNIT: Cantilever T-tail, of glassfibre/epoxy resin construction. Tailplane has pushrod-operated elevators with spring trim. No tabs. Fin is integral with fuselage and has internally-mounted VHF aerial.

LANDING GEAR: Mechanically-retractable monowheel (Stomil, size 350 × 135, tyre pressure 3·43 bars; 49·8 lb/sq in) and fixed 200 mm diameter tailwheel. Disc brake on main wheel.

SZD-9bis Bocian 1E tandem two-seat general-purpose sailplane

SZD-30A Pirat single-seat Standard Class sailplane, which has been in production since 1967

SZD-36A Cobra 15 single-seat high-performance Standard Class sailplane

SZD-38 Jantar-1 Open Class sailplane, first series-built all-plastics Polish sailplane

SZD-41 Jantar Standard single-seat Standard Class sailplane

SZD-42 Jantar-2 single-seat high-performance sailplane, built for the 1976 World Open Class Championships in Finland

ACCOMMODATION AND EQUIPMENT: Single semi-reclining seat under two-piece fully-transparent removable canopy. Normal cockpit instrumentation, plus VHF transceiver, oxygen equipment and artificial horizon. Headrest and backrest adjustable on ground; rudder pedals adjustable during flight.

DIMENSIONS, EXTERNAL:
- Wing span: 19·00 m (62 ft 4 in)
- Wing chord at root: 0·90 m (2 ft 11½ in)
- Wing chord at tip: 0·377 m (1 ft 2¾ in)
- Wing chord (mean): 0·74 m (2 ft 5¼ in)
- Wing aspect ratio: 27
- Length overall: 7·11 m (23 ft 4 in)
- Height over tail: 1·60 m (5 ft 3 in)
- Tailplane span: 2·60 m (8 ft 6¼ in)

AREAS:
- Wings, gross: 13·38 m² (144·0 sq ft)
- Ailerons (total): 0·88 m² (9·47 sq ft)
- Trailing-edge flaps (total): 1·40 m² (15·07 sq ft)
- Airbrakes (total): 0·69 m² (7·43 sq ft)
- Fin: 0·60 m² (6·46 sq ft)
- Rudder: 0·50 m² (5·38 sq ft)
- Tailplane: 1·35 m² (14·53 sq ft)
- Elevators (total): 0·812 m² (8·74 sq ft)

WEIGHTS AND LOADINGS:
- Weight empty, equipped: 290 kg (639 lb)
- Max T-O weight:
  - with water ballast: 520 kg (1,146 lb)
  - without water ballast: 420 kg (926 lb)
- Max wing loading:
  - with water ballast: 38·9 kg/m² (7·97 lb/sq ft)
  - without water ballast: 31·4 kg/m² (6·43 lb/sq ft)

PERFORMANCE (without water ballast):
- Best glide ratio at 52·5 knots (97 km/h; 60·5 mph): 47
- Min sinking speed at 40·5 knots (75 km/h; 46·5 mph): 0·50 m (1·64 ft)/sec
- Stalling speed: 35 knots (65 km/h; 40·5 mph)
- Max speed (smooth air): 135 knots (250 km/h; 155 mph)
- Max speed (rough air): 89 knots (165 km/h; 102·5 mph)
- Max aero-tow speed: 81 knots (150 km/h; 93 mph)
- *g* limits: +5·3; −2·65

SZD-41A JANTAR STANDARD (AMBER)

This high-performance single-seat Standard Class sailplane was designed by Ing W. Okarmus on the basis of the prototype Open Class SZD-38 Jantar-1. The fuselage and tail unit are the same for both types; the wings of the SZD-41A are designed to OSTIV Standard Class requirements.

The Jantar Standard was flown for the first time at Bielsko-Biala on 3 October 1973, piloted by A. Zientek. Polish pilots flying the SZD-41 in the 1974 World Championships at Waikerie, Australia, gained third and seventh places in the Standard Class competition. Jantar Standards gained fourth, sixth and eighteenth places in the 1976 World Championships at Räyskälä, Finland in June 1976. The current production version is designated SZD-41A.

A total of 17 Jantar Standards had been built by the beginning of 1976.

The description of the SZD-38 applies also to the SZD-41A, except in the following details:

TYPE: Single-seat Standard Class sailplane.

WINGS: Cantilever shoulder-wing monoplane. Wing section NN-8. Dihedral 1° 30′. Sweepback 0° 30′ at quarter-chord. Single-spar ribless structure with foam-filled glassfibre/epoxy resin sandwich skin. Multi-hinged ailerons, hinged on the upper surface, are actuated by pushrods. No flaps. Provision for 80 kg (176 lb) water ballast.

DIMENSIONS, EXTERNAL:
- Wing span: 15·00 m (49 ft 2½ in)
- Wing chord at root: 0·95 m (3 ft 1½ in)
- Wing chord at tip: 0·45 m (1 ft 5¼ in)
- Wing chord (mean): 0·71 m (2 ft 4 in)
- Wing aspect ratio: 21·1

AREAS:
- Wings, gross: 10·66 m² (114·7 sq ft)
- Ailerons (total): 0·53 m² (5·70 sq ft)
- Airbrakes (total): 0·478 m² (5·15 sq ft)

WEIGHTS AND LOADINGS:
- Weight empty, equipped: 250 kg (551 lb)
- Max T-O weight:
  - with water ballast: 440 kg (970 lb)
  - without water ballast: 360 kg (793 lb)
- Max wing loading:
  - with water ballast: 41·3 kg/m² (8·46 lb/sq ft)
  - without water ballast: 33·8 kg/m² (6·92 lb/sq ft)

PERFORMANCE (without water ballast):
- Best glide ratio at 56·5 knots (105 km/h; 65 mph): 40
- Min sinking speed at 42 knots (78 km/h; 48·5 mph): 0·62 m (2·03 ft)/sec
- Stalling speed: 37 knots (68 km/h; 42·5 mph)
- Max speed (smooth air): 135 knots (250 km/h; 155 mph)
- Max speed (rough air): 86 knots (160 km/h; 99 mph)
- Max aero-tow speed: 81 knots (150 km/h; 93 mph)
- *g* limits: +5·3; −2·65

SZD-42 JANTAR-2 (AMBER)

This high-performance single-seat Open Class sailplane was developed by Dipl-Ing Adam Kurbiel from the series-built all-plastics SZD-38A Jantar-1, and flew for the first time on 2 February 1976.

The first two examples were flown by the Polish team in the 1976 World Championships at Räyskälä, Finland, where they gained second and third places in the Open Class; these two aircraft have wings built in two sections. The series-built version, to which the following description applies, have four-piece wings; one of these aircraft, flown by R. Johnson of the USA, gained seventh place in the 1976 World Championships.

TYPE: Single-seat high-performance Open Class sailplane.

WINGS: Generally similar to SZD-38A, but built in four parts and having 2° dihedral, light alloy DFS-type airbrakes and provision for 130 litres (28·6 Imp gallons) of water ballast. Flap/aileron pushrods carried in ball-bearings.

FUSELAGE: As described for SZD-38A.

TAIL UNIT: Cantilever cruciform structure of glass-fibre/epoxy resin. Fin, integral with fuselage, carries internally-mounted VHF aerial. Pushrod-operated elevator, with spring trim.

LANDING GEAR: As described for SZD-38A, except for monowheel tyre pressure of 2·96 bars (43 lb/sq in) and tailskid instead of tailwheel.

ACCOMMODATION AND EQUIPMENT: As described for SZD-38A.

DIMENSIONS, EXTERNAL: As for SZD-38A, except:
- Wing span: 20·50 m (67 ft 3 in)
- Wing chord at tip: 0·395 m (1 ft 3½ in)
- Wing chord (mean): 0·731 m (2 ft 4¾ in)
- Wing aspect ratio: 29·2
- Height over tail: 1·76 m (5 ft 9¼ in)

AREAS: As for SZD-38A, except:
- Wings, gross: 14·25 m² (153·4 sq ft)
- Ailerons (total): 1·15 m² (12·38 sq ft)
- Trailing-edge flaps (total): 1·38 m² (14·85 sq ft)
- Fin: 0·72 m² (7·75 sq ft)
- Rudder: 0·48 m² (5·17 sq ft)
- Elevators (total): 0·38 m² (4·09 sq ft)

WEIGHTS AND LOADINGS:
- Weight empty, equipped: 343 kg (756 lb)
- Max T-O weight:
  - with water ballast: 593 kg (1,307 lb)
  - without water ballast: 463 kg (1,020 lb)
- Max wing loading:
  - with water ballast: 41·6 kg/m² (8·52 lb/sq ft)
  - without water ballast: 32·5 kg/m² (6·66 lb/sq ft)

PERFORMANCE:
- Best glide ratio:
  - with water ballast, at 55 knots (102 km/h; 63·5 mph): 47
  - without water ballast, at 47·5 knots (88 km/h; 54·5 mph): 47
- Min sinking speed:
  - with water ballast, at 47 knots (87 km/h; 54 mph): 0·54 m (1·77 ft)/sec
  - without water ballast, at 40·5 knots (75 km/h; 46·5 mph): 0·46 m (1·51 ft)/sec
- Stalling speed:
  - with water ballast: 43·5 knots (80 km/h; 50 mph)
  - without water ballast: 35·5 knots (65 km/h; 40·5 mph)
- Max speed (smooth air):
  - flaps up: 135 knots (250 km/h; 155 mph)
  - flaps down: 89 knots (165 km/h; 102·5 mph)
- Max speed (rough air):
  - flaps up: 86 knots (160 km/h; 99 mph)
  - flaps down: 75·5 knots (140 km/h; 87 mph)
- Max aero-tow speed: 75·5 knots (140 km/h; 87 mph)
- g limits:
  - with water ballast: +4·0; −1·5
  - without water ballast: +5·3; −2·65

SZD-45A OGAR (GREYHOUND)

This two-seat school and training powered sailplane was designed by Dipl Ing Tadeusz Labuc, originally with a 33·5 kW (45 hp) Stamo engine and later with a 50·7 kW (68 hp) Sportavia Limbach engine. The first prototype (SP-0001) made its first flight on 29 May 1973, and was described in the 1973-74 *Jane's*.

Production of 16 Ogars, intended for service with the Polish Aero Club, had been completed by the beginning of 1976.

TYPE: Two-seat powered school and training sailplane.

WINGS: Cantilever shoulder-wing monoplane. Wortmann wing sections: FX-61-168 at root, FX-60-1261 at tip. Dihedral 1° 30′. Incidence 1°. Sweepback 1° at quarter-chord. Single-spar wooden structure, with moulded plywood stressed skin covered with glassfibre. Slotless ailerons of glassfibre sandwich construction, controlled by pushrods. Airbrake above and below each wing.

FUSELAGE: Pod and boom type. Main nacelle structure is a glassfibre/epoxy resin shell on two strong wooden frames, to carry the wings, engine mounting, fuel tank, and the tubular duralumin boom which supports the tail unit.

TAIL UNIT: Cantilever T-tail, the fin being integral with the metal tailboom.

LANDING GEAR: Semi-retractable monowheel (size 400 × 150), with shock-absorber and disc-type brake. Fully-castoring tailwheel. For school use, outrigger legs and wheels are mounted adjacent to the wingtips.

POWER PLANT: One 50·7 kW (68 hp) Limbach SL 1700 EC four-cylinder four-stroke aircooled engine, mounted behind the cabin and driving a two-blade Hoffmann pusher propeller. Fuel capacity 22 kg (48·5 lb).

ACCOMMODATION AND EQUIPMENT: Seats for two persons side by side under two-piece fully-transparent upward-hinged cockpit canopy. Dual controls standard. Backrests adjustable on ground, rudder pedals adjustable during flight. Engine controls are on a pedestal, mounted centrally beneath the panel, between the seats, with navigation instruments to the left and engine instruments to the right.

DIMENSIONS, EXTERNAL:
- Wing span: 17·53 m (57 ft 6¼ in)
- Wing chord at root: 1·578 m (5 ft 2 in)
- Wing chord at tip: 0·531 m (1 ft 8¾ in)
- Wing chord (mean): 1·089 m (3 ft 6¾ in)
- Wing aspect ratio: 16·2
- Length overall: 7·95 m (26 ft 1 in)
- Height over cabin roof: 1·15 m (3 ft 9¼ in)
- Height over tail: 1·72 m (5 ft 7¾ in)
- Tailplane span: 3·60 m (11 ft 9¾ in)
- Propeller diameter: 1·50 m (4 ft 11 in)

AREAS:
- Wings, gross: 19·10 m² (205·6 sq ft)
- Ailerons (total): 0·94 m² (10·12 sq ft)
- Fin: 0·96 m² (10·33 sq ft)
- Rudder: 0·82 m² (8·83 sq ft)
- Tailplane: 2·70 m² (29·06 sq ft)
- Elevators (total): 0·78 m² (8·40 sq ft)

WEIGHTS AND LOADINGS:
- Weight empty: 473 kg (1,042 lb)
- Max T-O weight: 700 kg (1,543 lb)
- Max wing loading: 36·6 kg/m² (7·5 lb/sq ft)
- Max power loading: 13·81 kg/kW (22·71 lb/hp)

PERFORMANCE (at max T-O weight, powered):
- Never-exceed speed: 121 knots (225 km/h; 139·5 mph)
- Max level speed at S/L: 97·5 knots (180 km/h; 112 mph)
- Cruising speed at S/L: 75·5 knots (140 km/h; 87 mph)
- Stalling speed: 37 knots (68 km/h; 42·5 mph)
- Max rate of climb at S/L: 159 m (522 ft)/min
- Service ceiling: 3,100 m (10,175 ft)
- T-O run: 200 m (656 ft)
- T-O to 15 m (50 ft): 576 m (1,890 ft)
- Range: 296 nm (550 km; 341 miles)

PERFORMANCE (at max T-O weight, unpowered):
- Best glide ratio at 51 knots (95 km/h; 59 mph): 22·6
- Min sinking speed at 43·5 knots (80 km/h; 50 mph): 1·10 m (3·61 ft)/sec

ROMANIA

ICA-BRASOV
INTREPRINDEREA DE CONSTRUCTII AERONAUTICE (Aircraft Construction Factory)

HEAD OFFICE AND WORKS: Brasov

As detailed in the Aircraft section, the current activities of the Romanian aircraft industry are divided between two industrial centres, IRMA in Bucharest and ICA at Brasov. In addition to its work on powered aircraft, the ICA is responsible for all sailplane development and production previously undertaken by URMV-3 (up to 1959) and IIL (Ghimbav) up to 1968. The principal Romanian designer of sailplanes is Dipl Ing Iosif Silimon, whose designs are prefixed with the letters IS. Details of his earlier designs have appeared in the 1961-62, 1965-66 and 1972-73 *Jane's*. Descriptions follow of those currently in production or under development:

IS-28

This tandem two-seat school and training sailplane, designed by a team led by Dipl Ing Iosif Silimon, flew for the first time in August 1970. It was certificated in December 1971, but is no longer in production.

A description and illustration can be found in the 1975-76 *Jane's*.

IS-28B

Despite its similar designation, this high-performance training sailplane represents a considerable advance over the IS-28 (1975-76 *Jane's*), from which it differs principally in having 17 m span all-metal wings, a longer and more slender fuselage, and reduced wing and tailplane dihedral. Design began in the Autumn of 1971, and the first IS-28B made its first flight on 26 April 1973.

There have been two versions, as follows:

IS-28B1. Version without wing flaps. Not now in production.

IS-28B2. Current (1976) production version, with Schempp-Hirth type (instead of DFS type) airbrakes and trailing-edge split flaps.

TYPE: Tandem two-seat high-performance training sailplane.

WINGS: Cantilever mid-wing monoplane. Wortmann wing sections: FX-61-163 at root, FX-61-126 at tip. Forward-swept all-metal wings, attached to fuselage by two adjustable tapered bolts at leading-edge and two fixed tapered bolts at trailing-edge. L-section main spar booms and dural web, dural auxiliary spar, and dural ribs. DFS type (B1) or Schempp-Hirth type (B2) metal airbrake above and below each wing. Ailerons and (B2 only) split trailing-edge flaps are fabric-covered, except for metal leading-edges. No tabs.

FUSELAGE: All-metal semi-monocoque structure, of oval cross-section, made up of frames, stringers and duralumin skin. Rear fuselage is a duralumin tubular monocoque.

TAIL UNIT: Cantilever T-tail, of generally similar design to that of IS-28 but with single-spar fin and less tailplane dihedral. Elevator trailing-edges and rudder fabric-covered. Trim tab in each elevator.

LANDING GEAR: Semi-retractable monowheel with oleo-pneumatic shock-absorber and disc brake. Manual retraction. Tyre size 330 × 135. Sprung tailskid, with rubber-block shock-absorber.

ACCOMMODATION: Seats for two persons in tandem under one-piece Plexiglas canopy, which hinges sideways (to starboard) and can be jettisoned in flight. Small sliding window and additional air inlets for ventilation. Dual controls standard.

EQUIPMENT: Conventional stick and rudder-pedal controls. Lever-controlled airbrakes. Monowheel retraction and extension controlled by lever on right-hand side of front cockpit. Nose towing hook, with Tost cable release, is standard; CG towing hook optional.

DIMENSIONS, EXTERNAL:
- Wing span: 17·00 m (55 ft 9¼ in)
- Wing aspect ratio: 16
- Length overall: 8·17 m (26 ft 9½ in)
- Height over tail: 1·80 m (5 ft 10¾ in)

AREA:
- Wings, gross: 18·24 m² (196·3 sq ft)

WEIGHTS AND LOADINGS:
- Weight empty:
  - B1: 340 kg (749 lb)
  - B2: 360 kg (793 lb)
- Max T-O weight (B1, B2):
  - single-seat: 500 kg (1,102 lb)
  - two-seat: 590 kg (1,300 lb)
- Max wing loading (B1, B2):
  - single-seat: 27·41 kg/m² (5·61 lb/sq ft)
  - two-seat: 32·35 kg/m² (6·62 lb/sq ft)

PERFORMANCE (B1 single-seat at 440 kg; 970 lb and two-seat at 530 kg; 1,168 lb AUW; B2 single-seat at 480 kg; 1,058 lb and two-seat at 540 kg; 1,190 lb AUW):
- Best glide ratio:
  - single-seat at 51 knots (94 km/h; 58 mph):
    - B1: 33
    - B2: 34
  - two-seat at 54 knots (100 km/h; 62 mph):
    - B1: 33
    - B1: 34
- Min sinking speed:
  - single-seat at 38 knots (70 km/h; 43·5 mph):
    - B1: 0·60 m (1·97 ft)/sec
    - B2: 0·62 m (2·03 ft)/sec
  - two-seat at 39 knots (72 km/h; 45 mph):
    - B1: 0·67 m (2·20 ft)/sec
    - B2: 0·69 m (2·26 ft)/sec
- Stalling speed:
  - single-seat:
    - B1: 34 knots (63 km/h; 39·5 mph)
    - B2: 33·5 knots (62 km/h; 39 mph)
  - two-seat:
    - B1: 38 knots (70 km/h; 43·5 mph)
    - B2: 39 knots (72 km/h; 45 mph)
- Never-exceed speed (both versions, single- or two-seat): 143 knots (266 km/h; 165 mph)
- g limits:
  - single-seat: +6·5; −4·0
  - two-seat: +5·3; −2·65

IS-28M

Two powered sailplane versions of the IS-28B2 have been developed. The rear fuselage, tail unit and main wings are virtually unchanged from the IS-28B2, but the powered versions are of low-wing configuration and have redesigned forward fuselages, cockpit canopies and main

SZD-45A Ogar side-by-side two-seat powered training sailplane

IS-28M2 side-by-side two-seat powered sailplane *(J. M. G. Gradidge)*

ICA-Brasov IS-29E single-seat Open Class sailplane

IS-28B2 tandem two-seat high-performance training sailplane

IS-29D2 single-seat Standard Class sailplane produced by ICA-Brasov

IS-29G, club version of this design by Dipl Ing Silimon, with 16·5 m span all-metal wings

landing gear. The wings of both versions can be folded from a point immediately inboard of the ailerons.

The two versions are designated as follows:

IS-28M1. Tandem two-seater, with main landing gear comprising a retractable central monowheel, with balloon tyre, and underwing outrigger wheels. Otherwise generally similar to IS-28M2, except that wing span is increased to 18·00 m (59 ft 0¾ in). Prototype under construction in Autumn 1976.

IS-28M2. Side-by-side two-seater, as described in detail, with main landing gear comprising two retractable wheels side by side under fuselage centre-section. Prototype (YR-1013) flew for the first time on 26 June 1976, and had completed approx 23 hr flying before arriving in UK on 2 September 1976. On 3 September it was reassembled at Lasham (assembly time 35 min), and two days later it made its international public debut at the Farnborough International air show. The first 10 aircraft of this type have been allocated to the UK, where they are marketed by Morisonics Ltd of The Parade, Frimley, Surrey.

The following description applies to the IS-28M2:

Type: Two-seat powered sailplane.

Wings: Cantilever low-wing monoplane, of mainly metal construction. Wortmann wing sections: FX-61-163 at root, FX-60-126 at tip. Dihedral 2°. Sweepforward 2° 30′ at quarter-chord. Single-spar structure, with aluminium ribs and skin, attached to fuselage by aluminium fittings, double panels and fairings. Fabric-covered metal ailerons and (optionally) all-metal split flaps on trailing edges. Flaps can be set to a negative position to aid high-speed gliding between thermals. All-metal two-section Hütter airbrakes on upper surfaces. Flaps, ailerons and airbrakes actuated by push-rods. Optional folding of outer wing panels, inward over inboard panels.

Fuselage: Conventional structure, in three parts: oval-section front portion, built up on two longerons and cross-frames and of metal construction except for glassfibre fairings and engine cowling panels; aluminium alloy monocoque centre portion; and rear portion of aluminium alloy frames and skin.

Tail Unit: Cantilever T-tail, with dihedral on horizontal surfaces. Single-spar fin and tailplane, entirely of aluminium alloy. Rudder and elevators also of aluminium alloy except for trailing-edges, which are fabric-covered. All tail control surfaces cable-operated. Trim tab in each elevator.

Landing Gear: Semi-retractable two-wheel main gear, mounted at inboard ends of wings and integral with fuselage. Mechanical extension and retraction. Rubber-ring shock-absorbers. Mechanically-operated main-wheel brakes. Steerable, non-retractable tail-wheel. Tyre sizes 340 × 125 mm on main units, 200 × 50 mm on tailwheel. Tyre pressure (all units) 2·45 bars (35·6 lb/sq in).

Power Plant: One 50·7/53·7 kW (68/72 hp) Limbach SL 1700 EI flat-four engine, driving a Hoffmann HO-V-62R two-blade variable-pitch fully-feathering propeller. Single fuel tank aft of cockpit, capacity (all usable) 36 litres (7·9 Imp gallons).

Accommodation: Reclinable seats for two persons side by side, with safety harnesses and full dual controls. Rearward-sliding Plexiglas jettisonable canopy, which can be opened in flight for cockpit ventilation. Adjustable vent for ram-air ventilation.

Systems and Equipment: Electrical system for fuel pump, alternator and engine starter motor, supplied by a Varta 12V 15Ah 65A battery and a Ducellier 7532 12V 22A generator. VHF antenna standard; provision for radio and navigation equipment to customer's requirements. Audible warning system, linked to airbrake selection lever, operates if main gear is not down for landing.

Dimensions, external:	
Wing span	17·00 m (55 ft 9¼ in)
Wing chord (mean)	1·17 m (3 ft 10 in)
Wing aspect ratio	15·8
Width, wings folded	9·00 m (29 ft 6¼ in)
Length overall	7·50 m (24 ft 7¼ in)
Fuselage: Max width	1·10 m (3 ft 7¼ in)
Max depth	1·10 m (3 ft 7¼ in)
Height overall	2·15 m (7 ft 0¾ in)
Tailplane span	3·478 m (11 ft 5 in)
Wheel track	1·36 m (4 ft 5½ in)
Wheelbase	5·40 m (17 ft 8½ in)
Propeller diameter	1·60 m (5 ft 3 in)
Propeller ground clearance (min)	0·15 m (6 in)
Dimensions, internal:	
Cockpit: Length	1·435 m (4 ft 8½ in)
Max width	1·10 m (3 ft 7¼ in)
Max height	0·975 m (3 ft 2½ in)
Areas:	
Wings, gross	18·24 m² (196·3 sq ft)
Ailerons (total)	2·574 m² (27·71 sq ft)
Flaps (optional, total)	1·936 m² (20·84 sq ft)
Airbrakes (total)	0·885 m² (9·53 sq ft)
Fin	0·90 m² (9·69 sq ft)
Rudder	0·60 m² (6·46 sq ft)
Tailplane	1·365 m² (14·69 sq ft)
Elevators (total, incl tabs)	1·365 m² (14·69 sq ft)
Weights and Loadings:	
Manufacturer's bare weight	500 kg (1,102 lb)
Weight empty, equipped	530 kg (1,168 lb)
Max T-O weight	730 kg (1,609 lb)
Max wing loading	40·0 kg/m² (8·19 lb/sq ft)
Power loading (50·7 kW; 68 hp)	14·4 kg/kW (23·7 lb/hp)

PERFORMANCE (at max T-O weight, powered):
Never-exceed speed
114·2 knots (211·7 km/h; 131·5 mph)
Max level speed 108 knots (200 km/h; 124 mph)
Design manoeuvring speed
93·6 knots (173·6 km/h; 107·9 mph)
Max cruising speed 92 knots (170 km/h; 106 mph)
Econ cruising speed 89 knots (165 km/h; 102·5 mph)
Fuel consumption at 81 knots (150 km/h; 93 mph)
11·5 litres (2·5 Imp gallons)/hr
Stalling speed 35·5 knots (65 km/h; 40·5 mph)
Max rate of climb at S/L 186 m (610 ft)/min
Service ceiling 5,000 m (16,400 ft)
T-O run 160 m (525 ft)
Landing run 65 m (213 ft)
Range 242 nm (450 km; 280 miles)
g limits +5·3; −2·65

PERFORMANCE (at max T-O weight, unpowered):
Best glide ratio at 54 knots (100 km/h; 62 mph) 30
Min sinking speed at 43 knots (80 km/h; 50 mph)
0·87 m (2·85 ft)/sec
Stalling speed, flaps up 40 knots (74 km/h; 46 mph)
Stalling speed, flaps down 38 knots (70 km/h; 43·5 mph)

IS-29

The IS-29, designed under the leadership of Dipl Ing Iosif Silimon, can be adapted to suit a variety of requirements or weather conditions. All versions have an identical fuselage and tail unit, and a choice of wings is available. The versions so far announced are as follows:

IS-29B. Standard Class version, with all-wooden wings of 15 m span. First flown April 1970; certificated September 1970. Not now in production. Described in 1975-76 *Jane's*.

IS-29D. Standard Class version, with all-metal 15 m span wings. First flown in November 1970; certificated in 1971. Approx 30 built by 1974. Described in 1975-76 *Jane's*. Current production version (1976), designated **IS-29D2**, has improved cockpit and controls, Hütter type airbrakes, separate tailplane and elevator for better longitudinal stability, and improved rigging system.

IS-29E. High-performance Open Class version, having increased-span wings, fitted with ailerons, flaps, Schempp-Hirth type airbrakes and integral water ballast tanks. First flown in August 1971 with 17·60 m (57 ft 9 in) span wings, as described in 1975-76 *Jane's*. Current production version (1976) is the **IS-29E3**, with 20 m span wings; a 19 m version, the **IS-29E2**, is under development for introduction in 1977.

IS-29G. Club version, with all-metal 16·5 m span wings. Prototype completed in 1972. Described in 1975-76 *Jane's*.

The following description applies to all current versions, except where a specific model is indicated:

TYPE: Single-seat sailplane.

WINGS: Cantilever shoulder-wing monoplane, with constant taper from root to tip. Wortmann wing sections: FX-61-165 at root, FX-61-124 at tip on D2; FX-K-170 (root) and FX-K-150 (tip) on E2 and E3. All-metal structure, with I-section main spar and false rear spar and riveted dural skin. Full-span trailing-edge flaps and ailerons, which are coupled to operate in unison but can be disconnected for separate operation during landing. Airbrakes (see under model listings for type) in upper and lower surface of each wing (upper surface only on D2). No tabs.

FUSELAGE: All-metal semi-monocoque structure of frames, stringers and dural skin, identical on all versions except for local variations at wing attachment points. Detachable glassfibre nosecap.

TAIL UNIT: Cantilever all-metal T-tail, with full-span elevator. Two-spar fin, integral with fuselage, and single-spar rudder.

LANDING GEAR: Manually-retractable monowheel, with brake, and non-retractable tailwheel.

ACCOMMODATION: Single adjustable seat under two-piece Plexiglas canopy, the rear portion of which hinges sideways to starboard and can be jettisoned in flight. Sliding window and additional inlet for cockpit ventilation.

EQUIPMENT: Conventional stick and rudder-pedal controls, consisting of levers, bars, cables and torsion axes beneath seat and cabin floor. Airbrakes, flaps, trim tabs and wheel brake are controlled by levers on left-hand side of cockpit, monowheel extension and retraction by a lever on the right. Towing hooks in nose (flush) and under fuselage.

DIMENSIONS, EXTERNAL:
Wing span:
D2 15·00 m (49 ft 2½ in)
E2 19·00 m (62 ft 4 in)
E3 20·00 m (65 ft 7½ in)
Wing aspect ratio:
D2 21·5
Length overall:
D2, E2, E3 7·38 m (24 ft 2½ in)
Height over tail (all versions) 1·68 m (5 ft 6¼ in)

AREAS:
Wings, gross:
D2 10·40 m² (111·9 sq ft)
E2 13·60 m² (146·4 sq ft)
E3 14·68 m² (158·0 sq ft)

WEIGHTS AND LOADINGS:
Weight empty:
D2 235 kg (518 lb)
E2 320 kg (705 lb)
E3 350 kg (771 lb)
Max T-O weight:
D2 360 kg (794 lb)
E2 (without water ballast) 440 kg (970 lb)
E3 (with water ballast) 530 kg (1,168 lb)
Max wing loading:
D2 34·62 kg/m² (7·09 lb/sq ft)
E2 (without water ballast) 32·35 kg/m² (6·62 lb/sq ft)
E3 (with water ballast) 36·10 kg/m² (7·39 lb/sq ft)

PERFORMANCE:
Best glide ratio:
D2 and E3 at 50 knots (93 km/h; 58 mph) 48
E2 at 48·5 knots (90 km/h; 56 mph) 46·5
Min sinking speed:
D2 and E3 at 43 knots (80 km/h; 50 mph)
0·43 m (1·41 ft)/sec
E2 at 42 knots (78 km/h; 48·5 mph)
0·50 m (1·64 ft)/sec
Stalling speed:
D2, E3 36·5 knots (67 km/h; 42 mph)
E2 33·5 knots (62 km/h; 39 mph)
Never-exceed speed:
D2, E3 135 knots (250 km/h; 155 mph)
E2 136 knots (253 km/h; 157 mph)
g limits:
D2, E2, E3 +5·3; −2·65

SWITZERLAND

NEUKOM
ALBERT NEUKOM SEGELFLUGZEUBAU

ADDRESS:
Flugplatz Schmerlat, CH-8213 Neuenkirch

Latest sailplanes built by Mr Neukom and three fellow-workers are the Standard Elfe 15 and 17 and Super-Elfe AN-66C, of which all available details follow:

NEUKOM S-4A ELFE 15

The S-4A Elfe 15 is a developed version of the Standard Class S-3 (see 1974-75 *Jane's*), from which it differs principally by having a two-piece strengthened wing with Schempp-Hirth airbrakes and a more roomy forward fuselage of all-plastics construction and better aerodynamic form. The prototype flew for the first time in 1970, and 10 had been built by early 1973. Production continues, though at a relatively slow rate. The Elfe 15 is also available in kit form for amateur construction.

The Elfe 15 is a cantilever high-wing monoplane, utilising Wortmann FX-61-163 and FX-60-126 wing root and tip sections. The wing, which is built in two parts, has an aluminium alloy main spar, is of plywood and foam sandwich construction, and is fitted with Schempp-Hirth airbrakes. Fuselage and tail unit are of glassfibre and plywood/foam sandwich construction. The landing gear comprises a rubber-sprung retractable monowheel, size 330 × 130, with brake. The cockpit is fitted with a removable transparent canopy.

DIMENSIONS, EXTERNAL:
Wing span 15·00 m (49 ft 2½ in)
Wing aspect ratio 19
Length overall 7·10 m (23 ft 3½ in)
Height over tail 1·50 m (4 ft 11 in)
Tailplane span 2·90 m (9 ft 6¼ in)

AREAS:
Wings, gross 11·80 m² (127·0 sq ft)
Ailerons (total) 0·85 m² (9·15 sq ft)
Airbrakes (total) 1·15 m² (12·38 sq ft)

WEIGHTS AND LOADING:
Weight empty 230 kg (507 lb)
Max T-O weight 350 kg (771 lb)
Max wing loading 29·6 kg/m² (6·06 lb/sq ft)

PERFORMANCE:
Best glide ratio at 48·5 knots (90 km/h; 56 mph) 37
Min sinking speed at 38 knots (70 km/h; 43·5 mph)
0·60 m (1·97 ft)/sec
Stalling speed
approx 35·5 knots (65 km/h; 40·5 mph)
Max speed (rough or smooth air)
113 knots (210 km/h; 130 mph)
Max aero-tow speed 75·5 knots (140 km/h; 87 mph)
Max winch-launching speed
54 knots (100 km/h; 62 mph)

NEUKOM ELFE 17

The Elfe 17 is a 17 metre Open Class version of the S-4A, employing the same fuselage, but having a two-piece wing of increased span. One water ballast tank in each wing leading-edge to contain a total of 60 kg (132 lb) water. A braking parachute is carried on this version.

A total of 10 Elfe 17s had been built by the Spring of 1973. Production continues, though at a relatively slow rate. The Elfe 17 is available also in kit form for amateur construction.

DIMENSIONS, EXTERNAL:
Wing span 17·00 m (55 ft 9¼ in)
Wing aspect ratio 21·8
Length overall 7·10 m (23 ft 3½ in)

AREA:
Wings, gross 13·20 m² (142 sq ft)

WEIGHTS AND LOADING:
Weight empty 255 kg (562 lb)
Max T-O weight 380 kg (837 lb)
Max wing loading 28·8 kg/m² (5·90 lb/sq ft)

PERFORMANCE:
Best glide ratio at 48·5 knots (90 km/h; 56 mph) 39
Min sinking speed at 40·5 knots (75 km/h; 46·5 mph)
0·56 m (1·80 ft)/sec
Stalling speed
approx 35·5 knots (65 km/h; 40·5 mph)
Max speed (rough or smooth air)
113 knots (210 km/h; 130 mph)
Max aero-tow speed 75·5 knots (140 km/h; 87 mph)
Max winch-launching speed
54 knots (100 km/h; 62 mph)

NEUKOM SUPER-ELFE AN-66C

This is a development of the AN-66-2 (see 1972-73 *Jane's*) having the same fuselage, but with an entirely new wing of Eppler 562/569 section and conventional instead of V-type tail surfaces. The wing is of plywood/balsa/-plywood sandwich construction, with a single duralumin spar, and is built in three parts. The centre-section is 6·50 m (21 ft 4 in) in length, the two outer panels each 8·25 m (27 ft 0¾ in) in length. A water ballast tank to contain 60 kg (132 lb) of water is situated in the leading-edge of each wing, and Schempp-Hirth airbrakes are fitted. Main interest lies in a newly-designed aerofoil flap which is able to increase the wing area by about 20%.

Flight testing of a prototype began on 11 September 1973 at Butzweilerhof in Germany.

FUSELAGE: Forward portion is a glassfibre sandwich structure, rear portion a wooden semi-monocoque.

LANDING GEAR: Retractable monowheel and tailskid.

ACCOMMODATION: Single seat under long flush transparent canopy.

DIMENSIONS, EXTERNAL:
Wing span 23·00 m (75 ft 5½ in)
Wing aspect ratio:
flaps in 33·1
flaps out 27·6
Length overall 8·10 m (26 ft 6¾ in)
Height over tail 1·85 m (6 ft 0¾ in)

AREAS:
Wings, gross:
flaps in 16·00 m² (172 sq ft)
flaps out 19·20 m² (207 sq ft)

WEIGHTS AND LOADINGS:
Weight empty 420 kg (926 lb)
Normal T-O weight 530 kg (1,168 lb)
Max T-O weight with ballast 650 kg (1,433 lb)
Wing loading at normal T-O weight:
flaps in 33·1 kg/m² (6·8 lb/sq ft)
flaps out 27·6 kg/m² (5·65 lb/sq ft)
Wing loading at max T-O weight:
flaps in 40·6 kg/m² (8·3 lb/sq ft)
flaps out 33·8 kg/m² (6·9 lb/sq ft)

PERFORMANCE:
Best glide ratio at 48·5 knots (90 km/h; 56 mph) 48
Min sinking speed at 40·5 knots (75 km/h; 46·5 mph)
0·50 m (1·64 ft)/sec
Stalling speed 33 knots (60 km/h; 37·5 mph)
Max speed (smooth air)
145·5 knots (270 km/h; 168 mph)

Neukom Elfe 17 single-seat Open Class sailplane *(Pio dalla Valle)*

Neukom Super-Elfe AN-66C high-performance sailplane *(Pio dalla Valle)*

PILATUS
PILATUS FLUGZEUGWERKE AG

Head Office and Works:
CH-6370 Stans, near Lucerne
Telephone: (041) 61 14 46
Telex: 78 329
Officers: See Aircraft section

PILATUS B4-PC11

Pilatus designed and built this single-seat Standard Class sailplane for multi-purpose training and aerobatics.

Swiss certification was granted on 12 June 1972, and the first delivery was made shortly afterwards. By 1976, more than 250 had been delivered to customers in 20 countries in all parts of the world. Certification for full aerobatic manoeuvres was granted at the end of January 1975. The current version is designated **B4-PC11AF**.

Type: Single-seat Standard Class sailplane.

Wings: Cantilever shoulder-wing monoplane. Wing section NACA 64_3-618. Dihedral 1°. Incidence 1° 30′. No sweepback. Conventional light alloy structure with PVC ribs. Conventional ailerons of similar construction. Mid-chord light alloy spoilers on wing upper surfaces. No tabs.

Fuselage: Semi-monocoque light alloy structure.

Tail Unit: Cantilever light alloy T tail with PVC ribs. Fixed-incidence tailplane. Elevator spring trim.

Landing Gear: Retractable (optionally non-retractable) main wheel, and fixed tailwheel, in tandem. Tost main wheel and tyre size 5·50-5. No shock-absorbers. Mechanical drum brake. Main wheel faired by doors when retracted.

Accommodation: Single semi-reclining seat under transparent cockpit canopy. Canopy is hinged for access to cockpit and is jettisonable in flight. Cockpit is ventilated. Battery, radio and oxygen system optional.

Dimensions, external:
Wing span 15·00 m (49 ft 2½ in)
Wing chord at root 1·07 m (3 ft 6 in)
Wing chord at tip 0·43 m (1 ft 5 in)
Wing chord (mean) 0·94 m (3 ft 1 in)
Wing aspect ratio 16
Length overall 6·57 m (21 ft 6¾ in)
Height over tail 1·57 m (5 ft 1¾ in)

Areas:
Wings, gross 14·04 m² (151·1 sq ft)
Ailerons (total) 1·15 m² (12·38 sq ft)
Fin 0·90 m² (9·69 sq ft)
Rudder 0·46 m² (4·95 sq ft)
Tailplane 1·15 m² (12·38 sq ft)

Weights and Loading:
Weight empty, equipped 230 kg (507 lb)
Max T-O weight 350 kg (771 lb)
Max wing loading 25·0 kg/m² (5·1 lb/sq ft)

Performance (at max T-O weight):
Best glide ratio at 46 knots (85 km/h; 53 mph) 35
Min sinking speed at 40·5 knots (75 km/h; 47 mph) 0·64 m (2·1 ft)/sec
Never-exceed speed (rough or smooth air) 129 knots (240 km/h; 149 mph)
Max manoeuvring and max aero-tow speed 87·5 knots (163 km/h; 101 mph)
Max winch-launching speed 70 knots (130 km/h; 80·5 mph)
Stalling speed 33 knots (61 km/h; 38 mph)
g limits:
Normal +6·32; −4·32
Aerobatic +7·0; −4·79

THE UNITED KINGDOM

HOLMES
KENNETH HOLMES

Enquiries to:
H. A. Torode, 16 Gravel Walk, Emberton, Olney, Buckinghamshire

HOLMES KH-1

The KH-1 high-performance sailplane was designed and built by Mr Kenneth Holmes, a meteorologist, and flew for the first time on 24 November 1971. A description and illustration of the first KH-1 appeared in the 1974-75 *Jane's*.

Following some 80 hr of flying, this prototype is now in storage. A second example, with new front fuselage contours, modified centre-section structure and simplified wing construction, is being built by Mr J. Halford at Moreton-in-the-Marsh, Gloucestershire, and was expected to be completed during 1976.

SLINGSBY
VICKERS-SLINGSBY (A Division of Vickers Limited Offshore Engineering Group)

Head Office and Works:
Kirkbymoorside, Yorkshire YO6 6EZ
Telephone: Kirkbymoorside (0751) 31751
Telex: 57911
Managing Director: G. E. Burton, BSc, ARCS
Technical Director:
J. S. Tucker, Dip Tech(Eng), BSc(Eng), CEng, MRAeS
Production Manager:
P. Smith

Vickers-Slingsby, a division of Vickers Ltd Offshore Engineering Group, was formed from the assets of the former Slingsby Aircraft Company, which went into liquidation in July 1969.

Present aircraft production is devoted to the Slingsby T.59H 22 metre Kestrel, which is a development of the Glasflügel 17 metre Kestrel described under the GHH heading in the German section. Production of the T.59D Kestrel 19 (1975-76 *Jane's*) has ended.

The company has a Ministry of Defence development contract for a new, glassfibre wing spar for the T.61 Falke; this was expected to lead in 1976 to a production contract for a batch of 15 of these aircraft for the Air Training Corps. A prototype is under construction, and sample spars and wings are being built for structural testing.

A total of 310 people were employed by Slingsby in January 1976; most of these were engaged in the manufacture of high technology glassfibre products for the Ministry of Defence (Navy) and for North Sea oil exploration.

Vickers-Slingsby has also publicised a 15 m Standard Class sailplane known as the **Vega,** to be available for delivery from June 1977, but no details of this were received for publication.

SLINGSBY T. 59H (GLASFLÜGEL) KESTREL 22 Series 2

Production of Slingsby-built Kestrels began in June 1971, and 102 had been delivered by October 1975, of which 96 were of the 19 metre version. The total also includes five 17 metre Kestrels, and a prototype 22 metre version produced in 1974 by manufacturing two 1·50 m (4 ft 11 in) stub wings to insert into the span of the 19 metre Kestrel.

Slingsby also completed a special 19 metre Kestrel, designated T.59C and having a carbon-fibre main spar. This aircraft made its first flight on 7 May 1971.

In 1973 Kestrel 19 m sailplanes achieved first place in the Finnish national championships and shared, with the German-built Kestrel 17 m, second to ninth places (inclusive) in the British championships. They also secured six of the first 28 places in the 1974 World Championships at Waikerie, Australia.

A description of the standard 17 metre Kestrel appears under the GHH heading in the German section; the following description applies to the Slingsby T.59H 22 metre version:

Type: Single-seat high-performance Open Class sailplane.

Wings: Four-piece wing, jointed at flap/aileron junction. Carbon-fibre main spar. Wortmann wing sections: FX-67-K-170 at root, FX-67-K-150 at tip. Dihedral 3° 45′. Incidence 0° 30′. Schempp-Hirth airbrake on each upper and lower surface.

Fuselage: Similar to 17 metre Kestrel up to just aft of canopy, beyond which a 750 mm (29·5 in) long additional section is inserted which considerably reduces the 'waisting' of the 17 metre version.

Tail Unit: Cantilever T tail. Tailplane same as for 17 metre version; fin and rudder area increased by approx 25%. Rudder lightened by fabric-covered cutout sections.

Landing Gear: Retractable Gerdes monowheel, size 5·00 × 5, tyre pressure 3·79 bars (55 lb/sq in). No shock-absorption. Disc brake, actuated hydraulically via master cylinder and cable from airbrake cross-shaft. Fixed tailwheel.

Accommodation: As 17 metre version.

Dimensions, external:
Wing span 22·00 m (72 ft 2¼ in)
Wing chord (mean) 0·702 m (2 ft 3½ in)
Wing aspect ratio 31·35
Length overall 7·55 m (24 ft 9¼ in)
Height over tail 1·94 m (6 ft 4¼ in)
Tailplane span 2·85 m (9 ft 4¼ in)

Areas:
Wings, gross (incl fuselage section) 15·44 m² (166·2 sq ft)
Ailerons (total) 0·78 m² (8·40 sq ft)
Flaps (total) 1·72 m² (18·51 sq ft)
Airbrakes (total frontal area, incl gap) 1·238 m² (13·33 sq ft)
Fin 0·713 m² (7·67 sq ft)
Rudder 0·578 m² (6·22 sq ft)
Tailplane 1·30 m² (14·00 sq ft)
Elevators 0·32 m² (3·45 sq ft)

Weights and Loading:
Weight empty, equipped (with instruments) 390 kg (860 lb)
Max T-O weight (with ballast) 659 kg (1,453 lb)
Max wing loading 42·68 kg/m² (8·74 lb/sq ft)

Performance (at max T-O weight):
Best glide ratio at 56 knots (104 km/h; 64·5 mph) 51·5
Min sinking speed at 46 knots (85 km/h; 53 mph) 0·48 m (1·57 ft)/sec
Max speed (smooth air) 135 knots (250 km/h; 155·5 mph)
Max speed (rough air) 105 knots (194·5 km/h; 121 mph)
Max aero-tow speed 81 knots (150 km/h; 93 mph)
Max winch-launching speed 70 knots (129·5 km/h; 80·5 mph)
g limits +5·3; −2·65

SLINGSBY T.61 (SCHEIBE SF-25B) FALKE

As previously recorded, Slingsby began producing the Scheibe SF-25B Falke powered sailplane under licence in 1970, and completed a total of 35. Production by Slingsby then ended, but as indicated in the introductory copy it was thought likely that manufacture of a further batch, with glassfibre main spars and able to carry a greater payload, would begin in November 1976.

Pilatus B4-PC11 single-seat fully-aerobatic high-performance Standard Class sailplanes

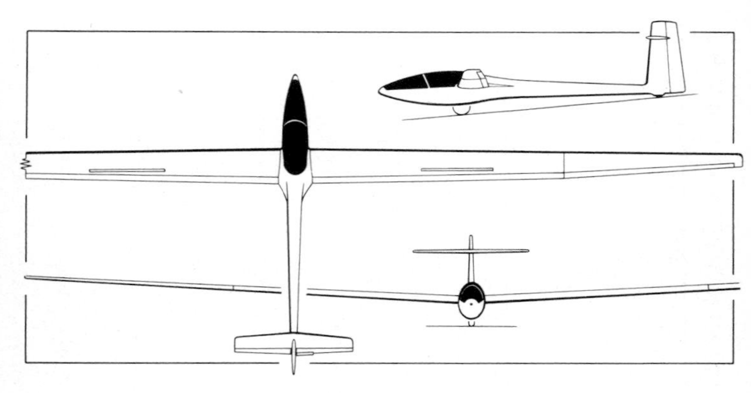

Slingsby T.59H 22 metre version of the Glasflügel Kestrel *(Michael A. Badrocke)*

Prototype Slingsby Kestrel 22, with carbon-fibre main spar

Ahrens AR 124 single-seat 13 metre sailplane

SWALES

SWALES SAILPLANES

ADDRESS:
Long Street, Thirsk, Yorkshire NR
Telephone: Thirsk 23096
OFFICERS:
Roy Swales
G. L. Kemp (Chief Inspector)

This company is producing two versions of the SD3 single-seat sailplane. This sailplane is not, as was reported in the 1975-76 *Jane's,* a development of either the Birmingham Guild BG 135 or the Yorkshire Sailplanes YS 55 Consort, neither of which continues in production. Swales Sailplanes is spares stockist for both of these earlier types, and also for the YS 53 Sovereign.

SWALES SD3-15T and SD-15V

Construction of the prototype of this single-seat sailplane started in September 1974, and it flew for the first time in March 1975. This aircraft was designated SD3-13V. Production aircraft are designated SD3-15V (first flight July 1975) and SD3-15T (first flight scheduled for June 1976), the suffix letter in each case denoting the tail configuration.

TYPE: Single-seat sailplane.

WINGS: Cantilever mid-wing monoplane. Wortmann FX-61-168 wing section. Dihedral 3°. Incidence 0°. No sweepback. Construction mainly of metal, with metal and polystyrene ribs and wingtips of glassfibre-reinforced plastics. All-metal trailing-edge flaps/airbrakes. Metal plain ailerons, with foam ribs. No spoilers.

FUSELAGE: Semi-monocoque structure of four longerons, metal stressed skin, and nosecone of glassfibre-reinforced plastics.

TAIL UNIT: Cantilever V (SD3-15V) or T (SD3-15T) tail, of metal construction with 50% foam ribs. T-tail version has full-span elevator with spring trim; V-tail version has all-moving surfaces with anti-balance tab.

LANDING GEAR: Non-retractable monowheel, size 5·00-5 × 6 in, with internally-expanding brake. Tyre pressure 2·41 bars (35 lb/sq in).

ACCOMMODATION: Single seat under one-piece flush-fitting transparent canopy which opens sideways to starboard.

DIMENSIONS, EXTERNAL:
Wing span 15·00 m (49 ft 2½ in)
Wing chord, mean 0·71 m (2 ft 4 in)
Wing aspect ratio 24
Length overall 6·10 m (20 ft 0 in)
Height over tail 1·30 m (4 ft 3 in)
Tailplane span 2·49 m (8 ft 2 in)

AREAS:
Wings, gross 9·48 m² (102·00 sq ft)
Fin 0·37 m² (4·00 sq ft)
Rudder 0·33 m² (3·50 sq ft)
Tailplane 0·56 m² (6·00 sq ft)

WEIGHTS AND LOADING:
Weight empty, equipped 218 kg (480 lb)
Max T-O weight 331 kg (730 lb)
Max wing loading 34·96 kg/m² (7·16 lb/sq ft)

PERFORMANCE (at max T-O weight):
Best glide ratio at 48 knots (88·5 km/h; 55 mph) 36
Min sinking speed at 42 knots (78 km/h; 48·5 mph) 0·73 m (2·40 ft)/sec
Stalling speed 34 knots (63·5 km/h; 39·5 mph)
Max speed (smooth air) 109 knots (201 km/h; 125 mph)
Max speed (rough air) 86 knots (159 km/h; 99 mph)
Max aero-tow speed 78 knots (145 km/h; 90 mph)
Max winch-launching speed 65 knots (121 km/h; 75 mph)
g limits +3·5; −1

THE UNITED STATES OF AMERICA

AHRENS

AHRENS AIRCRAFT CORPORATION

ADDRESS:
2800 Teal Club Road, Oxnard, California 93030
Telephone: (805) 985-2000
Telex: 65 9240
VICE-PRESIDENTS:
Kim Ahrens (Engineering)
Peter Ahrens

This company has built three prototypes of a single-seat sailplane, the AR 124. Orders for 20 production AR 124s had been received by Summer 1975.

AHRENS AR 124

Design and prototype construction of the AR 124 began in September 1974, and the V1 first prototype flew for the first time in December 1974. A photograph in the Addenda to the 1975-76 *Jane's* illustrated the V2 second prototype; the V3 third prototype, flown in August 1975, represents the production version, to which the following description applies:

TYPE: Single-seat sailplane.

WINGS: Cantilever mid-wing monoplane. Wing section NACA 64_3-618. Dihedral 2°. Constant-chord, all-aluminium square-tipped wings, with ailerons and upper-surface spoilers. No flaps or tabs.

FUSELAGE: All-aluminium elliptical-section flush-riveted monocoque structure.

TAIL UNIT: Cantilever T-tail, with sweptback vertical surfaces and constant-chord non-swept tailplane and one-piece elevator. No tabs.

LANDING GEAR: Non-retractable unsprung Gerdes monowheel, with Gerdes hydraulic disc brake, and tailwheel.

ACCOMMODATION: Single seat under one-piece moulded canopy. Basic flight instrumentation only.

DIMENSIONS, EXTERNAL:
Wing span 13·00 m (42 ft 7¾ in)
Wing chord (constant) 0·97 m (3 ft 2 in)
Wing aspect ratio 13
Length overall 6·10 m (20 ft 0 in)
Height over tail 1·57 m (5 ft 2 in)
Tailplane span 2·36 m (7 ft 9 in)

AREAS:
Wings, gross 12·54 m² (135·0 sq ft)
Spoilers (total) 0·37 m² (4·0 sq ft)

WEIGHTS AND LOADING:
Weight empty 158 kg (350 lb)
Max T-O weight 290 kg (640 lb)
Max wing loading 23·13 kg/m² (4·74 lb/sq ft)

PERFORMANCE:
Best glide ratio at 60 knots (111 km/h; 69 mph) 26
Stalling speed 30 knots (55·5 km/h; 34·5 mph)
Max speed (smooth air) 130 knots (241 km/h; 150 mph)
Max speed (rough air) 80 knots (148 km/h; 92 mph)
Max aero-tow speed 90 knots (166 km/h; 103 mph)
g limit +4·5

AmEAGLE

AmEAGLE CORPORATION

ADDRESS:
841 Winslow Court, Muskegon, Michigan 49441
Telephone: (616) 744-4220 (daytime); (616) 780-4680 (night)
PRESIDENT: Larry Haig

This company was formed by Mr Larry Haig, a 1958 graduate engineer from Wayne State University, following seven years experience in gas turbine engine development with the research laboratories of General Motors; a further seven years in automotive and diesel engine design and application work for the Allison Division of General Motors and Teledyne Continental Motors; and two years experience with composite laminate structures with Brunswick Corporation.

Its first product is a powered sailplane known as the American Eaglet.

AmEAGLE AMERICAN EAGLET

Design of the American Eaglet began in March 1969. Construction of the prototype started on 1 May 1975, and this aircraft (N101EA) made its first flight on 19 November of that year. See photograph on next page.

The aircraft is intended for amateur construction, and 66 sets of plans had been ordered by the Spring of 1976.

TYPE: Single-seat powered sailplane.

WINGS: Shoulder-wing monoplane, with single aluminium tube bracing strut on each side. Wing section Wortmann FX-61-184. Dihedral 1° from roots. Incidence 3°. No sweepback. Two-spar structure (Douglas fir) with urethane foam core, moulded glassfibre leading-edges and wingtips, and epoxy-bonded pre-curved glassfibre skin. Top-hinged ailerons of similar sandwich construction. Aluminium spoiler at 30% chord on each upper surface. No flaps or tabs.

FUSELAGE: Pod-and-boom structure, the forward portion comprising two pre-formed glassfibre half-shells pop-riveted to a framework of pre-formed tubular aluminium longerons. Tailboom is a thin-wall 6061-T6 aluminium tube with a moulded glassfibre tailcone. Combined pitot tube/lifting handle in nose.

TAIL UNIT: Cantilever inverted-Vee tailplane and elevators, of Wortmann FX-71-L-150/30 section, suspended from rear of tailboom.

LANDING GEAR: Manually-retractable nylon monowheel, with external friction-pad brake, forward of CG. Tyre size 10·5 × 3·5 × 5 in, pressure 0·83 bars (12 lb/sq in). Tailskid under tip of each half-tailplane.

POWER PLANT: One 9 kW (12·2 hp at 8,000 rpm) McCulloch 101B flat-twin two-stroke engine, installed aft of cockpit and driving a two-blade fixed-pitch pusher propeller with folding nylon blades. Recoil starting. Engine is for T-O and self-recovery only, and not for cross-country continuous operation. Fuel tank capacity 2 litres (0·5 US gallons). Normal fuel load of approx 1·81 kg (4 lb) (consumption at max power is 5·5 kg/hr; 12 lb/hr) is sufficient for one T-O and climb to 610 m (2,000 ft) and three airborne restarts and climbs from 150 m (500 ft) to 610 m (2,000 ft).

ACCOMMODATION: Single canvas sling seat under one-piece flush-fitting Plexiglas canopy.

EQUIPMENT: To builder's requirements: not supplied by AmEagle.

DIMENSIONS, EXTERNAL:
- Wing span: 10·97 m (36 ft 0 in)
- Wing chord at root: 0·76 m (2 ft 6 in)
- Wing chord at tip: 0·46 m (1 ft 6 in)
- Wing aspect ratio: 18
- Length overall: 5·03 m (16 ft 6 in)
- Height over tail: 0·86 m (2 ft 10 in)
- Tailplane span: 1·63 m (5 ft 4·3 in)
- Propeller diameter: 0·61 m (2 ft 0 in)
- Propeller ground clearance:
  - tail down: 76 mm (3 in)
  - tailboom horizontal: 178 mm (7 in)

AREAS:
- Wings, gross: 6·69 m² (72·00 sq ft)
- Ailerons (total): 0·65 m² (7·00 sq ft)
- Spoilers (total): 0·14 m² (1·50 sq ft)
- Tailplane: 0·645 m² (6·94 sq ft)
- Elevators (total): 0·28 m² (2·98 sq ft)

WEIGHTS AND LOADING:
- Weight empty: 68 kg (150 lb)
- Max T-O weight: 163 kg (360 lb)
- Max wing loading: 24·41 kg/m² (5·0 lb/sq ft)

PERFORMANCE:
- Best glide ratio at 52 knots (96·5 km/h; 60 mph): 24
- Min sinking speed at 39 knots (72·5 km/h; 45 mph): 0·95 m (3·1 ft)/sec
- Stalling speed:
  - in free air: 33 knots (61·5 km/h; 38 mph)
  - in ground effect: 29·5 knots (55 km/h; 34 mph)
- Max speed (smooth air): 100 knots (185 km/h; 115 mph)
- Max speed (rough air): 80 knots (148 km/h; 92 mph)
- Max aero-tow speed: 69·5 knots (129 km/h; 80 mph)
- T-O run: 230-305 m (750-1,000 ft)
- Landing run: 60-90 m (200-300 ft)
- *g* limits: +4·4; −2·3 (safety factor 1·5)

BEDE

BEDE AIRCRAFT INC

ADDRESS:
Newton Municipal Airport, Newton, Kansas 67114
Telephone: (316) 283-8870

PRESIDENT:
James R. Bede

BEDE BD-5S

The BD-5S was announced in 1975 as a sailplane version of the BD-5 Micro, from which it differs principally in having extended-span wings, modified landing gear, and a revised cockpit layout. A prototype was flown in 1975. Like other members of the BD-5 family, the BD-5S is intended for sale in plan and kit form for amateur construction.

A description and illustration of the prototype BD-5S appeared in the 1975-76 *Jane's*. Further development has been discontinued for the time being, pending completion of work on the BD-5 and BD-5G aircraft programmes (see Homebuilt Aircraft section).

BERKSHIRE

BERKSHIRE MANUFACTURING CORPORATION

ADDRESS:
Berkshire Valley Road, Oak Ridge, New Jersey 07438
Telephone: (201) 697-2020

CONCEPT 70

Concept 70 is a single-seat Standard Class sailplane, which first flew in 1970, a year after the start of design work. Construction is almost entirely of glassfibre.

Production began in 1973, and 16 Concept 70s had been built by Spring 1974, at which time FAA certification was nearing completion. No news of the company has been received since that time.

TYPE: Single-seat Standard Class sailplane.

WINGS: Cantilever shoulder-wing monoplane. Eppler-Wortmann wing sections, with thickness/chord ratio of 15%. Dihedral 2°. Incidence 2°. Glassfibre/PVC foam sandwich structure, with constant-chord centre-section and tapered outer panels. Aluminium 90° flap on each trailing-edge, between wing root and aileron. Ten-position trim. All control surfaces actuated by push/pull rods with ball-bearings.

FUSELAGE: Glassfibre monocoque structure, reinforced with steel tube frame in centre and cockpit sections.

TAIL UNIT: Cantilever type, of similar construction to wings. Fixed-incidence tailplane. Cable-actuated rudder; elevators actuated by push/pull rods with ball-bearings. Rudder pedals adjustable in flight.

LANDING GEAR: Manually-retractable Tost monowheel (size 4·00-4, tyre pressure 2·41 bars; 35 lb/sq in) and breakaway tailskid. Drum brake.

ACCOMMODATION: Single semi-reclining seat, recessed for American-type parachute; backrest adjustable in flight. One-piece flush Plexiglas canopy is hinged and jettisonable, and is fitted with sliding window and turnout air vent.

EQUIPMENT: Standard equipment includes basic cockpit instrumentation; retractable tow coupling; wave-trap antenna built into fin, with co-axial cable to panel; quickly-detachable instrument cover; oxygen bottle mounts; and two baggage compartments. Optional equipment includes wingtip protective wheels; 91 kg (200 lb) water ballast system; wheel-up warning system; radio (to customer's choice); tinted canopy; and Althaus venturi.

DIMENSIONS, EXTERNAL:
- Wing span: 15·00 m (49 ft 2½ in)
- Wing chord (centre-section, constant): 0·91 m (3 ft 0 in)
- Wing chord at tip: 0·43 m (1 ft 5 in)
- Wing aspect ratio: 20
- Length overall: 7·315 m (24 ft 0 in)
- Height over tail: 1·83 m (6 ft 0 in)
- Tailplane span: 2·565 m (8 ft 5 in)

AREA:
- Wings, gross: 11·52 m² (124·0 sq ft)

WEIGHTS AND LOADING:
- Weight empty, equipped: 226 kg (500 lb)
- Max T-O weight with ballast: 396 kg (875 lb)
- Max wing loading: 35·14 kg/m² (7·20 lb/sq ft)

PERFORMANCE (at max T-O weight):
- Best glide ratio at 52 knots (96·5 km/h; 60 mph): 40
- Min sinking speed at 43·5 knots (80·5 km/h; 50 mph): 0·62 m (2·03 ft)/sec
- Stalling speed, flaps up: 31·5 knots (58 km/h; 36 mph)
- Max speed (rough or smooth air): 105 knots (194·5 km/h; 121 mph)
- Max aero-tow speed: 78 knots (144·5 km/h; 90 mph)
- *g* limit: +6

BRIEGLEB

SAILPLANE CORPORATION OF AMERICA

ADDRESS:
El Mirage Rt, Box 101, Adelanto, California 92301
Telephone: (714) 388-4343

PRESIDENT: William G. Briegleb

Mr William G. Briegleb formed the Sailplane Corporation of America (see 1967-68 *Jane's*) to market gliders of his own design.

The BG 12 series of sailplanes are available as kits or plans for amateur construction. By early 1976, kits and/or plans had been purchased for 256 BG 12s and 43 BG 12-16s, and more than 95 aircraft were known to have been completed.

The company has also re-introduced, as a plans-only service at present, the 1940s-designed BG 6 and BG 7.

BRIEGLEB BG 6 and BG 7

The Briegleb BG 6 was originally type certificated in 1941, and was used both by civilians and by the US Army Air Corps. It was designed for home construction, student training, aero-tow and local soaring, and is built basically of steel tubing, wood and fabric. It can be built and licensed under FAR Pt 1 as a homebuilt aircraft, or licensed in the Standard category when built, inspected and test flown in accordance with the requirements of its type certificate.

The BG 7, although of similar construction, has an extended-span wing, with upper-surface spoilers, mated to the BG 6 fuselage and tail unit, and is a more advanced aircraft designed for higher performance and cross-country soaring. Although not possessing a type certificate, it was designed to the earlier CAR Pt 05; many of the required static tests were completed, and approx 90% of the stress data approved, when the second World War halted the certification programme. Today, the BG 7 may be built and licensed in the Amateur category of FAR Pt 1.

The general appearance of the BG 6 and 7 is shown in the accompanying illustration. At present, Sailplane Corporation of America is marketing only sets of plans for their construction.

DIMENSIONS, EXTERNAL:
- Wing span:
  - BG 6: 9·83 m (32 ft 3 in)
  - BG 7: 12·27 m (40 ft 3 in)
- Wing aspect ratio:
  - BG 6: 8·9
  - BG 7: 13·1
- Wing aerofoil section (both): NACA 4412
- Length overall: 5·58 m (18 ft 3⁷⁄₁₆ in)
- Height over wing: 1·30 m (4 ft 3 in)
- Tailplane span: 2·13 m (7 ft 0 in)

AREAS:
- Wings, gross:
  - BG 6: 10·87 m² (117·0 sq ft)
  - BG 7: 11·43 m² (123·0 sq ft)
- Ailerons (total):
  - BG 6: 1·85 m² (19·9 sq ft)
  - BG 7: 1·74 m² (18·7 sq ft)
- Vertical tail surfaces (total): 0·82 m² (8·8 sq ft)
- Horizontal tail surfaces (total): 1·47 m² (15·8 sq ft)

WEIGHTS AND LOADINGS:
- Weight empty:
  - BG 6: 106 kg (235 lb)
  - BG 7: 113 kg (250 lb)
- Max T-O weight:
  - BG 6: 193 kg (425 lb)
  - BG 7: 211 kg (465 lb)
- Max wing loading:
  - BG 6: 17·72 kg/m² (3·63 lb/sq ft)
  - BG 7: 18·45 kg/m² (3·78 lb/sq ft)

PERFORMANCE (at max T-O weight):
- Best glide ratio:
  - BG 6: 16
  - BG 7: 20
- Sinking speed:
  - BG 6: 0·91 m (3·00 ft)/sec
  - BG 7: 0·85 m (2·80 ft)/sec
- Stalling speed:
  - BG 6: 28 knots (52 km/h; 32 mph)
  - BG 7: 30 knots (55 km/h; 34 mph)
- Max towing speed:
  - BG 6: 62·5 knots (116 km/h; 72 mph)
  - BG 7: 70 knots (130 km/h; 81 mph)

BRIEGLEB BG 12BD and BG 12-16

The BG 12 series are single-seat high-performance sailplanes, the prototype of which flew for the first time in 1956. All ribs and bulkheads are cut from plywood and construction is similar to that of a model aeroplane. Standard wing span of 15·24 m (50 ft 0 in) may be clipped to 15·00 m (49 ft 2½ in) at builder's option.

AmEagle American Eaglet single-seat powered sailplane

Berkshire Concept 70 single-seat glassfibre Standard Class sailplane

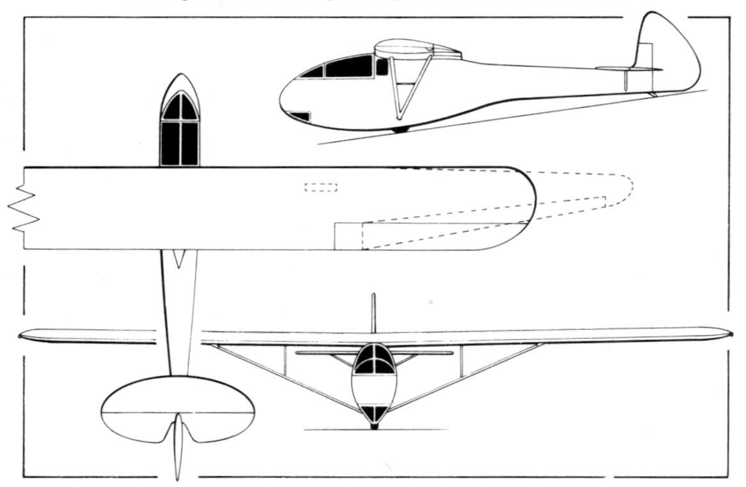

Briegleb BG 6 sailplane; BG 7 wing planform shown in dotted line *(Michael A. Badrocke)*

Briegleb BG 12-16 single-seat high-performance sailplane

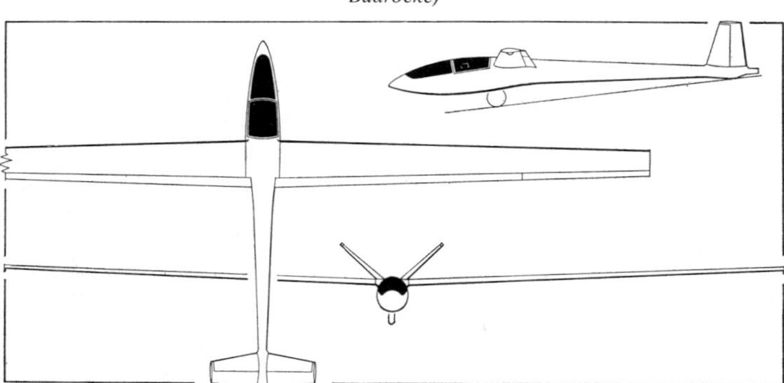

Bryan (Schreder) RS-15 single-seat Standard Class sailplane *(Michael A. Badrocke)*

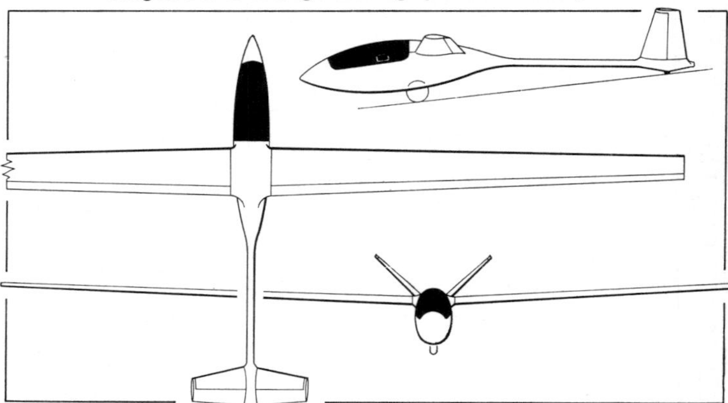

Bryan (Schreder) HP-18 sailplane, developed from the RS-15 *(Michael A. Badrocke)*

The earlier BG 12, BG 12A and BG 12B models were described in the 1972-73 *Jane's*. The two models currently available are:

BG 12BD. Two-piece wing with thicker section at root. Welded control system. First BG 12B production model flew in July 1973. The BD is similar except that there is no change in wing incidence between root and tip. The control system has an input action at the low speed range that deflects the ailerons upward to give the same effect as 'twist'. Kits and plans have been sold to customers in Australia, Belgium, Canada, Iceland, Japan, New Zealand, Peru, the UK and the USA.

BG 12-16. Development of BG 12BD having a slimmer fuselage, longer flaps and all-moving horizontal tail surfaces, which give a glide ratio up to two points better than the BD model. First flown in June 1969. Design modifications, including T-tail and greater use of glassfibre-reinforced plastics components, under consideration in 1976.

The following description applies to the BG 12BD, except where indicated:

Type: Single-seat high-performance sailplane.

Wings: Cantilever high-wing monoplane. Wing section NACA 4415R at root, NACA 4406R at tip. Dihedral 1°. Incidence 4° from root to tip. All-wood structure, with plywood covering. Plywood-covered wooden ailerons and trailing-edge flaps. Flaps used as airbrakes at speeds up to 112 knots (210 km/h; 130 mph).

Fuselage: Cutout plywood bulkheads, spruce longerons, plywood-covered.

Tail Unit (BG 12BD): Cantilever wooden structure. Ground-adjustable tailplane incidence. No tabs.

Tail Unit (BG 12-16): Cantilever structure, with swept-forward fin and plywood-covered rudder. All-moving horizontal surfaces of glassfibre-covered metal construction, built in two halves and fitted with anti-servo tabs.

Landing Gear: Shock-mounted nose-skid and sprung tailwheel or tailskid. Non-retractable unsprung monowheel with tyre size 10·50 × 4 or (BG 12-16) 12 × 4 × 5, manufactured by General Tire & Rubber Co. Tyre pressure 2·41 bars (35 lb/sq in). Briegleb circumferential brake.

Accommodation: Single seat under large moulded Plexiglas canopy. Adjustable seat and rudder pedals.

Dimensions, external:
- Wing span: 15·24 m (50 ft 0 in) *or* 15·00 m (49 ft 2½ in)
- Wing chord at root: 1·14 m (3 ft 9 in)
- Wing chord at tip: 0·31 m (1 ft 0¼ in)
- Wing aspect ratio: 17·9
- Length overall:
  - 12BD: 6·68 m (21 ft 11 in)
  - 12-16: 7·32 m (24 ft 0 in)
- Height over tail:
  - 12BD: 1·22 m (4 ft 0 in)
  - 12-16: 1·27 m (4 ft 2 in)

Areas:
- Wings, gross (full span): 13·10 m² (141·0 sq ft)
- Ailerons (total): 1·30 m² (14·00 sq ft)
- Trailing-edge flaps (total):
  - 12BD: 1·25 m² (13·45 sq ft)
  - 12-16: 1·30 m² (14·00 sq ft)
- Fin:
  - 12BD: 0·19 m² (2·05 sq ft)
  - 12-16: 0·42 m² (4·50 sq ft)
- Rudder:
  - 12BD: 0·65 m² (7·00 sq ft)
  - 12-16: 0·62 m² (6·70 sq ft)
- Tailplane (12BD): 1·00 m² (10·75 sq ft)
- Elevators (12BD): 0·57 m² (6·15 sq ft)
- Horizontal tail surfaces (12-16, total incl tabs): 0·86 m² (9·25 sq ft)

Weights and Loadings:
- Weight empty: 227-238 kg (500-525 lb)
- Max T-O weight:
  - 12BD: 340 kg (750 lb)
  - 12-16: 385 kg (850 lb)
- Max wing loading:
  - 12BD: 26·0 kg/m² (5·32 lb/sq ft)
  - 12-16: 29·4 kg/m² (6·03 lb/sq ft)

Performance:
- Best glide ratio:
  - 12BD at 45 knots (84 km/h; 52 mph): 33-34
  - 12-16 at 48·5 knots (90 km/h; 56 mph): 34-36
- Min sinking speed:
  - 12BD at 41 knots (76 km/h; 47 mph): 0·69 m (2·26 ft)/sec
  - 12-16 at 42 knots (77·5 km/h; 48 mph): 0·685 m (2·25 ft)/sec
- Stalling speed, flaps up:
  - 12BD: 33 knots (61 km/h; 38 mph)
  - 12-16: 34 knots (63 km/h; 39 mph)
- Stalling speed, flaps down:
  - 12BD: 29 knots (53 km/h; 33 mph)
  - 12-16: 30 knots (55 km/h; 34 mph)
- Max speed (smooth air):
  - 12BD, 12-16: 121 knots (225 km/h; 140 mph)
- Max speed (rough air) and max aero-tow speed:
  - 12BD, 12-16: 112 knots (210 km/h; 130 mph)
- Max winch-launching speed:
  - 12BD, 12-16: 65 knots (121 km/h; 75 mph)
- *g* limit: +10

BRYAN

BRYAN AIRCRAFT INC

HEAD OFFICE AND WORKS:
Williams County Airport, PO Box 488, Bryan, Ohio 43506
Telephone: (419) 636-1340
DIRECTOR:
R. E. Schreder

Details of earlier designs by Mr Schreder can be found under the Airmate heading in the 1965-66 *Jane's* and under the Bryan heading in the 1973-74 edition.

The company markets plans and kits for the RS-15 high-performance Standard Class sailplane, and since 1974 has completed a prototype of the HP-18, a more recent design. No later news of this aircraft has been received.

BRYAN (SCHREDER) RS-15

This 15 metre sailplane was designed by Mr Schreder to meet current OSTIV Standard Class specifications. It is designed for simple, rapid assembly by the homebuilder and is licensed in the amateur-built Experimental category. No jigs are required, and most major components are prefabricated, to reduce assembly time to approx 500 man-hours for a builder with average mechanical aptitude.

TYPE: Single-seat high-performance Standard Class sailplane.

WINGS: Cantilever shoulder-wing monoplane. Wing section Wortmann FX-67-150. Dihedral 2° 18′. Incidence 2°. All-metal structure except for polyurethane foam plastics ribs spaced at 102 mm (4 in) centres. Main wing spar caps pre-machined from solid aluminium plate stock. Water ballast carried inside wing box spars. Plain ailerons. Optional trailing-edge flaps/airbrakes, of aluminium sheet bonded to foam ribs, which can be linked with ailerons.

FUSELAGE: Monocoque structure. Prefabricated glassfibre forward pod, complete with bulkheads, floorboards and finish; 152 mm (6 in) diameter aluminium tube tail-boom.

TAIL UNIT: All-metal V tail which can be folded upward for towing or storage.

LANDING GEAR: Manually-retractable monowheel, size 500 × 5, and non-retractable steerable tailwheel. Hydraulic shock-absorbers on main wheel and tailwheel. Hydraulic brake on main wheel.

ACCOMMODATION: Single seat under one-piece Plexiglas canopy. Retractable tow hitch.

DIMENSIONS, EXTERNAL:
Wing span 15·00 m (49 ft 2½ in)
Wing aspect ratio 21·4
Length overall 6·71 m (22 ft 0 in)
Height over tail (extended) 1·17 m (3 ft 10 in)
Height over tail (folded) 1·52 m (5 ft 0 in)

DIMENSIONS, INTERNAL:
Cockpit: Length 1·68 m (5 ft 6 in)
Depth 0·91 m (3 ft 0 in)
Width 0·61 m (2 ft 0 in)

AREAS:
Wings, gross 10·5 m^2 (113·0 sq ft)
Trailing-edge flaps (total) 1·90 m^2 (20·5 sq ft)
Fixed tail surfaces (total) 0·79 m^2 (8·5 sq ft)
Movable tail surfaces (total) 0·65 m^2 (7·0 sq ft)

WEIGHTS AND LOADINGS:
Weight empty 200 kg (440 lb)
Normal T-O weight 335 kg (740 lb)
Max T-O weight with water ballast 426 kg (940 lb)
Normal wing loading 31·7 kg /m^2 (6·5 lb/sq ft)
Max wing loading 40·5 kg/m^2 (8·3 lb/sq ft)

PERFORMANCE:
Best glide ratio 38
Min sinking speed at 43·5 knots (80·5 km/h; 50 mph), AUW of 284 kg (626 lb) 0·64 m (2·1 ft)/sec
Stalling speed, flaps up, AUW of 335 kg (740 lb) 40 knots (74 km/h; 46 mph)
Stalling speed, flaps down, AUW of 335 kg (740 lb) 32·5 knots (60·5 km/h; 37·5 mph)
Max speed (smooth air) 130 knots (241 km/h; 150 mph)
Max speed (rough air) and max aero-tow speed 104 knots (193 km/h; 120 mph)
Max winch-launching speed 78 knots (145 km/h; 90 mph)

BRYAN (SCHREDER) HP-18

The HP-18 is a high-performance single-seat 15 metre sailplane, designed to meet current OSTIV Standard Class requirements. Generally similar in appearance to the RS-15, it has a slightly longer fuselage, with circular instead of oval section, and other features designed to reduce drag and produce a superior competition aircraft. These include better gap seals, new wingtips, removable tailwheel and better streamlining; the Tost monowheel has a tyre pressure of 2·07 bars (30 lb/sq in) and is fitted with a mechanically-expanding brake. Construction of a prototype began in December 1973, and was completed in 1975.

The structural description of the RS-15 applies generally also to the HP-18; specification of the HP-18 is as follows:

DIMENSIONS, EXTERNAL:
Wing span 15·00 m (49 ft 2½ in)
Wing chord at root 0·91 m (3 ft 0 in)
Wing chord at tip 0·46 m (1 ft 6 in)
Wing mean aerodynamic chord 0·69 m (2 ft 3 in)
Wing aspect ratio 21·1
Length overall 7·16 m (23 ft 6 in)
Height over tail (extended) 1·22 m (4 ft 0 in)
Height over tail (folded) 1·57 m (5 ft 2 in)

DIMENSIONS, INTERNAL:
Cockpit: Length 1·78 m (5 ft 10 in)
Depth 0·69 m (2 ft 3 in)
Width 0·61 m (2 ft 0 in)

AREAS:
Wings, gross 10·66 m^2 (114·7 sq ft)
Ailerons (total) 0·58 m^2 (6·2 sq ft)
Trailing-edge flaps (total) 1·64 m^2 (17·7 sq ft)
Fixed tail surfaces (total) 0·79 m^2 (8·5 sq ft)
Movable tail surfaces (total) 0·65 m^2 (7·0 sq ft)

WEIGHTS AND LOADINGS:
Weight empty 191 kg (420 lb)
Normal T-O weight 326 kg (720 lb)
Max T-O weight with water ballast 417 kg (920 lb)
Normal wing loading 30·8 kg/m^2 (6·3 lb/sq ft)
Max wing loading 39·1 kg/m^2 (8·0 lb/sq ft)

PERFORMANCE:
Best glide ratio 40
Min sinking speed at 39 knots (73 km/h; 45 mph), AUW of 275 kg (606 lb) 0·52 m (1·7 ft)/sec
Stalling speed, flaps up, AUW of 326 kg (720 lb) 35 knots (64·5 km/h; 40 mph)
Stalling speed, 60° flap, AUW of 326 kg (720 lb) 30·5 knots (57 km/h; 35 mph)
Max speed (smooth air) 130 knots (241 km/h; 150 mph)
Max speed (rough air) and max aero-tow speed 104 knots (193 km/h; 120 mph)
Max winch-launching speed 78 knots (145 km/h; 90 mph)
g limits ±12

DSK

DSK AIRCRAFT CORPORATION

ADDRESS:
11031 Glenoaks Boulevard, Pacoima, California 91331
Telephone: (213) 899-1016
EXECUTIVES:
James Maupin
Norman Barnhart

DSK BJ-1 DUSTER

About 150 Duster single-seat sailplanes are reportedly under construction, from kits supplied by DSK (Duster Sailplane Kits), and several of these have flown.

Plans of the Duster (but not kits) are also marketed by California Sailplanes. See next page for illustration.

The Duster sailplane was designed by Mr Ben Jansson, and the following details apply to the standard BJ-1:

WINGS: Cantilever shoulder-wing monoplane. NACA 4415 (modified) section. Flaps deflect 90° to serve as airbrakes.

LANDING GEAR: Non-retractable monowheel and tail-wheel.

DIMENSIONS, EXTERNAL:
Wing span 13·00 m (42 ft 7¾ in)
Wing aspect ratio 17·7
Length overall 6·10 m (20 ft 0 in)

AREA:
Wings, gross 9·60 m^2 (103·3 sq ft)

PERFORMANCE:
Best glide ratio at 47 knots (87 km/h; 54 mph) 29
Min sinking speed at 67·5 knots (125 km/h; 77·5 mph) 1·80 m (5·91 ft)/sec
g limit +6

EXPLORER

EXPLORER AIRCRAFT COMPANY

ADDRESS:
PO Box 6555, Reno, Nevada 89513
Telephone: (702) 322-1421

Explorer Aircraft Company is marketing plans of a biplane glider named the Aqua Glider. Designed by Colonel William L. Skliar, USAF (Ret'd), it is intended for tethered gliding by unlicensed pilots, towed behind any motor boat able to attain a speed of 30 knots (56 km/h; 35 mph). If the pilot has the necessary licence, he can cast off from the motor boat when airborne and make a free flight before landing back on the water.

Design of the Aqua Glider originated in September 1958. Construction of the prototype began in January 1959, and was completed in June 1959; the first flight was made during the following month. Plans are available to amateur constructors and approx 1,000 sets have been sold in Argentina, Australia, the Bahamas, Brazil, Canada, France, Japan, Malaysia, Mexico, the Philippines, Portugal, Rhodesia, Singapore, South Africa, Spain, Sri Lanka, Sweden, Switzerland, the UK, the USA and South Vietnam. About 12 completed Aqua Gliders are known to have flown, in the USA, the Bahamas, Brazil and Japan, and about 200 more are under construction.

The prototype, after making about 1,000 flights and being flown by about 60 pilots, was donated to the Experimental Aircraft Association Museum for permanent display.

EXPLORER PG-1 AQUA GLIDER

TYPE: Single-seat homebuilt waterborne glider.

WINGS: Forward-stagger single-bay biplane with N interplane and parallel centre-section struts. Wing section (both) NACA 4412. Dihedral 2° 30′ on lower wing only. Incidence 4°. No sweepback. Conventional single-spar wooden structure with fabric covering. Spoiler-type light alloy ailerons on lower wing only, immediately aft of main spar. No flaps or tabs. Balancer floats at lower wingtips.

HULL: Unstepped watertight wooden structure of spruce with 1/16 in mahogany plywood bow, bottom skins and sides. Plywood is glassfibre-covered below waterline. Tow hook on nose.

TAIL UNIT: Wire-braced spruce structure, with plywood and fabric covering, carried on welded steel tube or wire-braced wooden boom. All-moving one-piece tailplane with bungee trim. Conventional rudder.

LANDING GEAR: Standard jumper skis, 1·83 m (6 ft) in length, attached to small wire-braced struts below hull.

ACCOMMODATION: Single seat in open cockpit in hull, forward of wings.

DIMENSIONS, EXTERNAL:
Wing span:
upper 4·88 m (16 ft 0 in)
lower 4·57 m (15 ft 0 in)
Wing chord, constant (both) 0·91 m (3 ft 0 in)
Wing aspect ratio 5
Length overall 4·17 m (13 ft 8 in)
Height overall 1·52 m (5 ft 0 in)
Tailplane span 2·13 m (7 ft 0 in)

AREAS:
Wings, gross 8·7 m^2 (94 sq ft)
Ailerons/spoilers (total) 0·19 m^2 (2 sq ft)
Fin approx 0·37 m^2 (4 sq ft)
Rudder approx 0·28 m^2 (3 sq ft)
Tailplane approx 1·67 m^2 (18 sq ft)

WEIGHTS AND LOADING:
Weight empty 81 kg (180 lb)
Max T-O weight 181 kg (400 lb)
Max wing loading 20·5 kg/m^2 (4·2 lb/sq ft)

PERFORMANCE:
Best glide ratio at 39 knots (72·5 km/h; 45 mph) 6·5
Min airspeed, at 149 kg (330 lb) gross weight 30·5 knots (56·5 km/h; 35 mph)
Max airspeed 56·5 knots (104·5 km/h; 65 mph)
g limit +4

FLIGHT DYNAMICS

FLIGHT DYNAMICS INC

ADDRESS:
PO Box 5070, State College Station, Raleigh, North Carolina 27607
Telephone: (919) 834-6806
PRESIDENT:
Thomas H. Purcell Jr

FLIGHT DYNAMICS SEASPRITE

As indicated under this company's heading in the Homebuilts section, the Seasprite is the unpowered Stage I version of the aircraft which, in its fully-developed powered form, is known as the Flightsail VII. The Seasprite is of more simplified construction, and is available in plans form to amateur sailplane constructors.

The fuselage, of aerofoil/hydrofoil cross-section, is built on a framework of aluminium tubing. The sides and top are covered with polyethylene foil, except for the upper part of the nose section, which has a transparent covering of polyester film and Plexiglas. The pilot sits in the centre-section, to the right of the slender aluminium boom which carries the simple tail unit, his weight being balanced by a counterweight at the port wingtip if a second occupant is not carried. The underside of the fuselage/hull has catamaran-type twin floats, built of plastics with a plywood

DSK Duster single-seat homebuilt sailplane *(A. J. Smith)*

Explorer PG-1 Aqua Glider built by Mr Leslie G. Higgins of Suffield, Ohio. **See previous page for description of PG-1**

Flight Dynamics Seasprite, Stage I version of the Flightsail VII, with sailwing

Laister LP-15 Nugget single-seat Standard Class sailplane

covering. For only a minimal weight increase, the structure can be covered instead with aluminium foil, which has a much longer life.

The wings are of triangular planform, and are of the sailwing type, with an aluminium tube forming the leading-edge, a wire trailing-edge, and a polyethylene covering. They are braced to the fuselage sides, and can be folded rearwards when not in use. The wingtips pivot about their leading-edges, to provide roll control.

The fin/tailplane is built as a single unit, pivoting only in the vertical plane for control in pitch.

DIMENSIONS, EXTERNAL:
- Wing span: 10·36 m (34 ft 0 in)
- Length overall: 6·10 m (20 ft 0 in)
- Height overall: 2·44 m (8 ft 0 in)

WEIGHTS:
- Weight empty: 75 kg (165 lb)
- Max T-O weight:
  - single-seat: 165 kg (365 lb)
  - two-seat: 227 kg (500 lb)

PERFORMANCE:
- Best glide ratio: 6
- T-O speed:
  - single-seat: 22 knots (40 km/h; 25 mph)
  - two-seat: 28 knots (51·5 km/h; 32 mph)
- Max towing speed in flight: 30 knots (56 km/h; 35 mph)

HALL
STANLEY HALL
HALL VECTOR I

Completion of this single-seat foot-launched ultra-light glider was reported in the Autumn of 1975. Of wood and fabric construction, it has strut-braced wings, and a cantilever tail unit with an all-moving tailplane. The pilot is seated on a bicycle saddle on the pod-shaped fuselage, and is totally enclosed by two clamshell doors after 'retracting' his legs following take-off. Flight testing began in 1975, and it was hoped to achieve a best glide ratio of 18.

The Vector I is designed to the FAA limit load factor for gliders (5·33). It incorporates normal sailplane control surfaces, including spoilers, and has a stick-type control column. Gross weight is more than 68 kg (150 lb).

LAISTER
LAISTER SAILPLANES INC

ADDRESS:
2714 Chico Avenue, South El Monte, California 91733
Telephone: (213) 442-4945
PRESIDENT:
Jack Laister

This company is marketing, in kit form only, a Standard Class 15 m sailplane designated LP-49, of which a description follows. Also described is the LP-15 Nugget, a Standard Class 15 m sailplane which is being produced, as the company's prime product, in factory-built form only.

LAISTER LP-49

Available in kit form only, the LP-49 is a 15 m Standard Class sailplane approved to 1970 FAI specifications and type certificated by the FAA. More than 50 kits had been sold by early 1976, and about 35 of these had been completed and flown.

TYPE: Single-seat Standard Class sailplane for amateur construction.

WINGS: Cantilever high-wing monoplane, with constant-chord centre-section and tapered outer panels. Extruded aluminium main spar booms, curved in chordwise direction to follow aerofoil section. Roll-contoured sheet aluminium skin, butted and flush-riveted with blind pop rivets. Glassfibre wingtip fairings. All-metal statically-balanced ailerons, with automatic hookup to prevent take-off without aileron control. Flap on each inboard trailing-edge, also with automatic control hookup.

FUSELAGE: Semi-monocoque structure, consisting of two pre-moulded glassfibre halves reinforced with aluminium bulkheads and fittings. Self-retracting tow-hook in nosecone.

TAIL UNIT: Cantilever aluminium structure with swept-back fin and rudder. Glassfibre tip fairings on tailplane, elevator and rudder. Rudder tip fairing houses rudder static balance; that for the elevator operates internally in rear end of fuselage.

LANDING GEAR: Retractable main wheel with brake, glassfibre nose skid with steel shoe, and non-retractable shrouded tailwheel. Main-wheel shock-absorber incorporates coil spring and damper. Landing gear door rotates up into fuselage when main wheel is extended, to avoid damage.

ACCOMMODATION: Single seat (contoured seat optional) under one-piece moulded transparent canopy. Radio and oxygen system available.

DIMENSIONS, EXTERNAL:
- Wing span: 15·00 m (49 ft 2½ in)
- Wing aspect ratio: 17
- Length overall: 6·28 m (20 ft 7·2 in)

DIMENSION, INTERNAL:
- Cockpit: Max width: 0·58 m (1 ft 11 in)

AREA:
- Wings, gross: 13·28 m² (143·0 sq ft)

WEIGHTS:
- Weight empty: 208 kg (460 lb)
- Max T-O weight: 408 kg (900 lb)

PERFORMANCE:
- Best glide ratio at 50 knots (92·5 km/h; 57·5 mph): 36·5
- Min sinking speed at 43 knots (79·5 km/h; 49·5 mph): 0·63 m (2·07 ft)/sec
- Never-exceed speed: 135 knots (249 km/h; 155 mph)
- Stalling speed, flaps down: 30·5 knots (56·5 km/h; 35 mph)
- Max speed (rough or smooth air): 117 knots (217 km/h; 135 mph)

LAISTER LP-15 NUGGET

Design and construction of the LP-15 prototype began in February 1971, and this aircraft (N15LP) made its first flight in June 1971. Certification was expected in mid-1975, and an initial quantity of 30 is in production. Five of these had flown by early 1975. The LP-15 is supplied only as a factory-built aircraft.

TYPE: Single-seat Standard Class sailplane.

WINGS: Cantilever shoulder-wing monoplane. Wortmann wing sections. Thickness/chord ratio 17% at root, 15% at tip. Dihedral 3° from centre-section. Incidence 1°. Sweepback at quarter-chord approx 1°. Chem-Weld bonded aluminium alloy structure. Long-span trailing-edge flaps with negative travel (up) for high-speed flight, 8° positive travel (down) for thermalling and 85° positive for landing or use as airbrakes. Top-hinged plain ailerons of similar construction. Automatic aileron/flap interlock available optionally. All control surfaces have internal static balances. No spoilers or tabs. Provision for 84 kg (185 lb) of water ballast, carried in centre-section, with fillback through dump valve.

FUSELAGE: Semi-monocoque structure. Forward portion of moulded glassfibre, rear portion of bonded aluminium alloy.

TAIL UNIT: Cantilever bonded aluminium alloy T-tail, with slightly-swept fin and rudder. Adjustable-incidence tailplane and one-piece elevator. Rudder and elevator have internal static balances. No tabs. Rudder pedals adjustable in flight.

LANDING GEAR: Manually-retractable monowheel (size 4·00-5, tyre pressure 2·41 bars; 35 lb/sq in) and tailskid. Monowheel has air-oil shock-absorption, with coil spring, and is fitted with a mechanically-operated internal brake.

ACCOMMODATION: Single semi-reclining seat under fully-transparent two-piece canopy with removable hood and sliding ventilation panel. Equipment, including oxygen and radio, is to customer's specification.

DIMENSIONS, EXTERNAL:
- Wing span: 15·00 m (49 ft 2½ in)
- Wing chord at root: 0·91 m (3 ft 0 in)
- Wing chord at tip: 0·35 m (1 ft 2 in)
- Wing aspect ratio: 22·2
- Length overall: 6·10 m (20 ft 0 in)

Height over tail 1·27 m (4 ft 2 in)
Tailplane span 2·39 m (7 ft 10 in)
DIMENSION, INTERNAL:
Cockpit: Max width 0·61 m (2 ft 0 in)
AREA:
Wings, gross 10·13 m² (109·0 sq ft)
WEIGHTS AND LOADING:
Weight empty, equipped 210 kg (465 lb)
Max T-O weight (with water ballast) 408 kg (900 lb)
Max wing loading (with water ballast) 40·33 kg/m² (8·26 lb/sq ft)
PERFORMANCE (at max T-O weight):
Best glide ratio at 48 knots (89 km/h; 55 mph) 36·5
Never-exceed speed 135 knots (249 km/h; 155 mph)
Stalling speed, flaps up 39 knots (73 km/h; 45 mph)
Max speed (rough or smooth air) 126 knots (233 km/h; 145 mph)
Max aero-tow speed 86·5 knots (160 km/h; 100 mph)
Max winch-launching speed 60·5 knots (112·5 km/h; 70 mph)
Min landing speed, full flap 30·5 knots (56·5 km/h; 35 mph)
g limit +12

MARSKE

MARSKE AIRCRAFT CORPORATION

ADDRESS:
130 Crestwood Drive, Michigan City, Indiana 46360
Telephone: (219) 879-7039

MARSKE MONARCH

This single-seat ultra-light glider was designed and built by Mr Jim Marske in 1974. Plans and kits are available to amateur constructors.

Mr Marske has successfully test-flown the Monarch prototype with a 10 kW (12 hp) McCulloch engine installed behind the pilot's seat, driving a 0·635 m (2 ft 1 in) diameter pusher propeller. It is considered, however, that an engine of approx 15 kW (20 hp) should be fitted to homebuilt aircraft.

The following description applies to the standard 1976 glider version, as illustrated on the following page:

WINGS: Braced high-wing monoplane, with single steel strut each side. Wing section NACA 43112. Moulded glassfibre leading-edge, with glassfibre spar web, foam plastics ribs and wooden rear spar and trailing-edge. Ailerons (outboard) and elevators (inboard) have single wooden spars with foam plastics ribs. All surfaces Dacron-covered. Spoilers above each wing. All control surfaces cable-actuated.

FUSELAGE: Simple minimal beam-type structure of laminated glassfibre, moulded in two halves and joined at centreline. Forward section supports pilot's seat, with nose fairing over instrument panel; rear section forms integral fin leading-edge. Tow hooks on nose.

TAIL UNIT: Fin and rudder only, above and below level of wings. Glassfibre leading-edge (fin), wooden trailing-edge (rudder), foam ribs and Dacron covering.

LANDING GEAR: Reinforced underfuselage landing skid. Single landing wheel below pilot's seat.

ACCOMMODATION: Single open seat. Overhead control stick.

DIMENSIONS, EXTERNAL:
Wing span 11·89 m (39 ft 0 in)
Wing aspect ratio 8·6
Length overall 3·51 m (11 ft 6 in)
AREA:
Wings, gross 16·26 m² (175·0 sq ft)
WEIGHTS AND LOADING:
Weight empty 100 kg (220 lb)
Max T-O weight 204 kg (450 lb)
Max wing loading 12·55 kg/m² (2·57 lb/sq ft)
PERFORMANCE:
Best glide ratio at 35 knots (64 km/h; 40 mph) 18
Min sinking speed at 28 knots (51·5 km/h; 32 mph) 0·91 m (3·0 ft)/sec
Sinking speed at 46 knots (85 km/h; 53 mph) 1·98 m (6·5 ft)/sec
Stalling speed 21 knots (39 km/h; 24 mph)
Cruising speed 46 knots (85 km/h; 53 mph)
Never-exceed speed 65 knots (120 km/h; 75 mph)
Assembly time 15 min
g limits +4·0; +8·0 (ultimate)

MILLER

W. TERRY MILLER

ADDRESS:
Box 570, RR1, Furlong, Pennsylvania 18925

Mr Miller is the designer of the Tern and Tern II sailplanes, and of a low-powered high-altitude sport aircraft designated WM-2. All of these aircraft were designed for amateur construction. A description of the WM-2 can be found in the Homebuilt Aircraft section of this edition.

The prototype Tern sailplane, constructed by the designer, first flew in September 1965. The prototype for the Tern II first flew in August 1968.

By the beginning of 1976 a total of 28 Tern and Tern II sailplanes had been built and flown in the United States, Canada and Australia. Illustrations of both types appear on the following page.

MILLER TERN

TYPE: Single-seat high-performance sailplane.

WINGS: Cantilever shoulder-wing monoplane. Wing section Wortmann FX-61-184 at root, Wortmann FX-61-163 at tip. Dihedral 2°. Incidence 2°. No sweep at 50% chord. Two-piece two-spar spruce structure, all plywood-covered. Plain all-wood ailerons. No flaps or tabs. Lower-surface wooden airbrakes used for glide-slope control.

FUSELAGE: Semi-monocoque wooden structure. Plastics-reinforced glassfibre nose. Plywood skin aft of cockpit.

TAIL UNIT: Cantilever all-wood structure, with modified NACA laminar-flow sections. Special hingeline contouring to reduce drag and increase control effectiveness at large deflections. All control surfaces 60% mass-balanced. No tabs.

LANDING GEAR: Non-retractable monowheel forward of CG, in streamlined pod. Skids under nose and tail. Wheel and tyre size 5·00-4. Brake lever applies pressure directly to tyre.

ACCOMMODATION: Single partially-reclining seat under transparent canopy, centre portion of which hinges sideways for access to cockpit. Standard seven-instrument panel. Space for radio.

DIMENSIONS, EXTERNAL:
Wing span 15·54 m (51 ft 0 in)
Wing chord at root 1·02 m (3 ft 4 in)
Wing chord at tip 0·56 m (1 ft 10 in)
Wing aspect ratio 20
Length overall 6·49 m (21 ft 3½ in)
Height over tail 1·52 m (5 ft 0 in)
Tailplane span 2·59 m (8 ft 6 in)
AREAS:
Wings, gross 12·08 m² (130 sq ft)
Ailerons (total) 0·71 m² (7·60 sq ft)
Airbrakes (total) 0·48 m² (5·20 sq ft)
Fin 0·43 m² (4·66 sq ft)
Rudder 0·46 m² (5·00 sq ft)
Tailplane 0·73 m² (7·82 sq ft)
Elevators 0·61 m² (6·56 sq ft)
WEIGHTS AND LOADING:
Weight empty, equipped 215 kg (475 lb)
Max T-O weight 318 kg (700 lb)
Max wing loading 26·36 kg/m² (5·4 lb/sq ft)
PERFORMANCE:
Best glide ratio at 50·4 knots (93 km/h; 58 mph) 36
Min sinking speed at 41 knots (76 km/h; 47 mph) 0·64 m (2·1 ft)/sec
Stalling speed 34 knots (63 km/h; 39 mph)
Max speed (smooth air) 104 knots (193 km/h; 120 mph)
Max speed (rough air) and max aero-tow speed 78 knots (145 km/h; 90 mph)
Max winch-launching speed 61 knots (112 km/h; 70 mph)

MILLER TERN II

This developed version of the Tern differs from the original model in having a larger wing span and a drogue parachute in the base of the rudder. All other details are as for the Tern except:

DIMENSIONS, EXTERNAL:
Wing span 18·29 m (55 ft 6 in)
Wing chord at tip 0·51 m (1 ft 8 in)
Wing aspect ratio 22
AREAS:
Wings, gross 13·01 m² (140 sq ft)
Ailerons (total) 0·86 m² (9·3 sq ft)
WEIGHTS AND LOADING:
Weight empty, equipped 249 kg (550 lb)
Max T-O weight 363 kg (800 lb)
Max wing loading 28 kg/m² (5·72 lb/sq ft)
PERFORMANCE (nominal):
Best glide ratio at 52 knots (97 km/h; 60 mph) 38
Min sinking speed at 41·7 knots (77 km/h; 48 mph) 0·59 m (1·95 ft)/sec
Stalling speed 35 knots (64 km/h; 40 mph)
Max speed (rough air) amd max aero-tow speed 76 knots (142 km/h; 88 mph)

RYSON

RYSON AVIATION CORPORATION

ADDRESS:
548 San Fernando Street, San Diego, California 92106
Telephone: (714) 291-7311, extn 201 (T. Claude Ryan) or 413 (William Wagner)
PRESIDENT:
T. Claude Ryan
EXECUTIVE VICE-PRESIDENT AND OPERATIONS MANAGER:
Jerome D. Ryan
CHIEF ENGINEER:
L. Pazmany

Ryson Aviation Corporation was founded by Mr T. Claude Ryan, until 1969 Chairman and Chief Executive Officer of the Ryan Aeronautical Company (now Teledyne Ryan Aeronautical). Other executives include his son, Mr Jerry Ryan, and Mr Ladislao Pazmany. The aim of the Corporation is to develop aeronautical products and make them available for manufacture by other companies.

First new design to emanate from the Ryson company was a two-seat powered cruising sailplane designated STP-1 Swallow, details of which appeared in the 1973-74 *Jane's*.

Since early 1975 the company has been working on the new project mentioned briefly in the 1974-75 *Jane's*: this is a powered sailplane of completely new design, and is designated ST-100 (Soaring/Touring—100 hp).

RYSON ST-100

Believed to be the first American powered sailplane designed for production, the ST-100 has the traditional tractor-propeller configuration of the most popular European powered sailplanes. Its design, which was started on 18 March 1974, incorporates many basic advances for this class of aircraft, such as all-metal construction, fully-folding wings, power-operated flaps for air braking, and ailerons which operate with the flaps. It is designed to be aerobatic and to meet the new FAR Pt 23 gust load requirements (15·25 m; 50 ft/sec); is capable of performing as an aero-tow aircraft for unpowered sailplanes; and can be used as a conventional powered aircraft for cross-country flying.

Construction of the ST-100 prototype began on 11 July 1974, and this aircraft was expected to make its first flight in July 1976.

TYPE: Two-seat powered sailplane.

WINGS: Cantilever low-wing monoplane. Wing section Wortmann FX-67-K-170/17. Dihedral 4° from roots. Incidence 0°. No sweepback. All-metal safe-life structure, with some fail-safe features, built of 2024-T4 and 2-24-T6 extruded aluminium, with 2024-T3 sheet aluminium skin. Aluminium plain ailerons and plain trailing-edge flaps, with foam core, piano-hinged at bottom skin. Flaps, powered by electrically-operated screwjack, have 12° upward travel, and can be deflected downward to 72° for use as airbrakes. Ailerons operate with flaps, upward to 12° and downward to 8°. No spoilers or tabs. Wings fold back alongside fuselage for stowage or transportation.

FUSELAGE: Conventional all-metal semi-monocoque structure, with 2024-T4 extruded aluminium longerons, and sheet metal frames, bulkheads and skins.

TAIL UNIT: Cantilever T tail, with sweptback fin and rudder and non-swept horizontal surfaces, the latter comprising a fixed-incidence tailplane and a one-piece balanced elevator. Aerofoil sections: horizontal surfaces NACA 63_1A012, vertical surfaces NACA 64_2A015/A019·5. Rudder and elevator have riveted sheet aluminium primary ribs, bonded intermediate foam ribs, and bonded sheet aluminium skin. Elevator tips removable for transportation or storage. Anti-servo and trim tab, of similar construction, in centre of elevator.

LANDING GEAR: Two-wheel main gear and steerable tailwheel, all non-retractable. Ryson oleo-pneumatic shock-absorbers. Cleveland 5·00-5 main wheels, with 5·00-5 (Type III) tyres, size 14·5 × 5·2 × 6·5. Goodyear M3336 tailwheel, size 6 × 1·7, with solid rubber tyre. Non-cooled Cleveland hydraulic disc brakes on main units. Streamlined glassfibre fairings on main gear legs, main wheels and tailwheel.

POWER PLANT: One 74·5 kW (100 hp) Continental O-200 flat-four engine, driving a Hoffmann HO-V-62 three-position propeller with two composite blades. Fuel in two integral tanks in leading-edges of wing centre-section, with combined capacity of 121 litres (26·6 Imp gallons; 32 US gallons). Oil capacity 5·7 litres (1·25 Imp gallons; 1·5 US gallons).

ACCOMMODATION: Seats for two persons in tandem under fully-transparent one-piece framed canopy which opens sideways to starboard. Baggage space aft of rear seat. Cockpit heated and ventilated.

SYSTEMS: Hydraulics for main-wheel brakes only. Dual electrical system, using two 12V 12Ah Wisco JB-7A motor-cycle batteries. Oxygen equipment for occupants.

DIMENSIONS, EXTERNAL:
Wing span 17·58 m (57 ft 8 in)
Wing chord at root 1·63 m (5 ft 4·3 in)
Wing chord at tip 0·57 m (1 ft 10·48 in)
Wing aspect ratio 15·61

Marske Monarch single-seat ultra-light glider *(Michael A. Badrocke)*

Miller Tern single-seat sailplane built by Mr Alex Stuart of Phoenix, Arizona

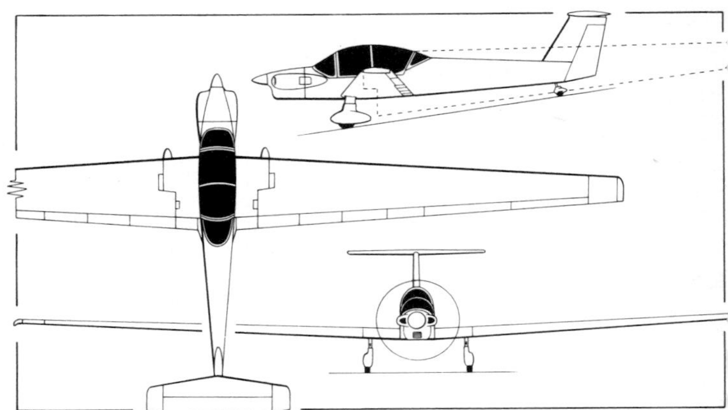

Ryson ST-100 two-seat powered sailplane *(Michael A. Badrocke)*

Miller Tern II built by Mr Charles Wieland of Huntington Beach, California

Schweizer SGS 1-26E single-seat medium-performance sailplane

Schweizer SGS 2-32 two/three-seat high-performance sailplane

Length overall (tail up) 7·78 m (25 ft 6·4 in)
Length overall (tail up, wings folded) 10·26 m (33 ft 8 in)
Width, wings folded 2·44 m (8 ft 0 in)
Fuselage max width 0·76 m (2 ft 6 in)
Height overall (tail down) 1·78 m (5 ft 10 in)
Tailplane span:
overall (ie elevator span) 3·05 m (10 ft 0 in)
elevator tips removed 2·44 m (8 ft 0 in)
Wheel track 2·18 m (7 ft 2 in)
Wheelbase 4·72 m (15 ft 6 in)
Propeller diameter 1·93 m (6 ft 4 in)
Propeller ground clearance 0·25 m (10 in)

Dimensions, internal:
Cockpit (firewall to rear seat back):
Length 2·03 m (6 ft 8 in)
Max width 0·66 m (2 ft 2 in)
Max height 1·11 m (3 ft 7·7 in)
Floor area 0·74 m² (8·00 sq ft)
Volume 1·36 m³ (48·00 cu ft)
Baggage compartment volume 0·093 m³ (3·30 cu ft)

Areas:
Wings, gross 19·79 m² (213·00 sq ft)
Ailerons (total) 1·05 m² (11·32 sq ft)
Trailing-edge flaps (total) 1·92 m² (20·67 sq ft)
Fin 0·573 m² (6·17 sq ft)
Rudder 0·70 m² (7·55 sq ft)
Tailplane 0·95 m² (10·25 sq ft)
Elevator, incl tab 1·37 m² (14·75 sq ft)

Weights and Loadings:
Weight empty, equipped 513 kg (1,130 lb)
Max T-O weight 726 kg (1,600 lb)
Max wing loading 36·67 kg/m² (7·51 lb/sq ft)
Max power loading 9·74 kg/kW (16·0 lb/hp)

Performance (estimated, at max T-O weight, powered):
Never-exceed speed 139 knots (257 km/h; 160 mph)
Max level speed at S/L 123 knots (228 km/h; 142 mph)
Max cruising speed (75% power) at 1,980 m (6,500 ft) 113 knots (209 km/h; 130 mph)
Econ cruising speed 56·5 knots (105 km/h; 65 mph)
Stalling speed at S/L, flaps up 43 knots (79·5 km/h; 49·4 mph)
Stalling speed at S/L, flaps down 37 knots (68·5 km/h; 42·5 mph)
Max rate of climb at S/L 273 m (895 ft)/min
Service ceiling 7,315 m (24,000 ft)
T-O run 174 m (570 ft)
T-O to 15 m (50 ft) 290 m (950 ft)
Landing from 15 m (50 ft) 244 m (800 ft)
Landing run 122 m (400 ft)
Range:
max payload and 45·4 litres (12 US gallons) fuel 195 nm (362 km; 225 miles)
max fuel at max cruising speed 521 nm (965 km; 600 miles)
max fuel at econ cruising speed 781 nm (1,448 km; 900 miles)

Performance (estimated, at max T-O weight, unpowered):
Best glide ratio, propeller feathered 28
Min sinking speed 0·89 m (2·93 ft)/sec

SCHWEIZER

SCHWEIZER AIRCRAFT CORPORATION

Head Office and Works:
Box 147, Elmira, New York 14902
Telephone: (607) 739-3821
Telex: 932459

President and Chief Engineer:
Ernest Schweizer

Vice-President and General Manager:
Paul A. Schweizer

Vice-President in Charge of Manufacturing:
William Schweizer

Sales Manager: W. E. Doherty Jr

Asst Treasurer: Joseph Kroczynski

Secretary: Kenneth Tifft

Schweizer Aircraft Corporation is the leading American designer and manufacturer of sailplanes. Its current products include the SGS 2-33A two-seat training sailplane, the one-design SGS 1-26E single-seat sailplane, the SGS 2-32 two/three-seat high-performance sailplane, the SGS 1-34B Standard Class single-seat high-performance sailplane, and the SGS 1-35 high-performance competition Standard Class sailplane.

By agreement with Grumman American, Schweizer also manufactures the Super Ag-Cat agricultural biplane, a description of which can be found under the Schweizer heading in the Aircraft section. Other subcontract work includes production of fuselage assemblies for Piper Aircraft Corporation and major structures for Bell Helicopter Textron.

Schweizer has also purchased design and production rights in the Thurston Teal two/three-seat light amphibian, and a description of this appears under the Schweizer heading in the Aircraft section.

About 400 people were employed by Schweizer in 1976.

SCHWEIZER SGS 1-26

This relatively small sailplane was developed for one-design class activities. It was designed to be produced both as a complete sailplane and in kit form for the homebuilder. The prototype 1-26 was first flown in January 1954, and with award of the FAA Type Certificate production of both complete sailplanes and kits began in November of that year.

The current production version, the **SGS 1-26E**, introduced an all-metal semi-monocoque fuselage. More than 645 SGS 1-26 sailplanes had been produced by January 1976, of which approximately 200 were in kit form.

TYPE: Single-seat medium-performance sailplane.

WINGS: Cantilever mid-wing monoplane. Dihedral 3° 30′. All-metal structure of aluminium alloy, with metal skin. Fabric-covered ailerons. Balanced airbrakes immediately aft of spar on each wing.

FUSELAGE: Production SGS 1-26E has an all-metal semi-monocoque fuselage.

TAIL UNIT: Cantilever aluminium alloy structure, covered with fabric.

LANDING GEAR: Non-retractable unsprung monowheel, size 4-80-4, with Schweizer brake, aft of rubber-sprung nose-skid. Small solid rubber tailwheel.

ACCOMMODATION: Single seat under blown Plexiglas canopy.

EQUIPMENT: Standard equipment includes fresh air vent, wheel cover, seat belt and shoulder harness, airspeed indicator and instrument panel. Provision for radio forward of pilot.

DIMENSIONS, EXTERNAL:
Wing span 12·19 m (40 ft 0 in)
Wing aspect ratio 10
Length overall 6·57 m (21 ft 6½ in)
Height over tail 2·21 m (7 ft 2½ in)
Tailplane span 2·29 m (7 ft 6 in)

AREAS:
Wings, gross 14·87 m² (160 sq ft)
Airbrakes (total) 0·26 m² (2·78 sq ft)

WEIGHTS AND LOADING:
Weight empty 195 kg (430 lb)
Max T-O weight 317 kg (700 lb)
Max wing loading 21·34 kg/m² (4·37 lb/sq ft)

PERFORMANCE:
Best glide ratio 23
Min sinking speed 0·79 m (2·60 ft)/sec
Stalling speed 29 knots (54 km/h; 33 mph)
Max permissible speed and max aero-tow speed
99 knots (183 km/h; 114 mph)
Max winch-launching speed
55 knots (101 km/h; 63 mph)

SCHWEIZER SGS 2-32

The SGS 2-32 has an unusually large cabin capable of carrying one very large or two average-sized passengers in addition to the pilot. The prototype flew for the first time on 3 July 1962. FAA Type Approval was received in June 1964 and production began immediately. More than 88 had been built by January 1976.

A number of world and national gliding records have been gained by pilots using this sailplane. A new world height record for women, in Class D2, was set on 5 March 1975 by Babs (Mary L.) Nott and Hannah Duncan (USA), with a flight to 10,809 m (35,462 ft).

The SGS 2-32 sailplane was chosen to form the basic airframe of the Lockheed YO-3A quiet observation aircraft and E-Systems L450F (see RPVs and Targets section). The Bede BD-2 was also a 2-32 development, and numerous powered adaptations of the 2-32 have been built by other companies.

TYPE: Two/three-seat high-performance and utility sailplane.

WINGS: Cantilever mid-wing monoplane. Wing section NACA 63_3618 at root, NACA 43012A at tip. Dihedral 3° 30′. All-metal single-spar structure with metal covering. Fabric-covered metal ailerons. Speed-limiting airbrakes on upper and lower surfaces.

FUSELAGE: All-metal monocoque structure.

TAIL UNIT: Cantilever metal structure, with all-moving tailplane. Fin metal-covered; control surfaces fabric-covered. Adjustable trim tab in tailplane.

LANDING GEAR: Non-retractable unsprung monowheel, wheel and tyre size 6·00-6, with hydraulic brake, and skid.

ACCOMMODATION: Pilot at front. Seat for one or two persons at rear. Dual controls. Rear control column removable for passenger comfort. Sideways-opening blown Perspex canopy.

EQUIPMENT: Lined cockpit, front and rear cabin air vents, seat belts and shoulder harnesses standard. Optional items include electrical and oxygen systems, radio, special instrumentation, canopy locks, cushions, map cases and wingtip wheels.

DIMENSIONS, EXTERNAL:
Wing span 17·40 m (57 ft 1 in)
Wing chord at root 1·45 m (4 ft 9 in)
Wing chord at tip 0·48 m (1 ft 7 in)
Wing aspect ratio 18·05
Length overall 8·15 m (26 ft 9 in)
Fuselage width at cockpit 0·81 m (2 ft 8 in)
Tailplane span 3·20 m (10 ft 6 in)

AREAS:
Wings, gross 16·70 m² (180 sq ft)
Ailerons (total) 1·37 m² (14·74 sq ft)
Airbrakes (total) 0·91 m² (9·76 sq ft)
Fin 0·73 m² (7·86 sq ft)
Rudder 0·67 m² (7·23 sq ft)
Tailplane 2·03 m² (21·88 sq ft)

WEIGHTS AND LOADINGS (H: high-performance category; U: utility category):
Weight empty, equipped:
H, U 385 kg (850 lb)
Max T-O weight:
H 608 kg (1,340 lb)
U 649 kg (1,430 lb)
Max wing loading:
H 36·32 kg/m² (7·44 lb/sq ft)
U 38·77 kg/m² (7·94 lb/sq ft)

PERFORMANCE (H: high-performance category, U: utility category; at AUW of 544 kg; 1,200 lb):
Best glide ratio at 51 knots (95 km/h; 59 mph):
H, U 34
Min sinking speed at 44 knots (80 km/h; 50 mph):
H, U 0·72 m (2·38 ft)/sec
Stalling speed:
H 42 knots (78 km/h; 48 mph)
U 44 knots (81 km/h; 50 mph)
Max speed (smooth air), airbrakes extended:
H 130 knots (241 km/h; 150 mph)
U 122 knots (225 km/h; 140 mph)
Max aero-tow speed 96 knots (177 km/h; 110 mph)

SCHWEIZER SGS 2-33A

The SGS 2-33 was developed to meet the demand for a medium-priced two-seat sailplane for training and general family soaring. The prototype was first flown in the Autumn of 1966 and received FAA Type Approval in February 1967. Production began in January 1967, and more than 430 SGS 2-33 sailplanes had been built by January 1976. The 2-33 is also available in kit form.

TYPE: Two-seat training sailplane.

WINGS: Strut-braced high-wing monoplane. Aluminium alloy structure with metal skin and all-metal ailerons. Airbrakes fitted.

FUSELAGE: Welded chrome-molybdenum steel tube structure. Nose covered with glassfibre, remainder with Ceconite fabric.

TAIL UNIT: Braced steel tube structure, covered with Ceconite fabric.

LANDING GEAR: Non-retractable Cleveland monowheel and 6·00-6 tyre, immediately aft of nose-skid. Rubber-block shock-absorption for skid. Wingtip wheels.

ACCOMMODATION: Two seats in tandem in completely lined cockpit, with dual controls. One-piece canopy. Rear door and window. Standard equipment includes air vent, seat belts, shoulder harnesses and airspeed indicator.

DIMENSIONS, EXTERNAL:
Wing span 15·54 m (51 ft 0 in)
Wing aspect ratio 11·85
Length overall 7·85 m (25 ft 9 in)
Height over tail 2·83 m (9 ft 3½ in)

AREAS:
Wings, gross 20·39 m² (219·48 sq ft)
Ailerons (total) 1·69 m² (18·24 sq ft)

WEIGHTS AND LOADING:
Weight empty 272 kg (600 lb)
Max T-O weight 472 kg (1,040 lb)
Max wing loading 23·14 kg/m² (4·74 lb/sq ft)

PERFORMANCE:
Best glide ratio 22·25
Min sinking speed:
solo 0·79 m (2·6 ft)/sec
dual 0·91 m (3·0 ft)/sec
Stalling speed:
solo 27 knots (50 km/h; 31 mph)
dual 30·5 knots (57 km/h; 35 mph)
Max speed (smooth air) and max aero-tow speed
85 knots (158 km/h; 98 mph)
Max winch-launching speed
60 knots (111 km/h; 69 mph)

SCHWEIZER SGS 1-34B

Design of this single-seat high-performance Standard Class sailplane, intended to replace the 1-23 series described in the 1967-68 *Jane's,* began in 1967 and construction of the prototype started in the following year. This flew for the first time in the Spring of 1969 and FAA type certification was awarded in September 1969. More than 90 production models had been completed by January 1976.

TYPE: Single-seat Standard Class club and syndicate sailplane.

WINGS: Cantilever shoulder-wing monoplane. Wing section Wortmann FX-61-163 at root, Wortmann FX-60-126 at tip. Dihedral 3° 30′. Incidence 1° at root, 0° at tip. No sweepback. All-metal aluminium alloy structure. Plain all-metal differential ailerons. Double-flap speed-limiting airbrakes, above and below wing.

FUSELAGE: All-metal aluminium alloy semi-monocoque structure.

TAIL UNIT: Cantilever all-metal aluminium alloy structure with swept vertical surfaces. Fixed-incidence tailplane. No trim tabs.

LANDING GEAR: Non-retractable monowheel, with forward skid and auxiliary tailwheel. Retractable monowheel optional. Cleveland wheel size 5·00-5, Type III 4-ply tyre. Cleveland wheel brake.

ACCOMMODATION: Single seat under bubble canopy.

DIMENSIONS, EXTERNAL:
Wing span 15·00 m (49 ft 2½ in)
Wing chord at root 1·31 m (4 ft 3·63 in)
Wing chord at tip 0·56 m (1 ft 10·13 in)
Wing aspect ratio 16
Length overall 7·85 m (25 ft 9 in)
Height over tail 2·29 m (7 ft 6 in)
Tailplane span 2·59 m (8 ft 6 in)

AREAS:
Wings, gross 14·03 m² (151 sq ft)
Ailerons (total) 1·01 m² (10·90 sq ft)
Fin 0·51 m² (5·51 sq ft)
Rudder 0·48 m² (5·18 sq ft)
Tailplane 1·23 m² (13·20 sq ft)
Elevators 0·55 m² (5·88 sq ft)

WEIGHTS AND LOADING:
Weight empty, equipped 249 kg (550 lb)
Max T-O weight 362 kg (800 lb)
Max wing loading 25·9 kg/m² (5·3 lb/sq ft)

PERFORMANCE:
Best glide ratio at 45 knots (84 km/h; 52 mph) 34
Min sinking speed at 40 knots (74 km/h; 46 mph)
0·64 m (2·1 ft)/sec
Max speed (smooth air)
117 knots (217 km/h; 135 mph)
Max aero-tow speed 100 knots (185 km/h; 115 mph)
g limits +8·33; −5·33

SCHWEIZER SGS 1-35

The SGS 1-35 is an-metal single-seat high-performance Standard Class sailplane, the prototype of which was flown for the first time in April 1973. The FAA certification programme was completed in Spring 1974.

The 1-35 has the widest wing loading range in the FAI Standard Class (28·22 to 43·75 kg/m²; 5·78 to 8·96 lb/sq ft), and carries 146·5 kg (323 lb) of water ballast. It can thus compete effectively in both light and strong soaring conditions. The 80° landing flaps are used instead of the more conventional airbrakes.

A total of 55 SGS 1-35s had been produced by January 1976.

TYPE: Single-seat high-performance Standard Class sailplane.

WINGS: Cantilever shoulder-wing monoplane. Wortmann wing sections: FX-67-K-170 at root, FX-67-K-150 at tip. Dihedral 3° 30′. Incidence 1° 30′. No sweep at quarter-chord. Aluminium stressed-skin and stringer construction. Bottom-hinged trailing-edge flaps and top-hinged ailerons of aluminium torque cell construction. No airbrakes or spoilers.

FUSELAGE: All-aluminium monocoque construction.

TAIL UNIT: Cantilever aluminium T-tail, with fixed-incidence tailplane and fabric-covered elevator. No tabs.

LANDING GEAR: Mechanically-retractable unsprung monowheel, tyre size 4·00-4, pressure 2·41 bars (35 lb/sq in). Hydraulic brake. Nose-skid and tailwheel or tailskid.

ACCOMMODATION: Single semi-reclining seat under one-piece detachable canopy.

DIMENSIONS, EXTERNAL:
Wing span 15·00 m (49 ft 2½ in)
Wing chord at aircraft c/l 0·91 m (3 ft 0 in)
Wing chord at tip 0·37 m (1 ft 2·67 in)
Wing aspect ratio 23·3
Length overall 5·84 m (19 ft 2 in)
Height over tail 1·35 m (4 ft 5 in)
Tailplane span 2·03 m (6 ft 8 in)

AREAS:
Wings, gross 9·64 m² (103·8 sq ft)
Ailerons (total) 0·475 m² (5·11 sq ft)
Trailing-edge flaps (total) 1·04 m² (11·20 sq ft)
Fin 0·53 m² (5·72 sq ft)
Rudder 0·44 m² (4·72 sq ft)
Tailplane 0·564 m² (6·07 sq ft)
Elevator 0·442 m² (4·76 sq ft)

WEIGHTS AND LOADING:
Weight empty, equipped 181 kg (400 lb)
Max T-O weight 422 kg (930 lb)
Max wing loading 43·75 kg/m² (8·96 lb/sq ft)

PERFORMANCE:
Best glide ratio 39
Stalling speed, 80° flap
28 knots (51·5 km/h; 32 mph)
Max speed (rough or smooth air) and max aero-tow speed 121 knots (223 km/h; 139 mph)
g limits +8·33; −5·33

Schweizer SGS 2-33 two-seat general-purpose sailplane

Schweizer SGS 1-34 single-seat high-performance Standard Class sailplane

Schweizer SGS 1-35 single-seat high-performance 15 metre all-metal sailplane
(Neil A. Macdougall)

Zanner OZ-5 single-seat Standard Class sailplane
(Howard Levy)

ZANNER
OTTO ZANNER

ADDRESS:
Vineland, New Jersey

ZANNER OZ-5

Mr Zanner has previously built, from kits and/or plans, a Schweizer SGS 1-26, a Thorp T-18, a Briegleb BG 12 and a Bryan (Schreder) HP-14. His fifth aircraft construction, the OZ-5, combines a fuselage and tail unit of his own design with the wings from a Bryan (Schreder) HP-15, and is completed to Standard Class specifications.

The fuselage/cockpit pod is a glassfibre structure to about one-third back on the tailboom; the remainder of the tailboom and the cantilever T-tail are all-metal structures. The monowheel landing gear can be fully retracted by a push/pull lever on the left-hand side of the single-seat cockpit. Flight testing was due to begin in 1975.

DIMENSIONS, EXTERNAL:
Wing span 15·00 m (49 ft 2½ in)
Length overall 6·71 m (22 ft 0 in)
Height over tail 1·22 m (4 ft 0 in)
Tailplane span 1·85 m (6 ft 1 in)

WEIGHTS:
Max pilot weight 106 kg (234 lb
Max T-O weight 303·5 kg (669·5 lb

PERFORMANCE (estimated):
Max speed (smooth air)
156 knots (290 km/h; 180 mph)
Max speed (rough air)
130 knots (241 km/h; 150 mph)
Max aero-tow speed 104 knots (193 km/h; 120 mph)
Max winch-launching speed
78 knots (145 km/h; 90 mph)

THE UNION OF SOVIET SOCIALIST REPUBLICS

LIETUVA

Known designs of this factory include the LAK-6 tandem two-seat motor glider, and the LAK-9 Open Class sailplane.

LIETUVA LAK-9

This single-seat sailplane was designed under the leadership of Balys Karvyalis of Vilna, Lithuania, and first flew in 1972; at that time it was referred to by the designation **BK-7**. It had a Wortmann FX-67-K-170 wing section, and a retractable monowheel landing gear, with tailskid.

The BK-7, which was of glassfibre construction, was subsequently reported to have entered series production, and the Polish publication *Skrzydlata Polska* has referred to a **BK-7A** version with a wing span of 20·00 m (65 ft 7½ in), length of 7·27 m (23 ft 10¼ in), gross weight of 380 kg (837 lb), best glide ratio of 46, and max speed of 113 knots (210 km/h; 130·5 mph).

An almost identical aircraft, registered CCCP-6301, was flown by O. Pasetsnik in the Open Class of the June 1976 World Championships at Räyskälä, Finland, and was referred to as the Lietuva **LAK-9.** It was not placed among the first 25 aircraft on this occasion, and it was reported that the LAK-9 had not, at that time, fully completed flight testing. The LAK-9 is the first Soviet sailplane to compete in the World Championships since 1968. It has water ballast tanks in the wings, and is fitted with trailing-edge flaps. Span is 19·00 m (62 ft 4 in) and gross weight 580 kg (1,278 lb).

The following specification data refer to the BK-7:

DIMENSIONS, EXTERNAL:
Wing span 17·80 m (58 ft 4¾ in)
Wing chord at root 0·96 m (3 ft 1¾ in)
Wing chord at tip 0·36 m (1 ft 2¼ in)
Wing aspect ratio 25·5
Length overall 7·20 m (23 ft 7½ in
Height over tail 1·40 m (4 ft 7 in)

AREAS:
Wings, gross 12·30 m² (132·4 sq ft
Ailerons (total) 0·66 m² (7·10 sq ft
Vertical tail surfaces (total) 1·38 m² (14·85 sq ft
Horizontal tail surfaces (total) 1·35 m² (14·53 sq ft

WEIGHTS AND LOADING:
Weight empty 290 kg (639 lb)
Normal T-O weight 380 kg (837 lb)
Max T-O weight (with ballast) 480 kg (1,058 lb)
Max wing loading (with ballast)
38·5 kg/m² (7·88 lb/sq ft)

PERFORMANCE:
Best glide ratio 43
Min sinking speed 0·52 m (1·71 ft)/sec
Stalling speed 40·5 knots (75 km/h; 47 mph)
Max permissible speed 121 knots (225 km/h; 139 mph)

OSHKINIS
BRONIS OSHKINIS

BRO-16

Details appeared in the Polish and Soviet press in 1974 of a single-seat basic training biplane glider designated BRO-16. Designed by Bronis Oshkinis of the Litovsk SSR, it made its first flight in August 1973 and is a seaplane development of his earlier land monoplane BRO-11 which was completed in 1954 and was subsequently produced for use by the DOSAAF. A modified version of the latter type, designated BRO-11-M, appeared in 1969 and was illustrated in the 1975-76 *Jane's*.

The general appearance of the BRO-16 can be seen in the accompanying illustration; all known details follow:

TYPE: Single-seat waterborne basic training sailplane.

WINGS: Strut-braced unequal-span biplane. Wing section TsAGI R-II, thickness/chord ratio 14%. Dihedral 3° 24′, incidence 3°, on both wings. Single-spar structures, upper wing being fabric-covered and lower wings plywood-covered. Fabric-covered wooden ailerons behind and below upper wing, over approx 87% of span, serve also as flaps and are actuated by pushrods. Aileron section TsAGI R-III, thickness/chord ratio 16%.

FUSELAGE: Nacelle-type hull, with flat planing bottom, of wooden construction with laminated plastics reinforcement. Strut framework, partially fabric-covered, carries the tail unit. Towing hook in nose.

TAIL UNIT: Fabric-covered wooden structure. Small fixed fin, large rudder, braced tailplane, and elevators. Rudder and elevators cable-controlled.

ACCOMMODATION: Single seat, in open cockpit with windscreen.

DIMENSIONS, EXTERNAL:
Wing span:
upper 7·80 m (25 ft 7 in)
lower 5·80 m (19 ft 0¼ in)
Wing chord (upper, constant) 1·00 m (3 ft 3¼ in)

VTC-Vrsac SSV-17 two-seat powered sailplane (Franklin 2A-120-A engine)

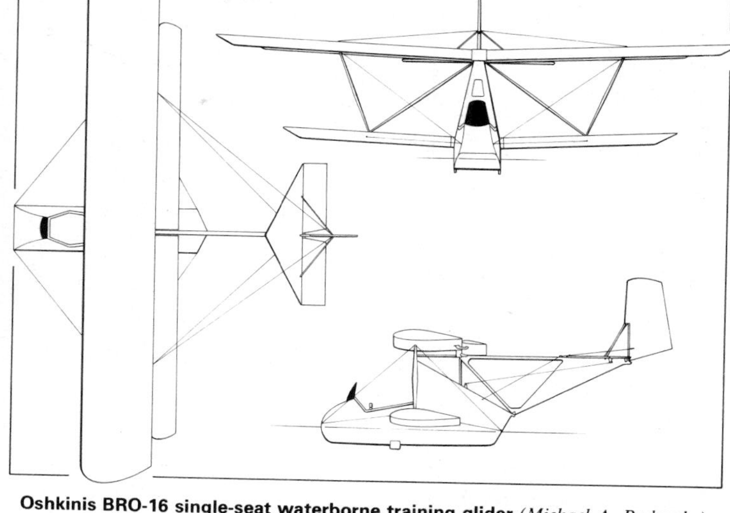

Oshkinis BRO-16 single-seat waterborne training glider *(Michael A. Badrocke)*

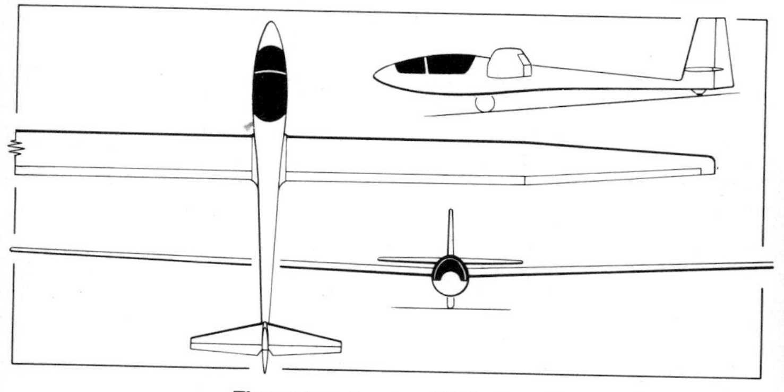

Three-view drawing *(Michael A. Badrocke)* and photograph of the Lietuva LAK-9 single-seat Open Class sailplane

Wing aspect ratio (upper)	3·8
Aileron span	6·80 m (22 ft 3¾ in)
Aileron chord (constant)	0·40 m (1 ft 3¾ in)
Length overall	5·24 m (17 ft 2¼ in)
Length of fuselage nacelle	3·70 m (12 ft 1¾ in)
Width of fuselage	0·70 m (2 ft 3½ in)
Height overall	3·65 m (11 ft 11¾ in)
Tailplane span	2·20 m (7 ft 2¾ in)
AREAS:	
Wings:	
upper	8·20 m² (88·26 sq ft)
lower	6·00 m² (64·58 sq ft)
Ailerons (total)	2·30 m² (24·76 sq ft)
Tailplane	0·65 m² (7·00 sq ft)
Elevator (total)	0·70 m² (7·53 sq ft)
WEIGHTS AND LOADING:	
Weight empty	129 kg (284 lb)
Max T-O weight	204 kg (450 lb)
Max wing loading	12·3 kg/m² (2·52 lb/sq ft)

PERFORMANCE: No details known

BRO-17U UTOCHKA

From illustrations published in the Soviet press, the BRO-17U appears to be essentially a shorter-span version of the BRO-16, having square-cut constant-chord upper wings with endplate surfaces at the tips. The fuselage nacelle and horizontal tail surfaces are also slightly modified in shape. What is unusual, however, is that each half of the upper wing appears to consist of three full-span slats, each of the first two overlapping the one behind it. Since the specification refers to an 'aileron area', the rear slat on each side presumably fulfils this function. The Soviet report did not make it clear whether the BRO-17U had yet been built, but gave the following specification:

DIMENSIONS, EXTERNAL:	
Wing span:	
upper	7·00 m (23 ft 0 in)
lower	4·50 m (14 ft 9¼ in)
Wing chord (upper, constant)	1·45 m (4 ft 9 in)
Wing aspect ratio (upper)	5·5
Wing dihedral (lower)	1°
Length overall	5·80 m (19 ft 0¼ in)
Length of fuselage nacelle	2·90 m (9 ft 6¼ in)
Height overall	2·60 m (8 ft 6½ in)
Elevator span	2·70 m (8 ft 10¼ in)
AREAS:	
Wings (upper)	10·50 m² (113·0 sq ft)
Ailerons (total)	2·30 m² (24·76 sq ft)
Fin	0·07 m² (0·753 sq ft)
Rudder	1·20 m² (12·92 sq ft)
Tailplane	0·50 m² (5·38 sq ft)
Elevator	1·35 m² (14·53 sq ft)
WEIGHTS AND LOADING:	
Weight empty	85 kg (187 lb)
Max T-O weight	160 kg (353 lb)
Max wing loading	15·0 kg/m² (3·07 lb/sq ft)
PERFORMANCE:	
Min sinking speed	1·20 m (3·94 ft)/sec
Max aero-tow speed	43 knots (80 km/h; 49·5 mph)
*Cruising speed (sic)	22 knots (40 km/h; 25 mph)
Landing speed	19 knots (35 km/h; 22 mph)

**Probably refers to towing speed on water*

SA

SPORTIVNAYA AVIATSIYA

SA-8T IDEL

This single-seat Standard Class sailplane, which first flew in August 1972, was designed by A. I. Osokin. It is believed to have been evaluated by the Central Aeroclub of the USSR, for possible aeroclub use by the DOSAAF flying training organisation, and production was under consideration in 1973. No further news of the SA-8T has been received since that time. A brief description and three-view drawing can be found in the 1975-76 *Jane's*.

YUGOSLAVIA

VTC

VAZDUHOPLOVNO TEHNICKI CENTAR—VRSAC

ADDRESS:
Vrsac, 29 Novembra b.b. Gudaricki put

Telephone: 80-111

DIRECTOR:
Veselinovic Zivota

CHIEF DESIGNER:
Dipl Ing Ivan Sostaric

The VTC at Vrsac is producing the Schempp-Hirth Cirrus under licence, manufacture in Germany by the parent company having ended in late 1971. The first production aircraft built at Vrsac was delivered in late 1971.

In collaboration with Sigmund Flugtechnik of Germany (see 1973-74 *Jane's*), the VTC developed and flew a two-seat powered sailplane designated SSV-17.

VTC (SCHEMPP-HIRTH) CIRRUS

This single-seat high-performance sailplane was designed in Germany by Dipl-Ing Klaus Holighaus, who was one of the co-designers of the Akaflieg Darmstadt D-36 Circe before joining Schempp-Hirth in 1965. He utilised a thick Wortmann series wing section, without flaps, to achieve good low-speed and climb characteristics. Provision for ballast overcomes the slight disadvantage this section has when compared with thinner, flapped profiles. Stalling characteristics are good with a thicker wing and it is claimed that weight is saved by comparison with a flapped wing of similar span and aspect ratio.

The first prototype Cirrus flew for the first time in January 1967 with a V tail unit. The second prototype had a conventional tail unit, as fitted to production models.

Schempp-Hirth built a total of 120 Cirrus sailplanes before ending production in late 1971. Since that time the Cirrus has been produced under licence by VTC in Yugoslavia, which had completed about 60 by early 1975.

TYPE: Single-seat high-performance sailplane.

WINGS: Cantilever mid-wing monoplane. Wortmann FX-66 series section. Thickness/chord ratio 19·6% at root, 16% at tip. Dihedral 3° at spar centreline. Incidence 2°. No sweep at spar centreline. Wing shell is a glassfibre/foam sandwich structure, with an all-glassfibre box spar. Hinged ailerons of glassfibre/balsa sandwich. No flaps. Schempp-Hirth aluminium alloy airbrakes.

FUSELAGE: Glassfibre shell 1·5 mm thick, stiffened with foam rings, secured with resin.

TAIL UNIT: Cantilever structure of glassfibre/foam sandwich. Tailplane mounted part-way up fin.

LANDING GEAR: Retractable monowheel type. Manual retraction. Annular rubber-spring shock-absorber. Tost wheel with Dunlop tyre size 3·50 × 5, pressure 3·38 bars (49 lb/sq in). Tost drum brake.

ACCOMMODATION: Single semi-reclining adjustable seat. Adjustable rudder pedals. Long flush Plexiglas canopy.

DIMENSIONS, EXTERNAL:	
Wing span	17·74 m (58 ft 2½ in)
Wing chord at root	0·90 m (2 ft 11½ in)
Wing chord at tip	0·36 m (1 ft 2¼ in)
Wing aspect ratio	25
Length overall	7·20 m (23 ft 7¼ in)

Height over tail 1·56 m (5 ft 0 in)
Tailplane span 2·50 m (8 ft 2½ in)

AREAS:
Wings, gross 12·6 m² (135·6 sq ft)
Ailerons (total) 1·04 m² (11·2 sq ft)
Fin 0·63 m² (6·8 q ft)
Rudder 0·52 m² (5·6 sq ft)
Tailplane 0·90 m² (9·7 sq ft)
Elevators 0·15 m² (1·6 sq ft)

WEIGHTS AND LOADING:
Weight empty, equipped 260 kg (573 lb)
Max T-O weight 400 kg (882 lb)
Max wing loading 31·7 kg/m² (6·5 lb/sq ft)

PERFORMANCE (at AUW of 360 kg; 793 lb):
Best glide ratio at 46 knots (85 km/h; 53 mph) 44
Min sinking speed at 39 knots (73 km/h; 45 mph) 0·50 m (1·64 ft)/sec
Stalling speed 33·5 knots (62 km/h; 39 mph)
Max speed (rough or smooth air) 119 knots (220 km/h; 137 mph)
Max aero-tow speed 76 knots (140 km/h; 87 mph)
Max winch-launching speed 59 knots (110 km/h; 68 mph)

Schempp-Hirth-built Cirrus single-seat high-performance sailplane

VTC SSV-17

The SSV-17 two-seat powered sailplane was developed jointly by the VTC and Sigmund Flugtechnik of Germany, under the design guidance of Dipl-Ings Alfred Vogt and Ivan Sostaric. The prototype (YU-M6009), flown by VTC test pilot A. Stanojevic, made its first flight on 24 June 1972. It was certificated in May 1973, and by the end of that year 10 had been ordered. No further news of the aircraft has been received since that time.

VTC TRENER

Believed to be in production by VTC for the Yugoslav Aero Club, the Trener is a modified version of the LIBIS-18 single-seat Standard Class sailplane which first flew on 20 October 1964 and was last described in the 1971-72 *Jane's*. An initial batch of 50 is believed to have been ordered.

Of all-wood construction (beech and Swedish plywood), the Trener has Wortmann wing sections with a thickness/chord ratio of 18% at the root, 3° dihedral and 1° 30′ sweepback on the leading-edges. The fuselage is a wooden monocoque, with a non-retractable monowheel. Tail unit comprises a sweptback fin and rudder and an all-moving tailplane of NACA 63-012 section. The tailplane is mounted slightly higher than on the LIBIS-18. Cockpit layout also has been improved.

DIMENSIONS, EXTERNAL:
Wing span 15·00 m (49 ft 2½ in)
Length overall 6·90 m (22 ft 7¾ in)

AREA:
Wings, gross 12·97 m² (139·6 sq ft)

PERFORMANCE:
Best glide ratio at 46 knots (85 km/h; 53 mph) 31
Min sinking speed at 42 knots (78 km/h; 48·5 mph) 0·72 m (2·36 ft)/sec
Never-exceed speed 118·5 knots (220 km/h; 136·5 mph)
Max speed, airbrakes extended 100 knots (185 km/h; 115 mph)
Max aero-tow speed (smooth air) 81 knots (150 km/h; 93 mph)
Max aero-tow speed (rough air) 65 knots (120 km/h; 74·5 mph)

HANG GLIDERS

In view of the increasing popularity of hang gliding or skysurfing as a sport, the Sailplanes section of *Jane's* began in 1973-74 to include some of the more popular craft of this type and the companies which manufacture them in complete or component form.

In 1975 the FAI set up a commission for hang-gliding, known as the CIVL (Commission International de Vol Libre), which held its first meeting in June 1975. It defined a hang glider as "a heavier-than-air, fixed-wing (not rotating) glider, capable of being carried, foot-launched and landed solely by the energy and use of the pilot's legs". Future tasks of the commission were to include safety, competition, classification of craft, and pilot ratings.

Details follow of the principal known national organisations governing the manufacture or operation of hang-gliders:

GERMANY (Federal Republic)

ALPINE DRACHENFLIEGER e.V.

ADDRESS:
8 München 60, Bauseweinallee 100
Telephone: (089) 95 12 13

NEW ZEALAND

NEW ZEALAND HANG GLIDING ASSOCIATION

SECRETARY AND TREASURER: PO Box 44117, Vic, Lower Hutt, Wellington

UNITED KINGDOM

BRITISH HANG GLIDER MANUFACTURERS' FEDERATION

ADDRESS:
'Mallows', Forest Drive, Kingswood, Tadworth, Surrey
Telephone: Mogador 2873
CHAIRMAN: R. Spooner
SECRETARY: Jillian Handley
TREASURER: L. Gabriels

BRITISH HANG GLIDING ASSOCIATION (Affiliated to the FAI)

ADDRESS:
Monksilver, Taunton, Somerset
Telephone: Taunton 88140
PRESIDENT: Mrs Ann Welch, OBE
CHAIRMAN: Martin Hunt
SECRETARY: Chris Corston
TRAINING OFFICER: Alvin Russell
PUBLICATION: *Wings!:* Editor, A. R. Fuell, 74 Eldred Avenue, Brighton, Sussex

UNITED STATES OF AMERICA

HANG GLIDER MANUFACTURERS' ASSOCIATION

ADDRESS:
c/o Peter Brock, UP Inc, PO Box 582, Rancho Temecula, California 92390
Telephone: (714) 676-5652
PRESIDENT: P. Brock

UNITED STATES HANG GLIDING ASSOCIATION

ADDRESS:
11312½ Venice Boulevard, PO Box 66306, Los Angeles, California 90066
Telephone: (213) 390-3065
PUBLICATION:
Ground Skimmer (monthly)

The former Self-Soar Association and United States Skysurfing Association, listed in the 1975-76 *Jane's*, are no longer operative. The publication *Hang Glider Weekly*, referred to under the former heading, continues in being as a separate entity, under the title *Hang Glider Magazine Newsweekly*. All enquiries to the Editor, Joseph P. Faust, Box 1860, Santa Monica, California 90406; telephone (213) 396-4241.

TYPES OF HANG GLIDER

As this edition was being prepared for press, international regulations governing the classification and operation of hang gliders were undergoing considerable revision, and for this reason it was considered advisable to omit any detailed listing of individual types from the current edition. It is hoped to restore this section in the 1977-78 *Jane's*.

In the meantime, the list given in the 1975-76 edition may be taken as a general guide to the principal manufacturers of hang gliders, several of whom have introduced new or improved models for 1976 and will be able to advise enquirers directly regarding the extent to which their current range of models meets the new regulations. Of those manufacturers who had replied to *Jane's* before this edition closed for press, the following had indicated a change of name and/or address from that given in the 1975-76 edition:

CANADA

FREE FLIGHT OF CANADA

PO Box number closed; presumed no longer operating.

KAUSCHE KITES

Now renamed **Windsports Industries Ltd;** address 5319 4th Street SW, Calgary T2R 0G9, Alberta; telephone (403) 255-4280.

SWITZERLAND

DR K. SCHLEUNIGER & CO

Address is now: Schöngrünstrasse 27, CH-4500 Solothurn; telephone (065) 220321.

UNITED KINGDOM

MILES HANDLEY

Now renamed **Miles Wings (Engineers) Ltd;** address Unidev Works, Croydon Road, Elmers End, Beckenham, Kent BR3 4BP; telephone number unchanged.

ULTRA LIGHT GLIDERS LTD

Now renamed **Aerosoar Gliders;** address unchanged.

UNITED STATES OF AMERICA

BILL BENNETT'S DELTA WING KITES AND GLIDERS INC

Address is now: PO Box 483, Van Nuys, California 91408.

MAN-FLIGHT SYSTEMS INC

No longer in business.

SKY SPORTS INC

Address is now: 394 Somers Road (Route 83), Ellington, Connecticut 06029; telephone (203) 872-7317.

LIGHTER-THAN-AIR: AIRSHIPS

GERMANY
(FEDERAL REPUBLIC)

WDL
WESTDEUTSCHE LUFTWERBUNG, THEODOR WÜLLENKEMPER KG

ADDRESS:
433 Mülheim/Ruhr, Flughafen
Telephone: (0 21 33) 31009/31000
Telex: 856810 WDL

In 1969 Herr Theodor Wüllenkemper founded the WDL airship works in Essen-Mülheim, with the objective of designing and building a new generation of non-rigid airships that would take advantage of present-day materials and technology.

The first three designs, WDL 1, 2 and 3, are all intended as experimental craft, increasing in size progressively and each intended to explore new concepts to simplify operation of such craft. Subject to successful development, it is planned to embody their basic constructional, control and handling concepts in a larger standard production airship, designated WDL 4.

WDL 2 is intended to explore special configurations of gondola to permit the carriage of international standard freight containers. These will take the form of interchangeable passenger/cargo gondolas. Power plant mountings, rotatable through 360°, will allow the propeller thrust to be used not only for fore and aft propulsion, but also in vertical modes to enhance lift or assist in landing procedures.

The standard production airship, WDL 4, will have a gross volume of 64,000 m³ (2,260,140 cu ft) and payload of 30,000 kg (66,140 lb). It is expected to have a heating system, from the exhaust, to raise the temperature of the helium lifting gas.

WDL 1

Construction of the WDL 1 serial number 100 was completed in mid-1972 and the first flight, piloted by Konrad Hess, was made on 12 August 1972.

This airship, named *The Flying Musketeer,* was severely damaged during a storm on 13 November 1972. It was repaired during the Winter months and began flying again on 28 April 1973, at which time it was renamed *Wicküler.*

WDL 1 airship owned by Orient Lease Co of Tokyo and decorated by Japanese artist Taro Okamoto *(Kazuo Miyazaki)*

The envelope is made of synthetic fabric and is helium-filled. An automatic pressure control system has been developed to maintain effective control of pressure during flight as well as on the ground, to preserve the shape and stability of the craft at all times. Two panels, each 40·0 m (131 ft 3 in) by 8·0 m (26 ft 3 in), one on each side of the envelope, together provide 10,000 electric bulbs for the display of advertising slogans. The power plant consists of two Rolls-Royce Continental aero-engines, each of 134 kW (180 hp); and all fuel for the engines is contained in cells mounted within the envelope. This arrangement has been adopted to provide maximum capacity within the gondola.

Construction of WDL 1 serial number 101 was completed in the late Summer of 1972 and a first flight was made during October. The envelope was damaged when the airship's inflatable hangar collapsed during a storm in November. Following repairs and air test, this airship was shipped to Japan, where it is being operated by Orient Lease Co Ltd of Tokyo. A German airship pilot was responsible for the training of a Japanese pilot during the Summer of 1973.

DIMENSIONS, ENVELOPE:

Length overall	55·0 m (180 ft 5 in)
Max diameter	14·5 m (47 ft 6¾ in)
Volume, gross	6,000 m³ (211,888 cu ft)

WEIGHTS:

Gross weight	6,300 kg (13,889 lb)
Envelope weight	1,600 kg (3,527 lb)
Payload	1,500 kg (3,307 lb)

PERFORMANCE (estimated):

Max speed	54 knots (100 km/h; 62 mph)
Operational radius	215 nm (400 km; 248 miles)

JAPAN

FUJI
FUJI MANUFACTURING CO LTD

HEAD OFFICE:
16 Hotoku-cho, Kita-ku, Nagoya
Telephone: (052) 991-8171
PRESIDENT: Kikuo Koizumi

In the Autumn of 1974 Fuji Manufacturing Co began construction of a small remotely controlled pilotless research airship. This had been conceived by the President of the company, and the work was put under the leadership of Mr Daisaku Okamoto. The first flight of this airship (construction number 2) was made successfully on 23 December 1974. (It is believed that No. 1 was the airship built by members of the Japan Experimental Aircraft Association and described under the JEAA heading in the 1975-76 *Jane's.)*

Airship No. 3 flew on 28 July 1975, this being a modified version of No. 2. No details are known of Nos 4, 5 and 6. No. 7, which first flew on 21 November 1975, has an overall length of 6·0 m (19 ft 7½ in) and was supplied to Tohoku University Geographical Laboratory. No. 8 was under construction in early 1976, for the Taisei Construction Company Ltd.

FUJI MODEL 500 AERO-SHIP

The word Aero-Ship, which forms a part of the designation of the Fuji Model 500, is an allusion to the unique aeroplane-type horizontal lifting surface which is attached to the lower structure of the gondola. The envelope is helium filled, and it is claimed that the wing not only provides additional lift but, in conjunction with the fairly large tail surfaces, makes the vessel highly manoeuvrable. Flown under radio control, it is suggested that the Aero-Ship could be utilised for such applications as aerial photography, various forms of meteorological and pollution measurement, seed and fertiliser distribution, and advertising.

The envelope of the Model 500 is of plastics, the gondola of Japanese cypress, balsa and plywood, silk-covered. Tail surfaces are of similar construction; tailplane incidence is normally 3°, but is ground adjustable. Trim tabs in elevators. The wing has Japanese cypress ribs, and unspecified balsa and plywood components, with silk covering. Wing section is Göttingen Gö 535 and the wing is normally mounted without incidence. Handling gear, comprising four wheels with tyres of 365 mm (14·4 in) diameter, has two wheels mounted beneath the wing, one at the aft end of the gondola and one beneath the lower fin. Power plant comprises two ENYA 60 IIIB aircooled engines, each developing 0·97 kW (1·3 hp), mounted at the rear of the gondola, at each end of an outrigger. The forward part of the gondola is able to house a camera or other specialised equipment. Radio control equipment and batteries are accommodated in the central area, and a fuel tank in the aft section.

Fuji Model 500 Aero-Ship, a radio-controlled research airship

Limited range of the radio control equipment used to date has prevented full evaluation of the Aero-Ship's performance, but a speed range of 5·4-43 knots (10-80 km/h; 6·2-50 mph) and ceiling of 1,000 m (3,280 ft) have been demonstrated.

DIMENSIONS:

Length overall	8·0 m (26 ft 3 in)
Wing span	3·26 m (10 ft 8½ in)
Wing chord, constant	0·68 m (2 ft 2¾ in)
Tailplane span	2·25 m (7 ft 4½ in)

AREAS:

Wings, gross	1·01 m² (10·87 sq ft)
Rudders (total)	0·43 m² (4·63 sq ft)
Tailplane	0·86 m² (9·26 sq ft)
Elevators (incl tabs)	0·49 m² (5·27 sq ft)

WEIGHTS AND LOADING:

Weight empty, equipped	22·7 kg (50 lb)
Weight, helium filled	4·0 kg (8·8 lb)
Max T-O weight	6·0 kg (13·2 lb)
Max wing loading	5·96 kg/m² (1·22 lb/sq ft)

THE UNITED KINGDOM

AEROSPACE DEVELOPMENTS
AEROSPACE DEVELOPMENTS

ADDRESS:
19-21 Newbury Street, London EC1A 7HU
Telephone: (01) 606-5981/3
Telex: 884985 MTAAD

Aerospace Developments has designed and is building the first of 22 non-rigid airships which its associated marketing organisation, Multimodal Transport Analysis Ltd, has contracted to supply to a Venezuelan customer, Aerovision. Two contracts are involved: the first covers the supply of a single airship in 1977; the second provides for a further 21 airships to be delivered over the ensuing ten years.

Five types of airships have been proposed, designated Types A to E, with payload capability respectively of 0·5, 2, 5, 10 and 20-25 tons. It is understood that other South American countries, in particular Argentina and Peru, are interested in the potential of airships for transport to and from undeveloped and otherwise inaccessible areas. It is suggested that by using such vessels also in a night-time advertising role, the cost of transport usage can be offset to some extent.

The airship currently under construction, and scheduled for roll-out in December 1976, is of the Type B configuration. It will be flown initially from Cardington, Bedfordshire, and the company has arranged for a US airship pilot to be responsible for the test and certification programme. This is expected to take from three to six months, after which the airship will be crated for shipment to Venezuela. It is anticipated that Aerovision will operate the vessel for some months before finalising the specification of follow-on airships. Following the assembly and testing of several airships in Britain, the remainder are expected to be manufactured in Venezuela, initially by the assembly of British-made components.

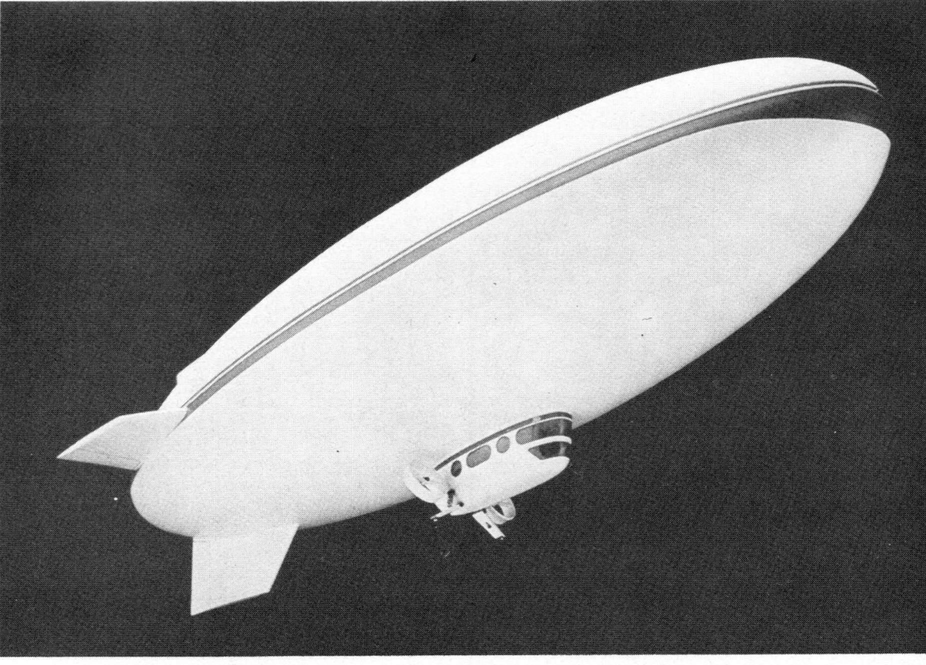

Model of Aerospace Developments Type B non-rigid airship

AEROSPACE DEVELOPMENTS TYPE B

The envelope of the Type B is being manufactured from a Dacron type of synthetic fabric, with a double layer of Mylar plastic film coating on the inner surface to reduce porosity and a single outer surface covering of Tevlar for weather resistance. Helium is to be used as the lifting gas.

The gondola is a one-piece structure of glassfibre with carbon fibre reinforcement, and will carry the crew, passengers, water ballast, advertising display control equipment and the two engines, one on each side. Fuel for the engines is contained in a tank in the leading-edge of the lower fin. The gondola serves also as mounting for the two fully-castoring main landing gear units, which are supplemented by a tailwheel in the base of the lower fin.

Power plant will comprise two Porsche six-cylinder air-cooled engines, each developing 149 kW (200 hp) at 6,000 rpm, and each driving via reversible reduction gear and shaft a five-blade adjustable-pitch ducted fan. The ducted fans can be swivelled to provide thrust-vectoring to supplement the gaseous lift, allowing the airship to be flown off in a 'heavy' condition. Conversely, they can be used to assist landing in the 'light' condition without the need to valve-off costly helium. The lower fin was selected for fuel storage because at the beginning of a flight the vessel's CG will be aft, causing the adoption of a nose-up attitude which will generate lift. At the end of a flight the CG will have moved forward to create a nose-down condition, helping to offset some of the airship's natural buoyancy.

DIMENSIONS, ENVELOPE:
- Length overall: 50·29 m (165 ft 0 in
- Max diameter: 13·93 m (45 ft 8½ in
- Volume, gross: 5,097 m^3 (180,000 cu ft

DIMENSIONS, GONDOLA:
- Length overall: 9·24 m (30 ft 3⅝ in
- Height: 2·26 m (7 ft 4⅞ in
- Width: 2·41 m (7 ft 10⅞ in

WEIGHTS (estimated):
- Weight empty: 2,522 kg (5,560 lb
- Max design gross weight (ISA + 30°C): 4,252 kg (9,375 lb

PERFORMANCE (estimated at 305 m; 1,000 ft, ISA):
- Max level speed: 80 knots (148 km/h; 92 mph
- Max cruising speed: 70 knots (130 km/h; 80 mph
- Normal cruising speed: 60 knots (111 km/h; 69 mph

CAMERON
CAMERON BALLOONS LTD

ADDRESS: 1 Cotham Park, Bristol BS6 6BZ
Telephone: (0272) 41455

DIRECTORS:
D. A. Cameron, BSc (Aero Eng), MIE
Kim Cameron
Tom Sage, ARPS, AIIP

Cameron Balloons Ltd, which designs and manufactures a wide range of hot-air balloons (which see), also designed and built the world's first hot-air airship. In addition to its use as a purely sporting vehicle, this type of aircraft is expected to have useful applications for a range of commercial tasks such as aerial photography, survey and advertising, and production versions are now available.

CAMERON D96 HOT-AIR AIRSHIP

First flight of the prototype (G-BAMK) was made at Wantage, Berkshire, on 7 January 1973 and was honoured by the award of the Royal Aero Club's silver medal and the American Lighter Than Air Society's 1973 Achievement Award. Since that time a considerable amount of work has been carried out to improve and develop the airship. The prototype had initially a single vertical stabiliser at the aft end, which provided marginal directional stability; the current production model has in addition two upper stabilisers, one on each side. An improved method of suspending the gondola has also been evolved, eliminating the distortion of the envelope which occurred in the early stages.

Production Cameron D96 hot-air airship for a customer in the USA

The envelope, like those of Cameron hot-air balloons, is made from a light but high-strength nylon fabric. The tubular-metal gondola installed on the prototype was built by a specialist engineering firm at Oxford, headed by Dr E. T. Hall. A new, lighter gondola has now been developed and this carries the propane burner, gas supply, pilot, passenger and power plant.

Power plant is a 33·56 kW (45 hp) 1,600 cc Volkswagen modified motor car engine, driving a large-diameter semi-shrouded pusher propeller. That installed in the prototype ran conventionally on motor gasoline, but has been converted to operate on propane gas, as used to supply the burner; production models also use propane.

The first production airship has been completed for a customer in the United States; two more were under construction in the Summer of 1976 for customers in Belgium and France.

DIMENSIONS, ENVELOPE:
- Length overall: 30·48 m (100 ft 0 in)
- Max diameter: 13·72 m (45 ft 0 in
- Volume, gross: 2,718 m^3 (96,000 cu ft

PERFORMANCE:
- Max speed: 13 knots (24 km/h; 15 mph
- Turning radius at 8·7 knots (16 km/h; 10 mph): 30·5 m (100 ft
- Endurance: 2 hour

GLOSTER—*see Addenda*

SMITH

ANTHONY SMITH

SMITH SANTOS-DUMONT

Mr Smith has constructed a small non-rigid airship (G-BAWL) which has the name *Santos-Dumont.* It flew for the first time in May 1974 and logged a total of six flying hours during flights in May and June, while it was inflated with hydrogen. Flying was resumed in 1975, by which time the envelope was inflated with helium.

The envelope is of standard configuration, with fore and aft ballonets. At the aft end of the envelope, lightweight tubular structures are attached to it by patches. Two of these structures, one on each side on the centreline of the envelope, carry fixed horizontal surfaces, to the trailing-edges of which are attached controllable elevators. Twin ventral surfaces are carried in a similar manner, and to the trailing-edge of each of these surfaces is attached a controllable rudder.

The gondola is also a lightweight tubular structure, at the rear of which are mounted two 14·91 kW (20 hp) Wankel-type engines, each driving a small ducted propeller of the kind used in hovercraft. Airscoops for inflation of the ballonets are mounted in the slipstream from the propellers. Fuel is contained in a single tank of 22·7 litres (5 Imp gallons) capacity. Ballast totalling 45 kg (100 lb) is carried.

When the airship is deflated and dismantled, its fabric parts will pack easily inside a station wagon carrying four persons, with the gondola and control surface structures on a roof rack.

General view of Anthony Smith's 934·46 m^3 (33,000 cu ft) airship

DIMENSIONS, ENVELOPE:	
Length overall	23·16 m (76 ft 0 in)
Max diameter	8·84 m (29 ft 0 in)
Propeller diameter (each)	0·71 m (2 ft 4 in)
Volume, gross	934·46 m^3 (33,000 cu ft)
Volume, ballonets (total)	170 m^3 (6,000 cu ft)
DIMENSION, GONDOLA:	
Length	3·66 m (12 ft 0 in)
WEIGHT:	
Weight empty	approx 481 kg (1,060 lb)
PERFORMANCE:	
Max speed	26 knots (48 km/h; 30 mph)
Pressure height	approx 1,825 m (6,000 ft)

THE UNITED STATES OF AMERICA

B AND G

B AND G AIRSHIPS LTD

BG-1 DOLPHIN

Limited details of a new semi-rigid airship have been received, the prototype of which was reported to be under construction in the Summer of 1976. Designated BG-1 Dolphin, it appears to be of conventional semi-rigid configuration and construction. The horizontal tail surfaces have elevators at their trailing-edges. The dorsal vertical surface is a fixed fin; the ventral surface has a rudder.

A gondola, beneath the envelope, is mounted slightly forward of the fore and aft centreline, and has accommodation for a crew of two. The 60 kW (80 hp) Barker CE 2000 engine is mounted at the aft end of the gondola, driving a fully-feathering and reversible pusher propeller. A 91 litre (24 US gallon) fuel tank is carried.

The envelope contains a single ballonet, and an air scoop for its inflation is sited in the propeller's slipstream. To simplify ground handling a fully-castoring wheel is mounted beneath the gondola, and a fixed wheel is mounted below the ventral tail surface.

DIMENSIONS, ENVELOPE:	
Length	29·63 m (97 ft 2½ in)
Max diameter	8·23 m (27 ft 0 in)
Height overall, incl gondola and fin	11·13 m (36 ft 6 in)
Propeller diameter	1·37 m (4 ft 6 in)
Fineness ratio	3·6
Volume, gross	1,075·6 m^3 (37,983 cu ft)
Volume, ballonet	169·9 m^3 (6,000 cu ft)
DIMENSIONS, GONDOLA:	
Length overall	2·97 m (9 ft 9 in)
Height	1·61 m (5 ft 3½ in)
Width	1·07 m (3 ft 6 in)
WEIGHTS (estimated):	
Weight empty	683 kg (1,506 lb)
Gross weight	930 kg (2,050 lb)
PERFORMANCE (estimated):	
Max level speed	30 knots (55 km/h; 34 mph)
Cruising speed	26 knots (48 km/h; 30 mph)
Service ceiling	1,524 m (5,000 ft)
Range at cruising speed	156 nm (289 km; 180 miles)

BOLAND

BRIAN J. BOLAND

ADDRESS: Pine Drive, RFD2, Burlington, Connecticut 06013

BOLAND ALBATROSS

Mr Brian Boland, an art and photographic teacher at Farmington High School, Connecticut, has, with his wife Kathy, designed and built a hot-air airship named *The Albatross.* Construction began in early 1975 and it was flown for the first time on 11 October 1975. By the end of that year a total of approximately 12 flying hours had been accumulated.

The envelope is made from 32 separate strips of polyurethane-coated rip-stop nylon, sewn together in just over three weeks. The gondola has a basic structure of light alloy, around which wicker sidewalls have been woven. It has oak landing skids on its undersurface, and is mounted on a wheeled trailer for easy ground handling, this being left on the ground on lift-off. The gondola is suspended from the envelope by twenty flexible steel cables which run through its light alloy tubular framework and are attached to two catenary suspension curtains. The tubular framework of the gondola serves also as mounting for three 8 million BTU propane burners. The fuel supply comprises six low-pressure cylinders, with a combined capacity of 227 litres (60 US gallons) of liquid propane, and a 22·7 litre (6 US gallon) tank of fuel/oil mixture for the propulsion engine. This is a 29·8 kW (40 hp) Rockwell JLO two-stroke engine, driving a Banks Maxwell two-blade ducted pusher propeller.

On the basis of flight experience with *The Albatross* in 1975, Mr Boland planned to introduce a number of improvements during 1976.

The Boland Albatross hot-air airship under test in 1975

DIMENSIONS, OVERALL:	
Length	34·14 m (112 ft 0 in)
Height	19·81 m (65 ft 0 in)
Width	15·24 m (50 ft 0 in)
Volume	3,965 m^3 (140,000 cu ft)
DIMENSIONS, GONDOLA:	
Length overall	3·35 m (11 ft 0 in)
Height	3·05 m (10 ft 0 in)
Width	1·22 m (4 ft 0 in)
WEIGHTS:	
Flight weight, fuelled	658 kg (1,450 lb)
Max design gross weight	1,089 kg (2,400 lb)
PERFORMANCE:	
Max speed	approx 10 knots (18·5 km/h; 8·5 mph)

CONRAD
CONRAD AIRSHIP CORPORATION

This company was founded by Mr Conrad and his son, and was reported in 1975 to be building a medium-size rigid airship in the open air near Phoenix, Arizona.

The hull is constructed of light alloy tubing, with 10 ring formers and 12 longitudinal stringers. Lift will be provided by 10 gas bags fabricated from laminated Mylar. Liquid ballast is to be carried, and will be distributed and/or dumped by high-volume transfer pumps. The power plant is said to consist of two 373 kW (500 hp) liquid-cooled engines, with the drive to the three propellers (one on the axis of the airship, at the tail) via two hydraulic units, each weighing 38·5 kg (85 lb). No other information was available by June 1976.

DIMENSIONS, ENVELOPE:
- Length overall: 68·58 m (225 ft 0 in)
- Max diameter: 15·24 m (50 ft 0 in)
- Volume, gross: 7,079 m³ (250,000 cu ft)

PERFORMANCE (estimated):
- Cruising speed: 74 knots (137 km/h; 85 mph)
- Range: 2,605 nm (4,828 km; 3,000 miles)

GOODYEAR
GOODYEAR TIRE & RUBBER COMPANY, GOODYEAR INTERNATIONAL CORPORATION—AIRSHIP OPERATIONS

ADDRESS:
1144 East Market Street, Akron, Ohio 44316
Telephone: (216) 794-2995
Telex: 098-6413

CHAIRMAN OF THE BOARD AND CHIEF EXECUTIVE OFFICER:
Charles J. Pilliod Jr

DIRECTOR INTERNATIONAL PUBLIC RELATIONS:
M. F. O'Reilly

Since 1917, Goodyear has built a total of 301 airships, more than any other company in the world. Of these, 244 were constructed under contract for the US Army and Navy, and included the USS *Akron* and USS *Macon,* the largest rigid airships constructed in the USA. The remaining 57 have been commercial airships, of which the first was the Pilgrim, launched in 1925.

In February 1968, Goodyear initiated a $5 million expansion and improvement programme for its airship operations, which included the provision of a new base at Houston, Texas, modernisation of the Mayflower III and Columbia II, based at Miami and Los Angeles respectively, and construction of a new airship, named America. This latter airship, completed in early 1969, was built at Wingfoot Lake, near Akron, from where it made its first flight on 25 April 1969. It is based currently at Houston, Texas. Columbia II and Mayflower III were described in the 1964-65 *Jane's,* the former having identical dimensions to the America, which was described briefly in the 1970-71 *Jane's.*

In 1971, a decision was made to build a fourth non-rigid airship to begin public relations operations in Europe in 1972. On 9 December 1971 components for its construction were flown from Akron Municipal Airport to the Royal Aircraft Establishment at Cardington, Bedfordshire, in an Aero Spacelines B-377MG Mini Guppy; and this airship, named Europa, made its first flight from Cardington on 8 March 2.

During the Winter months of each year the Europa operates from a base which has been established at Capena, Italy, north of Rome. To accommodate the Europa and ground support vehicles, and to provide office space, crew quarters and adequate facilities for complete maintenance of the airship, Goodyear has erected a hangar at Capena. This concrete, steel and glass structure is 76·20 m (250 ft) in length, 49·00 m (160 ft 9 in) wide and 27·43 m (90 ft) in height. During the Spring and Summer months the airship tours western Europe.

In 1975 a new airship was constructed, this being given the name Columbia III. Upon its completion the Columbia II was retired from service.

All available details of the Europa, which is basically similar to the America and Columbia III, are given below:

Europa about to touch down after one of its frequent publicity flights

EUROPA

The envelope of the Europa, which has a surface area of 2,006 m² (21,600 sq ft), is made of two-ply Neoprene-coated Dacron and, like that of its sister 'ships, is helium-filled. On each side of the envelope is a four-colour sign 32·00 m (105 ft 0 in) long and 7·47 m (24 ft 6 in) high, containing 3,780 lamps to flash static or animated messages. These can be read at a distance of 1·6 km (1 mile) when the airship is cruising at a height of 305 m (1,000 ft).

A turbojet APU, mounted in a removable pod on the undersurface of the 'ship's gondola, drives a 500A 28V generator to supply electrical power for the signs and their control equipment. The turbojet is designed to operate without developing any appreciable amount of forward thrust for the airship.

The gondola, attached to the undersurface of the envelope, has accommodation for a pilot and six passengers, and has a single non-retractable landing wheel mounted beneath it.

Power plant consists of two 156·6 kW (210 hp) Continental IO-360-D flat-six engines, each driving a Hartzell two-blade metal reversible-pitch pusher propeller. Standard tankage is provided for 527 litres (138 US gallons) of fuel, and auxiliary tankage for 598 litres (158 US gallons). Total available fuel capacity is thus 1,125 litres (296 US gallons).

By 1 January 1976 the Europa had flown 5,898 hours and carried more than 29,750 passengers since its first flight in 1972.

DIMENSIONS, OVERALL:
- Length: 58·67 m (192 ft 6 in)
- Width: 15·24 m (50 ft 0 in)
- Height: 18·14 m (59 ft 6 in)

DIMENSIONS, ENVELOPE:
- Length: 58·00 m (190 ft 3½ in)
- Max diameter: 14·00 m (45 ft 11 in)
- Fineness ratio: 14·4
- Volume, gross: 5,379·9 m³ (202,700 cu ft)
- Volume, ballonets: 1,662·2 m³ (58,700 cu ft)

DIMENSIONS, GONDOLA:
- Length overall: 6·93 m (22 ft 9 in)
- Height: 2·47 m (8 ft 1¼ in)
- Height, incl landing gear: 3·59 m (11 ft 9½ in)
- Width at ceiling: 2·13 m (7 ft 0 in)
- Width at floor: 1·31 m (4 ft 3½ in)

WEIGHTS:
- Weight empty: 4,252 kg (9,375 lb)
- Max design gross weight: 5,824 kg (12,840 lb)

PERFORMANCE:
- Max speed: 43·5 knots (80 km/h; 50 mph)
- Normal cruising speed: 30-35 knots (56-64 km/h; 35-40 mph)
- Max rate of climb at S/L: 732 m (2,400 ft)/min
- Max rate of descent: 427 m (1,400 ft)/min
- Normal operational altitude: 305-915 m (1,000-3,000 ft)
- Service ceiling: 2,285 m (7,500 ft)
- Endurance at cruising speed, standard fuel: approx 10 hours
- Endurance at cruising speed, max fuel: approx 23 hours

RAVEN
RAVEN INDUSTRIES INC

ADDRESS:
PO Box 1007, Sioux Falls, South Dakota 57105
Telephone: (605) 336-2750
Telex: TWX 910-660-0306

RAVEN STAR

During 1974 the Applied Technology Division of Raven Industries designed and built a hot-air airship named Raven STAR (Small Thermal Airship by Raven) which flew for the first time, in the USA, on 7 January 1975. It was subsequently shipped to Switzerland where, on 13 May 1975, it was named *Verkehrshaus Luzern.*

Slightly bigger than the British Cameron D96, it has an envelope of urethane coated Dacron and cruciform inflated tail surfaces. The aft portion of the lower vertical surface is movable and serves as a rudder, and the aft section of each horizontal tail surface is also movable, serving as an elevator to assist pitch control. A catenary suspension system carries the gondola, which has tubular metal skids and accommodates a crew of two as well as two

STAR (Small Thermal Airship by Raven) in flight

11 million BTU propane burners, 303 litres (80 US gallons) of liquid propane, 45 kg (100 lb) baggage and the power plant. This comprises one 48·5 kW (65 hp) Revmaster conversion of a Volkswagen motor car engine, driving a 1·47 m (4 ft 10 in) diameter shrouded propeller manufactured by A & T Engineering. The pitch of this propeller is ground-adjustable. A single fuel tank is installed for the engine, this having a capacity of 76 litres (20 US gallons). There is also a four-stroke APU which is used to maintain the pressure within the envelope at ·004 bars (0·05 lb/sq in).

Raven Industries believes there are many applications for low-cost vehicles of this type, their ability to fly low and slowly, and to loiter, making them ideal for tasks such as traffic observation, power-line inspection and advertising. The company intends, if there appears to be sufficient market interest, to develop the basic design of STAR for larger models made of improved materials which would offer increased lift and duration. Proposed design improvements include a clutch-operated propeller, variable-pitch envelope pressurisation fan and omni-directional landing gear.

DIMENSIONS, ENVELOPE:
Length overall 36·58 m (120 ft 0 in)
Max diameter 14·63 m (48 ft 0 in)
Volume, gross 3,965 m³ (140,000 cu ft)

WEIGHTS:
Weight of envelope 272 kg (600 lb)
Weight of gondola, with max fuel 590 kg (1,300 lb)
Gross lift 1,089 kg (2,400 lb)
Payload 227 kg (500 lb)

PERFORMANCE:
Max level speed 22 knots (40 km/h; 25 mph)
Design max altitude 1,525 m (5,000 ft)
Endurance 3 hr

TUCKER
TUCKER AIRSHIP COMPANY

ADDRESS:
13218 Lake Street, Los Angeles, California 90066
Telephone: (213) 398-6907; (213) 393-5254

The Tucker Airship Company was founded in 1973 by Curtis Tucker Jr, with the object of building airships that would benefit from current technology and the latest materials.

Production of three prototype airships has been planned as follows:

TX-1. An experimental flight vehicle to validate the basic design, test materials and mooring system, and subsequently be used for research, pilot training and advertising. Scheduled for completion in late 1976.

Proposal 751. A non-rigid airship for law enforcement, aerial photography and advertising, with a crew of two plus one passenger. Low noise and vibration levels make such a vehicle an excellent platform for surveillance and photography, providing, by comparison with a helicopter, long range and improved crew comfort. Early studies indicate that operating costs will be lower than those of a helicopter, although initial capital cost may be greater. Design parameters include length 41·15 m (135 ft), volume 2,265 m³ (80,000 cu ft), max speed 87 knots (160 km/h; 100 mph), ceiling 3,050 m (10,000 ft), range of approx 430 nm (800 km; 500 miles).

Proposal 761. A semi-rigid airship for geological exploration or as an experimental airship for naval use. In the former configuration this vessel would have quarters, facilities and laboratories for a crew of four and six passengers and would be able to remain airborne for periods of five to six days for exploration of remote areas. Initial applications in a naval role would be to appraise its potential for sea control and anti-submarine warfare. Reduced accommodation and the elimination of laboratories would provide for advanced electronic equipment to be carried over an extended range. Special features envisaged for this proposal include the provision of a small helipad on the upper surface of the envelope, artificial superheating and chilling to assist vessel trimming, ducted propellers, electronic control of subsystems and internal control runs for the operation of aerodynamic surfaces. Design parameters include length 106·68 m (350 ft), volume 28,317 m³ (1,000,000 cu ft), max speed 78 knots (145 km/h; 90 mph), ceiling 3,660 m (12,000 ft), pressure height 2,440 m (8,000 ft), range approx 4,165 nm (7,720 km; 4,800 miles).

TUCKER TX-1

Design of the TX-1 originated in October 1972 and construction began in May 1973, with completion and first flight scheduled for late 1977. It is very much an experimental vessel, intended to provide experience in design, construction and operation, and will be used to test new design ideas, materials and ground handling concepts. Should the future airship proposals develop, it is envisaged that the TX-1 type would serve as a low-cost primary trainer. It is intended to be available for research use by scientific, military and industrial organisations upon request.

A semi-rigid design was chosen to provide for multi-mission roles, permitting the carriage of external loads; all of the superstructure, including the tail control surfaces, is located beneath the envelope. Oversize ballast and trim tanks are provided to keep the airship level irrespective of load disposition, and to provide additional safety during pilot training. A unique feature of the design is its modular construction. The TX-1 breaks down into twelve units for transportation or storage, and can be assembled and inflated in an open field. A nose spindle and mast are provided for mooring in normal weather conditions. When broken down for transit, the entire vessel and its ground support equipment can be accommodated in a standard-width trailer, 4·57 m (15 ft) in length.

Working on the rigid structure of the Tucker TX-1, before assembly to the envelope, inflated at rear

The forward-mounted gondola provides accommodation for the pilot and one passenger, seated in tandem. Excellent visibility is provided by a wide and deep windscreen, plus large sliding side windows. Standard airship controls are provided and equipment includes VFR instrumentation and communication radio. A monowheel is mounted beneath the gondola. Tail control surfaces comprise a large fin and rudder, with a wide-span horizontal tail surface and elevator. Power plant is a 67 kW (90 hp) McCulloch two-stroke four-cylinder engine, mounted beneath the keel structure at approximately its mid-point, and driving an AAF tractor propeller 1·12 m (3 ft 8 in) in diameter. Total fuel capacity is 227 litres (60 US gallons).

Three holds are provided within the keel structure, each with a capacity of 90·7 kg (200 lb), and are available for equipment or baggage. Hardpoints are provided for the carriage of external loads.

The envelope, which contains a single ballonet, is manufactured from a plastic laminate, coated with a special sealant developed by the company, and is intended to be hydrogen filled for short test flights. For operational use the envelope would be inflated with helium. Special equipment includes pneumatic ballast controls and a one-man mooring system, and there are provisions for full IFR instrumentation.

DIMENSIONS, OVERALL:
Length 27·74 m (91 ft 0 in)
Height 7·92 m (26 ft 0 in)

DIMENSIONS, ENVELOPE:
Length 27·74 m (91 ft 0 in)
Max diameter 6·10 m (20 ft 0 in)
Length of constant-diameter section 13·11 m (43 ft 0 in)
Volume 1,903·9 m³ (20,500 cu ft)

DIMENSIONS, GONDOLA:
Length overall 2·74 m (9 ft 0 in)
Height 1·37 m (4 ft 6 in)
Max width 0·91 m (3 ft 0 in)

AREAS:
Fin 6·97 m² (75·0 sq ft)
Rudder 2·93 m² (31·5 sq ft)
Tailplane 7·43 m² (80·0 sq ft)
Elevator 5·39 m² (58·0 sq ft)

WEIGHTS (estimated):
Weight empty 295 kg (650 lb)
Max design gross weight 635 kg (1,400 lb)

PERFORMANCE (estimated):
Max speed 48 knots (88 km/h; 55 mph)
Service ceiling 1,370 m (4,500 ft)
Range with max fuel 434 nm (804 km; 500 miles)
Range with max payload 43 nm (80 km; 50 miles)

LIGHTER-THAN-AIR: BALLOONS

FRANCE

CHAIZE
MAURICE CHAIZE

ADDRESS:
48 rue Balay, 42000 Saint-Étienne
Telephone: (77) 33-43-76

Maurice Chaize was manufacturing two hot-air balloons in the Summer of 1976, one in the FAI AX-6 class, the other in the AX-7 class. The envelopes are made from high-strength nylon, with the lower panels in the throat made from Nomex flame-resistant fabric. They are supplied in a ready-to-fly state, complete with basket, burners, and gas cylinders, but without instruments.

The company holds the French Certificat de Navigabilité de Type for both models, details of which can be found in the accompanying tables.

THE UNITED KINGDOM

CAMERON
CAMERON BALLOONS LTD

ADDRESS:
1 Cotham Park, Bristol BS6 6BZ
Telephone: (0272) 41455

DIRECTORS:
D. A. Cameron, BSc (Aero Eng), MIE
Tom Sage ARPS, AIIP
Kim Cameron

Cameron Balloons has been manufacturing hot-air balloons in Bristol since 1968. The company claims to be the second largest manufacturer of hot-air balloons in the world, and has built the world's largest hot-air balloon, the *Gerard A. Heineken* of 14,158 m³ (500,000 cu ft). Cameron balloons hold the absolute altitude record for

Maurice Chaize balloon in the FAI AX-7 class

Cameron Balloons' economy balloon, the Viva-56, in tethered flight

The world's largest balloon, the Cameron-built *Gerard A. Heineken*

Cameron Balloons' Model O-42

hot-air balloons of 13,971 m (45,836 ft) and the absolute duration record of 18 hr 56 min. The company holds CAA, FAA, Certificat de Navigabilité de Type and the West German Musterzulassungsschein type certificates for its balloons. It constructed also the world's first hot-air airship (see Airships section).

Envelopes are made from rip-stop nylon which has been treated with a polyurethane sealant to reduce porosity. The lower section of the envelope is made from Nomex flame-resistant fabric. The envelopes are available with either a Velcro seal or parachute deflation system, and those with the Velcro seal have also a vertical flap-type vent. Cameron offers a wide range of balloons, details of which can be found in the accompanying table. Each is supplied complete with willow and cane basket, burner and gas cylinders and is ready to fly when delivered. Optional items include instrument pack, envelope thermometer, pressure scoop, Nomex skirt, rigid-suspension basket, trail rope and additional gas cylinders.

THUNDER

THUNDER BALLOONS LTD

ADDRESS:
75 Leonard Street, London EC2A 4QS

Telephone: (01) 729-0231

CHAIRMAN:
K. F. Simonds

DIRECTORS:
T. M. Donnelly (Managing)
A. R. Wirth

Thunder Balloons manufacture a range of ready-to-fly hot-air balloons, details of which can be found in the accompanying tables. They are of two basic types, one having a parachute-type rip panel, the other a Velcro circular rip panel.

Envelopes are made from rip-stop nylon, treated with polyurethane sealant to reduce porosity, and with the lower panels of Nomex flame-resistant fabric. The balloons are supplied complete with aluminium fuel cylinders, burner, and willow and rattan basket, which is equipped with a fire extinguisher and fire blanket as standard. Optional items include instrument pack, Nomex skirt, metal framed basket, tether/trail rope, quick-release strap, master and slave cylinder and cylinder jackets.

THE UNITED STATES OF AMERICA

ADAMS

MIKE ADAMS' BALLOON LOFT

ADDRESS:

PO Box 12168, Atlanta, Georgia 30305

Telephone: (404) 261-5818

Mike Adams' Balloon Loft is manufacturing a range of hot-air balloons, of which details can be found in the accompanying tables.

Envelopes are made from rip-stop nylon, which may be coated optionally with a polyurethane or aluminised sealant to reduce porosity. Nomex flame-resistant fabric is used to protect the envelope throat. Two manoeuvring vents are located in the crown of the balloon and there is a duffle bag type of deflation system.

Adams balloons are supplied complete with a woven rattan basket, burner, aluminium fuel cylinders, and instruments which include a sensitive altimeter, rate-of-climb indicator, envelope temperature gauge, bubble type compass and fuel gauges. Optional items include tank-mounted pressure gauge, Nomex skirt, inflator fan and additional fuel cylinders.

AVIAN

AVIAN BALLOON

ADDRESS:

South 4323 Locust Road, Spokane, Washington 99206

In the Spring of 1976, Avian Balloon had completed the FAA certification programme of a hot-air balloon designated Falcon II. The company plans to obtain certification of two other balloons which will be available in 1977.

The balloon envelopes are made from rip-stop nylon, which can be treated with a sealant to reduce porosity, or can optionally be made from aluminised balloon fabric. A feature of the design is the incorporation of a mechanical rapid deflation system. The basket is of wicker, and in addition to the burner and gas cylinders, an instrument panel that includes an altimeter, rate-of-climb indicator, envelope temperature gauge and compass is standard. Options include an inflater fan and extra gas cylinders. Details of Avian's balloons are given in the accompanying tables.

Thunder Balloons' *Highjack*, an American-owned hot-air balloon in the AX-7 class

Thunder Balloon's Model 77 in the AX-7 class

Left: One of the Mike Adams' Balloon Loft range of hot-air balloons

Right: Basket, burner and instrument panel for the family of Avian balloons

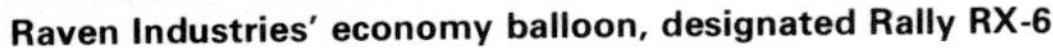

Raven Industries' economy balloon, designated Rally RX-6

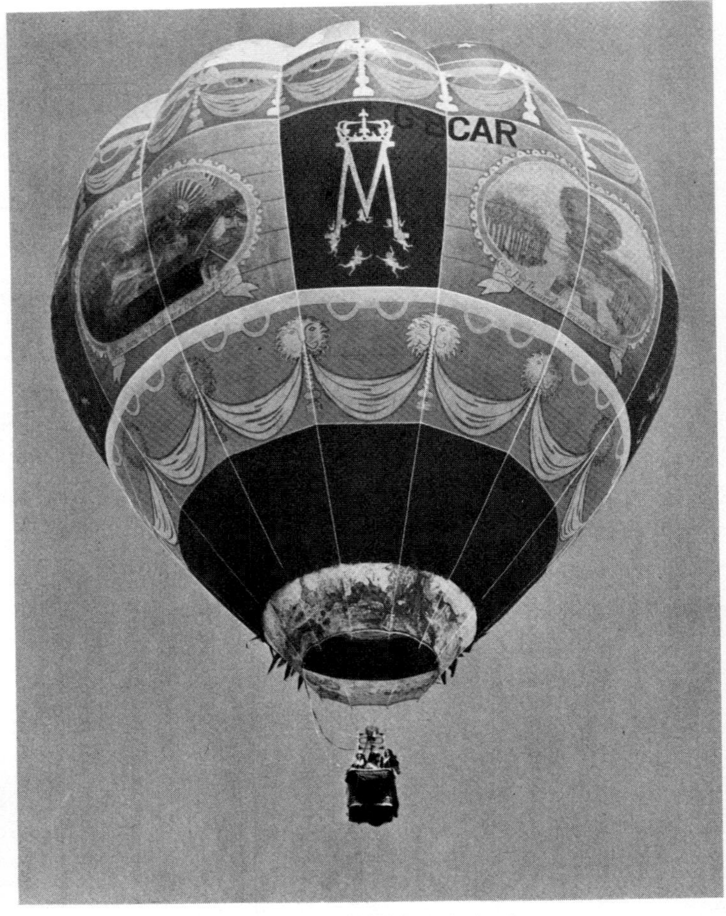

Thunder Balloons' AX-7 *Marie Antoinette*

BALLOON WORKS
THE BALLOON WORKS

Address:
Rhyne Aerodrome—RFD2, Statesville, North Carolina 28677
Telephone: (704) 873-0503
Directors:
Tracy Barnes
Dodds Meddock
Karl Stefan

The Balloon Works is currently marketing four hot-air balloons, details of which are given in the accompanying tables.

The envelopes of this company's balloons are made of a polyester fabric, which tests have shown to be less affected by heat exposure than other materials, and are coated with a urethane sealant to reduce porosity. Rapid deflation is made possible by use of a wide-diameter envelope valve, which is self-sealing immediately the valve line is released. Wicker baskets of both triangular and rectangular shape are available. A Nomex flame-resistant skirt is standard.

Standard equipment of the Firefly balloons includes an electronic instrument to give fast readings of air and envelope temperature, sensitive altimeter and climb and descent indicator. The Dragonfly has for instruments a sensitive altimeter, rate-of-climb indicator and envelope thermometer. An advanced type of burner and gas cylinders are standard; additional cylinders are available optionally.

HARE
HARE BALLOONS

Address:
1056 Laguna Avenue, Los Angeles, California 90026

Hare Balloons is marketing six different sizes of hot-air balloons in kit form for construction by amateur enthusiasts. The kits are complete with all materials, burner system and gas cylinders, and all but the smallest balloon have a glassfibre gondola. Woven wicker baskets are available optionally. The smallest AX-4 balloon is supplied with a special harness and quick release in lieu of a gondola/basket.

It is claimed that an enthusiastic constructor should be able to make the largest in the series in approximately 330 hours. Details of the Hare balloons are given in the accompanying tables.

RAVEN
RAVEN INDUSTRIES INC

Address:
Box 1007, Sioux Falls, South Dakota 57101
Telephone: (605) 336-2750

Raven Industries has been manufacturing hot-air balloons since the early 1960s, and has also built the STAR hot-air airship of which details are given in the Airships section.

Raven is marketing currently an economy balloon designated Rally RX-6, and three other balloons which are available either as tethered balloons for training purposes, or as free balloons. Details of these may be found in the accompanying tables.

Raven envelopes are made from rip-stop nylon treated with ultra-violet inhibitors to prolong fabric life. They embody a manoeuvring vent and a Velcro rapid deflation port. The balloons are available with lightweight aluminium and glassfibre gondolas or optional woven rattan baskets. Standard equipment includes the burner, gas cylinders and instrument panel which includes a sensitive altimeter, rate-of-climb indicator and envelope temperature gauge. A climb and descent indicator is available optionally, as are stainless steel or additional aluminium gas cylinders and flame-resistant skirt and envelope throat lining.

SEMCO
SEMCO BALLOON

Address:
Route 3, PO Box 514, Aerodrome Way, Griffin, Georgia 30223
Telephone: (404) 228-4005

Semco Balloon is marketing currently four standard hot-air balloons, and details of these are given in the accompanying tables.

Envelopes are made from rip-stop nylon and have a rapid deflation system and manoeuvring vent. Aluminised interior coating and Nomex fire-resistant envelope throat lining optional. Burners and gas cylinders are standard, but models are available with either an aluminium gondola or woven wicker basket. An instrument panel is standard, and this includes a sensitive altimeter, rate-of-climb indicator, compass and envelope thermometer. A ground inflation fan, climb and descent indicator and additional gas cylinders are available as options.

HOT-AIR BALLOON DATA

Country	Company	Model	Volume m^3/cu ft	Diameter m/ft	Height m/ft	Crew	Basket	Fuel Cylinders	Burners	Lift, S/L ISA kg/lb
FAI CLASS AX-4										
United Kingdom	Cameron Balloons	O-31	890/31,430	—	—	1	Willow/Cane	2	1	—
United Kingdom	Thunder Balloons	31	890/31,430	—	—	1	Willow/Rattan	2	1	—
United States	Hare Balloons	AX-4	900/31,779	12·50/41	15·24/50	1	Harness Only	1	1	240/529
United States	Semco Balloon	30-AL	850/30,000	12·19/40	18·29/60	1	Aluminium Chair	2	1	283/625
FAI CLASS AX-5										
United Kingdom	Cameron Balloons	O-42	1,190/42,024	—	—	2	Willow/Cane	2	1	—
United Kingdom	Thunder Balloons	42	1,190/42,024	—	—	2	Willow/Rattan	2	1	—
United States	Hare Balloons	AX-5	1,210/42,372	13·72/45	16·46/54	2	Glassfibre/Aluminium	2	1	325/716
FAI CLASS AX-6										378/834
France	M. Chaize	CS. 1600	1,600/56,503	—	—	3	—	2	—	—
United Kingdom	Cameron Balloons	O-56	1,590/56,150	—	—	3	Willow/Cane	2	1	—
United Kingdom	Cameron Balloons	Viva 56	1,590/56,150	—	—	3	Willow/Cane	2	1	—
United Kingdom	Cameron Balloons	M-53	1,500/52,972	—	—	3	Willow/Cane	2	1	—
United Kingdom	Thunder Balloons	56	1,590/56,150	—	—	3	Willow/Rattan	2	1	—
United Kingdom	Thunder Balloons	56A	1,590/56,150	—	—	3	Willow/Rattan	2	1	453/998
United States	Adams Balloon Loft	A50	1,557/54,985	—	—	2	Rattan	4	1	453/998
United States	Adams Balloon Loft	A50S	1,755/61,977	—	—	3	Rattan	4	1	—
United States	The Balloon Works	Firefly 6	1,585/56,000	—	—	3	Wicker	3	1	—
United States	The Balloon Works	Dragonfly	1,585/56,000	—	—	3	Wicker	3	1	—
United States	Hare Balloons	AX-6	1,600/56,496	15·24/50	17·68/58	3	Glassfibre/Aluminium	2	1	—
United States	Raven Industries	S-50A	1,597/56,400	15·24/50	17·68/58	2-4	Glassfibre/Aluminium	3	1	504/1,112
United States	Raven Industries	RX-6	1,597/56,400	15·24/50	17·68/58	2-3	Moulded	2	1	635/1,400
United States	Semco Balloon	Model T	1,585/56,000	15·24/50	20·73/68	2	Aluminium or Wicker	2	2	—
FAI CLASS AX-7										—
France	M. Chaize	CS. 2000	2,000/70,629	—	—	3-4	—	3	—	—
United Kingdom	Cameron Balloons	O-65	1,840/64,980	—	—	3	Willow/Cane	3	1	—
United Kingdom	Cameron Balloons	O-77	2,190/77,339	—	—	4	Willow/Cane	4	1	—
United Kingdom	Cameron Balloons	M-77	2,190/77,339	—	—	4	Willow/Cane	4	1	—
United Kingdom	Thunder Balloons	65	1,840/64,980	—	—	3	Willow/Rattan	3	1	503/1,108
United Kingdom	Thunder Balloons	77	2,180/76,986	—	—	4	Willow/Rattan	4	1	594/1,310
United Kingdom	Thunder Balloons	77A	2,180/76,986	—	—	4	Willow/Rattan	4	1	594/1,310
United States	Adams Balloon Loft	A55	2,123/74,973	—	—	3	Rattan	4	1	—
United States	Adams Balloon Loft	A55S	2,350/82,990	—	—	4	Rattan	4	1	—
United States	Avian Balloon	Falcon II	1,700/60,000	15·39/50·5	20·42/67	1-2	Wicker	2	1	376/830
United States	The Balloon Works	Firefly 7	2,180/77,000	—	—	4	Wicker	4	1	—
United States	Hare Balloons	AX-7	2,200/77,682	17·07/56	19·20/63	4	Glassfibre/Aluminium	2	1	693/1,529
United States	Raven Industries	S-55A	2,195/77,500	16·76/55	19·20/63	3-4	Glassfibre/Aluminium	3	1	657/1,450
United States	Semco Balloon	Challenger	2,145/75,750	16·00/52·5	21·34/70	3	Wicker or Aluminium	2	3	—
FAI CLASS AX-8										
United Kingdom	Cameron Balloons	O-84	2,380/84,049	—	—	4	Willow/Cane	4	1	—
United Kingdom	Cameron Balloons	A-105	2,970/104,884	—	—	6	Willow/Cane	6	2	—
United Kingdom	Thunder Balloons	105	2,970/104,884	—	—	6	Willow/Rattan	6	2	812/1,790
United States	Avian Balloon	Skyhawk	2,265/80,000	16·76/55	21·34/70	3-4	Wicker	3	1	517/1,140
United States	Avian Balloon	Skyhawk XL	2,973/105,000	18·29/60	22·56/74	4-5	Wicker	4	1	680/1,500
United States	The Balloon Works	Firefly 8	2,973/105,000	—	—	4-5	Wicker	4	1	—
United States	Hare Balloons	AX-8	3,000/105,930	18·90/62	21·03/69	4-5	Glassfibre/Aluminium	3	1	—
United States	Raven Industries	S-60A	2,973/105,000	18·29/60	21·03/69	3-4	Glassfibre/Aluminium	3	1	680/1,500
United States	Semco Balloon	TC-4	2,577/91,000	16·76/55	24·99/82	4	Birch/Aluminium	4	3	—
FAI CLASS AX-9										
United Kingdom	Cameron Balloons	A-140	3,960/139,846	—	—	8	Willow/Cane	6	2	—
United Kingdom	Thunder Balloons	140	3,965/140,022	—	—	8	Willow/Rattan	6	2	1,082/2,385
United States	Hare Balloons	AX-9	4,000/141,240	20·73/68	23·16/76	6-7	Glassfibre/Aluminium	3	1	1,261/2,781

RPVs AND TARGETS

AUSTRALIA

GOVERNMENT OF AUSTRALIA

DEPARTMENT OF INDUSTRY AND COMMERCE

ADDRESS: Anzac Park West Building, Constitution Avenue, Parkes, Canberra ACT 2600
Telephone: 48-2111
Telex: 62063
OFFICERS: See Aircraft section

Government Aircraft Factories

HEADQUARTERS:
Fishermen's Bend, Melbourne, Victoria 3207
AIRFIELD AND FINAL ASSEMBLY WORKSHOPS:
Beach Road Lara, Avalon, Victoria 3207
MANAGER: J. H. Dolphin

The Government Aircraft Factories are units of the Defence Production facilities owned by the Government of Australia and operated by the Department of Industry and Commerce.

Current products include the Jindivik weapons target and a target drone version of the Ikara anti-submarine missile, known as the Turana.

GAF JINDIVIK Mk 3B

GAF Jindivik Mk 3B target drone, with high-altitude wing extensions, jettisoning its take-off trolley

The Jindivik continues to be a standard weapons target in Australia and Great Britain.

Design began in March 1948 and construction of the prototype Jindivik Mk 1 started in December 1950. This prototype flew for the first time on 28 August 1952.

A total of 466 Jindiviks had been ordered by 1 January 1976, including 226 for the UK, 163 for a joint UK/Australia Weapons Project for use on the Woomera range, 42 for the US Navy, 25 for the Royal Australian Navy and 10 for Sweden. Of these, 430 had been delivered, including 14 Mk 1s, 111 Mk 2s (first flight 11 December 1953), 3 Mk 2As (first flight 18 September 1958), 76 Mk 2Bs (first flight 8 October 1959), 9 Mk 3s (first flight 12 May 1961), 147 Mk 3As (first flight 10 November 1961) and 70 Mk 3Bs (first flight 22 January 1970). Those used in the UK are assembled and equipped to British operational standards by the Guided Weapons Division of BAC. Operationally, their uses include the towing of targets for air-to-air missile firing practice, at the Missile Practice Camp at RAF Valley, Anglesey, by RAF Strike Command Lightnings (armed with Red Top and Firestreak) and Phantoms (Sparrows and Sidewinders).

Only the Mk 3B version of the Jindivik is still in production. Deliveries of this version began in April 1969. Customers include the UK (82 Mk 103B), Royal Australian Navy (15 Mk 203B) and Weapons Research Establishment, Woomera.

The Mk 3B was designed to cater for low-level trials at speeds in excess of 500 knots (925 km/h; 575 mph). Following miniaturisation and updating of some of the basic flight equipment, the front fuselage and equipment bay were redesigned to give greater volume for special trials equipment.

One Jindivik was delivered to Bell in 1973 for prototype installation of an ACLS (air cushion landing system), and in January 1975 negotiations were finalised between the Australian Dept of Industry and Commerce and the USAF to modify one Jindivik in Australia to carry out air cushion ground tests. The Royal Australian Navy made available a Mk 203A Jindivik for this purpose, and an accompanying photograph shows this aircraft fitted with a US-supplied ACLS under contract to the USAF's Flight Dynamics Laboratory. Taxying trials were in progress in February 1976 at the GAF's airfield at Avalon, Victoria. At that time negotiations were in progress between the DIC and the USAF to carry out flight testing of the vehicle after successful completion of the taxying trials. Flight testing was expected to take place, at the Royal Australian Navy missile range at Jervis Bay, NSW, in late 1976 or early 1977.

For low-altitude work, the standard-span Jindivik can be fitted either with Mk 5 wingtip camera pods, or with larger Mk 8 pods each containing a camera, a Luneberg lens and a small amount of fuel. For high-altitude work (also with a choice of Mk 5 or Mk 8 wing pods), constant-chord wing extension panels can be added outboard of the pods. For extra high altitude flying (with Mk 5 pods only), these panels can be replaced by increased-span panels, tapered on the leading-edge. A ventral tail fin is also fitted in this configuration.

Up to 1 January 1976, Jindiviks had flown 3,137 sorties at the RAE, Llanbedr, North Wales; one particular Mk 3A drone (WRE 418) was eventually destroyed after successfully completing 285 sorties at Woomera. Total sorties by all Jindiviks at that date amounted to well over 5,600.

TYPE: Pilotless target drone.

WINGS: Cantilever low/mid-wing monoplane. Wing section NACA 64A-106 with modified trailing-edge. Dihedral 2° 30'. Incidence 1°. Multi-spar box structure of aluminium alloy. Spars Araldite bonded to preformed heavy-gauge skins to form interspar torsion box which is utilised as integral fuel tankage. Leading-edge attached to front spar and rebated skins by Araldite. Trailing-edge box is an Araldite-bonded structure and is riveted to main spar box structure, which houses aileron control system of rods and bellcranks. Aluminium alloy monocoque flaps and ailerons hinged to this structure, with continuous piano hinges and three-point pin hinges respectively. Ailerons fitted with inset geared tab and driven by GAF-designed twin-motor servo motor. Flaps operated pneumatically.

Royal Australian Navy Jindivik Mk 203A fitted with air cushion landing systems

FUSELAGE: Aluminium alloy semi-monocoque structure, built in front, centre and rear sections. Front fuselage carries all control equipment, autopilot amd telemetry equipment on three removable trays. Pitot head and wave-guide boom mounted on permanent nose probe. A moulded glassfibre canopy, which lifts off for access to the equipment, forms the ram-type air intake. Rear end of front fuselage and front end of centre fuselage form bay in which all special trials equipment is carried. Centre fuselage also houses landing skid. Rear fuselage carries engines and jetpipe. Optional airbrake under rear fuselage, which may be used as dive brake to reduce descent time from high altitude.

TAIL UNIT: Cantilever multi-spar tailplane of light alloy bonded with Araldite. Elevators, formed of single wrapped skins stiffened by chordwise flutes and carried on piano hinges, driven by GAF-designed twin-motor servo motor. Inset geared tabs. Fin of light alloy skin bonded to two spars, stabilised by metal honeycomb filling. No rudder; but provision for an in-flight variable trim tab in fin. Ventral fin on extra high altitude version.

LANDING GEAR: Pneumatically-extended, manually retracted (on ground) central skid. Pneumatic jack acts as shock-absorber. Steel auxiliary skids at wingtips. See later paragraph on 'Take-off and Landing'.

POWER PLANT: One Rolls-Royce Viper Mk 201 turbojet engine, rated at 11·1 kN (2,500 lb st). Engine relight capability available in the event of flameout. Flexible rubber main fuselage fuel tank, capacity 291 litres (64 Imp gallons), and two integral wing tanks, total capacity 173 litres (38 Imp gallons). Single refuelling point in centre fuselage for all tanks. Wing tanks pressurised from engine compressor to feed fuselage tank from where fuel is pumped to engine. Pressurised fuel recuperator fitted in this delivery line to cope with negative *g* conditions. Compressed air for starting and throttle, operated by electric actuator through a cam switch box, allowing several fixed rpm engine conditions to be selected in addition to throttle 'beep' demands. Oil capacity 4·5 litres (1 Imp gallon).

CREW: Normal operating ground crew of four: flight commander, navigator, pilot and batsman.

SYSTEMS: No hydraulic system. Non-regenerative pneumatic system: air stored at 138 bars (2,000 lb/sq in) in power pack which supplies air to the flaps, airbrake and landing skid reduced to 39·6 bars (575 lb/sq in). If airbrake is to be used for speed control when two aircraft are formating, a separate air supply system may be fitted as optional equipment. Engine-driven 11·5kVA alternator delivers 208V AC 3-phase electrical supply at 300-550Hz. In the event of alternator failure, a 24V DC battery provides limited power for essential control functions. Automatic orbit and/or destruct systems provided, consistent with range safety requirements.

REMOTE-CONTROL EQUIPMENT: Aircraft is remotely controlled from a ground station. Radio control equipment comprises two receiver/selectors, the second of which may be used as standby or destruct, and GAF relay receiving set. Telemetry equipment consists of NIC transducers and Australian-designed transmitter and junction box.

TRIALS EQUIPMENT: Transponders and microwave reflectors for trials of active, semi-active or beam-riding missiles. Heat sources, including infra-red flare packs mounted in rear of fuselage, can be fitted to provide low-frequency IR output. Transponders in the X, S and C bands can be fitted for target acquisition and to enable the Jindivik to be tracked at greater range. Provision for a 'Tonic' winch pylon under each wing, on which can be carried a recoverable active radar or infra-red 'Tonic' CG tow 'bird' or an expendable tow bird with infra-red-augmented nose. These can be towed at 15-150 m (50-500 ft) behind aircraft; recovery is by electric winch. Other types of recoverable towed target may be fitted, carrying augmentation in the form of transponders, microwave reflectors or infra-red.

RECORDING EQUIPMENT: Cameras fitted with wide-angle lenses are carried in wingtip pods, with all-round viewing capability, and are used to film and record the approach path and proximity of missiles fired against the target. Variants are the Mk 5 pod with cameras only and the Mk 8 with cameras, fuel and provision for fitment of microwave reflectors in leading-edge and trailing-edge radomes. Rearward-facing cameras, and a Luneberg lens reflector mounted in fairing above jetpipe, may be used to film and record operations when towed targets are used.

TAKE-OFF AND LANDING: Jindivik is mounted on a tubular-framed tricycle take-off trolley. Aircraft/trolley steering is achieved by a servo-controlled nosewheel which responds to signals from the aircraft's autopilot. The aircraft/trolley combination accelerates under normal jet power with flaps retracted and with the aircraft set at a negative incidence. When unstick speed (125 knots; 231 km/h; 144 mph) is reached the aircraft is rotated to take-off incidence and flaps are lowered rapidly. Rotation of the aircraft initiates the trolley release system and the aircraft climbs away. At the same time trolley brakes equipped with Dunlop anti-skid devices are applied. When Jindivik is in the approach run, the flaps and skid are selected down for landing. On touchdown, at approx 120 knots (222 km/h; 138 mph), a 'sting' extended below the main skid rotates on impact and initiates rapid retraction of flaps. Fuel supply is terminated by radio command.

DIMENSIONS, EXTERNAL:
- Wing span:
  - short span, low altitude — 6·32 m (20 ft 8·99 in)
  - extended, high altitude — 7·92 m (26 ft 6 in)
  - extended, extra high altitude — 9·78 m (32 ft 1·4 in)
- Wing chord (constant) — 1·22 m (4 ft 0 in)
- Wing aspect ratio:
  - short span — 5·67
  - extended, high altitude — 6·88
  - extended, extra high altitude — 8·97
- Length overall:
  - incl nose probe — 8·15 m (26 ft 8¾ in)
  - excl nose probe — 7·11 m (23 ft 3¾ in)
- Height overall, skid extended — 2·08 m (6 ft 9·85 in)
- Tailplane span — 1·98 m (6 ft 6 in)

DIMENSIONS, INTERNAL:
- Equipment bays:
  - front fuselage — 0·32 m³ (11·46 cu ft)
  - centre fuselage — 0·15 m³ (5·24 cu ft)

AREAS:
- Wings, gross:
  - short span — 7·06 m² (76·0 sq ft)
  - extended span, high altitude — 9·48 m² (102·0 sq ft)
  - extended span, extra high altitude — 10·68 m² (115·0 sq ft)
- Ailerons (total) — 0·42 m² (4·5 sq ft)
- Flaps (total) — 1·07 m² (11·5 sq ft)
- Fin — 0·67 m² (7·2 sq ft)
- Tailplane — 1·36 m² (14·6 sq ft)
- Elevators — 0·39 m² (4·2 sq ft)

WEIGHTS:
- Weight empty, equipped (min) — 1,315 kg (2,900 lb)
- Max payload:
  - short-span version — 249 kg (550 lb)
  - extended-span versions — 181 kg (400 lb)
- Max T-O weight:
  - short span, Mk 5 wing pods — 1,451 kg (3,200 lb)
  - short span, Mk 8 wing pods — 1,655 kg (3,650 lb)
  - high altitude, Mk 5 wing pods — 1,474 kg (3,250 lb)
  - high altitude, Mk 8 wing pods — 1,655 kg (3,650 lb)
  - extra high altitude, Mk 5 wing pods — 1,496 kg (3,300 lb)

PERFORMANCE (A: short span, Mk 5 pods; B: short span, Mk 8 pods; C: high altitude, Mk 5 pods; D: high altitude, Mk 8 pods; E: extra high altitude, Mk 5 pods):
- Max level speed at max operational ceiling:
  - A, B, C, D — Mach 0·86 (490 knots; 908 km/h; 564 mph)
  - E — Mach 0·82 (470 knots; 871 km/h; 541 mph)
- Min operating height:
  - A, B — 15 m (50 ft)
- Max operational ceiling:
  - A — 17,375 m (57,000 ft)
  - B — 16,460 m (54,000 ft)
  - C — 19,200 m (63,000 ft)
  - D — 18,595 m (61,000 ft)
  - E — 20,420 m (67,000 ft)
- Time to max operational ceiling:
  - A — 26 min
  - B, C — 30 min
  - D, E — 34 min
- T-O run — 305 m (1,000 ft)
- T-O to 15 m (50 ft) — 670 m (2,200 ft)
- Landing from 15 m (50 ft) — 823 m (2,700 ft)
- Landing run — 457 m (1,500 ft)
- Typical max on-station endurance:
  - A, C — 1 hr 6 min
  - B — 1 hr 38 min
  - D — 1 hr 52 min
  - E — 1 hr 3 min
- Max range:
  - A — 430 nm (796 km; 495 miles)
  - B — 670 nm (1,240 km; 771 miles)
  - C — 540 nm (1,000 km; 621 miles)
  - D — 900 nm (1,667 km; 1,036 miles)
  - E — 700 nm (1,297 km; 806 miles)
- Max banking angle (all) — 60°
- Max diving angle (all) — 40°
- Max *g* pull-up at 15 m (50 ft):
  - A — +3·7
  - B — +6·0

GAF Turana turbojet-powered target drone

GAF TURANA

The Turana is a target drone based on the Ikara missile. It was developed by the Dept of Industry and Commerce initially to meet a Royal Australian Navy Staff Requirement for a modern gunnery and guided weapons target. The development programme was approved and funded in August 1969. First flight was made on 12 March 1971, and during subsequent test flying at the WRE Woomera Range and the RAN's missile range at Jervis Bay an altitude of 6,100 m (20,000 ft) and a speed of more than 390 knots (724 km/h; 450 mph) were attained in level flight. Speeds in excess of 500 knots (925 km/h; 575 mph) were achieved in shallow dives. The Turana is capable of demand turns of up to 3*g*, customer-specified manoeuvres of more than 10*g*, and demand heading changes of up to 26°. Construction ensures an average life of at least 10 flights per drone. Equipment is designed for at least 20 complete missions, including sea-water immersion.

Government Aircraft Factories are the co-ordinating design authority and prime contractor responsible for the airframe, autopilot, engine and fuel system. Guidance and rocket propulsion aspects of the project are subcontracted to the Weapons Research Establishment/EMI Electronics (Australia) and Weapons Research Establishment/Ordnance Factory and Explosives Factory, Maribyrnong, respectively.

An initial order for 12 Turanas, including spares and ancillary ground and shipboard equipment, was placed by the Royal Australian Navy, and deliveries of equipment were nearing completion in February 1976. Further orders are anticipated.

At that time a Concurrent Evaluation Phase (CEP) was proceeding, to ensure the target's readiness for entry into RAN service. Sixteen flights in this series had then been flown at the Jervis Bay missile range, and further flights were scheduled during 1976. Performance aspects demonstrated during the CEP have included sea-skimming gunnery presentations from 25 nm (46 km; 28·8 miles) at 15·25 m (50 ft), low-altitude flight at 11·25 m (37 ft), and precision control during gunnery engagements by two ships. One Turana has flown 10 times.

TYPE: Pilotless target drone.

WINGS: Cantilever mid-wing tail-less monoplane of cropped delta planform. Wing section modified NACA 64010. Spindle attachment to fuselage, the spindle being also the main spar. Two metal ribs and metal rear spar, the remainder of the wing being of foam-filled glassfibre. Full-span elevons are operated via a differential mechanical linkage by two electric actuators, one for pitch and one for roll control. The wings are quickly detachable from the fuselage.

FUSELAGE: Aluminium alloy torsion box, with removable glassfibre fairings, housing the autopilot, fuel tank, engine and various miss-distance equipments and transponders. Main structural member consists of an H-section structure of chemically-milled side skins, bulkheads and forgings, and a bottom diaphragm. Interface attachments are identical with those of Ikara.

TAIL UNIT: Single vertical fin, of NACA 64010 section, made of chemically-milled aluminium skins and aluminium ribs and spars, with glassfibre tip and trailing-edge.

POWER PLANT: One Microturbo Couguar 022 turbojet engine of 0·78 kN (176 lb st). Stainless steel/airbag fuel tank of 51·8 litres (11·4 Imp gallons) capacity. Airbag is pressurised from engine compressor. Compressed air is used for starting and the engine speed is controlled by an electronic unit, which forms part of a speed demand loop.

BOOST: Single-nozzle PMD41 solid-propellant booster motor specially developed for Turana, attached to fuselage by swivelling links and two explosive bolts. Nominal burning time 2 seconds; nominal thrust 26·7 kN (6,000 lb). Boost motor is jettisoned at the end of the boost phase. Launching is from the standard Ikara ship launcher, or from a lightweight portable launcher, and is made at a fixed elevation of 22° 30′.

SYSTEMS: All electric. Rechargeable silver/zinc battery pack, in compartment immediately aft of ejectable nose section, provides power for all services for more than one hour. Pyrotechnic charge for ejection of nose section.

CONTROL SYSTEM AND AUTOPILOT: Elevons on the wings are operated symmetrically or differentially by electric actuators. The autopilot includes a displacement gyro sensing roll and pitch, a rate gyro sensing yaw, an air data unit with airspeed and altitude transducers, signal summing and shaping networks and drive amplifiers for the servo system. A radar altimeter option is available, to allow automatic controlled flight at low altitudes. A pitot-static tube is fitted to the tip of the vertical fin. Height lock and speed lock loops are provided within the drone.

GUIDANCE: Drone is designed to be used initially in conjunction with an adaptation of the Ikara missile guidance system on board RAN ships. Navigation is by means of the Ikara tracking receiver and ship's plotting facilities. If required, a modified guidance equipment can be provided which is independent of ship facilities; alternative tracking and return data links are also possible. Guidance equipment is housed in the vertical fin.

Recovery: Command parachute recovery, the parachute being housed in the ejectable nose of the drone. The parachute recovery sequence is activated automatically in the event of either an engine, electrical power system or command link failure. Short or long time delays can be included in the command link failure mode to allow for momentary signal fades. Initial versions of Turana are designed for water recovery. For use on land, a larger parachute and land recovery system could be provided.

Trials Equipment: Drone is intended initially mainly for gunnery practice, for which purpose visual augmentation is provided in the form of a smoke release system. Forward-looking passive radar augmentation is provided by a 190 mm (7½ in) Luneberg lens in the nose. Space and large weight-carrying capacity are available for active augmentation or for other special equipment such as a 465MHz 48-channel telemetry system. Pyrotechnic flares can be accommodated to meet customer requirements. An acoustic miss-distance indication system is available. A short-tow system, to carry infra-red flares or a Luneberg lens radar reflector, can be developed.

Operations: Typical operation as a service target will begin by boosted launch from a ship and climb (or descent) to the required operational altitude. Control of the drone is effected by open loop demands from the ground and by a closed loop sensing and autocontrol system in the drone. This system obviates the necessity for either telemetered flight information or highly-skilled flight controllers. Altitude may be locked to the commanded height, which may be changed at will by the controller. Speed is also locked to a variable command datum, thus giving the controller freedom to use any section of the flight envelope, and control in azimuth can be exercised to provide a number of presentations, crossing, approaching or receding, for the exercise of guns, short-range or medium/long-range guided weapons. The number of presentations is dependent upon the chosen operating speed and altitude. The flight is concluded by descent to any required low altitude, and return to the neighbourhood of the ship, where the parachute is deployed and recovery from the sea is effected.

Dimensions, external:

Wing span	1·53 m (5 ft 0·2 in)
Wing chord at root	1·07 m (3 ft 6 in)
Wing chord at tip	0·41 m (1 ft 4 in)
Length overall	3·37 m (11 ft 0½ in)
Height (less boost motor)	1·02 m (3 ft 4 in)
Height (with boost motor)	1·19 m (3 ft 10·8 in)

Dimension, internal:

Special equipment capacity	0·028 m³ (1 cu ft)

Areas:

Wings, gross	1·23 m² (13·2 sq ft)
Elevons (total)	0·19 m² (2·04 sq ft)
Fin	0·32 m² (3·4 sq ft)

Weights (less special equipment):

Weight dry	165 kg (364 lb)
Weight at launch	279 kg (615 lb)
Weight less booster	251 kg (555 lb)
Weight at recovery (empty)	172 kg (380 lb)
Fuel	41 kg (91 lb)
Special equipment weight	more than 45 kg (100 lb)

Performance:

Max level speed	390 knots (724 km(h; 450 mph)
Boost acceleration (nominal)	10g
End-of-boost speed	170 m (560 ft)/sec
Service ceiling	10,000 m (33,000 ft)
Max rate of climb at S/L	1,219 m (4,000 ft)/min
Range	325 nm (602 km; 374 miles)
Endurance	55 min

HAWKER DE HAVILLAND

HAWKER DE HAVILLAND AUSTRALIA PTY LIMITED

Head Office:
PO Box 78, Lidcombe, NSW 2141
Telephone: (03) 649 0111
Telex: 20214

Officers: See Aircraft section

Hawker de Havilland has developed, in response to a Royal Australian Army tender, an RPV to provide an aiming and scoring system for that Service's Redeye surface-to-air missiles. Known as Enmoth, the complete system consists of a delta-shaped flying-wing drone, a transport container, and a self-contained ground support package. A description of the target drone follows:

HAWKER DE HAVILLAND HDH-10 ENMOTH TARGET AIRCRAFT

The HDH-10 Enmoth is a mini-RPV, having a cropped-delta wing configuration. Its aerofoil section ensures low drag and good low-speed characteristics, particularly pitching moment, while maintaining a generous thickness/chord ratio. The blended wing/body design helps in providing the required silhouette of the aircraft simulated, and the thick wing provides both generous storage space and a substantial leading-edge with good impact-absorbing properties. A generous main-wheel track ensures good ground handling characteristics, and the ground attitude of 2° nose down prevents excessive bounce and float after landing. The large internal volume available for payload gives the Enmoth considerable development potential.

Wings: Cantilever monoplane of cropped-delta planform. Wing section NACA 23012. Elevon on each trailing-edge, with Futaba flight controls and actuators. Built of plywood-covered polystyrene foam in prototype; production versions will have wings of glassfibre-covered foam.

Fuselage: Circular-section structure, blending into wings at point of maximum wing thickness.

Tail Unit: Cropped-delta fin only (rudder considered unnecessary, as aircraft is virtually unspinnable). No horizontal tail surfaces.

Landing Gear: Non-retractable tricycle type, with wide-track main gear and steerable nosewheel.

Power Plant: One 1·5 kW (2 hp) K & B 61 piston engine, driving a pusher propeller. Four 369 g (13 oz) plastics fuel tanks, pressurised by engine, disposed symmetrically on each side of CG to eliminate trim changes during flight. Max fuel load 1·47 kg (3·25 lb).

Launch and Recovery: Normally by conventional T-O and landing. Parachute recovery system available optionally.

Guidance and Control: Radio command guidance system. Aerodynamic control (roll and pitch) by elevons.

Special Equipment: Infra-red source, with all-round 'view', on top of fin. Radio control receiver and payload housed in fuselage. Futaba flight controls and actuators used throughout the RPV.

Dimensions, external:

Wing span	1·52 m (5 ft 0 in)
Length overall	1·57 m (5 ft 2 in)
Height overall	0·56 m (1 ft 10 in)
Wheel track	0·61 m (2 ft 0 in)
Wheelbase	0·58 m (1 ft 11 in)
Propeller diameter	0·32 m (1 ft 0½ in)

Weights:

Weight empty	6·8 kg (15 lb)
Max normal fuel	1·5 kg (3·25 lb)
*Max normal payload	3·0 kg (6·75 lb)
Max normal T-O weight	11·3 kg (25 lb)

**Payload capability at a min speed of 9 m (30 ft)/sec is over 3 kg (6·75 lb), and at a min speed of 15·24 m (50 ft)/sec is over 5 kg (11 lb)*

Performance (at max T-O weight. A: at max power, B: at 50% power):

Max level speed:	
A	70·5 knots (130·5 km/h; 81 mph)
B	57 knots (106 km/h; 66 mph)
Stalling speed:	
A	20 knots (37 km/h; 23 mph)
B	30 knots (56 km/h; 34·5 mph)
Endurance:	
A	35 min
B	52 min

Hawker de Havilland HDH-10 target aircraft

HAWKER DE HAVILLAND HDH-11 BEEMOTH

This is a projected development of the Enmoth, having a wing span of 2·29 m (7 ft 6 in), length of 2·44 m (8 ft 0 in), AUW of 38 kg (84 lb), and a useful load (payload plus fuel) of 15·5 kg (34 lb).

BELGIUM

MBLE

MANUFACTURE BELGE DE LAMPES ET DE MATÉRIEL ELECTRONIQUE SA

Address:
Rue des Deux Gares 80, B-1070 Brussels
Telephone: (02) 523 00 00
Telex: 21.420

Deputy Manager:
Albert Colpaert

This company, which employs 5,300 people, has contributed to a number of European aerospace programmes, including those of the F-104G Starfighter, Hawk missile and ELDO space research projects. It has also developed, built and is producing a battlefield surveillance system named Épervier. A version known as the Asmodée, for battlefield surveillance and target acquisition, has been advertised.

MBLE ÉPERVIER (SPARROWHAWK) SYSTEM

The Épervier is a battlefield reconnaissance system developed and built entirely in Belgium. It comprises, essentially, a drone vehicle known as the X-5, with its sensors; a short ramp launcher; and a drone control centre (DCC). The X-5 is a small, unmanned vehicle powered by a 0·51 kN (114 lb st) Lucas CT3201 turbojet engine. The airframe, of glass-reinforced plastics, is manufactured under subcontract by Fairey SA at Gosselies. Brief details of the earlier X-1 to X-4 drone prototypes have appeared in previous editions of *Jane's*.

The X-5 can carry Omera 5 in or Omera or Oude-Delft 70 mm day or night cameras, and SAT Cyclope infra-red line-scanning equipment with real-time transmission. The launcher equipment comprises a short orientatable ramp and an associated checkout device. The DCC contains all the necessary electronic equipment for guiding and tracking elements. A mobile unit for photographic processing and interpretation is part of the system.

Initially, the Épervier system was developed to meet NATO specifications. The Belgian Ministry of Economic Affairs decided on 11 July 1969 to support the programme financially. In early 1971, a co-operation agreement was signed between MBLE and the Belgian Ministry of Defence, and development continued to more advanced technical characteristics and operational specifications on behalf of the Belgian Army.

MBLE Épervier X-5 battlefield surveillance RPV

In late 1972 and early 1973, the Épervier successfully underwent an official military operational evaluation. During more than 80 flights, the system proved to be a flexible, easy-to-operate and accurate means of reconnaissance, capable of photographing pinpoint targets or large areas of terrain during guided and/or programmed flights up to more than 38 nm (70 km; 43·5 miles) from its base. Photoflash gear can be carried for night operation.

An initial order for the Belgian Army, announced in 1974, led to the start of production, in two batches, of 43 Éperviers and associated ground equipment.

DIMENSIONS (X-5):
- Wing span: 1·72 m (5 ft 7¾ in)
- Length overall: 2·38 m (7 ft 9¾ in)
- Height overall: 0·92 m (3 ft 0¼ in)

WEIGHTS (X-5):
- Payload: approx 20 kg (45 lb)
- Max launching weight: 142 kg (313 lb)

PERFORMANCE (X-5):
- Cruising speed: 270 knots (500 km/h; 310 mph)
- Operating height limits: 305-1,830 m (1,000-6,000 ft)
- Max mission radius: 40 nm (75 km; 47 miles)
- Endurance: more than 25 min

CANADA

CANADAIR

CANADAIR LTD (Subsidiary of General Dynamics Corporation)

HEAD OFFICE AND WORKS:
PO Box 6087, Montreal 101, Quebec
Telephone: (514) 744-1511
OFFICERS: See Aircraft section

In addition to its work on manned aircraft, Canadair is developing or producing various drone systems, of which all available details follow:

CANADAIR CL-89

NATO designation: AN/USD-501

The Canadair CL-89 (AN/USD-501) airborne surveillance drone system evolved from a need of the western Allied armed forces for an intelligence-gathering device for battlefield commanders. Canadair, in co-operation with the Canadian government and Canadian Armed Forces, conducted a study to identify the ideal unmanned airborne surveillance system. It was agreed that the system should be simple to operate and maintain; have a high survivability in a sophisticated enemy air defence environment; be capable of detecting and recording enemy formations and weapons of tactical significance in all weathers, by day or night; and provide accurate information in time for the recipient to react.

Development was started in 1961 by Canadair Ltd on a shared-cost basis with the Canadian Department of Industry, Trade and Commerce.

With a very high probability of survival against all known air defence systems, the CL-89 can acquire timely and accurate battlefield intelligence using its photographic and infra-red line-scanning equipment.

The system, consisting of the air vehicles plus the related ground support and operational maintenance equipment, is totally integrated, mobile, and independent of such external services as electrical power supplies.

Details of the flight and operational evaluation programmes have been given in previous editions of *Jane's*.

The CL-89 system has been produced for Canada, the Federal Republic of Germany, Italy (jointly with Meteor SpA, which see) and the UK. More than 500 drones had been manufactured by early 1976, with production continuing.

TYPE: Recoverable airborne surveillance drone system.

WINGS: Four rectangular single-spar stub wings at rear of drone body, in a cruciform arrangement at 45° to the horizontal and vertical centrelines. Upper pair fold out of the way when the landing airbags are inflated. Ailerons on port upper and starboard lower wings.

FOREPLANES: Two pairs of canard foreplane surfaces, of cropped-delta planform, aft of nosecone on horizontal and vertical centrelines, for pitch and yaw trim respectively.

BODY: Cylindrical metal sturucture, with curved nosecone and tapering tailcone. Three detachable dorsal packs for forward and rear landing bags and flare container, and two detachable ventral packs for sensor equipment and parachute recovery system.

POWER PLANT: One 0·56 kN (125 lb st) Williams Research WR2-6 turbojet engine, with variable exhaust nozzle, installed in tailcone aft of wings. Air intake duct on each side of fuselage, forward of wings. Fuel and oil tanks in central body compartment, forward of air intakes.

BOOST: One 20·29 kN (4,550 lb st) average thrust Bristol Aerojet Wagtail rocket motor, with electrical ignition, in a helically-welded steel case which is fitted with detachable rectangular cruciform tail fins and is attached to the body of the drone by three Vee-shaped thrust arms and a cable. After 2·5 seconds of flight the cable is cut automatically, freeing the thrust arms and allowing the booster assembly to fall away.

SYSTEMS: Engine-driven alternator for electrical power during flight.

GUIDANCE: The flight path, altitude and sensor on/off commands are controlled by a preset programmer which receives information from an Air Distance Measuring Unit (ADMU) and combines this with the preset programme to control the flight path. A ground homing beacon positions the drone in its final stages of flight to ensure the accuracy of the landing.

RECOVERY: After final positioning by the ground homing beacon, the drone's drogue parachute deploys to slow it down until the main parachute is deployed. The drone is inverted; the forward and rear landing bags are then automatically inflated and deployed to absorb the landing shock.

Launch of a Canadair CL-89 airborne surveillance/target acquisition drone system

EQUIPMENT: Air Distance Measuring Unit (ADMU) in nose probe. Nosecone houses programmer, static power converter and homing receiver. Compartment aft of nosecone houses shaping amplifier, flash detector, directional and vertical gyros, transponder antenna, forward landing bag container and air bottle to inflate both landing bags. Ventral sensor pack is immediately aft of this compartment. Two sensor systems are currently in use: the Carl Zeiss KRb 8/24 camera system and the Hawker Siddeley Dynamics Type 201 infra-red linescan system. Linescan is a reconnaissance technique in which the terrain overflown is scanned at high speed in narrow strips at right angles to the flight path. Forward motion builds up a continuous picture of radiation from the ground below. In an infra-red linescan the radiation is collected by an optical scanner and focussed on to an infra-red detector. Variations in the radiations received cause corresponding fluctuations of the signal output from the detector. The detector output is processed electronically and is used to modulate the intensity of a light source which exposes a photographic film. This equipment greatly enhances the night performance of the CL-89, since the infra-red linescan can produce continuous imagery on the darkest night without the use of illuminating flares. It is also possible to detect 'hot' targets, such as military vehicles which are under camouflage, by virtue of the difference in temperature from their surroundings. Aft of the sensor pack is the compartment for the fuel and oil tanks. This compartment also has a ventral forward-hinged door providing access to the engine start air connector, and a dorsal pack containing 12 photoflares just forward of the rear landing bag container. The final cylindrical compartment houses the rear landing bag container itself and the parachute recovery pack, between the dorsal and ventral pairs of wings respectively. The sustainer engine is mounted in the tailcone.

DIMENSIONS, EXTERNAL:
- Length overall, excl nose probe:
  - with booster: 3·73 m (12 ft 3 in)
  - without booster: 2·60 m (8 ft 6½ in)
- Body diameter: 0·33 m (1 ft 1 in)
- Span of wings: 0·94 m (3 ft 1 in)
- Span of foreplanes: 0·48 m (1 ft 7 in)

WEIGHTS:
- Weight dry (less fuel, oil and payload): 78·2 kg (172·4 lb)
- Payload: 15·1 kg (33·3 lb)
- Max launching weight:
  - with booster: 156 kg (343 lb)
  - without booster: 108 kg (238 lb)

Parachute recovery of Canadair CL-89 drone vehicle

PERFORMANCE:
- Max speed: 400 knots (741 km/h; 460 mph)

CANADAIR CL-289

NATO designation: AN/USD-502

The CL-289 is a joint Canadair/Dornier programme, based on AN/USD-501 (CL-89) technology, to meet a German military technical requirement for a longer-range drone.

Resembling a scaled-up CL-89, the CL-289 will have greater payload capacity and will be fitted with both photographic and infra-red line-scanning (IRLS) sensors and a real-time data transmission system for IRLS imagery. Launch and recovery systems are basically similar to those of the CL-89. Range will be approx 215 nm (400 km; 248 miles).

CANADAIR CL-227

The CL-227 has been designed to meet a requirement for a medium-range surveillance, target acquisition and fire adjustment system. It is a rotating-wing vehicle, handling real-time data transmission.

It has an 'hour-glass' shaped body, with a power plant in the upper bulged portion driving two co-axial counter-rotating rotors amidships to provide VTOL capability. The lower bulged portion will contain the payload, which can include various types of electro-optical sensors such as TV, low light level TV and thermal imaging systems.

Shown in model form in the accompanying photograph, the CL-227 has a radio command guidance system. The RPV is designed to take off and land using a small, simple platform.

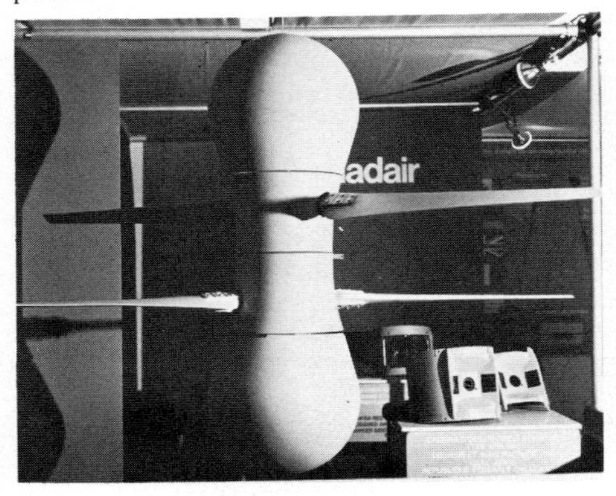

Model of the Canadair CL-227 *(Brian M. Service)*

Artist's impression of the Canadair CL-289 long-range surveillance drone

FRANCE

AÉROSPATIALE
SOCIÉTÉ NATIONALE INDUSTRIELLE AÉROSPATIALE

HEAD OFFICE:
37 Bd de Montmorency, 75781 Paris-cédex 16

OFFICERS: See Aircraft section

Drones in current production by Aérospatiale (Division Engins Tactiques) are the CT.20 target and the R.20 battlefield reconnaissance system.

Under development is a target for close-range air-to-air missiles, designated C.20. A sea-skimming target designated CM.38, which is in the project stage, is designed to present a radar image similar to that of the Exocet missile. Development of a high-altitude, long-endurance reconnaissance vehicle has also been reported.

AÉROSPATIALE CT.20

The CT.20 is a turbojet-powered radio-controlled target of medium performance, which is also used as a tug for a towed target. Series production began in 1958; by 1 January 1976 a total of 1,221 had been built, including 281 for export to NATO and non-NATO countries such as Italy and Sweden.

It is standard equipment for training military units in the use of air-to-air and surface-to-air missiles, in particular the Hawk.

The CT.20, fitted with a Thompson-CSF active homing radar, and with a warhead in place of the recovery system, also formed the basis of the M.20 anti-ship missile designed by Aérospatiale for the Swedish Navy, of which 68 were built by Saab-Scania as the RB08.

The CT.20 is launched by booster rockets from a nearly zero-length ramp, and attains a speed of 329 knots (610 km/h; 379 mph) by the time it reaches its maximum acceleration. The drone then continues to fly under the control of a radio operator located on the ground or in a 'mother' aircraft.

Nine signals can be transmitted: turns to right and left, nose up and down, increase and reduce power, trace smoke, operate cameras, and land. The turning signal controls bank and the turns are executed without reverse yaw. The pitch signals act on the elevators via the autopilot. When the landing signal is transmitted, the engine is stopped, the brake parachute opens and, at the end of a delay period, the recovery parachute is released. The descent is made in a level attitude and the impact with the ground is cushioned by an airbag forward of the centre-section. In the case of radio control failure the landing sequence occurs automatically; and the drone can be recovered from the sea, as it is designed to float.

Low-altitude flight, under barostatic altitude control, can be programmed currently at 90-120 m (300-400 ft). A version known as the CT.20 TBA, with very low altitude capability (about 30·5 m; 100 ft) is operational. This has a TRT AVH-6 radio altimeter and an improved remote guidance system.

Aérospatiale CT.20 drone, with underwing towed target

The CT.20 can be used to tow a Dornier target 1·20 m (3 ft 11¼ in) long, with a fin span of 400 mm (1 ft 3¾ in) and body diameter of 250 mm (9·85 in), pylon-mounted under the starboard wing of the CT.20 at launch. When towed in flight on an 800-1,200 m (2,625-3,950 ft cable, it has negligible effect on the CT.20's speed, and flight duration is reduced by no more than 15%.

Alternatively, the CT.20 can be used with a trailed target system developed for the Centre d'Essais des Landes and used primarily for training purposes in connection with air-to-air missiles with electro-magnetic or infra-red homing systems. This system exists in two versions: the BEY electro-magnetic version, with one or two Luneberg lenses; and the EMIR infra-red version, equipped with four infra-red sources and a telecommand receiver. The targets can be trailed at up to 800 m (2,625 ft) behind the CT.20.

The R.20 reconnaissance drone, developed from the CT.20, is described separately.

The following description applies to the CT.20, which is currently being built in two forms: a standard version, and an extended version with greater endurance.

TYPE: Turbojet-powered radio-controlled target.

WINGS: Cantilever mid-wing monoplane with medium sweepback. Each wing is a light alloy conventional structure. Lateral control spoilers at wingtips.

FUSELAGE: In three main sections. Forward section, of aluminium alloy, contains command guidance, autopilot, batteries and principal recovery parachute. Cenral section consists of a structural steel tank divided into two parts, one for fuel and the other containing chemicals for the tracking smoke. Rear fuselage, of aluminium alloy, contains the engine and carries the tail unit. A braking parachute is housed in a cone above the jet nozzle.

TAIL UNIT: Vee tail of aluminium alloy, comprising two

Aérospatiale D 15 low-cost artillery target

elevator surfaces controlled simultaneously by a single jack.

POWER PLANT: One Turboméca Marboré II (3·92 kN; 880 lb st) turbojet engine in CT.20 Version IV, or Marboré VI (4·7 kN; 1,056 lb st) turbojet engine in CT.20 Version VII.

DIMENSIONS, EXTERNAL (A: standard version; B: extended version with trailed target):

Wing span	3·16 m (10 ft 4½ in)
Length overall:	
A	5·45 m (17 ft 10½ in)
B	5·60 m (18 ft 4½ in)
Body diameter (max)	0·66 m (2 ft 2 in)

AREA:

Wings, gross	3·20 m² (33·34 sq ft)

WEIGHTS (A: standard version; B: extended version with trailed target):

Weight empty:	
A	490 kg (1,080 lb)
B	610 kg (1,344 lb)
Max launching weight:	
A	660 kg (1,455 lb)
B	800 kg (1,763 lb)

PERFORMANCE (A: standard version; B: extended version with trailed target):

Max speed at 10,000 m (32,800 ft):	
Marboré II	485 knots (900 km/h; 560 mph)
Marboré VI	512 knots (950 km/h; 590 mph)
Max Mach number (A, B)	Mach 0·85
Service ceiling:	
Marboré II	12,000 m (39,375 ft)
Marboré VI	15,000 m (49,200 ft)
Max operating height:	
A	14,000 m (45,925 ft)
B	13,000 m (42,650 ft)
Time to max operating height:	
A, B	15 min
Time to 10,000 m (32,800 ft)	6 min
Endurance at max operating height:	
A	50 min
B	1 hr 10 min
Min operating height:	
A, B	30 m (100 ft)
Practical range of command guidance and tracking system	135 nm (250 km; 155 miles)
Endurance at min operating height:	
A	15 min
B	21 min
Average endurance at 10,000 m (32,800 ft)	45 min

AÉROSPATIALE R.20

The Aérospatiale R.20 battlefield reconnaissance drone is developed from the CT.20 target, to which it is externally similar. It is powered by a Marboré IID turbojet and launched with the aid of two solid-propellant booster rockets from a short ramp on a standard Berliet GB-C8-KT Army lorry. Standard NATO cameras or other surveillance equipment, including an SAT Cyclope infra-red line-scanning device, are carried in its nose and in interchangeable wingtip containers. Two other standard vehicles are used to carry support equipment, including radio control equipment and the antenna system.

Aérospatiale R.20 battlefield reconnaissance drone, developed from the CT.20 target

When close to its launch-post the R.20 is controlled directly from the ground. Over longer distances, the drone is controlled automatically by a gyroscopic platform and an electronic programmer, enabling it to follow a pre-arranged flight plan.

It is claimed to offer an over-target accuracy of within 300 m (985 ft) at a distance of 54 nm (100 km; 62 miles) from its launch site. Average operating height is 1,000 m (3,300 ft), but it can be set to fly higher or lower, as required. It can photograph more than 200 km² (77 square miles) of territory during a single low-altitude sortie, using three synchronised Omera 114 × 114 mm cameras. Data can be sent back during flight by radio link. Flares can be carried for night photography.

After initial testing in 1963, the R.20 was tested under operational condidtions in February 1964. It is in operational use in the French Army and performs routine flights over French training grounds. Production had reached a total of 62 by 1 January 1976, and was continuing to fulfil outstanding orders for the French Army. A test programme, involving three firings, was successfully completed by the 7th Artillery Regiment of the French Army at Landes in Spring 1973 to evaluate the R.20's reconnaissance system.

DIMENSIONS, EXTERNAL:

Wing span	3·72 m (12 ft 2½ in)
Width, wings folded	1·35 m (4 ft 5 in)
Length overall	5·71 m (18 ft 9 in)
Body diameter (max)	0·66 m (2 ft 2 in)
Wingtip containers:	
Length	1·90 m (6 ft 3 in)
Diameter	0·40 m (1 ft 3¾ in)

WEIGHTS:

T-O weight of drone	850 kg (1,875 lb)
T-O weight with booster	1,100 kg (2,425 lb)
Payload	150 kg (330 lb)

PERFORMANCE:

Operating speed	Mach 0·65
Operating height range	200 to 10,000 m (660 to 32,800 ft)
Operating radius at low altitude	86 nm (160 km; 100 miles)

AÉROSPATIALE D 15

Aérospatiale developed the D15 as a small, inexpensive target drone for anti-aircraft artillery training. It made its first flight on 10 April 1975, powered by two 1·27 kW (1·7 hp) engines, and has radio command guidance.

The D 15 is of high-wing monoplane configuration, with a single fin and rudder and horizontal tail surfaces. Twin main wheels are fitted, with a third wheel to the rear, under the centre of the fuselage.

Structure is extremely simple, mostly of plastics materials, and utilises many components and accessories already in production. Assembly takes 5 minutes, filling of the 2 litre (3·5 Imp pint) fuel tank another 2 minutes, and starting of the engines another 1 or 2 minutes. The fuel capacity is sufficient for 1 hour's flight at full power, and the D 15 can simulate a low-level strike at speeds of up to 108 knots (200 km/h; 124 mph) or diving attacks at up to 162 knots (300 km/h; 186 mph).

DIMENSIONS, EXTERNAL:

Wing span	2·40 m (7 ft 10½ in)
Length overall	2·20 m (7 ft 2½ in)
Wing area	1·00 m² (10·76 sq ft)

WEIGHTS:

Payload	6 kg (13·2 lb)
T-O weight	18 kg (39·7 lb)

PERFORMANCE:

No details known

DORAND

GIRAVIONS DORAND

ADDRESS:
5 rue Jean Macé, 92153 Suresnes, BP 3-30
Telephone: 772.18.20
Telex: Iteser 28823 F Serv 458

MANAGING DIRECTOR:
P. de Guillenchmidt

Under contract to the Direction des Recherches et Moyens d'Essais (DRME), Dorand built a DS.7 test example of a gyro-glider which can be operated under pre-programmed controls to recover loads dropped from aircraft. Its purpose is to provide the load to be recovered with the means of making an autorotative descent, terminating in a flareout which reduces its vertical rate of descent to less than 2 m (6 ft)/sec at touchdown.

All available details of the DS.7 were given in the 1974-75 *Jane's*.

The initial phase of capability testing of the DS.7 gyro-glider model was followed in 1974 by a second phase concerned with rotor head and control system simplification prior to the start of production studies. These tests were conducted with half-scale working models of the DS.7; one of these is shown in the accompanying photograph, waiting to be lifted for a drop test.

Applications foreseen for this system, if tests prove successful, include the recovery of loads at very low altitudes or under high wind conditions; recovery of the Aérospatiale R.20 reconnaissance vehicle (which see); steering of a dropped load towards a predetermined point under remote control; automatic recovery of a helicopter following an engine failure; ejection seat recovery; and use as a personnel gyro-glider. Among the advantages claimed for the recovery system are its light weight (only 5 to 7 per cent of the recovered load), its very small volume when folded, fast rotor spin-up, high disc loading and ability to soft-land delicate loads. Payloads ranging from 100 kg (220 lb) to 15,000 kg (33,070 lb) could be recovered with a system of this type, depending on rotor size, and release altitude could be as low as 15 m (50 ft).

Dorand DS.7 half-scale gyro-glider test model

FRANCE-ENGINS

SOCIÉTÉ ANONYME FRANCE-ENGINS

ADDRESS:
c/o Microturbo SA, Chemin du Pont-de-Rupé, 31019 Toulouse Cédex
Telephone: (61) 47.63.26

PRESIDENT-DIRECTOR GENERAL:
André de Boysson

This company was formed, in early 1975, by Microturbo SA (see Aero-Engines section) and Société Soulé. In June 1975 it displayed publicly for the first time, at Satory, the Mitsoubac decoy and target drone.

FRANCE-ENGINS MITSOUBAC

Of simple, streamlined shape, the Mitsoubac has an airframe of moulded laminated polyester, with small low-mounted clipped-delta wings and a sweptback fin. There are no horizontal tail surfaces. The drone can be pre-programmed or radio controlled in flight, via an on-board light aircraft autopilot with a vertical gyro. Elevons are actuated by pneumatic jacks.

The Mitsoubac is powered by a Microturbo TRS 18-056 turbojet, rated at 0·98 kN (220 lb st), derived from the engine that powers the Bede BD-5J piloted aircraft. The design permits parachute recovery, catapult launch, or air launch from a helicopter or fixed-wing aircraft; in the latter case the Mitsoubac is carried inverted on the launch aircraft's external stores pylons.

The Mitsoubac is envisaged in several versions: (1) as an expendable target drone for pilot training, capable of flying for 15 minutes at 9,145 m (30,000 ft) and Mach 0·85; (2) as a recoverable target drone for anti-aircraft firing practice, flying between 45 m (150 ft) and 275 m (900 ft) at Mach 0·6-0·75, with a range of 108-162 nm (200-300 km; 124-186 miles); and (3) as a decoy, flying at low altitude and Mach 0·7, with a range of 162 nm (300 km; 186 miles). In each version, payload is 35 kg (77 lb), or 25 kg (55 lb) if a parachute recovery system is fitted.

Flight testing, begun in 1975 with air launch from a light helicopter, was continuing satisfactorily in the Spring of 1976.

DIMENSIONS, EXTERNAL:

Wing span	1·35 m (4 ft 5¼ in)
Length overall	2·86 m (9 ft 4½ in)
Height overall	0·72 m (2 ft 4½ in)

WEIGHTS:

Weight empty	83 kg (183 lb)
T-O weight:	
version 1	100 kg (220·5 lb)
version 2	120 kg (264·5 lb)
version 3	135 kg (297·5 lb)

France-Engins Mitsoubac beneath a Bell 47 helicopter

MARCHETTI
SA CHARLES MARCHETTI

HEAD OFFICE:
80 avenue de la Grande Armée, 75017 Paris
Telephone: 380.17.69
PRESIDENT: Charles Marchetti

This company has undertaken design studies of a variety of rotating-wing aircraft in recent years, under contract from the French Direction des Recherches et Moyens d'Essais (DRME).

MARCHETTI HELISCOPE

The Heliscope is an electrically-powered flying platform, consisting basically of two three-blade light alloy fixed-pitch rotors of co-axial contra-rotating design. These rotors are attached respectively to the rotor and stator of an electric motor located between their hubs.

The power supply for this motor is provided by a generator mounted on the vehicle which serves as the mobile ground station, and is fed through a cable which runs to the centre of the base of the Heliscope.

Marchetti has used test results to project a platform using three Heliscopes, each with a four-blade rotor of 1·60 m (5 ft 3 in) diameter. Such a vehicle could support a useful load of 100 kg (220 lb) with a simplified stabilisation system at an altitude of 230 m (755 ft) above the ground.

The company is continuing its studies of Heliscopes of higher performance and new electrical free-flying platforms.

Motorised mockup of a flying platform with three Marchetti Heliscopes

Right: Marchetti Heliscope flying platform

GERMANY
(FEDERAL REPUBLIC)

DORNIER
DORNIER GmbH

ADDRESS:
Postfach 2160, 8000 München 66
Telephone: München 8715480
Telex: 05-23543
OFFICERS: See Aircraft section

Current activities of Dornier GmbH include the development of drones, reconnaissance systems, missiles and air target systems.

The Dornier Aerial Target System (DATS) can be adapted to a wide variety of aircraft and target drones, and was described briefly in the 1975-76 *Jane's*.

The aircraft/helicopter division of Dornier GmbH has developed a mobile drone system, designated Do 34 Kiebitz, based on the Do 32 K Experimental Kiebitz; and an unmanned VTOL research vehicle known as the Aerodyne.

Jointly with Canadair (which see), Dornier is developing the CL-289, a growth version of the CL-89 with increased power and longer range.

DORNIER Do 32 K EXPERIMENTAL KIEBITZ (PEEWIT)

The Experimental Kiebitz was developed, under contract from the Federal Ministry of Defence, to evaluate the feasibility of an automatically-stabilised tethered rotor platform, capable of utilising sensors for reconnaissance, communications and ECM purposes at an altitude which considerably increases their effectiveness.

The complete system consists of a tethered rotating-wing platform (first demonstrated in flight in late 1970) and a mobile ground vehicle which serves as transporter, take-off and landing ramp, and power supply station. The drone is reeled in and out by a winch mounted on the ground vehicle.

Five experimental rotating-wing vehicles were built, and these completed approx 100 hours of successful test flying, as described in earlier editions of *Jane's*. The basic flight test programme was completed in 1974, and has been followed by tests designed to evaluate the detectability of the Kiebitz. In addition to measurement of the vehicle's infra-red radiation and radar profile, details have also been obtained of the distances at which the rotor platform is discernible acoustically and optically. Also, so far as possible within the experimental vehicle's payload capacity, tests have been performed with applications sensors. These include an RDF system, developed by Dornier, which used the Kiebitz rotor as an integral part of the D/F system by means of a blade-tip antenna; and a Grundig television camera, with oscillation-damping suspension and Dynalens lens-stabilisation, to determine the feasibility of such an installation for reconnaissance and surveillance purposes. The experimental programme, particularly with regard to payload and sensor investigations, was still active in 1975.

DIMENSIONS, EXTERNAL:
- Diameter of rotor — 7·50 m (24 ft 7¼ in)
- Height overall — 1·60 m (5 ft 3 in)
- Body diameter at bottom edge — 0·75 m (2 ft 5½ in)

WEIGHTS:
- Weight without tether or payload — 200 kg (440 lb)
- Max payload to 200 m (660 ft), ISA — 50 kg (110 lb)

PERFORMANCE:
- Reel-in/reel-out speed — 1·5 m (4·9 ft)/sec
- Operational ceiling — 200 m (660 ft)

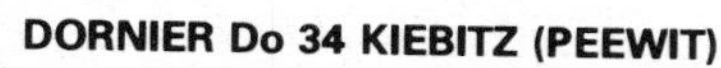

DORNIER Do 34 KIEBITZ (PEEWIT)

In August 1972 Dornier announced the receipt of a DM 7 million contract from the Federal Ministry of Defence for the first development phase of a prototype of an operational Kiebitz system, to be used for reconnaissance, fire control, communications and traffic monitoring duties. This followed the completion in mid-1972 of the design of a Do 34 operational model as a sensor platform with a payload of 140 kg (308 lb). Full-scale mockups of the Do 34 have been completed, and work is proceeding on the construction of two prototype flight vehicles and two ground stations. First flight was anticipated in 1976.

The complete system is vehicle housed, and consists of a landing platform, winch system, guidance and control post, flight vehicle and sensor, checkout system, fuel tank for 12 hours' operation, and auxiliary equipment. The flight vehicle has a cone-shaped airframe to reduce radar reflectivity, and the payload compartment is located on the underside of this, enabling sensors to be changed quickly and allowing space for a large-volume radome. The rotor's twin blades are attached by straps and driven by cold air expended through the blade-tip nozzles, a principle which ensures that no torque is produced which could act on the platform. Air for the rotor blades is supplied by a radial compressor and an Allison 250-C20 turboshaft engine, the latter being installed on the slant to ensure a good intake position.

After arrival on site, the drone can be in position at an operational height of 300 m (985 ft) in 8 minutes. Limiting factors in the guidance and control system are a wind speed of 14 m (46 ft)/sec ±8 m (26 ft)/sec; the available thrust reserves; and the requirements of the various sensors. The control system aligns the Kiebitz according to airframe attitude and position in relation to the ground. An electromagnetic sensor measures any drift from the desired position.

Under a bilateral agreement between the German and French governments, one Kiebitz flight vehicle will be fitted with an advanced version of the French LCT Orphée radar, to define, integrate and test a new battlefield reconnaissance system known as **Argus** (see International section).

Operational trials will be carried out with the other Kiebitz prototype fitted with passive ECM in the form of a Decca RDL-2 surveillance radar, operating in the 0·5-18CHz band to provide immediate determination of azimuth, frequency measurement and signal analysis. Another project under discussion is for a Seekiebitz, carrying a Ferranti Seaspray search radar, which could enhance the strike range of medium-range naval missile systems by elevating the radar well above the height of existing masthead aerials. A similar application might be for the detection of low-flying aircraft flying in the shadows of hills.

The following details apply to the basic Do 34 flight vehicle:

Dornier Experimental Kiebitz and mobile ground control station

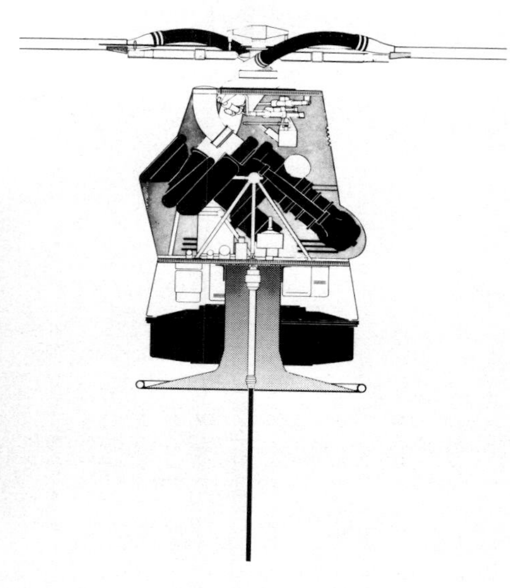

Diagram of the Do 34 Kiebitz flying platform for electronic surveillance: lower portion contains Decca RDL-2 passive ECM detector

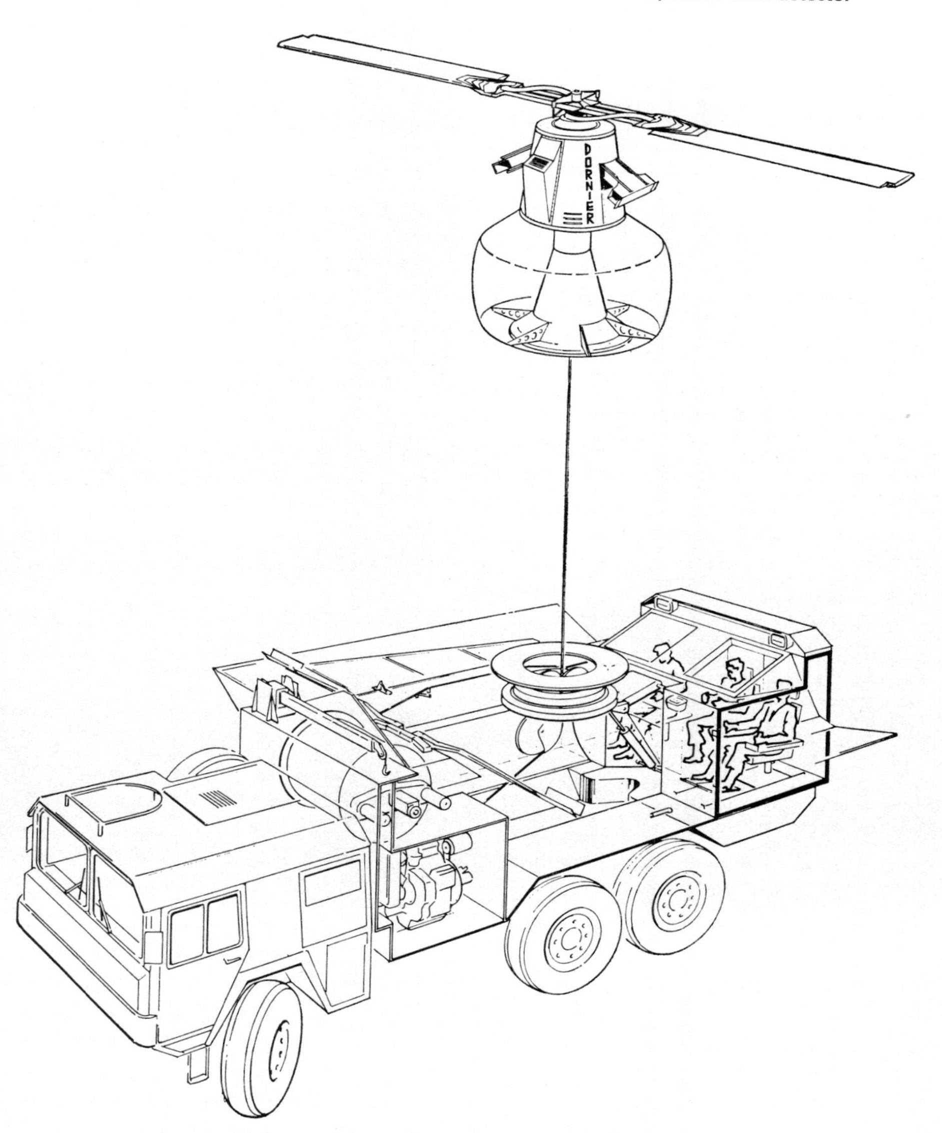

Dornier Do 34 Operational Kiebitz tethered rotor platform

DIMENSIONS, EXTERNAL:
- Diameter of rotor — 8·00 m (26 ft 3 in)
- Height overall — 1·45 m (4 ft 9 in)
- Body diameter at bottom edge — 1·05 m (3 ft 5¼ in)

WEIGHTS:
- Weight without tether or payload — 280 kg (617 lb)
- Max payload to 300 m (985 ft), ISA — 140 kg (308 lb)

PERFORMANCE:
- Reel-in/reel-out speed — 3 m (9·8 ft)/sec
- Operational ceiling — 300 m (985 ft)
- Mission endurance — 24 hr

DORNIER UKF

The UKF (Unbemanntes Kampfflugzeug: unmanned combat aircraft) has been proposed by Dornier, under the Federal Defence Ministry's Experimental Component Programme, as a means of examining the basic problems of target acquisition, armament and flight guidance for ground support RPVs. VFW-Fokker is assisting in the programme.

An Aeritalia (Fiat) G91 aircraft is serving as the testbed in this programme; in 1975 Dornier was engaged in integrating the necessary test equipment in this aircraft.

DORNIER AERODYNE E 1

Development of the Aerodyne wingless high-speed VTOL aircraft was begun by the late Dr A. M. Lippisch in the USA. It has been continued since 1967 by Dornier GmbH, under an experimental programme for the Federal Ministry of Defence, within the framework of which the feasibility of the Aerodyne design is demonstrated by means of flight tests with an unmanned experimental vehicle.

The Aerodyne principle is marked by the combination of the means of lift and propulsion in a single unit, namely a shrouded propeller, the slipstream from which is deflected downward by cascade-type vanes for vertical take-off and landing. Control is by deflection of the turboshaft exhaust, which emerges at the end of the tailboom, and vanes in the propeller slipstream.

A radio-controlled prototype vehicle, known as the Aerodyne E 1, was built in 1971 and made its first flight, tethered, on 18 September 1972. Hovering tests were concluded successfully on 30 November 1972, after 74 flights totalling almost 1½ hours in the air. The last of these was, unintentionally, a free flight, from which the E 1 was successfully brought down from the hover to a normal landing.

In 1974, studies were carried out to define operational versions of the Aerodyne for various tasks. These studies confirmed that the Aerodyne, due to its unique launch and recovery characteristics, would be a suitable V/STOL RPV for a variety of naval applications, and proposals for such applications were submitted to NATO.

Work on the Aerodyne, under Federal Defence Ministry contract, continued in 1975-76 and involves detailed investigation of a ship-based application of an operational vehicle for over-the-horizon fire control missions. A description of a twin-fan Aerodyne proposal for a maritime RPV, evolved jointly by Dornier and Hawker Siddeley Dynamics, is given separately.

Test flights of the Aerodyne E 1 from a moving platform were to be carried out in the Autumn of 1976. For these tests a tail unit, reinforced landing gear and a more powerful engine (a Lycoming LTS 101-650C) were to be installed.

TYPE: Radio-controlled VTOL experimental RPV.

AIRFRAME: Annular fan shroud and deflection chamber, with bullet-shaped pod attached centrally in mouth of fan duct. In E 1 prototype, this pod houses electronics system and control unit; in a production version it would house the mission payload. The power plant is mounted on top of the fan shroud, with a hollow tailboom to the rear. Non-retractable landing gear, designed for vertical landings, consists of two main legs and a shorter leg under the tailboom. Shock-absorption by interchangeable aluminium honeycomb damper in each leg.

POWER PLANT: One 276 kW (370 shp) MTU 6022A-3 turboshaft engine, driving a five-blade variable-pitch Hoffmann fan via a Zahnradfabrik/Dornier GmbH Z-type reduction gear. (To be replaced in 1976 by Lycoming LTS 101-650C turboshaft.) Fuel in two tanks located between inner and outer walls of deflection chamber.

FLIGHT CONTROLS: The fan shroud and deflection chamber form an inner flow channel, at the end of which the airflow is deflected by cascade-type vanes. In the E 1 prototype, this deflection amounts to approx 60° during hovering flight and it is intended to be increased to approx 75° during future development. The cascades are not moved during hovering. The engine's exhaust is directed aft through the tailboom and provides pitch and yaw control during hovering. For this purpose the jet can be deflected up and down as well as sideways by hydraulically-operated cascades. (For pitch and yaw control during forward flight, conventional tail control surfaces, not present during hover tests, will be fitted to the tailboom.) Roll control is effected during both hovering and forward flight by a vertical keel flap, located immediately aft of the main cascade flaps, which supplies a roll moment when deflected, due to its arrangement below the CG. This keel flap and the fan pitch control mechanism are operated hydraulically.

COMMAND AND GUIDANCE: Radio command, developed by Dornier GmbH. During hovering flight, the remote control pilot commands the flight vehicle's attitude via a small control stick. Bodenseewerk on-board attitude control system, which ensures that the attitude command is kept within very small limits, incorporates a vertical gyro, three rate gyros and three accelerometers as sensors. Telemetry system, with more than 50 channels, for transmitting data to ground station. Power supply system incorporates battery for emergency power in the event of a generator failure.

DIMENSIONS, EXTERNAL (approx):

Max diameter	1·90 m (6 ft 2¾ in)
Fan diameter	1·10 m (3 ft 7¼ in)
Length overall	5·50 m (18 ft 0½ in)

WEIGHT:

Max T-O weight	435 kg (959 lb)

Dornier Aerodyne E 1 prototype VTOL RPV in tethered flight

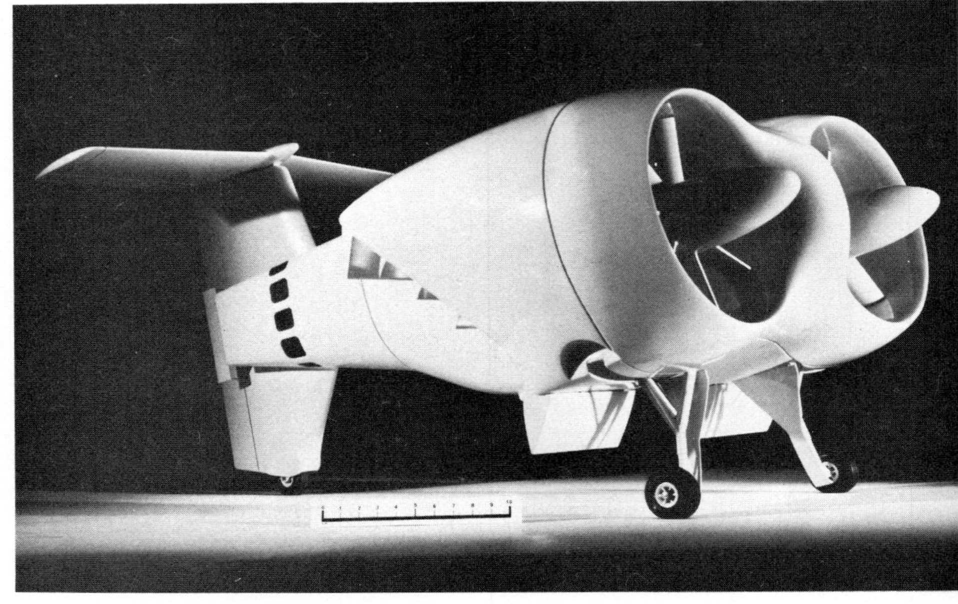

Model of HSD/Dornier proposal for a maritime version of the Aerodyne

MARITIME AERODYNE

Applications of the Aerodyne principle to operational RPV systems are being studied under a 1971 agreement between Dornier and Hawker Siddeley Dynamics, and include the naval RPV proposal shown in the accompanying illustration.The Maritime Aerodyne is suited to a number of naval applications, including reconnaissance weapon control and delivery, ASW, ECM and in-fligh missile guidance. The version illustrated has twin ductec fans and could carry two homing torpedoes.

DIMENSIONS, EXTERNAL:

Width overall	3·00 m (9 ft 10 in
Length overall	4·90 m (16 ft 9¼ in
Height overall	2·60 m (8 ft 6¼ in

WEIGHT:

Max T-O weight	1,360 kg (3,000 lb

PERFORMANCE (estimated):

Max level speed	329 knots (610 km/h; 379 mph
Service ceiling	6,700 m (22,000 ft
Hovering ceiling	1,200 m (3,940 ft
Range at max level speed with 100 kg (220·5 lb payload	809 nm (1,500 km; 932 miles
Endurance: with 100 kg (220·5 lb) payload	3 hr 20 mi
with 400 kg (882 lb) payload	1 hr 20 mi

INDIA

ADE
AERONAUTICAL DEVELOPMENT ESTABLISHMENT

ADDRESS:
c/o Scientific Adviser's Department, The Defence Ministry, South Block, Central Secretariat, New Delhi 2

DIRECTOR:
Vivek R. Sinha

It was reported in the Spring of 1974 that flight testin was imminent of a supersonic target drone developed b the ADE and apparently of indigenous design.

No further details are available.

INTERNATIONAL PROGRAMMES

ARGUS

PARTICIPATING COMPANIES:

Dornier GmbH, Postfach 2160, 8000 München 66, German Federal Republic
Telephone: München 8715480
Telex: 05-23543

LCT (Laboratoire Central de Télécommunications), 18-20 Rue Grange-Dame-Rose, 78140 Vélizy-Villacoublay, BP 40, France
Telephone: 946-96-15
Telex: 69892

ARGUS

Argus (Autonomes Radar Gefechtsfeld Uberwachungs System) is a battlefield surveillance system being developed under a German-French government agreement signed on 6 March 1974. Dornier has the main contract for co-ordinating, integrating and testing the system, and the programme is administered by a board of directors representing the two governments.

The Argus prototype consists of a Dornier Do 34 Kiebitz tethered rotor platform in which is installed an advanced version of the LCT Orphée radar as the primary sensor. The complete Argus system also incorporates a mobile ground station and a tether cable. The ground station is installed in a container mounted on a cross-country 7 ton truck and provided with all equipment necessary for transportation and a 12 hr operation of the system. Operating height of the flight vehicle is 300 m (985 ft).

A description of the Kiebitz appears under the Dornier heading in the German section of this edition. The Argus prototype flight vehicle was scheduled to begin flight testing in 1976.

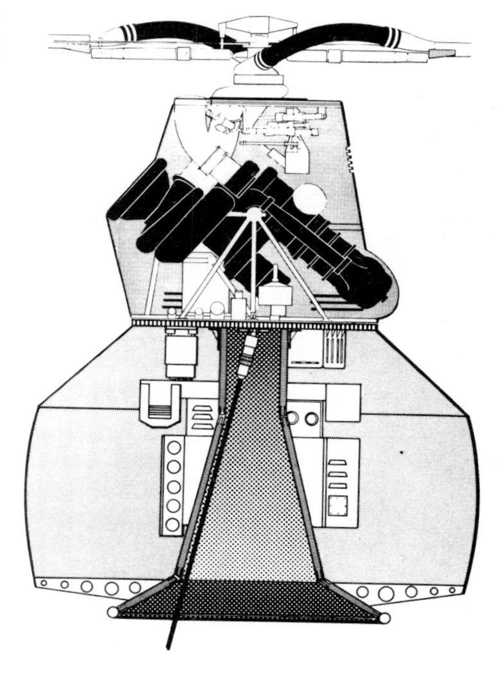

Dornier Do 34 Kiebitz in Argus configuration: tilted engine shown in solid black; central shaded area indicates LCT radar location

ITALY

METEOR

METEOR SpA COSTRUZIONI AERONAUTICHE ED ELETTRONICHE

HEAD OFFICE:
146 Via Nomentana, 00162 Rome
Telephone: (06) 8380232 and 8392145
Telex: 68136 Meteorom

WORKS:
PRODUCTION:
34074 Monfalcone, Stazione F.S. Ronchi Nord
Telephone: (0481) 778001
Telex: 46288 Meteormo

FLIGHT UNIT:
34074 Monfalcone, Aeroporto Giuliano
Telephone: (0481) 77441
Telex: 46288 Meteormo

TEST CENTRE:
09040 Villaputzu S.S. 125 Km 78
Telephone: (070) 99937-99946
Telex: 79076 Avielsar

PRESIDENT AND GENERAL MANAGER:
Dott Furio Lauri

CHAIRMAN OF AUDITORS' COMMITTEE:
Dott Luciano Davanzo

MANAGERS:
Mrs Luciana Bortolotti (Manager, General and Administrative Affairs Service)
Gen SA (r) Roberto Fassi (Chairman, Military Exports Committee)
Antonio Castelli (Monfalcone Plant Manager)
Guido Borsari (Manager, Marketing Co-ordination Planning Service)
Carlo Spano (Manager, Technical and Operational Service)

Meteor was established in Trieste as a joint stock company in 1947. Its present head office is in Rome, supported by facilities at Monfalcone (Trieste) and Cagliari (Sardinia). The former is a production factory, the latter being equipped for flight operations and for technical assistance to the users of the tri-service range at Salto di Quirra in southern Sardinia.

Meteor has developed and is producing for the Italian and foreign armed forces a range of propeller-driven and turbojet-powered radio-controlled drones covering a speed range from 323 knots (600 km/h; 372 mph) up to Mach 2·8, and altitudes from 10 m (35 ft) to 25,000 m (82,000 ft) above sea level. It also co-produces the Northrop-Ventura Meteor 1 and 2, Northrop-Ventura Meteor/USD-1, Aérospatiale Meteor 20 (CT.20) target drones and the BM-1 (Beech Meteor 1) missile target system. Under licence from Canadair, Meteor produces 50% of the AN/USD-501 (CL-89) reconnaissance systems ordered by Italy.

Naval version of the Meteor P.1 target drone on zero-length launcher

METEOR P.1

The Meteor P.1 is a subsonic target drone, used for training with directed anti-aircraft batteries of medium and large calibre and with ground-to-air missiles. It is currently available in two versions, both of which can be used for in-sight and out-of-sight operation. The first version is fitted with a 74·5 kW (100 hp) Meteor Alfa 1 four-cylinder X-type two-stroke aircooled engine for operation at heights up to 8,000 m (26,250 ft). The second version has a Meteor Alfa 1AQ engine, giving 89·5 kW (120 hp) constant up to 6,500 m (21,325 ft) and permitting operation at heights up to 13,000 m (42,000 ft).

The P.1 is made largely of glassfibre-reinforced polyester resin. Its engine is manufactured from anti-corrosive and special steel and aluminium, with extensive chromium plating. This permits recovery from salt water and re-use. Flotation is ensured by the use of blocks of expanded resin inside the structure.

All Meteor targets are launched normally with the engine running at peak rpm, with the aid of jettisonable solid-propellant rockets. Alternatively, a catapult can be used, or the targets can be air-launched.

For out-of-sight radio control, over ranges up to 54-86 nm (100-160 km; 62-100 miles), the operator uses a series of control levers linked to a UHF ground transmitter which emits a five-tone modulated carrier signal. The receiver in the target transforms the signals into seven distinct control operations. Two tones control the elevator, two control the ailerons, and the fifth is used to stop the engine and open the recovery parachute at the end of a flight. The ailerons and elevators are operated by electrical servo controls, those for the ailerons being combined with a gyro which stabilises the target laterally. Electronic equipment in the target also includes a two-axis automatic stabilisation system. The target's track and altitude are plotted normally by the radar of the gun or missile battery using it, and wingtip reflectors can be fitted to amplify the echoes from the target. However, a UHF tracking and tele-control system can be used, in conjunction with a transponder in the target weighing only 2 kg (4·4 lb).

DIMENSIONS, EXTERNAL:
Wing span without wingtip containers — 3·68 m (12 ft 1 in)
Length overall — 3·39 m (11 ft 1½ in)
Height overall — 0·65 m (2 ft 1½ in)

WEIGHTS (out-of-sight versions. A: 100 hp, B: 120 hp):
Weight without fuel and electronics:
A, B — 133 kg (293 lb)
Electronics and gyro guidance equipment:
A, B — 25 kg (55 lb)
Launching weight (one-hour flight):
A — 195 kg (425 lb)
B — 202 kg (444 lb)
Max launching weight:
A — 220 kg (484 lb)
B — 225 kg (495 lb)

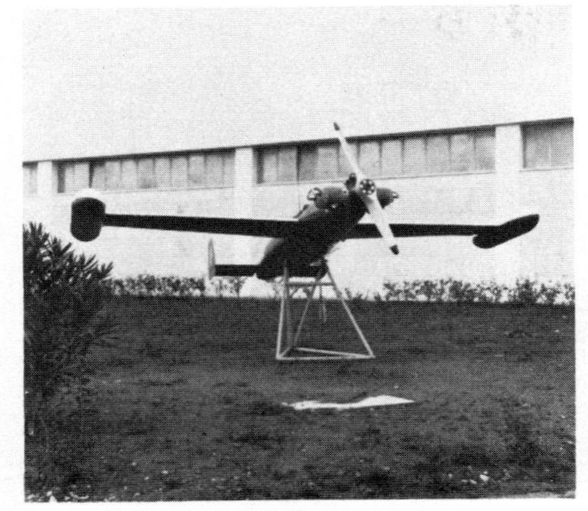

Meteor P.X radio-controlled target drone

Meteor Gufo system control truck and Gufone drone on towed launcher

PERFORMANCE:
- Max level speed at 6,500 m (21,325 ft):
  - A: 178 knots (330 km/h; 202 mph)
  - B: 296 knots (550 km/h; 342 mph)
- Stalling speed:
  - A, B: 62 knots (115 km/h; 71 mph)
- Max rate of climb at S/L:
  - A: 900 m (2,950 ft)/min
  - B: 1,200 m (3,940 ft)/min
- Time to 6,100 m (20,000 ft):
  - A: 10 min
  - B: 5 min

METEOR P.X

The Meteor P.X is generally similar in layout to other Meteor drones, but is powered by a 53·7 kW (72 hp) flat-four engine. It is designed for zero-length launching, with the assistance of a Meteor 8785 solid-propellant booster rocket.

Guidance is by radio control, via a Meteor RSS 529 fully-transistorised two-axis autopilot, with radar tracking of the drone's position. Recovery is by parachute, deployed automatically or on receipt of a signal from the ground.

A number of P.X systems have been produced for the armed forces of Italy and other countries, and production was continuing in 1976.

DIMENSIONS, EXTERNAL:
- Wing span without wingtip containers: 3·56 m (11 ft 8 in)
- Length overall: 3·46 m (11 ft 4¼ in)
- Diameter of fuselage: 0·40 m (1 ft 3¾ in)
- Span of tail unit: 1·21 m (3 ft 11½ in)

WEIGHT:
- Max launching weight: 165 kg (363 lb)

PERFORMANCE (at max launching weight):
- Max level speed: 194 knots (360 km/h; 224 mph)
- Time to reach 6,100 m (20,000 ft): 10 min
- Service ceiling: 8,000 m (26,250 ft)
- Radius of action: 86 nm (160 km; 100 miles)
- Endurance: 1 hr

METEOR GUFO

Meteor evolved the Gufo tactical reconnaissance system to meet anticipated military requirements during the 1970s. The system is claimed to be particularly suitable for use in mountainous country. In operational form it would enable up to 27 kg (60 lb) of sensors to be carried at 400 knots (740 km/h; 460 mph) to target areas up to 110 nm (200 km; 125 miles) from the launch site. Recovery can be within a radius of 100 m (330 ft) from a predetermined spot.

The drone part of the system, known as the **Gufone (Owl)**, is proposed in three versions, the standard operational vehicle being based on the American Northrop Chukar I (MQM-74A) target drone modified by Meteor to carry new guidance equipment and sensors, together with inflatable bags to cushion the landing shock. Details of the three versions are shown in the table. Equipment for day and night operations can include a variety of infra-red sensors and cameras using 50, 70 or 75 mm film to photograph a strip of terrain more than 55 nm (102 km; 63 miles) long and, respectively, 1,000 m (3,280 ft), 2,000 m (6,560 ft) or 3,000 m (9,840 ft) wide, respectively from altitudes of 305 m (1,000 ft), 610 m (2,000 ft) or 915 m (3,000 ft). For night operations, the Gufone can carry 14 wingtip flares which are dropped at preselected time intervals. At take-off, the 0·54 kN (121 lb st) turbojet engine is supplemented by two Meteor 8785/CNS solid-propellant jettisonable boosters, providing a total thrust of 25·5 kN (5,730 lb) for 0·7 seconds.

A military unit deploying the Gufo system would comprise a launching section; a guidance and control section; a sensor recovery, interpretation and headquarters section; and a vehicle recovery and preparation section.

The Gufo system is designed to overcome the problem of poor accuracy that sometimes mars results when pre-programmed drones are used over ranges of more than 30 nm (55 km; 35 miles). The Gufone can be launched in any direction and normally makes the first part of its flight under guidance over friendly territory. This permits the effects of factors such as wind and engine performance to be calculated, so that the Gufone can be directed very precisely on to the first stage of its programmed flight to the target. Once it has been put on course, it becomes 'deaf' to all friendly or enemy electronic signals until it approaches the end of its return flight and comes under command guidance for recovery.

Provision is made for an intermediate pre-programmed guidance phase between the guided and 'deaf' phases. In this case, the drone will accept only specially-coded commands of very short duration, for the sole purpose of correcting its course.

The Gufo system has been evaluated successfully by the Italian armed forces; production of a version for use in an Italian operational environment was expected to be authorised during 1976.

DIMENSIONS (Gufone):
- Wing span: 1·69 m (5 ft 6¾ in)
- Length overall: 3·61 m (11 ft 10 in)
- Height overall: 0·73 m (2 ft 5 in)

WEIGHTS (Gufone):
- Sensors (max): 27 kg (60 lb)

PROPOSED VERSIONS OF GUFONE DRONE

	Gufone	A	B
Payload	30 kg (66 lb)	60 kg (132 lb)	30 kg (66 lb)
Range penetration from FEBA at S/L	29 nm (55 km; 34 miles)	29 nm (55 km; 34 miles)	48 nm (90 km; 56 miles)
Range penetration from FEBA at 10,000 m (32,800 ft)	72 nm (135 km; 83 miles)	72 nm (135 km; 83 miles)	113 nm (210 km; 130 miles)
Flight altitude	100-10,000 m (328-32,800 ft)		
Cruising speed	378-405 knots (700-750 km/h; 435-466 mph)		
Landing accuracy	150 m (492 ft) CEP		
Wind during launch	20 m (65 ft)/sec		
Pictures	Improved film by day and night		
Recovery	By parachute		
Launch	By booster rocket from a jeep		
Radar reflectivity	0·2 m² (2·15 sq ft)		

Mirach mini-RPV component of Meteor Andromeda system *(see next page)*

Max launching weight 136 kg (300 lb)

PERFORMANCE (Gufone):

Max level speed 400 knots (740 km/h; 460 mph)

Max cruising height 10,670 m (35,000 ft)

METEOR ANDROMEDA SYSTEM

Meteor is concentrating most of its current design effort on a multi-role RPV system known as Andromeda, which consists of the following elements:

Mirach. RPV component of Andromeda system, described separately.

Alamak. Training version of the electronic ground and airborne subsystem for guidance, tracking, plotting, and telemetry and/or navigation missions. Already in production and service with other systems.

Sirak. Combat version of Alamak, for use in high-jamming areas. Design is at the optimisation stage.

Ground support subsystem, suitable for both training and combat missions. Already in production and service with other systems.

METEOR MIRACH

The Mirach mini-RPV is one of the three basic elements of Meteor's Andromeda multi-role RPV system. It is of modular construction, using mostly materials of European manufacture so as to facilitate the interchangeability of components between versions produced for different roles. The general appearance of the Mirach can be seen in the accompanying photograph. It is ground-launched from a zero-length launcher, and is self-recoverable.

When fitted with a 2·94 kN (660 lb st) turbojet engine, the Mirach has a maximum speed of approx 540 knots (1,000 km/h; 621 mph). The following details refer to the basic version with a 1·47 kN (330 lb st) turbojet engine:

DIMENSIONS, EXTERNAL:

Wing span 2·40 m (7 ft 10½ in)

Wing leading-edge sweepback 45°

Length overall 4·20 m (13 ft 9¼ in)

DIMENSION, INTERNAL:

Payload volume 0·06 m³ (2·12 cu ft)

WEIGHTS:

Weight empty 182 kg (401 lb)

Fuel and lubricant 104 kg (229·5 lb)

Internal payload 58 kg (128 lb)

External payload 100 kg (220·5 lb)

Max T-O weight:

internal payload only 344 kg (758·5 lb)

internal and external payload 444 kg (979 lb)

PERFORMANCE:

Max level speed at S/L 432 knots (800 km/h; 497 mph)

Max level speed at 9,000 m (29,525 ft) 442 knots (820 km/h; 509 mph)

Econ cruising speed 372 knots (690 km/h; 429 mph)

Service ceiling 12,000 m (39,375 ft)

Out-and-back range at 9,000 m (29,525 ft) 256 nm (475 km; 295 miles)

Endurance at max speed at S/L 30 min

Endurance at econ cruising speed at 9,000 m (29,525 ft) 1 hr 24 min

JAPAN

FUJI

FUJI JUKOGYO KABUSHIKI KAISHA (Fuji Heavy Industries Ltd)

HEAD OFFICE:

Subaru Building, 7-2, 1-chome, Nishi-shinjuku, Shinjuku-ku, Tokyo

Telephone: Tokyo (03) 347-2505

Telex: 0-232-2268

OFFICERS: See Aircraft section

Under contract from the Japan Defence Agency, Fuji is building the Teledyne Ryan BQM-34A Firebee I subsonic target drone (see US section) for use in the training of Tartar missile and gunnery crews and for the evaluation of air-to-air missile systems.

It was expected that applications of Fuji-built BQM-34As during 1976 would include the use of some of these drones fitted with RALACS (Radar Altimeter Low Altitude Control System: see Teledyne Ryan entry in US section), for training the crews of defensive weapons against attack by anti-shipping missiles.

Thirteen Fuji-built BQM-34As had been completed and delivered to the Japan Maritime Self-Defence Force, and two to the Technical Research Institute of the Japan Defence Agency, by the end of FY 1975. A further seven were scheduled to be completed during FY 1977.

Fuji has also received a Japan Defence Agency contract to build the prototype of a ramjet-powered vehicle, possibly for reconnaissance applications. No other details have been received of this vehicle, which made a successful first flight in the Autumn of 1975.

During the past three years, at the request of the JASDF, Fuji has carried out a feasibility study for the conversion of JASDF F-86F Sabre fighters to QF-86F target drone configuration. If this programme is approved, Fuji would develop the remote control system and flight control system for the drone version, and Mitsubishi would undertake the necessary airframe modifications.

Fuji-built Firebee I target drone of the JMSDF

NEC

NIPPON ELECTRIC CO, LTD (Nippon Denki Kabushiki Kaisha)

HEAD OFFICE:

33-1 Shiba Gochome, Minato-ku, Tokyo 108

Telephone: Tokyo (03) 454-1111

Telex: NECTOK A J22686

PRESIDENT:

Koji Kobayashi

MANAGER, OVERSEAS RELATIONS:

Kiyoshi Yamauchi

Under a technical aid agreement with Northrop Corporation, USA, NEC is responsible for production and repair of Northrop Shelduck target drones (MQM-36) for the Japan Maritime Self-Defence Force and Ground Self-Defence Force. Delivery of these drones to the JMSDF and JGSDF began in 1961, and a total of 228 had been delivered by the end of 1975. Production was expected to continue at the rate of approx 15 per year.

NEC-built Northrop Shelduck target drone

THE UNITED KINGDOM

AEL
AERO ELECTRONICS (AEL) LTD

ADDRESS:
Gatwick House, Horley, Surrey RH6 9SU
Telephone: 02 934-5353
Telex: 877362 and 87116
DIRECTORS:
H. K. Hughes, DFC
J. F. Hughes
M. H. Nicholas (Managing)

This company manufactures small, low-cost radio- and optically-guided systems for target drone, reconnaissance or other RPV applications.

Complete Snipe system, including trailer, launcher, two aircraft, towed target and optical tracking gear

AEL4111 SNIPE

The AEL4111 Snipe target system is an international standard target device for the training of anti-aircraft weapon and missile operators.

The following description applies to the Snipe target-towing aircraft:

AIRFRAME: Of high-wing monoplane configuration, constructed of veneer-covered polystyrene foam.

POWER PLANT: One 56 cc single-cylinder piston engine, with glo-plug ignition, driving a two-blade wooden propeller. Methanol/oil fuel mixture, tank capacity 2·3 litres (0·5 Imp gallons). Additional fuel tanks available optionally.

LAUNCH AND RECOVERY: Launched by catapult from trailer-mounted ramp. Recovery by parachute (stored in nose compartment) or by orthodox landing. Fail-safe system shuts down engine and deploys parachute in event of radio interference or loss of control signal.

GUIDANCE AND CONTROL: Radio command guidance system. Each Snipe customer must specify two control frequencies in the 27MHz or 70MHz waveband which are clear of all other traffic. Each system operates in a frequency slot with an approx bandwidth of 25KHz and with an ERP maximum of 5W. If it is intended to fly more than one aircraft simultaneously, at the same site or within radio frequency control range of each other, individual control frequencies must be specified for each set. If such coincidental operation is not intended, the customer's total order can be supplied to operate on the same control and reserve frequencies.

SPECIAL EQUIPMENT: The complete Snipe system, which can be operated by a two-man crew, comprises two miniature aircraft, including their engines, airborne guidance equipment, visual enhancement high-intensity strobe lights and radio-triggered recovery parachutes; tripod-mounted ground control equipment, complete with optics; a ground support unit, including engine starter, fuel container, accumulator and battery chargers; radio monitor; a field spares pack, including a set of key replacement subassemblies and components; a complete target drogue assembly, including a 3·05 m (10 ft) × 0·76 m (2 ft 6 in) banner and 152 m (500 ft) towline; five spare drogues; and a lockable trailer for containing, storing and transporting the complete system. The system catapult launcher forms an integral part of the trailer. Optional extras include additional fuel tanks or an additional 4·5 kg (10 lb) of payload, such as radar or infra-red enhancement packages. A Polaroid camera installation was expected to be available from mid-1976. Other optional packages under development.

DIMENSIONS, EXTERNAL:
Wing span	2·51 m (8 ft 3 in)
Length overall	2·13 m (7 ft 0 in)

WEIGHTS:
Payload	4·5 kg (10 lb)
Max launching weight	18 kg (40 lb)

PERFORMANCE:
Max speed	111 knots (206 km/h; 128 mph)
Max controllable range	2·7 nm (5 km; 3·1 miles)
*Average flight endurance	45 min

**extendable if additional fuel tanks fitted*

The Snipe target towing aircraft in flight

BAC
BRITISH AIRCRAFT CORPORATION LTD (GUIDED WEAPONS DIVISION)

ADDRESS:
Six Hills Way, Stevenage, Hertfordshire SG1 2DA
Telephone: Stevenage (0438) 2422
Telex: 825125/6
OFFICERS: See Air-Launched Missiles section

BAC/AJEP TAILWIND (DRONE VERSION)

As a company-funded venture, BAC's Guided Weapons Division is developing, in conjunction with AJEP Ltd (see UK Aircraft section) a drone version of the latter company's Series 2 Tailwind aircraft. The purpose of this programme is to provide missile operators with a realistic but inexpensive full-size target vehicle; if trials are successful, the drone Tailwind may become an integral part of BAC's Rapier air defence system in the future.

For initial trials, BAC has acquired the original Series 2 Tailwind prototype G-AYDU, which was scheduled to make its first flight under remote control at the end of 1976.

A description of the manned Tailwind Series 2 can be found on page 170 of this edition. The only changes made in the drone version are the installation of an autopilot and radio command link equipment; all manual controls are retained, so that the Tailwind can be flown on normal duties when not in use as a drone.

AJEP Tailwind Series 2 prototype, converted by BAC for unmanned operation

RCS
RCS GUIDED WEAPONS SYSTEMS (Division of Radio Control Specialists Ltd)

ADDRESS:
National Works, Bath Road, Hounslow, Middlesex TW4 7EE
Telephone: 01-572 0933 and 0934
DIRECTORS:
E. A. Falkner, CEng, AMRAeS, FIED
P. R. Conway, BSc (Eng), CEng, MICE
K. E. Mackley, CEng, MIERE

This company manufactures a range of small, low-cost radio- and optically-guided systems for target drone, reconnaissance or other RPV applications, brief details of which follow. Of these, the Merlin, Falcon, Heron, Heron HS, Mossette, Training Drone and SATS were previously listed erroneously in *Jane's* as products of AEL (which see), and were said to have been superseded. This is not so, although AEL is no longer the marketing agent for these systems, which continue to be available direct from RCS Guided Weapons Systems.

The standard systems offer payload capabilities of up to 13·6 kg (30 lb) and a visual control range of up to 2·7 nm (5 km; 3·1 miles). Non-standard systems can be produced with payload capabilities of up to 22·7 kg (50 lb). To simplify control, an autostabiliser is available, with options of height lock and heading lock. An autopilot is under development to permit automatic out-and-back sorties

RCS4020 Merlin composite target/camera system

RCS Falcon mini-RPV, developed from the Merlin

beyond, and returning to, visual control. All systems are controlled using a standard tripod-mounted transmitter, and each system operates in a slot with an approximate bandwidth of 25KHz and an ERP maximum of 10W.

RCS MERLIN

The Merlin system consists of a delta-shaped flying-wing drone, built of polystyrene foam with a plywood skin, powered by a 10 cc piston engine and fitted with a tricycle landing gear. A six-function pulse width modulated VHF guidance system with RF output of 5W is used.

Three versions are available, as follows:

RCS4018 Target System. For training air defence weapon crews to acquire and track target aircraft. Capable of simulating a full-size aircraft, with a mission response time of less than 4 minutes. For surveillance radar response, can carry X- and S-band radar enhancement devices. A 2·13 m (7 ft) drogue target can be towed on a 152 m (500 ft) cable for live firing target work.

RCS4019 Camera System. For short-range aerial reconnaissance with standard accessories. Radio-triggered Robot 35 mm camera, with f/2·8 lens, giving 18 high-definition photographs per loading at altitudes of up to 915 m (3,000 ft). Optional accessory kit for processing and printing film in approx 5 min from removal from aircraft.

RCS4020 Composite Target/Camera System. Total facilities equivalent to those of RCS4018 and 4019 combined.

LAUNCH AND RECOVERY: Orthodox T-O from ground or launch from system launcher; recovery by orthodox landing.

DIMENSIONS, EXTERNAL:
Wing span 1·45 m (4 ft 9 in)
Length overall 1·22 m (4 ft 0 in)

WEIGHTS:
Payload 1·4 kg (3 lb)
Max launching weight 4·8 kg (10·5 lb)

PERFORMANCE:
Max speed 78 knots (145 km/h; 90 mph)
Stalling speed 20 knots (37 km/h; 23 mph)
Max controllable range (optical) 1·6 nm (3 km; 1·9 miles)
Endurance 30 min

RCS FALCON

The Falcon is of generally similar appearance and construction to the Merlin, but is larger, has a greater payload and longer endurance, and is powered by a 3·7 kW (5 hp) engine. Guidance, launch and recovery are the same as for the Merlin system.

Three versions are available, as follows:

RCS4012 Target System. For training air defence weapon crews to acquire and track low-level high-speed aircraft. Mission response time less than 4 minutes. Can be equipped with X- and S-band radar enhancement devices. Facility for trailing a 3·05 m (10 ft) towed target on a 152 m (500 ft) cable. Supplied to Libyan government in 1973-74.

RCS4015 Camera System. For short-range aerial reconnaissance, for which it has same equipment as Merlin, including optional processing/printing kit. Target location by photo/map comparison. Supplied to Libyan government 1973-74.

RCS4017 Composite Target/Camera System. Combines total facilities of RCS 4012 and 4015.

DIMENSIONS, EXTERNAL:
Wing span 2·13 m (7 ft 0 in)
Length overall 1·98 m (6 ft 6 in)

WEIGHT:
Max launching weight 16·3 kg (36 lb)

PERFORMANCE:
Max speed 110 knots (204 km/h; 126·5 mph)
Stalling speed 18 knots (33 km/h; 21 mph)
Max controllable range (optical) 2·7 nm (5 km; 3·1 miles)
Endurance 45 min

Right: RCS4020 Heron target drone, ordered by the British Army

RCS4024 HERON

This RPV is of high-wing monoplane configuration, of simple and robust construction and powered by a 3·7 kW (5 hp) engine. Fuel capacity is 4·5 litres (1 Imp gallon). Although designed specifically for target work, its payload capability and 0·03 m³ (1 cu ft) payload bay make it equally suitable for other applications, and up to 5 kg (11 lb) of the total payload may consist of pylon-mounted underwing stores such as flares or small bombs. A good X-band radar signature is provided for tracking. Guidance, launch and recovery are the same as for the Merlin system. The Heron is controllable in winds gusting up to 30 knots (56 km/h; 34·5 mph).

A number of Herons have been ordered by the British Army.

DIMENSIONS, EXTERNAL:
Wing span 3·00 m (9 ft 10 in)
Wing chord (constant) 0·432 m (1 ft 5 in)
Length overall 2·13 m (7 ft 0 in)
Fuselage: Max width 0·254 m (10 in)
Tailplane span 1·00 m (3 ft 3¼ in)

WEIGHTS:
Max payload:
catapult launch 6·8 kg (15 lb)
ground T-O 13·6 kg (30 lb)
Max weight:
catapult launch 15·4 kg (34 lb)
ground T-O 24 kg (53 lb)

PERFORMANCE:
Max speed 104 knots (193 km/h; 120 mph)
Max controllable range (optical) 2·7 nm (5 km; 3·1 miles)
Endurance 30 min

RCS HERON HS

This is a high-speed version of the standard Heron, improved aerodynamically by reducing the wing thickness and area, introducing linear taper and eliminating wing struts and pylons. The wheeled landing gear is replaced by a sprung metal skid and catapult launching points, although a take-off dolly can be provided if a clear area is available, permitting a 25% increase in payload.

Dimensions are the same as for the standard Heron, except for a 2·44 m (8 ft 0 in) wing span. Max payload is 4·5 kg (10 lb) and max speed 121·5 knots (225 km/h; 140 mph).

RCS4041 MOSSETTE

The Mossette is a re-usable sweptwing target drone with an in-flight profile similar to that of wire-guided anti-tank missiles in current use, and is intended for training operators of helicopter- and vehicle-launched missiles of that type. It is powered by a 6 cc engine. At the bottom of the underfin is fitted a smokeless flare to enable the trainee operator to align the Mossette with the target, thus realistically simulating the operational missile. Immediately in front of the target is erected the system catching net in which the Mossette is retrieved for re-use. The nosecone, which is expendable, is replaced prior to the next exercise. Guidance system as for Merlin and other RCS RPVs.

High-speed RCS Heron HS

RCS Mossette sweptwing target drone

LAUNCH AND RECOVERY: Launch from system catapult; recovery with catching net.

DIMENSIONS, EXTERNAL:
Wing span 1·22 m (4 ft 0 in)
Length overall 1·07 m (3 ft 6 in)

WEIGHT:
Max launching weight 1·8 kg (4 lb)

PERFORMANCE:

Max speed	69 knots (128 km/h; 80 mph)
Stalling speed	18 knots (33 km/h; 21 mph)
Max controllable range (optical)	1·6 nm (3 km; 1·9 miles)
Endurance	20 min

RCS4030 TRAINING DRONE

Designed to train operators of operational drone/RPV systems, the RCS4030 is of broadly similar external appearance to the Heron RPV, but is slightly smaller and is powered by a 10 cc engine. It is fitted with a 6-channel proportional radio control designed for use with the standard tripod-mounted ground control system for the Merlin and other RCS RPVs.

DIMENSIONS, EXTERNAL:

Wing span	2·13 m (7 ft 0 in)
Length overall	1·83 m (6 ft 0 in)

WEIGHTS:

Payload	1·8 kg (4 lb)
Weight dry	5 kg (11 lb)

PERFORMANCE:

Max speed	52·1 knots (96·6 km/h; 60 mph)
Stalling speed	13 knots (24·1 km/h; 15 mph)
Endurance	35 min

RCS SATS

SATS (Small Arms Target System) is used as a live target for infantry anti-aircraft firing practice with rifles and sub-machine-guns. Controlled with a hand-held transmitter, it has an optical range fully satisfying its design operational needs.

DIMENSIONS, EXTERNAL:

Wing span	2·21 m (7 ft 3 in)
Wing chord at root	0·38 m (1 ft 3 in)
Wing chord at tip	0·25 m (10 in)
Length overall	1·98 m (6 ft 6 in)

WEIGHT:

T-O weight	4·1 kg (9 lb)

PERFORMANCE:

Max speed	74 knots (137 km/h; 85 mph)
Endurance	40 min

RCS SWIFT

The Swift is a photographic or target drone, of swept-wing aircraft configuration. It has a glassfibre fuselage, and plywood-covered wings with a styrofoam core.

LAUNCH AND RECOVERY: Launched from specially designed toggle action catapult, and flown along a fixed heading at a fixed angle of climb, using an optical tracking system.

GUIDANCE AND CONTROL: Basic system includes a tripod-mounted high-power radio transmitter (range 5·4 nm; 10 km; 6·2 miles) and optical sight. Drone can also be operated with 6-channel hand-held transmitter.

SPECIAL EQUIPMENT: 35 mm camera, barometric trip switch and marker flash.

DIMENSION, EXTERNAL:

Wing span	1·83 m (6 ft 0 in)

WEIGHTS:

Payload	0·9 kg (2 lb)
Max T-O weight	4·5 kg (10 lb)

PERFORMANCE:

Max speed	100-120 knots (185-222 km/h; 115-138 mph)
Range	4·3 nm (8 km; 5 miles)
Endurance	30 min

RCS SMALL ROTARY-WING DRONE

This small rigid-rotor drone is intended primarily as a target with which to train operators of radar-controlled missiles, and is powered by a 0·9 kW (1·2 hp) 10 cc piston engine. Fuel capacity 0·5 litres (0·1 Imp gallons).

LAUNCH AND RECOVERY: Conventional helicopter T-O and landing, using non-retractable tricycle landing gear.

GUIDANCE AND CONTROL: Six-channel proportional radio control system (four for pitch, roll, yaw and throttle control, and two for camera controls).

SPECIAL EQUIPMENT: Choice of half-frame (f/2·8) or full-frame (f/2·5) 35 mm camera, each with 18 exposures per wind and automatic film advance. Heavy-duty 12V electric starter optional.

DIMENSIONS, EXTERNAL:

Main rotor diameter	1·63 m (5 ft 4 in)
Tail rotor diameter	0·36 m (1 ft 2 in)
Length overall	1·83 m (6 ft 0 in)

WEIGHTS:

Payload	0·9 kg (2 lb)
Weight dry	5 kg (11 lb)

PERFORMANCE:

Max speed	61 knots (113 km/h; 70 mph)
Endurance	20 min

RCS LARGE ROTARY-WING RPV

This surveillance mini-RPV is powered by a 3·7 kW (5 hp) two-stroke piston engine, and the following details apply to this version. A twin-engined version, capable of carrying an 18 kg (40 lb) payload, was under test in early 1976.

LAUNCH AND RECOVERY: Conventional helicopter T-O and landing, using wide-track tubular skid landing gear.

RCS4030 Training Drone

RCS Small Arms Target System (SATS)

The RCS small rotary-wing drone, of mini-helicopter configuration (10 cc piston engine)

Single-engined version of RCS's large rotary-wing RPV, designed for surveillance use

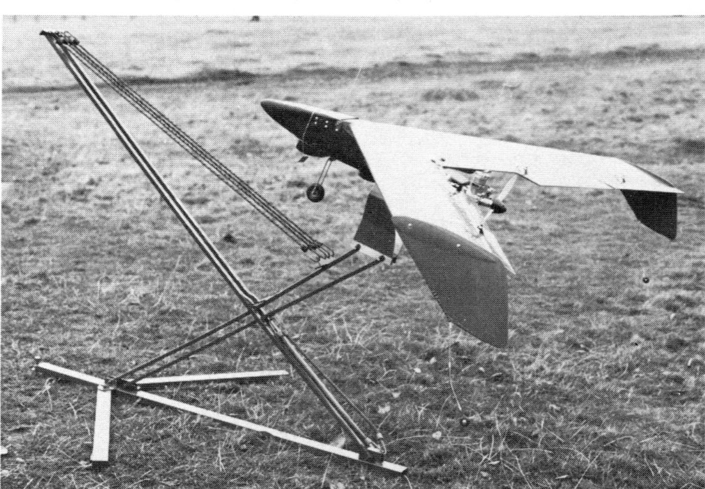

Right: RCS Swift sweptwing photo-reconnaissance or target drone, on its catapult launcher

SPECIAL EQUIPMENT: Standard payload comprises ground surveillance TV system, cameras, sensors and other equipment. Gyroscopic stabilisation optional.

DIMENSION, EXTERNAL:

Main rotor diameter	2·29 m (7 ft 6 in)

WEIGHTS:

Payload	8·5 kg (18·7 lb)
Max T-O weight	20 kg (44 lb)

PERFORMANCE:

Max speed	32 knots (60 km/h; 37 mph)
Guidance radius	2·7 nm (5 km; 3·1 miles)

SHORTS

SHORT BROTHERS & HARLAND LTD

ADDRESS:
PO Box 241, Airport Road, Belfast BT3 9DZ, Northern Ireland
Telephone: 0232 58444
Telex: 74688
OFFICERS: See Aircraft section

SHORTS SKYSPY

The Skyspy is a pilotless remotely controlled VTOL aerial reconnaissance vehicle, developed by Shorts as a private venture and suitable for use in military or naval applications in battlefield areas, other high-risk zones, or at sea. It is very small (see dimensions of the prototype vehicle which follow), structurally and mechanically simple, relatively inexpensive, and easily transportable. Its small size and low power also confer a number of operational advantages, such as low radar and infra-red signatures, low noise level, low gust response, and a very small visual silhouette, all of which contribute to a low damage risk under operational conditions. Surveillance can be carried out at all angles of attack between conventional forward flight and the hover mode.

A wide variety of applications is envisaged, including army reconnaissance; naval over-the-horizon viewing; weapon control and delivery, including the capability of providing a command link for over-the-horizon weapon control systems; target spotting; coastguard surveillance; border patrol and police duties; fishery protection; search and rescue operations; forest fire spotting; and emergency relief and medical support service. It can also be used for electronic countermeasures (ECM) and, mounting an airborne laser, as a guidance facility for missiles. As a substitute for full-size strike aircraft in the anti-armour role, a Skyspy system could pinpoint its target, accurately deliver a warhead, and immediately relay the results to its base without hazard to skilled personnel or expensive equipment. Similarly, in a naval application, it could deliver weapons against submarines or surface targets, or serve as a substitute for reconnaissance helicopters.

A development programme involving tethered hovering and forward flight trials began in 1975, and by the Spring of 1976 the prototype had completed some 70 hr of test flying, including a series of vertical take-offs and landings. The technique being developed by Shorts involves the use of a launching arm which supports the vehicle in the T-O attitude until it achieves the required vertical thrust and lifts free. Although the Skyspy can be landed vertically on level ground, the conditions likely to be met in service demand a technique that will cater for rough terrain. To meet this requirement, a 6·1 × 6·1 m (20 × 20 ft) landing net is employed, which can be transported easily and set up in virtually any type of terrain or on board ship. The vehicle is brought to the hover and then descends smoothly under power into the net, to be retrieved and refuelled for its next mission.

In its simplest form, one Skyspy and a complete ground control system can be carried on a Land-Rover with a two-man crew. A larger system, for main battlefield support duties, will consist of a suitable transporter, accommodating two Skyspys and spare fuel tanks, and mounting a lightweight crane to serve both for launching and for recovery after landing. A second transporter will provide a ground control centre to operate the RPV and to monitor the information fed back by it.

Having successfully completed the VTOL flight test phase, Skyspy was continuing its test programme in 1976, in which the safety restraints will be gradually eliminated in the progress towards unrestrained flight trials.

TYPE: Remotely controlled VTOL aerial reconnaissance and surveillance drone.

AIRFRAME: The basic vehicle consists of a centrebody carrying the engine, fuel, and control and stabilisation actuators; a low-pressure fan; and an axially symmetrical duct connected to the centrebody by an engine mounting spider and by stators. Aerofoil surfaces, for pitch, roll/yaw, and rotational stabilisation and control, are set across the duct exit and integrate in part with low aspect ratio wings located on the exterior of the duct. The centrebody comprises the major part of the vehicle weight, the duct being a simple, light but rigid structure. An equipment/payload pod fairing is located on the outer surface of the duct. The autopilot and power supply equipment are at the rear, in the wall of the duct.

POWER PLANT: Lift and propulsive thrust are obtained by vectoring the gross thrust output of a single-stage multi-blade fixed-pitch low-pressure ducted fan, powered by a small piston engine (in prototype, a two-stroke-in-line engine of 48·4 kW; 65 hp) and augmented by aerodynamic force components generated on the duct surfaces and intake lip.

OPERATIONAL EQUIPMENT: Operational payload can include TV camera, sensors, automatic data-gathering and other equipment, to operator's requirements, installed in pod fairing on the outer forward surface of the duct. The Skyspy is intended to be flown under remote control to the chosen surveillance area, where it can hover over a stationary target (or track a moving one) and relay positional details of the target in real time, using secure data links, to a ground- or ship-based controller.

TYPICAL DIMENSIONS AND WEIGHTS (48·4 kW; 65 hp engine):

Overall diameter	1·08 m (3 ft 6½ in)
Fan diameter	0·85 m (2 ft 9½ in)
Height overall	1·37 m (4 ft 6 in)
Weight empty	85 kg (187 lb)
Payload	20 kg (44 lb)
Fuel	25 kg (55 lb)
Max T-O weight	130 kg (286 lb)

PERFORMANCE:
Dependent upon operational requirements and payload. A vehicle of the size quoted could attain a max speed of 102 knots (190 km(h; 118 mph) and could operate at altitudes of up to 1,825 m (6,000 ft); typically, it could carry a 20 kg (44 lb) payload for a sortie of 1½ hours' duration.

Shorts Skyspy surveillance and general purpose RPV

SHORTS MATS-B

MATS (Model Aircraft Target System) has been developed under Ministry of Defence contract for use in practice firings of Shorts Blowpipe and other small anti-aircraft weapons.

AIRFRAME: Cantilever high-wing monoplane configuration, built of metal and glassfibre-reinforced plastics. Polyurethane foam-filled wings, fin and tailplane.

POWER PLANT: One 9·3 kW (12·5 hp) 123 cc single-cylinder piston engine, driving a two-blade propeller. Fuel tank in centre of fuselage.

LAUNCH AND RECOVERY: Launched pneumatically from vehicle-mounted rail launcher. Recovery by triple parachute or by conventional landing, using flares housing as a skid. Launching can be carried out into head-winds of up to 25 knots (46 km/h; 29 mph).

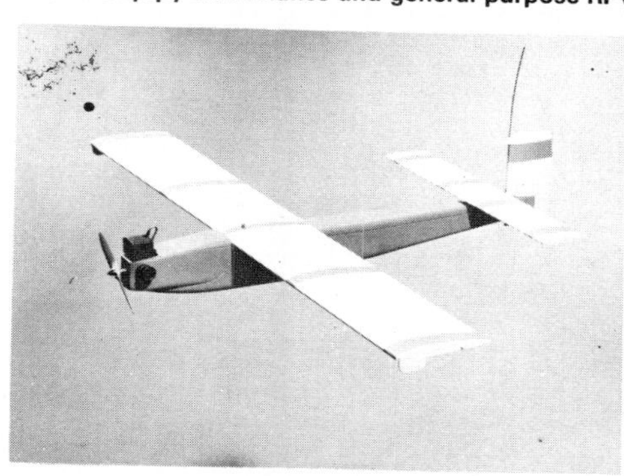
Shorts MATS-B miniature target aircraft

GUIDANCE AND CONTROL: Seven-channel radio command guidance system (hand-held transmitter), in the 68MHz band. Aerodynamic control by servo-assisted ailerons and all-moving tailplane. Roll stabilisation and automatic (barometric) height control standard.

SPECIAL EQUIPMENT: Tracking flares or smoke generators in underfuselage payload compartment. Radar reflector in fuselage. Stabilisation sensors in each wing, near tip. Parachute compartment in forward fuselage.

DIMENSIONS, EXTERNAL:

Wing span	3·35 m (11 ft 0 in)
Length overall	2·44 m (8 ft 0 in)

WEIGHTS:

Payload	6·8 kg (15 lb)
Max launching weight	45 kg (100 lb)

PERFORMANCE:

Max level speed	125 knots (231 km/h; 144 mph)
Min endurance	50 min

WESTLAND

WESTLAND HELICOPTERS LTD

HEAD OFFICE, WORKS AND AERODROME:
Yeovil, Somerset BA20 2YB
Telephone: Yeovil (0935) 5222
Telex: 46277
OFFICERS: See Aircraft section

Westland Helicopters Ltd began work on RPHs (remotely piloted helicopters) in 1968, with a series of studies covering configuration, electronics, vehicle performance, control systems and vulnerability. Experimental work began in 1972 using models for wind tunnel work and the measurement of radar, optical and infra-red signatures. At the same time simulated flights were made using a computer to check the stability and controllability of the RPH.

Much of this ground work had been completed by 1974, and was followed in mid-1975 by the start of a flight development programme, using a small flying testbed aircraft known as the Mote to prove the aerodynamic characteristics and to continue development work on the control system. This led to a Ministry of Defence development contract for the Wisp, a small RPH for the British Army.

By the Spring of 1976 Westland had been selected as the winner of the competition, referred to briefly under the Ministry of Defence heading in previous editions of *Jane's,* to develop a second RPH, known as Wideye, for battlefield reconnaissance for the British Army. This RPH is also being developed under contract.

WESTLAND WISP

A small number of Wisps are being built for trials pur-

Left: Third of the small batch of Westland Wisp remotely piloted helicopters being built for trials purposes; right: model showing the general configuration of the larger Wideye RPH

poses, and the third of these was displayed publicly for the first time at the Farnborough International air show in September 1976, when first details of the Wisp were also released. Flight testing was scheduled to begin later that month.

As the accompanying photograph shows, the Wisp has a small, flattened-sphere-shaped body, which is of glassfibre construction and contains the power plant and mission payload. A fixed, four-legged landing gear is fitted to the underside of the body. The power plant drives a pair of two-blade co-axial counter-rotating rotors. Directional control of the RPH is by differential changing of the collective pitch of the rotors. The Wisp can be carried in a standard Land-Rover vehicle, and the modular construction of the body enables a variety of payloads to be accommodated. These payloads can be removed or installed very quickly when a change of role is required.

DIMENSIONS, EXTERNAL:

Rotor diameter (each)	1·52 m (5 ft 0 in)
Body diameter	0·61 m (2 ft 0 in)
Body depth	0·41 m (1 ft 4 in)

WESTLAND WIDEYE

Wideye, which is at the design definition stage, is a larger and more advanced RPH, capable of carrying out a wide variety of operational roles. Like the Wisp, it has a co-axial rotor system, a similar remote control system, and low radar, noise and infra-red signatures.

The Electro-Optical Systems Division of Marconi-Elliott Avionics Systems Ltd has been appointed by the MoD as electronics systems contractor, to work in conjunction with Westland Helicopters on Wideye for the British Army. This work will draw substantially upon MEASL's experience of various electronic surveillance systems, including daylight and low-light sensors, stabilised sensor mountings, signal processors, secure data and command links, specialised displays, vehicle tracking and target location systems, and ground control stations.

THE UNITED STATES OF AMERICA

AERONUTRONIC FORD
AERONUTRONIC FORD CORPORATION
(Subsidiary of Ford Motor Company)

HEAD OFFICE:
728 Parklane Towers East, One Parklane Boulevard, Dearborn, Michigan 48126
Telephone: (313) 594-0698

AERONUTRONIC DIVISION:
Ford Road, Newport Beach, California 92663
Telephone: (714) 640-1500
Telex: 678470

OFFICERS:
See Air-Launched Missiles section

DIRECTOR OF PUBLIC AFFAIRS:
Donald E. Flamm

In addition to its work on guided missiles, Aeronutronic Ford (formerly Philco-Ford) developed and manufactured the low-cost rocket-powered LOCAT air target, of which details were given in the 1972-73 *Jane's*. The change to the present company name was made in the Spring of 1975.

Under various Department of Defense contracts beginning in March 1972, Aeronutronic Ford has designed, developed and produced small, low-cost mini-RPVs for battlefield reconnaissance, target acquisition and laser designation. Both day and night versions have been and continue to be developed, and details of the earlier Praeire and Calere programmes appeared in the 1974-75 *Jane's*.

Currently, Aeronutronic Ford's Aeronutronic Division at Newport Beach, California, is engaged in a general programme, on behalf of the US Air Force Aeronautical Systems Division, of advanced studies for small RPV airframes, engines and payloads.

Aeronutronic Ford Praeire IIB mini-RPV, which has demonstrated successful anti-tank operation in conjunction with the US Army's CLGP (Cannon Launched Guided Projectile)

AERONUTRONIC FORD PRAEIRE IIB

Praeire IIB (the name is Latin for 'leading soldier') is the current daytime version of Aeronutronic Ford's mini-RPV, and features all-glassfibre/polyurethane foam construction. The conventional wing/tail/body design utilises a mid-fuselage pusher engine configuration to clear the nose area for the TV sensor payload. Power plant is a Kolbo Korp 6 kW (8 hp) flat-twin engine. Guidance is by radio command. An Earth's electric field autopilot provides attitude control when out of sight of the pilot.

The TV sensor package, also developed by Aeronutronic Ford, contains a high resolution camera with 4 : 1 zoom capability sightline. Pointing and stabilisation are provided by a coelostat mirror optical system which provides 360° azimuth and +5°/−75° evelation coverage. An International Laser Systems miniature laser designator provides target ranging and designation through a common optical path with the TV. Total weight of the sensor system is 11·3 kg (25 lb). The Praeire IIB carries the sensor system in a nose module which rotates 180° after runway or catapult take-off of the aircraft to place the optical dome underneath for viewing. The dome is made of plastics.

In 1975, Praeire IIB successfully acquired and designated a tank target at the White Sands Missile Range, which then received a direct hit from a US Army CLGP (Cannon Launched Guided Projectile). Praeire IIB loitered near the tank at a height of approx 457 m (1,500 ft); the CLGP was fired from a 155 mm howitzer located 4·3 nm (8 km; 5 miles) from the target.

Aeronutronic Ford has also developed an electronic warfare payload for the Praeire for tactical communications interception and jamming.

DIMENSIONS, EXTERNAL:

Wing span	3·96 m (13 ft 0 in)
Length overall	3·35 m (11 ft 0 in)

WEIGHT:

Max T-O weight	61·2 kg (135 lb)

PERFORMANCE:

Max level speed at 1,525 m (5,000 ft)	74 knots (137 km/h; 85 mph)
Stalling speed at 1,525 m (5,000 ft)	33 knots (62 km/h; 38 mph)
Ceiling	3,050 m (10,000 ft)
Mission radius	Radio line-of-sight
Mission endurance	4 hr

AERONUTRONIC FORD CALERE

Night versions of Praeire are called Calere and utilise FLIR for full night-time capability. To meet the needs of future mini-RPVs, Aeronutronic Ford has developed a miniature FLIR sensor which weighs only 2·7 kg (6 lb) and includes three fields of view.

AERONUTRONIC FORD CALERE III

This mini-RPV is being developed by Aeronutronic Ford under an ARPA contract for a LANDSS (Lightweight Advanced Night/Day Surveillance System) vehicle. Two prototypes have been ordered, each powered by a 3·7 kW (5 hp) Kolbo Korp engine; one prototype will be equipped to carry a 9·1 kg (20 lb) payload which will include a lightweight FLIR (forward-looking infra-red) package, optical surveillance devices, and an International Laser Systems laser target dsignator. Calere III has a wing span of 2·74 m (9 ft 0 in) and AUW of 27·2 kg (60 lb). Flight testing was due to begin in 1976.

AMR—*See Addenda*

APL
APPLIED PHYSICS LABORATORY, JOHNS HOPKINS UNIVERSITY

ADDRESS:
8621 Georgia Avenue, Silver Spring, Maryland 20910

APL RPD2

This drone, developed from an earlier US Army programme vehicle by the Applied Physics Laboratory of Johns Hopkins University, underwent testing in 1975 by the US Naval Surface Weapons Center as a miniature target for the evaluation of improved shipboard fire control systems.

POWER PLANT: One 9 kW (12 hp) McCulloch MC-101A single-cylinder piston engine.

LAUNCH AND RECOVERY: Launch from small trolley, by rubber cord, or from larger catapult system. Parachute recovery system.

GUIDANCE AND CONTROL: System includes autopilot which uses Earth's electrostatic field for pitch and roll stabilisation.

SPECIAL EQUIPMENT: Includes passive radar augmentation devices housed in nose and in fairing on fin trailing-edge.

WEIGHTS:

Payload	10 kg (22 lb)
Max launching weight	36·3 kg (80 lb)

PERFORMANCE:

Max level speed	150 knots (278 km/h; 173 mph)
Cruising speed	65 knots (120 km/h; 75 mph)
Service ceiling	5,500 m (18,000 ft)
Endurance	1 hr 30 min

BEECHCRAFT
BEECH AIRCRAFT CORPORATION

HEAD OFFICE AND MAIN WORKS:
Wichita, Kansas 67201
Telephone: (316) 689-7111

OFFICERS: See Aircraft section

In addition to manufacturing piloted aircraft, Beech has been designing and producing pilotless target drones of various types since 1955. By 1 February 1976 it had built a total of 5,285 target drones. This figure includes more than 2,200 Model 1001s, of which more than 1,360 were MQM-61As for the US Army. Production of the Models 1001 and 1025 (1972-73 *Jane's*) has ended, but targets on hand continue to support various military services.

Beech has also carried out design studies to examine the

suitability of its range of light aircraft for operation under remote control.

BEECHCRAFT MODELS 1019, 1072, 1088, 1094 and 1095

US military designation: AQM-37A (Model 1019)
UK designations: SD.2 Stiletto (Model 1072) and Model 1095
Italian designation: Model 1088
French designation: Vanneau (Model 1094)

Winner of a 1959 US Navy/Air Force design competition, the Beechcraft Model 1019 (US Navy AQM-37A, formerly KD2B-1) target system is designed to simulate aircraft and missile threats, and to provide defence weapon system evaluation and operational crew training.

The complete target system includes a launcher which is adaptable to a variety of fighter aircraft, test and checkout equipment, handling and servicing equipment and launch aircraft controls, as well as the target vehicle itself. The target is expendable and thus requires no recovery support.

The target provides both active and passive radar augmentation for radar acquisition and lock-on. A chemical flare is provided for missions which require infra-red augmentation. Two optional miss-distance indication systems are available.

Flight termination is normally through aerodynamic means, but an explosive destructor system is available, to provide additional range safety and operational flexibility. The only procedures required to ready the target for flight are decanning, battery servicing, pre-flight checking out, pressure cartridge inserting and nitrogen pressurising.

The AQM-37A was launched successfully for the first time on 31 May 1961, at the Naval Missile Center, Point Mugu, California. In subsequent development tests, after being launched at 10,050 m (33,000 ft) from an F-3B Demon, it flew higher and faster than any previous drone developed for target duties. During weapon system training operations at Point Mugu in the Spring of 1965, an AQM-37A, launched from an F-4B at a speed of Mach 1·3 at 14,300 m (47,000 ft), climbed to 27,750 m (91,000 ft) and maintained a speed of Mach 2·8.

The AQM-37A has been operational since 1963 from shore installations and aircraft carriers and is being launched at present from three types of US Navy aircraft, the F-4, F-8 and A-4. Beech was awarded a follow-on contract from the US Navy on 19 December 1972 for a further 202 AQM-37A targets, and production of these was completed in 1974.

In the Spring of 1968, Beech received contracts to modify 10 AQM-37As for evaluation by the USAF and three for the US Army, making their electronics and destruct systems compatible with current advanced weapon systems and range requirements of these services.

In late 1974 the US Naval Air Systems Command awarded Beech a contract to modify 10 AQM-37A missile targets to perform low-altitude flight missions. This classified Navy programme, designated **Sea Skipper** (formerly Sea Skimmer), called for development and modification of missile targets capable of being air-launched over the sea at 366 m (1,200 ft) altitude and then flown as low as 15 m (50 ft) over predetermined points. The purpose of the programme is to demonstrate the feasibility of modifying AQM-37A targets for use in the evaluation of ship defence systems and the training of crews. By the end of 1975, Beech had delivered all 10 for flight evaluation at the Naval Missile Center, Point Mugu, California, and the Navy had exercised an option to purchase a further 10, of which deliveries were scheduled to begin in March 1976.

In 1975 the US Army contracted with Beech to modify 10 government-furnished AQM-37As for weapon systems evaluation. Deliveries of these also were due to begin in March 1976.

In early 1976 the US Army ordered 28 recoverable AQM-37As (the first with such capability), fitted with a two-stage parachute system. Under this contract Beech will produce two versions of the AQM-37A. One will be a supersonic high-altitude target capable of operating at up to 1,181 knots (2,188 km/h; 1,360 mph) and 21,335 m (70,000 ft) altitude; the other a low-altitude modification which can be flown to within 55 m (180 ft) of the terrain.

Deliveries were scheduled to begin in November 1976.

In addition to the AQM-37A variants built for the US services, four other models have been announced:

Model 1072 (Shorts SD.2 Stiletto). Version for UK, substantially re-engineered by Beech and Short Bros & Harland to meet British requirements, including virtually complete replacement of the radio and radar systems and control system changes. Total of 75 ordered, fitted by Beech with a single-chamber rocket motor. Deliveries of the last 20 of these targets, the contract for which was announced in August 1974, were scheduled for completion in January 1976. The Stiletto is launched from Canberra PR Mk 3 aircraft. In a successful first test flight at Llanbedr on 1 August 1968, the drone was released at 16,750 m (55,000 ft) and flew for more than 28 nm (52 km; 32 miles) at an average speed of Mach 1·4 before the flight was terminated by a commanded explosive detonation.

Principal modifications made by Shorts were the incorporation of a British EMI T44/1 telemetry system; provision of additional 15V flight break-up system (WREBUS); installation of radioactive miss-distance indicator (RAMDI), with associated radio link; introduction of Plessey IR 112A/IR 310 telecommand system and heading and turns command circuitry; modification to propulsion system to give Mach 2 performance at 18,300 m (60,000 ft); and changes in the radar augmentation system. The current Stiletto system consists of the basic target vehicle plus a number of optional mission kits which can be either installed by Shorts or delivered separately for customer installation.

Beechcraft Model 1019 AQM-37A supersonic target drone

Model 1094 Vanneau version of the Beechcraft AQM-37A, modified by Matra for the French Air Force

Model 1088. Manufacturer's designation of five targets supplied to Italy. Intended for air-launch from F-104S aircraft of the Italian Air Force.

Model 1094 (Matra Vanneau: Lapwing). Fifteen Model 1094 targets ordered in mid-1973 by the French Air Force, deliveries of which began about a year later. A follow-on contract for 30 was announced in August 1974; these are being modified to French Air Force requirements by Matra. The first of this second batch was launched from Cazaux on 22 January 1975, and the first interception of a Vanneau by a Matra Super 530 missile was made on 28 May 1975.

Model 1095. The British Ministry of Defence in August 1974 ordered 10 Model 1095 targets. These are being modified by Shorts to MoD specifications and will be used for crew training exercises on the Hebridges range.

Production of the drone was transferred from Wichita to Beech's Aerospace Division at Boulder, Colorado, in 1968. Total deliveries have been made to date of more than 2,800 targets of all versions.

TYPE: Supersonic air-launched expendable or recoverable target drone.

WINGS: Cantilever mid-wing monoplane of cropped-delta planform, mounted at rear of fuselage. Modified double-wedge wing section. No dihedral or incidence. Sweepback on leading-edge 76°. Full-span ailerons.

FUSELAGE: Cylindrical centre-fuselage, with ogival nose section and tapering rear section over rocket chambers. Underbelly tunnel for rocket-engine cartridge-operated start valves, plumbing, infra-red flare and miss-distance scoring system antenna.

TAIL UNIT: Fixed endplate fins on each wingtip. Canard foreplane control surfaces of modified double-wedge section.

POWER PLANT: One Rocketdyne/AMF LR64 P-4 two-chamber liquid-propellant rocket engine (2·81 kN; 631 lb st). Three propellant tanks, for nitrogen pressurant, mixed amine fuel (MAF-4) and IRFNA oxidiser, form integral part of centre-fuselage.

GUIDANCE: Programmed guidance system.

DIMENSIONS, EXTERNAL:	
Wing span	1·00 m (3 ft 3½ in)
Wing chord at root	1·98 m (6 ft 6 in)
Wing chord at tip	0·53 m (1 ft 9 in)
Length overall	3·82 m (12 ft 6½ in)
Height overall	0·51 m (1 ft 8 in)
Diameter of fuselage	0·33 m (1 ft 1 in)
AREAS:	
Wings (exposed)	0·87 m² (9·35 sq ft)
Ailerons (total)	0·088 m² (0·95 sq ft)
Fins (total)	0·39 m² (4·20 sq ft)
Foreplanes (total, exposed)	0·071 m² (0·76 sq ft)
WEIGHT:	
Max launching weight	256 kg (565 lb)
PERFORMANCE (rated):	
Operating speed	Mach 0·4 to Mach 3·0
Operating height	300-24,385 m (1,000-80,000 ft)
Endurance	5-15 min
Range	more than 100 nm (185 km; 115 miles)

BEECHCRAFT MODEL 1070 HAST

Beech Aircraft is continuing a development programme for the USAF Armament Laboratory at Eglin AFB, Florida, to provide a high-performance air-launched aerial target system for use by the three Services of the US Department of Defense. The HAST (High Altitude Supersonic Target) is a continuation of the former Sandpiper project, which concluded a successful flight test programme in 1968 and was described in the 1970-71 *Jane's*. A hybrid propulsion system and a command manoeuvring system were demonstrated in that flight test programme. In 1971 Beech received authority to build 12 flight test units of the HAST and 13 refurbishment kits, and delivery of the former began in 1972. A recovery system development programme, utilising an inert HAST vehicle, has been completed. This programme has proved the recovery system, which uses a 14 m (45 ft) diameter parachute, with both water and aerial retrieval of the vehicle.

In August 1974, under a $2·9 million contract, Beech and the US Air Force initiated a 28-month programme to demonstrate flight performance. By the end of 1975, flight had been demonstrated successfully at up to Mach 3·0 at 24,385 m (80,000 ft). Subsequent missions in 1976 were planned to demonstrate flight at up to Mach 4·0 and 27,432 m (90,000 ft).

The flight performance envelope of the HAST covers a range from Mach 1·2 at 12,200 m (40,000 ft) to Mach 4·0 at 30,500 m (100,000 ft). The target is designed to be air-launched at speeds of Mach 1·2 to 2·5. Manoeuvres of between 5*g* at 10,670 m (35,000 ft) and 1·15*g* at 27,400 m (90,000 ft) are to be performed. The vehicle is to be capable of performing 'S' and 180° turns in the horizontal plane and altitude changes in the vertical plane. Manoeuvres can be pre-programmed or initiated via ground radio link.

Modular payloads with a wide variety of options will be available for accurate simulation of aircraft or missile threats. Payloads will include various radar and infra-red augmentation devices, as well as a flare/chaff dispenser. Vector miss-distance scoring systems will also be included.

The modular recovery system has been developed as an optional feature for mid-air retrieval of HAST, or for land or water recovery. Refurbishment of recovered targets will permit their re-use.

TYPE: Supersonic air-launched target drone.

WINGS: Clipped delta planform, with a sweepback of 75° and a constant thickness except for the leading-edge. Full-span aileron on trailing-edge of each wing.

FOREPLANES: Arrow planform canards, mounted on forward portion of fuselage, for longitudinal control.

FUSELAGE: Cylindrical body, with a 3·5 calibre von Kármán nose section and a conical boat-tail section.

TAIL UNIT: Fixed endplate vertical stabiliser on each wingtip.

POWER PLANT: Hybrid rocket engine developed by Chemical Systems Division of UTC. Propellant is polybutadiene and polymethyl-methacrylate, with inhibited red fuming nitric acid oxdiser. The system is inherently safe, since the propellants will not burn unless external ignition is applied. The engine is throttleable, with thrust variable from 0·53 to 5·34 kN (120 to 1,200 lb st). The 0·33 m (13 in) thrust chamber forms an integral part of the fuselage assembly. Oxidiser pressurisation and electrical power are provided by a ducted power unit. This unit, developed by the Marquardt Company, is powered by a ram-air turbine with air intake and exit on the lower side of the fuselage mid-section. A free siphon device has been developed by Beech to provide uninterrupted oxidiser flow during manoeuvres. Manoeuvring requirements dictate a positive expulsion system for the oxidiser.

GUIDANCE: Pre-programmed, with ground command interface.

DIMENSIONS, EXTERNAL:

Wing span	1·02 m (3 ft 4 in)
Length	5·08 m (16 ft 8 in)
Height (stabiliser)	0·66 m (2 ft 2 in)
Body diameter	0·33 m (1 ft 1 in)

AREAS:

Wings (total exposed)	0·97 m² (10·44 sq ft)
Foreplanes (total exposed)	0·12 m² (1·28 sq ft)
Stabilisers (each)	0·31 m² (3·38 sq ft)

VOLUME:

Payload volume	0·041 m² (2,500 cu in)

WEIGHTS:

Launching weight	519 kg (1,145 lb)
Propellant	297 kg (655 lb)
Payload	38 kg (85 lb)

PERFORMANCE:

Endurance at Mach 3	5 min

Model of the Beechcraft Model 1070 High Altitude Supersonic Target (HAST)

Beechcraft MQM-107 variable-speed training target for the US Army's Missile Command

BEECHCRAFT MODEL 1089 VSTT

US Army designation: MQM-107A Streaker

Beech Aircraft Corporation took part, with Northrop Ventura (which see), in a 'price and performance' competition to design and develop a Variable-Speed Training Target (VSTT) for the US Army's Missile Command. The programme consisted basically of the following three phases: (1) design and manufacture of hardware; (2) contractor flight tests; and (3) Army evaluation tests. Beech was announced the winner of this competition in the Spring of 1975.

The principal function of the VSTT is to tow a variety of tow targets for missile training and evaluation. Two TA-8 radar augmentation or infra-red augmentation targets can be carried on each mission and towed separately up to 2,440 m (8,000 ft) behind the VSTT.

The VSTT will serve as an aerial target for air defence systems such as Chaparral, Redeye, Hawk and Stinger, and is expected to become the primary subsonic missile training target for the US Army. It will be capable of operating at altitudes from 90 m (300 ft) to 12,200 m (40,000 ft) and at speeds of more than 500 knots (926 km/h; 575 mph).

The Beech MQM-107A is powered by a Teledyne CAE 372-2 (J402-CA-700) turbojet engine of 2·85 kN (640 lb st) and carries 242 litres (64 US gallons) of fuel. It is launched from the ground with a JATO booster and has a drogue and main parachute command recovery system.

A total of 49 flights in the contractor development test programme and US Army evaluation programme were completed in 1974. In these flights, high altitude, speed, flight control and towing capability were demonstrated, as well as operation of the recovery system. Recoveries have been made with minimal damage to the crushable, impact-absorbing nose.

On 4 April 1975 Beech announced the award of a $7·7 million initial contract for production and contractor operation of its winning design, and these are being built at Beech's Boulder, Colorado, Division. First flight of a production MQM-107A was made in April 1976. It is estimated that 317 of these targets will eventually be required by the US Army, and by September 1976 Beech had received contracts totalling more than $18 million for the production of 176 MQM-107As, with deliveries extending to December 1977.

The modular design employed throughout the system provides for ease of manufacture and economy in operation and maintenance. One unique design feature is the flat aerofoil section of the wing and tail surfaces. The fuselage is cylindrical. Low-cost bonded honeycomb is used for the non-moving aerofoil surfaces; control surfaces have an aluminium skin and are foam-filled. Support equipment is minimal and lightweight, providing for easy transportability and deployment. The checkout and launch equipment is accommodated in two suitcase-size containers.

The guidance and control system provides for both ground control and pre-programmed flight. The flight control operator is provided with all pertinent flight information by radio link from sensors located in the vehicle, can command both manoeuvre and recovery of the ve cle. In flight the guidance and control system automatic stabilises about the roll, yaw and pitch attitudes and p vides altitude and speed hold modes.

GUIDANCE: Radio command.

DIMENSIONS, EXTERNAL:

Wing span	3·00 m (9 ft 10
Length	5·13 m (16 ft 10
Height (total)	1·47 m (4 ft 10
Body diameter	0·38 m (1 ft 3

AREAS:

Wings (total projected)	2·52 m² (27·16 sc
Horizontal tail surfaces	0·79 m² (8·48 sc
Vertical tail surfaces	0·43 m² (4·66 sc

WEIGHTS:

Launching weight (incl booster)	460 kg (1,014
Usable fuel	173 kg (382

PERFORMANCE:

Operating speed range
247-499 knots (459-925 km/h; 285-575 m

Operating height range	S/L to 12,200 m (40,00
Endurance	more than

BEECHCRAFT MODEL 1089E TEDS

Beech is one of two US companies selected by USAF to develop and produce flight demonstr examples of a new Tactical Expendable Drone Sy (TEDS). The contract, announced in June 1975, is w $1·7 million, and covers a 21-month validation p ramme which will include four months of flight testing definition of the system in its tactical role.

A brief account of the TEDS programme is given under the US Air Force heading in this section. Its primary goals are the maximum use of effective, existing RPVs at the lowest possible life-cycle costs, and long storage life. In the USAF role, TEDS would be deployed in large numbers on one-way missions in support of tactical forces.

Basis of the Beech entry is the Model 1089 VSTT drone vehicle, with a Teledyne CAE turbojet power plant and a payload developed by Sanders Associates. Prototypes of a competitive design have been ordered from Northrop (which see), and at completion of the fly-off phase the USAF will decide whether to proceed with full-scale engineering development by a single contractor, after which a production decision will be taken.

Modified Model 1089E being entered by Beech for the USAF's TEDS programme

BOEING

BOEING AEROSPACE COMPANY

ADDRESS:
PO Box 3999, Seattle, Washington 98124

COMPASS COPE PROGRAMME DIRECTOR:
Donald B. Jacobs

Under the Compass Cope programme sponsored by the US Air Force (Aeronautical Systems Division) and the National Security Agency, Teledyne Ryan (which see) and Boeing Aerospace Company received contracts to build prototype RPVs for competitive evaluation. The Boeing vehicle has the USAF designation YQM-94A.

Design studies have been or are being conducted by Boeing of other RPV configurations, among which have been reported a low-level multi-purpose penetration vehicle, for photographic reconnaissance or weapons delivery, and a high-altitude reconnaissance vehicle having high aspect ratio wings reinforced with carbon fibre or other composite materials.

BOEING B-GULL (COMPASS COPE B)

USAF designation: YQM-94A

Together with the Teledyne Ryan Model 235/YQM-98A, the Boeing YQM-94A competed in the US Air Force's Compass Cope programme to select a high-altitude, long-endurance RPV. The Boeing design was selected for continued (pre-production) development on 27 August 1976 (see Addenda).

Performance goals set for the Compass Cope flight demonstration programme included an altitude of more than 16,765 m (55,000 ft), endurance of more than 20 hr, and payload of 317·5-680 kg (700-1,500 lb). These parameters were to allow an operational vehicle to perform the following types of mission: battlefield reconnaissance; signal intelligence gathering; communications relay; photographic reconnaissance; ocean surveillance; and atmospheric sampling. The USAF has the eventual Compass Cope vehicle under consideration as a data relay platform for manned strike aircraft, as part of its Precision Location Strike System (PLSS). The US Navy also is monitoring the Compass Cope programme, with a view to its possible application to ocean surveillance.

The production Compass Cope may also be used by the USAF to replace the RB-57 in its Pave Nickel programme for monitoring radar emissions along the western borders of the German Democratic Republic. Another typical application which has been quoted is that of patrolling areas of the Arctic Ocean to monitor firings from the northern missile test site of the USSR, a task at present carried out by Boeing RC-135 manned aircraft flying from Elmendorf AFB, Alaska.

In mid-1976 Boeing, in association with Dornier GmbH, was co-operating with the Federal German government to assess the role which Compass Cope might play in that country's overall defence system.

Boeing Aerospace design studies for a vehicle of this type began in September 1970, and a prototype contract was awarded by the USAF on 15 July 1971. Two YQM-94A prototypes were ordered, of which the first made a successful first flight on 28 July 1973. On 4 August 1973, on its second flight, it was destroyed in a crash following a failure of the lateral accelerometer which caused the ground-based pilot to lose control. The second aircraft was flown on 2 November 1974. During its second flight, on 23/24 November 1974, this aircraft completed an endurance trial of 17 hr 24 min over the southern California desert, attaining altitudes of more than 16,765 m (55,000 ft) and meeting most major objectives of the flight test programme. During high-altitude manoeuvres it encountered and survived a temperature of −70·6°C.

Completion of four demonstration flights was announced by the USAF on 27 November 1974. Systems engineering studies were then planned to determine design changes necessary to produce an operational Compass Cope RPV.

Both Compass Cope contenders take off and land using conventional runway techniques. An all-weather automatic landing system may become part of the future Compass Cope system.

The following description applies to the YQM-94A prototypes:

Second prototype of the Boeing Compass Cope B high-altitude strategic RPV

Artist's impression of Compass Cope B, showing nose TV camera

TYPE: Experimental high-altitude long-endurance strategic RPV.

WINGS: Cantilever shoulder-wing monoplane. Constant-chord centre-section, slight sweepback on outer wing leading-edges. Aluminium skin, with bonded glassfibre honeycomb core. Airbrakes and ailerons on each trailing-edge; no trailing-edge flaps or leading-edge lift devices.

FUSELAGE: Semi-monocoque structure of basically circular section, tapering towards rear. Aluminium longerons, glassfibre bulkheads and glassfibre honeycomb skin.

TAIL UNIT: Cantilever unit, of similar construction to wings. Tailplane indexed in line with wings. Twin endplate fins and rudders, the former having small fore-and-aft pointed fairings at the base. Full-span elevator, with tabs.

LANDING GEAR: Retractable tricycle type, basically that of an Aero Commander, with single wheel on each unit. All units retract rearward, the main units into fairings which project aft of the wing trailing-edges.

POWER PLANT: Prototypes each powered by one General Electric J97-GE-100 non-afterburning turbojet engine, rated at 23·4 kN (5,270 lb st), installed in a pylon-mounted pod above the fuselage, in line with the wings. Alternative power plants under consideration include turbofan engines of Avco Lycoming, Garrett-AiResearch or General Electric (TF34) design. Fuel in integral tanks occupying full span of wings. Provision for restarting engine in flight.

GUIDANCE AND CONTROL: Electronics module, located in fuselage forward of wings, is removable as a complete unit. On-board instrumentation, developed by Sperry Flight Systems, includes an integrated flight control system with internally-generated ILS (and, on proposed production version, automatic detection and correction of flight control malfunctions); a redundant stabilisation system; and an APW-26 airborne transceiver and other data link equipment. A TV camera is mounted in the undernose fairing; in the second protoype a heater is provided to prevent frosting of the camera glass. Antenna bay in rear lower portion of fuselage. Ground control via a command module which embodies standard cockpit instrumentation, TV screen and navigation display; a data transmission system, to permit the return of video signals; and a TPW-2A X-band radar van.

OPERATIONAL EQUIPMENT: Apart from the nose-mounted TV camera, details of other operational equipment are classified. All sensors and antennae are housed in the lower half of the fuselage, the payload being located just forward of the electronics module.

DIMENSIONS, EXTERNAL:

Wing span	27·43 m (90 ft 0 in)
Wing aspect ratio	16·7
Length overall (excl nose probe)	12·19 m (40 ft 0 in)
Wheel track	6·40 m (21 ft 0 in)

WEIGHTS:

Weight empty	2,494 kg (5,500 lb)

Payload for 24 hr mission	317·5 kg (700 lb)
Max T-O weight	6,531 kg (14,400 lb)

PERFORMANCE (at max T-O weight):

Cruising speed at altitudes from 15,240 to 21,340 m (50,000 to 70,000 ft)	Mach 0·5 to 0·6

BOEING ADVANCED RPV

In April 1975, Boeing was one of three aerospace companies to be awarded a one-year fixed-price study contract by the US Air Force to design an advanced RPV (ARPV) capable of carrying out reconnaissance, electronic warfare and srike missions, for service in the 1980s. The Boeing contract, worth $646,750, covers the complete system, including the RPV, ground-operated controls, recovery elements and support systems. Objectives of the study include improved cost-effectiveness and rapid missior turnaround.

CELESCO
CELESCO INDUSTRIES INC

ADDRESS:
3333 Harbour Boulevard, Costa Mesa, California 92626
Telephone: (714) 546-8030

On behalf of the US Air Force's Avionics Laboratory, Celesco developed a small decoy glider known as the Maxi-Decoy, and a larger rocket-powered development known as Propelled Decoy, to be carried by US strike aircraft. The decoys, which are carried on the aircraft's external stores points, have small, spring-loaded wings, with some 30° of sweepback when extended, which are opened in flight after the decoy has been released in the target area.

No answer has been received to enquiries made during the past two years, but the following reported descriptions are believed to be substantially correct:

CELESCO MAXI-DECOY

Unpowered decoy glider, having a square-section body and flip-out sweptback wings. Capable of high subsonic speed. Under development by USAF primarily as an active or passive ECM carrier, with payload test flights scheduled during 1975, using an F-4 Phantom carrier aircraft. Three Maxis can be accommodated in each of the F-4's four underfuselage Sparrow missile recesses, and two can be carried on each 340 kg (750 lb) underwing hardpoint; several decoys can be deployed simultaneously.

Electronic warfare payloads are designed and manufactured by the Electromagnetic and Aviation Systems Division of RCA. Ten payloads scheduled for testing at Eglin AFB during 1975. Typical payloads include a C-band jammer; or a G-band jammer combined with radar signature augmentation.

DIMENSIONS:

Wing span	approx 0·91 m (3 ft 0 in)
Body width	0·13 m (5 in)
Length	1·14 m (3 ft 9 in)
Payload volume	0·008 m³ (500 cu in)

WEIGHT:

Launch weight	59 kg (130 lb)

PERFORMANCE:

Max speed	Mach 0·8-0·9

CELESCO PROPELLED DECOY

Larger than Maxi-Decoy, with circular-section body, flip-out sweptback wings and cruciform sweptback tai surfaces; powered by an Atlantic Research solid propellant rocket motor. Flight testing began in 1974 Two-axis (roll and heading/altitude) autopilot allows th decoy to manoeuvre at high cruising speeds as well a performing in glide modes; unlike the Maxi, the Propelle Decoy can cruise at high subsonic speeds. It was buil under a $1·5 million USAF contract, and six successfu launchings from an F-4 carrier aircraft were made at Eglir AFB in 1974. One Propelled Decoy can be carried on eacl of the Phantom's 340 kg (750 lb) underwing hardpoints Payload definition studies by RCA were under way i early 1975, primarily in the field of active and passiv ECM.

DIMENSIONS:

Wing span	1·40 m (4 ft 7 in
Body diameter	0·25 m (10 in
Length	2·24 m (7 ft 4 ir
Payload volume	0·03 m³ (1,800 cu ir

WEIGHT:

Launch weight	136 kg (300 lt

PERFORMANCE:

Burn time	5 mi
Max speed	Mach 0·8-0·

DSI
DEVELOPMENTAL SCIENCES INC

ADDRESS:
15747 East Valley Boulevard, PO Box 1264, City of Industry, California 91749
Telephone: (213) 330-6865
PRESIDENT: Dr Gerald R. Seemann

DSI designed and built, under contract to NASA (which see), a prototype yawed-wing research aircraft, which made its first flight in mid-1976.

The company has also designed and built the Sky Eye and Scout mini-RPVs, and is designing and developing, for LMSC, the Aquila, based on the Sky Eye.

DSI/NASA OBLIQUE WING REMOTELY PILOTED RESEARCH AIRCRAFT

As indicated under the NASA heading, the design and development of this RPRA (Remotely Piloted Research Aircraft) was sponsored by NASA's Ames Research Center; detail design and development was subcontracted to DSI.

Design guidelines called for a flying-wing type of vehicle, with control achieved principally by use of the elevons, able to fly at wing yaw angles between 0° and 45°. A horizontal tail was added to provide longitudinal stability. The aircraft can be flown in the zero-to-moderate yaw configuration without the horizontal tail, but is statically unstable without it in the 45° yaw configuration.

Detail design started on 5 September 1972, and construction of the prototype by DSI began on 16 February 1973. Taxying tests began in the late Autumn of 1974; additional telemetry equipment was then added, and modifications made to the nosewheel unit and the on-board TV scanner. In February 1975 it was tested in the Ames Research Center 12·2 × 24·4 m (40 ft × 80 ft) wind tunnel; after additional tests in the tunnel in August 1975, further modifications were made to the propeller and duct; the tailboom was extended by 0·86 m (2 ft 10 in); and a Cessna 150 nosewheel and strut were incorporated. First flight took place at NASA's Dryden Flight Research Center at Edwards AFB, California, on 6 August 1976.

WINGS: Cantilever mid-wing monoplane of elliptical planform. Special reflexed aerofoil section, of 21% thickness/chord ratio, is constant over the entire wing. No dihedral, anhedral, incidence or aerodynamic twist. Sweepback at quarter-chord 5° 30′. Wing is capable of in-flight sweeps of 15°, 30° and 45°. Glassfibre and epoxy resin construction. Outboard leading-edge is largely non-structural, and is removable for installation and removal of payload. The wing is 'dished' near the centre, producing a flared spanwise thickness distribution, to permit yawing of the fuselage with respect to the wing without breaking the contour of the fuselage/wing mating surfaces. The trailing-edge is fitted with a vertical stabiliser, ahead of the normal fin and rudder, to provide 'weathercock' stability at the larger wing yaw angles. The wingtips and stabiliser tip are frangible to minimise damage during recovery. Control is effected primarily by wide-span elevons on the trailing-edge, operable differentially (as ailerons) or in unison (as elevators). They are of glassfibre, epoxy resin and Ceconite construction, and are actuated electrically by directly-coupled servos based on lightweight gear motors by Globe Industries Division of TRW.

Prototype DSI/NASA Oblique Wing RPRA, shown without the wing-mounted vertical stabiliser

FUSELAGE: Basically cylindrical structure of conventional skin/stringer construction, with moulded glassfibre/epoxy resin and aluminium skin. Transparent (Lucite) removable hemispherical nose-cap. Central portion, housing the engine, parachute and other equipment, is described as the 'cookie', since in essence it is a circular disc of constant (173 mm; 6·9 in) thickness. A spanwise slit through the cookie accommodates the propeller disc.

TAIL UNIT: Cruciform surfaces, carried on a large-diameter tubular boom designed to accommodate the wing yawing motion and braced by two additional struts forming a Vee from the propeller duct to the tail. These struts are carried through the duct and continued upstream to the fuselage nose. Angular sweptback fin and rudder and elliptical one-piece all-moving tailplane. The latter is constructed of a thin layer of foam-filled glassfibre/epoxy material, and is frangible (as is the fin-tip) to minimise recovery damage. The horizontal tail is actuated by a separate servo commanded by the same circuit as the elevons' elevator functions. Thus, the tailplane can be removed and the aircraft flown as a flying wing while still utilising the elevator function of the elevons.

LANDING GEAR: Non-retractable tricycle gear, with aluminium leaf-spring shock-absorption, fitted for initial taxi and flight testing. Nosewheel and strut are from a Cessna 150. Main wheel tyres size 5·00-4, nosewheel tyre size 4·00-8. No brakes. Streamlined fairings over main wheels. Paraform 'chute (diameter 14·17 m; 46 ft 6 in) in rear of 'cookie', which deploys automatically if the command signal is lost. The aircraft can also be recovered deliberately by this method in response to a command signal from the ground.

POWER PLANT: One 67 kW (90 hp) McCulloch 4318B Model O-100-1 flat-four engine, driving a Brock two-blade fixed-pitch wooden pusher propeller turning within an annular duct. The duct, which is of symme cal section (18% thickness/chord ratio), improves st thrust and thrust at low speeds, permitting a high cr speed propeller to be used while still achieving satis tory T-O and climb performance. The duct also red propeller tip noise and protects operators from the t ing propeller. A tubular supporting strut connects fuselage nose to the base of the duct, serving als deflect the arresting cable in the case of a low appr when using a horizontal-cable snag recovery syst Fuel tank in underside of forward fuselage, capacit kg (60 lb).

SYSTEMS: No hydraulic or pneumatic systems. Elect system (1kW Delco alternator and battery) powers c trols, TV, radio and autopilot.

ELECTRONICS AND EQUIPMENT: Proline command recei special FM data transmitter, Green Ray TV transmi ELT and radar transponder. The nose section cont the TV scanner and related zoom and tilt mechanis flight instrumentation and instrumentation cam command receiver; yaw, roll and pitch gyros; TV si transmitters; UHF tracking; battery; and vac pump.

DIMENSIONS, EXTERNAL:

Wing span	6·81 m (22 ft
Wing chord at root	1·73 m (5 ft 8·
Wing aspect ratio	
Length overall	5·89 m (19 ft 1
Fuselage length	1·61 m (5 ft 3½
Fuselage diameter	0·46 m (1 ft
Height overall	2·03 m (6 ft
Tailplane span	2·29 m (7 ft
Wheel track	2·60 m (8 ft 6½
Wheelbase	1·33 m (4 ft 4½
Propeller diameter	1·23 m (4 ft 0½
Propeller ground clearance	0·35 m (1 ft

DIMENSION, INTERNAL:
Payload volume — 0·25 m^3 (9 cu ft)
AREAS:
Wings, gross — 9·29 m^2 (100·00 sq ft)
Elevons (total) — 1·07 m^2 (11·50 sq ft)
Fin — 0·52 m^2 (5·56 sq ft)
Rudder — 0·12 m^2 (1·33 sq ft)
Tailplane — 1·05 m^2 (11·3 sq ft)
WEIGHTS AND LOADINGS:
Operating weight empty — 412 kg (908 lb)
Max payload — 68 kg (150 lb)
Max T-O weight — 499 kg (1,100 lb)
Max wing loading — 53·7 kg/m^2 (11·0 lb/sq ft)
Max power loading — 7·45 kg/kW (12·0 lb/hp)
PERFORMANCE (estimated, at max T-O weight):
Never-exceed speed — 250 knots (463 km/h; 288 mph)
Max level speed at 1,525 m (5,000 ft) — 150 knots (278 km/h; 173 mph)
Max cruising speed — 100 knots (185 km/h; 115 mph)
Econ cruising speed — 80 knots (148 km/h; 92 mph)
Stalling speed — 60 knots (111 km/h; 69 mph)
Max rate of climb at S/L — 305 m (1,000 ft)/min
Service ceiling — 3,050 m (10,000 ft)
Endurance — 1 hr

DSI RPA-12 SKY EYE

Design of the prototype Sky Eye was started on 28 February 1973; construction was completed on 19 April, and the aircraft flew for the first time on 26 April 1973. Originally of sweptwing tail-less configuration (Sky Eye I-A), it had a pusher propeller mounted in an annular duct aft of the wing, as illustrated in the 1973-74 *Jane's*.

It was later modified to Sky Eye I-B configuration, with the propeller duct removed, twin fins mounted outboard near the wing trailing-edges, and a ciné camera attached externally to the fuselage in addition to the on-board TV camera. This version was still in use in 1976 as a testbed for instrumentation and sensors.

Current versions of the RPV developed by DSI are the expendable **Sky Eye II-E** and the recoverable **Sky Eye II-R**, the latter being 6% larger than the II-E and having endplate fins at the wingtips and a lower radar cross-section.

A derivative of the Sky Eye, known as Aquila, is being designed and built for Lockheed under a US Army contract; this is described under the LMSC entry.

The following description applies to Sky Eye I-B:

WINGS: Cantilever high-wing monoplane. Wing section NACA 23015 (modified). No dihedral or anhedral. Incidence 5°. Sweepback 27° at quarter-chord. Plywood skins, 1/64 in (0·4 mm) thick, with Styrofoam core. Elevon on each trailing-edge, of similar construction, functioning both as aileron and elevator and actuated by modified radio-control servos. No flaps or other high-lift devices.

FUSELAGE: Cylindrical structure, of wooden bulkheads with epoxy-glass skin. Transparent hemispherical nose-cap.

TAIL UNIT: Twin fins, attached to wing trailing-edge near tips.

LANDING GEAR: Non-retractable tricycle gear, mounted on spring struts. Available optionally without landing gear for use with alternative launching system and with recovery by parachute or net.

POWER PLANT: One 9 kW (12 hp) McCulloch 101A piston engine, modified to reduce radio frequency interference, mounted in rear of fuselage. Two-blade wooden fixed-pitch pusher propeller. One fuselage fuel tank (petrol/oil mixture), capacity 3·8 litres (1 US gallon), with provision also for internal wing tanks to raise total capacity to 11·4 litres (3 US gallons).

SYSTEMS AND EQUIPMENT: Total-loss battery system. Alternator can be added for flights of more than 1 hr duration. VHF command system, based on radio control equipment with 15W command output.

DIMENSIONS, EXTERNAL:
Wing span — 3·50 m (11 ft 6 in)
Wing chord at root — 1·03 m (3 ft 4 in)
Wing chord at tip — 0·64 m (2 ft 1 in)
Wing aspect ratio — 4·1
Length overall — 1·70 m (5 ft 7 in)
Length of fuselage — 1·50 m (4 ft 11 in)
Height overall — 0·89 m (2 ft 11 in)
Propeller diameter — 0·61 m (2 ft 0 in)
Wheel track — 0·81 m (2 ft 8 in)
Wheelbase — 0·58 m (1 ft 11 in)
AREAS:
Wings, gross — 3·02 m^2 (32·5 sq ft)
Elevons (total) — 0·12 m^2 (1·3 sq ft)
WEIGHTS AND LOADINGS (A: conventional T-O and landing; B: launch T-O, parachute recovery):
Weight empty, equipped:
A — 25 kg (55 lb)
B — 27·2 kg (60 lb)
Max payload:
A — 20·4 kg (45 lb)
B — 18 kg (40 lb)
Max T-O and landing weight:
A, B — 56·5 kg (125 lb)
Max wing loading:
A, B — 18·79 kg/m^2 (3·85 lb/sq ft)
Max power loading:
A, B — 6·28 kg/kW (10·4 lb/hp)
PERFORMANCE (at max T-O weight):
Never-exceed speed (structural):
A, B — 200 knots (370 km/h; 230 mph)
Max level speed:
A — 90 knots (167 km/h; 103·5 mph)
B — 120 knots (222 km/h; 138 mph)
Stalling speed:
A, B — 35 knots (65 km/h; 40·5 mph)
Max rate of climb at S/L:
A — 548 m (1,800 ft)/min
B — 670 m (2,200 ft)/min
Service ceiling:
A — 3,960 m (13,000 ft)
B — 4,570 m (15,000 ft)
Control-limit range:
A, B — 43 nm (80 km; 50 miles)
Max endurance at 55 knots (102 km/h; 63 mph):
A — 6 hr
B — 9 hr

DSI (LOCKHEED) AQUILA

A derivative of the Sky Eye (which see), known as Aquila, has been designed and is being manufactured by DSI for Lockheed Missiles and Space Co under US Army contract. The programme calls for DSI to manufacture 30 of these vehicles, of which 13 had been delivered to LMSC by the end of January 1976. A description of the Aquila appears under the LMSC entry in this section.

DSI RPA-9 SCOUT

For Northrop, under a US Air Force contract, DSI designed, built and flight tested an electrically-propelled mini-RPV named Scout.

Development is now being continued by DSI as a company-funded programme, with modifications to the RPV which include a new engine and autopilot. By January 1976, DSI had flown 25 missions with the Scout; for the majority of these a tricycle landing gear was fitted.

WINGS: Cantilever shoulder-wing monoplane. Wing section NACA 63-415 (modified). Dihedral 2° 30′. Incidence 5°. Twist 5°. Sweepback 27° 45′ at quarter-chord. Kevlar skins, from 0·011 in to 0·045 in thick, with partial urethane foam support. Elevon on each trailing-edge, of similar construction, functions both as aileron and elevator and is actuated by modified radio-control servos. No flaps or other high-lift devices.

FUSELAGE: Of super-elliptical cross-section, tapered down towards engine, and built of Kevlar skins and bulkheads. Removable nose-cap.

TAIL UNIT: Twin endplate fins, attached to wingtips.

POWER PLANT: One K & B 60 piston engine, driving a two-blade pusher propeller.

SYSTEMS AND EQUIPMENT: Eight channels of telemetry to monitor speed, altitude, autopilot functions and on-board sensors. VHF command system with 15W command output. DSI-PSA-WL-01 stability augmentation autopilot.

DIMENSIONS, EXTERNAL:
Wing span — 2·715 m (8 ft 10·92 in)
Wing chord at root — 0·61 m (2 ft 0 in)
Wing chord at tip — 0·328 m (1 ft 0·96 in)
Wing aspect ratio — 5·35
Length overall — 0·91 m (3 ft 0 in)
Height overall — 0·20 m (8·04 in)
Fuselage:
Max width — 0·36 m (1 ft 2 in)
Max height — 0·15 m (6 in)
Propeller diameter — 0·305 m (1 ft 0 in)

DSI Sky Eye I-B mini-RPV

Model of DSI Sky Eye II mini-RPV

DSI RPA-9 Scout mini-RPV, with tricycle landing gear

AREAS:
Wings, gross — 1·38 m^2 (14·86 sq ft)
Elevons (total) — 0·095 m^2 (1·02 sq ft)
WEIGHTS AND LOADING:
Weight empty, equipped — 15·4 kg (34 lb)
Max T-O weight — 25 kg (55 lb)
Max wing loading — 18·1 kg/m^2 (3·7 lb/sq ft)
PERFORMANCE (at max T-O weight):
Max level speed — 60 knots (111 km/h; 69 mph)
Stalling speed — 32·5 knots (60 km/h; 37 mph)
Max rate of climb at S/L — 91 m (300 ft)/min

DSI RPMB

The RPMB (Remotely Piloted Mini-Blimp) is a small, helium-filled experimental drone airship. It is 4·88 m (16 ft) long and has an envelope max diameter of 1·22 m (4 ft). No other details are known, except that it carries on-board electronics and a ciné camera in addition to the power plant.

DSI Remotely Piloted Mini-Blimp (RPMB) experimental airship drone

E-SYSTEMS

E-SYSTEMS INC

Melpar Division:
7700 Arlington Boulevard, Falls Church, Virginia 22046
Telephone: (703) 560-5000
Director, Development Planning (Melpar Division):
D. R. Gibbs

E-Systems Inc's activities lie predominantly in aerospace systems development and manufacture, in addition to which it has carried out specialised conversion work on nearly 400 C-135 and KC-135 series aircraft for the US Air Force. Its Greenville Division designed and built the L450F, a single-engined high-altitude aircraft capable of manned or unmanned operation. Melpar Division has designed a number of mini-RPVs, of which details follow.

E-SYSTEMS E-45 and E-55

Built by the Melpar Division of E-Systems Inc, the original E-45 was developed under an ARPA contract and was one of four types of mini-RPV evaluated in 1973-74 in the US Army RPAODS programme.

Since that time Melpar has tested the E-45 extensively, using a gimballed TV camera for surveillance purposes. In this role the E-45 can locate a 9·14 m (30 ft) target while operating at an altitude of approx 1,525 m (5,000 ft). A modified version, designated **E-55,** is under development as a successor to the E-45. This has a more powerful engine and increased payload capability. Both models are capable of live TV reconnaissance, jamming, targeting or homing missions.

The following description applies to the E-45, of which 12 examples are believed to have been built:

Wings: Cantilever constant-chord high-wing monoplane, of NACA 4415 section. Horizontal centre-section; 15° dihedral on outer panels. Incidence 5° at root. Provision to extend span by addition of no-dihedral tip panels.

Fuselage: Pod-shaped central nacelle.

Tail Unit: 51 mm (2 in) diameter twin tailbooms, supporting twin endplate fins and rudders and enclosed tailplane with full-span elevator.

Power Plant: One 1·5 kW (2 hp) single-cylinder two-stroke engine, mounted in rear of fuselage nacelle and driving a two-blade pusher propeller.

Launch and Recovery: Launch by vehicle-mounted launcher, using jettisonable take-off dolly, or by catapult. Recovery by conventional landing, on skid beneath central nacelle, or by net.

Guidance and Control: RPV is controlled either by a real-time tracking radar mounted on top of a two-man mobile van, or by a standard non-tracking radio control link. The radar system, which transmits command data and receives telemetry replies at frequencies between 5·4 and 5·9GHz, can control the RPV effectively for approx 43·5 nm (80·5 km; 50 miles). This control can be extended to a range of 130 nm (241 km; 150 miles) by pre-programming the aircraft's flight and using an airborne relay for video downlink. The aircraft can be controlled either manually or by a combination of pre-programmed flight and manual landing. Aerodynamic control is by use of elevator and rudders.

Special Equipment: All versions of E-45, and E-55, are fitted with a Melpar-designed five-axis autopilot weighing 0·9 kg (2 lb), which can be used for dead reckoning or Omega navigation. As a payload for the E-45, Melpar has designed and is flight testing a low-cost TV surveillance system which provides the means for remote manual control, target identification, and navigational assistance away from the launch site. This utilises a modified sub-miniature commercial TV camera mounted on a Melpar-developed gimbal, the camera having a standard 'C' mount lens adaptable to multiple fields of view and operating with a video bandwidth of 6MHz.

Dimensions, external:	
Wing span	2·41 m (7 ft 11 in)
Wing chord (constant)	0·29 m (11·2 in)
Length overall	2·36 m (7 ft 9 in)
Height overall	0·51 m (1 ft 8 in)
Width over tailbooms	0·71 m (2 ft 4 in)
Tailplane span inside booms	0·60 m (1 ft 11½ in)
Fuselage diameter	0·20 m (8 in)
Propeller diameter	0·46 m (1 ft 6 in)
Area:	
Wings, gross	0·68 m² (7·36 sq ft)
Weights:	
Weight empty, equipped	11 kg (24 lb)
Payload	6·8 kg (15 lb)
Fuel	2·7 kg (6 lb)
T-O weight	20·5 kg (45 lb)
Performance (at 18 kg; 40 lb T-O weight):	
Cruising speed	43·5 knots (80 km/h; 50 mph)
Stalling speed	35 knots (65 km/h; 40 mph)
Rate of climb at S/L	229 m (750 ft)/min
Service ceiling	3,050 m (10,000 ft)
Endurance	5 hr

E-Systems E-55, more powerful version of the E-45 mini-RPV, mounted on its launcher

E-Systems E-100X mini-RPV, an enlarged development of the E-45

E-SYSTEMS AXILLARY

This mini-RPV, a modified version of the Melpar Division E-45, was being evaluated by the US Air Force in 1976 as a potential very-low-cost harassment vehicle (VLCHV). Five are reported to have been ordered. The vehicle has a radar homing capability which would enable it to fulfil a defence suppression role if fitted with a small explosive charge and directed at an enemy target. It is similar in external appearance to the standard E-45, but has wings with neither dihedral nor anhedral, and non-swept fins and rudders with equal area above and below the tailbooms.

Other aspects of the VLCHV programme under consideration include unarmed vehicles, designed to loiter for up to 6 hr over enemy defences and so impose radar silence upon them; and alternative sensor payloads such as forward-looking infra-red, radiation detectors and signals jammers.

E-SYSTEMS E-100X

Essentially, the E-100X is an enlarged development of the E-45, having extended-span wings as standard, greater engine power, and increased payload volume. Like the E-45, it consists of three major assemblies (fuselage pod, wing, and booms and tail unit), all designed for quick and simple assembly or dismantling. The wing assembly contains a sealed fuel tank. The pod section is a sealed unit accommodating the on-board electronics packages and the power plant.

Some 0·085 m³ (3 cu ft) of volume and more than 22·7 kg (50 lb) of weight can be devoted to possible payloads. Potential payloads for the E-100X include a low-cost miniature TV reconnaissance system, similar to that developed for the E-45, or packages for photographic reconnaissance, electronic jamming, tactical strike support, communications relay, and remote communications interception.

Power Plant: One 6·7 kW (9 hp) Rosspower flat-four piston engine, driving a two-blade pusher propeller. Fuel in integral tank in wings; fuel load 8·2 kg (18 lb), or approx 11·4 litres (3 US gallons).

Airframe: Similar to E-45.

Launch, Recovery, Guidance and Control: Similar to E-45.

Special Equipment: According to mission (see introductory copy).

Dimensions, external (typical):	
Wing span	3·28 m (10 ft 9 in)
Length overall	2·74 m (9 ft 0 in)
Height overall	0·57 m (1 ft 10½ in)
Fuselage diameter	0·27 m (10¾ in)
Tailplane span (c/l of tailbooms)	0·66 m (2 ft 2 in)
Propeller diameter	0·56 m (1 ft 10 in)
Area:	
Wings, gross	1·33 m² (14·33 sq ft)
Weights (typical):	
Payload	22·7-23·6 kg (50-55 lb)
Fuel	8·2 kg (18 lb)
T-O weight	45·4 kg (100 lb)
Performance (typical):	
Max level speed	95 knots (175 km/h; 109 mph)
Cruising speed	75 knots (139 km/h; 86 mph)
Stalling speed	42 knots (78 km/h; 48·5 mph)
Rate of climb at S/L	305 m (1,000 ft)/min
Service ceiling	above 3,050 m (10,000 ft)
Endurance	more than 5 hr

E-SYSTEMS L450F

USAF designation: XQM-93

The Greenville Division of E-Systems built a single-seat monoplane designated L450F, powered by a turboprop engine, capable of manned or unmanned operation and of carrying data-gathering equipment or electronic relay equipment (similar to that of a communications satellite) to a height of more than 13,715 m (45,000 ft).

The first prototype made its first flight, with a pilot, during February 1970. On 24-25 January 1972 the modified second prototype, converted to unmanned configuration for USAF evaluation and designated XQM-93 (serial number 70-1287), made a nonstop flight of more than 21 hours at Edwards AFB, California. The flight was part of the USAF's Compass Dwell programme in which the XQM-93 was flown competitively with the Martin Marietta Model 845A. The evaluation was completed in early 1972, neither vehicle being selected for production.

On 23 and 27 March 1972 the L450F, flown by test pilot Don Wilson, established six new international altitude records and ten time-to-height records at Majors Field, Greenville, Texas.

A description and illustration of the L450F can be found in the 1973-74 and subsequent editions of *Jane's*. The study of potential military and civil applications, especially the latter, was continuing in 1975. It is not known if this programme is still active.

EGLEN

EGLEN HOVERCRAFT INC

ADDRESS:
801 Poplar, Terre Haute, Indiana 47807
Telephone: (812) 234-4307
PRESIDENT:
Jan Alan Eglen

This company has rights to manufacture the USAF's Falcon mini-RPV, designed by the AF Flight Dynamics Laboratory (erroneously attributed to Meridian Corporation in the 1975-76 *Jane's*). It has also built a prototype of the USAF's Eagle RPV. All known details of these two vehicles follow.

Eglen Hovercraft Inc manufactures other drone aircraft of various sizes, and undertakes custom-building of RPRVs for NASA and other agencies.

EGLEN (USAF) FALCON

The Falcon is a mini-RPV which is in service with NATO and other military and civil authorities for testing sensor and weapon systems and for weather monitoring. It was designed at the USAF Flight Dynamics Laboratory, by a team led by Donald Lowe.

Other potential applications include reconnaissance, surveillance, photography, search missions, flood or forestry survey, air sampling, traffic observation, infra-red or laser target designation, and air testing of autopilots, flight computers and other equipment. It is available also in a target version.

AIRFRAME: Of high-wing monoplane configuration, built mainly of glassfibre, foam plastics and plywood, with epoxy paint finish.

LANDING GEAR: Normally fitted with non-retractable tricycle landing gear, with steerable nosewheel. This gear can be replaced by skis for operation in snow or desert conditions. Falcon is also suitable for shipboard operation, using a catapult launch and net recovery system.

POWER PLANT: One Homelite two-stroke piston engine, of approx 2·24 kW (3 hp), driving a two-blade pusher propeller.

LAUNCH AND RECOVERY: Conventional T-O and landing; or catapult launch and net recovery.

GUIDANCE AND CONTROL: Radio command guidance system. Aerodynamic control by ailerons, elevators and rudder.

SPECIAL EQUIPMENT: Six-channel Proline radio control system standard (four channels for flight control and two for payload operation). Eight-channel system available optionally. Standard airborne TV payload package includes Conic 8W CTM UHF 408V S-band TV transmitter with antenna; Sony AVC 1400 TV camera and lens (modified for airborne use); and battery pack. Each TV payload package is modified to accept the microwave transmission system utilised, and is fully systems-tested.

DIMENSIONS, EXTERNAL:
Wing span — 2·44 m (8 ft 0 in) *or* 3·05 m (10 ft 0 in)
Fuselage: Length — 1·83 m (6 ft 0 in)
Max width — 0·165 m (6½ in)
Max depth — 0·235 m (9¼ in)
Height over fin — 0·64 m (2 ft 1 in)
Wheel track — 0·84 m (2 ft 9 in)

AREA:
Wings, gross — 1·11 m² (12·00 sq ft) *or* 1·39 m² (15·00 sq ft)

WEIGHTS AND LOADING:
Weight empty — 11·3 kg (25 lb)
Max payload — 11·3 kg (25 lb)
Balanced T-O weight — 15·4 kg (34 lb)
Max T-O weight — 22·7 kg (50 lb)
Max wing loading — 24·4-34·2 kg/m² (5-7 lb/sq ft)

PERFORMANCE:
Max level speed — 100 knots (185 km/h; 115 mph)
Service ceiling — approx 4,575 m (15,000 ft)
Max endurance — 5 hr

EGLEN (USAF) EAGLE

This larger RPV has a 9·7 kW (13 hp) piston engine, and carries a fully-gimballed TV relay system in the nose, behind a transparent nose-cap. Like the Falcon, it was designed at the US Air Force Flight Dynamics Laboratory, which built the first prototype shown in an accompanying photograph. A second prototype was under construction by Eglen in early 1976.

DIMENSIONS, EXTERNAL:
Wing span — 4·88 m (16 ft 0 in)
Propeller diameter — 0·61 m (2 ft 0 in)

WEIGHTS:
Payload — 36·3-40·8 kg (80-90 lb)
Max T-O weight — 91 kg (200 lb)

PERFORMANCE:
No details received for publication

Eglen (USAF) Falcon multi-purpose mini-RPV

First prototype Eagle mini-RPV, built by the USAF

FAIRCHILD

FAIRCHILD REPUBLIC COMPANY

DIVISIONAL OFFICE AND WORKS:
Farmingdale, Long Island, New York 11735
OFFICERS: See Aircraft section

Fairchild Space and Electronics Company, one of the divisions of Fairchild Industries, began to be actively engaged in RPV development in 1968. A family of low-cost RPV systems in the low subsonic speed range was designed, built and extensively tested. This family of RPVs incorporated a unique sailwing folding wing design, developed at Princeton University under the sponsorship of Fairchild Industries, and has been described in recent editions of *Jane's*.

During 1975 these RPV programmes were transferred from Fairchild Space & Electronics Company to Fairchild Republic Company; but their development has now ended.

FAIRCHILD AIRCRAFT SERVICE DIVISION

ADDRESS:
PO Box 1177, Crestview, Florida 32536
Telephone: (904) 682-2746

Fairchild Aircraft Service Division received in 1975 a subcontract for the conversion of 29 Convair F-102A Delta Dagger fighters to PQM-102 drone configuration.

This programme follows earlier contracts under which Fairchild converted eight F-102As to remotely piloted prototype vehicles, and was scheduled to complete the conversion of a further 31 F-102As in June 1976. Five of the aircraft so far converted have been configured to carry a human pilot, to monitor the aircraft's response to remote commands; these manned test aircraft are designated QF-102. The conversions are being carried out under subcontract to Sperry Flight Systems Division, under whose entry further details of this programme can be found.

FSI

FLIGHT SYSTEMS INC

ADDRESS:
4000 Westerly Place, PO Box 2400, Newport Beach, California 92663
Telephone: (714) 833-9661
DIRECTOR, TEST PROGRAMMES:
E. T. Binckley
DIRECTOR, QF-86E PROGRAMME:
S. C. Warrick

FSI QF-86E SABRE

FSI has produced for US Army evaluation two remotely controlled examples of the North American F-86 Sabre sweptwing jet fighter, the converted aircraft actually being Canadair-built Sabre 5s, structurally similar to the US-built F-86E version. These were demonstrated to the US Army's Missile Command in early 1975, at the White Sands Missile Range. The QF-86Es, which retain provision for a monitoring human pilot in the cockpit, are controlled remotely by a Vega Precision Laboratories ground control system and are capable of a full range of pre-programmed flight manoeuvres including take-off and landing. Manoeuvres of up to 7*g* can be performed, and the QF-86Es can also deploy stores, initiate jamming and provide other countermeasures, all under remote control.

Flight Systems Inc QF-86E drone conversion of the Canadair Sabre 5

A description of the basic F-86 airframe can be found in the 1959-60 *Jane's;* the following details apply to the FSI QF-86E/Sabre 5:

POWER PLANT: One Orenda 10 turbojet engine, rated at 28·15 kN (6,325 lb st).

LAUNCH AND RECOVERY: Conventional runway T-O and landing, on retractable tricycle landing gear. Aircraft is programmed to come to a halt, or to continue and make a safe T-O and climb-out, if the ground control link should be lost during the T-O run.

GUIDANCE AND CONTROL: Radio command guidance system. Primary mode of operation is NOLO (No Local Operator), but provision for an on-board human pilot is retained (eg for manned practice presentations, or maintenance or ferry flights). Remote control exercised from one fixed and one mobile ground station, both manufactured by Vega Corporation. These stations are the same as those used in operation of the Sperry/Convair QF/PQM-102 (which see). Aerodynamic control of QF-86E via the normal aircraft control surfaces.

SPECIAL EQUIPMENT: On-board electronics and instrumentation comprises three basic installations: autopilot, with FSI electronics; Vega Precision Laboratories command/telemetry data system; and an FSI interface coupler for processing uplink command and downlink telemetry data to and from the drone aircraft. Radar altimeter optional, for simulated low-level attack presentations. Ancillary equipment, according to mission, may include scoring gear, infra-red flare dispenser or ECM pod.

DIMENSIONS, EXTERNAL:
- Wing span: 11·31 m (37 ft 1·2 in)
- Length overall: 11·43 m (37 ft 6 in)
- Height overall: 4·48 m (14 ft 8·4 in)

WEIGHTS:
- Basic weight empty: 4,921 kg (10,850 lb)
- T-O weight 'clean': 6,123 kg (13,500 lb)
- T-O weight with two 454 litre (120 US gallon) drop-tanks: 6,894 kg (15,200 lb)

PERFORMANCE:
- Max level speed above 11,000 m (36,000 ft): Mach 0·92 (527 knots; 977 km/h; 607 mph)
- Service ceiling: 13,715 m (45,000 ft)
- *g* limit: +7·0

KAMAN
KAMAN AEROSPACE CORPORATION

ADDRESS:
Old Windsor Road, Bloomfield, Connecticut 06002
Telephone: (203) 242-4461
OFFICERS: See Aircraft section

Kaman is conducting a number of independent design studies and mission analyses for rotating-wing RPVs, one of which, designated K-244, competed in 1974 for the US Army's RPAODS competition, since won by the LMSC/DSI Aquila.

Kaman is also investigating the feasibility of using an unmanned, tethered, rotating-wing aerial platform for battlefield surveillance and target acquisition.

In the hardware stage, for the US Navy, is a similar type of vehicle known as STAPL.

Kaman Ship Tethered Aerial Platform (STAPL) first prototype

KAMAN STAPL

Under contract to the US Office of Naval Research, Kaman designed, built and is flight testing two prototypes of a Ship Tethered Aerial Platform (STAPL), preliminary design of which was carried out in 1970 under a previous ONR contract. Testing of the two aircraft and their mobile launch and recovery platform and control system began in 1974 and is planned to continue into 1977.

The prototype aircraft are of autogyro configuration, and are equipped with automatic flight control systems and data recording equipment. Modifications have been made to the second prototype, but details of these have not been received for publication. The self-contained AFCS, engineered by Kaman, provides three-axis stabilisation and automatic flight path control, and incorporates redundancy for mission reliability.

LMSC
LOCKHEED MISSILES AND SPACE COMPANY INC

ADDRESS:
1111 Lockheed Way, Sunnyvale, California 94088
Telephone: (408) 742-4321
PUBLIC RELATIONS:
Paul J. Binder

As part of a general programme of investigation into the field of tactical RPVs, Lockheed Missiles and Space Co built a testbed aircraft, the RTV-2, with which to conduct preliminary flight testing. This was described briefly in the 1974-75 *Jane's*.

Lockheed Missiles and Space Co's Aquila was in 1974 declared the winner of the US Army's competition for a battlefield surveillance mini-RPV.

LMSC is also developing a folding-wing mini-RPV known as Aequare, for flight test during 1976.

LMSC AQUILA

US Army designation: XMQM-105

This mini-RPV is being developed under a US Army contract, as a system technology demonstrator, and in 1974 replaced that service's RPAODS programme. It was formerly known as Little 'r'; its present name is the Latin for 'eagle'. Thirty XMQM-105s, and four ground stations, are being built under this contract, and a half-scale test model was flown for the first time on 14 July 1975. First flight of a full-size Aquila was made in December 1975, when a tricycle landing gear was fitted.

Design and development of the Aquila was undertaken under subcontract to LMSC by DSI (which see), and the RPV is derived from the latter company's Sky Eye. A Sky Eye was used in 1974 in preliminary tests of the Aquila launch and recovery system. By the end of January 1976, DSI had delivered 13 Aquilas to LMSC.

The Aquila has a lower wing, of near-delta planform, a larger fuselage, and its construction, entirely of Kevlar 49 honeycomb material, gives it a smaller radar cross-section than the Sky Eye. The airframe, which dismantles into four major components (centrebody, two wings, and propeller duct), has six access panels and a quick-disconnect bladder fuel system. Power plant is an 8·2 kW (11 hp) McCulloch MC-101M/C piston engine, driving a two-blade pusher propeller mounted in an annular duct. The Aquila will normally have no landing gear, being launched pneumatically and recovered in a transportable arrester/net system developed jointly by LMSC and the All American Engineering Co.

On-board equipment includes a Lockheed autopilot, Aacom data link, Honeywell laser designator and Perkin-Elmer reconnaissance camera. Five different

Above: LMSC Aquila, prior to first test flight, with tricycle landing gear

Right: Aquila mini-RPV with fuselage access panels removed

Aquila on truck-mounted pneumatic rail launcher

payload configurations are to be evaluated: for real-time surveillance, photographic reconnaissance, target acquisition, laser target designation, and target location and artillery adjustment.

Evaluation was to take place at the US Army artillery school at Fort Sill, Oklahoma, and at Fort Huachuca, Arizona.

DIMENSIONS, EXTERNAL:
- Wing span: 3·63 m (11 ft 10·8 in)
- Length: 1·83 m (6 ft 0 in)

WEIGHTS:
- Payload: 13·6 kg (30 lb)
- Launch weight: 54·4 kg (120 lb)

PERFORMANCE (approx):
- Max speed: 120 knots (222 km/h; 138 mph)
- Min speed to become airborne: 44 knots (81·5 km/h; 50·5 mph)
- Service ceiling: above 6,100 m (20,000 ft)
- Max endurance: 3 hr

LMSC AEQUARE

First flown in mid-1975, Aequare (Latin for 'to equalise') is an expendable reconnaissance and laser target designation mini-RPV, produced as a feasibility vehicle for evaluation by the US Air Force. It has a glassfibre airframe manufactured by Windecker Industries Inc.

Tests scheduled for 1976 included use in connection with launch of 500 lb Mk 82 laser-guided bombs.

POWER PLANT: One 9·3 kW (12·5 hp) McCulloch MC-101 single-cylinder piston engine, driving a three-blade ducted pusher propeller.

LAUNCH AND RECOVERY: Air-launched at 7,620 m (25,000 ft) from fighter aircraft (F-4 Phantom in tests), under starboard wing of which it is carried with wings folded inside a modified SUU-42 flare dispenser pod. A CTV-2 pod containing the command/telemetry/video data link is carried under the fighter's port wing. Upon launch, end cap of carrier pod is released by explosive charge, deploying a drogue parachute on which pod descends to 4,270 m (14,000 ft). Barometric device then releases main 'chute, which pulls pod clear of RPV. Wings of Aequare then pivot forward into flight position, and engine is started. Provision also made for ground launch by rocket. RPV airframe is expendable, but parachute system is provided for recovery of payload package.

GUIDANCE AND CONTROL: After launch of pod, carrier aircraft climbs to 10,670 m (35,000 ft) and acts as relay between RPV and ground control station. On release from pod, RPV is pre-programmed to enter level, powered flight and to set up data link with ground controller. Remainder of flight is normally controlled from ground station, but can be pre-programmed if desired.

SPECIAL EQUIPMENT: Recoverable payload compartment in lower fuselage contains Aeronutronic Ford stabilised TV (non-stabilised in initial tests) and laser target designator, Aacom data link equipment, Lockheed autopilot, and X- and C-band beacons.

DIMENSIONS, EXTERNAL:
- Wing span: approx 3·66 m (12 ft 0 in)
- Length overall: approx 2·44 m (8 ft 0 in)

WEIGHT:
- Max launching weight: 68 kg (150 lb)

PERFORMANCE:
- Range: nearly 174 nm (322 km; 200 miles)

LOCKHEED

LOCKHEED-GEORGIA COMPANY

ADDRESS:
86 South Cobb Drive, Marietta, Georgia 30063

OFFICERS: See Aircraft section

LOCKHEED (NASA) ACT

NASA's Dryden Flight Research Center has awarded a $92,000 contract to Lockheed-Georgia to build a flying model to demonstrate the potential benefits of Active Control Technology (ACT). The model is a remotely controlled one-seventh flying scale model of the Center's JetStar business aircraft, and is provided with two alternative sets of wings and tail surfaces. One of these sets represents the basic JetStar design, the other conforms to the ACT JetStar, based upon the principle of using control surfaces for more than one function. The object of this is to permit the use of smaller control surfaces and to make possible more efficient flight and reduced fuel consumption.

The ACT remotely piloted JetStar built by Lockheed for NASA, seen here with modified engine pods and a T tailplane

MCDONNELL DOUGLAS

MCDONNELL DOUGLAS CORPORATION

HEAD OFFICE AND WORKS:
PO Box 516, Saint Louis, Missouri 63166
Telephone: (314) 232-0232

OFFICERS: See Aircraft section

McDonnell Douglas Corporation is involved in various activities concerning RPVs, and some contracts and/or IRAD projects are still active at McDonnell Douglas Astronautics Company (see below), McDonnell Aircraft Company and Actron. Much of this work is classified and is concerned with payloads and control subsystems rather than with the flight vehicles themselves.

Conversion of McDonnell Douglas F-4B Phantom aircraft to QF-4B configuration for target drone purposes is described under the US Navy heading in this section.

MCDONNELL DOUGLAS ASTRONAUTICS COMPANY

Under a company-funded programme, MDAC developed a mini-RPV. Originally conceived to meet DARPA (Defense Advanced Research Projects Agency) and US Army requirements for a small, low-cost battlefield reconnaissance and laser target designation vehicle, many other mission possibilities were investigated, together with appropriate payloads.

Under contract to the USAF, and funded by DARPA, the MDAC vehicle was flown against simulated and real defences at Eglin AFB, Florida, in 1975. All contract requirements were completed successfully.

Brief details of an earlier feasibility engineering model, of which two examples were built, appeared in the 1975-76 *Jane's*.

The mini-RPV programme at MDAC is currently inactive.

NASA

NATIONAL AERONAUTICS AND SPACE ADMINISTRATION

HEADQUARTERS:
1520 H Street NW, Washington, DC 20546

Ames Research Center

ADDRESS:
Moffett Field, California 94035

NASA OBLIQUE WING REMOTELY PILOTED RESEARCH AIRCRAFT

NASA's Ames Research Center is sponsoring the design and development of an advanced-technology RPRA (Remotely Piloted Research Aircraft) incorporating the concept of the oblique wing. Both military and civil applications of such a vehicle are being studied.

Engineering models have been tested in flight and in the wind tunnel at Ames. The full-size RPRA has been designed and built under contract by DSI (which see), and is described under that company's heading in this section. It flew for the first time on 6 August 1976.

In addition, the Dryden Flight Research Center of NASA has awarded a contract to Teledyne Ryan (which see) to investigate the possibility of using a Firebee II supersonic target drone to carry out flight research with the oblique wing concept.

NASA/DSI Oblique Wing RPRA, with wing at maximum yaw angle of 45°

HUGH L. DRYDEN FLIGHT RESEARCH CENTER

ADDRESS:
PO Box 273, Edwards, California 93523
Telephone: (805) 258-3311

MINI-SNIFFER PROJECT MANAGER:
Robert D. Reed

NASA MINI-SNIFFER II

The Mini-Sniffer is a simple, low-cost vehicle platform, designed for high-altitude atmospheric research in remote areas, to measure air turbulence and atmospheric trace gas and to determine the constituents of particles in the atmosphere.

Payload instrumentation is being developed by the Wallops Flight Center and the National Oceanic and Atmospheric Administration to measure nitric oxide, chlorine, Freon 11 and 12, free oxygen, and the hydroxyl radical. Langley Research Center plans to use the low-speed Mini-Sniffer to make accurate measurements of fine turbulence which cannot be obtained with faster aircraft. Equipped with lightweight wire impact particle grab samplers developed by Ames Research Center, and a real-time telemetered particulate density sensor, the Mini-Sniffer will be able to make low-speed manouevres through aerospace vehicle wakes or other suspected particle concentrations. Radiation sensor payloads being developed by the Atomic Energy Commission may allow the multiple

NASA Mini-Sniffer II, with variable-dihedral wingtips

NASA Mini-Sniffer I prototype (background) and Mini-Sniffer II atmospheric sampling RPRVs

Teledyne Ryan Firebee II used as an RPRV by NASA's Dryden Flight Research Center

and prompt monitoring of nuclear activity.

The Dryden Flight Research Center began by test-flying prototype vehicles to acquire airframe, aerodynamic, and guidance and control data, and to develop launch and recovery techniques. The first two of these prototype vehicles, designated Mini-Sniffer I, were described and illustrated in the 1975-76 *Jane's*.

Design of the Mini-Sniffer I began in October 1973, and the first flight was made in August 1974, with an interim air-breathing engine. With this engine, 15 flights were made at altitudes up to 6,100 m (20,000 ft).

A third prototype, with a twin-tailboom configuration and designated Mini-Sniffer II, made its first flight in the Summer of 1975; the following details apply to this aircraft, which in 1976 was fitted with a hydrazine-burning engine to permit flights to altitudes of up to 30,480 m (100,000 ft):

TYPE: Remotely piloted research vehicle.

WINGS: Cantilever mid-wing monoplane. Mississippi State University MO6-13-128 wing section with 13% thickness/chord ratio. Variable dihedral on outer panels. Sweepback 20° at quarter-chord. Structure of foam, glassfibre and wood, fitted with elevons. Modified Kraft Radio electric servo system.

FUSELAGE: Conventional structure, built of same materials as wings.

TAIL UNIT: Twin sweptback fins and rudders, carried on slender tailbooms extending from wing trailing-edges. Built of same materials as wings.

LANDING GEAR: Non-retractable tricycle type. Cantilever tapered spring steel legs, and aluminium wheels (of 127 mm; 5 in diameter) without tyres.

POWER PLANT (mission vehicle): One 3-22·4 kW (4-30 hp) Akkerman 235 piston engine, driving a two-blade constant-speed variable-pitch feathering propeller. Single aluminium integral fuel tank in fuselage, for 26·5 litres (7 US gallons) of hydrazine monopropellant. Oil capacity 0·95 litres (1 US quart).

SYSTEMS AND EQUIPMENT (mission vehicle): Alternator on Akkerman 235 engine provides electrical power during climb and cruise; 28V nickel-cadmium battery for use during unpowered flight. Kraft (in prototypes) or Resdel (in mission vehicle) uplink control. SCI-680-PCM 16-channel data telemetry system. Vega radar transponder. NASA autopilot. Sun and magnetic sensors for wing levelling, airspeed hold and heading hold.

DIMENSIONS, EXTERNAL:
- Wing span 5·49 m (18 ft 0 in)
- Wheel track 1·14 m (3 ft 9 in)
- Wheelbase 1·27 m (4 ft 2 in)
- Propeller diameter 1·83 m (6 ft 0 in)

DIMENSIONS, INTERNAL:
- Payload compartment:
  - Length 1·03 m (3 ft 4 in)
  - Width 0·25 m (10 in)
  - Volume 0·033 m³ (1·16 cu ft)

WEIGHTS AND LOADING (mission vehicle):
- Max payload 22·7 kg (50 lb)
- Max T-O weight 65·8 kg (145 lb)
- Max wing loading 19·9 kg/m² (4·08 lb/sq ft)

PERFORMANCE (mission vehicle, estimated):
- Max level speed at 21,340 m (70,000 ft) 160 knots (296 km/h; 184 mph)
- Stalling speed 26 knots (49 km/h; 30 mph) IAS
- Max rate of climb at S/L 792 m (2,600 ft)/min
- Service ceiling 27,430 m (90,000 ft)
- Range with max payload 434 nm (805 km; 500 miles)

NASA HiMAT

Details of this programme can be found under the Rockwell International heading in this section.

NASA ACT

Details of this programme can be found under the Lockheed heading in this section.

NASA (TELEDYNE RYAN) FIREBEE II RPRV

NASA's Dryden Flight Research Center is conducting flight research, using Teledyne Ryan BQM-34F Firebee II supersonic target drones modified to operate as RPRVs (Remotely Piloted Research Vehicles), in a programme with applications to military and civil aircraft designed to operate in the transonic and low supersonic speed ranges.

The programme includes research into aerodynamics, structures and unique configuration characteristics, and the Firebee will be modified to have new wings, tail surfaces and other features. Data from sensors, instrumentation and related hardware on board the drone will be telemetered back to the Dryden RPRV control facility. One such application will be evaluation of the oblique wing concept.

The Firebees will be air-launched; recovery will be initially by mid-air retrieval system (MARS) and later by conventional landing on the Rogers Dry Lake bed.

LANGLEY RESEARCH CENTER

ADDRESS:
Hampton, Virginia

NASA PROJECT DAST

Project DAST (Drones for Aerolastic and Structural Testing) is a programme under which NASA's Langley and Dryden Flight Research Centers plan to pursue further the use of RPRVs for investigating unusual aerodynamic configurations, aerodynamic load problems and other techniques.

Some preliminary work has already been done, using Teledyne Ryan BQM-34E Firebee II supersonic drones acquired from the US Navy. Subsequently, NASA obtained its own Firebee II vehicles with which to continue work under this programme, as described under the Dryden Flight Research Center heading. In addition, the Firebees used in the Himat programme will also eventually be used in the DAST programme, in which they will be fitted with new wings to evaluate flutter suppression and other phenomena. The first set of DAST wings is being funded and built by Langley Research Center.

NORTHROP
NORTHROP CORPORATION—VENTURA DIVISION

DIVISION HEAD OFFICE:
1515 Rancho Conejo Boulevard, Newbury Park, California 91320
Telephone: 498-3131
Telex: 659 220

GENERAL MANAGER:
W. E. Woolwine

VICE-PRESIDENTS:
Walter Dedrick (Marine Systems)
John E. Evans (Manufacturing)
George C. Grogan Jr (Marketing)
V. W. Howard (Engineering)
W. F. Sternadel (Finance)

Northrop's Ventura Division designs and manufactures pilotless target aircraft and related equipment. It also manufactures glassfibre wing fairings for the Boeing 747 transport aircraft. It has diversified into the marine systems field and is producing an unmanned underwater vehicle (MK-30) as a target for US Navy anti-submarine warfare training. The division also produces torpedo modification kits which convert the MK-37 torpedo to the NT-37C torpedo now used by a number of navies throughout the world.

Northrop Ventura (formerly Radioplane) undertook the design, development and construction of its first radio controlled target drone in the mid-thirties. Since then it has become a leader in the field of pilotless aircraft. More than 70,000 drones have been delivered to the US military services and 25 allied nations.

Parachute landing systems designed and produced by Northrop Ventura returned safely to Earth all the US astronauts who accomplished space flight under the Mercury, Gemini, Apollo and Skylab programmes.

Northrop also developed a Variable-Speed Training Target (VSTT), which was evaluated in competition with an entry from Beech Aircraft Corporation to meet the requirements of the US Army's Missile Command. This was described and illustrated in the 1974-75 and 1975-76 *Jane's*.

Components of a Northrop MQM-74A are used in the VATOL (vertical attitude take-off and landing) RPV described under the US Navy heading in this section.

NORTHROP SHELDUCK

This target drone is currently in use by the armed forces of 18 countries as a training device for ground-to-air gunnery and is used as a training target for surface-to-air missiles such as Seacat, Tigercat, Redeye, Blowpipe, Sparrow, Chaparral, Hawk, Sidewinder and Nike.

Design of the drone was started in 1946 and the prototype flew for the first time in 1947. Since then more than 60,000 of this type, including early KD2R versions, have been built and production continues.

The target is surface-launched from land or ship, either by rotary or zero launcher. Radio control is utilised, the target being tracked visually or by radar. After completion of its mission, it is recovered by the use of a radio-command-released parachute. In the event of serious damage by gunfire or loss of radio control or electrical power, the parachute is deployed automatically. The target is designed to be repaired easily if damaged by gunfire.

TYPE: Remotely controlled aerial target.

WINGS: Cantilever high-wing monoplane. Wing section NACA 23012 at root, NACA 4412 at tip. No dihedral. Incidence 1° at root, −2° at tip. Conventional aircraft aluminium alloy construction. Conventional ailerons servo-operated by type D-9 actuators.

FUSELAGE: Semi-monocoque aluminium alloy structure, with integral steel fuel tank.

TAIL UNIT: Cantilever aluminium alloy structure. Fixed-incidence tailplane. Elevator servo-operated by type D-9 actuator.

POWER PLANT: One 67 kW (90 hp) Northrop O-100-3 flat-four engine, driving a two-blade fixed-pitch wooden propeller. Steel fuel tank in mid-fuselage, capacity 44 litres (11·6 US gallons). Refuelling point in fuselage forward of wing.

SYSTEMS: Electrical power only, from 28V battery.

ELECTRONICS AND EQUIPMENT: AN/ARW-79 remote flight control system with automatic altitude hold control. Radar or FM type tracking systems or equivalent. L-band tracking system, smoke generating, infra-red tow target, night light kits, tow banner and many other accessories available to customer's requirements. For radar appearance augmentation, two wingtip reflector pods are optional.

DIMENSIONS, EXTERNAL:

Wing span	3·50 m (11 ft 6 in)
Wing chord (mean)	0·51 m (1 ft 8·13 in)
Wing aspect ratio	7
Length overall	3·85 m (12 ft 7½ in)
Height overall	0·76 m (2 ft 6 in)
Tailplane span	1·27 m (4 ft 2 in)
Propeller diameter	1·10 m (3 ft 8 in)

AREAS:

Wings, gross	1·74 m² (18·7 sq ft)
Ailerons (total)	0·12 m² (1·3 sq ft)
Fin	0·17 m² (1·8 sq ft)
Tailplane	0·28 m² (3·0 sq ft)
Elevators	0·13 m² (1·4 sq ft)

WEIGHTS:

Weight empty	123 kg (271 lb)
Max launching weight	163 kg (360 lb)
Max zero-fuel weight	133 kg (292 lb)
Max landing weight	154 kg (340 lb)

PERFORMANCE (at max launching weight):

Never-exceed speed	250 knots (463 km/h; 288 mph)
Max level speed at S/L and max cruising speed	194 knots (359 km/h; 223 mph)
Stalling speed	58 knots (108 km/h; 67 mph)
Max rate of climb at S/L	1,341 m (4,400 ft)/min
Service ceiling	8,230 m (27,000 ft)
Range with max fuel	173 nm (320 km; 199 miles)

NORTHROP NV-128

First flown in May 1974, the NV-128 was an MQM-74A modified to accommodate additional payload volume and weight. It carried a CAI 70 mm frame camera as a primary sensor and a forward oblique motion camera. An infra-red line-scanner or panoramic camera was interchangeable with the still camera system. Navigation was by an on-board Litton AMECON Loran system which provided automatic commands to the Northrop autopilot. Overall length was increased to 4·08 m (13 ft 4¾ in), and the NV-128 was capable of speeds up to 450 knots (834 km/h; 518 mph) and altitudes of more than 10,670 m (35,000 ft).

Two NV-128s were completed, and were evaluated in 1974 as a part of the tri-service Solid Shield amphibious landing exercises held at Camp Lejeune, North Carolina. The NV-128 was reportedly a contender in 1975 for a US Army requirement for a tactical high-speed surveillance RPV.

NORTHROP CHUKAR II

US military designation: MQM-74C

The MQM-74C is an improved version of the MQM-74A (1974-75 *Jane's*), evolved via an MQM-74B developmental model to meet requirements for a 500 knot (926 km/h; 576 mph) target. Production is under way of 718 MQM-74Cs for the US Navy; delivery of these began in early 1974, and the 400th was completed in October 1975.

Modified versions of the MQM-74C have been tested as remotely piloted vehicles, fitted with a Motorola radiation seeker and a warhead, under the US Navy's Persistent Arm programme. Three others were to be used in 1976 as flight demonstration vehicles in the US Air Force's TEDS programme.

Northrop Shelduck basic training target drone, a type produced for 18 countries

Shipboard launch of a Northrop NV-128 experimental surveillance RPV

Northrop MQM-74C Chukar II target drones (Williams WR24-7 turbojet engine)

The standard Chukar II/MQM-74C target aircraft was designed to meet requirements for a small, lightweight target for anti-aircraft gunnery, surface-to-air missile training and weapon system evaluation. Chukar II is used at the NATO Missile Firing Installation (NAMFI) in the Mediterranean, and at the US Army's White Sands missile range. Meteor of Italy (which see) has been selected by NATO's Hawk management office to provide Chukar II services at the Salto di Quirra range in Sardinia, and the Chukar II is used also in the Persian Gulf, to provide crew training and weapon system evaluation for the Imperial Iranian Navy.

TYPE: Turbojet-powered radio-controlled recoverable target drone.

WINGS: Cantilever shoulder-wing monoplane. No dihedral. Detachable aluminium wings, each with electrically-actuated aileron.

FUSELAGE: Aluminium semi-monocoque structure housing all equipment, power plant and fuel tankage. Nose and tail skins removable for access to electronic components and power plant. Underslung engine air intake duct.

TAIL UNIT: Cantilever aluminium structure of inverted Y form, comprising fixed vertical fin, fixed tailplane halves and two electrically-actuated elevators. Tailplane anhedral 30°.

POWER PLANT: One Williams Research Corporation Model WR24-7 (J400-WR-401) turbojet engine, rated at 0·78 kN (176 lb st). Fuel tank in centre of fuselage.

SYSTEMS: Electrical power only, from engine-driven alternator through a rectifier-regulator. 28V nickel-cadmium battery secondary power source used during glide.

CONTROL SYSTEM: Out-of-sight control by automatic stabilisation and command, with radar tracking; in-sight control with visual acquisition aids. Proportional feedback stabilisation and control system for pitch and bank. Engine throttle position, altitude hold initiation and recovery system initiation controlled by audio tone signals. Components include Motorola AN/DKW-1 integrated target control system, aileron and elevator servos, and altitude hold pressure transducer. Command control antenna in upper forward fuselage.

EQUIPMENT: On-board acquisition and tracking aids include fore and aft Luneberg lenses for passive radar augmentation, four wingtip-mounted MK-28 Mod 3 infra-red flares, pyrotechnic infra-red plume augmentors, active L-band augmentation, and a chaff or a pyrotechnic flash and smoke system, designed to improve visual augmentation. Main payload compartment is in front fuselage between control equipment bay and fuel tank.

LAUNCH AND RECOVERY: Zero-length launching by means of two Mk 91 Mod 0 JATO rockets and a ZL-5 launcher. Two modes of command recovery are utilised. Normal method consists of automatic drone pull-up followed by main parachute deployment, and is initiated automatically in emergencies such as interruption of continuous radio signal or loss of parachute command channel. Alternative mode consists of direct main parachute deployment and is initiated automatically on loss of electrical power. Main parachute, housed in fuselage immediately aft of wing, is a 9·14 m (30 ft) diameter extended-skirt nylon canopy, with automatic disconnect and impact.

DIMENSIONS, EXTERNAL:

Wing span	1·73 m (5 ft 8 in)
Length overall	3·87 m (12 ft 8½ in)
Body diameter	0·36 m (1 ft 2 in)
Height overall	0·71 m (2 ft 4 in)

WEIGHTS:

Weight empty	122 kg (269 lb)
Max launching weight	196·6 kg (433·5 lb)

PERFORMANCE:

Max level speed at 6,100 m (20,000 ft)	515 knots (954 km/h; 593 mph)
Max level speed at S/L	475 knots (880 km/h; 547 mph)
Econ cruising speed at S/L	250 knots (463 km/h; 288 mph)
Max rate of climb at S/L with full fuel	1,780 m (5,840 ft)/min
Service ceiling	12,200 m (40,000 ft)
Range at max speed at S/L	195 nm (360 km; 224 miles)
Range at max speed at 6,100 m (20,000 ft)	315 nm (584 km; 363 miles)
Range at econ cruising speed at S/L	230 nm (426 km; 265 miles)

NV-130 'Nutcracker', Northrop's contender in the USAF's TEDS competition

Quarter-scale model of Northrop ARPV

NORTHROP NV-130 'NUTCRACKER'

US military designation: Tactical Expendable Drone System (TEDS)

Northrop and Beech Aircraft Corporation (which see) have been selected by the US Air Force to develop and produce flight demonstration examples of a new Tactical Expendable Drone System (TEDS). The Northrop contract, announced in June 1975, is worth $1·5 million, and covers an 18-month validation and development programme which will include flight testing, systems definition, and computer simulations of operational effectiveness. First flight of a Northrop prototype, which has the manufacturer's designation NV-130, was made on 22 January 1976. After completion of the validation programme the USAF will decide whether to proceed with full-scale development, after which a production decision will be taken. The drone system is intended for use as a credible decoy, to increase the survivability of manned strike aircraft, or for an ECM role in support of manned aircraft.

Northrop's TEDS, named 'Nutcracker' by the company, has a high degree of commonality with the MQM-74C target drone from which it is derived. Three MQM-74C vehicles were modified for the flight test programme, and an additional vehicle was used for extensive ground tests and payload performance measurements. The basic fuselage, wings and engine are common with the MQM-74C, except that the expendable mission permits an uprating of the engine thrust; electrical power has been increased; the recovery system is replaced with an auxiliary fuel tank; the lower aft fuselage fairing has been enlarged; and wing pods for payload equipment have been added.

TYPE: Turbojet-powered expendable ECM drone.

WINGS: As MQM-74C.

FUSELAGE: Of same construction as MQM-74C. Entire Chukar II nose shell, with increased nose payload volume, is removable for rapid access to mission payload. Aft fairing enlarged to accommodate flight electronics. In operational configuration, the recovery system is replaced by an auxiliary fuel tank.

POWER PLANT: One Williams Research Corporation WR24-7 (J400-WR-401) turbojet engine, uprated for short-life expendable application to 0·87 kN (195 lb st).

LAUNCH AND RECOVERY: JATO-assisted zero-length launch. No recovery system (drone programmed to crash-impact).

SYSTEMS: Electrical power from engine-driven alternator: 28V DC through converter-regulator, and three-phase 200V AC.

CONTROL SYSTEM: Two-axis attitude controller (pitch and roll). Roll-to-turn manoeuvres are performed by using the ailerons. Altitude control is closed around the pitch loop. Vehicle airspeed is controlled by sensing engine rpm and advancing or retarding throttle position. Control of the vehicle in altitude, heading and airspeed is through the introduction of flight profile commands from the programmer to the autopilot. Navigation is pre-programmed, using predicted airspeed and an on-board heading reference sensor. Component include programmer/autopilot, directional gyro, vertical gyro, altitude transducer, rpm sensor and contro surface actuators.

EQUIPMENT: Penetration aids.

DIMENSIONS, EXTERNAL:

Wing span	2·04 m (6 ft 8½ in
Wing aspect ratio	4·(
Length overall	4·04 m (13 ft 3¼ in
Body diameter (max)	0·36 m (1 ft 2 in
Height overall	0·71 m (2 ft 4⅛ in

AREA:

Wings, gross	0·74 m² (8·0 sq ft

WEIGHTS:

Max launching weight:	
incl boosters	265 kg (585 lb
without boosters	233·5 kg (515 lb

NORTHROP ADVANCED RPV

In April 1975, Northrop was one of three aerospace companies to be awarded a one-year fixed-price study contract by the US Air Force to design an advanced RPV (ARPV) capable of carrying out reconnaissance, electronic warfare and strike missions, for service in the 1980s. Northrop, whose contract is worth $499,614, is assisted in the study by Texas Instruments and General Research Corporation.

Previous related studies performed by Northrop for the USAF include the multi-mission RPV system and the drone control and data retrieval system study undertaken with RCA to design a system for data links and control of multiple RPVs.

Northrop's ARPV contender is powered by a General Electric J85 turbojet engine, is 9·14 m (30 ft) long, and has a wing span of 4·57 m (15 ft). It can be air or ground launched, is recoverable, and can carry both external stores and an internal payload.

ROCKWELL INTERNATIONAL

ROCKWELL INTERNATIONAL CORPORATION COLUMBUS AIRCRAFT DIVISION

ADDRESS:
4300 East Fifth Avenue, Columbus, Ohio 43216
MANAGER, PUBLIC RELATIONS:
Dent Williams

Columbus Aircraft Division has a number of company-funded activities concerning RPV development. It also has programme contracts from the US Air Force and US Army which include an ARPV study (described hereafter); and studies for advanced remotely piloted modular aircraft, RPV man/machine interface, and RPV operator task loading simulation.

ROCKWELL INTERNATIONAL ADVANCED RPV

In April 1975, Rockwell International was one of three aerospace companies to be awarded a one-year fixed-price study contract by the US Air Force to design an advanced remotely piloted vehicle (ARPV) capable of carrying out reconnaissance, electronic warfare and strike missions, for service in the 1980s. The Rockwell contract is worth $699,684.

As shown in the accompanying illustration, the Rockwell contender is of sweptwing, Vee-tailed configuration, with the power plant mounted in a dorsal pod. It is expected also to have a retractable tricycle landing gear, and to be able to carry mission payloads in wingtip-mounted pods.

Further details of the ARPV programme can be found under the US Air Force heading in this section.

Artist's impression of the Rockwell International entry in the US Air Force Advanced RPV competition

LOS ANGELES AIRCRAFT DIVISION

ADDRESS:
International Airport, Los Angeles, California 90009

ROCKWELL INTERNATIONAL/NASA HiMAT

HiMAT (Highly Manoeuvrable Aircraft Technology) is a programme evolved by NASA's Dryden Flight Research Center at Edwards AFB, California. Its basic purposes are to speed up the progress of advanced design technology into the flight test phase; to assist designers in taking larger technological steps forward between generations of aircraft; and, more specifically, to provide a low-cost, low-risk means of testing the advanced manoeuvring capability of future aircraft.

After receipt of programme proposals from Grumman Aerospace Corporation, McDonnell Aircraft Company and Rockwell International, NASA announced in October 1975 the award of an $11·8 million contract to Rockwell for the design and construction of two prototype HiMAT remotely piloted research vehicles (RPRVs). The general appearance of these can be seen in the accompanying three-view drawing. In a 30-month programme, of which some 18 months will comprise the flight test phase, NASA will evaluate a number of advanced design technology features by means of the HiMAT vehicles.

To meet the requirements of the programme, HiMAT's design consists of a basic core vehicle, with a life of 100 hours, which will include the engine and all essential subsystems. To the core vehicle will be added, as modular units, the main wings, canard surfaces, tail surfaces, and engine intake and afterburner/exhaust structures. In this way the modular components can be replaced during the programme, at minimum cost, with others of alternative design. Among these is expected to be a so-called '2D' vectored-thrust exhaust nozzle; other features to be tested include advanced supercritical wings; variable-camber wings; deformable, self-trimming outer wings; CCV (control configured vehicle) techniques; a digital fly-by-wire system; and a variable-thrust engine control system.

The two HiMAT prototype vehicles are due to be delivered to NASA in October and November 1977, and to make their first flights in February and April/May 1978. They will be air-launched at about 13,720 m (45,000 ft) from a B-52 carrier aircraft, and their performance monitored by TV, telemetry and radar.

Meanwhile, in 1975 NASA began evolving control techniques for the HiMAT programme, first by using one of the three-eighths-scale unpowered glassfibre models of the F-15 fighter used in that aircraft's development programme, and later with two modified US Air Force BQM-34F Firebee II target drones to evaluate control in powered, supersonic flight.

The following description applies to the HiMAT prototypes as envisaged in 1976:

TYPE: Remotely piloted research vehicle.

WINGS: Blended wing/body design, of roughly double-delta configuration. Cantilever mid-wing, with sharply-swept main wings and canard forebody surfaces. Main wings have neither dihedral nor anhedral, and have ailerons and flaps/airbrakes on the trailing-edges. Canard surfaces have marked dihedral, are fitted with elevators, and have ogival-curve leading-edge strakes.

FUSELAGE: Blended wing/body, with area ruling.

TAIL UNIT: Twin, swept vertical tail fins at extremities of main wings; and twin, swept, outward-canted fins and rudders on short booms extending from trailing-edges of main wings at approx mid-span.

LANDING GEAR: Retractable tricycle gear, of skid type for landing on dry lake bed at Edwards AFB. All units retract rearward, main units to form continuous fairing with wing/tail booms, nose unit into underside of engine air intake trunk.

POWER PLANT: One General Electric J85-GE-21 turbojet engine (15·6 kN; 3,500 lb st dry and 22·25 kN; 5,000 lb st with afterburning), mounted centrally under fuselage.

GUIDANCE AND CONTROL: Primary control from ground console, by TV, telemetry, and radar link with on-board systems. If ground control is lost, backup inputs from the RPV will be relayed to a TF-104G chase-plane. On occasions when the TF-104G is out of control range, the HiMAT has an on-board self-righting system that will bring the RPRV into straight and level subsonic flight until the former can resume control.

EQUIPMENT: On-board equipment includes TV camera in cockpit, radar altimeter under nose, and angle of attack sensor in nose probe.

DIMENSIONS, EXTERNAL:

Wing span	4·63 m (15 ft 2·4 in)
Length overall, incl probe	6·43 m (21 ft 1·2 in)
Height overall	1·31 m (4 ft 3·6 in)

WEIGHTS:

Weight empty	1,270 kg (2,800 lb)
Max air-launching weight	1,542 kg (3,400 lb)
Thrust/weight ratio	approx 1

PERFORMANCE (estimated):

Max level speed (for 3 min at 12,200 m; 40,000 ft)	Mach 1·4
g limit	+8

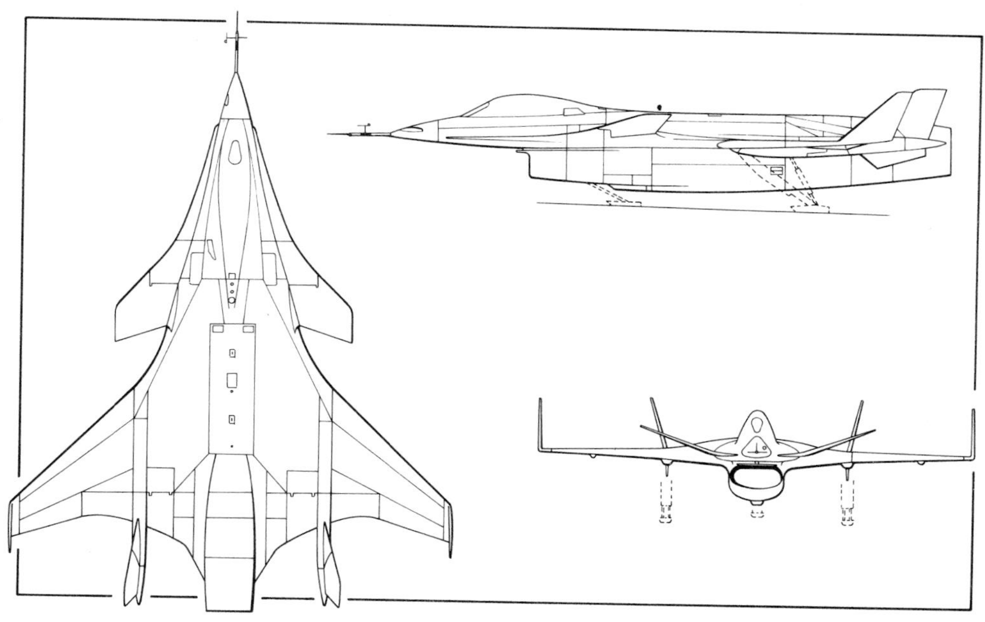

Rockwell International/NASA HiMAT advanced technology RPRV *(Michael A. Badrocke)*

Artist's impression of HiMAT being launched from B-52 carrier aircraft

SPERRY

SPERRY FLIGHT SYSTEMS DIVISION, SPERRY RAND CORPORATION

ADDRESS:
P.O Box 21111, Phoenix, Arizona 85036
Telephone: (602) 942-2311

SPERRY (CONVAIR/GENERAL DYNAMICS) F-102A DELTA DAGGER

USAF designations: QF-102A and PQM-102

Under a $5·5 million contract awarded on 31 March 1973, Sperry Flight Systems Division undertook conversion for the USAF of eight Convair F-102A interceptors to drone configuration, to provide up-to-date 'threat simulation' targets for US Air Force air-to-air weapons tests.

Two of the aircraft, designated QF-102A, retain normal cockpit controls and can be flown by monitoring pilots. The others of the development batch were designated PQM-102 and were designed and equipped entirely for unmanned operation.

The PQM-102 is the first-ever fighter aircraft converted for drone duties with no provision whatever for manned flight, and cannot be flown except under remote control. It was developed under sponsorship of the Armament Development and Test Center of AF Systems Command at Eglin AFB, Florida, and has greater speed and manoeuvrability than the manned F-102A, resulting from lower operating weight. The PQM-102 is ground controlled through a remote data link, and has a pre-programmed manoeuvring capability that can be initiated or terminated by the ground controller.

The first PQM-102 was flown for the first time on 13 August 1974, and 41 'Nullo' (unmanned) flights had been made by November 1975. These flights involved the firing of 53 missiles, including air-to-air and surface-to-air types. First operational flight of a PQM-102 was made on 25 June 1975, and nine PQM-102 targets had been expended by November 1975. Flight testing, conducted by the 6585th Test Group of the Air Force Special Weapons Center at Holloman AFB, New Mexico, was carried out over the White Sands Missile Range. Follow-on contracts to convert F-102As to PQM-102s have been awarded to Sperry Flight Systems, which has subcontracted airframe modification work to Fairchild Industries' Aircraft Service Division (which see), and certain control electronics to Vega Precision Labs, Vienna, Virginia. A total of 68 targets have been ordered, of a total planned USAF procurement of 128 PQM-102s over a six-year period. Of these, five are QF-102As with provision for a human pilot. The US Army will receive 24 of these for use in evaluating ground-to-air missiles. Most flights will take place over the Gulf of Mexico from Tyndall AFB, Florida, site of Aerospace Defense Command's Air Defense Weapons Center.

The following details apply to the PQM-102:

POWER PLANT: One Pratt & Whitney J57-P-23A turbojet engine, rated at 52 kN (11,700 lb st) dry and 76·5 kN (17,200 lb st) with afterburning.

PQM-102, converted from an F-102A by Sperry Flight Systems and the USAF's first converted-fighter target drone to have no provision for manned operation

DIMENSIONS, EXTERNAL:
- Wing span: 11·62 m (38 ft 1½ in)
- Length overall: 20·84 m (68 ft 4⅔ in)
- Height overall: 6·46 m (21 ft 2½ in)

AREA:
- Wings, gross: 64·57 m² (695·0 sq ft)

PERFORMANCE:
- Max speed: Mach 1·2 (688 knots; 1,274 km/h; 792 mph)
- Operating height range: 61 m (200 ft) to 16,760 m (55,000 ft)
- *g* limit: +8

TELEDYNE RYAN

TELEDYNE RYAN AERONAUTICAL

HEAD OFFICE AND WORKS:
2701 Harbor Drive, San Diego, California 92112
Telephone: (714) 291-7311

CHAIRMAN:
Robert C. Jackson

PRESIDENT:
Barry J. Shillito

VICE-PRESIDENTS:
D. L. Arney (Industrial Relations)
H. D. Drake (Electronic & Space Systems)
R. D. Fields (Finance and Controller)
T. E. Flannigan (Washington DC Office)
E. C. Oemcke (Aerospace Systems)
W. J. Wiley (Plant Operations)

PUBLIC RELATIONS AND COMMUNICATIONS MANAGER:
Robert B. Morrisey

The former Ryan Aeronautical Company was an indirect successor to Ryan Airlines Inc, which produced the aeroplane in which Charles A. Lindbergh made the first nonstop flight from New York to Paris in 1927. In February 1969, Ryan Aeronautical became a wholly-owned subsidiary of Teledyne Inc, and in December 1969 the company was renamed Teledyne Ryan Aeronautical.

The current activities of the company fall into two major categories, under the headings of Aerospace Systems and Electronic and Space Systems. The former group is concerned principally with the design, production and field operation of high-performance aerial jet targets and RPV systems.

The Electronic and Space Systems group is responsible for design and production of radar equipment for landing spacecraft on Mars in the Viking programme; electronic navigation and positioning equipment for rotating-wing and fixed-wing aircraft; remote sensors for Earth resources studies; electronic warfare systems; and microwave antennae.

The Teledyne Ryan AN/APN-200 Doppler velocity sensor, installed in the Lockheed S-3A ASW patrol aircraft, incorporates electronic techniques perfected by Teledyne Ryan in earlier Dopplers and in its Apollo Moon-landing radars; it was the first major subsystem to work with the S-3A's on-board central computer. In combination with the computer and an inertial system, the APN-200 can be used in accurate point-to-point navigation and in the critical localisation procedures to pinpoint enemy submarines.

Teledyne Ryan continues to maintain a strong technical interest in vertical take-off and landing aircraft, but this makes only a minor contribution to the volume of the company's business.

Major production items at Teledyne Ryan's plant for many years have been the Firebee jet-powered targets and special-purpose vehicles (pre-programmed and remotely piloted) for various types of reconnaissance mission.

The company has other important contracts in the missile and space fields, including design and fabrication of radar altimeters, precision antennae and structures for advanced space vehicles.

Teledyne Ryan Model 124 Firebee I subsonic target drone of the US Air Force

As described under the NASA entry, Teledyne Ryan BQM-34E Firebee II supersonic drones are being used in remotely piloted research vehicle (RPRV) test programmes at the Administration's Dryden and Langley Research Centers. Teledyne Ryan has itself conducted, under contract to NASA, a feasibility study involving the fitting of various different wing planforms and aerofoil sections to the Firebee II for test purposes. These included three types of supercritical wing, three types of supersonic wing, and a laminar-flow wing.

Among other activities, Teledyne Ryan is studying a number of concepts for mini-RPVs, a field in which the US services are particularly interested at the present time.

TELEDYNE RYAN MODEL 124 FIREBEE I

USAF/US Navy designations: BQM-34A and BQM-34S

US Army designation: MQM-34D

The Firebee I is a remotely piloted high-speed turbojet-powered vehicle which was developed as a joint US Air Force/Army/Navy project, with the USAF Air Research and Development Command having technical cognisance of its development.

Glide flight tests of the original version of the Firebee were begun in March 1951, and the first powered flights were made that Summer at the USAF Holloman Air Development Center, Alamogordo, New Mexico. A total of 1,280 of these early Q-2A and KDA versions were built eventually for all three US services and for the RCAF, and full details of these can be found in previous editions of *Jane's*.

Development of the current **BQM-34A** (originally Q-2C) Firebee began on 25 February 1958, with the object of obtaining a much-improved all-round performance. Construction of the prototype started on 1 May and it flew for the first time on 19 December 1958. The first production model flew on 25 January 1960.

Firebee I descending by recovery parachute

By February 1976, a total of 5,637 Firebee Is (including

more than 4,350 BQM-34A/S and MQM-34D targets) had been produced. The latest contracts, for an additional 318 Firebees, extend production through the 1977 calendar year. They include a contract from the Japan Defence Agency for Firebees for use in training missile and gunnery crews. These are being built by Fuji (which see), which had completed 21 by the end of 1975 and was scheduled to manufacture five more in 1977.

In 1971 Teledyne Ryan announced that a BQM-34A Firebee drone had successfully performed a simulated dogfight against a US Navy Phantom fighter aircraft over the Pacific Missile Test Range using a company-developed flight control system known as MASTACS. This development has evolved into a full three-axis flight control system capable of executing co-ordinated flight high-*g* manoeuvres, which has been produced to satisfy high-*g* manoeuvring performance requirements.

The Firebee targets currently being manufactured to incorporate a Motorola integrated track and control system (ITCS) or Vega Track and Control System (VTCS) have the designation **BQM-34S**.

Up to the end of January 1976 such targets had provided more than 24,670 flights in support of weapon system and target research, development, test, evaluation, quality assurance, training and annual service practices conducted by the US Army, Navy and Air Force, and certain foreign governments.

On 16 and 30 January 1974 a 'stretched' version of a US Navy Firebee I target made two flights over the Atlantic Fleet Weapons Range at Cabras Island, Puerto Rico, carrying 181 kg (400 lb) more fuel than the standard model. The flights lasted 2 hr 6 min and 2 hr 17 min respectively, compared with the normal average endurance, with standard full load, of approx 1 hr.

RPVs (remotely piloted vehicles) using airframes developed from that of the Firebee I include those bearing the Teledyne Ryan Model numbers 147, 234 and 239; these are described separately.

The following details refer to the standard BQM-34A target vehicle:

TYPE: Remotely piloted jet target vehicle.

WINGS: Cantilever mid-wing monoplane. Wing section from leading-edge to 0·264 chord NACA 0009·932; from 0·264 chord to trailing-edge NACA 63A014·63. Thickness/chord ratio 14%. No dihedral or incidence. Sweepback at quarter-chord 45°. Three-spar aluminium alloy semi-monocoque structure, incorporating leading-edge droop. Single-spar ailerons of magnesium, aluminium and stainless steel, operated by Lear servo-actuators. Wingtips detachable to minimise damage on landing. Provision for wingtip extensions to increase span.

FUSELAGE: Conventional semi-monocoque structure of aluminium alloy, with chemical-etched components to save weight and simplify subassemblies. Glassfibre tailcone and nose section. Keel under central portion, to absorb landing impact.

TAIL UNIT: Single assembly attached to fuselage by four bolts. All surfaces swept 45° at quarter-chord. Fin is multi-spar aluminium alloy monocoque structure. Trim rudder is operated electrically by Bendix actuator. Single-spar aluminium alloy monocoque tailplane. Magnesium elevators powered by Lear servo. Glassfibre fin-tip houses telemetry antenna. Glassfibre tailplane tips house radar echo enhancing antennae. Ventral fin under tailcone, aft of main tail unit.

POWER PLANT: One 7·56 kN (1,700 lb st) Teledyne CAE J69-T-29 turbojet engine. Fuel tank integral within forward section of fuselage, capacity 378 litres (100 US gallons). Provision for one 94·5 litre (25 US gallon) auxiliary fuselage tank and one 378 litre (100 US gallon) drop-tank under each wing. Refuelling point above forward fuselage. Oil capacity 5·75 litres (1·5 US gallons).

SYSTEMS: Electrical power only. Primary power furnished by a 28V DC engine-driven generator of 200A capacity. Power for control systems furnished by a 400Hz 115V 250W AC inverter; a 28V 12·5Ah lead-acid battery provides power for the electrical devices of the recovery system and for control during the pre-landing glide phase.

ELECTRONICS AND EQUIPMENT: AN/DRW-29 radio receiver with Dorsett TM-4-31A telemetry system, or DKW-2 guidance transponder. A/A37G-3 or A/A37G-8 flight control system.

LAUNCH AND RECOVERY: Either air-launching, from a suitably-modified aircraft, or ground-launching, using a 50·3 kN (11,300 lb st) (nominal) solid-propellant JATO bottle, can be used; the US Navy has also launched BQM-34A Firebees from ships under way at up to 15 knots (27·5 km/h; 17 mph). The two-stage parachute recovery system operates automatically in the event of a target hit, loss of radio wave carrier from the remote control station, engine failure, or upon command by the remote control operator. To prevent damage by dragging, the recovery system incorporates a disconnect which releases the parachute from the Firebee on contact with the ground or water.

GUIDANCE AND CONTROL: Remote control methods for Firebee I include a choice of radar, radio, active seeker and automatic navigator, all developed and designed by Teledyne Ryan. Remote control is normally accomplished through a UHF radio link using an AN/FRW-2 or SRW-4 ground transmitter and an AN/DRW-29 airborne receiver. The target can be controlled either from a manned aircraft or from a surface station. Remote command includes activation of special scoring and augmentation equipment in the target. A beacon in the Firebee facilitates radar tracking from the remote control station; there is provision to install a telemetry system to relay pertinent flight data to the controller if required. Basic commands consist primarily of on/off functions which are received by the on-board radio receiver and relayed to the appropriate subsystem for action. A Motorola ITCS (integrated tracking and control system) is fitted in the BQM-34S version. Other types of remote command and tracking systems can include a microwave command and guidance system which can control the Firebee beyond line-of-sight from a ground station through an airborne relay station. This equipment operates with coded impulses which reduce the possibility of interference from other electronic signals. Operational Firebees can be equipped with an increased manoeuvrability three-axis flight control system for tactical air combat simulation which gives the target the capability to perform 4, 5 or 6*g* manoeuvres. The Firebee I can also be equipped with active and passive radar augmentation systems as well as afterburning plume devices to provide realistic threat simulation for training personnel in the firing of air-to-air and surface-to-air weapon systems. A Radar Altimeter Low Altitude Control System (RALACS), when added to the Firebee I control system, permits precision low-altitude flights at 15 m (50 ft) over water and 30 m (100 ft) over land.

OPERATIONAL EQUIPMENT: Wide range of possible 'building block' equipment combinations, including visual or radar-reflecting banner targets; radar or infra-red Towbee towed targets or tow target Doppler 'bird'; two underwing drop-tanks, 500 lb bombs or bomblet dispensers; AN/ALE-33 or other ECM containers; wingtip tow launchers, camera pods, scoring equipment, flares or other forms of infra-red augmentation, or reflector pods for radar augmentation. The BQM-34A can be equipped with adjustable travelling wave tube amplifiers for use as radar echo enhancers in the L, S, X and C frequency bands. These devices provide realistic radar appearances for all-size targets from the smallest fighter to the largest bomber aircraft.

BQM-34A Firebee I target drone dropping away from the launch pylon of its carrier aircraft

DIMENSIONS, EXTERNAL (BQM-34A):
- Wing span 3·93 m (12 ft 10·8 in)
- Wing chord (streamwise, constant) 0·85 m (2 ft 9·4 in)
- Wing aspect ratio 4·632
- Length overall 6·98 m (22 ft 10·8 in)
- Body diameter 0·94 m (3 ft 1·2 in)
- Height overall 2·04 m (6 ft 8·4 in)
- Tailplane span 2·26 m (7 ft 5 in)
- Tailplane chord (streamwise, constant) 0·69 m (2 ft 3 in)
- Fin chord at tip 0·56 m (1 ft 10 in)

AREAS:
- Wings, gross 3·34 m² (36·0 sq ft)
- Ailerons (total) 0·39 m² (4·16 sq ft)
- Fin 1·05 m² (11·28 sq ft)
- Ventral fin 0·13 m² (1·43 sq ft)
- Rudder 0·043 m² (0·46 sq ft)
- Tailplane 1·55 m² (16·69 sq ft)
- Elevators 0·64 m² (6·84 sq ft)

WEIGHTS AND LOADING:
- Weight empty 680 kg (1,500 lb)
- Basic gross weight 934 kg (2,060 lb)
- Max launching weight 1,134 kg (2,500 lb)
- Max wing loading 338·3 kg/m² (69·3 lb/sq ft)

PERFORMANCE:
- Never-exceed speed Mach 0·96 (635 knots; 1,176 km/h; 731 mph at 15,240 m; 50,000 ft)
- Max level speed at 1,980 m (6,500 ft) 600 knots (1,112 km/h; 690 mph)
- Max cruising speed at 15,240 m (50,000 ft) at 816 kg (1,800 lb) AUW 547 knots (1,015 km/h; 630 mph)
- Stalling speed, power on, at 816 kg (1,800 lb) AUW 101 knots (187 km/h; 116 mph)
- Max rate of climb at S/L at 1,000 kg (2,200 lb) AUW 4,875 m (16,000 ft)/min
- Operating height range 15 m-18,300 m (50 ft to more than 60,000 ft)
- Endurance at 15,240 m (50,000 ft), incl 2 min 40 sec glide after fuel expended 75 min 30 sec
- Max range 692 nm (1,282 km; 796 miles)
- Flotation time with 25% fuel 24 hr

TELEDYNE RYAN MODEL 147

USAF designations: in AQM-34 series, and XQM-103

The Model number 147, and the basic USAF designation AQM-34, encompass a large family of surveillance, reconnaissance and ECM RPVs evolved from the subsonic BQM-34A/MQM-34D Firebee I target. They are air-launched from DC-130A or E Hercules motherplanes which combine the functions of command, tracking and data relay aircraft. The original Model 147A, which was little more than a modified Firebee I with a new guidance system and increased fuel capacity, was developed in 1962.

Details of the early Model 147s, and their uses from 1964-74, can be found in the 1975-76 *Jane's*.

By 1974, many hundreds of Model 147s had been delivered by Teledyne Ryan for operational use, and in September 1972 the company was permitted to identify no fewer than 24 members of this large family. Details of the most recent variants follow, together with the appropriate USAF designation where applicable:

Modified Model 147G (XQM-103). Research RPV, highly modified from standard Model 147G under a programme managed by AF Systems Command's Flight Dynamics Laboratory at Wright-Patterson AFB, Ohio. First of six captive flights made on 19 January 1974; first free flight made in early 1975. Differs structurally from standard 147G in having wings and fuselage strengthened to withstand a sustained symmetrical load factor of 10*g*. The XQM-103 is air-launched from a DC-130 and air-recovered by CH-53 helicopter in the normal fashion. It was expected to make 24 flights during 1975-76, to provide data for use in other near-term RPV programmes using similar vehicles and to begin to answer some of the long-term questions regarding future RPVs. Among areas to be studied are the relationship between the RPV and its remote pilot; high-*g* manoeuvring; target acquisition under co-operative conditions; remote pilot techniques; and subsystem evaluations. An important part of the programme is the remote pilot station, one of which has been specially built at Edwards AFB and one at Wright-Patterson AFB.

Model 147H (AQM-34N). Medium-altitude reconnaissance version. Details in 1975-76 *Jane's*.

Model 124I. Designation of a hybrid Model 124/147-type vehicle produced for export, about a dozen of which are said to have been supplied to Israel in 1971 and later used for high-altitude photographic reconnaissance overflights of Egyptian (and possibly other Arab) territory. Other press reports, also unconfirmed, have suggested that Israel has used Firebee I-type drones (not necessarily of the same type) to deliver Israeli-built air-to-ground missiles against Egyptian missile sites. Teledyne Ryan programme and command guidance system. Photographs released in early 1975 reveal this RPV, which was used for day and night reconnaissance during the October 1973 war, to have endplate auxiliary fins and an elongated nose similar to that of the AQM-34L and M versions used by the USAF. Israeli RPVs were also used during the 1973 war to decoy Arab SAM missiles; one vehicle successfully drew the fire of no fewer than 32 SAMs and still returned to its base.

Model 147NA (AQM-34G). Medium-altitude ECM version. Has extended-span wings, strengthened to carry active (jamming) or passive (chaff dispensing) ECM pods externally (one pylon under each wing); small antenna fairing on top of fin. Teledyne Ryan programme and command guidance system. Built under USAF Compass Bin and Combat Angel programmes. In service with 11th Tactical Drone Squadron of USAF at Davis-Monthan AFB, Arizona.

Model 147NC (AQM-34H). Medium-altitude ECM version, similar to 147NA but with higher launching weight and twin cropped-delta-shape endplate auxiliary fins. Underwing ECM (ALE-2 chaff dispenser) pods. Teledyne Ryan programme and command guidance system. Other equipment includes Sperry Univac APW-25 or 26 transponder. Built under USAF Compass Bin and Combat Angel programmes. Developed for use in Vietnam, but not used operationally. In service since 1969 with 11th Tactical Drone Squadron, Davis-Monthan AFB. Illustrated in 1973-74 *Jane's*. Has been used to evaluate a prototype multiple drone control (MDC) system installed in a DC-130A launch/director aircraft and able to control up to eight drones simultaneously. This system is to be modified to enable it similarly to handle multiple drones of the improved AQM-34V type (see next entry).

Teledyne Ryan Model 147NC (AQM-34V) electronic warfare RPV *(US Air Force)*

Model 147NC (AQM-34V). Update of AQM-34H Combat Angel RPVs by Teledyne Ryan, E-Systems (Melpar Division) and Sperry, to increase active and passive ECM jamming. Also improved are flight controls, launch, recovery, and multiple drone control capabilities. Power plant as AQM-34H; AUW increased. New equipment includes E-Systems (Melpar Division) modular noise jammers, containing five jammers covering three bands, and MB Associates ALE-38 improved bulk chaff dispenser pods on the underwing hardpoints. The jammers are contained in a 1·70 m (5 ft 7 in) modular nose section; ALE-38 pods weigh 227 kg (500 lb) each. Prominent airscoop on top of nose. Thirty-four existing AQM-34H and J drones involved in conversion programme under $11·95 million USAF contract awarded in September 1974. Follow-on contract for $6·07 million announced in Spring 1975, covering 16 new-build AQM-34Vs. First flight of an AQM-34V was made in the Summer of 1976, when it was air-launched at 4,570 m (15,000 ft) from a DC-130E carrier aircraft of the US Air Force's 6514th Test Squadron at Hill AFB, Utah. The RPV climbed to 7,620 m (25,000 ft) before initiation of the MARS (mid-air retrieval system); all objectives of the flight were achieved successfully. First deliveries, to 11th Tactical Drone Squadron, due in 1976.

Model 147NC (M-1) (AQM-34J). Medium-altitude training version of 147NC, to which it is externally similar except for the absence of underwing pylons. Built under USAF Compass Bin and Combat Angel programmes. In service for reconnaissance RPV training with 11th Tactical Drone Squadron.

Models 147NP, NQ, NRE, NX, SA and SB. For details see 1975-76 *Jane's*.

Model 147SC (AQM-34L). Low-altitude photographic reconnaissance version. Endplate auxiliary fins. Teledyne CAE J69-T-41A engine, with 517 kg (1,140 lb) fuel load, pre-programmed navigation system, utilising a Doppler navigator and digital programmer, and remote control capability via a microwave command guidance system or radio link from an airborne or ground station. Built under USAF Compass Bin and Buffalo Hunter programmes. After the cessation of bombing on 15 January 1973 this RPV and the Lockheed SR-71 manned strategic reconnaissance aircraft were the only USAF reconnaissance types permitted to overfly North Vietnam. Three AQM-34Ls, modified by Teledyne Ryan as prototypes and redesignated **AQM-34L (TV-1),** were used in Vietnam for real-time navigation and reconnaissance through forward- and down-looking TV camera. Details of AQM-34L sensors in 1973-74 *Jane's*. One AQM-34L used in 1975 to flight test a new TV target detection system with a dual recording capability and a stop/instant replay action similar to that used in domestic television sports broadcasts. Three AQM-34Ls modified as prototypes for BGM-34C (Model 239) RPV (which see).

Model 147SC (YAQM-34U). Designation of six AQM-34Ls in Lear Siegler Update programme (see 1973-74 *Jane's*). Five later modified as prototypes for Model 239 (BGM-34C) (which see).

Model 147SD (AQM-34M). Low-altitude photographic reconnaissance version, generally similar to AQM-34L. Radar altimeter fitted. Seventy-eight ordered, including eight for flight testing. Built under USAF Compass Bin and Buffalo Hunter programmes. Some fitted in 1972 with Teledyne Systems Co Loran receivers and

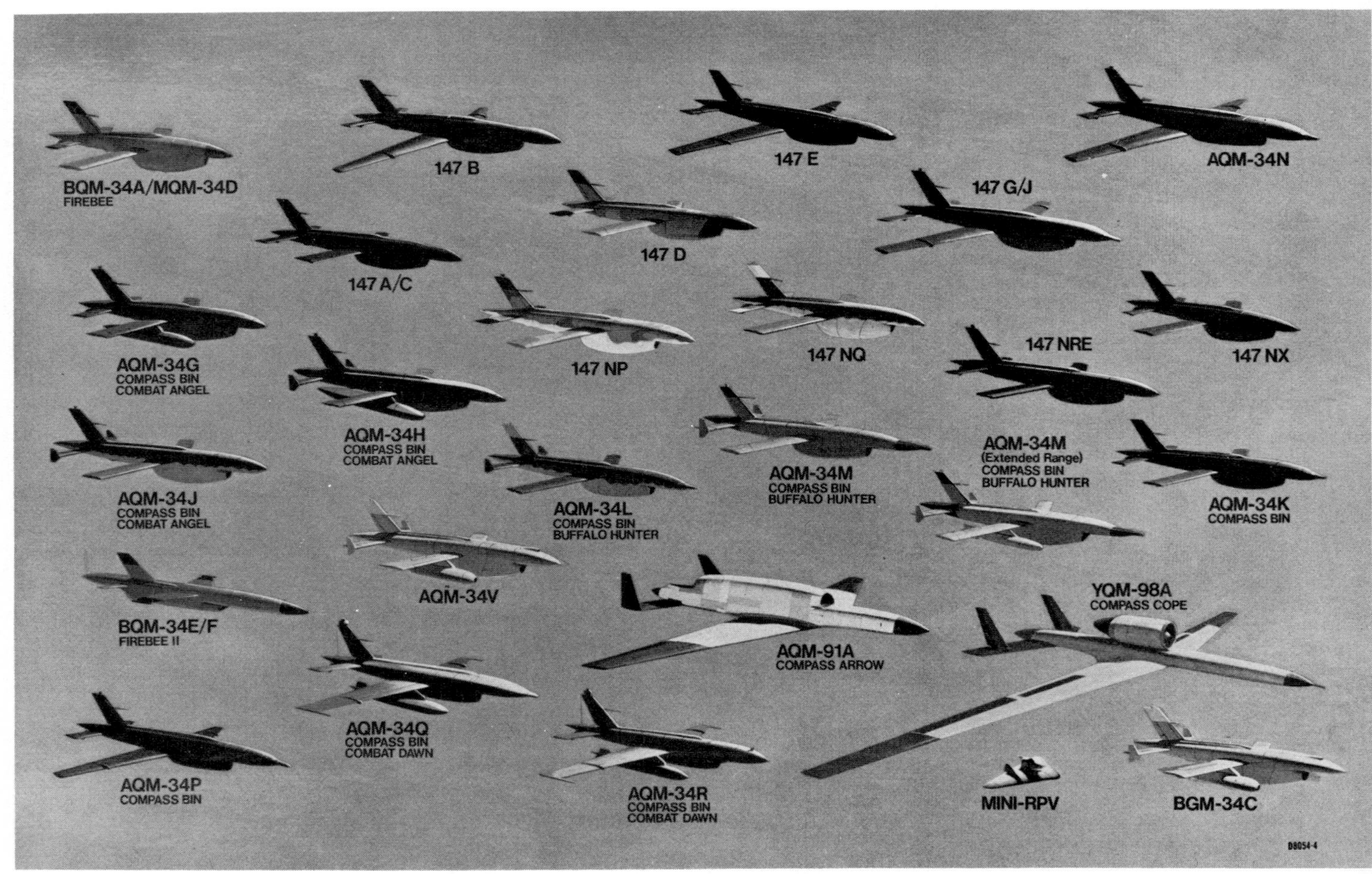

Currently-identifiable range of drones and RPVs designed by Teledyne Ryan, showing USAF designations and programme associations

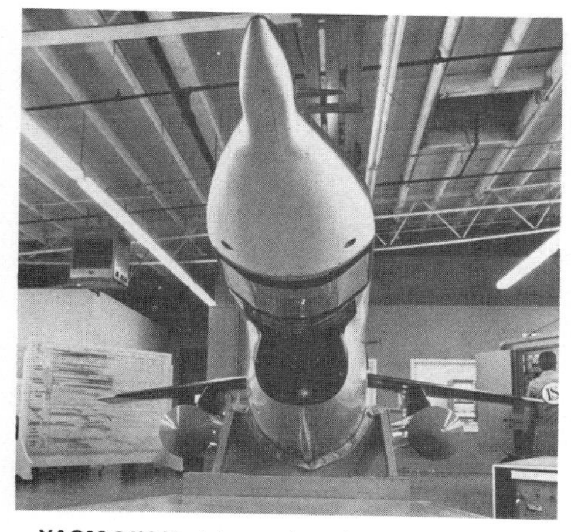

YAQM-34U Update version of the Model 147SC

redesignated **Model 147SD Loran/AQM-34M(L)**. Extended-range version, with underwing drop-tanks, also built. One AQM-34M was allocated for flight testing at Edwards AFB, California, under USAF Compass Robin programme (see 1975-76 *Jane's* for details).

Model 147SK. Version evaluated by US Navy (see 1975-76 *Jane's*).

Model 147SRE (AQM-34K). Low-altitude night reconnaissance version, similar to other 147S models except for slightly shorter fuselage. No antenna fairing on top of fin. Built under USAF Compass Bin programme.

Model 147T (AQM-34P). High-altitude surveillance version, with much-increased wing span of the 147H and Teledyne Ryan programme and command guidance system. Built under USAF Compass Bin programme. Damaged airframes displayed in Peking and by North Vietnam included one example of this version (airframe No. 17).

Model 147TE (AQM-34Q). Medium-altitude surveillance version. Teledyne Ryan programme and command guidance system and two underwing drop-tanks. Built under USAF Compass Bin programme, and with 147TF was also one of two models which formed part of USAF's Combat Dawn programme. Extended-span wings, with fences, have a slender tubular weight fairing at each wingtip and are of increased chord compared with other medium/high-altitude versions. Large bulbous data link antenna fairing on top of fin, and other aerials projecting from fuselage.

Model 147TF (AQM-34R). Built under USAF Compass Bin programme. Was also the second Combat Dawn version for Strategic Air Command: twenty reported to have been ordered in 1971. Airframe externally similar to 147TE, but probe-type fairing on nose is omitted and there is a small blade antenna on top of the nosecone. Provision for underwing stores. Said to be capable of cruising above 18,300 m (60,000 ft) at a speed of 420 knots (780 km/h; 485 mph).

Power Plants: All Model 147s have a single Teledyne CAE turbojet engine as follows:
- NA, NC, NC(M-1): 7·56 kN (1,700 lb st) J69-T-29.
- SC, SD, SD Loran, SRE, 124I: 8·54 kN (1,920 lb st) J69-T-41A.
- T, TE, TF: 12 kN (2,700 lb st) J100-CA-100.

Dimensions, external:

Wing span:

SC	3·96 m (13 ft 0 in)
NA, NC, NC(M-1), SD, SD Loran, SRE and 124I	4·42 m (14 ft 6 in)
T, TE and TF	9·75 m (32 ft 0 in)

Length overall:

NA, NC and NC(M-1)	7·92 m (26 ft 0 in)
SRE	8·84 m (29 ft 0 in)
SC, SD, SD Loran, T, TE and TF	9·14 m (30 ft 0 in)
124I	9·45 m (31 ft 0 in)

Body diameter:

NA, NC, NC(M-1), SC, SD, SD Loran, SRE and 124I	0·94 m (3 ft 1·2 in)
T, TE and TF	1·01 m (3 ft 3·6 in)

Weights:

Max launching weight:

NA	1,671 kg (3,684 lb)
NC (AQM-34H)	1,700 kg (3,749 lb)
NC (AQM-34V)	2,041 kg (4,500 lb)
NC(M-1)	1,299 kg (2,865 lb)
SC	1,390 kg (3,065 lb)
SC (TV-1)	1,429 kg (3,150 lb)
SD and SD Loran	1,412 kg (3,113 lb)
SRE	1,527 kg (3,367 lb)
T	1,720 kg (3,792 lb)
TE	1,755 kg (3,870 lb)
TF:	
without underwing stores	1,859 kg (4,100 lb)
with underwing stores	2,812 kg (6,200 lb)
124I	1,474 kg (3,250 lb)

Teledyne Ryan 147SD (AQM-34M) RPV under the wing of a DC-130 Hercules director aircraft

Teledyne Ryan Model 147TE remotely piloted vehicle (USAF designation AQM-34Q). Other versions of Model 147 are suspended from pylons of DC-130E launch aircraft in background

Teledyne Ryan AQM-34R (Model 147TF) being launched from a USAF Lockheed DC-130E control aircraft

TELEDYNE RYAN MODEL 166 FIREBEE II

US Navy designations: BQM-34E and BQM-34T
USAF designation: BQM-34F

Under contract to the US Navy and USAF, Teledyne Ryan Aeronautical is producing the Model 166 Firebee II supersonic target vehicle, an advanced development of the BQM-34A Firebee I which can provide aerial target presentations at above 18,300 m (60,000 ft) at a supersonic dash speed of Mach 1·5 for a period of 14 minutes.

Fourteen XBQM-34E development Firebee IIs were built under US Navy contract, and these underwent a successful operational test and evaluation programme in 1968-69. One static test airframe was also completed.

Three versions of the Firebee II have been built in quantity, as follows:

BQM-34E. For US Navy: total of 116 built. First operational flight 1 June 1972 at Pacific Missile Test Range. In service from 1973 with US Navy Squadron VC-8 in Puerto Rico. Production completed.

BQM-34F. For US Air Force: total of 99 built. Slightly heavier than BQM-34E, with corresponding adjustment of performance, due to different augmentation and scoring systems and addition of recovery parachute assembly required for mid-air retrieval system (MARS). Under-

Teledyne Ryan Model 166 Firebee II supersonic target, in production for the US Navy

US Air Force BQM-34F Firebee II supersonic target drone, with ventral fuel pod still attached

went operational testing and evaluation in 1973 with the USAF's 6514th Test Squadron at Edwards AFB, California. Production completed.

BQM-34T. Current production version for US Navy, incorporating a Motorola integrated track and control system (ITCS). Total of 54 ordered; deliveries began in mid-1974.

Under an agreement announced in September 1968, the Guided Weapons Division of British Aircraft Corporation has licence rights, covering the UK, Australia, Norway and Denmark, to manufacture and sell the Firebee II.

To preserve the supersonic configuration, protuberances beyond the basic airframe lines are avoided by designing the external attachments and antennae to be flush. A nose radome, similar to that of the subsonic Firebee I, houses the radar augmentation system antenna and passive augmentation. Directly behind this is the equipment compartment containing electrical, electronic and scoring systems, followed by the central fuselage, consisting of the fuel tank and structure for supporting the wing. The inlet and oil tank assembly is slung under the equipment compartment. The inlet duct passes from the inlet opening through the fuel tank to the engine, which is installed in a fuselage half-shell integral with the central fuselage structure. The entire aft portion of the fuselage is a removable subassembly which forms the upper shell, covering the engine.

Among the unusual design characteristics of the supersonic Firebee are its 'clean' wings. No ailerons are used, roll control being achieved by differential deflection of the all-moving horizontal tail surfaces.

As noted under the NASA entry, Firebee II drones feature in a number of experimental programmes, including the Rockwell/NASA HiMAT and Project DAST. Teledyne Ryan also has a $41,000 NASA contract to study the possibility of using the Firebee II for flight research into the oblique wing concept. Five Air Force BQM-34Fs have been supplied to NASA for these programmes.

Type: Remotely piloted supersonic jet target vehicle.

Wings: Cantilever shoulder-wing monoplane. Sweepback 53° at leading-edge. Thickness/chord ratio 3%. Basic structure consists of aluminium honeycomb core, steel skins tapering from 2·5 to 0·3 mm (0·10 to 0·012 in) in thickness, steel leading-edge, machined aluminium trailing-edge and detachable aluminium wingtips. No ailerons.

Fuselage: Conventional aluminium semi-monocoque structure. Shear and torsional forces are carried by the skin. Longitudinal members such as side longerons, keel, riser, trough and skins carry the fuselage bending loads. Frames, bulkheads and formers shape and hold the skin to its contour. Glassfibre nose radome.

Tail Unit: Sweptback (45°) all-moving horizontal surfaces and sweptback (53°) tapered fin and rudder. The horizontal tail surfaces are used for both roll and pitch control. The control surfaces are actuated by an electro-hydraulic actuator unit; this is a self-contained package with two output shafts for the horizontal tail surfaces and one for the rudder, which is used for directional trim and yaw damping. Aluminium honeycomb cores, with steel skins tapering from 2·5-0·4 mm (0·10 to 0·016 in) in the horizontal tail surfaces, with machined aluminium leading- and trailing-edges and a steel machined attachment fitting. Fin consists of aluminium honeycomb core, steel skins tapering from 3·7-0·9 mm (0·145 to 0·035 in) in the central section, 0·38 mm (0·015 in) steel forward and aft skins, and aluminium leading-and trailing-edges. The fin tip is a glassfibre housing for antennae.

Power Plant: One 8·54 kN (1,920 lb st) Teledyne CAE J69-T-406 turbojet engine. Wing centre-section and main fuselage total fuel tank capacity 119 kg (263 lb) in BQM-34T. External fuel pod capacity 181 kg (400 lb); weight of fuel plus tank, 210 kg (463 lb). With all tanks, the target will perform subsonic flight missions with similar performance capability, endurance and range to those of the subsonic BQM-34A. For supersonic flights, the external pod is jettisoned. Oil tank capacity 5·75 litres (1·5 US gallons). Provision for 50·3 kN (11,300 lb st) (nominal) solid-propellant JATO bottle.

Systems: Electrical power only. Primary power furnished by a 28V 200A DC engine-driven starter/generator. Power conversion by means of a 250VA 400Hz 115V AC static inverter. Power for recovery system and for drone control during glide phase furnished by a 28V 10Ah nickel-cadmium battery.

Electronics and Equipment (BQM-34E and T): AN/DLQ-3 ECM equipment, AN/DRQ-4 missile scoring system, AN(APX-71 L-band beacon, special low-altitude radar altimeter kit for 15 m (50 ft) altitude, X- or C-band tracking beacons, AN/DRW-29 radio control receiver, Dorsett AN/AKT-21 telemetry system. A Motorola ITCS (integrated track and control system) is fitted in the BQM-34T version and performs the functions of the tracking beacon, radio control receiver and telemetry system. Radar augmentation includes travelling wave tube (TWT) in S-, C- and X-band and a nose-mounted Luneberg lens passive radar reflector. Infra-red augmentation is by wingtip-mounted MK 37 Mod 0 IR flares.

Launch and Recovery: The Firebee II can be launched from either a ground or shipborne launcher, or air-launched from a modified DP-2E Neptune or DC-130A or E Hercules aircraft, in an essentially similar manner to the subsonic Firebee I. The DP-2E can carry two Firebees underwing, and the DC-130 four; these may be launched from altitudes up to 5,485 m (18,000 ft) at approx 200 knots (370 km/h; 230 mph). Recovery is by a two-stage parachute system similar to that fitted to the Firebee I, and the Firebee II also can be recovered after landing on water or (BQM-34F only) by helicopter mid-air retrieval system (MARS). In a MARS recovery, the helicopter snares a 5·72 m (18 ft 9 in) diameter engagement parachute which extends above the 24·08 m (79 ft 0 in) main parachute. Once engaged, the main parachute is released automatically to allow the helicopter to reel in the Firebee II and transport it back to the target operations area. Provision is also made for emergency recovery. In all versions, the recovery parachutes are housed in the fuselage tailcone.

Guidance and Control: Remote control of the BQM-34E or F, including activation of the recovery system, is accomplished by a frequency-modulated UHF radio guidance system, with 20 separate command channels, utilising an AN/DRW-29 on-board radio receiver and a compatible transmitter at the remote control station. This receiver is installed in the equipment compartment, and the antenna is located in the glassfibre fin-tip. A Motorola ITCS (integrated track and control system) is fitted in the BQM-34T version. The Firebee II has an automatic flight control system (AFCS) consisting of six elements: a three-axis rate gyro, vertical gyro, air data computer, flight control box, low-altitude control box, and three-axis electro-hydraulic actuator assembly. Positioning data during a flight is provided to the remote control station by an on-board radar tracking beacon, the antennae for which are located on top of the nose compartment and in the lower aft portion of the fuselage. To relay data to the remote controller, a 10-channel telemetry system is used which comprises data collection, conversion and FM/FM transmitting equipment in the drone and receiving and data display units in the remote control station. When engine or generator power is shut down at high altitude, either by remote command or because of fuel depletion, the flight control system continues to operate on battery power. The recovery sequence is preceded, at altitudes above 4,570 m (15,000 ft), by a power-off glide, and can be initiated by remote command at any time during the glide. When necessary to gain altitude and reduce speed for safe parachute deployment and recovery, a power-off climb is initiated automatically below 15,000 ft either by normal recovery command or if there is a loss of engine power or a generator failure.

Operational Equipment: In general, Firebee II can be equipped with active and passive radar augmenters, electronic and photographic scoring systems, ECM, low-altitude radar sensing systems and infra-red flares or pods. Augmentation equipment, for weapon systems evaluation and personnel training, includes provisions for target identification, GCI tracking, variable radar-image size and augmented infra-red radiation. There is a smoke system to aid long-distance visual identification at altitudes of 6,100 m (20,000 ft) and above; a low-altitude smoke generator is under development. Positive electronic identification is provided by the L-band IFF beacon. A travelling wave tube (TWT) system is employed to provide radar echo augmentation in various patterns, by the use of specially-designed antenna systems, to represent various sizes of target.

Dimensions, external:
- Wing span:
  - BQM-34E and T — 2·71 m (8 ft 10·8 in)
  - BQM-34F — 2·95 m (9 ft 7·9 in)
- Length overall:
  - BQM-34E and T — 8·73 m (28 ft 8 in)
  - BQM-34F — 8·89 m (29 ft 1·9 in)
- Body diameter — 0·61 m (2 ft 0 in)
- Height overall — 1·71 m (5 ft 7·2 in)
- Tailplane span — 1·46 m (4 ft 9·6 in)

Weights (A: BQM-34E; B: BQM-34F):
- Weight empty:
  - A — 658 kg (1,452 lb)
  - B — 780·8 kg (1,721·4 lb)
- Max launching weight:
  - A, air launch — 855·7 kg (1,886·5 lb)
  - B, air launch — 951 kg (2,097 lb)
  - A, ground launch — 1,027·2 kg (2,264·5 lb)
  - B, ground launch — 1,122·5 kg (2,475 lb)

Performance:
- Max speed:
  - at S/L — Mach 1·1
  - at 13,715 m (45,000 ft):
    - A — Mach 1·8
    - B — Mach 1·78
  - above 18,300 m (60,000 ft) — Mach 1·5
- Operating height range — 15 m (50 ft) to 18,300 m (60,000 ft)
- Service ceiling:
  - A — 18,300 m (60,000 ft)
  - B — 16,765 m (55,000 ft)
- Control range — 200 nm (370 km; 230 miles)
- Typical range, external tank on:
  - low-altitude, subsonic cruise/transonic dash — 221 nm (409 km; 254 miles)
  - high-altitude, subsonic cruise/supersonic dash — 606 nm (1,123 km; 698 miles)

high-altitude, subsonic cruise throughout
617 nm (1,142 km; 710 miles)

Max range	774 nm (1,434 km; 891 miles)
Endurance (total time)	1 hr 14 min
Flotation time	24 hr

TELEDYNE RYAN MODELS 234 and 239

USAF designation: BGM-34

Although this RPV shares the Firebee I parentage of the Model 147/AQM-34, it has been developed primarily for tactical strike and other defence suppression roles, reflecting plans to evolve combat drones for a variety of missions which at present require manned aircraft.

Details have been announced of three versions, as follows:

BGM-34A (Model 234). Four built to evaluate the feasibility of using RPVs to deliver defence suppression weapons by day, under real-time control. Details in 1975-76 *Jane's*.

BGM-34B (Model 234A). Generally similar to BGM-34A, but with 8·54 kN (1,920 lb st) J69-T-41A engine, modified tail unit, enlarged control surfaces, and added operational capability. Eight ordered. Sperry Univac radio command guidance system. One or two fitted with extended, modified nose containing a low light level TV camera and an Aeronutronic Ford stabilised laser designator/receiver system. This installation enables the BGM-34B to act in a pathfinder role, locating and locking on to a target and signalling its position to other RPVs, carrying weapons, in the force. One other BGM-34B was fitted with a Hughes high-resolution FLIR (forward-looking infra-red) nose sensor instead of the TV installation.

Further details of these installations can be found in the 1975-76 *Jane's*.

BGM-34C (Model 239). Multi-mission validation RPV, with modular nose sections for reconnaissance, electronic warfare and air-to-ground strike missions. Eight ordered in November 1974, under a $15·68 million USAF contract, to conduct development, integration and test beginning in 1976. This vehicle combines, in interchangeable modular noses, the capabilities of the AQM-34V electronic warfare, AQM-34M reconnaissance and BGM-34B TV-guided Maverick strike RPVs. The Lear Siegler YAQM-34U electronics, with computer-based quick-turnaround automatic guidance electronics (AGE), are integrated in the BGM-34C vehicle. Flight testing was scheduled to begin in September 1976 and involve 32 flights, ending in April 1977. Five vehicles are being used in this programme, using three 'reconnaissance' noses, two 'strike' noses and one 'electronic warfare payload' nose. Prototypes are converted from Model 147SC (three AQM-34L and five YAQM-34U) RPVs, with structure strengthened for ground launch. Control aircraft is a specially-modified DC-130H, able to control up to eight drones simultaneously.

DIMENSIONS, EXTERNAL:

Wing span	4·42 m (14 ft 6 in)
Length overall:	
BGM-34A	7·45 m (24 ft 5·2 in)
BGM-34B	8·35 m (27 ft 4·6 in)
BGM-34C	8·69 m (28 ft 6·2 in)
Body diameter	0·94 m (3 ft 1·2 in)

WEIGHTS:

Max launching weight:	
BGM-34B	1,465 kg (3,230 lb)
BGM-34C	2,721 kg (6,000 lb)

TELEDYNE RYAN MODEL 235 R-TERN (COMPASS COPE R)

USAF designation: YQM-98A

This aircraft was ordered by the USAF in 1972 for evaluation in its Compass Cope programme for a high-altitude sensor platform RPV.

Representing a third-generation vehicle to follow the Model 147H/AQM-34N and Model 154/AQM-91A, the Teledyne Ryan Model 235 has extremely high aspect ratio wings and an overfuselage pod mounting for its power plant. Design features of the Garrett ATF 3 engine configuration are such as to produce a low infra-red signature, low radar reflectivity, very low smoke and noise emissions, and a capability for very high altitude operation.

Two YQM-98A protoypes were ordered in 1972, and construction began in February 1973. Both prototypes (72-01871 and '872) were rolled out on 4 January 1974.

Delivery was made to Edwards AFB on 10 April 1974, and the first flight, lasting 1 hr 50 min, took place on 17 August 1974. During the flight demonstration programme, which was completed successfully at Edwards AFB on 27 November 1974, the YQM-98A made five flights totalling more than 36 hr. On its fifth flight, on 3/4 November, it set an unofficial endurance record for RPVs by remaining airborne for well over 24 hr and reached an altitude in excess of 16,765 m (55,000 ft).

Flight testing of the YQM-98A was resumed on 8 May 1975, from Cape Canaveral AFB, over the USAF's Eastern Test Range in Florida, as part of a joint USAF/Ryan programme of system engineering studies to determine any design changes necessary to produce an operational Compass Cope RPV. A total of 17 flights had been made by September 1975. One prototype was lost in a landing accident in the Autumn of 1975.

On 27 August 1976, the Boeing YQM-94A vehicle was selected for continued (pre-production) development, but as this section closed for press it was understood that Teledyne Ryan had officially contested this decision.

The following description applies to the YQM-98A prototypes:

Rollout of first Teledyne Ryan BGM-34C multi-mission RPV for the USAF, August 1976

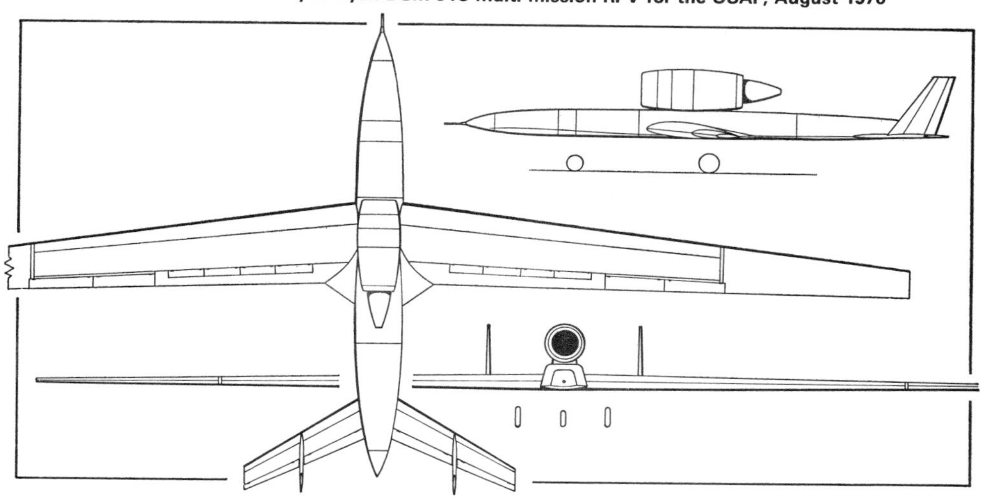

Teledyne Ryan YQM-98A high-altitude long-endurance strategic RPV *(Roy J. Grainge)*

Second prototype of the Teledyne Ryan YQM-98A Compass Cope R long-endurance high-altitude strategic RPV

TYPE: Prototype high-altitude long-endurance strategic RPV.

WINGS: Cantilever low-wing monoplane, of very high aspect ratio. Approx 1° anhedral from roots. Sweepback approx 7° at quarter-chord. Two-section ailerons on each trailing-edge, inboard of outer panels. Four-section spoilers on each upper surface. Triangular fillet on each trailing-edge at root. Conventional semi-monocoque structure, with selected use of composite materials. Detachable 4·27 m (14 ft) outer panels have an aluminium core and graphite composite skin; trailing-edges and fairings make extensive use of DuPont PRD 49 glassfibre.

FUSELAGE: Semi-monocoque structure, of approximately rectangular cross-section with rounded upper edges, tapering towards front and rear. Undersurface flared and flattened to reduce radar reflectivity. Nose, tailcone and parts of fuselage are components from AQM-91A. Nosecone, forward fuselage and tailcone are detachable for transportation.

TAIL UNIT: Cantilever low-set swept tailplane, with twin sweptback fins and overhanging rudders at approx half-span. Sweepback approx 32° on tailplane leading-edge, approx 34° on fin leading-edges. Full-span elevators. Part of tailplane is component from AQM-91A.

LANDING GEAR: Retractable tricycle type. Single-wheel main units, modified from those of a Cessna A-37, retract inward into wings; nosewheel unit, from a Canadair CF-5, retracts rearward.

POWER PLANT: One Garrett-AiResearch ATF 3 (YF104-GA-100) turbofan engine, rated at 18 kN (4,050 lb st), mounted in a pod on a shallow pylon on top of the fuselage, in line with the wings. Design output of this engine is 22·2 kN (5,000 lb st). Fully-automatic electronic fuel control system. Fuel tank in each inboard wing panel, each with refuelling point on wing upper surface, and in fuselage.

SYSTEMS AND EQUIPMENT: Equipment compartments in nose and tail portions of fuselage. Main mission payload compartment in lower forward fuselage, with provision for additional payload to be carried in rear fuselage. Singer Kearfott Talar IV tactical airborne landing approach radar. Details of other systems and equipment are not available officially, but these are expected to include an integrated flight control system.

GUIDANCE AND CONTROL: Teledyne Ryan radio command guidance system.

DIMENSIONS, EXTERNAL:

Wing span	24·75 m (81 ft 2½ in)
Wing aspect ratio	19
Length overall	approx 11·68 m (38 ft 4 in)
Length of fuselage	11·38 m (37 ft 4 in)
Fuselage: Max depth	0·79 m (2 ft 7 in)
Height overall	2·44 m (8 ft 0 in)
Tailplane span	6·53 m (21 ft 5 in)
Wheel track	approx 2·29 m (7 ft 6 in)
Wheelbase	3·07 m (10 ft 1 in)

AREA:

Wings, gross	32·24 m² (347 sq ft)

WEIGHTS AND LOADINGS (approx):

Weight empty	2,540 kg (5,600 lb)
Payload for 24 hr mission	317·5 kg (700 lb)
Max T-O weight	6,490 kg (14,310 lb)
Min wing loading	78 kg/m² (16 lb/sq ft)
Max wing loading	200 kg/m² (41 lb/sq ft)

Teledyne Ryan Model 262 STAR mini-RPV being launched in April 1976, during initial tests by the US Navy's Naval Surface Weapons Center

PERFORMANCE (estimated):

Cruising speed at altitudes from 15,240 to 21,340 m (50,000 to 70,000 ft)	Mach 0·5 to 0·6
T-O field length	1,067 m (3,500 ft)
Max endurance	30 hr

TELEDYNE RYAN MODEL 262

The Model 262 mini-RPV, developed originally as a contender in the US Army's RPAODS competition, was designed and built under contract to Teledyne Ryan by AMR. A number of AMR Ryan-Benson RPV-007 half-scale test vehicles were built to flight test the basic design, power plant and systems.

An accompanying illustration depicts a radar cross-section mockup, or 'feasibility version', of the Model 262, which underwent testing in 1974. Wing span is about 2·29 m (7 ft 6 in); length about 1·5 m (5 ft); average speed in flight approx 120 knots (222 km/h; 138 mph). Later, the original engine was replaced by an 18·6 kW (25 hp) McCulloch flat-twin engine, driving a five-blade ducted propeller mounted centrally on top of the craft. The engine and fuel tank are mounted entirely within the air-craft, the rotor being driven by a complex system of belts and gears.

The unique delta configuration is claimed to contribute towards low cost; very low radar cross-section and low visual, acoustic and infra-red signatures; and towards minimum logistic, maintenance and technical support requirements for operational use.

The full-size Model 262, shown in another photograph, has a wing span of 2·29 m (7 ft 6 in), and, fully fuelled, weighs approx 75 kg (165 lb), depending upon its mission. It is fitted with a hybrid autopilot, employing analogue techniques for the inner control loop and digital techniques for the outer control loop.

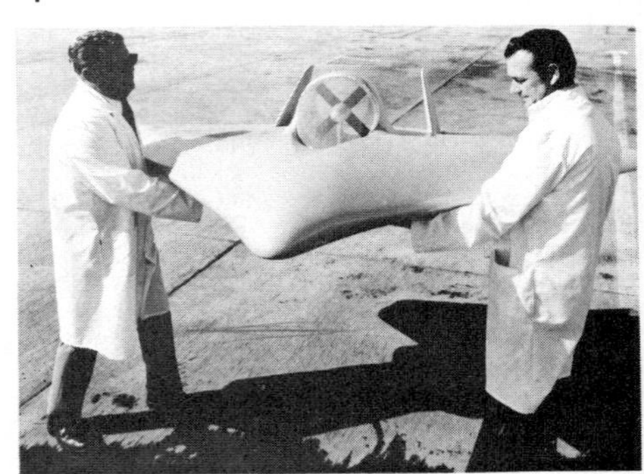

Radar cross-section mockup of the Teledyne Ryan Model 262 mini-RPV

Deployment in many tactical applications, at sea and ashore, would involve launch from a trainable compressed-air rail launcher and net recovery; but other versions could be fitted with a conventional landing gear, such as the retractable tricycle type designed by AMR.

Teledyne Ryan has a contract worth more than $1 million to develop the Model 262 as a competitor in the US Navy's STAR (Ship Tactical Airborne RPV) programme, and three were built for evaluation. First shipboard tests were held in 1976. The Model 262 was unsuccessful in the US Army competition, in which the LMSC Aquila was selected.

US DEPARTMENT OF DEFENSE

The Director of the Defense Advanced Research Projects Agency (ARPA) told the Senate Armed Services Committee in the Spring of 1975 of some major areas in which the ARPA planned to concentrate RPV research using FY 1976 funds. These included the testing of a missile- or aircraft-launched mini-RPV; development and test of shipboard launch and recovery techniques; battery- and solar-powered RPVs; miniature infra-red and laser designator and rangefinder systems that will enable a mini-RPV to find targets and guide weapons at night; and data links resistant to hostile jamming. The use of mini-RPVs for hostile weapon location is foreseen, using the air vehicle as a platform equipped with a lightweight high-performance radar, or a laser line-scanner, and a direct line-of-sight flash detector. A small, low-cost system for locating a mortar projectile is being designed, and investigation of radar options for projectile tracking is in progress.

In addition to drone/RPV programmes for which hardware contracts have been awarded (described under the appropriate contractor's heading in this section), the following are among the more important programmes which are or were recently receiving attention by the three US services:

US AIR FORCE

ADDRESS:

Office of Information, Air Force Systems Command, Andrews AFB, Maryland 20334

Telephone: (301) 981-4137/8

ARPV (Advanced RPV). Intended successor to the Teledyne Ryan BGM-34C, for service in the early 1980s; formerly known as AMMR (Advanced Multi-Mission RPV). Designs, invited in 1975 from three competing companies, will incorporate a drone control and data retrieval system (CDRS) and data relay by Compass Cope-type RPVs. The CDRS, aimed at demonstrating simultaneous control of up to 20 RPVs in the face of intensive jamming, entered the prototype stage in late 1974. Industry briefing took place in late 1974; one-year study contracts were awarded in April 1975 to Boeing Aerospace, Northrop Corporation and Rockwell International (which see).

Constant Angel. Electronic warfare programme involving two types of ground-launched RPV: a high-subsonic tactical expendable drone system (TEDS, which see), to act as a decoy, and an orbiting jamming vehicle.

LAMPS (Low-Altitude Multi-Purpose System). Programme to develop a low-altitude long-range RPV; not to be confused with US Navy's Light Airborne Multi-Purpose System involving manned helicopters.

Recovery Systems. Several alternative recovery system studies are under way, both for the improvement of the present mid-air retrieval system (MARS) and for alternatives to this system. Concepts being examined include ACLS (air cushion landing systems), ALS (automatic take-off and landing systems), VTOL vehicles, recovery and storage by a large launch aircraft such as the Boeing 747, arrester gear systems and steerable recovery parachutes. An ALS is probable for any production version of the Compass Cope RPV; the E-Systems L450F (1975-76 *Jane's*) has been mentioned as a possible ALS test vehicle.

TEDS (Tactical Expendable Drone System). Original plans to issue RFPs (Requests For Proposals) for a TEDS vehicle to two competing contractors in October 1972 were deferred when proposed funding was deleted from the FY 1974 budget request. The TEDS programme was reintroduced in FY 1975, and a flight demonstration and validation phase began in early 1976. This is using, as recoverable demonstrators, three Beechcraft Model 1089 VSTT prototypes and three Northrop NV-130s, modified to TEDS configuration; these are described under their manufacturers' entries in this section. Contractor proposals were submitted to AF Systems Command on 24 February 1975, and contracts were placed in May and June 1975. The Beech and Northrop vehicles are intended as proof-of-concept demonstrators only, and are not necessarily contenders for a production TEDS. The latter will, of course, be non-recoverable, and is intended primarily for mobile ground launch, although air-launch capability may be included as an option. Payload will be in the order of 23-45 kg (50-100 lb), and primary electrical power source 1kW or less. TEDS is seen as having two basic functions: as a straightforward decoy, carrying active or passive ECM, or to accompany strike aircraft and provide them with ECM support. Luneberg lenses will provide the primary means of radar augmentation.

VLCHV (Very Low Cost Harassment Vehicle). Mini-RPV programme to develop a small vehicle (possibly pre-programmed rather than remotely piloted) able to loiter over hostile air defences for up to 4 hr, either in an unarmed capacity to draw the defences' fire and provoke radar silence, or armed with homing devices and an explosive charge to destroy a hostile radar. About a dozen US aerospace companies were invited to complete paper studies in early 1975; system studies, by E-Systems' Melpar Division and Lockheed Missiles and Space Co, were submitted in mid-1975. Northrop is also a contender. Field tests have been carried out with the E-Systems Axillary mini-RPV (which see), which has already demonstrated its radar homing capability. Another possible VLCHV contender is a 3·05 m (10 ft) long vehicle with a 4·27 m (14 ft) wing span, able to carry a 34 kg (75 lb) payload and two 2·75 in rockets. The US Army has a broadly similar programme for a small 'kamikaze' RPV able to fly a warhead into a target.

US ARMY

ADDRESS:

Office of Information, Army Aviation Systems Command, Mart Building, PO Box 209, St Louis, Missouri 63166

Electronic Warfare RPV. Project for a 68-91 kg (150-200 lb) mini-RPV able to carry electronic jamming equipment.

'Kamikaze' RPV. Broadly similar programme to USAF's VLCHV (which see) for a small (about 27 kg; 60 lb) expendable RPV to fly a warhead into a target.

Little Scout. Target designating RPV for service entry in about 1980, to be evolved from LMSC Aquila (which see).

RPAODS. This programme (see 1975-76 *Jane's*) was phased out in FY 1975, and replaced by the LMSC Aquila (which see). During 1975, an Aeronutronic Ford RPAODS vehicle, equipped with a laser designator, took part in a successful test of the Cannon Launched Guided Projectile.

Tactical High-Speed Surveillance RPV. The US Army plans to issue RFPs for three systems, each comprising 12 RPVs and one ground control station; procurement plans for these have been deferred for the time being. These are to be off-the-shelf vehicles; contenders are understood to include the Canadair CL-89 and the Northrop NV-128 (which see).

US NAVY

ADDRESS:

Office of Information, Department of the Navy, Main Navy Building, the Pentagon, Washington, DC 20301

Naval Air Development Center QF-4B. For use by the Naval Missile Center, the NADC at Warminster, Pennsylvania, converted 44 McDonnell Douglas F-4B Phantoms to QF-4B configuration. The first aircraft was delivered in the Spring of 1972 to the NMC at Point Mugu, California, for further flight development and subsequent use as a supersonic manoeuvring target for new missile development.

Flight testing was completed in January 1974, and the target certificated for use when the prototype QF-4B was flown as a drone for a missile system test. Vought DF-8L Crusaders are used as drone control aircraft.

Naval Ship Research and Development Center VATOL. This US Navy programme is aimed at demonstrating VATOL (vertical attitude take-off and landing). Tethered flight tests, at the Pacific Missile Range in late 1975, were to be followed by free flights from a runway. The VATOL test vehicle is modified from a Northrop MQM-74A target, the major changes involving the fitting of croppped-delta main wings, of NACA 64A008 section, with elevons; fixed canard foreplane surfaces; a much larger fin and rudder; and a non-retractable tricycle landing gear. In this airframe are installed the power plant (a 2·94 kN; 660 lb st Teledyne CAE J402-CA-400 turbojet engine fitted with a vane-type thrust vector control system) and the midcourse guidance system of a McDonnell Douglas AGM-84A Harpoon air-to-surface missile. Fuel load is 42·6 kg (94 lb), and an emergency recovery parachute is fitted.

One of the 29 F-86H Sabre fighters converted to QF-86H drone configuration by the US Naval Weapons Center at China Lake, California *(Henry Artof)*

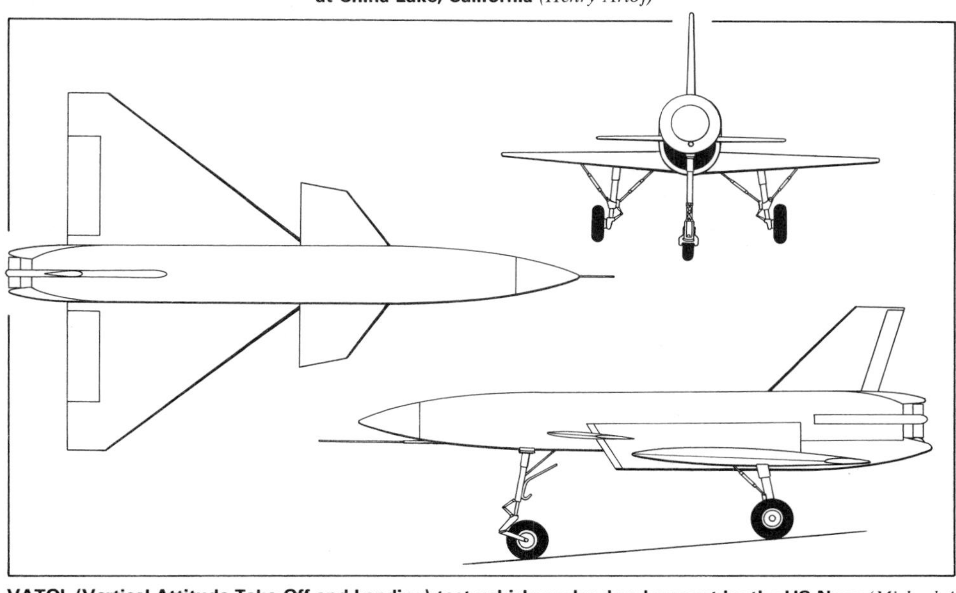

VATOL (Vertical Attitude Take-Off and Landing) test vehicle under development by the US Navy *(Michael A. Badrocke)*

DIMENSIONS, EXTERNAL:

Main wing span	2·21 m (7 ft 3 in)
Main wing chord (mean)	1·09 m (3 ft 7 in)
Main wing area	2·02 m² (21·74 sq ft)
Foreplane span	1·10 m (3 ft 7½ in)
Body diameter	0·36 m (1 ft 2 in)
Length overall	approx 3·58 m (11 ft 9 in)

PERFORMANCE:

Max cruising speed	147 knots (272 km/h; 169 mph)
Time to 1,525 m (5,000 ft)	8·8 sec
Endurance	approx 30 min

Naval Surface Weapons Center HASPA. HASPA (High Altitude Superpressure Powered Aerostat) is a programme for a remotely piloted airship for high altitude (20,725 m; 68,000 ft) ocean surveillance and communications relay, sponsored by Electronic Systems Command of the USN and managed by the NSWC. Martin Marietta is a prime contractor. HASPA will be helium-filled, with an electric propulsion system, will be 101·50 m (333 ft) long, and will have a diameter of 20·42 m (67 ft). An unpowered prototype, built by Sheldahl Inc, was due to fly in late 1975 or early 1976.

Naval Weapons Center QF-86H. Twenty-nine North American F-86H Sabres began undergoing conversion in 1974, at the Naval Weapons Center, China Lake, California, into QF-86H pilotless target aircraft. The QF-86H is an interim, all-attitude manoeuvring target pending the development of a new, highly-manoeuvrable drone known as **NAT** (Navy Agile Target). Modifications in the QF-86H include an on-board TV camera, Sperry autopilot, a non-redundant proportional control system and a telemetry system.

Naval Weapons Center QT-38. The NWC at China Lake, California, is converting four ex-USAF T-38 Talon supersonic trainers to drone configuration. A Sperry control system is fitted.

THE UNION OF SOVIET SOCIALIST REPUBLICS

SOVIET TARGET DRONES

Soviet films shown to invited audiences in the West have depicted two types of target drone in current use for training surface-to-air missile crews and the pilots of interceptor fighter aircraft. Both are shown in accompanying illustrations.

LAVOCHKIN La-17

This small radio-controlled target has a slim cylindrical fuselage, with unswept wings of constant chord mid-set on the fuselage at about mid-length. The conventional tail surfaces are also unswept and of constant chord, with the tailplane mounted partway up the fin to avoid the exhaust from two jettisonable solid-propellant booster rockets which are fitted under the wings for take-off. The La-17, which was developed in the early 1950s, has an all-metal airframe. Cruise propulsion is provided by a turbojet engine mounted in a pod under the fuselage. It has been suggested that this may be a modified version of the 8·83 kN (1,985 lb st) RU-19-300 engine which is used as an APU in the Antonov An-24/26/30 series of aircraft. Equipment is reported to include a parachute recovery system and a Luneberg lens radar reflector.

DIMENSIONS (estimated):

Wing span	6·95 m (22 ft 9½ in)
Length overall	7·10 m (23 ft 3½ in)
Body diameter	0·50 m (1 ft 7¾ in)
Height overall	2·53 m (8 ft 3½ in)

WEIGHTS (estimated):

Weight empty	1,200 kg (2,645 lb)
Launching weight	1,800 kg (3,968 lb)

PERFORMANCE (estimated):

Max level speed:	
at high altitude	Mach 0·85
at low altitude	Mach 0·70
Service ceiling	14,000 m (45,925 ft)
Max endurance	1 hr

YAKOVLEV Yak-25RD (?)

Shown as a target for an SA-2 *(Guideline)* surface-to-air missile in a film made in the late 1960s, this Soviet target aircraft appeared to be almost identical with published impressions of the Yakovlev high-altitude reconnaissance aircraft known to NATO as *Mandrake.* This suggests that the *Mandrakes* may have been adapted into high-flying pilotless targets when retired from active service in their original role.

Mandrake was a development of the Yak-25, with an existing fuselage married to a new straight wing of extended span. This made possible the high-altitude performance needed for 'U-2 type' missions, with a minimum of new design effort. The fuselage was believed to be generally similar to that of the Yak-25R/26, with a single-seat cockpit and a new 'solid' nose to house reconnaissance equipment. The zero-track main landing gear of the Yak-25 was retained, with single nosewheel and twin-wheel rear unit under the centre-fuselage, requiring the provision of housings for the outrigger wheels on the tips of the new wings. The engines are believed to have been Tumansky R-9 turbojets, each rated at 25·5 kN (5,730 lb st) dry and 39·2 kN (8,818 lb st) with afterburning. Photographs of a generally similar aircraft, referred to as the Yak-25RD, have appeared in the Eastern European press; this is almost certainly the 'Yakovlev RV' which set a number of payload-to-height records in July 1959 and a women's closed-circuit distance record in September 1967.

DIMENSIONS (estimated):

Wing span	21·50 m (70 ft 6 in)
Length overall	15·665 m (51 ft 4¾ in)
Wing area	50·00 m² (538·2 sq ft)

Lavochkin La-17 jet-powered target drone taking off with the aid of booster rockets

Yakovlev high-altitude special reconnaissance aircraft (NATO reporting name *Mandrake*) adapted for use as an unpiloted target

AIR-LAUNCHED MISSILES

FRANCE

AÉROSPATIALE

SOCIÉTÉ NATIONALE INDUSTRIELLE AÉROSPATIALE

HEAD OFFICE:
37 boulevard de Montmorency, 75781-Paris Cédex 16
Telephone: 524.43.21

OFFICERS:
See Aircraft section

Division Engins Tactiques

EXECUTIVE OFFICE:
2 rue Béranger, 92320-Châtillon
Telephone: 655.54.00
Telex: 25881

When Nord-Aviation became part of Aérospatiale, responsibility for its guided missiles, target vehicles and test vehicles was allocated to Aérospatiale's Tactical Missiles Division. Subsequently, this Division developed and put into production a new generation of short-range battlefield weapons, in association with Messerschmitt-Bölkow-Blohm of Germany (see Euromissile entry in International section).

A total of 396,080 tactical missiles and 1,794 target and reconnaissance RPVs had been ordered from the Division and its overseas partners by the beginning of 1976. Employees totalled 5,695 at that time.

AÉROSPATIALE AS.11

The AS.11 is an air-to-surface version of the SS.11 line-of sight wire-guided battlefield missile, which was developed for use from vehicles of all kinds. It has a cylindrical body and swept cruciform wings, and is powered by a two-stage solid-propellant rocket motor. Directional control is achieved by varying the sustainer efflux through two side-nozzles.

In action, with a typical helicopter installation, the operator acquires the target by means of a stabilised and magnifying optical sight. As soon as the missile enters his field of vision after launch, he passes to it the signals needed to align it with the target, while keeping it above the terrain until impact. The signals are given by the operator by means of a control stick which makes it possible to send simultaneously up or down and port or starboard commands. The signals are transmitted to the missile over wires, wound on twin spools. Tracer flares are installed on the rear of the missile for visual reference.

Since 1962, the AS.11 B.1 version, using transistorised firing equipment, has been in production. It is available with a variety of different warheads, including an inert type for practice, the Type 140AC anti-tank warhead capable of perforating 60 cm (24 in) of armour plate, the Type 140AP02 explosive warhead (2·6 kg; 5·72 lb of explosive) which will penetrate an armoured steel plate 1 cm (0·4 in) thick at a range of 3,000 m (9,800 ft) and explode about 2·1 m (7 ft) behind the point of impact, and the Type 140AP59 high-fragmentation anti-personnel type with contact fuse.

The AS.11 B.1 and the basic surface-launched SS.11 B.1 have been supplied to all three French services and the armed forces of 27 other countries, including the USA and UK. Orders totalled 166,337 by the beginning of 1976, including surface-to-surface Harpons with automatic infra-red command guidance. Of these, 58% were exported. Licence manufacture was undertaken in Federal Germany and India.

DIMENSIONS:
Length overall — 1·20 m (3 ft 11 in)
Body diameter — 0·164 m (6½ in)
Wing span — 0·50 m (1 ft 7½ in)

WEIGHT:
Launching weight — 29·9 kg (66 lb)

PERFORMANCE:
Average cruising speed — 313 knots (580 km/h; 360 mph)
Time of flight (propelled) — 20-21 sec
Min turning radius — approx 1,000 m (3,300 ft)
Range — 500-3,000 m (1,650-9,840 ft)

AÉROSPATIALE AS.12

The wire-guided air-to-surface AS.12 (and similar surface-to-surface SS.12) is a spin-stabilised missile derived from the AS.11/SS.11 series but with a warhead weighing 28·4 kg (62·6 lb), about four times as much as that of the latter, making it suitable for use against fortifications as well as tanks, ships and other vehicles. The current OP.3C warhead can pierce 40 mm (1·5 in) of armour and explode on the other side.

The AS.12 has a cylindrical body, cruciform wings and two-stage solid-propellant rocket motor. As in the AS.11, its wires are wound on twin spools. A command to line-of-sight guidance system is used; this missile is not available with the automatic infra-red command guidance system used with some versions of the AS.30.

The missile can be air-launched at speeds up to about 200 knots (370 km/h; 230 mph), in conjunction with an APX 260 or APX 334 gyro-stabilised sight. It can be used at night with target illuminating equipment.

The AS.12 arms Breguet Atlantic and P-2 Neptune aircraft of the Royal Netherlands Navy, and is also carried by the Breguet Alizé and Hawker Siddeley Nimrod. It is operational on helicopters of the French Navy, the Royal Navy and several other naval air arms.

A total of 7,079 SS.12s and AS.12s had been ordered by the beginning of 1976, when manufacture was continuing.

DIMENSIONS:
Length overall — 1·87 m (6 ft 1·9 in)
Body diameter — 0·18 m (7 in)
Warhead diameter — 0·21 m (8·25 in)
Wing span — 0·65 m (2 ft 1½ in)

WEIGHT:
Launching weight — 77 kg (170 lb)

PERFORMANCE:
Speed at impact:
AS.12 (fired at 200 knots; 370 km/h; 230 mph) — 180 knots (335 km/h; 210 mph)
Time of flight — 32 sec
Max range:
AS.12 in relation to surface — approx 8,000 m (26,250 ft)
AS.12 in relation to aircraft — approx 5,500 m (18,000 ft)

AÉROSPATIALE AS.30

This tactical air-to-surface missile has a cylindrical body, canted cruciform sweptback wings indexed in line with cruciform 'flip-out' tailfins, a two-stage solid-propellant power plant and radio command guidance system. It can be fitted with a general-purpose warhead or a semi armour-piercing type with thicker casing.

The requirements to which the AS.30 was designed included an initial launch range of at least 5·4 nm (10 km; 6·2 miles), with the provision that the launching aircraft should not approach to within 1·6 nm (3 km; 1·8 miles) of the target. A CEP (circular error probability) of less than 10 m (33 ft) was specified and this standard of accuracy has been exceeded by the AS.30. Minimum launching speed is approximately Mach 0·45; there is no limitation for launching at supersonic speeds.

As an alternative to the original manual 'steering' system, the AS.30 can utilise the TCA infra-red automatic guidance system evolved by the company's Tactical Missiles Division. With this, the pilot has no other task to perform than conventional aiming with his weapon-sight, keeping the target centred in the sight during the flight of the missile. An infra-red tracker is trained constantly on an IR flare on the missile. An axial gyroscope compensates for sight movement. From data provided by the two devices, deviations of the missile from the correct flight path are detected and corrected by command signals radioed to the missile. Accuracy is claimed to be as great as that achieved by a fully-trained operator controlling the missile by hand.

This automatic guidance system is operational in the French Air Force, which utilises the AS.30 with a 230 kg (510 lb) HE warhead and alternative delay or non-delay fuses.

Other customers for the AS.30 have included the French Navy and the German, Swiss and South African air forces.

A total of 3,863 AS.30s had been produced by the beginning of 1976, of which 78% were exported to six countries. A laser-guided version is being developed jointly by Aérospatiale and Thomson-CSF, as armament for Jaguar aircraft.

Aérospatiale AS.30 air-to-surface missile on Mirage III-E underfuselage launcher

Aérospatiale AS.11 air-to-surface missiles on a British Army Scout helicopter *(Peter R. March)*

Aérospatiale AS.12 air-to-surface missile installation on a Lynx helicopter

DIMENSIONS:

Length overall:	
with X35 warhead	3·885 m (12 ft 9 in)
with X12 warhead	3·839 m (12 ft 7 in)
Body diameter	0·34 m (1 ft 1½ in)
Wing span	1·00 m (3 ft 3¼ in)
WEIGHT:	
Launching weight	520 kg (1,146 lb)
PERFORMANCE:	
Speed at impact	450-500 m/sec (1,475-1,640 ft/sec)
Range (average)	5·9-6·5 nm (11-12 km; 6·8-7·5 miles)

AÉROSPATIALE AM39 EXOCET

Exocet is a missile that was devised originally to provide warships with all-weather attack capability against other surface vessels. In its basic MM38 and longer-range MM40 surface-to-surface forms it can be fitted in all classes of surface warships, including fast patrol boats and hydrofoil craft, and offers an economical means of defence against missiles like the Soviet *Styx* by attacking the launching vessels rather than attempting to intercept the missiles after launch. An air-to-surface version is designated AM39.

The Exocet missile is in the form of a cylindrical body fitted with cruciform sweptback wings and cruciform tail control surfaces indexed in line with the wings. Propulsion is provided by a tandem two-stage solid-propellant motor, and the highly-destructive warhead is described as being in the same order as that of a torpedo. The missile is stored in a container which also serves as a launcher during surface-to-surface use, and is usually installed in a fixed position.

The AM39 air-to-surface version has a reduced launch weight and a new rocket motor that burns for 150 seconds, giving an increased range. Another modification to the propulsion system by comparison with the MM38 ensures a one-second delay in ignition after launch, allowing the missile to drop clear of the launch aircraft.

The AM39 has been selected for carriage by the French Navy's Atlantic and Super Etendard aircraft and by the Mirage F1. It is suitable for operation from helicopters in the class of the Super Frelon and Sea King.

The missile's high subsonic flight profile consists of a pre-guidance phase, during which it travels towards the target, whose range and bearing have been determined by an airborne radar or ship's fire control computer and set up in the missile pre-guidance circuits before launch, and a final guidance phase during which the missile flies directly towards the target under the control of its active homing head. Throughout the flight the missile is maintained at a very low altitude (reported to be 2 to 3 metres; 6·5 to 10 ft) by an FM radio altimeter. Its homing head is reported to pick up the target over a range of up to 6·5 nm (12 km; 7·5 miles).

Exocet is intended to operate efficiently in an ECM (electronic countermeasures) environment.

No further details are available officially, but Exocet is reported to have a warhead weighing 160 kg (352 lb). It uses a modified version of the inertial low-level guidance system developed for the Kormoran missile and a sustainer motor similar to that of the Martel missile.

Subcontractors to Aérospatiale include Electronique Marcel Dassault for the ADAC homing head, TRT for the AHV-7 radio altimeter, SNPE for the solid propellants, SERAT for the explosive charge, SFENA and SAGEM for accelerometers and gyroscopes, Jaeger for control surface actuators, and ECAN-Ruelle for servo control equipment. Following the placing of British and German contracts, participating companies include BAC, Hawker Siddeley Dynamics, Morfax, ROF Patricroft (Royal Ordnance Factory), Vosper-Thornycroft, Newton, Sperry and Smiths in the UK, and MBB of Germany.

Super Frelon helicopter carrying two Exocet air-to-surface missiles

The firing installation supplied by Aérospatiale for the surface-to-surface versions is compatible with the full range of existing shipboard fire control systems, such as the Thomson Vega and systems manufactured by Marconi and HSD.

In August 1971, The Boeing Company acquired a licence to manufacture and market Exocet in the USA.

Firing trials of MM38 Exocet missiles fitted with guidance systems began in mid-1971. The manufacturer's tests were concluded on 4 July 1972, with a direct hit of a missile with a live warhead on a 300-ton hull, which was sunk. Evaluation trials by the French Navy, with the participation of the Royal Navy and the West German Navy, began in October 1972. Other customers for the MM38 surface-to-surface version include the navies of Brazil, Chile, Ecuador, Greece, Malaysia and Peru. Production has been under way at the rate of 15 missiles per month since January 1973.

Launch tests from a Super Frelon helicopter began in April 1973. The type of missile employed was an intermediate version of Exocet designated AM38, which consisted of an MM38 fitted with the one-second ignition delay that is embodied in the AM39. Inert missiles were dropped in the first three tests at forward speeds of 80 knots (148 km/h; 92 mph), 100 knots (185 km/h; 115 mph) and 120 knots (222 km/h; 138 mph) respectively. Two launches of missiles with live power plants but no guidance followed on 21 and 25 June 1973, with the helicopter flying at 100 knots and 125 knots (185 km/h; 115 mph and 232 km/h; 144 mph) respectively at a height of 500 m (1,650 ft).

In each of the powered launches, the missile dropped about 10 m (33 ft) in a stable, horizontal attitude from its pylon before ignition, despite rotor downwash.

Carrying two AM39 Exocets and 5,000 litres (1,100 Imp gallons) of fuel, a Super Frelon could carry out a 6-hour patrol near its shore or ship base, or attack a specific target up to 350 nm (650 km; 400 miles) from base. Equipment added to the helicopter consists of two launch pylons, an operator's console, an Omera ORB-31D X-band search and tracking radar capable of acquiring targets the size of a fast patrol boat over a range of 28·5 nm (53 km; 33 miles) in sea state 4 or 5, an EMD Alto-2 Doppler radar altimeter, a SFIM CV-153 vertical gyro for a compass system, a compensated airspeed indicator and a Crouzet 70 computer.

Manufacture of the AM39 for the French Navy was expected to begin in 1976. All 75 production AM39s built by January 1976 were exported.

Several more countries are contemplating deployment of the AM39 air-launched version of Exocet from rotating-wing and fixed-wing aircraft. Use of the MM38 version in a shore-to-ship role and from submarines is also anticipated.

DIMENSIONS (MM38):

Length	5·20 m (17 ft 0¾ in)
Body diameter	0·348 m (1 ft 1¾ in)
Span of wings	1·04 m (3 ft 5 in)
Span of fins	0·758 m (2 ft 5¾ in)
WEIGHTS:	
Launching weight:	
MM38	730 kg (1,609 lb)
AM39	less than 650 kg (1,430 lb)
PERFORMANCE:	
Max level speed	Mach 0·93

Range (estimated):

AM39, launched from a helicopter at 60 knots (110 km/h; 69 mph) at a height of 100 m (330 ft) 28 nm (52 km; 32·25 miles)

AM39, launched from an Atlantic aircraft at heights between 300 and 5,000 m (1,000 ft and 16,400 ft) 29-32·25 nm (54-60 km; 33·5-37·25 miles)

AM39, launched from a Super Etendard aircraft at heights between 100 m (330 ft) and 10,000 m (33,000 ft) 32-37 nm (60-70 km; 37-43 miles)

NEW MISSILE

It was announced by M Robert Galley, the French Defence Minister, on 24 January 1974, that a new air-to-surface missile with a nuclear warhead was to be developed for future use from fighter aircraft engaged on penetration missions. The missile will have a range of 43-80 nm (80-150 km; 50-93 miles). Its warhead will have a yield of 500-600 kilotons.

MATRA
SA ENGINS MATRA

HEAD OFFICE:
4 rue de Presbourg, 75116-Paris

MANAGEMENT AND WORKS:
37 avenue Louis Breguet, 78140-Vélizy

POSTAL ADDRESS:
BP No. 1, 78140-Vélizy

Telephone: 946.96.00

Telex: ENMATRA 69.077F

OTHER WORKS:
rue de la Convention, 41300-Salbris

Telephone: (39) 83.02.50

CHAIRMAN: Marcel Chassagny

GENERAL MANAGER:
Jean-Luc Lagardere

PUBLIC RELATIONS DIRECTOR:
Philippe Chassagny

Since 1948, Matra has been engaged in extensive research and experimental work in the guided missile, propulsion and guidance fields.

After prolonged testing on Meteor and Canberra aircraft, the company's type R.510 air-to-air weapon went into small series production for training purposes. Described in the 1958-59 *Jane's,* it was superseded by the fully-developed type R.511 missile, of which approximately 1,000 were manufactured as standard weapons of the French Air Force and continue in service on Mirage III-C interceptors.

The R.511 was followed in turn by the R.530, and further improved weapons are now in production or being developed as armament for the interceptors of 1980 onward.

In September 1964 an Anglo-French government agreement was signed, providing for the joint development and production by Matra and Hawker Siddeley of an air-to-surface guided weapon that was subsequently designated AS.37/AJ.168 Martel. Details of this can be found in the International section.

Matra is also designing and producing on its own, or in collaboration with foreign companies, other missiles to meet current tactical requirements, including the surface-to-air Crotale and a variant of Crotale, known as Cactus, for operation by South Africa.

Matra is responsible for the development and quantity production of launchers for unguided solid-propellant rockets of 37 mm, 68 mm and 100 mm calibre, which form standard armament on French and British fighter and bomber aircraft, and combat aircraft and helicopters of other nations.

Since 1960, Matra has also been working on retardation systems for bombs, under the direction of the Service Technique de l'Aéronautique. The system in series production and service comprises a cruciform parachute, mechanism to check the release parameters, and nose and tail fuses. It has been used with French SAMP 250 kg and 400 kg bombs since 1966 and is in service in some ten foreign countries, including West Germany, in certain cases on standard British or US manufactured bombs. It can be used at heights down to 30 m (100 ft).

Under development is the Durandal rocket-propelled penetration bomb for attacking runways. Other Matra weapon systems include dispersal armaments; launchers for marine markers, sonobuoys and flares; and pods for DEFA 30 mm guns, 20 mm guns and 12·7 mm machine-guns for use on all categories of combat aircraft, including light aircraft and helicopters.

MATRA R.530

Since 1963, more than 3,000 R.530 air-to-air missiles have been ordered as standard armament for Vautour and Mirage III and F1 interceptors of the French Air Force, Mirages of the South African, Israeli, Australian and Brazilian Air Forces, and F-8E (FN) Crusaders of the French Navy. Production continued in 1976.

Suitable for use at heights from sea level to 21,000 m (69,000 ft), the R.530 is an all-weather missile, with inter-

Versions of the Matra R.530 air-to-air missile with semi-active radar guidance (left) and infra-red homing head (right)

changeable EMD AD-26 semi-active radar and infra-red homing heads. It can be fired at the target from any relative direction, its homing head being sufficiently sensitive not to require firing from astern of the enemy aircraft.

The R.530 has a cylindrical body, with cruciform delta wings, two of which are fitted with ailerons, and cruciform tail control surfaces. It is powered by a two-stage Hotchkiss-Brandt solid-propellant rocket motor, rated at 83·36 kN (18,740 lb st).

There are two types of Hotchkiss-Brandt high-explosive warhead, each weighing 27 kg (60 lb) and fitted with a proximity fuse. The latest types of anti-jamming ECM devices are fitted to the AD-26 head.

DIMENSIONS:
- Length overall: 3·28 m (10 ft 9¼ in)
- Body diameter: 0·26 m (10¼ in)
- Wing span: 1·10 m (3 ft 7¼ in)

WEIGHT:
- Launching weight: 195 kg (430 lb)

PERFORMANCE:
- Max speed: Mach 2·7
- Range: 9·5 nm (18 km; 11 miles)

Matra Super 530 high-performance air-to-air missile under wing of Mirage F1

MATRA SUPER 530

This new high-performance air-to-air weapon system is intended initially as armament for the Dassault Mirage F1 interceptors of the French Air Force, offering ranges compatible with the Cyrano IV fire control system of these aircraft. Compared with the R.530, it has much improved aerodynamics, structure, radome, electronics and power plant, doubling both the acquisition distance and the effective range. It is an all-weather and all-sector weapon, with the ability to attack targets flying at an altitude more than 7,600 m (25,000 ft) higher or lower than that of the launch aircraft.

The general appearance of the Super 530 is shown in the accompanying photograph of a test round. It has a Super AD-26 semi-active radar homing head, developed and built by Electronique Marcel Dassault. Its rocket motor, supplied by Thomson/Brandt, utilises Butalane propellant, with a much higher specific impulse than the motor of the R.530, raising the missile's max speed to an estimated Mach 4·5. The warhead is supplied by Thomson/Brandt, with a proximity fuse by Thomson-CSF; and the latest ECM anti-jamming circuits are fitted.

Ramp-launch tests of the Super 530 began in January 1971. Flight trials of the homing head were started in September 1972, and the first controlled model was fired successfully on 27 February 1973. Development and evaluation launches, performed under the auspices of the Centre d'Expériences Aériennes Militaires, Mont-de-Marsan, will continue until 1977. Many complete missiles had been fired successfully against airborne targets by January 1976, when the first launch from a Mirage F1 was conducted successfully.

DIMENSIONS:
- Length overall: 3·54 m (11 ft 7¼ in)
- Body diameter: 0·26 m (10¼ in)
- Wing span: 0·64 m (2 ft 1¼ in)
- Fin span: 0·90 m (2 ft 11½ in)

WEIGHT:
- Launching weight: 227 kg (500 lb)

PERFORMANCE (estimated):
- Max speed: Mach 4·5
- Operational ceiling: above 21,350 m (70,000 ft)
- Range: 16-19 nm (30-35 km; 18·5-21·75 miles)

Matra R.550 Magic short/medium-range air-to-air missile on wingtip of Dassault Mirage F1

MATRA R.550 MAGIC

This air-to-air weapon system was intended initially to meet a French Air Force requirement for a highly-manoeuvrable short/medium-range 'dogfight' missile. Development is reported to have been started in 1967 as a private venture and to have continued under official contract since 1969.

The accompanying illustration shows the unique 'double-canard' configuration of the R.550 Magic, with a set of movable foreplane control surfaces immediately behind and indexed in line with cruciform fixed surfaces. The missile has a solid-propellant motor and an infra-red homing head with a cooled cell, manufactured by SAT. It can utilise the same aircraft launcher as the Sidewinder which it will replace.

The first air-launch of an R.550 complete with guidance equipment, against a target, took place at the Landes Test Centre on 11 January 1972, from a Gloster Meteor test aircraft. The first full test of the missile's manoeuvrability was conducted on 30 November 1973, with a launch from a Mirage III against a CT.20 target drone.

The Magic is reported to be effective over ranges from about 500 m (1,650 ft) to nearly 3·25 nm (6 km; 3·75 miles), to be able to accept load factors of up to 7*g*, and to have high acceleration. Delivery of the first production missiles for operational testing and evaluation by the French Air Force began in 1974. Full series production began in 1975 and more than 1,000 Magics had been delivered by early 1976. Adaptation to the Mirage III, Mirage F1, Jaguar and French Navy Crusader fighter has been completed; adaptation to the Super Étendard was underway in the Spring of 1976. In addition, large export orders have been received from nine countries, requiring adaptation to other aircraft such as the Aermacchi M.B. 326K.

DIMENSIONS:
- Length overall: 2·80 m (9 ft 2¼ in)
- Body diameter: 0·15 m (6 in)
- Spans of fins: 0·65 m (2 ft 1½ in)

WEIGHT:
- Launching weight: 88 kg (194 lb)

MATRA/HSD AS.37/AJ.168 MARTEL

See International section

GERMANY
(FEDERAL REPUBLIC)

MBB
MESSERSCHMITT-BÖLKOW-BLOHM GmbH

HEAD OFFICE AND WORKS:
Ottobrunn bei München, 8 München 80, Postfach 801220

Telephone: (089) 60 00 25 90

Defence Technology Division

WORKS:
Nabern and Schrobenhausen

OFFICERS:
See Aircraft section

In addition to its manufacture of piloted aircraft, MBB has engaged in the development and manufacture of guided weapons for many years. It has achieved considerable success with its short-range surface-to-surface and surface-to-air battlefield missiles, some of which result from partnership with Aérospatiale of France in the Euromissile team.

Among weapons under current development or entering production are an air-launched anti-shipping missile known as the Kormoran, and a TV-guided air-to-surface missile named Jumbo. MBB has also produced the Aérospatiale AS.20 air-to-surface missile under licence since 1971, for carriage by F-104G aircraft engaged on anti-shipping duties.

MBB is engaged on the development of research rockets and satellites (see Spaceflight section).

MBB KORMORAN

This roll-stabilised air-to-surface anti-shipping missile results from a joint development programme by MBB and Aérospatiale of France, under the design leadership of MBB and financed by the Federal German government.

Kormoran is an all-weather weapon which can be carried by any aircraft able to maintain a speed between Mach 0·6 and 0·95 during the attack, and which is equipped with target detection radar and an autonomous navigation system such as an inertial platform or Doppler. The aircraft's electronics must be supplemented by a position and homing indicator (PHI) and Kormoran firing equipment. Launch information is obtained from the aircraft's target detection radar, navigation system, and a navigation support device which adapts the signals representing velocity for the Kormoran airborne computer. In addition, if the missile is operated in an optical mode, instead of through target detection radar, the PHI has to be used in conjunction with computer co-ordinated data (CCD) and a vector addition unit (VAD).

The general appearance of Kormoran is shown in the accompanying illustration. It has a cylindrical body, fitted with cropped-delta cruciform wings and with cruciform tail control surfaces indexed in line with the wings. Two built-in boosters accelerate the missile to high subsonic cruising speed, which is then maintained by the solid-propellant sustainer motor. The guidance system employs pre-guidance and homing phases, using a Thomson-CSF radar terminal seeker. The new-type high-energy warhead, developed at MBB's Schrobenhausen works, weighs 160 kg (352 lb) and can penetrate 70-90 mm (2·75-3·5 in) of steel plate, making it effective against ships up to the size of a destroyer.

The launch aircraft can approach the target area at very low level, to escape detection by the ship's radar, by means of the PHI, with radar switched off. With the radar then turned on, the target is acquired and its position is fed into the PHI; the radar is again switched off while the aircraft takes up the optimum attack position. The radar is switched on, locked on to the target, and initial data are transmitted automatically to the missile. As soon as the missile has been launched, the aircraft breaks away, outside the range of enemy anti-aircraft defences. The missile descends to its programmed flight level and locks on to the target, guided by inertial navigation aided by its radar altimeter. At a prescribed distance, the inertial system releases the active radar seeker. After this has locked on target, the inertial guidance is slaved and corrected in azimuth and range by the seeker head. At a short distance from the target the missile descends to its terminal flight level in order to hit the target just above the waterline.

The secondary optical firing mode, in which the pilot aims the missile by means of his optical weapon sight, is used against surprise targets, in the presence of extreme enemy ECM, or when the target is a small vessel, such as a fast patrol boat or minesweeper. Kormoran itself is immune to all known kinds of ECM.

In-plant testing of Kormoran was completed by early 1974, at which time official tests from F-104G Starfighter aircraft were under way. An initial production contract to equip F-104Gs of the German naval air arm was placed by early 1975, with very large follow-up orders planned. Deliveries began in 1976.

DIMENSIONS:

Length overall	4·40 m (14 ft 5 in)
Body diameter	0·34 m (1 ft 1½ in)
Wing span	1·00 m (3 ft 3½ in)

WEIGHT:

Launching weight	600 kg (1,320 lb)

MBB-built AS.20 missile under wing of F-104G Starfighter of Marinefliegergeschwader I
(Ing Hans Redemann)

MBB Kormoran missile under wing of F-104G Starfighter

Jumbo television-guided air-to-surface missile, under development by MBB

MBB/AÉROSPATIALE HOT

See International section.

MBB JUMBO

Jumbo is a television-guided air-to-surface missile, with interchangeable nuclear and conventional warheads, which is being developed primarily as a weapon for the Panavia Tornado aircraft. MBB claims that the missile will offer improved ECM and ECCM capabilities by comparison with current types, and will have a long range.

Development was continuing in 1976 with German government approval.

As can be seen in the accompanying illustration, Jumbo has a configuration very like that of Kormoran, but is larger and weighs nearly twice as much, suggesting the possibility of a warhead as heavy as 500 kg (1,100 lb). It is powered by a solid-propellant rocket motor, and carries its TV optics between two of the swept wings, in the front of an underbelly fairing duct. To cope with European weather conditions, the eventual development of a dual-mode radar/TV seeker seems likely.

MBB has stated that Jumbo is intended for use against point and area targets, with launch at virtually any altitude. An inertial guidance system and radar altimeter are expected to offer a variety of flight paths to the target. Video/command link equipment will be carried in a pylon-mounted pod on the launch aircraft, which will include F-4s of the Luftwaffe.

DIMENSIONS:

Length overall	5·24 m (17 ft 2½ in)
Body diameter	0·50 m (1 ft 7¾ in)
Wing span	1·25 m (4 ft 1¼ in)

WEIGHT:

Launching weight	1,150 kg (2,535 lb)

INDIA

BHARAT DYNAMICS

BHARAT DYNAMICS LTD

ADDRESS:
10-3-310 Masab Tank, Hyderabad-500 028

MANAGING DIRECTOR:
Brig J. P. Anthony

FACTORY MANAGER:
Wing Cdr V. M. Chitale

FINANCIAL CONTROLLER:
S. A. Rahman

This company was formed by the Government of India to establish a national guided missile industry. It occupied a new factory at Hyderabad in 1971 and, as a first step, undertook the manufacture of SS.11 anti-tank missiles under licence from Aérospatiale of France. The first batch of missiles, produced from subassemblies supplied by Aérospatiale, came off the Indian assembly line and passed acceptance tests only 12 months after signature of the licensing agreement. By late 1973, the SS.11 was being manufactured entirely in India. It is planned to extend production to other types of weapon, including several of the second-generation missiles developed by Euromissile (which see).

The Soviet-designed K-13A (NATO *Atoll*) infra-red homing air-to-air missile is also reported to be produced in India, for carriage by the MiG-21 fighters of the IAF.

INTERNATIONAL PROGRAMMES

EUROMISSILE

EUROMISSILE GROUPEMENT D'INTÉRÊT ÉCONOMIQUE

ADDRESS:
7 rue Béranger, BP 84, 92320-Châtillon, France
Telephone: 657.12.44
Telex: EUROM 204691 F

PRESIDENT:
Marcel Morer

VICE-PRESIDENT:
Friedemann Striegel

DIRECTOR OF FINANCE AND CONTRACTS:
Fritz Ramjoue

SALES DIRECTORS:
Pierre Chaboureau
Emil-Otto Wittmann

Euromissile is a Groupement d'Intérêt Economique formed by Aérospatiale of France and MBB of Germany. It enables the industrial management and marketing of missiles which the two parent companies have developed jointly to be handled by a single organisation, responsible for the entire programme.

First products of the Euromissile team are three short-range battlefield weapons known as Milan, Hot and Roland, developed initially for the armed services of France and Federal Germany. Of these, Hot is the only one that has an air-launch capability.

Six Hot missile launchers mounted on an MBB BO 105 helicopter

HOT

The Hot (High-subsonic, optically-guided, tube-launched) is a tube-launched wire-guided anti-tank missile suitable for both surface-to-surface and air-to-surface use. It has fins which fold down against the body when it is in its launching tube, and open out to spin-stabilise it in flight. The power plant is a two-stage solid-propellant rocket motor. Because of its comparatively high speed, the time of flight to a target is about half that for the earlier Aérospatiale SS.11. Guidance is by means of the TCA type of automatic optical sighting/infra-red system in one of three forms: a type made up of separate components for adaptation to the turrets of vehicles like the French AMX 10; a periscopic type for turret-less vehicles such as the German Jagdpanzer-Rakete; a stabilised-sight type for use on helicopters such as the BO 105, Gazelle and Bell UH-1D. A jet vane control system is used.

The HLVS (Hot, stabilised localiser-sight) system offers magnifications of 2·8 × 22° field and 11·2 × 5·6° field. Sight limits are ±120° in bearing and +28°/−20° in elevation. To engage a target, the aimer maintains a sighting cross on the target, switches on the firing installation and selects a missile on the control box.

Development of the automatic guidance system for use on helicopters began in early 1972, with the first installation on an Alouette III. The results of initial tests were as good as those achieved earlier on the ground in terms of accuracy and reliability, over ranges up to 4,000 m (13,125 ft) during hovering and manoeuvring flight.

Flight testing of an operational Hot installation on an Aérospatiale/Westland SA 341 Gazelle helicopter, in conjunction with the HLVS stabilised sight system, was performed in 1973. With the target hidden from the sight of the pilot and missile operator before take-off, the helicopter proved able to acquire the target and launch a missile within 4 seconds of lift-off. Hit probability proved to be 90% over ranges varying from a minimum 400 m (1,310 ft) to a maximum of 4,000 m (13,125 ft) in hovering flight and 4,600 m (15,000 ft) in translational flight at speeds up to 108 knots (200 km/h; 124 mph). After launch, the helicopter was able to take evasive action at a turning speed of up to 6° per second.

Launching a Hot missile from a Gazelle helicopter

Further helicopter trials, by the official services of France and Germany, took place during 1974, from a Gazelle and a BO 105 respectively. Both nations have since decided to adopt Hot systems as standard combat installations on these helicopters.

The Hot missile, mountings and guidance units are common for all surface or airborne installations. Tests already conducted have confirmed that the guidance accuracy is high enough to allow the missile to be used for defence against attack by low-flying aircraft and helicopters.

DIMENSIONS:

Length overall	1·30 m (4 ft 3¼ in)
Body diameter	0·175 m (6¾ in)
Fin span	0·31 m (1 ft 0¼ in)

WEIGHTS (helicopter installation):

Launching weight of missile	22 kg (48·5 lb)
Warhead	6 kg (13·2 lb)
Missile and container	32 kg (70·5 lb)
Sight/localiser group	33·8 kg (74·5 lb)
Command/guidance group	30·3 kg (66·8 lb)
Launcher group:	
Gazelle, 4 launchers	39·8 kg (87·8 lb)
BO 105, 6 launchers	125 kg (275 lb)

PERFORMANCE (air-to-surface):

Max speed	545 knots (1,010 km/h; 625 mph)
Flight times:	
2,000 m (6,560 ft)	8·7 sec
3,000 m (9,840 ft)	12·5 sec
4,000 m (13,125 ft)	16·3 sec
Range	400-4,000 m (1,310-13,125 ft)

MATRA/HSD

SA ENGINS MATRA, avenue Louis Breguet, 78140 Vélizy-Villacoublay, France
Telephone: 946-96-00

HAWKER SIDDELEY DYNAMICS LTD, Manor Road, Hatfield, Herts, England
Telephone: Hatfield 62300

Matra and HSD have developed jointly and are manufacturing two versions of an air-to-surface precision tactical strike missile known as Martel, to meet the requirements of the British and French armed services.

MATRA/HSD AS.37/AJ.168 MARTEL

Martel (Missile Anti-Radar and TELevision) is a guided air-to-surface missile which is operational in two versions, one using passive radar homing and the other TV guidance. The two versions have maximum commonality of structure and systems.

The anti-radar version of Martel (AS.37), produced by Engins Matra in France, offers all-weather attack capability against radar antennae in several frequency bands. Depending on the mission profiles, it can be launched at very low, medium or high altitudes. It then flies a homing trajectory into the emitting target source. This is done without further information or control from the parent aircraft, which can return to its base immediately after launch.

The television version of Martel (AJ.168), produced in the UK by Hawker Siddeley Dynamics, follows a pre-programmed course immediately after launch, but the final impact on target is effected by the weapon operator, who is given a direct visual picture of the target on a high-brightness monitor. Command instructions are sent back from the aircraft to the missiles if changes are required to the missile flight path, in either elevation or azimuth. Control signals generated within the missile alter the flight path to bring the axis of the missile into line with that of the television camera, once the target has been selected.

A wide variety of initial flight patterns can be produced to ensure that the TV system has the best possible opportunity of identifying the target, with the least possible danger to the launching aircraft.

Other major companies associated with Martel are the Marconi Company, which provides the TV and radio link equipment that forms part of the guidance system of the TV version; Electronique Marcel Dassault, which provides the AD-37 passive homing head of the anti-radar version; and BAC, which supplies missile radomes and high-precision gyroscopes through its Reinforced and Microwave Plastics Group and Precision Products Group respectively. The solid-propellant motors are produced by Hotchkiss-Brandt and Aérospatiale.

No other details of Martel may yet be published. The general appearance of the two versions is shown in the accompanying illustrations.

The first simulated firings and mockup launchings were made in the Summer of 1964 and prototypes of both versions were completed in 1965-66. Development was completed in 1968 with a highly successful series of firings. Evaluation, which began in 1969, proved very satisfactory. Delivery of production missiles and equipment began in 1972; the first firing of a production Martel took place on 10 December 1973, when an anti-radar missile was launched from a Breguet Atlantic ASW aircraft of the French Navy, off Landes Flight Test Centre. The Martel impacted on a target radar mounted on a ship.

Both versions of Martel continued in production in 1976, as armament for a variety of aircraft, including the Hawker Siddeley Buccaneer operated by the British services, and the Dassault Mirage III-E, SEPECAT Jaguar and Breguet Atlantic operated by the French services.

DIMENSIONS:	
Length overall:	
AS.37	4·12 m (13 ft 6¼ in)
AJ.168	3·87 m (12 ft 8½ in)
Span of wings	1·20 m (3 ft 11¼ in)
Body diameter	0·40 m (1 ft 3¾ in)
WEIGHTS:	
Launching weight:	
AS.37	530 kg (1,168 lb)
AJ.168	550 kg (1,213 lb)
Warhead	150 kg (330 lb)

AS.37 Martel missile on the underwing launcher of a Buccaneer aircraft

Buccaneer carrying three Martels and a Martel systems pod

MATRA / OTO MELARA

SA ENGINS MATRA

avenue Louis Breguet, 78140 Vélizy-Villacoublay, France

OTO MELARA SpA

Via Valdilocchi 15, CP 337, 19100 La Spezia, Italy

Telephone: 502.005 and 504.041
Telex: 27368 OTO

These two companies have developed jointly an anti-shipping missile known as Otomat, which is turbojet-powered and has an unusually long range for this class of weapon. Its name is a contraction of **Oto** Melara and **Matra**.

The basic ship-launched version of Otomat is in production for three countries and was described in the 1974-75 *Jane's*. A coastal defence variant is being developed at the request of several potential operators, and an air-launched version has been projected, for installation on helicopters or fixed-wing aircraft, but has not yet been adopted as a production weapon.

ISRAEL

RAFAEL

RAFAEL ARMAMENT DEVELOPMENT AUTHORITY

ADDRESS:
Ministry of Defence, POB 2082, Haifa
Telephone: 04-714168

The duty of Rafael Armament Development Authority is to develop and supply advanced weapons and weapon systems for use by the Israeli Defence Forces. As well as meeting urgent requirements for complete weapons, it is responsible for research, development and manufacture of propellants, aircraft armament, fuses, explosives, small computers, electronic systems, communications systems and other products. To make this possible, it possesses a variety of structural testing, environmental testing and other laboratories and facilities.

One of the weapon systems developed and manufactured by Rafael is the Shafrir air-to-air missile.

SHAFRIR

Shafrir is a short-range air-to-air dogfight missile developed for use against aircraft at heights up to 18,000 m (60,000 ft). Its development was completed by the late 'sixties and production was then started. Many rounds have been fired in air combat against enemy aircraft since 1969, with considerable success.

This was particularly evident after the Yom Kippur War of October 1973, when Israel is reported to have destroyed about 335 Arab aircraft in air combat for the loss of only six of its own aircraft. Nearly 200 of the Arab aircraft were shot down by infra-red homing missiles,

Shafrir under wing of a Kfir fighter

some Sidewinders but mostly Shafrirs, according to the US *Armed Forces Journal*. In 70 other successful combats, the weapon responsible for destroying the target was not determined.

Relatively small in size and simple in conception, Shafrir has a slim cylindrical body, with an infra-red seeker head and cruciform canard control surfaces indexed in line with cruciform fixed wings mounted at the tail. A rolleron is inset in the tip of each wing to help stabilise the missile in flight.

Shafrir has a solid-propellant rocket motor and is a solid-state weapon, fully transistorised and with all components built to strict military specifications. The foreplanes are actuated pneumatically. Electronic circuitry is kept to a minimum, with no computers. Guidance is by proportional navigation, for optimum results against manoeuvring targets.

The missile and its launcher are mounted under the wing of the aircraft on a specially-designed adapter which is capable of carrying other types of weapon as an alternative to Shafrir. Attachment is mechanical and the missile requires no support from the aircraft except for the firing circuit. When a target is detected within firing range, an audio signal is heard and a light is switched on automatically on the pilot's control panel as an indication that the firing button should be pressed. After launch, the missile tracks the target entirely automatically, and the warhead is detonated either on impact or by the proximity fuse within optimum distance of the target.

Sales are reported to have been made to several overseas customers, including Taiwan.

DIMENSION:
Length overall 2·50 m (8 ft 2½ in)
WEIGHTS:
Warhead 11 kg (24·25 lb)
Launching weight 93 kg (205 lb)
PERFORMANCE:
Max range 2·7 nm (5 km; 3·1 miles)

ITALY

SELENIA

SELENIA INDUSTRIE ELETTRONICHE ASSOCIATE SpA

HEAD OFFICE AND WORKS:
via Tiburtina km. 12.400, 00131 Rome
POSTAL ADDRESS:
PO Box 7083, 00100 Rome
Telephone: 43601
Telex: 61106 Seleniat
PUBLIC RELATIONS:
Dott Ing Paolo De Gaetano

This major Italian company produces missiles, missile components and equipment, in addition to a wide range of radar, telecommunications, automation and other electronic products.

ASPIDE

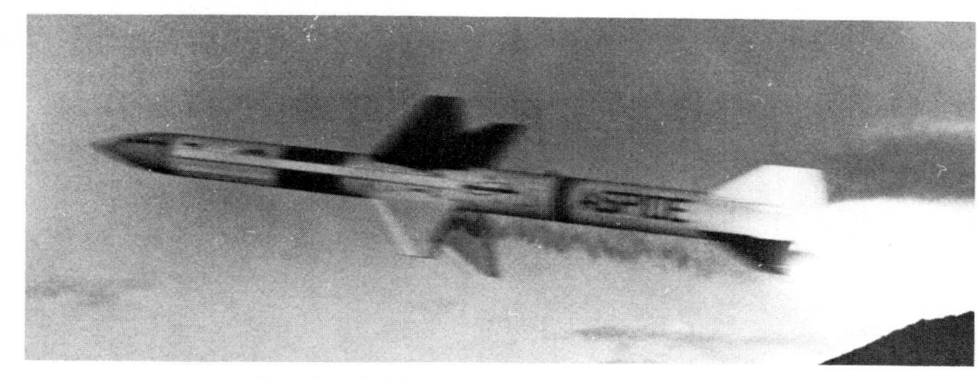

Ground-to-air firing of an Aspide missile in Sardinia

This missile is under development as an all-weather all-aspect weapon for high-performance interceptors, for use in the Albatros naval air defence system, and for use in the Spada ground-based low-altitude air defence system. It will improve the effectiveness of each of these weapon systems in terms of maximum missile range, operation at very low altitudes, multiple target engagement and resistance to advanced ECM. In particular, it is expected to enhance the dogfight and shootdown capabilities of the F-104S interceptors of the Italian Air Force.

The air-to-air version of Aspide has more-tapered delta-shape wings and fins, of greater span, than those of the surface-to-air version. A single-stage solid-propellant rocket motor, supplied by SNIA Viscosa, gives it a speed described as being "well in the hypersonic field". Its guidance system is of the semi-active radar type, developed by Selenia. The warhead, with an estimated weight of 33 kg (73 lb), is a fragmentation type produced by SNIA Viscosa. It is located immediately forward of the pivoted wings, this compartment being used for telemetry equipment in the development rounds.

The missile's final configuration, almost identical to that of the US Sparrow, was determined by the Spring of 1974, and captive flight testing of prototypes began on an F-104S in July of that year. Qualification tests and construction of pre-series prototypes for engineering evaluation were under way by that time. Ground-to-air firings of Aspide, using a prototype of the Spada system missile launcher, began at the Salto di Quirra test range, Sardinia, in May 1975. First-phase firing tests were concluded on 17 December 1975, as were the captive flight trials which had been carried out in parallel. Both test programmes achieved fully all design goals, and the first Aspide production missiles will therefore leave the production line on schedule in 1977.

DIMENSIONS:
Length overall 3·70 m (12 ft 1½ in)
Body diameter 20·3 cm (8 in)
Wing span 1·00 m (3 ft 3¼ in)
Fin span 0·80 m (2 ft 7½ in)
WEIGHT:
Launching weight 220 kg (485 lb)

SISTEL

SISTEL—SISTEMI ELETTRONICI SpA

HEAD OFFICE:
via Tiburtina 1210, 00131 Rome
Telephone: 415841 and 414651
Telex: 68112 Sistelro
MANAGING DIRECTOR:
Prof Ing Giovanni Malaman

Sistel—Sistemi Elettronici SpA—was formed jointly by Montecatini-Edison, Contraves Italiana, Fiat, Finmeccanica and SNIA in late 1967, to develop new products in the missile field. It embodies the former missiles and space equipment branch of Contraves Italiana.

Sistel's current products include the Indigo short-range surface-to-air missile, the Sea Killer Mk 1 and Mk 2 surface-to-surface missiles, and the Marte helicopter-launched version of Sea Killer Mks 1 and 2. Under development is the Sea Killer Mk 3. FM/FM telemetry packages and telemetry receiving stations are also designed and manufactured by Sistel.

MARTE

In 1967, the Italian Navy initiated a development programme, known as Project Marte, to enhance the capabilities of shore-based or shipborne helicopters by arming them with anti-shipping missiles. Eventually, Sistel's Sea Killer family of missiles was chosen as most suitable for this application, because of the weapons' all-weather operability, automatic guidance system, inherent insensibility to ECM and sea-skimming flight profile.

Sistel was appointed prime contractor and supplier of Sea Killer Mk 1 and Mk 2 missiles, described individually under this entry. Agusta manufactures under licence the Sikorsky SH-3D and Bell Model 204 helicopters which were envisaged as carriers of the Marte weapon system, with airborne radars supplied by SMA. The system has, however, been studied for installation on a wide range of Agusta helicopters, with take-off weights varying from 3,000 kg to more than 10,000 kg (6,600-22,050 lb), dependent on the number of missiles to be carried and the possible requirement for simultaneous anti-submarine and surface strike capability.

Equipment added to the Marte helicopter is lightweight and easily operated. The radar performs navigation, search and target tracking as well as guidance of the missile in azimuth. Thus an SH-3D helicopter, fully equipped for surface strike and anti-submarine duties, and with an autopilot for instrument flying, can be assigned a 4¼-hour maritime patrol at a search speed of 100 knots (185 km/h; 115 mph), carrying a crew of four. Total weight of the system is 1,165 kg (2,568 lb), made up of 400 kg (882 lb) for the missile launching equipment, 600 kg (1,323 lb) for two Sea Killer Mk 2 missiles, 143 kg (315 lb) for sonar equipment, and 22 kg (48 lb) for an optical sight. The operator has a missile control console inside the main cabin of the aircraft. The radar antenna is mounted inside a shallow radome beneath the hull, under the flight deck.

In action, it can be assumed that the helicopter will locate a target in a few seconds of radar operation at the limit of radar range. To reduce the possibility of enemy recognition of the helicopter's radar interrogation, the airborne radar is then switched off and the aircraft descends in order to fly toward the target at the lowest practical height above the water. At an estimated distance just beyond missile range, the helicopter climbs again to missile launching height, re-acquires the target and launches a Sea Killer, which takes slightly more than one minute to reach the enemy ship.

The Sistel radar altimeter in the missile can be pre-set before launch to control the cruising height at values down to 2 m (6 ft 7 in) or less, depending on factors such as sea state. The altitude can be changed by command signal during flight, if required. Control in azimuth can be achieved either by automatic radar mode in all-weather conditions or by a standby fair-weather system, using an optical sight and joystick controller.

Test firing a Marte missile from an AB 204AS helicopter in flight

Initial firing trials of the Marte system in flight, from an AB 204 helicopter, had been made successfully by the Spring of 1973. Subsequently the Italian Navy awarded a pre-production contract for Marte systems utilising Sea Killer Mk 2 missiles, for installation on SH-3D helicopters.

SISTEL SEA KILLER Mk 1

This fully-qualified short-range surface-to-surface ship-based guided weapon, known for a time as Nettuno, is carried in a five-round multiple launcher on board the fast patrol boat *Saetta* of the Italian Navy.

Sea Killer Mk 1 is also one of the sea-skimming missiles utilised in the Marte air-to-surface weapon system for helicopters.

WINGS: Movable cruciform control surfaces at centre of missile.

BODY: Cylindrical light alloy structure.

TAIL SURFACES: Cruciform stabilising fins.

POWER PLANT: Solid-propellant rocket motor (19·6 kN; 4,410 lb st).

GUIDANCE: Alternative all-weather beam-rider/command/radio altimeter guidance, or optical radio command/radio altimeter guidance.

CONTROL: Via movable wings.

WARHEAD: High-explosive fragmentation type, with impact/proximity fuse; weight 35 kg (77 lb).

DIMENSIONS:

Length overall	3·73 m (12 ft 3 in)
Body diameter	0·20 m (7·87 in)
Span of wings	0·85 m (2 ft 9½ in)

WEIGHTS:

Launching weight	168 kg (370 lb)
Weight at burnout	118 kg (260 lb)

PERFORMANCE:

Speed at burnout	Mach 1·9
Max effective range	5·4 nm (10 km; 6·2 miles)

SISTEL SEA KILLER Mk 2

Known for a time as Vulcano, this two-stage surface-to-surface guided missile is a development of the Sea Killer Mk 1 with a heavier warhead and extended range. First flight trials of Sea Killer Mk 2 missiles were made in mid-1969. Qualification trials were carried out in mid-1971, and the weapon is operational on board four Vosper Mk 5 frigates of the Imperial Iranian Navy. It is also one of the sea-skimming missiles utilised in the Marte air-to-surface weapon system for helicopters.

Sistel Sea Killer Mk 2 missile, as used in Marte air-to-surface weapon system

WINGS, BODY, TAIL SURFACES: As for Sea Killer Mk 1. Booster also has stabilising fins.

POWER PLANT: SEP 299 solid-propellant booster, rated at 43·16 kN (9,702 lb st) for 1·6 sec. SEP 300 solid-propellant sustainer, rated at 0·98 kN (220·5 lb st) for 73 sec.

GUIDANCE AND CONTROL: As for Sea Killer Mk 1.

WARHEAD: High-explosive semi-armour-piercing with impact/proximity fuse; weight 70 kg (154 lb).

DIMENSIONS:

Length overall	4·70 m (15 ft 5 in)
Body diameter	0·20 m (7·87 in)
Span of wings	1·00 m (3 ft 3½ in)

WEIGHT:

Launching weight	300 kg (660 lb)

PERFORMANCE:

Cruising speed	Mach 0·7
Max effective range	over 13·5 nm (25 km; 15·5 miles)

JAPAN

MITSUBISHI
MITSUBISHI DENKI KABUSHIKI KAISHA (Mitsubishi Electric Corporation)

HEAD OFFICE:
2-3, Marunouchi 2-chome, Chiyoda-ku, Tokyo
Telephone: 218-2111
PRESIDENT: Sadakazu Shindo

Since 1921, Mitsubishi Electric has been responsible for production of a wide range of electrical and electronic products as a sister company of Mitsubishi Heavy Industries. Its recent aerospace products include radar systems, computers, fire control systems, radio equipment and ADF installations.

In 1968, the company was awarded a contract to produce Hawk surface-to-air missiles for the Ground Self-Defence Force, under licence from Raytheon (USA). The missiles are being delivered to the JGSDF in the period 1968-77.

Subsequently, Mitsubishi Electric was named prime contractor for licence production of Sparrow air-to-air missiles to arm F-4EJ fighters of the JASDF. This work began in FY 1972, and about 600 Sparrow IIIs are being delivered, for air-to-air use, between 1974 and 1978. In addition, 16 Sea Sparrow ship-to-air missiles were ordered by the JMSDF in FY 1975, to arm frigates.

MITSUBISHI JUKOGYO KABUSHIKI KAISHA (Mitsubishi Heavy Industries, Ltd)

HEAD OFFICE:
5-1, Marunouchi, 2-chome, Chiyoda-ku, Tokyo 100
OFFICERS: See Aircraft section

Mitsubishi Heavy Industries is developing and producing air-to-air missiles for the JASDF. It is also manufacturing the Nike-Hercules surface-to-air missile under licence from McDonnell Douglas Corporation, and has Japanese government contracts to develop an air-to-surface anti-shipping missile.

Details of its research rockets can be found in the Spaceflight section of this edition.

MITSUBISHI AAM-1

The AAM-1 is an infra-red homing air-to-air missile which Mitsubishi developed and produced for the Japan Defence Agency. Deliveries began in November 1970, to replace the Sidewinder on the F-86F and F-104J interceptors of the JASDF; total planned production of 330 missiles was completed by late 1971. No details are available.

MITSUBISHI AAM-2

Mitsubishi is developing this missile as a replacement for the AAM-1. It is a collision-course weapon, whereas the AAM-1 is limited to pursuit-course attack. Few details are available, except that Mitsubishi, as prime contractor, awarded a contract to Nihon Electric Company to cover manufacture of experimental infra-red homing devices for the AAM-2.

A contract awarded in March 1973 covered the manufacture of a second batch of 21 prototype AAM-2s for continued testing, at a cost of about $4·36 million. Air-launches of 40 pre-production missiles were scheduled to begin in the Spring of 1975, and the AAM-2 was expected to enter service later that year.

MITSUBISHI ASM-1

The Japanese government approved the appropriation of $1 million in FY 1973 and $5 million in FY 1974 to cover the initial development phase of an air-to-surface rocket-powered anti-shipping missile to be carried by Mitsubishi FS-T2-KAI attack aircraft of the JASDF. In November 1973 Mitsubishi Heavy Industries was selected as prime contractor for this missile, which will also be suitable for ship or ground launching when fitted with a rocket booster.

Known only as the ASM-1 at present, the new missile will travel to the target at a low altitude, using an inertial system for mid-course guidance and an active radar seeker for terminal guidance.

DIMENSIONS (provisional):

Length overall	4·00 m (13 ft 1½ in)
Body diameter	0·50 m (1 ft 7¾ in)

WEIGHTS (provisional):

Warhead	150 kg (330 lb)
Launching weight	500 kg (1,102 lb)

PERFORMANCE (provisional):

Cruising speed	Mach 1 class
Max range	20 nm (37 km; 23 miles)

NORWAY

KONGSBERG
A/S KONGSBERG VAAPENFABRIKK

HEAD OFFICE AND WORKS:
Postboks 25, N-3601 Kongsberg
Telephone: (034) 33250
Telex: 11491

Defence Products Division

SALES MANAGER, DEFENCE EQUIPMENT:
T. Engebakken
PUBLIC RELATIONS MANAGER: E. Frisvaag

This government-owned company is the only armament manufacturer in Norway. Its products include small arms, guns, rockets and missiles, proximity fuses, fire control equipment and weapon systems.

A/S Kongsberg Vaapenfabrikk was prime contractor for European production of the Bullpup air-to-surface missile and is the manufacturer of the Norwegian-developed Terne anti-submarine system and Penguin anti-ship missile system.

PENGUIN

This anti-ship missile system was developed by the Norwegian Defence Research Establishment, with assistance from the US and Federal German navies. It is in quantity production by Kongsberg Vaapenfabrikk, and can be installed on ships, helicopters and other platforms.

In its ship-to-ship version, Penguin is delivered in a container with integral launch-rail and can utilise most existing types of shipboard fire-control systems. It embodies a two-stage solid-propellant rocket motor and an inertial guidance system with infra-red terminal homing. Its warhead is said to weigh 120 kg (264 lb) and to be similar to that of Bullpup, with a contact fuse.

Each of the 20 *Storm* class gunboats of the Royal Norwegian Navy carries six individual Penguin launchers on its rear deck. The six torpedo boats of the *Snogg* class are each fitted with four launchers; others are installed on the five *Oslo* class frigates and on four Turkish *Kartal* class fast attack craft.

An improved version, the **Penguin Mk 2,** is being developed jointly for the Norwegian and Swedish Navies.

The air-launched version of Penguin Mk 2, which is being studied for carriage by jet fighters, would have reduced wing span and would not need the booster stage of the rocket motor. Its primary potential application is on the General Dynamics F-16 aircraft ordered for service with NATO air forces in Europe.

The following details apply to the ship-launched versions of Penguin Mk 1 and Mk 2:

DIMENSIONS:
- Length overall: 3·05 m (10 ft 0 in)
- Body diameter: 28 cm (11 in)
- Wing span: 1·40 m (4 ft 7 in)

WEIGHTS:
- Launching weight: 330 kg (727 lb)
- Weight with container/launcher: 640 kg (1,410 lb)

PERFORMANCE:
- Cruising speed: Mach 0·8
- Max range:
  - Mk 1: 10 nm (18·5 km; 11·5 miles)
  - Mk 2: 15 nm (27 km; 17 miles)

Mockup installation of Penguin missile on F-104G aircraft

SOUTH AFRICA

First official news of guided weapon development in South Africa was given by the Minister of Defence, Mr P. W. Botha, on 2 May 1969. He announced that both surface-to-air and air-to-air missiles were then under development.

WHIPLASH

Few details of this missile are available except that it is a purely South African venture, has "some unique characteristics" and had already been tested successfully on a range at St Lucia at the time of Mr Botha's original announcement of the project on 2 May 1969. Development is said to have begun in 1966. In September 1971, it was stated that a Mach 2 target had been hit by such a missile three seconds after the missile was launched from a Mirage III aircraft of the South African Air Force. Initial production was underway in late 1972. Reports in the UK press have suggested that the missile is named Whiplash.

SWEDEN

BOFORS
AB BOFORS

HEAD OFFICE AND WORKS:
PO Box 500, S-690 20 Bofors
Telephone: 0586-360 00
Telex: 732 10

MANAGING DIRECTOR: C. U. Winberg
ORDNANCE DIVISION: L. Pålsson
PUBLIC RELATIONS MANAGER: U. Carlström

This world-famous Swedish armament manufacturing concern has been producing unguided air-to-air and air-to-surface rockets for many years. As its first project in the guided missile field it developed as a private venture the Bantam wire-guided anti-tank weapon.

BOFORS BANTAM
Swedish military designation: RB53

The Bantam is a small wire-guided anti-tank missile, designed originally for operation by a single infantry soldier and since adapted for air-to-surface use also.

The cylindrical body and cruciform wings are made largely of glassfibre-reinforced plastics. Control is by vibrating spoilers, in each wing trailing-edge; and features of the guidance system, such as automatic control of the spoiler frequency, make Bantam easier to guide than other missiles of the so-called first generation. A two-stage solid-propellant rocket motor is used. The high-explosive warhead weighs 1·9 kg (4·1 lb); and hit probability is claimed to be 95 to 98 per cent over ranges between 800 and 2,000 m (2,625 and 6,600 ft).

The wings fold at mid-span, making possible the use of a very small carrying container. When the missile is fired from the container the wings unfold and their bent rear corners then cause the missile to rotate in flight.

The total weight of the entire weapon system in its basic infantry-operated form, including launcher, carrying rack, cable and control unit, is 20 kg (44 lb).

Bantam missile being fired from a Saab Supporter light attack aircraft

The Bantam continues in large-scale production for surface-to-surface use by the Swedish and Swiss Armies, with which it is standard equipment. It can be installed on and fired from vehicles of most types, including light combat aircraft and helicopters, and has been ordered to equip SK 61 (Scottish Aviation Bulldog) aircraft of the Swedish Air Force. Current contracts will maintain series production throughout the 1970s.

DIMENSIONS:
- Length overall: 0·85 m (2 ft 9½ in)
- Body diameter: 0·11 m (4·3 in)
- Wing span: 0·40 m (1 ft 3¾ in)

WEIGHT: Launching weight: 7·5 kg (16·5 lb)

PERFORMANCE:
- Cruising speed: 165 knots (306 km/h; 190 mph)
- Range: 250-2,000 m (820-6,600 ft)

SAAB-SCANIA
SAAB-SCANIA AKTIEBOLAG

HEAD OFFICE AND WORKS:
S-581 88 Linköping
Telephone: 013-12 90 20
Telex: 50040 saablgs

Aerospace Division

OFFICERS: See Aircraft section

In addition to its work on piloted aircraft, Saab-Scania is developing a modernised version of the Swedish 04 air-to-ship missile and is producing a new air-to-surface weapon system, designated Saab-05A.

Saab-Scania is also prime contractor for licence manufacture of two versions of the Hughes Aircraft Company's Falcon air-to-air missile. These are the **RB27** (Hughes designation HM-55), with semi-active radar homing, and the **RB28** (HM-58), with infra-red homing. Both versions are carried by Saab J 35F Draken interceptors of the Swedish Air Force, and Saab-Scania is offering the RB28 for export under the designation **M-58.** Its Hughes L-24 launcher is adaptable to most types of combat aircraft.

A new long-range air-to-air missile, designated Saab 372, is under development as armament for the Saab JA 37 Viggen. Under study is a surface-launched anti-ship missile with advanced characteristics, to meet future requirements for coastal and naval defence.

SAAB-04E

Saab-Scania is prime contractor for development and production of this modernised version of the 04 anti-shipping missile, on behalf of the Swedish Air Force. Three Saab-04Es constitute the most important of the various alternative weapon loads that can be carried by the Saab AJ 37 Viggen attack aircraft. Release is reported to be possible at aircraft speeds between Mach 0·4 and near-sonic speeds, after which the missile flies a pre-programmed descent to low level for attack.

Details of the development and characteristics of earlier versions of the 04 can be found under the 'Vapenavdelningen' heading in previous editions of *Jane's*. The modernisation undertaken by Saab-Scania on the 04E involves changes to the missile's structure and guidance system, to increase reliability and target hit capability.

Deliveries of the 04E were continuing in 1976. The weapon is operational with certain AJ 37 Viggen squadrons of the Swedish Air Force.

TYPE: Air-to-ship guided weapon.

WINGS: Cantilever mid-wing monoplane, mounted at rear of weapon. Ailerons in trailing-edges. Fixed fins at tips.

Body: Circular-section all-metal structure.
Tail Surfaces: Tail-first design. Cruciform control surfaces on nose. Fixed fins at wingtips.
Power Plant: One IMI (Summerfield Research) solid-propellant rocket motor.
Guidance: Radar homing, with Philips (Sweden) active seeker head.
Control: Autopilot with pneumatically-driven gyros and pneumatic control surface servos.
Warhead: High-explosive with proximity fuse. Weight approx 300 kg (660 lb).
Dimensions:
Length overall 4·47 m (14 ft 8 in)
Body diameter 0·50 m (1 ft 7¾ in)
Wing span 1·97 m (6 ft 5½ in)
Weight:
Launching weight 616 kg (1,358 lb)
Performance:
High subsonic cruising speed

SAAB-05A

The Saab-05A is a manually-guided supersonic air-to-surface tactical missile for use against targets at sea or on land. It can also be used against aerial targets.

Known originally under the project designation Saab 305A, its development began in 1960, when an initial contract was received from what is now the Air Materiel Department of the Defence Materiel Administration of the Armed Forces.

The 05A is in service with the Swedish Air Force as one of the alternative weapons carried by the AJ 37 Viggen. It can, because of its simplicity, be adapted readily for carriage by other types of aircraft.

The airframe of the 05A is made of conventional aircraft materials and consists of a pointed cylindrical body with long-chord cruciform wings and aft-mounted cruciform control surfaces. The VR-35 pre-packaged liquid-propellant smoke-free rocket motor, supplied by Volvo Flygmotor AB, is centrally mounted and is fitted with a tailpipe which passes through the rear of the body.

The armament system is located in the nose and most of the control equipment at the rear of the missile.

After launching, the missile moves automatically into the centre of the pilot's line of sight, and is then guided by command signals from a pilot-operated joystick. These are transmitted over a microwave radio link which is highly resistant to jamming and permits full control at low altitudes over all kinds of terrain. The high precision of the guidance system and high manoeuvrability of the missile make it possible to attack targets which are at considerable offset angles to either side of the aircraft's course.

Guidance signals received by the missile are converted by an autopilot to control surface deflections through the medium of four gas-driven actuators, supplied by a solid-fuel gas generator. The autopilot also takes care of roll stabilisation.

The very effective proximity-fused armament system is of a special design developed by the Swedish Research Institute of National Defence, and is manufactured under subcontract by Forenade Fabriksverken.

Equipment for assembly, testing and servicing forms part of the weapon system, as do the special launching rack on which the missile is carried and the control equipment in the aircraft. The missiles are stored fully assembled in containers, where they can be kept for three years without maintenance or testing.

Training of pilots to utilise the 05A missile involves the use of different kinds of ground-based and airborne simulators, and training missiles.

Dimensions:
Length 3·60 m (11 ft 10 in)
Body diameter 0·30 m (1 ft 0 in)
Wing span 0·80 m (2 ft 8 in)
Weight:
Launching weight approx 305 kg (675 lb)

SAAB-05B

Announced in May 1975, the Saab-05B was intended to complement the 05A missiles in service with AJ 37 Viggen squadrons of the Swedish Air Force. Unlike the manually-guided 05A, it would be guided to the target by its electro-optical target seeker, which the pilot would lock on to the selected target before launch. This would improve the weapon's stand-off capability, by leaving the pilot free to take evasive action after launch.

As can be deduced from the accompanying illustration, the 05B was intended to retain many features of the 05A. Its electro-optical seeker was evolved from the same Saab-Scania research which produced the TVT-300 ship or ground based TV tracking system ordered by a number of armed services.

In the Spring of 1976, the Swedish government decided to order Maverick missiles from the USA in preference to 05Bs; but Saab has continued development as a private venture.

Dimensions:
As for Saab-05A, except:
Length overall 3·64 m (11 ft 11¼ in)
Weight:
Launching weight approx 330 kg (728 lb)

Saab-04E air-to-surface guided missile mounted under the wing of a Saab AJ 37 Viggen

Two Saab-05A missiles on the inboard underwing pylons of a Saab 105

The proposed Saab-05B air-to-surface missile, with TV target seeker

SAAB 372

Saab is developing this new air-to-air missile to equip the JA 37 interceptor version of the Viggen. Few details are yet available, except that it is an infra-red homing weapon and can be carried on any or all of the stores pylons of the JA 37.

Saab claims that the medium-range 372 offers a close-range capability and great tolerance to launch errors and target manoeuvres, enabling it to satisfy both dogfight and intercept missile requirements. A Swedish government decision to undertake full development of the missile was made in July 1975.

DIMENSIONS:

Length overall	2·345 m (7 ft 8¼ in)
Body diameter	0·175 m (6·9 in)
Wing span	0·62 m (2 ft 0·4 in)

WEIGHT:

Launching weight	approx 110 kg (243 lb)

Saab 372 infra-red air-to-air missile on the underwing pylon of a Saab AJ 37 Viggen

SAAB SKA

Announced in the Spring of 1975, the SKA is intended for both surface-to-surface and air-to-surface use, by coastal artillery units, from torpedo boats and flotilla leaders of the Swedish Navy, and from the next generation of attack aircraft operated by the Swedish Air Force. It is described as being of the 'launch and leave' type, with a homing head capable of tracking both sea and land targets.

Saab claims that the SKA would be highly resistant to countermeasures, and would have a long range despite relatively small dimensions and weight. No other details are available for publication.

THE UNITED KINGDOM

BAC

BRITISH AIRCRAFT CORPORATION (GUIDED WEAPONS) LTD

DIRECTORS:

G. R. Jefferson, CBE, BSc, CEng, MIMechE, FRAeS (Chairman and Managing Director)
E. L. Beverley, DFC, CEng, FRAeS (Commercial)
J. Cattanach, BSc, CEng, MIEE (Design)
T. G. Kent, CEng, MIMechE, MRAeS (Deputy Managing Director)
Lt Col H. Lacy, MBE, BSc (London Director)
H. Metcalfe, OBE, BSc, ARCS, CEng, FRAeS (Chief Executive, Stevenage)
R. J. Parkhouse, MSc, BSc, CEng, FIProdE, MRAeS (Guided Weapons Projects Director)
R. J. Raff, FCA, ACMA (Financial)
D. Rowley, MA, CEng, FRAeS (Assistant Managing Director, Electronic and Space Systems)
L. A. Sanson, OBE (Sales and Service)
A. T. Slator, MBE, MA (Vice-Chairman)
S. A. Smith, JP, MA, CEng, MRAeS (Chief Executive, Bristol)
J. McG. Sowerby, OBE, BA, CEng, FIEE, SMIEEE (Divisional Technical Director)

SPECIAL DIRECTORS:

E. M. Dowlen, DLC, CEng, MSc, FRAeS, AFAIAA (Guided Weapon Research Projects)
G. E. King, OBE, BA (Divisional Quality Manager)
J. A. Leitch, MSc, CEng, (Divisional Production Manager)
R. N. Settle, BSc, CEng, MIEE, AInstP (Overseas Projects)
I. D. Woodhead, MBE (Divisional Product Support Manager)

SECRETARY:

M. W. Plimley, LLB

On 1 January 1964 the entire aircraft interests of BAC, together with its Guided Weapons Division, were integrated into a wholly-owned subsidiary company named British Aircraft Corporation Ltd, as described in the Aircraft section. The Guided Weapons Division remains under the management of British Aircraft Corporation (Guided Weapons) Ltd.

In addition to its work on the guided weapons described below, and the Rapier and Seawolf surface-to-air missiles, this Division is playing a leading part in the design and construction of British satellites, and also produces the highly successful Skylark research rocket. Details of these activities can be found in the Spaceflight section.

BAC announced in December 1972 that it had been awarded a Ministry of Defence (Procurement Executive) contract for evaluation and demonstration of an electro-optical seeker, as a possible air-to-surface weapon guidance system. The seeker, manufactured by Rockwell International, is already fitted to 'smart bombs' used by the USAF. Its ability to home automatically on a target, by means of a TV seeker, is considered to be of significance to British forces.

Evaluation of various aspects of TV and laser guidance for air-to-surface weapon systems is continuing, with extensive flight and laboratory trials of different equipments in progress.

BAC (Guided Weapons) Ltd has a total of 9,000 employees at its Stevenage and Filton works.

BAC SWINGFIRE AND HAWKSWING

Swingfire is a wire-guided anti-tank weapon system, designed to defeat the heaviest battle tanks at ranges from 140 to 4,000 m (460-13,125 ft). The first formal reference to the weapon was made by the UK Minister of Defence on 10 August 1962, when he announced that it was being developed for the British Army, with British Aircraft Corporation as the prime contractor, under a Ministry of Defence contract.

Swingfire was first installed in the FV432 armoured personnel carrier, which went into service as the FV438 in 1969. Missiles are launched from a bin at a QE of 35°, and are controlled by an operator who views the engagement with a periscopic sight from within the vehicle. Alternatively, the operator can take up a position with a portable sight anywhere within a 100 m (328 ft) radius circle from the vehicle. The field of fire is a sector of 4,000 m (13,125 ft) radius, which subtends 90° at the launching site without the need to traverse the launcher. The elevation coverage is ±20° with respect to the vehicle datum in the direct fire mode, when the operator is in the vehicle, and +20°/−10° relative to the horizontal in the separated mode. Horizontal separation distances up to 100 m (328 ft) and vertical separations up to 23 m (75 ft) are within the system's capability.

The missile, which is stored in and launched from an hermetically-sealed disposable container, carries a hollow-charge warhead designed to defeat the heaviest battle tanks at angles of attack up to 70°. This warhead and its safety and arming mechanisms are mounted in front of the solid-propellant motor. The rear body houses the autopilot and the wire spool, and supports four stabilising fins which unfold and lock into position after launch. Lateral control is achieved by thrust vector control (jetavator) under autopilot command. By this means, and with a 6g launch acceleration, the missile is swung automatically on to the gathering trajectory until it enters the field of view of the operator's sight, hence the name 'Swingfire'.

Commands for the gathering trajectory are generated by the programme unit, which is pre-set with the parameters of the engagement's geometry (i.e. azimuth and elevation angles, location of operator and local crest clearance angle). This entirely automatic gathering ends with a steady-state signal, which guides the missile in the direction of the target. The operator's signals are added to the steady-state signal, to bring the missile accurately on to the line of sight and keep it there until impact. This operator task is greatly helped by use of the velocity control system. Joystick movements demand change of heading under autopilot control (i.e., under steady-state conditions the missile will fly in the straight line which corresponds to the position of the joystick plus the steady-state signal). Two of the many advantages of this system are the ease with which it can be installed in vehicles, because there is no need for a traversing turret, and the tactical freedom to fire from behind solid cover.

The system described is fitted currently to the FV438 APC, in which two missiles are ready to fire and twelve are stowed and ready for internal reloading; in the FV712 Ferret Mk 5 scout car, which carries four missiles ready to fire and two stowed; and in the Alvis Striker CVRT (combat vehicle reconnaissance tracked) which carries five missiles ready to fire and five missiles internally stowed.

Swingfire anti-tank missile leaving its launch container on a helicopter during air-to-surface trials

Striker is in production for the British and Belgian armies.

A palletised application known as **Infantry Swingfire** has also been developed, and can be fitted to vehicles as small as the ¾-ton Land-Rover. Missiles may be launched from this equipment when it is either on the vehicle or on the ground.

The main development programme has been completed also for an airborne anti-tank version, known as **Hawkswing**, which is claimed to offer greater lethality, range and tactical flexibility than any other contemporary system of its kind, as well as considerable development potential. It has a range of 4,000 m (13,125 ft), with a large angle-off and crest clearance capability, and the missile can be launched and controlled to the target while the helicopter is undergoing all normal manoeuvres and evasive flying. The hollow-charge warhead will defeat all known combinations of armour and will destroy the heaviest battle tanks.

The Hawkswing system requires only limited maintenance and includes an airborne trainer. The missile is a round of ammunition, highly reliable and easily transportable. It has been designed as an easily-removable weapon system to fit the Lynx helicopter, which can carry six missiles, and change of role can be effected quickly.

By the beginning of 1975 the trials development programme had been completed, including highly successful demonstrations. The system can be fitted easily to most other helicopters currently in service or projected.

DIMENSIONS:

Length overall	1·17 m (3 ft 8 in)
Max body diameter	0·17 m (6·7 in)
Wing span	0·40 m (15·73 in)

BAC SEA SKUA

This lightweight all-weather sea-skimming anti-ship missile is designed for operation from the Royal Navy's Lynx helicopter, to meet the threat from missile-firing fast patrol craft and to provide a good stand-off capability for the helicopter, with consequent protection from counter-attack. The complete helicopter-borne weapon system includes surveillance and tracking radar, fire control equipment, launching arrangements and the semi-active homing missile. Main contractors are BAC for the missile and fire control, Ferranti for the Seaspray radar, and Westland for the Lynx helicopter.

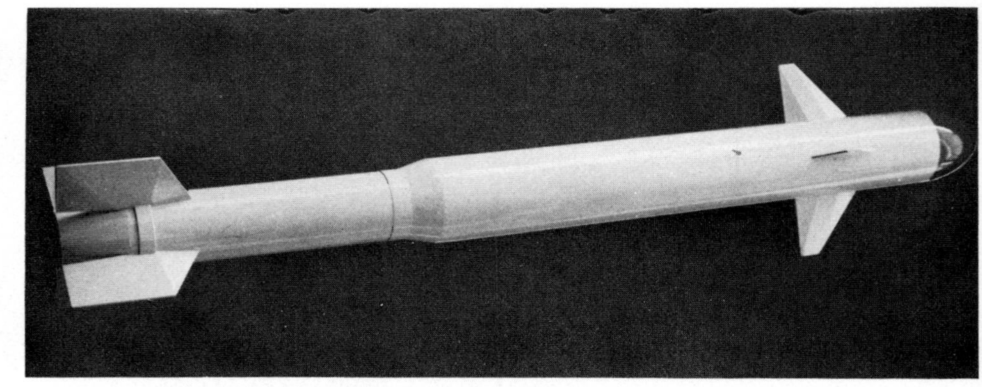

Model of the Sea Skua lightweight air-launched anti-surface-vessel missile

In addition to its ability to engage fast patrol boat targets, Sea Skua has a significant effectiveness against larger warships. It offers high lethality, with the capability of firing salvoes. The radar provides continuous surveillance, and includes facilities for weapon control and target identification.

Sea Skua is a quick-reaction system, at immediate readiness throughout the helicopter sortie, with a high degree of system reliability. Wherever possible, it embodies existing, proven techniques and components. The missile has four fixed stabilising tail-fins, indexed at 45° to the cruciform movable wing control surfaces. The solid-propellant boost and sustainer motor provides a high subsonic speed. Sea-skimming capability is conferred by a radio altimeter, with semi-active radar for terminal homing and a high-efficiency warhead actuated by a direct-action fuse. Simple go/no-go testing is all that is required on board ship before the missile is mounted on the helicopter.

The weapon system will enter service with the Royal Navy in the late 1970s and will be deployed widely on Lynx helicopters. It can be fitted to other medium-size helicopters and long-range maritime patrol aircraft.

HSD

HAWKER SIDDELEY DYNAMICS LTD

HEADQUARTERS:
Manor Road, Hatfield, Hertfordshire AL10 9LL
Telephone: Hatfield (30) 62300
Telex: 22324

DIRECTORS:
Sir John Lidbury, FRAeS (Chairman)
Capt E. D. G. Lewin, CB, CBE, DSO, DSC, RN(ret'd) (Managing Director)
Dr G. H. Hough, PhD, CBE, BSc, FRAeS, FIEE (Deputy Managing Director)
J. A. Airey, BSc, ARCS, CEng, FRAeS (Director in Charge, Stevenage)
J. D. Crane, FCA (Commercial)
R. G. Dancey, BA, TD (Administration)
L. G. Evans, MA, CEng, FRAeS
P. R. Franks, BSc, MSc, CEng, FRAeS, MIEE (Technical & Quality Assurance)
J. McGregor Smith, ACMA (Financial)
W. T. Neill, MBE, CEng, FIPE, FRAeS, FBIM, MTI (Director in Charge, Lostock)

SECRETARY:
J. Creed, FCA

EXECUTIVE DIRECTORS:
R. Birch, FCA (Finance)
P. L. V. Hickman (Space)
H. R. Leather, CEng, FRAeS (Mechanical Equipment & Systems)
T. H. Lettice (Contracts)
P. C. F. Morgan, FInstPS, CRAeS (Purchasing)

PUBLIC RELATIONS MANAGER:
Michael K. Hird

This subsidiary of the Hawker Siddeley Group came into being as a consequence of the reorganisation of the Group on 1 July 1963. It is responsible for the design, development and production of all Hawker Siddeley guided weapons, space-launch vehicles, satellites, propellers and air-conditioning systems.

The company is prime contractor for the Sea Dart ship-to-air/ship-to-surface missile, and has developed jointly with Engins Matra of France the Martel long-range air-to-surface weapon (see International section).

Development of the company's own SRAAM highly-manoeuvrable close-range air-to-air weapon is continuing at technology demonstration level, which includes a number of development firings. A new medium-range all-weather air-to-air missile known as Sky Flash, based on the American Sparrow, is in full production for the Royal Air Force. Hawker Siddeley Dynamics is also responsible for development and manufacture of an infra-red linescan system which equips the CL-89 surveillance drone, produced under a joint British, Canadian and West German programme.

Hawker Siddeley Dynamics is the British 'daughter firm' for the Australian-developed Ikara long-range anti-submarine weapon for the Royal Navy.

HSD SRAAM

Combat experience in Vietnam and other war theatres has emphasised the need for an air-to-air missile able to cope with the high accelerations encountered in dogfight situations. The design of SRAAM (Short Range Air-to-Air Missile) is such that it can operate effectively down to a very short minimum range, even with the target manoeuvring to its design limits. At the same time, its maximum range is comparable with that of existing air-to-air weapons.

The simplicity of the fire control system would enable SRAAM to be carried by almost any type of interceptor or air superiority fighter, strike or reconnaissance aircraft, without requiring aircraft modification. Missiles have folding fins to allow them to be carried in launch tubes, attached to a launch beam which contains all the fire control system.

SRAAM is visually aimed, with a wide aiming tolerance, and is guided by a passive infra-red homing system. Steering is by thrust-vector control, by means of semaphores which project into the motor efflux, and would enable SRAAM to outmanoeuvre any aircraft, including crossing targets at very short range. The high-explosive warhead is designed to be detonated by proximity or contact fuse.

A technological development programme, which includes a limited series of air firings from a Hunter aircraft, is in progress to prove the SRAAM concept.

DIMENSIONS:

Length overall	2·73 m (8 ft 11¼ in)
Body diameter	0·168 m (6½ in)

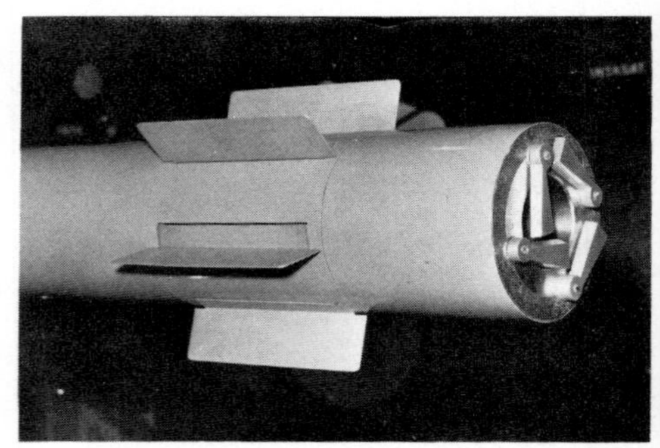

Tail of SRAAM air-to-air missile, showing steering semaphores and folding fins *(Flight International)*

Hawker Siddeley Dynamics SRAAM air-to-air missile

SRAAM launchers under wings of a two-seat Hawker Siddeley Harrier V/STOL aircraft

SKY FLASH (XJ521)

Swedish Air Force designation: RB71

On 17 April 1973 Hawker Siddeley Dynamics announced that it had received a prime contract from the Ministry of Defence for project definition and pre-development studies of a new medium-range all-weather air-to-air missile based on the American AIM-7E Sparrow, which is in service with both the RAF and Royal Navy. Raytheon, the US Sparrow prime contractor, is participating in the programme through cross-licensing agreements with Hawker Siddeley Dynamics and Marconi Space and Defence Systems Ltd.

Project definition was completed in 1973, and it was announced on 26 February 1974 that Hawker Siddeley Dynamics had received instructions from the Ministry of Defence to proceed with full development and initial production. A contract for full production was awarded in 1975. In October 1976 it was announced that Sky Flash was to be evaluated as armament for the Swedish Air Force's Saab JA37 Viggen all-weather fighters.

The 'boost and coast' missile, known as Sky Flash (XJ521), has the same general configuration and dimensions as the Raytheon AIM-7E, but is fitted with a semi-active radar homing head developed by Marconi under a Ministry of Defence contract. This has been designed as a completely self-contained unit with modular construction, which provides considerable flexibility in the overall design to cater for a wide range of different applications. Microstrip circuitry and the latest types of solid-state technology and micro-circuits have offered reliability, reduced size and high performance. The complete homing head is only 283 mm (11 in) long and can be fitted in a missile body of 180 mm (7 in) internal diameter.

The new, advanced, fuse system is designed by EMI Electronics Ltd, and is claimed to offer a high single-shot kill capability against targets at high, medium and low altitudes. Hawker Siddeley Dynamics has updated the autopilot and power systems, and is responsible for building the missile structure, as well as for assembly and test.

Sky Flash medium-range all-weather air-to-air missiles

The first five firing trials of Sky Flash missiles, from an F-4J Phantom of the US Navy at Point Mugu, California, were all successful. The first and second achieved a grazing pass and a direct hit respectively against 'glinting' targets. The third scored a direct hit on a manoeuvring target. The fourth produced a lethal close pass in a 'tail chase' situation. The fifth obtained a hit when Sky Flash was launched from medium altitude against a near sea level target.

HSD/MATRA AJ.168/AS.37 MARTEL

See International section.

THE UNITED STATES OF AMERICA

BOEING

BOEING AEROSPACE COMPANY

Head Office and Works:
Seattle, Washington 98124

Current responsibilities of the Boeing Aerospace Company include development of the E-3A, E-4 and YC-14 aircraft, and YQM-94A Compass Cope RPV for the USAF; B-1 bomber electronics integration; operational support for the AGM-69A short-range attack missile (SRAM); modernisation of the Minuteman ICBM force; military and commercial hydrofoil programmes; and most of The Boeing Company's diversification programmes, including personal rapid transit systems, agri-business development, water resources projects and electronics programmes.

SRAM

US military designation: AGM-69A (WS-140A)

This supersonic air-to-surface defence suppression and primary attack missile is carried by the General Dynamics FB-111 and the B-52G and H versions of the Boeing Stratofortress, and has been designated as primary armament for the Rockwell International B-1 supersonic strategic bomber, now under development. SRAM is intended to be capable of penetrating advanced enemy defence systems and has nuclear capability.

Details of the early history of SRAM development can be found in the 1975-76 *Jane's*. Production was authorised by the USAF on 12 January 1971, and deployment by Strategic Air Command began on 4 August 1972, when the B-52Gs of the 42nd Heavy Bombardment Wing became operational with SRAM at Loring AFB, Maine. USAF contracts covered production of 1,500 missiles, to equip 17 B-52 wings and two FB-111 wings, at 18 SAC bases.

The 1,000th SRAM was delivered on 14 August 1974, and deliveries of the authorised total of 1,500 missiles were completed in July 1975.

As system integration contractor, Boeing was responsible for overall SRAM system performance. The task of marrying the SRAM system to the carrier aircraft was its responsibility, but it was assisted in the flight test programme by General Dynamics/Fort Worth and Boeing-Wichita, in the respective roles of associate and subcontractor. The programme was managed by the Aeronautical Systems Division of the Air Force Systems Command.

Subcontractors in the SRAM programme were Lockheed Propulsion Company for the LPC-415 restartable solid-propellant two-pulse rocket motor; Kearfott Division of The Singer Company for the guidance subsystem (inertial with terrain avoidance capability), Guidance and Controls Division of Litton Industries for the B-52 inertial measurement unit; Autonetics Division of Rockwell International for the FB-111/B-52 aeroplane computer; Delco Electronics for the missile computer; Stewart-Warner Electronics Division for the radar receiver/transmitter; and Unidynamics Phoenix Division of Universal Match Corporation for the missile safe-arm-fuse subsystem. In addition, International Business Machines received a subcontract for modifying the bomb/navigation system of the B-52.

B-52 bomber, with four three-round SRAM clusters visible under its wings

Each B-52G and H can carry 20 SRAMs, twelve in three-round underwing clusters and eight on a rotary launcher in the aft bomb bay, together with up to four Mk 28 thermonuclear weapons. Alternatively, the rotary launcher can be carried simultaneously with two underwing AGM-28B Hound Dogs and decoy missiles. An FB-111 can carry six SRAMS, four on swivelling underwing pylons and two internally.

Range of each missile is reported to be slightly more than 90 nm (160 km; 100 miles) in the 'high mode', 30 nm (55 km; 35 miles) in the 'low mode'. It is able to fly 'dog-leg' courses and its radar signature is said to be no larger than that of a machine-gun bullet.

When SRAM is carried externally, a tailcone, 0·56 m (22·2 in) long, is added for aerodynamic reasons.

The Department of Defense's FY 1977 budget requests include $16 million for continued development and testing of a new SRAM motor, to replace the current LPC-415 which was designed for a five-year service life, and $21 million to procure new SRAMs for the B-1. Use of the latter sum is contingent upon a B-1 production decision.

DIMENSIONS:

Length overall	4·27 m (14 ft 0 in)
Body diameter	44·5 cm (1 ft 5½ in)

WEIGHT (approx):

Launching weight	1,010 kg (2,230 lb)

ALCM

USAF designation: AGM-86A

The Air-Launched Cruise Missile (ALCM) is a small unmanned winged air vehicle capable of sustained subsonic flight following launch from an airborne carrier aircraft. It is propelled by a turbofan engine, incorporates a nuclear warhead, and is programmed to strike a predetermined surface target. Guidance is by a combination of inertial and terrain comparison techniques. In the terrain comparison mode the missile is kept on course by a computer which compares pre-programmed geographical features on its flight plan with the geography seen by its sensors during actual flight.

The ALCM design makes maximum use of the air vehicle and small turbofan engine developed under the former AGM-86A Subsonic Cruise Armed Decoy (SCAD) programme, for which Boeing was prime contractor.

When launched in large numbers, each of the missiles would have to be countered, making defence against them both difficult and costly. The missiles are intended to dilute defences and improve the ability of manned aircraft to penetrate to major targets. Small radar signature and low-level flight capability will enhance their effectiveness.

The ALCM is intended to have a 'hard target' kill capability, and is being considered also for decoy, reconnaissance and non-nuclear applications in Europe. It is similar in overall dimensions to SRAM, and is suitable for carriage on the rotary launcher developed for this latter weapon, with wings and tail folded and engine air intake retracted. A B-52 will be able to carry 12 ALCMs externally and 8 internally. A B-1 will be able to carry 24, all internally. When carried externally, ALCM will be able to have an underbelly auxiliary fuel tank fitted to increase its range.

The ALCM programme is directed by Air Force Systems Command's Aeronautical Systems Division at Wright-Patterson AFB, Ohio. The missile, designated AGM-86A, is continuing in advanced development with a funding request of $79 million for fiscal year 1977 before Congress. In accordance with a directive from the US Secretary of Defense, the ALCM and the US Navy's Sea-Launched Cruise Missile (SLCM) are to be continued in advanced development until concept demonstration has been completed. Meanwhile, the two programmes are being restructured to ensure maximum commonality.

The Boeing Company is responsible for the ALCM airframe, which differs in configuration from that of the SLCM. Williams Research Corporation is supplying F107-WR-100 turbofan engines for the advanced development phase ALCMs, with guidance by McDonnell Douglas; but any future production ALCMs and SLCMs are intended to utilise a common engine and common guidance system, and the contracts for these will be subject to industry competition.

Cost of the ALCM advanced development programme is estimated at $125 million, including two jettison tests which were made successfully from a B-52 in June and July 1975 and are being followed by seven powered flights of prototype ALCMs. The first of these took place successfully on 5 March 1976, the missile impacting 70 nm (130 km; 80 miles) downrange at White Sands, New Mexico, after a 10 min 40 sec flight at Mach 0·65. The second and third prototypes, with inertial guidance platform installed, were to explore the complete flight envelope, in 30 min flights; the last four were to have complete guidance packages.

Simultaneously, captive ALCMs are being 'flown' electronically while mounted on the B-52. Full-scale development could begin in early 1977, followed by a full production decision in 1979, after a number of flights. Total programme cost should then be less than $1,500 million.

DIMENSIONS:

Length overall	4·27 m (14 ft 0 in)
Wing span, deployed	2·90 m (9 ft 6 in)
Diameter	0·64 m (2 ft 1 in)

WEIGHT:

Launching weight:	
with belly tank	1,088 kg (2,400 lb)
without belly tank	860 kg (1,900 lb)

AIR-LAUNCHED MINUTEMAN/MX

To counter the possibility of future vulnerability of America's force of 1,000 silo-based Minuteman ICBMs, the Department of Defense has undertaken an Advanced ICBM Technology Programme. One mode of deployment proposed for the new missile, which is known by the designation MX, is as an air-mobile system on board modified wide-bodied jet transports or new low-cost aircraft that might maintain an airborne alert. Weight of the MX missile, which would be cold-launched from a canister, is estimated at about 68,000 kg (150,000 lb), double the weight of a Minuteman I. To ensure navigation accuracy

Test models of the SRAM air-to-surface short-range attack missile

SRAMs on rotary launcher being prepared for uploading into B-52

Prototype AGM-86A Air Launched Cruise Missiles (ALCMs) undergoing final inspection

Minuteman I ICBM leaving hold of C-5A during air-launch trial on 24 October 1974

for the launch aircraft, its crew might utilise the Navstar satellite global positioning system now in an early development stage.

As an initial evaluation of the air-mobile concept, the USAF has conducted a series of test drops from a C-5A Galaxy transport aircraft. These comprised three 'bath-tub' drops of concrete slabs of increasing size and weight; three 'mass simulation' drops to investigate missile shape stability; and one drop each of an inert but instrumented Minuteman I, a fuelled but unfired Minuteman I, and finally, on 24 October 1974, a 'short-burn' Minuteman I.

In this last test, the missile was pulled by drogue parachutes out of the rear door of the C-5A. Held upright by parachutes, it fell to a height of about 2,500 m (8,000 ft) before its motor ignited. It then climbed to more than 6,100 m (20,000 ft) during 10 seconds of full thrust, before falling into the sea at the Western Test Range, Vandenberg AFB, California.

GENERAL DYNAMICS

GENERAL DYNAMICS CORPORATION

HEAD OFFICE:
Pierre Laclede Center, St Louis, Missouri 63105

Convair Division

HEADQUARTERS:
5001 Kearny Villa Road, San Diego, California 92123

VICE PRESIDENT AND GENERAL MANAGER:
Grant L. Hansen

Pomona Division

HEADQUARTERS:
PO Box 2507, Pomona, California 91766

WORKS:
Pomona and Window Rock, Arizona

VICE PRESIDENT AND GENERAL MANAGER:
Dr Leonard F. Buchanan

The Pomona Division of General Dynamics is responsile for development and production of tactical weapon systems for use by the US Navy, US Army, US Air Force and friendly foreign powers. In current production are the Standard Missile-1, more than 7,000 of which have been produced since 1967 to replace the Terrier and Tartar naval ship-launched weapons, primarily for air defence but also for use against selected surface targets; and the the Standard ARM anti-radiation missile. The US Naval Air Systems Command has awarded a contract to Pomona for second-source production of the AIM-7F Sparrow air-to-air missile (described under Raytheon entry). Under development and nearing the production phase in early 1976 were the Stinger advanced man-portable air defence weapon and Phalanx, the Navy's all-weather automatic gun fire-control system. Also under development is the longer-range Standard Missile-2.

Convair Division was selected in 1976 to develop the US Navy's Sea Launched Cruise Missile, under the designation BGM-109 Tomahawk. An air-launched version is also under test. Convair's activities in the aircraft and spaceflight fields are described in the relevant sections of this edition.

AGM-78D version of Standard ARM under the wing of an F-105 Thunderchief

STANDARD ARM

US Navy designations: AGM-78 and RGM-66D

It was announced in September 1966 that the US Naval Air Systems Command had awarded a $7·5 million letter contract to Pomona Division for initial development of an air-launched weapon that would home on radiation emitted by a ground radar set and destroy the installation with its explosive warhead. Known as Standard ARM (anti-radiation missile), this weapon utilises an adaptation of the Navy's RIM-66A medium-range ship-to-air Standard Missile-1 propulsion system. It was intended to provide a significant increase in capability over earlier weapons in countering the threat of enemy radar-controlled anti-aircraft guided missiles and guns. 'Launch and leave' omni-directional attack from outside the lethal radius of enemy surface-to-air missiles is an inherent feature of the system.

Successful operation of the weapon system has led to development of several advanced versions, some of them highly classified. The initial version used the passive-homing target-seeking head of a Shrike missile. Current models have improved seeker heads and electronics for better target selection, more effective operation against target countermeasures and still greater attack range. A capability to counter hostile search and ground control intercept radars is inherent in these missiles.

Associated with Pomona Division in the Standard ARM development programme have been Texas Instruments for the Shrike seeker of the original AGM-78A version, Maxson Electronics for the initial improved seeker head of the AGM-78B version (a subsequent, further improved seeker is being developed by GD/Pomona), IBM and Bendix for improved electronics, Aerojet-General for the Mk 27 Mod 4 dual-thrust solid-propellant rocket motor, Grumman (A-6 launch aircraft), Fairchild Republic Division of Fairchild Industries (F-105 launch aircraft), and Convair Aerospace Division of General Dynamics for aircraft modification.

The general appearance of Standard ARM is shown in an accompanying illustration. The basic airframe consists of a cylindrical body with pointed ogival nose, and cruciform long-chord narrow-span wings, with forward portions of much reduced span, indexed in line with cruciform tail control surfaces.

Equipment carried by the launch aircraft includes a Target Identification and Acquisition System (TIAS), which is able to determine and pass to the missile specific target parameters.

Production continues of a surface-to-surface version of Standard ARM, designated RGM-66D. Its in-service air-launched counterpart is the AGM-78D.

DIMENSIONS:

Length overall	4·57 m (15 ft 0 in)
Body diameter	0·343 m (1 ft 1½ in)

WEIGHT:

Launching weight, basic version	615 kg (1,356 lb)

PERFORMANCE:

Max speed	supersonic

TOMAHAWK

US Navy designation: BGM-109

In December 1972 the US Navy awarded study contracts for a Sea Launched Cruise Missile (SLCM) to General Dynamics, Vought, Lockheed, Boeing and McDonnell Douglas. The first two of these companies were selected by Naval Air Systems Command as competitive cruise missile airframe contractors in January 1974.

As early as November 1972, during preliminary studies, GD had selected a unique capsule/missile design concept for its version of the SLCM when deployed in its primary role, for underwater launching through submarine torpedo tubes. The missile, designated YBGM-109 in its development form, was delivered and stored inside a protective stainless steel capsule, with wings and tail fins folded and engine air intake retracted into the body. By jettisoning this capsule at launch, GD was able to concentrate all protective features into the disposable capsule instead of penalising the flight performance of the missile with the added weight.

Underwater launches of inert test vehicles began at the Naval Undersea Center, San Clemente Island, California, in July 1974. Tests from nuclear submarines followed, and by 13 February 1976 the GD missile had become the first US SLCM to complete successfully wind tunnel, underwater launch and boost-to-glide tests. On 17 March, the Secretary of the Navy announced that Convair Division of

General Dynamics had been selected as prime contractor for the SLCM, which had been named Tomahawk. The contract, funded at $34·8 million, called for a flight of the first prototype missile in April 1976. The guidance systems, under development by McDonnell Douglas Astronautics Company—East, were next to be integrated into the prototype, and flown as a complete weapon system in the early Autumn of 1976. A decision on whether or not to proceed to full-scale development is scheduled for January 1977.

There are two basic versions of Tomahawk, as follows:

Strategic version: Designed for use over ranges of more than 1,500 nm, this version has a McDonnell Douglas inertial navigation system with Terrain Contour Matching (Tercom). For launch from submarines, surface ships and land platforms it is intended to be powered by an Atlantic Research Corporation solid-propellant booster motor, in the 31·1 kN (7,000 lb) thrust class, and a Williams Research Corporation F107-WR-100 cruise turbofan rated at more than 1·33 kN (300 lb st). The boost motor is not fitted for air-to-surface launch from strategic aircraft. This version has a nuclear warhead, and is intended to fly at low altitudes in all weathers.

Tactical version. Intended for use over ranges of more than 300 nm, this version differs from the strategic version basically in having a non-nuclear warhead, reduced fuel, and modified forms of the 1·8 kN (400 lb st) Teledyne turbojet engine and McDonnell Douglas guidance system fitted to the Harpoon missile.

The airframe is basically common to both versions, with a cylindrical torpedo-shape body made of aluminium, tapering at the tail. Two narrow-chord wings, positioned approximately midway between the nose and tail, fold inside the body when the missile is in its launch capsule and extend forward for cruising flight in the atmosphere. Cruciform folding tail surfaces are indexed in line with the wings. The ventral air intake for the cruise engine is retractable. Design emphasis is on low radar cross-section and minimum visual and infra-red signatures.

First air launch of a tactical Tomahawk was made from beneath the wing of a US Navy A-6 Intruder, at the Pacific Missile Test Center, on 28 March 1976. After being launched at a height of 9,600 m (31,500 ft), the missile made turns to port and starboard and changed altitude, as programmed. The turbojet shut down prematurely, but the flight was terminated successfully with wing retraction and deployment of parachutes and flotation bags to permit recovery and re-use of the test round.

Tactical Tomahawk under the wing of an A-6 Intruder test aircraft

The first fully guided Tomahawk flight after launch from an A-6 aircraft, 5 June 1976

DIMENSIONS:
- Length overall:
  - with boost motor: 6·40 m (21 ft 0 in)
  - without boost motor: 5·49 m (18 ft 0 in)
- Body diameter: 0·53 m (1 ft 9 in)

WEIGHT (approx):
- Handling weight, strategic version 1,814 kg (4,000 lb)

PERFORMANCE (estimated):
- Cruising speed, strategic version: 478 knots (885 km/h; 550 mph)
- Range:
  - strategic: over 1,500 nm (2,779 km; 1,727 miles)
  - tactical: over 300 nm (555 km; 345 miles)

HUGHES
HUGHES AIRCRAFT COMPANY

HEAD OFFICE:
Culver City, California 90230

Missile Systems Group

VICE-PRESIDENT AND GROUP EXECUTIVE:
William F. Eicher

Hughes Aircraft began developing an air-to-air missile for the USAF in 1950, and this weapon has been in squadron use for many years as the Falcon. Production in the USA ended after a total of some 45,000 missiles had been delivered.

Currently, Hughes is producing an air-to-air missile named Phoenix for the US Navy, an air-to-surface missile named Maverick for the USAF, and a wire-guided anti-tank missile named TOW for the US Army and other customers. It is also assisting the US Navy with development of an air-to-air anti-radiation missile known as Brazo.

Hughes manufactures fire-control systems for the majority of USAF interceptors, and is engaged extensively in space research and communications satellite programmes (see Spaceflight section).

Assembling the long-chord wings of an AIM-54A Phoenix air-to-air missile

PHOENIX

US Navy designation: AIM-54A (formerly AAM-N-11)

Intended originally to form the primary armament of the abandoned F-111B, Phoenix arms the Grumman F-14A Tomcat two-seat carrier-based fighter and is claimed to have capabilities exceeding those of any other air-to-air system yet operational.

Configuration is similar to that of earlier Hughes air-to-air missiles, with a cylindrical body and long-chord cruciform wings; but cruciform tail control surfaces replace the wing trailing-edge surfaces of the earlier weapons.

Phoenix has a Mk 47 Mod 0 solid-propellant rocket motor produced by Rocketdyne, and is fitted with a large proximity-fused high-explosive warhead. The weapon system consists of the missile itself, AWG-9 airborne weapon control system, and LAU-93A missile launcher. Hughes is prime contractor for the entire system, including support equipment. Control Data Corporation supplies the central processing portion of the missile control system computer.

The launch aircraft's AWG-9 weapon control system is able to lock on to an enemy aircraft, surface-launched cruise missile or air-to-surface missile, at high or low altitude, in any kind of weather, and launch the Phoenix missile. The missile then takes over to intercept the target.

The data from the radar is processed by a solid-state, high-speed general-purpose digital computer, the output of which is displayed to the missile control officer in the launch aircraft on two displays: a 254 mm (10 in) cathode-ray tube and a 127 mm (5 in) multi-mode storage tube.

The long-range high-power pulse Doppler radar has a planar array antenna, providing a 'look-down' capability that enables it to pick out moving targets from the ground clutter that normally obscures them in a conventional radar. The AWG-9 also incorporates an infra-red subsystem that provides an independent search and track sensor, and has a track-while-scan radar mode that makes it possible to launch up to six missiles and keep them on course while searching for other possible targets, all in the presence of sophisticated enemy countermeasures.

The first AWG-9 reconfigured for the F-14A fighter was delivered to the US Navy in February 1970, just one year after design began. Extensive use of hybrid circuits and new packaging techniques had enabled its weight to be reduced to less than 635 kg (1,400 lb), compared with 907 kg (2,000 lb) for earlier, less efficient versions. The subsequent addition of air combat manoeuvre modes provides improved 'dogfight' capability, and the AWG-9 is able to launch other naval air-to-air weapons, including Sparrow and Sidewinder missiles, as well as Phoenix. It can also direct the firing of the M61 Vulcan 20 mm cannon.

Contractor testing was concluded on 13 June 1973 with a launch from an F-14A against a BQM-34E Firebee target drone over a record range of 110 nm (204 km; 126 miles). The drone, augmented by radar signal to look as large as a bomber and equipped with an on/off blinking noise jammer, approached at an altitude of 15,850 m (52,000 ft) and speed of Mach 1·55. The F-14A, flying at 13,700 m (45,000 ft) and Mach 1·45, began tracking the Firebee at very long range, locked on and launched a single Phoenix at 110 nm. During flight, the Phoenix reached a high point in its trajectory of more than 30,500 m (100,000 ft) and then passed the drone within the lethal distance of its warhead.

Of the 56 Phoenix missiles launched from various aircraft during contractor tests, 43 were scored as hits. The success rate for the 17 missiles launched from the F-14A was 88%. Only a few additional tests remained to be completed by USN crews prior to fleet introduction of Phoenix aboard F-14s at NAS Miramar, California, in the Autumn of 1973.

In one of these USN tests, on 21 November 1973, six Phoenix missiles were launched from an F-14, from Point Mugu, and were guided simultaneously against six target drones simulating an attack force of enemy aircraft. All six missiles were launched within 37 seconds, two of them in a 3½-second period, at a distance of 43 nm (80 km; 50 miles). Two Lockheed QT-33 drones were destroyed by the unarmed missiles. Two others, a subsonic Teledyne Ryan BQM-34A and a supersonic BQM-34E, were damaged by direct hits. One missile was unable to intercept its target because the BQM-34 veered off course, leaving a weakened radar signature that could not be tracked at such a long range. The sixth missile apparently suffered a hardware failure about one-third of the way through its flight.

Effectiveness of Phoenix against low-flying anti-ship cruise missiles was demonstrated in August 1974, when the crew of a US Navy F-14 launched a Phoenix from a range of 19 nm (35 km; 22 miles) against a BQM-34 drone flying only 15 m (50 ft) above the sea to simulate such a missile. The unarmed Phoenix homed on the target and scored a hit by passing within the lethal distance of the warhead that is fitted to operational rounds.

All F-14s built for the US Navy and Imperial Iranian Air Force are equipped to carry Phoenix. Production was initiated in October 1970, when the US Navy awarded a $145 million contract to Hughes for production of AWG-9 systems at the company's El Segundo, California, electronics manufacturing facility. In December 1970, the Navy announced the award of a letter contract of $40 million for production of Phoenix missiles at Tucson, Arizona. The production contracts for FY 1973 totalled $120 million for AWG-9 systems and $62·3 million for 180 Phoenix missiles, plus spares, support and test equipment, field service and technical manuals. FY 1974 funding provided $130 million for 54 AWG-9 systems, and $100 million for Phoenix missiles. Budget requests for FY 1975 and 1976 each included the procurement of a further 340 missiles, with 240 requested in FY 1977 and additional purchases envisaged in FY 1978 and 1979.

Phoenix entered first-line service in September 1974, when two squadrons of US Navy F-14s were deployed on board the nuclear-powered USS *Enterprise* for a six-month tour of duty in the Western Pacific. It went into service on the USS *John F. Kennedy* in the Atlantic in mid-1975.

DIMENSIONS (approx):

Length overall	3·95 m (13 ft 0 in)
Body diameter	0·38 m (1 ft 3 in)
Span of wings	0·91 m (3 ft 0 in)

WEIGHT:

Launching weight	447 kg (985 lb)

BRAZO/PAVE ARM

Under a joint project, known to the US Navy as Brazo (Spanish for 'arm') and to the USAF as Pave Arm, Hughes Aircraft is assisting in feasibility demonstrations of an air-to-air anti-radiation missile (ARM). Its main task has been to design and manufacture the guidance subsystems, based on a broad band receiver designed by the Naval Electronics Laboratory Center, and integrate these subsystems into each of the eight modified Raytheon Sparrow missiles that were allocated to initial flight tests, under an 18-month demonstration programme. Hughes also modified the LAU-17/A launcher and fabricated the cockpit control panel for the F-4D aircraft provided for Brazo flight trials.

The Navy is responsible for design and development of the missile, with the USAF responsible for testing and evaluation at the Air Force Special Weapons Center, Holloman AFB, New Mexico. It was expected that Varian Associates would receive a contract for missile electronics, with an antenna from AIL Division of Cutler-Hammer. The missile radome was to be chosen from competitive tenders by Avco, Brunswick and Texas Instruments.

The aim is to evolve a weapon for use against advanced all-weather fighters in the class of the Soviet MiG-25. Brazo is intended to home on fire control radar emissions from enemy aircraft, and the development programme has the appropriate name of ERASE (Electromagnetic Radiation Source Elimination).

First development firing took place on 16 April 1974, when a Brazo fired from an F-4D intercepted a BQM-34 Firebee drone in a look-down tail attack at Holloman AFB. Two further hits were scored in the next two launches, which involved long-range look-down nose intercepts. This completed the initial demonstration phase of the programme.

In early 1976, design changes were being made to improve the missile's low-altitude capability. A contract for 12 more missiles for continued flight trials was expected to be awarded in the Autumn of 1976.

TOW

US Army designation: BGM-71A

Hughes Aircraft was prime contractor for development of this high-performance surface-to-surface and air-to-surface anti-tank missile, the basic characteristics of which are indicated by its name, as TOW is an acronym for Tube-launched, Optically-tracked, Wire-guided.

The basic ground-fired TOW system consists of a glassfibre launch tube, a tripod, a traversing and sighting unit, an electronic package, and missiles encased in shipping containers. Total weight of the entire weapon system, including missile, is approximately 91 kg (200 lb), but the launcher and electronics can be broken down into four units for carrying by infantry.

The missile has low aspect ratio wings and tail control surfaces that remain folded while in the launcher and flick open as the missile leaves the launch tube. The wings flick forward during extension, the tail surfaces rearward.

TOW is inserted into the rear end of the tube in its container, which forms an extension of the tube. Electrical and mechanical connections to the missile are made automatically during this operation. The Hercules K-41 solid-propellant motor gives two separate boost periods. It fires first to propel the missile from the launcher. To

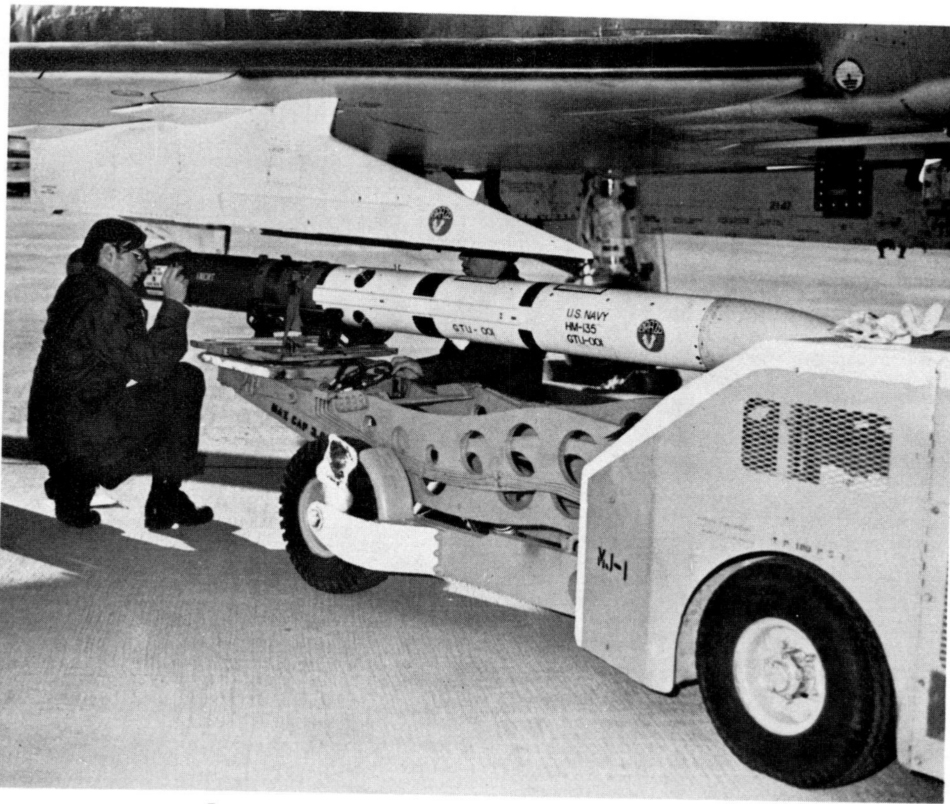

Brazo air-to-air anti-radiation missile, based on the Sparrow

BGM-71A TOW missile undergoing final inspection

ensure safety for the operator, the missile then coasts for a period after leaving the mouth of the tube, before the second stage of the booster fires.

The operator guides the missile by keeping the target centred in a telescopic sight. Movement of the sight generates electronic signals to correct the missile's course, the signals being passed through two wires. The tail control surfaces are actuated by a Chandler Evans CACS-2 system, using high-pressure stored helium gas to operate four differential piston actuators in matched pairs to control yaw, pitch and roll. The warhead is a high-explosive shaped charge, developed under the Army Munitions Command, Picatinny Arsenal, New Jersey.

TOW is intended as a heavy assault weapon for use against tanks, armoured vehicles and gun emplacements over ranges of more than one mile. In its surface-to-surface form, it can be mounted on a variety of ground vehicles, including the M-113 armoured personnel carrier.

To permit use of TOW from helicopters, Hughes developed under US Army contract a gyro-stabilised sight that would eliminate the effects of aircraft vibration and manoeuvres. As part of the XM-26 missile/launcher/sight subsystem, this was installed on a UH-1B helicopter, from which air-to-surface firing tests of TOW missiles were then made successfully at Redstone Arsenal, Alabama, with hits on moving tank targets over ranges of more than one mile.

Following such tests, the US Army Aviation Systems Command awarded Bell Helicopter Company a contract in March 1972 to modify eight AH-1G HueyCobras to carry eight TOW missiles each, under the Improved Cobra Armament Program (ICAP). Helmet sights for weapon aiming were supplied by Univac.

Experience showed that, in fact, this helicopter could carry only two to six TOWs, depending on temperature and altitude, in addition to its other armament and fuel load.

Some of the initial batch of eight modified HueyCobras were shipped to Vietnam on 24 April 1972. After a short training period, during which each pilot-gunner fired one TOW missile for the first time, the aircraft were committed at Kontum to meet an expected armour threat. By 27 June, in 77 combat launches, they had scored 62 hits on point targets and had destroyed 39 armoured vehicles, trucks and howitzers. None of the helicopters had been hit by hostile fire.

As a result of these successes, the US Army embarked on a programme to convert 290 AH-1G HueyCobras to AH-1Q Cobra/TOW standard. All of these aircraft were scheduled to be converted further to AH-1S standard, with uprated engine, gearbox and transmission, that would enable them to carry a full load of fuel, TOW missiles and other standard ordnance. TOW is also specified as armament of the US Army's new Advanced Attack Helicopter.

TOW has been in production for the US Army since 29 November 1968, with Chrysler Corporation as second-source supplier under a $33·3 million contract covering production until January 1973. Since that date, Hughes has been the sole producer of TOW. Orders totalled more than 217,000 missiles by January 1976, of which 122,500 were scheduled for delivery by the beginning of May 1976.

The missiles are manufactured in Hughes' missile factory at Tucson, Arizona, the launchers at the company's El Segundo, California, plant. Deployment to US troops in the USA and Europe began in November 1970, to replace the 106 mm recoilless rifle and the Entac and SS.11 missiles. Subsequently, TOW was ordered by 17 other countries, including Canada, Federal Germany, Italy, the Netherlands, Luxembourg, Turkey, Denmark, Norway and Iran.

The Department of Defense purchased in FY 1974 a total of 18,000 TOW missiles and 1,518 launchers for the US Army, and an initial quantity of 5,425 missiles and 100 launchers for the Marine Corps. The FY 1975 request was for $149 million, covering the purchase of 30,319 missiles and 1,041 launchers for the two services, and continued development of a night sight for the weapon system. Requests for FY 1976 and FY 1977 are intended to provide a total of about 46,500 TOW missiles, 2,809 launchers and 75 initial production night sights for the Army; 3,700 missiles and 51 launchers for the Marine Corps.

An extended-range version has been demonstrated in helicopter firings at ranges up to 3,750 m (12,300 ft), an increase of 25 per cent over the basic missile to which the following details apply:

DIMENSIONS OF MISSILE:
Length overall: 1·16 m (3 ft 9·7 in)
Diameter: 15 cm (5·9 in)
WEIGHT OF MISSILE:
Launching weight: 22 kg (48 lb)
PERFORMANCE:
Max range: over 3,000 m (9,850 ft)
Min range: under 65 m (215 ft)

Production line of AGM-65A Maverick TV-guided missiles at Hughes' Tuscon, Arizona, works

MAVERICK

USAF designation: AGM-65

Development of this air-to-surface missile began in 1966, when Hughes and North American each received a project definition contract to verify preliminary design and engineering studies of the projected weapon and to provide information for development and production contracts.

After evaluation of the results of these contracts, in July 1968, the USAF awarded Hughes a $95 million fixed-price incentive contract to cover development, test and evaluation of Maverick over a three-year period, with options for follow-on production of up to 17,000 missiles. All of these had been delivered by November 1975, and production continues under subsequent contracts.

The basic AGM-65A version of Maverick has a cylindrical body, with rounded glass nose and long-chord delta wings, indexed in line with cruciform tail control surfaces mounted close to their trailing-edges. It is powered by a Thiokol TX-481 solid-propellant motor and is television-guided.

It can be carried by the A-7D, A-10A, F-4D and F-4E, normally in three-round underwing clusters, and has a high-penetration conical shaped-charge warhead, intended for use against pinpoint targets such as tanks and columns of vehicles. Maverick has also been carried by Teledyne Ryan BGM-34 RPVs.

Unlike earlier TV-guided missiles, the AGM-65A Maverick is self-homing. The pilot of the launch aircraft first selects the desired weapon station, and a timed indicator light signals completion of the required gyro run-up time. After visually detecting a target, the pilot depresses the uncage switch which removes the protective dome cover from the nose of the missile and activates the video circuitry. The scene viewed by the Maverick TV seeker appears on a high-brightness TV screen in the cockpit. The pilot then manoeuvres his aircraft to place his optical sight reticle on the target, or slews the missile seeker head. After depressing the track switch, he waits until the cross-hairs are positioned over the target, then effects lock-on by releasing the track switch and launches the round. Maverick is homed on the target by an electro-optical device in its nose.

The first unguided air launch of the Maverick missile was conducted successfully at Edwards AFB, California, on 15 September 1969. The test was the first of 15 air launches conducted by McDonnell Douglas, with Hughes support, to prove the safe separation of the missile from an F-4 Phantom II throughout the aircraft's flight envelope. On 18 December 1969, a Maverick, complete except for warhead, was launched at medium range from an F-4D in a diving attack against a stripped-down M-41 tank at Holloman AFB, New Mexico. It scored a direct hit in this, its first guided test flight.

During subsequent tests at Fort Reilly, Kansas, Maverick was launched at distances ranging from a few thousand feet to many miles from the target, and from high altitudes down to treetop level, against manoeuvring Army tanks.

As a result of the success of the first 27 flight tests, the USAF decided in early 1972 to cancel the 13 further launches that had been planned. Production of the AGM-65A was initiated in 1971, under a $69·9 million contract, and the USAF formally accepted the first of the initial quantity of 2,000 production Mavericks at Hughes' Tucson, Arizona, plant on 30 August 1972.

By October 1973, the number of Mavericks ordered by the USAF had increased to 11,000 in four batches. In that month, the missile was used operationally by the Israeli Air Force during the Yom Kippur war, with results that were described as "Quite impressive, although the conditions there were much more favourable for such electro-optical weapons than would be the case in Europe". Other Mavericks have been delivered to Iran.

A major disadvantage of any conventional TV-guided missile is that it is restricted to use in clear-visibility conditions. Considerable research is being devoted to improving the versatility of Maverick, and by the Spring of 1975 four different versions were known to exist, as follows:

AGM-65A. Initial TV-guided version, as already described. Deliveries, under contracts totalling about $400 million, reached the originally planned programme total of 17,000 on 21 November 1975. Production of a further 2,000 then started.

AGM-65B. Basically similar to AGM-65A, but with a modified 'scene magnification' TV seeker, making it possible for the pilot to identify and lock on to the target from a longer range or to attack smaller targets. Engineering development completed by January 1975. Order for 4,000 announced in August 1975; deliveries began in December 1975.

AGM-65C. Laser-guided version intended specifically for close air support against designated targets. Capable of day and night use in conjunction with airborne or ground designator. Ten missiles fired during development at Eglin AFB in June-October 1973, using Long Knife airborne designator pod and ILS-NT200 ground designator. Batch of 100 requested under FY 1977 budget.

AGM-65D. Version with imaging infra-red seeker (IIR), which produces an image by sensing small differences in infra-red heat radiated by objects in view, so being suitable for day and night use, even through haze. In early tests, targets such as an offshore drilling rig and oil storage area were detected by a forward-looking infra-red (FLIR) acquisition aid fitted to an F-4D. Handover was then made to the IIR seeker in the missile, for a homing attack. In the Summer of 1974, Hughes received a $10·2 million contract to install eight IIR seekers on Maverick missiles for free flight trials in the Autumn of 1975. The first firing of an IIR Maverick scored a direct hit on the tank target.

Later development will include adaptation of Maverick to carry the 113 kg (250 lb) MK-19 warhead for possible USAF/US Navy use against larger hardened targets such as command bunkers or ships.

DIMENSIONS (AGM-65A):
Length overall: 2·46 m (8 ft 1 in)
Body diameter: 30 cm (12 in)
Wing span: 0·71 m (2 ft 4 in)
WEIGHT (AGM-65A, approx):
Launching weight: 210 kg (462 lb)

MDAC

MCDONNELL DOUGLAS ASTRONAUTICS COMPANY (A Division of McDonnell Douglas Corporation)

ADDRESS:
5301 Bolsa Avenue, Huntington Beach, California 92647
Telephone: (714) 896-3311

PRESIDENT:
Robert L. Johnson

EXECUTIVE VICE-PRESIDENT:
Ben G. Bromberg

VICE-PRESIDENTS:
C. James Dorrenbacher (Engineering)
John L. Sigrist (Contracts and Pricing)
Paul L. Smith (Marketing)
W. H. Peter Drummond (Competitive Cost Project)
Theodore D. Smith (Delta Programmes)
Charles S. Perry (Deputy Programme Manager, BMD Sys. Tech. Programme)
Ned T. Weiler (Programme Manager, Ballistic Missile Defence Programmes)
Theodore D. Dunn (Chief Counsel)

FIELD AND TEST CENTERS:
Vandenberg Test Center, Vandenberg AFB, California 93436
Kwajalein Test Center
Florida Test Center, Cocoa Beach, Florida 32931

VICE-PRESIDENT:
Raymond D. Hill (Director)

TI-CO DIVISION:
PO Box 600, State Road 405, Titusville, Florida 32780

VICE-PRESIDENT:
Thurman W. Stephens (General Manager)

McDonnell Douglas Astronautics Company-East

ADDRESS:
Box 516, St Louis, Missouri 63166
Telephone: (314) 232-0232

VICE-PRESIDENTS:
Erwin F. Branah (General Manager)
Raymond A. Pepping (Space Programmes)
Clifford D. Marks (Engineering)
Harry W. Oldeg (Fiscal Management)
R. Wayne Lowe (Missile Programmes)

This company was formed on 26 June 1968, by merging the former Douglas Missile and Space Systems Division and the McDonnell Astronautics Company into a single management structure.

In the missile field, McDonnell Douglas Astronautics Company is developing and building the Harpoon anti-shipping missile for the US Navy and other customers, producing the Dragon anti-tank missile for the US Army and working on other projects of a classified nature. Its important space programmes are described in the Spaceflight section.

HARPOON

US Navy designations: AGM-84A and RGM-84A-1

Harpoon is a US Navy missile suitable for launching from aircraft, ships and submarines against shipping targets, from extended stand-off ranges. It is an all-weather weapon, with an air-breathing propulsion system.

Five major US aerospace companies responded to requests for proposals issued by Naval Air Systems Command on 22 January 1971, at the end of three years of research and study. This had included seeker flight and ground tests, propulsion studies, aerodynamic testing and analytical investigations of a number of possible configurations.

In May 1971, General Dynamics and McDonnell Douglas were asked to submit additional technical and financial data on their proposals, which were then evaluated by representatives of Naval Air Systems Command, Naval Ordnance Systems Command and selected Navy field activities. These evaluations considered all aspects of the proposals, including the design approach, the extent of modifications required to existing launch systems (eg, the Asroc launcher) to accommodate the new missile, the technical risk and projected costs.

As a result of this review, the Navy selected McDonnell Douglas as prime contractor for development of the Harpoon missile, in June 1971, under the programme management of Naval Air Systems Command and with major support from Naval Ordnance Systems Command. The work was allocated to McDonnell Douglas Astronautics Company-East, with Texas Instruments Inc and Sperry Systems Management as major subcontractors.

The initial contract, valued at $66 million, awarded to MDAC on 21 June 1971, covered the development and demonstration of a number of engineering-model missiles over a two-year period. This phase of the programme was completed successfully on schedule, and was followed by a

30-month final development phase, funded at $110 million, in which more than 30 missiles were launched from Harpoon-designated aircraft, ships and submarines. Operational evaluation was under way in early 1976, utilising pilot production missiles. Initial production was authorised in July 1975 for approximately 300 missiles; the US Navy requested $178·3 million for 350 in FY 1977.

Teledyne CAE and Garrett AiResearch were awarded contracts for initial development of the Harpoon's turbine propulsion system. The J402-CA-400 turbojet proposed by the former company was eventually selected for this application.

The general configuration of the **AGM-84A** air-launched version of Harpoon is shown in the accompanying illustration. It is a torpedo-shape missile, with cruciform wings indexed in line with cruciform tail surfaces. The turbojet power plant is housed in the rear of the body, with a ventral flush air intake.

Prior to launch, targeting data for Harpoon are provided by the command and launch subsystem, which interfaces with on-board systems. The Harpoon data processer, a general-purpose digital computer, receives targeting and attitude data from existing systems, and computes the necessary missile and launcher orders. After launch, guidance is provided by a midcourse guidance system consisting of a strapdown attitude reference assembly and digital computer. No inputs from the launch platform are required by the missile after launch. Cruise altitude is monitored by a radar altimeter, enabling the flight to the target to be made at low altitude, so offering both optimum target acquisition capability through reduction of clutter effects, and the ability to penetrate enemy defences. Control is exercised via the tail-fins, which are driven by electro-mechanical actuators. Offset launch capability is provided for all launch modes.

When the target comes within the search area of the active radar seeker, the high-resolution system detects and locks on to the target, even in rain and high sea states. Seeker lock-on is maintained until impact. Capability to perform high-*g* manoeuvres throughout flight permits successful operation against fast manoeuvring targets. A terminal 'pop-up' manoeuvre counters close-in defences and offers maximum warhead effectiveness. Counter-countermeasures devices are installed. The warhead is a high-explosive blast type.

Addition of a short solid-propellant tandem booster, with cruciform fins, converts the Harpoon into the **RGM-84A-1** ship-launched version. This is launched at a low elevation angle and follows a ballistic trajectory until booster separation. The sustainer engine then starts and the missile descends to cruise altitude, following the same trajectory as that for air-launch. Launchers can include standard Tartar, Terrier, Asroc and Mk 26 types.

The Harpoon canister launcher provides a lightweight means of adapting Harpoon to nearly any surface launch application, including land-based shore defence systems. Under development for Patrol Hydrofoil Missile (PHM) ships, the launcher, weighing approximately 907 kg (2,000 lb), consists of a cluster of four Harpoon canisters and associated support structure. Each canister contains one Harpoon. The missile configuration is the same as that for ship launch, except that the aerodynamic surfaces are modified to permit folding within the canister diameter.

The submarine-launch Harpoon configuration is the same as for the canister launcher, except that the missile is installed in a buoyant capsule, which is fired from the submarine's torpedo tubes. As the capsule rises toward the surface, aft-mounted control fins unfold to maintain the required attitude. Upon broaching the surface, the ends of the capsule separate automatically and the missile's solid-propellant booster ignites, launching it into the same trajectory as for surface launches.

AGM-84A Harpoon air-to-surface anti-shipping missile on launcher

The first drop-test of an AGM-84A development round was made in May 1972 from a P-3 Orion aircraft flying at 6,100 m (20,000 ft) over the Pacific. The test was designed to demonstrate separation characteristics and release mechanisms. Powered flights began in July 1972.

First launch of a complete missile against a target was made on 20 December 1972, when a Harpoon fired from a P-3 followed a full operational flight path, including terminal manoeuvre, before scoring a direct hit on the target ship, USS *Ingersoll,* moored off Point Mugu, California. During 1974 and 1975, a total of 34 missiles were launched from P-3A and P-3C aircraft, DE-1052, DLG and DDG ships, and attack submarines. Launch conditions were varied to include the full scope of anticipated operational conditions. These tests, including warhead shots, proved that Harpoon met all specified performance requirements.

The AGM-84A is intended to be carried by a variety of aircraft, including the P-3 Orion, S-3 Viking, A-7E Corsair II and A-6E Intruder.

DIMENSIONS:
Length overall:
AGM-84A — 3·84 m (12 ft 7 in)
RGM-84A-1 — 4·57 m (15 ft 0 in)
Body diameter — 0·34 m (1 ft 1½ in)

WEIGHTS:
Launching weight:
AGM-84A — 526 kg (1,160 lb)
RGM-84A-1 — 662 kg (1,460 lb)

PERFORMANCE:
Cruising speed — high subsonic

ASALM

Under USAF contracts, McDonnell Douglas and Martin Marietta are engaged on concept formulation studies and advanced technology development for an Advanced Strategic Air-Launched Missile (ASALM) to replace both the SRAM and the ALCM subsonic cruise missile in the 1980s. An artist's impression of the Martin Marietta proposal is reproduced on this page. It is expected to offer a somewhat higher speed than SRAM and about half the ALCM's range of more than 1,040 nm (1,930 km; 1,200 miles). Size will be comparable with that of SRAM, since the new missile must be compatible with the rotary launcher utilised for the current weapon. ASALM must also be capable of both high-level and low-level penetration of enemy airspace.

Artist's impression of ASALM (Advanced Strategic Air-Launched Missile)

Further contracts, covering design and ground testing of ASALM propulsion systems, have been awarded to the Marquardt Company and Chemical Systems Division of UTC. The results of this aspect of the programme are expected to be generally applicable to other projects, including air-launched tactical missiles, surface-launched defensive missiles and a variety of air-to-air missiles.

The type of propulsion system proposed is an integral rocket-ramjet, involving both solid rocket and liquid fuel ramjet technology. It will operate as a solid rocket booster until it reaches supersonic speeds. At that point, through a series of mechanical changes that take place in flight, it will become a ramjet. These changes involve the opening of air-inlets, nozzle changes and a switch to the burning of liquid fuel and air in the combustion chamber within a common system. Prior to the development of this concept, a ramjet required an external solid rocket booster to propel it to its supersonic velocity. The spent booster was then jettisoned.

NWC
NAVAL WEAPONS CENTER

HEADQUARTERS:
China Lake, California 93555
Telephone: (7xr) o3o-3ttt

COMMANDER:
Rear Admiral W. J. Moran

TECHNICAL DIRECTOR:
H. G. Wilson

HEAD, PUBLIC AFFAIRS OFFICE:
J. H. McGlothlin

The Naval Weapons Center (formerly US Naval Ordnance Test Station) is located 250 km (155 miles) north-east of Los Angeles, on the Mojave Desert, and comes under the command of the Chief of Naval Material, Department of the Navy.

The mission of the Center is to conduct a programme of warfare analysis, research, development, test, evaluation, systems integration and fleet engineering support in naval weapons systems, principally for air warfare, and to conduct investigations into related fields of science and technology.

Chief ordnance developments of the Center are guided missiles, rockets, and aircraft fire-control and bomb-directing systems. Weapons developed at the Center include the Mighty Mouse 2·75 in folding-fin aircraft rocket, the 11·75 in Tiny Tim rocket, the Zuni 5 in folding-fin aircraft rocket, the Sidewinder air-to-air guided missile, the Snakeye 250/500 lb bomb with folding dive-brake retardation system to avoid fragmentation damage to the launch aircraft during low-level strikes, the Shrike and Standard ARM air-to-surface anti-radar missiles and the Walleye glide bomb. Production of many of these weapons was entrusted to commercial companies under whose entries they are, or were, described in *Jane's.*

SIDEWINDER

US military designation: AIM-9

The **AIM-9A** prototype version of the Sidewinder air-to-air missile was developed by the NWC and was first fired successfully on 11 September 1953. The first-generation production version, redesignated **AIM-9B,** was manufactured in very large numbers by Philco (now Aeronutronic Ford Corporation) and General Electric for the US Navy and the USAF, and was supplied to many foreign services, including the air forces of nine NATO countries, Australia, Nationalist China, Japan, the Philippines, Spain and Sweden, and the Royal Navy, Royal Canadian and Royal Netherlands Navies.

Production in the USA, from 1955, totalled more than 80,000 AIM-9Bs. The number of heat-seeking heads manufactured for these missiles exceeded 100,000, and delivery of improved versions was continuing in 1975. Many countries still deploy the basic AIM-9B, of which licence manufacture was also undertaken in Germany by Bodenseewerk, in association with subcontractors in the Netherlands, Denmark, Norway, Greece, Portugal and Turkey. The first of some 9,000 licence-built rounds was delivered in November 1961. Bodenseewerk subsequently conducted a Sidewinder improvement programme.

When the AIM-9B first entered service, it was described as having fewer than two dozen moving parts and no more electronic components than a domestic radio, making it one of the simplest and cheapest guided missiles yet produced in quantity. It has a cylindrical aluminium body, cruciform control surfaces at the nose and cruciform tail fins. Power plant is a Naval Propellant Plant solid-propellant rocket motor. Guidance is by infra-red homing, and an 11·4 kg (25 lb) high-explosive warhead is fitted.

Eight developed versions of Sidewinder have been produced for air-to-air use, to enhance and update the missile's capabilities, as follows:

AIM-9C. Only version of Sidewinder not fitted with

AIM-9L advanced version of Sidewinder

infra-red guidance. Motorola semi-active radar guidance, to provide the US Navy's F-8 fighters with an all-weather weapon. Production completed.

AIM-9D. First advanced model of the Sidewinder with higher speed and greater range capability. Developed by NWC at China Lake; produced in large numbers by Raytheon Company for the US Navy and UK. External changes include a tapering nose, longer-chord nose fins and greater sweepback on the leading-edge of the tail fins. Rocketdyne Mk 36 Mod 5 solid-propellant motor. New guidance unit, fuse and warhead. Production completed.

AIM-9E. Development of AIM-9B, with improved guidance and control; produced by Philco for the USAF, by modification of AIM-9Bs. Production completed.

AIM-9G. Similar to AIM-9D but with improved target acquisition and lock-on. Production by Raytheon for US Navy and USAF completed in FY 1970.

AIM-9H. Version for US Navy with improved close-range 'dogfight' capability. Basically similar to AIM-9G, but with solid-state guidance for improved reliability and maintainability, decreased minimum ranges and faster angle tracking rates. Off-boresight acquisition/launch capability. Lead bias function moves missile impact point forward to a more vulnerable area on the target aircraft. Total of 4,120 acquired between FY 1971 and FY 1975. Only version of Sidewinder in production in early 1975. Operational on F-14 Tomcat and other aircraft. USAF requested 800 under FY 1976 budget.

AIM-9J. Advanced version of AIM-9E, evolved to enhance 'dogfight' capability by improved manoeuvring ability. Being produced for USAF by Aeronutronic Ford Corporation to equip the F-15 Eagle and other types, by modification of remaining 590 AIM-9Bs in USAF inventory and a further 1,410 AIM-9Bs acquired from the US Navy. Deliveries scheduled to begin in 1977 and to be completed in 1978.

AIM-9J+ (J-3). Further improvement of AIM-9J, with solid-state electronics and same fuse as AIM-9L. Will have an all-aspect performance equivalent to that of AIM-9L. To be produced by conversion of existing AIM-9Es and AIM-9Js, with deliveries scheduled for FY 1978-1980.

AIM-9L. Third-generation version of Sidewinder, under development jointly by US Navy and USAF. Research and development completed by early 1975, when user evaluation was under way. Mk 36 Mod 6 solid-propellant motor. Double-delta nose fins for improved inner boundary performance and better manoeuvrability. AM-FM conical scan for increased seeker sensitivity and improved tracking stability. Rate bias and active optical fuse for increased lethality and low susceptibility to countermeasures. Annular blast fragmentation warhead. Series of 20 joint service technical evaluation firings completed on 1 March 1975, including successful firings against PQM-102 targets, and with particular emphasis on difficult shots on the beam or forward quarter of the target. Planned procurement between FY 1976 and FY 1980 totals 3,550 for US Navy and 4,810 for the USAF. In FY 1976, Raytheon was expected to deliver 1,300 AIM-9Ls and Aeronutronic Ford the remaining 210 missiles to qualify them for full production.

DIMENSIONS:	
Length overall:	
AIM-9B	2·83 m (9 ft 3½ in)
AIM-9D	2·91 m (9 ft 6½ in)
Body diameter	0·13 m (5 in)
Fin span:	
AIM-9B	0·56 m (1 ft 10 in)
AIM-9D	0·64 m (2 ft 1 in)
LAUNCHING WEIGHT:	
AIM-9B	72 kg (159 lb)
AIM-9D	84 kg (185 lb)
PERFORMANCE:	
Speed:	
AIM-9B	Mach 2·5
Range:	
AIM-9B	1·75 nm (3·35 km; 2 miles)
AIM-9D	over 1·75 nm (3·35 km; 2 miles)

AGILE/AIMVAL

US Navy designation: AIM-95

Development of this close-range 'dogfight' missile was initiated by the Naval Weapons Center to arm the Grumman F-14 Tomcat fighter.

Further development of Agile, and of the USAF's projected CLAW 'dogfight' missile, has been abandoned i favour of a common weapon for both services, as follow-on to the AIM-9L Sidewinder. The characteristi of this new missile were to be determined as a result of test programme known as 'Aimval', conducted durin 1976 on the ACMR facilities at Yuma and Las Vegas. considerable number of F-14, F-15, A-7E and F-5 'aggressor' aircraft were assigned to this programme.

SHRIKE

US Navy designation: AGM-45A

Known originally as ARM (anti-radar missile), th Shrike is a supersonic air-to-surface weapon which hom on to enemy radar installations.

Texas Instruments and Sperry Univac produce guidanc and control assemblies for Shrike, which has a Rocketdyr Mk 39 Mod 7 or Aerojet Mk 53 solid-propellant mot and is armed with a proximity-fused high-explosive fra mentation warhead, weighing 66 kg (145 lb). Its gener configuration is shown in the accompanying illustratior

Delivery to carrier-based attack squadrons of the U Navy began in 1964. It can be carried by all Navy attac aircraft, and by F-105G and most F-4 aircraft of th USAF, and was used operationally in Vietnam from 196 Many improvements have increased Shrike's effectivene since it first became operational, leading to a series twelve versions designated AGM-45-1, -1A, -2, -3, -3A 3B, -4, -6, -7, -7A, -9 and -10. These differ primarily in th frequency coverage of the front end detachable seek sections.

Between 1965 and FY 1974, the US Navy procure 11,167 Shrikes; the USAF procured 9,638, followed b 270 more in FY 1975. Only the USAF planned continue procurement in FY 1976, with a projected purchase 1,318 of the AGM-45-7A, -9 and -10 models, followed b 1,637 more by the end of FY 1977. Aircraft carrying the latest models will include the 'Wild Weasel' F-4Gs.

DIMENSIONS:	
Length overall	3·05 m (10 ft 0 i
Body diameter	20·3 cm (8 i
Wing span	0·91 m (3 ft 0 i
WEIGHT:	
Launching weight	182 kg (400 l
PERFORMANCE (estimated):	
Range	over 2·6 nm (5 km; 3 mile

Shrike anti-radar missile on underwing rack of a McDonnell Douglas A-4E Skyhawk

RAYTHEON

RAYTHEON COMPANY

HEAD OFFICE:
141 Spring Street, Lexington, Massachusetts 02173
Telephone: (617) 862-6600
Telex: 92-3455

CHAIRMAN OF THE BOARD AND CHIEF EXECUTIVE OFFICER:
Thomas L. Phillips

PRESIDENT:
D. Brainerd Holmes

SENIOR VICE-PRESIDENT:
Aldo R. Miccioli (General Manager, Missile Systems Division)

VICE-PRESIDENTS:
Justin M. Margolskee (General Manager Operations, Missile Systems Division)
M. W. Fossier (Asst General Manager, Technical)
Floyd Wimberly (Advanced Systems)
Joseph Glasser (Manufacturing Manager)
Richard P. Axten (Public and Financial Relations)

Raytheon Company's Missile Systems Division is prime contractor for the US Army's surface-to-air Hawk and US Navy/USAF Sidewinder (described under NWC heading) and Sparrow air-to-air weapon systems. It has developed a surface-to-air version of Sparrow as the RIM-7H Sea Sparrow and is prime contractor for development of the important Patriot surface-to-air missile.

Missile Systems Division is also active in the fields of microminiaturised computers, advanced electronics systems, phased array radars and lasers.

Raytheon Company's Equipment Division is prime contractor for the NATO Sea Sparrow system, under development for the US Navy, the Royal Norwegian Navy, the Royal Danish Navy, the Royal Netherlands Navy, the Royal Belgian Navy and the Italian Navy.

SPARROW

US military designations: AIM-7 and RIM-7H

The AIM-7 Sparrow, developed by Raytheon Company for the US Naval Air Systems Command, is a radar-homing air-to-air missile with all-weather all-altitude operational capability. It can also be used against shipping targets from aircraft or ships.

Engineering of the Sparrow weapon system is done at Raytheon's laboratories at Bedford, Mass, and the company's engineering proving site at Oxnard, California. Production of the missile and its associated systems is centred in plants at South Lowell, Mass, and Bristol, Ten In addition, Mitsubishi is manufacturing more than 600 Japan for the Japan Air Self-Defence Force, under licen from Raytheon.

The Sparrow equips McDonnell Douglas F-4 Phanto II aircraft of the US Navy, USAF, Royal Navy, Royal A Force, Imperial Iranian Air Force, Israeli Air Forc Republic of Korea Air Force and Spanish Air Force. Th Lockheed F-104S fighters licence-built in Italy a Sparrow-armed, and both the USAF's McDonnell Dou las F-15 Eagle and the US Navy's Grumman F-14 Tomc employ this missile.

Without change, the Sparrow is used on US Navy shi as a surface-to-air and anti-shipping weapon in the Bas Point Defense Surface Missile System. It is also used Canadian ships in the Close Range Missile System.

Sparrow has a cylindrical body, pivoted cruciform win and cruciform tail fins in line with the wings. The curren advanced, AIM-7F version is powered by a larger He cules Mk 58 Mod 0 solid-propellant motor and has Raytheon semi-active Doppler radar homing system, wi a smaller solid-state seeker. The heavier continuous-r warhead is actuated by a proximity fuse or contact fus and is mounted forward of the wings, instead of aft as o ɜarlier versions. Manoeuvrability has been improved, and Sparrow is now considered a good dogfight missile as well ıs a good medium-range weapon with all-aspect capabili-ty.

The AIM-7F is the fourth operational version of the Sparrow. Production of the earlier AIM-7C, D and E nodels totalled 34,000 missiles, and the AIM-7F is now in ɔroduction for the US Navy and USAF. FY 1976 budget ·equests, on behalf of both services, included $140·9 mill-on for a further 980 missiles, $2·1 million for spares, and $6·9 million for research, development, test and evalua-ion. A further $156·4 million, for 1,530 missiles, RDT & E and spares, was requested in FY 1977. General Dynamics is to be brought in as second-source contractor. Total procurement under the AIM-7F programme is expected to provide 6,864 missiles for the US Navy and 5,415 for the USAF.

Development of a monopulse seeker for the AIM-7F was started in 1975, with the aims of reduced cost and ncreased performance in the ECM and lookdown/clutter reas. Initial operational capability is planned for 1981.

The ship-launched RIM-7H, now operational, has fold-ng wings to reduce the size of its launcher. A new ship-aunched version is planned, based on the AIM-7F, with onger range than the RIM-7H.

A new missile known as Sky Flash, evolved from Spar-ow, is in production for the Royal Air Force by Hawker iddeley Dynamics of the UK (which see).

Sparrow being launched from a McDonnell Douglas F-15A Eagle air superiority fighter

The following details apply specifically to the AIM-7F:

DIMENSIONS:

Length overall	3·66 m (12 ft 0 in)
Body diameter	0·20 m (8 in)
Wing span	1·02 m (3 ft 4 in)

WEIGHT:

Launching weight	227 kg (500 lb)

PERFORMANCE (estimated):

Speed	over Mach 3·5
Range	24 nm (44 km; 28 miles)

ROCKWELL INTERNATIONAL

ROCKWELL INTERNATIONAL CORPORATION

EXECUTIVE OFFICES:

1700 East Imperial Highway, El Segundo, California 90246

600 Grant Street, Pittsburgh, Pennsylvania 15219

OFFICERS: See Aircraft section

Missile Systems Division

4300 East Fifth Avenue, Columbus, Ohio 43216

MANAGER, PUBLIC RELATIONS:

Dent Williams

Rockwell International's Columbus (Ohio) Missile Sys-ɜms Division is playing a major role in several important US weapon programmes. Brief details are given below of ıe Condor and HOBOS air-to-surface weapons and the Air Defense Suppression Missile which the Missile Sys-ɜms Division is developing and/or producing for the US Army, Navy and Air Force.

This Division is also engaged on extensive work involv-ıg the applications of laser technology. In particular, it has ıpplied hardware for the US Army Missile Command's ıser terminal homing guidance demonstration program-ıe, which is leading to development of the Hellfire anti-rmour weapon system.

CONDOR

US Navy designation: AGM-53B

Condor, developed by the Rockwell International Mis-le Systems Division for the Naval Air Systems Com-and; is an advanced automatic-homing air-to-surface issile.

Layout is conventional for a supersonic missile, with a ylindrical body, rounded nose, cruciform delta wings and ruciform tail control surfaces indexed in line with the ings. The guidance package in the nose contains an ectro-optical contrast tracker, switchable wide and nar-ow field-of-view optics, a high resolution camera, and wo-axis stabilised platform. A full three-axis autopilot ffers heading, altitude and pitch hold modes, with titude and rate stabilisation. Immediately forward of the ings is the warhead section, consisting of a 285 kg (630) 12-point linear shaped charge, with aerodynamic arm-g, and instantaneous or delayed fusing. To the rear of the ocketdyne long-burning solid-propellant rocket motor, ith offset blast tube/nozzle, are the secure data link, flight ntrols, a silver zinc primary battery and electromechani-l actuators.

Prior to missile launch the weapon system operator ommunicates with the missile through an umbilical cable; fter launch, all communication between missile and air-raft is via the data link. Prior to launch, the launch-point, ırget co-ordinates and missile cruise altitude are entered n the control panel.

Immediately after launch, the aircraft data link acquires he missile data link and goes into the track mode. All equencing of the data link from this point to target impact s completely automatic.

The fuse mechanism in the warhead is armed automati-ally after 8 seconds of flight. The autopilot sends a motor gnition signal to the rocket motor within 3 seconds after ıe missile has reached the pre-selected cruise altitude.

The computer in the control system pod on the aircraft ccepts navigation inputs from the aircraft, and missile osition (bearing and range) inputs from the data link. With this data and pre-selected target coordinates, the omputer generates midcourse steering commands which re transmitted automatically to the missile via the data nk. Using this automatic midcourse guidance system, the missile can be flown to the target area in all weather conditions. As the range readout shows that the missile is approaching the target, the operator uses the video to observe terrain features which identify the target area. When he acquires the target area, he switches the TV from wide to narrow field-of-view to facilitate target acquisi-tion. When the operator locates the target, he slews the seeker to position the cross-hairs on the desired aiming point and enters the terminal phase.

The terminal phase has two modes of operation: missile follow and lock-on. In the missile follow mode the operator can control the missile manually, to fly it to the aiming point. In the lock-on mode, the tracker automati-cally tracks the final aiming point. In both modes the seeker provides a tracking rate command to the autopilot and the missile flies a proportional homing trajectory to impact the aiming point.

Condor made its first, successful, air launching test from an F-4 fighter on 31 March 1970. Direct hits on land and ship targets have been achieved in subsequent tests, using both inert and live warheads, and a live warhead strike against a ship target (USS *Vammen)* on 4 February 1971 resulted in sinking of the vessel. During one subsequent test, a Condor hit a target 47 nm (87 km; 54 miles) from the launch aircraft. Weather capability was demonstrated when one particular test missile penetrated significant weather, including icing conditions, during its midcourse flight to the target area, after which the target was acquired and a direct hit achieved.

Operational evaluation was completed in July 1975, by which time deliveries totalled 53 missiles, 7 Grumman A-6E launch aircraft and 6 of the 3·66 m (12 ft) long, 306 kg (675 lb) airborne missile control system (AMCS) pods that are carried under the belly of the aircraft. The prog-ramme for FY 1976/77T was authorised to provide 215 Condors, 100 A-6E aircraft installation kits and 8 more AMCS pods. A further 40 Condors were included in FY 1977 budget requests. However, the US House-Senate conferees terminated the Condor programme when approving the FY 1977 Defense Department appropria-tion in the Summer of 1976. This may result in abandon-ment of further development and production.

Being of modular design, Condor is capable of accepting alternative guidance, propulsion, data link and other sub-systems to provide all-weather terminal guidance and longer range. A variant known as YAGM-53A-1 Turbo Condor, with a Garrett AiResearch J401 turbojet replac-ing the normal rocket motor, has been flight tested. It hit a manoeuvring target that was 104 nm (195 km; 120 miles) distant from its launching point.

The following details apply to the basic AGM-53B Condor:

DIMENSIONS:

Length overall	4·22 m (13 ft 10 in)
Body diameter	0·43 m (1 ft 5 in)
Wing span	1·35 m (4 ft 5 in)

WEIGHT:

Launching weight	966 kg (2,130 lb)

PERFORMANCE:

Max range	60 nm (111 km; 69 miles)

AGM-53B Condor electro-optically guided air-to-surface missile under the wing of an F-14A Tomcat

ADSM

In early 1973, the US Army Missile Command awarded Rockwell's Missile Systems Division a contract to design and manufacture a dual-mode anti-radiation homing/infra-red (ARH/IR) Air Defense Suppression Missile (ADSM) for Army evaluation.

The ADSM utilises the basic 17·8 cm (7 in) diameter airframe of Rockwell's Hornet missile, described in the 1973-74 *Jane's*. Both vehicle and dual-mode seeker are designed, fabricated and integrated by its Missile Systems Division.

By early 1974, direct hits on the target had been achieved in three consecutive firings of the ADSM. The third launch was performed by a Rockwell crew flying a UH-1 helicopter at the US Army Missile Command's Redstone Arsenal, Alabama. The missile was fired over a range of more than 2 nm (4 km; 2·5 miles) and from an altitude of 305 m (1,000 ft), and struck the centre of the turret of a stationary tank target. The latter was equipped with radar to represent an armoured vehicle fitted with radar-directed multiple anti-aircraft guns.

ADSM development round on a Bell UH-1 helicopter

HELLFIRE

The US Army requested $11·1 million under the 1974 Fiscal Year budget proposals for a programme to determine the feasibility of a helicopter-launched anti-armour missile known as Hellfire. The initial version was intended to use a laser homing guidance system, and early tests were made at the Army Missile Command headquarters in Huntsville, Alabama, using Rockwell International Hornet missile airframes and Rockwell laser seekers.

The programme advanced a stage further in the early Summer of 1974, when Rockwell International and Hughes Aircraft Company were selected as competing contractors in a one-year advanced development programme aimed at producing a modular Hellfire able to accept semi-active laser, dual-mode ARH/IR, TV and imaging IR homing heads.

In November 1974, two test missiles fitted with laser seekers, supplied by Rockwell International, scored direct hits on two tank targets in the first successful ripple fire demonstration of such missiles at Redstone Arsenal. A US Army crew, flying an AH-1G HueyCobra helicopter, 'popped up' from cover and fired the first test missile at a tank illuminated by a ground-based laser designator. A quarter of a second later, the second missile was launched against a tank situated about 20 m (65 ft) from the first target and illuminated by a stabilised platform airborne laser on board a second AH-1G.

The FY 1976/77T Military Appropriations Act provided funding necessary to conduct a competition for development of the laser Hellfire system and to initiate one follow-on contract in 1976. It was announced in October 1976 that Rockwell had won the competition and had been awarded a $66·7 million contract for full-scale engineering development.

Launch weight of Hellfire is expected to be in the region of 41 kg (90 lb).

Hellfire development rounds on the wingtip pylons of a Bell AH-1G HueyCobra

HOBOS (GBU-15)

HOBOS (HOming BOmb System) is a modular weapon system developed under the USAF's electro-optical guided bomb (EOGB) programme. It consists of a KMU-353A/B or KMU-390/B (dependent on bomb size) guidance and control kit designed for easy installation on MK 84 (2,000 lb) and M118E1 (3,000 lb) general-purpose bombs. Kit installation does not alter the conventional bomb suspension, release or jettison functions, thus permitting the conversion of standard bombs into highly-accurate guided weapon systems.

Each kit consists of a forward guidance section, the warhead or interconnect section (including the bomb), and the aft control section.

The guidance or nose section consists of target seeker optics and sensor mounted on a gyro-stabilised platform, and the associated electronics which include a camera, platform and tracker electronics, and associated power supplies. The guidance section also furnishes reference signals to the autopilot located in the control section.

The warhead section consists of the MK 84 or M118E1 bomb with an interconnect assembly made up of four strakes, external electrical conduit, umbilical receptacle, and strake and umbilical receptacle attachment bands. The strakes are indexed in line with the wings to provide aerodynamic stability. The conduit transmits electrical signals between the guidance and control sections. The umbilical receptacle provides the necessary aircraft signal interface. Fusing of the HOBOS can be accomplished either electrically or mechanically.

The control or aft section consists of a cylindrical body with cruciform wings fitted with trailing-edge flap control surfaces for flight manoeuvring. The weapon system batteries and flight control system components are located within this section, including the autopilot which collects pitch and yaw steering data from the guidance section and converts this information into signals which drive the pneumatic control surface actuators.

The effectiveness of the HOBOS modular weapon system has been demonstrated fully in a large number of successful air drops, notably against targets such as aircraft in revetments, jungle roads and bridges in Vietnam.

Left to right: **TV-guided HOBOS based on 2,000 lb MK 84 bomb; IR-guided HOBOS based on 2,000 lb MK 84 bomb; TV-guided HOBOS based on 3,000 lb M118 bomb**

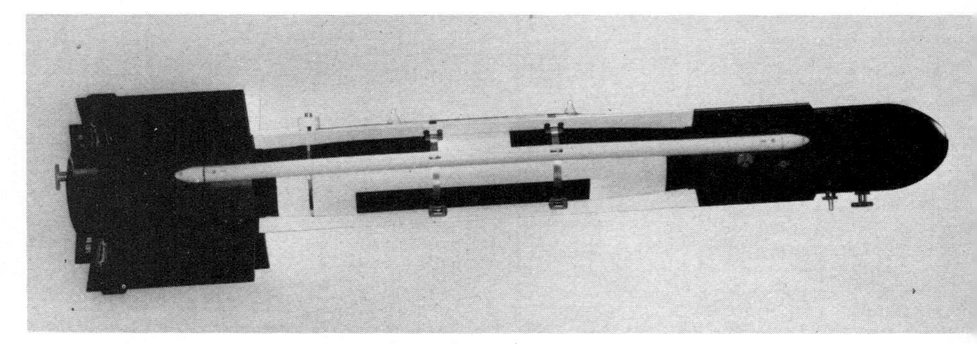

MK 84 HOBOS with new midcourse guidance, including DME, for increased accuracy

Missile Systems Division is under contract for production of the version built around the MK 84 bomb, for Greece and Israel, as well as for the US services.

USAF designation for this weapon system in its lates form is GBU-15(V), covering a family of modular glid weapons. Adaptation to other types of warhead, such as an SUU-54 cluster munition dispenser, is possible; and several alternative types of guidance system may be utilised to provide day and night all-weather capability, including a laser seeker, imaging infra-red (IIR), and a midcourse system that includes distance measuring equipment (DME) for increased accuracy. Each system provides a glide bomb with self-contained guidance, high accuracy and moderate/long-range stand-off capability. The laser and IIR seekers are the same as those developed for the Maverick missile.

Further options available on the GBU-15(V) are planar and cruciform wing modules, as alternatives to the basic small wing/strake module. One accompanying illustration shows an MK 84 HOBOS electro-optically guided round fitted with a planar wing module, which attaches above the weapon to increase its range when launched at high altitude. This 907 kg (2,000 lb) weapon carries a data link to allow controllability at stand-off ranges. The wings extend after launch.

A further winged variant is fitted with large cruciform rear-mounted wings and small cruciform foreplanes, giving increased manoeuvrability after launch at low altitude. This direct attack GBU-15(V) is expected to precede the planar wing/DME version into service.

DIMENSIONS:

Length overall:	
MK 84	3·78 m (12 ft 5 in)
M118E1	3·71 m (12 ft 2 in)
Body diameter:	
MK 84	0·46 m (1 ft 6 in)
M118E1	0·61 m (2 ft 0 in)
Wing span:	
MK 84	1·12 m (3 ft 8 in)
M118E1	1·32 m (4 ft 4 in)

WEIGHT:

Launching weight:	
MK 84	1,016 kg (2,240 lb)
M118E1	1,544 kg (3,404 lb)

GBU-15 PAVE STRIKE

As a result of the successful HOBOS programme, the USAF has selected Rockwell International's Missile Systems Division to serve as Principal Associate Contractor for integrating all system elements of modular guided weapon systems covered by its GBU-15 Pave Strike project. The Division's responsibilities, under terms of the contract, include the integration of the DME system from IBM, the range extension system from Celesco, data link equipment from Hughes Aircraft Company, the SUU-54 cluster warhead from Martin, fuses, proximity sensors, and future guidance systems and other system elements currently under development.

Concurrently with its integration management of GBU-15 Pave Strike, the Division has initiated engineering development of the Pave Strike improved version of HOBOS, to provide the USAF with a weapon offering greater stand-off range and generally improved performance. This will embody features to which reference is made under the HOBOS (GBU-15) entry.

Modular glide bomb version of the electro-optically guided MK 84 HOBOS, with wings to increase range

Modular guided weapon developed under GBU-15 Pave Strike programme

TEXAS INSTRUMENTS

TEXAS INSTRUMENTS, INC

CORPORATE OFFICES:

North Building, 13500 North Central Expressway, Dallas, Texas 75222

Texas Instruments has played a major part in many US missile programmes. The latest missile for which it is prime contractor is the AGM-88A HARM, of which brief details follow:

HARM

US Navy designation: AGM-88A

On 24 May 1974, Texas Instruments was named as prime contractor for HARM (High-speed Anti-Radiation Missile) by the US Naval Air Systems Command. The initial phase of the contract, valued at $1·4 million, covered four months of basic design co-ordination. Latest budget request is $33·5 million for continued RDT&E in FY 1977. The main four-phase programme is expected to involve four years' work. This will be centred in the company's Dallas facilities, where production of the Shrike anti-radiation missile has taken place for more than a decade.

Few details of HARM may yet be published. Its basic configuration is conventional, with a slim cylindrical body, ogival head, cruciform double-delta wings at mid-length for simplified roll control, and cruciform tail fins indexed in line with the wings. The fixed antenna for proportional navigation is in the extreme nose, with the seeker and other equipment. The solid-propellant rocket motor will be provided by Thiokol.

The emphasis on high speed reflects experience gained in Vietnam, where Soviet surface-to-air missile radar systems sometimes detected the approach of US anti-radiation missiles such as the first-generation Shrike and ceased operation before the missile could lock on to them. HARM is intended to have a sufficiently high performance to lock on to any target radar before it can shut down. It will cover a wide range of frequency spectra, so enabling one basic missile to be used against a variety of different radars. Aircraft likely to carry HARM, if its development proves successful, include the A-4M, A-6E, A-7, F-4, F-14, P-3 and S-3. Procurement of a total of 5,000 missiles is envisaged, with initial operational capability in 1979.

THE UNION OF SOVIET SOCIALIST REPUBLICS

AIR-TO-AIR MISSILES

AA-1

NATO reporting name: *Alkali*

This first-generation Soviet air-to-air missile was produced as standard armament for the Sukhoi Su-9 and of the all-weather versions of the MiG-19, which continue in service with the air forces of the Soviet Union and some of its allies.

Alkali is a solid-propellant missile, with large delta cruciform wings at the rear and small cruciform foreplanes indexed in line with the wings. There appear to be control surfaces in the trailing-edges of the wings and *Alkali* is believed to employ semi-active radar homing. The warhead is carried immediately aft of the foreplanes.

DIMENSIONS:

Length	1·88 m (6 ft 2 in)
Body diameter	178 mm (7 in)
Wing span	0·58 m (1 ft 10¾ in)

PERFORMANCE:

Range	3·2-4·3 nm (6-8 km; 3·7-5 miles)

AA-2

NATO reporting name: *Atoll*

This missile has been seen under the wings of a variety of Soviet aircraft and is standard equipment on home and export versions of the MiG-21. It is almost identical to the first-generation American Sidewinder (AIM-9B) in size and configuration and appears to have a similar infra-red guidance system.

The body is cylindrical with cruciform control surfaces near the nose, indexed in line with the fixed cruciform tail-fins. There are no external cable or control conduits.

The triangular control surfaces have a compound sweep averaging about 60° on the leading-edge and 10° on the trailing-edge. They are linked in opposite pairs, with a maximum movement of 20-30°.

Leading-edge sweep on the tail-fins is about 40°, with straight trailing-edges. A small gyroscopically-controlled tab is inset in the trailing-edge of each fin, at the tip, presumably for anti-roll stabilisation but possibly with an added control function.

Nozzle diameter of the solid-propellant motor is 8 cm

(3⅛ in). Weight and performance of *Atoll* should be very similar to those of Sidewinder.

DIMENSIONS:

Length overall	2·80 m (9 ft 2 in)
Body diameter	120 mm (4·72 in)
Span of control surfaces	0·45 m (1 ft 5¾ in)
Span of tail-fins	0·53 m (1 ft 8¾ in)

WEIGHT AND PERFORMANCE:
No details available except:
Range approx 2·5-3·5 nm (5-6·5 km; 3-4 miles)

Advanced Atoll

The latest versions of the MiG-21 carry a mix of standard infra-red *Atolls* and a new version of this weapon with a radar homing head. For convenience, the radar version is known at present as *Advanced Atoll.*

AA-3

NATO reporting name: *Anab*

First seen as underwing armament on the Yakovlev Yak-28P fighter *(Firebar)* in the 1961 Soviet Aviation Day display, *Anab* is a standard air-to-air missile in the Soviet Air Force. It was carried by Yak-28, Sukhoi Su-11 and Sukhoi Su-15 interceptors taking part in the 1967 air display at Domodedovo.

Anab has a cylindrical body, with small cruciform canard control surfaces indexed in line with very large cruciform tail-fins. Both infra-red and semi-active radar homing versions are operational.

DIMENSIONS (estimated):

Length:	
IR version	4·1 m (13 ft 5 in)
radar-homing version	4·0 m (13 ft 1 in)
Body diameter	280 mm (11 in)
Wing span	1·30 m (4 ft 3 in)

PERFORMANCE:
Range 4·3-5·4 nm (8-10 km; 5-6·2 miles)

AA-5

NATO reporting name: *Ash*

Ash is the large air-to-air missile shown under the wings of Tupolev Tu-28P *(Fiddler)* fighters on photographs in the Aircraft section of this edition. It has cruciform wings and tail surfaces indexed in line, and is operational in two versions, which have infra-red and semi-active (or active) radar homing heads respectively.

DIMENSIONS (estimated):

Length:	
IR version	5·5 m (18 ft 0 in)
radar homing version	5·2 m (17 ft 0 in)

AA-6

NATO reporting name: *Acrid*

This air-to-air missile was identified during 1975 as standard armament of the *Foxbat-A* interceptor version of the MiG-25. Its configuration is similar to that of *Anab* but it is considerably larger. Photographs suggest that the version of *Acrid* with an infra-red homing head is normally carried on each inboard underwing pylon, with a radar-homing version on each outer pylon. The wingtip fairings on the fighter, different in shape from those of *Foxbat-B,* are thought to house continuous-wave target illuminating equipment for the radar-homing missiles.

DIMENSION (estimated):
Length: radar-homing version 6·10 m (20 ft 0 in)

AA-7

NATO reporting name: *Apex*

This long-range air-to-air missile is one of two types of missile known to be carried as standard armament by interceptor versions of the MiG-23. No details are yet available, except that *Apex* has a solid-propellant rocket motor.

PERFORMANCE (estimated):
Range 15 nm (27 km; 17 miles)

AA-8

NATO reporting name: *Aphid*

Second type of air-to-air missile carried by the MiG-23, *Aphid* is a close-range solid-propellant weapon.

PERFORMANCE (estimated):
Range 3-4 nm (5·5-8 km; 3·5-5 miles)

HELICOPTER MISSILE

NATO reporting name: *Swatter*

No photographs of the Mil Mi-24 *(Hind-A)* assault helicopter have yet shown missiles mounted on its wingtip launchers. However, there is no evidence to suggest that it carries wire-guided missiles. The only standard Soviet anti-tank missile known to operate without wire guidance is *Swatter,* shown in surface-to-surface use in an accompanying illustration.

Swatter is controlled by elevons on the trailing-edges of its rear-mounted cruciform wings. The two small canard surfaces at the nose are also movable. The motor appears to exhaust through two vents diametrically opposed between the wings. Two more tubes, projecting rearward from opposite wings, probably house tracking flares. The blunt nose of *Swatter* suggests the likelihood of a terminal homing system operating via the canard foreplanes.

DIMENSIONS:

Length	1·12 m (3 ft 8 in)
Wing span	0·65 m (2 ft 2 in)

MiG-19 with four of the missiles known to NATO as *Alkali* **on underwing launchers** *(Tass)*

Atoll **air-to-air missile under wing of Indian MiG-21. Of interest also is the GP-9 underbelly gun pack**

Swatter **anti-tank missiles on BRDM vehicle. Missiles of this type are believed to arm the Mil Mi-24 helicopter** *(Tass)*

AIR-TO-SURFACE MISSILES

AS-2

NATO reporting name: *Kipper*

The missile carried by the Tu-16 *(Badger)* in the 1961 Aviation Day display, and described as an anti-shipping weapon, has a conventional sweptwing aeroplane layout, with an underslung power plant which is almost certainly a turbojet. It appears to be about 9·5 m (31 ft) long, with a wing span of 4·88 m (16 ft). Its performance is likely to be much inferior to that of the more refined and larger American Hound Dog air-to-surface missile of somewhat similar configuration, with an estimated range of 115 nm (213 km; 132 miles) and cruising speed of Mach 1·2. Radar is carried in the nose of the Tu-16 carrier aircraft.

AS-3

NATO reporting name: *Kangaroo*

Largest of the air-to-surface missiles first seen in the Soviet Aviation Day display at Tushino in 1961, and known to be operational, was that carried by the Tu-95 *(Bear)* and given the NATO reporting name *Kangaroo.* It is a winged missile with an airframe similar in size and shape to a sweptwing turbojet-powered fighter aircraft. The tail unit is conventional, with sweepback on all surfaces. The vertical surfaces, concealed inside the launch-aircraft until the missile is dropped, are of rhomboid form.

What was believed originally to be a radome on the missile's nose was identified subsequently as either a duct through which air can be fed to start the missile's turbojet engine prior to launching or a fairing over the air intake. Radar guidance equipment is carried in the nose of the Tu-95 carrier aircraft. Length of this missile is reported to be 14·9 m (48 ft 11 in), with a span of 9·15 m (30 ft), speed of Mach 2 and range of 350 nm (650 km; 400 miles).

AS-4

NATO reporting name: *Kitchen*

The air-to-surface missile carried semi-submerged in the fuselage of the Tupolev Tu-22 bomber *(Blinder)*, and under the wings of the variable-geometry *Backfire-B,* looks considerably more advanced than the weapons already described. It appears to have stubby delta wings and cruciform tail surfaces, and is believed to be powered by a liquid-propellant rocket engine. The speed of *Kitchen* is probably high. It is about 11·3 m (37 ft) long and has an estimated range of 160 nm (300 km; 185 miles) at low altitude.

AS-5

NATO reporting name: *Kelt*

In September 1968 a photograph released officially in Moscow showed this air-to-surface missile under the port wing of a Tu-16 bomber. It is externally similar to the earlier turbojet-powered *Kennel* missile, which looked rather like a scaled-down unpiloted version of the MiG-15 with a hemispherical radome above its air intake. On *Kelt,* both the ram-air intake and radome are replaced by a hemispherical nose fairing, probably housing a larger radar. This implies that *Kelt* is rocket-powered and it may be significant that, unlike *Kennel,* it has an underbelly fairing of the kind seen on the rocket-powered *Styx* missile. *Kelt* also appears to be longer than *Kennel* and its underwing carrier is much larger.

According to Israeli reports, about 25 *Kelts* were launched against Israeli targets by Tu-16s from Egypt during the Arab-Israeli War of October 1973. Twenty of the missiles were claimed as destroyed by the air and ground defences; the others hit two radar sites and a supply centre in Sinai.

DIMENSIONS (estimated):

Length overall	9·45 m (31 ft 0 in)
Span	4·57 m (15 ft 0 in)

PERFORMANCE (estimated):

Range	over 175 nm (320 km; 200 miles)

AS-6

Little is known about this new missile, which is reported to be one of the weapons carried by the Tupolev supersonic 'swing-wing' bomber known to NATO as *Backfire.* Unofficial reports have suggested that it can be fitted with alternative 450 kg (1,000 lb) nuclear or high-explosive warheads. Propulsion is said to be by liquid-propellant rocket motor, with inertial midcourse guidance and active radar terminal homing. Range is said to vary from 135 nm (250 km; 155 miles) at low altitude to 375-430 nm (700-800 km; 435-500 miles) at high altitude.

AS-7

NATO reporting name: *Kerry*

This tactical air-to-surface missile is reported to be carried by the Su-17, Su-20 and MiG-23 close support aircraft.

Kennel **anti-shipping missile under the wing of a Tu-16 bomber**

Two views of Tupolev Tu-95 *(Bear-B)* **with AS-3** *Kangaroo* **air-to-surface missile**

The *Badger-G* **version of the Tupolev Tu-16, with two** *Kelt* **missiles on underwing launchers**

SPACEFLIGHT
AND
RESEARCH ROCKETS

AUSTRALIA

GOVERNMENT OF AUSTRALIA
DEPARTMENT OF DEFENCE

ADDRESS:
Canberra ACT 2600
Telephone: 659111
Telex: 62053
SECRETARY: Sir Arthur Tange, CBE

Weapons Research Establishment:
Salisbury, South Australia 5108
DIRECTOR: Dr M. W. Woods

Current types of upper atmosphere sounding rockets which have been developed by the Weapons Research Establishment and built in Australia are as follows:

COCKATOO

This sounding rocket was developed to replace the HAD vehicle described in the 1970-71 *Jane's*. Cockatoos have apogees in the 125-145 km (77·5-90 miles) range. They are used for falling sphere, lithium trail and ultra-violet investigations of the upper atmosphere.

CORELLA

The Corella Mk 1 rocket is used for 'chemical seeding' experiments involving grenade releases, lithium and TMA trail releases to determine the physical and chemical properties of the upper atmosphere.

DIMENSION:
Length — 810 cm (26 ft 7 in)
Diameter of 1st stage — 25·4 cm (10 in)
Diameter of 2nd stage — 17·8 cm (7 in)

WEIGHTS:
Launching weight — 448-466 kg (988-1,028 lb)
Payload — 38-56 kg (85-125 lb)

PERFORMANCE (85° launch angle):
Ceiling:
with 38 kg; 85 lb payload — 240 km (149 miles)
with 56 kg; 125 lb payload — 195 km (121 miles)

KOOKABURRA

Kookaburra Mk 1 is used to release a parachute-borne dropsonde for transmitting synoptic information on atmospheric temperature and ozone concentrations to Australian meteorological stations, as well as for measuring wind profiles. It uses two Australian solid-propellant rocket motors, the Lupus 1 and Musca, with a total impulse of 37,800 Ns and 19,100 Ns respectively. Payload capacity of the Kookaburra Mk 1 is 5·9 kg (13 lb) which can be carried to a height of 80 km (50 miles).

A Kookaburra Mk 2 version, with an apogee of 125 km (78 miles) with a 4·8 kg (10·6 lb) payload has been developed for use in inflatable falling sphere experiments and ionospheric experiments in the D and E region. The first-stage Lupus 2 motor of this rocket has a total impulse of 46,200 Ns. Kookaburra Mk 3 is also used in falling sphere experiments, with a Lupus 3 first-stage motor having a total impulse of 60,900 Ns.

DIMENSIONS:
Length:
Mk 1 — 353 cm (11 ft 7 in)
Mk 2 — 358 cm (11 ft 9 in)
Mk 3 — 390 cm (12 ft 9 in)
Diameter of 1st stage — 12·7 cm (4·95 in)
Diameter of 2nd stage — 8·9 cm (3·50 in)

WEIGHTS:
Launching weight:
Mk 1, 2 — 54 kg (120 lb)
Mk 3 — 64 kg (142 lb)

Lorikeet sounding rocket

Payload:
Mk 1 — 5·9 kg (13 lb)
Mk 2 — 4·8-5·9 kg (10·6-13 lb)
Mk 3 — 5·7 kg (12·5 lb)

PERFORMANCE (85° launch angle):
Ceiling:
Mk 1 — 80 km (50 miles)
Mk 2:
with 4·8 kg; 10·6 lb payload — 125 km (78 miles)
with 5·9 kg; 13 lb payload — 110 km (68 miles)
Mk 3 — 135 km (84 miles)

LORIKEET

The Lorikeet sounding rocket was developed to replace the Cockatoo. It utilises two Australian solid-propellant rocket motors, the Dorado (172,000 Ns total impulse) and Lupus. Lorikeet Mk 1, with a Lupus 1 second-stage motor (37,800 Ns total impulse), is used for large parachute borne dropsondes; Lorikeet Mk 2, with a Lupus 3 second-stage motor (60,900 Ns total impulse), is used for dispensing Lithium trails, ionospheric experiments and ultra-violet investigations of the upper atmosphere.

DIMENSIONS:
Length:
Mk 1 — 568 cm (18 ft 7¾ in)

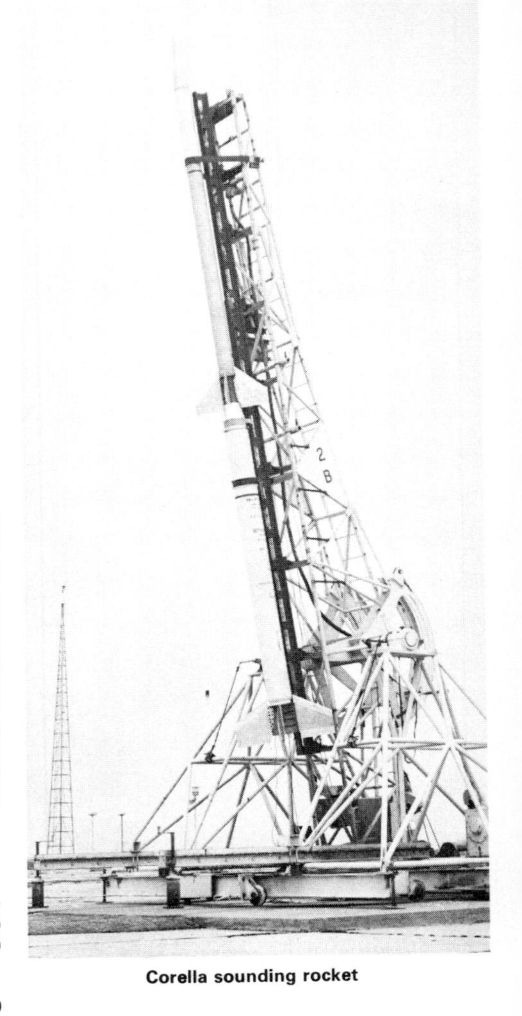

Corella sounding rocket

Mk 2 — 610 cm (20 ft 0 in)
Diameter of 1st stage — 17·8 cm (7 in)
Diameter of 2nd stage — 12·7 cm (4·95 in)

WEIGHTS:
Launching weight:
Mk 1 — 211 kg (465 lb)
Mk 2 — 195-202 kg (431-446 lb)
Payload:
Mk 1 — 21 kg (47 lb)
Mk 2 — 16-23 kg (35-50 lb)

PERFORMANCE (85° launch angle):
Ceiling:
Mk 1 — 85 km(53 miles)
Mk 2:
with 16 kg; 35 lb payload — 160 km (100 miles)
with 23 kg; 50 lb payload — 138 km (86 miles)

BRAZIL

AVIBRÁS
AVIBRÁS INDÚSTRIA AEROESPACIAL SA

HEAD OFFICE AND WORKS:
Antiga Estrada de Paraibuna, km 118 (CP 229), São José dos Campos, São Paulo State
Telephone: (0123) 21-7433
MANAGING DIRECTOR:
Eng João Verdi Carvalho Leite
SALES DIRECTOR:
Hely Adilson de Oliveira

This company is engaged in research and development of rockets, propellants and ancillary systems for both civil and military purposes. In the military field, it is producing a variety of rocket pods and is developing an air-to-air missile (MAA-1) and an air-to-surface missile (MAS-1), as well as other types of guided airborne weapons. Brief details of its current research rocket programmes follow:

AVIBRÁS SONDA 1/A

The Sonda 1/A vehicle is a two-stage solid-propellant sounding rocket, designed to reach heights of up to 100 km (62 miles) with a 4·5 kg (10 lb) payload package. Developed and built in series by Avibrás under contract to the Brazilian Ministry of Aeronautics, it is now operational at the Ministry's Barreira do Inferno Range in Natal, Rio Grande do Norte State, where it has replaced US-built types.

The radio-sonde payload package is ejected when the rocket reaches its ceiling, and descends by parachute. Telemetry data are transmitted at a frequency of 403MHz, and the package is tracked by radar.

The booster is a 127 mm (5 in) calibre military aircraft rocket, developed and manufactured by Avibrás for the Brazilian Air Force.

PERFORMANCE (85° launch angle):
Max speed at 2nd stage burnout — 2,720 knots (1,400 m/sec; 3,135 mph)
Nominal ceiling — 54 nm (100 km; 62 miles)
Peak altitude achieved — 66 nm (123 km; 76 miles)

ABIVRÁS SONDA 1/B

This two-stage solid-propellant sounding rocket differs from the Sonda 1/A in having a second stage of reduced length. Max payload is 5·5 kg (12 lb).

PERFORMANCE:
Max speed at 2nd stage burnout — 2,430 knots (1,250 m/sec; 2,800 mph)
Nominal ceiling — 40·5 nm (75 km; 46·5 miles)
Peak altitude achieved — 45 nm (83 km; 51 miles)

IAE/EDE
CENTRO TÉCNICO AEROESPACIAL INSTITUTO DE ATIVIDADES ESPACIAIS (IAE)
EDE—Divisão de Engenhos Espaciais

ADDRESS:
12.200 São José dos Campos, São Paulo State
DIRECTOR OF THE IAE:
Col Hugo de Oliveira Piva
HEAD OF THE EDE:
Lt Col Abner Maciel de Castro

The EDE—Divisão de Engenhos Espaciais (Space Vehicles Division) is the branch of IAE—Instituto de Atividades Espaciais (Space Activities Institute) responsible for research, development and prototype construction of rockets for scientific and military use. Several of the latter (of 37 mm, 70 mm and 127 mm calibres) are in production by private industry under Ministry of Aeronautics contracts. Its major scientific programmes concern the Sonda II and Sonda III sounding rockets.

IAE/EDE SONDA II

Sonda II is a single-stage fin-stabilised solid-propellant rocket, designed to carry a max payload of 62 kg (136 lb) to altitudes of up to 180 km (115 miles).

The rocket is produced in three versions, Types A, B and C, the major difference being in the length of the motor: Type A 2·40 m (7 ft 10½ in); Type B 3·68 m (12 ft 0¾ in); Type C 3·04 m (9 ft 11¾ in).

Up to January 1976, a total of 39 firings had been made for evaluation and development of the vehicle.

IAE/EDE SONDA III

This fin-stabilised two-stage solid-propellant rocket is designed to carry a 50 kg (110 lb) payload to a height of 500 km (310 miles). The second stage is a modified Sonda II motor. It is being flight tested for qualification of the vehicle.

DIMENSIONS:
- Length overall: 8·00 m (26 ft 3 in)
- Diameter of 1st stage: 0·56 m (1 ft 10 in)
- Diameter of 2nd stage: 0·30 m (11·8 in)

WEIGHT:
- Launching weight: 1,562 kg (3,443 lb)

INPE

CONSELHO NACIONAL DE PESQUISAS (CNPq) INSTITUTO DE PESQUISAS ESPACIAIS (INPE)

ADDRESS:
São José dos Campos, São Paulo State

GENERAL DIRECTOR:
Fernando de Mendonça

Brazil began its space research programme officially in 1961, by creating, within the National Research Council, a Group for the Organisation of a National Commission for Space Activities. Ten years later, in April 1971, this Comissão Nacional de Atividades Espaciais (CNAE) became a permanent organisation and was renamed the Instituto de Pesquisas Espaciais (INPE). Today INPE is the principal Brazilian agency for civilian space research.

From the start, INPE concentrated on research in two sectors: Pure Research and Appplied Research. The first sector is concerned with scientific programmes such as Project MATE, a study of the geomagnetic field conducted in collaboration with US and German organisations; Project MIRO, an analysis of the upper atmosphere, including the development of dye-lasers; Project EXAMETNET, a meteorological investigation using sounding rockets; and other projects concerned with ionospheric and atmospheric observations. Together, these programmes absorb about 20 to 30 per cent of the Institute's resources.

The second sector, demanding the larger share of the resources, is aimed at making early and important contributions to national development. Programmes include Project MESA, which consists of meteorological studies based on photographs taken by satellite and transmitted automatically to several receiving stations throughout the country, built to INPE specifications by the Brazilian electronics industry; and Project SERE, which utilises remote sensing for surveys of the natural resources existing in the vast territory of Brazil. For this work, INPE has its own Brazilian-developed Embraer Bandeirante aircraft, fitted with remote sensing equipment, and uses also photographs taken by NASA aircraft and the Landsat spacecraft. The most ambitious programme is Project SACI, (Sátelite Avançado de Communicaçoes Interdisciplinares), intended to evaluate the potential ability of a Brazilian geo-stationary satellite to enhance the nation's educational facilities and so help in achieving the highly-prized goal of universal education for some 24 million students in Brazil's primary schools.

It includes a comprehensive educational experiment in northeastern Brazil (Rio Grande do Norte State), involving use of instructional broadcasts via the NASA-Fairchild ATS-6 geostationary spacecraft to some 500 schools with an estimated 20,000 pupils; and feasibility study of a national educational satellite that would assist provision of first-class education throughout Brazil.

Further details were given in the 1975-76 *Jane's*.

CANADA

BAL

BRISTOL AEROSPACE LIMITED

HEAD OFFICE AND WORKS:
Winnipeg International Airport, PO Box 874, Winnipeg, Manitoba R3C 2S4
Telephone: (204) 775-8331
Telex: 07-57774

PRESIDENT AND GENERAL MANAGER: W. M. Auld

VICE-PRESIDENT MARKETING: R. H. May

VICE-PRESIDENT ROCKET AND SPACE DIVISION:
A. W. Fia

MARKETING MANAGER: R. J. Bevis

Bristol Aerospace Limited (BAL) has developed and is manufacturing a series of seven high-altitude sounding rockets known as Black Brants. These vehicles have been designed for upper atmosphere research and have been used successfully by Canadian, United States and European research institutes.

Black Brant rockets have solid-propellant motors, and are designed for simplicity of operation, enabling them to be launched from ranges offering only minimum support facilities.

BAL is able to offer a completely comprehensive service in the sounding rocket field, including propellants, vehicle hardware, recovery systems, electronic components and payload system design and buildup, scientific experiment integration and checkout facilities. Experienced launch crews are available to conduct or support launches at any rocket range.

In addition to the Black Brant vehicles, BAL has developed a new meteorological rocket.

BLACK BRANT IIIA

Black Brant IIIA is a single-stage fin-stabilised rocket, powered by a 9KS11000 solid-propellant motor. The payload compartment is a 5·3 : 1 cone with 25·4 cm (10 in) cylindrical section as standard. The cylindrical section can be lengthened to 127 cm (50 in) if required.

Black Brant IIIA is launched in an overslung position from a Nike launcher or underslung from a boom launcher. Low impact dispersion characteristics suit it for launch from ranges with minimum impact area.

DIMENSIONS:
- Length overall (standard configuration): 5·54 m (18 ft 2¼ in)
- Body diameter: 0·26 m (10¼ in)

WEIGHT:
- Launching weight (less payload): 281 kg (618 lb)

NOMINAL PERFORMANCE:
- Altitude with 40 kg (88 lb) gross payload* (85° launch angle): 100 nm (185 km; 115 miles)
- Max acceleration: 27*g*

**gross payload includes payload compartment and igniter housing*

BLACK BRANT IIIB

This vehicle offers a 40% increase in performance by comparison with Black Brant IIIA when carrying similar payloads. It is speciallly suited for use at remote sites, where ease of operation is a major consideration.

Vehicle hardware and dimensions are identical with those of Black Brant IIIA, with the exception of the 12KS10000 motor. The same alternative launchers can be used.

DIMENSIONS:
- Length overall, standard configuration: 5·54 m (18 ft 2¼ in)
- Body diameter: 0·26 m (10¼ in)

WEIGHT:
- Launching weight (less payload): 308 kg (678 lb)

NOMINAL PERFORMANCE:
- Altitude with 51 kg (112 lb) gross payload* (85° launch angle): 127 nm (235 km; 146 miles)
- Max acceleration: 27*g*

**gross payload includes payload compartment and igniter housing*

BLACK BRANT IVA

The first stage of this vehicle utilises the 15KS25000 engine used in the single-stage Black Brant VA; the second stage is generally similar to the production version of Black Brant IIIA, with the same payload capacity.

Black Brant IVA can be assembled in a few hours and may be held at instant launch readiness for many days. It is launched in an underslung position from a rail launcher.

DIMENSION:
- Length overall (standard configuration): 11·33 m (37 ft 2 in)

WEIGHT:
- Launching weight (less payload): 1,388 kg (3,059 lb)

NOMINAL PERFORMANCE:
- Altitude with 38 kg (84 lb) gross payload* (85° launch angle): 499 nm (925 km; 575 miles)
- Max acceleration: 38*g*

**gross payload includes payload compartment and igniter housing*

BLACK BRANT IVB

This is an uprated version of the Black Brant IVA. Vehicle hardware and dimensions are identical with those of the IVA, except that a 12KS10000 solid-propellant motor is used in the second stage. Three low aspect ratio fins may be added to the conical stabiliser to increase stability and so allow payloads up to 153 cm (60 in) long to be flown.

WEIGHT:
- Launching weight (less payload): 1,414 kg (3,116 lb)

NOMINAL PERFORMANCE:
- Altitude with 38 kg (84 lb) gross payload* (85° launch angle): 572 nm (1,060 km; 660 miles)
- Max acceleration: 38*g*

**gross payload includes payload compartment and igniter housing*

BLACK BRANT VA

Black Brant VA is a single-stage high-altitude research rocket, powered by a 15KS25000 solid-propellant motor. It is suitable for carrying payloads of up to 280 kg (615 lb) to the 67 nm (125 km; 77 miles) region.

Black Brant VA carries its payload in a 4·3 : 1 ogival fairing 1·87 m (6 ft 1 in) long, or in a 5·1 : 1 cone 2·18 m (7 ft 2 in) long. Cylindrical extensions up to 1·78 m (5 ft 10 in) long may be added. It is launched in an underslung position from a rail.

DIMENSIONS:
- Length overall (standard configuration): 8·13 m (26 ft 8 in)
- Body diameter: 0·44 m (17·2 in)

WEIGHT:
- Launching weight (less payload): 1,123 kg (2,473 lb)

NOMINAL PERFORMANCE:
- Altitude with 140 kg (308 lb) gross payload* (85° launch angle): 97 nm (180 km; 112 miles)
- Max acceleration: 16*g*

**gross payload includes payload compartment and igniter housing*

BLACK BRANT VB

The Black Brant VB single-stage research rocket has the same dimensions and payload space as the Black Brant VA but utilises a different solid-propellant motor, the 26KS20000.

DIMENSIONS:
- Length overall (standard configuration): 8·13 m (26 ft 8 in)
- Body diameter: 0·44 m (17·2 in)

WEIGHT:
- Launching weight (less payload): 1,294 kg (2,849 lb)

NOMINAL PERFORMANCE:
- Altitude with 140 kg (308 lb) gross payload* (85° launch angle): 202 nm (375 km; 233 miles)
- Max acceleration: 15*g*

**gross payload includes payload compartment and igniter housing*

BLACK BRANT VC

Black Brant VC employs the same motor as Black Brant VB, but is modified to make the vehicle compatible with launch towers at White Sands Missile Range, New Mexico, and Wallops Island, Virginia.

WEIGHT:
- Launching weight (less payload): 1,311 kg (2,888 lb)

NOMINAL PERFORMANCE:
- Altitude with 140 kg (308 lb) gross payload* (85° launch angle): 190 nm (325 km; 218 miles)
- Max acceleration: 15*g*

**gross payload includes payload compartment and igniter housing*

METEOROLOGICAL ROCKET

This low-cost rocket is designed to carry a 3·2 kg (7 lb) payload, including parachute, to a height of 72 km (45 miles). The rocket is designed for worldwide use and can operate in any climatic and weather conditions. It can be launched from existing permanent missile ranges or from semi-permanent remote sites. A complete system, including launcher and firing control equipment, can be supplied.

DIMENSIONS:
- Length overall (standard configuration): 2·82 m (9 ft 3 in)
- Body diameter: 0·12 m (4¾ in)

WEIGHT:
- Launching weight (incl payload and parachute): 52·4 kg (115 lb)
- Max acceleration: 47*g*

NOMINAL PERFORMANCE:
- Altitude (with payload) (80° launch angle): 40 nm (72 km; 45 miles)

SPAR
SPAR AEROSPACE PRODUCTS LIMITED

HEAD OFFICE AND WORKS:
825 Caledonia Road, Toronto, Ontario M6B 3X8
Telephone: (416) 781-1571
Telex: 02-2054 Sparcal Tor
CHAIRMAN OF THE BOARD: L. D. Clarke
PRESIDENT: R. D. Richmond
DIRECTORS:
C. H. Barrett
D. S. Beatty
R. B. Dodwell
W. H. Jackson
Dr P. A. Lapp
R. A. Perigoe
D. A. B. Steele
VICE-PRESIDENTS:
G. J. Aubrey (Finance)
G. B. Gnomes (Contracts)
J. E. Lockyer (Engineering)
J. D. MacNaughton (Remote Manipulator System)
R. E. Marcille (Marketing)
G. R. Rutledge (Manufacturing)
MANAGER, PUBLIC AND GOVERNMENT RELATIONS: John F. Walker

Spar Aerospace Products Limited was responsible for design and fabrication of the Alouette I and II and ISIS I and II satellite structures. All of these spacecraft used Spar extendible STEM devices as long sounder antennae. Other STEM space applications include gravity gradient booms; magnetometer, transponder and spectrometer booms; astronaut mechanical aids; and extendible solar array actuators. Since 1960, more than 500 STEM devices have been flown successfully in Canadian, US, Soviet and European space programmes.

Details of the Alouette and ISIS spacecraft can be found in earlier editions of *Jane's*. Spar has since manufactured the primary structures of three Anik communications satellites for Canadian domestic use, under contract from Hughes Aircraft Company, the prime contractor to Telesat Canada. The company also provided technical support to Hughes throughout the design and development phases of the programme, and was selected by Hughes to manufacture similar structures for a minimum of 15 satellites which Hughes expects to sell in world markets. (Anik is described under the Hughes entry in the US section).

Under contract from the Communications Research Centre, Spar was responsible for design and manufacture of the major subsystems of the Communications Technology Satellite (CTS).

In association with NASA, Spar has been conducting detailed studies on the special-purpose manipulator system (SPMS) for the Earth observatory satellite (EOS) series of spacecraft to be launched by the Space Shuttle.

In 1975 Spar, in association with other Canadian companies, was awarded contracts totalling $21·7 million by the Canadian Department of Supply and Services to build a Remote Manipulator Simulator facility and to design the Remote Manipulator System (RMS) to be installed on the Shuttle Orbiter for the purposes of deploying and retrieving payloads.

The RMS consists of a 15 m (50 ft) long mechanical arm with joints simulating a human shoulder, elbow and wrist. Canada is to supply at its own expense the first RMS flight unit which will be a simplified version of the production manipulators.

Space Shuttle Remote Manipulator System (RMS) under development by Spar

Communications Technology Satellite

COMMUNICATIONS TECHNOLOGY SATELLITE (CTS)

Launched on 17 January 1976 by a Delta 2914 vehicle from the Kennedy Space Center, Florida, into a synchronous orbit, the Communications Technology Satellite (CTS) is a joint Canadian/US space venture to demonstrate the operation of satellite communications at power levels significantly greater than those provided by earlier spacecraft. In orbit, CTS is stationed over the equator at 116° W longitude, just west of South America and in line with Calgary, Alberta.

The objective is to evolve a new technology that will make possible high-quality television reception and two-way voice communication with the use of small and simple user-operated ground terminals.

The CTS is the second such satellite. The first is the Applications Technology Satellite-6 (ATS-6) launched in May 1974. ATS-6 achieves its high power by use of a 9 m (30 ft) reflector antenna, CTS by means of a very high power (200 W) transmitter. ATS-6's effective radiated power is 53 dbW (decibel watt), CTS's 59 dbW. The CTS transmitting power level is 10 to 20 times higher than that of current commercial communications satellites.

The CTS programme is expected to provide communications relative to education, health care, news and cultural events to segments of the population that cannot be reached by either terestial or existing satellite communications systems.

The basic spacecraft structure consists of a central thrust tube, housing the apogee motor, forward and aft equipment platforms, and equipment panels on opposite sides of the body. Curved panels on the other two sides of the spacecraft carry the majority of the body-mounted solar cells that provided power until the main solar arrays were deployed. Since the forward platform carries the super-high frequency antennae and earth sensors, which must be very accurately aligned, it is isolated thermally from the rest of the spacecraft.

At launch CTS weighed 675 kg (1,485 lb) and had an overall height and outside diameter of 188 cm (74 in) and 183 cm (72 in) respectively. In orbit, the deployable solar array extended to span 16·07 m (52 ft 9 in). The array has a total of 25,272 solar cells and is designed to produce 1,257 W of useful power.

The spacecraft's high-efficiency 200W transmitter was built to a NASA Lewis Research Center specification for operation in the 12GHz frequency band, one of the new satellite frequency bands allocated by the International Telecommunications Union. The transmitter embodies a NASA-built travelling-wave-tube amplifier of novel design, having an efficiency of 50 per cent, at the power output of 200W.

In addition to the high-technology solar arrays and amplifier, the spacecraft embodies an advanced three-axis stabilisation system. This employs a fixed momentum wheel and hydrazine gas thrusters, to maintain antenna boresight pointing accuracy to within ±0·2° in pitch and roll, and ±1° in yaw. The purpose of the stabilisation system is to keep the satellite antennae pointing accurately toward the centre of selected target areas, while the solar arrays always face the sun. Conventional communications satellites are stabilised by spinning, which means that, with the solar power cells mounted on the outer circumference, about two-thirds of the cells are always in darkness; such power systems are thus only one-third efficient.

The United States and Canada are sharing equally in experiment time during the satellite's expected two-year life. The Office of Applications, NASA Headquarters, has responsibility for the United States portion of the CTS programme, and has assigned project management to the Lewis Research Center. The Canadian Department of Communications, which has overall responsibility for the Canadian portion of the CTS programme, has assigned programme responsibility to the Communications Research Centre.

CHINA
(PEOPLE'S REPUBLIC)

China launched her first satellite (Norad designation *Chicom 1*) on 24 April 1970, and thus became the fifth country to orbit a payload using national resources, following the USSR, USA, France and Japan. Few details were released; a New China News Agency statement gave the weight of the spacecraft as 172 kg (380 lb), and said that it was injected into a 2,382 × 439 km (1,480 × 273 mile) orbit at an inclination of 68·5°. It is believed to have been launched from the main Chinese rocket centre near

Shuang Cheng Tsu, 800 km (500 miles) east of the nuclear test establishment of Lop Nor.

No details of any experiments are available, but as a prototype it is probable that only basic vehicle development instrumentation was installed.

A second satellite (Norad *Chicom 2*) was launched on 3 March 1971 but was not announced by the Chinese authorities until 16 March. This spacecraft was described as being purely scientific, with a weight of 221 kg (487 lb). The data issued by Peking included an initial apogee of 1,825 km (1,134 miles), perigee of 266 km (165 miles) and orbital period of 106 minutes. Clear radio signals were transmitted on 20.008MHz.

A third satellite was launched on 26 July 1975. According to the New China News Agency this satellite was to support China's 'preparedness against war'. Launched into a relatively low 464 × 186 km (288 × 115 mile) orbit, it may thus have been a reconnaissance vehicle.

A fourth satellite, of undisclosed weight, was launched on 26 November 1975 into a similar orbit, and six days later returned to Earth. as programmed, according to the New China News Agency. It is believed that the reference to retrieval concerns a package ejected in orbit rather than to the complete spacecraft.

China's fifth satellite (Norad *Chicom 5*) was launched on 16 December 1975, and was placed in a 388 × 185 km (241 × 115 mile) orbit. This spacecraft was designated China 5 by the New China News Agency.

It is widely surmised that the development of the satellites and their launch vehicle was managed by Dr Tsien Hsue-Shen, a scientist who was employed by the Jet Propulsion Laboratory in California as a member of a rocket design team during the second World War. After the war the Doctor worked at the Massachusetts Institute of Technology. He returned to China in 1955.

FRANCE

AÉROSPATIALE

SOCIÉTÉ NATIONALE INDUSTRIELLE AÉROSPATIALE

Division des Systèmes Balistiques et Spatiaux

MANAGEMENT:
BP 96, 78130-Les Mureaux
Telephone: 474-72-13

CANNES:
BP 52, 06322 Cannes-la-Bocca
Telephone: 47-06-52

AQUITAINE:
BP 11, 33160-St Médard en Jalles
Telephone: 90-92-13

LES MUREAUX:
BP 2, 78130-Les Mureaux
Telephone: 474-72-11

DIRECTOR: P. M. Usunier
ASST DIRECTOR: Louis Marnay

The Space and Ballistic Systems Division of Aérospatiale handles research, development, testing and engineering for entire missile and space vehicle systems. It produces all kinds of major aerospace components, power stages, equipment modules, re-entry vehicles and fully-equipped satellite structures. The Division also handles integration and assembly, ground and in-flight testing, study and development of ground support facilities, and command and control subsystems.

Details of the Division's current launch vehicle, research rocket and spacecraft programmes are given below.

As a member of the CIFAS, CESAR and COSMOS international consortia, Aérospatiale is participating in other satellite programmes (see International section).

AÉROSPATIALE SOUNDING ROCKETS

Aérospatiale has developed and produces a range of solid-propellant sounding rockets which are fully proven in operation and are being improved continually in terms of performance. All four currently-available types of rocket are stabilised by fixed fins and need no guidance or control systems, which keeps their weight to a minimum. More than 350 rockets had been built by early 1976, and launches have taken place from 18 bases throughout the world.

The main characteristics and performance of these rockets are shown in the accompanying table.

DIAMANT B P-4

Following the final launch of the original Diamant B satellite launch vehicle, the programme is continuing with an improved version of the vehicle, using a new second stage. Aérospatiale again provides all three stages, under CNES sponsorship. Details are as follows:

L-17

This first stage has a metal structure housing 12,129 kg (26,740 lb) of nitrogen tetroxide and 5,887 kg (12,978 lb) of UDMH liquid propellants. Its rocket engine develops 349·12 kN (78,500 lb) of thrust, with a specific impulse of 221·1 seconds.

P-4

New second stage, developed originally for the M-1 version of the MSBS submarine-launched ballistic missile. Rita I motor contains 4,000 kg (8,820 lb) of solid propellant, in a wound glassfibre casing, and gives approximately 178 kN (40,000 lb) of thrust.

P-06

Third-stage solid propellant motor of 45·36 kN (10,200 lb) thrust, in a wound glassfibre structure.

Diamant B P-4 can put an increased payload of approximately 150 kg (330 lb) into a 500 km (310 mile) circular orbit.

Artist's impression of Ariane three-stage launch vehicle

Diamant B P-4 successfully launched the Starlette satellite on 6 February 1975, D 5-A (Castor) and D 5-B (Pollux) on 17 May 1975, and the D 2-B (Aura) satellite on 27 September 1975.

DIMENSIONS:
Length:
L-17 — 14·08 m (46 ft 2½ in)
P-4 — 2·60 m (8 ft 6 in)
P-06 — 2·07 m (6 ft 9½ in)
Complete vehicle — 18·94 m (62 ft 2 in)
Body diameter:
L-17 — 1·40 m (4 ft 7 in)
P-4 — 1·50 m (4 ft 11 in)
P-06 — 0·80 m (2 ft 7 in)

WEIGHT:
Complete vehicle at lift-off — 27,492 kg (60,610 lb)

ARIANE

This heavy three-stage launch vehicle is being developed in Europe as a co-operative project, with the French space agency CNES as prime contractor. Payload potential will be 1,500 kg (3,300 lb) into a transfer orbit, or applications satellites of up to 750 kg (1,650 lb) weight into synchronous orbit.

Diamant B P-4 satellite launch vehicle

The first test launch is scheduled for 1979, and the vehicle is expected to be operational in late 1980.

Aérospatiale is responsible for the design of Ariane and for integration of the entire vehicle. It is producing major structural elements of the first stage and is responsible for the final delivery of all three stages and the fairings.

SEP is developing the engines and associated subsystems for all three stages. Matra is building the vehicle equipment bay and checkout facilities. Air Liquide is

Name of rocket	Type and weight of propellant	Diameter	Max overall length•	Payload	Corresponding altitude
Bélier III	Isolane: 230 kg (506 lb)	30·5 cm (12 in)	5·112 m (16 ft 9 in)	30-120 kg (66-264 lb)	135-60 km (84-37 miles)
Dauphin	Plastolane: 687 kg (1,515 lb)	56 cm (22 in)	6·23 m (20 ft 5 in)	100-230 kg (220-507 lb)	158-100 km (98-62 miles)
Dragon III	Plastolane: 687 kg (1,515 lb) Isolane: 230 kg (506 lb)	56 cm (22 in) 30·5 cm (12 in)	8·5 m (27 ft 10½ in)	30-120 kg (66-264 lb)	700-390 km (435-242 miles)
Eridan	Plastolane: 687 kg (1,515 lb) Plastolane: 687 kg (1,515 lb)	56 cm (22 in) 56 cm (22 in)	10·2 m (33 ft 5 in)	100-420 kg (220-926 lb)	460-200 km (286-124 miles)

developing the third-stage structures, and tanks for the liquid hydrogen and liquid oxygen propellants. All four French companies are subcontracting work extensively throughout Europe to manufacturers in the countries supporting the programme.

Details of the individual stages are as follows:

L 140

This first stage is made up of two identical steel tanks for the 141,000 kg (310,850 lb) of UDMH and N_2O_4 propellants, linked together by means of a cylindrical skirt. It is powered by four Viking engines (each 588·4 kN; 132,275 lb st), carried on a cylindrical thrust frame and protected by fairings with fins. Burn time is 139 seconds.

A truncated cone-shaped interstage skirt joins the first and second stages.

L 33

The second stage comprises two light alloy tanks for 33,000 kg (72,750 lb) of UDMH and N_2O_4 propellants, separated by a common bulkhead. The single Viking engine 686·5 kN; 154,325 lb st in vacuum) is linked to the stage by a conical thrust frame. Burn time is 131 seconds.

A cylindrical interstage skirt connects the second and third stages.

H 8

The third stage is made of light alloy and houses 8,800 kg (19,400 lb) of liquid hydrogen and liquid oxygen propellants for its single HM7 cryogenic engine (64·73 kN; 14,550 lb st). Burn time is 563 seconds.

A cylindrical vehicle equipment bay is situated above the third stage. The equipment platform is an annular plate, the inside flange of which supports the payload attachment fittings. Aluminium payload fairings, made up of two half-shells, are attached to the outside flange of the equipment platform.

DIMENSIONS:

Length:	
L 140	18·40 m (60 ft 4½ in)
Interstage structure	4·80 m (15 ft 9 in)
L 33	11·60 m (38 ft 0½ in)
Interstage structure	2·65 m (8 ft 8½ in)
H 8	8·50 m (27 ft 10½ in)
Equipment bay	0·30 m (1 ft 0 in)
Payload fairings	8·60 m (28 ft 2½ in)
Overall length	48·0 m (157 ft 6 in)
Body diameter:	
L 140	3·80 m (12 ft 5½ in)
L 33, H 8, equipment bay	2·60 m (8 ft 6½ in)
Payload fairings	3·20 m (10 ft 6 in)

WEIGHTS:

L 140	154,000 kg (339,510 lb)
L 33	36,300 kg (80,025 lb)
H 8	9,400 kg (20,725 lb)
Payload fairings	810 kg (1,785 lb)
Total weight at lift-off	more than 200,000 kg (440,920 lb)

D2-B (AURA)

As in the case of the D2-A Tournesol scientific satellite, launched on 15 April 1971, Aérospatiale was responsible for the satellite structure, including the solar panels and their deployment mechanism, the antennae, the equipment module with the de-spin and separation devices, the cold-gas stabilisation system, and the gilding of the cover of the generally similar D2-B. Matra was prime contractor.

The D2-B is basically cylindrical, with four solar panels which deployed after orbit had been achieved. Weighing 106 kg (234 lb) the D2-B was launched on 27 September 1975. In orbit the spacecraft is also known as Aura.

D5

The first flight model of this composite technological satellite was lost as a result of the failure of its launcher. Details of the D5-A and D5-B components can be found in the 1972-73 *Jane's*.

A second pair, D5-A (Castor) and D5-B (Pollux), was launched successfully on 17 May 1975. Purpose of the mission was to test a hydrazine motor developed by SEP and a micro-accelerometer developed by ONERA. Aérospatiale was responsible for the satellite structures, the equipment module, the antennae and the separation mechanisms.

Display model of SLT communications satellite

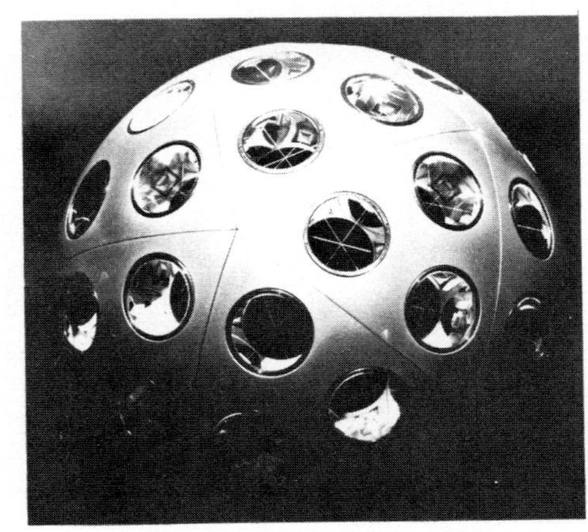
Starlette geodesic satellite

D5 satellite undergoing vibration tests

STARLETTE

This passive geodesic satellite was launched on 6 February 1975, by a Diamant B P-4 vehicle, from the CNES facility at Kourou, French Guiana. Weighing 47 kg (104 lb), the satellite was intended for analysis of the Earth's gravity and of the Earth as a deformable solid body. Of spheroid shape, it is 24 cm (9·45 in) in diameter. Its core is made of Uranium 238, and its surface is covered with 60 laser reflectors made by Aérospatiale. The satellite is being observed by the Smithsonian Astrophysical Observatory (SAO) and French laser stations.

SLT

SLT is a project for a multi-purpose modular spacecraft, intended for geostationary telecommunications. The 1,500 kg (3,300 lb) spacecraft could be launched by either an Ariane or an Atlas Centaur vehicle. It has a rectangular box-shaped body with three main elements.

A service module, containing support equipment and the apogee motor.

A central platform, housing the orbital and attitude control systems.

A payload module, including the transporters and antennae.

European partners for the project include MBB (West Germany) and ETCA (Belgium).

MATRA
SA ENGINS MATRA

HEAD OFFICE:
4 rue de Presbourg, 75116-Paris

MANAGEMENT AND WORKS:
37 avenue Louis Breguet, 78140-Vélizy

Telephone: 946.96.00

Telex: ENMATRA 69.077F

OFFICERS: See Missiles section

For the CNES, Matra was prime contractor for the D2-B (Aura) satellite, and is developing the telescope and radiometer for the Meteosat satellite. The Matra radiometer will be the biggest such instrument developed in Europe for carriage by a satellite. Its purpose is to determine wind speeds and sea surface temperature. Two flight models have already been delivered; one is undergoing qualification tests in vibration and acoustic noise at IABG Munich, and the other thermal vacuum tests at SOPEMA Toulouse.

Matra is also working on the Command and Data Management System for Spacelab, as contractor for ESA, and on the Ariane European launcher, for which it is supplying a digital computer. It is developing the attitude and orbital control system for the Orbital Test Satellite (OTS), for which Hawker Siddeley Dynamics is prime MESH contractor under an ESA programme.

D2-B (Aura)

Under contract to the CNES, Matra was prime contractor for the D2-B (Aura) scientific satellite, which was launched by a Diamant B P-4 vehicle, from Kourou, on 27 September 1975.

This satellite is described briefly under the Aérospatiale entry. As prime contractor, Matra was responsible for design and development of the scientific experiments, and for qualification and acceptance testing of the satellite equipment and subsystems.

SIGNE 3

Under contract to CNES Matra is prime contractor for the SIGNE 3 astronomical satellite which is to be launched by a Soviet rocket in 1977. This satellite has a cylindrical body with four deployable solar panels, and is of similar configuration to the D2-B satellite.

SIGNE 3 astronomical satellite

ONERA
OFFICE NATIONAL D'ÉTUDES ET DE RECHERCHES AÉROSPATIALES

ADDRESS:
29 avenue de la Division-Leclerc, 92320-Châtillon
Telephone: 657.11.60
DIRECTOR: Pierre-Louis Contensou
HEAD, INFORMATION AND PUBLIC RELATIONS DEPT:
Serge Baume

In addition to its work as a national research and experimental centre in the service of the entire French aircraft and missile industry, ONERA has designed, built and launched various types of experimental vehicle to obtain basic data in fields such as propulsion and kinetic heating and in connection with scientific research programmes sponsored by the Centre National d'Etudes Spatiales (CNES).

ONERA designed and built the CACTUS (Capteur Accelerometrique Capacitif Triaxial Ultra-Sensible) experiment carried on board the D5-A Castor satellite launched on 17 May 1975. This is an ultra-sensitive triaxial capacitive accelerometric detector, designed to measure accelerations between one thousand millionth and one hundred thousandth (10^{-9} to 10^{-5}) of the gravity acceleration at the surface of the Earth.

The instrument is basically a gold-plated rhodanised platinum sphere, 4 cm (1·57 in) in diameter, housed in a spherical cage of slightly larger diameter, in which its position at the centre is maintained by electrostatic forces. The unit weighs about 10 kg (22 lb) and the power requirement of 1·6W, is met by the satellite's solar cells. Although designed for a life of six months, the instrument was still providing data after 10 months in orbit.

Details of rockets developed earlier by ONERA can be found in the 1969-70 and 1975-76 *Jane's*.

THOMSON-BRANDT
COMPAGNIE FRANÇAISE THOMSON HOUSTON—HOTCHKISS-BRANDT
Armament and General Engineering Branch

ADDRESS:
52 avenue des Champs-Elysées, 75008-Paris
Telephone: 359.18.87
Telex: 29966 Brantarm-Paris
WORKS:
Saint Denis, La Ferte St Aubin (Loiret), Tulle (Corrèze)

Since 1964, the Armament Division of Thomson-Brandt has been prime contractor to the SECT (Firing-Range Equipment Department) for the design and production of sounding rockets, utilising rocket motors of French design in mass production for military purposes. Some of these were described briefly in the 1970-71 and 1975-76 editions of *Jane's*.

GERMANY
(FEDERAL REPUBLIC)

DORNIER
DORNIER SYSTEM GmbH

HEAD OFFICE AND WORKS:
7759 Immenstaad/Bodensee (near Friedrichshafen)

POSTAL ADDRESS:
Postfach 6136048, 7990 Friedrichshafen

Telephone: Immenstaad (07545) 81
Telex: 0734359

BONN OFFICE:
Allianzplatz, 5300 Bonn

GENERAL MANAGERS:
Dipl-Ing Dr Jr Karl-Wilhelm Schäfer
Dr Ing Bernhard Schmidt
Dr Ing Helmut Ulke
Dipl-Kfm Klaus-Peter Thomé
PUBLIC RELATIONS:
Gerhard Patt

Dornier System, a member company of the Dornier group, employs about 1,200 engineers and scientists in activities involving space flight, new technologies, electronics and research. It is a major European contractor for the design, development and integration of equipment on scientific satellites.

Currently, Dornier System is leading the STAR consortium for development of the international Sun/Earth explorer ISEE-B within the co-operative ESA/NASA programme. This satellite is due to be launched at the end of 1977. Dornier System is also developing the environmental control and life support sub-system (ECLSS) and the instrument pointing system (IPS) for the Spacelab to be carried into orbit by the Space Shuttle.

For Ariane, the European launch vehicle, Dornier is engaged on design and production of the second-stage tank structure.

ERNO
ERNO Raumfahrttechnik GmbH

ADDRESS:
Hünefeldster. 1-5, P.O.B. 10 59 09, 2800 Bremen 1
Telephone: (0421) 5391
Telex: 024 5548
DIRECTORS:
Dipl-Ing Hans E. W. Hoffmann
R. A. Kosegarten

ERNO is an affiliated company of the Zentralgesellschaft VFW-FOKKER GmbH, Düsseldorf. Its current space activities include work on the Spacelab to be carried into orbit by the Space Shuttle, the second stage of Ariane, the European launch vehicle, satellites and space probes. In particular, ERNO has participated in the programmes for Azur, Intelsat III, TD-1A, Aeros and the Helios solar probes, and is currently involved with the Orbital Test Satellite (OTS) and the Maritime OTS (MAROTS). In addition to these projects, ERNO is engaged in propulsion technology, hydrazine technology, electronics, nuclear technology, oceanology, protection of the environment and planning technology.

SPACELAB

On 5 June 1974, a ten-nation consortium, led by VFW-Fokker/ERNO, was awarded by ESA a contract valued at about £98 million (1974 basis) to design and develop the Spacelab space laboratory which represents Europe's contribution to the US Post Apollo Programme of re-usable Space Shuttles.

Largest European space project initiated in the present decade, the work will extend over a period of six years, leading to delivery of one hard mock-up, two engineering models, one flight unit, three sets of ground support equipment and initial spares. The Spacelab flight unit, fully qualified and ready for the installation of experiments, will be delivered to NASA in early 1979.

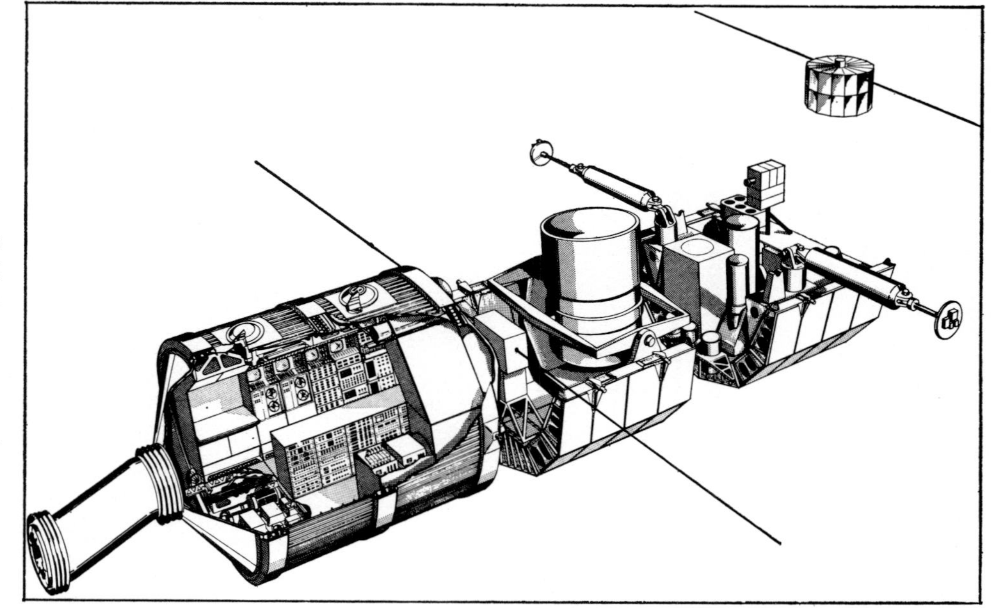

Diagram of Spacelab, including two experiment pallets. The entire spacecraft will remain connected to the Space Shuttle Orbiter vehicle during operation

Spacelab will be the only manned payload for the Space Shuttle. A typical payload will comprise a large module and two pallets with an overall length of 13·80 m (45 ft 3 in) and a shell diameter of 4·06 m (13 ft 4 in).

The first launch of the Space Shuttle with a Spacelab is scheduled for 1980. Four payload specialists will work in two shifts at an orbital height of 200-500 km (124-310 miles) for periods of 7 to 30 days. Spacelab will remain attached to the cargo bay of the Orbiter during the mission and will be brought back to Earth when the Orbiter lands. The payload specialists will be able to work under normal atmospheric conditions in the pressurised module. For work on the pallet section pressure suits will be necessary.

MBB
MESSERSCHMITT-BÖLKOW-BLOHM GmbH
Space Division

HEAD OFFICE AND WORKS:
8012 Ottobrunn bei München
Telephone: (089) 60 00 25 90

In addition to its manufacture of piloted aircraft, Messerschmitt-Bölkow-Blohm is engaged in the development of guided weapons (see Missiles section), research rockets and satellites.

MBB was prime contractor for the Helios solar probe, of which details follow, and is a member of several of the international teams working on current satellite programmes. They include the CESAR consortium which is developing the COS B scientific satellite; the CIFAS consortium responsible for the Franco-German Symphonie experimental communications satellite; and the COSMOS consortium entrusted with development of ESA's Meteosat (see International section).

HELIOS

Helios is the largest German space probe project to date and, at the time of its inception, was the largest bilateral space project in which NASA had participated. Contracts for the German share of the programme totalled approximately 180 million DM; a further 60 million DM were provided for experiments. The contracting agency was the German Federal Ministry of Research and Technology, through its department responsible for space research, the Gesellschaft für Weltraumforschung mbH. NASA supplied the launch vehicles, and the US contribution to the project cost approximately 180 million DM.

As prime contractor, the Space Division of MBB was made responsible for about 45% of the work performed under the German part of the programme. In particular, it was responsible for project management, system management, integration and test, as well as development and manufacture of the attitude measurement, attitude control, electrical distribution, and antenna subsystems. Subcontractors for other major subsystems were ERNO, Bremen (structure and thermal control); AEG-Telefunken, Ulm (telecommunications receiver); SEL, Stuttgart (data handling); Thomson-Houston CSF, France (transmitter); AEG-Telefunken, Hamburg (solar generator); and ETCA, Belgium (power conditioning).

Helios is an automatically-functioning solar probe. The basic structure is spool-shaped, with a cylindrical central experiment compartment 1·75 m (5 ft 8 in) in diameter by 0·55 m (1 ft 9½ in) long. A conical solar array is attached to each end, completing the spool shape.

Helios is intended to fly closer to the Sun than any previous spacecraft. At its nearest point, it passes within 45 million km (28 million miles) of the Sun's surface, and is subjected to temperatures of up to 370°C, which is hot enough to melt lead; so the dissipation of heat is a major problem. Heat from the central compartment is radiated to space, mainly in an axial direction, from radiating areas on the top and bottom, via louvre systems. In addition, optical surface reflectors cover half of the outer surface, and it was anticipated that their mirror finish would reflect up to 90% of all incident solar energy. Internally, several layers of insulation are fitted between the reflectors and the payload. In addition, the spacecraft spins once every second to distribute evenly the heat coming from the Sun. It was anticipated that these measures would maintain the spacecraft's internal temperature below 30°C.

Above the central body, within and protruding above the upper solar array, is the telecommunications antenna system. This includes a narrow-beam high-gain antenna (23 dB) with a mechanically de-spun reflector. Above the high-gain antenna is a medium-gain antenna (torroidal pattern, 7 dB); on top of the antenna mast is a third antenna system, with a quasisotropic pattern.

Two deployable double-hinged booms are fitted, to carry magnetometer experiments. Two other deployable booms, diametrically opposed, are used as antennae for a radio-wave experiment.

Ten experiments are carried, seven designed by institutes and scientists in Germany, and three in the USA; all are designed to obtain new information on interplanetary space in the region close to the Sun. Each experiment represents a complete research project, including measurement of the solar wind, magnetic fields, solar and galactic cosmic rays, electromagnetic waves, micro-meteoroids and the zodiacal light.

Helios 1 was launched by NASA, using a Titan-Centaur, from the Kennedy Space Center, Florida, on 10 December 1974. Controlled from the German space operations centre near Munich, data transmitted by the spacecraft are correlated with that obtained from the IMP Explorers 47 and 50 in Earth orbit, the Pioneer spacecraft orbiting the Sun at about 150 million km (93 million miles), and from Pioneers 10 and 11 in the outer solar system. By March 1976, Helios 1 had passed three perihelia (closest point to the Sun) and was still functioning well.

Helios 2, identical to Helios 1, was launched on 15 January 1976.

DIMENSIONS:

Height overall	4·20 m (13 ft 8 in)
Diameter, central compartment	1·75 m (5 ft 8 in)
Diameter, over solar arrays	2·77 m (9 ft 1 in)
Width, with boom extended	32·0 m (105 ft 0 in)

Helios solar probe

WEIGHTS:

Total	370 kg (815 lb)
Experiments	72 kg (158 lb)

INDIA

ISRO
INDIAN SPACE RESEARCH ORGANISATION

ADDRESS:
c/o Department of Space, "F" Block, Cauvery Bhavan, District Office Road, Bangalore 560 009
Telephone: 29822 and 28215
Telex: BG-499 and BG-326

ISRO functions under the Department of Space (DOS) of the Government of India. It is responsible for the planning, execution and management of space research activities and space applications programmes that have been undertaken by the Department of Space. It provides rockets and laboratory facilities to the scientists belonging to different organisations in India, for the purpose of conducting approved space science experiments. It operates and maintains the UN-sponsored Thumba Equatorial Rocket Launching Station (TERLS) and encourages international collaboration in space research experiments using rockets. ISRO is also the sponsoring agency for Indian scientists participating in research programmes supported by foreign space agencies. These include laboratory experiments and ground-based observations in support of space programmes, rocket experiments and satellite experiments.

The programmes and activities of ISRO can be divided into three broad areas:

Space Sciences. Programmes in aeronomy, X-ray and gamma ray astronomy, cosmic rays, studies of lunar material, meteorites, etc.

Space Technology. All aspects of work related to the development of rockets, satellites, propellants, scientific payloads and ground and space instrumentation.

Space Applications. Translation of the knowledge gained in space research and technology into practical applications in the fields of satellite communications, satellite television, meteorology, geodesy and Earth resources survey.

The activities of ISRO, with its Headquarters at Bangalore, are at present carried out at the following three centres:

Vikram Sarabhai Space Centre (VSSC)
ISRO PO, Trivandrun 695 022
Telephone: 4671; 3451 and 3561
Telex: Space TV 048-201
DIRECTOR: Dr Brahm Prakash

VSSC serves as the main research and development centre for space technology. It is engaged in the development of satellites, sounding rockets, satellite launch vehicles and related systems.

Space Applications Centre
Post Box No. 11, Jodhpur Tekra, Ahmedabad 380053
Telephone: 837714
Telex: ANTARIX-012-239
DIRECTOR: Prof Yash Pal

This Centre is concerned with space applications programmes of ISRO in the areas of satellite communications, TV broadcasting via satellites, survey of Earth resources and meteorological parameters from space/aerial platforms using remote sensing techniques.

SHAR Centre
Sriharikota PO 524124, via Sulurupeta, Nellore District (Andhra Pradesh)
Telephone: Sulurupeta 225
Telex: 041-7353 SHAR
PROJECT ENGINEER: Dr Y. J. Rao

This centre contains facilities for launching and flight testing large multi-stage sounding rockets and satellite launch vehicles.

The current ISRO programmes are:

ARYABHATA SATELLITE

India's first satellite, Aryabhata, named after an Indian astronomer and mathematician of the fifth century, was launched from within the Soviet Union by a Soviet rocket on 19 April 1975, under an agreement signed between the two countries in 1972.

Fabrication of the satellite, which involved both government organisations and private industry, took place at Peenya, near Bangalore. More than 90 per cent of the structure was manufactured in India, but many of the electronic components were imported. Assembly and integration of these units into the various sub-systems were accomplished by Indian scientists, working under the guidance of Soviet technicians.

ISSP Indian Aryabhata scientific satellite

Model of SLV-3 launch vehicle *(Théo Pirard)*

Weighing 360 kg (794 lb) the satellite was placed successfully into a 600 km (370 mile) near-circular orbit. The spin-stabilised, truncated-octagonal spacecraft was 116 cm (45·6 in) high × 147 cm (57·9 in) in diameter. Electrical power was provided by a total of 18,500 solar cells mounted on its 26 flat faces. Design lifetime in orbit was 6 months.

On board were three scientific experiments: one for investigations in X-ray astronomy; one for measuring solar neutron and gamma emissions; and one for measuring ionospheric parameters. Owing to a failure in the power supply system, these had to be switched off during the fifth day in orbit. Useful data was, however, obtained during the five days of operation, and this is being published.

Apart from the expriments, the other subsystems on board the satellite functioned well. Signals from the satellite were received at the ground station at the Shar Centre, which also transmitted commands to the satellite. Data was received at a ground station set up at Bear's Lake, near Moscow, for a period of six months.

SATELLITE INSTRUCTIONAL TELEVISION EXPERIMENT (SITE)

This experiment is aimed at bringing televised educational programmes to the villages and remote areas of India through the medium of a satellite. NASA's ATS-6 Applications Technology Satellite was used for this purpose, the SITE programme being inaugurated on 1 August 1975 by the Prime Minister Smt. Indira Gandhi. The instructional TV programmes were produced mainly by All India Radio, and a few by ISRO. Development of the necessary hardware for the ground segment, as well as work relating to the software aspect of the instructional programmes, were the responsibility of ISRO. The TV programmes were beamed to the satellite from the Experimental Satellite Communications Earth Station (ESCES) of SAC at Ahmedabad, which served as the primary ground station for the experiment. A second Earth Station, built by ISRO, operated from Delhi.

SATELLITE LAUNCH VEHICLE (SLV-3)

This VSSC project has the objective of developing an

indigenous satellite launch vehicle capable of putting a 40 kg (88 lb) satellite into a 400 km (250 mile) near-circular Earth orbit. The four-stage solid-propellant vehicle, designated SLV-3, will have a maximum length of 23 m (75 ft 6 in) and will weigh 17,000 kg (37,480 lb) at lift-off. Thrust of the first stage will be 43,000 kg (95,000 lb). Inertial control and guidance will be employed.

The first launch is planned to take place in mid-1978.

SEO SATTELLITE

This Satellite for Earth Observations (SEO) is currently under development by ISRO, and is scheduled to be launched from the Soviet Union in 1978.

INTERNATIONAL PROGRAMMES

CESAR

BRITISH AIRCRAFT CORPORATION LTD (BAC)
100 Pall Mall, London SW1Y 5HR, England

ÉTUDES TECHNIQUES ET CONSTRUCTIONS AÉROSPATIALES SA (ETCA)
BP 97, 6000 Charleroi, Belgium

MESSERSCHMITT-BÖLKOW-BLOHM GmbH (MBB)
Space Division: 8012 Ottobrunn bei München, Germany

SELENIA-INDUSTRIE ELETTRONICHE ASSOCIATE SpA
CP 7083, 00100 Rome, Italy

SOCIÉTÉ NATIONALE INDUSTRIELLE AÉROSPATIALE
37 boulevard de Montmorency, 75781-Paris-cédex 16, France

Under the auspices of ESA, this international consortium developed and built a scientific satellite known as COS B, of which brief details follow:

COS B

The COS B satellite was developed and produced by the CESAR consortium for ESA, with MBB as prime contractor. Its primary purpose was to measure gamma radiation from the galaxy, with particular regard to energy spectrum, the direction and intensity of incoming radiation, and changes as a function of time. Of special interest is the behaviour of pulsars, discovered as recently as 1967 and believed to be composed of highly-compressed nuclear matter.

Aérospatiale's Cannes facility was responsible for supplying the satellite structure and thermal control system. The scientific instrument payload was developed as a European co-operative venture. It weighs approximately 120 kg (265 lb) and can be considered as an astronomical observatory designed specifically to measure extra-terrestrial gamma radiation.

BAC was involved in major subsystem technology developments, including a unique attitude measurement sensor and electronics system, electronics and pneumatics for the spark chamber gas flushing subsystem, electronics and pneumatics for the attitude control subsystems, and the satellite's solar arrays.

The COS B contract covered the design and manufacture of two mockup satellites, a prototype, and two flight models which were to be delivered in 1973-74.

The COS B satellite is cylindrical, with two sensor units and the spin-rate adjustment thrusters mounted approximately midway along the curved cylindrical surface. Four monopole antennae protrude below the bottom of the spacecraft. A total of 9,080 solar cells, divided among 12 panels, are mounted over the outer curved surface.

The 275 kg (606 lb) spacecraft was launched on 9 August 1975 by NASA, for ESA, on a Delta 2913 vehicle from the Western Test Range, California, and was placed successfully in a highly elliptical orbit of 99,800 × 340 km (62,000 × 212 miles), with an inclination of 90° to the equator. It was expected to transmit scientific data to Earth for a period of two years.

Artist's impression of COS B gamma ray research satellite

DIMENSIONS:

Height without antenna	1·20 m (3 ft 11 in)
Height overall	1·71 m (5 ft 7 in)
Body diameter	1·40 m (4 ft 7 in)

CIFAS

CONSORTIUM INDUSTRIEL FRANCO-ALLEMAND POUR LE SATELLITE SYMPHONIE (DEUTSCH-FRANZÖSICHES INDUSTRIE-KONSORTIUM FÜR DEN SATELLITEN SYMPHONIE)

ADDRESS FOR CORRESPONDENCE:
BP 62, 78130-Les Mureaux, France
Telephone: 474.72.13
Telex: AIRSPA 60159F

ADMINSTRATOR:
P. M. Usunier

This consortium was formed on 25 April 1968 in response to a request from the French and German governments for proposals to develop a telecommunications satellite named Symphonie. Members of the consortium, under the technical, industrial and commercial direction of Société Nationale Industrielle Aérospatiale (Division des Systèmes Balistiques et Spatiaux), are Messerschmitt-Bölkow-Blohm, Thomson-CSF, SAT, Siemens-AG and AEG-Telefunken.

SYMPHONIE

The first operational Symphonie telephone/television satellite was launched by a Delta vehicle on 19 December 1974, and reached its geostationary position over the South Atlantic, at longitude 11·5°W, on 8 January 1975, giving coverage of two zones through two elliptical beams, each of 13° × 8°. The first beam covers Europe, North and Central Africa; the second beam covers South America and the eastern parts of North and Central America.

The second Symphonie was launched on 26 August 1975, and is also positioned at longitude 11·5°W. It was used operationally for the first time, by the French television service, for the transmission of film of President Giscard d'Estaing's trip to the Soviet Union in October 1975.

First flight model of Symphonie communications satellite

The main body of the Symphonie satellite is a shallow six-sided box, with three deployable panels of solar cells, giving a minimum total power of 170W. Three parabolic and horn telecommunications antennae are mounted above the body. Two transponders have a transmission capacity of 600 two-way telephone channels and provide coverage of the two Earth receiving zones in various combinations. A three-axis stabilisation system and a built-in MBB liquid rocket apogee motor (40 kg; 88 lb st for 20 minutes) are fitted. The spacecraft weighed 402 kg (886 lb) at launch, and weighs 240 kg (530 lb) in orbit.

Symphonie is intended for 24-hour use over a five-year period. The basic programme calls for three flight units, plus the necessary spares. Additional satellites can be built to meet the requirements of potential users.

COSMOS

ÉTUDES TECHNIQUES ET CONSTRUCTIONS AÉROSPATIALES SA (ETCA)
BP 97, 6000-Charleroi 1, Belgium

SOCIÉTÉ NATIONALE INDUSTRIELLE AÉROSPATIALE
BP 96, 78130-Les Mureaux, France

SOCIÉTÉ ANONYME DE TÉLÉCOMMUNICATIONS (SAT)
41 rue Cantagrel, 75624-Paris-cédex 13, France

MESSERSCHMITT-BÖLKOW-BLOHM GmbH (MBB)
8 München 80, Postfach 801 169, Germany

SIEMENS AG
8 München 70, Hofmannstrasse 51, Germany

MARCONI SPACE & DEFENCE SYSTEMS LTD
The Grove, Stanmore, England

SELENIA SpA
CP 7083, 00100 Rome, Italy

Announcement of the formation of this consortium by the seven major European companies listed above was made on 3 November 1970.

In addition to many study contracts carried out for ESA and their respective governments, COSMOS companies are now involved in several major space programmes, notably Meteosat, of which details follow:

METEOSAT

Aérospatiale of France, acting as prime contractor for the COSMOS consortium, was awarded a $56·76 million contract by ESA for development of this geosynchronous meteorological satellite in December 1973. The contract covers work over a period of four years, culminating in launch of the first Meteosat flight model by a Delta 2914 vehicle in mid-1977.

Aeronutronic Ford assists Aérospatiale's Cannes establishment in co-ordinating and integrating the project. Of the other COSMOS members, MBB is responsible for the spacecraft structure; Marconi Space & Defence Systems will supply the attitude and orbit control system and checkout equipment, and is involved with Siemens on the satellite's main communications package; ETCA is responsible for electric power supply; Selenia for data transmission and service telecommunications; and SAT for the telemetry encoder and solar cells. Associated sub-contractors include Terma of Denmark, Crouzet of France, SRA of Sweden and CIR of Switzerland.

Meteosat represents Europe's contribution to a worldwide Global Atmospheric Research Programme (GARP) which will utilise also US, Japanese and Soviet satellites to provide up-to-the-minute data for the World Weather Watch Organisation. Its three primary functions will be to take photographs of the Earth and its cloud cover by day and, in infra-red, at night; collection of meteorological data obtained by numerous ground and sea stations, and by satellites in low polar orbits; transmission of unprocessed pictures and meteorological data to Earth, and relay of processed photographs to users.

A key sensor will be the radiometer, used to determine wind speeds and sea surface temperatures. Development of this device was begun under a special contract by Matra, assisted by Marconi, and is being continued by Aérospatiale.

The general appearance of Meteosat is shown in the accompanying photograph. It will be spin stabilised in orbit, when the solar cells covering its sides will give a minimum of 200W. The telescope housed within the cylindrical body will be provided with an aperture in the side of the spacecraft.

DIMENSIONS:
Height overall 1·45 m (4 ft 9 in)
Diameter 2·10 m (6 ft 10½ in)

WEIGHT:
Initial weight in orbit 300 kg (661 lb)

Meteosat geostationary meteorological satellite

MESH

SA ENGINS MATRA
BP No 1, 78140-Vélizy, France

ERNO RAUMFAHRTTECHNIK GmbH
28 Bremen 1, Hünefeldstrasse 15, Postfach 1199, Germany

SAAB-SCANIA AKTIEBOLAG
S-581 88 Linköping, Sweden

HAWKER SIDDELEY DYNAMICS LTD
Gunnels Wood Road, Stevenage SG1 2AS, Herts, England

AERITALIA SpA
Piazzale V. Tecchio 51/A, 80125 Napoli, Italy

This international consortium was formed in October 1966, the word "MESH" being an acronym of the initial letters of the four original founder members. They agreed not only to collaborate in manufacturing programmes, but to submit joint tenders to specifications for space vehicles and satellites issued within and outside Europe.

As a result of one such tender, MESH received in March 1967 a contract to design and develop the TD-1A satellite for the European Space Research Organisation (ESRO). Largest and most advanced satellite built in Europe at that period, it was launched on 12 March 1972. After a subsequent period of "hibernation", it was triggered into normal operation once more on 12 February 1973, without difficulty. Further details can be found in the 1972-73 *Jane's*.

In December 1969, Fiat SpA of Italy, through its Aviation Division (now embodied in Aeritalia SpA), joined MESH as a full member. Having a direct technical exchange agreement with TRW of America, its membership strengthened the team considerably.

In addition to the OTS satellite, described here, MESH is also involved in the maritime communications satellite, MAROTS, which employs the same basic OTS technology (see under MSDS in the UK section).

OTS

This Orbital Test Satellite is being produced under a three-year contract, with Hawker Siddeley Dynamics as prime contractor, responsible for both development and launch support. The satellite will mark completion of the technological and development phase of the European Communications Satellite programme.

OTS will be placed in geostationary orbit in 1977, in order to demonstrate the operational capabilities of on-board high-technology communications payloads and spacecraft systems. It will also test experimental advanced communications concepts and propagation assumptions in space.

The primary aim of the OTS programme is to develop a space communication system which will make facilities available to European postal and telecommunications authorities grouped in CEPT (Conference Européenne des Postes et Télécommunications). These facilities will offer satellite links for a significant portion of inter-European telephony, telegraphy and telex traffic in the 1980s, and will satisfy the requirements of the EBU (European Broadcasting Union) or Eurovision relay.

MAROTS

An adaptation of OTS, the Maritime Orbital Test Satellite (MAROTS) is under development for the European Space Agency and is scheduled for launch in late 1977. MAROTS will be the forerunner of future worldwide maritime communications services.

Prime contractor for the spacecraft is Hawker Siddeley Dynamics, leading MESH. The communications payload is the responsibility of Marconi Space and Defence Systems, leading a consortium of communications companies.

The 462 kg (1,018 lb) attitude-stabilised MAROTS is being designed to the technical requirements of the Intergovernmental Maritime Consultative Organisation (IMCO). Much of the structure and equipment is common to that of OTS. Since MAROTS is an experimental system, designed to evaluate the potential market as well as the technology, its communication equipment will be highly versatile. Voice, high-speed data, teleprinter and telex channels will all go through a single wide-band transponder. Because of the small diameter and low gain of a ship's aerial, and the relatively low performance of its receiving equipment, the shore-to-ship service, in which there can be up to 60 voice channels, will make by far the heaviest power demands on the satellite.

The design life of the satellite is seven years, with a reliability of 80 per cent after three years.

Artist's impression of OTS Satellite in orbit

STAR

BRITISH AIRCRAFT CORPORATION LTD (BAC)
100 Pall Mall, London SW1Y 5HR, England

CONTRAVES AG
Schaffhauserstrasse 580, CH-8052 Zurich 11, Switzerland

CGE-FIAR
via G.B. Grassi 93, 20157 Milano, Italy

DORNIER SYSTEM GmbH
Postfach 648, 799 Friedrichshafen/Bodensee, Germany

TELEFONAKTIEBOLAGET L. M. ERICSSON
126 25 Stockholm, Sweden

MONTEDEL (MONTECATINI EDISON ELETTRONICA SpA)
Via E Bassini 15, 20133 Milano, Italy

SENER SA
Guzman el Bucuo 121, Madrid 3, Spain

SOCIÉTÉ EUROPÉENNE DE PROPULSION
Tour Roussel-Nobel, cédex 3, 92080-Paris Défense, France

THOMSON-CSF
173 boulevard Hausmann, 75360-Paris, cédex 08, France

The STAR (Satellites for Telecommunications, Applications and Research) consortium was formed in December 1970 by industrial companies from eight of the member states of the European Space Research Organisation (ESA), to respond to tenders issued by ESA for both applications and scientific satellites. Additional companies have since joined the consortium and some have left.

Since its formation, the STAR consortium has completed a number of important studies on behalf of ESA, and has been awarded the main development contract for the GEOS satellite, with BAC as prime contractor and Hughes Aircraft Company, USA, as consultant.

GEOS satellite in the BAC electromagnetically clean test facility at Bristol

GEOS

GEOS is the first European scientific geostationary satellite. It will probe the Earth's magnetospheric environment from a synchronous orbit 36,000 km (22,300 miles) above the equator, carrying scientific experiments devised by 11 scientific groups from seven countries, to study the electric, magnetic and particle fields at a constant geographical position in the Earth's outer magnetosphere during an operational lifetime of two years. In doing so, it will be capable of being moved along its orbit between the limits of 35° West and 30° East as required.

Artist's impression of ISEE-B Explorer satellite

One of the major technological problems that had to be overcome in the development of GEOS, was the need to minimise the fields generated by the electronic equipment within the spacecraft. The sensitivity of the spacecraft's equipment necessitated the building of a large welded steel-plate screened test facility especially for electromagnetic cleanliness (EMC) testing.

GEOS was scheduled to be launched by a Delta vehicle from the Eastern Test Range in the USA in 1976.

DIMENSIONS:
Height overall 110 cm (43·3 in)
Diameter 162 cm (63·8 in)

WEIGHT:
Launching weight 549 kg (1,210 lb)

ISEE-B EXPLORER SATELLITE

Dornier System is prime contractor for the International Sun Earth Explorer (ISEE-B) satellite, on behalf of the STAR consortium, with BAC responsible for subsystems.

BAC's work involves the spacecraft's attitude and orbit control subsystems, sensors, hinged booms for sensor deployment and mechanical ground support equipment.

This is a joint NASA/ESA programme, involving three satellites. Two of these (ISEE-A and ISEE-C) are being developed under NASA auspices; ISEE-B is the responsibility of the European Space Agency.

ISEE-A and B will be launched by a single Delta vehicle in late 1977 and ISEE-C will be launched in mid-1978. The missions will investigate solar-terrestrial relationships at the outermost boundaries of the Earth's magnetosphere.

ITALY

AERITALIA
SPACE, AVIONICS AND INSTRUMENTS GROUP

HEADQUARTERS:
Piazzale Vincenzo Tecchio 51, 80125 Naples
Telephone: (081) 619522
Telex: 71370 (AERIT)

TURIN AREA:
Caselle Works, Turin Airport, 10100 Turin
Telephone: (011) 991363
Telex: 31602 (FIATVOLI)

Caselle Works is responsible for the Aeritalia Group's work on avionics, for the design of aerospace systems such as satellite attitude control devices, and for the repair of space equipment.

This Group also designed the structure and thermal control system of the Spacelab, which is to be carried into orbit by the NASA Space Shuttle. The structure will be manufactured by the Aircraft Group (which see).

CIA
COMPAGNIA INDUSTRIALE AEROSPAZIALE

ADDRESS:
Viale di Villa Grazioli 23, Rome

This consortium was formed in 1965 for the design and development of aerospace systems, and currently comprises CGE of Milan, Fiat of Turin, Finmeccanica of Rome, Montedison of Milan, Selenia of Rome and SNIA Viscosa of Milan.

CIA is involved primarily in design and development of the Sirio satellite project, including integration tests and launch procedures, Activities also include studies of how Sirio satellites can be used for educational TV broadcasts for African and Latin-American countries, and for newspaper facsimile transmissions. Future activities will include Earth resources remote sensing, ecological and energy conversion studies.

SIRIO

The Sirio (Satellite Italiano per Ricerche Industriali Orientate) developed and built under the sponsorship of the Consiglio Nazionale Ricerche (CNR) of Rome. It was intended to be put into geostationary orbit by means of a Delta launch vehicle from Cape Canaveral, in late 1976.

Main Sirio programme objective is to perform propagation and communications experiments in the 12-18GHz band from geostationary orbit.

Sirio is basically cylindrical in shape, and is to be spin-stabilised at 90 rpm in orbit. Its apogee motor has a specific impulse of 280 sec. An auxiliary propulsion system utilises hydrazine as a monopropellant. The required 100W electrical supply throughout the planned two-year lifetime of the satellite will be provided by solar cells.

DIMENSIONS:
Height overall, incl antennae and apogee motor 1·99 m (6 ft 6 in)
Height of body 0·95 m (3 ft 1½ in)
Diameter 1·45 m (4 ft 9 in)

WEIGHTS:
Launching weight 398 kg (877 lb)
Weight in orbit 198 kg (437 lb)

Sirio geostationary research satellite *(Théo Pirard)*

CSTM
CENTRO STUDI TRASPORTI MISSILISTICI (Missile Transport Research Centre)

HEAD OFFICE:
Via Squarcialupo 19-A, 00162 Rome
Telephone: 423.833
PRESIDENT: Glauco A. Partel

CSTM has terminated development of the Bora-Sound sounding rocket, of which details are given in the 1975-76 *Jane's*.

JAPAN

MITSUBISHI
MITSUBISHI JUKOGYO KABUSHIKI KAISHA (Mitsubishi Heavy Industries Ltd)

HEAD OFFICE:
5-1, Marunouchi, 2-chome, Chiyoda-ku, Tokyo 100
OFFICERS: See "Aircraft" section
MANAGER, SPACE SYSTEMS DEPARTMENT:
Y. Kato

On 1 June 1964, Shin Mitsubishi Jukogyo, Mitsubishi Zosen (Shipbuilding) and Mitsubishi Nihon Jukogyo were amalgamated as Mitsubishi Jukogyo Kabushiki Kaisha (Mitsubishi Heavy Industries Ltd). The units known formerly as Shin Mitsubishi and Mitsubishi Zosen, together with Mitsubishi Denki (Electric Machinery) have been engaged since 1955 in the development of missiles and the production of sounding rockets, of which all available details are given below.

Within the framework of Japan's national space programme, Mitsubishi is collaborating with the Science and Technology Agency, the National Space Development Agency, and the Institute of Space and Aeronautical Science, University of Tokyo. It has manufactured the rocket chambers of the Kappa, Lambda and Mu rockets, the attitude control motors of Lambda and Mu, and the launching and assembly tower for the Mu.

For the Japan Defence Agency, Mitsubishi is developing and producing air-to-air missiles and is manufacturing the Nike-Hercules surface-to-air missile under licence.

MITSUBISHI S-B, S-C, LS-A and LS-C

Mitsubishi has been awarded Science and Technology Agency contracts to develop and test a series of research rockets. First of these were for the S-B and S-C meteorological sounding rockets for observation of the upper atmosphere.

The S-B and S-C are each fitted with a single-stage solid-propellant rocket motor in a fibre-reinforced plastics casing. The S-B is capable of carrying a payload of 2·3 kg (5·1 lb) to a height of more than 70 km (44 miles); the S-C is designed to carry a 2·5 kg (5·5 lb) payload to a height of more than 80 km (50 miles). At the peak altitude, a radio telemetry and parachute recovery package is released to measure the atmospheric temperature, wind direction and strength.

First launching trials of the S-B were made in 1964. The S-C was launched for the first time in 1969.

First launching of the two-stage LS-A (solid-propellant first stage, liquid-propellant second stage) was made in 1964. Subsequently, Mitsubishi received a follow-on contract for the larger LS-C, the second stage of which has been developed by the National Space Development Agency of Japan as part of the development programme for the N launch vehicles that orbited the Kiku Engineering Test Satellite (ETS-1) on 9 September 1975, and the Ionosphere Sounding Satellite (ISS-1) on 29 February 1976.

DIMENSIONS:
Overall length:
S-B 2·78 m (9 ft 1½ in)
S-C 2·85 m (9 ft 4 in)
LS-A 7·55 m (24 ft 9 in)
LS-C 11·14 m (36 ft 6 in)
Body diameter:
S-B 0·16 m (0 ft 6 in)
S-C 0·17 m (6·5 in)
LS-A, second stage 0·30 m (11·8 in)
LS-C, second stage 0·60 m (23·6 in)

WEIGHTS:
Launching weight:
S-B 67·6 kg (149 lb)
S-C 77 kg (170 lb)
LS-A 760 kg (1,675 lb)
LS-C 2,536 kg (5,591 lb)

PERFORMANCE:
Ceiling (75° launch angle):
S-B 75 km (47 miles)
S-C 85 km (53 miles)
LS-A 110 km (68 miles)
LS-C 59 km (37 miles)

NASDA
NATIONAL SPACE DEVELOPMENT AGENCY OF JAPAN

ADDRESS:
2-4-1, Hamamatsu-cho, Minato-ku, Tokyo

PRESIDENT: Hideo Shima

DIRECTOR, SYSTEMS PLANNING DEPARTMENT:
Yasuhiro Kuroda

The National Space Development Agency was established on 1 October 1969 to take over functions of the National Space Development Center of the Science and Technology Agency, as well as the Ionosphere Sounding Satellite Development Division of the Radio Research Laboratories of the Ministry of Posts and Telecommunications. The major duties of the Agency, as the central organisation for promoting space activities in Japan, include the development of satellites, launch vehicles, tracking systems, and associated facilities and equipment.

The work is performed in accordance with the space development programme approved by the Japanese Prime Minister, on the basis of recommendations by the Space Activities Commission.

Engineering Test Satellite (ETS-1)

Ionosphere Sounding Satellite (ISS)

Medium-capacity Communications Satellite (CS)

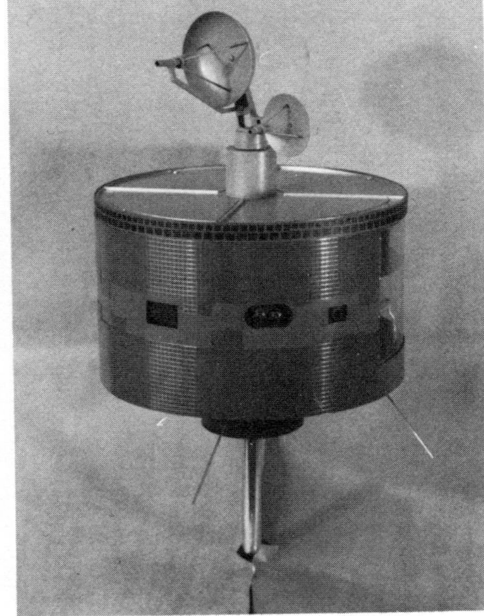

Experimental Communications Satellite (ECS)

Engineering Test Satellite (ETS-2)

Geostationary Meteorological Satellite (GMS)

Broadcasting Satellite Experimental (BSE)

In addition to the programme of satellites to be launched by the N vehicle (which see), NASDA has another programme for launching three geostationary satellites—Geostationary Meteorological Satellite (GMS), the medium-capacity geostationary Communications Satellite (CS) for experimental purposes, and the medium-scale Broadcasting Satellite (BSE) for experimental purposes by means of US Delta launch vehicles during 1977. Each of these three satellites will weigh about 350 kg (772 lb) in orbit. The GMS is to be launched as part of the Global Atmospheric Research Programme (GARP) of the World Weather Watch (WWW) plan, and the CS and BSE are to be launched for domestic pruposes.

N LAUNCH VEHICLE

The N launch vehicle three-stage rocket is capable of putting a 920 kg (2,028 lb) satellite into a 200 km (124 mile) orbit or a 130 kg (287 lb) satellite into a geostationary orbit. The first stage embodies a liquid-propellant rocket (liquid oxygen/kerosene) with Nissan solid-propellant strap-on boosters (approx 1,471 kN; 330,600 lb st); the second stage is powered by a storable liquid rocket (NTO/A-50), and the third stage by a solid rocket. The guidance system is of the strap-down radio command type.

Using the N vehicle, NASDA successfully launched the 85 kg (187 lb) Kiku Engineering Test Satellite (ETS-1) into a circular orbit of approximately 1,000 km (620 miles), at an inclination of 47°, on 9 September 1975, and the 135 kg (298 lb) Ume Ionosphere Sounding Satellite (ISS) into a similar orbit with an inclination of 70°, on 29 February 1976. With the N vehicle, NASDA plans to launch a second 254 kg (560 lb) Engineering Test Satellite (ETS-2) into a synchronous orbit in 1976, and the 130 kg (287 lb) Experimental Communications Satellite (ECS) into a geostationary orbit in 1978. The N vehicle is also expected to be utilised for launching of other satellites later.

Studies are in hand to uprate the performance of the rocket.

The main parameters of the current N vehicle are as follows:

DIMENSIONS (approx):
- Length overall: 32·6 m (106 ft 11 in)
- Diameter of first stage: 2·4 m (8 ft 2½ in)

WEIGHT (approx): 90,000 kg (198,400 lb)

First launching of N rocket

NISSAN
NISSAN JIDOSHA KABUSHIKI KAISHA (Nissan Motor Co Ltd)

HEAD OFFICE:
17-1, 6-chome, Ginza, Chuo-ku, Tokyo

AERONAUTICAL AND SPACE DIVISION:
5-1, 3-chome, Momoi, Suginami-ku, Tokyo

Telephone: Tokyo (390) 1111

Telex: 0-232-2271

CHAIRMAN:
Katsuji Kawamata

PRESIDENT:
Tadahiro Iwakoshi

MANAGING DIRECTOR AND VICE-PRESIDENT:
Takashi Ishihara

DIRECTOR, MANAGER OF AERONAUTICAL AND SPACE DIVISION:
Yasuakira Toda

Nissan Motor Co Ltd is the oldest automobile manufacturer in Japan, having been founded in 1933 under the name of Jidosha Seizo Co Ltd. It adopted its present name in 1934 and, following its merger with the former Prince Motors Ltd, on 1 August 1966, is Japan's largest automobile company. The merger also gave Nissan a leading position in the national aerospace industry, as the Space and Aeronautical Division of Prince Motors became a Division of Nissan Motor Co.

Known before and during World War 2 as Nakajima Aircraft Company, Prince formed its Space and Aeronautical Division as an offshoot of its primary post-war business of automobile manufacture. Today, the Aeronautical and Space Division is responsible for the major share of work on rockets and missiles in Japan.

In the field of military missiles and rockets, the company has been engaged in extensive research and development since 1953. In association with the Technical Research and Development Institute of the Japan Defence Agency, it has developed many types of air-to-air, air-to-surface and surface-to-surface solid-propellant unguided rockets, several of which are in production, including the "30-rocket", the largest surface-to-surface type in service in Japan.

In addition, this Department has produced many sounding rockets such as Pencil, Baby, Kappa, Lambda and Mu for the Institute of Space and Aeronautical Science, University of Tokyo, and has produced many other research rockets for the Japanese Aerospace Laboratory, Meteorological Agency and the Antarctic Research Centre.

Under contracts from the National Space Development Agency, NASDA, Nissan has developed a number of rocket motors of the kind used in the NASDA N satellite launch vehicle. The first-stage strap-on boosters are produced by Nissan under licence from Thiokol Corporation, USA, and the third-stage solid-propellant motor are procured and serviced by Nissan.

Taiyo Solar Radiation and Thermospheric Structure (SRATS) satellite

NISSAN SOUNDING ROCKETS

Nissan has developed a range of sounding rockets, and seven different types are now available. Over 350 of these rockets have been launched, including 23 in the Antarctic. Their main characteristics are detailed in the accompanying table.

Mu-3C-2 launch vehicle for Taiyo (SRATS) satellite

NISSAN SOUNDING ROCKETS

Designation	Stages	Length m (ft)	Diameter mm (in)	Total Weight kg (lb)	Max Alt. km (miles)	Payload kg (lb)
MT-135P	1	3·3 (10·9)	135 (5·3)	69 (152)	60 (38)	12 (26)
S-160 JA	1	3·9 (12·9)	160 (6·3)	110 (243)	90 (56)	15 (33)
S-210 JA	1	5·3 (17·5)	210 (8·3)	260 (573)	130 (81)	23 (51)
S-310	1	6·8 (22·4)	310 (12·2)	700 (1,543)	180 (113)	40 (88)
K-9M	2	11·2 (37)	420 (16·5)	1,400 (3,086)	350 (219)	80 (176)
K-10	2	10·0 (33)	420 (16·5)	1,800 (3,968)	250 (156)	240 (529)
L-3H	3	16·3 (53·8)	735 (28·9)	9,500 (21,000)	2,100 (1,313)	270 (595)

Mu ROCKET

The planned series of satellite launches by Mu-4S vehicles ended in August 1972, when an Mu-4S successfully put into orbit Japan's fourth satellite, known as Dempa (Radio Exploration Satellite). Earlier, the satellite Ohsumi had been launched by a Lambda rocket in 1970, and Tansei and Shinsei by Mu-4S rockets in 1971.

The Mu-4S was then superseded by the advanced Mu-3C, with three stages and control systems, which had been under development by Nissan and other manufacturers since 1972, under the supervision of the Institute of Space and Aeronautical Science, University of Tokyo. Only the first-stage M10 and SB engines have been retained from the Mu-4S. The second- and third-stage engines, structures and control systems are all newly developed.

Mu-3C-1 was launched on 16 February 1974, at the Uchinoura facilities of the Institute of Space and Aeronautical Science. It put Japan's fifth satellite, Tansei-2, into the planned orbit with a perigee of 273 km (169 miles) and apogee of 3,180 km (1,976 miles).

An Mu-3C-2 successfully put into orbit Japan's third scientific satellite, known as the Taiyo SRATS (Solar Radiation and Thermospheric Structure satellite) on 24 February 1975. Next will come CORSA (Cosmic Radiation Satellite), launched by Mu-3C, and a series of EXOS (Exospheric Satellites), launched by Mu-3H vehicles, which will be advanced versions of the Mu-3C, in 1978.

The following details apply to the Mu-3C:

TYPE: Solid-propellant three-stage satellite launch vehicle.

POWER PLANT: M10 first stage gives 866 kN (194,670 lb st) at sea level, and is supplemented by eight SB strap-on boosters, giving a total of 136·8 kN (30,750 lb st). The M22 second stage gives 360 kN (80,910 lb st) in vacuo and is fitted with liquid-injection thrust vector control. The M3A third stage gives 73·55 kN (16,535 lb st) in vacuo and has a similar control system.

DIMENSIONS:

Length overall:	
M10	11·92 m (39 ft 1 in)
M22	4·88 m (16 ft 0 in)
M3A	1·54 m (5 ft 1 in)
Diameter:	
M10	1·50 m (4 ft 11 in)
M22	1·41 m (4 ft 7½ in)
M3A	1·14 m (3 ft 9 in)

WEIGHT:

Launching weight:	
M10	25,140 kg (55,423 lb)
SB (8)	3,825 kg (8,432 lb)
M22	8,635 kg (19,038 lb)
M3A	1,240 kg (2,736 lb)

NETHERLANDS

ANS
INDUSTRIEEL CONSORTIUM ASTRONOMISCHE NEDERLANDSE SATELLIET

ADDRESS: PO Box 7600, Schiphol-Oost

Telephone: 020 5449111

Telex: FOSP 15707

Project work on the first Dutch satellite was started in January 1970. Known as the Astronomical Netherlands Satellite (ANS), it was developed and built by a consortium formed by Fokker-VFW and Philips Gloeilampen Fabrieken, in collaboration with astronomers of the Universities of Groningen and Utrecht and with NASA. The project was supervised by the NIVR (Netherlands Agency for Aerospace Programmes) on behalf of the Netherlands government.

ANS

The ANS small astronomical satellite was launched by a NASA Scout vehicle from Vandenberg Air Force Base, California, on 30 August 1974, into a Sun-synchronous polar orbit of 1,173 × 257 km (729 × 160 miles). This is somewhat different from the planned near-circular orbit of 560 × 510 km (348 × 317 miles), but did not affect adversely the X-ray and ultra-violet investigations into young hot stars.

The X-ray and ultra-violet regions of the spectrum, most wavelengths of which are opaque to the Earth's atmosphere, are among the least explored. Unexpected discoveries made by earlier satellites, notably NASA's Small Astronomy Satellite (Explorer 42/UHURU) and Orbiting Astronomical Observatory 3 (OAO-3), provided strong motivation for carrying out an extensive survey of the sky in these wavelength regions.

In orbit ANS performed well, and excited astronomers by discovering a new phenomenon of X-ray bursts—a discovery of the first rank scientifically and considered by some to be comparable in importance to the discovery of radio pulsars. For more than a decade it has been speculated that special processes in the nuclei of galaxies may lead to the formation of "black holes" millions of times more massive and quite different from those formed through the ordinary processes of stellar evolution. ANS data provided the first direct evidence of black holes of this magnitude.

Although ANS operations had ceased, the potential importance of the discovery is so great that the spacecraft was re-activated for the period 1 March to 19 April 1976 to investigate bursts on X-ray sources in the galactic center.

Further details of ANS can be found in the 1975-76 *Jane's*.

POLAND

IMGW
INSTYTUT METEOROLOGII I GOSPODARKI WODNEJ

HEAD OFFICE: 00-967 Warszawa, ul Podlesna 61
Telephone: 34-16-51/59
Telex: 814331
DIRECTOR: Jan Zielinski, DSc eng, Associate Professor
CHIEF OF ROCKET RESEARCH SECTION:
Dr J. Walczewski

Rocket Sounding Laboratory:
Kraków, ul P. Borowego 9

All meteorological research projects in Poland are now undertaken exclusively by the Institute of Meteorology and Water Economy (IMGW), which is conducting experiments in accordance with the recommendations of COSPAR and the World Meteorological Organisation.

Rockets used by the IMGW include the Meteor-1E and 3E, evolved from the Meteor-1 and 3 designed at the Instytut Lotnictwa and described briefly in the 1973-74 *Jane's*. Use of the IMGW's own Rasko-2 rocket for the artificial modification of clouds has ended, but a new version known as Rasko-2R has been developed.

IMGW RASKO-2R

This simple unguided solid-propellant rocket has a cylindrical metal body and cruciform tail surfaces. It is basically similar to the Rasko-2, described in the 1971-72 *Jane's,* but its payload comprises a radar transponder instead of chemicals.

A prototype of the Rasko-2R has been completed. It is designed for use as a test rocket for training radar operators and for radar checkout in preparation for the use of larger meteorological rockets.

METEOR-1E and 3E

The original Meteor-1 and Meteor-3 meteorological sounding rockets designed at the IL are no longer in use. Experimental versions known as Meteor-1E and Meteor-3E have been evolved, for use as test vehicles for a new dart-sonde system named Grot-Somit developed at the IMGW in 1972-73.

Meteor-1E is a two-stage vehicle, comprising the same rocket booster as that used in the original Meteor-1 plus an unpowered dart, known as Grot E-50/11. This has a plastic body, containing a modified version of a standard parachute sonde for temperature and wind measurements. The sonde transmits on a frequency of 1,700MHz. The ground telemetry and tracking equipment consists of the Meteorit station used with the RKZ balloon sonde.

Meteor-3E differs from the 1E in having an additional Meteor-1 motor in tandem, making it a three-stage vehicle.

DIMENSIONS:

Length overall:	
1E	2·86 m (9 ft 4½ in)
3E	4·66 m (15 ft 3½ in)
Length of dart	1·16 m (3 ft 9½ in)
Diameter of motor stages	12 cm (4·7 in)
Diameter of dart	5·1 cm (2·0 in)

WEIGHTS:

Launching weight:	
1E	35·5 kg (78 lb)
3E	68·0 kg (150 lb)
Dart	7·2 kg (16 lb)

PERFORMANCE:

Ceiling:	
1E	21,000 m (68,900 ft)
3E	44,000 m (144,350 ft)

Meteor-1E experimental sounding rocket

SPAIN

CASA
CONSTRUCCIONES AERONAUTICAS SA

HEAD OFFICE:
Rey Francisco 4, Madrid-8

WORKS AND OFFICERS: See "Aircraft" section

In addition to its work on aircraft design, manufacture and overhaul, CASA is engaged on space research contracts and spacecraft fabrication in co-operation with CONIE (Comision Nacional de Investigacion del Espacio) and ESA (European Space Agency).

CASA participated in production of the structure of the HEOS-A2 satellite, under subcontract to MBB, and the structure and thermal control equipment for the COS B satellite, under subcontract to Aérospatiale.

Under subcontract to CONIE, CASA contributed to development of the first Spanish satellite, Intasat. It is developing the inter tanks and forward skirts of the Ariane launch vehicle, and is setting up an automated battery test centre for the evaluation of battery cells for ESTEC.

CONIE
COMISION NACIONAL DE INVESTIGACION DEL ESPACIO

ADDRESS:
Paseo del Pintor Rosales 34, Madrid-8

INTA
INSTITUTO NACIONAL DE TECNICA AEROESPACIAL

Space research in Spain is conducted under the sponsorship and control of CONIE, of which INTA is the Technological Centre. Annually, CONIE recommends a programme of space activities, within the framework of a six-year programme approved by the Spanish government and dating from 1968. It uses as much as possible the facilities offered by research institutions in Spain and the services of Spanish manufacturers. The national space facilities and laboratories, including the El Arenosillo sounding rocket launch station on the Atlantic coast of Mazagon (Huelva), are operated by INTA.

In partnership with NASA, the Institut d'Aeronomic Spatiale de Belgique, and the Max Planck Institut für Ionosphere, CONIE is making ionospheric studies. It is also carrying out astrophysical and meteorological observations.

INTA was prime contractor for Intasat-1, the first Spanish satellite (see 1975-76 *Jane's*). It is now developing a sounding rocket, of which brief details follow:

INTA-300

The INTA-300 is a two-stage sounding rocket, with motors and payload package of constant diameter. The first-stage booster comprises a standard INTA-255 single-stage rocket, which has been used in large quantities to carry scientific payloads. The second stage embodies a new high-performance motor which combines low thrust with long burning time in order to limit kinetic heating of the vehicle's skin and to reach peak speed at a low-density altitude.

DIMENSIONS:
- Length overall: 7·27 m (23 ft 10 in)
- Diameter (constant): 0·26 m (10 in)
- Max fin span: 1·10 m (3 ft 7 in)
- Length of payload section: 0·53 m (1 ft 9 in)

WEIGHTS:
- Launching weight: 508 kg (1,120 lb)
- Payload, incl telemetry systems: 44 kg (97 lb)

PERFORMANCE:
- Max speed: Mach 8
- Ceiling with current payload: 327 km (203 miles)
- Future ceiling with 10 kg (22 lb) payload: 500 km (310 miles)
- Range (85° launch angle): 147 km (91 miles)
- Max acceleration: 29*g*

SENER

HEAD OFFICES:
Madrid and Bilbao

CHIEF EXECUTIVE:
Dr Manual Sendagorta

Sener has been awarded a £2·3 million contract for work on the ESRO Spacelab by the Instituto Nacional de Tecnica Aeroespacial (INTA), the main ESRO contractor for Spain which is one of the nine European countries participating in the NASA Space Shuttle Spacelab programme. The work involves design and development of the equipment needed to transport Spacelab and to position the laboratory on board the Space Shuttle Orbiter, as well as special equipment needed for emergencies including those which could occur when the Orbiter is in the vertical position ready for launching.

Sener has also received a contract to design and manufacture the launching platform and the umbilical service tower for the European Ariane launch vehicle.

THE UNITED KINGDOM

BAC
BRITISH AIRCRAFT CORPORATION
Guided Weapons Division
Electronic and Space Systems (ESS)

STEVENAGE WORKS:
Six Hills Way, Stevenage, Herts SG1 2DA

BRISTOL WORKS:
PO Box 77, Filton House, Bristol BS99 7AR

ASSISTANT MANAGING DIRECTOR: D. Rowley
SALES DIRECTOR: L. A. Sanson, OBE
MANAGER SPACE SYSTEMS AND SOUNDING ROCKETS: R. G. T. Munday
PUBLICITY MANAGER: R. F. Bailey
PUBLICITY OFFICER (Bristol): T. C. Bickerton

BAC has played a significant part in British and international space projects for more than 10 years, involving 26 different spacecraft and over 340 research rockets. Such activities are handled by its Electronic and Space Systems organisation, comprising the Electronic and Space Systems Group at Bristol and the Reinforced and Microwave Plastics and Precision Products Groups at Stevenage.

BAC was prime contractor for Britain's Ariel IV satellite (see 1972-73 *Jane's*); it participated in both of Europe's HEOS satellite programmes and has manufactured major subsystems for Europe's COS B satellite. As leader of the STAR consortium of companies (see International section), BAC is prime contractor for the GEOS geostationary scientific satellite and is also developing systems for the International Sun Earth Explorer Satellite (ISEE-B).

As main overseas contractor to Hughes Aircraft Company for the Intelsat IV series of communications satellites, six of which are in service around the Earth, BAC was the first company outside the USA to participate in this advanced and competitive field. It is currently manufacturing subsystems for the larger-capacity Intelsat IVA satellites, two of which have been launched, and is also supplying four sets of satellite hardware for American Telephone and Telegraph Company's Comstar 1 communications satellite system under contract to Hughes Aircraft Company.

BAC has been selected as the design authority for the structure and the deployable booms of Britain's sixth satellite, UK6. Prime contractor is Marconi Space and Defence Systems (which see).

BAC has in hand a variety of space technology contracts from the European Space Agency.

Other BAC Electronics and Space Systems activities concern high-speed data handling systems for aircraft, mathematical modelling of complex systems, HF notch antenna tuning units, microwave phased arrays, precision gyroscopes and inertial guidance equipment for the services and industry, and a wide range of reinforced and microwave plastics components for the marine, electronics and aerospace industries, including radomes, radar reflectors, antennae and structural components.

BAC SKYLARK

The Skylark high-altitude sounding rocket was developed originally by the Royal Aircraft Establishment and used in a programme of upper atmosphere research associated with the International Geophysical Year 1957-58. The basic concept of the rocket proved so sound and flexible that it became Europe's most successful upper atmosphere research rocket. By early 1976 a total of 345 Skylarks had been launched from ranges around the world.

Skylark is powered by a Raven main stage, with an optional Cuckoo or Goldfinch first stage or booster. These are all solid-propellant radial-burning rocket motors.

Five versions of Skylark give a wide range of altitude/payload performance. Typically, a Cuckoo-boosted Raven 6 (Skylark type 3) will lift a 150 kg (330 lb) payload to about 250 km (155 miles) altitude, while the combination of Raven 6 and Goldfinch (Skylark type 5) will lift a 235 kg (520 lb) payload to the same height.

A range of three-axis attitude control units is made by GEC-Elliott Electronics, using nitrogen gas-jets to point the payload with great accuracy at either the Sun, Moon or a star. The Sun-pointing ACU will acquire and complete its payload stabilisation in less than 30 seconds.

Stabilisation of the Skylark payload permits observations and experiments in solar spectroscopy UV and stellar UV, and solar and stellar X-ray photography. Results to date have included the discovery of a Nova-like strong variable X-ray source in the constellation Centaurus. Another notable achievement was the identification in 1971 of X-ray source GX3+1, accurately pinpointed by experiments in two Skylark rockets launched within seconds of this X-ray star's occlusion by the Moon.

Other systems available include a timer-intiated roll control unit, and a boost-phase spin system using small rocket thrusters and canted fins to provide roll rates up to 3Hz for improved aerodynamic stability. A yo-yo system is used to de-spin the payload before ACU acquisition begins.

A payload recovery system is available and is initiated at a height of about 4 km (2·5 miles), with a gun-released pilot chute which slows down the payload before the main 6 m (20 ft) canopy opens.

An additional use for Skylark is in the new technology of space processing. Skylark can subject experiments to zero *g* conditions for periods of up to 5 minutes followed by payload recovery.

More extensive details of Skylark can be found in the 1972-73 *Jane's*.

DIMENSIONS:
- Length:
  - nominal (Raven plus average payload): 9·0 m (30 ft)
  - with Cuckoo booster: 10·5 m (32 ft)
  - with Goldfinch booster: 11·5 m (35 ft)
- Body diameter: 43·8 cm (17·25 in)

LAUNCHING WEIGHTS:
- Typical boosted Skylark: 1,850 kg (4,080 lb)
- Payload limits: 70 kg (175 lb) to 310 kg (680 lb)

PERFORMANCE:
- Payload to height:
  - Type 1 (Raven 8, unboosted): 100 kg (220 lb) to 170 km (106 miles)
  - Type 2 (Raven 8, Cuckoo boosted): 150 kg (330 lb) to 200 km (125 miles)
  - Type 3 (Raven 6, Cuckoo boosted): 215 kg (475 lb) to 200 km (125 miles)
  - Type 4 (Raven 8, Goldfinch boosted): 150 kg (330 lb) to 270 km (168 miles)
  - Type 5 (Raven 6, Goldfinch boosted): 250 kg (550 lb) to 240 km (150 miles)
  - Type 7 (Raven 11, Goldfinch boosted): 300 kg (660 lb) to 275 km (172 miles)
  - Type 9 (Stonechat single stage): 700 kg (1,543 lb) to 250 km (156 miles)
  - Type 12 (Raven 11, Goldfinch boosted and Cuckoo IV): 150 kg (330 lb) to 770 km (480 miles)
- Time above 100 km (60 miles):
  - Raven 8: 4 minutes
  - Raven 8/Cuckoo: 5 minutes
  - Raven 6/Goldfinch: 6 minutes

BRISTOL AEROJET
BRISTOL AEROJET LTD

HEAD OFFICE:
Banwell, Weston-super-Mare, Somerset BS24 8PD
Telephone: Banwell (0934-82) 2251
Telex: 44259

OFFICERS: See Aero-Engines section

Bristol Aerojet pioneered the development of metal and fibre-wound pressure shells, including rocket motors formed from helically welded steel strip up to 914 mm (36 in) in diameter. In addition, the Company specialises in the design, manufacture and firing of research rockets for a wide variety of upper atmosphere investigations and Earth resources work.

The Company's research rockets have been used by the British Meteorological Office and for British University experiments since 1964. They have also been used by similar organisations in Canada, Germany, Spain, Australia, Pakistan, Sweden and France. About 1,000 of these rockets had been fired by early 1976.

The high performance and relative cheapness of the Skua and Petrel rockets makes them suitable also for use as ballistic targets, both for development work and for service trials on weapon systems. A wide variety of potential threats can be represented by simple adjustment of the target performance, and the Petrel target has been proved on extensive trails at velocities up to Mach 3.

BRISTOL AEROJET/INTA FLAMENCO

Described in earlier editions of *Jane's* as the Bristol Aerojet/INTA 300, this two-stage research rocket has been developed in collaboration with Instituto Nacional de Tecnica Aeroespacial Estoban Terrades of Spain. It is designed to lift a nominal payload of 35 kg/43 litres (77 lb/2,700 in³) to a height of 360 km (223 miles). Further details of performance are given in the accompanying table.

The Flamenco uses rail-type launchers already in general use.

The principle of high initial acceleration has been retained. Combined with the use of small fin-tip spin rockets, actuated automatically as the vehicle leaves the launcher, this ensures a high velocity/low dispersion feature which has already proved valuable on the Skua and Petrel.

Following successful proving flights, this rocket is now generally available. A further 12 launchings are currently contemplated by British and Spanish scientific groups.

The higher performance of the Flamenco, its greater payload capacity and the use of PCM telemetry, considerably enhance the capability of this rocket to fill a gap in the conventional range of sounding rockets.

BRISTOL AEROJET PETREL

This upper atmosphere research rocket is larger than Skua and lifts a nominal payload of 16 kg/11 litres (35 lb/670 in³) to a height of 152 km (94 miles). Two versions are available; the performance of each is shown in the accompanying table.

Petrel uses the same launch system as Skua and is fitted with a similar boost carriage system, to give a high velocity during the first 200 m (650 ft) of flight and relative insensitivity to side winds, with a consequent low dispersion error.

Petrel, like Skua, has a full-diameter instrument bay forward of the propulsion motor, and has accepted a wide variety of experiments and telemetry systems, including those used in the Skua and magnetometers, barium cloud payloads, mass spectrometers and electron density probes.

BRISTOL AEROJET SKUA

This low-cost rocket lifts a nominal payload of 5·5 kg/8·2 litres (12 lb/500 cu in) to a height of 100 km (62 miles). It is available in two versions, of which performance details are given in the accompanying table. The original Mark 1 and Mark 3 (extended temperature limits) versions are no longer in production.

A feature of Skua is its low dispersion area, achieved by the use of small boost rockets (Chicks), which give it a very high launch velocity. The number of boost rockets is variable to provide the required performance, and they are fitted in a recoverable carriage. The low-cost launch-tube can be mounted on a simple rail-type launcher, or on a ground or truck mounting.

Forward of the motor, the Skua can carry a wide variety of sondes, instruments, experiments or telemetry in a full-diameter instrument bay. Various ejection mechanisms can be fitted to expel the payload instruments at any predetermined time during the flight; a radar-reflective parachute can be ejected to lower the instruments slowly, and to provide a radar target determination of wind speed and height. A small 24-channel telemetry sender, working in the 432·5 to 450HZ band, can be fitted into the rocket nosecone section and can be ejected with the instruments.

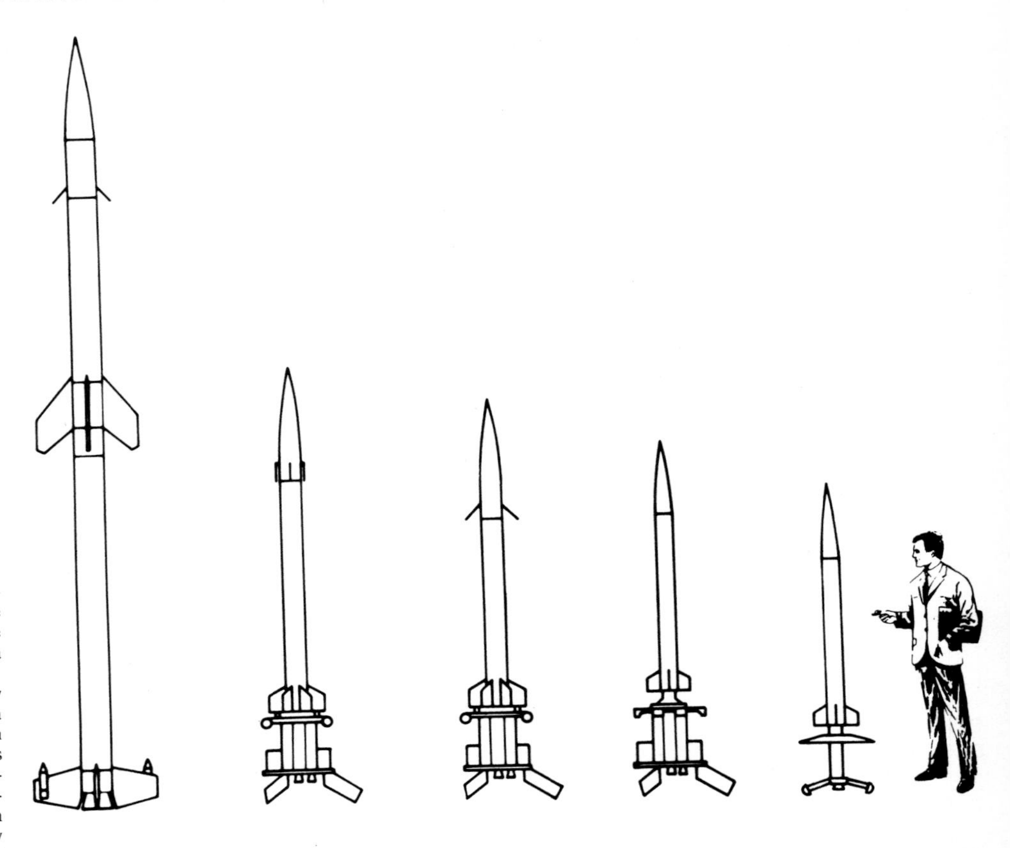

Range of Bristol Aerojet sounding rockets, to same relative scale. Left to right: Flamenco, Petrel 2, Petrel 1, Skua 4, Skua 2

Vehicle	Boost	Dia/length mm (in)	Launch Weight kg (lb)	Nominal Thrust kN (lb) Boost/Flight	Payload kg (lb)	Apogee km (miles)	Dispersion km/m/sec Wind Error	Remarks
Flamenco	Heron	260 × 7,270 (10·25 × 286·25)	450 (992)	115·7 (26,015)/16·7 (3,750)	35-50 (77-110)	360-280 (225-175)	20-15	In production
Petrel 2	3 Chicks	190 × 3,500 (7·5 × 137·75)	120 (265)	60·3 (13,560)/4·46 (1,003)	16-20 (35-44)	184-160 (114-100)	13·8-12·3	Uprated version of Petrel 1
Petrel 1	3 Chicks	190 × 3,340 (7·5 × 131·5)	114 (251)	60·3 (13,560)/4·46 (1,003)	16-20 (35-44)	145-130 (90-80)	10·2-9·3	Over 100 fired
Skua 4	4 Chicks	128 × 2,620 (5 × 103)	45 (99)	20·1 (4,520)/1·78 (401)	6-9 (13-20)	135-90 (84-55)	4·0-3·0	High precision version for firing from restricted ranges
Skua 2	1 Chick	128 × 2,420 (5 × 95·25)	45 (99)	20·1 (4,520)/1·78 (401)	5-8 (11-17·5)	125-84 (77·5-52)	12·2-8·4	Over 1,000 fired

HSD
HAWKER SIDDELEY DYNAMICS LTD

HEADQUARTERS:
Manor Road, Hatfield, Herts

SPACE DIVISION: Gunnels Wood Road, Stevenage SG1 2AS, Herts

OFFICERS: See Missiles section

In 1976 Hawker Siddeley Dynamics completed twenty-one years of work under major contracts in the space field. The company has been prime contractor for eight projects (Blue Streak, ESRO II, ESRO IV, X-4, S2 68 U/V telescope, Nimbus G radiometer, OTS and MAROTS) and a major subcontractor on four others (Intelsat III, TD-1A, Spacelab and Ariane). HSD is now involved in development of the OTS and MAROTS communications satellites, which are described in the MESH entry in the International section.

MSDS
MARCONI SPACE AND DEFENCE SYSTEMS LTD

HEAD OFFICE:
The Grove, Stanmore, Middlesex HA7 4LY
Telephone: (01) 954-2311
Telex: 22616

Marconi Space and Defence Systems, a member of GEC-Marconi Electronics, was prime contractor to the Ministry of Defence for Skynet II, the first operational communications satellite to be built outside the USA or USSR. It was also prime contractor for the Ariel 5 scientific satellite, and is involved in a number of major European collaborative programmes. Details of Skynet II and Ariel 5 can be found in the 1975-76 *Jane's*.

As a member of the COSMOS consortium, it is providing the full attitude control system and other equipment for the Meteosat satellite. Under subcontract to Aérospatiale, it is responsible for attitude and orbit control systems on the Ariane launch vehicle. It is also prime contractor for the MAROTS portion of the OTS satellite project. MSDS has also completed, as prime contractor, the initial design phase of the UK's next scientific satellite, UK 6.

MAROTS

The MAROTS satellite is being designed to provide ship-to-shore communications links vastly superior to the present highly-congested conventional high-frequency circuits. For the first time, merchant ships in the Atlantic and Western Indian Oceans should be able to establish contact with the shore almost instantaneously.

MAROTS will use the basic structure and control system which are being developed for OTS (see MESH entry in this section). On this will be mounted the complete communications package which is being developed by Marconi Space and Defence Systems under a contract worth over $26 million.

The launch of MAROTS is scheduled for the Autumn of 1977, when it will be put into a geostationary orbit over the Atlantic.

Marconi is also to study and define the basic parameters of the shipborne and shore-based terminals. The former are to be made as simple as possible, and to this end MAROTS will be equipped with a high-efficiency 'shaped beam' antenna coupled with a transistorised L-band power amplifier. Ships will use the L-band frequencies to communicate to and from the satellite. As shore terminals will be able to use more sophisticated equipment, the shore-to-satellite frequency will be in the 14GHz band while satellite-to-shore communication will use the 11GHz band.

A minimum three-year life span is intended for MAROTS, and success in this experimental and pre-operational phase being undertaken by ESA would enable the Inter-Governmental Maritime Consultative Organisation (IMCO) to envisage a worldwide maritime satellite communications system.

UK 6 (ARIEL) SATELLITE

UK 6, Britain's sixth scientific satellite in the Ariel series, is intended to undertake investigations in the field of high-energy astrophysics, and to this end it will carry three experiments. One will study the ultra-heavy component of cosmic radiation; two other experiments will make observations of X-ray sources and other features of the X-ray sky.

The satellite is funded by the Science Research Council and will be managed from the Council's Appleton Laboratory at Slough, from where it will be controlled in orbit.

UK 6 is designed to have an operational life of two years, and is scheduled to be launched by a Scout rocket from Wallops Island into a 550 km (342 mile) circular orbit in October 1977.

THE UNITED STATES OF AMERICA

AERONUTRONIC FORD
AERONUTRONIC FORD CORPORATION

AERONUTRONIC DIVISION:
Ford Road, Newport Beach, California 92663
Telephone: (714) 640-1500
WESTERN DEVELOPMENT LABORATORIES DIVISION:
3939 Fabian Way, Palo Alto, California 94303
Telephone: (415) 494-7400

The former Philco-Ford Corporation changed its name to Aeronutronic Ford Corporation on 31 March 1975. Details of its current space programmes follow:

SYNCHRONOUS METEOROLOGICAL SATELLITE (SMS)

Aeronutronic Ford Corporation's Western Development Laboratories Division was selected as prime contractor to NASA's Goddard Space Flight Center for the first two research and development SMS (Synchronous Meteorological Satellite) spacecraft. Designed to meet National Operational Meteorological Satellite System (NOMSS) requirements, as specified by the National Oceanic and Atmospheric Administration (NOAA), these spacecraft are providing continuous observation of the atmosphere on an operational basis. Their design provides for a day-and-night cloud cover viewing capability, for the collection and dissemination of meteorological data.

SMS-1, first of the 2·62 m (103 in) tall, 1·90 m (75 in) diameter satellites, was orbited by a Delta 2914 vehicle from the Eastern Test Range, Cape Kennedy, on 17 May 1974. The basic payload of the 627 kg (1,379 lb) spacecraft comprises a telescope radiometer, called the Visible Infra-red Spin-Scan Radiometer (VISSR), providing both infra-red (IR) and high resolution visible photography; a communications system for data collection and distribution; and a Space Environment Monitoring (SEM) subsystem.

The satellite transmits day and night 'full-disc' pictures of the western hemisphere every 30 minutes. The telescope is capable of producing 1·0 km (4 mile) resolution images in visible light, and 9 km (5 mile) resolution images during night-time, using the infra-red sensors.

SMS-1 was stationed initially over the eastern Atlantic Ocean to support tropical storm research, as part of the Global Atmospheric Research Programme, in the summer of 1974. During this experiment SMS-1 provided the first near-continuous day-night coverage of a major hurricane (Carmen) in September 1974. It was moved to its present station at longitude 75°W over the equator, south of Bogota, Colombia, in time to observe weather phenomena of the spring tornado season in the mid and eastern regions of the United States.

In addition to its main task of weather surveillance, the satellite is operating as the orbital component of the Geostationary Operational Environmental Satellite system planned by NASA, NOAA, and the US Department of Commerce. In this role the satellite receives data from up to 10,000 data-collection platforms, located around the USA on board ocean-going ships and buoys, and at manned and unmanned environmental stations.

The satellite is also being used to monitor solar flare activity, and the intensity of the Earth's magnetic field and of incoming X-radiation.

SMS-2 was launched on 6 February 1975, and was placed in a synchronous orbit over the equator at longitude 115°W, east of Hawaii. From this position it can 'view' the western half of the USA and Hawaii, while SMS-1 'views' the eastern USA from its position at longitude 75°W. This will enable the two spacecraft to be the first satellites used in the Global Atmospheric Research Programme (GARP), a worldwide weather watch sponsored by the United Nations and the International Council of Scientific Unions. When GARP is fully under way in 1977, the United States' SMS and GOES satellites will be joined by similar spacecraft from Europe (Meteosat), the USSR and Japan, to form a global weather monitoring network that should enable meteorologists to forecast weather a week or more in advance with extreme accuracy.

GEOSTATIONARY OPERATIONAL ENVIRONMENTAL SATELLITE (GOES 1)

Similar to the SMS satellites, but embodying improvements in communication systems performance and reliability, the 294 kg (647 lb) GOES 1 (Geostationary Operational Environmental Satellite) was built for the National Oceanic and Atmospheric Administration (NOAA) by Aeronutronic Ford's Western Development Laboratories Division, under a contract awarded through NASA's Goddard Space Flight Center.

GOES-1 was launched successfully on 16 October 1975. It will be followed by GOES-B and C, also to be built by Aeronutronic Ford.

NATO 3

In February 1973, Aeronutronic Ford received a $27·7 million contract to build a new generation of communications satellites for NATO. The contract was placed by the USAF's Space and Missile Systems Organisation on behalf of NATO's Integrated Communications System Management Agency. Costs of the project are being shared by participating NATO countries; both foreign and US subcontractors are being used. The fixed-price-incentive contract called for delivery of two NATO-3 satellites for launch in 1976.

NATO-3A was launched in April, 1976, and is now in synchronous orbit over the Atlantic at longitude 15·5°W, midway between Africa and South America. NATO-3B was due to be launched later in 1976, to serve as a backup to the first satellite.

These satellites offer considerably more power and wider frequency bands than the earlier NATO-2 satellites, also built by Aeronutronic Ford. Each cylindrical spacecraft carries more than 32,000 solar cells to provide energy for an estimated operating life of seven years. Two primary 'beams' of communications coverage operate simultaneously: one serving NATO countries throughout the northern hemisphere; the other covering only western Europe.

Each satellite has three channels for receiving, translating frequencies, amplifying and re-transmitting voice, telegraph, facsimile and wide-band data originated by ground stations in the multi-nation network.

DIMENSIONS:

Height overall	3·10 m (10 ft 2 in)
Height of body	2·24 m (7 ft 4 in)
Diameter	2·18 m (7 ft 2 in)

WEIGHTS:

At launch	720 kg (1,540 lb)
In orbit	310 kg (677 lb)

JAPANESE ENGINEERING TEST SATELLITES (ETS-2)

Aeronutronic Ford's Western Development Laboratories Division received a $20 million contract in March 1974 to develop Japan's Engineering Test Satellites (ETS-2), which will be used to flight test that country's 'N' launch vehicle. The contract was awarded by Mitsubishi Electric Corporation, which is responsible to Japan's National Space Development Agency for the ETS-2 programme.

Aeronutronic Ford will design, develop and build a flight-qualified ETS-2 prototype and an actual flight model. Although assembled in the United States, the spacecraft will be launched in early 1977 from Tanegashima Space Centre in Japan. ETS-2 will be the first payload placed in synchronous orbit by an 'N' launch vehicle. Equipment on board the satellite will monitor continuously the rocket's launch characteristics, including vibration, shock, acceleration and other parameters. Once in orbit, it will be used also to evaluate Japanese satellite control techniques and ground support systems.

The ETS-2 spacecraft will be spin-stabilised, with mechanically de-spun antennae. Preliminary designs call for the cylindrical satellites to be approximately 1·40 m (55 in) in diameter and 0·84 m (33 in) high, with a 0·76 m (30 in) antenna on top. Each will weigh about 254 kg (560 lb) at launch and 130 kg (285 lb) in orbit.

EXPERIMENTAL JAPANESE COMMUNICATIONS SATELLITES

Aeronutronic Ford's Western Development Laboratories Division received a contract valued at approximately $30 million in February 1974, from Mitsubishi Electric Corporation, to design two experimental communications satellites for Japan. Mitsubishi is under contract to Japan's National Space Development Agency to administer the programme.

Aeronutronic's contract calls for the company to deliver two medium-capacity spacecraft for launch from the United States in early 1977. The experimental satellites (a flight-qualified prototype and an actual flight model) are intended to pave the way for later launches of full-capacity operational spacecraft.

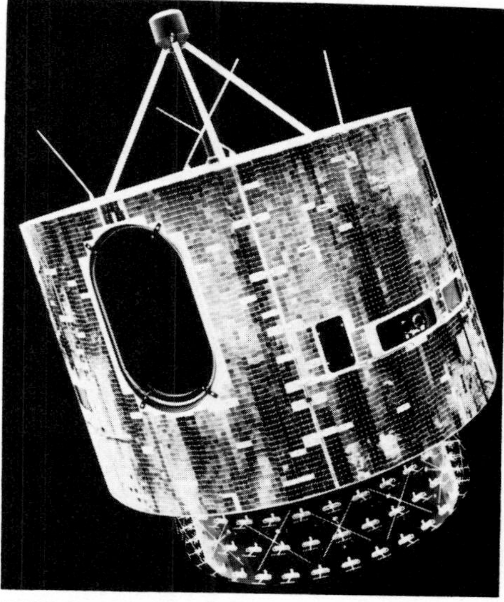

GOES/Geostationary Operational Environmental Satellite

Engineering Test Satellite (ETS-2)

Medium-capacity Communications Satellite

Early design plans call for spin-stabilised satellites with mechanically de-spun antennae, designed to operate from synchronous geostationary orbit. They will be used for telephone communications tests over the islands of Japan. Tests will include C-band (4 to 6GHz) and K-band (18 to 30GHz) frequencies.

ALRC
AEROJET LIQUID ROCKET COMPANY

ADDRESS:
Highway 50 and Aerojet Road, PO Box 13222, Sacramento, California 95813
Telephone: (916) 355-1000
PRESIDENT: J. L. Heckel

This operating company of Aerojet-General Corporation is responsible for production of the Aerobee series of upper atmosphere research rockets. These boosted single-stage vehicles are used by all the US services and by civilian and government organisations in connection with missile and satellite programmes.

Aerobee vehicles are available with a number of accessory subsystems. A fully flight-tested attitude control system, which utilises residual pressurisation gas, can be used for vehicle de-spin, orientation and up to ten manoeuvres after sustainer burnout. A yo-yo de-spin system is also available and provides a simple and effective means for reducing spin rate. Fully-proven payload recovery systems are available for use over land or water.

Also in production, to supplement some versions of Aerobee, is a family of space-probe vehicles known as Astrobees. They include two research vehicles known as Astrobee D and F, utilising a newly-developed high-energy long-burning solid propellant.

AEROBEE 150 (AEROBEE-HI)

Structural and power plant details of this boosted single-stage liquid-propellant research rocket can be found in the 1970-71 *Jane's*.

PAYLOAD: 54-181 kg (120-400 lb) in 0·38 m (15 in) diameter, 2·21 m (87 in) ogive, plus 1·52 m (60 in) maximum cylindrical extension.

DIMENSIONS:
Length 9·41 m (30 ft 11 in)
Body diameter 0·38 m (1 ft 3 in)

WEIGHT:
Firing weight, with booster, less payload 865 kg (1,908 lb)

PERFORMANCE:
Ceiling with 56·7 kg (125 lb) payload 300 km (186 miles)
Max acceleration 11·1g

AEROBEE 170

This higher-performance version of the Aerobee 150 utilises a Nike-E5 solid-propellant motor as a booster. It made a successful first flight on 26 October 1968, carrying a 119 kg (263 lb) stellar spectra experiment payload to a height of 257 km (160 miles). The rocket was fitted with yo-yo de-spin, attitude control and recovery systems.

PAYLOAD COMPARTMENT: Either a 5:1 ogive or 3:1 cone may be used, with up to 1·52 m (60 in) of cylindrical section.

DIMENSIONS:
Length with booster 12·47 m (41 ft 0 in)
Body diameter 0·38 m (1 ft 3 in)

WEIGHT:
Firing weight, with booster 1,370 kg (3,010 lb)

PERFORMANCE (with 113 kg; 250 lb payload):
Speed at burnout 1,975 m (6,489 ft)/sec
Peak altitude 267 km (166 miles)

AEROBEE 200

The improved propellant used in this version extends the burn time of the sustainer motor to 59 seconds.

PERFORMANCE (with 113 kg; 250 lb payload):
Peak altitude 317 km (197 miles)

AEROBEE 300 (SPAEROBEE)

Structural details of this two-stage boosted liquid-propellant research rocket can be found in the 1970-71 *Jane's*. A developed version, for NASA, is the Aerobee 300A which differs in having four fins instead of three.

PAYLOAD: Payload assembly, fabricated of aluminium, consists of a 10° half-angle nosecone extending back to a cylinder, and was designed to provide 0·028 m³ (1 cu ft) of usable volume for a nominal 22·7 kg (50 lb) payload. Net payload capabilities are 11·3-59·5 kg (25-120 lb).

DIMENSIONS:
Overall length, with booster 10·54 m (34 ft 7 in)
Body diameter, sustainer and booster 0·38 m (1 ft 3 in)
Body diameter, second-stage assembly 0·2 m (8 in)

WEIGHT:
Gross weight, with booster, less payload 939 kg (2,070 lb)

PERFORMANCE:
Ceiling with 22·7 kg (50 lb) payload 454 km (282 miles)
Max acceleration 63·8g

AEROBEE 350

Structural and power plant details of this boosted single-stage liquid-propellant sounding rocket can be found in the 1970-71 *Jane's*.

DIMENSIONS:
Length with booster 15·34 m (50 ft 4 in)
Body diameter 0·56 m (1 ft 10 in)

WEIGHT:
Launching weight, with booster 3,040 kg (6,700 lb)

PERFORMANCE:
Speed at burnout Mach 9·0
Max ceiling 480 km (300 miles)
Ceiling with 227 kg (500 lb) payload 330 km (205 miles)

ASTROBEE 250

The Astrobee 250 is a single-stage solid-propellant vehicle boosted by two Thiokol 1·5KS-35,000 Recruits, which are mounted on the sides of the main motor.

The payload compartment has a diameter of 0·61-0·79 m (24-31 in) and volume of 1·13 m³ (40 cu ft), accommodating a payload of 182-682 kg (400-1,500 lb).

DIMENSION:
Length 10·41 m (34 ft 2 in)

WEIGHT (less payload):
Firing weight, with boosters 4,580 kg (10,100 lb)

PERFORMANCE:
Ceiling with max payload 206 km (128 miles)
Ceiling with min payload 326 km (203 miles)
Max acceleration 15g

ASTROBEE 1500

Largest in the current series of Astrobees, the Astrobee 1500 is a two-stage solid-propellant rocket, with two Thiokol 1·5KS-35,000 Recruits mounted on the sides of the first stage as boosters.

The payload compartment has a diameter of 0·51 m (20 in) and volume of 0·104 m³ (3·67 cu ft) accommodating a payload of 22-136 kg (50-300 lb).

After first-stage burnout the second stage is spin-stabilised prior to separation by four MARC 49A1 (·5KS-180) spin motors. The second-stage casing and payload do not separate after burnout. A yo-yo de-spin unit is available for payloads requiring low spin rates.

DIMENSION:
Length, with boosters 10·41 m (34 ft 2 in)

WEIGHT (less payload):
Firing weight, with boosters 5,240 kg (11,541 lb)

PERFORMANCE (85° launch angle):
Ceiling with 136 kg (300 lb) payload 1,300 km (805 miles)
Ceiling with 22 kg (50 lb) payload 2,970 km (1,840 miles)
Max acceleration 41·4g

ASTROBEE D and F

Astrobee D and F are the first of a family of research vehicles which utilise a newly developed high-energy long-burning solid propellant known as Hydroxyl Terminated Polybutadiene (HTPB). This allows optimum delivery of impulse while also providing a more moderate acceleration environment for the scientific payload.

Following a completely successful flight test programme, Astrobee D was put into operational use by NASA and AFCRL for meteorological and D region physics experiments.

Astrobee F completed its static testing and flight tests in 1972, and production models entered the NASA inventory in 1974. During the same year, a Nike-boosted version of Astrobee F was expected to be flown.

DIMENSIONS:
Length overall:
Astrobee D 3·56 m (11 ft 8 in)
Astrobee F 6·86 m (22 ft 6 in)
Body diameter:
Astrobee D 0·15 m (6 in)
Astrobee F 0·28 m (15 in)

LAUNCHING WEIGHTS:
Astrobee D (no payload) 88 kg (193 lb)
Astrobee F 1,210 kg (2,670 lb)

PERFORMANCE (Astrobee F data estimated):
Ceiling:
Astrobee D with 4·5 kg (10 lb) payload 145 km (90 miles)
Astrobee F with 114 kg (250 lb) payload 356 km (221 miles)

SUPER CHIEF

The Super Chief is a rail-launched boosted solid-propellant stabilised vehicle for high-altitude experimentation. The booster is powered by a Talos motor, and there are four fixed fins on both the booster and sustainer.

Two versions are available, as follows:

Super Chief I. With Sergeant sustainer motor. Able to carry a 454 kg (1,000 lb) payload, contained within a 2·03 m (80 in) long extension, to a height in excess of 201 km (125 miles).

Super Chief II. With Castor sustainer motor. Able to carry a 500 kg (1,100 lb) payload to a height in excess of 341 km (212 miles). A 2·36 m (93 in) long nosecone may be used for the payload.

DIMENSIONS:
Length overall 13·41 m (44 ft 0 in)
Body diameter 0·79 m (2 ft 7 in)

WEIGHT:
Launching weight 6,026 kg (13,284 lb)

NIRO

Niro is a two-stage solid-propellant vehicle for small to medium payloads (18-82 kg; 40-180 lb), which it can carry to the 'F' region of the ionosphere. It was developed to meet a USAF requirement for a small and economical vehicle which could provide roll control and good structural stability.

Standard diameter (7·75 in) and oversize (9 in) payloads have been flown successfully. Usable volume can be varied up to 0·113 m³ (4 cu ft). Both land and water recovery systems are available.

DIMENSIONS:
Length overall, without payload 6·5 m (21 ft 5 in)
Body diameter:
first stage 0·42 m (16·5 in)
second stage 0·20 m 7·75 in)

WEIGHT (less payload):
Launching weight 723 kg (1,591 lb)

PERFORMANCE (85° launch angle):
Ceiling with 18 kg (40 lb) payload 287 km (178 miles)
Ceiling with 82 kg (180 lb) payload 113 km (70 miles)

BENDIX
THE BENDIX CORPORATION

HEAD OFFICE:
Bendix Center, Southfield, Michigan 48075

AEROSPACE SYSTEMS DIVISION

Ann Arbor, Michigan

The Aerospace Systems Division of Bendix is responsible for the Corporation's primary missile and space activities. Products include systems for the USAF's Samos reconnaissance satellites, and the stable platform for the inertial guidance system of the Saturn launch vehicle. The division also built the Lageos satellite for the NASA Marshall Space Flight Center.

LAGEOS

The Laser Geodynamic Satellite (Lageos) was launched on 4 May 1976 by a Delta 2913 from the Western Test Range. The programmed near-circular polar orbit was 5,790 km (3,600 miles) high.

Looking like a large golf ball, Lageos is designed to provide a stable point in the sky from which very precise measurements of specific locations on the surface of the Earth can be made. By timing the return of reflected pulses of laser light to an accuracy of about one ten-thousand-millionth of a second, scientists expect to measure the relative location of participating ground stations. During the first four years of the programme it is estimated that measurements will be accurate to within 0·1 m (4 in); eventually, when the measurement techniques have been perfected and more laser stations become operational, an accuracy of 20 mm (0·8 in) is hoped for. Using such data, scientists expect to gain important information regarding Earth motions and movement along geological faults. These movements, within the surface of the Earth, are the fundamental cause of natural disasters such as earthquakes. The likelihood and magnitude of earthquakes, and other natural disasters, may eventually be forecast.

Lageos is 0·61 m (2 ft) in diameter, weighs 411 kg (906 lb) and carries 426 special prisms designed to return laser pulses to their exact point of origin on the Earth. Known as cube-corner retro-reflectors, the prisms are three-dimensional units that reflect light back to its source regardless of the angle at which it is received. The retro-reflectors are made of high-quality fused silica, a synthetic quartz.

Structurally the satellite consists of two aluminium hemispheres bolted together around a heavy solid brass core. Key to the extremely accurate measurements anticipated is its highly stable orbit. Because it is so small and yet has so much mass, the drag effects of solar radiation, traces of the Earth's atmosphere, or variation in the Earth's gravity field will be minimised.

Sealed inside the satellite is a message for the future. Prepared by Dr Carl Sagan, Cornell University, the message, etched on stainless steel, displays the numbers one to ten in binary arithmetic, a drawing of the Earth in orbit round the Sun, and three maps of the Earth's surface. One map shows the continents close together, as they were about 225 million years ago; the centre map shows their position today; the third map shows the estimated movements after 8·4 million years, the estimated lifetime of the satellite.

The Lageos project is part of the Earth and Ocean Dynamics Application Programme (EODAP) being conducted by the NASA Office of Applications and is managed by the Marshall Space Flight Center.

See NASA entry for illustration (page 694).

BOEING
THE BOEING COMPANY

HEAD OFFICE AND WORKS:
Seattle, Washington 98124

Boeing Aerospace Company

OFFICERS: See Aircraft section

Details of the Burner upper-stage boosters, which are no longer in production, are given in the 1975-76 *Jane's*.

CELESCO

CELESCO INDUSTRIES INC

ADDRESS: Costa Mesa, California 92626

This company markets the range of sounding rockets and research vehicles listed under the Susquehanna Corporation entry in the 1972-73 *Jane's*.

ARGO B-13 NIKE-APACHE

This two-stage rocket, like its predecessor the Nike-Cajun, is used worldwide for scientific research in the ionosphere. As an example of the vehicle's versatility, it has been adapted successfully as a ballistic target for tactical surface-to-air missile development. The solid-propellant motors consist of a Nike M-5 first stage and Apache TE307 second stage.

The B-13 Nike-Apache is designed to carry a payload of approximately 27 kg (60 lb), which it will lift to a height of 250 km (155 miles).

ARGO D-4 JAVELIN

The Argo D-4 is a research vehicle which is able to carry an instrument payload of 22·7 kg (50 lb) to a height of 965 km (600 miles). It is a four-stage vehicle, comprising an Honest John M6, followed by two Nike-Ajax M5 boosters and a final stage designed by the Allegany Ballistic Laboratory and designated X-248. With a payload of 50 kg (110 lb) gross, 28 kg (62 lb) net experiment, a speed of about Mach 13 can be attained.

Argo D-4 is used by NASA, the USAF and other customers.

DIMENSIONS:
Length overall 14·83 m (48 ft 8 in)
Max body diameter 0·58 m (1 ft 10·8 in)
WEIGHT:
Launching weight 3,355 kg (7,400 lb)

SWIK MOD A, B, C & D

The basic SWIK vehicle (MOD A) is a two-stage solid-propellant rocket composed of an XM33-E8 Castor, assisted by two auxiliary XM-19 Recruits, as the first stage and an X-254 Antares second stage.

The SWIK MOD B uses the same first stage, but the X-254 is replaced with the X-259.

The SWIK MOD C & D are also two-stage solid-propellant rockets, each with a TX-354 Castor II, assisted by two auxiliary XM-19 Recruits, as the first stage. An X-254 is used as the second stage of the MOD C, with the X-259 used as the second stage in the MOD D configuration.

All SWIK vehicles follow a ballistic trajectory and can carry net payloads of 45 kg (100 lb) to an altitude in excess of 1,850 km (1,150 miles) or 227 kg (500 lb) net payload to an altitude in excess of 1,295 km (805 miles).

DIMENSIONS:
Length overall 10·86 m (35 ft 8 in)
Max body diameter 0·79 m (2 ft 7 in)
WEIGHT:
Launching weight 5,817 kg (12,824 lb)

TRAILBLAZER II

Trailblazer II is a four-stage solid-propellant re-entry test vehicle. Two stages are fired upward, two stages downward to achieve a re-entry velocity of 6,700 m/sec (22,000 ft/sec) with a 9 kg (20 lb) net payload.

DIMENSIONS:
Length overall 15·24 m (50 ft 0 in)
Max body diameter 0·79 m (2 ft 7 in)
WEIGHT:
Launching weight 6,044 kg (13,324 lb)

ATHENA

Athena has been used to impact experimental re-entry vehicle payloads on the White Sands Missile Range under the advanced ballistic re-entry systems (ABRES) programme. It is being launched currently at Wallops Island, on behalf of the Defense Nuclear Agency.

Work on the Athena project began in 1962, under a USAF contract. Technical acceptance was received in December 1963 and the first launch was made on 10 February 1964. By the end of 1971, a total of 132 launchings had been made, with further launches planned in the period to the end of 1975.

Depending on the requirement, Athena can have a three- or four-stage configuration, with payload capacity ranging from 23-113 kg (50-250 lb). It utilises standard solid-propellant rocket motors, including the Thiokol Castor, Thiokol Recruit, Hercules X-259, Thiokol 261 and 23KS11,000. Cruciform fins, indexed in line, are fitted to the first and second stages. Size of the payload compartment is 2·16 m × 0·56 m (85 in long by 22 in in diameter).

Athena H re-entry research vehicle

Initial Athena launcher offset is adjusted as directed by the output of a ground-based meteorological computer loop. The inputs to this loop are vehicle dynamic characteristics and measured wind profiles to 61,000 m (200,000 ft). The vehicle is spin-stabilised during ascent boost. Ground commands, based upon radar-derived trajectory dispersion data, are generated by a ground-based computer and adjust the pre-set re-entry angle resulting from the control system manoeuvre. The vehicle is re-spun to provide stability during re-entry boost.

Midcourse attitude correction is provided by a Honeywell DHG 138A attitude controller employing two two-degree-of-freedom attitude gyros in a COG orientation.

The velocity package is fired after Athena reaches its apogee of either 259,000 m (850,000 ft) for high-angle re-entry (IRBM) simulation or 182,900 m (600,000 ft) for lower-angle (ICBM) simulation.

DIMENSIONS:
Length overall 15·74 m (51 ft 8 in)
Body diameter, second stage 0·79 m (2 ft 7 in)
WEIGHT:
Max launching weight 7,260 kg (16,000 lb)
PERFORMANCE:
Max speed at re-entry test altitude (nominal) 7,010 m (23,000 ft)/sec
Range (based on current range use) 780 km (485 miles)

ATHENA H

This new and much-enlarged version of the Athena re-entry research vehicle became available for use in 1971. It is large enough to carry a full-scale military re-entry body if required, its max payload being about 454 kg (1,000 lb).

Athena H is available in both two-stage and three-stage configurations, comprising a booster stage and single-stage or two-stage velocity package. After booster burn-out, it will coast to a peak altitude of nearly 305,000 m (1,000,000 ft). During this phase, the velocity package will be re-orientated and re-ignited at the altitude calculated to produce the desired re-entry angle.

DIMENSION:
Length overall 18·52 m (60 ft 9 in)

Percheron sounding rockets on launchers

WEIGHT (approx):
Max launching weight 14,515 kg (32,000 lb)
PERFORMANCE (estimated):
Max speed at re-entry test altitude:
45 kg (100 lb) payload 7,620 m (25,000 ft)/sec
90 kg (200 lb) payload 7,160 m (23,500 ft)/sec

PERCHERON A and B

The basic Percheron is a single-stage solid-propellant sounding rocket, which utilises an XM-33-E8 Castor I (Percheron A) as the main stage, assisted by two TE-M-29 Recruit auxiliary motors. Percheron B utilises a TX-354-4 Castor II as the main stage, also assisted by two Recruit auxiliary motors.

These vehicles fly a ballistic trajectory and can carry a gross payload of 454 kg (1,000 lb) to an altitude of 259 km (160 miles) in the case of Percheron A, and 370 km (230 miles) in the case of Percheron B.

DIMENSIONS:
Length overall 10·56 m (34 ft 8 in)
Max body diameter 0·79 m (2 ft 7 in)
WEIGHTS:
Launching weight (incl gross payload of 454 kg; 1,000 lb)
Percheron A 5,064 kg (11,180 lb)
Percheron B 5,409 kg (11,942 lb)

PERCHERON C

Percheron C is a single-stage solid-propellant sounding rocket which utilises a TX-526 Castor IV motor as the main stage, assisted by four TE-M-29 Recruit auxiliary motors. It follows a ballistic trajectory and can carry a payload of 1,814 kg (4,000 lb) to an altitude of 400 km (250 miles).

DIMENSIONS:
Length overall 14·2 m (46 ft 9½ in)
Max body diameter 1·0 m (3 ft 3¼ in)
WEIGHT:
Launching weight (incl gross payload of 1,814 kg; 4,000 lb) 13,300 kg (29,350 lb)

FAIRCHILD INDUSTRIES

FAIRCHILD INDUSTRIES INC

EXECUTIVE OFFICE:
Germantown, Maryland 20767
OFFICERS AND DIVISIONS:
See Aircraft section

Fairchild Space and Electronics Company was prime contractor for ATS-6, NASA's second-generation Applications Technology Satellite.

ATS-6

The most complex, versatile and powerful communications satellite yet built, ATS-6 (ATS-F), the sixth of NASA's Applications Technology Satellite series, was launched into a 35,760-35,792 km (22,220-22,240 mile) near-synchronous orbit on 30 May 1974, by a Titan IIIC from the Eastern Test Range, Cape Kennedy.

Weighing 1,348 kg (2,972 lb), the spacecraft is 8 m (26 ft) high. It consists basically of a rectangular Earth Viewing Module (EVM), housing controls and Earth-orientated experiments, connected by a tubular support structure to a 9 m (30 ft) diameter deployable reflector antenna. Two arms, each supporting a semi-cylindrical solar array, extend from the hub that supports the antenna. Mounted on top of the hub is an Environmental Measuring Experiments (EME) package. Overall width of the satellite at the solar array is about 16 m (52 ft).

The spacecraft embodies several major technology innovations for communications satellites, including a large parabolic antenna, a digital computer for attitude control, the use of graphite composite material for primary structure, heat pipes for primary thermal control, monopulse tracking for attitude control, and an offset pointing capability.

Heart of the spacecraft is the EVM. Containing high-power electronic units capable of operating in many hundreds of modes, it requires sophisticated cooling to prevent overheating. This is provided by the thermal control system, consisting of thermal louvres, heat pipes and highly efficient insulation. The louvres, mounted in the 'north' and 'south' faces of the EVM, open and close as required, to dispel or retain heat and so maintain a temperature within 15 to 20°C.

Located initially over the Galapagos Islands, the satellite, in conjunction with ATS-1 and ATS-2, was to be used in a Health Education Telecommunications experiment covering remote and poorly served rural areas in the Appalachian and Rocky Mountain States, and the States of Washington and Alaska. The experiment demonstrated the feasibility of remote medical consultation and diagnosis, and general education.

The satellite was repositioned over Lake Victoria in Kenya, East Africa, in July 1975, from where it was used by the Indian government to conduct its Satellite Instructional Television Experiment, involving the broadcasting of programmes dealing with agriculture, family planning, hygiene and occupational training to some 5,000 remote villages equipped with low-cost receivers. While in this position, the satellite was also used to track and relay data from the docked Apollo-Soyuz Test Project spacecraft.

The satellite is equipped to perform 20 other experiments, including evaluation of a high-resolution radiometer for meteorological work, and of a caesium-ion-thruster attitude stabilisation system; and investigations of the radiation environment at synchronous altitude, and of the phenomenon of mutual interference between spacecraft and ground communications in the 6GHz band. It is performing well in orbit, and was to be moved back to the western hemisphere during August 1976.

ATS-6 applications technology satellite

GENERAL DYNAMICS

GENERAL DYNAMICS CORPORATION

HEAD OFFICE:
Pierre Laclede Center, St Louis, Missouri 63105

Convair Division
San Diego, California 92138

OFFICERS: See Aircraft section

In June 1974, the Convair Aerospace Division was reorganised into two separate divisions of General Dynamics Corporation—the Convair Division in San Diego, California, and the Fort Worth Division in Texas.

Convair Division devotes a major part of its activity to production and launch of Atlas and Centaur space launch vehicles, as well as production of DC-10 fuselages. Convair is also producing Cruise Missile prototypes for the US Navy, and the Space Shuttle Orbiter mid-fuselage for NASA; and is tendering for the Interim Upper Stage (IUS) and Space Tug programmes for the Shuttle.

ATLAS E and F

Atlas E and F series rockets were formerly ICBMs deployed at Strategic Air Command bases as part of the US strategic missile force. Phase-out of the Atlas ICBM force in 1965 made them available for conversion into launch vehicles. The series E and F vehicles are essentially identical, the primary difference being in their method of operational deployment.

The missiles are stored at Norton AFB, California, until they enter the refurbishment and launch programme. Refurbishment from missile to launch vehicle is undertaken by Convair Division at Vandenberg AFB, California, utilising some of the personnel who conduct the launchings. This on-site activity was made possible by a reduction in planned launch rates, and led to a substantial reduction in the overall 'launched cost' of each vehicle.

In January 1976, nine Atlas E and F missiles remained available to be launched. All of these are now assigned to missions. During 1975 NASA selected Atlas E and F to launch the next generation of meteorological satellites and the first Seasat.

Atlas E and F launch vehicles are used for both orbital and sub-orbital missions, and have been integrated with a variety of upper stages and payload delivery systems. The addition of a launch site in South Vandenberg has substantially increased polar orbit capability, previously accomplished by launching westerly and then 'doglegging' south during powered flight. Sub-orbital flights are launched normally toward the Kwajalein Missile Range in the Pacific Ocean. Such launches continue to be of particular value in experiments conducted by the USAF Space and Missile Systems Organization and the US Army Advanced Ballistic Missile Defense Agency.

The vehicle descriptions of Atlas E and F are generally similar to that of the Atlas SLV, described separately.

DIMENSIONS:
Length overall without payload 21·72 m (71 ft 3 in)
Body diameter 3·05 m (10 ft 0 in)

WEIGHTS:
Launching weight:
Atlas E 122,470 kg (270,000 lb)
Atlas F 122,000 kg (269,000 lb)

PERFORMANCE:
Ballistic:
Range with 2,720 kg (6,000 lb) payload 9,265 km (5,750 miles)
Orbital (Atlas alone):
907 kg (2,000 lb) payload in 185 km (115 mile) Earth polar orbit
Orbital (with upper stages):
Atlas E or F with upper stages such as OV1 or Burner II can insert varying number of payloads into various orbits: eg Atlas 107F, with three OV1s, inserted 27 separate experiments into four different orbits on the same launch

Atlas F launch vehicle

ATLAS SLV

The Atlas Standardised Launch Vehicle (SLV) had its inception in Atlas (above), the United States' first intercontinental ballistic missile (ICBM). Continuous technological improvements have been embodied in the SLV series, of which two versions are currently in service:

SLV-3A, for use with Agena or OV1 upper stages.

SLV-3D, for use with the Centaur D-1A high-energy upper stage.

These vehicles differ from their immediate predecessor, the SLV-3, mainly in increased tank length and rocket engine thrust.

Atlas/Centaur AC-36, launch vehicle for an Intelsat IVA communications satellite

Atlas is a 'stage-and-a-half' vehicle, consisting of sid booster and central sustainer sections. The sustainer sec tion includes the propellant tanks and a single rocke engine. The booster engines receive fuel from the sus tainer tanks and are jettisoned midway into flight.

The engine system is the Rocketdyne MA-5, usin liquid oxygen and RP-1 propellants. Total thrus developed is 1,917·5 kN (431,040 lb), including 1,646 kN (370,000 lb) total from the two boosters, 266·9 kN (60,000 lb) from the sustainer, and 4·6 kN (1,040 lb) tota axial thrust from the two vernier rockets. All engines are ignited at lift-off.

Propellant tanks are pressurised and made of thin-wall stainless steel, with an intermediate bulkhead separating the oxidiser and fuel. One noticeable difference between the SLV-3D and the SLV-3A is in the forward tank structure. For the SLV-3D the tank is a constant 3·05 m (10 ft) diameter up to the adapter for the upper stage. The forward tank of the SLV-3A tapers to a 1·78 m (5 ft 10 in) diameter at the adapter attach ring. Total propellants carried by the SLV-3D weigh 121,582 kg (268,040 lb); those of the SLV-3A weigh 134,055 kg (295,540 lb).

Most of the electronic command and control functions for the SLV-3D are generated by its Centaur D-1A upper-stage astrionics system. The SLV-3A has its own systems, independent of the upper stage, including radio guidance.

By the beginning of 1976, Atlas had been the booster for 162 space launches, with many space 'firsts' to its credit. The world's first communications satellite was launched by Atlas in 1958, as were the first US manned orbital flights (Mercury), starting in 1962. All US planetary spacecraft have been launched by Atlas or Centaur. In 1972 Pioneer 10 was started on its flight path to Jupiter with the highest velocity ever imparted to a spacecraft. Launch vehicle was an Atlas/Centaur with an additional TE-M-364-4 solid-propellant rocket motor.

By January 1976 Atlas had amassed a total of 424 space and ballistic launches. During a 5½-year period, it had one unbroken succession of 53 successful space launches.

DIMENSIONS:

Diameter	3·05 m (10 ft 0 in)
Length:	
SLV-3A	21·6 m (71 ft 0 in)
SLV-3A/Agena	36·0 m (118 ft 0 in)
SLV-3D	18·6 m (61 ft 0 in)
SLV-3D/Centaur	39·9 m (131 ft 0 in)

PERFORMANCE:

Atlas/Centaur: See Centaur entry

Atlas SLV-3A/Agena:

3,992 kg (8,800 lb) into 185 km (115 mile) circular orbit

1,325 kg (2,920 lb) into synchronous transfer orbit

Atlas/OV1: See OV1 entry

CENTAUR

Centaur was the first US high-energy upper stage and the first to utilise liquid hydrogen as a propellant. The latest version, Centaur D-1, is combined with either the Atlas SLV-3D or the Titan IIIE, providing for a wide range of applications and capability.

The original contracts for Centaur were awarded by the Advanced Research Projects Agency (ARPA) in 1958. The programme was transferred from ARPA to NASA in 1959. NASA's Lewis Research Center provides overall management and integration of the Centaur programme.

The 21st operational launch of Atlas/Centaur, and the final launch of the Centaur D series, occurred in August 1972 with the successful launch of an Orbiting Astronomical Observatory. Other payloads launched successfully by Atlas/Centaur D include the Surveyor series of seven lunar spacecraft, Applications Technology Satellite, the Mariner 7 and 9 Mars spacecraft, four Intelsat IV communications satellites and Pioneer spacecraft.

In April 1973, the first Atlas/Centaur D-1A launched Pioneer 11 on a Jupiter fly-by mission; the same spacecraft is scheduled to pass Saturn in 1979. Three more Intelsat IVs, Mariner 10 and two Intelsat IVAs have also been launched successfully by this vehicle. In early 1976, Atlas/Centaur D-1A had been assigned future missions extending into 1981. They include further launches of Intelsat IVAs and Vs, Comstars (domestic communication satellites leased to AT and T by Comsat), HEAOs, Pioneer missions to orbit Venus and probe its atmosphere, and the US Navy's Fleetsatcom.

The second and third operational missions of Titan III E/Centaur D-IT launched successfully two Viking Mars orbiter/landers in 1975. In early 1976 the Helios 2 solar probe was launched successfully. This mission was also designed to prove further Centaur's multi-burn and extended coast capability. After completing two burns and separating from the Helios spacecraft, Centaur went on to perform five additional burns separated by zero-gravity coasts of up to 5½ hours. The first operational use of this new capability will occur during the 1977 Mariner Jupiter/Saturn missions.

Centaur D-1 retains the same propulsion and structural features as its predecessor, Centaur D, with stainless steel pressurised tanks and two 66·72 kN (15,000 lb st) Pratt & Whitney RL10A liquid oxygen/liquid hydrogen rocket engines. Specific impulse with this propellant combination is 445 sec, the highest of any current space vehicle. Total propellants carried weigh 13,950 kg (30,750 lb). Attitude control is achieved by gimballing the two main engines or by clusters of small hydrogen peroxide rocket motors.

Several of the astrionics components have been redesigned or repackaged for Centaur D-1. The most significant addition is a 16,000 word capacity Teledyne digital computer. Navigation, guidance, vehicle stability, tank pressurisation, propellant management, telemetry formats and transmission, and event initiation are all controlled by the computer. Guidance, control and sequencing for the Atlas booster are provided by the Centaur D-1A astrionics system. The Centaur D-1T also provides guidance for its Titan booster.

Payloads are carried on adapters which mount to the forward end of the Centaur. A 3·05 m (10 ft) diameter fairing protects payloads for Centaur D-1A. For Titan/Centaur a 4·27 m (14 ft) shroud encloses both the payload and the Centaur D-1T.

DIMENSIONS:

Centaur length	9·14 m (30 ft 0 in)
Centaur diameter	3·05 m (10 ft 0 in)
Atlas/Centaur length	39·9 m (131 ft 0 in)

PERFORMANCE:

Atlas/Centaur:

5,080 kg (11,200 lb) into 185 km (115 mile) circular orbit

1,860 kg (4,100 lb) into synchronous transfer orbit

590 kg (1,300 lb) to near planet

Titan/Centaur:

15,400 kg (34,000 lb) into 185 km (115 mile) circular orbit

3,310 kg (7,300 lb) into synchronous equatorial orbit

3,720 kg (8,200 lb) to near planet

GENERAL ELECTRIC

GENERAL ELECTRIC COMPANY SPACE DIVISION

HEAD OFFICE:

Valley Forge Space Center, PO Box 8555, Philadelphia, Pennsylvania 19101

General Electric's Space Division is vehicle contractor for NASA's Landsat programme (formerly Earth Resources Technology Satellite, ERTS). It is also responsible for design and manufacture of the stabilisation and control system for NASA's Orbiting Astronomical Observatory (OAO).

LANDSAT (ERTS)

Designed to study the resources of Earth from space, ERTS-1 (Earth Resources Technology Satellite) was launched by NASA, using a Delta vehicle from Vandenberg, on 23 July 1972, into a near-circular 901 × 933 km (560 × 580 mile) polar orbit. It was subsequently renamed Landsat-1.

Based on the Nimbus meteorological satellite, Landsat is 3·05 m (10 ft) high, not including the solar paddles, and weighs 891 kg (1,965 lb). The solar array consists of two panels measuring 1·2 × 2·4 m (4 × 8 ft); additional solar cells are mounted on the transition sections of the array.

Two main groups of sensors are carried: a multispectral scanner subsystem (MSS) and a return beam vidicon (RBV) subsystem. The two sensors repetitively take photo-like images of the Earth, while a data collection system (DCS) collects environmental data of various types from ground-based remote platforms. A high-performance recorder, known as a wide-band video tape recorder (WBVTR), stores photo images from the RBV and MSS, as needed, when the satellite is out of range of a direct-readout data acquisition station.

The RBV camera system, developed by the Astro Electronics Division of RCA, comprises three 2 in TV-type 4,125-line cameras, taking photographs in green, red and near-infra-red bands, and with each picture covering a 100 nm (185 km; 115 mile) square with a resolution of about 100 m (330 ft). Shallow ocean areas stand out in the green band, as do sedimentation and pollution. Man-made structures appear brightly in the red band, where vegetation appears very dark, to give information on land-use mapping. Water areas stand out as dark areas on the infra-red photographs.

The MSS, built by Hughes Aircraft Co, is an optical-mechanical sensing system which simultaneously detects optical energy in the green, red and two infra-red spectral bands, including near-infra-red. The system scans the same 100 nm (185 km; 115 mile) wide flight path as that observed by the RBV camera array. Scanning is achieved by means of a mechanically oscillating flat mirror, which flip-flops from side to side about 13 times each second. MSS images are photograph-like in appearance and of excellent quality, with a resolution of 70 m (230 ft).

Both RBV and MSS produce black and white photographs; those from the different bands can be put together to form false-colour photographs.

The DCS provides users of the space data with near real-time environmental information, collected from ground-based sensor instruments measuring soil, water, air and other parameters, to assist in the interpretation of data from the two sensors.

Owing to a switch problem in the satellite's power source, the RBV system was de-activated 14 days after launch. In March 1973 the video tape recorder, which had exceeded its 500-hour designed lifetime, developed sporadic bursts of noise which degraded the transmission of pictures and consequently was turned off for investigation. Photographs transmitted live were still excellent.

By early 1976 Landsat-1 had returned more than 150,000 images, covering the entire USA and many other parts of the globe. It takes 18 days to complete a cycle over the whole Earth, beginning again on the 19th day. Analysis of these images has produced valuable results, including more accurate estimates of the acreage of agricultural crops and of forest timber volumes; more rapid provision of maps for land use; more accurate estimates of areas covered by water, of flood areas and of water run-off from melting snow; identification of previously unknown Earth structural features, important for the detection of minerals amd potential earthquake zones; provision of coastal zone data for prediction of the dispersal of riverborne sediments, and industrial wastes and sewage; and classification of strip-mined areas for the estimation of mined acreage and reclaimed regions.

In 1976 images from the spacecraft were used to help Alaskan Indians select thousands of acres of potential commercial timber land and promising areas for mineral exploration from vast tracts of wilderness offered by the US federal government to settle native claims going back to the US purchase of Alaska from Russia in 1867.

Landsat-2 (known formerly as ERTS-B) was launched on 22 January 1975, into an orbit similar to that of Landsat-1, and was then manoeuvred 180° out of phase from it to provide repetitive coverage of all portions of the Earth every nine days, instead of once every 18 days.

Landsat-1 Earth resources technology satellite

Landsat-2 is similar to Landsat-1 and weighs 953 kg (2,100 lb). Day-to-day operation of the spacecraft is, however, controlled by a unique 14 kg (30 lb) digital computer with a 4,096 word memory, the most advanced of its type ever flown on an unmanned NASA satellite. The computer can handle 55 separate commands from ground stations, to carry out routine operations on board the satellite for periods of up to 24 hours. Such remote, computer-directed operations are essential, as the spacecraft is out of tracking station range about 80 per cent of the time. By early 1976 Landsat-2 had returned more than 50,000 images.

A third spacecraft in the series, Landsat-C, is scheduled to be launched late in 1977. This will carry improved sensors to collect more detailed data than its predecessors.

HUGHES

HUGHES AIRCRAFT COMPANY

HEAD OFFICE:

Culver City, California 90230

Telephone: (213) 391-0711

Current space programmes for which Hughes Aircraft is prime contractor include development and manufacture of NASA's new-generation Orbiting Solar Observatory (OSO) satellites. The company is also engaged on development and manufacture of classified military satellite systems, and on the investigation and flight testing of experimental ion engines for advanced spacecraft.

Hughes has an $800,000 contract from the Indonesian government to define a proposed satellite system that that country wished to have in service by the end of 1976. The system calls for two satellites and 50 ground stations, located in key areas of Indonesia's scattered 13,000-island archipelago. The entire system is expected to cost about $100 million.

INTELSAT IV

Under a $72 million contract awarded on 18 October 1968, Hughes Aircraft Company became prime contractor for the Intelsat IV satellite. The initial order for one prototype and four flight satellites was placed by Communications Satellite Corporation (Comsat) on behalf of the International Telecommunications Satellite Consortium (Intelsat). Contracts for a second series of four satellites were negotiated in the Autumn of 1970.

Larger than any previous communications satellite, Intelsat IV is the fifth-generation Hughes satellite to see service since the tiny Syncom 2 was launched in 1963. In its design and manufacture, Hughes was assisted by subcontractors in nine member-nations of Intelsat. Northern Electric in Canada was also a subcontractor.

Intelsat IV was designed to offer communications facilities 25 times greater than those of any satellite previously put into service. It is able to carry 6,000 two-way telephone calls, transmit 12 simultaneous colour television broadcasts, or handle an infinite variety of different kinds of communications signals. Design lifetime in orbit is seven years.

A unique feature is the ability to focus power into two 'spotlight' beams, 4½° wide, and direct them at any selected areas, thus providing a stronger signal and increased number of available channels in areas of heaviest communication traffic. This is made possible by mounting on the satellite two steerable dish antennae, each 1·27 m (50 in) in diameter, controlled by command signals from Earth. Two horn antennae, with 17° beams, provide coverage outside the areas encompassed by the spotlight beams. Electronic switching enables ground controllers to adjust the amount of power going into each of the two antennae systems. Two Earth-coverage horns (one as a backup) are used for reception.

Intelsat IV has 12 broad-band communications channels. Each has a bandwidth of 40MHz, providing capacity for some 750 communications circuits. Four of the repeaters serve the Earth-coverage antennae, the other eight are intended to be switched as required to either Earth-coverage or spot-beam antennae.

A total of 45,012 solar cells provide 569W of power initially. Effective isotropic radiated power (EIRP) at beam centre is 33·7 dbw for the spot-beam antennae and 22 dbw for the Earth-coverage beams. A total of 24 output TWTs is used (12 for redundancy), each with an ouput of about 7·5W.

The satellite is basically drum-shape, with an Aerojet-General SVM-4A apogee motor containing 686 kg (1,514 lb) of solid propellant. Positioning and orientation control are provided by a redundant hydrazine system, with 122 kg (273 lb) of fuel.

The first flight Intelsat IV and its subsystems were built and tested at the Hughes space facilities, El Segundo, California, with the member-nation subcontractors participating directly. The second was assembled and tested by Hughes, but most of its subsystems were built by the participating subcontractors. The third spacecraft was assembled by British Aircraft Corporation at Bristol, England, using subsystems furnished by subcontractors. Similar arrangements were made for the second series of four satellites.

Subcontractors to Hughes in the Intelsat IV programme were Thomson-CSF (France), for telemetry and command antennae, telemetry horn and telemetry and command equipment; Nippon Electric Co (Japan) for repeater F-2; AEG-Telefunken (Germany) for repeater F-3; Northern Electric Co (Canada) for repeater F-4; Kolster Iberica SA (Spain) for TWT power supply converters for drivers; Etudes Techniques et Constructions Aérospatiales (Belgium) for battery controller and relay; Svenska Radio AB (Sweden) for solenoid and squib drivers; Contraves AG (Switzerland) for antennae positioning electronics and de-spin control electronics; Selenia SpA (Italy) for Earth coverage—transmit and receive antennae, and spot-beam communications antennae; British Aircraft Corporation (United Kingdom) for nutation damper, positioning and orientation subsystem, battery pack, structure and harness, Sun sensor, and solar panel, with solar cells supplied by Société Anonyme de Télécommunications (France), Ferranti Ltd (United Kingdom) and AEG-Telefunken (Germany).

Other participants in the programme included Etudes Techniques et Constructions Aérospatiales (Belgium) for digital portion of systems, test equipment and ground control equipment; Svenska Radio AB (Sweden) for RF portion systems test equipment; and British Aircraft Corporation for handling equipment and spacecraft integration and test.

Intelsat IV-A was launched from Cape Kennedy on 25 January 1971 by an Atlas-Centaur vehicle. It was put into a synchronous orbit and allowed to drift toward its operational station at 24·5°W, where it has been in use since 28 March 1971, linking countries with Earth stations ringing the Atlantic basin.

Intelsat-B was launched at 8.10 pm EST on 19 December 1971 and allowed to drift eastward toward its station over the Atlantic, where it entered commercial service at the end of February 1972. The British-assembled Intelsat-C was launched at 7.12 pm EST on 22 January 1972 and was placed in synchronous orbit two days later. It was on station over the Pacific in time to transmit TV and press coverage of President Nixon's visit to Peking on 21-28 February.

Intelsat IV-D (F-5) was launched on 13 June 1972 and has been positioned on the equator over the Indian Ocean, at longitude 62°E, to complete the first global system of Intelsat IVs. This spacecraft added 12 television channels to those now available between the USA and other nations, and can also carry 5,000 to 6,000 two-way telephone conversations under average conditions.

Intelsat F-7 was launched on 23 August 1973, and was placed in service over the Atlantic at longitude 29·8°W on 21 November 1973. Intelsat F-8, launched on 21 November 1974, was stationed over the Pacific Ocean at longitude 174°E to join the earlier Intelsat launched in January 1972. Commercial traffic was transferred to the new spacecraft on 14 December 1974.

Intelsat F-6 (the seventh in the series) was launched on 20 February 1975, but failed to go into orbit due to a malfunction of the Atlas/Centaur launching vehicle. This spacecraft was to have joined and replaced over the Indian Ocean the operational Intelsat IV satellite, which is suffering from a fall-off in performance.

The standby satellite for F-6 was an Intelsat III; but a contingency plan arranged after the failure substituted the first Intelsat IV to be built, the F-1, which was not used because of an assembly accident which was subsequently rectified. This F-1 standby spacecraft was launched successfully on 22 May 1975 and is now on station over the Indian Ocean. It completed the originally-planned worldwide network of seven Intelsat IV spacecraft, comprising three over the Atlantic, two over the Pacific and two over the Indian Ocean.

In addition to these seven spacecraft, the Intelsat IV system consists of 111 antennae at 88 Earth stations in 54 countries. The system has accommodated an increase in transoceanic telephone traffic over the past 10 years from 3 million calls a year to over 50 million.

DIMENSIONS (F-8):

Diameter	2·38 m (7 ft 10 in)
Height of solar drum	2·82 m (9 ft 3 in)
Height overall	5·28 m (17 ft 4 in)

WEIGHTS:

At launch	1,406 kg (3,100 lb)
In orbit	700 kg (1,544 lb)

INTELSAT IVA

Pending the availability of Intelsat V, Hughes is building six improved Intelsat IVA spacecraft, with BAC as a principal subcontractor, to handle the world's mounting telecommunications traffic up to 1979. These new spacecraft will have 20 transponders (individual receiver-transmitters) compared with the 12 on Intelsat IV. A new advanced antenna system concentrates signal beams like spotlights into the communication and business centres on both sides of an ocean, via 16 of the 20 transponders. Using a technique called beam diversity on 'frequency re-use', the spacecraft casts separate transmitting and receiving spot beams. These permit an Earth station in one beam to receive and transmit on the same frequencies as an Earth station in the other beam without interference. The result is an increase in capacity to 11,000 two-way telephone conversations or 20 colour TV channels, almost twice that of its predecessors.

The first Intelsat IVA, F-1, was launched successfully on 26 September 1975, and was stationed in synchronous orbit over the Atlantic at longitude 25°W on 9 December 1975. It became operational on 1 January 1976. It was later joined by Intelsat IVA F-2, launched on 29 January 1976, and now on station over the Atlantic at longitude 29·5°W, where it is serving as a backup to the F-1 spacecraft.

A third Intelsat IVA was due to be launched in late 1976.

DIMENSIONS:

Diameter	2·36 m (7 ft 9 in)
Height overall	7·00 m (22 ft 11 in)

WEIGHT:

At launch	1,495 kg (3,300 lb)

ANIK

Under a $31 million contract from Telesat Canada, Hughes Aircraft Company built three spacecraft to provide a domestic satellite communications system in Canada. Major subcontractors were Northern Electric Company of Montreal, which provided the complete electronics system, including the entire communications package; and Spar Aerospace Products of Malton, Ontario, which provided the spacecraft structures and engineering support services. These two subcontractors will provide the electronics packages and structures for up to 15 additional spacecraft of similar type which Hughes expects to sell in worldwide markets.

The Telesat contract called for delivery of the first satellite, to be known as Anik-1, by October 1972, followed by the second and third at four-month intervals thereafter. Each satellite was equipped to provide 12 radio frequency channels, ten of which are available for commercial use with the remaining two on standby. Each channel carries one colour TV signal or its equivalent in message traffic. This can be as high as 960 one-way voice channels.

Solar cells, arranged around the body of each satellite, deliver 250W of power. The satellite's antenna is a 1·52 m (60 in) parabolic dish, providing a beam width of 3° by 8°.

Anik-1 (Telesat-A) was launched on 9 November 1972, by the first 'Straight-Eight' Delta launch vehicle, from Cape Kennedy. An on-board solid-propellant rocket motor was fired during the apogee of the transfer orbit to 'kick' the satellite into synchronous orbit above the equator at longitude 114°W, a point due south of Gallup, New Mexico, on 24 November. Small on-board jets, supplied by 45 kg (100 lb) of hydrazine propellant, provide attitude and station-keeping control of the satellite while on station for a period of at least seven years. Anik-1 became operational on 11 January 1973, when the first commercial telephone call was relayed between Resolute, in Queen Elizabeth Islands, and Ottawa. Full-scale commercial operation followed the orbiting of Anik-2, which was launched on 20 April 1973 and was also positioned subsequently over the equator at longitude 109°W. A third satellite was built as a spare, but was launched on 7 May 1975 to serve initially as a backup, and will enter active service as a replacement or when an increase in business requires the use of all three satellites. Afte

Intelsat IVA communications satellite

Canada's Anik communications satellite

aunch Anik-3 was stationed at 104°W longitude over the Pacific Ocean almost due south of Carlsbad, New Mexico. All three Anik satellites were performing satisfactorily in mid-1976.

DIMENSIONS (approx):

Height overall	3·3 m (11 ft)
Diameter	just over 1·83 m (6 ft)

WEIGHTS:

Launching weight	approx 562 kg (1,240 lb)
In orbit	270 kg (600 lb)

ORBITING SOLAR OBSERVATORY (OSO)

Built by Hughes Aircraft Company, the second-generation Orbiting Solar Observatory satellite OSO-I was launched on 21 June 1975 at 11.43 GMT by a two-stage Delta from Cape Canaveral into a 560 × 544 km (348 × 338 mile) orbit.

Larger, heavier (OSO-I 1,064 kg; 2,346 lb. OSO-1 200 kg; 440 lb) and more advanced than earlier OSOs (see previous editions of *Jane's)*, the new spacecraft is designed to continue investigations into how the Sun, with a 'relatively cool' surface temperature of 10,000°F, heats its 16,000 km (10,000 mile) wide corona to 4,000,000°F. Understanding the Sun is important not only because it is the ultimate source of energy that supports all life on Earth, but also because it is the closest nearby star for astronomers and physicists to study in detail. A secondary objective is to investigate celestial sources of X-rays in the Milky Way galaxy and beyond.

Originally, when the contract for the spacecraft was awarded by NASA in September 1971, it was planned to produce three of the new satellites, OSO-I, J and K. Through budget restrictions, however, OSO-J and K have been eliminated from the programme.

The spacecraft structure consists of a spinning base section, called the 'wheel', and an upper portion, or 'sail'. The wheel carries experiments such as cosmic X-ray telescopes and associated spacecraft subsystems not requiring sustained solar pointing. The sail carries those experiments such as a spectrometer and the associated subsystems that must point continuously at the Sun. It also mounts the solar cell array.

Eight experiments are carried, six in the wheel and two in the sail. The wheel experiments include: high sensitivity crystal spectrometer and polarimeter; mapping X-ray heliometer; investigation of soft X-ray background radiation; cosmic X-ray spectroscopy; high energy celestial X-ray experiment; and an extreme ultraviolet radiations from earth and space experiment. The two sail experiments are the high resolution ultraviolet spectrometer, from the University of Colorado, and the chromosphere fine structure experiment, from the Centre National de la Recherche Scientifique (CNRS), Paris, France.

The sail has a pointing accuracy of 1 arc second (1/3,600°), equivalent to a rifle marksman keeping a 3 m (10 ft) target in his sights at a distance of 640 km (400 miles). Earlier OSOs pointed with an accuracy of 60 arc seconds (1/60°), scanning a solar area of 43,000 km (27,000 miles). The improved pointing accuracy of OSO-I enables its instruments to scan areas of the Sun's rim in 725 km (450 mile) segments, an area considered small as the Sun's diameter is approx 1,390,000 km (864,000 miles).

The experiments are commanded automatically by a memory system programming itself. In the case of earlier OSOs, ground controllers had to wait up to 1½ hours for completion of an orbit before they could send commands to the satellite.

The OSO-I programme is under the overall management of NASA's Office of Space Science. Project management is the responsibility of the Goddard Space Flight Center, which is also responsible for tracking and data acquisition.

DIMENSIONS:

Wheel:	
Diameter	1·52 m (60 in)
Height	0·72 m (28 in)
Sail:	
Height	2·35 m (93 in)
Width	2·10 m (82 in)

WEIGHT:

At launch	1,064 kg (2,346 lb)

Orbiting Solar Observatory satellite OSO-I

MARISAT

Built by Hughes Aircraft Company for the Comsat General Corporation, Marisat A was launched by NASA using a Delta 2914 vehicle from Cape Canaveral on 19 February 1976 at 22·32 GMT.

The spacecraft, the world's first commercial maritime telecommunications satellite, is designed to provide rapid, high quality communications between ships at sea and shore offices. Telephone and telex messages may be exchanged without fear of interference or delay due to severe weather or ionospheric disturbances that might disrupt radio traffic. The satellite is expected to improve significantly the communication of distress, safety, search and rescue, and weather reports.

After launch, Marisat A was manoeuvred into a synchronous orbit and is now on station over the Atlantic at longitude 15° W, where it covers a total of 155 million square kilometres (60 million square miles) of busy sea areas.

A second satellite, Marisat B, was launched in May 1976 and is now stationed over the Pacific at longitude 176° W, to serve the commercial maritime industry in this area. A third satellite has been built as a reserve.

Based upon the proven technology of the Intelsat IV and IVA satellites, also built by Hughes, the 655 kg (1,445 lb) spacecraft is drum-shaped, with the antenna system mounted on one end. Earth-pointing capability of the antenna system is achieved through a non-contacting three-channel co-axial rotary joint, which couples the de-spun antenna 'farm' to the rotating platform containing satellite subsystems, including the communications repeaters. Electrical power of 330W is generated mainly by approximately 7,000 solar cells on the outer surface of the cylindrical body.

Three communications repeaters are carried. One of these, the UHF (240-400MHz) repeater, contains one wideband and two narrowband channels, all three of which can be turned on and off by ground command, and initially will be used exclusively by the US Navy. The remaining two repeaters provide ship-to-shore and shore-to-ship civil maritime communications. These two latter channels, both approx 4MHz wide, operate in the L and C bands. One channel translates shore-to-ship signals from 6GHz to 1·5GHz; the other translates ship-to-shore signals from 1·6GHz to 4GHz.

Earth stations operating with Marisat have 12·8 m (42 ft) diameter antennas, but ship-based terminals employ above-deck antennae of only 1·22 m (4 ft) diameter, and can be either bought or leased by users.

The US Navy's use of the UHF channels will be for at least two years, pending the introduction of its own Tac-

Marisat maritime communications satellite

scan satellites. The remaining channels will be used by the Maritime Administration for demonstration of the system's use for commercial shipping. When the Navy contract is complete, the Marisat system will be made available for a variety of commercial users during the remainder of its five-year planned life.

DIMENSIONS:

Diameter	2·13 m (7 ft 1 in)
Height	3·81 m (12 ft 6 in)

WEIGHTS:

At launch	655 kg (1,445 lb)
In orbit	330 kg (728 lb)

USAF 711 SATELLITE

The US technical press reports that Hughes has developed for the USAF a new surveillance satellite to monitor foreign radar activity. Developed under the Air Force's 711 programme, it is intended to be launched by Titan III booster into a highly elliptical orbit and then to relay data to ground stations for analysis.

The new satellite is intended to supersede the Ferret type produced by Lockheed.

USAF 313 SATELLITE

Under the USAF's 313 programme, Hughes was contracted to develop a military satellite data relay system. This is intended to utilise satellites in polar orbit to relay directly to US ground stations reconnaissance photographs and early warning data from other satellites.

STRATEGIC COMSAT SYSTEM

A new strategic communications satellite system is being developed by Hughes for the USAF. Its purpose is to supplement the lower-latitude coverage that will be provided in the mid-seventies by the TRW Fltsatcom network, and provide high-latitude coverage for SAC air-to-ground communications.

Few details are available except that each satellite system is expected to weigh only 16 kg (35 lb), and to be carried by a USAF cloud-cover surveillance satellite in polar orbit.

LOCKHEED

LOCKHEED AIRCRAFT CORPORATION

HEAD OFFICE AND WORKS:
Burbank, California

OFFICERS: See Aircraft section

Lockheed Missiles & Space Company Inc

HEAD OFFICE:
1111 Lockheed Way, Sunnyvale, California 94088
Telephone: (408) 742-6688

OTHER FACILITIES:
Palo Alto and Santa Cruz, California

DIRECTOR, PUBLIC INFORMATION:
George Mulhern

Lockheed Missiles & Space Company is heavily engaged in both missile work and the design, development and production of satellites and space vehicles. Details of some of its current space programmes follow:

AGENA D

The Agena satellite, for which LMSC was named prime contractor after a design competition in 1956, is a versatile space vehicle which is used normally as the upper stage of a two-stage launcher, in combination with a Thor, Atlas, Thrust-Augmented Thor, Long-tank Thor or Titan IIIB. The current Agena D version consists of a cylindrical body containing a Bell Aerosystems Model 8096 (YLR81-BA-11) restartable liquid-propellant rocket engine (71·2 kN; 16,000 lb st) and propellant tanks, telemetry, instrumentation, guidance and attitude control systems. It has carried most types of power supply, including a nuclear reactor electric power supply and an ion engine. The payload section (nosecone) can accommodate a wide variety of Earth-orbiting and space probes weighing up to several hundred pounds. The Agena system and its attached payload have functioned for more than six months in some missions for the USAF.

Agena D differs from earlier versions in being able to accept a variety of payloads, whereas its predecessors had integrated payloads. The restartable engine permits the satellite to change its orbit in space.

In the period 1959-71, well over 300 Agena spacecraft applications were announced, in terms of launches attempted, of which all but about 30 had achieved success, with payload injected into orbit. By the same date, more than 45 Agenas had been used as the upper stages of launch vehicles for other spacecraft, with only two recorded failures. No Agena had failed to achieve orbit since April 1967. Among many significant achievements, Agena spacecraft were first to achieve a circular orbit, to achieve a polar orbit, to be stabilised in all three axes in orbit, to be controlled in orbit by ground command, to return a man-made object from space, to propel themselves from one orbit to another, to propel spacecraft on successful Mars and Venus flyby missions, to achieve a rendezvous and docking by spacecraft in orbit, and to provide propulsion power in space for another spacecraft. Agena also forms the basis of the new Seasat-A ocean survey satellite, described separately in this entry.

A high proportion of the unidentified US satellites included in the table at the end of this section can be assumed to be Agena payloads of various kinds.

The following details refer to Agena D:

DIMENSIONS:	
Length (typical)	7·09 m (23 ft 3 in)
Diameter	1·52 m (5 ft 0 in)
WEIGHTS (typical):	
Propellant weight	6,148 kg (13,553 lb)
Vehicle weight empty	673 kg (1,484 lb)
Weight in orbit, less payload	579 kg (1,277 lb)

SPACE TEST PROGRAM

Some details of the US Department of Defense's Space Test Program (formerly Space Experiment Support Program) were released on 17 October 1971, when a Lockheed Agena carrying multiple scientific experiments was launched into polar orbit from Vandenberg AFB, California, by a Thor booster.

The Space Test Program provides space flights for DoD-approved space research projects which are not allocated individual launches. Eligible projects include those which require space flight for completion, are part of a DoD development test and evaluation programme, or are sponsored by another US federal agency.

Largest of the experiments aboard this particular spacecraft was a 113 kg (250 lb) flexible solar array capable of producing 1,500W of electrical power from the Sun's energy. Developed by the Air Force Aero Propulsion Laboratory, the Hughes-built solar array was rolled up in a cylinder at launch and then unrolled in orbit like a window shade to a size of 9·75 m × 1·83 m (32 ft × 6 ft). The array had a pointing system to keep it facing the Sun for maximum power output.

Most complicated experiment aboard the Agena was built by the Lockheed Research Laboratory in Palo Alto, for the Office of Naval Research and the Defense Nuclear Agency. The experiment was in two parts: low- and high-energy particle detectors, and an Earth Reflecting Ionospheric Sounder (ERIS). It used 19 individual instruments to collect data on proton, alpha and electron particles that enter the upper atmosphere. In the polar regions, ERIS was designed to transmit high-frequency signals to the ground, from where they were reflected back to the satellite. Scientists expected the collected data to be useful in understanding the effects of solar storms on polar phenomena.

Other experiments concern satellite communications and celestial sphere measurements.

Total mission life for the Agena and its cargo was expected to be at least six months. Such long life is made possible by Agena's ability to respond to ground command and to make necessary adjustments to maintain its orbit.

Scientific data gathered in orbit by this particular Agena were transmitted to tracking stations located around the world. Equipped with 11 antennae, the Agena used 800 telemetry channels to return data to the experimenters. It was capable of responding to 248 different real-time ground commands.

SEASAT SATELLITE

Lockheed Missiles & Space Company has been awarded a contract by the Jet Propulsion Laboratory, on behalf of NASA, to develop an experimental Seasat ocean survey satellite. The project represents another application of the company's versatile Agena spacecraft.

To be launched in mid-1978 from Vandenberg Air Force Base, California, Seasat-A is designed to send back information on surface winds and temperatures, currents, wave heights, ice conditions, ocean topography and coastal storm activity. In a near-circular polar orbit of about 800 km (430 miles), it will circle the earth 14 times daily and cover 96% of the world's oceans every 36 hours.

The 1,815 kg (4,000 lb) satellite will comprise three major elements: a standard bus; a sensor module support structure; and a sensor module with the sensors and antennae.

The bus will be basically an Agena spacecraft. After separation from an Atlas launcher, the Agena engine will provide power for the orbital insertion and circularisation manoeuvres. In orbit, the Agena will supply 626W average and 1,180W peak power, and will provide stabilisation, control and guidance, as well as reaction control for orbital trim. Orbit attitude control will be achieved through a momentum-biased system of momentum wheels, attitude sensors and electronics required to provide a constant pointing accuracy of 0·5°, with pointing knowledge to be better than 0·2°.

The satellite will be equipped with five special sensors, comprising three active radars and two passive radars, to

Agena satellite and experimental payload for Department of Defense Space Test Program. Copper discs o front of vehicle *(left)* are communications antennae. Long cylindrical object over heads of engineers is a flexibl solar array which unrolls like a window blind in space

make its oceanographic measurements, as follows:

Synthetic aperture radar: to provide all-weather imagery of open oceans and ice conditions, with 25 m (82 ft) resolution.

Radar altimeter: to measure the geoid (ocean topography) to ±10 cm (34 in) and wave height to ±1·0 m (3·0 ft).

Radio scatterometer: to measure surface winds under 20 m/sec (66 ft/sec).

Scanning multifrequency microwave radiometer: to measure surface winds above 20 m/sec (66 ft/sec), ocean temperatures to within ±2°C, detect current changes, and map ice coverage.

Visible/infra-red radiometer: to provide imagery and correlation data, during day and night, and in both clear and cloudy weather, to corroborate inputs from the other sensors.

Seasat-A's mission will be to prove the feasibility of employing an operational, multiple-satellite Seasat network to monitor the world's oceans on a continuous, near-real-time basis. Such a system could provide ships at sea with detailed charts of routes, updated to show the latest weather conditions, sea state and hazards. It has been estimated that the potential economic value of such a system to the USA alone could approach $2,000 million by the year 2000.

LOCKHEED/USAF PR SATELLITES

Several different families of photographic reconnaissance satellites based on the Agena vehicle are known to be in regular use by the USAF, although no officially-released details are available. A summary of the reported characteristics of two classes of spacecraft. follows:

Photo/video type. Satellites of this type appear to have been orbited three or four times each year by Long-Tank Thrust-Augmented Thor-Agena D (Thorad) vehicles. Orbital inclination has usually been in the 75-88° bracket, with an orbital life of 22-28 days. Estimated weight of each satellite is about 1,815 kg (4,000 lb). It is used for basic 'seek-and-find' missions, photographs being processed on board the satellite and transmitted to Earth by radio link. On some missions a small **P-11** 'pick-a-back' satellite is launched simultaneously into a circular orbit of around 555 km (345 miles), presumably for electronic intelligence (elint) gathering.

Recoverable type. Once launched at approx quarterly intervals, these satellites are believed to take close-look high-resolution photographs of targets detected by the photo/video spacecraft. Satellite weight is 2,950-3,175 kg (6,500-7,000 lb), requiring a Titan IIIB launch vehicle for the second-stage Agena. Orbits are elliptical, with apogee of around 445 km (275 miles) and perigee in the order of 130 km (80 miles), maintained for up to 25 days by using Agena propulsion to restore any rapid degradation of the orbit. Inclination averages about 110°. The capsules are recovered eventually by the air-snatch technique.

The relationship between the above satellites and the Samos programme (see 1975-76 *Jane's*) is not known. Their use appears to be declining now that Big Bird is available.

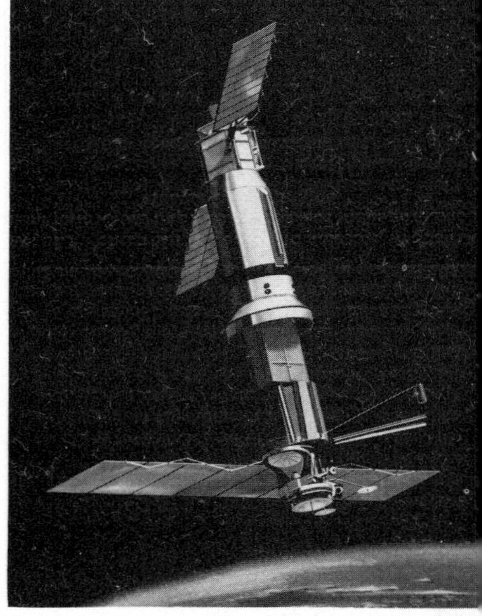

Artist's impression of Seasat-A ocean observatio satellite

LOCKHEED/USAF 467 BIG BIRD

First launched on 15 June 1971, from Pt Arguello, thi highly-advanced photographic reconnaissance satellite i reported to weigh about 11,340 kg (25,000 lb) and to b 15·25 m (50 ft) long. The first Big Bird was launched b Titan IIID into an orbit with an apogee of 299 km (18 miles), perigee of 183 km (114 miles) and inclination o 96·41°. Its capabilities clearly included the same kind o close-look high-resolution photography as that of th recoverable type of Agena vehicle; and it is reported t have ejected a series of capsules for air-snatch recovery.

Big Bird is believed to process photographs taken b cameras and transmit information to Earth in the form o digital data by radio link. Some reports suggest that it als carries infra-red mapping and side-looking radar equip ment. Its orbit takes its cameras within range of ever point on the Earth twice in each period of 24 hours.

Three more Big Birds were launched in 1972; other followed on 9 March, 13 July and 10 November 1973, o 10 April and 29 October 1974 and 8 June 1975, and o subsequent occasions as listed in *Jane's* satellite launc tables. It is expected that these spacecraft will continue t be launched at roughly four-to-six-monthly intervals t provide an increasing share of the total US satellite intel ligence. They have demonstrated an endurance of up t four months in orbit.

MARTIN MARIETTA

MARTIN MARIETTA CORPORATION

Corporate Headquarters:
11300 Rockville Pike, Rockville, Maryland 20852
Aerospace Headquarters:
1800 K Street NW, Washington, DC 20006
Telephone: (202) 833-1900
President and Chief Executive Officer, Martin Marietta Corporation:
J. Donald Rauth
Vice-President, Public Relations:
Roy Calvin
President, Martin Marietta Aerospace:
Thomas G. Pownall
Vice-Presidents, Aerospace Company:
Herman Pusin (Engineering and Research)
Robert J. Whalen (Advanced Systems and Planning)
Laurence J. Adams (Denver Division)
Howard W. Merrill (Baltimore Division)
Sidney Stark (Orlando Division)

Baltimore Division
103 Chesapeake Park Plaza, Baltimore, Maryland 21220

Denver Division
PO Box 179, Denver, Colorado 80201
Telephone: (303) 979-7000
Director, Public Relations:
John H. Boyd Jr

Orlando Division
PO Box 5837, Orlando, Florida 32805
Telephone: (305) 352-2000
Director, Public Relations:
Edward J. Cottrell

Current activities of Martin Marietta's Baltimore Division include manufacture of components for the Rockwell B-1 strategic bomber and the McDonnell Douglas DC-10 transport, and thrust reversers for General Electric CF6-6 and CF6-50 turbofan engines.

The Denver Division of Martin Marietta is responsible for the Titan III family of space launch boosters, the Viking Mars orbiter/landers, the Space Shuttle project, and for other spacecraft, their systems and related research.

Orlando Division is engaged in development of the SAM-D missile, the terminally-guided Pershing II missile, the Cannon-Launched Guided Projectile (CLGP), electro-optical guidance and fire-control systems, and a wide range of military and commercial communications and electronics equipment. This Division has extensive laboratories for research, development, test and evaluation of warheads, materials, special munitions, structures, propellants, lasers, guidance and control, fluidics, reconnaissance devices, digital communications and millimetre wave techniques.

TITAN III

Titan III is America's standard heavy-duty space 'workhorse' booster and is used for both military and non-military space launch missions. It provides a high frequency launch capability for a wide variety of manned and unmanned payloads, ranging from 15,875 kg (35,000 lb) in Earth orbit to 3,175 kg (7,000 lb) for planetary missions such as the exploration of Mars. The Space and Missile Systems Organisation of the Air Force Systems Command has executive management of the programme. Martin Marietta at Denver, Colorado, was named systems integrator for the industry/contractor team on 20 August 1962. Technical direction was assigned to Aerospace Corporation.

Martin Marietta, in addition to its role as systems integrating contractor, builds the airframe and liquid-propellant stages, supplies the flight control system, and is integrating contractor for facilities and launch operation at Cape Canaveral. Aerojet Liquid Rocket Co produces the liquid-propellant engines. UTC's Chemical Systems Division supplies the solid-propellant boosters used in the more powerful models. Guidance systems for the Titan IIIC, D and E are built by General Motors Corporation's Delco Division, Western Electric and Honeywell Corporation respectively.

The core section of the Titan III consists of elements which provide a high degree of commonality throughout all configurations. It consists of two booster stages evolved from the Titan II ICBM and an upper stage, known as Transtage, that can function both in the boost phase of flight and as a restartable space tug propulsion vehicle. All stages use storable liquid propellants and have gimbal-mounted thrust chambers for vehicle control.

Titan III exists in four configurations:

Titan IIIB. Basically the first two stages of the core section. It can accommodate a variety of specialised upper stages. First launched on 29 July 1966. Series of launches continued through 1975, all with Agena upper stages and classified USAF payloads, including reconnaissance satellites.

Titan IIIC. Consists of the core section of the main airframe, including the Transtage upper stage, with solid-propellant rocket motors attached to each side to function as a booster stage before ignition of main engines. Payloads include USAF and NASA unmanned military, scientific and communications satellites and spacecraft, including about 80% of all satellites placed into synchronous equatorial orbit from US launch sites.

Titan IIIE-Centaur launch of the Mars Viking I spacecraft 20 August 1975

Titan III-Centaur heavy duty launch vehicle

Titan IIID. Basically similar to IIIC but has only a two-stage liquid-propellant core (without Transtage) and radio guidance instead of the standard inertial guidance. Able to accept a variety of upper stages. Production order placed by USAF in November 1967. First used to orbit the first Lockheed Big Bird advanced photo-reconnaissance spacecraft, weighing about 11,340 kg (25,000 lb), from Pt Arguello on 15 June 1971.

Titan IIIE-Centaur. Basically a Titan IIID which has been modified to take a Centaur high-energy upper stage. The first launch, on 11 February 1974, was terminated by the range safety officer when the second-stage engine failed to ignite. First operational launch took place from the Eastern Test Range on 10 December 1974, carrying the German Helios Sun-probe spacecraft. Primary mission of Titan IIIE-Centaur was the launch in 1975 of two Viking spacecraft designed to land on the surface of Mars in 1976.

A Titan IIIC, with a 7·62 m (25 ft) long typical payload fairing and all stages mated, is 39·62 m (130 ft) in height. Future payloads will extend the overall height considerably; the Titan IIIE-Centaur launch vehicle for the Viking Mars exploration spacecraft, for example, had an overall height of 48·77 m (160 ft) when the shroud protecting the payload was in place.

The first stage of the main airframe (core vehicle) is 22·25 m (73 ft) long and 3·05 m (10 ft) in diameter. Its engines, which use a blend of hydrazine and unsymmetrical dimethylhydrazine (UDMH) for fuel, and nitrogen tetroxide as an oxidiser, have a 15 : 1 expansion ratio and are ignited at an altitude where efficiency is increased, giving a thrust of 2,339·6 kN (526,000 lb) in vacuo.

The second stage of the core vehicle is 7·10 m (23 ft 3½ in) tall and 3·05 m (10 ft) in diameter. Its engine uses the same propellants as the first stage and develops 453·7 kN (102,000 lb st).

The Transtage space propulsion vehicle is 4·57 m (15 ft) tall and 3·05 m (10 ft) in diameter and also uses UDMH/hydrazine and nitrogen tetroxide as propellants. The twin-chamber engine produces 71·17 kN (16,000 lb) of thrust and is capable of multiple restarts in space, which permits a wide variety of manoeuvres, including change of plane, change of orbit, and transfer to deep-space trajectory. Transtage also houses the control module for the entire vehicle, including the guidance system and segments of the flight control and vehicle safety systems.

Titan IIIC/D/E's solid-propellant booster motors are each 25·91 m (85 ft) long and 3·05 m (10 ft) in diameter. Each motor is built in five segments and develops more than 5,115·2 kN (1,150,000 lb st). Steering for the booster stage is accomplished through a thrust vector control system, which injects nitrogen tetroxide into the engine nozzle.

The Titan III 17-vehicle research and development flight testing programme involved launch of four IIIAs and 13 IIICs. The first launch of a Titan IIIA development vehicle occurred on 1 September 1964. Titan IIIC made its maiden flight on 18 June 1965 and had completed 19 successful flights by the beginning of 1974, putting a total of 37 satellites into synchronous orbit in the process.

Titan III vehicles had performed successfully in 86 consecutive launches in the twelve-year period to 1975. Additional contracts for various models have extended production to 1979.

Dimensions: See text

Launching Weights (approx):
Core vehicle 204,120 kg (450,000 lb)
635,030 kg (1,400,000 lb)

Performance (Titan IIIC, approx):
Speed at burnout:
Solid-propellant boosters
3,560 knots (6,600 km/h; 4,100 mph)
1st stage 8,860 knots (16,300 km/h; 10,200 mph)
2nd stage 14,850 knots (27,520 km/h; 17,100 mph)
Transtage 15,200 knots (28,160 km/h; 17,500 mph)

VIKING

Viking is the name of the programme under which NASA sent two unmanned spacecraft to orbit the planet Mars and to make soft landings there in 1976. It replaced

Impression of Viking spacecraft in orbit. The Lander Module was carried inside the saucer-shaped protective shield during the long journey to Mars

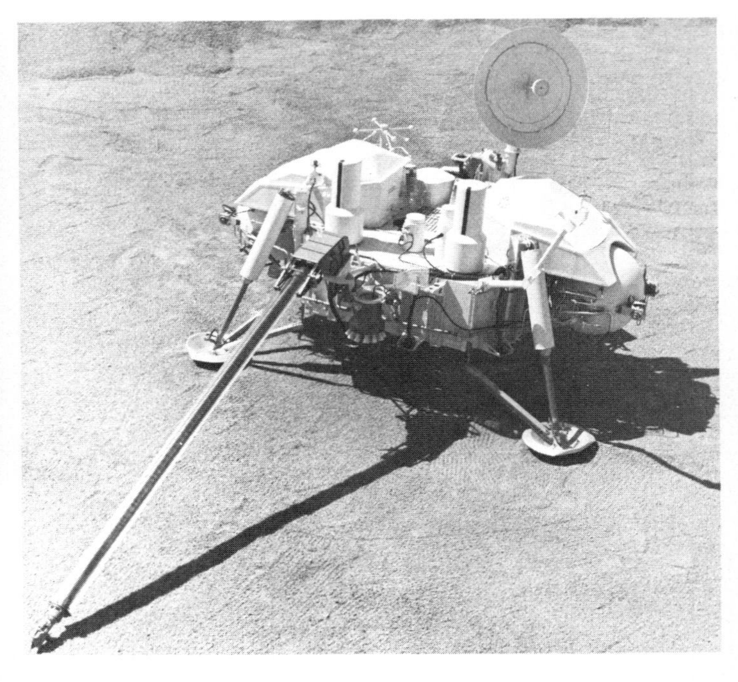

Engineering model of the Viking Lander Module

the earlier, more costly, Voyager project.

Launched on 20 August and 9 September 1975, by Titan IIIE-Centaur vehicles from Cape Canaveral, each of the Viking spacecraft consists of two modules, known as the Viking Orbiter and Viking Lander. Total weight at launch of each spacecraft was about 3,630 kg (8,000 lb).

Viking Orbiter module was built by the Jet Propulsion Laboratory, California, and is a large development of the JPL's earlier Mariner class of spacecraft, with a liquid-propellant retro-rocket engine to decelerate the craft into orbit on arrival at the planet.

The basic structure of the Orbiter module is an octagon, approximately 2·4 m (8 ft) across by 0·46 m (18 in) high. Overall the Orbiter is 3·3 m (10 ft 9½ in) high and 9·7 m (32 ft) across, with the solar panels extended. Its fuelled weight is 2,325 kg (5,125 lb).

Orbiter experiment instruments include two high-resolution television cameras, an infra-red spectrometer and an infra-red radiometer, all mounted on a scan platform to survey the same area of Mars.

Viking Lander was built by Martin Marietta's Denver Division. The module comprises five basic systems, the Lander body, bioshield cap and base, the aeroshell (an ablative heat shield), the base cover and parachute, and the Lander subsystems. Assembled, the Lander is about 3 m (10 ft) across and 2 m (7 ft) tall. It weighed about 1,090 kg (2,400 lb) at separation and approximately 576 kg (1,270 lb) at touchdown without fuel, including approximately 90 kg (200 lb) of instruments.

The Lander body is basically a platform for scientific instruments and operational subsystems. It is a hexagon-shaped box with three 1·09 m (43 in) side beams and three 0·56 m (22 in) short sides. The body is supported on three legs, 1·3 m (51 in) long, giving a ground clearance of 0·22 m (8·7 in). Three terminal descent engines provided attitude control and reduced the Lander's velocity after separation. The engines embody an advanced exhaust system designed to minimise disturbance to the landing site environment, an unusual grouping of 18 small nozzles on each engine spreading the engine exhaust over a wide angle.

Lander module equipment includes two television cameras, a surface sampler attached to an extendable boom, a biology experiment, a gas chromatograph-mass spectrometer and an X-ray fluorescence spectrometer, a meteorology instrument, and a three-axis seismometer.

After successful launch, the year-long, 815 million km (505 million mile) journey was completed without major incidents, except for indications during in-flight checkout tests that one of the three small ovens carried on each Lander may have failed on both Viking 1 and Viking 2. These ovens are designed to heat Martian soil samples to 500°C, to release constituents in the soil for analysis by the gas chromatomass spectrometer. Also, on 31 October 1975, attempts to charge the batteries in the Lander of Viking 2 using the primary charger were unsuccessful. Both Viking 1 and 2 were launched with those batteries essentially uncharged to prolong their lives. This potentially serious problem was overcome by using a backup charger.

Viking 1 was due to enter a Martian orbit about 18 June 1976 and Viking 2 about 7 August 1976. Lander 1 was expected to touch down around 4 July and Lander 2 about 9 September.

Four landing sites were selected for the Landers, two primary and two secondary. Prime site for Lander 1 is Chryse, a region at the north-east end of a 4,800 km (3,000 mile) long rift canyon discovered by Mariner 9. Lander 2's primary site is Cydonia, in the Mare Acidalium region at the edge of the southernmost reaches of the north polar hood.

Once in orbit, site acquisition of Viking 1 will be effected by trim manoeuvres to establish a synchronous orbit in which the spacecraft will pass daily near the site with a periapsis of 1,500 km (930 miles) and an apoapsis of 32,600 km (20,200 miles). Viking 2 will be inserted into a super-synchronous initial orbit with a period of about 28·7 hours.

After establishment of the required orbit, the Lander will then separate and leave orbit, by means of a retro-rocket. It will be protected by a heatshield as it decelerates in the Martian atmosphere. The aeroshell heatshield below the Lander. will be jettisoned about 6,400 m (21,000 ft) above the planet's surface. Simultaneously a parachute system will be deployed to decelerate the craft further. At 1,220 m (4,000 ft) over Mars, the parachute will jettison and the three terminal descent engines will fire to slow the Lander for a soft touchdown at 3 m/sec (6 mph). The engines will shut down as the Lander's footpads contact the surface.

While in orbit round Mars, the instruments on Viking Orbiter will photograph the planet and map its atmospheric water vapour and thermal properties.

The cameras on Viking Lander will take pictures of Mars from the surface, and the instruments will study the planet's biology, molecular structure, inorganic chemistry, meteorology and seismology, and physical and magnetic properties. Of special interest is the surface sampler, which will be extended to dig up soil samples for incubation and analysis inside the biology instrument's three metabolism and growth experiment chambers, in a search for living organisms.

By the time this entry was being prepared for printing both Vikings had landed safely. The first Viking Lander touched down on 20 July in the Chryse Planita basin, after a three-week search for an alternative to the originally-planned landing site, which proved unsuitable. Lander 2 touched down on 4 September in the Utopia Planita secondary site, on the opposite side of Mars. Both spacecraft have returned remarkably clear photographs of the rock-strewn Martian surface, and are continuing their investigations into the nature of the Martian soil, returning data that often appear to be conflicting.

MCDONNELL DOUGLAS
MCDONNELL DOUGLAS CORPORATION

Head Office and Works:
Box 516, St Louis, Missouri 63166

McDonnell Douglas Astronautics Company

Headquarters:
5301 Bolsa Avenue, Huntington Beach, California 92647

Officers: See Missiles section

This company was formed on 26 June 1968, by a merger of the former Douglas Missile and Space Systems Division and the McDonnell Astronautics Company into a single management structure. Details of its current space programmes follow:

THOR

Details of the development, and operational deployment by the RAF, of the Thor IRBM can be found in the 1962-63 *Jane's*.

The Thor force was disbanded during 1963 and all the missiles were flown back to the United States. All were subsequently converted into space boosters by McDonnell Douglas. The company is also continuing production of Thors for use as first-stage boosters for the various space launch vehicles described separately.

Thor has a circular-section aluminium body of lightweight integrally-stiffened design, providing integral tankage for its liquid oxygen and kerosene propellants. Propulsion is by a Rocketdyne liquid-propellant engine, the chamber of which is gimbal-mounted to provide directional control and stability. Two liquid-propellant vernier engines, on each side of the main engine, provide speed adjustment after main engine burnout, as required, plus roll stabilisation.

A total of 468 Thor IRBMs and their derivative space boosters had been launched by 1 January 1976.

LONG TANK THOR

Details of this launch vehicle can be found in previous editions of *Jane's*. It is no longer in use.

DSV-3 DELTA

In May 1959, Douglas was awarded a prime contract by NASA to develop a three-stage vehicle named Delta, capable of placing a 218 kg (480 lb) payload into a 480 km (300 mile) Earth orbit or of sending a 45 kg (100 lb) payload on deep space probes. The original orders for 12 and 14 Deltas respectively were followed by further contracts in 1963-75, bringing the total ordered, in many versions, to 149.

Satellites launched by Delta have included many of the Explorer series, Pioneer, Tiros and Nimbus weather satellites, Echo I, Orbiting Solar Observatory series, Ariel, HEOS A, Biosatellite, Landsat, and active communications satellites such as Telstar, Relay, Syncom, Early Bird, Intelsat, Skynet, Telesat (Anik), Westar, Symphonie and Satcom.

Details of early versions of the Delta can be found in the 1971-72 and 1972-73 *Jane's*. All vehicles prior to the DSV-3P have been launched. Production is now centred on the DSV-3P, of which details follow:

DSV-3P Extended Long Tank Delta (also known as 'Straight-Eight' or '2000 Series Delta'). This launch vehicle has a constant 2·44 m (8 ft) diameter from the base of the boat-tail to the conical nose section of the shroud. This provides an enlarged volume, to accommodate larger payloads. The second stage, with a TRW LMDE engine, is suspended within the 2·44 m (8 ft) diameter barrel section. The first-stage length is increased by 3·05 m (10 ft) by comparison with earlier long-tank versions, providing an increase of 13,600 kg (30,000 lb) in propellant capacity. In addition, the MB-3-III main engine is replaced by a Rocketdyne RS-27 of 911·84 kN (205,000 lb st). As an alternative to the TE-364-3 motor, a higher-performing motor, the TE-364-4, is available as a third stage. The two-stage capability is increased to 1,880 kg (4,150 lb) into a 370 km (230 mile) circular orbit. The three-stage capability is increased to 700 kg (1,550 lb) into a synchronous transfer orbit.

DSV-3P (Delta 3914). First launched on 12 December 1975, this uprated version of the standard 'Straight-Eight' Delta utilises nine Castor IV solid-propellant strap-on motors in place of the nine Castor IIs used on the Delta 2914. Development costs were borne by MDAC, and will be recovered through a user charge imposed on non-government users of the vehicle. The RCA-A Satcom was the first spacecraft launched by a Delta 3914. Vehicle dimensions are the same as for the standard DSV-3P Delta 2914, but firing weight has been increased to 191,400 kg (422,000 lb). Delta 3914 is capable of putting a 907 kg (2,000 lb) payload into a geosynchronous transfer orbit.

DIMENSIONS:

Length overall	35·15 m (115 ft 4 in)
Body diameter	2·44 m (8 ft 0 in)

WEIGHTS:

Firing weight of DSV-3P (2000 Series):

3 solid motors	104,330 kg (230,000 lb)
6 solid motors	117,930 kg (260,000 lb)
9 solid motors	131,540 kg (290,000 lb)

USAF 437 ANTI-SATELLITE SYSTEM

No details of this anti-satellite system are available, although the US technical press suggests that it has been operational since 1964. It is described as an unguided direct ascent anti-satellite vehicle, launched by Thrust Augmented Thor and fitted with a nuclear warhead.

Delta 3914 launching RCA Satcom satellite

NASA
NATIONAL AERONAUTICS AND SPACE ADMINISTRATION

HEADQUARTERS: Washington, DC 20546

ADMINISTRATOR: Dr James C. Fletcher

NASA is responsible for co-ordinating and conducting virtually all US non-military space projects. Its Office of Manned Space Flight is responsible for the Space Shuttle programme. NASA's Office of Space Science and its predecessor, the Office of Space Science and Applications, had launched over 300 spacecraft into Earth orbit or interplanetary space by 1976, and had launched more than 1,600 sounding rockets into near-Earth space. To launch automated spacecraft OSS has collaborated with private industry in developing a series of versatile launch vehicles, including Scout, Delta and Atlas-Centaur. Adaptation of the Titan III and Titan IIIE-Centaur for launching larger automated spacecraft is under way.

Details of most of these programmes appear under the entries for the respective industry prime contractors in this section of *Jane's*. Following are details of other current programmes for which NASA is responsible:

EXPLORER 54

Launched on 6 October 1975 by a Thor Delta from the Western Test Range, California, into an elliptical (3,817 × 154 km; 2,372 × 96 mile) polar orbit, Explorer 54 was the second in the series of three manoeuvrable spacecraft designed to explore an area of the Earth's outer atmosphere where important energy transfer and other processes take place and that are critical to the heat balance of the atmosphere.

The general configuration of the satellite is a 16-sided polyhedron. The drum-shaped spacecraft is 1·35 m (53·2 in) in diameter and 1·15 m (45 in) high. It weighs 675 kg (1,488 lb) of which 95 kg (210 lb) is instrumentation. Solar cells mounted on the top and sides of the outer shells supply electrical power for the spacecraft and experiments. Various sensors and probes project through the outer skin to collect data and provide spacecraft attitude control information. The spacecraft is equipped with hydrazine thrusters to provide an in-orbit manoeuvring capability.

Explorer 54 carries twelve scientific instruments: ultraviolet (nitric oxide) photometer; cylindrical electrostatic probe; atmosphere density accelerometer; photoelectron spectrometer; retarding potential analyser (measures ion and electron data); visual airglow photometer; solar extreme ultraviolet spectro-photometer; magnetic ion mass spectrometer; low energy electron spectrometer; open-source neutral mass spectrometer; and neutral atmosphere composition and temperature spectrometers.

In orbit the instruments made simultaneous measurements of incoming solar radiation and of the Earth's atmosphere to provide information on the physical processes that govern the composition of the lower thermosphere and the ionosphere, thus making possible study of the closely interlocking cause-and-effect relationships that control the Earth's near-space environment.

Explorer 54 satellite

The first spacecraft in the series, Explorer 51 (*Jane's* 1974-75) discovered that the weather in this region is constantly changing with winds up to ten times as severe as those normally present in the Earth's surface. Prior to Explorer 51, it was believed that the atmosphere in this region behaved predictably and was relatively stable. Now it is known to be very dynamic and unpredictable.

Although Explorer 51 completed its primary mission in early 1975, it was used early in the flight of Explorer 54 so that areas of interest at different altitudes could be sampled simultaneously.

Overall programme direction is the responsibility of NASA's Office of Space Science, Washington, with Goddard Space Flight Center providing the spacecraft and rocket direct management.

See also Explorer 55.

EXPLORER 55

Launched on 20 November 1975 by a Thor Delta from Cape Canaveral, Florida, into an elliptical (2,983 × 156 km; 1,854 × 97 mile) equatorial orbit, Explorer 55 is the third and last in the series of manoeuvrable spacecraft designed to explore an area of the Earth's outer atmosphere where important energy transfer processes take place that are critical to the heat balance of the atmosphere.

The other two satellites in the series are Explorer 51 (*Jane's* 1974-75) and Explorer 54, which see. Explorer 55 is generally similar to Explorer 54, but is heavier, weighing 720 kg (1,587 lb), including 107 kg (237 lb) of instrumentation.

In addition to the twelve scientific instruments for the primary mission, Explorer 55 also carries an instrument to measure the Earth's ozone layer between 20 degrees north and south. Called a backscatter ultraviolet spectrometer (BUV), it was added to the spacecraft's payload as part of NASA's programme to measure the atmospheric distribution of ozone on a global basis.

Information returned by this instrument in conjunction with the others could represent a major step in understanding the interaction of upper atmosphere constituents

Ozone sounding instrument being installed on Explorer 55

with solar ultraviolet light and the resulting impact on the ozone layer.

The data may form the basis on which decisions can be made regarding the desirability of banning the widespread use of aerosol cans using chlorofluoromethanes (Freon). These gases, expelled from aerosol cans, slowly over a period of about ten years, work their way up to the ozone layer where sunlight and resulting chemical reacions are believed to cause them to destroy ozone.

EXPLORERS 56 and 57

Explorer 56 and 57 were the designations to be allocated after they achieved orbit to two Dual Air Density (DAD) Explorer satellites launched by a single Scout rocket from the Western Test Range, California, on 5 December 1975.

Unfortunately, the fourth stage of the Scout rocket failed to ignite, and the two spacecraft did not go into orbit.

GEOS-3

Launched into a near-circular orbit of 843 km (524 miles) by a Delta rocket, from the Western Test Range, on 9 April 1975, GEOS-3 was intended to measure accurately the topography of the ocean surface, and the wave height, period and direction of the sea state. It was designated GEOS-3 (Geodynamics Experimental Ocean Satellite-C before launch), and is the third in a series of spacecraft designed to increase knowledge of the Earth's shape and dynamic behaviour. The first spacecraft in the series was Explorer 29, launched in November 1965; the second was Explorer 36, launched in January 1968.

The 1·32 m (52 in) diameter GEOS-3 comprises a densely packed eight-sided aluminium body topped by a truncated pyramid, and weighs 340 kg (750 lb). The outer surfaces of the body are covered with panels of solar cells, designed to provide maximum output and ensure minimum daily fluctuations in the satellite's exposure to sunlight as it orbits the Earth.

Digital solar attitude sensors, mounted below three equatorial solar cell panels, provide information on the spacecraft's orientation relative to the Sun; a three-axis vector magnetometer is installed for measuring the orientation with respect to the Earth's magnetic field. An electromagnet is carried, to stabilise the satellite magnetically.

The satellite is orientated in orbit with its antennae and experiments pointed towards the Earth at all times. This is achieved by means of an extremely precise gravity-gradient stabilisation system, employing a 6 m (19 ft 7 in) scissors-type boom with a 45 kg (100 lb) end mass; the boom was extended from a housing in the pyramid end of the body after the spacecraft had been magnetically stabilised. The boom can be retracted or extended by means of a motor inside the body. A constant-speed momentum wheel, like a gyroscope, augments the gravity-gradient boom, to provide full three-axis stabilisation.

Experiments installed in the spacecraft include a radar altimeter (the first to be carried on an unmanned spacecraft) to measure the sea state; two C-band transponders, to support the altimeter; one S-band transponder, for satellite-to-satellite tracking and for Earth tracking experiments; laser retro-reflectors, for measuring the satellite's range at optical frequencies; and a radio Doppler system that transmits on two coherent frequencies and is used to obtain precise satellite range rate data.

The orbit of GEOS-3 must be known precisely, so that its height above sea level, measured by the radio altimeter, can be calibrated accurately. This is achieved by tracking from the ground by laser, radio Doppler, C-band radar and S-band radar, as well as by tracking via the Applications Technology Satellite ATS-6, launched in May 1974. The satellite-to-satellite tracking experiment, the first of its kind, is expected to provide more precise orbit information on the observed satellite than has been obtainable by the less frequent observations of ground stations.

The schedule of altimeter calibrations started on 22 April and continued until 20 May, when ATS-6 began to move from its then-current position over South America towards a position over Africa, to support Indian educational programmes. Availability of the large communications satellite permitted substantial data gathering through the calibration area.

In 1975 a NASA-US Navy team spent a month in Newfoundland where the sea conditions were measured using aircraft and GEOS-3 simultaneously, to verify that satellite observations of the sea can be as reliable as those made directly on the surface.

The GEOS-3 programme is under the management of NASA's Office of Applications; the NASA Wallops Flight Center has project management responsibility. Mission operations are being managed by the Goddard Space Flight Center, which also managed the Delta launch vehicle. The spacecraft was designed and fabricated by the Applied Physics Laboratory of the Johns Hopkins University. Launch site operations were managed by the NASA Kennedy Space Center Unmanned Launch Operations Directorate.

MARINER

This is a NASA project covering the design and manufacture of a series of unmanned space-probes for missions to Mars and Venus. The general appearance of one of the latest probes, for which Jet Propulsion Laboratory holds the prime contract, is shown in an accompanying illustration.

Details of many successful missions by Mariner spacecraft can be found in the 1969-70 and 1973-74 *Jane's*. Mariner 10 (Venus/Mercury) was described in the 1975-76 *Jane's*.

MARINER-JUPITER-SATURN SPACECRAFT

Two Mariner-type spacecraft will be launched on a Jupiter-Saturn mission in 1977, one in August and the second in September. They will fly by both planets, and conduct exploratory investigations of the Jupiter and Saturn planetary systems and the interplanetary medium out to Saturn. Major investigations will include imaging, radio science, infra-red and ultraviolet spectroscopy, magnetometry, charged particles, cosmic rays, photopolarimetry, planetary radio astronomy, plasma and particulate matter.

MARINER-URANUS SPACECRAFT

NASA is considering the possibility of a mission to Uranus, the third farthest planet from the Sun, in 1979. Using a Mariner-type spacecraft, the mission could begin on 3 November 1979, the spacecraft arriving at Jupiter in April 1981. Using the gravity of Jupiter to accelerate the Mariner and redirect its trajectory, the spacecraft would proceed to a mid-1985 encounter with Uranus, possibly passing as close as 24,000 km (15,000 miles).

Launch vehicle for the mission would be the Titan IIIE/Centaur combination used for the 1976 Viking mission to Mars and the currently planned 1977 Mariner Jupiter-Saturn mission.

Mariner 10 spacecraft built for 1973-74 mission to Venus and Mercury

PIONEER VENUS 1978

NASA plans to send both an orbiter and a multiprobe spacecraft to Venus in 1978 to conduct a detailed scientific examination of the planet's atmosphere and weather. Both spacecraft are being built by Hughes Aircraft Co.

The orbiter will be launched in May and inserted into Venusian orbit in December; the multiprobe spacecraft will be launched in August and the probes will enter the Venusian atmosphere six days after arrival of the orbiter.

Weight of the orbiter will be about 567 kg (1,250 lb) and that of the multiprobe spacecraft about 885 kg (1,950 lb). Both will be about 2·44 m (8 ft) in diameter.

The orbiter, carrying 43 kg (95 lb) of instruments, is designed to study the Venusian atmosphere over one 243 day period. An elliptical orbit will bring the spacecraft to within 200 km (125 miles) of the surface, the maximum distance being 60,000 km (37,300 miles). Most of the data-gathering will occur when the craft is closest to the planet, about one hour a day.

The multiprobe spacecraft will carry four separate probes which will be released from the main body about 20 days before penetration of the atmosphere. Three of the probes will be small, weighing about 86 kg (189 lb), including 2·7 kg (6 lb) of instruments. Most of the probe's weight will be accounted for by the heatshields and pressure vessels. These probes will measure atmospheric pressure and temperature, and investigate the exchange of heat energy between the Sun and atmosphere.

The fourth probe is larger, weighing about 291 kg (642 lb) and carrying about 28 kg (62 lb) of instrumentation. Its payload includes a mass spectrometer and a gas chromatograph to provide details about the identity of components in the atmosphere.

All four probes will be targeted to different locations on the surface, and will be tracked during their 70 minute descent through the atmosphere, to gain information on winds and circulation patterns.

After release of the probes the main body will make a shallow entry into the atmosphere, obtaining measurements of the upper atmosphere until it burns up at an altitude of about 120 km (75 miles).

The Pioneer Venus mission will be managed for NASA's Office of Space Science by the Ames Research Center, California.

SKYLAB

Details of the Skylab 1 Orbital Workshop, and of the Skylab 2, 3 and 4 crew missions, were given in the 1974-75 *Jane's*. In mid-1976, Skylab 1 was in a 450 km (280 mile) high orbit, with a nominal lifetime of 10 years. It has been suggested that the space station could be refurbished and used for additional experiments when the Space Shuttle becomes operational.

SPACE SHUTTLE

The Space Shuttle will be the first re-usable space vehicle, consisting basically of two stages: a booster and an orbiter. The orbiter has a delta wing and looks very like a conventional aeroplane, but is powered by three Rocketdyne high-pressure rocket engines. The liquid oxygen/liquid hydrogen propellants for these engines will be carried in a large external jettisonable tank, on which the orbiter will be mounted at lift-off. Two large solid-propellant jettisonable boosters will be mounted on opposite sides of the propellant tank for lift-off.

Lageos satellite, built for NASA by Bendix (see pag 684)

Initial stage of the Space Shuttle project was a Phase feasibility study, which confirmed the engineering ar financial practicability of a re-usable orbiting vehicl Phase B called for a complete definition and costing of tl Shuttle system. Two large industrial groups were co tracted to prepare and submit competitive designs NASA, one headed by Rockwell International, the oth by McDonnell Douglas. A third industrial group, heade by Boeing and Grumman, was commissioned to re-che some Phase A work to ensure that the best basics had be embodied into the Phase B submissions.

In early 1972 the political and financial decision w taken to proceed to Phase C/D, involving design ar development of the system, and construction of the fi two flying vehicles.

On 26 July 1972, North American Rockwell (no Rockwell International) Corporation's Space Division Downey, California, was selected as prime contractor f design, development and production of the orbiter and i integration with all other elements of the Shuttle syster

On 16 August 1973, the Martin Marietta Corporatio Denver Division, was selected to design, develop, test a evaluate the Shuttle external tank.

The final major Shuttle contractor, Thiokol Chemic Corporation of Brigham City, Utah, was selected on November 1973 to design, develop, test and evaluate t solid-propellant rocket motors.

The orbiter will normally be operated by a crew of thre comprising pilot, co-pilot and a mission specialist. F multi-payload missions, up to four additional paylo specialists can be carried. Hatches in the top of the fus age give access to the payload compartment, which is 18 m (60 ft) long and 4·57 m (15 ft) in diameter, this lar space being made possible by the fact that the main prop lant tanks are external. The interior of the craft will pressurised, enabling the crew to work without spacesui and no astronaut training will be needed by passenger

In operation, the Shuttle will be launched vertical with all engines firing in both the boosters and orbiter. an altitude of about 40 km (25 miles), the booster stag will separate and descend into the sea by parachute, f

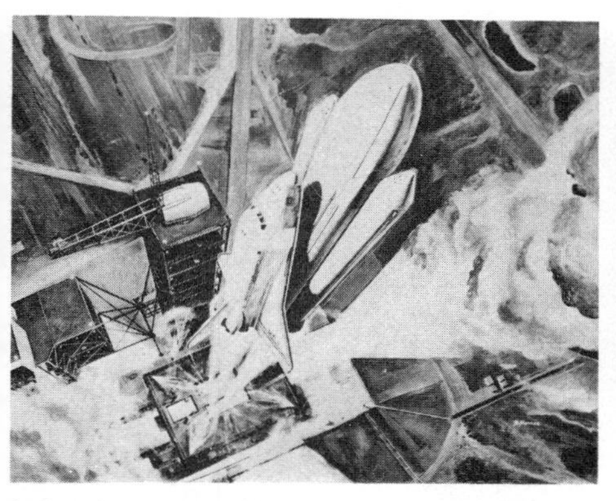

Artist's impression of Space Shuttle launch. The two solid propellant rocket boosters and the Orbiter main engine fire in parallel

The solid propellant rocket boosters are jettisoned after burnout and recovered by parachute

The large external tank is jettisoned before the orbiter goes into orbit

Typical in-orbit experimental payload. An impression of a mission utilising the remote manipulator system, which can be used to lift payloads out of the Orbiter, is shown in the Spar entry in the Canadian section

After the orbital operations de-orbiting manoeuvres are initiated and re-entry into the Earth's atmosphere at a high angle of attack mode

At low altitude the orbiter goes into horizontal flight for an aircraft-type approach and landing

recovery. The orbiter will continue under its own power, and will jettison its large underbelly propellant tank just before attaining orbit.

In space, the orbiter will manoeuvre by means of two smaller rocket engines, also mounted in the rear-fuselage propulsion cluster. For minor course corrections and adjustments of attitude, the orbiter will have a series of small thrusters.

The orbiter's main initial tasks are expected to be the placing of satellites into orbit, retrieval of satellites from orbit, and the repair and servicing of satellites in orbit. It could be used to put a propulsive stage and satellite into precise low Earth orbit, for subsequent transfer to synchronous orbit or to an 'escape' mission into space. It could also be used for short-duration scientific and applications missions, as an orbiting research laboratory or reconnaissance vehicle, for space rescue, as a tanker for space refuelling, and for support of orbiting space stations.

On some flights a pressurised Spacelab, being developed by nine European countries at their own expense, will be carried in the payload bay and will serve as a space laboratory. For the first time, scientists and engineers who are not astronauts will have an opportunity to accompany and conduct their experiments in space.

Spacelab will be the means by which man-associated experiments can be performed in the orbiter payload bay. Experiments can be assembled, checked out and mated in advance, and the Spacelab installed in the payload bay just prior to flight. The Spacelab will include a pressurised enclosure housing support equipment (to make it habitable) as well as the experimental equipment. When sensors require direct exposure to the space environment, a pallet will be used in association with the pressurised enclosure. On other types of missions, a pallet may be used alone, with control of the instruments being exercised from the orbiter cabin or even from the ground. The Spacelabs are being designed around a basic seven-day mission, which is extendable up to 30 days by trading payload weight and volume for the additional consumables necessary to accommodate the further time in orbit.

On conclusion of its mission, the orbiter will fly back into the atmosphere towards its land base, protected by a new form of heat shielding which will survive 100 missions, unlike current ablative-type heatshields. Once through the re-entry phase, the orbiter will be able to glide up to 950 nm (1,760 km; 1,100 miles) to its base, steered by aerodynamic controls.

Special equipment being developed for use on the Space Shuttle includes a new type of spacesuit and a rescue system. The shuttle suit being developed at NASA's Johnson Space Center, Houston, features an 'adjustable fit' concept, a departure from the Apollo programme in which the suits were customised for each astronaut, a long and costly process. The Shuttle suit, a two-piece combination of upper and lower torso, will be manufactured in small, medium and large sizes, to accommodate all crew members, including females. The suit will contain a life support system as an integral part of the rigid upper torso.

The rescue system, a Personal Rescue Enclosure, also under development at the Johnson Space Center, is a 0·86 m (34 in) diameter ball which contains its own short-term simplified life support and communication systems. The ball has three layers (Urethane, Kevlar and an outside thermal protective layer) and a small viewing port of tough Lexan.

Space Shuttle Approach and Landing Test (ALT), the initial flight test of the Shuttle Programme, is scheduled to begin in mid-1977. The ALT flights will be conducted at the NASA Dryden Flight Research Center in California. The Orbiter, OV-101, will be carried aloft to an altitude of about 7,620 m (25,000 ft) on top of a specially modified Boeing 747. It will then be released, allowing the crew to fly the Orbiter to the ground. Several unmanned and manned non-release flights will precede the initial 'free flight' of the Orbiter.

The second Orbiter, OV-102, is scheduled to make its first flight into space, from the Kennedy Space Center, on 1 April 1979. Three months later, on 1 July, it will be sent into orbit around the Earth, carrying the first fare-paying payload. NASA has a provisional list of payloads, known as the mission model, defining some 572 flights during 1980-1990, the first decade of Shuttle operation.

Basic dimensions and weights of the Shuttle are as follows:

DIMENSIONS, EXTERNAL:	
Wing span of orbiter	23·77 m (78 ft 0 in)
Length overall	56·08 m (184 ft 0 in)
Length of orbiter	37·19 m (122 ft 0 in)
DIMENSIONS, INTERNAL:	
Payload bay:	
Length	18·30 m (60 ft 0 in)
Diameter	4·57 m (15 ft 0 in)
WEIGHTS:	
Shuttle complete	1,998,500 kg (4,406,000 lb)
Orbiter empty	68,040 kg (150,000 lb)
Orbiter design landing weight	85,275 kg (188,000 lb)
Payload:	
due east	29,485 kg (65,000 lb)
at 104°	14,515 kg (32,000 lb)
THRUST:	
Total, at lift-off	28,135 kN (6,325,000 lb)
Orbiter, main engines (3), each	1,668 kN (375,000 lb)
Boosters (2), each	11,565·4 kN (2,600,000 lb)

APOLLO-SOYUZ TEST PROJECT (ASTP)

The Apollo-Soyuz Test Project, which involved the docking in Earth orbit of an American Apollo manned spacecraft and a Soviet Soyuz manned spacecraft, in July 1975, was described in the 1975-76 *Jane's*.

RCA

RCA CORPORATION

HEAD OFFICE:
30 Rockefeller Plaza, New York, NY 10020
Telephone: (212) 265-5900

CHAIRMAN OF THE BOARD AND CHIEF EXECUTIVE OFFICER: Robert W. Sarnoff

PRESIDENT: Anthony L. Conrad

EXECUTIVE VICE-PRESIDENTS:
Kenneth W. Bilby (Public Affairs)
Dr J. Hillier (Research and Engineering)
Irving K. Kessler (Government and Commercial Systems)

VICE-PRESIDENT, INTERNATIONAL NEWS AND INFORMATION:
Leslie Slote

RCA is prime contractor for a number of major defence programmes, including the US Navy's Aegis advanced ship-to-air missile system, the USAF's Block 5D and 417 satellites, and Tiros meteorological satellites.

BLOCK 5D SATELLITE

Block 5D is the name given to new advanced meteorological satellites forming part of the Defence Meteorological Satellite Programme (DMSP) managed by the Air Force Space and Missiles Systems Organisation (SAMSO).

Two DMSP satellites are normally in orbit at any one time providing weather data on a real-time basis to the Air Weather Service (AWS) and Navy ground and shipboard terminals located round the world. Data from the satellites is made available to civilian services through the Commerce Department's National Oceanic and Atmospheric Administration (NOAA).

Each spacecraft circles the world at an altitude of about 800 km (500 miles) in a near-polar, Sun-synchronous orbit. Each scans an area 2,960 km (1,800 miles) wide; thus each satellite can cover the entire Earth in about 12 hours.

Designed and built by RCA Astro-Electronics Division, Princeton, Block 5D spacecraft incorporate many changes which will increase their life span, support larger and more numerous sensors, and provide for a near constant resolution of visible and infra-red data.

Full title of the satellite is the Block 5D Integrated Spacecraft System, as it embodies a unique 'two-in-one' design approach in which the functions of the launch vehicle upper stage and the orbital satellite have been integrated into a single overall system.

The spacecraft comprises four major sections: a precision mounting platform for sensors and other equipment requiring precise alignment; an equipment module housing the bulk of the electronics; a reaction-control equipment support structure containing the spent third stage rocket motor and supporting the ascent phase reaction control equipment; and a solar cell array.

Primary sensor is the Operational Linescan System (OLS), made by Westinghouse Electronics Corporation, which provides visual and infra-red imagery. The flexible design of the spacecraft also allows the addition of special sensors which are of three basic types: temperature/moisture sounders, precipitating electron spectrometers for auroral detection, and upper atmosphere density sounders.

The major improvement in the Block 5D OLS over previous scanning radiometers is its ability to produce imagery with a near constant resolution across the scan. This is accomplished by using segmented visual and infra-red detectors, and employing a detector switching technique as the OLS scans away from the satellite sub-point (nadir) towards the edge-of-scan. This detector switching is combined with a basic sinusoidal scan pattern to produce the near constant resolution.

Three-axis stabilisation is achieved by the use of three reaction wheel assemblies, which provide a pointing accuracy of better than 0·1°. Power for operating the spacecraft subsystems is produced by a deployable, Sun-tracking solar array, covered by 10,560 silicon cells. The array produces 900W of power.

DIMENSIONS:
- Height: 5·18 m (17 ft)
- Width: 1·83 m (6 ft)

WEIGHT:
- At lift-off: 2,676 kg (5,900 lb)
- In orbit: 473 kg (1,043 lb)

Block 5D Air Force meteorological satellite

ITOS (NOAA) SATELLITE

Designed and built by the RCA Astro-Electronics Division, Princeton, for NASA's Goddard Space Flight Center, the first Improved Tiros Operational Satellite, also known as Tiros M, was launched from the Western Test Range, California, on 23 January 1970. A second-generation operational weather satellite, it represented a significant improvement over its predecessors (ten Tiros and nine ESSA spacecraft) in that it was capable of mapping the Earth's cloud cover at night as well as by day. A complete scan of the Earth was thus possible in 12 hours rather than just once a day. Other satellites of similar type have been built.

The ITOS satellite is box-shaped, 4·27 m (14 ft) wide with solar panels deployed, and weighs 309 kg (682 lb). Instead of spinning, as did ESSA satellites, it is stabilised so that it always faces the Earth. Employing a large spinning flywheel and appropriate electronic circuitry, this stabilisation system is called Stabilite.

Picture equipment includes two advanced vidicon cameras, with tape storage, and two automatic picture transmission (APT) cameras. Data from the latter can be picked up by the 500 or so relatively simple APT ground receiving stations located in over 50 countries. Other experiments include a solar proton monitor for solar flare warnings, and a radiometer to measure the Earth's heat balance.

ITOS-1 was launched into polar orbit, together with a small Australian tracking satellite, by a two-stage Delta-N, with six Castor solid-propellant strap-on boosters. It was followed on 11 December 1970 by **ITOS-2**. This craft, and the four remaining spacecraft in the series, are funded by the US Commerce Department. Once in orbit ITOS-2 was handed over to the National Oceanic and Atmospher Administration (formerly ESSA) and was given the desig nation **NOAA-1**.

Artist's impression of NOAA-4 experimental satellit in orbit

An attempt to put a second NOAA satellite into orb on 21 October 1971 was unsuccessful. The satellite re entered the atmosphere due to incorrect orientation at th time of injection. A replacement **NOAA-2** was launche successfully on 15 October 1972. The Delta launch vehicl carried as secondary payload the Oscar 6 amateur radi satellite.

The initial **NOAA-3** (ITOS-E) was also lost in a launc failure, in July 1973. Its replacement (ITOS-F) wa launched successfully from the Western Test Range on November 1973.

NOAA-4 was launched on 15 November 1974, th Delta launch vehicle also placing into orbit the Oscar amateur radio satellite and the Spanish Intasat-1 satellit NOAA-4 was the 24th RCA-built Tiros series spacecra the first of which was launched on 1 April 1960. Since the the spacecraft have returned more than 2·5 million view of the world's weather.

USAF 417 SATELLITE

This reconnaissance support satellite is reported to hav been operational since the mid-sixties. It weighs an est mated 270 kg (600 lb) and is launched by Thor/Burner Its purpose is to report cloud cover over areas that hav been selected for surveillance by photographi reconnaissance satellites, to prevent wastage of film. Orb tal height is about 800 km (500 miles).

ROCKWELL INTERNATIONAL
ROCKWELL INTERNATIONAL CORPORATION
Space Division

ADDRESS:
12214 Lakewood Boulevard, Downey, California 90241

Telephone: (213) 922-2111

OTHER FACILITIES:
Seal Beach, California; Cocoa Beach, Florida; Houston, Texas

PRESIDENT: G. W. Jeffs

The Space Division of Rockwell International was principal contractor for the building of Apollo Command and Service Modules and the second stage of the Saturn V launch vehicle used for the highly-successful 11-year Apollo lunar exploration programme. It was principal contractor for the Command and Service Modules which were used in 1973-74 to carry astronauts/scientists to the orbiting Skylab 1 laboratory. It built the Docking Module and docking system, and modified the Apollo Command and Service Modules for the Apollo-Soyuz Test Project of July 1975.

The Division is under contract to NASA to design, build and test the Space Shuttle Orbiter, and to act as integrator for the Space Shuttle transportation system. It is also under contract to the USAF Space and Missile Systems Organisation (SAMSO), to build the Navstar global positioning system satellites.

NAVSTAR

The Navstar global positioning system satellite is intended to satisfy future precise positioning and navigation needs of all the US military services, and to have potential civil applications. Navstar will provide suitably equipped users with highly accurate (to within 10 m; 30 ft) three-dimensional position and velocity information and a precise timing reference in real time. Envisaged applications include en-route navigation for space, air, land and sea craft; aircraft runway approach; photo mapping; geodetic surveys; aerial rendezvous; refuelling; and range instrumentation, safety, and search and rescue operations.

The 446 kg (980 lb) Navstar spacecraft will be placed in subsynchronous, 12-hour circular orbits of about 20,000 km (13,000 miles) in three orbital planes at 63° inclination, with eight satellites per ring, giving a total of 24 satellites.

Concept validation is expected in 1977, and the complete system should be operational by 1984.

Rockwell International has been awarded a $60 million contract by USAF Space and Missile Systems Organisation to develop spacecraft for the initial phase of Navstar. The contract covers the design and manufacture of five flight navigation satelllites and a qualification vehicle. Weight of each satellite will be about 742 kg (1,636 lb) at launch.

SPACE SHUTTLE TRANSPORTATION SYSTEM

The National Aeronautics and Space Administration's Space Shuttle will be the world's first re-usable space transportation system, and will be the keystone of America's space programme through this century. Rockwell International's Space Division is integrating the system and developing the Shuttle's payload-carrying orbiter stage for NASA, under a six-year $2,800 million contract awarded on 26 July 1972, following three years of study and a competitive proposal.

The Shuttle system includes the orbiter stage, capable of carrying up to 29,485 kg (65,000 lb) of varied cargo into Earth orbit; an external propellant tank; and two solid-propellant rocket boosters. The orbiter will lift off from Earth like a rocket, operate in orbit as a spacecraft, and return to land in a manner similar to that of a conventional aeroplane. Characteristics of the system are as follows:

ORBITER: A delta-winged aeroplane-like craft, 37·19 m (122 ft) long and 17·37 m (57 ft) high, with a 23·77 m (78 ft) wing span. Blended wing/fuselage design for optimum aerodynamic and manoeuvring characteristics. Structure of conventional aircraft design and fabrication, basically utilising aluminium, with outer thermal protective covering.

BOOSTER: Two solid-propellant rocket boosters, together with the orbiter's main engines, will power the orbiter from lift-off to approximately 43 km (27 miles) altitude. They will be jettisoned about two minutes into the flight, at an altitude of about 43,000 m (140,000 ft), dropped by parachute into the ocean and recovered for re-use. The boosters will each develop 11,565 kN (2,600,000 l st) and will be positioned under the wings of the orbite attached one on each side of the orbiter's external prop ellant tank. Supplied by Thiokol, each booster will b about 45·42 m (149 ft) long, with a diameter of 3·66 (12 ft).

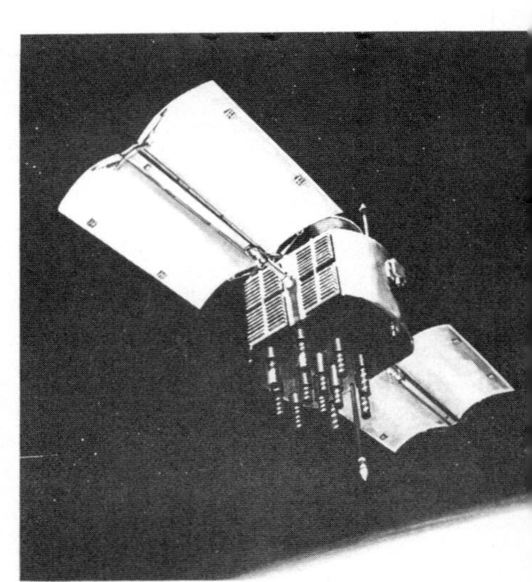
Artist's impression of NOAA-4 environmental satellite

EXTERNAL PROPELLANT TANK: Contains the liquid oxyge and liquid hydrogen main propellants for the orbite Aluminium monocoque construction, with foam exte nal insulation. Approximately 46·93 m (154 ft) long an 8·37 m (27 ft 6 in) in diameter. Under development b Denver Division of Martin Marietta Aerospace.

PAYLOAD BAY: 4·57 m (15 ft) in diameter by 18·29 m (6 ft) long, with manipulator arm equipped with televisio for deploying and retrieving one or more payloads.

CREW COMPARTMENT: Two-deck seating arrangement; tw

flight crewmen and two mission crewmen on upper deck, with room for a minimum of three scientists or specialists, and some living area. Dual flight controls for pilot and co-pilot. This compartment houses systems for controlling and operating the orbiter.

THERMAL PROTECTION: Silica fibre-based high-temperature and low-temperature re-usable surface insulation over a majority of the craft, with a reinforced carbon-carbon composite for the nose and wing leading-edges.

MAIN PROPULSION: Three high-pressure liquid oxygen/liquid hydrogen engines, each developing 2,090 kN (470,000 lb st) in space, provide the main propulsion for the orbiter. The engines are being developed for NASA under separate contract by Rockwell International's Rocketdyne Division.

ORBIT MANOEUVRING ENGINES: Two 26·7 kN (6,000 lb st) engines will be used for the orbiter's orbit manoeuvring subsystem (OMS). The engines will be housed in pods, one on each side of the orbiter's aft fuselage.

REACTION CONTROL ENGINES: The orbiter's reaction control subsystem (RCS) will utilise thirty-eight 3·87 kN (870 lb thrust) engines and six 0·11 kN (25 lb st) vernier thrusters. Fourteen of the RCS engines will be on the orbiter's nose and 24 will be on the aft end, 12 in each OMS pod.

ELECTRONICS: The electronics system consists of six subsystems: guidance, navigation and control; data processing and software; communications; displays and controls; flight instrumentation; and electrical power distribution and control. More than 50 per cent of the equipment for the overall system will be of 'mature design' or equipment already proven or easily available.

Space Shuttle Orbiter rollout September 1976

DIMENSIONS, EXTERNAL:

Length of complete vehicle at lift-off	56·08 m (184 ft 0 in)
Height of complete vehicle	23·16 m (76 ft 0 in)

See also Space Shuttle entry under NASA heading in this section.

P72-2 SATELLITE

The one-of-a-kind P72-2 multi-payload satellite was developed for the USAF Space and Missile Systems Organisation (SAMSO) by Rockwell's Space Division.

One of the most complex satellites produced under the Air Force Space Test Programme, it contained scientific instruments to measure the concentration and distribution of upper atmosphere aerosols (particles). When it was launched in 1975 the booster malfunctioned and the satellite was destroyed by the range officer.

TRW

TRW SYSTEMS GROUP

HEAD OFFICE:
1 Space Park, Redondo Beach, California 90278
Telephone: (213) 535-4321

VICE-PRESIDENT AND GENERAL MANAGER:
Dr George E. Solomon

VICE-PRESIDENT AND ASST GENERAL MANAGER:
Dr Edward B. Doll

PUBLIC AFFAIRS AND COMMUNICATIONS DIRECTOR:
Raymond Weil

Known formerly as TRW/Space Technology Laboratories (STL), and originally as the Guided Missiles Division of Ramo-Wooldridge, TRW Systems Group has provided technical direction and systems engineering for the USAF Atlas/Titan/Minuteman missile programme since 1954. The company has also designed and built more than 100 spacecraft and scores of major subsystems. Its recent and current military and civilian contracts have included prime contracts for the NASA OGO and Pioneer, USAF Vela and DSCS II, international communications satellites and the US Navy Fleet Communications Satellite.

The company built the variable-thrust descent engine for the Apollo Lunar Module (LM) and continues to produce a variety of low-thrust liquid-propellant and radioisotope thrusters. It also produced the LM abort guidance system.

TRW Systems provided NASA with mission analysis and spacecraft systems planning for the Apollo and Skylab projects, and furnishes systems integration and test support for the US Navy's anti-submarine warfare (ASW) programme.

DEFENSE SATELLITE COMMUNICATIONS SYSTEM, PHASE II (DSCS II)

The Phase II Defense Satellite Communications System (DSCS II) utilises synchronous-orbit, high-capacity, super high frequency communications satellites and surface terminals to provide reliable worldwide circuits for carrying essential military communications. The satellites are developed and produced by TRW Inc for the US Air Force, and supersede the DSCS I system described under the Philco-Ford entry in the 1972-73 *Jane's*.

Protected against interference, the satellites are each equipped with steerable narrow-beam antennae that focus a portion of the satellite's energy to areas 870 nm (1,600 km; 1,000 miles) in diameter. Within these specially illuminated areas, the narrow beam antennae allow small terminals to be used in place of more costly larger terminals. The narrow beams are designed to be steered in a matter of minutes to different locations on the Earth's surface, and the satellites are so designed that they can be moved in a matter of days to new synchronous orbital positions. In this way antenna coverage can be tailored to fit defence contingency communications all over the world.

The Phase II satellites each weigh 522 kg (1,150 lb), are 2·75 m (9 ft) in diameter and 3·95 m (13 ft) tall with antennae extended. Electrical power is supplied by solar arrays with an output of 535W at launch, decreasing to a minimum of 358W after five years. The X-band single-frequency conversion repeater weighs 81 kg (178 lb) and has 20W of power output from each of two travelling wave tubes. Its bandwidth is 410MHz. It has a capacity of 1,300 voice channels or up to 100 megabit per second of data.

The Earth-coverage antennae have a transmit beam-width of 18°, a gain of 16·8 dbi, and effective radiated power of 28 dbw. The narrow-coverage antennae have a beamwidth of 2·6°, a gain of 33 dbi and an effective radiated power of 43 dbw. They are steerable to ±10°.

First launch of two DSCS II satellites was on 3 November 1971, by Titan IIIC launch vehicle, from Cape Kennedy, and these are now serving the Atlantic and Pacific theatres. A second pair was launched on 14 December 1973, and a third pair in 1975, to give the DSCS a worldwide coverage. In 1974 contracts were placed for six additional DSCS II satellites to provide continuity in global coverage during the period 1977-80.

To support the early post-launch phase of the system, Aeronutronic Ford was contracted to design and build one heavy transportable ground terminal, one medium transportable terminal and one maintenance and supply van. These Phase 1 terminals have been upgraded and redeployed and are fully operational. In addition, two new heavy 18 m (60 ft) static terminals are operational.

DSCS II defence communications satellite

FLTSATCOM

The US Navy-sponsored Fltsatcom satellite system will provide worldwide high-priority UHF communications between naval aircraft, ships, submarines, ground stations, Strategic Air Command and the presidential command networks. The Air Force Systems Command's Space and Missile Systems Organisation (SAMSO) is the contracting agency.

Four of the three-axis stabilised satellites will be placed into geosynchronous equatorial orbit to provide complete Earth coverage, except for the polar regions. A fifth satellite will serve as a contingency spare. Fltsatcoms will be launched individually by Atlas-Centaur vehicles from Cape Kennedy, beginning in early 1977. Each will have a design life of five years and will provide more than 30 voice and 12 teletype channels.

The satellite consists of two major components, each with a basic 2·44 m (8 ft) hexagonal body:

The payload module contains UHF and X-band communications equipment and antennae. Each of its six side panels carries related communications components. The 4·88 m (16 ft) parabolic UHF antenna is made up of ribs and mesh, and opens like an umbrella.

The spacecraft module contains nearly all the remaining subsystem equipment, including the Earth sensors, attitude and velocity control, telemetry, tracking and command, and electrical power and distribution, as well as the buried, non-separable apogee kick motor. The solar array, never shadowed, is exposed to sunlight in both folded and deployed configurations. Each of the two panels measures 3·96 m × 2·13 m (13 ft × 7 ft). Together, they will provide at least 1,200W. Three nickel-cadmium batteries will provide power during eclipse.

WEIGHTS:

At lift-off	1,724 kg (3,800 lb)
In orbit	885 kg (1,950 lb)

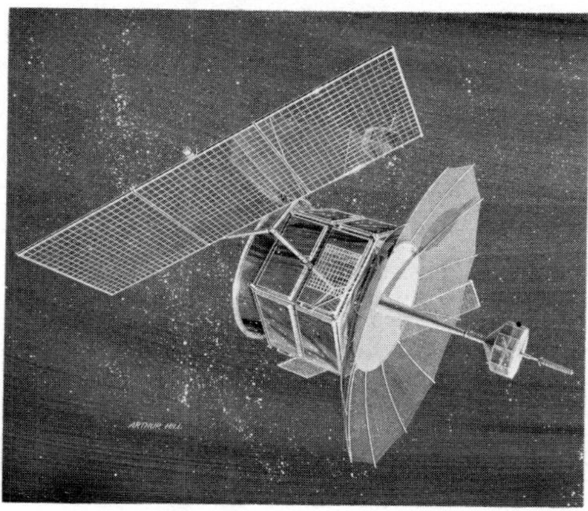

Artist's impression of Fltsatcom satellite

ERS

The ERS Environmental Research Satellite is a small spacecraft for conducting scientific and engineering research experiments in space. The satellite was developed by TRW Systems Group in 1961 and the first ERS was launched from Vandenberg AFB in September 1962. Since then, more than 30 ERS, all launched as piggyback payloads, have been orbited successfully.

An accompanying illustration shows the variety of configurations in which ERS satellites can be supplied. They range in weight from 0·7-36 kg (1½ to 80 lb) and generally carry a single experiment.

The most commonly flown ERS has been the octahedron, with eight sides on which triangular solar panels are fastened. The ERS can incorporate many different subsystems. Stabilisation can be either spin, passive magnetic, gravity gradient, or active magnetic. Electric power for the satellite is derived from solar cells, often supplemented by rechargeable batteries. VHF transmitters have been employed on all ERS flights for telemetry and tracking beacon signals, and command/receiver systems have been used. Satellite antenna subsystems have included dipole, crossed dipole and monopole. The satellite normally employs a passive thermal control subsystem.

The ERS has been used by NASA to check the manned spaceflight network (TETR-1, launched 13 December 1967, TETR-2, launched 8 November 1968, and TETR-3, launched 28 September 1971) and by the USAF to conduct experiments concerned with solar cell radiation damage, radiation measurements, surface contact bonding of materials in space, surface friction, zero gravity heat transfer, and solar X-ray and nuclear particle monitoring.

The ERS is one of the most reliable and least expensive active satellites yet produced. The latest configuration is the prism, which may be used as an Earth resources satellite.

ERS satellites are available to any prospective purchaser and can be made to meet any specified dimensions to house space instrumentation.

Two small satellites of the same generic type were developed to orbit the Moon as part of the Apollo programme. Known as Apollo particles and fields subsatellites, one was injected into lunar orbit from Apollo 15's Service Module during that mission, and returned data for more than six months. A second subsatellite was placed in orbit by Apollo 16, but because the spacecraft's lunar orbit had not been circularised, as the result of a guidance system gimbal lock alarm early in the mission, the satellite was ejected into a highly-elliptical orbit and crashed into the Moon after a few days of operation.

Examples of environmental research satellites developed by TRW for piggyback launching

IMEWS (USAF 647 SATELLITE)

The USAF's 647 series of satellites, for which TRW Systems is prime contractor, is intended to provide early warning of a hostile ballistic missile launch by detecting the infra-red emission from the missiles by means of an Aerojet-General infra-red 'telescope'. The programme is believed to be known now by the acronym IMEWS (Integrated Missile Early Warning Satellite) and to be the operational successor to the original Midas project.

A first launch was attempted on 6 November 1970, using a Titan IIIC vehicle. The intention was to place the 647 satellite into an initial synchronous orbit over the USA for checkout, then to move it westward to a position from where it could observe missile tests in China and firings down the Soviet Pacific test range. The propellants that would have been used for this re-location were exhausted in trying to correct a faulty initial orbit. Reports suggest that the infra-red telescope system was tested subsequently by observing US rocket launches.

The second launching, by a Titan IIIC from Cape Kennedy on 5 May 1971, is believed to have been successful. IMEWS-3 was launched on 1 March 1972, and IMEWS-4 on 12 June 1973. French reports suggest that the IMEWS satellite is cylindrical, weighs 820 kg (1,800 lb) at launch, has an inertial three-axis stabilisation system and measures approximately 3 m (9 ft 10 in) in diameter and 3 m (9 ft 10 in) in height; cruciform solar panels are said to span 7 m (23 ft). The same reports stated that the satellite carries high-resolution cameras, able to transmit photographs of any missiles that are located to a ground station 500 km (300 miles) north-west of Adelaide, Australia, for onward transmission via Programme 313 synchronous-orbit relay satellites to NORAD headquarters, Colorado Springs, USA.

PIONEER 6

Launched on 16 December 1965, Pioneer 6 (*Jane's* 1969-70) completed ten years' operation in space in December 1975, and continues to function satisfactorily, exceeding by a large margin its six-month design life.

Since launch, the 63 kg (140 lb) spacecraft has circled the Sun 12 times, covering over 9,656 million km (6,000 million miles). During this time the spin-stabilised solar-powered craft, one of a series of four injected into solar orbit to study the Sun, has been radioing back data continually, to a total of over 50,000 million bits of data.

The spacecraft has measured the Sun's corona, returned data on solar storms from the inaccessible, invisible side of the Sun, and measured the tail of the Comet Kohoutek. It has made new discoveries about the Sun, and helped to chart the solar wind, solar cosmic rays and the solar magnetic field, all three of which extend far beyond the orbit of Jupiter.

Five of the spacecraft's six scientific experiments continue to function; on the sixth, the magnetometer, a key part has worn out.

PIONEER 10

Details of Pioneer 10 were given in the 1974-75 *Jane's*. On 10 February 1976 the spacecraft crossed the orbit of Saturn, when it was 1,436 million km (892 million miles) from Earth. In mid-March 1976, sensors on the spacecraft indicated that it was passing through the magnetosphere of Jupiter, probably meaning that the magnetic 'tail' of the latter extends over 690 million km (430 million miles), spanning the distance between the orbits of Jupiter and Saturn.

Project officials hope that communication with the spacecraft will be possible at least up to the orbit of Uranus, 3,200 million km (2,000 million miles) from the Earth, which it will reach in 1979.

Pioneer 10 will leave the solar system altogether when it crosses the orbit of Pluto, some 6,000 million km (3,600 million miles) away, in 1987.

PIONEER SATURN (PIONEER 11)

Launched on 6 April 1973, Pioneer 11 is a backup spacecraft for Pioneer 10, full details of which were given in the 1974-75 *Jane's*. It was redesignated Pioneer Saturn in December 1974 after its successful encounter with the planet Jupiter. The spacecraft is virtually identical to Pioneer 10, except that a second magnetometer was added to measure high magnetic field close to Jupiter.

Like its predecessor, Pioneer 11 survived the long journey to Jupiter, involving the crossing of the asteroid belt beyond the orbit of Mars, and encountered the planet on 2 December 1974.

Had Pioneer 10 not been successful, Pioneer 11 would have served its primary purpose as a backup spacecraft and would have been guided by means of a midcourse correction for the same or a more conservative trajectory. However, in view of the success of the earlier spacecraft, Pioneer 11 was redirected for a close-in encounter which took it to within 41,000 km (25,500 miles) of the outer cloudtops.

The key course correction manoeuvre took place on 19 April 1974 when, at a distance of 676 million km (420 million miles) from Earth, thrusters on the spacecraft were commanded to fire for a duration of 42 minutes and 36 seconds. The burn added a velocity increment of 230 km/h (140 mph) to Pioneer's existing velocity of 46,200 km/h (28,700 mph). The spacecraft's speed then continued to fall gradually, until it entered Jupiter's sphere of influence. There the velocity increased to a maximum of 172,000 km/h (107,000 mph), the highest speed ever attained by a man-made object, as Pioneer passed the planet. The trajectory was such that, not only did it take Pioneer 11 three times closer to the planet than its predecessor, it also put the spacecraft on a Jupiter-gravity assisted course for the planet Saturn.

The spacecraft approached Jupiter ahead of and below the South Pole. Gravity then pulled the vehicle upward, towards the North Pole and around the planet opposite to its direction of rotation, in a corkscrew-like pattern.

During the Jupiter flyby, the spacecraft obtained pictures of the planet's poles, and its large inner moons, and carried out a full programme of observations of Jupiter's radiation belts, magnetosphere and atmosphere. The data confirmed that Jupiter is primarily a liquid planet, consisting mostly of hydrogen, and that it radiates more heat than it absorbs from the Sun. It also indicated that Jupiter is surrounded by a magnetic field very much more complex than was previously thought. The inner magnetic field is tilted with respect to Jupiter's spin axis, and displaced from the planetary centre.

While close to the planet, the spacecraft was subjected to intense bombardment by energetic proton radiation approaching rates of 150 million particles/cm²/sec. This caused anomalies in experiments and systems, the most notable being the loss of about 40% of the infra-red radiometer data on the outboard leg of the encounter. However, owing to the relatively short time of exposure, due to its very high speed, Pioneer emerged from the zone of peak danger in better overall shape than did Pioneer 10 after its encounter.

After passing Jupiter, and on course for Saturn, Pioneer 11 was redesignated Pioneer Saturn by NASA.

On 20 November 1975 the spacecraft, using its imaging photopolarimeter, took its first look at Saturn even though it was still over 1,300 million km (800 million miles) away.

Pioneer 11 undergoing final checkout before launch

The observation was from an 'off to the side' angle, showing the planet at a 'phase angle' (away from the sun-planet line) four times larger than the largest angle at which the planet can ever be seen from the Earth.

The spacecraft's course is such that, as it crosses the solar system, it will rise above the ecliptic plane until it reaches a high point of some 160 million km (100 million miles) in early 1977. It will then arc back to the ecliptic, intercepting it again for its encounter with Saturn, currently scheduled for 1 September 1979.

In mid-1976 the exact flyby course had not been selected, but a complicated course change manoeuvre was successfully completed on 18 December 1975 to give NASA flight officials two options. One would take the Pioneer between Saturn's rings and the planet, and the other would take the spacecraft in under the rings and then upward outside of them. The former flyby should enable Pioneer to get a close look at Titan, Saturn's sixth moon, and bigger than the planet Mercury.

The course change manoeuvre increased the spacecraft velocity by 108 km/hr (67 mph), and to effect this controllers at Ames Research Center, California, had to lose communication with the spacecraft, then 462 million km (287 million miles) from Earth, for several hours, and allow the spacecraft to command itself to change position, fire its thrusters and then reposition itself to point its antenna back to Earth in order to resume communications.

Should Pioneer Saturn survive the encounter, it ought to be able to continue to transmit its findings as it goes, before reaching the limit of communication somewhere near the orbit of Uranus, 3,200 million km (2,000 million miles) from the Earth.

VOUGHT

VOUGHT CORPORATION (Subsidiary of THE LTV CORPORATION)

Systems Division

HEADQUARTERS:
PO Box 5907, Dallas, Texas 75222

OFFICERS: See Aircraft section

Vought Systems Division, which resulted from a merger of the former Vought Aeronautics Company and Vought Missiles and Space Company in January 1973, is prime contractor for the NASA/DoD Scout launch vehicle. Its other products and capabilities relating to astronautics and space technology include a radiator system used in the Apollo spacecraft command module, and a manned aerospace flight simulator that was used by Apollo astronauts.

The Division is also engaged in the fields of advanced defence systems, ramjet propulsion systems and laser technology. Contracts have been awarded to Vought to develop the non-metallics needed to protect space shuttle vehicles during repeated re-entries from space. The materials involved are of the all-carbon type, known as reinforced carbon carbon (RCC).

A pioneer in the extra-vehicular manoeuvring unit field, it developed the USAF's Astronaut Manoeuvring Unit (AMU) and flight demonstration models of the unmanned, radio-controlled Remote Manoeuvring Unit (RMU). Under subcontract to Chrysler, this division produced the fuel and oxidiser containers for the first stage of the Saturn 1B space booster.

SCOUT (XRM-91)

Against competition from 12 other companies, Chance Vought (now part of Vought Corporation) won the major contract for the NASA/Department of Defense Scout four-stage solid-propellant space research vehicle in April 1959. In addition to being responsible for assembly of the overall vehicle, the company developed the nose section and airframe protecting the payload, the inter-stage sections between the various rocket engines, stage separation devices and the jet vanes and fin assemblies.

As prime vehicle contractor, Vought's Systems Division now performs duties extending from initial assembly and test at the Dallas plant to management and launch services. It also has built the launching towers for the rocket, including a type which permits horizontal checkout of the vehicle and erection to any desired position up to vertical for launch. Launching is possible at any angle from vertical to 20° from vertical.

Scout was designed to make possible space, orbital and re-entry research at comparatively low cost, using 'off-the-shelf' major components where possible. Its first stage is the 511·5 kN (115,000 lb st) Algol IIB (Aerojet Senior) by Aerojet-General, or the new Algol III (see below); the second stage is the 266·9 kN (60,000 lb st) Castor II by Thiokol; the third stage is the 93·41 kN (21,000 lb st) Antares II (X259) by Hercules Inc's Allegany Ballistics Laboratory; the fourth stage was originally the 13·34 kN (3,000 lb st) Altair (X248) or the more powerful Altair X258, but these have been superseded by a UTC stage (see below). Honeywell provides the simplified gyro guidance system. Spin stabilisation of the fourth stage is by Vought.

Final assembly of the Scout is done at NASA's Wallops Island facility, Virginia; at Vandenberg AFB, California, the west coast launch site for the vehicle; or at Italy's sea-based San Marco platform off the east coast of Africa, near Kenya.

On 16 February 1961, a Scout became the first solid-propellant vehicle ever to put a satellite into orbit when it was used to launch the Explorer 9 inflatable sphere.

An improved version, with FW-4S fourth stage (26·7 kN; 6,000 lb st) by United Technology Center, was launched for the first time on 10 August 1965, with complete success. In addition to increasing the payload capability to 145 kg (320 lb) in a 480 km (300 mile) orbit, this version can be manoeuvred in yaw and can send a 45 kg (100 lb) payload more than 25,750 km (16,000 miles) from the Earth.

A new heat shield, first used during a launch on 20 June 1971, provides a volume of 1·0 m³ (35·3 cu ft) and increases the diameter of the payload that Scout can carry from 0·8 m (2 ft 7½ in) to 1·0 m (3 ft 3½ in). The Algol III first-stage motor, available since 1971, provides a total impulse of 3,266,000 kg/sec (7,200,000 lb/sec) compared with about 2,476,000 kg/sec (5,450,000 lb/sec) for the Algol IIB. It increases the weight that Scout put into a 500 km (310 mile) easterly orbit from 150 kg (330 lb) to approximately 193 kg (425 lb).

A fifth-stage velocity package is under development, which will increase the Scout's hypersonic re-entry performance, make possible highly-elliptical deep-space orbits, and extend the vehicle's probe capabilities to the Sun.

Scouts have been used by NASA for a series of re-entry experiments. In one of these, in 1966, a special 0·43 m (17 in) spherical motor was used as a fifth stage to thrust back into the atmosphere at more than 29,000 km/h (18,000 mph) a payload designed to provide data on a heat-shield material for nose-caps. Total payload, including the motor, weighed 180 kg (400 lb) at lift-off.

In addition to its use by NASA and the Department of Defense, Scout is used for international programmes, including those of the United Kingdom, Italy, France, Germany, the Netherlands and the European Space Agency (ESA). On 26 April 1967, a Scout inaugurated use of a sea-based platform on the equator, off the east coast of Africa. As space booster for Italy's San Marco programme, it became the first vehicle to orbit a satellite from a launch site at sea.

Scout ended 1972 with its 28th consecutive successful launch, by orbiting Germany's Aeros satellite from Vandenberg in the first launch using a combination of both the Algol III first stage and the larger heat-shield.

By January 1976 a total of 93 Scouts had been launched (37 in a row without a failure until 5 December 1975, when a Scout failed to orbit NASA's Dual Air Density satellites). One which utilised an Alcyone IA fifth-stage motor to place NASA's Hawkeye satellite in a highly-elliptical orbit was the first Scout to incorporate a fifth stage as part of the launch vehicle rather than the payload.

NASA Scout 181 launch vehicle for Aeros satellite

DIMENSIONS:

Overall height	22·92 m (75 ft 2½ in)
Max body diameter	1·14 m (3 ft 9 in)

WEIGHT:

Launching weight	21,400 kg (47,185 lb)

THE UNION OF SOVIET SOCIALIST REPUBLICS

The information concerning recent research projects contained in the following notes is in most cases based on official Soviet news releases.

Full details of earlier Soviet satellites, spaceprobes and spacecraft have appeared in the 1959-60 and subsequent editions of *Jane's*, together with descriptions of the A-2 and A-3 research rockets and meteorological rockets which are standard vehicles in constant use.

COSMOS SATELLITES

This series of satellites is continuing the Soviet programme of research into physical phenomena in space and the Earth's upper atmosphere; into the technical problems involved in spaceflight and the development of spacecraft design and their systems; and of experiments of an applied nature of interest to science and the USSR economy.

The wide terms of reference mean that a Cosmos satellite can vary from a small uninstrumented device to a large spacecraft capable of life support, and military applications. The majority of the scientific satellites are of a basic standard design in which various experiment payloads can be accommodated. They are basically cylindrical in shape, approximately 1·83 m (6 ft) long by 1·05 m (3 ft 6 in) in diameter and weigh about 360 kg (800 lb).

Details of Cosmos 186, 188, 212 and 213 (automatic rendezvous and docking) and 215 (astronomical observatory) can be found in the 1970-71 *Jane's*.

Cosmos 605, launched on 31 October 1973, was a Soyuz-type spacecraft containing rats, tortoises, a mushroom bed, four beetles and living bacteriological spores. The mission lasted three weeks, and provided data on their reactions to prolonged weightlessness. Cosmos 690, launched on 22 October 1974, was also used for a biological space mission. The spacecraft contained animals and systems being tested for life support performance in space.

Cosmos 782, launched on 25 November 1975, inaugurated an international biological programme involving the co-operation of scientists of Czechoslovakia, France, Hungary, Poland, the Soviet Union and the United States. The satellite carried a centrifuge, for reproducing terrestral-type gravitation in conditions of weightlessness, and was used in experiments to study the influence of simulated gravity on different biological processes. The centrifuge comprised two main assemblies, one immovable and the other capable of rotating; each part contained identical groups of animals, plants and cells. Participating US scientists included those from Colorado State University, the State University of New York, the University of San Francisco and the Johnson Space Center.

Cosmos 637, launched on 26 March 1974, was the first satellite in the series to be placed in a synchronous orbit. Positioned over the Indian Ocean, it is assumed to be an early version of the Statsionar communications satellite announced by the Soviet authorities in 1970. Statsionar 1, the first operational satellite of this class, was launched on 22 December 1975. See the Raduga entry in this section.

MILITARY COSMOS

As the satellite tables on pages 704-5-6 indicate, Cosmos designations are given to Soviet reconnaissance satellites and other types of military spacecraft.

Cosmos reconnaissance satellites are launched from the bases at Plesetsk and Tyuratam, usually into orbits with inclinations of 52°, 65° or 72°. Most eject capsules after 8 days and these are presumably recovered. Some, such as Cosmos 228, have demonstrated the capacity for in-flight frequency changing; others, such as Cosmos 251, 264 and 280, were among the first to have a small manoeuvring capability for more precise target coverage.

Cosmos scientific satellite *(Maurice Allward)*

Cosmos 317 appeared to be the first of a new series of

operational reconnaissance satellites with an 11-13 day life, instead of the 8 days of early versions. The longer flights could imply that the craft carry a larger film package. If so, the same coverage will be obtained with fewer launches. An unusual characteristic of some of the longer-life craft is the ejection of a capsule just before recovery. The ejected capsule goes into a slightly lower orbit, where it remains for several days until it decays naturally. Many of the reconnaissance satellites use the on-board short-wave tracking beacon for Morse code type telemetry transmissions, but some of the later types use other equipment and are thus probably more sophisticated.

A number of test vehicles for the Soviet Union's Fractional Orbital Bombardment System have been given Cosmos designations.

Cosmos designations have also been given to a series of spacecraft which seem to have the capability of intercepting other satellites. Early 'interceptors' were Cosmos 249, 252, 374 and 375. Cosmos 397, launched on 25 February 1971, passed close to Cosmos 394 launched sixteen days earlier, and was subsequently destroyed. Cosmos 400, launched on 19 March 1971, was 'intercepted' by Cosmos 404 on the day of its launch, 3 April 1971. Cosmos 804, launched on 16 February 1976, is reported to have made several orbit plane changes to intercept, but without exploding, the Cosmos 803 'target', launched on 12 February 1976.

Cosmos 518, launched on 15 September 1972 from Plesetsk, and Cosmos 519, launched the following day from Tyuratam and involviing a rare high-inclination orbit from that site, were presumably launched to obtain supplementary data on NATO naval manoeuvres then being conducted in the North Sea. Specific launches are often made to observe such manoeuvres, and also to cover areas of special political concern. Thus in 1971 Cosmos 463 and 464 were used to obtain photographic coverage of the Indo-Pakistan War, and in 1973 Cosmos 597, 600 and 602 were used to survey the October Israeli-Arab Yom Kippur War. After fighting stopped, the orbits of Cosmos 609, 612, 616 and 625 were such as to indicate that they were monitoring the cease-fire, while Cosmos 630 monitored the withdrawal of Israeli forces.

Cosmos 629, launched from Plesetsk on 24 January 1974 into an orbit inclined at 62·8° to the equator, was the first recoverable reconnaissance satellite to employ this inclination. It has been used by the majority of reconnaissance satellites since then.

A number of Cosmos satellites have been launched into orbits suitable for ocean surveillance missions. The first of these was Cosmos 198, launched on 27 December 1967. Other satellites with similar orbits include Cosmos 209, launched 22 March 1968; Cosmos 367, launched 3 October 1970; Cosmos 402, launched 1 April 1971; Cosmos 469, launched 25 December 1971; Cosmos 516, launched 21 August 1972; Cosmos 626, launched 27 December 1973; Cosmos 651, launched 15 May 1974; Cosmos 654, launched 17 May 1974; Cosmos 723, launched 2 April 1975; and Cosmos 724, launched 7 April 1975. Cosmos 651 and subsequent spacecraft in this series appear to be bigger, with a length of about 12 m (40 ft) compared with 6 m (20 ft) of the earlier satellites. All the satellites were launched into an initially low orbit of about 250 km (155 miles) but after a few weeks moved into an approximately circular orbit of 1,000 km (620 miles).

COSMOS LAUNCHERS

Two categories of launch vehicle appear to be used for Cosmos and Intercosmos satellites, and other Soviet spacecraft. One category is based on the structures and power plants of standard missiles, such as the SS-4 (*Sandal*), SS-5 (*Skean*) and SS-9 (*Scarp*), with additional upper stages as required. The other combines the basic core vehicle developed originally for the Vostok manned spacecraft with a variety of upper stages. Examples are as follows:

SS-4 + Cosmos stage. First stage powered by 706·34 kN (158,800 lb st) RD-214 four-chamber liquid-propellant rocket engine, burning nitric acid and kerosene. Second stage powered by RD-119 single-chamber engine, burning liquid oxygen and dimethylhydrazine, and giving 107·87 kN (24,250 lb st) in vacuum. Typical launch, on 26 June 1974, orbited Cosmos 662, a 408 kg (900 lb) ellipsoid, 1·83 m (6 ft) long with a diameter of 1·22 m (4 ft), intended for scientific research.

SS-5 + Restart stage. A typical application for the SS-5 is to orbit satellites like Cosmos 655 and 661. Shaped as cylinders, 1·83 m (6 ft) long and 0·91 m (3 ft) in diameter, with paddle-type solar panels, these are thought to have navigation and/or electronic intelligence missions.

SS-9 + FOBS stage. Frequent launches of this vehicle are expected to contribute to continued development of Fractional Orbital Bombardment System techniques and/or to ocean surveillance missions. Satellites like Cosmos 651 and 654 normally remain in low parking orbit for two months, then split and move into a 104 min orbit.

Vostok core + Venus stage. This standard launch vehicle has many applications. It is used with an escape stage to orbit the 1,250 kg (2,750 lb) uprated Molniya 2 communications satellites. Typical military payloads were Cosmos 639, a manoeuvrable reconnaissance satellite intended probably to study the breakup of Arctic pack ice; and Cosmos 658, a reconnaissance satellite in the form of a four-ton sphere-cylinder, 5·0 m (16 ft 6 in) long, which remained in orbit for 12 days.

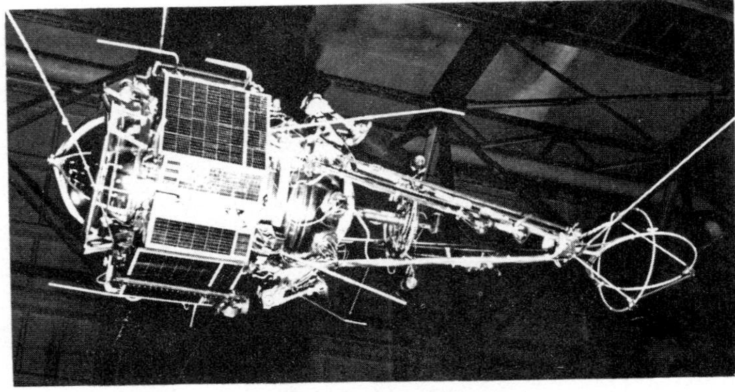

Intercosmos 14 satellite *(Tass)*

INTERCOSMOS SATELLITES

Launched on 14 October 1969, Intercosmos 1 involved the co-operation of seven Socialist countries: Bulgaria, Czechoslovakia, Germany (Democratic Republic), Hungary, Poland, Romania and the USSR. The satellite carried scientific instruments developed and made in Czechoslovakia and Germany as well as in the USSR. Its programme was devoted mainly to research connected with the Sun, and involved simultaneous observations in the seven participating countries into radio-astronomical, ionospheric and optical phenomena.

Further details of Intercosmos 1 and the later Intercosmos 2 can be found in the 1970-71 *Jane's*. Intercosmos 3 and 4 were described in the 1971-72 edition. Intercosmos 5 was launched on 2 December 1971. Intercosmos 6, launched on 7 April 1972, was the first of a new series, with recoverable payloads, and weighed 1,070 kg (2,360 lb).

Intercosmos 9, a Soviet-Polish satellite, launched on 19 April 1973, was named Intercosmos Copernicus 500 to mark the 500th anniversary of the birth of the Polish scientist. This spacecraft carried a Polish-built receiver to detect solar radiation and to study the effects of solar radiation on the Earth's ionosphere. A feature in a Polish magazine indicated that Intercosmos Copernicus 500 was launched by a rocket housed in a silo. The launching vehicle, although marked Kosmos, appears to be the Soviet IRBM designated SS-5 by NATO, which has been used to launch other Intercosmos satellites. For space launch missions an upper second stage is added. Before launch the top of this stage, mounting the payload, protrudes above the ground, facilitating pre-launch adjustments and system checks.

Intercosmos 10 was launched on 30 October 1973 and Intercosmos 11 on 17 May 1974. Intercosmos 12, launched on 31 October 1974, carried experiments provided by Bulgaria, Czechoslovakia, Germany (Democratic Republic), Hungary, Romania and the USSR. Intercosmos 13, launched on 27 March 1975, was designed to conduct studies of the upper atmosphere. The series of launches is continuing (see Satellites table).

LUNA SPACECRAFT

Descriptions of the soft-landing Luna 9 and 13 and lunar-orbiting Luna 10, 12 and 14 spacecraft can be found in the 1968-69 *Jane's*.

Reference to the Luna 15 spacecraft, sent to the Moon at the time of America's Apollo 11 mission, appears in the 1969-70 edition. The 1971-72 edition contains details of Luna 16, which brought samples of Moon rock back to Earth, and Luna 17, the transport vehicle for the Lunokhod 1 automatic lunar roving vehicle. The mission of Luna 20, the second to return with a lunar soil sample, is described in the 1972-73 edition.

Luna 21, which carried the Lunokhod 2 lunar roving vehicle to the Moon, is described in the 1973-74 *Jane's*. **Luna 22** was launched on 29 May 1974 and was placed into lunar orbit on 2 June for a programme of near-Moon space research. On 11 November the orbit was changed to 1,437 × 171 km (893 × 106 miles). On 2 April 1975, following further changes to its trajectory, the orbit was 1,409 × 200 km (875 × 124 miles) and the spacecraft had completed 2,842 revolutions of the Moon.

Luna 23 was launched into Earth orbit on 28 October 1974, and subsequently, on 2 November, was placed in a lunar orbit of 94 × 104 km (58 × 65 miles). After orbital manoeuvres on 4 and 5 November, the station was soft-landed in the target area, in the southern part of the Sea of Crises. On touchdown, the impact damaged some equipment, including a drill designed to obtain samples from a depth of up to 2·5 m (8 ft) below the surface, presumably for return to Earth.

An emergency research programme was adopted, and was conducted for three days before contact with the station ceased completely.

Luna 24, launched on 9 August 1976, was put into a lunar orbit on 14 August. The trajectory was adjusted on 16 and 17 August, to place the spacecraft into a 120 × 12 km (75 × 7 mile) elliptical orbit around the Moon, and it was soft-landed successfully on the lunar surface on 18 August, in the south-east corner of the Sea of Crises. After landing, on command from the Earth, an on-board drill extracted small samples of the lunar soil from a succession of pre-set depths of up to 2 m (7 ft). These samples were placed in a hermetically sealed capsule which, on 19 August, was launched from the lunar surface, the main structure of the spacecraft serving as the launch pad. The sealed capsule was landed successfully in the Soviet Union on 23 August, about 200 km (124 miles) south-east of Surgut in Western Siberia. The lunar samples, powdery and including small grains more than 5 mm (0·20 in) across, resemble outwardly those brought back to Earth by Luna 16 in 1970.

METEOR

Meteor is the name given to the Soviet series of first-generation operational meteorological satellites, developed from a number of Cosmos prototypes (described in previous editions of *Jane's*).

The satellites, which provide information about the state of the atmosphere both on the 'daylight' and 'night' sides of the Earth, are stabilised so that the camera lenses and infra-red instruments always point towards the Earth.

Information received from Meteors is supplied to the Soviet hydro-meteorological service and to the World Meteorological Service. Cloud cover picture charts are transmitted to Washington, Geneva, Tokyo, Sydney and other foreign weather services.

Meteor 1 was launched on 26 March 1969; and launchings have continued at a current rate of three or four per year, with Meteor 21 launched from Plesetsk on 1 April 1975. It is a cylinder with two solar panels, about 4·88 m (16 ft) long, with a diameter of 1·5 m (5 ft).

The Meteor system consists of three satellites in orbital planes at 90° and 180° to each other, so that they pass over a given area of the Earth on the northbound pass at intervals of about 6 hours and 12 hours, and again during the southbound pass.

MOLNIYA

Molniya 1 (Lightning), launched on 23 April 1965, was a communications satellite placed in a highly elliptical orbit designed to provide the longest possible communications sessions between Moscow and Vladivostok. It was the first Soviet communications satellite, and was in the form of a cylinder with conical ends, one end containing the correcting engine and a system of micro-jets, and the other end containing solar and Earth-orientation sensors. Six solar panels and two parabolic aerials were mounted on the central body. Also attached were the radiation surfaces of the temperature-control system comprising a radiation/refrigerator and a heating panel in the form of a flat ring, which also accommodated solar batteries.

During flight the satellite was orientated with its solar panels facing the Sun and one of its aerials was directed simultaneously towards the Earth. Signals were transmitted in a relatively narrow beam, ensuring a strong reception at the surface of the Earth. The other aerial was in reserve, and for this to be used the satellite had to be rotated 180° longitudinally.

Molniya can handle a television programme, a large number of telephone conversations, still pictures, telegraphic information and other forms of information.

Details of early Molniya 1s can be found in previous editions of *Jane's*.

On 24 November 1971, **Molniya 2A**, the first of a new uprated version, was launched from Plesetsk. It is believed to have a central structure 4·10 m (13 ft 6 in) tall, with a diameter of 1·52 m (5 ft), and to weigh about 1,250 kg (2,750 lb).

Launches of both the original and improved versions are continuing, and the current operational system consists of four pairs of each type of Molniya circling the Earth in orbital planes spaced at 90° intervals. The planes rotate at

Prognoz 2 solar research spacecraft *(Tass)*

Full-scale model of Molniya 2 communications satellite
(Brian M. Service)

the rate of about 1° a day, in order to maintain stationary ground tracks over the Earth throughout the year, thus providing predictable communications coverage over the whole of the USSR.

Molniya 3 is the latest version of the long-established series of Molniya communications satellites, and embodies improved communications equipment able to accommodate colour television and new communications frequencies. The first of the new spacecraft was launched on 21 November 1974, and was placed in a normal Molniya-type 12 hr 17 min orbit, with its 650 km (400 mile) perigee over the southern hemisphere and its apogee of 40,690 km (25,283 miles) over the northern hemisphere.

The spacecraft is intended to "ensure long-distance and cable radio communication in the USSR to points within the Orbita system, and for international co-operation".

PROGNOZ

Launched on 14 April 1972, Prognoz 1 (Forecast) was the first of a new series of satellites designed specifically to study processes of solar activity, and their influence on the interplanetary medium and the Earth's magnetosphere.

Weighing 857 kg (1,890 lb), the spacecraft was injected into a highly-elliptical orbit of 950 × 200,000 km (600 × 124,000 miles) from an initial Earth parking orbit.

Prognoz 1 is basically sphere-shaped, with four cruciform solar panels extended to provide power via an internal rechargeable chemical battery. The sphere, filled with an inert gas, contains telemetry equipment, temperature control components and the electrical system. Experiments on board are intended to measure: electromagnetic solar radiation generated simultaneously with solar flares; solar cosmic radiation; solar wind plasma; and radio waves.

Prognoz 2 was launched on 29 June 1972, to begin a programme of joint experiments with Prognoz 1. In addition to Soviet equipment, the second spacecraft carried instruments developed by French scientists. Prognoz 3, launched on 15 February 1973, operated in conjunction with Lunokhod 2 in attempts to determine the cause of variations in the local lunar magnetic fields. It was still functioning in February 1974, returning data three times a week on solar flares, X-ray activity and the effect of solar radiation on the Earth's ionosphere. Prognoz 4 was launched on 22 December 1975.

RADUGA

Raduga (Rainbow) is the name of a new Soviet system of geostationary-orbit communications satellites, the first operational version of which, Statsionar 1, was launched on 22 December 1975.

The satellite, launched at 1300 GMT, is now on station over the Indian Ocean at longitude 80°E. Russia stated that it would be joined by Statsionar 2 and 3 in 1976, which would be located at longitude 35°E and 85°E respectively. The Soviet authorities have also given notice of their plans to launch a further seven satellites in the series, Statsionars 4 to 10, over the Atlantic, Pacific and Indian Oceans to establish a global communication network operating at conventional, internationally agreed C-band frequencies (4GHz to 6GHz).

Statsionars are three-axis stabilised, and transmit at higher power levels than comparable Intelsat IV spacecraft. This enables them to communicate with smaller and therefore cheaper ground terminals. Currently, the spacecraft will communicate with the Orbita network of ground stations set up for the Molniya programme.

Existence of Statsionar was initially revealed on 3 February 1969 in an official Soviet communication to the ITU, and further details of the spacecraft were registered with that organisation on 1 December 1970. The first test spacecraft in the system was designated Cosmos 637, launched on 26 March 1974, and this was followed on 29 July 1974 by what appears to be an engineering test model of Statsionar but designated Molniya 1S. The long delay between the initial Soviet announcement and the launch of Cosmos 637 could mean that serious technical difficulties were encountered during the development of the new spacecraft.

One worrying aspect of the complete Raduga system is its potential for interference with other established systems: for example, the Indonesian system and Intelsat. One Statsionar, planned for longitude 58°E, will be only 2° away from an operating Intelsat IV.

SALYUT

Salyut (Salute) spacecraft have served as orbital scientific stations for the crews of Soyuz manned spacecraft. Basically, the station is a stepped cylinder 12·2 m (40 ft) long, from 2·13 m (7 ft) to 4·0 m (13 ft) in diameter, with a weight of 18½ tons.

Details of Salyut 1 and 2 can be found in the 1973-74 *Jane's*. Salyut 3 was described in the 1975-76 edition.

Salyut 4 was launched on 26 December 1974 into an initial 270 × 219 km (167 × 136 mile) orbit. The purpose of this mission was to test modifications embodied as a result of experience gained with Salyut 3, and to provide the means for additional experiments in space.

Externally, Salyut 4 differs little from Salyut 3, but significant changes were made internally to accommodate the programme of experiments. A major new piece of equipment is an Orbital Sun Telescope, OST-1, made at the Crimean Astrophysical Observatory. The telescope has a main mirror of 0·25 m (10 in) diameter, and embodies a 'shortwave diffraction spectrometer', which registers photo-electrically the far ultraviolet range of 850-1,350 Angstroms. Earth photographic equipment is also carried, which can produce imagery of the ground with 'varying degrees of surface detail and in various spectral ranges'. Other equipment includes an infra-red telescope spectrometer and X-ray telescopes.

Soyuz 17, crewed by Alexei Gubarev and Georgi Grechko, docked with Salyut 4 on 12 January 1975, for a 30-day programme of research. Experiments included lengthy observations of the Sun, using OST-1, and spectrographic measurements of atomic oxygen, which is the most Sun-sensitive element of the Earth's ionosphere. In a 'vegetable garden' peas sprouted, causing Grechko to comment that it was good to see green plants in view of the cosmic cold outside the space station.

A major event, on 4 February, was the 'repairing' of the Sun telescope by spraying a fresh reflecting surface on to its mirror to counter the dulling effect of three weeks'

Full-scale representation of the latest type of Salyut space station, with a third solar panel on top *(Tass)*

exposure to space micro-debris. On Earth, work of this nature has to be carried out in a vacuum chamber; the Salyut cosmonauts operated in the vacuum of space. If this experiment proves to have been entirely successful, Soviet engineers will be encouraged to develop second-generation telescopes with mirror diameters between 1 and 3 m (3 ft 3 in and 10 ft).

Of great importance were the elaborate exercises undertaken by the cosmonauts, to minimise the physiological deterioration that can result from prolonged exposure to weightlessness. Equipment included an integrated physical trainer, embodying an endless-belt running track with associated elasticised gear; a bicycle ergometer, the pedals of which were operable by either hands or feet; and two types of load-generating suits. One suit, a *g*-type, is designed to load the body to prevent blood pooling in the lower limbs; the other embodies load-generating straps linked at the shoulders, waist and shoes, so that the wearer has to exert considerable effort to stand erect.

Physical training exercises occupied the cosmonauts for 2·5 hours daily, which can be compared with the 0·5 hours of the Skylab 2 crew, 1·0 hour of the Skylab 3 crew, and 1·5 hours expended daily by the Skylab 4 crew during their record 84 days in orbit.

The Soyuz 17 undocked and returned safely to Earth on 9 February 1975. Initial medical examinations made at the landing site indicated that both cosmonauts had stood up well to their long stay in orbit. Soyuz 20, launched on 17 November 1975, docked with Salyut 4 on 19 November. This spacecraft (described separately in this section) was unmanned, the purpose of the mission being to gain experience in the supplying of space stations by such spacecraft.

Salyut 5, launched on 22 June 1976, is a modest development of previous Salyuts, the major change being the embodiment of an additional docking port so that more than one Soyuz spacecraft can be docked with it at the same time.

Salyut 5 is reported to be designed for manned missions lasting up to 90 days, although the basic mission will be between two and three months. The space station is capable of being used for longer missions, but the Soviet cosmonauts, Vitali Sevastyanov and Pyotr Klimuk, who piloted the Salyut 4-Soyuz 18 mission which established a new Soviet manned space flight record of 63 days in orbit, reported that they were relatively tired after their two months in orbit.

With its two docking ports Salyut 5 can support a four-man crew brought to the station by two-man Soyuz spacecraft. If the Soyuz are re-configured back to their original three-man capability, then two vehicles could both launch and return a six-man crew.

New spacesuits have been developed for use on Salyut 5 and will be used during extensive extravehicular activities which will include both engineering and scientific experiments.

The unmanned automatic supplies spacecraft technique, tested when Soyuz 20 docked with Salyut 4, is reported to become operational with Salyut 5. Such craft will be used to replenish consumables such as air, food, film and scientific equipment, while the space station is operational.

On 7 July 1976 Soyuz 21 (which see) docked with the space station.

SOYUZ SPACECRAFT

Developed for the Russian Earth-orbital space station programme, Soyuz spacecraft are equipped for missions of up to 30 days duration.

Each spacecraft comprises three basic sections or modules: a laboratory-cum-rest compartment (orbital module), a descent compartment (landing module) and a propulsion and instrument section (service module). The orbital module is mounted on the extreme nose of the craft, and communicates with the landing module via a hermetically-sealed hatch. The orbital and landing modules are pressurised to 1·01 bars (14·7 lb/sq in), have a combined internal volume of 9 m³ (318 cu ft) and can accommodate up to four cosmonauts.

The service module contains the main systems for orbital flight, together with a liquid-propellant propulsion system embodying two motors (one a standby) each with a thrust of 3·92 kN (880 lb). These allow midcourse manoeuvres, up to heights of 1,300 km (800 miles), and are used for the de-orbit manoeuvre. Another system provides attitude control. Attached to the service module is a solar-cell array having an area of about 14 m² (150 sq ft).

The landing module contains the parachutes and landing rockets. A backup parachute system is available in case of failure. The main parachute, preceded by a pilot 'chute, is deployed at 8,000 m (27,000 ft). Retro-rockets, operating at a height of about 1 m (3 ft) above the ground, ensure a landing velocity not exceeding 3 m/sec (10 ft/sec). The aerodynamic design of the landing module permits landing loads to be kept within 3-4*g*, although ballistic re-entries involving loads of 8-10*g*, can be made if required. The overall length of the craft is about 9 m (30 ft), the diameter of the crew compartments about 2·1 m (7 ft) and the all-up

Soyuz launch vehicle

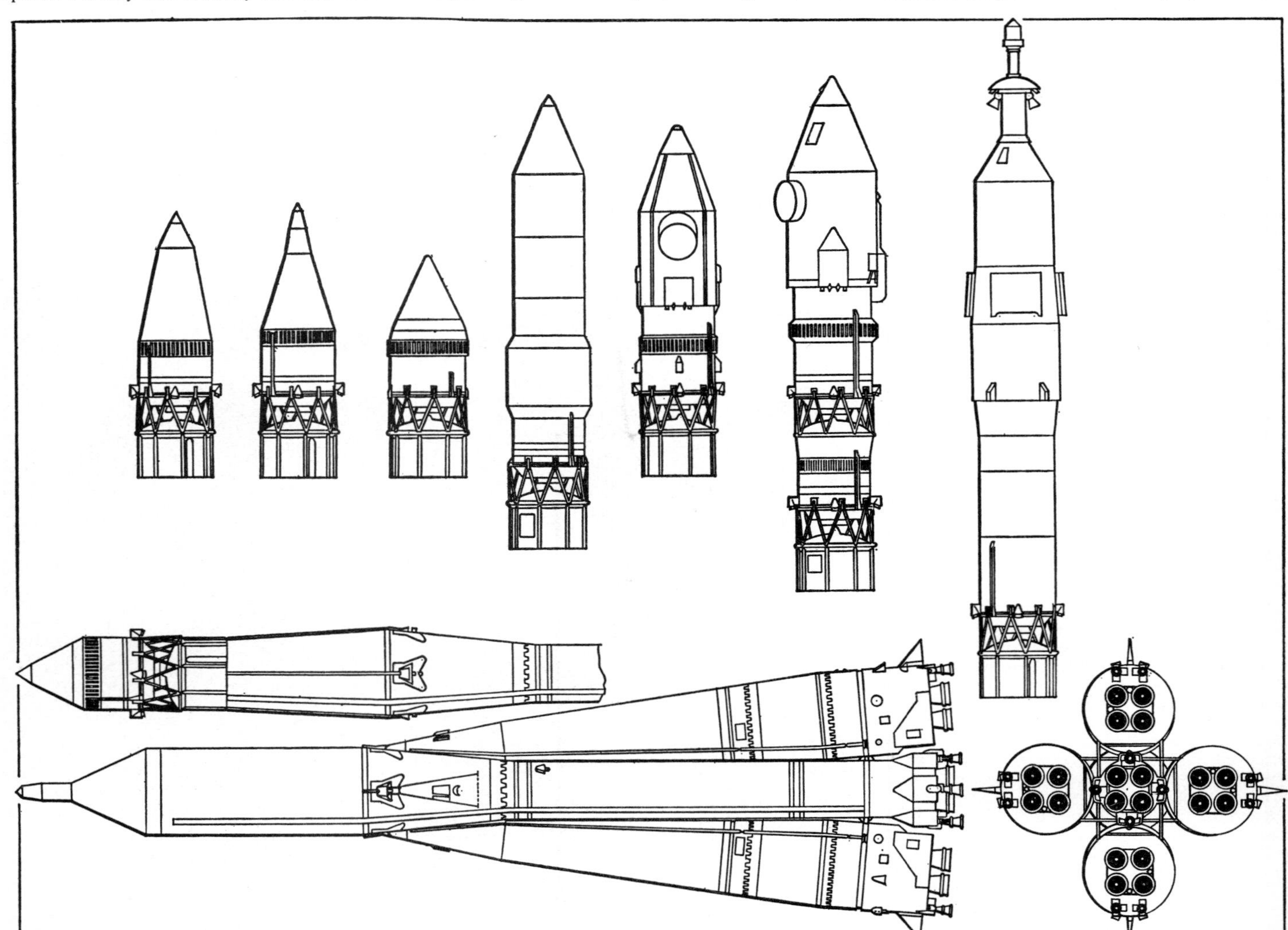

Soviet launch vehicles. The two bottom views are a side elevation and end view of the nozzle arrangement of the basic vehicle, made up of the core and four boosters in development test configuration. Immediately above, *left,* is a scrap view of the top of the core stage with the additional stage used to orbit Sputnik 1 on 4 October 1957. Top *(left to right)* are adaptations of the vehicle used to launch Sputnik 2, Sputnik 3, Luna 1 and 2, Mars 1 and Venus 1, Vostok 1, Voskhod 1 and 2, and Soyuz. Overall length of the Vostok launcher was 38 m (124 ft 8 in) *(Sherwood Designs Ltd, with acknowledgement to Skrzydlata Polska)*

weight about 6,000 kg (13,000 lb).

The Soyuz craft are equipped with an automatic control system for approach and docking manoeuvres, the technique and external aerials being similar to those employed on the Cosmos spacecraft 186 and 188, and 212 and 213.

Soyuz 1 to 5 were described in the 1969-70 *Jane's*, Soyuz 6, 7 and 8 in the 1970-71 edition, Soyuz 9, 10 and 11 in the 1971-72 edition, and Soyuz 12 and 13 in the 1974-75 edition. Soyuz 14, 15, 16, 17 and 18 were described in the 1975-76 edition.

Soyuz 19 was the designation given to the spacecraft used for the successful joint USSR-USA Apollo-Soyuz Test Project (ASTP) in July 1975, as described under the NASA entry in the 1975-76 *Jane's*.

Soyuz 20. This unmanned spacecraft was launched on 17 November 1975 and docked with the Salyut 4 space station two days later. Salyut 4, launched on 26 December 1974, had by then been functioning automatically since the crew, ferried up in Soyuz 18 (described in the 1975-76 edition), had left on 26 July 1975. After docking, the Soyuz-Salyut complex was in a 367 × 343 km (228 × 213 mile) orbit.

Soyuz 20 contained tortoises, Drosophila flies, cacti, gladioli corms, vegetable seeds, maize and leguminous plants. Experiments with the tortoises were intended to help predict changes in the bodies of animals during spaceflight, while experiments with the flies will assist in studying the influence of weightlessness on their growth and development. In addition some 20 species of higher plants were used to study changes in the molecular structure of plants and the manner in which they are influenced by cosmic radiation.

The results of these experiments will assist in designing new life support systems.

The prime object of the Soyuz 20 mission, however, was described as to gain experience in the supply by unmanned spacecraft of scientific equipment or food reserves to manned space stations. Having accomplished this task the spacecraft could then be used by the crew of the space station as an auxiliary transport vehicle.

An additional capability of such a mission would be the use of the unmanned spacecraft as an ambulance to bring back to Earth sick or injured crew members.

Soyuz 21. Launched on 6 July 1976 into a 252 × 193 km (157 × 120 mile) orbit, Soyuz 21 was crewed by cosmonauts Air Force Col Boris Volynov and Army Lt Col Vitaly Zholobov. Prior to the spacecraft's fourteenth revolution the orbit was raised to 278 × 251 km (173 × 156 miles).

On 7 July 1976 the spacecraft docked with the space station Salyut 5 (which see) to start a mission involving, according to official Russian sources, 'research into the effects of weightlessness, and a survey of geological features on the Earth's surface and weather phenomena'.

Soyuz 22 was launched into a low Earth orbit on 15 September 1976, from the manned spaceflight centre at Baikonur, with Valery Bykovsky and Vladimir Aksenor as the crew. The spacecraft was stated to be part of a programme of space co-operation among Communist countries, and carried East German photographic equipment. The main aim of the mission was stated to be to develop methods of studying geography and geology from very high altitudes.

Soyuz 21 landed safely on 24 August 1976, near the town of Kokchetov in Kazakhstan, after 48 days in orbit. During the mission the crew carried out an extensive programme of biological, metallurgical and Earth resources experiments. Although not a prime scientific task, the cosmonauts also made numerous meteorological observations which regularly complemented data from unmanned spacecraft, to produce a more complete world weather picture. A week before the return, the Russian newspaper *Izvestia* reported that music was played to the crew in an effort to ease the effects of prolonged isolation. After landing, the crew were reported to be suffering from 'sensory deprivation'.

SOYUZ LAUNCHER

The vehicle used for launching Soyuz spacecraft appears to be a development of the booster used for launching the original Vostok spacecraft, with some 11·8 m (36 ft) of additional upper staging and structures. To cater for the increased weight and bending moment the inter-stage truss is strengthened. During launch, the Soyuz vehicle is surmounted by an escape tower with three rows of rocket nozzles. Under the projecting domed fairing is a ring of eleven or twelve main nozzles, surmounted by four small vernier nozzles. At the base of the cylindrical section at the top is a ring of still-smaller nozzles of the kind seen around the tail of the *Frog-7* missile.

It is not possible to identify the current engines in the launch vehicle, or give their individual ratings. However, official Soviet reports have stated that the vehicle has a total thrust of around 60 million horsepower, which is three times the power quoted for the original Vostok launcher. The basic configuration has not changed. Thus, the first stage consists of a central core, powered by an engine with four primary nozzles and four verniers. This is surrounded by four wrap-round boosters, each with four primary nozzles and two verniers, so that 32 rocket chambers are fired simultaneously during lift-off.

VENUS

Venus (Venera in Russian) is the name given to a series of spacecraft launched to explore the planet Venus.

Descriptions of Venus 1 to 7 are given in earlier editions. Venus 8, which successfully soft-landed on the surface of the planet in July 1972, is described in the 1973-74 edition.

Venus 9 was launched on 9 June 1975, followed on 14 June by **Venus 10,** to soft-land television cameras and other instruments on Venus. Each spacecraft consisted of an orbiter module and a lander module. The Tass announcement of the launchings stated that the two spacecraft were of a new type, and embodied radically new features compared with previous Venus probes.

A major difference was a reduction in the ratio between the spacecraft's weight and their cross-sectional area, so that the first stage of deceleration would be completed more quickly than previously.

Other design changes included a new parachute system, and two new devices to help the spacecraft make a successful landing. One device was a circular metallic disc designed to act as an airbrake to ensure that the craft would reach the surface rapidly but safely after the main parachute had been released, and the other was a crushable metallic ring, to help absorb the impact of landing. The reason for a rapid descent was to minimise the exposure to high temperature.

A new heatshield enabled a number of instruments to be installed on the outside of the body of the Venus lander, yet be protected during the descent through the dense and turbulent atmosphere. The shield also embodied a shock-absorbing device. The heatshield had to be light and yet be able to withstand a temperature of 2000°C, and its release mechanism is reported to have presented severe design problems.

The two spacecraft completed the 370 million km (230 million mile), 136-day journey to Venus safely, and went into orbit round the planet in October 1975.

The orbiter modules of previous Venus spacecraft were designed only to explore space on the way to the planet and to eject a landing capsule. The Venus 8 and 9 orbiters, after releasing the lander modules, some two days before reaching the planet went into highly elliptical orbits (Venus 9: 1,300 × 112,000 km; 800 × 69,600 miles; Venus 10: 1,400 × 114,000 km; 870 × 70,840 miles). From these orbits the orbiters first relayed to Earth data from the landers, and then continued to transmit data on the Venusian cloud cover and magnetic field.

Venus 9 lander entered the atmosphere at 06·58 Moscow time on 22 October at a speed of 10·7 km/sec (6·65 miles/sec). Aerodynamic braking reduced the speed to 250 m/sec (820 ft/sec). The main parachute then deployed, at which point the lander started to transmit data on the Venus cloud layers. The cloud layer was found to extend vertically for 30-40 km (19-25 miles) with its base lying 30-35 km (19-22 miles) above the surface. At a height of 50 km (31 miles) the parachute was released and the lander continued its descent retarded only by the circular airbrake, to land at 08·13 Moscow time. Final descent velocity was about 7 m/sec (23 ft/sec). Before landing the craft was cooled to −10°C in order to increase its operational life.

Replica of the lander module of the Venus 9 and 10 spacecraft

After impact the lander transmitted, via Venus 9 orbiter, data on the environment for 53 minutes. It also transmitted the first spectacular picture of the surface of this planet. The pictures were so good they were released for publication without intermediate processing. Contrary to some expectation, they clearly showed a scatter of large rocks, typical of a young mountainscape.

Venus 10 lander separated from its orbiter on 23 October and touched down on 25 October at 05·17 Moscow time after a descent time of 75 minutes. It landed 2,200 km (1,370 miles) from its twin. On the surface it operated for 65 minutes, during which time it measured a surface temperature of 465°C, a pressure of 92 Earth atmospheres and a wind speed of 3·5 m/sec (8 mph). Venus 10 lander also transmitted spectacular pictures showing a rocky landscape typical of old mountain formations.

To utilise fully the limited post-entry life of the landers, the flight plans were designed so that, after their separation, the orbiter and lander were arranged to approach one another from opposite sides of the planet. This technique put the orbiters in the ideal position to receive signals from the lander for up to two hours after entry of the latter.

By any standards the Venus 9 and 10 missions were completely successful.

VENUS LAUNCHER

Although basically similar in configuration to the launchers used for the Soviet manned spacecraft, the launcher for the Venus probes (see 1971-72 *Jane's*) appears to have elongated first and second stages, giving a considerable increase in propellant tankage.

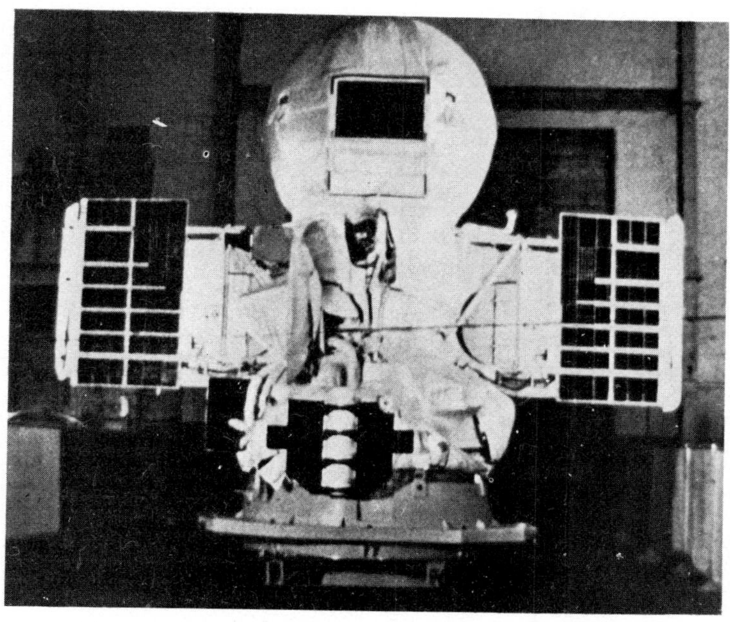

Venus spacecraft under test prior to launch in June 1975 *(Tass)*

SATELLITES AND SPACECRAFT LAUNCHED IN 1975

Note: Both the USA and the USSR have withheld information on some launchings and this list may be incomplete.
Data in italics are approximate or estimated

Date	Origin	Name	Total weight kg	lb	Launch vehicle	Apogee km	miles	Perigee km	miles	Inclin-ation °	Lifetime	Remarks
10 Jan	USSR	Soyuz 17	—	—	—	349	217	336	209	51·58	29½ days	Crew: Gubarov and Grechko. Docked with Salyut 4 space station.
17 Jan	USSR	Cosmos 702	*6,000*	13,000	—	313	195	204	127	71·33	12 days	*Reconnaissance satellite.* Recovered.
21 Jan	USSR	Cosmos 703	408	900	—	1,518	943	198	123	81·96	*303 days*	Continued Cosmos programme.
22 Jan	USA	Landsat 2	953	2,100	Uprated Delta	919	571	906	563	99·09	*100 years*	Second Earth Resources Technology Satellite.
23 Jan	USSR	Cosmos 704	*6,000*	*13,000*	—	329	204	213	132	72·86	14 days	*Manoeuvrable reconnaissance satellite.* Recovered.
28 Jan	USSR	Cosmos 705	408	900	—	502	312	272	169	70·97	*294 days*	Continued Cosmos programme.
30 Jan	USSR	Cosmos 706	*1,270*	*2,800*	—	39,783	24,720	623	387	62·85	*10 years*	*Military communications satellite.*
5 Feb	USSR	Cosmos 707	—	—	—	550	342	505	313	74·03	*10 years*	*Military satellite.*
6 Feb	USSR	Molniya 2M	1,250	2,750	—	39,750	24,700	602	374	62·81	*11 years*	Communications satellite.
6 Feb	France	Starlette 1	47	104	Diamant B P-4	1,107	688	805	500	49·82	*2,000 years*	Geodetic satellite; with laser reflectors.
6 Feb	USA	SMS 2	627	1,379	Uprated Delta	36,693	22,800	35,680	22,170	1·10	Unlimited	Second Synchronous Meteorological Satellite.
12 Feb	USSR	Cosmos 708	—	—	—	1,423	884	1,387	862	69·23	*6,000 years*	*Military satellite.*
12 Feb	USSR	Cosmos 709	*6,000*	*13,000*	—	333	207	188	117	62·83	13 days	*Manoeuvrable reconnaissance satellite.* Recovered.
24 Feb	Japan	Taiyo (SRATS)	—	—	Mu-3C-2	3,132	1,946	249	155	31·54	*5 years*	Solar Radiation And Thermospheric Satellite.
26 Feb	USSR	Cosmos 710	*6,000*	*13,000*	—	355	221	180	112	64·99	14 days	*Manoeuvrable reconnaissance satellite.* Recovered.
28 Feb	USSR	Cosmos 711 to 718	—	—	—	1,530	951	1,449	900	74·00	*10,000 years*	*Military communications satellites.*
10 Mar	USA	SDS 1	—	—	Titan IIIB-Agena D	39,300	24,420	295	183	63·5	*10 years*	Military data relay satellite.
12 Mar	USSR	Cosmos 719	*6,000*	*13,000*	—	329	204	182	113	64·98	13 days	*Manoeuvrable reconnaissance satellite.* Recovered.
21 Mar	USSR	Cosmos 720	*6,000*	*13,000*	—	1,280	795	223	139	62·81	12 days	*Reconnaissance satellite.* Recovered.
26 Mar	USSR	Cosmos 721	*6,000*	*13,000*	—	241	150	210	130	81·33	12 days	*Reconnaissance satellite.* Recovered.
27 Mar	USSR	Cosmos 722	*6,000*	*13,000*	—	359	223	210	130	71·35	13 days	*Manoeuvrable reconnaissance satellite.* Recovered.
27 Mar	USSR	Intercosmos 13	408	900	—	1,690	1,050	285	177	82·95	*4 years*	Scientific satellite.
1 April	USSR	Meteor 21	—	—	—	906	563	877	545	81·21	*500 years*	Weather satellite.
2 April	USSR	Cosmos 723	—	—	—	277	172	256	159	65·02	*600 years*	*Ocean Survey Satellite.*
7 April	USSR	Cosmos 724	—	—	—	276	171	258	160	64·97	*600 years*	*Ocean Survey Satellite.*
8 April	USSR	Cosmos 725	408	900	—	508	316	283	176	70·99	273 days	Continued Cosmos programme.
9 April	USA	GEOS 3	340	750	Uprated Delta	853	530	838	521	114·96	*200 years*	Geodynamic Experimental Ocean Satellite.
11 April	USA	Cosmos 726	—	—	—	1,008	626	972	604	82·99	*1,200 years*	Purpose not known.
14 April	USSR	Molniya 3B	*3,300*	*1,500*	—	40,660	25,265	636	395	62·86	*12 years*	Communications satellite.
16 April	USSR	Cosmos 727	*6,000*	*13,000*	—	358	222	180	112	64·98	12 days	*Reconnaissance satellite.* Recovered.
18 April	USSR	Cosmos 728	*6,000*	*13,000*	—	350	217	211	131	72·83	12 days	*Reconnaissance satellite.* Recovered.
18 April	USA	USAF	*3,000*	*6,600*	Titan IIIB-Agena D	401	249	134	83	110·54	48 days	Reconnaissance satellite.
19 April	India	Aryabhata	360	194	—	610	379	570	354	50·68	*10 years*	First Indian research satellite.
22 April	USSR	Cosmos 729	—	—	—	1,011	628	980	609	82·97	*1,200 years*	Purpose not known.
24 April	USSR	Cosmos 730	*6,000*	*13,000*	—	269	167	169	105	81·33	12 days	*Reconnaissance satellite.* Recovered.
29 April	USSR	Molniya 1AE	—	—	—	40,830	25,370	430	267	62·83	*100 years*	Communications satellite.
7 May	USA	Explorer 53	195	430	Scout	508	316	499	310	2·99	*15 years*	Third Small Astronomy Satellite.
7 May	Canada	Anik-3 (Telesat-3)	600	270	Uprated Delta	35,920	22,320	35,212	21,880	0·05	Unlimited	Third Canadian domestic communications satellite.
17 May	France	Pollux (D-5B)	—	—	Diamant B P-4	1,284	798	269	167	29·96	*Two years*	Carries microrocket experiment.
		Castor (D-5A)	—	—		1,268	788	272	169	29·95	*Two years*	Carries accelerometer experiment.
20 May	USA	DSCS 5 & 6	—	*608*	Titan IIIC	249	155	150	93	28·58	6 days	Defence communications twin satellites, failed to achieve synchronous orbit.
21 May	USSR	Cosmos 731	*6,000*	*13,000*	—	296	184	203	126	64·97	12 days	*Reconnaissance satellite.* Recovered.
22 May	USA	Intelsat IV-G (F-1)	1,406	3,100	Atlas-Centaur	35,790	22,240	35,775	22,230	0·4	Unlimited	International communications satellite.
24 May	USA	DMSP	—	—	Thor-Burner II	892	554	814	506	98·93	*80 years*	Data Meteorological Satellite.
24 May	USSR	Soyuz 18	—	—	—	349	217	338	210	51·59	63 days	Ferried Cosmonauts Klimuk and Sevastianov to Salyut 4.
28 May	USSR	Cosmos 732-739	—	—	—	1,529	950	1,448	900	74·00	10,000 years	Probably navigation or communications satellites.
28 May	USSR	Cosmos 740	*6,000*	*13,000*	—	327	203	172	107	64·97	13 days	*Reconnaissance satellite.* Recovered.
30 May	USSR	Cosmos 741	*6,000*	*13,000*	—	231	144	211	131	81·34	12 days	*Reconnaissance satellite.* Recovered.
3 June	USSR	Cosmos 742	—	—	—	375	233	189	117	62·82	12 days	*Reconnaissance satellite.* Recovered.
5 June	USSR	Molniya 1AF	—	—	—	40,845	25,380	434	270	62·82	*100 years*	Communications satellite.
	France	SRET 2 (MAS 2)	29·6	62	—	40,813	25,360	513	319	62·83	*12 years*	Research satellite.
8 June	USSR	Venus 9	—	—	—	—	—	—	—	—	—	Venus probe.
8 June	USA	USAF	—	25,000	Titan IIID	269	167	154	96	96·38	*5 months*	*Big Bird* advanced reconnaissance satellite.
		SSU 1	—	—	—	1,402	871	1,389	863	95·09	*10,000 years*	Pickaback satellite with own motor.
12 June	USA	Nimbus 6	827	1,823	Uprated Delta	1,104	686	1,091	678	99·96	1,600 years	Improved weather satellite.
12 June	USSR	Cosmos 743	*6,000*	*13,000*	—	355	220	190	118	62·81	13 days	*Reconnaissance satellite.* Recovered.
14 June	USSR	Venus 10	—	—	—	—	—	—	—	—	—	Venus probe.
18 June	USA	USAF	—	—	—	*35,820*	*22,260*	*35,650*	*22,150*	0·2	Unlimited	Purpose not known.
20 June	USSR	Cosmos 744	—	—	—	650	404	612	380	81·25	*60 years*	Purpose not known.
21 June	USA	OSO I (Eye)	1,064	2,346	Uprated Delta	560	348	544	338	32·94	*20 years*	Orbiting Solar Observatory.
24 June	USSR	Cosmos 745	408	900	—	540	335	274	170	71·00	*6 months*	Continued Cosmos programme.
23 June	USSR	Cosmos 746	*6,000*	*13,000*	—	346	215	188	117	62·80	13 days	*Reconnaissance satellite.* Recovered.
27 June	USSR	Cosmos 747	*6,000*	*13,000*	—	309	192	197	122	62·83	12 days	*Reconnaissance satellite.* Recovered.

Date	Origin	Name	Total weight kg	lb	Launch vehicle	Apogee km	miles	Perigee km	miles	Inclination °	Lifetime	Remarks
3 July	USSR	Cosmos 748	6,000	13,000	—	356	221	179	111	62·81	13 days	Manoeuvrable reconnaissance satellite. Recovered.
4 July	USSR	Cosmos 749	—	—	—	550	342	508	316	74·04	10 years	Military satellite.
8 July	USSR	Molniya 2N	1,250	2,750	—	39,976	24,840	459	285	62·89	100 years	Communications satellite.
11 July	USSR	Meteor 201	—	—	—	890	553	858	533	81·29	500 years	Weather satellite.
15 July	USSR	Soyuz 19 (ASTP)	6,690	14,750	—	232	144	217	135	51·76	6 days	Crew: Leonov and Kubasov. Docked with Apollo 18 in Apollo-Soyuz Test Project.
15 July	USA	Apollo 18 (ASTP)	16,830 (Incl Docking Module)	37,100	Saturn IB (SA 210)	232	144	217	135	51·75	7 days	Crew: Stafford, Slayton and Brand. Docked with Soyuz 19 in Apollo-Soyuz Test Project.
17 July	USSR	Cosmos 750	408	900	—	803	499	272	169	71·04	Two years	Scientific satellite.
23 July	USSR	Cosmos 751	6,000	13,000	—	314	195	198	123	62·82	12 days	Reconnaissance satellite.
24 July	USSR	Cosmos 752	—	—	—	515	320	481	299	65·85	10 years	Military satellite.
26 July	China	China 3	3,600	8,000	—	460	286	183	114	69·02	50 days	Scientific satellite.
31 July	USSR	Cosmos 753	6,000	13,000	—	303	188	171	106	62·83	16 days	Manoeuvrable reconnaissance satellite. Recovered.
9 Aug	USA	Cos. B	275	606	Delta	99,800	62,000	340	212	90·13	5 years	European Celestial Observation Satellite.
13 Aug	USSR	Cosmos 754	6,000	13,000	—	345	214	210	130	71·37	13 days	Manoeuvrable reconnaissance satellite. Recovered.
14 Aug	USSR	Cosmos 755	—	—	—	1,025	637	991	616	82·90	1,200 years	Navigation satellite.
20 Aug	USA	Viking 1	3,625	8,000	Titan IIIE-Centaur (TC-3)	—	—	—	—	51·0	Unlimited	Mars probe. Landed 20 July 1976 and photographed planet's suface.
22 Aug	USSR	Cosmos 756	—	—	—	649	403	627	390	81·24	60 years	Military satellite.
27 Aug	USA	Symphonie 2	402	886	Delta	35,856	22,280	35,357	21,970	0·0	Unlimited	French-German communications satellite.
27 Aug	USSR	Cosmos 757	6,000	13,000	—	337	209	190	118	62·82	15 days	Manoeuvrable reconnaissance satellite. Recovered.
2 Sept	USSR	Molniya 1AG	1,000	2,204	—	40,681	25,278	639	397	62·87	10 years	Communications satellite.
5 Sept	USSR	Cosmos 758	6,000	13,000	—	351	218	181	112	67·14	1 day	Reconnaissance satellite. Exploded in orbit.
9 Sept	USSR	Molniya 2P	1,250	2,750	—	39,880	24,780	449	279	62·92	15 years	Communications satellite.
9 Sept	Japan	Kiku (ETS-1)	85	187	'N' rocket	1,102	685	975	606	46·99	800 years	Engineering Test Satellite.
9 Sept	USA	Viking 2	3,625	8,000	Titan IIIE-Centaur (TC-4)	—	—	—	—	51·0	Unlimited	Mars probe. Landed 4 Sept 1976
12 Sept	USSR	Cosmos 759	6,000	13,000	—	281	175	234	145	62·80	12 days	Reconnaissance satellite. Recovered.
16 Sept	USSR	Cosmos 760	6,000	13,000	—	355	220	181	112	64·96	14 days	Manoeuvrable reconnaissance satellite. Recovered.
17 Sept	USSR	Cosmos 761-768	40	88	—	1,553	965	1,480	920	74·00	10,000 years	Military communications satellite.
18 Sept	USSR	Meteor 22	—	—	—	901	560	837	520	81·26	500 years	Weather satellite.
23 Sept	USSR	Cosmos 769	6,000	13,000	—	331	206	211	131	72·83	12 days	Reconnaissance satellite.
24 Sept	USSR	Cosmos 770	—	—	—	1,222	759	1,188	738	82·94	3,000 years	Military satellite.
25 Sept	USSR	Cosmos 771	6,000	13,000	—	247	153	219	136	81·33	13 days	Manoeuvrable reconnaissance satellite. Recovered.
26 Sept	USA	Intelsat IV-A (F-1)	1,495	3,300	Atlas-Centaur	35,775	22,230	35,373	21,980	0·5	Unlimited	Advanced communications satellite with 6,000 telephone circuits or 20 TV channels.
27 Sept	France	Aura (D-2B)	106	234	Diamant B P-4	723	449	499	310	37·13	15 years	Scientific ultra-violet radiation satellite.
29 Sept	USSR	Cosmos 772	6,500	14,330	—	320	199	201	125	51·79	4 days	Soyuz-type ferry craft. Recovered.
30 Sept	USSR	Cosmos 773	—	—	—	828	514	791	492	74·06	120 years	Military satellite.
1 Oct	USSR	Cosmos 774	6,000	13,000	—	333	207	212	132	71·35	14 days	Manoeuvrable reconnaissance satellite. Recovered.
6 Oct	USA	Explorer 54	675	1,488	Delta	3,817	2,372	154	96	90·10	Three years	Second Atmospheric Explorer satellite.
8 Oct	USSR	Cosmos 775	1,250	2,756	—	35,900	22,307	35,900	22,307	0·03	Unlimited	Purpose undisclosed.
9 Oct	USA	USAF	3,000	6,613	Titan IIIB-Agena D	356	221	125	78	96·41	52 days	Manoeuvrable reconnaissance satellite.
12 Oct	USA	Triad	94	207	Scout	704	438	362	225	90·74	Four years	Transit Improvement Programme satellite.
16 Oct	USA	GOES 1	294	647	Uprated Delta	35,800	22,245	35,775	22,230	1·00	Unlimited	Synchronous meteorological satellite.
17 Oct	USSR	Cosmos 776	6,000	13,000	—	310	192	203	126	62·82	12 days	Reconnaissance satelite. Recovered.
29 Oct	USSR	Cosmos 777	—	—	—	456	283	437	272	65·02	Five years	Purpose undisclosed.
4 Nov	USSR	Cosmos 778	—	—	—	1,018	632	989	615	82·96	1,200 years	Navigational satellite.
4 Nov	USSR	Cosmos 779	6,000	13,000	—	334	208	188	117	62·80	14 days	Manoeuvrable reconnaissance satellite. Recovered.
14 Nov	USSR	Molniya 3C	1,500	3,300	—	39,847	24,760	483	300	62·80	12 years	Communications satellite.
17 Nov	USSR	Soyuz 20	—	—	—	367	228	343	213	51·59	—	Unmanned spacecraft. Docked with Salyut 4 in space station supply test.
20 Nov	USA	Explorer 55	720	1,587	Delta	2,983	1,854	156	97	19·70	3 years	Third Atmospheric Explorer satellite.
21 Nov	USSR	Cosmos 780	6,000	13,000	—	298	185	206	128	65·01	12 days	Reconnaissance satellite. Recovered.
21 Nov	USSR	Cosmos 781	—	—	—	557	346	508	316	74·03	10 years	Purpose not known.
25 Nov	USSR	Cosmos 782	—	—	—	405	251	227	141	62·81	11 weeks	Co-operative biological satellite, with Soviet and US experiments
26 Nov	China	China 4	—	—	—	478	297	179	111	62·95	6 days	Reconnaissance satellite. Recovered.
28 Nov	USSR	Cosmos 783	—	—	—	838	520	797	495	74·06	120 years	Purpose not known.
3 dec	USSR	Cosmos 784	6,000	13,000	—	252	156	216	134	81·33	12 days	Reconnaissance satellite. Recovered.
4 Dec	USA	USAF	—	—	Titan IIID	233	145	156	97	92·27	5 months	
		USAF	—	—	—	1,558	968	236	147	96·28	Two years	Pickaback satellite, purpose not known.
11 Dec	USSR	Intercosmos 14	544	1,200	—	1,707	1,060	345	214	73·99	5 years	Scientific satellite.
12 Dec	USSR	Cosmos 785	—	—	—	1,020	634	898	558	65·07	600 years	Purpose not known. Moved into high orbit.
13 Dec	USA	Satcom 1	—	—	Delta	35,775	22,230	35,743	22,210	1·0	Unlimited	Radio Corporation of America commercial communications satellite.
14 Dec	USA	IMEWS 5	—	—	Titan IIIC	35,856	22,280	35,600	22,120	0·5	Unlimited	Integrated Missile Early Warning Satellite.

Date	Origin	Name	Total weight kg	lb	Launch vehicle	Apogee km	miles	Perigee km	miles	Inclin-ation °	Lifetime	Remarks
16 Dec	China	China 5	—	—	—	388	241	185	115	69·00	*Six weeks*	Purpose not known.
16 Dec	USSR	Cosmos 786	*6,000*	*13,000*	—	347	215	180	112	65·00	13 days	*Reconnaissance satellite.* Recovered.
17 Dec	USSR	Moliniya 2Q	1,250	2,750	—	39,928	24,810	436	271	62·86	*10 years*	Communications satellite.
22 Dec	USSR	Prognoz 4	857	1,890	—	199,237	123,800	634	394	65·00	10 years	Particles and radiation belt studies satellite.
22 Dec	USSR	Statsionar 1 *(Rainbow)*	—	—	—	35,775	22,230	35,775	22,230	0·1	Unlimited	Synchronous communications satellite.
25 Dec	USSR	Meteor 23	—	—	—	901	560	842	523	81·26	*500 years*	Weather satellite.
27 Dec	USSR	Molniya 3D	*1,500*	*3,300*	—	40,800	25,352	470	292	62·81	*10 years*	Communications satellite.

AERO-ENGINES

THE ARGENTINE REPUBLIC

CICARÉ
CICARÉ AERONAUTICA SC

ADDRESS:
Ave Ibañez Frocham s/n, CC24, Saladillo, Provincia de Buenos Aires

Telephone: (Comodoro Mantel, in Buenos Aires) 41-5260

OFFICERS: See "Aircraft" section.

Several types of Cicaré automotive engine have been produced in quantity for surface applications. Cicaré Aeronautica, formed in 1972, is developing light helicopters (see "Aircraft" section) and light-aircraft engines. The first engine is the 4C2T (four-cylinder two-stroke).

CICARÉ 4C2T

This is a light, low-cost engine intended to be used in a wide range of aeroplanes, motor-gliders and helicopters. It is intended in particular to replace the many Continental A65 and similar engines used in the past in Argentina and other countries as a standard power unit for light aviation. The engine first ran in October 1973.

In 1974 a power of 50 kW (67 hp) was reached at only 3,500 rpm using a fixed-pitch wooden propeller. New aerobatic carburettors are now being tested. The first series of production engines were being built in 1976.

TYPE: Four-cylinder horizontally-opposed air-cooled two-stroke piston engine.

CYLINDERS: Bore 74 mm (2·91 in). Stroke 76 mm (3·00 in). Swept volume 1,314 cc (80·18 cu in). Compression ratio 8·0 : 1. Cylinders on each side cast in light alloy as single unit complete with half crankcase. Steel liners.

PISTONS: Aluminium alloy castings, each with two compression rings and one scraper ring.

CONNECTING RODS: Forged steel, with needle-roller bearings in both big and small ends.

CRANKSHAFT: Four-throw steel forging carried in three ball bearings.

CRANKCASE: Divided on vertical centreline; each half cast complete with pair of cylinders.

INDUCTION: Twin carburettors, each equipped with hot-air anti-icing. Mixture passed through crankcase and thence through inlet and exhaust ports in cylinders.

FUEL: Mixture of 40 parts 80-octane gasoline to 1 part SAE.40 oil.

IGNITION: Dual magnetos, serving two Champion UK10, Autolite BT3 or PVI BT3 plugs per cylinder.

LUBRICATION: See under "Fuel".

ACCESSORIES: Rear pads for 12V alternator and starter.

DIMENSIONS:
Length 810 mm (31·89 in)
Width 660 mm (25·98 in)
Height 550 mm (21·65 in)

WEIGHT, DRY:
Estimated, with accessories 68·0 kg (150 lb)

Prototype Cicaré 4C2T showing propeller shaft driven by gearbox at rear. Rated at 52·2 kW (70 hp)

PERFORMANCE RATING:
Max T-O 52·2 kW (70 hp) at 3,500 rpm

SPECIFIC CONSUMPTION:
Fuel/oil mix at optimum rating
111 μg/J (0·66 lb/hr/hp)

AUSTRALIA

CAC
COMMONWEALTH AIRCRAFT CORPORATION LTD

HEAD OFFICE AND WORKS:
304 Lorimer Street, Port Melbourne, Victoria 3207

Telephone: 64-0771

OFFICERS: See "Aircraft" section

Recent engine activities of the Commonwealth Aircraft Corporation have been centred on licensed manufacture of the SNECMA Atar 9C and Rolls-Royce Viper 11 turbojets. These engines power Australian-built Mirage III-O ground attack aircraft and Aermacchi M.B. 326H training aircraft, respectively.

Their production, amounting to 140 Atars and 112 Vipers, is now complete; but overhaul of the Atar, Viper and Rolls-Royce Avon Mks 1, 26 and 109 turbojets continues to be undertaken for the RAAF, together with the manufacture of spares.

Commonwealth Aircraft-built SNECMA Atar 9C turbojet, rated at 62·8 kN (14,110 lb st)

BELGIUM

FN
FABRIQUE NATIONALE HERSTAL SA

HEAD OFFICE AND WORKS:
B-4400 Herstal

Telephone: (04) 64 08 00

FN has completed licensed production of the SNECMA Atar 9C turbojet for Belgian-built Mirage 5-B aircraft in collaboration with SNECMA and Fairey SA. FN has gained sub-contracts on the GRTS Larzac 04 and the CFM56.

Participation in Larzac production was one of the offsets included in the adoption of the Alpha Jet trainer by the Belgian Air Force. FN's share in the CFM56 project will be a part of the 50 per cent work-split due to SNECMA of France. FN also secured participation in the SNECMA M53, which would take effect if this engine found a customer.

FN is also a member of the consortium which made Rolls-Royce Tyne 21 and 22 turboprops in co-operation with Rolls-Royce, MTU and SNECMA (Hispano-Suiza) to power Breguet Atlantic maritime reconnaissance aircraft and Transall C-160 transports. Additional engines probably will be needed to supply the re-opened Transall line.

FN maintains and repairs General Electric J79-GE-11A turbojets which the company produced in association with MAN Turbo (now MTU) in Germany and Fiat in Italy to power F-104G fighters. Other repair and overhaul work concerns Turboméca Marboré IIF turbojets and engine controls and accessories.

CANADA

ORENDA
HAWKER SIDDELEY CANADA LTD, Orenda Division

HEAD OFFICE AND WORKS:
Mississauga, Ontario

Telephone: (416) 677-3250

POSTAL ADDRESS:
Box 6001, Toronto, AMF, Ontario L5P 1B3

GENERAL MANAGER:
R. F. Tanner

DIRECTOR OF ENGINEERING: B. A. Avery

DIRECTOR OF OPERATIONS: P. K. Peterson

DIRECTOR OF MARKETING: D. J. Caple

DIRECTOR OF FINANCE: K. R. Church

Since late August 1973, complete ownership of Orenda has been restored to Hawker Siddeley Canada Ltd. Main activities of the division are the manufacture of aircraft turbine engines and components, and the design and manufacture of industrial gas turbines.

Orenda's main activity is repair, overhaul and spares manufacture of J85-CAN-15 turbojet engines for Canadian-built Northrop CF-5s for the Canadian Armed Forces and NF-5s for the Royal Netherlands Air Force, under licence from General Electric. The engine, similar to the J85-GE-15 described under General Electric on a later page, was in production at Orenda from 1967 until 1974.

Orenda is also overhauling the J85-CAN-40, used in the Canadair CL-41 Tutor, and the J79-OEL-7 and Orenda turbojet engines. It supplies parts to Canada, West Germany, the Netherlands, Pakistan, Belgium, Norway, Italy, Venezuela and the United States.

The J79-OEL-7, used in the Canadair CF-104, most nearly resembles the J79-GE-11A. It has a four-strut front frame, cartridge-operated emergency nozzle closure system and other, minor differences.

P&WC
PRATT & WHITNEY AIRCRAFT OF CANADA LTD

HEAD OFFICE AND WORKS:
PO Box 10, Longueuil, Quebec J4K 4X9

Telephone: (514) 677-9411

CHAIRMAN, BOARD OF DIRECTORS:
T. E. Stephenson

PRESIDENT AND CHIEF EXECUTIVE OFFICER:
D. C. Lowe

VICE-PRESIDENTS:
R. H. Guthrie (Industrial & Marine Division)
A. L. Tontini (Personnel)
R. G. Raven (Helicopter & Systems Division)
E. L. Smith (Operations)
E. H. Schweitzer (Product Support)
K. H. Sullivan (Marketing)
V. W. Tryon (Finance)

Pratt & Whitney Aircraft of Canada is a major subsidiary of the United Technologies Corporation, Connecticut, USA, and was formed originally to manufacture and overhaul reciprocating engines and spare parts designed by UTC's Pratt & Whitney Aircraft Division. In 1957 its activities were enlarged to include design and development of turbine engines. Today P&WC is the UTC company responsible for engines for general aviation. It is also the prime source of spare parts production for all Pratt & Whitney reciprocating engines.

Original turbine work by the company was initiated by the concept and preliminary design of the JT12 (J60) turbojet, development and manufacture of which were taken over subsequently by Pratt & Whitney Aircraft. Design, development and manufacture of the PT6, ST6, PT6T and JT15D series of small turbine aero-engines represents over 70 per cent of the company's activities.

Since 1974 P&WC has been eager to launch the larger JT25D turbofan. Funds have not yet been made available, and United Technologies has discussed instead the possibility of a joint programme with Rolls-Royce (1971) Ltd of the UK on the latter's RB.401 in the same category.

P&WC is owned 90 per cent by the United Technologies Corporation. Approximately 18 per cent of its sales are linked to defence requirements. It occupies more than 139,400 m² (1·5 million sq ft) of space in four plants and employs more than 5,350 persons.

In 1975 a total of 2,397 new engines were delivered. The peak rate was reached in September 1975 when shipments reached 244.

Exterior *(right)* **and cutaway drawing** *(below)* **of the Pratt & Whitney (Canada) JT15D-1 twin-spool turbofan (9·8 kN; 2,200 lb st)**

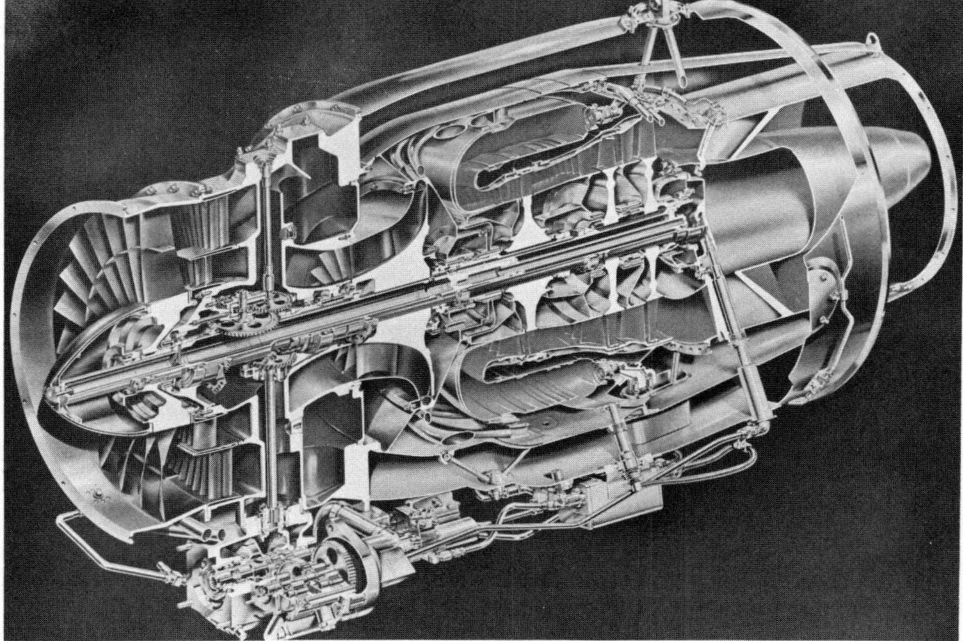

P&WC JT15D

Following a comprehensive performance study of small turbofan engines carried out by P&WC during 1965, detail design of a definitive engine, the 8·9-11·12 kN (2,000 lb to 2,500 lb) JT15D was initiated in June 1966. First run of the new turbofan was on 23 September 1967, within eight days of the target set at the start of design. The engine exceeded its rated thrust on its second build in November 1967 and achieved its guaranteed sfc in May 1968. Flight testing of the JT15D in a nacelle under a modified Avro CF-100 testbed aircraft started on 22 August 1968.

In May 1971 the JT15D-1 received full FAA/MoT certification. The engine has flown to 14,360 m (48,000 ft) and Mach 0·8, and has undergone testing for bleed-air purity, anti-icing, noise certification, inlet distortion, altitude starting to 12,200 m (40,000 ft), intake, vortex, accessory gearbox 270 hours' full rating, cold-weather starting, foreign-object ingestion including birds, numerous 1,000-cycle tests at elevated turbine temperatures and thrust levels, and other testing.

In 1969 the T-O thrust was raised from an initial 8·9 to 9·8 kN (2,000 to 2,200 lb), with an sfc of 0·540. The first growth version, the JT15D-4, was run successfully ahead of schedule in January 1972, developing a rated thrust of 10·28 kN (2,310 lb) to 25°C ambient at the rated sfc of 0·554.

Intended to power business aircraft, small transports and counter-insurgency combat aircraft in the 8,000 lb to 12,500 lb AUW category, the JT15D is an advanced technology two-spool front-fan engine having a minimum number of aerodynamic components. Major design objectives were a significant improvement in sfc, and simplicity of construction to ensure low first cost and maintenance costs. Other objectives were low noise levels, ease of handling, and the attainment of airline standards of reliability.

Initial application for the JT15D was the twin-engined Cessna Citation 500. Flightworthy prototype engines were delivered in August 1969 for the Citation's first flight in mid-September.

By March 1976 more than 715 JT15D-1s had been delivered to Cessna for the Citation. Certificated to FAR Part 36 noise specifications, the JT15D-1-powered Citation is claimed to be the quietest executive jet flying. TBO of the D-1 engine is 2,400 hr.

A second application for the engine is the Aérospatiale SN 601 Corvette twin-engined business jet. The first JT15D-4 prototype engines were delivered to Aérospatiale in August 1972 for the Corvette programme. Certification of the JT15D-4 and production deliveries occurred three months ahead of schedule in September 1973. The TBO of the JT15D-4 is 1,500 hours; deliveries totalled 89 by March 1976.

In March 1976 total operating hours of all JT15D engines amounted to 606,600. Engines were installed in 318 aircraft of 210 operators in 31 countries.

The following description relates to the JT15D-1 (JT15D-4 features in brackets):

TYPE: Two-shaft turbofan.

AIR INTAKE: Direct pitot intake without inlet guide vanes. Hot-air anti-icing for nose bullet.

FAN: Single-stage axial fan, aerodynamically related to that of the JT9D but on a much smaller scale. Forged disc fitted with 28 solid titanium blades secured by dovetail fixings riveted to disc. Blades have part-span shrouds. Casing, which forms the engine air intake, of forged stainless steel. Circular splitter ring behind fan, held between two rows of 33 inner wrapped-sheet stators and single row of 66 outer stator blades. Total air mass flow, 34 kg (75 lb)/sec; by-pass ratio about 3·3:1; by-pass flow typically 26 kg (57·5 lb)/sec; primary core air flow 8 kg (17·5 lb)/sec; fan pressure ratio 1·5:1.

COMPRESSOR: Primary airflow enters eye of single-stage titanium centrifugal compressor. Single-sided impeller, with 16 full vanes and 16 splitter vanes, secured to shaft by special bolt and key-washer. Two-piece casing with diffuser in form of pipes containing straightening vanes. Overall pressure ratio almost 10:1. (JT15D-4 compressor airflow augmented by axial boost stage between fan and compressor.)

COMBUSTION CHAMBER: Annular reverse-flow type. Outer casing of heat-resistant steel; flame tube of nickel alloy, supported on low-pressure turbine stator assembly. Spark igniters at 5 and 7 o'clock positions (viewed from rear).

FUEL SYSTEM: Engine-driven sandwich-mounted pump delivering through FCU, flow divider and dual manifolds at 44·8 bars (650 lb/sq in); DP-LI pneumatic control unit mounted on pump.

FUEL GRADES: JP-1, JP-4, JP-5 conforming to PWA Spec. 522.

NOZZLE GUIDE VANES: High-pressure ring of 15, air-cooled, integrally cast in cobalt alloy.

TURBINE: Single-stage HP turbine with 71 solid blades held in fir-tree roots in thick-hub disc of refractory alloy; two-stage LP turbine with nickel alloy discs, first stage being cast integrally with 61 blades and second stage carrying 55 blades in fir-tree roots. LP fan shaft drives fan, with ball thrust bearing behind fan and roller gear and intershaft bearings; HP shaft drives centrifugal compressor, with front ball thrust bearing and rear roller bearing. Gas temperature 960°C before turbine, 562°C after turbine.

JET PIPE: Nickel alloy cone and sheet-metal pipe. Provision made for adjusting the area to match engines and to trim performance.

ACCESSORY DRIVES: Package under front of engine driven by power offtake from front of HP shaft.

LUBRICATION SYSTEM: Integral oil system, with gear-type pump delivering at up to 5·52 bars (80 lb/sq in). Capacity, 9·0 litres (2·4 US gallons; 2·0 Imp gallons).

OIL SPECIFICATION: PWA521 Type II.

MOUNTING: Hard or soft, according to customers' choice. Four main pads on front casing, arranged two on each side at 30° above and below horizontal. One rear mount at top or on either side of centreline.

STARTING: Air-turbine starter or electric starter/generator.

DIMENSIONS:

Diameter:	
JT15D-1	691 mm (27·2 in)
JT15D-4	686 mm (27·0 in)
Length overall:	
JT15D-1	1,506 mm (59·3 in)
JT15D-4	1,600 mm (63·0 in)
Frontal area	0·37 m² (4 sq ft)

WEIGHT, EQUIPPED:

JT15D-1	231 kg (509 lb)
JT15D-4	253 kg (557 lb)

PERFORMANCE RATINGS:

T-O:	
JT15D-1	9·8 kN (2,200 lb st)
JT15D-4	11·12 kN (2,500 lb st)
Max continuous:	
JT15D-1	9·3 kN (2,090 lb st)
JT15D-4	10·56 kN (2,375 lb st)

SPECIFIC FUEL CONSUMPTION (T-O):

JT15D-1	15·30 mg/Ns (0·540 lb/hr/lb st)
JT15D-4	15·92 mg/Ns (0·562 lb/hr/lb st)

P&WC PT6A

The PT6A is a free-turbine turboprop, built in many versions.

An experimental PT6 ran for the first time in November 1959 and flight trials in the nose of a Beech 18 began in May 1961. Civil certification of the first production model, the 578 ehp PT6A-6, was granted in late 1963. Progressively higher rated versions have followed to power a wide variety of aircraft, and deliveries have for many years been made at rates in excess of 100 engines a month.

Related series of engines include the T74 military turboprop, the PT6B commercial turboshaft, PT6T coupled turboshaft, T400 military coupled engine, and the ST6 series of APU, industrial and marine engines. Technology from the PT6/ST6 family has also been embodied in P&WC's JT15D turbofan.

Principal versions of the PT6A are as follow:

PT6A-6. Flat rated at 431 ekW; 410 kW (578 ehp; 550 shp) at 2,200 propeller rpm to 21°C, this version received civil certification in December 1963. A total of 350 PT6A-6s were built between then and November 1965. Among aircraft powered by the PT6A-6 are the de Havilland Canada Turbo-Beaver and early DHC-6 Twin Otter Series 100.

PT6A-20. Flat rated at 432 ekW; 410 kW (579 ehp; 550 shp) at 2,200 propeller rpm to 21°C, the -20 offered improved reliability and increases in max continuous, max climb and max cruise power ratings over the PT6A-6. The PT6A-20 was certificated in October 1965. Between then and 1974 approximately 2,400 were built to power the Beech King Air B90, Beech Model 99 Commuter Liner, prototypes of the EMBRAER EMB-110 Bandeirante, de Havilland Canada DHC-6 Twin Otter Series 100 and 200, James Aviation (Fletcher FU-24) conversion, Marshall of Cambridge (Grumman) Goose conversion, McKinnon G-21C and G-21D Turbo-Goose (Grumman Goose) conversions, Pilatus PC-6/B1-H2 Turbo-Porter, Pilatus PC-7 Turbo-Trainer and the Swearingen Merlin IIA (which can be re-engined with the PT6A-27).

PT6A-20A. Similar to A-20; fitted to early Beech King Air C90.

PT6A-21. Flat rated at 432·5 ekW; 410 kW (580 ehp; 550 shp) at 2,200 propeller rpm to 21°C the A-21 offers improved fuel consumption and reliability, mainly by mating the A-27 power unit with the A-20A gearbox. Certificated on 10 December 1974. Fitted to current Beech King Air C90.

PT6A-25. Flat rated at 432·5 ekW; 410 kW (580 ehp;550 shp) at 2,200 propeller rpm to 33°C. Uprated A-21 with inverted-flight capability. Major castings in magnesium offered as option. Certificated May 1976. Fitted to Beech T-34C.

PT6A-27. Flat rated at 553 ekW; 507 kW (715 ehp; 680 shp) at 2,200 propeller rpm to 22°C, attained by 12½ per cent increase in mass flow provided by larger-diameter compressor, at lower turbine temperatures than in PT6A-20. Production began in November 1967 and 1,000 engines had been delivered by March 1975. Production continues. Applications include the Hamilton Westwind II/III (Beech 18) conversions, Beech Model 99, Beech Model 99A Commuter Liner, Beech U-21A and U-21D, de Havilland Canada DHC-6 Twin Otter Series 300, Pilatus/Fairchild Industries PC-6/B2-H2 Porter, Frakes Aviation (Grumman) Mallard conversion, IAI Arava, Let L-410A Turbolet, Saunders Aircraft ST-27A (de Havilland Heron) conversion and EMBRAER EMB-110 Bandeirante.

PT6A-28. Similar to the PT6A-27 and with the same T-O and max continuous ratings, this version has an additional normal cruise rating of 562 ehp available up to 21°C corresponding to the max cruise rating conditions of the 27. In addition the max cruise rating of the -28 gives 652 ehp up to the higher ambient of 33°C. This model continues in production, with more than 800 delivered by February 1975. Applications are Beech King Air E90 and A100 and Piper Cheyenne.

PT6A-34. Flat rated at 584 ekW; 559 kW (783 ehp; 750 shp) at 2,200 propeller rpm to 31°C, this version has aircooled nozzle guide vanes to allow operation at higher turbine entry temperatures. More than 80 had been delivered by February 1975, for the IAI Arava, Saunders ST-28, Frakes Aviation (Grumman) Mallard conversion, Jetstream conversion and EMBRAER EMB-111.

PT6A-38. Derated A-41, flat rated at 597 ekW; 559 kW (801 ehp; 750 shp) to 39°C. Certificated May 1975. Installed in Beech C-12A.

PT6A-41. This higher mass flow version embodies air-cooled stage-one turbine nozzle guide vanes and a two-stage free turbine (in place of the previous single-stage unit) to give improved power absorption. Length is thus increased by 101 mm (4 in) to 1,676 mm (66 in). The -41 has a T-O rating of 673 ekW; 634 kW (903 ehp; 850 shp) at 2,000 propeller rpm, available up to 41°C. Thermodynamic power is 812 ekW (1,089 ehp). By March 1976 more than 340 engines had been delivered for Beech Super King Air 200.

PT6A-45A. Similar to PT6A-41 but with redesigned gearbox to transmit higher powers at reduced propeller speeds. Rated at 875·5 ekW; 835 kW (1,174 ehp; 1,120 shp) at 1,620-1,700 rpm to 15°C, or to 28°C with water injection. Certificated February 1976. Powers Shorts SD3-30 and Mohawk 298.

PT6A-50. Similar to PT6A-41 with a longer, higher-ratio reduction gear to give lower propeller tip speed for quieter operation at T-O. Length is consequently increased to 2,134 mm (84 in). Rating at T-O is 875·5 ekW; 835 kW (1,174 ehp; 1,120 shp) available with water injection up to 34°C at 1,210 propeller rpm. Certification was due in Summer 1976. Powers de Havilland Canada DHC-7.

By March 1976 the total flight time of PT6A engines amounted to 15,053,363 hr. Engines were installed in 2,582 aircraft, of 40 different types, used by 1,250 operators in 98 countries.

The following data apply generally to the PT6A series:

TYPE: Free-turbine axial-plus-centrifugal turboprop engine.

The 553 ekW (715 ehp) P&WC PT6A-27 free-turbine turboprop

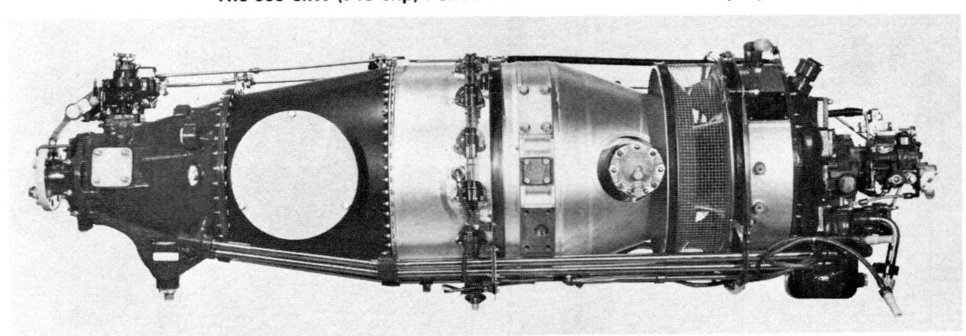

The 875·5 kW (1,174 ehp) P&WC PT6A-45A free-turbine turboprop

PROPELLER DRIVE (all models up to and including PT6A-41): Two-stage planetary gear train. Ratio 15 : 1. Rotation clockwise when viewed from rear. Drive from free turbine. Flanged propeller shaft. Plain bearings. Higher-ratio reduction gears developed for PT6A-45A and -50.

AIR INTAKE: Annular air intake at rear of engine, with intake screen. Aircraft-supplied alcohol anti-icing system or inertial separation anti-icing system.

COMPRESSOR: Three axial-flow stages, plus single centrifugal stage. Single-sided centrifugal compressor, with 26 vanes, made from titanium forging. Axial rotor of disc-drum type, with stainless steel stator and rotor blades. The stator vanes (44 first-stage, 44 second-stage, 40 third-stage) are brazed to casing. The rotor blades (16 first-stage, 32 second-stage and 32 third-stage) are dovetailed to discs. Discs through-bolted, with centrifugal compressor, to shaft. Fabricated one-piece stainless steel casing and radial diffuser. PT6A-27: compression ratio 6·7 : 1, air mass flow 3·1 kg/sec (6·8 lb/sec).

COMBUSTION CHAMBER: Annular reverse-flow type of stainless steel construction, with 14 simplex burners around periphery of chamber. All versions up to A-34 have two glow plug igniters with option of two spark igniters; A-38 onwards, two spark igniters. PT6A-27 has one plug at 64° on starboard side of vertical centreline and one at 90° on port side.

FUEL SYSTEM: Bendix DP-F2 pneumatic automatic fuel control system. Pneumatic computing section, with automatic inlet air temperature compensation, fuel metering and regulating section, gas generator governor and free turbine governor. Primary and secondary flow manifolds with seven nozzles per manifold.

FUEL GRADE: Commercial jet fuels JP-1, JP-4, JP-5, MIL-J-5624. Use of aviation gasolines (MIL-G-5572) grades 80/87, 91/98, 100/130 and 115/145 permitted for a period of up to 150 hours during any overhaul period.

NOZZLE GUIDE VANES: 29 nozzle guide vanes; A-34 onward, 14 aircooled HP vanes.

TURBINES: Models up to A-34 have two single-stage axial; HP turbine (with 58 blades) drives compressor, and LP turbine (with 41 shrouded blades) drives output shaft. PT6A-38 onward have two-stage LP turbine. All blades have fir-tree root fixings.

BEARINGS: Each main rotor (gas generator and free turbine) supported by one ball and one roller anti-friction bearing.

JET PIPE: Collector duct surrounding free-turbine shaft, exhaust through two ports on horizontal centreline.

ACCESSORIES: Mounting pads on accessory case (rear of engine) for starter/generator, hydraulic pump, aircraft accessory drive, vacuum pump and tachometer-generator. Mounting pad on the shaft-turbine reduction gear case for propeller overspeed governor, propeller constant-speed control unit and tachometer generator. All accessories mounted on the ends of the engine and do not protrude beyond the major diameter.

LUBRICATION SYSTEM: One pressure and four scavenge elements in the pump stacks. All are gear type and are driven by the gas generator rotor. Engine has an integral oil tank with a capacity of 8·75 litres (2·3 US gallons). Oil supply pressure is 4·48 bars (65 lb/sq in).

OIL SPECIFICATION: CPW202, PWA522 Type II (7·5 cs vis) (MIL-L-23699, MIL-L-7808 for military engines).

MOUNTING: Up to A-34, three-point ring suspension. A-38 onward, four-point mounting.

STARTING: Electric starter/generator on accessory case.

DIMENSIONS:
- Max diameter: 483 mm (19 in)
- Length, less accessories:
  - PT6A-6 to -34: 1,575 mm (62 in)
  - PT6A-38, -41: 1,701 mm (67 in)
  - PT6A-45A: 1,829 mm (72 in)
  - PT6A-50: 2,133 mm (84 in)
- Frontal area: 0·18 m² (1·95 sq ft)

WEIGHT DRY:
- PT6A-20: 130 kg (286 lb)
- PT6A-21, -27, -28: 136 kg (300 lb)
- PT6A-25: 144 kg (317 lb)
- PT6A-34: 141 kg (311 lb)
- PT6A-38, -41: 168 kg (371 lb)
- PT6A-45A: 192 kg (423 lb)
- PT6A-50: 263 kg (580 lb)

PERFORMANCE RATINGS:
- T-O rating:
  - See under model listings
- Max continuous rating:
  - PT6A-20: 432 ekW; 410 kW (579 ehp; 550 shp) at 2,200 rpm (to 22°C)
  - PT6A-21: 432·5 ekW; 410 kW (580 ehp; 550 shp) at 2,200 rpm (to 33°C)
  - PT6A-25: 432·5 ekW; 410 kW (580 ehp; 550 shp) at 2,200 rpm (to 33°C)
  - PT6A-27, -28: 553 ekW; 507 kW (715 ehp; 680 shp) at 2,200 rpm (to 22°C)
  - PT6A-34: 584 ekW; 559 kW (783 ehp; 750 shp) at 2,200 rpm (to 30°C)
  - PT6A-38: 597 ekW; 559 kW (801 ehp; 750 shp) at 2,200 rpm (to 39°C)
  - PT6A-41: 673 ekW; 634 kW (903 ehp; 850 shp) at 2,000 rpm (to 41°C)
  - PT6A-45A: 798 ekW; 761 kW (1,070 ehp; 1,020 shp) at 1,620-1,700 rpm (to 27°C)
  - PT6A-50: 762 ekW; 725·5 kW (1,022 ehp; 973 shp) at 1,210 rpm (to 32°C)
- Max climb rating:
  - PT6A-20: 422 ekW; 401 kW (566 ehp; 538 shp) at 2,200 rpm
  - PT6A-21: 423·5 ekW; 401 kW (568 ehp; 538 shp) at 2,200 rpm (to 15°C)
  - PT6A-25: 432·5 ekW; 410 kW (580 ehp; 550 shp) at 2,200 rpm (to 33°C)
  - PT6A-27, -28: 486 ekW; 462 kW (652 ehp; 620 shp) at 2,200 rpm (to 21°C)
  - PT6A-34: 545 ekW; 522 kW (731 ehp; 700 shp) at 2,200 rpm (to 28°C)
  - PT6A-38: 597 ekW; 559 kW (801 ehp; 750 shp) at 2,000 rpm (to 27°C)
  - PT6A-41: 673 ekW; 634 kW (903 ehp; 850 shp) at 2,000 rpm (to 28°C)
  - PT6A-45A: 749 ekW; 713 kW (1,004 ehp; 956 shp) at 1,425 rpm (to 15°C)
  - PT6A-50: 706 ekW; 671 kW (947 ehp; 900 shp) at 1,020-1,160 rpm (to 23°C)

Max cruise rating:

PT6A-20	389 ekW; 369 kW (522 ehp; 495 shp) at 2,200 rpm
PT6A-21	390 ekW; 309 kW (523 ehp; 495 shp) at 2,200 rpm (to 15°C)
PT6A-25	432·5 ekW; 410 kW (580 ehp; 550 shp) at 2,200 rpm (to 33°C)
PT6A-27	486 ekW; 462 kW (652 ehp; 620 shp) at 2,200 rpm (to 21°C)
PT6A-28	486 ekW; 462 kW (652 ehp; 620 shp) at 2,200 rpm (to 33°C)
PT6A-34	545 ekW; 522 kW (731 ehp; 700 shp) at 2,200 rpm (to 19°C)
PT6A-38	597 ekW; 559 kW (801 ehp; 750 shp) at 2,000 rpm (to 27°C)
PT6A-41	673 ekW; 634 kW (903 ehp; 850 shp) at 2,000 rpm (to 28°C)
PT6A-45A	749 ekW; 713 kW (1,004 ehp; 956 shp) at 1,425 rpm (to 15°C)
PT6A-50	706 ekW; 671 kW (947 ehp; 900 shp) at 1,020-1,160 rpm (to 23°C)

SPECIFIC FUEL CONSUMPTION:

At T-O rating:

PT6A-20	109·7 μg/J (0·649 lb/hr/ehp)
PT6A-21, 25	106·5 μg/J (0·630 lb/hr/ehp)
PT6A-27, -28	101·7 μg/J (0·602 lb/hr/ehp)
PT6A-34	100·6 μg/J (0·595 lb/hr/ehp)
PT6A-38	106·3 μg/J (0·629 lb/hr/ehp)
PT6A-41	99·9 μg/J (0·591 lb/hr/ehp)
PT6A-45A, -50	94·6 μg/J (0·560 lb/hr/ehp)

At max continuous rating:

PT6A-20	109·7 μg/J (0·649 lb/hr/ehp)
PT6A-21, -25	106·5 μg/J (0·630 lb/hr/ehp)
PT6A-27, -28	101·7 μg/J (0·602 lb/hr/ehp)
PT6A-34	100·6 μg/J (0·595 lb/hr/ehp)
PT6A-38	106·3 μg/J (0·629 lb/hr/ehp)
PT6A-41	99·9 μg/J (0·591 lb/hr/ehp)
PT6A-45A	96·8 μg/J (0·573 lb/hr/ehp)
PT6A-50	97·7 μg/J (0·578 lb/hr/ehp)

At max cruise rating:

PT6A-20	113·2 μg/J (0·670 lb/hr/ehp)
PT6A-21	109·7 μg/J (0·649 lb/hr/ehp)
PT6A-25	106·5 μg/J (0·630 lb/hr/ehp)
PT6A-27, -28	103·4 μg/J (0·612 lb/hr/ehp)
PT6A-34	102·1 μg/J (0·604 lb/hr/ehp)
PT6A-38	106·3 μg/J (0·629 lb/hr/ehp)
PT6A-41	99·9 μg/J (0·591 lb/hr/ehp)
PT6A-45A	97·7 μg/J (0·578 lb/hr/ehp)
PT6A-50	98·5 μg/J (0·583 lb/hr/ehp)

OIL CONSUMPTION:

Max 0·091 kg (0·20 lb)/hr

P&WC T74

T74 is the US designation for military versions of the PT6A turboprop and PT6B turboshaft. The T74 turboprop is of the same configuration as the PT6A. Military versions are:

T74-CP-700. US Army counterpart of the PT6A-20. More than 300 T74-CP-700s have been delivered to Beech for 129 U-21A aircraft. Inertial separator system developed under Army contract to protect against sand and dust ingestion.

T74-CP-702. Rated at 778 ehp and retrofitted in Beech U-21 aircraft engaged in US Project Crazydog electronic countermeasures.

A further application of the military T74/PT6A is the Helio Stallion Model 550, turbine version of the single-engined Helio U-10, for which FAA certification has been completed. Deliveries to the USAF for the Credible Chase mission were made in 1971.

P&WC PT6B/PT6C

The PT6B is the commercial turboshaft version of the PT6A and has a lower-ratio reduction gear. Past applications include the Lockheed XH-51A and Model 286 rigid-rotor helicopters.

Principal versions of the PT6B are:

PT6B-9. Rated at 550 shp at 6,230 rpm available to 25°C. Civil certification received in May 1965. Production complete.

PT6B-34. Similar to B-16 except based on PT6A-34. T-O rating 750 shp at 6,188 rpm to 35°C (2½ min contingency 900 shp to 15°C). Certification due Summer 1976.

PT6C. This series of engines provides direct drive from the power turbine, with no reduction gearing.

DIMENSIONS:

Max diameter	483 mm (19 in)
Length, less accessories, PT6B-34	1,499 mm (59·0 in)
Frontal area	0·18 m² (1·95 sq ft)

WEIGHT, DRY:

PT6B-9	111 kg (245 lb)
PT6B-34	139 kg (306 lb)

PERFORMANCE RATINGS:

T-O:

See under model listings

Max continuous:

PT6B-9 373 kW (500 shp) at 6,230 rpm (to 22°C)
PT6B-34 559 kW (750 shp) at 6,188 rpm (to 35°C)

Max climb:

PT6B-9 373 kW (500 shp) at 6,230 rpm (to 22°C)

Max cruise:

PT6B-9 362 kW (485 shp) at 6,230 rpm
PT6B-34 466 kW (625 shp) at 6,188 rpm (to 15°C)

SPECIFIC FUEL CONSUMPTION:

At T-O rating:

PT6B-9	112·4 μg/J (0·665 lb/hr/shp)
PT6B-34	104·6 μg/J (0·619 lb/hr/shp)

At max continuous rating:

PT6B-9	115·8 μg/J (0·685 lb/hr/shp)
PT6B-34	104·6 μg/J (0·619 lb/hr/shp)

At max cruise rating:

PT6B-9	116·6 μg/J (0·69 lb/hr/shp)
PT6B-34	110·4 μg/J (0·653 lb/hr/shp)

OIL CONSUMPTION:

Max μ0·091 kg (0·20 lb)/hr

The 1,342 kW (1,800 shp) P&WC PT6T-3 Twin Pac coupled free-turbine turboshaft

P&WC PT6T TWIN PAC

First run in July 1968, the PT6T Twin Pac comprises two PT6 turboshaft engines mounted side by side and driving into a combining gearbox to provide a single output drive. The engine was launched as a coupled power unit for a family of twin-engined helicopters based on the Bell Helicopter UH-1 series. First of these, jointly financed by Bell, P&WC and the Canadian government, was the 15-seat Bell Model 212, an improved version of the 205A commercial helicopter which first flew with the PT6T-3 in April 1969.

Installation of the 1,342 kW (1,800 shp) PT6T-3 in the Model 212, in addition to offering true engine-out capability, provides an additional 300 shp over the single-engine 205A and gives enhanced hot-day and high altitude performance. Qualified PT6T-3s became available in the third quarter of 1970 coincident with certification of the Model 212. Deliveries of the helicopter started in early 1971. The Bell 212 is also produced under licence by Agusta in Italy.

Another application of the PT6T-3 engine is for conversion from piston engine to turbine power of the Sikorsky S-58. The prototype S-58T flew in August 1970 and certification was received in April 1971. It provides increased operating economy, twin-engine reliability and improved work capability over the S-58. The additional power of the PT6T-3 increases the altitude and hot-day capability due to flat rating, and improves the payload by reducing the empty weight.

In these two helicopter applications, total shaft-power output is limited by the helicopter transmission. In the Model 212 the 1,342 kW (1,800 shp) PT6T-3 is restricted to a T-O rating of 962 kW (1,290 shp) and 843 kW (1,130 shp) for continuous power. In the S-58T the limits are 1,122 kW (1,505 shp) at T-O and 935 kW (1,254 shp) for continuous operation. The PT6T-3 is easily adapted to such power requirements by a simple setting of its torque control. In the event of a power-section failure, torquemeters in the combining gearbox signal the other power section to maximum power. A single-engine 30-minute rating is included for use, at pilot discretion, in such contingencies.

By March 1976, a total of 612 T-3 engines had been delivered and had logged 857,666 hours in 373 helicopters.

An uprated Twin Pac, the PT6T-6, was certificated in December 1974. The higher power is achieved by material and aerodynamic improvements to the compressor-turbine nozzle guide vanes and rotor blades. By February 1974, more than 120 of these engines had been ordered for installation in Agusta-Bell 212 and Sikorsky S-58T helicopters. Deliveries began in early 1975; by March 1976 a total of 92 T-6 engines had logged 11,369 hr in 57 helicopters. A military variant is the T400-WV-402, described under the next entry.

The following details describe the main features differing from those of the standard PT6 single-engine configuration:

TYPE: Coupled free-turbine turboshaft.

SHAFT DRIVE: Combining gearbox comprises three separate gear trains, two input and one output, each contained within an individual sealed compartment and all interconnected by drive shafts. Overall reduction ratio 5 : 1. Input gear train comprising three spur gears provides speed reduction between power sections and output gearbox. The two drives into the output gearbox are via Formsprag fully-phased overrunning clutches with input third gear forming outer member of clutch, and interconnect shaft forming inner, overunning member. Output gear train comprises three helical spur gears, i.e. two input pinions meshing with single output gear. Output shaft drives forward between gas generators. Rotation clockwise viewed from front of engine. Hydro-mechanical torquemeter (of PT6 design concept) provided in each interconnected drive shaft, measuring power transmitted by each gas generator as a hydraulic pressure used to control torque balancing (between gas generators) and limiting.

AIR INTAKES: Additional inertial particle separator fitted upstream of engine to reduce sand and dust ingestion. High frequency compressor noise suppressed.

FUEL SYSTEM: As PT6 with manual backup system, and dual manifold for cool starts. Automatic power sharing and torque limiting. Torquemeters provide signals to Bendix fuel system metering valves to maintain power at level set by pilot's selective-collective control. Fuel heaters.

FUEL GRADES: JP-1, JP-4 and JP-5.

JET PIPE: Single upward-facing exhaust port on each gas generator.

ACCESSORIES: Starter/generator and tacho-generator mounted on accessory drive case at front of each power section. Other accessory drives on combining gearbox, including individual power turbine speed governors and tacho-generators, and provision for blowers and aircraft accessories.

LUBRICATION SYSTEM: Independent lubrication system on each power section for maximum safety during single-engine operation. Integral oil tanks. Separate oil system for output section of combining gearbox.

OIL SPECIFICATION: PWA Spec 521. For military engines, MIL-L-7808 and -23699.

STARTING: Electrical, with cold weather starting down to −54°C.

DIMENSIONS:

Height	838 mm (33·0 in)
Width	1,118 mm (44·0 in)
Length	1,702 mm (67·0 in)

WEIGHT, DRY (standard equipment):

PT6T-3	288 kg (635 lb)
PT6T-6	291 kg (642 lb)

PERFORMANCE RATINGS:

T-O (5 min):

Total output, at 6,600 rpm:

PT6T-3	1,342 kW (1,800 shp)
PT6T-6	1,398 kW (1,875 shp) (to 21°C)

Single power section only, at 6,600 rpm:

PT6T-3	671 kW (900 shp)
PT6T-6 (2½ min)	764 kW (1,025 shp)

30 minute power (single power section only), at 6,600 rpm 723 kW (970 shp)

Max continuous:

Total output, at 6,600 rpm:

PT6T-3	1,193 kW (1,600 shp)
PT6T-6	1,249 kW (1,675 shp) (to 19°C)

Single power section only, at 6,600 rpm:

PT6T-3	596·5 kW (800 shp)
PT6T-6	615 kW (825 shp) (to 19°C)

Cruise A:
  Total output, at 6,600 rpm:
    PT6T-3 932 kW (1,250 shp)
    PT6T-6 1,014 kW (1,360 shp)
  Single power section only, at 6,600 rpm:
    PT6T-3 466 kW (625 shp)
    PT6T-6 500 kW (670 shp)
Cruise B:
  Total output, at 6,600 rpm:
    PT6T-3 820 kW (1,100 shp)
    PT6T-6 891 kW (1,195 shp)
  Single power section only, at 6,600 rpm:
    PT6T-3 410 kW (550 shp)
    PT6T-6 440 kW (590 shp)
Ground idle, at 2,200 rpm 44·7 kW (60 shp) max

Specific Fuel Consumption:
  At T-O and 30 minute ratings (total output):
    PT6T-3 100·6 μg/J (0·595 lb/hr/shp)
    PT6T-6 100·0 μg/J (0·592 lb/hr/shp)
  At max continuous rating (total output):
    PT6T-3 101·2 μg/J (0·599 lb/hr/shp)
    PT6T-6 101·9 μg/J (0·603 lb/hr/shp)
  At Cruise A rating (total output):
    PT6T-3 106·1 μg/J (0·628 lb/hr/shp)
    PT6T-6 108·7 μg/J (0·643 lb/hr/shp)
  At Cruise B rating (total output):
    PT6T-3 110·4 μg/J (0·653 lb/hr/shp)
    PT6T-6 114·4 μg/J (0·677 lb/hr/shp)

Oil Consumption:
  Max (for both gas generators) 0·18 kg (0·4 lb)/hr

P&WC T400

Military version of the PT6T Twin Pac, the T400-CP-400 coupled turboshaft was the first US Navy turboshaft (or turboprop) to be designated under the new US military aircraft turbine engine designation system. The T400-CP-400 and PT6T-3 helicopter engines have the same performance and are similar externally, the major difference being that in the T400 aluminium castings replace magnesium. For military roles, P&WC describes the T400 as producing a minimum infra-red signature. Military Qualification Tests (MQT) were completed in March 1970, and production deliveries started in the same month.

The T400 is used in the US Air Force and Navy Bell UH-1N (military version of the Model 212), the US Marine Corps Bell AH-1J, and the Canadian Armed Forces Bell CH-135. T400 field operations started in the middle of 1970.

TBO on the T400-CP-400 is 2,000 hours on both the power section and reduction gearbox. By March 1976, 496 engines had been delivered.

The T400-WV-402 is the military counterpart of the PT6T-6 and has similar ratings (see PT6T entry). The WV-402 is used in the improved AH-1 helicopter for Iran. By March 1976, 247 engines had been delivered. Total flight time of T400 engines then stood at 904,861 hours, in 486 helicopters.

Dimensions (CP-400 and WV-402):
  Height 828 mm (32·6 in)
  Width 1,115 mm (43·5 in)
  Length 1,659 mm (65·3 in)

Weight, Dry:
  T400-CP-400 324 kg (714 lb)
  T400-WV-402 338 kg (745 lb)

Performance Ratings:
  Intermediate:
    T400-CP-400 1,342 kW (1,800 shp) at 6,600 rpm
    T400-WV-402 1,469 kW (1,970 shp) at 6,600 rpm
  Max continuous:
    T400-CP-400 1,141 kW (1,530 shp) at 6,600 rpm
    T400-WV-402 1,248 kW (1,673 shp) at 6,600 rpm

Specific Fuel Consumption (Intermediate rating):
    T400-WV-402 99·9 μg/J (0·591 lb/hr/shp)
    T400-CP-400 100·4 μg/J (0·594 lb/hr/shp)

CHINA
(PEOPLE'S REPUBLIC)

NATIONAL AIRCRAFT ENGINE FACTORY

Main Location: Shenyang

Although Chinese central and regional governments made attempts to build up an aircraft industry as early as 1913 (see "Aircraft" section), and several western engines were imported during the 1920s (the most important being the Bristol Jupiter), it was not until the Japanese invaded Manchuria in 1932 and set up the puppet state of Manchukuo that a self-sufficient aviation industry existed. The main Japanese-managed plant was located at Mukden, today the city of Shenyang. During the second World War there were at least three airframe or engine factories in Manchuria, producing nearly 2,200 fighter and trainer aircraft and about the same number of engines, the latter all being aircooled radials of basically Japanese design.

In 1945 the Soviet Union dismantled the industries in Manchuria; but following the Communist Chinese takeover in 1949 it provided assistance in re-establishing a Chinese aircraft industry, helping to build up fully-equipped airframe and engine factories, mainly on the original sites in Manchuria. The first product was the Yak-18 primary trainer, and it is believed that a licence to construct the aircraft and its M-11FR engine was signed in Moscow in November 1952. This 119 kW (160 hp) five-cylinder radial (see USSR entry in this section) was first produced at Shenyang in 1956, and is believed to have been built under licence in quantity. It is unlikely that new M-11 engines are still being made.

In 1958 licences were obtained by the 2nd Ministry of Machine Building for two additional Soviet aircooled radial engines, the 194 kW (260 hp) Ivchenko AI-14R and 746 kW (1,000 hp) Shvetsov ASh-62IR (both described under Poland), fitted respectively to the locally-built Chinko No. 1 (Yak-12) and Fong Shou No. 2 (An-2). Both of these aircraft and their engines are believed to have been built in large numbers. By 1959 the Manchurian plants were licence-building the Soviet Mi-4 helicopter and the Czech Super Aero 45 light twin. It is thought that in each case the engine (respectively the 1,268 kW (1,700 shp) Shvetsov ASh-82V 14-cylinder radial and the 104·4 kW (140 hp) M 332 four-in-line, the latter last described in the 1971-72 *Jane's*) was also produced either at Shenyang or at one of the other national factories. One possibility is that the Czech engine was made at the works at Harbin, because it was there that the Chinese version of the Super Aero 45 was produced. Harbin may also have taken over the M-11FR programme, because from 1959-60 that factory manufactured the M-11-powered Hai Lun-kiang No. 1, a locally designed liaison aircraft resembling the Yak-12.

GAS TURBINE ENGINES

During the Korean War (1950-53) large numbers of MiG-15 fighters were ferried through Manchuria. Chinese technicians became familiar with the aircraft and its Klimov RD-45 (Rolls-Royce Nene derivative) engine. In 1955 a licence for the manufacture of the MiG-15 fighter and MiG-15UTI trainer was signed in Moscow, and from 1958 several hundred of the latter were produced, powered by the RD-45 of 24·24 kN (5,450 lb st). The MiG-15 fighter was apparently not built in China, but in 1959 the first Chinese F-4, a licence-built MiG-17, began a production run of well over 1,000 aircraft, all probably powered by Chinese-built Klimov VK-1 turbojets rated at 26·47 kN (5,950 lb st).

In February 1959 the Chinese signed a licence agreement for the manufacture of the MiG-19 supersonic fighter, powered by RD-9 turbojets. Soon afterwards the relationship with the Soviet Union was severed; but the Chinese, working alone, managed to fly a locally-built F-6 (MiG-19) in 1961, and subsequently constructed a number estimated to reach 1,500. Thus, probably more than 4,000 RD-9 engines have been made at Shenyang. A subsequent production programme concerns the MiG-21. As described in the Aircraft section, this fighter and its 42·26 kN (9,500 lb st) R-11 axial turbojet were put into production in China without a licence or any Soviet help. Deliveries of the R-11 from Shenyang are thought to have begun in 1965.

Chinese versions or developments of the RD-9 are likely to have been used in locally produced military prototypes. One of these is the F-9 twin-engined strike fighter.

In 1975, after detailed negotiations, the Chinese government signed a preliminary contract with Rolls-Royce (1971) Ltd for the licensed manufacture of a supersonic afterburning Spey turbofan generally similar to the Spey 202/203 of the British F-4 Phantom. The value of British contracts on this programme, which includes a new production plant, is tentatively put at £100 million. It is assumed that the engines will power a military aircraft, probably of Chinese design, and that the Chinese objective is total self-sufficiency in their production and operation.

NATIONAL CHINESE ACADEMY OF SCIENCE

Location: Peking

While the national aviation industry has made remarkable progress in building engines to foreign designs, the Chinese have, as a long-term undertaking, sought to establish a national capability to design aircraft engines on a basis of self-sufficiency. The Chinese Academy of Science's Institute of Mechanics was charged with this task, and under the direction of Dr Chien Hsuehshen a 12-year plan was begun in 1956. This plan was presumably succeeded by another plan in 1968, but nothing has been disclosed concerning any indigenous Chinese engines.

CZECHOSLOVAKIA

AVIA
AVIA NP

Address: Letňany, Prague 9
Telephone: Prague 89-5231

Originally a member of the Czechoslovak Aviation Industry Group, Avia National Corporation was transferred to the Czechoslovak Automotive Industry (CAZ) Group in 1960. The company is at present engaged in series production of the M 137, M 337 and M 462 types of piston engine, as well as propeller and spare parts manufacture.

AVIA M 137

Designed to power light aerobatic, training, and single-engined and multi-engined sports aircraft, the 134 kW (180 hp) M 137A piston engine is a modification of the M 337 with fuel and oil systems for aerobatic operation and without a supercharger. It powers the Zlin 42 M and Z 526 F. The M 137AZ is a modified version, with the air intake port at the rear so that a dust filter can be incorporated. Details are as for M 337, with the following differences:

Crankshaft: No oil holes for propeller control.

Fuel System: Type LUN 5150 pump; system designed for sustained aerobatics.

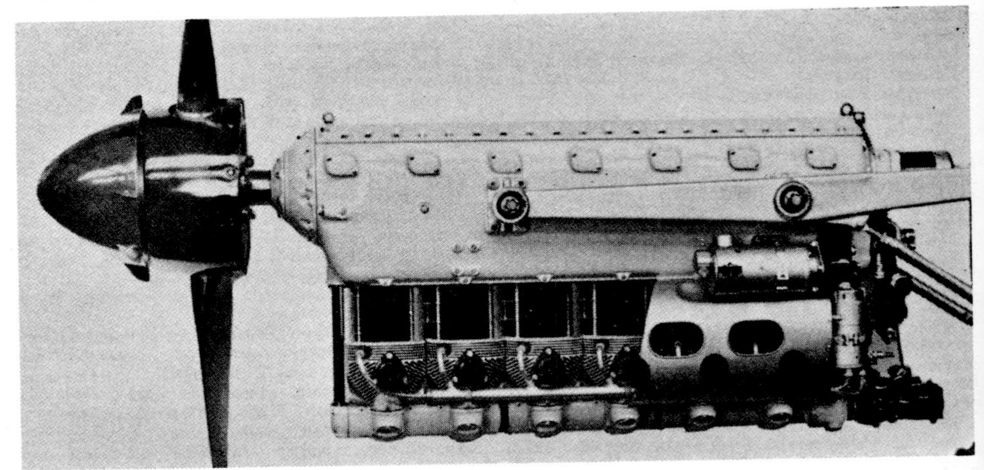

The 134 kW (180 hp) Avia M 137A six-cylinder aircooled piston engine

STARTER: LUN 2131 electric.
DIMENSIONS:
Length 1,344 mm (52·9 in)
Width 443 mm (17·44 in)
Height 630 mm (24·80 in)
WEIGHT (including starter): 141·5 kg (312 lb)
PERFORMANCE RATINGS:
T-O 134 kW (180 hp) at 2,750 rpm
Max continuous 119 kW (160 hp) at 2,680 rpm
Max cruising 104·5 kW (140 hp) at 2,580 rpm
SPECIFIC FUEL CONSUMPTION:
At T-O rating 91·26 μg/J (0·540 lb/hr/hp)
At max cruise rating 81·96 μg/J (0·485 lb/hr/hp)

AVIA M 337

The basic M 337 six-cylinder aircooled supercharged engine powers the Morava L-200D light aircraft. The M 337A, with modified fuel system, powers the Zlin 43. The M 337AK has fuel and oil systems designed for fully aerobatic operation; it is designed to power several aircraft, including the Zlin 726K.

TYPE: Six-cylinder inverted in-line aircooled, ungeared, supercharged and with direct fuel injection.

CYLINDERS: Bore 105 mm (4·13 in). Stroke 115 mm (4·53 in). Swept volume 5·97 litres (364·31 cu in). Compression ratio 6·3 : 1. Steel cylinders with cooling fins machined from solid. Cylinder bores nitrided. Detachable cylinder heads are aluminium alloy castings. Cylinder and head assembly attached to crankcase by four studs. Valve seats of special steel. Valve guides and sparking plug bushes of bronze.

PISTONS: Aluminium alloy stampings with graphited surfaces. Two compression rings and two knife-shaped scraper rings in common groove above gudgeon-pin. Gudgeon-pins secured by spring-circlips.

CONNECTING RODS: H-section aluminium alloy forgings. Two split big-ends bolted together by two bolts. Steel two-piece liner, lead-bronze plated.

CRANKSHAFT: Forged from special chrome-vanadium steel, machined all over. Nitrided crank-pins. Carried in seven steel-backed lead-bronze plated slide bearings which are lightly lead-lined, and in one ball thrust bearing at the front. Terminating in a wedge-shaped cone for the propeller hub mounting.

CRANKCASE: Heat-treated magnesium alloy (Elektron) casting, with top and front covers. Deep-sunk bearing covers forged from aluminium alloy, with double cross webs.

VALVE GEAR: Camshaft on the cylinder heads actuates the valves by means of rocker arms. Camshaft driven by vertical shaft and bevel gears. One inlet valve of heat-treated steel, one sodium-filled exhaust valve of austenitic steel with stellite seat. Nitrided valve stems.

IGNITION: Shielded type. Two vertical PAL-LUN 2221.13 magnetos with automatic sparking advance, driven by bevel gears. Two PAL L 22-62 sparking plugs per cylinder, 12 × 1·25 mm.

LUBRICATION: Dry sump pressure-feed type. Double gear-type oil pump with pressure and scavenge stages mounted on rear wall of crankcase. Oil from tank passes through triple filter into pressure stage of oil pump and then into main channel drilled in crankcase. Pressure control valve adjusted to 3·5-4 atm. Inlet union of main channel provided with oil-pressure gauge connecting pipe. Oil returned from sump by scavenge stage of oil pump to tank. Special gear-type oil pump draws oil from cam box and forces it into crankcase from where it flows into sump.

SUPERCHARGER: Centrifugal type mounted on engine rear flange. Driven through a damping rubber coupling from crankshaft. Planetary gear, ratio 7·4 : 1, engaged via band friction clutch. Force feed lubrication of supercharger from main engine lubrication system.

FUEL SYSTEM: Low-pressure injection system. LUN 5152 pump driven from camshaft. Fuel injection nozzles located in front of intake valves. Automatic control in relation to engine manifold pressure. Fuel supplied to injection pump by fuel pressure pump located in common body with injection pump. (The M 337A has a unified fuel injection pump, type LUN 5150, and other minor changes. Specific fuel consumption is slightly higher.)

FUEL GRADE: Minimum 72-78 octane, with maximum TEL 0·06 per cent (volume).

COOLING: Airscoop on port side, designed to provide easy access to sparking plugs and easy removal of scoop and baffles.

STARTING: Electric starter combined with supercharger. Electric motor rotates the starter dog which is engaged by an electromagnet. Gears and clutch of supercharger serve the starter also.

ACCESSORIES: One 600W 28V dynamo. Electric rpm transmitter, drive 1 : 1. Propeller control unit. Mechanical tachometer on oil pump, drive 1 : 2. High-pressure hydraulic pump type P 6121A.

MOUNTING: Four engine-bearer feet with rubber dampers.

PROPELLER DRIVE: Direct left-hand tractor.

DIMENSIONS:
Overall length, without propeller boss 1,410 mm (55·51 in)
Width 472 mm (18·58 in)

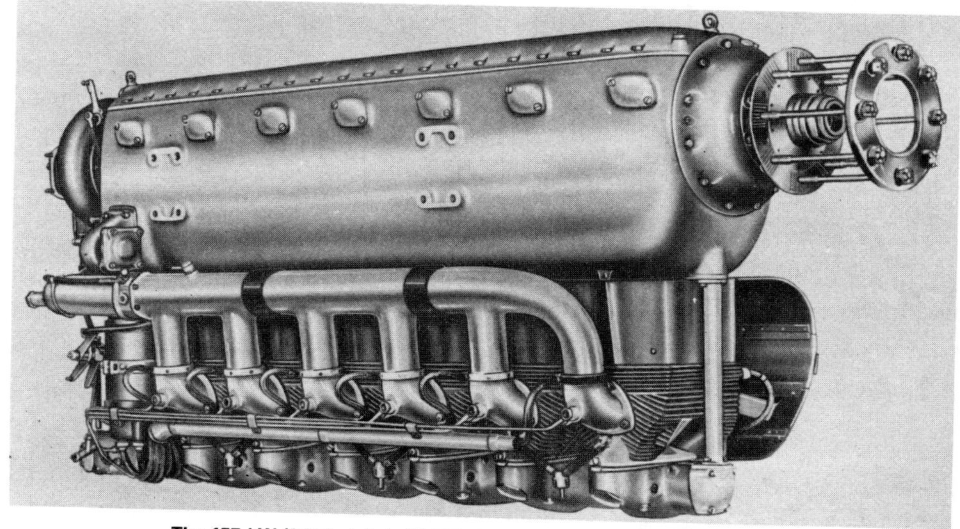

The 157 kW (210 hp) Avia M 337 six-cylinder aircooled piston engine

The 235 kW (315 hp) Avia M 462RF nine-cylinder geared supercharged radial engine

Height 628 mm (24·72 in)
Frontal area 0·20 m² (2·15 sq ft)
WEIGHT, DRY: 148 kg (326·3 lb)
PERFORMANCE RATINGS:
T-O rating 157 kW (210 hp) at 2,750 rpm
Max continuous power 127 kW (170 hp) at 2,600 rpm
Max cruising power at 1,200 m (3,940 ft) 112 kW (150 hp) at 2,400 rpm
SPECIFIC FUEL CONSUMPTION:
At T-O rating 100·6 μg/J (0·595 lb/hr/hp)
At max cruising power at 1,200 m (3,940 ft) 72·7 μg/J (0·430 lb/hr/hp)

AVIA M 462RF

This engine is a modification of the Soviet Ivchenko AI-14RF. It powers the Z-37A Cmelák agricultural monoplane. The major parts are imported from the Soviet Union.

TYPE: Nine-cylinder aircooled radial engine.

CYLINDERS: Bore 105 mm (4·13 in). Stroke 130 mm (5·12 in). Swept volume 10·16 litres (620 cu in). Compression ratio 6·2 : 1. Cylinders forged from steel. All surfaces machined. Cylinder bores nitrided. Each cylinder attached to crankcase by eight studs. Cylinder heads are aluminium alloy castings, screwed on to the barrels.

PISTONS: Aluminium alloy forgings. Three compression rings and two scraper rings. First compression ring trapezoidal in cross-section, with sliding surfaces chrome-plated. Gudgeon-pin case-hardened and quenched, with conical lightening hole.

CONNECTING RODS: Heat-treated chrome-nickel steel forgings with polished surfaces. Articulated rod ends and master rod gudgeon-pin end have bronze bushes. Master rod main bearing of steel with centrifugally-cast bronze coating of 0·5 mm thickness.

CRANKSHAFT: Single-throw type of chrome-nickel steel. All surfaces machined, heat-treated and polished. Carried in two type 2213 anti-friction bearings. Front bearing is thrust-bearing, rear one free.

CRANKCASE: Aluminium alloy forging in two sections.

VALVE GEAR: One inlet and one exhaust valve per cylinder. Both valves of austenitic steel; exhaust valve sodium-cooled. Valve seats of austenitic steel, pressed in and rolled. Cam disc situated in front section of crankcase and driven by spur gears.

INDUCTION SYSTEM: Mixture fed from supercharger. Floatless carburettor type AK-14RF fed by pump type 702 ML.

SUPERCHARGER: Single-stage centrifugal type. Spring-loaded, driven by spur gearing. Gear ratio 8·65.

FUEL: 78 octane minimum.

IGNITION: Four-spark screened ignition system. Two magnetos mounted on accessory drive housing and driven by spur gearing. Two SD-49SMM spark plugs per cylinder.

LUBRICATION: Dry-sump pressure-feed type. Gear-type oil pump with pressure and scavenge stages.

REDUCTION GEAR: Planetary reduction gearing, ratio 0·787 : 1. Hollow shaft for oil supply to hydraulically-actuated variable-pitch propeller.

ACCESSORIES: One 1,500W generator, fuel pump, electric tachometer drive, type AK 50M air compressor, and drive for agricultural equipment (see "Aircraft" section.)

STARTING: Compressed-air starting.

MOUNTING: Engine bolted to mounting at eight points through rubber dampers.

DIMENSIONS:
Length overall 1,020 mm (40·15 in)
Diameter 1,000 mm (39·37 in)
Frontal area 0·755 m² (8·03 sq ft)
PERFORMANCE RATINGS:
T-O (5 min) 235 kW (315 hp) at 2,450 rpm
Max continuous 209 kW (280 hp) at 2,200 rpm
Max rated 183 kW (245 hp) at 2,000 rpm
Max cruise 145·5 kW (195 hp) at 1,900 rpm
SPECIFIC FUEL CONSUMPTION:
At T-O rating 108 μg/J (0·639 lb/hr/hp)
At max cruise power 83·8 μg/J (0·496 lb/hr/hp)

MOTORLET
MOTORLET NC, ZÁVOD JANA SVERMY

ADDRESS: Prague-Jinonice
Telephone: Prague 522241
GENERAL MANAGER:
Zdenek Horcík
ASSISTANTS TO GENERAL MANAGER:
TECHNICAL DIRECTOR: Ing Z. Pisařík
ECONOMIC DIRECTOR: Ing Josef Svoboda
PRODUCTION DIRECTOR: Bohumil Hamerník
HEAD OF DESIGN DEVELOPMENT:
Ing Vladimír Pospísil

Motorlet National Corporation operates the main aero-engine establishment in Czechoslovakia, based on the former Walter factory at Jinonice, previously well known for its radial and in-line piston engines. Today, the Walter name continues in use only as a trade-mark for Motorlet piston and turbine engines.

Motorlet started turbine engine manufacture in 1952 with licensed production of the Russian RD-45 centrifugal turbojet for MiG-15 fighters. Present production activities concern the small M 701 centrifugal turbojet, Czechoslovakia's first indigenous turbine, for the L-29 Delfin trainer, and the M 601 turboprop.

In addition Motorlet is manufacturing hydraulic instruments, precision castings, and non-aero gas-turbine components.

The Walter M601 free-turbine turboshaft, rated at 552 ekW (740 ehp)

Cutaway Motorlet M 701-c500 turbojet, rated at 8·73 kN (1,962 lb st)

WALTER M 601

Second of Czechoslovakia's small turbine engines to enter production, the M 601 is a free-turbine turboprop having a combined axial-and-centrifugal compressor. Designed to power the Czech twin-engined L-410 light transport aircraft, it is rated at 740 ehp and drives a VJE-508 constantdspeed three-blade propeller with hydraulically varible pitch.

The first version of the M 601, rated at 550 ehp, ran in October 1967. Development of a revised 522 ekW (700 ehp) version, of increased diameter, started during 1968. Until this higher-powered model is certificated, the L-410A has Pratt & Whitney Aircraft of Canada PT6A-27 turboprops of 533 ekW (715 ehp).

TYPE: Free-turbine combined axial-and-centrifugal turboprop.
PROPELLER DRIVE: Reduction gear at front of engine with drive from free-turbine. Reduction ratio 14·9 : 1.
AIR INTAKE: Annular intake at rear of engine, with debris screen, feeds air to compressor plenum chamber.
COMPRESSOR: Two axial stages plus single centrifugal stage. Pressure ratio 6·4 : 1 at 36,660 rpm gas generator speed. Air mass flow 3·25 kg (7·17 lb)/sec.
COMBUSTION CHAMBER: Annular combustor with rotary fuel injection and low-voltage ignition.
COMPRESSOR TURBINE: Single-stage.
POWER TURBINE: Single-stage.
FUEL SYSTEM: Low-pressure LUN 6590 system, with two-lever control providing gas-generator and power-turbine speed controls.
FUEL GRADE: PL4, PL5 kerosene.
JET PIPE: Collector duct surrounding power turbine shaft. Exhaust through two ports on horizontal centreline.
ACCESSORIES: Mounting pads on accessory case at rear of engine. Propeller controls mounted on reduction gear case at front of engine.
LUBRICATION SYSTEM: Pressure gear-pump circulation. Integral oil tank and cooler.
OIL SPECIFICATION: B3V synthetic oil.
MOUNTING: Three elastically-supported pins on compressor casing.
STARTING: Electric.
DIMENSIONS:
Diameter — 420 mm (16·54 in)
Length — 1,675 mm (65·95 in)
WEIGHT, DRY: — 178 kg (392·5 lb)
PERFORMANCE RATINGS:
T-O rating — 552 ekW (740 ehp)
Cont. rating — 485 ekW (650 ehp) to 18°C
SPECIFIC FUEL CONSUMPTION:
At T-O rating: — 109·55 μg/J (0·648 lb/ehp/hr)

MOTORLET M 701

The M 701 turbojet powers the L-29 and L-29A Delfin trainers. Production started in 1961 and by the spring of 1969 over 4,500 M 701s had been built. Production of the engine was continuing in 1975.

All models of the M 701 have the same ratings and differ mainly with regard to TBO, as indicated by their individual designations. The TBOs for the M 701-b150, M 701-c250, M 701-c400 and M 701-c500 are respectively 150, 250, 400 and 500 hr. The M 701-c250 introduced flame tube and turbine improvements, and the M 701-c400 has further improvements in turbine design.

TYPE: Single-shaft centrifugal turbojet.
AIR INTAKE: Annular air intake, with central bullet fairing, at front of engine. De-icing by hot engine bleed air.
COMPRESSOR: Single-stage centrifugal type.
IMPELLER: Single-sided. Blade-type diffuser. Pressure ratio 4·3 : 1. Air mass flow 16·9 kg/sec (37·25 lb/sec) at 15,400 rpm.
COMBUSTION CHAMBER: Seven straight-flow chambers, interconnected by flame channels. Two igniter plugs in Nos. 2 and 7 chambers.
FUEL SYSTEM: Fuel pump of the LUN 6201.03 multi-plunger type. Barometric pressure control acts on servomechanism to vary fuel delivery according to altitude and speed. High-pressure shut-off cock. Max fuel pressure 0·49-0·98 bars (7·1-14·2 lb/sq in) behind fuel filter, 82·74 bars (1,200 lb/sq in) behind fuel pump.
FUEL GRADE: PL-4 to TPD-33d01960 standard, T-1 to GOST-4138-49 standard, or other similar fuels.
TURBINE: Single-stage axial-flow type, with 61 blades. Gas temperature after turbine 680-700°C.
JET PIPE: Fixed-cone type.
ACCESSORY DRIVES: Drives on engine front casing to fuel pump, 28V generator, hydraulic pump and tachometer. One spare drive.
LUBRICATION SYSTEM: Wet sump type. Sump at bottom of front case. One three-stage gear-type pump. Sump capacity 3·5 litres (0·75 Imp gallons). Normal oil supply pressure 2-2·5 kg/cm² (28·5-35·5 lb/sq in).
OIL SPECIFICATION: OLE-TO to TP 200/074-59 standard, or GOST 982-53.
STARTING: Electric starter.
DIMENSIONS:
Max width — 896 mm (35·28 in)
Max height — 928 mm (36·53 in)
Length overall — 2,067 mm (81·38 in)
WEIGHT, DRY: — 330 kg (728 lb) + 2·5%
PERFORMANCE RATINGS:
Max T-O — 8·73 kN (1,926 lb st) at 15,400 rpm
Rated power — 7·85 kN (1,764 lb st) at 14,950 rpm
Max cruise rating — 7·06 kN (1,587 lb st) at 14,500 rpm
Idling — 0·69 kN (154 lb st) at 5,400 rpm
SPECIFIC FUEL CONSUMPTION:
At rated power — 32·3 mg/Ns (1·14 lb/hr/lb st)

OMNIPOL
OMNIPOL FOREIGN TRADE CORPORATION

ADDRESS:
Washingtonova 11, Prague 1
Telephone: 2126

Omnipol is responsible for exporting products of the Czech aviation industry and for supplying information on those products which are available for export.

FRANCE

G2P
GROUPEMENT POUR LES GROS PROPULSEURS À POUDRE

HEAD OFFICE:
3 avenue du Général de Gaulle, 92080-Puteaux
Telephone: 772 12 12
MAIN ESTABLISHMENT:
Bordeaux-St Aubin de Médoc
ADMINISTRATOR:
Roger Guernon

On 1 October 1972 SEP and SNPE (both listed in this section) pooled their interests in a group called G2P (Groupement pour la Propulsion à Poudre) to ensure the close co-ordination of their activities and to act as prime contractor in the field of solid-propellant propulsion.

On 1 August 1974 the name of the group was changed to Groupement pour les Gros Propulseurs à Poudre (grouping for large solid motors), with the same initials. Its activities are centred upon the motors of strategic missiles (MSBS, SSBS) and large tactical missiles (Pluton).

MICROTURBO
MICROTURBO SA

Head Office and Works:
Chemin du Pont de Rupé, 31019-Toulouse Cédex
Telephone: (61) 47 63 26
Telex: 510835
Directorate:
G. Bayard
L. Pech (Commercial Director)
P. Calmels (Chief Engineer)

Microturbo was established in 1960 for the production of small gas turbines. The initial product was the Noelle 60290 free-turbine starter for the SNECMA Atar turbojet, and from this a wide range of units has been evolved. Current production and projects are:

Propulsion units: Couguar 022 for GAF Turana drone; TRS 18 for BD-5J aircraft and Caproni A-21J sailplane; TRI 60 in development for RPVs.

Auxiliary power units: Saphir series, for Dassault Mystère 20/Falcon; IAI Westwind; Hawker Siddeley HS.125; Westland Commando. Gevaudan series, for Nord 2501 and Frégate, Aérospatiale Caravelle 6R. 170 series air-transportable packs for helicopters, etc. An APU derived from the Dragon 021 air producer is under development.

Starting systems: Noelle 002 and 015 for Dassault Mirage and Etendard aircraft; Jaguar 007 for Sepecat Jaguar aircraft; 047MK2 for Hawker Siddeley Hawk aircraft, Dragon 021 for SNECMA M53-engined aircraft.

Other products include turbochargers, heat exchangers, regenerators, various aviation components and special test equipment for in-situ and shop testing of Microturbo units and their accessories.

Jointly with Soc. Soulé the company has formed SA France-Engins (see "RPVs and Targets" section).

MICROTURBO TRS 18-046

This single-shaft turbojet was designed for installation in gliders, to impart a self-launch and climb capability, but has since been adapted for ultra light aeroplanes. It is in production for the Bede BD-5J jet.

The TRS 18—046 is of modular construction. The forward module incorporates the air intake, gearbox, electronic governing and protection unit and the start sequencing and indication unit. The 28V 600W starter/generator is located in the nose bullet. The oil tank, with submerged pump, is on the underside, and includes provision for inverted flight. The HP oil filter and pressure transducer are on the top of this module. Adjacent to the compressor are the probes for engine speed and air temperature.

The turbine module comprises: the one-piece centrifugal compressor, with diffuser and straightener vanes; the axial turbine rotor and nozzle diaphragm; and the main frame, carrying the rotor assembly on two ball bearings between the compressor and turbine. The aft module comprises: the turbine casing backplate, carrying the annular folded combustion chamber liner, exhaust cone and nozzle; 10 spill-type burners; two igniter plugs, used only during starting; and the jet pipe with thermocouple.

The fuel pump is driven electrically. The lubrication system is a closed circuit, with pressure supply to the rotor and gearbox bearings. The engine can be shut down and restarted in flight, and incorporates automatic fault and protection systems.

Dimensions:
Length 650 mm (25·59 in)
Width 325 mm (12·797 in)
Height 350 mm (13·78 in)
Weight, Dry:
Basic 32·0 kg (70·5 lb)
With igniter and voltage regulator 33·4 kg (73·63 lb)
Performance Rating (ISA, S/L):
T-O and max continuous
0·898 kN (202 lb st) at 44,000 rpm
Specific Fuel Consumption:
At above condition 36 mg/Ns (1·27 lb/hr/lb st)

Microturbo TRS 18-046 single-shaft turbojet, rated at 0·898 kN (202 lb st)

Microturbo TRS 18-056 single-shaft turbojet, rated at 0·98 kN (220·5 lb st)

MICROTURBO COUGUAR 022

This single-shaft turbojet was developed as a powerplant for small aircraft and RPVs. The 022 version powers the Australian GAF Turana target drone.

The installed engine, which incorporates only the equipment needed for flight, was designed to withstand the high mechanical and thermal stresses imposed by accelerated launch, telemetered shutdown and subsequent immersion in sea-water. Equipment for defuelling, refurbishing, refuelling, starting, ground running and pre-flight checkout is all accommodated in an associated servicing rig connected via an umbilical which is disconnected immediately prior to flight. Starting is accomplished by an air-impingement nozzle integral with the engine impeller shroud, the air supply being controlled from the servicing rig.

Dimensions:
Length (less jet pipe) 628 mm (24·725 in)
Length overall 853 mm (33·58 in)
Width 282 mm (11·1 in)
Height 386 mm (15·2 in)
Weight, dry:
Basic, less jet pipe 26·5 kg (58·4 lb)
With control box 28·3 kg (62·4 lb)
Performance Ratings (ISA, S/L):
T-O and max continuous
0·79 kN (178 lb st) at 48,500 rpm
Idle 0·138 kN (31 lb st) at 28,000 rpm
Specific Fuel Consumption (ISA, S/L, static):
T-O and max continuous 35·4 mg/Ns (1·25 lb/hr/lb st)

MICROTURBO TRS 18-056

A simplified version of the TRS 18-046, the TRS 18-056 retains only the gas-generator section. It has been developed to power the France-Engins Mitsoubac and other RPVs. The lubrication system is of the total-loss type. During starting, the engine is cranked either by impingement with air supplied from a bottle or by windmilling, according to the RPV in which it is installed. Ignition for starting is provided by an electrically fired cartridge. At shutdown, two independent systems shut off the fuel supply: one is an instantaneously actuated shut-off valve and the other is releasing the gas pressure from the fuel tank.

Microturbo Couguar 022 single-shaft turbojet rated at 0·79 kN (178 lb st)

Microturbo TRI 60 single-shaft turbojet, rated at 3·43 kN (772 lb st)

Dimensions:
Length 600 mm (23·6 in)
Width 305 mm (12·0 in)
Height 345 mm (13·6 in)
Weight, Dry:
Basic, no jet pipe 23·0 kg (50·7 lb)
Performance Rating (ISA, S/L):
T-O and max continuous
0·98 kN (220·5 lb st) at 45,000 rpm
Specific Fuel Consumption:
At above condition 36 mg/Ns (1·27 lb/hr/lb st)

MICROTURBO TRI 60

Representing a significant French development in the propulsion of cruise-type unmanned vehicles, the TRI 60 was designed under a contract from the Direction des Recherches et Moyens d'Essais. It is an extremely simple single-shaft turbojet for use in subsonic missiles and RPVs. The design has been biased towards minimal cost and absence of any maintenance or overhaul, though engine design life exceeds 20 hr.

The annular intake contains the accessory gearbox in the central bullet, together with an alternator or starter/generator; the struts house fuel and oil pipes. The simple axial compressor operates at a pressure ratio of about 4 : 1, with airflow of 5·6 kg (12·3 lb)/sec, and is carried between front and rear bearings with labyrinth seals. The smokeless combustor is of the axial type, with multiple spray burners fed by a peripheral manifold. The axial turbine is overhung behind the rear bearing on the central diffuser housing.

An air bleed provides up to 1·5 per cent of total airflow. There is an engine-driven fuel pump, but lubrication is by either pre-lubricated bearings or a total-loss system from a pressurised reservoir. Speed control can be mechanical, electronic, fluidic or pneumatic, according to installation. Starting can be by impingement, electrical, cartridge or other means.

Dimensions:
Length overall 882 mm (34·72 in)
Envelope diameter 310 mm (12·20 in)
Weight, Dry: 45 kg (99·2 lb)
Performance Rating (ISA, S/L):
Max T-O 3·43 kN (772 lb st) at 28,500 rpm
Specific Fuel Consumption (as above):
35·4 mg/Ns (1·25 lb/hr/lb)

RECTIMO
RECTIMO AVIATION SA

Offices and Works:
Aérodrome de Chambéry, 73420-Savoie
Telephone: (79) 63 40 06
Director:
André Rosselot

Rectimo has manufactured over 500 Type 4 AR 1200 single-ignition derivatives of the Volkswagen four-cylinder aircooled car engine, which together with the larger 4 AR 1600 are used in the Sportavia RF4D powered glider and various ultra-light aircraft. The 30 kW (40 hp) 4 AR 1200 engine has a 1,192 cc cubic capacity, 7 : 1 compression ratio and weighs 61·5 kg (136 lb). Fuel consumption under cruise conditions is 11 litres (2·4 Imp gal)/hr. The 4 AR 1600 produces 45·5 kW (61 hp) at T-O and has a cubic capacity of 1,600 cc and an 8 : 1 compression ratio. Weight is 64 kg (141 lb). Both engines have a maximum speed of 3,600 rpm.

Rectimo 4 AR 1200 piston engine of 30 kW (40 hp)

SEP
SOCIÉTÉ EUROPÉENNE DE PROPULSION

Head Office:
3 avenue du Générale de Gaulle, Tour Nobel, Cédex 3, 92080-Paris La Défense
Telephone: 772 12 12
Telex: 63906 Putau
Centres and Establishments:
Bordeaux-Le Haillan, Bordeaux-Blanquefort, Vernon, Melun-Villaroche and Istres
President Director General:
P. Soufflet

SEP specialises in the design and development of all categories of propulsion systems and engines for aircraft, missiles, space launchers and satellites, and it possesses the most important rocket facilities in Europe. Two-thirds of its 2,200 personnel are engineers and technicians specialising in research, development and testing.

SEP produces a wide range of solid- and liquid-propellant motors as sustainers and boosters for French and European guided and unguided missiles and space launchers. Some 60 different types of motor have been designed since 1950, including the three stages of the Diamant A and B and BP 4 French space launchers, the second stage and perigee motor of the ELDO Europa II launcher, and motors for the SSBS, MSBS and Pluton nuclear-warhead missiles. The company has acquired great experience in cryogenic-propellant rockets through developing the HM4 and HM7 engines.

Centre National d'Etudes Spatiales has entrusted SEP with the entire propulsion systems of the three stages of the L 3 S Ariane launch vehicle. A major engine for manned aircraft is the SEP 844 which provides thrust boost for Dassault Mirage III fighters in service with the French Air Force and other air forces.

Other SEP developments include engines using hybrid propellants, fluorine and fluorine compounds, monopropellants, compressed gases, as well as electric "thrusters". The company is now applying its missile and space technology to oceanology.

SEP FAON, ELAN

In April 1972 SEP announced the Faon, later joined by the larger Elan. These are extremely simple and reliable small motors, intended primarily for young experimenters in aerospace clubs. The Faon was designed within the framework of these clubs, and under CNES sponsorship, to carry a 3 kg (6·6 lb) payload to a height of 3,500 m (11,500 ft).

Basic Data (F: Faon, E: Elan):

Length (excluding igniter):	
F	185 mm (7·28 in)
E	340 mm (13·4 in)
Max diameter:	
F	62 mm (2·44 in)
E	120 mm (4·72 in)
Weight, loaded:	
F	1·1 kg (2·43 lb)
E	5·24 kg (11·55 lb)
Mean thrust:	
F	0·19 kN (42·7 lb)
E	1·5 kN (337·2 lb)
Burn time:	
F	5 sec
E	3·77 sec

SEP HM4

The HM4 is the smaller of two upper-stage liquid-propellant rocket engines currently being developed by SEP.

Type: Liquid-propellant rocket engine.

Propellants: Liquid oxygen and liquid hydrogen.

Thrust Chamber Assembly: Four-chamber unit of 42 : 1 nozzle area ratio, regeneratively cooled, and with double-wall machined casing in stainless steel and Inconel X750. Operating sequence initiated by hydrogen pre-cooling and pre-opening of hydrogen injection valve. Concentric-tube propellant injection system with central oxygen flow. Pyrotechnic ignition. Combustion pressure 23·29 bars (337·8 lb/sq in) and temperature 2,627°C.

Thrust Chamber Mounting: Chambers hinged around axis concentric with engine axis.

Propellant Pumps: Axial-plus-centrifugal pumps, co-axial.

Propellant Flow: Liquid hydrogen flow rate 1·67 kg (3·67 lb)/sec at 40 bars (580 lb/sq in). Liquid oxygen flow rate 8·33 kg (18·32 lb)/sec at 36 bars (522 lb/sq in).

Turbine: Two-stage axial-flow impulse unit in Inconel X750. Gas inlet temperature 617°C.

Gas Generator: Liquid hydrogen flow rate 0·088 kg (0·19 lb)/sec. Liquid oxygen flow rate 0·079 kg (0·17 lb)/sec. Pyrotechnic ignition.

Lubrication System: Uses tributyl phosphate spray into gaseous hydrogen.

Starting: Solid-grain primer.

Thrust Control: Thrust held constant by regulation of turbopump speed via control of gas generator propellant supply.

Dimensions:

Height overall	1,170 mm (45·6 in)
Diameter overall	1,220 mm (46·5 in)

SEP HM4 liquid oxygen/liquid hydrogen rocket engine for upper-stage propulsion (40·4 kN; 9,080 lb)

Weight, Dry:	174 kg (382·8 lb)
Performance:	
Max thrust	40·4 kN (9,080 lb)
Overall propellant mixture ratio	5 : 1
Specific impulse	412

SEP P4

The P4 solid-propellant propulsion system was designed as the second stage of the French Navy's MSBS (underwater-to-surface ballistic missile), a weapon now equipped with the Rita II engine.

The P4 has a case of wound glass filament, weighing only 320 kg (705 lb). It is loaded with a grain of Isolane propellant, based on polyurethane binder, of high specific impulse. Ignition is by a solid-propellant microrocket on the front closure, which is also provided with controllable thrust-termination ports. The fixed nozzle is of carbon-fibre laminate with a graphite throat. Pitch and yaw control is effected by freon injection through four electro-valves. Roll is controlled by two separate steerable motors.

P4 is to replace Topaze as the second stage of the B P 4 version of the Diamant launch vehicle.

Dimensions:	
Length	2,500 mm (98·4 in)
Diameter	1,500 mm (59 in)
Weight, Loaded:	about 4,000 kg (8,820 lb)
Performance:	
Vacuum thrust	176·5 kN (39,680 lb)
Constant-thrust burn	55 sec

SEP VIKING

The Viking series of turbopump-fed rocket engines was designed for simplicity and low cost. The thrust chamber, fed with unsymmetrical dimethyl hydrazine (UDMH) and nitrogen tetroxide (N_2O_4), has a single wall of HS 25 steel, coated with zirconium oxide, fuel-film cooled. The nozzle throat is of graphite. The light-alloy injector is of the radial type, with alternate doublets.

Mounted directly on the chamber, the turbopump has a two-stage Curtiss turbine, driven by propellant gases cooled by water. The turbine shaft carries impellers for UDMH, N_2O_4 and water. The gas-generator also provides on-board power and pressurises the tanks. Combustion pressure is regulated against a reference by varying turbine speed. Mixture ratio is maintained by controlling the flow of N_2O_4.

In 1972 SEP ran Viking I qualification tests of 150 sec duration. The L 3 S Ariane vehicle will have a first stage, Lilo, with four Viking II engines now under development. Testing of the cluster is to begin in 1976. The second-stage Ariane engine is Viking IV, a Viking II tuned for vacuum operation, with two-axis thrust vectoring. The nozzle has a bell shape, fabricated by welding rolled and Flo-turned steel sheet. Testing of the Viking IV was scheduled to begin in June 1976.

Weight, Dry:	
Viking II	776 kg (1,735 lb)
Viking IV	850 kg (1,876 lb)
Performance:	
Nominal S/L thrust:	
Viking II	588·4 kN (132,275 lb)
Vacuum thrust:	
Viking II	686·5 kN (154,323 lb)
Viking IV	717·9 kN (161,378 lb)
Chamber pressure:	
Viking II	55·5 bars (805 lb/sq in)
Vacuum specific impulse:	
Viking II	278·6
Viking IV	292·3

SEP HYDRAZINE THRUSTER

SEP has for a long period been engaged in the development of small monopropellant thrusters for satellite attitude and orbit control. Most of this work has been based on hydrazine, decomposed through a catalyst to serve as a monopropellant. The present SEP hydrazine propulsion system uses CNESRO 1 catalyst, developed jointly by SEP and the Faculté des Sciences of Paris.

SEP P4 solid-propellant propulsion system for Diamant B P 4

Propulsion bay of Lilo, first stage of the L 3 S Ariane launch vehicle, showing the four SEP Viking II engines. Total S/L thrust is 2,353·6 kN (529,100 lb)

SEP hydrazine thruster system for D-5A satellite, shown with conical fairing removed

SEP delivered to CNES (Space Centre of Toulouse) a flight model of its hydrazine micropropulsion system, comprising: surface-tension tank, engine, European CNESRO catalyst, hydrazine electrovalve, sensor and on-board electronics. This micropropulsion system was assembled on the D-5A satellite, launched by Diamant B P-4. During endurance testing a D-5A thruster operated continuously for 145,000 seconds, without attention to the catalyst bed.

SEP has now developed a larger thruster for mounting on the GEOS satellite. During qualification testing this thruster operated for 8,600 seconds (specified time 7,800 sec) in 34,000 impulses, and for 3,400 seconds in continuous operation (specification 3,000 sec). In the following data this thruster is referred to as GM, the earlier unit being referred to as D-5A.

WEIGHT, DRY:
- D-5A: 0·18 kg (0·397 lb)
- GM: 0·35 kg (0·771 lb)

CHAMBER PRESSURE:
- D-5A: 15·3-30·5 bars (222-442 lb/sq in)
- GM: 10-30 bars (145-435 lb/sq in)

ELECTRICAL LOAD:
- D-5A: 5W
- GM: 6W

THRUST:
- D-5A: 0·0028-0·0016 kN (0·629-0·359 lb)
- GM: 0·014-0·006 kN (3·15-1·35 lb)

SPECIFIC IMPULSE (Vacuum):
- D-5A, GM: 215-230

SEP ION THRUSTER

Like other electric thrusters that generate a jet of charged ions, this device uses caesium as the working fluid. Caesium from the supply tank is vaporised and ionised between the hollow cathode and inner anode in the presence of a powerful magnetic field. The ions are accelerated through an electrostatic grid and then neutralised by a secondary hollow cathode immediately downstream. Its useful life is to amount to 10,000 hours of cumulative operation, enabling the engine to perform missions ranging from 7 to 10 years in orbit.

WEIGHT, DRY: 1·5 kg (3·3 lb)

PERFORMANCE:
- Thrust: 6·7 mN or about 0·00067 kg
- Flow-rate of caesium: 0·5 gr (0·0011 lb)/hr
- Beam orientation: ±20°
- Power input: 250W
- Specific impulse: 5,000

SEP HM7

Under development for upper-stage propulsion, the HM7 is a 70·06 kN (15,750 lb) liquid oxygen/liquid hydrogen engine. It is also referred to in the entry for MBB of West Germany.

TYPE: Liquid-propellant rocket engine.

PROPELLANTS: Liquid oxygen and liquid hydrogen.

SEP ion thruster, using caesium ions to give very small thrust at high specific impulse

THRUST CHAMBER ASSEMBLY: Single-chamber unit of 48 : 1 nozzle area ratio, regeneratively cooled, and of stainless steel tube construction. Operating sequence initiated by hydrogen pre-cooling and pre-opening of hydrogen injection valve. Concentric-tube propellant injection system with central oxygen flow. Pyrotechnic ignition. Combustion pressure 35 bars (507·5 lb/sq in) and temperature 2,727°C.

THRUST CHAMBER MOUNTING: Gimballed assembly, turbopump integral with chamber.

PROPELLANT PUMPS: Axial-plus-centrifugal pumps, coaxial.

PROPELLANT FLOWS: Liquid hydrogen flow rate 2·76 kg (6·07 lb)/sec at 65 bars (942·5 lb/sq in). Liquid oxygen flow rate 14·21 kg (31·26 lb)/sec at 52 bars (754 lb/sq in).

TURBINE: Two-stage axial-flow impulse unit in Inconel X 750. Gas inlet temperature 617°C.

GAS GENERATOR: Liquid hydrogen flow rate 0·133 kg (0·29 lb)/sec. Liquid oxygen flow rate 0·12 kg (0·26 lb)/sec. Pyrotechnic ignition.

LUBRICATION SYSTEM: Uses tributyl phosphate spray into gaseous hydrogen.

STARTING: Solid grain primer.

THRUST CONTROL: Thrust held constant by regulation of turbopump speed via control of gas generator propellant supply.

SEP HM7 high-energy upper-stage rocket engine (vacuum thrust 70·06 kN; 15,750 lb)

DIMENSIONS:
- Height overall: 1,617 mm (63·06 in)
- Diameter overall: 847 mm (33·03 in)

WEIGHT, DRY: 145 kg (319 lb)

PERFORMANCE:
- Max thrust in vacuo: 70·06 kN (15,750 lb)
- Overall propellant mixture ratio: 5·15 : 1
- Specific impulse: 425

TYPICAL SEP SOLID-PROPELLANT ROCKET ENGINES

Type	Thrust at S/L	Duration of Thrust (sec)	Total Weight	Length	Diameter
SEP 163 Sioule	1·4 kN (315 lb)	37	40 kg (88 lb)	1,745 mm (68·7 in)	141 mm (5·5 in)
SEP 6854 Odet	37 kN (8,320 lb)	4	124 kg (273 lb)	2,560 mm (101 in)	226 mm (8·9 in)
SEP 7382 Yonne	94 kN (21,130 lb)	20	1,250 kg (2,756 lb)	4,062 mm (160 in)	584 mm (23 in)
SEP 7392 Rance	162·8 kN (36,520 lb)	17·5	1,565 kg (3,450 lb)	4,605 mm (181·3 in)	578 mm (22·7 in)
SEP 7342 Vienne	293 kN (65,860 lb)	4·6	846 kg (1,870 lb)	3,195 mm (125·8 in)	590 mm (23·2 in)
SEP 299 Arz	43·15 kN (9,702 lb)	1·6	57·5 kg (126·78 lb)	1,076 mm (42·2 in)	250 mm (9·8 in)
SEP 300 Drac	0·96 kN (215·8 lb)	73	52·5 kg (115·76 lb)	1,100 mm (43·3 in)	207 mm (8·1 in)
SEP Dropt	40 kN (9,000 lb)	45	751 kg (1,655·9 lb)	1,500 mm (59 in)	800 mm (31·5 in)
SEP Ball	4·2 kN (944 lb)	4·3	23·1 kg (50·9 lb)	430 mm (16·9 in)	221 mm (8·7 in)
SEP Trap	Start 8·05 kN (1,809 lb) Cruise 1·78 kN (402 lb)	Start 3·8 Cruise 10·2	45·2 kg (99·75 lb)	835 mm (32·87 in)	221 mm (8·7 in)
SEP Arc	Start 1·90 kN (20,233 lb) Start 2·53 kN (11,915 lb) Cruise 14·1 kN (3,170 lb)	Start 10·1 Start 20·2 Cruise 3·5	57 kg (125·7 lb)	2,306 mm (90·8 in)	157 mm (6·2 in)
SEP Ciron (GEOS)	16·4 kN (3,687 lb)	45	300 kg (661 lb)	1,070 mm (42·1 in)	684 mm (26·9 in)
SEP Mage	30·4 kN (6,835 lb)	40	345 kg (761 lb)	1,170 mm (46·1 in)	764 mm (30·1 in)

SNECMA

SOCIÉTÉ NATIONALE D'ÉTUDE ET DE CONSTRUCTION DE MOTEURS D'AVIATION

HEAD OFFICE:
150 Boulevard Haussmann, 75361 Paris Cedex 08
Telephone: 227 33 94
Telex: 65383 Motavia Paris

CHAIRMAN AND MANAGING DIRECTOR:
René Ravaud

DEPUTY MANAGING DIRECTOR FOR AERO-ENGINE PROGRAMMES AND MARKETING:
Jean Péquignot

DEPUTY MANAGING DIRECTOR FOR AERO-ENGINE DEVELOPMENT AND PRODUCTION:
Joseph Millara

DEPUTY MANAGING DIRECTOR FOR SUBSIDIARY COMPANIES AND DIVISIONS:
François Carrier

GENERAL SECRETARY:
Bernard Denis

PERSONNEL DIRECTOR:
Philippe Sappey

FINANCE AND ECONOMICS DIRECTOR:
Guy Zarrouati

PERSONAL ASSISTANT TO THE CHAIRMAN:
Roger Abel

DIRECTOR FOR INTERNATIONAL RELATIONS:
Jean Crépin

TECHNICAL DIRECTOR:
Michel Garnier

TECHNICAL MANAGER:
Jean Devriese

MANAGER, ENGINE PRODUCTION:
Michel Viret

PUBLIC RELATIONS DEPARTMENT:
Alexandre Barbé

Villaroche Centre:
77550-Moissy-Cramayel

Design, development and ground test centre. A subsidiary establishment for flight and noise tests is located at Istres.

Evry-Corbeil:
RN 7, 91 Evry, BP 81-91003 Evry Cedex

Engine production, quality control, service, procurement and laboratories for research and development.

Gennevilliers:
291 Avenue d'Argenteuil, BP 30-92234 Gennevilliers

Forging and casting production, complete machining of mechanical parts.

Suresnes—ELECMA Division:
22 Quai Galliéni, 92150-Suresnes

Design, development and production of electronic devices, especially electronic control systems for the aircraft industry.

Bois-Colombes—Hispano-Suiza Division:
Rue du Capitaine Guynemer, 92270-Bois-Colombes

Design, development and production of industrial gas turbines, nuclear equipment, and production of parts for jet engines and aeronautical equipment.

SNECMA (Société Nationale d'Etude et de Construction de Moteurs d'Aviation) was born on 29 August 1945 from the merger of several aero-engine companies: Gnome et Rhône, Société Anonyme des Moteurs Renault pour l'Aviation, Société Générale de Mécanique et d'Aviation (former Moteurs Lorraine), and Groupe d'Etudes des Moteurs à Huile Lourde.

These companies already had a long aeronautical tradition and SNECMA has always devoted its main activity to aero-engines. More than 4,500 Atar turbojets have been ordered; they have played a significant part in the worldwide success of Mirage fighters. SNECMA is developing the M53 turbojet for fighters of the next decade.

SNECMA is also participating in the following international collaborative programmes:

The Olympus 593 for Concorde, developed and produced with Rolls -Royce; the M45H, also developed and produced with Rolls-Royce which has now taken over this programme; the CF6-50 for the Airbus A300, for which engine SNECMA and MTU in Germany are partners with General Electric; the CFM56, which SNECMA shares equally with General Electric, but with other European engine manufacturers being associated within SNECMA's share; and the Larzac, produced in co-operation with Turboméca and with production also including the German companies MTU and KHD, and FN in Belgium.

These engines are discussed in the International section, apart from the Larzac which appears under Turboméca-SNECMA, and the M45H now listed under Rolls-Royce.

The merger in 1968 with the Hispano-Suiza company, which is now a division of SNECMA, brought in production of Tyne engines for Transall and Breguet Atlantic aircraft; this engine was manufactured under a Rolls-Royce licence within a consortium including SNECMA, Rolls-Royce, MTU and FN in Belgium. Spare parts production is continuing.

Other activities of the SNECMA group include: electronic control systems for engines and miscellaneous equipment for aerospace vehicles produced by the ELECMA Division; Martin-Baker ejection seats produced under licence by the Hispano-Suiza Division for Mirage, Jaguar and Alpha Jet aircraft; repair and overhaul of aero-engines and components, carried out in the Billancourt and Chatellerault facilities of the subsidiary company SOCHATA-SNECMA.

Another subsidiary, Messier-Hispano, is the main French producer of aircraft landing gears, wheels, brakes and associated hydraulic systems. A Messier-Hispano subsidiary is the Bugatti company at Molsheim. The Messier-Hispano/Bugatti group equips practically all French military and commercial aircraft and participates in most European collaborative projects. Outside the aviation market, industrial gas turbines, turbochargers, nuclear equipment (such as compressors, specialised pumps and miscellaneous equipment) are produced by the Hispano-Suiza Division. CNMP-Berthiez produces armaments and high quality machine-tools. This company, a SNECMA subsidiary, was born from the merger in 1968 of CNMP (Compagnie Normande de Mécanique de Précision) which took over in 1964 a state-owned factory in Le Havre, and the "Anciens Etablissements Berthiez" company located in Givors.

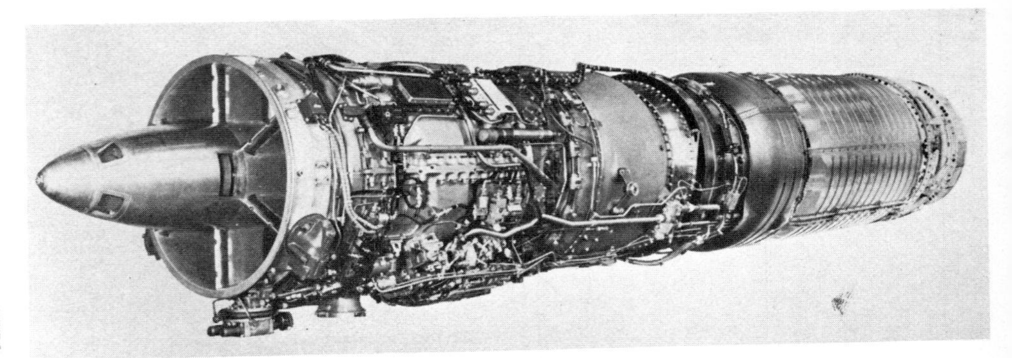

SNECMA Atar 9K-50 turbojet of 70·6 kN (15,870 lb st) with afterburner

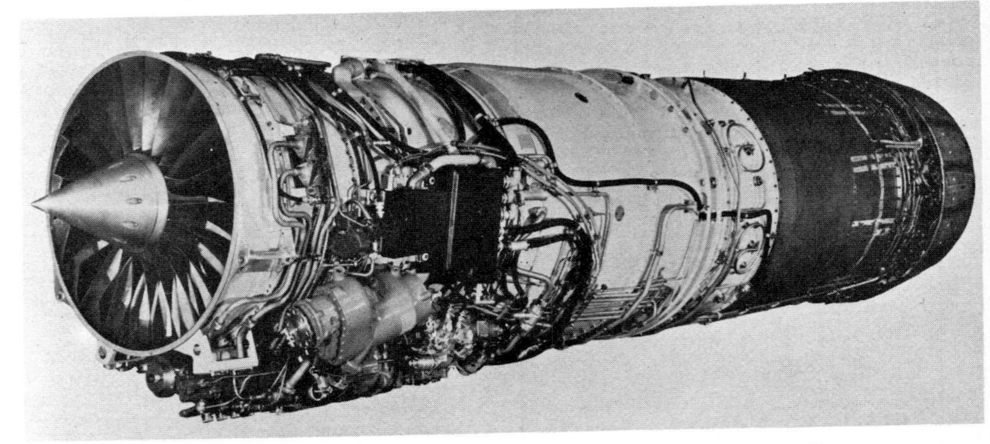

SNECMA M53 augmented by-pass turbojet of 83·4 kN (18,740 lb st)

SNECMA ATAR

The Atar is a single-shaft military turbojet first run in 1946 and since greatly developed and cleared for flight at Mach numbers greater than 2. Major versions in 1976 are:

Atar 9C. Compared with the earlier 9B this introduced a new compressor, a self-contained starter and an improved overspeed which comes into operation automatically when the aircraft reaches Mach 1·4 giving power equivalent to a sea level thrust of 62·76 kN (14,110 lb st). The compressor rotor has steel blades on stages 1, 2, 7, 8 and 9 and light alloy blades on stages 3-6: the stator has steel blades on stages 1 and 2 and light alloy blades on stages 3-8. Air mass flow 68 kg (150 lb)/sec. Pressure ratio 5·5 : 1. Equips most Mirage III and 5, and has been produced under licence in Switzerland and Australia. It is also assembled in Belgium, with a part-Belgian content, for Belgian Mirage 5 aircraft.

Atar 9K-50. Derived from the Atar 9C and fitted as initial power plant of the Mirage F1 and prototype Mirage G8. Designed to offer improved subsonic specific fuel consumption, increased thrust for supersonic acceleration and improved overhaul life. The main improvements are in an entirely redesigned turbine with blades not forged but cast and coated with refractory metal from the vapour phase. This wholly new turbine section includes a section of engine carcase, exit cone and fixed vanes. Stages 1 and 8 of the compressor have been redesigned, resulting in pressure ratio raised from 6 : 1 to 6·15 : 1, coupled with slightly augmented mass flow. The intake section has been revised to accommodate a rearranged accessory-drive system, and the control and electronic equipment have been revised and extended to improve the security of single-engined aircraft. The 9K-50 has done extensive ground and flight testing in Mirage III and F-1 aircraft and was homologated for service use following its final 150 hour test at the CEP, Saclay, between 8 September and 24 November 1969. The 9K-50 is the power plant of the Mirage F1-C fighters of the Armée de l'Air and of Mirage F1 aircraft of all overseas customers at the time of going to press with this edition. In addition, the 9K-50 equips certain export versions of the Mirage III, especially late models for the Republic of South Africa.

Atar 8K-50. This is essentially the 9K-50, the latest variant in production, re-engineered to have a simple unaugmented jet-pipe and fixed nozzle, for the Super Etendard. The original Etendard was powered by the Atar 8C, compared with which the 8K-50 has 16 per cent higher thrust and 50 per cent longer time between overhauls. All parts are protected against sea corrosion. Production 8K-50 engines have been ordered for Super Etendards of the Aéronavale, and the engine completed its 150 hr official type test in May 1975.

Dimensions:
- Diameter: 1,020 mm (40·2 in)
- Length overall:
  - Atar 8K-50: 3,936 mm (155 in)
  - Atar 9C, 9K-50: 5,944 mm (234 in)

Weights:
- Dry, complete with all accessories:
  - Atar 8K-50: 1,155 kg (2,546 lb)
  - Atar 9C: 1,450 kg (3,197 lb)
  - Atar 9K-50: 1,587 kg (3,500 lb)

Performance Ratings:
- Max with afterburner:
  - Atar 9C: 58·8 kN (13,200 lb st) at 8,400 rpm
  - Atar 9K-50: 70·6 kN (15,870 lb st) at 8,400 rpm
- Max without afterburner:
  - Atar 8K-50: 49 kN (11,025 lb st) at 8,700 rpm
  - Atar 9C: 42 kN (9,430 lb st) at 8,400 rpm
  - Atar 9K-50: 49·2 kN (11,055 lb) st

Specific Fuel Consumption:
- At max rating with afterburner:
  - Atar 9C: 57·5 mg/Ns (2·03 lb/hr/lb st)
  - Atar 9K-50: 55·5 mg/Ns (1·96 lb/hr/lb st)
- At max rating without afterburner:
  - Atar 8K-50: 27·5 mg/Ns (0·97 lb/hr/lb st)
  - Atar 9C: 28·6 mg/Ns (1·01 lb/hr/lb st)
  - Atar 9K-50: 27·5 mg/Ns (0·97 lb/hr/lb st)

Oil Consumption: 1·5 litres (2·64 Imp pints)/hr

SNECMA M53

Design started in 1967 on the M53 to provide an engine of superior performance to present series Atar engines but of simpler and less costly design than the SNECMA TF 306 turbofan. The result is a single-shaft turbofan—more strictly a continuous-bleed turbojet or by-pass turbojet—having the capability of propelling fighter aircraft at high altitude initially at Mach 2·5.

The M53 is intended to power the Delta Mirage 2000 and the twin-engined Delta Super Mirage. It is of modular construction.

The single shaft comprises a three-stage fan and five-stage compressor driven by a two-stage turbine designed for operation at high gas temperature. There are no inlet guide vanes. Max airflow is 84 kg (185 lb)/sec. Between the fan and compressor is a mid-frame incorporating accessory drives and front roller bearing and ball thrust bearing. The annular combustion chamber is designed for smoke-free operation. The turbine delivery casing incorporates the third bearing. Fuel to the combustion chamber and reheat system, and the multi-flap nozzle, are controlled by a fuel system monitored by an ELECMA electronic computer.

The first prototype engine was tested in February 1970, 18 months after programme go-ahead. The second ran in August, and achieved maximum rpm after 30 minutes of operation. Military rating (51 kN; 11,466 lb) was reached in October, three months ahead of contractual commitment. Testing with afterburner began in November 1970. The first official test took place at 54·92 kN (12,346 lb) in May 1971. In the same month a 50 hr test was run at afterburning thrusts up to 81·39 kN (18,298 lb), 98 per cent of the nominal rating. In September thrust exceeded the nominal figure, at 83·4 kN (18,740 lb), and in December simulated altitude trials began at Saclay. By March 1972 three prototype M53 engines were running. Flight trials began on 18 July 1973 in the starboard pod of

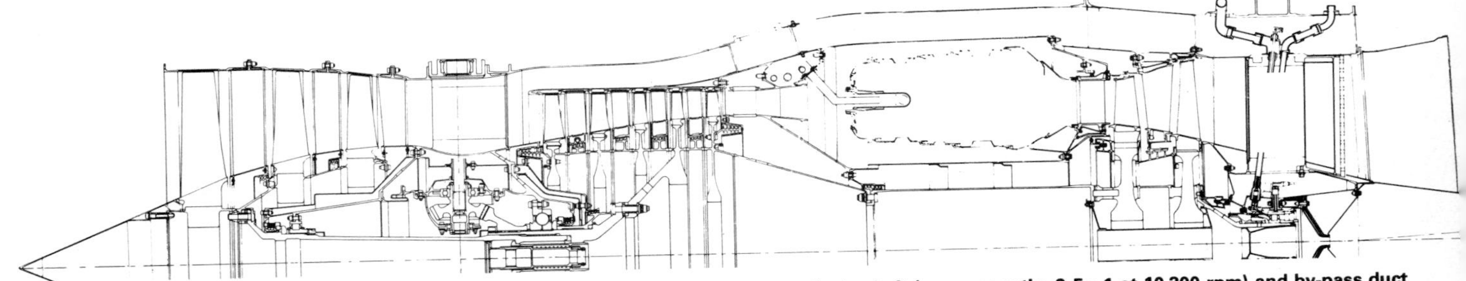

Longitudinal section through the SNECMA M53 showing LP and HP sections on single shaft (pressure ratio, 8·5 : 1 at 10,200 rpm) and by-pass duct

a Caravelle testbed. The supersonic flight envelope is being explored with the Mirage F1-M 53 flying testbed, which first flew in December 1974. In December 1975, M 53 running time totalled over 4,500 hours.

DIMENSIONS:
Overall length 4,850 mm (190·5 in)
Max diameter 1,040 mm (40·9 in)
WEIGHT, DRY: 1,420 kg (3,130 lb)
PERFORMANCE RATINGS:
Max thrust, with afterburner 83·4 kN (18,740 lb)
Max thrust, cold 55 kN (12,350 lb)
SPECIFIC FUEL CONSUMPTION:
At max cold rating 24·64 mg/Ns (0·87 lb/hr/lb st)

SNECMA/GE CFM 56

After carefully exploring the world market, and the prospects for collaborating with other large engine companies while retaining design leadership, SNECMA decided in the Autumn of 1971 to develop the CFM 56, a subsonic turbofan in the ten-tonne-thrust (22,000 lb) class, in cooperation with General Electric. The French and US companies are splitting all the work on a 50/50 basis. The project is covered more fully in the International section.

GE/SNECMA/MTU CF6-50

To provide engines for the Airbus Industrie A300B2 and B4 programmes, SNECMA and MTU of Federal Germany participate in a co-production programme with General Electric to make the CF6-50 turbofan. SNECMA manufactures parts totalling 21 per cent of the cost of each engine, and MTU 10·5 per cent. GE supplies 61·3 per cent. Assembly is performed by SNECMA at Corbeil, and static testing done at SNECMA Villaroche. These tasks account for 6 per cent of the cost of each engine. SNECMA and MTU also supply sets of parts to GE for engines to power the DC-10 Series 30. By the end of 1975 Airbus Industrie had ordered 142 engines, and SNECMA's co-production programme was extended to include four engines for a Boeing 747-200F for Air France.

SNECMA/ROLLS-ROYCE OLYMPUS 593

SNECMA is collaborating with Rolls-Royce in the design, development and manufacture of the Olympus 593 turbojet for the Concorde. A description of the engine appears under 'Rolls-Royce/SNECMA' in the International section.

SNECMA/TURBOMÉCA LARZAC

This joint design of a 13·23 kN (2,976 lb st) turbofan by SNECMA and Turboméca is being developed to power business jets, military liaison aircraft and trainers. A description of the Larzac is given under the entry for Turboméca-SNECMA GRTS, in this section.

SNPE
SOCIÉTÉ NATIONALE DES POUDRES ET EXPLOSIFS

HEADQUARTERS:
12 Quai Henri IV, 75181 Paris Cédex 04
Telephone: 277 15 70
Telex: 22356 Poudres Paris

USINE DE SAINT MÉDARD

St Médard-en-Jalles (Gironde)
Telephone: (56) 44 21 25

Established by Royal Decree in 1679, the former Poudrerie Nationale de St Médard was one of the largest establishments in the Service des Poudres. In October 1971 the SDP was taken over by the SNPE, a new national company responsible for all solid propellant charges from pistol ammunition to an ICBM.

St Médard has a payroll of almost 1,300 of whom 1,000 are skilled groups of civil and military engineers, working in more than 500 buildings dispersed in the pine forest north-west of Bordeaux.

Brief details of some current production motors are given in the table. Composite propellants include polybutadiene, polychlorates of vinyl or polyurethane, ammonium perchlorate (with additions such as dispersed aluminium) and numerous other composite propellants, often with 12-15 separate ingredients, as well as the more traditional double-base extruded propellants (type SD) derived from cordite and similar gun propellants which are made cheaply in sizes up to 200 mm (7·9 in) diameter.

SD motors are used mainly for such applications as anti-tank missiles and take-off boosters. The plastic/plastolane composites are cast in free blocks, whereas the isolite/isolane/butalite/butalane series are case-bonded permanently into the metal vehicle stage which is also sometimes fabricated at St Médard.

The establishment has test-fired a 255 kN (57,320 lb) motor of 2 m (78·8 in) diameter, and at the 1969 Paris Salon exhibited an inert segment weighing approximately 10,000 kg (22,000 lb) and of 3 m (118 in) diameter to demonstrate its capacity to make segments suitable for Titan IIIC.

Since 1959 the establishment has been manufacturing the rocket motor of the US Hawk missile under licence, and in turn has licensed the governments of India and Pakistan to make the plastolane motors used in Aérospatiale sounding rockets.

An original way of forming the charge for large motors has been developed which, instead of using a star-centred filling, uses a retractable boring tool (with an extraction duct for the swarf) to machine annular bleed slots around the cast charge. This technique, used on the P.4 and P.6 missile stages, reduces the time taken in fabricating the internal profile of the charge to two days, compared with 15 days for methods involving a retractable or fusible mandrel or former.

PRINCIPAL SNPE MOTORS

Motor	Application	Total Impulse kN-sec (lb-sec)	Duration (sec)	Diameter mm (in)	Charge weight kg (lb)
Isolane propellant					
P.16 (Type 902)	SSBS first stage	40,993 (9,215,350)	76	1,500 (59)	16,000 (35,275)
P.10 (Type 903)	SSBS second stage	—	—	1,500 (59)	10,000 (22,050)
P.10 (Type 904)	MSBS first stage	—	—	1,500 (59)	10,000 (22,050)
P.4 (Rita)	MSBS second stage	9,709 (2,182,580)	55	1,500 (59)	4,000 (8,820)
P.6					
Diamant	Diamant third stage	1,314 (295,420)	45	650 (25·5)	640 (1,410)
Soleil VE.111	Diamant second stage	4,609 (1,036,170)	36·7	800 (31·5)	2,260 (4,982)
Polka	Masurca boost	1,569 (352,740)	4·6	560 (22)	690 (1,521)
Jacée	Masurca cruise	618 (138,900)	26	400 (15·75)	320 (705)
Plastolite/plastolane					
Marie-Antoinette	Matra R. 530	78·46 (17,635)	9	203 (8)	42·5 (93·7)
Vénus	Malafon boost	167 (37,480)	2·9	275 (10·8)	92 (203)
Épervier	AS.30 boost	105 (23,600)	2	330 (13)	57 (126)
Mammouth	VE.110 test vehicle	3,609 (811,300)	18·2	800 (31·5)	1,910 (4,210)
Stromboli	Aérospatiale Dragon	1,422 (319,675)	16·5	550 (21·6)	685 (1,510)
SD					
ACRA	ACRA missile	—	—	—	—
Entac	Entac missile	—	18·7	62 (2·45)	1·47 (3·24)
Mk 43	2·75 in rocket	—	1·5	62 (2·45)	2·65 (5·84)
CT.20	CT.20 boost trolley	—	2	92 (3·62)	90 (198)

TURBOMÉCA
SOCIÉTÉ TURBOMÉCA

HEAD OFFICE AND WORKS:
Bordes, 64320 Bizanos
Telephone: (15-59) 32 84 37
Telex: 560928
OTHER WORKS:
Mézières S/Seine (Yvelines) and Tarnos (near Bayonne)
PARIS OFFICE:
1, Rue Beaujon, Paris 8e
PRESIDENT AND DIRECTOR-GENERAL:
J. R. Szydlowski

The Société Turboméca was originally formed in 1938 by M Szydlowski and M Planiol to develop blowers, compressors and turbines for aeronautical use.

Turboméca is the leading European manufacturer of small turbine aero-engines. Since it first started development of gas turbines in 1947, the company has developed about 50 different types of power plant of which some 15 have entered production and ten types have been manufactured under licence in five countries.

By 1 January 1976, 15,000 Turboméca engines for fixed and rotary-wing applications and aircraft auxiliary duties had been delivered to customers in 94 countries, including France. Approximately 12,000 more engines have been built under licence by what are today Rolls-Royce (1971) Ltd in the UK, Teledyne CAE in the US, ENMASA in Spain, Hindustan Aeronautics Ltd in India, Bet-Shemesh in Israel and state factories in Romania and Yugoslavia. Present production rate by Turboméca totals some 130 new and overhauled engines per month.

A new 12,077 m² (130,000 sq ft) extension to the company's factory at Tarnos was commissioned in September 1968, bringing the total covered floor area for Turboméca's three plants at Bordes, Mézières and Tarnos to 125,352 m² (1,347,000 sq ft). At 1 January 1976 the company employed a total of 4,668 people.

In addition Turboméca has a 51 per cent holding in Bet-Shemesh Engines, an aero-engine factory built in Israel, near Jerusalem, in conjunction with the Israeli government. The first section of the factory, based on the same layout as Turboméca's Tarnos plant, was officially opened in January 1969. Details of Bet-Shemesh Engines are given in the Israeli entry in this section.

A high degree of interchangeability exists among the range of Turboméca turbine engines. Most of them have been described in previous editions of *Jane's* and the entries which follow are concerned with only the more important current types.

Two important turbofans of part-Turboméca design and manufacture are the Adour (shared with Rolls-Royce) and the Larzac (jointly developed with SNECMA and produced by a consortium including MTU and KHD of West Germany).

ROLLS-ROYCE TURBOMÉCA RB.172/T260 ADOUR

This turbofan was developed jointly by Rolls-Royce and Turboméca for the SEPECAT Jaguar tactical strike fighter and advanced trainer. Other versions have been developed subsequently for the Hawker Siddeley Hawk and Mitsubishi T-2. A brief description of the Adour is given in the International section.

TURBOMÉCA ASTAFAN

The Astafan is a low-consumption lightweight turbofan of high by-pass ratio, low noise level design which made its first run during the Summer of 1969. Comprising an Astazou turboprop power section, operating at constant speed, driving a single-stage variable-pitch fan via reduction gearing, the Astafan is being developed in several versions corresponding to different development stages of the Astazou. All have the high by-pass ratio of 7 : 1 and are characterised by very low specific fuel consumption. Variable-pitch blading facilitates constant-speed operation and permits off-loading of the engine during starting.

The following are the first two versions under development:

Astafan III. Derived from the Astazou XVIII, with air-cooled turbine.

Astafan IV. Derived from the Astazou XX, with three-stage axial high-pressure compressor.

Flight development of prototype Astafan engines began in a Hawk Commander on 8 April 1971. Two podded underwing Astafans replace the original piston engines, conferring a substantial improvement in flight performance and a reduction in noise and vibration.

On 24 January 1976 an Aero Commander 690 flew at Pau for the first time powered by two Astafan IVF6 engines, similar to the basic Astafan IV but rated at 10·49 kN (2,359 lb) take-off thrust and with even lower noise and fuel consumption.

TYPE: Single-shaft turbofan with geared fan.

ENTRY CASING: Annular light alloy entry cowl and fan duct supported on double row of air straightener vanes downstream of fan rotor. Annular intake to gas generator section located at exit to fan duct. Rear casing of secondary intake carries accessories and accessory drives (using arrangement similar to Astazou XVIII intake).

FAN: Single-stage fan with variable-incidence rotor blading overhung at front without entry guide vanes. Drive from gas generator section is via two-stage epicyclic gear train housed in cylindrical casing forming inner wall of fan duct. Astafan IV has fan of increased diameter (700 mm; 27·5 in).

COMPRESSOR, COMBUSTION SYSTEM AND TURBINE: Same as Astazou XVIII. Normal gas-generator operating speed, 43,000 rpm. (Astafan IV has Astazou XX compressor, running at 42,000 rpm.)

JET PIPE: Fixed type with straight frustum inner cone. Extension jet pipe to convergent propulsive nozzle and ejector nozzle at rear of engine pod casing.

ACCESSORIES: Mounted on casing forming rear of secondary air intake.

FUEL SYSTEM: Independent control systems for starting and normal operation. Fuel regulator maintains speed constant with pilot operating single lever controlling fan blade pitch to vary thrust output. Turboméca "thermic" load limiter controls turbine entry temperature between set limits (using principle of operation similar to that on Astazou XVIII).

FUEL GRADES: AIR 3404A, 3405 or 3407A.

LUBRICATION SYSTEM: Pressure lubrication to bearings and reduction gear, with annular engine-mounted oil tank.

STARTING: Automatic electrical starting with compressor blow-off valve and fan in minimum pitch.

DIMENSIONS:
Length overall:
Astafan III — 2,030 mm (80·0 in)
Astafan IV — 2,218 mm (87·5 in)
Max diameter over fan cowl:
Astafan III — 665 mm (26·2 in)
Astafan IV — 780 mm (30·7 in)

WEIGHTS:
Astafan III bare engine — 210 kg (462 lb)
Equipped Astafan III — approx 230 kg (507 lb)
Astafan IV bare engine — 220 kg (485 lb)

PERFORMANCE RATINGS:
T-O, wet:
Astafan III — 8·34 kN (1,870 lb st)
Astafan IV — 12·06 kN (2,710 lb st)
T-O, dry:
Astafan III — 7·75 kN (1,740 lb st)
Astafan IV — 11·28 kN (2,530 lb st)

SPECIFIC FUEL CONSUMPTION:
At T-O rating:
Astafan III — 10·34 mg/Ns (0·365 lb/hr/lb st)
Astafan IV — 8·78 mg/Ns (0·310 lb/hr/lb st)
At max continuous rating:
Astafan III — 10·17 mg/Ns (0·359 lb/hr/lb st)
Astafan IV — 8·64 mg/Ns (0·305 lb/hr/lb st)

TURBOMÉCA-SNECMA LARZAC

This small turbofan has been developed jointly by Turboméca and SNECMA to power military trainers and other small aircraft. A description of the Larzac is given under Turboméca-SNECMA GRTS (which follows this entry).

TURBOMÉCA MARBORÉ

The Marboré turbojet is the most widely used of Turboméca's range of gas turbines. By the beginning of 1975, a total of 4,270 Marboré II engines of 3·92 kN (880 lb) st had been delivered by Turboméca and a further 10,000 by Continental Aviation and Teledyne CAE (see US section) as the J69. Production of the Marboré IID continues for the Aérospatiale CT.20 target drone.

This initial version of the engine was joined in production by the 4·71 kN (1,058 lb st) Marboré VI with receipt of type approval in June 1962. By the beginning of 1975, a total of 959 Marboré VI turbojets had been built and production continues under a large French government order awarded in December 1968. Four versions, each with differing accessory arrangements, have been delivered; the Marboré VIC for the Morane-Saulnier Paris II, the Marboré VID for the Aérospatiale M.20 drone, the Marboré VIF for the CM.170 Super Magister, and the Marboré VIJ for the Morane-Saulnier Paris IA. During 1968, the TBO for the Marboré VIF2 was increased to 1,000 hr.

The Marboré VI was also built under licence in Spain by ENMASA as the Marboré M21.

The following particulars relate to the Marboré VI series:

TYPE: Single-shaft centrifugal-flow turbojet.

AIR INTAKE: Annular sheet metal nose intake bolted to front of light alloy compressor casing.

COMPRESSOR: Single-sided impeller machined from two alloy forgings, shrunk on steel shaft and locked and dowelled to maintain alignment. Externally-finned light alloy compressor casing supports front ball-bearing for rotating assembly in a central housing supported by three streamlined struts. This housing also contains gears for accessory drives. Pressure ratio 3·84 : 1. Air mass flow 9·8 kg (21·6 lb)/sec.

COMBUSTION CHAMBER: Composed of inner and outer sheet metal casings, forming annular flame tube. Air from compressor passes through both radial and axial diffuser vanes and divides into three main flows, two primary for combustion and one secondary. Two primary flows enter combustion zone from opposite ends of chamber, the rear stream through turbine nozzle guide vanes which it cools. Secondary flow enters through outer casing for dilution and cooling of combustion gases. Two torch igniters.

FUEL SYSTEM: Fuel, pumped through hollow impeller shaft, is fed to combustion zone by rotating injector disc around periphery of which are number of vents which act as nozzles. Fuel is vented by centrifugal force, being atomised in the process. Fuel delivery at low thrust settings regulated by by-pass valve.

FUEL GRADE: AIR 3405 (JP-1).

NOZZLE GUIDE VANES: Twenty-five hollow sheet steel guide vanes cooled by part of primary combustion air.

TURBINE: Single-stage turbine with thirty-seven blades with fir-tree root fittings in steel disc. Bolted to main shaft and tail shaft, latter supported by rear roller bearing for rotating assembly. Gas temperature 613°C at 21,500 rpm.

JET PIPE: Inner and outer sheet metal casings, latter supported by three hollow struts. Inner tapered casing extends beyond end of outer casing to induce airflow through struts to cool rear main bearing and inner casing.

ACCESSORY DRIVES: Gear casing in central compressor housing with drives for fuel and oil pumps. Connecting shaft to underside of accessories gear case above compressor casing. Accessories include tachometer generator and electric starter. Take-off (4 hp continuous) for remotely-driven accessory box.

LUBRICATION SYSTEM: Pressure type. Single gear-type pump serves front gear casing, two main bearings and rpm governor. Three scavenge pumps return bearing oil to tank via cooler. Normal oil pressure 2·76 bars (40 lb/sq ft).

OIL SPECIFICATION: AIR 3512 (mineral) or AIR 3513A (synthetic).

MOUNTING: Four points, with Silentbloc rubber mountings, two at front and two at rear.

STARTING: Air Equipement 24V electric starter or compressed air starter. Two Turboméca igniter plugs.

DIMENSIONS:
Length with exhaust cone but without tailpipe — 1,416 mm (55·74 in)
Width — 593 mm (23·35 in)
Height — 631 mm (24·82 in)

WEIGHT, DRY:
Equipped — 140 kg (309 lb)

PERFORMANCE RATINGS:
T-O — 4·71 kN (1,058 lb st) at 21,500 rpm
Cruising — 4·21 kN (925 lb st) at 20,500 rpm

SPECIFIC FUEL CONSUMPTION:
At T-O rating — 30·87 mg/Ns (1·09 lb/hr/lb st)
At cruising rating — 30·31 mg/Ns (1·07 lb/hr/lb st)

TURBOMÉCA ARBIZON IIIB (TR 281)

Announced in 1970, the Arbizon IIIB is a simple single-shaft turbojet with minimum overall dimensions and weight. The main rotating assembly comprises an axial and a centrifugal compressor driven by a single-stage turbine. General design is similar to that of other Turboméca engines, with an annular combustion chamber supplied with centrifugally-injected fuel. At the front the axial intake opens out into a four-lobed bell-mouth to provide a large front face for the electric starter and other engine driven accessories.

Mass flow is 6 kg (13·2 lb)/sec, and pressure ratio 5·5. The Arbizon III has already been produced in small numbers (16 by 1 January 1975) to power the Otomat guided missile. Originally designated TR281, it was derived from the Turmo IIIC$_3$.

DIMENSIONS:
Diameter of front face — 410 mm (16·14 in)
Diameter of combustion chamber — 405 mm (15·95 in)
Overall length, with accessories — 1,361 mm (53·58 in)

WEIGHT, DRY: 115 kg (253 lb)

PERFORMANCE RATINGS:
T-O — 3·73 kN (836 lb st) at 33,000 rpm
Max continuous — 3·24 kN (727 lb st) at 32,000 rpm

Turboméca Astafan III geared variable-pitch turbofan on testbed, giving 8·34 kN (1,870 lb st)

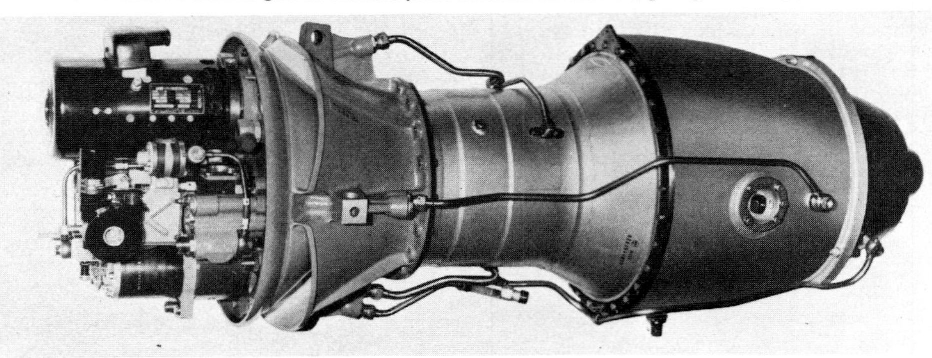

Turboméca Arbizon III expendable turbojet, rated at 3·73 kN (836 lb st)

ᴾECIFIC FUEL CONSUMPTION:
At T-O rating 31·87 mg/Ns (1·125 lb/hr/lb st)
At max continuous rating
31·44 mg/Ns (1·11 lb/hr/lb st)

TURBOMÉCA ARRIEL

This new turboshaft engine is intended initially for use two Aérospatiale helicopters, the single-engined AS 50 Ecureuil and twin-engined SA 365 Dauphin. It could so power a future version of the SA 341 Gazelle.

The Arriel is intended to have low first cost, low mainte- nce cost and low specific weight. It is characterised by odular construction, and is expected eventually to form e basis for a single-shaft turboprop and a turbofan in the 90 kN (1,100 lb) class. The gas generator ran in 1973. ne first complete engine ran on the bench on 7 August)74. Flight development began on 17 December 1974 in e SA 341-02 Gazelle, which had been converted for rriel development by Aérospatiale and the CGTM. The vin-engined SA 365 first flew on 24 January 1975. Pro- ction engine deliveries were scheduled for late 1976.

ʏPE: Single-shaft axial-plus-centrifugal free-turbine turboshaft.

ɪʀ INTAKE: Direct pitot entry to axial compressor.

ᴏMPRESSOR: Single-stage axial compressor, machined from titanium forging, cantilevered ahead of shaft running in two ball bearings and attached by axial lock to turbine shaft at centrifugal rotor. Supersonic centrifugal stage also machined from titanium, and connected to turbine shaft by central bolt, with drive by curvic coupling. Downstream are radial and then axial stators. High rotational speed for maximum attainable pressure ratio (9:1).

ᴏMBUSTION CHAMBER: Annular chamber, with flow radially outwards and then inwards. Centrifugal fuel injection without central tube.

ᴀS-GENERATOR TURBINE: Two integral cast axial stages with solid blades. Assembled by curvic couplings and central bolt. Shaft supported by axial-compressor bearings at front and roller bearing at rear. Turbine shield capable of disc containment.

ᴏWER TURBINE: Single axial stage with inserted blades. Cantilevered ahead of roller bearing; rear of shaft held in ball bearing, cages of both bearings being secured to exhaust diffuser arms.

ᴛ PIPE: Exhaust diffuser fabricated by welding, with central portion around output shaft and flared outer wall.

ᴇDUCTION GEAR: Light alloy gearbox, containing two stages of helical gears, giving drive at 6,000 rpm to output shaft extending whole length of engine, with drive connections to both front and rear. Hydraulic torquemeter.

ᴄCESSORY DRIVES: Two bevel gears and radial quill shaft drive accessory gearbox at front end, carried between compressor case and output shaft. Main pad provides for optional 12,000 rpm alternator; other drives for oil pumps, tachometer generator, governor and starter.

ᴜBRICATION SYSTEM: Independent circuit. Oil from tank passes through gear pump and metallic-cartridge filter. Return from engine via three gear scavenge pumps. Temperature probe and pressure switch to verify operation.

ʟ SPECIFICATION: AIR 3512 (mineral) or AIR 3513A (synthetic).

ᴏUNTING: Multi-point flanges allow easy mounting in single or twin installation.

ᴀRTING: Electric starter or starter/generator.

MENSIONS:

Length, excl accessories	1,090 mm (42·91 in)
Height overall	569 mm (22·40 in)
Width	430 mm (16·93 in)

ᴇIGHT, DRY:

With all engine accessories	90 kg (198 lb)

ʀFORMANCE RATINGS:

Max contingency, initial	508 kW (681 shp)
Max contingency, later	544 kW (730 shp)
Take-off and intermediate contingency	478 kW (641 shp)
Max continuous	441·5 kW (592 shp)

ᴇCIFIC FUEL CONSUMPTION:

Max contingency	93·1 μg/J (0·551 lb/hr/shp)
Intermediate continegency	96·8 μg/J (0·573 lb/hr/shp)

TURBOMÉCA ASTAZOU

The Astazou is the major turboprop in the Turboméca nge and is in production in its 636 kW (853 ehp) stazou XIVC and 760·6 kW (1,020 ehp) Astazou XVI rsions to power a number of different aircraft. These rsions are also marketed by Rolls-Royce Turboméca ternational Ltd under the designations AZ14 and Z16.

The Astazou XIV was certificated by the French airworiness authorities in October 1968, followed by RB/FAA certification of the Astazou XIVC and C1 in arch 1969. The Handley Page (now Scottish Aviation) tstream received ARB/FAA cerification during the following month. On 15 November 1968, a Pilatus Turbo- rter STOL aircraft powered by a 585 hp Astazou XIVE hieved a world altitude record for C-1-c class aircraft with a flight to 13,485 m (44,242 ft).

Turboméca Arriel free-turbine turboshaft, with initial ratings up to 508 kW (681 shp)

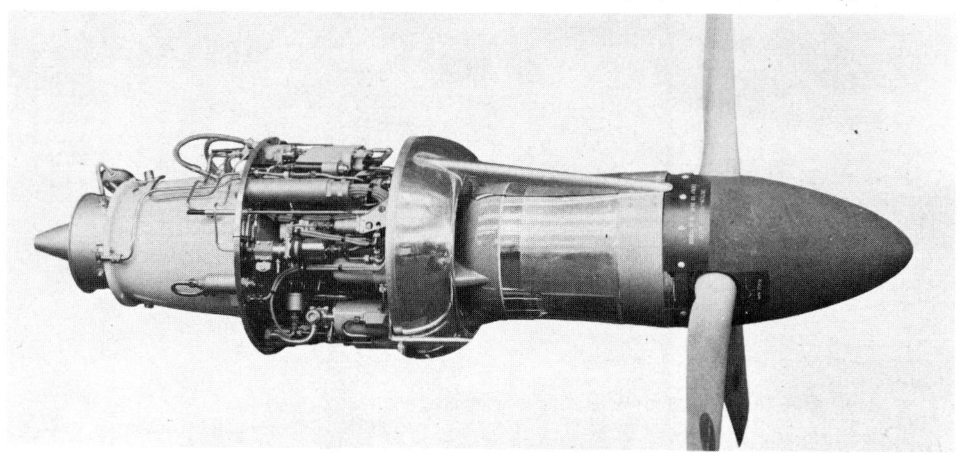

Turboméca Astazou XIVC turboprop engine, rated at 636 kW (853 ehp)

Current versions of the Astazou are:

Astazou XII. Powered Shorts Skyvan Srs 2 at 690 shp and Pilatus Turbo-Porter PC-6/A1-H2 at 700 ehp.

Astazou XIV (alias AZ14). Developed from Astazou XII. Powers early Jetstream business aircraft at 853 ehp.

Astazou XVI (alias AZ16). Higher rated version of Astazou XIV and first engine to enter production with new Turboméca aircooled turbine. Completed French official endurance tests in November 1968 at 800 kW (1,073 ehp). Fully flight tested by CGTM in modified Nord 260, following initial testing in a Morane-Saulnier MS 1500 Épervier. Further endurance testing carried out with distilled water injection to provide flat rating performance. The XVID, without starter/generator. powers the Jetstream; the XVIZ powers the Nord 260A. On offer as alternative power plant for IAI Arava STOL transport. The Astazou XVIG, equipped for sustained inverted flight, was certificated by the Services Officiels Français on 30 April 1971; it powers the Argentinian Pucará combat aircraft. By the end of 1974, 99 Astazou XVI engines had been built.

Astazou XVIII. Higher rated version of Astazou XVI which first ran in early 1969 with T-O rating of 860·5 kW (1,154 ehp). Potential application in Astazou XVI installations.

Astazou XX. Under development. This engine has two transonic axial compressor stages in titanium, in addition to the centrifugal stage machined in steel. Maximum T-O rating is 1,077·5 kW (1,445 ehp).

The following description relates to versions from Astazou XIVC to XX. Details of the Astazou series of turboshafts are given separately.

TYPE: Single-shaft axial-plus-centrifugal turboprop.

PROPELLER: Hamilton Standard three-blade single-acting, counterweight type with hydraulically-actuated variable pitch from full negative through to feather position. Emergency hydraulic feathering provided. Diameter to suit individual applications.

REDUCTION GEAR: Mounted in tapered cylindrical casing at front of engine, with two-stage epicyclic reduction gear having helical primary gears and straight secondary gears. Reduction ratio 24·115 : 1 (XVIG, 21·8 : 1). Driven from front of compressor and mounted on ball and roller bearings. (Astazou XX gearbox incorporates torquemeter.)

AIR INTAKE: Annular air intake at rear of reduction gear casing. Hot-air de-icing.

COMPRESSOR: Two-stage axial followed by single-stage centrifugal with single-sided impeller. Two rows of stator blades aft of each axial rotor. Centrifugal stage has radial and axial diffusers. Axial stages have steel discs with integral steel blades. Discs and two-piece steel impeller located on compressor shaft by radial lugs. Shaft carried on ball bearings ahead of stage-one axial disc and ahead of impeller. (Astazou XX has a four-stage compressor, comprising three axial followed by one centrifugal.)

COMBUSTION SYSTEM: Reverse-flow annular type with centrifugal fuel injector using rotary atomiser disc. Ignition by two ventilated torch igniters.

TURBINE: Three-stage axial with blades integral with discs. (Air cooling provided for Astazou XVI, XVIII and XX.) Discs attached by curvic couplings and through-bolts. Rotor carried on compressor rear ball bearing and roller bearing aft of stage-three disc supported by struts across turbine exhaust.

JET PIPE: Fixed type with curved inner cone.

ACCESSORIES: Mounted on casing forming rear of air intake. Drive pads provided for starter/generator, oil pump, fuel pump and speed governor, tacho-generator, AC generator (optional) and hydraulic pump (optional).

MOUNTING: Trunnion located on each side of turbine casing front flange, plus third trunnion on underside of turbine casing.

FUEL SYSTEM: Automatic constant-speed system with propeller Beta-control and Turboméca "thermic" load limiter and speed governor.

FUEL GRADES: AIR 3404, 3405, or 3407.

LUBRICATION SYSTEM: Pressure lubrication to bearings and reduction gear, with 8 litre (14 pint) oil tank mounted at front of engine.

OIL SPECIFICATION: AIR 3515 or synthetic AIR 3573.

STARTING: Electric.

DIMENSIONS:

Diameter over intake cowl	546 mm (21·5 in)
Overall length, incl propeller	2,047 mm (80·6 in)

WEIGHTS:

With accessories:

Astazou XIV	approx 206 kg (454 lb)
Astazou XVID	205 kg (452 lb)
Astazou XVIG	228 kg (502 lb)
Astazou XVIZ	213 kg (468 lb)
Astazou XVIII	approx 205 kg (452 lb)
Astazou XX	220 kg (484 lb)

PERFORMANCE RATINGS:

T-O:

Astazou XIV 636 ekW; 596·5 kW (853 ehp; 800 shp) at 43,000 rpm
Astazou XVID 722·5 ekW; 681 kW (969 ehp; 913 shp) at 43,089 rpm
Astazou XVIG, XVIZ 760·5 ekW; 720 kW (1,020 ehp; 965 shp) at 43,000 rpm
Astazou XVIII 860·5 ekW; 813 kW (1,154 ehp; 1,090 shp) at 43,000 rpm
Astazou XX 1,075 ekW; 1,030 kW (1,442 ehp; 1,381 shp) at 42,000 rpm

Max continuous:

Astazou XIV 574 ekW; 537 kW (770 ehp; 720 shp) at 43,000 rpm
Astazou XVID 626·5 ekW; 586 kW (840 ehp; 786 shp) at 43,089 rpm
Astazou XVIG, XVIZ 696·5 ekW; 654 kW (934 ehp; 877 shp) at 43,000 rpm
Astazou XVIII 809 ekW; 768 kW (1,085 ehp; 1,030 shp) at 43,000 rpm
Astazou XX 951·5 ekW; 907·5 kW (1,276 ehp; 1,217 shp) at 42,000 rpm

SPECIFIC FUEL CONSUMPTION:

At T-O rating:

Astazou XIV 92·4 μg/J (0·547 lb/hr/shp)
Astazou XVI (all versions) 88·7 μg/J (0·525 lb/hr/shp)
Astazou XVIII 85·3 μg/J (0·505 lb/hr/ehp)
Astazou XX 75·9 μg/J (0·449 lb/hr/ehp)

At max continuous rating:

Astazou XIV 93·6 μg/J (0·554 lb/hr/shp)
Astazou XVI (all versions) 90·2 μg/J (0·534 lb/hr/shp)
Astazou XVIII 84·5 μg/J (0·50 lb/hr/ehp)
Astazou XX 78 μg/J (0·462 lb/hr/ehp)

TURBOMÉCA ASTAZOU TURBOSHAFT

This turboshaft series of the Astazou family is derived from the early second-generation Astazou II turboprop fitted to the Mitsubishi MU-2 and Pilatus Turbo-Porter. Variants are as follows:

Astazou IIA. Rated at 390 kW (523 shp) and powers the Aérospatiale SA 318C Alouette II Astazou helicopter.

Astazou IIN. Original version of turboshaft, powering the prototype Aérospatiale/Westland SA 341.

Astazou IIIN. Definitive turboshaft for Anglo-French helicopter programme for production SA 341 Gazelle. Derived from Astazou IIA but with revised profile of turbine, using higher temperature alloy to match power needs of SA 341. Produced jointly by Turboméca and Rolls-Royce (1971) Ltd.

Astazou XIVB and XIVF. In production for the SA 319B Alouette III; XIVB is civil and XIVF military. Flat rated to 441 kW (591 shp) (one hour) up to 4,000 m (13,125 ft) or +55°C.

Astazou XIVH. In production for SA 341 Gazelle, with much increased power. Flat-rated to transmission limit, to remove all altitude and temperature limitations. Certificated October 1974.

Astazou XVIIIA. Further increase in power gained by improved turbine, allowing higher gas temperature. Powers SA 360 Dauphin.

The following description relates to the Astazou IIIN except where indicated:

TYPE: Single-shaft axial-plus-centrifugal turboshaft.

REDUCTION GEAR: Similar to Astazou XIV turboprop. Reduction ratio 7·039 : 1 (Astazou XIVB/F, 7·345; XVIA, XVIIIA, 7·375).

AIR INTAKE: Annular air intake at rear of reduction gear casing.

COMPRESSOR: Single-stage axial (IIA, IIN, IIIN) or two-stage axial (all subsequent versions) followed by single-stage centrifugal with single-sided impeller. Two rows of stator blades aft of axial rotor. Otherwise similar to Astazou XIV compressor. Air mass flow 2·5 kg/sec (5·5 lb/sec).

COMBUSTION SYSTEM: Similar to combustor on Astazou XIV.

TURBINE: Similar to Astazou XIV.

JET PIPE: Similar to Astazou XIV.

ACCESSORIES: Mounted on casing forming rear of air intake. Drive pads provided for starter/generator, oil pump, fuel pump and governor, tacho-generator, AC generator (optional).

MOUNTING: At front by flange located at power take-off section, and at rear by two lugs on accessory mounting pad section.

FUEL SYSTEM: Automatic constant speed control with speed governor.

LUBRICATION SYSTEM: Pressure type with gear type pumps. Oil tank of 8 litre (14 pint) capacity mounted at front of engine.

STARTING: Electrical, automatic.

DIMENSIONS:

Height: Astazou IIA 458 mm (18 in)
Astazou IIIN, XIVH 460 mm (18·1 in)
Astazou XVIA, XVIIIA 698 mm (27·48 in)
Width: Astazou IIA 480 mm (18·8 in)
Astazou IIIN, XIVH 460 mm (18·1 in)

Turboméca Astazou XVIIIA turboshaft engine, flat rated at 651 kW (873 shp)

The 780 kW (1,046 shp) Turboméca Bastan VII single-shaft turboprop powering the Frégate

Length overall: Astazou IIA 1,272 mm (50·0 in)
Astazou IIIN, XIVB/F 1,433 mm (56·3 in)
Astazou XVIA, XVIIIA 1,327 mm (52·2 in)
Astazou XIVH 1,470 mm (57·9 in)

WEIGHTS:

Bare engine: Astazou IIA 113 kg (249 lb)
Astazou IIIN 115 kg (253 lb)
Equipped: Astazou XIVB/F 166 kg (366 lb)
Astazou XVIA, XVIIIA 155 kg (341 lb)
Astazou XIVH 160 kg (353 lb)

PERFORMANCE RATINGS:

Max power: Astazou IIA 390 kW (523 shp)
Astazou IIIN 441·5 kW (592 shp)
One hour: Astazou XIVB/F 440·7 kW (591 shp)
Astazou XVIA 651 kW (873 shp) maintained at sea level to 30°C
Astazou XVIIIA 651 kW (873 shp) maintained at sea level to 40°C
Max continuous: Astazou IIA 352·7 kW (473 shp)
Astazou IIIN 390 kW (523 shp)
Astazou XIVB/F 405 kW (543 shp)
Astazou XVIA, XVIIIA 600 kW (805 shp)
Astazou XIVH thermodynamic power 640 kW (858 shp) flat-rated in SA 341 to 440·7 kW (591 shp) to 55°C or 4,000 m (13,125 ft)

SPECIFIC FUEL CONSUMPTION:

At max power rating:

Astazou IIA 105·3 μg/J (0·623 lb/hr/shp)
Astazou IIIN 108·7 μg/J (0·643 lb/hr/shp)
Astazou XIVB/F 105·5 μg/J (0·624 lb/hr/shp)
Astazou XVIA 93·8 μg/J (0·555 lb/hr/shp)
Astazou XVIIIA 91·3 μg/J (0·540 lb/hr/shp)

At max continuous rating:

Astazou IIA 107·1 μg/J (0·634 lb/hr/shp)
Astazou IIIN 111·4 μg/J (0·659 lb/hr/shp)
Astazou XIVB/F 109·5 μg/J (0·648 lb/hr/shp)
Astazou XVIA 95·8 μg/J (0·567 lb/hr/shp)
Astazou XVIIIA 92·1 μg/J (0·545 lb/hr/shp)
Astazou XIVH 89·6 μg/J (0·53 lb/hr/shp)

TURBOMÉCA BASTAN

The Bastan turboprop is one of the second generation of Turboméca engines, characterised by their two-stage axial-centrifugal compressor. The Bastan VIC rated at 786·7 kW (1,055 ehp) powers the Aérospatiale N 262 and was certificated by the Services Officiels Français and the FAA in 1964. The 1,000 ehp Bastan VID powers the Argentinian GII.

A second version, the Bastan VII, flat rated at 780 kW (1,046 shp), is derived from the Bastan VI and powers the Aérospatiale Frégate. The Bastan VII was certificated b the Services Officiels Français on 3 August 1970.

The following description relates to both the Bastan V and VII, except for the differences indicated:

TYPE: Single-shaft axial-plus-centrifugal turboprop.

REDUCTION GEAR: Two-stage epicyclic type, insid tapered cylindrical casing at front of engine. Ratio 1 21·0957. Propeller shaft carried in ball bearing at front.

AIR INTAKE: Annular intake at rear of reduction gea casing. Outer wall of intake, of triangular cross-sectior provides mounting for accessories. Front ball bearin for compressor shaft carried by air intake assembly.

COMPRESSOR CASING: Central portion carries rear ba bearing for compressor shaft.

COMPRESSOR: Single axial stage for Bastan VIC, and tw axial stages for Bastan VII, followed by single cen trifugal stage. Two rows of diffuser vanes between axi stages and two more aft of the centrifugal stage, of whic the first is radial and the second axial. On Bastan V first axial rotor blades are titanium and pin-mounted i disc, and second axial rotor blades are light allo integral with disc. Bastan VIC pressure ratio 5·83 : and air mass flow 4·5 kg (10 lb)/sec. Bastan VII pressur ratio 6·68 : 1 and mass flow 5·9 kg (13·1 lb)/se Water-methanol injection in Bastan VIC.

COMBUSTION CHAMBER: Direct-flow annular type. Usu Turboméca rotary atomiser fuel injection system. Tw torch igniters. Gas temperature before turbine 870°C

TURBINE CASING: Houses combustion chamber and tu bine nozzle assembly. Supports engine rear roller bear ing at rear end.

TURBINE: Three-stage axial-flow turbine with separat discs. Each turbine preceded by axial-flow nozzle guid vane assembly.

JET PIPE: Annular welded sheet assembly comprisin cylindrical outer casing and central bullet fairing.

ACCESSORY DRIVE: Upper pinion train drives dynamo star ter, propeller governor and fuel pump with fuel mete ing device. Lower gear drives electric tachomete transmitter, fuel pump, 20kVA alternator and landin gear pump. All accessories mounted on intake casin

MOUNTING: Three attachment points, two lateral, one bottom of engine.

ENGINE CONTROL: By two governors. One adjusts fuel flo entering engine so that it is maintained at the value se by the power control lever, as a function of the varia tions of pressure and temperature at the engine a intake. The second governor maintains the propelle rpm at the value set by the rpm control lever, by varyin propeller pitch.

STARTING: Automatic starter/generator on Bastan VII.

DIMENSIONS:

Height: Bastan VIC 775·5 mm (30·53 in)
Width: Bastan VIC 685 mm (26·97 in)
Diameter: Bastan VII 550 mm (21·7 in)
Length: Bastan VIC 1,548·6 mm (60·95 in)
Bastan VII 1,911 mm (75·2 in)

WEIGHT, DRY:

Fully equipped:
Bastan VIC 322 kg (710 lb)
Bastan VII approx 370 kg (816 lb)

PERFORMANCE RATINGS (S/L, ISA):

Bastan VIC:
T-O and max continuous 700·4 ekW; 595 kW (1,060 ehp; 798 shp) at 33,500 rpm

Bastan VII
T-O 1,089 ekW; 780 kW (1,460 ehp; 1,046 shp) at 32,000 rpm maintained up to 40°C or to 3,650 m (11,975 ft)
Max continuous 1,048 ekW; 780 kW (1,405 ehp; 1,046 shp) at 32,000 rpm maintained up to 36°C

SPECIFIC FUEL CONSUMPTION:

Bastan VIC, T-O 98·4 μg/J (0·582 lb/hr/ehp)
Bastan VII, T-O 88·7 μg/J (0·525 lb/hr/ehp)

TURBOMÉCA ARTOUSTE III

The Artouste IIIB is a single-shaft turboshaft derived from the Artouste II. It is a member of the second generation of Turboméca engines with two-stage axial-centrifugal compressor and three-stage turbine. The Artouste IIIB has a pressure ratio of 5·2 : 1. Air mass flow is 4·3 kg/sec (9·5 lb/sec) at 33,300 rpm.

Type approval at the rating given below was received on 25 May 1961, following completion of a 150-hour official type test. Production at Turboméca continues. In addition, Artouste IIIBs are being built under licence in India by Hindustan Aeronautics Ltd.

The Artouste IIIB, which powers the Aérospatiale SA 316B Alouette III, obtained FAA certification in March 1962 and in August 1968 similar certification of the Artouste IIC1, C2, C5 and C6, powering the SE 3130 and 313B Alouette II Artouste, was also obtained.

An uprated version, the Artouste IIID, was certificated on 30 April 1971. It differs in having a reduction gear giving 5,864 rpm at the driveshaft (instead of 5,773 rpm) and in slightly revised equipment. The IIID powers a late version of the Alouette III.

DIMENSIONS:

Length 1,815 mm (71·46 in)
Height 627 mm (24·68 in)
Width 507 mm (19·96 in)

WEIGHT, DRY:

Fully equipped:
Artouste IIIB 182 kg (400 lb)
Artouste IIID 178 kg (392 lb)

PERFORMANCE RATINGS (maintained up to 55°C at S/L or up to approximately 4,000 m; 13,150 ft):

T-O:
Artouste IIIB 420 kW (563 shp)
Artouste IIID 440 kW (590 shp)
Max continuous (both) 405 kW (543 shp)

SPECIFIC FUEL CONSUMPTION:

T-O:
Artouste IIIB 128·8 μg/J (0·762 lb/hr/shp)
Artouste IIID 126·2 μg/J (0·747 lb/hr/shp)
Max continuous (both) 130·1 μg/J (0·77 lb/hr/shp)

TURBOMÉCA TURMO

The Turmo is a free-turbine engine which is in service in both turboshaft and turboprop versions. Each has a gas generator section comprising a single-stage axial plus single-stage centrifugal compressor, annular combustor, and two-stage turbine. The power turbine and transmission system vary according to the engine series.

The main variants of the Turmo are as follows:

Turmo IIIC$_3$. This was the original power plant of the triple-engined SA 321 Super Frelon helicopter. Maximum contingency rating is 1,104 kW (1,480 shp).

Turmo IIIC$_4$. Developed from Turmo IIIC$_3$ and with a maximum contingency rating of 1,032 kW (1,384 shp), this all-weather version is manufactured jointly by Turboméca and Rolls-Royce to power SA 330 Puma twin-engined helicopters under the Franco-British helicopter agreement of October 1967. Certificated by the Services Officiels Français on 9 October 1970.

Turmo IIIC$_5$, IIIC$_6$, IIIC$_7$. Similar to Turmo IIIC$_3$ but with different ratings. The SA 321F and 321J Super Frelons powered by these engines obtained French certification in June 1968.

Turmo IIID. Turboprop version, similar in basic construction to Turmo IIIC series but with output speed limited to 6,000 rpm. Gas generator section is mounted beneath output shaft which is driven by free-turbine. The overhung forward drive leads through a freewheel and dog clutch to the propeller reduction gearbox, which gives a final drive at 1,200 rpm. A drive pad at the rear of the primary gearbox enables the engines of a multi-engined aircraft to be coupled together by spanwise shafting, as employed with the four Turmo IIID$_6$ turboprops powering the Breguet Br 941S STOL transport.

The 420 kW (563 shp) Turboméca Artouste IIIB single-shaft helicopter turboshaft

The 1,032 kW (1,384 shp) Turboméca Turmo IIIC$_4$ turboshaft which powers the SA 330 Puma helicopter

Turmo IIIE$_3$. Similar to Turmo IIIC$_3$ but with different ratings. In production for SA 321 Super Frelon.

Turmo IIIE$_6$. Similar except for material of gas-generator turbine, which is improved to allow higher gas temperatures.

Turmo IV. The Turmo IVA is a civil engine derived from the IIIC$_4$, with a maximum contingency rating of 1,057 kW (1,417 shp). The IVB is a military version having the same ratings as the IIIC$_4$.

Turmo XII. Version tailored to rail-traction use.

The following description applies generally to the Turmo IIIC$_3$, C$_4$, C$_5$ and E, except where indicated:

TYPE: Free-turbine axial-plus-centrifugal turboshaft.

REDUCTION GEAR: Turmo IIIC$_3$, C$_5$ and E$_3$ fitted with rear-mounted reduction gear mounted in bifurcated exhaust duct with rear-facing power take-off shaft. Output shaft from free-turbine drives into high-speed gear of simple helical spur train of 3·53 : 1 reduction ratio. Output shaft also drives reduction gear driving oil cooler fan mounted on front of main reduction gear case. Turmo IIIC$_4$ is a direct-drive engine.

AIR INTAKE: Annular forward-facing intake, with de-icing in Turmo IIIC$_4$ and C$_5$. Centre housing contains forward ball bearing for compressor shaft and bevel gear drive to accessories mounted above and below intake casing.

COMPRESSOR: Single-stage axial followed by single-stage centrifugal with single-sided impeller. Two rows of light alloy stator blades aft of axial stage. Centrifugal stage has steel radial and axial diffusers; impeller located by lugs on turbine shaft. Axial rotor blades, titanium in Turmo IIIC$_3$, C$_5$ and E$_3$ and steel in Turmo IIIC$_4$, pin-mounted in steel disc with integral shaft. Pressure ratio 5·9 : 1 on Turmo IIIC$_3$. Air mass flow 5·9 kg (13 lb)/sec. Axial rotor carried on ball bearing ahead of disc and roller bearing aft of disc. Also, ball bearing ahead of impeller.

COMBUSTION SYSTEM: Reverse-flow annular type with centrifugal fuel injector using rotary atomiser disc. Ignition by two ventilated torch igniters.

GAS GENERATOR TURBINE: Two-stage axial unit with integral rotor blades. Discs with curvic couplings through-bolted to compressor shaft. Carried on roller bearing at rear of second-stage disc.

POWER TURBINE: Two-stage axial unit in Turmo IIIC$_3$, C$_5$ and E$_3$, and single-stage in Turmo IIIC$_4$. Blades carried in discs by fir-tree roots. Rotor overhung from rear on through-bolted output shaft. Shaft carried on roller bearing at front (at rear of turbine disc) and ball bearing at rear (at input to reduction gear). In all advanced production engines of IIIC$_4$ derivation the power turbine speed is 22,840 rpm under all high-power conditions.

JET PIPE: Fixed type with lateral bifurcated exhaust duct in Turmo IIIC$_3$, C$_5$ and E$_3$, and single lateral duct on Turmo IIIC$_4$.

ACCESSORIES: Mounted above and below intake casing with drive pads for oil pump, fuel control unit, electric starter, tacho-generator and, on Turmo IIIC$_4$, oil cooler fan. Control unit remote drive also provided on Turmo IIIC$_4$ from bevel gear drive on power turbine output shaft.

MOUNTING: Two lateral supports fitted to lower part of turbine casing at rear flange output shaft protection tube. On Turmo IIIC$_4$, also on reduction gear case.

FUEL SYSTEM: Fuel control unit for gas generator on Turmo IIIC$_3$, C$_5$ and E$_3$, with speed limiter for power turbine also fitted on E$_3$. Constant-speed system fitted on Turmo IIIC$_4$ power turbine, with speed limiter also fitted on gas generator.

FUEL GRADE: AIR 3405 for Turmo IIIC$_4$.

LUBRICATION SYSTEM: Pressure type with oil cooler and 13 litre (23 Imp pint) tank at front of engine on Turmo IIIC$_4$, with oil tank only around intake casing on Turmo IIIC$_3$, C$_5$ and E$_3$, and by intake accessory drive gear on Turmo IIIC$_4$.

OIL SPECIFICATION: AIR 3155A, or synthetic AIR 3513, for Turmo IIIC$_4$.

STARTING: Automatic system with electric starter motor.

DIMENSIONS:

Height:
Turmo IIIC$_3$, C$_5$ and E$_3$ 716·5 mm (28·2 in)
Turmo IIIC$_4$ 719 mm (28·3 in)
Turmo IIID$_3$ 926 mm (36·5 in)

Width:
Turmo IIIC$_3$, C$_5$ and E$_3$ 693 mm (27·3 in)
Turmo IIIC$_4$ 637 mm (25·1 in)
Turmo IIID$_3$ 934 mm (36·8 in)

Length:
Turmo IIIC$_3$, C$_5$ and E$_3$ 1,975·7 mm (78·0 in)
Turmo IIIC$_4$ 2,184 mm (85·5 in)
Turmo IIID$_3$ 1,868 mm (73·6 in)

WEIGHT, DRY:

Turmo IIIC$_3$ and E$_3$, fully equipped 297 kg (655 lb)
Turmo IIIC$_5$, III$_6$ and IIIC$_7$ 325 kg (716 lb)
Turmo IIIC$_4$, equipped engine 225 kg (496 lb)
Turmo IIID$_3$, basic engine 365 kg (805 lb)

PERFORMANCE RATINGS:

T-O: Turmo IIIC$_3$, D$_3$ and E$_3$ 1,104 kW (1,480 shp)
Turmo IIIE$_6$ 1,181 kW (1,584 shp)

Max contingency:
Turmo IIIC$_4$ at 33,800 gas-generator rpm 1,032 kW (1,384 shp)
Turmo IIIC$_6$ at 33,550 gas-generator rpm 1,156 kW (1,550 shp)
Turmo IIIC$_7$ at 33,800 gas-generator rpm 1,201 kW (1,610 shp)
Turmo IVA at 33,950 gas-generator rpm 1,057 kW (1,417 shp)
Turmo IVC at 33,800 gas-generator rpm 1,163 kW (1,560 shp)

T-O and intermediate contingency:
Turmo IIIC$_4$ at 33,450 gas-generator rpm 978 kW (1,312 shp)
Turmo IIIC$_5$ 1,050 kW (1,408 shp)

Max continuous:
Turmo IIIC$_3$ and E$_3$ 956 kW (1,282 shp)

Turmo IIIC4 872·5 kW (1,170 shp)
Turmo IIIC5, C6, C7 951 kW (1,275 shp)
Turmo IIID3 956 kW (1,282 shp)
Turmo IIIE6 951·5 kW (1,276 shp)
Turmo IVA at 32,800 gas-generator rpm 872·5 kW (1,170 shp)
Turmo IVC at 32,400 gas-generator rpm 939·5 kW (1,260 shp)

SPECIFIC FUEL CONSUMPTION:
At T-O rating:
Turmo IIIC3 and E3 101·9 μg/J (0·603 lb/hr/shp)
Turmo IIID3 104·1 μg/J (0·616 lb/hr/shp)
At max contingency rating:
Turmo IIIC4, C5, C6, C7 and IV 106·8 μg/J (0·632 lb/hr/shp)
Turmo IVA 106·3 μg/J (0·629 lb/hr/shp)
At T-O and intermediate contingency rating:
Turmo IIIC4 108·1 μg/J (0·640 lb/hr/shp)
At max continuous rating:
Turmo IIIC3 and E4 (108·1 μg/J (0·640 lb/hr/shp)
Turmo IIIC4, C5, C6, C7 and IVC 110·9 μg/J (0·656 lb/hr/shp)
Turmo IIID3 108·1 μg/J (0·640 lb/hr/shp)

TURBOMÉCA MAKILA

This new turboshaft engine, rated at an initial 1,323·5 kW (1,775 shp) for take-off and intermediate contingency, is under development to power the Aérospatiale SA 331 Super Puma helicopter. Derived partly from the Turmo family, it incorporates all the latest features of the company's advanced engines, including: rapid-strip modular construction; three axial stages of compression plus one centrifugal; later fuel inlet to centrifugal atomiser; two-stage gas-generator turbine (probably with cooled blades); two-stage free power turbine; and lateral exhaust. A provisional section drawing appeared in the 1975-76 edition of *Jane's*.

During 1974 this engine was confirmed as partner to the Arriel in laying the foundation for the company's marketing in the next 15 years. The world market for this size of engine is put at 10,000 units. The first engine was delivered for bench test in 1976, with production deliveries scheduled to start within three years.

The following data are provisional:

DIMENSIONS:
Length, intake face to rear face 1,395 mm (54·94 in)
Width 530 mm (20·9 in)
Height 514 mm (20·25 in)
WEIGHT, DRY: 210 kg (463 lb)
PERFORMANCE RATINGS (ISA, S/L):
Max contingency
1,424 kW (1,910 shp) at 36,300 gas-generator rpm
T-O and intermediate
1,323·5 kW (1,775 shp) at 35,500 gas-generator rpm
Max continuous
1,215·5 kW (1,630 shp) at 34,750 gas-generator rpm
SPECIFIC FUEL CONSUMPTION:
Max contingency 80·4 μg/J (0·476 lb/hr/shp)
T-O and intermediate 80·8 μg/J (0·478 lb/hr/shp)
Max continuous 82·3 μg/J (0·487 lb/hr/shp)

Turboméca Makila free-turbine turboshaft, with initial ratings up to 1,425 kW (1,910 shp). This photograph depicts a mock-up; a longitudinal section drawing appeared in the 1975-76 *Jane's*

The Turboméca Turmastazou XIV free-turbine turboshaft, rated at 663 kW (889 shp)

TURBOMÉCA TURMASTAZOU

The Turmastazou turboshaft engine comprises the Astazou single-seat turboprop with the addition of a free-turbine (Turmo is an abbreviation for "turbine motoriste" and by implication refers to a free-turbine engine). Development is proceeding on the five engines built by March 1969. The Turmastazou is intended for helicopter applications and is also in service with the Orléans-type Bertin Aérotrain.

TYPE: Free-turbine axial-plus-centrifugal turboshaft.

POWER DRIVE: Direct at rear of engine. No reduction gear fitted.

AIR INTAKE: Annular forward-facing intake at front of engine, feeding direct to compressor inlet.

COMPRESSOR AND COMBUSTION SYSTEM: Similar to Astazou.

COMPRESSOR TURBINE: Two-stage axial unit with rotor blades integral with turbine discs. Discs through-bolted with curvic couplings.

POWER TURBINE: Two-stage axial unit with rotor blades integral with turbine discs. Discs through-bolted with curvic couplings. Constant output speed 29,000 rpm.

JET PIPE: None fitted as standard.

ACCESSORIES: Mounted on compressor casing behind oil tank. Drive pads fitted for oil pump, tacho-generator, fuel control unit, starter/generator, AC generator and hydraulic pump (optional).

FUEL SYSTEM: Automatic control system with constant speed control of free-turbine.

LUBRICATION SYSTEM: Pressure type system with gear pump. Oil tank mounted around front of engine.

STARTING: Automatic electrical starting.

DIMENSIONS:
Height 552 mm (21·7 in)
Width 440 mm (17·3 in)
Length overall 1,332 mm (52·4 in)
WEIGHT:
Complete with accessories 160 kg (352 lb)
PERFORMANCE RATINGS:
Turmastazou XIV:
T-O 663 kW (889 shp)
Max continuous 590·5 kW (792 shp)
Turmastazou XVI:
T-O 735 kW (986 shp)
Max continuous 684·5 kW (918 shp)
SPECIFIC FUEL CONSUMPTION:
Turmastazou XIV and XVI:
T-O 86·9 μg/J (0·507 lb/hr/shp)
Max continuous 85·7 μg/J (0·514 lb/hr/shp)

TURBOMÉCA DOUBLE TURMASTAZOU

The Double Turmastazou free-turbine coupled turboshaft comprises two Turmastazou turboshafts coupled by a combining gearbox to drive a common output shaft. The engine is intended for twin-engined helicopter installations. It is specified for a new helicopter by Agusta, to which company an engine has been delivered.

The Double Turmastazou XIV comprises two Turmastazou XIV engines and is under development at 1,775 shp. A higher-powered model, the 2,071 shp Double Turmastazou XVI, is derived from the Astazou XVI. In both cases the usual output shaft speed is 6,600 rpm.

TURBOMÉCA-SNECMA

GROUPEMENT TURBOMÉCA-SNECMA (GRTS)

(Groupement d'Interêt Economique under the law of 23 September 1967)

OFFICES:
1 Rue Beaujon, BP 37-08, 75362 Paris Cedex 08
Telephone: 924-18-61

ADMINISTRATORS:
R. Florentini
E. Delfour

MANAGEMENT CONTROL COMMITTEE:
R. Martin
L. Henrion

FINANCIAL COMMISSARY:
C. Hirt

Announced in March 1969, Groupement Turboméca-SNECMA is a company formed jointly by Société Turboméca and SNECMA to be responsible for the design, development, manufacture, sales and service support of the Larzac all-axial small turbofan launched in 1968 as a joint venture by the two companies. Groupement Turboméca-SNECMA has no capital at present and primarily comprises a joint management organisation to produce the new engine.

TURBOMÉCA-SNECMA LARZAC

Originally this small turbofan was planned for a wide range of applications, and the first prototype was a 9·8 kN (2,200 lb) engine aimed at the commercial market. This type of engine ran in May 1969 and began flight development in a pod carried by a Constellation in March 1971. By this time the main immediate market had shifted to military trainers, and GRTS designed the Larzac 04 for this purpose.

In February 1972 the Larzac 04 was selected for a joint Franco-German programme to provide propulsion for the Alpha Jet trainer (see International entry in "Aircraft" section). In addition to the two French partners in GRTS, two German companies, MTU and KHD, were added to the programme. Both have played a part in the manufacture of prototype engines and the achievement of endurance tests. All four companies are sharing in production, and complete engines will be assembled in both countries for the Alpha Jet programme.

Following the adoption of the Alpha Jet by Belgium, the Belgian engine company FN is also expected to participate in the Larzac 04 programme. In January 1976 420 engines were ordered for Alpha Jets for France and Germany, with first delivery scheduled in January 1978. The work split, fixed by intergovernmental agreement and achieved in practice, is France 56·5 per cent, Germany 43·5.

Bench testing of the Larzac 04 began in May 1972. Flight development with the Constellation testbed began in March 1973, and with a Falcon 10 in July 1973. The first Alpha Jet flew on 26 October 1973, and qualification of the Larzac 04 was accomplished on schedule in May 1975. The rating given is at a turbine entry temperature of 1,130°C; growth thrust potential greater than 15 per cent is forecast without dimensional change.

GRTS states that the Larzac 04 is the subject of several evaluation requests from French and foreign aircraft manufacturers for military training and liaison aircraft and civil executive jet aircraft. It is claimed that the engine can meet all these uses without significant modification.

In September 1972 the French Services Officiels approved an agreement between GRTS and Teledyne CAE covering the production, marketing and after-sales support of the Larzac in the United States and Canada. As described in the Teledyne CAE entry, the US engine will be designated Model 490-04.

The Larzac 04 has a two-stage fan, four-stage HP compressor, annular combustion chamber with vaporising burners, single-stage HP turbine with cooled blades and single-stage LP turbine. Maximum airflow is 26·6 kg (60·8 lb)/sec, pressure ratio 10·6 and by-pass ratio 1·13. A single fixed-area jet pipe is used. All accessories are driven by the HP spool and grouped under the fan case. The engine is mounted by an isostatic suspension on either side of the centre of gravity. The engine is of modular design and is intended to produce minimum noise and smoke.

DIMENSIONS:
- Overall length of basic engine: 1,150 mm (45·3 in)
- Overall diameter: 600 mm (23·6 in)

WEIGHT, DRY:
- Larzac 04: 290 kg (640 lb)

T-O THRUST (S/L, static):
- Larzac 04: 13·24 kN (2,976 lb)

SPECIFIC FUEL CONSUMPTION:
- Larzac 04: 20·1 mg/Ns (0·71 lb/hr/lb)

Turboméca-SNECMA Larzac 04 two-shaft turbofan, rated at 13·24 kN (2,976 lb) st

GERMANY
(FEDERAL REPUBLIC)

KHD
KLÖCKNER-HUMBOLDT-DEUTZ AG

ADDRESS:
5 Köln (Cologne) 80, Postfach 80 05 09
Telephone: Cologne (0221) 8221

KHD is a leading manufacturer of diesel engines and trucks. In its subsidiary at Oberursel it concentrates its activities in the gas-turbine field. It designs, manufactures and services gas turbines, power transmissions and complete power systems for aircraft, road vehicles, marine craft and industrial applications. Its most important gas turbine is the T312, the compact lightweight APU (auxiliary power unit) of the Panavia MRCA, providing full shaft power 114 kW (153 hp) or full air bleed of 0·22 kg (0·49 lb)/sec or a combination. KHD also supplies the complete MRCA secondary power system linking the T312 to a cross-shaft coupled to each main engine.

KHD is also participating in the production of the T64 turboshaft (see MTU in this section) and Larzac turbofan (see Turboméca-SNECMA in this section).

LIMBACH
LIMBACH MOTORENBAU

HEAD OFFICE AND WORKS:
Kotthausener Str 5, D-533 Königswinter 21, Sassenberg
Telephone: (02244) 2322
PRESIDENT: P. Limbach

This company manufactures four-stroke piston engines for ultra-light aeroplanes and powered gliders. All are of similar basic design, though one sub-type has a greater cylinder stroke, and a new range about to go into production has substantially larger capacity.

LIMBACH SL 1700

Several variants of this engine have been certificated by the Luftfahrt-Bundesamt (Federal Office of Civil Aviation). Apart from the first sub-type listed below all are four-stroke (Otto) engines:

Limbach SL 1700D. Dual-ignition. Not certificated. Fitted to Sportavia-Pützer RF7.

Sportavia-Limbach SL 1700E. Basic engine of the current range. Fitted to Sportavia/Pützer RF5 and RF5B.

Limbach SL 1700EA. Differs in having front-end starter and different induction system. Fitted to Scheibe SF-25C Falke.

Limbach SL 1700EAI. Similar to EA except equipped to drive Hoffmann variable-pitch propeller. Fitted to Scheibe SF-28.

Limbach SL 1700EB. Similar to E except for having increased cylinder stroke and twin carburettors.

Limbach SL 1700EBI. Similar to EB except equipped to drive Hoffmann variable-pitch propeller. Fitted to Schleicher ASK 16.

Limbach SL 1700EC. Similar to E except for having a carburettor intake heating box.

Limbach SL 1700 ECI. Similar to EC except equipped to drive Hoffmann variable-pitch propeller.

Sportavia-Limbach SL 1700EI. Similar to E except equipped to drive Hoffmann variable-pitch propeller. Optional for RF5B.

Unless otherwise stated, the following description refers to the SL 1700E:

TYPE: Four-cylinder horizontally-opposed air-cooled piston engine.

CYLINDERS: Bore 88 mm (3·46 in). Stroke 69 mm (2·71 in) (EB, EBI, 74 mm; 2·87 in). Swept volume 1,680 cc (102·51 cu in) (EB, EBI, 1,800 cc; 108·56 cu in). Compression ratio 8 : 1.

INDUCTION: Stromberg-Zenith 150CD carburettor (two in EB, EBI). (EA, EAI, one Zenith 28 RX2).

FUEL GRADE: 90 octane.

IGNITION: Single Slick 4030 magneto feeding one Bosch WB 240 ERT 1 plug in each cylinder.

STARTING: One Fiat 0·5 hp starter (EA, EAI, one Bosch 0·3 kW; 0·4 hp).

Limbach SL 1700D flat-four two-stroke engine, rated at 48·5 kW (65 hp)

ACCESSORIES: Ducellier 250W alternator (EA, EAI, 150W Ducati); APG 17.09.001 fuel pump (EA,EAI, 17.09.001A).

DIMENSIONS:
- Length overall:
  - SL 1700D: 649 mm (25·6 in)
  - SL 1700EA, EAI: 558 mm (22·0 in)
  - SL 1700E, EI, EC, ECI: 618 mm (24·33 in)
  - other variants: 580 mm (22·8 in)
- Width overall:
  - SL 1700D: 800 mm (31·5 in)
  - SL 1700EA, EAI: 770 mm (30·3 in)
  - other variants: 764 mm (30·1 in)
- Height overall:
  - SL 1700D: 451 mm (17·8 in)
  - SL 1700EA, EAI: 392 mm (15·4 in)
  - other variants: 368 mm (14·5 in)

WEIGHT, DRY:
- SL 1700E, EI: 73 kg (161 lb)
- SL 1700EA, EAI: 70 kg (154 lb)
- SL 1700EB, EBI, EC, ECI: 74 kg (164 lb)

PERFORMANCE RATINGS:
- T-O:
  - SL 1700D: 48·5 kW (65 hp) at 3,600 rpm
  - SL 1700E, EI, EC, ECI: 51 kW (68 hp) at 3,600 rpm
  - SL 1700EA, EAI: 44·7 kW (60 hp) at 3,550 rpm
  - SL 1700EB, EBI: 53·7 kW (72 hp) at 3,600 rpm
- Continuous:
  - SL 1700E, EI, EC, ECI: 45·5 kW (61 hp) at 3,200 rpm
  - SL 1700EA, EAI: 41·7 kW (56 hp) at 3,300 rpm
  - SL 1700EB, EBI: 49·2 kW (66 hp) at 3,200 rpm

LIMBACH L 2000-2600

Instead of planning a single new engine—the SL 2400 introduced in the 1975-76 *Jane's*—Limbach is developing similar engines in three sizes. The smallest is the L 2000 of just under 2,000 cc capacity; the intermediate L 2400 has larger bore, and the largest, the L 2600, combines the larger bore with longer stroke. They will be available with propeller flanges for a solid propeller or, with suffix D-I, a Hoffmann variable-pitch propeller. Later these engines will be available with reduction gear. In 1976 development was proceeding towards certification (FAR Pt. 33).

Sportavia-Limbach SL 1700E flat-four four-stroke engine, rated at 51 kW (68 hp)

Limbach SL 1700EA flat-four four-stroke engine, rated at 44·7 kW (60 hp)

TYPE: Four-cylinder horizontally-opposed aircooled four-stroke piston engine.

CYLINDERS: Bore (L 2000) 94 mm (3·70 in), (L 2400, 2600) 103 mm (4·06 in). Stroke (L 2000, 2400) 71 mm (2·79 in), (L 2600) 78·5 mm (3·09 in). Swept volume (L 2000) 1,970 cc (120·2 cu in), (L 2400) 2,368 cc (144·49 cu in), (L 2600) 2,616 cc (159·62 cu in). Compression ratio 8 : 1.

INDUCTION: One Marvel-Schebler 3 PA carburettor fed by APG 20.09.001 fuel pump.

FUEL GRADE: 90 octane.

IGNITION: Two Slick 4001 magnetos, with Slick high-temperature harness feeding two Bosch WB 240 ERT 1 plugs in each cylinder.

STARTER: Bosch 0·3 kW (0·4 hp) electric.
ACCESSORIES: Limbach 250 W alternator; oil capacity 4 l (4·2 qt).
DIMENSIONS (L 2400):

Length	640 mm (25·20 in)
Width	790 mm (31·10 in)
Height	378 mm (14·88 in)

WEIGHT, DRY:

L 2000 D-I	91 kg (201 lb)
L 2400 D-I	92 kg (203·5 lb)
L 2600 D-I	93 kg (205 lb)

PERFORMANCE RATING (T-O and max cont):

L 2000 D-I	52·2 kW (70 hp) at 3,000 rpm
L 2400 D-I	59·6 kW (80 hp) at 3,000 rpm
L 2600 D-I	67 kW (90 hp) at 3,000 rpm

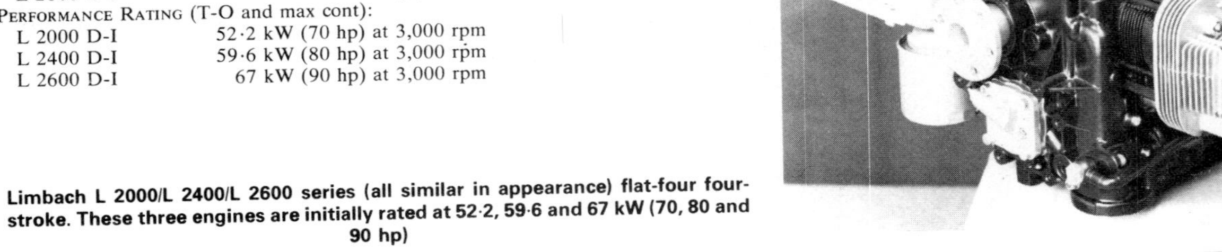

Limbach L 2000/L 2400/L 2600 series (all similar in appearance) flat-four four-stroke. These three engines are initially rated at 52·2, 59·6 and 67 kW (70, 80 and 90 hp)

MBB
MESSERSCHMITT-BÖLKOW-BLOHM GmbH

ADDRESS:
8 Munich 80, POB 801 220
Telephone: (0811) 6 00 01
DEVELOPMENT AND PRODUCTION CENTRES:
Ottobrunn bei München; Lampoldhausen; Hamburg; and Bölkow-Apparatebau GmbH at Nabern and Schrobenhausen
OFFICERS: See "Aircraft" section

As noted in the Aircraft section, Messerschmitt-Bölkow and Hamburger Flugzeugbau merged in 1969 to form the MBB group. The former Bölkow element of this group is engaged in the design and development of a wide variety of medium- and high-energy rocket engines and motors. These include liquid bipropellant and mono-propellant engines, liquid- and solid-propellant air-augmented rockets, solid-propellant motors for small and medium-size missiles, and hybrid engines. Other activities include thrust augmentation by afterburning, engine casing design, propellant insulation and mounting, and preparatory work towards series production and reliability. Other recent development of liquid-propellant engines by MBB has embraced advanced thermodynamic combustion engines, injection and cooling of engines of low thrust level and engines with high- and medium-energy storable propellants. The following are some of the major programmes:

MBB solid-propellant ram rocket on combustion test in a wind tunnel

MBB/SEP HM7 high-energy rocket engine for Ariane third-stage propulsion

MBB MONOPROPELLANT ENGINES

MBB has developed N_2H_4 engines for thrust levels of 0·01/0·8/2·58/3·65/5·56/6·45 kN (2/180/580/820/1,250/1,450 lb). The smallest unit was developed from 1965 to 1968 for a satellite attitude-control system. The programme ended with a flight prototype optimised for the Shell 405 catalyst. The 2·58 kN (580 lb) engine was designed for a meteorological sounding rocket and was used for actual rocket flights in 1968. The 6·45 kN (1,450 lb) engine was the most powerful N_2H_4 monopropellant engine ever tested in Europe.

MBB MONOPROPELLANT GAS-GENERATORS

MBB has produced 13 different types of gas-generator, with flow rates from 0·00018 kg (0·0004 lb) to 8·2 kg (18·2 lb)/sec and using the propellants N_2H_4, $N_2H_4 + H_2O$, and $H_2O_2 + H_2O$. The bigger gas-generators are operating with catalytic and thermic decomposition. The smallest generator supplies a cold-gas satellite attitude-control system, while the bigger types are for propellant-tank pressurisation. Some 4,500 generators had been produced and delivered to industrial customers by the end of 1975.

MBB STORABLE ROCKET ENGINES

This engine family includes motors of 0·01/0·03/0·05/0·08/0·3/0·4/0·5/97·9 kN (2·2/6·6/11/18/66/88/110/22,000 lb) thrust. Except for the 0·01 kN (2·2 lb) engine, which uses MMH/N_2O_4, all other engines run on $AZ50/N_2O_4$. The 0·01 kN (2·2 lb) engine is in use as the attitude-control thruster for the Franco-German Symphonie communications satellite; it is designed for steady-state and pulse-mode operation. By the end of 1975 a total of 74 had been produced. The 0·4 kN (88 lb) engine was the vernier for the German third stage of the ELDO-A launcher, and in modified form it was to be used as the apogee motor for ELDO-II. This engine is highly qualified and in several ELDO-A launches has operated in space without failure. A total of 135 was produced. In 1973 a new tactical missile engine was developed, operating at thrust levels of 23·5 kN (5,290 lb); 11·77 kN (2,646 lb); 7·85 kN (1,764 lb). The 98 kN (22,000 lb) engine was developed as the prototype power plant for an artillery rocket.

MBB HIGH-ENERGY ROCKET ENGINES

MBB has worked with the cryogenic propellant combination H_2/F_2 since 1962. After the installation of a fluorine liquefaction facility and a technology programme for a 0·3 kN (66 lb) engine, the development of a 4·9 kN (1,100 lb) engine started in 1967-68. It was the aim to use a cluster of two engines of this size as a "kick stage" for space probes. The programme was discontinued after the first series of tests with regeneratively-cooled chambers had showed high combustion efficiency. The propellant combination H_2/O_2 was used from 1962 in three engine projects at the thrust levels of 0·3/6·67/9·34 kN (66/1,500/2,100 lb). The 66 lb H_2/O_2 engine was presented in 1966 as a flight prototype, the first cryogenic rocket engine in Germany. The engine incorporated integral propellant valves and an ignition system. In altitude-simulation tests a specific impulse of 415 sec was achieved and multiple restart capability and throttleability were demonstrated. For possible post-Apollo programme participation, the 1,500 lb and 2,100 lb engines were developed in 1970-72. Both are regeneratively cooled, electrically ignited, restartable and pulsable (10Hz). They have high performance and a combustion efficiency of more than 98%. In 1972-73 an LH_2/LO_2 engine rated at 5·0 kN (1,125 lb) in vacuum conditions was developed for multiple-restart use in upper stages. This engine can be throttled over a range of 8 : 1 (possibly 10 : 1).

Since 1973, in partnership with SEP of France, MBB has developed the HM7 engine to power the third stage of the European Ariane space launch vehicle. This LH_2/LO_2 engine has a vacuum thrust of 57·87 kN (13,009 lb). Its regeneratively cooled thrust chamber of milled copper operates at a pressure of 28·6 bars (420 lb/sq in), and the

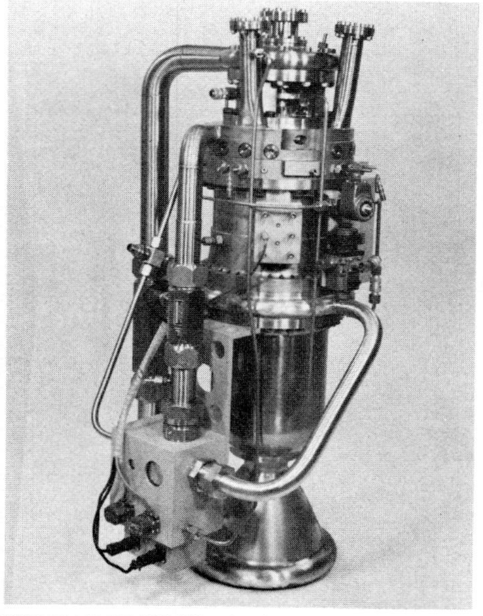

MBB 49 kN (11,000 lb) st lox/kerosene high-pressure topping cycle rocket engine

specific impulse is 435 sec. Total weight, excluding the turbopump, is 63 kg (138·9 lb). The nozzle extension, with an area-ratio of 61, is dump-cooled with LH_2. By the start of 1976, a total of 150 firings had been completed on MBB's high-pressure test facility under sea-level and simulated altitude conditions. SEP is Ariane propulsion prime contractor. MBB's role is to supply the thrust chamber and main propellant valves.

MBB HIGH-PRESSURE ENGINES (TOPPING CYCLE)

In the course of development of high-pressure liquid-rocket engines MBB developed a turbopump engine utilising the topping cycle in 1963. This 49 kN (11,000 lb) lox/kerosene engine was the first integrated (autonomous) topping cycle engine to run in the western world. It was designed for 96·5 bars (1,400 lb/sq in) abs and was throttleable at 14 : 1, and remains the first and most powerful turbopump rocket engine developed in Federal Germany since 1945.

For the storable propellants $UDMH/N_2O_4$ a 9·8 kN (2,200 lb) pressure-fed engine was developed and tested at chamber pressures of more than 193 bars (2,800 lb/sq in) abs to demonstrate high-performance injection systems and electroformed thrust-chamber technology for corrosive propellants. During the past year the company has developed completely electroformed integral chambers entirely of nickel which, in comparison with a copper-nickel chamber of the same size and thrust, are approximately 30 per cent lighter.

MBB developed and fabricated some 133·44 kN (30,000 lb) H_2/O_2 engines designed for 207 bars (3,000 lb/sq in) abs. These were tested at Rocketdyne's facilities in California up to chamber pressures of 276 bars (4,000 lb/sq in) abs without failure. The electroforming thrust-chamber technology demonstrated in this programme is now the baseline for the Space Shuttle Main Engine of Rocketdyne (which see).

MBB valveless pulsejet shown on bench test

MBB RAM ROCKETS

Since 1965 MBB has been developing ram rocket engines with liquid and solid propellants, and liquid-fuelled ramjet engines, for the propulsion of missiles with supersonic cruise speed and long range capability.

MBB has studied high-energy boron propellants as well as medium-energy composite grains, UDMH, MMH and kerosene. For missile applications, studies were carried out with semi-integrated and fully integrated engines with many configurations of air inlet. These studies were supported by half-scale burning tests (net thrust 4·0 kN; 899 lb) and aerodynamic tests in wind-tunnels.

Solid-propellant ram rockets are inherently simple in structure and have high reliability because of hypergolic ram combustion.

MBB PULSEJET

Since 1968 MBB has been developing the first of a series of valveless pulsejet engines intended for a wide range of applications, including missiles, drones and remotely piloted vehicles. The unit has a bellmouth inlet, connected with the combustion chamber by a cylindrical tube. Downstream the combustion chamber leads through another cylindrical tube, to a divergent nozzle. Ignition is by a spark plug; subsequent operation is automatic.

Static testing of an experimental model is being conducted in collaboration with the DFVLR. Considerably more than 250 hours have been logged.

DIMENSIONS:
Length overall 3,000 mm (118 in)
WEIGHT, DRY: 10 kg (22·0 lb)
PERFORMANCE RATING: 0·353 kN (79·4 lb st)

MTU
MOTOREN-UND-TURBINEN-UNION MÜNCHEN GmbH

HEAD OFFICE AND WORKS:
München-Allach, Dachauer Str 655 (postal address, 8 München 50, Postfach 50 06 40)
Telephone: (0811) 1 48 91

DIRECTORS:
Dr Ing Karl Schott (Chairman of the Board)
Rolf Breuning and Hugo Berthold Saemann (Speakers for the management)
Dr Hans Dinger
Dr Karl Adolf Müller
Werner Niefer
Dr Ernst Zimmermann

MTU München has a nominal capital of DM100 million, and is owned half by Maschinenfabrik Augsburg-Nürnberg AG (MAN) and half by Daimler-Benz AG. This company now manages all the aircraft engine programmes formerly managed by MAN Turbo and Daimler-Benz. These programmes are being concentrated at Munich. At the beginning of 1976 total employment was 5,700, including a development staff of 1,000. Covered floor area is 452,000 m² (4,865,300 sq ft).

MTU München holds 40 per cent of the shares of Turbo-Union Ltd, an equal percentage being owned by Rolls-Royce and 20 per cent by Fiat (see "International" section).

By far the largest of MTU München's programmes is its participation, through Turbo-Union, in the RB.199 programme for the Panavia MRCA. With Rolls-Royce and Fiat of Italy, it is participating in the Pratt & Whitney JT10D programme (see US section).

MTU is a member of the consortium of companies producing the Larzac 04 turbofan for the Alpha Jet (see Turboméca-SNECMA under France).

The company's own main development activities concern the MTU 6022 turboshaft (no longer used in aircraft propulsion), and a derivative of this unit under development for MAN AG and Daimler-Benz for use in heavy road vehicles. Production programmes include licence manufacture of General Electric J79-MTU-17A engines for F-4E Phantoms and T64 engines for CH-53G helicopters. It has concluded a technical support contract with Detroit Diesel Allison covering manufacture, sale and support of the 250-B17 turboprop and 250-C20 turboshaft and future uprated versions.

MTU also manufactures parts of the Spey turbofan under subcontract to Rolls-Royce and is a member of the consortium which made the Rolls-Royce Tyne. MTU shares in the manufacture of the CF6-50 engine for the Airbus Industrie A300. Servicing and overhaul activities include work on the J79-11A, -17A and -J1K, T64-MTU-7, Tyne, Solar Titan turbine (CH-53G APU) and Lycoming piston engines.

The MTU J79-MTU-J1K turbojet with afterburner

MTU/GE CF6

Under the terms of a licence agreement signed with the International General Electric Company, MTU shares in the manufacture of the CF6-50 engine for the Airbus A300, together with SNECMA. In addition MTU supplies parts to GE for the DC-10 programme. The main task of MTU is production of the complete HP turbine, with electrolytically stem-drilled film-cooled blades, vanes, impeller spacer and heat shield. A description of the CF6 appears under General Electric in the US section.

MTU J79

MAN Turbo (now MTU) took over the German licence rights to the General Electric J79 afterburning turbojet from BMW Triebwerkbau which had previously participated in the manufacture of considerable numbers of the engine for European-built Lockheed F-104G Starfighters. In addition to providing an overhaul, repair and maintenance service for licence-built J79-GE-11As, MTU also developed a slightly improved version of this model under the designation J79-MTU-J1K. The company is modifying J79-GE-11As in service with the German Air Force to J1K standard to increase performance, economy and reliability. MTU is also producing J79 components for McDonnell Douglas RF-4E Phantom fighters for the German Air Force and has taken over complete technical and logistic support for the J79-GE-17 engines involved. In 1972 the company went into production with 448 new J79-MTU-17A engines for the 175 F-4E fighters for the Luftwaffe.

A description of the J79 is given under the General Electric entry in the US section of this edition. Manufacturer's details of the J79-MTU-J1K include the use of a steel compressor rotor and magnesium and steel casings; flame tubes fabricated in Hastelloy; turbine blades in Udimet 700; jet pipe fabricated in Inconel; fuel and oil specification MIL-L-5624 and MIL-7808 respectively; and use of an air supply starter. Technical data for the-J1K include: air mass flow 74·4 kg (164 lb)/sec; compressor pressure ratio 12·4; diameter 995 mm (39·2 in); length 5,291 mm (208 in); and weight 1,685 kg (3,715 lb). Max thrust is raised from 44·5 kN (10,000 lb) dry to 46·5 kN (10,460 lb) and from 70·3 kN (15,800 lb) with reheat to 71 kN (15,950 lb), all at 7,460 rpm. MTU reports the sfc at military rating as 0·84, and 2·07 with afterburning. In addition to improving performance, the MTU afterburner doubles the engine overhaul period from 400 to 800 hours.

MTU T64

MTU München has a licence to manufacture the General Electric T64 free-turbine turboshaft/turboprop. Between 1972 and 1974 the company produced the 2,927 kW (3,925 shp) T64-MTU-7 turboshaft to power the Sikorsky CH-53G medium helicopter chosen by the German Army, in co-operation with GE and Klöckner-Humboldt-Deutz AG. MTU is now supporting these engines in se[rvice]. A description of the T64 is given under the Gener[al] Electric entry in the US section.

RFB
RHEIN-FLUGZEUGBAU GmbH

ADDRESS and other details: See "Aircraft" section

Though not a constructor of aircraft engines, Rhein-Flugzeugbau is marketing a Fan Pod as a complete unit for fitment to power-assisted sailplanes and ultra-light aircraft, as well as to surface vehicles. In 1976 main markets were seen as: motor-gliders, aerofoil boats, air-cushion vehicles, RPVs and water or snow vehicles.

RFB FAN POD SG 85

This is a fixed-geometry pod marketed as a complete unit for installation by the purchaser. It was run on the bench in 1974, with Fichtel & Sachs Wankel engines, and was subsequently flight tested in a Blanik sailplane. In certain types of sailplane with side-by-side seating it should be possible to make the installation retractable. This is desirable, because the windmilling drag of the pod is high.

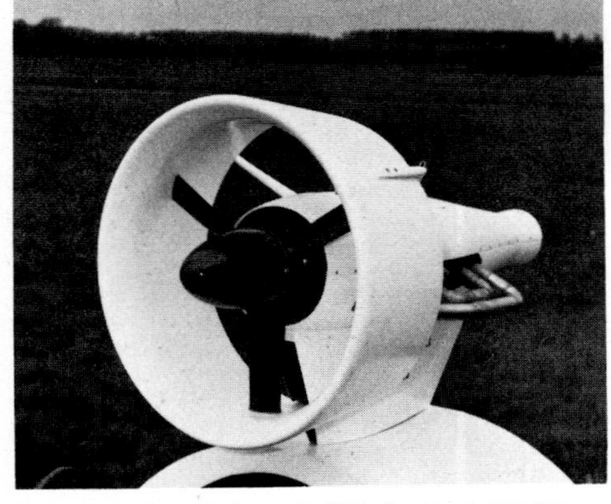

Prototype installation of RFB fan pod SG 85, as mounted for flight testing on Blanik sailplane (0·932 kN, 209·5 lb st at S/L)

Omnipol L-13 Blanik in sustained flight with RFB SG 85 fan pod

ENGINE: Dual rotating-combustion (Wankel-type) engines mounted in tandem. Fichtel & Sachs KM 914/2 V-85 engines are regarded as standard, with a combined rating of 50 hp. Engines are equipped with 12V starter and generator, and with silenced exhaust system.

FAN ROTOR: Three-blade fan moulded in Friedel-Krafts reinforced plastics material, with hard erosion-resistant strip along each leading-edge. Direct drive from engines without separate bearings. Rotating spinner.

FAN DUCT: Single shell in FK sandwich reinforced plastics with PUR foam filling.

PYLON: Light alloy construction, with downward projecting spigot attachment 200 mm (8 in) long mating with airframe.

FUEL GRADE: Automotive gasoline and oil mixture in rat[io] 30 : 1.

DIMENSIONS:
- Length: 1,200 mm (47·24 i[n])
- Fan duct external diameter: 750 mm (29·53 i[n])
- Height, incl aircraft spigot: 1,000 mm (39·37 i[n])

WEIGHT, DRY: 56 kg (123·5 l[b])

PERFORMANCE RATING:
- Full throttle, S/L: 0·932 kN (209·5 lb)

FUEL CONSUMPTION:
- At full throttle (5,500 rpm), S/L 15 l (3·3 Imp gal)/[h]
- At cruise at 5,000 rpm, S/L 11·5 l (2·5 Imp gal)/[h]

NOISE (full throttle, S/L, 305 m; 1,000 ft) 57 dB

PIEPER
PIEPER MOTORENBAU GmbH

ADDRESS:
495 Minden/Westf, Postfach 1229
Telephone: (0571) 34088

Pieper is manufacturing the 45 hp Stamo MS 1500-1 modified Volkswagen four-cylinder aircooled piston engine for the Scheibe SF-25B Falke two-seat powered glider. The capacity of this is 1,500 cc, compression ratio 7·2 : 1, length 640 mm (25 in), width 745 mm (29·3 in), height 395 mm (15·5 in), and dry weight 52 kg (115 lb). The MS 1500-1 operates on either 80/86 or 90 octane fuel, and is started by a pull-cord. Production has now also begun of the MS 1500-2, with electric starter and generator. This increases overall height to 450 mm (17·7 in) and dry weight to 60 kg (132 lb). By 1976 well over 500 engines had been delivered, and the period between complete overhauls had been extended to 1,000 hr.

Pieper MS 1500-1 four-cylinder four-stroke engine, rated at 33·6 kW (45 hp)

INDIA

HAL
HINDUSTAN AERONAUTICS LTD

HEAD OFFICE:
Indian Express Building, Vidhana Veedhi, PO Box 5150, Bangalore 1
Telephone: 75004/5/6

OFFICERS: See "Aircraft" section

The Bangalore and Koraput Engine Divisions of HAL comprise the main aero-engine design, development and manufacturing elements of the Indian aircraft industry.

BANGALORE COMPLEX (Engine Division)

This Division is engaged on a variety of gas-turbine and piston engines of both indigenous and licensed designs. These include the HJE-2500 turbojet designed by HAL; the Orpheus 701 and 703 turbojets being built under licence from Rolls-Royce to power the Gnat and HF-24 Mk I fighters; the Viper 11, made under Rolls-Royce licence to power the HJT-16 Mk I Kiran trainer; the Dart 531 turboprop, built under licence from Rolls-Royce to power the HS 748 transport; the Artouste IIIB turboshaft, built under licence from Turboméca to power Alouette III and SA 315 helicopters; and the PE 90 and HPE-4 piston engines designed by HAL. The Engine Division also overhauls and repairs Rolls-Royce Avon and HAL Orpheus turbojets, and various types of piston engines, for the Indian Air Force.

HAL HJE-2500

The HJE-2500 is a small turbojet engine which HAL is developing as a potential power plant for production versions of its HJT-16 Kiran basic training aircraft. It is a single-spool design, with a seven-stage compressor driven by a single-stage turbine. All components except the fuel system were manufactured at Bangalore.

The HJE-2500 ran for the first time on the test bed on 30 December 1966. This prototype engine has now achieved its full design performance.

TYPE: Single-shaft turbojet.

AIR INTAKE: Made of aluminium-magnesium alloy with three radial struts supporting the main thrust bearing and starter. Inlet guide vanes are fixed type.

COMPRESSOR CASING: Two-piece aluminium-magnesium alloy casing with seven rows of aluminium alloy blades and one row of flow straightener vanes.

COMPRESSOR: Built up of seven steel discs mounted on a common drum supported on rear roller bearing and front ball thrust bearing. 410 aluminium alloy rotor blades. Two-piece aluminium alloy casing with 606 dovetailed stator blades. Air mass flow 20·4 kg (45 lb)/sec at 12,500 rpm. Pressure ratio 4·2 : 1.

COMBUSTION CHAMBER: Cannular type with seven flame tubes and duplex burners. High-energy ignition system with plugs in tubes 2 and 5.

FUEL SYSTEM: Positive displacement pump with pressure, fuel/air ratio and acceleration control.

FUEL GRADE: DEngRD. 2494.

TURBINE CASING: Nimocast with 65 solid nozzle gui[de] vanes of Nimonic 90.

TURBINE: Single stage with fir-tree root fitting for 1[..] blades of FV.448. Rear bearing ahead of turbine whee[l].

JET PIPE: Fixed nozzle.

CONTROL SYSTEM: Single lever master control.

LUBRICATION: Return-flow system for front bearing a[nd] gears. Total loss for centre and rear bearings.

MOUNTING: Spherical joints on sides of delivery casi[ng] with steady on compressor casing.

STARTING: Electric starter in nose bullet.

DIMENSIONS:
- Length, flange to exit cone: 2,160 mm (85 i[n])
- Diameter: 660 mm (26 i[n])
- Frontal area: 0·34 m² (3·66 sq [ft])

WEIGHT, DRY: 265 kg (585 [lb])

PERFORMANCE RATING:
- T-O: 11·12 kN (2,500 lb st) at 12,500 r[pm]

SPECIFIC FUEL CONSUMPTION: 27·8 mg/Ns (0·98 lb/hr/lb [st])

HAL HPE-4

Replacing the six-cylinder HPE-2, the HPE-4 is to [be] an eight-cylinder piston engine intended for the Bas[ant] agricultural aircraft. It will be unsupercharged, with f[uel] injection and a direct drive. Cubic capacity will be 11·8 litres (724·92 cu in) and compression ratio 8·7 : 1. T

HPE-4 is to have wet-sump pressure-feed lubrication and will run on 100/130 octane fuel. By 1976 two prototype HPE-4 engines had been constructed, and development was progressing satisfactorily.

DIMENSIONS:
Height 524 mm (20·63 in)
Width 870 mm (34·252 in)
Length 1,170 mm (46·063 in)

WEIGHT, DRY (estimated): 255 kg (562·28 lb)
PERFORMANCE, RATING:
T-O at S/L at 2,650 rpm 298 kW (400 hp)

KORAPUT DIVISION

This Division of HAL is located at Koraput in Orissa. It was established to manufacture under Soviet Government licence the Tumansky R-11 and R-13 afterburning turbojets for HAL-built MiG-21 fighters. With help from the Soviet Union, the first engine was run on the bench (which it was used to calibrate) in early 1969. HAL Koraput is already considerably larger than the original Bangalore factory, and in 1970 had cost Rs 350m (£19·4 m). Production has been centered on the R-11-F2-300 for the MiG-21FL and R-13-F2S-300 for the MiG-21M (see under Tumansky, Soviet Union, in this section).

INTERNATIONAL PROGRAMMES

CFM INTERNATIONAL
CFM INTERNATIONAL SA

ADDRESS:
4 Rue de Penthièvre, 75008 Paris, France
Telephone: 265 97 72

CHAIRMAN AND CHIEF EXECUTIVE:
J. Sollier

MEMBERS OF THE BOARD:
R. Ravaud (SNECMA)
G. Neumann (GE)
J. C. Malroux (SNECMA)
J. Crepin (SNECMA)
F. Carrier (SNECMA)
J. W. Sack (GE)
B. H. Rowe (GE)
J. I. Hope (GE)
N. Burgess (GE)

CFM International, a joint company, was formed by General Electric and SNECMA in early 1974 to provide overall programme management for the CFM56 engine and a single customer interface for sales and service. Owned and managed on a 50/50 share basis, the company has been staffed with experienced people from the two parent companies.

In addition to the CFM International management team which directs the overall programme, the parent companies have their own CFM56 programme managers. For SNECMA, this position is held by Jean Bagneux. His GE counterpart is Jack I. Hope. The two men contribute to overall direction in conjunction with the Chief Executive. In their roles as programme managers they also ensure that the resources of the parent organisations are properly applied.

The SNECMA/GE agreement is not a profit-sharing arrangement. Responsibilities for hardware design, development and production are assigned through CFM International on an equal basis to the parent companies. Each company then assumes responsibility and funding for its assigned task throughout the life of the programme. This is a unique concept among international co-operative ventures in the aerospace field. SNECMA has from the outset been agreeable to participation of other European engine companies within its 50 per cent share. FN of Belgium has joined SNECMA for certain low-pressure system components.

GE is responsible for design integration, the core engine and the main engine control. The core engine is that of the F101 turbofan developed for the B-1 bomber for the USAF. SNECMA is responsible for the low-pressure system, reverser, gearbox and accessory integration and engine installation.

Each company will continue to be responsible for its assigned hardware from design through development, production and product improvement. CFM International will provide the planning and integration of the product support programme.

CFM INTERNATIONAL CFM56

In the late 1960s General Electric and SNECMA made independent studies of the market requirement for the next generation of high by-pass ratio engines. The GE studies were centered around an engine designated GE13, the core of which is now being used in the F101. SNECMA's studies were based on an engine designated M56. Each company concluded that a large market existed for a high by-pass ratio engine in the "ten tonne class" (97·86-106·76 kN; 22,000-24,000 lb st), with low noise, low emissions and low fuel consumption, coupled with ease of maintenance and low operating costs.

In April 1971 SNECMA began a search for possible partners to undertake development of a commercial engine in this class. By December 1971 SNECMA had chosen GE as its partner and, after obtaining French government approval, detailed design activity began. Work was stopped, however, in September 1972 when the export licence restriction applied by the US State Department Office of Munitions Control made the programme untenable for the two companies. It was not resumed on a joint basis until September 1973, when negotiations led to a workable licensing agreement which also protected US technology. Since that time, a working agreement and management structure have been defined and four engines have been built and put on test.

CFM56 second prototype (97·86 kN; 22,000 lb to 30°C), fully instrumented for running at Melun

The first CFM56 demonstrator engine ran at the GE Evendale Plant on 20 June 1974, as scheduled almost one year previously. This engine reached its full rated 97·86 kN (22,000 lb st) within 10 hours of running, with fuel consumption lower than specification. Since then three further engines have entered the test programme, and running time amounted to 1,500 hours in April 1976. Cross-wind, bird ingestion and icing tests have provided early confirmation of design integrity. Noise tests indicate that predicted levels will be achieved.

Two more engines were scheduled to run in 1976, with hardware being accumulated for four additional engines. Flight testing of one engine mounted in a Caravelle is scheduled to begin in early 1977, with engine certification due in the third quarter of 1978. The test programme has been designed to allow this schedule to be varied to suit market demands.

Another engine is to fly in 1977 in a USAF McDonnell Douglas YC-15 as part of the AMST programme.

CFM International is studying growth versions of the CFM56. It is estimated a rating of 122·3 kN (27,500 lb) st can be achieved with the existing fan diameter.

Potential applications for the CFM56 are numerous. The engine size is ideal for derivative or re-engined versions of existing commercial aircraft including the twin-engined BAC One-Eleven, DC-9 and Mercure; the three-engined Boeing 727 and Trident; and the four-engined A300 derivative, Boeing 707 and DC-8-60 series. Several airframe companies in the United States and Europe are studying use of the CFM56 for new commercial transport aircraft in the two-, three- and four-engined categories, for service in the 1980s and 1990s. In addition to this potentially large commercial market, an extensive military market for engines in this category is foreseen, for military transports, tankers and long-duration patrol and reconnaissance missions. Design targets include low noise characteristics (FAR 36 minus 8-10 EPNdB), low emissions (1979 EPA standards as a goal) and low fuel consumption (15-25 per cent improvement in fuel burned per seat-mile), coupled with design simplicity.

TYPE: Two-shaft turbofan for subsonic applications.

AIR INTAKE: Direct pitot entry, without inlet guide vanes.

FAN: Single-stage axial. Forged titanium disc holding 44 inserted titanium blades, each with a tip shroud to form a continuous peripheral ring. Fan and the attached LP compressor (booster) run in front roller bearing and rear ball bearing. Pointed conical spinner rotates with fan. Alloy steel fan frame of continuous ring construction, carried by 12 radial struts of low thickness/chord ratio well downstream of fan. Max airflow 354 kg (780 lb)/sec. By-pass ratio 6 : 1.

LP COMPRESSOR: Three axial stages on titanium drum bolted to fan disc serve as booster to supercharge core. Downstream flow curves sharply inwards to match diameter of HP compressor. In this section is main fan frame and sumps and bearings for front end of both shafts. Ring of bleed doors allows core airflow to escape into fan duct at low power settings. Bleed doors are closed at all normal flight power settings.

HP COMPRESSOR: Nine-stage axial with tapering tip diameter. Rotor of high-strength corrosion-resistant alloy, with blades of titanium (to stage 4) or steel. First four stators variable. Split titanium casing. Based upon HP compressor of F101, with minor modifications. Overall pressure ratio in 25 : 1 class.

COMBUSTION CHAMBER: Fully annular with advanced film cooling. Based upon F101 combustor but modified for long-life operation at reduced turbine temperature with minimal emissions. Level of pollution from the core is claimed to be below that of any engine at present in airline service.

HP TURBINE: Single-stage axial with aircooled stator and rotor blades. Entry gas temperature in 1,260°C class. High stage loading. HP system carried in only two bearings.

LP TURBINE: Four-stage axial.

EXHAUST UNIT (FAN): Constant-diameter duct of sound-absorbent construction. Outer cowl and engine cowl form convergent plug nozzle, with airframe-mounted reverser.

EXHAUST UNIT (CORE): Fixed-area type with convergent plug nozzle. Sound-absorbent construction.

ACCESSORY DRIVE: Gearbox in front sump transmits drive from front of HP spool, via radial shaft in fan frame, to transfer gearbox mounted on underside of fan case. Drive faces on both front and rear sides. Air starter at transfer gearbox.

FUEL SYSTEM: Hydromechanical with electronic trim.

LUBRICATION: Non-pressure-regulated system.

DIMENSIONS:
Front flange diameter 1,814 mm (71·4 in)
Length, excl spinner 2,280 mm (89·8 in)

WEIGHT, DRY: 1,886 kg (4,150 lb)

PERFORMANCE RATINGS:
Max T-O
97·86 kN (22,000 lb st) flat rated to 30°C
Cruise at 9,144 m (30,000 ft) at Mach 0·80
25·35 kN (5,700 lb st)

SPECIFIC FUEL CONSUMPTION:
Cruise rating, as above 18·7 mg/Ns (0·66 lb/hr/lb st)

ROLLS-ROYCE/ALLISON
ROLLS-ROYCE (1971) LIMITED

HEAD OFFICE:
Norfolk House, St James's Square, London SW1Y 4JR, England

THE DETROIT DIESEL ALLISON DIVISION, GENERAL MOTORS CORPORATION

HEAD OFFICE:
Detroit, Michigan, USA

Co-operation between Rolls-Royce and Allison started in November 1958, when the two companies began work on the design and development of high-performance jet engines for commercial and military applications. Under the terms of a Memorandum of Understanding signed by the US and Britsh governments in October 1965, they developed jointly a version of the Rolls-Royce Spey turbofan, under the designation Allison TF41, to power advanced versions of the Vought A-7 Corsair II attack aircraft.

The two partners were also involved in a joint demonstrator programme for an advanced lift-jet engine for V/STOL fighter aircraft. This programme successfully demonstrated all design goals and was concluded in June 1971. Both the US and British governments reported the engine ready for full-scale emergency development.

ALLISON/ROLLS-ROYCE TF41

Manufacturers' designations: Rolls-Royce Spey RB. 168-62 and -66, Allison Model 912-B3 and -B14

In August 1966 Allison and Rolls-Royce were awarded a joint $200 m contract by USAF Systems Command for the development and production of an advanced version of the RB.168-25 Spey turbofan, to power Vought A-7D Corsair II fighter-bomber aircraft for the USAF. The requirement was to provide an engine offering maximum thrust increase over the TF30-P-6 powering USN A-7As. The amount of the contract was increased to $230 m in December 1966, Rolls-Royce's share being about $100 m.

Development and production were undertaken jointly by Rolls-Royce and Allison, with Rolls-Royce supplying parts common to existing Spey variants and Allison, which is manufacturing under licence, being responsible for items peculiar to the TF41. This provided an approximately 50/50 division of manufacturing effort, but with Allison also undertaking assembly, test and delivery.

Design of the RB.168-62 started in June 1966 and, following the award of the USAF contract, the engine was given the USAF designation TF41-A-1. Major change compared with the RB.168-25 is the move forward of the by-pass flow split into the LP compressor, to give a larger three-stage fan followed by a two-stage IP compressor, all five stages being driven by the two-stage LP turbine. The number of HP compressor stages is reduced from 12 to 11, the HP turbine remaining at two stages. These modifications raise the mass flow and the by-pass ratio (from 0·7 : 1 to 0·76 : 1). No afterburner is fitted.

Other design changes compared with the RB.168-25 include omission of the fan inlet guide vanes, the first rotor stage being overhung on a bearing supported by the first-stage stator vanes. The fan and IP compressor are of more modern aerodynamic design, and the HP and LP turbine nozzle throat areas have been increased to pass the additional flow. The HP turbine is of modified aerodynamic design, and an annular exhaust mixer replaces the RB.168-25's chuted design.

First run of the TF41-A-1/RB.168-62 was at Rolls-Royce, Derby, in October 1967, the first Allison engine following at Indianapolis in March 1968. Development continued ahead of schedule, delivery of the first production TF41-A-1 being made in June 1968.

A second version of the TF41 is the A-2, developed for the US Navy and ordered in 1968 to power the Vought A-7E Corsair. Differences are slight, although the thrust rating is appreciably increased by raising the engine speed. This required re-stressing the disc of the LP turbine and HP compressor. Mass flow is slightly increased, the by-pass ratio being 0·74 : 1. The engine has additional protection against corrosion.

By 1976 more than 1,500 engines had been delivered, and production of the TF41 for the A-7 is expected to continue into the 1980s. In combat service both versions of the TF41 have shown outstanding reliability. The exceptional overhaul life (for a combat engine) of 1,200 hours was reached after four years of service.

Proposals have been made by Detroit Diesel Allison for TF41 developments offering substantially greater thrust for future versions of the Corsair. In general these proposals envisage a growth in thrust of up to 20 per cent within the same installation envelope. An industrial TF41, designated Spey Mk 1900, has been developed by the Rolls-Royce Industrial and Marine Division.

The two current production versions of the TF41 are known to Rolls-Royce as the RB.168-62 and RB.168-66; the corresponding Allison designations are Model 912-B3 and 912-B14. The following description refers basically to the TF41-A-1; where the A-2 differs, the data for that engine are given in brackets.

TYPE: Military turbofan.

AIR INTAKE: Direct entry, fixed, without intake guide vanes.

COMPRESSOR: Two-shaft axial. 3 fan stages, 2 intermediate stages on same shaft and 11 high-pressure stages. All rotor blades carried on separate discs. Fan and LP rotor blades of titanium, held by dovetail roots in slots broached in discs which are bolted together through curvic couplings and similarly attached to the stubshafts. HP rotor blades also of titanium, except stages 9, 10 and 11 of stainless steel, the first HP stage being pinned and the remainder being dovetailed into broached slots; discs similarly bolted together but driven through splined shaft coupling. LP rotor carried in 3 roller bearings and HP by 2, with central ball location bearing and intershaft ball bearing. LP casing of steel and aluminium; HP casing of stainless steel, both split at horizontal centreline. Stainless steel LP stator blades slotted laterally into casing, intermediate stators welded to inner casing sub-assembly rings. HP stator blades of stainless steel, slotted laterally into casing. Overall pressure ratio 20 : 1 (A-2, 21·4 : 1); mass flow 117 kg/sec (258 lb/sec) (A-2, 119 kg/sec; 263 lb/sec). HP compressor pressure ratio, 6·2 : 1; mass flow, 67 kg/sec (148 lb/sec).

COMBUSTION CHAMBER: Tubo-annular, with 10 interconnected Ni-Co alloy flame tubes in steel outer casing. Duple spray atomising burner at head of each chamber. High-energy 12-joule igniter plug in chambers 4 and 8.

FUEL SYSTEM: Hydromechanical HP system with automatic acceleration and speed control and emergency manual override. Variable-stroke dual fuel pump.

FUEL GRADE: JP-4 (A-2, JP-5).

NOZZLE GUIDE VANES: Two HP stages with air cooling; two LP stages uncooled, but 1st LP vanes are hollow and contain air pipes to cool LP rotor. All stators precision cast in Ni-Co alloy.

TURBINE: Impulse-reaction axial type, two HP stages and two LP. All blades forged in Ni-Co alloy; first HP stage blades cooled internally by HP compressor air; remainder have solid discs of Inco 901, LP discs of steel (A-2, Inco 901). All discs bolted to drive shafts.

JET PIPE: Fixed, heat-resistant steel.

ACCESSORY DRIVES: External gearbox driven by radial shaft from HP system; provision for starter, fuel boost pump, two hydraulic pumps, HP fuel pump, fuel control, HP tachometer, CSD and alternator, permanent-magnet generator, LP fuel pump and oil pumps. Additional low-speed (LS) gearbox, driven from LP shaft, serving LP rotor governor and tachometer.

LUBRICATION SYSTEM: Self-contained, with engine-mounted tank, fuel/oil heat exchanger and gear type pump; pressure 3·45 bars (50 lb/sq in). Tank capacity: A-1, 4·5 litres (1·0 Imp gal); A-2, 10·3 litres (2·27 Imp gal).

MOUNTING: Main ball-type trunnions on compressor intermediate casing; rear tangential steady-type at rear of by-pass duct.

STARTING: Integral gas turbine (air turbine):

DIMENSIONS:

Length overall	2,610 mm (102·6 in)
Intake diameter	953 mm (37·5 in)
Height overall	1,026 mm (40 in)

WEIGHT, DRY:

A-1	1,353 kg (2,980 lb)
A-2	1,370 kg (3,018 lb)

PERFORMANCE RATING (Max T-O):

A-1	64·5 kN (14,500 lb) to ISA +8°C
A-2	66·7 kN (15,000 lb) to ISA

SPECIFIC FUEL CONSUMPTION (Max T-O):

A-1	17·93 mg/Ns (0·633 lb/hr/lb st)
A-2	18·33 mg/Ns (0·647 lb/hr/lb st)

Cutaway drawing of the Allison/Rolls-Royce TF41-A-1 (Spey RB.168-62) turbofan (64·5 kN; 14,500 lb st)

ROLLS-ROYCE/SNECMA

ROLLS-ROYCE (1971) LTD

Head Office:
Norfolk House, St James's Square, London SWIY 4JR, England

Telephone: 01-839-7888

SOCIÉTÉ NATIONALE D'ÉTUDE ET DE CONSTRUCTION DE MOTEURS D'AVIATION

Head Office:
150 Boulevard Haussmann, 75361-Paris cédex 08, France

Telephone: 227-33-94

Rolls-Royce (1971) Ltd and SNECMA are jointly responsible for development and production of the Olympus turbojet engine to power the Anglo-French Concorde supersonic transport. They are also developing M45S series variable-stator turbofan engines for both civil and military use.

ROLLS-ROYCE/SNECMA M45SD-02 (RB.410D-2)

Ultra-quiet demonstrator engine based on M45H but driving Dowty Rotol variable-pitch fan. Joint programme by RR (1971), SNECMA and Dowty Rotol, with supplementary funding by the British government.

The geared variable-pitch fan is a new principle which enables a considerable reduction to be made in engine noise while, at the same time, maintaining engine performance. By the introduction of a reduction gear it is possible to run the fan turbine at high speed and, in consequence, at high efficiency, yet still use a large, low-speed fan to give high thrust with low noise levels. The M45SD-02 is designed to generate source noise not exceeding 95 PNdB at 152 m (500 ft).

The large fan, giving a high by-pass ratio, has variable-pitch blades, which help to reduce even further the part-thrust noise of this basically quiet engine by allowing the fan to run at optimum pitch. Moreover, the variable-pitch feature offers improved handling qualities important to airline safety.

These characteristics are achieved with variable pitch because:

It provides a fundamentally faster thrust-response rate with rapid pitch changes and the core remaining at relatively high regime. This will become of increased importance for thrust control on steep approach paths being studied for minimum noise and improved air traffic control, and also for baulked landing, and for effective selection of reverse thrust after touchdown. Moreover, it enables reverse thrust to be operated down to zero aircraft speed without hot-air ingestion problems and so can be used for aircraft ground manoeuvring purposes.

It enables very low thrust levels to be achieved without thrust spoiling in order to obtain quiet, steep descent paths on the approach and from cruising altitude.

Thrust can be optimised at all forward speeds and in all climatic conditions, by using variable pitch to provide the best possible engine matching.

It allows the independent selection of engine speed for any given part-thrust condition in order to minimise noise.

The demonstrator engine, designated M45SD-02 and with the Rolls-Royce project number of RB.410D-2, was assembled at the Bristol Engine Division plant at Parkside (Coventry) and ran at the government engine test station at Aston Down, Glos, in April 1975. The results of the programme will be relevant to many classes of aircraft, mainly of an R/STOL nature. Data from this technology demonstrator will be used to refine the design of a definitive production engine, the **M45S-11 (RB.410-11)**. This has the following specification: basic weight, 1,309 kg (2,885 lb); by-pass ratio, 8·8 : 1; total fan airflow, 286 kg (630 lb)/sec; overall pressure ratio 17·8 : 1; T-O (S/L, ISA, static), 63·9 kN (14,370 lb st); corresponding specific fuel consumption, 0·295; noise at 152 m (500 ft) horizontal distance, 96 PNdB.

The following description relates to the M45SD-02 (RB.410D-2):

Type: Geared variable-pitch fan demonstrator engine.

Air Intake: Direct pitot, with no struts or inlet guide vanes.

Fan: Single stage, of aluminium alloy construction throughout. Fourteen variable-pitch blades held in rotor hub by nuts. Rotor runs in overhung ball and plain bearings, with hub bolted to tubular shaft driven through epicyclic gearbox of 2·38 : 1 ratio by LP turbine shaft. Fan cowl and duct of aluminium acoustic panels. Fan hub shrouded by large anti-iced rotating spinner. Rear of spinner is large tubular fixed shroud over reduction gear, carrying fan duct on 34 aluminium-alloy stator vanes. Max airflow 219 kg (484 lb)/sec. Bypass ratio 8·73 : 1.

Core Engine: Basically identical to M45H-01 except for added LP turbine stage. Overall engine pressure ratio 13·5 : 1.

LP Turbine: Four-stage axial, with fourth stage comprising 48 stators and 62-blade rotor held by tie-bolts to M45H-01 turbine.

Dimensions and Weights: Not finalised.

Performance Rating:
T-O, S/L, ISA 44·48 kN (10,000 lb st)

Final assembly (before adding fan duct) of M45SD-02 quiet variable-pitch turbofan at the Parkside (Coventry) factory of Rolls-Royce (1971) Ltd. This engine has since run at the Aston Down test site

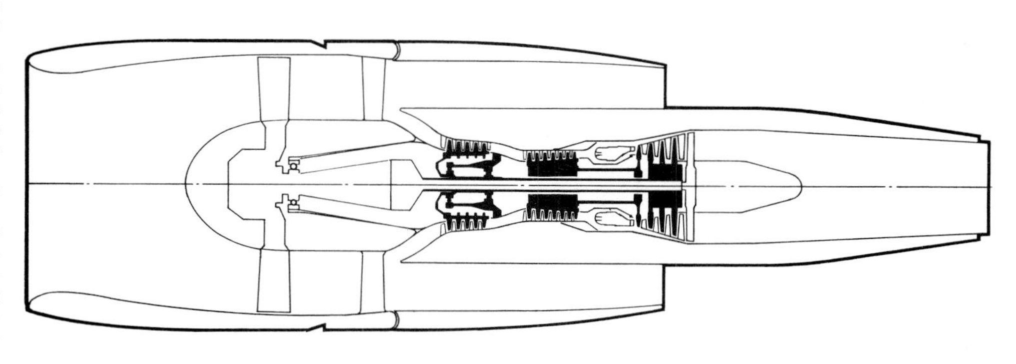

Provisional cross-section of Rolls-Royce/SNECMA M45SD-02 geared variable-pitch turbofan (44·48 kN; 10,000 lb st). Parts common to the M45H-01 are shown in solid black

Specific Fuel Consumption:
T-O, S/L, ISA 9·07 mg/Ns (0·32 lb/hr/lb st)

ROLLS-ROYCE/SNECMA OLYMPUS 593

The Olympus 593 is manufactured by Rolls-Royce (1971) Ltd in Britain and SNECMA in France as the power plant for the Concorde supersonic airliner. The work is shared on a 60%/40% basis, respectively. Rolls-Royce is producing the gas generator and SNECMA the convergent/divergent exhaust nozzle, thrust reverser and afterburner system.

Pre-flight Olympus development engines, designated 593D, were used for bench testing from mid-1964. The first of the Olympus 593 flight-type engines made its initial test run in November 1965. Thrusts ranging up to 178 kN (40,000 lb) have been obtained with afterburning in operation. A Vulcan testbed, with a single Olympus 593 mounted beneath its fuselage in a representative Concorde half-nacelle, assisted flight development from September 1966 to July 1971. Concordes have been flying since February 1969.

In March 1974 a production standard engine, the Olympus 593 Mk 610, successfully completed an official 150 hr type test. Full certification was achieved in September 1975, when total running time exceeded 50,000 hours. In January 1976, at the start of scheduled services by British Airways and Air France, more than 54,000 hours had been run, on the bench, in the Vulcan, and in seven Concordes. This far exceeded the time logged by any other engine at the start of service (and does not include the much greater time logged by earlier Olympus engines).

Industrial and marine versions of this engine are being considered.

Type: Two-spool turbojet with partial afterburning.

Air Intake: Fabricated titanium casing, with zero-swirl five-spoke support for front LP compressor bearing. In the Concorde the engine is installed downstream of an intake duct incorporating auxiliary intake and exit door systems and a throat of variable profile and cross-section (see "Aircraft" section).

LP Compressor: Seven-stage, with titanium blading and discs. Single-piece casing machined from stainless steel forging, electro-chemically machined.

Intermediate Casing: Titanium, with vanes supporting LP and HP thrust bearings. Incorporates drive for engine-mounted aircraft and engine gearboxes.

HP Compressor: Seven-stage axial. First three stages of blades titanium alloy. Remaining stages refractory alloy. Stainless steel single-piece casing. Mass flow 186 kg (410 lb)/sec. Overall pressure ratio 15·5 : 1.

Delivery Casing: Electro-chemically machined. Incorporates burner manifold and main support trunnions.

Combustion Chamber: Annular, cantilever-mounted from rear. Fabricated as single unit from nickel alloy, with all joints butt-welded. Total of 16 vaporising burners, each with twin outlets, welded directly into chamber head. Fuel injectors are simple pipes which enter each vaporiser intake with no physical contact. Combustion leaves virtually no visible smoke.

Fuel Grades: DERD. 2494 Issue 7, AIR 3405B (3rd edition, amendment 1), ASTM D-1655-71 (Jet A) and ASTM D-1655-71 (Jet A1).

Fuel System: Lucas system, incorporating mechanically driven first-stage pump and a second-stage pump driven by an air turbine, which is shut down at altitude cruise conditions as fuel-pressure requirements can be met by the first-stage pump alone. The first-stage pump also supplies afterburner fuel. Fuel-cooled oil cooler. Ultra electronic system, with integrated-circuit amplifier, provides combined control of fuel flow and primary nozzle area. Afterburner fuel controlled by ELECMA unit.

HP Turbine: Single-stage turbine, with cooled stator and rotor blading.

LP Turbine: Single-stage, with cooled rotor blades. LP drive-shaft co-axial with HP shaft.

Primary Nozzle Assembly: Straight jet pipe and pneumatically-actuated variable primary convergent nozzle which permits maximum LP-spool speed and turbine-entry temperature to be achieved simultaneously over wide range of compressor-inlet temperatures.

Afterburner: Single-ring sprayer with programmed fuel control as function of main engine fuel flow.

Secondary Nozzle Assembly: Monobloc structure with each twin nacelle manufactured from Stresskin panels. Each power plant terminates in a pair of "eyelids" which form a variable-area secondary divergent nozzle and thrust reverser. Eyelid position is programmed to maintain optimum power plant efficiency through all the flight regimes. When completely closed they act as thrust reversers.

Accessory Drives: Two gearboxes, mechanically driven off HP shaft (LP shaft has a pulse-probe signal source and provision for hand or mechanical turning). LH

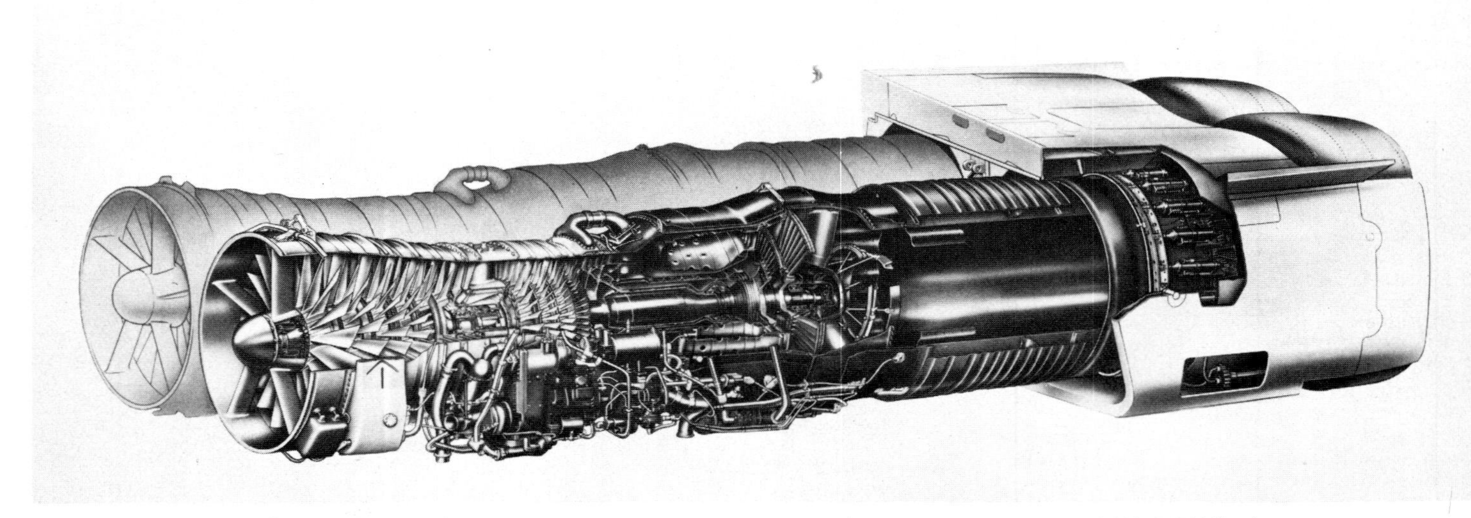

Cutaway drawing of installed pair of Rolls-Royce/SNECMA Olympus 593 turbojets (each 167·7 kN; 37,700 lb st)

gearbox drives main engine oil pressure/scavenge pumps and first-stage fuel pump. RH gearbox drives aircraft hydraulic pumps and IDG (Integrated Drive Generator).

LUBRICATION SYSTEM: Closed system, using oil to specification DERD.2497, MIL-L-9236B. Pressure pump, multiple scavenge pumps and return through Serck fuel/oil heat exchanger.

STARTING SYSTEM: SEMCA air-turbine starter drives HP spool. Dual high-energy ignition system.

MOUNTING: Main trunnions on horizontal centreline of delivery casing. Front stay from roof of nacelle picks up top of intake casing.

DIMENSIONS:
- Max diameter at intake: 1,206 mm (47·5 in)
- Length, flange-to-flange: 3,810 mm 149·5 in)
- Intake flange to final nozzle: 7,112 mm (280 in)

WEIGHT, DRY:
- With exhaust system: 3,386 kg (7,465 lb)
- Without exhaust system: 3,075 kg (6,780 lb)

PERFORMANCE RATINGS:
- T-O (S/L, ISA): 167·7 kN (37,700 lb st
- Contingency (S/L, ISA): 173·0 kN (38,900 lb st
- Cruise (M2·0 at 16,100 m; 53,000 ft, ISA + 5° C) 44·61 kN (10,030 lb

SPECIFIC FUEL CONSUMPTION:
- Cruise (as above): 33·7 mg/Ns (1·190 lb/hr/lb st

ROLLS-ROYCE TURBOMÉCA
ROLLS-ROYCE TURBOMÉCA LIMITED

ADDRESS:
4/5 Grosvenor Place, London SW1, England
Telephone: 01-235-3641

This company was formed jointly by Rolls-Royce and Turboméca in June 1966 to control the design, development and production programmes for the Adour two-shaft turbofan. The main function of the company is to receive contracts from the British Ministry of Defence on the Adour for both the British and French governments. The company can also enter into commercial contracts for the sale of the Adour to customers other than the British and French governments and grant licences for its manufacture.

ROLLS-ROYCE TURBOMÉCA ADOUR

The Adour was originally designed for the SEPECAT Jaguar strike/trainer aircraft to meet requirements laid down by the British and French joint air and naval staffs. It is a two-shaft turbofan engine fitted with an integral, modulated afterburner of advanced design. The whole engine is simple and robust and contains features permitting modular exchange, major engine sections being replaceable in the field, avoiding the need for return to an overhaul base.

The complete propulsion unit has been designed for an overhaul life of 1,000 hours. The engine temperatures and rotational speeds are moderate and a thrust growth of the order of 40 per cent within the confines of existing installations is envisaged.

Rolls-Royce Turboméca Adour 151 for Hawker Siddeley Hawk T.1, rated at 23·13 kN; 5,200 lb st

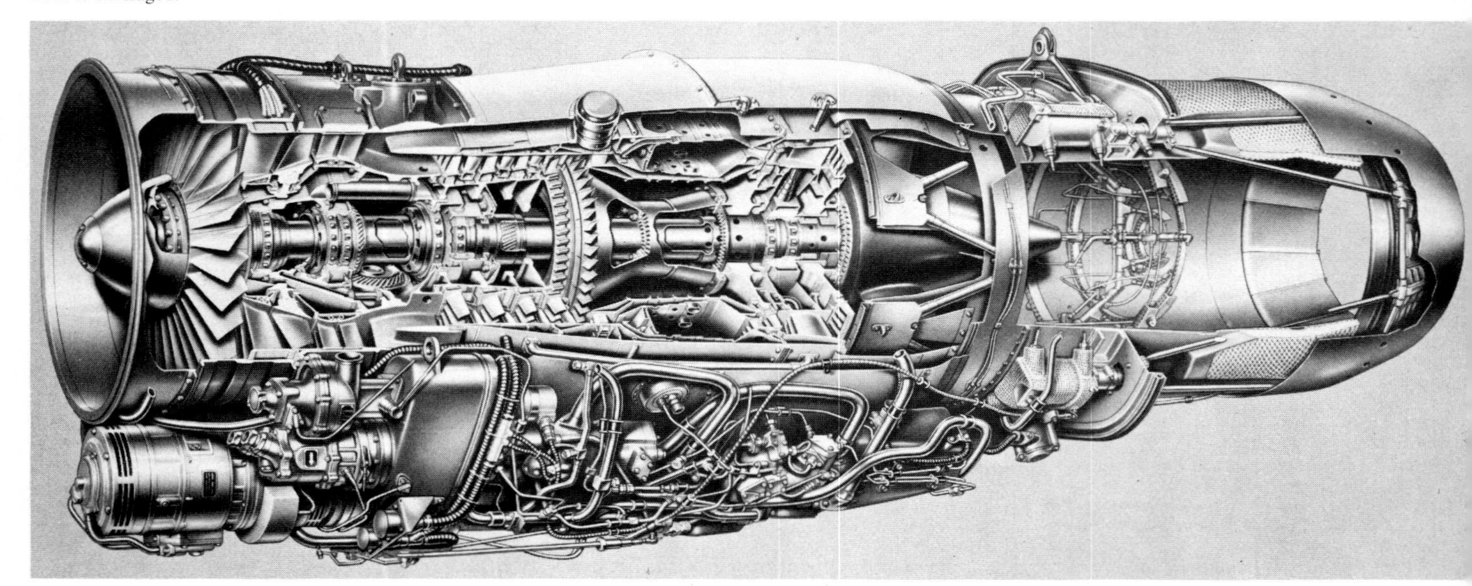

Cutaway drawing of the Rolls-Royce Turboméca Adour Mk 102 augmented turbofan (32·5 kN; 7,305 lb st)

Bench testing began at Derby on 9 May 1967. Engines re assembled at Derby (R-R) and Tarnos (Turboméca) om parts made at single sources in Britain and France. 'urboméca makes the compressors, casings and external ipework (to preserve Anglo-French parity the after-urner is subcontracted to SNECMA); Rolls-Royce akes the remainder. By mid-1976 total running time xceeded 200,000 hours.

The Adour has been selected for the Japanese T-2 ainer and FS-T2-KAI fighter/support aircraft, and since 970 Ishikawajima-Harima Heavy Industries has been roducing the Adour under a licence agreement. In 1972 a on-afterburning Adour was selected to power the lawker Siddeley Hawk advanced trainer. Further appli-ations of the Adour are under consideration, in military ircraft, business jets and small commercial transports.

Current versions of the engine are as follows:

Adour Mk 102. Standard production engine for Jaguars service with RAF and Armée de l'Air. Qualified in 972.

Mk 801A. Japanese designation TF40-IHI-801A. tandard production engine for Mitsubishi T-2 and S-T2-KAI. Qualified in 1972. (See Ishikawajima-arima in Japanese section).

Mk 151. Non-afterburning version for Hawk. Internal omponents and certification temperatures identical to lk 102 and Mk 801A. Qualified in 1975.

Mk 804. Uprated engine for Jaguar International. nstallationally interchangeable with Mk 102. General crease in thrust of some 10 per cent, with greater increase at high forward speeds (rating with full afterburner at Mach 0·9 at S/L, ISA increased by 27 per cent) giving improved aircraft performance throughout flight envelope. Qualified in 1976.

Further versions of the Adour, giving increased thrust and improved specific fuel consumption, are under development.

The following description refers to the Adour Mk 102:

TYPE: Two-shaft turbofan for subsonic and, with augmentation, supersonic aircraft.

INTAKE: Formed by forward extension of fan casing. No radial struts or inlet guide vanes.

FAN: Two-stage. Rotating spinner, anti-iced by turbine-bearing cooling air, on front of first-stage disc. Individually replaceable blades with no part-span or tip shrouds. Fixed stators and exit vanes. Whole unit overhung ahead of front LP roller bearing of squeeze-film type. Full-length by-pass duct leading to afterburner. By-pass ratio, 0·81 : 1.

COMPRESSOR: Five-stage compressor on HP shaft. Large-diameter double-conical shaft for rigidity with bolted curvic couplings. Wide-chord blades of titanium. Steel stator blades. Overall pressure ratio 11 : 1.

COMBUSTION CHAMBER: Annular, with straight-through flow. Fitted with 18 air-spray fuel nozzles and two igniter plugs. Engine fuel system by Lucas GTE.

HP TURBINE: Single-stage, aircooled.

LP TURBINE: Single-stage. Both turbine bearings of squeeze-film type.

JET PIPE: Fully modulated afterburner of compact, short-length design incorporating four concentric but staggered spray rings and vapour gutters. Catalytic igniters between inner gutters. Variable nozzle has eight master and eight slave petals positioned by eight-sided frame moved axially by four fuel-operated nozzle rams. Afterburner fuel flow and nozzle system by Dowty Fuel Systems, with vapour-core pump.

DIMENSIONS:
Inlet diameter (all) 559 mm (22 in)
Max width (all) 762 mm (30 in)
Max height (all) 1,041 mm (41 in)
Length:
Mks 102, 801A, 804 2,970 mm (117 in)
Mk 151 1,956 mm (77 in)

WEIGHT, DRY:
Mks 102, 801A 706 kg (1,556 lb)
Mk 151 535 kg (1,220 lb)
Mk 804 715 kg (1,576 lb)

PERFORMANCE RATINGS:
T-O dry:
Mks 102, 801A 22·75 kN (5,115 lb)
Mk 151 23·12 kN (5,200 lb)
Mk 804 23·4 kN (5,260 lb)
T-O with afterburner:
Mks 102, 801A 32·5 kN (7,305 lb)
Mk 804 35·6 kN (8,000 lb)

SPECIFIC FUEL CONSUMPTION (Mk 102):
S/L static, dry 21 mg/Ns (0·74 lb/hr/lb st)
Mach 0·8, 11,890 m (39,000 ft) 27 mg/Ns (0·955 lb/hr/lb st)

TURBO-UNION

TURBO-UNION LTD

ADDRESS:
PO Box 3, Filton, Bristol BS12 7QE, England
Telephone: 0272-693871

Formed in October 1969, this international company vas established to manage the entire programme for the RB.199 engine for the Panavia Tornado multi-role com-at aircraft. Shares are held in the ratio Fiat SpA, 20 per ent; MTU München GmbH, 40 per cent; Rolls-Royce 1971) Ltd, 40 per cent. The overall work-share of the RB.199 programme, involving design, development and roduction, is shared on the basis of the engine modules in he ratio: Fiat 15 per cent, MTU 42·5 per cent, RR 42·5 er cent.

TURBO-UNION RB.199

Designed originally by Rolls-Royce Bristol Engine Division, this engine competed with the Pratt & Whitney TF16 for propulsion of the Panavia Tornado (which see) nd was announced as the winner on 4 September 1969.

A three-shaft augmented turbofan of extremely dvanced design, the RB.199 has a lightweight fan based n that of the Pegasus on a reduced scale, an advanced fterburner designed for an extremely wide operating nvelope, and an integral thrust-reverser. The RB.199 ffers low fuel consumption for long-range dry cruise, ven at sea level, and approximately 100 per cent thrust ugmentation with full afterburner for combat manoeuvre nd supersonic acceleration. Further design goals included ninimal weight and frontal area, moderate first cost and conomic maintenance and overhaul. Strip, inspection nd rebuild are facilitated by modular construction, and by he use of electron-beam welding to reduce the number of eparate components.

The first prototype RB.199 ran at Bristol in September 971. The second ran in April 1972 on a newly built pen-air testbed at MTU Manching, near Munich, equip-ed with oil injection for flow visualisation in the reverse-hrust regime. In May 1973 the bench development pro-ramme of 15 engines was completed by the start of run-ing on a new enclosed test cell built by Fiat at Sangone, ear Turin. High-altitude testing under simulated Mach umbers exceeding 2 has been proceeding at Bristol, Derby and the National Gas Turbine Establishment at 'yestock. Flight testing began in April 1973 with an ngine installed in a dummy portion of the Tornado, com-lete with inlet and duct system and 27 mm Mauser gun, arried beneath a Vulcan. This subsonic flight testbed is nvestigating in-flight relighting, operation at high angles f attack and yaw, windmilling, gun firing, extraction of air leed to supply airframe services, and response to pilot emand. A further important development tool is an HP pool (core) rig, used for research on the crucial HP tur-ine rotor blades, which are among many components that ave benefited considerably from the optimisation of nanufacturing techniques.

Various problems encountered in 1973, centred on the IP turbine rotor blades and IP turbine disc, delayed flight learance from February to August 1974, though these roblems were completely solved in the course of 1974. Ground running in Tornado P.01 began in March 1974 nd this aircraft flew at Manching on 14 August 1974. Aircraft P.02 flew at Warton on 30 October 1974. By that ime more than 2,000 hours had been logged in static esting and 320 hours of subsonic flight with the Vulcan. Afterburner light-up has consistently been exemplary, nd as early as January 1974 main-engine relight had been

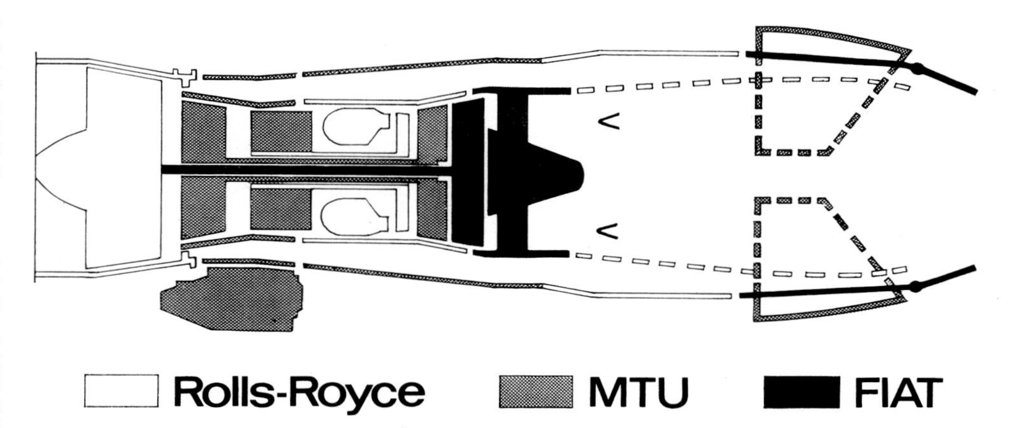

Simplified section through Turbo-Union RB.199 showing portions for which each company is responsible

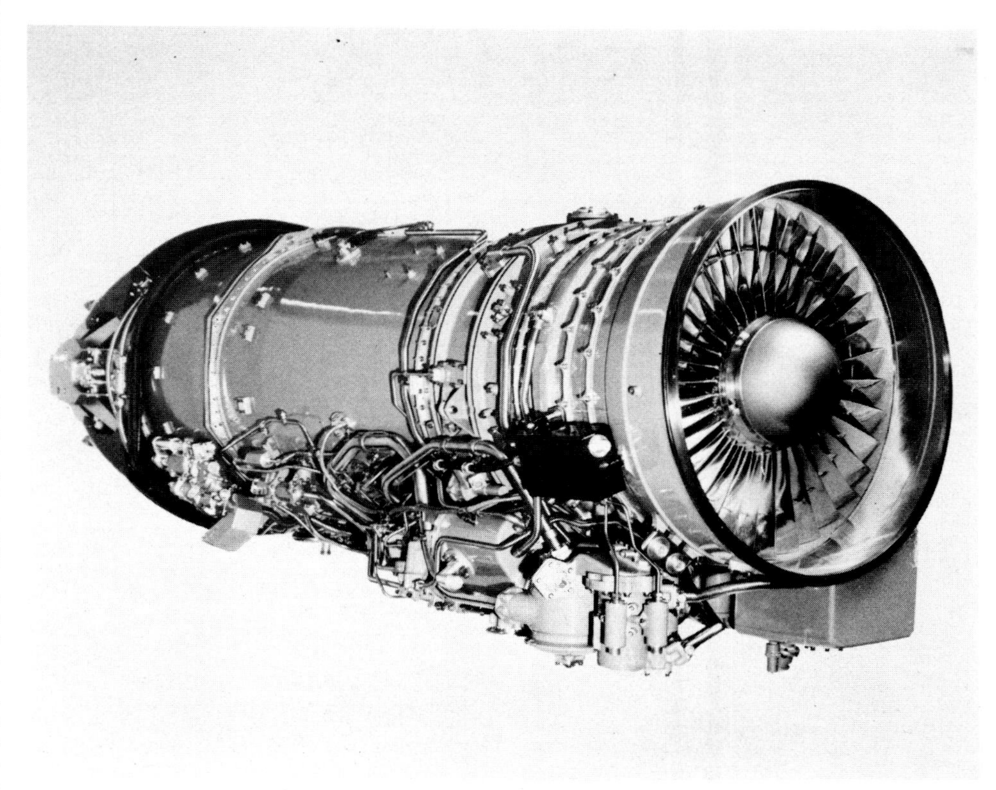

Turbo-Union RB.199-34R three-shaft augmented turbofan (photograph shows mockup/display exhibit)

demonstrated 1,525 m (5,000 ft) higher than demanded by specification for entry into service, and 3,050 m (10,000 ft) higher than specification for initial flying.

The following are principal existing or projected versions of the RB.199 which have been announced officially or unofficially:

RB.199-34R. Designation of the basic engine adopted to power the Tornado. The official figures for performance given later apply to this engine.

RB.199-01. Initial engine used in flight development. For prototype flight clearance a reduced performance was accepted, dry thrust being 11 per cent below that required for FQT (Formal Qualification Testing) and maximum augmented thrust 19 per cent below; both figures were well within the allowed limits. This engine has since been superseded by the -02, and all Tornado flying was being accomplished on -02 engines by late 1975.

RB.199-02. Has an increased annulus area in IP and LP turbines, achieved by minor redesign of the blades within the existing turbine casing. In addition to improving the development potential of the engine, the immediate effect was to enable the engine to deliver increased thrust at existing temperatures. A flight clearance test was completed in late 1974, and one engine was fitted to Tornado P.02. Despite delays, which by April 1975 had led to a shortage of engines and delayed the flight of P.03, a second RB.199-02 was fitted to P.02; by April this aircraft had conducted valuable flight development with the improved engine and explored relight boundaries. The -02 engines were in April 1975 being installed in both P.03 and P.04, and retrofitted to P.01 as it became due for engine change. A bench engine built to the -02 standard had demonstrated final FQT specification thrust, both dry and with afterburner at lower turbine gas temperatures than had previously been achieved. It had demonstrated an sfc 5·0 per cent better than that achieved with any -01 engine, and an afterburning sfc better than that required for FQT.

RB.199-03. Embodies significant improvements to fan and afterburner, giving increased performance. Tornado flying with -03 engines was scheduled to begin in Autumn 1976.

RB.199 Mk 101. This designation applies to the initial production-standard engine, as ordered in July 1976. It corresponds to the -04 development standard.

RB.199 uprated. Turbo-Union is reported to have studied an advanced RB.199 rated at 79·6 kN (17,900 lb) st with maximum augmentation. Such an engine could power later Tornado versions or the projected Hindustan HF-73. A third potential application is in developments of the McDonnell Douglas/Northrop F-18.

The following description refers to the RB.199-34R in 02 form:

TYPE: Three-shaft augmented turbofan with integral afterburner and reverser.

INTAKE: Direct pitot type; details restricted.

FAN: Three-stage axial, aerodynamically derived from that of Pegasus. Discs and blades of titanium alloy. Maximum air mass flow approx 70 kg (154 lb)/sec. By-pass ratio more than 1 : 1.

IP COMPRESSOR: Three-stage axial. Rotor mainly of titanium alloy. Casing copy-milled in titanium.

HP COMPRESSOR: Six-stage axial. Overall pressure ratio greater than 20 : 1.

COMBUSTION CHAMBER: Annular, with vaporising burners. Combustion without visible smoke.

HP TURBINE: Highly loaded single stage with advanced cooling techniques. Entry gas temperature over 1,277°C.

IP TURBINE: Single-stage axial, with cooled rotor blades.

LP TURBINE: Two-stage axial.

JET PIPE: Close-coupled afterburner, giving fully modulated augmentation. Maximum gas temperature over 1,627°C. Convergent/divergent nozzle of variable profile and area, matched to wide range of flight conditions. Outer nozzle flaps actuated by four screwjacks driven by an air motor, moving translating ring and in turn driving segments of primary nozzle.

REVERSER: Target type, with upper and lower clamshells driven by left and right actuators.

ACCESSORIES: Radial shaft drives packaged external gearbox carrying hydromechanical portion of main and afterburner fuel system. Fuel control of advanced electronic type. Engine gearbox mechanically connected to aircraft gearbox and KHD starter gas-turbine.

DIMENSIONS:
- Max diameter: 870 mm (34·25 in)
- Overall length: approx 3,250 mm (128 in)

WEIGHT, DRY: approx 8,165 kg (1,800 lb)

PERFORMANCE RATINGS:
- T-O dry: over 31·1 kN (7,000 lb st)
- T-O with max afterburning: over 62·2 kN (14,000 lb st)

ISRAEL

BET-SHEMESH
BET-SHEMESH ENGINES LTD

OFFICE AND WORKS:
Bet-Shemesh (between Tel Aviv and Jerusalem)

The first section of a 12,077 m² (130,000 sq ft) new Israeli aero-engine factory, Bet-Shemesh Engines Ltd, was officially inaugurated on 15 January 1969. The company is owned 49 per cent by the Israeli government and 51 per cent by Turboméca SA and the manufacturing plant is based on the Turboméca factory at Tarnos. The payroll is planned to rise to around 1,100. Initially Bet-Shemesh manufactured turboprop components on behalf of Turboméca. By 1973 complete Marboré VI turbojets for CM 170 Super Magister trainers were being produced. The company also manufactures parts of the Marboré II, Artouse II and III, Turmo II and Astazou II. A non-aviation product is the M2T1 industrial gas turbine.

An associate company of Bet-Shemesh Engines has also been established to produce many of the special small-scale cast items for Israeli-manufactured Turboméca engines and components. Known as Misco-Bet-Shemesh Ltd, the company is owned 25 per cent each by the Israeli government and Turboméca, and 50 per cent by the Howmet Corp of Muskegon, USA.

ITALY

ALFA-ROMEO
SOCIETÀ PER AZIONI ALFA-ROMEO

HEAD OFFICE:
Via Gattamelata 45, 20149 Milan
Telephone: Milan 3977

AVIATION WORKS:
80038 Pomigliano D'Arco, Naples
Telephone: 8841 344

This company, famous as an automotive manufacturer, entered the aero-engine industry in Italy in 1925 by acquiring licences for the Jupiter engine from the Bristol Aeroplane Co Ltd, and the Lynx engine from Armstrong Siddeley Motors Ltd. In 1930 the company produced its first engine of original design and remained an important manufacturer of piston engines for Italian aircraft until 1956.

For the following few years, Alfa-Romeo restricted its aviation work to overhaul of Curtiss-Wright R-1820 and R-3350 piston engines and Wright J65, and Rolls-Royce Dart, Avon and Conway turbine engines. Subsequently, overhaul has also been undertaken of Rolls-Royce Gnome turboshafts, later marks (514, 527 and 528) of the Dart turboprop, General Electric J85-GE-13A turbojets and T58 turboshafts, Pratt & Whitney JT3D and JT8D turbofans and PT6 turboprops, and Allison T56 turboprops.

Alfa-Romeo resumed its manufacturing activities by participating in the European production programme for General Electric J79-GE-11A turbojets to power Lockheed F-104G Starfighters built in Europe. It is collaborating with Rolls-Royce in the manufacture and overhaul of Gnome H.1000, H.1200 and H.1400 turboshafts. A licence agreement has also been signed for the manufacture of General Electric T58 turboshafts.

Alfa-Romeo is prime contractor for the manufacture, under General Electric licence, of the J85-GE-13A turbojet to power the G91Y aircraft. It is also manufacturing the "hot" section of the J79-GE-19 turbojet engines for the F-104S Starfighters that are being produced in Italy for the Italian Air Force, as well as many parts for the J79-11B, 17 and -1K. Alfa-Romeo manufactures CF6 combustors under subcontract to General Electric; assembles the kits and overhauls P&WC PT6T engines for the AB 212 helicopter; under GE licence, it is responsible for the hot section of the T64-P4D turboprop co-produced with Fiat; and Alfa-Romeo participates in development and hot-section component manufacture for the Turbo-Union RB.199 for the Panavia MRCA and Pratt & Whitney JT10D. It is continuing to study a gas turbine engine in the 373 kW (500 hp) class for general aviation use.

Alfa-Romeo is a member of the Finmeccanica-IRI group of companies.

FIAT
FIAT, SOCIETÀ PER AZIONI

HEADQUARTERS:
Corso Marconi 10, 10100 Turin
Telephone: (011) 65651

CHAIRMAN: G. Agnelli
VICE CHAIRMAN: F. Rota
MANAGING DIRECTOR: U. Agnelli
GENERAL DIRECTOR: N. Gioia

Aviation Division:
Via Nizza 312, 10100 Turin
Telephone: (011) 6399

DIRECTOR: G. C. Boffetta
DESIGN AND DEVELOPMENT DIRECTOR: G. Maoli
PRODUCTION DIRECTOR: P. G. Covati

TEST CENTRE:
Officine del Sangone, Strada del Drosso 145, 10100 Turin

The Fiat company was incorporated in 1899 and built its first aero-engines in 1908.

Fiat entered the gas-turbine field after the second World War, by undertaking licence production of the de Havilland Ghost centrifugal-flow turbojet engine and then, in 1953, the manufacture of parts and assemblies of the Allison J35 turbojet engine for the USAF.

In 1960, Fiat began manufacture of the Orpheus 803 turbojet engine, and overhaul of the Orpheus 801 and 803, under licence from Bristol Siddeley Engines Ltd. In collaboration with FN of Belgium, BMW of Germany and Alfa-Romeo of Italy, Fiat also produced components and carried out the assembly of General Electric J79 turbojet engines for F-104G Starfighters built in Europe.

Fiat's main aircraft engine programmes now concern the Turbo-Union RB.199, Rolls-Royce Viper 600, Pratt & Whitney JT10D and General Electric J79, J85, CF6 and T64, all of which are referred to below. In addition the company overhauls many types of engine, including the J79 and Orpheus, and the R-2800 piston engine. It is engaged in the design, development and production of the LM 2500 marine gas turbine, in collaboration with General Electric, and the design and production of main gearboxes for Aérospatiale helicopters.

TURBO-UNION RB.199

Fiat holds 20 per cent of the shares of Turbo-Union Ltd, the joint company set up to produce the RB.199-34R engine for the MRCA. Fiat's responsibility is the LP turbine and shaft, exhaust diffuser, jetpipe and nozzle. The programme is described under Turbo-Union in the International part of this section.

PRATT & WHITNEY JT10D

Since 1974 Fiat has been responsible for design and development of the accessory drive gearbox for the Pratt & Whitney JT10D civil turbofan engine. The programme is described under Pratt & Whitney in the US part of this section.

GENERAL ELECTRIC CF6

Fiat is engaged in the manufacture of components for the CF6 civil turbofan for both General Electric and SNECMA. For GE the company supplies complete accessory gearboxes, transfer gearboxes, inlet gearboxes and shafts. SNECMA is supplied with various gearbox components and shafts for CF6-50 engines for the Airbus A300.

GENERAL ELECTRIC T64-P4D

This free-turbine turboprop powers the Aeritalia G 222 military transport aircraft. Under a licence agreement between the General Electric Company and the Italian government, the engine is being manufactured in Italy under the leadership of Fiat as prime contractor, with Alfa-Romeo a major subcontractor.

GENERAL ELECTRIC J79-19

This afterburning turbojet powers the Lockheed F-104S fighters of the Italian Air Force. Fiat, as prime contractor for Italy, is producing the engine under a licence agreement between the General Electric Company and the Italian government.

GENERAL ELECTRIC J85-13

With Alfa-Romeo as prime contractor, Fiat participates in production under licence of the J85-13A, power plant of the Aeritalia G91Y aircraft. Fiat manufactures the compressor rotor assembly, power takeoff, accessory gearbox, afterburner casing and other components amounting to approximately 40 per cent of the engine.

ROLLS-ROYCE VIPER 600

The development and production of this turbojet for business aircraft and military trainers is a joint venture between Rolls-Royce (1971) and Fiat. Fiat's responsibility extends to all components rearward of the compressor housing (except turbine discs and blades), together with the exhaust assembly and silencer designed to meet noise legislation. The Viper 600 is described in the Rolls-Royce entry in this section.

PIAGGIO
INDUSTRIE AERONAUTICHE E MECCANICHE RINALDO PIAGGIO SpA

Head Office:
Viale Brigata Bisagno 14, 16129 Genoa (426)
Telephone: 540 521

Works and Officers:
See "Aircraft" section

The Aero Engine Division of Piaggio is currently manufacturing the following engines under various licence agreements: Rolls-Royce Bristol Viper 11, 526 and 540 turbojets to power the Aermacchi M.B. 326 and the Piaggio PD-808 (a sublicence for the manufacture of the Viper 11 and 540 was issued to Atlas Aircraft to power South African-built M.B. 326 aircraft); Avco Lycoming T53-L-13A and L-13B and T55-L-11A turboshaft engines ordered by the Italian government for installation in the Bell 204B/205 and CH-47C Chinook helicopters, respectively; and the Lycoming VO-435 and GSO-480 piston engines.

SNIA VISCOSA
DEFENCE AND AEROSPACE DIVISION

Address:
via Lombardia 31, 00187 Rome
Telephone: (06) 4680
Telex: 61114 SNIA
Factories:
Colleferro (Rome) and Ceccano (Frosinone)
Division Manager: Ing E. Svizzeretto
Chief of Rocket, Missiles and Space R & D Department: Ing P. Laurienzo

The SNIA Viscosa Defence and Aerospace Division is engaged in the research, design and production of solid propellants, solid-propellant rocket motors, complete rockets and missiles.

The company possesses all the necessary installations for the production of double-base solid propellants, from the nitration of glycerine and cotton to the production of grains by extrusion or casting. Composite solid-propellant grains and motors incorporating polyurethane or polybutadiene polymers as binder are also produced in all sizes, with high specific impulses and outstanding physical and mechanical characteristics.

Complete propulsion units of up to 600 mm (23·6 in) diameter, with combustion times ranging from a fraction of a second to about 1 minute, are manufactured by SNIA Viscosa's Defence and Aerospace Division for use in military rockets and missiles. A major production item in this field is the SNIA Type ARF/8M 2 in folding-fin air-to-surface rocket.

SNIA's Defence and Aerospace Division also produced the motors for the Italian Sparrow air-to-air missile programme, and is currently involved in the development of many weapon systems in co-operation with companies such as Messerschmitt-Bölkow-Blohm, Selenia, Breda Meccanica Bresciana and Oto Melara.

In the aerospace field, in addition to various single- and two-stage sounding rocket prototypes which have been tested successfully many times, the apogee motor for the ELDO Europa II programme was brought to the final qualification stage by SNIA. For the Italian Sirio space programme a new amagnetic high-performance apogee motor has been developed, with a titanium alloy case and improved polybutadiene propellant. Other apogee motors are being designed for ESA programmes: for the GEOS satellite a glassfibre apogee motor is being developed with the co-operation of SEP of France; for OTS, Meteosat and Aerosat a high-performance apogee motor with carbon

SNIA Viscosa Sirio apogee boost motor

fibre case is being developed by a consortium formed by SNIA, SEP and MAN. In early 1974 SNIA was assigned by CNES the responsibility for design and manufacture of retro and ullage motors for the separation of all three stages of the Ariane vehicle.

JAPAN

IHI
ISHIKAWAJIMA-HARIMA JUKOGYO KABUSHIKI KAISHA (Ishikawajima-Harima Heavy Industries Co Ltd)

Head Office:
No 2-1, 2 chome, Ote-Machi, Chiyoda-ku, Tokyo
Aero Engine and Space Development Group:
3-5-1, Mukodai-cho, Tanashi-shi, Tokyo 188
Telephone: (0424) 66-1252
President: Dr Hisashi Shinto
Executive Vice-President:
Dr Osamu Nagano
Managing Director and General Manager, AESD Group:
Dr Kaneichiro Imai

On 1 December 1960, Ishikawajima Heavy Industries Co was merged with Harima Shipbuilding Co and has since operated as Ishikawajima-Harima Heavy Industries Co.

Its work on turbojet engines dates from 19 June 1956, when the Japanese government approved a licence agreement with the General Electric Company of Cincinnati, Ohio, USA, under which Ishikawajima has been producing spares for the J47-GE-27 turbojet engines which power F-86F Sabre fighter aircraft of the Japan Air Self-Defence Force.

In February 1960 IHI began the licence production of General Electric J79-IHI-11A turbojet engines for Japanese-built Lockheed F-104J Starfighters. In 1967 the company built a J79 gas generator for use in an 8,000 kW generating set.

Under further licensing agreements with General Electric, IHI is producing the J79-IHI-17 turbojet for the McDonnell Douglas F-4EJ, the T58 turboshaft for helicopters and other applications, including the propulsion of air cushion vehicles and hydrofoil boats, and the T64-IHI-10 turboprop engine to power the JMSDF's PS-1 anti-submarine flying -boat and US-1 search and rescue amphibian, and the Kawasaki P-2J maritime patrol aircraft. In addition the T64 has performed well in a Japanese Navy torpedo boat. By December 1975 deliveries totalled 505 T58s, 286 T64s and 481 J79s.

The manufacture of the Rolls-Royce Turboméca Adour augmented turbofan in Japan under licence agreement received government approval in September 1970. Its production began in early 1973, under the Japanese designation TF40-IHI-801A.

IHI undertakes overhaul and repair of Pratt & Whitney JT8D and Rolls-Royce RB.211 commercial turbofans, and General Electric J79, T58 and T64, and R-R Turboméca TF40 military engines, and Turboméca Artouste and Astazou turboshafts.

Ishikawajima-Harima J3-IHI-7C turbojet engine (13·7 kN; 3,080 lb st)

Prior to the start of licence production, in April 1959, IHI had been responsible for the J3 turbojet engine which had been under development by the Nippon Jet-Engine Company since 1956. The J3-IHI-7 version is installed in Fuji T-1B intermediate jet trainer and Kawasaki P-2J aircraft.

IHI participated in developing the XJ11 liftjet, as well as the JR100, JR200 and JR300 built under supervision of the NAL. In addition, in collaboration with Mitsubishi and Kawasaki, IHI made the prototypes of the FJR710 turbofan. The JR100, JR200 and FJR710 are described in the NAL entry.

IHI is co-operating with JDA in the development of small turbofans in the 11·8 kN (2,645 lb) thrust class.

IHI J3-IHI-7C

The J3-IHI-7C is a derivative of the J3-1, of which a description appeared in the 1959-60 *Jane's,* under the entry for 'Nippon Jet-Engine Company'. It is installed in the Kawasaki P-2J aircraft currently in service with the JMSDF, and is in production. By December 1975 177 had been delivered, with about 80 remaining to be built.

Studies are also underway for converting the J3-IHI-7 to an augmented turbofan. Major modifications include the introduction of a shorter combustion chamber, provision for air bleed for aircraft BLC purposes, revised turbine blading, and the addition of an aft-fan and afterburner. An experimental version with augmentation reached a thrust of 20·2 kN (4,542 lb) during bench tests in December 1972.

The following data apply to the J3-IHI-7C:
TYPE: Axial-flow turbojet.
AIR INTAKE: Annular nose air intake. Anti-icing system for front support struts.
COMPRESSOR: Eight-stage axial-flow type, built of Ni-Cr-Mo steel. Rotor consists of a series of discs and spacers bolted on to shaft. Rotor and stator blades of AISI 403 steel. Stator blades brazed on to fabricated base which is fixed in casing with circumferential T-groove. Rotor blades dovetailed to discs. Light alloy casing in upper and lower sections, flange-jointed together. Pressure ratio 4·5 : 1. Air mass flow 25·4 kg (56 lb)/sec.
COMBUSTION CHAMBER: Annular type. AISI 321 steel outer casing. L 605 steel flame tube. Thirty fuel supply pipes located in combustion chamber outer casing and 30 vaporiser tubes located at front of flame tube. Ignition by low-voltage high-energy spark plug in each side of combustion chamber.
FUEL SYSTEM: Hydromechanical, with IHI FC-2 fuel control.
FUEL GRADE: JP-4.
NOZZLE GUIDE VANES: Single row of aircooled fabricated vanes.
TURBINE: Single-stage axial-flow type. Disc bolted to shaft. Precision-forged blades.
BEARINGS: Rotating assembly carried in front (double ball) and rear (roller) compressor rotor bearings and rear (roller) turbine shaft bearing.
JET PIPE: Fixed-area type.
ACCESSORY DRIVES: On gearbox under compressor front casing.
LUBRICATION SYSTEM: Forced-feed system for main bearings and gear case. Dry sump. Vane-type positive displacement supply and scavenge pump.
OIL SPECIFICATION: MIL-L-7808.
MOUNTING: Three-point suspension, with one pickup by a pin on starboard side of compressor front casing and a trunnion on each side of the compressor rear casing.
STARTING: Electrical starter in intake bullet fairing.
DIMENSIONS:
Length, less tailpipe 1,661 mm (65·4 in)
Length overall with rear cone 1,994 mm (78·5 in)
Diameter overall 627 mm (24·7 in)
Frontal area 0·28 m² (3·01 sq ft)
WEIGHT, DRY:
Bare 380 kg (838 lb)
With accessories 430 kg (948 lb)
PERFORMANCE RATING:
T-O 13·7 kN (3,080 lb st)
SPECIFIC FUEL CONSUMPTION:
At T-O rating 29·74 mg/Ns (1·05 lb/hr/lb st)
OIL CONSUMPTION:
At normal rating (max) 0·60 litres (1·06 Imp pints)/hr

IHI J3-IHI-8

This engine is an uprated version of the J3-IHI-7 and has completed military qualification testing. All details for the J3-IHI-7 apply equally to the J3-IHI-8, with the following exception:
PERFORMANCE RATING:
T-O 15·2 kN (3,415 lb) st at 13,000 rpm

IHI XJ11

The XJ11 is an engine which IHI is developing to follow the J3 series. Design objectives include an especially lightweight structural design, highly-loaded main engine components, simplified fuel and lubrication systems, and thrust/weight ratio of 20 : 1.

IHI TF40-IHI-801A

This is the Rolls-Royce Turboméca Adour augmented turbofan engine, as built under licence by IHI. TF40 engines are now being delivered with Japanese-made parts in every major assembly. Total deliveries by November 1975 were 44. For details see Rolls-Royce Turboméca entry in International section.

KAWASAKI
KAWASAKI JUKOGYO KABUSHIKI KAISHA (Kawasaki Heavy Industries Ltd)

HEAD OFFICE:
2-16-1 Nakamachi-Dori, Ikuta-ku, Kobe
Telephone: Kobe (078) 341-7731
WORKS:
Gifu and Akashi
OFFICERS: See "Aircraft" section

Kawasaki's factory at Akashi started repair overhaul and component manufacturing for aircraft engines, on behalf of the US armed forces and the Japan Defence Agency, in 1953. Since then, Kawasaki has overhauled more than 11,000 engines, mainly of the Allison J33, General Electric J47, Rolls-Royce Orpheus, Westinghouse J34, and Kawasaki KT5311A and KT5313B series.

In 1967, under a licence agreement with Avco Lycoming, Kawasaki started manufacturing T5311A turboshaft engines (Kawasaki designation KT5311A), followed in 1972 by the T53-L-13B (T53-K-13B) and T5313B (KT5313B), which are installed in Bell UH-1B/H helicopters manufactured by Fuji in Japan. Deliveries of KT5311A, KT5313B and T53-K-13B engines had reached 197 units by February 1976. About 20 are being manufactured annually.

In collaboration with MHI and IHI, Kawasaki participates in the Ministry of International Trade and Industry's development programme for the FJR710 high by-pass ratio turbofan engine (see NAL entry).

Kawasaki shares in parts manufacturing for the Pratt & Whitney JT8D engines produced by MHI for the Kawasaki C-1 medium-range military transport aircraft, and also for the Rolls-Royce Turboméca Adour produced by IHI for the Mitsubishi T-2 supersonic trainer and FS-T2-KAI.

MITSUBISHI
MITSUBISHI JUKOGYO KABUSHIKI KAISHA (Mitsubishi Heavy Industries Ltd)

HEAD OFFICE:
5-1, Marunouchi 2 chome, Chiyoda-ku, Tokyo 100
ENGINE WORKS:
Daiko Plant, Nagoya Aircraft Works, 1-1, Daiko-cho, Higashi-ku, Nagoya 455
Telephone: Nagoya (052) 721-3111
Komaki North Plant, Nagoya Aircraft Works, 1200, Higashi-Tanaka, Komaki-Shi, Aichi 485
Telephone: Komaki (0568) 79-2111
OFFICERS:
See "Aircraft" section

Since 1952 Mitsubishi has been responsible for the repair and overhaul of engines of the Japan Defence Agency, domestic and foreign airlines and the US Air Force.

In 1967 Mitsubishi resumed its activity in the aviation gas-turbine field by undertaking manufacture of the CT63 turboshaft engine to power Hughes 369HM helicopters of the JGSDF under a licence agreement with Allison. A total of 179 engines had been delivered to the Japan Defence Agency by January 1976, and the eventual number to be manufactured in Japan is expected to be 300. In 1972, under licence agreement with Pratt & Whitney Aircraft, Mitsubishi began the manufacture of the JT8D-M-9 turbofan. The first was delivered in January 1973. By 1977 a total of 70 engines are to be delivered to the Japan Defence Agency for use in the Kawasaki C-1 military transport. Deliveries in January 1976 had reached 31 engines, on schedule.

Mitsubishi is engaged in the series production of a gas-turbine compressor set of its own design. Designated GCM-1, this is built into the power pack of the starter trolley which supports the fighter aircraft of the Japan Air Self-Defence Force.

NAL
NATIONAL AEROSPACE LABORATORY

ADDRESS:
1880 Jindaiji-machi, Chofu City, Tokyo
Telephone: 0422-47-5911
DIRECTOR: Masao Yamanouchi
HEAD OF AERO-ENGINE DIVISION:
Masakatsu Matsuki

The National Aerospace Laboratory (NAL) is a government establishment responsible for research and development in the field of aeronautical and space science. Since 1962 it has extended its activity to include V/STOL techniques. The decision was made in that year to initiate development of an engine to fulfil the requirement for a lightweight lift-jet power plant for VTOL aircraft.

The lift-jet engine was designated NAL/IHI JR100 series, and features a thrust-to-weight ratio of 10. The first version was completed in 1964. Two sets of NAL/IHI JR100Fs were installed in the NAL Flying Test Bed (see 1972-73 *Jane's),* and altogether IHI produced six lift-jets of the JR100 series.

The more advanced NAL/IHI JR200 was developed in 1966, the NAL/IHI JR220 was completed in 1971, and the latest development programme concerns the JR300.

In 1971 the Agency of Industrial Science and Technology, Ministry of International Trade and Industry (MITI), funded a high by-pass ratio turbofan engine (FJR710) development programme. NAL has completed the basic design of this engine, and many component tests are being made at NAL. Industrial companies, chiefly IHI, are making development engines for this programme.

NAL/IHI JR100

As a part of a V/STOL research and development programme, NAL designed and developed a simple 10 : 1 thrust/weight ratio lift-jet. The prototype engine, manufactured by IHI, was completed during 1964, and in the course of 150 hours of testing, including an endurance test in March 1969, some 1,300 starts have been completed. Engines were formerly being employed in testing a height control system for a "soft" landing, using the NAL/IHI JR 100H version, and in a Flying Test Bed for stability studies, using the NAL/IHI JR100F. This research was complete by mid-1971.

An interim version, the JR100H2, was to be tested in 1975. Later the operational JR100V will be built, five being planned for NAL VTOL research aircraft. A JR100H is also being used for noise research.
TYPE: Single-shaft axial-flow lift-jet.
AIR INTAKE: Forward-facing annular type.
COMPRESSOR: Six-stage axial unit of 3·9 : 1 pressure ratio and 27·5 kg (60·6 lb)/sec air mass flow.
COMBUSTION CHAMBER: Annular type.
TURBINE: Single-stage axial unit with 850°C entry temperature.
JET PIPE: Fixed area.
FUEL SYSTEM: Hydromechanical system with single master control.
LUBRICATION SYSTEM: Non-return, intermittent oil supply system.
DIMENSIONS:
Diameter 600 mm (23·6 in)
Length overall 975 mm (38·4 in)
WEIGHT, DRY: 156 kg (347·5 lb)
PERFORMANCE RATINGS:
Max T-O:
JR100F 14 kN (3,150 lb st)
JR100H 14·9 kN (3,360 lb st)
JR100V 14·2 kN (3,200 lb st)
SPECIFIC FUEL CONSUMPTION:
At max T-O rating:
JR100F 32·6 mg/Ns (1·15 lb/hr/lb)
JR100H 32 mg/Ns (1·13 lb/hr/lb)

NAL/IHI JR100 lift-jet (14-14·9 kN; 3,150-3,360 lb st)

NAL/IHI JR200 and JR220

Following work on the NAL/IHI JR100, NAL designed and developed the higher-thrust NAL/IHI JR200 of improved thrust/weight ratio, and this was manufactured by IHI. The design objectives of the NAL/IHI JR200 are the same as for the NAL/IHI JR100, but use is made of a higher air mass flow, smaller combustion chamber and more extensive lightweight materials. The prototype NAL/IHI JR200 was completed in the Summer of 1966.

An improved version, the NAL/IHI JR220 with higher pressure ratio and higher turbine entry temperature, is now under initial development. The prototype NAL/IHI JR220 has achieved its performance at NAL's test cell.

The following details relate to the JR200:
TYPE: Single-shaft axial lift-jet
AIR INTAKE: Forward-facing annular type.

MITI/NAL FJR710 two-shaft turbofan (49 kN; 11,025 lb st)

NAL/IHI JR200 lift-jet (20·4 kN; 4,585 lb st)

COMPRESSOR: Five-stage axial unit of 4 : 1 pressure ratio and 37·2 kg (82 lb)/sec air mass flow at 12,450 rpm. Air bleed 3 kg (6·6 lb)/sec.
COMBUSTION CHAMBER: Annular type.
TURBINE: Single-stage axial unit with 850°C entry temperature.
JET PIPE: Fixed area.
WEIGHT, DRY: 127 kg (280 lb)
PERFORMANCE RATINGS:
Max, without air bleed 20·4 kN (4,585 lb)
Max, with air bleed 17·8 kN (4,012 lb)
SPECIFIC FUEL CONSUMPTION:
At max rating, without air bleed 1·13
At max rating, with air bleed 1·17

MITI/NAL FJR710

In the late 1960s the Japanese government and industry, seeking an engine programme that might remain competitive for many years, decided to embark on the design of a subsonic turbofan of high by-pass ratio. After a preliminary study by the NAL, funding was provided by the Ministry of International Trade and Industry in 1971 for a prototype demonstrator and test programme.

NAL has managed the design of the resulting FJR710. Manufacture of the prototype and development engines was subcontracted to IHI, Kawasaki and Mitsubishi. The first engine made its first run in May 1973. By the end of 1975 six engines (three FJR710/10 and three FJR710/20 with small changes) had run a total of 200 hours. One engine had completed a 60-hour endurance test.

The current test programme is hoped to lead to a production engine in the 44·5-66·8 kN (10,000-15,000 lb st) class. It is planned that there should also be a second major stage of development, probably several years hence, which would raise thrust to at least 89 kN (20,000 lb).

The following description applies to the first prototype engine, and is provisional:

TYPE: Two-shaft high by-pass ratio turbofan for subsonic commercial or military aircraft.
AIR INTAKE: Direct annular entry around fan spinner.
FAN: Single-stage fan, with rotating spinner and inserted titanium blades with part-span shrouds. Metal fan duct held by eight aerofoil struts, preceded by ring of flow-straightening vanes. By-pass ratio 6·5 : 1.
COMPRESSOR: Mechanically independent HP compressor. Multi-stage axial assembly with inserted blades of titanium and, at delivery end, high-nickel alloy. Several rows of variable stator blades held in upper and lower half-casings and operated by peripheral rings scheduled by hydraulic ram.
COMBUSTION CHAMBER: High-intensity smokeless annular type.
TURBINE: Two-stage HP gas-generator turbine with cooled blades. Multi-stage LP fan turbine. Entry temperature 1,150°C.
JET PIPE: Fixed area.
DIMENSIONS (approx):
Length 3,300 mm (130 in)
Diameter 1,520 mm (60 in)
WEIGHT, DRY: 1,247 kg (2,750 lb)
PERFORMANCE RATINGS (ISA):
T-O 49 kN (11,025 lb)
Cruise at 6,100 m (20,000 ft) at Mach 0·7 16·7 kN (3,748 lb)
SPECIFIC FUEL CONSUMPTION:
T-O 9·63 mg/Ns (0·34 lb/hr/lb st)
Cruise, as above 17·3 mg/Ns (0·61 lb/hr/lb st)

POLAND

BORZECKI
JOZEF BORZECKI

ADDRESS:
Wroclaw, ul. Sernicka 20/4

BORZECKI 2RB

The 2RB is a small piston engine designed and built by J. Borzecki for use in motor gliders. The prototype was built in 1970 and in the same year underwent bench tests. A flat-four, it may be used with tractor or pusher propeller. The engine began flight testing in early 1972 in the Borzecki Alto-Stratus motor glider. Development is continuing. The engine is based on the Riedel engine used to start German turbojets in World War 2.

TYPE: Four-cylinder two-stroke horizontally-opposed air-cooled piston engine.
CYLINDERS: Bore 70 mm (2·76 in).Stroke 35 mm (1·38 in). Swept volume 540 cc (33 cu in). Compression ratio 7·2 : 1. Aluminium alloy cylinders and heads attached to crankcase by four studs. Steel liners.
PISTONS: Aluminium alloy, with two compression rings.
CONNECTING RODS: Milled from steel.
CRANKSHAFT: Steel shaft, supported in four ball bearings and one ball-thrust bearing at the front.
CRANKCASE: Aluminium alloy case, divided at the vertical longitudinal and transverse centre-lines with aft cover.
INDUCTION: Jawa 250 motor-cycle carburettor.
FUEL: Mixture of 76 octane petrol and oil.
IGNITION: Battery ignition. One M14-250 14 mm (0·55 in) sparking plug per cylinder.
MOUNTING: Four rubber dampers at rear of crankcase or four studs at cylinder heads.
PROPELLER DRIVE: Direct tractor or pusher.
DIMENSIONS:
Length overall, with propeller boss 450 mm (17·7 in)
Width 250 mm (9·8 in)
Height, with carburettor 260 mm (10·2 in)
WEIGHT, DRY: 15 kg (33 lb)
PERFORMANCE RATINGS:
T-O 17·9 kW (24 hp) at 6,000 rpm
Continuous 12 kW (16 hp) at 4,500 rpm

The Borzecki 2RB light piston engine

SPECIFIC FUEL CONSUMPTION:
At max T-O rating 145·3 μg/J (0·86 lb/hr/shp)
At continuous rating 126·75 μg/J (0·75 lb/hr/shp)

IL
INSTYTUT LOTNICTWA (Aeronautical Institute)

HEADQUARTERS:
Al. Krakowska 110/114, 02-256 Warsaw-Okecie
Telephone: Warsaw 460993
MANAGING DIRECTOR:
Ing Zbigniew Pawlak
SCIENTIFIC DIRECTOR:
Dr Czeslaw Skoczylas
DEPUTY DIRECTOR:
Dipl Ing Jerzy Grzegorzewski
CHIEF OF TECHNICAL INFORMATION DIVISION:
Dipl Ing Andrzej Glass

The Aeronautical Institute is an establishment concerned with aeronautical research, aerodynamic tests, strength tests, test flights of aeroplanes, helicopters and gliders, aviation equipment, materials, technical information and standardisation. The Institute has a special manufacturing plant responsible for constructing prototypes to its own design.

IL SO-1

The Aeronautical Institute designed the SO-1 turbojet to power the Polish TS-11 Iskra (Spark) jet basic trainer. It was designed to permit the full range of aerobatics, including inverted flight. Guaranteed overhaul life is 200 hours. Production was handled by the WSK-Rzeszów, as noted in that organisation's entry.

TYPE: Single-shaft axial-flow turbojet.
AIR INTAKE: Annular intake casing manufactured as a cast shell. Fixed inlet guide vanes.
COMPRESSOR CASING: Manufactured as a cast shell in two parts, split along horizontal centreline, in aluminium alloy.
COMPRESSOR: Seven-stage axial-flow compressor. Drum-type rotor built up of disc assemblies, with constant

diameter over tips of rotor blades. Carried in ball bearing at front and roller bearing at rear. Steel stator blades bonded with resinous compound into slots in carrier rings. Rotor of steel and duralumin, with first three blade rows of steel and remainder of aluminium alloy. Pressure ratio 4·8.

COMBUSTION CHAMBER: Annular type with 24 integral vaporisers. Outer casing made of welded steel.

FUEL SYSTEM: Two independent systems supplied by one pump. Starting system consists of six injectors, with direct injection. Main system consists of twelve twin injectors with outlets towards the vaporisers.

FUEL SPECIFICATION: Kerosene P-2 or TS-1.

TURBINE: Single-stage axial-flow type. Blades attached to disc by fir-tree roots. Supported in roller bearing at rear.

JET PIPE: Outer tapered casing and central cone connected by streamlined struts. Nozzle area adjusted by exchangeable inserts.

LUBRICATION SYSTEM: Open type for rear compressor and turbine bearings, supplied by separate pumps. Closed type for all other lubrication points, fed by separate pumps.

OIL SPECIFICATION: Type AP-26 (synthetic).

ACCESSORY DRIVES: Gearbox mounted at bottom of air intake casing and driven by bevel gear shaft from front of compressor.

STARTING: 27V starter/generator and bevel gear shaft, driven by aircraft battery or ground power unit, mounted on air intake casing.

DIMENSIONS:

Length overall	2,151 mm (84·7 in)
Width	707 mm (27·8 in)
Height	764 mm (30·1 in)

IL SO-1 turbojet (9·8 kN; 2,205 lb st), initial power plant of the TS-11 Iskra (Spark) trainer *(BIIL)*

WEIGHT, DRY:	303 kg (668 lb)
PERFORMANCE RATINGS:	
T-O	9·8 kN (2,205 lb st) at 15,600 rpm
Max cont	8·7 kN (1,958 lb st) at 15,100 rpm
SPECIFIC FUEL CONSUMPTION:	
At T-O rating	29·6 mg/Ns (1·045 lb/hr/lb st)
OIL CONSUMPTION:	0·8 litres (1·4 Imp pints)/hr

IL SO-3

This improved version of the SO-1 replaced the earlier type in production at the WSK-Rzeszów. The SO-3 is intended for tropical use and incorporates minor changes in compressor, combustion chamber and turbine, data remaining the same as for the SO-1. It is fitted in the current production TS-11 Iskra jet trainer. TBO has been doubled, to 400 hours.

JANOWSKI
JAROSLAW JANOWSKI

ADDRESS:
Lodz 11, ul. Nowomiejska 2/29

JANOWSKI SATURN 500

The Saturn 500 was designed by Mr Jaroslaw Janowski and built by Mr S. Polawski for the Janowski J-1 ultra-light amateur-built aircraft (see "Aircraft" section). The prototype Saturn 500 was built in 1969. This two-cylinder two-stroke engine may be used with tractor or pusher propeller, and is intended for ultra-light aircraft built by amateurs.

In 1972 work began on a new version of the Saturn 500, with new cylinder heads, improved crankshaft and dual ignition. Its rating (max T-O) is increased to 30 hp; dry weight is believed to be about 25 kg (55 lb).

The following description applies to the initial 25 hp version:

TYPE: Two-cylinder two-stroke horizontally-opposed aircooled piston engine.

CYLINDERS: Bore 70 mm (2·76 in). Stroke 65 mm (2·56 in). Swept volume 500 cc (30·5 cu in). Compression ratio 8·5 : 1. Steel barrels with aluminium alloy cylinder heads. Cylinder and head assembly attached to crankcase by four studs.

PISTONS: Of aluminium alloy. Two compression rings and one oil scraper ring.

CONNECTING RODS: Steel forgings.

CRANKSHAFT: Steel counterbalanced shaft, supported in two lead-bronze plain bearings and one ball-thrust bearing at the front.

CRANKCASE: Aluminium alloy case, split in the vertical plane, with front and aft covers.

INDUCTION: Two BVF 28N1 carburettors.

FUEL: Petrol/oil mixture using aviation 90 octane.

IGNITION: Two magnetos. One M14-250 14 mm (0·55 in) sparking plug per cylinder.

MOUNTING: Four rubber dampers at rear of crankcase.

PROPELLER DRIVE: Direct tractor or pusher.

DIMENSIONS:

Length overall, with propeller boss	430 mm (16·93 in)
Width, without sparking plugs	515 mm (20·27 in)
WEIGHT, DRY:	27 kg (59·5 lb)
PERFORMANCE RATING:	
T-O	18·65 kW (25 hp) at 4,000 rpm
SPECIFIC FUEL CONSUMPTION:	
Max T-O rating	118·23 μg/J (0·70 lb/hr/hp)
Normal cruising power	111·48 μg/J (0·66 lb/hr/hp)

Saturn 500 two-cylinder two-stroke engine designed by Jaroslaw Janowski

PZL
POLSKIE ZAKLADY LOTNICZE

HEADQUARTERS:
ul. Miodowa 5, 00251 Warsaw

SALES REPRESENTATIVE:
Pezetel, ul. Przemyslowa 26, 00450 Warsaw
Telephone: Warsaw 285071

The entire Polish aircraft industry is subordinate to the Zjednoczenie Przemyslu Lotniczego i Silnikowego PZL (Aircraft and Engine Industry Union). Pezetel handles all export sales of Polish aeronautical material.

WSK-PZL-KALISZ
WYTWÓRNIA SPRZETU KOMUNIKACYJNEGO-PZL-KALISZ

HEAD OFFICE AND WORKS:
ul. Czestochowska 140, 62800 Kalisz
Telephone: 4081-3

GENERAL MANAGER:
Dipl Ing Zbigniew Girulski

In 1952 the Soviet Union transferred responsibility for manufacture and service support of Russian aircooled radial piston engines to the WSK (transport equipment manufacturing centre) at Kalisz. No longer in production, except as spares, are the 125 hp M-11D, used in the Po-2 and CSS-13, and the 160 hp M-11FR, used in the Junak-2 and -3. A Polish team under Dipl Ing Wiktor Narkiewicz designed the 330 hp WN-3 radial for the TS-8 Bies, and this was produced by WSK-Kalisz. All these engines have been described in earlier editions of *Jane's*.

Current production is centred on: the 1,000 hp ASh-62IR (Polish designation ASz-62IR) for all versions of the An-2; and the 260 hp AI-14RA for the PZL-104 Wilga (see Aircraft section). Production of the ASz-62IR exceeds 10,000 engines. WSK-Kalisz has developed a version of the AI-14 with electric starter, the AI-14RC.

IVCHENKO AI-14

The original 260 hp AI-14R version of this nine-cylinder aircooled radial engine has been produced in very large quantities, in both the Soviet Union and Poland. Since 1960 several later versions have gone into production at Kalisz for fixed-wing aircraft and helicopters. The following versions can be listed:

AI-14R. Basic production engine, rated at 194 kW (260 hp); fitted to Yak-12, Yak-18, PZL-101A Gawron and PZL-104 Wilga 3.

AI-14RC. Version fitted with electric starter. Developed at Kalisz.

AI-14RF. Modified by I. M. Vedeneev to power An-14 light STOL transport. Take-off rating 224 kW (300 hp).

AI-14ChR. Further modified by Vedeneev to give take-off power of 261 kW (350 hp) at higher rpm.

M-14VF. Helicopter engine, rated at 209 kW (280 hp); fitted to Ka-15 and Ka-18.

M-14V-26. Improved helicopter engine developed in Soviet Union by Vedeneev, with higher running speed, fan cooling and numerous refinements. Powers Ka-26.

The following description refers to the AI-14R and AI-14RF:

TYPE: Nine-cylinder single-row aircooled radial, geared and supercharged.

CYLINDERS: Bore 105 mm (4·125 in). Stroke 130 mm (5·125 in). Displacement 10·16 litres (620 cu in). Nitrided steel barrel; cast light alloy head incorporating air starting valve. Two spark plugs and single inlet and exhaust valves. Compression ratio 5·9 (AI-14RF, 6·2).

PISTONS: Aluminium forgings, each with two chromium-plated compression rings, upper oil control ring with slot and, on skirt, tapered scraper ring.

CONNECTING RODS: One master rod, with lead-bronze big-end bearing; eight link rods articulated by steel cemented knuckle pins fixed in the master-rod cheeks against rotation and secured laterally by retaining strips.

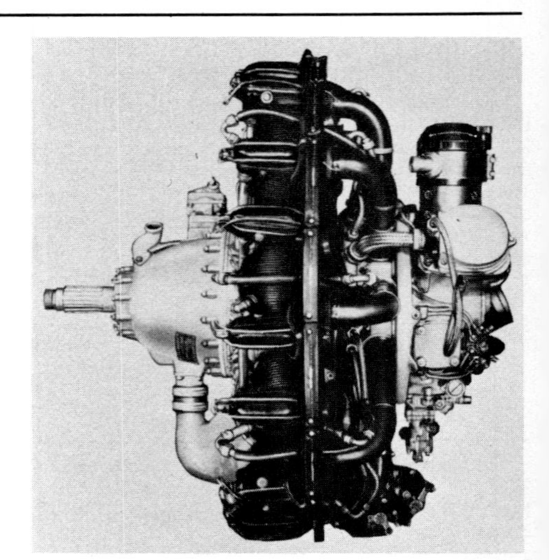

Polish-built AI-14R piston engine

CRANKSHAFT: Heat-treated steel in two parts, the front portion being gripped in the split cheek of the rear portion and held by a pinch bolt. Both portions carry a counterweight, the rear counterweight being pendulous type which balances inertia forces and also serves as vibration damper. Shaft held in two main roller bearings and ball thrust bearing in crankcase front cover.

RANKCASE: Front and rear forgings in heat-treated light alloy, joined by nine bolts. At the rear a mixture collector receives the mixture from the supercharger.

ALVE GEAR: Each valve is opened by push/pull rod, from a cam plate geared to rotate in opposition to the crankshaft.

DUCTION SYSTEM: Carburettor type, with choke tube upstream of supercharger.

JEL GRADE: B-70 gasoline to specification GOST 1012-54, 73 octane to DERD.2485 or 80 grade non-ethylated to US MIL-G-5572C. (AI-14RF, SB-78, not under 78 octane.)

JPERCHARGER: Aluminium forged impeller geared up from crankshaft to ratio of 7·105 (AI-14RF, 8·65). Magnesium cast diffuser.

JBRICATION: Gear-type pressure and scavenge pumps.

NITION: Left magneto serves front plugs, right serves rear; automatic timing and fully screened.

ROPELLER DRIVE: Three planetary gears, ratio 0·787.

ARTING: By compressed air from airborne or ground compressor or bottle-fed through distributor to cylinder-head valves (and to purge lower cylinders of drained liquid).

IMENSIONS:

Diameter	985 mm (38·78 in)
Length:	
AI-14R	956 mm (37·63 in)
AI-14RF	1,002 mm (39·45 in)
M-14V-26	1,145 mm (45·08 in)

EIGHT, DRY:

AI-14R	197 kg (434 lb)
AI-14RF	230 kg (507 lb)
M-14V-26	245 kg (540 lb)

RFORMANCE RATINGS:

Max T-O:	
AI-14R	194 kW (260 hp) at 2,350 rpm
AI-14RF	224 kW (300 hp) at 2,400 rpm
M-14V-26	242·4 kW (325 hp) at 2,800 rpm
AI-14R	164 kW (220 hp) at 2,050 rpm
AI-14RF	212·5 kW (285 hp) at 2,300 rpm
Cruise (60% max continuous):	
AI-14R	98·4 kW (132 hp) at 1,730 rpm
AI-14RF	112 kW (150 hp) at 1,730 rpm
M-14V-26	141·7 kW (190 hp) at 2,350 rpm

SPECIFIC FUEL CONSUMPTION:

Max T-O:		
AI-14R	95-104·3	μg/J (0·562-0·617 lb/hr/shp)
AI-14RF	98·7-108	μg/J (0·584-0·639 lb/hr/shp)
Cruise:		
AI-14R	76·4-83·8	μg/J (0·452-0·496 lb/hr/shp)
AI-14RF	78·2-85·7	μg/J (0·463-0·507 lb/hr/shp)

SHVETSOV ASz-62

Power plant of the An-2 transport biplane, the ASz-62 is a 1,000 hp nine-cylinder aircooled radial engine. It was developed from the Wright Cyclone R-1820 by Arkadiya Shvetsov's bureau in the Soviet Union, as the ASh-62. Several variants have been built, including the ASz-62IR/TK driving a turbo-compressor to maintain 850 hp up to a height of 9,500 m (31,000 ft).

There are two current versions, as follows:

ASz-62IR. Standard power plant of the Li-2 (Soviet DC-3) and all versions of the An-2 except for the An-2M. Transferred to WSK-PZL-Kalisz about 1955.

ASz-62M. Developed by Vedeneev from ASh-62IR to power An-2M. Has a power take-off shaft, rated at up to 58 hp, to drive agricultural spraying or dustjng gear, two electrical generators and the flight deck air-conditioning equipment. Weight of the take-off package is 113 kg (249 lb).

Both versions have a cylinder bore of 155·5 mm (6⅛ in), swept volume of 29·87 litres (1,823 cu in) and compression ratio of 6·4 : 1.

The planetary reduction gear of the ASh-62M has a ratio of 0·637 : 1.

DIMENSIONS (ASz-62M):

Length overall, without power take-off	1,130 mm (44·50 in)
Diameter	1,375 mm (54·13 in)

WEIGHT, DRY (ASz-62M):

Without power take-off	567 kg (1,250 lb)

PERFORMANCE RATINGS (ASz-62M: A, with power take-off driving generators and air-conditioning equipment; B, driving also agricultural equipment):

Polish-built ASz-62IR piston engine

T-O: A	1,000 hp at 2,200 rpm
Rated power:	
A	820 hp at 2,100 rpm
B (75% power)	615 hp at 1,910 rpm
Cruise:	
B (50% power)	410 hp at 1,670 rpm

SPECIFIC FUEL CONSUMPTION (corresponding to above ratings):

T-O: A	112 μg/J (0·661 lb/hp/hr)
Rated power:	
A	104 μg/J (0·617 lb/hr/hp)
B	89 μg/J (0·529 lb/hr/hp)
Cruise:	
B	80 μg/J (0·474 lb/hr/hp)

WSK-PZL-RZESZÓW
WYTWÓRNIA SPRZETU KOMUNIKACYJNEGO-PZL-RZESZÓW

HEAD OFFICE AND WORKS:
ul. Obronców Stalingradu 120, 35078 Rzeszów, Postbox 340

Telex: 83411

GENERAL MANAGER:
Dipl Ing Jozef Rokoszak

WSK-Rzeszów, founded in 1938 as PZL-Rzeszów, at first produced Bristol Pegasus and Walter Junior and minor engines under licence. After World War 2 the works expanded considerably. The first product was the Soviet M-11D, followed by series production of the ASz-62IR and Lit-3 piston engines and HO-10 and SO-1 turbojets.

Current production is centred on: the Soviet-designed GTD-350 turboshaft, together with WR-2 reduction gear for Mi-2 helicopters; the tropicalized SO-3 turbojet for the TS-11 Iskra trainer; and the PZL-3S piston engine for agricultural aircraft. The SO-3 is described (with the SO-1) on an earlier page under IL.

Isotov GTD-350 turboshaft engine (322 kW; 431 shp)

GTD-350

The GTD-350 is a free-turbine helicopter power plant. In the version used in the twin-engined Mi-2, the drive is taken from the rear, with the twin jet pipes of each engine exhausting to port (port engine) and starboard (starboard engine). The GTD-350 can be supplied with downward-facing jet pipe and with drive from the front, if required. Although developed and initially produced by the Isotov bureau in the Soviet Union, it is now in production only in Poland. WSK-Rzeszów is developing a version rated at 400-700 shp to be used in a proposed single-engined WSK helicopter. In 1970, time between overhauls for the original version was 500 hours; by 1973 this time had been extended to 1,000 hours, and in 1975 selected engines were being run to 2,000 hours.

TYPE: Axial/centrifugal-flow free-turbine turboshaft engine.

AIR INTAKE: Annular intake casing and inlet guide vanes of stainless steel. Automatic de-icing of inlet guide vanes and central bullet by air bleed from compressor.

COMPRESSOR: Seven axial stages and one centrifugal stage, all of steel, connected together with a tie-bolt. Discs shrunk-fitted to shaft. Blades of axial stages have dovetail roots. Shaft carried in front roller bearing and rear ball bearing. Pressure ratio 5·9 : 1. Air mass flow 2·19 kg (4·83 lb)/sec at 45,000 rpm.

COMPRESSOR CASING: Horizontally-split aluminium alloy casing, with stator blades brazed to semi-rings. No diffuser blades.

COMBUSTION CHAMBER: Reverse-flow type with air supply through two tubes. Centrifugal duplex single-nozzle burner. Ignition system comprises burner and semi-conductor spark-plug. Eight thermocouples at gas outlet.

FUEL SYSTEM: Includes NR-40T pump governor with shut-off cock, which feeds fuel to burner, controls gas-generator rpm and limits max output; RO-40T power turbine rpm governor, DS-40 signal transmitter controlling bleed valves; and electromagnetic valve to provide fuel for starting.

FUEL GRADE: TS-1 or TS-2.

COMPRESSOR TURBINE: Single-stage turbine with aircooled disc. Shrouded blades with fir-tree roots. Precision-cast fixed guide vanes. Turbine casing has metal-graphite insert in plane of blades. Shaft supported in ball bearing at rear. Temperature before turbine 970°C.

POWER TURBINE: Two-stage constant-speed type (24,000 rpm). Shrouded blades with fir-tree roots. Discs bolted together. Stator blades welded to rings. Airflow is again reversed aft of power turbine.

JET PIPES: Two fixed-area jet pipes.

REDUCTION GEARING: Two sets of gears, with ratio of 0·246 : 1, in cast magnesium alloy casing. Output shaft speed, 5,900 rpm.

LUBRICATION SYSTEM: Closed type. Gear-type pump with one pressure and four scavenge units. Nominal oil pressure 2·94±0·5 bars (43 lb/sq in). Oil cooler and oil tank, capacity 12·5 litres (2·75 Imp gallons), fitted to airframe.

OIL GRADE: B3-W (synthetic).

ACCESSORIES: STG3 3kW starter/generator, NR-40T governor pump, D1 tachometer generator and oil pumps mounted on reduction gear casing and driven by gas-generator. RO-40T rotating speed governor, D1 tachometer generator and centrifugal breather, also mounted on reduction gear casing, driven by power turbine.

STARTING: STG3 starter/generator suitable for operation at up to 4,000 m (13,125 ft) altitude.

DIMENSIONS:

Length overall	1,385 mm (54·53 in)
Max width	520 mm (20·47 in)
Width (with jet pipes)	626 mm (24·65 in)
Max height	630 mm (24·80 in)
Height (with jet pipes)	760 mm (29·9 in)

WEIGHT, DRY:

Less jet pipes and accessories	135 kg (298 lb)

PERFORMANCE RATINGS:

Max contingency
322 kW (431 shp) at 97% max gas-generator rpm

T-O rating (6 min)
295 kW (395 shp) at 94% max gas-generator rpm

Nominal rating (1 hr)
235 kW (315 shp) at 89% gas-generator rpm

Cruising rating (II)
210 kW (281 shp) at 86·5% gas-generator rpm

Cruising rating (I)
173 kW (232 shp) at 83·5% gas-generator rpm

Specific Fuel Consumption:

T-O	136 μg/J (0·805 lb/hr/shp)
Nominal	146 μg/J (0·861 lb/hr/shp)
Cruising (II)	154 μg/J (0·913 lb/hr/shp)
Cruising (I)	165 μg/J (0·978 lb/hr/shp)

PZL-3S

One of the diminishing number of new radial engines, the PZL-3S has been developed from the LiT-3 (in turn based on the Soviet AI-26W) and benefits from the total time of nearly nine million hours logged by the earlier engine. It is claimed to be simple and reliable, with easy starting in the most severe climatic conditions.

Type: Seven-cylinder aircooled radial.

Cylinders: Bore 155·5 mm (6·12 in). Stroke 155 mm (6·1 in). Swept volume 20·6 litres (1,265 cu in). Compression ratio 6·4 : 1.

Pistons: Forged aluminium.

Induction System: Float-type carburettor. Mechanically driven supercharger.

Fuel Grade: B-91/115 gasoline or 91/96 grade.

Lubrication: Gear-type oil pump. Oil grade Mk-20, MS-22 or 100.

Propeller Drive: Direct. Provision for constant-speed propeller Type US-129000. Propeller governor type 21322-AO3.

Accessories: GSK 1,500W generator, AK-50 air compressor. Provision on rear cover for auxiliary accessories.

Starting: Electrical.

PZL-3S seven-cylinder radial

Dimensions:

Diameter	1,267 mm (49·88 in)
Length	1,065 mm (41·9 in)

Weight, Dry: 400 kg (882 lb)

Performance Ratings:

Max T-O	448 kW (600 hp) at 2,200 rpm
Max continuous	410 kW (550 hp) at 2,050 rpm
Cruise (75 per cent)	310 kW (415 hp) at 2,000 rpm

PZL-Franklin engines (the name appears in this form on the valve-gear covers): above right, 2A-120, showing electric starter and generator; above left, 4A-235B; below, 6AS-350A with turbocharger

Specific Fuel Consumption:

T-O, max cont	112 μg/J (0·66 lb/hr/hp)
Cruise	86 μg/J (0·51 lb/hr/hp)

PZL-FRANKLIN ENGINES

In 1975 Pezetel acquired rights to manufacture and market the entire range of aircooled piston engines formerly produced by the Franklin Engine Company (Aircooled Motors) of the United States. These engines will be produced in Poland for light aircraft of all kinds, including helicopters and motor-gliders. The minimum of engineer ing changes will be effected, and batch production wa beginning at the WSK-PZL-Rzeszów as this edition wen to press. The first models to enter production are th PZL-Franklin 2A-120, 4A-235B, 6A-350C, 6AS-350A and 6A-350D.

All these engines are of the horizontally-opposed type with cylinders of 117·48 mm (4·625 in) bore and 88·9 mm (3·5 in) stroke. All have direct drive and operate on 100/130 grade fuel. Other details are tabulated:

PZL-FRANKLIN ENGINES

Engine model	Cylinder arrangement	Capacity cc (cu in)	Compression ratio	Max T-O rating at S/L (kW, hp at rpm)	Overall dimensions mm (in) length	width	height	Weight dry kg (lb)	Remarks
2A-120	2 horiz	1,916 (117)	8·5	45 (60) at 3,200	581 (22·9)	795 (31·3)	515 (20·3)	75·8 (167)	
4A-235B	4 horiz	3,850 (235)	8·5	93 (125) at 2,800	774 (30·5)	795 (31·3)	637 (25·1)	117·6 (259)	
6A-350C	6 horiz	5,735 (350)	10·5	164 (220) at 2,800	952 (37·5)	795 (31·3)	641 (25·25)	167 (367)	
6AS-350A	6 horiz	5,735 (350)	7·4	186·4 (250) at 2,800*	1,097 (43·2)	868 (34·2)	983 (38·7)	189 (417)	turbocharged
6A-350D	6 horiz	5,735 (350)	10·5	175 (235) at 3,200	825 (32·5)	795 (31·3)	642 (25·3)	145 (320)	helicopter engine

*Rated at 175 kW (235 hp) at 2,800 rpm at 5,000 m (16,400 ft).

SOUTH AFRICA

ATLAS

ATLAS AIRCRAFT CORPORATION OF SOUTH AFRICA (PTY) LTD

Address and Officers: See Aircraft section

Atlas is manufacturing the Rolls-Royce Viper 540 turbojet under sublicence from Piaggio of Italy, for use i Atlas Impala attack trainers.

SPAIN

CASA

CONSTRUCCIONES AERONÁUTICAS SA

Head Office and Officers:
See Aircraft section

DIVISIÓN DE MOTORES

Office and Works:
Carretara de Ajalvir, Km 3.5, Apdo 111 Torrejón de Ardoz, Madrid
Telephone: 407 34 66 and 407 37 66

Managing Director:
Antonio Barrón Medrano

Production Manager:
Florencio Manteca Martinez

Engineering Manager:
Anselmo Andrés y Andrés

Sales Manager:
César Francisco Natal Fernández

The Division de Motores of CASA was formed in June 1973. It was formerly the Empresa Nacional de Motores de Aviación (ENMASA), as described in previous editions of *Jane's*.

The first turbojet engine produced by ENMASA wa the Marboré II, which was built under licence from th French Turboméca company as the Marboré M21.

The Division's current aeronautical activities includ overhaul and repair of General Electric J79 and J85 SNECMA Atar 9C and Turboméca Marboré II and V turbojet engines, Garrett-AiResearch TPE 331 turbo props, and Lycoming T53 turboshafts, in the plant a Ajalvir. The same factory also overhauls and repairs air craft and engine accessories.

SWEDEN

FLYGMOTOR
VOLVO FLYGMOTOR AB

HEAD OFFICE AND WORKS: S-461 01 Trollhättan
Telephone: 0520-301-00

This company was founded in 1930 and began by building under licence the Bristol Pegasus I aero-engine, under the designation My VI. Since 1970 Volvo Flygmotor has been a wholly owned subsidiary of AB Volvo.

Volvo Flygmotor AB holds a licence to build Rolls-Royce Avon engines and is also engaged in research and development work on turbojet engines, ramjet engines and rocket engines. Its major current programme involves the production of two versions of the RM8 supersonic turbofan developed from the Pratt & Whitney JT8D to power different sub-types of the Saab 37 Viggen combat aircraft.

It is also engaged in the development of experimental hybrid and liquid-fuel rocket engines. The company has a technical collaboration agreement on ramjet development with Rolls-Royce (1971) Ltd.

FLYGMOTOR RM6C AVON

RM6C is the designation of the licence-built version of the Rolls-Royce Avon 300-Series turbojet which Flygmotor manufactured to power the Saab 35D, E and F Draken fighters. It is rated at approximately 74·7 kN (16,800 lb) st with its Swedish-developed afterburner in use.

Production of the RM6C was completed in late 1974. Manufacture of spares, and general service support continue for a large number of Avon engines, including earlier versions.

FLYGMOTOR RM8

The RM8 is a Swedish military version of the Pratt & Whitney JT8D civil subsonic turbofan which Flygmotor has developed to power the Saab 37 Viggen supersonic multi-purpose combat aircraft.

The programme is being undertaken with the assistance of Pratt & Whitney, which supplied Flygmotor initially with three of the first production JT8D-1 engines. The first was put on test on 6 August 1964, and was being run with afterburning by the Spring of 1965. Approximately 4,000 hr of bench and flight testing had been completed by March 1974.

Development of the engine, and manufacture of 195 RM8As for the Viggen, is covered by a 689 million kroner contract from the Swedish government, the largest single order ever received by Flygmotor. Production, which will absorb 595 million kroner, started in mid-1968 and was continuing in the mid-1970s.

The RM8 is fitted with an afterburner with fully-variable exit nozzle, which gives a 70% increase in thrust at sea level.

Flygmotor has so far devoted almost 7,500,000 man-hours to the development of the RM8. Though based on the JT8D, the RM8 is made of new parts in different materials and having changed dimensions.

Until 1970 the main effort was devoted to development and manufacture of the RM8A version for the AJ 37 strike version of the Viggen, and all engines delivered by the end of 1973 were of this type. In late 1970 a substantially modified version, the **RM8B,** was planned to meet the propulsion requirements of the fighter Viggen, the JA 37. Research at Pratt & Whitney and Flygmotor showed that a changed design could improve the reliability of operations at high altitudes and in severe manoeuvres, as well as increase thrust in all regimes. In collaboration with Pratt & Whitney the design of the RM8B was completed in late 1971. The major change to improve functional reliability at high altitude has been to take the first stage off the LP compressor and add it to the fan, giving a three-stage fan and three-stage LP spool, both having a revised aerodynamic configuration according to the latest P & W research. To increase thrust the RM8B has a four-nozzle burner combustion system and a new HP turbine.

The first RM8B LP compressor was tested in 1971 in the P & W Willgoos high-altitude laboratory and showed very good performance and stability. Five complete RM8B prototype engines have now also been built and extensively tested at Trollhättan. The flight development programme, which began in a Viggen testbed on 13 September 1974, indicates that the RM8B will meet all requirements for the JA 37.

The following description refers to the RM8A:

TYPE: Axial-flow two-spool turbofan with modulated afterburner.

AIR INTAKE: Annular, with 19 fixed inlet guide vanes.

FAN: Two-stage front fan. Titanium blades.

LP COMPRESSOR: Four-stage axial-flow, integral with fan stages, on inner of two concentric shafts. Blades of titanium. Steel casing.

HP COMPRESSOR: Seven-stage axial-flow on outer hollow shaft. Blades made of special high-temperature alloys of type used for turbine blading. Overall pressure ratio 16·5 : 1. By-pass ratio approximately 1 : 1. Total air mass flow 145 kg (320 lb)/sec.

COMBUSTION CHAMBER: Cannular type with nine cylindrical flame tubes, each downstream of a single Duplex burner and discharging into a single annular nozzle. Two high-energy spark plugs, each with its own igniter box.

Volvo Flygmotor RM8B augmented turbofan for JA 37 fighter version of Viggen, rated at 125 kN (28,110 lb) with maximum augmentation

HP TURBINE: Single-stage axial-flow, with cast aircooled blades.

LP TURBINE: Three-stage axial-flow, with cast blades. Exit guide vanes after turbine.

AFTERBURNER: Double-skinned to provide duct for cooling air. Outer skin of titanium. Inner skin of special alloys. One hot-streak igniter. Hydraulically-actuated fully-variable nozzle, using fuel as the operating fluid.

BEARINGS: Main shafts run in total of six bearings.

CONTROL SYSTEMS: There are two systems. The main system for the gas-generator comprises a Bendix hydromechanical control of advanced design. A further Bendix unit controls the fuel flow to the afterburner and the nozzle area. A single power lever controls thrust from maximum afterburner down to idle; further movement below idle actuates the fuel cut-off valve.

MOUNTING: Three-point. Main mountings on each side of compressor casing; one under turbine casing.

ACCESSORY DRIVE: Via gearbox, under engine, driven from HP turbine shaft.

DIMENSIONS:	
Length overall:	
RM8A	6,160 mm (242·5 in)
RM8B	6,230 mm (245·25 in)
Max diameter (both versions)	1,397 mm (55 in)
Inlet diameter (both)	1,030 mm (40·55 in)
WEIGHT, DRY:	
RM8A	2,100 kg (4,630 lb)
RM8B	2,350 kg (5,181 lb)
PERFORMANCE RATINGS (ISA, S/L):	
Max T-O, augmented:	
RM8A	115·6 kN (25,990 lb)
RM8B	125 kN (28,110 lb)
Max T-O, dry:	
RM8A	65·6 kN (14,750 lb)
RM8B	72 kN (16,200 lb)
SPECIFIC FUEL CONSUMPTION:	
Max augmented:	
RM8A	70·0 mg/Ns (2·47 lb/hr/lb st)
RM8B	71·4 mg/Ns (2·52 lb/hr/lb st)
Max dry:	
RM8A	17·8 mg/Ns (0·63 lb/hr/lb st)
RM8B	18·1 mg/Ns (0·64 lb/hr/lb st)
Max continuous (both)	17·3 mg/Ns (0·61 lb/hr/lb st)

FLYGMOTOR VR35

The VR35 is a prepackaged liquid-propellant rocket engine with positive expulsion of the storable inhibited red fuming nitric acid (IRFNA) and Hydyne propellants. A solid-propellant gas generator delivers the expelling gas and programmes the thrust into a boost phase followed by a sustain blow-down period.

The positive expulsion is accomplished by a gas-pressurised piston for the fuel, contained in a central tank, and an inward-collapsible aluminium bladder for the oxidiser in its concentric tank. This expulsion system enables the engine to be fired under any acceleration direction, a capability which is essential for a missile engine with a thrust programme that includes the sustain phase.

Another advantage with this type of engine is the completely smoke-free exhaust, leaving no signature at launch or during flight.

The VR35 is now in production for the Saab-Scania RB05 air-to-surface missile, which is one of the main weapons for the AJ 37 Viggen (see "Air-Launched Missiles" section).

DIMENSIONS:	
Length overall	1,770 mm (69·7 in)
Max diameter	300 mm (11·8 in)
WEIGHT:	127 kg (280 lb)
PROPELLANT MASS FRACTION:	0·59
OPERATING TEMPERATURE RANGE:	−50°C to +65°C

Volvo Flygmotor VR35 prepackaged liquid rocket on static test at dive angle

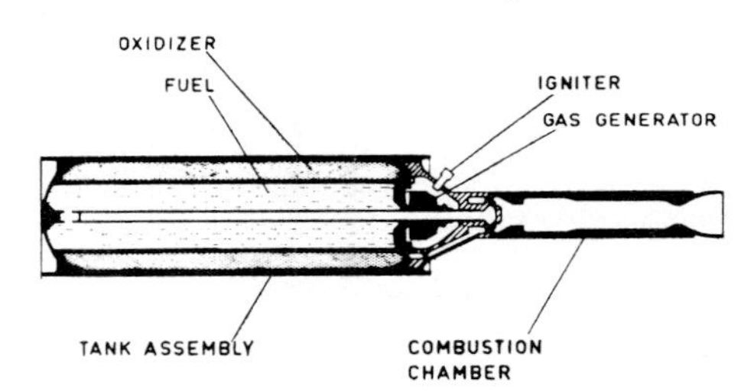

VR35 longitudinal cross-section

FLYGMOTOR RAMJETS

Ramjet engine research and development have been under way at Volvo Flygmotor since 1952, when the company's underground high-pressure air magazine was completed. The stored air drives the combustion test rigs and wind tunnels necessary for this kind of work.

The engine research culminated in several flight tests with the RR2, designed for a Mach number of 2·85 and intended for pod-mounting on a ground-to-air missile. The RR2 engine has a diameter of 260 mm (10·25 in) and an empty weight 42·0 kg (92·5 lb). The kerosene-type fuel is stored in the missile main body and is delivered to the ramjet combustion chamber by means of an air-driven turbopump integrated into the engine forebody.

The main research effort is now directed towards an integral rocket-ramjet engine. This new concept is a ramjet engine containing a solid- propellant grain in its combustion chamber. The rocket is fired to launch and accelerate the vehicle. After completion of the boost phase the chamber is used for ramjet combustion. The transition between the boost and the ramjet-sustain phase is accomplished by suddenly increasing the nozzle area, opening the air inlet and injecting the liquid fuel.

Two basic types of rocket-ramjet have been investigated and tested. The **RRX1** has a circumferential nose air intake and is designed for a Mach number of 5·0. The body diameter is 190 mm (7·5 in). The **RRX5** has four side-mounted air intakes to free the nose for a missile warhead and guidance. It is intended for a Mach-number range of 1·8-3·5, with a diameter varying between 300-450 mm (11·8 and 17·7 in).

THE UNITED KINGDOM

BRISTOL AEROJET
BRISTOL AEROJET LIMITED

HEAD OFFICE AND WORKS:
Banwell, Weston super Mare, Somerset BS24 8PD
Telephone: Banwell (0934-82) 2251
Telex: 44259
DIRECTORS:
Dr F. Liewellyn Smith, CBE (Chairman)
R. M. Howarth (Managing Director)
G. A. Harrison
E. N. Vidler
W. K. Bachelder (USA)
M. G. Hatley (USA)
E. A. Lowe (USA)
J. D. Nichols (USA)
R. J. Mill (USA) Alternate

In 1952 a team was formed to meet the rocket-motor requirements of early British guided missile projects. This team constituted the nucleus of Bristol Aerojet, formed in 1958 by the Bristol Aeroplane Company and the Aerojet-General Corporation of California.

The company specialises in the design, development and production of solid-propellant and packaged liquid-propellant rocket motors for missiles and for satellite launch and research rockets, working in collaboration with the Rocket Propulsion Establishment, Ministry of Defence, and others.

The company provided motors for Bloodhound 1 and 2, Seaslug 1 and 2, and Thunderbird 1 and 2. Production vehicles which use Bristol Aerojet motors include the Seawolf, AN/USD-501 (CL-89), Skylark, Rapier, Petrel, Seacat, Tigercat, Skua, Jaguar/Jabiru, INTA 300 and INTA 255.

As well as being the primary supplier of rocket motors in the UK, the company is also a supplier to NATO, Australia, Belgium, Canada, France, Germany, India, Italy, Pakistan, Spain and Sweden.

Bristol Aerojet rocket-motor manufacturing processes have been licenced to Breda Meccanica Bresciana in Italy and to Instituto Nacional de Tecnica Aerospatial (INTA) in Spain, and are used under different arrangements in Canada and Australia. Under an agreement with the British government the company is supplying British "plastic" propellant manufacturing and filling technology and equipment for a rocket-motor filling facility set up by INTA, and is able to supply a similar service to other approved organisations.

Bristol Aerojet produces rocket motors to customer requirements, undertakes launcher design, and provides technical service for assembly and firing of experiment rockets at British and overseas ranges. The company als produces research rockets and ballistic targets.

BRISTOL AEROJET LIQUID MOTORS

Fully-integrated rocket motors of this type have bee successfully developed in collaboration with the Rock Propulsion Establishment, Ministry of Defence. Fligh proven motors give complex thrust/time programmes o command, and have a shut-down and re-start capabilit The simple propellants are smoke-free and offer insta readiness after long storage.

BRISTOL AEROJET SOLID MOTORS

The following details may be published of the princip types of solid rocket motor manufactured by Bristol Aer jet Ltd:

Motor Name	Diam mm (in)	Length mm (in)	Burn Time (sec)	Total impulse (kN-sec)	Application
Bantam	125 (4·92)	1,473 (58)	33·6	52·0	Skua
Chick	68 (2·68)	554 (21·8)	0·18	4·45	Skua and Petrel boost
Cuckoo	431·8 (17)	1,312 (51·65)	4·1	360·0	Skylark boost
Goldfinch	431·8 (17)	2,224 (87·56)	3·6	701·0	Skylark boost
Lapwing	176·8 (6·96)	1,826 (71·9)	25·0	153·0	Petrel
Raven	431·8 (17)	5,232 (206)	30·0	1,510·0	Skylark
Siskin	141·2 (5·56)	600·0 (23·62)	3·58	16·5	Black Arrow
Waxwing	712 (28)	(spherical)	55·0	845·5	Black Arrow apogee

CLUTTON
ERIC CLUTTON

ADDRESS:
92 Newlands Street, Shelton, Stoke-on-Trent, Staffs ST4 2RF

In order to provide propulsion for his FRED (see 1971-72 *Jane's*) and Easy Too (see "Aircraft" section) Mr Clutton has been engaged in independent modification of the Volkswagen 1,500 cc engine and intends ultimately to market plans and a conversion kit.

His first conversion involved a geared drive to the propeller by means of a toothed belt. The ratio was 0·5 : 1 and, with the 1,829 mm × 1,118 mm (72 in × 44 in) ex-Cirrus Minor propeller used on FRED Srs 2, the engine speed was held to 3,750 rpm.

In 1968 development began on a true geared conversion, and two types were described (and one illustrated) in the 1972-73 *Jane's*. For reasons of cost Mr Clutton h now returned to 2 : 1 multi-vee-belt drive, but this diffe from all earlier types in that it is taken off the bell-housir end and, in its prototype form, uses off-the-shelf materi and parts. There are no castings, and only simple machi ing is needed on the two shafts. The prototype geare engine weighs 95·2 kg (210 lb), but is expected to be 13 kg (30 lb) lighter when more light alloy parts are used

LUCAS
LUCAS AEROSPACE LTD

HEAD OFFICE:
Monkspath, Shirley, Solihull, Warwickshire
Telephone: 021-744-8522
CHAIRMAN: B. F. W. Scott

Lucas Aerospace was formed in 1971 to bring together Lucas Gas Turbine Equipment Ltd and Rotax Ltd, both Lucas companies engaged in aerospace work. The assets and goodwill of the Rover Gas Turbine Co were acquired from Alvis Ltd in December 1972. The Electrical Group is now responsible for the CT 3201 (formerly designated TJ.125) turbojet manufactured previously by Rover Gas Turbines.

LUCAS CT 3201

This small 0·51 kN (114 lb) st turbojet engine utilises the gas-generator section of the Lucas 2S/150A engine and has been developed as a power plant for surveillance RPVs.

Initial deliveries of CT 3201 engines have been made to MBLE to power the Belgian company's Epervier surveillance RPV. Discussions are in progress with other drone manufacturers.

TYPE: Single-shaft turbojet.
AIR INTAKE: Aluminium alloy bifurcated.
COMPRESSOR: Single-stage centrifugal. Cast aluminium with steel rotating guide vanes. Rotor hydraulically pressed to shaft. Mass flow 0·93 kg (2·05 lb)/sec.
COMBUSTION CHAMBER: Annular reverse-flow. Six spray burners. High-energy 2 Joule igniter.
FUEL SYSTEM: Double-datum piston type.
FUEL GRADE: JP-4, JP-5 and equivalents.
TURBINE: Single, radial-flow, 12 blades integral with steel disc. Retained by sleeve/bolt on shaft, hung behind needle and roller bearings. Inlet gas temperature 917°C.
ACCESSORY DRIVES: Fuel pump and alternator between intakes driven by toothed belts.
LUBRICATION SYSTEM: Self-contained, 10 cc. SAE.5 below 25°C, SAE.10 above.
DIMENSIONS:
Length overall — 584 mm (23·0 in)
Max width — 269 mm (10·6 in)
Max depth — 311 mm (12·25 in)
WEIGHT, DRY: — 18·6 kg (41·0 lb)

Lucas CT 3201 turbojet of 0·51 kN (114 lb st) as fitt to the MBLE X-5 Épervier RPV

PERFORMANCE RATINGS:
Max thrust at ISA — 0·51 kN (114
Reduced thrust — 0·32 kN (73
SPECIFIC FUEL CONSUMPTION:
At max rating — 37·1 mg/Ns (1·31 lb/hr/lb

NPT
NOEL PENNY TURBINES LTD

HEAD OFFICE:
Siskin Drive, Toll Bar End, Coventry CV3 4FE
Telephone: 0203-301528
DIRECTORS:
E. R. Ponsford (Chairman)
R. N. Penny (Managing Director)
The Earl of Minto
S. Penny (Secretary)

Noel Penny Turbines was incorporated in 1972 with registered offices at Solihull (Birmingham area), and factories at Coventry, Harwich and Southam. The Coventry factory is extensively equipped with research, design, analytical, test and manufacturing facilities. The Harwich factory is equipped for the manufacture and test of gas turbines. Southam concentrates upon the manufacture of small high-precision items and group training.

NPT is planning new research and test facilities at Coventry, and is currently negotiating for an adjacent site to expand its manufacturing facilities. Similar activities are taking place at Harwich.

NPT's basic objectives are to produce advanced-concept power plants for world markets, to exploit new low-cost materials, especially in high temperature environments, and to expand its services to industry and the batch manufacture of engineering components and sub-assemblies. The main products continue to be small gas turbines, but NPT is active in solid-state electronics. It also produces such equipment as dynamic balancing machines, engine and process controllers, electronic tachometers and high-energy ignition systems.

The engines described are of modular design, developed for use as turbojets, gas producers or sources of shaft power. More advanced units are under development for various applications including aircraft propulsion.

The NPT 100 engine is currently used in a four-engined air-cushion unit designed by NPT to relieve road loads during the transport of heavy equipment and machinery. The NPT 400 is the chosen power plant for an advanced drone system now being developed.

NPT 100, 400

These are the first NPT gas-turbine products. Their simplest forms are the 101 and 401 turbojets, but other versions including shaft drive can be provided. The following description refers to the NPT 101, with NPT 401 differences noted in brackets.

TYPE: Single-shaft turbojet.
AIR INTAKE: Direct pitot.
COMPRESSOR: Single-stage centrifugal (two axial stages, one centrifugal). Forged aluminium-alloy rotor (cast steel) in cast aluminium-alloy casing. Maximum airflow 1·5 kg (3·3 lb)/sec (2·28 kg; 5·03 lb/sec).
COMBUSTION CHAMBER: Reverse-flow annular with va orising burners. HE spark igniter.
FUEL SYSTEM: Gear pump. Electronic control system.
FUEL GRADE: Kerosene, No 2 diesel.
TURBINE: Two-stage axial. Cast Nimonic disc with integr blades. Bolted to shaft with roller bearing. Inlet g temperature 877°C (933°C).
ACCESSORY DRIVES: Two bevel-driven shafts on compre sor casing.
LUBRICATION: Closed circuit with cooling.
MOUNTING: Flange on compressor casing.
STARTING: Air or electric.
DIMENSIONS:
Diameter — 381 mm (15·0 i
Length (NPT 101) — 889 mm (35·0 i
(NPT 401) — 940 mm (37·0 i
WEIGHT, DRY:
NPT 101 — 34·5 kg (76·0 l
NPT 401 — 38·6 kg (85·0 l
PERFORMANCE RATING:
NPT 101 — 0·756 kN (170 l
NPT 401 — 1·335 kN (300 l
SPECIFIC FUEL CONSUMPTION:
NPT 101 — 35·69 mg/Ns (1·26 lb/hr/l
NPT 401 — 31·16 mg/Ns (1·10 lb/hr/l

ROLLASON

ROLLASON AIRCRAFT AND ENGINES LTD

Head Office and Works:
Brighton, Hove and Worthing Joint Municipal Airport, Shoreham-by-Sea, Sussex BN4, 5FJ
Telephone: Shoreham-by-Sea 62680
Officers: See "Aircraft" section

In support of its manufacture (now ended) of the Druine Turbulent light aeroplane, Rollason Aircraft and Engines Ltd is undertaking the conversion of Ardem 4CO2 power plants for this aircraft, from motor car engines.

Rollason has developed several versions of the Ardem engine with capacities of 1,500 or 1,600 cc. Any of these Ardem engines can be installed in Nipper aircraft. The Mk X is also used in the Australian Corby Starlet and several other homebuilts.

ARDEM 1,500 cc

The standard model has a compression ratio of 7·8 : 1 and gives 33·6 kW (45 hp) at 3,300 rpm for take-off, with a fuel consumption of 17 litres (3·75 Imp gallons)/hr at max rating. In addition, a high-compression version is available, as follows:

Cylinders: Bore 83 mm (3·27 in). Stroke 69 mm (2·72 in). Cast steel barrels, light alloy heads. Compression ratio 8·5 : 1.
Pistons: Aluminium alloy high-compression pistons, each with two compression rings and one scraper ring. Floating gudgeon pins.
Connecting Rods: White metal bearings in big-end. Bronze bearings in little-end.
Crankshaft: Runs in four white metal bearings.
Crankcase: Magnesium case.
Valve Gear: Two valves per cylinder. Camshaft geared to crankshaft.
Induction: Zenith 32 KL P10 carburettor.
Fuel Grade: 100 octane.
Ignition: Lucas SR4 magneto mounted below engine, with chain drive. Two Lodge LH spark plugs per cylinder.
Lubrication: Wet sump type, with single gear-type pump.
Oil Specification: Shell W80.
Propeller Drive: Direct drive.
Accessories: SEV 46C fuel pump.
Dimensions:
Length 426 mm (16·75 in)
Width 750 mm (29·50 in)
Height 559 mm (22·00 in)
Performance Rating:
Max 39·5 kW (53 hp) at 3,600 rpm
Fuel Consumption:
At max rating 18·2 litres (4·0 Imp gallons)/hr

ARDEM 1,600 cc

After extensive testing this engine is now approved and is produced as the Ardem Mk XI. It differs from the Mk X in having cylinders of 85·5 mm (3·365 in) bore and dual ignition.

Weight:
With accessories 71·6 kg (158 lb)
Performance Ratings:
Max 41 kW (55 hp) at 3,300 rpm
Cruise 26·5 kW (35·5 hp) at 2,500 rpm
Fuel Consumption:
At cruise rating 12·5 litres (2·75 Imp gallons)/hr

ROLLS-ROYCE

ROLLS-ROYCE (1971) LIMITED

Head Office:
Norfolk House, St James's Square, London SW1Y 4JR
Telephone: 01-839-7888
Main Locations:
PO Box 3, Filton, Bristol BS12 7QE
Telephone: 0272-693871
PO Box 31, Moor Lane, Derby DE2 8BJ
Telephone: 0332-42424
PO Box 72, Ansty, Coventry CV7 9JR
Telephone: 0203-32-2311
Leavesden, Watford WD2 7BZ
Telephone: 47-7400
Scottish group of factories mainly to the south of Glasgow
Chairman and Chief Executive:
Sir Kenneth Keith
Deputy Chairman:
Sir William Nield
Vice Chairman:
Marshal of the Royal Air Force Sir Denis Spotswood
Technical Director:
Sir Stanley Hooker
Projects Director:
Sir William Cook
Financial Director:
J. E. M. Gardner
Managing Director, Aero Division:
D. A. Head
Executive Office Director:
D. J. Pepper
Board Members:
Sir St John Elstub
Sir Arthur Knight
R. T. Whitfield
Company Secretary:
H. E. Trevan-Hawke

Rolls-Royce (1971) Ltd, which produces a range of gas turbines and ramjets, retains the experience in aircraft engines built up over more than 60 years by the predecessor companies that it incorporated. The company also represents British experience in lightweight high-power gas turbines for industrial and marine purposes, since such engines were first derived from aircraft gas turbines more than a decade ago.

More than 190 million hours of operating experience have been accumulated with Rolls-Royce civil and military gas turbines, which are used by 224 airlines and 80 armed forces.

In addition to the products designed, developed and manufactured solely in Britain, the company works with partners abroad on a number of joint civil and military aircraft engine programmes. Licences for the manufacture of Rolls-Royce engines or components are also held by many countries throughout the world.

The company was registered on 23 February 1971, and at present the British government is the sole shareholder. On 22 May of the same year the company was vested with assets which were those concerned with the gas turbine business of Rolls-Royce Ltd.

The main activities are at Derby, Glasgow, Bristol, Coventry and Leavesden, where aircraft gas turbines are produced, and at Ansty, where aircraft gas turbine techniques are applied to industrial and marine uses. A total of 63,550 people are employed.

One of the most advanced products is the RB.211 three-shaft turbofan which is in service in the Lockheed TriStar airliner. Development and production continue jointly with the French company SNECMA on the Olympus two-shaft turbojet which powers the Concorde supersonic airliner (see International section). Other Rolls-Royce engines which are in commercial service include the Spey, M45 and Conway turbofans, Avon and Viper turbojets, and Dart and Tyne turboprops.

Rolls-Royce (1971) is a member of the Turbo-Union company (see International section), in which it is associated with MTU of Germany and Fiat of Italy on the RB.199 engine for the MRCA multi-role combat aircraft. In addition to the Olympus, the company is also associated with SNECMA on the M45SD-02, with Detroit Diesel Allison Division of General Motors on the TF41 and with Turboméca of France on the Adour, and is expecting to complete agreements with United Technologies Corporation for the JT10D and possibly other engines.

A new turboshaft, the Gem, is being produced by Rolls-Royce for the Westland/Aérospatiale Lynx multi-purpose helicopter. Together with the Turmo and Astazou, on which production is shared with Turboméca, it is part of the Anglo-French helicopter programme, designed to meet the medium and light helicopter requirements of the two countries.

In May 1973 Rolls-Royce (1971) and General Electric's Aircraft Engine group announced that an agreement signed by the two companies in 1971, under which GE is exploring opportunities for the Gem (RS.360) in the United States, would be continued for the current year.

Versions of several Rolls-Royce aircraft gas turbines are used as power units for large and small ships, hydrofoils, air cushion vehicles, electricity-generating sets, gas and oil pumping equipment and for other industrial uses.

ROLLS-ROYCE TURBOMÉCA ADOUR

This turbofan was designed by Rolls-Royce and Turboméca to power the Jaguar, and is in production on a 50 : 50 basis by the two companies for this aircraft. A similar version is made under licence by IHI of Japan for the T-2, and an unaugmented version powers the HS Hawk. The Adour is described in the International part of this edition.

ROLLS-ROYCE RB.162

The RB.162 is a simple ultra-lightweight turbojet engine which was developed initially to meet the requirements of the aircraft industries of Britain, France and Federal Germany for a lift-jet unit for V/STOL aircraft. Bench testing began in 1962. The RB.162-1/4 series was described in the 1969-70 edition of *Jane's,* together with the more powerful RB.162-30 series and RB.162-81.

The current version is as follows:

RB.162-86. Developed as take-off booster for transport aircraft. In February 1971 received full Transport Category C of A after completing 550 operating cycles with some 19 out of the 50 hours of running time being at take-off rating. Installed in HS Trident 3B in service with British Airways and CAA of China.

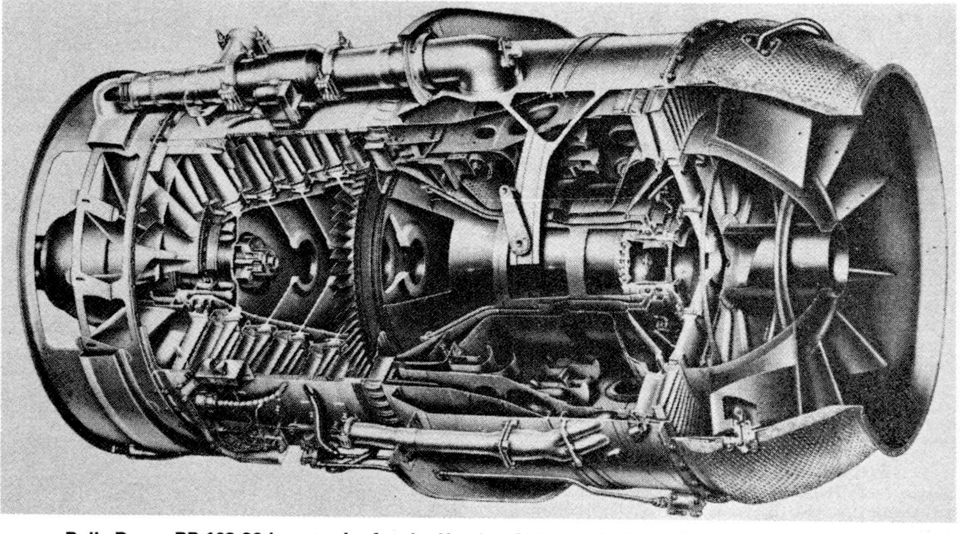

Rolls-Royce RB.162-86 booster jet for the Hawker Siddeley Trident 3B (23·35 kN; 5,250 lb st)

Type: Simple turbojet for take-off boost.
Air Intake: Direct pitot. Steel with anti-iced fixed inlet guide vanes. Two vertical struts integral with nose bullet and annulus house fuel, oil and air pipes. Front cover of these struts removable.
Compressor Casing: In two halves, in GRC (glass-reinforced composite). Mountings for accessories, including throttle, igniter box, oil tank, electrical disconnect bracket, anti-icing valve and Firewire. Five stages of stator vanes, and shroud boxes, bonded to form integral part of each half-casing. Steel front flange bonded to each casing. Five introscope bosses permit viewing of all rotor stages.
Compressor Rotor: Six-stage axial. Aluminium drum and shaft formed from individual discs and spacers electron-beam-welded to form single unit. Location for front bearing as part of 2nd-stage disc. Curvic coupling incorporated at rear of shaft to connect with turbine disc. Stage 1 blades, aluminium, clappered at ¼ span and pin-retained. Stage 2 aluminium, and stages 3-6 all GRC, retained by dovetails. Fixed hollow steel diffuser vanes aft of compressor. Pressure ratio 4·5 : 1. Mass flow 38·5 kg (85 lb)/sec.
Combustion Chamber: Annular type, inside welded sheet steel casing. Outer flame tube is continuous drum with perforations for secondary air. Inner flame tube is Nimonic drum carrying 18 equally spaced burners, two starting atomisers and two high-energy igniters 30° each side of vertical section plane.
Fuel System: Self-contained system in four units. One unit, with aluminium body, is housed in nose bullet and driven at engine speed. This contains backing pump, HP gear pump, accelerator control unit and two datum governors; also incorporates oil pump. Second unit is combined throttle and flow control, again in aluminium, mounted on compressor casing; incorporates spill flow adjuster. Third unit is electrical actuator, on throttle, which controls position of throttle plate: Closed, Ground Idle, Flight Idle, Climb, Take-off. Fourth unit is pressurising and dump valve interspaced between throttle and main burner rail. Simple form of filtering included, but system designed to accept dirty or icy fuel without heating. Fuel grades: JP-1, JP-4.
Nozzle Guide Vanes: Hollow aircooled refractory alloy. Form integral part of combustion casing.
Turbine: Single-stage axial. Steel disc. Hollow aircooled blades of refractory alloy held by fir-tree roots.
Bearings: Only two ball bearings, one between inlet guide vanes and compressor, the other forward of turbine. Both are single-row, the turbine bearing being for location and the intake bearing making provision for axial expansion.

JET PIPE: Two parts. Exhaust unit comprises short outer cone, with inner cone supported by 12 radial aircooled struts. Exterior protected by insulation blankets. Second unit is bolt-on jet pipe with common diameter final nozzle, adjusted by trimmers. Pipe supports JPT thermocouples.
LUBRICATION SYSTEM: Total loss system. Oil supplied from 2·4 litre (6 pint) tank, supported by compressor casing, with electric contents indicator. Oil fed to both bearings during engine running by rotating piston-type pump between rotor and fuel pump. Oil scavenged from both bearings and ejected overboard.
MOUNTING: Four equally spaced spherical trunnions, of which any three used.
STARTING: Direct air impingement on turbine blades. Air can be bled from main propulsion engines.
ACCESSORIES: Following provide control parameters: engine-speed signal generated by phonic wheel and probe incorporated in fuel pump; contact thermocouples on front and rear bearings; for JPT, 12 thermocouples in series in jet pipe; two Sperry vibration pickups, one on compressor forward casing, one on turbine casing; Firewire detector around compressor casing (Zone 1); oil contents indicator in oil tank. Signals from all indicators fed to flight deck warning system via computer/amplifier.
DIMENSIONS:
Diameter 737 mm (29 in)
Length 2,464 mm (97 in)
WEIGHT, DRY: (with fluid systems): 283 kg (625 lb)
PERFORMANCE RATING:
Max T-O 23·35 kN (5,250 lb st)

ROLLS-ROYCE RB.199

This advanced augmented turbofan is the power unit for the Panavia Tornado. The RB.199 programme is managed by Turbo-Union (see International part of this section).

ROLLS-ROYCE RB.211

The RB.211 is an advanced technology three-shaft turbofan of high by-pass ratio, high pressure ratio design in the 178 kN (40,000 lb) to 235·75 kN (53,000 lb st) bracket. The engine was selected by Lockheed in March 1968 to power the L-1011 TriStar, and later by Boeing as an alternative option on the 747.

Rolls-Royce initiated design studies of three-shaft turbofans in 1961 and a twin-spool engine, the RB.178, was tested in 1967 to provide relevant component and gas generator experience. Among the advantages afforded by a three-shaft layout are its ready use of a high pressure ratio with fewer compressor and turbine stages whilst maintaining excellent handling. The need for compressor variable stator mechanisms can also be minimised and the rotating assemblies can be made relatively short and rigid while preserving light construction. As a result the RB.211 has demonstrated outstandingly low seal and aerofoil wear, thus maintaining a high level of performance throughout engine life.

For the Lockheed TriStar, Rolls-Royce developed and now manufactures the complete propulsion system, comprising the engine, fan airflow reverser, pod cowlings and related systems, and noise attenuation for the intake, fan cowl and turbine exhaust duct.

The engine is divided into seven modules. This permits rapid change of engine parts, and enables service life to be set up individually for each module. It also facilitates rapid repair, as a damaged or time-expired module can be replaced with the engine installed in the aircraft. Maximum provision is made for in-service monitoring of engine condition and visual inspection on the ground of all engine sections.

The RB.211 combustion chamber is of annular design, giving significant advantages over tubo-annular systems in terms of reduced cost, weight and length, and improved efficiency. The reduced length makes a two-bearing HP system possible, with both bearings located away from the high temperatures of the combustion area. Detailed design of the combustion liner, fuel injection nozzles and fuel control system has been aimed at reducing exhaust contaminants to a minimum. As a result the smoke level of the RB.211 shows a significant reduction compared with turbofans in service previously.

The HP turbine blades are single-pass convection cooled, and have film cooling at the leading edge and on both faces. Cooling air is fed to the turbine disc via preswirl nozzles which accelerate the air in the direction of disc rotation, thus reducing its temperature before entry to the blade root.

The fan air stream of the RB.211 flows through a three-quarter length duct. With high by-pass ratio engines the required reverse thrust can be achieved by reversing the fan air stream only. Fan stream reversing is achieved by translating rearwards a section of fan cowling, thus uncovering sets of cascades and at the same time closing the fan duct downstream by hinged blocker doors.

In order to minimise fan noise, the RB.211 has no inlet guide vanes; the distance between the single-stage fan and its outlet guide vanes is optimised for minimum noise generation, as is the numerical ratio of the two rows of blading. Additional noise attenuation is achieved by the use of acoustic lining material in the intake, fan and turbine exhaust ducts. The RB.211 achieved FAA noise certification with comfortable margins. The FAA described the TriStar as "quieter than any other large wide-body airliner".

Rolls-Royce RB.211-524 three-shaft turbofans (222·4 kN; 50,000 lb st)

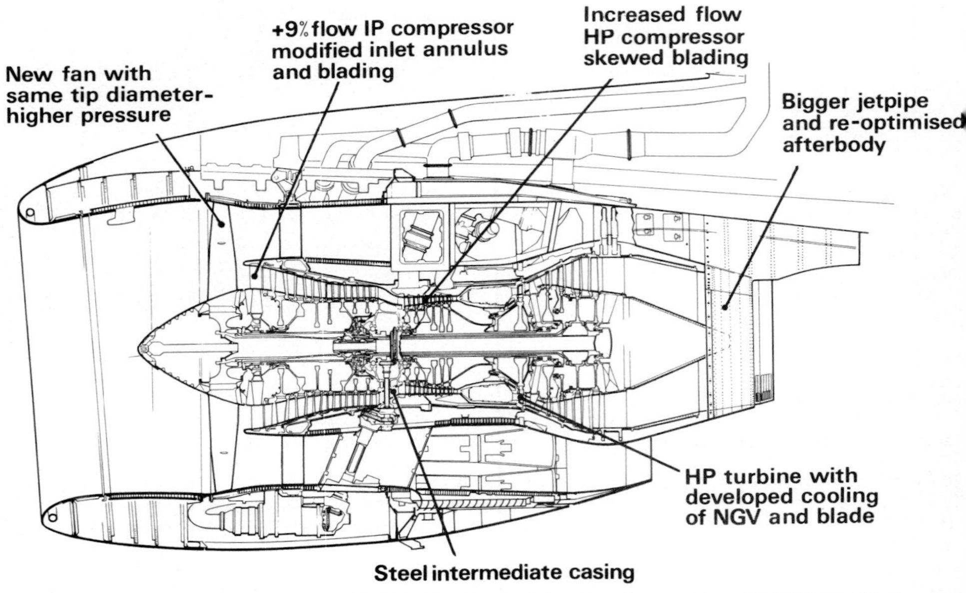

Longitudinal cross-section of complete RB.211-524 wing pod showing changes from RB.211-22B. This drawing is not the same as that published last year, the main alteration being the new afterbody

The **RB.211-22B,** the standard engine of the L-1011-1 TriStar, is flat rated at 187 kN (42,000 lb st) to 28·9°C. This engine was certificated in February 1973 by the CAA and in April 1973 by the FAA. By the spring of 1976 more than 530 engines had been delivered, and engine flight hours in service were close to 2 million.

The **RB.211-524** represents the initial step in the RB.211's programme of growth. It preserves maximum commonality with the -22B but has increased core airflow, slightly higher turbine gas temperature and improved component efficiency. First run took place on 1 October 1973. A few days later the full take-off thrust was achieved. In early 1974 a -524 at the National Gas Turbine Establishment demonstrated maximum cruise and climb sfc guarantees.

By April 1976 nine RB.211-524 prototype engines had run 3,000 hours, including a series of cycles specifically aimed at testing the engine severely. The first production RB.211-524 was run in December 1975. Deliveries for aircraft certification began in March 1976; deliveries for airline service were due in late 1976.

Engines for the extended-range TriStar are rated at 213·5 kN (48,000 lb) to 28·9°C and are designated RB.211-524. Engines for B.747 aircraft have identical configuration but have been certificated at 222·4 kN (50,000 lb) to 28·9°C and are designated **RB.211-524B.** The designation **RB.211-524E** has been reserved for a derated engine, flat-rated at 193·5 kN (43,500 lb). Without major change, the -524 could be uprated to 235·75 kN (53,000 lb st), and development to higher ratings is under consideration.

On 20 June the British Department of Industry announced support for the Boeing 747 with RB.211-524B engines, rated at 222·4 kN (50,000 lb st) initially and later at 235·75 kN (53,000 lb). The government will pay approximately half the programme launch cost, recovering the investment through a levy on sales. Production-standard engines were due to be delivered to Boeing in the Spring of 1976, for service entry one year later. Rolls-Royce calculates the RB.211-powered 747 will "use less fuel and be cheaper to operate" than earlier versions. An RB.211-524 completed a declared type test at 50,000 lb in June 1975, five weeks ahead of schedule. Full certification was achieved in December 1975.

The following description relates to the RB.211-22B:
TYPE: Three-shaft axial turbofan.
AIR INTAKE: Forward-facing pitot.
LP FAN: Single-stage overhung fan driven by LP turbine, the whole rotor assembly being supported on three bearings. Front bearing is large roller, squeeze-film supported behind fan. Axial location of rotor is by inter-shaft ball bearing in rear end of IP compressor drum. LP turbine supported on roller bearing, squeeze-film mounted in exhaust cone panel. Rotating spinner supported from fan rotor disc and hot-air anti-iced via central feed-tube within shaft. Titanium alloy used for 33 fan rotor blades, and steel for 70 fan outlet guide vanes. Titanium fan disc bolted with curvic coupling to LP shaft. Aluminium fan casing. Total fan airflow (T-O rating), 626 kg (1,380 lb)/sec (-524, 657·7 kg; 1,450 lb). By-pass ratio 5 : 1 (-524, 4·4 : 1.)
IP COMPRESSOR: Seven-stage compressor rotor driven by IP turbine and supported on three bearings located directly in support panels. Front squeeze-film bearing is roller. Mid bearing at rear of IP compressor is ball bearing providing axial location for IP rotor. Rear bearing is roller, squeeze-film supported in panel between HP and IP turbines. Two drums, one of titanium discs welded together and the other of welded steel discs, are bolted to form one rotor, carrying titanium rotor blades. Aluminium and steel casings carry aluminium and stator steel blades. Single-stage titanium variable inlet guide vanes.

Rolls-Royce RB.211-22B three-shaft turbofan of 187 kN (42,000 lb st) for the Lockheed TriStar

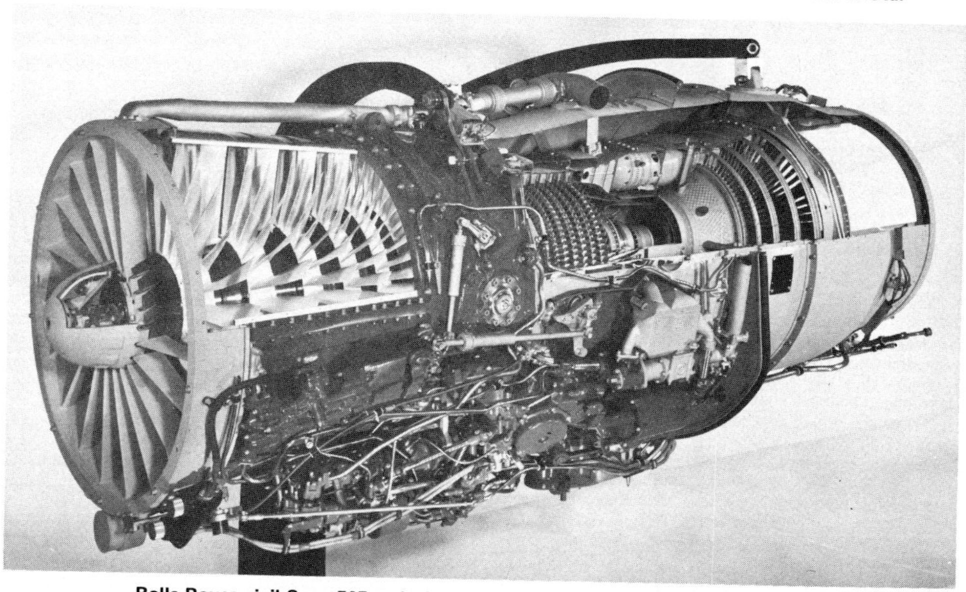

Rolls-Royce civil Spey 505 turbofan engine, cut away to show internal details

HP Compressor: Six-stage compressor rotor driven by HP turbine connected by large-diameter shaft and carried on ball location bearing at front and roller bearing squeeze-film mounted in panel behind HP turbine disc. Welded titanium discs, a single steel disc and welded nickel alloy discs are bolted together to form the rotor, carrying titanium, steel and nickel alloy blades. Steel casing carries steel and Nimonic stator blades. Overall pressure ratio 25 : 1.

Combustion Chamber: Fully annular, with steel outer casings and Nimonic flame tube. Downstream fuel injection by 18 airspray burners with annular atomisers. Ignition by starting atomisers and high-energy igniter plugs in Nos 8 and 12 burners.

HP Turbine: Single-stage axial unit with Nimonic nozzle guide vanes and Nimonic rotor blades, both rows air-cooled. Convection and film-cooled blades mounted in Nimonic disc by fir-tree roots.

IP Turbine: Single-stage axial unit with Nimonic nozzle guide vanes and Nimonic rotor blades. NGVs aircooled. Rotor blades fir-tree mounted in Nimonic disc.

LP Turbine: Three-stage axial unit with Nimonic rotor blades fir-tree mounted in steel disc.

Jet Pipe: Steel jet pipe without spoiler.

Accessory Drives: Radial drive from HP shaft to gearbox on fan casing. Accessories driven include integrated-drive generator and aircraft hydraulic pumps.

Lubrication System: Continuous circulation "dry sump" system with single gear-type pressure pump and multiple gear-type scavenge pumps. Oil tank 21 litres (37 Imp pints) capacity integral with gearbox.

Mounting: Two-point mounting system. Front mount on fan casing takes thrust, vertical and side-loads. Rear link mount on exhaust casing takes torsional, side and vertical loads. Both mounts are fail-safe and allow for carcase expansion.

Dimensions:
- Length overall: 3,033 mm (119·4 in)
- Intake diameter: 2,172 mm (85·5 in)

Weight, Dry:
- Ship set of three installed propulsion systems:
  - RB.211-22B: 17,464 kg (38,500 lb)
  - RB.211-524: 17,962 kg (39,600 lb)

Performance Ratings:
- T-O, flat rated to 28·9°C:
  - RB.211-22B: 187 kN (42,000 lb st)
  - RB.211-524: 213·5 kN (48,000 lb st)
  - RB.211-524B: 222·4 kN (50,000 lb st)
  - RB.211-524E: 193·5 kN (43,500 lb st)
- Cruise at 10,670 m (35,000 ft) and Mach 0·85:
  - RB.211-22B: 41·8 kN (9,400 lb st)
  - RB.211-524 and -524B: 48·8 kN (10,970 lb st)

Specific Fuel Consumption:
- At cruise rating, as above:
  - RB.211-22B: 18·1 mg/Ns (0·640 lb/hr/lb st)
  - RB.211-524: 18·6 mg/Ns (0·657 lb/hr/lb st)
  - RB.211-524B: 18·6 mg/Ns (0·657 lb/hr/lb st)
  - RB.211-524E: 18·0 mg/Ns (0·638 lb/hr/lb st)

ROLLS-ROYCE RB.163 CIVIL SPEY

Design of the Spey RB.163 began in September 1959, and the first engine ran at the end of December 1960.

Flight testing of two Speys in a Vulcan began on 12 October 1961, and prototype flight trials of the Spey-engined Hawker Siddeley Trident began on 9 January 1962. In July 1962 the ARB issued special category approval of the Spey in the Trident pod, which involved completion of a 150 hour type test to combine UK/US schedule. Civil Speys are in service in the Trident, BAC One-Eleven, Grumman Gulfstream II and F.28 Fellowship, and in the C-8A augmentor-wing research aircraft of NASA.

The following versions of the civil Spey are in service:

Mk 505-5. T-O rating of 43·8 kN (9,850 lb st) at 12,490 rpm, for Hawker Siddeley Trident 1 fleet of British Airways.

Mk 506-14. T-O rating of 46·3 kN (10,410 lb st) at 12,530 rpm, for BAC One-Eleven.

Mk 506-14AW. As 506-14, but with water injection to maintain rating to 35°C.

Mk 511-5, 511-8 and 511-14. T-O rating of 50·7 kN (11,400 lb st) at 12,390 rpm. Mk 511-5 for trident, Mk 511-8 for Gulfstream II and Mk 511-14 for One-Eleven.

Mk 511-5W and 511-14W. As 511-5 and 511-14 but with water injection to maintain rating to 35°C.

Mk 512-5 and 512-14. T-O rating 53·2 kN (11,960 lb st) in Mk 512-5 for Trident, and 53·4 kN (12,000 lb st) in Mk 512-14 for One-Eleven, in each case at 12,390 rpm. Both available with water injection.

Mk 512DW. T-O rating 55·8 kN (12,550 lb st). Similar to Mk 512 but with T-O rating increased by increasing limiting compressor delivery pressure at T-O, with turbine entry temperature restrained by water injection.

Mk 555-15. Lightened and simplified version, with T-O rating of 43·8 kN (9,850 lb). For F28 Fellowship. Has changed company project number of **RB.183.**

Mk 606. Redesigned low-noise engine, described separately (Spey 67C).

The military versions are described separately.

According to Fokker-VFW a near-relative of the Spey 67C is the RB.183-72, a refanned version of the simplified RB.183 Spey 555 which powers existing F28 Fellowship versions. To be rated in the 64·5-71·2 kN (14,500-16,000 lb) st class, the Spey-72 is seen by Fokker-VFW as a prime candidate engine for the proposed F28-2.

The following details refer specifically to the Spey Mk 512-14DW, as fitted to the BAC One-Eleven Series 500, except where indicated:

Type: Two-spool axial-flow turbofan engine.

Air Intake: Annular, with bleed air thermal anti-icing.

Compressor: Two spools. Five-stage (four-stage on Mks 505, 506 and 555) low-pressure (LP) and 12-stage high-pressure (HP). First-stage HP stator vanes variable-incidence. LP compressor steel drum type, pinned to shaft. HP compressor is of the steel disc type, first stage bolted to shaft, remaining stages splined. HP stator blades steel; LP stators aluminium. LP rotor blades aluminium (Mk 512 titanium 1st stage); HP blades steel and titanium. Stators slotted into casing; rotor blades pinned or dovetailed. LP casing two-piece aluminium. HP casing two-piece steel. Pressure ratio 21·2 : 1 (15·0 on Mk 505 and 555, 17·2 on Mk 506, 18·9 on Mk 510 and 511). Air mass flow 94·4 kg (208 lb)/sec (90·27 kg; 200 lb on Mk 505 and 555, 92 kg; 203 lb on Mk 506, 92·5 kg; 204 lb on Mk 510 and 511). By-pass ratio 0·64 : 1 (1·0 on Mk 505, 555 and 506).

Combustion Chamber: Tubo-annular with 10 Nimonic sheet liners. Duplex downstream burners, one per chamber. High energy igniters in chambers 4 and 8.

Fuel System: Plessey LP pump feeding through fuel-cooled oil cooler and Marston Excelsior fuel heater to LP filter at inlet to Lucas GD pump. HP metered by Lucas regulator, embodying combined speed and acceleration control and fed through Lucas LP governor and shut-off valve to Duple spray nozzles. Maximum pressure 124 bars (1,800 lb/sq in).

Fuel Grade: DERD.2482 or 2486.

Water Injection System: (engines bearing "W" suffix): Water supplied by Lucas air turbopump through engine-mounted automatic shut-off valve to injector passages in fuel spray nozzles (water sprays into primary airflow through flame tube swirlers).

Nozzle Guide Vanes: Hollow cast in nickel-based alloy HP aircooled.

Turbines: Two two-stage. First HP aircooled. HP discs nickel-based alloy, bolted to shaft (HP discs steel on Mks 505, 506 and 555). LP discs creep-resisting ferritic steel. Nickel-based alloy blades attached by fir-tree roots.

Bearings: LP compressor supported in roller bearings, plus ball thrust bearing. HP compressor has front roller bearing and ball thrust bearing. Turbine bearings all roller type flexibly mounted.

Jet Pipe: Fixed-area stainless steel sheet.

Reverser and Suppressor: Normally internal clamshell (Gulfstream II target type, not Rolls-Royce supplied, and F28 has no reverser). Five- or six-chute silencing nozzles available.

Accessory Drives: Port gearbox, driven from LP rotor, carries LP governor and LP tacho. Starboard gearbox, driven from HP rotor, carries LP and HP fuel pumps, fuel regulator, main oil pumps, airflow control rpm signal transmitter, starter and HP tacho. Provision in starboard gearbox for aircraft ancillaries.

Lubrication System: Self-contained continuous circulation. Single pressure pump feeds oil from tank through fuel-cooled cooler and HP filter to gearboxes and shaft bearings. Five main scavenge pumps. Tank capacity 6·8 litres (12 Imp pints). Usable oil 5·1 litres (9 Imp pints). Normal pressure 2·41-3·45 bars (35-50 lb/sq in).

OIL SPECIFICATION: DERD.2487.

MOUNTING: Two trunnions, two saddle mountings and one rear mounting.

STARTING: Plessey 220 air-turbine starter. Rotax alternative on Mks 505, 511 and 512; AiResearch on Mk 555.

DIMENSIONS:

Length, less tailpipe:	
Mk 505, 506, 555	2,795 mm (110·0 in)
Mk 510, 511	2,911 mm (114·6 in)
Diameter:	
Mk 505, 506, 555	940 mm (37·0 in)
Mk 510, 511	942 mm (37·1 in)

WEIGHT, DRY:

Mk 505-5	998 kg (2,200 lb)
Mk 506-14	1,024 kg (2,257 lb)
Mk 506-14AW	1,038 kg (2,288 lb)
Mk 510-5, 511-5	1,049 kg (2,312 lb)
Mk 510-14, 511-14	1,058 kg (2,332 lb)
Mk 510-14W, 511-14W	1,188 kg (2,621 lb)
Mk 511-5W	1,050 kg (2,317 lb)
Mk 512	1,168 kg (2,574 lb)
Mk 555-15	995 kg (2,194 lb)

PERFORMANCE RATINGS:

Max T-O: See under series descriptions

Max continuous:

Mk 505	42 kN (9,450 lb st) at 12,260 rpm
Mk 506	44·4 kN (9,990 lb st) at 12,385 rpm
Mk 511	48·7 kN (10,940 lb st) at 12,240 rpm
Mk 512, Mk 512DW	51·5 kN (11,580 lb st) at 12,450 rpm
Mk 555-15	42·1 kN (9,470 lb st) at 11,900 rpm

Typical cruise rating at 450 knots (834 km/h; 518 mph) at 9,750 m (32,000 ft):

All versions 13·7 kN (3,070 lb st)

SPECIFIC FUEL CONSUMPTION:

At T-O rating:	
Mk 505	15·9 mg/Ns (0·560 lb/hr/lb st)
Mk 506	15·95 mg/Ns (0·563 lb/hr/lb st)
Mk 511	17·3 mg/Ns (0·612 lb/hr/lb st)
Mk 555-15	15·9 mg/Ns (0·560 lb/hr/lb st)
At typical cruise rating:	
Mks 505, 506	21·5 mg/Ns (0·760 lb/hr/lb st)
Mk 510, 511	22·3 mg/Ns (0·790 lb/hr/lb st)
Mks 512, 555	22·7 mg/Ns (0·800 lb/hr/lb st)

OIL CONSUMPTION:

Max (all Marks) 0·42 litres (0·75 Imp pints)/hr

ROLLS-ROYCE SPEY 606

The proposed Spey Mk 606 (-67C) is a high by-pass ratio turbofan derived from the Spey Mk 512. The engine is designed to provide low noise levels and improved fuel consumption to meet the requirements of the 1980s. It is expected that the 90 dB footprint for a typical twin-jet will be reduced to under 10·33 km² (4 sq miles) and that noise levels in the region of FAR Pt 36 minus 10 dB will be achieved.

TYPE: Two-spool axial turbofan.

COMPRESSOR: Single-stage fan and three-stage intermediate pressure (IP) compressor on low-speed shaft and 12-stage high-pressure (HP) compressor on high-speed shaft. Fan/IP compressor has impact-resistant steel and titanium blading and no inlet guide vanes. HP compressor of disc construction, with blades of steel and titanium, with stator vanes slotted into two-piece steel casing. Air mass flow 195 kg (429 lb)/sec. Pressure ratio 21·9. By-pass ratio 1·96.

COMBUSTION AND FUEL SYSTEMS: Similar to Spey Mk 512. No water-injection system.

NOZZLE GUIDE VANES: Hollow cast in nickel-based alloy. HP aircooled.

TURBINES: Two-stage HP and three-stage LP. Both HP stages cast aircooled. LP blades forged.

BEARINGS: Similar to Spey Mk 512.

JET PIPE: Fixed-area stainless sheet, with noise-suppression honeycomb lining.

REVERSER: Internal clamshell with noise-suppression.

ACCESSORY DRIVES: Gearbox driven from HP rotor carries fuel pumps, fuel regulator, oil pumps, starter and aircraft accessories.

DIMENSIONS:

Fan diameter 1,194 mm (47 in)

Length (inlet flange to exhaust flange) 2,412 mm (95 in)

DRY WEIGHT: Not available.

PERFORMANCE RATING:

Max T-O 75·1 kN (16,900 lb st)

SPECIFIC FUEL CONSUMPTION:

Max cruise (7,620 m , 25,000 ft; Mach 0·75, ISA) 22·4 mg/Ns (0·791 lb/hr/lb)

ROLLS-ROYCE RB.168 MILITARY SPEY

The military Spey RB.168 incorporates modifications to meet higher-duty conditions.

Design of the RB.168-1, Mk 101, started in November 1960; this mark powers the Hawker Siddeley Buccaneer strike aircraft.

The RB.168-25R (Mks 202/3) supersonic engine with afterburner has a dry rating of 54·5 kN (12,250 lb) plus a 70 per cent static augmentation and powers the McDonnell Douglas Phantom FG.1 and FGR.2. Design of this variant of the Spey started at the beginning of 1964 and the first run was in April 1965. Major change is the introduction of a robust shaft-and-disc LP compressor. As with Mk 101, use is made of HP compressor bleed air for aircraft BLC purposes. A Plessey gas-turbine starter is fitted. Augmentation is thrust-modulating from an initial boost, at sea level static, of six per cent. The afterburner incorporates three vee-gutter flame stabilisers, multi-bar fuel injection via four upstream manifolds, self-contained ignition, a fully-variable primary nozzle and a fixed secondary nozzle. Longitudinal movement of the divergent ejector nozzle by six hydraulic rams operates the primary nozzle flaps.

In December 1975 the chairman of Rolls-Royce signed contracts with the Chinese government for an initial supply of supersonic augmented Spey engines and for licence-production of such engines in China. The initial agreement is valued at over £80 million, and the eventual total, including plant and equipment, is expected to exceed £100 million.

The Spey RB.168-20 Mk 250, closely based on the commercial Spey, powers the Hawker Siddeley Nimrod. Embodying extensive anti-corrosion treatment, this variant provides a higher thrust than its civil counterpart, through operation at higher rpm and higher turbine entry temperature. Provision is made for driving a large alternator.

DIMENSIONS:

Diameter	825 mm (32·5 in)
Length:	
Mks 101, 250	2,985 mm (117·5 in)
Mks 202, 203	5,204 mm (204·9 in)

WEIGHT, DRY:

Mk 101	1,181 kg (2,603 lb)
Mk 250	1,227 kg (2,704 lb)
Mks 202, 203	1,857 kg (4,093 lb)

PERFORMANCE RATING:

Max T-O:

Mk 101	49·1 kN (11,030 lb)
Mks 202, 203	91·25 kN (20,515 lb)
Mk 250	53·4 kN (11,995 lb)

ROLLS-ROYCE/ALLISON SPEY TF41

Versions of the Spey are being produced jointly by Rolls-Royce and the Detroit Diesel Allison Division of General Motors to power the LTV A-7 Corsair II close support aircraft. These engines, the TF41-A-1 and TF41-A-2, are described in the International section.

ROLLS-ROYCE M45

Collaboration between Rolls-Royce Bristol and SNECMA on the M45 series of advanced turbofan engines began in late 1964, and a formal agreement was signed in February 1965. First of the series to be built was the M45F demonstrator engine, which ran for the first time in June 1966.

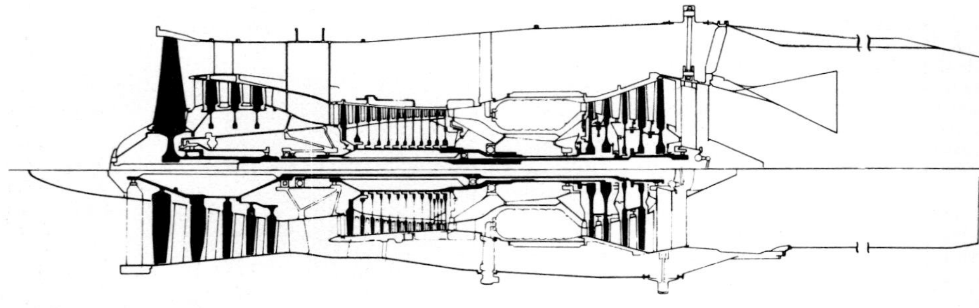

Comparative cross-sections of the Rolls-Royce Spey 512 and Spey 606, the latter to be rated at 71·2 kN; 16,90 lb st. The HP spool, combustor and first two LP turbine stages are common

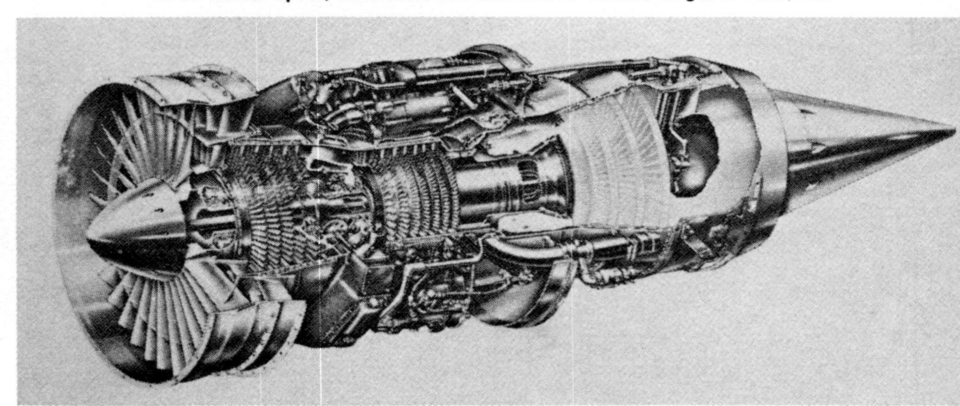

Cutaway drawing of Rolls-Royce M45H-01 civil turbofan, rated at 33·8 kN (7,600 lb st)

Under the collaborative agreement a subsonic civil ver sion, the M45H, was developed for the VFW 614 trans port. This engine was certificated in 1974 and since 197 has been in scheduled airline service. In 1976 the com panies mutually agreed that Rolls-Royce (1971) Lt should take over full responsibility for the M45H, which i now entirely a Rolls-Royce product.

The M45H is a series of twin-spool commercial engine in the 35·84 kN (8,000 lb) thrust class. The **M45H-01,** th first in this series, ran in January 1969 and has bee developed to power the twin-engined VFW 614 short haul airliner, on which it is mounted in unique overwin pods. The M45H-01, with a take-off thrust of 33·8 k (7,600 lb), passed its type approval test in August 1974 The VFW 614, which began its flight-test programme i July 1971, received its LBA Certificate of Airworthines in August 1974.

Modular construction, using ten modules, has bee incorporated to reduce strip time for repair and overhau Internal inspection for "on-condition" maintenance i possible by means of borescopes. In addition magneti chip-detectors, fine filtration in the oil scavenge line an vibration measuring devices assist in giving early warnin of incipient failure.

Low gas velocities result in noise levels substantiall lower than those of most current engines. The omission o fan inlet guide vanes, and the relatively large fa rotor/stator spacing, result in low compressor noise levels.

Derivatives of the M45H-01 under consideration a present include:

M45H-05. Uprated version of M45H-01, with nomina 15 per cent increase in cruise thrust.

Also under consideration are derivatives involving th application of fixed-pitch and variable-pitch geared fans which, with other modifications, will enable thrusts of u to 80·06 kN (18,000 lb) to be obtained.

The ultra-quiet demonstrator engine, designate **M45SD-02** (RB.401D-2), remains a collaborative prog ramme with SNECMA, and is described under the tw companies in the International part of this section.

The following particulars apply to the M45H-01:

TYPE: Twin-spool turbofan.

AIR INTAKE: Pitot, designed to integrate with VFW 614 short-cowl pod.

FAN: Single-stage axial unit integral with the LP system. Blades, in titanium, have snubbers at approximately two-thirds of blade height. Inlet guide vanes omitted t reduce icing problems, weight, cost and noise. Fan centrifuging effect and use of snubbers reduce risk of damage from foreign object ingestion. Mass flow 108 kg/se (238 lb/sec). By-pass ratio 3·0 : 1.

LP COMPRESSOR: Five-stage axial unit of constant roo diameter, driven by LP turbine. Rotor is of monoblo construction to give smooth running and long life. Fa and LP compressor rotor overhung on LP front shaf carried on bearings supported from compressor inter mediate casing. Front bearing is main ball thrust unit Rear bearing is roller providing radial location for inter connection of LP front and rear shafts. Fan and LP compressor casing are single ring assemblies giving stiff construction with high degree of circularity. This, together with use of abradable spacer coatings, gives enhanced compressor efficiency through small blade tip clearances. All compressor stators are punched and brazed into stator rings, eliminating stator wear and fretting. Low-pressure air bleed provided to ventilate accessory zone. No variable-geometry blading.

INTERMEDIATE CASING: One-piece casing carrying engine and aircraft accessories, and gearbox. Internally provides support for fan/LP compressor bearings and for HP compressor front bearing. Contains accessory drive gears from LP and HP shafts.

HP COMPRESSOR: Seven-stage axial unit of constant tip diameter, driven by HP turbine. Rotor uses multi-disc construction with through bolts and curvic couplings. Front stub shaft carried on main ball thrust bearing supported by compressor intermediate casing. Rear HP disc bolted to HP turbine shaft. HP compressor casing and stator mounting as for LP compressor. High-pressure air bleed provided for aircraft cabin air-conditioning and, when required, engine nose cowl anti-icing system. No variable-geometry blading. Overall pressure ratio 16 : 1.

COMBUSTION SYSTEM: Short-length annular chamber with vaporising burners.

HP TURBINE: Single-stage axial unit with air-cooled rotor blading. Rear of HP rotor carried on intershaft roller bearing downstream of turbine disc.

LP TURBINE: Three-stage axial unit of constant mean diameter. Roller bearing, supported by exhaust cone assembly, carries rear of LP turbine shaft adjacent to stage 3 disc.

BEARINGS: Mainshaft bearings, except intershaft, of squeeze-film type.

JET PIPE: Plug nozzle type.

FUEL SYSTEM: Dowty hydro-mechanical system. Fuel from first-stage centrifugal pump passes through fuel heater and filter, and thence to second-stage gear pump to main control system and acceleration control. Centrifugal governors on LP and HP shafts cause fuel spill-back in event of more than 2 per cent shaft overspeed.

ACCESSORIES: Engine accessories and engine-driven aircraft accessories are mounted in annulus between gas generator cowling and engine carcase. Accessories mounted on engine gearbox and driven from LP and HP compressor. Accessories include fuel heater, engine oil cooler, fuel control units, LP and HP governors, alternators, constant speed unit, oil tank, air starter, hydraulic and lubricating pumps.

LUBRICATION SYSTEM: Gear-type pressure pump feeds oil to all main bearings, LP and HP drives and accessory gearbox. Oil scavenged from front and rear compartments and from accessory gearbox passes through fuel-cooled oil cooler before returning to tank.

STARTING: Air starter on HP accessory gearbox.

DIMENSIONS:
- Fan intake diameter: 909 mm (35·8 in)
- Overall length: 2,795 mm (110·0 in)

WEIGHT:
- Max, with accessories: 708 kg (1,560 lb)

PERFORMANCE RATINGS:
- T-O:
  - S/L, ISA: 33·8 kN (7,600 lb st)
  - S/L, ISA+15°C: 32·3 kN (7,260 lb st)
- Cruise:
  - Mach 0·65 and 6,405 m (21,000 ft) ISA: 11·9 kN (2,670 lb st)

SPECIFIC FUEL CONSUMPTION:
- At T-O rating:
  - S/L, ISA: 13·03 mg/Ns (0·460 lb/hr/lb st)
  - S/L, ISA+15°C: 13·40 mg/Ns (0·473 lb/hr/lb st)
- At cruise rating:
  - Mach 0·65 and 6,405 m (21,000 ft) ISA: 21·25 mg/Ns (0·750 lb/hr/lb st)

ROLLS-ROYCE DART

Beginning life in 1945 at 738 kW (990 hp), this classic turboprop was developed to give 2,420 ekW (3,245 ehp) in military and 2,256 ekW (3,025 ehp) in civil versions. Current production of new engines is centred on the less powerful models listed in the table below. About 7,000 Dart engines have been delivered, and demand remains strong.

The 1,700 ekW (2,280 ehp) Rolls-Royce Dart 532 single-shaft turboprop, typical of the versions in production

TYPE: Single-shaft centrifugal-flow turboprop engine.

REDUCTION GEAR: Double helical high-speed train and final helical drive. Gear trains connected by three layshafts. High-speed pinion driven by inner shaft system bolted directly to turbine discs.

AIR INTAKE: Circular intake with annular duct leading to impeller eye of first-stage compressor. Oil tank cast integral with casing. Secondary air intake supplies air to oil cooler.

COMPRESSOR: Two-stage centrifugal. Each impeller has nineteen vanes and steel rotating guide vanes. Mass air flow at maximum rpm typically 10·66 kg (23·5 lb)/sec at 5·62 : 1 pressure ratio.

COMBUSTION CHAMBERS: Seven inclined flame tubes with atomisers for downstream injection. High-energy igniter plugs in Nos. 3 and 7 chambers.

FUEL SYSTEM: Variable-stroke pump delivers fuel to burners through flow control unit, which incorporates a filter, throttle valve, shut-off cock and barometric pressure control. Operation of control unit is function of intake pressure and throttle valve pressure drop, thus determining fuel/air ratio for all engine operating conditions. Fuel pressure at burners varies from 2·76 bars (40 lb/sq in) at idling speed to 82·75 bars (1,200 lb/sq in) at maximum power. Automatically-progressive injection of water/methanol to maintain take-off power under high ambient temperature. System linked with throttle lever to prohibit use except at take-off rpm. Fuel filter de-icing by hot air from compressor.

TURBINE: Three-stage axial. First and second stage discs bolted together by five bolts and all three by further five. Drive shaft divided, with inner shaft connecting turbine with reduction gear and outer shaft with compressor. Blades of Nimonic alloy secured by fir-tree roots.

EXHAUST UNIT: Inclined to suit installation. Outer shell supports inner cone on three struts enclosed in aerofoil-section fairings.

ACCESSORY DRIVES: Gearbox drive from main-shaft centre-coupling immediately behind compressor.

LUBRICATION: Integral oil tank (14 litres; 25 Imp pints) feeds via standpipe and feathering pump through tank base, to ensure feathering possible even after prolonged system oil leak. Gear pump supplies oil to all bearings and reduction-gear jets. Combined delivery from four scavenge pumps returned via oil-cooler. Pressure and scavenge pumps in single housing and driven by common shaft.

CONTROLS: Throttle interconnected with propeller controller and high-pressure cock linked with feathering controls. Propeller feathered by moving shut-off cock past closed position; depression of unfeathering button returns blades to fine pitch. On landing it is necessary to select removal of flight fine stop to permit blades to move down to zero pitch.

MOUNTING: Four feet at 90° on compressor casing, although only three need be used.

DIMENSIONS, WEIGHTS AND PERFORMANCE: See table.

ROLLS-ROYCE VIPER

This well established turbojet remains in quantity production for civil and military customers. Approximately 5,000 Vipers have been ordered by 36 countries, 27 of which have chosen the engine for trainer and light strike aircraft.

Current versions are as follows:

Viper 11 (Mk 200 Series). Single-shaft seven-stage axial-flow compressor driven by single-stage turbine. Air mass flow 20 kg/sec (44 lb/sec). Type-tested at 11·12 kN (2,500 lb st) and powers Jindivik Mk 3 drone, BAC Jet Provost T 4 and 5, Yugoslav Soko Galeb and Hindustan HJT-16 Kiran trainers.

A Viper 11 version, the 22-1, was built under licence in Italy by Piaggio for the Aermacchi M.B.326 trainer and by the Atlas Corporation of the Republic of South Africa and Commonwealth Aircraft Corporation of Australia for similar aircraft.

Viper 500 Series. Development with increased airflow, achieved by zero stage on compressor. Major applications early HS.125 (Mks 521, 522) and PD-808 executive aircraft (Mk 526) and BAC 167 Strikemaster (Mk 535), Aermacchi M.B.326GB (Mk 540) and Soko Jastreb (Mk 531) training and light combat aircraft. Mk 540 built under licence by Piaggio and Atlas.

Viper 600 Series. Eight-stage axial-flow compressor driven by two-stage turbine; annular vaporising combustion chamber. Take-off rating 16·7 kN (3,750 lb st) civil and 17·8 kN (4,000 lb st) military. Agreement signed with Fiat (Italy) in July 1969 for technical collaboration in design, development and production (see Fiat entry).

The civil Viper 601 powers the HS.125-600, and the military Viper 632 is fitted to the Aermacchi M.B.326K and M.B.339.

The following details apply to the Viper 600 series:

TYPE: Single-shaft axial turbojet.

AIR INTAKE: Direct pitot. Anti-icing by hot compressor air. No inlet guide vanes.

COMPRESSOR: Eight-stage. Steel drum-type rotor with disc assemblies. Magnesium alloy casing with blow-off valve. Stator blades mounted in carrier rings slotted into

ROLLS-ROYCE DART TURBOPROP ENGINES (current production)

Mark Number	Take-off Guaranteed Minimum Power	Cruising Specific Fuel Consumption (300 knots; 555 km/h; 345 mph at 6,100 m; 20,000 ft)	Maximum Basic Dry Weight	Gear Ratio	Length (without jet pipe)	Diameter	Remarks
Mk529 (RDa.7)	1,626 ekW; 1,484 kW (2,180 ehp; 1,990 shp) at 15,000 rpm	97·7 μg/J (0·578 lb/hr/ehp)	560 kg (1,235 lb)	0·093 : 1	2,480 mm (97·6 in)	963 mm (37·9 in)	Powers Grumman Gulfstream I and Fairchild FH-227. (3·50 m; 11 ft 6 in propeller).
Mk532 (RDa.7L)	1,700 ekW; 1,551 kW (2,280 ehp; 2,080 shp) at 15,000 rpm	97·7 μg/J (0·578 lb/hr/ehp)	561 kg (1,237 lb)	0·093 : 1	2,480 mm (97·6 in)	963 mm (37·9 in)	Uprated RDa.7, powers HS 748, Friendship and Argosy 222.
Mk550 (RDa.8)	1,827 ekW; 1,678 kW (2,450 ehp; 2,250 shp) at 15,000 rpm	97·7 μg/J (0·578 lb/hr/ehp)	561 kg (1,237 lb)	0·093 : 1	2,480 mm (97·6 in)	963 mm (37·9 in)	Powers HS 748 Model 228.

casing. All stator blades and 1st, 2nd and 8th stage rotor blades of steel; remainder aluminium alloy. Zero-stage and first-stage rotor blades attached by fir-tree roots; stages 3-8 riveted. Pressure ratio 5·8 : 1. Air mass flow 26·5 kg (58·4 lb)/sec.

Combustion Chamber: Short annular type with 24 vaporising burners and six starting atomisers. Electric ignition.

Fuel System: Hydromechanical, consisting primarily of fuel pump, barometric fuel control and air/fuel ratio control.

Fuel Grade: JP-1 or JP-4.

Turbine: Two-stage axial. Shrouded blades attached to discs by fir-tree roots and locking strips. Discs attached by Hirth couplings.

Bearings: Ball-thrust type at forward end of compressor, roller bearings at centre section and at rear end of combustion chamber inner casing.

Jet Pipe: Cone of heat-resisting steel rings butt-welded together. (Viper 601: eight-lobe convoluted nozzle to meet FAR Pt 36 noise requirements).

Accessory Drives: Gearbox driven from front of compressor by bevel gear.

Lubrication System: Self-contained. Recirculatory system supplying front bearing and gearbox, metered feed supplied to centre and rear bearings by micro-pumps. Military version fully aerobatic.

Oil Specification: Mobil Jet 2, Shell ASTO 500 and Castrol 580.

Mounting: Civil: cantilevered, side mounted, single spherical bearing in centre-section casing, with top and bottom links and attachment at intake casing. Military: trunnion mounted at centre-section with additional support at intake casing.

Starting: 24V starter/generator.

Dimensions:

Max casing diameter:

All versions	624 mm (24·55 in)

Length (flange to flange):

Viper 11	1,626 mm (64·0 in)
Viper 521, 522 (plus jet pipe and nozzle)	2,159 mm (85·0 in)
Viper 531, 535, 540, 632	1,806 mm (71·1 in)
Viper 601 (plus jet pipe)	2,270 mm (89·4 in)

Weights, Dry:

Viper 11	284 kg (625 lb)
Viper 521	336 kg (740 lb)
Viper 522, 531	345 kg (760 lb)
Viper 526	340 kg (750 lb)
Viper 535	331 kg (730 lb)
Viper 540	342 kg (755 lb)
Viper 601, 632	358 kg (790 lb)

Performance Ratings:

T-O:

Viper 11	11·12 kN (2,500 lb st)
Viper 521, 531	13·9 kN (3,120 lb st)
Viper 522, 526, 535, 540	14·9 kN (3,360 lb st)
Viper 601	16·7 kN (3,750 lb st)
Viper 632	17·8 kN (4,000 lb st)

Specific Fuel Consumption:

Viper 11	30·3 mg/Ns (1·07 lb/hr/lb st)
Viper 500 series	28·3 mg/Ns (1·00 lb/hr/lb st)
Viper 601	26·6 mg/Ns (0·94 lb/hr/lb st)
Viper 632	27·5 mg/Ns (0·97 lb/hr/lb st)

Oil Consumption:

All versions	0·57 litres (1 Imp pint)/hr

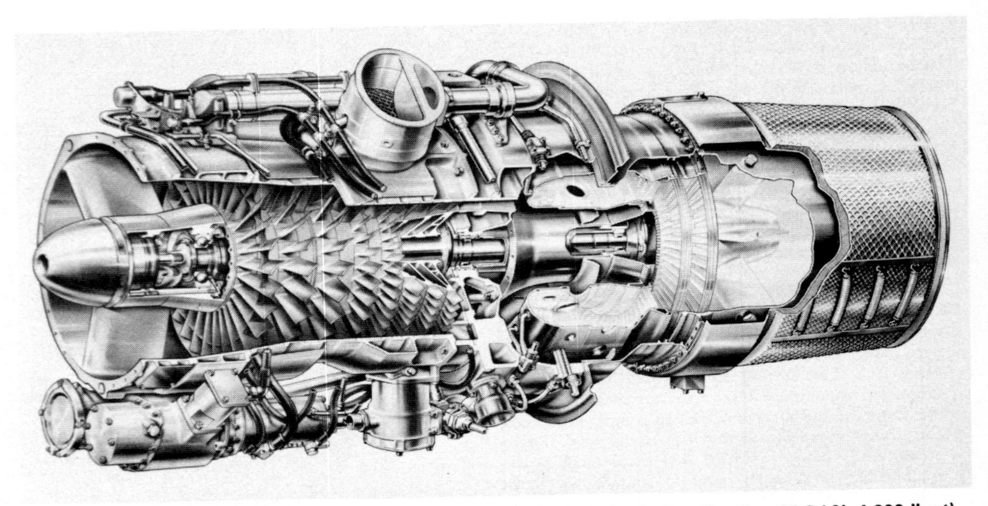

Cutaway drawing of the Rolls-Royce Viper 632 single-shaft turbojet (take-off rating 17·8 kN, 4,000 lb st)

Rolls-Royce RB.401-06 turbofan demonstrator engine, rated at 22·7 kN (5,100 lb st)

ROLLS-ROYCE RB.401

The RB.401 is a two-shaft, medium by-pass ratio, front-fan turbofan in the 24·5 kN (5,500 lb) thrust class. It is aimed principally at the business jet market, but could also have application to trainer/light attack aircraft.

A demonstrator engine, the RB.401-06, ran on the testbed on 21 December 1975, ahead of schedule, and exceeded its design thrust of 22·7 kN (5,100 lb) within the first 7 hours' running.

The RB.401 incorporates components already proven in company research programmes, with technology which reads across from the RB.211 and other advanced commercial engines. It has been designed to meet all foreseeable requirements regarding noise and exhaust pollution, in addition to having an extremely low specific fuel consumption.

The following data apply to the RB.401-07, the proposed initial production engine:

Type: Two-shaft subsonic turbofan.

Fan: Single-stage axial, based on latest RB.211 technology. Low aspect ratio titanium blades without snubbers. Maximum airflow 82·5 kg (182 lb)sec. By-pass ratio 4·2 : 1.

By-Pass Duct: Full-length, giving installational simplicity with optimum overall nacelle performance.

Intermediate Casing: One-piece magnesium casting with integral accessory gearbox and oil tank.

HP Compressor: Multi-stage axial, with early stages in titanium, later stages in Nimonic. Pressure ratio of HP spool alone, 11·5 : 1.

Combustion System: Annular, with vaporising burners.

Fuel System: Hydromechanical, with provision for electronic supervision of engine parameters.

Turbines: Single-stage HP turbine, with air-cooled rotor and stator blades. Two-stage LP turbine, with uncooled fully-shrouded blades.

Lubrication System: Self-contained, including tank, pumps, filter, scavenge strainer, tubes, fittings, and cooler.

Dimensions:

Length, flange to flange	1,525 mm (60·0 in)
Diameter, fan casing	823 mm (32·4 in)

Weight, Dry:	477 kg (1,050 lb)

Performance Ratings:

T-O:

S/L, ISA	24·7 kN (5,540 lb) st
S/L, ISA+15°C	24·0 kN (5,400 lb) st

Cruise:

Mach 0·7 and 12,000 m (40,000 ft) ISA	5·0 kN (1,130 lb)

Specific Fuel Consumption:

T-O (S/L ISA)	12·72 mg/Ns (0·449 lb/hr/lb st)
T-O (S/L ISA+15°C)	13·05 mg/Ns (0·461 lb/hr/lb st)
Cruise (as above)	20·0 mg/Ns (0·707 lb/hr/lb st)

ROLLS-ROYCE ORPHEUS

This single-spool turbojet was initiated in December 1953 as a private venture. The Orpheus 701 was type-tested in November 1956 at 20·1 kN (4,520 lb st) and is used in the Gnat fighter; a modified version powers the Indian HAL Ajeet.

The Mk 703, rated at 21·6 kN (4,850 lb st), is in service with the Indian Air Force in the Hindustan HF-24 Marut fighter. For all Indian applications the engine is built under licence at Bangalore by Hindustan Aeronautics Limited (HAL).

Other versions of the Orpheus power the Fiat G91, Fuji T1F2 and Gnat T.1 trainer. A description of the G91 engine appeared in the 1967-68 *Jane's*.

ROLLS-ROYCE OLYMPUS

The Olympus was the first British two-spool turbojet. It entered production in 1953 as the Mk 101 with a T-O thrust of 49 kN (11,000 lb) for the Hawker Siddeley Vulcan B.1 bomber. The engines currently in service with the Vulcan B.2 are the Mk 202 (75·6 kN; 17,000 lb) and Mk 301 (89 kN; 20,000 lb). From these engines has been derived a range of Industrial Olympus engines in worldwide use for power generation and gas pumping, producing 25,200/32,800 egkW (33,800/44,000 eghp). Marine Olympus derivatives of 22,800 egkW (28,000 bhp) are in use as propulsion units for medium and large warships of twelve navies.

Details of the more powerful Olympus 593 for the Concorde are given in the International section.

ROLLS-ROYCE PEGASUS

The Pegasus is a turbofan for V/STOL applications. It has two main rotating systems which are mechanically independent and rotate in opposite directions, thus minimising gyroscopic effects. It has a three-stage axial-flow fan of transonic design and an eight-stage high-pressure compressor, each driven by a two-stage turbine. Thrust vectoring is achieved by four rotatable nozzles simultaneously operated and symmetrically positioned on each side of the engine. The front nozzles discharge by-pass air, whilst the rear nozzles discharge the turbine efflux. The total thrust is divided between the four nozzles, and the resultant thrust passes through a fixed point irrespective of nozzle angle, thus minimising aircraft control problems. HP bleed air is used for aircraft stabilisation. The Pegasus was the first engine designed with an overhung fan, without inlet guide vanes, requiring no anti-icing.

By varying the angle of the nozzles, an aircraft powered by the Pegasus can take off vertically or with a short run or with a conventional long run. The reserve of power in the normal runway take-off allows for an appreciable increase in payload. The engine is stressed for operation up to Mach 2, ISA tropopause, at maximum thrust.

The Pegasus ran in August 1959, and flight trials in the Hawker Siddeley P.1127 prototypes began in October 1960. The **Pegasus 3,** which powered prototype P.1127 aircraft, was rated at 60 kN (13,500 lb st). The **Pegasus 5,** which powered the tripartite V/STOL evaluation Kestrels and was fitted to the Dornier Do 31 V/STOL transport, was rated at 69 kN (15,500 lb st). The **Pegasus 6,** with the production designation **Mk 101,** had a maximum rating of 84·5 kN (19,000 lb), and entered service with the RAF in the Hawker Siddeley Harrier GR.1 and T.2 in April 1969. All Mk 101 engines have been converted to Mk 103 standard. The **Pegasus 10 Mk 102,** a Pegasus 6 version uprated to 91·2 kN (20,500 lb), obtained type approval in March 1970; it entered service in 1971 with Harriers of the RAF and AV-8As of the US Marine Corps. All Mk 102 engines have been brought up to Mk 103 standard.

The **Pegasus 11 Mk 103,** at a rating of 95·6 kN (21,500 lb), was funded by the British government and completed its type approval in July 1971. The increase in thrust was obtained by aerodynamic redesign of the fan and increased turbine entry temperature. The Pegasus 11 is cleared to use VIFF (vectoring in forward flight) to enhance combat manoeuvrability. A variant of the Pegasus 11 is the **Mk 104** for maritime operation. The LP and intermediate casings are of corrosion-resistant materials, and the gearbox drive is strengthened for higher electrical power generation. The Mk 104, rated at 95·6 kN (21,500 lb st), will power the first 25 Sea Harriers ordered for the Royal Navy in May 1975. In US service the Pegasus 11 has the British designation **Mk 803** and the US designation **F402-RR-401.**

In October 1971 an agreement was signed with Pratt & Whitney Aircraft for joint development of the Pegasus; the US company also has an option to secure a manufacturing licence. Among the engine options being considered in April 1976 by the two companies for the AV-8B Advanced Harrier were: the existing Pegasus 11; the 11D, with aerodynamic improvements to give another 3·56 kN (800 lb st); and the 11+, an 11D with slight increase in rpm and turbine temperatures, to give a further 0·89 kN (200 lb st).

The following data apply specifically to the Pegasus 11:

TYPE: Two-shaft vectored-thrust turbofan.

AIR INTAKE CASING: One-piece casting in ZRE magnesium-zirconium alloy.

FAN: Three-stage axial, overhung ahead of front bearing. Titanium blades with part-span snubbers. Maximum airflow 196 kg (432 lb)/sec. Pressure ratio 2·3 : 1. By-pass ratio 1·4 : 1.

INTERMEDIATE CASING: Houses front fan bearing, accessory drives and HP compressor front bearing. All engine-driven accessories mounted above this casing.

HP COMPRESSOR: Eight-stage with titanium rotor blades. Overall pressure ratio 14 : 1.

COMBUSTION SYSTEM: Annular, with low-pressure vaporising burner system.

FUEL SYSTEM: Hydromechanical, comprising centrifugal backing pump, gear-type pressure pump, HP shut-off cock and overspeed governors and emergency manual control.

TURBINES: Two-stage HP turbine and two-stage LP turbine. First-stage HP blades precision cast. Remaining three rotor stages have forged blades. Both HP stages aircooled.

THRUST NOZZLES: Two steel cold front-thrust nozzles and two Nimonic hot thrust nozzles, actuated by duplicated air motors through shafts and chains. Vectored-thrust control by pilot command lever.

LUBRICATION SYSTEM: Self-contained, comprising pressure pump and three scavenge pumps, with fuel-cooled oil cooler.

MOUNTING: Four-point suspension, with main trunnions on each side of delivery casing and tie link at rear of turbines.

STARTING: Gas-turbine starter/APU on intermediate casing.

DIMENSIONS:

Length, without nozzles	2,510 mm (98·84 in)
Diameter, fan casing	1,220 mm (48·05 in)
Length, with nozzles	3,480 mm (137 in)

WEIGHT, DRY:

Without nozzles	1,392 kg (3,077 lb)

PERFORMANCE RATINGS:
See under sub-type descriptions.

ROLLS-ROYCE ODIN

The Odin ramjet powers the Hawker Siddeley Sea Dart, which has been developed to meet a Royal Navy and NATO requirement for a medium-range guided weapon system for small warships. It is in service with HMS *Bristol* and will equip Type 42 destroyers of the Royal and Argentine Navies and also the through-deck cruiser HMS *Invincible*. The ramjet forms an integral part of the missile body, and gives a longer range and better performance characteristics against fast-manoeuvring targets than a solid-rocket-powered missile. Details of the Odin are classified.

ROLLS-ROYCE GEM

The Gem, originally designated BS.360, was designed and developed at the Helicopter Engine Group at Leavesden, and is now in production, for the Westland/Aérospatiale Lynx helicopter. Although this is an element of the Anglo-French helicopter programme, the Gem development programme has been funded entirely by the British government, and pre-production engines have been wholly British built. Production engine deliveries, with a proportion of parts supplied by Turboméca of France, began in late 1975.

The Gem gas generator was run in July 1969 and the complete engine in September 1969. Engine ground running in the Lynx rotor rig began in September 1970, and the first flight of the Lynx took place in March 1971.

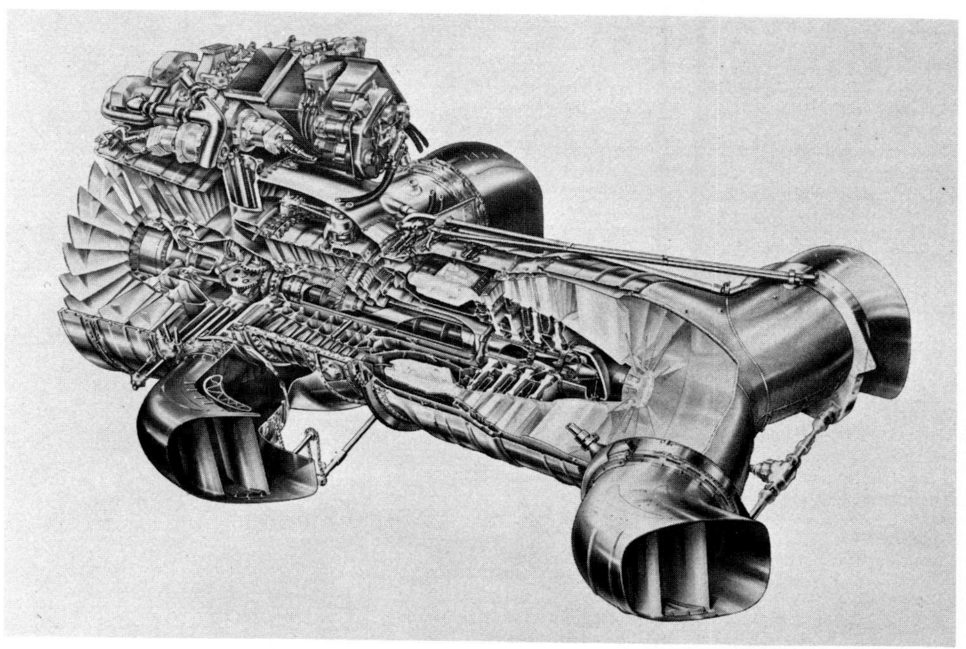

Cutaway drawing of the Rolls-Royce Pegasus 11 turbofan engine (95·6 kN; 21,500 lb st)

Cutaway drawing of the Rolls-Royce Gem free-turbine turboshaft, with initial service rating of 671 kW (900 shp)

A total of 51 pre-production engines were delivered to the bench, rotor rig and flight development programmes. Type approval was imminent as this edition went to press.

Choice of a two-spool gas generator gives fast response to power demand without the need for a complex control system. Conservative stressing and thermo-dynamic loading, and use of proven design and manufacturing techniques, are features which experience has shown to contribute to engine reliability.

The design concept of the engine is based upon seven major modules, each capable of being assembled, tested and released as interchangeable units for service use in the interest of reducing the operator's product support commitments.

The nine main bearings each have labyrinth seals pressurised by LP compressor air which also cools the bearings and minimises heat transfer to the oil, and oil cooler and fan requirements.

Provision is made for in-flight and on-ground condition monitoring systems. Features include access ports for intrascope inspection of each LP compressor stage, HP compressor, combustor, LP turbine and power turbine, and mountings for vibration pickups.

To date four versions of the Gem have been announced:

Gem Mk 10001. Engine-change unit for Lynx, rated at 671 kW (900 shp). In quantity production, shared with Turboméca. Scheduled to enter service in 1976 with RN Intensive Trials Unit.

Gem 2. Export military engine, rated at 900 shp. In production for export naval and multi-role versions of Lynx ordered by several foreign governments.

Gem Mk 501. Civil version of Gem 2. To be rated at 900 shp and certificated to BCAR to match availability of Westland 606, for which this engine has been selected.

Gem 4. First stage of uprating programme; rated at 1,050 shp. Based on Gem 2 with modified LP compressor to increase mass flow by approximately 10 per cent, together with small increase in turbine entry temperature. Selected for 4,762 kg (10,500 lb) Lynx for military service in 1979.

The following description relates to the Gem Mk 10001; other versions are generally similar:

TYPE: Free-turbine turboshaft, with two-spool gas generator.

AIR INTAKE: Annular forward facing.

SHAFT DRIVE: Compact single-stage double-helical reduction gear with rotating planet cage carried by ball bearing at front and roller bearing at rear. Reduction gear mounted within intake casing and driven by power turbine shaft. Gearbox comprises No. 1 module, and power turbine shaft No. 2 module. Gearbox provides governed output speed of 6,000 rpm. No. 2 module provides for signal to phase-displacement torquemeter (customer option).

LP COMPRESSOR: Four-stage axial rotor carried by roller bearing at rear. Stator blades mounted directly in casing. Air intake casing and forward end of compressor casing supported by conical outer casing mounted off compressor intermediate casing. LP compressor and intake case comprise No. 3 module.

INTERMEDIATE CASING: Cast casing forming junction between LP and HP compressors. Carries accessory drive and wheelcase, and provides support for LP ball bearings and HP ball and roller bearings.

HP Compressor: Single-stage centrifugal impeller having alternate inducer and radial vanes. Combined radial-and-axial diffuser feeds compressor delivery air to annular combustor. Overall pressure ratio 12·0 : 1.

Combustion Chamber: Fully annular reverse-flow with air-atomiser fuel sprays supplied by external fuel manifold. High-energy ignition box mounted on power turbine/jet pipe case.

HP Turbine: Single-stage axial close-coupled to HP impeller. Rotor blades and aircooled nozzle guide vanes based on R-R Dart technology. Roller bearing downstream of turbine disc carries rear of HP spool. Bearing supported by structure inboard of hollow LP nozzle guide vanes. HP spool with compressor intermediate casing and combustor comprise No. 4 module.

LP Turbine: Single-stage axial with shrouded rotor blades, drives LP compressor. Roller bearing downstream of turbine disc carries rear of LP rotor. This bearing, together with power turbine upstream roller bearing, is supported by structure inboard of hollow power turbine stage-one nozzle guide vanes. LP turbine and main shaft comprise No. 6 module.

Power Turbine: Two-stage axial with shrouded rotor blades. Thick-section discs have integral stub shafts which abut with centre tie-bolt forward to long, small-diameter drive shaft. Discs carried on upstream and downstream roller bearings, latter being supported by four cruciform struts in exhaust duct. Rear of power drive shaft drives output speed governor and overspeed fuel cut-off trip mechanism via spur and bevel gear train in exhaust cone. Power turbine and jet pipe form No. 7 module.

Jet Pipe: Short-length duct with casing extending forward to combustor rear casing. Four cruciform struts integral with exhaust cone.

Accessory Drives: Bevel gear on front of HP compressor shaft drives accessory shaft extending through compressor intermediate casing to spiral bevel gear drive to accessory wheelcase mounted atop intermediate casing. Drives provided for starter/generator, fuel pump, oil cooler fan and other accessories. Wheelcase forms No. 5 module.

Fuel System: Plessey fuel system with fluidics circuit providing fully automatic control, and power matching for multi-engine installation. Also automatic restoration of power from "good" engine in event of single engine failure. Incorporates fuel filter.

Lubrication System: Engine-mounted oil tank and cooler to provide self-contained system. Magnetic chip detectors fitted in each scavenge line. Oil filter incorporated in accessory wheelcase.

Dimensions:

Height overall	596 mm (23·5 in)
Width overall	575 mm (22·6 in)
Length overall	1,095 mm (43·1 in)

Weight, Dry:

Gem 10001, 2, 501	150 kg (330 lb)
Gem 4	156 kg (343 lb)

Performance Ratings:

Max contingency (2½ min):	
Gem 10001, 2, 501	671 kW (900 shp)
Gem 4	783 kW (1,050 shp)
Intermediate contingency (1 hr)/max T-O (5 min):	
Gem 10001, 2, 501	619 kW (830 shp)
Gem 4	734·5 kW (985 shp)
Max continuous:	
Gem 10001, 2, 501	559 kW (750 shp)
Gem 4	645 kW (865 shp)

Specific Fuel Consumption:

Max T-O:	
Gem 10001, 2, 501	89·57 μg/J (0·53 lb/hr/shp)
Gem 4	86·19 μg/J (0·51 lb/hr/shp)
50 per cent max T-O:	
All versions	111·5 μg/J (0·66 lb/hr/shp)

ROLLS-ROYCE GNOME

Gnome is the name given to the versions of the General Electric T58 turboshaft which Rolls-Royce manufactures in the UK. The first British-built engine ran on 5 June 1959.

The major difference between the Gnome and T58 lies in the replacement of the Hamilton Standard fuel control system by a Lucas system controlled by an electrical computer by Hawker Siddeley Dynamics.

By January 1976 approximately 1,700 Gnome engines had been delivered. To date, four versions have been announced:

H.1000. Initial version, rated at 783 kW (1,050 shp). Power plant for military Whirlwind HAR.Mk 9, HAR.Mk 10 and HCC.Mk 12, civil S-55 Series 3 and Agusta-Bell 204B.

H.1200. Rated at 932 kW (1,250 shp). Used in Agusta-Bell 204B, Boeing Vertol 107 and some Kawasaki KV-107/II-5s. Coupled version for Wessex Mks 2, 5, 50 and 60 series comprises two H.1200s driving through a coupling gearbox designed and manufactured by Rolls-Royce.

H.1400. Rated at 1,044 kW (1,400 shp). Based on the H.1200, with modified compressor to increase airflow. Turbine diaphragm cooling redesigned to increase temperature capacity and life. Dimensions unchanged. In production for Westland Sea King and Commando.

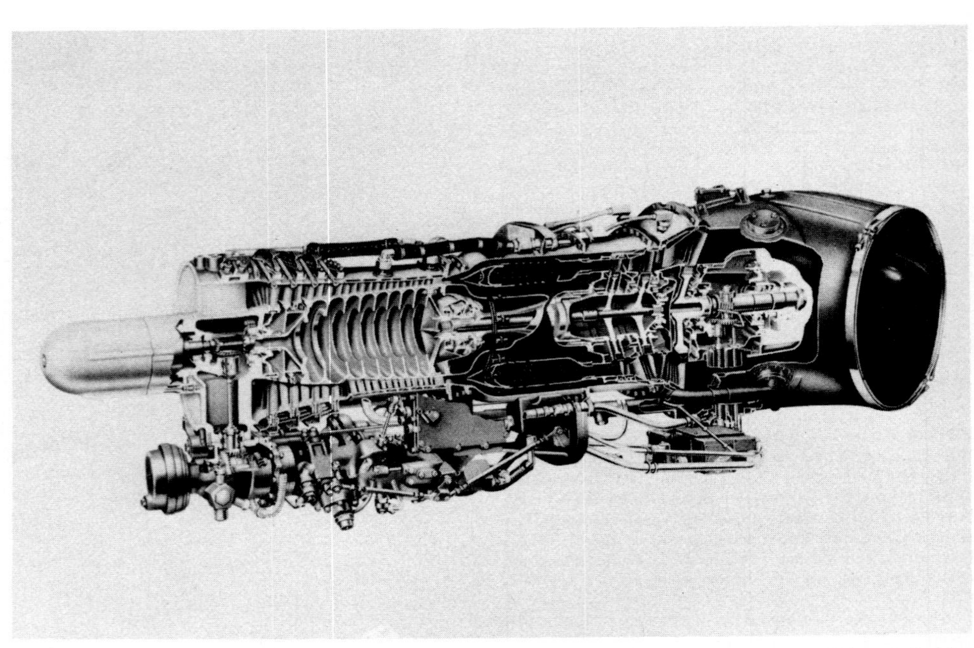

Cutaway drawing of the Rolls-Royce Gnome H.1400-1 free-turbine turboshaft, rated at 1,238 kW (1,660 shp) for 2½ min; one-hour rating is 1,145 kW (1,535 shp)

H.1400-1. Rated at 1,145 kW (1,535 shp). Uprated from H.1400, without change in size or weight, by increasing gas-generator speed and using improved gas-generator turbine-blade material allowing increased temperature. In production for Westland Sea King and Commando.

In addition a new two-stage power turbine permits further increases in power. The Gnome **H.1400-3.** is proposed to be rated at 1,287 kW (1,725 shp), a conservative figure for this engine. Further developments are under consideration for using this power turbine at ratings of 1,376 kW (1,845 shp) and 1,547 kW (2,075 shp).

The following description refers specifically to the H.1400 turboshaft version:

Type: Axial-flow free-turbine turboshaft engine.

Air Intake: Annular forward-facing. Centre housing carrying front main bearing supported by four radial struts. Struts and inlet guide vanes anti-iced with hot compressor bleed air and oil drainage.

Compressor: Ten-stage axial. Controlled variable incidence for inlet guide vanes and first three rows of stator blades. Integral spool-type rotor assembly with rotor blades secured in dovetail root fittings. Rotor splined to shaft which is carried on roller bearings at front and ball bearing at rear. Main steel casing split along horizontal centreline, with stator blades brazed in carrier rings. Pressure ratio 8·4 (H.1200, 8·12). Air mass flow 6·22 kg (13·7 lb)/sec (H.1200, 5·70 kg (12·55 lb/sec). A short-length casing interposed between compressor and combustor has radial vanes across compressor outlet to carry main centre bearing.

Combustion System: Straight-through annular chamber with outer casing split along horizontal centreline. Sixteen Simplex-type fuel injectors, eight on each of two sets of manifolds. One Lodge capacitor-discharge high-energy igniter plug.

Fuel System: Lucas hydromechanical units, comprising variable-stroke multi-plunger pump, flow control unit and throttle controlled by HSD electrical control computer and throttle actuator.

Fuel Grade: DERD.2453, 2454, 2486, 2494 and 2498 (NATO F34, F40, F35 and F44).

Gas-Producer Turbine: Two-stage, coupled to compressor shaft by conical shaft. Extended-root blading with fir-tree attachments. A short-length intermediate casing interposed between gas-producer and power turbines carries power-turbine nozzle guide vanes.

Power Turbine: Single-stage free turbine. Extended-root blading with fir-tree attachments. Rotor disc integral with output shaft and overhung from rear on roller bearing on downstream face of disc and ball bearing at rear of shaft. Complete assembly mounted inside exhaust ducting.

Exhaust System: Curved exhaust ducting arranged to suit individual applications.

Reduction Gear: Optional double-helical gear providing reduction from nominal 19,500 rpm power turbine speed to 6,600 rpm at output shaft. Provision for power take-off to left or right.

Accessory Drives: Quill shaft drive through lower intake strut. Fuel and lubrication systems mounted beneath compressor casing. Power take-off shaft up to 100 shp on primary reduction gear casing for separate accessories gearbox.

Lubrication: Fully scavenged gear pumps. Serck oil cooler.

Oil Specification: Military, DEngRD 2487 and 2493, Castrol 205 GTO and Esso Turbo Oil 2380. Commercial Aero Shell Turbine Oil 750, Esso Extra Turbo Oil 274, Castrol 98, Castrol 205 GTO and Esso Turbo Oil 2380.

Mounting: Three forward mounting faces on intake casing. Two rear mounting faces on upper portion of primary gear casing. When no reduction gear fitted, rear mounting face on engine centreline at power-turbine output shaft housing.

Starting: Rotax electric starter in nose bullet.

Dimensions:

Length:	
H.1000, H.1200, H.1400-1 (all ungeared)	1,392 mm (54·8 in)
H.1400-3	1,598 mm (62·9 in)
Coupled H.1200 (Wessex)	1,747 mm (68·8 in)
Max height:	
H.1000, H.1200, H.1400-1 (all ungeared)	549 mm (21·6 in)
H.1400-3	614 mm (24·2 in)
Coupled H.1200 (Wessex)	1,031 mm (40·6 in)
Max width:	
H.1000, H.1200 (ungeared)	462 mm (18·2 in)
H.1400-1 (ungeared)	577 mm (22·7 in)
H.1400-3	599 mm (23·6 in)
Coupled H.1200 (Wessex)	1,059 mm (41·7 in)

Weight, Dry:

H.1000 (ungeared)	134 kg (296 lb)
H.1200 (ungeared)	142 kg (314 lb)
H.1400-1 (ungeared)	151 kg (334 lb)
H.1400-3 (ungeared)	185 kg (408 lb)
Reduction gearbox	52·6 kg (116 lb)
Coupled H.1200 with coupling gearbox: for Wessex	422 kg (930 lb)

Performance Ratings (at power-turbine shaft):

Max contingency (2½ min; multi-engine aircraft only):	
H.1200	1,007 kW (1,350 shp)
H.1400-1	1,238 kW (1,660 shp)
H.1400-3	1,435 kW (1,925 shp)
Max one-hour (single engine):	
H.1000	783 kW (1,050 shp)
H.1200	932 kW (1,250 shp)
H.1400-1	1,145 kW (1,535 shp)
H.1400-3	1,286 kW (1,725 shp)
Max continuous:	
H.1000	671 kW (900 shp)
H.1200	783 kW (1,050 shp)
H.1400-1	932 kW (1,250 shp)
H.1400-3	1,044 kW (1,400 shp)

Specific Fuel Consumption:

At max contingency rating:	
H.1200	104·4 μg/J (0·618 lb/hr/shp)
H.1400-1	102·75 μg/J (0·608 lb/hr/shp)
H.1400-3	88·9 μg/J (0·526 lb/hr/shp)
At max one-hour rating:	
H.1000	109·9 μg/J (0·650 lb/hr/shp)
H.1200	105·5 μg/J (0·624 lb/hr/shp)
H.1400-1	102·75 μg/J (0·608 lb/hr/shp)
H.1400-3	90·6 μg/J (0·536 lb/hr/shp)
At max continuous rating:	
H.1000	113·2 μg/J (0·670 lb/hr/shp)
H.1200	108·5 μg/J (0·642 lb/hr/shp)
H.1400-1	106 μg/J (0·627 lb/hr/shp)
H.1400-3	96 μg/J (0·568 lb/hr/shp)

ROLLS-ROYCE MOTORS

ROLLS-ROYCE MOTORS LTD

HEAD OFFICE:
Crewe, Cheshire CW1 3PL
Telephone: 0270-55155

MAIN LOCATIONS:
Crewe (Car Division, Specialist and Light Aircraft Engine Division, Investment Foundry Division)
Shrewsbury, Shropshire SY1 4DP (Diesel Division, Military Engine Division)
London NW (Mulliner Park Ward Division)

DIRECTORS:
I. J. Fraser (Chairman)
D. A. S. Plastow (Managing Director)
C. S. Aston
T. P. Barlow
H. P. N. Benson
L. W. Harris (Commercial Director)
T. Neville (Financial Director)
H. Wuttke

DIRECTOR OF PUBLICITY:
W. D. J. Roscoe

Rolls-Royce Motors, which began operations on 24 April 1971, comprises the businesses formerly carried on by the Motor Car and Oil Engine Divisions of the defunct Rolls-Royce Ltd. The company has four factories employing 8,500 people; annual turnover is about £48 million. It is unconnected with Rolls-Royce (1971) Ltd.

Although primarily concerned with the design, development and manufacture of products in the automotive field, Rolls-Royce Motors also has a major interest in aviation through the production of light-aircraft engines, gas turbine investment castings, machined parts and sheet-metal fabrications.

Under the terms of a licence agreement signed in 1960 with Teledyne Continental Motors of the United States, Rolls-Royce Motors markets Continental light-aircraft engines and spare parts throughout the world, with the exception of North and South America and certain Far Eastern countries. Five models from the Continental range are manufactured by the company's Specialist and Light Aircraft Engine Division at Crewe, and these are described hereafter.

ROLLS-ROYCE CONTINENTAL C90

This is the 70·8 kW (95 hp) Continental C90 four-cylinder horizontally-opposed aircooled engine built under licence. Further details are given under the "Teledyne Continental" heading in the US section.

ROLLS-ROYCE CONTINENTAL O-200-A

This is the 74·5 kW (100 hp) Continental O-200-A four-cylinder horizontally-opposed aircooled engine, built under licence. Details are given under the "Teledyne Continental" heading in the US section.

ROLLS-ROYCE CONTINENTAL O-240-A

The O-240-A is the first light-aircraft engine to be developed by Rolls-Royce Motors in conjunction with Teledyne Continental. The engine had its origins in the United States but, soon after Continental had built and run a prototype, a changed commercial situation and the need to divert engineering resources to other work, especially the new Tiara range, caused Teledyne Continental to cease its active development. Rolls-Royce took over the programme in 1968 with a view to developing the engine specifically for the European market. Certification was completed in January 1970 and FAA Type Validation was awarded in February 1971. Production deliveries began in mid-1970, initially for the Reims Aviation Aerobat and Rollason Condor. The O-240 is being marketed in the United States by Teledyne Continental Motors.

TYPE: Four-cylinder, horizontally-opposed, air-cooled, carburetted, unsupercharged.

CYLINDERS: Bore 112·5 mm (4·438 in). Stroke 98·4 mm (3·875 in). Capacity 3,933 cc (240 cu in). Compression ratio 8·5 : 1. Cast aluminium alloy finned heads are screwed and shrunk on to forged steel barrels.

PISTONS: Heat-treated aluminium alloy. Two compression rings and one oil control ring above the gudgeon pin, one scraper below. Fully floating, ground steel tube gudgeon pins with pressed-in aluminium end plugs.

CONNECTING RODS: Forged steel I-section. Big-end bearings are thin steel backed overlay plated copper lead. Little-end bearings are roller bronze bushings.

CRANKSHAFT: Alloy steel forgings, nitrided all over for greater fatigue strength, having three journals running in thin steel backed overlay plated copper lead bearings.

CRANKCASE: Cast aluminium alloy, split along the vertical centreline.

VALVE GEAR: Two valves per cylinder. Steel inlet valves with hardened tips. Steel exhaust valves with hardened tips faced with Stellite "F". Valve seats shrunk into position. Camshaft, in centre of crankcase beneath the crankshaft, driven by gear from the crankshaft.

INDUCTION: Float-type carburettor with a manual mixture control.

FUEL GRADE: 100/130 octane minimum.

IGNITION: Two Slick type 4001 or two Bendix Scintilla S4LN-21 magnetos on rear of crankcase driven by gears from camshaft. Two Champion REM 38EC, REM 38W or Lodge RSE 23/3R 18 mm spark plugs per cylinder.

LUBRICATION SYSTEM: Wet sump. Magnesium crankcase cover houses the engine-driven gear type oil pump. An oil pressure relief valve is mounted in the cover. Provision is made for an airframe-mounted oil cooler and optional full-flow filter.

PROPELLER DRIVE: Direct drive, clockwise when viewed from rear. ARP 502 Type 1 flange.

ACCESSORIES: Ford 15V 60A alternator. Mechanical tachometer drive from oil pump at rear of engine. Fuel pump is operated from an eccentric on the camshaft at front of engine. An AND 20,000 accessory drive pad is provided at the front of the crankcase.

STARTING: Prestolite EO 19508 12V starter.

MOUNTING: Four rear-mounted ring type mounting brackets to which vibration isolators can be attached.

DIMENSIONS:

Length	826 mm (32·5 in)
Width	798 mm (31·4 in)
Height	633 mm (24·9 in)

WEIGHT, DRY: incl accessories 112 kg (246 lb)

PERFORMANCE RATINGS:

Take-off 97 kW (130 hp) at 2,800 rpm
Maximum recommended cruise 72·7 kW (97·5 hp) at 2,540 rpm

SPECIFIC FUEL CONSUMPTION:

Max rich	81·1 μg/J (0·48 lb/hr/hp)
Max lean	71·0 μg/J (0·42 lb/hr/hp)

OIL CONSUMPTION:

Maximum 2·53 μg/J (0·015 lb/hr/hp)

ROLLS-ROYCE CONTINENTAL O-300

This is the 108 kW (145 hp) Continental O-300 six-cylinder horizontally-opposed aircooled engine, built under licence.

Versions currently available include the O-300-C and D, of which full details can be found under the "Teledyne Continental" heading in the US section.

ROLLS-ROYCE CONTINENTAL IO-360

This is the 156·6 kW (210 hp) Continental IO-360 six-cylinder horizontally-opposed aircooled engine built under licence. The turbocharged TSIO-360 of 156·6 or 168 kW (210 or 225 hp) is also produced by Rolls-Royce Motors.

olls-Royce Continental piston engines; above, O-200, rated at 74·5 kW (100 hp); bove right, O-240, rated at 97 kW (130 hp); right, IO-360, rated at 156·6 kW (210 hp)

THE UNITED STATES OF AMERICA

AEROJET

AEROJET-GENERAL CORPORATION (Subsidiary of The General Tire & Rubber Company)

CORPORATE EXECUTIVE OFFICES:
9100 East Flair Drive, El Monte, California 91734
Telephone: (213) 572-6000
CHAIRMAN OF THE BOARD:
M. G. O'Neil
PRESIDENT:
J. H. Vollbrecht
Aerojet Solid Propulsion Company
PRESIDENT: George H. Hage
Aerojet Liquid Rocket Company
PRESIDENT: Jack L. Heckel

Aerojet-General Corporation has activities in five major areas of business: chemicals, electronics, engineering and construction, mechanical systems and metal products, research and development facility management. In the chemicals area, Aerojet Solid Propulsion Company develops, produces and tests solid-propellant rocket motors for aerospace and defence programmes. In the mechanical systems area, Aerojet Liquid Rocket Company is active in research, development, testing and production of liquid-propellant rocket engines and sounding rockets for defence and aerospace programmes, and waterjet propulsion systems for US Navy craft and Army amphibious vehicles. Aerojet also has 50 per cent ownership and is the manager of Bristol Aerojet which develops and produces rocket motor systems for the British Ministry of Defence (see UK section).

Aerojet is a wholly owned subsidiary of The General Tire & Rubber Company, Akron, Ohio, and had 11,000 employees in December 1973.

Applications of the Corporation's rocket technology include:

Aerojet Solid Propulsion Company (ASPC), Sacramento, California. Development and manufacture of the second-stage motors for the US Air Force Minuteman ICBM, the motors for Hawk, Sparrow and Standard ARM, and the first stage of the Scout launch vehicle.

Aerojet Liquid Rocket Company (ALRC), Sacramento, California. Development and manufacture of all liquid-fuel stages for the US Air Force Titan family of vehicles, and Aerobee and Astrobee sounding rockets.

AEROJET APOLLO SPS

Aerojet Liquid Rocket Company produced the engine used to propel the Apollo spacecraft's Service Module. Known as the Service Propulsion System (SPS), the engine was designed to steer the module to the Moon, place it in lunar orbit, eject it from that orbit and bring it back to Earth. In the Skylab programme it was used to de-orbit the spacecraft and return it to Earth.

The SPS engine, which utilises storable liquid propellants, produces 89 kN (20,000 lb st) and is 4 m (13 ft 4 in) high. It is designed to operate repeatedly for a total of 12·5 minutes, with a maximum single burn of 10·5 minutes, and is the largest and most powerful ablatively-cooled rocket engine yet developed in the USA.

The SPS rocket engine has fired as programmed on each mission. Firings ranged from a minimum of 0·5 sec duration to the longest duration of 7 min 25 sec during flight.

AEROJET TITAN III ENGINES

The production of Titan III first, second and Transtage engines for use as booster propulsion on the Titan family of vehicles has been under way continuously since 1962 by Aerojet Liquid Rocket Company and its predecessor Aerojet organisations. These engines, utilising storable propellants, develop 2,313 kN (520,000 lb), 445 kN (100,000 lb) and 71·2 kN (16,000 lb) thrust respectively. Their flight reliability is in the 90-99 per cent class. The nominal weights of these engines are 1,977, 564 and 196 kg (4,360, 1,245 and 432 lb) respectively. These weights are below those of other comparable liquid rocket engines.

In August 1974 ALRC was awarded a $32·5 million contract by the USAF Space and Missile Organisation for 14 additional Titan propulsion systems. This will extend production through October 1977.

AEROJET HAWK MOTOR

The single-chamber solid-propellant rocket motor of the Hawk surface-to-air missile was the first dual-thrust dual-grain motor to be mass-produced.

Within its single propellant mass, the motor has an inner core of propellant constituting a short-duration booster grain which launches and accelerates the missile to supersonic speed. When this inner core is consumed, a slower-burning outer core, forming the sustainer portion of the propellant, takes over and keeps the missile at the required velocity.

The Hawk missile has been in production for more than 18 years, and has been licensed to be manufactured by several foreign firms. It serves as a primary air defence system for the US Army and Marine Corps and a number of foreign countries. The rocket motor has demonstrated a high degree of reliability and a shelf life capability in excess of 12 years.

An Improved Hawk propulsion system, using an upgraded and higher-impulse polyurethane propellant, passed its qualification testing at the US Army Test and Evaluation Command, White Sands Missile Range. Several hundred Improved Hawk motors have been test fired and flight tested without failure, demonstrating the same high degree of reliability and long life as the Basic Hawk motor. The Improved Hawk motor is now in production at Sacramento. Aerojet Solid Propulsion Company has signed agreements with several foreign firms for its production. ASPC is also offering an uprated motor with the pintle nozzle which has been demonstrated successfully by both the Air Force and the Army. This would double the range of the Hawk at a slight increase in cost.

AEROJET MINUTEMAN MOTORS

Aerojet Solid Propulsion Company produces the second stage of LGM-30G Minuteman III. This motor has polybutadiene/ammonium perchlorate propellant packaged in a titanium case, and the single submerged nozzle has liquid injection thrust-vector control. Loaded weight is 7,076 kg (15,600 lb) and average thrust 269·5 kN (60,000 lb). More than 3,000 motors had been delivered by the end of 1974.

AEROJET SPACE VEHICLE MOTORS (SVM)

The Aerojet family of space vehicle motors provides a wide range of impulse for synchronous orbit insertion, retrograde or upper-stage propulsion applications in communications, meteorological and research satellites. Communications and meteorological satellites have been placed in synchronous orbit by SVM-1, SVM-2, SVM-4A, SVM-5 and SVM-7 motors. The motors differ mainly in size (diameter) and use glass-filament-wound cases, advanced propellant and modern nozzle materials. Mechanical-electrical safe-and-arming devices are included in the basic configuration. Impulse flexibility for new applications can be obtained by minor variations in case length and/or propellant loading.

SVM-1. SVM-1 was developed and qualified to place the Intelsat II communications satellite in synchronous orbit and was first flown in 1966. The motor has a case diameter of 457 mm (18·0 in), an overall length of 836 mm (32·9 in) and contains 73·9 kg (163·0 lb) of propellant. Total motor weight is 87·3 kg (192·1 lb).

SVM-2. This motor is 889 mm (35 in) long overall, with a 565 mm (22·25 in) diameter case, and weighs a total of 159·2 kg (350·2 lb). Maximum propellant weight in the baseline (basic) configuration is 143·2 kg (315·0 lb), and motors with up to 10 per cent propellant off-load have been static tested under space conditions. The SVM-2 was originally developed as the apogee motor for synchronous orbit injection of the Intelsat III communications satellite. It is in production for the Japanese ETS-II programme.

SVM-4A. By 1976 eight Intelsat IV satellites had been placed on station by the SVM-4A motor. This 932 mm (36·7 in) diameter motor is 1,532 mm (60·3 in) long overall, with 643·6 kg (1,416 lb) of propellant; it weighs a total of 707·7 kg (1,557 lb).

SVM-5. The SVM-5 is a 762 mm (30·0 in) diameter motor developed and qualified for NASA's Synchronous Meteorological Satellite. This motor has a propellant weight of 287·8 kg (633·2 lb) and a total weight of 319·2 kg (702·2 lb). It is 902 mm (35·5 in) long overall.

SVM-6. This motor was qualified for the NATO III communications satellite programme and is compatible with Thor/Delta 2914 launch vehicle applications.

SVM-7. This motor was designed for synchronous-orbit injection of Thor/Delta 3914 payloads. SVM-7 has a 762 mm (30·0 in) diameter case and an overall length of 1,445 mm (56·9 in), and features a lightweight nozzle of advanced carbon-carbon material. Propellant weight is 409·1 kg (900 lb) and total motor weight is 440·9 kg (970 lb). The SVM-7 was qualified in mid-1975 and flown successfully in December 1975.

AEROJET ALGOL II B

The Algol II B is used for the first stage of the Scout launch vehicle. This motor produces an average thrust of 381·5 kN (85,742 lb) for 47·3 seconds. Earlier versions of the motor, known as the Algol I, were clustered in the Little Joe II booster for development flight testing of the Apollo spacecraft.

Approximately 100 Algol motors have been successfully used in flight tests. Nozzles with an adjustable cant angle from 0° to 14° have been used on the Algol I. This experience can be used to adapt the motor to strap-on booster applications.

AEROJET ALCOR 1B

The Alcor 1B is a high-performance Aerojet Solid Propulsion Company motor featuring a high-specific-impulse polybutadiene propellant and extremely light-weight inert components. The nozzle is a unique combination of laminated reinforced-plastics materials, and the chamber is a very thin, welded, high-strength titanium 6Al-4V alloy structure. This highly efficient chamber has demonstrated excellent resistance to external flight loads in structural tests.

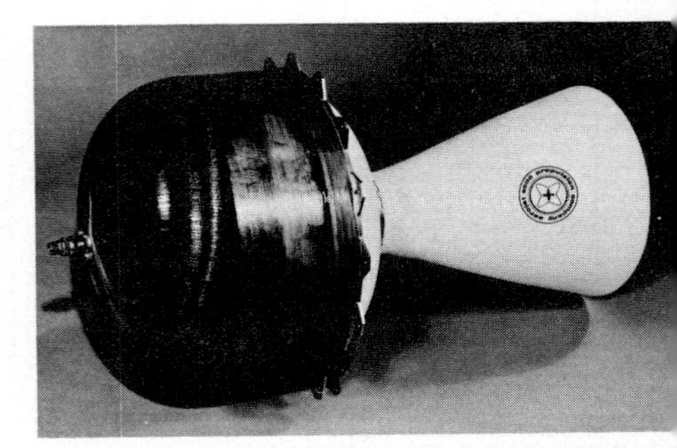

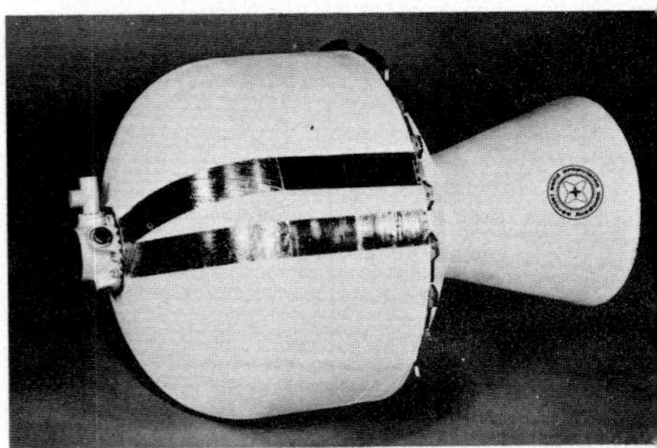

ASPC space-vehicle motors: from top, SVM-1, SVM-2, SVM-4A, SVM-5, SVM-6 and SVM-7

The Alcor 1B is used as third stage on the Athena test vehicle and is the second-stage propulsion system on the Astrobee 1500 launch vehicle. It can be applied as an upper-stage sounding rocket motor, a small component test vehicle, a synchronous orbit injection motor, or an upper-stage booster for low and medium orbits. The motor can be spin-stabilised and is fabricated to a very small thrust-misalignment tolerance (0·0004 radians angular and 0·51 mm; 0·020 in linear). This motor has been used on more than 100 flights without a failure.

AEROJET ASTROBEE D

The Astrobee D is a multi-purpose sounding-rocket motor for operation with small payloads in the D-region of the ionosphere, and with meteorological sensing payloads at altitudes up to 164 km (102 miles).

Astrobee D was engineered to facilitate low-cost production by design simplicity. A unique one-piece moulded nozzle that features integral fin attachments was developed for this motor. The chamber insulation material permits low-cost automatic application techniques. The motor has hydroxyl-terminated polybutadiene (HTPB) propellant that is low in cost yet delivers high specific impulse at very low burning rates. The low burning rate extends the duration, thereby reducing drag, structural loads and aerodynamic heating and thus providing improved performance. Aerojet has designed dual-thrust characteristics into the motor using the HTPB propellant, allowing high off-the-launcher acceleration without the necessity for adding an auxiliary boost motor.

AEROJET ASTROBEE F

The Astrobee F is a dual-thrust sounding-rocket motor incorporating many of the design features of the successful Astrobee D. The boost thrust averages 169 kN (38,000 lb) for 3 seconds followed by a sustain thrust of 3,765 kg (8,300 lb) for 53 seconds. The 381 mm (15 in) diameter motor is designed to be compatible with the Aerobee payload systems and vehicle facilities. The initial Astrobee F test design is capable of delivering 90·7 kg (200 lb) payloads up to altitudes of 378 km (235 miles). Design refinements which will increase performance approximately 10 per cent are anticipated after production flight data are available.

This motor successfully completed its flight tests in 1972, and is in production under NASA contract.

AEROJET SOLID PROPULSION COMPANY MOTORS

Name	Designation	Fuel	Oxidiser	Average thrust kN (lb)	Max length m (in)	Max dia m (in)	Total weight kg (lb)	Remarks/Primary Application
Strategic Motors								
Minuteman 2nd stage	SR19-AJ-1	Polybutadiene	NH_4ClO_4	269·5 (60,000)	4·01 (162)	1·32 (52)	7,076 (15,600)	Minuteman LGM-30F second stage, titanium case, single submerged nozzle, liquid injection TVC.
Polaris 1st stage	A3P	Polyurethane	NH_4ClO_4	—	4·62 (182)	1·37 (54)	10,886 (24,000)	Polaris A3, glass case, four rotatable nozzles; nitroplasticiser additive.
Tactical Motors								
Phoenix	Mk 60 Mod 0	Polyurethane	NH_4ClO_4	—	1·78 (70)	0·38 (15)	199 (439)	Propulsion for Navy's fleet-defence air-to-air missile.
Sparrow III, AIM-7E	Mk 52 Mod 2	Polybutadiene	NH_4ClO_4	—	1·32 (52)	0·2 (8)	68·5 (151)	Propulsion for Navy's Sparrow air-to-air missile.
Sparrow III, AIM-7F	Mk 65 Mod 0	Polybutadiene	NH_4ClO_4	—	1·55 (61)	0·2 (8)	93·5 (206)	Propulsion for Navy's Advanced Sparrow air-to-air missile.
Shrike	Mk 53 Mod 2	Polybutadiene	NH_4ClO_4	—	1·32 (52)	0·2 (8)	71 (157)	Propulsion for Navy's AGM-45 anti-radiation air-to-surface missile.
Shrike, Improved	Mk 53 Mod 3	Polyurethane	NH_4ClO_4	—	1·30 (51)	0·2 (8)	78 (172)	Propulsion for anti-radiation air-to-surface missile.
Tartar	Mk 1 Mod 0	Polyurethane	NH_4ClO_4	—	2·62 (103)	0·34 (13·5)	345 (760)	Dual-thrust propulsion for Navy ship-to-air missile.
Tartar, Improved	Mk 27 Mod 2, 3	Polyurethane	NH_4ClO_4	—	2·62 (103)	0·34 (13·5)	354 (780)	Dual-thrust propulsion for Navy ship-to-air missile.
Standard Missile	Mk 56 Mod 0	Polybutadiene Polyurethane	NH_4ClO_4	—	2·62 (103)	0·34 (13·5)	411·5 (907)	Dual-thrust propulsion for Navy ship-to-air Missile Type 1, MR.
Standard ARM	Mk 27 Mod 4	Polybutadiene Polyurethane	NH_4ClO_4	—	2·62 (103)	0·34 (13·5)	358 (790)	Dual-thrust propulsion for air-launched anti-radiation version.
Harpoon	MX-(TBD)B446-2	Polyurethane	NH_4ClO_4	—	0·61 (24)	0·34 (13·5)	119 (262)	Booster for Navy Harpoon anti-ship missile.
2·75 in FFAR, Improved	SR105-AJ-1	Polyurethane	NH_4ClO_4	—	0·34 (33)	0·07 (2·75)	5·9 (13)	Air-launched forward firing.
Hawk	XM22E8	Polyurethane	NH_4ClO_4	—	2·77 (109)	0·36 (14)	388 (856)	Dual-thrust motor for Army's surface-to-air Hawk missile.
Hawk, Improved	XM112	Polyurethane	NH_4ClO_4	—	2·77 (109)	0·36 (14)	395 (870)	Dual-thrust motor for Army's surface-to-air Hawk missile.
Launch Vehicle Boosters and Space Motors								
Algol II B		Polyurethane	NH_4ClO_4	381·4 (85,742)	8·89 (358)	1·02 (40)	10,886 (24,000)	First-stage booster for Scout.
Alcor 1B		Polybutadiene	NH_4ClO_4	44·5 (10,000)	1·43 (76)	0·51 (20)	455·5 (1,004)	Third stage of Athena test vehicle and 2nd stage for Astrobee 1500 launch vehicle.
Astrobee D		Polybutadiene	NH_4ClO_4	16/8·9 (3,600/ 2,000)	2·79 (110)	0·15 (6)	82 (181)	Dual-thrust meteorological rocket (10 lb to 100 miles altitude).
Astrobee F		Polybutadiene	NH_4ClO_4	169/37 (38,000/ 8,300)	7·11 (280)	0·43 (15)	1,255 (2,768)	Dual-thrust sounding rocket (200 lb to 235 miles altitude).
SVM-1	SVM-1	Polybutadiene	NH_4ClO_4	—	0·84 (33)	0·51 (20)	87 (192)	Apogee-boost motor for Intelsat II synchronous communications satellite.
SVM-2	SVM-2	Polybutadiene	NH_4ClO_4	—	0·89 (35)	0·58 (23)	159 (350)	Apogee-boost motor for Intelsat III synchronous communications satellite.
SVM-4A	SVM-4A	Polybutadiene	NH_4ClO_4	—	1·52 (60)	0·94 (37)	706 (1,557)	Apogee-boost motor for Intelsat IV synchronous communications satellite.
SVM-5	SVM-5	Polybutadiene	NH_4ClO_4	21·8 (4,900)	0·91 (36)	0·76 (30)	318·5 (702)	Apogee-boost motor for NASA Synchronous Meteorological Satellite.
SVM-6	SVM-6	Polybutadiene	NH_4ClO_4	31·1 (6,850)	1·37 (54)	0·76 (30)	354 (780)	Apogee-boost motor for NATO III communications satellite.
SVM-7	SVM-7	Polybutadiene	NH_4ClO_4	43·2 (9,520)	1·45 (57)	0·76 (30)	440 (970)	Apogee-boost motor for T/D 3914 satellites.
Gas Generators and JATOs								
Sprint, Launch Eject		Polybutadiene	NH_4ClO_4	n.a.	0·56 (22)	0·68 (27)	392 (865)	Used to launch the Sprint missile from its silo.
Turbine Start Cartridge		Butyl Rubber	NH_4NO_3	n.a.	0·41 (16)	0·18 (7)	11 (24)	For Titan II first stage.
Turbine Start Cartridge		Butyl Rubber	NH_4NO_3	n.a.	0·48 (19)	0·10 (4)	5·5 (12)	For Titan II second stage.
Gas Gen Mk 46		Butyl Rubber	NH_4NO_3	n.a.	1·09 (43)	0·305 (12)	—	Prime power source for the Navy's Mk 46 Mod 0 torpedo.
Controllable Motors								
CCSRM		Polyurethane	NH_4ClO_4	—	0·38 (15)	0·51 (20)	415 (915)	Single-chamber controllable solid-rocket motor.
Air Launched (VTM)		Polybutadiene	NH_4ClO_4	—	1·37 (54)	0·43 (17)	185 (407)	Variable-thrust motor; has both thrust-magnitude and vector control.
Air-to-Air Controllable (ATAC)		Polybutadiene	NH_4ClO_4	—	1·70 (67)	0·20 (8)	95 (210)	Throttling over wide temperature range.
Stop-Start Motor (SSM)		Polyurethane	NH_4ClO_4	—	2·21 (87)	0·51 (20)	396·5 (874)	Stop/start operation on command.

ALLISON

DETROIT DIESEL ALLISON DIVISION, GENERAL MOTORS CORPORATION

HEAD OFFICE:
Detroit, Michigan
Telephone: (313) 531-7100
INDIANAPOLIS OPERATIONS:
PO Box 894, Indianapolis, Indiana 46206
Telephone: (317) 244-1t11
GENERAL MANAGER:
James E. Knott
DIVISIONAL COMPTROLLER:
Victor H. Laurie
GENERAL SALES MANAGER:
Chester B. Plum
MANAGER, INDIANAPOLIS OPERATIONS:
Edward B. Colby

The former Allison division of General Motors was in 1970 merged with the Detroit Diesel Division, but the aircraft gas turbine operations at Indianapolis remain generally unchanged by the merger. The aircraft engines still continue to be marketed under the single name "Allison". Detroit Diesel Allison's Gas Turbine Operations continue to produce T56 turboprop engines for military and commercial versions of the Lockheed C-130 Hercules transport, and for the Lockheed P-3 Orion anti-submarine aircraft, Grumman E-2 Hawkeye airborne early-warning aircraft and Grumman C-2A Greyhound transport. A commercial counterpart of the T56, the Model 501-D13, powers the Lockheed Electra and Convair 580 airliners (see 1970-71 *Jane's*). A turboshaft version, the T701, was selected to power the Boeing XCH-62A helicopter.

The Allison T63 small gas turbine, and its commercial counterpart, the Model 250, have been developed through many versions with numerous applications, as listed under each model heading. Production of T63 and Model 250 turboshaft engines had reached 11,000 by the end of 1975.

A turboprop version of the Model 250 was certificated in March 1969 for light fixed-wing aircraft applications. In 1966 Allison established a worldwide distributor organisation to provide local service and support for all Model 250-powered equipment. Main franchise-holder for Europe is Hants & Sussex Aviation; for SE Asia and Australasia it is Hawker de Havilland.

It was announced in January 1967 that Allison and Rolls-Royce of England would develop and produce jointly a version of the Rolls-Royce Spey turbofan engine, under the designation TF41, to power advanced versions of the LTV A-7 Corsair II aircraft. The TF41 remains in production for the USAF A-7D close-support attack aircraft and the US Navy's A-7E carrier-based attack bomber. Further details of the Rolls-Royce/Allison programmes can be found in the International section.

ALLISON MODEL 250

US military designation: T63

The Model 250 is a small turboshaft engine in which power is derived from a free power turbine and is delivered through an offset gearbox which includes all accessory drive pads.

A development contract for the T63 military version was received by Allison in June 1958 and the engine was first run in the Spring of 1959.

The original T63-A-5, rated at 186·5 kW (250 shp), completed a 50-hour preliminary flight rating test in March 1962, prior to the start of its flight test programme in a Bell UH-13R helicopter. The 150-hour military model qualification test was completed in September 1962, with simultaneous completion of tests required for FAA Type Approval. In December 1962, the T63-A-5 was awarded an Approved Type Certificate by the FAA and was accepted by the US Army in a ceremony at Allison.

The T63-A-5A engine, rated at 236 kW (317 shp), completed its qualification-certification tests in July 1965 and was awarded an FAA Type Certificate in September 1965. This engine powers the Hughes OH-6A and Bell OH-58A light observation helicopters. Delivery of production A-5A engines began in December 1965.

T63-A-700. Upgraded version, with T-O power of 317 shp. Fitted to Bell OH-58A Kiowa. Corresponds to commercial Model 250-C18 and B15.

T63-A-701. Further uprated version, rated at 400 shp. Corresponds to commercial C20 and B17.

T63-A-720. Hot-end improvements, increasing T-O rating to 313 kW (420 shp). Corresponds to commercial C20B and B17B. Specified for ASH (Advanced Scout Helicopter) helicopters, for US Army evaluation, by Hughes and Bell. Planned for the Bell OH-58C Interim Scout, a modified LOH.

250-C18. Derived directly from the military T63-A-5A the C18 was the initial commercial version of the Model 250. Rated for take-off at 236 kW (317 shp), it powers all the initial commercial versions of the Bell JetRanger, Hughes 500 and Fairchild FH-1100, as well as the Agusta-Bell 206A, US Navy Bell TH-57A SeaRanger, and the Kawasaki (Hughes) 369HS. Deliveries began in December 1965. Production by Allison is complete, but the C18 is the subject of licence agreements with MTU (West Germany) and Kawasaki (Japan).

The 313 kW (420 shp) Allison Model 250-C20B turboshaft engine

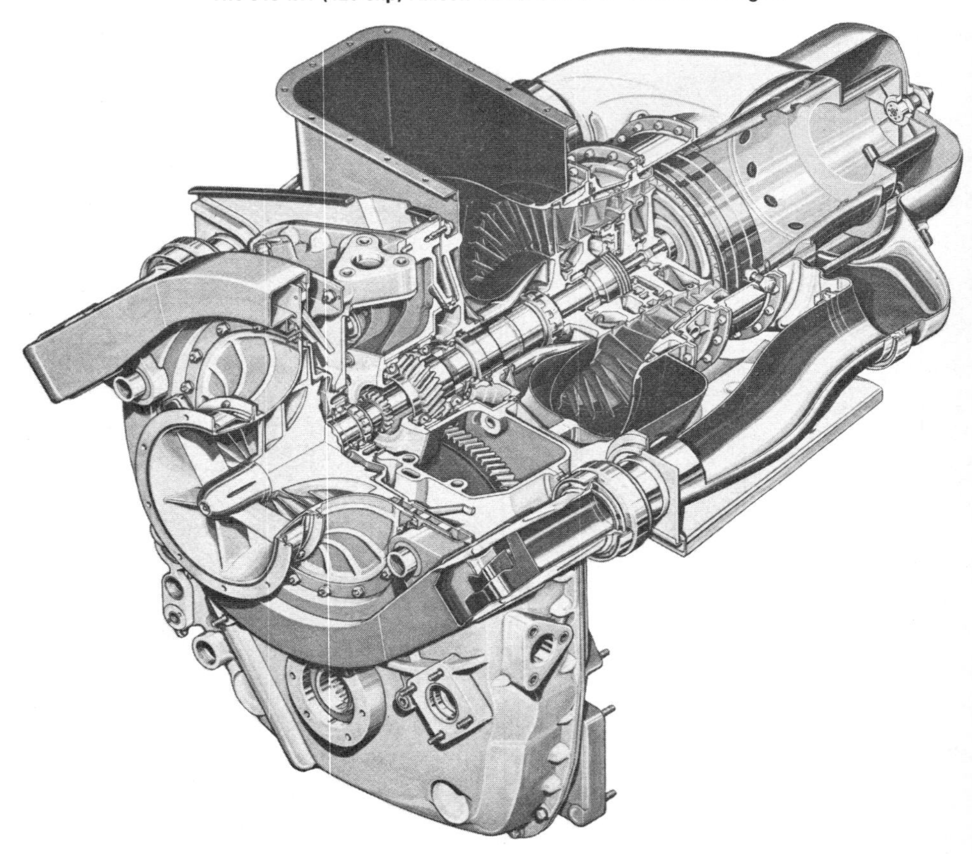

Cutaway drawing of Allison 250 Series III (250-C28) turboshaft, to be rated at 373 kW (500 shp)

250-B15. A direct conversion of the C18, the B15 was the original 236·5 kW (317 shp) turboprop version of the Model 250. The engine is essentially a C18 mounted in the inverted position, with compressor below the new propeller reduction gearbox and the twin jet pipes discharging obliquely downwards. The B15 was certificated in March 1969. Production engine, designated B15G, was fitted to the prototype SIAI-Marchetti SM.1019.

250-B17. Announced in 1972, the B17 is an uprated version of the B15 turboprop, corresponding to the C20 turboshaft. Rated at 298 kW; 311 ekW (400 shp; 417 ehp), it is fitted to the American Jet Industries Turbostar 402 (Cessna 402 conversion), the GAF Nomad and SIAI-Marchetti SM.1019E. In April 1974 Allison announced the B17B, operating at 17°C higher turbine gas temperature and with hot-end improvements similar to those of the C20B, which maintain full power at high ambient temperatures. The B17B entered production in September 1974. It has been selected for future Turbostar 402 conversions, for the Turbostar 414 and for possible future versions of the GAF Nomad.

250-C20. This is the most important current production version of the Model 250. Incorporating numerous improvements to increase airflow, component efficiency and turbine temperature, the C20 is rated at 298 kW (400 shp). Dry weight is increased by only 8·6 kg (19 lb) compared with the C18. Fully certificated for production delivery in February 1974, the C20 is fitted to the Bell 206B JetRanger II, MBB BO 105A, Agusta-Bell 206B and 206B-1, Agusta A 109, Dornier Do 34 Kiebitz RPV (with MTU power transmission), and Soloy conversion of the UH-12E. An uprated version, the 313 kW (420 shp) C20B, powers the Bell 206L LongRanger, advanced versions, of the Agusta A 109 and military versions of the Boeing Vertol (MBB) 105C.

250-C28. Representing Series III of the Model 250 evolutionary process, the 373 kW (500 shp) C28 is a near-total redesign. The axial multi-stage compressor, one of the primary features of the original design, has been eliminated. Instead a single-stage front-entry centrifugal impeller is used, handling a considerably increased airflow. The philosophy behind Series III was to reduce

The 3,661 ekW (4,910 ehp) Allison T56-A-15 turboprop engine which powers late versions of the Lockheed C-130 Hercules transport

ıoise and emissions, and despite the increase in power the ound pressure level of the bare engine has been reduced. Emissions are approximately halved by the completely ıew premix-swirl combustion chamber. The compressor acceleration bleed is eliminated, and the exhaust leaves hrough a single low-velocity stack which also has a minimal infra-red signature. The main gearbox has new gears with increased helix and decreased pressure angles. Flight-cleared prototype engines were first available in December 1975, Type Certification was scheduled for early 1976, and production engines were due in May 1976.

250-C30. Representing Series IV of the Model 250 growth programme, the C30 has a more advanced single-tage compressor, handling a higher mass flow at an ncreased pressure ratio. The engine has numerous new eatures, one of which is dual ignition to comply with FAR Pt 29. The C30 will have an initial T-O rating of 485 kW 650 shp), with growth potential to approximately 567 kW 760 shp). Prototype engines scheduled for September 976, FAA certification is due in March 1978 and production deliveries in May 1978.

GMA 500. Representing a projected Series V configuration, this may mature as a different engine not even nown as a Model 250. In 1975 GMA 500 was in the preliminary design stage. Features are likely to include: front-mounted gearbox, single-stage centrifugal compressor with 360° inlet, low-emission combustor, air-cooled turbine, rear exhaust and modular construction. Weight may be 136 kg (300 lb), T-O rating 559 kW (750 shp) and pecific fuel consumption less than 93 μg/J (0·55 b/hr/shp).

The following description applies to the 317 shp Series I ngines. Other series engines are described in brackets.

Type: Light turboshaft or turboprop engine.

Compressor: Axial/centrifugal compressor with six axial stages and one centrifugal. Axial stages of 17-4 PH cast as single units comprising integral wheels and blades. (Series III onwards, single centrifugal compressor only). Compressed air delivered through a vaned diffuser to a collector scroll and thence via two external tubes, one on each side of the engine, to the combustion chamber. Pressure ratio 6·2 : 1 (400 shp, 7·0 : 1; 420 shp, 7·2 : 1; C28, 7·08 : 1; C30, 8·5 : 1). Air mass flow 1·36 kg (3·0 lb)/sec (400 shp, 1·5 kg; 3·4 lb. 420 shp, 1·63 kg; 3·6 lb. C28, 1·96 kg; 4·33 lb. C30, 2·54 kg; 5·6 lb).

Combustion Chamber: Single can-type chamber at aft end of engine. Single duplex fuel nozzle in rear face of chamber. One igniter.

Turbines: Two-stage gas-producer turbine and two-stage "free" power turbine. Integrally-cast blades and wheels. Combustion gases after passing through turbines enter exhaust hood in middle of engine where they are collected and exhausted upward. Gas-producer turbine outlet temperature 750°C (B17B, C20B, 810°C; C28, 801°C; C30, 771°C). Turbine/combustor assembly, including exhaust collector, bolted to rear face of gearcase. Gas-producer rotor speed, B17B, C20B, 50,970 rpm; C28, 51,005 rpm; C30, 50,000 rpm. Power turbine rotor speed, B17B, C20B, 33,290 rpm; C28, C30, 33,420 rpm.

Gearcase: A magnesium casting which forms primary structure of engine and contains all power and accessory gear trains, torque sensor, oil pumps and engine main bearings. Compressor and combustor/turbine assemblies bolted to front and rear faces respectively. One spur gear train engages pinion driven by power turbine shaft and transmits output power to horizontal shaft on centreline of engine below (in turboprops, and optionally on turboshaft models, above) compressor turbine output shaft accessible on both front and rear faces of gearcase. Rated shp available at either front or rear spline, or any combination totalling rated power. Second spur gear train engages on gas generator turbine shaft and provides drive for engine accessory pads. Turboshaft version has output speed of 6,000 rpm. Turboprop has additional reduction gear to propeller shaft at top front of engine.

Control System: Pneumatic-mechanical system consisting essentially of fuel pump and filter assembly, gas producer fuel control and power turbine governor (B17B, hydromechanical; C20B, C28, C30, pneumatic-mechanical).

Fuel: Primary fuels are ASTM-A or A-1 (Model 250-C20, ASTM D-1655) and MIL-T-5624, JP-4, JP-5 and diesel fuel.

Lubrication: Dry sump.

Oil Specification: MIL-L-7808 and MIL-L-23699.

Dimensions:

Length:	
B17B	1,143 mm (45·0 in)
C20B	1,046 mm (40·8 in)
C28	1,042 mm (40·63 in)
C30	1,097 mm (43·2 in)
Width:	
B17B, C20B	483 mm (19·0 in)
C28, C30	557 mm (21·94 in)
Height:	
B17B	572 mm (22·5 in)
C20B	589 mm (23·2 in)
C28, C30	638 mm (25·13 in)

Weight, Dry:

B17B	88·4 kg (195 lb)
C20B	71·5 kg (158 lb)
C28	99·3 kg (219 lb)
C30	106·6 kg (235 lb)

Performance Ratings (S/L, ISA):

T-O:	
C20B (5 min)	313 kW (420 shp)
C28 (30 min)	373 kW (500 shp)
C30 (2½ min)	522 kW (700 shp) to 32·2°C
Max continuous:	
B17B	287 kW (385 shp)
C20B	298 kW (400 shp)
C28	356 kW (478 shp)
C30	485 kW (650 shp)
Cruise B (75 per cent):	
B17B	205 kW (275 shp)
C20B	207 kW (278 shp)
C28	244·5 kW (328 shp)
C30	312 kW (418 shp)

Specific Fuel Consumption:

At T-O rating:	
B17B	110·7 μg/J (0·655 lb/hr/shp)
C20B	110 μg/J (0·650 lb/hr/shp)
C28	108·2 μg/J (0·640 lb/lb/shp)
C30	100 μg/J (0·592 lb/hr/shp)
At cruise B rating:	
B17B	121 μg/J (0·716 lb/hr/shp)
C20B	120 μg/J (0·709 lb/hr/shp)
C28	120·5 μg/J (0·713 lb/hr/shp)
C30	111 μg/J (0·657 lb/hr/shp)

ALLISON GMA 300

This is the latest core engine developed under the USAF/USN Ategg (Advanced Turbine Engine Gas Generator) programme. Its first application is to provide the basis for the 501-M62 (T701), described separately. Projected applications include a wide range of turbofans, some with variable-pitch blades, and turboshaft engines for air and surface applications.

ALLISON GMA 500

This advanced 559 kW (750 shp) class core is included as Series V of the Model 250 described separately.

ALLISON T56

Current versions of the T56 are as follows:

T56-A-14. Rated at 3,661 ekW (4,910 ehp). Generally similar to T56-A-15, but seven-point suspension like T56-A-10W and detail changes. Powers the P-3B and C Orion.

T56-A-15. Rated at 3,661 ekW (4,910 ehp). Introduced aircooled turbine blades. Powers C-130H (all versions), C-130K, HC-130N, HC-130P and some AC-130s; specified for growth version of Aeritalia G222.

T56-A-422. Rated at 3,661 ekW (4,910 ehp). Powers Grumman E-2C Hawkeye and C-2A Greyhound.

T56-A-423. Rated at 3,661 ekW (4,910 ehp). Powers US Navy versions of the C-130.

Including the Model 501 commercial engines, production of these engines reached 11,000 by February 1976. Production is likely to continue into the early 1980s.

The following details apply to the T56-A-15:

Type: Axial-flow turboprop engine.

Propeller Drive: Combination spur/planetary gear type, primary step-down by spur, secondary by planetary. Overall gear ratio 13·54 : 1. Power section rpm 13,820. Cast magnesium reduction-gear housing. Gearbox assembly supported from power section by main drive shaft casing 711 mm (28 in) long and two inclined struts. Weight of gearbox assembly approximately 249 kg (550 lb) with pads on rear face for accessory mounting.

Air Intake: Circular duct on engine face. Thermal de-icing.

Compressor: Fourteen-stage axial-flow. Series of fourteen discs with rotor blades dovetailed in peripheries and locked by adjacent discs. Rotor assembly tie-bolted to shaft which runs on one ball and one roller type bearing. Fifteen rows of stator blades, welded in rings. Disc, rotor and stator blades and four-piece cast casing of stainless steel. Compressor inlet area 1,004 cm^2 (155·65 sq in). Pressure rato 9·5 : 1. Air mass flow 14·70 kg (32·4 lb)/sec.

Combustion Chamber: Six stainless steel cannular-type perforated combustion liners within one-piece stainless steel outer casing. Fuel nozzles in forward end of each combustor liner. Primary ignition by two igniters in diametrically-opposite combustors.

Fuel System: High-pressure type. Bendix control system. Water/alcohol augmentation system available.

Fuel Grade: MIL-J-5624, JP-4 or JP-5.

Nozzle Guide Vanes: Hollow aircooled blades of special high-temperature alloy.

Turbine: Four-stage. Rotor assembly consists of four stainless steel discs, with first stage having hollow air-cooled blades of special high-temperature alloy, secured in peripheries of discs by fir-tree roots. Discs splined to rotor shaft which runs on front and rear roller bearings. Steel outer turbine casing. Gas temperature before turbine 1,076°C.

Jet Pipe: Fixed. Stainless steel.

Accessory Drives: Accessory pads on rear face of reduction-gear housing at front end of engine.

Lubrication System: Low-pressure. Dry sump. Pesco dual-element oil pump. Normal oil supply pressure 3·8 bars (55 lb/sq in).

Oil Specification: MIL-L-7808.

Mounting: Three-point suspension.

Starting: Air turbine, gearbox-mounted.

Dimensions:

Length (all current versions)	3,708 mm (146 in)
Width:	
All versions	686 mm (27 in)
Height:	
A-15, A-422, A-423	991 mm (39 in)
A-14	1,118 mm (44 in)

Weight, Dry:

A-14	855 kg (1,885 lb)
A-15	828 kg (1,825 lb)
A-422	859 kg (1,984 lb)
A-423	836 kg (1,844 lb)

Performance Ratings (S/L, ISA, static):

T-O:

A-14, A-15, A-422, A-423

3,661 ekW; 3,424·kW (4,910 ehp; 4,591 shp) at 13,820 rpm

Normal:

A-14, A-15, A-422, A-423

3,255 ekW; 3,028 kW (4,365 ehp; 4,061 shp) at 13,820 rpm

SPECIFIC FUEL CONSUMPTION:
At max rating:
A-14, A-15 84·67 μg/J (0·501 lb/hr/ehp)
At normal rating:
A-14, A-15, A-422, A-423
87·4 μg/J (0·517 lb/hr/ehp)
OIL CONSUMPTION:
A-14, A-15 1·3 litres (0·35 US gallons)/hr

ALLISON 501-M62

US military designation: T701-AD-700

Developed from the T56, the T701 is a free-turbine turboshaft engine for helicopters and other applications. Though similar in size to the gas generator of the T56, it has a compressor, combustion chamber and turbine section of more advanced design, handling a substantially greater airflow (see Allison GMA 300).

In 1972 the T701 was selected to power the Boeing XCH-62A, the US Army heavy lift helicopter (HLH), and the engine was fully funded by the Army. The XT701 is the prototype preliminary flight rating test and safety test engine (PPFRT/SDT). Flight qualification was accomplished in early 1975 (but see "Aircraft" section for termination of XCH-62A programme).

TYPE: Free-turbine turboshaft engine.

INTAKE: Circular casting incorporating six aerofoil struts with thermal anti-icing. Accessory drive shafts at top and bottom, with main accessory gearbox on underside. This section carries front bearing for output shaft.

COMPRESSOR: Thirteen-stage axial, with variable inlet guide vanes and first five stator rows. Rotor built up from rings and discs, with large-diameter tubular central shaft rearwards from second stage. Rotor supported by front roller bearing and rear ball-thrust bearing. Longitudinally jointed casing incorporates large bleed manifold for 10 th stage air. Pressure ratio 12·8. Mass flow 20·1 kg (44·3 lb)/sec.

COMBUSTION CHAMBER: Annular, with 16 burners disposed around inner wall of flame tube fed with primary air through narrow annular gap at upstream end of snout section. Smoke-free combustion. Secondary air admitted through peripheral slits in flame tube giving film cooling.

FUEL GRADE: MIL-T-5624 grades JP-4, JP-5.

TURBINE: Gas-generator turbine has two axial stages with aircooled blades. Both stages cantilevered behind rear roller bearing. Power turbine has two axial stages assembled by row of bolts at first disc, carried between central ball thrust bearing on mid-frame through centre of combustion chamber and rear roller bearing.

JET PIPE: Fixed-area type, with truncated central bullet and tangential struts carrying rear bearing.

OUTPUT: Power turbine forward shaft is splined to central drive shaft carried in two ball bearings at front end and incorporating torque sensing assembly. Rotation clockwise, viewed from rear. Rated speed 11,500 rpm.

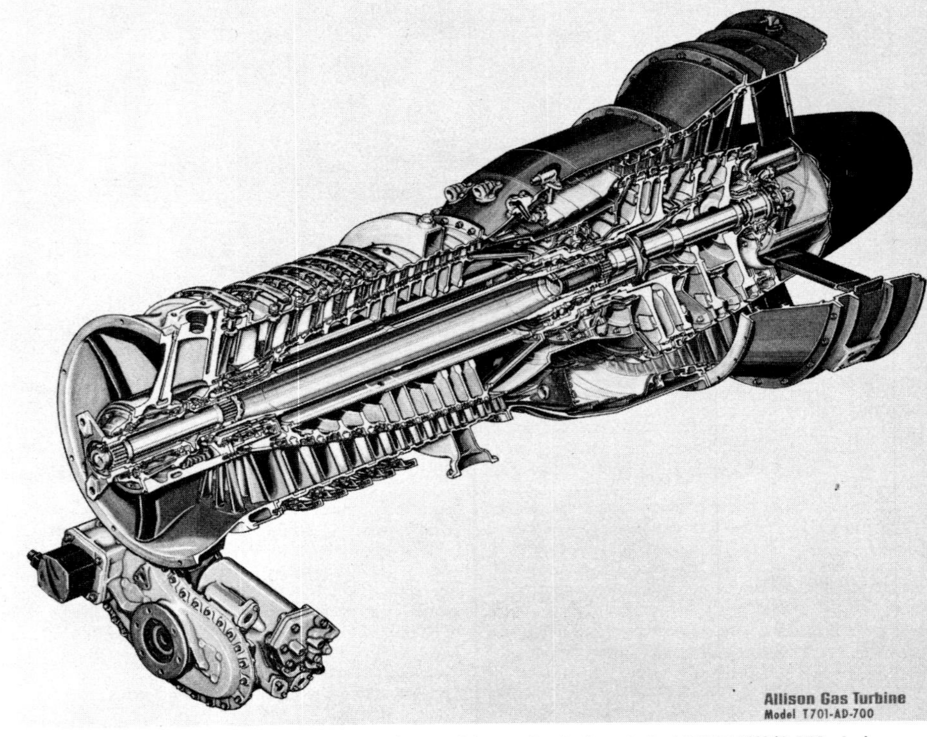

Cutaway drawing of Allison T701 free-turbine turboshaft, rated at 6,025 kW (8,079 shp)

ACCESSORY DRIVES: Main accessory gearcase beneath air intake section. Drives on front for starter/generator and tachometer, and on rear face for fuel pump and fuel control unit.

LUBRICATION SYSTEM: Self-contained integral oil system with external tank carried on left side of compressor casing.

OIL SPECIFICATION: MIL-L-23699, MIL-L-7808.

MOUNTING: Main suspension on each side of intake casing; seven possible mounting pads arranged around jet pipe casing.

DIMENSIONS:
Length (intake face to jet pipe exit)
1,633 mm (64·3 in)
Length overall 1,880 mm (74·0 in)
Intake diameter 516 mm (20·3 in)
Jet pipe diameter 714 mm (28·1 in)
Width 767 mm (30·2 in)
Height 935 mm (36·8 in)
WEIGHT, DRY: 534 kg (1,179 lb)
PERFORMANCE RATINGS (ISA, S/L, static) at gas generator speed of 15,049 rpm:
Intermediate (30 min) 6,025 kW (8,079 shp)
Max continuous 5,447 kW (7,305 shp)
75% 4,085 kW (5,478 shp)
50% 2,720 kW (3,648 shp)
25% 1,362 kW (1,827 shp)
SPECIFIC FUEL CONSUMPTION (conditions as above):
Intermediate 79·6 μg/J (0·471 lb/hr/shp)
Max continuous 78·1 μg/J (0·462 lb/hr/shp)
75% 79·1 μg/J (0·468 lb/hr/shp)
50% 85·5 μg/J (0·506 lb/hr/shp)
25% 107·7 μg/J (0·637 lb/hr/shp)

AVCO LYCOMING

AVCO LYCOMING STRATFORD DIVISION OF AVCO CORPORATION

HEAD OFFICE:
550, South Main Street, Stratford, Connecticut 06497
Telephone: (203) 378-8211
PRESIDENT OF AVCO CORPORATION:
George L. Hogeman
AVCO LYCOMING ENGINE GROUP EXECUTIVE:
John M. Ferris
VICE-PRESIDENTS:
J. S. Bartos (General Manager)
S. L. Rosenburg (Controller)
J. F. Shanley (Administration)
E. L. Wilkinson (Operations)
Michael S. Saboe (Engineering)
M. J. Leff (Marketing and Product Support)
Dr. F. Haber (International Operations)
L. H. Sample (Washington Operations)

Avco Lycoming Stratford is primarily the engine manufacturing division of Avco Corporation. It is producing two families of turbine engines, the T53 and T55 free-turbine units, of which turboshaft, turboprop, turbofan, industrial and marine versions are available. Development is continuing on the LTS 101 turboshaft and turboprop in the 447 kW (600 shp) class, the PLT-27 turboshaft in the 1,491 kW (2,000 shp) class, and on the advanced technology LTC4V series.

Avco Lycoming is also engaged in the production of test systems for gas turbine engines, and in the development and production of mechanical constant-speed transmissions. It has manufactured re-entry vehicles for the Minuteman ICBM and components for the Titan III space launch vehicle, and has also done missile-tube machining.

AVCO LYCOMING ALF SERIES

Announced in 1970, these high by-pass ratio turbofan engines were private venture developments until late 1971. They succeeded the earlier PLF1A-2 and PLF1C-1 turbofans which began bench running during the Winter of 1963-64.

All the Avco Lycoming turbofans utilise proven parts, and techniques from the earlier free-turbine turboshaft engines, and all have a high by-pass ratio suited to the propulsion requirements of subsonic aircraft. The chief models at present on offer are decribed below.

AVCO LYCOMING ALF 502

This turbofan family is aimed primarily at the commercial light transport and executive market. It is derived from the T55 turboshaft engine, like the discontinued ALF 501 of which a cutaway drawing appeared in the 1971-72 *Jane's.* The first two models are the 502A, rated at 32 kN (7,200 lb) st, and the 502D with a basic rating of 29 kN (6,500 lb) st.

There are many potential applications of the ALF 502. The main effort concerns the Learstar 600 with two ALF 502Ds. The projected, heavier Dassault Breguet Atlantic 2A and 2B patrol aircraft are likely to have two ALF 502 booster pods. Both the Fokker-VFW F27 and HS 748 have been projected in twin-ALF 502 forms. Smaller transport projects in the executive class include the Falcon 40 and projects by Rockwell, all with two ALF 502s. At least four other US aircraft companies have ALF 502-powered projects, including the Grumman G-159 (re-engined Gulfstream I).

TYPE: Two-shaft, high by-pass ratio, geared turbofan.

AIR INTAKE: Annular, around fan spinner and fan gearbox. The front frame includes main engine-mounting provisions and ducting for by-pass fan and core-engine airflows. Hollow struts provide ducts for necessary services across the fan and core-engine flow streams. Accessory drive and power take-off from the compressor rotor is transmitted through a bevel gear assembly in the front frame which is externally mounted for optimum accessibility to the accessory gearbox. Starting torque is transmitted through the reverse path.

FAN: The fan rotor includes a single-stage fan with an additional core-engine supercharging compressor cantilevered aft from the fan wheel. Fan blades have both base shrouds and mid-span shrouds for vibration and impact damping. The fan rotor is mounted on a conical support by means of a thrust bearing and a roller bearing, separated sufficiently to minimize moment loads during flight manoeuvres. A single-stage planetary helical reduction gear transmits power from the core-engine to the fan rotor. Opposing thrusts on the fan rotor and the fan drive turbine are partially balanced by reaction loads on the helical gear teeth, thereby reducing thrust bearing loads. Fan by-pass ratio approximately 6 : 1. Air mass flow 108·9 kg (240 lb)/sec.

COMPRESSOR: The compressor rotor comprises a seven-stage axial spool in tandem with a single-stage centrifugal compressor. Compressor casing halves are individually removable for stator or blade replacement in the field.

COMBUSTION CHAMBER: The combustor is a folded annular atomising burner with the turbine parts packaged concentrically within it. This concept provides a shorter, more compact engine, minimising moment loads on the mounting structure. It has the further advantages of reduced casing temperatures and improved blade containment. The design arrangement permits fan-drive turbine removal, either separately or in combination with the combustor, with single-flange disassembly.

TURBINE: The compressor turbine is a two-stage aircooled axial turbine directly coupled to the compressor shaft. Blades for both stages are base-shrouded for reduced attachment temperatures and improved vibration damping. The fan-drive turbine is a two-stage uncooled axial turbine with the drive extending forward concentrically through the hollow compressor shaft to the fan reduction gear. The fan-drive turbine blades are tip-shrouded. The fourth-stage turbine nozzle has long-chord hollow vanes also serving as the fan-drive turbine bearing support.

EXHAUST UNIT: Separate discharge of fan air and core gas. Fan air expelled past row of straightener vanes and eight hollow aerofoil struts joining inner and outer walls of fan duct. Core jet pipe of minimum length and fixed area. Provision will be made for fan airflow reversal.

ACCESSORIES: Driven accessories are externally mounted on the accessory gearbox, carried on the fan casing, for optimum accessibility and reduced environmental temperatures.

DIMENSIONS:
Max diameter 1,067 mm (42·0 in)
Length overall 1,443 mm (56·8 in)
WEIGHT, DRY: 565 kg (1,245 lb)

PERFORMANCE RATINGS:
T-O, S/L static:
ALF 502D 29 kN (6,500 lb)
SPECIFIC FUEL CONSUMPTION (T-O rating):
ALF 502D 11·9 mg/Ns (0·42 lb/hr/lb st)

AVCO LYCOMING F102

This military turbofan matches the gas-generator of the T55 with a "newly designed fan package". The cutaway drawing (1975-76 *Jane's*) shows many detail differences compared with the ALF 502, illustrated on the right. Its first application, as the YF102-LD-100, was in the Northrop A-9A, the unsuccessful contender in the US Air Force's AX programme.

In 1974 the F102 was chosen to power the NASA Ames QSRA (Quiet Short-haul Research Aircraft), which will be a conversion of a de Havilland Canada C-8A Buffalo from the National Science Foundation. Four F102 engines will provide low-emission hybrid propulsion and lift, with upper-surface flap blowing.

Lycoming's F102 turbofan programme includes extensive testing at the Stratford facility in test cells as well as in flight. An official Preliminary Flight Rating Test was conducted in March 1972. In order to obtain early flight experience Lycoming purchased a former Navy carrier-based bomber, the North American AJ-2. This has been fitted with a retractable support system on which the test turbofan engine is deployed through the bomb bay into the airstream while in flight. In June 1972 testing had been completed at altitudes up to 5,334 m (17,500 ft). The first engine delivery to Northrop was made in late February 1972 and preparations for ground testing in the A-9A started soon afterwards. Prototype development flight tests began on 30 May 1972, leading to delivery of two A-9A aircraft to the USAF for flight evaluation in the Autumn of 1972. Research flying with the QSRA is due to begin in late 1976.

DIMENSIONS:
Diameter 1,067 mm (42·0 in)
Length overall 1,488 mm (58·6 in)
WEIGHT, DRY: 565 kg (1,245 lb)
PERFORMANCE RATING:
T-O, S/L static 35 kN (7,860 lb)

AVCO LYCOMING LTS 101

US military designation: YT702-LD-700

This new turboshaft engine is the smallest of the company's aircraft gas turbines, yet it achieves the remarkable power/weight ratio of 0·184 kg (0·405 lb)/shp. It is a simple, robust unit, intended to be sold at the lowest possible price for commercial applications (hopefully, less than $25,000), and several single- and twin-engined installations are being pursued.

It has an axial compressor stage followed by a centrifugal stage, reverse-flow annular combustion chamber and single-stage compressor and power turbines. A prominent particle separator and scroll surrounds the air intake. The LTS 101 can deliver power front or rear, but the initial version has a front-drive offset gearbox. Mass flow is 2·27 kg (5·0 lb)/sec and pressure ratio 8·5 : 1.

The first version to find a customer was the LTS 101-650C, which has been delivered for Bell 222 production since late 1975. Another version, under development since 1974, is the LTS 101-600B/650A, which is a dual-ignition engine for twin-engine installation. This engine, rated at 592 shp, has been selected for the Kawasaki KH-7 helicopter. Altogether, seven variants were certificated by the FAA in October 1975.

DIMENSIONS:
Length 785 mm (30·9 in)
Width 406 mm (16·0 in)
Height:
LTS 101-650C 483 mm (19·0 in)
LTS 101-600B/650A 584 mm (23·0 in)
WEIGHT, DRY:
LTS 101-650C 105 kg (232 lb)
LTS 101-600B/650A 109 kg (240 lb)
PERFORMANCE RATINGS (T-O, S/L):
LTS 101-650C 447 kW (600 shp)
LTS 101-600B/650A 441·5 kW (592 shp)
SPECIFIC FUEL CONSUMPTION (T-O rating):
LTS 101-650C 98 μg/J (0·58 lb/hr/shp)
LTS 101-600B/650A 99·7 μg/J (0·59 lb/hr/shp)

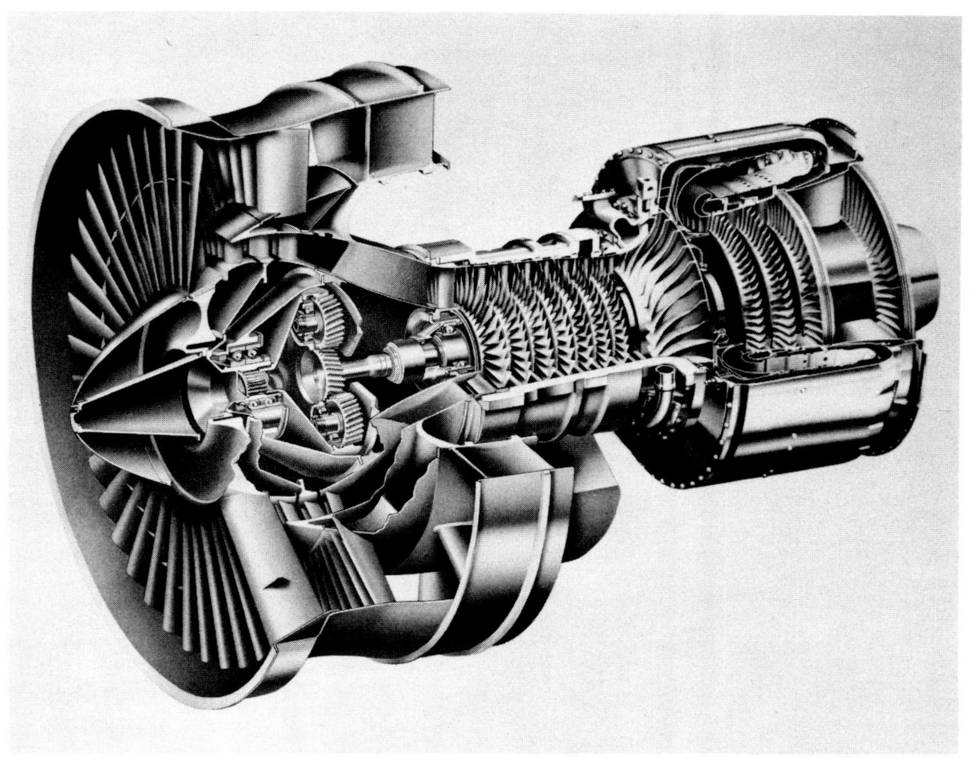

Avco Lycoming ALF 502D geared turbofan rated at 29 kN (6,500 lb st)

AVCO LYCOMING LTP 101

At the 1973 Paris Air Show Avco Lycoming presented both the LTS 101 turboshaft and the previously unannounced LTP 101 turboprop version. The preliminary design of the LTP 101 is dated "1972-73". It is designed for tractor or pusher operation, above or below the wing in a multi-engine aircraft, and is of modular design, with the same gas-generator as the LTS 101. The additional single-stage propeller/gearbox module raises the output shaft to just below the gas-generator axis. The output-shaft speed can lie in the range 1,700-2,000 rpm, and typical propellers are given as Hartzell 2·36 m (93 in) diameter models with three blades (241 m; 790 ft/sec) or five blades (210 m; 690 ft/sec).

The LTP 101 is being developed for both new aircraft and for retrofitting piston-engine aircraft. Proposed applications (including some that are well into development) include the Piaggio P.168, the Britten-Norman Turbo Islander and various agricultural aircraft designs. The Piaggio and Britten-Norman programmes were scheduled to reach first-flight status during 1976, with new aircraft as well as retrofit kit-offerings planned. Other retrofit

AVCO LYCOMING GAS TURBINE ENGINES

Manufacturer's and civil designation	Military designation	Type*	T-O Rating kN (lb) or max kW (hp)	SFC μg/J ‡mg/Ns (lb/hr/hp; ‡lb/hr/lb st)	Weight dry less tailpipe kg (lb)	Max diam mm (in)	Length overall mm (in)	Remarks
T5313B	—	ACFS	1,044 kW (1,400 shp)	98 (0·58)	245 (540)	584 (23)	1,209 (47·6)	Powers Bell 205A
T5317A	—	ACFS	1,119 kW (1,500 shp)	99·7 (0·59)	256 (564)	584 (23)	1,209 (47·6)	Based on T5319A
T5319A	—	ACFS	1,342 kW (1,800 shp)	96·3 (0·57)	256 (564)	584 (23)	1,209 (47·6)	Awaiting FAA certification
—	T53-L-11	ACFS	820 kW (1,100 shp)	115 (0·68)	225 (496)	584 (23)	1,209 (47·6)	Bell UH-1B, D, F; Kaman H-43
T5311A	—	ACFS	820 kW (1,100 shp)	115 (0·68)	225 (496)	584 (23)	1,209 (47·6)	Bell 204B
—	T53-L-13B	ACFS	1,044 kW (1,400 shp)	98 (0·58)	245 (540)	584 (23)	1,209 (47·6)	Advanced UH-1s and AH-1G
—	T53-L-702	ACFS	1,417 kW (1,900 shp)	94·6 (0·56)	254 (561)	584 (23)	1,209 (47·6)	Military T5319A
—	T53-L-703	ACFS	1,155 kW (1,549 shp)	101·4 (0·60)	247 (545)	584 (23)	1,209 (47·6)	Bell AH-1Q, AH-1S TOWCobra
LTC1K-4C	—	ACFS	1,417 kW (1,500 shp)	98 (0·58)	247 (545)	584 (23)	1,209 (47·6)	Canadair CL-84
T5321A	—	ACFP	1,393 ekW (1,868 ehp)	96·3 (0·57)	306 (675)	584 (23)	1,656 (65·2)	Turboprop T5319A
—	T53-L-15	ACFP	897 ekW (1,203 ehp)	101·4 (0·60)	274 (605)	584 (23)	1,483 (58·4	Grumman OV-1D
LTC4R-1	—	ACFP	2,837 ekW (3,804 ehp†)	87·9 (0·52)	422 (930)	615 (24·2)	1,580 (62·2)	Turboprop T55-L-11A
—	T55-L-7B	ACFS	1,976 kW (2,650 shp)	104·8 (0·62)	263 (580)	615 (24·2)	1,118 (44)	Boeing CH-47A, Bell HueyTug
—	T55-L-7C	ACFS	2,125 kW (2,850 shp)	101·4 (0·60)	267 (590)	615 (24·2)	1,118 (44)	Boeing CH-47B, Bell KingCobra and Bell 214A
T5508D (LTC4B-8D)	—	ACFS	1,750 kW (2,347 shp) (flat-rated)	106·5 (0·63)	274 (605)	610 (24)	1,118 (44)	Bell 214A, 214B
—	T55-L-11A	ACFS	2,796 kW (3,750 shp)	89·6 (0·53)	322 (710)	615 (24·2)	1,181 (46·5)	Boeing CH-47C
LTC4B-12	—	ACFS	3,430 kW (4,600 shp)	86·2 (0·51)	329 (725)	615 (24·2)	1,118 (44)	Improved T55-L-11A
LTC4V-1	—	AFS	3,729 kW (5,000 shp)	62·3 (0·41)	267 (590)	559 (22)	1,062 (41·8)	In development
LTS 101	YT702-LD-700	ACFS	441·5 kW (592 shp)	98 (0·58)	109 (239)	584 (23)	785 (30·9)	Bell 222, Dornier Aerodyne, Kawasaki KH-7
LTP 101	—	ACFP	482 ekW (646 ehp)	93 (0·55)	145 (320)	533 (21)	914 (36)	B-N Turbo-Islander, Piaggio P.168
PLT-27	—	ACFS	1,529 kW (2,050 shp)	76·1 (0·45)	145 (320)	437 (17·2)	965 (38)	In development
ALF 301B	—	ACFF	12·9 kN (2,894 lb)	‡12·46 (‡0·44)	286 (630)	826 (32·5)	1,209 (47·6)	In development
ALF 502D	—	ACFF	29 kN (6,500 lb)	‡11·61 (‡0·41)	565 (1,245)	1,257 (49·51)	1,443 (56·8)	Learstar 600
—	F102-LD-100	ACFF	35 kN (7,860 lb)	—	500 (1,100)	1,041 (41)	1,422 (56)	Northrop A-9A, QSRA (converted NASA Buffalo)

*ACFS = axial plus centrifugal, free-turbine shaft; ACFP = axial plus centrifugal, free-turbine propeller; AFS = axial, free-turbine shaft; ACFF = axial plus centrifugal, free-turbine fan

†2,752 kW (3,690 shp); also has military rating of 2,574 ekW/2,494 kW (3,452 ehp/3,344 shp).

Avco Lycoming turbine engines in the 600 shp class (from left to right): LTS 101-600B/650A 441·5 kW (592 shp), LTS 101-650C 447 kW (600 shp) and LTP 101 462 ekW (620 ehp)

orders are being negotiated in the business twin and utility STOL segments of the turboprop market.

DIMENSIONS:
- Length overall (propeller flange to jet pipe connection) 940 mm (37·0 in)
- Height overall 483 mm (19·0 in)

WEIGHT, DRY: 145 kg (320 lb)

PERFORMANCE RATINGS:
- Take-off (S/L, ISA) 462 ekW; 447 kW (620 ehp; 599 shp)
- Take-off (S/L, 32°C) 409 ekW; 393 kW 584 ehp; 527 shp)
- Max cruise (S/L, ISA) 429 ekW; 414 kW (575 ehp; 555 shp)
- Max cruise (1,525 m; 5,000 ft, ISA, 180 knots; 334 km/h; 207 mph) 399 ekW; 392 kW (535 ehp; 526 shp)
- Max cruise (4,575 m; 15,000 ft, ISA, 180 knots; 334 km/hr; 207 mph) 306 ekW; 300 kW (410 ehp; 402 shp)

SPECIFIC FUEL CONSUMPTION:
- T-O (S/L, ISA) 96 μg/J (0·568 lb/hr/shp)
- T-O (S/L, 32°C) 99·5 μg/J (0·589 lb/hr/shp)
- Max cruise (S/L, ISA) 96·8 μg/J (0·573 lb/hr/shp)
- Max cruise (1,525 m; 5,000 ft, 180 knots; 334 km/h; 207 mph) 92·1 μg/J (0·545 lb/hr/shp)
- Max cruise (4,575 m; 15,000 ft, ISA, 180 knots; 334 km/h; 207 mph) 90·4 μg/J (0·535 lb/hr/shp)

AVCO LYCOMING LTC1

US military designation: T53

The T53 is a turboshaft with a free power turbine, which was developed under a joint USAF/US Army contract. A total of 17,000 units had logged over 26 million hours of operation, with every US armed service and in 29 other countries, by January 1975.

Licences for manufacture of the T53 are held by Klöckner-Humboldt-Deutz in Germany, Piaggio in Italy and Kawasaki in Japan. In general these involve the supply of kits of "hot end" parts from Lycoming, with initial production in all three countries being centred on the T53-L-11.

Versions currently in production or under development are as follows:

T53-L-13. Uprated version of L-11, which it superseded in production in August 1966. Redesigned "hot end" and initial stages of compressor section to provide substantially increased power for hot-day and high-altitude performance. Four turbine stages, compared with two in earlier models, and variable-incidence inlet guide vanes combined with redesigned first two compressor stages, permit greater airflow and lower turbine temperatures. This version has atomising combustor to facilitate operation on a wider range of fuels. Powers Bell UH-1C and UH-1D and CH-118 Iroquois and AH-1G HueyCobra. A specially-modified version of the L-13B powers the Bell XV-15 tilt-rotor VTOL aircraft being built under NASA/Army contract (see "Aircraft" section). The engines are located in pivoted wingtip pods, which carry the directly-driven rotors. The **T5313A** commercial version of the T53-L-13 received FAA type certification in Spring 1968 and powers Bell 205A helicopters. Marine and industrial versions are the 1,150 shp **TF12A** and 1,400 shp **TF14B.**

T53-L-701. Turboprop version of the L-13 incorporating the Lycoming "split-power" propeller reduction gear. Produced for Grumman OV-1D previously powered by T53-L-15, and specified for stretched version of Air-Metal AM-C111. In production for T-CH-1 (Taiwan). Earlier L-7 powers Swiss FFA C-3605.

T53-L-703. Turboshaft engine similar to L-13. Flat rated for AH-1Q and AH-1S HueyCobra.

LTC1K-4C. Generally similar to the T53-L-13, but incorporating special seals to allow operation in the attitude range from 105° nose up to 90° nose down. Has 10-minute rating of 1,500 shp and was produced in limited quantity for use in the Canadair CL-84-1 VTOL aircraft.

T5319A. Latest growth version of T53 turboshaft family. Improvements over L-13 include new gearing, improved cooling of first gas producer turbine nozzle plus aircooled blades in first turbine rotor. Also incorporates new materials in other turbine stages. Rated at 1,800 shp at take-off.

T5317A. Lower-powered version of -19A with take-off rating limited to 1,500 shp by use of standard L-13 reduction gear.

T5321A. Turboprop version of -19A with "split-power" gear. Uses standard SBAC No. 4 propeller shaft with through-the-shaft oil provisions.

The following details apply to the T53-L-13 and L-701:

TYPE: Free-turbine turboshaft engine.

AIR INTAKE: Annular casing of magnesium alloy, with 6 struts supporting reduction gearbox and front main bearings. Anti-icing by hot air tapped from engine.

COMPRESSOR: Five axial stages followed by a single centrifugal stage. Four-piece magnesium alloy casing with one row of variable-incidence inlet guide vanes and five rows of steel stator blades, bolted to one-piece steel alloy diffuser casing with tangential outlet to combustion chamber. Rotor comprises one stainless steel and four aluminium alloy discs with stainless steel blades and one titanium impeller mounted on shaft supported in forward ball thrust and rear roller bearings. Compression ratio 7·4 : 1. Air mass flow 4·85 kg/sec (10·7 lb/sec) at 25,240 gas producer rpm.

COMBUSTION CHAMBER: Annular reverse-flow type, with one-piece sheet steel outer shell and annular liner. Twenty-two atomising fuel injectors.

FUEL CONTROL SYSTEM: Hydromechanical controls for gas generator and for power sections. Chandler Evans TA-2S system with one dual fuel pump. Pump pressure 41·4 bars (600 lb/sq in). Main and emergency flow controls. Separate interstage air bleed control.

FUEL GRADE: ASTM A-1, MIL-J-5624, MIL-F-26005A, JP-1, JP-4, JP-5, CITE.

TURBINE: Four axial-flow turbine stages. Casing fabricated from sheet steel. First two stages, driving compressor, use hollow aircooled stator vanes and cored-out cast steel rotor blades and are mounted on outer co-axial shaft to gas producer. Second two stages, driving reduction gearing, have solid steel blades, and are spline-mounted to shaft.

EXHAUST UNIT: Fixed-area nozzle. Steel outer casing and inner cone, supported by four radial struts.

ACCESSORIES: Electric starter or starter/generator (not furnished). Bendix-Scintilla TGLN high-energy ignition unit. Two igniter plugs.

LUBRICATION: Recirculating system, with gear pump with one pressure and one scavenge unit. Filter. Pump pressure 4·83 bars (70 lb/sq in).

OIL GRADE: MIL-L-7808, MIL-L-23699.

DIMENSIONS:
- Length overall:
  - L-13 1,209 mm (47·6 in)
  - L-701 1,483 mm (58·4 in)
  - T5321A 1,656 mm (65·2 in)
- Diameter:
  - All versions 584 mm (23·0 in)

WEIGHT, DRY:
- Less tailpipe:
  - L-13 249 kg (549 lb)
  - T5317A, T5319A 256 kg (564 lb)
  - T5321A 306 kg (657 lb)
  - LTC1K-4C, L-703 247 kg (545 lb)
  - L-701 312 kg (688 lb)

PERFORMANCE RATINGS:
- Max at S/L:
  - L-13 1,044 kW (1,400 shp)
  - L-701 1,082 ekW; 1,044 kW (1,451 ehp; 1,400 shp plus 0·57 kN (128 lb) at 20,430 rpm
  - LTC1K-4C, T5317A 1,119 kW (1,500 shp)
  - T5319A 1,342 kW (1,800 shp)
  - T5321A 1,393 ekW; 1,342 kW (1,868 ehp; 1,800 shp)
  - L-703 1,107 kW (1,485 shp) to 28°C

SPECIFIC FUEL CONSUMPTION:
- At max rating:
  - L-13, LTC1K-4C 98 μg/J (0·58 lb/hr/shp)
  - L-701, T5319A, T5321A 96·3 μg/J (0·57 lb/hr/shp)
  - T5317A 99·7 μg/J (0·59 lb/hr/shp)
  - L-703 101·4 μg/J (0·60 lb/hr/shp)

OIL CONSUMPTION:
- All versions 450 gr (1·0 lb)/hr

AVCO LYCOMING LTC4

US military designation: T55

This engine is based on the T53 design concept but with higher mass flow. It was developed under a joint USAF/US Army contract. Total operating time by early 1976 was 3·4 million hours. Most of this time has been

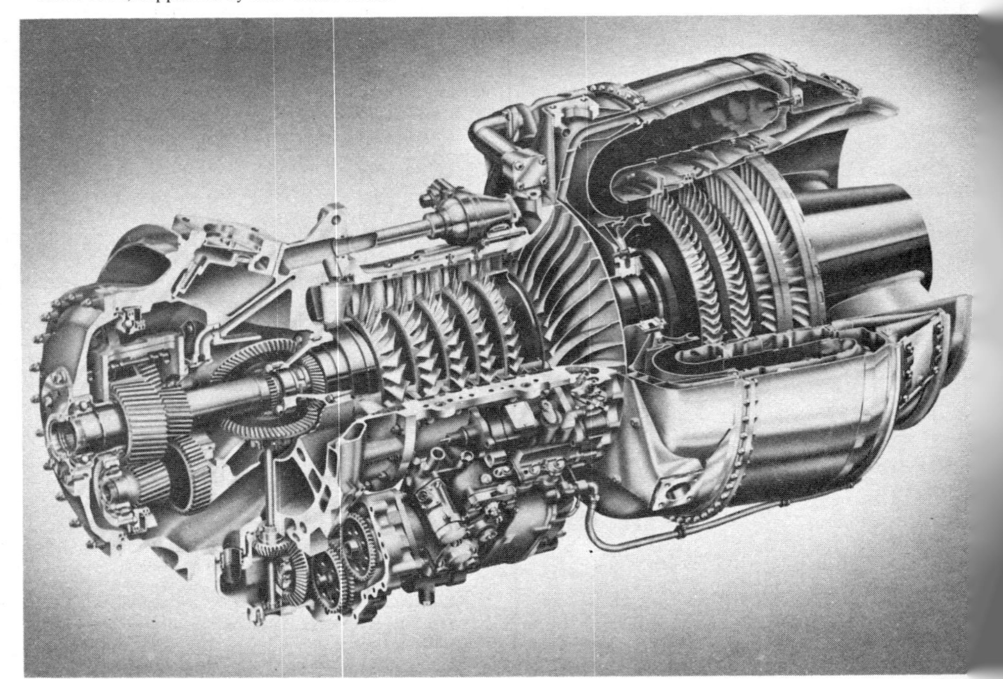

Cutaway drawing of the 1,400 shp Avco Lycoming T53-L-13 turboshaft engine

logged by L-7 versions which power the CH-47A and CH-47B Chinook.

Current production and development versions are as follows:

T55-LTC4B-8D. Modified version of the T55-L-7C. Powers Bell 214A and 214C utility helicopters for Iran, flat rated to transmission limit of 1,678 kW (2,250 shp).

T5508D. Commercial version of LTC4B-8D. Powers Bell 214B.

YT55-L-9. Turboprop version of LTC4B-8D. Chosen for Piper Enforcer programme with reduced flat rating of 1,886 kW (2,529 ehp).

T55-L-11 (LTC4B-11B). Uprated and redesigned version of L-7, with a second stage added to the compressor turbine, and variable-incidence inlet guide vanes ahead of the compressor. First two compressor stages transonic. New atomising fuel nozzles. Powers CH-47C Chinook, first deliveries having been made in August 1968.

T55-L-712. Known as RAM-D, this engine is planned to reduce overall cost of the US Army's modernised Boeing Vertol CH-47C. The basic engine is the T55-L-11D, with wide-chord compressor blades and welded rotor. Improvements in design and materials are aimed at achieving a time between overhauls of 2,500 hr. Military qualification is due in mid-1978.

LTC4B-12. Growth version with 3,430 kW (4,600 shp) maximum power rating, 3,258 kW (4,370 shp) on hot day. Higher turbine entry temperature and increased turbine cooling.

LTC4R-1. Turboprop version of L-11 with Lycoming split-power reduction gear.

TF25, TF35. Industrial and marine versions of the LTC4; TF25C is rated at 1,976 kW (2,650 shp) and TF35C is rated at 2,461 kW (3,300 shp). The advanced TF40 version is rated at 2,722 kW (3,650 shp).

The following description applies to the T55-L-11 and LTC4R-1:

Type: Free-turbine turboshaft engine.

Air Intake: Annular type casing of magnesium alloy with four struts supporting reduction gearbox and front main bearings. Anti-icing by hot air tapped from engine. Provision for intake screens.

Compressor: Seven axial stages followed by a single centrifugal stage. Two-piece magnesium alloy stator casing with one row of variable inlet guide vanes and seven rows of steel stator blades, bolted to steel alloy diffuser casing to which combustion chamber casing is attached. Rotor comprises seven stainless steel discs and one titanium impeller mounted on shaft supported in forward ball-thrust bearing and rear roller bearing. Pressure ratio 8·2 : 1. Air mass flow 12·25 kg (27 lb)/sec.

Combustion Chamber: Annular reverse-flow type. Steel outer shell and inner liner. Twenty-eight fuel burners with downstream injection.

Fuel System: Hamilton Standard JFC 31 fuel control system. Gear-type fuel pump, with gas producer and power shaft governors, flow control with altitude compensation and shut-off valve.

Fuel Grade: MIL-J-5624 grade JP-4, JP-5, MIL-F-46005A or CITE.

Turbine: Two mechanically independent axial turbines. Gas-generator turbine has single stage (two stages on T55-L-11 series) with cored-out cast steel blades having inner cooling airflow. Disc flange-bolted to drive shaft. Hollow stator vanes. Two-stage power turbine has solid steel blades.

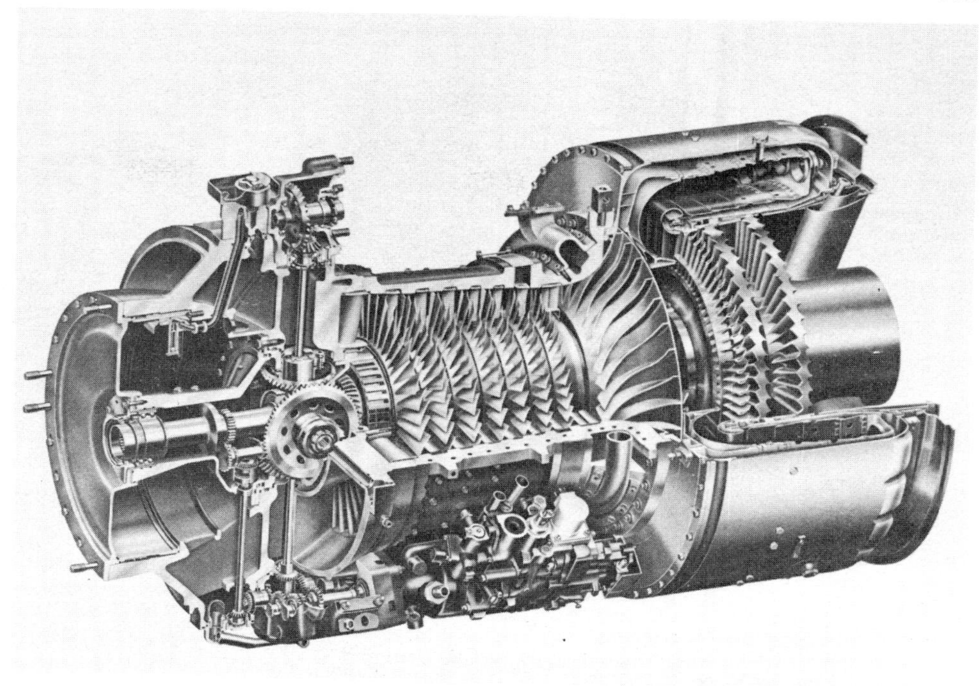

Avco Lycoming T5508D (LTC4B-8D) turboshaft, flat-rated at 1,750 kW (2,250 shp)

Exhaust Unit: Fixed-area nozzle, with inner cone, supported by six radial struts.

Accessories: Electric starter or starter/generator, or air or hydraulic starter. Bendix-Scintilla TGLN high-energy ignition unit. Four igniter plugs.

Lubrication: Recirculating type. Integral oil tank and cooler on L-11, external tank for 4R-1.

Oil Grade: MIL-L-7808, MIL-L-23699.

Dimensions:

Diameter (all versions) 616 mm (24·25 in)

Length overall:

L-7, LTC4B-8D, T5508D, LTC4B-12 1,119 mm (44·03 in)

L-9, LTC4R-1 1,580 mm (62·2 in)

L-11A 1,181 mm (46·5 in)

Weight, Dry:

L-7B 263 kg (580 lb)

L-9, LTC4R-1 422 kg (930 lb)

L-11A 322 kg (710 lb)

LTC4B-8D, T5508D 274 kg (605 lb)

LTC4B-12 329 kg (725 lb)

Performance Ratings (T-O, S/L):

L-7B 1,976 kW (2,650 shp)

L-9, LTC4R-1 2,837 ekW; 2,752 kW (3,804 ehp; 3,690 shp plus 1·27 kN;285 lb)

LTC4B-8D, T5508D 2,200 kW (2,950 shp) flat rated to 1,687 kW (2,250 shp)

LTC4B-12 3,430 kW (4,600 shp)

Specific Fuel Consumption (T-O, S/L):

L-7B 104·8 μg/J (0·62 lb/hr/shp)

L-9, LTC4R-1, L-11A 89·6 μg/J (0·53 lb/hr/shp)

LTC4B-8D, T5508D (flat rated) 106·5 μg/J (0·63 lb/hr/shp)

LTC4B-12 86·2 μg/J (0·51 lb/hr/shp)

AVCO LYCOMING PLT 27A

A completely new design (developed in parallel with an engine for armoured vehicles), this three-shaft engine was originally planned to meet US Army helicopter needs, in competition with the GE12 and Pratt &Whitney ST9. Development has been continued and broadened as an advanced technology engine in the 2,000 hp class. A demonstrator engine first ran in 1972, and current development is concentrated upon two turboshaft versions, the PLT 27A/B and C/D. Both have multi-stage axial and centrifugal compressors, and folded annular combustion chambers.

Dimensions:

Max diameter 437 mm (17·2 in)

Length overall 970 mm (38·2 in)

Weight, Dry:

PLT 27A/B 145/163 kg (320/360 lb)

PLT 27C/D 165/186 kg (365/410 lb)

Performance Ratings (T-O, S/L):

PLT 27A/B 1,529 kW (2,050 shp)

PLT 27C/D 1,864 kW (2,500 shp)

Specific Fuel Consumption (T-O, S/L):

Both versions 72·7 μg/J (0·43 lb/hr/shp)

Two versions of the Avco Lycoming PLT 27 advanced turboshaft: left, PLT 27A/B 1,529 kW (2,050 shp); right, PLT 27C/D 1,864 kW (2,500 shp)

AVCO LYCOMING WILLIAMSPORT DIVISION

Williamsport, Pennsylvania 17701

Telephone: (717) 323-6181

Vice-Presidents:

Peter J. Goodwin (General Manager)

A. E. Light (Engineering)

L. J. Anderson (Sales and Service)

W. R. Bower (Operations)

H. A. Schuck (Controller)

Williamsport Division is engaged primarily in the production of well-known Lycoming series of horizontally-opposed aircooled reciprocating engines ranging from 115 to 450 hp. Turbocharging is being offered on additional models and the horsepower, in some cases, has been increased. Development efforts are being directed to improvements resulting in lower cost of manufacturing and longer time between overhauls, to help offset increasing labour and material costs. During recent years FAA approval has been received for several turbocharged six-cylinder engines of both direct-drive and geared types. The turbocharger provides air for cabin pressurisation, and the engines have provision for a freon compressor for cabin cooling. Williamsport is studying the market for gas turbines for light aircraft.

AVCO LYCOMING O-235 and O-290 SERIES

The version of the O-290 Series engine in current production is the O-290-D2C, which differs from the preceding O-290-D2B by having retard breaker magnetos.

Type: Four-cylinder horizontally-opposed aircooled.

Cylinders: Bore (O-235-C1B) 111 mm (4⅜ in), (O-290-D2C) 123·7 mm (4⅞ in). Stroke (both) 98·4 mm (3⅞ in). Aluminium alloy head screwed and shrunk on

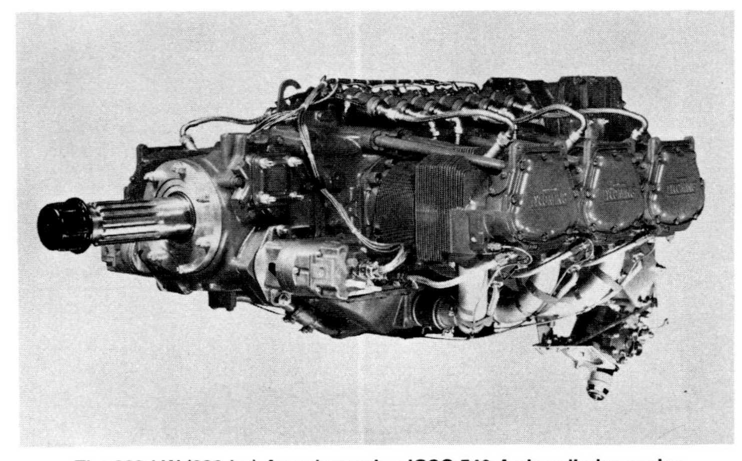

The 283 kW (380 hp) Avco Lycoming IGSO-540-A six-cylinder engine

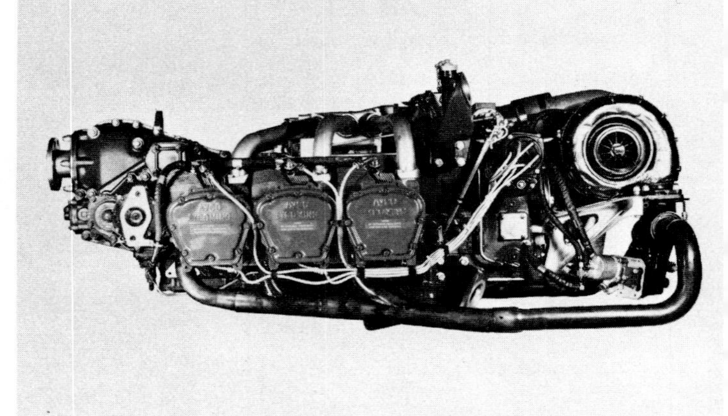

The 317 kW (425 hp) Avco Lycoming TIGO-541-E six-cylinder engine

to steel barrel. Cylinder assemblies attached to crankcase by studs and nuts.

PISTONS: Machined from aluminium alloy forgings. O-235 piston has four rings: two compression, an oil regulator and an oil scraper. O-290 has three rings: two compression and one oil regulating. Fully-floating gudgeon-pins with aluminium alloy retaining plugs.

CONNECTING RODS: Forged steel. Copper-lead steel-backed precision type bearings. Bronze bushed little-ends.

CRANKSHAFT: One-piece forged chrome nickel molybdenum steel four-throw shaft on four nitrided bearings.

CRANKCASE: Aluminium alloy casting split on vertical centreline. Four precision copper-lead steel-backed main bearings.

VALVE GEAR: Two valves per cylinder. Inlet valves of Silchrome No. 1, exhaust valves of AMS 5682 with Stellite-faced heads. Valve seats of AMS 5700 shrunk into head.

INDUCTION: Marvel-Schebler MA-3A and MA-3S1A carburettor with manual altitude control and idle cut-off. Centre zone distribution chamber in oil sump.

IGNITION: Two Bendix Scintilla S4LN magnetos, incorporating a retard breaker.

LUBRICATION: Full pressure wet sump type.

ACCESSORIES: Starter, generator and tachometer drive. Optional drives for fuel pump and vacuum pump can be supplied.

DIMENSIONS, WEIGHTS AND PERFORMANCE: See table.

AVCO LYCOMING O-320 SERIES

The O-320 is basically the same as the O-290-D2C except for an increase in cylinder bore to 130 mm (5⅛ in), with a corresponding increase in swept volume to 5·2 litres (319·8 cu in), and use of a Marvel-Schebler MA-4SPA carburettor.

It is available in low-compression and high-compression versions for use with 80/87 or 100/130 octane fuels respectively.

For other details see table.

AVCO LYCOMING IO-320 SERIES

The O-320 engines are available as fuel-injected models with both high and low compression. For further details see table.

AVCO LYCOMING O-340 and O-360 SERIES

The O-340 is basically the same as the O-320 except for an increase in stroke to 105 mm (4⅛ in), with a corresponding increase in swept volume to 5·58 litres (340·4 cu in), and use of the larger Marvel-Schebler MA-4-5 carburettor. The O-360, the same as the O-340 except for a further increase in stroke to 111 mm (4⅜ in), has a corresponding increase in swept volume to 5·92 litres (361 cu in). Both engines are available in low- and high- compression versions for use with 80/87 or 100/130 octane fuel respectively.

The VO-360-B1A is the helicopter version of the O-360, arranged for installation with the crankshaft vertical.

The IMO-360-B1B is a fuel-injected single-ignition version of the O-360 for use in unmanned aircraft.

For further details see table.

AVCO LYCOMING IO-360 SERIES

The O-360 is built in two fuel-injection versions: the IO-360-A series with tuned injection, tuned induction and high-output cylinders, and the IO-360-B series with continuous-flow port injection and standard cylinders.

AVCO LYCOMING O-435 SERIES

The O-435 Series includes direct-drive, geared, and geared and supercharged models, details of which will be found in the table. The VO-435-A1F, TVO-435-A1A and TVO-435-B1A are helicopter engines for vertical installation. The TVO-435 engines are equipped with an AiResearch exhaust-driven turbocharger which allows them to maintain rated power to 6,100 m (20,000 ft). The GO-435-C2B2-6 is a geared-drive wet sump engine with a propeller governor drive, mounted on the left side of the propeller reduction-gear housing.

TYPE: Six-cylinder horizontally-opposed aircooled, incorporating major components of the O-290.

CYLINDERS: Bore 123·7 mm (4⅞ in). Stroke 98·4 mm (3⅞ in).

PISTONS: Aluminium alloy pistons with two compression and two oil control rings.

CONNECTING RODS: H-section steel forgings with replaceable bearings inserts in big-ends and split bronze bushings in little-ends.

CRANKSHAFT: Machined from chrome nickel molybdenum steel forging. All bearing surfaces nitrided.

CRANKCASE: Aluminium alloy casting split on the vertical centreline. Additional ball-thrust bearing at forward end of case.

INDUCTION: Marvel-Schebler MA-4-5 or Stromberg PS-5BD single-barrel carburettor attached to bottom of oil sump casting. The distributing zone is submerged in oil. Separate induction pipes lead to inlet valves.

IGNITION: Two Bendix-Scintilla magnetos driven by spur gears from the timing gear.

LUBRICATION: Full pressure type, including valve mechanism. Crankshaft equipped with centrifugal sludge-removers. Pistons, gudgeon-pins and accessory drive gears lubricated by splash.

ACCESSORY HOUSING: Aluminium alloy casting bolted to rear of crankcase and top rear of oil sump. Houses oil pump and geared accessory drives, and provides mounting for starter and generator, fuel pump, tachometer drive and magnetos. Vacuum pump drive optional equipment.

STARTING: Delco-Rémy 12V automotive type starter. Starter torque applied to crankshaft gear through Bendix-type starter drive.

DIMENSIONS, WEIGHTS AND PERFORMANCE: See table.

AVCO LYCOMING O-480 SERIES

The O-480 Series is basically the same as the O-435 Series except for an increase in cylinder bore to 130 mm (5⅛ in) and in swept volume to 7·8 litres (479·7 cu in). The geared and normally aspirated engines are available in low- and high-compression versions for use with 80/87or 100/130 minimum octane fuels respectively. The geared and supercharged GSO-480-B Series have a supercharger drive ratio of 11·27 : 1, providing rated power to 2,440 m (8,000 ft) on 100/130 minimum octane fuel. The IGSO-480-A1F6 is similar to the GSO-480 except that it is fitted with direct fuel injection into the eye of the supercharger. High-compression and supercharged engines are provided with internal oil cooling of the pistons as standard equipment. For other details see table.

AVCO LYCOMING O-540 SERIES

The O-540 is basically a direct-drive six-cylinder version of the four-cylinder O-360, with the same bore and stroke and a swept volume of 8·86 litres (541·5 cu in). It is currently available in a high-compression configuration for use with 100/130 minimum octane fuel. A low-

Right: **The 201 kW (270 hp) Avco Lycoming TVO-435-B1A turbocharged vertical helicopter engine**

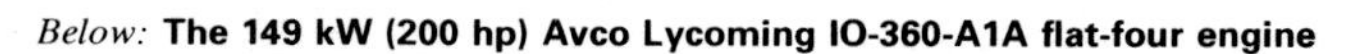

Below: **The 149 kW (200 hp) Avco Lycoming IO-360-A1A flat-four engine**

AVCO LYCOMING HORIZONTALLY-OPPOSED PISTON ENGINES

Engine Model	No. of Cylinders	Rated output at Sea Level kW (hp) at rpm*	Capacity litres (cu in)	Compression Ratio	Fuel Grade	Weight Dry kg (lb)	Length Overall mm (in)	Width Overall mm (in)	Height Overall mm (in)	Gear Ratio†
O-235-C1B	4	86 (115) at 2,800	3·85 (233)	6·75	80	109 (240)	757 (29·81)	812 (32·00)	569 (22·40)	D
O-235-L2A	4	88 (118) at 2,800	3·85 (233)	8·5	100/130	100 (222)	751 (29·56)	812 (32·00)	569 (22·40)	D
O-290-D2C	4	101 (135) at 2,600 T-O 104 (140) at 2,800	4·75 (289)	7·0	80/87	107 (235)	757 (29·81)	819 (32·24)	576 (22·68)	D
O-320-E2D	4	112 (150) at 2,700	5·2 (319·8)	7·0	80/87	113 (249)	738 (29·05)	819 (32·24)	584 (22·99)	D
O-320-H2AD	4	119 (160) at 2,700	5·2 (319·8)	9·0	100/130	114 (252)	795 (31·31)	830 (32·68)	621 (24·46)	D
IO-320-B1A	4	119 (160) at 2,700	5·2 (319·8)	8·5	100/130	117 (259)	853 (33·59)	819 (32·24)	488 (19·22)	D
IO-320-C1A	4	119 (160) at 2,700	5·2 (319·8)	8·5	100/130	122 (268)	853 (33·59)	819 (32·24)	488 (19·22)	D
LIO-320-B1A	4	119 (160) at 2,700	5·2 (319·8)	8·5	100/130	119 (262)	853 (33·59)	819 (32·24)	488 (19·22)	D
AEIO-320-E1B	4	112 (150) at 2,700	5·2 (319·8)	7·0	80/87	117 (258)	738 (29·05)	819 (32·24)	589 (23·18)	D
O-360-A1D	4	134 (180) at 2,700	5·92 (361)	8·5	100/130	116 (256)	757 (29·81)	848 (33·37)	625 (24·59)	D
O-360-A1F6	4	134 (180) at 2,700	5·92 (361)	8·5	100/130	120 (265)	780 (30·70)	848 (33·37)	625 (24·59)	D
O-360-A4A	4	134 (180) at 2,700	5·92 (361)	8·5	100/130	121 (267)	751 (29·56)	848 (33·37)	625 (24·59)	D
AEIO-360-A1B6	4	149 (200) at 2,700	5·92 (361)	8·7	100/130	139 (307)	780 (30·70)	870 (34·25)	492 (19·35)	D
HIO-360-A1A	4	134 (180) at 2,900 to 915 m (3,000 ft)	5·92 (361)	8·7	100/130	128 (283)	855 (33·65)	870 (34·25)	492 (19·35)	D
HIO-360-C1A	4	153 (205) at 2,900	5·92 (361)	8·7	100/130	133 (293)	791 (31·14)	870 (34·25)	495 (19·48)	D
HIO-360-D1A	4	142 (190) at 3,200 to 1,280 m (4,200 ft)	5·92 (361)	10·0	100/130	132 (290)	894 (35·28)	870 (34·25)	495 (19·48)	D
IO-360-A1A	4	149 (200) at 2,700	5·92 (361)	8·7	100/130	133 (293)	757 (29·81)	870 (34·25)	549 (21·61)	D
IO-360-B1B	4	134 (180) at 2,700	5·92 (361)	8·5	100/130	121 (267)	757 (29·81)	848 (33·37)	633 (24·91)	D
IO-360-F1A	4	134 (180) at 2,700	5·92 (361)	8·5	100/130	123 (272)	815 (32·09)	848 (33·37)	526 (20·70)	D
IVO-360-A1A	4	134 (180) at 2,900	5·92 (361)	8·5	100/130	124 (274)	762 (30·00)	848 (33·37)	583 (22·95)	D V
TIO-360-A1B	4	149 (200) at 2,575 to 4,575 m (15,000 ft)	5·92 (361)	7·3	100/130	161 (355)	1,153 (45·41)	870 (34·25)	506 (19·92)	D
VO-435-A1F	6	186 (250) at 3,200 T-O 194 (260) at 3,400	7·1 (434)	7·3	80/87	181 (399)	882 (34·73)	853 (33·58)	612 (24·13)	DV
TVO-435-B1A	6	164 (220) at 3,200 to 6,100 m (20,000 ft) T-O 201 (270) at 3,200	7·1 (434)	7·3	100/130	217 (478)	882 (34·73)	853 (33·58)	905 (35·65)	D V
TVO-435-G1A	6	164 (220) at 3,200 to 6,100 m (20,000 ft) T-O 209 (280) at 3,200 to 5,486 m (18,000 ft)	7·1 (434)	7·3	100/130	211 (465)	882 (34·73)	853 (33·58)	906 (35·67)	D V
IGO-480-A1A6	6	209 (280) at 3,000 T-O 220 (295) at 3,400	7·8 (479·7)	8·7	100/130	206 (455)	1,036 (40·76)	842 (33·12)	712 (28·02)	0·642
GO-480-B1D	6	194 (260) at 3,000 T-O 201 (270) at 3,400	7·8 (479·7)	7·3	80/87	196 (432)	981 (38·64)	842 (33·12)	712 (28·02)	0·642
GO-480-G1D6	6	209 (280) at 3,000 T-O 220 (295) at 3,400	7·8 (479·7)	8·7	100/130	201 (444)	981 (38·64)	842 (33·12)	712 (28·02)	0·642
IGSO-480-A1F6	6	239 (320) at 3,200 to 3,350 m (11,000 ft) T-O 254 (340) at 3,400	7·8 (479·7)	7·3	100/130	233 (513)	1,208 (47·56)	842 (33·12)	570 (22·44)	0·642
O-540-A1A5	6	186 (250) at 2,575	8·86 (541·5)	8·5	100/130	166 (367)	945 (37·22)	848 (33·37)	624 (24·56)	D
O-540-B2B5	6	175 (235) at 2,575	8·86 (541·5)	7·2	80/87	166 (366)	945 (37·22)	858 (33·77)	624 (24·56)	D
O-540-J1A5D	6	175 (235) at 2,400	8·86 (541·5)	8·5	100/130	161·5 (356)	989 (38·93)	848 (33·37)	624 (24·56)	D
VO-540-9	6	227 (305) at 3,200 to 915 m (3,000 ft)	8·86 (541·5)	8·7	100/130	205 (452)	882 (34·73)	880 (34·70)	649 (25·57)	D V
IO-540-C4B5	6	186 (250) at 2,575	8·86 (541·5)	8·5	100/130	170 (375)	976 (38·42)	848 (33·37)	622 (24·46)	D
IO-540-D4A5	6	194 (260) at 2,700	8·86 (541·5)	8·5	100/130	170 (375)	976 (38·42)	848 (33·37)	622 (24·46)	D
AEIO-540-D4B5	6	194 (260) at 2,700	8·86 (541·5)	8·5	100/130	175 (388)	999 (39·34)	848 (33·37)	622 (24·46)	D
IO-540-J4A5	6	186 (250) at 2,575	8·86 (541·5)	8·5	100/130	172 (380)	999 (39·34)	848 (33·37)	622 (24·46)	D
IO-540-K1A5	6	224 (300) at 2,700	8·86 (541·5)	8·7	100/130	201 (443)	999 (39·34)	870 (34·25)	498 (19·60)	D
IO-540-P1A5	6	216 (290) at 2,575	8·86 (541·5)	8·7	100/130	192 (424)	999 (39·34)	870 (34·25)	498 (19·60)	D
HIO-540-A1A	6	216 (290) at 2,575	8·86 (541·5)	8·7	100/130	201 (443)	999 (39·34)	870 (34·25)	498 (19·60)	D
IGO-540-B1C	6	242 (325) at 3,000 T-O 261 (350) at 3,400	8·86 (541·5)	8·7	100/130	227 (500)	1,178 (46·38)	870 (34·25)	550 (21·66)	0·642
IGSO-540-A1D	6	268 (360) at 3,200 to 3,200 m (10,500 ft) T-O 283 (380) at 3,400	8·86 (541·5)	7·3	100/130	245 (540)	1,223 (48·15)	870 (34·25)	722 (28·44)	0·642
VO-540-B1B3	6	227 (305) at 3,200	8·86 (541·5)	7·3	80/87	201 (444)	882 (34·73)	880 (34·70)	617 (24·29)	D V
VO-540-B2D	6	Max continuous 227 (305) at 3,200	8·86 (541·5)	7·3	80/87	200 (442)	882 (34·73)	880 (34·70)	649 (25·57)	D V
VO-540-C1A	6	227 (305) at 3,200 to 915 m (3,000 ft)	8·86 (541·5)	8·7	100/130	200 (441)	882 (34·73)	880 (34·70)	649 (25·57)	D V
IVO-540-A2C	6	227 (305) at 3,200 to 915 m (3,000 ft)	8·86 (541·5)	8·7	100/130	197 (435)	882 (34·73)	880 (34·70)	615 (24·22)	D V
TIO-540-A1A	6	231 (310) at 2,575 to 4,575 m (15,000 ft)	8·86 (541·5)	7·3	100/130	230 (506)	1,304 (51·34)	870 (34·25)	577 (22·71)	D
TIO-540-F2BD	6	242 (325) at 2,575 to 4,575 m (15,000 ft)	8·86 (541·5)	7·3	100/130	232 (511)	1,304 (51·34)	870 (34·25)	570 (22·42)	D
TIO-540-J2BD	6	261 (350) at 2,575 to 4,575 m (15,000 ft)	8·86 (541·5)	7·3	100/130	234 (518)	1,308 (51·50)	870 (34·25)	573 (22·56)	D
TIVO-540-A2A	6	227 (305) at 3,200 to 4,575 m (15,000 ft)	8·86 (541·5)	7·3	100/130	230 (507)	882 (34·73)	880 (34·70)	914 (36·00)	D V
TIO-541-A1A	6	231 (310) at 2,575 to 4,575 m (15,000 ft)	8·86 (541·5)	7·3	100/130	249 (549)	1,247 (49·09)	870 (34·25)	543 (21·38)	D
TIO-541-E1C4	6	283 (380) at 2,900 to 4,575 m (15,000 ft)	8·86 (541·5)	7·3	100/130	270 (596)	1,272 (50·07)	905 (35·66)	640 (25·17)	D
TIGO-541-C1A	6	298 (400) at 3,200 to 4,575 m (15,000 ft)	8·86 (541·5)	7·3	100/130	319 (703)	1,462 (57·57)	885 (34·86)	575 (22·65)	0·667
TIGO-541-D1A	6	336 (450) at 3,200 to 4,575 m (15,000 ft)	8·86 (541·5)	7·3	100/130	318 (701)	1,462 (57·57)	885 (34·86)	575 (22·65)	0·667
TIGO-541-E1A	6	317 (425) at 3,200 to 4,575 m (15,000 ft)	8·86 (541·5)	7·3	100/130	318 (701)	1,462 (57·57)	885 (34·86)	575 (22·65)	0·667
IO-720-A1B	8	298 (400) at 2,650	11·84 (722)	8·7	100/130	257 (567)	1,179 (46·41)	870 (34·25)	573 (22·53)	D

*Note: horsepower is the figure in brackets; †D, Direct drive; V, Vertical mounting

compression model for use with 80/87 fuel can be provided and the vertically-mounted VO-540 is in production as a helicopter power plant. The TIVO-540-A1A, equipped with an AiResearch turbocharger, maintains rated power to 5,180 m (17,000 ft).

During 1968-70 computer analysis of engine service records enabled time between overhauls to be increased by up to 50 per cent. The O-540-B (Cherokee 235) and O-540-E (Comanche and Cherokee SIX) are both now cleared at 1,800 hours. For other details see table.

AVCO LYCOMING IO-540 SERIES

A fuel-injection, tuned induction version of the O-540 with high-output cylinders, model IO-540-B1A5, is rated at 216 kW (290 hp) at 2,575 rpm. A geared version of this engine, model IGO-540-B1A, is rated at 261 kW (350 hp) at 3,400 rpm for take-off. The geared and supercharged model IGSO-540-B1A has a supercharger ratio of 11·27 : 1 and provides take-off power to 3,350 m (11,000 ft) altitude on 100/130 minimum octane fuel. The IO-540-J (Turbo Aztec) has since 1970 been cleared to operate 1,500 hours between overhauls, and the IO-540-C (Aztec), IO-540-D (Comanche), IO-540-K (Cherokee SIX) and IO-540-M and TIO-540-A and -C series (Navajo) have been cleared to 1,800 hours. The latest addition to this series is the TIO-540-J, a 261 kW (350 hp) top-exhaust, direct-drive, turbocharged engine used in the Piper Navajo Chieftain. For further details see table.

AVCO LYCOMING TIO-541 SERIES

First engine in this turbocharged six-cylinder series was the TIO-541-A1A, which gives 310 hp to 4,570 m (15,000 ft) and 172 kW (230 hp) to 7,620 m (25,000 ft). Lycoming has now extended this family with the 283 kW (380 hp) TIO-541-E series and the 317 kW (425 hp) geared TIGO-541-E. These have an overhaul life of 1,200 hours. A double-scroll blower, to provide cabin pressurisation also, is available on all these turbocharged engines.

AVCO LYCOMING IO-720 SERIES

This eight-cylinder version of the IO-540 engine is available and has a rating of 298 kW (400 hp) at 2,650 rpm. Time between overhauls is 1,500 hours. For further details see table.

AVCO LYCOMING NEW RANGE

Avco Lycoming, Williamsport, is continuing the development of a new series of high-performance normally-aspirated and turbocharged flat opposed piston engines. Research into gas turbines in the same power category as Lycoming piston engines has been moved to Stratford.

The 231 kW (310 hp) Avco Lycoming TIO-540 six-cylinder engine fitted to the Piper Turbo Navajo

The 298 kW (400 hp) eight-cylinder Avco Lycoming IO-720-A1A engine

BELL

BELL AEROSPACE COMPANY DIVISION OF TEXTRON INC

HEAD OFFICE AND WORKS:
Buffalo, New York 14240
Telephone: (716) 297-1000
OFFICERS: See "Aircraft" section

Bell has been engaged in the design, development and production of liquid-propellant rocket engines since 1946.

A single-chamber engine which Bell developed originally to power a nuclear weapon pod to be carried by the B-58 Hustler bomber is now being used in modified form in the Agena space vehicle and provides upper-stage propulsion for numerous spacecraft. The latest version is able to offer multiple re-start capability and was used for the Gemini rendezvous programme. The Agena vehicles used in this programme were also fitted with a Bell-produced secondary propulsion system of 0·071 kN (16 lb) and 0·88 kN (200 lb) st radiation-cooled rocket motors for fine adjustment of their velocity prior to the docking manoeuvre.

On 20 February 1974, Bell completed 15 years of space use of the Agena series of engines. In that time the Models 8096, described hereunder, and 8247 have been used in more than 300 missions by the US Air Force and NASA.

Bell provided the post-boost propulsion system (PBPS) for the Minuteman III ICBM.

BELL MODEL 8096 AGENA ENGINE

This engine was first developed as the power plant for one of the weapon pods that was to be carried by the B-58 Hustler supersonic bomber. It is used in modified form, with gimballed chamber, as the power unit of the Lockheed-built Agena vehicle, forming the second stage of the Thor-Agena, Atlas-Agena and other space vehicles.

During its development the engine has undergone five major modifications, each resulting in an improvement in specific impulse. The present version, designated Model 8096, has a specific impulse of nearly 300 sec, which is more than 10% better than the original version of 1959, an increase equivalent to 225 kg (500 lb) in payload for Earth orbital missions.

The Model 8096 engine is a single-chamber pump-fed engine, running on red fuming nitric acid and unsymmetrical dimethyl-hydrazine (UDMH) hypergolic propellants.

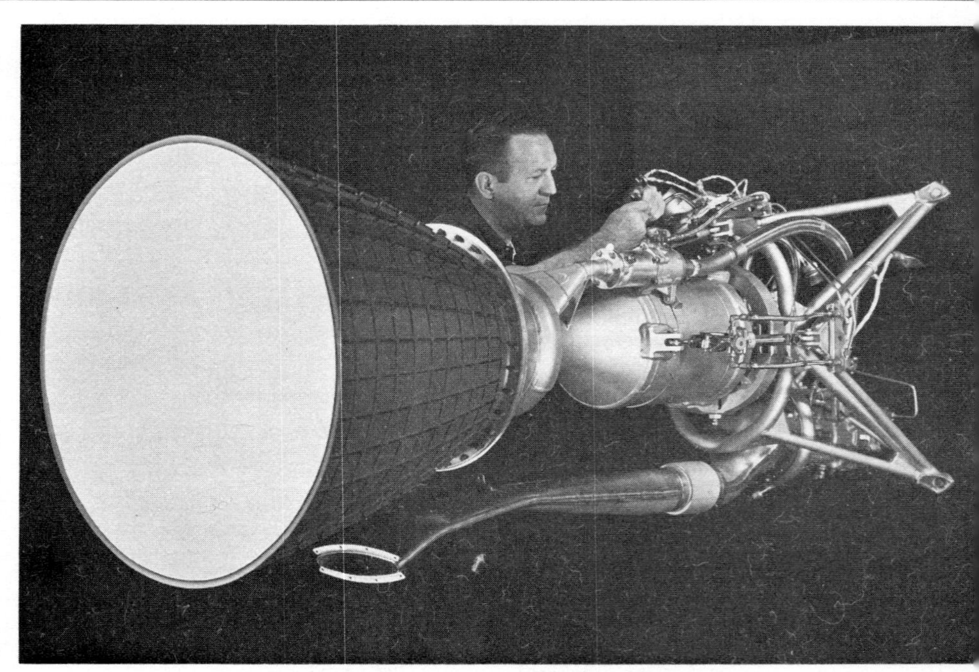

Bell Agena single-chamber liquid-propellant rocket engine

It gives 71 kN (16,000 lb) st and has re-start capability in space. This feature can be used, for example, to change from a circular to an elliptical orbit.

The Model 8096 engine has the ability to be re-started twice in space. It powers the Agena vehicles used in many US Air Force and NASA programmes, including Ranger, Mariner, Nimbus, Echo 2, Alouette, OGO, POGO, AOSO and OAO.

Bell has qualified the Agena engine to run on high-density acid (HDA), which burns at increased temperature to give increased thrust and efficiency; it requires a silicone additive to protect the thrust chamber. A higher-performance baffled injector has been developed and qualified for the current production engine. The company is investigating future propellant combinations and potential Agena applications in the Space Shuttle programme.

DIMENSIONS:
- Length overall: approx 2,134 mm (84 in)
- Nozzle diameter: 825·5 mm (32·5 in)

WEIGHT: approx 132 kg (290 lb)

PERFORMANCE:
- Thrust: 71 kN (16,000 lb)
- Chamber pressure approx: 34·5 bars (500 lb/sq in)
- Specific impulse: approx 300 sec

CSD

CHEMICAL SYSTEMS DIVISION

(A division of United Technologies Corporation)

HEADQUARTERS:
Sunnyvale, California 94088
Telephone: (408) 739-4880
DIVISION PRESIDENT:
Barnet R. Adelman
DIVISION VICE-PRESIDENTS:
Dr David Altman
Eugene Roberts

On 1 May 1975 the former United Technology Corporation (UTC) was renamed Chemical Systems Division (CSD). Simultaneously its parent, the former United Aircraft Corporation, was renamed United Technologies Corporation.

CSD is currently engaged in research, development and production of rockets, rocket propellants and advanced propulsion systems, as well as a range of "spin-off" by-products. Since its founding in the late 1950s it has conducted a continuous programme of rocket technology and is now producing a variety of advanced space propulsion systems for both the Department of Defense and the National Aeronautics and Space Administration.

CSD's largest programme is the production of 3,048 mm (120 in) diameter, segmented solid-propellant booster rockets. Other programmes described hereafter include the Algol III, FW-4 and FW-5 rocket motors, all of which are fully operational.

Testing of CSD's patented Techroll fluid bearing continues. This bearing, considered to be the first major improvement in rocket steering methods in almost a decade, is being applied primarily to rocket motor technology for programmes being funded by the US Air Force, Navy and Army. Of major importance was the award in mid-1974 of a three-year Air Force contract calling for the demonstration of a booster stage movable nozzle system applicable to an Advanced Intercontinental Ballistic Missile. This system is to be a high deflection, low torque component, which will provide high performance, long service life and resistance to high environmental stress levels. It will also make use of advanced carbon composite nozzle materials. Other uses for the Techroll bearing now being researched include a main rotor bearing for helicopters and a joint in high-pressure fluid-transmission lines.

CSD 3,048 mm (120 in) SEGMENTED MOTOR

Mass-produced at CSD's Coyote, California, rocket production facility, this 3,048 mm (120 in) diameter solid-propellant rocket motor centre segment is interchangeable with any other centre segment. Its cylindrical metal case is manufactured from high-strength steel (D6AC) and heat treated to an ultimate strength of 13,445 bars (195,000 lb/sq in). Less than 12·7 mm (0·5 in) thick, each case is equipped with clevis-type end joints with holes for cylindrical fastening pins. Loaded with propellant, a synthetic rubber (polybutadiene acrylonitrile) with aluminium additives as fuel and ammonium perchlorate as oxidiser, a segment has an overall length of 3,200 mm (10 ft 6 in) and weighs 35,830 kg (79,000 lb). Designed as the basic building block for large solid-propellant booster rockets, the segmented motor can be assembled in from one- to eight-segment configurations with thrusts ranging from 1,112-6,673 kN (250,000 to 1·5 million pounds). A fully configured motor contains a destruct system, a forward-end ignition system a liquid-injection thrust-vector control system for steering and has its own staging sequence capability.

CSD has made progress in the development of more efficient and economical means of fabricating the steel motor cases used in the 120 in programme, through production processes known as internal roll extrusion and shear forming. Both techniques reduce production time and result in a more reliable and economical product.

A five-segment configuration of this motor is used in pairs as the zero (launch) stage of the US Air Force Titan IIIC and D space launch vehicles and the new Titan IIIE of NASA, first flown on 11 February 1974. Operating together, the 26·12 m (86 ft) tall motors produce a thrust of 10,675 kN (2,400,000 lb). Launched for the first time on 18 June 1965, the boosters have flown 39 successful missions in as many launches, using 78 motors.

Production continued in 1975, following a $59 million US Air Force contract for 16 five-segment motors for delivery in 1976 and 1977. This raises total purchase by the USAF Space and Missile Systems Organisation to 116.

Essentially a lengthened version of the five-segment motor, a seven-segment motor has been statically test-fired four times. Producing a record 6,227 kN (1·4 million pounds) of thrust, the motors operated perfectly while testing an advanced steering system, called an ullage blow-down, which is now used on all flights of Titan IIIC, IIID and IIIE vehicles.

CSD ALGOL III

CSD is in full production of Algol III, a more powerful solid-propellant first stage for NASA's Scout space launch vehicle. 9·14 m (30 ft) tall and 1·14 m (45 in) in diameter, the improved solid booster permits an increase of 40-54 per cent in the Scout's payload/weight capability. When loaded with its propellant (PBAN with aluminium additives and ammonium perchlorate oxidiser) Algol III weighs 13,605 kg (30,000 lb). Its nozzle is fabricated from steel with a graphite cloth-phenolic and silica cloth-phenolic cone liner. Ignition is by a small nose-mounted solid rocket. Producing 623 kN (140,000 lb) of lift-off thrust, Algol III burns for 75 seconds to boost the Scout to an altitude of 48 km (30 miles) before burnout. The motor flew for the first time in August 1972 and at the end of 1975 had completed 12 missions.

CSD Algol III solid-propellant motor on vertical (inverted) static test, generating thrust of 622·5 kN (140,000 lb)

CSD FW-4, the upper-stage motor of a Scout at Vandenberg AFB, with the payload above and one of the two nose shrouds in place

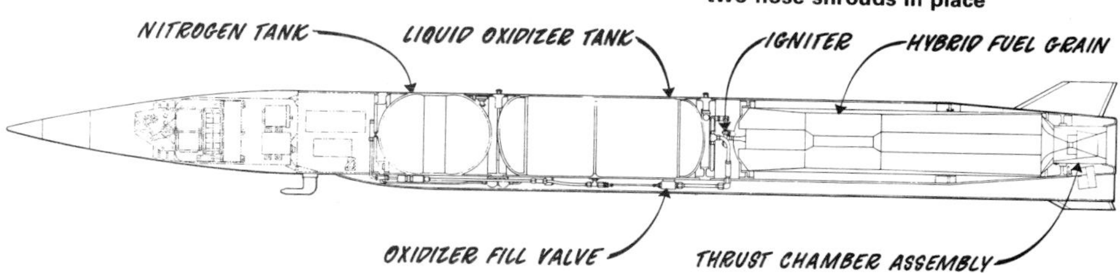

Longitudinal section of supersonic target vehicle with CSD hybrid propulsion, showing disposition of solid fuel grain and liquid propellants

CSD FW-4

The FW-4 solid motor was designed to be interchangeable with existing upper-stage motors on launch vehicles with orbital, probe or re-entry missions, as well as for retro-rocket propulsion in space vehicles. It is used in both NASA and military programmes, flying as the top stage on Air Force Atlas and Thor boosters and NASA Scout and TAD vehicles. By the end of 1975 FW-4 motors had placed 66 satellites in precise orbit.

The FW-4 is 498 mm (19·6 in) in diameter and has an overall length of 1,484 mm (58·43 in). Its case is fabricated from glassfibre and epoxy resin, with a cylindrical wall thickness of 2·03 mm (0·08 in) and domes with geodesic contours with a nominal wall thickness of 0·76 mm (0·03 in). The igniter and nozzle attachment fittings are fabricated from aluminium, and the aluminium interstage flanges are bonded and riveted to the integrally-wound glassfibre skirt of the motor case. The standard FW-4 rocket motor has an inert weight of 25·2 kg (55·5 lb) and a propellant weight of 274 kg (605 lb), resulting in a motor mass fraction of 0·92. The average motor thrust is 25 kN (5,620 lb), with a burn time of 30·5 seconds.

In January 1975 CSD demonstrated an advanced version of this motor with increased performance for an LTV-NASA group.

CSD FW-5

Similar to the company's FW-4 motor, the FW-5 is likewise designed to be interchangeable with existing upper-stage motors. While its primary application is that of an apogee motor, it can be used for orbital, probe or re-entry missions, as well as retro-rocket propulsion for space vehicles.

The FW-5 is 691 mm (27·2 in) diameter and has an overall length of 1,123 mm (44·2 in). Its case is fabricated from glassfibre and epoxy resin, with a cylindrical wall thickness of 3·3 mm (0·13 in) and end domes with solar isotensoid contours with a nominal wall thickness of 15·2 mm (0·60 in). The igniter and nozzle attachment fittings are fabricated from aluminium, and the aluminium interstage flange is bonded and riveted to the integrally-wound glassfibre skirt of the motor case.

The standard FW-5 has an inert weight of 29·7 kg (65·6 lb) and a propellant weight of 260 kg (575 lb), resulting in a motor mass fraction of 0·89. The average motor thrust is 25·04 kN (5,630 lb) with a burn time of 42·1 seconds.

FW-5 operated perfectly on its first mission in November 1972 when it placed Telesat Canada's Anik communications satellite into final orbit. The second Anik was put on station by an FW-5 in April 1973, followed by a third in May 1975. In addition, CSD supplied six FW-5 motors to Hughes Aircraft, two of which were used in 1974 to orbit Westar I and II, the first US domestic communications satellite system.

CSD HAST HYBRID PROPULSION

CSD's hybrid rocket for the Department of Defense High Altitude Supersonic Target (HAST) drone (see under Beech heading in the RPVs and Targets section) is 330 mm (13 in) in diameter, has an overall length of 3,556 mm (140 in), and weighs 390 kg (860 lb). It uses a polybutadiene and methyl methacrylate fuel grain, and nitric acid as its liquid oxidiser. Ablative-cooled, its thrust ranges from 0·88-5·34 kN (200 to 1,200 lb) at altitudes between 10,670 and 30,480 m (35,000 and 100,000 ft). The HAST successfully completed all its initial flight-readiness test-firings in 1973 and had two successful flight tests at Edwards AFB in 1975.

CSD LIQUID ROCKET ENGINES

The company's liquid rockets range in size from 1,321 mm (52 in) long and 660 mm (26 in) in diameter and weighing 32·6 kg (72 lb) to 1,918 mm (75·5 in) long, 1,219 mm (48 in) in diameter and weighing 83·9 kg (185 lb). Their propellant is 50/50 hydrazine and unsymmetrical dimethyl hydrazine and nitrogen tetroxide. The engine nozzles are a composite structure with a glassfibre shell and silica-phenolic liner. Ignition is hypergolic.

CSD RAMJET PROPULSION

In early 1973, CSD was selected to spearhead the efforts of its parent, United Technologies Corporation, in the research and development of ramjet propulsion systems, with the support of UTC's Hamilton Standard Division and United Technologies Research Center. In mid-1973 CSD was awarded a contract by the US Navy to research, design and develop a Modern Ramjet Engine (MRE). Called an integral rocket/ramjet, the propulsion device involves both solid rocket and liquid fuel ramjet technology. It will operate as a solid rocket booster until it reaches supersonic speeds. At that point, through a series of mechanical changes that take place in flight, it becomes a ramjet. These changes involve the opening of air inlets, an increase in the nozzle diameter and a switch to the burning of liquid fuel and air in the combustion chamber within a common system. The propulsion technology acquired in carrying out this project will be applied to the development of an advanced air-to-air missile system.

In mid-1974 CSD was awarded another ramjet programme by the US Air Force, to design and ground test an advanced integral rocket/ramjet propulsion system which would satisfy the requirements of an advanced strategic air-launched missile (ASALM). The Air Force said that, while this new three-year programme was designed to provide propulsion information specifically applicable to ASALM, it would be generally applicable to other missions including air-launched tactical missiles, surface-launched defensive missiles and a variety of air-to-air missiles. Fuels and propellants are now being evaluated, and components being designed and fabricated for early static tests for both of these programmes.

Shortly after the Navy's MRE project got underway, the Air Force and Navy jointly awarded a programme to CSD to research and develop the technology for a solid fuel integral rocket/ramjet. In this concept the solid rocket would use two types of propellants in the same combustion chamber. An inner layer would burn through the boost phase, and an outer layer would provide the fuel for the ramjet phase.

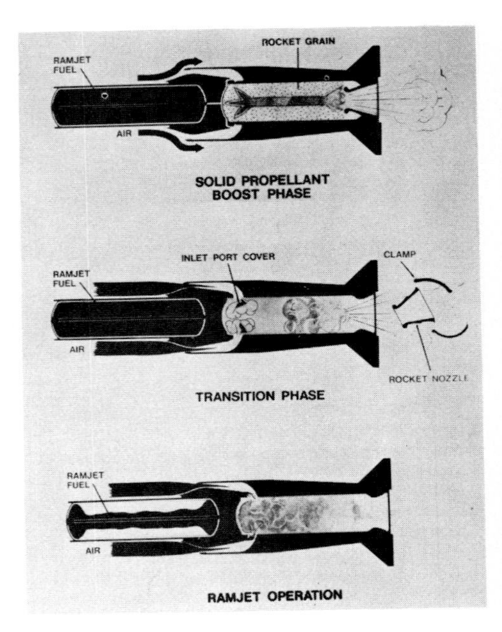

Simplified sequence drawings showing operation of CSD integral rocket/ramjet engine. In the final (lowest) illustration ramjet fuel is being forced out of its collapsible cell by gas pressure

CURTISS-WRIGHT

CURTISS-WRIGHT CORPORATION, WOOD-RIDGE FACILITY

HEAD OFFICE AND WORKS:
One Passaic St, Wood-Ridge, New Jersey 07075
Telephone: (201) 777-2900

CHAIRMAN OF THE BOARD AND PRESIDENT:
T. Roland Berner

SENIOR VICE-PRESIDENTS:
Charles E. Ehringer
Seymour S. Bitterman
Richard P. Sprigle

VICE-PRESIDENT, ENGINEERING:
A. F. Kossar

VICE-PRESIDENTS:
D. Lasky
W. Figart

The Wood-Ridge facility of Curtiss-Wright Corporation is engaged in the research, development and licensing of Wankel-type rotary engines, manufacture of engine parts, aircraft and industrial engine overhaul, electric power generation systems and advanced marine and turbine technology.

Wright engines continue in worldwide service in fixed-wing aircraft and helicopters.

These engines have been described in previous editions of *Jane's*.

Cessna Cardinal testbed flying with Curtiss-Wright RC2-60 engine, rated at 149 kW; 200 hp

CURTISS-WRIGHT SETE

In competition with Garrett-AiResearch and Pratt & Whitney, Curtiss-Wright is participating in the SETE (Supersonic Expendable Turbine Engine) programme of the US Navy. The objective is the cheapest possible jet engine capable of unfailingly-reliable instant starting and flight propulsion, under severe conditions of manoeuvre, over a wide band of speeds and heights, on a single flight of a missile. Details are restricted.

CURTISS-WRIGHT RC ENGINES

In 1958 Curtiss-Wright Corporation obtained a licence for the NSU-Wankel type of rotating-combustion (RC) engine and embarked on a major programme of independent development of a range of such engines aimed at a wide spectrum of applications. At first the company concentrated on large engines in the power range around 500 hp for aircraft use, but during the past decade much smaller engines have dominated the hardware test and development programme, some of which has been funded by US military agencies, including the Naval Air Systems Command.

Most research has been carried out on versions of the **RC2-60** (twin rotors each of about 0·983 litre; 60 cu in capacity), rated at up to 149 kW (200 hp) at 5,500 rpm and with possible future potential to reach twice this rotational speed in view of the near-perfect balance. One of these engines powered the Lockheed Q-Star acoustic research aircraft, specially designed for miniumum noise level (see 1971-72 *Jane's*). Further testing has been completed successfully in the Cessna Cardinal illustrated. Under the sponsorship of the US Army Aviation Command and the Hughes Helicopters company, the RC2-60 successfully completed flight evaluation in a TH-55 training helicopter, illustrated in the 1974-75 *Jane's*.

In 1965 the 310 hp **RC2-90** was run, with helicopter applications in mind, and the 1·47 litre (90 cu in) rotor has since been used in extensive development of stratified charge engines capable of operating on a range of fuels including JP-4 and JP-5 gas turbine kerosenes.

Curtiss-Wright RC2-75, rated at 254 kW (340 hp), on bench with three-blade propeller

The **RC2-75-Y3** is one of a very important family of engines, regarded as optimally sized for a wide range of general aviation aircraft. As the designation indicates, it is based on two rotors each of nominal 1·23 litres (75 cu in) capacity, and has liquid cooling and a geared drive. Engine development had in early 1975 progressed to running engines with propellers on static testbeds. The compactness and favourable power/weight ratio are evident from the data.

DIMENSIONS:
- Length: 798 mm (31·4 in)
- Width: 602 mm (23·7 in)
- Height: 546 mm (21·5 in)

WEIGHTS:
- Basic, dry: 127 kg (280 lb)
- Installed, with starter, oil cooler, oil tank, coolant and radiator and mounting brackets: 167 kg (368 lb)

PERFORMANCE RATING:
- Max T-O: 254 kW (340 hp) at 7,000 rpm

DREHER

DREHER ENGINEERING COMPANY

ADDRESS:
933 5th Street, Santa Monica, California 90403
Telephone: (213) 395-6510

Mr Max Dreher, an aeronautical engineer, has built a series of small turbojet engines over a period of 24 years. One of them, known as the TJD-76 Baby Mamba, was mounted on his Prue 215A all-metal 12·0 m sailplane as an auxiliary turbojet.

Development of the TJD-76C has been virtually completed, apart from the recent addition of an automatic fuel control device which is operating well and will later be applied to the larger engines. During the past year it has become evident that the main market favours the TJD-76E, and a suitable production manufacturer is being sought.

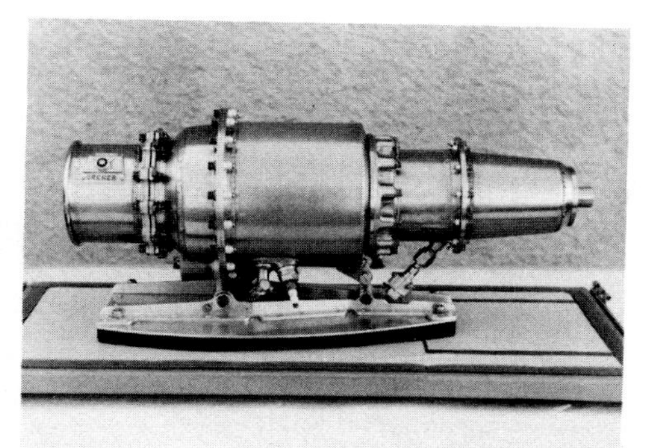

Two views of the Dreher Baby Mamba single-shaft turbojet: (left) TJD-76C on mount; (right) TJD-76C Jet Pack installed on sailplane. Thrust rating is 0·245 kN (55 lb st)

DREHER TJD-76A BABY MAMBA

This very small turbojet engine has a single-stage centrifugal compressor, straight-through-flow annular combustion chamber, with six injectors, and single-stage axial-flow turbine. The shaft runs on two ball bearings. Starting is by compressed air.

Testbed for the Baby Mamba is a Prue 215A all-metal sailplane, with a span of 12·00 m (39 ft 4½ in) and T-O weight of 275 kg (605 lb). Glide ratio with engine fitted is 28 : 1.

DIMENSIONS:

Length overall	414 mm (16·3 in)
Diameter	152 mm (6 in)

WEIGHTS:

Bare turbojet	7·7 kg (17 lb)
Complete power plant package	11·35 kg (25 lb)

DREHER TJD-76C BABY MAMBA

This new version of the Baby Mamba is lighter and introduces several mechanical and aerodynamic improvements, including a tachometer generator for direct rpm reading.

TYPE: Single-shaft turbojet.

AIR INTAKE: At front. Air flow 0·50 kg (1·1 lb)/sec.

COMPRESSOR: Single-stage mixed-flow. Single 17-4 PH stainless steel impeller with sixteen vanes. Splined to shaft and supported in two ball bearings. Mixed-flow two-stage diffuser of 347 stainless steel. Pressure ratio 2·8 : 1.

COMPRESSOR CASING: Of 2024 aluminium alloy and 347 stainless steel.

COMBUSTION CHAMBER: Annular type with Hastelloy X outer casing and flame tube. Vaporising system with fuel/air pre-mix. One spark plug in flame tube.

FUEL SYSTEM: Manual with pressurised fuel supply, or electrically-driven fuel pump. Fuel pressure 5·52 bars (80 lb/sq in). Automatic system for drone applications.

FUEL GRADE: Kerosene or petrol.

NOZZLE GUIDE VANES: Single axial stage, with sixteen investment-cast vanes in Stellite 31.

TURBINE: Single-stage axial-flow, with nineteen integrally-cast blades, of Inconel 713 LC. Gas temperature 770°C before turbine, 675°C after turbine, at continuous cruising power.

JET PIPE: Fixed type, with jet pipe and cone of Hastelloy X.

LUBRICATION: Air/oil mist system with total loss, using bleed air equivalent to 2·5 per cent of total mass flow. Capacity 1 litre (2 US pints).

OIL GRADE: MIL-L-7808E (Turbo 15).

MOUNTING: Two rigid connections on diffuser section and one flexible connection on turbine section.

STARTING: Compressed air 10·34 bars (150 lb/sq in), via three nozzles driving turbine wheel.

DIMENSIONS:

Length overall	416 mm (16·38 in)
Diameter	151 mm (5·94 in)

WEIGHTS:

Dry	6·4 kg (14·1 lb)
Complete with fuel tank	10·0 kg (22 lb)

PERFORMANCE RATINGS:

Max	0·245 kN (55 lb)
Continuous	0·20 kN (45 lb)

SPECIFIC FUEL CONSUMPTION:

at max rating	1·5

OIL CONSUMPTION:

at max rating	25 cc/min

DREHER TJD-76D and E

These versions were derived from the TJD-76C in 1972 to meet a need for a very low-cost short-life unit for the propulsion of small drones and other expendable vehicles. Both have similar performance to the TJD-76C but weigh approximately 4·5 kg (9·9 lb).

DSI

DEVELOPMENTAL SCIENCES INC

ADDRESS:
15747 East Valley Boulevard, City of Industry, California 91749
Telephone: (213) 330-6865
PRESIDENT: Dr Gerald R. Seemann

DSI RESONATING RAMJET

This simple engine consists of an inlet diffuser, inlet with turning vanes, fuel injectors, flameholder, combustion chamber, transition section, tailpipe and nozzle. It is fabricated by welding from 321 stainless steel. Certain area ratios are critical to successful operation at approximately 300Hz. Flight Mach number ranges from 0·5 to 0·95.

DSI has both flown and tunnel-tested this engine. In its basic form it is a very low-cost power plant. Recently experiments have been made with an internal ram-air turbine for on-board shaft power.

Among suitable fuels are propane (gas or liquid), gasoline (petrol), JP-4 and JP-5. Fuel is fed by a pump to multiple injector nozzles, with fuel scheduling regulated by a Mach number feedback. For cold starting the combustion chamber is fitted with a 5 sec Holex igniter. The engine is available in sizes smaller and larger than that described below.

DIMENSIONS:

Length overall	1,270 mm (50 in)
Width (or height)	197 mm (7·75 in)
Height (or width)	279 mm (11·0 in)

WEIGHT, DRY: 13·6 kg (30 lb)

PERFORMANCE RATING (S/L):

Max thrust at Mach 0·7	0·89 kN (200 lb)

SPECIFIC FUEL CONSUMPTION: 3·5-4·8

Development Sciences resonating ramjet in 0·89 kN (200 lb) thrust size

FRANKLIN

FRANKLIN ENGINE COMPANY, INC

(Subsidiary of Audi SA)

HEAD OFFICE AND WORKS:
PO Box 8, Syracuse, New York 13208
Telephone: (313) 457-2200

Known formerly as Aircooled Motors Inc, this company produced the first of its light horizontally-opposed air-cooled engines in 1938. By 1941 it had placed on the market engines of four and six cylinders ranging in output from 65 to 150 hp.

In 1961 the company was resurrected by a new owner, Aero Industries Inc, and began trading as the Franklin Engine Co, Inc. In 1973 it ran into financial problems, ceased manufacturing in 1974, and in 1975 was bought by Audi SA, of Sao Paulo, Brazil. It is now being restructured. In 1975 Pezetel, the Polish national aerospace marketing organisation, bought rights to the complete range of Franklin engines (details in 1975-76 *Jane's*) and hopes to manufacture selected models in Poland. This is reported under WSK-Rzeszów, Poland (which see).

GARRETT-AIRESEARCH

THE GARRETT CORPORATION (one of The Signal Companies)

HEAD OFFICE:
9851 Sepulveda Boulevard, Los Angeles, California 90009
Telephone: (213) 776-1010
PRESIDENT:
Harry H. Wetzel
EXECUTIVE VICE-PRESIDENT, SALES AND SERVICE:
William J. Pattison
GROUP VICE-PRESIDENT:
Ivan E. Speer

AIRESEARCH MANUFACTURING COMPANY of Arizona (a division of The Garrett Corporation)

HEAD OFFICE AND WORKS:
Sky Harbor Airport, 402 South 36th Street, Phoenix, Arizona 85034
Telephone: (602) 267-3011
VICE-PRESIDENT AND MANAGER:
John A. Teske
SALES MANAGER:
Malcolm E. Craig

The Garrett Corporation's AiResearch Manufacturing Company of Arizona, at Phoenix, has been called the world's largest producer of small gas turbines. Development of the first AiResearch small turbines began in 1946 and the division has since produced over 70% of the total of gas-turbine units with power ratings from 60 to 2,500 hp built in the United States and Europe.

The first use of AiResearch turbines as prime movers occurred in 1957 when the McDonnell Aircraft Corporation used three GTC85 turbo-compressors to power the Model 120 pressure-jet helicopter. The GTC85 was followed by the Model 331, the first AirResearch engine designed as an aircraft prime mover.

AiResearch began development of the Model 331, as a private venture, in December 1959, with the object of producing an engine suitable for use as a turboshaft for helicopters and as a turboprop for fixed-wing aircraft.

The first version, the 373 kW (500 shp) TSE 331 turboshaft, was assembled and ready for initial testing by December 1960. Flight tests in a Republic Lark (licence-built Alouette II) helicopter began on 12 October 1961. From this engine evolved a family of commercial TPE 331 and military T76 turboprop versions.

Today Garrett is producing the TFE 731 and ATF 3 series of turbofan engines for the executive, light military and commuter market. Garrett also has development programmes in the low-cost, expendable engine field for propulsion applications in missiles and remotely piloted vehicles. It has discontinued the TSE 36 and TSE 231 helicopter engines described in the 1975-76 *Jane's*.

GARRETT-AIRESEARCH ATF3

US military designation: F104-GA-100

The Garrett-AiResearch ATF3 is a turbofan of unusual layout in the 17·8 to 26·7 kN (4,000 lb to 6,000 lb) thrust range. It is designed to provide, through its low fuel consumption, extended range for subsonic business or military aircraft, and through its relatively high overall pressure ratio, flexible operation in high-altitude applications.

Considered to be the first three-spool engine to run in the United States, it is the first engine in the world to combine the three-spool features with a reverse-flow combustion system and turbines, and mixed-flow exhaust.

The arrangement of components allows the fan design to be determined largely independently of the gas-generator compressor requirements, and permits operation at optimum fan speed. Omission of fan inlet guide vanes, mixing of the gas-generator exhaust with the fan airflow, and double reversal of the internal airflow enable the ATF3 to offer significant reductions in overall noise generation.

Other design considerations include reliability, maintainability and elimination of visible smoke. The accessories are revealed by removing the tailcone fairing, and their positioning at the rear of the engine is claimed to reduce installed drag.

All design and early development of the ATF3 took place at the AiResearch Torrance (Los Angeles) facility. The conceptual design was completed in early 1966, and testing of demonstrator engines was initiated in May 1968. Under US Air Force contract, the ATF3 successfully completed preliminary flight rating tests in 1972 at 18 kN (4,050 lb) thrust. Both the aerodynamic and mechanical design criteria were established around sea-level, ISA +15°C take-off rating of 22·24 kN (5,000 lb).

In 1973 the YF104 was chosen to power fly-off models of the Ryan Compass Cope RPV. The engine is especially well suited to such an application in view of its low infra-red and noise signatures, capability for operation at high altitude, and the ease with which the control system can be integrated with microcircuit guidance and telemetry. The Compass Cope R engine is flat rated at 18 kN (4,050 lb), though considerably greater thrust could be obtained. The reduced thrust is matched with large shaft-power extraction and other special features. Extensive USAF altitude testing of this engine brought total YF104 and ATF3 running time by January 1974 to 2,500 hours. By 1976 this time had more than doubled, and the YF104 had set turbofan altitude and unrefuelled endurance records in Compass Cope RPVs (over 16,770 m; 55,000 ft, and over 24 hr).

After extensive further development Garrett announced the ATF3-6 commercial version in October 1975. Closely similar to the YF104, it has already demonstrated sea-level brochure performance at Garrett's Torrance plant, while altitude cruise guarantees have been met during testing at the NASA Lewis Research Center. It is predicted that the ATF3-6 will meet 1979 emissions standards.

In May 1976 it was announced that the ATF3-6 had been selected by Dassault-Breguet to power the Falcon 20G business jet, with US coast-to-coast range. Flight-test engines are due in May 1977 and commercial certification is scheduled for early 1978, after a total of 11,000 hours of ATF3 testing. The ATF3-6 is offered under a retrofit programme for existing Falcon 20 aircraft.

TYPE: Three-shaft axial-flow turbofan.

INTAKE: Direct pitot, fixed type. No inlet vanes or struts. Total airflow 73·5 kg/sec (162 lb/sec).

LOW-PRESSURE (FAN) SYSTEM: Single-stage titanium fan, driven by three-stage IP turbine. One thrust bearing and one roller bearing support independent LP shaft. Bypass ratio 2·8 at take-off.

INTERMEDIATE-PRESSURE SYSTEM: Five-stage titanium axial IP compressor, each stage having a separate disc, driven by two-stage LP turbine. Airflow is then delivered to rearward-facing HP compressor via eight tubes feeding into an annular duct concentric with the by-pass duct. One thrust bearing and one roller bearing support independent IP shaft. Core airflow 18·15 kg (40 lb)/sec.

HIGH-PRESSURE SYSTEM: Single-stage titanium centrifugal compressor, driven by single-stage HP turbine. IP airflow enters the single-sided impeller from the rear. One thrust bearing and one roller bearing support the independent HP shaft. Overall pressure ratio (T-O) 21, (high-altitude cruise) 25.

COMBUSTION SYSTEM: Reverse-flow annular type.

TURBINES: Single-stage HP, three-stage IP and two-stage LP turbines drive, respectively, the HP, fan (LP) and IP compressors. IP and LP turbines have fully shrouded blades. Aircooled first-stage nozzle vanes and HP rotor blades. Exhaust gases turned 180° through eight sets of cascades to mix with fan by-pass flow.

FUEL SYSTEM: Electromechanical, incorporating solid-state computer. Manual emergency backup system.

ACCESSORY DRIVES: Three drive pads on rear-mounted gearbox driven by HP shaft, providing for hydraulic pump, starter/generator and one spare. Accessory cooling by fan discharge air which is exhausted through a nozzle at the tip of the fairing.

EXHAUST SYSTEM: Mixed fan and turbine exhaust discharged to atmosphere through annular nozzle surrounding combustion section.

LUBRICATION SYSTEM: Self-contained hot-tank type; tank integral with gearbox.

MOUNTING: Two-plane pickup system.

STARTING: Electrical or pneumatic.

DIMENSIONS:

YF104:

Length	2,495 mm (98·25 in)
Max diameter	833 mm (32·78 in)

ATF3-6:

Length	2,316 mm (91·2 in)
Max diameter	853 mm (33·6 in)

WEIGHT, DRY:

YF104, bare	408 kg (900 lb)
ATF3-6	431 kg (950 lb)

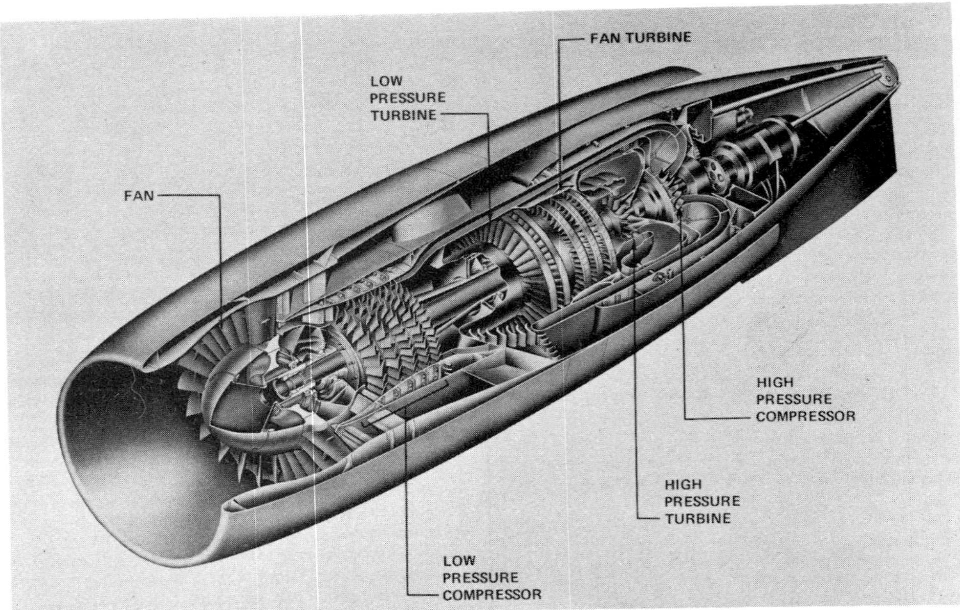

Cutaway of the Garrett-AiResearch ATF3-6 three-shaft turbofan with double flow-reversal (22·46 kN; 5,050 lb st)

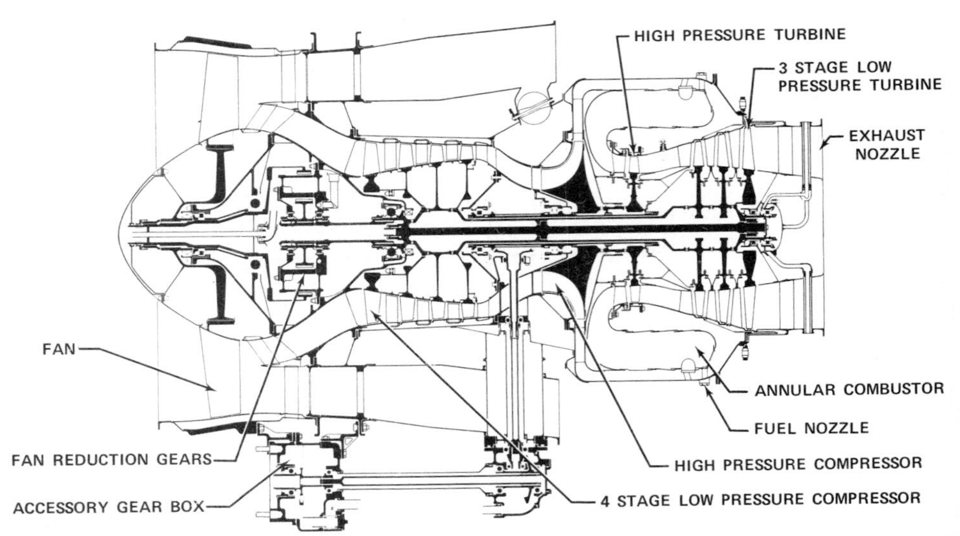

Cross-section *(top)* **and external view** *(above)* **of the 15·57 kN (3,500 lb st) Garrett-AiResearch TFE731 geared front-fan engine**

PERFORMANCE RATINGS (Uninstalled):
T-O (S/L, static):
YF104 18 kN (4,050 lb), ISA
ATF3-6 22·46 kN (5,050 lb), ISA+15°C
Cruise (12,200 m; 40,000 ft at Mach 0·8):
ATF3-6 4·5 kN (1,012 lb)
SPECIFIC FUEL CONSUMPTION:
At T-O rating (S/L, ISA static)
13·6 mg/Ns (0·48 lb/hr/lb st)
At cruise (as above) 22·38 mg/Ns (0·79 lb/hr/lb st)

GARRETT-AIRESEARCH TFE731

Announced in April 1969, the TFE731 is a two-spool geared turbofan designed to confer US coast-to-coast range upon business jet aircraft. Use of a geared fan confers flexibility in operation and yields optimum performance both at low altitudes and at up to 15,250 m (50,000 ft).

The LP spool is made up of a three-stage turbine driving a geared fan and four-stage compressor. The HP spool consists of a single-stage turbine driving a single-stage centrifugal compressor derived from that of the TPE331/T76 turboprop.

Component testing began in March 1969. The first engine ran in September 1970, and was tested at Phoenix in a Learjet 25 (illustrated in 1972-73 *Jane's*). FAA certification and first production deliveries to Dassault for the Falcon 10 took place in August 1972.

In October 1972 it was stated that the Lockheed JetStar would be re-engined with the TFE731-3, flat rated at 16·46 kN (3,700 lb) st by a modest increase in turbine inlet temperature. The TFE731-3 was certificated in 1974. The modified aircraft is designated JetStar II, and first flew in July 1974. AiResearch Aviation is converting a number of JetStar I aircraft to have TFE 731-3 power.

Since 1973 the TFE731, in various sub-types, has been selected for five further business jets: the Gates Learjet 35/36, Cessna 700 Citation III (tri-jet), Dassault-Breguet Falcon 50 (tri-jet), Hawker Siddeley) 125-700 and IAI 1124 Westwind. The engine has also been selected for the CASA C-101 trainer and light attack aircraft, and proposals have been made for RPV and other short-life programmes.

In October 1975 Garrett announced the more powerful TFE731-4, with five-stage LP compressor. Design specification was to be fixed in July 1976, with certification possible in late 1978.

By January 1976 deliveries of TFE731-2 and -3 engines had reached 380. Output had risen to 30 per month, with expansion planned to rise to 70 per month to meet market demand.

TYPE: Turbofan with two shafts and geared front fan.

AIR INTAKE: Direct pitot, fixed, without guide vanes.

FAN: Single-stage axial titanium fan, with inserted blades. Mounted on a simple shaft supported by a roller bearing, located under the fan disc, and by a ball-thrust bearing. The fan shaft is connected directly to the planetary gearbox ring gear. Max fan airflow, sea level static, 51·25 kg (113 lb)/sec (-3, 53·7 kg; 118·3 lb/sec; -4, 54·6 kg; 120·4 lb/sec). By-pass ratio 2·66 (-3, 2·80; -4, 2·28).

COMPRESSOR: Low-pressure compressor has four stages (TFE731-4, five stages), each with a separate disc. Rotors and stators have inserted blades and vanes. High-pressure compressor, carried on a separate shaft running at higher speeds, is centrifugal. Overall pressure ratio (S/L, static): -2, 14·0; -3, 14·6; -4, 17·5.

COMBUSTION CHAMBER: Annular combustion chamber of reverse-flow type, with 12 fuel nozzles inserted radially and injecting fuel tangentially.

FUEL SYSTEM: Hydro-electronic, with single-lever control to mechanical and electronic elements.

TURBINES: High-pressure turbine has a single axial stage with inserted blades. Low-pressure turbine has three axial stages, all with inserted blades. Average inlet gas temperature to HP turbine, S/L, max T-O thrust, 1,010°C (-3 and -4, higher).

SHAFTING: High-pressure spool consists of HP turbine and HP compressor, mounted on shaft supported by one roller bearing and one ball bearing. This spool drives the accessory gearbox through a tower shaft transfer gearbox system. Low-pressure spool consists of the LP turbine and the LP compressor. It is composed of separate components interconnected by curvic couplings and simply supported on one ball bearing at the compressor end and one roller bearing at the turbine end. LP spool drives the fan shaft through a quill shaft and a planetary gear reduction system. Overall gear ratio is 0·555 : 1.

JET PIPE: Short fan duct with cool discharge around remainder of engine, facilitating installation of fan reverser. Hot gas pipe at rear, with fixed nozzle of minimum length.

ACCESSORY DRIVES: Accessories driven from HP spool are grouped around underside of the forward section of the fan duct. Pads are provided on the front side of the accessory gearbox for the airframe-type accessories: hydraulic pump, starter/generator or starter motor and alternators. Pads on the back side of the gearbox drive the engine accessories: fuel control unit and oil pump.

DIMENSIONS:
Intake diameter 716 mm (28·2 in)
Length overall (-2, -3) 1,263 mm (49·73 in)
Length overall (-4) 1,397 mm (55·0 in)
Width 869 mm (34·20 in)
Height overall 992 mm (39·07 in)

WEIGHT, DRY:
TFE731-2, -3 329 kg (725 lb)
TFE731-4 349 kg (770 lb)

PERFORMANCE RATINGS:
Max T-O (S/L, 24·4°C):
TFE731-2 15·57 kN (3,500 lb st)
TFE731-3 16·46 kN (3,700 lb st)
TFE731-4 17·61 kN (3,959 lb st)
Cruise (12,200 m; 40,000 ft at Mach 0·8):
TFE731-2 3·36 kN (755 lb)
TFE731-3 3·92 kN (881 lb)
TFE731-4 4·48 kN (1,008 lb)

SPECIFIC FUEL CONSUMPTION:
Max T-O (as above):
TFE731-2 13·88 mg/Ns (0·49 lb/hr/lb)
TFE731-3 14·59 mg/Ns (0·515 lb/hr/lb)
TFE731-4 15·52 mg/Ns (0·548 lb/hr/lb)
Cruise (as above):
TFE731-2 23·08 mg/Ns (0·815 lb/hr/lb)
TFE731-3 23·65 mg/Ns (0·835 lb/hr/lb)
TFE731-4 24·13 mg/Ns (0·852 lb/hr/lb)

GARRETT-AIRESEARCH ETJ 131

Evolving from Garrett's long turbocharger experience, the ETJ131 turbojet is designed for military applications such as low-cost decoys and target vehicles. Although details are not available, it is generally known that the design uses turbocharger components mated to a single-can combustor. In the 0·445 kN (100 lb) thrust class, the engine is described as "low-cost, with very simplified controls".

GARRETT-AIRESEARCH TJE 341

In the 4·45 kN (1,000 lb) thrust class, the TJE341 turbojet is derived from earlier Garrett low-cost expendable engine technology and demonstration programmes. Initial application aimed at is in medium-range RPVs now under study. The TJE341 has a re-usable rating for recoverable RPVs, yet is expected to retain the overall simplicity and potentially low production cost that were design goals for earlier Garrett turbojets.

GARRETT-AIRESEARCH TPE331

US military designation: T76

Originally based upon extensive experience with APUs, this was the first AiResearch engine for aircraft propulsion; it has since been the main product which has helped to provide funding for later engines. The military T76 has achieved only modest sales, but the civil TPE331 has been most successful and developed in a series of versions. By 1976 all production models had achieved a TBO of at least 3,000 hours. Deliveries in January 1976 were at the rate of more than 40 engines per month, and total deliveries then exceeded 4,500. Flight time exceeded 8,500,000 hours in more than 1,450 aircraft of 44 types owned by 960 operators.

The following are major versions:

TPE331 Series I, II. Initial production version, FAA certificated in February 1965. Rated at 451 ekW; 429 kW plus 0·33 kN (605 ehp; 575 shp plus 75 lb st). Redesignated **TPE331-25/61** and **-25/71** and produced until 1970. Powers Mitsubishi MU-2 (A to E models), Fairchild Industries/Pilatus Porter, Carstedt Jet Liner, Volpar Super Turbo 18, Aerospace FU-24, Rockwell International Hawk Commander and 680 and DHC-2 Turbo Beaver.

TPE331-1 series. Certificated December 1967 at 526 ekW; 496 kW plus 0·44 kN (705 ehp; 665 shp plus 100 lb st). Powers Mitsubishi MU-2 (F and G), Pilatus Turbo-Porter and Fairchild Industries AU-23A Peacemaker, Texas Airplane CJ600, Volpar Turboliner, Conroy Stolifter, Interceptor 400, Rockwell International Turbo Commander and (customer option) Thrush Commander, Swearingen Merlin IIB and Aerospace Fletcher 1284.

TPE331-2 series. The -201 was certificated in December 1967 at 563 ekW; 533 kW plus 0·45 kN (755 ehp; 715 shp plus 102 lb st). A variant, the -251, was certificated in March 1970; this incorporates the 626 kW (840 shp) gas generator of the -3 series but retains the 715 shp gearbox and is flat rated at 533 kW (715 shp) to 2,135 m (7,000 ft). Powers Shorts Skyvan 3 (-201), Beech King Air B100 (-251), CASA 212 Aviocar (-251) and Rockwell International Turbo Commander 690A (-250).

TPE331-3 series. Certificated in March 1970 at 674 ekW; 626 kW plus 0·71 kN (904 ehp; 840 shp plus 159 lb st). Uprated gas generator with increased airflow and pressure ratio, but same turbine temperature as in original TPE 331. Selected first for Handley Page C-10A (terminated). Powers Swearingen Merlin III, IV and Metro, Air-Metal AM-C 111 and FMA IA 58 (prototype).

TPE331-8. Matches compressor and gearbox of -251 with new turbine section. Thermodynamic power of 676 ekW; 645 kW (865 shp plus 47·7 kg, 105 lb thrust), but flat rated at 533 kW (715 shp) to 36°C. Certification was due in September 1976. Powers Cessna 441.

TPE331-9, -10, -11. New models under development with ratings 533 kW (715 shp) to 746 kW (1,000 shp).

T76. Military engine, with gas generator similar to TPE331-1 series but with front end inverted, to give inlet above instead of below spinner. Two versions, originally designated T76-G-10 and G-12 and restyled G-410 and G-411, respectively giving clockwise and anticlockwise propeller rotation (seen from rear). Development near completion of revised T76-G-420/421 rated at 776 kW (1,040 shp), with certification in September 1976. All

Garrett-AiResearch ETJ131 low-cost turbojet for short-life applications (0·445 kN, 100 lb st class). The air inlet is on the left, the combustor at upper right and the jet pipe behind at lower right

Garrett AiResearch TPE331 series commercial turboprop engine

models power Rockwell International OV-10 Bronco.

The TPE331 and T76 are of similar frame size, and the following data apply generally to both models:

TYPE: Single-shaft turboprop engine with integral gearbox.

PROPELLER DRIVE: Two-stage reduction gear, one helical spur and one planetary, with overall ratio of 20·865 : 1 or 26·3 : 1. Shaft, driven from single-spool compressor, is carried in ball and roller bearings. Rotation clockwise or anticlockwise, as required.

AIR INTAKE: Single scoop intake duct at top (T76) or bottom of engine, at front. Provision for bleed air de-icing.

COMPRESSOR: Tandem two-stage centrifugal type. Each impeller is single-sided, and is made from titanium. Impellers attached to shaft by curvic couplings. First-stage casing of magnesium, with aluminium diffuser. Second-stage casing and diffuser of stainless steel. Mass flow, 2·61 kg (5·78 lb)/sec for 25/61, 25/71, 2·81 kg (6·2 lb)/sec for -1, 2·80 kg (6·17 lb)/sec for -2 and T76, 3·52 kg (7·75 lb)/sec for -251 and 3·54 kg (7·8 lb)/sec for -3. Pressure ratio 8·0 for 25/61, 25/71, 8·34 for -1, 8·54 for -2 and T76, 10·37 for -251 and -3.

COMBUSTION CHAMBER: Annular type of high-temperature alloy. High-energy capacitor discharge ignition. Igniter plug on turbine plenum.

FUEL SYSTEM: Woodward or Bendix control system for use with Beta propeller governing control system. Five radial primary nozzles in continuous operation. Ten axial simplex nozzles. Max fuel pressure 41·4 bars (600 lb/sq in).

FUEL GRADE: (TPE331): Aviation turbine fuels ASTM designation D1655-64T types Jet A, Jet B and Jet A-1; MIL-F-5616-1, Grade JP-1.

FUEL GRADE: (T76): MIL-L-5624F(2), Grades JP-4 and JP-5; MIL-G-5572, Grade 115/145.

NOZZLE GUIDE VANES: Axial vanes made from Inco 713C castings.

TURBINE: Three-stage axial-flow type. Discs of first two stages of Inco 100, third stage of Inco 713C, attached to shaft by curvic couplings. Blades cast integrally with disc. Turbine inlet gas temperature, 987°C for 25/61, 25/71, 993°C for T76, 1,005°C for all other models.

BEARINGS: One ball bearing at compressor end of shaft, one roller bearing at turbine end.

JET PIPE: Fixed type. Cone and jet pipe both of stainless steel.

ACCESSORIES: AND 20005 Type XV-B tachometer generator, AND 20002 Type XII-D starter/generator, AND 20010 Type XX-A propeller governor and AND 20001 Type XI-B hydraulic pump, all mounted on aft face of accessories case.

LUBRICATION SYSTEM: Medium-pressure dry sump system. Gerotor internal gear-type pressure and scavenge pumps. Normal oil supply pressure 6·90 bars (100 lb/sq in). Provision for automatic fuel filter anti-icing.

OIL SPECIFICATION: MIL-L-23699-(1) or MIL-L-7808.

MOUNTING: Five-point suspension. Three pads on aft face of accessory case, two pads at aft end of turbine plenum.

STARTING: Pad for 399A starter/generator on aft face of accessory case.

DIMENSIONS (approx):

Length overall:

TPE331	1,092 to 1,168 mm (43-46 in)
T76	1,118 mm (44 in)

Width:

TPE331	533 mm (21 in)
T76	483 mm (19 in)

Height:

TPE331	660 mm (26 in)
T76	686 mm (27 in)

WEIGHTS, DRY:

TPE331-25/61, 71	152 kg (335 lb)
TPE331-1, -2	152·5 kg (336 lb)
T76	155 kg (341 lb)
TPE331-251	163 kg (360 lb)
TPE331-3	161 kg (355 lb)

PERFORMANCE RATINGS:

T-O: see under model listings

Military (30 min):

T76-G-410/411: 533 kW; 563 ekW (715 shp; 755 ehp)

Normal:

T76-G-410/411: 485 kW; 514·5 ekW (650 shp; 690 ehp)

Max cruise (ISA, 3,050 m; 10,000 ft and 250 kt; 463 km/h; 288 mph):

TPE331-25/61, 71	332 kW (445 shp)
TPE331-1	404 kW (542 shp)
TPE331-2, T76	430 kW (577 shp)
TPE331-251, -3	529·5 kW (710 shp)

SPECIFIC FUEL CONSUMPTION:

At T-O rating:

TPE331-25/61, 71	111·5 μg/J (0·66 lb/hr/shp)
TPE331-1	107·0 μg/J (0·633 lb/hr/shp)
TPE331-2	99·4 μg/J (0·588 lb/hr/shp)
TPE331-251	105·8 μg/J (0·626 lb/hr/shp)
TPE331-3	99·7 μg/J (0·59 lb/hr/shp)
T76-G-410/411	101·4 μg/J (0·60 lb/hr/shp)

OIL CONSUMPTION:

Max: 0·009 kg (0·02 lb)/hr

Garrett-AiResearch TPE331 series commercial turboprop engine

GARRETT-AIRESEARCH TSE331-3U

This turboshaft, derived from the TPE331-3 turboprop, was certificated in April 1970. It powers the Sikorsky S-55T helicopter conversion certificated by Aviation Specialties Inc of Mesa, Arizona, flat rated at 522 kW (700 shp). In early 1976 over 40 conversions had been completed.

TYPE: Single-shaft turboshaft with front end drive.

POWER DRIVE: Two-stage reduction gear, one helical spur and one planetary, with overall ratio of 16·410 : 1. Output shaft rotation is clockwise, looking forward from rear of engine, and has a bolted flange attachment.

AIR INTAKE: Single scoop intake duct at top front of engine. Provision for bleed air anti-icing.

COMPRESSOR: Two-stage centrifugal. Tandem, single-sided titanium impellers are attached to the shaft by curvic couplings. First-stage casing of magnesium with aluminium diffusers. Second-stage casing and diffusers of stainless steel. Pressure ratio 10·32 : 1.

COMBUSTION CHAMBER: Annular type of high-temperature alloy. High-energy capacitor discharge ignition. Igniter plug on turbine plenum.

FUEL SYSTEM: Woodward fuel control for automatic speed control and fuel metering to match engine power to rotor load. Fuel filter, fuel shut-off valve, fuel-flow divider and manifold drain valve, fuel manifold and nozzle assemblies (five primary, ten secondary), start-fuel system and fuel anti-ice system.

FUEL GRADE: Aviation turbine fuels ASTM designation D1655-68T, Types Jet A, A-1, and B, MIL-T-5624G-1, Grades JP-4 and JP-5, MIL-F-5161-1, Grade JP-1.

NOZZLE GUIDE VANES: Axial vanes made from Inco 713C castings.

TURBINE: Three-stage axial-flow type. Discs of first two stages of Inco 100, third stage of Inco 713C, attached to shaft by curvic couplings. Blades cast integrally with disc.

BEARINGS: One ball bearing at compressor end of shaft, one roller bearing at turbine end.

EXHAUST DUCT: Fixed type. Cone and jet pipe both of stainless steel.

ACCESSORIES: AND 20005 Type XV-B tachometer generator, AND 20002 Type XII-D starter/generator, AND 20001 Type XI-B hydraulic pump, all mounted on aft face of accessories case.

LUBRICATION SYSTEM: Medium-pressure dry sump system. Gerotor internal gear type pressure and scavenge pumps. Normal oil supply pressure 6·90 bars (100 lb/sq in). Provision for automatic fuel filter anti-icing.

OIL SPECIFICATION: MIL-L-23699A or MIL-L-7808D.

MOUNTING: Five-point suspension. Three pads on aft face of accessory case, two pads at aft end of turbine plenum.

STARTING: Pad for 300A starter/generator on aft face of accessory case.

DIMENSIONS (approx):

Length overall	1,118 mm (44 in)
Width	533 mm (21 in)
Height	686 mm (27 in)

WEIGHT, DRY: 161 kg (355 lb)

PERFORMANCE RATINGS:

T-O	596·5 kW (800 shp)
Max continuous	522 kW (700 shp)

SPECIFIC FUEL CONSUMPTION:

T-O: 99·7 μg/J (0·59 lb/hr/shp)

OIL CONSUMPTION:

Max: 0·009 kg (0·02 lb)/hr

GENERAL ELECTRIC

GENERAL ELECTRIC COMPANY AIRCRAFT ENGINE GROUP

HEADQUARTERS:
1000 Western Avenue, West Lynn, Massachusetts 01905

Telephone: (617) 594-0100

GROUP LOCATIONS:
Lynn and Everett, Massachusetts; Cincinnati, Ohio; Rutland and Ludlow, Vermont; Hooksett, New Hampshire; and Albuquerque, New Mexico. Also test facilities at Edwards Air Force Base, California, and Peebles, Ohio. Further facilities at Seattle, Washington; Arkansas City, Kansas; Ontario, California.

VICE-PRESIDENT AND GROUP EXECUTIVE:
Gerhard Neumann

Military Engine Division:
VICE-PRESIDENT AND GENERAL MANAGER:
James E. Worsham

Airline Programs Division:
VICE-PRESIDENT AND GENERAL MANAGER:
Brian H. Rowe

Commercial Engine Projects Division:
VICE-PRESIDENT AND GENERAL MANAGER:
Robert H. Goldsmith

Group Engineering Division:
VICE-PRESIDENT AND GENERAL MANAGER:
Edward Woll

Group Manufacturing Division:
VICE-PRESIDENT AND GENERAL MANAGER:
Raymond E. Letts

Marine and Industrial Project Department:
GENERAL MANAGER: O. R. Bonner

CFM56 Program Department:
GENERAL MANAGER: Jack I. Hope

Group Strategic Planning Operation:
VICE-PRESIDENT: Fred O. MacFee Jr

Group Product Quality Operation:
GENERAL MANAGER: Paul C. Setze

Group Finance and Management Support Operation:
MANAGER: Robert D. Desrochers

Group Legal Operation:
GROUP COUNSEL: James W. Sack

The General Electric Company entered the gas-turbine field in about 1895. Years of pioneering effort by the late Dr Sanford A. Moss produced the aircraft turbosupercharger, successfully tested at height in 1918 and mass-produced in the second World War for US fighters and bombers.

The company built its first aircraft gas turbine in 1941, when it began development of Whittle-type turbojets, under an arrangement between the British and American governments.

Current products of the Aircraft Engine Group include the F101, F103, F404, J79, J85, T58, T64, T700 and TF34 for military use, and the CF6, CF34, CF700, CJ610, CT58 and CT64 for the commercial and general aviation market. In partnership with SNECMA of France a company has been formed to develop and market the CFM56 turbofan, as described in the International part of this section under CFM International.

In January 1968, as part of a series of major changes to the corporate structure, General Electric's Flight Propulsion Division (one of four divisions forming the company's previous Aerospace and Defence Group), which hitherto had been responsible for all GE's aero-engine work, was promoted to become one of the nine operating groups now comprising General Electric. This change in organisation was aimed at strengthening GE's civil and military aircraft engine activities in the domestic US and international markets.

GENERAL ELECTRIC J79

Development of the J79, America's first high-compression variable-stator turbojet, began in 1952. It was flight tested for the first time in 1955 and became the first production Mach 2 engine when it was selected to power the General Dynamics B-58 Hustler bomber. In addition to production by General Electric, versions of the J79 have been or are being manufactured by Orenda of Canada to power the Canadair CF-104/F-104G (MAP), by Ishikawajima-Harima in Japan for the licence-built F-104DJ, and by MTU of Germany, Fiat of Italy and FN of Belgium for the European-built F-104G. The Italian production team, including Alfa-Romeo, is now producing the J79-GE-19, an improved engine similar to the

J79-GE-17 but configured for the F-104S Starfighter.

Overall, the International Technical Assistance Programme has been responsible for assembly of more than 2,500 J79 turbojets for the F-104 and F-4. A total of more than 16,000 J79s had been built by GE and licensees by January 1976.

Derivatives of the J79 have been the CJ805-3 turbojet and CJ805-23 turbofan, powering the Convair 880 and 990 Coronado, respectively, as well as the LM1500 industrial and marine gas turbine.

Versions of the J79 in service are as follows:

J79-GE-7A. Powers the Lockheed F-104C and D Starfighters. Built under licence by Orenda (as J79-OEL-7) for Canadair CF-104.

J79-GE-8. For production versions of McDonnell Douglas F-4B and RF-4B Phantom II and North American (Rockwell) RA-5C Vigilante. Air mass flow 76·5 kg (169 lb)/sec. Pressure ratio 12·9 : 1.

J79-GE-10. Advanced version powering RA-5C and F-4J. Entered production in June 1966, superseding the J79-GE-8. Pressure ratio 13·5 : 1.

J79-GE-11A. For US-built Lockheed F-104G Starfighters. Built under licence in Japan (as J79-IHI-11A), Germany, Italy, Belgium and Canada.

J79-GE-15. Powers F-4C, F-4D and RF-4C for USAF. Similar to J79-GE-8 except for self-contained starting.

J79-GE-17. Similar to J79-GE-10, but for F-4E, F-4F and F-4EJ.

J79-GE-19. Advanced version designed to supersede J79-GE-11A in F-104. Used in F-104S and F-104A. Differs from J79-GE-10/17 only in external characteristics. Guided expansion jet nozzle derived from nozzles of J79-GE-5C and YJ93. Afterburner system provides continuous thrust modulation. Fuel flow can be modulated from 1,225 kg (2,700 lb)/hr to 15,420 kg (34,000 lb)/hr.

The following details cover the basic features of all J79 variants except where otherwise indicated:

Type: Variable-stator single-shaft axial-flow turbojet.

Air Intake: Annular type, surrounding central bullet fairing. Struts and inlet guide vanes anti-iced with compressor discharge air. First-stage stator anti-icing on J79-GE-8, -10 and -15.

Compressor: Seventeen-stage axial-flow. First six stator stages and the inlet guide vanes have variable incidence. Setting of variable-incidence vanes adjusted by dual actuators moved by engine fuel to achieve optimum airflow angles for each stage at all engine speeds. Rotor, which runs on two bearings, is made from Lapelloy, B5F5 and titanium. All engines have type 403 stainless steel blades and vanes except J79-GE-7A which has A286 stator vanes at stages 7 to 17 inclusive. Total of 1,260 stator vanes and 1,271 rotor blades. Variable stator vanes have a platform, trunnion and threaded stem arrangement for external attachment to the actuation system linkage. Fixed stator vanes are inserted into T-slots on rear casing. All rotor blades have dovetail roots. Front compressor stator casing is made from a magnesium-thorium casting or Chromolloy forging, depending on engine model. On those engines requiring an intermediate compressor casing this is made of either A286 or 321 SS. All models have a forged and machined rear compressor stator casing, constructed in two halves for ease of assembly and dis-assembly.

Combustion Chamber: Cannular type consisting of 10 combustion cans. Outer casing of Chromolloy, flame tube of Hastelloy. J79-GE-7A, -11A, -15, -17 and -19 have dual igniters in cans 4 and 5. J79-GE-2, -8 and-10 have single igniter in can 4.

Fuel System: Hydromechanical range-governing control system composed of two separate and distinct systems, the main fuel system and afterburner fuel system. Main system is controlled by main fuel control, which is a flow-controlling unit. The afterburner system is controlled by an independent control, also of the flow-controlling type. Automatic acceleration control with exhaust temperature limiting. Gear-type main fuel pump. Engine-driven centrifugal afterburner fuel pump.

Fuel Grade: JP-4 or JP-5.

Nozzle Guide Vanes: Three-stage; first with 58 vanes of R41, second with 62 vanes of Hastelloy R235 and R41, third with 44 vanes of A286.

Turbine: Three-stage axial-flow type. Stages 1 and 2 bolted to shaft, stage 3 integral with aft shaft. J79-GE-8 and -15 have first- and second-stage wheels of V57 and third-stage wheel of A286. All three stages of J79-GE-10, -17 and -19 have intermediate aged V57. Other models have all stages of A286. J79-10, -17 and -19 first stage has 148 blades of Udimet 700 or René 80, second stage has 114 blades of Udimet 500, third stage 84 blades of M252. All blades attached by fir-tree roots. Lightweight casing of fabricated A286 in two easily-removable halves.

Bearings: Three only. Roller in front frame, ball (main thrust) in compressor frame, roller in turbine frame.

Jet Pipe: Liner of N155 and L605 with ceramic coating. Jet pipe of A286.

Afterburner: Short type (max 1,985°C) with fully-variable nozzle of "petal" type. Actuation by hydraulic rams utilising engine lubricating oil. Three-ring, quadrant-burning on all models except J79-GE-8, -10, 15, -17 and -19, which have core annulus burning with radial spraybars.

General Electric J79-GE-17 turbojet (79·6 kN; 17,900 lb st with afterburning)

Accessory Drives: All engine controls and accessories, aircraft hydraulic pumps, generators, alternators and constant-speed drives (as required) are driven by two gearboxes on bottom of engine and a nose inlet gearbox.

Lubrication: Dry-sump system. Vane-type pumps. Sump pressure provided from compressor. Oil cooling from fuel. Sump capacity ranges from 15 to 19 litres (4-5 US gallons). Average normal oil supply pressure 3·45 bars (50 lb/sq in).

Oil Specification: MIL-L-7808, MIL-L-23699.

Mounting: Pads provided on front frame and turbine frame for a variety of mounting arrangements, depending on airframe requirements.

Starting: J79-7A, -11A and -19 have pneumatic turbine starter mounted on front frame of inlet gearbox. J79-GE-8 and -10 have turbine air impingement starter. J79-GE-15 and -17 have combination cartridge/pneumatic starter on transfer gearbox.

Dimensions:

Length overall:

J79-GE-7A, 11A	5,283 mm (207·96 in)
J79-GE-8	5,295 mm (208·45 in)
J79-GE-10, 17, 19	5,301 mm (208·69 in)

Diameter at compressor:

J79-GE 7A, 8, 11A, 15	973 mm (38·3 in)
J79-GE-10, 17, 19	992 mm (39·06 in)

Weight, Dry:

J79-GE-7A	1,622 kg (3,575 lb)
J79-GE-8	1,666 kg (3,672 lb)
J79-GE-10	1,749 kg (3,855 lb)
J79-GE-11A	1,615 kg (3,560 lb)
J79-GE-15	1,672 kg (3,685 lb)
J79-GE-17, 19	1,740 kg (3,835 lb)

Performance Ratings:

T-O, with afterburning:

J79-GE-7A, 11A	70·3 kN (15,800 lb st)
J79-GE-8, 15	75·6 kN (17,000 lb st)
J79-GE-10, 17, 19	79·6 kN (17,900 lb st)

Military:

J79-GE-7A, 11A	44·5 kN (10,000 lb st)
J79-GE-8, 15	48·5 kN (10,900 lb st)
J79-GE-10, 17, 19	52·8 kN (11,870 lb st)

Cruise:

J79-GE-7A, 11A	11·8 kN (2,650 lb st)
J79-GE-8, 10, 15, 17, 19	11·6 kN (2,600 lb st)

Specific Fuel Consumption:

At T-O rating:

J79-GE-7A, 11A	55·8 mg/Ns (1·97 lb/hr/lb st)
J79-GE-8	54·67 mg/Ns (1·93 lb/hr/lb st)
J79-GE-15	55·1 mg/Ns (1·945 lb/hr/lb st)
J79-GE-10, 17, 19	55·66 mg/Ns (1·965 lb/hr/lb st)

At military rating:

J79-GE-8, 15	24·36 mg/Ns (0·86 lb/hr/lb st)
J79-GE-7A, 10, 11A, 17, 19	23·79 mg/Ns (0·84 lb/hr/lb st)

At cruise rating:

J79-GE-7A, 8, 11A, 15	29·74 mg/Ns (1·05 lb/hr/lb st)
J79-GE-10, 17, 19	26·91 mg/Ns (0·95 lb/hr/lb st)

GENERAL ELECTRIC J85

The following are major versions of the J85 small military turbojet, the -21 being the main production version. By 1976 approximately 11,000 J85 engines had been delivered to air forces in 26 nations.

J85-4A. Powers the Rockwell International T-2C Buckeye trainer.

J85-5. Afterburning version with 6·6 : 1 thrust-to-weight ratio; powers Northrop T-38 Talon supersonic trainer.

J85-13. Developed from J85-5, with increased turbine inlet temperature for Northrop F-5A/B supersonic fighter. As the J85-13A, licence-built by Alfa-Romeo, also powers the Aeritalia G91Y.

J85-15. Version of J85-13 with improved turbine and hydraulically actuated exhaust nozzle to power CF-5 and NF-5. Manufactured under licence in Canada by Orenda.

J85-17A/B. Powers Saab 105G attack/reconnaissance aircraft and Cessna A-37B attack aircraft. Also used as take-off and climb booster for Fairchild C-123K and AC-119K.

J85-21. Higher airflow version with zero stage to give total of nine compressor stages. Equipped with afterburner for supersonic aircraft. Powers Northrop F-5E/F Tiger II.

General Electric J85-21 turbojet (22·2 kN; 5,000 lb st with afterburning)

General Electric CJ610-4 (12·68 kN; 2,850 lb st) and *(right)* CJ610-5 (13·1 kN; 2,950 lb st) turbojet engines

J85/J1. Non-afterburning derivative with nine-stage compressor.

Civil version of the J85 is the CJ610 turbojet, to which the aft-fan CF700 turbofan is closely related. Both are described separately.

The following data refer specifically to the J85-5 and 13, except where otherwise stated:

TYPE: Single-shaft turbojet.

AIR INTAKE: Annular type, surrounding central bullet fairing. Variable-incidence inlet guide vanes, with hot-air anti-icing.

COMPRESSOR: Eight-stage axial-flow type, with variable inlet guid vanes and automatically controlled bleed valves (-21 and /J1, nine stages, no bleed valves but first three stator stages variable-incidence). Titanium rotor blades, first stage (first two in -21 and /J1) having part-span shrouds. Discs joined at periphery. Casing in upper and lower halves. Pressure ratio approximately 7 : 1 (8·3 : 1 in -21 and /J1). Air mass flow 20 kg/sec (44 lb/sec) (24·0 kg/sec; 53·0 lb in -21 and /J1).

COMBUSTION CHAMBER: Annular type with perforated liner. Twelve duplex fuel injectors. Ports in outer casing facilitate inspection of liner.

TURBINE: Two-stage axial-flow type. Casing is in halves, split horizontally. Turbine inlet temperature (-21) 977°C.

AFTERBURNER (J85-5, -13, -15, -21): Consists of a diffuser and a combustor. A pilot burner with four spraybars and a main burner of 12 spraybars are located in the diffuser section. Combustion is initiated by a single igniter plug and is then self-sustained. Nozzle position governs exit area and is regulated automatically by the afterburner control system as a function of turbine exit temperature and throttle lever position.

LUBRICATION: Positive displacement, pressurised recirculating type.

STARTING: Air impingement starter on afterburning engines. Provision for starter/generator on non-afterburning engines.

DIMENSIONS:
Length overall:
J85-4 1,029 mm (40·50 in)
J85-5 with afterburner 2,657 mm (104·6 in)
J85-13, -15 with afterburner 2,682 mm (105·6 in)
J85-17 1,039 mm (40·5 in)
J85-21 with afterburner 2,858 mm (112·5 in)
J85/J1 1,186 mm (46·7 in)
Max diameter:
J85-4 450 mm (17·7 in)
J85-5, -13, -21 533 mm (21·0 in)
J85-17 450 mm (17·7 in)
J85/J1 508 mm (20·0 in)

WEIGHT, DRY:
J85-4 188 kg (415 lb)
J85-5 265 kg (584 lb)
J85-13 271 kg (597 lb)
J85-15 279 kg (615 lb)
J85-17 181 kg (398 lb)
J85-21 310 kg (684 lb)
J85/J1 208 kg (458 lb)

PERFORMANCE RATINGS:
Max rating, with afterburner:
J85-5 17·1 kN (3,850 lb st)
J85-13 18·1 kN (4,080 lb st)
J85-15 19·1 kN (4,300 lb st)
J85-21 22·2 kN (5,000 lb st)
Military rating, without afterburner:
J85-4 13·1 kN (2,950 lb st)
J85-5 11·9 kN (2,680 lb st)
J85-13 12·1 kN (2,720 lb st)
J85-17 12·7 kN (2,850 lb st)
J85-21 15·6 kN (3,500 lb st)
J85/J1 16·3 kN (3,670 lb st)

SPECIFIC FUEL CONSUMPTION:
At max rating, with afterburner:
J85-5 62·3 mg/Ns (2·20 lb/hr/lb st)
J85-13 62·9 mg/Ns (2·22 lb/hr/lb st)
J85-15 61·75 mg/Ns (2·18 lb/hr/lb st)
J85-21 60·3 mg/Ns (2·13 lb/hr/lb st)
At military rating, without afterburner:
J85-4 27·9 mg/Ns (0·98 lb/hr/lb st)
J85-5, -13, -15 29·2 mg/Ns (1·03 lb/hr/lb st)
J85-17 28·1 mg/Ns (0·99 lb/hr/lb st)
J85-21 28·3 mg/Ns (1·00 lb/hr/lb st)
J85/J1 27·8 mg/Ns (0·98 lb/hr/lb st)

General Electric CF700-2D turbofan (18·9 kN; 4,250 lb st)

GENERAL ELECTRIC CJ610

Announced in May 1960, the CJ610 is a power plant tailored for commercial and executive aircraft of 5,700-7,500 kg (12,500-16,500 lb) gross weight. It is essentially similar to the basic J85 turbojet, without afterburner, and incorporates an eight-stage axial-flow compressor, annular combustion chamber, two-stage reaction turbine, fixed-area concentric exhaust section and integrated control system. Air mass flow is 20 kg (44 lb)/sec.

By December 1975, a total of 1,600 CJ610s had accumulated more than 3,300,000 flying hours in more than 660 aircraft. TBO reached 3,000 hr in 1974.

There are six versions:

CJ610-1, CJ610-4. Initial production versions, differing only in accessory gearbox location.

CJ610-5, CJ610-6. Developed versions of -1 and -4 respectively, providing increased T-O thrust. Power Gates Learjet 24D, 25B and 25C, Hansa and IAI Westwind 1121.

CJ610-8, CJ610-9. Developed for production deliveries beginning in 1969. Power Hansa, IAI Westwind 1123 and NAL (Japan) experimental VTOL.

DIMENSIONS:
Length overall:
CJ610-1, -5, -9 1,298 mm (51·1 in)
CJ610-4, -6, -8 1,153 mm (45·4 in)
Max flange diameter 449 mm (17·7 in)

WEIGHT, DRY:
CJ610-1 181 kg (399 lb)
CJ610-4 176 kg (389 lb)
CJ610-5 183 kg (402 lb)
CJ610-6 180 kg (396 lb)
CJ610-8 185 kg (407 lb)
CJ610-9 191 kg (421 lb)

PERFORMANCE RATINGS (guaranteed):
T-O:
CJ610-1, -4 12·7 kN (2,850 lb st)
CJ610-5, -6 13·1 kN (2,950 lb st)
CJ610-8, -9 13·8 kN (3,100 lb st)
Max continuous:
CJ610-1, -4 12 kN (2,700 lb st)
CJ610-5, -6 12·4 kN (2,780 lb st)
CJ610-8, -9 13 kN (2,925 lb st)

SPECIFIC FUEL CONSUMPTION:
At T-O rating:
CJ610-1, -4 28·05 mg/Ns (0·99 lb/hr/lb st
CJ610-5, -6, -8, -9 27·75 mg/Ns (0·98 lb/hr/lb st
At max continuous rating:
CJ610-1, -4 27·5 mg/Ns (0·97 lb/hr/lb st
CJ610-5, -6 27·2 mg/Ns (0·96 lb/hr/lb st
CJ610-8, -9 27·2 mg/Ns (0·96 lb/hr/lb st

GENERAL ELECTRIC CF700

Like the CJ610 turbojet, the CF700 is also derived fro the J85 engine. Utilizing the same gas generator, it is a aft-fan turbofan suitable for military and commercial ai craft. Since it can be tilted while in steady-state operatio and operate vertically, it affords lift/cruise capability i VTOL aircraft.

FAA certification of the original version was receive on 1 July 1964. The uprated CF700-2D was certificated i early 1968. The CF700-2D has an improved compress turbine with higher thermodynamic efficiency. Th CF700-2D2 incorporates a new design of tailpipe. In Ma 1975 the FAA approved an uprating of the CF700-2D take-off thrust from 4,315 to 4,500 lb.

CF700 engines power the Dassault-Breguet Falcon 2 and Rockwell Sabre 75A executive transports. By 197 the TBO had reached 3,000 hr. By September 1975 ove 900 CF700s had flown nearly 2,000,000 hr in more tha 330 aircraft.

The general description of the J85 turbojet applies als to the CF700, with the following additional assembly:

AFT FAN: Single-stage free-floating fan. By-pass ratio 1·6 1. Mass air flow through fan 39·9 kg (88·0 lb)/sec.

DIMENSIONS:
Overall length, compressor nose to tailcone tip 1,912 mm (75·57 ir
Length, flange to flange 1,361 mm (53·6 ir
Max diameter 840·4 mm (33·1 ir
Max diameter less fan 447 mm (17·6 ir

WEIGHT, DRY:
CF700-2C 330 kg (725 lt
CF700-2D, -2D2 334 kg (737 lt

PERFORMANCE RATINGS:
Max T-O (flat-rated to 30°C):
CF700-2C 18·35 kN (4,125 lb st
CF700-2D, -2D2 18·90 kN (4,250 lb st
Max continuous:
CF700-2C 17·8 kN (4,000 lb s
CF700-2D, 2D2 18·3 kN (4,120 lb s

SPECIFIC FUEL CONSUMPTION:
Max T-O:
CF700-2C, -2D, -2D2 18·4 mg/Ns (0·65 lb/hr/lb s

Max continuous:

CF700-2C, -2D	18·4 mg/Ns (0·65 lb/hr/lb st)
CF700-2D2	18·1 mg/Ns (0·64 lb/hr/lb st)

GENERAL ELECTRIC J97

Directly derived from the GE1 (the basic core engine used for several current engines, as described in the 1973-74 *Jane's*), the J97 is an advanced, lightweight turbojet. During a five-year development programme it has run for more than 2,000 engine hours, including altitude performance demonstration, 60-hour endurance, −51°C starting, water-ingestion tolerance and other MIL-5007C tests. In 1972 the J97-100 version successfully completed the USAF 60-hour preliminary flight rating test. The initial production J97-GE-3 powered the Teledyne Ryan 154 RPV, and the -GE-100 version powers the Boeing YQM-94A Compass Cope RPV. The J97 is a candidate engine for the McDonnell Douglas 260 lift/cruise demonstrator and for other projected aircraft.

The advanced technology demonstrated is adaptable to many applications. An early use will be in the NASA V/STOL research vehicle, with a lift-fan system offering high thrust/weight ratio for optimum lift/cruise fan power. In afterburning and unaugmented forms it is suitable for advanced trainers, light tactical fighters and drones and remotely piloted vehicles. Other versions have been planned as turboshafts for large helicopters and as turboprops for tilt-wing transports. Components related to the J97-100 have been utilised on the TF39 and F101 engines.

TYPE: Single-shaft turbojet or gas-generator.

COMPRESSOR: 14-stage axial with six variable stator rows. Max mass flow 31·75 kg (70 lb)/sec. Pressure ratio 13·8.

COMBUSTION CHAMBER: Annular, straight-through type with 16 vaporising burners.

FUEL GRADE: JP-4, JP-5.

TURBINE: Axial, two stages, with aircooled nozzle guide vanes. Turbine entry gas temperature, over 1,095°C.

JET PIPE: Simple fixed-area type on J97-100; other versions incorporate deflector valves, free turbine or an advanced film-cooled afterburner and variable nozzle.

DIMENSIONS:

Diameter	620 mm (24·4 in)
Length	2,781 mm (109·5 in)

WEIGHT, DRY: 314 kg (694 lb)

PERFORMANCE RATINGS:

Unaugmented	23·4 kN (5,270 lb st)
Augmented	31·1-44·5 kN (7,000-10,000 lb st)

SPECIFIC FUEL CONSUMPTION:

Unaugmented	25·9 mg/Ns (0·915 lb/hr/lb st)

GENERAL ELECTRIC F404

The F404 is an advanced technology augmented turbofan described as "in the 16,000 lb thrust class". It is the US Navy derivative of the successful YJ101 engine flown in the USAF YF-17 aircraft (see *Jane's* 1975-76). The changed designation from J to F (turbofan) reflects a higher by-pass ratio, and the number in the 400-series indicates funding by the US Navy.

In contrast, the YJ101 was funded by the US Air Force, in April 1972, to power the twin-engined Northrop YF-17 Air Combat Fighter. Seven engines in two YF-17 prototypes logged 719 hr in 302 flights. These test flights explored a large part of the flight envelope, and a maximum of Mach 2·05 was reached at 12,500 m (41,000 ft). All engine commitments were achieved during the seven-month YF-17 flight programme.

In May 1975 the US Navy selected the McDonnell Douglas/Northrop team to develop its Navy Air Combat Fighter (NACF), designated F-18. The F-18 will be a derivative of the YF-17, powered by two F404 engines.

Compared with that of the J101, the F404 fan diameter is increased less than 25·4 mm (1·0 in) while increasing the by-pass ratio from 0·20 to 0·34. The fan is driven by a slightly larger LP turbine. The technology, and the core, comprising the HP compressor, combustion chamber and HP turbine, remain the same.

First F404 engine test is scheduled for early 1977, and engine qualification for early 1979. The F-18 is scheduled to make its first flight in mid-1978, and to become operational with the US fleet in 1982.

TYPE: Two-shaft augmented low-ratio turbofan (turbojet with continuous by-pass bleed).

AIR INTAKE: Plain annular. Fixed central bullet, fixed and variable inlet vanes.

FAN: Three-stage axial. Outer flow diverted to by-pass duct. By-pass ratio 0·34.

HP COMPRESSOR: Seven-stage axial. Overall pressure ratio, 25 : 1 class.

COMBUSTION CHAMBER: Single-piece annular.

HP TURBINE: Single-stage axial. Highly loaded air-cooled blades.

LP TURBINE: Single-stage axial.

EXHAUST SYSTEM: Close-coupled high-augmentation afterburner with combustion in both core and by-pass flows. Convergent-divergent exhaust nozzle with hydraulic actuation.

CONTROL SYSTEM: Electrical-hydromechanical.

DIMENSIONS:

Length overall	4,030 mm (158·8 in)
Max diameter	880 mm (34·8 in)

WEIGHT, DRY: approx 908 kg (2,000 lb)

PERFORMANCE RATING:

Max T-O	71·2 kN (16,000 lb st) class

General Electric J97-GE-100 turbojet of 23·4 kN; 5,270 lb st without afterburning

General Electric TF34-GE-400A turbofan of 41·3 kN; 9,275 lb st

GENERAL ELECTRIC TF34

It was announced in April 1968 that the US Naval Air Systems Command had awarded General Electric a contract for development of the TF34. This high by-pass ratio turbofan had won a 1965 US Navy competition aimed at providing a tailor-made engine in the 40 kN (9,000 lb) st category for the VS(X) aircraft by 1972 within a budget of $96 million. In August 1972 the **TF34-GE-2,** the initial variant for this application (now called the Lockheed S-3A Viking), completed its Model Qualification Test (MQT) and subsequently entered production. GE, working under a Naval Air Systems Command fixed-price incentive-fee contract, completed the development within the price, time and major technical goals laid down in 1965.

The contract to develop the TF34 engine was awarded to General Electric in March 1968. The first engine ran in May 1969 and development GE-2 engines successfully completed the Preliminary Flight Rating Test (PFRT) on 28 February 1971, earning General Electric full incentive by completing this milestone two months ahead of schedule. This test established the flightworthiness of the engine, and subsequently 28 YTF34-GE-2 engines were delivered on time to Lockheed for S-3A flight testing. The TF34-GE-2 has top mountings and a short fan duct. It was qualified for production in August 1972. By January 1975, a total of 180 GE-2 engines had been shipped, and operating time exceeded 45,000 hours. The S-3A entered fleet service in February 1974, and GE and the US Navy have defined a 4,000 hour TBO extension programme.

In January 1975 GE began shipment of the **TF34-GE-400A,** which replaces the GE-2 as S-3A engine. The new model incorporates various improvements, with changed external piping, an adaptive control system for optimising accessory power extraction, and a simplified rocket gas ingestion system.

In 1970 the TF34 was selected to power the twin-engined Fairchild Republic A-10A attack aircraft to compete in the AX competition. The A-10A application led in July 1972 to an Air Force contract for development of the **TF34-GE-100.** This was re-engineered to minimise unit price. It has a long fan duct and side mountings. The GE-100 flew in the first A-10A in May 1972. The A-10A won the AX competition, and the TF34-GE-100 was formally qualified for production in October 1974.

In 1974 a third version of the TF34, most nearly resembling the GE-2, was selected to provide auxiliary (thrust) power for the Sikorsky S-72 RSRA (Rotor System Research Aircraft) under development for NASA and the US Army.

The TF34 engine has undergone a wide variety of development test including climatic, altitude, overspeed, overtemperature, inlet distortion and corrosion susceptibility. Other tests included noise, smoke, infra-red measurements and operation while ingesting water, steam, sand and rocket gases. The TF34 programme has also included over 400 hours of operation in a B-47 test aircraft and another 450 hours in the altitude and climatic test chambers at the Naval Air Propulsion Test Center. In a further test a GE-100 engine was subjected to low-cycle fatigue testing equivalent to 5,000 mission hours in an A-10A. Desired time between overhauls for the GE-2 and GE-100 is 3,500-4,000 hours.

TYPE: Two-shaft high by-pass ratio turbofan for subsonic aircraft.

AIR INTAKE: Plain annular intake. No fixed inlet struts or

guide vanes. Small spinner rotates with fan.

FAN: Single-stage fan has blades forged in titanium, without part-span shrouds. Blades replaceable with engine installed. Performance at max S/L rating, mass flow 153 kg (338 lb)/sec at 7,365 rpm with pressure ratio 1·5. By-pass ratio 6·2.

COMPRESSOR: 14-stage axial on HP shaft. Inlet guide vanes and first five stators variable. First nine rotor stages titanium, remainder high-nickel alloy. Blades and vanes through stage 5 individually replaceable. Split casing provides bleeds for engine cooling, seal pressurisation, anti-icing and airframe use. Performance at max S/L rating, core airflow 21·3 kg (47 lb)/sec at 17,900 rpm with pressure ratio 14 : 1, overall engine pressure ratio 21.

COMBUSTION CHAMBER: Annular chamber designed for highly efficient and complete combustion with near-zero smoke. Hastelloy chamberliner and front dome, providing ports for primer nozzles, igniters and 18 carburetting burners.

TURBINE: Two-stage HP gas generator turbine with convection-cooled rotor blades and stator vanes, the first-stage nozzle vanes having film and impingement cooling. Four-stage LP fan turbine with tip-shrouded blades; LP blades and stators replaceable on installed engine. Turbine entry gas temperature 1,225°C maximum.

FUEL SYSTEM: Contamination-resistant, carburetting type. Integrated hydromechanical control unit with electronic amplifier. Fuel grade JP-4 or JP-5.

ACCESSORY DRIVES: Engine and customer accessories mounted around horseshoe-shaped gearbox, fitting closely around lower half of compressor casing. Radial shaft drive from front of HP shaft. Fan airflow passes outside accessories through optimised duct.

LUBRICATION: Enclosed, pressurised, dual system with vent along centre shaft.

DIMENSIONS:
Max diameter:

TF34-GE-400A	1,326 mm (52·2 in)
TF34-GE-100	1,259 mm (48·6 in)
Bsasic length (both)	2,540 mm (100·0 in)

WEIGHT, DRY:

TF34-GE-400A	661 kg (1,458 lb)
TF34-GE-100	647 kg (1,427 lb)

PERFORMANC RATINGS:
Max T-O (S/L, static):

TF34-GE-400A	41·3 kN (9,275 lb st)
TF34-GE-100	40·3 kN (9,065 lb st)

SPECIFIC FUEL CONSUMPTION:
Max T-O, S/L static:

TF34-GE-400A	10·3 mg/Ns (0·363 lb/hr/lb st)
TF34-GE-100	10·5 mg/Ns (0·370 lb/hr/lb st)

Cruise, Mach 0·6, 9,150 m (30,000 ft):

TF34-GE-400A	17·2 mg/Ns (0·607 lb/hr/lb st)
TF34-GE-100	17·1 mg/Ns (0·604 lb/hr/lb st)

GENERAL ELECTRIC CF34

In April 1976 General Electric's General Aviation Engine Department, at Lynn, announced the CF34 as a new turbofan in the 31-36 kN (7,000-8,000 lb st) class for business and commercial aircraft. A natural derivative of the military TF34 at Cincinnati, the CF34 will be closely similar to the military engine to reap maximum advantage of its wide experience (in April 1976 in excess of 100,000 hours), but will be rated at lower thrust levels. FAA certification was in progress in early 1976.

GENERAL ELECTRIC TF39

General Electric produced the 182·8 kN (41,100 lb st) TF39 turbofan for the Lockheed C-5 Galaxy heavy logistics transport aircraft.

By the end of 1975 the TF39 had completed over 900,000 hours of running.

Full details can be found in the 1973-74 *Jane's*.

GENERAL ELECTRIC CF6

US military designation (CF6-50E): F103-GE-100

On 11 September 1967 General Electric announced the endorsement and commitment of corporate funding for development of the CF6 turbofan for the then-forthcoming generation of wide-body transports. From the initial family of 142 to 160 kN (32,000 lb to 36,000 lb st) CF6 two-shaft engines announced in September 1967 to cover the anticipated thrust requirements of the Lockheed and McDonnell Douglas airbus projects, the CF6 evolved through a series of variants to the CF6-6D, flat rated at 178 kN (40,000 lb) to 31°C and tailored to the McDonnell Douglas DC-10 Series 10 intermediate-range transport. Announcement that this engine had been selected by United Air Lines and American Airlines was made on 25 April 1968. Further orders have since been placed by many airlines for the CF6-6 and -50 series.

Basic configuration of the CF6-6 comprises a "1¼-stage" fan driven by a five-stage LP turbine energised by a slightly modified TF39 core engine, consisting of a 16-stage HP compressor, annular combustor and two-stage turbine. Modifications have been introduced to enable the accessory systems to suit airline installation requirements, while other changes are aimed at enhancing reliability, durability and maintainability.

The construction is modular, featuring easily-removable components that are interchangeable to enable airlines to minimise spare-parts holdings and facilitate sectional overhaul procedures. Provisions have been made for mounting sensors and detection devices to monitor engines during flight. Borescope ports are provided at every compressor and turbine stage, and around the combustion chamber, enabling engine checks to be made without disassembly.

Two views of the General Electric CF6 turbofan *(above)*, **CF6-6D, 178 kN (40,000 lb) thrust class;** *(below)* **CF6-50A, 218 kN (49,000 lb) thrust class**

The CF6 fan is designed for low noise output and a 30,000 hr operational life. It offers high resistance to erosion and foreign-object damage, and provides inherent material separation capability. Rather than entering the HP compressor inlet, ingested foreign objects are centrifuged into the fan and emerge via the fan nozzle. The fan rotor is designed to meet FAA reliability criteria and has substantial speed and stress margins. A blade containment system and automatic engine shutdown system are also provided to enhance safety.

Particular attention has been paid to noise suppression and combustor smoke reduction. A 25·4 mm (1 in) thick glassfibre sandwich structure developed by GE is incorporated along the outer walls of the fan duct, and the inner walls are of bonded aluminium honeycomb. The CF6-6D and -50A installations in the DC-10-10 and -30 have met all FAR 36 noise limitation requirements. The TF39 combustor in the core engine has been modified to introduce axial swirlers, directing more air through the dome to the burning zone, and smoke level is well below the visible range.

CF6-6D. Initial 178 kN (40,000 lb st) version of engine in production for intermediate-range DC-10 Series 10. First ran on 21 October 1968 and 18 days later attained 203·5 kN (45,750 lb st). Following a series of successful factory and outdoor tests, engine was released for production in February 1969. The second CF6-6D, built to the production configuration, first ran in May 1969. By December 1970 a total of 30 engines had been shipped and flight testing with a single engine hung on the starboard inner pylon of a B-52 had extended to 15,250 m (50,000 ft), Mach 0·896 and 420 knots (779 km/h; 484 mph) indicated airspeed. Delivery of flight test engines to McDonnell Douglas started in late 1969, with aircraft first flight following in September 1970. Certification of the CF6-6D for commercial service was granted by the FAA in September 1970, and the engine entered airline service in the DC-10 Series 10 in August 1971.

CF6-6D1. In August 1971 this growth version was FAA certificated and offered to take advantage of the demonstrated margin of the -6D. The D1 rating is increased by 1,000 lb to 182·4 kN (41,000 lb st) at 28·9°C. By the end of 1974 more than 400 6-6D and 6-6D1 engines had been shipped.

CF6-6R. De-rated version, simplified for low-cost operation. Flat-rated at 166·8 kN (37,500 lb st) to 30°C. Not yet certificated.

CF6-45A. Economical de-rated version of CF6-50 series (described later) giving flat rating of 206·8 kN (46,500 lb st) to 36·1°C. Not yet certificated.

CF6-50A. Announced by GE in January 1969, the 218 kN (49,000 lb st) CF6-50A is a growth version of the CF6-6 to power the DC-10 Series 30, the Airbus Industrie A300 and Boeing 747-200. The increased thrust is achieved by increased flow through the core engine (reducing the by-pass ratio from 5·9 to 4·4) at slightly decreased turbine entry temperature. A major change is the introduction of two additional booster stages behind the single-stage LP compressor of the CF6-6, with no change in the turbofan's external dimensions. To provide for flow matching between the two rotors, variable by-pass doors are incorporated between the LP and HP compressors. A 41 per cent scale model fan with three-stage compressor and variable by-pass doors started testing in January 1969. In October 1970 a CF6-50A attained a thrust of 258 kN (58,000 lb) in a test cell at 5·6°C. FAA certification testing was completed in March 1972. The CF6-50A entered airline service in December 1972 in the

DC-10 Series 30. The CF6-50 series also powers the Airbus Industrie A300, which first flew in October 1972 and entered scheduled service in May 1974; and the Boeing 747, which first flew with the GE engine in June 1973 and entered scheduled service in November 1975.

CF6-50C. The CF6-50C is rated at 226·8 kN (51,000 lb st) up to 30°C. Higher thrust is provided by an increase in turbine temperature, with improved cooling of hot-section components. Certificated November 1973.

CF6-50D. This engine is flat-rated at 226·8 kN (51,000 lb) up to 25°C. It received FAA certification in November 1972. Engines began flight testing in the Boeing 747 in June 1973, leading toward aircraft certification in 1974. No longer offered commercially but used as F103 in YC-14 (see below).

CF6-50E. (Military designation F103-GE-100). This engine is rated to give 233·5 kN (52,500 lb st) up to 26°C. Certificated November 1973. Powers Boeing 747-200, E-4 and YC-14 (see below).

CF6-50L. Most powerful version offered commercially. Detail engineering and material improvements permit rating of 240·2 kN (54,000 lb). To be available in December 1977. See "Addenda" section.

The first military application of the F103-GE-100, the military CF6, is the Boeing YC-14 AMST (see "Aircraft" section of this edition). The two F103 engines, each rated at 226·8 kN (51,000 lb st), are mounted ahead of and immediately above the inner part of the wing, to provide upper-surface blowing. With flaps extended, this will give a lift coefficient exceeding 4·0.

In April 1974 GE announced that the two Boeing E-4A/B Advanced Airborne Command Post aircraft then being built would be powered by the F103 engine, and that similar engines would later be retrofitted, at no cost to the USAF, to the two E-4As delivered in 1973.

By January 1976 the first 397 CF6-6 engines had accumulated nearly 2,300,000 engine hours. Engine-attributable three-month unscheduled removal rate was 0·47 per 1,000 hours, and engine-attributable three-month in-flight shut-down rate only 0·11 per 1,000 hours. During the same time period 521 CF6-50 engines had accumulated over 1,600,000 hours with a three-month engine-attributable unscheduled removal rate of 0·29 per 1,000 hours and engine-attributable in-flight shutdown rate of 0·03 per 1,000 hours.

The following data relate to the CF6-6D, with the differing features of the CF6-50 series also detailed.

TYPE: Two-shaft high by-pass ratio commercial turbofan.

AIR INTAKE: Single forward-facing annular configuration.

FAN: Single-stage fan with integrally-mounted single-stage LP compressor (described together as a 1¼-stage fan), both driven by LP turbine. Fan has rotating spinner and omits inlet guide vanes. Blade-containment shroud provided against possible blade failure. The 38 fan rotor blades are individually removable from the thick-section disc bolted to forward conical extension of LP shaft system. Blade aerofoil has anti-vibration shrouds at two-thirds span. Fan exit airflow split between LP compressor and fan slipstream. Fan front frame has 12 radial struts across fan slipstream exit. Front frame provides support for LP and HP rotor front bearings, fan being overhung ahead of large-diameter ball-thrust bearing with rear roller bearing ahead of core engine. Blades, discs, spool of titanium; exit guide vanes of aluminium; fan frame and shaft of steel; spinner and fan case of aluminium alloy. Total airflow 593 kg/sec (1,307 lb/sec), by-pass ratio 5·9 : 1. Configuration of CF6-50 is similar but with two added LP stages and by-pass doors (described below). Total airflow 654 kg (1,439 lb)/sec; by-pass ratio 4·4 : 1.

LP COMPRESSOR: Single-stage compressor acting as booster to airflow into core engine. Rotor blades carried on rear rim of tapered drum bolted to rear of fan disc. Stators cantilevered off short-chord shroud ring, supported by radial outer struts and radial/tangential inner struts located on fan front frame. Compressor exit flow free to balance between core engine and fan slipstream exit. Configuration of CF6-50 modified to three compressor booster stages carried on flanged rotor drum. Continuous shroud extends to fan front frame with 12 integral by-pass doors located between canted radial struts in fan exit inner casing. These doors maintain proper flow matching between the fan/LP system and core by opening at low power settings to permit LP supercharged flow to bleed into the fan airstream. The doors are closed during take-off and cruise.

HP COMPRESSOR: Sixteen-stage compressor of near-constant tip diameter, with inlet guide vanes and first six stator rows having variable incidence. Provision for interstage air bleed for airframe use and engine cooling. Rotor is of combined drum-and-disc construction with front stage and rear three stages overhung on conical shaft providing location on HP front bearing and HP main shaft. All rotor blades held in rabbeted discs and individually replaceable without rotor disassembly. Stages 1-14 blades forged titanium, 15-16 steel. Stages 1-10 disc titanium, 11-16 and aft casing Inconel 718. Casing split on horizontal centreline: stator vanes held in dovetail slots and replaceable individually. Stages 1-2 stators titanium, 3-15 steel; inlet guide vanes titanium, outlet guide vanes steel. Double-skin inner casing shrouds the LP main shaft. Outlet frame contains compressor diffuser and incorporates support structure for HP rotor mid-bearings. Overall pressure ratio (T-O), 24·2 (6D), 24·7 (6D1), 22·4 (6R). Core airflow 86 kg (190 lb)/sec. CF6-50A has 15th and 16th stages removed to pass greater core airflow of 121 kg (267 lb/sec) and reduce pressure and temperature of air entering combustion chamber. Improved materials and strengthened structure in later stages. Overall pressure ratio (T-O), 27·3 (45A), 28·6 (50A), 29·5 (50C), 30·3 (50E), 31·4 (50L).

COMBUSTOR: Fully annular with comprehensive film-cooling. Separate snout, dome and inner/outer skirts, with nozzles, igniter, leads and manifold externally removable. Dome contains ports for two igniters and axial swirler cups for 30 fuel nozzles. Igniters of high-voltage surface-gap type with energy level of 2·0 joules, each igniter operated independently. Forged steel nozzles with liner of Hastelloy X. Nozzle and dome designed to minimise smoke, and entrance diffuser has gradual profile to assure low temperature gradient to turbine under all flight conditions. CF6-50A combustor is shorter, of improved material (HS 18-8), and can be removed with fuel nozzles in place.

HP TURBINE: Two-stage aircooled turbine with 1,290°C entry temperature. Rotor blades are film and convection cooled. Rotor blades cast from René 80; discs and forward and rear shafts of Inconel 718. First-stage nozzle guide vanes supported at inner and outer ends, second-stage cantilevered from outer ends, with inner ends carrying interstage labyrinth seals. First-stage vanes cast from X40 and film cooled by compressor discharge pressure. Second-stage vanes are cast from René 80 material and are convection cooled. Vanes are welded into pairs to decrease number of gas leakage paths. Thin-section discs with heavy-section centreless hubs are bolted to front and rear conical shafts, including conical and arched inter-disc diaphragms. Configuration for CF6-50 is similar but introduces improved materials and cooling, and vanes are not Siamesed but individual.

LP TURBINE: Five-stage constant tip-diameter turbine with nominal 871°C inlet temperature. Rotor blades tip-shrouded and cast in René 77, not aircooled. Forward and rear shafts, case and discs of Inconel 718. First-stage nozzle guide vanes supported at inner and outer ends, remaining stages are cantilevered from outer ends, with inner ends carrying inter-stage labyrinth seals. Stages 1-3 guide vanes cast in six-vane segments in René 77, stages 4 and 5 cast in pairs in René 41. Vanes held in slots machined in the two half-stator casing. Drum and centreless disc construction, located on LP rotor by front and rear conical diaphragms attached to third- and fourth-stage discs. Front diaphragm attached to LP main shaft, rear diaphragm to rear stub shaft. Drive to rotor by means of long fan midshaft. On CF6-50 a four-stage LP turbine is used, all stages being modified in geometry and cooled by 7th HP-stage compressor air instead of 9th.

EXHAUST UNIT (FAN): Fixed-area annular duct with outer cowl and engine cowl forming convergent plug nozzle for fan slipstream.

EXHAUST UNIT (TURBINE): Short-length fixed-area exhaust duct with convergent plug nozzle. Provision for exhaust thrust reverser.

THRUST REVERSER (FAN): Annular cascade reverser with blocker doors across fan duct. For reverse thrust, rear portion of fan outer cowl translates aft on rotating ballscrews to uncover cascade vanes. Blocker doors (16 off) flush-mounted in cowl on link arms hinged in inner cowl, rotate inwards to expose cascade vanes and block fan duct. Reverser hinged at top to open in L/R halves for access to HP casing and combustor.

THRUST REVERSER (TURBINE): Post nozzle exit, cascade type. Two cascade screens are mounted in vertical plane on fixed pivot aft of turbine exhaust and are enclosed in fairing forming aerofoil-shaped plug. Aft translation of fairing uncovers cascades which open across nozzle exit and divert turbine exhaust radially outward and slightly forward in horizontal plane. Configuration for CF6-50 similar to fan thrust reverser with nine blocker doors, but not split (CF6-50 also available with fixed nozzle). Acoustic treatment is provided in the nozzle flow path.

ROTOR SUPPORT SYSTEM: Eight bearings (four for each rotor) at seven locations. Fan and LP compressor carried on ball thrust bearing (1) behind fan disc and roller bearing (2) at front of LP main shaft; both bearings mounted in fan front frame structure, which also supports HP compressor front roller bearing (3). LP turbine carried on roller bearings at front and rear of turbine rotor assembly—rear bearing (7) being mounted in spider structure across turbine exit, and front bearing (6) on major spider structure between HP and LP turbines. HP compressor carried at rear on adjacent roller bearing (4R) and ball-thrust bearing (4B) at interconnection with HP turbine front conical shaft, both bearings being mounted on support structure integral with compressor outlet diffuser. A roller bearing (5), mounted in the inter-turbine structure, carries the aft HP conical shaft.

ACCESSORY DRIVE: This consists of the inlet gearbox, radial gearbox, radial driveshaft, transfer gearbox, horizontal driveshaft and accessory gearbox. The inlet gearbox is located in the forward sump of the engine. The gearbox transfers energy from the core-engine (HP) rotor to the radial driveshaft located in a housing aft of the bottom vertical strut of the fan frame. The transfer gearbox is mounted on the bottom of the fan frame. Accessory mounting pads are provided on both the forward and aft faces of the gearbox. The engine accessories mounted on the gearbox are starter, fuel pump, main engine control, lubrication pump and tachometer. Pads are also provided for mounting the aircraft hydraulic pumps, constant-speed drive and alternator.

FUEL SYSTEM: Hydromechanical fuel control system regulates steady-state fuel flow and schedules acceleration and deceleration fuel flow. It also schedules and powers variable-stator vane position. A governor in the Woodward control provides core-engine speed stability during steady-state operation. During transient operation, core-engine fuel flow is scheduled on the basis of throttle position, compressor inlet temperature, compressor discharge pressure and core-engine speed. The fuel control and fuel pump are mounted in the accessory package as an integrated unit which avoids interconnecting high-pressure fuel lines and potential leakage points (they are separable for change or maintenance). This configuration provides a single drive mounting flange. The filter, fuel/oil heat exchanger and control pressurising valve may be removed individually without removing the entire assembly. The fuel manifold is double-wall constructed for safety and mounted on the exterior of the engine. For CF6-50, fuel control is modified to provide scheduling function for LP compressor by-pass doors.

FUEL GRADES: Fuels conforming to ASTM-1655-65T, Jet A, Jet A1 and Jet B, and MIL-T-5624G2 grades JP-4 or JP-5 are authorised, but Jet A is primary specification.

LUBRICATION SYSTEM: Dry-sump centre-vented system in which oil is pressure-fed to each engine component requiring lubrication. Oil is removed from the sump areas by scavenge pumps, passed through a fuel/oil heat exchanger and filter to the engine tank. Nominal lubrication system pressure is 2·07-6·21 bars (30-90 lb/sq in) above sump reference pressure. All pressure and scavenge pumps and filters are located in the lubrication centre on the forward side of the gearbox.

OIL SPECIFICATION: Conforming to General Electric specification D50TFI classes A & B, equivalent to MIL-L-7808 or MIL-L-23699A.

MOUNTING: Main thrust mount located on the inner fan frame; aft flight mount located on the turbine mid-frame.

STARTING: Air-turbine starter mounted on the front of the accessory gearbox at the through shaft.

NOISE SUPPRESSION EQUIPMENT: Acoustic panels integrated with fan casing, fan front frame and thrust reverser.

DIMENSIONS:

Fan tip diameter	2,195 mm (86·4 in)
Max width (cold)	2,390 mm (94·1 in)
Max height (over gearbox)	2,675 mm (105·3 in)
Length overall (cold):	
CF6-6D	4,775 mm (188 in)
CF6-50 series	4,648 mm (183 in)

WEIGHT, DRY (basic engine):

CF6-6D, -6D1	3,523 kg (7,765 lb)
CF6-6R	3,509 kg (7,730 lb)
CF6-45A	3,829 kg (8,435 lb)
CF6-50A, -50C	3,793 kg (8,355 lb)
CF6-50E	3,799 kg (8,375 lb)
CF6-50L	3,962 kg (8,735 lb)
Fan and turbine reverser:	
CF6-6D, -6D1, -6R	914 kg (2,016 lb)
CF6-50A, -50C	939 kg (2,069 lb)
CF6-50E	971 kg (2,141 lb)
CF6-50L	984 kg (2,167 lb)

PERFORMANCE RATINGS:

Max T-O, uninstalled ideal nozzle:

CF6-6D: 177·9 kN (40,000 lb st) at 9,800 core engine rpm and 3,500 fan rpm, flat rated to 31·1°C

CF6-6D1: 182·4 kN (41,000 lb st), flat rated to 28·9°C

CF6-6R: 166·8 kN (37,500 lb st), flat rated to 30°C

CF6-45A: 206·8 kN (46,500 lb st), flat rated to 36·1°C

CF6-50A: 218 kN (49,000 lb st) at 10,200 core engine rpm and 3,800 fan rpm, flat rated to 30·6°C

CF6-50C: 226·8 kN (51,000 lb st), flat rated to 30°C

CF6-50E: 233·5 kN (52,500 lb st), flat rated to 26°C

CF6-50L: 240·2 kN (54,000 lb st), flat rated to 30°C

Max altitude and Mach No:

CF6-6 and -50: 13,700 m (45,000 ft) at Mach 1·0

Max cruise at 10,670 m (35,000 ft) and Mach 0·85, flat rated to ISA + 10°C, uninstalled, real nozzle:

CF6-6D	40·3 kN (9,060 lb st)
CF6-6D1	40·9 kN (9,200 lb st)
CF6-6R	39·45 kN (8,870 lb st)
CF6-50A, C	48 kN (10,800 lb st)
CF6-50E	50·3 kN (11,300 lb st)
CF6-50L	52·8 kN (11,870 lb st)

SPECIFIC FUEL CONSUMPTION:

At T-O thrust, as above:

CF6-6D	9·86 mg/Ns (0·348 lb/hr/lb st)
CF6-6D1	9·91 mg/Ns (0·350 lb/hr/lb st)
CF6-6R	9·74 mg/Ns (0·344 lb/hr/lb st)
CF6-45A	10·65 mg/Ns (0·376 lb/hr/lb st)
CF6-50A	10·88 mg/Ns (0·384 lb/hr/lb st)
CF6-50C	11·05 mg/Ns (0·390 lb/hr/lb st)
CF6-50E	11·16 mg/Ns (0·394 lb/hr/lb st)
CF6-50L	11·33 mg/Ns (0·400 lb/hr/lb st)

At max cruise thrust, as above:

CF6-6D	17·87 mg/Ns (0·631 lb/hr/lb st)
CF6-6D1	17·93 mg/Ns (0·633 lb/hr/lb st)
CF6-6R	17·82 mg/Ns (0·629 lb/hr/lb st)
CF6-50A, -50C	18·52 mg/Ns (0·654 lb/hr/lb st)
CF6-50E	18·72 mg/Ns (0·661 lb/hr/lb st)
CF6-50L	18·24 mg/Ns (0·644 lb/hr/lb st)

OIL CONSUMPTION: 0·9 kg (2·0 lb)/hr

GENERAL ELECTRIC F101

The F101-GE-100 is the engine designed by General Electric for the Rockwell International B-1 strategic bomber.

The F101, four of which power each B-1, is an augmented turbofan in the 133 kN (30,000 lb) thrust class which represents a considerable advance in the state of the art. Compared with the latest augmented J79, it is of approximately the same weight and bulk, with a reduced overall length, yet gives twice the thrust with a greatly reduced specific fuel consumption. In the B-1 the engine is required to give reliable flight performance at high subsonic speed at sea level and up to Mach 2 at extreme altitude.

The complete engine with afterburner ran in early 1972. In March 1974 the official PFRT (Preliminary Flight Rating Test) was successfully demonstrated. Shipment of flight engines began the following month.

The first B-1 flight took place on 23 December 1974. Engine qualification was scheduled for mid-1976, and a production decision is expected before the end of the year. GE's contract funding through FY76 is $457 million, and covers development of the engine and shipment of 25 engines (2 ground and 23 flight) to the airframe contractor. Funding is being released for each stage of development only when the preceding stage has been demonstrably accomplished.

AIR INTAKE: Direct pitot with 20 fixed radial vanes with variable trailing flaps.

FAN: Two axial stages. Solid titanium blades with tip shrouds for improved clearance control. Inlet guide vanes and for vanes installed in horizontally split titanium casing which permits blades and vanes to be individually replaceable. Pressure ratio, over 2. Airflow, approximately 159 kg (350 lb)/sec.

COMPRESSOR: The high stage-loading technology developed in the GE1 series of engines (see 1973-74 *Jane's*) has been applied to the F101 axial compressor to obtain, in nine stages, a pressure ratio exceeding 11. Rotor constructed by inertia-welding separate discs into drum. First six rotor stages titanium alloy, stages 7-9 Inconel 718. Inlet guide vanes and stator vanes 1-3 steel, with variable incidence; remainder Inconel 718, fixed. Repairability and maintenance enhanced by split compressor casing, with front section of titanium and rear of steel, allowing blades and vanes to be individually replaceable. Borescope inspection ports permit visual inspection of vanes, blades and clearances.

COMBUSTION CHAMBER: Very short, annular machined Hastelloy X. Carburettor system feeding vaporiser tubes to inject fuel vapour into dome area. Smokeless combustion downstream gives a uniform temperature profile at HP turbine nozzle.

HP TURBINE: Air-cooled, single-stage, high-energy-extraction design. Light weight achieved by use of advanced materials and manufacturing processes. Blades and vanes are hollow airfoils in René 125, convective- and film-cooled. Stationary shroud is segmented and cooled, giving growth characteristics compatible with rotor to provide tip-clearance control.

LP TURBINE: Two-stage assembly. First stators of thoria-dispersed nickel-chrome with René 95 rotor disc and René 125 blades. Second stage of René 80. Shaft of Inconel 718. All blades uncooled. Blades individually replaceable; vanes replaceable in segmented groups.

AFTERBURNER: Fully modulated augmentation. Fan and core flows each supplied through 28 augmentor chutes and mixed downstream of turbine. Reheat fuel supplied initially to core flow only, giving automatic light-up with minimal jump in thrust. Further increase in augmentor fuel-flow causes additional outer spraybars and circular flameholder gutters to come into operation, until at maximum augmentation combustion is complete across the whole engine flow. Fan air at about 121°C used to provide film-cooling of augmentor duct wall of Inconel 625.

PROPELLING NOZZLE: Exhaust duct of welded titanium. Variable-area, variable-profile primary and secondary nozzles, each of multiple-flap type. Eight primary rams drive translating ring, cams and links to secure optimum independent profile of both nozzles over wide range, with con-di operation in supersonic cruise at all levels.

General Electric F101 augmented turbofan, rated in the 133 kN (30,000 lb st) class

The 1,394 kW (1,870 shp) General Electric T58-GE-16 turboshaft

ACCESSORY DRIVES: Main accessory gearbox self-contained module at 6 o'clock on fan casing. Drives main and augmentor fuel pumps, fuel boost pump, fuel control, lube pressure and scavenge pumps, hydraulic pump, alternator and power take-off shaft. Integral lube and hydraulic tanks.

DIMENSIONS:

Length overall	4,596 mm (181 in)
Intake diameter	1,397 mm (55 in)

WEIGHT, DRY: about 1,814 kg (4,000 lb)

PERFORMANCE RATINGS:

Cold	about 75·6 kN (17,000 lb st)
Full augmentation	over 133·4 kN (30,000 lb st)

GENERAL ELECTRIC T58

The T58 is a small free-turbine turboshaft engine which was developed originally for the US Navy Bureau of Weapons. A civil version, the CT58, was awarded a Type Certificate by the FAA on 1 July 1959 and is described separately.

The engine is intended primarily as a power unit for helicopters. It has also been developed for marine and industrial use.

Hydromechanical constant-speed control system featured in the T58 maintains essentially constant rotor speed by regulating the engine power automatically, so eliminating the need for speed adjustment by the pilot during normal operation.

Initial flight tests of the T58 were made in a Sikorsky SH-34H, which flew for the first time with two T58s in a nose installation on 30 January 1957.

Rolls-Royce (1971) Ltd produces modified versions of the T58 under licence in Great Britain as the Gnome. The T58 is also licensed for manufacture in Italy and Japan. Industrial and marine version of the T58 is the LM100. By January 1976 more than 7,000 T58 engines had been manufactured.

Versions currently in service or in production are as follows:

T58-GE-3. Five-minute rating of 988 kW (1,325 shp). Powers Bell UH-1F.

T58-GE-5. Five-minute rating of 1,119 kW (1,500 shp). Powers Sikorsky CH-3E, HH-3E/F and NASA RSRA (Sikorsky S-72).

T58-GE-8E, F. Rated at 1,007 kW (1,350 shp). Powers Boeing Vertol CH-46A, Kaman SH-2, Sikorsky SH-3A/G and HH-52A.

T58-GE-10. Rated at 1,044 kW (1,400 shp). Powers Sikorsky SH-3D/H, and Boeing Vertol CH-46D/F.

T58-GE-16. Rated at 1,394 kW (1,870 shp). US military qualified. Aircooled gas-generator turbine and two-stage power turbine. Powers Boeing CH-46E.

T58-GE-100. Uprated T58-GE-5. Ten-minute rating 1,119 kW (1,500 shp) to 15°C or 1,100 kW (1,475 shp) to 26°C. Will power selected CH/HH-3E. Qualification due mid-1976.

TYPE: Free-turbine turboshaft.

AIR INTAKE: Annular intake casing with four hollow radial struts supporting central housing for starter drive clutch and front main roller bearing. Casing and struts anti-iced by air bled from compressor.

COMPRESSOR: Ten-stage axial-flow. Variable-incidence inlet guide vanes. First three of the eleven rows of stator blades also have variable incidence. One-piece steel construction for last eight stages of rotor hub. Casing divided into upper and lower halves. Pressure ratio 8·4 : 1. Air mass flow 5·62 kg (12·4 lb)sec in T58-GE-3 and 8E, 6·21 kg (13·7 lb)/sec in T58-GE-5 and -10, 6·30 kg (13·9 lb)/sec in T58-GE-16.

COMBUSTION CHAMBER: Annular type. Sixteen fuel nozzles (eight on each of two manifolds) mounted on front of inner liner. Dual capacitor discharge ignition unit. Outer casing in two halves to facilitate inspection.

Gas Generator Turbine: Two-stage short-chord axial-flow type, coupled directly to compressor by hollow conical shaft. Centre ball thrust bearing, rear roller bearing. Cooling by air bled from compressor. T58-GE-16 has aircooled first-stage turbine nozzle and blades and second-stage nozzle.

Power Turbine: Single-stage (two-stage in T58-GE-16) axial-flow type, mechanically independent of gas generator turbine. Operated nominally at 19,500 rpm. Engines with single-stage power turbine can have reduction gear giving output at 6,000 rpm. Power turbine accessory drive unit and flexible feed-back cable provide a speed signal to the control.

Torque Sensor Speed Decreaser Gearbox (optional): Gearbox with integral lubrication system. Reduces power speed to 6,000 rpm. Assembly includes an integral torque sensing system.

Jet Exhaust: Two positions (90° left or right) on all versions. T58-GE-16 can also be supplied with downward-ejecting or multiple-position exhaust.

Controls (except T58-GE-10 and -16): Free turbine constant-speed control. Hydro-mechanical controls.

Controls (T58-GE-10, -16): Integrated hydro-mechanical/electrical power control system for isochronous speed governing and twin-engine load sharing.

Accessory Drives: Engine accessories driven from compressor shaft. Airframe accessories mounted on free-turbine reduction gearbox or rotor hub.

Dimensions:
- Max width:
  - except T58-GE-16: 526 mm (20·7 in)
  - T58-GE-16: 607 mm (23·9 in)
- Length overall:
  - except T58-GE-16: 1,499 mm (59·0 in)
  - T58-GE-16: 1,626 mm (64·0 in)

Weight, Dry:
- T58-GE-3: 140 kg (309 lb)
- T58-GE-5, -100: 152 kg (335 lb)
- T58-GE-8E, F: 138 kg (305 lb)
- T58-GE-10: 159 kg (350 lb)
- T58-GE-16: 201 kg (443 lb)

Performance Ratings:
- Five-minute:
  - See under model listings
- Military:
  - T58-GE-3: 988 kW (1,325 shp) at 20,960 rpm
  - T58-GE-5, 10: 1,044 kW (1,400 shp) at 19,500 rpm
  - T58-GE-8E, F: 1,007 kW (1,350 shp) at 19,500 rpm
  - T58-GE-16: 1,394 kW (1,870 shp) at 19,500 rpm
  - T58-GE-100: 1,119 kW (1,500 shp) at 19,500 rpm
- Cruise:
  - T58-GE-3: 798 kW (1,070 shp)
  - T58-GE-5, 10: 932 kW (1,250 shp)
  - T58-GE-8E, F: 857·5 kW (1,150 shp)
  - T58-GE-16: 1,320 kW (1,770 shp)
  - T58-GE-100: 1,015 kW (1,360 shp) at 19,500 rpm

Specific Fuel Consumption:
- At military rating:
  - T58-GE-3: 103 μg/J (0·61 lb/hr/shp)
  - T58-GE-5, 8E/F, 10, 100: 101 μg/J (0·60 lb/hr/shp)
  - T58-GE-16: 89·5 μg/J (0·53 lb/hr/shp)
- At cruise rating:
  - T58-GE-3: 106·5 μg/J (0·63 lb/hr/shp)
  - T58-GE-5, 100: 103 μg/J (0·61 lb/hr/shp)
  - T58-GE-8E, F: 105 μg/J (0·62 lb/hr/shp)
  - T58-GE-10: 105 μg/J (0·62 lb/hr/shp)
  - T58-GE-16: 91 μg/J (0·54 lb/hr/shp)

GENERAL ELECTRIC CT58

The commercial version of the T58 is designated CT58 and was the first US helicopter turbine to receive FAA certification.

Current versions are as follows:

CT58-110. Rated at 932 kW; 1,250 shp (1,007 kW; 1,350 shp for 2½ min) at 19,500 rpm. Air mass flow 5·67 kg (12·7 lb)/sec. Pressure ratio 8·2 : 1.

CT58-140. Rated at 1,044 kW; 1,400 shp (1,119 kW; 1,500 shp for 2½ min) at 19,500 rpm. Air mass flow 6·21 kg (13·7 lb)/sec. Pressure ratio 8·4 : 1.

The CT58 powers the Sikorsky S-61 and S-62 and Boeing Vertol 107 Model II.

Dimensions:
- Max width: 406 mm (16·0 in)
- Length overall: 1,500 mm (59·0 in)

Weight, Dry:
- CT58-110: 143 kg (315 lb)
- CT58-140: 154 kg (340 lb)

Performance Ratings:
- 2½ min and normal T-O:
  - See under model listings
- Cruise:
  - CT58-110: 783 kW (1,050 shp)
  - CT58-140: 932 kW (1,250 shp)

Specific Fuel Consumption:
- At normal T-O rating: 103 μg/J (0·61 lb/hr/shp)
- At cruise rating:
  - CT58-110: 108 μg/J (0·64 lb/hr/shp)
  - CT58-140: 105 μg/J (0·62 lb/hr/shp)

The 2,928 kW (3,925 shp) General Electric T64-GE-413A turboshaft

GENERAL ELECTRIC T64

The T64 is a versatile aircraft gas turbine engine which was initially developed for the US Navy. The basic T64 turboshaft engine becomes a turboprop with the addition of a two-part speed-reduction gearbox.

Current versions include:

T64-GE-7. Direct-drive turboshaft rated at 2,928 kW (3,925 shp). Powers US Air Force CH-53B, CH-53C and HH-53C. Also powers VFW-built CH-53D/G. Produced under licence by MTU (see entry under Germany, Federal Republic).

T64-GE-7A. Direct-drive turboshaft flat-rated at 2,935 kW (3,936 shp) to 28°C. Powers Sikorsky S-65.

T64-GE-10. Turboprop engine with propeller gearbox above centreline. Rated at 2,215 kW (2,970 shp). Produced under licence by Ishikawajima Harima Heavy Industries in Japan for Shin Meiwa PS-1 flying-boat (four engines) and Kawasaki P-2J patrol aircraft (two engines).

T64-GE-413A. Direct-drive turboshaft rated at 2,927 kW (3,925 shp). Two engines power the US Navy CH-53D and RH-53D.

T64-GE-415. Growth version with improved combustion liner and turbine cooling. Max rating 3,266 kW (4,380 shp). Powers Sikorsky RH-53D and three-engined CH-53E helicopters.

T64-GE-820. CT64-820-1, -2 and -3 turboprops based on early turboshaft versions power DHC-5C Buffalo and prototype Aeritalia G222. FAA certificated CT64-820-4 has improved components of T64-415 and is flat-rated at 2,336 kW (3,133 shp) to 38°C, in production for DHC-5D Buffalo.

T64-P4D. Turboprop version flat rated at 2,535 kW (3,400 shp) to 45°C. Two P4D engines power the production Aeritalia G222 transport. Production by Fiat, supported by Alfa-Romeo, with deliveries beginning in 1975.

All T64s are qualified to operate from 100° nose-up to 45° nose-down. The T64 was designed for extensive growth: current production engines rated at 3,266 kW (4,380 shp) are a result of growth made possibly largely by aircooling of the first-stage gas generator turbine rotor and stator. The addition of aircooling to the second turbine stage provides further horsepower growth beyond 3,729 kW (5,000 shp) without significant change in external dimensions.

By January 1975 a total of over 1,900 T64 engines of all kinds had been delivered by GE to customers in nine countries. In addition licences to produce the T64 are held by MTU in Germany, Fiat in Italy, Rolls-Royce (1971) in Britain and IHI in Japan.

Type: Free-turbine turboshaft/turboprop engine.

Compressor: Fourteen-stage axial-flow. Single-spool steel rotor for -10 and -820-1/2/3. Titanium and steel compressor for -7A, -413A, -415, -P4D and CT64-820-4. Inlet guide vanes and first four stages of stator blades variable. Compressor blades can be removed individually without rotor disassembly. Casing flanged along centreline. Stator blades removable. Air mass flow per second: -10, 11·6 kg (25·5 lb); -7A, -413A, 12·8 kg (28·3 lb); -415, 13·3 kg (29·4 lb); -820-4, 11·9 kg (26·2 lb); P4D, 12·2 kg (27·0 lb). Pressure ratio: 10, -820-4, 12·5; -7A, -413A, 14·1; -415, 14·8; P4D, 13·0.

Combustion Chamber: Annular type. Double fuel manifold feeds twelve duplex-type fuel nozzles with external flow divider. Nozzles mounted on outer diffuser wall of compressor rear frame.

Gas Generator Turbine: Two-stage axial-flow type, coupled directly to compressor rotor by spline connection.

Power Turbine: Two-stage axial-flow type, mechanically independent of gas generator turbine.

Reduction Gear: Remotely-mounted basic reduction gear for turboprop versions is offset and accessible for inspection and replacement. Gear-driven by power turbine, using co-axial shafting through the compressor. Propeller gear ratio 13·44 : 1.

Starting: Mechanical, airframe supplied.

Dimensions:
- Length:
  - T64-GE-7A, 413A, -415: 2,006 mm (79 in)
  - T64-GE-10, -820-4, P4D: 2,793 mm (110 in)
- Width:
  - T64-GE-7A, -413A, -415: 660 mm (26·0 in)
  - T64-GE-10, -P4D, -820: 683 mm (26·9 in)
- Height:
  - T64-GE-7A, 413A, -415: 825 mm (32·5 in)
  - T64-GE-10, -P4D: 1,168 mm (46 in)
  - CT64-820: 1,026 mm (40·4 in)

Weight, Dry:
- T64-GE-413A: 325 kg (716 lb)
- T64-GE-7A, -415: 327 kg (720 lb)
- T64-GE-10: 529 kg (1,167 lb)
- CT64-820-1: 513 kg (1,130 lb)
- CT64-820-2, -4: 520 kg (1,145 lb)
- CT64-820-3: 518 kg (1,140 lb)
- T64-P4D: 538 kg (1,188 lb)

Performance Ratings:
- Max rating (sea level):
  - T64-GE-10: 2,215 kW (2,970 shp) at 1,160 output rpm
  - T64-GE-7A, -413A: 2,935 kW (3,936 shp)
  - CT64-820-4: 2,336 kW (3,133 shp)
  - T64-P4D: 2,535 kW (3,400 shp)
  - T64-415: 3,266 kW (4,380 shp)

Specific Fuel Consumption (S/L):
- At max rating:
  - T64-GE-7A, -415: 79·4 μg/J (0·47 lb/hr/shp)
  - T64-GE-10: 84·5 μg/J (0·50 lb/hr/shp)
  - T64-GE-413A, P4D: 81 μg/J (0·48 lb/hr/shp)
  - CT64-820-4: 83 μg/J (0·49 lb/hr/shp)

GENERAL ELECTRIC T700

Under a contract awarded by the US Army Aviation Materiel Laboratories Propulsion Division, General Electric initiated a two-year demonstration programme in August 1967 to design, build and test a new 1,118·5 kW (1,500 shp) advanced technology turboshaft. Designated GE12, the engine was developed by the Aircraft Engine Group's Military Engine Division. During the latter part of 1968 the basic contracts for the GE12 and competing Pratt & Whitney ST9 demonstrator turboshaft were extended to the end of 1969 and additional work was also funded to carry the GE12 programme through to 30 September 1971. The GE12 demonstrator was illustrated in the 1971-72 *Jane's*.

A new competition was conducted in 1971 to provide the power plant for the Army's projected utility tactical transport system (UTTAS) proposed as a replacement for the present Bell UH-1 family of Army helicopters. In early 1972 it was announced that the winner was the GE engine, and that a contract for a production version, the T700-GE-700, was being negotiated with the US Army Aviation Systems Command.

The first T700 engine was tested in February 1973. Shipment of ground test engines was accomplished on schedule in February 1974, and flight qualification testing was completed ahead of schedule in August 1974. Shipments for UTTAS flight aircraft began immediately, with first flights of the Boeing YUH-61A and Sikorsky YUH-60A following during the autumn of 1974. Identical T700 engines power the two advanced attack helicopter (AAH) contenders, the Bell YAH-63A and Hughes YAH-64A.

Basic development of the T700-GE-700 was completed by the successful inspection of two engines after Model Qualification Test (MQT) in March 1976. At that time 19,000 hours had been run, including 3,300 flight hours in 12 helicopters. A total of 86 engines had been shipped and had logged up to 1,000 hours field operation without unscheduled removal.

The T700 has been designed to be compatible with the Army's special operating and environmental conditions, and embodies high reliability, simplicity of maintenance, low vulnerability to combat damage, and high performance combined with compact dimensions. Use is made of higher pressure ratios and turbine entry temperatures than with existing small turboshafts to assist in reducing size and weight. Specific fuel consumption of the T700 is 25 to 30% lower than that of present turboshafts, and weight is some 33% lower than that of current helicopter engines of the same power category.

To reduce vulnerability, all external lines and leads are short in length and are grouped compactly for minimum exposure. Self-contained electrical and lubrication systems are fitted. Multiple mounting points allow for ease of installation and the necessary airframe connections have been minimised and are located close to the engine centreline. The whole engine is of modular construction for swift field maintenance or section replacement without special tools (for example, complete hot-section inspection and replacement by two men in 75 minutes using standard toolbox).

GE has announced studies for derivatives, including a family of turbojets for RPVs and other fixed-wing aircraft. Other possibilities include turboprops and high by-pass ratio turbofans, while front and rear drive modifications make surface applications possible.

A programme of growth has been started to meet the widening spectrum of applications. As envisioned, 10 per cent greater power will be available from the existing basic engine. Addition of an LP booster (zero stage) will then provide 20-30 per cent growth, depending on the increase in turbine temperature chosen. These growth versions are planned to be interchangeable with the present T700-GE-700 aircraft installation.

Material changes and coatings have been defined for "navalisation" of the T700 to power ship-based aircraft, such as US Navy LAMPS Mk III helicopter.

TYPE: Ungeared free-turbine turboshaft engine.

INTAKE: Annular type, with anti-iced integral inlet particle separator containing no moving parts yet designed to remove 95 per cent of sand, dust and foreign-object ingestion. Extracted matter discharged by separator blower driven from accessory gearbox.

COMPRESSOR: Combined axial/centrifugal. Five axial stages and single centrifugal stage mounted on same shaft. Each axial stage is one-piece "blisk" (blades plus disc) in AM355 steel highly resistant to erosion. Inlet guide vanes and first two stator stages are variable. Pressure ratio, about 15 : 1. Airflow about 4·5 kg (10 lb)/sec.

General Electric T700-GE-700 turboshaft engine, showing (light coloured duct rising at mid-section) the extractor for solid particles removed by the inlet separator. Intermediate (T-O) rating, 1,151 kW (1,543 shp)

COMBUSTION CHAMBER: Fully annular. Compact short-length configuration, designed for maximum reliability and long life. Central fuel injection to maximise acceptance of contaminated fuel and give minimal smoke generation and uniform temperature profile into the turbine. Flame tube is machined ring in Hastelloy X. Ignition system obtains power from separate winding on engine-nounted alternator and serves dual plugs.

TURBINE: Two-stage gas-generator (HP) turbine operates at gas temperatures exceeding 1,100°C. First-stage nozzle investment-cast in X40. Second-stage nozzle investment-cast in two-vane segments in R80. Discs, cooling plates, and blades of both stages clamped by five short tiebolts; five larger bolts then tighten turbine to shaft, driving via curvic joint. Rated shaft speed (S/L, ISA, max T-O power), 44,720 rpm. Two-stage free power turbine, designed for high efficiency at part-power levels (especially 30 and 60% of military power), with tip-shrouded blades and segmented nozzles. Power turbine inlet temperature at intermediate power, 827°C. Nozzle guide vanes René 77, rotor discs Inco 718, rotor blades René 80 uncooled. Output speed, 20,000 rpm. Power output shaft for front drive.

CONTROLS: Hydromechanical control can be replaced in less than 12 minutes and requires no adjustment or lockwire. Electrical control, coupled with hydromechanical control, provides twin-engine speed and torque matching.

ACCESSORIES: Grouped at top of engine, together with engine control system, for maximum simplicity, accessibility and combat survivability. Integral lubrication supply tank, plus 6-minute emergency supply of mist lubrication following total loss of main supply. Torque sensor provides signal to electrical control system.

DIMENSIONS:
- Length overall: 1,181 mm (46·5 in)
- Width: 635 mm (25 in)
- Height overall: 584 mm (23 in)

WEIGHT, DRY (with particle separator): 188 kg (415 lb)

PERFORMANCE RATINGS:
- Intermediate (30 min): 1,151 kW (1,543 shp)
- Max continuous: 933 kW (1,251 shp)
- 72% max continuous: 671 kW (900 shp)

SPECIFIC FUEL CONSUMPTION:
- Intermediate: 79·25 μg/J (0·469 lb/hr/shp)
- Max continuous: 80·6 μg/J (0·477 lb/hr/shp)
- 72%: 88 μg/J (0·520 lb/hr/shp)

GESCHWENDER
GESCHWENDER AEROMOTIVE

ADDRESS:
Box 5152, Lincoln, Nebraska 68505
Telephone: (402) 464-2434

Fred L. Geschwender has now established a range of engines which were originally aimed primarily at homebuilt aircraft which are replicas of World War 1 aircraft, or scaled replicas of World War 2 fighters, and other aircraft having in-line or vee-type engines. The generally available certificated engines preclude an amateur constructor from achieving realism.

His search narrowed down possible engines to vee-type automotive units and, finally, to the "small block" Ford engines of 289, 302, 351 and 400 cu in capacity. These are V-8 engines, but some V-6 units can also be used. Geschwender's present policy is to design, build and market hardware for conversion kits by which stock car engines can be fitted for unrestricted aerial use.

The kit normally includes: 2 : 1 or 2·66 : 1 reduction gear, using precision chain drive; new ignition system with single plugs but dual magnetos and leads (one set being normally disconnected but available for immediate use); new drive with torsion-damper and electric-starter ring; various carburettor and accessory options; and cooling system (generally very similar to that of a modern liquid-cooled car) tailored to the aircraft.

Since 1974 it has become evident that by far the largest and most urgent market is the agricultural one. Increasingly the company's efforts are devoted to replacing World War II radial engines in agricultural aviation.

Geschwender Ford V-8 conversion undergoing flight testing in a Funk 23B

In 1974-75 development began on the Ford Capri and Mustang 2 V-6 engines of 159 and 171 cu in, rated at 160 hp before conversion, and the Ford Pinto and Mustang 2 engines of 122 and 140 cu in, rated originally at 100 hp. Engines are installed in a Beech T-34 (Ford V-8) and the company's Funk 23B and Piper Pawnee Brave; a Grumman Ag-Cat and Cessna AGtruck were expected to fly in mid-1976. Turbocharged engines rated at up to 600 kW (800 hp) are planned.

Work is in hand on a four-cylinder engine rated at 67 kW (90 hp) for advanced homebuilts such as the Bede BD-5. An aircooled dual-overhead-cam engine, this will be developed later to 97 kW (130 hp) unblown, and will also be available with reduction gear.

View from above of the Ford V-8 conversion, rated at up to 448 kW (600 hp)

The following are basic data for existing models:

DIMENSIONS (length/width/height, mm; in):
- 302 cu in: 1,358; 53·5/559; 22·0/616; 24·25
- 351 cu in (351C): 1,358; 53·5/584; 23·0/622; 24·5
- 400 cu in: 1,358; 53·5/584; 23·0/635; 25·0

WEIGHT, DRY:
- 302 cu in: 241 kg (530 lb)
- 351C: 272·2 kg (600 lb)
- 400 cu in: 318 kg (700 lb)

MAX POWER (S/L):
- 302 cu in: 224 kW (300 hp) at 4,800 rpm (crankshaft)
- 351C: 246-335·5 kW (330-450 hp) at 5,400 rpm (crankshaft)
- 400 cu in: 194 kW (260 hp) at 4,400 rpm to 448 kW (600 hp) at 5,500 rpm (crankshaft)

GLUHAREFF
EMG ENGINEERING COMPANY

ADDRESS:
18518J South Broadway Avenue, Gardena, California 90248
Telephone: (213) 321-8699

Eugene M. Gluhareff, a pioneer of ultra-light rotorcraft, has been developing a unique type of air-breathing jet engine which he considers to offer notable advantages over all other systems for rotor tip-drive. The first model is now out of the development stage and is on sale in three sizes in various forms for tip-drive or sailplane auxiliary propulsion. It has also been used for surface application in a go-kart, and in numerous static rigs sold to universities and other organisations. Production has been hard pressed to keep up with demand, a fast-growing market being radio-controlled flight vehicles. Testing of a 0·36 kN (80 lb st) engine began in 1974.

GLUHAREFF G8-2

Although extremely simple, the G8 series corresponds to no prior jet system. The design is based on propane, a readily-available volatile fuel. The pressure of the liquid propane in the tank delivers the fuel, via a needle valve serving as the throttle, to the burner unit. The pipe enters the burner duct and is immediately vaporised in a hot heat-exchanger. Vapour then passes back down an insulated pipe to the injector where its residual pressure is converted to kinetic energy. The high-velocity gas jet induces air through three "supercharger" intakes, each synchronised to the internal flow, which gives the correct final fuel/air ratio to the mixture entering the combustion chamber. Here the mixture is initially ignited by a spark plug and thereafter burns continuously. The intake ducts are tuned to each other to create one-way flow. Resonance in the tailpipe is undesirable and is prevented by making the propelling nozzle of fishtail shape.

Once started, by opening the throttle valve and closing the induction-coil igniter circuit, the engine takes up to ten seconds to warm up and thereafter can be swiftly cycled between minimum and maximum thrust, which varies approximately linearly with fuel flow. Static thrust is about 90 per cent of the value attainable dynamically with maximum ram-air augmentation. There is no visible flame or smoke, though the tailpipe is short enough not to bend under centrifugal load when hot. Noise is considered "low in comparison with other types of jet engine". Life is dictated solely by the slow oxidation of the stainless steel parts.

At present production is concentrated on two sizes of jet unit. The G8-2-15 is 914 mm (36 in) long; has a tailpipe diameter of 89 mm (3·5 in) and weighs 2·5 kg (5·5 lb). The equivalent horsepower at 644 km/h (400 mph) is 19·2. The G8-2-40 is 978 mm (38·5 in) long, has a tailpipe diameter of 127 mm (5·0 in) and weighs 5·2 kg (11·5 lb). The equivalent horsepower at 644 km/h (400 mph) is 46·0.

Static thrust (S/L, ISA) for the smaller unit, the G8-2-15, ranges from 0·004 kN (1 lb) at nozzle pressure of 1·38 bars (20 lb/sq in) to 0·067 kN (15 lb) at about 8·96 bars (130 lb/sq in). At the lower end the specific fuel consumption is over 10, but at higher powers levels off close to 6. Addition of a ram-air scoop reduces static performance (max about 0·047 kN (10·5 lb), but raises dynamic thrust at 91·5 m (300 ft)/sec from 0·071 kN (15·9 lb) to 0·078 kN (17·5 lb). Altitude performance has been explored statically, on a mountain; more than 0·053 kN (12 lb) thrust was available at 2,455 m (8,050 ft), the greatest height reached.

An installation comprises tank (on aircraft CG), tank valve, needle throttle valve, fuel lines, jet unit, ignition system and pressure gauges for the tank and fuel nozzle. Care must be taken to install the unit so that radiant heat from the main duct and from the hot vapour line (up to 540°C) cannot harm the vehicle structure or rotor blade. Light sailplanes have flown successfully on a single G8-2-15, taking off unaided into a 24-32 km/h (15-20 mph) wind.

The units are manufactured by Gluhareff's subsidiary EMG Engineering Co. Customers have the option of buying plans only, a construction package, partly prefabricated or an assembly kit or finished engine.

Gluhareff G8-2-15 jet unit, rated at 0·067 kN (15 lb st), without ram-air scoop

JACOBS

PAGE INDUSTRIES OF OKLAHOMA INC

Head Office:
Cimarron Airport, PO Box 191, Yukon, Oklahoma 73099
Telephone: (405) 354-5385
President: O. J. Butts
Vice-President and Treasurer:
Currey Smith
Engineering Manager:
Merrill H. Bumbaugh

Page Industries of Oklahoma is the successor to the Jacobs Aircraft Engine Co, which had manufactured air-cooled radial piston engines since 1929. Current production is centred upon the various versions of the R-755 described below. Page is holder of the FAA Approved Parts Inspection System (APIS) for production of new Jacobs engine parts.

JACOBS R-755

This long-established seven-cylinder radial is being manufactured in several versions. The basic engine is the **R-755A** which is the subject of the detailed description. The current engines in this series are the R-755A2, and the A2M with dual ignition. The **R-755B** has a lower power rating, and is available as the R-755B1, driving a fixed-pitch propeller, and the B2 which can have a control valve for a two-position propeller or provision for a hydraulically-operated constant-speed propeller. The **R-755S** is a turbocharged version, available with dual ignition as the R-755SM.

Type: Seven-cylinder aircooled radial piston engine.

Cylinders: Bore 133 mm (5·25 in). Stroke 127 mm (5 in). Swept volume 12·3 litres (757 cu in). Barrels machined from steel forgings with close-spaced fins. Aluminium alloy heads screwed and shrunk on. Aluminium-bronze valve-seats shrunk into heads.

Pistons: Forged aluminium alloy. Three compression rings and one scraper ring above gudgeon-pin (piston-pin) and two scraper rings below. Fully-floating nitrided gudgeon-pins.

Connecting Rods: One-piece steel master rod. Forged aluminium alloy link rods bearing directly upon nitrided-steel pins.

Crankshaft: Two-piece clamp type, machined from forgings in chrome-nickel-molybdenum steel.

Crankcase: Assembled from five parts. Aluminium alloy front case carries ball thrust bearing and valve-operating gear. Front half of main case is aluminium alloy casting supporting crankshaft front roller bearing. Rear half of main case is aluminium alloy casting supporting crankshaft rear roller bearing and incorporates ring-type intake manifold. Aluminium alloy rear plate carries additional crankshaft ball bearing and supports accessory drives. Aluminium alloy rear case carries accessories.

Valve Gear: Cam ring, drive gears, tappets and pushrods all in nose section, with all moving parts enclosed. Tulip-type inlet valves and sodium-cooled exhaust valves, each with two springs.

Induction: Single updraught Stromberg NA-R7A carburettor. R-755S and SM have this carburettor fed at up to 35 in Hg by AiResearch exhaust-gas turbocharger between Nos. 2 and 3 cylinders. System set to maintain power to 5,945 m (19,500 ft), with aneroid relief valve to prevent overboosting beyond 36 in Hg. Manifold pressure controlled by throttle.

Fuel: 80 octane (755S, SM, 80/87 octane).

Ignition: One Scintilla magneto and battery distributor, incorporating automatic advance (engines with "M" suffix, two magnetos).

Lubrication: Single unit comprising gear pressure pump and two scavenge pumps feeds all plain bearings. Dry sump. Automatic valve lubrication. Optional provision to operate adjustable-pitch or constant-speed propeller.

Propeller Drive: Direct, RH tractor, SAE 20 spline.

Mounting: Choice of eight locations incorporated around rear main crankcase.

Dimensions:

Diameter	1,118 mm (44 in)
Length overall	1,020 m (40 7/32 in)

Jacobs R-755A seven-cylinder radial piston engine, rated at 224 kW (300 hp)

Weight, Dry:

R-755A, B1	229 kg (505 lb)
R-755B2	232 kg (511 lb)
R-755S	261 kg (576 lb)
R-755SM	264 kg (583 lb)

Performance Ratings:
Rated power:
R-755A 224 kW (300 hp) at 2,200 rpm
R-755B 205 kW (275 hp) at 2,200 rpm
R-755S, SM:
rated power 205 kW (275 hp) at 2,030 rpm maintained to 5,945 m (19,500 ft); take-off (1 min limit) 261 kW (350 hp) at 2,200 rpm

LPC

LOCKHEED PROPULSION COMPANY

Headquarters:
PO Box 111, Redlands, California 92373
Telephone: (714) 794-5111
President:
G. Graham Whipple
Executive Vice-President, Operations:
N. B. Chase
Vice-President:
A. H. Von Der Esch (Technical, Marketing)

Lockheed Propulsion Company is the former Grand Central Rocket Company, which was founded in 1952 as the first major US company devoted entirely to the advancement of solid-propellant rocket development. It became a wholly-owned subsidiary of the Lockheed Aircraft Corporation in late 1961, and a division of Lockheed in February 1963.

Current programmes include production of the solid-propellant pulse motor for the USAF Short Range Attack Missile (SRAM), several alternative thrust vector control systems for large solid-propellant motors, and small, variable-thrust motors for attitude control, orbital ejection and re-entry. LPC has also developed and test-fired more than 300 hybrid rockets utilising a solid-propellant fuel and liquid oxidiser, with diameters of up to 0·48 m (19 in). This work was done for the US Army.

LPC developed a "self-eject" launch technique. In this, the solid rocket vehicle ejects itself from a launch tube using low-pressure gas flow from its own first-stage motor. Once out of the tube the motor comes to full thrust automatically. A vehicle weighing more than 136,078 kg (300,000 lb) has been launched successfully by this method.

The company has developed a high-energy smokeless propellant for use in tactical missiles and is studying techniques for launching rockets from guns. One such concept keeps the rocket vehicle intact while being driven from the barrel by a powder charge. Once out of the nozzle, and while travelling at approximately 915 m (3,000 ft)/sec, the rocket's own motor fires. Such techniques offer a considerable extension of a missile's range.

LPC has continued to develop solid propellants called polycarbutenes. They are based on polybutadiene prepolymer binder systems (PBAA, PBAN, CTPB, HTPB) and have been characterised to meet a broad spectrum of operational requirements. These propellant systems have been fired with grains weighing up to 165,108 kg (364,000 lb). Solid-propulsion units have successfully completed temperature cycling from minus 100 to plus 93°C, random vibration from 20 to 2,000 cycles per second, 10·5*g* acceleration, 20*g* drop shock, ageing in vacuum and ambient ageing up to seven years. The ballistics of these propellant systems have been tailored to meet broad burning rate requirements, maximum impulse, and low to high pressure exponents for an exceedingly wide range of solid propulsion design requirements. Physical properties meet complex structural integrity requirements.

Recent development programmes include advanced pulse motors which permit sequential or simultaneous burning of pulse grains to produce a wide range of thrust levels, and a unique grain-retention system which produces minimum propellant strains over a wide temperature and vibration environment. A third new development is a grain design concept utilising embedded fuses which, as they burn, form grain ports. This grain configuration combines the high volumetric efficiency of an end-burner with the lower burning rate and greater ballistic flexibility of a centre-perforated grain.

LPC has developed and is producing flexible seals, called Lockseals, for use on movable-nozzle thrust-vector control systems. The Lockseal movable joint is composed of alternate layers of rubber and metal in a laminated structure. Lockseal joints also are being produced for commercial applications. The company is conducting propellant structural-integrity programmes under USAF contracts, including a programme to design and fabricate a highly instrumented inert rocket motor for measurement of missile environments for air-launched rocket applications.

Additionally, and in support of its work on advanced upper-stage rockets, LPC has developed a family of nitrocellulose-base composite propellants (nitroplastisols) of very high specific impulse. Since 1968 research has also been conducted on an advanced monopropellant motor.

LPC is also engaged in fabrication of hardware for the aerospace industry and components for nuclear reactors.

Available details of motors which have aircraft, missile or space vehicle applications are given hereafter:

LPC-415

US military designation: SR75-LP-1

This is the propulsion system for the SRAM (Short Range Attack Missile) in service with the USAF as the AGM-69A. LPC developed, qualified, and produced the motor under sub-contract to Boeing. It is an advanced, two-pulse motor which provides for multiple flight trajectories. It is a 444 mm (17·5 in) diameter, high-pressure/long-duration end-burner, with variable inter-pulse delay.

LPC MINI MOTOR (LP-117)

The Mini Motor is used to impart a spinning motion to research satellites. Total vacuum impulse can be varied between 7·7 and 15·9 kg-sec (17 and 35 lb-sec) by pre-selecting grain bore diameter and length. Cartridge-loading facilitates this impulse selection. The low solids content of the exhaust minimises particle impingement on the host vehicle.

DIMENSIONS:
Length overall 153 mm (9·58 in)
Diameter 38 mm (1·51 in)
WEIGHT: 0·23 kg (0·51 lb)
PERFORMANCE:
Average thrust in vacuum 0·11 kN (24·8 lb)
Total operating time 1·39 sec
Specific impulse 254 sec

LPC R/B EJECT MOTOR (LP-118)

This is a very small motor used to eject and spin a re-entry body. The motor provides a high degree of ballistic reproducibility and a low content of alkaline elements in the exhaust. The internal-burning grain is cartridge-loaded.

DIMENSIONS:
Length overall 92 mm (3·62 in)
Diameter 32 mm (1·25 in)
WEIGHT: 0·315 kg (0·694 lb)
PERFORMANCE:
Average thrust in vacuum 0·22 kN (49·4 lb)
Action time 0·288 sec
Specific impulse 249 sec

LPC LARGE ORBITAL BOOST MOTOR (LP-119)

This motor is used in pairs for spacecraft orbit adjustment. It is a high-performance, relatively high-mass-fraction motor utilising a case-bonded, end-burning propellant grain and a high-strength aluminium case. Total impulse can be varied from 2,948 to 6,146 kg-sec (6,500 to 13,550 lb-sec) by selection of grain length.

DIMENSIONS:
Length overall 330 mm (13·0 in)
Diameter 282 mm (11·1 in)
WEIGHT: 31·2 kg (68·7 lb)
PERFORMANCE:
Average thrust in vacuum 2·63 kN (592 lb)
Action time 22·5 sec
Specific impulse 248 sec

LPC HIGH-THRUST MOTOR

This high-thrust, short-duration motor was developed and qualified for use in ejecting special payloads from missiles or space vehicles. Similar motors of smaller diameter have been used in tube-launched, shoulder-fired tactical weapons. The high-thrust motor contains a rubber-base PBAN non-aluminised composite fuel with ammonium perchlorate oxidiser, within a 7075 aluminium chamber with a steel nozzle throat. The motor is fired by a low voltage squib triggered by an electrical pulse; maximum chamber pressure is 275 bars (4,000 lb/sq in).

Static test-firing of the LPC-415 solid-propellant pulse rocket motor which was mass-produced by LPC for the US Air Force's AGM-69A short-range attack missile (SRAM)

DIMENSIONS:
Length overall 255 mm (10·03 in)
Diameter 174 mm (6·86 in)
WEIGHT: 6·57 kg (14·5 lb)
PERFORMANCE:
Max thrust at S/L 133·4 kN (30,000 lb)
Operating time 0·024 sec
Specific impulse 230 sec

SPIN/THRUST MOTOR

Avco Corporation, prime contractors for advanced re-entry vehicles for late-model LGM-30 Minuteman missiles, subcontracted to LPC the task of providing a reliable rocket motor capable of imparting either forward thrust or rotational spin (the motors are probably installed either in separate multiple warheads or in decoys or other penetration aids). The Spin/Thrust motor weighs 1·18 kg (2·6 lb) and imparts a thrust variable from 0·169-3·56 kN (38 to 800 lb) over a burn time of 0·15-6·2 sec. The main charge fires through a central nozzle, while spin rates of 630-1,170 rpm (10·5-19·5 rps) are imparted by three canted tangential nozzles around the base. To suit the requirements of particular missions the propellant division between thrust and spin can be altered over a wide range, replacing propellant by inert filling if necessary to preserve motor weight and centre of gravity position.

LPC LSM 156-5

When this three-segment solid-propellant motor was fired in 1965 it was the largest flight prototype solid motor tested at the time. It developed about 13,344 kN (3,000,000 lb) thrust and consumed some 317,500 kg (700,000 lb) of polycarbutene propellant in one minute of forced-draught burning. It was the first rocket of this size to be fitted with all major components of a steering system suitable for use in an actual flight.

The LSM 156-5 was made up of a 150 ton 6·70 m (22 ft) long central segment and forward and aft segments of almost equal size. Liquid nitrogen tetroxide, under high pressure, was sprayed through selected injectors in the nozzle expansion cone to create a shock-wave and also to react chemically with the exhaust gases to generate side forces for steering, at right angles to the direction of main gas flow.

This was the fifth consecutive successful firing in the USAF's Large Solid Motor Program, and the third by LPC.

DIMENSIONS:
Length overall 24·4 m (80 ft 0 in)
Diameter 3·96 m (156 in)

LPC LSM 156-5 flight prototype 156 in solid-propellant motor

WEIGHT: 385,500 kg (850,000 lb)
PERFORMANCE:
Max thrust at S/L 13,344 kN (3,000,000 lb)

LPC LSM 156-6

This monolithic (single-segment) 3·96 m (156 in) solid-propellant motor was the fourth fired by LPC under the USAF's Large Solid Motor Program. It could be considered a potential second stage of a space launch vehicle using a motor like the LSM 156-5, described above, as the first stage. It was filled with a polycarbutene propellant.

DIMENSIONS:
Length overall 10·36 m (34 ft 0 in)
Diameter 3·96 m (156 in)
WEIGHT: 147,400 kg (325,000 lb)
PERFORMANCE:
Max thrust at S/L 4,448 kN (1,000,000 lb)

LTV—*See 'Vought'*

MARQUARDT

THE MARQUARDT COMPANY (a division of CCI Corporation)

HEAD OFFICE:
16555 Saticoy Street, Van Nuys, California 91409
Telephone: (213) 781-2121
PRESIDENT:
G. H. Hanauer
VICE-PRESIDENT:
K. E. Woodgrift (Engineering)
A. L. Sorensen (Operations)
DIRECTORS:
B. E. Huston (Finance)
T. Linton (Rocket Systems)
O. B. McCutcheon (Contracts)
J. Matthews (Administration)

The Marquardt Corporation was formed in November 1944 to undertake research and development of ramjet engines, and it produced the first American subsonic ramjet in 1945. Since that time the company has diversified into production of aerospace controls and accessories, space rocketry, ordnance components, environmental systems and industrial products.

Marquardt's main engineering business continues to be advanced aerospace propulsion and the supply of ram-air turbine power systems. A major portion of sales is currently associated with manufacture of sophisticated structures and components for the aerospace industry, and the production of clustered munition ordnance.

Marquardt is currently developing security-classified types of composite rocket/air-breathing propulsion systems for the US Air Force and Navy. These are regarded by the company as likely to lead to a new generation of power plants for supersonic strategic missiles and expendable tactical vehicles.

In the field of precision control rockets, Marquardt provided manoeuvring, stabilisation and control propulsion for the Apollo Service and Lunar Modules, and is now engaged in developing reaction control engines for the Space Shuttle (the R-40A described separately). The company's precision control rockets also served as main propulsion for the five Lunar Orbiters. Recently, the control rocket work has been expanded to include development of a monopropellant rocket system for a classified programme, and advanced development activities for a hydrogen/oxygen water electrolysis rocketry system. A complete range of monopropellant and bipropellant rockets, precision valves and rocketry components is now being marketed.

Marquardt maintains extensive test facilities at Van Nuys for research and testing of air-breathing rocket engines, and controls and accessories. Total land area occupied exceeds 56 acres, and covered buildings exceed 46,400 m² (500,000 sq ft); employment exceeds 800.

MARQUARDT R-4D

This liquid bipropellant rocket reaction control engine was developed and produced by Marquardt for the Apollo spacecraft Service Module and Lunar Module. After the completion of the Apollo flights, the engines were used in the Skylab and Apollo-Soyuz programmes.

TYPE: Liquid-propellant reaction control rocket.

PROPELLANTS: Nitrogen tetroxide and monomethyl hydrazine.

THRUST CHAMBER ASSEMBLY: Single chamber. Area ratio 40. Made of aluminium, steel and molybdenum. Radiation cooling. Started by electrical signal to on/off solenoid valves. Hypergolic ignition. Combustion pressure 6·21 bars (90 lb/sq in). Combustion temperature 2,870°C.

THRUST CHAMBER MOUNTING: Flange bolt circle on injector head.

DIMENSIONS:
- Length overall: 343 mm (13·5 in)
- Height: 168 mm (6·6 in)
- Width: 152 mm (6·0 in)

WEIGHT, DRY: 2·27 kg (5·0 lb)

PERFORMANCE RATINGS:
- Max rating at S/L: 0·28 kN (63 lb st)
- Max rating in vacuum: 0·45 kN (100 lb st)

MARQUARDT R-40A

This precision control rocket is being developed and qualified for the Space Shuttle orbiter vehicle.

TYPE: Liquid-propellant reaction control rocket.

PROPELLANTS: Nitrogen tetroxide and monomethyl hydrazine.

THRUST CHAMBER ASSEMBLY: Single chamber. Area ratio 20. Made of welded coated columbium, with welded-on orthogonal and scarfed nozzle extension in same material. Internal film cooling. Exterior insulated for buried installation. Started by electrical signal to on/off solenoid valve. Multiple doublet injector with hypergolic ignition.

THRUST CHAMBER MOUNTING: Flange bolt circle on injector head.

PROPELLANT FEED SYSTEM: Pressurised tanks, with feed of 526 gr (1·16 lb) fuel and 838 gr (1·85 lb) oxidant per second at 16·4 bars (238 lb/sq in abs).

DIMENSIONS:
- Length overall: 457 mm (18·0 in)
- Nozzle exit diameter: 267 mm (10·5 in)

WEIGHT, DRY: 6·3 kg (14·0 lb)

PERFORMANCE RATINGS:
- Max thrust (vacuum): 3·87 kN (870 lb)
- Chamber pressure: 10·5 bars (152 lb/sq in)
- Specific impulse: 286

MARQUARDT R-30

This precision control rocket has been qualified for use on the kick stage of the USAF Defense Meteorological Satellite and NASA Tiros-No.

TYPE: Liquid monopropellant precision control rocket.

PROPELLANT: Hydrazine.

THRUST CHAMBER: Single chamber. Area ratio 22. Fabricated in L605 stainless steel. Shell 405 catalyst bed in combustion chamber. Multiple element injection.

THRUST CHAMBER MOUNTING: Fixed, by flange bolt circle on injector head.

PROPELLANT FEED SYSTEM: Pressurised tank. Flow rate 0·297 kg (0·657 lb)/sec at 31 bars (450 lb/sq in abs).

DIMENSIONS:
- Length overall: 273 mm (10·75 in)
- Diameter: 103 mm (4·0 in)

WEIGHT, DRY: 1·36 kg (3·0 lb)

PERFORMANCE RATINGS:
- Max thrust (vacuum): 0·69 kN (155 lb)
- Chamber pressure: 15·5 bars (225 lb/sq in)
- Specific impulse: 236

MARQUARDT R-1E

This small high-performance rocket was qualified for use on the reaction control subsystem of the subsequently-abandoned USAF Manned Orbital Laboratory, and now being qualified as the vernier engine for the Space Shuttle Orbiter.

TYPE: Liquid bipropellant rocket for use in space.

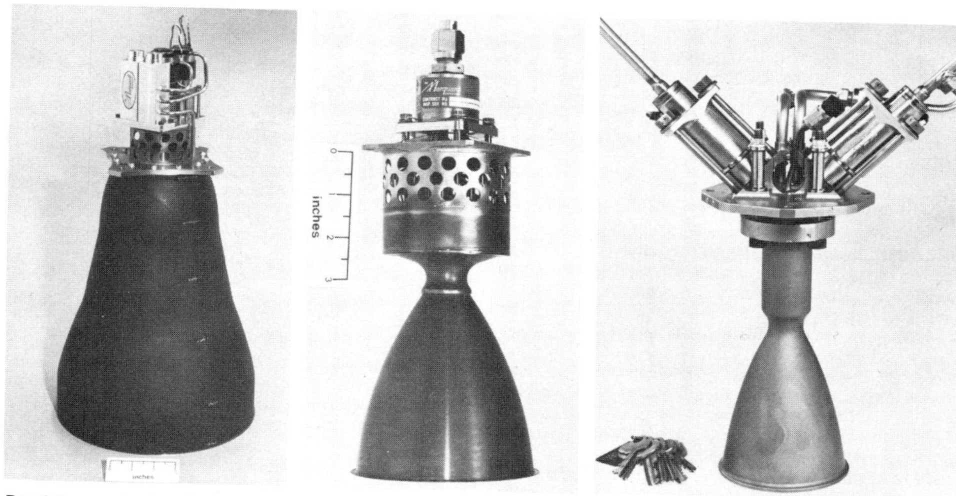

Precision control rocket engines by Marquardt *(from left to right)*: **R-40A 3·87 kN (870 lb), R-30 0·69 kN (155 lb), R-1E 0·11 kN (25·0 lb), all ratings being in vacuum**

Marquardt MA212-XAA conical-inlet supersonic ramjet (net thrust 6·05 kN; 1,360 lb at Mach 2·5 at 12,200 m; 40,000 ft)

PROPELLANTS: Nitrogen tetroxide and monomethyl hydrazine.

THRUST CHAMBER: Single chamber. Minimum area ratio 26 with orthogonal and scarfed nozzles. Made of metal-coated columbium. Insulated for buried installation. Started by electrical signal to on/off solenoid valve. Single doublet injector with hypergolic ignition.

THRUST CHAMBER MOUNTING: Fixed, by flange bolt circle on injector head.

PROPELLANT FEED SYSTEM: Pressurised tank. Flow rate 0·016 kg (0·0354 lb)/sec fuel and 0·0256 kg (0·565 lb)/sec oxidant.

DIMENSIONS:
- Length overall: 241 mm (9·5 in)
- Width: 145 mm (5·7 in)
- Height (depth): 103 mm (4·0 in)

WEIGHT, DRY: 1·0 kg (2·20 lb)

PERFORMANCE RATINGS:
- Max thrust (vacuum): 11·3 kg (25·0 lb)
- Chamber pressure: 7·45 bars (108 lb/sq in)
- Specific impulse (area ratio 40): 275

MARQUARDT MA210/212

Two versions of this low-cost ramjet engine have been developed. The pitot-inlet MA210-XAA is designed primarily for the high-subsonic speed regime of Mach 0·7 to 0·9, although it has been tested at Mach 2·0. The conical-inlet version, MA212-XAA, provides higher thrust and better cruise fuel consumption in supersonic flight conditions. The engines use the same combustor system with an interchangeable inlet-diffuser assembly selected for either subsonic or supersonic flight operation. Both versions are of single-wall all-steel construction with two-point mounting to the vehicle pylon. Ignition is by pyrotechnic flare.

DIMENSIONS:
- Length overall:
  - MA210-XAA: 2,070 mm (79 in)
  - MA212-XAA: 2,134 mm (84 in)
- Diameter: 381 mm (15 in)

Marquardt R-4D reaction control engines

WEIGHT, DRY:
- MA210-XAA: 31·75 kg (70 lb)
- MA212-XAA: 35·2 kg (77·5 lb)

PERFORMANCE:
- Operating envelope:
  - MA210-XAA: S/L to 9,145 m (30,000 ft), Mach 0·7-1·5
  - MA212-XAA: S/L to 18,300 m (60,000 ft), Mach 1·0-2·5
- Design thrust (net):
  - MA210-XAA: 2·94 kN (660 lb), S/L, Mach 0·9
  - MA212-XAA: 6·05 kN (1,360 lb), 12,200 m (40,000 ft), Mach 2·5

NELSON

NELSON AIRCRAFT CORPORATION

HEAD OFFICE:
PO Box 454, Irwin, Pennsylvania 15642
Telephone: (412) 863-5900

PRESIDENT:
Charles R. Rhoades

VICE-PRESIDENT:
Lawrence J. Rhoades

Nelson Aircraft Corporation, among its many industrial activities, produces to order the Nelson H-63 four-cylinder two-cycle aircooled engine, which has been certificated by the FAA as a power unit for single-seat helicopters, and is now available also as a power plant for propeller-driven aircraft. All these engines are capable of sustained inverted flight. Recommended overhaul period is 800 hours.

In 1975 Nelson completed testing the magneto and speed-reducing (M and R) versions, but these are not yet certificated. The liquid-cooled engines described are new.

NELSON FOUR-STROKE RANGE

Nelson Aircraft is developing a new range of lightweight four-stroke engines. They will be compact, and will have liquid cooling. The three sizes have outputs of 44·7 kW (60 hp), 89·5 kW (120 hp) and 112 kW (150 hp). Dry weights are, respectively, 39·95 kg (88 lb), 57·2 kg (126 lb) and 70·37 kg (155 lb).

NELSON H-63

US military designation: YO-65

Developed originally as a power unit for single-seat helicopters, the H-63 is now available in two versions, as follows:

H-63C. Basic helicopter power unit for vertical installation. Battery ignition and direct drive. Certificated by FAA. Supplied as complete power package, including clutch, cooling fan and shroud.

H-63CP. Basically as H-63C, but without clutch, fan and shroud. Intended primarily for installation in horizontal position, with direct drive to propeller. FAA certificated.

H-63CPM. Magneto ignition (see "Ignition" in detail description).

H-63CPMR, CPR. Versions with speed-reducing propeller drive (see "Power Take-off" in detail description).

Nelson has developed a 1·07 m (42 in) wooden propeller with glassfibre covering for use with the H-63. It is suitable for either tractor or pusher installation.

TYPE: Four-cylinder horizontally-opposed aircooled, two-stroke.

CYLINDERS: Bore 68·3 mm (2¹¹/₁₆ in). Stroke 70 mm (2¾ in). Total capacity 1·03 litres (63 cu in). Compression ratio 8 : 1. Each complete cylinder is machined from an aluminium alloy casting, the bore being porous-chrome plated for wear resistance. Cylinders bolted to and detachable from crankcase.

PISTONS: Aluminium alloy casting. Two piston rings. Two needle roller bearings pressed in boss. Piston (gudgeon) pin pressed into small end of connecting rod.

CONNECTING RODS: Alloy steel forging. Caged roller bearing at big-end.

CRANKSHAFT: Four-throw. Nitralloy shaft on ball and roller bearings.

CRANKCASE: Two-piece case divided on horizontal centreline. Each half is a magnesium alloy casting.

INDUCTION: Nelson diaphragm-type all-angle fuel control carburettor. Hot-air anti-icing. Fuel/oil mixture valves from crankcase through specially-designed rotary valve driven by crankshaft. Intake to and exhaust from cylinders through ports. Exhaust stacks are of aluminium alloy.

FUEL: 80/87 octane gasoline and SAE 30 kerosene-base oil in 8 : 1 mixture for fuel and lubrication.

IGNITION (except M models): Battery-type dual-ignition with automatic retard for starting. (M models): Two Slick magnetoes. Two Champion D-9 or 5 COM spark plugs per cylinder.

LUBRICATION: See under "Fuel".

POWER TAKE-OFF: (H-63C): Hollow shaft extension from Salisbury centrifugal clutch output drive. (R models): Hy-Vo high-speed chain drive with ratio of 8 : 5 (0·0625).

STARTING: 12V DC Autolite electric motor and Bendix drive.

COOLING: (H-63C): Centrifugal aluminium fan and two-piece glassfibre shrouding designed to maintain all temperatures within acceptable limits on an FAA hot day of 37·8°C, S/L. (Other versions): by propeller slipstream.

MOUNTING: Four Lord-type mounts, two on each half of crankcase.

DIMENSIONS (H-63C):

Length	508 mm (20·0 in)
Height	376 mm (14·8 in)
Width	605 mm (23·8 in)

WEIGHT, DRY:

H-63C, with accessories	34·5 kg (76 lb)
H-63CP, with accessories	30·8 kg (68 lb)
H-63CPM	30·9 kg (68·0 lb)
H-63CPR, H-63CPMR	38·6 kg (85·0 lb)

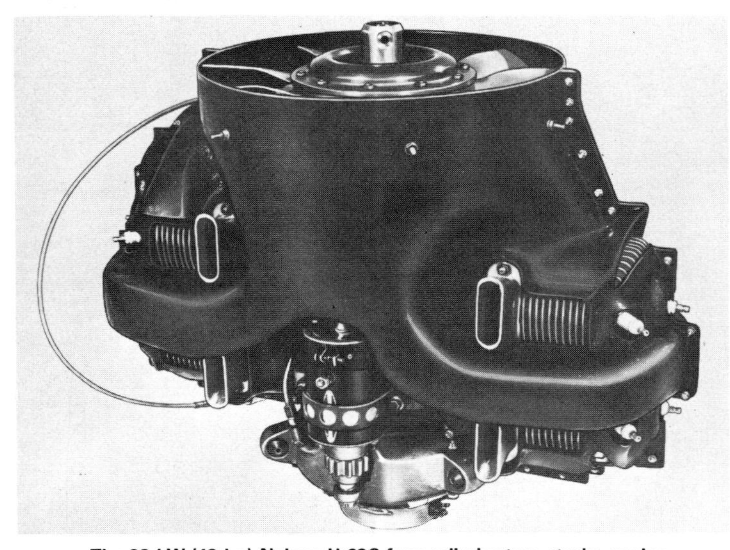

The 32 kW (43 hp) Nelson H-63C four-cylinder two-stroke engine

The 36 kW (48 hp) Nelson H-63CP for fixed-wing aircraft

First photograph of the Nelson H-63CPR with speed-reducing chain drive. Maximum rating will be 35·4 kW (47·5 hp)

POWER RATINGS:

T-O:

H-63C	32 kW (43 hp) at 4,000 rpm
H-63CP, H-63CPM	35·8 kW (48 hp) at 4,400 rpm
H-63CPR, H-63CPMR	35·4 kW (47·5 hp) at 2,750 output rpm

Max continuous:

H-63C	32 kW (43 hp) at 4,000 rpm
H-63CP, H-63CPM	33·6 kW (45 hp) at 4,000 rpm
H-63CPR, H-63CPMR	33·25 kW (44·6 hp) at 2,500 output rpm

CONSUMPTION:

Direct-drive

24 litres (6·3 US gal; 5·2 Imp gal)/hr at 4,000 rpm

R models

16 litres (4·2 US gal; 3·47 Imp gal)/hr at 2,500 output rpm

NORTHROP

NORTHROP CORPORATION, VENTURA DIVISION

HEAD OFFICE AND MAIN PLANT:
1515 Rancho Conejo Boulevard, Newbury Park, California 91320

Telephone: (805) 498-3131

In 1972 Northrop Corporation acquired the rights to this engine from McCulloch Corporation (see 1972-73 *Jane's*). The 4318 series continues in production at Ventura Division, to power the MQM-36 Shelduck (see "RPVs" section).

NORTHROP MODEL 4318F

US military designation: O-100-3

TYPE: Four-cylinder horizontally-opposed air-cooled two-stroke.

CYLINDERS: Bore 80·8 mm (3³/₁₆). Stroke 79·4 mm (3⅛ in). Displacement 1·6 litres (100 cu in). Compression ratio 7·8 : 1. Heat-treated die-cast aluminium with integral head and hard chrome plated walls. Self-locking nuts secure cylinders to crankcase studs.

PISTONS: Heat-treated cast aluminium. Two rings above pins. Piston (gudgeon) pins of case-hardened steel.

CONNECTING RODS: Forged steel. "Free-roll" silver-plated bearings at big-end. Small-end carries needle bearing. Lateral position of rod controlled by thrust washers between piston pin bosses and small-end.

CRANKSHAFT: Four-throw one-piece steel forging on four anti-friction bearings, two ball and two needle, one with split race for centre main bearing.

CRANKCASE: One-piece heat-treated permanent-mould aluminium casting, closed at rear end with cast aluminium cover which provides mounting for magneto.

VALVE GEAR: Fuel mixture for scavenging and power stroke introduced to cylinders through crankshaft-driven rotary valves and ported cylinders.

INDUCTION: Crankcase pumping type. Diaphragm-type carburettor with adjustable jet.

FUEL SPECIFICATION: Grade 100/130 aviation fuel mixed in the ratio 20 parts fuel with one part 40SAE two-cycle outboard motor oil (or 30 parts fuel to one part Super Red oil).

IGNITION: Single magneto and distributor. Directly connected to crankshaft through impulse coupling for easy starting. Radio noise suppressor included. BG type RB 916S, AC type 83P or Champion REM-38R spark plugs. Complete radio shielding.

LUBRICATION: Oil mixed with fuel as in conventional two-stroke engines.

PROPELLER DRIVE: RH tractor. Keyed taper shaft.

STARTING: By separate portable hydraulic starter.

MOUNTING: Three mounting lugs provided with socket for rubber mounting bushings.

DIMENSIONS:

Length	686 mm (27·0 in)
Width	711 mm (28·0 in)
Height	381 mm (15·0 in)

WEIGHT, DRY:

Less propeller hub	34·9 kg (77 lb)

POWER RATING:

Rated output 62·6-71·6 kW (84-96 hp) at 4,100 rpm

Production of Northrop Ventura RPVs powered by O-100-3 piston engines rated at 62·6-71·6 kW (84-96 hp)

SPECIFIC CONSUMPTION:

Fuel/oil mixture	152·1 g/J (0·90 lb/hr/shp)

PRATT & WHITNEY

THE PRATT & WHITNEY AIRCRAFT GROUP OF UNITED TECHNOLOGIES CORPORATION

GROUP HEADQUARTERS:
East Hartford, Connecticut 06108
Telephone: (203) 565-4321
GROUP PRESIDENT:
Bruce N. Torell
GROUP EXECUTIVE VICE-PRESIDENT, TECHNOLOGY AND STRATEGIC PLANNING:
Richard J. Coar
GROUP MANAGER PUBLIC RELATIONS:
Robert H. Zaiman

Commercial Products Division
East Hartford, Connecticut
DIVISION PRESIDENT:
David J. Hines

Government Products Division
West Palm Beach, Florida
DIVISION PRESIDENT:
Edmund V. Marshall

Manufacturing Division
DIVISION PRESIDENT:
Donald Nigro

Pratt & Whitney Aircraft of Canada
See separate entry under Canada.

Pratt & Whitney Aircraft was formed in 1925 and rapidly became a world leader in aircraft piston engines. In 1947, when it had become the largest single division in United Aircraft Corporation, it built its first turboprop, and went into production with an Americanised Rolls-Royce turbojet. Since then Pratt & Whitney has become the world's largest producer of gas-turbine engines.

On 1 May 1975 United Aircraft changed its name to United Technologies Corporation (UTC), reflecting its diversified products. The former UTC (United Technology Center) was renamed Chemical Systems Division (see under CSD on an earlier page). On 23 April 1976 Pratt & Whitney was itself restructured as the Pratt & Whitney Aircraft Group, composed of four divisions listed above.

Commercial Products Division is responsible for commercial aircraft engines. Government Products Division is responsible for military engines. Manufacturing Division provides plant and facilities for making the products of the CPD and GPD. Pratt & Whitney Aircraft of Canada (P&WC), which has its own entry under that country on an earlier page, is responsible for engines for general aviation.

The Pratt & Whitney Aircraft Group has approximately 33,000 employees at locations in Connecticut, and a further 10,000 in Florida and Canada.

Excluding P&WC, the divisions of the Group had by 1976 manufactured 57,000 gas-turbine engines, nearly all of them for aircraft. These engines had accumulated over 380 million flight hours in military and commercial service. Most of this time has been logged by the JT3D, JT8D and JT9D turbofans on which the major part of the world's air transport is based.

In addition to its production series of engines, Pratt & Whitney was in February 1970 awarded a contract to develop two versions of its JTF22 demonstrator engine, an augmented turbofan. The F401 version powers the US Navy F-14B fighter; the F100 powers the Air Force F-15 and YF-16 fighters. Further important demonstrator work includes the Advanced Turbine Engine Gas Generator programme. The ATEGG programme, funded by the US Air Force, has been active for several years and is continuing to demonstrate the overall, integrated performance of advanced component technology in a full-scale engine.

In 1973 Pratt & Whitney launched a programme to develop the JT10D turbofan in the 89-133·5 kN (20,000-30,000 lb) st class. By 1974 agreement had been reached for the limited participation of MTU of West Germany and Fiat of Italy in the programme. As this edition went to press an announcement was awaited of a collaboration agreement with Rolls-Royce (1971) Ltd to join in the JT10D programme.

In August 1975 Pratt & Whitney and Rolls-Royce (1971) signed an agreement for joint further development of the Pegasus vectored-thrust turbofan, which includes an option to take up a manufacturing licence for the British engine. In 1971 Mitsubishi Heavy Industries of Japan was licensed to manufacture parts and assemble JT8D engines for the C-1 military transport.

The Royal Swedish Air Board is licensed to manufacture afterburning versions of the JT8D turbofan engine and its parts. This work is being handled by Volvo Flygmotor (which see).

Licence agreements for the manufacture of certain of the Division's piston engines or piston-engine parts continue with Commonwealth Aircraft Corporation of Australia; Fiat of Italy; and Metal Leve SA of Brazil.

Pratt & Whitney J52-P-408 two-shaft turbojet rated at 49·7 kN (11,200 lb) st

Pratt & Whitney J58 (JT11D-20B) turbojet with afterburner for the Lockheed SR-71

PRATT & WHITNEY JT12

US military designation: J60

The J60 is a small high-performance turbojet engine which has a nine-stage axial-flow compressor, cannular type combustion section with eight flame tubes, and a two-stage turbine. The rotor runs in three bearings. Pressure ratio is 6·5 : 1.

Design studies began in July 1957 and the prototype ran in May 1958. The first prototype (50 hour engine) was delivered in July 1959, with T-O rating of 13 kN (2,900 lb). Delivery of production engines, rated at up to 14·7 kN (3,300 lb), began in October 1960, and over 2,900 have been shipped.

Versions of the J60/JT12 power the Lockheed JetStar, and the Rockwell International T-39 Sabreliner and Sabre 75, and T-2B Buckeye trainer. A turboshaft version is the JFTD12, described later.

DIMENSIONS:
Diameter 556 mm (21·9 in)
Length 1,981 mm (78·0 in)
WEIGHT, DRY: 212 kg (468 lb)
PERFORMANCE RATINGS: see above

PRATT & WHITNEY JT8

US military designation: J52

The J52 is a medium-sized turbojet which was designed under the auspices of the US Navy Bureau of Weapons. It powers all versions of the Grumman A-6 Intruder/Prowler attack and ECM aircraft, current versions of the McDonnell Douglas A-4 Skyhawk, and the North American Rockwell AGM-28 Hound Dog air-to-surface missile, as listed hereafter:

J52-P-3. Rated at 33·4 kN (7,500 lb). Powers AGM-28 Hound Dog.

J52-P-6A, 6B, 8A, 8B. Rated at 37·8 kN (8,500 lb) (6A, 6B) or 41·4 kN (9,300 lb) (8A, 8B). Powers A-4E, A-4F, TA-4F, TA-4J, A-6A, A-6B, A-6C, A-6E.

J52-P-408. Rated at 50 kN (11,200 lb). Powers A-4F, A-4M, some export A-4 versions, EA-6B.

The J52 is a two-spool turbojet, with total of 12 compressor stages, a "cannular" type combustion system fed by 36 dual-orifice injectors and independent high-pressure and low-pressure single-stage turbines. Pressure ratio is 14·5 : 1. Several advanced design features are incorporated to achieve the rating increases with a minimum change in engine envelope and weight compared to previous JT8 (J52) models. These include two-position inlet guide vanes and aircooled first-stage turbine vanes and blades. In addition, the burner cans include features for reduced smoke.

DIMENSIONS:
Diameter 766 mm (30·15 in)
Length 3,018 mm (118·5 in)
WEIGHT, DRY:
J52-P-6A, B 933 kg (2,056 lb)
J52-P-8A, B 961 kg (2,118 lb)
J52-P-408 1,052 kg (2,318 lb)
PERFORMANCE RATINGS:
See under model listings

PRATT & WHITNEY JT11

US military designation: J58

The J58 is an advanced single-spool turbojet engine designed for operation at flight speeds in excess of Mach 3·0, and at very high altitudes. It is rated in the 133·5 kN (30,000 lb) class.

The general configuration of the J58 is shown in the accompanying illustration. It is fitted with an advanced control system which governs automatically the variable intake, fuel supply and variable-area nozzle.

Two J58s (JT11D-20) form the power plant of the YF-12A and SR-71 versions of the Lockheed A-11 military aircraft, giving them a Mach 3 cruising performance. The power plant installation was described in the entry for the Lockheed YF-12A and SR-71 in the 1974-75 *Jane's*.

PRATT & WHITNEY JT3D

US military designation: TF33

The JT3D is a turbofan version of the J57 turbojet, handling almost 2·5 times more air than the J57 and with pressure ratio ranging from 13 : 1 on the JT3D-1 to 16·1 : 1 on the JT3D-8A (TF33-P-7).

Evolution from the J57 involved removal of the first three stages of the J57 compressor and replacement by two fan stages. Of considerably larger diameter than the compressor, the fan extends well outside the compressor casing. The third-stage turbine on the J57 was enlarged and a fourth stage added to provide the power necessary to drive the low-pressure compressor rotor and integral fan. A new short discharge duct was designed to exhaust the fan air well forward on the engine nacelle just after it has passed through the fan. Some JT3D and military TF33 installations have a full-length by-pass duct to a single reverser and nozzle handling the whole mass flow.

The JT3D produces 50% more take-off and 27% more cruise thrust than the J57, while giving a 13% better cruising specific fuel consumption.

Flight trials in a B-52 Stratofortress bomber and Boeing 707 and DC-8 transports began in 1960. The JT3D powers all late versions of these aircraft. The Lockheed C-141A StarLifter military transport uses the TF33-P-7 version, with an additional stage of compression. In January 1973 the -7 engine, modified to incorporate additional accessory drives, was selected to power the Boeing E-3A (AWACS) aircraft. Designation of the E-3A engine is TF33-PW-100A (JT3D-8B).

More than 8,300 JT3D turbofans, including converted JT3C engines, had been delivered by the end of 1975. Additional engines remain to be delivered.

DIMENSIONS:
Diameter:
JT3D-3B 1,350 mm (53·14 in)
TF33-PW-100A 1,373 mm (54·06 in)

Length:
JT3D-3B 3,479 mm (137 in)
TF33-PW-100A 3,607 mm (142 in)
WEIGHT, DRY:
JT3D-3B 1,969 kg (4,340 lb)
TF33-PW-100A 2,173 kg (4,790 lb)
PERFORMANCE RATINGS (T-O, S/L, static):
JT3D-3B 80 kN (18,000 lb)
TF33-PW-100A 93·4 kN (21,000 lb)
SPECIFIC FUEL CONSUMPTION (T-O rating):
JT3D-3B 15·5 mg/Ns (0·535 lb/hr/lb st)
TF33-PW-100A 15·86 mg/Ns (0·560 lb/hr/lb st)

PRATT & WHITNEY JT8D

This turbofan engine was developed as a company-sponsored project to power short/medium-range transport aircraft, including the Boeing 727. United Aircraft's French licensee, SNECMA, designed the complete power plant nacelle, with thrust reverser, used in the Caravelle Super B airliner, and later developed and produced the advanced quiet plug nozzle for the -15 engine for the Mercure. In March 1971 Mitsubishi was licensed to manufacture the -9 engine for the Kawasaki C-1 transport.

Construction of the JT8D is largely of steel and titanium. An annular by-pass duct runs the full length of the engine, with balanced mixing of the hot and cold air streams in the tailpipe.

Manufacture of prototype engnes began in November 1960 and the first engine run was made in April 1961. Flight testing was carried out initially under a B-45 testbed aircraft. Prototype engines for airframe testing were delivered to Boeing in 1962. Production deliveries began in the first half of 1963. By the end of 1975 more than 7,700 had been delivered and flight time exceeded 96 million hours with more than 140 operators.

The following are the basic versions:

JT8D-1, -1A, -1B. Initial version rated at 62·3 kN (14,000 lb st). Powers Boeing 727-100, -100C and -200, McDonnell Douglas DC-9-10 and -10F, and Aérospatiale Caravelle 10R.

JT8D-7, -7A, -7B. Develops 62·3 kN (14,000 lb st) to 28·9 °C at S/L. Specified for Boeing 727-100, -100C and 200, Boeing 737-100 and -200, McDonnell Douglas DC-9-10, -30 and -30F, Aérospatiale Caravelle 10R and 11R.

JT8D-9, -9A. Develops 64·5 kN (14,500 lb st) to 28·9°C at S/L. Specified for Boeing 727-100, -100C and 200, 737-100, -200, -200C and T-43A, McDonnell Douglas DC-9-20, -30, -40, C-9A and C-9B, Aérospatiale Caravelle 12 and Kawasaki C-1. Deliveries began in July 1967.

JT8D-11. Develops 66·7 kN (15,000 lb st) to 28·9°C at S/L. Specified for McDonnell Douglas DC-9-20, -30 and 40 series aircraft and Boeing 727-200. Deliveries began in November 1968.

JT8D-15. Develops 69 kN (15,500 lb st) to 28·9°C. FAA certification was recieved and deliveries began in April 1971. Selected for Dassault Mercure, Boeing Advanced 727 and 737, and DC-9. The -15 has completed an endurance test of 6,000 take-off cycles, equivalent to 2-3 years of normal operation.

JT8D-17. Develops 71·2 kN (16,000 lb st) to 28·9°C. Certificated on 1 February 1974 (tenth anniversary of JT8D-1 entry to airline service). Entered service July 1974. More than 400 ordered by December 1975 for advanced versions of Boeing 727 and 737, DC-9 and McDonnell Douglas YC-15 prototypes.

JT8D-17R. Latest version for Advanced 727, 737 and DC-9. New rating senses any significant thrust loss in any other engine and automatically provides ATT (Alternate Take-off Thrust) of 77·40 kN (17,400 lb) to 25°C. Certification and first deliveries were due in Spring 1976.

JT8D-100 Series. Under NASA contract, P&WA has designed, built and tested refanned versions of the JT8D which offer increased thrust and reduced noise. The new fan has a single stage, and its diameter is increased by 220 mm (8·7 in). On 9 January 1975 flight testing started with a DC-9 re-engined with two JT8D-109 engines (converted JT8D-9) each rated at 71·2 kN (16,000 lb st). On the same day, running began of a Boeing 727 rear section fitted with a JT8D-115 engine (converted JT8D-15) at Boardman, Oregon.

Pratt & Whitney TF33-P-7 turbofan engine rated at 93·4 kN (21,000 lb) st

Pratt & Whitney JT8D-17R two-shaft turbofan rated at 77·39 kN (17,400 lb) st to 25°C

JT8D-200 Series. The first members of this series are the JT8D-209, rated at 80 kN (18,000 lb st) and the JT8D-217, rated at 84·5 kN (19,000 lb st). These reduced-noise derivatives of the JT8D family are substantially redesigned. They combine the HP compressor, HP turbine spool and combustion section of the existing JT8D (the JT8D-209 and -217 use JT8D-9 and -17 components, respectively) with advanced LP technology derived from the NASA JT8D Refan Programme and other recently developed P&WA engines. The 200 Series offers substantially increased thrust with reduced noise and specific fuel consumption, together with the established reliability and low maintenance cost of the JT8D HP spool. The new single-stage fan has increased diameter. The new six-stage LP compressor, integral with the fan, offers increased pressure ratio. The LP turbine has 20 per cent greater annular area and achieves a higher efficiency. Surrounding the engine is a new by-pass duct. A JT8D-209 is to be test flown on one of the McDonnell Douglas YC-15 prototype transports (which see).

Since January 1970 all new JT8D engines have incorporated smoke-reduction hardware, and conversion kits are available for in-service engines. Two noise-reduction options are also available for all JT8D models. Maximum TBO for the JT8D is 16,800 hours.

A supersonic military version of the JT8D with afterburning was developed and is manufactured under licence in Sweden by Volvo Flygmotor AB. The Swedish engine, designated RM8, powers the Mach 2 Saab 37 Viggen multi-purpose combat aircraft. The first production engine was shipped on 28 October 1970.

TYPE: Axial-flow two-spool turbofan.

AIR INTAKE: Annular with 19 fixed inlet guide vanes.

FAN: Two-stage front fan (200 Series, single stage). First stage has 30 titanium blades dovetailed into discs. First-stage blades have integral shroud at about 61% span. Airflow: -1, -1A, -7, -7A, 143 kg (315 lb)/sec; -9, 9A, 144 kg (318 lb)/sec; -11, 146 kg (321 lb)/sec; -15, 147 kg (324 lb)/sec; -17, 148 kg (327 lb)/sec; -209, 217, 213 kg (470 lb)/sec. By-pass ratio: -1, -1A, -7, 7A, 1·10; -9, -9A, 1·03; -11, 1·01; -15, -17, 0·99; -209, 217, 1·65.

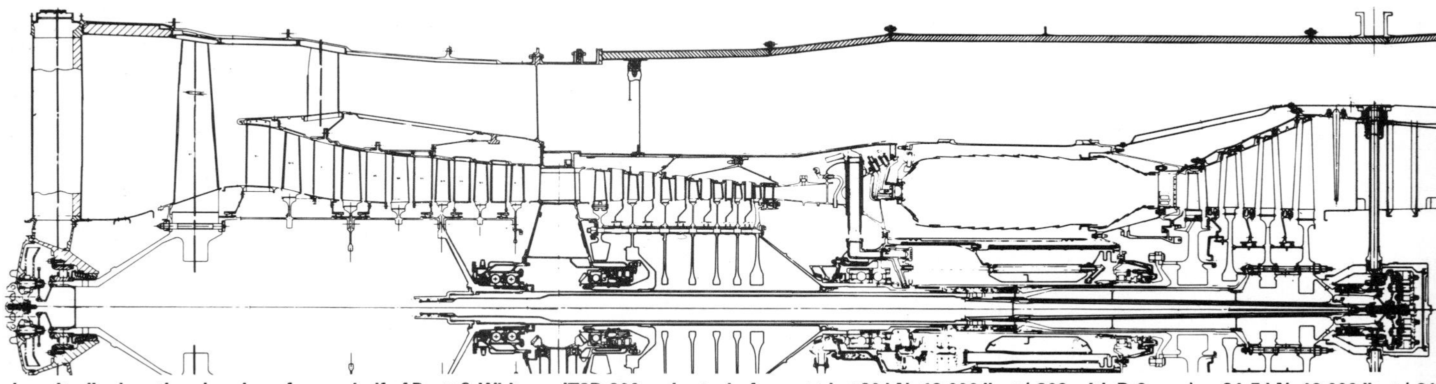

Longitudinal section drawing of upper half of Pratt & Whitney JT8D-200 series turbofan, rated at 80 kN; 18,000 lb st (-209 with D-9 core) or 84·5 kN; 19,000 lb st (-217 with D-17 core)

LP Compressor: Six-stage axial-flow, integral with fan stages, on inner of two concentric shafts. Blades made of titanium. Shaft carried in double ball bearings, either half of each bearing being able to handle the complete loading.
HP Compressor: Seven-stage axial-flow on outer hollow shaft which, like the inner shaft, is carried in double ball bearings. One-piece casing. Blades made of steel or titanium. Pressure ratio: -1, -1A, -7, -7A, 16·2; -9, -9A, 16·9; -11, 17·5; -15, 18; -17, 17·6; -209, -217, 18.
Combustion Chamber: Cannular type with nine cylindrical flame-tubes, each downstream of a single Duplex burner and discharging into a single annular nozzle.
HP Turbine: Single-stage axial-flow. Solid blades in -1 to 9, aircooled in -11 and later; guide vanes hollow and aircooled in all models.
LP Turbine: Three-stage axial-flow. Solid blades and guide vanes.
Dimensions:
Diameter:
-1 to -17 1,080 mm (42·5 in)
-209, -217 1,431 mm (56·34 in)
Length:
-1 to -17 3,048 mm (120·0 in)
-209, -217 3,282 mm (129·2 in)
Weight, Dry:
JT8D-1, -1A, -1B 1,431 kg (3,155 lb)
JT8D-7, -7A, -7B 1,454 kg (3,205 lb)
JT8D-9, -9A 1,475 kg (3,252 lb)
JT8D-11, -15 1,501 kg (3,309 lb)
JT8D-17 1,510 kg (3,330 lb)
JT8D-209 1,860 kg (4,100 lb)
JT8D-217 1,896 kg (4,180 lb)
Performance Ratings:
T-O thrust (S/L, static): see model descriptions
Max cruise thrust (10,665 m; 35,000 ft at Mach 0·8):
JT8D-1, -1A 15·7 kN (3,520 lb)
JT8D-7, -7A 16·1 kN (3,630 lb)
JT8D-9, -9A 16·7 kN (3,760 lb)
JT8D-11 17·6 kN (3,950 lb)
JT8D-15 18·3 kN (4,120 lb)
JT8D-17 18·9 kN (4,240 lb)
(at 9,145 m; 30,000 ft at Mach 0·8)
JT8D-209 23·6 kN (5,290 lb)
JT8D-217 26·3 kN (5,890 lb)
Specific Fuel Consumption:
T-O rating:
JT8D-1, -1A, -7, -7A
16·57 mg/Ns (0·585 lb/hr/lb st)
JT8D-9, 9A 16·85 mg/Ns (0·595 lb/hr/lb st)
JT8D-11 17·56 mg/Ns (0·620 lb/hr/lb st)
JT8D-15 17·84 mg/Ns (0·630 lb/hr/lb st)
JT8D-17 18·27 mg/Ns (0·645 lb/hr/lb st)
Max cruise rating, as above:
JT8D-1, -1A 22·24 mg/Ns (0·785 lb/hr/lb st)
JT8D-7, -7A 22·38 mg/Ns (0·790 lb/hr/lb st)
JT8D-9, -9A 22·86 mg/Ns (0·807 lb/hr/lb st)
JT8D-11 23·14 mg/Ns (0·817 lb/hr/lb st)
JT8D-15 23·45 mg/Ns (0·828 lb/hr/lb st)
JT8D-17 23·62 mg/Ns (0·834 lb/hr/lb st)

PRATT & WHITNEY JT9D

US military designation (JT9D-7): F105-PW-100

Based on technology stemming from the USAF heavy freighter propulsion of 1961-63, the JT9D was the first of the new era of very large, high by-pass ratio turbofans on which the design of the present generation of wide-body commercial transports rests.

The main advances in the JT9D are: (1) improved fan design to achieve the desired pressure ratio at high efficiency from a single stage with no inlet guide vanes; (2) improved compressor to attain a pressure ratio of 24 : 1 in 15 stages, compared with 14 : 1 in 16 stages for the JT3D; (3) improved combustion chamber to give greater temperature rise in appreciably shorter length than in previous engines, with lower pressure loss and better exit temperature distribution (and able to use smoke-reduction technology from the outset); (4) new high-temperature materials and cooling systems to allow a substantial rise in turbine gas temperature; (5) a controlled-vortex turbine design, allowing much higher stage-loadings (effectively eliminating two turbine stages); and (6) design features which enable thrust to be more than doubled with considerably less noise.

In its basic design the JT9D is compact, being shorter than the JT3D, and has two shafts, each supported in two bearings. In cruising flight the installed sfc is 22-23% lower than for the JT3D or JT8D. Careful attention has been paid to maintenance. The engine is made in ten modules which can be individually removed in short times, as demonstrated in numerous tests. By adding brackets and twin rails to the airframe the engine can be dismantled on the airframe. There are 21 borescope ports for inspecting all stages of blading and the combustion section; and provision is made to facilitate chip detectors, eddy current, ultrasonic and radioisotope inspection.

Company investment in facilities for the JT9D programme exceeds $100 million for production and $38 million for engineering development.

First run of the JT9D was in December 1966, and first engine flight test, with the engine mounted on the starboard inboard pylon of a Boeing B-52E, was in June 1968. The first flight of the Boeing 747 occurred on 9 February 1969. The DC-10-40 flew on 28 February 1972.

Pratt & Whitney JT9D-59A turbofan, rated at 236 kN (53,000 lb st)

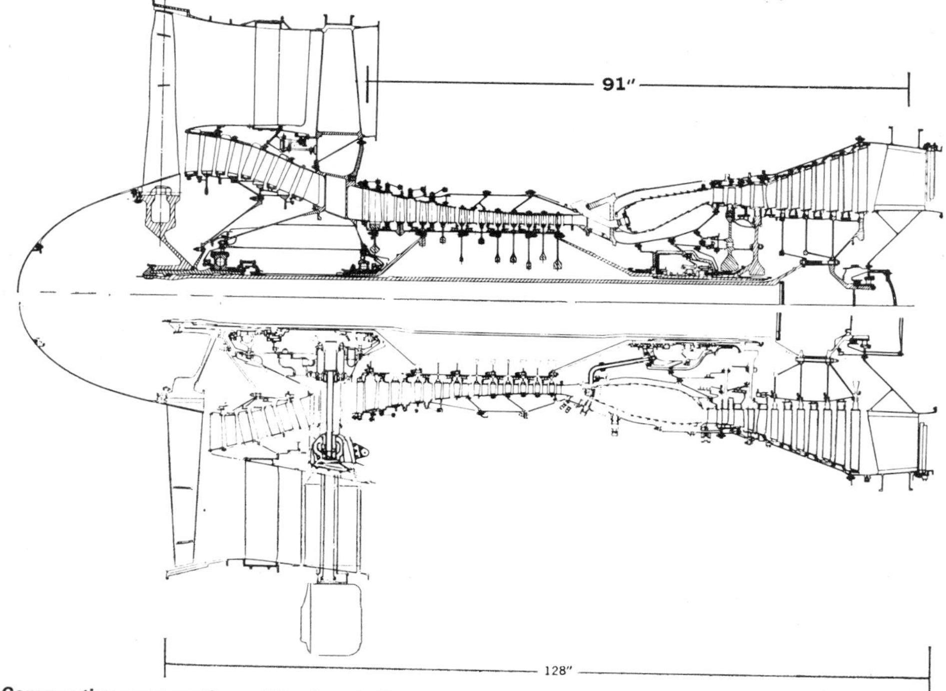

Comparative cross-sections of the Pratt & Whitney JT9D-20 *(lower half)* **and JT9D-59 turbofans, showing the latter's redesigned fan, LP compressor, combustion chamber and turbine**

Versions of the JT9D include:

JT9D-3. The initial production model, rated at 193·5 kN (43,500 lb) to 26·7°C. Fitted to first production Boeing 747. Engines delivered from April 1969 and certificated the following month.

JT9D-3A. Incorporates water injection for wet rating of 200 kN (45,000 lb) to 26·7°C. Powers Boeing 747-100 and -200B. Engines delivered from December 1969 and certificated on 9 January 1970.

JT9D-7. This engine incorporates improvements resulting from -3A service experience. The LP compressor has blades and vanes sloped back perpendicular to the inclined core airflow for increased stability and life; pylon-matched fan exit vanes reduce sfc; HP compressor discs have a longer life, and the stators are driven through a low-friction mechanism; a short-cone hooded burner increases durability and reduces smoke emission far below the visible level; changes to HP and LP turbines increase life, and improved HP disc sealing improves performance. The -7 was certificated in June 1971. It powers the 747-200B, C, F and SR, raising the certificated take-off weight from 322,050 to 351,530 kg (710,000 to 775,000 lb). On 30 November 1971 the 747-200 was certificated at full weight and thrust, and with a fixed-inlet cowl, quieter than the original type with blow-in doors (104 EPNdB "traded" compared with 112).

In 1972, Pratt & Whitney stated that, while late JT9D-3A engines were suffering only half as many engine-caused unscheduled removals as the first engines, the more powerful -7 engine had "exhibited a removal rate four times better than these improved 3As", with only two removals attributable to engine problems in its first 75,000 hours— "a record unmatched by any engine".

JT9D-7A. This incorporates a number of aerodynamic improvements which provide higher component efficiencies. The result is an increased thrust capability at the same turbine temperature. This has been reflected in rating increases over the JT9D-7. A version of the -7A, with improved component efficiencies to reduce specific fuel consumption, powers the 747SP.

JT9D-7F. Aerodynamically identical to the JT9D-7A, the -7F has first- and second-stage turbine rotor blades and second-stage stator vanes of directionally solidified material, allowing a rise in turbine gas temperature. This is reflected in further increase in thrust. The -7F permits operation at the Boeing 747 basic structural limit of 362,870 kg (800,000 lb). The -7F was certificated in September 1974; first deliveries were made in March 1975.

JT9D-20. This engine, which replaced the JT9D-15 and JT9D-25 (mentioned in the 1972-73 *Jane's* and originally intended for installation in the DC-10-20), has the same ratings as the JT9D-15 except that the take-off rating with water injection has been increased 220 kN; 49,400 lb to 30°C. With this engine, the DC-10-20 was redesignated DC-10-40 and certificated at a gross weight of 530,000 lb for dry operation. At the wet rating **(JT9D-20W)** it is certificated at 251,745 kg (555,000 lb). The D-20 is similar to the D-7, except for external configuration changes such as accessory-gearbox location, thrust-transmitting points and plumbing hardware locations. The gearbox is under the fan exit casing, and the new mounting has enabled the "Thrust frame" yoke (added to earlier engines to prevent ovalising of the casing) to be eliminated. The D-20 was certificated in October 1972.

JT9D-59A. This engine is the first member of the family of growth versions to be selected to power the DC-10. It evolved from an intensive component development programme begun in 1970, which led to the running of a complete experimental engine at 276 kN (62,000 lb st). The Dash 59A differs from earlier JT9D engines mainly in the following respects: the fan has a diameter approximately 25·4 mm (one inch) larger and reprofiled blades of higher efficiency; the low-pressure compressor has a zero (fourth) stage and is completely redesigned and the whole hot end is entirely redesigned. The burners are recontoured, an HP turbine carbon seal is added, the HP turbine rotor blades are of directionally solidified PWA 1422 superalloy, the HP turbine annulus is of greater area, and the LP turbine is mechanically and aerodynamically redesigned. The carcase of the engine is stressed for 249 kN (56,000 lb st). With a dry rating of 236 kN (53,000 lb) the JT9D-59A was certificated on 12 December 1974; production deliveries began in January 1975. These engines are configured for installation in a common nacelle, developed jointly by P&WA and Rohr Industries, for both the 747 and DC-10. The growth potential of this size of engine is predicted at 267 kN (60,000 lb).

JT9D-70A. This is the corresponding growth version of the JT9D for the Boeing 747. The engine was certificated on 12 December 1974, and first deliveries were made in January 1975 at a rating of 236 kN (53,000 lb). First application is the 747F certificated at 362,870 kg (800,000 lb).

Since entry to service on 21 January 1970 the JT9D has gained experience in the 747 more rapidly than any previous engine. Within one year 653 engines had been delivered, and early in 1973 the total exceeded 1,132. Rate of delivery has since slowed but the total now exceeds 1,550 and flight time in early 1976 was in excess of 14 million hours.

The following description applies to early versions of the JT9D, with data for later models given in parentheses:

TYPE: Two-shaft turbofan of high by-pass ratio.

INTAKE: Direct pitot, annular fixed geometry (except that airframe inlet on early 747 aircraft has blow-in side doors around periphery). No inlet guide vanes ahead of fan. Airflow improved by rotating spinner.

FAN: Single stage, with 46 titanium blades of 4·6 aspect ratio and two part-span shrouds held by dovetails in steel LP rotor. Downstream are 108 aluminium alloy exit guide vanes, followed by nine discharge-case radial struts. Fan case of stainless steel and aluminium alloy, designed to contain fan blades. Discharge case lined with perforated acoustic material. Nominal airflow 684 kg (1,509 lb)/sec at 3,650 rpm (-7, 698 kg; 1,540 lb/sec at 3,750 rpm; -59A, -70A, 734 kg; 1,619 lb/sec). Pressure ratio; typically 1·6 : 1. By-pass ratio: -3A, 5·17 : 1; 7, 5·15 : 1; -59A, -70A, 4·9 : 1.

LP COMPRESSOR: Three stages (JT9D-59A and -70A four different stages), rotating with fan. Rotor made up of rings, spacers and conical disc splined to steel LP shaft and held by lock-nut ahead of fan and overhung ahead of main LP ball thrust bearing. Hydraulically opened bleed ring at LP exit to increase flight-idle stall margin and excess air during deceleration. Rotor stages have 124, 132 and 130 dovetailed blades of titanium alloy. First stator stage anti-iced by 9th stage bleed air. Stator stages have 88, 128 and 126 titanium vanes and 120 (4th stage) nickel alloy vanes, all riveted to outer rings. Casing of aluminium alloy. Core airflow typically 118 kg (260 lb)/sec (all versions).

HP COMPRESSOR: Eleven stages. All stages have rings or centreless discs with integral spacers carried on conical discs at 3rd and 11th stages on HP shaft of titanium alloy (front) and high-nickel alloy (rear), bolted at rear hub. Rotor stages have 60, 84, 102, 100, 110, 108, 104, 94 and 100 dovetailed titanium blades and 102 and 90 nickel alloy blades. Stator has 76, 70, 80, 106, 100 and 112 titanium alloy vanes and 126, 146, 154, 158 and 92 vanes of nickel alloy, all brazed to inner and outer rings. First four stator stages are variable, positioned by hydraulic actuator to provide adequate stall margin for starting, acceleration and part-power operation. Casing of titanium alloys (last two stages, nickel alloy) has bleed ports supplying 8th-stage air for airframe requirements. Max HP speed: -3A, 7,580 rpm; -7, 8,000 rpm. Overall engine pressure ratio: -3A, 21·5 : 1; 7, 22·3 : 1; -59A, -70A, 24 : 1.

COMBUSTION CHAMBER: The diffuser case, which extends from the HP compressor to the midpoint of the combustion section, incorporates two sets of bleed ports for 15th-stage air for airframe requirements. The forward set takes air from the outside case via an integral manifold and the rear set bleeds air from the inner diameter via four of the ten radial struts. The combustor itself is fabricated in nickel alloy and is annular, with the forward end of the liner extended in 20 conical primary zones held in 20 burners fed from external fuel manifolds. In early models (-3A, -7 and -20), the outer casing can be slid forward over the diffuser for access to the HP turbine. Ignition by dual AC 4-joule capacitor system serving two plugs just above chamber centreline on each side.

FUEL SYSTEM: Pressure type with hydraulic control system operating at up to 76 bars (1,100 lb/sq in). Main components are fuel control, pump, fuel/air heater and fuel/oil heat exchanger. Provision for water injection, as customer option, with regulator, piping and spray nozzles, adds 18·1 kg (40 lb) to engine weight.

FUEL GRADE: P&W specification PWA 522.

HP TURBINE: Two stages. Both have high-nickel discs splined to HP shaft, secured by lock-nut, carrying high-nickel blades in fir-tree roots; first stage has 116 aircooled blades and second has 138 solid blades. Stators have 66 and 90 high-nickel alloy vanes, both rows aircooled. Turbine inlet temperature (-3A, max T-O), typically 1,243°C (-59A, -70A, 1,350-1,370°C).

LP TURBINE: Four stages. Stages have 108, 126, 122 and 116 solid nickel alloy blades held in fir-tree roots in discs of nickel alloy (last disc, iron alloy). Stators have 122, 120, 110 and 102 solid nickel alloy vanes. Exhaust gas temperature after turbine, typically 452°C (-3A) or 482°C (-7, -20).

JET PIPE: Fixed Inconel assembly, with large central plug cone.

REVERSER: Fan duct reverser comprises a translating sleeve (the rearmost portion of fan duct) which moves aft, causing long links to close the blocker doors and simultaneously pulling aft the cascade vanes. Primary (core) reverser, largely of Inconel 625, uses fixed cascades which are uncovered by aft movement of translating sleeves to which are hinged blocker doors pulled by links against the central nozzle plug. No primary reverser is used on -59A and -70A.

ACCESSORY DRIVES: Main accessory gearbox driven by tower bevel shaft from front of HP spool and mounted under central diffuser case (-20, -59A, -70A, under fan discharge case). Main driven accessories include CSD, fuel pump and control, starter, hydraulic pump, alternator and N_2 tachometer; Boeing 747 includes primary reverser motor and the DC-10-40 a second hydraulic pump and a fuel boost pump (747 has electric tank pump). The box also includes numerous lubrication system items, and provides for hand-turning the HP spool during borescope inspection.

LUBRICATION SYSTEM: Pressure feed through fuel/oil cooler to four main bearings and return through scavenge pumps (-20 also centrifugal scavenge) to 18·8-37·6 litre (5-10 US gal; 4·16-8·32 Imp gal) tank.

Pratt & Whitney JT8D refanned (JT8D-209 and -217 are similar externally) rated at 80 or 84·5 kN (18,000 or 19,000 lb st)

Pratt & Whitney JT9D-7 turbofan, rated at 202·5 kN (45,500 lb st)

OIL GRADE: PWA 521 (blend of synthetic and/or mineral oils).

MOUNTING: From above, in two planes. Front mount (-3A, -7) is double flange at top of fan discharge case, absorbing vertical and side loads. On -20, -59A, -70A the mount is rectangular block above intermediate case, taking vertical and side loads, and thrust brackets at 40° each side of vertical on intermediate-case outer flange. Rear mount (-3A, -7) in double flange above casing in plane of turbine LP bearing, to which engine thrust is transmitted via Y-shaped thrust frame from intermediate case in arrangement that prevents thrust loads reaching (and distorting) the turbine exhaust case. On 20, -59A, -70A, the frame is eliminated and rear mount takes only vertical, side and torsional loads.

STARTING: Pneumatic, by HamStan PS 700 or AiResearch ATS100-384 (DC-10, PS 700 only). Supplied at 2·76-3·10 bars (40-45 lb/sq in) from APU, ground cart or cross-bleed.

DIMENSIONS:

JT9D-3A, -7, -7A, -7F, -20:

Diameter	2,428 mm (95·6 in)
Length (:ange to flange)	3,256 mm (128·2 in)

JT9D-59A, -70A:

Diameter	2,466 mm (97·7 in)
Length	3,358 mm (132·2 in)

WEIGHT, DRY:

Guaranteed, including standard equipment:

JT9D-3A	3,905 kg (8,608 lb)
JT9D-7, -7A	3,982 kg (8,780 lb)
JT9D-7F	4,036 kg (8,900 lb)
JT9D-20	3,833 kg (8,450 lb)
JT9D-59A	4,116 kg (9,075 lb)
JT9D-70A	4,123 kg (9,090 lb)

PERFORMANCE RATINGS:

T-O, dry:

JT9D-3A	193·5 kN (43,500 lb)
JT9D-7	202·5 kN (45,500 lb)
JT9D-7A	205·25 kN (46,150 lb) to 26·7°C
JT9D-7F*	213·5 kN (48,000 lb) to 26·7°C
JT9D-20*	206 kN (46,300 lb)
JT9D-59A, -70A*	206 kN (53,000 lb) to 30°C

T-O, wet:

JT9D-3A	200 kN (45,000 lb) to 26·7°C
JT9D-7	209 kN (47,000 lb)
JT9D-7A	212 kN (47,670 lb) to 30°C
JT9D-7F*	222·5 kN (50,000 lb) to 30°C
JT9D-20*	220 kN (49,400 lb) to 30°C

Max cruise performance, 10,665 m (35,000 ft) at Mach 0·85:

JT9D-3A, -7	44·3 kN (9,950 lb)
JT9D-7A	46·3 kN (10,400 lb)
JT9D-7F	49·2 kN (11,050 lb)
JT9D-20*	47·5 kN (10,680 lb)
JT9D-59A*, -70A*	53·2 kN (11,950 lb)

SPECIFIC FUEL CONSUMPTION:

At dry T-O rating, S/L static, ISA:

JT9D-3A	9·80 mg/Ns (0·346 lb/hr/lb st)
JT9D-7	10·08 mg/Ns (0·356 lb/hr/lb st)
JT9D-7A	10·11 mg/Ns (0·357 lb/hr/lb st)
JT9D-7F	10·34 mg/Ns (0·365 lb/hr/lb st)
JT9D-20*	9·89 mg/Ns (0·349 lb/hr/lb st)
JT9D-59A, -70A*	10·57 mg/Ns (0·373 lb/hr/lb st)

Cruise, Mach 0·85, 10,665 m (35,000 ft):

JT9D-3A	17·84 mg/Ns (0·630 lb/hr/lb st)
JT9D-7	18·01 mg/Ns (0·636 lb/hr/lb st)
JT9D-7A	18·16 mg/Ns (0·641 lb/hr/lb st)
JT9D-7F	18·55 mg/Ns (0·655 lb/hr/lb st)
JT9D-20*	17·67 mg/Ns (0·624 lb/hr/lb st)
JT9D-59A*, -70A*	17·93 mg/Ns (0·633 lb/hr/lb st)

Ideal nozzles

PRATT & WHITNEY JT10D

When Pratt & Whitney Aircraft explored the design and marketing possibilities for an engine transferring JT9D technology to a lower range of thrusts, studies hardened on the range 89-133·5 kN (20,000-30,000 lb st), to meet future requirements of commercial transport aircraft. After more than two and a half years of design and component testing, the first JT10D flight-weight demonstrator engine ran on 9 August 1974. On 19 August, it achieved a thrust of 102·3 kN (23,000 lb).

Pratt & Whitney JT10D demonstrator, run at 102·3 kN (23,000 lb) st

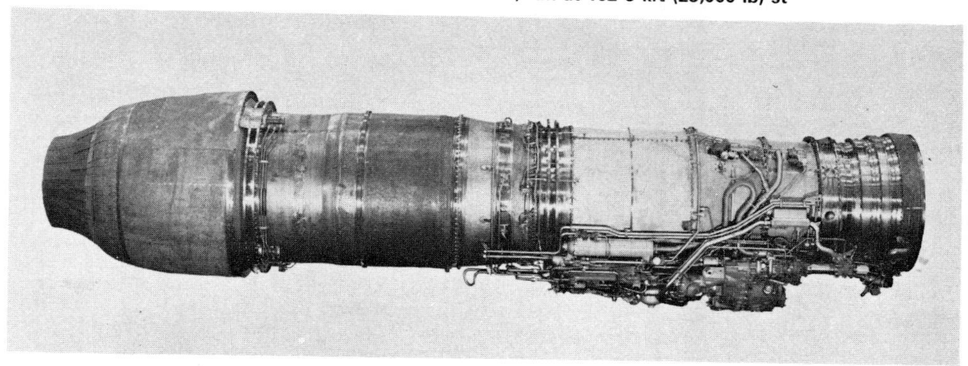

Pratt & Whitney TF30-P-412 afterburning turbofan, rated at 93 kN (20,900 lb) st

P&WA has been working with airframe manufacturers and airlines to develop an engine configuration that will result in minimum trip fuel consumption with low emission levels, and will reduce the man-hours required for engine installation and flight-line maintenance. The JT10D uses the JT9D modular maintenance concept, but with fewer parts per module. It consists of seven modules, each individually replaceable. No trim balancing on the engine of the rotating elements is required, since each is prebalanced in its case. In addition, the individual parts in the modules are designed to be readily accessible, and easily removed and replaced. The engine incorporates low-noise features of the JT9D, including the use of a high by-pass ratio, a single-stage fan without inlet guide vanes, wide axial separation between the blade and vane rows, and a moderate fan tip speed. The engine is designed to be compatible with acoustically treated nacelles to achieve noise levels below FAR.36 requirements.

P&WA is continuing to test the JT10D demonstrator engine. By early 1975 design work had been initiated on a derivative of the demonstrator, designated JT10D-2. This engine with a take-off thrust of 109 kN (24,500 lb), will be designed to meet requirements for the following potential applications: derivatives of the Boeing 707 and McDonnell Douglas DC-8 or DC-9, the Boeing 7X7 family, Lockheed Advanced Transport Aircraft (ATA), Japanese YX and European CAST aircraft. In addition, the engine offers improved performance in potential military applications such as the AMST, re-engined KC-135, and a re-engined or stretched C-141. Target date for JT10D-2 certification is November 1979.

Participating with P&WA in the development of the

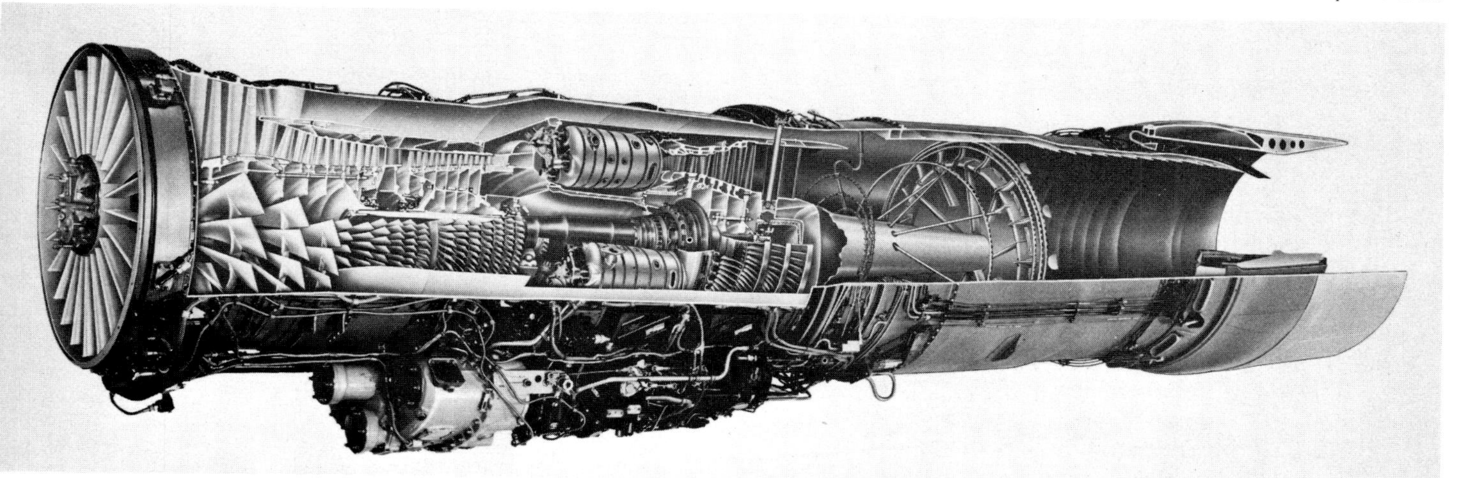

Cutaway drawing of Pratt & Whitney TF30-P-100 afterburning turbofan, rated at 111·7 kN (25,100 lb) st

JT10D are Rolls-Royce (1971) Ltd of Britain, Motoren- und Turbinen-Union GmbH (MTU) of Federal Germany and Fiat SpA of Italy. Rolls-Royce (1971) Ltd is likely to be a major collaborator, bearing about 34 per cent of the programme; P&WA is expected to bear 54 per cent, MTU 10 per cent and Fiat 2 per cent. Rolls-Royce engineers have been participating in work at Hartford since September 1975, but a formal agreement had not been announced in March 1976.

The following details refer to the JT10D-2:

TYPE: Two-shaft turbofan of high by-pass ratio.

AIR INTAKE: Direct front entry. No inlet guide vanes or anti-icing.

FAN: Single stage. Titanium forged hub with 40 inserted titanium alloy blades with part-span shrouds. Downstream are single row of exit guide vanes and radial struts supporting fan case. Rotating fan spinner. By-pass ratio 5·65 : 1.

HP COMPRESSOR: Twelve stages. Overall engine pressure ratio 28 : 1.

COMBUSTION CHAMBER: Annular, with flame tube fabricated in nickel alloy.

HP TURBINE: Two-stage axial with aircooled blades.

LP TURBINE: Four-stage axial.

WEIGHT, DRY:	2,177 kg (4,800 lb)
PERFORMANCE RATING:	
T-O (S/L, static)	109 kN (24,500 lb)

PRATT & WHITNEY JTF10A

US military designation: TF30

Development of this high-compression two-spool turbofan was begun in 1958 as a private venture, and resulted in testing of the first turbofan with afterburning. It was chosen subsequently as the power plant for the General Dynamics F-111.

The version used initially in the F-111 was designated TF30-P-1 (JTF10A-20) which provides 82·3 kN (18,500 lb st) with afterburning. It was superseded in the F-111A by the TF30-P-3 (JTF10A-21) which provides the same thrust with reduced sea level supersonic specific fuel consumption. The F-111D is powered by the TF30-P-9 (JTF10A-36) engine with afterburning, rated at 87·2 kN (19,600 lb st). The FB-111 bomber is equipped with the TF30-P-7 (JTF10A-27D) engine, which is in the 89 kN (20,000 lb) thrust class with afterburning. The F-111F is equipped with the TF30-P-100 (JTF10A-32C) engine, an advanced version with higher thrust. The Vought A-7A and A-7B Corsair II tactical attack aircraft are powered by the TF30-P-6 (JTF10A-8) and TF30-P-8 (JTF10A-9), these being simplified versions without afterburning and rated at 50·5 kN (11,350 lb) and 54·3 kN (12,200 lb st) respectively. TF30-P-8 engines are being converted to TF30-P-408 (JTF10A-16A) standard, with a thrust rating of 59·6 kN (13,400 lb st).

In July 1965, the TF30-P-1 completed successfully its official ground tests for military qualification, involving two 150-hour tests, with 12½ hours of full-power operation in simulated Mach 1·2 flight at sea level. In November 1966, the TF30-P-3 successfully completed a 150-hour military qualification test, with 56·25 hours of simulated Mach 2·2, and 12·5 hours of simulated Mach 1·2 flight at sea level.

Most recent application of the TF30 is the US Navy's Grumman F-14A Tomcat fighter, powered by the **TF30-P-412,** a modified version of the TF30-P-12, with a revised form of afterburning nozzle. The P-412 has an afterburning rating of 93 kN (20,900 lb st).

The most advanced TF30 production version was the USAF TF30-P-100, qualified in January 1971, in which weight is held below 1,815 kg (4,000 lb) while increasing thrust to the 111 kN (25,000 lb) class, with reduced fuel consumption.

Time between overhauls of the early models (P-3 with afterburner and P-6 without) reached 1,000 hr in 1972. The TBO of the P-100 was then at 450 hr, and planned to climb in stages, reaching 1,000 hr in 1975. A similar TBO progression is planned for the -412, currently at 450 hr.

TYPE: Two-shaft axial-flow turbofan.

INTAKE: Direct pitot annular type with 23 fixed inlet guide vanes (19 on P-8 and P-408). Hollow vanes pass anti-icing air.

FAN: Three stages (two on P-8 and P-408). Rotor and stator and casings all of titanium. Three rotor stages have 28 (with part-span shrouds), 36 and 36 blades, all dovetailed; stator stages have 44, 44 and 48 blades, all rivet-retained. Pressure ratio 2·14 : 1. Mass flow typically 112 kg (247 lb)/sec (P-100 118 kg; 260 lb/sec).

LP COMPRESSOR: Six stages (seven on P-8 and P-408), constructed integrally with fan to form nine-stage spool. Wholly of titanium construction, except stator blades of steel.

HP COMPRESSOR: Seven stages, constructed mainly of nickel-based alloy.

COMBUSTION CHAMBER: Can-annular, with steel casing and eight Hastelloy X flame cans each held at the front by four dual-orifice burners. Spark igniters in chambers 4 and 5.

FUEL SYSTEM: HP system (above 69 bars; 1,000 lb/sq in), with conventional hydromechanical control. Main elements comprise fuel pump, filter, heater, fuel control, P & D valve and nozzles. Separate afterburner system for A/B engines. No water injection.

FUEL GRADE: JP-4, JP-5.

HP TURBINE: Single stage, with film-cooled nozzle guide vanes (stators) and aircooled rotor blades of cobalt-based alloy (P-100 vanes and blades of directionally solidified alloy). Max gas temperature, early models 1,137°C, P-100 1,316°C.

LP TURBINE: Three stages of nickel-based alloys. Rotor stages have 88, 86 and 72 fir-tree root blades. Gas temperature after turbine, typically 550°C.

JET PIPE (non-A/B engine): Simple steel pipe where fan airflow and core gas mix before passing through fixed nozzle.

AFTERBURNER: Diffuser leads to combustion section comprising double-wall outer duct and inner liner carrying five-zone combustion system. Ignition by auxiliary squirt in A/B diffuser, coupled with main squirt in No. 4 burner can which produces hot-streak of fuel through the turbine (P-100 engine, fully modulated light-up by 4-joule electrical ignition system). Max gas temperature 1,490°C.

NOZZLE (A/B engines): Primary nozzle has variable area, with six hinged segments actuated by engine-fuel rams (P-100, 18 iris segments translated along curved profile by six long-stroke rams). Ejector nozzle has six blow-in doors with free tail-feathers.

ACCESSORY DRIVES: Main gearbox under compressor, driven by bevel shaft from HP spool. Contains major elements of lubrication and breather systems. Drive pads at front and rear for main and A/B fuel pumps, main oil pump, N_2 tachometer, starter, fluid power pumps and power take-off.

LUBRICATION SYSTEM: Self-contained dry-sump hot-tank system. Accessory gearbox housing forms 15 litre (4 US gal; 3·3 Imp gal) tank. Oil circulated at 3·10 bars (45 lb/sq in) through pump, filter, coolers (air/oil on airframe, fuel/oil on engine and A/B fuel/oil cooler) and three main bearing components; returned by scavenge pumps and de-aerator.

OIL GRADE: MIL-L-7808.

MOUNTING: Two-planar. Front peripheral pair of flanges absorb vertical, side and thrust loads; rear pair of peripheral flanges (in line with No. 6 bearing behind LP turbine) absorb vertical and side loads.

STARTING: Air-turbine starter on left forward drive pad of accessory gearbox.

DIMENSIONS:	
Max diameter:	
TF30-P-412A	1,293 mm (50·9 in)
TF30-P-100	1,242 mm (48·88 in)
Length overall:	
TF30-P-412A	5,987 mm (235·7 in)
TF30-P-100	6,139 mm (241·7 in)
WEIGHT, DRY:	
TF30-P-412A	1,800 kg (3,969 lb)
TF30-P-100	1,807 kg (3,985 lb)
PERFORMANCE RATINGS (T-O, S/L):	
TF30-P-412A	93 kN (20,900 lb)
TF30-P-100	111·7 kN (25,100 lb)
SPECIFIC FUEL CONSUMPTION (T-O):	
TF30-P-100	69·40 mg/Ns (2·450 lb/hr/lb st)

PRATT & WHITNEY JTF22

US military designations: F100 and F401

Stemming partly from the JTF16 demonstrator engine designed in 1965-66, the JTF22 is an advanced-technology military turbofan with afterburner for highly supersonic applications. Basic development has been funded as a demonstrator programme for the US Air Force. In February 1970 the decision was taken to use the JTF22 core engine as the basis of two highly refined power units: the **F100-PW-100** (JTF22A-25A) for the twin-engined McDonnell Douglas F-15 Eagle fighter for the US Air Force, and the **F401-PW-400** (JTF22A-24A) for the twin-engined Grumman F-14B Tomcat fighter prototype for the US Navy. Subsequently, the F100 was adopted for the single-engined General Dynamics F-16. The F-16 engine is identical with the F-15 engine.

Some 3,000 hours of development testing were accomplished between 1968 and the 60 hr PFRT (preliminary flight rating test) in February 1972. The 150 hr QT (qualification test) was scheduled to be completed in early 1973, but very severe development difficulties resulted in this test not being passed until October 1973. Some of the problems involved catastrophic mechanical failure of the compressor and turbine, and the US Air Force set up a special F100 board of enquiry. Flight development, on the other hand, has gone well. In February 1976, out of a total running time of some 70,000 hours, more than 20,000 had been during development flying of the F-15, TF-15 and

Pratt & Whitney F100-PW-100 augmented turbofan for McDonnell Douglas F-15, rated at 106 kN (23,810 lb) st with full afterburning and shown also below in longitudinal cross-section

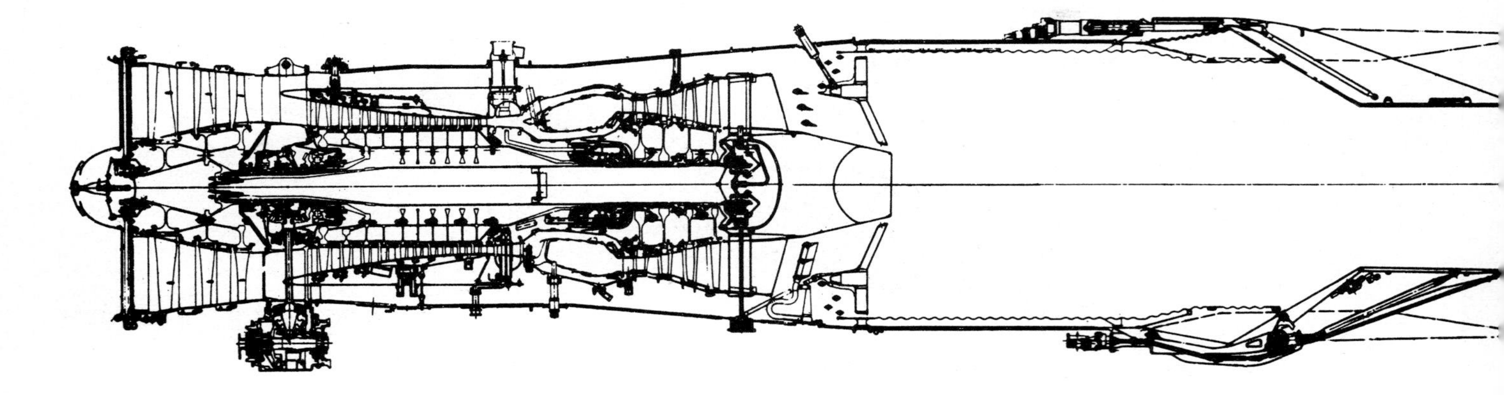

YF-16, and pilot opinions were generally highly complimentary. Production of an initial batch of 72 engines, for 30 F-15 and TF-15 aircraft for inventory service, was completed between June 1974 and April 1975. A second batch of 145 engines was delivered between May 1975 and April 1976. Delivery of operational aircraft to the US Air Force began in November 1974. By early 1976 more than 240 engines had been delivered; in mid-1976 output was to rise from 15 to 25 per month.

The prototype F401 ran in September 1972 and two flight-cleared engines flew in an F-14B development aircraft in June 1973. Though considerably more powerful and more efficient than the F100, the F401 is not a fully funded programme and the F-14B may not go into production, the stumbling-block being financial.

Type: Two-shaft turbofan with high-augmentation afterburner.

Intake: Direct pitot type. Fabricated titanium, with fixed nose bullet. Single row of 21 inlet guide vanes, with hot-air anti-iced leading-edges and variable-camber trailing-edge flaps.

Fan: Three stages (3½ in F401 which has added single-stage IP compressor downstream of fan to supercharge core). Fan blades have part-span shrouds. Discs of titanium 6-2-4-6, blades titanium 8-1-1. Entry diameter 928 mm (36·5 in) (F401 1,079 mm; 42·5 in). By-pass ratio 0·6 (F401 1·0).

Compressor: Ten-stage axial, on HP shaft. First three stages have variable stators. Discs 1-2, forged Ti 6-2-4-6; 3, forged Ti 8-1-1; 4, forged PWA 1016; 5, 7 and 9, PWA 1027; 6, 8 and 10, Gatorised (isothermal squeeze forging) IN-100. Blades 1-3, Ti 8-1-1; 4, 5, Ti 6-2-4-6; 6-9 Incoloy 901; 10, PWA 1005. Pressure ratio 8 : 1. Overall engine pressure ratio 23 : 1 (F401 26·9 : 1).

Combustion Chamber: Annular. Fabricated in nickel alloy with film cooling throughout. Large-diameter duplex fuel nozzles. Primary atomising stage used at all times, with secondary air-blast stage operative at high fuel flow rates. Capacitor-discharge ignition.

HP Turbine: Two stages. Discs forged IN-100. Blades PWA 1422 directionally solidified alloy with PWA 73 coating; first rotor impingement cooled, second with convective (HP bleed air) only. Maximum gas temperature 1,399°C (both engines). Maximum speed 14,650 rpm (F401 14,600 rpm).

LP Turbine: Two stages. Discs forged IN-100. Blades, uncooled, cast in IN-100 with PWA 73 coating. Modified LP turbine in F401. Maximum speed 9,600 rpm (both engines).

Afterburner: Four concentric spray rings in flow from core engine; three slightly further downstream in by-pass airflow. Flameholder assembly downstream of spray nozzles, with high-energy electrical ignition to give modulated light-up. Carcase, like by-pass duct and other major portions, fabricated in Stresskin stainless-steel sandwich. Interior liner of refractory material.

Nozzle: Multi-flap balanced-beam articulated nozzle giving very wide range in area and profile.

Control System: Unified hydromechanical fuel and nozzle-area control, with electronic engine control.

Dimensions:
- Overall diameter:
  - F100-PW-100: 1,180 mm (46·5 in)
  - F401: 1,283 mm (50·5 in)
- Intake diameter:
  - F100-PW-100: 884 mm (34·8 in)
  - F401: 1,079 mm (42·5 in)
- Length, excl bullet:
  - F100-PW-100: 4,851 mm (191·0 in)

Weight, Dry:
- F100-PW-100: 1,371 kg (3,020 lb)
- F401: 1,655 kg (3,649 lb)

Performance Ratings (S/L, ISA):
- Max T-O, dry:
  - F100-PW-100: 64 kN (14,375 lb)
  - F401: 73 kN (16,400 lb)
- Max T-O, augmented:
  - F100-PW-100: 106 kN (23,810 lb)
  - F401: 125 kN (28,090 lb)

Specific Fuel Consumption (S/L, ISA):
- Max T-O, dry:
  - F100-PW-100: 19·26 mg/Ns (0·68 lb/hr/lb st)
  - F401: 17·56 mg/Ns (0·62 lb/hr/lb st)
- Max T-O, augmented:
  - F100-PW-100: 72·23 mg/Ns (2·55 lb/hr/lb st)
  - F401: 69·40 mg/Ns (2·45 lb/hr/lb st)

PRATT & WHITNEY JFTD12

US military designation: T73

This free-turbine turboshaft engine consists basically of the gas generator of a JT12 turbojet with a two-stage free-turbine added downstream to provide a rear drive. The exhaust is taken out to one side, and in the case of the installation on the Sikorsky S-64 Skycrane helicopter one engine exhausts to port, the other to starboard. Ratings are 3,355 and 3,579 kW (4,500 and 4,800 shp). Production was completed at 351 engines. Further details were given in the 1975-76 *Jane's*.

PRATT & WHITNEY RL10

The RL10 rocket engine, for the propulsion of space vehicle upper stages, is a regeneratively cooled, turbopump-fed engine with a single chamber. The current RL10A-3-3 production version is rated at 66·7 kN (15,000 lb) thrust at an altitude of 61,000 m (200,000 ft), and has a nominal specific impuse of 444 seconds. Propellants are liquid oxygen and liquid hydrogen, injected at a nominal oxidiser-to-fuel mixture ratio of 5·0 : 1. Rated engine thrust is achieved at a nominal design chamber pressure of 27·6 bars (400 lb/sq in) absolute, with a nominal nozzle area ratio of 57 : 1. The engine can be used for multi-engine installation on an interchangeable basis and is capable of multiple starts after extended coast periods.

Pratt & Whitney RL10A-3-3 rocket engine

First deliveries were made in August 1960 for use in NASA's Centaur stage of the Atlas-Centaur rocket, which is powered by two RL10 engines. A six-engine cluster of RL10A-3 engines powered the S-IV stage of the Saturn I, achieving a perfect performance record for the entire launch programme. Over 9,000 RL10 firings have been accomplished, and 114 engines have flown on operational Saturn and Centaur vehicles, accomplishing 168 successful in-flight starts. In the Titan-Centaur 5 mission two RL10 engines accomplished a record seven in-flight starts, five of them after spacecraft separation. Between two starts there was a 5·25 hr coast at zero-*g*.

Advanced versions of the RL10 have been tested at the Pratt & Whitney Aircraft Florida Research and Development Center. These tests include variable-thrust operation, low idle operation, pumped idle operation, operation with a 205: 1 ratio nozzle extension, operation on fluorine/hydrogen, lox/propane and flox/methane propellants. NASA has ordered additional RL10 engines to power Centaur missions during the late 1970s.

ROCKETDYNE

ROCKETDYNE DIVISION OF ROCKWELL INTERNATIONAL

Headquarters:
6633 Canoga Ave, Canoga Park, California 91304
Telephone: (213) 884-4000
Other Facilities:
McGregor, Texas
Santa Susana, California

President:
N. J. Ryker

Executive Vice-President:
N. C. Reuel

Vice-Presidents:
E. B. Monteath (Advanced Programmes)
D. J. Sanchini (SSME Programme Manager)
S. J. Domokos (Laser Programmes)
O. I. Thorsen (Solid Rocket Division)
M. C. Ek (Engineering and Test)
J. C. McMillen (Finance and Administration)

Rocketdyne is a division of Rockwell International, devoted primarily to the design and manufacture of rocket engines for the US Air Force, Army and Navy and the National Aeronautics and Space Administration. It was established as a separate division on 8 November 1955.

Rocketdyne's work on liquid-propulsion engines is centred at Canoga Park, California, and work on solid motors at McGregor, Texas.

Rocketdyne liquid-propellant engines power more than three-quarters of all large US space vehicle stages, and powered all three stages of the Saturn V used in the Apollo programme and the Saturn IB used for Skylab launches in 1973 and 1974.

Current products of Rocketdyne's McGregor plant include propulsion systems for the US Navy's Sparrow III, Shrike, AIM-9C/D Sidewinder, Condor and Phoenix missiles, the Army's Chaparral, and miscellaneous turbine starters and gas generators.

Many Rocketdyne solid motors are filled with Flexadyne, a high-performance composite propellant unaffected by long-term storage at −60 to 76°C. By early 1975 more than 43,000 Flexadyne motors had been delivered. During 1973-74 Flexadyne technology transfer activity was accomplished with Federal Germany and Norway.

ROCKETDYNE SSME

On 13 July 1971, the Rocketdyne Division of Rockwell International was selected by the US National Aeronautics and Space Administration to design and develop the main engine for the orbiter stage of the US Space Shuttle. Three of these engines will provide a total of 6,272 kN (1,410,000 lb) vacuum thrust.

Two large solid-propellant boosters will be strapped on the sides of the orbiter's expendable propellant tank which will carry the liquid oxygen and liquid hydrogen for the three main engines in the oribiter. The orbiter rides piggyback on the propellant tank in a parallel configuration. The solid motors and the three Space Shuttle Main Engines (SSME) will produce 28,469 kN (6,400,000 lb st) to lift the vehicle from the pad in a conventional vertical flight path. The solid motors will burn out at about 40 km (25 miles) altitude, separate from the orbiter stage, and be lowered by parachutes into the ocean for recovery. The three main engines will continue to power the vehicle to near orbit; the external tank will then separate and be de-orbited and disposed in a safe area of the ocean. After mission completion the orbiter will re-enter the Earth's atmosphere and manoeuvre to a landing site for an unpowered horizontal landing similar to that of a conventional jet aircraft.

In overall configuration, the SSME is slightly smaller in size than the F-1 engine used in the Saturn V vehicle first stage. It burns liquid oxygen and liquid hydrogen propellants and has been designed for high reliability, reusability, multiple re-start capability and low cost. It will be capable of 7½ hours of burn time, accrued during 55 flights. Modified airline maintenance procedures will be used to service the engine between flights without removing it from the vehicle.

The design combines the merits of high-chamber-pressure operation, an optimum-performance contoured bell-shaped nozzle, and a regeneratively-cooled thrust chamber, capable of 11° gimballing, for maximum performance and long life. The chamber wall is cooled so efficiently that it is at 567°C, although the combustion temperature is about 3,300°C. No propellants are wasted in the cooling process. The combustion chamber wall is made of slotted metal, rather than tubes, using Rocketdyne-developed NaRloy-Z, a copper alloy that is easily machined, has higher strength than pure copper, and has very high thermal conductivity. Tubes are incorporated in the lower nozzle section.

The SSME is controlled by a unique system incorporating dual-redundant digital computers. This system monitors engine parameters such as pressure and temperature and the engine is automatically adjusted to operate at the required thrust and mixture ratio. The system also develops a record of engine operating history for maintenance purposes to improve serviceability and extend total engine life.

Rocketdyne moved into the hardware development phase in 1972, supported by two principal subcontractors: Honeywell Inc is designing the engine controller; Hydraulic Research and Manufacturing Co is designing the hydraulic actuators.

Combustion Chamber: Channel-wall construction with regenerative cooling by the hydrogen fuel. Concentric-element injector.

Turbopumps: Two low-pressure pumps boost the inlet pressures for two high-pressure pumps. Dual pre-burners provide turbine-drive gases to power the high-pressure pumps. Hydrogen-pump discharge pressure is 478·3 bars (6,937 lb/sq in) at 37,250 rpm; it develops 56,554 kW (75,840 hp).

Controller: Honeywell digital computer controller provides closed-loop engine control, in addition to data processing and signal conditioning for control, check-out, monitoring engine status, and maintenance data acquisition.

Controls: A hydraulic-actuation control system is used. The dual-redundant self-monitoring servo-actuators respond to signals from the controller to position the ball valves. A pneumatic system provides backup for the hydraulic system for engine cut-off.

Maintenance: Engine to be maintained using airline-type maintenance procedure for on-the-vehicle servicing. Time between overhauls is 55 flights or 7·5 hours of cumulative operation.

DIMENSIONS:

Length	4,242 mm (13 ft 11 in)
Diameter at nozzle exit	2,388 mm (7 ft 10 in)

PERFORMANCE:

S/L thrust (one engine)	1,668 kN (375,000 lb)
Vacuum thrust	2,091 kN (470,000 lb)
Specific impulse	455 sec
Chamber pressure	207 bars (3,000 lb/sq in)
Throttling ratio	2 : 1

ROCKETDYNE RS-27

The RS-27 power plant consists of an RS2701A main engine and two LR101-NA-11 vernier engines. The verniers provide vehicle control during flight and vehicle stabilisation prior to stage separation. The RS-27 is used as the booster propulsion system for the Delta launch vehicle, replacing the Rocketdyne MB-3 (USAF designation LR79) propulsion system.

The RS2701A is a single-chamber bipropellant fixed-thrust gimballed engine. It utilises liquid oxygen and RP-1 propellants at a nominal mixture ratio of 2·245 : 1. Its rated thrust is 93,000 kg (205,000 lb) at sea level, with a maximum duration of 242 seconds. The thrust and mixture ratio are controlled by fixed orifices. The engine is a hybrid design which utilises the turbopump, turbine, gas generator, valves and thrust chamber of the H-1 engine, and the control system, start system and component-packaging arrangement of the MB-3 engine.

DIMENSIONS:

Overall length	3,607 mm (11 ft 10 in)
Envelope max diameter	1,900 mm (6 ft 4 in)
WEIGHT, DRY (approx):	1,025 kg (2,261 lb)

ROCKETDYNE MA-5

USAF designations: YLR89-NA-7 booster and YLR105-NA-7 sustainer

The MA-5 propulsion system consists of a dual-chamber liquid-propellant booster engine, a single-chamber liquid-propellant sustainer engine, and two vernier engines to control vehicle roll and to trim final velocity and directional control after burnout of the sustainer. This propulsion system powers the Atlas-Agena and Atlas-Centaur launch vehicles. It is developed from the MA-2 system last described in the 1966-67 *Jane's*.

The design consists of two gimballed tubular-wall booster chambers, with twin-turbopump feed for the liquid oxygen and RP-1 propellants, and a single gimballed tubular-wall sustainer chamber, with similar feed. Ignition of both boosters and the sustainer engine takes place shortly before the vehicle is launched. Each YLR89-NA-7 booster is rated at 1,646 kN (370,000 lb) but can be derated to 1,495 kN (336,000 lb). The YLR105-NA-7 is rated at 267 kN (60,000 lb) but can be derated to 254 kN (57,000 lb).

Current production is continuing, and the MA-5 is being used to boost launches scheduled for the period 1976-81.

DIMENSIONS:

Length	2,490 mm (8 ft 2 in)
Diameter, nozzle exit	1,219 mm (4 ft)

WEIGHT, DRY:

Booster	1,372 kg (3,024 lb)
Sustainer	427 kg (941 lb)

ROCKETDYNE MODEL 16NS-1,000

The 16NS-1,000 was developed as a standard JATO unit for the USAF, but has wider applications. It consists of a steel cylinder, closed at the forward end. The igniter is located on the forward end, with the exhaust nozzle and pressure release diaphragm at the aft end. Thrust is transmitted to the aircraft attachment fittings through three mounting lugs welded on the cylinder.

DIMENSIONS:

Length	890 mm (2 ft 11 in)
Diameter	267 mm (10·5 in)

WEIGHTS:

Without propellant	48·4 kg (106·6 lb)
Complete	89·2 kg (196·9 lb)

PERFORMANCE RATING:

4·45 kN (1,000 lb) for 16 seconds

ROCKETDYNE MK 25 JATO

The Mk 25 was developed as a standard JATO unit for the US Navy and is used in three forms: Mod 0, with 30° canted nozzle, for launching the A-3 series aircraft; Mod 1, with 15° cant, for boosting the A-4 series; and Mod 2, with straight nozzle, for sled applications. The case is of 4130 steel and the RDS-135 solid propellant burns along inner and outer radii in the form of a sponge-supported cylindrical grain. Performance varies greatly with ambient temperature, a hot day giving much higher thrust for a shorter burn; at 15°C action-time thrust is 19·4 kN (4,360 lb) for 5·41 seconds. The 1,371 mm (54 in) long motor weighs 94 kg (208 lb) filled and 37·6 kg (83 lb) after firing.

ROCKETDYNE RS2101C

The Rocketdyne RS2101C is the Viking Orbiter engine which provided trajectory-correction and orbit-insertion for the Mars orbiter (with lander) which landed on that planet in 1976. The engine is derived from a family of similar 'football sized' engines (RS-21, described in 1972-73 *Jane's*) which provide gimballed axial thrust for space vehicles. An earlier model (RS2101) provided manoeuvring control for the Mars-Mariner 9, which was launched on 30 May 1971 and placed in orbit about Mars on 13 November 1971. The engine is powered by storable nitrogen tetroxide (oxidiser) and monomethyl-hydrazine (fuel), fed by gas pressure, and stabilised by acoustic-cavity damping. The RS2101 is the first rocket engine to have a chamber cooled by fuel to provide internal regenerative (interegen) cooling; its nozzle is radiation cooled. The engine can be gimballed ±9° in any direction through a throat-plane gimbal. A mechanically-linked torque-motor-operated bipropellant valve is integrally pinned above the aluminium multi-element unlike-double injector for impulse control. The injector is joined by the beryllium chamber to the L605 expansion nozzle. Designed to be fired further from Earth than any previous man-made engine, the RS2101C also has the unusually long continuous burn time of 45 minutes.

The first Rocketdyne SSME (Space Shuttle Main Engine) completed its installation in April 1975 at NASA's National Space Technology Laboratories at Bay St Louis, Miss. It has since been extensively tested (sea-level thrust 1,667 kN, 375,000 lb)

DIMENSIONS:

Length	554 mm (21·8 in)
Nozzle exit diameter	272 mm (10·72 in)
WEIGHT:	8·2 kg (18·0 lb)

PERFORMANCE (vacuum):

Thrust	1·33 kN (300 lb)
Specific impulse	292 sec
Chamber pressure	7·98 bars (115·7 lb/sq in)
Area ratio	60
Oxidiser flow rate	0·281 kg (0·62 lb)/sec
Fuel flow rate	0·186 kg (0·41 lb)/sec

Rocketdyne RS2101C rocket engine for Viking Orbiter

ROCKETDYNE P8E-9

Rocketdyne has been under contract to the US Army Missile Command since August 1971 to deliver production quantities of the P8E-9 rocket engine to power the Lance surface-to-surface missile; previous Lance propulsion had been procured through LTV, the prime contractor. The P8E-9 operates on storable bipropellants (IRFNA, inhibited red fuming nitric acid, and UDMH, unsymmetrical dimethyl hydrazine). The engine uses dual concentric thrust chambers for booster and sustainer operation, with propellants fed from pressurised tanks. Ignition is hypergolic, and chamber and nozzle cooling of both chambers is ablative.

Start of the system is initiated by ignition of the solid-propellant gas generator, which pressurises the propellant tanks, provides engine start, and provides a gas flow through side-mounted spin nozzles for missile rotation. Booster duration is 6 seconds, with both booster and sustainer engines operating; sustainer operation continues for 120 seconds with variable thrust. Booster thrust is in

Rocketdyne P8E-9 packaged liquid rocket engine (thrust chamber illustrated) for Lance, rated at 222·4-22·24 kN (50,000/5,000 lb)

the 222·4 kN (50,000 lb) class and sustainer thrust range is from 22·24 kN (5,000 lb) to zero. Area ratio values are 5·7 (booster) and 4·0 (sustainer). Mixture ratio is 3·4, chamber pressure is 65·5 bars (950 lb/sq in), and the motor measures 492·8 mm (19·4 in) long by 338·5 mm (21·2 in) diameter.

ROCKETDYNE MK 38/39

The Mk 38 solid-propellant rocket motor developed and produced by Rocketdyne for the Sparrow III air-to-air missile was the first to combine a special free-standing propellant charge (grain) with the company's Flexadyne propellant. Based on a carboxy-terminated linear polybutadiene fuel-binder, the new propellant provides a substantial increase in missile performance and has superior physical properties which give it resistance to cracking or tearing at extremely low temperatures. The motor has a diameter of 203 mm (8 in) and length of 1,316 mm (51·8 in).

The development contract for the motor was placed in 1961 and flight tests began successfully 12 months later. Development and qualification of the motor were completed in 22 months. The McGregor plant shortly thereafter, in July 1963, began manufacturing the Mk 39 motor for the AGM-45A Shrike anti-radar missile. This is similar in design and ballistic performance. Improved designs of these motors have been qualified, incorporating a case-bonded Flexadyne grain to provide 50 per cent more power with a corresponding increase in missile range.

ROCKETDYNE MK 36 MOD 5

The McGregor, Texas, plant of Rocketdyne received a development contract for motors for the AIM-9C and AIM-9D advanced versions of the Sidewinder air-to-air missile in 1963. Standard Sidewinder cases were loaded with the company's Flexadyne propellant and tested under temperature extremes ranging from sub-zero to over 150°C. They showed perfect reliability in over 200 firings during development and operational evaluation, and first production contracts were awarded in 1964.

Designated Mk 36 Mod 6, the Rocketdyne Sidewinder motor is approximately 1,830 mm (72 in) long, 127 mm (5 in) in diameter and contains 27 kg (60 lb) of Flexadyne propellant.

ROCKETDYNE MK 47

The Mk 47 Mod 0 solid-propulsion system for the Navy's AIM-54A Phoenix missile was developed at the McGregor plant since 1963. The first powered flight was in April 1966, two months after completion of the propulsion system development programme, in the course of which over 60 motors were subjected to such tests as multiple-temperature cycling, shock tests simulating catapult and arrested landings, and extensive vibration tests. Motor dimensions are 381 mm (15 in) diameter by 1,775 mm (69·9 in) long.

The Mk 47 motor utilises an improved version of Flexadyne, particularly adaptable to Phoenix missile requirements of high volumetric loading, high total impulse and long burning time, to provide the long-range missile operational capability required. The propellant has excellent ballistic properties, a 5-10 year shelf life and exhaust characteristics that minimise radar attenuation. Rocketdyne has successfully test fired similar propellant at −60°C in a large research motor after two complete temperature cyclings between −60 and 77°C. Rocketdyne is now in its fifth production run with this motor, under US Navy contract.

Rocketdyne Mk 47 Mod 0 solid-propellant motor for Hughes AIM-54A Phoenix air-to-air missile

ROCKETDYNE CONDOR MK 70 MOD 0

The Mk 70 Mod 0 rocket motor developed and produced by Rocketdyne-McGregor for the Condor air-ground missile utilises a Flexadyne solid-propellant grain, case-bonded with a stress-relieving liner. The liner allows the motor to perform over a wide range of severe temperature and dynamic conditions. Motor dimensions are 432 mm (17 in) diameter and 2,642 mm (104 in) long.

The development contract for the Mk 70 motor was placed in late 1969 and qualification was completed in late 1971. The motor has a record of 100 per cent reliability in missile flight tests. The all-up Condor missile assembly line began at the McGregor plant in 1974.

ROTORWAY
ROTORWAY INC

ADDRESS:
14805 S. Interstate 10, Tempe, Arizona 85284
Telephone: (602) 963-6652
GENERAL MANAGER: Al Newell

RotorWay Inc is a builder of small helicopters (see entry in "Aircraft" section). For a considerable period it has been developing its own power plant, for these and for other light aircraft. The following is preliminary information:

ROTORWAY FLAT-FOUR

As yet lacking a published designation, the engine designed and being developed by RotorWay is matched to the Scorpion Too light helicopter, but is expected to find a wide market. It has been designed with the following objectives in mind: high power/weight ratio; improved fuel economy; reduced noise and emissions; smooth operation; and long, reliable life.

TYPE: Horizontally-opposed, vertical-crankshaft, water-cooled four-stroke piston-engine.

CYLINDERS: Offset left and right for plain connecting rods side-by-side. Swept volume 2·19 litres (134 cu in). Compression ratio 10 : 1.

INDUCTION: Through circular air cleaner to single downdraught carburettor with fixed main jet and adjustable idle jet.

IGNITION: Aircraft-type magneto.

LUBRICATION: Oil temperature 82°-99°C. Oil pressure 2·72-4·1 bars (40-60 lb/sq in).

COOLING: Closed water system, operating temperature 82°C.

WEIGHT, DRY (with starter): 75 kg (165 lb)

PERFORMANCE Not disclosed, except that maximum torque is generated at operating speed of 4,500 rpm.

RotorWay water-cooled flat-four helicopter engine (power not yet disclosed)

TELEDYNE CAE
TELEDYNE CAE DIVISION OF TELEDYNE INC

HEAD OFFICE:
1330 Laskey Road, Toledo, Ohio 43697
Telephone: (419) 470-3000
PRESIDENT: James L. Murray
VICE-PRESIDENTS:
Henry C. Maskey (Engineering)
Eugene R. Sullivan (Finance)
Robert P. Schiller (Marketing)
Richard A. Myers (Operations)
David A. McQuillian (Industrial Relations)

Teledyne CAE (formerly Continental Aviation and Engineering) became a division of Teledyne Inc during 1969. Teledyne CAE has long experience in the design, development and production of gas-turbine engines, and is now devoted exclusively to turbine engine work.

The headquarters for management, marketing, finance, engineering and production is the Toledo, Ohio, facility of over 32,500 m² (350,000 sq ft).

From 1951 until 1960 almost all development was based on Turboméca designs. By far the most important of these was the Marboré, from which stemmed the J69 series of turbojets on which the manufacturing programme has depended. Since 1960 Teledyne CAE has embarked on an in-house development programme on a large scale. To a considerable degree the newer Teledyne CAE engines are aimed at target drones, unmanned reconnaissance aircraft and cruise-type guided missiles. The company has long claimed to be the largest maker of engines for unmanned applications.

TELEDYNE CAE 352 and 356
US military designation: J69

The J69 was originally the Turboméca Marboré, which has been developed to meet American requirements. Four versions are currently available as follows:

J69-T-25 (Teledyne CAE Model 352-5A). Long-life version, which powers the Cessna T-37B trainer and is FAA certificated as the Model CJ69-1025. Its air mass flow is 9 kg (19·8 lb)/sec. Operational ceiling is 13,720 m (45,000 ft).

J69-T-29 (Teledyne CAE Model 356-7A). Powers the Teledyne Ryan BQM-34A subsonic target drone. Operational ceiling is 18,300 m (60,000 ft). This is the Teledyne CAE counterpart to the Turboméca Gourdon turbojet, comprising a Marboré II with the addition of a single-stage transonic axial compressor supercharging the centrifugal stage.

J69-T-41A (Teledyne CAE Model 356-29A). Transonic axial compressor and revised centrifugal stage handling airflow of 13·5 kg (29·8 lb)/sec with pressure ratio of 5·45 : 1. Operational ceiling in excess of 21,030 m (69,000 ft). In production as improved version of J69-T-29 powering special-purpose subsonic RPVs.

YJ69-T-406 (Teledyne CAE Model 356-34A). In production for the US Navy's BQM-34E and the USAF's BQM-34F supersonic target drones. Initial qualification testing was completed during 1967 and deliveries of production engines began in 1970. The T-406 engine can propel the BQM-34E to Mach 1·5 at 18,300 m (60,000 ft) altitude. Future development of the T-406 engine involves addition of an advanced axial compressor stage and an

Teledyne CAE J69-T-29 turbojet of 7·56 kN (1,700 lb st)

Teledyne CAE 356-28E turbojet for unmanned high-altitude applications, rated at 18·7 kN (4,200 lb st)

afterburner for Mach 2·5 drone performance; another project would have an aircooled turbine.

The J69-T-29, YJ69-T-406 and J69-T-41A have a single-stage axial compressor ahead of the standard centrifugal compressor. Combustion system and turbine arrangements are basically the same as on the J69-T-25.

DIMENSIONS (nominal):
Length overall:
J69-T-25 — 899 mm (35·39 in)
YJ69-T-406, J69-T-41A and J69-T-29 — 1,138 mm (44·8 in)
Width:
J69-T-25 — 566 mm (22·30 in)
J69-T-41A, J69-T-29 — 568 mm (22·36 in)
YJ69-T-406 — 572 mm (22·52 in)

WEIGHT, DRY:
J69-T-25 — 165 kg (364 lb)
J69-T-29 — 154 kg (341 lb)
J69-T-41A — 159 kg (350 lb)
YJ69-T-406 — 163 kg (360 lb)

PERFORMANCE RATINGS:
Max rating:
J69-T-25 — 4·56 kN (1,025 lb) at 21,730 rpm
J69-T-29 — 7·56 kN (1,700 lb) at 22,000 rpm
J69-T-41A — 8·54 kN (1,920 lb) at 22,000 rpm
YJ69-T-406 — 8·54 kN (1,920 lb) at 22,150 rpm
Normal rating:
J69-T-25 — 3·91 kN (880 lb) at 20,700 rpm
J69-T-29 — 6·12 kN (1,375 lb) at 20,790 rpm
J69-T-41A — 7·34 kN (1,650 lb) at 20,900 rpm
YJ69-T-406 — 7·65 kN (1,719 lb) at 21,450 rpm

SPECIFIC FUEL CONSUMPTION:
At max rating:
J69-T-25 — 32·30 mg/Ns (1·14 lb/hr/lb st)
J69-T-41A, J69-T-29 — 31·16 mg/Ns (1·10 lb/hr/lb st)
YJ69-T-406 — 31·44 mg/Ns (1·11 lb/hr/lb st)
At normal rating:
J69-T-25 — 31·72 mg/Ns (1·12 lb/hr/lb st)
J69-T-29 — 30·73 mg/Ns (1·085 lb/hr/lb st)
J69-T-41A — 30·87 mg/Ns (1·09 lb/hr/lb st)
YJ69-T-406 — 31·16 mg/Ns (1·10 lb/hr/lb st)

TELEDYNE CAE 356-28A

US military designation: J100-CA-100

The Model 356-28A has been developed by Teledyne CAE as a power plant for RPVs and other unmanned aircraft. The engine is derived from the J69 family but has no parts in common with the J69 family. It has a two-stage transonic axial compressor ahead of the centrifugal stage, handling a mass flow of 20·4 kg (44·9 lb)/sec with a pressure ratio of 6·3 : 1. The combustion chamber is annular with centrifugal fuel injection. The turbine has two axial stages, each fitted with replaceable blades. Fixed geometry is used throughout, although the engine is at present operating at altitudes in excess of 22,860 m (75,000 ft).

The J100-CA-100 completed a 108 hour qualification test in June 1969. Applications include the Teledyne Ryan 147TE and 147TF medium-altitude intelligence-collection RPVs.

DIMENSIONS:
Length, intake flange to jet pipe flange — 1,225 mm (48·21 in)
Max width — 629 mm (24·75 in)
Max height — 663 mm (26·10 in)

WEIGHT, DRY: 195 kg (430 lb)

PERFORMANCE RATINGS:
Max — 12·01 kN (2,700 lb) at 20,700 rpm
Normal — 10·81 kN (2,430 lb) at 20,120 rpm

SPECIFIC FUEL CONSUMPTION:
At max rating — 31·16 mg/Ns (1·10 lb/hr/lb st)
At normal rating — 30·60 mg/Ns (1·08 lb/hr/lb st)

Teledyne CAE YJ69-T-406 turbojet of 8·54 kN (1,920 lb st)

Teledyne CAE J100-CA-100 turbojet of 12·01 kN (2,700 lb st)

TELEDYNE CAE 356-28E

Derived directly from the J100 (Model 356-28A), the Model 356-28E has a geared zero-stage which raises pressure ratio to about 8·1 : 1. The higher pressure is an essential stepping stone to further increase in operating altitude, one of the urgent demands on the company in meeting propulsion requirements for future unmanned aircraft. The 356-28E has a mass flow of 29·4 kg (65 lb)/sec. This engine has a centrifugal compressor, combustion chamber and turbine basically similar to the J100. Operating ceiling exceeds 27,430 m (90,000 ft).

DIMENSIONS:
Length, intake flange to jet pipe flange — 1,641 mm (64·6 in)
Basic overall diameter — 648 mm (25·5 in)

WEIGHT, DRY: 220 kg (485 l
PERFORMANCE RATING (T-O): 18·7 kN (4,200 l

TELEDYNE CAE 356-28F

US military designation: J100-CA-101

This version of the J100 has been optimised for lo altitude performance with minimal cost. Changes inclu a slight increase in shaft speed, revised radial-diffuser va angle and reduced turbine inlet nozzle area. Applicati has not been disclosed.

DIMENSIONS:
Length overall — 1,234 mm (48·6
Max width — 627 mm (24·7
Max height — 643 mm (25·3

WEIGHT, DRY: 195 kg (430

PERFORMANCE RATINGS:
Military S/L static 13·6 kN (3,050 lb)
Military S/L Mach 0·95 13·3 kN (3,000 lb)
SPECIFIC FUEL CONSUMPTION:
Military S/L static 31·16 mg/Ns (1·10 lb/hr/lb st)
Military S/L Mach 0·95 38·52 mg/Ns (1·36 lb/hr/lb st)

TELEDYNE CAE 365

US military designation: LJ95

This family of engines had its inception in a lift-jet, the Model 365-7, developed for the US Air Force as the XLJ95-T-1. Details remain classified, except that the engine is in the 22·24 kN (5,000 lb st) class, has an above-average turbine gas temperature and offers a ratio of thrust to weight exceeding 20 : 1, yet is intended for propulsion of manned aircraft. From the 365-7 unit Teledyne CAE has projected various cruise turbojets, of which one of the most important could be the 365-20, a possible candidate for the propulsion of very-high-performance (up to Mach 3 at above 22,860 m; 75,000 ft) target drones for the training of fighter pilots.

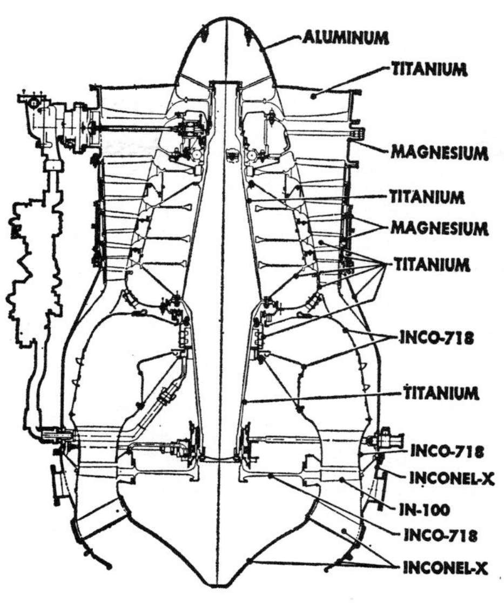

Cross-section drawing of XLJ95 lift-jet, detailing materials

Teledyne CAE XLJ95-T-1 lift-jet

TELEDYNE CAE 370

US military designation: J402-CA-400

This low-cost expendable engine was designed for the propulsion of cruise-type missiles and is in production for the US Navy AGM-84A and RGM-84A Harpoon missiles. The J402 is noteworthy for its compact component and accessory disposition, giving minimum frontal area. Though the entire design minimises production time and cost, high reliability was a prime requirement. Flight limits are 12,200 m (40,000 ft) and Mach 0·9 continuous or Mach 1·1 for limited periods. Engine life is reported unofficially to be 1 hr.

In 1974 the J402-CA-400 was selected as cruise power plant for the tactical versions of the US Navy Tomahawk Sea Launched Cruise Missile. In partnership with the competing SLCM contractors, Vought Systems Division and GD Convair, Teledyne CAE supported extensive testing, leading to selection of the GD tactical SLCM in mid-1976.

TYPE: Single-shaft turbojet.
INTAKE: Direct pitot inlet with four struts.
COMPRESSOR: Single transonic axial compressor with precision cast construction. Single centrifugal compressor with precision cast construction. Max airflow 4·35 kg (9·6 lb)/sec. Pressure ratio 5·8.
COMBUSTION CHAMBER: Annular type.
FUEL SYSTEM: Low-pressure supply to centrifugal injection nozzles in compressor shaft. Electronic control system with automatic sequencing and regulation to meet demands of missile flight profile.
TURBINE: Single-stage axial.
JET PIPE: Fixed-area.
ACCESSORIES: Pyrotechnic starting and ignition systems. Optional integral alternator and alternator regulator to give 6 kW of DC power.
MOUNTING: Four main mountings disposed radially around main (compressor diffuser) frame.
DIMENSIONS:
Length (excl bullet) 748 mm (29·44 in)
Overall diameter 318 mm (12·52 in)
WEIGHT, DRY: 45·36 kg (100 lb)
PERFORMANCE RATING:
Max S/L static 2·94 kN (660 lb) at 41,200 rpm
SPECIFIC FUEL CONSUMPTION (S/L, static):
34·0 mg/Ns (1·20 lb/hr/lb st)

Teledyne CAE J402-CA-400 expendable low-cost turbojet of 2·94 kN (660 lb st)

TELEDYNE CAE 372-2

US military designation: J402-CA-700

This turbojet is in production for the Beech MQM-107 variable-speed training target. It is based on the Model 370 (J402) but differs in detail engineering and equipment, reflecting the need for repeated missions of extended duration. The electronic fuel control governs engine operation throughout the starting cycle and over the whole operating range. A shaft-mounted high-speed alternator provides 1·2 kW of DC power. Engine life is unofficially reported to be 15 hr.

DIMENSIONS:
Length (excl bullet) 753 mm (29·65 in)
Overall diameter 317 mm (12·50 in)
WEIGHT, DRY: 52 kg (115 lb)
PERFORMANCE RATING:
Max S/L static 2·85 kN (640 lb) at 40,400 rpm
SPECIFIC FUEL CONSUMPTION (S/L, static):
33·71 mg/Ns (1·19 lb/hr/lb st)

TELEDYNE CAE 440/555

A possible basis for a wide family of advanced small engines for the period after 1975, the 440 and 555 core engines developed as a result of the company's participation in the US Air Force Advanced Turbine Engine Gas Generator programme. Like ATEGG studies by other companies, the Models 440 and 555 have design parameters (pressure ratio, turbine entry temperature and specific fuel consumption) similar to those of the most advanced large engines. Most likely applications of these engine cores would be in turbofans in the 13·3-22·24 kN (3,000-5,000 lb) thrust class for piloted aircraft or high-performance RPVs.

Teledyne CAE J402-CA-700 turbojet of 2·85 kN (640 lb st) for MQM-107 variable-speed training target

TELEDYNE CAE 471

Not yet allotted a US military designation, the Model 471-11D two-shaft turbofan was a candidate engine for the US Navy Tomahawk Sea Launched Cruise Missile in its strategic version. Teledyne CAE was awarded a contract for 471-11D development by Vought Systems Division, unsuccessful contender in the programme; so the future of the engine is uncertain.

The core engine comprises three axial compressor stages plus one centrifugal stage; an annular combustion chamber with slinger fuel injection; and a two-stage axial turbine. The fan has two stages and is driven by a two-stage LP turbine. Beyond the fact that the 471 is in the 2·67 kN (600 lb) st class, all other design and performance details are classified. It is unofficially reported that the 471 is being considered in various modified forms, with ratings in the range 1·78-4·45 kN (400-1,000 lb) for future RPV programmes.

TELEDYNE CAE 490

This is the French-designed Turboméca-SNECMA Larzac (see Turboméca-SNECMA GRTS in French section), an 'exclusive agreement' for which was announced by Teledyne CAE in January 1973. The American company will "market, manufacture and service" the European turbofan for the United States and Canada.

The president of Teledyne CAE said that the Larzac, the initial US version of which is designated Model 490-4, "provides a valuable new source of flight-ready jet engines for strike and trainer aircraft, missiles and remotely piloted vehicles". A commercial Model 490 was planned to be available in 1976. A Model 490-4 demonstrator engine began running at Toledo in March 1973.

Mockup of Teledyne CAE 471-11Dx two-shaft turbofan, described as "in the 2·67 kN/600 lb st class"

Display mockup of Teledyne CAE 490-4 (Larzac) turbofan in the 13·3 kN (3,000 lb) thrust class

TELEDYNE CONTINENTAL
TELEDYNE CONTINENTAL MOTORS
Aircraft Products Division

ADDRESS:
PO Box 90, Mobile, Alabama 36601
Telephone: (205) 438-3411

PRESIDENT:
D. G. Bigler

VICE-PRESIDENTS:
L. Waters (Aeronautical Engineering)
D. Rauch (Controller)
H. Moffett (Operations)
S. Levey (Personnel and Industrial Relations)
R. Hillard (Quality Assurance)

DIRECTOR, MARKETING:
W. K. Danhof

DIRECTOR, PROCUREMENT:
B. Carroll

LEGAL COUNSEL:
J. Bales

DIRECTOR, COMMUNICATIONS:
Don Fairchilds

In 1928, the former Continental Motors Corporation, one of the largest automobile engine manufacturers in the world, produced its first aero-engine, a sleeve-valve air-cooled radial incorporating the Argyll (Burt-McCollum) patents, which had been purchased by the Corporation from the British Argyll Company in 1925.

In 1931 the 38 hp A40 flat-four was put on the market. This was followed by the A50, A65, A75 and A80 engines.

The current range of Teledyne Continental light aircraft engines includes horizontally-opposed four- and six-cylinder engines, some with fuel injection, rated between 100 and 435 hp. The first of the new-generation Tiara engines, the 6-285-B, is also in production.

In October 1960 it was announced that Rolls-Royce Ltd of England had acquired the licence to manufacture and sell certain engines from the complete range of Continental piston engines throughout the world, apart from the Americas and certain countries in the Far East. (See Rolls-Royce Motors Ltd in UK section.)

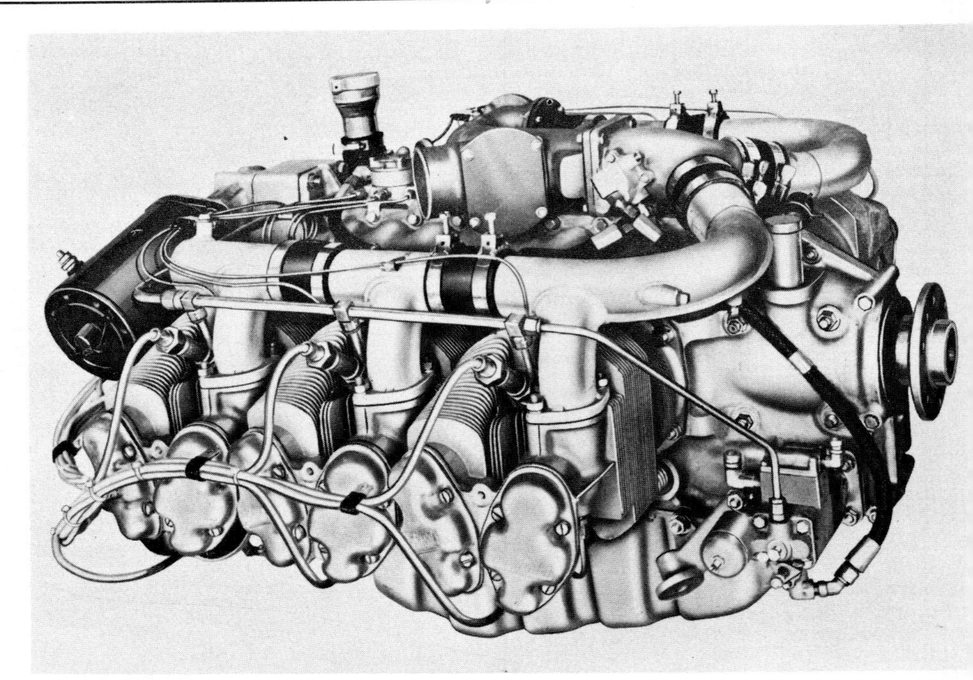

The 168 kW (225 hp) Teledyne Continental TSIO-360-C

CONTINENTAL C90 SERIES

The C90 Series includes the C90-8F which has a flanged crankshaft but does not have provisions for installing either a starter or generator; the C90-12F and -14F which have a flanged crankshaft and starter and generator; and the -16 which has a vacuum pump adaptor.

C90 Series engines have an approved take-off rating of 95 hp at 2,625 rpm.

TYPE: Four-cylinder horizontally-opposed aircooled.

CYLINDERS: Bore 103·2 mm ($4\frac{1}{16}$ in). Stroke 98·4 mm ($3\frac{7}{8}$ in). Capacity 3·28 litres (201 cu in). Compression ratio 7 : 1. Externally-finned aluminium alloy head castings, screwed and shrunk permanently on externally-finned steel barrels.

PISTONS: Cam ground aluminium castings. Three compression rings above pin. Top ring chrome-faced. Oil control ring below pin. Holes in groove provided for interior drain. Pins are full floating ground steel tubes with ground aluminium end plugs.

CONNECTING RODS: "I" beam-type, split bronze pin bushings, identical precision inserts (same as main bearings, steel-backed, lead alloy lined).

CRANKCASE: Aluminium alloy.

CRANKSHAFT: Steel alloy forging, with nitrided journals and crankpins for greater strength.

CAMSHAFT: Steel alloy forging.

VALVE GEAR: Exhaust valves are Stellite-faced and stem tips are hardened. Bronze valve guides.

GEAR TRAIN: Torque is transmitted to engine components from the crankshaft via gears machined from alloy steel forgings, conforming to SAE specifications.

INDUCTION: Small float-type carburettor with a simplified manual mixture control.

FUEL GRADE: 80/87 octane minimum.

IGNITION: Radio shielded, impulse couples, small or standard size magnetos optional.

LUBRICATION SYSTEM: Magnesium crankcase cover houses the engine-driven gear-type oil pump. An oil pressure relief valve and oil screen are also mounted in the cover.

ACCESSORIES: 12V 20 or 35A alternator standard (12V 60A optional).

For other details, see table.

CONTINENTAL O-200 SERIES

The O-200-A engine is generally similar to the C9 Series engines. It is fitted with a single updraught carburettor, dual magnetos and starter and generator.

The O-200-B is similar to the O-200-A, but is designe for pusher installation.

For other details see table.

CONTINENTAL IO-360 SERIES

The IO-360 is a six-cylinder horizontally-opposed air cooled engine with fuel injection. Design and materials ar generally similar to those of IO-346-A (1970-71 *Jane's*) except for number and size of cylinders. Accessorie include oil cooler, two magnetos, propeller governo drive, vacuum pump and 24V alternator. The IO-360 ha a sandcast crankcase, with the accessory case mounted a the rear. The cylinders are shell-moulded.

The IO-360-C has dual accessory drive. The TSIO 360-A, B and C have a turbocharger pressurised inductio system, revised fuel system, starter and accessory driv scavenge pump and full-flow oil cooler. These engine power the Cessna T337 Skymaster. The TSIO-360-E i equipped with a complete exhaust system, and the tu bocharger is engine mounted. The LTSIO-360-E is ider tical except that the crankshaft rotates in the opposi direction; it powers the Piper Seneca II.

For further details see table.

CONTINENTAL O-470 SERIES

Engines in the O-470 series (including the E-185 and E-225) are all basically similar. Engines prefixed "IO" have direct fuel-injection.

The O-470 family of engines are manufactured in four power ranges, from 168 kW to 194 kW (225 hp to 260 hp) as follows:

168 kW (225 hp) IO-470-K
172 kW (230 hp) O-470-R
186 kW (250 hp) IO-470-C, G
194 kW (260 hp) IO-470-D, E, F, L, N, V

The 168 kW and 172 kW (225 hp and 230 hp) models have a compression ratio of 7 : 1, the 186 kW (250 hp) models a ratio of 8 : 1, and the 194 kW (260 hp) models a ratio of 8·6 : 1.

The following description refers specifically to the O-470-R, but is generally applicable to all versions:

TYPE: Six-cylinder horizontally-opposed aircooled.

CYLINDERS: Bore 127 mm (5 in). Stroke 101·6 mm (4 in). Swept volume 7·5 litres (471 cu in). Compression ratio 7 : 1. Forged steel barrels with integral cooling fins. Heat-treated cast aluminium alloy heads screwed and shrunk on to barrels.

PISTONS: Aluminium. Four rings, two compression and one oil control above pin and scraper ring below. Steel gudgeon pins with permanently forged-in aluminium end plugs.

CONNECTING RODS: Forged steel. Trimetal bronze replaceable type big-end bearings, bronze bushing little-ends.

CRANKSHAFT: One-piece six-throw chrome-nickel-molybdenum steel forging. Outer surfaces nitrided. One 5th and one 6th order counterweights attached to shaft. Five bearings of replaceable shell type.

CRANKCASE: Two-piece heat-treated aluminium casting divided at vertical lengthwise plane through crankshaft, with integral cast accessory section. The O-470-S and IO-470-VO have squirt nozzles installed in the crankcase to provide oil cooling to the piston inner dome.

VALVE GEAR: Two poppet-type valves per cylinder: one steel inlet and one steel exhaust with Stellite seat. Camshaft gear-driven from crankshaft in lower part of crankcase.

INDUCTION: Updraught gravity-feed carburettor.

FUEL: 80/87 octane.

IGNITION: Two magnetos on top of accessory section. Two spark plugs per cylinder. Shielded ignition harness.

LUBRICATION: Pressure type. Oil cooler on front of crankcase. Oil filter in crankcase. One impeller type pump. Oil pressure 2·07-4·14 bars (30-60 lb/sq in).

PROPELLER DRIVE: RH drive. Direct. Flanged propeller shaft. Provision for constant-speed propeller.

ACCESSORIES: Generator on accessory section. Drives for vacuum pump and tachometer.

STARTING: Electric starter.

MOUNTING: Four mounting points, one at each lower corner of crankcase.

DIMENSIONS, WEIGHTS AND PERFORMANCE: See table.

CONTINENTAL IO-520 SERIES

These engines are basically similar to the IO-470, but with cylinders of larger bore. They are fitted with an alternator driven either by a belt or by a face gear on the crankshaft. All IO-520 series engines are rated at 213 kW (285 hp) except for the IO-520-D, -E and -F which have a take-off rating of 224 kW (300 hp). IO-520 engines power the Beechcraft Baron and Bonanza, Navion and Cessna 210. New in 1970 were the generally similar IO-520-J, -K and -L, also rated at 213 kW (285 hp) (-K and -L are cleared to 224 kW (300 hp) at 2,850 rpm at take-off). The IO-520-M was developed in 1975 for use in the Cessna 310, replacing the IO-470-V.

The TSIO-520 series are turbocharged. Take-off rating is 213 kW (285 hp) except for the -E and -G, rated at 224 kW (300 hp), and the TSIO-520-J rated at 231 kW (310 hp) and equipped with an intercooler and provision for an overboost valve. These engines power the Cessna 414, 320D, T210 and 210F, and turbocharged Bonanza. The TSIO-520-L was developed for use in the Beech Pressurised Baron. It develops 231 kW (310 hp) at 2,700 rpm, has a complete exhaust system and an engine mounted turbocharger. The TSIO-520-N is used in the Cessna 340A and 414.

For other details see table.

CONTINENTAL GTSIO-520

This is similar to the TSIO-520 range but is geared and uprated. The -C model, rated at 254 kW (340 hp) at 3,200 rpm, powers the Cessna 411. The GTSIO-520-D, rated at 280 kW (375 hp) at 3,400 rpm, powers the Cessna 421. The -K has an integral turbocharger and complete exhaust system; the most powerful Continental engine in production, used in the Rockwell Commander 685. The -G is used in a military application, the -H powers the Cessna 421A Golden Eagle. The -L is used in the Cessna 421C and the -M in the Cessna 404.

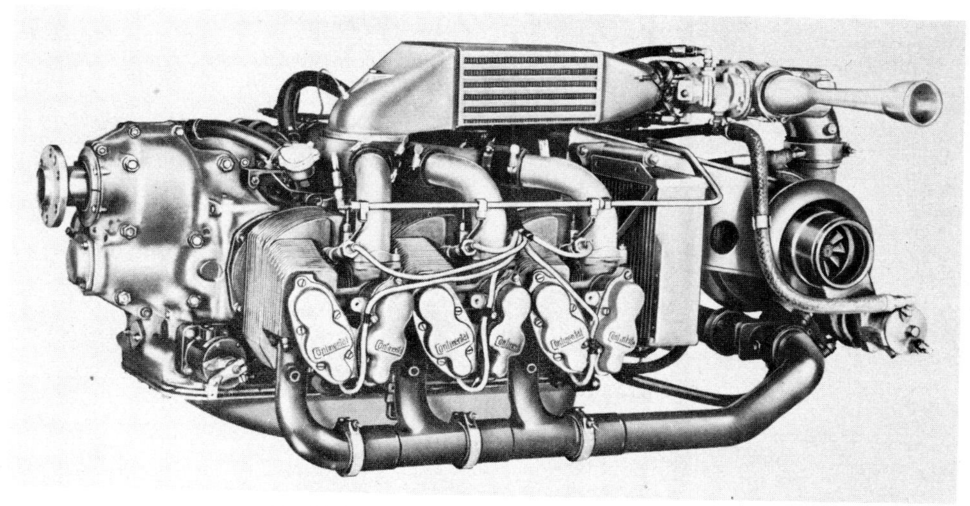

The 435 hp Teledyne Continental GTSIO-520-K

Above: **The 213 kW (285 hp) Teledyne Continental Tiara 6-285-B six-cylinder engine, and** *(below)* **a cutaway drawing of the same engine**

CONTINENTAL TIARA SERIES

In 1972 Teledyne Continental Motors initiated production of a comprehensive new range of general aviation piston engines of aircooled horizontally-opposed geared design, known as the Tiara family. Design parameters considered in engineering the new series were reduction in weight to horsepower ratio, improved crankshaft versus propeller speed efficiency and easier servicing. Significant achievements have been made on all counts.

All Tiara engines are reduction geared at 2 : 1 ratio, thus creating a favourable high crankshaft speed to relatively low propeller speed. The transfer of torque from low to high speed is controlled by a VTC (vibratory torque control) unit. In addition to providing a smooth transition from low to high speeds, this eliminates the need for crankshaft pendulum dampers. The VTC unit hydraulically locks the system into a torsionally stiff configuration at low speeds, and unlocks as speed builds up, allowing the engine to operate on a flexible quill shaft system at higher rpm. A major reduction in vibratory torque is thus achieved in propeller gearing, propeller and accessory systems, permitting the use of lighter, slower and quieter propellers. Lower component stresses allow the use of lighter parts.

The cylinder barrel and head are of lighter design than hitherto, offering enhanced cooling, lower manufacturing cost and ease of servicing and replacement. The head is a shell-moulded casting, providing greater uniformity of dimensions and permitting the use of thinner fins. As a result, the Tiara cylinder design requires approximately a quarter less cooling air pressure than contemporary engines.

As a consequence of operating at higher engine rpm, the camshaft rpm is compatible with that of the propeller. A single pair of spur gears of 2 : 1 ratio drives both the propeller and camshaft. This novel arrangement eliminates the two gears normally required to drive the camshaft. Only seven gears instead of the normal 13 are required for the accessory gearing. The train drives seven side-mounted accessories including the oil pump and, in the turbocharged models, an additional scavenge pump. Bevel gears provide a right-angle drive from the rear end of the camshaft, facilitating vertical positioning of the accessory section to control overall height.

Common features of all models include a cylinder bore and stroke of 123·8 mm (4⅞ in) and 92·08 mm (3⅝ in) respectively, overhead valves, dual ignition, a Teledyne Continental fuel injection system, a minimum fuel grade of 100/130 aviation gasoline, and a lubrication system having a wet sump with full flow filtering. Accessory drives are provided for a tachometer, two magnetos, a starter, a belt-driven alternator, and a propeller governor. In addition there are two spare drives. Some Tiara variants announced to date are as follows:

Model 6-285-B, C. Six-cylinder naturally-aspirated engine of 213 kW (285 hp). Has completed all required development, endurance, type and flight tests, and has been granted FAA Type Certificate. Initial production engine in Tiara series, for Piper Pawnee Brave.

Model T6-320. Six-cylinder turbocharged engine of 239 kW (320 hp).

REPRESENTATIVE TELEDYNE CONTINENTAL HORIZONTALLY-OPPOSED ENGINES

Engine Model	No. of Cylinders	Bore and Stroke mm (in)	Capacity litres (cu in)	Power Ratings kW (hp) at rpm		Comp. Ratio	Dry Weight* kg (lb)	Dimensions			Octane Rating
				Take-off	M.E.T.O.			Length mm (in)	Width mm (in)	Height mm (in)	
C90-16F	4	103·2 × 98·4 (4 1/16×3⅞)	3·28 (201)	71 (95) at 2,625	67 (90) at 2,475	7·0	84·4 (186)	794 (31·25)	800 (31·5)	615 (24·2)	80/87
O-200-A	4	103·2×98·4 (4 1/16×3⅞)	3·28 (201)	74·5 (100) at 2,750	74·5 (100) at 2,750	7·0	99·8 (220)	725 (28·53)	802 (31·56)	589 (23·18)	80/87
IO-360-D	6	112·5×98·4 (4 7/16×3⅞)	5·9 (360)	157 (210) at 2,800	157 (210) at 2,800	8·5	148·3 (327)	877 (34·53)	798 (31·40)	618 (24·33)	100/130
TSIO-360-C, D	6	112·5×98·4 (4 7/16×3⅞)	5·9 (360)	168 (225) at 2,800	168 (225) at 2,800	7·5	136 (300)	910† (35·84)	838 (33·03)	603 (23·75)	100/130
IO-470-H	6	127×101·6 (5×4)	7·7 (471)	194 (260) at 2,625	194 (260) at 2,625	8·6	202·5 (446·5)	1,100 (43·31)	852 (33·56)	502 (19·75)	100/130
O-470-R, S	6	127×101·6 (5×4)	7·7 (471)	172 (230) at 2,600	172 (230) at 2,600	7·0	193·2 (426)	915 (36·03)	852 (33·56)	723 (28·42)	80/87
TSIO-470-D	6	127×101·6 (5×4)	7·7 (471)	194 (260) at 2,600	194 (260) at 2,600	7·5	231·8 (511)	1,465 (58·07)	852 (33·56)	514 (20·25)	100/130
IO-520-A	6	133×101·6 (5¼×4)	8·5 (520)	213 (285) at 2,700	213 (285) at 2,700	8·5	215·9 (476)	1,053 (41·41)	852 (33·56)	502 (19·75)	100/130
IO-520-BA	6	133×101·6 (5¼×4)	8·5 (520)	213 (285) at 2,700	213 (285) at 2,700	8·5	207·3 (457)	1,009 (39·71)	853 (33·58)	678 (26·71)	100/130
IO-520-D	6	133×101·6 (5¼×4)	8·5 (520)	224 (300) at 2,850	213 (285) at 2,700	8·5	208·2 (459)	949 (37·36)	901 (35·46)	604 (23·79)	100/130
IO-520-M	6	133×101·6 (5¼×4)	8·5 (520)	213 (285) at 2,700	213 (285) at 2,700	8·5	188 (415)	1,189 (46·80)	852 (33·56)	518 (20·41)	100/130
TSIO-520-B	6	133×101·6 (5¼×4)	8·5 (520)	213 (285) at 2,700	213 (285) at 2,700	8·5	219 (483)	1,490 (58·67)	852 (33·56)	516 (20·32)	100/130
TSIO-520-C	6	133×101·6 (5¼×4)	8·5 (520)	213 (285) at 2,700	213 (285) at 2,700	7·5	208 (458)	1,040† (40·91)	852 (33·56)	509 (20·04)	100/130
TSIO-520-E	6	133×101·6 (5¼×4)	8·5 (520)	224 (300) at 2,700	224 (300) at 2,700	7·5	219 (483)	1,010† (39·75)	852 (33·56)	527 (20·74)	100/130
TSIO-520-J, N	6	133×101·6 (5¼×4)	8·5 (520)	231 (310) at 2,700	231 (310) at 2,700	7·5	221·3 (487·8)	997 (39·25)	852 (33·56)	516 (20·32)	100/130
TSIO-520-L	6	133×101·6 (5¼×4)	8·5 (520)	231 (310) at 2,700	231 (310) at 2,700	7·5	244·5 (539)	1,286 (50·62)	852 (33·56)	508 (20·02)	100/130
GTSIO-520-C	6	133×101·6 (5¼×4)	8·5 (520)	254 (340) at 3,200	254 (340) at 3,200	7·5	252·7 (557)	1,081 (42·56)	880 (34·04)	587 (23·1)	100/130
GTSIO-520-F, K	6	133×101·6 (5¼×4)	8·5 (520)	324 (435) at 3,400	324 (435) at 3,400	7·5	290·3 (640)	1,426 (56·12)	880 (34·04)	664 (26·15)	100/130
GTSIO-520-H, L‡, M‡	6	133×101·6 (5¼×4)	8·5 (520)	280 (375) at 3,400	280 (375) at 3,400	7·5	250 (550·37)	1,081 (42·56)	880 (34·04)	680 (26·78)	100/130
Tiara Series											
6-285-B, C	6	123·8×92·08 (4⅞×3⅝)	6·65 (406)	213 (285) at 4,000	213 (285) at 4,000	9·0	185 (409)	1,019 (40·11)	836 (32·91)	615 (24·22)	100/130
T6-320	6	123·8×92·08 (4⅞×3⅝)	6·65 (406)	239 (320) at 4,400	239 (320) at 4,400	9·6	185 (409)	1,019 (40·11)	836 (32·91)	615 (24·22)	100/130

*With accessories; †Not including turbocharger;
‡Similar to -H except length 1,114 mm (43·87 in) and weight 248 kg (547 lb) (-M, height 662 mm; 26·08 in)

THERMO-JET

THERMO-JET STANDARD INC

HEAD OFFICE:
PO Box 55976, Houston, Texas 77055

Telephone: (713) 683-9177

MANAGER:
John A. Melenric

This company specialises in the design and manufacture of valveless pulse-jet units for remotely piloted vehicles and the homebuilt aircraft markets. These engines are devoid of moving parts and are characterised by multiple reverse-flow air inlets to a combustion chamber in which is burned propane, butane or compressed natural gas, obviating the need for a fuel pump. Intermittent combustion and expulsion takes place at a cycle frequency determined by the chamber size and geometry and combustion pressure.

At present Thermo-Jet is offering four sizes of unit, described separately below. Each has a structure fabricated in Type 321 stainless steel, with Type 304 stainless fuel piping and 2024 T4 aluminium fuel nozzles. The throttle regulates fuel flow only. To start, ignition is provided by a hand-cranked magneto connected through a plug and socket to a Champion CJ-6 sparking plug. The throttle is then advanced to the vapour mode and the engine lights up. Once running, the throttle is switched to the liquid mode for normal operation. Combustion chamber mean temperature is 1,150°C in each unit.

All Thermo-Jet engines are now equipped with one annular reverse-flow vane mounted at the end of each air-inlet tube. The vane, in conjunction with the air-inlet bellmouth, forms an essentially constant-area duct, which recovers total pressure with minimum loss as the engine moves through the air at high velocity. Test-stand data indicate gross thrust increases of 100 per cent and specific fuel consumption decreases of 50 per cent can be had at a Mach number of 0·5 with the new vanes. An exhaust diffuser is available on special order which permits engine operation in the Mach 0·8 range. Each diffuser is designed for a specific Mach number.

THERMO-JET J7-300

Smallest of the company's units, this has three air inlets to a duct terminating in a straight-pipe exhaust tube.

DIMENSIONS:
- Diameter: 178 mm (7 in)
- Length: 1,245 mm (49·0 in)

WEIGHT, DRY: 3·63 kg (8·0 lb)

PERFORMANCE:
(for max power):
- Thrust at S/L: 0·09 kN (21 lb)
- Specific impulse: 735 sec
- Mean effective pressure: 0·21 bars (3 lb/sq in)
- Cycle frequency: 119 cps

(for max endurance):
- Thrust at S/L: 0·031 kN (7 lb)
- Specific impulse: 950 sec
- Mean effective pressure: 0·07 bars (1 lb/sq in)
- Cycle frequency: 119 cps

SPECIFIC FUEL CONSUMPTION:
- Max power: 139 mg/Ns (4·9 lb/hr/lb st)
- Max endurance: 108 mg/Ns (3·8 lb/hr/lb st)

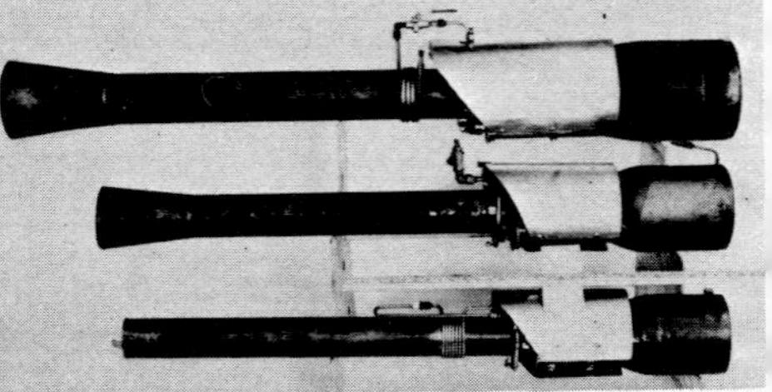

From the top: Thermo-Jet J7-300, J8-200 and J10-200

THERMO-JET J8-200

This larger unit has two air intakes and an exhaust tube with flared end.

DIMENSIONS:
- Diameter: 203 mm (8 in)
- Length: 1,422 mm (56 in)

WEIGHT, DRY: 4·99 kg (11·0 lb)

PERFORMANCE:
- (for max power):
  - Thrust at S/L: 0·133 kN (30 lb)
  - Specific impulse: 720 sec
  - Mean effective pressure: 0·19 bars (2·8 lb/sq in)
  - Cycle frequency: 98 cps
- (for max endurance):
  - Thrust at S/L: 0·044 kN (10 lb)
  - Specific impulse: 1,000 sec
  - Mean effective pressure: 0·07 bars (1 lb/sq in)
  - Cycle frequency: 91 cps

SPECIFIC FUEL CONSUMPTION:
- Max power: 141·8 mg/Ns (5·0 lb/hr/lb st)
- Max endurance: 102 mg/Ns (3·6 lb/hr/lb st)

THERMO-JET J10-200

This further enlarged unit has two air inlets to a duct having a flared exhaust tube nozzle.

DIMENSIONS:
- Diameter: 254 mm (10 in)
- Length: 1,778 mm (70 in)

WEIGHT, DRY: 8·8 kg (19·5 lb)

PERFORMANCE:
- (for max power):
  - Thrust at S/L: 0·244 kN (55 lb)
  - Specific impulse: 692 sec
  - Mean effective pressure: 0·186 bars (2·7 lb/sq in)
  - Cycle frequency: 69 cps
- (for max endurance):
  - Thrust at S/L: 0·08 kN (18 lb)
  - Specific impulse: 945 sec
  - Mean effective pressure: 0·07 bars (1 lb/sq in)
  - Cycle frequency: 65 cps

SPECIFIC FUEL CONSUMPTION:
- Max power: 147 mg/Ns (5·2 lb/hr/lb st)
- Max endurance: 108 mg/Ns (3·8 lb/hr/lb st)

THERMO-JET J13-202

Largest and newest of the present range, this unit has two air inlets to a duct having a flared exhaust tube nozzle.

DIMENSIONS:
- Diameter: 330 mm (13 in)
- Length: 2,311 mm (91 in)

Thermo-Jet J13-202 valveless pulse-jet (0·40 kN; 90 lb st)

WEIGHT, DRY: 15 kg (33 lb)

PERFORMANCE:
- (for max power):
  - Thrust at S/L: 0·40 kN (90 lb)
  - Specific impulse: 655 sec
  - Mean effective pressure: 0·19 bars (2·8 lb/sq in)
  - Cycle frequency: 58 cps
- (for max endurance):
  - Thrust at S/L: 0·142 kN (32 lb)
  - Specific impulse: 900 sec
  - Mean effective pressure: 0·07 bars (1 lb/sq in)
  - Cycle frequency: 56 cps

SPECIFIC FUEL CONSUMPTION:
- Max power: 156 mg/Ns (5·5 lb/hr/lb st)
- Max endurance: 113 mg/Ns (4·0 lb/hr/lb st)

THIOKOL
THIOKOL CORPORATION

CORPORATE OFFICE:
PO Box 1000 Newtown, Pennsylvania 18940
Telephone: (215) 968-5911
Cable: Thiokol Ntow

GOVERNMENT SYSTEMS GROUP:
PO Box 9258 Ogden, Utah 84409
Telephone: (801) 863-3511

SOLID PROPELLANT ROCKET MOTOR PLANTS:
Elkton, Maryland
Huntsville, Alabama
Marshall, Texas
Brigham City, Utah

PYROTECHNIC AND ORDNANCE PLANT:
Shreveport, Louisiana

CHAIRMAN OF THE BOARD:
Dr H. W. Ritchey

PRESIDENT AND CHIEF EXECUTIVE OFFICER:
R. E. Davis

VICE-PRESIDENT AND TREASURER:
A. P. Roeper

GROUP VICE-PRESIDENTS:
R. D. Evans (Fibres)
J. S. Jorczak (Chemical)
James M. Stone (Government Systems)
E. R. Kearney (General Products)

Organized in 1929, Thiokol Chemical Corporation produced and marketed the first synthetic rubber manufactured in the United States. In 1943, the discovery by Thiokol of liquid polymer, a new type of synthetic rubber, paved the way for the practical development of the "case-bonded" principle of rocket power plant design. The company's polysulphide liquid polymer proved to be the catalyst for the first mass production of efficient solid-propellant rocket motors, as well as for the development of large solid-propellant motors. The firm's operations have now been organised into separate groups to serve widening areas of related products. Reflecting this diversity, the name has been changed to Thiokol Corporation.

Details of some of the more important solid-propellant rocket motors used in missiles, sounding rockets, spacecraft and space launch vehicles are given below. Important current rocket motor activity, of which details cannot be reported, includes: recent completion of production of the motors for the three stages of the Spartan missile, production of gas generators for the Poseidon missile; recent completion of production of the TX-481 Maverick motor, and engineering development of the TX-486 motor for the SAM-D air-defence missile.

In 1974 Thiokol Corporation was awarded a contract to develop the solid rocket motors for the NASA Space Shuttle.

THIOKOL TRIDENT MOTORS

The three stages of propulsion for the Trident C-4 fleet ballistic missile are being developed jointly by Thiokol and Hercules Inc. All three stages have advanced solid-propellant rocket motors, but details are classified.

THIOKOL POSEIDON FIRST-STAGE MOTOR

The motor for the first stage of the Poseidon C-3 fleet ballistic missile is manufactured jointly by Thiokol and Hercules Incorporated.

THIOKOL MINUTEMAN THIRD-STAGE MOTOR

US military designation: SR73-AJ-1

The SR73-AJ-1 is the third-stage rocket motor for the Minuteman III missile. It is 2·36 m (93 in) long, 1·32 m (52 in) in diameter, and weighs 3,649 kg (8,046 lb). The motor supplies 155·1 kN (34,876 lb) of thrust over a 59·6 second burn time.

The cylindrical case for the SR73-AJ-1 is prefabricated of glassfibre filaments, wound on a soluble mandrel and then cured. The solid-propellant fuel uses a polybutadiene acrylonitrile polymer binder with an ammonium perchlorate oxidiser and aluminium additive. The motor has a single nozzle and a liquid-injection thrust-vector control system.

Thiokol TU-122, the stage-1 Minuteman motor, on an in-plant transporter at Brigham City

THIOKOL TU-122

US military designation: M-55

The TU-122 is the first-stage rocket motor for the solid-propellant Minuteman ICBM. The motor is approximately 7·62 m (25 ft) in length, 1·68 m (5 ft 6 in) in diameter, and weighs about 22,680 kg (50,000 lb). It produces approximately 889·5 kN (200,000 lb) thrust during a 60 second firing time.

The motor case is manufactured from D6AC steel. The composite solid propellant employs a polybutadiene acrylic acid polymer binder with ammonium perchlorate oxidiser and aluminium powder additives. Thrust vector control is achieved through the use of four movable nozzles. Advanced versions continue in production for the LGM-30G; over 1,000 of earlier versions have been delivered.

IMPROVED SRAM MOTOR

In late 1974 Thiokol was awarded a contract, valued at about $1 million, from Boeing for the study of an improved motor for the US Air Force AGM-69A Short-Range Attack Missile (SRAM). Thiokol will use "a propellant with longer life than that used in missiles now being produced", and will also investigate the life of the propellant of the present motor (made by LPC). Boeing stated that, with the US Air Force, it would use the results to optimise the design of "a new, more effective motor for the missiles to be built in the future for the supersonic B-1 bomber."

THIOKOL TX-174

The TX-174 is the first stage of the Pershing tactical weapon system. The motor case is a thin-wall flight design fabricated from modified AISI H-11 or type D6AC steel. The contoured nozzle has an expansion ratio for sea-level operation.

The TX-174 has an overall length of 102·61 in and an outside diameter of 1·02 m (40 in). The nominal propellant weight is 2,019 kg (4,451 lb). The nominal total weight of the TX-174 is 2,270 kg (5,004 lb). It provides an average thrust of 117 kN (26,290 lb) and a total impulse of 1,017,200 lb-sec. The web burning time is 38·30 sec. The TX-174 utilizes a cylindrical-core propellant configuration.

THIOKOL TX-175

The TX-175 is the sustainer stage for the Pershing. The motor case is a thin-wall flight design fabricated from modified AISI H-11 or type D6AC steel. The forward dome of the case has three ports for impulse control. The contoured nozzle is sized for altitude operation.

The TX-175 has an overall length of 2·46 m (96·72 in) and an outside diameter of 1·02 m (40·0 in). The nominal propellant weight is 1,263 kg (2,785 lb). The nominal total weight of the TX-175 is 1,471 kg (3,244 lb). At vacuum conditions, it provides an average thrust of 85·5 kN (19,220 lb) and a total impulse of 757,200 lb-sec. The web burning time is 39·0 sec. The TX-175 utilizes a cylindrical-core propellant configuration.

THIOKOL TX-30

The TX-30 is the sustainer of the Nike-Hercules missile. Overall length including blast tube is 4·43 m (174·4 in), and diameter 0·72 m (28·4 in). Total weight is 1,288 kg (2,840 lb) with a propellant weight of 985 kg (2,172 lb). Average thrust is 63·2 kN (14,200 lb) for a burning time of 26·6 sec. Total impulse is 399,000 lb-sec.

THIOKOL TX-261

The TX-261 is a PBAA-propellant version of the Nike-Hercules sustainer, with an altitude expansion nozzle instead of a blast tube/nozzle as on the TX-30. It is used in a variety of applications, including Pershing Drop Boost Test Vehicle and Defense Nuclear Agency Project Flame for re-entry vehicle nosecone evaluations. It is also used as the second stage of the ARPAT (Advanced Research Projects Agency Terminal) test vehicle. The motor case is 2·92 m (114·89 in) long and 0·72 m (28·44 in) in diameter. Its total loaded weight is 1,301 kg (2,870 lb). Propellant weight is 1,054 kg (2,324 lb). Sea-level thrust is 254·5 kN (57,200 lb) over a total burning time of 8·92 sec for a total impulse of 546,800 lb-sec.

THIOKOL TX-306

The TX-306 is one of a group of motors that can be used separately, or in combination, for a wide variety of applications. It has a spherical configuration. Its contoured nozzle provides 20 : 1 expansion ratio for altitude applications. It was used in conjunction with TX-33, TX-261 and TE-M-29 motors to provide propulsion for the ARPAT target vehicle. As the third stage of that vehicle it accelerated simulated warheads during high-velocity intercept tests. The TX-306 is 29·11 in in diameter and with nozzle it is 1·20 m (47·31 in) long. Its case-bonded, composite propellant weighs 312 kg (688 lb), and its total weight is 383 kg (845 lb). It provides an average thrust of 59·6 kN (13,400 lb). Its total burning time is approximately 11 sec and its total impulse is 158,000 lb-sec at sea level.

THIOKOL TE-260G

The TE-260G rocket motor for the Subroc missile incorporates both directional control and thrust reversal. It consists of a cylindrical case containing the Thiokol propellant charge, a dual forward bulkhead design incorporating a thrust-reversal system, and an aft bulkhead containing four nozzles, each of which is equipped with a jetevator thrust vector control system. Six cable-carrying conduits run the full length of the case on the inside wall to allow guidance signals to be transmitted to the aft vector control system.

The TE-260G employs a composite solid propellant which is bonded to the case wall. Its principal constituents are polyurethane fuel binder and ammonium perchlorate oxidiser in a propellant system designed to provide a high specific impulse and a low burning rate. Thiokol developed the polyurethane system specifically to meet the Subroc requirements.

THIOKOL TU-289

US military designation: SR49-TC-1

The TU-289 solid-propellant rocket motor powers the AIR-2A Genie unguided air-to-air missile. It replaces an earlier motor, and has an improved propellant which increases the storage life and permits the missile to be deployed in a wide range of environmental temperatures. This motor remains in production.

THIOKOL TX-354

TX-354 motors are used in a variety of applications such as first and second stages, and as strap-on boosters for launch vehicles. The TX-354-3 has a high-altitude nozzle and is used as the second stage of the Scout vehicle. TX-354-4 has a sea-level straight nozzle and is used as a first stage for the Stripi vehicle with Recruit strap-on motors. TX-354-5 is the strap-on booster for the Delta vehicle. It has an 11° canted sea-level nozzle and is used in groups of three, six, or nine. TX-354 weighs 4,320 to 4,410 kg (9,525 to 9,743 lb). The case without nozzle is 5,130 mm (202 in) long and has a diameter of 787 mm (31 in). Sea-level thrust is 232 kN (52,150 lb) over a total burning time of 39 sec for a total impulse of 886,043 kg-sec (1,953,400 lb-sec).

THIOKOL TX-581

The TX-581 (Castor II-X) is an extended-length version of the TX-354 (Castor II) motor and was developed as a strap-on booster with an 11° canted nozzle. It can also be used in first- or second-stage applications with a straight nozzle. The motor is 787·4 mm (31 in) in diameter, 8,000 mm (315 inches) long and weighs about 5,806 kg (12,800 lb). Sea-level performance provides an average thrust of approximately 339·4 kN (76,300 lb) over a total burning time of 35·8 sec, giving a total impulse of 1,195,164 kg-sec (2,634,900 lb-sec), a 35 per cent increase over Castor II.

THIOKOL TX-526

TX-526 motors are used in a variety of booster/strap-on applications. The TX-526-0 motor, combined with four Recruit strap-on motors, is the booster for the Athena H vehicle. The TX-526-1 with 7° canted nozzle is qualified for strap-on booster applications. The TX-526-2 with 11° canted nozzle is used as a strap-on booster for the Delta 3914 vehicle, with each vehicle having nine motors. The 9,093 mm (358 in) long, 1,016 mm (40 in) diameter solid-propellant motor weighs 10,550 kg (23,250 lb). It provides an average thrust of 379·3 kN (85,270 lb), with a total burning time of approximately 58 sec for a total impulse of 2,159,544 kg-sec (4,760,900 lb-sec) at sea-level.

THIOKOL TE-M-416 TOMAHAWK

The Tomahawk is a high-performance motor designed specifically for use in sounding rocket systems. It is used in the Tomahawk vehicle as a single stage and in several other vehicles, such as Nike-Tomahawk and Terrier-Tomahawk, as the second stage. The motor is 3,607 mm (142 in) long and has a diameter of 229 mm (9 in). It weighs 220 kg (486 lb) and produces 48·9 kN (11,000 lb) thrust at sea level.

Thiokol TU-289 motor for the Genie air-to-air nuclear missile, being prepared for static firing at the company's Wasatch Division at Brigham City

Three Thiokol Castor I and three Castor II strap-on booster rockets attached to Thrust-Augmented Thor launch vehicle

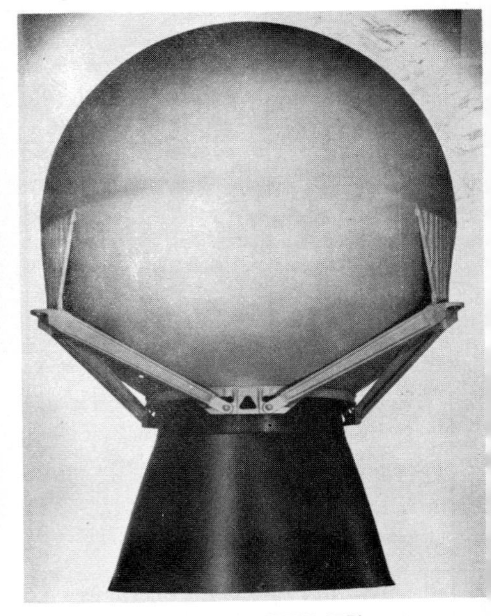

Thiokol TE-M-364-2 (STAR-37B) motor

THIOKOL TE-M-29 (RECRUIT)

The Recruit was developed for the X-17 re-entry test vehicle. It is especially useful for sounding rockets, sleds and auxiliary boost applications because of its high overall performance. The TE-M-29-1 version was used in the X-17, Project Argus, Project Farside, Trailblazer and Stripi programmes. Its total loaded weight is 164 kg (361·5 lb); it is 2,674 mm (105·28 in) long and 229 mm (9 in) in diameter. Used with the TE-P-372 pyrogen or TE-I-436 pyrotechnic igniter, it is supplied with 6°, 6·5°, 9°, or 9·5° standard canted adaptors. The TE-M-29-2, used in the Project Farside, Little Joe I and II and Squirt programmes, uses a 4·26 : 1 expansion ratio nozzle. The TE-M-29-3 provides boost to the Athena first stage; it has a 5° 56′ canted adaptor. The TE-M-29-4 is the retro motor for the S-IVB stage of Saturn V, and the TE-M-29-5 is the retro motor for the S-IV stage of Saturn IB.

THIOKOL TE-M-29-8

Super Recruit, TE-M-29-8, is an improved version of previous Recruit motor configurations. The burn-time motor total impulse is increased from 55,000 to 62,000 lb-sec, while keeping burn time unchanged. Weight is increased by about 6 per cent.

THIOKOL TE-M-424 (SATURN S-IC RETRO)

This motor was developed to serve as retrograde propulsion on the S-IC Saturn V first stage. Eight motors are employed for each launch. The TE-M-424 is man-rated because it was used in the Apollo programme. Motor length is 2,141 mm (84·3 in) and diameter is 386 mm (15·2 in). The motor produces 390·5 kN (87,800 lb) thrust for 0·633 seconds. Total motor weight is 228·8 kg (504·5 lb).

THIOKOL TE-M-307 (APACHE)

The TE-M-307-3 rocket motor was designed for second-stage applications, and therefore includes a 20 second delay igniter. It is 2,741 mm (107·91 in) long, 174·2 mm (6·86 in) in diameter and is used both as a sounding rocket and as a target missile. The TE-M-307-4 version was designed for single-stage applications. It uses the same loaded case and headcap assembly as the TE-M-307-3, with a 3·32 : 1 expansion ratio nozzle and an instantaneous TE-P-415 pyrogen. It is also used as a sounding rocket and as a target missile.

THIOKOL TE-M-364 (STAR-37)

The TE-M-364-2 (STAR-37B) is a 939·8 mm (37 in) diameter spherical main retro-rocket designed for the Surveyor and modified for use on the Burner II stage. Modifications consisted of increasing propellant loading to 653 kg (1,440 lb) and strengthening the attachment structure to accommodate higher inertial loads.

The TE-M-364-3 (STAR-37D) is a Surveyor main retro-rocket modified for use as third stage propulsion on the Improved Delta vehicle. Modifications consisted of again increasing propellant load, to 1,440 lb, redesigning the attachment structure to mate with the Delta launch vehicle and changing the diameter to 952·3 mm (37·49 in).

The TE-M-364-4 (STAR-37E) is an elongated version of the Delta motor, the AP/hydrocarbon/Al propellant grain being increased in mass from 653 kg (1,440 lb) to 1,040 kg (2,290 lb) by adding a 355 mm (14 in) cylinder to the case. Average thrust is 68·8 kN (15,472 lb) for a burn time of 41·96 sec. This motor provides third-stage propulsion on Improved Delta.

The TE-M-364-19 (STAR-37F) is a shorter version of the Delta motor, accommodating 845 kg (1,836 lb) of propellant. The body is of composite asbestos, glass and graphite phenolic structure and has a 178 mm (7 in) cylindrical section. Average thrust is 55·9 kN (12,470 lb) for a burn time of 42·55 sec. It provides the impulse to circularise the orbit of the FltSatCom satellite at the apogee of the launch orbit.

The TE-M-364-11 (STAR-37G) is a very similar extended Delta motor, likewise used for Improved Delta third-stage propulsion. Average thrust is 62·9 kN (14,14[illegible] lb) for a burn time of 45·48 sec.

The TE-M-364-14 (STAR-37N) is a version having a propellant loading of 557·9 kg (1,230 lb). Average thrust is 38·4 kN (8,634 lb) for a burn time of 37·7 sec. This motor provides third-stage propulsion on the Japanese [illegible] vehicle.

The TE-M-364-15 (STAR-37S) is a titanium (6Al 4V) spherical-cased version with modified attachment and propellant loading of 657·7 kg (1,450 lb). Average thrust is 43·5 kN (9,790 lb) for a burn time of 42·2 sec.

Thiokol TE-M-442-1 (STAR-26B) motor

provides the third-stage propulsion for weather satellites flown on the USAF Thor vehicle.

THIOKOL TE-M-442-1 (STAR-26B)

This motor is spherical, 663 mm (26·1 in) in diameter and 839 mm (33·05 in) long; propellant weight is 238 kg (525 lb) and total motor weight is 261 kg (576 lb). The TE-M-442-1 was developed from the TE-M-442 of 1965 and features a case of titanium instead of steel. It flies as an additional stage to the standard Burner II launch vehicle, atop the TE-M-364-2 second stage.

THIOKOL TE-M-479 (STAR-17)

The TE-M-479 is a 442 mm (17·4 in) spherical rocket motor developed for NASA's Radio Astronomy Explorer satellite programme. The motor is 687 mm (27·06 in) long and serves as the apogee kick stage which makes the orbit of the spacecraft truly circular. Total motor weight is 78·8 kg (173·8 lb); propellant weight is 69·4 kg (153 lb). High mass-fraction and excellent performance reproducibility characterise this motor for space systems application. The TE-M-479 was first flown in July 1968.

THIOKOL TE-M-521 (STAR-17A)

This 444 mm (17·5 in) diameter and 980 mm (38·6 in) long motor was developed by adding a 175 mm (6·9 in) straight section to the spherical TE-M-479 (RAE) motor. The TE-M-521 has a propellant weight of 112 kg (247 lb) and a total weight of 123·9 kg (273·2 lb). It served to "circularise" the orbit of the Skynet I, NATO I and IMP-H and -J satellites. The motor has a titanium case and flight-proven propellant.

THIOKOL TE-M-541/542 (STAR-6)

This small glassfibre motor measures 157 mm (6·2 in) in diameter and 356 mm (14 in) long and serves in a classified space application. Using the same hardware, with minor insulation changes, the motor is loaded to either of two configurations: 4·85 kg (10·7 lb), 1,395 kg-sec (3,075 lb-sec) total impulse, 5·99 kg (13·2 lb) total weight (TE-M-541); and 3·27 kg (7·2 lb), 930 kg-sec (2,050 lb-sec) total impulse, 4·8 kg (10·6 lb) total weight (TE-M-542). These motors have an extensive flight history.

THIOKOL TX-280

This motor was developed to serve as the ullage control

Thiokol TE-M-521 (STAR-17A) motor

motor for the S-IV and S-IVB stages of the Saturn vehicles. Four motors are used on the S-IV stage of the Saturn I, Block II vehicle, three are used on the S-IVB stage of the Saturn IB, and two are used when the S-IVB is on the Saturn V. The TX-280 is man-rated and was used for the Apollo programme. It measures 922 mm (36·28 in) in length and is 211 mm (8·29 in) in diameter. Its loaded weight is 40 kg (88·4 lb). Propellant weight is 27 kg (58·7 lb). The TX-280 achieves a sea-level thrust of 15·2 kN (3,420 lb) over a burn time of 3·81 sec, resulting in total impulse of 7,045 kg-sec (15,500 lb-sec).

THIOKOL TE-M-473 (SANDHAWK)

The Sandhawk TE-M-473 is a high-performance 330 mm (13 in) diameter, 5,105 mm (201 in) long rocket motor designed for sounding rocket use. It features a regressive thrust-time trace, which results in near-constant vehicle acceleration during its 15 second burn time and provides an extremely smooth flight environment. This motor is suited for use in single-stage, two-stage and three-stage vehicle configurations.

THIOKOL TE-M-236 (SARV RETRO)

This is a retrograde motor for an unmanned satellite. It uses an internal-burning case-bonded grain weighing 18·3 kg (40·34 lb) in a case of 4130 steel, with a re-entrant conical rear closure to keep overall length to only 324 mm (12·76 in). Burn-time (7·5 sec) average thrust is 5·6 kN (1,250 lb).

THIOKOL TE-M-640 ALTAIR III (STAR-20)

This motor was developed as the fourth stage of the Scout launch vehicle. The 274·1 kg (604·3 lb) CTPB propellant grain is contained in a filament-wound glassfibre case. The lightweight external nozzle is a composite of graphite, plastics and steel. Total loaded weight, including consumable pyrogen igniter, is 301·4 kg (664·5 lb). Average thrust is 27·5 kN (6,175 lb) for a burn time of 27·4 sec.

THIOKOL TE-M-458 (STAR-13)

This is a deceleration motor used in the Anchored Interplanetary Monitoring Platform (AIMP) programme. The 31 kg (68·3 lb) charge of AP/Al/ urethane is con-

Thiokol TE-M-541 (STAR-6) motor

tained in a spherical case of 6Al-4V titanium, with graphite/vitreous silica phenolic nozzle. Loaded weight is 35·65 kg (78·6 lb) and average thrust 3·8 kN (850 lb) for a burn time of 21·8 sec.

THIOKOL TE-M-516 (STAR-13A)

This apogee-boost motor is made by mating the propellant and nozzle of the TE-M-444 with the case and igniter of the TE-M-458. Average thrust is 5·8 kN (1,309 lb) for a burn time of 15·3 sec. The motor is used as an injection stage of the Thor Burner II carrying two satellites: Secor and Aurora.

THIOKOL TE-M-604 (STAR-24)

This apogee-boost motor has a 199·5 kg (439·8 lb) charge of AP/hydrocarbon/Al propellant contained in a spherical case of 6Al-4V titanium with graphite/carbon phenolic nozzle. Average thrust is about 20 kN (4,500 lb) for a burn time of 30·21 sec. The motor was used as the apogee kick stage on the UK Skynet II and USAF Space Test Platform; it is being modified for use on the NASA IUE and LAGEOS spacecraft.

THIOKOL TE-M-616 (STAR-27)

This apogee-boost motor has a 332·9 kg (733·9 lb) charge of AP/hydrocarbon/Al propellant in a case (spherical with cylindrical centre portion) of 6Al-4V titanium with graphite/carbon phenolic nozzle. Average thrust is 26·7 kN (5,996 lb) for a burn time of 34·84 sec. First customer was the Canadian Communications Research Centre, for the Communications Technology Satellite. This motor will also be used for Delta Vehicle 2914 apogee kick for the Japanese Broadcast and Geo-Meteorological satellites, and for the Atlas-launched USAF Global Positioning Satellite.

THIOKOL TE-M-700

This apogee-boost motor has a 463 kg (1,020 lb) charge of AP/hydrocarbon/Al propellant in a case of 6Al-4V titanium with a nozzle of graphite/carbon-carbon. The motor total weight is 491 kg (1,082 lb) and has mass fraction of 0·943. Average thrust is 26·4 kN (5,930 lb) over a burn time of 55 sec. This motor was developed for apogee-boost assignments on Delta 3914 and Atlas.

TRW

TRW SYSTEMS

HEAD OFFICE:
One Space Park, Redondo Beach, California 90278
Telephone: (213) 535-4321

TRW developed, built and launched the first monopropellant hydrazine propulsion system to enter and be started in space.

TRW is testing a wide variety of chemical propulsion engines. One of these, the man-rated Lunar Module Descent Engine, landed Astronauts on the Moon. Another engine was built to provide midcourse trajectory corrections for the Mariner '69 missions to Mars and the Mariner 10 (Venus-Mercury) missions. TRW provided the monopropellant hydrazine orbit-adjust propulsion system for three NASA Atmospheric Explorer Satellites launched in 1973, 1974 and 1975.

TRW's propulsion research programmes include low-thrust monopropellant, bipropellant, colloid, ion, radio-isotope and electro-thermal engines. In addition, an active research programme in low-cost propulsion technology is being continued. Tests of TRW rocket engines are conducted at the company's test site at San Juan Capistrano, California.

TRW TR-201 (DELTA)

A bipropellant engine designed for vacuum operation, the TR-201 serves as propulsion of the second stage of the NASA/McDonnell Douglas Delta launch vehicle. The first eight launches using this engine were: Delta 101, Westar-1, April 1974; 102, SMS-1, May 1974; 103, Westar-2, October 1974; 104, ITOS-7, November 1974; 105, Skynet IIB, November 1974; 106, Symphonie I, December 1974; 107, ERTS-2, January 1975; 108, SMS-2, February 1975.

TYPE: Liquid bipropellant rocket engine.

PROPELLANTS: Nitrogen tetroxide and 50/50 mix of hydrazine and UDMH.

THRUST CHAMBER: Single chamber. Area ratio 43. Chamber of quartz phenolic construction with ablative cooling. Nozzle of columbium, with radiation cooling. Co-axial injector with hypergolic ignition. Starting by 28V electrical signal to on/off solenoid valves.

THRUST CHAMBER MOUNTING: Gimbal attachment above injector.

PROPELLANT FEED SYSTEM: Pressure feed system by McDonnell Douglas Astronautics. Gas pressure 15·51 bars (225 lb/sq in). Flow rate 5·62 kg (12·4 lb) fuel and 8·92 kg (19·7 lb) oxidant per second.

TRW TR-201 Delta engine, with vacuum thrust of 43·6 kN (9,800 lb)

DIMENSIONS:
- Length overall: 2,156 mm (84·9 in)
- Nozzle diameter: 922 mm (36·3 in)

WEIGHT, DRY: 113 kg (250 lb)

PERFORMANCE RATING:
- Max thrust (vacuum): 43·6 kN (9,800 lb)
- Combustion pressure: 7·03 bars (102 lb/sq in)
- Combustion temperature: 2,700°C
- Specific impulse: 302

VOUGHT

VOUGHT CORPORATION

HEAD OFFICE:
PO Box 5907, Dallas, Texas 75222
Telephone: (214) 266-2011
OFFICERS: See "Aircraft" section

VOUGHT LVRJ

Vought Corporation, a division of LTV Aerospace (itself a subsidiary of The LTV Corporation) is developing, under a US Navy contract, a new air-breathing propulsion system designed for future missiles and other unmanned vehicles. Handled by the Vought facility at Grand Prairie, Texas, the system is known as the Low-Volume Ram Jet (LVRJ). The propulsion method is described as a 'stepping stone toward missile systems of the future'. While performance details have not been released, the company announces that in broad terms the propulsion system will have several times the specific impulse of current systems and will represent a quantum jump in propulsion effectiveness for advanced stand-off tactical missile system applications.

The LVRJ features an integral rocket-ramjet propulsion system using a common motor case for compact design. Boosted to high speed by the solid-propellant motor, the vehicle then uses the empty motor case as a combustion chamber for ramjet fuel to provide high-speed long-duration flight. Four inlets on the vehicle's aft section provide ram air for combustion. Test vehicles scheduled for the flight demonstration programme are 4,547 mm (179 in) long and 381 mm (15 in) in diameter, and weigh 680 kg (1,500 lb), but smaller or larger models can be designed for air-to-air or other applications. Funding of the development and flight test programme is under contract from the Naval Air Systems Command.

The vehicle made its first flight at the Pacific Missile Test Center at Point Magu, California, on 2 December 1974. Test officials described its performance as excellent. Launched from an A-7 attack aircraft, the vehicle flew a distance of more than 56 km (35 miles) and attained speeds well above Mach 2 (2,334 km/h; 1,450 mph), precisely as programmed. On its second flight in November 1975 the LVRJ flew more than 64 km (40 miles), exceeded 2,736 km/h (1,700 mph) and demonstrated high-*g* manoeuvres. Still higher performance was planned during five additional test flights in 1976.

Vought LVRJ carried by A-7E Corsair

WILLIAMS

WILLIAMS RESEARCH CORPORATION

ADDRESS:
2280 W Maple Road, Walled Lake, Michigan 48088
Telephone: (313) 624-5200

Sam Williams believed in 1956 that gas-turbine technology could be extended down to very small sizes, and that if a small turbojet were made available it would find a market. Accordingly this company began a development programme on the WR2, described below, on a scale of effort and funding reflecting its very limited resources.

The WR2 first ran at a design thrust of 0·31 kN (70 lb) in 1962 and has since been developed into the WR2-6 and WR24-6. The more advanced WR19 uses an aerodynamically similar core and Williams Research is also building a range of shaft-drive engines, all characterised by their very simple design and aerodynamic similarity of their centrifugal compressors, axial compressor and power turbines.

Industrial and automotive engines of 75, 150 and 500 shp have been produced and the first of these, based on the WR2, was also developed for one-man helicopters. Williams Research is now developing, under contract from a private company, an automotive engine with regenerator and, for another company, a shaft-drive engine rated at significantly above 500 shp.

The company is now strong and experienced enough to mount a planned attack on the manned-aircraft market and believes that, as has happened with drone engines, aircraft markets will appear when reliable turbojet and turbofan engines sized between the 1·91 kN (430 lb) thrust of the WR19 and an upper limit of 8·90 kN (2,000 lb st) become available.

WILLIAMS WR2 and WR24

US military designation (WR24): J400

Although simple in design, almost to the point of appearing crude, this single-shaft turbojet has shown itself to be an effective power plant for high-subsonic drone aircraft and a suitable base for development of more advanced engines.

Air enters at the eye of a single-sided light alloy centrifugal compressor which handles an air mass flow of 1 kg (2·2 lb)/sec at a pressure ratio of 4·1 : 1. After passing through the diffuser which provides the structural basis for the engine the air divides, part of it flowing radially inwards as primary combustion airflow and the main bulk entering the short outward-radial annular combustor, through dilution apertures around the outer and rear face of the flame tube.

Fuel is sprayed centrifugally through a group of fine holes in the main compressor drive shaft. Surrounding the fuel pipe along the centreline of the main drive shaft is a cool airflow bled from the diffuser, which escapes through holes in the drive shaft to cool the combustion flames and reduce metal shaft and bearing temperatures, the main bearing being behind the compressor. A single igniter is mounted in the chamber at 12 o'clock. The hot gas, at about 955°C, then turns inwards and exits rearwards through the single-stage axial turbine and simple jet pipe.

The first production versions of the WR2 are the WR2-6, fitted to the Canadair AN/USD-501 high-performance battlefield reconnaissance vehicle; and the WR24-6 and -7 (YJ400-WR-400 and J400-WR-401) which power, respectively, the Northrop Chukar I and II target drones. The WR2-6 has a variable-area exhaust nozzle with translating central bullet, and drives a DC generator. The WR24 family have a minimal fixed-area jet pipe and drive a 4,000Hz alternator. The WR24-7 runs at higher temperature than the WR24-6 and incorporates detail modifications. The WR24-17, not yet in production but used in the Northrop variable-speed training target (VSTT), is further uprated though similar externally.

DIMENSIONS:
- Overall length:
  - WR2-6: 566 mm (22·3 in)
  - WR24-6: 490 mm (19·3 in)
  - WR24-7, WR24-17: about 635 mm (25 in)
- Max diameter:
  - WR2-6, WR24-6: 274 mm (10·8 in)
  - WR24-7, WR24-17: about 305 mm (12 in)

WEIGHT, DRY:
- WR2-6, WR24-6: about 13·6 kg (30 lb

MAXIMUM RATINGS (S/L, static):
- WR2-6: 0·56 kN (125 lb) at 60,000 rpm
- WR24-6: 0·54 kN (121 lb) at 60,000 rpm
- WR24-7: 0·76 kN (170 lb
- WR24-17: 0·89 kN (200 lb

SPECIFIC FUEL CONSUMPTION:
- WR2-6, WR24-6: 35·41 mg/Ns (1·25 lb/hr/lb st

WILLIAMS WR19

To produce this two-shaft turbofan Williams Researc used the WR2 as core and added an additional fan, axi compressor and drive turbine on a separate shaft, togethe with a by-pass duct. The LP turbine is related to thos developed for the company's shaft-drive engines.

The WR19 is the power plant used in the Bell Aerosy tems Flying Belt. It has also been used in the Willian Aerial Systems Platform (WASP) and Kaman Stowab Aircrew VEhicle Rotoseat (SAVER). From it has bee derived the US Air Force/Navy F107, described separate ly.

In early 1970 the company received a $1,400,000 co tract from the USAF for further development of a turb fan for future decoys. The company is making great effor to increase the maximum gas temperature, particularly the WR19 and derived engines. At present the temper ture actually used is about 955°C, with potential of tl present materials (Haynes 31 cobalt-base alloy for inl

Left: Williams Research WR2-6 turbojet, for the Canadair CL-89 (AN/USD-501) reconnaissance drone 0·56 kN (125 lb st). Right: Williams Research WR24-7 turbojet 0·76 kN (170 lb st); the 0·89 kN (200 lb st) WR24-17 is visually identical

guide vanes, Inco 100 for first-stage turbine blades and Inco 713 for other hot parts) limited to about 1,010°C.

Despite the mechanical difficulty of working on such small components, with turbine rotor discs and blades cast as single units, Williams is experimenting with aircooled turbine rotor blades and expects soon to be able to operate at gas temperatures higher than 1,100°C. The WR19 would be the first engine offered with cooled blades, and it also continues the company philosophy of using specially developed alternators, governors and other accessories capable of running at the full 60,000 rpm of the main shaft.

AIR INTAKE: Direct pitot type with four struts but no fixed inlet guide vanes. Unlike most WR2 engines the WR19 has a plain annular entry instead of a side intake downstream of an alternator or generator on the nose of the main shaft.

COMPRESSOR: Two-stage metal fan and two-stage axial IP compressor on common shaft leading to HP centrifugal compressor, handed to rotate in opposite direction to minimise gyroscopic couple. Total air mass flow, about 2 kg (4·4 lb)/sec; overall pressure ratio, 8·1; by-pass ratio, approximately 1 : 1.

COMBUSTION CHAMBER: Folded annular type, with fuel sprayed from revolving slinger on HP shaft. Dilution airflow admitted through perforated liner; cooling air injected through two sets of holes in HP shaft. Single igniter mounted diagonally on engine upper centre-line.

FUEL SYSTEM: Fuel fed at low pressure through transfer seal into pipe in HP shaft and ejected at high centrifugally-induced pressure, through calibrated fine orifices drilled radially through HP shaft in line with combustion chamber.

TURBINE: Single-stage axial-flow HP turbine, with Haynes 31 nozzle guide vanes and rotor wheel cast as single unit in Inco IN 100. Two-stage LP turbine, again with both wheels cast as single units, in Inco 713. Provision to be made for aircooling to raise entry gas temperature from 955°C to above 1,100°C.

JET PIPE: Mixer unit immediately downstram of LP turbine allows by-pass flow to merge with core gas flow to pass through plain propelling nozzle.

ACCESSORIES: Fuel and control system, filters, oil pump, tacho-generator and optional other accessories grouped into flat packages around upper part of fan/IP compressor casing. Starting system, depending on application, drives HP spool.

MOUNTING: Depending on application, main mounting above centrifugal diffuser casing with two double-lug pickups on horizontal centre-line at LP turbine casing.

DIMENSIONS:
- Length overall: 610 mm (24 in)
- Envelope diameter: 305 mm (12 in)

WEIGHT:
- Dry, depending on equipment: 27·6-30·8 kg (61-68 lb)

PERFORMANCE RATING:
- Max thrust, S/L static: 1·91 kN (430 lb)

SPECIFIC FUEL CONSUMPTION:
- Max thrust, S/L static: 19·83 mg/Ns (0·7 lb/hr/lb st)

Williams Research Corporation WR19 two-shaft turbofan engine 1·91 kN (430 lb st). Believed to be the smallest turbofan in the world, the WR19 weighs 30·4 kg (67 lb) as shown

WILLIAMS WR3

US military designation: F107

The F107-WR-100 turbofan is an advanced and uprated WR19 designed to propel the US Air Force ALCM (Air Launched Cruise Missile). Rated in the 2·67 kN (600 lb st) class, the F107 also powers the General Dynamics Convair Tomahawk Sea Launched Cruise Missile. Other prospective applications include a number of RPVs and light aircraft.

The F107 core is similar to that of the WR2, 19 and 24. The engine is an extremely simple two-shaft unit, with a high-altitude starting system and mixed-flow exhaust. Preliminary Flight Rating Test (PFRT) running at Arnold Engineering Development Center was completed in October 1975, and PFRT approval was granted in November. ALCM flight testing began in February 1976.

Williams Research F107-WR-100 two-shaft turbofan, developed for the Air-Launched Cruise Missile. Rating is "in the 2·67 kN/600 lb st class"

DIMENSIONS:
- Length overall: 800 mm (31·5 in)
- Envelope diameter: 305 mm (12 in)

WEIGHT, DRY: 58·7 kg (130 lb)

PERFORMANCE RATING: 2·67 kN (600 lb st) class

THE UNION OF SOVIET SOCIALIST REPUBLICS

GDL

ADDRESS: Leningrad

The Gas Dynamics Laboratory was founded in Leningrad in 1929. Since that time it has completed a very large number of rocket-engine programmes, mainly using liquid propellants. The laboratory has occupied many facilities, but is still headquartered in Leningrad. Three of its best-known engines are described below.

GDL RD-107

This four-chamber liquid-propellant rocket engine was developed during 1954 to 1957. The RD-107 and its derivatives have been in use for many years as launch vehicle first-stage engines for Soviet satellites to the Sun and Moon, and automatic stations launched to the Moon, Venus and Mars. They also powered the Vostok and Voskhod manned launch vehicles.

TYPE: Four-chamber liquid-propellant rocket engine.

PROPELLANTS: Liquid oxygen and kerosene.

THRUST CHAMBERS: Four primary thrust chambers of double-wall construction, with fabricated corrugations between walls and inner walls of copper or copper-rich alloy. Conical nozzles. All-welded heads. Flat-plate injectors, with concentric rings of tubes in which propellants are pre-mixed before injection. Estimated diameters: throat 150-165 mm (6-6·5 in), nozzle 685 mm (27 in). Combustion pressure 60·8 bars (882 lb/sq in).

VERNIER CHAMBERS: Two chambers of double-wall construction, with finning between walls. Estimated diameters: throat 75 mm (3 in), nozzle 305 mm (12 in).

TURBOPUMP: One single-shaft turbopump mounted in tubular thrust frame and feeding all chambers. Assembly comprises turbine exhaust hood containing coiled heat exchanger, single-sided shrouded centrifugal kerosene pump, double-sided shrouded centrifugal liquid oxygen pump, gearbox, and two auxiliary centrifugal pumps, one of which supplies the monopropellant gas generator. Fuel lines to main chambers pass through common valve.

PERFORMANCE (in vacuum):
- Rated thrust: 1,000 kN (224,870 lb)
- Specific impulse: 314 sec

GDL RD-119

This more modern single-chamber liquid-propellant engine, which has been in use since 1962, forms the second-stage engine of a launch vehicle for Cosmos research satellites. Many hundreds of these satellites have been launched, using two-, three- and four-stage launch vehicles of various types and having lifting capacities ranging from hundreds of pounds to 7·6 tonnes (7·5 tons).

TYPE: Single-chamber liquid-propellant rocket engine.

PROPELLANTS: Liquid oxygen and dimethyl-hydrazine.

THRUST CHAMBER: Single fixed chamber, possibly of tubular-wall construction, with fuel entry above base of nozzle. Estimated diameters: throat 100 mm (4 in), nozzle 940 mm (37 in). Combustion pressure 81·1 bars (1,176 lb/sq in).

TURBOPUMP: One single-shaft turbopump, driven by

Exhibition model of the four-chamber RD-107 first-stage rocket engine of the Vostok space launch vehicle, with two-chamber vernier engine in front
(TAM Air et Cosmos)

monopropellant (hydrazine) gas generator. Exhaust from gas generator taken to multiple auxiliary nozzles for control in roll, pitch and yaw.

PERFORMANCE (in vacuum):

Rated thrust	108 kN (24,250 lb)
Specific impulse	352 sec

GDL RD-214

This neat liquid-propellant rocket engine, developed at GDL in 1952-57, has been adopted as the standard first-stage propulsion for launching the Cosmos series of Soviet satellites. It has four thrust chambers, burning nitric acid and kerosene, each chamber being rated at 176·6 kN (39,700 lb) thrust at sea level. Vacuum rating of the engine is 725·7 kN (163,142 lb). The propellants are fed by a single large turbopump group mounted above the chamber group, the fuel being supplied straight to a bolted connection on the welded chamber heads and the nitric acid passing through part-flexible pipes to the regeneratively cooled chamber nozzle and throat. Chamber pressure is given by GDL as 3·1 bars (45 kg/cm²) and specific impulse as 246. The four chambers are rigidly fixed to the Cosmos launcher first stage, vehicle control being accomplished by four refractory deflector vanes, one per chamber, mounted on the vehicle skirt control packages and projecting into the rocket exhaust. Several hundred RD-214 engines have been made and flown from Kapustin Yar on non-recoverable Cosmos missions.

The RD-219 powers the second stage of launchers; a nitric acid/UDMH engine, its twin chambers have vacuum thrust of 883 kN (198,441 lb)

The 706·4 kN (158,800 lb) thrust GDL RD-214 rocket engine used to power the first stage of the Cosmos launcher. Vacuum rating 725·7 kN (163,142 lb)

GLUSHENKOV

This design bureau became known in 1969 when it was revealed as that responsible for the TVD-10 engines fitted in the Beriev Be-30 STOL transport. The TVD-10 is a small turboprop in the 950 shp class, supplied to Beriev as a complete cowled unit with a reversing propeller produced specially for this application. The engine air intake is positioned below the propeller axis, but no details of the TVD-10 have yet been made available. It has been reported that it was mainly due to difficulties with the TVD-10 that the Be-30 programme was terminated in 1973.

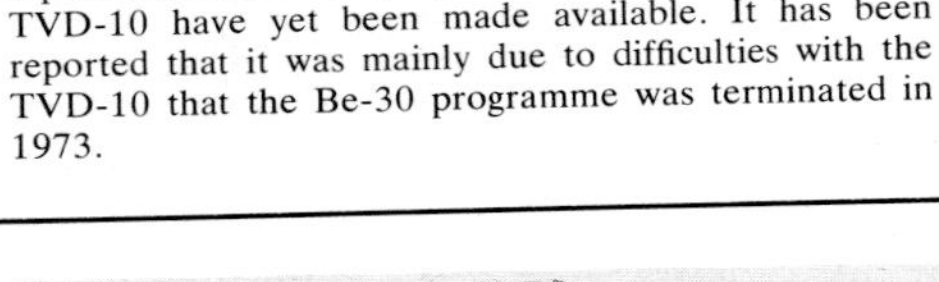

Glushenkov is also responsible for the 671 kW (900 shp) GTD-3 turboshaft engine, two of which power the Kamov Ka-25 helicopter family. This engine and the TVD-10 appear to have a generally similar layout, with a free power turbine, and it is possible that they use a common gas generator.

ISOTOV

GENERAL DESIGNER IN CHARGE OF BUREAU:
Sergei Pietrovich Isotov

This bureau was responsible for the GTD-350 and TV2-117A turboshaft engines which power the Mil Mi-2 and Mi-8 helicopters respectively. The former is in production in Poland, and is described under the WSK-PZL-Rzeszów entry in this edition. Isotov is also responsible for the TVD-850 turboprop engines which power the Antonov An-28 transport aircraft (see Aircraft section).

ISOTOV TV2-117A

The power plant of the Mi-8 comprises two TV2-117A engines coupled through a VR-8A gearbox. As is common with modern Soviet helicopters, the engines and gearbox are delivered and thereafter treated as a single unit. The complete package incorporates a control system (separate from the control system of each gas generator) which maintains desired rotor speed, synchronises the power of both engines, and increases the power of the remaining engine if the other should fail.

The TV2 engine is of conservative design, being biased in favour of long and trouble-free life rather than attempting to rival the small size and weight of some Western engines in the same power class.

TYPE: Free-turbine helicopter turboshaft engine.

AIR INTAKE: Direct pitot, with main front casing providing vertical upper and lower drive-shafts to accessory package. Main accessory group above the engine projects ahead of intake face. Casing incorporates variable-incidence inlet vanes.

COMPRESSOR: Ten-stage axial. Construction principally in titanium to reduce weight in comparison with the steel that would otherwise be used. Inlet guide vanes and stators of stages 1, 2 and 3 are of variable incidence to facilitate starting and increase compressor efficiency over a wide speed range; for the same reasons the casing incorporates automatic blow-off valves. Pressure ratio 6·6 : 1 at 21,200 rpm.

COMBUSTION CHAMBER: Annular, with eight burner cones. Fabricated from inner and outer diffuser casings, flame tube, casing, burners, and anti-icing bleed air pipe.

FUEL GRADE: T-1 or TS-1 to GOST 10227-62 specification (Western equivalents, DERD.2494, MIL-F-5616).

Isotov TV2-117A turboshaft engine (1,118 kW; 1,500 shp)

TURBINE: Two-stage axial compressor turbine bolted to rear of splined shaft with front extension to drive accessories. Solid rotor blades, held by fir-tree roots in discs cooled by bleed air (first disc 10th-stage air, all other discs 8th-stage). First- and second-stage stators have 51 and 47 inserted blades respectively. Free power turbine of similar two-stage design; its rotors have 43 and 37 blades respectively.

EXHAUST UNIT: Large fixed-area duct which deflects the gas out at 60°. It comprises a pipe, pipe shroud and tie-band, shroud connector links and exhaust pipe attachments. The exhaust pipe and shroud together form a double-wall assembly which minimises heat transfer into the power plant nacelle, the pipe being cooled by air circulating in the double wall.

OUTPUT SHAFT: The main drive-shaft is an extension of the power turbine rotor shaft. It conveys torque from the free turbine to the overrunning clutch of the helicopter main gearbox (VR-8A) and is also coupled to the speed governor of the free-turbine rotor. Max output speed 12,000 rpm; main rotor speed 192 rpm.

ACCESSORIES: Mounted on the main drive box above the intake casing, in which a train of bevel and spur gears provides drives for airframe and engine accessories. The engine automatic control system includes a fuel system, hydraulic system, anti-icing system, gas temperature restriction system, engine electric supply and starting system, and monitoring instruments. The hydraulic system positions the variable stators according to a preset programme, depending on compressor speed and air temperature at the inlet; it also sends electrical signals to control the starter/generator system, close the startin bleed air valves and restrict peak gas temperature t 600°C. Air up to 1·8 per cent of the total mass flow ca be used to heat the intake and other parts liable to icin Fire extinguishant can be released manually by the pilo upon receipt of a fire warning, through a series of spra rings and pipes.

LUBRICATION: Pressure circulation type. Oil is supplied b the upper pump and scavenged from the five main bea ings by the lower pump, returned through th helicopter-mounted air/oil heat exchanger and then to the helicopter tank. The oil seals and air/oil labyrin seals are connected to a centrifugal breathing systen

OIL GRADE: Synthetic, permitting operation at oil tempe atures above 200°C, combined with easy starting minus 40°C without heating the oil. Grade B-3V MRTU 38-1-157-65 (nearest foreign substitute Castr 98 to DERD.2487). Consumption, not over 0·5 lit per hour per engine.

STARTING: Electrical, fuel, and ignition systems are inte rated. The SP3-15 system comprises DC sta ter/generator, six storage batteries, control pan ground supply receptacle, and control switches a relays; of these all are airframe mounted except for t GS-18TP starter/generator which cranks the compre sor during the starting cycle. The ignition unit compris a control box, two semiconductor plugs, solenoid val and switch. The starting fuel system comprises automatic starting unit on the NR-40V fuel regulati pump, constant-pressure valve, and two igniters.

DIMENSIONS:
Length, with accessories and exhaust pipe 2,835 mm (111·5 in)
Length, intake face to rotor gearbox connection 2,391 mm (94·25 in)
Width 547 mm (21·5 in)
Height 745 mm (29·25 in)
WEIGHT, DRY:
Engine, without generator, transducers, etc 330 kg (727 lb)
VR-8A gearbox, less entrapped oil 745 kg (1,642 lb)
PERFORMANCE RATINGS:
T-O (S/L, static) 1,118 kW (1,500 shp)
Cruise (122 knots; 225 km/h; 140 mph at 500 m; 1,640 ft) 746 kW (1,000 shp)
SPECIFIC FUEL CONSUMPTION:
T-O, as above 102·4 μg/J (0·606 lb/hr/shp)
Cruise, as above 115·4 μg/J (0·683 lb/hr/shp)

ISOTOV TVD-850

Power plant of early Antonov An-28 STOL transports, the TVD-850 turboprop has a rating of 810 shp. The cowled engine looks almost identical to the Turboméca Astazou, and the reduction gearbox must be housed in the centre of an annular air intake section. No details are yet available.

IVCHENKO

The design team headed by general designer Alexander G. Ivchenko until his death in June 1968 was based in a factory at Zaparozhye in the Ukraine, where all prototypes and pre-production engines bearing the 'AI' prefix were developed and built. The bureau is now headed by General Designer Lotarev, who has his own entry on a later page.

The first engine with which Ivchenko was associated officially was the 41 kW (55 hp) AI-4G piston engine used in the Kamov Ka-10 ultra-light helicopter. He progressed ultimately, via the widely used AI-14 and AI-26 piston engines, to become one of the Soviet Union's leading designers of gas turbine engines. Since 1952 all Soviet piston engines have been assigned to Poland (see under WSK-PZL-Kalisz in Polish section).

IVCHENKO AI-20

The Ivchenko bureau was responsible for the AI-20 turboprop engine which powers the Antonov An-12 and Ilyushin Il-18 transport aircraft, Il-38 maritime patrol aircraft and Beriev M-12 Tchaika amphibian.

Prototype engines probably ran in 1954-55. Series production started in 1957 with four variants of the basic AI-20. They were followed by two major production versions:

AI-20K. Rated at 2,942 kW (3,945 ehp). Used in Il-18V, An-10A and An-12.

AI-20M. T-O rating of 3,124 kW (4,190 ehp). Used in Il-18D/E, An-10A and An-12. Probably fitted to Il-38 and Beriev M-12. Capable of operation on a wide range of fuels and lubricating oils. Later the power was increased to 3,169 kW (4,250 ehp), and this is the engine featured in the detailed description below.

The AI-20 was designed to operate reliably in all temperatures from −60°C to +55°C at heights up to 10,000 m (33,000 ft). It is a constant-speed engine, the rotor speed being maintained at 12,300 rpm by automatic variation of propeller pitch. Gas temperature after turbine is 560°C in both current versions. TBO of the AI-20K was 4,000 hours in the Spring of 1966; the same life was reached by the -20M in 1968.

In the Il-18 installation, the AI-20 turboprop is supplied as a complete power plant with cowling, mounting and automatically-feathering reversible-pitch four-blade propeller.

TYPE: Single-shaft turboprop.

AIR INTAKE: Inner and outer cones connected by six radial struts. Outer casing carries accessories and front mountings. Centre casing carries reduction gear.

COMPRESSOR: Ten-stage assembly of discs running in roller bearing in front casing, joined to first disc by tubular extension shaft, and ball-thrust bearing in combustion chamber casing on through-bolted rear shaft. Magnesium alloy stator casing in upper and lower halves, bolted together. Pressure ratio 9·2 under altitude cruise conditions. Air mass flow 20·7 kg (45·6 lb)/sec.

COMBUSTION CHAMBER: Annular chamber with ten burner cones welded to front ring, and separate inner and outer shrouds. Burners anchored by flanges on chamber casing. Pilot burners and ignition plugs at top of casing.

FUEL GRADE: T-1 or TS-1 to GOST-10227-62 (DERD.2492, JP-1 to MIL-F-5616).

TURBINE: Three stages overhung on cantilevered shaft running in roller bearing in tapered cone of combustion chamber casing and splined to compressor drive-shaft. Rotor blades shrouded at inner and outer ends and installed in pairs in slots in aircooled discs. Stator blades secured in grooves in casing, first stage being aircooled and second stage being hollow to ensure uniform heating.

JET PIPE: Fixed-area type with five radial struts. Nozzle area 0·225 m² (2·42 sq ft).

REDUCTION GEAR: Planetary type, incorporating six-cylinder torquemeter and negative-thrust transmitter (type IKM), with self-checking device, for autofeathering AV-681 propeller. Ratio 0·08732 (input speed 12,300 rpm except ground-idle 10,400 rpm).

LUBRICATION: Pressure-feed type with full re-circulation; hourly consumption not over 0·8 litres (0·175 Imp pints).

OIL GRADE: Mixture 75% transformer oil GOST 982-56 or MK-8 to GOST 6457-66 (equivalent to DERD.2490 or MIL-O-6081B) and 25% MS-20 or MK-22 to GOST 1013-49 (DERD.2472 or MIL-O-6082B).

ACCESSORIES: Engine and airframe accessories driven off compressor front extension shaft, via radial shafts at 6 and 12 o'clock. Full ice-protection and fire-extinguishing systems.

Ivchenko AI-20M turboprop of 3,169 ekW (4,250 ehp) *(courtesy of Aviation Magazine International, Paris)*

The 1,875 ekW (2,515 ehp) Ivchenko AI-24A turboprop *(courtesy of Aviation Magazine International, Paris)*

STARTING: Two electric starter/generators, Type STG-12 TMO-1000, supplied from ground source or from APU Type TG-16.
DIMENSIONS:
Length 3,096 mm (121·89 in)
Width 842 mm (33·15 in)
Height 1,180 mm (46·46 in)
WEIGHT, DRY: 1,040 kg (2,292 lb)
PERFORMANCE RATINGS:
T-O 3,169 kW (4,250 ehp)
Cruise (350 knots; 650 km/h; 404 mph at 8,000 m; 26,000 ft) 2,013 kW (2,700 ehp)
SPECIFIC FUEL CONSUMPTION:
T-O 104·3 μg/J (0·617 lb/hr/shp)
Cruise, as above 73·3 μg/J (0·434 lb/hr/shp)

IVCHENKO AI-24

This turboprop powers the An-24 and its derivatives. Production began in 1960 and the following data refer to engines of the second series, which were in production by the Spring of 1966.

The AI-24 of 1,875 kW (2,515 ehp) powered the An-24V Series I, and was followed by the AI-24A with provision for water injection, in the main production version of that aircraft.

The more powerful AI-24T of 2,103 kW (2,820 ehp) with water injection is used in the An-26. This engine has in-flight vibration monitoring, automatic relief of power overloads and gas temperature behind the turbine, and auto-shutdown and feathering.

The AI-24 is a constant-speed engine, maintained at 15,100 rpm by automatic variation of propeller pitch. The engine is flat-rated to maintain its nominal output to 3,500 m (11,500 ft). TBO was 3,000 hours in the Spring of 1966; by 1968 the later AI-24T had reached 4,000 hours.

TYPE: Single-shaft turboprop.

AIR INTAKE: Large magnesium alloy casting, comprising inner and outer cones joined by four radial struts. Carries accessories, reduction gear, front mountings and compressor inlet guide vanes.

COMPRESSOR: Ten-stage axial. Stainless steel rotor, comprising rigidly-connected discs carrying dovetailed blades. Front shaft runs in roller bearing and is bolted to propeller drive-shaft of reduction gear; rear shaft runs in ball-thrust bearing and is splined to turbine shaft. Welded steel casing in bolted left and right halves, with welded front and rear connecting flanges. Pressure ratio (max continuous, 6,000 m; 18,300 ft, 272 knots; 505 km/h; 314 mph) 7·85 : 1. Air mass flow 14·4 kg (31·7 lb)/sec.

COMBUSTION CHAMBER: Annular, of spot-welded heat-resistant steel, with eight simplex burners inserted into swirl-vane heads. Contains two starting units, each comprising a body, pilot burner and igniter plug.

FUEL GRADE: T-1, TS-1 to GOST 10227-62 (DERD.2494 or MIL-F-5616).

TURBINE: Three-stage axial. Three discs carry solid blades in fir-tree roots, and are automatically centred on each other when connected by stay-bolts to the extended flange at the rear of the turbine shaft. Shaft splined to compressor rear shaft and held by tie-rod; runs in roller bearing ahead of first turbine disc. Three stator diaphragms through-bolted together and to combustion-chamber casing. First nozzle diaphragm cooled by secondary air from combustion chamber. Rotor/stator sealing effected by soft inserts mounted in grooves in nozzle assemblies. Peak exhaust temperature during starting 750°C.

JET PIPE: Fixed-area type. Inner and outer rings connected by three hollow struts carrying 12 thermocouples.

REDUCTION GEAR: Planetary type, incorporating hydraulic torquemeter and electromagnetic negative-thrust transmitter for propeller auto-feathering. Magnesium alloy casing. Front flange of propeller shaft has end splines and 12 stud holes for type AV-72 propeller (AI-24T drives AV-72T propeller). Ratio 0·08255.

LUBRICATION: Pressure circulation system; hourly consumption not over 850 gr (1·87 lb).

OIL GRADE: Mixture of 75% transformer oil GOST 982-56 or MK-8 (DERD.2490 or MIL-O-6081B) and 25% MS-20 or MK-22 (DERD.2472 or MIL-O-6082B).

ACCESSORIES: Mounted on front casing are starter/generator, alternator, propeller speed governor and centrifugal breather. Below casing are oil unit, air separator and removable box containing LP and HP fuel pumps and drives to hydraulic pump and tachometer generators. Also on front casing are an aerodynamic probe, ice detector, and negative-thrust feathering valve, torque transmitter and oil filter.

STARTING: Electric STG-18TMO starter-generator supplied from ground power or from TG-16 APU.

DIMENSIONS:

Length overall	2,346 mm (92·36 in)
Width	677 mm (26·65 in)
Height	1,075 mm (42·32 in)

WEIGHT, DRY: 600 kg (1,323 lb)

PERFORMANCE RATINGS:

T-O:

AI-24A	1,875 kW (2,515 ehp)
AI-24T	2,103 kW (2,820 ehp)

Cruise rating at 243 knots (450 km/h; 280 mph) at 6,000 m (18,300 ft):

AI-24A	1,156 kW (1,550 ehp)
AI-24T	1,178 kW (1,580 ehp)

SPECIFIC FUEL CONSUMPTION:

At cruise rating:

AI-24A	91·3 μg/J (0·540 lb/hr/shp)
AI-24T	90·1 μg/J (0·533 lb/hr/shp)

OIL CONSUMPTION: 0·85 kg (1·87 lb)/hr

IVCHENKO AI-25

This turbofan powers the three-engined Yakovlev Yak-40 STOL transport. The Aeroflot Yak-40 has the basic AI-25 engine, but the Yak-40B of the Soviet Air Force has an AI-25 with an aircooled HP turbine and rating of 17·13 kN (3,850 lb st). The AI-25 also powers the Czech L-39 trainer, and it had been intended that it should be manufactured by Motorlet, Czechoslovakia, as the Walter Titan. The emergence of the Polish WSK/PZL/Mielec M-15 agricultural aircraft has transferred AI-25 activity to Poland, and mass production is now likely to take place in that country to provide engines for an M-15 programme estimated at 3,000 aircraft for the Soviet Union alone.

The Ivchenko bureau planned the engine for small transports, trainers and business jet aircraft. It is claimed to have an exceptional margin of flow stability and to be unusually robust and simple.

TYPE: Two-shaft turbofan.

AIR INTAKE: Fabricated from titanium sheet. Central bullet and intake leading-edges anti-iced by hot bleed air.

FAN: Three-stage axial. Drum/disc construction with pin-jointed blades. Casing and fan duct of magnesium alloy. Peak pressure ratio, 1·695 at 10,750 rpm. By-pass ratio 2.

Ivchenko AI-25 turbofan (14·68 kN; 3,300 lb st)

COMPRESSOR: Eight-stage axial. Drum/disc construction of titanium, with aluminium and magnesium casing. Dovetailed blades. Peak pressure ratio, 4·68 at 16,640 rpm. Overall pressure ratio, 8.

COMBUSTION CHAMBER: Annular. Inner and outer casings joined upstream to 12 burner heads with stabilisers.

FUEL GRADE: T-1, TS-1 to GOST 10227-62 (DERD.2494, MIL-F-5616).

TURBINE: Single-stage HP turbine; two-stage LP turbine. Shrouded solid rotor blades held by fir-tree roots in cooled discs.

JET PIPE: Plain fixed convergent nozzles for core and by-pass airflow. No mixer.

LUBRICATION: Self-contained, pressure circulating.

OIL GRADE: MK-8 to GOST 6457-66 or MK-6 to GOST 10328-63 (Western equivalents, DERD.2490 or MIL-O-6081B). Consumption 0·3 litres (0·53 Imp pints)/hr.

ACCESSORIES: All shaft-driven accessories mounted on gearbox on underside of engine and driven off HP spool. Equipment includes automatic fire extinguishing (agent can be supplied into oil-contacted labyrinth cavities), ice protection, automatic starting and control system, oil-system chip detector and casing vibration monitor.

STARTING: Pneumatic. Air starter type SV-25 is supplied from ground hose coupling or from APU type AI-9 or an operating engine bleed. System claimed to develop high torque for rapid start in any climatic condition, and to be cleared for exceptional number of starts in given overhaul life.

DIMENSIONS:

Length overall	1,993 mm (78·46 in
Width overall	820 mm (32·28 in
Height overall	895 mm (35·24 in

WEIGHT, DRY:

Without accessories	290 kg (639 lb

PERFORMANCE RATINGS:

T-O: 14·68 kN (3,300 lb st

Long-range cruise rating, 6,000 m (20,000 ft) and 29 knots (550 km/h; 342 mph) 3·49 kN (785 lb st

SPECIFIC FUEL CONSUMPTION:

T-O	15·86 mg/Ns (0·56 lb/hr/lb st
Cruise, as above	23·71 mg/Ns (0·837 lb/hr/lb st

KLIMOV

GENERAL DESIGNER IN CHARGE OF BUREAU:

Not known.

Led by the late V. Ya. Klimov, this bureau was in 1947-48 assigned the task of developing the RD-45, the Soviet copy of the Rolls-Royce Nene which had been imported from Britain and put into mass production without a licence. From this engine stemmed technology which assisted the important axial engines used in the next generation of Soviet fighters. It is not known whether the bureau is still operative, or whether it led in turn to the Tumansky bureau (which see). The RD-9 is believed to have been designed in the Klimov era.

KLIMOV VK-1

This simple centrifugal turbojet was an increased-airflow development of the RD-45 (Nene), similar in many respects to the developments by Rolls-Royce, Pratt & Whitney and Hispano-Suiza. The original VK-1 powered the MiG-15bis, S-103, LiM-2, MiG-15UTI (as a retrofit in many cases), the later Tu-12 versions and the Il-28. The VK-1A powered the MiG-17. The VK-1F powers the MiG-17P, -17F and -17PFU, F-4, S-104 and all LiM-5 versions. Most VK-1 engines were delivered from Engine Plant No 45, near Moscow, in 1950-56. As noted on an earlier page, the VK-1F is also believed to have been produced in China, for the F-4 and Chinese-built Il-28.

PERFORMANCE RATINGS (T-O):

VK-1	26·47 kN (5,950 lb st)
VK-1A	26·87 kN (6,040 lb st)
VK-1F with afterburner	33·15 kN (7,450 lb s

KLIMOV AXIAL ENGINES

The engine which appears to link the Klimov an Tumansky bureau is the single-shaft axial turbojet know most generally as the RD-9, but also referred to as th VK-5, VK-7, VK-9, M-109 and by other designation Until after 1965 this was usually ascribed to the Klimo bureau (as the VK designation would appear to indicat but today it is agreed that the RD-9 is a Tumansky engin and it appears under that heading on a later page. Th inference is that Tumansky was promoted to becom General Designer in charge of the bureau that had prev ously borne Klimov's name.

KOLIESOV

During an official tour of the Soviet aircraft industry in mid-1973, a representative of *Air Force Magazine* was told of the existence of the hitherto-unreported Koliesov Engine Design Bureau. It is developing an alternative engine to the Kuznetsov NK-144, the power plant used currently in the Tu-144 and, probably, in the Tupolev bomber known to NATO as *Backfire*.

The new Koliesov engine is described as a variable-geometry, variable by-pass ratio engine which functions as a turbojet in supersonic flight and as a turbofan in the subsonic regime. No such advanced design exists in th West.

Soviet engine designers are reported to be experimen ing also with hypersonic vehicles powered by scramj propulsion systems and capable of operating in the Mach to Mach 7 range.

KUZNETSOV

GENERAL DESIGNER IN CHARGE OF BUREAU:

Nikolai Dmitrievich Kuznetsov

This bureau appears to have devoted its entire energy to large transport gas turbine engines. Though the formidable NK-12 was a surprising success, the bureau has not been entirely able to overcome technical problems, and its NK-8 turbofan has been replaced by engines from the Soloviev bureau in the Il-62M long-range airliner. Kuznetsov himself led a delegation of engineers to visit Rolls-Royce in late 1973 to discuss collaboration with the British company.

KUZNETSOV NK-8

One of the first Russian civil turbofans, the NK-8 has been developed through a number of variants, the most powerful of which is the NK-144 supersonic augmented engine for the Tupolev Tu-144 supersonic transport. Basic versions are the 99·1 kN (22,273 lb st) **NK-8-4** which originally powered the Ilyushin Il-62 four-engined transport, and the further-developed 93·2 kN (20,950 lb) **NK-8-2** which was the original engine of the Tupolev Tu-154. At one time it was planned to replace the NK-8-2 by the Soloviev D-30K, which replaced the NK-8-4 in the Il-62M. It is believed that all current Tu-154A/B aircraft are powered by the **NK-8-2U** of greater thrust (details unknown in the West). The NK-8-4 remains in service with several Il-62 (not Il-62M) operators, including LOT.

TYPE: Two-shaft turbofan.

AIR INTAKE: Fabricated from outer ring, inner splitter and welded stator blades (15 in core airflow, 30 ahead of fan). Hot-air ice-protection.

FAN: Two-stage axial, with anti-flutter sweptback blades on first rotor stage. Pressure ratio 2·15 at 5,350 rpm. By-pass ratio 1·02 (NK-8-2, 1·00).

COMPRESSOR: Two IP stages on fan shaft. Six-stage HP compressor. Construction of both rotors and stators, including blading, almost wholly of titanium alloy. Core pressure ratio, 10·8 at 6,950 HP rpm (NK-8-2, 10 at 6,835 rpm).

COMBUSTION CHAMBER: Annular, with 139 burners. Claimed to produce no visible smoke.

FUEL GRADE: T-1 and TS-1 to GOST 10227-62 or T-7 to GOST 12308-66 (equivalent to Avtur 50 to DERD.2494 or MIL-F-5616).

TURBINE: Single-stage HP turbine, two-stage LP turbine, all with shrouded rotor blades, aircooled discs and hollow nozzle blades (stators). All shafting carried between shock-absorbing bearings at each end, with labyrinth and contact (rubbing) graphite seals to prevent gas leakage. Gas temperature, not over 870°C (1,143 ahead of turbine, not over 670°C (NK-8-2, 650 downstream, both values sea level, static.

JET PIPE: Mixer leads by-pass flow into common jet p which may be fitted with blocker/cascade-type rever giving up to 48% (NK-8-2, 45%) reverse thrust, a noise suppressor.

LUBRICATION: Continuous pressure feed and recirculati Oil consumption not over 1·3 kg (2·87 lb)/hr. Press not less than 2·28 bars (33 lb/sq in).

OIL GRADE: Mineral oil MK-8 or MK-8P to GOST 64 66 (DERD.2490 or MIL-O-6081B). External tank left side of front casing.

ACCESSORIES: These include automatic flight-deck wa ing of vibration exceeding permissible limit, ice and All accessories grouped beneath fan duct casing. Eng claimed to need no attention for long periods, ot than inspection of fuel and oil filters. RTA-26-9-1 bine gas temperature controller by Smiths Industri

STARTING: HP spool driven by constant-speed drive PPO-62M, or started pneumatically by air from A type TA-6, from ground hose or by air from ano engine (NK-8-2, pneumatic starter only). Time to id speed not over 80 sec. Engine can be windmill-starte the air under all conditions, up to altitudes of 11,00 (36,000 ft).

DIMENSIONS:
NK-8-4:
Length, no reverser 5,100 mm (201 in)
NK-8-2:
Length, with reverser 5,288 mm (208·19 in)
Length, without reverser 4,762 mm (187·48 in)
Diameter 1,442 mm (56·8 in)
WEIGHT, DRY:
NK-8-4:
No reverser 2,100 kg (4,629 lb)
With reverser 2,400 kg (5,291 lb)
NK-8-2:
No reverser 2,100 kg (4,629 lb) max
With reverser 2,350 kg (5,180 lb) max
PERFORMANCE RATINGS:
NK-8-4:
T-O rating 99·1 kN (22,273 lb st)
Cruise rating at 11,000 m (36,000 ft) and 458 knots (850 km/h; 530 mph) 27 kN (6,063 lb st)
NK-8-2:
T-O rating 93·2 kN (20,950 lb st)
Cruise (as above) 17·65 kN (3,968 lb st)
SPECIFIC FUEL CONSUMPTION:
At cruise rating at 11,000 m (36,000 ft) and 458 knots (850 km/h; 530 mph):
NK-8-4 22·1 mg/Ns (0·78 lb/hr/lb st)
NK-8-2 21·53 mg/Ns (0·76 lb/hr/lb st)

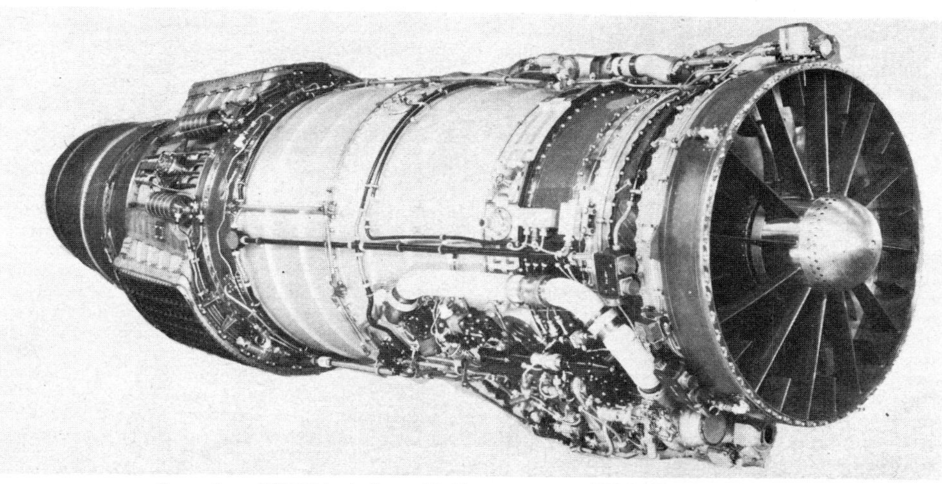

Kuznetsov NK-8-2 turbofan with thrust reverser (93·2 kN; 20,950 lb st)

KUZNETSOV NK-144

This is the two-spool augmented turbofan developed for the Soviet Union's first supersonic transport aircraft, the Tu-144. It is a development of the NK-8 and the first five pre-production NK-144s completed some 1,500 hours of bench-testing by October 1965. The engine flew in at least one airborne test aircraft before the start of the Tu-144 flight programme in 1968.

Since 1972 the afterburner augmentation has been increased, raising maximum rating from 171·6 kN (38,580 lb) to the figure given below. A version of the NK-144 is believed to be the engine of at least the first sub-type of the Tupolev supersonic bomber known to NATO as *Backfire*.

The NK-144 is reported to have a two-stage titanium fan, three-stage IP compressor, eleven-stage HP compressor, annular combustion chamber, single-stage HP turbine and two-stage LP turbine. Aircooled blades are used in the HP turbine, and titanium is used extensively in construction of the engine. By-pass ratio is reported to be 1 : 1, maximum mass flow 250 kg (551 lb)/sec, and pressure ratio 15 : 1. The jet pipe incorporates an afterburner, with hydraulically-actuated variable-area nozzle. Gas temperature at turbine entry is 1,050°C.

DIMENSIONS:
Length overall 5,200 mm (204·7 in)
Diameter 1,500 mm (59 in)
WEIGHT:
Without jet pipe, but with afterburner 2,850 kg (6,283 lb)
PERFORMANCE RATINGS:
Max, without afterburning 127·5 kN (28,660 lb st)
Max, with afterburning 196·1 kN (44,090 lb st)

Kuznetsov NK-12MV single-shaft turboprop of 11,033 ekW (14,795 ehp)
(courtesy of Aviation Magazine International, Paris)

KUZNETSOV NK-12M

Designed at Kuibishev under the leadership of N. D. Kuznetsov and former German engineers, the NK-12M is the most powerful turboprop engine in the world. In its original form as the NK-12M it developed 8,948 kW (12,000 ehp). The later NK-12MV is rated at 11,033 kW (14,795 ehp) and powers the Tupolev Tu-114 transport, driving four-blade contra-rotating propellers of 6·5 m (18 ft 4 in) diameter. As the NK-12MA, rated at 11,185 kW (15,000 shp), it powers the Antonov An-22 military transport, with propellers of 6·2 m (20 ft 4 in) diameter. A third application is in the Tupolev Tu-95 bomber and its derivatives, and Tu-126 'AWACS'.

The NK-12M has a single 14-stage axial-flow compressor. Compression ratio varies from 9 : 1 to 13 : 1 according to altitude, and variable inlet guide vanes and blow-off valves are necessary. A cannular-type combustion system is used: each flame tube is mounted centrally on a downstream injector, but all tubes merge at their maximum diameter to form an annular secondary region. The single turbine is a five-stage axial. Mass flow is 65 kg (143 lb)/sec.

The casing is made in four portions, from sheet steel, precision welded. An electric control for variation of propeller pitch is incorporated, to maintain constant engine speed.

DIMENSIONS:
Length 6,000 mm (236·2 in)
Diameter 1,150 mm (45·3 in)
WEIGHT, DRY: 2,350 kg (5,181 lb)
PERFORMANCE RATINGS:
T-O 11,033 kW (14,795 ehp)
Nominal power 8,826 kW (11,836 ehp) at 8,300 rpm
Idling speed 6,600 rpm

LOTAREV

GENERAL DESIGNER IN CHARGE OF BUREAU:
Vladimir Lotarev
ADDRESS:
Zaparozhye

LOTAREV D-36

As successor to Alexander Ivchenko at Zaparozhye, Vladimir Lotarev has developed the turbofan engine that powers the Yakovlev Yak-42. With a by-pass ratio of 5·34 : 1 the D-36 is the first avowed turbofan—as distinct from a by-pass turbojet—to emerge in the Soviet Union. Bench testing was in progress in September 1973, and flights in a pod carried beneath a Tu-16 testbed aircraft were scheduled to begin at the end of 1973.

No details had been disclosed as this edition went to press, though Western engineers visiting the Soviet Union report that the D-36 is a more advanced and modern engine than other Russian commercial gas turbines. Of three-shaft design, it is in many respects similar to a scaled RB.211, and its behaviour during development and certification running is said to have been encouraging.

WEIGHT, DRY: 1,080 kg (2,480 lb)

PERFORMANCE RATINGS:
T-O rating 62·76 kN (14,110 lb st)
Cruise rating at 8,000 m (26,200 ft) at Mach 0·84 15·6 kN (3,505 lb st)

SPECIFIC FUEL CONSUMPTION:
At T-O rating 10·76 mg/Ns (0·38 lb/hr/lb st)
At cruise rating as above 18·7 mg/Ns (0·66 lb/hr/lb st)

LYULKA

GENERAL DESIGNER IN CHARGE OF BUREAU:
Arkhip Mikhailovich Lyulka

During the late 1930s Arkhip Lyulka worked on the design of an axial turbojet that became an early war casualty. In 1942 he planned a more advanced engine that finally materialised as the TR-1, of 12·75 kN (2,866 lb st), run on the bench in 1944 and used in the Ilyushin Il-22 four-jet bomber and Sukhoi Su-11 twin-jet fighter prototypes, both of 1947. Ultimately, in 1948, this pioneer Soviet-designed turbojet was developed to give 14·71 kN (3,307 lb st).

In 1946 Lyulka began the design of a very ambitious axial engine to give a thrust of 44·13 kN (9,920 lb), and in 1950 this began bench trials under the designation AL-5. Although of basically simple, single-shaft configuration, with a seven-stage compressor and single-stage turbine, the AL-5 was more powerful than all Western engines apart from the prototype Olympus and J57. By 1951 it was rated at 45·1 kN (10,140 lb st) and flew in the prototype Ilyushin Il-30 twin-jet bomber; later in 1951-52 uprated AL-5 engines, giving a static thrust of 49·1 kN (11,032 lb), powered the Il-46 twin-jet bomber and the transonic Lavochkin La-190 and Yakovlev Yak-1000 fighters. An advanced civil version of the same engine, the AL-5 rated at 53·93 kN (12,125 lb st) powered the Tu-110 four-engined derivative of the Tu-104 airliner that never went into production (at the time, in 1959, this engine was reported in the West as the 'Lu-4').

By the time the AL-5 was running, Lyulka had conducted extensive research with axial compressors having supersonic airflow through some or all of the stages. It was clear that, if problems of flow breakdown and inefficiency could be resolved, such a compressor would enable turbojets to be made much smaller and lighter for a given thrust and with greater thrust per unit frontal area, and thus much better suited to the propulsion of supersonic fighters. By 1952 a supersonic-compressor engine had been designed and built. This, the AL-7, became Lyulka's first major success.

LYULKA AL-7

The first AL-7 ran on the bench in late 1952 and the first production version was cleared for use in 1954 at a design rating of 63·74 kN (14,330 lb st). Its initial application was on the Il-54, yet another Ilyushin twin-jet bomber that failed to see production, despite the fact that its speed at low altitude of 620 knots (1,150 km/h; 714 mph) was probably unrivalled by any other bomber in 1955. In the same year the Sukhoi Su-7 single-seat ground-attack fighter was designed around the AL-7F afterburning version of this engine, with thrust increased by about 40 per cent (see data below). By 1956 the Su-7 was flying, and the AL-7F had also been chosen for the basically similar Su-9 all-weather fighter. Subsequently the -7F was also produced for the Su-11 and -11U.

By 1958 a further developed version of the basic unaugmented engine, the AL-7PB, had been chosen by Beriev for the Be-10 reconnaissance flying-boat, which—apart from being the only pure-jet flying-boat ever to go into service anywhere—set up a number of world records for speed, load-carrying and altitude. Other versions of the AL-7, in both cases of the -7F afterburning family, powered the unsuccessful Tu-98 bomber and La-250 strike fighter of 1956.

The AL-7F, or a development of it, has been persistently reported to be the power plant of the Tu-28P twin-engined interceptor. This is not confirmed. The Tu-28P is considered to need greater thrust, greater even than provided by the AL-21 which was not available when the Tu-28 entered service in the early 1960s.

Type: Single-shaft axial-flow turbojet, available with or without afterburner.

Air Intake: Central bullet fairing and 14 fixed aerofoil struts anti-iced by compressor bleed air.

Compressor: Nine-stage axial (probably eight stages in original AL-7 design). First two stages widely separated axially, with variable stators ahead of second stage. Each stage has blades inserted in centreless disc held by peripheral spacers at correct distance from adjacent discs, the whole being coupled together finally by the central drive-shaft in tension. Pressure ratio probably about 8 : 1.

Combustion Chamber: Annular type with perforated inner flame tube. Multiple downstream fuel injectors inserted through cups in forward face of liner. Liner outer casing provided with multiple inward secondary-air injection ducts.

Turbine: Two-stage axial-flow type. Both wheels overhung behind rear bearing; front disc bolted to flange on hollow tubular driveshaft which, in turn, is splined to rear of compressor shaft running in main centre bearing which locates compressor axially against end loads.

Afterburner (AL-7F): Comprises upstream diffuser and downstream combustion section. Pilot combustor on turbine exit cone includes single nozzle ring and flame-holder; main spray ring and gutter flame-holder assembly located further downstream at greater radius. Refractory liner in combustion section. Variable-area nozzle, with multiple hinged flaps which govern nozzle size and profile according to signals from reheat control system based on turbine exit temperature and throttle lever position.

Accessories: Fuel pump and control unit, oil pumps, hydraulic pump, electric generator, tachometer and other items grouped into quickly replaceable packages beneath compressor casing.

Performance Ratings:

Max rating:

AL-7F, unaugmented	68·64 kN (15,432 lb st)
AL-7F, afterburning	98·1 kN (22,046 lb st)
AL-7PB	63·74 kN (14,330 lb st)

LYULKA AL-21

When continued development of the Su-7 family by the Sukhoi bureau was matched with increased thrust it was logical to suppose that the engine would be a derivative of the AL-7. It is now reported that this is the case, and the designation AL-21 has been given. Compared with the AL-7 the -21 is closely similar, and may be installationally interchangeable, but has significant improvements to the compressor and other components. The first production version was the AL-21F-3, for the Su-17 and export Su-20 variable-geometry tactical aircraft. The same engine is reported to power the twin-engined Su-15 interceptor. Sukhoi's latest production aircraft, the very important twin-engined Su-19, has completely different (turbofan) engines, though in the same thrust class and possibly from the Lyulka bureau.

Performance Ratings:

Max S/L, unaugmented	80·1 kN (18,000 lb st)
Max S/L, afterburning	109 kN (24,500 lb st)

MIKULIN

General Designer in Charge of Bureau:
Alexander Alexandrovich Mikulin

Eighty years old on 16 February 1975, Mikulin has been engaged in aircraft engine design since 1916. Notable Vee-12 water-cooled engines from his designs were the AM-13, AM-34, AM-38 (used in the Il-2 and Il-10 Stormovik) and AM-42.

Very large numbers of M-11 five-cylinder radial engines have been built in the Soviet Union and (from 1949) Poland to power a variety of light aircraft and helicopters. Best-known variants are the 93 kW (125 hp) M-11D and the 119 kW (160 hp) M-11FR which powers the Yak-18 primary trainer. Brief descriptions appeared in the 1975-76 *Jane's*.

The large turbojet described below was designed immediately after the second World War. Though a version was fitted to M-4 four-engined prototypes, it is not known if Mikulin's bureau developed the much more powerful 'D-15' engine fitted to later M-4 aircraft.

MIKULIN RD-3M-500

The basic RD-3M (or AM-3M) single-spool axial-flow turbojet was developed under the design leadership of P. F. Zubets from the original Mikulin M-209 (civil RD-3 or AM-3) engine which powers the Tu-16 and early M-4 bombers and was adapted for Russia's first jet transport, the Tu-104.

The RD-3M-500 was evolved, in turn, from the RD-3M and powers the Tu-104A and Tu-104B commercial transports. It has a simple basic configuration, with an eight-stage axial-flow compressor, annular-type combustion system with 14 flame tubes, and a two-stage turbine. The compressor casing is made in front, centre and rear portions, the front casing housing a row of inlet guide vanes. A bullet fairing mounted centrally in the annular ram-air intake houses a type S-300M gas-turbine starter, developing 90-100 hp at 31,000-35,000 rpm. The jet pipe consists of a central cone and fixed nozzle with an orifice diameter of approximately 840 mm (33 in). Pressure ratio is 6·4 : 1; temperature after turbine 720°C.

Dimensions:

Length overall	5,340 mm (210·23 in)
Diameter	1,400 mm (55·12 in)

Performance Rating:

T-O	93·15 kN (20,940 lb st)

MIKULIN RD-5

Though no longer in service, it is worth recording that this was the original engine of the MiG-19 and Yak-25 twin-engined fighters. Designed in 1949-52 as the AM-5, it was a single-shaft axial turbojet of simple design, used dry in the Yakovlev fighter and as the RD-5F (AM-5B) with afterburner in the original MiG-19.

Performance Rating (T-O):

AM-5	21·35 kN (4,800 lb st)
AM-5F	29·8 kN (6,700 lb st)

SHVETSOV

Founder of Bureau:
Arkadiya Dmitrievich Shvetsov

Arkadiya Shvetsov had meteoric rise to fame with his aircooled radial engines based on original US designs, which transformed Soviet fighters, bombers and transport aircraft during and after the second World War. The most important, and most powerful, of these engines was the ASh-82 series, derived from the Pratt & Whitney Twin Wasp, which at ratings up to 1,491 kW (2,000 hp) powered Lavochkin fighters, the Tu-2 and Tu-4 bombers, Il-12 and -14 transports and Mi-4 and Yak-24 helicopters (see 1975-76 *Jane's*). The ASh-62, originally based on the Wright Cyclone, was made in large numbers at ratings in the 750 kW (1,000 hp) class for the Li-2 and An-2. The ASh-62M agricultural version, developed by Vedeneev, and all An-2 engines after 1952, were transferred with all other Soviet aircooled radials to Poland (see WSK-PZL-Kalisz and WSK-PZL-Rzeszów in Polish part of this section). Another important Shvetsov piston engine was the 545 kW (730 hp) ASh-21, fitted to the Yak-11. All these engines have been described in earlier editions.

SOLOVIEV

General Designer in Charge of Bureau:
P. A. Soloviev

Engines for which Soloviev's design team is responsible include the turbofans fitted in the Ilyushin Il-62M and Il-76 and Tupolev Tu-124 and Tu-134 transport aircraft, and the turboshafts which power the Mi-6, Mi-10 and V-12 helicopters.

SOLOVIEV D-15

This engine was first reported, in 1959, as that fitted to the four-engined Type 201-M aircraft which gained a number of world records for speed and altitude. Over the years it has become apparent that the aircraft was a special Myasishchev M-4, and that the engine is standard in the later service versions of this long-range reconnaissance and ESM carrier. Details of the D-15 are still unknown in the West, but it is probably safe to deduce that it is a two-shaft by-pass turbojet (low by-pass ratio turbofan). It is likely that it laid the foundation upon which Soloviev's bureau produced the civil D-20 and D-30, both of which are considerably smaller engines.

Performance Ratings:

T-O	128·6 kN (28,660 lb st)

Soloviev D-20P turbofan (52·96 kN; 11,905 lb st)

SOLOVIEV D-20P

The D-20P is a two-spool turbofan fitted to the Tupolev Tu-124 twin-engined passenger transport. Of conservative design, it underwent prolonged testing before entering service. It was designed for maximum economy and reliability over the range of ambient temperatures between −40°C and 40°C.

Type: Two-shaft turbofan (by-pass turbojet).

Air Intake: Eight radial struts and central bullet fairing, de-iced by hot bleed air from fourth HP stage (from final stage at low rpm).

Fan: Three-stage axial, with supersonic blading in first stage. Mass flow 113 kg (249 lb)/sec at 8,550 rpm. Pressure ratio (S/L, static at max cont 7,900 rpm) 2·4 : 1. By-pass ratio 1 : 1.

Compressor: Eight-stage axial. Automatically-controlled flap valves downstream of the third and fourth stages bleed air into the fan duct to stabilise behaviour. Pressure ratio (at max continuous, 11,170 rpm) 5 : 1; overall pressure ratio 13 : 1.

Combustion Chamber: Can-annular, with 12 flame tubes each fitted with duplex burner.

Fuel Grade: T-1, TS-1 to GOST 10227-62 (Avtur-50 to DERD.2494, MIL-F-5616).

Turbine: Single-stage HP turbine with cast blades; stato

blades and both sides of disc cooled by bleed air. Two-stage LP turbine with forged blades. Max gas temperature downstream of turbine 650°C.

JET PIPE: Concentric pipes for fan airflow and core gas, terminating in supersonic nozzles of fixed-area type.

LUBRICATION: Open type, with oil returned to tank. Consumption in flight, not over 1 kg (2·2 lb)/hr. Typical pressure 3·4-4·5 kg/cm² (50-64 lb/sq in).

OIL GRADE: Mineral oil MK-8 or MK-8P to GOST 6457-66 (DERD.2490 or MIL-O-6081B).

ACCESSORIES: Two gearboxes provide drives for starter/generator, tachometer, air compressors, hydraulic pump, oil pump and other controls and instruments. For re-starting in flight, an altitude sensing device meters fuel flow appropriate to height. An automatic fire extinguishing system is fitted. De-icing of the air intake and inlet guide vanes is controlled automatically. The engine also has oil chip detectors, vibration monitors and turbine gas temperature limiters.

STARTING: Electric (DC) system, incorporating STG-18TM starter/generator.

DIMENSIONS:
Length overall 3,304 mm (130 in)
Diameter, bare 976 mm (38·3 in)

WEIGHT, DRY: 1,468 kg (3,236 lb)

PERFORMANCE RATINGS:
Max T-O rating 52·96 kN (11,905 lb st)
Long-range cruise, Mach 0·75, 11,000 m (36,000 ft) 10·79 kN (2,425 lb st)

SPECIFIC FUEL CONSUMPTION:
Max T-O 20·4 mg/Ns (0·72 lb/hr/lb st)
Long-range cruise, as above 25·5 mg/Ns (0·90 lb/hr/lb st)

SOLOVIEV D-25V

D-25V is the Soloviev bureau designation for the free-turbine turboshaft which powers the Mil Mi-6, Mi-10 and V-12 helicopters and was also fitted to the Kamov Ka-22 experimental convertiplane. It is usually referred to by its official designation of **TV-2BM.**

The complete helicopter power plant comprises two D-25V engines, identical except for handed jet pipes, and an R-7 gearbox. The latter has four stages of large gearwheels providing an overall ratio of 69·2 : 1. The R-7 is 2,795 mm (110·04 in) high, 1,551 mm (61·06 in) wide and 1,852 mm (72·91 in) long. Its dry weight is 3,200 kg (7,054 lb), more than that of the pair of engines.

The D-25V is flat rated to maintain rated power to 3,000 m (10,000 ft) or to temperatures up to 40°C at sea level.

The D-25VF turboshafts fitted to the Mil V-12 helicopter are uprated to 6,500 shp. These engines are believed to incorporate a zero stage on the compressor and to operate at higher turbine gas temperatures. The following details apply to the basic D-25V:

TYPE: Single-shaft turboshaft with free power turbine.

AIR INTAKE: Six hollow radial struts, the two vertical struts housing splined shafts driving upper and lower accessory drive boxes. Vertical struts de-iced by oil drained from upper drive box; four inclined struts and bullet fairing de-iced by hot oil returned from engine to tank.

COMPRESSOR: Nine-stage axial. Comprises fixed inlet guide vane assembly, first-stage stator ring, upper and lower casings with dovetailed stator blades, ninth-stage stator ring and exit vanes, rotor, and air blow-off valves. Pressure ratio 5·6 at T-O power, 10,530 rpm.

COMBUSTION CHAMBER: Can-annular. Assembled from diffuser (the structural basis of the engine), inner shroud, 12 flame tubes with transition liners, diaphragm and compressor-shaft shroud.

FUEL GRADE: T-1, TS-1 to GOST 10227-62 (DERD.2494, MIL-F-5616).

TURBINE: Single-stage compressor turbine, overhung behind rear roller bearing. Two-stage power turbine, overhung on end of rear output shaft. Both turbines rotate counter-clockwise, seen from the rear. Normal power turbine rpm, 7,800-8,300; maximum 9,000. Transmission shaft in three universally-jointed sections, allowing for 10 mm (4 in) misalignment between engine and gearbox.

JET PIPE: Large fabricated assembly in heat-resistant steel, curved out to side to allow rotor transmission to pass through duct wall in aircooled protecting pipe.

LUBRICATION: Pressure circulation at 3·45-4·41 bars (50-64 lb/sq in). Separate systems for gas-generator and for power turbine, transmission and gearbox.

OIL GRADE: Gas-generator, MK-8 to GOST 6457-66 or transformer oil to GOST 982-56. Power turbine and gearbox, mixture (75-25 Summer, 50-50 Winter) of MK-22 or MS-20 to GOST 1013-49 and MK-8 or transformer oil. Hourly oil consumption, gas-generator not over 1 kg (2·2 lb), power turbine and transmission not over 2 kg (4·4 lb).

ACCESSORIES: SP3-12TV electric supply and starting system; fuel supply to separate LP and HP systems; airframe accessories driven off upper and lower gearboxes on inlet casing.

STARTING: The SP3-12TV system starts both engines and also generates electric current. It comprises an STG-12TM starter/generator on each engine, igniter unit, two spark plugs with cooling shrouds, two switch-over

The 5,500 shp Soloviev D-25V turboshaft

Soloviev D-30 turbofan (66·68 kN; 14,990 lb st) with thrust reverser

Soloviev D-30KU turbofan (107·9 kN; 24,250 lb st)

contactors, solenoid air valve, pressure warning, PSG-12V control panel and electro-hydraulic cutout switch of the TsP-23A centrifugal governor. In the starter mode the system draws current from a ground supply receptacle or from batteries.

DIMENSIONS:
Length overall, bare 2,737 mm (107·75 in)
Length overall with transmission shaft 5,537 mm (218·0 in)
Width 1,086 mm (42·76 in)
Height 1,158 mm (45·59 in)

WEIGHT, DRY:
With engine-mounted accessories 1,325 kg (2,921 lb)

PERFORMANCE RATINGS:
T-O 4,101 kW (5,500 shp)
Rated power 3,504 kW (4,700 shp)
Cruise (1,000 m; 3,280 ft, 135 knots; 250 km/h; 155 mph) 2,983 kW (4,000 shp)

SPECIFIC FUEL CONSUMPTION:
T-O, as above 108 μg/J (0·639 lb/hr/shp)
Cruise, as above 118·1 μg/J (0·699 lb/hr/shp)

SOLOVIEV D-30

This two-spool turbofan powers the Tu-134 twin-engined airliner and is derived from the D-20. Major portions of the core and carcase are similar, but the complete power plant is larger than the D-20, and more powerful and efficient.

In turn the D-30 has been developed into the considerably larger D-30K, described separately. They showed a continuing allegiance to the form of power plant pioneered by Rolls-Royce as the 'by-pass turbojet', a term still used in the Soviet Union to describe these engines, in which the LP system comprises several stages and the by-pass ratio is not greater than unity (in the West most turbofans have single-stage fans with a by-pass ratio of from 3 to 8).

TYPE: Two-shaft turbofan (by-pass turbojet).

AIR INTAKE: Titanium alloy assembly, incorporating air bleed anti-icing of centre bullet and radial struts.

FAN: Four-stage axial (LP compressor). First stage has shrouded titanium blades held in disc by pinned joints. Pressure ratio (T-O rating, 7,700 rpm, S/L, static), 2·65 : 1. Mass flow 125 kg (265 lb)/sec. By-pass ratio 1 : 1.

COMPRESSOR: Ten-stage axial (HP compressor). Drum and disc construction, largely of titanium. Pressure ratio (T-O rating, 11,600 rpm, S/L, static), 7·1 : 1. Overall pressure ratio, 17·4 : 1.

COMBUSTION CHAMBER: Can-annular, with 12 flame tubes fitted with duplex burners.

FUEL GRADE: T-1 and TS-1 to GOST 10227-62 (equivalent to DERD.2494 or MIL-F-5616).

TURBINE: Two-stage HP turbine. First stage has cooled blades in both stator and rotor. LP turbine also has two stages. All discs aircooled on both sides, and all blades shrouded to improve efficiency and reduce vibration. All shaft bearings shock-mounted.

JET PIPE: Subsonic fixed-area type, incorporating main and by-pass flow mixer with curvilinear ducts of optimum shape. D-30-2 engine of Tu-134A fitted with twin-clamshell (Rolls-type) reverser.

LUBRICATION: Open type, with oil returned to tank.

OIL GRADE: Mineral oil MK-8 or MK-8P to GOST 6457-66 (equivalent to DERD.2490 or MIL-O-6081B). Consumption in flight not over 1·0 kg (2·2 lb)/hr.

ACCESSORIES: Automatic ice-protection system, fire extinguishing for core and by-pass flows, vibration detectors on casings, oil chip detectors and automatic limitation of exhaust gas temperature to 620°C at take-off or when starting and to 630°C in flight (5 min limit). Shaft-driven accessories driven via radial bevel-gear shafts in centre casing, mainly off HP spool, accessory gearboxes being provided above and below centre casing and fan duct. D-30-2 carries constant-speed drives for alternators.

STARTING: Electric DC starting system incorporating STG-12TVMO starter/generators.

DIMENSIONS:
Overall length 3,983 mm (156·8 in)
Base diameter of inlet casing 1,050 mm (41·3 in)

WEIGHT, DRY: 1,550 kg (3,417 lb)

PERFORMANCE RATINGS:
T-O 66·68 kN (14,990 lb st)
Long-range cruise rating, 11,000 m (36,000 ft) and Mach 0·75 12·75 kN (2,866 lb st)

SPECIFIC FUEL CONSUMPTION:
T-O 17·56 mg/Ns (0·62 lb/hr/lb st)
Cruise, as above 21·81 mg/Ns (0·77 lb/hr/lb st)

SOLOVIEV D-30K

Despite its designation, this turbofan is very different from the D-30 described previously. It is larger, has a much higher rating, has a by-pass ratio considerably higher, and very few parts (in the core) common to the D-30.

The basic **D-30KU** version, to which the specification details below apply, replaced the Kuznetsov NK-8 as power plant of the Ilyushin Il-62M long-range transport. The more powerful **D-30KP,** rated at 117·7 kN (26,455 lb st), powers the Ilyushin Il-76 freight transport and Il-86 passenger transport. Clamshell-type thrust reversers are fitted to all four engines of both aircraft, and to the outer engines of the Il-62M. These reversers are not an integral part of the engine but are airframe assemblies incorporated in the nacelle.

TYPE: Two-shaft turbofan, with integral flow mixer and reverser.

AIR INTAKE: Fabricated from titanium alloy. Fixed spinner and 26 cambered inlet guide vanes anti-iced by air bled from sixth or eleventh stage of HP compressor (depending on rpm). Integral front roller bearing for LP shaft.

FAN (LP Compressor): Three stages, mainly of titanium alloy. First-stage rotor blades held in dovetail slots, with part-span anti-vibration snubbers. Other two stages have pinned rotor blades. Spool rotates between front roller bearing and rear ball bearing, with additional roller bearing behind LP turbine. Drum/disc construction, coupled with tie-bolt and driven by splined shaft connection. Mass flow at take-off, 269 kg (593 lb)/sec at 4,730 rpm (87·9 per cent), with by-pass ratio of 2·42.

DIVISION CASING: Linking the LP and HP compressors, this is the main structural attachment band to the aircraft. Magnesium-alloy casting, held by front and rear rows of peripheral bolts. Carries LP mid bearing and HP front bearing and incorporates vertical radial drive to front drive box for accessories on underside.

HP COMPRESSOR: Eleven stages. Drum/disc rotor, with discs centred on shaft by rectangular splines. Rotor blades held in dovetail slots, first two stages having part-span snubbers. Construction of titanium alloys, except for shafts, rear casing, and rotor blades and discs of stages 9-11, and stator vanes of stages 10-11, which are steel. To reduce blade vibration inlet guide vanes are turned through up to 30° according to preset programme over speed range of 7,900-9,600 rpm, while air is bled from fifth and sixth stages under transient conditions; in addition a closed peripheral chamber with perforated walls surrounds the first-stage rotor blades. HP shaft supported in front roller bearing in division casing, ball thrust bearing at rear of compressor spool and roller bearing ahead of turbine. Casing split horizontally. Overall pressure ratio (S/L, static) 20 at HP speed of 10,460 rpm (96 per cent).

COMBUSTION CHAMBER: Cannular type with 12 flame tubes in annular chamber. Each tube comprises hemispherical head and eight short sections welded with gaps for dilution air. Single swirl-type main/pilot burner centred in each tube. Igniter plugs in two tubes. Outer casing and duct shroud provided with longitudinal joints for access to flame tubes and HP turbine nozzle ring.

FUEL GRADE: T-1, TS-1, GOST-10227-62, A-1 (D1655/63t), DERD.2494 or 2498, Air 3405/B or 3-GP-23e.

TURBINES: Two-stage HP turbine with first-stage nozzles, part of second-stage nozzles and both sets of discs and rotor blades cooled by HP bleed air. Second-stage rotor blades tip-shrouded. Both discs interchangeable. Take-off inlet gas temperature 1,122°C. Four-stage LP turbine with uncooled shrouded rotor blades carried in four identical discs cooled by by-pass air.

JET PIPE: Downstream of LP turbine a rear support frame serves as the rear structural band attaching the engine to the aircraft. This frame incorporates the rear LP shaft roller bearing and 12 thermocouples, and also includes the 16-chute mixer for the core and by-pass flows.

LUBRICATION: Closed type, with oil returned to tank. Incorporates fuel/oil heat exchanger and centrifugal air separator with metal-particle warning unit.

OIL GRADE: MK-8 or MK-8P to GOST 6467-66 (mineral) or BNII NP-50-1-4F to GOST 13076-67 (synthetic) or Western equivalents.

ACCESSORIES: Front and rear drive boxes underneath engine carry all shaft-driven accessories. Differential constant-speed drive to alternator and air-turbine starter.

STARTING: Pneumatic air-turbine starter fed by ground supply, APU or cross-bleed from running engine. Start cycle time to idling rpm, 40-80 sec depending on ambient temperature (limits, −60° to +50°C). In-flight starting up to 9,000 m (27,430 ft) by windmilling.

DIMENSIONS:

Length with reverser	5,700 mm (224 in)
Inlet diameter	1,464 mm (57·6 in)
Maximum diameter of casing	1,560 mm (61·4 in)

WEIGHT, DRY:

With reverser	2,650 kg (5,842 lb)
Without reverser	2,300 kg (5,071 lb)

PERFORMANCE RATINGS (ISA):

T-O 108 kN (24,250 lb st) to 21°C

Cruise at 11,000 m (36,000 ft) and Mach 0·8 27 kN (6,063 lb st)

SPECIFIC FUEL CONSUMPTION:

At T-O rating	13·88 mg/Ns (0·49 lb/hr/lb st)
Cruise, as above	19·83 mg/Ns (0·70 lb/hr/lb st)

TUMANSKY

GENERAL DESIGNER IN CHARGE OF BUREAU:
Not known

Academician Sergei K. Tumansky, who died in 1973, left a legacy of turbojet and by-pass jet engines which were the propulsion basis on which the MiG bureau created the MiG-21, in worldwide service, and the extremely high-performance MiG-25. Tumansky is believed also to have designed the engine of the MiG-23. Previously the bureau had provided the engine used in the Yak-30 and Yak-32 jet trainer prototypes.

TUMANSKY RD-9

This simple axial turbojet was almost certainly designed under the leadership of V. Ya. Klimov; it was originally known as the VK-7 (and by other designations) before becoming generally referred to as the Klimov RD-9. After the death of Klimov it is probable that his bureau became led by Tumansky, and most recent reports describe the RD-9 as a Tumansky engine. The original engine may, in turn, have owed something to the earlier RD-5 (AM-5) credited to the Mikulin bureau (which see). In the mid-1950s the RD-5 was replaced by the RD-9 as the standard engine of the MiG-19 and Yak-25. In the Yak night fighter the engine was the RD-9, while (so far as is known) all versions of the MiG-19 have afterburning engines designated RD-9B.

PERFORMANCE RATINGS (T-O):

RD-9	25·5 kN (5,732 lb st)
RD-9B with afterburner	31·9 kN (7,165 lb st)

TUMANSKY R-11 and R-13

Designated TRD Mk R37F (turbojet R37F) by the Soviet armed forces, the R-11/R-13 family have been built in very large numbers, and two versions have been licence-built by HAL in India. At least one version is produced, without a licence, in China.

A single-shaft turbojet with afterburner, the R-11 entered production in 1956 with dry and afterburning ratings of 38·25 kN (8,600 lb) and 50 kN (11,240 lb) respectively. In 1959 a world speed record was set by the E-66, powered by an R-11-F2-300, with the same dry rating but a new afterburner giving a maximum thrust of 58·36 kN (13,120 lb). This engine also powered the production variant of the E-66, the MiG-21F, as well as the MiG-21PF, FM, FL and possibly other versions.

A very similar engine is the R-11-F2S-300, which powers the Indian-built MiG-21M. It is believed that versions of the R-11 power all known versions of the twin-engined Yak-28.

The R-13-300 incorporates major changes to handle an increased airflow, and has dry and afterburning ratings of 50 kN (11,240 lb) and 64·72 kN (14,550 lb) respectively. It powers the latest known MiG-21 versions, including the MF. It is believed not to be installationally interchangeable with the R-11 series.

TUMANSKY MiG-25 ENGINE

In about 1961 the Tumansky bureau began the design of a jet engine with afterburner considerably more powerful than the R-11 and intended for flight at speeds higher than Mach 3. This engine was adopted by the MiG bureau for the MiG-25 twin-engined fighter. Existence of this Mach 3·2 aircraft was disclosed in April 1965 when its first world record was announced. At that time the aircraft was referred to as the E-266, and its engine as the TRD Mk 31. Thrust rating was given as 98·1 kN (22,046 lb). It is believed that production MiG-25 fighters, MiG-25R reconnaissance aircraft and MiG-25U trainers are powered by an advanced version of this engine with maximum ratings of 82·29 kN (18,500 lb st) dry and 138 kN (31,000 lb st) with maximum afterburning. No details of the engine are known in the West, but it is believed to be a two-shaft by-pass jet (turbofan of by-pass ratio less than unity). Recent record climb submissions call the engine 'RD-F'. The large afterburner exhausts through an advanced con-di nozzle, and is almost certainly fully modulated. It is believed that a methanol/water spray system is used at Mach numbers over 3 to cool the air in the inlet duct.

TUMANSKY MiG-23 ENGINE

Despite its importance, the engine of the MiG-23 is unknown in the West, though generally attributed to the Tumansky bureau. Basic design of the single-engined variable-geometry MiG-23 dates from 1962-63. Prototypes were flying by 1967, and the MiG-23B was in service by 1970. This has an engine designed for STOL and low-level operation, with large LP compressor (fan) and simple afterburner. Dry and augmented ratings are estimated at 66·72 kN (15,000 lb) and 88·96 kN (20,000 lb) respectively. The engine is installed with simple fixed inlets and has a short two-position nozzle. The MiG-23S all-weather fighter has an engine sharing a common core but having a smaller LP compressor and large afterburner. This engine is designed for supersonic operation at Mach 1·8-2·1, and is fed by fully-variable supersonic inlets and discharges through a large fully modulated primary and secondary nozzle of variable profile. Dry and augmented ratings are estimated at 62·28 kN (14,000 lb) and 107 kN (24,000 lb) respectively. As far as can be judged, the same engine is fitted to the MiG-23U two-seat trainer.

VEDENEEV

GENERAL DESIGNER IN CHARGE OF BUREAU:
Ivan M. Vedeneev

This designer was responsible for the improvement and development of certain models of the AI-14 piston engine designed by the Ivchenko bureau. He also developed the ASh-62M, produced in Poland as the ASz-62M, from Shvetsov's ASh-62IR nine-cylinder radial. He now heads his own bureau.

M-14V-26

Derived from the Ivchenko AI-14 family of engines for fixed-wing aircraft, the M-14V-26 powers the Kamov Ka-26 helicopter. In this installation the stub-wing carries an engine on each tip. Beneath the rotor an R-26 gearbox combines the power of the engines and distributes it equally between the two co-axial main rotors turning in opposite directions.

It is said to incorporate all the experience gained in many years of developing engines in this class, and shows numerous areas of refinement compared with the AI-14 series. The engine has forced cooling by an axial fan driven via a friction clutch and extension shaft ahead of the main output bevel box at 1·452 times crankshaft speed. The engine planetary gearbox has a ratio of 0·309 and incorporates friction and ratchet clutches. The central R-26 gearbox has a ratio of 0·34; it also drives the generator, hydraulic pump, oil pump and tachometer generator.

DIMENSIONS:

Diameter	985 mm (38·78 in)
Length	1,145 mm (45·08 in)

WEIGHT, DRY: 245 kg (540 lb)

PERFORMANCE RATINGS:

T-O	242 kW; 325 hp at 2,800 rpm
Max continuous I	205 kW; 275 hp at 2,450 rpm
Max continuous II	142 kW; 190 hp at 2,350 rpm
Cruise I	142 kW; 190 hp at 2,350 rpm
Cruise II	108 kW; 145 hp at 2,350 rpm

SPECIFIC FUEL CONSUMPTION:

At cruise ratings 77·7 μg/J (0·46 lb/hr/hp)

VEDENEEV M-14P

With this engine Vedeneev has reverted to the original fixed-wing application, apparently independently of the Ivchenko bureau. The M-14P is used with direct drive to a fixed-pitch two-blade propeller in the Moscow Aviation Institute OSKB-1-3PM and the Yak-18T.

In the former aircraft the T-O rating is given as 242 kW (325 hp), and in the Yak-18T as 269 kW (360 hp).

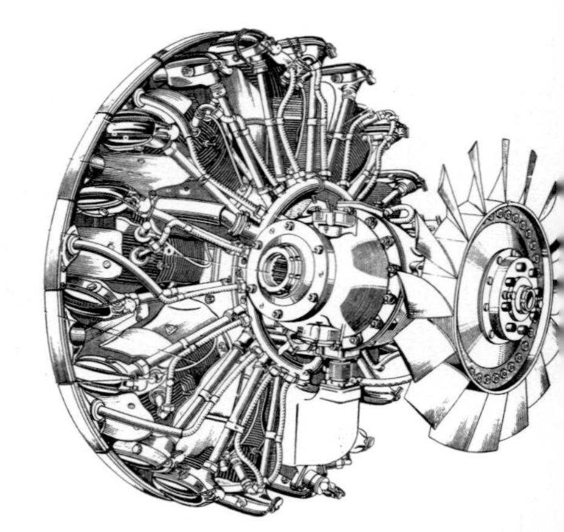

Vedeneev M-14V-26 radial piston engine, with cooling fan, for Kamov Ka-26 helicopter

ENGINES OF UNKNOWN DESIGN

1. **GRD Mk U2.** The rocket engine fitted to the Mikoyan E-66A aircraft which set up a world height record in April 1961 carries this designation. It develops 29·4 kN (6,615 lb) thrust, presumably at sea level.

2. **TRD Mk P.166/TRD 31.** P.166 is the designation given to the 98·1 kN (22,046 lb st) turbojet engine fitted in the Mikoyan E-166 aircraft that held the world's absolute speed record. The same rating is quoted for the TRD 31 turbojet fitted in the T-431 aircraft, implying that the designation is related to the aircraft in which the engine is fitted; and the TRD Mk P.166 and TRD 31 are almost certainly versions of the same turbojet. Afterburning p

vides a 50 per cent thrust boost, static.

3. **RU 19-300.** Auxiliary turbojet of 8·83 kN (1,985 lb st) for which the only known applications are in the Antonov An-24/26/30 transports. Mounted in the rear of the starboard nacelle in place of the TG-16 APU, the RU 19-300 provides additional take-off thrust and also drives an integrally-mounted generator to relieve the aircraft's AI-24T turboprops of supplying electrical power during take-off. This arrangement increases the An-24RV's take-off performance under hot and high conditions, and improves single-engine handling and stability. After take-off, the auxiliary turbojet is shut down and the AI-24Ts are coupled mechanically to the engine-mounted generators. In this dual role the RU 19-300 provides 2·16 kN (485 lb st) for take-off. During flight the auxiliary turbojet is available for use as an APU. The version installed in the An-30 is designated RU 19A-300.

4. **Lift-Jet.** Soviet design bureaux conducted extensive research into jet lift from the late 1950s, and purpose-designed lift-jets were probably flying by 1962 (possibly much earlier). At the 1967 air show at Domodedovo three lift-jet V/STOL research aircraft were displayed. A MiG aircraft based on the MiG-21 had two lift-jets in the mid-fuselage, with a single large inlet door above. A larger MiG, with fuselage and tail closely similar to the eventual MiG-23 but with a small delta wing, had a similar arrangement, differing in detail design of the dorsal door. A Sukhoi aircraft, similar to the twin-engined Su-15, had three lift engines fed by two upper doors. All these installations had large open slots in the dorsal doors (which were hinged at the rear) and transverse louvres filling the ventral jet aperture (probably hinged and under pilot control to give variable forward thrust). Photographs of these aircraft last appeared in the 1971-72 *Jane's*. The two lift-jets used in the V/STOL aircraft carried on board the carrier *Kiev* are probably developments of the same engines, though details are not yet known. They could be turbojets or turbofans. The inlet door has open louvres, and it is safe to assume pilot-controlled exit cascades. Thrust estimates for these engines are agreed at around 26·7 kN (6,000 lb) each.

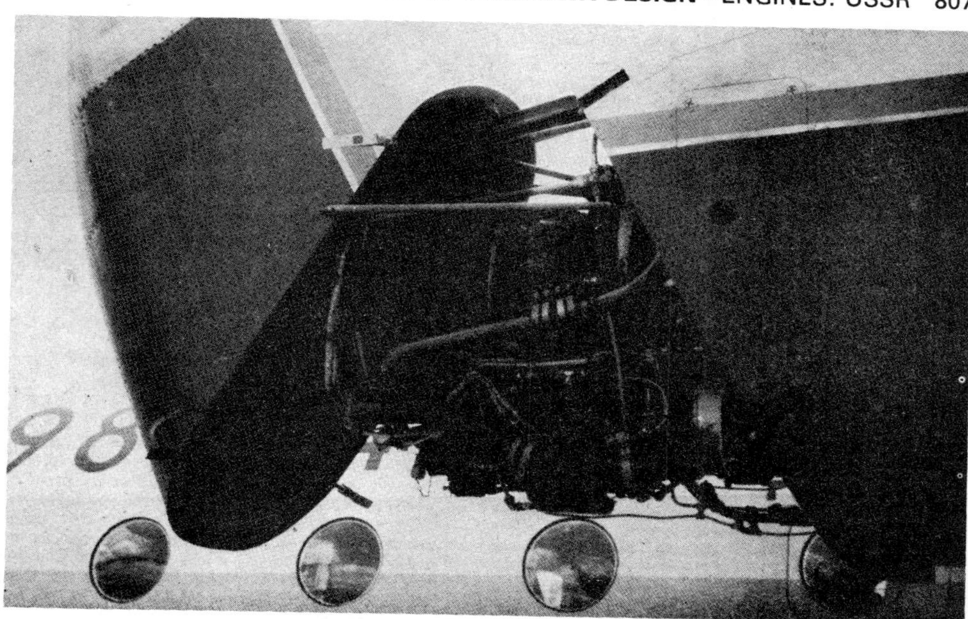

RU 19-300 auxiliary turbojet (8·83 kN; 1,985 lb st) installed in Antonov An-24RV to provide take-off boost and electrical supply

5. **Lift/Cruise Engine.** The Yakovlev V/STOL research aircraft (NATO *Freehand*) demonstrated at the 1967 Domodedovo air show was powered by a twin-nozzle vectored-thrust propulsion system. The use of a large bifurcated nose intake and only two exhaust nozzles suggests that the installation may have used two standard turbojets or turbofans, each fitted with a single swivelling nozzle. On balance, however, it is likely that the *Freehand* research aircraft used the same large single engine as the V/STOL combat aircraft carried on board the *Kiev*. This is probably a turbofan, similar in essentials to the British Pegasus and Anglo-German RB.193 (used in the VAK 191B), but with the entire efflux discharged through left and right rear nozzles. The HP spool must be bled to provide air for the reaction control jets at nose, tail and wingtips. There is not thought to be any afterburning in this engine, and preliminary estimates in September 1976 put its T-O thrust in the 89 kN (20,000 lb st) class.

ADDENDA

ADDENDA

BRAZIL

EMBRAER

EMBRAER EMB-111 BANDEIRANTE (Page 12)

The Brazilian Air Force has awarded a multi-million dollar contract to the AIL Division of Cutler-Hammer for fourteen AN/APS-128 lightweight sea patrol airborne search radar (SPAR-1) systems, for installation in the EMB-111 maritime patrol version of the Bandeirante. The entire installation weighs less than 79 kg (175 lb). Deliveries were scheduled to begin in mid-1976.

The AIL radar will be fully integrated with the EMB-111's on-board inertial navigation, high-powered searchlight, signal cartridge launcher, and camera systems to provide operational flexibility over a variety of missions including surveillance, search and rescue. The Brazilian Air Force will utilise the EMB-111 for both military and civil missions, including operations with naval vessels, sonar searches, shipping surveillance, anti-smuggling patrol, and transport of cargo and personnel.

The principal feature of the AN/APS-128, which is designed to operate in numerous roles and modes, is its ability to detect small targets over large areas under varying sea conditions. The 360°-of-rotation antenna assembly, which weighs only 17 kg (38 lb), is mounted in the EMB-111's nose radome, and provides more than 240° of azimuth coverage. A tilt adjustment of ±15° permits various depression angles, and automatic roll and pitch stabilisation of the antenna compensates for the varying effects of aircraft attitudes up to ±20° from straight and level flight.

Inside the EMB-111, a 178 mm (7 in) operator's PPI will display adjustable range scales of 25, 50 and 125 nm (46, 92 and 232 km; 29, 58 and 144 miles), with 5, 10 and 25 nm (9, 18 and 46 km; 5·75, 11·5 and 29 mile) markers. The pilot's PPI can be used for navigation as well as for weather and terrain avoidance.

First prototype of the EMBRAER EMB-121 Xingu 6/9-passenger transport, rolled out on 19 August 1976 (see page 12)

CANADA

CANADAIR/LEAR AVIATION

CANADAIR LIMITED (Page 17)

Head Office and Works:
Cartierville Airport, St Laurent, Montreal, Quebec H3C 3G9

In July 1974 Mr William P. Lear, designer of the original Learjet business aircraft, initiated the development of a new turbifan light transport to which he gave the name LearStar 600. A feature of this aircraft from the start was use of a supercritical wing, making the LearStar 600 the first commercial transport designed to take advantage of technology resulting from the wing aerodynamic research conducted by NASA's Richard T. Whitcomb in the 1960s. It was envisaged at first that the aircraft would have a three-engine layout, but a change was made to two rear-mounted turbofans as the design progressed.

Lear Aviation Corporation planned to begin construction of a prototype in July 1975, and announced that a new company would be formed to produce the LearStar 600 if it attracted sufficient interest from prospective operators. But in April 1976 Canadair Ltd acquired from Lear Aviation an option for exclusive rights to manufacture and market the aircraft worldwide, since when a number of important design changes have been made. In particular the wing span has been increased, the fuselage now has a wide-body section, and the original tailplane anhedral has been eliminated.

With an estimated worldwide requirement for more than 1,000 business aircraft in the category of the LearStar 600 in the decade from 1978 to 1988, Canadair believes that this aircraft could capture some 40% of the market. The company is reported to require confirmation of 40-50 orders before making a production decision. Its provisional pre-production and certification schedule envisages the start of detail fabrication for three prototypes before the end of 1976, first flight around the turn of the year 1977/78, and certification by the end of February 1979. Production deliveries could then begin in the second quarter of 1979, to total 20 aircraft in that year, building up to a rate of 56 aircraft a year by 1981.

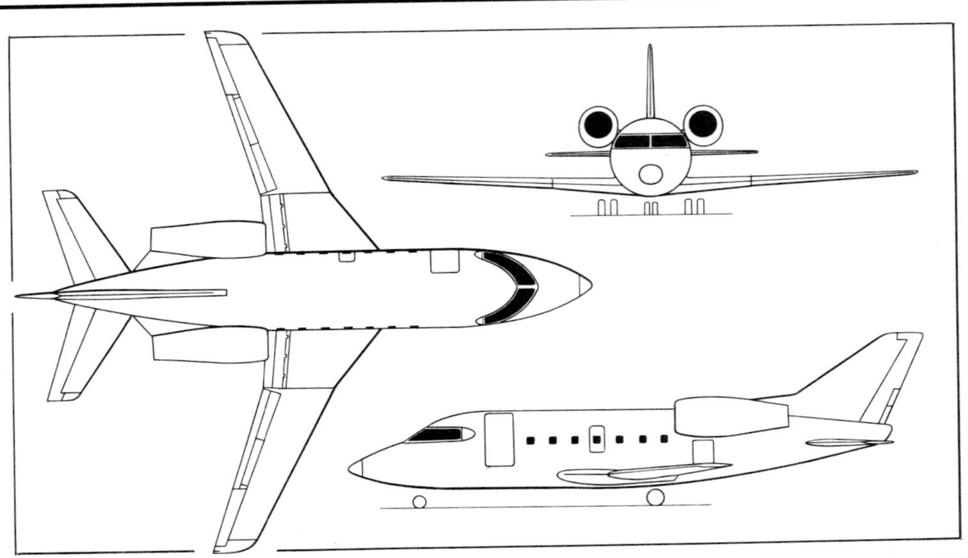

Canadair LearStar 600 twin-turbofan business, priority air cargo, airline, and commuter passenger transport *(Pilot Press)*

CANADAIR LEARSTAR 600

Type: Twin-turbofan business, priority air cargo, airline, and commuter passenger transport.

Wings: Cantilever low-wing monoplane, built in one piece. Supercritical wing section. Sweepback at quarter-chord 25°. Two-spar structure, primarily of light alloy; spars covered with skin-stringer panels to form rigid torsion box. Replaceable wingtips. Trailing-edge flaps over 75% of span. Hydraulically-powered all-metal ailerons and outboard roll-control spoilers. Dual inboard spoilers for descent control and ground lift dumping. Trim tabs in ailerons.

Fuselage: Light alloy fail-safe semi-monocoque structure of circular cross-section, with clad frames, stringers, and chemically-milled skins. Nose radome of glassfibre honeycomb.

Tail Unit: Cantilever light alloy structure, with swept vertical and horizontal surfaces. Fin and tailplane of multi-spar construction, with ribs and spanwise stiffened skin panels. Tailplane incidence adjusted by irreversible drive from the flap gearbox to trim the aircraft as a function of flap position. Control surfaces mechanically operated. Trim tab in rudder. All-metal honeycomb trim tabs in elevator. Tailplane leading-edge anti-iced by engine bleed air.

Landing Gear: Hydraulically-retractable tricycle type, with twin wheels on each unit. Main wheels retract inward into fuselage, nose unit forward. Nose unit steerable. Hydraulically-operated multiple-disc brakes. Fully-modulated anti-skid system. Provision for emergency extension of landing gear.

Power Plant: Two 33·36 kN (7,500 lb st) Avco Lycoming ALF 502 high by-pass ratio turbofan engines, pylon-mounted on each side of the rear fuselage, and fitted with cascade-type fan-air thrust reversers. Integral fuel tank in each wing; total capacity 8,244 litres (2,178 US gallons). Single-point pressure fuelling at up to 945 litres (250 US gallons)/min, with a supply pressure of 3·45 bars (50 lb/sq in). Provision for overwing gravity fuelling to 100% capacity.

Accommodation: Pilot and co-pilot side by side on flight deck with dual controls. Blind-flying instrumentation standard. Door on port side, forward of wing, on all versions, with built-in airstairs. Typical 11-passenger executive layout has wardrobe forward of entrance and cabinet aft of entrance on port side, with crew locker, buffet and bar, and cabinet opposite on starboard side; four swivelling armchairs in pairs, separated by tables, in centre of cabin; a three-place settee on the port side at the rear, with two pairs of facing seats, separated by a table, opposite; separate lavatory compartment and wardrobe to rear of cabin; and rear baggage compartment with internal access and external baggage door on port side. Typical 10-passenger executive configuration has wardrobe (port), and crew locker and lavatory (starboard) at front; two pairs of swivelling armchairs separated by tables on port side of cabin; one armchair and twin-seat separated by table, and four-place settee on starboard side; with buffet (port), bar cabinet (starboard), and separate baggage compartment with internal and external access at rear. Thirty-passenger commuter version has washroom, toilet, and stewardess seat forward of door on port side, with wardrobe and electronics bay opposite; seven pairs of seats on port side of cabin, eight pairs on starboard side, with centre aisle; and rear baggage compartment with external door on port side. All passenger versions have an overwing type III emergency exit on each side of cabin; commuter transport has a third emergency exit opposite door. Air cargo version has a toilet and wardrobe at front of cabin, with type III exit opposite door; upward-hinged cargo door, also forward of wing, on starboard side; completely clear cabin space, able to house five containers with a total volume of 21·24 m³ (750 cu ft), or up to 3,400 kg (7,500 lb) of general freight. Overhead exit panel above flight deck optional. Windscreen anti-iced electrically.

Systems: Pressurisation and air-conditioning by engine bleed air, with max pressure differential of 0·69 bars (10 lb/sq in). Backup cabin pressure control system standard. Dual independent hydraulic systems, pressure 207 bars (3,000 lb/sq in), with variable-displacement pump on each engine, using synthetic phosphate ester fluid. Emergency hydraulic system. DC electrical system includes two 28V generators, one on each engine, and two standby storage batteries; AC power supplied by static inverters. DC external power receptacle. Emergency oxygen system, pressure 124 bars (1,800 lb/sq in), with automatic demand regulators on flight deck. Provision for passenger emergency oxygen system. Structural provisions for APU in rear fuselage. Engine fire detection system and two-shot extinguisher system to suppress a fire in either nacelle. Stability augmentation system, operating in conjunction with autopilot, has Mach trim compensation.

Electronics: Standard items include dual VHF-20A com transceivers, dual VIR-30A nav receivers, dual TDR-

90 ATC transponders, dual FD 109Z flight directors, APS-80 autopilot, dual DME-40 DMEs, AH-55 radio altimeter, dual MC-103 compasses, ADF-60 ADF, dual 346B-3 audio systems, ADS-80 air data computer, weather radar, ground proximity warning system and associated antennae, including HF. Provisions for HF com, second ADF, third VHF com, VLF nav system, inertial nav system, voice recorder and flight recorder.

EQUIPMENT: Standard items include navigation, anti-collision, wing ice inspection, landing and taxi lights; duplicated lighting system for flight deck; gust locks for all control surfaces, with a safety interlock to the engine throttle linkage; electrically heated pitot systems; capacitance type fuel gauges; cabin fire extinguishers; first aid kit; smoke masks and oxygen cylinders; emergency exit lights; and emergency battery pack.

DIMENSIONS, EXTERNAL:

Wing span	18·21 m (59 ft 9 in)
Wing aspect ratio	8·5
Length overall	19·45 m (63 ft 10 in)
Height overall	6·01 m (19 ft 8½ in)
Tailplane span	7·29 m (23 ft 11 in)
Wheel track	2·95 m (9 ft 8 in)
Wheelbase	8·00 m (26 ft 3 in)
Cabin door: Width	1·01 m (3 ft 4 in)
Optional cargo door: Width	1·37 m (4 ft 6 in)

DIMENSIONS, INTERNAL:

Cabin:	
Length, excl flight deck	8·61 m (28 ft 3 in)
Max width	2·49 m (8 ft 2 in)
Max height	1·85 m (6 ft 1 in)
Volume, excl flight deck	28·09 m³ (992 cu ft)
Baggage compartment volume	4·47 m³ (158 cu ft)

AREA:

Wings, gross	39·02 m² (420 sq ft)

WEIGHTS AND LOADINGS (estimated):

Weight empty, equipped	6,656 kg (14,675 lb)
Operating weight empty	7,665 kg (16,900 lb)
Max fuel	6,717 kg (14,810 lb)
Payload with max fuel	426 kg (940 lb)
Max payload	3,400 kg (7,500 lb)
Max T-O weight	14,742 kg (32,500 lb)
Max ramp weight	14,810 kg (32,650 lb)
Max landing weight	14,061 kg (31,000 lb)
Max zero-fuel weight	12,250 kg (27,000 lb)
Max wing loading	377·8 kg/m² (77·38 lb/sq ft)
Max power loading	221 kg/kN (2·17 lb/lb st)

PERFORMANCE (estimated):

Max level speed at S/L
300 knots (555 km/h; 345 mph) EAS

Max operating speed above 3,050 m (10,000 ft)
375 knots (695 km/h; 432 mph) EAS

Max operating speed above 7,160 m (23,500 ft)
Mach 0·90

Max cruising speed at 11,000 m (36,000 ft), ISA, at AUW of 11,793 kg (26,000 lb) Mach 0·88

Normal cruising speed at 12,000 m (39,000 ft), ISA, at AUW of 14,061 kg (31,000 lb) Mach 0·85

Long-range cruising speed at 13,100 m (43,000 ft) at AUW of 14,061 kg (31,000 lb) Mach 0·80

Max certificated ceiling 14,935 m (49,000 ft)

Max range with 8 passengers at long-range cruising speed 4,030 nm (7,468 km; 4,640 miles)

OPERATIONAL NOISE CHARACTERISTICS (FAR 36, estimated):

T-O	78 EPNdB
Approach	90 EPNdB
Sideline	87 EPNdB

DE HAVILLAND CANADA

DHC-6 TWIN OTTER (Page 20)

The 500th Twin Otter, a Series 300 for Metro Airlines of Houston, Texas, was delivered on 15 July 1976. By September 1976 orders for all versions had reached a total of 518.

The 500th de Havilland Canada DHC-6 Twin Otter in the insignia of Metro Airlines of Houston, Texas

DHC-7R RANGER

First details of this maritime reconnaissance version of the Dash 7 were given on 5 September 1976, on the opening day of the Farnborough International air show in the UK.

Principal differences from the standard Dash 7 airliner, described on pages 22-23 of this edition, are increased fuel tankage, to provide approx 10-12 hour endurance at normal patrol speeds; two observers' stations in the fuselage, with bubble windows; Litton LASR-2 search radar in an underfuselage radome; and on-board electronics and equipment for a range of maritime surveillance duties including day and night photography. The Ranger can be converted easily to a standard 50-passenger transport configuration, and retains the capacity to carry up to 26 passengers without removal of the reconnaissance installation.

In addition, the advantages inherent in the basic DHC-7 design include multi-engine safety; low fuel consumption; quiet operation, with low interior vibration and noise levels; and the ability to use short, semi-prepared airstrips close to the reconnaissance area.

Engineering design work on the Ranger is under way, and a prototype is scheduled to fly in the Autumn of 1978.

TYPE: Four-turboprop maritime reconnaissance aircraft.

WINGS: As for DHC-7.

FUSELAGE: Generally as for DHC-7, except for addition of ventral radome.

TAIL UNIT AND LANDING GEAR: As for DHC-7.

POWER PLANT: Four 835 kW (1,120 shp) Pratt & Whitney Aircraft of Canada PT6A-50 turboprop engines, as in DHC-7, each driving a Hamilton Standard 24PF series constant-speed fully-feathering reversible-pitch slow-turning (1,210 rpm) propeller with four glassfibre blades. Fuel load increased from 4,626 kg (10,200 lb) in DHC-7 to 7,734 kg (17,050 lb) in DHC-7R, equivalent to increase in total tank capacity from 5,602 litres (1,480 US gallons; 1,232 Imp gallons) to 9,350 litres (2,470 US gallons; 2,056 Imp gallons).

ACCOMMODATION: Pilot and co-pilot on flight deck. Work stations in forward part of cabin for two observers (one each side), with swivelling seats and 180° bubble windows, and for navigator/tactical co-ordinator aft of starboard observer's station. Fully-equipped galley and toilet/washroom at rear of cabin. All reconnaissance installations are of modular design, permitting quick and easy removal to make entire interior available for use in transport role. Alternatives to primary reconnaissance layout include 50-passenger transport, with reconnaissance installation removed; seating for up to 26 passengers in rear of cabin without removal of reconnaissance installation at front; or mixed passenger/cargo layout with reconnaissance installation removed, freight loading door and movable cabin bulkhead added (typical load, three standard freight pallets and 18 passengers). With all of these layouts, toilet and buffet provisions at rear of cabin, and 6·8 m³ (240 cu ft) of baggage space, are standard.

SYSTEMS: Generally as described for DHC-7, including cabin pressurisation at 0·294 bars (4·26 lb/sq in); dual hydraulic systems, each of 207 bars (3,000 lb/sq in); and 115/200V AC and 28V DC electrical systems. Adequate electrical power is provided to allow mission to be completed in event of an engine shutdown.

ELECTRONICS: Standard electronics comprise VHF/FM (maritime), dual VHF and HF/SSB communications; Canadian Marconi CMA-734 Omega VLF navigation system; dual VHF nav with glideslope; marker beacon; DME; ATC transponder; ADF with remote magnetic indicator; radar altimeter; gyro magnetic compass system; autopilot; dual flight director system; two air data computers; flight data recorder; flight compartment voice recorder; integrated audio system; and emergency locator beacon. Optional electronics include UHF com; UHF/DF receiver; Litton LTN-72 inertial navigation system; Doppler navigation system; and Ontrac III VLF navigation system.

Artist's impression of the DHC-7R Ranger four-turboprop maritime patrol aircraft

OPERATIONAL EQUIPMENT: Electronics racks on port side near front of cabin, just aft of observer's station, with flare stowage and flare launcher (for night photography) to rear of these racks. On centreline of cabin are the 360° scan Litton LASR-2 search radar, between electronics racks and navigator's station, and a vertical camera installation and floor window opposite the flare stowage racks. Photo annotation system records on the film the appropriate position data obtained from the aircraft's navigation system. Main camera can be supplemented by hand-held cameras at the two observers' stations. Six-man life raft at front of cabin, adjacent to starboard observer's station. Nose-mounted weather radar is optional. A range of specialised sensing equipment can be installed, to customer's requirements, for resource surveillance.

DIMENSIONS, EXTERNAL: As for DHC-7

DIMENSIONS, INTERNAL:

Cabin, excl flight deck:	
Length	12·04 m (39 ft 6 in)
Max width	2·596 m (8 ft 6·2 in)
Floor width	2·13 m (7 ft 0 in)
Max height	1·94 m (6 ft 4½ in)
Height under wing	1·85 m (6 ft 1 in)
Volume	54·1 m³ (1,910 cu ft)
Baggage compartment volume	6·8 m³ (240 cu ft)

WEIGHTS AND LOADINGS (estimated):

Basic weight empty (standard)	11,282 kg (24,874 lb)
Operating weight empty (standard)	12,927 kg (28,500 lb)
Max payload	4,332 kg (9,550 lb)
Max fuel (standard tanks)	7,734 kg (17,050 lb)
Max T-O weight	20,411 kg (45,000 lb)
Max zero-fuel weight	17,690 kg (39,000 lb)
Max landing weight	18,597 kg (41,000 lb)
Max wing loading	255·5 kg/m² (52·3 lb/sq ft)
Max power loading	6·11 kg/kW (10·04 lb/shp)

PERFORMANCE (estimated, at max T-O weight except where indicated):

Max cruising speed at S/L:	
ISA	233 knots (432 km/h; 268 mph)
ISA +15°C	230 knots (426 km/h; 265 mph)
Service ceiling:	
ISA	6,705 m (22,000 ft)
ISA +15°C	6,100 m (20,000 ft)
Service ceiling, one engine out:	
ISA	5,030 m (16,500 ft)
ISA +15°C	4,420 m (14,500 ft)
T-O run at S/L:	
ISA	787 m (2,580 ft)
ISA +15°C	860 m (2,820 ft)
Landing run at S/L at max landing weight:	
ISA and ISA +15°C	677 m (2,220 ft)

Typical mission profile, incl radar search at 1,525 m (5,000 ft) and 30 min inspection at 305 m (1,000 ft), at 800 nm (1,480 km; 920 miles) from base, reserves for 45 min hold at 1,525 m (5,000 ft):

time on search, out and back at 3,050 m (10,000 ft)
2 hr 30 min

time on search, out and back at optimum altitude
3 hr 40 min

total mission time, out and back at 3,050 m (10,000 ft) 11 hr 0 min
total mission time, out and back at optimum altitude 12 hr 0 min
Typical patrol endurance, cruising at 80% power at 4,570 m (15,000 ft), reserves as above:
total mission time 9 hr 30 min
Range with max fuel, cruising at 80% power at 4,570 m (15,000 ft), reserves as above:
with reconnaissance installation and 26 passengers 1,430 nm (2,650 km; 1,646 miles)
with 50 passengers, reconnaissance installation removed 800 nm (1,482 km; 921 miles)

TEAL AIRCRAFT CORPORATION

HEAD OFFICE AND WORKS:
Buttonville Airport, Markham, Ontario M3P 3J9
Telephone: (416) 297-3027
Telex: 06219564
PRESIDENT: R. F. English

In the Spring of 1976, this company acquired from Schweizer Aircraft Corporation the complete production tooling and fixtures, and all rights in the **Teal amphibian,** designed by Mr David B. Thurston and described on page 380 of this edition. Teal Aircraft Corporation has set up a US subsidiary to hold the aircraft's FAA Type Certificate, but remains the parent company for the Teal programme and hopes to have the aircraft back in production during 1977. It is adapting the airframe to take a 134 kW (180 hp) engine.

CZECHOSLOVAKIA

LET

L-410 TURBOLET

The first example of the L-410M version (see page 29) has been completed and flown. Initial data for this version are as follows:

POWER PLANT: Two 544 kW (730 ehp) Walter M 601 A turboprop engines, each driving an Avia V 508 three-blade hydraulically-adjustable reversible-pitch constant-speed fully-feathering propeller of 2·50 m (8 ft 2½ in) diameter. Fuel and oil capacities as for L-410A.

WEIGHTS AND LOADINGS:
Basic weight empty (17 seats, full IFR equipment) 3,720 kg (8,201 lb)
Max payload 1,410 kg (3,108 lb)
Max fuel 1,020 kg (2,248 lb)
Max T-O weight 5,700 kg (12,566 lb)
Max landing weight 5,500 kg (12,125 lb)
Max wing loading 173·46 kg/m² (35·53 lb/sq ft)
Max power loading 5·23 kg/kW (8·61 lb/ehp)

PERFORMANCE (at max T-O weight, ISA):
Cruising speed at 3,000 m (9,850 ft) 197 knots (365 km/h; 227 mph) TAS
Stalling speed, flaps down 62 knots (114 km/h; 71 mph) EAS
Max rate of climb at S/L 450 m (1,476 ft)/min
Rate of climb at S/L, one engine out 84 m (275 ft)/min
Max operating altitude 6,000 m (19,685 ft)
Service ceiling, one engine out 2,800 m (9,190 ft)
T-O to 10 m (33 ft) 630 m (2,067 ft)
Landing from 15 m (50 ft) 670 m (2,198 ft)
Max range with 760 kg (1,675 lb) payload, 30 min fuel reserves 625 nm (1,160 km; 720 miles)

L-410M version of the Let Turbolet transport aircraft (two M 601 A turboprop engines)

FRANCE

AÉROSPATIALE

Aérospatiale SA 330Z (page 41), fitted with large 'fenestron' tail rotor and T tailplane as part of the Super Puma development programme

DASSAULT-BREGUET

Artist's impression of the Dassault Delta Mirage 2000 (SNECMA M53 turbofan engine), brief details of which are given on page 51

DASSAULT-BREGUET FALCON 20G (Page 53)

The Falcon 20G will differ from the current 20F in having two Garrett AiResearch ATF 3-6 turbofan engines, each rated at 22·47 kN (5,050 lb st). The new power plant, complete with nacelles and thrust reversers, will be offered initially as a retrofit for existing Falcon/Mystère 20 aircraft, with full production of the 20G scheduled for a later date.

Carrying six passengers under ISA conditions, the Falcon 20G will offer a maximum range of 2,240 nm (4,147 km; 2,577 miles), with 45 min reserves, representing an increase of 30% compared with the 20F. At max T-O weight of 13,000 kg (28,660 lb), balanced field length will be 1,495 m (4,900 ft) at 30°C. Time to climb to cruising height is expected to be reduced by 26%.

It was announced in the Autumn of 1976 that a tender by Falcon Jet Corporation to supply 41 Falcon 20Gs to meet a US Coast Guard requirement for medium-range surveillance aircraft had been announced as the lowest bid. If ordered, the aircraft will be assembled in the USA and fitted with Collins electronic equipment.

Model of the Dassault-Breguet Falcon 20G, with Garrett AiResearch ATF 3-6 turbofan engines

DASSAULT-BREGUET FALCON 50 (Page 55)

The prototype Falcon 50 (F-WAMD) made its first flight on 7 November 1976.

DASSAULT-BREGUET MERCURE 200/ASMR

Under the title of Advanced Short-to-Medium Range (ASMR) commercial jet transport, a developed version of the Mercure 200 was being offered in the Autumn of 1976 under the joint sponsorship of Dassault-Breguet, Aérospatiale and McDonnell Douglas. The general configuration of the ASMR remains generally similar to that described under the Mercure 200 entry on page 57, except that wing span and area are increased to 32·85 m (107 ft 9 in) and 125 m² (1,345 sq ft) respectively, and max T-O weight is increased to 71,668 kg (158,000 lb). Typically, the ASMR version would be able to carry 160 passengers over a range of 1,750 nm (3,240 km; 2,015 miles).

Prototype Dassault-Breguet Falcon 50, rolled out at Bordeaux-Mérignac on 4 September 1976

ROBIN (Page 60)

ROBIN R 2160

This two-seat all-metal light training aircraft was announced in October 1976, as the first of a new series of aircraft designated R for Robin. The R 2160 flew for the first time in July 1976. It is derived from the current HR 200, which it will replace, and embodies many components of that aircraft, including the basic fuselage and tail fin. The wing is new and of increased chord, with an NACA 23015 section and incidence of 3° throughout the span. Each trailing-edge comprises an inboard slotted flap and a fully-balanced slotted aileron, both of increased chord compared with those of the HR 200. The rudder is enlarged, and there is an additional small ventral fin to improve spinning characteristics. Greater comfort for the two occupants has been ensured by moving back each seat position by 50 mm (2 in).

Power plant of the R 2160 prototype is a 119 kW (160 hp) Lycoming flat-four engine. The R 2108 will have an 80·5 kW (108 hp) engine, and a version with an 89 kW (120 hp) engine is projected.

The Robin R 2160, evolved from the HR 200 two-seat trainer

GERMANY (FEDERAL REPUBLIC)

MASTER PORTER FLUGZEUGENTWICKLUNGSGESELLSCHAFT mbH & CO KG

HEAD OFFICE:
8000 München 90, Pfalzer-Waldstrasse 70, Postfach 900 566
Telephone: (089) 682001/681001
Telex: 05-23887
PRESIDENT: E. Stoeckl
CHIEF ENGINEER: E. F. Moehring

Formed in 1971, this company was known originally as Poligrat-Development GmbH & Co KG. In 1974 it announced that its first aircraft programme would involve development and manufacture of a twin-engined cargo and passenger transport named PD-01 Master Porter, of which details were given in the 1975-76 *Jane's*. Subsequent market research suggested that the major potential markets for such an aircraft, in Africa and South America, require a slightly larger design. As a result, development is now being concentrated on the Master Porter PD-02, with increased overall dimensions, more powerful engines, more than doubled fuel capacity and provision for air-conditioning.

MASTER PORTER PD-02

The Master Porter PD-02 is a twin-turboprop QSTOL transport aircraft, intended for third-level passenger and/or cargo operation. It has been designed to meet FAR 25 and CAB Part 298 standards; and the ultimate objective of its manufacturer is to market the aircraft as a product for assembly by approved foreign licensees.

Under contract to Master Porter, Pilatus in Switzerland (assisted by Eidgenössische Flugzeugwerke, Emmen, the Flug- & Fahrzeugwerke Altenrhein and others) is building two prototypes and a static test airframe.

TYPE: Twin-turboprop transport aircraft.

WINGS: Cantilever high-wing monoplane of light alloy construction, built in three sections. Wing section NACA 23018. Constant chord. No dihedral. Incidence 2°. Ailerons and electrically-operated double-slotted trailing-edge flaps of light alloy construction. Balance tab in starboard aileron; trim and balance tab in port aileron. Pneumatic de-icing boots on wing leading-edges.

FUSELAGE: Conventional all-metal semi-monocoque fail-safe structure of basically rectangular section.

TAIL UNIT: Cantilever light alloy two-spar structure, with dorsal fairing forward of fin. Electrical and manual adjustment of variable-incidence tailplane. Trim and balance tab in rudder; balance tab in each elevator. Pneumatic de-icing boots on leading-edges of fin and tailplane.

LANDING GEAR: Hydraulically-retractable and steerable twin-wheel nose unit, retracting forward. Non-retractable single main wheels, mounted in stub fairings attached to base of fuselage. Menasco oleo-pneumatic shock-absorbers. Dunlop tyres size 36 × 13·00-12 on main wheels. Dunlop tyres size 7·00-6 on nosewheels. Menasco hydraulic brakes. Optional float installation.

POWER PLANT: Two 835 kW (1,120 shp) Pratt & Whitney Aircraft of Canada PT6A-50 turboprop engines, each driving a Hamilton Standard four-blade metal propeller. Integral fuel tanks in wings, with standard capacity of 2,660 litres (585 Imp gallons) and max optional capacity of 3,880 litres (853 Imp gallons). Single pressure refuelling point in starboard main landing gear fairing; gravity refuelling points on upper surface of wings. Military versions can carry optional underwing fuel tanks.

ACCOMMODATION: Crew of two on flight deck. Three-abreast seating for 24 passengers, or four-abreast for 30 passengers in high-density layout, with provision for toilet at front and baggage compartments. Quick-change (30 min) conversion capability to all-cargo configuration, including provision for folding and stowing passenger seats if required. Combined passenger/cargo and paratroop layouts available. Ambulance version can accommodate 16 stretcher patients and six attendants. Passenger door, with integral steps, ahead of wing on port side. Large rear-loading door, which can be lowered to serve as a ramp or opened upward and inward. Fuselage cross-section can accept standard 2·24 × 2·24 m (88 × 88 in) pallets or LD-1, -3 or -7 containers. Roller conveyor system and crash net available at customer's option. Cabin air-conditioned.

SYSTEMS: Electrical system supplied by two 28V DC engine-driven generators. Two 25Ah batteries. Inverters for AC supply. Hydraulic system for nosewheel steering and retraction, with duplicated system for brakes. Air-conditioning system by Hamilton Standard. Oxygen system for flight crew. De-icing system uses electrical heating for engine air intakes, propellers and pitot heads, and engine bleed air for wing and tail unit de-icing boots.

EQUIPMENT: Standard equipment includes communications radio and cockpit voice recorders. Blind-flying instrumentation standard.

DIMENSIONS, EXTERNAL:

Wing span	21·10 m (69 ft 8¾ in)
Wing chord, constant	2·18 m (7 ft 1¾ in)
Wing aspect ratio	9·68
Length overall	15·12 m (49 ft 7¼ in)
Height overall	6·30 m (20 ft 8 in)
Tailplane span	8·70 m (28 ft 6½ in)
Wheel track	3·50 m (11 ft 4¾ in)
Wheelbase	4·80 m (15 ft 9 in)
Propeller diameter	3·43 m (11 ft 3 in)
Propeller ground clearance	0·80 m (9⅝ in)

Passenger door (port, fwd):
- Height 1·76 m (5 ft 9¼ in)
- Width 0·68 m (2 ft 2¾ in)
- Height to sill 0·73 m (2 ft 4¾ in)

Passenger door (rear):
- Height 1·90 m (6 ft 2¾ in)
- Width 0·70 m (2 ft 3½ in)
- Height to sill 0·73 m (2 ft 4¾ in)

Emergency exits (3):
- Height 0·92 m (3 ft 0¼ in)
- Width 0·51 m (1 ft 8 in)

DIMENSIONS, INTERNAL:
- Cabin: Length 7·20 m (23 ft 4½ in)
- Max width 2·30 m (7 ft 6⅝ in)
- Max height 2·00 m (6 ft 6⅝ in)
- Floor area 16·56 m² (178·25 sq ft)
- Volume 33·45 m³ (1,181·3 cu ft)
- Baggage holds (2), each 3·32 m³ (117·24 cu ft)

AREAS:
- Wings, gross 46·00 m² (495·2 sq ft)
- Ailerons (total) 3·45 m² (37·14 sq ft)
- Trailing-edge flaps (total) 6·53 m² (70·29 sq ft)
- Vertical tail surfaces, incl tab 7·80 m² (83·96 sq ft)
- Horizontal tail surfaces, incl tabs 13·50 m² (145·3 sq ft)

WEIGHTS AND LOADINGS (estimated. A: civil; B: military overload):
- Basic operating weight:
  - A 5,060 kg (11,155 lb)
  - B 5,120 kg (11,288 lb)
- Max T-O weight:
  - A 9,000 kg (19,841 lb)
  - B 9,670 kg (21,319 lb)
- Max ramp weight:
  - A 9,130 kg (20,128 lb)
  - B 9,850 kg (21,716 lb)
- Max zero-fuel weight:
  - A 8,260 kg (18,210 lb)
  - B 8,120 kg (17,902 lb)
- Max landing weight:
  - A, B 8,900 kg (19,621 lb)
- Max wing loading:
  - A 196 kg/m² (40·14 lb/sq ft)
  - B 210 kg/m² (43·01 lb/sq ft)
- Max power loading:
  - A 5·39 kg/kW (8·86 lb/shp)
  - B 5·79 kg/kW (9·52 lb/shp)

PERFORMANCE (estimated, at max T-O weight):
- Never-exceed speed 275 knots (510 km/h; 317 mph) EAS
- Max cruising speed at 3,050 m (10,000 ft) 222 knots (412 km/h; 256 mph)
- Econ cruising speed at 3,050 m (10,000 ft) 193 knots (358 km/h; 222 mph)
- Stalling speed, flaps down, power off 71 knots (132 km/h; 82 mph)
- Stalling speed, flaps down, power on 53 knots (98 km/h; 61 mph)
- Max rate of climb at S/L 690 m (2,264 ft)/min
- Service ceiling 7,500 m (24,600 ft)
- Service ceiling, one engine out 4,650 m (15,255 ft)
- T-O run (FAA field length) 720 m (2,362 ft)
- Landing run (FAA field length) 675 m (2,215 ft)
- Range with max fuel, 45 min reserve, plus 10% en-route fuel 970 nm (1,800 km; 1,118 miles)
- Range with max payload (passenger version), 45 min reserve, plus 10% en-route fuel 280 nm (520 km; 323 miles)
- Ferry range, with 300 kg (661 lb) payload at 4,570 m (15,000 ft), 30 min holding plus 10% en-route fuel 2,125 nm (3,940 km; 2,448 miles)

MBB

MBB BO 105 (Page 71)

By September 1976 a total of 270 BO 105s were in operation, in 17 countries.

MBB BO 107 (Page 73)

MBB announced in September 1976 that this seven/nine-seat enlarged development of the BO 105 helicopter was in the definition phase as a joint venture between MBB and Kawasaki of Japan.

MBB HFB 320 HANSA

Four HFB 320s are being specially equipped for the Luftwaffe by MBB's Hamburger Flugzeugbau Division, for training duties in connection with the West German air defence system. Development of this version, which is understood to carry ECM equipment, began in the Spring of 1974, and the first aircraft was delivered on 31 August 1976. The HFB 320 was last described in the 1973-74 *Jane's*. Altogether, 46 were built, of which the Luftwaffe has acquired twelve.

RFB (Page 74)

RFB/GRUMMAN AMERICAN FANLINER

An accompanying illustration shows the second prototype of the Fanliner (D-EBFL), which flew for the first time on 4 September 1976. Powered by an RFB-modified Audi/NSU Wankel-type rotary piston engine, giving 112 kW (150 hp) and driving a Dowty Rotol integral ducted fan, this prototype has a considerably refined airframe compared with the original Fanliner.

The flush cabin has been styled by industrial designer Luigi Colani and now provides an exceptional field of view. The wings and tailplane are those of the Grumman American Cheetah. Landing gear fairings and rear fuselage lines have been improved, and all radiators are aft of the cabin.

A total of 420 flights had been logged by Fanliner 001 by early September 1976. A production decision will be taken in 1977, leading to series production in 1978 if the go-ahead is given.

Second prototype of the RFB/Grumman American Fanliner

SPORTAVIA

SPORTAVIA FOURNIER RF6 SPORTSMAN

The description given on page 75 is not, as stated, that of the Sportavia production version of the Sportsman, the correct designation of which is **RF6-180.**

The RF6-180, of which a prototype (D-EHYO) first flew on 28 April 1976, has a more powerful engine, increased cabin volume, and an outer skin of glassfibre-reinforced plastics (GRP). Details are as follows:

TYPE: 2 + 2-seat lightweight sporting aircraft.

WINGS: Cantilever low-wing monoplane. Wing section NACA 63-218 at root, NACA 63-215 at tip. Dihedral 5° from roots. Incidence 0°. One-piece single spar of laminated beech, birch plywood and pine ribs, mahogany plywood covering (upper surfaces only), and GRP outer skin. All-metal split flap on each trailing-edge, with electrical actuation. Fabric-covered wooden Frise-type ailerons, actuated by pushrods. No tabs. Turned-down wingtips.

FUSELAGE: Conventional wooden structure of frames and longerons, with mahogany plywood covering and GRP outer skin.

TAIL UNIT: Cantilever wooden structure, with GRP-covered fixed surfaces and fabric-covered control surfaces. Fixed-incidence tailplane. Rudder and elevators actuated by pushrods. Flettner tab in port elevator.

LANDING GEAR: Non-retractable tricycle type. Main units have hydraulic shock-absorption. Steerable nosewheel, with oleo-pneumatic shock-absorber. Single wheel on each unit. Hydraulic disc brakes. Streamline fairings over all three wheels.

POWER PLANT: One 134 kW (180 hp) Lycoming O-360-A1F6D flat-four engine, driving a Hoffmann three-blade constant-speed propeller with spinner. Fuel tank in each wing, combined capacity 220 litres (48·5 Imp gallons) in four-seat version, 150 litres (33 Imp gallons) in three-seat version. Overwing refuelling point above each tank. Oil capacity 6·5 litres (1·5 Imp gallons).

ACCOMMODATION: Two front seats side by side under rearward-sliding Perspex 'bubble' canopy, and one or two 'occasional' rear seats. 'Solid' canopy roof is planned for production aircraft. Cabin heated and ventilated.

SYSTEMS: Hydraulic system for brakes. Electrical power provided by 12V 60A engine-driven generator.

Prototype Sportavia RF6-180 Sportsman 2 + 2-seat light aircraft

ELECTRONICS AND EQUIPMENT: Radio, radio navigation equipment and full blind-flying instrumentation to customer's requirements.

DIMENSIONS, EXTERNAL:
- Wing span 10·50 m (34 ft 5½ in)
- Wing aspect ratio 7·6
- Wing taper ratio 0·65
- Length overall 7·15 m (23 ft 5½ in)
- Height overall 2·56 m (8 ft 4¾ in)
- Propeller diameter 1·72 m (5 ft 7¾ in)

AREA:
- Wings, gross 14·50 m² (156·08 sq ft)

WEIGHTS AND LOADINGS:
- Weight empty 595 kg (1,311 lb)
- Max T-O weight:
  - 3-seat 1,000 kg (2,204 lb)
  - 4-seat 1,100 kg (2,425 lb)
- Max wing loading:
  - 3-seat 68·97 kg/m² (14·13 lb/sq ft)
  - 4-seat 75·86 kg/m² (15·54 lb/sq ft)
- Max power loading:
  - 3-seat 7·46 kg/kW (12·24 lb/hp)
  - 4-seat 8·19 kg/kW (13·47 lb/hp)

PERFORMANCE (at max T-O weight; A: 3-seat, B: 4-seat):
- Never-exceed speed:
  - A, B 194 knots (360 km/h; 223 mph)
- Max level speed at S/L:
  - A, B 156 knots (290 km/h; 180 mph)
- Cruising speed (75% power) at 2,590 m (8,500 ft):
  - A, B 146 knots (270 km/h; 168 mph)
- Cruising speed (65% power) at 2,590 m (8,500 ft):
  - A, B 134 knots (248 km/h; 154 mph)
- Cruising speed (55% power) at 2,590 m (8,500 ft):
  - A, B 123 knots (228 km/h; 142 mph)
- Stalling speed, flaps up:
  - B 51 knots (94 km/h; 58·5 mph)
- Stalling speed, flaps down:
  - A 43·5 knots (80 km/h; 50 mph)
  - B 44·5 knots (82 km/h; 51 mph)
- Max rate of climb at S/L:
  - A 366 m (1,200 ft)/min
  - B 324 m (1,063 ft)/min
- Service ceiling:
  - A 6,500 m (21,325 ft)
  - B 5,400 m (17,725 ft)
- T-O run (S/L, zero wind, ISA + 15°C):
  - A 200 m (656 ft)
  - B 235 m (771 ft)
- T-O to 15 m (50 ft) (S/L, zero wind, ISA + 15°C):
  - A 380 m (1,247 ft)
  - B 435 m (1,427 ft)
- Max range at 2,000 m (6,560 ft), 52% power, no reserves:
  - A 518 nm (960 km; 596 miles)
  - B 815 nm (1,510 km; 938 miles)

INTERNATIONAL

VTI/CIAR

ORAO (IAR-93) (Page 96)

The following expanded structural description of the Orao is based upon further study of available photographs, and the specification data upon reports appearing in the international press during 1976. The Orao has the Romanian designation IAR-93.

TYPE: Single-seat ground attack fighter.

WINGS: Cantilever shoulder-wing monoplane, of low aspect ratio. Anhedral approx 4° from roots. Sweepback approx 43° on leading-edges. Leading-edge slats. Wide-chord Fowler-type trailing-edge flaps. Trim tab on each aileron.

FUSELAGE: All-metal semi-monocoque structure. Door-type perforated airbrake under each side of lower front fuselage, forward of main-wheel bays. 'Pen-nib' fairing above exhaust nozzles. Space provision in nose for ranging radar.

TAIL UNIT: Cantilever metal structure, with sweepback on all surfaces. Low-set all-moving tailplane, with tip-mounted anti-flutter weights which project forward of leading-edges. Fin has a small dorsal fairing. Trim tab in rudder. Auxiliary ventral fin on each side beneath rear fuselage.

LANDING GEAR: Messier-Hispano retractable tricycle type, with single-wheel nose unit and twin-wheel main units. All units have oleo-pneumatic shock-absorbers. Hydraulic actuation, all units retracting forward into fuselage. Braking parachute in bullet fairing at base of rudder.

POWER PLANT: Two 17·8 kN (4,000 lb st) Rolls-Royce Viper Mk 632 non-afterburning turbojet engines in prototypes, mounted side by side in fuselage, with lateral air intakes and twin exhaust nozzles. Internal fuel load approx 2,500 kg (5,510 lb). Production aircraft to be fitted with Rolls-Royce-developed afterburners, increasing power of each engine to approx 26·5 kN (5,950 lb st).

ACCOMMODATION: Pilot only, on ejection seat beneath rear-hinged, upward-opening canopy. Production aircraft expected to include tandem two-seat operational training version.

SYSTEMS AND EQUIPMENT: Graviner Firewire and BCF fire detection and extinguishing systems. Fairey Hydraulics filters and sampling valves. Landing light under nose. Ram-air scoop aft of cockpit on each side; smaller air-scoops on top of fuselage aft of canopy and at front of dorsal fin, and below rear fuselage.

ARMAMENT: Two 30 mm cannon in lower front fuselage, aft of nosewheel bay; one underfuselage and four underwing stations for external stores. Max external load approx 2,000 kg (4,410 lb) on prototypes, approx 3,000 kg (6,615 lb) on production version.

DIMENSIONS, EXTERNAL (estimated):
- Wing span: 7·56 m (24 ft 9¾ in)
- Wing aspect ratio: 3·22
- Wing area, gross: 18·00 m² (193·75 sq ft)
- Length overall: 12·90 m (42 ft 3¾ in)
- Height overall: 3·78 m (12 ft 4¾ in)

WEIGHTS AND LOADINGS (estimated; A: prototypes, B: production version):
- Weight empty, equipped:
  - A: 4,300 kg (9,480 lb)
  - B: 4,700 kg (10,360 lb)
- T-O weight 'clean':
  - A: 7,000 kg (15,430 lb)
  - B: 7,300 kg (16,095 lb)
- Max T-O weight with external stores:
  - A: 9,000 kg (19,840 lb)
  - B: 10,300 kg (22,700 lb)
- Wing loading:
  - A at T-O weight 'clean': 388·8 kg/m² (79·6 lb/sq ft)
  - A at max T-O weight: 500·0 kg/m² (102·4 lb/sq ft)
  - B at T-O weight 'clean': 405·5 kg/m² (83·0 lb/sq ft)
  - B at max T-O weight: 572·2 kg/m² (117·2 lb/sq ft)
- Power loading:
  - A at T-O weight 'clean': 196·6 kg/kN (1·93 lb/lb st)
  - A at max T-O weight: 252·8 kg/kN (2·48 lb/lb st)
  - B at T-O weight 'clean': 137·7 kg/kN (1·35 lb/lb st)
  - B at max T-O weight: 194·3 kg/kN (1·91 lb/lb st)

PERFORMANCE (estimated; A: prototypes, without afterburning; B: production aircraft with afterburning):
- Max level speed at low level:
  - A: Mach 0·92 (609 knots; 1,128 km/h; 701 mph)
  - B: Mach 1·0 (662 knots; 1,226 km/h; 762 mph)
- Max level speed at high altitude:
  - A: Mach 0·95 (544 knots; 1,009 km/h; 627 mph)
  - B: Mach 1·6 (917 knots; 1,699 km/h; 1,056 mph)
- Landing speed:
  - A, B: 121 knots (225 km/h; 140 mph)
- Max rate of climb at S/L:
  - A: 5,520 m (18,110 ft)/min
  - B: 12,000 m (39,370 ft)/min
- Time to 11,000 m (36,000 ft):
  - A: 5 min 0 sec
  - B: 1 min 36 sec
- Service ceiling:
  - A: 14,000 m (45,925 ft)
  - B: 16,000 m (52,500 ft)
- T-O run:
  - A at 8,500 kg (18,740 lb) AUW: 925 m (3,035 ft)
  - B at max T-O weight: 1,000 m (3,280 ft)
- Landing run:
  - A at 8,500 kg (18,740 lb) AUW: 1,000 m (3,280 ft)
  - B at max T-O weight: 1,000 m (3,280 ft)
- Combat radius with 2,000 kg (4,410 lb) external stores:
  - A, lo-lo-lo: 108 nm (200 km; 124 miles)
  - B, lo-lo-lo: 175 nm (325 km; 202 miles)
  - A, hi-lo-hi: 216 nm (400 km; 248 miles)
  - B, hi-lo-hi: 350 nm (650 km; 404 miles)
- *g* limits:
  - A: +6·8
  - B: +7·5

ISRAEL

IAI

IAI KFIR (LION CUB) (Page 97)

The Kfir utilises a basic airframe similar to that of the Dassault Mirage 5, the main changes being a shorter but larger-diameter rear fuselage, to accommodate the J79 engine; an enlarged and flattened undersurface to the forward portion of the fuselage; introduction of a dorsal airscoop, in place of the triangular dorsal fin, to provide cooling air for the afterburner; and a strengthened landing gear, with longer-stroke oleos. Several internal changes have also been made, including a redesigned cockpit layout, addition of a considerable amount of Israeli-built electronics equipment, and increased internal fuel tankage compared with the Mirage 5. Intended for both air defence and ground attack roles, the Kfir retains the standard Mirage fixed armament of two 30 mm DEFA cannon, and can carry a variety of external weapons including the Rafael Shafrir air-to-air missile. It has demonstrated stall-free gun firing throughout the flight envelope

On 20 July 1976, at the Israeli Air Force base at Hatzerim, in the Negev, IAI gave the first public demonstration of a modified version known as the **Kfir-C2**. This has a number of changes from the previous model, the most significant of which are the addition of non-retractable, sweptback canard surfaces just aft of the engine air intakes; a small strake on each side of the extreme nose; and an extended wing leading-edge, created by increasing the chord on approximately the outer 40% of each half-span.

These changes, which add some 85 kg (187 lb) to the structural weight, recall the retractable 'moustaches' fitted by Dassault to the experimental Milan version of the Mirage and described in earlier editions of *Jane's*. The canard surfaces of the Kfir-C2, however, are much larger in area than those of the Milan, and by virtue of their different location they eliminate two of the principal criticisms made of the Milan installation: impairment of the pilot's view forward and downward, and the creation of adverse wake effects in the engine air intakes. The Kfir-C2 installation is, perhaps, more analogous with that of the Saab 37 Viggen.

The Kfir-C2 is expected to become the principal production version, both for the Israeli Air Force and for export. The new modifications, which can be retrofitted to existing Kfirs, were designed to improve the aircraft's dogfighting manoeuvrability at the lower end of the speed range and to enhance take-off and landing performance. It is claimed that, in particular, they give a better sustained turning performance, with improved lateral, longitudinal and directional control; contribute to a very low gust response at all operational altitudes, especially at very low level; offer improved handling qualities at all angles of attack, high *g* loadings, and low speeds; reduce take-off and landing distances, and landing speeds; and permit a more stable (and, if required, a steeper) approach, with a flatter angle of approach and touchdown.

TYPE: Single-seat interceptor, long range patrol fighter and ground attack aircraft.

The IAI Kfir-C2 single-seat multi-purpose combat aircraft, with fixed canard surfaces and other airframe refinements

WINGS: Cantilever low-wing monoplane of delta planform, with conical camber. Thickness/chord ratio 4·5% to 3·5%. Anhedral 1°. Incidence 1°. Sweepback on leading-edges 60° 34'. All-metal torsion-box structure, with stressed skin of machined panels with integral stiffeners. Two-section elevons on each trailing-edge, with smaller elevator/trim flap inboard of inner elevon. Elevons powered by hydraulic jacks; trim flaps are servo-assisted. Small, hinged plate-type airbrake above and below each wing, near leading-edge. Kfir-C2 has additional modifications which include extended chord on outer leading-edges, and sweptback canard fixed surfaces above and forward of wings, near top lip of each engine air intake. Metal Resources Inc of Gardena, California, has an IAI subcontract to manufacture replacement wing components for Israeli Mirages.

FUSELAGE: All-metal semi-monocoque structure, 'waisted' in accordance with area rule. Cross-section of forward fuselage has a wider and flatter undersurface than that of Mirage 5. Nosecone built of locally-developed composite materials, with (on Kfir-C2) a small horizontal strake or 'body fence' on each side near the tip. UHF antenna under front of fuselage, forward of nosewheel door. Enlarged-diameter rear fuselage, compared with Mirage 5, with approx 0·61 m (2 ft) shorter tailpipe. Ventral fairing under rear fuselage.

TAIL UNIT: Cantilever all-metal fin; rudder powered by hydraulic jack, with servo-assisted trim. UHF antenna in tip of fin. Triangular-section dorsal airscoop forward of fin, to provide cold air for afterburner cooling. No horizontal tail surfaces.

LANDING GEAR: Retractable tricycle type, with single wheel on each unit. Hydraulic retraction, nose unit rearward, main units inward into fuselage. Longer-stroke oleos than on Mirage 5, and all units strengthened to permit higher operating weights. Main-gear leg fairings shorter than on Mirage; inner portion of each main-leg door is integral with fuselage-mounted wheel door. Steerable nosewheel, with anti-shimmy damper. Oleo-pneumatic shock-absorbers and disc brakes. Braking parachute in bullet fairing below rudder.

POWER PLANT: One General Electric J79 turbojet engine (modified GE-17), with variable-area nozzle, rated at 52·8 kN (11,870 lb st) dry and 79·62 kN (17,900 lb st) with afterburning. Air intakes enlarged, compared with Mirage 5, to allow for higher mass flow. Adjustable half-cone centrebody in each air intake. Internal fuel in five fuselage and four integral wing tanks. Total internal capacity is probably in the order of 4,000 litres (880 Imp gallons), perhaps slightly more. There is a refuelling point on top of the fuselage, above the forward upper tank. In addition, there are wet-points for the carriage of one or two drop-tanks beneath each wing, and one under the fuselage; these tanks may be of 500, 600, 1,300 or 1,700 litres (110, 132, 286 or 374 Imp gallons) capacity. External capability should be comparable to that of the Mirage 5, which can carry up to 4,700 litres (1,034 Imp gallons) of auxiliary fuel in external drop-tanks, or 1,000 litres (220 Imp gallons) in combination with 4,000 kg (8,820 lb) of ordnance.

ACCOMMODATION: Pilot only, on Martin-Baker JM.6 zero-zero ejection seat, under rearward-hinged upward-opening canopy. Revised cockpit layout compared with Mirage 5. Cockpit pressurised, heated and air-conditioned.

SYSTEMS: Two separate environmental control systems (ECS), one for cockpit heating, pressurisation and air-conditioning, and one for electronics compartments. Two independent hydraulic systems, probably of 207 bars (3,000 lb/sq in) pressure. No. 1 system actuates flying control surfaces and landing gear; No. 2 actuates flying controls, airbrakes, landing gear, wheel brakes and utilities. Constant-speed drive unit (CSD) for essential services. Electrical system probably similar to that of Mirage 5, with DC power provided by two 24V 40Ah batteries and a 26·5V 9kW generator, and AC power by a 125VA (200V 400Hz) transformer-rectifier, a 9kVA (115/200V 400Hz) alternator and a static inverter. Oxygen system for pilot.

ELECTRONICS AND EQUIPMENT: MBT Weapons Systems twin-computer fly-by-wire flight control system, with integrated memory unit (IMU), two-axis gyro and standby compass, autopilot, radar altimeter, angle of attack transmitter and indicator, and accelerometer indicator. Elta Electronics multi-mode computer-based navigation and weapon delivery system, with Tacan, Doppler radar, IFF/SIF and nose-mounted fire control radar. Israeli-built head-up display and automatic gun-sight. Duplicated UHF radio. Twin landing lights on nosewheel leg; anti-collision light in fin leading-edge.

ARMAMENT: Fixed armament of one IAI-built 30 mm DEFA cannon in underside of each engine air intake (125 rds/gun on Mirage 5). Seven hardpoints (three under fuselage and two under each wing) for external stores. For interception duties, one Rafael Shafrir infra-red homing air-to-air missile can be carried under each outer wing. Ground attack version can carry two 1,000 lb bombs or an air-to-surface missile under the fuselage, and two 1,000 lb or four 500 lb bombs (conventional or 'concrete dibber' type) under the wings. Alternative external stores may include rocket pods; napalm; Shrike, Maverick or Hobos air-to-surface missiles; ECM pods; or drop-tanks.

DIMENSIONS, EXTERNAL:
- Wing span: 8·22 m (26 ft 11½ in)
- Wing aspect ratio: 1·94
- Foreplane span (estimated): 3·50 m (11 ft 6 in)
- Length overall: 15·55 m (51 ft 0¼ in)
- Height overall: 4·25 m (13 ft 11¼ in)
- Wheel track: 3·15 m (10 ft 4 in)
- Wheelbase: 4·87 m (15 ft 11¾ in)

WEIGHTS:
- Weight empty (interceptor, estimated):
  - Kfir: 7,200 kg (15,873 lb)
  - Kfir-C2: 7,285 kg (16,060 lb)
- Typical combat weight (interceptor), 50% internal fuel and two Shafrir missiles:
  - Kfir: 9,305 kg (20,514 lb)
  - Kfir-C2: 9,390 kg (20,701 lb)
- Max combat T-O weight (all versions): 14,600 kg (32,188 lb)

PERFORMANCE (estimated):
- Max level speed above 11,000 m (36,100 ft):
  - Kfir: over Mach 2·2 (1,260 knots; 2,335 km/h; 1,450 mph)
  - Kfir-C2: over Mach 2·3 (1,317 knots; 2,440 km/h; 1,516 mph)
- Max rate of climb at S/L: 14,000 m (45,950 ft)/min
- Time to 11,000 m (36,100 ft): 1 min 45 sec
- Stabilised ceiling (combat configuration): above 15,240 m (50,000 ft)
- T-O run at 11,000 kg (24,250 lb) AUW (Kfir): 700 m (2,300 ft)
- Landing run at 9,000 kg (19,840 lb) AUW (Kfir): 450 m (1,475 ft)
- Combat radius:
  - interceptor, two 600 litre drop-tanks: 200-288 nm (370-535 km; 230-332 miles)
  - ground attack, lo-lo-lo: 351 nm (650 km; 404 miles)
  - ground attack, hi-lo-hi: 700 nm (1,300 km; 807 miles)

ITALY

AERMACCHI

AERMACCHI M.B. 326 (Page 104)

Orders announced in the Autumn of 1976 included one from the Ghana Air Force, for six M.B. 326Ks, and one from the Tunisian Republican Air Force, for four M.B. 326GBs and six M.B. 326Ks.

PIAGGIO

The twin-turboprop Piaggio P.166-DL3, which made its first flight on 3 July 1976. A brief description of this version of the P.166 appears on page 115

NETHERLANDS

FOKKER-VFW

FOKKER-VFW F27 MARITIME

The following details are additional to those given on pages 135-6:

POWER PLANT: Total fuel capacity (including pylon tanks) 9,310 litres (2,460 US gallons; 2,048 Imp gallons).

WEIGHTS:
- *Typical operating weight empty: 12,430 kg (27,403 lb)
- Max internal fuel load: 5,870 kg (12,941 lb)
- Max internal and external fuel load: 7,395 kg (16,303 lb)
- Max T-O weight: 20,410 kg (45,000 lb)
- Max landing weight: 19,730 kg (43,500 lb)
- *Typical mission zero-fuel weight: 12,745 kg (28,100 lb)
- Max zero-fuel weight: 17,920 kg (39,500 lb)

**dependent upon customer requirements*

PERFORMANCE (with pylon tanks, at max T-O weight except where indicated):
- Cruising speed at 6,100 m (20,000 ft), AUW of 18,150 kg (40,000 lb): 230 knots (427 km/h; 265 mph)
- Typical search speed at 610 m (2,000 ft): 145 knots (270 km/h; 168 mph)
- Service ceiling: 7,070 m (23,200 ft)
- T-O run:
  - S/L, ISA: 975 m (3,200 ft)
  - S/L, ISA +20°C: 1,080 m (3,545 ft)
- Landing distance (unfactored, ISA at S/L):
  - at 19,730 kg (43,500 lb) AUW: 610 m (2,000 ft)
  - at 13,620 kg (30,000 lb) AUW: 530 m (1,740 ft)
- Transport range at 6,100 m (20,000 ft) with 4,536 kg (10,000 lb) payload, 30 min loiter and 5% reserves: 1,000 nm (1,850 km; 1,150 miles)
- Max range at 6,100 m (20,000 ft), 30 min loiter and 5% reserves: 2,215 nm (4,100 km; 2,550 miles)

FOKKER-VFW F28 FELLOWSHIP (Page 136)

Linjeflyg of Sweden has increased its order for the Mk 4000 version to 10, with options on a further five. The Mk 4000 made its first flight on 20 October 1976.

UNITED KINGDOM

BAC

BAC ONE-ELEVEN (Page 170)

Cyprus Airways has ordered two One-Eleven Series 500s, bringing the total number of all versions in service or on order to 222.

BRITTEN-NORMAN

BRITTEN-NORMAN BN-2A ISLANDER (Page 173)

Britten-Norman announced at Farnborough 1976 the availability of two new versions of the BN-2A Islander: an Agricultural Islander and an Islander Firefighter.

The agricultural version comprises the standard BN-2A-26 version with a specially-designed liquid/powder hopper mounted beneath each wing. This system has been evolved so that the cabin area is kept clean and uncontaminated by chemical deposit or odour, and can be converted rapidly for conventional passenger or military usage. High volume liquid spraying can be carried out by the use of a Transland BoomMaster pump and spraybar applied to each hopper; low and ultra-low volume spraying is achieved by the use of one or two Micronair rotary atomisers per hopper; and solid material can be distributed by fitting a standard Transland gate box to each hopper, with electrical agitators or dust doors as required.

Each underwing hopper has an internal volume of 0·57 m³(20 cu ft), allowing for the carriage of 783 kg (1,726 lb) of liquid chemicals, 790 kg (1,742 lb) of liquid chemicals for distribution by Micronair ultra low volume units, or 776 kg (1,711 lb) of dry chemicals.

The firefighter version is equipped with four specially designed interconnected liquid tanks which are mounted on a 9*g* restraint structure attached to cabin floor pickups for basic cabin seating. Each of the four containers has an outlet positioned approximately on the aircraft's centreline, and the contents of the tanks can be dumped simultaneously or on a two-shot basis. During either method of discharge, changes to the aircraft's CG and trim are negligible.

Total capacity of the system is 800 litres (176 Imp gallons) which, subject to the availability of suitable ground-based pumping equipment, can be charged with water or fire-retardant in approximately two minutes. Optimum speed for discharge is 65 knots (120 km/h; 75 mph) at an altitude of 61 m (200 ft), the load being distributed over an area of 100 m by 15 m (330 ft by 50 ft) in 2·5 seconds.

FLIGHT INVERT

FLIGHT INVERT LTD

ADDRESS:
College of Aeronautics, Cranfield Institute of Technology, Cranfield, Bedford MK43 0AL
Telephone: Bedford (0234) 51551
Telex: 825072

CHIEF DESIGNER: Prof D. Howe

Flight Invert Ltd is a non-profit-making company formed by the Cranfield Institute of Technology (Cranfield R and D Ltd) and George House (Holdings) Ltd to manage the A1 programme.

FLIGHT INVERT CRANFIELD A1

The A1 (G-BCIT) is an aerobatic aircraft produced to the requirements of Neil Williams, for use by the British aerobatic team. Design began in 1968 and construction of the prototype was initiated in 1971. Lack of finance stopped any major work until 1975, when the project was assisted by Mr Alan Curtis. The A1 is stressed to maximum load factors of $+13{\cdot}5g$ and $-9g$ in aerobatic configuration.

The A-1 made its first flight on 23 August 1976.

TYPE: Single-seat aerobatic aircraft; two seats for training and ferrying.

WINGS: Cantilever low-wing monoplane. NACA 23 series wing section. Max thickness/chord ratio 15%. Dihedral 3°. Incidence ground-adjustable. Sweepback at quarter-chord 9° 36′. One-piece wing of light alloy skin-stringer construction, with machined extrusions for centre of front spar. Some fail-safe features. Light alloy ailerons with fluted skins. No flaps or spoilers.

FUSELAGE: Welded steel tube structure, with light alloy floor and wooden formers, fabric-covered.

TAIL UNIT: Light alloy construction, similar to wings, except for rudder which is fabric-covered. Trim tab in elevators. Ground-adjustable tailplane incidence.

LANDING GEAR: Non-retractable tailwheel type from Chipmunk. Main-wheel tyres of 0·36 m (14 in) diameter, pressure 1·72 bars (25 lb/sq in). Tailwheel tyre of 0·127 m (5 in) diameter, pressure 2·76 bars (40 lb/sq in). Hydraulic brakes.

POWER PLANT: One specially-prepared 156·5 kW (210 hp) Rolls-Royce Continental IO-360-D flat-six engine, with dry sump and separate negative *g* oil tank for unlimited inverted flight. Two-blade Hartzell variable-pitch propeller. Main fuel tank in fuselage; two auxiliary tanks in wing leading-edges.

ACCOMMODATION: Single seat under sideways-opening bubble canopy in aerobatic configuration. Two seats for training and ferrying.

SYSTEMS: Electrical system, supplied by 28V DC engine-driven alternator and optional lead-acid batteries.

DIMENSIONS, EXTERNAL:

Wing span	10·00 m (32 ft 10 in)
Wing chord at root	2·08 m (6 ft 10 in)
Wing chord at tip	0·91 m (3 ft 0 in)
Wing area	15 m² (161·5 sq ft)
Wing aspect ratio	6·7
Length overall	8·05 m (26 ft 5 in)
Tailplane span	3·11 m (10 ft 2½ in)
Propeller diameter	1·90 m (6 ft 2¾ in)

WEIGHT:

Max T-O weight, two seats	920 kg (2,028 lb)

Flight Invert Cranfield A1 single-seat aerobatic aircraft

HAWKER SIDDELEY

HAWKER SIDDELEY 748 COASTGUARDER (Page 180)

Further details of this new version of the HS 748 are now available:

Its development was initiated by the need for an economic and versatile aircraft to protect vulnerable offshore energy and fishery resources and, at the same time, to provide a vehicle suitable for anti-smuggling, search and rescue, and general maritime reconnaissance missions. The airframe is generally similar to that of the standard HS 748 civil and military transport; but there is crew accommodation for two pilots, two beam observers, and a tactical navigator, to enable the Coastguarder to fulfil its primary roles. An 0·30 m (1 ft 0 in) diameter chute is mounted in the aft fuselage for the air launch of five-man rescue dinghies, and smoke or flame floats. The standard radio, radar and navigation equipment has been expanded to cover the normal naval radio frequencies, and to provide adequate navigation aids for long overwater flights.

The tactical navigator's station is situated midway down the cabin, on the starboard side, and is equipped with an MEL MAREC radar display and plotting board, Decca 72 Doppler, and a Decca 9447 TANS computer/display. The MAREC radar was chosen as standard on the basis of experience gained in previous ASV, ASW and SAR applications. It has an underfuselage antenna, an 0·43 m (1 ft 5 in) diameter main display and plotting board, with an 0·13 m (5 in) repeat display for the pilot. Used in conjunction with the Doppler, TANS computer and Marconi Omega VLF navigation system, the resulting tactical navigation system can, in addition to satisfying all normal search and navigation requirements, provide effective tactical plotting to control an exercise involving a group of friendly vessels and other radar targets, including aircraft. MAREC provides up to 200 nm (370 km; 230 miles) display range in all directions for the tactical navigator and up to 250 nm (460 km; 285 miles) for the pilot's repeater display. A choice of presentation scale between 0·5 and 25 nm (0·9-46 km; 0·6-29 miles) per inch allows enlargement of any selected part of the display.

To provide the additional range required for a maritime reconnaissance role, the fuel tankage has been increased to a maximum of 9,956 litres (2,190 Imp gallons). The standard Coastguarder may be used for a number of maritime and other roles without any change to the basic configuration. Additional passengers can be accommodated by fitting seats to the standard rails which run the full length of the cabin. The optional rear freight door provides an air-dropping capability, allowing the despatch of large dinghies or supplies in an air/sea rescue role. As many as twelve 30-man dinghies can be transported and dropped for the rescue of a large number of aircraft/ship survivors. The Coastguarder can also be converted easily for cargo carrying, by removal of the tactical navigator's station and other equipment.

The description of the standard HS 748 Series 2A applies also to the Coastguarder, except as follows:

TYPE: Twin-turboprop maritime patrol aircraft.

POWER PLANT: Fuel in integral wing tanks and fuselage tanks with a max combined capacity of 9,956 litres (2,190 Imp gallons).

ACCOMMODATION: Standard Coastguarder layout has two pilots on flight deck; two beam observers seated at forward end of cabin, one each side, with domed windows; and tactical navigator approximately midway down cabin at tactical station on starboard side. Toilet on starboard side at aft end of cabin, with galley opposite. Main door on port side at rear of cabin; smaller door for emergency exit on starboard side. Crew door on port side at front of cabin. Large rear freight door optional. Four airline-type seats, forward of tactical navigator's station, on starboard side, serve as crew rest area.

ELECTRONICS AND EQUIPMENT: Electronics include dual Collins 618M-3 VHF com transceivers, dual Collins 51RV-2B VHF nav receivers, Collins AN/ARC-159 UHF com transceiver, Collins 51Z-4 marker beacon receiver, dual Collins DF 206 ADF, Collins DF 301E UHF D/F, Collins 618T-3 HF transceiver, Collins 346D-1 address system, Ultra UA 60 interphone, Sperry RN 200 radio navigation display, Honeywell AN/APN-171 radio altimeter, MEL MAREC radar with 0·43 m (1 ft 5 in) main display and 0·13 m (5 in) pilot's repeat display, Marconi AD 1800 Omega VLF nav system, Decca 72 Doppler, and Decca 9447 TANS computer/display. Attitude stabilised antenna, size 0·91 m × 0·53 m (3 ft 0 in × 1 ft 9 in), in underfuselage radome provides 360° azimuth viewing, plus selected sector scan facilities. Provisions for optional ATC transponder, DME, and height encoding altimeter. Standard equipment includes an 0·30 m (1 ft 0 in) launch chute for five-man rescue dinghies and smoke or flame floats. Optional equipment includes large rear freight door, additional passenger seats, large dinghies, and other rescue equipment.

WEIGHTS AND LOADINGS:

Weight empty	10,354 kg (22,827 lb)
Basic operating weight	11,971 kg (26,393 lb)
Typical sortie T-O weight	20,446 kg (45,076 lb)
Normal max T-O weight	21,092 kg (46,500 lb)
Normal zero-fuel weight	17,460 kg (38,500 lb)
Normal landing weight	19,500 kg (43,000 lb)
Overload max T-O weight	23,135 kg (51,000 lb)
Overload zero-fuel weight	19,500 kg (43,000 lb)
Overload max landing weight	21,545 kg (47,500 lb)
Max wing loading	307·09 kg/m² (62·9 lb/sq ft)
Max power loading	6·80 kg/kW (11·1 lb/ehp)

PERFORMANCE (estimated, at max normal T-O weight, unless indicated otherwise):

Max cruising speed at 18,145 kg (40,000 lb) AUW
242 knots (448 km/h; 278 mph)

Max rate of climb at S/L at 18,145 kg (40,000 lb) AUW
402 m (1,320 ft)/min

Service ceiling	7,620 m (25,000 ft)
Min ground turning radius	11·89 m (39 ft 0 in)
T-O run	1,030 m (3,380 ft)
T-O to 15 m (50 ft)	1,237 m (4,060 ft)

Landing from 15 m (50 ft) at normal landing weight
620 m (2,035 ft)

Landing run at normal landing weight 390 m (1,280 ft)

HAWKER SIDDELEY HAWK (Page 184)

On 4 November it was announced that the Hawk had been selected for service with the Finnish Air Force, against competition from Czechoslovakia, France, Germany, Italy and Sweden. A firm order for up to 50 aircraft is expected to be signed in mid-1977.

On the same day, the first delivery of an RAF Hawk was made to No. 4 Flying Training School at RAF Valley in Anglesey, North Wales.

NORMAN

N. D. NORMAN

ADDRESS:
The Barn, Kingates Farm, Whitwell, Isle of Wight
Telephone: Niton 730116
Telex: 21120

Mr Desmond Norman, CBE, a founder and former Joint Managing Director of Britten-Norman Ltd, has designed and is building the prototype of an aircraft named Firecracker, of which the following brief details may be published:

NORMAN NDN-1 FIRECRACKER

The Firecracker is a tandem two-seat light aircraft powered by a 194 kW (260 hp) Lycoming IO-540 flat-six engine. It is intended primarily for use by government or civilian flying training schools. The military version will have provision for hardpoints to carry weapons or other stores.

It is expected that the Firecracker will be offered in kit form, on a technology transfer basis, to countries interested in establishing their own aircraft industry on the basis of a modern, high-performance multi-purpose type.

UNITED STATES OF AMERICA

AJI

AJI HUSTLER MODEL 400 (Page 207)

Revised specification data for the Hustler have been received, as follows:

WINGS: Fowler flaps are now double-slotted. Pneumatic de-icing boots.

TAIL UNIT: Pneumatic de-icing boots.

POWER PLANT: Turbojet is now a Teledyne CAE J402-CA-700, rated at 2·94 kN (660 lb st).

SYSTEMS: Pressurisation differential 0·55 bars (8·0 lb/sq in). Propeller and windscreen de-iced electrically.

DIMENSIONS, EXTERNAL:

Wing span	9·94 m (32 ft 7½ in)
Length overall	11·71 m (38 ft 5 in)
Height overall	3·30 m (10 ft 10 in)
Tailplane span	4·06 m (13 ft 4 in)

AREA:

Wings, gross	16·82 m² (181·15 sq ft)

WEIGHTS AND LOADING:

Weight empty	1,678 kg (3,700 lb)
Max fuel	925 kg (2,040 lb)
Max T-O and landing weight	2,950 kg (6,500 lb)
Max wing loading	175·2 kg/m² (35·88 lb/sq ft)

PERFORMANCE (estimated, at max T-O weight; front engine only except where indicated):

Max cruising speed at 6,100 m (20,000 ft)
330 knots (611 km/h; 380 mph) TAS

Econ cruising speed at 10,650 m (35,000 ft)
286 knots (531 km/h; 330 mph) TAS

T-O speed	61 knots (113 km/h; 70 mph) CAS
Approach speed	74 knots (137 km/h; 85 mph) IAS
Max rate of climb at S/L	915 m (3,000 ft)/min
Service ceiling:	
front engine	11,033 m (36,200 ft)
rear engine	4,115 m (13,500 ft)
T-O run	205 m (670 ft)
T-O to 15 m (50 ft)	387 m (1,270 ft)
Landing from 15 m (50 ft)	305 m (1,000 ft)
Landing run	153 m (500 ft)

Range at max cruising speed, 30 min reserves
1,804 nm (3,340 km; 2,076 miles)

Max range at econ cruising speed, 30 min reserves
2,580 nm (4,780 km; 2,970 miles)

AMR

AERO-MARINE RESEARCH

ADDRESS:
PO Box 5194, San Bernardino, California 92408
Telephone: (714) 864-0072

PRESIDENT:
Cdr William Benson, BSc, AeEng, MInstPI

In addition to its work on remotely piloted vehicles (see under RPVs and Targets in this Addenda), AMR has flown or is building a number of manned aircraft. All available details follow:

BENSON XR 207 ROTORPLANE

The prototype for the Rotorplane was a small two-seat autogyro, described briefly in the 1975-76 *Jane's*. Four of these were built, of which the last has now been redesignated **XR 207-2** and is serving as the pre-production aircraft for a tandem two-seat production version.

An accompanying photograph shows this aircraft, which has been strengthened structurally to enable it to be fitted with a Benson delta wing and an enclosed cockpit. In this form, it was due to begin flight testing during the latter half of 1976.

A larger, five-seat production version, designated **XR 207-5,** is also planned. This will be powered by two 74·5 kW (100 hp) or 112 kW (150 hp) Rolls-Royce Continental flat-four engines each driving an externally-mounted Rotorduct ducted propeller, and will have a three-blade rigid rotor of 8·05 m (26 ft 5 in) diameter, with spin-up device for vertical take-off. Accommodation will be for a pilot and either four passengers (one beside the pilot and three on a bench seat at the rear) or three stretchers and a medical attendant.

The following details apply to the XR 207-5:

DIMENSIONS, INTERNAL:
Cabin: Length 3·05 m (10 ft 0 in)
Max width 1·83 m (6 ft 0 in)
Max height 1·52 m (5 ft 0 in)

WEIGHTS (estimated):
Weight empty 567 kg (1,250 lb)
Max T-O and landing weight 1,814 kg (4,000 lb)

PERFORMANCE (estimated):
Max level speed 226 knots (417 km/h; 260 mph)
Max rate of climb at S/L 610 m (2,000 ft)/min
Range with max payload 347 nm (641 km; 400 miles)

BENSON 110 NOVA

As indicated briefly in the 1975-76 edition of *Jane's,* AMR is developing non-rotary-winged aircraft based on the Benson delta wing configuration and having a Rotorduct propulsion system. Plans current in mid-1976 were for three versions, all to the same general configuration, but of differing sizes. These are:

Nova Trainer. Tandem two-seat trainer, with instructor in the rear seat, which is slightly elevated. Single external Rotorduct.

Nova Sport. Four-seat private and sporting STOL aircraft. Single external Rotorduct.

Nova Executive. Ten-seat business transport, as shown in the accompanying artist's impression. Twin 'buried' Rotorducts.

No construction of full-size prototypes of any of the Nova series had been undertaken by mid-1976, but the essential design features are being, or were shortly due to be, tested upon various XD 110 scale research aircraft; details of these are given in the RPVs & Targets section of this Addenda.

XR 207-2 pre-production prototype for two-seat Benson Rotorplane

XD 110A scale research model showing wing planform proposed for Rotorplane

Artist's impression of five-seat Benson XR 207-5 Rotorplane

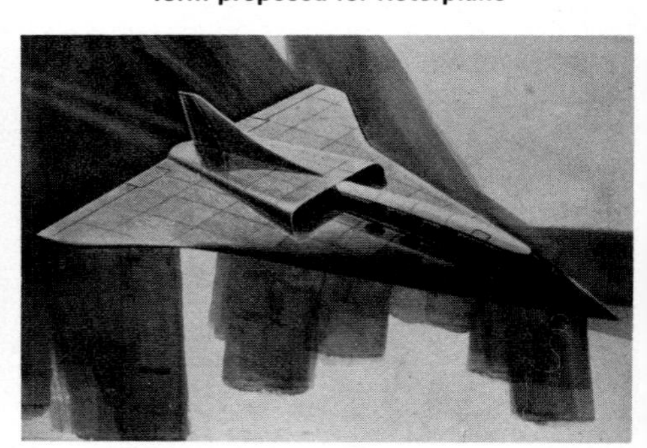

Artist's impression of Benson 110 Nova Executive ten-seat transport

BEECH AIRCRAFT CORPORATION (Page 209)

Beech has designed and produced a wingtip-mounted in-flight refuelling system for installation on the Boeing 707-3J9C tanker-transport aircraft operated by the Imperial Iranian Air Force. Given the Beech Model number 1080, the system is of the hose-and-drogue type, and was first displayed publicly at the Farnborough International air show in September 1976. The IIAF Boeings have a fuel capacity of 109,020 litres (28,800 US gallons), and with the Beech system installed can transfer fuel to two aircraft simultaneously. They also have an underfuselage Boeing 'flying boom' installation for refuelling aircraft equipped with receptacles for this system.

BEECHCRAFT TURBO MENTOR (Page 209)

Fourteen T-34C Turbo Mentors were ordered by the Ecuadorean Air Force during 1976.

BEECHCRAFT BARON 58TC

The following details are additional to those given on page 216:

PERFORMANCE (at 2,767 kg; 6,100 lb max T-O weight except where indicated):
Max level speed 249 knots (461 km/h; 287 mph)
Max cruising speed at approx 81% power and average cruise weight:
at 3,050 m (10,000 ft) 212 knots (393 km/h; 244 mph)
at 4,575 m (15,000 ft) 222 knots (411 km/h; 256 mph)
at 6,100 m (20,000 ft) 232 knots (430 km/h; 267 mph)
Cruising speed at approx 74% power and average cruise weight:
at 3,050 m (10,000 ft) 203 knots (376 km/h; 234 mph)
at 4,575 m (15,000 ft) 214 knots (396 km/h; 246 mph)
at 6,100 m (20,000 ft) 223 knots (413 km/h; 257 mph)
Cruising speed at approx 64% power and average cruise weight:
at 3,050 m (10,000 ft) 190 knots (352 km/h; 219 mph)
at 4,575 m (15,000 ft) 201 knots (372 km/h; 231 mph)
at 6,100 m (20,000 ft) 210 knots (389 km/h; 242 mph)
Cruising speed at approx 55% power and average cruise weight:
at 3,050 m (10,000 ft) 175 knots (324 km/h; 201 mph)
at 4,575 m (15,000 ft) 186 knots (345 km/h; 214 mph)
at 6,100 m (20,000 ft) 194 knots (359 km/h; 223 mph)
Stalling speed, power off:
flaps up 83 knots (154 km/h; 96 mph)
flaps down 79 knots (147 km/h; 91 mph)
Max rate of climb:
at S/L 445 m (1,461 ft)/min
at 4,575 m (15,000 ft) 407 m (1,337 ft)/min
Rate of climb, one engine out:
at S/L 62 m (204 ft)/min
at 1,525 m (5,000 ft) 59 m (194 ft)/min
Service ceiling above 7,620 m (25,000 ft)
Service ceiling, one engine out 4,390 m (14,400 ft)
T-O run 475 m (1,556 ft)
T-O to 15 m (50 ft) 760 m (2,495 ft)
Landing from 15 m (50 ft) 761 m (2,498 ft)
Landing run 450 m (1,471 ft)
Range at average cruise weight with 719 litres (190 US gallons) usable fuel, allowances for start, taxi, T-O, climb and 45 min reserves:
at approx 81% power:
at 3,050 m (10,000 ft) 890 nm (1,649 km; 1,025 miles)

Imperial Iranian Air Force Boeing 707 tanker-transport fitted with Beechcraft Model 1080 twin hose-and-drogue in-flight refuelling system

Beechcraft Baron 58TC twin-engined business light transport aircraft

at 4,575 m (15,000 ft)
924 nm (1,712 km; 1,064 miles)
at 6,100 m (20,000 ft)
968 nm (1,794 km; 1,115 miles)
at approx 74% power:
at 3,050 m (10,000 ft)
964 nm (1,786 km; 1,110 miles)
at 4,575 m (15,000 ft)
992 nm (1,838 km; 1,142 miles)
at 6,100 m (20,000 ft)
1,032 nm (1,912 km; 1,188 miles)
at approx 64% power:
at 3,050 m (10,000 ft)
1,086 nm (2,012 km; 1,250 miles)
at 4,575 m (15,000 ft)
1,107 nm (2,051 km; 1,275 miles)
at 6,100 m (20,000 ft)
1,131 nm (2,096 km; 1,302 miles)
at approx 55% power:
at 3,050 m (10,000 ft)
1,189 nm (2,203 km; 1,369 miles)
at 4,575 m (15,000 ft)
1,210 nm (2,242 km; 1,393 miles)
at 6,100 m (20,000 ft)
1,231 nm (2,281 km; 1,417 miles)

BEECHCRAFT MODEL 76

The following details are additional to those given on pages 224-5:

DIMENSIONS, EXTERNAL:

Wing span	11·58 m (38 ft 0 in)
Length overall	8·84 m (29 ft 0 in)
Height overall	2·72 m (8 ft 11 in)
Wheel track	3·23 m (10 ft 7 in)
Wheelbase	2·13 m (7 ft 0 in)
Propeller diameter	2·24 m (7 ft 4 in)

BEECHCRAFT MODEL 77

Beech has decided to go ahead with Preliminary Design PD 285 (see pages 210-11), under the production Model number 77. Deliveries will begin in 1978. With an 85·7 kW (115 hp) Lycoming O-235 series flat-four engine, the gross weight of the Model 77 is 748 kg (1,650 lb). The following details have also been released:

DIMENSIONS, EXTERNAL:

Wing span	9·14 m (30 ft 0 in)
Length overall	7·28 m (23 ft 10¾ in)
Height overall	2·30 m (7 ft 6½ in)
Tailplane span	3·00 m (9 ft 10 in)
Wheel track	2·54 m (8 ft 4 in)
Wheelbase	1·50 m (4 ft 11¼ in)
Propeller diameter	1·75 m (5 ft 9 in)

BOEING COMMERCIAL AIRPLANE COMPANY

Additional orders announced up to mid-October 1976 included the following:

BOEING 707 (Page 238)

Nigeria Airways
1 707-3F9C (making total sales 920)

BOEING 727 (Page 239)

Total sales increased to 1,347 by:
American Airlines
10 Advanced 727-223 (making 16 in all)
Braniff International
2 Advanced 727-227
Nigeria Airways
2 Advanced 727-2F9
United Air Lines
28 727-222

BOEING 747 (Page 242)

Total sales increased to 313 (of which 289 delivered) by:
Avianca
1 747-159
SAS
1 747-283B (plus 1 on option)
Singapore Airlines
1 747-212B (making 6 in all)

The first Boeing 747 powered by four Rolls-Royce RB.211-524B turbofans made its first flight, lasting 1 hr 51 min, from Everett on 3 September 1976. Certification is scheduled for April 1977. Meanwhile, under officially observed conditions, this aircraft has set a world record by taking off at a greater weight than that at which any other aeroplane has flown. Taking off at a weight of 381,244 kg (840,500 lb), it climbed to a height of more than 2,000 m (6,562 ft), claiming a record for the greatest mass lifted to that height.

CESSNA AIRCRAFT COMPANY

CESSNA MODEL 150 AEROBAT (Page 254)

Fifteen Aerobats have been ordered by the Zaïre Air Force.

Cessna announced on 14 September 1976 the development of three business jet aircraft, the first of which, the Citation I, would be available in December 1976. The other two aircraft have the designations Citation II and Citation III, the former being scheduled for delivery from February 1978.

CESSNA CITATION I

This model differs from the original Citation, first introduced in late 1971, by having a wing of increased span and Pratt & Whitney Aircraft of Canada JT15D-1A turbofan engines, giving an improved rate of climb and higher cruising speeds.

The description of the Citation 500 Series (pages 274-5) applies also to the Citation I, except as follows:

POWER PLANT: As for Citation 500 except for installation of Pratt & Whitney Aircraft of Canada JT15D-1A turbofan engines, each rated at 9·77 kN (2,200 lb st).

DIMENSIONS, EXTERNAL: As for Citation 500, except:

Wing span	14·35 m (47 ft 1 in)

WEIGHTS:

Weight empty (incl electronics)	2,932 kg (6,464 lb)
Max ramp weight	5,443 kg (12,000 lb)
Max T-O weight	5,375 kg (11,850 lb)
Max landing weight	5,148 kg (11,350 lb)
Max zero-fuel weight	3,810 kg (8,400 lb)
Optional max zero-fuel weight	4,309 kg (9,500 lb)

PERFORMANCE (at max T-O weight, except where indicated):
Cruising speed at average cruising weight
351 knots (649 km/h; 403 mph) TAS
Stalling speed at max landing weight
83 knots (154 km/h; 95·5 mph) CAS
Max rate of climb at S/L 817 m (2,680 ft)/min
Rate of climb at S/L, one engine out
244 m (800 ft)/min
Max certificated altitude 12,495 m (41,000 ft)
Service ceiling, one engine out 6,400 m (21,000 ft)
T-O to 10·7 m (35 ft) 838 m (2,750 ft)
Landing run at max landing weight 701 m (2,300 ft)
Range with 6 passengers, 45 min reserve
1,333 nm (2,470 km; 1,535 miles)

CESSNA CITATION II

This version of the Citation will introduce several new features, including a fuselage lengthened by 1·07 m (3 ft 6 in), an increased-span high aspect ratio wing, increased fuel and baggage capacity, and installation of Pratt & Whitney Aircraft of Canada JT15D-4 turbofans.

The description of the Citation 500 Series (pages 274-5) applies basically to the Citation II, except as follows:

POWER PLANT: Two Pratt & Whitney Aircraft of Canada JT15D-4 turbofan engines, each rated at 11·12 kN (2,500 lb st) for take-off, mounted in pod on each side of rear fuselage. Integral fuel tanks in wings, with usable capacity of 2,702 litres (714 US gallons).

ACCOMMODATION: As for Citation 500 Series, except seating for 8-10 passengers in main cabin, with toilet and increased baggage capacity.

DIMENSIONS, EXTERNAL:

Wing span	15·75 m (51 ft 8 in)
Wing aspect ratio	8·3
Length overall	14·40 m (47 ft 3 in)
Height overall	4·55 m (14 ft 11 in)
Tailplane span	5·74 m (18 ft 10 in)
Wheel track	5·36 m (17 ft 7 in)
Wheelbase	5·54 m (18 ft 2 in)

DIMENSIONS, INTERNAL:

Cabin:	
Length, front to rear bulkhead	6·38 m (20 ft 11 in)
Max width	1·50 m (4 ft 11 in)
Max height	1·45 m (4 ft 9 in)
Baggage capacity	2·21 m³ (78 cu ft)

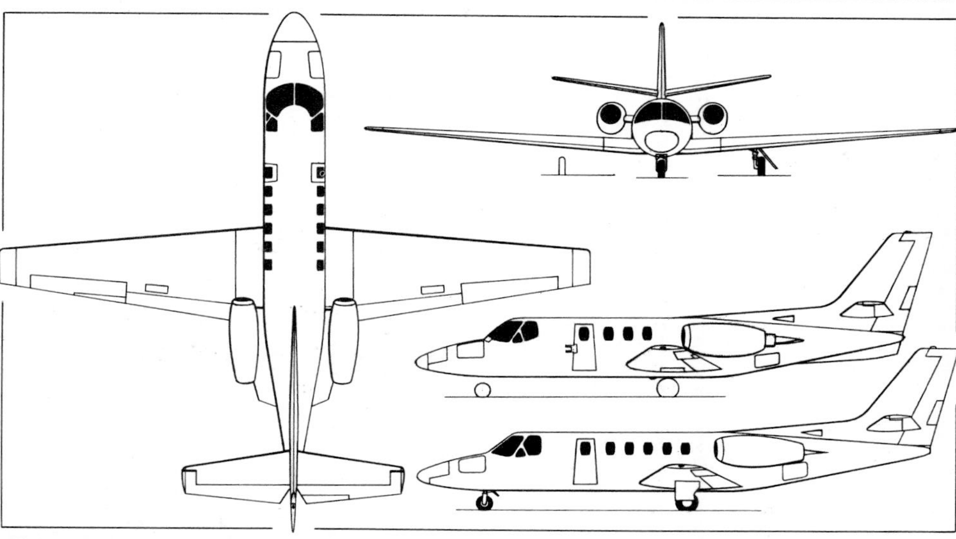

Three-view drawing of the Cessna Citation II, with additional side view (centre) of the shorter Citation I *(Pilot Press)*

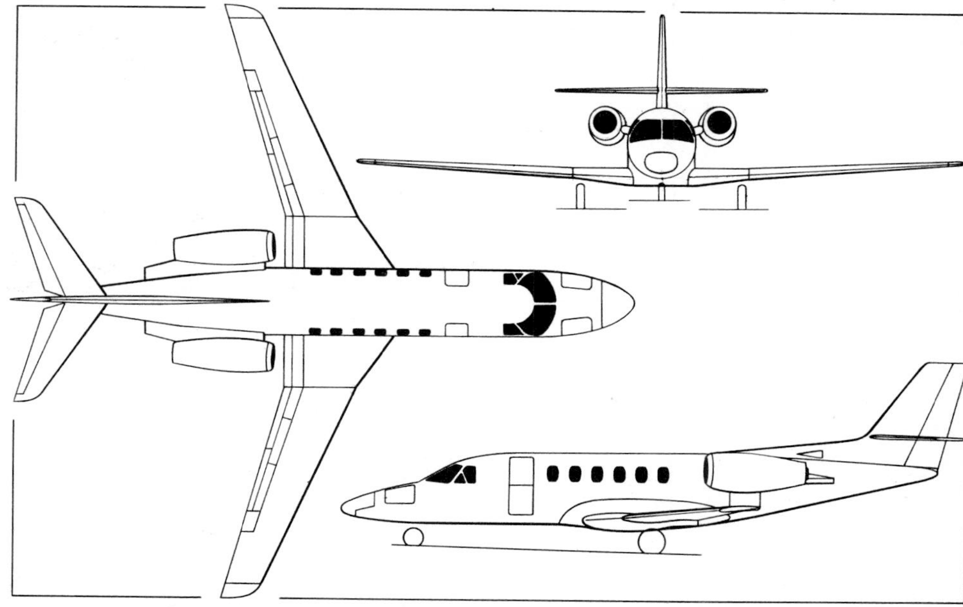

Cessna Citation III twin-turbofan sweptwing business aircraft *(Pilot Press)*

WEIGHTS:
Weight empty (incl electronics) 3,157 kg (6,960 lb)
Max ramp weight 5,761 kg (12,700 lb)
Max T-O weight 5,670 kg (12,500 lb)
Max landing weight 5,443 kg (12,000 lb)
Max zero-fuel weight 4,763 kg (10,500 lb)
Optional max zero-fuel weight 4,989 kg (11,000 lb)
PERFORMANCE (estimated, at max T-O weight, except where indicated):
Cruising speed at average cruising weight
365 knots (676 km/h; 420 mph) TAS
Stalling speed at max landing weight
81 knots (150 km/h; 93 mph) CAS
Max rate of climb at S/L 1,067 m (3,500 ft)/min
Rate of climb at S/L, one engine out
311 m (1,020 ft)/min
Max certificated altitude 13,105 m (43,000 ft)
Service ceiling, one engine out 8,075 m (26,500 ft)
T-O to 10·7 m (35 ft) 732 m (2,400 ft)
Landing run at max landing weight 698 m (2,290 ft)
Range with 10 passengers, 45 min reserve
1,806 nm (3,347 km; 2,080 miles)

CESSNA CITATION III

While retaining some general similarity to earlier members of the Citation family, the Citation III, which is scheduled for initial delivery in early 1980, is a very different aeroplane. Only brief details were released on 14 September 1976, but the Citation III is to have a supercritical swept wing. It will be larger in size, powered by a version of the Garrett AiResearch TFE 731 turbofan engine, and will be produced in two versions, a 10/15-seat Transcontinental and an 8/13-seat Intercontinental, the latter with increased fuel capacity. Cabin pressurisation will be at 0·61 bars (8·9 lb/sq in), to permit operation at a max certificated altitude of 13,715 m (45,000 ft).

DIMENSIONS, EXTERNAL:
Wing span 15·42 m (50 ft 7 in)
Length overall 15·72 m (51 ft 7 in)
Height overall 5·18 m (17 ft 0 in)
Tailplane span 5·33 m (17 ft 6 in)
Wheel track 4·06 m (13 ft 4 in)
Wheelbase 6·02 m (19 ft 9 in)
DIMENSIONS, INTERNAL (A: Transcontinental; B: Intercontinental):
Cabin:
Length, front to rear bulkhead:
A 7·01 m (23 ft 0 in)
B 6·45 m (21 ft 2 in)
Max width 1·63 m (5 ft 4 in)
Max height 1·68 m (5 ft 6 in)
Baggage capacity 2·27 m³ (80 cu ft)
WEIGHTS (estimated, A: Transcontinental; B: Intercontinental):
Weight empty, equipped:
A 4,282 kg (9,441 lb)
B 4,364 kg (9,621 lb)
Max T-O weight:
A 7,779 kg (17,150 lb)
B 8,664 kg (19,100 lb)
Max landing weight:
A, B 7,121 kg (15,700 lb)
Max ramp weight:
A 7,870 kg (17,350 lb)
B 8,754 kg (19,300 lb)
Max zero-fuel weight:
A, B 5,897 kg (13,000 lb
Max fuel capacity:
A 3,030 kg (6,680 lb
B 3,833 kg (8,450 lb
PERFORMANCE (estimated at max T-O weight, excep where indicated otherwise. A: Transcontinental; B Intercontinental):
Cruising speed at average cruising weight:
A, B 470 knots (871 km/h; 541 mph) TA
Stalling speed at max landing weight:
A, B 93 knots (172 km/h; 107 mph) CA
Max rate of climb at S/L:
A 1,623 m (5,325 ft)/mi
B 1,430 m (4,690 ft)/mi
Rate of climb at S/L, one engine out:
A 497 m (1,630 ft)/mi
B 418 m (1,370 ft)/mi
Max certificated altitude:
A, B 13,715 m (45,000 ft
Service ceiling, one engine out:
A 9,050 m (29,700 ft
B 8,230 m (27,000 ft
FAA T-O field length:
A 1,216 m (3,990 ft
B 1,466 m (4,810 ft
Landing run at max landing weight:
A, B 1,036 m (3,400 ft
Range, 45 min reserves:
A 2,397 nm (4,442 km; 2,760 miles
B 2,996 nm (5,552 km; 3,450 miles

GATES LEARJET

A Gates Learjet Model 24 modified by Swedair of Sweden for towing sleeve-type or dart-type targets. It is shown fitted with a belly-mounted ECM equipment pod and a dart-type tow 'bird' stowed beneath the rear fuselage

GENERAL DYNAMICS

GENERAL DYNAMICS F-16 (Page 287)

It was announced on 27 October 1976 that the government of Iran had signed a letter of intent for the purchase of up to 160 General Dynamics F-16 air combat fighters, with deliveries to begin in the early 1980s. Iran has a requirement for a total of 300 F-16s.

GENERAL DYNAMICS CCV YF-16

Under a $6 million contract awarded by the USAF, General Dynamics Fort Worth Division has modified the first prototype of the F-16 air combat fighter as a testbed for a control configured vehicle (CCV) programme. This is being directed by the Air Force Systems Command's Air Force Flight Dynamics Laboratory at Wright-Patterson AFB, Ohio.

The YF-16 was selected for this programme because, in place of conventional flying controls, it has a quadruple-redundant fly-by-wire control system, in which electrical circuits replace the usual mechanical linkages between the pilot's controls and the related aerofoil control surfaces. This system was integrated into the F-16 design to exploit, from the outset, the total capabilities of flight control system technology through the CCV principle. In the F-16, the application of CCV technology was concerned with the relationship of aircraft balance to static longitudinal stability, allowing the CG to be moved further aft than is normally possible with an aircraft of conventional configuration and control. It results in a significant reduction of trim drag at high load factors and at supersonic speeds. The effect is to reduce overall drag, which includes both the tail drag and the change in drag on the wing resulting from changes in wing lift required to balance the down load on the tail.

Until the advent of the YF-16 prototypes, CCV technology was a mainly-theoretical concept which aerodynamicists had exploited by the conventional approach of analysis, models, and wind tunnel research. With the availability of CCV basic hardware, in the form of the highly successful YF-16 which had given some hint of CCV potential, further exploration of the principle was inevitable.

The only external change now visible on the YF-16 prototype is the addition of two canard surfaces, each 0·74 m² (8 sq ft) in area, mounted one each side of the engine air intake duct and operated by hydraulic actuators. Other changes include means of isolating port and starboard wing fuel tanks from the forward and aft fuselage fuel cells, providing a manual means of varying the aircraft's CG position; modifications to the flight control system to permit the use of wing trailing-edge flaperons (flaps/ailerons) in combination with the all-moving tailplane to provide direct lift control; and similarly to use the new canards and the conventional rudder in conjunction to give direct side-force control.

First prototype YF-16, modified by General Dynamics for CCV (Control Configured Vehicle) research

The expected result of these changes is to give the aircraft radically different performance characteristics. Thus the CCV should be able to point its nose in any direction without changing its flight path; or rise, descend, and move sideways without changing its nose direction. Movement of the nose to port or starboard, sideways movement without bank or roll, and wings-level turns are executed by using the canards and rudder in conjunction. Pitching movements of the nose, and climb and descent, are effected by co-ordinated movements of the wing flaperons and all-moving tailplane.

These additional control freedoms may prove invaluable for better target tracking performance and weapon delivery accuracy. Instead of having only a fleeting moment for aim and weapon discharge, it would seem possible to fly a CCV aircraft with the nose crabbed a few degrees to port or starboard, or similarly pitched up or down a few degrees, or a combination of both, allowing the target to be aligned accurately in the gunsight. Another advantage of the CCV results from the very rapid response of the flight control system, which senses, reacts, and damps out gust effects before the pilot is aware of them.

The CCV/YF-16 flew for the first time with its canard surfaces operative on 24 March 1976, and by 8 April had already accumulated 14 hours of flight testing. The programme was scheduled for a period of seven months during which time the Flight Dynamics Laboratory hoped to demonstrate that these additional control surfaces would make for better performance, ease the pilot's workload, and make possible completely new combat manoeuvres. The end result may be an aircraft that is smaller, lighter, and less expensive, with the ability to combine better payload/range and combat kill potential than previous fighters.

GRUMMAN AMERICAN

GRUMMAN AMERICAN GULFSTREAM III

On 10 November 1976, Grumman American announced that it was developing an executive jet that would fly faster and further, carrying more passengers than any other business aircraft in the world. Named Gulfstream III, it is planned as an improved version of the current Gulfstream II (page 296), with production to begin in 1980, and will be certificated for 19 passengers and crew.

Key to the performance of the Gulfstream III will be its use of a NASA/Grumman supercritical wing, fitted with 'winglets' at the tips to give a further improvement in cruising fuel consumption. Other features will include an enlarged cabin, larger cockpit with improved field of view, and a more streamlined nose for improved aerodynamics and a quieter flight deck environment. There will also be increased baggage space and improvements to systems. The Gulfstream II power plant, comprising two Rolls-Royce Spey Mk 511-8 turbofans, will be unchanged in the new aircraft, although T-O thrust of each engine will be adjusted to less than 40 kN (9,000 lb) to meet environmental standards.

The following data are provisional:

DIMENSIONS, EXTERNAL:
Wing span 25·85 m (84 ft 9½ in)
Length overall 26·61 m (87 ft 3½ in)
Height overall 7·25 m (23 ft 9½ in)
Fuselage depth 2·99 m (9 ft 9½ in)
Fuselage width 2·40 m (7 ft 10¾ in)

WEIGHTS:
Max ramp weight 30,570 kg (67,400 lb)
Max T-O weight 30,345 kg (66,900 lb)
Max landing weight 28,890 kg (63,700 lb)

PERFORMANCE (estimated):
Max cruising speed Mach 0·88
Long-range cruising speed Mach 0·84
Operating height 15,550 m (51,000 ft)
Approach speed 130 knots (240 km/h; 150 mph)
Time to initial cruising height of 13,100 m (43,000 ft) 23 min
FAA T-O distance (ISA) 1,800 m (5,900 ft)
FAA landing distance 1,070 m (3,500 ft)
Range with max fuel, 30 min reserves 4,400 nm (8,150 km; 5,060 miles)
NBAA/IFR range 4,000 nm (7,400 km; 4,600 miles)

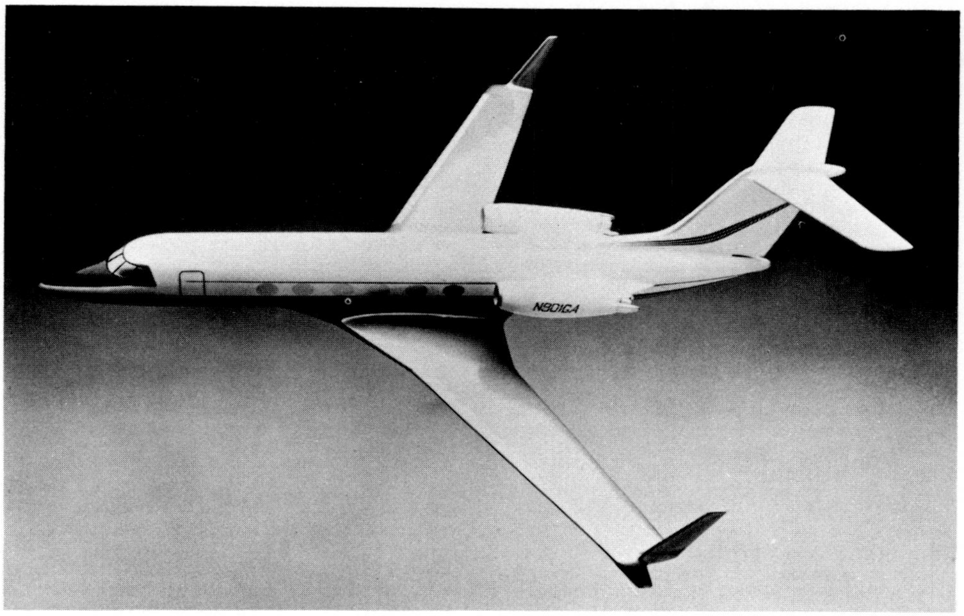

Model of the projected Grumman American Gulfstream III, with supercritical wings and 'winglets'

SCHWEIZER (GRUMMAN) SUPER AG-CAT

Grumman American is evaluating an Ag-Cat (see page 381) which has been re-engined with a Polish-built 447 kW (600 hp) PZL-3S radial engine, driving a Dowty Rotol R289 three-blade metal propeller. This power plant is being assessed for potential use in new-production Ag-Cats.

LOCKHEED-CALIFORNIA COMPANY

LOCKHEED CP-140 AURORA (Pages 313-4)

The purchase of 18 special variants of the Lockheed P-3 Orion maritime patrol aircraft for the Canadian Armed Forces was announced by the Hon James Richardson, Canadian Minister of National Defence, on 21 July 1976. This marked the terminal phase of a procurement programme that originated in 1972, when Air Specification 15-14 defined the Canadian government's requirements for a Long-Range Patrol Aircraft (LRPA) to replace the CP-107 Argus maritime reconnaissance aircraft serving currently with the CAF. Following two years of concept definition by five aerospace companies, Lockheed and The Boeing Company were selected for final contract definition; in late 1975 Lockheed's 'P-3 LRPA' design was announced as the winner.

Designated subsequently as the CP-140 Aurora, this new aircraft combines the P-3 Orion's airframe, power plant and basic aircraft systems with the electronic systems and data processing capability of the carrier-based Lockheed S-3A Viking. Able to perform missions involving a range of more than 4,000 nm (7,400 km; 4,600 miles), or flights of up to 17 hours' duration, the CP-140 will be deployed initially for ASW duties; national sovereignty patrols; shipping, fisheries and Arctic surveillance; ice reconnaissance; and search and rescue. By the addition of a weapons bay sensors canister at a later date, the CP-140 will be able to undertake additional civilian tasks such as resources location, pollution control and aerial survey.

The cabin interior of the P-3C has been changed extensively to meet Canadian requirements: immediately aft of the flight deck are an observer's station on the port side and crew rest bunks on the starboard side. Moving aft, the tactical compartment comes next, with accommodation for the Tactical Navigator (TACNAV), Navigator/Communicator (NAVCOM), two Acoustic Sensor Operators (ASO), and two Non-Acoustic Sensor Operators (NASO), all on the port side. Aft of the tactical compartment is the search stores and camera bay, with two more observer stations, one on each side. At the rear of the cabin are a galley, on the port side, a dinette area, and an airborne maintenance station on the starboard side. A toilet is located on the port side of the cabin, immediately aft of the forward observer's position.

On the flight deck, an ASA-82 Multi-Purpose Display (MPD) provides the pilots with a real-time presentation of the tactical situation and sensor information; directions from the TACNAV and NAVCOM are fed through the computer for display on both the MPD and the Flight Director Indicators (FDIs). Cues and alerts, indicating required sequences of action, are displayed on the periphery of the MPD.

An AJN-15 Flight Director system supplies attitude, heading and fly-to-point references. For long-range navigation, data from the Horizontal Situation Indicator are normally adequate. For precise, close-in tactical manoeuvring the FDI is used, and the automatic flight control system includes full-time attitude control and proportional control-wheel steering.

The three observer stations each have a fully-swivelling seat and are provided with intercom. Each of the observation windows gives full hemispherical view, and there are power and storage provisions for a hand-held camera. Each position is provided with isolation curtains, to screen observer and window from cabin lighting during night visual search. A fourth station can be made available on the starboard side, by removal of the crew rest bunks.

The TACNAV has a console which includes an ASA-82 MPD, an ASQ-147 keyset and trackball, and armament controls. With his keyset the TACNAV can control, via the computer, the Sonobuoy Reference System (SRS), and can call up and display FLIR and other radar data on his MPD. The NAVCOM also has an ASA-82 and ASQ-147, plus HF, VHF (FM) and UHF transceivers; inertial, VLF (Omega) and Doppler navigation sets; LF and UHF ADF, and VHF homer; a high-speed teleprinter and teletype keyboard; provisions for Tactical Satellite Communications (TACSATCOM); data link; control of reconnaissance photography; provisions for control of survey photography; and provisions for secure communications. The NAVCOM's MPD and keyset serve as a backup for the TACNAV in the event of equipment failure.

The two ASOs share a dual console and each has an ASA-82 MPD and ASQ-147 keyset and trackball. They share also an ASA-82 Auxiliary Readout Unit (ARU), a time code generator, and an AN/ASH-27 28-track tape recorder. Their MPDs can display acoustic data or the tactical plot, but the ARU is a dedicated acoustic display. The acoustic functions of receiving, processing, display and recording are controlled by the keysets through the computer.

The two NASOs also have a dual console, each with an ASA-82 MPD and ASQ-147 keyset and trackball, the keysets being used to control radar, electronic support measures (ESM) and FLIR through the computer. Principal controls shared by these two operators, or available to only one of them, include ASQ-501 MAD, OA-5150/ASQ(FACS II) MAD compensator, video tape recorder, SIF and provisions for SLAR.

The heart of the entire control system is a Univac AN/AYK-10 Navigation/Tactical computer. Its two central processors function independently; both have coordinated access to a core memory of 65,536 words. There is growth capacity for an additional 32,000 words, and space has been allocated for a 127,000-word auxiliary memory in the acoustic system processor for the computer.

The search stores and camera bay has stowage for 'A' size sonobuoys, large and small marine markers, Signals Underwater Sound (SUS) and flares. Intercom controls and an ordnance status panel are provided for the ordnance crew member. The computer-controlled electrically-fired cartridge-actuated A-size launchers can all be operated with the aircraft pressurised. They comprise 36 underfloor launchers, loadable only on the ground, and three which can be loaded from the cabin with the aircraft pressurised or unpressurised. A C-size chute, just aft of the three cabin launch tubes, allows free-fall launch (with the aircraft unpressurised) of flares, small marine markers, SUS and mail, and air drops to remote ships or stations.

A KA-107A day/night reconnaissance camera is installed beneath the floor in this area, and is accessible in flight through a floor hatch. The illuminator for night reconnaissance photography is located beneath the floor of the in-flight maintenance station. This position has a bench with 28V DC and 115V 400Hz AC power outlets, and there are provisions for a microfiche reader.

Aircraft operational support equipment for the CP-140 includes the ground-based Data Interpretation and Analysis Center (DIAC), and a Ground Support Computer Complex (GSCC). The former provides operational support for the operating squadrons; the latter provides technical support for the operational software, and maintains software configuration records.

The aircraft's weapon bay, which has a maximum capacity of 2,177 kg (4,800 lb) on eight stations, can accommodate and drop the Canadian SKAD/BR search and rescue kit, as well as a variety of ordnance. There are ten underwing hardpoints, with an individual capacity ranging from 277 kg (611 lb) to 1,111 kg (2,450 lb).

The CP-140 Aurora is designed primarily to carry out military tasks essential to North American and NATO defence, and to provide long-range surveillance of Canada's coastal waters. It is scheduled to enter service in 1980 and, because of the growth potential of its equipment, is expected to serve into the next century.

TYPE: Four-turboprop long-range ASW and maritime patrol aircraft.

WINGS: Cantilever low-wing monoplane. Wing section NACA 0014 (modified) at root, NACA 0012 (modified) at tip. Dihedral 6°. Incidence 3° at root, 0° 30′ at tip. Fail-safe box beam structure of extruded integrally-stiffened aluminium alloy. Lockheed-Fowler trailing-edge flaps. Aluminium alloy ailerons operated by dual hydraulic boosters supplied from two independent hydraulic systems. Trim tabs in ailerons. Anti-icing by engine bleed air ducted into leading-edges.

FUSELAGE: Conventional aluminium alloy semi-monocoque fail-safe structure.

TAIL UNIT: Cantilever aluminium alloy structure with dihedral tailplane and dorsal fin. Fixed-incidence tailplane. Rudder and elevators each operated by dual hydraulic boosters, supplied from two independent hydraulic systems. Trim tabs in elevators and rudder. Electrical anti-icing system for leading-edges of fin and tailplane.

LANDING GEAR: Hydraulically-retractable tricycle type with twin wheels on each unit. All units retract forward, main wheels into inner engine nacelles. Oleo-pneumatic shock-absorbers. All units can free-fall to the down and locked position in emergency. Hydraulically-powered steerable nose unit, controlled by handwheel on the pilot's side console. Hydraulically-operated dual segmented-disc brakes. Pneumatic emergency braking system.

POWER PLANT: Four 3,661 kW (4,910 ehp) Allison T56-A-14 turboprop engines, each driving a four-blade metal constant-speed fully-feathering and reversible propeller. Fuel in one fuselage and four wing integral tanks, with total usable capacity of 34,826 litres (9,200 US gallons). Single-point pressure refuelling, and four overwing gravity refuelling points, are provided. Fuel dump system. Propeller blade cuffs and spinners de-iced by electrical heating.

ACCOMMODATION: Normal eleven-man crew, with seating for five additional passengers. Dual controls standard. Flight deck has wide-vision windows, and circular windows for up to four observers are provided in the main cabin, each bulged to give 180° visibility. Main cabin fitted out as detailed in introductory paragraphs. Door on port side, aft of wing. Overwing emergency exit on each side of cabin; others in side and ceiling of flight deck. De-fogging and anti-icing of windscreens by electrical heating; windscreens have mechanical wipers, a washing system for the removal of salt deposits, and a rain-repellent spray system. Stowage for clothing, life jackets and parachute harness. Four floor tie-down areas have a combined baggage/cargo capacity of 442 kg (975 lb).

SYSTEMS: Air-conditioning and pressurisation system supplied by two engine-driven compressors, maintaining cabin temperatures between 15·6°C and 26·7°C (60°F and 80°F), and a cabin altitude of 2,440 m (8,000 ft) to a height of 9,145 m (30,000 ft). Two independent hydraulic systems, each at a pressure of 207 bars (3,000 lb/sq in), are powered by three interchangeable electrically-driven pumps, any two of which can maintain full hydraulic services. Pneumatic system at pressure of 207 bars (3,000 lb/sq in) for emergency braking. Electrical system of 120/208V 400Hz AC supplied by three 60/90kVA engine-driven generators, any one of which can maintain full normal load. DC power supplied by three 200A 24V transformer-rectifiers and one 31Ah storage battery. APU drives a 60/90kVA generator and provides power and bleed air for ground air-conditioning, weapons bay heating and engine starting; it can also provide emergency electrical power in flight. Oxygen system for crew of three on flight deck with 3·5 hour capacity. Individual portable chemical oxygen generators for emergency use by all crew members. Automatic flight control system (AFCS) with dual-channel fail-safe autopilot; includes tactical and airways nav modes and proportional control wheel steering.

ELECTRONICS AND EQUIPMENT: Univac AN/AYK-10 navigation/tactical computer; digital magnetic tape units; teleprinter; display generator units; APS-116 search radar; OR-89/AA (modified) FLIR; video recorder for FLIR imagery; ARS-2 sonobuoy reference system; OL-82 (modified) acoustics data processor; RD-348, ASQ-147 and ASA-82 displays; LN-33 inertial navigation system; APN-208 Doppler; ARN-115 Omega; Tacan; revised airways/approach nav aids; dual VOR/ILS; communications sets comprising HF, UHF, VHF (AM), VHF guard receiver, VHF(FM); HF SIMOPS filters; RCVR homing; USH 502 crash position indicator/flight data recorder; ASW-31 AFCS; ALR-47 ESM; AN/ASH-27 28-track tape recorder; ASQ-501 MAD; OA-5150/ASQ (FACS II) MAD compensator; SLAR provisions; IFF; data link: Airborne Radiation Thermometer (ART) provisions; and time coding generator. Equipment includes KA-107A day/night reconnaissance camera and night illuminator; provisions for civil sensors canister; galley with refrigerator and sink; white edge lighting for all console-mounted control panels; white cabin lighting; reading lights at all crew positions; white overhead lights; and aisle lights.

PERFORMANCE (with mission payload of 2,540 kg; 5,600 lb except where stated otherwise):

Max transit speed at optimum altitude
395 knots (732 km/h; 455 mph)

Max level speed below cruise ceiling
375 knots (695 km/h; 432 mph)

FAR balanced field length 2,408 m (7,900 ft)

T-O to 15 m (50 ft) 1,829 m (6,000 ft)

Landing from 15 m (50 ft) at 51,714 kg (114,000 lb) landing weight 975 m (3,200 ft)

LOCKHEED-GEORGIA

LOCKHEED C-130 HERCULES

Portugal has ordered two Hercules for delivery in August 1977.

LOCKHEED MODEL L-100 SERIES COMMERCIAL HERCULES (Page 320)

Lockheed-Georgia announced on 5 September 1976 that it had completed preliminary design work on a new high-capacity version of the Hercules, designated L-100-50. Extra capacity results from 'stretching' the L-100 basic airframe by an additional 10·67 m (35 ft 0 in), with a plug 6·10 m (20 ft 0 in) long forward of the wing and another 4·57 m (15 ft 0 in) long aft of the wing. This would provide a cargo compartment length of 26·03 m (85 ft 4¾ in), offering a bulk capacity of 229·4 m³ (8,100 cu ft). To cater for the proposed maximum take-off weight of 77,110 kg (170,000 lb), strengthening of the wing and landing gear structure is included in the design.

This new version of the commercial L-100 Hercules offers a maximum payload of 27,215 kg (60,000 lb), which could be carried over a range of 1,240 nm (2,300 km; 1,430 miles). Range with a 22,680 kg (50,000 lb) payload would be 1,910 nm (3,540 km; 2,200 miles). For palletised cargo operations this represents a 40 per cent improvement in revenue payload capability by comparison with the current production L-100-30 Hercules. The cargo compartment would be able to accommodate ten 2·24 m × 3·00 m (7 ft 4 in × 9 ft 10 in) or ten 2·44 m × 3·00 m (8 ft 0 in × 9 ft 10 in) pallets, a mix of 2·44 m × 2·44 m (8 ft 0 in × 8 ft 0 in) containers 3·05, 6·10 or 12·20 m (10, 20 or 40 ft) long, or 20 LD-3 containers.

Artist's impression of the L-100-50 'stretched' version of the commercial Hercules, with 10·67 m (35 ft) longer fuselage

MARSH AVIATION COMPANY

ADDRESS:
5060 East Falcon Drive, Mesa, Arizona 85205
Telephone: (602) 832-3770

MARSH/ROCKWELL TURBO THRUSH

Marsh Aviation Company has converted a piston-engined Rockwell Thrush Commander to turbine power by the installation of an AiResearch TPE 331-1-101 turboprop engine. Derated to 447 kW (600 shp) for this conversion, the full 580 kW (778 shp) output of the TPE 331 is available in emergency. The empty weight of the Turbo Thrush is 227 kg (500 lb) less than that of the Rockwell Thrush Commander, providing increased payload capability and improved speed and performance. For agricultural operators working in remote areas the TPE 331 installation has the advantage that ordinary automotive diesel fuel can be used if jet fuel is not available.

Following more than 600 hours of flight by the prototype, an FAA Supplemental Type Certificate has been issued. The first production conversion was handed over in September 1976, and other domestic and foreign deliveries were scheduled before the end of the year.

Marsh Turbo Thrush, a turbine-engined conversion of the Rockwell International Thrush Commander

DIMENSIONS, EXTERNAL:
As for Thrush Commander except:
Length overall 9·27 m (30 ft 5 in)

WEIGHTS AND LOADINGS:
As for Thrush Commander except:
Weight empty 1,633 kg (3,600 lb)

PERFORMANCE (at 2,721 kg; 6,000 lb T-O weight, except where indicated):

Cruising speed, 50% power
127 knots (235 km/h; 146 mph)

Working speed, 50% power
108·5 knots (201 km/h; 125 mph)

Stalling speed, flaps up 41·5 knots (77 km/h; 48 mph)

Stalling speed, flaps down 38 knots (71 km/h; 44 mph)

Stalling speed, flaps up at normal landing weight
39 knots (72·5 km/h; 45 mph)

Stalling speed, flaps down at normal landing weight
37 knots (69 km/h; 43 mph)

Max rate of climb at S/L 915 m (3,000 ft)/min

Service ceiling 7,620 m (25,000 ft)

T-O run 183 m (600 ft)

Landing run 91 m (300 ft)

Ferry range, at 60% power
521 nm (966 km; 600 miles)

MARTIN MARIETTA

MARTIN MARIETTA MODEL 845A

Martin Marietta designed around the basic Schweizer SGS 1-34 Standard Class sailplane a remotely piloted powered aircraft under the designation Martin Marietta Model 845A. This aircraft was described in the RPVs & Targets section of the 1973-74 and 1974-75 *Jane's*.

The aircraft illustrated, Serial No. 845A-1, has since been converted to a single-seat aircraft with conventional controls, and has been used to carry out air sampling sorties for the US government prior to the launch of space vehicles from Cape Canaveral, Florida.

First prototype Martin Marietta Model 845A, a former RPV now converted to manned operation for high altitude air sampling tests *(Jean Seele)*

MCDONNELL DOUGLAS CORPORATION

MCDONNELL DOUGLAS F-18 (Page 329)

Subcontracts awarded by McDonnell Douglas during October 1976 included those for the head-up display (to Kaiser Aerospace and Electronics Corporation), intercom (SCI Systems Inc), wing leading-edge flap mechanical drive units (AiResearch Manufacturing Co), and rudder hydraulic servo-cylinders (Ronson Hydraulic Units Corporation).

MCDONNELL DOUGLAS DC-8 CARGO CONVERSION

McDonnell Douglas is modifying two DC-8-43 passenger transports to an all-cargo configuration. The modification includes the removal of all passenger facilities, replacing the floor with the seven-track cargo floor used on DC-8 freighters, replacing cabin windows with metal plugs, installation of a main deck cargo door 2·16 m (7 ft 1 in) high and 3·56 m (11 ft 8 in) wide, provision of a 9*g* cargo barrier net, cabin interior cargo liner and a smoke detection system. Also included in the modification is all work necessary to qualify the aircraft for increased operating weights and FAA certification.

Should the customer so desire, the Rolls-Royce Conway engines of the DC-8-40 series can be replaced by Pratt & Whitney JT3D turbofans, reclassifying them as DC-8-50 series aircraft. This is being done in the initial two conversions for Frederick B. Ayer and Associates.

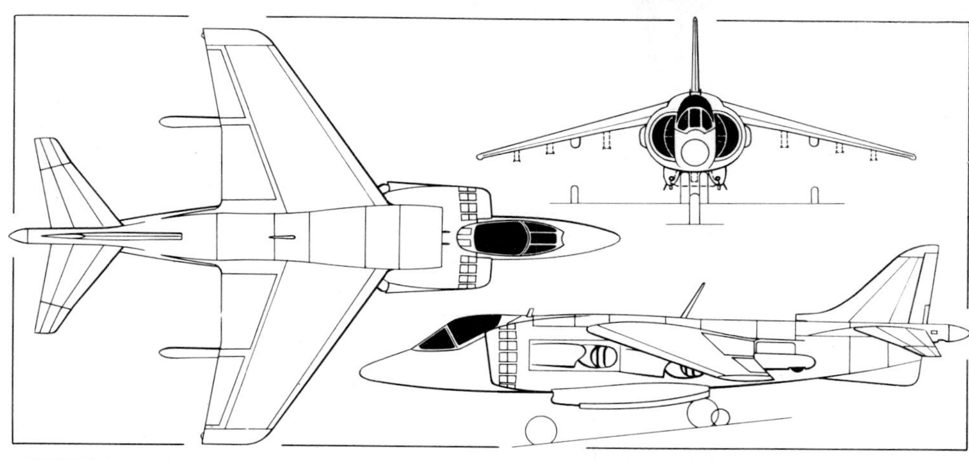

AV-8B Advanced Harrier under development by McDonnell Douglas for the US Marine Corps *(Pilot Press)*

NASA

Artist's impression of the QSRA (Quiet Short-Haul Research Aircraft), being developed by Boeing Commercial under an approx $ 20 million NASA contract to investigate the reduction of airport congestion and noise. Embodying NASA 'quiet propulsive lift' technology, the aircraft, a modified de Havilland Canada C-8A Buffalo, will be powered by four Avco Lycoming YF102 overwing turbofans and will have a similar wing lift system to the Boeing YC-14. Testing will be done at NASA's Ames Research Center, with first flight scheduled for the second half of 1978

ROCKWELL INTERNATIONAL CORPORATION

ROCKWELL INTERNATIONAL OV-10 BRONCO (Pages 369-70)

The following additional version has been announced:

OV-10G. Version for Korean Air Force; 24 ordered in 1976, for delivery in 1977.

ROCKWELL INTERNATIONAL SABRELINER 65A and 80A

Rockwell announced on 8 September 1976 that the company was planning to introduce two new Sabreliner business jets in 1978/79, with retrofit programmes available for existing aircraft. Both will incorporate advanced technology improvements developed by the Raisbeck Group (which see), and which include increased wing chord, leading-edge slats replaced by a blunt cambered leading-edge, and introduction of Fowler-type trailing-edge flaps. This new wing can accommodate approximately 450 kg (990 lb) of additional fuel.

The first to be certificated in early 1978 will be the Sabreliner 80A, with the new wing and powered by two 20·0 kN (4,500 lb st) General Electric CF700-2D-2 turbofans as used in the Sabreliner 75A. Approximately a year later it is planned to certificate a Sabreliner 65A, which will use the new wing and be powered by two 16·5 kN (3,700 lb st) Garrett AiResearch TFE 731-3 turbofan engines; this version will have transcontinental range.

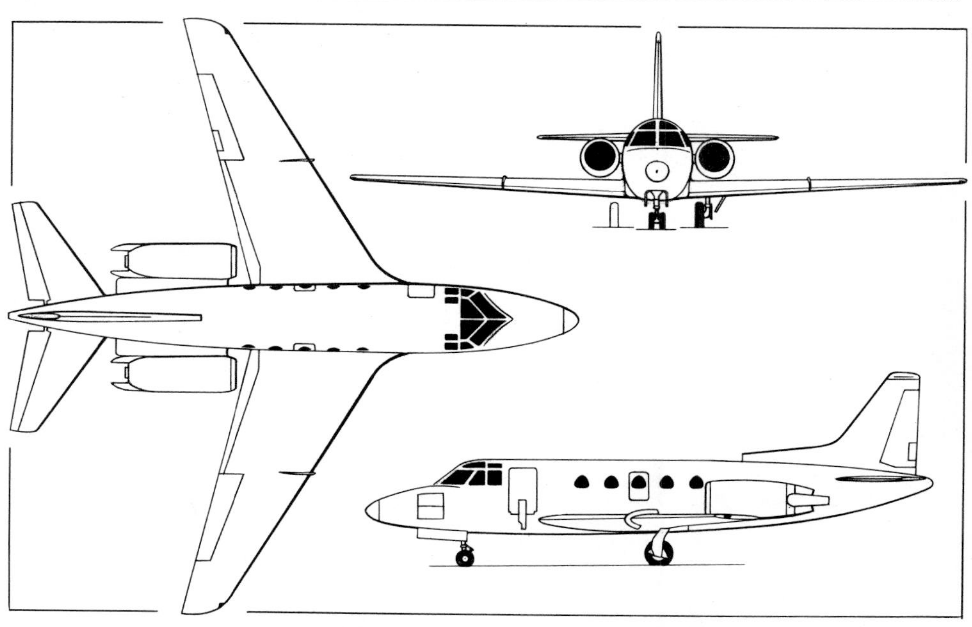

Rockwell International Sabreliner 65A, one of two new versions announced in 1976 *(Pilot Press)*

Rockwell Thrush Commander fitted in Poland with a PZL-3S aircooled radial engine *(BIIL)*

SPITFIRE

SPITFIRE HELICOPTER COMPANY LTD

ADDRESS:
PO Box 61, Media, Pennsylvania 19063
Telephone: (215) 565-2986
PRESIDENT: Jack Fetsko
VICE-PRESIDENT: Mike Meger

Spitfire Helicopter Company has completed the prototype of a new lightweight turbine-powered helicopter which it has designated as the Spitfire Mark I. Evolved from the basic design of the Enstrom F-28A, the new Spitfire has a turbine power plant replacing the Lycoming piston engine of the F-28A. The drive from this passes via a reduction gear instead of through multiple Vee belts. This installation not only provides a weight saving of 91 kg (200 lb), but frees space which can be allocated to additional cargo or auxiliary fuel. The other important feature of the Spitfire is the provision of tricycle landing gear, the resulting STOL capability offering significant fuel economy.

SPITFIRE MARK I

TYPE: Three-seat light turbine-powered helicopter.

ROTOR SYSTEM: Fully-articulated metal three-blade main rotor. Blades of bonded light alloy construction, each attached to rotor hub by retention pin and drag link. Two-blade teetering tail rotor. Blades do not fold.

ROTOR DRIVE: Shaft drive to both main and tail rotors through conventional reduction gear.

FUSELAGE: Glassfibre and light alloy cab structure, with welded steel tube centre section. Semi-monocoque light alloy tailcone structure.

LANDING GEAR: Non-retractable tricycle type. Single wheel on each unit, with wheel fairing.

POWER PLANT: One 313 kW (420 shp) Allison 250-C20B turboshaft engine, derated to 179 kW (240 shp) for take-off. Total standard fuel capacity of 265 litres (70 US gallons).

ACCOMMODATION: Pilot and two passengers, side by side on bench seat. Door on each side of cabin. Baggage space forward of engine compartment, with external access door on each side of fuselage: helicopter can be operated with these doors removed to permit loading of outsize cargo. Cabin heated and ventilated.

SYSTEMS: Electrical power provided by engine-driven alternator.

ELECTRONICS AND EQUIPMENT: Various nav/com systems available to customer's requirements. Cargo hook and litters optional. It is intended that survey and agricultural equipment will also be available.

DIMENSIONS, EXTERNAL:
Diameter of main rotor 9·75 m (32 ft 0 in)
Diameter of tail rotor 1·42 m (4 ft 8 in)
Length overall 8·96 m (29 ft 4¾ in)
Height overall 2·79 m (9 ft 2 in)
Width over wheel fairings 2·44 m (8 ft 0 in)

AREAS:
Main rotor disc 74·7 m² (804 sq ft)
Tail rotor disc 1·59 m² (17·1 sq ft)

WEIGHTS AND LOADINGS (estimated):
Weight empty 567 kg (1,250 lb)
Max T-O weight 1,043 kg (2,300 lb)
Max disc loading 13·97 kg/m² (2·86 lb/sq ft)
Max power loading 5·83 kg/kW (9·58 lb/shp)

PERFORMANCE (estimated, at max T-O weight):
Max level speed 112 knots (208 km/h; 129 mph)
Cruising speed 95·5 knots (177 km/h; 110 mph)
Max rate of climb at S/L 472 m (1,550 ft)/min
Service ceiling 4,570 m (15,000 ft)
Hovering ceiling in ground effect 4,085 m (13,400 ft)
Hovering ceiling out of ground effect 2,440 m (8,000 ft)
Max endurance 4 hr

VOLPAR (Page 398)

VOLPAR/SCOTTISH AVIATION CENTURY JETSTREAM

In mid-1976 Volpar was involved in the prototype installation of two 597 kW (800 shp) Garrett AiResearch TPE 331-3U-303 turboprop engines in a Scottish Aviation Jetstream owned by Century Aircraft of Santa Barbara. These engines replace the Turboméca Astazous which have a comparatively low TBO. Scottish Aviation assisted Volpar in this project by supplying installation drawings of the Garrett-engined Jetstream developed by Handley Page for the USAF under the designation C-10A. Simultaneously with the installation of these engines, fuselage ice guards were being manufactured and fitted. It was planned to convert the remaining six Jetstreams owned by Century Aircraft following completion of the prototype's certification programme.

UNION OF SOVIET SOCIALIST REPUBLICS

TUPOLEV

TUPOLEV VARIABLE-GEOMETRY BOMBER (Page 439)

The following data concerning *Backfire-B* have been released officially in the USA:

DIMENSIONS, EXTERNAL:
Wing span:
fully spread 34·45 m (113 ft)
fully swept 26·21 m (86 ft)
Length overall 40·23 m (132 ft)
Height overall 10·06 m (33 ft)

WEIGHTS:
Nominal weapon load 7,935 kg (17,500 lb)
Max T-O weight 122,500 kg (270,000 lb)

HOMEBUILT AIRCRAFT

FRANCE

STARCK

M André Starck has been flying since 1927, when he first left the ground in a Chanute-type hang glider. He next built a Mignet Pou-du-Ciel, followed by the first of his own designs, the AS-10. This was a tandem two-seat biplane, and led to the AS-20, first flown on 23 October 1942, with the sharply staggered narrow-gap biplane wing arrangement first conceived by an aerodynamicist named Nenadovitch.

Five fairly conventional monoplane designs were next, designated AS-70 Jac, AS-71, AS-57, AS-80 Holiday and AS-90 New-Look. In the new AS-27 Starcky and AS-37, described briefly below, M Starck has reverted to the narrow-gap staggered biplane formula.

STARCK AS-27

First flown in the Summer of 1975, the AS-27 is a small single-seat racing aircraft powered by a 78 kW (105 hp) Potez flat-four engine. It embodies the narrow-gap staggered biplane configuration favoured by M Starck for his current designs, but is otherwise conventional, with cantilever non-retractable main landing gear and a steerable tailwheel. No details are available except that it was built by M Claude Chevassut and first flown at Chavenay by M Robert Buisson.

STARCK AS-37

Starck AS-37 two-seat homebuilt biplane *(Deutsches Museum)*

As can be seen in the accompanying illustration, this latest design by M André Starck also embodies the narrow-gap, sharply staggered biplane wing configuration that has characterised several of his earlier products. The power plant is also unusual, as the 44·5 kW (60 hp) Citroen GS aircooled four-cylinder engine is mounted behind the side-by-side two-seat cabin and drives two pusher propellers mounted in the gap between the wings, on each side. This is claimed to enhance, by means of the propeller slipstream, the slot effect produced by the wing arrangement. Wingtip 'curtains', inclined at 45° to join the wingtips, stiffen the overall structure, make dihedral unnecessary on the main wings, and are claimed to improve stall characteristics and lateral control; the ailerons are attached to their trailing-edges. Construction of the aircraft is all-wooden. Disc brakes are fitted to the main wheels of the non-retractable tricycle landing gear.

DIMENSIONS, EXTERNAL:
Wing span 6·30 m (20 ft 8 in)
Wing area, gross 13·60 m² (146·4 sq ft)
Length overall 6·00 m (19 ft 8 in)
Height overall 1·60 m (5 ft 3 in)

WEIGHTS:
Weight empty 400 kg (882 lb)
Max T-O weight 620 kg (1,366 lb)

PERFORMANCE (estimated):
Max level speed 100 knots (185 km/h; 115 mph)
Cruising speed 91 knots (170 km/h; 105 mph)
T-O speed 38 knots (70 km/h; 43·5 mph)
Rate of climb at S/L 210 m (690 ft)/min
Service ceiling 4,500 m (14,750 m)
T-O run 140 m (460 ft)
Range with 90 litres (19·75 Imp gallons) fuel 810 nm (1,500 km; 930 miles)

UNITED KINGDOM

WHITTAKER

WHITTAKER EXCALIBUR

A letter from Mr C. M. Robertson, on 3 October 1976, reported that the prototype Excalibur had completed a successful 20 minute first flight at St Mawgan, Cornwall about six weeks earlier. With an 89 kg (14 stone) pilot, full 36 litres (8 Imp gallons) of fuel, battery, radio and all electrics, take-off run into a light 8-10 knot wind was approximately 122 m (400 ft), initial rate of climb just over 183 m (600 ft)/min and cruising speed 82 knots (153 km/h; 95 mph), all at 4,100 rpm, against the ducted fan's design speed of 4,500 rpm.

All control responses were reported to be normal, the ailerons slightly heavy and rudder response, although positive, required rather more movement than the pilot considered desirable.

The prototype is more than 45 kg (100 lb) overweight, as a result of taking the glassfibre fuselage shell from a plug, to avoid the expense of making a mould; and of utilising a duct weighing 28 kg (62 lb) instead of the planned 10 kg (22 lb).

Before resuming flight tests, Mr Robertson's company has experimented with modifications to the power plant to increase the output and rpm. Following Mr Whittaker's departure from the company, another engineer is redesigning the aircraft as the production-type Excalibur 2, of which the first five examples are scheduled for completion by the late Summer of 1978.

Excalibur 2 is expected to have no dihedral on the centre wing, to simplify the wing root joints and boom-to-wing attachments. Tail areas will be increased; wheel track reduced from 2·44 m (8 ft) to 1·83 m (6 ft) to improve handling on uneven grass fields; wheel fairings fitted; and the duct redesigned to permit manufacture from rolled aluminium, with only the leading-edge of glassfibre. To avoid adverse effects on fan efficiency and noise levels, due to disturbed airflow around the cylinder heads and exhaust stacks of the present Volkswagen, a different engine will be fitted. Under evaluation are the Dawes and Scot two-stroke engines and an unspecified liquid-cooled in-line engine. It is anticipated that the modifications will permit a 91 m (300 ft) take-off run, 244 m (800 ft)/min rate of climb and cruising speed of over 104 knots (193 km/h; 120 mph).

UNITED STATES OF AMERICA

KICENIUK
TARAS KICENIUK JR

ADDRESS:
Palomar Observatory, Palomar Mountain, California 92060
Telephone: (714) 742-3933

KICENIUK HPA 1

Mr Kiceniuk, whose Icarus hang gliders are described in the 1975-76 *Jane's,* has designed and built a one-man-powered aircraft with the assistance of Mr Bill Watson. Known as the HPA 1 (for Human Powered Aircraft), it underwent initial trials at El Mirage dry lake and at Van Nuys airport, California, in the Summer of 1976. By mid-August it had made several towed tests, confirming its low-drag and good control characteristics, and had also flown for a short distance in free flight.

The general appearance of the HPA 1 can be seen in the accompanying illustration. The power system uses conventional bicycle pedals, with a chain drive to a pusher propeller mounted aft of the fuselage nacelle. Propeller/pedal rpm ratio is 4:1.

Kiceniuk HPA 1 man-powered aircraft prototype

DIMENSIONS, EXTERNAL:
- Wing span: 12·50 m (41 ft 0 in)
- Wing area: 22·85 m² (246·00 sq ft)
- Length overall: 6·86 m (22 ft 6 in)
- Elevator span: 3·15 m (10 ft 4 in)
- Elevator area: 2·88 m² (31·00 sq ft)
- Propeller diameter: 1·91 m (6 ft 3 in)

WEIGHT:
- Weight empty: 66 kg (145 lb)

RED BARON
RED BARON FLYING SERVICE

ADDRESS:
PO Box 497, Idaho Falls, Idaho 83401
Telephone: (208) 523-1812

Red Baron Flying Service, a company which provides flight instruction, maintenance, sales and charter service, sponsored a modified version of the North American P-51D which, currently designated Red Baron P-51, is flown in competition by Roy McClain. It is powered by an extensively modified Rolls-Royce Griffon 57 engine, driving de Havilland three-blade metal contra-rotating constant-speed propellers.

The company is sponsoring also Darryl Greenamyer's modified version of the Lockheed F-104 Starfighter with which he plans to attack the current low-level world speed record of 784 knots (1,452 km/h; 902 mph) as well as the world absolute height record of 36,240 m (118,898 ft) held currently by a MiG-25 of the Soviet Union. Designated RB-104, this aircraft has been undergoing modification for some years. It has a special water injection system to provide maximum power from its turbojet engine, and reaction controls in the fuselage nose are designed to assist the pilot to maintain the very critical low-level altitude at such high speed.

It was anticipated that the attempt on these records would be made from Edwards AFB, California, in the Autumn of 1976.

Red Baron RB-104, modified version of the Lockheed F-104, for attempt on the low-level speed record

AIRSHIPS

UNITED KINGDOM

SKYSHIPS LTD

ADDRESS:
Camelot, Waverley Avenue, Fleet, Hampshire
Telephone: Fleet 29914
MANAGING DIRECTOR: R. Rogers

The last two issues of *Jane's* have given very brief details of an airship referred to as the Gloster. It is now known that the name of this vessel was changed to *Skyship 1* and that the persons concerned in the development and construction of this prototype have incorporated Skyships Ltd to continue its development and to design and build a four-seat production version. The company also intends to investigate the design of a larger airship.

Skyship 1 was nearing completion in mid-October 1976, at which time it was awaiting final assembly and gas inflation before the beginning of flight trials. The airship is already registered with the CAA, and has been allocated the British registration G-AWVR.

An experimental feature is the inclusion of two suspension systems for the vessel's gondola. An internal system suspends the gondola from two catenary curtains attached to the top of the envelope, and the external system comprises a large diaphragm affixed to the undersurface of the envelope and laced to the perimeter of the gondola roof. Each system is capable of supporting the gondola on its own, with an adequate safety factor. By adjusting the length of the internal suspension straps the load can be divided between the two systems, or can be transferred to one or the other so that the characteristics of each can be explored.

The envelope is made of single-ply polyurethane-proofed nylon fabric with taped and cemented seams, and has been reinforced by the application of an abrasion-resistant urethane coating. Fins and control surfaces are of the thick aerofoil section inflated type, maintained at a pressure double that of the envelope by an electric blower mounted within the rear ballonet. Tubular light alloy nose stiffeners are laced to the envelope and support a metal nosecone which incorporates a fitting for mooring to a mast.

The gondola is an enclosed stressed-skin structure which contains two seats in tandem, controls, instruments, ballast tank and a fuel tank of 68 litres (15 Imp gallons) capacity. A triangular steel tube truss is attached by rubber anti-vibration mountings to the aft end of the gondola, forming the attachment point for the outriggers to which the engines are mounted. The upper struts of these outrigger structures have wing-shaped fairings which incorporate the ballonet airscoops, and through which pass the engine controls, fuel pipes and electric cables. The undersurface fairing of the gondola is filled with plastics foam.

Power plant comprises two 19·4 kW (26 hp) Hirth F10.1B flat-four two-stroke engines, each driving a Hoffmann two-blade metal fixed-pitch propeller.

DIMENSIONS, OVERALL:
- Length: 25·0 m (82 ft 0 in)
- Width: 7·62 m (25 ft 0 in)
- Height: 9·75 m (32 ft 0 in)

DIMENSIONS, ENVELOPE:
- Length: 25·0 m (82 ft 0 in)
- Max diameter: 7·62 m (25 ft 0 in)
- Volume, gross: 679·6 m³ (24,000 cu ft)
- Volume, ballonet: 169·9 m³ (6,000 cu ft)

DIMENSIONS, GONDOLA:
- Length overall: 3·66 m (12 ft 0 in)
- Width: 1·22 m (4 ft 0 in)
- Width over engines: 3·66 m (12 ft 0 in)
- Propeller diameter: 1·47 m (4 ft 10 in)

PERFORMANCE (estimated):
- Max speed: 43 knots (80 km/h; 50 mph)
- Max operating altitude: 1,830 m (6,000 ft)
- Max endurance at 21 knots (39 km/h; 24 mph): 10 hr

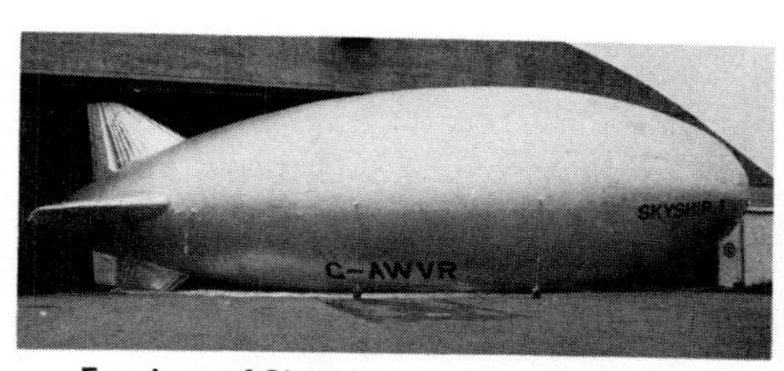

Envelope of Skyship 1 during inflation test

Gondola of Skyship 1, showing port engine

RPVs & TARGETS

UNITED STATES OF AMERICA

AMR
AERO-MARINE RESEARCH

ADDRESS:
PO Box 5194, San Bernardino, California 92408
Telephone: (714) 864-0072
PRESIDENT:
Cdr William Benson, BSc, AeEng, MInst PI
OPERATIONS MANAGER:
Col Peters

Aero-Marine Research was founded in 1953, by Cdr W. Benson, as a patent-holding and R & D company. It has since become involved in flight testing, military and aerospace research and consultancy, and the design, testing and marketing of its own products.

The most important of these, which is in use by several major aerospace companies, is the high mass flow Benson Rotorduct system; with patents issued and pending, this system is making a major breakthrough in ducted propulsion and lifting power units.

Also the subject of patented technology is the Benson Delta, a 'pure' wing, the basic design of which was first flown in October 1946 (in the form of a ¼-scale controlled research model) and which incorporates several new aerodynamic features. Continuous development since that time resulted in a light aircraft research machine (the XD 109) and, more recently, the XD 110 series of unmanned research aircraft of which descriptions follow. The application of the Benson Delta and Rotorduct to manned aircraft programmes is mentioned briefly in the Aircraft section of this Addenda.

The RPV versions of the XD 110 series have attracted considerable interest from the aerospace industry and from military authorities in the USA and overseas. Several are involved in current research programmes, and the advanced technology derived from these deltas has been incorporated into several designs produced for or by other companies.

AMR has also demonstrated successfully the launching of some of its RPVs from air cushion vehicles.

AMR RPV-004

This small RPV continues to serve a useful purpose. It has been demonstrated to the US Army and US Navy, including bad-weather flying; has air-tested autopilot systems; and, in one demonstration, took off under ground remote control, was handed over to helicopter airborne control while in flight, and returned to ground control for landing.

As shown in an accompanying photograph, the only aerodynamic control surfaces are an elevator in the centre of the wing trailing-edge, outboard of which is an elevon on each side. The RPV is highly aerobatic, being able to sustain high positive and negative *g* factors. It is powered by a K & B engine, driving a two-blade tractor propeller, and has a max speed of 113 knots (209 km/h; 130 mph) and a max rate of climb of 366 m (1,200 ft)/min.

The RPV-004 served as a systems testbed for the half-scale RPVs built by AMR as feasibility models for the Teledyne Ryan mini-RPV (which see).

AMR RYAN-BENSON RPV-007

The RPV-007 was designed, built and flown by AMR under contract from Teledyne Ryan as a half-scale testbed for the latter company's mini-RPV (which see), latterly with an approx 1·6 kW (2·2 hp) K & B engine. It was described in the 1975-76 *Jane's*. The radio control equipment and autopilot were first flight-tested in the RPV-004.

AMR RPV-004 (left) and RPV-007 mini-RPVs. Electronics bay of the 004 is located beneath the striped hatch in the centre of the wing. The 007 has a dorsally-mounted Rotorduct

XD 110A scale research model for two-seat Benson Nova Trainer

AMR XD 110

Built originally as a scale test model for a two-seat STOL advanced trainer, the third prototype **XD 110A** became in 1969 the first mini-RPV to be demonstrated in flight. In this initial version, it is powered by a 4 hp Ross flat-six engine, and has demonstrated minimum and maximum flying speeds of 11·03 knots (20·4 km/h; 12·7 mph) and 228·2 knots (422·9 km/h; 262·8 mph) respectively. It has a patented Benson double-delta wing, which will not stall and will maintain full control in all axes during a mush below minimum flying speed. All control surfaces are sealed. The wing has a span of 1·98 m (6 ft 6 in) and a centreline chord of 2·82 m (9 ft 3 in). A single Benson Rotorduct is fitted at the rear of the fuselage, and is driven by an extension shaft, the engine being mounted amidships. The Rotorduct has a 0·25/0·27 m (10/10½ in) diameter moulded shroud, and can employ propellers (rotors) of composite material with any number of blades from two to eight and with fixed or variable pitch. A retractable tricycle landing gear is fitted, with electrically-actuated wheel brakes.

The XD 110A can carry a 22·7 kg (50 lb) max payload, and continues to prove a useful testbed for RPV engines, wind tunnel work for the XD 110 programme, remote control flight, and RPV systems proving. In particular, it was serving in 1976 as a research test model for the full-size tandem two-seat Benson 110 Nova Trainer (see Aircraft section of this Addenda).

Successful test flights have been carried out in conditions ranging from 52·8°C at 915 m (3,000 m) above S/L to sub-zero temperatures at 6,100 m (20,000 m) above S/L. The 'buried' engine, with the exhaust emitting and cooled through the external duct, results in virtually no infra-red signature; and the glassfibre and plastics construction gives the XD 110 an extremely low radar signature.

AMR (Benson) XD 110B research RPV with engine hatch removed; one of the two Rotorducts is shown to the right

The fourth prototype, designated **XD 110B**, was undergoing flight test as an RPV in the Spring of 1976. Slightly larger than the XD 110A, this has twin buried Rotorducts in the rear fuselage, as shown in an accompanying photograph. If desired, the buried-duct configuration of the XD 110B permits low-speed recovery by net, instead of a conventional wheeled landing. Flight demonstrations have been requested by the US Air Force, the US Marine Corps, Israel and Japan.

The XD 110B has a wing span of 2·13 m (7 ft 0 in) and a centreline chord of 3·05 m (10 ft 0 in); it can carry a 36·3 kg (80 lb) payload and reach a max speed of 260 knots (483 km/h; 300 mph). A patented Benson Delta Flow System enables the RPV to fly at a minimum speed as low as 9·03 knots (16·7 km/h; 10·4 mph).

BOEING AEROSPACE COMPANY

BOEING B-GULL (COMPASS COPE B) (Page 621)

As mentioned briefly in the main RPVs & Targets section, the USAF Aeronautical Systems Division announced on 27 August 1976 that Boeing Aerospace had been selected as single contractor for pre-production development of the Compass Cope HALE (high altitude, long endurance) RPV system. The initial funding of $2·75 million covers the design, development, manufacture and testing of three pre-production prototypes, including an associated ground command and control facility, support data and spares. Overall cost of the programme is expected to amount to $77·2 million.

Requests for proposal (RFPs) were issued to Boeing Aerospace and Teledyne Ryan on 21 April 1976, and source selection began in June. The pre-production phase will last for 52 months, and is intended to result in an RPV which can take off and land automatically on conventional runways, and perform a number of operational missions with a minimum of configuration changes. The major changes in the pre-production vehicles will be the adoption of an in-production turbofan engine (the General Electric TF34), and a new dual digital flight control system capable of automatic flight malfunction detection and switching (the YQM-94A prototypes were equipped with an improvised analogue system). The work will be done at Boeing's Developmental Center south of Seattle, and will employ, at peak, about 300 people.

One reason for the selection of a more powerful engine is the decision to increase the payload capacity of the Compass Cope RPV from 317-340 kg (700-750 lb) to 544-907 kg (1,200-2,000 lb), though even so the TF34 engine will be derated in this application from its normal 40·03 kN (9,000 lb st) to about 33·36 kN (7,500 lb st). Boeing has subcontracted Honeywell to develop a 'brass-board' version of the digital flight control system; other subsystems will include a Sperry Univac continuous-wave frequency-modulated command guidance system, and a Bendix automatic landing system.

Although the two programmes are not specifically linked, the principal application of Compass Cope is expected to be in the US Air Force's Precision Location Strike System (PLSS), in which event up to 100 of these RPVs may be required, some 50-60 for operational use and the remainder to cover attrition and other contingencies. Alternative applications, possibly requiring fewer Compass Cope vehicles, include side-looking airborne radar (SLAR) surveillance, communications relay, signature intelligence (sigint), and (for the US Navy) ocean surveillance.

ENGINES

FRANCE

ONERA/AÉROSPATIALE

ONERA (see Spaceflight & Research Rockets section) and Aérospatiale announced on 23 September 1976 that they had jointly been conducting flight experiments with ram rockets operating on solid fuel. Under the supervision of the Direction Technique des Engins (engin = missile) the two organisations are to continue this research, with flight trials at the Landes Test Centre. The fuel is a gas formed by the decomposition of a solid grain—developed by ONERA and with production handled by SNPE (see main Engines section)—having a low oxidiser content.

INTERNATIONAL

ROLLS-ROYCE/ALFA-ROMEO RB.318

Under the project number RB.318, Rolls-Royce and Alfa-Romeo are reported to be studying a turboprop in the 450 kW (600 shp) class. The proposal stems from the EPM600 of 1974 (see 1974-75 *Jane's*) in which MTU was also associated. Elimination of a turboshaft version is hoped to enable the price to be reduced, by allowing the RB.318 to utilise a single-shaft layout.

JAPAN

XENOAH
XENOAH COMPANY

Head Office:
2-3-6 Akasaka, Minato-Ku, Tokyo

This company produces a range of small high-speed two-stroke piston engines which have gained wide acceptance for the propulsion of many kinds of vehicles, and for other applications. They are being used in a number of ultra-light aircraft of both homebuilt and factory-built types. In the United States, Xenoah engines are marketed by Bede Aircraft (see Aircraft section).

Type: Three-cylinder aircooled two-stroke piston engine.
Cylinders: Bore 72·0 mm (2·835 in). Stroke 59·5 mm (2·342 in). Swept volume 726·7 cc (44·34 cu in). Compression ratio (Model G72C-C) 6·2:1; (Model G72C-N) 7·1:1.
Induction: Three Mikuni VM34SS float-type carburettors with mixture control. Two Mikuni DF52 fuel pumps.
Fuel Grade: Mixture of 100/130 octane gasoline and lubricating oil. Engine can be run on pre-mixed fuel or have an oil injection pump as a customer option.
Ignition: Dual capacitor-discharge ignition system, with RF shielding.
Exhaust System: Optional 3:1 manifold and tuned silencer.
Starting: Hitachi electric starter, 12V, 1kW.
Accessories: Optional Mitsubishi 14V 35A alternator and mounting bracket for instrument vacuum pump.

Typical Xenoah three-cylinder engine for light aircraft applications

Dimensions:
Length overall 550 mm (21·65 in)
Width overall (incl carb's) 384 mm (15·1 in)
Height overall (excl plugs) 358 mm (14·1 in)
Weight, Dry:
Bare 56·6 kg (125 lb)
With manifold, silencer, starter, alternator and vacuum pump 63·5 kg (140 lb)
Performance Rating:
Max continuous 52·2 kW (70 hp) at 6,250 rpm
Specific Fuel Consumption:
Max at 75% power (49 kW; 65 hp) 135 μg/J (0·8 lb/hr/hp)
Fuel Consumption:
Max at 75% power 24·5 litres (5·4 Imp gallons; 6·5 US gallons)/hr

UNITED KINGDOM

BONNER
AERO BONNER LTD

Address:
Shoreham Airport, Sussex BN4 5FJ
Telephone: 079-17-5764

This company has been developing an aircraft conversion of the British Ford V-6 motor car engine. In 1976 the first flight prototype engine was installed in a Chipmunk (actually a former winner of the King's Cup race) and was exhibited statically at the SBAC show at Farnborough. The start of the flight test programme was imminent in October 1976.

BONNER SUPER SAPPHIRE

This engine for general aviation is a watercooled development of a standard production car engine. Early bench development has been most successful, and the Super Sapphire is expected to set 'a new level of cheap, lightweight, compact, quiet and smooth power' for aircraft requiring engines in the 149 kW (200 hp) class.

Type: Six-cylinder four-stroke piston engine, turbocharged and with geared drive.
Cylinders: V-6 configuration. Watercooled. Swept volume 3,000 cc (183 cu in).
Induction: Turbocharged. Direct fuel injection.
Ignition: Dual, with one magneto and one coil.

De Havilland Chipmunk exhibited at Farnborough in 1976 with a Bonner Super Sapphire V-6 engine
(Air Portraits)

Lubrication: Dry sump.
Propeller Drive: Reduction gear with quill shaft, 0·5 ratio.
Dimensions:
Length overall 914 mm (36·0 in)
Width overall 457 mm (18·0 in)
Height overall 609 mm (24·0 in)
Weight, Dry: 149 kg (328 lb)
Performance Rating:
Max T-O 149 kW (200 hp) at 5,500 rpm
Fuel Consumption:
At cruise (74·5 kW; 100 hp) 34 litres (7·5 Imp gal)/hr

ROLLS-ROYCE

ROLLS-ROYCE RB.211

Rolls-Royce announced in September 1976 that it was studying the design of a lower-thrust version of the RB.211, designated RB.211-535, with the core of the 22B but a smaller fan. Thrust would be 142 kN (32,000 lb), flat-rated to 20°C.

UNITED STATES OF AMERICA

GENERAL ELECTRIC

GENERAL ELECTRIC CF6

In September 1976 General Electric's Aircraft Engine Group announced FAA certification of the CF6-50C1, an uprated version of the -50C flat-rated at 233·52 kN (52,500 lb st) to 30°C. Later in the same month it announced the CF6-32, to be rated initially at 133·4 kN (30,000 lb st). This will use the core of the CF6-6, operating at reduced temperature, matched with a new front fan and LP turbine. Like the RB.211-535 mentioned above, the CF6-32 is aimed at proposed new medium-range transports.

In October 1976 GE announced the most powerful of all CF6 versions, the CF6-50M, to be rated at 247 kN (55,500 lb st) to 30°C. The -50M is scheduled for certification in 1979.

GENERAL ELECTRIC CT7

Under this designation General Electric's Aircraft Engine Group announced on 14 September 1976 a commercial version of the T700 turboshaft engine. Preliminary specifications are virtually identical to those of the military (US Army) engine, except that dry weight is increased to 195 kg (430 lb) and length to 1,194 mm (47 in). Some external piping has been fireproofed and locking/fastening devices conform to commercial standards. FAA certification of the CT7 is due in the second quarter of 1977, with deliveries following in mid-1978. GE expect to sell "from 1,000 to 3,000 units over the next decade."

INDEXES

GENERAL INDEX OF AIRCRAFT

Items printed in this type refer to this edition — *Items printed in italics refer to the ten previous editions*

A

B

C

D

E

F

G

H

I

J

Q

R

S

T

U

SAILPLANES

RPVs AND TARGETS

AIR-LAUNCHED MISSILES, SPACEFLIGHT & RESEARCH ROCKETS

B

C

D

AERO ENGINES

N

O

P

Q

R

S

T

U

V

W